LOURDES COLLEGE LIBRARY
3 0379 1003 4138 9

D0850027

GRAY'S

Manual of Botany

GRAY'S
Manual of Botany

EIGHTH (*CENTENNIAL*) EDITION—ILLUSTRATED

A HANDBOOK OF THE FLOWERING PLANTS AND FERNS OF THE CENTRAL AND NORTHEASTERN UNITED STATES AND ADJACENT CANADA

Largely rewritten and expanded by

Merritt Lyndon Fernald

FISHER PROFESSOR OF NATURAL HISTORY, EMERITUS,
AND FORMER DIRECTOR OF THE GRAY HERBARIUM, HARVARD UNIVERSITY

WITH ASSISTANCE OF SPECIALISTS IN SOME GROUPS

CORRECTIONS SUPPLIED BY R. C. ROLLINS

Biosystematics, Floristic & Phylogeny Series,
Volume 2
Theodore R. Dudley, Ph.D., General Editor

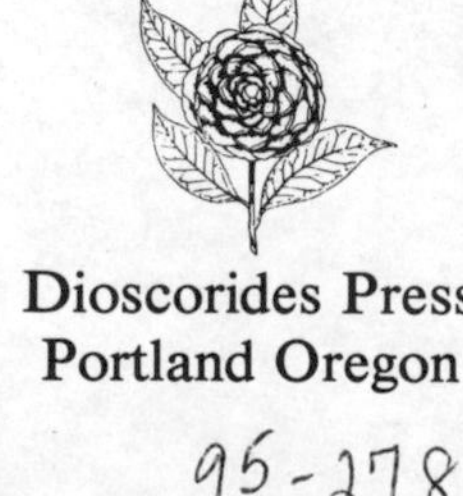

Dioscorides Press
Portland Oregon

Reprint published 1987 by
Dioscorides Press (an imprint of Timber Press, Inc.)
9999 S.W. Wilshire
Portland, Oregon 97225

Fourth Printing 1993

ISBN 0-931146-09-7

Printed in Hong Kong

Contents

The Eighth Edition of Gray's Manual of Botany was completed and published in 1950. Not long after, the author, Professor Merritt Lyndon Fernald, died. Those of us closest to the scene during the period of tremendous and dedicated effort that resulted in this revision were relieved that the author lived to see his beloved book in print. We were aware that Fernald, for a period of forty years, had directed his botanical activities toward this revision, which was to mark the centennial of its beginning. First he explored the northeastern portion of what was known to him and his associates as the "manual range." And he extended this range over that of the previous edition to include an area he thought was of particular botanical interest, the Straits of Belle Isle in eastern Canada. Then, as he said, when his energies and health no longer permitted him to do field work in the more rigorous areas of the north, he turned to southern borders of the "range" to fulfill his thirst for field work on the plants to be included in the Manual. Most of Fernald's contemporaries would know in detail of his findings and experiences, for he wrote of them extensively for the pages of *Rhodora,* which he edited. Thus he had full license to relate in great detail his eye-opening encounters with the plants he came to know intimately. Every student of Fernald knew of the keenness of his observations and his ability to retain these in his mind, then to explicate the nature of the taxa with which he dealt.

The high quality of Gray's Manual can be attributed to tradition in part, but the Eighth Edition has a special quality that has to be associated with the author's particular extensive acquaintance in the field with the plants about which he wrote. He knew the plants on their home ground. From this comes a resonance that is evident in nearly every page of the book. And this aspect will persist as long as the book does. But no book is perfect. Errors, where so many entries of specifics are involved, are bound to creep into the final copy. Fortunately, in his introduction, Fernald invited the botanical public to send to the Gray Herbarium any errors detected. Finding errors in Gray's Manual, such as mm. for cm., became a game for friendly colleagues, and corrections for the manual accumulated over time. In one case, the lost leg of a key was a serious omission. Teachers, their students, and others doing floristic research were a rich source of needed additions, ommissions, and corrections. These we tried to accommodate in a corrected printing, published in 1973. At that time, there was no opportunity to thank those who helped to improve the accuracy of Gray's Manual, and I cannot name them individually now. But I do appreciate their help and hereby tender them my grateful thanks.

The reissue of a book, now 37 years beyond its original publication, will be questioned by some. Ideally, it should be revised. But that possibility is not open at the present time. Yet we know the content endures and there is a demand for Gray's Manual. To satisfy that demand, this reissue is being made. It is as simple as that.

Reed C. Rollins
Asa Gray Professor
of Systematic Botany, Emeritus

Preface

Earlier Editions. — One hundred years before the introductory paragraphs of this Preface were written, Asa Gray issued the first edition of his Manual of the Botany of the Northern United States (1848), a book of 710 pages, covering the flowering plants, vascular cryptogams, mosses and liverworts, the treatment of the two latter groups by William Starling Sullivant. The range covered the area from New England to (and including) Wisconsin, and southward took in Pennsylvania and Ohio. It contained descriptions of 2213 named species and varieties of flowering plants, ferns and fern-allies. In the second edition (1856) the range was extended to include all of Virginia and Kentucky, thence northward to cover all states east of the Mississippi. Isaac Sprague supplied 14 plates, illustrating the genera of cryptogams and the volume contained 739 pages, with descriptions of 2426 species. By the fifth edition (1867) the mosses and liverworts were excluded but there were plates by Sprague of analytical drawings of the genera of grasses and sedges. This was the last edition actually prepared by Gray, although in subsequent issues of it there were addenda by him. In the sixth edition (1890), revised by Sereno Watson and John Merle Coulter, the liverworts, with new illustrations, the treatment of them prepared by Lucien Marcus Underwood, were restored, although the mosses were omitted, and the range extended westward to the 100th meridian. In this edition 3138 vascular plants were included. In the last or seventh edition (1908), prepared by Benjamin Lincoln Robinson and the present writer, "adjacent Canada" (meaning from Prince Edward Island and northern New Brunswick to southwestern Ontario) was added at the north and the western limit was reduced to follow the western boundary of Minnesota and Iowa, thence southward along the 96th meridian, thus taking in easternmost Nebraska and Kansas but excluding a large number of plants of the Great Plains, which do not extend eastward into the more forested country. The liverworts and the former plates were excluded but 1036 very accurate text-figures, chiefly by F. Schuyler Mathews (some of the grasses by J. Franklin Collins) included, primarily in the more technical groups. A total of 4885 species, varieties and named forms were described, the volume containing 926 pages.

The longest interval between editions of the past was 23 years. By the time the present edition is off the press 42 years will have elapsed since the publication of the seventh edition. This epoch has been one of fundamental changes, both in our understanding of our plants and in the requirements of the International Rules of Botanical Nomenclature. Incidentally, the previous editions were prepared with only slight familiarity with the growing plants. As was well recognized, Gray and Watson were not primarily active field-students of our flora. Gray was almost overwhelmed by the endless collections sent him for critical study from all parts of the continent, as well as from eastern Asia, by his active (largely unassisted) part in teaching and in the development of the then appreciated and fruitful Botanic Garden, his philosophical and editorial writing and his trenchant reviews and appreciative biographical notes. He depended on others, rarely more than one or two from each of a few states, for material from the "Manual-range". Watson's field-work had been on the western side of the continent during the survey by the Clarence King Exploring Expedition. When he became Curator of the Gray Herbarium (then the Harvard Herbarium, including all groups of plants), his extra-curatorial activities were little concerned with the plants growing about him, but with detailed bibliographic work and monographic studies covering groups of broad geographic areas. Coulter's brief period of field-work, likewise, had been chiefly in an area far removed from the range covered in the Manual. Robinson took over from Watson extensive study of the Mexican flora, based on collections of others, the continuation and editing for a time of unfinished monographic studies by Gray and, later, monographic studies of large genera of Mexico or South America. Physically unable to carry on extended field-work, his personal familiarity in the field with most plants of the Northeast was necessarily limited.

Recent Activity in Field-work. — In the past 40 years, however, a host of zealous and highly intelligent workers, both amateur and professional, has discovered that the flora of the temperate eastern half of North America is full of absorbing and largely still unsolved problems; and the knowledge of our own region has vastly multiplied, all because the so-called "closet-botanist" of the past has recently come out of his closet to the fields and woods and is more and more doing so. In 1829 Amos Eaton (*Eat.* of our abbreviations) stated that there are "not, probably, 50 undescribed species of Phenogamous plants in the United States — perhaps not one species east of the Mississippi". If everyone had scrupulously hidden in his "closet" we might not now smile at Eaton's naïveté. But the insistent lure of the field has led to search for unexplored areas, resulting, during the last half-century, in the botanical recognition of many hundreds of species new to our range at the southeastern, southwestern and northwestern corners of the area. Furthermore, the intense botanical interest of the Gaspé Peninsula of Quebec, the Magdalen Islands and Newfoundland, regions with amazingly isolated or relict species or endemic representatives of geographically remote allies, has been tardily recognized. Since topographically these areas are merely outliers of the rest of temperate eastern Canada, the extension of the Manual to include them has seemed quite logical. Consequently, the northern limit of range now includes the area south of the Straits of Belle Isle and from Anticosti Island westward along the 49th parallel of latitude in Quebec to the northwestern corner of Minnesota. The western and southern limits are unchanged. When, in 1932, the proposal was made to extend the limits eastward to include Gaspé, Anticosti and Newfoundland, the objection was raised, that the number of additional species was forbidding. As it proves, however, intensive exploration going on at the southeastern, southwestern and northwestern corners of the old area (Virginia, Missouri and the region from northern Lake Huron across Minnesota) has already added three or four times as many species as from the new northeastern extension. In fact, so rapidly are discoveries being made that, in order to prevent perpetual upsetting of analytical keys and the numbering of species in large genera, it has been necessary to refrain from including some of the latest additions being made by ardent helpers. When practicable these most recent additions are included as memoranda but the final setting-up of the text has to be the deadline!

Asa Gray, in his first edition, specially acknowledged the help of three active botanists of the period (Carey, Oakes and Olney) who had supplied him with material and technical aid in treating the vascular plants. Hundreds of enthusiastic coöperators, about 400 of them by direct correspondence and by supplying critical specimens, have stimulated the present work. Some have sent most helpful notes on technical matters, others liberal material from their special regions, others selected specimens of additions, novelties or puzzles. To all of these most helpful friends I am deeply indebted; if significant points in their correspondence and notes have been overlooked, the endless mass of detail and the thousands of items to be coordinated must be the excuse. Space naturally forbids enumeration of all those who have supplied such data, but a few whose help in such matters is outstanding must be noted, although the scores of active workers in New England and the Maritime Provinces may be taken for granted. Special appreciation may be expressed for the unlimited encouragement of the late AGNES M. AYRE (Newfoundland); Professor E. LUCY BRAUN (Kentucky); the late Professor FREDERIC K. BUTTERS (Minnesota); Dr. C. C. DEAM (Indiana); the late CHARLES K. DODGE (Michigan); Professor NORMAN C. FASSETT (Wisconsin); Professor RUSKIN S. FREER (Virginia); Professor RAY C. FRIESNER (Indiana); Professor FRANK C. GATES (Kansas); Professor ADA HAYDEN (Iowa); NEIL HOTCHKISS (aquatics); the late Dr. HOMER D. HOUSE (New York); FRANCIS WELLES HUNNEWELL (specialties, Virginia); BAYARD LONG (New Jersey, Pennsylvania, etc.); Père LOUIS-MARIE (Quebec); Professor FRANK T. MCFARLAND (Kentucky); Dr. JOHN W. MOORE (Minnesota); Professor JACQUES ROUSSEAU and associates (Quebec); Professor C. OTTO ROSENDAHL (Minnesota); Professor ORIN A. STEVENS (western Minnesota); Dr. JULIAN A. STEYERMARK (Missouri); Dr. ROBERT R. TATNALL (Delaware Peninsula); Professor THOMAS M. C. TAYLOR and associates (Ontario); the late Frère MARIE-VICTORIN and associates (Quebec and Ontario); SAMUEL C. WADMOND (Wisconsin and Minnesota); the late Professor KARL M. WIEGAND (New York and Virginia). All this recent and current activity accounts for the inclusion in the present largely rewritten edition of more than 8000 species, varieties, forms and named hybrids.

SYSTEM AND CATEGORIES ADOPTED. — In the sequence of families it has seemed most practical to follow, as in the seventh edition, the system of ENGLER & PRANTL, since that (as epitomized in the *Genera Siphonogamarum* of Dalla Torre & Harms) is the system most generally followed (except in Britain) by taxonomists; and it is the only widely recognized one for which treatments of essentially all families and genera of vascular plants of the whole world have been provided. Several other systems or partial systems, departing from it in many respects but not so fully elaborated, are obviously better in some details. However, a handbook of the flora of a limited region being primarily a presentation of a ready means for the identification of plants now living, not the elaboration of hypotheses as to the possible derivation of the groups from theoretical ancestors of past geological time, adherence to the most familiar and fully elaborated general system seems the suitable course.

The somewhat general groupings under the genera are followed: subgenera, sections (indicated by the sign §) and series, the two latter categories not clearly differentiated in current practice and often treated as interchangeable. Until their exact use has been more clearly defined than at present and the priority of names in such subordinate categories carefully established, the terms used by the best recent monographers are here accepted. Much more definite (at least in most groups except the actively hybridizing and asexually reproducing *apomicts*, like the Brambles, the ubiquitous weedy Dandelion or the adventive European Hawkweeds) are the terms species, variety and form. The SPECIES is conceived as a series of individuals (usually numberless) occupying, until disturbed by man's activity, a natural geographic area and having essentially identical morphological characters of flower, fruit or reproductive structure, somewhat exemplifying the biblical definition, "It is by their fruits ye shall know them", for most critical taxonomic study starts, when possible, with flower, fruit, seed or spore. As here used the term VARIETY, *i.e.*, VARIETAS (as opposed to any sort of variation or fluctuation) refers to a *geographic variety*, a strongly fixed variation of a species with the essential reproductive parts unchanged but showing somewhat constant departures in size of parts, shape of leaves, or modification of the less fundamental parts of the flower, etc., and occupying a somewhat segregated geographic area: for example, *Abies balsamea*, widely ranging across much of northern Canada, coming south to cool woods of the northernmost states and along the higher mountains southward, its var. *phanerolepis* differing in smaller cones and in having the bracts in maturity with exserted (instead of included) awns, the variety in bleaker habitats northeastward and unknown south of the higher mountains of northern New England; or

Acer rubrum, with samaras 1.5–2.5 cm. long, widely ranging from Newfoundland to Manitoba and south beyond our limits, but var. *Drummondii*, with larger samaras (2.7–5 cm. long), the leaves and flowers as in typical *A. rubrum*, occurring on the southern Coastal Plain and pushing northward to New Jersey, southern Indiana, southern Illinois and Missouri. The FORMS (FORMA, pl. FORMAE) are minor departures in vegetative characters (dwarfing, depressing, etc.), indument, multiplication or loss of parts, change of color, etc., striking variations but without fundamental or constant trend over large areas, occurring here-and-there in the broad range of the species. The term SUBSPECIES, too often substituted for variety, is not here accepted in that sense. A good illustration of two true subspecies is *Stachys palustris:* the Eurasian subspecies having many gland-tipped trichomes on the closely viscid-pilose calyx, this subspecies with several varieties which are adventive with us; the indigenous North American plants lacking the glands on the long-hirsute or long-pilose calyx and this subspecies having several geographically limited varieties. HYBRIDS of two usually distinct species frequently occur and in some very plastic and unstable groups, like many genera of the *Rosaceae*, or in *Oenothera*, they seem just now to be endless. These become the bases of important cultivated strains (like the cultivated apples, blackberries, etc.), but in undisturbed Nature they introduce utmost confusion into any system of classification. Their characteristics are usually in some ways a combination of those of the parents but the plants are often more vigorous and floriferous and their recognition is the result of experience and trained judgment. In most cases they are not specially described in this volume but merely mentioned, though in a few recurrent, abundant and tolerably stable plants of presumable hybrid-origin, such as × *Lysimachia producta*, they are admitted to full standing with the sign × preceding the binomial. Others, less fixed but having received binomials, are mostly not described and their status is indicated by an × before the trivial name, while those which have received no binomial are indicated by the same sign between the specific names of the supposed parents.

INDICATION OF STATUS IN OUR FLORA. — In the Descriptive Flora (except in keys) all maintained names of genera (or their initials) accepted as in our flora are in bold-face type; the generic synonyms, at the close of the paragraph, in Roman capitals. The names of indigenous (*Indig.* or *Nat.*) species, varieties, forms and named hybrids are in bold-face type, those of plants of foreign origin in Roman type. Of the latter we have two main groups: INTRODUCED (abbreviated *Introd.*), those plants, like Japanese Honeysuckle or the cultivated *Primula veris*, which were intentionally brought from outside our area; ADVENTIVE (abbreviated *Adv.*), coming uninvited from outside areas as vagrants or weeds, only rarely as desirable elements in the flora. As members of our wild flora species of either of these groups may have become NATURALIZED (*Natzd.*) or firmly established, as the too common Sheep-Sorrel or the Orange Hawkweed, or they may be only locally or casually established or adventive. In so large an area as that here covered, plants indigenous in one section may be only adventive or naturalized in another. In such cases an attempt is made to indicate this dual status. *Synonyms* within the species are indicated by italics and when the synonym differs only in the generic name the latter only is frequently (but not always) stated. By no means all synonyms are given; usually only those in current works of similar character are included.

RANGES AND FLORISTIC AREAS. — In general the ranges are given from east to west across the North, then to the south (*Pinus Strobus* for example), although in case of many plants (*Pinus australis* for instance) primarily southern and passing out northward the stated range attempts to indicate that fact. Similarly, plants which come from a broad range west of our area often have that fact indicated in the statement. The occurrence of species on other continents or in areas wholly remote from our own is usually indicated in parentheses. The "life-zones", often much emphasized, are not here generally taken up. In the eastern United States and Canada the Arctic-alpine habitat (for plants of bleak areas and chiefly isolated from an arctic area), the Coastal Plain, the Great Plains and, of course, maritime areas are fairly distinctive. Beyond that, most of the floristic areas often cited become too confluent or indistinct. Their use has always to be tempered by too many apologies. Unless otherwise stated the dates under the species indicate the flowering period.

KEYS. — In general (except in the SYNOPSIS OF ORDERS AND FAMILIES) the synoptic keys so much used in previous editions are not here adopted. Instead, regularly dichotomous keys are employed. The primary or contrasting divisions are shown by regularity of indention, and in the longer keys by identical introductory letters of an alphabet, the differing alphabets in the longest keys distinguished by the type used. A letter of any of the alphabets is used only twice: for the contrasting divisions of the key. The keys have been made with the hope that plants in different stages of development, whether in flower or fruit, can be reached; they are, in fact, brief diagnoses. It is hoped that by their use greater satisfaction will be secured than by the "one-character" key which so often fails to turn in the lock. In many descriptions of species and varieties *italics* are used for emphasis, the italicized phrases, read continously, constituting a supplementary key.

MEANINGS OF GENERIC AND SPECIFIC NAMES. — Many scientists say "A name is a name; that is all we need". Most names, however, have a background and there are many users of a handbook who desire to know something of their original meanings. The present author admits his own curiosity about such matters. Generic names like *Commelina* and *Tillandsia*, products of the fanciful imagination of Linnaeus, *Andromeda* and *Dirca* from his knowledge of Greek mythology, and *Aristolochia*, *Paronychia* and *Saxifraga* from ancient medical beliefs, add, when their origin is understood, vastly more interest to study of the plants than do such unpoetic or unimaginative generic names of this "practical" era as *Pseudoëspostoa* or *Haageocactus*. So far as practicable but always as a secondary feature, the meanings (or sources) of generic, specific, varietal and formal names have been given; and even such simple trivials as *alba* or *albiflora* have been translated for the benefit of those without some background of Latin. This may seem unnecessary but my constant adviser on such matters, ARTHUR STANLEY PEASE, Pope Professor of the Latin Language and Literature at Harvard University, assures me that no understanding of the simplest Latin words can nowadays be assumed.

In case of generic names commemorating persons, a great many, like *Linnaea*, *Burmannia*, *Jussiaea*, *Mertensia*, *Tradescantia* or *Ruppia*, call to mind early great founders of our science; while active explorers of the North American flora gave distinction to such names as *Claytonia*, *Mitchella*, *Kalmia*, *Bartonia*, *Menziesia* or *Lesquerella*. The commemorative names below the rank of genus are very numerous. When possible, without long search of local archives and birth-certificates, the full name and date of birth (or death) has been entered. Often only the general period of the subject's activity is given, and when local plants have been dedicated merely to "Mr. Brown" or "the late Mrs. Smith", the search for the desired data has not been prolonged.

In capitalization of specific names and those beneath the species the recommendation of the International Rules of Botanical Nomenclature and the best practice of taxonomists for two centuries is followed. Old generic names (such as in *Euphorbia Peplus*) taken over as specific epithets, aboriginal names (as in *Euphorbia Ipecacuanhae*) similarly applied, and personal genitives (such as in *Euphorbia Geyeri*) are written with capital initials. This practice of the great founders and leaders of our science is, however, with the present tendency to "standardize" everything and to avoid the mental effort of trying to understand the meanings of names, in danger of becoming overthrown by a hasty and superficial majority. If that time comes *Silybum Marianum*, named for the Virgin Mary, will to many Americans indicate "coming from Maryland"; the specific name of *Zea Mays* (from the aboriginal name now rendered as Maize) will be untranslatable; and *Fumaria Bastardii* (honoring the distinguished French botanist, Toussaint Bastard) will seem to mean "of a bastard". The retention of the capital initial has much value; its rejection is too often a source of misunderstanding.

COLLOQUIAL NAMES. — An effort has been made to record the colloquial names under which the more conspicuous plants are known by the layman. These names are, of course, very variable and often very local, while such a name as "Mayflower" is used in so many senses as to be essentially meaningless outside restricted areas. In general, popular usage, as indicated in many local records and including those of the French population of Quebec, has been accepted. Such library-made names as are mere translations of the Latin name and not colloquially used are not admitted, the translation of the specific epithet being sufficient; neither are names arbitrarily invented by enthusiasts in such manufacture, to displace those which are more generally known to the layman; and the alteration of such long-familiar names as TREE-OF-HEAVEN to "TREEOFHEAVEN" is too absurd for acceptance.

ILLUSTRATIONS. — Many of the figures by Matthews and Collins, originally drawn for the seventh edition, have been retained. Others have given way to somewhat more detailed illustrations; and, of course, supplementary figures under genera already well-illustrated are by a younger group of artists. It has seemed unnecessary to retain figures in many groups which are much-illustrated in popular works (*Botrychium* and most Orchids, for instance) but rather to concentrate, so far as draftsmen have been available, upon large or technical genera. At one period, when the present author spent his "out-of-hours" time in preparing a very detailed work on a more restricted area, he employed students and friends who could make technical drawings to prepare many plates illustrating genera of the limited area. With the increasing demand for the rewriting of Gray's Manual the other manuscript was set aside. Its details are incorporated in the present work and some of the illustrations taken over. Plates 1 and 2 are undisturbed plates of the series. Some other plates, with the drawings much interlocked, have been divided to show essentials (as in many figures in *Panicum*). These newer drawings by different illustrators are quite unlike in style and technique, thus showing how differently several illustrators proceed. Most of the new drawings are by five artists, whose individual patterns will be readily recognized through a few examples each: the late ARTHUR MONRAD JOHNSON (plates 1 and 2 and supplementary or all drawings in *Eleocharis, Scirpus, Juncus, Luzula,* etc.); RUTH PEABODY ROSSBACH (*Panicum, Ranunculus, Antennaria, Solidago,* etc.); SHIRLEY GALE — now CROSS (*Rhynchospora, Ulmus, Hypericum, Myriophyllum, Pyrola, Bidens,* etc.); SARA R. STEWART — now HINCKLEY (*Potentilla, Geum, Astragalus, Polygala, Verbena,* etc.); WILLIAM HOLLAND DRURY, JR. (supplementary figures in *Carex*, such as figs. 522–524, 560–562 and 645–647, *Salix* and *Quercus*). Other figures taken from *Contributions from the Gray Herbarium* or from the journal, Rhodora, are noted under their respective genera. Since measurements of parts are recorded in the specific descriptions the enlargements or reductions in the illustrations can readily be estimated. In general, therefore, no statement of these very variable proportions is given.

PRONUNCIATION AND ACCENTUATION. — The general principles followed in the seventh edition are here accepted. In these chiefly English-speaking countries it is presumed that the English sounds of letters will be used in pronunciation. Accentuation, however, follows the classical Latin model except in cases of names derived from those of persons, which might be too much distorted for recognition by the regular accentuation in Latin. In these cases an attempt has been made to throw the accent where it would not seriously disguise the name of the person honored: *Michauxiàna*, *Pálmeri*, etc. Two accents are used, both indicating the syllable to be stressed: the grave (`) to indicate the long English sound of the vowel; the acute (′) to show the shortened or otherwise modified sound.

OTHER ACKNOWLEDGMENTS. — Many helpers and contributors to the making of this volume have been noted. Some, like Professor Pease, who has been ever ready to advise on many points and who has most generously scrutinized both copy and proof, deserve a second word of thanks and appreciation. Similarly, Professor ERNEST ROULEAU's scrutiny of much of the manuscript and his most generous reading of proof-sheets have been aids of the greatest importance and are thankfully acknowledged.

The necessity to understand the TYPES or original specimens, described by European botanists to whom colonial Americans sent material, has imposed a large burden upon several officers of European herbaria where these irreplaceable types have been preserved. Without their most kindly coöperation in sending or in aiding in the making of photographs of hundreds of such types many misidentifications would have been perpetuated. Among those who have thus largely helped in our securing these significant aids to accuracy in the identification of our species the following should be mentioned: the late Professor LUDWIG DIELS of the Botanischer Garten und Botanisches Museum, Berlin-Dahlem; Drs. B. P. G. HOCHREUTINER and CHARLES BAEHNI of the Conservatoire et Jardin Botaniques, Genève; Dr. H. HUMBERT and M. R. METMAN of the Múseum National d'Histoire Naturelle, Paris; Mr. SPENCER SAVAGE of the Linnean Society of London; Drs. JOHN RAMSBOTTOM, GEORGE TAYLOR and others of the British Museum (Natural History), London; the late Sir ARTHUR HILL and Dr. W. B. TURRILL, the present Keeper of the Herbarium, of the Royal Botanic Gardens, Kew, and their associates.

Naturally, the members of the staff of the Gray Herbarium have been under constant call in bibliographic and technical matters. To the late CHARLES ALFRED WEATHERBY I am specially indebted for his patient checking of details in nomenclature and presentation. It was originally hoped that he would act as coauthor in the writing of the text but his modesty and the crowding of his more special interests made that hope a vanishing one, greatly to the regret of the author remaining. The treatments of many genera or groups have often been prepared with aid from monographic or revisional studies by others. When their studies have been significantly the bases of the present treatments acknowledgment is given in the footnotes. In three cases the treatments of genera have been contributed by specialists upon them: *Juglans* and *Carya* largely revised from the text of the present author by Dr. WAYNE E. MANNING; the treatment of the intricate genus *Crataegus* prepared by the only student who professes to understand it, ERNEST JESSE PALMER; *Desmodium* by the recognized authority on that genus, Dr. BERNICE G. SCHUBERT. Finally, especially since the impairment of his eyesight, the author has had the most loyal, conscientious and unlimited aid of Dr. SCHUBERT in the exacting details of coordinating usages in the text, editing the manuscript for the printer, checking and double-checking the citations of figures, and in the scores of other details necessary in the book as it goes to press. My appreciation of all her helpfulness can not be adequately expressed. — M. L. F.

Cambridge, June, 1949.

Synopsis of the Orders and Families of Vascular Plants Included in this Volume

DIVISION I. PTERIDÓPHYTA (Vascular Cryptogams)

Rush-like, fern-like, moss-like or quill-leaved plants without true flowers. Reproduction by spores (without embryos).

Class I. ARTICULÀTAE

Represented with us only by

Order I. EQUISETÀLES. Stems and branches rush-like, jointed, with nodes covered by sheaths composed of more or less united scarious leaves, otherwise leafless; sporangia borne on inner surface of peltate scales of terminal spike-like cones. A single family:

Fam. 1. Equisetaceae.

Class II. LYCOPODIÌNAE

Stem elongate, covered by persistent small leaves; sporangia in axils of or upon upper surfaces of leaves or in terminal bracted spikes or strobiles.

Order II. LYCOPODIÀLES. Leaves without a ligule, in 4–16 ranks; sporangia axillary, bearing uniform spores. A single family:

Fam. 2. Lycopodiaceae.

Order III. SELAGINELLÀLES. Leaves ligulate, in 4–6 ranks; sporangia axillary or on upper surfaces near bases of leaves, of 2 kinds, some bearing minute (male) microspores, others with few large (female) macrospores. Represented with us only by

Fam. 3. Selaginellaceae.

Class III. ISOËTÌNAE

Stem short and thick, corm-like, bearing elongate-subulate or thickish-linear broad-based ligulate, rosulate leaves; sporangia immersed in the base of the leaf, some filled with minute (male) microspores, others with larger (female) macrospores. A single order:

Order IV. ISOËTÀLES. A single family:

Fam. 4. Isoëtaceae.

Class IV. FILICÌNAE, Ferns and Fern-allies

Leaves (fronds) not closely imbricated or, if narrow and slightly imbricated, without axillary sporangia.

Subclass I. EUSPORANGIÀTAE

Walls of sporangia of several layers; sporangia without a transverse ring. Represented with us only by

Order V. OPHIOGLOSSÀLES. Fertile and sterile halves of frond very dissimilar. A single family:

Fam. 5. Ophioglossaceae.

Subclass II. LEPTOSPORANGIÀTAE

Wall of sporangium of a single layer of cells; the sporangia with or without a transverse or oblique or vertical ring.

Order VI. EUFILICÀLES. Sporangia borne on foliaceous or modified fronds.

Fam. 6. Osmundaceae. Fertile fronds or fertile portions of frond unlike the sterile ones; sterile fronds pinnate; sporangia globose, opening longitudinally; the transverse ring obsolete or nearly so, when present apical.

Fam. 7. SCHIZAEACEAE. Differing from Fam. 6 in its palmate or simple sterile frond and ovoid sporangia with complete apical transverse ring.

Fam. 8. HYMENOPHYLLACEAE. Sporangia sessile at base of a bristle-like receptacle, with oblique ring; frond of a single layer of cells.

Fam. 9. POLYPODIACEAE. Sporangia stalked, without bristle-like receptacle, the stalk passing into a vertical jointed incomplete ring which, by straightening transversely, ruptures the sporangium.

Order VII. HYDROPTERIDÀLES. Sporangia inclosed in basal sporocarps (inclosing involucres) borne along the creeping stem or rhizome.

Fam. 10. MARSILEACEAE. Slenderly rhizomatous plants with long-petioled leaves (fronds) 4-foliolate with us; sporocarps crustaceous, on ascending peduncles.

Fam. 11. SALVINIACEAE. Small moss-like floating plants with 2-ranked simple fronds; sporocarps soft, sessile or nearly so, borne on under side of axis.

DIVISION II. SPERMATÓPHYTA (SEED-PLANTS, FLOWERING PLANTS, PHANEROGAMS)

Plants with or without green leaves, with true flowers bearing stamens or carpels (pistils) or both. Male generative cells (pollen-grains), with rare extralimital exceptions, passive, developing an elongate tube. Reproduction (except in unusual sterile and vegetatively propagating forms) normally by seeds containing an embryo or minute plant.

SUBDIVISION I. GYMNOSPÉRMAE (GYMNOSPERMS)

Trees or shrubs with needle-like, scale-like or broad and firm leaves; wood often pitchy or gummy; flowers mostly unisexual, rarely with a perianth; ovules naked, not inclosed in an ovary. Represented with us only by the

Order VIII. CONÍFERAE (CONIFERÀLES). Leaves slender, narrow or scale-like, chiefly evergreen; fertilization by passive sperm-cells; fruit a cone or berry- or drupe-like. Two families:

Fam. 12. TAXACEAE. Flowers solitary, axillary; the globular staminate ones composed of few naked stamens; the pistillate ones with a single plump and erect ovule which becomes a bony-coated seed partly surrounded by a juicy disk.

Fam. 13. PINACEAE. Flowers ament-like or borne in aments; the staminate mostly elongate; the pistillate scaly and 2–several-flowered and ripening into cones or becoming drupe-like.

SUBDIVISION II. ANGIOSPÉRMAE (ANGIOSPERMS)

Trees, shrubs or herbs with leaves of diverse form and texture, or true leaves wanting. Flowers with or without a perianth. Ovule or ovules in a closed ovary which in maturity becomes the fruit.

CLASS I. MONOCOTYLEDÒNEAE (MONOCOTYLEDONS)

Stems without central pith or annular layers, but having the woody fibers distributed through them (a transverse slice showing the fibers as dots scattered through the cellular tissue). Embryo with a single cotyledon, the early leaves alternate. Parts of flower in twos, threes, fours or sixes, not in fives. Leaves chiefly parallel-veined. Our representatives herbaceous or rarely shrubby or (in *Smilax*) with woody flexible stems.

a. *Carpels* 1 *or, if more, becoming distinct when mature or forming an indehiscent inferior ovary; cycles of flower often of unequal number.*

b. *Flowers not in the axils of regularly imbricated scales nor on a fleshy or frondose axis.*

Order IX. PANDANÀLES. Perianth primitive or rudimentary, of bristles or dry scales; flowers monoecious, in heads or dense spikes; seed with abundant endosperm. Two families with us:

Fam. 14. TYPHACEAE. Flowers densely crowded in a terminal spike; perianth represented by slender bristles.

Fam. 15. SPARGANIACEAE. Flowers radiating in mostly scattered heads; perianth represented by flattish scales.

Order X. HELÒBIAE (NAJADÀLES). Perianth herbaceous, petaloid or none; seed without endosperm.

c. *Carpels distinct or becoming distinct in maturity.*

d. *Perianth wanting or of a single herbaceous series.*

Fam. 16. ZOSTERACEAE (POTAMOGETONACEAE). Flowers clustered or in spikes; carpels free from the first; aquatic branching plants with flat leaves.

Fam. 17. NAJADACEAE. Flowers solitary in axils of sheathing leaves; branching aquatics.

Fam. 18. JUNCAGINACEAE. Carpels at first united, separating in maturity; perianth herbaceous; paludal plants with simple ascending stems and terete leaves.

d. *Perianth of 2 series, the outer of 3 herbaceous sepals, the inner of 3 white to pink petals; carpels many, forming disks or rings or covering the receptacle.*

Fam. 19. ALISMATACEAE. Carpels usually 1-ovulate, forming achenes; embryo campylotropous.

Fam. 20. BUTOMACEAE. Carpels many-ovulate, forming follicles; embryo orthotropous.

c. *Carpels united into an obscurely or hardly 3-locular ovary.*

Fam. 21. HYDROCHARITACEAE. Ovary inferior, indehiscent or irregularly rupturing; flowers often subtended by a spathe, with 3 sepals and 3 petals; aquatic; flowers unisexual.

b. *Flowers in the axils of regularly imbricated scales or upon a fleshy or frondose axis.*

Order XI. GLUMIFLÒRAE (GRAMINÀLES). Flowers in the axils of regularly imbricated scales which form dry spikes or spikelets; fruit indehiscent, with copious endosperm back of the straight embryo, forming a grain or achene; grasses and sedges.

Fam. 22. GRAMINEAE (Grasses). Stem (culm) hollow or with readily removed pith, or solid, cylindric or compressed, the closed nodes usually hard; cauline leaves 2-ranked, with sheath usually open (rarely closed) on the side opposite the blade; scales of spikelets usually paired; perianth represented by 2 or 3 inconspicuous rudiments (lodicules) at base of floret; anthers versatile, attached above the sagittate base to the filaments; fruit (grain) a hard seed usually inseparable from the thin pericarp (caryopsis), a utricle or a naked achene.

Fam. 23. CYPERACEAE (Sedges). Culm usually solid and 3-angled, with soft nodes; cauline leaves (when present) 3-ranked; sheaths closed; scales mostly single; perianth wanting or of bristles or scale-like sepals; anthers basifixed; achene naked (but sometimes inclosed in a spathe or pouch), not fused to pericarp.

Order XII. SPATHIFLÒRAE (ARÀLES). Flowers crowded on a fleshy or spongy axis (spadix), this often subtended by a colored spathe, or borne upon tiny floating fronds.

Fam. 24. ARACEAE. Plant leafy, rooting in soil; flowers closely crowded on a spadix, with or without a spathe; perianth wanting or of 4 or 6 sepals; fruit berry-like.

Fam. 25. LEMNACEAE. Plant tiny, floating, consisting of free or colonial flattish, lenticular, strap- or rod-shaped, ellipsoid or globose fronds, with rare flowers projecting from margin or surface.

a. *Carpels 3 or rarely 2, united into a compound ovary; cycles of flower usually of same or multiples of same number.*

e. *Seeds with endosperm (albumen stored within embryo-sac).*

Order XIII. FARINÒSAE (XYRIDÀLES). Endosperm mealy or granular.

Fam. 26. XYRIDACEAE. Scapose plants with narrow basal leaves; flowers in dense conical or cylindric to globose heads, wholly or nearly covered by firm bracts, perfect; corolla of 3 yellow evanescent petals; ovary 1-locular, many-seeded.

Fam. 27. ERIOCAULACEAE. Scapose, with soft basal and slender leaves; flowers crowded on top of receptacle, forming button-like heads, monoecious or dioecious; perianth chaffy, white to drab or fuscous; ovary 2- or 3-locular, each locule with 1 ovule.

Fam. 28. BROMELIACEAE. Scurfy plants; ours a gray elongate slender branching epiphyte hanging from branches of trees; its leaves slender; seeds clavate, raised on a long hairy-tufted stalk.

Fam. 29. COMMELINACEAE. Green-leaved terrestrial rather fleshy herbs; sepals green; petals brightly colored; ovules and seeds orthotropous.

Fam. 30. PONTEDERIACEAE. Green-leaved, terrestrial or aquatic; sepals and petals all petaloid, all colored; ovules and seeds anatropous.

Order XIV. LILIIFLÒRAE (LILIÀLES). Endosperm horny or fleshy.

f. *Ovary superior.*

Fam. 31. JUNCACEAE. Sepals and petals scale-like or glume-like; leaves narrowly lanceolate to filiform, often in basal tussocks, or represented by sheaths; stem unbranched; the rushes.

Fam. 32. LILIACEAE. Sepals (herbaceous only in *Trillium*) colored and more or less petaloid like the petals; plants of various habit, with simple or branching stems or scapose, rarely rush-like.

f. *Ovary partly or wholly inferior.*

Fam. 33. HAEMODORACEAE. Perianth woolly on the outside; stamens 3.

Fam. 34. DIOSCOREACEAE. Flowers small, not brightly colored, smooth or glabrous, chiefly dioecious; slender-stemmed, from stout roots or rhizomes; leaves broad, strongly ribbed and netted-veined; fruit a 3-angled or -winged capsule.

Fam. 35. AMARYLLIDACEAE. Flowers often showy, perfect, the sepals and petals similarly petaloid; stamens 6; anthers introrse; leaves narrow, parallel-veined.

Fam. 36. IRIDACEAE. Similar to Fam. 35 but stamens 3; anthers extrorse.

e. *Seeds without or essentially without endosperm (except in Fam. 38, tiny orchid-like plants with minute seeds with loose testa and undifferentiated embryo); ovary inferior.*

Order XV. SCITAMÍNEAE. Leaves pinnately veined; perianth very irregular, petaloid; fertile stamen 1, the others changed to petaloid staminodia; anther 1-locular; seed with copious perisperm (albumen stored outside embryo-sac). Represented with us only by

Fam. 37. MARANTACEAE.

Order XVI. MICROSPÉRMAE (ORCHIDÀLES). Fertile stamens 1–3; seeds minute, without perisperm, with loose testa and undifferentiated embryo; leaves parallel-veined or wanting.

Fam. 38. BURMANNIACEAE. Flower regular; fertile stamens 3, distinct; style distinct; seeds with endosperm.

Fam. 39. ORCHIDACEAE. Flowers zygomorphic (irregular), with 1 petal (lip) unlike the others; fertile stamens 1 or 2, their filaments fused to the style; seeds without endosperm.

CLASS II. DICOTYLEDÒNEAE (DICOTYLEDONS)

Stem commonly formed of bark, wood and pith; the wood forming a zone between the other two and increasing, when the stem continues from year to year, by addition of a new layer to the outside, next the bark. Leaves mostly netted-veined. Embryo with a pair of opposite cotyledons (or cotyledons very rarely fused or seemingly single). Cycles of flowers mostly in twos, threes, fours or fives or their multiples.

Subclass I. ARCHICHLAMÝDEAE (CHORIPÉTALAE, including APÉTALAE and POLYPÉTALAE)

Petals wanting or distinct (or, if rarely some of them united at base, with the anthers not opening by terminal pores). For Subclass II see p. xxvi.

g. *Petals wanting or rudimentary.*

h. *Flowers chiefly staminate or pistillate, commonly in aments, those of either one or both sexes without a perianth or with an inconspicuous bract-like perianth.*

i. *Herbs; flowers perfect, in a dense spike, of petaloid texture.*

Order XVII. PIPERÀLES. Herbs (rarely shrubs) with jointed stems, alternate entire leaves, mostly perfect flowers without perianth and in close spikes, each flower in the axil of a small bract; carpels 1 or, if more, united; ovules orthotropous, mostly 1.

Fam. 40. Saururaceae. Erect, with broad cordate leaves with converging ribs; flowers white, with 3–5 basally united indehiscent few-ovulate carpels.

i. Trees or shrubs; flowers of one or both sexes in aments, not petaloid.

j. Carpels 2, united, forming a many-seeded dehiscent capsule; seeds hairy-tufted. One order:

Order XVIII. SALICÀLES. One family:

Fam. 41. Salicaceae.

j. Carpel with 1–several ovules, in fruit forming 1-seeded nuts, samaras or drupes; seeds not hairy-tufted.

k. Ovary superior or naked; staminate flowers without a perianth; fruit a drupe or 1-seeded dry nut.

Order XIX. MYRICÀLES. Style forking or with 2 stigmas; fruit a wax- or resin-coated nut with an orthotropous ovule; wood close and hard. One family:

Fam. 42. Myricaceae. Resiniferous or aromatic shrubs and trees.

Order XX. LEITNERIÀLES. Style simple; fruit a drupe with an amphitropous ovule; wood very light and soft. One family:

Fam. 43. Leitneriaceae. Non-aromatic large shrubs or small trees; the wood lighter than cork.

k. Ovary wholly or partly inferior; staminate flowers with or without a perianth; fruit a fleshy-seeded nut or a samara.

Order XXI. JUGLANDALES. Leaves odd-pinnate; ovary with 1 basal orthotropous ovule; nut with invaginated and forking shell, surrounded by an indehiscent or dehiscent and deciduous fibrous-fleshy epicarp (husk). One family:

Fam. 44. Juglandaceae.

Order XXII. FAGÀLES. Leaves simple; ovary with 2–6 pendulous ovules; fruit a 1-seeded nut or samara, without adherent pericarp.

Fam. 45. Corylaceae (Betulaceae). Styles 2; ovary 2-locular, each locule with 2 pendulous ovules; pistillate flowers and fruits in elongate aments or short clusters, subtended by foliaceous involucres or simple to 5-lobed bracts.

Fam. 46. Fagaceae. Styles 3; ovary 3–7-locular at base, each locule with 1 or 2 ovules; pistillate flowers solitary or clustered; fruits subtended by or inclosed in involucres made up of innumerable imbricated bracts.

h. Flowers perfect or unisexual, rarely in aments, with or without a calyx.

l. Seeds without albumen or nearly so.

Order XXIII. URTICÀLES. Ovary superior; fruit indehiscent; leaves with stipules.

Fam. 47. Ulmaceae. Trees or shrubs with alternate leaves; flowers polygamous or polygamo-monoecious; fruit a samara, nut or drupe with pendulous ovule, solitary or in axillary fascicles or racemes; sap watery.

Fam. 48. Moraceae. Trees or shrubs, alternate-leaved; flowers unisexual, the staminate ones in racemes or spikes, the pistillate crowded in heads or spikes; fruit an achene embraced by the finally juicy calyx; sap milky.

Fam. 49. Cannabinaceae. Herbs with mostly opposite dissected or lobed leaves; flowers unisexual; calyx of pistillate flower a single sepal embracing the ovary; filaments erect in bud; stigmas 2, elongate; fruit an achene with pendulous ovule; embryo coiled or curved.

Fam. 50. Urticaceae. Herbs (with us) with opposite or alternate unlobed leaves; calyx of pistillate flowers tubular or cup-shaped, or of 2–5 lobes or sepals; filaments inflexed in bud; style or stigma 1; achene with erect ovule; embryo straight.

l. Seed with copious albumen (except in some halophytic members of Fam. 55 with fleshy exstipulate subterete or reduced leaves, depressed utricle, or flowers immersed in stem, and spiral embryo).

m. Ovary wholly or partly inferior.

Order XXIV. SANTALÀLES. Fruit indehiscent, a drupe, nut or berry; ovary 1-locular; seed without loose testa; plants parasitic.

Fam. 51. SANTALACEAE. Plants terrestrial, their roots attaching to those of the host; leaves green, not heavily coriaceous; ovule or ovules suspended.

Fam. 52. LORANTHACEAE. Plants parasitic on trunks or branches of trees; leaves green to brown, heavily coriaceous or reduced to connate scales; ovule erect.

Order XXV. ARISTOLOCHIÀLES. Fruit dehiscent, a 6-locular capsule; calyx enlarged, often petaloid. One family with us:

Fam. 53. ARISTOLOCHIACEAE. Calyx tubular or campanulate with 3 lobes; seeds numerous with endosperm and often perisperm; leaves broad and entire, often cordate.

m. *Ovary superior.*

Order XXVI. POLYGONÀLES. Fruit a triangular or lenticular achene; embryo straight or curved but not spiral nor annular; stipules sheathing (at least when young). One family:

Fam. 54. POLYGONACEAE.

Order XXVII a. CENTROSPÉRMAE (CHENOPODIÀLES). Fruit a capsule, utricle or berry; embryo crescentic, spiral or annular.

n. *Ovary or its locules each with 1 ovule; fruit indehiscent or circumscissile.*

o. *Parts of flower spirally arranged; perianth herbaceous or scarious; fruit a utricle or rarely a drupe; leaves chiefly alternate, exstipulate.*

Fam. 55. CHENOPODIACEAE. Flowers without scarious bracts; calyx herbaceous.

Fam. 56. AMARANTHACEAE. Flowers subtended by scarious bracts; calyx dry and scarious.

o. *Parts of flower chiefly in whorls or cycles.*

Fam. 57. NYCTAGINACEAE. Calyx tubular or funnelform, petaloid and showy, its persistent and hardened base embracing the single achene; leaves opposite, exstipulate.

Fam. 58. PHYTOLACCACEAE. Calyx with free unmodified petaloid sepals; ovary a ring of several united carpels, in fruit becoming a many-scalloped berry; leaves alternate, exstipulate.

(Fam. 61. CARYOPHYLLACEAE). Sepals herbaceous or scarious, in fruit surrounding the 1-seeded utricle; leaves mostly opposite, often with scarious stipules.

n. *Ovary many-ovulate; fruit a capsule.*

Fam. 59. AIZOACEAE. Parts of flower cyclic; leaves (in ours) mostly opposite or pseudoverticillate.

g. *Petals commonly present, only exceptionally wanting or rudimentary or in some genera of evident relationship represented by showy petaloid sepals; flowers prevailingly perfect.*

p. *Ovary prevailingly superior (the carpels exceptionally on an invaginated receptacle or surrounded by a fleshy disc in some* Rosaceae).

q. *Stamens mostly hypogynous (on the receptacle).*

r. *Ovary single, compound.*

Order XXVII b. CENTROSPÉRMAE (CARYOPHYLLÀLES). With 1 (rarely 3–5) locule; ovules and seeds several to very many, attached to base or central column of capsule; embryo usually coiled or curved. Leading to polypetalous members of

Fam. 60. PORTULACACEAE. Sepals or calyx-lobes 2; stamens 5–∞, when of same number as petals borne opposite them; plants succulent or fleshy.

Fam. 61. CARYOPHYLLACEAE. Sepals or calyx-lobes or -teeth 4 or 5; stamens 4–10; when of same number as petals borne opposite the sepals; plants of dryish texture, rarely succulent.

r. *Ovary simple or, if compound, of more than 1 locule; ovules not curved.*

Order XXVIII. RANÀLES (RANUNCULÀLES). Carpels solitary or distinct, often numerous, spirally arranged; stamens numerous, usually at least twice as many as sepals, spirally or cyclically arranged.

s. *Flowers without true perianth; seed exalbuminous.*

Fam. 62. CERATOPHYLLACEAE. Flowers minute, with green calyx-like involucre; fruit a large achene; brittle aquatics with whorled and dissected foliage.

s. *Flowers with regular calyx and usually a corolla; seeds mostly with albumen.*

t. *Leaves centrally peltate or, if not so, the fruit with locules ovule-bearing over the whole inner surface.*

Fam. 63. NYMPHAEACEAE. Aquatic herbs.

t. *Leaves not centrally peltate; carpels forming achenes, follicles, berries or drupes; ovules on restricted placentae.*

u. *Non-aromatic herbs or shrubs without oil-tubes, mostly acrid, acid, bitter or with poisonous alkaloids.*

Fam. 64. RANUNCULACEAE. Anthers not opening by uplifted lids; fruit an achene, capsule or berry with non-arillate seed; flowers mostly perfect; stem not twining.

Fam. 65. BERBERIDACEAE. Differing from Fam. 64 in having anthers opening by subapical uplifted lids or, if not, with the seeds of the large berry inclosed in a pulpy aril; fruit a capsule or berry.

Fam. 66. MENISPERMACEAE. Fruit a drupe with large often crescentic seed; flowers small, dioecious; flexible stems twining; leaves palmate or marginally peltate.

u. *Aromatic trees or shrubs with oil-tubes in the parenchyma.*

v. *Stamens spirally arranged, very many; anthers not opening by lids.*

Fam. 67. MAGNOLIACEAE. Sepals 3; pistils on an elongate axis, forming a cone-like mass of follicles or samaras; seed with solid albumen; leaves alternate.

Fam. 68. CALYCANTHACEAE. Sepals numerous; pistils inserted on base and inner surface of calyx-tube, forming achenes without endosperm; leaves opposite.

Fam. 69. ANNONACEAE. Sepals 3; petals 3 or 6; pistils few, ripening into pulpy several-seeded fruits; seeds with deeply ruminated or furrowed albumen; leaves alternate.

v. *Stamens in whorls, 6–12; anthers opening by 2 or 4 uplifted valves.*

Fam. 70. LAURACEAE. Fruit a drupe; seed without albumen; leaves alternate, entire or lobed.

r. *Carpels 2–∞, united into a compound ovary, with mostly nerve-like parietal placentae; parts of flower in whorls.**

Order XXIX. RHOEADÀLES (PAPAVERALES). Plants with ordinary foliage, not insectivorous; the leaves neither tubular nor ordinarily covered with stalked glands; sepals mostly 2 or 4 (–7 in Fam. 74).

w. *Sepals 2; embryo minute, at base of fleshy albumen. One family:*

Fam. 71. PAPAVERACEAE (incl. FUMARIACEAE). Herbs with the 2 sepals large and fugacious, the flowers regular, the juice milky or colored; or the 2 sepals scale-like, the flower irregular, the juice watery.

w. *Sepals 4–7; embryo curved or folded, completely filling seed; albumen none.*

Fam. 72. CAPPARIDACEAE. Sepals and petals 4, regular; stamens 6 or more, nearly equal; capsule 1-locular, 2-valved, without transverse partition; leaves stipulate; inflorescence with bracts.

Fam. 73. CRUCIFERAE. Sepals and petals 4, regular; stamens 6, tetradynamous (2 shorter and inserted lower than the other 4); capsule 2-locular by a transverse partition, or indehiscent or transversely jointed; stipules and usually bracts wanting.

* Note that the index-letter r is here used for the third time.

Fam. 74. RESEDACEAE. Sepals and petals 4–7, irregular; stamens indefinite or on a hypogynous disk; capsule 3–6-lobed, subcylindric or barrel-shaped, opening at top.

Order XXX. SARRACENIÀLES. Plants insectivorous; leaves all radical. either tubular or flat and covered with long gland-tipped hairs; sepals 5, persistent.

Fam. 75. SARRACENIACEAE. Leaves hollow, hooded; ovary and capsule 5-locular, with axile placentae; style 1, expanded above into a broad umbrella-like limb.

Fam. 76. DROSERACEAE. Leaves flat to filiform, circinate, the blade bearing many long glands; ovary 1-locular, with parietal placentae; styles 3 or 5, each slenderly 2-parted.

q. *Stamens mostly perigynous (adnate to perianth), epigynous (on the ovary) or inserted on or at the base of a hypogynous disk.*

x. *Carpels solitary or several and distinct (rarely united); stamens commonly perigynous; sepals generally united at base or confluent with receptacle.*

Order XXXI. PODOSTEMONÀLES. Plants thallus-like, attached to inundated rock by haptera (nonvascular attachments), resembling liverworts, algae or lichens; flowers subtended by spathe-like involucres; mostly without perianth, or perianth reduced. A single family:

Fam. 77. PODOSTEMACEAE. A single species with us.

Order XXXII. ROSÀLES. Plants with roots, stems and leaves; the perianth usually of both calyx and corolla.

y. *Endosperm usually copious and fleshy (if scanty, the plant succulent).*

Fam. 78. CRASSULACEAE. Carpels of same number as calyx-segments; endosperm scanty; plant fleshy or succulent.

Fam. 79. SAXIFRAGACEAE. Carpels fewer than calyx-segments, mostly 2; ovules and seeds 2–∞ in each locule of capsule or berry; plants nonsucculent herbs or shrubs.

Fam. 80. HAMAMELIDACEAE. Differing from Fam. 79 in woody 2-locular capsule with 1 seed in each locule; trees or shrubs.

y. *Endosperm none or barely developed (see also Fam. 78 above).*

Fam. 81. PLATANACEAE. Trees with large palmately lobed leaves and tiny monoecious flowers in dry spherical heads; achene slenderly obconic, surrounded by long bristles, with 1 orthotropous pendulous ovule.

Fam. 82. ROSACEAE. Herbs, shrubs or trees with perfect or rarely dioecious regular flowers; corolla usually petaloid and colored, rarely wanting; pistils 1–∞, distinct, or the carpels united and inclosed by calyx-tube (pome) or 1 and forming a drupe; ovules 1–∞, amphitropous or anatropous.

Fam. 83. LEGUMINOSAE. Differing from Fam. 82 in its 1 pistil which forms a legume with 1–several seeds borne in 2 alternating rows along the ventral suture, the legume dehiscent along both sutures; or fruit a loment, with 1-seeded mostly indehiscent segments (articles); corolla regular or (with us) more often irregular.

x. *Carpels united into a compound ovary; sepals mostly distinct.*

z. *Stamens few, rarely more than twice as many as sepals; stamen-bearing disk usually well developed.*

a. Stamens 1–12, rarely more, if of same number as sepals opposite them.

Order XXXIII. GERANIÀLES. Ovules pendulous, with the raphe toward axis of ovary or ascending and with raphe dorsal; stamens mostly borne on base of perianth.

b. Stamens more than 1; perianth or perianth-like structures present; plants terrestrial.

c. Flowers perfect, polygamous or dioecious, with normal perianth; carpels 2, 4, 5, 6, 8 or 10 (if 3 the plants with pellucid-punctate foliage and aromatic oil).

d. Flowers regular.

e. Herbs (with us); flowers perfect.

Fam. 84. LINACEAE. Capsule 5-locular or, by false septa dividing the carpels, becoming 10-locular; fertile stamens as many as petals, united below into a tube; leaves narrow, simple.

Fam. 85. OXALIDACEAE. Capsule splitting along dorsal suture of each of the 5 carpels; fertile stamens twice or thrice the number of petals, radiating at different levels from the tube of united filaments; leaves (in ours) 3-foliolate.

Fam. 86. GERANIACEAE. Carpels 5, slender-beaked, splitting apart at maturity from the central axis and coiling; the body of each dehiscent carpel 1-seeded; leaves palmately lobed or imparipinnate or pinnatisect.

Fam. 87. ZYGOPHYLLACEAE. Carpels 5, separating in maturity and often indehiscent, 1–5-seeded, prickly or muricate; leaves in ours paripinnate.

e. Trees, shrubs or half-shrubs; flowers perfect, polygamous or unisexual.

Fam. 88. RUTACEAE. Leaves punctate-dotted, aromatic or pungently fragrant; hypogynous disk prominent.

Fam. 89. SIMAROUBACEAE. Flowers (in ours) unisexual or polygamous, in terminal panicles; filaments free; fruit an elongate samara; leaves not punctate, simply pinnate.

Fam. 90. MELIACEAE. Flowers (in ours) perfect, in axillary panicles; filaments united into a tube; fruit a globose drupe; leaves not punctate, pinnately decompound.

d. Flowers very irregular.

Fam. 91. POLYGALACEAE. Flowers showy, somewhat papilionaceous; petals 3; stamens commonly 8.

c. Flowers mostly unisexual, apetalous, rarely polypetalous, or modified and with somewhat petal-like colored glands borne on the cup-like involucre; ovary of 3 carpels; leaves not punctate. A single family:

Fam. 92. EUPHORBIACEAE. As defined in last paragraph.

b. Stamen 1; flowers naked or inclosed by membranous bracts.

Fam. 93. CALLITRICHACEAE. Flowers in axils of opposite leaves; the pistillate ones 4-locular and 4-lobed, separating when ripe into the locules; small, often aquatic, herbs.

Order XXXIV. SAPINDÀLES. Ovules pendulous, with raphe away from axis of ovary, or ascending and with raphe ventral; stamens mostly borne on or at base of a hypogynous disk.

f. Flowers regular or essentially so; fruit not elastically dehiscent.

g. Petals wanting (sepals sometimes petaloid); flowers unisexual or polygamous; low evergreen shrubs.

Fam. 94. BUXACEAE. Succulent-stemmed; leaves broad; flowers racemose; fruit (in ours) a 3-horned dehiscent capsule.

Fam. 95. EMPETRACEAE. Stems slender, woody; leaves linear or narrowly oblong, firm, crowded, heath-like; flowers axillary or in terminal heads; fruit a fleshy or dryish berry.

g. Petals normally present; flowers perfect, polygamous or unisexual.

Fam. 96. LIMNANTHACEAE. Weak annual herbs with pinnate leaves; flowers 3-merous, the 3 carpels ripening into achenes.

Fam. 97. ANACARDIACEAE. Trees or shrubs with 3 styles or stigmas; ovary 1-locular, forming a drupe.

Fam. 98. CYRILLACEAE. Trees or shrubs with alternate pinnately veined simple leaves; flowers in elongate slender or spiciform racemes, perfect; ovary 2-locular, each locule 2-ovulate; fruit a fleshy loculicidally dehiscent capsule with slender seeds.

Fam. 99. AQUIFOLIACEAE. Differing from Fam. 98 in its axillary dioecious or polygamous flowers; fruit a berry-like drupe with plump crustaceous seeds.

Fam. 100. Celastraceae. **Seed with a brightly colored aril; fruit a plump capsule; upright or depressed opposite-leaved or twining alternate-leaved shrubs.**

Fam. 101. Staphyleaceae. **Shrubs or small trees, with pinnate 3- or 5-foliolate opposite leaves;** flowers perfect, on jointed pedicels; ovary 3-carpellate, the 3-locular capsule bladdery-inflated; embryo straight.

Fam. 102. Aceraceae. **Trees or large shrubs, with palmately lobed or (in 1 species) pinnate (with 3–7 leaflets) leaves; flowers dioecious or polygamous, on continuous pedicels; ovary 2-locular, 2-lobed, maturing into a pair of finally separating 1-seeded compressed and winged samaras; embryo coiled or folded.**

Fam. 103. Hippocastanaceae. **Trees or shrubs with mostly digitate opposite leaves of 3–9 elongate leaflets; flowers showy, slightly irregular, in a thyrse or panicle, polygamous or staminate; fruit a leathery capsule with large chestnut-like seeds.**

Fam. 104. Sapindaceae. **Trees, shrubs or climbers (rarely herbaceous), with alternate pinnate or bipinnate many-foliolate leaves; fruit a drupe or bladdery capsule.**

f. Flowers very irregular; fruit elastically dehiscent.

Fam. 105. Balsaminaceae. **Posterior sepal enlarged and saccate; valves of dehisced capsule coiling; watery-stemmed herbs.**

a. Stamens as many as sepals and alternate with them; ovules erect.

Order XXXV. RHAMNÀLES. **Flowers cyclic; carpels 2–5, each with 1 or 2 ascending ovules; shrubs (often climbing) or trees.**

Fam. 106. Rhamnaceae. **Shrubs, trees or twining vines with pinnately veined simple leaves; fruit a drupe or capsule.**

Fam. 107. Vitaceae. **Tendril-bearing vines (with us), with palmately veined or divided leaves; fruit a berry.**

z. Stamens many to very many (few in Elatinaceae, *paludal or aquatic annuals, and* 5 *in* Violaceae *with irregular flower and with adnate anthers connivent over the pistil); hypogynous disk scarcely or not at all developed.*

Order XXXVI. MALVÀLES. **Sepals or calyx-lobes valvate; placentae united in the axis of the ovary; pubescence mostly stellate; mucilaginous herbs, shrubs or trees.**

Fam. 108. Tiliaceae. **Sepals deciduous; stamens in several brush-like fascicles; anthers 2-locular; embryo straight; trees, rarely herbs.**

Fam. 109. Malvaceae. **Sepals persistent; stamens monadelphous (filaments united into a tube); anthers 1-locular; embryo curved; mostly herbs and shrubs.**

Order XXXVII. PARIETÀLES (VIOLÀLES). Sepals 5, imbricated or convolute in bud; stamens as many as petals or commonly more numerous, borne on perianth or just below petals; carpels 2, 3 or 5, united; placentae usually parietal, sometimes basal or axile.

h. Styles 3 *or* 5, *distinct or only rarely united above base; placentae mostly basal or axile; seed without or essentially without endosperm; flowers regular.*

Fam. 110. Theaceae (Ternstroemiaceae). Ours trees or shrubs with alternate exstipulate leaves; flowers terminal or axillary, with 1 or 2 basal bracts; flowers with 5 or 6 sepals imbricate in bud and showy silky-backed petals; stamens very many; ovary 5-locular; styles 5, free or united; capsule woody, loculicidally dehiscent.

Fam. 111. Guttiferae (incl. Hypericaceae). Ours opposite-leaved herbs or shrubs; the leaves commonly pellucid-punctate, without stipules; the usually 5 sepals and petals convolute in bud; stamens few–∞, often in fascicles; ovary 3- or 5-locular; styles free or united at base; capsule not woody, septicidally dehiscent.

Fam. 112. Elatinaceae. Small aquatic or paludal herbs with opposite or whorled stipulate leaves; flowers axillary; inconspicuous sepals, petals and stamens persistent in fruit; the capsule 2–5-locular.

Fam. 113. TAMARICACEAE. Heath-like shrubs with small exstipulate alternate leaves; ebracteate 5-merous flowers in slender racemes; capsule 1-locular, with basal or axillary placentae; seed with long terminal tuft of hairs.

h.Styles united nearly or quite to summit or 1 *and simple; placentae parietal; seed with copious endosperm.*

Fam. 114. CISTACEAE. Herbs or half-shrubs with us; upper or all leaves alternate; 2 outer sepals much smaller than inner 3, often bract-like; inner sepals and petals convolute in bud; stamens mostly borne on a thickened zone beneath the ovary; style 1 and undivided or stigma sessile; endosperm copious.

Fam. 115. VIOLACEAE. Herbs with us, with alternate stipulate leaves; flower very irregular, commonly spurred at base; sepals uniform, convolute in bud; anthers connivent about and nearly covering the united (often upwardly enlarged) styles; capsule 3-locular, the three valves with the margins approximate after discharge of the hard and plump seeds.

Fam. 116. PASSIFLORACEAE. Herbaceous vines (with us) with tendrils in the axils of alternate often palmately lobed leaves; the 5 filaments united into a tube embracing the long stalk of the ovary, many sterile filaments between the corolla and the tube of fertile stamens; fruit a 1-locular many-seeded berry.

Fam. 117. LOASACEAE. Herbs with harsh or adhesive pubescence and alternate exstipulate leaves; calyx-tube adherent to the 1-locular ovary and capsule; calyx-lobes and petals convolute in bud, deciduous; stamens borne on throat of calyx; 3 styles somewhat united; capsule dehiscent at summit.

p.Ovary inferior or inclosed in a tubular or campanulate calyx-tube.

Order XXXVIII. OPUNTIÀLES. Fleshy and usually prickly plants with enlarged joints, mostly leafless; sepals and petals imbricated in several series, their bases adherent to summit of 1-locular ovary; the many stamens borne on inside of tube formed by union of sepals and petals. A single family:

Fam. 118. CACTACEAE. As above.

Order XXXIX. MYRTIFLÒRAE (MYRTÀLES). Trees, shrubs or herbs with flowers variously disposed but not in umbels, nor small and very numerous in corymbs; stamens in two whorls, in ours borne on the perianth; style 1, simple (except in the aquatic Fam. 126); placentae axile; endosperm none or scanty.

i.Ovary inclosed in the tubular or campanulate persistent calyx.

Fam. 119. THYMELAEACEAE. Shrubs with nonscurfy leaves; perfect flowers with petaloid calyx and usually no petals; ovary 1-locular, with 1 pendulous ovule, ripening into a drupe.

Fam. 120. ELAEAGNACEAE. Shrubs or trees with scurfy leaves; perfect or dioecious flowers with petaloid calyx; ovary 1-locular, with 1 erect ovule, ripening into an achene surrounded by the pulpy calyx, thus forming a false drupe.

Fam. 121. LYTHRACEAE. Herbs or shrubs (or trees) with mostly opposite or whorled leaves; calyx surrounding the membranous 2–6-locular and 2–∞-ovulate ovary; petals usually present; flowers perfect or dimorphous or trimorphous.

i.Ovary truly inferior or adherent at least to base of calyx.

Fam. 122. NYSSACEAE. Trees with alternate exstipulate leaves; dioecious or polygamous inconspicuous greenish flowers in small terminal (or axillary) heads; staminate flowers with 5-toothed or nearly obsolete calyx, pistillate with small calyx fused to 1–2-locular ovary with solitary pendulous anatropous ovule; fruit a drupe, crowned by remains of calyx; endosperm abundant.

Fam. 123. MELASTOMATACEAE. Herbs (in the Tropics often arborescent) with prominently ribbed opposite leaves; flowers perfect, with showy petals; ovary adherent to base of calyx-tube; anthers elongate, opening by apical pores; capsule 4-locular, with many-seeded placentae projecting from central axis.

Fam. 124. HYDROCARYACEAE. Aquatics with opposite or whorled leaves; calyx-tube short, inclosing base of ovary, its limb 4-parted, the segments becoming spinescent in fruit; fruit a large indehiscent horned nut.

Fam. 125. ONAGRACEAE (OENOTHERACEAE). Herbs (with us) with 2–6-merous symmetrical flowers; tube of calyx closely adherent to the 2–4-locular ovary, its lobes valvate in bud or obsolete; stamens as many or twice as many as petals, inserted at summit of calyx-tube; petals (sometimes wanting) convolute in bud; small anatropous seeds without albumen.

Fam. 126. HALORAGACEAE. Aquatic or paludal herbs (with us), with the immersed leaves mostly dissected; flowers perfect, polygamous or monoecious, with or without 4 petals; stamens 3, 4 or 8; stigmas 3 or 4; fruit indehiscent, 3- or 4-locular, forming a nutlet or nutlets; endosperm none.

Fam. 127. HIPPURIDACEAE. Differing from Fam. 126 in its undivided whorled leaves; perfect or polygamous flowers with entire calyx; stamen and slender style 1; 1-seeded fruit 1-locular.

Order XL. UMBELLIFLÒRAE (UMBELLÀLES). Flowers in umbels or many-flowered corymbs, mostly perfect, with 1 whorl of epigynous stamens; sepals small or much reduced; carpels and styles (or stigmas) 1–5; ovule 1 in each locule, pendulous, anatropous; fruit a 1–∞-seeded drupe or of 2 dry finally separating mericarps; endosperm copious.

Fam. 128. ARALIACEAE. Herbs, shrubs or trees with flowers in umbels; leaves (in ours) compound; flowers perfect, polygamous or dioecious, in umbels or panicles or racemes of small umbels, the 5 small petals epigynous; stamens epigynous; carpels 2–5, forming several-seeded berry-like drupes.

Fam. 129. UMBELLIFERAE. Herbs with mostly hollow internodes and compound or simple leaves (or phyllocladia); flowers in umbels or heads; calyx 5-toothed or entire, its tube fused to the 2-locular and 2-ovulate ovary; the 5 petals and 5 stamens borne on a disk crowning the ovary and surrounding the base of the 2 styles; fruit of 2 seed-like dry carpels (mericarps) cohering by their inner faces but separating in maturity, commonly bearing longitudinal canals filled with aromatic oil.

Fam. 130. CORNACEAE. Trees, shrubs or herbs, with mostly opposite simple leaves; flowers in corymbs (rarely panicles), perfect in ours; calyx minutely 4-toothed; petals 4; stamens 4, borne at base or margin of disk capping the 2-locular 2-ovulate ovary; style 1; fruit a drupe with 2 stone-like seeds.

Subclass II. METACHLAMÝDEAE (GAMOPÉTALAE or SYMPÉTALAE)

Petals more or less united (distinct in the early families of Order *Ericales* in which the 2-locular anthers open by terminal pores).

j. Ovary superior (or, if inferior, with anthers opening by apical pores or chinks).

Order XLI. ERICÀLES. Herbs, shrubs or trees with exstipulate usually alternate simple leaves; flowers perfect, regular or slightly irregular; stamens free or nearly free from base of corolla, as many as and alternate with corolla-lobes or petals, or twice as many; anthers opening by terminal pores or chinks. Ovary 2–10-locular, mostly with axile placentae, the style 1; fruit a capsule, berry or drupe; endosperm abundant.

Fam. 131. CLETHRACEAE. Shrubs or trees with stellate pubescence; flowers in panicles or racemes; petals 5, distinct; stamens 10, the anthers becoming inverted; pollen-grains simple; ovary superior, 3-locular; style 3-cleft; capsule 3-valved, the valves 2-cleft, the many-seeded porrect placentae remaining attached to the columella; seeds with loose transparent cellular coat.

Fam. 132. PYROLACEAE. Herbs or half-shrubs with evergreen leaves, or saprophytic and without green leaves; petals distinct or united; mature anthers usually resupinate; pollen-grains simple or compound; ovary and capsule 5 (or 4)-locular; placentae much as in the last; seed-coat loose and cellular.

Fam. 133. ERICACEAE. Shrubs or trees; calyx free from ovary and the fruit capsular, or united to it and the fruit baccate or drupaceous and crowned by calyx-lobes; corolla gamopetalous (rarely polypetalous); hypogynous disk well-developed; anthers upright; pollen-grains compound (4 united); ovary 2–10-locular, forming a capsule or berry-like drupe; seed-coat close or loose.

Order XLII. DIAPENSIÀLES. Herbaceous or suffruticose, differing from *Ericales* in the 3-locular ovary and absence of stamineal disk; stamens borne on the corolla or connate into a tube, those opposite the corolla-lobes (when present) reduced to staminodia; anther-locules dehiscent by longitudinal, oblique or transverse slits. One family:

Fam. 134. DIAPENSIACEAE.

Order XLIII. PRIMULÀLES. Herbs, shrubs or trees with exstipulate simple leaves; flowers regular, perfect, mostly gamopetalous (in succulent halophytic *Glaux* apetalous but with brightly colored petaloid calyx); sepals distinct or united, mostly herbaceous; stamens of same number as corolla-lobes and inserted on tube or throat opposite them; ovary unilocular, with simple style, the several ovules with 2 integuments and sessile on central placenta; fruit a capsule; embryo surrounded by horny or fleshy albumen. One herbaceous family with us:

Fam. 135. PRIMULACEAE.

Order XLIV. PLUMBAGINÀLES. Differing from *Primulales* in bracteolate inflorescence; persistent calyx membranaceous or scarious; ovary with 1 basal anatropous ovule; styles 5; fruit an achene or utricle; endosperm mealy. A single family:

Fam. 136. PLUMBAGINACEAE. Ours halophytic or xerophytic, with firm rosulate basal leaves.

Order XLV. EBENÀLES. Trees or shrubs with simple alternate leaves; flowers regular, perfect, polygamous or dioecious; ovary 2–∞-locular; seed with hard coat.

Fam. 137. SAPOTACEAE. Trunks, branches and leaves with milky juice (latex); flowers perfect, in axillary or supra-axillary clusters; sepals strongly imbricated; corolla-lobes imbricated in bud; staminodia commonly alternating with the true corolla-lobes; anthers often versatile; ovary superior; style single; fruit baccate or drupaceous, by abortion 1-seeded, the seed with a single bony coat.

Fam. 138. EBENACEAE. Latex wanting; flowers dioecious or polygamous, the staminate with many stamens, the pistillate with some imperfect stamens; ovary superior, its styles as many or half as many as the 2–16 locules, each locule with 1 or 2 pendulous anatropous 2-coated ovules; fruit a berry, with large bony-coated seed or seeds.

Fam. 139. SYMPLOCACEAE. Pubescence simple; flowers axillary, perfect; calyx-lobes imbricated in bud; stamens very many, usually a cluster adnate to base of each petal or corolla-lobe; anthers short and innate; ovary strongly inferior, completely septate, 2–5-locular, the style and stigma entire; fruit a dryish drupe or nut, 1-locular and 1-seeded; embryo terete.

Fam. 140. STYRACACEAE. Differing from Fam. 139 in its stellate or scurfy pubescence; flowers in simple or compound racemes; calyx-teeth very short or obsolete, open in bud; stamens in a single series, the filaments forming a tube borne from base of corolla; anthers elongate, introrse, adnate; ovary septate only below middle; fruit a dry drupe or nut or 3-valved; embryo flat or leaf-like.

Order XLVI. OLEÀLES. Trees or shrubs with opposite simple or pinnate leaves; inflorescences often racemose or paniculate; flowers perfect or unisexual, regular, with calyx small or almost wanting; petals free, united or absent; stamens usually 2, alternate with the 2 2- or 4-ovulate carpels, the ovules pendulous or basal; style 1 or 0; seed with straight embryo and usually fleshy albumen. Represented with us by

Fam. 141. OLEACEAE. Fruit a drupe, samara or tough capsule.

Order XLVII. CONTÓRTAE (GENTIANÀLES). Herbs or shrubs (with us); leaves most often opposite; flowers regular, with 4 or 5 sepals or calyx-segments; corolla of 4 or 5 segments, these mostly convolute in bud; stamens 4 or 5, inserted low on corolla; carpels 2, united, or distinct below the united anthers; ovules with 1 integument and with endosperm.

Fam. 142. LOGANIACEAE. Opposite leaves with stipules or stipular lines connecting their bases; flower 4- or 5-merous except for the 2-locular ovary of 2 wholly united carpels; nectariferous disk wanting or very small; stamens free, pollen of simple grains; style 1, with the terminal stigma capitate or 2- or 4-lobed; fruit a capsule; endosperm copious; juice watery.

Fam. 143. GENTIANACEAE. Herbs differing from Fam. 142 in exstipulate leaves; ovary 1-locular but with 2 introflexed parietal placentae; style 1 or with lobes stigmatose down their inner faces.

Fam. 144. APOCYNACEAE. Plants with abundant latex; leaves exstipulate; flowers 5-merous except for ovary; filaments very short, distinct; anthers introrsely dehiscent, distinct but closely connivent around stigma; carpels 2, distinct or united into an ovary with 2 placentae; style 1, the stigmatic surface on a ring underneath the apex of the single stigma; pollen loose, of simple grains; fruit a follicle; seeds with or without a coma.

Fam. 145. ASCLEPIADACEAE. Differing from *Apocynaceae* in monadelphous stamens (the tube of filaments forming the gynostegium or column), the anthers permanently attached to a large stigmatic body; pollen combined in waxy pollinia or granular masses; carpels united only by the common stigmatic mass.

Order XLVIII. TUBIFLÒRAE (POLEMONIÀLES). Ovary 1, compound (except in *Phrymaceae*), with 2–5 locules or placentae; flowers regular or irregular; stamens distinct, mostly adnate to middle or upper part of corolla.

k. Corolla regular (sometimes slightly zygomorphic in Fam. 149), *the stamens as many as its divisions*, 4 *or* 5.

Fam. 146. CONVOLVULACEAE. Stem commonly twining or trailing, sometimes erect, frequently with milky juice, with green alternate petioled leaves (in parasitic *Cuscuta* the plant without chlorophyll, the leaves reduced to scales); flowers axillary, peduncled or cymose-glomerulate; sepals imbricated; corolla with entire or 4- or 5-lobed margin, plicate; ovary (2 ovaries in *Dichondra*) 2- or 3-locular, each locule with 2 erect anatropous ovules, these becoming large seeds with smooth or hairy testa; embryo incurved, with ample foliaceous plaited and crumpled cotyledons (in *Cuscuta* the long spiral embryo without cotyledons), with little or no surrounding albumen, radicle inferior.

Fam. 147. POLEMONIACEAE. Herbs or half-shrubs with opposite or alternate entire or divided exstipulate leaves and colorless juice; ovary 3-carpellate, with 3 locules, becoming a loculicidal capsule with seeds borne on thick axile placentae; corolla convolute in bud, not plicate, 5-lobed; style 3-cleft or 3-lobed at summit; stigmas introrse; embryo straight, in sparse endosperm.

Fam. 148. HYDROPHYLLACEAE. Herbs with mostly alternate simple, lobed or compound leaves, differing from the *Polemoniaceae* in the ovary 1-locular with 2 parietal placentae or becoming 2-locular by union of the placentae in the axis; styles 2 or 2-cleft; capsule 2-valved; seeds reticulate or pitted; albumen solid; inflorescence often scorpioid.

Fam. 149. BORAGINACEAE. Mostly herbs (with us), with alternate chiefly entire and often harsh leaves; inflorescence cymose, often scorpioid, the mostly uniparous or biparous cymes evolute into unilateral false spikes or racemes; ovary 2-carpellate but often appearing 4-carpellate from the 4 lobes (2-lobed carpels) around the base of the central style; the 4 mature nutlets separate or separable; or ovary not lobed, containing or splitting into nutlets; seed 1, with mostly straight embryo and little or no albumen.

k. Corolla irregular, usually somewhat bilabiate, sometimes barely irregular or even regular; stamens usually fewer than corolla-lobes, 2 *or* 4 (*or of same number,* 5, *in Fam.* 152)*; style undivided; stigma entire,* 2-*lobed or bilamellar; ovary of* 2 *carpels,* 2- *or* 4-*locular, rarely* 3–5-*locular.*

l. Carpels 1- *or* 2-*seeded, separating as nutlets, or drupaceous.*

Fam. 150. VERBENACEAE. Ovary 2- or 4-locular, in fruit separating into 2 or 4 nutlets, the style apical; stamens 4 and didynamous or 2; hypogynous disk rarely developed; flowers axillary or in racemes, spikes or heads, often with bracts; fruit dry or a berry-like drupe; ovule erect, from base of each locule; endosperm scanty or none; herbage mostly not aromatic.

Fam. 151. LABIATAE. Differing from *Verbenaceae* in the ovary deeply 4-lobed around the style, the lobes separating as seed-like nutlets; hypogynous disk often prominent; stem usually 4-angled; flowers often in cymes, the pair of cymes in opposite leaf-axils forming false whorls; herbage usually aromatic, containing immersed oil-glands.

l. Carpels several-seeded and forming berries or capsules, not separating as nutlets, or 1-*seeded and forming an achene.*

m. Carpels several-seeded, forming capsules or berries.

Fam. 152. SOLANACEAE. Herbs or shrubs (with us), with alternate or unequally paired exstipulate leaves and inflorescences not truly axillary; flower slightly zygomorphic or regular; corolla-lobes or margin induplicate-plicate (sometimes imbricate) in the bud; stamens of same number as corolla-lobes or one of them rudimentary; anthers 2-locular, opening by slits or pores; ovary 2-locular or falsely 3–5-locular, the many ovules on thickened axile placentae; style simple, stigma entire or 2-lamellar; seeds mostly with incurved or coiled embryo in fleshy albumen.

Fam. 153. SCROPHULARIACEAE. Herbs or half-shrubs (rarely trees) with exstipulate leaves and axillary flowers or cymose or racemose inflorescences; flower strongly to but slightly zygomorphic; calyx 4- or 5-parted; corolla imbricate or convolute (not plicate) in bud; ovary and capsule completely 2-locular, the placentae median on the partition; seeds often very small and abundant, with straight or slightly curved embryo in copious fleshy albumen; cotyledons rarely broader than radicle.

Fam. 154. BIGNONIACEAE. Trees or shrubs with mostly opposite decussate leaves and showy zygomorphic campanulate or funnelform corollas; ovary and capsule 2-locular by the extension of a partition beyond the two parietal placentae or sometimes 1-locular; seeds large, mostly winged, transverse, filled by the horizontal embryo, without albumen; the broad cotyledons foliaceous, plane, emarginate at both ends, the basal notch including the short radicle.

Fam. 155. MARTYNIACEAE. Annual herbs with rounded petioled opposite leaves; calyx bracteolate, inflated, deciduous; anthers gland-tipped, their locules divergent; ovary 1-locular, with 2 parietal placentae meeting in the axis, thence diverging into lamellae bearing 1 or 2 rows of ovules; fruit beaked, fleshy-drupaceous, the fleshy exocarp finally 2-valved and deciduous, the fibrous-woody endocarp 2-valved through the beak, indehiscent below.

Fam. 156. OROBANCHACEAE. Root-parasites, essentially without chlorophyll, the leaves reduced to colored scales, the flowers axillary or in terminal racemes; ovary and capsule 1-locular, with 2 or 4 parietal many-ovuled placentae; seeds with fleshy albumen, the minute embryo without differentiation of parts.

Fam. 157. LENTIBULARIACEAE. Aquatic or paludal insectivorous or carnivorous plants with scapes or scapiform peduncles; bilabiate corolla with prominent palate (personate) and usually a spur; stamens 2, the anthers confluently 1-locular; ovary 1-locular, with a free central many-ovulate placenta; capsule subglobose, usually irregularly bursting; seeds without endosperm, filled by the solid embryo.

Fam. 158. ACANTHACEAE. Herbs or shrubs (rarely trees), with opposite exstipulate decussate leaves and usually with cystoliths in epidermis or parenchyma of leaves and stems; anthers often hairy or spurred, their halves frequently separated and borne at different heights; ovary 2-locular, with axile placentae, with 2–10 ovules in each locule; capsule loculicidal, firm, elastically dehiscent; seed without endosperm, often mucilaginous (when moistened), ours flat and roundish, clasped by retinacula.

m.Carpel 1, *maturing as an achene.*

Fam. 159. PHRYMACEAE. Opposite-leaved herbs; flowers abruptly reflexed in the spike after anthesis; calyx slender, constricted at throat, with subulate teeth, the 3 upper teeth prolonged as hooks; carpel 1, maturing as an achene; seed without albumen; radicle superior, the broad cotyledons convolute around their axis.

Order XLIX. PLANTAGINÀLES. Corolla scarious and nerveless, its lobes imbricated in the bud; flowers tetramerous, regular; calyx imbricated; stamens 4 or fewer; style entire; ovary and circumscissile capsule (pyxidium) 1- or 2-locular, the locules sometimes with false septa (thus doubling apparent locules); seeds mostly amphitropous and peltate, the straight embryo in firm fleshy albumen. One family:

Fam. 160. PLANTAGINACEAE. Acaulescent or caulescent herbs with rosulate basal leaves often strongly ribbed; the flowers chiefly in axillary peduncled spikes, or in one genus flowers axillary.

j.Ovary inferior or half-inferior.

Order L. RUBIÀLES. Herbs, shrubs or trees with opposite or whorled leaves; stamens of same number as or fewer than corolla-lobes (or in *Adoxaceae* stamens doubled), borne on the corolla and alternate with the lobes; style or styles simple or cleft above; ovule or ovules anatropous; albumen usually abundant.

Fam. 161. RUBIACEAE. Flowers in ours regular, 5- or 4-merous, except for the frequently 2 carpels; leaves simple, entire, with stipules between them or at bases of petioles or with whorled leaves (the extra leaves considered as replacing stipules); calyx-segments open in aestivation; corolla funnelform to rotate, with stamens borne on the tube; anthers introrse; a disk commonly developed on summit of ovary; fruit a loculicidal capsule or a drupe or berry.

Fam. 162. CAPRIFOLIACEAE. Differing from Fam. 161 in exstipulate opposite frequently toothed leaves (or stipular growths at bases of petioles in pinnate-leaved *Sambucus*); corolla regular or zygomorphic; fruit indehiscent or rarely a septicidal capsule.

Fam. 163. ADOXACEAE. Herbs with ternately compound radical exstipulate leaves; inflorescence capitate; stamens a pair below each sinus of the rotate 4–6-cleft corolla, each with a peltate anther.

Fam. 164. VALERIANACEAE. Herbs (rarely shrubs) with exstipulate simple or pinnate leaves; flowers irregular or unsymmetrical; stamens 1–4, fewer than corolla-lobes; ovary with 1 locule containing a single suspended ovule with exalbuminous seed and usually two empty locules; calyx modified to a ring or developing into pappus-like plumes; fruit achene-like.

Fam. 165. DIPSACACEAE. Herbs (rarely shrubs) with exstipulate leaves; flowers in involucrate heads, each embraced by a pair of united bracteoles forming an epicalyx; stamens 4 or some of them aborted; the dry indehiscent 1-seeded fruits inclosed by the enlarged epicalyx.

Order LI. CUCURBITÀLES. Mostly tendril-bearing succulent herbs, with long-petioled exstipulate alternate palmately nerved leaves; flowers unisexual, regular, 5-merous except for the 1–3-locular ovary; stamens usually united by anthers or filaments; fruit fleshy or membranaceous, indehiscent; seeds large, filled by the leaf-like cotyledons. One family:

Fam. 166. CUCURBITACEAE.

Order LII. CAMPANULÀTAE (CAMPANULÀLES). Exstipulate herbs or shrubs (or trees) with epigynous flowers; calyx of sepals or sepaloid lobes in families with exinvolucrate inflorescences, changed to pappus in involucrate plants; anthers free or connate into a tube around the style; fruit a capsule or modified achene (cypsela).

Fam. 167. CAMPANULACEAE (incl. LOBELIACEAE). Herbs (with us) with alternate leaves and often with latex; flowers solitary or in racemes, panicles or cymes, not in involucrate heads; calyx normal; corolla regular or zygomorphic; anthers free or united into a tube around the brush-like summit of the style; ovary 2–5 (–10)-locular with many anatropous ovules on axile placentae; fruit a capsule; embryo straight, in fleshy endosperm.

Fam. 168. COMPOSITAE. Herbs or shrubs (with us); flowers aggregated on the receptacle, forming heads subtended by an involucre; calyx-teeth changed to a pappus of bristles, scales or awns; corolla tubular or campanulate to strap-shaped (ligulate); anthers often forming a tube around the style (sometimes free); fruit an achene covered by the closely adherent calyx-tube (a cypsela).

Artificial Analytical Key to the Families

(CARRIED OUT, IN SOME CASES, TO SUBFAMILIES, GENERA OR SPECIES)

DIVISION I. PTERIDÓPHYTA (VASCULAR CRYPTOGAMS)

Rush-like, fern-like, moss-like or quill-leaved plants without true flowers. Reproduction chiefly by spores (without embryos).

*a.*Terrestrial or submersed plants rooting in the ground (not free-floating). . . *b.*

*b.*Stems conspicuously jointed, their nodes covered by sheaths composed of basally united scarious leaves, otherwise leafless; sporangia borne on inner surface of peltate scales of terminal spike-like cones. . . . . EQUISETACEAE, p. 3

*b.*Stems not conspicuously jointed, bearing green leaves or leaf-like fronds. . . *c.*

*c.*Leaves small, very numerous and imbricated or quill-like and crowning a short corm-like stem; sporangia sessile, axillary or nearly so.

Stem elongate, covered by persistent small leaves; sporangia near base of upper leaf-surfaces or in terminal bracted (when present) spikes or strobiles.

Leaves without a ligule, in 4–16 ranks; strobiles terete; spores uniform, minute. LYCOPODIACEAE, p. 10

Leaves ligulate, in 4–6 ranks; strobiles (in ours) 4-sided; sporangia of two kinds, some containing very many minute microspores (male), others with fewer and larger macrospores (female). . . . . . SELAGINELLACEAE, p. 15

Stem short, thick and corm-like, crowned by a rosette of quill-like leaves; spores of two kinds. . . . . ISOËTACEAE, p. 16

*c.*Leaves (fronds) not closely imbricated or, if very narrow and slightly imbricated, without axillary sporangia. . . *d.*

*d.*Fronds simple or variously cleft, not 4-foliolate; sporangia not inclosed in basal sporocarps. . . *e.*

*e.*Fertile fronds or fertile portions of fronds conspicuously unlike the sterile. . . *f.*

*f.*Sterile fronds linear-filiform, tortuous, crowded on a short crown; fertile frond filiform, tipped by a one-sided short fruiting portion with few obliquely ascending crowded finger-like pinnae 2–4 mm. long. . *Schizaea*, p. 25

*f.*Sterile fronds or portions of fronds dilated, not crowded; fertile frond or portion of frond larger than in preceding genus. . . *g.*

*g.*Stem twining, filiform; leaves alternately in pairs, the blades palmate. *Lygodium*, p. 25

*g.*Stem not twining nor filiform; the leaves or portions of frond not alternately paired. . . *h.*

*h.*Sterile half of frond simple, the fertile a long-stalked simple spike of 2 rows of coherent sporangia. . . . . *Ophioglossum*, p. 23

*h.*Sterile frond or portion of frond usually divided; fertile frond or portion of frond variously divided. . . *i.*

*i.*Sporangia naked; sterile frond or portion of frond divided (if simple, the small and weak plant with single frond from a scarcely rhizomatous base); the fertile variously divided and with distinct globose sporangia.

Rhizome scarcely developed, the 1 or 2 fronds arising from above the fleshy roots; frond consisting of a sterile lower half and a fertile upper panicle or spike; sporangia 2-ranked, their walls of several layers. . . . . *Botrychium*, p. 20

Rhizome stout, the fronds clustered at its summit; sporangia not 2-ranked, with a single layer of cells. . . . . OSMUNDACEAE, p. 24

*i.*Sporangia partly or wholly covered by the rolled-up modified pinnules, forming globular berry-like divisions of the stiff fertile frond; rhizome elongate.

Fronds in vase-like clumps, the tall sterile ones lanceolate and regularly pinnate, with free veins, surrounding the simple pinnate fertile ones. . . . . *Pteretis*, p. 30

Fronds solitary or scattered, the sterile ones coarsely pinnatifid and deltoid-ovate, with veins forming a mesh, the fertile bipinnate. *Onoclea*, p. 31

*e.*Fertile fronds or fertile portions of fronds similar to (though sometimes of different size from) the sterile.

Sporangia sessile at base of a bristle-like receptacle, with an oblique ring; frond filmy, of a single layer of cells. . . . . . . HYMENOPHYLLACEAE, p. 25

Sporangia stalked, without bristle-like receptacle, with an incomplete jointed vertical ring; frond of more than 1 layer of cells. . POLYPODIACEAE, p. 26

*d.*Fronds long-petioled, 4-foliolate; sporangia inclosed in crustaceous sporocarps borne on the slender rhizome. . . . . MARSILEACEAE, p. 50

*a.*Floating moss-like small plants with very small 2-ranked simple fronds (leaves); sporangia inclosed in soft sessile or subsessile sporocarps borne on under side of axis. SALVINIACEAE, p. 51

DIVISION II. SPERMATÓPHYTA (SEED-PLANTS, FLOWERING PLANTS)

Plants ordinarily with true flowers containing stamens or pistils or both. Reproduction typically by seeds containing an embryo or minute plant (exceptions, with sterile flowers or reproduction vegetative, noted in the key).

SUBDIVISION I. GYMNOSPÉRMAE (GYMNOSPERMS)

Woody plants, ours with needle-like, slender or scale-like leaves; wood often pitchy or gummy; flowers mostly unisexual, usually without a perianth; ovules naked, not inclosed in an ovary.

Leaves linear, spreading in 2 ranks, yellowish-green beneath, without resin-ducts; staminate flowers globular, stalked, of a few naked stamens with 5–9 pollen-sacs; pistillate flower of imbricate scales, the uppermost scale subtending 1 ovule above a disk; fruit a red pulpy disk with a central hard ovoid seed. . . . . TAXACEAE, p. 52

Leaves linear, needle-like or scale-like, with resin-ducts; flowers cone-like or in cones; staminate scales with 2–6 2–5-locular anthers; pistillate flower cone-like, with 1 or 2 ovules at upper side of each scale; fruit a cone, or drupe-like and bluish and consisting of coalescent thick scales. . . . . PINACEAE, p. 52

SUBDIVISION II. ANGIOSPÉRMAE (ANGIOSPERMS)

Woody or herbaceous, with leaves of various forms or true leaves wanting; flowers with or without a perianth; ovule or ovules in a closed ovary which in maturity becomes the fruit.

CLASS I. MONOCOTYLEDÒNEAE (MONOCOTYLEDONS)

Stems without central pith or annular layers, but having the woody fibers distributed through them (a transverse section showing the fibers as dots scattered through the cellular tissue). Parts of flower in twos, threes, fours or sixes, not in fives. Leaves chiefly parallel-veined. Ours herbaceous, or rarely shrubby (in *Smilax*) with flexible woody stems.

*a.*Plants free-swimming or floating, small or minute, distinct or adhering in mats or colonies, the fronds flattish or lenticular, globose or strap- or rod-like; rare flowers borne in the margin or on the surface. . . . . LEMNACEAE, p. 385

*a.*Plants with leaves and stems, rooting in the substratum. . . *b.*

*b.*Flowers normally developed; reproduction by seeds. . . *c.*

*c.*Carpels 1 or, if more, becoming distinct or separated when mature; cycles of flower often of unequal number. . . *d.*

*d.*Flowers not in the axils of regularly imbricated scales nor on a fleshy axis. . . *e.*

*e.*Flowers unisexual, monoecious, in dense spikes or heads; perianth of hair-like bristles or of dry scales; seed containing endosperm.

Flowers densely crowded in a terminal spike, the lower half pistillate, the summit staminate; perianth of slender bristles. . . . . . TYPHACEAE, p. 60

Flowers in globose heads, the upper heads staminate, the lower ones pistillate; perianth of flat scales. . . . . SPARGANIACEAE, p. 61

*e.*Flowers not densely spicate nor in heads, if loosely spicate perfect; perianth herbaceous, petaloid or wanting; seed without endosperm. . . *f.*

*f.*Carpels distinct or separating at maturity, superior. . . *g.*
*g.*Perianth wanting or of a single herbaceous series.
Aquatic branching plants with flat to linear-filiform leaves; carpels free from the first.
Flowers clustered or in spikes. . . . . ZOSTERACEAE, p. 64
Flowers solitary in axils of sheathing leaves. . . . . NAJADACEAE, p. 81
Paludal plants with simple ascending stem and terete or quill-like leaves; perianth herbaceous. . . . . . JUNCAGINACEAE, p. 82
*g.*Perianth of 2 series, the outer of 3 herbaceous sepals, the inner of white to pink petals; carpels many, forming disks or rings or covering the receptacle.
Carpels mostly 1-ovulate, forming achenes (indehiscent). ALISMATACEAE, p. 83
Carpels many-ovulate, forming follicles (dehiscent). . . BUTOMACEAE, p. 92
*f.*Carpels at first united into a superior ovary, 3–6, becoming distinct at maturity; flowers perfect, in racemes or spikes; leaves terete. JUNCAGINACEAE, p. 82
*d.*Flowers in the axils of regularly imbricated scales or upon a fleshy axis. . . *h.*
*h.*Flowers in the axils of regularly imbricated dry scales forming spikes or spikelets; fruit an achene or modified achene (grain or caryopsis) with abundant endosperm; leaves narrow, parallel-veined.
Stem (culm) commonly hollow or with readily removed pith, cylindric or compressed, with closed hard nodes; cauline leaves 2-ranked, the sheath usually open (sometimes closed); scales of spikelet usually paired; perianth obsolete; anthers versatile; fruit usually a caryopsis (achene inseparable from a thin pericarp). . . . . GRAMINEAE, p. 94
Stem (culm) usually solid and 3-angled, with soft nodes; cauline leaves (when present) 3-ranked, with closed sheath; perianth wanting or of bristles or scales; anthers basifixed; achene free, not fused to pericarp. . CYPERACEAE, p. 236
*h.*Flowers crowded upon a fleshy or spongy axis (spadix), this often subtended by a spathe; fruit a berry or berry-like; leaves narrow to broad, parallel- or netted-veined. . . . . ARACEAE, p. 381
*c.*Carpels 3 *or rarely 2*, united into a compound ovary; cycles of flower usually of same or multiples of same number. . . *i.*
*i.*Ovary superior. . . *j.*
*j.*Perianth of dry white to drab or fuscous chaff-like scales; flowers crowded on summit of receptacle and forming small button-like heads. ERIOCAULACEAE, p. 390
*j.*Perianth green or colored; flowers not in button-like heads. . . *k.*
*k.*Stamens all alike and fertile. . . *l.*
*l.*Plant moss-like, elongate, loosely branching, epiphytic, hanging from trees, gray-scurfy; seeds clavate, raised on a long hairy-tufted stalk. BROMELIACEAE, p. 391
*l.*Plant neither moss-like nor scurfy. . . *m.*
*m.*Plant a rush, with slender leaves or leaves reduced to sheaths; perianth dry, scarious to firm, green to brown or purplish. . . . JUNCACEAE, p. 397
*m.*Plants of various habit; perianth herbaceous or petaloid. . . *n.*
*n.*Inflorescence a cone-like terminal head, the flowers covered by the hard and dry closely imbricated bracts; sepals narrow and glume-like; petals yellow, fugacious; leaves basal; scape naked. XYRIDACEAE, p. 387
*n.*Inflorescence various, not cone-like; sepals like petals or herbaceous.
Sepals green; petals brightly colored.
Stem bearing alternate parallel-veined leaves; flowers several or many; fruit a capsule. . . . . COMMELINACEAE, p. 392
Stem with a whorl of usually 3 netted-veined leaves near summit, these subtending the solitary peduncled or sessile flower; fruit a berry. . . . . *Trillium*, p. 443
Sepals colored like the petals . . . . LILIACEAE, p. 420
*k.*Stamens dissimilar, some of them reduced, modified or sterile.
Sepals green, petals brightly colored; plants terrestrial. . COMMELINACEAE, p. 392
Sepals and petals similar and similarly colored; plants aquatic or subaquatic. . . . . PONTEDERIACEAE, p. 395
*i.*Ovary wholly or partly inferior. . . *o.*
*o.*Aquatic, with solitary unisexual flowers borne from spathes; carpels united into an indehiscent or finally rupturing leathery capsule; seeds without endosperm. . . . . HYDROCHARITACEAE, p. 93
*o.*Terrestrial or paludal plants with mostly perfect or clustered flowers; fruit a dehiscent capsule or a berry. . . *p.*
*p.*Stamens 3 or more; the flower regular or nearly so. . . *q.*
*q.*Leaves broad and netted-veined; flowers numerous, in panicles or racemes, unisexual, the staminate ones very small, the pistillate ripening into 3-angled capsules; stem erect or twining. . . . . . DIOSCOREACEAE, p. 451

*q.*Leaves narrow, parallel-veined; flowers perfect. . *r.*
*r.*Perianth densely woolly outside, the flowers in terminal corymbs.
Stamens 3; anthers introrse; juice of plant red. . . HAEMODORACEAE, p. 451
Stamens 6; anthers introrse; juice not red. . . . *Lophiola*, p. 456
*r.*Perianth not woolly outside.
Stamens 6; anthers introrse. . . . . AMARYLLIDACEAE, p. 452
Stamens 3.
Perennials with narrow 2-ranked equitant leaves; perianth dropping from summit of capsule; seeds rather large, firm, with close testa. IRIDACEAE, p. 456
Annual tiny epigaean herbs or subterranean perennial saprophytes with tiny scale-like leaves; seeds minute, soft, with loose testa. BURMANNIACEAE, p. 463
*p.*Stamen (at least the fertile) 1, or in *Cypripedium* 2; flowers very irregular.
Anther 1-locular; seed 1, large, with copious albumen around the embryo. MARANTACEAE, p. 462
Anther 2-locular; seeds innumerable, tiny, without endosperm. ORCHIDACEAE, p. 463
*b.*Flowers undeveloped or aborted; reproduction by vegetative buds, bulblets or sprouts (distortions due to aphids, nematodes, fungi, etc., not included). . *s.*
*s.*Leaves or sheaths narrow and elongate; plant not twining. *t.*
*t.*Inflorescence or reproductive bodies compact or a compact cluster at summit of naked or few-leaved stem. *u.*
*u.*Plant with bulbous base.
Plant with strong odor of onion; bulb with thin coats; inflorescence an umbel of small bulbs; onions. . . . . . . *Allium*, p. 429
Plant without onion-odor; bulbs crowded, hard and tough, with the flat leaves forming turf; inflorescence a panicle of narrow-leaved buds; a grass. *Poa bulbosa*, p. 120
*u.*Plant not bulbous-based. *v.*
*v.*Scape terminated by a single small cone of firm imbricated broad scales, these subtending tufts of small leaves. *Xyris caroliniana*, form, p. 389
*v.*Scape or stem terminated by a branching modified inflorescence or by several flowers.
Flowers of 3 green sepals and 3 white or pink petals; carpels numerous, covering summit of receptacle, aborting in northern half of range; reproduction by subterranean tubers. . *Sagittaria graminea* and *S. Eatoni*, p. 89
Flowers (aborted) often replaced by hard elongate corms or leafy tufts in a terminal branching cluster.
Modified inflorescence an umbel subtended by leafy bracts; flat scaly spikelets like 2-edged combs often present. . . . . *Cyperus dentatus*, p. 244
Modified inflorescence a panicle, with scaly bracts at base of leafy tufts; narrow-leaved grasses. . . . . . . *Festuca*, p. 104
*t.*Reproductive bodies not in compact clusters at the summit of a scape.
Aquatic with elongate submersed stems, flat leaves with stipulate bases and terminal or axillary hardened tufts of leaves and stipules (winter-buds). *Potamogeton*, p. 65
Terrestrial or aquatic, with slender leaves consisting of 2 or more tubes; diffuse panicles of remote tiny flowers, these often replaced by subulate bulblets. *Juncus pelocarpus*, etc., p. 415
*s.*Leaves broad, cordate, netted-veined; stem twining; rounded bulbs or clusters of bulbs in leaf-axils. . . . . *Dioscorea Batatas*, p. 452

CLASS II. DICOTYLEDÒNEAE (DICOTYLEDONS)

Stem commonly formed of bark, wood and pith; the wood forming a zone between the other two and increasing, when the stem continues from year to year, by additions of a new layer to the outside, next the bark. Leaves mostly netted-veined. Embryo with a pair of opposite cotyledons (or cotyledons rarely fused or seemingly single). Cycles of flowers mostly in twos, threes, fours or fives or their multiples.

A.Flowers normally developed, with stamens or carpels or both; reproduction by seed. . B.
B.Flowers with only 1 floral envelope (calyx or sometimes corolla) or none. . . C.
C.Flowers unisexual, either staminate or pistillate (groups with perfect or hermaphrodite and unisexual flowers mixed or on the same inflorescence not here included). . . D.
D.Parasitic on branches of trees; the stem and branches jointed; fruit a pulpy or leathery berry. . . . . LORANTHACEAE, p. 562

D. Not parasitic on trees, rooting in the ground (rarely free-swimming). . . . E.

E. Trees, shrubs or woody climbers. . F.

F. Leaves pinnate or deeply pinnatifid (excluding oaks). . G.

G. Leaves merely pinnatifid with rounded lobes, very aromatic, with semicordate stipules; monoecious, with staminate flowers in slender aments; pistillate flowers in globular bur-like heads; ovary and small shining nut overtopped by 8 linear-bristleform soft bracts; small shrub. . *Comptonia*, p. 525

G. Leaves pinnate; large shrubs or trees.

Leaves alternate, commonly aromatic.

Branches and petioles not prickly; leaves not punctate-dotted; monoecious; staminate flowers in pendulous aments; pistillate flowers in short racemes or spikes, each surrounded by a 3- or 4-lobed involucre; fruit a large hard-shelled nut surrounded by a husk, the fleshy kernel lobed; trees. . . . . JUGLANDACEAE, p. 525

Branches prickly or spiny; leaves punctate-dotted; flowers all in panicles or corymbs, not involucrate; fruit 2-valved, the thin outer coat (exocarp) soon separating and exposing the 2 lustrous black large seeds; shrub or small tree. . . . . *Xanthoxylum*, p. 951

Leaves opposite, not pellucid-punctate nor aromatic; sepals or calyx-lobes 4 or 5 or obsolete; ovary 1- or 2-locular; styles or stigmas 2; fruit a samara.

Leaflets 3 or 5; staminate inflorescences loose fascicles from below leafy tip, the flowers on pendulous hairy pedicels; pistillate inflorescence a raceme; sepals distinct; ovary deeply 2-lobed, with 2 long styles; fruit of 2 inequilateral 1-winged samaras. . . ACERACEAE, p. 984

Leaflets mostly 5–11; staminate inflorescences dense panicles from axils of last year's leaves, pistillate a panicle; sepals united below or obsolete; ovary not lobed, tapering to a long 2-cleft style; fruit equilateral, 2 (or 3)-winged. . . . . *Fraxinus*, p. 1147

F. Leaves simple or merely lobed. . . H.

H. Leaves dilated, not closely imbricated, mostly more than 1 cm. long. . . I.

I. Ascending or upright trees and shrubs, not twining. . . J.

J. Both staminate and pistillate flowers in aments, heads, dense spikes or cone-like structures. . . K.

K. Pistillate flowers without a calyx. . . L.

L. Pistillate inflorescence a short-stalked or sessile ament, glomerule or woody cone; staminate inflorescence an ament; leaves not palmately lobed. . . M.

M. Bracts of ament unlobed, free and deciduous, or none.

Ovary and capsule 2–4-valved; seeds very many, hairy-tufted; staminate flowers subtended by 1 or more glands or by a cup- or saucer-like disk; filaments simple, with 2-locular anther; young leaves stipulate. . . SALICACEAE, p. 487

Ovary and fruit 1-ovulate, ripening into a drupe or plump to globose nut subtended by basal bractlets; staminate flowers without basal gland or disk.

Pistils ripening into crowded wax-coated or resin-dotted nuts with entire basal bractlets; filaments united at base; leaves short-petioled, aromatic; overwintering buds not tomentose. . . . . *Myrica*, p. 523

Pistils ripening into few remote smooth elongate drupes subtended by glandular-ciliate bractlets; filaments free; leaves long-petioled, not aromatic; overwintering buds tomentose. . . . . LEITNERIACEAE, p. 525

M. Bracts of pistillate ament upwardly dilated, usually 3- or 5-lobed at summit, free and deciduous or persistent and forming woody cone-like aments; carpel flat, without basal bractlets, forming a flattish usually thin winged nutlet; staminate aments pendulous, dry, dark-colored, the filaments unforked and with 2-locular anther or forking and with each fork with 1 locule of the anther. . . CORYLACEAE, p. 530

L. Pistillate inflorescence a long-stalked pendulous spherical head of confluent 2-beaked ovaries, these in fruit hard and filled with few perfect winged seeds and many aborted sawdust-like ones; staminate heads in a raceme; long-petioled leaf palmately lobed. . . . . *Liquidambar*, p. 752

K. Pistillate flowers with a calyx.

Pistillate flowers at base of slender moniliform stiffish staminate ament, within an involucre which becomes a prickly bur; their calyces slender tubes overtopping ovary, the summit lobed, the throat bearing sterile stamens; nuts large, lenticular. *Castanea*, p. 540

Pistillate flowers in independent aments, spikes or heads.

Pistillate inflorescence an ament with foliaceous or bladdery bracts; the calyx adnate to ovary and merely short-toothed at summit; fruit a hard-shelled nut; staminate ament pendulous, slender, brown-bracted, the flowers apetalous; each fork of filament with 1 locule of the anther. . CORYLACEAE, p. 530

Inflorescence a dense spike or globose head of crowded flowers with basal calyx of 4 unequal sepals; stamens 4, with inflexed filaments. . . . . MORACEAE, p. 554

J. Pistillate flowers not in aments, very dense heads or very dense spikes. . . N.

N. Calyx wanting.

Pistillate inflorescence a few-flowered head-like scale-covered small cluster, with 2-cleft red styles protruding; fruit a nut covered by a foliaceous toothed involucre; staminate aments drab, dry, pendulous, each bract with 4–8 bifurcate stamens; leaves and scars alternate. . . . . *Corylus*, p. 530

Pistillate and staminate inflorescences small panicles or racemes, the pistillate of naked pedicelled flowers, the staminate in close bracted groups; style simple, stigma 2-lobed; fruit a drupe; stamens 2–4, simple; leaves opposite. . . . *Forestiera*, p. 1150

N. Calyx present, at least in flowers.

Calyx adnate to ovary; pistillate flowers without rudimentary stamens; seed with copious albumen; bark non-aromatic.

Styles 3–5; ovary subtended or covered by an accrescent involucre which in fruit becomes a scaly cup or 4-parted bristly bur; fruit a nut; staminate flowers in drooping moniliform or globular aments. . . . . . FAGACEAE, p. 539

Style 1; fruit a drupe, not covered by or embraced by a bur or scaly cup; staminate flowers not in aments.

Inflorescence a spike and leaves alternate, or pistillate flowers solitary and short-stalked in axils of opposite leaves; calyx-teeth or -lobes definite at summit of adherent calyx in both flower and fruit. . SANTALACEAE, p. 560

Inflorescence of 1–few crowded sessile flowers at tip of slender spreading to drooping peduncles from axils of alternate leaves; calyx-teeth not evident at summit of drupe. . . . . . . NYSSACEAE, p. 1048

Calyx free, of broad sepals or lobes.

Leaves not scurfy, alternate, punctate-dotted, aromatic; sepals free or united at base, petal-like; pistillate flowers bearing rudimentary stamens; fruit a drupe, not overtopped by the calyx; anthers opening by pores and lids. . . LAURACEAE, p. 677

Leaves scurfy, opposite; calyx urceolate, 4-lobed at summit, the throat with an 8-lobed crown, loosely embracing the nut-like ovary and overtopping it, its tube becoming fleshy and red or orange in maturity; anthers without pores or lids. *Shepherdia*, p. 1045

I. Slender twining vine with often palmately lobed long-petioled leaves and long-peduncled panicles; calyx of 6 petal-like sepals; 3 upright pistils capped by radiating star-like or fimbriate stigmas; stone of drupe somewhat cup-like or lunate; stamens numerous. MENISPERMACEAE, p. 674

H. Leaves narrowly linear to linear-oblong, evergreen, rather crowded, 2–8 mm. long; flowers axillary or in terminal small heads, without calyx but subtended by bracts or with 3 petal-like sepals; fruit a 3–9 seeded drupe; anthers versatile. . . . . EMPETRACEAE, p. 974

E. Herbs. . O.

O. Leaves compound.

Frail and brittle aquatic with whorled and finely dissected leaves; solitary involucrate flowers sessile in leaf-axils. . . . . CERATOPHYLLACEAE, p. 636

Firm-stemmed terrestrial upright plants with alternate petioled ternately decompound leaves with distinct leaflets; flowers with petaloid or herbaceous sepals, in panicles or corymbs. . . . . . . *Thalictrum*, p. 656

O. Leaves simple, unlobed to deeply divided. . P.

P. Leaves mostly deeply palmate-lobed or -cleft.

Leaves peltate; coarse glaucous-stemmed smooth annual; staminate flowers with 3–5 colored sepals, the slender filaments forking; pistils 3-locular, forming weakly spiny 3-valved large capsules with mottled seeds. . . . . *Ricinus*, p. 962

Leaves not peltate; filaments not forking.

Plants without long stinging hairs; flowers paniculate; sepals distinct, greenish; stamens 5, short; pistillate flowers sessile in spikes, 1-locular, with 2 simple styles; fruit an achene. . CANNABINACEAE, p. 555

Plant with long stinging bristles; flowers loosely cymose; calyx of staminate flowers corolla-like, tubular, with rotate white limb; stamens numerous, the filiform filaments in several lengths; pistillate flowers with 3 forking styles; fruit a 2-valved capsule. *Cnidoscolus*, p. 959

P. Leaves not obviously lobed or merely hastate. . . Q.

Q. Nodes of stem and panicled racemes covered by tubular sheaths; calyx reddish; stigmas 3, plumose; achene 3-angled, surrounded at least below by reddish sepals; foliage acid.

***Rumex*, subgen. *Acetosella* and *Acetosa*, p. 571**

Q. Nodes without tubular sheaths. . R.

R. Flowers without a calyx.

Weak aquatic or terrestrial soft herbs with opposite leaves; 4-lobed ovaries in lower axils and single stamens in upper axils.

CALLITRICHACEAE, p. 972

Firm terrestrial plants; flowers borne from base of calyx-like involucre with colored glands near summit; the stalked 3-locular pistillate flower central; the numerous staminate flowers each subtended by a bract and having but 1 stamen. . . . *Euphorbia*, p. 963

R. Flowers (at least the staminate) with a typical calyx. . . S.

S. Staminate (or all) flowers in spikes, heads or groups of spikes. . . T.

T. Spikes or heads of flowers interrupted, moniliform or with remote glomerules or flowers.

Leaves opposite.

Inflorescence a terminal panicle; flowers alternate on spiciform branches. . . . . *Iresine*, p. 605

Inflorescences axillary, the flowers in remote glomerules.

Spikes overtopped by the subtending leaves, often leafy-tufted at apex; calyx 4-parted, with a rudimentary 1-locular ovary. . . . . *Boehmeria*, p. 559

Spikes overtopping leaves, not leafy-tufted above; calyx of 3 sepals, without rudimentary ovary. . . *Mercurialis*, p. 960

Leaves alternate.

Flowers in glomerules, all of a spike staminate.

Individual flowers bracted at base, the bracts and sepals scarious. . . . . AMARANTHACEAE, p. 601

Individual flowers bractless; sepals herbaceous or fleshy.

CHENOPODIACEAE, p. 590

Flowers not glomerulate, those of upper part of spike staminate, lowest flowers pistillate.

Inflorescences terminal or from upper axils; anthers short, not long-exserted; lower pistillate flowers with roundish carpels. . . . . EUPHORBIACEAE, p. 958

Inflorescences basal, overtopped by the naked stem and broad terminal leaves; anthers elongate, greatly exserted; pistillate flowers with 3 horn-like carpels.

BUXACEAE, p. 974

T. Spikes or heads of flowers close or continuous.

Leaves mostly opposite; calyx equally 4-parted and with 4 stamens in staminate flowers, unequally parted and with young achene bearing sessile often stellate stigma in pistillate flowers. . . . . URTICACEAE, p. 556

Leaves alternate.

Staminate and pistillate flowers in sessile axillary glomerules; calyx deeply 4-parted or of 4 sepals; the 4 stamens inflexed. . . . . *Parietaria*, p. 559

Staminate spikes exserted, the flowers not in axillary glomerules.

Inflorescence a panicle; the staminate spikes terminating the otherwise pistillate flowering branches, these bearing 1-ovulate green flowers. . . . . *Axyris*, p. 598

Inflorescence not paniculate; staminate spikes separate or borne below (rarely above) the few pistillate flowers. EUPHORBIACEAE, p. 958

S. Staminate flowers not spicate.

Flowers paired, axillary, 1 staminate and 1 pistillate in each axil; calyx 5- or 6-parted; stamens 3, mostly united at base. *Phyllanthus*, p. 963

Flowers numerous, in loose compound axillary cymes; calyx 5-parted; stamens 5, free; broad-leaved nettle with stinging hairs. . . . . . *Laportea*, p. 558

C. Flowers (or at least some in each inflorescence) perfect or with both stamens and fertile pistil. . . U.

U. Woody; shrubs, trees or climbers. . . V.

V. Twining, or climbing by tendrils or leaf-stalks.

Calyx tubular, elongate, curved; ovary single; leaves simple.

Twining; leaves deeply cordate; flowers solitary on long axillary peduncles; perianth sigmoid or curved like a Dutch pipe, with 3 divergent broad lobes. . . . . *Aristolochia*, p. 564

Climbing by terminal tendrils; leaves not cordate; flowers in racemes, slightly arching, with a broad wing extending down one side and along petiole, the summit with 5 narrow erect lobes. . . . *Brunnichia*, p. 590

Calyx of large free blue or purple petal-like sepals; ovaries numerous in a head, each with a long usually plumose style; leaves with 3 or more leaflets, their petioles twisting around supports. . *Clematis*, p. 663

V. Not twining nor climbing. . . W.

W. Leaves pinnate, with thin coarsely toothed or cleft leaflets, forming a tuft above the numerous simple or forking pendulous racemes; 5 petaloid sepals brownish-purple; pistils several, forming 1-seeded capsules; low shrub. *Xanthorhiza*, p. 672

W. Leaves simple; coarser shrubs or trees.

Foliage scurfy with stellate hairs or scales.

Leaves coarsely toothed, stipulate; flowers in dense terminal spikes or heads; top of calyx extending half-way to summit of ovary and fruit; stamens about 24, the long-exserted filaments dilated upward; styles 2; capsule woody, 2-valved. . . . . . *Fothergilla*, p. 752

Leaves entire, exstipulate; flowers few in axils; calyx surrounding and overtopping 1-locular stone-like ovary, becoming pulpy, the tube bearing a disk; stamens 4–8, not long-exserted. . . ELAEAGNACEAE, p. 1044

Foliage not scurfy.

Style 1, simple or cleft, or stigma sessile; fruit a smooth drupe.

Exstipulate leaves and bark spicy-aromatic; sepals 6, in 2 whorls of 3 each; stamens in 3 or 4 lengths, some of them changed to staminodia; anthers opening by valves; drupe 1-seeded. . . LAURACEAE, p. 677

Exstipulate or stipulate leaves and bark not aromatic; lobes of campanulate or tubular calyx 4 or 5; stamens in 1 or 2 series; anthers without valves.

Leaves stipulate; stamens 4 or 5 in 1 series; style 2–4-cleft or subsessile, stigma lobed; drupe with 2–4 stones. . . . *Rhamnus*, p. 992

Leaves exstipulate; stamens 8 or 10 in 2 series; stigma capitate or long style simple; drupe 1-seeded. . . . THYMELAEACEAE, p. 1044

Styles 2; fruit a drupe or a samara.

Leaves alternate, unlobed; ovary unlobed; fruit symmetrical, a rounded samara or a drupe. . . . . ULMACEAE, p. 553

Leaves opposite, angulate-lobed; ovary 2-lobed; fruit a pair of unsymmetrical samaras, each winged on one side. . . . . ACERACEAE, p. 984

U. Herbs. . . X.

X. Plant thalloid (without true stem and leaves), resembling a seaweed or liverwort, adhering to inundated rock by non-vascular attachments; flowers solitary, nearly sessile in a sac-like involucre, without perianth. PODOSTEMACEAE, p. 731

X. Plants with root, stem and leaves. . . Y.

Y. Leaves (at least the lower) deeply lobed, divided or compound. . . Z.

Z. Ovary solitary.

Aquatics with at least the submersed leaves finely and pinnately dissected; calyx adnate to ovary.

Upper subrosulate floating leaves with inflated petioles and rhombic blades; peduncled flowers solitary, maturing into a large coriaceous nut with spines formed by calyx-lobes. . . HYDROCARYACEAE, p. 1050

Upper leaves not rosulate, narrow; flowers sessile in leaf-axils, all perfect, or some imperfect and with the uppermost staminate; fruit small, 4-lobed or 3-angled. . . . . . HALORAGACEAE, p. 1071

Terrestrial, with broad segments or leaflets; calyx free from ovary.
Leaves merely pinnately lobulate or pinnatifid; sepals not petaloid.
Cauline leaves lobulate; flowers glomerulate, axillary or in panicles; calyx fleshy; fruit a utricle, with ring-like embryo. CHENOPODIACEAE, p. 590
Cauline leaves (at least the upper) uncleft; flowers racemose; sepals thin; fruit a flat rounded notched silicle, with 2 boat-shaped valves; embryo not circular; plant with mustard-flavor. *Lepidium*, p. 701
Leaves with distinct leaflets or palmate; sepals petaloid.
Leaves decompound; flowers pedicelled, in elongate racemes (simple or panicled); sepals distinct, caducous; 2-horned modified stamens in place of petals; fruit a follicle. . . . . . *Cimicifuga*, p. 671
Leaves simply pinnate or palmately divided; flowers in dense heads or spikes; calyx with top-shaped tube and 4 lobes, in fruit surrounding the achene. . . . *Alchemilla*, p. 864 and *Sanguisorba*, p. 867
Z.Ovaries 2–∞. . . . . RANUNCULACEAE, p. 642
Y.Leaves simple, not strongly lobed nor deeply dissected. . . a.
a.Calyx (when present) free from ovary. . . b.
b.Calyx wanting; flowers white, crowded in a long terminal slender raceme; stem succulent, jointed, with petioled, cordate alternate leaves; fruit of 3 or 4 indehiscent, wrinkled, basally united, beaked carpels. . . . . SAURURACEAE, p. 487
b.Calyx present. . . c.
c.Nodes of stem and inflorescence covered by mostly tubular sheaths; calyx variously colored or petaloid; styles or stigmas 3 or 2; achene 3-angled or lenticular. . . . . . . POLYGONACEAE, p. 565
c.Nodes not covered by tubular sheaths; fruits mostly with seeds. . . d.
d.Stem very succulent, jointed, with leaves reduced to short mostly buried scales; the flowers deeply immersed in hollows of the joints. . . . . *Salicornia*, p. 599
d.Stems only slightly, if at all, succulent; leaves of normal development; flowers not deeply immersed. . . e.
e.Leaves, at least of main axis, opposite or whorled.
Flowers clustered in a colored corolla-like involucre; calyx-limb deciduous, the tube forming a pericarp around the unilocular 1-seeded nut-like fruit. . . . NYCTAGINACEAE, p. 605
Flowers exinvolucrate.
Calyx-limb or sepals persistent; capsule or utricle naked.
Calyx not corolla-like; styles or stigmas 2–5; embryo strongly curved.
Leaves whorled or opposite; stamens inserted at base of calyx; capsule 1–5-locular. . . . AIZOACEAE, p. 607
Leaves opposite; stamens borne on receptacle; capsule 1-locular. . . . . . . CARYOPHYLLACEAE, p. 611
Calyces campanulate to globose, nearly to quite covering ovary, sessile in axils of narrow leaves; style 1; stamens attached to base of calyx; capsule subglobose or ovoid; seed not strongly curved.
Calyx green or greenish, not corolloid, 4-toothed at summit; stamens 4 or 8; capsule indehiscent or irregularly bursting. . . . . . . LYTHRACEAE, p. 1045
Calyx roseate, corolloid, deeply 5-lobed; stamens 5; capsule dehiscent, 5-valved. . . . . *Glaux*, p. 1143
Calyx of free and deciduous large brightly colored petaloid sepals; flower solitary, terminal, with many free carpels ripening into long-tailed achenes. . . . *Clematis*, p. 663
e.Leaves alternate.
Inflorescences elongate racemes borne from side of stem back of leaf-bases; sepals petal-like, white or pink; ovary a ring of 10 (in ours) united carpels ripening into a 10-scalloped berry. . . . . PHYTOLACCACEAE, p. 606
Inflorescences axillary or terminal.
Flowers axillary or in terminal panicles of spiciform branches; sepals not petaloid, 2–5, distinct; stamens 1–5; styles or stigmas 2 or 3; fruit a 1-locular and 1-seeded utricle.
Flowers without scarious bracts; calyx herbaceous or fleshy. . . . . CHENOPODIACEAE, p. 590
Flowers with scarious bracts; calyx scarious. AMARANTHACEAE, p. 601

Flowers in terminal cymes with secund branches; calyx 5- or 6-parted; stamens 10 or 12; carpels 5 or 6, united half-way up, forming a 5- or 6-beaked capsule with many seeds. *Penthorum*, p. 736

a. Calyx adnate to ovary.

Leaves whorled, narrow; flowers axillary; calyx entire; stamen 1, borne from margin of calyx, embracing the wholly stigmatic simple style in a groove; nutlet 1-seeded. . . . . . . HIPPURIDACEAE, p. 1076

Leaves not whorled; calyx lobed at summit; stamens 4–12; fruit drupe-like or capsular.

Calyx large, campanulate to sigmoid, lurid to purple-brown, with 3 broad lobes, fused at base to the 6-locular ovary; stamens 6 or 12, often united to style, with anthers adnate to stigma; capsules 6-locular; seeds with large fleshy aril. . . . . ARISTOLOCHIACEAE, p. 562

Calyx small and straight, with 4 or 5 lobes or teeth; the tube adnate half-way or to summit of 1- or 4-locular ovary; anthers free from stigma; seed non-arillate.

Terrestrial, erect alternate-leaved root-parasites; flowers in single or numerous small umbels or cymes; calyx covering ovary and drupe- or nut-like fruit, with a lobed disk at summit of tube, the 5 stamens alternating with lobes of disk; style 1. . SANTALACEAE, p. 560

Aquatic or subaquatic, with prostrate or weak stems and opposite (rarely alternate) leaves; styles 2 or subsessile stigmas 4.

Leaves roundish to reniform; flowers solitary or in irregular leafy cymes; calyx adnate to lower half of ovary; stamens 4–10, inserted on a disk; styles 2; capsule depressed, 2-lobed, 2-valved. . . . . *Chrysosplenium*, p. 744

Leaves oval or ovate; flowers axillary, sessile; campanulate calyx-tube extending to summit of ovary, with 4 spreading lobes; stamens 4; ovary and 4-angled capsule 4-locular, capped by 4-lobed stigma; capsule opening by chinks along the angles. *Ludwigia*, p. 1052

B. Flowers with 2 floral envelopes, calyx and corolla (or corolla-like structures). . . f.

f. Petals or petal-like parts distinct or barely united at base. . . g.

g. Flowers all unisexual, staminate and pistillate ones on different plants or in separate inflorescences. . . h.

h. Trees or shrubs. . . i.

i. Leaves pinnate.

Leaves bipinnate; flowers in terminal panicles or racemes, the pistillate inflorescence smaller and denser than the staminate; calyx tubular, 5-lobed at summit; 5 erect petals borne at summit of calyx-tube; style 1; stamens 10; fruit a heavy and woody legume. . . . . . *Gymnocladus*, p. 884

Leaves simply pinnate; flowers cymose or paniculate; calyx cleft to base; styles or stigmas 2–5; fruit not a legume.

Flowers in simple or forking cymes or open panicles; pistils with united or connivent styles; fruit 2-valved, the thin outer coat (exocarp) soon separating and exposing the 2 lustrous black large seeds; branches prickly; leaves punctate-dotted. . . . . *Xanthoxylum*, p. 951

Flowers paniculate; styles free or somewhat united; fruit indehiscent; branches not prickly; leaves not punctate.

Panicle loose and open; petals valvate; stamens 10, from base of 10-lobed disk; ovary deeply 2–5-lobed; fruit an elongate twisted samara. . . . . SIMAROUBACEAE, p. 952

Panicle dense; petals imbricate; stamens 5; ovary unlobed, 1-ovulate; fruit a red dryish drupe. . . . . . . ANACARDIACEAE, p. 976

i. Leaves simple. . . j.

j. Twining, with alternate leaves.

Slender vines with palmately veined leaves and axillary panicles; stamens 6–24, not on a disk; carpels 3 or 6; fruit a drupe with 1 curved or cup-like stone. . . . . MENISPERMACEAE, p. 674

Coarser and strongly ligneous, with pinnately veined leaves and terminal or axillary panicles or cymes; stamens 5, at angles of a disk; carpel 1; fruit a 3-valved drupe-like capsule with 1 or 2 pulpy-arillate brightly colored seeds in each locule. . . . . . *Celastrus*, p. 984

j. Not twining. . . k.

k. Flowers clustered (rarely solitary) in a head terminating a long axillary peduncle.

Head globose, with closely packed dry flowers, terminating a pendulous peduncle; overwintering buds with 1 scale; palmately angulate leaves with sheathing stipules; staminate flowers with 3–8 stamens; pistillate ones 3–8-carpellate, with slender styles; slenderly 1-seeded nutlets surrounded by bristles. . . . PLATANACEAE, p. 753

Head lobulate, few-flowered, terminating a stiffly divergent peduncle; bud of imbricated scales; entire or merely toothed leaves exstipulate; staminate flowers with disk- or cup-like calyx and 5–12 stamens; pistillate flowers with campanulate calyx, 1-carpellate, with obsolete style; drupe smooth. . . . NYSSACEAE, p. 1048

k. Flowers not in heads.

Leaves palmately lobed, opposite; stamens mostly 8, on summit of disk; ovary deeply 2-lobed, with 2 long styles; fruit 2 inequilateral samaras winged down 1 side. . . . . ACERACEAE, p. 984

Leaves not palmately lobed; fruit not a samara.

Ovary 3-locular, the 3 slender styles 2-cleft; staminate flowers with 5 or 6 petals and stamens, pistillate flowers with petals reduced; fruit a capsule splitting into 3 2-valved carpels, each valve 1-seeded. . . . *Andrachne*, p. 963

Ovary 2–5-locular, with a single style or a sessile stigma; stamens and petals 4 or 5; fruit drupaceous, 2–5-locular, with 1 bony nutlet in each locule, or the drupe 1-seeded.

Leaves alternate or opposite; inflorescence not pendulous; petals small, white to green or yellowish; stamens free from petals; anthers blunt; drupe red, yellow or black, with 2–5 nutlets.

Sepals united below; calyx with no disk; stigma sessile; seed with copious albumen. . . . AQUIFOLIACEAE, p. 980

Sepals distinct; calyx with a lobed disk; style definite; seed with little or no albumen. . . . RHAMNACEAE, p. 991

Leaves opposite; inflorescence a loose pendulous panicle; corolla-segments greatly prolonged, linear, white, the stamens borne at their bases; anthers long-beaked; drupe blue, with 1 stone. *Chionanthus*, p. 1147

h. Herbs. . . l.

l. Leaves (or many of them) compound.

Leaves twice- or thrice-pinnate; slender spikes in a large panicle; pistillate flowers reflexed in fruit, the 3 or 4 carpels splitting along the ventral suture; staminate flowers with the stamens borne on the upper (or inner) surface of a discoid expansion of floral axis. . . *Aruncus*, p. 756

Leaves simply pinnate or palmate.

Acaulescent, with 3-foliolate stipulate basal leaves and usually superficial stolons; corymb or raceme peduncled; calyx rotate, bractlets alternating with calyx-lobes; pistillate flower with many 1-seeded small carpels borne on a fleshy finally juicy plump receptacle; staminate flower similar but larger, the receptacle reduced. . . *Fragaria*, p. 802

Caulescent, without superficial stolons; leaves not basal; flowers axillary and sessile or in a roundish terminal umbel; receptacle not fleshy.

Aquatic, with finely dissected pinnate leaves and axillary flowers; pistillate and perfect flowers with 4 nutlet-like carpels, the staminate with 4–8 stamens. . . . . *Myriophyllum*, p. 1071

Terrestrial, with a whorl of 3 palmately 3–5-foliolate leaves at summit, these subtending a long-peduncled subglobose umbel; fruit a several-seeded drupe; staminate flowers with 5 stamens at border of sterile ovary. . . . *Panax*, p. 1077

l. Leaves simple. . . m.

m. Pistillate flowers in the axils or forks of stem or at bases of staminate spikes; styles 2 or 3, forked; fruit an achene or 1–3-locular capsule; staminate flowers with 5–15 stamens alternating with glands or in 2 or 3 cycles borne on a slender column; plant silvery or scurfy with stellate scales . . . EUPHORBIACEAE, p. 958

m. Pistillate flowers variously disposed; style simple; plant not scurfy. . n.

n. Leaves opposite; styles 3 or 5; capsule 1-locular; stamens 10. CARYOPHYLLACEAE, p. 611

n. Leaves alternate or spirally arranged; carpels several or many or ovary 8–10-locular. . o.

o. Low (up to 4 dm. high); unilocular distinct carpels and their single styles free.

Stoloniferous, with slender upright 1–3-leaved stems with single terminal white rose-like flower and a group of 1-seeded indehiscent carpels.

Stem simple, erect, with 1–3 reniform lobulate leaves and 1 terminal flower; pistillate flower with many carpels and maturing as a raspberry-like fruit; staminate flower similar, without carpels. *Rubus*, subgen. *Chamaemorus*, p. 819

Stem prostrate, branching, the roundish leaves not lobed; petaliferous flowers mostly staminate; pistillate flowers apetalous, on recurving basal pedicels, with a few dryish drupes. *Dalibarda*, p. 864

Cespitose, with heavy base and fleshy many-leaved stem, with terminal flattish corymb of small yellow or greenish staminate and yellowish or purplish pistillate flowers; leaves oval or narrower, fleshy; fruit a many-seeded follicle. . . . . . *Sedum*, p. 731

o.Tall (1–2 m. or more), with palmately divided leaves; flowers in a terminal panicle; tube of many antherless filaments surrounding the 8–10 styles in the pistillate, or tube antheriferous in the staminate ones; fruit depressed, separating into 8–10 kidney-shaped carpels. . . . . . . *Napaea*, p. 1003

g.Flowers all, or at least some in each inflorescence, perfect or essentially so. . . p.

p.Trees or shrubs. . . q.

q.Twining or climbing. . . r.

r.Climbing by tendrils, twining petioles or aërial rootlets.

Tendril-bearing; inflorescence an open cyme or a thyrsoid panicle, terminal or opposite an alternate long-petioled leaf; stamens opposite the 4 or 5 valvate petals; style 1 or obsolete; ovary 2-locular; fruit a 1–4 -seeded berry. . . . VITACEAE, p. 994

Without tendrils.

Climbing by aërial roots along stem and branches.

Leaves opposite, not lobed; white flowers in terminal corymbs; sepals and petals 7–10; stamens 20–30; styles united into a column, with 7–10-lobed stigma; capsule urceolate, opening between ribs, with many fine seeds. . . . . . . *Decumaria*, p. 747

Leaves alternate, those of sterile branches 3- or 5-lobed, evergreen; terminal umbels nearly globular, single or in racemes; petals and stamens 5; styles 5, united; drupe 3–5-seeded. . . . *Hedera*, p. 1078

Climbing by twining petioles of the compound opposite leaves; flowers solitary and terminal or corymbose-paniculate; ovaries many, ripening into achenes with long usually plumose style. *Clematis*, p. 663

r.Twining, without aërial roots or tendrils.

Leaves pinnate; pendulous racemes with showy blue or violet papilionaceous flowers; all but 1 of the 10 stamens united; fruit a tough legume. *Wisteria*, p. 903

Leaves simple, with straight parallel veins; greenish-white regular flowers in small terminal panicles and axillary clusters; the 5 stamens free; fruit a drupe. . . . . . *Berchemia*, p. 992

q.Not climbing nor twining. . . s.

s.Flower papilionaceous (as in the pea), with the stamens forming a simple or divided tube around the style or 10 and distinct; fruit a legume. LEGUMINOSAE, p. 879

s.Flowers not papilionaceous. . . t.

t.Leaves compound. . . u.

u.Leaves bipinnate.

Stamens 10 or more (if fewer the tree thorny); carpel 1, forming a legume. . . . . LEGUMINOSAE, p. 879

Stamens 5–12; ovary 4- or 5-locular, forming a capsule or drupe.

Slender and low, suffruticose, rarely 1 m. high; flowers in terminal open cymes; petals fringed; stamens borne on receptacle, the alternate ones shorter; ovary and capsule 4- or 5-lobed. . *Ruta*, p. 952

Strong trees or large shrubs; flowers in panicles or panicles of umbels.

Prickleless; showy lilac or lavender flowers in loose panicles; filaments forming a 10- or 12-lobulate tube, the anthers between the lobes; ovary superior, forming a large drupe. MELIACEAE, p. 953

Prickly; small whitish flowers in large panicles of many roundish umbels; stamens 5, on margin of disk; ovary inferior, forming a berry-like drupe. . . . . . . *Aralia*, p. 1077

u.Leaves simply pinnate or palmate. . . v.

v.Stamens borne on the calyx in 1 or more cycles around the ovary or ovaries, without a hypogynous disk; pistils distinct or the carpels united and surrounded by the fleshy calyx-tube. ROSACEAE, p. 753

v. Stamens borne on or near margin of a hypogynous disk. . . w.

w. Leaves pinnate.

Leaflets 3, punctate-dotted and aromatic; flowers from axils of the preceding year's leaves or in terminal cymes; ovary 2–8-locular, with 1 style, forming a round samara or an orange-like berry. . . . . RUTACEAE, p. 951

Leaflets 3–41, not punctate; flowers in terminal or axillary panicles or racemes.

Leaves alternate.

Leaflets 13–41; panicle large and open, terminal; stamens 10; carpels united by their styles or stigmas; fruit an oblong twisted samara with round seed near middle. SIMAROUBACEAE, p. 952

Leaflets 3–31; panicles axillary or, if terminal, a dense thyrse; stamens 5; 3 styles or stigmas free; ovary 1-locular, forming a drupe. . . . ANACARDIACEAE, p. 976

Leaves opposite, with 3 or 5 leaflets; white pendulous flowers on jointed pedicels; ovary 3-carpellate, forming a large bladdery-inflated capsule. . . . . STAPHYLEACEAE, p. 984

w. Leaves opposite, digitate, of 5–9 elongate leaflets; flowers irregular, showy, in terminal thyrse or panicle; fruit a large leathery subglobose capsule, with a large chestnut-like seed. HIPPOCASTANACEAE, p. 988

t. Leaves simple or represented by small scales. . . x.

x. Ovary wholly superior or not surrounded by prolonged calyx-tube. . . y.

y. Stamens opening by pores, slits or valves.

Woody shrubs or trees.

Prickly shrub with small yellow flowers in racemes or corymbs; wood yellow; sprout-leaves altered to sharp 3-pronged or branching prickles, the flowering shoots of the next season from their axils; sepals 6; petals and stamens 6; fruit a berry with 1–few crustaceous seeds. . . . . *Berberis*, p. 674

Not prickly.

Evergreen, mostly with aromatic bark and leaves.

Medium-sized tree; flowers paniculate or cymose; sepals (or sepals and petals) in 2 similar whorls of 3 each; the 12 stamens in successive whorls of 3 each, the innermost ones sterile; anthers opening by lids; fruit a blue drupe with 1 stone. . . . . *Persea*, p. 678

Small shrubs with umbelliform corymbs; calyx-lobes or sepals and petals 5; stamens 5–8, similar; anthers opening by pores or chinks; fruit a 5-valved many-seeded capsule. . . . . ERICACEAE, p. 1114

Deciduous; only the flowers aromatic, these in slender racemes; sepals 5; petals 5; stamens 10, similar; capsule 3-locular, with minute seeds. . . . . . CLETHRACEAE, p. 1108

Only weakly suffrutescent, through persistence of the evergreen leaves and slender axis, nearly herbaceous; flowers in racemes or corymbs; sepals and concave petals 5; stamens 10; anthers extrorse in bud, becoming inverted and introrse in flower, opening by pores at base; capsule 5-lobed, 5-valved. PYROLACEAE, p. 1108

y. Stamens not opening by lids, pores or chinks. . . z.

z. Stamens borne on the receptacle, without hypogynous disk.

Leaves alternate; sepals 3, colored like petals.

Flowers white or whitish to roseate; petals 6–15; carpels 1- or 2-seeded, forming a cone of follicles or samaras. MAGNOLIACEAE, p. 675

Flowers lurid or purplish, the sepals 3; the petals in 2 whorls of 3 each, enlarging after expansion, the 3 outer with spreading tips, the 3 inner erect; carpels many-seeded, forming a pulpy berry. . . . . ANNONACEAE, p. 677

Leaves opposite; bark aromatic; flowers at tips of lateral branches; sepals and petals undifferentiated, spirally arranged, very many; carpels numerous, lining base of a prolonged receptacle, forming achenes. . CALYCANTHACEAE, p. 676

z. Stamens inserted at base of perianth or on a hypogynous disk. . . A.

A. Leaves opposite.

Leaves palmately veined and shallowly lobulate; ovary 2-locular, 2-lobed, forming 2 separable laterally 1–2-winged and 1-seeded samaras. . . . . ACERACEAE, p. 984

Leaves pinnately veined, unlobed; fruit a capsule or drupe.

Stamens 4–∞; petals not elongate-linear, yellow, greenish or purplish; fruit a capsule.

Leaves not punctate; flowers not conspicuous, greenish or purplish; stamens 4 or 5, at margin of large disk; fruit a fleshy capsule, with few large arillate seeds. CELASTRACEAE, p. 982

Leaves mostly pellucid-punctate; flowers with showy yellow petals; stamens very numerous, often in brushes; no hypogynous disk; capsule membranaceous, with many tiny seeds without aril. GUTTIFERAE, p. 1007

Stamens 2, anthers nearly sessile; panicle pendulous; petals or segments of corolla elongate-linear, white; fruit a 1-seeded blue drupe. . . . . *Chionanthus*, p. 1150

A. Leaves alternate. *B.*

B. Carpels 1–∞, all distinct; stamens inserted on calyx; calyx with sepals united at base or confluent with receptacle.

Flower regular; fruit a capsule, berry or drupe. ROSACEAE, p. 753

Flower papilionaceous, roseate; leaf broadly rounded, palmate-veined; stamens 10; fruit a legume. . . *Cercis*, p. 886

B. Carpels united into a compound ovary; sepals distinct or united only near base. . . *C.*

C. Pedicels of many sterile caducous flowers of the large open terminal panicle becoming flowerless plumes; fertile flowers yellowish; 5 stamens opposite sepals, at margin of large disk; fruit a 1-sided drupe, with the 3 styles about midway on one side. . . . . *Cotinus*, p. 976

C. Pedicels all with fertile flowers, not plumose. . . . *D.*

D. Stamens and sepals or calyx-lobes of same number.

Foliage of scarious-margined scales, suggesting heaths; flowers ebracteate, in slender racemes; capsule 1-locular; seed with long terminal tuft of hairs. TAMARICACEAE, p. 1016

Foliage of dilated herbaceous leaves; stamens 4–8, borne near margin of hypogynous disk; calyx-segments or sepals valvate; seeds not hairy-tufted.

Stamens opposite sepals or calyx-segments.

Flowers in slender elongate spiciform racemes borne just below new leafy shoots of the season; fruit a fleshy yellow loculicidal 2-valved capsule with small slender seeds. . . . CYRILLACEAE, p. 979

Flowers solitary or clustered in leaf-axils; fruit a red or black (rarely yellow) berry-like drupe with plump crustaceous seeds. AQUIFOLIACEAE, p. 980

Stamens alternate with calyx-segments; 2–5 carpels each 1- or 2-ovulate; fruit a drupe or capsule. RHAMNACEAE, p. 991

D. Stamens very numerous; hypogynous disk scarcely, if at all, developed.

Sepals or calyx-lobes valvate; placentae united in axis of ovary; pubescence often stellate.

Flowers in axillary cymes, the peduncle united to a ligulate bract; sepals deciduous; small creamy petals imbricated in bud; stamens in several fascicles; anthers 2-locular; globular 1-locular and 1- or 2-seeded fruit nut-like. . . TILIACEAE, p. 999

Flowers axillary, on free jointed peduncles; persistent calyx-segments subtended by an involucre; large roseate to white corolla convolute in bud; stamens united into a column, the anthers 1-locular; ovary a ring of 5 loculicidal many-seeded carpels. . . . . *Hibiscus*, p. 1005

Sepals 5 or 6, imbricated in bud; the 5 or 6 showy petals silky on back; ovary 5-locular, forming a large woody capsule. . . . . . THEACEAE, p. 1007

x.Ovary partly or wholly inferior (adnate to calyx) or covered by or inclosed in calyx-tube. *E.*

*E.*Stamens free from the ovary. *F.*

*F.*Plant fleshy, without leaves or merely scaly, bearing tufts of stinging bristles or prickles; sepals and petals all brightly colored, imbricated in several series; stamens ∞; fruit a berry. CACTACEAE, p. 1043

*F.*Plant not fleshy, with typical stems and leaves; calyx and petals dissimilar. . . *G.*

*G.*Leaves opposite or verticillate; capsule 2–5-locular.

Leaves opposite; calyx equally 4- or 5-cleft; stamens 8–∞, similar; tube of calyx adnate to ovary; shrubs, not tip-rooting. . . . . SAXIFRAGACEAE, p. 735

Leaves verticillate or opposite; calyx with alternating short and broad and long and slender teeth, the enveloping tube free from the ovary; stamens 8 or 10, alternately long and short; tip-rooting shrub. . . . . *Decodon*, p. 1047

*G.*Leaves alternate.

Petals distinct.

Carpels 1- or 2-locular; styles or stigmas 2; seed with copious albumen.

Seeds small and soft, 2–∞ in each locule of the membranous 2-beaked capsule or the juicy berry. SAXIFRAGACEAE, p. 735

Seeds large and hard, 1 in each locule of the woody abruptly exploding capsule. . . . HAMAMELIDACEAE, p. 751

Carpels each 1-locular and with 1 style each, the 3–5 carpels adnate to the fleshy calyx, or the very numerous distinct carpels borne within the fleshy calyx-tube. ROSACEAE, p. 753

Petals free except at the very short base of the corolla; calyx adnate to ovary.

Pubescence of simple hairs; flowers in dense axillary clusters; stamens clustered at bases of petals; ovary completely septate. . . . . . SYMPLOCACEAE, p. 1146

Pubescence stellate or scurfy; flowers in simple or compound racemes; stamens in a single tube; ovary septate only below middle. . . . . STYRACACEAE, p. 1146

*E.*Stamens borne on disk at summit of ovary.

Very prickly; alternate leaves lobulate, palmately veined; inflorescence a large panicle of umbels; styles 2 . . . *Oplopanax*, p. 1078

Prickleless; opposite leaves unlobed, pinnately veined; inflorescence a corymb or panicle; style 1. . . . CORNACEAE, p. 1105

p.Herbs. *H.*

*H.*Anthers opening by apical or subapical pores or lids.

Leaves compound or deeply lobed, not evergreen; anthers opening by uplifted lids; ovary 1-locular. . . . . BERBERIDACEAE, p. 672

Leaves simple, not lobed; anthers opening by pores; ovary 4- or 5-locular.

Leaves in a basal cluster or paired along the stem, evergreen, not prominently ribbed; or plant saprophytic and without chlorophyll, the leaves reduced to scales; anthers becoming resupinate, opening by pores at the bottom; capsule free from calyx. . . PYROLACEAE, p. 1108

Leaves in pairs along stem, prominently ribbed, deciduous; anthers upcurved, the pores at summit; capsule surrounded by and partly adnate to urn-shaped calyx. . . . MELASTOMATACEAE, p. 1049

*H.*Anthers not opening by apical lids or pores. . *I.*

*I.*Ovary superior or not inclosed by or covered by tube of calyx. . . *J.*

*J.*Stamens borne on the receptacle (hypogynous) or on a disk only in *Resedaceae* (Mignonettes). . . *K.*

*K.*Ovary 1-locular but compound, with 2–8 styles; ovules and seeds attached to base or central column of capsule; embryo coiled or curved.

Sepals or calyx-lobes 2; stamens 5–∞, when of same number as petals borne opposite them; plants succulent or fleshy. PORTULACACEAE, p. 608

Sepals or calyx-lobes or -teeth 4 or 5; stamens 4–10, when of same number as petals alternate with them; plants mostly of dryish texture. . . . . . CARYOPHYLLACEAE, p. 611

*K.*Ovary simple or, if compound, of more than 1 locule; ovules not curved. . *L.*

L. Carpels solitary, or few to many, distinct and spirally arranged; stamens usually at least twice to many times as many as sepals, spirally or cyclically arranged.

Aquatic plants with leaves centrally peltate or, if not with peltate leaves, having the locules of ovary and fruit with ovules or seeds over the whole inner surface. . . . NYMPHAEACEAE, p. 637

Terrestrial (a few aquatic), the leaves only rarely centrally peltate; ovules on restricted placentae; carpels forming achenes, follicles, berries or drupes.

Fruit an achene, capsule or acrid or unpalatable berry; seeds non-arillate; leaves compound or simple and variously lobed, cleft or entire. . . . . . RANUNCULACEAE, p. 642

Fruit a large insipidly sweet berry with arillate seeds; large white (or roseate) flower in fork of upright stem, beneath a large peltate and deeply lobed umbrella-like leaf. *Podophyllum*, p. 672

L. Carpels 2–∞, united into a compound ovary, with mostly nerve-like parietal placentae; parts of flower in whorls.

Plants with ordinary or typical leaves, not insectivorous; sepals 2 or 4 (–7 in *Resedaceae*).

Sepals 2; flower regular or irregular; embryo minute, at base of fleshy albumen. . . . PAPAVERACEAE, p. 679

Sepals 4–7; embryo curved or folded, completely filling the seed; albumen none.

Petals and sepals 4, regular; stamens not on a hypogynous disk; capsule 1- or 2-locular, closed, longitudinally dehiscent, or indehiscent and disarticulating.

Stamens 6 or more, nearly equal; capsules 1-locular, 2-valved, without transverse partitions; leaves stipulate, palmate; raceme with bracts at base of flowers. CAPPARIDACEAE, p. 684

Stamens 6, in 2 lengths, 2 shorter and inserted lower down than the other 4 (tetradynamous); capsule 2-locular by a transverse partition or indehiscent and transversely jointed; stipules and usually bracts wanting. CRUCIFERAE, p. 685

Petals and sepals 4–7, irregular; stamens indefinite in number or on a hypogynous disk; capsule 3–6-lobed, barrel-shaped or subcylindric, opening at summit before maturity of seed. . . . . RESEDACEAE, p. 728

Plants insectivorous; the leaves all radical, tubular, or flat to filiform and bearing long stipitate glands; sepals and petals usually 5, persistent.

Leaves hollow, hooded; ovary and capsule 5-locular, with axile placentae; style 1, expanding above into a broad and umbrella-like limb. . . . . SARRACENIACEAE, p. 728

Leaves flat to filiform, uncoiling from base to apex (circinate), the blade covered with long glands; ovary 1-locular, with parietal placentae; styles 3 or 5, each slenderly 2-parted. DROSERACEAE, p. 729

J. Stamens mostly perigynous (adnate to perianth) or inserted on or at base of a hypogynous disk. . . *M.*

M. Carpels solitary or several and distinct (rarely united); stamens perigynous; sepals often united at base or confluent with the receptacle.

Endosperm copious (if wanting, the plants very fleshy or succulent); flowers regular.

Carpels 2-∞, distinct or united, 1–5-locular, the few to many ovules and seeds sessile or nearly so, with fleshy endosperm; stamens borne on calyx; sepals or calyx-lobes not conspicuously petaloid; leaves basal or cauline.

Plant succulent or fleshy; carpels the same number as calyx-segments; endosperm scanty. . . . . CRASSULACEAE, p. 731

Plant hardly succulent; carpels fewer than calyx-segments, mostly 2; ovules and seeds 2–∞ in each locule of capsule; endosperm copious. . . . . . . SAXIFRAGACEAE, p. 735

Carpel 1, 1-locular, 1-ovulate, becoming an achene or utricle; ovule pendulous on a slender funiculus arising from base of carpel; endosperm mealy; stamens borne on bases of petals; calyx petaloid, brightly colored. . PLUMBAGINACEAE, p. 1144

Endosperm none or rarely developed; sepals united at base or confluent with the receptacle.

Flower regular; carpels 2–∞, forming achenes, follicles or drupelets; stamens not forming brushes or tubes. . . . ROSACEAE, p. 753

Flowers irregular and papilionaceous or regular; carpel 1, forming a legume with the seeds in 2 alternate rows along the ventral suture, or a loment with 1-seeded indehiscent segments; stamens monadelphous or diadelphous (forming a simple or divided tube around the ovary) or distinct, 5, or more often 10. LEGUMINOSAE, p. 879

M. Carpels united into a compound ovary; sepals mostly distinct or separate nearly to base. . . *N.*

N. Stamens few, rarely more than twice as many as sepals; stamen-bearing hypogynous disk usually well developed. . . *O.*

O. Ovules pendulous, with raphe toward axis of ovary, or ascending and with raphe dorsal; stamens mostly borne on base of perianth.

Flowers regular; petals 4 or 5; stamens 4, 5 or 10; ovary 4-, 5- or 10-locular.

Leaves simple, entire, narrow; fertile stamens as many as petals, united below into a tube; capsule with false septa, each dividing its locule into two. . . . . . LINACEAE, p. 940

Leaves deeply lobed or compound; fertile stamens of same number as or twice or thrice the number of petals.

Leaves 3-foliolate, with obcordate leaflets (acid to taste); petals yellow, violet, roseate or whitish; filaments united into a tube, the antheriferous tips radiating from it at different levels; capsule splitting into many-seeded segments. . . . . OXALIDACEAE, p. 943

Leaves lobulate, or pinnate with more than 3 leaflets; filaments distinct or barely united at base; mature separated carpels 1–few-seeded.

Styles united below into a beak, 5-parted at summit; the 5 smooth carpels separating at maturity, the bodies 1-seeded, the long beak coiling; petals roseate. GERANIACEAE, p. 946

Style 1, undivided; the 5 carpels prickly or muricate, 1–5-seeded; petals yellow. . . . ZYGOPHYLLACEAE, p. 950

Flowers very irregular, somewhat papilionaceous, showy; petals 3; stamens commonly 8; ovary 2-locular, 2-ovulate. POLYGALACEAE, p. 953

O. Ovules pendulous, with raphe away from axis of ovary, or ascending and with raphe ventral; stamens mostly borne on or at base of a hypogynous disk.

Leaves pinnate, with 3 or 5 leaflets; flowers regular; sepals 3, larger than the 3 petals; stamens 6; ovaries 3, 1-locular, forming fleshy achenes; small and weak annual. LIMNANTHACEAE, p. 975

Leaves simple; flowers very irregular, showy; sepals 5, the 2 anterior ones united, the posterior large and saccate; petals 2, 2-lobed; stamens 5, with 5 appendages of filaments connivent over the stigma; ovary 1, 5-locular; capsule elastically dehiscent, its freed valves coiling; watery-stemmed annuals. BALSAMINACEAE, p. 990

N. Stamens few to many; hypogynous disk scarcely or not at all developed. . . *P.*

P. Sepals or calyx-lobes valvate, persistent; stamens monadelphous; the many anthers 1-locular; placentae united in axis of ovary; embryo curved; pubescence mostly stellate; juice mucilaginous. . . . . MALVACEAE, p. 1000

P. Sepals imbricate or valvate, 5; stamens as many as petals or more numerous, borne on perianth or just below petals; carpels 2, 3 or 5, united; placentae parietal or basal, exceptionally axile. . . *Q.*

Q. Styles 3 or 5, distinct or only rarely united above base; seed without endosperm; flowers regular; leaves opposite.

Leaves punctate-dotted, exstipulate; stamens few–∞, often in fascicles; inflorescences terminal corymbs, with or without secondary axillary ones; sepals 4 or 5; petals mostly showy, yellow or roseate, deciduous. GUTTIFERAE, p. 1007

Leaves not punctate, stipulate; flowers axillary, inconspicuous; stamens 2–10, not fascicled; sepals 2–5; petals persistent in fruit. . . . ELATINACEAE, p. 1015

Q. Styles united nearly or quite to summit or 1 and simple; seed with copious endosperm; upper or all leaves alternate or leaves basal.

Depressed or ascending, not climbing nor tendril-bearing; fruit a capsule.

Leaves basal, long-petioled, cordate-rotund, lustrous, evergreen; naked scape terminated by a spiciform raceme of bracted small white flowers; 10-toothed tube of filaments adnate to bases of petals, 5 teeth naked, the alternate 5 shorter and bearing 1-locular anthers opening across the top. . . . . *Galax*, p. 1136

Leaves (at least the upper) alternate or, if basal, not evergreen.

Suffrutescent or tough-stemmed narrow-leaved herbs; the leaves either marcescent and scale-like or deciduous, nearly entire; flowers nearly regular; the 2 outer sepals much smaller or larger than the inner 3, often bract-like; stamens on a thickened zone beneath the ovary, 3 or ∞, free. . . . CISTACEAE, p. 1016

Herbs with dilated soft and toothed or divided leaves of the season; flowers very irregular, spurred at base; the 5 sepals uniform; petals mostly violet, white or yellow; anthers connivent about and nearly covering the united styles. . . . . VIOLACEAE, p. 1022

Vines with tendrils in axils of palmately divided leaves; 5 filaments forming a tube embracing the long stalk of the ovary, many sterile filaments between corolla and tube of fertile stamens; fruit a 1-locular many-seeded berry.
PASSIFLORACEAE, p. 1042

I. Ovary inferior or adnate to or inclosed by the tubular or campanulate calyx-tube. . *R.*

R. Stamens inserted on the calyx; flowers not in umbels.

Ovary and capsule surrounded by calyx-tube but free from it.
LYTHRACEAE, p. 1045

Ovary and capsule fused to calyx-tube; leaves alternate or opposite.

Aquatics with finely dissected pinnate leaves and axillary flowers; perfect flowers with 4 nutlet-like carpels and 4–8 stamens.
Myriophyllum, p. 1071

Terrestrial or subaquatic, with simple or broadly foliolate pinnate leaves.

Leaves interruptedly pinnate; inflorescences slender spiciform racemes of yellow flowers; calyx and fruit top-shaped to subglobose, indurated, the margin covered with hooked prickles, inclosing 2 achenes. . . . . *Agrimonia*, p. 865

Leaves simple, at most pinnatifid.

Foliage harsh with adhesive pubescence; stamens 20–200 or more; petals 5–10; styles 3, variously united. . . LOASACEAE, p. 1042

Foliage not adhesive; stamens 2–12 of same number or twice as many as the 2–6 (mostly 4 or 2) petals; styles united; stigmas capitate or clavate, 4–6-lobed. . . . ONAGRACEAE, p. 1051

R. Stamens borne on or at margin of disk at summit of ovary; flowers in umbels or in a dense cyme subtended by 4 petaloid bracts.

Flowers in simple or variously clustered umbels, without showy petaloid bracts; leaves alternate or whorled, compound or simple.

Styles 2–5; ovary 2–5-locular, maturing into a 2–5-seeded berry or drupe. . . . . ARALIACEAE, p. 1077

Styles 2; ovary 2-locular, 2-ovulate; fruit 2 seed-like dry carpels coherent by their inner faces but usually separating in maturity.
UMBELLIFERAE, p. 1078

Flowers in a dense head-like cyme, subtended by 4 (usually) white or pink petal-like spreading bracts; petals and stamens 4; ovary 1-locular, 2-ovulate; style 1; fruit a red drupe with 2 stones; leaves opposite or crowded in a false whorl, simple. . . . CORNACEAE, p. 1105

f. Petals united into a gamopetalous corolla. *S.*

S. Stamens more numerous than lobes or segments of corolla. . *T.*

T. Ovary 1-locular, at least above middle.

Ovary superior; placentae parietal; fruit not drupaceous; leaves stipulate or, if exstipulate, compound, not stellate-pubescent.

Leaves exstipulate, compound, finely dissected; flowers irregular or with dimerous regularity; sepals 2; petals in 2 pairs; stamens 2 sets of 3 each; fruit indehiscent or capsular; watery-stemmed herbs. PAPAVERACEAE, subfam. FUMARIOIDEAE, p. 681

Leaves stipulate, simple or compound; flowers regular or papilionaceous; calyx 4- or 5-toothed; petals 3– usually 5; stamens 3–∞ and mostly distinct, or 10 and monadelphous or diadelphous; fruit a legume or loment; trees, shrubs or herbs. . . . LEGUMINOSAE, p. 879

Ovary inferior, forming a drupe-like fruit, 1-locular above middle, 3-locular below; placentae basal; calyx-teeth short or obsolete; fruit drupaceous; simple leaves exstipulate, scurfy or stellate-pubescent. . STYRACACEAE, p. 1146

*T.*Ovary 2–∞-locular. . *U.*

*U.*Ovary 2-locular, locules 1-ovulate; flowers very irregular, somewhat papilionaceous; petals 3; stamens commonly 8; simple-leaved herbs. POLYGALACEAE, p. 953

*U.*Ovary 3–∞-locular; flowers regular; petals or corolla-lobes 4–8. . . *V.*

*V.*Stamens free from the corolla.

Style 1; leaves simple; anthers opening by apical pores; plant woody or suffrutescent. . . . . ERICACEAE, p. 1114

Styles 5; leaves 3-foliolate; anthers not opening by pores; deciduous-leaved herbs. . . . . OXALIDACEAE, p. 943

*V.*Stamens attached to base or tube of corolla.

Saprophytic herbs without green leaves. . . . PYROLACEAE, p. 1108

Not saprophytic; leaves green. . . *W.*

*W.*Trees, shrubs or undershrubs; anthers mostly 2-locular.

Filaments united into 1–5 groups.

Ovary superior; flowers solitary, axillary or subterminal; large petals silky on back; styles 5; woody capsule 5-valved. THEACEAE, p. 1007

Ovary inferior; flowers clustered in axils or forming leafy-bracted racemes; petals not silky on back; style 1; fruit a 1-seeded drupe or 3-valved drupe-like capsule.

Pubescence simple; calyx-lobes imbricated in bud; a cluster of stamens on base of each petal. . . SYMPLOCACEAE, p. 1146

Pubescence scurfy or stellate; calyx open in bud, its teeth short or obsolete; stamens in a single tube from base of corolla. STYRACACEAE, p. 1146

Filaments free.

Style 1; anthers opening by terminal pores; flowers perfect, with 4–10 fertile stamens; fruit capsular, if a berry the ovary inferior and the seeds small. . . . . . . ERICACEAE, p. 1114

Styles 4; anthers not opening by pores; flowers polygamous, the male with 16 fertile stamens, the female with 8 sterile ones; fruit a large berry with 4–8 large and hard seeds. . . . EBENACEAE, p. 1145

*W.*Herbs; anthers 1-locular.

Leaves alternate, stipulate; sepals 5, united at base, free from ovary; petals mostly showy; stamens many, monadelphous in a column; pistils several in a ring or forming a several-locular capsule; relatively coarse plants. . . . . . MALVACEAE, p. 1000

Leaves basal or 1 pair of opposite 3-lobed cauline exstipulate ones; calyx with hemispherical tube adnate to ovary, its limb 3-toothed; stamens a pair below each sinus of small rotate corolla; fruit a 2–5-seeded drupe; tiny weak herb. . . . . . ADOXACEAE, p. 1342

*S.*Stamens not more numerous than lobes of corolla. . *X.*

*X.*Stamens of same number as corolla-lobes and opposite them.

Corolla appendaged with petaloid scales inside; ovary 5-locular; drupe with 1 stone; trees or shrubs. . . . . . SAPOTACEAE, p. 1144

Corolla without scales inside; ovary 1-locular; herbs.

Style 1; leaves cauline or basal, herbaceous; calyx ebracteolate, herbaceous; fruit a several-seeded capsule. . . . . PRIMULACEAE, p. 1136

Styles 5; leaves at base of scape, coriaceous, persistent; calyx bracteolate, scarious or corolloid, brightly colored; fruit a 1-seeded utricle. PLUMBAGINACEAE, p. 1144

*X.*Stamens alternate with corolla-lobes or fewer. . . *Y.*

*Y.*Ovary free from calyx-tube (superior). . . *Z.*

*Z.*Corolla regular or nearly so. . . **a.**

a.Stamens as many as corolla-lobes. . . **b.**

b.Ovary more than 1 or, if 1, deeply lobed. . . **c.**

c.Ovaries 2 or, if only 1, 2-horned.

Filaments united into a tube, the anthers permanently attached to a large stigmatic body; carpels united only by the common stigmatic mass; follicles many-seeded; seeds usually with a coma. . . . . . ASCLEPIADACEAE, p. 1169

Filaments distinct.

Stipules or stipular membrane or line between the opposite leaves; ovary 2-horned. . . . . . . LOGANIACEAE, p. 1151

Stipules none; ovaries 2.

Leaves reniform, alternate; slender creeping plant with 1-flowered axillary peduncles; the campanulate corolla not exceeding calyx; utricular capsules 2, each 1- or 2-seeded. . . . . . *Dichondra*, p. 1178

Leaves not reniform, opposite or, if alternate, with the stem upright and with terminal cymes; ovaries distinct but styles or stigmas united; fruit a several–∞-seeded follicle. APOCYNACEAE, p. 1166

c. Ovaries deeply 4-lobed around the central style.

Leaves alternate, mostly entire; stamens 5; flowers mostly in scorpioid racemes or on racemose branches of cymes, unrolling from base to apex. . . . . . . BORAGINACEAE, p. 1195

Leaves opposite, toothed; stamens 4; flowers in axillary glomerules or the glomerules forming spiciform racemes. LABIATAE, p. 1213

b. Ovary 1, not deeply lobed. . d.

d. Ovary 1-locular.

Seed 1; corolla scarious and veinless; small aquatic or subaquatic herb with monoecious flowers, the staminate one terminating the scape and with long-exserted filaments, the 2 pistillate ones basal. . . . . *Littorella*, p. 1318

Seeds several–∞; corolla colored and veined.

Leaves opposite, entire, mostly sessile and herbaceous or scale-like and often alternate, mostly glabrous; corolla convolute in bud; the 2 parietal placentae many-seeded; style 1, simple or, if 2-forked, the branches stigmatose down the inner face; capsule dehiscent through the placentae. . GENTIANACEAE, p. 1153

Leaves alternate, commonly petioled, if opposite deeply cleft, lobed or pinnate.

Leaves alternate, long-petioled, simple, round-cordate and repand or 3-foliolate, glabrous; inflorescence not scorpioid; corolla valvate in bud; capsule irregularly bursting or only tardily dehiscent; aquatic or paludal plants.
GENTIANACEAE, subfam. MENYANTHOIDEAE, p. 1166

Leaves alternate, rarely opposite, deeply cleft, lobed or pinnate, commonly pubescent; inflorescence often scorpioid; 2 styles distinct, with terminal stigmas; corolla convolute or imbricated in bud. . . . . . . HYDROPHYLLACEAE, p. 1191

d. Ovary 2–10-locular. . . e.

e. Leafless twining parasites without chlorophyll. . . *Cuscuta*, p. 1182

e. Leafy, with chlorophyll. . . f.

f. Leaves opposite, their bases connected by a stipular line. LOGANIACEAE, p. 1151

f. Leaves alternate or, if opposite, with no trace of stipules. . . g.

g. Stamens free from corolla or nearly so.

Style 1. . . . . ERICACEAE, p. 1114

Style none. . . . . AQUIFOLIACEAE, p. 980

g. Stamens on tube or summit of corolla. . h.

h. Stamens at notches of corolla; low suffruticose leathery-leaved evergreens. . . . . . . DIAPENSIACEAE, p. 1135

h. Stamens on corolla-tube. . . i.

i. Stamens 4; leaves opposite or in basal rosettes.

Leafy-stemmed; corolla petaloid, brightly colored, veiny; fruit 4 nutlets or a drupe. VERBENACEAE, p. 1208

Acaulescent or leafy-stemmed; corolla scarious, veinless; capsule circumscissile. . . PLANTAGINACEAE, p. 1313

i. Stamens 5 or rarely more. . . . j.

j. Fruit indehiscent, drupaceous or of dry nutlets.

Herbs with mostly scorpioid inflorescences; fruit of 2 or 4 seed-like nutlets. . . . BORAGINACEAE, p. 1195

Shrubs with small axillary flowers or clusters; fruit a drupe. . . . . . AQUIFOLIACEAE, p. 980

j. Fruit a few–∞-seeded capsule.
Styles 2 (exceptionally 3).
Plant usually twining or trailing; corolla plicate; a pair of erect ovules borne at base of each locule. CONVOLVULACEAE, p. 1177
Plant not twining; corolla not plicate; 4–∞ ovules borne on parietal placentae. HYDROPHYLLACEAE, p. 1191
Style 1, often branched.
Branches of style or lobes of stigma 3.
Plant twining; corolla plicate. CONVOLVULACEAE, p. 1177
Not twining; corolla not plicate. POLEMONIACEAE, p. 1185
Branches of style or lobes of stigma 2 or 4.
Corolla convolute or twisted in bud; fruit a 4-seeded capsule. CONVOLVULACEAE, p. 1177
Corolla valvate or imbricate in bud; fruit a berry or many-seeded capsule. SOLANACEAE, p. 1251

a. Stamens fewer than corolla-lobes.
Anthers 4, in pairs.
Corolla convolute in bud; capsule elastically or explosively dehiscent, throwing the discoid seeds from their clasping retinacula (hook-like processes from placentae); leaves bearing cystoliths. ACANTHACEAE, p. 1308
Corolla imbricated in bud; fruit 2 or 4 bony nutlets or drupaceous. VERBENACEAE, p. 1208
Anthers 2, rarely 3.
Herbs.
Flowers glomerulate in axils of opposite leaves; ovary 4-lobed, forming nutlets. . . . . LABIATAE, p. 1213
Flowers not glomerulate; fruit a capsule.
Acaulescent or leafy-stemmed; flowers crowded in dense spikes; corolla scarious, nerveless; capsule elongate, not flattened, circumscissile. PLANTAGINACEAE, p. 1313
Caulescent, with opposite leaves; flowers not densely spicate; corolla petaloid, veined; capsule flattened, broad, not circumscissile. . . . . *Veronica*, p. 1280
Trees and shrubs. OLEACEAE, p. 1147

Z. Corolla irregular. . . . **k.**

k. Anthers 5.
Stamens free from corolla; anther-locules opening by terminal pores; shrubs. . . . . *Rhododendron*, p. 1116
Stamens inserted on corolla.
Ovary deeply 4-lobed around style, separating as 4 1-seeded nutlets. . . . . *Echium*, p. 1200
Ovary not deeply lobed, many-ovulate, forming a capsule.
Calyx 5-parted; some or all filaments woolly; capsule dehiscent, 2-valved. . . . . *Verbascum*, p. 1263
Calyx tubular-urceolate, closely surrounding ovary; filaments not woolly; circumscissile top of capsule coming off as a lid. *Hyoscyamus*, p. 1259

k. Anthers 2 or 4. . . . **l.**

l. Ovules 1 in each of the 1–4 locules.
Ovary 4-lobed around the style; plants with aromatic oil. LABIATAE, p. 1213
Ovary not lobed, the style apical; plant not aromatic.
Ovary 1-locular, forming an achene at bottom of the tightly closed unguiculate-toothed calyx; the mature flowers reflexed in the spike. . . . . PHRYMACEAE, p. 1313
Ovary 2–4-locular; flowers not becoming reflexed. VERBENACEAE, p. 1208

l. Ovules 2–many in each locule. . . . **m.**

m. Ovary imperfectly or falsely 4-locular by intrusion of the 2 parietal placentae into the axis, they there diverging as broad lamellae bearing 1 or 2 rows of ovules; large fruit with strong incurved beak. . . . . MARTYNIACEAE, p. 1301

m. Ovary definitely 1- or 2-locular.
Locule 1.
Parasitic on roots of terrestrial hosts, without chlorophyll or true leaves; stamens 4. . . . . OROBANCHACEAE, p. 1302
Not parasitic; insectivorous or carnivorous, either aquatic

or paludal, with traps (bladders) borne on leaves or branches, or with margins of simple rosulate leaves infolded; stamens 2. LENTIBULARIACEAE, p. 1304

Locules 2.

Trees or woody climbers; placentae parietal. . BIGNONIACEAE, p. 1300

Herbs (1 genus of trees); placentae axile.

Seeds mostly numerous, not embraced by retinacula (clasping hook-like processes from placentae); capsule not explosive; leaves without cystoliths. SCROPHULARIACEAE, p. 1261

Seeds 2–12, discoid, clasped by retinacula and thrown by abrupt bursting of the 2-valved firm capsule; leaves bearing cystoliths. . . . . ACANTHACEAE, p. 1308

Y.Ovary inferior, adherent to calyx-tube. . n.

n.Tendril-bearing herbs; anthers often united. . . . . . CUCURBITACEAE, p. 1348

n.Tendrils none. . o.

o.Stamens separate.

Stamens free from the corolla or nearly so, as many as its lobes; leaves alternate; stipules none. . . . . . CAMPANULACEAE, p. 1350

Stamens inserted on the corolla; leaves opposite or whorled.

Stamens 1–3, always fewer than corolla-lobes; limb of calyx nearly or quite obsolete or setiform and developing into pappus; ovary with 1 1-ovulate locule and 2 empty or abortive ones. VALERIANACEAE, p. 1343

Stamens 4 or 5; limb of calyx (when present) not becoming pappus-like.

Ovary 2–5-locular.

Leaves opposite or perfoliate, neither whorled nor with true stipules. . . . . CAPRIFOLIACEAE, p. 1330

Leaves either opposite and stipulate or whorled and exstipulate. RUBIACEAE, p. 1318

Ovary 1-locular.

Shrubs or trees; flowers in compound cymes; fruit a 1-seeded drupe. . . . . *Viburnum*, p. 1338

Herbs; flowers in dense involucrate heads; fruit an achene. DIPSACACEAE, p. 1347

o.Stamens united by their anthers, these forming a ring or tube around the style.

Flowers separate or densely clustered, mostly not involucrate; corolla regular or irregular; fruit a capsule opening near top; seeds numerous. . . . . CAMPANULACEAE, p. 1350

Flowers crowded on a common receptacle, forming a "head" surrounded or subtended by an involucre, or rarely singly involucrate flowers in dense heads; stamens inserted on corolla, the anthers more or less united and forming a tube; calyx-tube adnate to the 1-locular ovary, its limb (the *pappus*) modified into bristles, hairs, awns, teeth, scales, etc., or obsolete; fruit seed-like (*achene*), dry. COMPOSITAE, p. 1357

A.Flowers aborted, asexual or wanting; reproduction vegetative.

Reproductive buds, bulblets or leafy tufts in terminal clusters; leaves mostly basal.

Blades of rosette-leaves covered with stipitate glands; inflorescence changed to a glomerule of small glandular-leaved rosettes. . . . . *Drosera*, p. 729

Blades without stipitate glands.

Leaves coriaceous, unlobed; inflorescence a spike with sterile flowers and hard bulbils or wholly of bulbils. . . . . *Polygonum*, p. 572

Leaves thin, palmately lobed; inflorescence more open, with or without sterile flowers and small leafy tufts or bulbils. . . . *Saxifraga*, p. 737

Reproductive buds moniliform, borne in axils of mostly opposite punctate-dotted leaves on a leafy stem. . . . . *Lysimachia*, p. 1139

Explanations of Abbreviated Names of Authors

(The first name in a combination is indicated only by its initial (or initials) if it appears in full directly above.)

Abrom. — ABROMEIT, Johannes.
Adans. — ADANSON, Michel.
Ait. — AITON, William.
Ait. f. — AITON, William Townsend.
Alef. — ALEFELD, Friedrich.
Alex., E. J. — ALEXANDER, Edward Johnston.
All. — ALLIONI, Carlo.
Almq. — ALMQUIST, Sigfrid.
Anders., E. — ANDERSON, Edgar.
Anders. & Woodson. — A., E., and WOODSON, Robert Everard, Jr.
Anderss. — ANDERSSON, Nils Johan.
Andr. — ANDREWS, Henry C.
Andrz. — ANDRZEJOWSKI, Anton Lukianowicz.
Ångstr. — ÅNGSTRÖM, Johan.
Ard. — ARDUINO, Pietro.
Arn. — ARNOTT, George Arnold Walker.
Arv.-Touv. — ARVET-TOUVET, Casimir.
Aschers. — ASCHERSON, Paul (Friedrich August).
Aschers. & Graebn. — A. and GRAEBNER, Paul.
Aschers. & Magnus. — A. and MAGNUS, Paul.
Aschers. & Prantl. — A. and PRANTL, Karl von.
Aubl. — AUBLET, Jean Baptiste Christophe Fusée.
Aust. — AUSTIN, Coe Finch.
B. & H. — BENTHAM, George, and HOOKER, Joseph Dalton.
Bab. — BABINGTON, Charles Cardale.
Baill. — BAILLON, Henri Ernest.
Balb. — BALBIS, Giovanni Battista.
Baldw. — BALDWIN, William.
Barn. — BARNEOUD, F. Marius.
Barnh. — BARNHART, John Hendley.
Bart. — BARTON, Benjamin Smith.
Bartl. — BARTLING, Friedrich Gottlieb.
Bartr. — BARTRAM, William.
Baumg. — BAUMGARTEN, Johann Christian Gottlob.
Beadle & F. E. Boynt. — BEADLE, Chauncey Delos, and BOYNTON, Frank Ellis.
Beauv. — BEAUVOIS, Ambroise Marie François Joseph Palisot de.
Beck, G. — BECK VON MANNAGETTA, Günther.
Beckh. — BECKHAUS, K.
Benn., Ar. — BENNETT, Arthur.
Benth. — BENTHAM, George.
Berl. — BERLANDIER, Jean Louis.
Bernh. — BERNHARDI, Johann Jacob.
Bertol. — BERTOLONI, Antonio.
Bess. — BESSER, Wilibald Swibert Josef Gottlieb.
Beurl. — BEURLING, Pehr Johan.
Bickn. — BICKNELL, Eugene Pintard.
Bieb. — BIEBERSTEIN, Friedrich August, Marschall von.
Bigel. — BIGELOW, Jacob.
Bisch. — BISCHOFF, Gottlieb Wilhelm.
Biv. — BIVONA-BERNARDI, Antonio.
Bjornstr. — BJORNSTRÖM, Friedrich Johann.
Blake or Blake, S. F. — BLAKE, Sidney Fay. [S. F. Blake used only in *Cyperaceae*, where in competition with S. T. Blake (Austral.)]
Blanch. — BLANCHARD, William Henry.
Boeckl. — BOECKELER, Otto.
Boehm. — BOEHMER, Georg Rudolf.
Boenn. — BOENNINGHAUSEN, Clemens Maria Friedrich von.
Bogenh. — BOGENHARD, Karl.
Boiss. — BOISSIER, Edmond.
Boiss. & Reut. — B. and REUTER, Georges François.
Bong. — BONGARD, Heinrich Gustav.
Borkh. — BORKHAUSEN, Moritz Balthasar.
Boynt., C. L. & Beadle. BOYNTON, Charles Lawrence, and BEADLE, Chauncey Delos.
Boynt., F. E. — BOYNTON, Frank Ellis.
Br., A. — BRAUN, Alexander.
Br., E. L. — BRAUN, Emma Lucy.
Br., P. — BROWNE, Patrick.
Br., R. — BROWN, Robert.
Brain. — BRAINERD, Ezra.
Briq. — BRIQUET, John (-Isaac).
Briq. & Cavill. — B. and CAVILLIER, François.
Britt. — BRITTON, Nathaniel Lord.
Britt. & Rose. — B., N. L., and ROSE, Joseph Nelson.
Britt., E. G. — BRITTON, Elizabeth Gertrude.
Brongn. — BRONGNIART, Adolphe Theodore.
BSP. — BRITTON, Nathaniel Lord, STERNS, Emerson Ellick, and POGGENBERG, Justus Ferdinand.
Buchenau & Fern. — BUCHENAU, Franz (Georg Philipp), and FERNALD, Merritt Lyndon.
Buckl. — BUCKLEY, Samuel Botsford.
Burgsd. — BURGSDORFF, Friedrich August Ludwig von.

Burm. f. — BURMANN, Nikolaus Laurens.
Butt. & Abbe. — BUTTERS, Frederic King, and ABBE, Ernst Cleveland.
Butt. & St. John. — B. and ST. JOHN, Harold.
C. & R. — COULTER, John Merle, and ROSE, Joseph Nelson.
C. & S. — CHAMISSO, Adalbert von, and SCHLECHTENDAL, Diedrich Franz Leonhard von.
Camb. — CAMBESSEDES, Jacques.
Campd. — CAMPDERA, François.
Carr. — CARRIÈRE, Élie Abel.
Casp. — CASPARY, Robert.
Cass. — CASSINI, Henri.
Çav. — CAVANILLES, Antonio José.
Čelak. — ČELAKOVSKY, Ladislav.
Cerv. — CERVANTES, Vicente.
Cham. — CHAMISSO, Adalbert von.
Chapm. — CHAPMAN, Alvan Wentworth.
Chev. — CHEVALLIER, Auguste.
Chois. — CHOISY, Jacques-Denis.
Christens. — CHRISTENSEN, Carl (Fredrik Albert).
Clairv. — CLAIRVILLE, Joseph Philippe de.
Clayt. — CLAYTON, John.
Coss. & Germ. — COSSON, Ernest, and GERMAIN DE SAINT PIERRE, Ernest.
Coult. — COULTER, John Merle.
Cov. & Britt. — COVILLE, Frederick Vernon, and BRITTON, Nathaniel Lord.
Cutler & Anders. — CUTLER, Hugh, and ANDERSON, Edgar.
Cyrill. — CIRILLO, Domenico [Italian form of name].
Darl. — DARLINGTON, William.
Davenp. — DAVENPORT, George Edward.
DC. — DECANDOLLE, Augustin Pyramus.
DC., A. — DECANDOLLE, Alphonse (Louis Pierre Pyramus).
Dcne. DECAISNE, Joseph.
Desf. — DESFONTAINES, Réné Louiche.
Desmaz. — DESMAZIÈRES, Jean Baptiste Henri Joseph.
Desr. — DESROUSSEAUX, Louis Auguste Joseph.
Desv. — DESVAUX, Augustin Nicaise.
Dew. — DEWEY, Chester.
Dietr. — DIETRICH, Albert.
Döll & Aschers. — DÖLL, Johann Christof, and ASCHERSON, Paul.
Dougl. — DOUGLAS, David.
Drej. — DREJER, Salomon Thomas Nicolai.
Dudl. — DUDLEY, William Russell.
Dufr. — DUFRESNE, Pierre.
Duham. — DUHAMEL DE MONCEAU, Henri Louis.
Dumont. — DUMONT DE COURSET, Georges Louis Marie.
D'Urv. — DUMONT D'URVILLE, Jules Sébastien César.
Dumort. — DUMORTIER, Barthélemy Charles.
Dur. — DURIEU DE MAISONNEUVE, M. C.
Eat. — EATON, Amos.
Eat., D. C. — EATON, Daniel Cady.
Eat., H. H. — EATON, Hezekiah Hulbert.
Eat., R. J. — EATON, Richard Jefferson.
Eat. & Grisc. — E., R. J., and GRISCOM, Ludlow.
Egglest. — EGGLESTON, Willard Webster.
Egglest. & Sheld. — E. and SHELDON, Edmund Perry.
Ehrh. — EHRHART, Friedrich.
Ehrh., B. — EHRHART, Balthasar.
Ell. — ELLIOTT, Stephen.
Endl. — ENDLICHER, Stephan Ladislaus.
Engelm. — ENGELMANN, George.
Engelm. & Gray. — E. and GRAY, Asa.
Epl. — EPLING, Carl Clawson.
Esch. — ESCHSCHOLTZ, Johann Friedrich.
Farw. — FARWELL, Oliver Atkins.
Fern. — FERNALD, Merritt Lyndon.
Fern. & Bissell. — F. and BISSELL, Charles Humphrey.
Fern. & Brack. — F. and BRACKETT, Amelia Ellen.
Fern. & Eames. — F. and EAMES, Arthur Johnson.
Fern. & Gale. — F. and GALE (CROSS), Shirley.
Fern. & Grisc. — F. and GRISCOM, Ludlow.
Fern. & Harris. — F. and HARRIS, Stuart Kimball.
Fern. & Hodgdon. — F. and HODGDON, Albion Reed.
Fern. & Knowlt. — F. and KNOWLTON, Clarence Hinckley.
Fern. & Long. — F. and LONG, Bayard.
Fern. & Rehd. — F. and REHDER, Alfred.
Fern. & St. John. — F. and ST. JOHN, Harold.
Fern. & Schub. — F. and SCHUBERT, Bernice Giduz.
Fern. & Weath. — F. and WEATHERBY, Charles Alfred.
Fern. & Wieg. — F. and WIEGAND, Karl McKay.
Fisch. — FISCHER, Friedrich Ernst Ludwig von.
Fisch. & Lall. — F. and AVE-LALLEMANT, Julius Edward Leopold.
Fisch. & Mey. — F. and MEYER, Karl Anton.
Forsk. — FORSKÅL, Pehr.
Forst. — FORSTER, Johann Reinhold and Georg.
Forst., T. F. — FORSTER, Thomas Furly.
Foug. — FOUGEROUX, Auguste Denis.
Fourn. — FOURNIER, Eugène.
Fourr. — FOURREAU, Jules.
Franch. — FRANCHET, Adrien.
Fresn. — FRESENIUS, Johann Baptist Georg Wolfgang.
Froel. — FROELICH, Joseph Aloys.
Gaertn. — GAERTNER, Joseph.

Gaertn., Mey. & Scherb. — GAERTNER, Philipp Gottfried, MEYER, Bernhard, and SCHERBIUS, Johannes.
Gal. — GALEOTTI, Henri.
Gaud. — GAUDICHAUD-BEAUPRÉ, Charles.
Gay & Dur. — GAY, Jacques, and DURIEU DE MAISONNEUVE, M. C.
Gilib. — GILIBERT, Jean Emmanuel.
Gmel. — GMELIN, Samuel Gottlieb.
Gmel., J. F. — GMELIN, Johann Friedrich.
Gmel., J. G. — GMELIN, Johann Georg.
Gmel., K. C. — GMELIN, Karl Christian.
Godr. — GODRON, Dominique Alexandre.
Good. — GOODENOUGH, Samuel.
Grab. — GRABOWSKI, Heinrich Emanuel.
Graebn. — GRAEBNER, Paul.
Grant & Epl. — GRANT, Elizabeth, and EPLING, Carl Clawson.
Greenm. — GREENMAN, Jesse More.
Greenm. & Larisey. — G. and LARISEY, Mary Maxine.
Gren. — GRENIER, Charles.
Grev. — GREVILLE, Robert Kaye.
Grev. & Hook. — G. and HOOKER, William Jackson.
Griseb. — GRISEBACH, Heinrich Rudolph August.
Gronov. — GRONOVIUS, Jan Fredrik.
Guenth., Grab. & Wimm. — GUENTHER, Karl Christian, GRABOWSKI, Heinrich, and WIMMER, Friedrich.
Gunn. — GUNNERUS, Johann Ernst.
Guss. — GUSSONI, Giovanni.
H. & A. — HOOKER, William Jackson, and ARNOTT, George Arnold Walker.
Hack. — HACKEL, Eduard.
Hagstr. — HAGSTRÖM, Johan Oskar.
Hall. — HALLER, Albrecht.
Hall. f. — HALLER, Gottlieb Emmanuel von.
Hand.-Maz. — HANDEL-MAZZETTI, Heinrich.
Hartm. — HARTMAN, Carl Johan.
Hassk. — HASSKARL, Justus Carl.
Haussk. — HAUSSKNECHT, Carl.
Haw. — HAWORTH, Adrian Hardy.
HBK. — HUMBOLDT, Friedrich Wilhelm Heinrich Alexander von, BONPLAND, Aimé, and KUNTH, Carl Sigismund.
Hedl. — HEDLUND, Johan Theodor.
Hedw. f. — HEDWIG, Romanus Adolph.
Hegelm. — HEGELMAIER, Friedrich.
Hegetschw. — HEGETSCHWEILER, Johann.
Herb. — HERBERT, William.
Herm., F. J. — HERMANN, Frederick Joseph.
Herrm. — HERRMANN, W.
Heynh. — HEYNHOLD, Gustav.
Hieron. — HIERONYMUS, Georg.
Hitchc. — HITCHCOCK, Albert Spear.
Hitchc., E. — HITCHCOCK, Edward.
Hochr. — HOCHREUTINER, Bénédict Pierre Georges.
Hoffm. — HOFFMANN, Georg Franz.
Hoffm., R. — HOFFMANN, Ralph.
Hoffmgg. — HOFFMANNSEGG, Johann Centurius, Graf von.
Hoffmgg. & Link. — H. and LINK, Heinrich Friedrich.
Hollick & Britt. — HOLLICK, Arthur, and BRITTON, Nathaniel Lord.
Holmb. — HOLMBERG, Otto Rudolph.
Holz. — HOLZINGER, John M.
Honck. — HONCKENY, Gerhard August.
Hook. — HOOKER, William Jackson.
Hook. f. — HOOKER, Joseph Dalton.
Hornem. — HORNEMANN, Jens Wilken.
Houtt. — HOUTTUYN, Martin.
Huds. — HUDSON, William.
Humb. — HUMBOLDT, Friedrich Wilhelm Heinrich Alexander von.
Humb. & Bonpl. — H. and BONPLAND, Aimé.
Jacq. — JACQUIN, Nicolaus Joseph.
Jaeg. — JAEGER, Hermann.
Jancz. — JANCZEWSKI VON GLINKA, Edward.
Jord. — JORDAN, Alexis.
Juss. — JUSSIEU, Antoine Laurent de.
Juss., B. — JUSSIEU, Bernard de.
Karst. — KARSTEN, Hermann.
Ker. — BELLENDEN, John, or more correctly BELLENDEN, John Ker; before 1804 GAWLER, John.
Kern. — KERNER, Johann Simon von.
Kihlm. — KIHLMAN, Alfred Oswald.
Kindb. — KINDBERG, Nils Conrad.
Kirch. — KIRCHNER, Georg.
Kit. — KITAIBEL, Paul.
Koch, A. — KOCH, Alfred.
Koch, H. — KOCH, (Otto Wilhelm) Heinrich.
Koch, J. F. W. — KOCH, Johann Friedrich Wilhelm.
Koch, K. — KOCH, Karl.
Koch, M. — KOCH, Max.
Koch, W. D. J. — KOCH, Wilhelm Daniel Joseph.
Koel. — KOELER, Georg Ludwig.
Kostel. — KOSTELETZKY, Vincenz Franz.
Krecz. — KRECZETOWICZ, V.
Krock. — KROCKER, Anton Johann.
Kronf. — KRONFELD, Moriz.
Ktze. — KUNTZE, Otto.
Kükenth. — KÜKENTHAL, Georg.
L. — LINNAEUS, Carolus, or LINNÉ, Carl von.
L. f. — LINNÉ, Carl von (the son).
Lacks. — LACKSCHEWITZ, Paul.
Laestad. — LAESTADIUS, Lars Levi.
Lag. — LAGASCA, Mariano.
Lag. & Rodr. — L. and RODRIGUEZ, José Demetrio.
Lak. — LAKELA, Olga.
Lall. — AVE-LALLEMANT, Julius Edward Leopold.
Lam. — LAMARCK, Jean Baptiste Pierre Antoine de Monet de.
Lamb. — LAMBERT, Aylmer Bourke.

La Pey. — La Peyrouse, Philippe Picot, Baron de.
Lat. — Latourette, Marc Antoine Louis Claret de.
Laxm. — Laxmann, Eric.
Leavenw. — Leavenworth, Melines Conklin.
Ledeb. — Ledebour, Carl Friedrich von.
Lehm. — Lehmann, Johann Georg Christian.
Lej. & Court. — Lejeune, Louis Simon Alexander, and Courtois, Richard.
Lesp. & Thév. — Lespinasse, Gustave, and Théveneau, A.
Less. — Lessing, Christian Friedrich.
Lestib. f. — Lestiboudois, Thémistocle.
Lévl. & Vaniot. — Léveillé, August Abel Hector, and Vaniot, Eugène.
Leyss. — Leysser, Friedrich Wilhelm von.
L'Hér. — L'Héritier de Brutelle, Charles Louis.
Liebm. — Liebmann, Frederik Michael.
Lightf. — Lightfoot, John.
Lilj. — Liljeblad, Samuel.
Lindb. f. — Lindberg, Harald.
Lindbl. — Lindblom, Alexis Eduard.
Lindl. — Lindley, John.
Lindm. f. — Lindman, Carl Axel Magnus.
Lloyd & Underw. — Lloyd, Francis Ernest, and Underwood, Lucien Marcus.
Lodd. — Loddiges, Conrad.
Loefl. — Loefling, Pehr.
Loisel. — Loiseleur-Deslongchamps, Jean Louis Auguste.
Loud. — Loudon, John Claudius.
Lour. — Loureiro, Juan.
Luerss. — Luerssen, Christian.
Macbr. — Macbride, James Francis.
Macfad. — Macfadyen, James.
Mackenz. — Mackenzie, Kenneth Kent.
Mackenz. & Bush. — M. and Bush, Benjamin Franklin.
MacM. — MacMillan, Conway.
Mansf. — Mansfeld, Rudolph.
Marsh. — Marshall, Humphrey.
Mart. — Martius, Karl Friedrich Philipp von.
Mart. & Schrad. — M. and Schrader, Heinrich Adolph.
Mast. — Masters, Maxwell Tylden.
Math. & Const. — Mathias, Mildred Esther, and Constance, Lincoln.
Mattf. & Kükenth. — Mattfeld, Johannes, and Kükenthal, Georg.
Maxim. — Maximowicz, Carl Johann.
Medic. — Medicus, Friedrich Casimir.
Meerb. — Meerburgh, Nicolaas.
Meisn. — Meisner, Carl Friedrich.
Merr. — Merrill, Elmer Drew.
Mert. — Mertens, Franz Carl.
Mert. & Koch. — M. and Koch, Wilhelm Daniel Joseph.
Mett. — Mettenius, Georg Heinrich.
Mey. — Meyer, Ernst Heinrich Friedrich.
Mey., C. A. — Meyer, Carl Anton.
Mey. & Bunge. — M., C. A., and Bunge, Alexander von.
Mey., G. F. W. — Meyer, Georg Friedrich Wilhelm.
Michx. — Michaux, André.
Michx. f. — Michaux, François André.
Mill. — Miller, Philip.
Mill., E. S. — Miller, Elihu Sanford.
Mill., G. S. — Miller, Gerritt Smith.
Mill. & Standl. — M., G. S., and Standley, Paul Carpenter.
Millsp. — Millspaugh, Charles Frederick.
Miq. — Miquel, Friedrich Anton Wilhelm.
Moc. & Sessé — Mociño, José Mariano, and Sessé, Martín.
Moq. — Moquin-Tandon, Alfred.
Mort., C. V. — Morton, Conrad Vernon.
Muell. Arg. — Mueller, Jean (of Aargau).
Muell., O. F. — Mueller, Otto Frederik.
Muell., P. J. — Mueller, Philipp Jakob.
Muenchh. — Muenchhausen, Otto, Freiherr von.
Muhl. — Muhlenberg, Gotthilf Henry Ernest.
Murb. — Murbeck, Svante.
Murr. — Murray, Johann Andreas.
Nannf. — Nannfeldt, J. A.
Neck. — Necker, Noel Joseph de.
Nees. — Nees von Esenbeck, Christian Gottfried.
Nees & Eberm. — N. and Ebermaier, Karl Heinrich.
Neilr. — Neilreich, August.
Nels. — Nelson, Aven.
Nels. & Cockerell. — N. and Cockerell, Theodore Dru Allison.
Nels. & Macbr. — N. and Macbride, James Francis.
Neum. — Neuman, Leopold Martin.
Newm. — Newman, Edward.
Nieuwl. — Nieuwland, Julius Arthur.
Nilss., A. — Nilsson, (Lars) Albert.
Nilss., Hj. — Nilsson, Nils Hjalmar.
Nutt. — Nuttall, Thomas.
Nym. — Nyman, Carl Frederik.
Ostenf. — Ostenfeld, Carl Emil Hansen.
Pall. — Pallas, Peter Simon.
Palmer & Steyerm. — Palmer, Ernest Jesse, and Steyermark, Julian Alfred.
Parl. — Parlatore, Filippo.
Pax & Hoffm. — Pax, Ferdinand, and Hoffmann, Käthe.
Perry & Fern. — Perry, Lily May, and Fernald, Merritt Lyndon.
Peterm. — Petermann, Wilhelm Ludwig.
Pfeiff., N. — Pfeiffer, Norma.
Phil. — Philippi, Rudolf Amandus.
Planch. — Planchon, Jules Émile.
Poir. — Poiret, Jean Louis Marie.

Poll. — Pollich, Johann Adam.
R. & P. — Ruiz Lopez, Hipolito, and Pavon, Josef.
R. & S. — Roemer, Johann Jakob, and Schultes, Joseph August.
Rabenh. — Rabenhorst, Ludwig.
Raf. — Rafinesque-Schmaltz, Constantine Samuel.
Raunk. — Raunkiaer, Christen.
Rech. f. — Rechinger, Karl.
Rehd. — Rehder, Alfred.
Rehd. & Wilson. — R. and Wilson, Ernest Henry.
Reichenb. — Reichenbach, Heinrich Gottlieb Ludwig.
Reichenb. f. — Reichenbach, Heinrich Gustav.
Retz. — Retzius, Anders Johan.
Reut. — Reuter, Georges-François.
Richards. — Richardson, John.
Robins. — Robinson, Benjamin Lincoln.
Robins. & Schrenk. — R. and Schrenk, Hermann von.
Robins., J. — Robinson, John.
Roem. — Roemer, Max J.
Rohrb. Rohrbach, Paul.
Rosend. — Rosendahl, Carl Otto.
Rosend. & Butt. — R. and Butters, Frederic King.
Rosend., Butt. & Lak. — R., B. and Lakela, Olga.
Rosend. & Moore. — R. and Moore, John William.
Rostk. — Rostkovius, Friedrich Wilhelm Gottlieb.
Rostk. & Schmidt. — R. and Schmidt, Wilhelm Ludwig Ewald.
Rottb. — Rottboell, Christen Fries.
Roxb. — Roxburgh, William.
Rupr. — Ruprecht, Franz Joseph.
Rydb. — Rydberg, Per Axel.
St. Hil. — St. Hilaire, Auguste de.
Salisb. — Salisbury, Richard Anthony.
Sam. — Samuelsson, Gunnar.
Sarg. — Sargent, Charles Sprague.
Savat. — Savatier, Ludwig.
Sch. Bip. — Schultz, Karl Heinrich (distinguished as **Bipontinus**, *i.e.*, of Zweibrücken).
Schaffn. — Schaffner, John Henry.
Schindl. — Schindler, Anton Karl.
Schinz & Thell. — Schinz, Hans, and Thellung, Albert.
Schleich. — Schleicher, Johann Christoph.
Schleid. — Schleiden, Matthias Jacob.
Schneid. — Schneider, Camillo Karl.
Schnizl. — Schnizlein, Adalbert.
Schrad. — Schrader, Heinrich Adolph.
Schreb. — Schreber, Johann Daniel Christian von.
Schub. — Schubert, Bernice Giduz.
Schum. — Schumann, Karl.
Schweig. & Koerte. — Schweigger, August Friedrich, and Koerte, Franz.
Schwein. — Schweinitz, Lewis David de.
Schwein. & Torr. — S. and Torrey, John.
Scop. — Scopoli, Johann Anton.
Scribn. — Lamson-Scribner, Frank.
Scribn. & C. R. Ball. — L.-S. and Ball, Carleton Roy.
Scribn. & Merr. — L.-S. and Merrill, Elmer Drew.
Scribn. & Sm. — L.-S. and Smith, Jared Gage.
Ser. — Seringe, Nicolas Charles.
Sheld. — Sheldon, Edmund Perry.
Shuttlw. — Shuttleworth, Robert.
Sibth. — Sibthorp, John.
Sibth. & Sm. — S. and Smith, James Edward.
Sieb. — Siebold, Philipp Franz von.
Sieb. & Zucc. — S. and Zuccarini, Joseph Gerhard.
Sincl. Sinclair, George.
Sm. — Smith, James Edward.
Sm., H. Smith, Harald.
Sm., J. — Smith, John.
Sm., J. D. — Smith, John Donnell.
Sm., J. G. — Smith, Jared Gage.
Sm., L. B. — Smith, Lyman Bradford.
Sm. & Bush. — Smith, Jared Gage, and Bush, Benjamin Franklin.
Sobol. — Sobolewski, Gregor.
Soland. — Solander, Daniel.
Spegaz. — Spegazzini, Carlos.
Spreng. — Sprengel, Kurt (Polykarp Joachim).
Standl. — Standley, Paul Carpenter.
Sternb. — Sternberg, Caspar M. von.
Sternb. & Hoppe. — S. and Hoppe, David Heinrich.
Steud. — Steudel, Ernst Gottlieb.
Stev. — Steven, Christian.
Steyerm. — Steyermark, Julian Alfred.
Sudw. — Sudworth, George Bishop.
Suksd. — Suksdorf, Wilhelm Nicholaus.
Sulliv. — Sullivant, William Starling.
Sw. — Swartz, Olaf.
T. & G. — Torrey, John, and Gray, Asa.
Ten. — Tenore, Michele.
Thed. — Thedenius, Knut Fredrik.
Thell. — Thellung, Albert.
Thuill. — Thuiller, Jean Louis.
Thunb. — Thunberg, Carl Pehr.
Thurb. — Thurber, George.
Tidestr. — Tidestrom, Ivar.
Torr. — Torrey, John.
Torr. & Hook. — T. and Hooker, William Jackson.
Torr., G. S. — Torrey, George Safford.
Tratt. — Trattinnick, Leopold.
Trautv. — Trautvetter, Ernst Rudolf von.
Trel. — Trelease, William.
Trev. — Treviranus, Christian Ludolf.

Trin. — TRINIUS, Karl Bernhard von.
Trin. & Rupr. — T. and RUPRECHT, Franz Josef.
Tuckerm. — TUCKERMAN, Edward.
Turcz. — TURCZANINOW, Nicolaus.
Underw. — UNDERWOOD, Lucien Marcus.
Vent. — VENTENAT, Étienne Pierre.
Vict. — Frère MARIE-VICTORIN (Conrad Kirouac).
Vict. & Rousseau. — M.-V. and ROUSSEAU, Jacques.
Vierh. — VIERHAPPER, Friedrich.
Vill. — VILLARS, Dominique.
Vis. — VISIANI, Roberto de.
Vitm. — VITMAN, Fulgenzio.
Wahlb. — WAHLBERG, Pehr Fredrik.
Wahlenb. — WAHLENBERG, Georg.
Waldst. & Kit. — WALDSTEIN, Franz de Paula Adam, Graf von, and KITAIBEL, Paul.
Wallm. — WALLMANN, Johan Hacquinus.
Wallr. — WALLROTH, Carl Friedrich Wilhelm.
Walp. — WALPERS, Wilhelm Gerhard.
Walt. — WALTER, Thomas.
Wang. — WANGENHEIM, Friedrich Adam Julius von.
Warnst. — WARNSTORF, Carl.
Wats., E. E. — WATSON, Elba Emanuel.
Wats., P. W. — WATSON, Peter William.
Wats., S. — WATSON, Sereno.
Wats. & Coult. — W., S., and COULTER, John Merle.
Weath. — WEATHERBY, Charles Alfred.
Weath. & Adams. — W. and ADAMS, John.
Web. — WEBER, Friedrich.
Webb. & Moq. — WEBB, Philipp Barker, and MOQUIN-TANDON, Alfred.
Wedd. — WEDDELL, Hugh Algernon.
Wettst. — WETTSTEIN, Richard von.
Wieg. — WIEGAND, Karl McKay.
Wiesb. — WIESBAUER, Johann Baptist.
Willd. — WILLDENOW, Karl Ludwig.
Wimm. — WIMMER, Friedrich.
Wimm. & Grab. — W. and GRABOWSKI, Heinrich.
Wimm., E. — WIMMER, (Franz) Elfried.
With. — WITHERING, William.
Wolfg. — WOLFGANG, Johann Friedreich.
Wormsk. — WORMSKIOLD, M.
Wulf. — WULFEN, Franz Xavier.
Wulf. & Schreb. — W. and SCHREBER, Johann Christian Daniel von.
Zimm., F. — ZIMMERMANN, Ferdinand Joseph von.

Other Abbreviations and Signs

Abund., abundant.
Adj., adjacent.
Adv., adventive, *i.e.*, as yet only casual and sporadic.
Afr., Africa.
Ala., Alabama.
Alta., Alberta.
Alt., altitude.
Am., America or American.
Arct., Arctic.
Ariz., Arizona.
Ark., Arkansas.
Atl., Atlantic.
Austral., Australia.
Auth., authors.
B.C., British Columbia.
Calif., California.
Can., Canada.
C.B., Cape Breton I., Nova Scotia.
Centr., central.
Cleist. fls., cleistogamous flowers.
Cm., centimeter (or centimeters), the hundredth part of a meter = about ⅖ of an inch.
Colo., Colorado.
Cosmop., cosmopolitan.
Ct., Connecticut.
Ct. val., valley of the Connecticut River, New England.
D.C., District of Columbia.
Del., Delaware.
Dm., decimeter (or decimeters), the tenth part of a meter = about 4 inches.
E., east or eastern.
Eastw., eastward.
Esc., escaping or escaped.
Estab., established.
Eu., Europe.
F., filius, son, or the younger (when following the name of an author).
F., forma; *ff.*, formae.
Fla., Florida.
Fls., flowers; *stam. fls.*, staminate flowers; *pist. fls.*, pistillate flowers.
Fr., fruit.
Freq., frequent.
Ga., Georgia.
Greenl., Greenland.
I. or *Id.*, island.
Ia., Iowa.
Ida., Idaho.
Ill., Illinois.
Ind., Indiana.
Indig., indigenous.
Infreq., infrequent.
Introd., introduced, *i.e.*, brought in intentionally, as through horticulture, etc.
Kans., Kansas.
Ky., Kentucky.
L., lake.
L.I., Long Island, New York.
La., Louisiana.
Lab., Labrador.
L. Sup., Lake Superior.
M., meter (or meters) = about 39⅓ inches.
Mackenz., Mackenzie District, Canada.
M.I., Magdalen Islands, Quebec, Canada.
Man., Manitoba.
Mass., Massachusetts.
Md., Maryland.
Me., Maine.
Mediterr. reg., Mediterranean region.
Mex., Mexico.
Mich., Michigan.
Minn., Minnesota.
Miss., Mississippi.
Mm., millimeter (or millimeters) = about 1/25 of an inch.
Mo., Missouri.
Mont., Montana.
Mt., mts., mountain, mountains.
N., north or northern.
N. Am., North America or North American.
N.B., New Brunswick.
N.C., North Carolina.
N.D., North Dakota.
N.E., New England.
N.H., New Hampshire.
N.J., New Jersey.
N.M., New Mexico.
N.S., Nova Scotia.
N.Y., New York.
Natzd., naturalized, *i.e.*, thoroughly established but not native.
Ne., northeast or northeastern.
Neb., Nebraska.
Nev., Nevada.
Nfld., Newfoundland.
No., number.
Northw., northward.
Nw., northwest.
O., Ohio.
Okla., Oklahoma.
Ont., Ontario.

Oreg., Oregon.
Pa., Pennsylvania.
P.E.I., Prince Edward Island.
Pen., peninsula.
Que., province of Quebec.
R., river.
R.I., Rhode Island.
Reg., region or regions.
S., south or southern.
S. Am., South America or South American.
S.C., South Carolina.
S.D., South Dakota.
Sask., Saskatchewan.
Scotl., Scotland.
Sd., Sound.
Ser., Series.
Southw., southward.
Southwestw., southwestward.
St. P. et Miq., St. Pierre et Miquelon, south of Newfoundland.
Subgen., subgenus.
Subtrop., subtropical.
Sw., southwest.
Subsp., subspecies.
Syst., system.
Temp., temperate.
Tenn., Tennessee.
Tex., Texas.
Trop., tropics or tropical.
Ung., Ungava District, Canada.
Va., Virginia.
Val., valley.
Vt., Vermont.
W., west or western.
W.I., West Indies.
W. Va., West Virginia.
Wash., Washington.
Westw., westward.
Wisc., Wisconsin.
Wyo., Wyoming.
Yuk., Yukon Terr.

Figures or words connected by the short dash indicate the extremes of variation, as "5–12 mm. long, few-many-flowered", *i.e.*, varying from five to twelve millimeters in length and from few- to many-flowered.

∞ Of indefinite number, usually many.

μ Micron, *pl.*, micra (pronounced mu). A micron, the millionth part of a meter, a measure used in microscopic studies.

§ Section.

? Indicates doubt.

♀ Bearing pistils or archegonia but neither stamens nor antheridia.

♂ Bearing stamens or antheridia but neither pistils nor archegonia.

× Crossed with, the symbol for a hybrid.

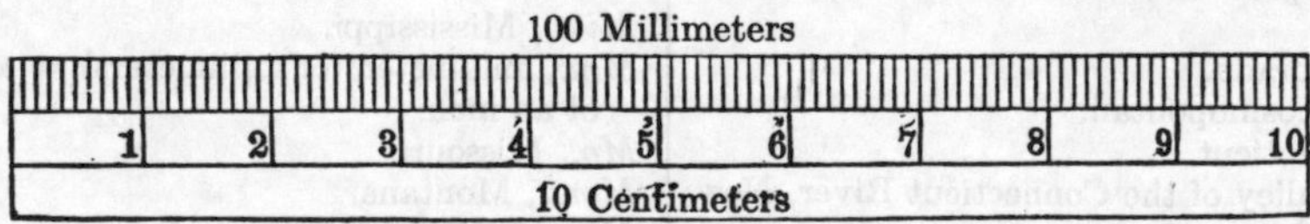

⅒ Meter, or 1 Decimeter

Families, etc.	Genera: Native	Genera: Introd., Adventive or Natzd.	Species: Native	Species: Introd., Adventive or Natzd.	Varieties: Native	Varieties: Introd., Adventive or Natzd.	Forms and Named Hybrids: Native	Forms and Named Hybrids: Introd., Adventive or Natzd.
Division I. PTERIDOPHYTA								
Fam. 1. Equisetaceae	1		11		4		34	
2. Lycopodiaceae	1		13		13		5	
3. Selaginellaceae	1		3					
4. Isoëtaceae	1		11		3		2	
5. Ophioglossaceae	2		9		7		9	
6. Osmundaceae	1		3		1		14	
7. Schizaeaceae	2		2					
8. Hymenophyllaceae	1		1					
9. Polypodiaceae	19		64	1	17		61	
10. Marsileaceae	1		1	1				
11. Salviniaceae	2		2	1				
Division II. SPERMATOPHYTA								
Subdivision I. Gymnospermae								
Fam. 12. Taxaceae	1		1					
13. Pinaceae	9		25	4	5		12	
Subdivision II. Angiospermae								
Class I. *MONOCOTYLEDONEAE*								
Fam. 14. Typhaceae	1		4				1	
15. Sparganiaceae	1		10		1			
16. Zosteraceae	4		40	1	23		10	
17. Najadaceae	1		6					
18. Juncaginaceae	2		4					
19. Alismataceae	4		24		4		11	
20. Butomaceae		1		1				
21. Hydrocharitaceae	3		4	1				
22. Gramineae	75	35	389	98	131	21	46	18
23. Cyperaceae	17		451	22	123	2	85	
24. Araceae	6	1	10	1			11	
25. Lemnaceae	4		10		1			
26. Xyridaceae	1		12		1		2	
27. Eriocaulaceae	2		5					
28. Bromeliaceae	1		1					
29. Commelinaceae	3		14	1	3	1	4	
30. Pontederiaceae	2	1	5	1			5	
31. Juncaceae	2		65	2	28		13	
32. Liliaceae	28	7	85	25	24	2	29	2
33. Haemodoraceae	1		1					
34. Dioscoreaceae	1		3	1			1	
35. Amaryllidaceae	6	3	11	6			2	
36. Iridaceae	3	1	19	4	4		4	
37. Marantaceae	1		1					
38. Burmanniaceae	2		2					
39. Orchidaceae	20	1	74	1	19		25	
Class II. *DICOTYLEDONEAE*								
Subclass I. Archichlamydeae								
Fam. 40. Saururaceae	1		1					
41. Salicaceae	2		52	13	31	3	24	2
42. Myricaceae	2		6		2		1	
43. Leitneriaceae	1		1					
44. Juglandaceae	2		12		9		5	
45. Corylaceae	5		26	4	14		18	1
46. Fagaceae	3		30	1	9		63	
47. Ulmaceae	3		9	3	5		3	

Families, etc.	Genera: Native	Genera: Introd., Adventive or Natzd.	Species: Native	Species: Introd., Adventive or Natzd.	Varieties: Native	Varieties: Introd., Adventive or Natzd.	Forms and Named Hybrids: Native	Forms and Named Hybrids: Introd., Adventive or Natzd.
Fam. 48. Moraceae	2	2	2	3				
49. Cannabinaceae	1	1	1	2				
50. Urticaceae	5		11	3	2			2
51. Santalaceae	5		7					
52. Loranthaceae	2		2					
53. Aristolochiaceae	2		10	1	4		1	
54. Polygonaceae	7	1	48	26	28	9	14	4
55. Chenopodiaceae	8	5	27	24	6	3		
56. Amaranthaceae	4	2	11	14	4			
57. Nyctaginaceae	1		4	1				
58. Phytolaccaceae	1		1					
59. Aizoaceae	2	5	2	5				
60. Portulacaceae	4		12	3			2	
61. Caryophyllaceae	8	10	56	44	24	4	7	3
62. Ceratophyllaceae	1		2					
63. Nymphaeaçeae	5		14	2	2		3	
64. Ranunculaceae	16	3	88	15	34	9	27	1
65. Berberidaceae	5		5	2			3	1
66. Menispermaceae	3		4					
67. Magnoliaceae	2		6		1			
68. Calycanthaceae	1		2					
69. Annonaceae	1		2					
70. Lauraceae	4		5		1		3	
71. Papaveraceae	6	6	14	11	2		3	1
72. Capparidaceae	3		4	1				
73. Cruciferae	20	29	71	76	33	8	5	7
74. Resedaceae		1		3				
75. Sarraceniaceae	1		2		1		2	
76. Droseraceae	1		7		1		4	
77. Podostemaceae	1		1					
78. Crassulaceae	2	1	7	11	1			
79. Saxifragaceae	15	2	59	9	17		17	
80. Hamamelidaceae	3		4		1		1	
81. Platanaceae	1		1		1			
82. Rosaceae	21	6	489	62	135	1	49	
83. Leguminosae	43	25	168	69	68	13	40	2
84. Linaceae	1	1	9	3	2			
85. Oxalidaceae	1		8	2	7		8	
86. Geraniaceae	1	1	5	10	2		1	
87. Zygophyllaceae	1	1	1	2				
88. Rutaceae	2	3	3	3	1		1	
89. Simaroubaceae		1		1				
90. Meliaceae		1		1				
91. Polygalaceae	1		16		7		7	
92. Euphorbiaceae	9	2	44	13	8		1	
93. Callitrichaceae	1		6	1				
94. Buxaceae	1		1	1				
95. Empetraceae	2		4				2	
96. Limnanthaceae	1		1					
97. Anacardiaceae	2		8	1	7		10	
98. Cyrillaceae	1		1		1			
99. Aquifoliaceae	2		10		4		6	
100. Celastraceae	3		5	4				
101. Staphyleaceae	1		1					
102. Aceraceae	1		9	5	7		6	
103. Hippocastanaceae	1		5	1	2			
104. Sapindaceae	1	2	1	2				
105. Balsaminaceae	1		2	3			7	1

Families, etc.	Genera: Native	Genera: Introd., Adventive or Natzd.	Species: Native	Species: Introd., Adventive or Natzd.	Varieties: Native	Varieties: Introd., Adventive or Natzd.	Forms and Named Hybrids: Native	Forms and Named Hybrids: Introd., Adventive or Natzd.
Fam. 106. Rhamnaceae	3		7	2	4		1	
107. Vitaceae	4		18	2	6		5	
108. Tiliaceae	1		4	3				
109. Malvaceae	7	5	23	13	1	2	2	2
110. Theaceae	1		2					
111. Guttiferae	2		26		12		2	
112. Elatinaceae	2		5					
113. Tamaricaceae		1		1				
114. Cistaceae	3		14		9		1	
115. Violaceae	2		49	4	13		69	1
116. Passifloraceae	1		2		1			
117. Loasaceae	1		2					
118. Cactaceae	2		5					
119. Thymelaeaceae	1	1	1	1				
120. Elaeagnaceae	2		3	2			1	
121. Lythraceae	6	1	11	3	3	1		
122. Nyssaceae	1		2		3			
123. Melastomataceae	1		6		3			
124. Hydrocaryaceae		1		1				
125. Onagraceae	7		60	6	41		4	
126. Haloragaceae	2		13	1	3		2	
127. Hippuridaceae	1		2				1	
128. Araliaceae	3	1	7	1			1	
129. Umbelliferae	30	16	66	19	17	1	6	2
130. Cornaceae	1		12	1	1		12	
Subclass II. Metachlamydeae								
Fam. 131. Clethraceae	1		2				1	
132. Pyrolaceae	6		15		5		2	
133. Ericaceae	20	2	70	6	22		37	
134. Diapensiaceae	3		3					
135. Primulaceae	9	1	15	2	4		4	1
136. Plumbaginaceae	2		3		1		1	
137. Sapotaceae	1		3		1			
138. Ebenaceae	1		1		2			
139. Symplocaceae	1		1		1			
140. Styracaceae	2		3		1			
141. Oleaceae	4	3	9	7	6		5	
142. Loganiaceae	4	1	5	1				
143. Gentianaceae	10		47	3	7		17	
144. Apocynaceae	3	1	8	3	4			
145. Asclepiadaceae	5	2	34	3	6		7	
146. Convolvulaceae	6	1	20	12	7	2	2	3
147. Polemoniaceae	3	1	19	3	6		8	
148. Hydrophyllaceae	5		17		2		1	
149. Boraginaceae	8	9	25	34	1	1	1	1
150. Verbenaceae	3		14	2	2		2	
151. Labiatae	24	15	85	62	37	13	20	3
152. Solanaceae	3	7	25	23	3	3		
153. Scrophulariaceae	26	9	112	40	36		26	6
154. Bignoniaceae	3		3	2				
155. Martyniaceae	1		1					
156. Orobanchaceae	3		7	2			1	
157. Lentibulariaceae	2		15		1		2	
158. Acanthaceae	3		10		11		7	
159. Phrymaceae	1		1		1			
160. Plantaginaceae	2		11	6	4	4	1	4
161. Rubiaceae	7	3	41	13	18		5	1

Families, etc.	Genera Native	Genera Introd., Adventive or Natzd.	Species Native	Species Introd., Adventive or Natzd.	Varieties Native	Varieties Introd., Adventive or Natzd.	Forms and Named Hybrids Native	Forms and Named Hybrids Introd., Adventive or Natzd.
Fam. 162. Caprifoliaceae	7	1	33	16	12	1	7	
163. Adoxaceae	1		1					
164. Valerianaceae	2		13	2	3			
165. Dipsacaceae		3		6				
166. Cucurbitaceae	5	4	5	7				
167. Campanulaceae	3	1	21	11	4		6	1
168. Compositae	82	33	564	139	266	21	121	14
Totals	849	284	4425	1098	1487	125	1121	84

SUMMARY BY DIVISIONS, CLASSES, ETC.

Division, Class, etc.	Genera Native	Genera Introd.	Species Native	Species Introd.	Varieties Native	Varieties Introd.	Forms and Named Hybrids Native	Forms and Named Hybrids Introd.
Pteridophyta . . .	32		120	3	45		125	
Spermatophyta . . .	817	284	4305	1095	1442	125	996	84
Gymnospermae . . .	10		26	4	5		12	
Angiospermae . . .	807	284	4279	1091	1437	125	984	84
Monocotyledoneae .	191	50	1250	165	362	26	249	20
Dicotyledoneae . .	616	234	3029	926	1075	99	735	64
Archichlamydeae	344	137	1767	521	601	54	451	30
Metachlamydeae	272	97	1262	405	474	45	284	34

SUMMARY BY MINOR GROUPS

Families		168	
Genera	native	849	
	introduced	284	
	total	1133	
Species	native	4425	
	introduced	1098	
	total		5523
Varieties	native	1487	
	introduced	125	
	total		1612
Forms and Named Hybrids	native	1121	
	introduced	84	
	total		1205

Whole number of different plants (species, varieties, forms and named hybrids) treated in this work. 8340

GRAY'S
Manual of Botany

Descriptive Flora

DIVISION I. PTERIDÓPHYTA (Vascular Cryptogams)

Rush-like, fern-like, moss-like or quill-leaved plants with fibrovascular bundles but without flowers. Reproduction by spores (without embryos).

FAM. 1. EQUISETÀCEAE (Horsetail Family)

***Rush-like, often branching plants, with jointed and often hollow stems from creeping rhizomes, having sheaths at the joints and, when fertile, terminated by the conical or spike-like fructification composed of peltate stalked scales bearing the spore-cases beneath.* — A single genus.**

1. EQUISÈTUM L. Horsetail. Prêle (Que.)

Rhizomes perennial, jointed, branched, widely creeping, dark-colored, often tuber-bearing, the nodes provided with toothed, often felted sheaths; roots in verticils from the nodes, felted. Stem erect to depressed, simple to branched, cylindrical, the surface regularly striated, overlaid with teeth, dots, bands, rosettes or a smooth coat of silex; the stomata in the grooves in regular rows or broad bands; the internodes usually (with us except in *E. scirpoides*) with a large central air-cavity (centrum), a medium-sized one (vallecular) under each groove, with which the stomata connect, and a smaller one (carinal) under each ridge. Nodes closed and solid, each bearing a whorl of segments joined by their edges into cylindrical sheaths, their tips thinner and prolonged into persistent or deciduous teeth. Branches, when present, often in whorls from the nodes. Fruiting cone formed of regular verticils of stalked sporophylls, the 6 or 7 sporangia opening down the inner side and discharging many loose green spores, each provided with 4 elastic, hygroscopic, clavate bands. Prothallus found in damp places, dioecious, green, variously lobed. — Small, nearly cosmop. genus. (The ancient name from *equus, horse,* and *seta, bristle.*)*

a. Stems annual, with or without regular verticils of branches; cones rounded at summit; stomata scattered or in 1 or 2 broad bands in each groove, their surfaces overlaid with a silex-plate bearing a central vertical slit. . . *b.*

b. Fruiting stem succulent, whitish, pinkish-brown or fulvous, appearing before the green and branching sterile ones, promptly wilting. . . *c.*

c. Fertile stem without or with only tardily developed green branches; branches of sterile stem divergent to ascending, not strongly recurving at tips. *d.*

d. Fertile stem 0.5–3 dm. high, becoming rather slender, its sheaths 0.5–2.5 cm. long and with the 8–12 teeth all distinct; cone 0.5–3.5 cm. long; sterile stem at most 4–5 mm. thick and 10 dm. high, the teeth of the sheaths of its 3–4-angled branches 1-keeled.

Teeth of primary sheaths brown throughout; internodes of sterile stem smoothish, the silex occurring as low dots; sheaths of branches with lance-attenuate teeth. . . . . 1. *E. arvense.*

Teeth of primary sheaths with white margins; internodes of sterile stem scabrous with 3 rows of spinules on each ridge; sheaths of branches with deltoid teeth. . . . . 2. *E. pratense.*

d. Fertile stem 2.5–6 dm. high, 1–2.5 cm. thick, its sheaths 2–5 cm. long and with the 20–30 teeth united into 2- or 3-lobed groups; cone 4–9.5 cm. long; sterile stem 0.5–2 cm. thick, 0.5–3 m. high, the teeth of its 4–6-angled branches 2-keeled. . . . . 3. *E. Telmateia,* var. *Braunii.*

* The illustrations show cross-sections of stem and, in some cases, the epidermis.

c. Fertile stem bearing spreading green branches at fruiting time.

Sheaths of main stem cuff-shaped, close, pale, with white-margined free attenuate brown teeth; branches from nodes simple or nearly so, horizontally spreading, not arched-recurving, their sheaths with distinct 1-ribbed deltoid teeth; ridges of main sterile stem with 3 rows of broad silex-processes. . . . . 2. *E. pratense.*

Sheaths of main stem loosely ventricose, flaring upward, their membranous brown to fulvous teeth fused into 3 or 4 compound lobes; branches simple to profusely branched, at first with recurving tips, later less recurved to ascending; rameal sheaths with fused teeth forming prolonged several-ribbed lobes; ridges of main sterile stem with 2 rows of hooked spinules. . . . . 4. *E. sylvaticum.*

b. Fertile and sterile stems similar, simple or branching, green; the cones not precocious. . . . *e.*

e. Centrum (central cavity) of main stem slender, about one-sixth the diameter of the stem, the vallecular cavities nearly as large; sheath of main stem loose, expanded upward, with 10 or fewer usually white-margined teeth. . . . . 5. *E. palustre.*

e. Centrum about one-half to four-fifths diameter of main stem; vallecular cavities much smaller or wanting.

Centrum one-half to two-thirds diameter of main stem; vacuoles well developed; sheaths becoming loose, with their narrowly white-margined teeth often united in twos or threes; spores abortive. 6. *E. litorale.*

Centrum four-fifths diameter of main stem; vacuoles wanting or minute; sheaths tight, their 15–20 dark brown teeth distinct; spores fertile. 7. *E. fluviatile.*

a. Stems mostly evergreen (except in no. 8 and var. of no. 10), simple or sparingly branched; cone apiculate or blunt; stomata (in ours) a single regular row on each side of the groove, overlaid by the siliceous coat of the stem, having access to the air through an irregular hole. . . . *f.*

f. Teeth articulated to and mostly finally dropping from the sheath, mature cone 1–2 cm. long.

Stem annual, soft and easily crushed; cone rounded at summit, without firm and sharp tip. . . . . 8. *E. kansanum.*

Stem perennial, firm and resistant to slightly soft; cone tipped by a firm dark point. . . . . 9. *E. hyemale.*

f. Teeth not articulated at base, persistent; mature cone 2–10 mm. long.

Stem stiff and straight, 0.7–4.5 mm. thick, hollow at center, with several vallecular cavities; teeth of sheath elongate, much longer than broad; cone 5–10 mm. long. . . . . 10. *E. variegatum.*

Stem flexuous, angulate-filiform, 0.5–1 mm. thick, solid at center, with 0 or 3 vallecular cavities; teeth of sheath as broad as long; cone 2–5 mm. long. . . . . 11. *E. scirpoides.*

1. E. arvénse L. (of cultivated fields), Common or Field-H., Devil's-guts, Queue de renard (Que.). — Extensively creeping by slender and freely forking dark felted and tuber-bearing rhizomes; *fertile stems usually precocious,* appearing in early spring, *succulent, whitish, pinkish or fulvous,* 0.5–3 dm. high, elongating and *becoming rather slender, mostly without green branches* from the nodes; their *sheaths* 0.5–2.5 *cm. long, with* 8–12 *distinct teeth; cone* 0.5–3.5 *cm. long,* subcylindric, *rounded at summit; sterile stems* erect to prostrate, 1–4 (rarely –5) mm. thick, *at most* 10 *dm. high,* the *main stem* 10–14-*furrowed, with* silex in dots and *the centrum one-half to two-thirds the diameter of the stem; branches* 3–4-*angled,* solid, *spreading or ascending, not recurving, mostly simple; their sheaths with lance-attenuate teeth.*

1. E. arvense.

Fig. 1. — Polymorphous species. The most pronounced vars. and forms with us are the following:

a. Branches 4- or 5-angled; sheaths of simple branches mostly 4-toothed, of branchlets 3-toothed. . . . *b.*

b. Branches simple. . . . *c.*

c. Cones chiefly terminating succulent mostly unbranched early stems without chlorophyll, 1–3.5 cm. long; the green branching stems appearing later. . . . *d.*

d. Sterile stem erect or nearly so, its branches numerous in whorls.

Branches loosely ascending to widely divergent, mostly subequal and elongating to 1–3 dm. . . . . *E. arvense* (typical).

Branches strongly appressed-ascending, often unequal, rarely more than 3–5 cm. long. . . . . Forma *varium.*
*d.*Sterile stem depressed, usually short, with ascending branches. . Forma *alpestre.*
*c.*Cones terminating slender green branches of the otherwise sterile stems, only 0.5–1.5 cm. long. . . . . Forma *campestre.*
*b.*Branches forking.
Sterile stem erect or strongly ascending, up to 6 dm. high.
Stem 1–3.5 dm. high, rather bushy-branched; the internodes 1–2.5 cm. long, mostly hidden by the strongly ascending branches. . . . . Forma *ramulosum.*
Stem 3–6 dm. high; internodes mostly 3–6 cm. long; branches loosely divergent. . . . . Forma *pseudo-silvaticum.*
Sterile stem depressed or prostrate, 0.5–2 dm. long, with upright branches. Forma *diffusum.*
*a.*Branches 3-angled, simple or essentially so, their sheaths 3-toothed. . . . *e.*
*e.*Stem erect or nearly so, with distant nodes and whorls of branches.
Branches loosely ascending to spreading, elongating to 1–3 dm. . . . Var. *boreale.*
Branches appressed-ascending, becoming 3–5 cm. long. . . . . . Forma *pseudo-varium.*
*e.*Stem depressed, with crowded nodes and ascending branches. . . . . Forma *pseudo-alpestre.*

E. arvénse L. (typical). — Open low ground, sterile meadows, embankments, roadsides and damp open woods and thickets, generally common, throughout our range and beyond. *Fr.* March–June. (Eurasia) — Very lax and tall plants of woods and thickets (doubtless mere responses to shade and fertile soil), with prolonged internodes and loosely divergent branches 2–3 dm. long, are forma **nemorðsum** (A.Br.) Klinge (of woodland); forma **vàrium** (Milde) Klinge (variable), Greenl. to Alaska, s. to Nfld., N.E., Great Lakes States, Neb., Colo. and Calif. (Eurasia); forma **alpéstre** (Wahlenb.) Luerss. (of high mountains), Arct. reg., s. to open shores and damp rocks or sand of Nfld., N.S., n. N.E., n. N.Y., L. Superior reg., Wyo. and n. Calif. (Eurasia); forma **campéstre** (C. F. Schultz) Klinge (of low fields), local, Arct. e. Am., s. to Nfld., M.I., N.E., N.Y., Mich. and Man.; forma **ramulðsum** (Rupr.) Klinge (having many branchlets), damp shores and slopes, Anticosti to Yuk., s. to N.E., N.J., Ind., Mo. and Wash.; forma **pseùdo-silváticum** (Milde) Luerss. (simulating *E. sylvaticum;* from the forking branches), local, wooded slopes, often calcareous, Que. to Wash., s. to Va.; forma **diffùsum** (A. A. Eat.) Clute (diffuse) = forma *decumbens* ("G. F. W. Mey." sensu Milde) Klinge, not spielart *decumbens* G. F. W. Mey., basonym, open habitats, Lab. to Alaska, s. to Nfld., N.E., n. N.J., Mich., Wisc., Minn. and Wash. (Eurasia)

Var. **boreàle** (Bong.) Ledeb. (northern). — Lab. to Alaska, s. to Nfld., M.I., n. N.E., Mich., Wisc., Minn. and B.C. (N. Eurasia) — Forma **pseùdo-vàrium** Vict. (simulating forma *varium*), more open habitats; forma **pseùdo-alpéstre** Vict. (simulating forma *alpestre*), wet sands or marls or rocky shores.

2. E. praténse Ehrh. (of the meadows), MEADOW-H. — More slender than the erect forms of no. 1, *with whitish-green stiff* slender *sterile stems* 1.5–5.25 dm. high, and *with the* 8–20 *ridges beset with* 3 *rows of flat spinules, the centrum about one-sixth the diameter of stem;* sheaths pale, with white-margined slender brown *teeth*, those of fertile precocious and fleshy whitish to pink or brown axis 8–15 mm. long, of sterile firm axes only 3–8 mm. high, cupuliform; branches wide-spreading, whorled, very slender, 3-angled, becoming 4–15 cm. long, or in forma **nànum** Milde (dwarf) only 1–4 cm. long; *teeth of rameal sheaths deltoid;* fertile cone slenderly cylindric, pale, 1–2.2 cm. long. — Alluvial woods, thickets, mossy glades, rich slopes and calcareous meadows, Nfld. to Alaska, s. to N.B., n. and w. N.E., n. N.J., N.Y., Mich., Wisc., ne. Ia., N.D., Mont. and s. B.C. *Fr.* late April–early July. (Eurasia) — The rare forma **ramulðsum** Milde (with branchlets) has forking sterile branchlets; forma **serðtinum** Milde (late) has sterile branches tardily produced from nodes of fertile stems. FIGS. 2 and 3.

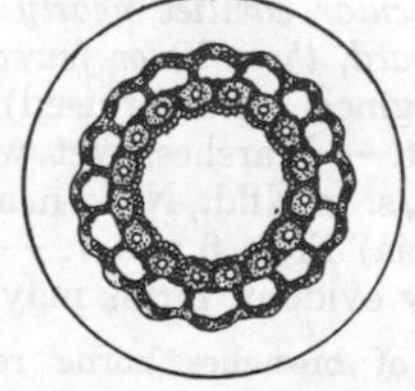
2. E. pratense.

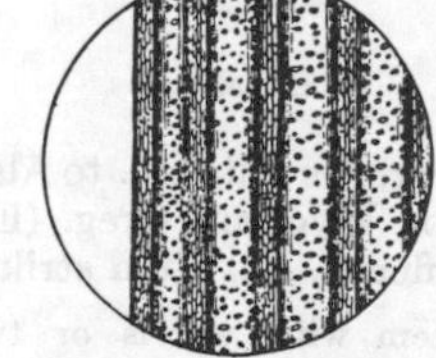
3. E. pratense.

3. E. Telmateìa Ehrh. (old name meaning muddy pool), var. **Braùnii** Milde (for ALEXANDER BRAUN, 1805–1877), GIANT OR IVORY H. — Very coarse; the *fertile* precocious whitish or pale *succulent stems* 1–2.5 *cm. thick and* 2.5–6 *dm. high; their whitish scarious sheaths* 4–9.5 *cm. long* *with* **20–30** *brown* lance-attenuate *teeth united into groups of* 2 *or* 3, the upper fully developed sheaths with teeth one-fourth to half as long as the tube, the uppermost sheath with a spreading nearly entire flange; cone 4–9.5 cm. long, 1–2.5 cm. thick; *sterile stem* 0.5–2 *cm. thick, becoming* 0.5–3 *m. high; its whitish-green* prominent nodes *scabrous* with siliceous teeth, *its* relatively short *pale sheaths with pale* slender *distinct teeth; whorled simple* divergent *scabrous* 4–6-*angled branches becoming* 1–2 *dm. long, their sheaths with* 2-*keeled teeth.* — Low woods or thickets or borders of streams, B.C. to s. Calif.; local, Keweenaw Pen., Mich. (*Farwell*) *Fr.* May, early June. — Typical *E. Telmateia* of Eu., sw. Asia and n. Afr. (*E. maximum* Lam., an illegitimate name) is lower, with sheaths of fertile stem with mostly quite distinct teeth, cone shorter, internodes of sterile stem smoother and with rounded ridges, etc.

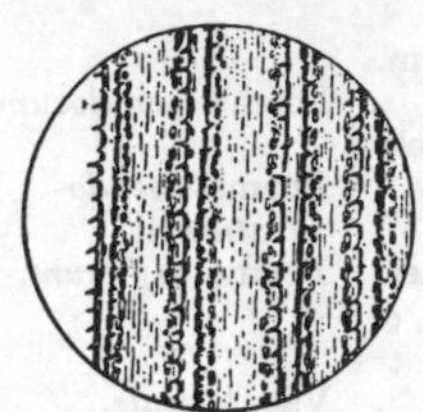

4. E. sylvaticum.

4. E. sylváticum L. (of woodland), Wood-H. — *Differing* from no. 1 *in the spreading and arched-recurving green branches borne from the nodes of the fertile precocious fleshy stems, the latter* 0.5– finally 4.5 dm. high, *subpersistent; sheaths* of the latter *loosely inflated and flaring upward,* 1–2.3 cm. long, *with the membranous brownish or fulvous teeth fused into* 3 *or* 4 *compound lobes; sterile stems* slender, 1.5–7.5 dm. high, with sheaths similar to those of fertile stems, with *the at first recurving slender* subsimple to freely forking *branches* finally spreading to ascending. Figs. 4 and 5. Variable species:

Green branches scabrous, especially along the lower internodes. . . . . *E. sylvaticum* (typical).

Green branches smooth.

Branches simple, subsimple or only slightly forking, the mature vegetative shoot 0.7–1.5 dm. across. . . . . Var. *pauciramosum.*

Branches copiously forking and reforking, the mature vegetative shoot 1–2 dm. across. . . . . Forma *multiramosum.*

E. sylváticum (typical). — Woodlands, thickets and openings, local, Que. to Alaska, s. to n. N.Y., Wisc., etc. (Eurasia)

Var. **pauciramòsum** Milde (few-branched) — Greenl. and Lab. to Alaska, s. to Nfld., N.S., N.E., N.Y., Ohio, Mich., Wisc., Minn., etc. — More often represented by forma **multiramòsum** Fern. (many-branched) = var. *multiramosum* (Fern.) Wherry, extending s. to s. N.E., n. N.J., Pa., O., Mich., Wisc., ne. Ia., S.D. and s. B.C. *Fr.* late Apr. (southw.)–June.

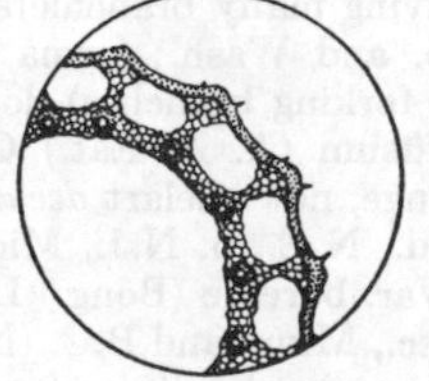

5. E. sylvaticum.

5. E. palústre L. (of marshes), Marsh- or Meadow-H. — Rhizome coarse, extensively and deeply creeping; sterile and fertile stems similar, green or greenish, very variable, from simple to sparsely or abundantly branched with simple (rarely forking) branches very short or up to 4 dm. long; primary stem deeply 5–10-grooved, with narrow elevated ridges; *centrum slender, about one-sixth the diameter of stem, the vallecular cavities nearly as large; sheaths of primary stem enlarged upward, their* 10 *or fewer* brownish to blackish *teeth* usually white-margined (until bruised); cone 1–2.5 cm. long, mature from June–Sept. — Marshes, wet woods, meadows, wet shores, etc., often in calcareous soil, Lab. to Alaska, s. to Nfld., N.B., n. and w. N.E., n. Pa., n. O., Mich., Ill., Minn., Neb., Wyo. and Oreg. (Eurasia) Figs. 6 and 7. — Extremely variable in habit; the following confluent, but often strikingly evident, forms may be recognized:

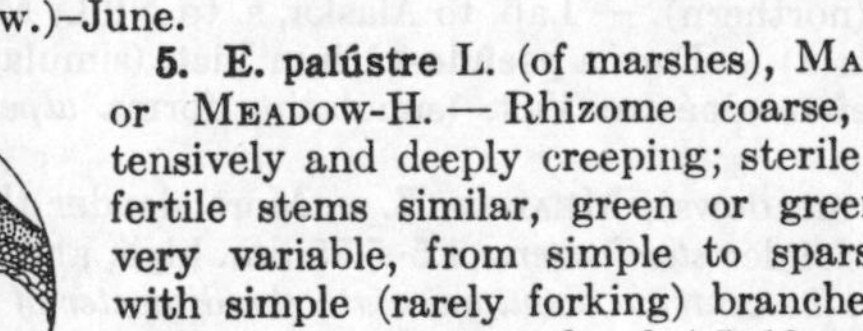

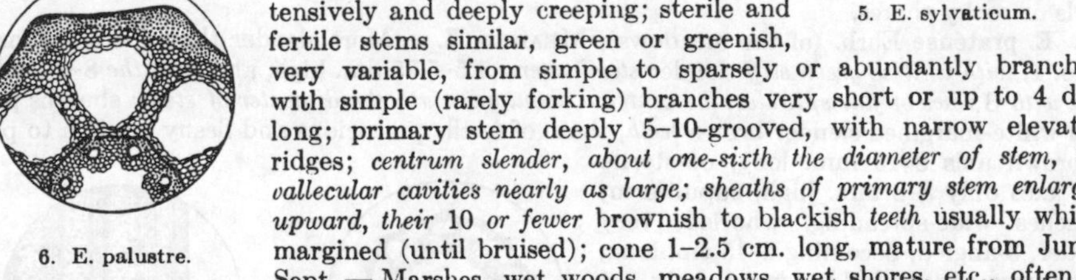

6. E. palustre.

*a.*Stem with whorls or tufts of branches borne regularly from the nodes. . . *b.*

*b.*Nodal branches all simple. . . *c.*

*c.*Stem erect or strongly ascending, with 3-many branches diverging symmetrically as whorls. . *d.*

*d.*Cone borne only from summit of stem. . . *e.*

*e.*Branches from median and upper nodes (those above submerged level) strongly arched-ascending to suberect. . . *f.*

*f.*Branches becoming 1.5–6 cm. long.

Teeth of primary sheaths white-margined. . . . . . . *E. palustre* (typical).

Teeth blackish throughout, without pale margin. . . . . . Forma *nigridens.*
f. Branches 0.7–3 dm. long, very strongly ascending. . . . . . Forma *verticillatum.*
e. Branches from median and upper nodes horizontally or subhorizontally divergent. . . . . Forma *arcuatum.*
d. Cones borne from tips of many ascending branches. . . . . . Forma *polystachion.*
c. Stem prostrate, greatly elongate, its elongate (up to 4 dm. long) strongly ascending or erect branches mostly in pairs. . . . . Forma *fluitans.*
b. Nodal branches (at least the lower ones) with short branchlets. . . . Forma *ramulosum.*
a. Branches only 1 or 2 (rarely 3), irregularly scattered or none.
Main or primary stem elongate, with many nodes, 2–5 mm. thick, with or without more slender secondary stems. . . . . Forma *simplex.*
Main axis nearly suppressed, the very numerous slender erect stems or branches nearly uniform and only 1–2 mm. thick. . . . . . . Forma *filiforme.*

E. palústre (typical). — Stem and axis 2–7.5 dm. high (var. *americanum* Vict.). — Throughout range; forma **nígridens** (St. John) Vict. (black-toothed), local, Côte Nord, Que.; forma **verticillàtum** Milde (whorled), up to 1 m. high, the internodes often hidden by the long ascending branches (forma *luxurians* in part of Vict.), the commonest form; forma **arcuàtum** Milde (arched like a bow), internodes clearly visible between the divergent whorls (incl. type of forma *luxurians* Vict.); forma **polystáchion** (Weigel) Duval-Jouve (having many spikes), many or most stiff and ascending whorled branches terminated by spikes 0.5–2 cm. long, local throughout; forma **flùitans** Vict. (floating), the prostrate or floating stems with the secund and very slender elongate ascending branches mostly paired (1's–6's), in or by water, local; forma **ramulòsum** (Milde) Klinge (having branchlets), similar to formae *verticillatum* or *arcuatum* but with branchlets 0.5–6 cm. long (incl. f. *ramosissimum* (Peck) Broun), local; forma **símplex** Milde (simple), primary stem simple or meagerly branched, 0.2–5 dm. high (f. *tenue* (Döll) Duval-Jouve), scattered through range; forma **filifórme** Lacks. (threadlike), with densely clustered erect filiform stems 1–4.5 dm. high, local.

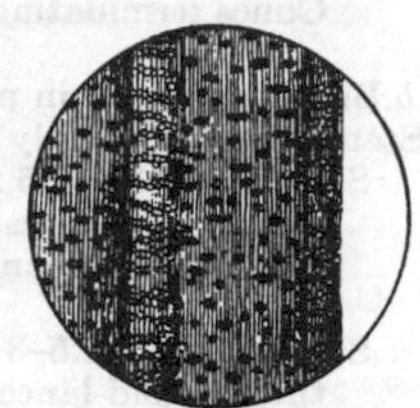
7. E. palustre.

6. × E. litoràle Kuhlewein (of the shore). — Combining in various degrees characters of nos. 1 and 7 and considered a recurrent or persistent hybrid of them; stem 6–18-grooved, the summits deeply so; *centrum one-half to two-thirds the diameter, the vacuoles well developed; sheaths loose,* slightly spreading, *the narrowly white-margined teeth often united in* 2's or 3's; branches 3–5-angled, winged, often solid as in no. 1; cones 0.7–2 cm. long, the *spores mostly aborted.* Figs. 8 and 9. — Several superficially dissimilar vegetative forms, of which the following are sometimes recognizable:

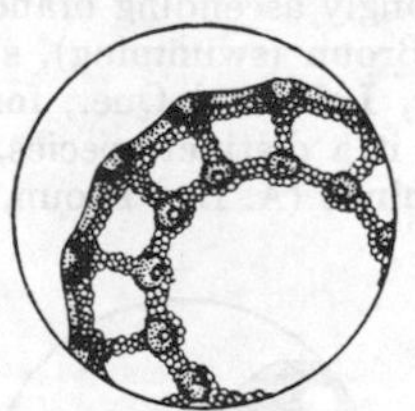
8. × E. litorale.

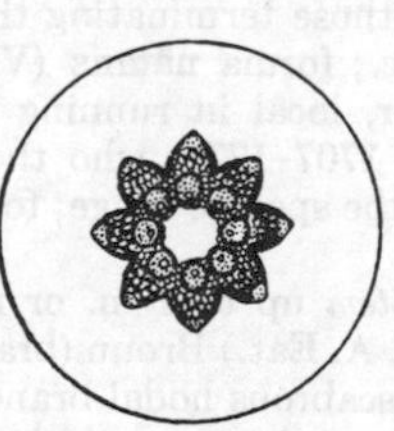
9. × E. litorale.

Branches regularly several and uniformly ascending or upwardly arching from most nodes.
Stem erect, up to 9 dm. high, its internodes clearly visible between the whorls of arched-ascending branches. . . . . *E. litorale* (typical).
Stem depressed or but slightly ascending, 0.5–3 dm. long, its internodes partly hidden by the strongly ascending and overlapping branches. . . . Forma *arvensiforme.*
Branches few and scattered or wanting.
Branches few, irregularly scattered . . . . Forma *humile.*
Branches wanting, stems very slender. . . . . Forma *gracile.*

E. litoràle (typical). — Shores, ditches, swales, etc., somewhat local, St. P. et Miq. and Que. to B. C., s. to N. S., N. E., N. J., Pa., Mich., Wisc., Minn. and Wash. (Eu.) — Forma **arvensifórme** (A. A. Eat.) Vict. (in form like *E. arvense*), local, easily confused with no. 1; forma **hùmile** (Milde) Vict. (humble), occasional; forma **grácile** (Milde) Vict. (slender), frequent.

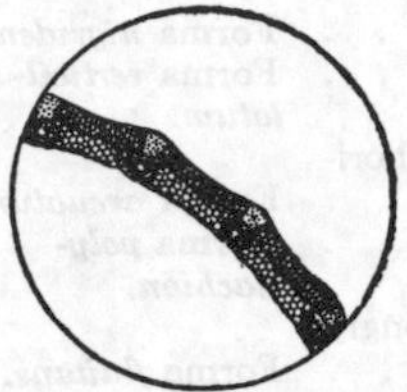
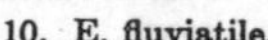
10. E. fluviatile.

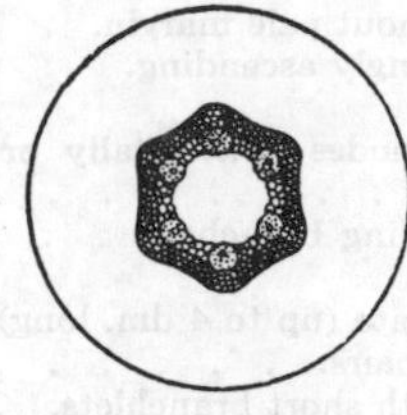
11. E. fluviatile.

7. E. fluviátile L. (of a river), WATER-H., PIPES. — Superficially resembling no. 5; *stems with* 10–30 *shallow grooves,* simple or branching, soft and brittle, nearly hollow; *centrum four-fifths the diameter of the main stem, the vacuoles wanting or minute* except near bases of coarsest stems; branches 4–6-angled, hollow, not winged; *sheaths mostly tightly appressed, with* 15–20 *dark narrow distinct teeth;* cone 0.7–3 cm. long; spores fertile, from May–Aug. FIGS. 10 and 11. — Polymorphous; the most pronounced forms with us are:

*a.*Stem with many essentially uniform and definite tufts or whorls of branches at many nodes. . . *b.*
*b.*Branches in regularly divergent whorls of 4–16, from nodes of erect or strongly ascending stem.
Cone single at summit of main axis (or wanting). . . . *E. fluviatile* (typical).
Cones terminating many stiffly ascending branches. . . . . . . Forma *polystachyum.*
*b.*Branches mostly in pairs, ascending from the prostrate or horizontal stem . Forma *natans.*
*a.*Stem simple or merely with a few scattered or solitary slender branches.
Stem stout, 3.5–7.5 mm. in diameter (in dried material); sheaths of mature primary stem usually closely appressed, their linear-lanceolate teeth mostly more than 2 mm. long and black throughout. . . Forma *Linnaeanum.*
Stem slender, 1.5–3 mm. in diameter (dry); sheaths usually rather loose, their deltoid-lanceolate teeth mostly less than 2 mm. long and pale toward base. . . . . Forma *minus.*

E. fluviátile (typical). — Stem up to 1.5 m. high, with whorls of ascending to spreading branches 0.2–2 dm. or more long. (*E. limosum,* forma *verticillatum* Döll). — Shallow water, wet shores and swales, Nfld. to Alaska, s. to N. S., N. E., Del., Va., O., Ind., Ill., Ia., Neb., Wyo. and Oreg. (Eurasia) — Forma **polystáchyum** (Brückn.) Broun (with many spikes), coarse, with cone of primary axis elongate, those terminating the strongly ascending branches much smaller, local, N. S., Me., Mich., etc.; forma **nàtans** (Vict.) Broun (swimming), stem floating or extending out in running water, local in running water, Lab. and Que.; forma **Linnaeànum** (Döll) Broun (for LINNAEUS, 1707–1778, who thought it a distinct species, *E. limosum* L.), Lab. to Alaska, s. throughout the specific range; forma **minus** (A. Br.) Broun, the simplest and slenderest extreme, occasional.

8. E. kansànum Schaffn. (Kansan). — *Stem* up to 1 m. or more tall, *annual,* simple, or in forma **ramòsum** (A. A. Eat.) Broun (branching) with slender and more or less elongate scabrous nodal branches, hollow, *relatively soft and easily compressed, smooth or smoothish,* with elongate internodes; the centrum three-fourths the diameter of the stem; *sheaths enlarged upward,* pale *green,* with very narrow blackish limb, the narrow attenuate blackish *teeth promptly deciduous; cone* plump, 1–2 cm. long, *with rounded summit.* (*E. laevigatum* sensu A. A. Eat., not A. Br. as to type) — Moist or dry siliceous or argillaceous banks, shores, bottoms, ditches, etc., Mich. to s. B. C., s. to O., Ind., Ill., Mo., Tex., n. Mex. and s. Calif. *Fr.* mid-May–July. FIG. 12.

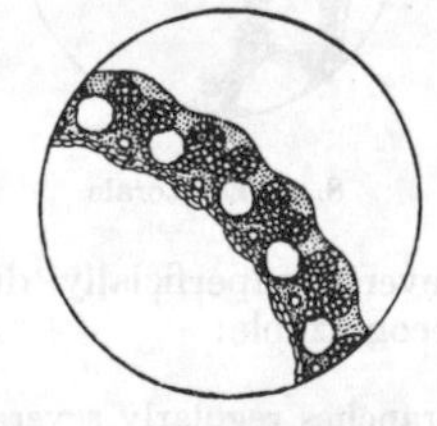
12. E. kansanum.

9. E. HYEMÀLE L. (of winter; because evergreen), SCOURING-RUSH, PRÊLE DES TOURNEURS (Que.). — *Stems* erect or ascending, mostly simple, up to 3 m. high, firm, scabrous, *evergreen,* hollow; *the centrum commonly two-thirds the diameter; the ridges* 2-angled or rounded on the back *with a row of tubercles on each side (in true Eurasian plant) or in ours with cross-bands of silex; sheaths* tight and cylindric or slightly expanded upward, *their narrow teeth mostly soon deciduous* (in one var. tardily so); *cone tipped by a firm point.* — Highly variable, the typical var. in Eurasia, with representatives in Pacific N. Am. Ours are mostly included in the following key:

*a.*Sheaths slightly broadened upward, green or greenish (except the oldest), narrowly dark-girdled at summit below the promptly disarticulating teeth, without or with a lower dark ring; long internodes of stem smooth or but slightly scabrous, easily compressed.
Stems simple or with few irregular sterile slender branches; cone solitary at

tip of main axis. . . . . Var. *intermedium.*

Stem with strongly ascending stiff branches terminating in a small cone. . Forma *proliferum.*

*a.*Sheaths cylindric, close and tight, drab to fuscous, usually with blackish rings toward base and at summit; stems firm, scabrous. . . . *b.*

*b.*Stem 2–10 dm. high, 2–12 mm. thick; the teeth of most of the sheaths promptly disarticulating. . . *c.*

*c.*Stems simple, with only a single terminal cone, at most sparsely and irregularly branching below broken or injured regions.

Stems 3–10 dm. high, stoutish (5–12 mm. thick). . . . . Var. *affine.*

Stem 2–4 dm. high, slender (2–4 mm. thick). . . . . Forma *pumilum.*

*c.*Stems short-branching above, some or all of the branches ending in a small cone. . . . . Forma *polystachyum.*

*b.*Stems 0.5–3 m. high, 6–18 mm. thick, simple, scabrous; teeth of sheaths subpersistent, tardily disarticulating. . . . . Var. *robustum.*

Var. intermèdium A. A. Eat. (intermediate; between var. *affine* and no. 8), **Smooth Scouring-rush.** — Stems mostly clustered, 0.2–1.3 m. high, 2–8 mm. thick; tubercles only slightly developed on the ridges. (*E. laevigatum* A. Br.) — Sandy shores, embankments, roadsides, etc., Anticosti I., Que., and s. Ont. to B. C., s. to w. N. E., Pa., W. Va., O., Ind., Mo., Tex., Ariz. and Calif. *Fr.* late May–Aug. Fig. 13. — Forma **proliferum** Haberer (bearing offspring), infrequent.

13. E. hyemale, v. intermedium.

Var. **affine** (Engelm.) A. A. Eat. (allied), Common Scouring-rush (*E. prealtum* Raf.). — Dry or moist sandy or argillaceous shores, embankments, roadsides, open woods, etc., Nfld. to s. Alaska, s. to N. S., N. E., Ga., Ala., La., Tex. and N. M. — Forma **pùmilum** (A. A. Eat.) Vict. (dwarfish), local; forma **polystáchyum** Prager (with many spikes), infrequent.

Var. **robústum** (A. Br.) A. A. Eat. (stout), *E. robustum* A. Br. — S. Ont. to Yuk., s. to Tenn., La., Tex., Mex. and Calif. (Asia) Figs. 14 and 15.

10. E. variegàtum Schleich. (variegated), Variegated H. — Slender, with tufted ascending *stems* 0.7–4.5 *mm. thick,* up to 6 dm. high, 5–15-*angled,* simple (or slender-branched below injured areas); *centrum one-third to two-thirds the diameter, with vallecular cavities;* sheaths dark, the upper becoming green to ashy, darkened at summit, *with lanceolate or lance-deltoid persistent white-margined or whitish teeth,* these with or without filiform tips; *cone strongly apiculate,* 5–10 *mm. long,* scarcely or barely exserted from the upper sheath, mature from late June–Oct. — Three vars.:

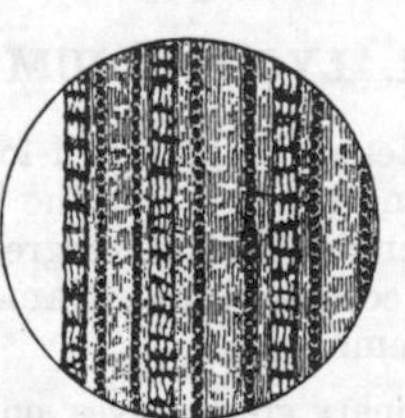

14. E. hyemale, v. robustum.

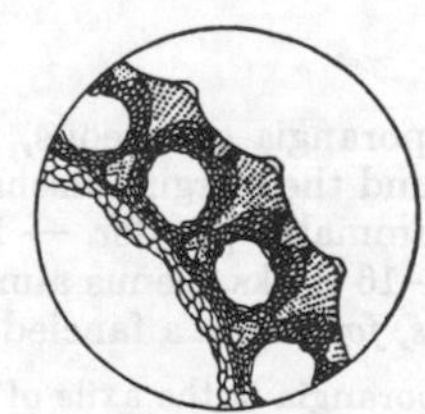

15. E. hyemale, v. robustum.

Ridges of perennial stems biangulate, with 2 rows of tubercles.

Stem 0.7–3 mm. thick, 1–2.5 dm. high, its 4–10 angles evident, the ridges deeply grooved; centrum one-third the diameter of the stem; sheaths slightly broadened upward, loose, promptly losing the bristle-tips; uppermost sheath (subtending cone), including whitish teeth, 3–5 mm. long; cone 5–8 mm. long. . . . . *E. variegatum* (typical).

Stem 2–4.5 mm. thick, up to 4.5 dm. high, with 10–15 slightly grooved rectangular ridges; centrum one- to two-thirds the diameter; sheaths tight, cylindric, their teeth retaining the bristle-tips; uppermost sheath (subtending cone), including dark but narrowly white-margined teeth, 5–10 mm. long; cone 7–10 mm. long. . . . . Var. *Jesupi.*

Ridges of mostly annual stems 8–12, rounded, with 1 row of tubercles, the stems 1.5–6 dm. high; centrum half diameter of stem; sheaths loose, slightly broadened upward; their dark teeth hyaline-margined, retaining the bristle-tips; cone 7–10 mm. long. . . . . Var. *Nelsoni.*

E. variegàtum (typical), including var. *anceps* Milde, the slenderest extreme. — Damp, often calcareous sands, shores, marly bogs, etc., Arct. reg., s. to Nfld., N. B., n. and w. N. E., n. N. J., nw. Pa., Mich., n. Ind., n. Ill., Minn., Neb., Colo. and Calif. (Eurasia)

Var. **Jésupi** A. A. Eat. (for HENRY GRISWOLD JESUP, 1826–1903). — Gaspé Pen., Que., to Mich., s. to N. B., n. and w. N. E. and n. Ind. — A possible hybrid of typical *E. variegatum* and *E. hyemale*, var. *affine*, but often dissociated from the latter; by Broun considered identical with European *E. trachyodon* A. Br., a very dubious identification; also called *E. hyemale*, var. *Jesupi* (A. A. Eat.) Vict., thus indicating its transitional traits.

Var. **Nélsoni** A. A. Eat. (named in 1904 for its discoverer, N. L. T. NELSON), *E. Nelsoni* (A. A. Eat.) Schaffn. — Ne. N. Y. to Mont., s. to Ct., n. Ind., n. Ill., Wyo. and Utah.

11. E. scirpoides Michx. (like *Scirpus; i.e.* in the sense of a small and very slender *Eleocharis*), DWARF SCOURING-RUSH. — *Densely cespitose, sending up many* (10–100 or more) prostrate to arched-ascending and *arched-recurving and flexuous or zigzag stems*, these *very slender*, 0.5–1 *mm. thick*, 0.3–2 (–3) dm. long, *solid at center, with* 6–8 *ridged* (*through the deep grooving of the* 3 or rarely 4) *angles*; vallecular cavities 0 or 3, tiny; *sheaths* loose, *with* the 3 or 4 *deltoid short teeth* white-margined and persistent or subpersistent, their subulate tips either persistent or soon breaking off; *cone* 2–5 *mm. long*, dark, barely exserted, mature in summer or holding over and expanding the following spring. — Woods, thickets, mossy knolls or springy banks, often partly buried in humus, Greenl. and Baffin I. to Alaska, s. to Nfld., N. S., Me., ne. Mass., w. N. E., N. Y., Mich., Wisc., Minn., S. D. and Wash. (N. Eu.) FIG. 16.

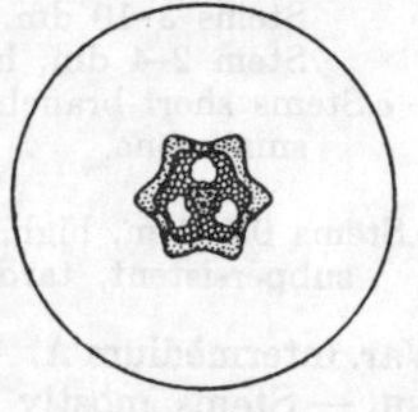

16. E. scirpoides.

FAM. 2. LYCOPODIÀCEAE (CLUB-MOSS FAMILY)

Low plants (*with us*), *usually with a coarsely moss-like aspect, with elongate simple to much branched stems covered with small lanceolate to oblong* (*rarely rounded*) *or subulate or linear evergreen entire or shallowly toothed leaves; sporangia* 1–3-*locular, solitary in axils of leaves or sporophylls, or on their upper surface, when ripe opening into two or three valves and shedding the numerous yellow spores, which are all of one kind.* — The family, as here defined, consists mainly of the large genus:

1. LYCOPÓDIUM L. CLUB-MOSS

Sporangia coriaceous, flattened to plump, 1-locular, 2-valved, mostly by a transverse line around the margin, discharging the spores in the form of a copious more or less sulphur-colored inflammable powder. — Perennials, with evergreen one-nerved leaves imbricated or crowded in 4–16 ranks. Genus almost cosmopolitan. (Name compounded of the Greek *lycos, a wolf*, and *pous, foot*, from a fancied resemblance.)

a. Sporangia in the axils of ordinary green leaves, not forming definite strobiles, splitting nearly to the base into 2 flat valves; broad and flattened gemmae or reproductive buds often borne in the upper axils. . *b.*

b. Plant with short slender rooting base 1–8 cm. long; leaves 3–8 mm. long.

Plant 0.3–3 dm. high; leaves plump and hollow at base, essentially uniform; arctic-alpine and boreal species. . . . . 1. *L. Selago.*

Plant 0.5–1.7 dm. high; leaves flat at base, in irregular longer and shorter alternating belts; southern species. . . . . 2. *L. porophilum.*

b. Plant with prostrate rooting marcescently leafy bases 1–4 dm. long; leaves in irregularly alternating longer and shorter belts, loosely divergent to reflexed, 6–15 mm. long, flat at base. . . . 3. *L. lucidulum.*

a. Sporangia in terminal scaly- or leafy-bracted strobiles; reproductive buds wanting. . . *c.*

c. Sterile branches creeping or recurved-procumbent, not definitely ascending; strobiles with green and leaf-like bracts essentially like foliage-leaves or, if with yellowish and scale-like bracts, the solitary strobile terminating slender branches springing directly from the creeping base. . . *d.*

d. Bracts of the strobile yellowish and firm, ovate-acuminate, strongly contrasting with the short subulate leaves of the peduncle-like fertile branch; sporangia reniform, splitting across the summit into 2 valves. 4. *L. carolinianum.*

d. **Bracts** of strobile green and leaf-like (yellowing only in extreme age), soft and elongate, with prolonged slender tips, nearly like the leaves of the fertile branches; sporangia ovoid to subglobose, finally opening at base.

Erect fertile branches, including the mostly bristly-ciliate leaves, 0.5–1.5 cm. thick; strobiles 1–2.5 cm. thick, their loosely ascending to horizontally spreading bracts lance-attenuate, scarcely dilated at base. . . . . 5. *L. alopecuroides.*

Erect fertile branches, including the mostly entire leaves, 2–8 mm. thick; strobiles 0.3–1.4 cm. thick, their spreading to appressed-ascending bracts with ovate bases abruptly narrowed to the linear-attenuate tip. . . . . 6. *L. inundatum.*

c. Sterile branches erect or strongly ascending from the creeping primary superficial or subterranean stems; strobiles with firm yellowish scale-like bracts, borne on erect leafy branches; sporangia reniform. . . . *e.*

e. Ascending branches simple or few-forked; longer leaves 2.5–11 mm. long, uniform, spirally arranged in several ranks, with slender-subulate to hair-like tips; strobile 0.6–11 cm. long, at least its lower erose to fimbriate bracts with subulate-aristate or hair-like tips.

Strobile 1, sessile, 0.6–4.5 cm. long; its bracts merely crenulate-erose, the lower ones(as well as the leaves) with subulate-aristate tips. . 7. *L. annotinum.*

Strobiles 1–6, on a leafy-bracted peduncle, 1–11 cm. long; its bracts fimbriate-erose, the lower ones (as well as the leaves) usually with a long soft hair-like tip. . . . . 8. *L. clavatum.*

e. Ascending branches freely forking, tree-like, bushy-branched or fan-like; free tips of leaves 0.5–5 mm. long; strobiles 0.6–8.5 cm. long, their entire or merely erose bracts acuminate to subulate-tipped. . . *f.*

f. Ascending branches (from subterranean creeping stem) erect and tree-like, 1–3 dm. high, with very numerous crowded ascending or recurving branchlets; leaves spirally to somewhat dorso-ventrally arranged, in 4 or 6 ranks, the free portion 3–5 mm. long; strobiles sessile at the tips of the branchlets, 1.5–8.5 cm. long, 3.5–6 mm. thick. . . . . 9. *L. obscurum.*

f. Ascending branches tufted, bushy or fan-like; the branchlets flattened or concave beneath and with 4-ranked minute leaves or, if with branchlets not strongly flattened, the free tips of the uniformly spiral and 4–5-ranked leaves only 1–3 mm. long and the commonly more or less peduncled strobiles only 2–3 mm. thick. . . *g.*

g. Sterile branchlets cylindric or merely compressed, not strongly flattened or concave beneath; leaves in 4 or 5 rows, uniform; their free slenderly subulate blades 1–3 mm. long, from nearly equaling to exceeding the adnate base; strobiles mostly solitary. . . . . 10. *L. sabinaefolium.*

g. Sterile branchlets flattened or concave on lower side; leaves in 4 rows, those of the under surface smaller than or unlike the marginal, the latter conspicuously decurrent and forming wings on the sterile branchlets with their free tips deltoid-ovate to lance-subulate and only 0.5–3 mm. long. . . *h.*

h. Strobiles sessile, 0.6–2 cm. long, at the tips of subcylindric leafy branches 0.4–1.4 dm. high; leaves of the glaucous sterile branchlets of 3 forms, the marginal ones with deltoid-ovate to -lanceolate, falcate free tips about equaling the decurrent base, the upper ones ovate-lanceolate, the lower ones trowel-shaped. . . 11. *L. alpinum.*

h. Strobiles or groups of strobiles usually peduncled, 1–5 cm. long; fertile branches 0.8–4 dm. high; marginal leaves of sterile branchlets with lance- or deltoid-subulate free tips much shorter than the decurrent base, leaves of lower side subulate and appressed.

The long running stems on or near the surface of the ground; sterile branches (excluding the free tips of the leaves) 2–4 mm. broad; leaves of lower surfaces of branchlets much smaller than the lateral ones; strobiles 1–5 cm. long. . . . . . . 12. *L. complanatum.*

The long creeping stems deep in the ground; sterile branches (excluding the free tips of the leaves) 1.2–1.8 mm. broad; leaves of lower surfaces of branchlets about equaling the lateral ones; strobiles 1–3 cm. long. . . . . 13. *L. tristachyum.*

1. L. Selàgo L. (old name for the group), Mountain- or Fir-C. — *Rooting or marcescent base of plant only* 1–8 *cm. long; stems tufted*, densely to loosely ascending, simple above base or more or less forking above, usually *forming dense flat-topped or corymbiform strongly ascending tufts* 0,3–3 dm. high; the stems, including the leaves, 0.3–1.8 cm. thick, densely clothed with lanceolate pungent and *lustrous* entire or denticulate *essentially uniform leaves* 3–8 *mm. long;* upper axils often bearing dilated gemmae or reproductive buds; *sporangia axillary,* reniform, splitting nearly to base into 2 flat valves; *spores with concave sides, the base with coarse papillae.* — Three ecological vars.:

Leaves ascending or appressed; stems 0.3–2 dm. long, densely tufted and forming flat-topped clumps.
Leaves lance-attenuate, 5–8 mm. long, ascending. . . . . *L. Selago* (typical).
Leaves mostly ovate-lanceolate, 3–5 mm. long, appressed or incurved. . . . Var. *appressum.*
Leaves strongly divergent, lance-attenuate, 5–8 mm. long; stems often 2–3 dm. long, loosely ascending. . . . . Var. *patens.*

L. Selàgo (typical). — Arct. reg., s. to damp or mossy rocks, barrens, cold woods or bare mts. to Nfld., coast of e. Me., n. N. H., s. Ct., w. N. Y., mts. to Va., n. Mich., n. Wisc., Mont. and Wash. *Fr.* July–Sept. (Eurasia; Mex.; S. Am.)

Var. **apprèssum** Desv. (appressed). — S. on mts., barrens and bogs to Nfld., n. N. E., centr. Mass., mts. to N. C., s. B. C. and Wash. (Eurasia)

Var. **pàtens** (Beauv.) Desv. (spreading). — Cold rocks and cliffs (often calcareous), Nfld. and Côte Nord, Que., to n. Wisc., s. to Gaspé Pen., Que., and Lamoille Co., Vt. (Alaska)

2. L. poròphilum Lloyd & Underw. (lover of soft stone). — Tufted and *short-based* as in no. 1, 0.5–1.7 *dm. high,* the crowded and ascending *branches mostly simple above the base; leaves* green, *scarcely lustrous,* linear-lanceolate, attenuate, *loosely ascending, spreading or reflexed,* 5–8 mm. long, *flat at base, in irregularly alternating longer and shorter belts;* spores much as in no. 1. (*L. lucidulum,* var. Clute) — Sandstone and limestone cliffs or cold woods, W. Va. and O. to Wisc., s. to Ala., Tenn. and Mo. *Fr.* July–Sept.

3. L. lucídulum Michx. (somewhat shining), Shining C. — *Stems sprawling and much elongate, loosely ascending, the rooting base* 1–4 *dm. long and covered with marcescent brown leaves,* the green ascending portions of the stems subsimple or loosely forking and 1–2.5 dm. high; *leaves in irregularly alternating belts, some longer, some shorter, loosely divergent, becoming reflexed, lustrous, more or less oblanceolate,* the larger often *broadest above the middle,* toothed, *the longer and larger ones* 0.6–1.5 *cm. long and* 0.8–2.6 *mm. broad;* upper axils often bearing dilated gemmae; *sporangia* reniform, splitting nearly to base into 2 flat valves, *borne chiefly in the axils of the shorter leaves;* spores as in no. 1, the basal papillae mostly smaller or obscure. — Cool woods, Nfld. to Ont., s. to N. S., N. E., se. Va., upland of S. C. and Tenn., Ind., Ill. and Mo. *Fr.* July–Sept. (Asia)

Var. **occidentàle** (Clute) L. R. Wilson (western). — *Leaves narrower, more attenuate and entire or nearly so.* — Nfld. to s. B. C., s. to n. N. E., N. Y., Ind., Wisc. and Wash.

4. L. caroliniànum L. (of Carolina). — *Creeping stem close to ground,* 0.2–1 dm. long, *with few short closely repent branches; their whitish-green lanceolate to narrowly ovate leaves slightly oblique,* 4–7 mm. long, the lateral ones *curved upward; fruiting branch solitary, scape- or peduncle-like, very slender,* erect, 0.5–3 dm. high, *bearing scattered subulate appressed leaves; strobile* solitary, 1–13 cm. long, 2.5–5 *mm. thick; bracts firm, yellowish-green, triangular, sharply acuminate,* diverging in age; sporangia reniform, splitting across the summit; *spores with* convex *winged sides,* the rounded *base with delicate reticulation.* — Wet pine-barrens or damp peats and sands of Coastal Plain, Fla. to La., n. to se. N. C.; very local, s. Greensville Co., Va., and Prince George Co., Md.; s. N. J.; very local, Suffolk Co., L. I. *Fr.* Aug.–Nov. (A very ancient and primitive species widely and very interruptedly dispersed in W. I., Centr. Am., e. S. Am., trop. and S. Afr., Mauritius, etc., Ceylon, Austral. and N. Z.)

5. L. alopecuroìdes L. (resembling *Alopecurus*), Fox-tail C. — *Sterile stem low-arching or recurved-procumbent,* 2–6 dm. long, *rooting at tip,* subsimple to loosely forking, *closely invested with soft slender loosely spreading to upcurving spinulose-ciliate* to entire *leaves; fertile branches* erect, 1–∞, *including the very crowded and loosely ascending to spreading linear-attenuate leaves,* 0.5–1.5 *cm. thick,* 0.6–4.5 dm. high; *strobile* 2–11 cm. long, 1–2.5 *cm. thick; its* very crowded herbaceous soft loosely ascending to widely divergent *bracts lance-attenuate, scarcely dilated at base, bristly-ciliate* (or very rarely entire); *sporangia ovoid to subglobose, finally opening at base; spores* rounded, *with coarsely reticulate base, the apical end with coarse papillae crowded in rows.* — Sandy or peaty shores, savannas and wet barrens, Fla. to La., n. mostly on Coastal Plain to L. I. and to Nantucket I., Mass. *Fr.* Aug.–Oct. (W. I.)

6. L. inundàtum L. (inundated), Bog-C. — *Stem* creeping, *rooting at tip,* simple or forking, 0.2–6 dm. long, *covered with soft linear- or lance-attenuate* usually upwardly curved *entire or sparingly spinulose-toothed leaves; fruiting branches, including the* spreading to appressed *leaves,* 2–8 *mm. thick,* 0.1–4 dm. high; *strobile* 0.8–15 cm. long, 0.3–1.4 *cm. thick; its bracts* spreading to appressed-ascending, *with narrowly ovate or lanceolate entire or slightly toothed bases,* narrowed to linear-attenuate tips resembling the foliage-leaves; sporangia much as in no. 5; *spores* similar to those of no. 5 but *the apical end with scattered and distinct small papillae.* — Highly variable, with the following confluent vars. (wholly drowned plants with atypically elongate branches and leaves can hardly be keyed):

Leading creeping terminal shoot elongating from 1.5–10 (when submerged –15) cm. beyond the last of the 1 or 2 (very rarely 3) fertile branches; fertile branches 0.5–7 (rarely –10) cm. high; strobile 0.8–4 cm. long, 6–10 mm. thick, its ovate- or lanceolate-based bracts with loosely ascending to spreading tips; transcontinental northern plant. . . . . *L. inundatum* (typical).

Leading creeping terminal shoot elongating from 0.7–4 dm. beyond the last of the mostly 2–10 fertile branches; the latter 0.5–4 dm. high; strobile (1.5–)2–15 cm. long, its bracts with lanceolate bases; southeastern and eastern coastwise plants.

Fertile branches stiffly erect (flexuous only when drowned), 2–5 mm. thick, densely covered with loosely ascending to spreading leaves mostly 5–8 mm. long; strobiles 7–14 mm. thick at base, including tips of loosely ascending to spreading elongate (5–8 mm.) bracts; sporangia mostly hidden under the bracts. Var. *robustum*.

Fertile branches erect or flexuous, 1.5–3 mm. thick, more openly covered with appressed or ascending leaves 3–5 mm. long; strobile 3–7 mm. thick, with closely appressed or ascending shorter (3–6 mm. long) bracts; outlines of sporangia (at least in dry specimens) often evident through the thin bracts. Var. *Bigelovii*.

L. inundàtum (typical). — Damp peaty or sandy shores, swamps or bogs, Nfld. to Alaska, s. to N. S., N. E., N. J., Pa., mts. to w. Va. and W. Va., n. O., n. Ind., n. Ill., Minn., Id. and Oreg. *Fr.* July–Oct. (Eurasia) — Older creeping bases, decaying, often leaving the peripheral new and derived small plants in circles or rings.

Var. **robústum** R. J. Eat. (stout). — Sandy or peaty soils and sphagnous bogs or savannas, e. Mass. to Fla. and La. — Stouter plants simulate more slender ones of no. 5. — Forma **furcàtum** Fern. (forking) has some or all fertile branches with 2–several erect branchlets and strobiles.

Var. **Bigelòvii** Tuckerm. (for its discoverer, Jacob Bigelow, 1787–1879). — The slenderest extreme. (Incl. var. *adpressum* Chapm., *L. adpressum* Lloyd & Underw. and *L. Chapmani* Underw.) — Bogs, savannas, wet shores, etc., Fla. to e. Tex., n. on or near Coastal Plain to se. N. H. and N. S., there freely passing into typical *L. inundatum.* — Forma **polyclavàtum** (McDonald) Fern. (with many clubs) has some or all of the fertile branches forking and bearing 2–several strobiles.

7. L. annótinum L. (a year old; from the clearly marked separation of each year's growth of the ascending branches), Stiff, Bristly or Interrupted C. — *Stem trailing* or creeping, up to 2 m. or more *long,* forking, usually shallowly covered by humus; *ascending branches* at first *simple, later slightly forking* and increasing annually from the summits, 0.6–3 dm. high, commonly arching; *leaves firm or stiff and hard,* linear-attenuate to lance-oblong or oblanceolate, from reflexed or divergent to appressed-ascending, 2.5–11 *mm. long, subulate-tipped; strobiles sessile, solitary,* 0.6–4.5 *cm. long,* their *ovate erose-crenate sharp-pointed bracts* straw-colored. — Four geographic vars.:

Leaves mostly spreading to reflexed, those of the fruiting branches 5.5–11 mm. long.

Leaves lanceolate, linear-oblong or oblanceolate, firm but not rigid, distinctly serrate. . . . . *L. annotinum* (typical).

Leaves linear- or lance-attenuate, firm and subrigid, entire or only obscurely serrate. . . . . Var. *acrifolium*.

Leaves strongly ascending to tightly appressed, those of fruiting branches 2.5–6 mm. long.

Leaves linear- or lance-attenuate, thick and hard, dorsally convex, entire. . . Var. *pungens*.

Leaves lanceolate to lance-oblong, flat, obscurely serrate. . . . . Var. *alpestre*.

L. annótinum (typical). — Woods and clearings, Lab. to Alaska, s. to Nfld., N.S., N.E., n. N.J., Pa., upland to w. Va. and W.Va., Mich., Wisc., Minn., Colo. and Oreg. *Fr.* late July–early Oct. (Eurasia)

Var. **acrifòlium** Fern. (sharp-leaved). — Similar habitats and range, s. to Nfld., N.S., Me., e. and centr. Mass., n. Ct., nw. N.J., Pa., mts. to Va. and W.Va., Mich., Wisc., Alta. and B.C. (Siberia)

Var. **púngens** (LaPylaie) Desv. (pungent). — Exposed rocky or peaty habitats, Greenl. and n. Lab. to Alaska, s. to Nfld., C.B., P.E.I., sw. N.B., se. Me., higher mts. of n. N.E. and n. N.Y., Bruce Pen., Ont., n. Wisc., n. Minn. and Sask. (Ne. Asia)

Var. **alpéstre** Hartm. (of high mountains). — Greenl., Baffin I. and Lab.; subalpine forest, Mt. Lafayette, N.H.; Yuk. and Alaska to n. Wyo. and s. B.C. (N. Eu.)

8. L. clavàtum L. (club-shaped), Common or Running C., Coral- or Staghorn-Evergreen, Buckhorn, Wolf's-claws, Courants verts (Que.). — *Stem creeping, much elongated and lying on the ground,* forking; *ascending branches* at first simple, *becoming* prolonged and *dichotomous,* including the strobile or strobiles 0.7–4 dm. high; fertile branches terminated by a *leafy-bracted peduncle* and 1–6 short-stalked to sessile strobiles; *foliage-leaves* linear-subulate, incurved-spreading to appressed-ascending, *usually tipped by a soft hair-like bristle; leaves of the peduncle much shorter, scattered, bristle-tipped; strobiles* 1–11 *cm. long,* their straw-colored firm *bracts fimbriate-erose,* at least the lower *bristle-tipped.* — An almost cosmop. species, with many geographic vars. (One of the chief sources of Lycopodium-powder, the spores of this and other large species).

a. Leaves tipped by a soft hair-like bristle. . . *b.*
- *b.* Peduncle bearing usually 2–6 strobiles; leaves of the ascending branches usually spreading. . . . . *L. clavatum* (typical).
- *b.* Peduncle normally bearing but 1 strobile; leaves ordinarily ascending or appressed. . . *c.*
  - *c.* Strobile slender-cylindric, at most 5 mm. thick.
    - Peduncle 4.5–15 cm. long; strobile 3.5–11 cm. long, 3.5–5 mm. thick. . Var. *megastachyon.*
    - Peduncle 0.5–2.5 cm. long; strobile 1.5–4 cm. long, 3–4 mm. thick. . Var. *monostachyon.*
  - *c.* Strobile slenderly ellipsoid or thick-cylindric, 1–2.5 cm. long, 5–7 mm. thick; peduncle 3.5–5.5 cm. long. . . . . Var. *brevispicatum.*

a. Leaves without terminal bristle. . . . . Var. *integerrimum.*

L. clavàtum, typical (incl. var. *tristachyum* Hook. and vars. *laurentianum* and *subremotum* Vict.). — Dry woods, thickets and clearings, Nfld. and Côte Nord, Que., to Alaska, s. to N.S., N.E., Del., Pa., mts. to N.C., n. O., Mich., Wisc., Minn., Sask., Mont., Id. and Wash. *Fr.* July–Sept. (Eurasia) — Among the numerous monstrosities the most striking is forma **stérilis** House (sterile) with filiform prolonged sterile peduncle or pedicels without strobiles.

Var. **megastáchyon** Fern. & Bissell (mammoth-spiked). — Centr. and s. Nfld. and Côte Nord, Que., to Thunder Bay Distr., Ont., s. to N.S., Me., Mass., Ct., N.Y., n. Mich. and n. Wisc. — Forma **furcàtum** (Luerss.) Vict. (forking) has some or all strobiles forking or double.

Var. **monostáchyon** Grev. & Hook. (one-spiked). — Exposed situations, Greenl. and Lab. to Alaska, s. to bleak hilltops and alpine crests of Nfld., Shickshock Mts., Que., White Mts., N.H., and to Man., Sask., Mont. and s. B.C.

Var. **brevispicàtum** Peck (short-spiked). — Summits of Adirondack Mts., N.Y. — Our most extreme var.

Var. **integérrimum** Spring (most entire). — Local, n. Wisc.; B.C. and Wash.

9. L. obscùrum L. (obscure; application unexplained, except that Linnaeus stated that the fruit was unknown to him), Tree-C., Flat-branch Ground-Pine, Bunch-Evergreen, Petits Pins (Que.). — Creeping stem subterranean, rhizome-like; *upright branches* scattered, often in rows, *tree-like,* 1–3 dm. high, *simple below, densely bushy-forking above; the crowded sterile branchlets spreading or recurving at tip, flattened or concave beneath; leaves in 4 or 6 ranks, linear-lanceolate, about* 1 *mm. broad, their free tips* 3–5 *mm. long,* their bases decurrent, *the lower* (and often the upper) *series usually appressed, the lateral series spreading; strobiles sessile at the tips of branchlets,* 1.5–6.5 cm. long, 4–6 *mm. thick,* erect, 1–15 in number; the firm yellowish bracts cordate, acuminate. — Woods, copses and clearings, s. Lab. to Alaska, s. to Nfld., N.S., N.E., L.I., Ga., Tenn. and Mont. *Fr.* July–Nov. (Asia)

Var. **dendroìdeum** (Michx.) D. C. Eat. (tree-like), Round-branch Ground-Pine. — *More compact,* the *branchlets more ascending or erect; leaves linear-attenuate,* mostly *less than* 1 *mm. wide, all incurved-ascending or uniformly slightly spreading; strobiles* 1–27, *more slender,* 2–8.5 cm. long and 3.5–4.5 *mm. thick.* — Open woods and clearings, similar range.

10. L. sabinaefòlium Willd. (savin-leaved), Ground-Fir, Heath-Cypress. — Creeping stems mostly beneath the surface of the ground, very slender; the *tufted* and *rather crowded ascending branches* (including the strobiles) 0.7–3 *dm. high, with* many *erect dorso-ventrally compressed or flattened sterile branchlets* 1.5–4 (–5) mm. broad; *leaves of erect branchlets 4-ranked, their free lance-subulate tips* 1–3 *mm. long, shorter than the adnate base, uniform; strobiles* solitary (rarely 2 or 3), slenderly cylindric, 1–4.5 *cm. long and* 2.5–3 *mm. thick, on peduncles* mostly 1–8 *cm. long* (rarely sessile); bracts firm, acuminate. — Woods, thickets and clearings, Nfld. and

Côte Nord, Que., to Algoma Distr., Ont., thence to Alaska, s. to N.S., n. N.E., N.Y., Wayne Co., Pa. and n. Mich. *Fr.* mid-July–Sept.

Var. **sitchénse** (Rupr.) Fern. (of Sitka), Sitkan C. — Smaller and lower, the *ascending branches* 0.3–1.3 *dm. high; branchlets cylindric, with the spirally arranged leaves in* 4 *or* 5 *ranks, the free tips usually longer than the decurrent base; strobiles* 0.6–3.3 cm. long, *sessile or on peduncles up to* 3 *cm. long.* (*L. sitchense* Rupr.) — Barrens, mt.-slopes and -summits, open thickets, rarely in woods, Lab. to Alaska, s. to Nfld., C.B., P.E.I., n. N.E., ne. N.Y., n. Ont., Sask., Mont. and Wash.

11. L. alpìnum L. (alpine), Alpine C., Ground-Fir, Heath-Cypress. — Creeping stem slightly beneath surface of ground; *tufted ascending branches glaucous,* 0.4–1.4 *dm. high,* crowded; fertile branchlets subcylindric; *the sterile branchlets flattened, often revolute-margined; leaves* 4-ranked, those of the sterile branchlets of 3 forms, *the marginal ones with deltoid-ovate to -falcate free tips* about equaling the decurrent base, the upper ovate-lanceolate, *the lower trowel-shaped; strobiles* 1–3, *sessile,* 0.6–2 cm. long. — Arct. reg., s. to cold mossy banks at 900–1250 m., Shickshock Mts., Gaspé Pen., Que.; Mont. and s. B.C. *Fr.* Aug., Sept. (Eurasia)

12. L. complanàtum L. (flattened), Running or Trailing Evergreen, Ground-Pine, Ground-Cedar, Christmas Green, Creeping Jenny, Courants verts (Que.). — Elongate stem creeping on or near surface of ground; *ascending or sprawling branches* (including strobiles) 0.3–4 *dm. high, with crowded to loosely and remotely forking branchlets; the latter often strongly constricted between the season's growths, flattened* and (excluding short free tips of marginal leaves) 2–4 *mm. broad; leaves* 4-ranked, *the free tips of the marginal ones lance- or deltoid-subulate and much shorter than the strongly decurrent base; leaves of lower side much smaller; peduncles* 0.5–6 cm. long, *slender and remotely bracted, with* 1–4 (–5) *slender strobiles* 1–3 *cm. long,* these blunt or often with slender sterile tips. (Incl. vars. *canadense* and *elongatum* Vict.) — Dry woods, thickets and clearings, Greenl. and n. Lab. to Alaska, s. to Nfld., C.B., P.E.I., N.B., n. N.E., Ont., n. Mich., Wisc., Minn., Mont., Id. and Wash. *Fr.* July–Sept. (Eurasia) — The species highly variable and extending in different but confluent geographic vars. s. into trop. Asia and through Mex. to S.Am. The typical var. passing directly into

Var. **flabellifórme** Fern. (fan-shaped). — Creeping stem superficial; *sterile branches erect or nearly so, their comparatively uniform branchlets arched into nearly horizontal fan-like platforms or slightly ascending broad funnels; constrictions between seasons' growths commonly less pronounced; peduncles* 3–11 *cm. long,* or in forma **brachýpodum** Fern. (short-stalked) only 5–15 mm. long, *with* 2–9 *strobiles* 1.5–5 cm. long, or in forma **Wíbbei** (Haberer) House (named for J. Herman Wibbe, 1839–1899) with 1 strict strobile. (*L. flabelliforme* (Fern.) Blanch.) — Dry woods and clearings, sw. Nfld. and Bonaventure Co., Que., to Ont. and Minn., s. to N.S., N.E., L.I., S.C., Ky. and Ia.

13. L. tristáchyum Pursh (three-spiked), Ground-Pine, Ground-Cedar. — *Creeping stem usually deep in the ground* (therefore *pale brown in dried specimens*); ascending branches (including strobiles) 1.5–4 dm. high, the crowded glaucous to green *sterile branchlets ascending and forming broad funnels* (in the open) *or loosely divergent* (in shade), 1.2–1.8 mm. broad (excluding leaf-tips); free tips of marginal leaves much shorter than the decurrent bases; *leaves on the convex to flat lower surface about equaling the lateral ones;* peduncles very slender, 2–12 cm. long, bearing 2–6 mostly pedicelled strobiles 1–3 cm. long. — Dry siliceous or acidic woods, thickets and clearings, Nfld. and Gaspé Pen., Que., to n. Alta., s. to N.S., N.E., N.C. and Tenn. *Fr.* June–Sept. (Eu.)

FAM. 3. SELAGINELLÀCEAE (Spikemoss Family)

Leafy plants, terrestrial or rooted in mud, never very large; stems branching; leaves small, in 4–6 rows; sporangia unilocular, solitary, axillary or borne on the upper surface of the leaf at its base and enwrapped in its margins, some containing large spores (macrospores) and others small spores (microspores). The macrospores are in the shape of a low triangular pyramid with a hemispherical base, and marked with elevated ribs along the angles. In germination they develop a minute prothallus which bears archegonia to be fertilized by antherozoids developed from the microspores.

1. SELAGINÉLLA Beauv. Spikemoss

Fructification of two kinds, namely, of minute and oblong or globular spore-cases, containing reddish or orange-colored powdery microspores; and of mostly 2-valved tumid larger ones, filled by 3 or 4 (rarely 1–6) much larger globose-angular macrospores; the former usually in the upper and the latter in the lower axils of the leafy 4-ranked sessile spike, but sometimes the two kinds on opposite sides all along the spike. (Name a diminutive of *Selago,* an ancient name

of a *Lycopodium*, from which this genus is separated and which the plants greatly resemble in habit and foliage.)

a.Delicate creeping herbaceous plant with filiform stem and branches: leaves membranaceous, lanceolate to ovate, not bristle-tipped; spikes subterete or obtusely angled.

Sterile branches 0.5–6 cm. long, their leaves and those of the ascending fertile branches uniform, lanceolate; sporophylls ciliate; boreal species. . . 1. *S. Selaginoides.*

Sterile stems and branches forming carpets 1–4 dm. across, their leaves dimorphic, without cilia, 4-ranked, 2 rows oblong to oval and spreading, 2 smaller and appressed; sporophylls not ciliate; more southern. . . . 2. *S. apoda.*

a.Tough evergreen tufted plant with densely imbricated narrow bristle-tipped leaves; spikes sharply 4-angled, with broad ovate sporophylls. . . . 3. *S. rupestris.*

1. S. Selaginoìdes (L.) Link (old generic name; resembling *Selago*).–Forming small mats rarely 1.5 dm. across, the *prostrate sterile branches filiform, weak, with uniform lanceolate sparsely ciliate spreading-ascending acute leaves 2–4 mm. long; fertile branches* ascending, *with similar leaves,* 0.5–8 cm. high, *the upper somewhat longer leaves passing into sporophylls of the loose subcylindric spike* (0.7–5 *cm. long*). — Damp shores, mossy banks, etc., often calcareous, Greenl. and Lab. to Alaska, s. to Nfld., C.B., M.I., n. Me., Bruce Pen., Ont., n. Mich., n. Minn., Colo. and s. B.C. *Fr.* July, Aug. (Eurasia)

2. S. ápoda (L.) Fern. (footless; from the sessile spikes). — Stems *weak, slender, freely forking,* prostrate, *repent, forming pale or whitish-green mats* 1–4 *dm. across; leaves pellucid-membranaceous, 4-ranked, the larger* narrowly ovate or oblong; *bluntish non-ciliate lateral ones spreading, the smaller stipule-like dorsal and ventral ones appressed and pointed; spikes sessile* along and at tips of branches, 3–12 *mm. long,* their sporophylls similar to foliage-leaves. (*S. apus* Spring) — Low woods, wet rocks, meadows, etc., Fla. to Tex., n. to s. Me., s. Que., s. Ont., Mich., Wisc. and Mo. *Fr.* May–Dec. (S.Am.)

3. S. rupéstris (L.) Spring (of rocks). — *Much branched in close tufts* 2–6 cm. high; *leaves densely appressed-ascending, imbricated, linear-lanceolate, firm,* convex and with a grooved keel, *minutely ciliate, with a whitish straight bristle at tip; spikes strongly quadrate,* terminal, 1–2.5 cm. long, *their ovate sporophylls dentate; megaspores reticulate.* — Dry rocks or packed sand, Rimouski and w. Saguenay Cos., Que., to s. Man., s. to N.S., N.E., L.I., w. Ga., n. Ala., Ark. and e. Okla. *Fr.* July–Oct. (Eurasia) — Grayish-green, suggesting a rigid moss.

S. TORTÍPILA A. Br. (with twisted hair), differing from no. 3 in having the terminal bristle of the leaf spirally twisted, the megaspores tuberculate toward base, occurs so near our border that it may be found in the mts. of sw. Va.

FAM. 4. ISOËTÀCEAE (QUILLWORT FAMILY)

Small aquatic, amphibious or terrestrial herbs suggesting in habit tufted grasses or rushes. Stem a short and thick corm, crowned with numerous elongate-subulate leaves with basal sporangia. Spores of two kinds, usually in different sporangia. — A single genus; the species similar in habit and aspect and to be distinguished with certainty only by the aid of the compound microscope.

1. ISOÈTES L. QUILLWORT

Corm fleshy, more or less depressed, the roots arising from the 2–5-lobed base, the flattened top bearing the leaves from a central bud or crown. Leaves dilated and imbricated at base, rounded or somewhat angular above, orbicular to quadrate or triangular in cross-section, with 4 transversely septate air-tubes or channels, bearing fibro-vascular central bundles, and with or without 4 or more peripheral bast-bundles. Stomata none or in narrow bands over the air-cavities. Ligule a small triangular or elongate delicate tissue extending slightly above the sporangium. Sporangia solitary in an excavation of the dilated leaf-base, more or less covered by the *velum* (formed from thin edges of the excavation), attached by their backs, plano-convex, with internal transverse threads; *microsporangia* (male) and *megasporangia* (female) usually borne in alternating cycles of leaf-bases. Spores dimorphic, those in *microsporangia* minute male *microspores,*

ligule
fovea
velum
sporangium

17. Isoëtes (diagrammatic). Inner surface of leaf-base.

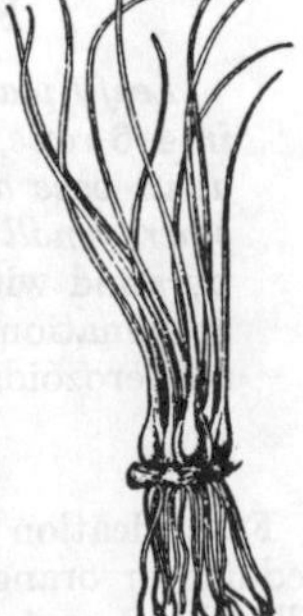
18. I. riparia, showing generic habit.

those in *megasporangia* commonly much larger female *megaspores*, the latter spherical and with an elevated median band (*equator*) and radiating from it to the summit of the upper hemisphere 3 shorter bands (*commissures*). (Name used by Pliny for a species of *Sedum*.)* Figs. 17 and 18.

a. Megaspores with tuberculate, pebbled, papillate or spinulose surface. . . . *b*.

b. Leaves with 4 bast-bundles; surface of megaspore minutely tuberculate, pebbled or papillate. § Tuberculatae.

Leaves 1.2–4.5 dm. long; sporangia 0.5–3 cm. long, brown-spotted, with narrow to broad velum; megaspores 280–440 μ in diameter; microspores 20–30 μ long, covered with spinules. . . . . 1. *I. melanopoda*.

Leaves 0.8–3 dm. long; sporangia 6–7 mm. long, usually brown-lineolate; megaspores mostly 440–650 μ in diameter; microspores 27–37 μ long.

Mature plant with 8–30 leaves 0.8–2.2 dm. long, these triangular in cross-section; velum wanting or very narrow; megaspores mostly 480–650 μ in diameter; microspores papillose; plant of Mississippi basin. . . . . 2. *I. Butleri*.

Mature plant with 8–15 leaves 1.5–3 dm. long, these quadrate in cross-section; velum covering one-fifth to one-third of sporangium; megaspores 440–470 μ in diameter; microspores covered with spinules; plant of Atlantic slope. . . . . 3. *I. virginica*.

b. Leaves without bast-bundles; surface of megaspore covered with spinules. § Echinatae. . . . . 4. *I. muricata*.

a. Megaspores with irregularly crested or reticulate surfaces. . . . *c*.

c. Surface of megaspore with irregular crests. § Cristatae. . . *d*.

d. Crests with crowded processes.

Leaves very numerous (20–200), usually with peripheral strands; sporangium 6–11 mm. long; faces of megaspore with labyrinthiform-convolute ridges. . . . . 5. *I. Eatoni*.

Leaves fewer (8–40), without peripheral strands; sporangium 3–5 mm. long; faces of megaspores with irregular interrupted granular or blunt-toothed ridges. . . . . 6. *I. saccharata*.

d. Crests remotely jagged or with isolated sharp peaks; leaves 10–30, with or without peripheral strands; sporangia 4–7 mm. long. . . . 7. *I. riparia*.

c. Surface of megaspore with regular or nearly regular reticulation. § Reticulatae. . . . *e*.

e. Megaspores mostly 600–800 (–1,100) μ in diameter; plants chiefly submersed, with stiff abruptly tipped leaves. . . . . . . 8. *I. macrospora*.

e. Megaspores mostly 360–570 (–600) μ in diameter; plants submersed or emerging, with leaves spirally spreading or recurving.

Leaves without peripheral strands (bast-bundles), 0.3–2 (–3) dm. long; sporangia 2–7 mm. long, often brown-spotted.

Sporangia 2–5 mm. long, unspotted or but slightly spotted, about one-third covered by the velum; upper segments of megaspore with nearly parallel and branching thin ridges, lower segment regularly reticulate. . . . . 9. *I. Tuckermani*.

Sporangia 4–7 mm. long, heavily spotted, with very narrow velum; megaspores covered all over by broad high ridges surrounding deep pits. . . . . 10. *I. foveolata*.

Leaves with peripheral strands (bast-bundles), 1–6 dm. long; sporangia 6–12 mm. long, pale and unspotted, one-fifth to two-thirds covered by the velum; megaspores deeply pitted, with narrow and sometimes crested reticulate ridges. . . . . 11. *I. Engelmanni*.

§ Tuberculàtae N. E. Pfeiff.

1. I. melanópoda Gay & Dur. (black-footed). — Plants mostly *polygamous*; *leaves* numerous (15–60), *strongly ascending* to stiffly erect, *castaneous or blackish and lustrous* at base, or in forma **pállida** (Engelm.) Fern. (pale) pale, 1.2–4.5 dm. high and 1.5–3.5 mm. broad, with 4 strong bast-bundles and several smaller peripheral ones; sporangia oblong or nearly so, 0.5–3 cm. long, brown-spotted, with velum very narrow to covering one-half the sporangium; ligule triangular-subulate; *megaspores* 280–440 *μ in diameter, with depressed tubercles crowded or confluent into low worm-like wrinkles or nearly smooth; microspores* 20–30 *μ long*, covered *with fine spinules*. — Shallow water,

19. I. melanopoda.

* The treatment by A. A. Eaton in ed. 7 modified to conform with the *Monograph of the Isoëtaceae* by Norma E. Pfeiffer in Ann. Mo. Bot. Gard. ix. 79–232 (1922). Under different species and vars. enlarged megaspores are shown.

shores, meadows, low woods, etc., Ill. and Ia., s. to Mo., Okla. and Tex. Fig. 19. — Engelmann noted some megasporangia bearing both megaspores and microspores.

2. I. Bútleri Engelm. (for its discoverer, George Dexter Butler, 1850–1910). — Superficially resembling pale-based form of no. 1, smaller, *dioecious; leaves* 8–30, *almost bristleform, with triangular cross-section,* 0.8–2.2 dm. long, 0.5–1.2 mm. broad, pale at base, with broad dissepiments, slender air-canals and 4 bast-bundles, the pale sheaths granular on the back; *sporangia* 6–7 *mm. long, commonly covered with brown lines, with velum wanting or very narrow;* ligule subulate, with the base cordate; *megaspores* (360–) 480–650 μ *in diameter,* covered *with* many *low and distinct* (sometimes confluent) *wart-like tubercles; microspores* 27–37 μ *long, covered with papillae.* — Rocky slopes, barrens, flats and depressions, Tenn., Mo. and e. Kans., s. into Ark. and Okla. Fig. 20.

20. I. Butleri.

3. I. virgínica N. E. Pfeiff. Differing from no. 2 in having only 8–15 *leaves* 1.5–3 *dm. long,* these *brown and shining at base, the slender blade nearly quadrangular in cross-section; velum well developed; megaspores* mostly 440–470 μ *in diameter, irregularly marked with points and short discontinuous ridges; microspores* 30–33 μ long, *spinulose.* — Mud and shallow water, interior Va. and W.Va., with a var. farther southw.

§ Echinàtae N. E. Pfeiff.

4. I. muricàta Dur. (with short hard points). — *Leaves loosely and often spirally ascending to recurving, without bast-bundles,* 5–75, deep green with pale sheaths, 0.4–3.5 (–5.5) dm. long, flaccid, with few or many stomata; *sporangia* 4–7 mm. long, often spotted, *completely or partially covered by the velum; megaspores* 420–580 μ in diameter, *covered with numerous* simple or toothed distinct or confluent *slender spinules; microspores* 23–33 μ long, *smooth or but slightly roughened.* (*I. Braunii* Dur., not Unger; *I. echinospora* Dur., vars. *muricata* (Dur.) Engelm. and *Braunii* (Dur.) Engelm.) — Shallow water or wet shores, s. Greenl. and Lab. to s. Alaska, s. to Nfld., N.S., N.E., N.J., Pa., O., Mich., Wisc., Minn., Colo., Utah and Calif. Fig. 21.

21. I. muricata.

§ Cristàtae N. E. Pfeiff.

5. I. Èatoni Dodge (for its discoverer, Alvah Augustus Eaton, 1865–1908). — Plant polygamous, with 20–200 rather coarse erect or arching to recurving deep green to fulvous *leaves* 1–7 dm. long and 2–4 *mm. in diameter,* these somewhat *flattened above,* with abundant stomata *and* usually *with bast-bundles; sporangia* 6–11 *mm. long, brown-spotted,* one-sixth to one-fourth covered by the velum; *megaspores* 300–520 μ *in diameter,* the upper half depressed, *with irregular commissural ridges, the faces covered with labyrinthiform-convolute ridges of crowded blunt processes; microspores very scarce, scattered among the megaspores,* 25–35 μ long, minutely tuberculate or smooth. — Fresh ponds, streams and shores or tidal flats, s. N.H. to N.J. — Forma **Gràvesii** (A. A. Eat.) Reed (for its discoverer, Charles Burr Graves, 1860–1936) of ponds and tidal pools and mud of s. Ct. has more strongly developed bast-bundles, the *macrospores averaging smaller and covered with short truncate single columns,* the *microspores* more abundant and *with high crests or tubercles* (*I. Gravesii* A. A. Eat.). Fig. 22.

22. I. Eatoni, f. Gravesii.

6. I. saccharàta Engelm. (sugary; because "macrospores as if sprinkled over with minute white grains of sugar"). — Corm much flattened, 2-lobed; *leaves* 8–30 (–40), olivaceous, arching, 0.3–2.5 dm. long, 1–2.5 mm. in diameter, with pale membranous base, flaccid, *without peripheral strands,* stomata abundant; *sporangia* 3–5 *mm. long, nearly as broad,* thickly spotted, one-fifth to one-third covered by the velum; *megaspores* 400–500 μ *in diameter, covered with very irregular and often discontinuous crowded ridges with granular or blunt processes; microspores* 22–32 μ long, *smooth or sparingly low-papillate.* — Fresh to slightly brackish tidal shores, Del. to D.C., s. to se. Va.

Var. **Àmesii** A. A. Eat. (in honor of Oakes Ames, 1874–). — *Corm mostly* 3–5-*lobed;* leaves up to 3 dm. long, very slender and quadrangular in cross-section with few stomata; *velum covering one-third to two-thirds of sporangium.* — Fresh ponds, streams and shores, s. N.E. and s. N.Y.

7. I. ripària Engelm. (occurring on banks of rivers). — Plant often polygamous; *leaves* 10–30, 0.9–3 *dm. long,* 1–3 mm. in diameter, *stiffly ascending but without bast-bundles;* ligule ovate with slender tip; *sporangia oblong,* 4–7 *mm. long,* brown-spotted, one-fourth to one-third covered by the velum; *megaspores* 440–660 μ *in diameter, bearing conspicuous jagged crests with sharp distinct or slightly confluent peaks; microspores* 25–33 μ long, *tuberculate or papillate.* — Fresh or tidal shores and shallow water, s. Me. to s. Ont., s. to s. N.E., Del. and Pa. Fig. 23.

23. I. riparia.

Var. **canadénsis** Engelm. (Canadian). — Usually *coarser;* the 15–75 *leaves* erect or recurving, 1–4.5 *dm. long,* without or with 2–4 bast-bundles; *sporangia* 5–8 *mm. long, one-fourth to one-half covered by velum; microspores* 27–37 μ long, *minutely roughened to spinulose.* (*I. Dodgei* A. A. Eat.) — Fresh ponds, streams, shores or tidal flats, s. Que. and s. Ont., s. to N.E., N.J. and Pa.

§ Reticulàtae N. E. Pfeiff.

8. I. macróspora Dur. (large-spored). — *Leaves* 10–40 or more, *stout, stiff, abruptly short-tipped, round in cross-section,* erect or recurving, 0.3–3 dm. long, with rather firm sheaths; *bast-bundles wanting;* stomata scarce or wanting; sporangia 3–5 mm. long, unspotted, with narrow velum; ligule short-triangular; *megaspores* 600–1,100 μ *in diameter, the* 3 *upper faces with thin more or less parallel-walled reticulation; rounded basal half with very coarse honeycomb-reticulation; microspores* 35–50 μ *long,* smooth or with short spinules. (Incl. var. *heterospora* A. A. Eat. and *I. Harveyi* A. A. Eat.) — Margins and shores of fresh pools, lakes and streams, ascending to alpine ponds, Nfld. to n. Minn., s. to N.S., n. N.E., ne. Mass., Catskill Mts., N.Y., Mich. and Wisc. Fig. 24. — Forma **hieroglýphica** (A. A. Eat.) N. E. Pfeiff. (with hieroglyphics) has the *megaspores only* 486–720 μ *in diameter,* their commissural faces with vermiform rounded or smooth ridges. (*I. hieroglyphica* A. A. Eat.) Fig. 25.

25. I. macrospora, f. hieroglyphica.

24. I. macrospora.

9. I. Tuckermáni A. Br. (for its discoverer, Edward Tuckerman, 1817–1886). — Corm 2- or 3-lobed; *leaves* 10–100, 0.3–2 (–3) *dm. long,* spirally spreading to recurving, gradually tapering to very slender tip, *without peripheral bast-bundles; sporangia* roundish or oblong, 2–5 *mm. long,* unspotted or only slightly spotted, about one-third covered by velum; *megaspores* 460–650 μ *in diameter, the commissural faces with nearly parallel and branching thin ridges, the round-based lower half regularly reticulate; microspores* 25–38 μ *long, smooth or minutely papillose,* rarely spinulose. (Incl. var. *borealis* A. A. Eat.) — Sandy, gravelly or muddy pond-margins and shores (frequently tidal), s. Lab. Pen. and Que., s. to Nfld., N.S., s. N.E. and N.Y. Fig. 26.

26. I. Tuckermani.

10. I. foveolàta A. A. Eat. (with small pits). — Differing from small plants of no. 9 in its larger *sporangia* (4–7 *mm. long) heavily dotted and with narrower velum; megaspores covered all over by broad high ridges surrounding smaller deep pits.* —Margins of ponds and streams, se. N.H. Fig. 27.

27. I. foveolata.

11. I. Engelmánni A. Br. (for its discoverer, George Engelmann, 1809–1884). — *Leaves* 15–100 or more, loosely spiraling and ascending to arching, 1–6 *dm. long, usually with bast-bundles; sporangia* 6–12 *mm. long, pale and unspotted, less than one-fourth covered by the velum; megaspores mostly* 400–615 μ *in diameter, with thin-walled honeycomb-reticulation; microspores* 21–30 μ *long, smooth or nearly so.* (Incl. vars. *valida* Engelm. and *fontana* A. A. Eat.) — Margins of ponds, pools and wet shores, or in muddy ditches, n. Fla. and Ala., n. to s. N.H., s. Vt., N.Y., s. Ind., s. Ill. and e. Mo. Fig. 28.

28. I. Engelmanni.

Var. **caroliniàna** A. A. Eat. (Carolinian). — Leaves 15–30, loosely spreading or sprawling, with 4 weak bast-bundles; *sporangia* 6–8 *mm. long, one-third to two-thirds covered by velum; megaspores* 400–530 μ *in diameter, the high honeycomb-ridges appearing spiny through the crisped and deeply cut margins; microspores spinulose.* — Bottomland-woods, shallow exsiccated pool-bottoms, pineland-streams, etc., se. Va.; and upland, N.C. to Ga.

FAM. 5. OPHIOGLOSSÀCEAE (Adder's-tongue Family)

Leafy and often somewhat fleshy plants; the leaves of the frond simple or branched, often fern-like in aspect, erect in vernation, developed from subterranean buds formed either inside the base of the old stalk or by the side of it, and bearing in specialized spikes or panicles rather large subcoriaceous bivalvular sporangia formed from the main tissue of the fruiting branches. Prothallus subterranean, not green, monoecious. — A small family, separated from the true Ferns by the different nature of the sporangia, the erect vernation, etc.

Sporangia in spikes or panicles; leafy branch divided or rarely simple, its veins free. . . . . 1. *Botrychium.*

Sporangia cohering in a simple 2-ranked spike; leafy branch in ours simple, its veins reticulated. . . . . 2. *Ophioglossum.*

1. BOTRÝCHIUM Sw. Moonwort. Grape-Fern

Rhizome short, erect, with clustered fleshy roots; base of stalk containing the bud for the next year's frond; frond with an anterior fertile and a posterior sterile segment or blade; the former mostly 1–3-pinnate, its contracted divisions bearing a double row of mostly sessile naked sporangia; these distinct, rather coriaceous, not reticulated, globular, without a ring, opening transversely into two valves. Sterile blade usually pinnately or ternately divided or compound; veins free. Spores copious, sulphur-colored. — Genus essentially cosmop. (Name a diminutive of the Greek *botrys, cluster of grapes*, from the appearance of the fruiting cluster.)*

a. Bud for the next year's frond completely enclosed by the sheathing base of the stalk. . . *b.*

b. Sterile blades ternately decompound, mostly on a long petiole (chiefly 1–17 cm. long) from near base of plant; the margin (seen under magnification) whitish, some or all blades overwintering; spores pitted, 25–35 μ in diameter, ripe in late summer or autumn. . . *c.*

c. Ultimate segments of sterile blade ovate, obovate, rhombic or rounded, with blunt or rounded tips, the chief terminal ones not greatly prolonged; sterile blades remaining green over winter.

Sterile blade heavily fleshy or coriaceous; ripe spores shed in August and September. . . . . 1. *B. multifidum.*

Sterile blade thinner, submembranaceous; spores shed from mid-September into October. . . . . 2. *B. dissectum,* forma *oneidense.*

c. Ultimate segments lanceolate to lance-ovate, acute or acutish, the chief terminal ones prolonged; sterile blades becoming bronze or purplish in autumn; ripe spores shed from mid-September into October. . . 2. *B. dissectum.*

b. Sterile blade pinnate or pinnatifid, borne high above base, if ternate and subbasal with petiole less than 1 cm. long; margin not white (as seen through lens); frond soft and shriveling in summer or autumn, not evergreen; spores tuberculate, 20–52 μ in diameter, ripening in spring or summer. . . *d.*

d. Sterile blade ternate, basal or subbasal. . . . . 4. *B. simplex.*

d. Sterile blade pinnate or pinnatifid, not ternate, from near base to summit of plant. . . *e.*

e. Sterile blade oblong to ovate, with oblong, obovate, or fan-shaped blunt or round-tipped segments; the expanding basal bud with both sterile and fertile blades erect, ascending or divergent, not both reflexed. . . *f.*

f. Sterile blade sessile or nearly so, with broadly fan-shaped or spoon-shaped pinnae or segments; spores 24–40 μ in diameter. . . . . 3. *B. Lunaria.*

f. Sterile blade petioled or, if sessile or subsessile, with oblong to oblong-obovate or fan-shaped segments; spores 20–52 μ in diameter.

Sterile blade petioled, borne from near base to near summit, simple or once-pinnate, the pinnate ones with fan-shaped to narrowly obovate segments or pinnae. . . . . 4. *B. simplex.*

Sterile blade sessile or very short-petioled, borne from above middle to summit of plant, pinnate or bipinnatifid, with oblong, oblong-ovate or narrowly obovate segments. . . . . 5. *B. matricariae-folium.*

e. Sterile blade deltoid, sessile at base of fruiting panicle; its segments lanceolate, acute; spores 24–42 μ in diameter; expanding basal bud with both fertile and sterile blades abruptly reflexed. . . . . . 6. *B. lanceolatum.*

a. Bud for the next year's frond exposed by the open sheathing base of the stalk; sterile blade broadly deltoid, more or less ternate, bi- or tripinnate, sessile below the long stalk of the fruiting panicle; spores 25–35 μ in diameter. . . 7. *B. virginianum.*

1. B. multífidum (Gmel.) Rupr. (much divided), Leathery G. — Plant 0.4–2.5 dm. high, *with* 1 *new green and often* 1 *old and overwintering green or yellowing very fleshy to heavily subcoriaceous basal ternate broadly triangular blade* (1–) 2–7 cm. long and 2–8(–10) cm. broad, on thickish petiole 1–8 dm. long, *its ovate to suborbicular round-tipped* often overlapping *segments entire or shallowly lobulate*, or in forma **dentàtum** Tryon (toothed) distinctly toothed; *fertile panicle* or panicles (rarely 2 or 3) obliquely thyrsoid-conical, 1.2–7 *cm. long;* sporangia 0.8–1 mm. in diameter; spores pitted, about 30 μ in diameter, mature in Aug. and Sept. (*B. ternatum,*

* Treatment derived in part from that of R. T. Clausen's *Monograph of Ophioglossaceae*, Mem. Torr. Bot. Cl., xix, no. 2 (1938); in part (with permission) from that of R. M. Tryon and others, Ferns and Fern Allies of Wisconsin (1940).

var. *rutaefolium* (A. Br.) D. C. Eat.) — Peaty, loamy or gravelly slopes, plains, thickets and clearings, Nfld. and Côte Nord, Que., to n. Alta. and B.C., s. to N.S., n. N.E., n. Mass., n. N.Y., n. Mich., Wisc. and Minn. (Eurasia) — Passing insensibly into

Var. **intermèdium** (D. C. Eat.) Farw. (intermediate). — *Coarser* throughout, 1–4.5 dm. high; sterile blade or blades 0.5–2 dm. long, 0.7–2.5 dm. broad, on heavy petioles 1–15 cm. long; *the more numerous segments ovate, obovate or oblong,* often not overlapping; *fruiting panicle* 0.3–1.8 *dm. long.* (Subsp. *silaifolium* (Presl) Clausen; *B. silaifolium* Presl) — Similar habitats, Que. to B.C., s. to N.S., N.E., n. N.J., e. Pa., Md., mts. of Va., n. O., n. Ind., n. Ill., ne. Ia., Mont. and Calif. — In the northern half of its range often growing with typical *B. multifidum* and seeming to be merely the larger and older individuals.

2. B. dissèctum Spreng. (dissected). — Similar to no. 1, usually less fleshy or coriaceous; *sterile blade usually becoming bronze or purple in autumn; its segments narrowly lanceolate to lance-oblong and acutish, the chief terminal ones two to four or five times as long as broad or, if shorter, broader and blunt, of thinnish texture;* spores ripe from mid-Sept. into Oct. or Nov. — Highly variable and through the last form passing insensibly into no. 1 but extending much farther south and not so far north.

*a.*Most segments of sterile blade narrowly lanceolate to lance-oblong, two to five times as long as broad, acute or bluntish, turning bronze or purple in autumn. . . *b.*
- *b.*Divisions of the twice pinnate blade deeply and incisely cleft into very many linear to oblong or quadrate emarginate or notched teeth or lobules; frond firm to subcoriaceous. . . . . *B. dissectum* (typical).
- *b.*Divisions blunt and shallowly toothed to nearly or quite entire. . . *c.*
  - *c.*Blade firm or subcoriaceous, bipinnate-pinnatifid, its 10–100 or more pinnules entire or minutely toothed.
    - Basal segments of basal pinnules of lowest pinnae not greatly exceeding the others. . . . . Forma *obliquum.*
    - Basal segments of basal pinnules of lowest pinnae prolonged to half the length of the pinnule. . . . . Forma *elongatum.*
  - *c.*Blade thin and submembranaceous, pinnate to bipinnate, its few (mostly 5–35) pinnules mostly unlobed or part of them only slightly lobed. . . . Var. *tenuifolium.*

*a.*Most segments broadly ovate, obovate or roundish, broadly rounded at summit, remaining green over winter. . . . . Forma *oneidense.*

B. dissèctum (typical), *B. obliquum,* var. *dissectum* (Spreng.) Clute — Pastures, clearings, old fields, open thickets or woods and sterile meadows, N.S. to ne. Me., w. to Minn., s. to N.C., Tenn. and Mo. — Northeastw. often passing freely into forma **oblìquum** (Muhl.) Fern. (oblique) (var. *obliquum* (Muhl.) Clute and *B. obliquum* Muhl.), extending s. to Ga., Ala., La. and e. Tex.; forma **elongàtum** (Gilbert & Haberer) Weath. (elongate; from the prolonged basal segments), scattered through range of last; forma **oneidénse** (Gilbert) Clute (of Oneida Co., N.Y.), embarrassingly transitional to species no. 1, chiefly in woods in range of no. 1 and, like it, with overwintering green blades, thus suggesting a late-fruiting shade-state of the latter, N.B. to Minn., s. to N.E., Va., upland of N.C., O. and Ind.

Var. **tenuifòlium** (Underw.) Farw. (thin-leaved). — Open woods, bottoms and damp clearings, Fla. to e. Tex., n. to Md., D.C., Ind. and Mo.

3. B. Lunària (L.) Sw. (old generic name), MOONWORT, HERBE À LA LUNE (Que.). — Succulent, *erect,* 0.3–2.5 dm. high; *basal bud with the expanding fertile and sterile blades ascending; the fertile panicle* (rarely 2 or 3) 0.1–1.5 *dm. long,* raised well above the *sterile blade; the latter sessile or essentially so, oblong to slightly oblong-ovate,* fleshy, 1–9 cm. long, 0.7–3.6 cm. broad, *with* (2–) 3–7 (–9) pairs of overlapping to slightly remote *broadly fan-shaped or semicircular pinnae; these pinnae as wide as or wider than long,* 0.4–2 *cm. broad, with low rounded-subtruncate summit and unequally concave bases,* entire or shallowly toothed, or in the rare forma **tripartìtum** (Moore) Weath. (three-parted) with the lower pinnae much elongated and pinnately divided, the blade thus 3-parted; *sporangia* 0.5–1 *mm. in diameter; spores* 25–35 *μ in diameter,* tuberculate, mature from June–Aug. — Open turfy, gravelly or ledgy slopes, shores and meadows, chiefly calcareous, Greenl. and Lab. to Alaska, s. to Nfld., e. Que., n. and e. Me., n. Mich., n. Wisc., n. Minn., Colo., Ariz. and Calif. (Eurasia, Australia, Tasmania and N.Z., with a var. in Patagonia) — Passing insensibly into forma **grácile** (Schur) Aschers. & Graebn. (slender), with narrowly oblong sterile relatively thin blade with remote roundish, oval or cuneate-obovate pinnae mostly longer than broad, 3–14 mm. broad (*B. onondagense* Underw. and var. *onondagense* and forma *onondagense* Clute), with the typical form (often in same colony), Nfld. and s. Lab. Pen. to Wash., s. to n. N.E., n. N.Y., Mich., etc. (Eurasia) Barely separable from the latter is forma **minganénse** (Vict.) Clute (of the Mingan Is., Que.) differing from forma *gracile* in fleshier texture, toothed or notched pinnae and spores averaging slightly larger, 30–40 μ in diameter (*B. minganense* Vict. or *B. Lunaria*, var. *minganense* Dole), with the typical form or forma *gracile* or by itself, Lab. to Alaska, s. to C.B., Vt., Wisc., etc. Other but hardly worthwhile forms have been segregated on the degree of toothing.

4. B. símplex E. Hitchc. (simple). — Weaker and more slender than no. 3, mostly 0.3–1.5 (rarely –2.5) dm. long or high, often arching or even reclining; *fruiting spike simple or but slightly compound,* 0.3–5 *cm. long,* stalked; *sporangia* 0.8–1.2 *mm. in diameter; spores* 33–52 μ *in diameter,* with reticulate, vermiculate or verrucose surface; *sterile blade* simple and unlobed, pinnately lobed or ternately divided, borne in the various forms from base nearly to summit of frond; in well-developed plants clearly petioled. — Very plastic, with the following recognizable vars. and forms, these too often seeming like responses to environment or to be stages of development:

*a.*Sterile blade borne from base to slightly above middle of plant; spores finely reticulate. . . *b.*
*b.*Blade simple or pinnately cleft. . . *c.*
*c.*Blade pinnately cleft, 0.7–4 cm. long, with subequal lateral lobes.
Lateral lobes 1–3 pairs, approximate, firm, broad-based; plants 0.3–1.6 dm. high. . . . . *B. simplex* (typical).
Lateral lobes 3–6 pairs, remote, at least the lower ones slender-based to petiolulate, thinnish; plants mostly 1–2 dm. high. . . . . . Forma *laxifolium.*
*c.*Blade entire, oval to obovate, 3–10 mm. long; plants 0.3–1.5 dm. high. . Forma *simplicissimum.*
*b.*Blade ternate or palmate, nearly basal, the long-stalked lowest divisions prolonged and often pinnately lobed, the blade 2–5 dm. broad but scarcely as long. . . . . Forma *compositum.*
*a.*Sterile blade borne near summit of plant, simple or pinnatifid with 1–5 pairs of remote and poorly developed lobes; spores with vermiculate or verrucose surface. . . . . Var. *tenebrosum.*

B. símplex (typical). — Meadows, pastures, open shores, etc., Nfld. to s. B.C., s. to N.S., N.E., L.I., N.J., Pa., Ind., Wisc., N.M. and Calif. Spores ripe in May and June. (Eurasia) — Forma **laxifòlium** (Clausen) Fern. (lax-leaved), var. *laxifolium* Clausen, in woods and thickets, N.E. to Wisc., s. to Pa.; forma **simplicíssimum** (Lasch) Milde (most simple), with the typical form, probably the juvenile state of it; forma **compósitum** (Lasch) Milde (compound), locally with the typical form or by itself.

Var. **tenebròsum** (A. A. Eat.) Clausen (in the dark). — Weak and lax, often maturing under fallen leaves. (*B. tenebrosum* A. A. Eat.) — Damp woods, N.E. and s. Que.

5. B. matricariaefòlium A. Br. (with leaves of *Matricaria*). — Differing from no. 3 in having the *basal bud, on expanding, with the tip of the sterile blade arched over and embracing the fertile one; mature plant glaucous,* up to 3 dm. high; fertile branch subsimple to paniculate, up to 8 cm. long; sporangia 0.6–1 mm. in diameter; spores 20–35 μ in diameter, muricate; *sterile blade* ovate, oblong or oblong-lanceolate, *subsessile or very short-petioled, borne from above middle to summit of plant,* 1–9 cm. long, 0.5–7 cm. broad, *pinnate, with the distant pinnae cut into oblong, oblong-ovate or narrowly obovate segments.* (*B. neglectum* Wood; *B. ramosum* of ed. 7, not *Osmunda ramosa* Roth, basonym) — Woods, thickets and dry to moist old fields, etc., Nfld. to s. B.C., s. to N.S., N.E., L.I., N.J., Pa., Md., w. Va., W.Va., O., Mich., Wisc., S.D. and Id. Spores ripe in June and July. (Eurasia; a var. in Patagonia) — Plants with pinnae entire are **forma grácile** (House) Weath. (slender); those with the sterile blade triangular-ovate, with 1 or 2 pairs of lower pinnae much elongate and pinnate are forma **palmàtum** Milde (palmate); plants with sterile blade with 3 equal parts, each part pinnatifid, are forma **compósitum** Milde (compound).

6. B. lanceolàtum (Gmel.) Ångstr. (lanceolate). — Plant stoutish, fleshy, 0.5–2.5 dm. high; *expanding basal bud with both fertile and sterile blades reflexed; fertile panicle rather dense,* 1–4.5 cm. long, with ascending branches; *sporangia* 1–1.5 *mm. in diameter,* crowded and scarcely immersed along the branches; spores averaging 35 μ in diameter; *sterile blade deltoid,* thick and fleshy, sessile at summit of stem, 0.8–4.5 cm. long, about as broad, *its lanceolate mostly pinnatifid pinnae subapproximate.* — Meadows, peaty slopes and clearings, Nfld. and

Côte Nord, Que., s. to Gaspé Pen., Que., and n. Me.; B.C. and s. Alaska, s. to Colo. and Wash. (alpine). *Fr.* late June–Aug. (Eurasia)

Var. **angustisegméntum** Pease & Moore (with narrow segments). — More slender, up to 3 dm. high; *fertile panicle more open*, the smaller *sporangia more deeply immersed; spores* 21–35 μ *in diameter; sterile blade thinner, with distant narrower segments.* (*B. angustisegmentum* Fern.) — Deciduous or mixed woods and openings, w. Nfld. to Wisc., s. to N.S., N.E., N.J., Pa., W.Va., w. Va. and O.

7. B. virginiànum (L.) Sw. (Virginian), RATTLESNAKE-FERN. — *Common stalk erect*, glabrous or glabrate, 0.5–4.5 dm. high, *bearing at summit the sessile broadly triangular ternately decompound sterile frond; the latter bi- or tripinnate*, membranaceous or slightly fleshy, and 0.5–4 dm. broad, with oblong-lanceolate toothed or lobed subremote ultimate segments; *fertile blade* pinnately decompound, 0.2–2 dm. long, *raised above the sterile frond on an erect stalk* 0.3–8 dm. long (rarely forking and bearing 2 or 3 panicles); sporangia 0.5–1 mm. in diameter, their valves widely spreading and recurved in dehiscence; spores 25–35 μ in diameter; *bud for the next year's frond exposed by the open sheathing base of the stalk.* — Rich deciduous or mixed woods, Fla. to Calif., n. to P.E.I., Que., Ont., Minn., S.D. and s. B.C. *Fr.* spring and early summer. (E. Asia)

Var. **europaèum** Ångstr. (European). — *Stiffer and firmer; sterile frond subcoriaceous*, its less toothed *ultimate segments approximate to overlapping; sporangia* up to 1.8 mm. in diameter, *their valves not widely spreading in dehiscence.* (Incl. var. *laurentianum* Butters) — Woods, thickets and damp openings, chiefly coniferous, s. Lab. to Alaska, s. to Nfld., N.B., n. N.E., n. Mich., n. Wisc., n. Minn., Colo. and Oreg. (N. Eu.)

2. OPHIOGLÓSSUM L. ADDER'S-TONGUE. HERBE SANS COUTURE (Que.)

Rhizome erect, fleshy and sometimes tuberous, with slender fleshy roots which are sometimes proliferous; bud situated at side of the base of the naked stalk of the frond; fronds with anterior fertile and posterior sterile branches or blades, the coriaceous sporangia connate and coherent in two ranks on the margins of a simple spike. Sterile segment or branch fleshy, simple in our species; its veins reticulate. Spores copious, sulphur-yellow. — Genus widely dispersed in all temp. and trop. reg. (Name from the Greek *ophis, a serpent*, and *glossa, tongue.*)

Sterile blade blunt or bluntish, its large areolae without smaller included areolae. 1. *O. vulgatum.*
Sterile blade apiculate, its large areolae surrounding a network of smaller areolae. 2. *O. Engelmanni.*

1. O. VULGÀTUM L. (common). — Rhizome subcylindric, erect, with numerous slenderly cord-like widely spreading roots (from which remote new fronds arise); *base of common stalk usually* (when not broken) *with a definite chartaceous or coriaceous blackish persistent cylindric sheath; sterile blade* oval to ovate, usually with rounded base, *blunt or rounded at summit*, pale or yellowish-green, dull, *its areolae* (seen by transmitted light) *without included smaller areolae;* fertile stalk elongate, the slender spike with a short sterile tip, the sporangia suborbicular or orbicular. — Eurasia; represented with us by two vars.:

Var. **pycnóstichum** Fern. (with crowded rows). — Habitally resembling the typical Eurasian plant, 1–3 dm. high; *common stalk* 4.5–13 cm. long, *usually with the firm dark tubular sheath persistent; sterile blade* fleshy, *deep green and lustrous, rounded, ovate or ovate-lanceolate, with rounded base,* 2.5–8.5 cm. long, 1–4 cm. broad; *sporangia crowded, transversely oblong.* — Rich, often calcareous, woods, marly pockets and bottomlands, S.C. and Tenn., n. to N.J., Md., W.Va., O. and Ind. *Fr.* April–early June. — Our nearest approach to the Eurasian typical plant.

Var. **pseudópodum** (Blake) Farw. (with a false foot; from the narrowed subpetiolar base of the sterile blade). — Common stalk usually naked at base (*usually without the hard and dark tubular basal sheath*); *sterile blade pale green, dull, oblanceolate, narrowly obovate, elliptic or lanceolate, usually tapering gradually to base,* 1–12 cm. long, 0.7–3.2 cm. broad; *sporangia suborbicular.* (Incl. *O. pusillum* Raf. and *O. arenarium* E. G. Britt.; by Clausen considered inseparable from the preceding var. and from the Eurasian plant.) — Peaty (acid) or grassy swales, wet thickets, shores, damp sands, sterile pastures, etc., P.E.I. and s. Que. to Wash., s. to N.S., N.E., L.I., Del., Pa., upland of Va., O., Ind., Ill., Neb., Ariz. and Mex. *Fr.* late May–Aug.

2. O. Engelmánni Prantl (in honor of GEORGE ENGELMANN, 1809-1884). — Differing from no. 1 in low stature (mostly 1–2 dm. high) and greater tendency to bear 2 or more leaves; the *sterile blade apiculate* at tip, elliptic to oblong and tapering at base, the *larger areolae* (seen by transmitted light) *embracing a network of smaller ones.* — Calcareous or argillaceous barrens, prairies, open woods, bluffs, etc., Fla. to Ariz. and Mex., n. to w. and n. Va., s. O., s. Ind., Mo., Kans. and Okla. *Fr.* spring and early summer.

FAM. 6. OSMUNDÀCEAE (Flowering Fern Family)

Leafy plants (ours herbaceous) *with creeping rhizomes. Sporangia naked, globose, mostly pedicelled, reticulated, with no ring or with mere traces of one near the apex, opening into two valves by a longitudinal slit.* Stipes winged at base.

1. OSMÚNDA L. Flowering Fern

Fertile fronds or fertile portions of the frond mostly lacking chlorophyll, much contracted and bearing on the margins of the narrow rachis-like divisions short-pedicelled and naked sporangia; these globular, thin and reticulated, large, opening by a longitudinal cleft into two valves and bearing near the apex a small patch of thickened oblong cells (the rudiment of a transverse ring); spores green. — Small genus of trop. and temp. reg. but absent from western N. Am. (Named, according to some writers, for *Osmunder*, the Saxon equivalent of the god *Thor.*)

Sterile fronds bipinnate, with simple pinnules; fertile fronds similar, terminated by the fruiting panicle. . . . . 1. *O. regalis.*

Sterile fronds once pinnate, the pinnae pinnatifid; fertile fronds separate and elongate or fertile pinnae borne on fronds similar to the sterile ones.

Fertile and sterile fronds similar, their mature pinnae not woolly at base; sterile pinnae borne above and below fertile ones, greenish, becoming blackish. . . . . 2. *O. Claytoniana.*

Fertile fronds separate from the sterile, shorter, dense, elongate, cinnamon-color, densely woolly, soon shriveling; bases of mature sterile pinnae woolly-tufted. . . . . 3. *O. cinnamomea.*

1. O. regàlis L. (royal), Flowering or Royal Fern. — Rhizome nearly superficial, the older portion soon decaying, the growing part covered with old stipe-bases and roots; fronds numerous, erect, commonly 1–3.5 m. high; stipes continuous with rhizome, at first reddish or glaucous, becoming green, about equaling blade, rounded on back, with broad stipule-like basal wings; *blade* lance-ovate, at first membranaceous, soon leathery, *bipinnate, with ascending pinnae, with oblong subentire or finely toothed blunt* nearly sessile *pinnules* rounded, subtruncate or auricled at base; *fertile fronds terminated by a* narrowly ovoid to oblong *panicle of* greenish, finally brownish, *racemose branches, the rachises bearing numerous rather persistent black hair-like scales.* — Eurasia; represented with us by

Var. **spectàbilis** (Willd.) Gray (showy). — *Lower*, commonly 0.6–1.8 m. high; *fronds* relatively broader, *oblong-ovate;* the *pinnules mostly oblong-oval to lance-oblong*, the longer ones 3–7 cm. long and 0.7–2 cm. broad (averaging about one-fourth as broad as long); *panicle more slender*, 0.7–3 dm. long, *without dark scales, rarely with a few slender axillary hairs.* — Low woods, peaty thickets, swales, shores, etc., Nfld. to Sask., s. to N.S., N.E., L.I., Fla., Ala., Miss., La., Tex. and Mex. *Fr.* spring and early summer. — Dwarf plants 1.5–3.5 dm. high, with largest pinnules elliptic to oblong-oval and only 1.5–3 cm. long and 0.8–1.1 cm. broad (averaging half as broad as long) and with fruiting panicles 5–10 cm. long, are forma **nàna** Fern. (dwarf) of alpine or bleak habitats in Nfld., Que. and mts. of n. N.E. — Aberrant forms have been designated as follows: forma **intercallàta** Dole (inserted between) with the median (instead of terminal) pinnae fertile (an aberration said to follow mowing or browsing); forma **anómala** (Farw.) Harris (anomalous) with some branches of fertile panicle foliaceous: forma **lineàris** Clute (linear) with the bases of the pinnules definitely auricled, the margins strongly undulate; forma **orbiculàta** Clute (round) with the few crowded sterile pinnules round to cordiform.

2. O. Claytoniàna L. (for John Clayton, pioneer botanist of Virginia, ?–1773), Interrupted Fern. — Rhizome heavy, creeping, covered with winged stipe-bases; *fronds once pinnate, with pinnae pinnatifid into blunt pinnules*, elliptic-oblong to oblong-oval, broadest at the middle, tapering to base and apex, *when expanding with promptly deciduous flocculent whitish-brown wool, soon quite glabrous;* stipes greenish, erect; *outer fronds usually sterile*, up to 1 m. tall and 3 dm. broad; *inner fronds* of crown *taller*, erect and *fertile; the fertile pinnae borne near middle of frond, greenish, becoming blackish.* — Moist woods and thickets, Nfld. to se. Man., s. to N.S., N.E., L.I., Va., upland to Ga., Ky. and n. Ark. *Fr.* spring and early summer. (A var. in e. and s.-centr. Asia) — Forma **dùbia** (Grout) Clute (doubtful) has remote pinnules, the outer ones elongate and deeply pinnatifid; forma **Mackiàna** Kittredge (named in 1922 for its discoverer, Clara McKenzie Mack) has the lanceolate blade with shortened upper pinnae with few irregularly spreading and often irregularly lobed somewhat triangular pinnules. — × **O. Rúggii** Tryon (named in 1940 for its discoverer, Harold Goddard Rugg), transitional in many characters, is a sterile hybrid of nos. 1 and 2.

3. O. cinnamòmea L. (cinnamon-colored), CINNAMON-FERN, BUCKHORN, FIDDLE-HEADS. — *Fronds dimorphic, the sterile oblong-lanceolate ones much taller than the narrow densely woolly and soon wilting fertile ones; stipes and unrolling fronds heavily clothed with cinnamon-brown wool, this soon deciduous from sterile blades but persisting as tufts at bases of pinnae;* sterile fronds up to 1.6 m. high, their lanceolate pinnae pinnatifid into broadly oblong obtuse divisions; fertile fronds appearing first, twice pinnate. — Swamps, low woods and thickets, Nfld. to Minn., s. to N.S., N.E., L.I. and the Gulf States. *Fr.* spring. (Vars. in trop. and subtrop. Am. and in e. Asia) — The following aberrant forms have been named: forma **auriculàta** (Hopkins) Kittredge (eared) with basal segments of lower side of each pinna greatly prolonged and acutely toothed; forma **bipinnatífida** Clute (twice-pinnatifid) with segments of sterile pinnae obtusely lobed or toothed; forma **incìsa** (J. W. Huntington) Gilbert (incised) like the last, but the teeth and lobes acute; forma **cornucopiaefòlia** Clute (with leaves bearing cornucopias) with the pinnae of the foliaceous fronds with rachis naked above (below leafy tip), many lower pinnules or segments bearing slender-stalked cornucopia-like clusters of sporangia from near the tip; forma **frondòsa** (T. & G.) Britt. (leafy) with fertile fronds partly leafy; forma **latipínnula** Blake (with broad pinnules) with segments few, triangular and about 2 cm. broad.

Var. **glandulòsa** Waters (glandular). — Segments and upper part of rachis glandular. — Locally in large and uniform colonies, on or near Coastal Plain, R.I. to e. Va.; Miss.

FAM. 7. SCHIZAEÀCEAE (CURLY-GRASS FAMILY)

Sterile fronds tufted and with us linear-filiform (Schizaea) or resembling a twining aerial stem with alternate paired palmately lobed leaves (Lygodium). Sporangia borne in double rows on narrow fertile segments, ovoid, sessile, having a complete transverse ring at the apex and opening by a longitudinal slit.

Plant not climbing, usually low and small; sterile fronds slenderly linear (often spiraling). . . . . 1. *Schizaea.*
Plant climbing, with pairs of stalked and alternate lobed leaves. . . . . . 2. *Lygodium.*

1. SCHIZAÈA Sm. CURLY-GRASS

Sporangia large, ovoid, with striate rays at apex, opening by a longitudinal cleft, naked, vertically sessile in a double row along the single vein of the narrow divisions of the pinnate or radiate fertile appendages to the slender and simply linear or (in some foreign species) flabelliform or dichotomously many-cleft fronds. — Characteristic ferns of S. Hemisph. and Tropics; our species isolated far to the north of the others. (Name from the Greek *schizo, to split.*)

1. S. pusílla Pursh (very small). — Forming dense tufts with crowded spiraling and curling slenderly linear sterile fronds; fertile fronds much longer, very slender (barely 0.5 mm. wide), erect, 1.5–12 cm. high, bearing at summit the one-sided fruiting portion consisting of 3–8 pairs of obliquely ascending crowded finger-like pinnae, each 1.5–4 mm. long. — Open damp peaty or sandy depressions, sphagnous bogs and low mossy open woods, or even in crevices of ledgy shores, tablelands and lowlands, Nfld. and N.S.; Pine Barrens of N.J. *Fr.* July–Sept.

2. LYGÒDIUM Sw. CLIMBING FERN

Fronds twining or climbing, bearing stalked and variously lobed (to very compound) blades alternately in pairs, with mostly free veins, bearing the fructification. Sporangia much as in *Schizaea,* fixed to the oblique veinlet by the inner side, next the base, one or rarely two covered by each scale-like indusium; indusia in groups on separate fructifications or divisions or spike-like lobes. — Several species in trop. and warm-temp. reg. (Name from the Greek *lygodes, flexible.*)

1. L. palmàtum (Bernh.) Sw. (palmate or hand-like). — Smooth, the stalk-like fronds slender, flexible and twining, 3–10 dm. or more long, from slender creeping rhizome; the short alternate petioles 2-forked, each fork bearing a rounded-cordate palmately 4–7-lobed blade 2–7 cm. broad; fertile blades uppermost, contracted and several times forked, forming a terminal panicle. — Moist acid soil of thickets, marshes and open woods, Ga. and Tenn., n., both in mts. and out to Coastal Plain, locally to s. N.H., Mass., se. and centr. N.Y., Pa., W.Va., se. O. and Ky. *Fr.* Aug., Sept.

FAM. 8. HYMENOPHYLLÀCEAE (FILMY FERN FAMILY)

Delicate ferns with slender often filiform creeping rootstocks. Fronds pellucid, of a single layer of cells. Sporangia sessile on a bristle-like receptacle within a cupuliform, tubular or bivalvular involucre, the ring transverse and complete. — Chiefly tropical, inhabiting damp places, often epiphytic. Fronds circinate in vernation.

1. TRICHÓMANES L. Filmy Fern

Involucre tubular-funnelform, its mouth nearly or quite truncate. Sporangia bursting vertically. — Small creeping ferns of most trop. and warm-temp. reg. (Greek name for some fern suggesting *Adiantum.*)

1. **T. Boschiànum** Sturm (for Roelof Benjamin van den Bosch, 1810–1862, student of this family). — Fronds oblong-lanceolate, 0.5–2 dm. long, 1.2–3.5 cm. broad, bipinnatifid; rachis narrowly winged; pinnae triangularly ovate, their divisions toothed or again lobed; receptacle capillary, much exserted. — Wet or dripping siliceous cliffs and slopes, Ala., n. to W.Va., Hocking Co., O., and Pope Co., Ill. *Fr.* July–Sept.

FAM. 9. POLYPODIÀCEAE (Fern Family)

Leafy plants (ours herbaceous) *with creeping rhizomes. Sporangia* (spore-cases) *collected in dots, lines or variously shaped clusters* (sori or fruit-dots) *on the back or margins of the frond or its divisions, cellular-reticulated, stalked; the stalk running into a vertical incomplete many-jointed ring, which by straightening at maturity ruptures the sporangium transversely on the inner side, discharging the spores. Fruit-dots often covered* (at least when young) *by a membrane called the indusium* (or less properly the *involucre*) *which grows either from the back or the margin of the frond.**

*a.*Indusium present, at least on young fruiting fronds. . . *b.*
*b.*Indusium dorsal on the frond or, if marginal, not formed by revolute margin of the frond. . *c.*
*c.*Indusium borne beneath the sorus, more or less surrounding it or its base as a finally ruptured, dissected or valved saucer- or cup-like structure. . . *d.*
*d.*Fertile and sterile fronds similar, flat or flattish, not rigid; sori flat or convex or, if globular, partly united with reflexed teeth of the frond-margin. . . . *e.*
*e.*Sori borne on backs of fronds.
Indusium borne symmetrically under the sorus, cup- or saucer-shaped and splitting into elongate segments. . . . 1. *Woodsia.*
Indusium hood-like or arched, attached by one side, partly under the sorus, opening at the summit. . . . . 2. *Cystopteris.*
*e.*Sori marginal, globular, inclosed in a cup-like indusium, opening at summit and partly adhering to reflexed teeth of the frond. . . 7. *Dennstaedtia.*
*d.*Fertile and sterile fronds very dissimilar, coarse and stiff; the sterile much larger than the erect rigid fertile ones; fertile segments subglobose and pod- or berry-like.
Fronds forming tall vase-like clumps, the stiffly feather-like simply pinnate fertile ones clustered, among or surrounded by the many times taller lanceolate regularly pinnate sterile ones, with veins free. 3. *Pteretis.*
Fronds scattered or solitary, the fertile ones bipinnate and the sterile ones coarsely pinnatifid and deltoid-ovate, with anastomosing veins. . . . . 4. *Onoclea.*
*c.*Indusium spreading from above or from one side over the sorus. . *f.*
*f.*Indusium peltate or attached at its center, orbicular to reniform.
Indusium reniform or with a deep sinus. . . . . 5. *Dryopteris.*
Indusium shield-shaped, without a sinus. . . . . . . 6. *Polystichum.*
*f.*Indusia linear or oblong to lunate, hooked or horseshoe-shaped, attached at margin. . . *g.*
*g.*Sori parallel to the oblique lateral veins. . . *h.*
*h.*Fronds herbaceous, annual, pinnate to tripinnate, with flattened, furrowed or angled stipes; indusia straight, curved or even horseshoe-shaped, along free veinlets. . . . . . . 8. *Athyrium.*
*h.*Fronds mostly evergreen, simple or, if pinnate, with slender nearly terete stipes; indusia straight or nearly so. . *i.*
*i.*Fronds simple, commonly auricled at base.
Fronds oblong, not attenuate to rooting tip; sori often paired along opposite sides of simple veins. . . . . . 9. *Phyllitis.*

* By Diels our 19 genera were placed among those of 5 tribes; by Christensen in 9 different subfamilies. The technical (often anatomical) distinctions are too erudite (and not agreed upon) for practical use in a local flora; the key to genera is, therefore, somewhat artificial.

Fronds lance-attenuate, with prolonged rooting tip; sori scattered or approximate along reticulate veins. . . 10. *Camptosorus.*

i. Fronds pinnate or pinnatifid, mostly firm or evergreen; indusia mostly separate, attached along inner side of free veins. . . 11. *Asplenium.*

g. Sori parallel to the midrib, in one or more chain-like rows on transverse anastomosing veinlets. . . . . . . 12. *Woodwardia.*

b. Indusium formed entirely or in part by the revolute margin of the frond or its segments. *j.*

j. Rhizome very short (except in no. 7); fronds mostly clustered (if scattered weak and fragile), the bases of their forks or pinnae without nectaries. . . *k.*

k. Sori soon confluent as a marginal band. *l.*

l. Pinnules and segments of frond articulated at base. . . . . 13. *Pellaea.*

l. Pinnules and segments not jointed at base. . . *m.*

m. Indusium, formed by reflexed margin of segments usually wanting; the lower surface of the triangular-ovate frond white and powdery beneath. . . . . 14. *Notholaena.*

m. Indusium well developed; lower surface glabrous or pubescent, not white-powdery.

Fertile and sterile fronds similar; reflexed indusial margins of segments whitish and membranaceous (if herbaceous, in species no. 2, the lanceolate frond hirsute). . . . . 15. *Cheilanthes.*

Fertile and sterile fronds very dissimilar, glabrous; reflexed indusial margins herbaceous. . . . . . 16. *Cryptogramma.*

k. Sori clearly distinct, short, mostly not confluent.

Indusia flattish, opening at side, transverse; stipe and frond glabrous; rhizome stout and short. . . . . 17. *Adiantum.*

Indusium cup-like, opening at summit, partly adherent to teeth of segments; stipe and finely dissected frond glandular-hairy; rhizome slender, elongate. . . . . 7. *Dennstaedtia.*

j. Rhizome coarse, elongate, forking and extensively creeping; fronds scattered, coarse and coriaceous, with nectaries at the lower forks; indusia continuous and transverse, the outer one conspicuous as a modified reflexed margin and an inner one less conspicuous. . . . 18. *Pteridium.*

a. Indusia wanting or nearly obsolete; sori roundish. . . *n.*

n. Fronds herbaceous, not evergreen, 2–4 times pinnatifid or pinnate, their stipes not articulated to rhizome.

Rhizome stout and short, forming a stipe-covered crown; fronds up to 1 m. high, in vase-like clumps, lanceolate. . . . . 8. *Athyrium.*

Rhizome very slender, elongate and forking, subterranean; fronds low, scattered or solitary, deltoid or broader. . . . 5. *Dryopteris.*

n. Fronds coriaceous, evergreen, simply pinnatifid (rarely bipinnatifid), the stipes articulated to the creeping superficial rhizome. . . . . . . 19. *Polypodium.*

1. WOÒDSIA R. Br. Woodsia

Sori round, borne on the back of simply forked free veins; the very thin and often evanescent indusium attached by its base all around the receptacle, *under* the sporangia, either small and open or else early bursting at the top into irregular lobes or segments. — Low and rather small tufted ferns with pinnately divided fronds. (Dedicated to *Joseph Woods*, 1776–1864, English botanist.)

a. Stipes jointed near the base, the joint appearing as a slightly thickened and darkened ring; the persistent old bases of stipes of essentially uniform length, mostly 1–3 cm. long; indusia with long curving, filamentous segments usually persistent and curving around or above the mature sporangia. . . *b.*

b. Stipes and at least the lower third of the rachises brown, firm, often chaffy at least at base, the old persistent stipe-bases 1–3 (–4) cm. high; fronds comparatively firm, linear to oblong-lanceolate, 0.5–5.7 cm. broad; indusia with 10–20 filaments conspicuously exceeding the sporangia.

Fronds permanently hairy and commonly chaffy beneath with whitish to rufescent chaff, lanceolate and stiff; stipes, at least when young, very chaffy. . . . . 1. *W. ilvensis.*

Fronds glabrous or promptly glabrate on both surfaces, without chaff, linear- to oblong-lanceolate, stiffish to submembranaceous; stipes without chaff or with few caducous scales. . . . . . 2. *W. alpina.*

b. Stipes and rachises green or stramineous, the latter chaffless, the former rarely chaffy above the joints; old stipe-bases 2–10 mm. high; fronds delicate, membranaceous, glabrous, linear or linear-lanceolate, 0.4–1.6 cm. broad; indusia with 5–8 filaments only slightly exceeding sporangia. . 3. *W. glabella.*

a. Stipes not jointed at base; their old denuded remnants often long and, including rachises, of irregular lengths; indusia hidden by the mature sporangia or splitting into few broad or jagged lobes or segments and forming a rosette about the sorus. . . *c.*

c. Indusia divided nearly to base into linear-lanceolate, linear or thread-like segments; northern species. . . *d.*

d. Frond and stipe glabrous or merely glandular, not hispidulous.

Frond lance-linear or -oblong, glabrous but sometimes glandular; pinnules of the principal pinnae approximate, broader than the separating sinuses; indusia with slenderly attenuate moniliform segments. . . . . 4. *W. oregana.*

Frond lanceolate, glandular-puberulent; pinnules of principal pinnae remote, scarcely broader than the wide separating sinuses; indusia with flat mostly 2-cleft segments. . . . . 5. *W. Cathcartiana.*

d. Frond and stipe hispidulous with white trichomes mixed with glands; indusia with flat lanceolate- to linear-attenuate segments. . . 6. *W. scopulina.*

c. Indusia cup-like, more or less cleft at summit into broadly oblong to ovate toothed segments; plants of southern or broad temperate range.

Frond and stipe hispidulous with white trichomes; indusia rarely cleft below middle; plant of Alleghany and Ozark Mts. . . . . . . 7. *W. appalachiana.*

Frond and stipe glabrous or merely glutinous except for chaff on stipe; indusia cleft nearly to base; wide-ranging. . . . . 8. *W. obtusa.*

1. W. ilvénsis (L.) R. Br. (of Island of Elba), Rusty or Fragrant W. — Rootstock ascending, usually freely forking, forming *dense tufts or tussocks covered with persistent brown old stipe-bases* 1–3 (–4) *cm. high;* new *stipes* firm, dark brown or blackish at the densely brown-chaffy base, *paler brown* above, *distinctly articulated* a third to half the way up to the lowest pinnae, *with the rachis more or less chaffy* (or with chaff-like trichomes); *frond lanceolate, firm,* dark green and glabrous or strigose above, *pilose and usually chaffy beneath* (chaff scanty or wanting in shade-specimens), 0.2–2 dm. long, 1–5.7 cm. broad, commonly aromatic, with 7–23 pairs of oblong-ovate to -lanceolate pinnatifid *pinnae* with oblong obtuse commonly *revolute and crenate lobes;* sori numerous, often confluent; *indusia with* 10–20 *long curving filamentous segments conspicuously overtopping the mature sporangia.* — Dry, mostly sterile rocks, cliffs and talus, frequently in exposed situations; Arct. reg., s. to Nfld., N.S., N.E., n. N.J., Pa., upland to N.C., Mich., n. Ill., n. Ia., Alta. and B.C. *Fr.* June–Oct. (Eurasia) — Var. **grácilis** Lawson (slender) or × *W. gracilis* (Lawson) Butters, with *elongate-lanceolate fronds with remote pinnae and reduced chaff,* is perhaps a hybrid with no. 2, growing with it northw.

2. W. alpìna (Bolton) S. F. Gray (alpine), Northern W. — Somewhat resembling no. 1; *frond* more delicate or less rigid, *linear- to narrowly oblong-lanceolate, glabrous or promptly glabrate, without chaff,* 0.1–1.5 dm. long, 0.5–2.5 *cm. broad, with* 5–18 pairs of suborbicular to oblong or lanceolate crenate to pinnatifid commonly remote *flat pinnae; sori nearly marginal,* distinct or confluent; indusia as in no. 1; stipes 0.5–1.5 mm. thick, without or with few caducous chaffy scales, flexuous and subcapillary in shade, stiff and coarser in exposed habitats. (*W. Belli* (Lawson) A. E. Porsild) — Arct. reg., s. to shaded or exposed, damp to dry slaty or calcareous rocky banks of Nfld., Gaspé Pen., Que., s. N.B., n. Me., n. Vt., ne. N.Y., Thunder Bay Distr., Ont., n. Mich. and n. Minn. *Fr.* late June–Aug. (Eurasia)

3. W. glabélla R. Br. (smooth), Smooth W. — Forming *small tufts* 0.5–3 (rarely –5) *cm. across at base; persistent old stipe-bases* 2–10 *mm. high; stipes soft and delicate, green or straw-colored, articulated near base,* usually *without chaff above the joint; frond linear or linear-lanceolate, glabrous throughout,* 1–16 cm. long, 4–16 *mm. broad,* with 6–23 pairs of *thin-membranaceous* (firmer in exposed situations) *suborbicular to ovate* (or uppermost lance-oblong) variously toothed, lobed, or cleft *divergent pinnae;* sori distinct or confluent; *indusia with* 5–8 *filamentous segments only slightly exceeding the sporangia.* — In thin moss or humus on calcareous rocks, often at crests of shaded cliffs, Arct. reg., s. to Nfld., Gaspé Pen., Que., n. N.E., Catskill Mts., N.Y., Thunder Bay Distr., Ont., and n. Minn. *Fr.* June–Aug.

4. W. oregàna D. C. Eat. (of Oregon), Oregon W. — *Glabrous* but sometimes glandular-glutinous, 0.5–3 dm. high; *fronds* bright green, *narrowly lance-oblong or -linear,* 1–3.5 cm. broad, bipinnatifid, with mostly distant pairs of triangular-oblong blunt pinnae; *pinnules of principal pinnae approximate,* blunt, *broader than the separating sinuses,* the marginal crenulate-serrate teeth often revolute; ***indusia with almost filiform conspicuously moniliform segments.*** —

Calcareous or siliceous ledges and cliffs, Alta. and B.C., s. to Neb., w. Okla. and N.M.; n. Ia. to n. Wisc., e. to Manitoulin I., Ont.; Rimouski Co., Que. *Fr.* June–Aug.

5. W. Cathcartiàna Robins. (named for Ellen Cathcart who discovered it in 1873), Cathcart's W. — *Finely glandular-puberulent;* fronds lanceolate, 0.6–2 dm. long, 1.5–6 cm. broad, much as in no. 4 but more open; *pinnules of principal distant oblong pinnae remote, scarcely broader than the wide separating sinuses,* often sharply denticulate; *indusia with flat linear often apically 2-cleft segments.* — Crevices and talus of rock (often trap), local, Algoma Distr., Ont., se. and s. to w. N.Y., n. Mich., n. Wisc. and n. Minn. *Fr.* July–Sept.

6. W. scopulìna D. C. Eat. (on rock), Rocky Mountain W. — *Loosely hispidulous with minute white hairs* mixed with glandular puberulence; frond 0.7–2.5 dm. long, 1.5–7 cm. broad, the distant pairs of pinnae oblong-lanceolate to -ovate; pinnules subapproximate, denticulate; *indusia with flat lanceolate- to linear-attenuate segments mostly hidden under the sorus.* — Rock-crevices (oftenest calcareous), Sask. to B.C., s. to Black Hills, S.D., N.M. and s. Calif.; locally in n. Minn., n. Wisc., n. Mich. and Algonquin Park, Ont.; n. Gaspé Co., Que. *Fr.* July–Sept. — Plant viscid and aromatic, the rachis and pinnae very brittle.

7. W. appalachiàna T. M. C. Taylor (of the Appalachians), Appalachian W. — Superficially resembling no. 6; *frond and stipe hispidulous with white trichomes; indusia somewhat cup-like, with coarsely lacerate margin or splitting into few broad segments.* — Shale or sandstone cliffs, mts. of w. Va. and W.Va.; Ozark Mts., Ark.

8. W. obtùsa (Spreng.) Torr. (obtuse), Blunt-lobed or Large W. — Tufts rather small, with persistent old fronds; *stipes straw-colored to pale brown, chaffy* at least when young; *fronds lanceolate,* firmly herbaceous, occasionally evergreen, 0.4–3.8 dm. long, 0.2–1 dm. broad, *minutely glandular or granular beneath and on the rachis;* pinnae mostly remote, 8–20 pairs, the lower deltoid-ovate and short, the median ovate-lanceolate to oblong and with 5–12 pairs of oblong obtuse deeply crenate or pinnatifid pinnules; *indusia when young subglobose and wrapping around the sorus, later splitting into about* 6 *broad oblong to ovate more or less radiating toothed segments.* — Rocky woods and ledges or dry wooded slopes, Ga. and Ala. to e. Tex., n. to s.-centr. Me., n.-centr. N.H., sw. Que., n. and w. N.Y., O., Mich., Wisc., Minn. and Neb. *Fr.* late May–Oct.

2. CYSTÓPTERIS Bernh. Bladder-fern

Sori roundish, borne on the back of a straight branch of the forking free veins; the delicate indusium hood-like or arched, attached by a broad base on the inner side (toward the midrib) partly under the sori, the indusium early opening at the outer side, which faces the apex of the lobe and is somewhat jagged, soon thrown back or withering away. — Delicate ferns with finely dissected pinnate fronds, nearly cosmop. in cool or temp. areas. (Name from the Greek *cystis, a bladder,* and *pteris, fern.*)*

Fronds lanceolate or lance-oblong, tufted toward the apex of the stoutish rhizome.

Fronds lanceolate or lance-oblong, broadest above the base, the lower pinnules of the pinnae decurrent on the rachis; veins mostly extending to the teeth; fertile and sterile fronds similar, not greatly prolonged above, without vegetative bulblets. . . . . 1. *C. fragilis.*

Fronds lanceolate, broadest at base, the fertile mostly narrower than the sterile and greatly prolonged to attenuate tips; lower pinnules not decurrent; veins mostly ending at the sinuses; back of frond often bearing roundish bulblets. . . . . 2. *C. bulbifera.*

Fronds distant, arising in rows from the slender cord-like much branched and prolonged rhizome, deltoid-ovate, ternately divided. . . . 3. *C. montana.*

1. C. frágilis (L.) Bernh. (fragile), Fragile Fern. — *Fronds tufted* at or toward the tip of the thickish rhizome; stipes slender, brittle, smooth, 10–25 cm. long; *blade* thin, lanceolate to lance-oblong, *broadest slightly above base,* 0.2–3 dm. long, 1–11.5 cm. broad, mostly twice- or thrice-pinnatifid; *pinnules decurrent on the margined rachis; indusia tapering to the free end,* soon deeply cleft, or more rounded and less cleft. — Highly variable semicosmop. species; the following vars. and forms recognized:

a. Rhizome oblique to horizontal, short and with short internodes or, if elongate, thickly beset with bases of old fronds; the growing point usually conspicuously chaffy, not prolonged horizontally beyond the fronds of the season. . . *b.*

b. Indusia relatively large, up to 1 mm. long, attenuate to slender tip when young, soon deeply cleft at apex. . . *c.*

* Treatment of no. 1 derived partly from that of C. A. Weatherby in Rhodora, xxxvii, 373–378 (1935).

c. Indusia glabrous; plant 0.3–4.5 dm. high; fronds 1–8 cm. wide. . *d.*
d. Basal secondary segments (pinnules) of larger pinnae nearly orbicular to ovate, deltoid or ovate-oblong, sessile, with broad or curving bases. . *e.*
e. Lower and larger pinnae lanceolate or narrowly lance-ovate, a third to half as broad as long.
Principal pinnae with pinnules distinct and separated to base . . *C. fragilis* (typical).
Principal pinnae merely lobulate-pinnatifid, mostly without distinct pinnules. . . . . Forma *dentata.*
e. Lower and larger pinnae deltoid-ovate, one-half to five-sixths as broad as long. . . . . Forma *simulans.*
d. Basal secondary segments (pinnules) of larger pinnae narrowly lanceolate, oblong, oblanceolate or spatulate, often cuneate to subpetiolar base, usually distant; pinnae lanceolate.
Pinnae tapering to unforked tips. . . . . Forma *angustata.*
Pinnae or some of them forking at tip. . . . . . . Forma *cristata.*
c. Indusia minutely glandular on back; plant 1.5–5 dm. high; frond 3–11.5 cm. broad. . . . . Var. *laurentiana.*
b. Indusia about 0.5 mm. long, round-ovate, not long-attenuate, nearly entire or shallowly lobed; plant 1–4.5 dm. high; pinnae lanceolate to lance-oblong or -ovate; lower pinnules of larger pinnae oblong to narrowly obovate, cuneate at base. . . . . Var. *Mackayii.*
a. Rhizome long-repent, with elongate internodes, the tuft of new fronds arising 2–4 cm. back from the prolonged horizontal scarcely chaffy growing tip; basal pinnules of larger pinnae broadly ovate to lanceolate or narrowly obovate, tapering to often petiolulate base, deeply cleft into oblong lobes; indusia as in var. *Mackayii*; fronds mostly 1–3 dm. long and 4–11.5 cm. broad. . . . . Var. *protrusa.*

C. frágilis (typical). — Arct. reg., s. on damp rocks, rocky slopes, in rich open woods, alluvium, etc., to Nfld., N.S., N.E., n. Pa., mts. of Va., n. O., n. Ind., n. Ill., Mo., Okla., Tex., N.M., Ariz. and Calif. *Fr.* June–Sept. (Eurasia) — Forma **dentàta** (Dickson) Clute (toothed) infrequent (Eu.); forma **símulans** Weath. (simulating; in this case the sterile fronds of no. 2), Ill., Tenn. and Mo. to Tex.; forma **angustàta** (Hoffm.) Clute (narrowed), Greenl. to Nfld., Que. and n. N.E., w. to Ont. and Mich. (Eu.); forma **cristàta** (Lowe) Weath. (crested) scattered eastward. (Eu.)

Var. **laurentiàna** Weath. (of the St. Lawrence region). — On chiefly calcareous rock or slopes, reg. of Gulf of St. Lawrence, w. Nfld. and Mingan Ids., Que., to Rimouski Co., Que., s. to C.B.; Algoma and Thunder Bay Districts, Ont., s. to Bruce Pen., Ont., Wisc. and Minn.

Var. **Mackaỳii** Lawson (for its discoverer, Alexander Howard MacKay, 1848–1929). — S. Que. to Minn. and S.D., s. to N.S., N.E., Del., Md., upland to N.C., Tenn. and Mo.

Var. **protrùsa** Weath. (pushing forward). — Moist rich (often calcareous) wooded slopes, rocky banks or alluvium, se. N.Y. to s. Minn., s. to Ala. and La.

2. C. bulbífera (L.) Bernh. (bearing bulbs), Bulblet-Fern. — Rhizome short and stout, terminated by a tuft of twice-pinnate fronds; *fronds* of two kinds, *broadest at base, most of the veins ending in the sinuses;* the usually *sterile fronds lanceolate,* without prolonged tips, these short fronds becoming fertile in forma **horizontàlis** (Lawson) Gilbert (horizontal); *the more generally fertile ones prolonged into very long tapering* ("flagelliform") *tips, these taller fronds* up to 9 dm. long and 2 dm. broad at base, *their rachises and the backs of their pinnae often bearing rounded bulblets* (soon dropping off and starting new plants); *pinnules* oblong, obtuse, *the distinct lower ones not decurrent; indusia short, truncate* on the free side. — Shaded ravines, rocky (chiefly calcareous) slopes and steep banks, Nfld. to Man., s. to N.S., n. and w. N.E., n. N.J., Pa., upland to Ga., Tenn., Ark., N.M. and Ariz. *Fr.* June–Sept.

3. C. montàna (Lam.) Bernh. (of mountains), Mountain B. — *Extensively creeping by slender cord-like branching prolonged smooth rhizomes, the fronds arising singly from them;* slender remotely chaffy stipes mostly much longer than the blade; *blade broadly triangular-ovate, with 3 divisions,* 0.5–2 *dm. long, nearly as broad,* the 2 lower divisions slightly narrower and shorter than the upper; pinnae 2–3-times pinnate; indusia roundish. — Greenl. to Alaska, s. to springy or damp calcareous slopes, thickets and woods of n. Nfld., Anticosti I., Gaspé Pen. and L. Mistassini, Que., Thunder Bay Distr., Ont., and mts. of Colo. and B.C. *Fr.* July, Aug. (Eurasia)

3. PTERÈTIS Raf. Ostrich-Fern

Sporangia somewhat as in *Onoclea,* normally subglobose and pod-like, borne on moniliform revolute-margined pinnules of the pinnate fertile frond. Fertile fronds borne in the midst of the vase-like crown of many times taller short-stalked broadly oblanceolate regularly pinnate sterile fronds, these with free veins. — Two species of N. Temp. reg. (Name from the Greek *pteris, a fern.*) MATTEUCCIA Todaro.

1. **P. pensylvánica** (Willd.) Fern. (Pennsylvanian). — Caudex stout, from widely creeping and forking rhizome, erect, heavily covered by persistent bases of stipes, sending out from the base stout scaly black stolons; STERILE FRONDS *in circles,* 0.5–2 (rarely –3) *m. high;* stipes green, 4-angled, deeply channeled on upper side, chaffy when young, the strongly flattened black bases covered outside with thin papery pale brown to cinnamon scales; *blade* 1.2–6 dm. broad, *very gradually narrowed at base, abruptly short-acuminate at tip; rachis grooved* on upper side, narrowly wing-margined, glabrous and lustrous, or in forma **pubéscens** (Terry) Fern. (pubescent) canescent-tomentose; the broadly linear acuminate spreading-ascending pinnae deeply pinnatifid into oblong blunt segments; FERTILE FRONDS erect, rigid, 2–6 dm. high, with lustrous brown stipe and with rachis rounded on back but broadly and shallowly channeled in front; the olivaceous (finally blackish) fronds 0.25–1 dm. broad, with crowded and often twisted or overlapping spreading-ascending moniliform pinnae, with pod-like pinnules, or in forma **obtusilobàta** (Clute) Fern. (obtusely lobed) pinnules flat, or in forma **foliàcea** (Farw.) Fern. (leafy) the lower half of fruiting frond flat and leaf-like, the upper with normal fruiting pinnules. (*Onoclea Struthiopteris* of ed. 7, not Hoffm.; *Struthiopteris germanica* of Am. auth., not Willd.; *Matteuccia nodulosa* (Michx.) Fern.; *P. nodulosa* (Michx.) Nieuwl.) — Rich or bottomland-thickets or woods or in alluvium, Nfld. and Côte Nord, Que., to s. Alaska, s. to N.S., N.E., n. Va., W.Va., O., Ind., Ill., Mo., S.D. and s. B.C. *Fr.* July–Oct.

4. ONOCLÈA L. SENSITIVE FERN

Sporangia borne on elevated receptacles, forming roundish sori imperfectly covered by very delicate hood-shaped indusia attached to the base of the receptacle. Fertile fronds stiffly erect, bipinnate; the pinnules ordinarily rolled up into globular berry-like divisions, at first completely concealing the sporangia, and at last, when dry and indurated, cracking open and allowing the spores to escape. — Rhizome creeping and forking, the solitary and scattered or few sterile fronds long-stalked, with deltoid-ovate deeply pinnatifid blade with anastomosing veins. A single species. (Name used by Dioscorides for some plant, certainly not this one.)

1. **O. sensíbilis** L. (sensitive; only to early frost).— Rhizome without chaff; stipes naked or with few scattered scales at base, 0.05–1.2 m. high, stiff and brittle; STERILE FRONDS firm, 0.15–5.5 dm. long, with rachis bordered by upwardly broadening wings; the 2–16 pairs of spreading-ascending lanceolate or lance-oblong segments entire, sinuate or coarsely pinnatifid (in larger plants), the lower segments contracted at base; FERTILE FRONDS (often not developed) becoming dark brown or blackish in age, lance-oblong; their erect pinnae with the pinnules rolled into tight subglobose bodies, or in the transitional forma **hemiphyllòdes** (Kiss & Kuemmerle) Weath. (like half-a-leaf) the one side of the lax frond fertile with the other half foliaceous, or in the frequent forma **obtusilobàta** (Schkuhr) Gilbert (obtusely lobate) the pinnules or many of them flat and more or less sterile, thus showing transition to the larger sterile fronds; sterile fronds blackening at first frost, the fertile ones persistent over winter. — Low open ground, alluvial thickets and low woods, most often fruiting in the open, Nfld. and s. Lab. Pen. to Man., s. to Fla., La. and Tex. *Fr.* June–Oct. (E. Asia)

5. DRYÓPTERIS Adans. SHIELD-FERN. WOOD-FERN

Sori round, centrally attached to the veins below or near their tips, naked or covered by a reniform to orbicular indusium with a marginal sinus, the fertile fronds scarcely modified. Stipe continuous with the rhizome. Fronds (in ours) variously divided, from once to thrice pinnate or pinnatifid to ternately compound. — Large and very complex world-wide genus, its sections by some considered to form genera. (Name from the Greek *drys, oak,* and *pteris, fern.*) ASPIDIUM, in part, of many auth., not Sw. THELYPTERIS Schmidel; NEPHRODIUM Richard; PHEGOPTERIS (Presl.) Fée

a. Rhizome slender and cord-like, freely forking and elongate, its branches 1–5 mm. thick; fronds thin, annual, scattered along the rhizome or only a few in tufts; stipe slender and fragile, naked or with chaffy scales only 1–6 mm. long; indusia 0.3–1.2 mm. in diameter or wanting. . . *b.*

b. Indusia present but soon shrivelling; fronds lanceolate, elongate, with 17–

46 pairs of pinnae; lowest pinnae shorter than to barely equaling the middle ones. (THELYPTERIS). *c.*

c. Lateral veins of the segments of the sterile fronds mostly forking, of the fertile simple or forked; lowest pinnae one-half to essentially as long as the middle ones; indusia glabrous or coarsely long-ciliate. . . . 1. *D. Thelypteris,* var. *pubescens.*

c. Lateral veins of the segments of both the sterile and fertile fronds simple; indusia finely glandular-ciliate.

Fronds with 18–31 pairs of pinnae; the lowest pair 1.7–7.5 cm. long, rarely less than a third as long as the middle ones; indusia 0.7–1.2 mm. broad. . . . . 2. *D. simulata.*

Fronds with 23–46 pairs of pinnae; the lowest pair 0.2–1.3 cm. long, many times shorter than the middle ones; indusia 0.3–0.8 mm. broad. . . . . 3. *D. noveboracensis.*

b. Indusia wanting from the first; fronds triangular to triangular-ovate or ternate, with 8–30 pairs of pinnae or segments; longest pinnae at or near the base. (PHEGOPTERIS). *d.*

d. Fronds more or less ternate, the 3 divisions and sometimes the 2nd pair of primary pinnae slender-petioled; rachis filiform, not winged; only the reduced upper pinnae or segments confluent.

Fronds membranaceous, glabrous (rarely sparsely glandular), the 2 lower divisions nearly as long as the terminal one; stipe and rachis lustrous, glabrous (rarely sparsely glandular). . . . . 4. *D. disjuncta.*

Fronds firm and stiffish, glandular-puberulent, the 2 lower divisions about half as long as the terminal one; stipe and rachis dull, puberulent. . . . . 5. *D. Robertiana.*

d. Fronds not ternate, bipinnatifid, triangular or triangular-ovate; rachis winged by the confluence of the lower segments of the sessile pinnae.

Wings of rachis not extending down to the lowest pinnae; frond narrower than long, commonly strigose-pubescent, the rachis and lower surface more or less chaffy with brown scales; united upper and lower segments of opposite pairs of pinnae not forming a fiddle-shaped wing. . . . . 6. *D. Phegopteris.*

Wings of rachis extending down to lowest pinnae; frond nearly or quite as broad as long, rarely strigose, minutely glandular-puberulent on rachis and veins beneath, at most with few white scales; united upper and lower basal segments of opposite pinnae forming a fiddle-shaped wing. . . . . 7. *D. hexagonoptera.*

a. Rhizome stout, closely covered with old stipe-bases; fronds forming terminal crowns, firm to nearly or quite evergreen; stipe comparatively stout, tough or brittle, with the abundant basal scales mostly 0.5–3.5 cm. long; indusia 0.3–2 mm. in diameter. Species often hybridizing. True DRYOPTERIS. *e.*

e. Fronds 0.1–1 m. or more long, 0.4–4 dm. wide, not especially aromatic; lowest pinnae mostly more than 2 cm. long and 1 cm. wide; indusia 0.4–2 mm. broad, not overlapping, their sinuses distinct; basal scales 0.5–3.5 cm. long; plants chiefly of woodlands and swamps. *f.*

f. Fronds tripinnate, tripinnatifid or bipinnate, evergreen or half-evergreen, firmly membranaceous but hardly coriaceous; their ultimate segments only 3–15 mm. long and 2–8 mm. broad, finely pinnate-cleft or incised, with their teeth ending in mucronate or short bristle-like tips; basal scales of stipe ovate, 0.5–1.5 cm. long; indusia 0.4–1.4 mm. in diameter.

Indusia glabrous or, if glandular, the frond finely tripinnatifid. . . 8. *D. spinulosa.*

Indusia glandular; frond bipinnate. . . . . 9. × *D. Boottii.*

f. Frond bipinnatifid or bipinnate (sometimes tripinnatifid near base only), firm to coriaceous, the teeth without bristle-tips; basal scales of stipe mostly longer; indusia 0.6–2 mm. in diameter. . *g.*

g. Sori not marginal; fronds firm-membranaceous to subcoriaceous. . *h.*

h. Fronds strongly dimorphic, the fertile much taller and narrower than the sterile, 0.6–1.8 dm. broad. their pinnae often turned at nearly right-angles to upper face of rachis. . . . . . . 10. *D. cristata.*

h. Fronds essentially alike, flat, the fertile ones 0.7–4 dm. broad. . . *i.*

i. Lowest pinnae of fertile fronds with 10–22 pairs of definite pinnules (excluding terminal teeth); fronds lanceolate to lance-oblong, the fertile ones one-fifth to half as broad as long; basal scales of stipe pale brown to fuscous, thin and scarious. . *j.*

j. Lowest pinnae obviously shorter than median ones; scales of stipe pale brown to cinnamon.
Basal scales of stipe ovate to ovate-lanceolate, 1–2 cm. long; lowest pinnae deltoid-ovate, the others broadly lanceolate; plant of swamps and wet woods, n. to s. Canada. . . . 10. *D. cristata*, var. *Clintoniana.*
Basal scales of stipe lance-linear and long-attenuate, 1.5–3 cm. long, mixed with shorter setiform ones; lowest pinnae lance-ovate or lanceolate, the others lance-linear; boreal upland plant. . . . . 11. *D. Filix-mas.*
j. Lowest pinnae nearly or quite as long as and like the median ones, all lanceolate; basal scales of stipe brown to fuscous, lance-attenuate to oblong-ovate, 0.7–1.5 cm. long; southeastern paludal species. . . . . 12. *D. celsa.*
i. Lowest pinnae of fertile frond with (15–) 20–31 pairs of definite pinnules; fronds ovate to ovate-oblong, one-half to five-sixths as broad as long; basal scales of stipe firm, castaneous to blackish, lustrous; plant of rich temperate woods. . . . . . . 13. *D. Goldiana.*
g. Sori marginal; fronds firm-membranaceous to coriaceous, blue- or gray-green. . . . . 14. *D. marginalis.*
e. Fronds 0.2–3 dm. long, 0.8–6 cm. broad, spicy-aromatic; lowest pinnae 0.2–2 cm. long, 0.2–1 cm. wide; indusia 1–2 mm. broad, often overlapping, the sinus often obscure; basal scales 3–15 mm. long; boreal xerophyte. . . . . 15. *D. fragrans.*

1. D. Thelýpteris (L.) Gray (old generic name), var. **pubéscens** (Lawson) Nakai (hairy), MARSH-FERN, MEADOW-FERN, SNUFFBOX-FERN. — Rhizome elongate, forking, blackish, slender, naked or nearly so; *stipes few or scattered,* chaffy when young, *becoming naked,* brittle, 1–7 dm. high; *fronds herbaceous, annual,* sensitive to first frost, *lanceolate to lance-oblong, strongly dimorphic, minutely pubescent, at least when young,* on both surfaces *and* especially *along rachis and lower side of midrib,* or in forma **suavèolens** (Clute) A. R. Prince (fragrant) glandular-aromatic when fresh; STERILE FRONDS very thin, 1–6 dm. long, 0.4–2 dm. wide, with 17–40 pairs of divergent or slightly recurving linear-lanceolate acute pinnae, or in forma **Púfferae** (A. A. Eat.) A. R. Prince (named in 1902 for MRS. J. J. PUFFER) with more or less crested forking tips; median pinnae with 8–25 pairs of oblong to oblong-ovate blunt slightly confluent *pinnules or segments, these with mostly forking lateral veins;* upper few pinnae abruptly shorter, simple or subsimple; *basal pair essentially as long as to half as long as median ones;* FERTILE FRONDS mostly taller, thicker and firmer, *their pinnules or segments* narrower and *commonly revolute-margined,* the pinnae (especially in sunny areas) often *falcate-recurved or contorted; lateral veins of segments simple or forking;* indusia glabrous or with long remote cilia, 0.3–1 mm. broad; sori often confluent. (*Aspidium Thelypteris* of ed. 7, not Sw.; *Thelypteris palustris* (Salisb.) Schott, var. Fern.) — Swamps, sterile meadows, bogs and low woods or thickets, s. Nfld. to se. Man., s. to N.S., N.E., L.I., Ga., Tenn. and Okla. *Fr.* June–Oct. (E. Asia) — Our representative of a wide-ranging species of N. Hemisph., Afr. and N.Z.

2. D. simulàta Davenp. (imitated; evidently intended for imitating), MASSACHUSETTS FERN. — Similar to nos. 1 and 3; *fronds* glabrous or sparsely pubescent, oblong-lanceolate, long-acuminate at tip, 1–4 dm. long, 0.35–1.7 dm. wide, *with* 18–38 *pairs of oblong-lanceolate* (in the fertile long-acuminate) *pinnae* and terminal segments; median pinnae with 10–28 pairs of oblong obtuse segments, *the lowermost pinnae 1.7–7.5 cm. long, rarely less than a third as long as the middle ones; fertile fronds* commonly *only slightly firmer than the sterile, their pinnules flat* (slightly revolute in exposed habitats); *lateral veins of segments of both fertile and sterile fronds simple; indusia 0.7–1.2 mm. broad, minutely glandular-ciliate; sori distinct* or only rarely confluent. (*Aspidium* Davenp.; *Thelypteris* Nieuwl.) — Boggy or swampy woods and thickets or on knolls in bogs, e. Va. and e. Pa. to Lincoln Co., Me. and sw. N.S.; inland locally to centr. N.H., e. Vt., sw. Que., ne. and centr. N.Y. and Md. *Fr.* July–Oct.

3. D. noveboracénsis (L.) Gray (of New York), NEW YORK FERN. — Rhizome slender, prolonged, forking; stipes slender, fragile, pale and lustrous, glabrous; *fronds elliptic to elliptic-lanceolate,* acuminate, *very gradually narrowed at base into remote and rapidly shortening pinnae,* the sterile and fertile similarly thin-membranaceous, *usually minutely hairy especially beneath,* or in forma **fràgrans** (Peck) Burnham (fragrant) glandular and aromatic, 2–6 dm. long, 5–12 cm. broad, *with* 23–46 *pairs of oblong-lanceolate flat pinnae;* median pinnae with 14–32 pairs of oblong blunt segments; *lowermost pinnae 0.2–1.3 cm. long, many times shorter than middle ones; lateral veins of segments of both sterile and fertile fronds simple; indusia* (very rarely wanting from the first) 0.3–0.8 *mm. broad, minutely glandular-ciliate,* sometimes hairy; sori rarely confluent. (*Aspidium* Sw.; *Thelypteris* Nieuwl.) — Dry to damp woods and thickets, Nfld. to s. Ont., Mich. and n. Ill., s. to N.S., N.E., L.I., Ga., Ala., Miss. and Ark. *Fr.* June–Sept.

4. D. disjúncta (Ledeb.) C. V. Mort. (disjoined), OAK-FERN. — Rhizome very slender, elongate, forking, at first chaffy with ovate papery scales, later naked, blackish, bearing scattered fronds; *stipe filiform,* slightly chaffy at base, otherwise *naked and with the rachis glabrous and lustrous* or rarely slightly glandular, 0.45–5 dm. high; *chaffy unrolling fronds with* 3 *drooping ball-like divisions; frond membranaceous or thin-herbaceous, glabrous, deltoid,* 0.4–3 dm. broad, *ternate, the* 3 *primary triangular divisions slender-stalked* as also sometimes the lower pinnae of the large median division; *the* 2 *lateral divisions divergent, nearly as long as the terminal one, asymmetrical,* the lower side with more prolonged pinnae; the 7–18 linear-oblong blunt sessile opposite pinnae of each division rapidly diminishing in length upward, the terminal ones confluent and merely toothed, the lower with oblong obtuse segments (or in large fronds again pinnatifid); *sori small, mostly nearly marginal, without indusia.* (*D. Linnaeana* Christens.; *Phegopteris Dryopteris* (L.) Fée; *Thelypteris Dryopteris* (L.) Slosson) — Greenl. and n. Lab. to Alaska, s. in cool mossy or rocky woods to Nfld., N.S., N.E., L.I., n. N.J., Md., upland to Va., n. O., Mich., Wisc., Mo., S.D., N.M., Ariz. and Oreg. *Fr.* June–Sept.

5. D. Robertiàna (Hoffm.) Christens. (on account of its glandular surface placed by early botanists with *Geranium Robertianum*), "LIMESTONE POLYPODY". — Somewhat suggesting no. 4; *stipe* 0.5–3 dm. high, *with the rachis minutely glandular-puberulent; uncoiling fronds crozier-like, not 3-divided; frond firmer and stiffer, minutely glandular-puberulent, more narrowly deltoid or deltoid-ovate,* 0.35–2 dm. long, 0.3–2 dm. broad at base, *the* 2 *lower divisions* (rarely not stalked) *about half as large as the terminal one and more nearly symmetrical than in no.* 4. (*Phegopteris* A. Br.; *Thelypteris* Slosson) — Damp to dry calcareous ledges, cliffs and talus, local, Nfld. to Alaska, s. to Bonaventure Co., Que., Restigouche R., N.B., Pa., Bruce Pen., Ont., n. Mich., Wisc., Ia. and Id. *Fr.* July–Sept. (Eurasia) — Fresh fronds aromatic.

6. D. Phegópteris (L.) Christens. (old generic name, Beech-Fern), LONG BEECH-FERN. — Slender forking rhizome at first chaffy, becoming naked; stipes few in tufts or scattered, slender and brittle, at first chaffy, becoming naked, pale, 0.4–4 dm. high; *frond triangular to triangular-ovate, long-acuminate, commonly about two-thirds as broad as long,* sometimes practically as broad, 0.6–3 dm. long, *herbaceous,* more or less *hairy on both surfaces and especially along rachis and midrib beneath, commonly with lanceolate brownish scales beneath; pinnae* and primary segments 13–30, *all but the lowest pair confluent at least at base; the lowest pinnae* somewhat *remote and projected downward and forward,* lance-acuminate or rarely blunt, 0.8–4 *cm. broad,* with oblong blunt oblique segments; *middle pinnae* similar, progressively narrower, mostly linear-lanceolate, *decurrent at base into semirhombic slenderly confluent wings;* upper pinnae reduced to closely confluent short lobes or segments; sori small, submarginal; *indusia wanting.* (*Phegopteris connectilis* (Michx.) Watt; *P. polypodioides* Fée; *Thelypteris* Slosson) — Woods, thickets and cool rocky banks, Greenl. and Lab. to Alaska, s. to Nfld., N.S., N.E., n. N.J., Pa., upland to N.C. and Tenn., O., Mich., Ia. and Oreg. *Fr.* June–Aug. (Eurasia)

7. D. hexagonóptera (Michx.) Christens. (six-cornered fern), BROAD BEECH-FERN. — Coarser than no. 6; stipe 1–6 dm. high; *frond commonly as broad as or broader than long* (rarely much narrower), 1–4 dm. long, *minutely glandular-puberulent* but rarely hairy *beneath,* especially on rachis and veins, *not chaffy, or merely with almost colorless slender scales; basal pinnae broadly lanceolate to narrowly rhombic, contracted at base,* long-acuminate, *ordinarily not deflexed,* 0.6–2 dm. long, 2–9 *cm. broad,* often asymmetrical, pinnatifid into oblong to lanceolate, crenate to again pinnatifid, blunt segments; *pinnae* above 2nd pair progressively shorter, narrower and more symmetrical, *the lower and median with their bases adnate to the rachis and forming fiddle-shaped slenderly confluent wings.* (*Phegopteris* Fée; *Thelypteris* Slosson) — Rich woods, Fla. to e. Tex., n. to N.E., sw. Que., s. Ont., O., Mich., Wisc., Minn. and e. Kans. *Fr.* June–Sept. — Forma **Simònii** Reed (named in 1945 for its discoverer, ANDREW SIMON) has the tips of the frond and of the pinnatifid pinnae forking and crested.

8. D. spinulòsa (O. F. Muell.) Watt (with minute spines), SPINULOSE WOOD-FERN, FANCY FERN, FLORIST'S FERN. — Rhizome stout, nearly horizontal, covered with old stipe-bases, terminated by a crown of several fronds; *stipes* 0.5–5 dm. high, pale, more or less *chaffy,* especially below, *with ovate papery to firm brown scales* 0.5–1.5 *cm. long; frond* lance-oblong to deltoid-ovate, *tripinnatifid, tripinnate or sometimes even quadripinnate,* evergreen or half-evergreen, *firmly membranaceous but hardly coriaceous,* 1–6.5 dm. long, 0.5–4 dm. broad, with 17–33 pairs of pinnae and primary upper segments; pinnae spreading or ascending, the lowest deltoid-ovate to -lanceolate, the others progressively narrower and more approximate; *pinnules* oblong, pinnate or pinnately cleft, *their ultimate segments or teeth with slenderly mucronate ("spinulose")* *tips;* sori small; the thinnish and quickly shriveling indusia 0.4–1.4 mm. broad, glabrous or glandular. — Highly variable species found in one or another of its geographic varieties over the forested N. Hemisph.; the following freely intergrading vars. with us:

a. Basal inferior and superior pinnules of lowermost pinnae subopposite, rarely more than 4 mm. apart; the inferior 1–6 cm. long, if more than twice as long as the superior not exceeding the 2nd inferior pinnule. . . *b.*

b. Frond glabrous, twice pinnate; the pinnae obliquely ascending, gradually tapering to apex; their pinnules rarely cleft nearly to the middle; basal inferior one usually longer than 2nd inferior one; indusia glabrous; scales of stipe pale brown or cinnamon-color. . . . . *D. spinulosa* (typical).

b. Frond commonly minutely glandular, especially on the rachis and racheolae, tripinnatifid or sometimes tripinnate; pinnae slightly ascending to divergent; basal inferior one shorter than to rarely exceeding the 2nd inferior one; indusia glandular; scales of stipe usually dark brown at base.

Mature indusia 0.8–1.4 mm. broad; pinnae gradually tapering to apex. Var. *fructuosa.*

Mature indusia 0.5–0.8 mm. broad; pinnae usually narrowed rather abruptly to prolonged lance-linear tips. . . . . . Var. *intermedia.*

a. Basal inferior and superior pinnules of lowest pinnae remote, 0.5–2 cm. apart; the inferior 3–10 cm. long, 2–4 times as long as the superior and commonly exceeding 2nd inferior pinnule.

Frond ovate to ovate-triangular, 1–6.5 dm. long, 1–4 dm. broad, tripinnatifid (basal pinnae sometimes tripinnate), not glandular; lower pinnules of lowermost pinnae with oblong obtuse sharply toothed or cleft segments 4–15 mm. long, 2–8 mm. wide; rachises of the spreading-ascending pinnae naked or with scattered linear- to lance-attenuate spreading scales; indusia glabrous. . . . . Var. *americana.*

Frond lanceolate, broad at base, gradually tapering from near middle to elongate tip, 2–6 dm. long, 1.2–2.5 dm. broad, tripinnate (basal pinnae sometimes quadripinnate), somewhat glandular beneath; ultimate divisions (of the 3rd or 4th order) of lowermost pinnae elliptic-lanceolate to narrowly rhombic, subpetiolulate or petiolulate, 2–8 mm. broad; rachises of spreading-ascending to falcate-recurving pinnae bearing ovate brown scales as well as bristle-like chaff; indusia glandular. . . Var. *concordiana.*

D. spinulòsa (typical), *Aspidium* Sw.; *Thelypteris* Nieuwl. — Low woods, thickets and swamps, less often in dry woods, s. Lab. to Alta., s. to Nfld., N.S., N.E., Va., O., Ind., Ill., Mo. and Id. *Fr.* June–Aug. (Eurasia)

Var. **fructuòsa** (Gilbert) Trudell (fruitful). — Fronds usually larger and less finely dissected than in the next, the pinnae more obliquely ascending. (*Aspidium spin.*, var. *dilatatum* of ed. 7, excl. form, not Gray) — Woods (wet or dry), common northw., Nfld. and Côte Nord, Que., to w. Ont., s. to N.S., N.E., Va., W.Va., O., Ind., Wisc. and Minn.

Var. **intermèdia** (Muhl.) Underw. (intermediate), *Aspidium spin.*, var. D.C. Eat.; *D. intermedia* Gray; *Thelypteris spin.*, var. Nieuwl. — Dry to wet woods and thickets, Nfld. to Thunder Bay Distr., Ont., s. to N.S., N.E., Va., upland to Ala. and Tenn., W.Va., O., Ind., Ill. and Ia.

Var. **americàna** (Fisch.) Fern. (American) = *Aspidium spin.*, var. *dilatatum*, forma *anadenium* Robins.; *Thelypteris spin.*, var. Weath.; and *D. campyloptera* (Kunze) Clarkson — Greenl. and n. Lab. to Alaska, s. in cool woods and thickets to Nfld., N.S., n. N.E., ne. and w. Mass., N.Y., mts. to N.C. and Tenn., n. Mich., n. Wisc., n. Minn., Id. and Wash., ascending to subalpine areas.

Var. **concordiàna** (Davenp.) Eastman (of Concord, Mass.), PURDIE'S FERN. — Local in low woods, Middlesex Co., Mass. — Wholly anomalous.

9. × D. Boóttii (Tuckerm.) Underw. (for its discoverer, WILLIAM BOOTT, 1805–1887), BOOTT'S WOOD-FERN. — Intermediate between *D. spinulosa*, var. *intermedia* and *D. cristata; fronds* firm-membranaceous, in outline intermediate, *bipinnate; indusium glandular.* (*Aspidium* Tuckerm.; *Thelypteris* Nieuwl.) — Probably a somewhat fertile hybrid, often abundant through coincident ranges of supposed parents.

10. D. cristàta (L.) Gray (crested), CRESTED WOOD-FERN. — Rhizome stout, creeping, heavily covered with old stipe-bases, terminated by a crown of *often strongly dimorphic fronds; fertile fronds much taller, on longer stipes and less evergreen than the sterile; stipes* erect, rather brittle, 0.5–5 dm. long, *with thin ovate to ovate-lanceolate cinnamon or pale brown scales* 0.5–2 *cm. long; fronds* firmly membranaceous or coriaceous, *at least the fertile erect, linear- to lance-oblong, slightly narrowed at base,* acuminate at tip, 1–7.5 dm. long, 0.6–3 dm. broad; *pinnae* 14–34, ascending or spreading, those of the fertile frond often twisted on the rachis, *the lowest deltoid-ovate, the others deltoid-oblong and deeply pinnatifid into coarse* serrate *oblong obtuse segments;*

the lowest with 4–18 *pairs of segments;* sori midway between margin and midvein; indusia 0.7–2 mm. broad, glabrous. — Very variable, with two vars. better marked than others:

Fronds linear-oblong to narrowly lance-oblong, 0.6–1.8 dm. broad; pinnae, at least of fertile frond, twisted on the rachis and more or less at right angles to it; lowest pinnae with 4–12 pairs of definite segments; median pinnae (7th from base) with 8–17 pairs. . . . . D. *cristata* (typical).

Fronds lance-oblong, 1.1–3 dm. broad, more uniform and flat; lowest pinnae with 12–18 pairs of definite segments; median pinnae with 14–22 pairs. . Var. *Clintoniana.*

D. cristàta (typical), *Aspidium* Sw.; *Thelypteris* Nieuwl. — Boggy or swampy open ground, thickets or wet woods, Nfld. to Alta., s. to N.S., N.E., L.I., Va., Tenn., n. La., Neb. and Id. *Fr.* June–Aug. (Eu.)

Var. Clintoniàna (D. C. Eat.) Underw. (for its discoverer, George William Clinton, 1807–1885), Clinton's Wood-Fern (*Aspidium cristatum*, var. D. C. Eat.; *Thelypteris cristata*, var. Weath.; *D. Clintoniana* Dowell). — Often in richer woods, w.-centr. Me. to sw. Ont. and Minn. s. to s. N.E., N.C., O., Ind. and Mo.

11. D. Fìlix-más (L.) Schott (old generic name, Male Fern), Male Fern. — Stout rhizome ascending, covered with long scales; *stipes stout and comparatively short,* 0.2–1 (or in deep shade –2.5) dm. high, *covered at base* and often throughout *by lance-linear and long-attenuate pale-brown papery scales* 1.5–2.5 *cm. long and numerous shorter setiform ones; frond lanceolate to lance-oblong, narrowed at base,* acuminate, dark green above, firm-membranaceous, rarely evergreen, 0.2–1 m. or more long, 0.7–4 dm. broad, *with* 21–52 *pairs of mostly lance-linear* spreading or spreading-ascending *pinnae; lowest pinnae short* and more ovate-lanceolate, *with* 7–20 pairs of crenate or serrate *obtuse oblong pinnules or segments; sori* midway between margin and midvein, *commonly confined to the upper half or third of the frond and to the lower three-fourths of each pinnule;* indusia firm, persistent, pale brown to whitish, 0.8–2 mm. broad. (*Aspidium* Sw.; *Thelypteris* Nieuwl.) — Rich woods, glades, upland pastures and rocky slopes (chiefly limestone, trap or slate), Nfld. to B.C., s. to C.B., York Co., N.B., Mt. Katahdin, Me., Vt., Niagara Falls to Bruce Pen., Ont., n. Mich., mts. to S.D., w. Okla., w. Tex., Mex., Ariz. and s. Calif. *Fr.* late June–early Sept. (Greenl.; Eurasia and n. Afr.)

12. D. célsa (Wm. Palmer) Small (high, exalted), Log-Fern. — *Rhizome* horizontal, *nearly superficial,* relatively slender (1.5–3 cm. thick), with few subterminal fronds; *stipe* pale, 1.5–4.5 dm. long, *covered with subpersistent scarious pale brown to fuscous lance-attenuate to oblong-ovate scales mostly* 0.7–1.5 *cm. long; frond firm-membranaceous,* 0.3–1.2 m. long, 1.5–3 dm. broad, *three-eighths to half as broad as long, lance-oblong or lanceolate, its upper third gradually narrowed and tapering; pinnae of lower two-thirds subequal,* these 9–11 pairs, *the basal practically as long as the median, all lance-attenuate* and 2–4 cm. broad, *the lowest with* 10–16 *pairs of* chiefly oblong blunt *pinnules;* sori near midvein; indusia about 1 mm. broad, glabrous, soon shriveling. (*Aspidium Goldianum*, var. *celsum* Robins.; *D. Goldiana*, subsp. *celsa* Wm. Palmer; *D. Goldiana*, forma *celsa* Clute) — Inundated acid swamps, cypress-swamps, cypress-knees and -logs, and wet woods, on or near Coastal Plain, La. to S.C., n. to se. Va. and locally to se. Pa. *Fr.* July–Oct.

13. D. Goldiàna (Hook.) Gray (for its discoverer, John Goldie, 1793–1886), Goldie's Fern. — Rhizome stout, ascending, covered with old stipe-bases and long scales and with a terminal crown of fronds; *stipes* pale brown, 1.5–5.5 dm. high, *covered at base with firm and lustrous lance-acuminate castaneous to blackish scales* 1.5–3.5 *cm. long; frond ovate or broadly ovate-oblong, abruptly short-acuminate at summit,* firm-membranaceous, 0.2–7.5 dm. long and 2–4 dm. broad, *one-half to five-sixths as broad as long, with* 20–38 *pairs of* spreading or slightly ascending *broad-lanceolate* deeply pinnatifid *pinnae; lowest pinnae* 3.5–7 *cm. broad, with* (15–) 20–31 *pairs of linear-oblong* blunt often subfalcate crenate or crenate-serrate *pinnules or segments; median pinnae of similar size; upper pinnae abruptly decreasing in size and segmentation; sori* near midvein, *commonly on all but the* 1–4 *lower pairs of pinnae and covering all but the upper third or fourth of the pinnule;* indusia 1–2 mm. broad. (*Aspidium* Hook.; *Thelypteris* Nieuwl.) — Rich, mostly calcareous, woods, w. N.B. and sw. Que. to Minn., s. to n. and w. N.E., N.C., Tenn. and Ia. *Fr.* June–Sept.

14. D. marginàlis (L.) Gray (marginal), Marginal Shield-Fern, Evergreen Wood-Fern. — Rhizome stout, ascending, covered with old stipe-bases and long brown chaffy scales, with a terminal spreading or ascending crown of *firm to coriaceous blue- or gray-green fronds; stipes* stout and brittle, 0.3–3.3 dm. long, *covered at base with thin papery cinnamon-colored lance-attenuate scales* 1–2.5 *cm. long,* chaffy or naked above; *frond lanceolate to oblong-ovate, evergreen,* 0.7–7 dm. long, 0.6–3 dm. broad, *with* 12–30 *pairs of* spreading or slightly ascending *lanceolate* deeply pinnatifid or pinnate or at base bipinnatifid *pinnae*, or in forma **élegans** (J. Robins.) F. W. Gray (elegant) all or nearly all pinnae bipinnatifid, or in forma **tripinnatífida** (Clute) Weath. (tripinnatifid) with the pinnules deeply toothed and very narrow to acute; *lowest pinnae* 0.3–1.5 dm. long, 1–6.5 cm. broad at base, *with* 7–17 *pairs of* oblong obtuse to acutish entire or toothed *pinnules or segments;* median pinnae slightly longer, the upper rapidly decreasing in size and segmentation; *sori marginal or nearly so* or at the sinuses of the teeth; indusia whitish to gray-brown, firm, 0.7–1.8 mm. broad. — Woods, clearings, and rocky slopes, Gaspé Pen., Que., to B.C., s. to N.S., N.E., L.I., Ga., Ala., Ark. and Okla. *Fr.* June–Oct. — Forma **Davenpórtii** (Floyd) A. R. Prince (for GEORGE EDWARD DAVENPORT, 1833–1907, American fern-specialist) has tips of some pinnae or of fronds forking.

15. D. FRÀGRANS (L.) Schott (fragrant), FRAGRANT CLIFF-FERN. — *Rhizome* stout, *covered with numerous persistent shrivelled and curled old fronds* and terminated by a crown of usually numerous spreading or ascending new fronds; *stipes* 1–15 cm. long, *glandular, chaffy* at base and often throughout *with lustrous brown or reddish often glandular-puberulent scales* 3–15 mm. long, the latter varying from lance-attenuate to ovate or suborbicular; *frond coriaceous,* 0.2–2 *dm. long,* 0.8–4 cm. broad, *glandular and aromatic, lanceolate to broadly linear, tapering from middle to base and apex, the overlapping and often inrolled pinnae* as well as stipe and rachis *heavily chaffy with cinnamon-brown to reddish scales, with* 17–40 *pairs* of oblong-lanceolate obtuse *pinnately incised or crenate pinnae* 0.2–2 *cm. long; indusia* whitish, becoming brown or fulvous, 1–2 mm. broad, often *densely crowded and overlapping,* with *sinus* commonly *obscure or very narrow,* the margin frequently toothed. (*Aspidium* Sw.; *Thelypteris* Nieuwl.) — Arct. reg., s. to Lab., Ung., Mackenz., Yuk. and Alaska. (Eu.) — Represented with us by

Var. **remotiúscula** Komarov (more remote). — Fronds submembranaceous, 0.7–3 dm. long, up to 6 cm. broad, sparsely scaly beneath; pinnae mostly flat and not overlapping. (*Thelypteris fragrans,* var. *Hookeriana* Fern.) — Dry cliffs and rocky banks (often calcareous), Lab. to Thunder Bay Distr., Ont., s. to Nfld., C.B., n. N.B., n. N.E., ne. N.Y., Mich., Wisc. and Minn., ascending to 1220 m. alt. *Fr.* June–Sept. (E. Asia)

6. POLÝSTICHUM Roth SHIELD-FERN

Fronds tufted at the end of a stout rhizome, chiefly of firm or leathery texture; stipes and rachises chaffy. Sori orbicular, opening on all sides of the circular peltate centrally attached indusium. — Genus nearly cosmop. (Name from the Greek *polys, many,* and *stichos, row,* the sori of some species being in many ranks.)

a. Sterile and fertile fronds similar, with 20–60 pairs of approximate or overlapping pinnae; lower fertile pinnae scarcely smaller than the upper sterile ones of the same frond; lowest pinnae of frond several times shorter than the median ones. . . *b.*

b. Fronds simply pinnate or with the pinnae merely pinnately lobed at base, coriaceous to rigid, evergreen, 1–6 cm. broad; pinnae crowded and overlapping; indusia 1–2 mm. in diameter.

Fronds dark green, rigid; pinnae acute, strongly auricled but not lobed at base, with strong pungent spreading teeth, chaffy on the back. . . . . 1. *P. Lonchitis.*

Fronds pale green, fleshy to coriaceous; pinnae obtuse, the lower and median ones pinnately lobed below the middle, with appressed or incurved short-tipped teeth, without chaff on back. . . . . . . 2. *P. mohrioides,* var. *scopulinum.*

b. Fronds bipinnate, only subcoriaceous, scarcely evergreen, 0.4–2.5 dm. broad; pinnae not overlapping, loosely chaffy on back; indusia 0.4–1 mm. in diameter. . . . . 3. *P. Braunii,* var. *Purshii.*

a. Sterile and fertile fronds dissimilar, with 10–35 pairs of simple pinnae; lower fertile pinnae often much shorter than the upper sterile ones of same frond, distant; lowest pinnae of fertile frond from half as long as to nearly equaling the median ones. . . . . 4. *P. acrostichoides.*

1. P. Lonchìtis (L.) Roth (*Lonchitis,* name used by Pliny for some plant with tongue-shaped leaf), HOLLY-FERN, TRIPE DE ROCHE (Que.). — Ascending rhizome stout, densely covered with old stipe-bases, decaying fronds and chaff, and terminated by a dense crown of spreading or ascending new fronds; *stipes* stout, 1–8 cm. long, densely *chaffy with lustrous* papery *red-brown to castaneous* ovate to linear mixed *scales; frond hard and rigidly coriaceous,* evergreen, dark green and lustrous above, *simply pinnate, linear-lanceolate, narrowed to base* and apex,

0.6–5.5 dm. long, 2–6 cm. broad, *with* 20–60 pairs of sessile or short-stalked *crowded* or overlapping *rigid spinulose-serrate acute pinnae;* rachis whitish, more or less chaffy; lowest pinnae symmetrically deltoid, 0.3–1 cm. long and nearly as broad; middle and upper pinnae obliquely lanceolate, upwardly falcate, strongly auricled on upper side, up to 1–3 cm. long; sori mostly confined to upper half of frond, at first distinct, in 2 single or double rows, later becoming confluent; indusia 1–2 mm. in diameter. — Calcareous cliffs and bushy talus, Nfld.; Gaspé Co. to Rimouski Co., Que.; C.B.; Niagara region, Ont., to n. Mich.; Yuk. and Alaska to Colo., Utah and s. Calif. *Fr.* June–Sept. (Greenl.; Eurasia)

2. P. mohrioìdes (Bory)Presl (resembling *Mohria*), var. **scopulìnum** (D. C. Eat.) Fern. (of rock or crag). — Rhizome with 1–∞ crowns of spreading or ascending new fronds; *stipes* whitish or stramineous above the castaneous base, 1–12 cm. long, *densely chaffy,* especially *at base, with cinnamon- or whitish-brown* ovate- to narrow-lanceolate erose-ciliate bristle-tipped *scales freely intermixed with* linear scales and *viscid trichomes; frond somewhat fleshy to coriaceous,* evergreen, *pale green,* pinnate, narrowly lanceolate, narrowed to base and apex, 0.5–3 dm. long, 1–5 cm. broad, often plicate or twisted, *with* 20–45 *pairs of* sessile or subsessile crowded and overlapping *obtuse pinnae;* rachis whitish or stramineous, broad, rounded on back, flattened and deeply furrowed above, more or less chaffy and viscid-pubescent; *pinnae* ovate to lance-oblong; *the lower and median ones more or less pinnately lobed below middle, coarsely serrate with* more or less spinulose *appressed teeth above;* the upper pinnae less cleft; *all smooth* or only slightly chaffy *on back and* somewhat deeply *pitted;* sori in 2 (rarely 4) median rows, at first distant, later often overlapping; indusia pale, 1.3–2 mm. in diameter; fresh plant strongly aromatic. (*P. scopulinum* (D. C. Eat.) Maxon) — Cliffs and talus of serpentine rock, steep ravines at 600–1000 m., Mt. Albert, Gaspé Co., Que.; very local, Mont., Id. and Utah; Wash. to Calif. *Fr.* July–Sept. — One of the geographic (and highly localized) vars. of *P. mohrioides* of subantarct. reg., thence n. along the Andes of S. Am.

3. P. Braùnii (Spenner) Fée (in honor of ALEXANDER BRAUN, 1805–1877), var. **Púrshii** Fern. (for its discoverer, FREDERICK TRAUGOTT PURSH, 1774–1820), BRAUN'S or PURSH'S HOLLY-FERN. — Stout rhizome terminated by a vase-like crown of ascending fronds; stipes pale brown above the castaneous base, stout, flattened and furrowed above, 0.2–2 dm. long, densely chaffy with thin pale brown intermixed ovate, lanceolate, linear and acicular scales; *fronds subcoriaceous, scarcely evergreen,* lustrous-green above, elliptic-lanceolate, *conspicuously tapering at base* and tip, 1.5–10 dm. long, 0.4–2.5 *dm. broad, bipinnate, with* 20–40 pairs of *approximate to slightly distant* lanceolate to narrowly oblong obtuse to acute *pinnae;* furrowed pale *rachis densely chaffy on back* with lanceolate and setiform scales; lowest pinnae oblong, obtuse, 1–4 cm. long; median pinnae narrowly oblong to lanceolate, 2–18 cm. long, with 6–18 pairs of pinnules; *pinnules* narrowly ovate to trapezoid-oblong, obtuse, nearly rectangular at base, slightly auricled at upper side, sharply serrate *with incurved bristle-tipped teeth and with slender chaff on lower surface;* sori in 2 rows near midrib; *indusia* 0.4–1 *mm. in diameter,* their margins often erose. — Rich woods, glades and shaded rock-slides and ravines, Nfld. and Anticosti I., Que., to Thunder Bay Distr., Ont., s. to N.S., s. N.B., n. N.E., nw. Mass., mts. of Pa. and n. Mich. *Fr.* June–Sept. — Differing from Eurasian and Alaskan *P. Braunii* in heavier and firmer fronds, thinner and longer-pointed large chaff of stipe, more scaly and less villous back of rachis, and shorter teeth of pinnules.

4. P. acrostichoìdes (Michx.) Schott (resembling *Acrostichum*), CHRISTMAS FERN, DAGGER-FERN, CANKER-BRAKE, FOUGÈRE-À-FAUCILLES (Que.). — Stout rhizome covered by old stipe-bases and wilted fronds and terminated by a crown of ascending to spreading new fronds; stipes greenish, with brown bases, rather slender, nearly terete, 0.5–2.5 dm. long, densely chaffy with intermixed ovate, lanceolate and almost hair-like thin slender-tipped brown scales; *fronds* firm, *dark green, thick-membranaceous to subcoriaceous, lustrous above, evergreen,* lanceolate, *only slightly narrowed at base,* 1–6 dm. long, 0.5–1.8 dm. wide, *simply pinnate* (usually); pinnae lanceolate to oblong, obtuse or acute, short-stalked, minutely serrulate with appressed bristle-tipped teeth; lowest sterile pinnae of the fertile fronds 0.5–1.5 cm. broad, opposite, spreading or reflexed, mostly with a strong auricle at base on upper side; *upper pinnae of fertile frond much reduced in size, remote,* the lowest 1.2–3.5 cm. long and 0.3–1 cm. broad (above the basal auricle); sori distinct or confluent; indusia 0.5–1.6 mm. in diameter. — Woods and rocky slopes, n. Fla. to e. Tex., n. to C.B., P.E.I., N.B., s. Que., s. Ont., Mich., Wisc., Ia. and e. Kans. *Fr.* June–Oct. Rarely hybridizes with no 3. — Passing into forma **incìsum** (Gray) Gilbert (incised), the fronds paler and often more coriaceous, with *coarsely serrate or incised pinnae,* the sterile pinnae of the fertile fronds up to 2.5 cm. broad, the fertile pinnae more gradually reduced in size and with some sori often near tips of long median or lower pinnae, the plant of similar range, sometimes seeming like a new growth after destruction of the early fronds. (Var. *Schweinitzii* (Beck) Small) — Among the endless teratological and taxonomically unimportant forms the following have been named: forma **crispum** Clute (crisped) with pinnae crisped and fluted at margin; forma **cristàtum** Clute (crested), with tip of frond more or less forking; forma **Gràvesii** Clute (named for its discoverer, JAMES ANSEL GRAVES, 1828–1909), with sterile pinnae truncate or emarginate, the midrib projecting; forma **lanceolàtum** Clute (lanceolate), with many of the sterile pinnae tapering to exauriculate base and to almost aristate tip, the marginal teeth slender and sharp; forma **multifidum** Clute (many-cleft), with few to many deeply pinnatifid pinnae; forma **orbiculàtum** Eames (orbicular), with many or all sterile pinnae reduced to orbicular to round-oblong or rounded-obovate short blades.

P. FALCÀTUM (L.f.) Diels (sickle-shaped), the cult. HOLLY-FERN of sect. *Cyrtomium*, sometimes called *C. falcatum* (L.f.) Presl, with short erect rhizome covered with dark ovate-acuminate scales, the simply pinnate tough frond with 3–10 pairs of ovate-falcate long-attenuate broad-based pinnae up to 3 or 4 cm. broad, the veins reticulate, the small sori scattered, is reported as spreading to walls, etc., n. to s. N.E. and N.J. (Introd. from e. Asia)

7. DENNSTAÈDTIA Bernh.

Sori small, globular, marginal, each borne on the apex of a free vein or fork; the sporangia borne on an elevated globular receptacle, inclosed in a membranaceous cupuliform indusium which is open at the top and, on the outer side, partly adherent to a reflexed toothlet of the frond. — Pubescent ferns with slender creeping rhizomes, found in trop. reg. and in e. Am. and Asia. (Named for *August Wilhelm Dennstedt*, who wrote on the flora of Weimar about 1800.) DICKSONIA in part of Am. auth., not L'Hér.

1. D. punctilóbula (Michx.) Moore (with dotted lobules), HAY-SCENTED FERN, BOULDER-FERN. — Rhizome slender, naked, extensively creeping and forking; fronds scattered, strongly sweet-scented in drying; stipe elongate, slender, chaffless, lustrous and brittle; lamina minutely glandular and pilose, 1.5–9 dm. long, lanceolate or narrowly lance-ovate, acuminate, very thin, pale green, mostly bipinnate; primary pinnae lanceolate, subopposite or subapproximate; pinnules pinnatifid into oblong and cut-toothed obtuse lobes; fruit-dots minute, each on a recurved toothlet, usually one at the upper margin of each lobe. (*Dicksonia* Gray) — Shady places, rocky open woods or pastures or damp slopes, in rather sterile soil, Ga. to Ark., n. to s. Nfld., C.B., M.I., N.B., sw. Que., Parry Sound Distr., Ont., Mich., s. Ill. and Mo. *Fr.* July–Oct. — Unusual forms are forma **schizophýlla** (Clute) Rugg (incised-leaved) with the ultimate segments deeply and sharply incised; forma **cristàta** (Maxon) Clute (crested) with the tips of many pinnae cristate-forked; forma **nàna** (Gilbert) Weath. (dwarf) with the pinnules of the pinnae of the usually small fronds crowded, overlapping and rounded toward tips of pinnae, very small and scattered toward its base; forma **Poýseri** Clute (named for WILLIAM ALDWORTH POYSER, 1882–1928, who had received a specimen from the discoverer) with blade bipinnate to bipinnatifid, with ovate pinnules and deeply lobed segments.

8. ATHÝRIUM Roth

Sori linear or oblong to lunate or arching into semicircular outline, either along one side of a free veinlet or curving above and crossing the veinlet; indusia thin, scarious, straight or curved, the free side usually toothed or glandular, or in no. 4 the indusium nearly or quite obsolete and represented by hidden shreds; scales of rhizome and stipe thin and scarious, delicate, thin-walled, fragile. — Herbaceous ferns with elongate pinnate to tripinnate or pinnatifid soft (often fragile) fronds; the green or greenish stipes furrowed or flattened above, with two bast-bundles below, these commonly united above into a 2-winged bundle. Genus semi-cosmop., mostly in N. Hemisph. (Name from the Greek *athyros*, *doorless*, the growth of the sporangia only tardily forcing back the outer margin of the indusia.) Often merged with *Asplenium*.

a. Fronds simply pinnate, the pinnae long-attenuate; sori straight, linear, 20–40 of them each side of midrib. . . . . 1. *A. pycnocarpon.*

a. Fronds bipinnatifid to bi- or tripinnate. . . *b.*

b. Fronds deeply bipinnatifid, each lobe of the fertile pinnae with 2–8 linear or linear-oblong straight or slightly curving sori each side of midrib, the sori not arching across the adjacent veinlet. . . . . 2. *A. thelypterioides.*

b. Fronds bi- to tripinnate; sori roundish, without apparent indusia, or (at least the lower ones) reniform, horseshoe-shaped or lunate, with the upper part often arching across the adjacent veinlet.

Indusium mostly crescent-shaped, broadly hooked or horseshoe-shaped; wide-ranging species. . . . . 3. *A. Filix-femina.*

Indusia obsolete or an obscure hidden rudiment; sori roundish; species of mts. of Nfld. and Gaspé Pen., Que. . . . . 4. *A. alpestre.*

1. A. pycnocárpon (Spreng.) Tidestr. (with crowded fruits), Glade-Fern, Narrow-leaved Spleenwort. — Rhizome nearly horizontal, bearing a subterminal tuft of erect fronds; stipes 1–7.5 dm. high, green to stramineous or pale brown; *fronds elongate-lanceolate, simply pinnate,* membranaceous; the sterile 2.5–7.5 dm. long, 1–1.8 dm. broad; *the numerous pinnae long-acuminate,* rounded to subtruncate or semi-hastate at base, 1–2 cm. broad, slightly undulate, the pale midrib evident beneath; *fertile fronds with narrow (lance-linear) pinnae; indusia linear, elongate, straight or barely falcate, 20–40 of them oblique to the midrib, borne on one side of the veinlets.* (*Asplenium* Spreng.; *Diplazium* Broun; *Asplenium angustifolium* Michx., not Jacq.; *Athyrium angust.* Milde; *Diplazium angust.* Butters) — Rich (mostly calcareous) wooded slopes, ravines, and bottoms, sw. Que. to Minn., s. to Mass., Ct., Va., upland to Ga., Ala., La. and e. Kans. *Fr.* Aug., Sept.

2. A. thelypterioides (Michx.) Desv. (resembling *Dryopteris Thelypteris*), Silvery Spleenwort. — Rhizome nearly horizontal; stipes pale, 1–7.5 dm. high, with deciduous pales; *frond* elliptic-lanceolate, narrowed to base and apex, 3.5–7.8 dm. long, 1–3 dm. broad across middle, *bipinnatifid; pinnae linear-lanceolate, deeply pinnatifid, the oblong obtuse lobes* minutely toothed and *bearing on the backs* of the fertile fronds *two rows of 2–8 whitish or silvery linear-oblong straight or slightly curving indusia, the sori not crossing over the adjacent nerve.* (*Asplenium* Michx.; *Diplazium* Presl) — Rich woods, bottomlands and shaded slopes, Ga. and Ala. to e. Mo., n. to C.B., Gaspé Pen. and s. Que., s. Ont., Mich., Wisc. and Minn. *Fr.* July–Sept. — Forma **acrostichoides** (Sw.) Gilbert (resembling *Acrostichum*) has the lobes more tapering to blunt or acutish tips and coarsely toothed (*Asplenium acrostichoides* Sw.; *Athyrium acrost.* Diels; *Diplazium acrost.* Butters).

3. A. Fìlix-fémina (L.) Roth (Lady-Fern, the Latin name originally applied to Bracken but transferred by Linnaeus to the present species, presumably on account of its fragility and delicate cutting as contrasted with the Male Fern; although, since *A. Filix-femina* is discouragingly variable, it has been suggested that Linnaeus had in mind the French proverb, "Souvent femme varie", etc.), Lady-Fern. — Rhizome strongly ascending to horizontal, chaffy, with a tuft of terminal or subterminal fronds; *stipes fragile, furrowed,* chaffy below, the chaff persistent or deciduous; *frond* membranaceous to subcoriaceous, herbaceous, narrowly to broadly lanceolate, *bipinnate* (rarely tripinnate), the *pinnules mostly lobed or toothed; indusia from merely arching to strongly arched at summit or even horseshoe-shaped or reniform, the summit of the larger ones arching across the adjacent vein.* — One of the most variable of species, consisting of many geographic vars. and innumerable minor vegetative forms. By Butters treated as being three species in our area; these, however, seeming to be too confluent for sharp specific differentiation from the Eurasian type. Only the following of the many proposed variations are here considered, the diagnostic characters partly from Butters in Rhodora, xix. 170–202 (1917) and from Weatherby in Am. Fern Journ. xxvi. 132–135 (1936).

a. Rhizome erect or ascending, the new growing tip in the center of the regular crown of stipes; frond tapering about equally from middle to apex and base, the lower pinnae only one-half to one-fourth as long as middle ones; larger indusia mostly less than 1 mm. long and more than half as broad, fringed (before injury) with long multicellular non-glandular cilia; spores yellow, warty. . . . . Var. *sitchense.*

a. Rhizome obliquely ascending to horizontal, with the growing tip projecting beyond the approximate series of living stipes; largest indusia mostly more than 1 mm. long but less than half as broad. . . *b.*

b. Rhizome compact, covered by persistent old bases of stipes, the whole usually 2–5 cm. thick; scales of stipe-base up to 1 cm. long and 1.5 mm. broad, brown to blackish, subpersistent; frond widest near middle; indusia toothed or ciliate, the cilia not gland-tipped; spores yellowish-brown, smooth or sparingly papillate. . . *c.*

c. Fronds dimorphic, the fertile ones more coriaceous and contracted or plicate; the sori at maturity confluent and covering lower side of fertile pinnae. . . *d.*

d. Longest pinnae of fertile frond 5–12 cm. long, their pinnules 4–12 mm. long and simple; sori mostly asplenioid (distal end not crossing the subtending vein); oblong obtuse pinnules of sterile frond only slightly toothed or lobed. . . . . Var. *Michauxii.*

*d.*Longest pinnae of fertile frond 1–2 dm. long, their pinnules 1.2–1.5 cm. long and pinnatifid; sori of lower segments often horseshoe-shaped; pinnules of sterile fronds oblong-lanceolate, strongly toothed or pinnatifid, acutish. . . . . Forma ***elatius.***

*c.*Fronds neither strongly dimorphic, coriaceous nor plicate; sori mostly not confluent at maturity. . . *e.*

*e.*Pinnules widely to perpendicularly divergent from rachis of pinnae. . *f.*

*f.*Pinnules regularly diminishing in size, oblong or linear-lanceolate, three to five times as long as broad, regularly and coarsely toothed or pinnatifid, the basal anterior segment usually largest. . . *g.*

*g.*Pinnules lanceolate, subacute, strongly toothed or pinnatifid, their segments toothed, their rachises scarcely winged. . . . . . Forma ***rubellum.***

*g.*Pinnules oblong, obtuse, only obscurely toothed, their rachises with strongly developed membranaceous wing. . . . . . Forma ***laurentianum.***

*f.*Pinnules irregularly varying from one another in size, irregularly lobed or toothed, joined by a broad membranaceous band, their broad lobes overlapping. . . . . Forma ***confertum.***

*e.*Pinnules oblique to rachis of pinna and prominently decurrent but not connected by a wing, their teeth acute. . . . . Forma ***elegans.***

*b.*Rhizome horizontally creeping, with few or no persistent old stipe-bases, the whole 1–1.5 cm. thick; scales of stipe-base few, soon deciduous, mostly less than 5 mm. long and 1 mm. broad, russet or brown; frond widest toward base; indusia ciliate, the cilia commonly gland-tipped; spores blackish, reticulate or wrinkled.

Pinnules less than 15 mm. long, oblong or oblong-linear, obtuse. . . . Var. ***asplenioides.***

Pinnules about 2 cm. long, triangular-lanceolate, pinnatifid. . . . . Forma ***subtripinnatum.***

Var. **sitchénse** Rupr. (of Sitka). — Fronds (except in dwarf colonies) 0.6–2 m. high, 1–3.5 dm. broad, subcoriaceous; sori round, distinct. — Damp subalpine slopes, brooksides and thickets, mts. of nw. Nfld. and Gaspé Pen., Que.; s. Alaska to Id. and Calif. *Fr.* Aug., Sept.

Var. **Michaùxii** (Spreng.) Farw. (for André Michaux, 1746–1802, who first described it), *A. angustum* (Willd.) Presl — Damp thickets, meadows and swamps, Que. to e. Man., s. to N.S., N.E., Pa., O., Wisc. and Ia.; southeastw. passing freely into var. *asplenioides. Fr.* through summer. — Our most intricate var., passing insensibly into many forms: forma **elàtius** (Link) Clute (tall), extending s. to Md., Ill., and Mo.; forma **cristàtum** (Hopkins) Clute (crested), like the last but with tips of blade and some pinnae forking, local; forma **rubéllum** (Gilbert) Farw. (somewhat red), with stipe and rachis and often the frond reddish or purplish or all of them green, in low woods, thickets and swamps, Lab. to Man., s. to Nfld., N.S., N.E., Va., O., Ind., Ill., Mo. and S.D.; forma **laurentiànum** (Butters) Fern. (of the region of the Gulf of St. Lawrence), in its compact rhizome, very short lower pinnae, winged rachises of pinnules, etc., closely simulating European forms of typical *A. Filix-femina*, Lab. and Nfld. to se. Me.; forma **confértum** (Butters) Fern. (crowded), local, Gaspé Pen., Que., to N.E.; forma **élegans** (Gilbert) Clute (elegant), Nfld. to Ont., s. to N.E. and N.Y.; forma **laciniàtum** (Butters) Fern. (slashed), somewhat suggesting forma *confertum* but with the pinnules deeply and irregularly slashed.

Var. **asplenioìdes** (Michx.) Farw. (resembling *Asplenium*), *A. asplenioides* (Michx.) Desv. — Wet woods and thickets, swamps and meadows, Fla. to Tex., n. to Mass., N.Y., O., Ind., Mo., and Okla. — Forma **subtripinnàtum** (Butters) Fern. (almost tripinnate), local.

4. A. alpéstre (Hoppe) Rylands (growing on mountains). — Rhizome heavy, forming a stout crown covered with coarse stipe-bases, the fronds in vase-like clumps; stipes brittle, with deciduous papery scales; *blade subcoriaceous, bipinnate or somewhat tripinnatifid*, elliptic-lanceolate to lance-ovate (rarely lance-oblong), mostly one-fourth to half as broad as long; *pinnules oblong-lanceolate, with the broad-based oblong ultimate lobes mostly approximate; sori round, without evident indusium (when very young sometimes with a rudimentary indusium of few filaments), median or submedian, the larger ones* 0.75–1.4 *mm. across;* spores blackish, reticulated. — Wet mossy or bushy slopes and stream-margins in quartzite rock, Highlands of St. John, nw. Nfld. *Fr.* July, Aug. (Iceland; Eurasia)

Var. **gaspénse** Fern. (of Gaspé Pen.). — *Fronds tripinnate or almost quadripinnate, the ultimate segments linear or linear-lanceolate and remote; sori* 0.3–0.8 *mm. across,* without evidence of indusia, *submarginal.* — Wet granitic rock, brooksides, meadows and subalpine woods, n. and e. slopes and tableland of Table-top Mt., Gaspé Co., Que. *Fr.* July, Aug.

9. PHYLLÌTIS Hill Hart's-tongue

Sori elongate, from oblong to narrowly linear, transverse, perpendicular or oblique to midrib, often longer and shorter ones alternating, approximate in opposite pairs, one on the upper side of one veinlet, the next on the lower side of the next superior veinlet, thus often appearing to have double indusia opening along the middle. — Fronds simple, unlobed or merely sinuate. Small genus, with localized or disjunct areas of occurrence in N. Hemisph. and S. Am. (*Phyllitis*, Greek name, supposed to be for this fern.)

1. P. Scolopéndrium (L.) Newm. (Greek name of some plant, taken over and applied to this), Hart's-tongue of Europe. — Fronds from a short caudex, oblong-lingulate or straplike, deeply cordate-auriculate at base, 1–6 (*av.* 3) *dm. long, the fruiting portion of the fertile fronds occupying one-third to the full length* (*av.* 73 *per cent*) *of the frond; longest indusia of each frond linear,* 0.17–3.3 (*av.* 1.7) *cm. long; foveolae* (*enlarged tips of veinlets*), as seen by transmitted light, *elliptic and nearly marginal; stipe with flat lance-attenuate scales mixed with much narrower or more slender ones.* (*Scolopendrium vulgare* Sm.) — Cult. and very locally natzd. in cool well-holes, etc. (Introd. from Eu.)

Var. **americàna** Fern. (American). — *Scales of stipe all narrow, curling, long-caudate; frond* narrower and *usually shorter,* 1–3.4 (*av.* 2.3) *dm. long, with midrib more promptly glabrate, mostly fruiting on upper half; longer indusia narrowly oblong,* 0.3–2.2 (*av.* 1.2) *cm. long; foveolae linear-oblanceolate, about* 1 *mm. from margin.* — Crevices and cool slopes or sink-holes of dolomite and other calcareous rock, highly localized, w. N.B. (perhaps extinct); centr. N.Y. (largely exterminated by quarrying and by federal "conservation"-activities); Dufferin and Durham Cos., Ont.; Grey and Bruce Cos., Ont., there frequent and abundant; Marion Co., Tenn. *Fr.* July–Sept.

10. CAMPTOSÒRUS Link Walking Leaf. Walking Fern

Sori oblong or linear, as in *Asplenium*, but irregularly scattered on either side of the *reticulated veins* of the simple frond, the sori next to the midrib single, the outer ones inclined to be approximate in pairs (so that their two indusia open face to face) or to become confluent at their ends, thus forming crooked lines (whence the name from the Greek *camptos, flexible,* and *soros, fruit-dot*). — Only two species, ours and one in e. Asia.

1. C. rhizophýllus (L.) Link (rooting leaf). — Fronds evergreen, subcoriaceous, growing in tufts, slender-petioled, arching or divergent; the mature blades 0.5–2 dm. long, lanceolate, entire or undulate, tapering to prolonged slender rooting tips, the cordate and round-auricled base 1–3 cm. broad. — Shaded rock (often calcareous), rarely on earth or tree-bases, Ga. to e. Okla., n. to w.-centr. Me. (rare), s. N.H., sw. Que., s. Ont., Mich., Wisc. and Minn. *Fr.* May–Sept. — The occasional forma **auriculàtus** R. Hoffm. (eared) has the basal auricles of some leaves divergent, acute and often long-tapering or even rooting. More infrequent are forma **angustàtus** F. W. Gray (narrowed) with blade almost linear-attenuate and about 5 mm. broad at base; forma **intermèdius** (Arthur) Clute (intermediate) with blade tapering or cuneate at base; and forma **Bòycei** C. L. Wilson (named in 1935 for its discoverer, Guy Boyce) with margin of blade cut to depth of 1–3 mm. into many wide- spreading teeth.

11. ASPLÈNIUM L. Spleenwort.

Sori oblong to linear, oblique, separate or becoming closely approximate; indusia straight or but slightly curved, attached lengthwise to the upper side of the fertile vein, the sori rarely double (then the indusia borne back to back); scales of rhizome or stipes narrow, firm, with thick-walled areolae. — Small often evergreen ferns of most reg., with slender to filiform stipes having the vascular bundles separate and peripheral or, if united toward the summit, forming a lunate bundle. (*Asplenon*, a name used by Dioscorides for some fern supposed to cure diseases of the spleen.)

a. Rachis plano-convex, flattened above, green and herbaceous. . . *b.*
b. Frond linear, herbaceous, rarely overwintering, simply pinnate, with 8–20 or more pairs of nearly uniform herbaceous rounded, rhombic or obovate small pinnae. . . . . 1. *A. viride.*
b. Frond lanceolate to ovate or deltoid, more or less evergreen, pinnate or pinnatifid, the lower pinnae or divisions longer than the others. . . *c.*
c. Frond lanceolate, very long-attenuate to a prolonged tail-like tip, simply pinnatifid or barely pinnate at base. . . . . 2. *A. pinnatifidum.*

c. Fronds broadly lanceolate to deltoid, 2- or 3-pinnate, not prolonged to tail-like tip.
Frond lanceolate or lance-ovate, tapering above, regularly pinnate, with 6–10 or more pinnae; stipe brown or purplish at base. . . . 3. *A. montanum.*
Frond deltoid-ovate, loosely divided into 2–4 pairs of remote alternate pinnae; stipe green. . . . . 4. *A. cryptolepis.*

a. Rachis subterete or flattened above, castaneous to purple-blackish and lustrous at least toward base. . . *d.*
d. Fronds all similar, most or all of them fertile and loosely ascending to spreading. . . *e.*
e. Fronds herbaceous, with rachis green at summit; pinnae 6– about 14 on each margin of rachis, the lower and median ones pinnatifid to pinnate. 5. *A. Bradleyi.*
e. Fronds firm to subrigid, with lustrous midrib dark to summit; pinnae mostly 15–35 pairs, usually not pinnatifid, the larger ones 3–13 mm. long.
Pinnae with 1 or 2 basal auricles, broadest at base; southern. . . . 6. *A. resiliens.*
Pinnae without basal auricles, broadest above base; wide-ranging. . . 7. *A. Trichomanes.*
d. Fronds of two types, the upright fertile ones much larger than the small spreading basal ones; pinnae of erect fronds alternate, auricled at base, the larger ones 1–6 cm. long. . . . . 8. *A. platyneuron.*

1. A. víride Huds. (green), Green S. — Rhizome short, forking into few to many crowded crowns, bearing blackish scales; *stipes slender, green above, merging into the herbaceous green rachis; frond linear to* linear-lanceolate, *herbaceous,* 2–13 cm. long, 5–15 mm. broad, simply *pinnate, with many round- or rhombic-ovate crenate pinnae;* sori borne toward the vague midrib. — Crevices and talus of calcareous rock, Nfld. and Côte Nord to L. Mistassini, Que., nw. to Alaska, s. to C.B., sw. N.B., Vt., ne. N.Y., Grey Co., Ont., Colo., Utah and Wash. *Fr.* June–Sept. (S. Greenl.; Eurasia)

2. A. pinnatífidum Nutt. (pinnatifid), Pinnatifid S. — Clustered, with a creeping thick rhizome; *stipes brown at base, green above, thence passing into the broad and pale green midrib* of the frond; *frond lanceolate, long-tapering into a prolonged tail-like tip, pinnatifid,* or rarely pinnate at base, 0.3–2.5 dm. long; lobes below the slender tip ovate, rounded at summit (rarely tapering to acuminate tips), entire or slightly crenate; the prolonged tip with lobes much reduced; sori irregular, those near the midrib often double. — Crevices of non-calcareous rock, centr. Ga. to Okla., n. to Pa., W.Va., O., s. and w.-centr. Ind., s. Ill. and Mo. *Fr.* summer. — Hybridizes with nos. 3 and 5; and with no. 8, producing × **A. Stótleri** Wherry (named in 1925 for one of its discoverers, T. C. Stotler) with distinct lobulate pinnae, the rachis brown at base but otherwise green.

3. A. montànum Willd. (montane), Mountain S. — Short rhizome with dark chaff at apex; fronds tufted, loosely spreading to ascending; *stipe* slender, *dark at base, green above, thence passing into the broad and flat green rachis; frond lanceolate to lance-oblong or -ovate,* subcoriaceous, *pinnate,* 3–12 cm. long, up to 5 cm. broad at base; the 6–10 or more *pinnae* ovate to ovate-oblong, *at least the lower ones cleft into cut-toothed pinnules.* — Shaded or sheltered crevices of chiefly non-calcareous rock, centr. Ga. and Ala., n. to s. Ct., w. Mass., se. N.Y., Pa., W.Va., e. and se. O.; n. Mich. *Fr.* May–Sept. — × **A. Trudélli** Wherry (named for one of its discoverers, Harry William Trudell, 1884–   ), reputed hybrid of nos. 3 and 2, seems to be intermediate between them.

4. A. cryptólepis Fern. (with hidden scales), Wall-Rue of N. America. — Tufted from a short rhizome, with persistent old stipe-bases; *scales of rhizome and stipe-bases hidden among rootlets* (not evident without removing rootlets), 1.5–4 *mm. long, firm, lanceolate, the cell-walls about as thick as the 3–6-seriate lumina; stipes green,* very slender, 1.5–6.5 cm. long, spreading; *frond deltoid-ovate, subcoriaceous,* 1–6.5 cm. long, *divided into* 2–4 *pairs of remotely alternate stalked pinnae; looser pinnae with* 3–7 *crenate-rhombic bluntly to subacutely dentate segments;* upper pinnae more simple, *their margins scarcely hyaline;* sori distinct or somewhat confluent; spores minutely rugulose. (*A. Ruta-muraria* sensu Am. auth., not L.; *A. Ruta-muraria,* var. *cryptolepis* Massey and subsp. *cryptolepis* Clausen & Wahl) — Calcareous cliffs and ledges, mostly local, Vt. to Algoma Distr., Ont., and n. Mich., s. to Ct., n. N.J., Pa., upland to N.C. and Ala., Tenn. and Ark. *Fr.* May–Sept. — Much confused with the European *A. Ruta-muraria,* which has the basal scales obvious, longer, and with narrower cell-walls; stipes mostly longer and chaffy or slenderly pubescent; teeth of segments more cartilaginous-bordered; indusia more crowded and spores with coarser ridges. — × **Asplenosórus inexpectàtus** E. L. Br. (unexpected) is a rare intergeneric hybrid of no. 4 and *Camptosorus rhizophyllus,* with slender-tipped but short deltoid-ovate coriaceous fronds more or less pinnate at base, once found with the parents in Adams Co., O.

Var. **ohiȍnis** Fern. (of Ohio). — Segments lanceolate, incised, long-attenuate. — **W. Pa.** to s. Ind., s. to w. Va.

5. A. Brádleyi D. C. Eat. (for its discoverer, FRANK H. BRADLEY, 1838–1879), BRADLEY'S S. — The short dark-scaly rhizome holding many old denuded stipes; new *stipes blackish and lustrous, the dark color passing to the lower half of the rachis; summit of rachis green; frond herbaceous or membranaceous but evergreen, oblong-lanceolate,* 0.4–2 dm. long, *pinnate, with* 6–14 *or more short-stalked* oblong-ovate obtuse incised to blunt-pinnatifid *pinnae each side of rachis,* the lowest pinnae longest. — Crevices of acidic rocks and cliffs, Ga. to e. Okla., n., locally, to se. N.Y., n. N.J., Pa., W.Va., s. O., Ky. and Mo. *Fr.* June–Sept. — × **A. Gràvesii** Maxon (named in 1918 for its discoverer, EDWARD W. GRAVES), a local hybrid of nos. 2 and 5 and growing with them, differs from the latter in simpler and less cut pinnae and slightly winged rachis, from the former in its sessile basal pinnae, merely acuminate fronds and dark stipe.

6. A. resíliens Kunze (recoiling), BLACK-STEM S. or LITTLE EBONY-S. — Short rhizome with rigid dull blackish scales; *fronds* evergreen, upright, 1–2.5 dm. high, *narrowly linear-oblanceolate; short stipe and rachis blackish and lustrous; frond pinnate, with numerous thickish and firm mostly opposite* nearly sessile often deflexed *oblong obtuse entire or crenulate pinnae broadest at base and auricled at base* on upper or both margins; *median pinnae* 6–13 *mm. long,* basal and terminal ones shorter. (*A. parvulum* Mart. & Gal., not Hook.) — Mostly shaded crevices of calcareous rock, Fla. to Ariz. and Mex., n. to s.-centr. Pa., W.Va., s. O., Ky., sw. Ill., Mo. and Kans. *Fr.* May–Sept. (W.I.; S. Am.)

7. A. Trichȍmanes L. (by pre-Linnaean botanists placed in the genus *Trichomanes*), MAIDEN-HAIR-S., DORADILLE CHEVELUE (Que.) — *Forming dense radiating or one-sided spreading tufts,* with many denuded old rachises; *slender firm stipe and rachis purple-brown and shining; frond linear,* dark green above, *firm,* usually 0.8–2.2 dm. long, *with* many pairs of *roundish-oblong or oval pinnae* 3–8 *mm. long; these inequilateral at* the cuneate to rounded *base, without auricles, broadest above the base.* — Shaded, often calcareous, rock-crevices, Rimouski Co., Que., to se. Man., s. to N.S., N.E., Va., upland to Ga. and Ala., Ark. and Okla.; s. Alta. and B.C., s. to Colo., Ariz. and Oreg. *Fr.* through summer. (Eurasia) — The rare forma **incìsum** (S. F. Gray) Clute (incised) has the pinnae sharply jagged-toothed.

8. A. platyneȗron (L.) Oakes (broad-nerved; inappropriate name, derived from an old figure with exaggeratedly broad rachis), EBONY-S. — *Dimorphic, with spreading and prostrate usually sterile short basal fronds with approximate* oblong *evergreen pinnae and much taller erect fertile fronds with wider sinuses and longer less evergreen pinnae; fertile fronds* 1–6 dm. high, 1.3–12 cm. broad, *on short chestnut-purple lustrous stipe and with similar rachis; the* subhorizontal *alternate pinnae narrowly ovate or oblong to lanceolate or linear* and *auricled at base,* firm to membranaceous, 0.5–6 cm. long, the median ones longest; overwintering fertile fronds sometimes producing vegetative buds or plantlets from some axils. (*A. ebeneum* Ait., more apt name) — Highly variable species, with vars. reputed to occur in W.I., S.Am., and s. Afr.; we have the following vars.:

a. Fertile fronds mostly 1–4 dm. high, 1.3–6 cm. broad, mostly with 25–50 pairs of pinnae. . *b.*

b. Pinnae obliquely and narrowly ovate to oblong or oblong-lanceolate, with rounded to merely subacute tips, the longer pinnae 0.5–2 (–2.5) cm. long.

Pinnae minutely crenulate, dentate or fine-serrulate. . . . . . *A. platyneuron* (typical).

Pinnae regularly and deeply pinnatifid, cleft half-way to midrib. . . . . Forma *Hortonae.*

b. Pinnae linear or linear-lanceolate, tapering to acute or acutish tips, more membranaceous; the longer ones 1.5–3 cm. long and with sharply serrate, incised or merely serrulate margins, the longest teeth minute to 3 mm. long. . . . . Var. ***incisum.***

a. Fertile fronds 3–6 dm. high, 6.5–12 cm. broad, with 45–70 pairs of thin-membranaceous pinnae; longest pinnae 3.2–6 cm. long, usually incised with irregular and unequal teeth or toothed lobes, the longest teeth or lobes 2–8 mm. long. . . . . Var. ***bacculum-rubrum.***

A. platyneȗron (typical). — Open woods, wooded slopes, rocky banks or crevices of ledges, Fla. to e. Tex., n. to w.-centr. Me., s. Que., s. Ont., O., s. Mich., s. Wisc., Ia. and Kans. *Fr.* late May–Sept. — Forma **Hórtonae** (Davenp.) L. B. Sm. (named in 1901 for its discoverer, FRANCES B. HORTON) rare. Plants with one or more forking pinnae are the trivial forma **furcàtum** Clute (forking). Plants with overwintering and partly dead fertile fronds producing vegetative buds or sprouts in the axils of one or more old pinnae, are by the enthusiast designated forma **prolíferum** (D. C. Eat.) Tanger (proliferating).

Var. **incìsum** (Howe) Robins. (incised; name appropriate for only extreme specimens). Incl. var. *serratum* (E. S. Mill.) BSP. or forma *serratum* (E. S. Mill.) R. Hoffm., and var. *euroaustrinum* Fern. in large part but not as to type. — Fla. to La., n. to Mass., centr. Vt., se. N.Y., Pa., W.Va. and Ky.

Var. **bâcculum-rûbrum** (Featherman) Fern. (Featherman's latinization of Baton Rouge). Including type of var. *euroaustrinum* Fern. — Rich woods, Fla. to La., n. on Coastal Plain to se. Va.

× **A. virgínicum** Maxon (Virginian) is a rare hybrid of typical no. 8 and no. 7, very slender, with linear fronds at most 1.8 cm. broad; pinnae 20–25 pairs, distant, deltoid-ovate, blunt, the lowest ones opposite, the others alternate, the largest (median) 6–9 mm. long.

× **Asplenosòrus** Wherry (intergeneric hybrid, the name derived from *Asplenium* and *Camptosorus*) has two members with us: (1) × **A. inexpectàtus** E. L. Br. (unexpected), *Asplenium cryptolepis* × *Camptosorus rhizophyllus* (see note under *Asplenium* no. 4); (2) × **A. ebenoìdes** (R. R. Scott) Wherry (like *Asplenium ebeneum*), Scott's Spleenwort, an occasional hybrid of *Asplenium platyneuron* and *Camptosorus* with fronds lanceolate and tapering to slender prolonged tip, the base pinnate or pinnatifid, the lobes round-ovate to acuminate, the brown stipes becoming green above, the rachis green.

12. WOODWÁRDIA Sm. Chain-Fern

Sori oblong or linear, arranged in one or more chain-like rows on transverse anastomosing veinlets parallel and near to the midrib. Indusium attached by its outer margin to the fruitful veinlet, free and opening on the side next the midrib. Veins more or less reticulated, free toward the margin of the frond. — Large ferns with widely creeping rhizomes, of temp. and trop. reg. of N. Hemisph., represented with us by one species of each of two sections. (Named for *Thomas Jenkinson Woodward*, English botanist, 1745–1820.)

§ Anchistèa (Presl) Hook.

1. W. virgínica (L.) Sm. (Virginian), Virginian C. — Rhizome blackish, stoutish, greatly elongate, the growing tips covered with dark slender scales; *sterile and fertile fronds similar*, erect, with tall lustrous dark-based stipes; *blade oblong-lanceolate, pinnate*, 2–8 dm. or more long, 0.7–3 dm. broad, the subdistant divergent *lanceolate pinnae pinnatifid; veins forming one row of areolae;* sori oblong, one on each areole, confluent when ripe. (*Anchistea* Presl) — Acid bogs, swamps, wooded bottoms, etc., Fla. to Tex., n. to N.S., sw. N.B., centr. Me., sw. Que., s. Ont., O. and s. Mich. *Fr.* July–Sept. (Bermuda) — Forma **fértilis** Farw. (fertile) has the normally sterile fronds with a few fertile upper pinnae.

§ Lorinsèria (Presl) Hook.

2. W. areolàta (L.) Moore (with areoles), Netted C. — Rhizome slender; *fronds strongly dimorphic, pinnatifid; sterile ones* with slender green stipes; *blades oblong-lanceolate to ovate*, fleshy-membranaceous, *their* lanceolate slightly toothed *divisions united at base by a broad wing, their veins forming many rows of areolae; fertile blades* on darker stipes, *taller and narrower, with narrowly linear almost distinct divisions*, their areoles and sori (6–10 mm. long) in a single row each side of the secondary midrib. (*Lorinseria* Presl) — Acid peat, boggy woods, swamps, etc., Fla. to Tex., n. on or near Coastal Plain to se. N.H. and sw. N.S., and inland n. to upland of N.C. and Tenn. and lowland of se. Mo. and Okla.; isolated in sw. Mich. *Fr.* July–Oct.

13. PELLAÈA Link Cliff-Brake

Sporangia in roundish or elongate clusters on the upper part of the slender free veins, distinct, or usually laterally confluent and thus forming a continuous submarginal line of fructification, this commonly covered by the broad membranaceous and usually continuous general indusium formed by the reflexed and altered margin of the fertile segment. — Small ferns of the S. Hemisph., the Malayan reg. and N. Am. with once–thrice-pinnate fronds, the pinnae and pinnules articulate at base; fertile and sterile fronds similar or the fertile taller and with narrower segments, the stipes dark-colored (whence the name from the Greek *pellos, dusky*).

1. P. atropurpùrea (L.) Link (blackish-purple), Purple C. — *Rhizome* short and stout, *bearing linear* rusty *scales; stipes* tufted, subrigid, dark brown or purplish to blackish, *scabrous with long appressed trichomes*, these extending to rachis, the stipes 0.1–2.3 dm. high; fronds coriaceous; *sterile fronds* much lower than the fertile, simply pinnate above, bipinnate toward base, *the upper pinnae and the* 1–3 *pairs of pinnules of the lower pinnae* round-ovate to ovate-oblong, *with pedicels up to 6 mm. long; fertile fronds* lanceolate to narrowly ovate, 0.5–3 *dm. long, with* remote pairs of pinnae, *the upper pinnae simple, the lower pinnately divided, all stalked except the upper pair;* the upper pinnae and the 1–5 pairs of pinnules of the lower linear, oblong, lanceolate or narrowly ovate, 1–5 cm. long. — Chiefly on exposed calcareous rocky slopes, nw. Fla. to Ariz., n. to w. N.E., n. N.Y., s. Ont., n. Mich., Wisc., Minn. and S.D. *Fr.* through summer. — Forma **cristàta** (Trel.) Clute (crested) is a rare form with dichotomously forked pinnae somewhat crowded toward summit of frond.

2. P. glabélla Mett. (smoothish), Smooth C. — Mostly smaller, the fertile fronds up to 2.5 dm. high; *stipe and rachis smooth and lustrous,* rarely with remote spreading hairs; *scales* at base of stipe *lanceolate,* often toothed; pinnae and pinnules smaller, broader, mostly sessile or very short-stalked. (*P. atropurpurea,* var. *Bushii* Mackenz. — perhaps the better name, and var. *glabella* (Mett.) Farw.) — Similar, often damper or more shaded habitats, Montmorency Co., Que., to B.C., s. through Vt. to Va. and to Tenn., Ark., Okla. and Colo. *Fr.* June–Sept.

14. NOTHOLAÈNA R. Br. Cloak-Fern

Sori rounded or oblong, borne near the ends of the veins, soon more or less confluent into an irregular marginal band, with no proper indusium. — Small mostly xerophytic ferns of warm-temp. and trop. reg., 1–4-pinnate, the lower surface almost always either hairy, woolly, chaffy or with a waxy white or yellow powder. (Name from the Greek *nothos, spurious,* and *chlaena, cloak;* the woolly coat of the original species forming a spurious covering to the sporangia, instead of an indusium as in related genera.)

1. N. dealbàta (Pursh) Kunze (whitened over). — Densely tufted, with many crowded and slender smooth stipes; *frond triangular-ovate,* 2–8 cm. long, 3–4-pinnate, *somewhat ternate;* rachis and branches shining, blackish; ultimate pinnules ovate-oblong, up to 2 mm. long, *white and powdery beneath.* — Clefts of dry calcareous rocks, Ark. to centr. Tex., n. to Mo., Neb. and e. Colo. *Fr.* May–Sept.

15. CHEILÁNTHES Sw. Lip-Fern

Sporangia borne on the thickened or clavate ends of the free veinlets, forming small and roundish distinct or nearly contiguous marginal sori, these covered by the reflexed mostly whitish and membranaceous (sometimes herbaceous) common indusia (or reflexed margins of the lobes or pinnae). — Low ferns of warm and temp. reg. (chiefly xerophytic), with mostly twice- or thrice-pinnate chaffy, pubescent or smoother fronds, the sterile and fertile ones often nearly alike, the divisions with the principal veins median. Some species with continuous indusia closely approach *Pellaea;* others with strongly dimorphic fronds as closely approach *Cryptogramma.* (Name from the Greek *cheilos, margin,* and *anthos, flower,* from the marginal sori.)

a. Sterile and fertile fronds essentially alike, all lanceolate to lance-oblong or ovate. . *b.*
  b. Fronds glabrous or hirsute.
    Fronds glabrous; indusium continuous, pale and firm. . . . . . . 1. *C. alabamensis.*
    Fronds villous-hirsute; ends of lobes reflexed and forming separate herbaceous involucres. 2. *C. vestita.*
  b. Fronds woolly or tomentose.
    Fronds mostly 1–3.5 dm. long, with densely matted wool on both surfaces; stipes densely long-woolly; involucre (hidden under wool) scarious and whitish. 3. *C. lanosa.*
    Fronds 2–12 cm. long, thinly pubescent above; stipes with scattered short flexuous trichomes; involucre green and herbaceous. . . . 4. *C. Feei.*
a. Sterile and fertile fronds somewhat dissimilar, deltoid- to rounded-ovate, commonly inrolled and cup-like, glabrous; the sterile smaller, on much shorter stipes and mostly soon decaying. . . . . 5. *C. siliquosa.*

1. C. alabaménsis (Buckl.) Kunze (of Alabama). — Rhizome chaffy; stipes blackish, nearly smooth; *fronds chartaceous, glabrous,* narrowly to broadly lanceolate, 0.7–3 dm. long, bipinnate, with numerous oblong-lanceolate pinnae; pinnules lanceolate to oblong-ovate, often lobed or auricled; *indusium continuous, pale and firm.* — Rocks and bluffs, chiefly calcareous, Ala. to Ariz. and Mex., n. to w. Va., Tenn., s. Mo. and Okla. *Fr.* June–Oct.

2. C. vestita (Spreng.) Sw. (clothed), Hairy L. — *Stipes* brown or castaneous, *sparsely villous-hirsute; fronds* herbaceous, rusty-green, linear-lanceolate to lance-oblong, 0.7–3 dm.

long, *villous-hirtellous*, especially beneath, *with straightish or slightly arching jointed rusty hairs*, bipinnate; pinnae or pairs of pinnae distant, triangular-ovate; pinnules oblong, crowded, more or less incised, the *ends of the roundish or oblong lobes reflexed and forming separate herbaceous involucres*, these pushed back by the ripening sporangia. (*C. lanosa* of ed. 7, not (Michx.) D.C. Eat.) — Rocks and cliffs, either siliceous or calcareous, Ga. to Tex., n. to s. Ct., se. N.Y., Pa., W.Va., s. Ind., s. Ill.; Mo. and Kans.; very local northw. *Fr.* June–Sept. — Fronds tightly curling during drought, quickly reviving after rain.

3. C. lanòsa (Michx.) D. C. Eat. (woolly), Woolly L. — Relatively stout; rhizome chaffy with brown or striate chaff; *stipes densely woolly; fronds* up to 5 dm. high, lanceolate, *tripinnate, gray-tomentose above, very densely woolly beneath with tangled white to brownish matted trichomes;* pinnae oblong to oblong-ovate; *ultimate pinnules distinct*, 1–2 *mm. long, roundish-obovate*, sessile or adnate-decurrent, the *reflexed narrow margins forming a continuous somewhat membranaceous indusium.* (*C. tomentosa* Link) — Cliffs and rocky slopes or gravelly banks, Ga. to Ariz. and Mex., n. to mts. of w. Va., W.Va. and Ky., and to Ark. and Okla. *Fr.* June–Sept. — Fronds curling during drought, revived by rain.

4. C. Feèi Moore (for Antoine Laurent Apollinaire Fée, 1789–1874, who first described it), Slender or Fée's L. — Dwarf, densely tufted, *very slender; stipes at first with scattered short flexible trichomes, finally glabrate, lustrous*, dark purple; *fronds* 2–12 *cm. long*, thinly pubescent above, *woolly beneath with soft whitish distinctly articulated flattened hairs*, twice or thrice pinnate; pinnae 8–12 mm. long, ovate, the lower distant, the upper contiguous; *pinnules crenately pinnatifid*, or mostly divided into roundish densely crowded segments 1–2 mm. long, *the herbaceous margin recurved and forming an almost continuous indusium.* — Dry rocks and cliffs, calcareous to circumneutral, w. Wisc. to s. B.C., s. to s. Ill., Ark., Tex., N.M., Ariz. and Calif. *Fr.* June–Sept.

5. C. siliquòsa Maxon (with siliques; from the silique-like fertile segments). — Densely tufted, *rigid;* stipes chestnut-brown to purplish, smooth; *fronds rigid, deltoid-ovate or rounded*, flattish after rain, more *often inrolled and cup-like; the sterile ones smaller and more dissected than the fertile, soon darkening or shriveling* (often discarded in collecting), *about half as high as the fertile; fertile fronds long-stalked*, 0.7–3 dm. high, tripinnate, the linear-lanceolate or -oblong segments with the revolute margins bearing a continuous revolute scarious-margined erose indusium. (*Cryptogramma densa* (Brack.) Diels) — Magnesian or calcareous rock, Mt. Albert, Gaspé Co., and region of Black Lake, Megantic Co., Que.; Grey Co., Ont.; s. B.C., s. to nw. Wyo., n. Utah and s. Calif. *Fr.* July–Sept. — By its dimorphic fronds and aspect about as well placed in *Cryptogramma*.

16. CRYPTOGRÁMMA R. Br. Rock-brake

Sori roundish or elongate and extending far down on the free forking veins. Margins of fertile segments herbaceous or barely scarious, at first reflexed and meeting at the midrib, at length opening out flat and exposing the confluent sporangia. — Low ferns of cool reg. of N. Hemisph., strongly dimorphic, the fertile fronds taller and with narrower divisions than the sterile. (Name from the Greek *cryptos, hidden,* and *gramme, a line,* alluding to the lines of sporangia at first concealed by the reflexed margin.)

1. C. críspa (L.) R. Br. (curled), Mountain-parsley. — *Tufted*, 1–3 dm. high, *vivid green*, with smoothish pale stipes; *sterile fronds* bi- or tripinnate, herbaceous or subcoriaceous, *deltoid-ovate*, the lower alternate pinnae largest, the pinnules ovate, with sharply toothed cuneate ultimate segments translucent, their veins not enlarged; *fertile fronds quadripinnate to bipinnate*, with linear pinnules 0.2–1 cm. long; *basal scales mostly concolorous, brown.* — Eu. and sw. Asia; represented with us by

Var. **acrostichoìdes** (R. Br.) C. B. Clarke (like *Acrostichum*). — Sterile *fronds* more *coriaceous*, opaque, the pinnules with oblong to narrowly elliptic crenate or incised segments, the nerve-tips (in dried material) obviously foveolate; fertile pinnules mostly 0.4–2 cm. long; *basal scales mostly with castaneous centers.* (*C. acrostichoides* R. Br.) — Rock-crevices and rocky slopes, Keewatin to Alaska and ne. Asia, s. locally to ids. of L. Huron, Ont., of L. Sup., Ont. and Mich., and along the mts. to N.M. and Calif. *Fr.* June–Aug.

2. C. Stélleri (Gmel.) Prantl (for its discoverer, Georg Wilhelm Steller, 1709–1746), Fragile R., Slender Cliff-brake. — *Frail, almost flaccid, the few scattered fronds arising singly from a horizontal whitish crisp rhizome; their fragile stipes almost chaffless, pale to purplish;* sterile *fronds* 0.3–2 dm. high, ovate to ovate-deltoid, *pale green*, with oblong, ovate or obovate flabelliform segments; fertile fronds stiffer, up to 3 dm. high, the segments linear, lanceolate or narrowly oblong. — Cool and shaded calcareous rock or springy slopes, se. Lab. Pen. to

Alaska, s. to Nfld. and locally to N.B., n. and w. N.E., n. Pa., W.Va., Mich., n. Ill., ne. Ia., Colo., Utah and Wash. *Fr.* late May–Sept. (Asia)

17. ADIÁNTUM L. Maidenhair. Capillaire (Que.)

Sporangia marginal, short, borne on the under side of a transversely oblong, lunate or roundish, more or less altered margin of a lobe or sinus of the pinnules, the margin reflexed to form an indusium; the sporangia attached to the approximate tips of the free-forking veins. — Ferns of rich and mesophytic to damp soils in trop. and temp. reg., with dark and often polished stipes, the main rib (costa) wanting (in ours) or at the lower margin. (The ancient name, meaning *unwetted*, the foliage shedding rain-drops.)

Frond palmately forking at the summit of the upright stipe. . . . . . . 1. *A. pedatum.*
Frond with a simple main rachis continuing the arching to pendulous stipe. . . . . 2. *A. Capillus-Veneris.*

1. A. pedàtum L. (palmately forking), Maidenhair-Fern. — *Stipes* deep red-brown to blackish, lustrous, from a slender rhizome, *erect or nearly so,* up to 6 dm. high, *forking at summit into* 2 (rarely 3) *widely divergent and arched-recurving branches* bearing on the upper side several spreading slender divisions, the expanded *frond* up to 5 dm. broad, *green, membranaceous* and very graceful; pinnules numerous, short-stalked and obliquely triangular-oblong, entire on lower margin, from which the veins all proceed, cleft and fruit-bearing on the other margin, with blunt lobes; indusia transversely linear or linear-oblong, mostly 2–5 mm. long and about 1 mm. wide. — Rich hardwoods, s. Que. to Minn., s. to N.S., N.E., Ga., Ala., Miss., La. and Okla. *Fr.* summer. — Forma **Bíllingsae** Kittredge (in honor of Elizabeth Billings, 1871–1944) is a rare form with most of the pinnules incurved and inrolled, thus appearing narrowly cuneate, but the upper ones broad and closely imbricated; forma **laciniàtum** (Hopkins) Weath. (slashed) is a rare form with some of the pinnules skeleton-like and finely dissected.

Var. **aleùticum** Rupr. (Aleutian), Aleutian M. — *Glaucous, stiffer,* the stipes more crowded on the stoutish knotty rhizome; *frond firm, blue-green, with merely spreading to strongly ascending branches* (or in exposed situations with cup-like fronds of twisted pinnules); pinnules smaller; indusia shorter and more lunate. — On serpentine or other magnesian rock or magnesian limestone, ascending to subalpine areas, local, Nfld.; Mt. Albert, Gaspé Co., and rocks of Megantic Co., Que., s. to n. Vt.; very local, w. Ont. and n. Wisc.; Alaska, s. to Utah and Calif. (Ne. Asia)

2. A. Capíllus-Véneris L. (old generic name, *Venus' hair*), Venus'-hair Fern. — *Fronds* 1–5 dm. long, *with a continuous main rachis, often pendulous* from a slender horizontal chaffy rhizome, the polished dark stipes 0.3–3 dm. long; the *blade ovate-lanceolate,* 2–3-pinnate at base, the upper third or half simply pinnate; pinnules thin-membranaceous, wedge-obovate or rhombic, mostly 1–3 cm. long, deeply incised, with blunt lobes bearing the transversely oblong to lunate indusia; veins flabellate-forking from base. — Moist, mostly shaded and calcareous rocks or steep banks, Fla. to Tex. and Mex., n. to sw. Va., Ky., Mo., S.D., Colo., Utah and Calif. *Fr.* summer. (Widely dispersed in warm-temp. and trop. reg.) — Our plant has longer and more slender rhizomes than the typical European plant; the various geographic vars. are not yet worked out.

18. PTERÍDIUM Gleditsch Bracken

Sori marginal, mostly continuous, the sporangia borne between the outer indusium (the modified margin of the segment) and a somewhat indefinite inner indusium (either a continuous membrane, an interrupted membrane or a few rudiments); receptacle a vascular strand connecting the ends of the veins, the inner indusium arising at its inner side. — A highly variable species found in most trop. and temp. reg., with extensively creeping and forking subterranean hairy rhizome, stiff upright stipes alternate upon the rhizome, and coarse and firm often tripinnate fronds, with revolute-margined segments, the lower pinnae with basal nectaries. (Name a diminutive of *Pteris, a wing,* with which genus *Pteridium* has often been merged.)*

1. P. aquilìnum (L.) Kuhn (of an eagle; from the wing-shaped fronds), Brake, Pasture-Brake, Hog-Brake, Grande Fougère (Que.), Fougère d'aigle (Que.). (*Pteris* L.) — The typical var. European and African; represented with us by 3 vars.:

Fertile and sterile indusia both ciliate and more or less pubescent on outer face; ultimate segments of frond usually pubescent beneath (between margin and midnerve); pinnules nearly at right angles to the costae. . . . . . Var. *pubescens.*

* Treatment taken from that of R. M. Tryon in Rhodora, xliii. 1–31 and 37–67 (1941).

Fertile and sterile indusia both glabrous; ultimate segments glabrous (rarely slightly pubescent) beneath (between margin and midnerve); pinnules oblique to the costae.
Margins of ultimate segments moderately pubescent; longest entire segment or entire portion of segment about four times as long as broad; terminal segments mostly 5–8 mm. wide. . . . . Var. *latiusculum.*
Margins of ultimate segments glabrous; longest entire segment or portion of segment six to fifteen times as long as broad; terminal segments mostly 2–4.5 mm. wide. . . . . Var. *pseudo-caudatum.*

Var. **pubéscens** Underw. (hairy). — Growing tip of rhizome with a tuft of dark hairs; frond 0.3–5 m. high, the taller fronds becoming scandent; tips of upper pinnae still unrolled when the lower pinnae are well expanded; blade usually 0.6–1 m. long, mostly ovate-triangular, not ternate but tripinnate or tripinnate-pinnatifid; rachis pubescent; segments mostly pubescent on upper surface, the lower surface densely so. (Var. *lanuginosum* (Bong.) Fern., not (Bory) Kuhn) — Open woods, thickets, burns and clearings, s. Alaska to nw. Mex., e. to w. S.D., Colo., N.M. and w. Tex.; isolated on Keweenaw Pen. and Mackinac I., n. Mich., Bruce Pen., Ont., and in Megantic Co., Que. *Fr.* summer. — Simulated by pubescent extremes of the next.

Var. **latiúsculum** (Desv.) Underw. (broadish). — Growing tip of rhizome naked or with few whitish (rarely dark) hairs; frond 0.3–1.5 m. high, the pinnae all unrolling at about the same time; blade 2–8 dm. long, usually broadly triangular (sometimes ovate), often ternate, usually tripinnate or tripinnate-pinnatifid, exceptionally only bipinnate-pinnatifid; rachis glabrate or glabrous; pinnules mostly oblique to the costae, glabrous or only sparsely (rarely densely) pubescent above and beneath; ultimate segments straight, with pubescent margin, the midnerve pubescent. — Dry sterile woods, clearings, burns, etc., Nfld. and Côte Nord, Que., to Minn., s. to N.S., N.E., L.I., N.J., Pa., D.C., upland to N.C., Tenn., Mo. and Okla.; isolated in Miss., w. S.D., Wyo. and Colo., and in ne. Mex. *Fr.* summer. (Eu.; e. Asia)

Var. **pseudocaudàtum** (Clute) Heller (imitating var. *caudatum*). — Growing tip of rhizome usually with a tuft of dark hairs; frond as in the last but usually smaller (2–7 dm. long); longest entire segments or entire parts of segment from six to fifteen times as long as broad, glabrous, sometimes pubescent on midnerve. — On or near Coastal Plain, Fla. to Tex., n. to se. Mass. and inland n. to s. O., s. Ill., s. Mo. and e. Okla.

19. POLYPÓDIUM L. Polypody

Sori round, naked, arranged on the back of the frond in one or more rows each side of the midrib, or irregularly scattered, each borne (in our species) on the end of a free veinlet. Rhizome creeping, branching, often covered with chaffy scales, bearing roundish knobs, to which the stipes are attached by a distinct articulation. — Large nearly worldwide genus, the plants mostly epiphytic or growing on rock. (Name from the Greek *polys, many,* and *pous, foot,* alluding to the branching rhizomes.)

Fronds green and glabrous beneath. . . . . 1. *P. virginianum.*
Fronds gray and scurfy beneath. . . . . 2. *P. polypodioides,* var. *Michauxianum.*

1. P. virginiànum L. (Virginian), Rock-P., Tripe de roche (Que.). — *Rhizome* spongy, rope-like, 2–7 mm. thick; *its scales* darkened on the back, loosely cellular with thick cell-walls, *cordate, with the sinus often closed,* 2–4.5 mm. long; stipe smooth, slender, 0.1–2 dm. long; *frond* evergreen, *glabrous,* 0.25–2.6 dm. long, 1.5–7 cm. broad, *green on both sides,* oblong-lanceolate to deltoid, usually deeply pinnatifid; the long pinnae linear-oblong or lanceolate, mostly entire or remotely and obscurely undulate-dentate, alternate, or the lowest subopposite and usually about as long as or slightly longer than the median ones, the latter mostly 2-8 mm. broad; *sori nearly marginal; sporangia mixed with long-stalked clavate simple or branching glands.* (*P. vulgare* of ed. 7, not L.) — The named forms are as follows, all but the typical form local or very exceptional.

a. Frond definitely and deeply pinnatifid into almost distinct elongate pinnae. . . *b.*
b. Frond lance-oblong, the basal pinnae not conspicuously longer than the median ones. . . *c.*
c. Frond firm to coriaceous, its pinnae, when simple, obtuse to round-tipped.
Pinnae all linear-oblong and simple, entire or shallowly undulate. . . *P. virginianum* (typical).

Pinnae upwardly dilated to broad lobulate summits, the terminal ones and apex of frond forking. . . . . Forma *chondroides.*

*c.*Frond submembranaceous to thin-coriaceous, the principal pinnae acutish to attenuate. *d.*

*d.*Pinnae entire or barely undulate.

Frond simple, gradually tapering to slender tip. . . . . . . Forma *acuminatum.*

Frond forking and crested at dilated apex. . . . . Forma *alato-multifidum.*

*d.*Pinnae or many of them dentate-pinnatifid. . . . . . Forma *bipinnatifidum.*

*b.*Frond deltoid, deltoid-lanceolate or broadly oblong-oval. *e.*

*e.*Frond deltoid-lanceolate, with unlobed blunt lower pinnae; upper half of frond abruptly contracted, narrow and merely undulate-lobulate. Forma *elongatum.*

*e.*Frond broader, with at least the lower pinnae pinnatifid or auricled.

Frond deltoid, gradually tapering to tip, its lower pinnae more or less hastate or auricled and often pinnatifid. . . . . . Forma *deltoideum.*

Frond broadly oblong or oblong-oval, abruptly short-tipped, its broad pinnae deeply cleft into many long often toothed segments. . . . Forma *cambricoides.*

*a.*Frond simple or with short and broad confluent pinnules, only 0.7–1.2 cm. broad.

Frond with the pinnules semicircular to broadly triangular, shallowly toothed and confluent. . . . . Forma *brachypteron.*

Frond simple or very shallowly undulate, narrowly lanceolate. . . . Forma *subsimplex.*

P. virginiànum (typical). — Frond 1.5–7 cm. broad. — On rocks, crests of ledges, bases of trees and rocky slopes, Nfld. to ne. B.C., s. to N.S., N.E., Va., upland to Ga. and Ala., Tenn. and Ark. *Fr.* mid-summer. (E. Asia) — Forma **chondroìdes** Fern. (like *Chondrus*), *P. vulgare* var. *bifido-multifidum* sensu Gilbert, not Druery, local in N.H. and N.Y.; forma **acuminàtum** (Gilbert) Fern. (gradually tapering to point), *P. vulgare* var. Gilbert, local in Pa.; forma **alàto-multífidum** (Gilbert) Fern. (dilated and much cleft), very local; forma **bipinnatífidum** Fern. (bipinnatifid), occasional from N.E. to W.Va.; forma **elongàtum** (Jewell) Fern. (elongate), *P. vulgare* var. Jewell, rare from Que. to W.Va.; forma **deltoìdeum** (Gilbert) Fern. (deltoid), *P. vulgare* ff. *deltoideum* and *hastatum* Gilbert, occasional from Que. and N.Y. to N.C.; forma **cambricoìdes** F. W. Gray (resembling *P. vulgare* var. *cambricum* of Wales), rare and most extreme of all the pinnatifid forms, from Que. and N.Y. to w. N.C.; forma **brachýpteron** (Ridlon) Fern. (short-winged), *P. vulgare* var. Ridlon, very rare in w. Vt.; forma **subsímplex** Fern. (almost simple), very rare in centr. N.H.

2. P. polypodioides (L.) Watt (resembling *Polypodium,* Linnaeus having thought it an *Acrostichum*), var. **Michauxiànum** Weath. (for André Michaux, 1746–1802, who first distinguished our plant), Resurrection-Fern. — Slender cord-like *rhizomes covered by* blackish but pale-margined *linear-subulate scales; stipes scurfy; fronds* coriaceous, *curling when very dry,* expanded in moister conditions, dark green and glabrous above, *brownish-, becoming grayish-scurfy beneath,* the blades 0.2–2 dm. long, 1–5 cm. broad, with blunt linear-oblong pinnae subopposite and divergent, the sinuses broad. — Extensively creeping in crevices of tree-trunks, along tree-branches, on old beams or on ledges or soil-covered knolls and banks, Fla. to Tex., s. to Guatemala, n. to Del., Va., Ky., s. Ill., Mo. and Okla. — The most northern var. of the wide-ranging *P. polypodioides* (*P. ceteraccinum* Michx.; *Marginaria polypodioides* Tidestr.) of trop. and warm-temp. N. and S. Am.

FAM. 10. MARSILEÀCEAE (Marsilea Family)

Perennial plants rooted in mud, having a slender creeping rhizome and either filiform or 4-parted long-petioled leaves; the somewhat crustaceous several-locular sporocarps borne on peduncles which rise from the rhizome near the leaf-stalks, or are more or less consolidated with the latter, and contain both macrospores and microspores.

1. MARSÍLEA L.

Submersed or emersed aquatic plants. Leaves 4-foliolate. Sporocarps with 2 teeth near the base, 2-locular vertically, splitting into 2 valves at maturity, and emitting an elastic cord or band of tissue, which carries the sporangia on a series of short branches or lobes. — Widely dispersed in trop. and temp. reg. (Named for *Luigi Fernando Conte Marsigli*, 1658–1730, Italian naturalist.)

1. M. QUADRIFÒLIA L. Leaflets broadly obovate-cuneate, glabrous; *sporocarps usually* 2 *or* 3 *on a short peduncle from near the base of the petioles,* pedicelled, glabrous or somewhat hairy; the basal teeth small, obtuse, or the upper one acute. — In water of lakes and quiet streams, local, originally cult., rapidly spreading where estab., N.E. to Ia. and Ky. *Fr.* June–Dec. (Introd. from Eu.)

2. M. mucronàta A. Br. (mucronate; from the upper of the basal teeth of the sporocarp). — *Leaflets* broadly cuneate, 5–15 *mm. long* and broad, *sparsely pubescent with short and broad appressed hairs;* petioles 2–11 cm. long; *peduncles free from the petiole,* shorter than the sporocarp; *sporocarp solitary,* obliquely oval, about 4 mm. long, *with the upper basal tooth longer than the lower,* the sinus between them rounded. (*M. vestita* of ed. 7, not Hook. & Grev.) — Shallow ponds, pools and wet shores, Fla. to Tex., n. Mex. and Ariz., n. to nw. Ia. and w. Minn., N.D., s. Sask. and s. Alta. *Fr.* Aug., Sept.

FAM. 11. SALVINIÀCEAE (SALVINIA FAMILY)

Floating plants of small size, having a more or less elongated and sometimes branching axis, bearing apparently distichous leaves; sporocarps (sori) very soft and thin-walled, two or more on a common stalk, 1-locular and having a central, often branched, receptacle which bears either macrosporangia containing solitary macrospores, or microsporangia with numerous microspores. — A small and interesting family of plants without close affinity to other groups.

Stems pinnately branching, covered by imbricated and lobed leaves. . . . 1. *Azolla.*
Stems simple, with the unlobed leaves appearing 2-ranked. . . . . . . 2. *Salvinia.*

1. AZÓLLA Lam. WATER-FERN

Small moss-like plants of trop. and warm-temp. reg.; the stems pinnately branched and covered with minute 2-lobed imbricated leaves, and emitting rootlets on the under side. Sporocarps in pairs beneath the stem; the smaller ones acorn-shaped, containing at the base a single macrospore with a few attached bodies of doubtful function above it; the larger ones globose and having a basal placenta which bears many pedicellate microsporangia which contain masses of microspores. (Name not satisfactorily explained.)*

1. A. caroliniàna Willd. (Carolinian). — *Plants* 0.5–1 *cm. in diameter,* dichotomously branched; divaricate *leaves* 0.5 *mm. long, nearly orbicular, smooth, not closely imbricated;* microsporangia 8–40 in an indusium. — Quiet waters, Fla. to La., n. to N.C., and locally to Mass., N.Y., O., Ind., Wisc., etc., often spread from cult. northw. (W.I.)

2. A. FILICULOÌDES Lam. (resembling *Filicula*). — *Plant elongate,* up to 6 cm. long; *the oblong to ovate papillose leaves about* 1 *mm. long, closely appressed and imbricated;* microsporangia 35–100 in an indusium. — Spreading from cult. to quiet waters. (Introd. from w. N. or S. Am.)

2. SALVÍNIA Adans. SALVINIA

Leaves apparently 2-ranked, horizontally floating or subaërial, a third series of foliar structures developed ventrally on the stem, these with the form of fascicles of root-like fibers. Sporocarps depressed-globose, formed from the tips of short basal divisions of the ventral leaves, clustered; a few of each cluster containing 10 or more macrosporangia, each of which contains a single macrospore; other sporocarps with numerous microsporangia, each with 64 microspores. (Named for *Antonio Maria Salvini*, 1633–1729, Florentine botanist.)

1. S. ROTUNDIFÒLIA Willd. (round-leaved). — Foliage-leaves suborbicular-oblong, thickish, mostly 1–1.5 cm. long, hairy or papillose on both sides, the lower surface commonly brownish or purplish. (*S. natans* of ed. 7, not All.) — Spreading from cult. to ponds, pools and marshes, local in the U.S. (Introd. from Mex. or S.Am.)

*** Descriptions of species taken directly from SVENSON in Am. Fern Journ. xxxiv. 74 (1944).**

DIVISION II. SPERMATÓPHYTA

(Seed-Plants, Phanerogams or Flowering Plants)

Male generative cells (with rare extra-limital exceptions) passive, developing an elongated tube. Flowers with stamens or pistils or both. Normal reproduction by seeds containing an embryo or minute plant.

Subdivision I. GYMNOSPÉRMAE (Gymnosperms)

FAM. 12. TAXÀCEAE (Yew Family)

Trees or shrubs, ours with evergreen linear leaves and dioecious (or more rarely monoecious) *flowers* (borne on short scaly peduncles), *the staminate globular, formed of a few naked stamens with anther-locules under a peltate somewhat lobed connective, the pistillate consisting of an erect or inverted ovule, which becomes a bony-coated seed more or less surrounded by a large fleshy disk* (or aril).

1. TÁXUS L. Yew. If (Que.)

Annular disk of the pistillate flowers cupuliform, globular, at length pulpy, red and berry-like. Cotyledons 2. — Leaves flat, mucronate, rigid, scattered, 2-ranked. Small genus of N. Hemisph. (The Greek *taxos*, *Yew-tree.*)

1. T. canadénsis Marsh. (of Canada), American Y., Ground Hemlock, Buis de Sapin (Que.). — A low straggling bush; stems diffuse (or rarely arborescent and 2 m. high); leaves linear, yellowish-green beneath. — Rich woods and thickets, Nfld. to Man., s. to N.S., N.E., upland to w. Va. and ne. Ky., w.-centr. Ind., n. Ill. and ne. Ia. Fig. 29. — Pulp sweet, edible; seeds and wilted foliage fatal to livestock.

29. T. canadensis.

FAM. 13. PINÀCEAE (Pine Family)

Trees and shrubs with resinous juice, mostly subulate, scale-like or linear entire leaves, and monoecious or rarely (in Juniperus*) dioecious flowers borne in or having the form of scaly aments, of which the pistillate become cones or berry-like. Ovules* 2 *or more at the base of each scale.* Mostly evergreen. In the following treatment the term *ament* is retained as the most convenient designation for the aggregates of scales bearing or inclosing either stamens or ovules. The morphology of the coniferous inflorescence is still doubtful. It seems probable that the staminate ament is a single flower, but paleophytological evidence suggests that the ovule-bearing cones are inflorescences.

a. Leaves spirally arranged (though sometimes turned into apparently 2 ranks), linear or needle-like; cones with several spirally arranged scales, mostly becoming woody. . . *b.*

b. Ovuliferous scales in the axils of persistent bracts; cones elongate, eventually opening, with basally attached scales; ovules inverted; seeds thin-winged; leaf-buds scaly. Subfam. I. Abietineae.

Leaves not in fascicles nor whorls, borne singly.

Leaves flattish, whitened along 2 lines beneath, often spreading into 1 plane, rarely pungent.

Cones erect, 2–8.5 cm. long (in ours), the scales falling away from the axis; leaves sessile. . . . . 1. *Abies.*

Cones pendulous, 1.2–3.5 cm. long, the scales persisting on the axis; leaves minutely petioled. . . . . 2. *Tsuga.*

Leaves tetragonal, mostly pungent, not turned into 1 plane, sessile on prominent persistent pulvini; cones pendulous or spreading. . . 3. *Picea.*

Leaves (of adult branches) in fascicles of 2–5 or clustered on short spurs.

Leaves many at the summits of divergent truncated spurs, deciduous. 4. *Larix.*

Leaves in sheathed (at least when young) fascicles of 2–5, evergreen. . 5. *Pinus.*

b. Ovuliferous scales peltate, without subtending bracts, forming globular to ovoid woody cones; ovules 2 or more, erect; seeds thick-winged; leaf-buds not scaly. Subfam. II. Taxodioideae.

Seeds 2 to each scale, 3-angled; leaves 2-ranked, deciduous. . . . . 6. *Taxodium.*

a. Leaves decussately opposite or ternate, usually scale-like and adnate (needle-like on young sprouts and in one species of no. 9); fertile scales few, decussately opposite or ternate, forming small closed or drupe-like cones; ovules erect; leaf-buds not scaly. Subfam. III. CUPRESSINEAE.

Monoecious; fruit a small cone with small winged or angled seeds; leaves opposite and somewhat 2-ranked.

Cones ellipsoid, drooping, with basally attached imbricated scales. . . 7. *Thuja.*

Cones angulate-globose, with peltate scales. . . . . 8. *Chamaecyparis.*

Dioecious (rarely monoecious); fruit drupe-like, with coalescent scales and ovoid, bony wingless seeds; leaves opposite or in 3's, not 2-ranked. . . 9. *Juniperus.*

Subfam. I. ABIETÍNEAE

1. ÁBIES Mill. FIR. SAPIN (Que.)

Staminate flowers pendulous from the axils of the previous year's leaves; anthers tipped by a knob, their locules bursting transversely; pollen as in *Pinus.* Cones erect on the upper sides of spreading branches, maturing the first year; their thin scales and bracts deciduous at maturity. Seeds and bark with balsam-bearing vesicles. — Leaves scattered, sessile, flat, with the midrib prominent on the whitened lower surface, on horizontal branches, appearing 2-ranked, leaving a circular scar. Trees of cool half of N. Hemisph. (The Latin name of an Old World species.)

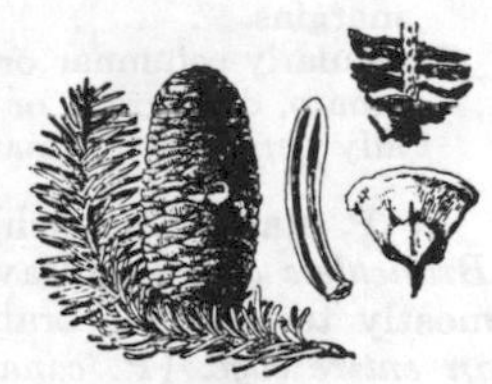

30. A. balsamea.

1. A. balsàmea (L.) Mill. (balsamic), BALSAM-F. or FIR-BALSAM. — Leaves narrowly linear, obtusely pointed or retuse, 1–3.2 cm. long; *cones subcylindric,* 3–8.5 cm. long, 2–3 cm. thick, at first violet-colored; *bracts obovate,* serrulate, abruptly slender-awned, *in maturity shorter than the scales.* — Woods, Lab. to Alta. s. to Nfld., N.S., N.E., mts. of Va. and W.Va., and to n. O., Mich., centr. Wisc. and ne. Ia. FIG. 30. — A slender tree; or in very exposed places a low or prostrate shrub with shorter and broader leaves, forma **hudsònia** (Jacques) Fern. & Weath. (of Hudson Bay).

31. A. balsamea, v. phanerolepis.

Var. **phanerólepis** Fern. (conspicuous-scaled). — Cones 2–5.5 cm. long, 1.5–2 cm. thick; *awns of the mature bracts exserted and spreading.* — Lab. to Ont., s. to Nfld., N.S., coast of Me., and high mts. of N.H. and Vt. FIG. 31.

2. A. Fràseri (Pursh) Poir. (for its discoverer, JOHN FRASER, 1750–1811), SOUTHERN F. or SHE-BALSAM. — Leaves narrowly linear, commonly retuse; *bracts* of the *ovoid cones dentate or erose-lacerate,* often emarginate and bearing a slender apical cusp, *the broad recurved tips much longer than the scales.* — Mts. of Va., N.C. and Tenn. — Forma **prostràta** Rehd. (prostrate) of exposed habitats, is depressed, with horizontal branches.

2. TSÙGA (Endl.) Carr. HEMLOCK. PRUCHE (Que.)

Staminate flowers subglobose clusters of stamens arising from the axils of the preceding year's leaves, the long stipe surrounded by numerous bud-scales; anthers tipped by a short spur or knob, their confluent locules opening transversely; pollen-grains simple. Cones on the ends of the preceding year's branchlets, maturing their first year, pendulous; their scales thin, persistent. — Leaves scattered, flat, whitened beneath, appearing 2-ranked, with a basal cushion. Few species of N.Am. and Asia. (The Japanese name of one of the species.)

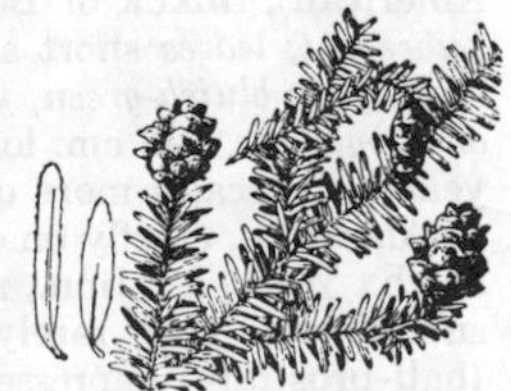

32. T. canadensis.

1. T. canadénsis (L.) Carr. (of Canada). — *Leaves* petioled, short-linear, obtuse, 8–13 *mm. long; cones* ovoid, 1.5–2.5 *cm. long,* the *scales suborbicular.* — Mostly hilly or rocky woods, N.B. and N.S. to Md. and e. Minn., and along the mts. to Ga. and Ala. FIG. 32. — A tall tree, with light and spreading spray, and delicate foliage bright green above, silvery beneath. — Forma **párvula** Vict. & Rousseau (dwarf), depressed, without ascending trunk, forming mats up to 1 m. high, Que. and n. N.E.

2. T. caroliniàna Engelm. (of Carolina). — *Leaves* petioled, linear, 15–18 *mm. long; cones* ovoid, 2–3.5 *cm. long; scales oblong,* in age loosely imbricated, widely and irregularly spreading. — Upland of w. Va. to Ga.

3. PÍCEA Dietr. SPRUCE. ÉPINETTE (Que.)

Staminate flowers on branchlets of the preceding year; anthers tipped by a rounded recurved appendage, their locules opening lengthwise. Cones maturing the first year, becoming pendulous; their scales thin, not thickened nor prickly-tipped, persistent. — Leaves somewhat spirally arranged on lower side of branchlet, needle-shaped and keeled above and below (4-sided), raised on peg-like pulvini. Trees of cold and cool-temp. areas of N. Hemisph. (The Latin name of some pine, from *pix, pitch.*)

Branchlets glabrous; bark pale brown; cones nearly cylindrical, becoming pale brown, their scales coriaceous; buds glabrous. . . . . 1. *P. glauca.*
Branchlets pubescent; bark reddish- to blackish-brown; cones ovoid to subglobose, reddish-brown, their scales becoming firm; buds pubescent.
Pyramidal tree; leaves acute or with firm, sharp tips, green, not glaucous; cones promptly deciduous, their scales entire or with barely denticulate margins. . . . . 2. *P. rubens.*
Irregularly columnar or slenderly pyramidal tree; leaves obtusish or emarginate, dark green or more or less glaucous; cones tardily deciduous, usually persistent for many years, their scales with erose margins. . . . 3. *P. mariana.*

1. P. glaùca (Moench) Voss (blue-green), WHITE OR CAT-S., ÉPINETTE BLANCHE (Que.). — *Branchlets glabrous;* leaves slender, pale or glaucous; *cones subcylindrical,* about 5 cm. long, mostly terminating branchlets, *deciduous; the thin, pale scales with an entire edge.* (*P. canadensis* of authors) — Woods, in good soils, Lab. to Alaska, s. to Nfld., N.S., n. N.E., ne. N.Y., n. Mich., Wisc., Minn., S.D. and Wyo. FIG. 33. — A handsome tree (to 45 m. high), in aspect somewhat resembling the BALSAM-FIR. In alpine and exposed habitats becoming the dwarf and depressed forma **párva** (Vict.) Fern. & Weath. (small).

33. P. glauca.

2. P. rùbens Sarg. (reddish), RED S., HE-BALSAM (southw.) — *Branchlets pubescent; leaves* mostly slender, 12–15 mm. long, usually acute or acutish, *dark green or yellowish-green,* usually upcurving; *cones elongated-ovoid,* mostly 3–4 cm. long, *clear brown or reddish-brown;* the scales rounded, entire or slightly erose. (*P. rubra* Dietr.) — Woods, St.P. et Miq. and P.E.I. to n. O., s. to N.S., N.E. and, chiefly on the uplands, to N.C. and Tenn. FIG. 34. — A valued conical timber-tree, to 35 m. high. Forma **virgàta** (Rehd.) Fern. & Weath. (wand-like), SNAKE-S., has long, slender branches almost without branchlets. — The tree of the southern mountains sometimes treated as a separate species, *P. australis* Small, but scarcely separable.

34. P. rubens.

3. P. mariàna (Mill.) BSP. (of Maryland; the name originally used by Miller as synonymous with North American), BLACK OR BOG-S., ÉPINETTE NOIRE (Que.). — *Branchlets pubescent; leaves* short and thickish, mostly 6–10 (rarely –13) mm. long, *pale bluish-green,* with strong whitish bloom; *cones short-ovoid or subglobose,* 2–3 cm. long, *dull grayish-brown,* persisting for several years; the scales more decidedly erose, rounded or often narrowed toward the apex. — Lab. to Alaska, s., chiefly on cool slopes and bogs, to Nfld., N.S., N.E., n. N.J., n. Pa., mts. to Va. Mich., Wisc., n. Minn., n. Man., n. Sask., Alta. and B.C. FIG. 35. — Chiefly a small irregularly subcylindric tree, rarely attaining 30 m. in height; the forma **semiprostràta** (Peck) Blake (half-prostrate) depressed, with leaves only 3–6 mm. long, chiefly alpine; forma **empetroìdes** Vict. & Rousseau (like *Empetrum*), similar but trunkless and trailing, mts. of Que.

35. P. mariana.

The NORWAY S., P. ÁBIES (L.) Karst., often cultivated as a shade tree, spreads slightly from planted trees northw. It has subglabrous branchlets, slender sharp-pointed dark green glossy leaves, and cones 1–1.5 dm. long. (Introd. from Eu.)

4. LÁRIX Mill. Larch. Mélèze (Que.)

Aments lateral, terminating short spurs on branches of a year's growth or more, short or globular, developed in early spring; the staminate from leafless buds; the pistillate mostly with leaves below. Anther-locules opening transversely. Pollen-grains simple, globular. Cone-scales persistent. — Leaves linear, soft, deciduous, very many in a circular cluster on the short spurs, developed in early spring from lateral scaly and globular buds, or scattered and spirally arranged along the long shoots of the season. Pistillate aments crimson or red, rarely greenish, in flower. Small genus of cold and cool-temp. half of N. Hemisph. (The classical name of the European species.)

36. L. laricina.

1. L. laricìna (DuRoi) K. Koch (larch-like; originally considered a pine), American or Black L., Tamarack, Hackmatack, Épinette rouge (Que.). — Leaves 1–2.5 cm. long; cones ovoid, 1.2–2 cm. long, of few rounded scales. — Lab. to Alaska, s., with us mostly in swamps, to Nfld., N.S., N.E., n. N.J., n. Pa., W.Va., n. O., n. Ind., n. Ill., Minn., Man., n. Sask., n. Alta. and ne. B.C. Fig. 36. — A tree (to 35 m. high), with hard and very resinous wood, often erroneously called "Juniper"; trunk dwarfed and branches prostrate in forma **depréssa** Rousseau (depressed).

L. decídua Mill. (deciduous) of Eu., with leaves 2.5–3 cm. long and cones 2–3.5 cm. long, occasionally spreads from cult. (Introd. from Eu.)

5. PÌNUS L. Pine. Pin (Que.)

Filaments short; connective scale-like; anther-locules 2, opening longitudinally. Pollen of 3 united locules, the 2 lateral ones empty. Fruit a cone formed of the imbricated mostly woody scales, which are persistent and spreading when ripe and dry; the 2 nut-like seeds partly sunk in excavations at the base of the scale. Cotyledons 3–12, linear. — Primary leaves thin and chaff-like, merely bud-scales, from the axils of which grow the secondary needle-shaped evergreen leaves; these in fascicles of 2–5 (with us), from slender buds; some thin scarious bud-scales sheathing the base of at least the young fascicle. Leaves when in pairs semicylindrical, becoming channeled; when more than 2 triangular, their edges in our species serrulate. Flowers developed in spring; the cones maturing in the second autumn. Genus of N. Hemisph. (The classical Latin name).

a. Leaves mostly 5 in a fascicle, the basal scales of the fascicle deciduous; leaves with 1 fibro-vascular bundle; cone-scales unarmed; wood soft, mostly without resin. § 1. Cembra (Soft Pines).
  Cones slenderly subcylindric, drooping on curved peduncles, scales thinnish. . . 1. *P. Strobus.*
a. Leaves 2 or 3 in a fascicle with persistent tubular sheath; fibro-vascular bundles 2; cone-scales thickened upward, armed or unarmed; wood hard, with bands of resin. § 2. Pinaster (Hard Pines). . . *b.*
 b. Cone-scales unarmed; leaves 2. . . *c.*
  c. Cones dehiscent at maturity, not lustrous; leaves 3–17 cm. long.
   Leaves full green, 7–17 cm. long; cone broadly ovoid-conic, reddish-brown, sessile, divergent. . . . . 2. *P. resinosa.*
   Leaves blue- or gray-green, 3–7 cm. long; cones slenderly conic, tawny or grayish, reflexed, on a short peduncle. . . . 3. *P. sylvestris.*
  c. Cones indehiscent for many years, oblong-conical, sessile, lustrous, yellowish-tawny, closely ascending to divergent; leaves 2–4 cm. long. . 11. *P. Banksiana.*
 b. Cone-scales armed with a sharp dorsal and subterminal spine or prickle; leaves 2 or 3. . . *d.*
  d. Cones 15–25 cm. long, cylindric or subcylindric; leaves 20–45 cm. long . 4. *P. australis.*
  d. Cones 3–12 cm. long, conic; leaves 2–20 (rarely –28) cm. long. . . *e.*
   e. Spine of cone-scales 1–3 mm. long. . . *f.*
    f. Leaves firm to subrigid, 0.7–1.5 mm. broad. . . *g.*
     g. Old cone when open subcylindric-ovoid, 6–12 cm. long; leaves 12–25 cm. long. . . . . 5. *P. Taeda.*
     g. Old cones when open broadly ovoid, 4–6 cm. long; leaves 4–13 cm. long.
      Spine of cone-scale minute, about 1 mm. long; leaves in 2's or 3's, 7–13 cm. long. . . . . 6. *P. echinata.*
      Spine of cone-scale 2–3 mm. long; leaves in 2's, 4–8 cm. long. . 7. *P. virginiana.*

f.Leaves rigid, 1.5–3 mm. broad. . . *h*.
h.Leaves in 3's, 3.5–28 cm. long; cones broadly ovoid-conic, symmetrical, deep-brown, divergent.
Leaves 3.5–14 cm. long; sheaths 4–14 mm. long; umbo of cone-scale pyramidal. . . . . 8. *P. rigida.*
Leaves 1.2–2.8 dm. long; sheaths 1–2.5 cm. long; umbo depressed or flattish. . . . . 9. *P. serotina.*
h.Leaves in 2's, 2–4 cm. long; cones oblong-conic, mostly asymmetrical, yellowish-tawny, ascending (rarely divergent). 11. *P. Banksiana.*
e.Spine of cone-scales 5–6 mm. long. . . . . 10. *P. pungens.*

Subgen. HAPLÓXYLON Koehne (SOFT PINES)

§ 1. CÉMBRA Spach

1. P. Stròbus L. (ancient name for some incense-bearing tree), WHITE P., PIN BLANC (Que.). — *Leaves in* 5's (rarely in 3's or 4's), very slender, glaucous; staminate flowers oval (8–10 mm. long), with 6–8 involucral scales at base; pistillate aments long-stalked, cylindrical; cones slender, subcylindrical, nodding, often curved, 0.6–2.5 dm. long; seed smooth; cotyledons 8–10. — Woods, Nfld. to Man., s. to N.S., N.E., e. Md., s. Pa., mts. of Ga. and Tenn., n. Ill. and centr. Ia. FIG. 37. — Important timber-tree to 50 (rarely to 75) m. high, *with bark of the branches smooth;* in bleak n. habitats sometimes depressed, with trailing branches, forma **prostràta** (Mast.) Fern. & Weath. (prostrate).

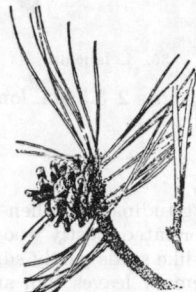

37. P. Strobus.

38. P. resinosa.

Subgen. DIPLÓXYLON Koehne (PITCH PINES)

§ 2. PINÁSTER W. D. J. Koch

2. P. resinòsa Ait. (resinous), RED or NORWAY P., PIN ROUGE (Que.). — *Leaves in 2's,* dark green, 7–17 cm. long, *their resin-ducts marginal; cones* ovoid-conical, smooth (*about 5 cm. long*), *their scales slightly thickened, pointless;* staminate flowers oblong-linear (12–18 mm. long), subtended by about 6 involucral scales which are early deciduous by an articulation above the base. — Dry woods, Nfld. to Man., s. to N.S., N.E., n. N.J., n. Pa., W.Va., Mich., Wisc. and Minn. FIG. 38. — A tall tree, up to 50 m. high, with reddish rather smooth bark and hard wood, not very resinous; forma **globòsa** Rehd. (globose), a rare form of dwarf growth, with globose crown.

P. nìgra Arnold (black), AUSTRIAN P., similar to no. 2 but the stiff leaves with median resin-ducts, the larger (5–8 cm. long) cones falling intact and their scales with a short prickle on the umbo, spreads slightly from cult. (Introd. from Eu.)

3. P. sylvéstris L. (of woodland), SCOTCH P., SCOTCH FIR. — *Leaves in 2's, bluish- or grayish-green,* 3–7 cm. long; *cones* 3–6 cm. long, slenderly conic, reflexed, *the thickened rhombic scales with central tubercle but not spinous.* — Much cult. and locally natzd. from N.E. to Ont., s. to N.J., O. and Ia. FIG. 39. — A valuable tree up to 40 m. high, with gray bark. (Natzd. from Eu.)

4. P. austràlis Michx. f. (southern), LONG-LEAF, YELLOW or GEORGIA P. — *Leaves* in 3's from long sheaths, *very long* (*20–45 cm.*), crowded at the summit of very scaly branches; staminate flowers 6–8 cm. long, rose-purple; *cones large* (*15–25 cm. long*), cylindrical or conical-cylindric, *the thick scales armed with a short recurved spine.* (*P. palustris* of many auth., not Mill.) — Sandy soil, Fla. to e. Tex., n. to se. Va. — A large tree (up to 40 m. high) with thin-scaled bark and exceedingly hard and resinous wood, the young trees forming columnar unbranched very leafy trunks.

39. P. sylvestris.

5. P. Taèda L. (ancient name for resinous pines), LOBLOLLY or OLDFIELD P. — *Leaves long* (*12–25 cm.*), in 3's or sometimes 2's, *with elongated sheaths,* light green; staminate flowers slender, 5 cm. long, usually with 10–13 involucral scales; *cones* slenderly conic, 6–12 cm. long, *their scales tipped with a stout triangular spine with concave sides;* seeds with 3 strong rough ridges on the under side. — Wet clay, or dry sandy soil, Fla. to e. Tex., n. to s. N.J., sw. Tenn., Ark. and Okla. — Tree up to 55 m. high, with light reddish bark breaking into large plates.

6. P. echinàta Mill. (spiny), YELLOW, LONG-TAG or SHORT-LEAF P. — *Leaves* dark green, in 2's or 3's, *slender*, 7–13 cm. long, with long sheaths; *cones* ovoid, 4–6 cm. long, *scales with a small weak prickle.* — Usually dry or sandy soil, n. Fla. to ne. Tex., n. to se. N.Y., N.J., W.Va., s. O., s. Ill., s. Mo. and e. Okla. FIG. 40. — Straight tree up to 40 m. high.

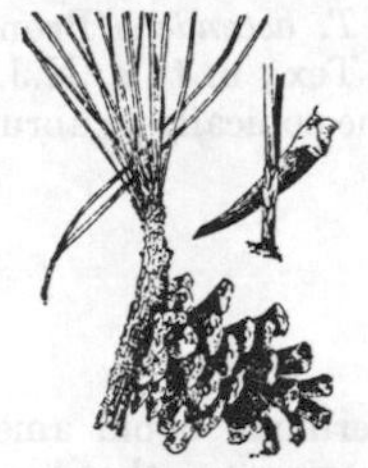
40. P. echinata.

7. P. virginiàna Mill. (Virginian), JERSEY, SPRUCE-, POVERTY- or SCRUB-P. — *Leaves short* (*4–8 cm. long*), in 2's; *cones* slenderly ovoid, 4–6 cm. long, *recurving, the scales tipped with a straight or recurved subulate prickle* 2–3 mm. long. — Barrens and sterile soil, Ga. to Ark., n. to se. N.Y., N.J., W.Va., centr. O. and s. Ind. FIG. 41. — An open-branched or straggling tree up to 12 (rarely 30) m. high, with spreading or drooping branchlets; young shoots with a purplish glaucous bloom.

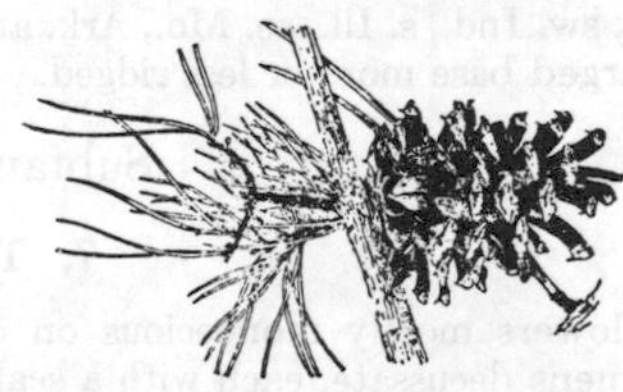
41. P. virginiana.

8. P. rígida Mill. (stiff), PITCH-P. — *Leaves* in 3's, 3.5–14 *cm. long, their sheaths* 4–14 *mm. long;* staminate flowers 1–2 cm. long; *cones* ovoid-conical to ovoid, 3–7 cm. long, often clustered, persistent, *mostly soon dehiscent; umbo of cone-scale prominent, pyramidal, terminated by a sharp prickle.* — Sandy or barren soil, se. Me. to e. Ont., w. N.Y., nw. Pa. and e. O., s. to Va. and mts. of Ga., e. Tenn. and Ky. FIG. 42. — Tree up to 25 m. high, with very rough dark bark and resinous hard wood. — Forma **globòsa** Allard (globular), compact, subglobose, with cones only 4.5–5.5 cm. long and about 5 cm. thick.

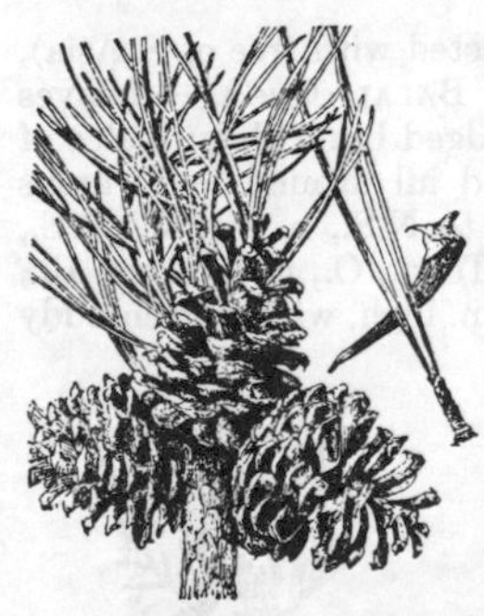
42. P. rigida.

9. P. seròtina Michx. f. (late), POND- or MARSH-P. — Similar to no. 8; *leaves* paler, 1.2–2.8 *dm. long,* tapering to more slender tips, their *sheaths* 1–2.5 *cm. long;* staminate flowers 1.5–3.5 cm. long; *cones indehiscent* or only tardily opening; *umbo of cone-scale depressed or flattish, the prickle delicate and soon deciduous.* — Pond-margins, swamps and sandy woods, Coastal Plain, Fla. and Ala., n. to s. N.J. — Slender tree up to 25 m. high.

10. P. púngens Lamb. (sharp-pointed), TABLE-MOUNTAIN or PRICKLY P. — *Leaves stout, short,* in 2's or 3's (3–7 cm. long), crowded, bluish; the sheath short (very short on old foliage); *cones* broadly ovoid, 5–9 cm. long, persistent, tardily dehiscent, *the scales armed with a strong hooked spine.* — Uplands, N.J., Pa. and W.Va., s. to Ga. and Tenn. — Rather small tree (to 18 m. high).

11. P. Banksiàna Lamb. (in honor of Sir JOSEPH BANKS, 1743–1820), GRAY, JACK- or SCRUB-P., CYPRÈS (Que.). — *Leaves* in 2's, very short and thick (2–4 cm. long), *oblique, divergent; cones* conical, *oblong, usually curved* (3–5 cm. long), smooth, *the scales pointless* or with a minute obsolescent prickle. — Barren, sandy, or rocky soil, n. Que. to Mackenz., s. to N.S., n. N.E., n. N.Y., Mich., n. Ind., n. Ill., Minn. Man., Sask. and Alta. FIG. 43. — Low tree, usually 5–10 (rarely 20) m. high.

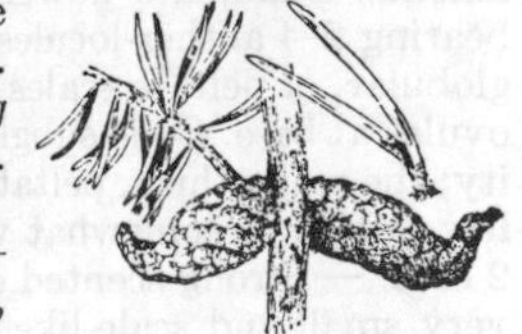
43. P. Banksiana.

Subfam. II. TAXODIOÌDEAE

6. TAXÒDIUM Richard BALD CYPRESS

Flowers monoecious, the two kinds on the same branches. Staminate flowers spiked-panicled, of few stamens; filaments scale-like, peltate, bearing 2–5 anther-locules. Pistillate aments ovoid, in small clusters, scaly, with a pair of ovules at the base of each scale. Cone globular, closed, composed of very thick and angular somewhat peltate scales bearing 2 3-angled or -winged seeds at their bases. Cotyledons 6–9. — Trees of se. U.S. and Mex., with light green deciduous leaves; a part of the slender leafy branchlets of the season also deciduous in autumn; commonly with erect unbranched columnar "knees" produced in areas of frequent flooding. (Name compounded of the Greek *taxos, the yew,* and *eidos, resemblance,* the leaves being yew-like.)

1. T. dístichum (L.) Richard (two-ranked). — Branches nearly horizontal to ascending (rarely drooping); leaves linear to linear-subulate, 0.5–2 cm. long, 2-ranked, at first often appressed or incurving, later widely spreading in 1 plane. (Including *T. ascendens* Brongn.) — Swamps and quiet waters, often overlying calcareous beds, Fla. to Tex., n. to s. N.J., w. Ky., sw. Ind., s. Ill., se. Mo., Ark. and Okla. — Tree up to 50 m. high, the conically to abruptly enlarged base more or less ridged.

Subfam. III. CUPRESSÍNEAE

7. THÙJA L. Arbor Vitae

Flowers mostly monoecious on different branches, in very small terminal ovoid aments. Stamens decussate, each with a scale-like filament or connective and bearing 4 anther-locules. Pistillate aments of few imbricated scales (fixed by the base), each bearing 2 erect ovules, the scales dry and spreading at maturity. Cotyledons 2. — Evergreen trees of N.Am. and e. Asia, with very flat 2-ranked spray, and closely imbricated small appressed persistent leaves; these of two sorts, on different or successive branchlets; one subulate; the other scale-like, blunt, short and adnate to the branch. (*Thyia* or *thya,* the ancient name of some resin-bearing evergreen.)

44. T. occidentalis.

1. T. occidentàlis L. (western; as contrasted with tree of e. Asia), Arbor Vitae, White Cedar, Cèdre or Balai (Que.). — Leaves appressed-imbricated in 4 rows on the 2-edged branchlets; scales of the cones pointless; seeds broadly winged all around. — Swamps and cool rocky banks, e. Que. to Sask. s. to N.S., n. and w. N.E., N.Y., limy areas among mts. to N.C. and Tenn., O., n. Ind., ne. Ill., Wisc. and Minn. Fig. 44. — Tree up to 20 m. high, with pale shreddy bark and light, soft but durable wood. — Forma **prostràta** Vict. & Rousseau (prostrate) making carpets rarely 1 m. high, summits and slopes, St. Lawrence R., Que.

8. CHAMAECÝPARIS Spach White Cedar. Cypress

Flowers monoecious on different branches, in terminal small aments. Staminate flowers composed of peltate scale-like filaments bearing 2–4 anther-locules under the lower margin. Pistillate aments globular, of peltate scales decussate in pairs, bearing few (1–4) erect ovules at base. Cone subglobose, firmly closed, but opening at maturity; the scales thick, peltate, pointed or bossed in the middle, with the few angled or somewhat winged seeds attached at base. Cotyledons 2 or 3. — Strong-scented evergreen trees of N.Am. and e. Asia, with very small and scale-like or some subulate closely appressed imbricated leaves, distichous branchlets and exceedingly durable wood. (From the Greek *chamai, on the ground,* and *cyparissos, cypress.*)

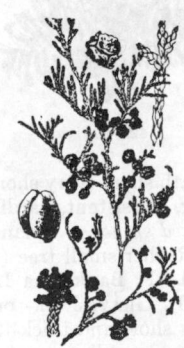

45. C. thyoides.

1. C. thyoìdes (L.) BSP. (like *Thya* or *Thuja*), White Cedar. — Leaves minute, pale, often with a small gland on the back, closely imbricated in 4 rows; cones small (6–9 mm. in diameter), of about 3 pairs of scales; seeds slightly winged. — Swamps, n. Fla. to Miss., n. to s. Me., s.-centr. N.H., centr. Mass. and se. N.Y. Fig. 45. — A tree to 25 m. high.

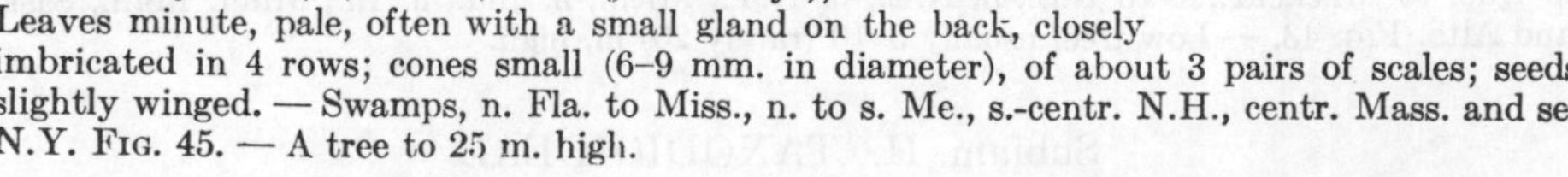

9. JUNÍPERUS L. Juniper. Genévrier (Que.)

Flowers dioecious or occasionally monoecious, in very small lateral aments. Anther-locules 3–6, attached to the lower edge of the peltate scale. Pistillate aments ovoid, of 3–6 fleshy coalescent scales, each 1-ovuled, in fruit forming a sort of berry, which is scaly-bracted underneath, bluish-black with white bloom. Seeds 1–12, ovoid, wingless, bony. Cotyledons 2. — Evergreen trees or shrubs of N. Hemisph. (The Latin name.)

Aments axillary; leaves in whorls of 3, free and jointed at base, linear or linear-subulate, without glands, with 2 white bands above. § 1. OXYCEDRUS.
  Leaves linear, concave above, the white bands confluent and broader than the green margins. . . . . 1. *J. communis.*
Aments terminal; leaves mostly opposite, sometimes subulate and loose, sometimes scale-like, appressed-imbricated and crowded, the scale-like ones usually with a dorsal gland. § 2. SABINA.
  Leaves entire; seeds 3–4.5 mm. long.
    Prostrate shrub, with trailing branches; fruits subglobose to oblate, often lobed, 6–10 mm. broad, on arched-recurving peduncles; seeds 3–5. . . 2. *J. horizontalis.*
    Upright tree (rarely depressed shrub); fruits subglobose to ovoid, 5–6 mm. broad, on straight or straightish ascending peduncles; seeds 1 or 2. . . 3. *J. virginiana.*
  Leaves minutely serrulate (under magnification); seeds 5–5.8 mm. long. . 4. *J. mexicana.*

§ 1. OXYCÉDRUS Endl.

1. J. commùnis L. (in clumps), COMMON J. — Arborescent, 2–12 m. high, with pyramidal or columnar form; leaves thin, straight, long and relatively narrow (12–21 mm. in length, 1.5 mm. broad at the base), widely spreading, grayish above, needle-pointed; fruit subglobose, 5–8 mm. in diameter. — Dry soil, rare or local, s. Me. to Man., s. to Md., mts. of Ga., Ind. and Ill. (Eu.) FIG. 46.

Var. **depréssa** Pursh (depressed), GROUND-J. — *Decumbent, forming large mats* up to 1.5 m. high and often several m. in diameter; *leaves* 8–18 *mm. long, straight or nearly so, sharp-pointed* and with a white stripe above; *fruits* 6–10 *mm. in diameter.* — Poor rocky soil, pastures, etc., Nfld. to Alta., s. to N.S., s. N.E., n. N.J., e. Va., n. O., n. Ind., n. Ill., and Minn., and on mts. to N.C. FIG. 47.

46. J. communis.

47. J. communis, v. depressa.

48. J. communis, v. saxatilis.

49. J. communis, v. megistocarpa.

Var. **saxátilis** Pallas (of rocks). — *Very depressed and trailing; leaves short and relatively broad, curved, subappressed,* 6–9 *mm. long,* 1.6–2 mm. broad, *short-pointed,* with a broad white stripe; fruits 6–9 mm. in diameter; seeds 4–6 mm. long. (Var. *montana* Ait.; *J. nana* Willd.; *J. sibirica* Burgsd.) — Exposed places, Greenl. to Alaska, s. to Nfld., coast of N.S., mts. of Que. and n. Me., Wisc., Wyo. and Calif. (Eurasia) FIG. 48.

Var. **megistocárpa** Fern. & St. John (largest-fruited). — Similar to the preceding; *fruits* 9–13 *mm. in diameter, with sweet, edible pulp;* seeds 5–7 mm. long. — Coasts, w. Nfld., M.I., e. N.B. and e. N.S.; Hudson Bay, Ung. FIG. 49.

§ 2. SABÌNA Endl.

2. J. horizontàlis Moench (lying flat), CREEPING SAVIN, SAVINIER (Que.). — *A procumbent, prostrate, or sometimes creeping shrub;* scale-like leaves acutely cuspidate; *fruit on short recurved peduncles,* 6–10 *mm. in diameter; seeds* 3–5 (commonly 4), often *chestnut-brown* and roughened. — Rocky or sandy banks, mossy bogs, etc., Nfld. to Alaska, s. to coast of N.S., Me. and N.H., and to sw. Vt., nw. N.Y., Mich., n. Ill., Minn., Neb. and Wyo. FIG. 50.

50. J. horizontalis.

3. J. virginiàna L. (Virginian), RED CEDAR or SAVIN, CÈDRE ROUGE (Que.). — Tree up to 30 m. high, when mature with ovoid to pyramidal crown, the lower branches often spreading, the branchlets and finer spray frequently drooping or pendulous; *leaves of adult branchlets tightly appressed, broadly deltoid, obtuse or merely subacute, entire; fruit on straightish erect peduncles,* 5–6 *mm. in diameter; seeds* 1 *or* 2, *their bases with conspicuous deep* *pits.* — Dry, rarely wet, open woods or rocky slopes and barrens, frequently calcareous, Fla. to Tex., n. to se. N.E., L.I., N.J., Ky. and Mo. — Passing into

51. J. virginiana, v. crebra.

Var. **crèbra** Fern. & Grisc. (frequent), NORTHERN RED CEDAR. — Tree usually more columnar or spire-like when mature, with *branchlets rarely drooping* (in wind-swept situations sometimes depressed); *leaves of adult branchlets less appressed, narrowly ovate, acute; seeds only shallowly pitted near base,* the flesh of the fruit often sweetish. — Extending northward on mostly circumneutral to acid soils to s. Me., N.H., sw. Que. and s. Ont. FIG. 51.

4. J. mexicàna Spreng. (Mexican), ONE-SEED J. — *Bushy tree, with several trunks* from one base, forming an *irregular or rounded head* 2–6 m. high; *scale-leaves minutely serrulate,* often with a conspicuous dorsal gland; *fruit* copper-color to dark blue, with a dense bloom, 6–8.5 *mm. long; seeds* 1 (sometimes 2 or 3), 5–5.8 *mm. long.* (*J. Ashei* Buchholz) — Rocky Mts., locally e. in mts. of Tex. and on limestones of the Ozark Mts., Ark. and sw. Mo. — Old bark gray; heart-wood pale brown.

SUBDIVISION II. ANGIOSPÈRMAE (ANGIOSPERMS)

CLASS I. MONOCOTYLEDÒNEAE, MONOCOTYLEDONS

FAM. 14. TYPHÀCEAE (CAT-TAIL FAMILY)

Paludal or aquatic herbs with nerved and linear sessile leaves and monoecious flowers on a spadix, destitute of proper floral envelopes. Ovary 1-locular, with usually persistent style and elongated 1-sided stigma; locule 1-ovuled. Fruit nut-like. Seed suspended, anatropous; embryo straight, in copious albumen. Root perennial.

1. TỲPHA L. CAT-TAIL FLAG. REED-MACE. MASSETTE or QUENOUILLE (Que.)

Flowers in a long and very dense cylindrical spike terminating the stem: the upper part (in the young condition consisting of several bracted confluent spikes) consisting of stamens inserted directly on the axis and intermixed with long hairs; the lower part consisting of stipitate 1-loculed ovaries, their stipes bearing slenderly clavate bristles, which form the copious down of the fruit. Nutlets minute, very long-stalked. — Spathes merely deciduous bracts or none. Rhizomes creeping. Leaves long, sheathing the base of the simple jointless stems, erect, thickish. Flowering in summer. Paludal or subaquatic upright herbs of trop. and temp. reg. (*Typhe,* the old Greek name.)*

a. Staminate and pistillate parts of spike usually contiguous; surface of pistillate (and ripening) portion appearing minutely pebbled or barely bristly under a lens; stigma lance-ovate; pollen-grains in 4's; leaves flat. . . . . . 1. *T. latifolia.*

a. Staminate and pistillate parts of spike usually separated by an interval; surface of pistillate portion appearing minutely bristly or curly; stigma linear or linear-lanceolate; pollen-grains single. . . . *b.*

b. Leaves fewer than 10, convex on the back, full green, the upper overtopping the reddish-brown spike; staminate half 0.7–2 dm. long; surface of pistillate half covered by crowded stigmas without conspicuous bractlets.

Plant 0.75–1.5 m. high; leaves 3–8 mm. wide; pistillate half of spike becoming 6–15 mm. in diameter; stigmas linear, with many dark brown bracts hidden beneath them among the hairs which surround abortive flowers; compound pedicels 0.5–0.7 mm. long. . . . . . 2. *T. angustifolia.*

Plant 2–3.5 m. high; leaves 7–10 mm. wide; pistillate half of spike becoming 1.6–2 cm. thick; stigmas linear-lanceolate, without or rarely with a few bracts hidden beneath them; compound pedicels 0.6–1.2 mm. long. . . . . 3. *T. glauca.*

b. Leaves 10 or more, flat, pale, overtopped by the whitish-brown spike; staminate half 2–4 dm. long; pistillate half covered by stigmas interspersed with ovate blades of bractlets. . . . . 4. *T. domingensis.*

1. T. latifòlia L. (broad-leaved), COMMON CAT-TAIL. — Stout and tall (1–2.7 m. high), the *flat* sheathing *pale or grayish-green leaves* 6–23 mm. broad, exceeding the stem; the *staminate*

* Treatment prepared with aid from Messrs. NEIL HOTCHKISS and HERBERT L. DOZIER.

and dark brown *pistillate parts of the spike usually contiguous*, the former 7–13 cm. long; the latter 2.5–20 cm. long, in fruit 1.2–3.5 cm. thick, its *surface* (under a lens) appearing *minutely pebbled with crowded persistent stigmas* and scarcely bristly; *pistillate flowers without bractlets among the bristles; stigma lance-ovate, fleshy, persistent; pollen-grains in fours;* denuded axis of old spike retaining slender pedicels 1–2 mm. long. — Marshes or shallow water, Nfld. to Alaska, s. through much of the U.S. into Mex. Late May–July. (Eurasia; n. Afr.) — Forma **ambígua** (Sonder) Kronf. (doubtful), with staminate and pistillate halves of spike separated, is often mistaken for the next.

2. T. angustifòlia L. (narrow-leaved). — Stem 0.75–1.5 m. high; *leaves few* (*less than* 10), somewhat *convex on the back, full green,* herbaceous, 3–8 mm. wide; *pistillate and staminate parts of spike usually separated by a short interval; the pistillate portion reddish-brown,* in fruit 6–15 mm. in diameter, 0.3–1.5 dm. long, *its surface minutely bristly with persistent linear stigmas; staminate half of spike* 0.7–2 *dm. long; pollen-grains simple; pistillate flowers with a linear fleshy stigma and usually with a hair-like bractlet with* dilated *blunt tips* among the bristles; denuded old axis covered with stout blunt *compound papillate pedicels* 0.5–0.7 *mm. long.* — Chiefly in basic or alkaline waters, N.S. and s. Me. to s. Que. and Ont., s. to S.C., W.Va., Ky., Mo. and Neb.; Calif. Late May–July. (Eurasia)

3. T. glaûca Godr. (glaucous). — *Differing from no. 2 in greater stature* (2–3.5 *m. high*)*; leaves* 7–10 *mm. broad; pistillate half of spike becoming* 1.6–2 *cm. thick and* 1.8–5 *dm. long; stigmas linear-lanceolate,* less fleshy, more persistent, *without or rarely with bractlets beneath; compound pedicels more bristly,* 0.6–1.2 *mm. long.* (*T. angustifolia*, var. *elongata* (Dudl.) Wieg.; *T. elongata* (Dudl.) Kronf.) — Centr. Me. to s. Ont., w. to Ia. and S.D., s. at low alts. to N.C. and Ala.; Calif. (Guatemala; Eu.)

4. T. domingénsis Pers. (of Santo Domingo). — Stem 2.5–4 m. high; *leaves* 10 *or more, flat, pale,* firm or *coriaceous,* 0.7–1.5 cm. wide, *much overtopped by the inflorescence; staminate half of spike* 2–4 *dm. long,* more or less *separated from the whitish-brown pistillate half;* surface of spike much as in nos. 2 and 3; *stigmas* linear, *interspersed with* many *apiculate-bladed bractlets, soon deciduous; compound pedicels* 0.5–0.8 *mm. long.* (*T. truxillensis* HBK.; *T. angustifolia*, var. *virginica* Tidestr.) — Brackish to fresh marshes and pools, Fla. to Tex. and s. Calif., n. along coast to Del. and e. Md., and inland n. to s.-centr. Kans., Utah, Nev. and n. Calif. June, July. (Trop. Am.)

FAM. 15. SPARGANIÀCEAE (Bur-reed Family)

Paludal or aquatic plants with alternate sessile linear 2-ranked leaves and monoecious flowers in globular sessile or pedunculate heads. Upper heads bearing sessile 3-androus naked flowers and minute scales irregularly interposed. The lower heads consisting of numerous sessile or shortly pedicelled pistillate flowers with a calyx-like perianth of 3–6 linear or spatulate scales. Ovary 1–2-locular. Fruit obovoid or fusiform, 1–2-seeded.

1. SPARGÀNIUM L. Bur-reed. Rubanier (Que.)

Heads scattered along the upper part of the simple or sparingly branched leafy stem, the bracts caducous or the lower persisting and leaf-like. — Perennials of cool and temp. areas of N. Hemisph. and of Austral. and N.Z., with fibrous roots and creeping horizontal rhizomes. Flowering through the summer. The pistillate heads becoming bur-like from the divergent beaks, but the pistils at maturity falling away separately in summer and autumn. (Name ancient, probably from the Greek *sparganion*, derivative of *sparganon, swaddling-band,* in allusion to the ribbon-like leaves.)*

*a.*Stigmas 2; fruits sessile, broadly cuneiform or obpyramidal, 4–8 mm. thick. . . 1. *S. eurycarpum.*

*a.*Stigma 1; fruits about equally narrowed to summit and to the more or less stipitate base, 1.2–3 mm. thick. . . *b.*

*b.*Staminate heads 2 (rarely 1)–20; fruiting heads 1.2–3.5 cm. in diameter; fruits 5.5–14 mm. long, with beak 1.5–6 mm. long. . . *c.*

*c.*Sepals borne chiefly at summit of stipe, half to two-thirds the length of the fruit; pericarp thin, readily removed; beak fragile, straight or curved; stigma linear or lanceolate, 0.6–4 mm. long; anthers 0.8–1.6 mm. long. . . *d.*

*d.*Heads or branches of inflorescence all axillary.

Leaves stiffish; bracts strongly ascending; branches bearing 3–8

* Treatment based primarily on that in Rhodora, xxiv. 26–34 (1922).

staminate and 0 (rarely 1 or 2) pistillate heads; stigma 2–4 mm. long; fruiting heads 2.5–3.5 cm. thick; receptacle fimbrillate-alveolate. . . . . 2. *S. androcladum.*

Leaves soft; bracts spreading or only slightly ascending; branches (when present) with 1–3 pistillate and 1–6 staminate heads; stigma 1–2 mm. long; fruiting heads 1.5–2.5 cm. thick; receptacle scarcely alveolate. . . . . 3. *S. americanum.*

*d.*Heads (or at least 1 of them) supra-axillary.

Staminate half of inflorescence 2–10 cm. long, of 4–9 mostly scattered heads (if shorter and with fewer heads, the plant very low and with erect lower bracts); fruit ribbed at summit between the 3 angles, its beak about equaling the body; tips of sepals appressed to fruit; plants erect and emersed. . . . . 4. *S. chlorocarpum.*

Staminate half of inflorescence 1–3 cm. long, of 1–4 (rarely –6) mostly crowded heads; fruit not strongly ribbed, its beak much shorter than the body; tips of sepals loosely ascending or spreading; leaves long and floating.

Leaves rounded on the back, 1.5–5 mm. wide, the strong nerves of the principal ones (seen on the under side) mostly 0.2–0.8 mm. apart; fruiting heads 1.2–2.2 cm. in diameter. . . . . . . 5. *S. angustifolium.*

Leaves flat, ribbon-like, 5–12 mm. wide, the strong nerves of the principal ones 0.8–2 mm. apart; fruiting heads 2–2.5 cm. in diameter. . . . . 6. *S. multipedunculatum.*

*c.*Sepals chiefly along the middle of the stipe, rarely reaching the middle of the fruit; pericarp closely investing the seed; fruit with a firm gladiate-falcate beak; stigma oblong to lance-ovate, 0.4–0.7 mm. long; anthers 0.4–0.7 mm. long. . . . . 7. *S. fluctuans.*

*b.*Staminate head 1 (very rarely 2); fruiting head 5–12 (–15) mm. in diameter; fruits 3–6.5 mm. long, with beak not exceeding 1.5 mm. or obsolete.

Pistillate heads all axillary; fruit tapering to a conical beak 0.5–1.5 mm. long. . . . . 8. *S. minimum.*

One or more heads supra-axillary.

Fruiting heads 1–3, distinct, 5–12 mm. in diameter; fruits 3.5–4.5 mm. long, rounded to beakless summit. . . . . 9. *S. hyperboreum.*

Fruiting heads 3–5, the upper densely crowded, 1–1.5 cm. in diameter; fruits 5–6.5 mm. long, fusiform, with slenderly conical beak. . . . 10. *S. glomeratum.*

1. S. eurycárpum Engelm. (broad-fruited). — Stout, 0.5–1.5 m. high; *leaves stiffish*, nearly flat, strongly ascending; *inflorescence forked*, branches usually with 1–3 pistillate and 2–20 staminate heads; *carpels* conical above, *with* 2 (rarely 1) *filiform stigmas* 2–3 *mm. long;* fruiting heads 2–3.5 cm. in diameter; *fruits sessile, broadly cuneate or obpyramidal*, often 2-seeded, 6–10 mm. long, 4–8 *mm. broad* at the abruptly beaked summit; receptacle shallowly alveolate, commonly bearing narrowly spatulate chaff. — Shallow water, chiefly in argillaceous or basic soils, M.I. to n. Alta. and s. B.C., s. to (?) Fla., O., Ind., Ill., Mo., Kans., Colo., Utah and Calif. FIG. 52.

52. S. eurycarpum.

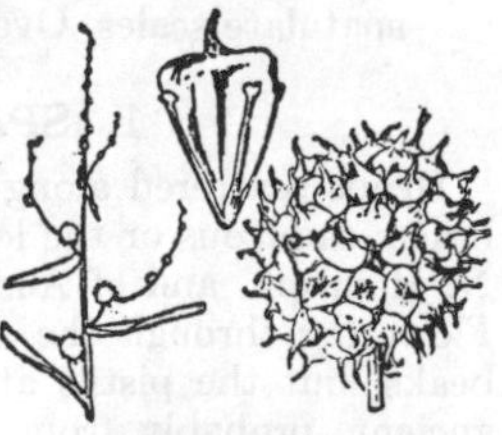

2. S. andrócladum (Engelm.) Morong (with staminate branches). — Stoutish, up to 1.2 m. high; *leaves stiffish*, very elongate, *strongly ascending, keeled*, 4–15 mm. wide; lower bracts similar, slightly scarious-margined at base; inflorescence simple or branched, the primary axis with 1–4 mostly sessile *axillary pistillate heads* and 4–10 staminate ones, the 1–3 filiform strongly arched geniculate *branches with* 3–8 *staminate heads and* rarely 1 *pistillate; filiform stigma* 2–4 *mm. long; fruiting heads* 2.5–3.5 *cm. in diameter; fruits lustrous*, 2.5–3 *mm. thick, the beak* 4.5–6 *mm. long; receptacle fimbrillate-alveolate;* anthers 1–1.6 mm. long. (*S. lucidum* Fern. & Eames) — Muddy or peaty shores, swamps or shallow water, Quebec Co., Que., to Minn., s. to s. Va., e. Ky., Ill., Mo. and Okla. FIG. 53.

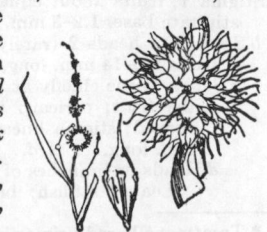

53. S. androcladum.

3. S. americánum Nutt. (American). — Stout to slender, up to 1 m. high; *leaves soft, thin, flat, translucent,* loosely ascending (sometimes floating), 4–20 mm. broad; lower *bract* similar, *spreading-ascending,* scarious-margined at base; inflorescence simple or branched, *heads or branches axillary,* the primary axis with 1–5 pistillate heads and 5–9 staminate ones, the *branches* (when present) *with* 1–6 *staminate heads and* 1–3 (rarely 0) *pistillate ones; stigma linear-oblong to lanceolate, 1–2 mm. long; fruiting heads* 1.5–2.5 *cm. in diameter; fruits opaque or but slightly lustrous, 2 mm. thick, the beak* 1.5–5 *mm. long; receptacle scarcely alveolate;* anthers 0.8–1.2 mm. long. (Incl. var. *androcladum* of ed. 7) — Muddy or peaty shores and shallow water, Nfld. to Muskoka Distr., Ont., Wisc., Minn. and N.D., s. to Fla., Ala. and Mo.; B.C. Fig. 54.

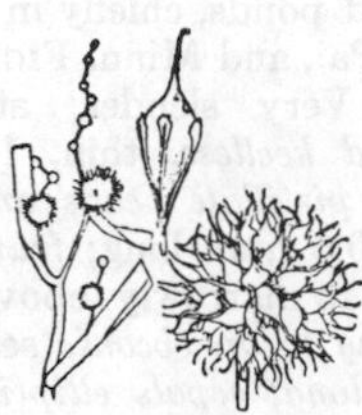
54. S. americanum.

4. S. chlorocárpum Rydb. (green-fruited). — Slender, up to 8.5 dm. high. *erect* (rarely submersed and flexuous); *leaves* erect or strongly ascending (rarely floating), *flat or slightly keeled, little if at all dilated at base* (except for the scarious margin), 2–12 mm. wide; the 3 or 4 ascending *bracts with broadly scarious-margined base;* inflorescence usually simple; 1 *or more* of the 1–4 remote or subremote *pistillate heads supra-axillary,* the lowest (sessile or short-peduncled) borne 1–6.5 dm. above the base of the stem; *staminate heads* 4–9, *mostly distant;* stigma lance-attenuate or linear-subulate, 0.8–1.7 mm. long; fruiting heads 1.5–2.7 cm. in diameter; *fruits* slightly lustrous, *ribbed at summit between the angles,* 1.2–2 mm. thick, the beak 2–4.3 mm. long; *sepals closely appressed-ascending,* cuneate-spatulate; anthers 0.8–1.2 mm. long. (*S. diversifolium* of Am. auth., not Graebn.) — Muddy or peaty soil or shallow water, Nfld. to Ung. and Algonquin Park, Ont., s. to N.S., N.E., n. N.J., Pa., Ind. and Ia. Fig. 55. — Passing into

55. S. chlorocarpum.

Var. **acaúle** (Beeby) Fern. (stemless). — *Pistillate heads* 1–3, *at least the upper approximate, in maturity* 1.2–2.2 *cm. in diameter, the lowest* 1–18 *cm. above the base of the plant; staminate heads* 2–5. (*S. acaule* Rydb.) — Similar range, but s. to uplands of Va., W.Va. and S.D.

5. S. angustifólium Michx. (narrow-leaved). — Slender *aquatic; the stems usually submersed,* up to 1.2 m. long; *leaves very elongate,* 1.5–5 *mm. wide, convex on the cellular-reticulate back and with the nerves only* 0.2–0.8 *mm. apart, opaque above;* middle and upper leaves and lower bracts *with dilated and subinflated base;* inflorescence usually simple; 1 or more of the 1–3 pistillate *heads supra-axillary,* all sessile or the lowest on peduncles up to 8.5 cm. long; *staminate heads* 1–4 (rarely –6), *mostly approximate; stigma* lance-attenuate, 0.6–1.5 *mm. long; fruiting heads* 1.2–2 *cm. in diameter;* fruits often reddish at base, 1.2–1.7 mm. thick, the ellipsoid body commonly constricted; the *beak* usually falcate, *about 2 mm. long; sepals loosely ascending, with narrow claw* and dilated often spreading blade; anthers 0.8–1.2 mm .long. (*S. affine* Schnizl.) — Deep or shallow water or wet shores, Lab. to Alaska, s. to Nfld., N.S., N.E., mts. of N.J. and Pa., Mich., n. Ill., Minn., Colo. and Calif. (Eurasia) Fig. 56.

56. S. angustifolium.

6. S. multipedunculátum (Morong) Rydb. (many-peduncled). — Similar to no 5. but coarser; *leaves flat,* ribbon-like, *scarcely dilated or inflated at base,* 5–12 *mm. wide; the strong nerves of the principal ones* (*beneath*) 0.8–2 *mm. apart; pistillate heads* 1–5, *in maturity* 2–2.5 *cm. in diameter,* sessile or the lower on peduncles up to 12 cm. long; stigma 1–1.8 mm. long. (*S. simplex* of ed. 7, not Huds.) — Lakes, ponds and pools, Nfld. and s. Lab. Pen. to Alaska, s. to e. N.S., n. N.E., s. Ont., Man., Sask., Colo. and Calif. Fig. 57.

57. S. multipedunculatum.

7. S. flúctuans (Morong) Robins. (fluctuating). — *Aquatic,* stems up to 1.5 m. long; *leaves flat and ribbon-like, thin, loosely cellular-reticulate,* 3–11 mm. broad, the middle and upper ones and the 2 foliaceous *obtuse bracts* dilated at base; *inflorescence branching, the main axis with* 0–3 *pistillate heads* and 4–6 staminate ones, the 1–3 arched-ascending or incurved *axillary branches with* 1 or 2 *sessile pistillate heads* and usually 1–4 staminate ones; *stigma* oblique, oblong or lance-ovate, 0.4–0.7 *mm. long;* fruiting heads 1.3–2.3 cm. in diameter; *fruit dark brown, opaque, of firm texture, with the pericarp closely investing the seed,* 2–2.2 mm. thick, tapering to a *strong gladiate-*

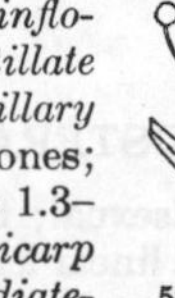
58. S. fluctuans.

falcate beak 2–3 mm. long; *sepals borne chiefly along the middle of the stipe, rarely reaching the middle of the fruit; anthers* ellipsoid, 0.4–0.7 *mm. long.* — Cold lakes and ponds, chiefly in siliceous regions, Nfld. to n. Que. and n. Alta., s. to N.S., N.E., mts. of n. Pa., and Minn. Fig. 58.

59. S. minimum.

8. S. mínimum (Hartm.) Fries (smallest). — Very slender; stems floating or suberect, 1–8 dm. long; *leaves flat and keelless,* thin, 1.5–8 mm. wide; inflorescence simple, *with 1–4 axillary pistillate heads and a remote staminate one; stigma ovate to lanceolate,* 0.3–0.6 mm. long; *fruiting heads 8–12 mm. in diameter; fruits* pale, ellipsoid to narrowly obovoid-fusiform, 3–3.5 *mm. long,* 1.8–2.5 *mm. thick, tapering to an obconic* sessile or substipitate *base and a conic beak* 0.5–1.5 *mm. long; sepals elliptic or cuneate-spatulate, one-half to two-thirds as long as the body of the fruit.* — Shallow pools, brooks, springs, etc., Nfld. to Alaska, s. to N.S., N.E., n. N.J., n. Pa., Mich., Wisc., Minn., Colo., Utah and Oreg. (Eurasia) Fig. 59.

9. S. hyperbòreum Laestad. (high-northern). — Resembling no. 8, 0.5–3 dm. long; leaves rather thick and opaque, 1–5 mm. wide; 1 *or more* of the 1–3 *pistillate heads supra-axillary;* staminate head 1 (rarely 2); *stigma nipple-shaped or oblong, sessile,* 0.2–0.5 *mm. long;* fruiting heads 5–12 mm. in diameter; *fruits* ellipsoid to obovoid, 3.5–4.5 *mm. long,* 1.6–2 *mm. thick, tapering or rounded to the beakless summit* and narrowed to a slender stipe; *sepals wanting or slenderly linear-spatulate,* and rarely reaching half the length of the fruit. — Greenl. and Lab. to Alaska, s. in peaty pools to Nfld., C.B., Anticosti I., nw. Que. and n. Man. (Eurasia) Fig. 60.

60. S. hyperboreum.

10. S. glomeràtum Laestad. (densely clustered). — Stem relatively stout, up to 6 dm. long; leaves normally floating, 0.3–2 cm. broad, thickish, with dilated sheaths; *inflorescence crowded, all but the lowest head usually overlapping;* the 1 or 2 staminate heads approximate to the 3–5 pistillate; *fruiting heads 1–1.5 cm. in diameter; fruits* pale, fusiform, 5–6.5 *mm. long, the slenderly conical beak about equaling the stipe;* sepals thin, linear, their tips extending above middle of carpel. — Shallow pools, s. Saguenay Co., Que., and n. Minn. (Eurasia)

FAM. 16. ZOSTERÀCEAE (Pondweed Family)

Aquatic herbs, mostly immersed or with floating upper leaves, with stems jointed and leafy; leaves sheathing at base or stipulate; flowers perfect or unisexual and often spathaceous, mostly in spikes or clustered; perianth none but in Potamogeton *the 4 anthers bearing from their connectives 4 valvate sepal-like outgrowths.* Stamens 1, 2 or 4, with extrorse anthers. Carpels 1–4 (rarely more), distinct, 1-locular, 1-ovuled, ripening into nutlets or dryish drupelets. Endosperm wanting. (Potamogetonaceae; Najadaceae of ed. 7 in large part)

Flowers in spikes or on an elongate axis.
- Spike completely sheathed in the leaf-like spathe, the unisexual flowers crowded on a flattened axis; stigmas 2, bristle-like; pollen filiform. Tribe I. Zostereae.
  - Stamens and pistils in 2 alternating rows on the axis. . . . . 1. *Zostera.*
- Spike when undeveloped covered by the stipule, in maturity exserted, simple; carpels 4; stigma 1, sessile; pollen globose or oblong and semi-lunate. Tribe II. Potamogetoneae.
  - Spike usually more than 2-flowered, in anthesis mostly raised above the stipule; anthers 4, sessile, each with a sepal-like outgrowth from the connective; carpels remaining sessile; pollen globose. . . . . 2. *Potamogeton.*
  - Spike 2-flowered, inclosed during anthesis in the sheathing leaf-base; anthers 2, without sepal-like outgrowths; carpels in fruit raised on slender stalks; pollen oblong, curved. . . . . 3. *Ruppia.*

Flowers axillary, unisexual, the pistillate borne in a small spathe, the staminate with a single stamen; stigma peltate or stelliform; pollen globose. Tribe III. Zannichellieae.
- Staminate flower naked, beside the group of 2–8 (mostly 4) pistillate ones; spathe of pistillate inflorescence cup-like. . . . . 4. *Zannichellia.*

Tribe I. Zostèreae Dumort.

1. ZOSTÈRA L. Grass-wrack. Eelgrass. Mousse de mer (Que.)

Flowers unisexual; the two kinds naked, sessile and alternate in two rows on the midrib of one side of a linear leaf-like spadix, which is hidden in a long sheath-like base of a leaf (spathe);

staminate flowers of a single ovate or oval 1-locular sessile anther, as large as the carpels, with filiform pollen; pistillate flowers of single ovate-oblong carpels attached near the apex, tapering to a subulate style, with a pendulous orthotropous ovule; stigmas 2, long and bristleform, deciduous. Pericarp bursting irregularly, inclosing a cylindric seed. Embryo short and thick (proper cotyledon almost obsolete), with an open chink or cleft its whole length, from which protrudes a doubly curved slender plumule. — Grass-like marine herbs of cooler halves of N. and S. Hemisph., growing wholly under high-water, from a creeping rhizome, the joints of the stem sheathed by bases of obtuse entire ribbon-like leaves (whence the name, from the Greek *zoster, a belt.*)

1. Z. marìna L. (of the sea). — Sterile simple stems with leaves (up to 1 m. long) broader than on the branching fertile ones; seed barrel-shaped, longitudinally ribbed, 3–4 mm. long. — The typical Eurasian plant has subtransparent leaves up to 12 mm. broad, mostly with 5–7 strong nerves besides the more numerous finer ones. Our plant is

Var. **stenophýlla** Aschers. & Graebn. (slender-leaved). — *Leaves* 1.5–6 *mm. wide, coriaceous, opaque, with* 3 (*rarely* 5) *strong nerves.* — Shallow sea-water, s. Greenl. and se. Lab. to N.C. (perhaps beyond); James Bay; Pacific N. Am. (Eurasia) — Better specimens and critical study may prove it a distinct species, *Z. stenophylla* Raf.

Tribe II. Potamogetòneae Reichenb.

2. POTAMOGÈTON L. Pondweed. Potamot or Herbes à Brochets (Que.)

Stamens 4, anthers 2-locular, the connectives bearing 4 valvate sepal-like outgrowths. Carpels 4 (rarely only 1), with an ascending campylotropous ovule; stigma sessile or on a short style. Fruit drupe-like when fresh, more or less compressed; endocarp (*seed*) crustaceous. Embryo hooked, annular or cochleate, the radicular end pointing downward. — Herbs of ponds and streams (a few in sea-margin), found in most regions, with jointed mostly rooting stems and 2-ranked leaves, which are usually alternate or imperfectly opposite; the submersed ones often pellucid, the floating ones often dilated and of a firmer texture. Stipules sheathing. Spikes sheathed by the stipules in the bud, mostly raised on a peduncle to the surface of the water. (An ancient name, composed of the Greek *potamos, a river,* and *geiton, a neighbor,* from the place of growth.) — By *fruit* the full-grown fresh or macerated fruit is intended; by *seed,* that with the dryish outer portion or epicarp removed. All measurements are from dried specimens. The month mentioned indicates the time of ripening of the fruit.*

a. Leaves all setaceous or linear-filiform, channeled, septate or divided their whole length by straight cross-partitions, with an elongate firm sheath consisting of the leaf-base and the adnate stipule; peduncles flexuous, filiform; spikes flexuous, moniliform; flowers in whorls; fruits plump, sessile. Subgen. I. Coleogeton. . . *b.*

b. Leaves retuse, blunt or merely short-apiculate; style almost wanting; beak of fruit wart-like and nearly central.

Sheaths tightly clasping, only slightly thicker than the stem, 0.4–2.2 cm. long, their margins connate below; spike with 2–5 unequally distant whorls (or upper whorls approximate), in maturity 0.5–5 cm. long; fruits 2–2.6 mm. long. . . . . 1. *P. filiformis.*

Sheaths (at least of the primary leaves) loose, subinflated, much thicker than the stem, 2–5 cm. long, open to base; spike with 5–12 nearly equidistant whorls, in maturity 3–8 cm. long; fruits 3–3.5 mm. long. . . 2. *P. vaginatus.*

b. Leaves of the branches sharply acute (the broader primary ones often obtuse); style slightly developed, in fruit forming a short recurving beak continuing the ventral margin. . . . . . . 3. *P. pectinatus.*

a. Leaves linear or terete to ovate or suborbicular, not regularly septate (though often with localized lacunae); peduncles and spikes stiff, the latter continuous or only slightly interrupted; stipules not adnate to leaf-base (except in no. 4 and slightly in nos. 20–23); fruits laterally compressed, often substipitate; wind-pollinated. Subgen. II. Eupotamogeton. . . *c.*

c. Leaf-blade auricled at base, stiff, linear to lanceolate, 20–60-nerved, with thin minutely serrulate cartilaginous margin, adnate to the lower half of the coarsely fibrous stipule (§ Adnati). . . . . 4. *P. Robbinsii.*

c. Leaf-blade not cartilaginous-margined, free from the stipule or, if adnate, delicate and very slender (§ Axillares). . . *d.*

---

* Treatments of species nos. 1–3 (Subgen. *Coleogeton*) derived largely from treatment of Harold St. John in Rhodora, xviii. 121–138 (1916); of species 6–25 from that of the writer in Mem. Am. Acad. Arts and Sci. xvii. pt. 1 (1932); of species 26–37 from the monograph of E. C. Ogden in Rhodora, xlv. 57–105, 119–163 and 171–214 (1943).

*d.*Leaves sharply serrulate, uniform, oblong to broadly linear; fruits 5–6 mm. long, subsagittate at base, the long-acuminate beak longer than the body; winter-buds bur-like, with strongly divergent indurated leaves 5 mm. or more broad and concave at base (Sub-§ CRISPI). . 5. *P. crispus.*

*d.*Leaves entire or merely undulate; fruits shorter (if rarely as long not subsagittate and the plant with dilated leaves), short-beaked or beakless; winter-buds, when present, with more ascending and narrower scarcely concave leaves. . . *e.*

*e.*Submersed leaves linear to linear-setaceous, 0.1–5 mm. broad, uniform; dilated upper leaves, if present, at most 4 cm. long and 2 cm. broad and with only 3–15 nerves and the plants with spikes at most 2.2 cm. long. . *f.*

*f.*Stipules all free from leaf-bases; spikes uniform, peduncled; fruits only slightly compressed, beaked, 1.6–5 mm. long; winter-buds frequently developed. . . *g.*

*g.*Leaves uniform, all linear to setaceous and submerged, the upper or involucral ones often with more developed lacunar system but not conspicuously broader. . *h.*

*h.*Stem flaccid, repeatedly forked; leaves flaccid, linear-setaceous, 0.1–0.5 mm. wide; stipules essentially nerveless, at most 12 mm. long; peduncle terminal, usually only 1, straight and erect, up to 2.4 dm. long; spike dense, with roundish light green fruits (Sub-§ MONTICOLI). . . . . 6. *P. confervoides.*

*h.*Stem firm; leaves firmer, mostly broader; stipules nerved, 0.3–3.5 cm. long; peduncles mostly in the upper forks, at most 9 cm. long; fruits obliquely obovoid to ovoid or quadrate, dark green, olivaceous or fulvous. *i.*

*i.*Stems strongly flattened, 0.7–3 mm. broad; principal leaves 2–5 mm. broad, with many (up to 35) nerves; peduncles stout, 1–1.6 mm. thick; spikes 1.5–3 cm. long, with 7–11 verticils; fruit quadrate, 4–5 mm. long (Sub-§ COMPRESSI). 7. *P. zosteriformis.*

*i.*Stems filiform or, if flattened, slender; leaves 1–9-nerved; peduncles slender; spikes 2–15 mm. long, with 1–5 verticils; fruits smaller or not quadrate (Sub-§ PUSILLI). *j.*

*j.*Stipules connate, forming cylinders with margins united two-thirds their length or more, in age rupturing and often shredded at tip (Series *Pusilli connati*). *k.*

*k.*Rootstock filiform, freely branching; leaves glandless and somewhat tapering at base; peduncles clavate, 3–10 (rarely –30) mm. long; spikes subcapitate, 2–5 mm. long; sepaloid connectives 0.6–1 mm. long; fruits with a thin or acute undulate to dentate dorsal keel. . . 8. *P. foliosus.*

*k.*Rootstock not elongate; leaves often with a pair of basal glands, scarcely tapering at base; peduncles filiform or with slightly broadened tip, if strongly clavate also flattened, 1–9 cm. long; spikes elongate, interrupted, 6–15 mm. long; sepaloid connectives 1.2–2.5 mm. long; fruits rounded on the back. . . *l.*

*l.*Stipules strongly fibrous, becoming whitish; peduncles broadened at tip; winter-buds covered by strongly fibrous white stipules.

Leaves thin and translucent, 5–7-nerved, 1.5–3.5 mm. broad, rounded at tip; stipules 7–11 mm. long; peduncles flattened, 1.5–5 cm. long; winter-buds on short lateral branches, their loosely ascending outer leaves not corrugated at base. . 9. *P. Friesii.*

Leaves firm, often revolute, 3 (rarely 5) -nerved, 0.5–2 (rarely –2.5) mm. broad, obtuse to sharply attenuate; stipules 0.8–2 cm. long; peduncles filiform-clavate, 1–9 cm. long; winter-buds terminating both long and short branches, their rigidly ascending outer leaves dilated and corrugated at base. . . . . 10. *P. strictifolius.*

*l.*Stipules scario-membranaceous, olivaceous; leaves 3-nerved; peduncles filiform, not enlarged at tip; winter-buds with scarious but not strongly fibrous stipules. . . . . 11. *P. pusillus.*

j. Stipules flat or convolute, the margins often inrolled but not connate (Series *Pusilli convoluti*). . . *m.*

m. Leaves 5–9-nerved, subrigid, attenuate to bristle-tips, the midrib not bordered by lacunae; stipules coarsely fibrous, 1.3–3 cm. long. . . . . 12. *P. longiligulatus.*

m. Leaves 1–3 (rarely –5) -nerved (lateral nerves obscure in slenderest plants), midrib bordered by lacunae; stipules delicate or, if coarsely fibrous, only 0.8–1.8 mm. long. . . *n.*

n. Leaves tapering to bristle-tips; stipules attenuate or sharply pointed.

Leaves linear-setaceous, 0.2–0.4 mm. wide, 1 (or obscurely 3) -nerved, with a single row of lacunae on each side; stipules subherbaceous, linear-attenuate; peduncles filiform, 1–3.5 cm. long; spike elongate, remotely flowered. . . . . 13. *P. gemmiparus.*

Leaves linear, 1–2.2 mm. wide, 3-nerved, midrib bordered on each side at base by 2 or more rows of lacunae; stipules coarsely fibrous, whitish; peduncles clavate, 0.5–1.5 cm. long; spike capitate, with crowded flowers. . . . . 14. *P. Hillii.*

n. Leaves obtuse to acute or mucronate, not bristle-tipped; stipules obtuse. *o.*

o. Plants fulvous; leaves 2–4 mm. broad; lateral nerves joining midrib at tip; fruits 3–4 mm. long, 2–2.6 mm. broad; bodies of winter-buds 1.5–4 cm. long, 2.5–7 mm. broad.

Stipules rather rigid, fibrous, finally fimbriate at apex; basal glands wanting or only 0.2–0.4 mm. broad; spike subcapitate, up to 7 mm. long . 15. *P. Porteri.*

Stipules delicately membranaceous; basal glands 0.6–1.2 mm. broad; spikes thick-cylindric, in fruit 0.6–1.3 cm. long. . . . . 16. *P. obtusifolius.*

o. Plant green (very rarely fulvous); leaves 0.3–2.4 mm. broad, lateral veins joining midrib well below tip or evanescent; glands at base of stipule minute; fruits 2–2.5 mm. long, 1.2–1.9 mm. broad; bodies of winter-buds 7–18 mm. long, 0.6–2.5 mm. broad. 17. *P. Berchtoldi.*

g. Leaves of 2 sorts, the submerged narrowly linear to linear-filiform; the floating ones dilated and coriaceous, 2–8 mm. wide; plants dimorphic, some with, others of the same colony without, floating leaves (Sub-§ Javanici).

Fruiting plants with floating leaves; sterile plants with only linear-filiform bristle-tipped leaves 0.1–0.5 mm. broad, and abundant small winter-buds; dilated leaves 3–8 mm. broad, on marginless petioles; fruits 1.6–2.2 mm. long, distinctly keeled on back. . . . . 18. *P. Vaseyi.*

Fruiting plants with only linear acutish leaves 0.4–1 mm. wide; sterile (but flowering) plants with floating leaves 2–4 mm. broad and tapering to margined petioles; fruit 2–2.4 mm. long, rounded on the back. . . . . 19. *P. lateralis.*

f. Stipules of some or all of submersed leaves adnate to base of leaf; inflorescences of 2 or 3 sorts, those in axils of linear leaves subglobose or few-flowered, those from axils of dilated leaves elongate; fruits suborbicular, strongly flattened, cochleate, beak wanting or a mere tooth; spiral curve of embryo clearly evident through the pericarp (Sub-§ Hybridi). . . *p.*

p. Submersed leaves linear, obtuse or acute, not bristle-tipped, 0.5–2 mm. wide; dilated leaves rounded at tip, the larger 5–15-nerved; lowest (subglobose) spikes subsessile or on ascending or arching peduncles only 1–4 mm. long.

Submersed leaves obtuse, usually rounded at tip; the connate sheath much longer than the free stipular tip; space between midrib and margin filled by lacunae; dilated leaves minutely emarginate; fruits 1.3–2.2 mm. in diameter, lateral keels scarcely developed, sides rounded. . . . . 20. *P. Spirillus.*

Submersed leaves subobtuse to acute, rarely lacunate; connate sheath half as long as free tip; dilated leaves round-tipped, not emarginate; fruits 1–1.5 mm. in diameter, lateral keels fine but definite, sides flat or depressed. . . . 21. *P. diversifolius.*

p. Submersed leaves setaceous, with bristle-tips, 0.1–0.6 mm. wide; dilated leaves acute or mucronulate, 3–7-nerved; lowest spikes on divergent or recurving peduncles 1.5–13 mm. long.

Fruits green or olivaceous, 1–1.5 mm. broad, the very narrow dorsal keel entire or minutely dentate; lateral keels very low, toothless or minutely toothed, sides flattish. . . . . . 22. *P. capillaceus.*

Fruits stramineous, 1.6–2.2 mm. broad, the broad and wing-like dorsal keel coarsely dentate; lateral keels wing-like, sinuate or coarsely dentate, sides broadly crateriform. . . . . . 23. *P. bicupulatus.*

e. Submersed leaves more broadly linear or greatly elongate to ovate or suborbicular, (0.1–) 0.5–7.5 cm. broad (if slenderly linear to subterete then with firm and lustrous floating leaves 2–12 cm. long and 1–6 cm. broad and fruiting spikes 1–5 cm. long and 7–12 mm. thick). . . *q.*

q. Submersed leaves linear or linear-filiform, 0.2–10 mm. broad; the median band each side of midrib largely composed of reticulate lacunae; fruit strongly laterally flattened, 3-keeled when dry, nearly orbicular, with short tooth-like submarginal beak; embryo subspiral, its tip strongly incurved above the base (Sub-§ NUTTALLIANI).

Stem compressed; submersed leaves ribbon-like, often strongly distichous, 1–10 mm. broad, 5–13-nerved, free from the ligule; dilated or floating leaves mostly opposite, 2–8 cm. long; fruits with depressed sides. . . . . 24. *P. epihydrus.*

Stem terete; submersed leaves linear-filiform, 0.2–1 mm. wide, adnate at base to ligule, 1- or obscurely 3-nerved; dilated leaves scattered, 2–4 cm. long; fruits with low-convex sides. . . . . 25. *P. tennesseensis.*

q. Stem terete; submersed leaves not distichous, linear-lanceolate to ovate or suborbicular, less lacunate; fruit with rounded sides, longer than broad, often more strongly beaked, the embryo making a partial ellipse. . *r.*

r. Submersed leaves petioled or tapering to sessile base, not strongly clasping; floating leaves often developed. . *s.*

s. Submersed leaves sessile, reddish or fulvous-olivaceous; floating leaves (when developed) delicate, translucent, tapering without sharp distinction into petiole; exocarp of fruit hard and smooth, tawny-olive (Sub-§ ALPINI). . . . . 26. *P. alpinus.*

s. Submersed leaves sessile or petioled; floating leaves coriaceous, dense or opaque, with sharp distinction between petiole and cordate- to cuneate-based blade; exocarp of fruit soft and porous, greenish, brownish or reddish. . . *t.*

t. Fruiting spike 5–7 mm. thick; fruit beakless or nearly so, 1.6–2.5 mm. long, with keel rounded or obsolete; endocarp-loop (within the fruit) usually with a central cavity; sepaloid connectives 1–2 mm. wide; floating or emersed leaves with rounded to subcordate bases; plant of Nfld. and Sable Id. (Sub-§ COLORATI). . . . . 27. *P. oblongus.*

t. Fruiting spike 6–15 mm. thick; fruit definitely beaked, mostly 2–5 mm. long; endocarp-loop solid; sepaloid connectives 1.2–3 mm. wide. . . *u.*

u. Submersed leaves broadly linear to ovate, less than thirty times as long as broad, with 3–37 definite nerves. . . *v.*

v. Margins of submersed leaves strictly entire, the nerves 11–37; floating leaves mostly with 21–51 nerves (Sub-§ AMPLIFOLII).

Stem not conspicuously dark-spotted; submersed leaves arcuate, 2.5–7.5 cm. broad; floating leaves cuneate to rounded at base, mostly with 30–51 nerves, about a quarter of the nerves more prominent than the others (seen by transmitted light); fruit cuneate at base, 3.5–5 mm. long. . . . . 28. *P. amplifolius.*

Stem usually with conspicuous blackish spots; submersed leaves not arcuate, 1–3.5 cm. broad; floating leaves cordate or broadly rounded at base, 21–35-nerved, the nerves essentially uniform; fruit rounded or lobed at base, 3–3.5 (–4) mm. long. . . . . . . 29. *P. pulcher.*

v. Margins of submersed leaves with fugacious 1-celled translucent denticles, the nerves 3–29; floating leaves with 9–21 (–29) nerves, tapering to rounded at base.

Submersed leaves with petioles 2–13 cm. long, acutish but neither sharp-pointed nor mucronate; fruits 3.5–4.3 mm. long, usually reddish, their keels usually muricate (Sub-§ Nodosi). . . . . 30. *P. nodosus.*

Submersed leaves sessile or on petioles rarely 4 cm. long, sharp-pointed or often mucronate; fruits 1.7–3.5 mm. long, mostly greenish to drab, with scarcely muricate keels (Sub-§ Lucentes).

Stem usually repeatedly branched, 0.5–1 mm. thick; submersed leaves 0.1–1(–1.5) cm. wide, five to thirty times as long as broad, sessile, their nerves mostly 3–9; floating leaves 1.5–5(–7) cm. long and 1–2(–3) cm. wide, their petioles mostly longer than the blades; stipules 0.5–3 cm. long, only faintly keeled; fruiting spikes 1–2.8 cm. long; sepaloid connectives usually 1.2–1.6 mm. broad; fruits 1.7–2.8 mm. broad. . . . . 31. *P. gramineus.*

Stem simple or but once branched, 1–5 mm. thick; submersed leaves 1.5–4.5 cm. wide, three to five times as long as broad, sessile or petiolate, their nerves mostly 9–17; floating leaves 4–19 cm. long and 2–6.5 cm. wide, their petioles mostly shorter than the blades; stipules 1–8 cm. long, prominently keeled; fruiting spikes 2.5–7 cm. long; sepaloid connectives 1.6–3 mm. broad; fruits 2.1–3 mm. wide. . . . . 32. *P. illinoensis.*

u. Submersed leaves slenderly linear to subterete, 0.3–2 mm. wide, fifty or more times as long as broad, with 3–5 obscure nerves (Sub-§ Natantes).

Stem 0.8–2 mm. thick; submersed leaves 0.8–2 mm. broad; blade of floating leaf 4–12 cm. long and 2.5–6.5 cm. wide, usually cordate, on petioles 1–2.5 mm. thick; stipules strongly keeled to tip, 4.5–11 cm. long; fruiting spike 9–12 mm. thick; mature fruits 2.5–3.5 mm. broad, scarcely keeled. . . . . 33. *P. natans.*

Stem 0.5–1 mm. thick; submersed leaves 0.3–1 mm. wide; blade of floating leaf 1.5–5.5 cm. long and 1–3 cm. wide, rounded to tapering at base, on petioles 0.2–1 mm. thick; stipules strongly keeled only at base, 1–5.5 cm. long; fruiting spike 7–9 mm. thick; mature fruits 1.6–2.4 mm. broad, prominently keeled. . . . . . 34. *P. Oakesianus.*

r. Submersed and upper leaves all cordate or rounded at base, clasping one-half to two-thirds the circumference of stem; no specialized floating leaves.

Rhizomes with rusty red spots; leaves mostly 1–2.5 dm. long, entire, often cucullate at apex; stipules conspicuous, rigid, 3–10 cm. long, usually persistent; peduncles 0.5–6 dm. long; bodies of fruits 4–5 mm. long and 3.2–4 mm. wide, with strong dorsal keel (Sub-§ Praelongi). . . . . 35. *P. praelongus.*

Rhizomes unspotted; leaves 1–10 cm. long, with marginal fugacious 1-celled denticles, flat-tipped; stipules 0.4–2 cm. long, delicate and fugacious or, if firm, soon disintegrating into fibers; peduncles 0.1–2.5 dm. long; bodies of fruits 2.3–3.5 mm. long and 1.7–3 mm. broad, keelless or only weakly keeled (Sub-§ Perfoliati).

Leaves narrowly lanceolate to lance-ovate, 1.5–10 cm. long, coarsely nerved; stipules coarse, soon disintegrating into strong white fibers; peduncles usually enlarged upward, 0.15–2.5 dm. long; fruits 2–3 mm. broad. . . . . . . 36. *P. Richardsonii.*

Leaves orbicular to ovate-lanceolate, the principal ones 1–6 cm. long, delicately nerved; stipules delicate, fugacious; peduncles slender to summit, 1–9 cm. long; fruits 2–2.3 mm. broad. . . . . 37. *P. perfoliatus.*

Subgen. 1. COLEOGÈTON (Reichenb.) Raunk.

§ CONNÀTI Hagstr.

1. P. filifórmis Pers. (thread-like). — Extensively creeping, *filiform stolons terminated by slender white tubers* 1–2 *cm. long;* stems filiform, bushy-branched or elongated to 9 dm.; *sheaths tightly clasping, only slightly thicker than the stem,* 0.4–2.2 *cm. long, margins connate below,* the free scarious tip 1–5 mm. long; *leaf-blades setaceous,* 0.2–0.5 mm. wide, blunt, 1–12 *cm. long; peduncle filiform, flexuous, up to* 1 *dm. long; spike moniliform,* 1.5–5 cm. long, *with* 2–5 *mostly remote whorls,* the upper whorls 3–12 mm., the lower 0.7–2.5 cm. apart; sepaloid connectives 0.5–1 mm. long; *fruits plump,* 2–2.6 *mm. long,* 1.6–2 *mm. broad,* strongly rounded on the back, *the short wart-like beak nearly central.* (*P. interior* Rydb.) — Calcareous or brackish waters, w. Nfld. to Alaska, s., very locally, to Gaspé Pen., Que., Vt., Minn., N.D. and Nev. July–Sept. (Greenl.; Eurasia)

Var. boreàlis (Raf.) St. John (northern). — Similar; leaves obtuse or retuse; spikes 0.5–2.5 cm. long, *the upper whorls crowded, the lower* 1–7 *mm. apart.* — Similar habitats, more common, Nfld. to Alaska, s. to C.B., P.E.I., ne. Me., Pa., Mich., Minn., Colo. and Utah. July–Sept. (Asia)

Var. Macoùnii Morong (for its discoverer, JOHN MACOUN, 1831–1920). — *Leaves slenderly linear,* 0.7–2 *mm. wide,* obtuse or short-apiculate; spikes as in the last. — M.I. and P.E.I.; n. Ont. to n. Alta., s. to n. Wisc., S.D., Colo., Utah, Nev. and Calif. (Sweden)

§ CONVOLÙTI Hagstr.

2. P. vaginàtus Turcz. (sheathed). — Coarser than no. 1; *stolons* 2–4 *mm. thick, tubers* 4–5 *cm. long; lower primary leaves with loose sheaths much thicker than the stem,* 2–5 *cm. long, open at base; leaf-blades* blunt, 1–2 mm. broad, 1–4.5 *dm. long;* upper leaves more slender, with closer sheaths; *spikes with* 5–12 *nearly equidistant whorls,* 3–8 cm. long; sepaloid connectives 1.5–2 mm. long; *fruits* 3–3.5 *mm. long,* 2–2.5 *mm. broad,* with nearly central wart-like beak. (*P. interior* in part, of ed. 7) — Deep fresh or brackish water, Lab. to Alaska, s. to w. Nfld., C.B., P.E.I., s. N.B., s. Que., n. N.J., w. N.Y., Mich., Wisc., Minn., N.D., Id. and Oreg. Aug.–Oct. (Eurasia)

3. P. pectinàtus L. (comb-like), "SAGO." — Rootstock very slender, tuber-bearing; *stems* filiform, up to 1 m. long, *abundantly branched and repeatedly forking above; leaves setaceous,* 0.2–1 mm. broad, *tapering to sharply pointed tips; sheaths* 2–5 *cm. long, the free tip less than half as long;* peduncle flexuous, up to 3 dm. long; *spike moniliform, with* 2–6 *unequally remote whorls,* 1.5–5 cm. long; *fruits* plump, 2.6–4.2 mm. long, *with a short slightly curved beak near the ventral margin.* (Including *P. interruptus* of ed. 7) — Saline, brackish or calcareous waters of coast and interior, Nfld. to Mackenz. and s. B.C., s. to Fla., W.Va., the Great Lakes States, Mo., Tex. and Mex. June–Sept. (S.Am.; Eurasia; Afr.)

Subgen. 2. EUPOTAMOGÈTON Raunk.

§ ADNÀTI Hagstr.

4. P. Robbínsii Oakes (in honor of JAMES WATSON ROBBINS, 1801–1879, pioneer student of the genus). — Rootstock scarcely developed, not tuberous; *stems rooting at lower nodes, the sterile* simple to freely and widely branching, *stiffly feather-like,* 1–9 dm. long, *closely invested by sheathing whitish stipules;* the flowering stems loosely elongating to 2 m.; *leaves stiffly 2-ranked,* those of the sterile stems *crowded, linear to lanceolate,* 3–8 mm. wide, *auricled at base, closely* 20–60-*nerved,* with a thin *cartilaginous minutely serrulate margin; free tip of the fibrous pale stipule* equaling or longer than its sheath, commonly *closely investing the internode above,* becoming fibrillose; leaves of elongating flowering stems remote and greatly reduced, their stipules with shorter tips; inflorescence branching; peduncles 1–26, straight; spikes stiff, interrupted, 0.7–2 cm. long; *fruits* (very rare) 4–5 *mm. long,* 2.7–3.3 *mm. broad, flattened, obliquely sigmoid-obovoid, prominently keeled, with nearly central beak* 1 *mm. long;* plant chiefly propagated by indurated branches and tips. — Muddy waters, N.B. to L. Mistassini, Que., w. to n. Ont., s. to N.E., Del., n. Ind., Ill. and Minn., s. Ala.; Wyo. to s. B.C., Wash. and Oreg. Aug., Sept.

§ Axillàres Hagstr.

Sub-§ Críspi Wallr.

5. P. críspus L. (crimped). — *Stem flattened, the broader flattened sides channelled,* freely branching above; *leaves sessile, oblong to broadly linear,* often reddish, *sharply serrulate,* in maturity often crisped and undulate; peduncles curved; spikes loosely flowered, up to 1.8 cm. long; *fruits obliquely ovate,* 5–6 *mm. long, flattened, with an elongate acuminate beak and subsagittate base; winter-buds bur-like,* 1–2.5 cm. long, *of strongly divergent and indurated leaves or their concave bases.* — Muddy, calcareous and brackish ponds and streams, locally aggressive, s. Que. and s. Ont. to Minn., s. to N.S., N.E., Va., W.Va. and Mo.; Calif. May–Sept. (Natzd. from Eu.)

Sub-§ Montícoli Hagstr.

6. P. confervoìdes Reichenb. (like *Conferva*). — *Rootstock filiform, extensively creeping,* bearing late in the season small *fusiform tubers; stems filiform, flaccid, repeatedly forking; leaves flaccid, linear-setaceous, attenuate to the slenderest of tips;* stipules delicate, almost nerveless; *peduncle at tip of main stem* (rarely on the branches or itself forking), straight, *erect, up to* 2.4 *dm. long; spike thick-cylindric or subglobose,* 0.3–1.2 cm. long; *fruits compressed, roundish, light green,* 2.2–3 mm. long, the dorsal keel sharp and often undulate; winter-buds (late in season) fusiform, 0.7–2 cm. long. — Sandy or peaty ponds and pools on mts., Nfld. to n. N.Y. and ne. Pa., and in the coastwise region to s. N.J.; also Mich. and Wisc. June–Aug.

Sub-§ Compréssi Fries

7. P. zosterifórmis Fern. (similar to *Zostera*), Flat-stem P. — *Stem* freely branching, *wing-flattened,* 0.7–3 *mm. wide,* constricted at the nodes; *leaves* linear and grass-like, *the primary ones* 1–2 *dm. long,* 2–5 mm. broad, obtuse to subacute, *with very many* (*up to* 35) *fine nerves and* 3 *larger ones; stipules firm, fibrous,* 1.5–3.5 *cm. long,* the lower obtuse, the upper acuminate; *spikes* cylindric, 1.5–3 *cm. long, with* 7–11 *subdistant whorls;* sepaloid connectives 2–2.6 mm. long; *fruits quadrate-oblong or -suborbicular,* 4–5 *mm. long,* 3–3.5 *mm. broad, the truncate base* 1.5–2.8 *mm. broad,* the dorsal keel wing-like and more or less dentate, the ventral margin usually umbonate at base; beak marginal, 0.6–1 mm. long; *winter-buds* 4–7.5 *cm. long,* the ascending leaf-tips usually exceeding the coarsely fibrous stipules. (*P. zosterifolius* of Am. auth., not Schum.; *P. compressus* of Am. auth., not L.) — Quiet waters, Que. to n. Alta. and s. B.C., s. to Va., W.Va., O., Ind., ne. Ill., Ia., Neb., nw. Mont. and n. Calif. July–Sept.

Sub-§ Pusílli (Graebn.) Hagstr.

8. P. foliòsus Raf. (leafy). — *Rootstock filiform,* rooting at the nodes; stems compressed-filiform, subsimple to loosely branched; *leaves* narrowly linear, *slightly tapering at base,* deep green to bronze, rather opaque, usually *without basal glands;* the primary ones 4–10 cm. long and 1.4–2.7 mm. broad, 3–5-nerved; the *midrib prominent,* compound below the middle, with 1–3 rows of lacunae on each side at base, these decreasing and disappearing toward the tip; the lateral nerves joining midrib 2 leaf-breadths below the tip; *stipules* at first *with connate margins, forming tubular delicately fibrous blunt sheaths* 0.7–1.8 cm. long, soon rupturing and deciduous; *peduncles* in upper forks, *slightly thickened upward,* 3–10 (rarely –30) *mm. long; spike subcapitate or thick-cylindric,* 2–5 *mm. thick, with* 2 *or* 3 *approximate whorls of* 2 *flowers each; sepaloid connectives* 0.6–1 *mm. long, brownish; fruits* fulvous or olivaceous, *obliquely suborbicular, strongly compressed,* 2–2.5 mm. long, the body orbicular to broadly obovate, the *dorsal keel thin and undulate or dentate,* beak 0.2–0.4 mm. long; winter-buds sessile in axils or terminating short branches; their hard bodies 1–1.6 cm. long, 1–2 mm. thick. (Var. *niagarensis* (Tuckerm.) Morong) — Fresh (often calcareous) or brackish waters, n. N.Y. to Wash., s. to the W.I. and Mex. July–Oct.

Var. **macéllus** Fern. (rather meagre). — *Smaller* throughout, mostly bushy-branched; *leaves bright green; the primary* ones 1–7 cm. long and 0.3–1.5 *mm. broad,* 1–3-nerved, *without lacunae or with* 1 *row* on each side *at base; stipules* 3–11 *mm. long;* connective green or brownish; *fruits* green, *obliquely obovoid,* 1.8–2.3 mm. long; *the body elliptical to obovate, longer than broad;* winter-buds (rare) mostly terminating elongate branches, the bodies 0.8–1.2 cm. long. (*P. foliosus* of ed. 7) — Gaspé Pen., Que., to Mackenz., s. to Fla., Ky., Mo., Kans., N.M. and Calif. July–Oct.

9. P. Frièsii Rupr. (for Elias Magnus Fries, 1794–1878). — Stem simple or subsimple below, subsimple or forking above, compressed; *leaves* linear, *bright green and translucent, the primary ones* 1.5–3.5 *mm. broad, obtuse or rounded at tip,* mucronate, 5–7-*nerved;* midrib usually bordered each side at base by a *slender row of lacunae; median nerves* fine but distinct, *joining* *midrib well below the tip* and with *frequent transverse connections; stipules strongly fibrous*, whitish, at first *connate* below, acute or acutish, 7–11 *mm. long, in age lacerate and 2-cleft*, usually with 2 *basal glands* 0.3–0.8 *mm. broad; peduncles flattened, broadened upward*, 1.5–5 *cm. long; spike* interrupted-cylindric, 0.7–1.5 *cm. long, with* 3–4 *remote whorls;* sepaloid connectives 1.5–2.5 mm. long; *fruits* obliquely ovoid or obovoid, 2–3 *mm. long*, 1.5–2 *mm. broad*, rounded on the back, the ventral face tapering to the short (0.2–0.8 mm.) beak; *winter-buds terminating short lateral branches; their oblong-lanceolate bodies* 1.2–2.5 *cm. long*, 2–4 *mm. broad, covered with coarsely fibrous pale stipules*, their loosely ascending leaves exserted 2–4 cm. — Calcareous or brackish waters, w. Nfld. to Mackenz., s., locally, to N.S., Mass., Ct., Va., O., Ind., Ill., Ia., S.D., s. Alta. and s. B.C. July–Sept. (Eu.)

10. P. strictifòlius Ar. Benn. (straight-leaved). — Stem compressed-filiform, simple or sparsely branched below, often forking and branching above, with *rigid branchlets; leaves mostly rigid, obtuse or abruptly mucronate, often revolute;* the primary ones 0.5–2.5 *mm. wide*, *3-nerved;* the *midrib rigid*, prominent, *without bordering lacunae* or with 1 row on each side at base, *rarely with conspicuous cross-reticulation; stipules connate* when young, *soon splitting, chartaceous, whitish, strongly fibrous*, 0.8–2 cm. long; *peduncles filiform, clavate at tip*, 1–9 *cm. long; spike interrupted*, cylindric, 0.8–1.5 cm. long, with 3–4 remote whorls; sepaloid connectives 1.3–1.8 mm. long; *fruits* obliquely obovoid to ovoid, 2–3 *mm. long*, 1.5–2 *mm. broad*, with marginal beak; *winter-buds* terminating elongate as well as short branches; the *fusiform body chartaceous, often whitish, fibrous*, 1–2.5 cm. long, the green *leaves with dilated or strongly corrugated bases* and prolonged or recurving tips. — Calcareous waters, Vt. to Sask., s. to w. Mass., n. and w. N.Y., n. Ind., Wisc., Minn. and Neb. July–Sept.

Var. **rutiloìdes** Fern. (like *P. rutilus*, yellowish-red). — *Leaves* firm, scarcely rigid, *very gradually tapering to a slender tip;* stipules less fibrous. (*P. rutilus* of ed. 7, not Wolfg.) — W. Que. to Mackenz., s. to w. Vt., nw. N.Y., nw. Pa., Ind., Mich., Minn., Neb. and Utah.

11. P. pusìllus L. (very small). — Stems capillary, slightly compressed, usually much branched; *leaves* linear, *firm*, usually *with* 2 *small translucent glands at base;* the primary ones 1–3 mm. wide, acute to obtuse, 3-nerved, light green; midrib prominent, usually not bordered by lacunae; *stipules scarious-membranaceous, slenderly tubular, with margins united to above the middle, finally rupturing and becoming lacerate (but hardly rigid) at summit*, 0.6–1.7 cm. long; *peduncles* axillary, *filiform*, 1.5–8 *cm. long; spikes elongate, strongly interrupted*, 6–12 *mm. long, of* 3–5 *distant whorls; fruits light olivaceous*, obliquely obovoid, often somewhat sigmoid, 1.9–2.8 mm. long, 1–1.8 mm. broad, *smooth but often deeply impressed on the somewhat flattened sides*, the back rounded, the ventral face arching to the prominent marginal beak; *winter-buds axillary* along the branches and terminal, their bodies 9–17 mm. long and 0.5–1.5 mm. broad. (*P. pusillus*, var. *capitatus* of ed. 7; *P. panormitanus* Biv. and var. *major* G. Fisch.) — Basic or alkaline waters, Gaspé Co., Que., to n. Alta. and s. B.C., s. to Va., Ala., La., Tex. and centr. and w. Mex. July–Sept. (W.I.; Azores; Eurasia)

Var. **mìnor** (Biv.) Fern. & Schub. (smaller). — Smaller; the larger or *primary leaves only* 0.3–1 *mm. wide*. — Mass. to n. Man. and s. B.C., s. to Md., Ala., La., Tex. and w. Mex. (Eurasia)

12. P. longiligulàtus Fern. (with long ligules). — *Stems* compressed-filiform, elongate, *sparsely branched; leaves* bright green, *subrigid*, often revolute, linear, *up to* 1.5 *dm. long*, 1.5–2 *mm. wide, attenuate to bristle-tips*, 5–9*-nerved*; midrib prominent, lateral nerves evanescent toward the tip; *stipules convolute, subrigid, strongly fibrous*, 1.3–3 *cm. long; peduncles* 1.5–3 *cm. long;* spikes cylindric, 1–1.3 cm. long, remotely flowered; sepaloid connectives firm, brownish, 2–2.3 mm. long; fruit unknown; *winter-buds subsessile or terminating short divergent branches*, 4–6.5 cm. long, the tips of the strongly ascending leaves much exceeding the fibrous stipules. — Calcareous waters, very local, nw. Nfld. to Minn., s. to nw. Ct., w. N.Y. and Mich. Aug., Sept.

13. P. gemmìparus Robbins (producing buds). — *Stems* filiform, *very freely branched, the long branches terminating in winter-buds; leaves linear-setaceous*, 0.2–0.4 *mm. broad, tapering to long bristle-tips; midrib bordered each side by a slender row of lacunae; stipules subherbaceous, linear-attenuate*, convolute, 0.5–2 cm. long, usually with 2 small basal glands; *peduncle filiform*, 1–3.5 cm. long; *spike interruptedly cylindric*, 5–8 mm. long, *of* 2–4 *remote whorls; fruits* (very rare) olive-green, *narrowly obovoid*, 2.2 *mm. long*, 1.5 mm. broad, rounded on the back, *the ventral face strongly rounded to the very short* (0.2 *mm.*) *wart-like nearly central broad beak; winter-buds terminating strongly ascending mostly prolonged* (up to 1.5 dm.) *branches;* their *lance-subulate bodies* 0.8–1.4 *cm. long* and 0.5–1.2 mm. broad, covered by subherbaceous stipules; *the divergent* or only slightly ascending *leaf-tips bristleform*, 1.5–3.5 *mm. long*. — Quiet waters, centr. Me. to s. Que. and e. Ct.

14. P. Híllii Morong (for its discoverer, Ellsworth Jerome Hill, 1833–1917). — Stem filiform, simple or with strongly ascending branches; *leaves* linear, pale green, the *primary ones 1–2.2 mm. broad, sharply acute and cuspidate, 3-nerved; midrib* usually *bordered* each side *by 1 or 2 rows of lacunae,* with additional lacunae at base; *stipules free, whitish,* 0.8–1.8 *cm. long, sharply acute, with strong persistent fibers; peduncles clavate,* 0.5–1.5 *cm. long,* often divergent or recurved; *spike capitate,* 1–4-flowered, 5–8 *mm. in diameter; fruits* compressed, obliquely obovate, 3–3.6 *mm. long,* 2.3–3 *mm. broad, dorsal keel prominent, ventral margin rounded to a* short *recurved beak;* winter-buds unknown. — Very local, e. Vt. to Mich., s. to w. Mass., Pa. and O. July, Aug.

15. P. Pórteri Fern. (for its discoverer, Thomas Conrad Porter, 1822–1901). — Stems compressed-filiform, with elongate branches; *leaves reddish-green,* linear, *obtuse, apiculate,* the primary ones 2–4 *mm. wide, 3-nerved; the prominent midrib bordered* each side *by 1 or 2 rows of lacunae and additional lacunae at base, the minute lateral nerves confluent with the midrib at tip; stipules* convolute, *rigid, fibrous,* 0.8–1.4 cm. long, in age with the apex lacerate and 2-parted, rarely with minute basal glands; *peduncles* remote, *barely 1 cm. long; spikes subcapitate,* few-flowered, in fruit 7 *mm. long; fruits* slightly compressed, obliquely obovoid, 3.4–4 *mm. long,* 2.3–2.6 *mm. broad,* the rounded back with a pronounced keel, the beak erect and 0.5 mm. long; *winter-buds* flabelliform, 2.5–3.5 *cm. long,* the strongly ascending leaf-tips much exceeding the fibrous stipules. — Streams, Dillerville Swamp, Lancaster, Pa. (*Porter*). Perhaps extinct, not collected since 1860. Sept., Oct.

16. P. obtusifòlius Mert. & Koch (blunt-leaved). — Stems much branched or in elongate plants slightly so; *leaves* linear, *warm green or reddish,* the primary ones 0.4–1 dm. long, 2–4 *mm. broad, rounded at tip,* subapiculate, *very translucent, with the broad compound midrib bordered* on each side, especially from base to above the middle, *by 2–4 bands of lacunae, the 2 very faint lateral nerves joining the midrib at tip; stipules* broad, obtuse, *delicately nerved,* 1.2–2 cm. long, with 2 *lustrous basal glands* 0.6–1.2 *mm. broad; peduncles slender, straight, ascending,* 0.8–2 (rarely –6) cm. long; *spikes* dense, *thick-cylindric,* 0.6–1.3 *cm. long; fruits* obliquely obovoid, fulvous, 3–4 *mm. long,* 2–2.3 *mm. broad,* with a low acutish keel on the rounded back, the ventral margin tapering to a short erect beak; *winter-buds* terminal, narrowly flabelliform, *their oblong scarious bodies* 2–4 *cm. long* and 3.5–7 *mm. broad,* much exceeded by the leaves. — Cold streams, springs and lakes, s. Lab. Pen. to Ont., s. to Nfld., C.B., N.E., n. N.J., ne. Pa., Mich., Wisc. and Minn.; s. B.C. July–Sept. (Eurasia)

17. P. Berchtòldi Fieber (for Friedrich, Graf von Berchtold, 1781–1876). — Stems capillary, subsimple to freely branching, the branches often terminated by winter-buds; *leaves linear to linear-setaceous,* usually *with a pair of small translucent glands at base,* 0.8–8.5 cm. long, 0.3–2.4 *mm. wide, 3-nerved* (lateral nerves obscure in narrowest or most translucent leaves), usually deep green and firm, sometimes translucent and flaccid; *lateral nerves* (when not evanescent) *joining the midrib well below the tip, midrib bordered by lacunae* (at least at base); *stipules flat or with inrolled but free margins,* hyaline or subherbaceous, faintly nerved, *obtuse* (often appearing acute from inrolling), 3–14 mm. long; *peduncles* from the upper axils, *filiform,* ascending, 0.35–3 (rarely –4.5) *cm. long; spikes subglobose, in fruit subcontinuous or but slightly interrupted,* 2–8 *mm. long,* of 1–3 *few-flowered whorls; fruits* dark olivaceous, obliquely obovoid, 2–2.5 mm. long, 1.2–1.9 mm. broad, *commonly rugulose on the surface* (when dry), the back rounded, the ventral face arching to a short beak; *winter-buds terminating the branches; their fusiform bodies olivaceous,* 7–18 *mm. long,* 0.6–2.5 *mm. broad,* the divergently ascending lower leaves 0.8–3.2 cm. long. — A highly variable species, represented with us by six varieties:

a. Leaf-tips mostly rounded or obtuse. . . *b.*

b. Midrib bordered on each side by 1 row (sometimes 2 at base) of lacunae; foliage mostly green.

Principal leaves 3–7 cm. long. . . . . *P. Berchtoldi* (typical).

Principal leaves 0.8–2.5 cm. long. . . . . Var. *polyphyllus*.

b. Midrib bordered on each side by 2–4 rows of lacunae; foliage fulvous. . . Var. *colpophilus*.

a. Leaf-tips subacute to sharply pointed. . . *c.*

c. Midrib of principal leaves (below involucral leaves) bordered on each side by 1 or 2 rows of lacunae.

Primary leaves (of principal stems) 0.5–1.5 mm. wide, with well-defined lacunae often in 2 rows each side of the midrib (at least in lower half). . Var. *acuminatus*.

Primary leaves 0.3–1 mm. wide, with 1 row of frequently evanescent lacunae each side of midrib. . . . . Var. *tenuissimus*.

c. Midrib of principal leaves bordered on each side by 3–5 rows of lacunae. . Var. *lacunatus*.

P. Berchtòldi (typical), *P. pusillus,* var. *mucronatus* (Fieber) Graebn.; *P. pusillus,* var.

Sturrockii of ed. 7, not Ar. Benn. — Greenl. and Nfld. to Alaska, s. to N.S., s. N.E., L.I., Del. Ind., Minn., Neb. and Mont. June–Sept. (Eurasia)

Var. **polyphýllus** (Morong) Fern. (many-leaved), *P. pusillus*, var. *polyphyllus* Morong — Shallow pools, e. Que. to n. Alta., s. to Me., Mass., N.Y., Wisc., Minn. and Neb.

Var. **colpóphilus** Fern. (lover of estuaries). — Brackish to fresh estuaries, Gaspé Co., Que.

Var. **acuminàtus** Fieber (tapering at tip), *P. pusillus* of ed. 7, not L. — E. Saguenay Co., Que., to Ont., s. to N.S., N.E., n. N.J., Pa. and La.; Yukon and Alaska to Oreg. (Eurasia)

Var. **tenuíssimus** (Mert. & Koch) Fern. (most slender). — Nfld. to Ont., s. to N.S., N.E., L.I., Va., Ky. and Okla.; Alaska to Calif. (Eurasia) — Hybridizes with var. of no. 36.

Var. **lacunàtus** (Hagstr.) Fern. (bearing lacunae). — Centr. Me. to Man., s. to s. N.E., L.I., Va. and Minn.; Vancouver I.

Sub-§ Javánici Graebn.

18. P. Vâseyi Robbins (for its discoverer, George Vasey, 1822–1893). — *Plants dimorphic, the flowering and fruiting form bearing dilated leaves* as well as occasional winter-buds; *the flowerless plants* (in the same or separate colonies) *without dilated leaves but with abundant winter-buds;* stems filiform, freely branched; *submersed leaves linear-filiform, almost bristle-tipped;* the primary ones 2.5–8 cm. long, 0.1–0.5 *mm. wide*, with a delicate midrib and (under high magnification) 2 lateral nerves; *their delicate stipules slender, attenuate,* 4–12 mm. long; *dilated leaves* elliptic or narrowly obovate, obtuse, 0.6–1.5 cm. long, 3–8 *mm. broad,* 5–9-nerved, *with slender petioles* often twice their length; *peduncles* 0.8–1.6 *cm. long,* at first straight and ascending, *in maturity recurving; spikes* at first emersed, *finally recurved and submersed, interrupted-cylindric,* 6–8 mm. long; *sepaloid connectives barely* 1 *mm. long; fruits compressed, with sides almost flat,* obliquely round-obovate, 1.6–2.2 mm. long, *with distinct dorsal keel and slender marginal recurved beak; winter-buds* (from early summer to autumn) *nearly sessile in the axils of submersed leaves on very short* (up to 1, rarely to 3, cm. long) *horizontal branchlets, their subulate bodies* 5–8 *mm. long* and 0.5–1.2 *mm. thick,* their loosely ascending leaves up to 2 cm. long. — Quiet muddy or calcareous waters, s. Que. and s. Ont., s. to N.B., Me., Mass., Pa., O., Ill. and Minn. July–Oct.

19. P. lateràlis Morong (one-sided). — *Dimorphic, fruiting form bearing only linear and submersed leaves, sterile* (but flowering) *form with some dilated leaves; submersed leaves narrowly linear,* 0.4–1 *mm. wide; the midrib prominent* to below the tip, with or without bordering lacunae; *dilated leaves* strongly divergent, often opposite in 1–3 pairs or single and paired with linear leaves; the blades oblong to spatulate-obovate, 8–14 mm. long and 2–4 *mm. broad, narrowed to margined petioles* 0.5–2 cm. long; *peduncles* subterminal or from the upper forks, 1–3 *cm. long,* in fruit often divergent; spikes slender-cylindric, 3.5–7 mm. long, remotely few-flowered; *sepaloid connectives* 1.5–2 *mm. long; fruits only slightly compressed,* 2–2.4 *mm. long, the strongly rounded back 2-grooved,* the *stigma nearly sessile; winter-buds* produced late in season *at the tips of proliferous upper branches, their bodies* 7–12 *mm. long* and 0.5–0.7 *mm. thick.* — Quiet waters, very local, e. Mass. and Ct. to Mich. July, Aug.

Sub-§ Hýbridi (Graebn.) Hagstr.

20. P. Spiríllus Tuckerm. (from *spira*, a coil). — Bushy-branched and bearing only submersed leaves, or more elongate and with some branches much prolonged and terminated by dilated and floating leaves; *submersed leaves* pale green, *linear,* straight or curving, *obtuse,* 1.5–4.5 cm. long and 0.6–2 *mm. wide, with the connate leaf-sheath much longer than the truncate or emarginate free stipular tip, the space between the midrib and delicate lateral nerves commonly filled by lacunae; dilated leaves* oblong to obovate, *rounded at tip and minutely emarginate,* 0.7–3 cm. long, 2–12 mm. broad, *with* 5–15 *nerves* deeply impressed beneath; petiole flat or channelled, at first connate with the hyaline stipule, later free; *spikes dimorphic or trimorphic; those from axils of the submersed linear leaves* 1–6*-flowered, subsessile or on erect peduncles rarely exserted beyond the sheathing leaf-base,* their flowers often without anthers; *spikes from axils of the coriaceous leaves* developed last, *cylindric,* 0.5–1.4 *cm. long, on clavate peduncles* 0.5–3 *cm. long;* sepaloid connectives 0.5–1 mm. long; *fruits greenish or yellowish;* those of the submersed spikes *lenticular, suborbicular or rounded-obovate,* 1.3–2.2 *mm. in diameter, with* the narrow to broad very unequally developed thin *dorsal keel toothless or* usually *with* 1–5 *remote low teeth, sides without definite keels but rounded and clearly conforming to the cochleate embryo* and with a deep central depression, beak obsolete. (*P. dimorphus*, in part, of ed. 7, not Raf.) — Quiet waters, Nfld. to e. Man., s. to N.S., s. N.E., Del., Va., W.Va., O., Wisc., Ia. and S.D. July–Nov.

21. P. diversifòlius Raf. (diverse-leaved). — Resembling no. 20; *submersed leaves subobtuse to acute,* 0.5–1.5 mm. wide, *the connate sheath about one-half the length of the free stipular tip;*

dilated leaves strongly rounded but scarcely emarginate at tip, 1–4 cm. long, 0.3–2 cm. broad; the *upper stipules* 0.6–3 cm. long, *in age becoming fibrous; upper spikes* 0.5–2 *cm. long, on peduncles* 0.2–1.5 *cm. long; fruits* greenish, suborbicular to semi-reniform, 3-*keeled,* 1–1.5 *mm. long, with very low and slender lateral keels* forming an entire or dentate border outside the embryo, *sides flat or but slightly depressed,* beak minute; winter-buds frequently formed in autumn in the upper axils. (*P. dimorphus,* in part, of ed. 7, not Raf.) — A southern species, extending from Mex. and the Gulf States n. to N.J., Pa., W.Va., O., Ind., Ill., Minn., S.D., Mont. and Oreg. June–Sept.

22. P. capillàceus Poir. (hair-like). — Habit of nos. 20 and 21, but extremely delicate; *submersed leaves flaccid, setaceous or setaceous-linear with setaceous tips,* 1.5–7 cm. long, 0.1–0.6 mm. broad, the short hyaline stipule either free and deciduous or partially adnate to the leaf-base; *dilated leaves* (when present) *lance- to oval-elliptic, acutely pointed or* if blunt at least *submucronate,* 0.7–2.6 cm. long, 1–10 mm. wide, *with* 3–7 *nerves* impressed beneath; petioles slender, free from the short (3–10 mm.) stipules; *submersed spikes* subglobose, usually *shorter than the slender and divergent or recurved peduncles* (3–13 *mm. long*); *upper spikes* subglobose to cylindrical, 0.3–1.5 cm. long, *on slender peduncles* 3–15 mm. long; *fruits greenish or olivaceous,* suborbicular, oblique, quadrate, strongly flattened, 3-keeled, 1–1.5 mm. long; the crest-like very *narrow dorsal keel* (0.1–0.3 mm. wide) *entire or minutely sinuate-dentate with* 3–12 *small teeth;* the very low and *slender lateral keels following the outer margin of the spiral,* toothless or toothed like the dorsal; *sides flattish and often cochleate-sulcate;* old plants (late in autumn) proliferating and producing branches of capillary leaves from the axils of the broad floating leaves. (*P. hybridus* of ed. 7, not Michx.; *P. diversifolius* Bart., not Raf.). — Quiet waters over sand, mud or peat, chiefly on or near the Coastal Plain, W.I. and Fla. to e. Tex., n. to s. Me., centr. N.H. and s. N.Y.; nw. Ind. to n. Wisc. June–Oct.

Var. **átripes** Fern. (black-footed). — *Rhizomes and stolons* rather *rigid, black; submersed leaves firm, ascending,* only 2–3 cm. long, *with 2 or 3 rows of lacunae each side of the midrib;* no globular submersed spikes developed. — Argillaceous spring-heads, se. Va.

23. P. bicupulàtus Fern. (having two cups). — Resembling no. 22; *fruits stramineous,* quadrate-rotund, 1.6–2 *mm. long; the dorsal keel broad and wing-like, coarsely and remotely* 5–12-*dentate; lateral keels high, sinuate or coarsely dentate, forming a deep cup or crater on each side of the fruit.* — Quiet waters, w. N.E. to mts. of Pa. and Tenn. July–Sept.

Sub-§ Nuttalliàni (Hagstr.) Fern.

24. P. epihỳdrus Raf. (upon the water). — Extensively creeping; *stem compressed,* simple or slightly branched below or sometimes forked at summit; *submersed leaves linear and ribbon-like,* 0.5–2.2 dm. long, 0.5–1 *cm. broad,* 7–13-nerved, *with at least the broad space within the inner nerves* (and frequently outside, rarely even to the margins) *loosely cellular-reticulate below the middle; floating leaves mostly opposite,* coriaceous, opaque, *elliptic to broadly oblong-oblanceolate,* bluntly cuspidate or merely rounded at tip, 3–8 cm. long, 1.5–3.5 cm. broad, 19–41-nerved, *tapering into flattened petioles;* transitional leaves often present; stipules of submersed leaves free, hyaline, obtuse, up to 4 cm. long, those of floating leaves firmer and often attenuate; peduncles 1.5–16 cm. long, thickish, from axils of floating leaves; spike in maturity 1.8–4 cm. long; sepaloid connectives flabelliform, 1.5–3 mm. long; *fruits laterally much flattened, pitted on the sides, round-obovate,* 3-keeled (when dry), 3–4.5 *mm. long,* 3–3.6 *mm. broad,* with dorsal keel 0.6–1.2 mm. wide; *beak a mere tooth at summit of ventral margin; embryo subspiral, its tip curved upward above the base.* (Var. *cayugensis* (Wieg.) Ar. Benn.) — Still or flowing water of lakes and streams, s. Que. to Minn., s. to N.B., Ct., N.J., Pa., W.Va., O., Ind., Ill. and Ia.; Id. and Wash. to n. Calif. July–Sept.

Var. **Nuttállii** (C. & S.) Fern. (for Thomas Nuttall, 1786–1859). — *Submersed leaves* (1–) 2–8 *mm. broad,* 3–7-nerved, *strongly distichous* and rather crowded *on sterile shoots; blade of floating leaf narrowly oblong or oblong-lanceolate,* rounded at tip, only rarely subacute, 2–7.5 cm. long, 0.4–2.5 cm. wide, 7–33-nerved; fruiting spike 0.8–3 cm. long; *fruits* 2.5–3.5 *mm. long* and 2–3 *mm. broad, the dorsal keel* 0.2–1 *mm. broad.* (*P. epihydrus* of ed. 7) — Quiet waters, Nfld. and se. Lab. Pen. to Man., s. to N.S., N.E., L.I., Va., and upland of n. Ga. and Tenn.; s. Alaska to Colo. and Calif. (Nw. Eu.) — Hybridizes, rarely, with nos. 29 and 30, the latter hybrid being × **P. subséssilis** Hagstr.

25. P. tennesseénsis Fern. (of Tennessee). — *Stem* very slender, *terete,* up to 1.5 mm. thick; *submersed leaves linear-filiform,* 0.2–1 *mm. broad,* 1 (or obscurely 3)-*nerved, adnate at base to the partially free ligule,* attenuate to tip, the space between margin and midrib filled with lacunae; *floating leaves* lanceolate or lance-oblong, acutish, 2–4 cm. long, 5–13 *mm. broad, scattered,* long-petioled, the 9–23 nerves impressed beneath; clavate peduncle 3–8 cm. long, ascending;

spike cylindric, 1–2.2 *cm. long,* in maturity 4.5–6 *mm. thick;* unguiculate connectives 2 mm. long, the oblate limb 1.5 mm. wide; *fruit* much as in no. 24, quadrate-orbicular, *with low-convex sides,* 2.5–3 mm. long, 2–2.5 *mm. broad,* the truncate base 0.8–1 mm. broad; the *acute dorsal keel* 0.5–0.8 *mm. broad,* entire or nearly so. — Still or flowing water, W.Va. to Tenn. Late May–Sept.

Sub-§ Alpìni (Graebn.) Hagstr.

26. P. alpìnus Balbis (alpine). — Slender rhizome sending up *simple* or upwardly rarely forked *stems with internodes much shorter than leaves; submersed leaves rufescent or warm olivaceous, pellucid,* narrowed to both ends, undulate, narrowly lanceolate to elliptic-oblong, *sessile,* up to 2.5 dm. long and 2.5 cm. broad; *the upper ones similar but firmer; floating leaves,* when developed, mostly alternate, lanceolate, elliptic or obovate, blunt, *tapering into the short petiole;* stipules hyaline, blunt; *spikes densely* flowered and *fruited,* much shorter than slender arching peduncle, the peduncle and rachis red; *fruit* obliquely obovoid, *plump, with hard smooth exocarp, tawny-olive, about* 2.5 *mm. long, the beak nearly central.* — Europe and w. Asia; represented with us by 2 vars.:

Var. tenuifòlius (Raf.) Ogden (thin-leaved). — *Submersed leaves oblong-linear to linear-lanceolate,* 0.7–2.5 dm. long, *usually eight times as long as broad, tapering to blunt or acutish tips;* flowering spike rather open, with 5–9 whorls of flowers, in fruit crowded and 1.5–3.5 cm. long, 0.8–1 cm. thick; sepaloid connectives orbicular to reniform; bodies of fruits 3–3.5 mm. long, 2–2.4 mm. broad, the beak lateral. (*P. alpinus* of ed. 7; *P. microstachys* Wolfg.; *P. tenuifolius* Raf.) — Chiefly in calcareous waters, sw. Greenl. and Lab. to Alaska, s. to Nfld., N.S., n. and w. N.E., e. Pa., N.Y., Mich., Wisc., Minn., Colo. and s. Calif. July–Sept. (E. Asia) — Crosses with no. 30 (producing × **P. subobtùsus** Hagstr.).

Var. subellípticus (Fern.) Ogden (nearly elliptic). — Similar to preceding var.; *submersed leaves oblong to oblong-ovate,* 4–10 cm. long, *usually less than eight times as long as broad, rounded and often slightly cucullate at apex.* — Calcareous waters, Nfld. to s. B.C., s. to C.B., M.I., N.B., n. N.E., n. N.Y., Mich., Wisc., Minn., Wyo. and Id. July–Sept.

Sub-§ Coloràti (Graebn.) Hagstr.

27. P. oblóngus Viviani (oblong). — Rootstock often spotted with cinnamon-color; stem simple, slender; submersed leaves (often wanting) lanceolate, 3–10 cm. long, 0.5–1.5 cm. broad, tapering to acutish apex and to petiole 1–3 cm. long, 7–11-nerved; the outer nerves marginal, with 2–4 rows of lacunae each side of midrib; *floating leaves* with coriaceous ovate blades tapering to obtuse tip, *rounded to slightly cordate at base,* 3–9 cm. long, 1–4 cm. broad, 11–21-nerved, the nerves all similar, petioles 1–15 cm. long; stipules 3 cm. long, practically keelless, subpersistent; peduncles 4–12 cm. long; flowering spike with about 10 whorls of flowers; *fruiting spike* 2–3 cm. long, 5–7 *mm. thick;* sepaloid connectives greenish, 1–2 mm. broad; *fruits* obovate to suborbicular, rounded at base, depressed on sides, 1.6–2.5 *mm. long,* 1.5–2.1 *mm. wide, with minute or obsolete beak, nearly or quite keelless, reddish; endocarp-loop with a large cavity.* (*P. polygonifolius* sensu most auth., not Pourret) — Shallow pools and muddy shores, e. Nfld., St. P. et Miq., and Sable I., N.S. Aug., Sept. (Eu.; N. Afr.; Madeira and Azores)

Sub-§ Amplifòlii Hagstr.

28. P. amplifòlius Tuckerm. (large-leaved). — Rhizome 2–4 mm. thick, with broadly obtuse black scales; stem 1–3.5 mm. thick, simple or tardily short-branched; *submersed leaves* (not the transitional ones) of 2 or more types; the lowest dark green lanceolate ones soon disappearing; *the median and upper ones* broadly lanceolate to ovate, with margins much longer than midrib, the leaf thus *arcuate,* 27–37-nerved, obtuse to acutish, tapering to petioles 1–6 cm. long; the blades 0.8–2 dm. long and 2.5–7.5 *cm. wide, entire,* with 3–6 rows of lacunae each side of midrib; *floating leaves* with gradual transition from submersed ones, coriaceous, opaque, ovate to elliptic, round-tipped or bluntly mucronate, *rounded to tapering at base,* 5–10 cm. long, 2.5–5 cm. broad, 21–51-*nerved, with about a quarter of the nerves stronger than the rest;* lacunae 0; petioles 0.8–2 dm. long; stipules of submersed leaves subpersistent, fibrous, becoming spirally stringy in age, 3.5–11 cm. long, those of floating leaves more definitely 2-keeled; peduncles usually thickened toward summit, 0.4–3 dm. long; *spikes with* 9–16 *whorls of flowers, in fruit* 4–8 cm. long and 1–1.5 *cm. thick; fruits* obovate, rounded on back, *cuneate at base,* the sides flattened, the body 3.5–5 *mm. long and* 2.5–3.7 *mm. broad,* with prominent beak, greenish, becoming reddish or orange-brown. — Lakes and streams, in rather deep water, s. Nfld. to N.D., s. to n. Va. and upland to nw. Ga. and ne. Ala., Ark. and Okla.; Mont. to s. B.C., s. to s. Id. and Calif. July–Sept. — Hybridizes with no. 32 (producing × **P. scoliophýllus** Hagstr.) and with nos. 35 and 36.

29. P. púlcher Tuckerm. (handsome). — Rhizome pale, often with purple spots; *stem usually dark-spotted;* lowest submersed leaves semiopaque, oblong, round-tipped; median and *upper submersed leaves* translucent, *lanceolate to lance-linear, not arcuate,* acutish, *tapering to short petioles or nearly sessile,* 0.8–1.8 dm. long, 1–3.5 *cm. broad,* entire, 9–21-nerved, with 4–8 rows of lacunae each side of midrib; *floating leaves* coriaceous, *ovate to roundish,* with bluntly mucronate to rounded tip, *cordate or broadly rounded at base,* 2–11 cm. long, 1.5–8.5 cm. broad, 19–35-*nerved, the nerves essentially uniform;* submersed stipules early decaying; those of floating leaves persistent, 2–5 cm. long, becoming acutish; peduncles 5–11 cm. long, of nearly uniform thickness; flowering *spike* with about 10 whorls, *in fruit* 2–3.5 cm. long and 8–11 *mm. thick; fruits* obliquely ovate, *rounded or lobed at base; the body* 3–4 *mm. long and* 2.6–3.4 *mm. broad,* acutely keeled, with dorsal keel often prominent, light brown to olive, the beak prominent and facial. — Peaty or muddy acid waters or shores, Fla. to e. Tex., n. to N.S. and s. N.H., s. N.E., s. N.Y., N.J., e. Pa., n. Md., s. O., nw. Ind., ne. Ill., s. Minn. and Okla. June–Sept. — Hybridizes with var. of no. 24.

Sub-§ Nodòsi Hagstr.

30. P. nodòsus Poir. (knotty). — Habitally similar to nos. 28 and 29; *submersed leaves* linear- to elliptic-lanceolate, 1–3.5 cm. wide, *tapering to petiole* 2–13 *cm. long,* and gradually to acutish tip, 7–15-nerved, with 2–5 rows of lacunae each side of midrib, the *young blades margined by fugacious translucent denticles; floating leaves long-petioled,* lance-oblong to lance-elliptic, *rounded to acutish at base and apex,* 3–11 cm. long, 1.5–4.5 cm. broad, 9–21-*nerved;* submersed stipules early decaying, linear, 3–9 cm. long, emersed ones similar but broader; peduncles usually stouter than stem, 1.5–2.3 mm. thick, 3–15 cm. long; spike becoming loose, of 10–17 whorls, in fruit 3–7 cm. long and 0.8–1 cm. thick; *bodies of fruits* obovate, 3.5–4.3 *mm. long,* 2.5–3 *mm. wide,* the dorsal keel strongly developed, the laterals often muricate, in maturity brownish or reddish, facial beak short. (*P. americanus* C. & S., incl. var. *novaeboracensis* (Morong) Ar. Benn.; *P. lonchites* Tuckerm.) — Ponds and streams, Ala. to Tex. and Mex., n. to s. N.B., s. Que., s. Ont., Mich., Minn., S.D., Mont. and s. B.C. Aug., Sept. (W.I.; S.Am.; Eurasia; Afr.) — Hybridizes with vars. of no. 24 and no. 26 and with no. 31 (producing × **P. argùtulus** Hagstr.) with no. 32 (producing × **P. Fáxoni** Morong) and with no. 36 (producing × **P. rectifòlius** Ar. Benn.)

Sub-§ Lucéntes Graebn.

31. P. gramíneus L. (grass-like). — *Stem* slender, 0.5–1 *mm. thick,* usually *much branched, the branches with small leaves; submersed leaves* linear, linear-lanceolate, lance-elliptic or sometimes oblanceolate, 1–13 cm. long, 0.1–1.5 *cm. wide, sessile,* acute and *usually sharp-pointed,* 3–11-*nerved,* with 1 or 2 rows of obscure lacunae each side of midrib, *the margins with fugacious 1-celled denticles; floating leaves* coriaceous, ovate to elliptic (rarely roundish), 1.5–7 *cm. long,* 1–3 *cm. broad,* obtuse, rounded to tapering at base, *mostly shorter than the petioles,* 13–23-nerved; *stipules* persistent, obtuse, slightly hooded at apex, 0.5–3 *cm. long, only faintly keeled;* peduncles 2–30 cm. long; spike compact, of 5–10 whorls, in fruit 1–2.8 cm. long; *sepaloid connectives* 0.7–2.3 *mm. broad;* bodies of *fruits* mostly obovate, 1.7–2.8 *mm. long,* 1.4–2.3 mm. wide, more or less keeled, the facial beak short and recurved. (*P. heterophyllus* sensu most auth., not Schreb.) — Our most variable species, with many vegetative responses to deep submergence or complete stranding, and tendency to cross with most species with which it occurs. The following varieties are the most stable.

Principal submersed leaves narrowly elliptic to oblanceolate, 1–13 cm. long, 0.2–1.5 cm. wide, five to ten times as long as broad (or, if longer, not less than 6 cm. long), the sides not parallel; nerves 5–11.
- Larger submersed leaves 1–6.5 cm. long, **0.2–0.8 cm.** wide, 5–7-nerved. . . *P. gramineus* (typical).
- Larger submersed leaves 3–13 cm. long, 0.6–1.5 cm. wide, 7–11-nerved. . . Var. *maximus.*

Principal submersed leaves linear, with essentially parallel sides, except for tapering apex, 1–5.5 cm. long, 1–3 mm. wide, ten to thirty times as long as broad, 3-nerved. . . . . Var. *myriophyllus.*

P. gramíneus (typical) = var. *graminifolius* Fries — Lakes, ponds and streams, Nfld. to Alaska, s. to N.S., N.E., n. N.J., N.Y., n. O., s. Ind., n. Ill., n. Ia., Neb., N.M., Ariz. and Calif. July–Sept. (Greenl.; Eurasia) — Hybridizes with var. of no. 26 and with nos. 30 and 32 (producing × **P. spathulaefórmis** (Robbins) Morong), with no. 36 (producing × **P. Hagstroèmii** Ar. Benn.) and very often with var. of no. 37 (producing × **P. subnitens** Hagstr.).

Var. **máximus** Morong (largest), *P. heterophyllus,* forma *maximus* Morong. — Commonly in deeper water, Lab. to Alaska, s. to Nfld., N.S., N.E., N.J., Pa., n. O., Wisc., n. Ia., N.D., N.M., Ariz. and Calif.

Var. **myriophýllus** Robbins (countless-leaved), *P. heterophyllus*, forma *myriophyllus* (Robbins) Morong — Shallow and quiet waters, local, N.H. to Minn., s. to Mass., R.I. and n. Ind.

32. P. illinoénsis Morong (of Illinois). — Rhizome stoutish; *stem simple or with simple branches,* 1–5 *mm. thick; submersed leaves* elliptic or elliptic-oblong or -ovate to lanceolate, or (by reduction of blade to midrib) even linear, 0.5–2 dm. long, (0.2–) 1.5–4.5 *cm. wide, sessile or tapering to petioles up to* 4 *cm. long,* acute, *usually mucronate,* 7–19-*nerved,* with 2–5 rows of lacunae along the midrib and larger nerves, *the margin with fugacious* 1-*celled denticles; floating leaves* (often wanting or merely transitions from submersed blades) with elliptic to ovate or oblong blades 4–19 *cm. long and* 2–6.5 *cm. wide,* obtuse but with a blunt mucro, rounded or tapering to base, *longer than petioles,* 13–29-nerved; *stipules* divergent, obtuse, 1–8 *cm. long, prominently* 2-*keeled;* peduncle thickish, 0.4–3 dm. long; *spikes* compact, of 8–15 whorls of flowers, *in fruit* 2.5–7 *cm. long* and 0.8–1 cm. thick; *sepaloid connectives* 1.6–3 *mm. broad; fruits* obovate, suborbicular or ovate, 2.5–3.5 mm. long, 2.1–3 *mm. broad,* laterally compressed, the acute dorsal keel prominent, the laterals less so but often with a projecting knob at base, the facial beak deltoid and about 0.5 mm. long. (*P. lucens* of Am. auth., not L.; *P. angustifolius* of Am. auth., not Berchtold & Presl) — Lakes and streams, chiefly calcareous, sw. Que. to s. B.C., s. to n. N.C., O., s. Ind., s. Ill., n. Ark., Neb., Colo., Utah and Calif.; Fla. to Tex. July–Sept. — Hybridizes with nos. 28, 30, 31, 36 and var. of 37.

Sub-§ NATÁNTES Graebn.

33. P. nàtans L. (swimming). — Rhizome with red spots; *stem* simple or slightly forking, 0.8–2 *mm. thick,* with transverse ridges; *submersed leaves firm, semiterete,* slenderly linear, 1–2 dm. long, 0.8–2 *mm. wide, with no differentiation of blade and petiole,* tapering to blunt tip, obscurely 3–5-nerved; *floating leaves* coriaceous, *the blade and long petiole* (1–2.5 *mm. thick*) *separated by a prolonged curving and flexible brownish joint,* the lustrous *blade* ovate to ovate-oblong, *cordate or rounded* (rarely tapering) *at base,* rounded or blunt at apex, 4–12 *cm. long,* 2.5–6.5 *cm. broad,* 13–37-nerved, with a third of the nerves prominent; *stipules* of submersed leaves clasping, whitish, fibrous, persistent, hooded at apex when young, soon splitting and ragged or spiraling, 4.5–11 *cm. long, with* 2 *strong keels;* stipules of floating leaves larger (5–12 mm. broad at base); peduncles thick, 3–8 cm. long; *spike* compact, of 8–14 whorls, *in fruit* 3–5 cm. long and 0.9–1.2 *cm. thick;* sepaloid connectives green, 1.8–2.8 mm. wide; *fruits* obovoid, 3–5 mm. long, 2.5–3.5 *mm. wide,* plump, *rounded on back* or slightly keeled, *the dry exocarp wrinkled,* beak short and broad. — Lakes and quiet streams, s. Greenl. and Nfld. to Alaska, s. to N.S., N.E., L.I., N.J., Pa., O., n. Ind., Ill., n. Ia., Neb., N.M., Ariz. and Calif. July–Sept. (Eurasia) — Sometimes without floating leaves or merely with narrow transitional blades.

34. P. Oakesiànus Robbins (for its discoverer, WILLIAM OAKES, 1799–1848). — Smaller throughout than no. 33; *stem* more often branched, 0.5–1 *mm. thick; submersed leaves delicate and flaccid,* 0.3–1 *mm. wide,* flat; *petioles of floating leaves* 0.2–1 *mm. thick; blades* oblong to narrowly ovate, *rounded or tapering at base,* 1.5–5.5 *cm. long* and 1–3 *cm. wide,* 7–25-nerved; *stipules delicately fibrous,* 1–5.5 *cm. long, keeled only at base; spikes* with 3–8 whorls, *in fruit* 1–3.5 cm. long and 7–9 *mm. thick; fruits* 2.5–3.7 mm. long, 1.6–2.4 *mm. broad, with acute dorsal keel* usually prominent, the greenish to buff *exocarp smoothish.* — Acid peaty-, sandy- or rocky-bottomed pools, Nfld. to w. Ont., s. to N.S., N.E., L.I., N.J., Pa., Mich. and Wisc. July–Sept.

Sub-§ PRAELÓNGI Hagstr.

35. P. praelóngus Wulfen (greatly prolonged), WHITE-STEM P. — *Rhizome* stoutish, *with rusty spots; stem* simple to branched, *white to olivaceous, zigzag,* 1.5–4 *mm. thick; leaves all submersed,* ovate- to lance-oblong, 0.5–3.6 *dm. long,* 1–3 *cm. broad,* 13–25-nerved, with 3–7 nerves stronger than others, *cordate to rounded at base and clasping a third to half the circumference of stem, hooded* (under pressure cracking) or rounded *at apex; stipules white, with strong nerves,* oblong to narrowly ovate, 3–10 *cm. long, keelless, usually persistent; peduncles* upwardly thickened, 0.5–6 *dm. long; spike* with 6–12 whorls, open in anthesis, in fruit 3–5 cm. long and 1.1–1.4 *cm. thick;* sepaloid connectives green, 1.7–2.9 mm. wide; *fruits* obovate, rounded on back, tapering to base, 4–5 *mm. long,* 3.2–4 *mm. wide,* with prominent thick beak, *the dorsal keel acute,* the exocarp green. — Cold waters of lakes and streams, usually deep, Lab. Pen. to Alaska, s. to Nfld., N.S., N.E., n. N.J., N.Y., n. O., n. Ind., ne. Ill., s. Wisc., n. Ia., Colo., Utah and Calif. June, July, usually withdrawing fruit to deep water to mature. (Eurasia) — Hybridizes with nos. 28 and 36.

Sub-§ Perfoliàti (Graebn.) Hagstr.

36. P. Richardsònii (Ar. Benn.) Rydb. (for its discoverer, Sir John Richardson, 1787–1865), Red-head P. — *Rhizome not spotted; stem* usually branched, 1–2.5 *mm. thick; leaves all submersed,* the lowest ovate to ovate-lanceolate, *the upper ovate-lanceolate to narrowly lanceolate,* 1.5–10 *cm. long,* 0.5–2 *cm. broad, prominently* 7–33-*nerved, with* 3–7 *nerves stronger than the others,* cordate-based, clasping one-half to three-fourths the circumference of stem, acutish to blunt, *the margin with fugacious more or less appressed denticles; stipules whitish, coarsely nerved,* ovate to lanceolate, obtuse, keelless, 1–2 *cm. long, soon disintegrating into stringy white fibers; peduncles* often *upwardly thickened,* 0.15–2.5 *dm. long; spikes* with 6–12 whorls, moniliform in anthesis, *in fruit* 1.5–4 *cm. long* and 1 *cm. thick;* bodies of grayish- to olive-green *fruits* obovate, rounded on back and at base, 2.5–3.5 mm. long, 2–3 *mm. wide,* the prominent beak up to 1 mm. long. — Lakes and rivers, frequently brackish or alkaline, Lab. to Alaska, s. to Gaspé Pen., Que., n. and w. N.E., e. Pa., N.Y., n. O., centr. Ind., ne. Ill., n. Ia., Neb., Colo., Utah, Nev. and Calif. July–Sept. — Hybridizes with nos. 28, 30, 31, 32, 35, and var. of 37.

37. P. perfoliàtus L. (with leaf clasping stem). — More slender than no. 36; *stem* 1–2 *mm. in diameter,* often much branched; *leaves delicate, membranaceous,* orbicular to ovate-lanceolate, *mostly ovate,* 1–6(–7) cm. long, 1.5–3 cm. wide, 11–21-*nerved, with* 5–7 *nerves more prominent than the others,* fulvous to olivaceous; *stipules delicate, translucent,* often appressed, *fugacious,* ovate-oblong, 0.4–2 cm. long; *peduncles not thickened,* 1–9 *cm. long; spikes* with 2–8 whorls, *in fruit* 1–2 *cm. long and* 8 *mm. thick;* bodies of fulvous to olive *fruits* 2.3–3 mm. long, 2–2.3 *mm. broad.* — Calcareous to brackish waters, s. Lab., Que. and N.B., local; reports from Minn.; not known with us to fruit. (Eurasia; n. Afr.; Austral.) Generally represented here by

Var. **bupleuroìdes** (Fern.) Farw. (resembling *Bupleurum*), Red-head P. — More slender, with *stem* 0.4–1.5 *mm. thick;* fulvous *leaves only* 0.5–2 *cm. wide, with* 7–17 *nerves.* (*P. bupleuroides* Fern.) — Calcareous or brackish waters, Nfld. to Ont., s. to N.S., N.E., L.I., coast of N.C., centr. Pa. and O.; nw. Fla. to La. July–Oct. — Hybridizes with var. *tenuissimus* of no. 17 (producing × **P. mýsticus** Morong) and with nos. 31, 32 and 36.

3. RÚPPIA L. Ditch-grass

Flowers 2, approximate on a slender spadix, which is at first inclosed in the sheathing spathe-like base of a leaf, consisting of 2 sessile stamens, each with 2 large and separate anther-locules, and 4 (rarely more) small sessile carpels with solitary campylotropous suspended ovules; stigma sessile, depressed. Fruits small obliquely ovoid pointed drupes, each on a slender stalk (*podogyne*) which develops after flowering, the spadix then raised on a filiform peduncle. Embryo ovoid, with a short and pointed plumule from the upper end, by the side of the short cotyledon. — Herbs of saline water, of semicosmop. range, with filiform forking stems, and almost capillary alternate leaves abruptly sheathing at the base. Flowers raised to the surface at the time of expansion. (Dedicated to *Heinrich Bernhard Ruppius,* a German botanist, 1689–1719.)*

Leaves scattered or but slightly tufted, their sheaths 0.2–1.2 cm. long, the blades 2–10 cm. long; fruits 1.5–3 mm. long. . . . . 1. *R. maritima.*
Leaves more flabellate-clustered, their sheaths 2–7 cm. long, the blades 1–2 dm. long; fruits 3–4 mm. long. . . . . 2. *R. occidentalis.*

1. R. marítima L. (of the sea). — Leaves linear-capillary, with *blades* 2–10 *cm. long,* their membranous *sheaths* 0.2–1.2 *cm. long; fruits* obliquely erect, 1.5–3 *mm. long. Fr.* July–Oct. — A highly variable nearly cosmopolitan species, represented with us by several striking but apparently confluent varieties, these sometimes treated as species:

a. Carpels ovoid, slightly oblique but not strongly excentric or curved, bluntish or not tapering to a conspicuous beak.
  Peduncles in maturity 1.5–6 cm. long; podogynes 0.6–2.5 cm. long; fruits 2–2.5 mm. long. . . . . Var. *obliqua.*
  Peduncles 2–10 mm. long; podogynes 1–6 mm. long.
    Podogynes distinctly longer than the fruits; the latter 2–2.5 mm. long. . Var. *intermedia.*
    Podogynes shorter than to about equaling the fruits; the latter 1.5–2 mm. long. . . . . Var. *brevirostris.*
a. Carpels strongly excentric and slenderly beaked or semilunate or curved. . . *b.*
  b. Mature carpels 2–3 mm. long, shorter than the podogynes.
    Mature peduncles 3–30 cm. long, spiraling or flexuous; podogynes mostly becoming 1–3.5 cm. long. . . . . Var. *longipes.*

* Treatment based upon that of Fernald and Wiegand in Rhodora, xvi. 119–127 (1914).

Mature peduncles at most 3 cm. long, not spiraling.
Peduncles usually 1.5–3 cm. long; podogynes 1–3.5 cm. long. . . . Var. *rostrata.*
Peduncles 0.5–1.5 cm. long; podogynes 1–6 mm. long. . . . . Var. *subcapitata.*
b. Mature carpels 1.5 mm. long, exceeding the very short podogynes. . . Var. *exigua.*

Var. **obliqua** (Schur) Aschers. & Graebn. (oblique), *R. obliqua* Schur — Saline pools and ditches, Nfld. to Rimouski Co., Que., s. to Me. (Eu.) Fig. 61.

Var. **intermèdia** (Thed.) Aschers. & Graebn. (intermediate), *R. intermedia* Thed. — Local, w. Nfld. to Rimouski Co., Que.; Wash. to Calif. (Eu.)

Var. **brevirόstris** Agardh (short-beaked), *R. brachypus* J. Gay — Local, w. Nfld. to Mingan Ids., Que. and M.I. (Eu., n. Afr.) Fig. 62.

61. R. maritima, v. obliqua.

62. R. maritima, v. brevirostris.

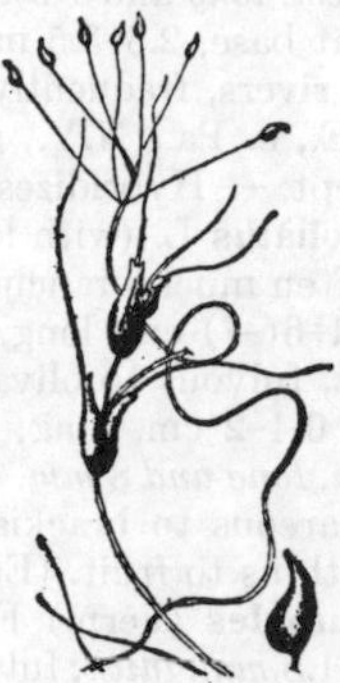

63. R. maritima, v. longipes.

Var. **lόngipes** Hagstr. (long-stalked). — Nfld. to Fla., W.I. and Mex. (Asia) Fig. 63.

Var. **rostràta** Agardh (beaked), *R. rostellata* Koch — Nfld. to lower St. Lawrence, Que., s. to Fla.; w. N.Y. to the Pacific, s. to Mo., Okla., etc. (W.I.; Mex.; S.Am.; Eurasia; Afr.) Fig. 64.

Var. **subcapitàta** Fern. & Wieg. (almost in a head). — E. Saguenay Co. to Temiscouata Co., Que., s. along coast to Ct. Fig. 65.

Var. **exígua** Fern. & Wieg. (little). — Tidal pools, Exploits R., Nfld. Fig. 66.

64. R. maritima, v. rostrata.

65. R. maritima, v. subcapitata.

66. R. maritima, v. exigua.

2. R. occidentàlis S. Wats. (western), Widgeon-grass. — Stouter than no. 1; *leaves often in fan-shaped groups, their blades* 1–2 *dm. long,* the *sheaths broadly scarious-margined and* 2–7 *cm. long,* with free tips; peduncles up to 5 dm. long, strongly spiraling; podogynes becoming 2–4 cm. long; *fruits ovoid,* short-beaked, only slightly inequilateral, 3–4 *mm. long.* — Alkaline waters, Alaska to B.C., eastw. to Sask., Minn. and Neb.

Tribe III. Zannichellìeae Kunth

4. ZANNICHÉLLIA L. Horned Pondweed

Flowers unisexual, sessile, naked, usually both kinds from the same axil; the staminate consisting of a single stamen, with a slender filament bearing a 2–4-locular anther; the pistillate of 2–8 (usually 4) carpels in the same cup-shaped involucre, forming obliquely oblong nutlets in fruit, beaked with a short style, which is tipped by an obliquely disk-shaped or stelliform stigma. Seed orthotropous, suspended, straight. Cotyledon tapering, bent and coiled. — Slender branching herbs, of most reg., growing under water, with mostly opposite long and linear filiform entire leaves, and sheathing membranous stipules. (Named in honor of *Gian Girolamo Zannichelli,* a Venetian botanist, 1662–1729.)

67. Z. palustris, v. major.

1. Z. palústris L. (of marshes). — Branches free-swimming or closely repent; *fruits sessile to short-stalked,* flattish, slightly incurving, *smooth or slightly dentate on the convex back, the body 2–3 mm. long, with beak* 0.8–1.5 *mm. long.* — Fresh, brackish or alkaline waters, e. Que. to Alaska, s. to w. N.Y., W.Va., Tenn., Mo., Tex. and Mex. July–Oct. (S.Am.; Eurasia; Afr.)

Var. **màjor** (Boenn.) W. D. J. Koch (larger). — *Fruits commonly pedicelled* and more or less *dentate* on the back, *the body* 3–3.6 *mm. long,* with *beak* 1–2 *mm. long.* (Var. *pedunculata* of ed. 7, not J. Gay) — Along the coast, Nfld. and e. Que. to Fla., extending slightly inland. (Eu.) Fig. 67.

FAM. 17. NAJADÀCEAE (Naiad Family)

Flowers unisexual, borne from the axils of branchlets and from sheaths of leaf-bases, the staminate a single terminal stamen in a sac-like perianth inclosed in a bottle-like spathe; pistillate flower consisting of a single naked ovary, with a basal anatropous ovule and terminated by 2 or 3 stigmas; seed covered by a thin pericarp closely conforming to the hard testa of the seed. Small branching aquatic annuals, with narrow opposite leaves having conspicuous basal sheaths. A single semicosmop. genus.

NÀJAS L. Naiad

The only genus, wide-spread in temp. and trop. reg., with characters of the family. Flowers through summer and autumn. (Greek *Naias, a water-nymph.*)

a. Leaves stiffish, sinuate, with broad-based coarse teeth, the basal sheath entire or only 1–2-toothed; plant dioecious; fruit plump, 1.5–3 mm. thick, 4–7.5 mm. long. Subgen. Eunajas. . . . . 1. *N. marina.*

a. Leaves flaccid to firm, finely toothed or entire, the basal sheath with 3–13 teeth or spinules on each shoulder; plant monoecious; fruits rather slender, rarely 2 mm. thick, 1.5–3.5 mm. long. Subgen. Caulinia. . . *b.*

b. Each margin of leaf with 20–60 spinules, the sheathing base herbaceous, with gradually sloping or rounded shoulders; staminate flowers 2.5–3.2 mm. long; fruit equilateral or nearly so; testa of seed with square to hexagonal areolae. . . *c.*

c. Leaves usually with inrolling margins and recurving tips, flaccid, pale green; anthers 1-locular; style and stigmas 0.8–2 mm. long; seed very lustrous, with 30–40 rows of hexagonal areolae. . . . . 2. *N. flexilis.*

c. Leaves flat or slightly crisped, straight, dark green, olivaceous or fulvous; anthers 4-locular; seed dull or opaque, with square or squarish areolae.

Style and stigmas 0.1–0.6 mm. long; seed with 10–20 longitudinal rows of areolae. . . . . 3. *N. guadalupensis.*

Style and stigmas 0.7–1.5 mm. long; seed with 50–60 rows of areolae. 4. *N. Muenscheri.*

b. Each margin of leaf with 6–20 minute spinules, the sheathing base subtruncate to conspicuously auricled at summit; staminate flowers 1–1.5 mm. long; seed slightly asymmetrical or curved, with elongate areolae.

Summit of leaf-sheath scarious, conspicuously auricled, with prolonged teeth; seed slightly falcate, 2.5–3.5 mm. long, with crowded rows of vertically elongate areolae about twice as long as broad. . . . 5. *N. gracillima.*

Summit of leaf-sheath herbaceous, subtruncate, with short teeth; seed barely asymmetrical, 2–3 mm. long, with (when dry) 12–18 ribs connected by transversely elongate ladder-like areolae. . . . . 6. *N. minor.*

Subgen. Eunàjas Aschers.

68. N. marina.

1. N. marìna L. (of the sea). — *Dioecious,* stoutish or slender, fragile, loosely branching up to 1 m. or more long, the internodes sometimes stoutly armed; *leaves stiffish,* linear to linear-oblong, 1–3 cm. long, 1–2.5 *mm. wide, each margin sinuate with* 3–12 *broad-based coarse sharp teeth,* the back sometimes toothed, the *basal sheath with the large rounded shoulder entire or with* 1–2 *teeth on each side; staminate flower* soon rupturing its spathe, 3–4 *mm. long,* anther 4-locular; pistillate flower about 3.5 mm. long, style and the 3 (or 2) stigmas 0.7–1.5 mm. long; *fruit* ellipsoid, 4–7.5 *mm. long,* 1.5–3 *mm. thick;* the pericarp at first inflated, becoming dry, tight and ruptured; seed opaque, closely alveolate. (Including vars. *gracilis* Morong and *recurvata* Dudl.) — Deep or shallow alkaline waters, centr. and w.N.Y.; Mich.; Minn.; N.D.; Utah to Calif. and Mex.; Fla. Aug.–Oct. (Tropics; Eurasia) Fig. 68.

Subgen. CAULÌNIA A. Br.

2. N. fléxilis (Willd.) Rostk. & Schmidt (flexible). — Monoecious, fragile, bushy-branched with crowded nodes or elongate and slender, light green to reddish or olive; *leaves linear* (slenderly to broadly) or narrowly lanceolate, *commonly with slightly inrolled margins and recurving tips, 1–4 cm. long,* 0.2–2 mm. wide, *each margin with* 20–40 *minute* 1-*celled spinules; each edge of the gradually tapering to rounded herbaceous sheath with* 6–13 *teeth;* staminate flowers 2.5–3.2 mm. long, borne near the tips of the fertile branches; *anther 1-locular;* pistillate flowers in the lower and middle axils, 1.6–2.5 mm. long; *style and* 2 commonly spinulose-based *stigmas* 0.8–2 *mm. long; fruit* slenderly to broadly ellipsoid, 2–3.5 *mm. long,* the *lustrous seed* (under magnification) *obscurely reticulate with* 30–40 *rows of hexagonal areolae,* closely invested by the yellowish to purplish pericarp. (Including var. *robusta* Morong) — Shallow fresh to brackish water, Nfld. to L. Mistassini, Que., w. to Minn., s. to Va., O., Ind., Ill. and Ia.; Alta. and B.C. to Oreg. July–Oct. (Nw. Eu.) FIG. 69.

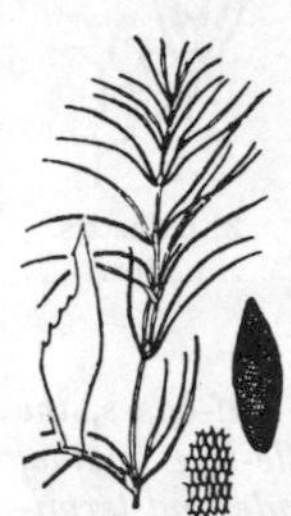

69. N. flexilis.

3. N. guadalupénsis (Spreng.) Magnus (of the island of Guadalupe). — Less fragile than no. 2, *rather firm,* deep green to purple; *leaves linear,* obtuse or acute, 0.5–2 *cm. long, flat* or *slightly crisped,* each margin of the gradually sloping base with 3–10 spinules; *anther 4-locular; style and* 2 or 3 *stigmas* 0.1–0.6 *mm. long; fruit* 1.5–2.5 *mm. long,* the *dull seed* clearly marked *with* 15–18 *rows of usually squarish areolae.* — Widely dispersed in trop. Am., extending n. to se. Mass., N.Y., sw. Que., nw. Pa., n. O., s. Mich., Minn., S.D., Id. and Oreg. Aug.–Oct. FIG. 70.

N. olivàcea Rosend. & Butt. (olive), a little known and usually sterile plant found from w.-centr. N.Y. to Minn., may prove distinct from no. 3, differing in stouter habit and seeds 2.3–2.5 mm. long, with smoother coats.

70. N. guadalupensis.

4. N. Muénscheri Clausen (for its discoverer, WALTER CONRAD LEOPOLD MUENSCHER, 1891–). — *Darker green than no.* 3, more slender and with longer internodes; *leaves about* 1 *cm. long,* 1 mm. wide, the broad bases somewhat lobed; *style and stigmas* 0.7–1.5 *mm. long; the seed with* 50–60 *rows of areolae, the raphe prominent.* — Fresh tidal margins of rivers, e. N.Y. to se. Va. Aug.–Oct.

5. N. gracíllima (A. Br.) Magnus (very slender). — *Very slender and delicate; the linear-setaceous leaves straight,* somewhat divergent from the *scarious sheathing base,* 1.5–3.5 cm. long, *rarely* 0.3 *mm. broad,* each leaf with 6–20 remote usually 3-*celled spinules* above the entire lower third; *the sheath with conspicuous rounded to truncate* 4–7-*toothed auricles; staminate flowers* 1.5 *mm. long; fruit* linear-oblong or slightly falcate, *with unequal sides,* 2.5–3.5 mm. long; *seed* dull or slightly lustrous, *bearing about* 24 *rows of longitudinal elongate quadrangular areolae.* — Muddy, peaty or sandy ponds, pools or shores, s. Me. to Minn., s. to Va., Ky. and Mo. July–Oct. FIG. 71.

71. N. gracillima.

6. N. MÌNOR All. (smaller). — In habit and *recurving leaf-tips* suggesting slender states of no. 2; *leaves* 0.3–0.5 mm. broad, each margin with 6–15 minute spinulose teeth, *the herbaceous sheath with subtruncate* 5–7-*toothed shoulders; fruits* 2–3 *mm. long,* about 0.6 mm. thick; *seed with* 12–18 *longitudinal ribs connected by transversely elongate ladder-like areolae,* the raphe prominent near the base. — Locally established in Hudson R. and other waters, N.Y.; Ohio R., W.Va. Aug.–Oct. (Natzd. from Old World)

FAM. 18. JUNCAGINÀCEAE (ARROW-GRASS FAMILY)

Paludal plants with terete bladeless leaves. Flowers perfect (in ours) or unisexual, in spikes or racemes, ours with herbaceous perianth of 3 greenish sepals and 3 similar petals. Carpels 3 to 6, at first more or less united, separating at maturity. Seeds anatropous; embryo straight, without endosperm. Carpels in fruit indehiscent and 1-seeded or follicular and 2-seeded. (SCHEUCHZERIACEAE)

Carpels 3–6, united until maturity; leaves radical, closed at tip; flowers bractless, in a spiciform raceme terminating the naked scape. . . . . 1. *Triglochin.*

Carpels nearly distinct, at length divergent; flowers bracteate, in a loose raceme on a leafy stem; leaves with a terminal pore. . . . . 2. *Scheuchzeria.*

1. TRIGLÒCHIN L. Arrow-grass. Troscart (Que.)

Sepals and petals concave, deciduous. Stamens 3–6; anthers oval, on very short filaments. Carpels united during anthesis, stigmas sessile; ovule solitary. Fruit splitting when ripe into 3–6 carpels, which separate from a persistent central axis. — Perennials of cosmop. or semi-cosmop. range, with rush-like fleshy leaves below, sheathing the base of the virgate naked and jointless scape. Flowers small, in a spiciform raceme, bractless. (Name composed of the Greek *treis, three*, and *glochis, point*, from the three points of the ripe fruit in no. 2 when dehiscent.)

Fruit longer than thick, 2.5–9 mm. long.
  Fruit (with 3–6 carpels) ovoid-prismatic, with 3–6 recurving beaks. . . . 1. *T. maritima.*
  Fruit (3-carpellate) clavate- or linear-prismatic, blunt. . . . . 2. *T. palustris.*
Fruit depressed-globose, 1.5–2 mm. long. . . . . 3. *T. striata.*

1. T. marítima L. (of the sea). — *Rootstock* covered with persistent whitish leaf-bases, *not stoloniferous;* scapes (up to 11.5 dm. high) and leaves thickish; *fruit ovoid or short-prismatic,* 1.3–3 mm. thick; *carpels* 3– (more often) 6, *rounded at base,* the edges acutish and reflexed when ripe, *the beaks recurving.* — Saline, brackish or fresh marshes and shores, Lab. to Alaska, s., on the coast to Del., and inland to Nfld., n. Me., s. Que., centr. and w. N.Y., nw. Pa., n. O., n. Ind., n. Ill., n. Ia., N.D., N.M., Ariz. and n. Mex. May–Aug. (Eurasia; n. Afr.; Patagonia) Fig. 72. — The smallest diffuse individuals with open raceme are *T. concinna* Davy, hardly a distinct species.

72. T. maritima.

2. T. palústris L. (of marshes). — *Rootstock* short, *emitting filiform* bulb-bearing *stolons;* scape (up to 7 dm. high) and leaves slender; *stamens* 6; *fruit linear-clavate,* 0.5–2 mm. thick, *blunt; carpels* when ripe separating from below upward, *awl-pointed at base.* — Damp brackish or calcareous places, Greenl. and Lab. to Alaska, s., on the coast to s. Me. and rarely to R.I., and inland to centr. Nfld., n. Me., s. N.Y., nw. Pa., n. O., Ill., n. Ia., Neb., N.M. and Calif. May–July. (Eurasia; s. S.Am.) Fig. 73.

73. T. palustris.

3. T. striàta R. & P. (streaked). — *Rootstock bearing stolons;* scape (up to 3.5 dm. high) and leaves slender; flowers very small; *sepals orbicular; stamens* 3; *fruit globose-triangular* or, when dry, 3-lobed, 1.5–2 *mm. long.* — Saline or brackish marshes, Fla. to La., n. very locally to Del. and Md.; Oreg. and Calif. May–Sept. (S.Am.; Afr.; Austral.)

2. SCHEUCHZÈRIA L.

Sepals and petals oblong, spreading, the latter narrower, persistent. Stamens 6, anthers linear. Carpels 3, globular or ovoid, slightly united at base, 2-ovuled, bearing flat sessile stigmas, in fruit forming 3 diverging and inflated 1–2-seeded follicles. — A low boreal bog-herb, with a creeping jointed rootstock, tapering into the ascending simple stem, which is zigzag, partly sheathed by the bases of the grass-like conduplicate leaves, and terminated by a loose raceme of a few flowers, with sheathing bracts, leaves tubular at the apex. (Named for *Johann Jacob* and *Johann Scheuchzer,* distinguished Swiss botanists, 1672–1733 and 1684–1738.)

74. S. palustris, v. americana.

1. S. palústris L. (of marshes). — The typical Eurasian plant has nearly beakless follicles 5–7 mm. long and seeds 3–4 mm. long. Ours is

Var. **americàna** Fern. (American). — Flowers 3–4 mm. long; follicles 7–10 mm. long, with a curving beak 0.5–1 mm. long, seeds narrowly ellipsoid, 4–5 mm. long, black. — Bogs, quagmires and peaty shores, Nfld. to Man. and Wash., s. to N.J., Pa., n. O., n. Ill., n. Ia., Neb., N.M. and Calif. May–July. Fig. 74.

FAM. 19. ALISMATÀCEAE (Water-plantain Family)

Paludal or aquatic herbs with scape-like stems, sheathing leaves and perfect or unisexual flowers; perianth of 3 *herbaceous mostly persistent sepals and* (*in ours*) *as many white or roseate deciduous petals, which are imbricate or involute in bud; stamens* 3 *or more, included; carpels numerous, distinct,* 1*-locular and mostly* 1*-ovuled, becoming achenes in fruit* (in our genera); *seeds erect, campylotropous, without endosperm.* — Roots fibrous; leaves radical, petiolate and strongly nerved with transverse veinlets, the earlier (and often all) without blade; flowers bracted, mostly verticillate in a raceme or panicle. (ALISMACEAE)

Carpels in a single ring, strongly flattened, with style on the ventral margin, distinctly below the tip; flowers all perfect; stamens 6. . . . 1. *Alisma.*
Carpels in dense heads, with an apical or subapical style (at summit of ventral margin); stamens more than 6.
Carpels plump; flowers all perfect. . . . . 2. *Echinodorus.*
Carpels flattened; upper flowers mostly staminate.
Lower flowers perfect; stamens 9–15. . . . . 3. *Lophotocarpus.*
Lower flowers pistillate or all flowers pistillate or all staminate; stamens very numerous. . . . . 4. *Sagittaria.*

1. ALÍSMA L. WATER- or MUD-PLANTAIN

Petals involute in the bud. Ovaries many in a simple circle on a flattened receptacle, forming flattened coriaceous achenes which are dilated and 2–3-keeled on the back. — Scape with whorled panicled branches. Flowers small, white or pale rose-color. Genus of warm or temp. reg. (The Greek name of a water-plant, of uncertain derivation.)

Leaves immersed and ribbon-like or, if emersed, with lanceolate to narrowly elliptic firm blades 2–10 cm. long, 0.5–2 cm. broad and gradually tapering at base; stamens barely equaling the ovaries. . . . . 1. *A. gramineum.*
Leaves essentially all emersed, with submembranaceous ovate to elliptic blades up to 2.5 dm. long and 1.5 dm. broad, in well-developed plants rounded to subcordate at base; stamens exceeding the ovaries.
Flowers 7–13 mm. broad; sepals with broad scarious margins, in anthesis 3–4 mm. long; petals 3.5–6 mm. long; styles about equaling ovaries; achenes 2.2–3 mm. long. . . . . 2. *A. triviale.*
Flowers 3–3.5 mm. broad; sepals not conspicuously scarious-margined, in anthesis 2–2.5 mm. long; petals 1–2 mm. long; styles about one-fourth as long as ovaries; achenes 1.5–2 mm. long. . . . . 3. *A. subcordatum.*

1. A. gramíneum K. C. Gmel. (grass-like). — *Submersed, with ribbon-like leaves* up to 1 m. long and 3–15 mm. broad, *or emersed with* slender-petioled firm leaves with *coriaceous lanceolate to acutely elliptic blades up to* 1 dm. long and 2 *cm. broad;* scapes from much shorter than to exceeding the leaves; inflorescence simple and loosely elongate, with 2 or more whorls, or forking, paniculate and up to 5 dm. long; *bracts ovate to ovate-lanceolate, scarious-margined;* pedicels 0.5–4 (–5) cm. long; flowers of submersed plants cleistogamous, of terrestrial plants expanded; sepals without broad scarious margins, 1.5–3 mm. long; *petals purplish or white,* 2–4 *mm. long; stamens barely equaling the ovaries;* anthers rounded, barely 0.5 mm. long; *style about half as long as the ovary;* achenes 2–2.7 mm. long, usually with narrow dorsal ridges. (*A. Geyeri* Torr.) — Calcareous to brackish waters and muddy shores, very local, sw. Que and n. N.Y.; Wisc. to Alta. and Wash., s. to S.D., Utah, Nev. and Oreg. Late July–Oct. (Eurasia; n. Afr.) PL. I, FIG. 1.

2. A. triviàle Pursh (ordinary). — *Emersed* perennial, with long-petioled ascending or erect leaves, *the ovate to elliptic submembranaceous blades* up to 2.5 dm. long and 1.5 dm. broad, in all but dwarfed individuals *rounded to subcordate at base;* scapes stiff, shorter than to much exceeding the leaves; inflorescence ovoid or pyramidal, 0.5–9 dm. long, much forked, the branches ascending; bracts ovate to ovate-lanceolate; *flowers* 7–13 *mm. broad; sepals* rounded-ovate, *broadly scarious-margined,* in anthesis 3–4 *mm. long; petals* white (rarely pink), 3.5–6 *mm. long; stamens twice as long as the ovaries;* anthers ovoid, 0.6–0.8 mm. long; *style about equaling the ovary; fruiting heads mostly* 4–7 *mm. in diameter; achenes* 2.2–3 *mm. long, with broadly rounded dorsal keels.* (*A. Plantago-aquatica* of ed. 7, in part, not L.; *A. brevipes* Greene) — Shallow water, muddy shores, ditches, etc., Que. to B.C., s. to N.S., N.E., Md., W.Va., Mich., Ia., Neb., N.M., Ariz. and n. Mex. June–early Sept. PL. I, FIG. 2.

3. A. subcordàtum Raf. (somewhat heart-shaped). — Similar to no. 2; panicle often more branched; bracts lanceolate; pedicels capillary; *flowers* 3–3.5 *mm. broad; sepals only narrowly margined,* in anthesis 2–2.5 *mm. long; petals* 1–2 *mm. long; stamens only slightly exceeding the ovaries; anthers spherical,* 0.3–0.5 *mm. long; style about a quarter as long as ovary; fruiting heads* mostly 3–4 *mm. broad; achenes* 1.5–2 *mm. long.* (*A. parviflorum* Pursh; *A. Plantago-aquatica,* var. *parviflorum* (Pursh) Farw.) — Similar habitats, Fla. to Tex. and Mex., n. to N.E., N.Y., s. Ont., Mich., Wisc., Minn. and Neb. Late June–early Sept.

2. ECHINÓDORUS Richard BURHEAD

Petals imbricated in the bud. Stamens 6–21 or more. — Mostly annuals or short-lived perennials of N. and S. Am., s. Eu. and Afr., with the habit of *Sagittaria,* the naked stems sparingly branched or simple, and the perfect flowers on rather short pedicels, in whorls of 3–6 or more, the achenes plump. Fl. summer and autumn. (Name from the Greek *echinus, rough husk,* and *doros, a leathern bottle,* applied to the ovary, which is in most species armed with the persistent style, forming a sort of prickly head of fruit.)

75. E. tenellus.

1. E. tenéllus (Mart.) Buchenau (very slender). — *Scapes* 1.5–10 *cm. high;* shoots often creeping and proliferous; submersed leaves lance-linear phyllodia; emersed *leaves* petiolate *with a lanceolate blade, acute* (1–3 *cm. long); umbel* single, 2–8-*flowered; pedicels reflexed* in fruit; *flower* 6 *mm. broad;* stamens 9; styles much shorter than the ovary; *achenes beakless,* 8-ribbed, reddish-brown, without glands. (*Helianthium tenellum* (Mart.) Britt.) — Sandy shores, Fla. to Tex., n., very locally, to se. S.C.; Del. and s. N.J.; w. L.I.; Middlesex Co., Mass.; sw. Ill. and e. Mo. July–Oct. (Mex.; W.I.; S.Am.) FIG. 75 and PL. I, FIG. 3.

2. E. rostràtus (Nutt.) Engelm. (beaked). — *Scape erect,* 1–6 *dm. high,* longer than the leaves; *leaves broadly ovate, cordate or truncate at base, obtuse,* the blade 2–11 cm. long, or lanceolate in forma **lanceolàtus** (Engelm.) Fern. (lanceolate); umbels proliferous, in a branched panicle; flowers 8–10 mm. broad; *stamens* 12; *styles longer than the ovary; achenes with a conspicuous erect beak.* (*E. cordifolius* of ed. 7, not (L.) Griseb.) — Muddy shores and bottoms, Fla. to Tex. and Mex., n. to O., Ill., Ia., Neb. and Calif. June–Oct. (Centr. Am.; W.I.) FIG. 76.

76. E. rostratus.

3. E. cordifòlius (L.) Griseb. (with heart-shaped leaf). — *Stems or scapes prostrate, creeping or arching, secund* (6–12 dm. long), proliferous, bearing many whorls of flowers; leaves somewhat truncately cordate, obtuse (5–20 cm. broad), long-petioled; flowers 12–20 mm. broad; *stamens about* 21; *styles shorter than the ovary; achenes with a short incurved beak, the keeled back denticulate.* (*E. radicans* (Nutt.) Engelm.) — About ponds and quiet streams, nw. Fla. to Tex. and Mex., n. to se. Va., s. Ind., Ill., Mo. and e. Kans. July–Oct. FIG. 77. — Roots sometimes bearing small fusiform to subcylindric tuber-like enlargements nearly their entire length.

77. E. cordifolius

3. LOPHOTOCÁRPUS Th. Durand

Sepals strongly concave, erect and appressed to the fruit. Lower flowers perfect. Stamens 9–15. — Perennials of Am., trop. Asia and Afr., with habit and carpels much as in *Sagittaria.* (Name from the Greek *lophotos, crested,* and *carpos, fruit,* not very applicable.)

1. L. calycìnus (Engelm.) J. G. Sm. (with large calyx). — Leaf-blades ovate, sagittate, up to 2 dm. broad, or in forma **depauperàtus** (Engelm.) Fern. (impoverished) unlobed or barely auricled and narrowly elliptic to spatulate, or in forma **máximus** (Engelm.) Fern. (largest) 3 dm. wide and much broader than long; raceme (rarely forking) with 2–7 whorls; *mature sepals rounded or oblate,* 7–13 *mm. broad, covering two-thirds of the fruiting head; heads of carpels* 0.7–1.6 cm. in diameter, *the beaks subappressed but readily visible; achenes* cuneate-oblanceolate to -obovate, 0.6–1.5 mm. broad, *their surfaces obscurely and finely reticulate,* the horizontal to oblique *beak nearly as long as the breadth of the achene, the margin* (without the wing) *conforming to the margin of the embryo.* — Sloughs and ponds, O. and Mich. to Minn. and S.D., s. to w. Va., Ala., La., Tex., and N.M. June–Oct. PL. I, FIG. 4.

2. L. spongiòsus (Engelm.) J. G. Sm. (spongy). — Leaves represented by thick spongy phyllodia, or in forma **laminàtus** Fern. (with leaf-blade) with lanceolate to ovate unlobed or sagittate blades; scapes 1–20 cm. long, much shorter than the leaves, simple (rarely forking), with 1–4 whorls of 1–3 long-pedicelled flowers; pedicels recurving in fruit; *mature sepals broadly ovate,* 3–8 *mm. broad, about half covering the fruiting head; heads of mature carpels* 7–10 mm. in diameter, *the beaks tightly appressed; achenes* narrowly cuneate-obovate, 1–1.5 mm. broad, *their surfaces coarsely spongy-reticulate,* the horizontal *beak barely half as long as the breadth of the achene; embryo about half filling the achene.* — On tidal mud of brackish estuaries, ne. N.B. to Va. July–Oct. PL. I, FIG. 5.

4. SAGITTÀRIA L. ARROWHEAD. SWAMP-POTATO. FLÈCHE D'EAU (Que.)

Sepals loosely spreading or reflexed in fruit. Petals imbricated in the bud. Carpels crowded in a spherical or somewhat triangular depressed head on a globular receptacle, in fruit forming

Plate I

Fig. 1, Alisma gramineum; fig. 2, A. triviale; fig. 3, Echinodorus tenellus; fig. 4, Lophotocarpus calycinus; fig. 5, L. spongiosus.

flat membranaceous winged achenes. Stamens mostly numerous. — Paludal or aquatic, mostly perennial, stoloniferous herbs of trop. and temp. reg., mostly American, with milky juice; the scapes sheathed at base by the bases of the long cellular petioles, of which the primary ones, and sometimes all, are destitute of any proper blade (*i.e.*, are phyllodia); blade, when present, unlobed or sagittate. Flowers produced all summer, whorled in threes, with membranous bracts. (Name from *sagitta*, *an arrow*, from a prevalent form of the leaves.)*

a. Pedicels of mature pistillate flowers arched-recurving.
- Leaves linear and bladeless or with blades up to 4 cm. long; only 1–3 of the lowest flowers pistillate; filaments glabrous; fruiting heads 4–6 mm. in diameter; achenes (rarely ripening) smooth, 0.7–1.4 mm. broad. . . . 1. *S. subulata.*
- Leaves with blades 0.5–2 dm. or more long; all flowers of the lower 1–4 whorls pistillate; filaments pubescent; fruiting heads 0.9–1.4 cm. in diameter; achenes spongy-rugulose, 1.4–2 mm. broad. . . . . . 2. *S. platyphylla.*

a. Pedicels of pistillate flowers spreading or ascending, not recurving. . . *b.*

b. Leaves not regularly sagittate or hastate, without basal lobes or at most with small basal auricles (if rarely sagittate with bracts connate, filaments dilated and pubescent, and achenes rugose, with subulate, not broad-based beak); bracts usually connate. . . *c.*

c. Scapes flexuous near the lowest whorl; pistillate flowers subsessile (very exceptionally pedicelled); head of carpels strongly echinate; beak of the rugose almost keelless achene 1–1.5 mm. long. . . . . . . 3. *S. rigida.*

c. Scapes straight; pistillate flowers slender-pedicelled; head of carpels not strongly echinate; beak of the keeled or winged achene only 0.1–0.6 mm. long. . . *d.*

d. Leaves with flattish phyllodia or definite blades. . . . *e.*

e. Fruiting heads 0.4–1.5 cm. in diameter; achenes 1.5–2.5 mm. long, 0.7–1.4 mm. broad, the faces plane or merely with low keels, beak 0.1–0.4 mm. long. . . *f.*

f. Stamens 20 or more; filaments slender, longer than the slender anthers; fruiting heads 1–1.5 cm. in diameter; beak at inner summit of achene.
- Bracts ovate, obtuse, strongly papillose, connate; filaments arachnoid; pistillate pedicels shorter than the staminate; achenes 2–2.5 mm. long, with ascending beak. . . . . . 4. *S. falcata.*
- Bracts lance-attenuate, glabrous or only slightly papillose, nearly free; filaments glabrous; pistillate pedicels longer than staminate; achene 1.5–2 mm. long, with horizontal or incurved beak. . . . . 5. *S. ambigua.*

f. Stamens fewer than 20; filaments dilated, shorter than the anthers; fruiting heads 4–10 mm. in diameter; beak usually borne below summit of achene (terminal in no. 8).
- Phyllodia 1–2.5 cm. broad, obtuse; fruiting pedicels 4–6.5 cm. long; bracts 5–10 mm. long; anthers linear-oblong, 2–2.5 mm. long; achenes 2–2.5 mm. long. . . . . 6. *S. Weatherbiana.*
- Phyllodia narrower, acuminate; pistillate or fruiting pedicels 0.5–3 cm. long; bracts 2–6 mm. long; anthers ellipsoid or rounded, 0.6–1 mm. long; achenes (rarely developed) 1.5–2 mm. long.
  - Emersed blades, when developed, firm and narrowly to broadly lanceolate; phyllodia thin; bracts 3–6 mm. long; pedicels of pistillate flowers 1–3 cm. long, about equaling those of staminate flowers; stamens about 18, the anthers and filaments subequal; beak of carpel borne well below summit; plant of fresh water and shores. . . . . . 7. *S. graminea.*
  - Emersed blades rare, linear; phyllodia spongy; bracts 2 mm. long; pedicels of pistillate flowers 5–8 mm. long, much shorter than those of staminate flowers; stamens 12, anther twice as long and broad as filament; beak of carpel terminal. 8. *S. Eatoni.*

e. Fruiting heads 1.5–2 cm. in diameter; achene 2.5–3 mm. long, 1.4–1.8 mm. broad, the faces and back with broad and thin wings, beak 0.4–0.6 mm. long. . . . . 9. *S. cristata.*

d. Leaves all represented by terete or quill-like phyllodia; stamens about 12, the dilated filaments shorter than the anthers; achene 2–2.7 mm. long, when ripe strongly rugose and several-keeled. . . . . . 10. *S. teres.*

* The illustrations show, in most cases, the outline of a characteristic leaf, the fruiting whorl and the achene.

b.Leaves all (or at least the latest) sagittate or hastate; bracts free or nearly so; stamens very numerous, the slender filaments mostly longer than the anthers. . *g*.

g.Achenes 2.3–4.5 mm. long, with broad-based marginal beak 0.5–2 mm. long; fruiting heads 1–3 cm. in diameter. . *h*.

h.Faces of achenes keeled, beak ascending to subhorizontal. *i*.

i.Inflorescence with 5–13 whorls; bracts firm, long-attenuate, mostly exceeding the pistillate pedicels; achenes 2.4–3.5 mm. long.

Pedicels of pistillate flowers 3–12 mm. long; fruiting heads 1–1.5 cm. in diameter. . . . . . 11. *S. australis.*

Pedicels of pistillate flowers 1–2 cm. long; fruiting heads 2–3 cm. in diameter. . . . . . 12. *S. brevirostra*

i.Inflorescence with 2–4 whorls; bracts scarious-margined, acute or subobtuse, mostly shorter than the fruiting pedicels (0.7–3 cm. long); achenes 3.6–4.5 mm. long. . . . 13. *S. Engelmanniana.*

h.Faces of achenes plane, beak subhorizontal to incurved; bracts subscarious, acute to obtuse.

Bracts boat-shaped or cucullate; pedicels terete; anthers linear or oblong, 1.5–2.3 mm. long; transcontinental species. 14. *S. latifolia.*

Bracts flat; pedicels flattened; anthers ovate or ovate-quadrate, 1–1.3 mm. long; southeastern. . . . 15. *S. planipes.*

g.Achenes 2–2.6 mm. long, with the slender (subulate) erect or incurved beak only 0.2–0.4 mm. long and borne from the summit well in from the margin; fruiting heads 0.8–1.3 cm. in diameter. . . . . . 16. *S. cuneata.*

1. S. subulàta (L.) Buchenau (awl-shaped). — Dwarf; *leaves linear, strap-shaped, obtuse or acutish,* 2–12 cm. long, 1–3 mm. wide, exceeding to shorter than the slender scape, rarely with a narrow blade (up to 2 cm. long and 4 mm. broad); *inflorescence* 1–4 *cm. long; pedicels* in 1–3 whorls, the 1 or 2 *fruiting ones stouter* and shorter than the others, *recurved,* 0.5–2 cm. long; *bracts scarious, connate or spathe-like* and oblique, obtuse or with prolonged tips, 3–5 *mm. long; filaments* 6–8, *glabrous; fruiting heads nodding,* 4–6 *mm. in diameter; achenes* obovate, 1.6–2.3 mm. long, 0.7–1.4 mm. broad, wing-margined, *with* slenderly keeled faces, the lateral to subterminal *subulate beak* 0.3–0.4 *mm. long.* — Fresh to brackish tidal mud, Fla. to Ala., n. to Mass. July–Sept. Fig. 78.

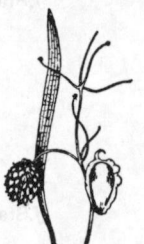

78. S. subulata.

Var. **nàtans** (Michx.) J. G. Sm. (floating). — *Leaves ribbon-like* or with dilated lanceolate to ovate blades up to 4 cm. long and 2 cm. broad; the obtuse bladeless leaves or phyllodia 1–3 *dm. long and* 3–6 *mm. broad,* often overtopped by the scape (1–4 dm. long); *inflorescence* 3–10 *cm. long;* the 1–3 recurving *pistillate pedicels* 0.5–3.5 *cm. long.* — Shallow pools along the coast, Fla. to se. Va. July–Sept.

Var. **gracíllima** (S. Wats.) J. G. Sm. (very slender). — *Very elongate* (up to 1 m. or more) and submerged; *leaves prolonged,* 1–3 *mm. wide;* prolonged scape flexuous; *inflorescence* 1–3 *dm. long,* with 2–4 very remote whorls; *bracts* (at least of the upper whorl) *subherbaceous, elongate, mostly caudate-tipped, nearly distinct,* 6–10 *mm. long; pedicels all elongate; the lower pistillate* arching or spreading, 0.3–2 *dm. long, not much thickened;* fruit unknown. — Deep water of streams, e. Mass. to s. Pa.; se. S.C.

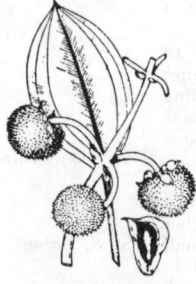

79. S. platyphylla.

2. S. platyphýlla (Engelm.) J. G. Sm. (broad-leaved). — *Leaves* erect, overtopping the scape, 0.3–1 m. high; the *blades lanceolate, elliptic or ovate,* unlobed or with short basal auricles; scape weak; whorls 3–7, *the* 1–4 *lower whorls pistillate, with soon recurving thickish pedicels* 1–2.5 cm. long; bracts ovate, scarious, strongly connate, 3–8 mm. long; filaments dilated, pubescent, longer than the anthers; *fruiting heads* 0.9–1.4 *cm. in diameter; achenes* cuneate-obovate, *spongy-rugulose,* 2.3–3 *mm. long,* 1.4–2 *mm. broad,* the dorsal keel rounded to the subtruncate summit, faces with 1–3 slender ridges; beak subulate, 0.1–0.4 mm. long, obliquely ascending, crowning the straight ventral margin. — Marshes and shores, s. Mo. and Kan. to Ala., La. and Tex. June–Aug. Fig. 79.

80. S. rigida.

3. S. rígida Pursh (stiff). — *Scape weak,* 1.5–8 dm. high or in deep water

many meters long, *flexuous,* sometimes procumbent, *usually bent near the lowest whorl,* shorter than the leaves; *leaves* linear to oval, *unlobed or with one or two narrow arching or divergent basal appendages; whorls* 2–7, *the lowest* 1 (rarely 2) *with pistillate subsessile* to short-pedicelled *flowers,* the staminate flowers on long pedicels; *bracts roundish, obtuse, connate; filaments pubescent,* broad at base, mostly longer than the broad anthers; *achenes narrowly cuneate-obovate, rugose,* 2.5–4 mm. long, 1.3–2 mm. broad, longitudinally costate, the convex back *with an obscure or very slender keel, the subulate erect or arching beak* 1–1.5 *mm. long.* (*S. heterophylla* Pursh, not Schreb.) — Calcareous or brackish mud or water, sw. Que. and s. Me. to Minn., s. to Va., Ky., Tenn., Mo. and Neb. July–Oct. (Introd. into Eu.) FIG. 80. — Foliage very variable: the typical form with linear-lanceolate to oblanceolate stiff acute blades long-attenuate to the tall petioles; forma **flùitans** (Engelm.) Fern. (swimming) with bladeless slender phyllodia; forma **ellíptica** (Engelm.) Fern. (elliptic) with broadly lanceolate, elliptic or oval blades, the unappendaged ones gradually curving to rounded at base.

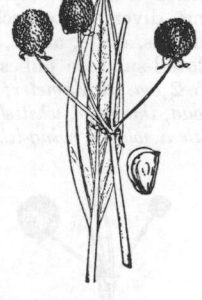

81. S. falcata.

4. S. falcàta Pursh (scythe-shaped). — Slender or stout, 0.8–1.5 m. high; *leaves* erect; the blades *linear-lanceolate to elliptic, unlobed, tapering to both ends, firm,* 1–4 dm. long; scapes simple or branching at lower nodes, the main axis with 6–10 whorls; the lower 1–4 whorls pistillate, with pedicels 0.5–2.5 cm. long; *staminate pedicels longer; bracts ovate, obtuse, strongly papillose,* 0.3–1 cm. long, connate below; sepals commonly papillose; *filaments* slender, *arachnoid,* longer than the slender anthers; fruiting heads about 1 cm. in diameter; *achenes cuneate-oblanceolate, falcate,* 2–2.5 mm. long, 0.7–1 *mm. broad,* narrowly winged, with one or two low facial ridges; *beak lance-subulate,* 0.2–0.4 *mm. long, ascending.* (*S. lancifolia* of ed. 7, not L.) — Swamps and shores (often brackish), Fla. to Tex., n. to Md. July–Oct. (E. Mex. and Guat.) FIG. 81.

82. S. ambigua.

5. S. ambígua J. G. Sm. (doubtful). — Similar to no. 4; scapes 3–9 dm. high, with 2–5 pistillate whorls; *pistillate pedicels* 1.5–3.5 cm. long, *longer than the staminate; bracts lance-attenuate,* slightly papillose, mostly 1–1.5 cm. long, nearly free; *filaments glabrous;* fruiting heads 1–1.5 cm. in diameter; *achenes cuneate-obovate,* 1.5–2 mm. long, 0.8–1.4 *mm. broad,* narrowly thin-winged, faces smooth or with a longitudinal thin keel; *beak minute,* 0.1–0.2 *mm. long, horizontal or incurved.* — Swamps and shores, sw. Mo., Kans. and Okla. May–July. FIG. 82.

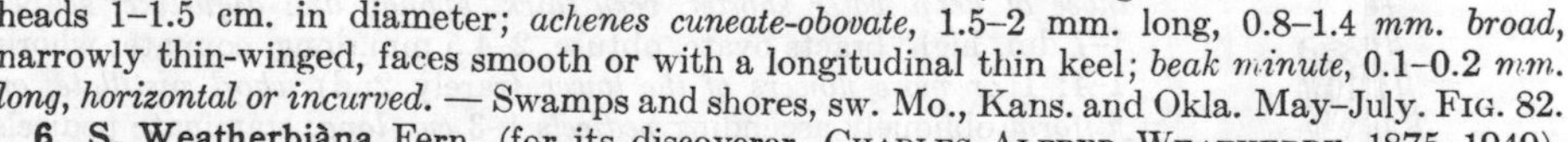

6. S. Weatherbiàna Fern. (for its discoverer, CHARLES ALFRED WEATHERBY, 1875–1949). — Plant 4–9 dm. high, with short thick rhizome and elongate stolons; *phyllodia membranous, dark green,* 1–2.5 *cm. broad, obtuse,* with loosely reticulate venation; *inner leaves dark green,* their thick *petioles* strongly *dilated below;* the lanceolate to elliptic *blade* acuminate at both ends, 1.2–2.5 *dm. long* and 2.5–7.5 *cm. broad, submembranaceous;* flowers in 3–6 whorls; *bracts* scarious, *ovate, acuminate,* united below the middle, 5–10 *mm. long; pedicels* filiform, 2.5–6.5 *cm. long;* sepals 6–9 mm. long; *stamens* 12–18, *with broadly dilated filaments and linear-oblong anthers* 2–2.5 *mm. long; achenes* flat, obliquely cuneate-obovate, 2–2.5 *mm. long,* 1–1.5 *mm. broad,* narrowly winged, the sides smooth or 1–3 ribbed. — Shallow water of wooded swamps, se. Va. to S.C. April, May. FIG. 83. — Fresh foliage developed in autumn and earliest spring.

83. S. Weatherbiana.

7. S. gramínea Michx. (grass-like). — Monoecious or dioecious, slender, 0.7–6 dm. high; *leaves* erect, *either* represented by thin *broadly linear-lanceolate* acute *phyllodia or with the slender petioles bladeless or with narrowly to broadly lanceolate tapering blades* 2–15 cm. long and up to 3.5 cm. broad; scapes simple, with 2–8 whorls; the lower one or two whorls of *pistillate flowers* (or all staminate), *on filiform spreading pedicels* 1–3 *cm. long,* staminate pedicels similar; *bracts ovate,* obtuse to acutish, *connate,* 3–6 *mm. long, scarious or membranaceous, not strongly ribbed;* petals white or pink; *filaments dilated, pubescent,* short; *an-*

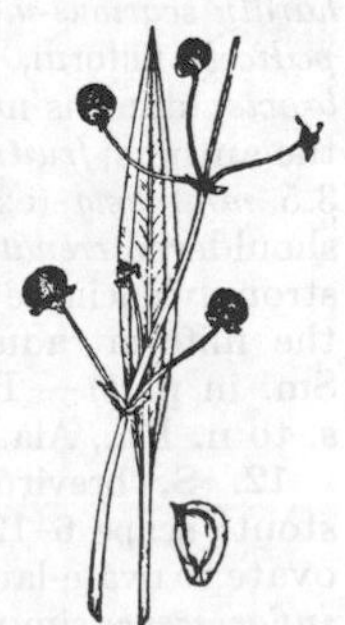

84. S. graminea.

thers ellipsoid to rounded, only 0.6–1 *mm. long, about equaling filaments;* fruiting heads (rarely developed northw.) 4–8 mm. in diameter; *achenes* narrowly obovate, rugulose, 1.5–2 *mm. long,* 0.8–1.2 *mm. broad,* the narrow-keeled back strongly rounded to a high shoulder, *the sides plane or with* 1 *or* 2 *slender ridges; the subulate beak minute* (0. 1–0.3 *mm. long*), terminating the straight or slightly curved ventral margin and *borne below the summit of the achene.* (Including *S. Edwardsiana* Clausen, which seems to be a submerged state.) — Wet sand and mud or in shallow water, Nfld. and s. Lab. to Ont., s. to Va. and "Fla.", W.Va., O., Ind., Mo., Ill. and Tex. Late May–Sept. Fig. 84.

8. S. Eàtoni J. G. Sm. (for its discoverer, Alvah Augustus Eaton, 1865–1908). — Similar to submersed states of no. 7; *phyllodia more fleshy;* petioles and blades rare, the latter linear and only 2–3 cm. long and 2–4 mm. wide; scape 0.5–1.2 dm. high, weak and slender, with 1 or 2 (rarely 3) whorls; *bracts* 2 *mm. long; pedicels of pistillate flowers* only 5–8 *mm. long,* much shorter than those of staminate flowers; petals roseate at base; *stamens* 12, *anthers much larger than filaments; carpels with terminal beaks;* fruit unknown. — Tidal mud and sands, s. N.H. to Va. — Rarely flowering, propagating by subterranean rhizomes.

9. S. cristàta Engelm. (crested). — Monoecious, slender, 2.5–7.5 dm. high; *leaves* erect, *either* represented by *broadly linear-lanceolate* and acuminate *spongy phyllodia or with* petioled *linear to elliptic-lanceolate thick blades* 6–10 cm. long and 0.3–2 cm. wide; scapes with 3–6 *whorls, the lowest one pistillate; filiform fruiting pedicels* 1.5–3 cm. long, divergent to arched-ascending; staminate pedicels similar; bracts subscarious, ovate, acute, 4–7 mm. long, connate below; stamens about 24, with dilated-subulate pubescent filaments longer than the anthers; *fruiting heads* 1.5–2 *cm. in diameter; achenes* cuneate-obovate, 2.5–3 *mm. long,* 1.4–1.8 *mm. broad, the broad whitish crenate dorsal wing strongly rounded,* the *faces* usually *with a whitish wing-like keel;* the subulate oblique or ascending *beak* 0.4–0.6 *mm. long,* terminating the straight ventral margin and borne below the summit of the achene. — Sandy margins, shores and bottoms of lakes and ponds, Bruce Pen., Ont., Wisc., Minn. and n. Ia. July, Aug. Fig. 85.

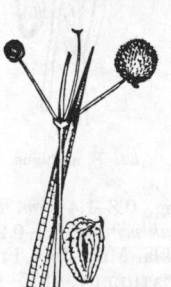

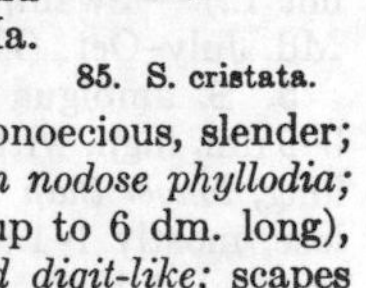

85. S. cristata.

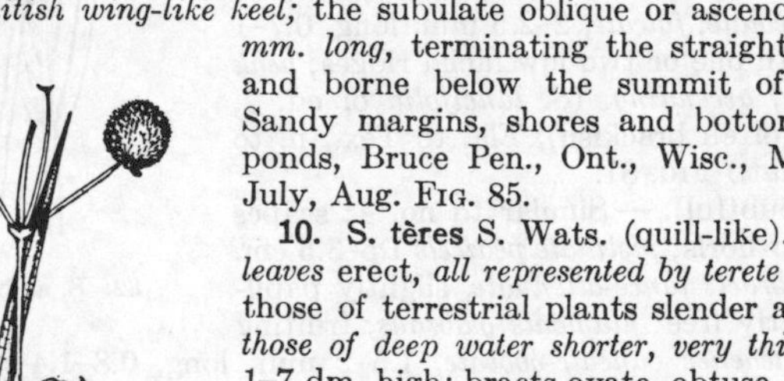

86. S. teres.

10. S. tères S. Wats. (quill-like). — Mostly monoecious, slender; *leaves* erect, *all represented by terete attenuate, often nodose phyllodia;* those of terrestrial plants slender and elongate (up to 6 dm. long), *those of deep water shorter, very thick, spongy and digit-like;* scapes 1–7 dm. high; bracts ovate, obtuse, 2–4.5 mm. long, connate; whorls 1–4; 1 *or more flowers of the lower* (rarely 2nd) *whorl pistillate, on filiform* obliquely ascending *pedicels* 1–3 *cm. long;* staminate pedicels mostly shorter; stamens about 12, the dilated pubescent filaments shorter than the anthers; *fruiting heads* 7–10 *mm. in diameter; achenes* cuneate-obovate, 2–2.7 *mm. long,* 1.3–1.8 *mm. broad, with* strongly rounded *crenate dorsal keel, the faces* (when fully ripe) *rugose and irregularly* 2–4(or more)-*keeled;* the thick-subulate obliquely ascending to horizontal *beak* 0.3–0.7 *mm. long,* crowning the slightly arching to straight ventral margin. — Sandy pond-margins, shores and swamps, e. Mass. and L.I. to e. Md. July–Sept. Fig. 86.

11. S. austràlis (J. G. Sm.) Small (southern). — Stoutish, 3–9 dm. high, monoecious; leaf-blades broadly ovate-oblong, obtusish, sagittate with broad basal lobes; *whorls* 5–11, *the pistillate* 2–4 (rarely only 1); *bracts* distinct, firm, *long-attenuate, hardly scarious-margined; pistillate pedicels* 3–12 *mm. long; staminate pedicels* filiform, 5–18 *mm. long, shorter than to barely equaling the bracts;* stamens numerous; filaments slender, equaling or longer than the anthers; *fruiting heads* 1–1.5 *cm. in diameter; achene* obovate, 2.8–3.5 *mm. long* (excluding beak), 2–3 mm. broad, *with* broad high-shouldered *crenate dorsal keel and lower facial keel;* the broad-based strongly uncinate beak ascending or strongly recurving, 1–2 mm. long; the filiform caducous style 1–1.5 mm. long. (*S. longirostra* J. G. Sm. in part) — By springs, rills and ponds, N.J. and Pa. to s. Ind., s. to n. Fla., Ala. and Mo. July–Oct. Fig. 87.

12. S. breviròstra Mackenz. & Bush (short-beaked). — Very stout; scape 6–12 dm. high; leaf-blades sagittate, with basal lobes ovate to ovate-lanceolate, acute, about equaling the terminal portion; *inflorescence* simple or branched at base, 2–5 *dm. long; the main axis with* 7–13 *whorls;* the *lower* 2–6 *whorls pistillate,* with pedicels 1–2

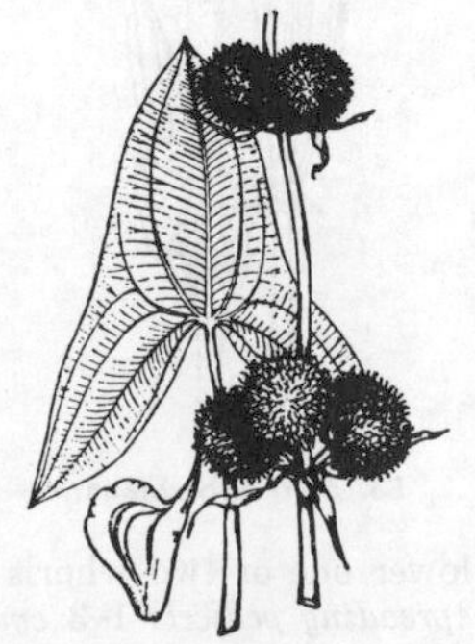

87. S. australis.

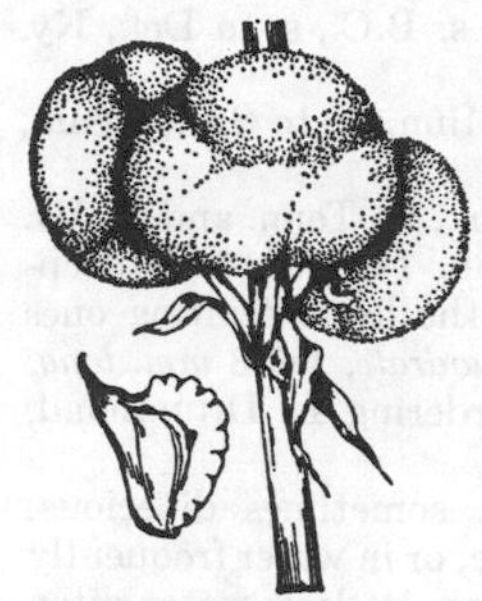
88. S. brevirostra.

cm. long, the staminate on slightly longer pedicels; *bracts firm, lanceolate, long-attenuate, the longer* 1.5–4 *cm. long;* filaments slender, glabrous, equaling the anthers; *fruiting heads depressed,* 2–3 *cm. in diameter; achenes* cuneate-obovate to quadrate, spongy-reticulate, 2.4–3 *mm. long,* 1.8–2.5 mm. broad, with broad *high-shouldered often dentate or serrate dorsal keel and one or two low ridges on the face; the broad-based beak obliquely ascending,* 0.5–1.5 *mm. long, terminating the straight ventral margin.* — Sloughs and wet shores, Ind. and Ill. to Kans. and Okla. July–Sept. Fig. 88.

13. S. Engelmanniàna J. G. Sm. (in honor of George Engelmann, 1809–1884). — Slender, 2.5–8 dm. high, mostly monoecious; blade and lobes of leaves narrowly linear to lance-attenuate, or the blades deltoid or ovate in forma **dilatàta** Fern. (enlarged); whorls 2–4, 1 (rarely 2–3) *whorl pistillate; bracts* distinct, acute to subobtuse, *scarious-margined; pistillate pedicels* 0.7–3 *cm. long; staminate pedicels* 1.3–5 cm. long, *much exceeding the bracts;* stamens numerous; filaments slender, equaling or longer than the anthers; fruiting heads 1.2–2.3 cm. in diameter, strongly echinate; *achenes* cuneate-obovate, 3.6–4.5 *mm. long* (excluding beak), 2–3.5 mm. broad, *with* ventral margin nearly straight, a rounded *entire broad dorsal wing and* 1 *facial wing and* 1 *or* 2 *lower facial keels;* the broad-based curving beak erect, ascending or horizontal, 1–2 mm. long. — Wet sand and peat, Mass. to S.C. July–Oct. Fig. 89.

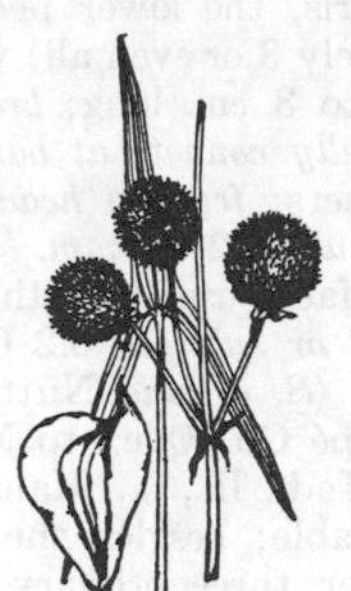
89. S. Engelmanniana.

14. S. latifòlia Willd. (broad-leaved), Wapato, Duck-potato. — Scape 1–15 dm. high, angled, with 1 or more of the lower whorls pistillate or all unisexual; *leaf-blades* ovate to linear, *mostly sagittate* (the earliest, and rarely all, without blades, the latest often of different outline); *the basal lobes* triangular-ovate to linear, *one-half as long as to longer than the body of the leaf;* bracts distinct or slightly connate, thin, subscarious, obtuse to acute; pedicels of pistillate flowers obvious, those of the staminate much longer; petals showy, white; filaments slender, mostly longer than the anthers; *achenes obovate,* 2.3–3.5 *mm. long,* 1.5–3 mm. broad, *with broad marginal wings but no facial keels,* the *subhorizontal to slightly incurved broad-based beak* 1–2 *mm. long.* Fig. 90. — In water or wet places, exceedingly variable, especially in leaf-outline. July–Sept. The more important variations are as follows:

90. S. latifolia.

a. Later leaf-blades acute or, if obtuse strongly narrowed from base to summit, the body or terminal lobe as broad as to many times narrower than long; basal lobes often arching or divergent; whorls of flowers 2–5 (rarely –10); heads of carpels rarely echinate; dorsal keel of achene continued around the summit as a high rounded shoulder, the filiform or subulate tip of the beak 0.3–0.8 mm. long. . . *b.*

b. Body of leaf ovate or ovate-deltoid, as broad as to two-thirds as broad as long, basal lobes triangular. . . . . *S. latifolia* (typical).

b. Body of leaf and basal lobes narrower.

Body of leaf narrowly triangular to oblong-lanceolate, acute or obtuse, two-fifths to one-seventh as broad as long. . . . . Forma *hastata.*

Body of leaf linear-lanceolate to narrowly linear, 0.2–1.5 cm. wide, one-eighth to one-sixteenth as broad as long. . . . . Forma *gracilis.*

a. Later (or all) leaf-blades obtuse or rounded at summit, the body nearly or quite as broad as long; the broad subacute to obtuse basal lobes parallel or only slightly divergent; whorls 4–10; head of carpels usually finely echinate; achene subtruncate at summit, the filiform tip of the beak 0.8–1.5 mm. long.

Bracts and calyx glabrous. . . . . Var. *obtusa.*

Bracts and calyx densely pubescent, the bracts comparatively short. . . . Var. *pubescens.*

S. latifòlia (typical). — N.B. to s. B.C., s. to S.C., Ala., La., Okla. and Calif. — Forma **hastàta** (Pursh) Robins. (halberd-shaped), Saguenay Co., Que., to s. B.C., s. to S.C., La. and

Kans.; Forma **grácilis** (Pursh) Robins. (slender), Gaspé Pen., Que., to s. B.C., s. to Del., Ky. and Neb.

Var. **obtùsa** (Muhl.) Wieg. (blunt), *S. esculenta* Howell — N.S. to Minn., s. to S.C., Tenn., La. and Okla.; Wash. to Calif.

Var. **pubéscens** (Muhl.) J. G. Sm. (hairy). — N.J. and Pa. to Fla., Ala., w. Tenn. and W.Va.

15. S. plánipes Fern. (flat-stalked). — Resembling var. *obtusa* of no. 14, with long deep-seated rhizome; *bracts flat; pedicels laterally compressed and flattened,* the lower fruiting ones erect or nearly so and up to 5 cm. long; *anthers broadly ovate to ovate-quadrate,* 1–1.3 *mm. long;* achenes similar to those of no. 14. — Deep wet argillaceous mud bordering L. Drummond, Great Dismal Swamp, Va. Aug.–Oct.

16. S. cuneàta Sheldon (wedge-shaped), WAPATO. — Monoecious, sometimes dioecious; scape 0.5–4 (or in submersed plants –8) dm. long, arching to erect, simple, or in water frequently branched; leaves in terrestrial forms with slender often curving petioles, in deep water often represented by broad ribbon-like phyllodia; *blades mostly sagittate;* the body linear-lanceolate to ovate, acute (rarely obtuse or with rounded tip); inflorescence with 2–5 whorls, the lower pedicels often replaced by branches, the lower 1 or 2 (rarely 3 or even all) whorls pistillate, these flowers subsessile or on pedicels up to 3 cm. long; *bracts* lanceolate or narrowly ovate, *acute to attenuate, usually connate at base;* filaments subulate, glabrous, about equaling the anthers; *fruiting heads* 0.8–1.3 *cm. in diameter; achenes* obovate, *spongy-reticulate,* 2–2.6 *mm. long,* 1.4–2.5 mm. broad, the wide dorsal keel rounded, the faces usually with a low slender ridge; the *subulate* usually curving *beak erect or suberect,* 0.2–0.4 *mm. long, terminating the strongly rounded ventral keel.* (*S. arifolia* Nutt.) — Calcareous or muddy shores and shallow water, Gaspé Co., Que., to Mackenz. and B.C., s. to N.S., n. and w. N.E., N.Y., O., Ind., Ill., Ia., Kans., N.M. and Calif. June–Sept. FIG. 91. — Excessively variable; besides the responses to submergence in and emergence from water, three primary forms may be noted: the typical form, with blades ovate to lanceolate, acute or subacute, the subparallel basal lobes one-fifth to one-half as long as the terminal lobe; forma **equíloba** Fern. (equally lobed) with more or less divergent sharp basal lobes about equaling the terminal one; forma **hemicỳcla** Fern. (half-circular) with the ovate blades strongly rounded at summit.

91. S. cuneata.

FAM. 20. BUTOMÀCEAE (FLOWERING RUSH FAMILY)

Paludal or aquatic herbs with scapes or long peduncles and perfect flowers; perianth of 3 *green or brightly colored persistent sepals, and* 3 *colored petals, which are imbricate or involute in bud; stamens* 9 (*in ours*) *or more; carpels* 6 *or more, forming many-seeded follicles; seeds straight, curved, or folded, without endosperm.* — Leaves radical or (in the frequently cult. yellow-flowered WATER-POPPY, *Hydrocleis nymphoides*) clustered at remote nodes of the floating stem, ensiform or with dilated blades; flowers showy, in umbels or solitary, with large bracts.

1. BÙTOMUS L. FLOWERING RUSH. JONC FLEURI (Que.)

Perianth with both sepals and petals corolloid and persistent. Stamens 9, with linear-subulate filaments and 2-locular basifixed short anthers with lateral dehiscence. Carpels 6, verticillate, barely united at base; style persistent as a beak; the stigma elongate, on the inner margin of the style. Follicles 6, coriaceous, united at base, slender-beaked, opening on the inner margin. Seeds straight, striate; embryo straight. — Single species, nat. of Eurasia. (From the Greek *bous, cow,* and *temno, to cut,* from the sword-like leaves.)

1. B. UMBELLÀTUS L. (with an umbel). — Tall, up to 1 m. or more high; rootstock fleshy, thick, late in the season bearing many freely deciduous erect grain-like tubers; leaves erect, ensiform, 3-angled at base; umbel before anthesis inclosed by 3 large scarious purple-tinged bracts; flowers very numerous, on long slender ascending pedicels; perianth 2–2.5 cm. broad, roseate, the sepals slightly tinged with green, the petals deeply colored outside; anthers red; young carpels pink; follicles inflated, long-beaked, about 1 cm. long. — Mud and shallow water of the St. Lawrence system, Que., s. to L. Champlain, Vt. and N.Y., and westw. along the Great Lakes to Mich. and O., rapidly spreading. June–Sept. (Natzd. from Eu.) — Forma VALLISNERIIFÒLIUS (Sagorski) Glück (with leaves of *Vallisneria*) has the submersed or finally floating leaves very elongate and thin, the plants sterile.

FAM. 21. HYDROCHARITÀCEAE (Frog's-bit Family)

Aquatic herbs with unisexual or polygamous mostly regular flowers sessile or on scape-like peduncles from a spathe, and with simple or double floral envelopes, which in the pistillate flowers are united into a tube and coherent with the 1–3-locular ovary. Stamens 3–12, distinct or monadelphous; anthers 2-locular. Stigmas 3 or 6. Fruit ripening under water, indehiscent, few–many-seeded.

Stems elongated, moss-like, with very numerous whorls of small linear to ovate sessile leaves. . . . . 1. *Elodea.*
Stems very short, or prolonged by stoloniferous branching; leaves ribbon-like, or long-petioled with dilated blades.
Leaves ribbon-like; pistillate spathes on long finally spiral scapes. . . . . 2. *Vallisneria.*
Leaves with ovate to reniform blades; spathes sessile or short-stalked. . . . 3. *Limnobium.*

1. ELODÈA Michx. Waterweed. Water-thyme. Ditch-moss

Flowers polygamodioecious, or perfect and with 3 stamens, from a sessile scarious axillary spathe. Staminate flowers with 3 sepals barely united at base, and usually 3 petals; filaments united into a short column; anthers 3–9. Pistillate or perfect flowers with 3 sepals and 3 often evanescent petals, raised to the surface by the greatly prolonged and slender calyx-tube or hypanthium. Ovary 1-locular, with 3 parietal placentae, each bearing a few orthotropous ovules; the capillary style coherent with the tube of the perianth; stigmas 3, large, 2-lobed or notched, exserted. Fruit coriaceous, few-seeded. — Perennial slender herbs of trop. and temp. reg., with veinless 1-ribbed sessile whorled or opposite leaves. (Name from the Greek *elodes, marshy.*)* Anacharis Richard, Philotria Raf. The name Elodea proposed for conservation.

a. Principal leaves 6–17 mm. long; staminate spathes 1-flowered, the petals wanting or up to 5 mm. long and 0.7 mm. wide.
Median and upper leaves 1–5 mm. wide, those of pistillate plants firm, dark green and closely imbricated above; spathe of pistillate flower with 2 broad apical teeth; flowers all with elongate filiform hypanthia, the staminate ones not liberated; sepals of pistillate flowers 2–2.2 mm. long; sepals of staminate flowers 3.5–5 mm. long, the slender-clawed petals 5 mm. long; anthers of staminate flowers 2–3.5 mm. long. . . . . . . 1. *E. canadensis.*
Median and upper leaves 0.3–1.5 mm. wide, flaccid, loosely divergent, not overlapping; spathe of pistillate flower with 2 acuminate teeth, the sepals about 1 mm. long; staminate flowers sessile, inclosed while developing within the spathe, then expelled and floating to surface of water, there expanding, their sepals about 2 mm. long, their petals wanting or minute (up to 0.5 mm. long); anthers 1–1.2 mm. long. . . . . 2. *E. Nuttallii.*
a. Principal leaves 2–3.3 cm. long; staminate spathes 2- or more-flowered, the petals about 1 cm. long and 6–9 mm. broad. . . . . 3. *E. densa.*

1. E. canadénsis Michx. (Canadian). — Plants dioecious; *pistillate plants* with slender dichotomously branched stems from creeping thread-like stolons; lower leaves opposite, ovate, small; *median and upper leaves* whorled in 3's, *oblong, oblong-ovate or ovate-lanceolate,* minutely serrulate, *firm, dark green, crowded and strongly imbricated toward summits of branches,* 6–13 mm. long and 1–5 *mm. broad; pistillate spathes* cylindric, in upper axils, *with* 2 *broad apical teeth;* pistillate flowers exserted by the thread-like base of the hypanthium prolonged to 2–15 cm., the dark-striate and *oblong-elliptic sepals* 2–2.2 *mm. long and* 1.1 *mm. broad,* the delicate broadly *elliptic-spatulate* white *petals* 2.6 *mm. long* and 1.3 *mm. broad,* the 3 acicular *staminodia* 0.7 *mm. long;* stigmas 3, broad, 2-cleft at apex, 4 mm. long; ovary lance-ovoid, 3 mm. long, with 3 or 4 erect ovules; *capsule* 6 mm. long, ovoid, *about* 3 *mm. thick;* seeds 4.5 mm. long, slenderly cylindric, glabrous. *Staminate plant* rare, with thin leaves linear to lance-oblong; *staminate spathes in upper axils, peduncular-based,* inflated and ellipsoid or ovoid above, 7 mm. long, 4 mm. thick, gaping at summit, *with* 2 *acute teeth; flowers with elongate thread-like base of hypanthium,* with dark-striate elliptic sepals 3.5–5 mm. long and 2–2.5 mm. broad, and delicate slender-clawed lanceolate petals about 5 mm. long and 0.3–0.7 mm. broad; stamens 9, the oblong-ellipsoid anthers 2–3.5 mm. long, the 3 inner elevated on a common (fused) stalk. (*Anacharis* Planch.; *Philotria* Britt.; *E. Planchonii* Caspary) — Quiet waters, often calcareous, Gaspé Pen., Que. to Ung., w. to Wash., s. to N.E., N.C., Ala., Ill., Ia., Okla., Colo., Utah and Calif. July–Sept. (Introd. into Eu.)

2. E. Nuttállii (Planch.) St. John (for Thomas Nuttall, 1786–1859). — More slender;

* Treatment made with aid of Harold St. John.

median and upper *leaves* in 3's or 4's, *linear or narrowly linear-lanceolate, pale green, flaccid,* 0.3–1.5 mm. broad, *loosely divergent, not appressed-imbricated; pistillate spathes* slenderly cylindric, slightly ovoid at base, *the* 2 *apical teeth acuminate; sepals obovate, about* 1 *mm. long and* 0.5 *mm. broad; petals broadly obovate,* 1.3 *mm. long,* 1 *mm. broad; staminodia* 0.5 *mm. long; capsule* 1.5–2 *mm. thick; staminate spathes in median axils, sessile,* ovoid, 2*-parted to below middle, the acuminate* often twisted *teeth forming an apiculation, the body* 2 *mm. long; flower sessile, toward maturity floating free and expanding at surface of water; sepals* ovate, *about* 2 *mm. long and* 1.5–1.7 *mm. broad,* often red-tinged; *petals wanting or ovate-lanceolate and only* 0.5 *mm. long; outer anthers* 1–1.2 *mm. long.* (*E. occidentalis* (Pursh) St. John, with illegitimate basonym; *E. minor* (Engelm.) Farw.; *Anacharis occidentalis* (Pursh) Vict.) — Shallow waters, either fresh or slightly brackish, s. Me. and s. Que. (often in tidal estuaries) to Id., s. to N.C., Ind., Ill., Mo., Kans. and Colo. July–Sept.

3. E. DÉNSA (Planch.) Caspary (dense). — Much coarser than nos. 1 and 2, the elongate branches with whorls of mostly 4 linear-lanceolate acuminate *leaves mostly* 2–3.3 *cm. long,* the lower whorls remote, the upper crowded; *staminate spathes with* 2*-several exserted flowers* with relatively showy *petals* 9–11 *mm. long and* 6–9 *mm. broad;* pistillate flowers not known with us. (*Philotria* Small; *Anacharis* Vict.) — Ponds, pools and quiet streams, originally cult., now rapidly spreading (vegetatively) from Fla. and Gulf States n. to N.E., N.Y., Ky., Neb. and presumably beyond. June–Oct. (Introd. and natzd. from S. Am.)

2. VALLISNÈRIA L. TAPEGRASS. EELGRASS

Dioecious; the staminate flowers crowded in a head, inclosed in an ovoid at length 3-valved spathe borne on a short scape; fertile stamens mostly 2. Pistillate flowers solitary and sessile in a tubular spathe on an exceedingly lengthened scape. Calyx 3-parted in the staminate flowers, 1 of the sepals smaller than the other 2; in the pistillate with a slender tube coherent with the 1-locular ovary, but not extended beyond it, 3-lobed (the lobes obovate). Petals 3, linear, small. Stigmas 3, large, nearly sessile, 2-lobed. Ovules very numerous, scattered over the walls, orthotropous. Fruit elongated, cylindrical. — Long linear leaves wholly submerged or their ends floating. The staminate flower-buds themselves, breaking from their short pedicels, float on the surface, where they gather around the fertile flowers, which are raised to the surface by sudden growth at the same time; afterward the filiform scapes coil up spirally, drawing the fruit under water to ripen. Small genus of trop. and temp. reg. (Named for *Antonio Vallisneri,* an Italian botanist, 1661–1730.)

1. V. americàna Michx. (American), WILD OR WATER-CELERY. — *Leaves* thin, up to 2 m. or more long, 0.5–2 *cm. broad,* obtuse, somewhat nerved and netted-veined; *staminate spathe bluntly acuminate,* 1–2 *cm. long, on thick clavate scapes* up to 5 cm. long; fruits 8–12 cm. long. (*V. spiralis* of ed. 7, not L.) — Quiet water, s. N.B. to N.D., s. to Fla. and Gulf States. July–Oct.

3. LIMNÓBIUM Richard AMERICAN FROG'S-BIT

Flowers dioecious, from sessile or somewhat peduncled spathes; the staminate spathe 1-leaved, producing about 3 long-stalked flowers; the pistillate 2-leaved, with a single short-stalked flower. Calyx 3-parted or -cleft; sepals oblong-oval. Petals 3, oblong-linear. Filaments in the staminate flowers entirely united into a central solid column, bearing 6–12 linear anthers at unequal heights; stamens in the pistillate flowers 3–6 subulate rudiments. Ovary 6–9-locular, with as many placentae in the axis, forming an ovoid many-seeded dry berry; stigmas as many as the locules, but 2-parted, subulate. — Floating in stagnant water and proliferating by runners; a single species. Leaves round-cordate or reniform, the young spongy-reticulated and purplish underneath. (Name from the Greek *limnobios, living in pools.*)

1. L. Spóngia (Bosc) Steud. (a sponge). — Leaf-blades 2–8 cm. long, faintly 5-nerved; peduncle of the staminate flower about 7.5 cm. long and filiform, of the pistillate only 2.5 cm. long and stout. — Stagnant water, Fla. to Tex., n. to se. Va. and very rarely to Del. and to Monroe Co., N.Y. (probably extinct); Union Co., Ill. and se. Mo. June–Sept. (Trop. Am.)

FAM. 22. GRAMÍNEAE (GRASS FAMILY)

Mostly herbs (in most BAMBUSEAE *woody-stemmed shrubs or trees) with the stem* (CULM) *often hollow or with readily removed pith (sometimes solid in the* PANICEAE *and the* ANDROPOGONEAE), *the nodes closed and usually hard, the branches, when developed, subtended by a* 2*-keeled* PROPHYLLUM *opposite the subtending leaf and by crowding of leaves often becoming flabelliform. Leaves alternate,* 2*-ranked, the* SHEATH *open at the often overlapping margins* *or closed, more or less embracing the internodes of the culm; the* BLADE *linear or lanceolate (rarely ovate) to acicular, parallel-veined in ours* (in some foreign genera netted-veined); *at the junction of blade and sheath a prolongation of the upper surface of the sheath forming the* LIGULE, *a variable structure (scarious, membranaceous or cartilaginous, prolonged or truncated, a simple blade or dissected into fringes or bristles, etc.). Inflorescence consisting of spikelets, these 1–∞-flowered and borne in panicles, racemes or spikes variously disposed.* SPIKELET *consisting of a series of distichous bracts: at base 2 empty ones, the 1st or lower and the 2nd or upper* GLUME *(in some groups the 1st glume reduced or obsolete, in the* ORYZEAE *and* ZIZANIEAE both often obsolete); the axis above the glumes in elongate spikelets called the RACHILLA; *above the 2nd glume 1–∞ similar alternate bracts, the* LEMMA *or* LEMMAS (modified in some genera); *each lemma normally subtending and opposite to a usually 1- or 2-nerved or -keeled prophyllum, the* PALEA, *this embracing the flower. At the base of the flower, between lemma and palea usually 2 (in the* BAMBUSEAE *and in* STIPA *often 3) mostly insignificant processes or scales, the* LODICULES, *representing a reduced perianth (becoming distended as the pollen and stigmas mature and by pressure throwing open the floret, usually at specific periods, then on drying allowing the lemma and glumes to close over and protect the developing fruit). Stamens 3 (rarely 2 or 1) or 6 (in some* BAMBUSEAE *as many as* 120), *the anther attached well above the sagittate base to the filament and thus appearing versatile. Ovary of 1 (perhaps 2 or 3 fused) carpel, with usually 2 styles or stigmas (style 1 in* NARDUS, *some* PANICEAE *and some* MAYDEAE, 3 *in some* BAMBUSEAE), *the stigma papillose, feathery or plumose. Fruit most commonly a* CARYOPSIS *(seed adherent to and inseparable from the thin pericarp) often closely embraced by palea and lemma, a naked achene (as in* ZIZANIOPSIS *and in most* BAMBUSEAE, *where juicy or baccate fruits sometimes occur) or a* UTRICLE *(with thin pericarp dehiscent or readily rupturing from the grain), as in* SPOROBOLUS; *the grain, except for the small embryo (at the base next the lemma), filled with starchy endosperm and bearing on the posterior side the linear or punctiform* HILUM. *Lemma, palea, lodicules, stamens and ovary or fruit collectively called the* FLORET, *the stalk of each spikelet the* PEDICEL, *the axis of the inflorescence or of its racemose or spiciform branches the* RACHIS. Divergences from typical spikelets frequent through reduction or modification of glumes or lemmas, change to unisexual or aborted florets, or development of involucres in some genera; these modifications the basis of tribal and generic differentiation. — Economically the most important family of plants, yielding most cereals, much forage and sugar, and other products; taxonomically one of the most technical and difficult groups, still poorly worked out as contrasted with the economically unimportant *Cyperaceae.*

Spikelet usually more or less laterally compressed to terete, the glumes and lemmas boat-shaped or with the 2 sides connivent, with the keel or midrib as an axis (spikelet dorsally compressed in *Milium*); florets 1–∞, when 1, often with the rachilla prolonged above the floret, when 2 or more, the rachilla with evident internodes; rachilla usually articulated above the glumes, finally disarticulating and leaving the glumes as persistent husks; if articulating below the glumes, the latter laterally compressed; rudimentary or imperfect floret, if present, usually borne at summit of spikelet. Subfam. I. POACOIDEAE (see p. 95).

Spikelet dorsally compressed, the glumes and perfect lemma somewhat flattened or merely arching; perfect floret 1, sometimes a 2nd or 3rd sterile, neutral or imperfect one or a rudiment borne below it; rachilla articulated below the glumes, the whole spikelet, including glumes, dropping in fruit. Subfam. II. PANICOIDEAE (see p. 188).

Subfam. I. POACOÏDEAE

a. Culms woody at least at base, perennial; leaf-blade petioled and articulated with the sheath, finally disjointing from it; spikelets (in ours) several-flowered; lodicules conspicuous. . . . . Tribe I. BAMBUSEAE (p. 96).

a. Culms chiefly herbaceous, mostly annual: leaf-blade not articulated to sheath. . . *b.*

b. Spikelets with one or more perfect flowers, if dioecious several-flowered, compressed; stamens mostly 3 (2 or 1). . . *c.*

c. Spikelets compressed but not flat, with 1 or 2 well-developed glumes; stamens 3 (or 2 or 1). . . *d.*

d. Spikelet with no sterile lemmas below the fertile floret; palea 1- or 2-nerved. . . *e.*

e. Spikelets in panicles, racemes or spikes, if in spikes or spiciform racemes not concentrated on one side of the rachis. . . *f.*

f.Spikelets with 2–∞ perfect flowers.
Spikelets pedicelled, in open or spiciform panicles or in racemes.
Glumes shorter than lowest lemma, not clasping the ripening series of lemmas; lemmas, if awned, with the awn terminal or from between terminal teeth. . . . . . Tribe II. Festuceae (p. 96).
Glumes mostly overtopping the lowest and sometimes all the lemmas, clasping the base or the entire series of ripening lemmas; awn of lemma, when present, dorsal. . Tribe IV. Aveneae (p. 142).
Spikelets (except the terminal one) sessile along opposite sides of the rachis, forming a spike. . . . . . Tribe III. Hordeae (p. 132).
f.Spikelets 1-flowered (only exceptionally 2-flowered), pedicelled, in open or spiciform panicles; glumes often longer than lower lemma. . . . . . Tribe V. Agrostideae (p. 151).
e.Spikelets in 2 rows concentrated along one side of rachis, forming simple 1-sided spikes or racemes or racemosely branched to digitate inflorescences. . . . . Tribe VI. Chlorideae (p. 178).
d.Spikelet with 2 sterile lemmas below and falling with the fertile floret or reduced to bristles or sometimes with a staminate flower; palea 1-nerved. . . . . Tribe VII. Phalarideae (p. 185).
c.Spikelets very flat, 1-flowered, with lemma and 1-nerved palea subequal, both strongly keeled; glumes (in ours) obsolete or nearly so; stamens 1–6; grain with long linear hilum. . . . . Tribe VIII. Oryzeae (p. 187).
b.Spikelets unisexual; the pistillate ones subulate or terete, with united styles; stamens 6. . . . . Tribe IX. Zizanieae (p. 187).

Tribe I. Bambùseae Nees

Represented with us only by

1. ARUNDINÀRIA Michx. Cane

Spikelets few–∞-flowered, perfect or the upper imperfect, laterally compressed, in racemes or panicles; glumes unequal, shorter than the lemmas, the first sometimes obsolete; lemmas firm, not keeled, many-nerved, attenuate-mucronate; paleas nearly as long as their lemmas, 2-keeled and several-nerved; lodicules 3, conspicuous; styles 2 or 3; grain free within the lemma and palea. — Woody or shrubby Am., Asiatic and Austral. grasses with terminal and lateral panicles of large spikelets, the petioled leaf-blade jointed to and deciduous from the sheath and somewhat tesselated. (Name from *arundo, a reed.*)

1. A. gigantèa (Walt.) Chapm. (very large), Large or Giant C. — *Culms arborescent*, 1.5–10 *m. high* and 1–7 cm. thick at base, rigid, simple the first year, branching the second, afterward fruiting at indefinite periods, then dying; leaves lanceolate, 2–5 cm. long, 1–3.5 cm. wide, smoothish or pubescent, the sheath ciliate on the margin and fimbriate at the summit; *panicle lateral*, composed of few simple unequal racemes, *on leafy branches;* spikelets 2.5–6 cm. long, 5–15-flowered, purplish or pale, erect. (*A. macrosperma* Michx.) — River-banks and swamps, Fla. to Tex., n. to s. Del., Md., s. O., s. Ind., s. Ill. and s. Mo., forming cane-brakes. Apr., May.

2. A. técta (Walt.) Muhl. (covered with thin cortex; from the large sheaths of the flowering shoots), Switch-C., Small C. — *Lower and more slender*, 0.6–4 *m. high*, branching above; leaves 8–20 cm. long, 0.8–3 cm. wide, more tapering at base; *panicles on shoots springing directly from the rhizome*, of few aggregated spikelets on long slender branches with rather loose sheaths, the blades very minute; spikelets 2.5–5 cm. long, 5–10-flowered. — Swamps, moist soil, or in water, Fla. to La., n. to Md. and Okla.; introd. and spreading n. to N.J. March–June. Fig. 92.

92. A. tecta.

Pseudosása Makino (false *Sasa*), a small Asiatic genus differing from *Arundinaria* in having the lower internodes of the stem unbranched, the summit of the leaf-sheath naked or with flexuous white smooth hairs, the paleas very unequal, is represented with us by the cult. **P. japónica** (Sieb. & Zucc.) Makino, a shrub up to 3.5 m. high, rapidly spreading by stolons, becoming natzd. in woods, on banks and along streams in se. Pa. and e. Md. (Introd. from e. Asia)

Tribe II. Festùceae Nees

a.Rachilla beardless or, if bearded, with hairs not overtopping lemmas; plants of small to medium stature, the culms relatively slender; leaves with blades relatively narrow; panicle not plumose. . . *b*.

b.Plants with perfect florets in each spikelet. . . *c*.
c.Spikelets uniform. . . *d*.
d.Callus or nerves of lemma glabrous, silky or cobwebby, not heavily bearded with divergent long beard. . . *e*.
e.Lemmas membranaceous to scarious or with scarious margins; paleas usually thin and soft. . . *f*.
f.Lemma 5-∞-nerved (or obscurely nerved in no. 10). . . *g*.
g.Panicle with branches spirally arranged or not strongly 1-sided. . . *h*.
h.Lemmas essentially uniform, the upper ones not forming a convolute mass. . . *i*.
i.Spikelets definitely longer than broad, with ascending florets (if florets widely divergent the plants perennial, with closed sheaths); lemmas definitely nerved, not cordate. . . *j*.
j.Lemmas convex on back or keeled only near tip. . . *k*.
k.Lemmas definitely 2-toothed at apex, often awned from the notch or from just below it.
Sheaths closed; ovary hairy at summit; stigmas sessile, plumose, borne laterally from below summit of ovary; grain adnate to palea. . . . . . 2. *Bromus*.
Sheaths closed; ovary glabrous; styles approximate, terminal; grain free. . . . . 3. *Schizachne*.
k.Lemmas not 2-lobed at apex; the awn, when present, terminal or terminating excurrent nerves. . . *l*.
l.Lemmas entire or erose, if awned with awn apical, the nerves not excurrent; callus and ovary glabrous.
Lateral nerves of lemma arched and converging to the midrib, terminal awn often present.
Perennials; florets not distended during anthesis, expanding; the plumose stigmas and the 3 anthers (on long filaments) projecting; grain ellipsoid or ovoid. . . . . . . 4. *Festuca*.
Annuals; florets distended above during anthesis, not opening, cleistogamous; the included short filament and 1 (rarely 3) anther appressed to palea or small included stigmas; grain linear-cylindric, attenuate at the ends. 5. *Vulpia*.
Lateral nerves of lemmas parallel, nearly straight, not arching to midrib, lemma awnless.
Sheaths open; lemmas faintly nerved, often minutely hairy at base; lodicules distinct; stigmas sessile; grain usually adherent to palea; annual or short-lived perennial halophytes. . . . . 6. *Puccinellia*.
Sheaths closed; lemmas mostly prominently nerved, glabrous or merely scabrous; lodicules united; styles definite; grains free; perennials of fresh habitats. . . . . 7. *Glyceria*.
l.Lemma somewhat toothed at summit, 1 or more lateral nerves excurrent as short subuli; callus villous; ovary pubescent. . . . . 8. *Scolochloa*.
j.Lemmas keeled, awnless, with 3 or 5 arching nerves often pubescent or cobwebby at base; grain tightly embraced; hilum punctiform. . . . . . . 9. *Poa*.
i.Spikelets nearly or quite as broad as long, with horizontally divergent inflated florets; lemmas faintly or obscurely nerved, cordate at base; ours annuals. . . . . . 10. *Briza*.
h.Lemmas dissimilar, the 2 or 3 uppermost empty and convolute into a single club-shaped mass; spikelets plump; tip of lemma bifid; sheaths closed. . . . . 11. *Melica*.
g.Panicle 1-sided or with 1-sided branches, the compressed and nearly sessile curved spikelets in crowded glomerules; glumes and lemmas sharp-keeled, at least the latter ciliate. . . . 12. *Dactylis*.
f.Lemmas 3- or 1-nerved. . . *m*.
m.Glumes and lemmas keeled.
Lemmas entire at tip, not awned.
Glumes and lemmas herbaceous, with scarious margins;

florets usually cobwebby at base; grain closely embraced, furrowed. . . . . 9. *Poa.*

Glumes and lemmas mostly scarious to chartaceous; florets not cobwebby; grain free from palea, not furrowed. . . 13. *Eragrostis.*

Lemmas 2-toothed at apex, mucronate or short-awned; grain free. . . . . 14. *Diplachne.*

*m.*Glumes and lemmas rounded on back, not keeled.

Spikelets strongly compressed.

Glumes and lemmas elongate, tapering, entire; callus densely bearded; tough erect involute-leaved stoloniferous perennial. . . . . 20. *Redfieldia.*

Glumes and lemmas short, subtruncate and erose; callus not bearded; soft depressed and matted flat- and broad-leaved plant. . . . . 15. *Catabrosa.*

Spikelets subterete, conical; grain closely embraced; tough, erect, tufted perennial. . . . . 16. *Molinia.*

*e.*Lemmas coriaceous, smooth and shining, without scarious margins; paleas firm to rigid.

Lemmas rounded on back, the lower fertile; palea 2-nerved; fruit beaked, as long as lemma. . . . . 17. *Diarrhena.*

Lemmas strongly flattened and keeled, the lower ones empty; palea with 2 wings; fruit much exceeded by lemma. . . . 18. *Uniola.*

*d.*Callus or nerves of lemma heavily bearded with long divergent hairs; tip of lemma toothed, lobed or emarginate. . . *n.*

*n.*Spikelets nearly terete; glumes spathiform, equaling or overtopping the crowded column of 5–9-nerved blunt-lobed lemmas; densely cespitose low perennial with short racemiform panicle suggesting that of *Danthonia;* plant of Nfld. and N.S. . . . . 19. *Sieglingia.*

*n.*Spikelets compressed; glumes narrower, shorter than column of mostly sharp-toothed 3-nerved lemmas. . . *o.*

*o.*Strong perennials, the culms not readily fracturing, ours with panicles 1–5 dm. long; no cleistogamous florets hidden in sheaths; palea, if bearded, hairy only below summit.

Nerves of lemma glabrous except for beard confined to callus and base of floret; palea narrow and acuminate, equaling lemma; plant with abundant slender rhizomes and stolons. . . . 20. *Redfieldia.*

Nerves, especially the lateral, bearded; palea broad, shorter than lemma; slender stolons wanting. . . . . 21. *Triodia.*

*o.*Slender wiry annuals, the mature culms disarticulating at the nodes, the fragments including sheaths and cleistogamous spikelets within them (at base); panicles 1–5 cm. long, simple or forking; palea strongly ciliate at summit. . . . . 22. *Triplasis.*

*c.*Spikelets of 2 sorts in a dense spike-like panicle, the terminal one of each fascicle terete, fertile, 2–3-flowered, sessile, nearly hidden by the lower sterile ones; these reduced to distichous fan-like groups of awns. . . 23. *Cynosurus.*

*b.*Plants dioecious; the staminate and pistillate spikelets dissimilar.

Depressed freely forking annual, forming mats; foliage pubescent; lemmas scario-membranaceous. . . . . 13. *Eragrostis.*

Erect perennial with simple rigid culms, with coarse rhizomes of papery white scales; foliage glabrous; lemmas coriaceous; halophytic. . . . 24. *Distichlis.*

*a.*Rachilla with long beard overtopping the lemmas; coarse and hard reeds, 1.5–4 m. tall, with thick culms, hard leaves 1–5 cm. broad and large plumose panicles. . . . . 25. *Phragmites.*

2. BRÒMUS L. Brome-Grass

Spikelets few–many-flowered; glumes unequal, 1–5-nerved; lemmas herbaceous, longer than the glumes, convex or sometimes keeled, 5–9-nerved, usually 2-toothed at the apex, awnless or awned from between the teeth or just below; palea equaling or a little shorter than the lemma, 2-keeled; stigmas widely separated, sessile, borne from below summit of ovary, emerging from sides of lemma; grain furrowed, adnate to the palea. — Annuals, biennials, or perennials of temp. reg., the flat leaves with closed sheaths, the terminal panicles of rather large spikelets. (An ancient Greek name for the oat, from *broma, food.*)

*a.*Tips of lemmas with teeth only 0.1–0.5 mm. long; perennials or coarse annuals or biennials. . . *b.*

b.Lower glume 1-nerved, 2nd glume 3-nerved (1st 3-nerved, 2nd 5-nerved only in no. 5 and no. 6, the former has convex densely pilose lemmas about 1 cm. long);

spikelets plump or subcompressed; the lemmas not strongly keeled, membranaceous to scarious. § ZERNA. . . *c.*

*c.*Branches of panicle finally arcuate or flexuous; lemmas awned, commonly pilose or silky at least at base or near the margins; anthers 1–5 mm. long. . . *d.*

*d.*Lower glume 1-nerved, 2nd 3-nerved; lemmas with awns 2–6 mm. long. . . *e.*

*e.*Most of the nodes covered by the leaf-sheaths; the latter closed or nearly so at summit, with a ruff-like horizontal chartaceous flange, in uninjured specimens these with uncinate basal auricles. . . . 1. *B. latiglumis.*

*e.*Most of the nodes exserted from the sheaths; the latter with a V-shaped orifice, without horizontal or auricled flange. . . *f.*

*f.*Tip of mature palea dilated, scarious or chartaceous and puckered, the dorsal face usually minutely pubescent; pubescence of lemma almost uniform over the back (rarely wanting), the lateral nerves obscure below the middle. . . . . 2. *B. purgans.*

*f.*Tip of mature palea rounded or tapering, flat, the dorsal face glabrous; pubescence of lemma concentrated in 2 marginal bands; lateral nerves distinct to base.

Rachilla exposed to view in mature spikelets; glumes strongly folded; lemmas strongly folded or becoming involute, lance-attenuate; palea linear. . . . . 3. *B. ciliatus.*

Rachilla usually hidden; glumes and lemmas flattish or convex, the latter oblong; palea oblong. . . . . 4. *B. Dudleyi.*

*d.*Lower glume (1-) 3-nerved, 2nd 5–7-nerved; back of lemma evenly pubescent.

Cauline blades 6–8; nodes mostly covered by sheaths; spikelets 1.8–4 cm. long; 2nd glume 5–7-nerved; awn of lemma 5–8 mm. long; southeastern species. . . . . 5. *B. nottowayanus.*

Cauline blades 3–5 (–6); nodes mostly exserted; spikelets 1.4–2.6 cm. long; 2nd glume 5-nerved; awn of lemma 1–3 mm. long; northern species. . . . . 6. *B. Kalmii.*

*c.*Branches of panicle ascending or stiffly spreading (loosely divergent in var. of no. 8); lemmas awnless or with awns up to 3 mm. long (or, if with awns more than 3 mm. long, not pilose); anthers 4–6 mm. long.

Cespitose; basal leaves slender and conduplicate, upper leaves flat; awns 5–6 mm. long. . . . . 7. *B. erectus.*

Stoloniferous; leaves all flat; awns wanting or at most 3 mm. long. . 8. *B. inermis.*

*b.*Lower glume 3–5-nerved, 2nd 5–9-nerved; spikelets strongly flattened, with keeled lemmas. § CERATOCHLOA.

Culms puberulent or pubescent, harsh; awns 4–7 mm. long; palea about equaling lemma; perennial. . . . . 9. *B. marginatus.*

Culms glabrous, smooth; awns wanting or up to 2 mm. long; palea one-half to three-fourths as long as lemma; annual or biennial. . . . . . 10. *B. catharticus.*

*a.*Tips of lemmas prolonged 0.6–2 mm. beyond base of awn (or in the awnless no. 17 lemmas broadly elliptical); annual or biennial weeds. . . *g.*

*g.*Lower glumes 3-nerved, 2nd glume 5–9-nerved; lemmas oblong, elliptical or oval. § ZEOBROMUS. . . *h.*

*h.*Spikelets subterete or only slightly compressed during anthesis, becoming compressed in fruit, 3–10 mm. broad; lemmas awned. . . *i.*

*i.*Pedicels all or nearly all about as long as or longer than the spikelets; lemmas somewhat coriaceous, their nerves not prominent. . . *j.*

*j.*Rachilla exposed to view by the inrolling of the lemmas in mature spikelets.

Branches of panicle ascending; awn straight or merely flexuous; ciliate tip of palea projecting beyond mature lemma. . . . 11. *B. secalinus.*

Branches of panicle divergent; awn straight or arching; palea not exserted. . . . . 12. *B. japonicus.*

*j.*Rachilla hidden, the lemmas not closely inrolling at maturity. . *k.*

*k.*Awns not strongly divergent; hyaline margin of lemma about 0.5 mm. broad; anthers 1.5–4 mm. long. . . *l.*

*l.*Lemmas (except terminal one) about equal, with prolonged acute teeth; anthers 4 mm. long. . . . . 13. *B. arvensis.*

*l.*Lemmas unequal; the lower much longer than the upper, with short blunt teeth; anthers 1.5–2.5 mm. long.

Lower lemmas about 7 mm. long; anthers 2–2.5 mm. long; branches of panicle solitary or in pairs. . . . . 14. *B. racemosus.*

Lower lemmas 9–11 mm. long; anthers 1.5–2 mm. long; branches (except in depauperate individuals) more numerous. . . . . 15. *B. commutatus.*

*k.*Awns strongly arching or divergent; hyaline margin of lemma 1 mm. broad; anthers 1 mm. long. . . . . 16. *B. squarrosus.*

*i.*Pedicels all or nearly all shorter than the spikelets; lemmas scarious or membranaceous, longitudinally plicate, with prominent nerves. . . 17. *B. mollis.*

*h.*Spikelets strongly flattened from the first, ovate, 0.8–1.3 cm. broad; lemmas awnless. . . . . 18. *B. brizaeformis.*

*g.*Lower glume 1-nerved, 2nd glume 3-nerved; lemmas narrow and elongate, tapering at tip. § Stenobromus. . *m.*

*m.*Spikelets (including awns) 6–10 cm. long; awns 3–4 cm. long, twice as long as body of lemma. . . . . 19. *B. rigidus.*

*m.*Spikelets (including awns) 2–6 cm. long; awns about equaling or slightly longer than body of lemma. . . *n.*

*n.*Panicle ellipsoid, with short ascending branches; spikelets 4–6 cm. long. . . . . 20. *B. madritensis.*

*n.*Panicle with loosely spreading or recurving lower branches.

Spikelets 4–5 cm. long, commonly solitary on the branches (1–3 on lower branches); awns 2–3 cm. long. . . . . 21 *B. sterilis.*

Spikelets 2–3.5 cm. long, often 2–6 on the branches; awns 1–1.7 cm. long. . . . . 22. *B. tectorum.*

§ Zérna (Panzer) Ledeb.

1. B. latiglùmis (Shear) Hitchc. (broad-glumed). — Cespitose or subcespitose perennial, 0.5–2 m. high; *cauline leaves* 10–20; their *sheaths mostly covering the nodes,* strongly costate, glabrous, or in forma **incànus** (Shear) Fern. (gray) strongly pilose; *the summit nearly or quite closed, with a brown cartilaginous horizontally divergent flange uncinate-auricled at base;* blades dark green, 5–17 mm. broad; the whitish midrib conspicuous, especially beneath; panicle becoming loosely ovoid, 1.5–3 dm. long, its elongate spreading to reflexed branches solitary or oftenest in pairs, with large basal pulvini; spikelets elliptic-lanceolate to -oblong, 1.5–3.5 cm. long, 5–9 mm. broad, loosely 3–8-flowered; glumes pilose or glabrous; lemmas membranaceous, strongly 5–7-nerved, 3–4 mm. broad, minutely sericeous toward base to almost glabrous, with awns 2–6 mm. long; palea with rounded plane tip; anthers 1.5–2.5 mm. long. (*B. altissimus* Pursh, not Gilib.) — Rich or alluvial thickets and woods, n. N.B. to Mont., s. to Md., w. N.C., W.Va., O., Ind., Ill., Mo., Tex. and N.M. Aug., Sept. Fig. 93.

93. B. latiglumis.

2. B. púrgans L. (purging; Linnaeus erroneously identifying it with *B. catharticus*). — Resembling no. 1, more slender; *sheaths mostly shorter than the internodes, with V-shaped orifice* and cartilaginous summit not horizontally spreading, the ligule prolonged, the sheaths and blades pilose, or in forma **laevivaginàtus** Wieg. (smooth-sheathed) glabrous; panicle 0.8–2 (rarely –3) dm. long, its branches with large basal pulvini; spikelets 2–4 cm. long, 3–12-flowered; glumes pilose or glabrous; *lemmas* 3–4 mm. broad, *tightly inrolled, finely sericeous-pilose over nearly the whole surface,* or glabrous in forma **glabriflòrus** Wieg. (smooth-flowered), the 5–7 *nerves obscure below the middle; palea dilated, scarious and finally puckered at tip,* usually minutely pubescent, nearly equaling the lemma; *anthers* 2.8–5 *mm. long.* — Rich woods and thickets, often on rocky slopes, s. N.H. to sw. Que., w. to Alta., s. to Fla., La. and Tex. June–early Aug. (rarely early Sept.). Fig. 94.

94. B. purgans.

3. B. ciliàtus L. (ciliate). — Perennial, often cespitose, 0.3–1.5 m. high; sheaths glabrous or sparsely hirtellous, with a narrow V-shaped orifice, mostly shorter than the internodes; blades sparsely hirtellous, scabrous, 3–16 mm. broad; panicle (except in dwarf plants) 1–3 dm. long, with arching and finally flexuous-spreading to drooping branches; *spikelets* 1.5–3 cm. long, 4–10 mm. broad, greenish, sometimes bronze- or purple-tinged, *loosely 3–9-flowered, at maturity displaying the rachilla; glumes strongly conduplicate, the upper narrowly lance-attenuate, with slender nerves; lemmas firm,* 2.5–3.5 mm. broad, *conduplicate or involute, lance-attenuate,* delicately nerved, *the marginal band pilose or sericeous; palea linear, usually closely embraced by the strongly folded lemma,* the green and ciliate *ribs marginal* *nearly their entire length*, the hyaline border abruptly folded toward the center; anthers 1–2.5 mm. long; caryopsis linear-lanceolate. — Thickets, shores and rocky slopes, Nfld. and s. Lab. Pen. to B.C., s. to N.S., Me., Mass., n. O., Ind., Wisc., Minn. and N.D.; Roan Mt., N.C.; the common plant northw. and in subalpine areas. July–early Oct. Fig. 95.

95. B. ciliatus.

Var. **intónsus** Fern. (unshaved). — Middle and upper sheaths villous or strongly retrorse-pilose; margins of lemmas finely sericeous to subglabrous. — Nfld. and s. Que. to s. Ont., s. to Pa., Ind. and Ill.; the common plant southw. and at low alt. July–Oct.

4. B. Dúdleyi Fern. (for its discoverer, William Russell Dudley, 1849–1911). — Perennial, the solitary or somewhat cespitose culms 0.4–1.2 m. high; leaf-sheaths mostly shorter than the internodes, glabrous or sometimes villous, with narrow V-shaped orifice; blades glabrous or sometimes pilose above, 4–12 mm. broad; panicles lanceolate to ovoid, loosely branched, 0.6–2 dm. long, the filiform branches ascending or spreading; *spikelets mostly purple- or bronze-tinged*, sometimes green, 2–2.5 cm. long, 5–9 mm. broad, 4–7-flowered, *the rachilla usually hidden; glumes flattish or dorsally convex, the 2nd coarsely brown- or purplish-ribbed; lemmas flattish, submembranaceous, narrowly oblong*, obtuse to subacute, *strongly 3–5-nerved*, 2.3–3 mm. broad, with awn 2–4 mm. long, *densely villous-hirsute outside the lateral nerves; palea oblong, flat, scarcely embraced by the lemma*, the ciliolate *nerves midway between the entire margin and the middle* except toward the tip (where marginal); anthers 1–2 mm. long; caryopsis oblong-lanceolate. — Boggy meadows, damp thickets and shores, Nfld. to B.C., s. to N.S., N.E., N.Y., mts. of W.Va., Mich., Minn. and Mont. June–Aug. Fig. 96.

96. B. Dudleyi.

5. B. nottowayànus Fern. (of Nottoway R., Va.). — Culms solitary or in small tufts, 0.6–1.5 m. high; *cauline leaves* 6–8, with retrorse-pilose *sheaths mostly covering the nodes*, the orifice V-shaped, the *ligule inconspicuous;* blades 0.6–1.3 cm. broad, weak, pilose on the upper surface; panicle lax, nodding, 0.5–2 dm. long, the *pulvini not strongly thickened; spikelets* 1.8–4 *cm. long*, 3–11-flowered; *lower glume* 1–3*-nerved, 2nd 5- or 7-nerved; lemmas evenly strigose-pilose on back, not inrolled*, with *awns* 5–8 *mm. long;* palea copiously pilose on back, with flat tip. — Bottomlands and rich woods, Nottoway Valley, se. Va. July, Aug.

6. B. Kálmii Gray (for its discoverer, Pehr Kalm, 1715–1779). — Perennial, the solitary or slightly cespitose culms 0.4–1.2 m. high; *leaf-blades* 3–5 (–6); the sheaths mostly shorter than internodes, with V-shaped orifice, villous or glabrous; *blades firm*, pubescent or glabrous, 4–10 mm. broad; panicle 5–15 cm. long, rather narrow, its short capillary branches in 2's, 3's or 4's; *spikelets* 1.4–2.6 cm. long, *closely* 5–11*-flowered; lower glume* 3*-nerved, 2nd 5-nerved, both pilose; lemmas convex, oblong to elliptic, densely hairy all over*, the *awns* 1–3 *mm. long;* palea oblong, flat, much shorter than the lemma; anthers about 2 mm. long. — Dry or moist open soil or thickets, often in calcareous areas, sw. Me. to se. Man., s. to w. Md., O., Ind., Ill., Ia. and N.D. Late June–Aug.

7. B. erectus Huds. (erect). — *Cespitose* perennial *bearing basal sterile tufts of slender plicate* pubescent *leaves;* culms 0.6–1.2 m. high; cauline leaves flat, with pubescent or glabrous sheaths; *panicle contracted, ellipsoid*, 0.7–1.5 dm. long, *with short erect branches;* spikelets lanceolate, 1.5–3 cm. long, 5–12-flowered; glumes acute, the 1st 1-nerved, 2nd 3-nerved; lemmas membranaceous, prominently nerved, glabrous or minutely scabrous, with awns 5–6 mm. long; anthers 4–6 mm. long. — Fields and roadsides, local, s. Me. to N.Y., Ont. and Wisc. May, June. (Adv. from Eu.)

8. B. inérmis Leyss. (unarmed or without awns), Awnless or Hungarian B. — *Loosely stoloniferous;* the culms mostly solitary, 0.4–1 m. or more high; *leaves all flat*, glabrous; *panicles* ellipsoid to ovoid, 1–3 dm. long, *with loosely ascending to suberect branches;* spikelets bronze- or purple-tinged, 1.5–3.5 cm. long, 5–10-flowered; glumes acute, the 1st 1-nerved, 2nd 3-nerved; *lemmas* membranaceous, with scarious margins, glabrous, or in forma villòsus (Mert. & Koch) Fern. (shaggy) pubescent, *blunt and awnless, or* in forma aristàtus (Schur) Fern. (awned) *with awns up to 3 mm. long.* — Roadsides and fields, esc. from cult., Nfld. to B.C., s. to Va., W.Va., Mo., e. Kans., etc. June, July. (Introd. and natzd. from Eu.) — In forma prolíferus Louis-Marie (bearing offspring) spikelets changed to bulblets or leafy tufts.

Var. DIVARICÀTUS Rohlena (divergent). — *Panicle* lax and open, *its loose and elongate branches widely spreading to reflexed.* — Local, Ct. to Mich. and Pa. (Introd. from Eu.)

§ CERATÓCHLOA (Beauv.) Griseb.

9. B. MARGINÀTUS Nees (margined). — Tufted *perennial; culms* up to 1 m. or more high, *puberulent or pubescent, harsh;* sheaths mostly shorter than the internodes, pilose to glabrous; *blades harsh; panicle erect,* narrow, 1–3 dm. long, with stiffly spreading to ascending branches; *spikelets strongly compressed,* oblong-lanceolate, 2.5–4 cm. long, 5–9-flowered; glumes scabrous, the 1st 3–5-nerved, the 2nd 5–7-nerved; *lemmas subcoriaceous,* pubescent, *with* 2 subacute hyaline apical teeth and a stout *awn* 4–7 *mm. long; palea about equaling the lemma.* — A species of western N. Am., locally natzd. eastw. to Ia. and casually farther eastw. June–Aug. (Natzd. from w. N.Am.)

10. B. CATHÁRTICUS Vahl (cathartic), RESCUE-GRASS, SCHRADER'S B. — Coarse *annual or biennial,* 2–8 dm. high, resembling no. 8; *culms glabrous, smooth;* glumes smooth or slightly scabrous; *lemmas* scabrous to almost glabrous, *awnless or with short awn* (*up to* 2 *mm. long*); *palea one-half to three-fourths as long as the lemma.* (*B. unioloides* (Willd.) HBK.) — Extensively cult. in warm reg., becoming natzd. n. to N.Y., Mo. and Kans. May–July. (Introd. and natzd. from Trop. Am.)

B. BREVIARISTÀTUS Buckl. (short-awned), a tufted perennial with pilose sheaths and blades, the latter becoming involute, and short-branched erect panicle with few spikelets 2–3 cm. long, the puberulent lemmas with awns 3–10 mm. long, a western species, is casual in w. Mo. (Adv. from farther west.)

§ ZEOBRÒMUS Griseb.

11. B. SECALÌNUS L. (rye-like), CHEAT or CHESS. — Annual, 0.1–1.3 m. high; *upper sheaths smooth,* strongly nerved; blades pilose above, harsh; *panicle* 0.3–2 dm. long (in starved plant reduced to 1–few spikelets), *branches ascending* or the lowest becoming divergent; *spikelets mostly on elongate pedicels,* 5–15-flowered, glabrous to scabrous, *in maturity lax, with the rachilla evident; lemmas spreading-ascending at maturity, strongly inrolling, firm,* subequal, *obscurely 7-nerved,* mostly 5–8 mm. long; awn straight or flexuous, 1–6 mm. long, deciduous, sometimes wanting; *palea equaling to exceeding the lemma, its ciliate tip slightly projecting in maturity;* anthers 1.5–1.8 mm. long. — Fields, waste places, etc., Que. to B.C., s. throughout much of the U.S. June–Sept. (Natzd. from Eu.) FIG. 97.

97. B. secalinus.

12. B. JAPÓNICUS Thunb. (Japanese). — Annual, 0.2–1 m. high; *sheaths soft-pubescent,* blades villous; *panicle* open, *secund, the divergent branches with drooping tips;* spikelets 2–2.5 cm. long, slender-pedicelled; 1st glume 3-nerved, acute, 2nd 5-nerved and obtuse; *lemmas glabrous,* 7–9 mm. long, obtuse, *firm, obscurely 9-nerved, the margins somewhat inrolling at maturity, with a twisted or divaricate awn* 8–12 *mm. long; palea distinctly shorter than its glume.* — Roadsides, waste places, etc., Mich. to Alta. and Wash., s. to Mo. and Neb.; casual in N.E. June–Aug. (Natzd. from Eurasia)

Var. PORRÉCTUS Hack. (directed forward). — Awns straight, directed forward. — Roadsides and waste places, Pa. to Calif., s. to N.C., Tenn., Ark. and Tex. June–Aug. (Natzd. from Eu.)

13. B. ARVÉNSIS L. (of cultivated land). — Annual or biennial, 3–9 dm. high; sheaths soft-pubescent, blades pubescent on both sides; *panicle large, open,* 1–3 dm. long (except in starved plants), the ascending branches often drooping at tip; *spikelets* mostly long-pedicelled, linear-lanceolate, 1.5–3 cm. long, 5–12-flowered, *becoming purple in maturity; lemmas about equal,* obtuse, glabrous or minutely scabrous, 7–8 mm. long, faintly 7-nerved, the hyaline margin *ending in prolonged acute teeth;* awn straight, 7–10 mm. long; *anthers about* 4 *mm. long.* — Fields, roadsides, etc., infrequent, Vt. to s. Ont. and Minn., s. to Fla., Mo. and Kans. June, July. (Adv. from Eu.)

14. B. RACEMÒSUS L. (racemose). — Annual, mostly 3–8 dm. high; sheaths and leaf-blades pubescent; *panicle slender, racemiform,* suberect; the ascending *branches solitary or in pairs, mostly shorter than their* 1 (rarely 2) *spikelet;* spikelets elliptic-lanceolate, lustrous, 6–10-flowered, 1–2 cm. long; *lower lemmas much longer than the upper,* 7–9 *mm. long,* obovate, faintly 7-nerved, with straightish awns and rounded apical teeth; *anthers* 2–2.5 *mm. long.* — Roadsides and waste places, local, N.S. to Minn., s. to N.C., Ky., Mo. and Kans. June–Aug. (Adv. from Eu.)

15. B. COMMUTÀTUS Schrad. (variable). — Similar to no. 14, usually stouter; *branches of panicle* (except in dwarfed plants) *mostly more numerous* (2–6), *the longer much exceeding their*

1–3 *spikelets; lower lemmas* 9–11 *mm. long; anthers* 1.5–2 *mm. long.* (*B. racemosus*, in large part, of ed. 7; *B. pratensis* Ehrh., not Lam.) — Dry roadsides and waste places, e. Que. to B.C., s. to Va., La., Kans., etc. June–early Aug. — Perhaps not specifically separable from no. 14. (Natzd. from Eu.)

16. B. SQUARRÒSUS L. (with spreading tips). — Annual, 2–6 dm. high; sheaths pubescent, blades pubescent or glabrous; *panicle lax, unilateral, with few capillary flexuous to nodding branches; spikelets* mostly solitary on the branches, *ellipsoid-lanceolate,* 2–4 cm. long, 5–10 *mm. broad, densely* 10–20-*flowered; lemmas* unequal, *oval,* obtuse, faintly 9-nerved, *with broad* (1 *mm.*) *hyaline margin;* the *awn* twisted and *strongly divergent,* 6–10 *mm. long; anthers* 1 *mm. long.* — Waste places, local, Ct. to Mich. June, July. (Adv. from Eu.)

17. B. MÓLLIS L. (soft to touch), SOFT CHESS. — Annual, 1–9 dm. high; *summits of culms* and leaves commonly *soft-pubescent; panicle erect and contracted,* ellipsoid, its longer branches bearing 2 or more spikelets; *pedicels mostly shorter than the lance-ovoid* 6–10-flowered soft-pubescent, or in forma LEIÓSTACHYS (Hartm.) Fern. (smooth-spiked) glabrous or glabrescent, *spikelets; glumes rather broad, the 2nd elliptic; lemmas membranaceous, strongly* 7–9-*nerved,* 7–9 mm. long, the awn 7–10 mm. long; anthers 1–2 mm. long. (*B. hordeaceus* of many auth., not L.) — Roadsides, old fields and waste places, Nfld. to B.C., s. to N.C., Ind., Ill., Mo., Kans., Utah and Calif. Late May–July. (Natzd. from Eu.)

18. B. BRIZAEFÓRMIS Fisch. & Mey. (resembling *Briza*), QUAKE-GRASS. — Annual, 2–8 dm. high; panicle lax, with long capillary spreading or drooping branches mostly with a single terminal *ellipsoid-ovate flat spikelet* 0.8–1.3 *cm. broad; lemmas rhombic-ovate, with broad hyaline margin, awnless.* — Roadsides and waste places, local (also cult. for ornament), Mass. to Wash., s. to Del., Ind., Ill., etc. Late May–July. (Introd. from Eu.)

§ STENOBRÒMUS Griseb.

19. B. RÍGIDUS Roth (stiff). — Annual, 0.3–1 m. high; *culms, leaves and panicle-branches pubescent or harsh;* panicle erect or ascending, the short branches terminated by 4–9-flowered *spikelets* 6–10 *cm. long including awns* (3–4 *cm. long*); 1st glume 1-nerved, 2nd 3-nerved, long-attenuate; lemmas unequal, lance-attenuate, *the lower with a long tooth* each side of the stout awn, *the upper tapering* to the beak, strongly ciliate; anthers about 1 mm. long. (*B. villosus* Forsk.) — Waste places, local, Md. and D.C., and rarely about ports n. to Mass.; Pacific States. May, June. (Adv. from Eu.)

20. B. MADRITÉNSIS L. (of Madrid). — Annual, 2–7 dm. high; leaves glabrous or pubescent; *panicle erect, ellipsoid,* 3–12 cm. long, *with short ascending to slightly spreading branches; spikelets* (*including awns*) 4–6 *cm. long,* 7–11-flowered; glumes lance-acuminate, the 1st 1-nerved, the 2nd 3-nerved; *lemmas* 1.5–2 *cm. long,* glabrous or scabrous, *the slightly arching awn of about the same length.* — Cult. for ornament; local in waste ground, Va., Mich., Ill., etc. June–Sept. (Introd. or adv. from Eu.)

21. B. STÉRILIS L. (sterile; early botanists thinking it a sterile oat). — Annual, 3–8 dm. high; culms glabrous; leaves pubescent; *panicle lax,* 1–2.5 dm. long, nodding, the *loosely spreading or recurving branches* usually *with* 1 (or 2 or 3 on the lowest) *spikelet* 4–5 *cm. long,* including *awn* (2–3 *cm. long*); lemmas strongly 7-nerved, scabrous; anthers 1–1.5 mm. long. — Waste places and roadsides, Mass. to s. Ont., s. to N.C., Ala., Ark. and Colo.; Pacific States. May–Sept. (Natzd. from Eu.)

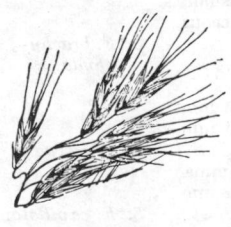

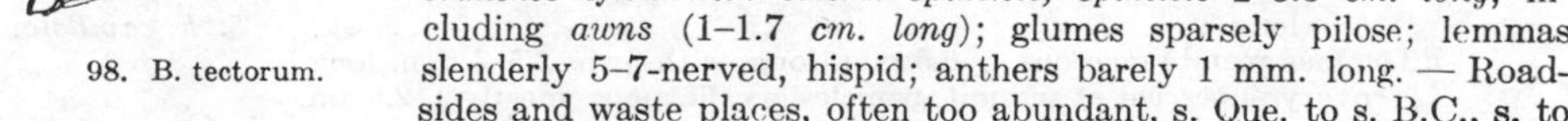
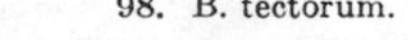
98. B. tectorum.

22. B. TECTÒRUM L. (of roofs, from its growing in thatch). — Annual, 0.2–1 m. high; leaves pubescent; *panicle rather dense,* 0.5–2 dm. long, *with spreading or recurving flexuous branches, the branches often with several spikelets; spikelets* 2–3.5 *cm. long,* including *awns* (1–1.7 *cm. long*); glumes sparsely pilose; lemmas slenderly 5–7-nerved, hispid; anthers barely 1 mm. long. — Roadsides and waste places, often too abundant, s. Que. to s. B.C., s. to Va., Miss., Mo., Kans., Colo. and Calif. May, June (rarely–Sept.). (Natzd. from Eu.) FIG. 98.

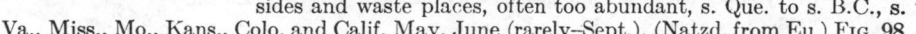

3. SCHIZÁCHNE Hack.

Spikelets 3–5-flowered, the glabrous rachilla articulate above the glumes; glumes unequal, 1st 3-nerved, 2nd 5-nerved, much shorter than the adjacent lemma; lemmas lanceolate, prominently 7-nerved, awned below the bifid apex, with a short bearded callus, the uppermost reduced; palea with very hairy submarginal keels, much shorter than lemma; ovary glabrous, with closely approximate styles; grain smooth and shining, free from palea. — Perennial

99. S. purpurascens.

herbs of N.Am. and e. Asia, the flat leaves with closed sheaths, the terminal panicle of rather large spikelets. (From the Greek *schizein, to split,* and *achne, chaff,* i.e. *lemma.*)

1. **S. purpuráscens** (Torr.) Swallen (purplish). — Loosely tufted; culms much exceeding the leaves, 0.3–1 m. high, very slender; panicle lax, slender, 0.5–1.5 dm. long; spikelets 1.3–2.3 cm. long, bronze to purplish, or pale in forma **álbicans** Fern. (whitish); glumes hyaline-margined; awns slightly divergent; grain dark-brown. (*Melica striata* (Michx.) Hitchc.) — Thickets and woods, Nfld. to s. Alaska, s. to Pa., W.Va., Wisc., Minn., S.D. and N.M. Mid-May–Aug. (northw.) Fig. 99.

4. FESTÙCA L. Fescue-Grass. Fétuque (Que.)

Spikelets 2 (rarely 1)–several-flowered; glumes mostly unequal, acute or acutish, the first 1-, the second 3-nerved and awnless or short-awned; lemma convex or subcarinate, 5-nerved, acute, obtuse or tapering to a straight awn; palea usually about equaling lemma; florets opening, exposing the nearly sessile plumose stigmas emerging from sides of lemma and the 3 free anthers raised on their filaments; grain ellipsoid or ovoid, with linear hilum, usually fused to lemma. — Perennials of frigid to temp. areas of all continents, with terminal panicles. (Ancient Latin name of some grass.)

a. Glumes firm and attenuate, opaque, much shorter than the spikelets; lemmas scarcely keeled; leaf-blades not obviously disarticulating from the old sheaths. . . *b.*

b. Basal leaves capillary or very slender, commonly involute or corrugated in drying (if flat not more than 3 mm. wide), usually densely tufted. § Ovinae. . . *c.*

c. Lower sheaths mostly whitish or drab, chartaceous, persistent, not soon disintegrating into fibers; new basal offshoots strongly ascending, from within the sheaths; anthers a quarter to half as long as the palea. . . *d.*

d. Lemmas coriaceous or membranaceous, with awns at most 3 mm. long (or wanting); ovaries glabrous; panicles strict, rarely 1 dm. long. . . *e.*

e. Awns 1.3–3 mm. long.

Panicles loosely open, with divergent short branches during anthesis, oblong to ellipsoid, mostly 3–10 cm. long; anthers 2.5–3 mm. long. . . . . 1. *F. ovina.*

Panicles spiciform and close or loosely linear-cylindric to loosely lanceolate; anthers less than 2 mm. long.

Panicle interruptedly linear-cylindric or loosely lanceolate, 2–10 cm. long; lemmas greenish, coriaceous, strongly involute; anthers 1.2–1.7 mm. long. . . . . 2. *F. saxi-montana.*

Panicle dense at least above, cylindric to lance-ovoid, usually 1–3 cm. long; lemmas purplish or bronze, membranaceous, tardily, if at all, involute; anthers 0.5–1 mm. long. . . . 3. *F. brachy-phylla.*

e. Awnless or with very short awns (up to 0.6 mm.).

Florets modified into leafy proliferous shoots; 2nd glume 3–5 mm. long; lemmas membranaceous, not strongly inrolled, 4–6 mm. long; stamens wanting. . . . . 4. *F. vivipara.*

Florets normal, perfect; 2nd glume 2–2.7 mm. long; lemmas coriaceous, tightly inrolled, 3–3.5 mm. long; anthers 1.5–2 mm. long. . . . . 5. *F. capillata.*

d. Lemmas membranaceous, 5–6.5 mm. long, with awns 2.5–7 mm. long; ovary pubescent at summit; panicles lax, flexuous, mostly 1–2.5 dm. long. . . . . 6. *F. occi-dentalis.*

c. Lower sheaths brown, red or purplish, membranaceous, friable, soon disintegrating into fibers; basal offshoots chiefly decumbent or divergent, their axes mostly projecting laterally from the old sheaths (strongly ascending and intravaginal in *F. rubra* var. *commutata*); anthers mostly one-half to two-thirds as long as the palea.

Spikelets (except in sporadic individuals) bearing only normal florets; lemmas (except in var. *mutica*) awned; anthers 2–4 mm. long. . . 7. *F. rubra.*

Spikelets with florets largely replaced by leafy tufts; lemmas awnless or merely acerose-attenuate; anthers (in rare fertile spikelets) 1.5–2 mm. long. . . . . 8. *F. prolifera.*

b. Basal and cauline leaves flat, 4–8 mm. wide, not densely tufted; lemmas mostly awnless. § Bovinae.
Panicle contracted, with short branches floriferous nearly to base; spikelets 8–12 mm. long; lemmas thin, scarious-margined. . . . . . . 9. ***F. elatior.***
Panicle diffuse, the long branches naked at base; spikelets 5–7 mm. long; lemmas firm.
Spikelets lanceolate, with glumes and lemmas appressed, mostly scattered at tips of branchlets. . . . . 10. ***F. obtusa.***
Spikelets ovate, with loosely ascending glumes and lemmas, approximate. . . . . 11. ***F. paradoxa.***
a. Glumes thin and scarious, ovate-lanceolate, lustrous, nearly as long as spikelet; lemmas keeled; leaf-blades disarticulating from the persistent basal sheaths. § Subbulbosae. . . . . 12. ***F. scabrella.***

§ Ovìnae Fries

1. F. ovìna L. (of sheep; from its growing on pastured hills), Sheep's-Fescue. — Densely tufted, forming strong tussocks with erect wiry culms 1.5–3 (–6) dm. high; *basal leaves* firm, capillary or strongly involute, 0.4–0.6 mm. in diameter, 5–7-nerved, *with chartaceous whitish non-fibrillose persistent sheaths;* the crowded *basal tufts erect or strongly ascending, projecting from the summits of the old sheaths;* cauline leaves 2, very short; *panicles* pale green or violet-tinged, *stiff, 2–5 cm. long, in anthesis loosely open and oblong or ellipsoid;* spikelets 5–7 mm. long, 3–6-flowered; *glumes* firm and attenuate, *the 2nd 2.5–4 mm. long; lemmas* lanceolate, *coriaceous, strongly involute, essentially nerveless,* glabrous, or hispid in forma hispídula (Hack.) Holmb. (with minute stiff hairs), the *lower 3–5 mm. long,* with *awns 1.3–3 mm. long; anthers 2.5–3 mm. long.* — Dry open soil or rocky slopes, local, Que. to Mich., s. to N.J., Pa. and W. Va. May, June. (Natzd. from Eurasia) Fig. 100.

100. F. ovina.

Var. duriúscula (L.) W. D. J. Koch (rather hard). — Coarser, up to 7 dm. high; leaves grooved, 0.7–1.2 mm. broad, 7–9-nerved; panicle 4–10 cm. long; spikelets 7–10 mm. long, 4–9-flowered; 2nd glume 3.5–5 mm. long; lowest lemma 4.5–6 mm. long. — More frequent, Nfld. to Minn. and Pa. (Natzd. from Eu.)

2. F. saximontàna Rydb. (of the Rocky Mountains). — Similar, 0.7–7 *dm. high;* leaves mostly shorter; *panicle interruptedly linear-cylindric to loosely lanceolate, 2–10 cm. long; spikelets* loosely flowered, *greenish;* glumes coriaceous, lance-subulate, 2nd 3–5 mm. long; *lemmas coriaceous,* almost nerveless, *strongly involute,* the longer 3–5 mm. long; *anthers 1.2–1.7 mm. long.* — Dry slopes, crests and open plains, w. Nfld.; e. Que.; n. Vt.; L. Huron, Ont., to Mackenz. and Alaska, s. to Mich., Wisc., Minn., Neb., Colo. and Utah. June, early July. Fig. 101.

101. F. saximontana.

3. F. brachyphýlla Schultes (short-leaved). — Habit and sheaths much as in nos. 1 and 2; culms capillary, 0.2–2.5 dm. high, mostly low; basal leaves capillary, 2–6 (–12) cm. long, rather soft; *panicles cylindric to lance-ovoid, compact, 1–3 cm. long; spikelets purplish or bronze; lemmas elliptic- to ovate-lanceolate, membranaceous,* usually *not strongly involute; anthers* 0.5–1 *mm. long.* (*F. ovina,* var. *brevifolia* (R. Br.) S. Wats.) — Arct. reg., s. to rocky summits and slopes of Nfld., Gaspé Pen. and L. Mistassini, Que., n. Mich., Colo., Utah, Nev. and Calif. Late June, July. (Arct. Eurasia) Fig. 102.

102. F. brachyphylla.

4. F. vivípara (L.) Sm. (viviparous, the florets altered to leafy tufts). — In habit, foliage and sheaths as in nos. 1 and 2; culms capillary, 0.5–2 (rarely –6) dm. high; *panicles* 2–10 cm. long, *strongly proliferous, the flowers represented by leafy tufts;* 2nd glume 3–5 mm. long; *lemmas membranaceous,* when not modified 4–6 *mm. long, awnless, not strongly inrolled;* axes and leaf-bases of proliferous tufts purple or green. (*F. ovina,* var. *vivipara* L.) — Greenl. and Lab. to Alaska, s. on exposed calcareous rock and peat to w. Nfld. and Anticosti I., Shickshock Mts. and L. Mistassini, Que. July, early Aug. (Eu.)

5. F. capillàta Lam. (hair-like). — Habit and sheaths of no. 1; basal *leaves delicately capillary,* 0.5–3 dm. long; culms capillary, 1–6 dm. high; panicle linear- to oblong-cylindric, rather dense, 1.5–7 cm. long; spikelets elliptical, 4–6 mm. long; *2nd glume 2–2.7 mm. long; lemmas* pale green, *coriaceous, tightly involute, the longer 3–3.5 mm. long, awnless or with a short mucro up to 0.6 mm. long; anthers 1.5–2 mm. long.* (*F. ovina,* var. *capillata* (Lam.) Alef.) — Dry open soil, Nfld. to Mich., Ill. and S.C.; apparently indigenous in Nfld., perhaps also in N.S.; introd. elsewhere. Late May–July. (Mostly natzd. from Eu.) Fig. 103.

103. F. capillata.

6. F. occidentàlis Hook. (western). — Densely tufted, the smooth and shining culms 5–9 dm. high; basal leaves capillary, 0.5–3 dm. long; the *basal offshoots erect, issuing from the summits of the membranaceous pale brown sheaths; panicle lax, subsecund,* flexuous, 1–2.5 *dm. long;* spikelets loosely 3–5-flowered, 6–10 mm. long; glumes narrow and acute, the 2nd 3.5–5 mm. long; *lemmas membranaceous,* 5–6.5 mm. long, with *awns* 2.5–7 *mm. long; ovary pubescent at summit.* — Woods and thickets, Bruce Pen., Ont., to n. Mich. and e. Wisc.; Mont. and B.C., s. to Wyo. and Calif. Late June, July. Fig. 104.

104. F. occidentalis.

7. F. rùbra L. (red). — *Loosely cespitose,* with matted and short or more or less creeping rootstocks, and culms geniculate at base to erect, 0.7–9 dm. high; *basal sheaths brown, red or purple, soon becoming loosely fibrous or disintegrating into fibrous shreds;* basal offshoots projecting laterally through the old sheaths or less often erect and projecting from the summits of the sheaths; basal leaves setaceous to plane, 0.5–3 mm. wide; cauline leaves similar or the upper flat; panicle 0.2–2.3 dm. long, with ascending branches or with lowest branches divergent; spikelets elliptic- to linear-lanceolate, stramineous, green or purplish, often glaucous, 0.7–2 cm. long, 3–10-flowered; glumes glabrous, unequal, 1st 1-nerved, 2nd longer and 3-nerved; lemmas lanceolate, mostly awn-tipped, obscurely or sometimes distinctly 3–5-nerved, narrowly scarious-margined, the lowest 4–8 mm. long (excluding awns); *anthers* 2–4 *mm. long, one-half to two-thirds the length of the palea;* ovary and grain glabrous. — A polymorphous species; the chief vars. and forms with us are the following:

a. Foliage comparatively soft, not strongly whitened (except in f. *glaucescens*). . . *b.*

b. Lemmas glabrous, scabrous or merely strigose-hirsute; panicles up to 2.3 dm. long, often with elongate branches. . . *c.*

c. Basal leaves setaceous or linear-involute, 0.5–1 mm. in diameter. . . *d.*

d. Basal offshoots all or in great part divergent or decumbent, forming loose mats. . . *e.*

e. Lemmas distinctly awned. . . *f.*

f. Lemmas glabrous. . . *g.*

g. Spikelets 7–10 mm. long, with 3–7 florets; 2nd glume 3–4.5 mm. long; body of 1st lemma 4–6 mm. long.

Foliage green. . . . . *F. rubra* (typical).

Foliage whitish. . . . . Forma *glaucescens.*

g. Spikelets 1–1.7 cm. long, with 6–10 florets; 2nd glume 4–6 mm. long; body of 1st lemma 6–8 mm. long. . . . . . Forma *megastachys.*

f. Lemmas puberulent or strigose-hirsute. . . . . Forma *squarrosa.*

e. Lemmas awnless or merely mucronate. . . . . Var. *mutica.*

d. Basal offshoots all erect, intravaginal, with leaves mostly elongate (1.5–6 dm.) and erect culms tall (2.5–9 dm.). . . . . Var. *commutata.*

c. Basal and cauline leaves all flat, 1.5–3 mm. broad; spikelets 1–1.7 cm. long, 6–10-flowered; body of 1st lemma 5.5–7 mm. long. . . . . Var. *multiflora.*

b. Lemmas densely villous to lanate; panicle narrow and spiciform or with short branches, 0.5–1.5 dm. long; spikelets 1–1.5 cm. long, ... Var. *arenaria.*

a. Foliage stiff, somewhat wiry, strongly whitened; spikelets 0.9–2 cm. long, 6–10-flowered; body of 1st lemma 5–7.5 mm. long, glabrous. . . . . Var. *juncea.*

F. rùbra (typical). — Sandy, rocky or peaty soils, Greenl. and Lab. to Alaska, s., most abundantly near the coast, to Va., mts. of N.C. and Tenn., Mo., and Calif., often introd. southw. June, July. — Forma **glaucéscens** (Hartm.) Holmb. (somewhat blue-green), occasional. Forma **squarròsa** (Fries) Holmb. (with recurving tips), var. *subvillosa* Mert. & Koch, occasional. Forma **megástachys** (Gaudin) Holmb. (with large spikelets), frequent. (Eurasia)

Var. **mùtica** Hartm. (without awns). — Greenl., s. on cool shores and mts. to Nfld. and e. Que. July, Aug. (Eu.)

Var. **commutàta** Gaudin (variable), var. *fallax* (Thuill.) Hack. Nfld. to Mich. and N.C. Late May–July. (Eu.)

Var. **multiflòra** (Hoffm.) Aschers. & Graebn. (many-flowered). — Local, Greenl. to e. Mass. July, Aug. (Eurasia)

Var. **arenària** (Osbeck) Fries (of sand). — Arct. reg., s. to Nfld., St. Paul I., N.S., Gaspé Pen., Que., Sask., Colo. and Oreg. July, Aug. (Eurasia)

Var. **júncea** (Hack.) Richter (rush-like). — Dunes, beaches and coastal headlands, Nfld. and e. Que. to se. Mass. Late June, July. (Eu.)

8. F. prolífera (Piper) Fern. (proliferous). — Intermediate between nos. 7 and 4; loosely stoloniferous to loosely cespitose; *basal sheaths reddish or purplish, membranaceous, coarsely ribbed, quickly disintegrating into loose fibers;* leaves soft, slender, linear-filiform; culms capillary, mostly geniculate at base, 1–4 dm. high; *inflorescence a simple flexuous raceme* (rarely branching and subpaniculate) *of* 3–13 *mostly proliferous glabrous spikelets;* glumes submembranaceous, the 2nd 4–6 mm. long; *lemmas membranaceous,* 5.5–8 mm. long, *muticous or merely pointed, not awned,* in many spikelets modified into leaves; *anthers* (rarely present) 1.5–2 *mm. long; florets mostly altered to leafy tufts* with purplish axes. (*F. rubra,* var. *prolifera* Piper) — Rocky or peaty soils, oftenest calcareous or neutral, w. Nfld.; mts. and valleys of Gaspé Pen., Que.; alpine regions of Mt. Katahdin, Me., and Mt. Washington, N.H. July–Sept. (Scotland)

Var. **lasiólepis** Fern. (hairy-scaled). — Lemmas pilose. — Calcareous cliffs and gravels, w. Nfld. and Anticosti I. and L. Mistassini, Que.

§ Bovìnae Fries

9. F. elàtior L. (taller), Taller or Meadow-Fescue.,— Loosely tufted, often with short creeping rootstocks; culms erect, 5–12 dm. high, smooth; blades 1–6 dm. long, 4–8 mm. wide, scabrous above; panicle erect, 0.5–3 dm. long, contracted after blooming, *branches spikelet-bearing nearly to the base; spikelets* 8–12 *mm. long;* glumes lanceolate; *lemmas* oblong-lanceolate, the scarious apex acute, *awnless,* or in forma aristàta Holmb. (awned) short-awned. — Meadows and roadsides, Nfld. to B.C., s. to Ala., La., Kans. and westw., often cult. June–Aug. (Natzd. from Eu.)

10. F. obtùsa Biehler (obtuse). — Culms solitary or few, erect, 4–12 dm. high; sheaths glabrous or pubescent; blades 1–3 dm. long, 4–11 mm. wide, scabrous, sometimes pubescent above; *panicle very loose,* 1–2.5 dm. long, usually subsecund and more or less nodding; *branches spikelet-bearing near the ends,* at first erect, *finally spreading; spikelets lanceolate to elliptic,* 1–5-flowered, 3.5–8 mm. long, 1.5–4 mm. broad, the *glumes and lemmas* usually *subappressed;* glumes firm, the 1st 2–3.5 mm., the 2nd 3–4.5 mm. long; *lemma* smooth, oblong-ovate, obtuse to subacute, 3–4.5 *mm. long,* the narrow margin hyaline. (*F. nutans* of ed. 7, not Moench nor Biehler) — Moist woods and copses, N.S. to s. Ont. and Man., s. to Fla., Miss. and Tex. June, July.

11. F. paradóxa Desv. (unusual). — Similar to no. 10, stiffer, up to 1.5 m. high; *panicle more compact, the ascending branches not divergent in age, floriferous from near the middle; the ovate spikelets* (5 *mm. broad*), 3–6-*flowered, approximate or subapproximate,* with glumes and lemmas loosely ascending; longer lemmas 4–5.5 mm. long. (*F. Shortii* Kunth) — Prairies, thickets and open woods, e. Md., se. Pa. and se. Va.; Ill. and Ky. to Minn., s. to Ga., Ala., Miss. and e. Tex. Late May–July.

§ Subbulbòsae Nym.

12. F. scabrélla Torr. (slightly harsh), Rough Fescue. — Densely cespitose, coarse, 3–9 dm. high; *old basal sheaths long-persistent, chartaceous, enlarged at base, their blades early disarticulating; basal leaves stiff, hard,* scabrous, often involute, erect, intravaginal; cauline only 1 or 2, the upper with long sheath and very short blade; panicle 0.5–2 dm. long, in anthesis ovate, with spreading branches, in fruit contracted; *spikelets* toward the tips of the branchlets, *elliptical or ovate,* 7–12 mm. long, 2–4(–6)-flowered; *glumes scarious, lustrous, broadly lanceolate to ovate,* the 1st 5–8 mm. long, the 2nd 5–9 mm. long; *lemmas* similar, *scario-membranaceous,* scabrous or puberulent, prominently 5-nerved, *the midnerve forming a keel;* palea about equaling the lemma; anthers 3 mm. long; ovary hirsute at summit. — Peaty or rocky meadows and barrens, mts. of w. Nfld.; Shickshock Mts. and serpentine-barrens of Megantic Co., Que.; Man. to s. B.C., s. to N.D., Colo. and Oreg. July–mid-Aug.

5. VÚLPIA K. C. Gmel.

Differing from *Festuca* in usually annual habit; first glume very small; lemmas compressed-subulate, slender-tipped or long-awned; florets rarely opening but upwardly distended in anthesis, close-pollinated; the 1 (rarely 3) anther on a very short filament, appressed to the palea or the small stigma, neither the anther nor stigma exserted; grain linear-cylindric, attenuate to both ends. — Chiefly annuals of temp. Eur., the Mediterr. reg. and temp. N. and S.Am., with open or spiciform panicles. (Name from *vulpes, fox,* from the many long awns of the panicle.) Festuca, subg. Vulpia (K. C. Gmel.) Hack.

1st glume one-fourth to one-third as long as the 2nd. . . . . 1. *V. myuros.*
1st glume two-thirds to three-fourths as long as the 2nd.
  Awns two to four times as long as the scarious-margined small (2–3.5 mm. long) lemmas. . . . . 2. *V. Elliotea.*

Awns much shorter than (or wanting) to but slightly longer than the firm and larger (the lower 3–5 mm. long) lemmas. . . . . 3. *V. octoflora.*

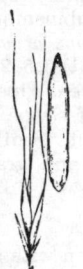

1. V. myùros (L.) K. C. Gmel. (mouse-tail). — Culms erect or geniculate at base, solitary or in small tufts, 2–7 dm. high; *sheaths* smooth, *overlapping;* blades smooth, linear, commonly involute; *panicle* 0.4–3 dm. long, slender, the branches soon appressed, *the tips flexuous or somewhat nodding; spikelets* 4–8-flowered, 8–11 *mm. long* (excluding awns); *glumes very unequal, the first* 1–1.5 *mm., the second* 4–5 *mm. long; lemma* linear-lanceolate, *scabrous above,* attenuate into a scabrous awn about twice its length. (*Festuca* L.) — Dry fields and waste places, s. Me. to Wisc., s. to Fla., La. and Tex.; Pacific States. Late May–July. (Natzd. from Eu.) Fig. 105.

2. V. Elliótea (Raf.) Fern. (for Stephen Elliott, its discoverer, 1771–1830). — Similar to no. 1, often lower; *panicle erect; spikelets* (excluding awns) 4–5 *mm. long; first glume* 2 *mm., second* 3.5 *mm. long; lemmas sparsely short-pubescent,* 2–3.5 *mm. long,* the awn two to four times as long. (*Festuca sciurea* Nutt.) — Sandy ground, Fla. to Tex., n. to N.J., Md., s. Mo. and Okla. May, June.

105. V. myuros.

3. V. octoflòra (Walt.) Rydb. (eight-flowered). — Culms slender, erect, often tufted, 0.5–7 dm. high; *sheaths shorter than the internodes;* blades narrowly linear, involute or rarely flat, soft, erect or ascending; *panicle* narrow, erect, 2–15 cm. long, *usually reduced to a more or less secund and open raceme;* spikelets 5–13-flowered, 5–12 mm. long; *glumes* subulate-lanceolate, *the lower* 3.5–4.5 *mm. long,* the upper one-third longer; *lemmas* lanceolate, *attenuate into a scabrous* straight *awn,* the *longer awns* 3.5–7 *mm. long.* (*Festuca* Walt.) — Dry sterile soil, Fla. to Tex., n. to s. N.J., s. Ill., centr. Mo. and Okla. May, June. Fig. 106.

Var. **tenélla** (Willd.) Fern. (very slender). — Smaller; inflorescence loosely spicate; *lower glumes* 2.3–4 *mm. long; awns* 1–3 *mm. long.* — Dry open soil, s. Me. to sw. Que., w. to s. B.C., s. to Ga., Ark., Tex., Colo. and Calif. Late May–early July.

106. V. octoflora.

Var. **glaúca** (Nutt.) Fern. (blue-green). — Inflorescence with *densely crowded spikelets; lower glume* 1.5–3 *mm. long; lemmas awnless or with awns up to* 2 *mm. long.* — W. Fla. to N.M., n. to Mo., S.D. and Wyo. May, June.

6. PUCCINÉLLIA Parl. Alkali-Grass. Goose-Grass

Spikelets much as in *Glyceria;* but lemmas usually firmer, rounded on the back, with 5 (rarely 7) obscure nerves, usually minutely hairy at base; palea broadly bidentate at apex, equaling or shorter than the lemma; lodicules separate; styles wanting; grain usually adherent to the palea. — Tufted or matted annuals or perennials chiefly of saline or alkaline habitats, in temp. and cold reg., with narrow or involute leaves with open or partially open sheaths. (Named for Prof. *Benedetto Puccinelli,* 1808–1850, an Italian botanist.)*

*a.*Anthers 1.5–2.2 mm. long; spikelets 0.5–1.2 cm. long, 4–11-flowered. . . . 1. *P. maritima.*
*a.*Anthers 0.5–1.2 mm. long; spikelets 3–7 (very rarely –9) mm. long, 2–6-flowered. . . *b.*
*b.*Grain 1–1.6 mm. long; lemmas serrulate or ciliolate. . . *c.*
*c.*Lower branches of short (0.2–1.6 dm.) panicle densely flowered to below the middle; lemmas thick, coriaceous, without a broad hyaline tip, the midrib usually excurrent as a sharp point. . . . . . . 2. *P. fasciculata.*
*c.*Lower branches floriferous chiefly above the middle or, if much below, loosely flowered and panicle 2–3 dm. long; lemmas thin and membranaceous (firm in no. 4) or at least with a broad hyaline tip; midrib not excurrent. . . *d.*
*d.*Glumes and lemmas truncate, erose-ciliolate (under magnification).
Mature panicle broadly ovoid, loose, its branches spreading or reflexed. . . . . 3. *P. distans.*
Mature panicle slender-cylindric to ellipsoid, dense, its branches erect or ascending. . . . . 4. *P. coarctata.*
*d.*Glumes and lemmas tapering to an acute or obtusish tip, not truncate, erose-ciliolate; panicle-branches erect or ascending, rarely spreading. . . *e.*
*e.*Panicle-branches smooth or nearly so; lemmas coriaceous, 2.3–2.8 mm. long. . . . . 5. *P. laurentiana.*

* Treatment based on that of Fernald and Weatherby in Rhodora, xviii. 1–23 (1916).

*e.*Panicle-branches strongly scabrous; lemmas thin or membranaceous.
Lemmas 2.5–3 mm. long; grain 1.4–1.6 mm. long. . . . . . . 6. *P. macra.*
Lemmas 1.5–2 mm. long; grain 1–1.2 mm. long. . . . . . . 7. *P. Nuttalliana.*
*b.*Grain 1.7–2.5 mm. long (if shorter, then with lemma not ciliolate); lemmas entire, serrulate or ciliolate. . . *f.*
*f.*Lemmas and glumes serrulate and ciliolate-erose (under magnification).
Gray-green; panicle-branches ascending or spreading, not reflexed, smooth or nearly so; lemmas coriaceous, 2.3–2.8 mm. long. . . . 5. *P. laurentiana.*
Bright green; panicle-branches at length reflexed, scabrous; lemmas thin and lustrous, 3–4 mm. long. . . . . 8. *P. lucida.*
*f.*Lemmas essentially entire, not ciliolate. . . . . 9. *P. paupercula.*

1. P. marítima (Huds.) Parl. (of the sea). — *Coarse perennial,* developing stiff repent stolons from autumn to spring; culms 1.5–10 dm. high, in dense clumps; leaves firm, usually involute in age, 0.2–1.7 dm. long; *panicle* 0.4–3 *dm. long,* the ascending or finally spreading *branches and the pedicels scabrous; spikelets* 0.5–1.2 *cm. long,* 4–11-*flowered;* 1st glume 2.2–3.4 mm. long, 3-nerved, acute; 2nd glume ovate, 3–4.5 mm. long, 3-nerved, abruptly tipped; *lemmas* 4–5 *mm. long, firm,* 5-nerved, the *midnerve prolonged* to the apex *and with the lateral nerves pubescent below;* palea 3.2–3.7 mm. long, lanceolate, ciliate on the nerves, the cilia longest below; *anthers* 1.5–2 *mm. long; grain* 2–2.2 *mm. long.* — Saline marshes and shores, Bonaventure Co., Que., to R.I., and doubtfully to Del. June, July (rarely –Oct.). (Eu.) Fig. 107.

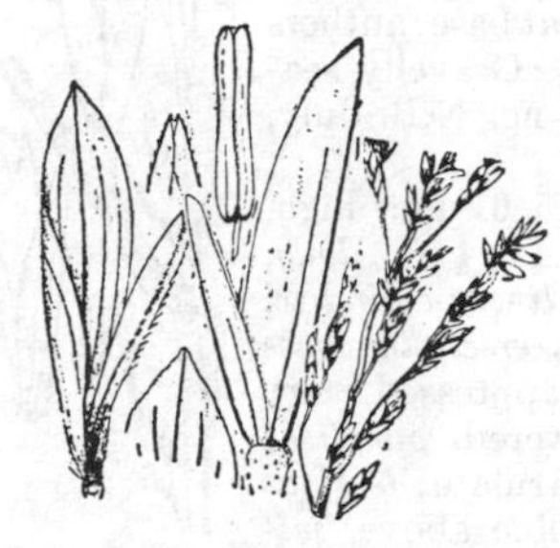
107. P. maritima.

2. P. fasciculàta (Torr.) Bickn. (with clusters). — Rather coarse *annual;* culms mostly geniculate, 1.5–8 dm. high; leaves 2–6 mm. broad, flat, or finally involute; *panicle* long-exserted, *ellipsoid to ovoid,* 0.2–1.6 dm. long, *contracted, the ascending or slightly spreading* sparingly scabrous *branches floriferous nearly to the base; spikelets* 3–4 *mm. long,* 2–5-*flowered; glumes* minutely serrulate; *the* 1*st* 0.75 *mm. long,* acutish, 1-nerved; the 2*nd* 1.5 *mm. long,* with 3 prominent nerves; *lemmas* 2–2.5 *mm. long, firm,* ovate, minutely serrulate at the bluntish tip, pubescent at base; palea slightly shorter, oblong, blunt and erose at apex, ciliate on the nerves; *anthers* 0.5–0.8 *mm. long; grain* 1.4 *mm. long.* (*P. Borreri* (Bab.) Hitchc.) — Sandy seashore, local, N.S. to Va.; saline shores, w. N.Y.; Utah. May–July. (Eu.) Fig. 108.

108. P. fasciculata.

3. P. distans (L.) Parl. (remote). — Annual, tufted; culms slender, 2–6 dm. high; leaves green, rarely glaucous, 3.5–10 cm. long; the cauline flat, 2–6 mm. wide; *panicle* green or violet-tinged, 0.8–2 dm. long, *ovoid, lax; the lower branches* 0.3–1.3 *dm. long, at first ascending but soon divergent and finally deflexed, flowering chiefly above the middle;* spikelets 4–5 mm. long, 4–6-flowered; *glumes erose-ciliolate,* the 1st 1 mm. long, 1-nerved, the 2nd 3-nerved with the lateral nerves faint; *lemmas* 2–2.5 mm. long, broadly ovate, faintly nerved, *obtuse or subtruncate, hyaline above and erose-ciliolate,* with a few hairs at base; palea shorter, ciliate in the upper half on the marginal nerves, erose-ciliolate at apex; anthers 0.7–0.9 mm. long; grain 1.4 mm. long. — Saline or calcareous waste ground and roadsides, coast of N.B. and N.S., s. to Del., occasionally inland to s. Ont. and Wisc. June–Oct. (Natzd. from Eu.) Fig. 109.

109. P. distans.

Var. angustifòlia (Blytt) Holmb. (narrow-leaved). — Cauline leaves 2 mm. or less wide, becoming involute; panicle 4–9 cm. long; lemmas 1.5–2 mm. long. — Local, s. N.B. to N.J., s. Ont. and w. N.Y., and occasional across the continent. (Natzd. from Eu.)

4. P. coarctàta Fern. & Weath. (contracted). — Cespitose *perennial* 0.8–3 dm. high, *glaucous;* cauline leaves flat; *panicles dense in anthesis,* 2–7 cm. long, finally loosening; the *lower branches* 1–3 *cm. long, ascending;* spikelets 3.5–5 mm. long, loosely 3–4-flowered; *glumes erose-ciliate,* the 1st 0.5–1 mm. long; the 2nd 3-nerved, 1.5–2 mm. long, broadly ovate, obtuse, sparsely

110. P. coarctata.

111. P. laurentiana.

hairy at base, hyaline and ciliolate above; palea subtruncate or bidentate at the broad apex, sparsely ciliate on the nerves. — Coasts of Lab., Nfld. and e. Saguenay Co., Que. July, Aug. FIG. 110.

5. P. laurentiàna Fern. & Weath. (of the Gulf of St. Lawrence). — Perennial; culms solitary or slightly tufted, 1–3 dm. high, *rigid, glaucous; cauline leaves with involute blades* 3–6 *cm. long; basal sheaths whitish; panicle* 0.6–1.3 *dm. long, its branches stiff, nearly glabrous, ascending or finally divergent* but not deflexed; *spikelets* 4–6.5 mm. long, 3–5-flowered, *whitish; glumes* erose-serrulate, acute; *the 2nd* 2–2.5 mm. long, broadly ovate, *abruptly narrowed to the apex; lemmas* ovate, hyaline above, 2.3–2.8 mm. long, *abruptly acuminate, erose-ciliolate,* very pubescent on the nerves below; palea shorter, lanceolate, with broad apex, scabrous above, ciliate at base; anthers 0.7–0.9 mm. long; *grain* 1.5–1.8 *mm. long.* — Gravelly seashores, Gulf of St. Lawrence, se. Que. and ne. N.B. July, early Aug. FIG. 111.

112. P. macra.

6. P. mácra Fern. & Weath. (large). — Cespitose, 4.5–6 dm. high, *green; cauline leaves with thin flat blades* 3–6.3 *mm. wide,* 0.4–1.4 *dm. long, the upper longer than the lower; basal sheaths purple; panicle linear-cylindric,* 2–3 *dm. long,* before and after anthesis *erect, the base included in the upper sheath; branches* appressed, *very scabrous; spikelets* 4–7 mm. long, 4–6-flowered, *purplish; 2nd glume obtuse,* 2.5 mm. long, erose-serrulate; *lemmas* 2.5–3 mm. long, ovate, *obtuse, thin,* hyaline above, *minutely erose-serrulate,* the nerves pubescent at base; palea 2.5–3 mm. long, oblanceolate, ciliate on the nerves, the lower cilia longer; anthers 0.9–1 mm. long; *grain* 1.4–1.6 *mm. long.* — Sea-cliffs and coastal sands, e. Gaspé Co., Que. Aug. FIG. 112.

113. P. Nuttalliana.

7. P. Nuttalliàna (Schultes) Hitchc. (for THOMAS NUTTALL, 1786–1859). Cespitose, green or slightly glaucous; culms slender, 2–9 dm. high; *cauline leaves* 1–2 *mm. wide,* 2–12 cm. long; *panicle* 0.4–3 dm. long, the *branches and pedicels strongly scabrous,* the lower branches (up to 1.2 dm. long) ascending to spreading; spikelets 3.5–7 mm. long, 3–6-flowered; *glumes thick,* erose-ciliolate, *the 1st* 1.2–1.5 *mm. long,* the *2nd* 1.5–2 *mm. long; lemmas* 1.5–2 *mm. long,* ovate, *rather abruptly contracted to a blunt or subacute apex,* not truncate, thin, hyaline above, erose-serrulate, the nerves pubescent below the middle; *anthers* 0.6–0.7 *mm. long; grain* 1–1.2 *mm. long.* (*P. airoides* (Nutt.) Wats. & Coult.) — Marshes (often saline) and prairies, Mackenz. and Yuk., s. to Minn., Tex., N.M. and Calif.; locally adv. eastw. to N.E. June, July. FIG. 113.

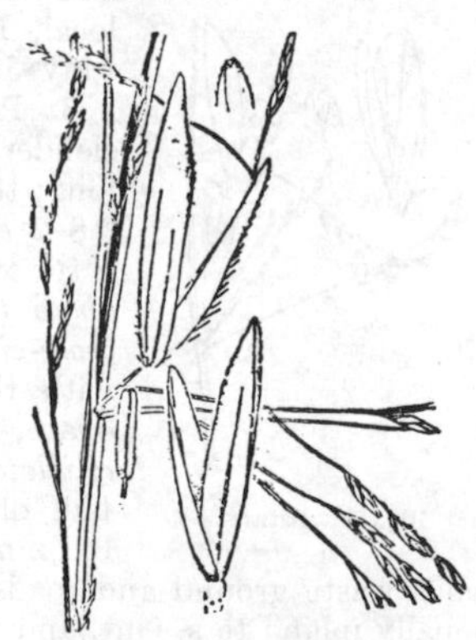
114. P. lucida.

8. P. lùcida Fern. & Weath. (shining). — Loosely cespitose, 1.5–7 dm. high, *green; cauline leaves with flaccid involute blades;* basal sheaths usually purplish; *panicle diffuse,* 1–2.5 *dm. long, the base commonly included in the upper sheath; branches filiform, scabrous, loosely divergent, finally deflexed;* spikelets 5–9 mm. long, 3–5-flowered, pale green; *glumes thin, lustrous;* 1st 2–3.5 *mm. long,* hyaline above, minutely serrulate, *the 2nd* 2.5–4 *mm. long; lemmas* 3–4 *mm. long,* broadly ovate, *acute,* erose-ciliolate, strongly pubescent toward the base with long hairs; *anthers* 1–1.2 *mm. long; grain* 1.8–2 *mm. long.* — Saline marshes and coastal sands, lower St. Lawrence R., Que.; also Wyo. to B.C. and Calif. July, Aug. FIG. 114.

9. P. paupércula (Holm) Fern. & Weath. (rather poor). — Densely cespitose or matted, sometimes (late in the season) with flagelliform stolons; leaves flat, soft; *panicle with smooth or smoothish* ascending, spreading or sometimes reflexed *branches;* spikelets 2–6-flowered; *2nd glume entire or only obscurely dentate, acutish or acuminate; lemmas entire, minutely pubescent at base only;* palea slightly scabrous on the nerves above, anthers 0.5–1 mm. long; *grain* 1.3–2.6 *mm. long,* readily separating from the palea. Three extreme vars.:

Panicle 0.1–1.8 dm. long; spikelets 3–7 mm. long; 1st glume 1–2 mm. long; 2nd glume 2–3 mm. long, 3-nerved; lemmas 2–4 mm. long, 5-nerved; grain 1.3–2.1 mm. long.

Culms 0.2–2 dm. high; cauline leaf-blades 1–4 cm. long; ligule less than 1 mm. long; panicle 1–6 cm. long; spikelets 3–5 mm. long; 1st glume 1–1.5 mm. long; lemma 2–2.5 mm. long; grain 1.3–2 mm. long. . . . . *P. paupercula* (typical).

Culms 0.5–4.5 dm. high; cauline leaf-blades up to 1.2 dm. long; ligule about 2 mm. long; panicle 0.2–1.8 dm. long; spikelets 4–7 mm. long; 1st glume 1.5–2 mm. long; lemma 2.5–4 mm. long; grain 1.7–2.1 mm. long. . . . Var. *alaskana.*

Panicle 1–2.5 dm. long; spikelets 0.6–1.2 cm. long; 1st glume 3–4 mm. long, 2nd 7–9 mm. long and 3–5-nerved; lemma 4.5–6 mm. long, 7-nerved; grain 2.2–2.6 mm. long. Var. *longiglumis.*

P. paupércula (typical). — Saline shores, sw. Greenl., n. Lab. and Ung., s. to Nfld., M.I. and Gulf of St. Lawrence, Que. July, Aug. (E. Asia) Fig. 115.

Var. **alaskàna** (Scribn. & Merr.) Fern. & Weath. (of Alaska), *P. angustata* of ed. 7; *P. pumila* (Vasey) Hitchc., based on a technically undescribed type. — Saline shores, Lab. and lower St. Lawrence, s. to Ct.; w. N.Y.; Alaska to B.C. (E. Asia) June–Sept. Fig. 116.

Var. **longiglùmis** Fern. & Weath. (with long glumes). — Marshes of Hillsborough R., P.E.I. July, Aug. Fig. 117.

115. P. paupercula (typical).

116. P. paupercula, v. alaskana.

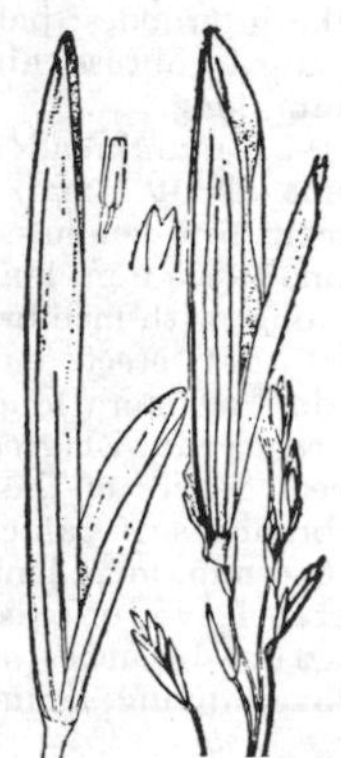

117. P. paupercula, v. longiglumis.

7. GLYCÈRIA R. Br. Manna-Grass

Spikelets few–many-flowered, subterete to compressed, in narrow or spreading panicles; glumes unequal, shorter than the florets; lemmas convex, firm, with a scarious margin or apex, and 5–9 usually strong parallel nerves; paleas equaling or a little longer than their lemmas, the strong nerves nearly marginal; lodicules united; styles distinct; grains free. — Paludal or aquatic perennials, chiefly of N. Am., Eurasia and Austral., with simple culms, closed or partially closed sheaths, flat blades and terminal panicles. (Name from the Greek *glyceros, sweet,* in allusion to the taste of the grain in true *Glyceria,* i.e., § *Fluitantes.*) Panicularia Heist.

a. Sheaths strongly compressed, ancipital; spikelets linear-subcylindric, mostly 1–4 cm. long. § Fluitantes. *b.*

b. Lemmas obtuse to subacute, 3–8 mm. long, about equaled to very slightly exceeded by the palea.

Spikelets 1–1.8 cm. long, those along the branchlets on slender pedicels; lemmas membranaceous, glabrous or only minutely scabrous except on the hirtellous nerves; anthers 0.6–1 mm. long; grain 1.2–1.6 mm. long. . 1. *G. borealis.*

Spikelets 1–4 cm. long, those along the branchlets sessile or on short upwardly broadened pedicels; lemmas coriaceous to submembranaceous, hirtellous or puberulent over the surface; anthers 0.5–3 mm. long; grain 1.5–3 mm. long.

Lemmas coriaceous, 3.6–5.5 mm. long, faintly nerved; upper glume 3.2–5 mm. long; anthers 1–2 mm. long; grain 1.8–2.5 mm. long. . 2. *G. septentrionalis.*

Lemmas membranaceous or submembranaceous, sharply nerved; upper glume 2.5–4 mm. long.

Lemmas membranaceous, 2.5–3 mm. long; anthers 0.5–0.8 mm. long; grain 1.5–2 mm. long. . . . . 3. *G. arkansana.*

Lemmas submembranaceous, 6–8 mm. long; anthers 1.8–3 mm. long; grain 3 mm. long. . . . . 4. *G. fluitans.*

*b.*Lemmas lance-attenuate, sharp-pointed, 6–10 mm. long, with the paleas prolonged 1.5–3 mm. beyond their tips. . . . . 5. *G. acutiflora.*

*a.*Sheaths terete or subterete; spikelets lanceolate, oblong or ovate, 2–9 mm. long. § Hydropoa. . . *c.*

*c.*Panicle contracted, with closely overlapping ascending branches.

Panicle linear-cylindric; spikelets about 4 mm. long; new cauline leaves (of the flowering culm) 7–9, soft and lax. . . . . 6. *G. melicaria.*

Panicle thick-cylindric to ellipsoid; spikelets 5–9 mm. long; new cauline leaves 3–6, firm, ascending. . . . . 7. *G. obtusa.*

*c.*Panicle open, lax, with loosely ascending to divergent branches. . . *d.*

*d.*Spikelets 3–5 mm. broad, if slightly narrower with obscurely nerved lemmas.

Cauline leaves of the season (new green leaves) 3–6, all but the lowest with sheaths shorter than internodes; panicle twice or thrice forked; lemmas prolonged 0.5–1 mm. beyond the paleas into hyaline acuminate tips, lower lemmas 3–4 mm. long. . . . . 8. *G. canadensis.*

Cauline leaves of the season 5–8, all but the upper with sheaths exceeding the internodes; panicle three to six times forked; lemmas firm, subacute to obtuse, about equaling to barely exceeding the paleas, 2–3 mm. long. . . . . 9. *G. laxa.*

*d.*Spikelets 1–2.5 mm. broad; lemmas strongly nerved. . . *e.*

*e.*2nd glume about 1 mm. long; spikelets ovate, 2–4.5 mm. long, with caducous firm lemmas. . . . . 10. *G. striata.*

*e.*2nd glume 2–2.5 mm. long; spikelets lanceolate or oblong, mostly 4–7 mm. long, with membranaceous lemmas. . . *f.*

*f.*Culms stout, erect, mostly 1 m. or more high; leaf-blades mostly 2 dm. or more long; panicle very compound; spikelets purple (rarely yellowish-green).

Green leaves of the season 3–6, with smooth sheaths; capillary branches of panicle smooth or smoothish; spikelets scattered, 5–6 mm. long; lemmas 2–2.7 mm. long. . . . . 11. *G. grandis.*

Green leaves of season 5–10, with scabrous sheaths; angular and coarse branches of panicle scabrous; spikelets 5–10 mm. long, overlapping; lemmas 2.5–4 mm. long. . . . . 12. *G. spectabilis.*

*f.*Culms slender, weak, decumbent or repent, rarely 1 m. high; leaf-blades rarely 2 dm. long; panicle weak, few-flowered; spikelets pale green.

Larger leaves 4–10 (–13) mm. wide, with ligules 5–8 mm. long; spikelets 5–7 mm. long; lemmas 2.5–3.5 mm. long; anthers elongate, 0.6–1.5 mm. long; grain 1.5 mm. long. . . . . . 13. *G. pallida.*

Larger leaves 1–3.5 mm. wide, with ligules 2–4 mm. long; spikelets 3–5 mm. long; lemmas 2–2.8 mm. long; anthers globose, 0.2–0.5 mm. long; grain 0.8 mm. long. . . . . 14. *G. Fernaldii.*

§ Fluitántes Anderss.

1. G. boreàlis (Nash) Batchelder (northern), Small Floating M., Float-Grass. — Culms somewhat flattened, erect from creeping or decumbent bases, 0.5–1.5 m. high; leaves smooth, with ancipital overlapping sheaths; blades very elongate, often floating, 2–5 mm. wide; ligules white, scarious, prolonged; panicles 1.5–5 dm. long, either simple or with few slender erect or spreading branches, *a pedicelled spikelet in each axil; spikelets numerous,* 7–13-flowered, 1–1.8 *cm. long, on slender pedicels one-fourth to two-thirds as long; glumes* obtuse or subacute, scarious-margined, *the upper* 2–3 *mm. long; lemmas thin, strongly* 7-*nerved,* 3–4 mm. long, *minutely scabrous or glabrous,* except on the nerves, with erose scarious summit, *slightly exceeding their paleas; anthers* 0.6–1 *mm. long; grain* 1.2–1.6 *mm. long.* — Wet places or shallow water, Nfld. to s. Alaska, s. to N.S., s. N.E., Pa., Ind., n. Ill., Ia., S.D., Colo., Ariz. and Calif. June–Aug. Fig. 118.

118. G. borealis.

119. G. septentrionalis.

2. G. septentrionàlis Hitchc. (northern), Floating M., Sweet Grass. — Usually coarser than no. 1; culms thick and soft; leaf-blades 1–2.5 dm. long, 2–12 mm. broad, obtuse; panicle 2–5 dm.

long, the subflexuous branches ascending, *a spikelet subsessile in each axil; spikelets* 1–2.5 cm. long, all but the terminal ones *subsessile or on short upwardly thickened pedicels; glumes* obtuse, *coriaceous* and hyaline, *the upper one* 3.2–5 *mm. long; lemmas coriaceous,* 3.6–5.5 *mm. long, faintly* 7-*nerved, hispidulous,* with a shining scarious erose-obtuse summit, slightly exceeded by the tip of the palea; *anthers* 1–2 *mm. long; grains* 1.8–2.5 *mm. long.* — Swamps, wet woods, ditches, etc., e. Mass. to s. Ont., s. to Ga., Ky., Mo. and Tex. May–July (rarely –Sept.). FIG. 119.

3. G. arkansàna Fern. (of Arkansas). — *Coarser* than no. 2; culms up to 1 cm. in diameter; *leaves flaccid,* 1–1.8 *cm. wide;* panicle 4–7 dm. long; branches at first ascending, finally spreading; *spikelets* 10–15-flowered, 1.5–2 *cm. long; upper glume* 2.5–3.5 *mm. long; lemmas* 2.5–3 *mm. long, delicately membranaceous, definitely hirtellous, sharply nerved; anthers* 0.5–0.8 *mm. long; grain* 1.5–2 *mm. long.* — swamps and wet bottomlands, w. N.Y. (*Sartwell*) to Ill., s. to Ark. and La.; se. Va. May–July. FIG. 120.

120. G. arkansana.

121. G. fluitans.

4. G. flùitans (L.) R. Br. (floating), FLOATING M., FLOAT-GRASS. — Resembling no. 2; blades 0.6–3 dm. long, 4–8 mm. broad; panicle 2.5–4 dm. long, with remote ascending, finally divergent, branches, *a spikelet subsessile in each axil; spikelets* 2–4 *cm. long,* all but the terminal ones nearly *sessile; glumes coriaceous,* shining, *the upper* 3–4 *mm. long; lemmas thin, submembranaceous, sharply* 7-*nerved, scabrous,* 6–8 *mm. long,* with a shining scarious margin and summit, *narrowed above* but obtuse, erose, the tip of the palea exserted; *anthers* 1.8–3 *mm. long; grain* 3 *mm. long.* — Shallow water, Nfld.; e. Gaspé Co., Que.; N.S.; Nantucket I., Mass. Late June–Aug. (Eurasia) FIG. 121.

5. G. acutiflòra Torr. (acute-flowered). — Culms flattened, weak and slender, 3–10 dm. high, from decumbent bases; sheaths overlapping, the uppermost inclosing the base of the panicle; blades 0.6–2 dm. long, 1–7 mm. wide, scabrous above; panicle simple or subsimple, 1.5–4 dm. long, the few stiff branches appressed or finally spreading; *spikelets subsessile,* 5–13-flowered, 1.5–4 *cm. long,* linear-cylindric; *lemmas* 6–10 *mm. long, sharply acute,* scabrous, *exceeded by* 1.5–3 *mm. by the long-acuminate bicuspidate paleas.* — Muddy pools and pond-margins, local, s. N.H. to Mich., s. to Del., W.Va., Tenn. and Mo. May–July, rarely –Aug. (Ne. Asia) FIG. 122.

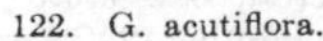

122. G. acutiflora.

§ HYDRÓPOA Dumort.

6. G. melicària (Michx.) F. T. Hubbard (similar to *Melica*). — Culms solitary or few, erect from a creeping base, 0.6–1.2 m. high; *new cauline leaves of the season* 7–9, with sheaths closed nearly to summit; *blades lax,* elongate, 2–8 mm. wide, scabrous; *panicle linear-cylindric, with closely appressed branches,* 1–3 dm. long, nodding at summit; *spikelets* appressed, 3–4-flowered, *about* 4 *mm. long;* stamens 2. (*G. Torreyana* (Spreng.) Hitchc.) — Wet woods, N.B. to O., s. to s. N.E., Md. and mts. of N.C. and Tenn. Late June–Aug. FIG. 123.

123. G. melicaria.

7. G. obtùsa (Muhl.) Trin. (obtuse). — *Culms stiff,* erect, 1.5–13 dm. high; *cauline leaves of the season* 3–6 (–7); sheaths closed except at summit; *blades firm, erect,* 2–8 mm. wide, smooth below, rough above; *panicle thick-cylindric to ellipsoid, dense,* 5–18 cm. long; *spikelets* 3–7-flowered, 5–9 *mm. long;* scarious apex of lemma often revolute. — Peaty and wet sandy soils, N.S. to centr. N.H., s. to N.C. Late July–Sept. FIG. 124.

124. G. obtusa.

8. G. canadénsis (Michx.) Trin. (of Canada), RATTLESNAKE-GRASS. — Culms solitary or few, erect, 3–10 dm. high; *cauline leaves of the season* 3–5 (–6); *all but the lowest sheaths shorter than the internodes;* blades firm, somewhat scabrous, 2–10 mm. wide; *panicle* 0.6–3 dm. long, *ovoid to pyramidal, very loose and open;* the capillary *branches mostly in* 1's–3's, spreading or drooping, naked below, *once or twice forked; spikelets* 4–10-flowered, *ovate,* in age broadly so, *tumid,* 4.5–7 *mm. long; lemmas clearly* 7-*nerved, the lower ones* 3–4 *mm. long, the hyaline acuminate tip prolonged* 0.5–1 *mm. beyond the palea; palea* (spread open) broadly elliptic, 1.2–2 *mm. broad;* stamens 2. — Bogs, meadows, damp

125. G. canadensis.

thickets and shores, Nfld. to L. Mistassini, Que., w. to s. Ont. and Minn., s. to N.S., s. N.E., Va., mts. of Tenn., Ind. and n. Ill. June–Aug. Fig. 125.

9. G. láxa Scribn. (loose). — Resembling no. 8, 0.7–1.6 m. high; *cauline leaves of the season* 5–8, very harsh, the *sheaths of all but the upper* 2 *or* 3 *overtopping the nodes; panicle* diffuse, 2–4 dm. long, pyramidal, *its principal branches in 3's–5's, two to five times forked; spikelets* 3–6-flowered, 3–5 *mm. long, oblong,* firm but not tumid; *lemmas firm, obscurely nerved, the lower ones* 2–3 *mm. long, subacute to obtuse, about equaling the paleas; palea* (spread open) oblong to elliptic, 0.8–1.4 *mm. broad.* (*G. canadensis,* var. *laxa* (Scribn.) Hitchc.) — Bogs and swamps, P.E.I. to Md. and W.Va. July, Aug. Fig. 126.

126. G. laxa.

10. G. striàta (Lam.) Hitchc. (with longitudinal lines), Fowl-meadow Grass. — Loosely to densely tufted, slender, 0.3–1.5 m. high; sheaths scabrous, closed nearly to summit; blades flat, 2–10 mm. wide, scabrous above, the uppermost 1–3 dm. long; *panicle lax, open, the loosely ascending capillary branches in age divergent or reflexed,* naked below; *spikelets* greenish or purplish, 3–7-flowered, 2–4 *mm. long, oblong-ovate; glumes minute, the 2nd about* 1 *mm. long, twice as long as the first; lemmas sharply 7-nerved,* barely if at all scarious-margined, *caducous;* stamens 3. (*G. nervata* (Willd.) Trin.) — Moist ground, common, Nfld. to Alta., s. to n. Fla., Ala. and Tex. June–Sept. (Natzd. in Eu.) Fig. 127.

127. G. striata (typical).

Var. **strícta** (Scribn.) Fern. (upright). — Lower, 2–9 dm. high, and stiffer; *leaves flat or conduplicate,* 1–5 mm. broad, the upper blade 0.3–2 dm. long; *panicle* 0.5–1.5 dm. long, *with ascending branches; spikelets* 3–4.5 *mm. long,* purple, rarely green; *lemmas with broad scarious tips.* — Lab. to Alaska, s. to Nfld., N.S., n. and w. N.E., L.I., Ill., Ia., S.D. and n. Mex. Fig. 128.

128. G. striata, v. stricta.

11. G. grándis S. Wats. (large), Reed-meadow Grass. — Culms clustered, stout, erect, 1–1.5 m. high; sheaths loose, the lower nodulose, overlapping; blades 6–15 mm. wide, smooth or slightly scabrous; *panicle* 2–4 *dm. long,* very compound, loose and open, nodding at the summit; *spikelets numerous, with purple florets,* or yellowish or pale in f. **palléscens** Fern. (becoming pale), *and whitish glumes,* 4–8-flowered, 5–6 *mm. long;* the palea nearly as long as the 7-nerved membranaceous lemma (2–2.7 mm. long). — Banks of streams, wet meadows, ditches, etc., Nfld. to Alaska, s. to Va., Tenn., Ia., N.M. and Oreg. June–Aug. Fig. 129.

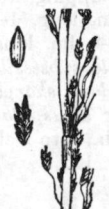

129. G. grandis.

12. G. spectábilis Mert. & Koch (showy). — Much coarser than no. 11; culms up to 2.2 m. high; *green leaves of the season* 5–10, their blades 0.8–2 cm. broad, *their sheaths scabrous; branches of panicle coarse, scabrous-angled; spikelets* 5–10 *mm. long, overlapping,* 4–9-flowered; *lemmas* 2.5–4 *mm. long; anthers* 1.5 *mm. long.* (*G. aquatica* (L.) Wahlenb., not J. & C. Presl) — Creeks and ditches, s. Ont. June, July. (Natzd. from Eurasia)

13. G. pállida (Torr.) Trin. (pale). — Culms loosely decumbent from creeping or floating bases, 3–10 dm. high; *leaves lax,* soft; *the larger blades* 0.5–2 dm. long, 4–10 (–13) *mm. wide, the lower sheaths usually divergent and free at summit; ligules* 5–8 *mm. long,* oblong, obtuse; *panicles weak,* 0.7–3 *dm. long,* the flexuous *branches ascending or somewhat divergent* and naked at base; *spikelets* lanceolate, pale green, loosely 4–9-flowered, 5–7 *mm. long;* glumes scarious, the upper obtuse; *lemmas* lanceolate, strongly 5–7-nerved, mostly dentate or erose at summit, 2.5–3.5 *mm. long; anthers elongate,* 0.6–1.5 *mm. long; grain* 1.5 *mm. long.* — Pools, sloughs and pond-margins, sw. N.S. and s. Me. to s. Ont., s. to Va., Tenn. and se. Mo. May–Aug. (E. Asia) Fig. 130.

130. G. pallida.

14. G. Fernáldii (Hitchc.) St. John (for the discoverer, 1873–). — Similar to no. 7, smaller, very slender; culms geniculate and loosely matted, 1–6 dm. high; *larger leaves* 4–10 cm. long, 1–3.5 *mm. wide, strongly ascending, the sheaths closely appressed; ligules* 2–4 *mm. long; panicles* 3–13 cm. long, *their lower branches strongly divergent or reflexed in age; spikelets* lanceolate, mostly 3–5-flowered, 3–5 *mm. long; lemmas* 2–2.8 *mm. long; anthers globose,* 0.2–0.5 *mm. long; grain* 0.8 *mm. long.* (*G. pallida,* var. *Fernaldii* Hitchc.; *G. neogaea* sensu Hitchc., not Steud.) — Shallow water and wet places, Nfld. to Minn., s. to N.S., s. N.E., L.I. and Pa. Late June–Sept. Fig. 131.

131. G. Fernaldii.

8. SCOLÓCHLOA Link

Spikelets 3–4-flowered; callus hairy; glumes acute; lemmas rigid, convex below, not keeled, the nerves unequal, one or more of them excurrent as slender teeth; apex of lemma often fimbriate; palea as long as its lemma or longer, 2-toothed; ovary hairy at the summit. — Tall perennials of n. temp. reg., with flat leaves and ample spreading panicles. (Name from the Greek *scolops, a prickle*, and *chloa, grass.*) FLUMINIA Fries

132. S. festucacea.

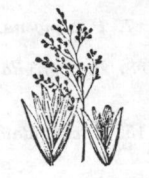

1. S. festucàcea (Willd.) Link (like *Festuca*), SPRANGLE-TOP. — Culms stout, erect, from thick soft rhizomes, 1–2 m. high; leaves 2–3 dm. long; panicles 1.5–3.5 dm. long, the fascicled branches spreading; spikelets 6–12 mm. long; glumes nearly as long as the florets, 3–5-nerved. — Marshes and shallow water, s. Mackenz. and B.C., s. to Ia., Neb. and Oreg. June, July. (Eurasia) FIG. 132.

9. PÒA L. MEADOW-GRASS. SPEARGRASS. PÂTURIN (Que.)

Spikelets 2–several-flowered, uppermost floret imperfect or rudimentary; glumes 1–3-nerved, keeled; lemmas herbaceous or membranaceous, mostly scarious-tipped, acute or obtuse, keeled, awnless, 3- or 5-nerved (the nerves arching, intermediate nerve sometimes very obscure), the dorsal or marginal nerves usually soft-hairy and often with a tuft of long cobwebby hairs at the base; palea 2-toothed; stigmas plumose, emerging from sides of lemma; grains tightly embraced; hilum punctiform. — Annuals or perennials of cool or temp. reg., with simple culms, narrow usually flat leaves ending in a cucullate tip, and terminal panicles. (An ancient Greek name for grass or fodder.)

a. 2nd glume 2–4.5 (rarely –6) mm. long, usually much overtopped by the upper lemmas; plant green, stramineous, purplish or but slightly glaucous; leaves weak to firm, mostly less than 5 mm. broad; culms slender, mostly less than 3 mm. in diameter. . . *b.*

b. Annuals or short-lived perennials, with soft foliage, the bases tufted but without long-persistent old leaves; culms weak, 0.3–5 dm. high; mature (closed) spikelets linear- or oblong-lanceolate.

Lemmas not cobwebby at base, distinctly 5-nerved; anthers 0.7–1 mm. long, exserted in anthesis. . . . . 1. *P. annua.*

Lemmas cobwebby at base, 3-nerved, intermediate nerves obscure; anthers 0.2 mm. long, included; flowers cleistogamous. . . . . 2. *P. Chapmaniana.*

b. Tufted or creeping perennials, the bases, when not thickened nor with obvious perennial offshoots, bearing persistent old leaves or dry sheaths; mature spikelets ovate to ovate-lanceolate. . . *c.*

c. Plants with elongate slender rhizomes or stolons or with loosely divergent, decumbent or geniculate basal offshoots. . . *d.*

d. Lemmas with faint or obscure intermediate nerves, obtuse; culms strongly flattened, with 3–7 leaves; panicles stiff, the branches floriferous mostly to below the middle. . . . . 3. *P. compressa.*

d. Lemmas prominently 5-nerved; culms not strongly flattened, with 2–4 new leaves; panicles with flexuous or delicate longer branches. . . *e.*

e. Branches of panicle with spikelets rather crowded from near the middle; lemmas lanceolate to lance-ovate, approximate, the lower internodes of the rachilla rarely prolonged; basal cobwebby tuft a third to half as long as lemma; anthers 1.2–1.8 mm. long. . . *f.*

f. Culms all or nearly all bearing erect or strongly ascending tufts of new green leaves from the basal sheaths.

Culms soft to firm, compressed at base and often geniculate, 2–3 mm. thick at base; basal leaves flat or flattish, as broad as thickness of culm; glumes broadly lanceolate to ovate, nearly straight. . . . . 4. *P. pratensis.*

Culms firm, terete to base, erect, 1–2 mm. in diameter at base; basal shoots with some filiform to involute blades much more slender than culm; glumes narrowly lanceolate, the 2nd arching. . . . . 5. *P. angustifolia.*

f. Culms chiefly arising from among old dried leaves at the tips of curving stolons, the new basal leaf-tufts all or nearly all on separate prolonged stolons or offsets.

Spikelets all or nearly all twice to many times length of pedicels; long pubescence, if any, confined to base of lemma.
2nd glume reaching nearly to tip of lemma above it; nerves of lemma glabrous. . . . . 6. *P. subcaerulea.*
2nd glume reaching only to middle of lemma above it; nerves of lemma minutely pubescent. . . . . . . 7. *P. alpigena.*
Spikelets nearly all on pedicels one-half to thrice their length; lemmas copiously villous to or above the middle. . . . . 8. *P. longipila.*
e.Branches with few scattered spikelets toward the tips; lemmas oblong, distant in anthesis, exposing the prolonged rachilla; basal web much shorter; anthers 2–3 mm. long. . . . . 15. *P. cuspidata.*
c.Plants tufted, without slender elongate rhizomes or stolons. . . *g*.
g.Lemmas distinctly 5-nerved. . . *h*.
h.Many or all branches of panicle floriferous to below the middle; the crowded spikelets subsessile or very short-pedicelled; ligules of upper leaves 5–8 mm. long; anthers 1.5–2 mm. long. . . . . . 9. *P. trivialis.*
h.Many or all branches loosely floriferous at tip; spikelets mostly on pedicels nearly or quite their own length; ligules of upper leaves 0.3–3 mm. long; anthers 0.6–3 mm. long. . . *i*.
i.Lemmas glabrous except for web at base.
Lemmas thin, acute, 2.5–4 mm. long; anthers 1–1.2 mm. long; upper ligules 0.3–1.5 mm. long. . . . . 10. *P. saltuensis.*
Lemmas coriaceous, obtuse, 2–3 mm. long; anthers 0.6–0.8 mm. long; upper ligules 2–3 mm. long. . . . . . . 11. *P languida.*
i.Lemmas pubescent, at least on the keel. . . *j*.
j.Lemmas without web at base. . . . . 12. *P. autumnalis.*
j.Lemmas webbed at base. . . *k*.
k.Spikelets 2.5–4 mm. long; lemmas 2–3 mm. long, firm, pilose between the nerves. . . . . 13. *P. sylvestris.*
k.Spikelets 5–8 mm. long; lemmas 4–5.5 mm. long, scario-membranaceous, glabrous between the nerves.
Lemmas acute, marginal nerves and keel villous; anthers 0.8–1.4 mm. long; leaves short-tipped. . . . . . 14. *P. Wolfii.*
Lemmas obtuse, marginal nerves and keel scabrous or hirtellous; anthers 2–3 mm. long; leaves cuspidate-tipped. 15. *P. cuspidata.*
g.Lemmas with only 3 distinct nerves, the intermediate nerves obscure. . . *l*.
l.Bases of leaf-tufts not definitely bulbous; spikelets normal. . . *m*.
m.Lemmas webbed at base. . . *n*.
n.Low, matted, alpine, with mature filiform flexuous or arching culms (0.5–3 dm. long) mostly leafless above the middle; panicle 1–6 cm. long, with few ascending branches. . . . 23. *P. Fernaldiana.*
n.Tall lowland or woodland plants with upright culms leafy above the middle; panicles mostly larger. . . *o*.
o.Lateral nerves of lemmas glabrous. . . . . . . 16. *P. alsodes.*
o.Lateral nerves pubescent. . . *p*.
p.Culms 1–few, filiform, mostly with 2 or 3 leaves 0.25–3 mm. wide; panicles 0.3–1.3 dm. long, their remote branches mostly in 1's or 2's; anthers 0.5–1 mm. long. . . . . 17. *P. paludigena.*
p.Culms stouter, numerous, mostly with about 6 broader leaves; panicles larger, with branches usually 3–5 at a node; anthers 0.8–1.5 mm. long.
Ligule truncate, 0.5–1 mm. long; anthers 1.2–1.5 mm. long. . . . . 18. *P. nemoralis.*
Ligule elongate, 2–5 mm. long; anthers 0.8–1 mm. long. 19. *P. palustris.*
m.Lemmas without web at base. . . *q*.
q.Glumes ovate; leaves broad; the lower cauline oblong-linear, blunt, 2–6 mm. wide.
Caudices stout, 0.5–1.5 cm. in diameter, heavily invested with whitish papery sheaths; glumes round-ovate, abruptly tipped, scario-membranaceous. . . . . . . 20. *P. alpina.*
Caudices slender, 2–4 mm. thick, loosely brown-sheathed; glumes ovate, acuminate, subcoriaceous. . . . . 21. *P. gaspensis.*
q.Glumes lanceolate or lance-ovate, acuminate; leaves slender, the lower cauline narrowly linear to involute, 0.3–2.5 mm. broad, with subulate-attenuate tips. . . *r*.
r.Marginal nerves of lemma pilose.

Stiff, more or less glaucous; culms wiry; glumes definitely shorter than lowest lemma. . . . . 22. *P. glauca.*

Lax, soft, green; culms capillary, flexuous or curving; glumes about equaling lowest lemma. . . . . . . 23. *P. Fernaldiana.*

*r.*Marginal nerves and keel of glabrous or merely scabrous-based scario-hyaline lemmas not pilose; basal tufts intravaginal, erect; leaves pruinose-glaucous, very narrow, often folded or convolute. . . . . 24. *P. Canbyi.*

*l.*Bases of erect leaf-tufts forming small hard bulbs; spikelets often proliferous. . . . . 25. *P. bulbosa.*

*a.*2nd glume 5.5–11 mm. long, not much exceeded by the lemmas; plant strongly whitened, freely stoloniferous; leaves hard, mostly 5–12 mm. broad; culms stout, mostly 3–9 mm. indiameter. . . . . 26. *P. eminens.*

1. P. ánnua L. (annual), Low Speargrass, Annual Bluegrass, Six-weeks Grass. — *Soft annual* or winter-annual, low (up to 5 dm. high), sometimes rooting at lower nodes; *sheaths loose; leaves very soft;* panicle pyramidal, 1–8 (–13) cm. long; *spikelets* crowded, 3–6-flowered, 3–7 mm. long, *in maturity linear- to oblong-lanceolate; lemma distinctly 5-nerved, not cobwebby at base,* the nerves hairy below; *anthers* 0.7–1 *mm. long.* — Cult. or waste ground and clearings, Nfld. to Alaska, s. to Fla., La. and Mex. Mar.–Dec., rarely all winter. (Natzd. from Eurasia)

Var. réptans Haussk. (creeping and rooting). — Shoots creeping, rooting at nodes, becoming perennial and producing leafy and flowering tufts at their tips. — Local, St. P. et Miq.; reported from Mich. (Adv. from Eu.)

2. P. Chapmaniàna Scribn. (in honor of Alvin Wentworth Chapman, 1809–1899). — Similar to no. 1, more strict, with terete culms; *sheaths close, mostly basal, reddish;* panicle narrower; *lemmas webbed at base, their intermediate nerves very obscure; flowers cleistogamous; anthers* 0.2 *mm. long.* — Dry soil, cult. fields, etc., Fla. to Tex., n. to Del., Ind., Ill., Ia. and e. Kans.; and locally adv. in waste and disturbed ground n. to Mass. Apr.–July.

3. P. comprēssa L. (flattened), Canada Bluegrass, Wiregrass. — Perennial, *with slender rhizomes, bluish-green,* 1–7 dm. high, stiff; *culms* geniculate-ascending, *wiry, strongly flattened; panicles* 1.5–20 cm. long, narrow, *stiff,* the usually short *branches in 2's, spikelet-bearing to below the middle; spikelets crowded, subsessile,* 3–6 (–9)-flowered, 3–8 mm. long; *lemmas obtuse, with faint or obscure intermediate nerves,* more or less bronzed at summit. — Dry soil, Nfld. to s. Alaska, s. to Ga. and Okla. May–Sept. (Natzd. from Eurasia)

P. Chaìxii Vill. (for Dominique Chaix, 1730–1799), Chaix's S., differing from no. 3 in its tall (up to 1 m. or more) and relatively soft culms, elongate pale flat leaves, loose open panicle up to 2.5 dm. long and with spikelets 8–9 mm. long, the lemmas acute, is reported from Minn. (Introd. or adv. from Eu.)

4. P. praténsis L. (of the meadows), Junegrass, Speargrass, Kentucky Bluegrass, Foin à vaches (Que.). — *Stoloniferous; culms arising from among crowded ascending tufts of leaves, soft to firm, compressed and* 2–3 *mm. wide at the often geniculate base,* 0.5–8 dm. high; *leaves flat or folded,* up to 6 mm. wide, *the basal as broad as the culms,* ligule 1.5 mm. long, the cauline blades short; panicle exserted, ellipsoid-ovoid to pyramidal, the lower slender spreading-ascending branches remote, the fascicles of 3–5, naked at base; spikelets crowded, 3–5-flowered, 4–5 mm. long; *glumes broadly lanceolate to ovate, nearly straight; lemmas* about 3 mm. long, *copiously webbed at base, intermediate nerves strong* and *glabrous.* — Moist slopes, shores, meadows and fields, indig. northw., introd. and cult. southw., Lab. to Alaska, s. beyond our range. May–Aug. (Eurasia) — Very variable; this and nos. 6–8 needing careful study. — A monstrous state induced by nematodes has the parts of the spikelet enlarged.

5. P. angustifòlia L. (slender-leaved). — Similar to no. 4; *culms* firm, *terete to base,* 1–2 *mm. thick,* erect; *tufts of capillary or closely involute very slender leaves borne at base, these much narrower than culms;* cauline leaves flat, 1–2 mm. broad; *glumes narrowly lanceolate, the 2nd arching.* — Dry open woods, clearings and rocky places, Nfld. to B.C., s. to N.C., Tenn., Ia., Neb., Colo. and Calif. May–July. (Eurasia)

6. P. subcaerùlea Sm. (bluish). — *Culms arising chiefly from among old dried leaves of the tips of curving stolons, the new basal leafy tufts all or nearly all on separate prolonged stolons or offsets; culms compressed at base,* 0.5–6 dm. high; leaves flat or folded, the basal 2–4 mm. broad; panicle ellipsoid to pyramidal, rather open, 3–7 cm. high; lower remote fascicles of 2–5 branches; *spikelets short-pedicelled,* borne from below the middle of the branches, 4.5–6.5 mm. long; *glumes lance-ovate, straight, the 2nd nearly reaching the tip of the lemma above it; lemmas with glabrous or merely papillate nerves.* (Incl. *P. irrigata* Lindm. f.) — Damp rocks, sands and open woods, Lab., s. to Nfld., M.I. and e. N.B. July, Aug. (Eu.)

P. árida Vasey (dry), Bunch- or Prairie-S., a *rigid densely tufted* but short-stoloniferous *species with culms arising from tufts of folded basal leaves; only* 1 *median cauline leaf; panicle dense, lance-ellipsoid, with appressed-ascending branches; spikelets very short-pedicelled and borne from near bases of branches,* 5–7 mm. long, the *lemma with densely villous nerves.* — Great Plains, Mont. to Ariz., e. to w. Minn., w. Ia., Kans., Okla. and n. Tex.

7. P. alpígena (Fries) Lindm. f. (born in alpine regions). — In habit like no. 6, usually *taller,* up to 7.5 dm. high; leaves firm; panicle up to 1.3 dm. high, its lax branches sometimes reflexed; spikelets 3–7 mm. long, rather crowded; *glumes* lance-ovate, *the 2nd reaching only to middle of lemma above it; nerves of* glabrous to scabrous *lemma minutely pubescent.* — Alpine meadows, wet slopes, bogs, wet shores, etc., Greenl. and Lab. to Yuk., s. to Nfld., M.I., P.E.I., n. Me. and alpine reg. of White Mts., N.H. June–Aug. (Eurasia) — May consist of more than one species.

8. P. longípila Nash (with long hair). — Habit of nos. 6 and 7; culms 2–6 dm. high; *panicle* lax and open, *with loosely divergent to reflexed branches,* 0.4–1.5 dm. long; *pedicels mostly* 0.3–3 *cm. long:* spikelets 5–8 mm. long, loosely flowered; 2nd glume 4–6 mm. long; *lemmas copiously villous to near the middle.* — Lab. to subalpine ravines and meadows of Shickshock Mts., Gaspé Pen., Que.; s. Alta. to Wyo. July, Aug. — Identification tentative.

9. P. triviàlis L. (ordinary), Rough-stalked Meadow-Grass. — Tufted, with decumbent or geniculate bases, 3–12 dm. high; culms weak, sometimes harsh below the panicle; sheaths and blades scabrous or smooth; *ligules of upper leaves* 5–8 *mm. long; panicle* 3–20 cm. long, *pyramidal, somewhat open,* the lower nodes *with* 1–11 (mostly 3–5) loosely spreading or ascending *branches, many of them floriferous to below the middle; spikelets crowded, subsessile or short-pedicelled,* 2–3-flowered, 2.7–4 mm. long, pale green, bronze or purplish; *lemma strongly nerved, glabrous except for the silky keel.* — Glades, spring-heads, brooksides, thickets and damp slopes, Nfld. to Ont., s. to Ga., La. and Kans. June–Aug. Indig. northw.; southw. perhaps only natzd. from Eu. Fig. 133.

133. P. trivialis.

10. P. saltuénsis Fern. & Wieg. (of bushy pastures). — Loosely to closely tufted, with long basal leaves; culms 2–8.5 dm. high, slender, with 2–4 remote cauline leaves 2–5 mm. wide, their blades shorter than the sheaths, their *ligules* 0.3–1.5 *mm. long; panicles lax, secund,* the principal ones 0.3–2 dm. long, their long and *slender branches* in 1's–3's and *floriferous at tip;* spikelets 3–5-flowered, 3.5–5.5 mm. long; glumes subequal, acute, about three-fourths as long as nearest lemma; *lemmas thin,* 3.2–4 *mm. long, acute, prominently 5-nerved, glabrous except for the webbed base,* green or purple-tinged, with narrowly hyaline margin; *anthers* 1–1.2 *mm. long;* grains 2–2.5 mm. long. — Open woods, thickets and recent clearings, e. Que. to w. Ont., s. to N.S., n. and w. N.E., Md., W.Va., Mich. and Minn. Late May–Aug. Fig. 134.

134. P. saltuensis.

Var. **micrólepis** Fern. & Wieg. (tiny-scaled). — Smaller throughout; spikelets 3–3.8 mm. long; lemmas 2.4–3 mm. long. — Woods and thickets, Nfld. and e. Que., s. to N.S. and n. N.E.; Minn.

11. P. lánguida Hitchc. (weak). — Resembling no. 9, 3–10 dm. high; sheaths compressed, much shorter than the internodes; *upper ligules* 2–3 *mm. long;* panicle nodding, 0.4–2 dm. long, the few (usually in 2's) long capillary branches ascending or spreading at the few-flowered tips; spikelets 2–4-flowered, 3–4 mm. long; *lemmas coriaceous, obtuse,* 2–3 *mm. long, distinctly 5-nerved, glabrous except the webbed base; anthers* 0.6–0.8 *mm. long.* (*P. debilis* Torr., not Thuill.) — Sandy or rocky woods and ridges, Vt. to Minn., s. to R.I., Ct., Pa., Ky., n. Ill. and Ia. Late May–early July. Fig. 135.

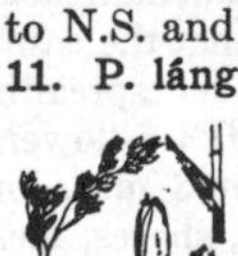

135. P. languida.

12. P. autumnàlis Muhl. (autumnal). — Tufted, with long soft basal leaves, 3–9 dm. high; ligules of cauline leaves 1–3 mm. long; panicle 0.5–2 dm. long, about as broad, the capillary flexuous spreading branches (mostly in 2's) with few spikelets at tips; *spikelets* 2–6-flowered, 5–8 *mm. long; lemmas oblong,* 3–4 mm. long, *pubescent below between the strong nerves, not webbed at base.* — Rich woods, N.J. and Pa. to s. Mich. and Ill., s. to n. Fla., Ala., La. and Tex. Late Mar.–early July, rarely late autumn.

13. P. sylvéstris Gray (of woodland). — Tufted, with long soft basal leaves; *culms subcompressed,* 3–12 dm. high; sheaths of cauline leaves shorter than internodes, their *ligules mostly* 1–2 *mm. long; panicle* 1–2 dm. long, oblong-pyramidal, silvery-green, *the short flexuous filiform branches* (*in 3's–8's*) *spreading or reflexed; spikelets* 2–4-flowered, 2.5–4 *mm. long;* 1st glume 1-, the 2nd 3-nerved; *lemmas firm, obtuse,* 2–3 *mm. long, webbed at base, pilose between the usually pubescent-based distinct nerves, the midnerve pubescent to*

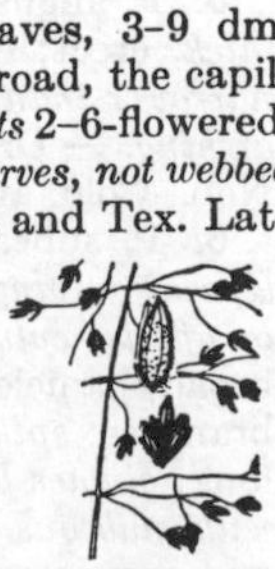

136. P. sylvestris.

summit; anthers 1.6–1.8 mm. long. — Rich woods, N.Y. to Minn. and Neb., s. to n. Fla., La. and Tex. Late Apr.–June. Fig. 136.

14. P. Wólfii Scribn. (for its discoverer, John Wolf, 1821–1897). — Resembling nos. 12 and 15; culms slender, 4–9 dm. high; *leaf-tips short;* panicle 8–15 cm. long; spikelets clustered at the ends of the ascending capillary branches, 2–4-flowered, 5–6 mm. long; *lemmas acute, 4 mm. long, webbed at base, strongly nerved, the marginal nerves and midnerve villous; anthers* 0.8–1.4 *mm. long.* — Woods, O. and n. Va. to Minn. and Mo., rare. Late Apr.–June.

15. P. cuspidàta Nutt. (tipped by a cusp). — Tufted, with long basal *cuspidate-tipped leaves and slender stolons;* culms soft, 2–6 dm. high, with 1–3 remote short leaves; panicle 6–15 cm. long, loosely broad-pyramidal; the few capillary branches mostly in 2's, spreading and spikelet-bearing at tips; *spikelets* 3–4-flowered, 5–8 *mm. long; lemmas oblong, obtuse, distant in anthesis, webbed at base, keel and marginal nerves scabrous or hirtellous, intermediate nerves prominent and naked; anthers* 2–3 *mm. long.* (*P. brachyphylla* Schultes) — Rich or rocky wooded slopes, N.Y. to Ill., s. to Ga. and Tenn. Mar.–June. Fig. 137.

137. P. cuspidata.

16. P. alsòdes Gray (of woods). — Tufted, with long soft basal *thin-sheathed leaves* 2–5 mm. broad; culms 3–8 dm. high; *upper leaves* with elongate sheaths *often inclosing the base of the panicle,* their ligules barely 1 mm. long; *panicle* 1–2.5 dm. long, diffuse, the filiform *branches mostly in 3's–5's,* finally spreading, or the lowest whorl ascending; spikelets 2–3-flowered, 4–6 mm. long; 1*st glume* 2.5 *mm. long,* 2*nd* 3–3.5 *mm. long; lemmas acute,* scarious-tipped, 4 *mm. long, cobwebby at base, intermediate nerves obscure, lateral nerves glabrous, keel villous below;* anthers 0.4–0.7 mm. long. — Rich woods and thickets, St. P. et Miq. to Ont., s. to N.S., N.E., N.C. and Tenn. May, June. Fig. 138.

138. P. alsodes.

17. P. paludígena Fern. & Wieg. (born of the marsh). — *Culms solitary or in very small tufts,* weak, compressed-filiform, 1.5–7 dm. high; *leaves* short, 0.25–3 *mm. broad, the cauline only* 2 *or* 3, *with ligules* 0.5–1.5 *mm. long; panicle* 3–13 cm. long, loose and open; the slender widely spreading *branches mostly* in 1*'s or* 2*'s,* spikelet-bearing above the middle; spikelets 2–5-flowered, 3–6 mm. long; *glumes about* 2 *mm. long; lemmas* 2.5–3.5 *mm. long,* acute or acutish, scarious-tipped, *scarcely webbed at base, marginal nerves and keel pilose at base, intermediate nerves obscure and glabrous;* anthers 0.5–1 mm. long. (*P. leptocoma* of Hitchc., as to our plant, not Trin.) — Sphagnum-bogs, tamarack-swamps, cold spring-heads, etc., centr. N.Y. to Wisc., s. to mts. of Pa., n. O., n. Ind. and n. Ill. Late May–early July. Fig. 139.

139. P. paludigena.

18. P. nemoràlis L. (of woodland), Foin à vaches (Que.). — Deep *green, in dense clumps* 0.1–1 m. high; *culms* slender, terete, *with about* 6 *spreading-ascending leaves* 1–3 mm. broad, mixed with crowded erect leafy shoots; *ligule truncate,* 0.5–1 *mm. long; panicle open and lax,* with branches usually 3–5 to a node, 0.5–1.5 dm. long; spikelets 3–6.5 mm. long, loosely disposed on the branches; *glumes* narrowly lanceolate, *long-acuminate,* straight; lemmas with few webby hairs at base, the marginal nerves and midnerve silky, the intermediate nerves obscure; *anthers* 1.2–1.5 *mm. long.* — Thickets, open woods, rocky slopes, open sands and shores, Greenl. and Lab. to Alaska, s. to Nfld., N.S., N.E., Pa., O., Mich., Wisc., Minn., Neb., etc. June–Sept. (Eurasia) — A polymorphic series in need of critical study.

19. P. palústris L. (of marshes), Fowl-meadow Grass. — In tussocks; *culms firm,* 0.3–1.5 *m. high, mostly with* 6 *nodes;* sheaths rather loose, blades 2–4 mm. wide; *ligules elongate,* 2–5 *mm. long; panicle* green or bronze to purple, 1–3 dm. long, *narrowly pyramidal to ellipsoid; the filiform spreading branches in remote fascicles of* 3–10, naked at base; spikelets 2–4-flowered, 2.8–4.5 mm. long; glumes 2–3 mm. long; lemmas 2–3.3 mm. long, *intermediate nerves obscure,* webby basal hairs copious. (*P. triflora* Gilib.; *P. crocata* Michx.) — Meadows, shores and thickets, Nfld. to Alaska, s. to Va., w. N.C., W.Va., O., n. Ind., n. Ill., Mo., Neb., Colo. and Calif. Late June–early Sept. (Eurasia; n. Afr.)

20. P. alpìna L. (alpine). — *Tufted, with stout ascending crowns* (0.5–1.5 *cm. thick) covered with papery whitish sheaths;* basal leaves firm, often distichous, broad; culms 0.5–6.5 dm. high; their *blades short, oblong-linear, blunt,* 2–6 *mm. broad;* panicle green, bronze or purple, pyramidal to ellipsoid, 1.5–11 cm. long; the filiform branches spreading or ascending, mostly naked at base; *spikelets* rather crowded, *broadly ovate,* 3–6-flowered, 4–8 *mm. long; glumes round-ovate, abruptly tipped, scario-membranaceous; lemmas* similar, webbed at base, *villous on midrib and margins, intermediate nerves obscure.* — Greenl.

140. P. alpina.

to Alaska, s., chiefly in calcareous areas, to rocky or peaty soils of Nfld., Gaspé Pen. and Rimouski Co., Que., Bruce Pen., Ont., n. Mich., Colo., Utah and Oreg. June–Aug. (Eurasia) FIG. 140.

21. P. gaspénsis Fern. (of the Gaspé Peninsula, Que.). — *Loosely cespitose, with slender caudices* 2–4 *mm. thick; basal leaves long and narrow* (1–4 *mm. wide*); culms slender, 1.5–5 dm. high, with 2 or 3 short leaves 2–5 mm. broad, their ligules 2–6 mm. long; panicle narrowly ovoid to subcylindric, 3–12 cm. long; its capillary ascending to spreading branches in 2's–4's, spikelet-bearing from near the middle; *spikelets narrowly ovate,* 3–4-flowered, 3–6 *mm. long,* pale green; *glumes ovate, acuminate, subcoriaceous,* lustrous, 2.8–4.5 mm. long; *lemmas* hyaline, white-margined, acute, 2.5–4.5 mm. long, *with keels long-pilose to above the middle, scabrous-ciliate at tip, lateral nerves long-pilose to above the middle, intermediate nerves pilose at base;* anthers 1.2–1.4 mm. long. — Rocky or gravelly shores and slopes, Gaspé Co., Que.; also Alaska. FIG. 141.

141. P. gaspensis.

22. P. glaùca Vahl (blue-green). — Resembling no. 18, in dense and mostly *stiff clumps* 1–7 dm. high, *blue-green to whitish-glaucous; cauline leaves about* 3, *stiff and slender,* 0.5–2.5 mm. broad, *the upper* somewhat spreading *from near middle of culm;* panicle lax and open to dense, 2–12 cm. long, with scabrous branches; *spikelets whitish-green to glaucous-purple,* narrowly ovate, 3.5–7 mm. long; glumes ovate, acute; *lemmas with no cobweb at base,* the margin and keel pilose, the intermediate nerves obscure. — Dry, often calcareous, rock, gravels and shores, Greenl. and Ellesmereland to Alaska, s. to Nfld., e. Que., mts. and headlands of n. N.E., Bruce Pen., Ont., n. Mich., n. Minn., Colo., Utah and Wash. June–Aug. (Eurasia) — Highly complex series, consisting of many unresolved trends; of these the recently proposed *P. scopulorum* Butt. & Abbe stands between nos. 22 and 23.

23. P. Fernaldiàna Nannf. (named for the present author, 1873–). — *Soft matted or tufted; culms filiform, flexuous or arching,* 0.5–3 *dm. long, mostly leafless above the middle; leaves soft,* 1–2 *mm. wide,* the few cauline with ligules 3–5 mm. long; *panicle* 1–6 *cm. long, simple or subsimple, often nodding,* green to bronze or purple, the filiform *branches ascending* and loosely spikelet-bearing at the ends; spikelets 2–4-flowered, 3–6 mm. long; *glumes about equaling the lowest lemmas; lemmas scario-membranaceous,* 3–4 mm. long, *pilose near the* webbed or webless *base on margins and keel,* intermediate nerves obscure; anthers 0.7–1 mm. long. (*P. laxa* of ed. 7, not Haenke) — N. Lab., s. to alpine reg. of Nfld., e. Que., Mt. Katahdin, Me., White Mts., N.H., Mt. Mansfield, Vt., and Mt. Marcy, N.Y. Late June–Aug. FIG. 142.

142. P. Fernaldiana.

24. P. Cánbyi (Scribn.) Piper (for its discoverer, WILLIAM MARRIOTT CANBY, 1831–1904). — In dense clumps, the *glaucous basal tufts erect, intravaginal;* culms 0.2–1 m. high, with 2–3 erect short *narrowly linear to convolute pruinose leaves* with prolonged ligules 3–5 mm. long; basal leaves pruinose, flat or convolute; *panicle slender, lanceolate,* 0.5–1.5 dm. long, *the short branches appressed; spikelets whitish-green* or purplish-tinged, 3–5-flowered, lanceolate, 5–6.5 mm. long; glumes hyaline, lance-ovate, the 1st 3.5, the 2nd 4.5–5 mm. long; *lemmas scarious, lustrous, rounded on the back, glabrous except for the scabrous base.* (Incl. *P. laevigata* Scribn.) — Dry calcareous ledges and cliffs, Gaspé Pen. and Rimouski Co., Que.; n. Mich. and ne. Minn.; Yuk. and B. C., s. to Neb., N.M. and Calif. July, Aug. FIG. 143.

143. P. Canbyi.

25. P. BULBÒSA L. (bulbous). — *Cespitose, the erect leafy tufts forming hard pale bulbs* 3–8 *mm. in diameter;* culms erect, stiff, 1.5–4 dm. high; *basal leaves short* (up to 1 dm. long), *erect,* 1–2 mm. broad; cauline leaves 2 or 3, with long sheaths and short flat blades, their *acute ligules* 2–3 *mm. long; panicle ellipsoid to slenderly ovoid, dense,* 3–5 cm. long, with short ascending branches; *spikelets* ovate, 3–4-flowered, 2.7–4 mm. long, or *often viviparous* (with lemmas, paleas, etc., modified to leafy tufts); lemmas scarious, the keel and marginal nerves pubescent. — Lawns and dry fields, L.I. to N.C. and Kans.; and on the Pacific coast. Mar.–June. (Adv. from Eurasia)

26. P. éminens J. S. Presl (conspicuous). — *Very glaucous,* glabrous, *freely stoloniferous,* stout; *culms coarse, mostly* 3–9 *mm. in diameter* at base, 3–10 dm. high, terete; *leaves thick and hard, mostly* (2–)5–12 *mm. broad,* with overlapping sheaths; *panicle heavy,* 0.8–3 dm. long, contracted (rarely loose); *spikelets* 3–5-flowered, 7–12 *mm. long,* pale or purple-tinged; 2*nd glume* 5.5–11 *mm. long, not much exceeded by the lemmas;* lemmas distinctly nerved, pubescent; anthers 1.7–2.5 mm. long. — Gravelly seashores, Lab., Nfld. and e. Que. (s. to M.I.). Late June–early Sept. (Alaska; ne. Asia)

10. BRÌZA L. Quaking Grass

Spikelets few–several-flowered, broad; florets crowded, almost horizontal, the uppermost usually imperfect; glumes subequal, firm-membranaceous, with broad scarious margins; lemmas 5–many-nerved (nerves often obscure), firm, subchartaceous with a scarious margin, navicular or ventricose, cordate at base; palea much smaller than its lemma; fruit dorsally compressed. — Annuals or perennials, nat. of Eurasia, n. Afr. and S. Am., with flat leaves and showy terminal panicles. (The Greek name of a kind of grain.)

144. B. media.

1. B. mèdia L. (intermediate). — *Perennial,* erect, 2.5–7 dm. high; *sheaths longer than the narrow blades;* panicle erect, the stiff capillary branches spreading; *spikelets* nodding, 5–9-flowered, 4–6 *mm. long,* about as broad, brown and shining; *lemmas naviculate.* — Roadsides, meadows and swamps, local, s. and w. N.E. to Ont. and Mich.; B.C.; often cult. June–Aug. (Introd. from Eu.) Fig. 144.

2. B. mìnor L. (smaller). — *Annual;* culms 1–6 dm. high, often branching at the base; *sheaths shorter than the blades;* panicle erect, its slender branches finally spreading, bearing fascicled branchlets; *spikelets* hardly nodding, 3–6-flowered, pale or plum-colored, broadly cordate, 3–5 *mm. long,* slightly broader; *lemmas strongly ventricose below.* — Waste places, se. N.Y. and southw., local; Pacific States; often cult. June (Introd. from Eu.)

3. B. máxima L. (greatest). — Annual; panicle with *few* (*up to* 8) *large ovate* 10–20-*flowered stramineous to brown spikelets* 1–2 *cm. long* and 1–1.5 *cm. broad.* — Roadsides, Mich., local; often cult. June. (Introd. from Eu.)

11. MÉLICA L. Melic-Grass

Spikelets 2–several-flowered, plump or but slightly compressed; rachilla prolonged beyond the fertile florets and bearing 2 or 3 gradually smaller empty lemmas convolute together or inclosing one another at the apex; glumes large, unequal, membranaceous or papery, scarious-margined, 3–5-nerved, little shorter than the florets; lemmas convex, 7–13-nerved, firm, with scarious margins, awnless or awned below the bifid apex; paleas shorter than their lemmas, the strong nerves nearly marginal; lodicule 1, entire or notched. — Perennials of temp. reg., with simple culms, closed sheaths, usually soft flat leaves and rather large spikelets in usually narrow panicles. (Classical name for some plant, taken up by Linnaeus for this genus.)

a. Glumes broad and papery; lemmas awnless. § Eumelica. . . *b.*
  b. Terminal sterile lemmas broad, hooded or truncate, quite unlike and overtopped by the 2 or 3 fertile ones.
    Cauline leaves 3 or 4, 2–5 mm. broad; glumes similar, subequal, nearly equaling the 2 fertile lemmas. . . . . 1. *M. mutica.*
    Cauline leaves 5–8, 5–12 mm. broad; glumes dissimilar, unequal, the lower about two-thirds as long as the lowest of the usually 3 fertile lemmas. . . . . 2. *M. nitens.*
  b. Terminal small sterile lemmas similar to and borne above the 3–5 fertile ones. 3. *M. Porteri.*
a. Glumes narrow, subherbaceous, with scarious margins; fertile lemmas awned, the small sterile ones similar. § Bromelica. . . . . 4. *M. Smithii.*

§ EuMélica Aschers.

1. M. mùtica Walt. (blunt, without slender tip). — Culms erect from knotted rhizomes, wiry, 4–9 dm. high; *sheaths* usually overlapping, *scabrous,* or pilose in forma **diffùsa** (Pursh) Fern. (loosely spreading); *blades* 3 *or* 4, the lower short, *the upper* 10–20 cm. long and 2–5 *mm. wide; panicle* 0.5–2.5 dm. long, simple, *with filiform ascending branches or reduced to a raceme;* spikelets 7–10 mm. long, pendulous on short pedicels; florets spreading, 6–8 mm. long; *lemmas* scabrous, *obtuse,* the intermediate nerves vanishing above; empty lemmas cucullate above, exceeded by the fertile ones. — Dry open woods and thickets, Md. to Ia., s. to Fla., Ala., La. and Tex. Apr.–June. Fig. 145.

145. M. mutica.

2. M. nitens Nutt. (shining). — *Culms* 0.7–1.5 *m. high,* erect from a short horizontal rhizome; *sheaths* overlapping, *glabrous; blades* 5–8, 1–2 dm. long, 5–12 *mm. wide;* panicle 1.5–2.5 dm. long; the *slender spreading branches* solitary or in pairs, simple or sparingly branched; spikelets numerous, 10–12 mm. long, usually 3-flowered, pendulous on short pedicels; *lemmas* 7–9 mm. long, scabrous, *acute;* empty lemmas 2, distant, broad at the summit, exceeded by the fertile ones. — Rocky woods, bluffs and dry clearings, Pa. to Minn., s. to Va., Ky., Ark. and Tex. May, June.

3. M. Pórteri Scribn. (for its discoverer, THOMAS CONRAD PORTER, 1822–1901). — Culms erect, slender, 4–10 dm. high; *sheaths* overlapping, *scabrous; blades* 5–7, 12–23 cm. long, 2–6 *mm. wide*, scabrous; panicle 1.5–2.5 dm. long; the *narrow spikelets* pendulous and *racemose along the slender ascending branches*, 4–6-*flowered*, 10–13 mm. long; lemmas 7–8 mm. long, subacute, scabrous; *empty lemmas like the fertile ones and exceeding them.* — Bluffs and stony hillsides, Tex. to Ariz., n. to Mo., Neb. and Colo. May–July.

§ BROMÉLICA Thurb.

4. M. Smíthii (Porter) Vasey (for its discoverer, CHARLES EASTWICK SMITH, 1820–1900). — Culms erect, slender, 0.7–1.5 m. high; sheaths scabrous; blades 1–3 dm. long, 6–15 mm. wide, lax, scabrous; panicle 1.2–4.3 dm. long, the solitary remote spreading branches spikelet-bearing toward the ends; *spikelets* 3–6-flowered, 1–2.4 *cm. long*, more or less tinged with purplish-chestnut; glumes acute; *lemmas glabrous, about* 10 *mm. long, excluding the awn, which is one-third to half as long.* (*Avena* Porter; *Bromelica* Farw.) — Rich woodlands, s. Ont. and Mich.; B.C. to Wyo. and Oreg. May–July. FIG. 146.

146. M. Smithii.

12. DÁCTYLIS L. ORCHARD-GRASS

Spikelets 2–6-flowered, compressed, nearly sessile and strongly overlapping, curved, in dense fascicles, these arranged in a 1-sided panicle; glumes unequal, ciliate or scabrous on the sharp keel, acute or mucronate; lemmas 5-nerved, hispid or ciliate-keeled, with short awn-points; paleas a little shorter than their lemmas. — Perennial, nat. of Eurasia and n. Afr., with flat leaves and glomerate panicles. (A name used by Pliny for a grass with digitate spikes, from the Greek *dactylos, a finger.*) FIG. 147.

1. D. GLOMERÀTA L. (gathered in bunches). — Coarse, tufted, glaucous, scabrous; culms erect, 4–15 dm. high; leaves broadly linear; panicle 0.3–2.5 dm. long, the few stiff branches naked below, contracted after flowering; spikelets crowded in dense one-sided clusters at the ends of the branches. The typical form has keels of glumes and lemmas long-ciliate, the backs glabrous. — Fields, roadsides and waste places, Nfld. to B.C., s. to Ala., N.M. and Calif. May–Sept. (Introd. and natzd. from Eu.) FIG. 148.

Var. CILIÀTA Peterm. (fringed). — Similar; glumes and lemmas pubescent on the back. — Less common. (Introd. and natzd. from Eu.) FIG. 149.

Var. DETÓNSA Fries (sheared-off). — Keels merely scabrous or short-hispid. — Occasional. (Introd. and natzd. from Eu.) FIG. 150.

147. Dactylis.

148. D. glomerata (typical).

149. D. glomerata, v. ciliata.

150. D. glomerata, v. detonsa.

13. ERAGRÓSTIS Beauv. LOVE-GRASS

Spikelets strongly compressed, 3–many-flowered; the uppermost floret sterile; rachilla articulated but sometimes not disjointing until after the fall of the glumes and lemmas with the grain; glumes keeled, much shorter than the spikelets; lemmas 3-nerved, broad, keeled; paleas shorter than their lemmas, often persistent after their fall, the strong nerves ciliate; grain globose to ovoid or oblong, not furrowed. — Annuals or perennials of warm-temp. and trop. reg., with loose or dense terminal panicles. (Name from the Greek, *Eros, god of love,* and *agrostis, a grass,* some Old World species long known as LOVE-GRASS.)

a. Culms extensively creeping, forming prostrate mats; panicles 1–6 cm. long.
  Flowers perfect; culms and sheaths glabrous; panicles all forking and open; lemmas glabrous; anthers 0.2–0.5 mm. long. . . . . 1. *E. hypnoides.*
  Dioecious; culms, sheaths, etc. pubescent; pistillate panicles subglobose, dense; lemmas pubescent; anthers about 2 mm. long. . . . . 2. *E. reptans.*

a.Culms erect or ascending, at most geniculate; primary panicles mostly larger. . . *b*.

b.Leaf-blades glandular or warty along the (often inrolled in drying) margin; panicles at most 2 dm. long, dense or, if open, only 2–7 cm. broad; spikelets 8–40-flowered, 5–17 mm. long; annuals with culms often geniculate and branching.

Spikelets 2–5.5 mm. broad; 2nd glume about 2 mm. long; lemmas closely imbricated, their bases not visible in the spikelet, 2–2.6 mm. long. . 3. *E. megastachya.*

Spikelets 1.5–2 mm. broad; 2nd glume 1.2–1.6 mm. long; lemmas more loosely imbricated, their bases becoming visible, less than 2 mm. long. 4. *E. poaeoides.*

b.Leaf-blades not glandular nor warty; panicles loose (or somewhat dense only in no. 9 with few-flowered spikelets less than 5 mm. long). . . *c*.

c.Culms (except in starved individuals) branching from middle or upper nodes; plants soft-based annuals; sheaths glabrous (except for ciliate orifice); primary or central panicle rarely 2 dm. long; 1st glume 0.5–1.2 mm. long; lower lemma 1–1.6 mm. long. . . *d*.

d.Auricles of upper leaf-sheaths long-ciliate; panicle loose. . . *e*.

e.Panicles ovoid or pyramidal; spikelets 5–15-flowered; lemmas 1.5–2.5 mm. long. . . *f*.

f.Branches of panicle spikelet-bearing from near base; spikelets 1.3–2.5 mm. broad, lance-oblong to narrowly ovate; glumes scabrous-nerved, the lower 1–2 mm. long, upper 1.5–2.3 mm. long; paleas very persistent.

Some or all spikelets on pedicels 5–8 mm. long, 1.8–2.5 mm. broad; upper glume and lower lemmas 1.8–2.3 mm. long.

Spikelets ovate-lanceolate, 2–2.5 mm. broad, stramineous or with purple tinge, 3–9-flowered; panicle lax and open, with long spreading pedicels. . . . . 5. *E. mexicana.*

Spikelets linear-lanceolate, lead-colored, 1.8–2 mm. broad, mostly 8–12-flowered; panicle with ascending branches, the ascending spikelets often imbricated. . . . . . . 6. *E. neomexicana.*

Some or all spikelets on pedicels only 1–5 mm. long, lead-colored, 1.3–1.7 mm. broad; upper glume and lower lemmas 1.5–1.7 mm. long. . . . . 7. *E. pectinacea.*

f.Branches of panicle spikelet-bearing one-half to two-thirds their length; spikelets 1 mm. wide; glumes glabrous, the lower 0.5 mm., upper 1 mm. long; lateral nerves of lemma inconspicuous; paleas soon deciduous. . . . . 8. *E. pilosa.*

e.Panicles slenderly ellipsoid; spikelets 2–7-flowered; lemmas 1.1–1.5 mm. long. . . . . 10. *E. Frankii.*

d.Auricles of upper leaf-sheaths naked; panicle rather dense. . . . 9. *E. multicaulis.*

c.Culms simple, firm to wiry (branching only at base in no. 11 which has pilose sheaths); panicle (or terminal one in no. 11) 1–9 dm. long; 1st glume 1–3 mm. long; lower lemma 1.3–3 mm. long; annuals or hard-based perennials. . . *g*.

g.Culms branching from base, capillary, soft; plant a soft annual; larger panicles 1–3 dm. long, 0.7–2 dm. broad, with glabrous pulvini; spikelets 1–5-flowered, 1.4–3 mm. long; lower lemma 1.3–1.8 mm. long. . . . . 11. *E. capillaris.*

g.Culms simple, firm to wiry; plants perennial, with hard bases, or firm-based annuals; panicles 2–9 dm. long; the pulvini, at least of the lower branches, usually ciliate or bearded; spikelets 3–25-flowered, 3–12 mm. long; lower lemma 1.5–3 mm. long. . . *h*.

h.Lateral as well as terminal spikelets distinctly pedicelled; lemmas ovate. . . *i*.

i.Panicles less than half as broad as long, elongate-ellipsoid; 1st glume 2.5–3 mm. long; lowest lemma 2.5–3 mm. long. . . . 12. *E. trichodes.*

i.Panicle half as broad as long or broader, broadly ellipsoid, ovoid or rhomboid; 1st glume 1–2 mm. long; lowest lemma 1.5–2.5 mm. long. . . *j*.

j.Culms few in tufts, 0.4–1.3 m. high; lemmas only obscurely nerved, lead-color or drab.

Leaf-blades 4–12 mm. broad; expanded panicle 3–9 dm. long, 2–4.5 dm. broad; spikelets 1–5-flowered, 3–4.5 mm. long; 1st glume 1.5 mm. long; lowest lemma 2–2.5 mm. long. . 13. *E. hirsuta.*

Leaf-blades 1–4 mm. broad; expanded panicle 1.5–3.5 dm. long, 1–2.5 dm. broad; spikelets 3–9-flowered, 3–10 mm.

long; 1st glume 1–1.2 mm. long; lowest lemma 1.8–2 mm. long. . . . . 14. *E. intermedia.*

*j.*Culms forming dense stools from subligneous horizontal rhizomes, 3–8 dm. high; lemmas prominently nerved, commonly purplish. . . . . 15. *E. spectabilis.*

*h.*Lateral spikelets sessile or essentially so, appressed; lemmas lance-acuminate. . . . . 16. *E. refracta.*

1. E. hypnoìdes (Lam.) BSP. (like *Hypnum*, or moss-like). — *Extensively creeping; culm* capillary, thickened and often rooting at the joints, *glabrous*, up to 5 dm. long, with short erect or ascending flowering branches 0.5–2 dm. high; leaves 1–4 cm. long, their *sheaths glabrous* or nearly so; *panicles simple or but slightly forked, usually open*, lanceolate, ellipsoid or slenderly ovoid, 1–6 cm. long; spikelets linear-lanceolate, 10–35-flowered, 5–15 mm. long, 1–2.5 mm. broad, the *flowers perfect and fertile; lemmas* very thin, acuminate, 1.5–2 *mm. long, smooth; anthers* 0.2–0.5 *mm. long.* — Gravelly or sandy shores, etc., centr. Me. and s. Que. to Wash., s. through much of U.S. to W.I., Mex. and S. Am. Late July–Nov. Fig. 151.

151. E. hypnoides.

2. E. réptans (Michx.) Nees (creeping). — Habit of no. 1, *coarser, dioecious; culms, peduncles, leaves, etc., harsh, pubescent; panicles on naked peduncles* 1.5–10 cm. long; staminate panicle open, with remote branches; *pistillate panicle dense, subglobose; spikelets* 2–4 *mm. broad; lemmas* 2–3.5 *mm. long, pubescent; anthers about* 2 *mm. long.* — Sandy shores and open places, w. Ky. and s. Ill. to S.D., s. to Tex. and Mex. July–Oct. (S. Am.) Fig. 152.

152. E. reptans.

3. E. megastáchya (Koel.) Link (large-spiked), Stink-, Snake- or Skunk-Grass, "Strong-scented Love-Grass". — Strong-scented (when fresh) annual, freely branching, with ascending or depressed and geniculate rather flaccid culms 2–9 dm. high; *leaves* 3–8 *mm. wide*, their margins (often inrolled) *bordered by wart-like glands;* panicles greenish-lead-color, handsome, 0.5–2 dm. long, dense, with elongate many-flowered spikelets, or open and with fewer-flowered short spikelets; *spikelets* 10–40-*flowered*, 5–17 mm. long, 2–3.5 *mm. wide, the florets closely imbricated;* pedicels and keels of the acute glumes and lemmas somewhat glandular; *lemmas* 2–2.6 *mm. long*, thin, *scabrous*, the lateral nerves prominent. (*E. cilianensis* sensu Vignolo-Lutati, apparently not *Poa cilianensis* All., the typonym; *E. major* Host) — Roadsides and waste places, Me. to Wash., s. through much of U.S. into W.I. and Mex. June–Oct. (Natzd. from Eu.) Fig. 153.

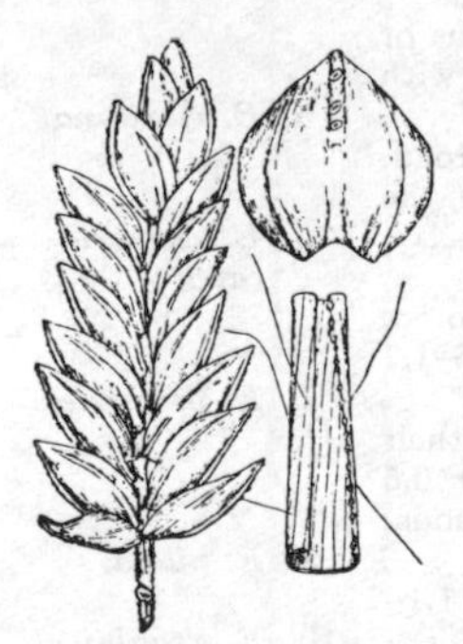
153. E. megastachya.

4. E. poaeoìdes Beauv. (like *Poa*). — Similar to no. 3, smaller, more slender; panicles open; *spikelets* 5–10 mm. long, 1.5–2 *mm. wide*, 8–20-flowered, *the florets less densely imbricated, the bases or rachilla-joints visible; lemmas* nearly smooth, *less than* 2 *mm. long.* (*E. minor* Host; *E. Eragrostis* (L.) Karst.) — Waste ground, less common than no. 3, through much of U.S. June–Oct. (Natzd. from Eu.)

5. E. mexicàna (Hornem.) Link (Mexican). — Annual, 1–7.5 dm. high, the slender culms simple or loosely branching; leaves pale, long-attenuate, 1–5 mm. broad, summit of sheath bristly-ciliate; *panicle* 0.4–2 dm. long, ellipsoid-ovoid, *very lax and open, the spreading or spreading-ascending branches with scattered ovate-lanceolate spikelets on spreading-ascending pedicels up to* 8 *mm. long; spikelets* 2–2.5 *mm. broad, stramineous or tinged with purple;* glumes acute, scabrous-keeled, the 1st about 1.5 mm. long, the 2nd 1.8–2 mm. long; lemmas 1.8–2.5 mm. long, scabrous on the keel. — Waste places and roadsides, local, Mex. and sw. states; adv. northeastw. to Ia. and Del. Aug.–Oct. (Adv. from the Southwest or from Mex.)

6. E. neomexicàna Vasey (of New Mexico). — Coarser and *darker green* than no. 5; culms 0.3–1 m. high; leaves 3–8 mm. broad; *terminal panicle* 1.5–4 *dm. long, with ascending branches with ascending spikelets mostly imbricated or overlapping; spikelets lead-colored, linear-lanceolate,* 1.8–2 *mm. broad*, 8–12-*flowered.* — Waste places and roadsides, locally ne. to Ia., Ind., Md. and Del. Aug.–Oct. (Adv. from the Southwest or from Mex.)

7. E. pectinàcea (Michx.) Nees (comb-like). — Slender annual, usually tufted (often densely so); the culms commonly geniculate at base, then ascending or spreading, 0.5–8 dm. high, branching from the lower nodes; sheaths shorter than the internodes, *the upper auricles*

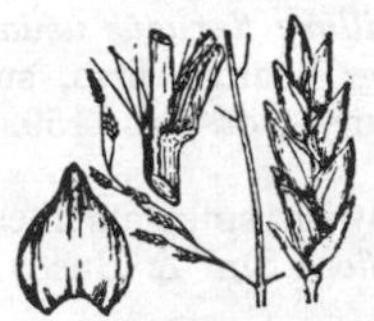

154. E. pectinacea.

with long cilia; panicles loose, up to 3 dm. long, the capillary *branches spikelet-bearing from near the base,* the axils naked or hairy, *lowest branches in 1's or 2's; spikelets ovate-lanceolate,* lead-color, 1.3–1.7 *mm. wide,* mostly more than 5 mm. long, the florets often 10–15; *glumes* acuminate, *scabrous on the keel, the lower* 1 *mm., the upper* 1.5 *mm. long; lemmas* mostly 1.5 mm. or more long, *their lateral nerves conspicuous; paleas very persistent* and tightly appressed to the rachilla; grain 0.7–0.8 mm. long. (*E. pilosa* in large part of ed. 7; *E. caroliniana* (Spreng.) Scribn.; *E. Purshii* Schrad.) — Sandy shores, ditches, roadsides, etc., sw. Que. to B.C., s. through most of U.S. to Mex. Mid-July–Oct. Fig. 154.

8. **E. pilòsa** (L.) Beauv. (hairy). — Similar to no. 7; differing in *lowest branches of panicle 2 or more; branches spikelet-bearing one-half to two-thirds their length; spikelets lance-linear, about* 1 *mm. wide,* their filiform *pedicels elongate; glumes glabrous, the lower* 0.5, *the upper* 1 *mm. long; lemmas* rarely more than 1.5 mm. long, *their lateral nerves inconspicuous; paleas soon deciduous.* — Roadsides, waste places and old fields, infrequent, Mass. to Colo., s. to Fla., Ala., La. and Tex. July–Oct. (Natzd. from Eu.) Fig. 155.

155. E. pilosa.

9. **E. multicaùlis** Steud. (many-stemmed). — Similar to no. 7, up to 4 dm. high; *auricles of upper sheaths naked; panicles dense, up to* 1.3 *dm. long; branches spikelet-bearing from near the base, the lowest solitary,* their axils naked; *spikelets ovate to ovate-oblong,* mostly 1.5 mm. wide and less than 5 mm. long and *with less than* 10 *florets, the pedicels very short; glumes glabrous,* the lower 0.5, the upper 1 mm. long; *lemmas* about 1.5 mm. long, *with inconspicuous lateral nerves; paleas promptly deciduous;* grain 0.5–0.6 mm. long. (*E. peregrina* Wieg.) — Roadsides, railroad-yards, etc., rapidly spreading, Me. to Wisc. and Va. July–Oct. (Semi-cosmop. weed; natzd. from e. Asia) Fig. 156.

156. E. multicaulis.

10. **E. Fránkii** C. A. Mey. (for its discoverer, Joseph C. Frank, 1782–1835). — *Annual,* erect from a decumbent base, or spreading, 1.5–5 dm. high, the *culms often branching at the nodes;* sheaths glabrous; ligule sparsely long-pilose; blades 5–12 cm. long, 2–4 mm. broad, scabrous above; *panicle slenderly ellipsoid, usually less than one-third the height of the plant,* 2–5 (–6) *cm. in diameter, the short branches diffusely spreading,* with many 2–6-flowered spikelets 2–4 mm. long; the *lateral pedicels mostly longer than the spikelets;* glumes and lemmas acute; *lemmas* faintly 3-nerved, 1.1–1.3 *mm. long; grain not grooved.* — River-silts and damp thin soil, sw. Que., s. Ont. and Minn., s. to Fla., Tenn., Miss., La. and e. Kans. Aug., Sept. Fig. 157.

157. E. Frankii.

Var. **brévipes** Fassett (short-stalked). — Panicle compact, 1.5–4.5 cm. thick; spikelets mostly 5–7-flowered, on shorter pedicels. — Wisc., Minn., Ill. and Ia.

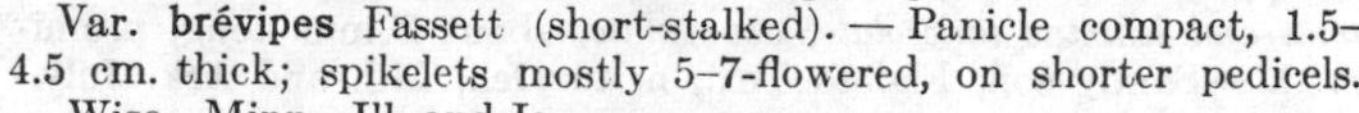

11. **E. capillàris** (L.) Nees (hair-like), Lace-Grass. — Slender erect commonly tufted *annual* 1.5–7 dm. high; *culms branching only at base, simple above;* sheaths overlapping, membranaceous, sparingly pilose or nearly glabrous, blades long and narrow; *panicle diffuse, much more than half the entire height of the plant,* ellipsoid-ovoid, *the larger* (of each plant) 1–3 *dm. long* and 0.7–2 *dm. in diameter; the lower* capillary *branches ascending,* finally divergent, *with small glabrous pulvini; spikelets* lead-color, 1–5-flowered, 1.4–3 mm. long, 1–2 mm. broad, *the lateral on divergent pedicels mostly* 0.5–1 *cm. long; glumes* acute, *the 2nd* 1.2–1.6 *mm. long; lemmas* acute, 1.3–1.8 *mm. long,* faintly 3-nerved; *grain longitudinally grooved.* — Dry sandy or rocky soil, s. Me. to Wisc., Ia. and Kans., s. to Ga., Tenn., Mo. and e. Tex. Late July–Oct. — Often lemon-scented. Fig. 158.

158. E. capillaris.

12. **E. trichòdes** (Nutt.) Wood (hair-like). — *Perennial,* with solitary or few erect culms 6–15 dm. high; *sheaths* overlapping, *smooth,* pilose at throat; blades 1–7 dm. long, 2–6 mm. wide, rather rigid, involute-taper-pointed; *panicle pale, purplish, lax,* ellipsoid, 3–6 *dm. long, less than half as broad,* lower axils sparingly pilose;

159. E. trichodes.

spikelets lanceolate to oblong-ovate, 3–6-flowered, 4–7 mm. long, *on capillary flexuous usually long (mostly* 6–30 *mm.) pedicels;* glumes sharply acute, 2.5–3 mm. long; lemmas firm, subacute, 2.5–3 mm. long. — Sandy soil, O. to Neb., s. to La. and Tex. July–Oct. Fig. 159. — Passing freely into

Var. **pilífera** (Scheele) Fern. (bearing hair). — Panicles paler, often yellowish; spikelets linear-oblong, 8–15-flowered, 8–12 mm. long. (*E. pilifera* Scheele; *E. grandiflora* Sm. & Bush) — Sandy barrens, Ill. to Neb. and Tex.

13. E. hirsùta (Michx.) Nees (hairy). — *Coarser* than no. 11, 0.5–1.3 *m. high, the stoutish culms unbranched* or merely with basal leafy offshoots; *leaves with subcoriaceous papillose-hispid sheaths and firm blades* 4–12 *mm. broad; panicle* 3–9 *dm. long,* 2–4.5 *dm. broad,* with the *coarse lower pulvini hirsute;* spikelets 3–4.5 mm. long; *lower lemmas* 2–2.5 *mm. long.* — Dry sandy soil, Fla. to Tex., n. to S.C. and sw. Mo.; adv. in e. Va. Aug.–Oct.

Var. **laevivaginàta** Fern. (with smooth sheaths). — *Sheaths glabrous* except for ciliate margin and summit. — Fla. to Md.

14. E. intermèdia Hitchc. (intermediate). — *Perennial* with few tufted wiry simple culms 3–9 dm. high; *sheaths subherbaceous,* glabrous or sparsely pilose, bearded at summit; *blades flat or involute,* 1–4 *mm. broad, with scabrous involute tips; panicle ovoid,* open, rather stiff, becoming somewhat diffuse, 1.5–3.5 *dm. long, two-thirds as broad,* the *lower pulvini usually pilose; spikelets all pedicelled,* the lateral spreading, *drab or brownish,* linear, *closely* 3–9-*flowered,* 3–10 *mm. long,* 1–1.5 *mm. broad;* 1*st glume* 1–1.2 *mm. long,* acute, 2nd slightly longer; *lemmas* ovate, *obscurely nerved, the lowest* 1.8–2 *mm. long.* (*E. lugens* of N. Am. auth., not Nees) — Prairies and dry slopes, Ga. to Ariz. and n. Mex., n. to Mo. and Kans.; casual in N.E. June–Oct.

15. E. spectábilis (Pursh) Steud. (showy), Tumble-Grass, Petticoat-climber. — *Hard tufted perennial,* with short stout rhizome, 3–8 dm. high; culms rigid, much shorter than the panicles; *sheaths* overlapping, glabrous or nearly so, *densely bearded at throat;* blades 4–8 mm. wide, often involute in drying; *panicles purple,* included at base, or slightly exserted at maturity, 2–4.5 dm. long, more than half as wide; *the branches becoming stiffly divergent to reflexed, pilose in the axils; spikelets* 3–15-*flowered,* 3–8 *mm. long,* 1.5–2 *mm. wide, on stiff pedicels; lemmas ovate, obtuse to subacute,* 1.5–2.5 mm. long, prominently nerved, *scabrous on the keel.* (*E. pectinacea,* var. *spectabilis* (Pursh) Gray) — Dry sand, Fla. to Tex., n. on coastal sands to se. Mass. and inland n. to O., Mich. and Minn. July–Oct.

Var. **sparsihirsùta** Farw. (remotely hairy). — Sheaths mostly pilose. (*E. pectinacea* of ed. 7, not (Michx.) Nees) — Dry sterile soil, generally common, s. Me. to S.D., s. to Fla., Ala., La., Tex. and N.M. July–Nov. — Panicles breaking off at maturity, forming tumbling, entangled masses.

16. E. refrácta (Muhl.) Scribn. (abruptly bent back). — Wiry perennial; culms more slender and taller (–9 dm.) than in no. 15; *sheaths overlapping, glabrous, sparingly villous at throat;* blades 2–4 mm. wide, nearly smooth; *panicle usually included at base,* 2–7 dm. long, two-thirds as broad, *the slender remote branches* sparsely pilose in the axils and *bearing few short-pedicelled or subsessile appressed* 6–25-*flowered spikelets* 6–12 *mm. long* and 1.3–2 mm. broad; *glumes and lemmas lanceolate, acuminate,* about 2 mm. long, prominently nerved. — Sandy soil, Fla. to Tex., n. to Del., Md. and Ark. July–Oct. Fig. 160.

160. E. refracta.

Several southwestern species, recorded as casual or about wool-waste and experimental plots, seem not to be established elements in our flora.

14. DIPLÁCHNE Beauv. Salt-meadow Grass

Spikelets several-flowered, narrow, erect and scattered along the slender rachis of long spike-like branches; flowers all perfect or the terminal staminate; glumes scario-membranaceous, keeled, acute, unequal; lemmas slightly longer, 1–3-nerved, keeled, 2-toothed at apex, mucronate or short-awned from between the teeth; stamens 3; styles distinct; grain 3-angled, free, unfurrowed. — Coarse grasses (ours annuals) of warm to trop. reg., with narrow flat leaves, the few–many spiciform branches forming terminal panicles. (From the Greek *diplous, double,* and *achne, chaff,* referring to the 2-toothed lemmas.)

2nd glume 1.7–3.5 mm. long; lemmas 2–4 mm. long.
  Lower branches of panicle 5–12 cm. long; 2nd glume acute; lemmas awned, acute. . . . . 1. *D. fascicularis.*
  Lower branches 2–6 cm. long; 2nd glume obtuse, emarginate; lemmas obtuse, often mucronate. . . . . 2. *D. Halei.*
2nd glume 4.5–7 mm. long; lemmas 5–8 mm. long.

Lemmas with awns about 1 mm. long. . . . . 3. *D. acuminata.*
Lemmas with awns 2.5–5 mm. long. . . . . 4. *D. maritima.*

1. D. fasciculàris (Lam.) Beauv. (in bundles or clusters). — Tufted annual 0.3–1.2 m. high; culms erect or decumbent and geniculate, branching; sheaths loose, subinflated, mostly exceeding the internodes; *ligules scarious, prolonged, 5–7 mm. long;* blades flat, 3–5 mm. wide, often becoming involute, scabrous; *panicle* 1–3.5 dm. long, its base included in the upper sheath, with suberect branches, the lower branches 5–12 cm. long, *axis and branches smooth or but slightly scabrous; spikelets* 5–10-flowered, linear-lanceolate, *mostly overlapping; glumes whitish, the 2nd 2–3.5 mm. long, acute,* with scabrous keel; *lemmas similar, pale green, 2–4 mm. long, with terminal awn* 0.5–1 *mm. long;* lateral nerves silky below, often excurrent as short lateral teeth. (*Leptochloa* Gray) — Brackish to fresh marshes and waste places, Fla. to Calif., n. in wet places to O., Ind., s. Wisc., Minn., S.D., Colo., Utah and Wash. July–Oct. Fig. 161. (W.I.; S.Am.)

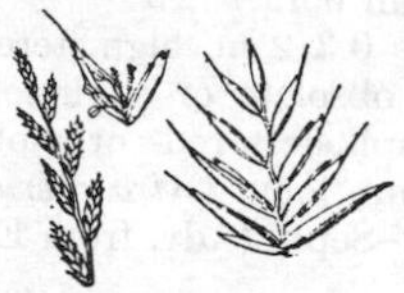
161. D. fascicularis.

2. D. Hàlei Nash (for its discoverer, Josiah Hale, ?1791–1856). — Resembling no. 1; *ligules 2–3 mm. long; panicle* oblong, 1–3 dm. long, the *lower branches 2–6 cm. long; spikelets ovate-lanceolate; 2nd glume ovate, obtuse, emarginate,* 1.7 mm. long; *lemmas* 2.5–3 mm. long, *obtuse,* often mucronate. (*Leptochloa floribunda* of Hitchcock's treatments, not Doell) — Marshes and flats, Fla. to Tex., very rarely inland to sw. Ind. Aug.–Oct.

3. D. acuminàta Nash (tapering at tip). — Similar to no. 1; leaf-blades very harsh; *panicle* up to 4 dm. long, *its axis and branches strongly scabrous; spikelets distant* or but slightly overlapping; *upper glume* 4.5–7 *mm. long,* subaristate, usually scabrous; *lemmas purplish or lead-color,* 5–8 *mm. long, acuminate,* only slightly toothed at apex, *the awn about* 1 *mm. long.* — Wet saline soil, w. Mo., Neb. (perhaps N.D.) and Colo., s. to La. and Tex. July–Sept.

4. D. marítima Bickn. (of the sea). — Depressed or ascending, 0.7–6 dm. high; *leaf-sheaths and -blades glabrous to barely scabrous,* the sheaths greatly distended; ligules 2–5 mm. long; *panicle* 0.5–2 dm. long, *its axis and branches smooth or but slightly scabrous* on the angles; spikelets distant or but slightly imbricated, commonly purplish or lead-color; *glumes long-attenuate, commonly awned, the upper* 5–7 *mm. long; lemmas* 5–8 *mm. long,* attenuate at tip, *with awn* 2.5–5 *mm. long.* — Brackish shores, local, coast of s. N.H. to S.C.; also Onondaga and Cayuga Lakes, N.Y. July–early Oct.

15. CATABRÒSA Beauv.

Spikelets usually 1–4-flowered; glumes unequal, shorter than the lemmas, erose at the broad summit; lemmas subcoriaceous, erose-truncate, strongly 3-nerved; palea as long as the lemma, the strong nerves near the margin. — A creeping perennial boreal aquatic with flat leaves and open panicles of small spikelets. (Name from the Greek *catabrosis, an eating,* referring to the erose or "nibbled" glumes.)

1. C. aquática (L.) Beauv. (aquatic), Water Hairgrass. — Smooth, soft, decumbent and rooting at lower nodes, the ascending culms finely corrugated; loose sheaths overlapping; cauline leaves with blades 0.3–2 dm. long, 4–15 mm. wide, broadly rounded at tip; panicle 1–2.5 dm. long, loose and open, with capillary spreading-ascending whorled branches up to 1 dm. long and floriferous from well above the base; spikelets 2–4-flowered, 3–5 mm. long. — Eurasia. Represented with us by

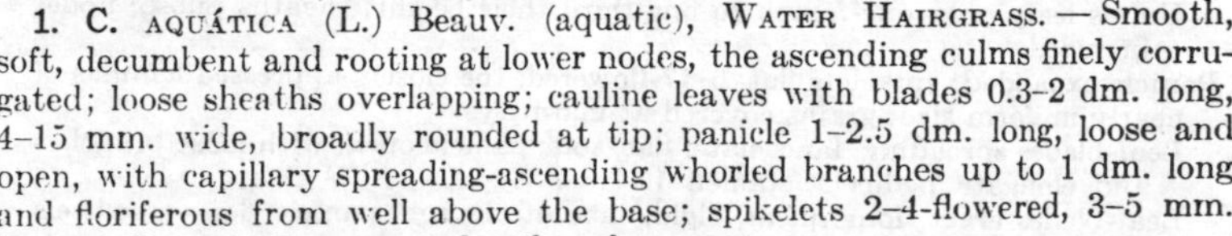
162. C. aquatica, v. laurentiana.

Var. **laurentiàna** Fern. (of the Gulf of St. Lawrence). — Ascending culms 1–8 dm. long, coarsely furrowed; cauline leaves 1–13 cm. long, 2–9 mm. broad, subacute to obtuse, callous-tipped; panicle lanceolate to narrowly ovoid, 2–16 cm. long, 1–8 cm. in diameter, the horizontal branches floriferous nearly to base; spikelets yellowish-brown to purplish, mostly 1 (rarely 2) -flowered; lemmas 2–3 mm. long. — Shallow fresh to brackish water or in marshes, coast of se. Lab., w. Nfld., se. Que., M.I., P.E.I. and ne. N.B. June–Aug. Fig. 162.

Var. **uniflòra** S. F. Gray (one-flowered). — Similar; with subascending branches floriferous chiefly from well above the base. — Rocky Mt. reg., eastw. to Man., Wisc. and Neb. (Eu.)

16. MOLÍNIA Schrank Moor-Grass

Spikelets 1–5-flowered, subterete, slenderly conical, upper floret imperfect; rachilla articulated; glumes convex, firm, 1-nerved, much shorter than the spikelet; lemmas cartilaginous,

remote, strongly 3-nerved, rounded on the back, acute; paleas about equaling the lemmas, obtuse, 2-keeled, with the lemma embracing the fruit. — Wiry hard-based tufted perennial, nat. of Eurasia, with elongate basal leaves, the cauline few and near the base of the seemingly nodeless (really with 1 low node) culms; the panicle elongate, contracted or diffuse. (Named for *Juan Ignazio Molina*, 1740–1829, early student of the Chilean flora.)

163. M. caerulea.

1. M. CAERÙLEA (L.) Moench (sky-blue). — Culms 0.3–2 m. high, terete; leaves rigid, 4–10 mm. broad, taper-pointed; ligule obsolete or a circle of hairs; panicle 0.3–3 dm. long, with erect branches; spikelets red- or violet-tinged or green; glumes 2–3 mm. long; lemmas 3–5 mm. long. (*Aira caerulea* L.) — Roadsides and fields, local, Nfld.; e. Me. to N.Y. and Pa. Aug.–Sept. (Adv. from Eu.) FIG. 163. — A striped-leaved form in cult.

17. DIARRHÈNA Beauv.

Spikelets 3–5-flowered, the uppermost florets sterile; glumes unequal, much shorter than the florets, lemmas broad, coriaceous, rigid, smooth and shining, convex below, 3-nerved, acuminate or mucronate-pointed; palea firm, 2-keeled; stamens 2, rarely 1; grain large, usually exceeding the lemma and palea, obliquely ovoid, obtusely beaked, with a shining coriaceous pericarp. — Nearly smooth perennials, ours and a Jap. species, with simple culms from a creeping rhizome, flat leaves and narrow few-flowered panicles. (Name composed of the Greek *dis*, *twice*, and *arrhen*, *male*, from the two stamens.)

164. D. americana.

1. D. americàna Beauv. (American). — Culms 6–10 dm. high; leaves nearly as long as the culm, 1–1.8 cm. wide; panicle very simple, 1–2.5 dm. long; spikelets short-pedicelled, 10–16 mm. long. (*D. diandra* Wood; *D. festucoides* (Raf.) Fern., not Raspail; *Korycarpus arundinaceus* Zea) — Shaded river-banks and woods, W.Va., O. and s. Mich. to S.D., s. to Ga., Ky., Ark., Okla. and Tex. July–Sept. FIG. 164.

18. UNÌOLA L. SPIKEGRASS. SPANGLEGRASS

Spikelets compressed, 3–many-flowered, the lower 1–4 lemmas empty; glumes compressed-keeled, acute or acuminate; lemmas firm-coriaceous or chartaceous, compressed-keeled, faintly many-nerved, the empty basal ones smaller; palea rigid; the keels broadly winged, nearly marginal; stamens 1 or 3. — Erect Am. perennials with simple culms, flat or involute leaves and terminal panicles. (Ancient name of some plant mentioned by Apuleius, diminutive of *unio, a kind of onion.*)

Panicle linear-filiform or slenderly virgate, simple or but slightly branched; spikelets subcompressed, 1–6-flowered, with the loosely ascending to divergent acuminate fertile lemmas strongly contrasted with the sterile basal ones; grains exposed at maturity.
- Culms leafy to the middle; sheaths glabrous; nodes mostly exserted. . . . 1. *U. laxa.*
- Culms leafy only one-fourth to one-third their height; sheaths pilose; nodes included. 2. *U. sessiliflora.*

Panicle expanded; spikelets flat, 6–17-flowered, the closely appressed lemmas of nearly uniform size; grains covered at maturity.
- Leaf-blades spreading, lanceolate, flat, soft; panicle open, with loose branches and elongate pedicels; stamen 1. . . . . 3. *U. latifolia.*
- Leaf-blades erect, long-linear, rigid, involute in age; panicle dense; pedicels short; stamens 3. . . . . 4. *U. paniculata.*

1. U. láxa (L.) BSP. (loose). — Tufted, from a knotty rhizome, 6–12 cm. high; *leaves extending half-way up the culm, long and narrow; sheaths smooth; nodes mostly exserted;* panicles linear-filiform, or slenderly virgate, 0.5–4.5 dm. long, simple or with slender erect branches; *spikelets rather plump*, short-pedicelled, 1–7-*flowered*, 3–9 mm. long; *lemmas rigid, greenish; the fertile* 3–5 *mm. long, acuminate, spreading and exposing the rugose grain at maturity;* palea shorter, arched; stamen 1. — Woods, meadows and swamps, Fla. to Tex., n. to L.I., N.J., e. Pa., Ky., Ark. and Okla. July–Sept. FIG. 165.

165. U. laxa.

2. U. sessiliflòra Poir. (sessile-flowered). — Resembling no. 1, coarser; *leaves mostly near the base, with strongly overlapping sheaths;* inflorescence a *stiffly moniliform* slender *spike-like raceme* 3–5.5 dm. long, with or without a few elongate branches; *fertile lemmas* 3.5–6.5 *mm. long.* (*U. longifolia* Scribn.) — Rich woods, Fla. to Tex., n. to se. Va., Tenn., Ark. and Okla. July–Sept.

3. U. latifòlia Michx. (broad-leaved), Wild Oats. — Culms simple or branching, 0.6–1.5 m. high; sheaths shorter than internodes; *ligule* 1 *mm. long, lacerate; blades lanceolate, soft, spreading,* 1–2.5 dm. long, 0.5–2.5 cm. broad; *panicle* 1–2.5 dm. long, *the filiform branches bearing a few pendulous flat broadly oval 6–17-flowered spikelets* 1.5–4 *cm. long* and 1–2 *cm. broad;* lemmas closely imbricated, the fertile 7–15 mm. long, hispidulous on the winged keel; *stamen* 1. — Shaded slopes and low thickets, N.J. to Ill. and Kans., s. to n. Fla., Ala., Miss., La. and e. Tex. July–Sept. Fig. 166.

166. U. latifolia.

4. U. paniculàta L. (with a panicle), Sea-Oats. — Stout, 1–2.5 m. high, with *numerous long rigid slender-tipped leaves involute in age;* ligule a ring of hairs about 1 mm. long; *panicles compact,* nodding, 2–6 dm. long, the stiff ascending branches bearing many *short-pedicelled stramineous flat* 8–16-flowered *spikelets* 1–2 *cm. long; lemmas* 8–10 *mm. long,* scabrous on the narrow keel; *stamens* 3. — Sandhills and drifting sands on coast, Va. to Tex., Mex. and S. Am. June, July.

19. SIEGLÍNGIA Bernh.

Spikelets terete, 3–5-flowered, the upper floret often imperfect, all the spikelets mostly cleistogamous; glumes subequal, ovate, herbaceous, with hyaline margins, nearly equaling to overtopping the florets, the upper 3-nerved; lemmas oval, coriaceous, rounded on the glabrous back, hairy-tufted at base, ciliate at the incurved margin, obscurely 5–9-nerved, closely imbricated, the tip with 2 blunt teeth and a short intermediate blunt mucro; palea lanceolate, with 2 ciliate nerves; grain closely embraced by lemma and palea. — Cespitose perennial with short strict simple panicles; late in the season often with 1-flowered cleistogenes within the bases of lower sheaths. (Named in 1800 for *Professor Siegling* of Erfurt.)

167. S. decumbens.

1. S. decúmbens (L.) Bernh. (reclining, with ascending tips), Heath-Grass. — Basal leaves tufted, firm, long and narrow, often convolute; ligule a tuft of hairs; culms slender, erect, 1–5 dm. high; panicle racemiform, 1.5–5.5 cm. long, with 3–15 narrowly ovoid spikelets 6–12 mm. long. — Peaty or turfy siliceous soils, indigenous in se. Nfld. and natzd. in sw. N.S. July, early Aug. (Eu.; n. Afr.) Fig. 167.

20. REDFIÈLDIA Vasey

Spikelets 3–4-flowered, compressed; glumes 1-nerved, acuminate, unequal; lemmas chartaceous, 3-nerved, the nerves glabrous except near the densely hairy callus; palea equaling the lemma; styles long, stigmas short; grain free, terete. — A single species, widely creeping by slender wiry rhizomes and stolons, the large panicle diffuse. (Named for *John Howard Redfield,* 1815–1895, distinguished botanist of Philadelphia.)

1. R. flexuòsa (Thurb.) Vasey (zigzag). — The only species; leaves and tall (6–10 dm.) culms stiff; panicle ellipsoid, 2–5 dm. long; spikelets 4–7 mm. long, the glumes whitish, the lemmas lead-colored. — Sandy plains and hills, Colo. and Ariz., e. to S.D., Neb., e. Kans. and Okla. July, Aug.

21. TRIÓDIA R. Br.

Spikelets 3–many-flowered in open or close panicles; florets perfect or the uppermost staminate; glumes thin or membranaceous, subequal, keeled; lemmas convex below, emarginate, toothed or lobed, 3-nerved, the nerves pubescent below and at least the middle one extending as a mucronate point between the teeth; palea broad, the nerves nearly marginal; grain convex on one face, concave on the other. — Perennials of Am. and Austral., with long narrow leaves and (in ours) terminal panicles. (Name from the Greek *triodous, three-toothed,* referring to the tip of the lemma.) Tridens R. & S.

Panicle loose and open, 2–4.5 dm. long; glumes shorter than lower lemmas.
  Branches of panicle loosely spreading-ascending, their bases villous only above; lateral spikelets appressed along the branchlets, subsessile or on pedicels up to 3 mm. long. . . . . 1. *T. flava.*
  Branches of panicle divergent, their bases villous all around; the divergent spikelets all on pedicels 0.3–2 cm. long. . . . . 2. *T. Chapmani.*
Panicle dense, spike-like, 1–3 dm. long; glumes much exceeding lower lemmas.
  Culms and sheaths smooth; panicles 1–2 cm. thick; spikelets compressed, 4–6 mm. long; glumes scabrous. . . . . 3. *T. stricta.*

Culms and sheaths harsh; panicles 3–6 mm. thick; spikelets subterete, 9–12 mm. long; glumes scabrous on keels. . . . . 4. *T. elongata.*

1. T. flàva (L.) Smyth (yellow), TALL RED-TOP. — *Culms* erect, 0.7–2 m. high, *viscid near summit* and on axes of panicle; sheaths bearded at summit, otherwise glabrous, as are the long flat (up to 1 cm. broad) or inrolling tapering blades; the showy *panicles* 2–4.5 dm. long, somewhat open, *the spreading-ascending* slender forking *branches* naked below, *bearded on the pulvini* (upper surface of base); *spikelets greenish to stramineous,* or in forma **cúprea** (Jacq.) Fosberg (coppery) purplish, 6–9 mm. long, 5–8-flowered, those terminating the branchlets distinctly pedicelled, *the several lateral ones appressed along the branchlets, subsessile or on pedicels up to 3 mm. long; glumes shorter than the lowest florets, obtuse, mucronate; the 3 nerves of the lemmas excurrent.* (*Tridens* Hitchc.) — Dry fields, roadsides, openings and borders of woods, s. N.H. to Minn. and Neb., s. beyond our range. Aug.–Oct. FIG. 168.

168. T. flava.

2. T. Chapmáni (Small) Bush (for its discoverer, ALVIN WENTWORTH CHAPMAN, 1809–1899). — More slender; leaf-blades 3–8 mm. broad, promptly involute; *panicle open* and skeleton-like, *with remote divergent branches; bases of primary branches long-villous all around; spikelets all long-pedicelled, the lateral on divergent pedicels* 0.3–2 *cm. long;* glumes more tapering. — Dry sandy woods and pinelands, Fla. to e. Tex., n. to se. Va., Cape May, N.J., Tenn. and s. Mo. Aug., Sept.

3. T. strícta (Nutt.) Benth. (upright). — Cespitose, 4.5–14 dm. high; *culms* stout, erect, *smooth;* leaves long and rigid, with *smooth sheaths; panicle* pale to purplish, *dense, spike-like,* 1–3 dm. long, 1–2 *cm. thick; spikelets compressed,* 4–6 *mm. long,* 4–10-flowered; *glumes* acute, *glabrous,* sharp-pointed, *exceeding lower* and sometimes all the *lemmas; lemmas* 2–3 *mm. long,* villous on the nerves below, the midnerve and often the laterals excurrent. (*Tridens* Nash) — Moist soil, Ala. to Tex., locally n. to Tenn., s. Mo. and se. Kans.; reported from se. Va. July–Oct.

4. T. elongàta (Buckl.) Scribn. (elongate). — Habit of no. 3, 3–8 dm. high, more slender; *culms scabrous; leaves harsh,* becoming involute; *panicle* linear-cylindric, 1.2–2.5 dm. long, 3–6 *mm. thick; spikelets subterete,* 9–12 *mm. long,* 10–12-flowered; *glumes* ovate, *scabrous on the keel,* the 2nd 3-nerved; *lemmas* 4–6 *mm. long,* the lateral very villous nerves evanescent below the tip. (*Tridens* Nash) — Limestone bluffs and barrens, sw. Mo. to Colo., s. to Ark., Tex. and Ariz. June–Oct.

22. TRÍPLASIS Beauv.

Spikelets 3–6-flowered, the florets remote, the lowest stipitate, perfect or the uppermost staminate; glumes unequal, keeled, shorter than the florets; lemmas 2-cleft, the 3 nerves strongly ciliate, the midnerve excurrent as a short awn between the lobes; palea shorter, broad, the nerves nearly marginal and densely long-ciliate from the middle to the apex. — Am. annuals or perennials with small nearly simple terminal panicles and cleistogamous spikelets or panicles within the enlarged sheaths; old culms with fruit-bearing sheaths disarticulating at the nodes. (Name from the Greek *triplasios, trifarious;* from the tip of the lemma.)

1. T. purpùrea (Walt.) Chapm. (purple), SAND-GRASS. — Culms tufted, widely spreading or ascending, wiry, 1–12 dm. long; nodes bearded; sheaths and the small rigid blades scabrous; terminal panicles 3–7 cm. long, the few stiff branches finally divergent; spikelets short-pedicelled, usually rose-purple, 5–8 mm. long; the awn of the lemma scarcely exceeding the truncate lobes. — In sand, on or near the coast, s. Me. to Fla. and Tex.; inland from w. N.Y. and s. Ont. to Minn. and Colo., s. to Ill., Mo. and Tex. Aug.–Oct. FIG. 169. — Plant acid to taste.

169. T. purpurea.

PAPPÓPHORUM Schreb. (bearing down), with the lemma fimbriate into many long bristle-like awns, is represented by P. MUCRONULÀTUM Nees (with short abrupt point), a tufted perennial up to 1 m. high, with a slender spiciform whitish panicle, the spikelets with 1 or 2 fertile and 2 or 3 sterile florets, the awns of the lemma up to 5 mm. long; and by P. BÌCOLOR Fourn. (two-colored), with more slender pink panicle; both casual on wool-waste. (Adv. from Mex. or our Southwest)

23. CYNOSÙRUS L. Dog's-tail

Spikelets dimorphous; the terminal terete fertile one in each fascicle 2–3-flowered, sessile, nearly hidden by the modified lower sterile ones, these reduced to a rigid fan-like distichous group of slender glumes and awned or pointed lemmas; glumes of fertile spikelet 2, rigid, shorter than the lemmas; fertile lemmas terete, 3-keeled, firm, mucronate; paleas with 2 ciliate keels; grain embraced by palea and lemma. — Tufted (ours perennial), nat. of Eurasia, with flat leaves, the fascicles of dimorphous spikelets crowded into close terminal spike-like panicles. (Name from the Greek *cynos, of a dog* and *oura, tail.*)

170. C. cristatus.

1. C. cristàtus L. (crested), Crested D. — Perennial, erect, 2–8 dm. high; the long basal leaves flat, soft, 1–3 mm. wide; the cauline 2–4, with long sheaths and short blades; panicle strict, 1-sided, 1.5–9 cm. long, 5–10 mm. thick. — Grasslands and roadsides, Nfld. to Ont., s. to N.S., N.E., N.C., W.Va. and O., local. June–Aug. (Natzd. from Eu.) Fig. 170.

C. echinàtus L. (prickly), with dense short ellipsoid panicle 2–3 cm. thick, and long awns, has appeared casually in waste places. (Adv. from Eu.)

24. DISTÍCHLIS Raf. Spike-Grass. Alkali-Grass

Dioecious; spikelets 8–16-flowered, compressed, the staminate soft and with the rachis continuous, the pistillate harder and with rachis articulated; glumes unequal, coriaceous, keeled, acute; lemmas coriaceous, rigid, faintly many-nerved, closely 2-ranked. — Rigid erect perennials of N. and S. Am. and Austral., with extensively creeping rhizomes covered with papery whitish scales, involute leaves and small crowded panicles of large smooth spikelets. (Name from the Greek *distichos, two-ranked.*)

1. D. spicàta (L.) Greene (spiked). — Pale or glaucous, 1 dm.–1.2 m. high; *leaves* 0.5–1.5 dm. long, spreading or ascending, flat to involute *with mostly smooth margins and bluntish tips;* ligule a ring of short hairs; panicle compact, subcylindric, 2–8 cm. long, mostly with 10–20 *spikelets* 5–14 *cm. long; 1st glume* 2–3.5 *mm. long, 2nd* 2.5–4 *mm. long; lemmas about* 3.5 *mm. long,* the staminate papery throughout, the pistillate with hyaline margin; paleas 3–4.5 mm. long, hyaline, with minutely ciliolate firm keels; *grain up to* 2 *mm. long, scarcely beaked;* anthers of staminate plants 2–3 mm. long, of pistillate plants reduced to minute globular to sagittate rudiments. — Saline marshes and alkaline soils, Fla. to Tex., locally inland to Mo.; n. along coastwise area to Gulf of St. Lawrence, C.B., P.E.I. and s. N.B.; Pacific Coast. Aug.–Oct. (S.Am.) Fig. 171.

171. D. spicata.

2. D. strícta (Torr.) Rydb. (straight). — Similar; *leaves usually strongly serrulate on margins and sharp tips; spikelets* 0.9–2.5 *cm. long,* the pistillate very firm; 1*st glume* 3.2–8 *mm. long, 2nd* 2–7 *mm. long; lemmas* 3.2–7.8 *mm. long,* the pistillate with conspicuous hyaline margin; *grain* 3–5 *mm. long, beaked;* anthers 2–4 mm. long. — Saline or alkaline reg., Minn. and Man. to B.C., s. to w. Mo., Okla., Tex. and Mex. Late May–July.

25. PHRAGMÌTES Trin. Reed. Roseau (Que.)

Spikelets loosely 3–7-flowered; rachilla clothed with long silky hairs; glumes unequal, lanceolate, acute; lemmas narrow, long-acuminate; that of the lowest floret somewhat longer, equaling the uppermost florets, empty or subtending a staminate flower, the other florets perfect; paleas one-half to two-thirds the length of their lemmas. — Tall semicosmop. perennials with stout leafy culms and large terminal panicles. (Name from the Greek *phragmites, growing in hedges,* apparently from its hedge-like growth along ditches.)

1. P. commùnis Trin. (growing in colonies). — Culms erect, stout, 1.5–4 (rarely –6) m. high, from long creeping rhizomes; sheaths overlapping; blades 1.5–6 dm. long; 1–6 cm. wide, flat, glabrous; panicle tawny, 1.5–4 dm. long; branches ascending, rather densely flowered; spikelets 10–17 mm. long; 1*st glume* 2.5–5 (*av.* 3.6), *2nd glume* 5–7 (*av.* 5.7) *mm. long;* the florets exceeded by the hairs of

172. P. communis, v. Berlandieri.

the rachilla. (*P. Phragmites* Karst.; *P. maxima* sensu Chiovenda, not *Arundo maxima* Forsk., the basonym) — Eurasia, and with its vars. nearly cosmop. Ours is

Var. **Berlandièri** (Fourn.) Fern. (for its discoverer, JEAN LOUIS BERLANDIER, 1805–1855). — 1*st glume* 4–6 (*av.* 4.6), 2*nd glume* 6–8.5 (*av.* 7.3) *mm. long*. (*P. maxima*, var. Moldenke) — Fresh to alkaline marshes, pond-margins, ditches, etc., e. Que. to B.C., s. to N.S., s. N.E., Md., O., Ind., Ill., La., Tex. and Mex. Late July–Sept. FIG. 172. — Rarely perfecting seed, spreading freely from the rhizomes, the leafy stolons often running on the surface of the ground for a distance of 5–10 m.

ARÚNDO (ancient name) DÓNAX L. (classical name), GIANT REED, a tall grass, 2–5 m. high, in clumps, with culms branching the 2nd year and with conspicuous 2-ranked long leaves (5–6 cm. broad) and erect panicle up to 6 dm. long, the rachilla glabrous, the lemmas densely villous, is cult. for ornament in the South; tending to escape northw. to Va. and Mo. Sept., Oct. (Introd. from the Mediterr. reg.)

Tribe III. HÓRDEAE Lindl. (see p. 96)

a. Spikelets solitary at each joint of the rachis. . . *b*.
- *b*. Spikes 1-sided; spikelets 1-flowered, alternating in 2 closely interlocking series; style simple, elongate. . . . . 26. *Nardus*.
- *b*. Spikes symmetrical, the spikelets on opposite sides of the rachis; stigmas 2. . . *c*.
  - *c*. Spikelets deeply imbedded in excavations of the very slender rachis, the 2 glumes thus pushed to front of the spikelet; mature spikelets falling with disarticulated internodes of rachis. . . . . 27. *Pholiurus*.
  - *c*. Spikelets not imbedded, the glumes normally placed. . . *d*.
    - *d*. Spikelets with broad side toward the rachis, with both glumes normally developed.
      - Spikelets compressed; lemmas with a definite callus; grain closely embraced by palea and lemma; rachis continuous or only tardily disarticulating, persistent after the falling or disintegration of the spikelets. . . . . 28. *Agropyron*.
      - Spikelets with inflated or subterete bases; lemmas without a callus; grain free; rachis soon disarticulating at base and finally at all joints, carrying the ripened spikelets. . . . . 29. *Aegilops*.
    - *d*. Spikelets with narrow side toward rachis, the latter with concave elongate pockets embracing the glumeless inner edge of lateral spikelets. . . . . 30. *Lolium*.

a. Spikelets commonly 2 or 3 at each joint of the rachis. . . *e*.
- *e*. Spikelets in 3's at each node of the flattened and disarticulating rachis, the lateral pair usually pedicelled and often reduced to awns, the central spikelet sessile and perfect. . . . . 31. *Hordeum*.
- *e*. Spikelets in 2's, all with perfect florets.
  - Rachis promptly disarticulating, the base of each segment forming a stipe-like process below the persistent spikelet above; glumes bristle-like, greatly prolonged. . . . . 32. *Sitanion*.
  - Rachis continuous; glumes, if prolonged, not bristle-like.
    - Spike dense or with ascending spikelets; glumes fully developed; grain adherent to lemma and palea. . . . . 33. *Elymus*.
    - Spike greatly interrupted, the spikelets widely divergent; glumes minute or obsolete; grain free within the lemma and palea. . . . 34. *Hystrix*.

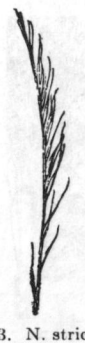

26. NÁRDUS L. MATGRASS

Spikelets 1-flowered, solitary and alternate in two closely interlocking series on one side of the slender continuous rachis; inner glume minute, adnate to the rachis, or wanting, outer glume wanting; lemma slender, awned, tightly inrolled around the palea, not appreciably loosening in anthesis; stamens 3; style simple, elongate. — Densely matted wiry perennial, with bristle-like leaves and terminal subulate spikelets in slender 1-sided spikes. (Name from the Greek *nardos, the spikenard*, the application here obscure.)

1. N. stricta L. (straight). — Rhizomes short, with crowded distichous, erect whitish-sheathed branches; culms bristle-like, naked or nearly so, 1–5 dm. high; spike 5–10 cm. long. — Sandy or peaty soil, seemingly indigenous, se. Nfld.; introd. from Eu. in poor grassland, local, N.S. to Mich. June–Sept. (Eurasia; Greenl.) FIG. 173.

173. N. stricta.

27. PHOLIÙRUS Trin. HARD GRASS. THIN-TAIL

Spikelets 1–2-flowered, awnless, solitary, alternate in excavations of the articulate rachis; glumes equal, placed edge to edge in front of the florets, except in the terminal spikelet, coriaceous rigid, 5-nerved, acute; lemma much smaller than the glumes, hyaline, keeled. — A low branching annual nat. of Eu. and adj. Asia, with slender cylindrical straight or curved terminal spikes which disarticulate at maturity, the joints falling with the appressed spikelets attached. (Name from the Greek *pholis, scale*, and *oura, tail* or *spike.*) LEPTURUS Am. auth., not R. Br.

1. P. INCÚRVUS (L.) Schinz & Thell. (curving-in). — Tufted, 1–2 dm. high, decumbent at base, glabrous; leaves short and narrow; spikes 3–10 dm. long, included at the base of the sheath; joints and spikelets 5 mm. long. (*Lepturus filiformis* Trin.) — Borders of brackish marshes, N.J. to Va.; and casual on ballast northw. May, June. (Adv. from Eu.) FIG. 174.

174. P. incurvus.

28. AGROPÝRON Gaertn.

Spikelets 3–many-flowered, solitary (rarely in pairs) in alternate notches of the continuous (rarely articulate) rachis, the side of the spikelet placed against the rachis; glumes subequal, opposite or edge to edge on the outer side of the spikelet, several-nerved, usually shorter than the florets, awnless or awned; lemmas convex or slightly keeled above, 5–7-nerved, awnless or awned from the apex; palea shorter than its lemma, bristly-ciliate on the keels; grain pubescent at the summit, usually adherent to the palea. — Ours perennials with simple culms and terminal spikes; a genus of temp. and cool reg. of both N. and S. Hemisph. (Name from the Greek *agrios, wild*, and *pyros, wheat.*) *

a. Spikelets at maturity readily disintegrating, individual florets promptly disarticulating, leaving empty glumes; anthers 1–3 mm. long; culms tufted or solitary, without elongate rootstocks. § GOULARDA. . . *b.*
- *b.* Anthers 1–2.5 mm. long; lemmas awnless or with essentially straight and ascending awns. . . . . 1. *A. trachycaulum.*
- *b.* Anthers 3 mm. long; lemmas with strongly arched-divergent long awns. . . 2. *A. spicatum.*

a. Spikelets at maturity dropping intact from rachis, individual florets not readily detached; anthers 3–7 mm. long; culms solitary or tufted; rootstocks very elongate and extensively creeping. § HOLOPYRON. . . *c.*
- *c.* Lemmas and rachillas (of spikelets) densely villous to lanate. . . . . . 3. *A. dasystachyum.*
- *c.* Lemmas and rachillas glabrous to merely scabrous. . . *d.*
  - *d.* Glumes linear-attenuate, tapering with straight margins from below the middle; spikelets 7–13-flowered; cartilaginous band of upper nodes of culm shorter than thick. . . . . 4. *A. Smithii.*
  - *d.* Glumes lanceolate to oblong, narrowed to tip from near the middle; spikelets 3–9 (in the maritime no. 5, –11)-flowered; cartilaginous band of upper nodes as long as thick.
    - Spike nearly square in section; internodes of rachis thick, usually 4-angled; spikelets 7–11 (very rarely only 3–5)-flowered; glumes coriaceous, with broadly rounded keel and ribs; leaves hard, very glaucous, mostly involute, with remote coarse ribs; top of culm solid (closely filled with pith). . . . . 5. *A. pungens.*
    - Spike usually not square; internodes of rachis mostly thin, rounded on the back, 2-edged; spikelets 2–9-flowered; glumes herbaceous, with slender keel and ribs; leaves soft, flat, with crowded fine ribs; top of culm hollow. . . . . 6 *A. repens.*

§ GOULÁRDA (Husnot) Holmb.

1. A. trachycaùlum (Link) Malte (rough-stemmed). — Perennial, green or glaucous, with tufted or non-stoloniferous bases; leaves flat (sometimes inrolled), 2–10 mm. broad; spikes 0.5–2.5 dm. long, cylindric to 1-sided, erect to arching; *spikelets not strongly compressed*, appressed, 2–7-flowered; glumes keeled, strongly 3–7-ribbed, tapering to pointed or awned tips; *lemmas* smooth or merely scabrous, *awnless or with straight ascending awns* up to 4 cm. long; *anthers 1–2.5 mm. long.* — Very variable; five varieties with us:

a. Awns wanting or at most half as long as body of lemma. . . *b.*

---

* Treatment based largely on that in Rhodora, xxxv. 161–185 (1933).

b. Glumes coriaceous or subcoriaceous, with hyaline margins 0.4–0.6 mm. broad; spikelets scarcely imbricated, the tips rarely reaching the bases of those above on the same side; internodes of rachis often quadrate, with all 4 sides concave and wing-margined, mostly 0.8–2 cm. long; rachilla usually scabrous or strigose. . . . . *A. trachycaulum* (typical).

b. Glumes subcoriaceous to herbaceous, the hyaline margin 0.1–0.4 mm. broad; spikelets mostly closely imbricated; internodes shorter, mostly convex on the back and 2-edged; rachilla usually villous.

Body of glume averaging 12.5 (10–16) mm. long; contracted (mature) spike averaging 7 (5-12) mm. thick. . . . . Var. *majus*.

Body of glume averaging 8 (7–10) mm. long; contracted spike averaging 5 (3–6) mm. thick. . . . . Var. *novae-angliae*.

a. Awns nearly equaling to much longer than body of lemma.

Body of glumes averaging 9 (6–12) mm. long; awn of lemma 7–20 mm. long; contracted fruiting spike (excluding awns) averaging 5 (3–10) mm. thick. . . . . Var. *glaucum*.

Body of glume averaging 13.5 (12–18) mm. long; awn of lemma 10–40 mm. long; contracted spike averaging 9 (6–13) mm. thick. . . . . . . Var. *unilaterale*.

A. trachycaùlum (typical). — Spikes up to 2.5 dm. long, very slender, with subremote spikelets. (*A. tenerum* Vasey; *A. pauciflorum* (Schwein.) Hitchc., not Schur) — Limy soils about the Gulf of St. Lawrence, e. Que.; prairies and open soils, L. Huron, Ont., to Alaska, s. to ne. Ill., Ia., Mo. (introd.), e. Kans., N.M. and Calif. July, Aug. Fig. 175. — *Elymus Macounii* is thought by some to be a hybrid of this with *Hordeum jubatum*.

Var. **màjus** (Vasey) Fern. (large). — Spikes dense, stout, 0.5–2 dm. long. (Including *A. biflorum* of ed. 7, not R. & S.; *A. pseudorepens* Scribn. & Sm.) — Gravelly soils, s. Lab. to s. B.C., s. to Nfld., N.S., coast of s. Me. and to n. N.E., w. Ont., Neb., Colo., Nev. and Oreg. July, Aug. Fig. 176.

Var. **nòvae-ángliae** (Scribn.) Fern. (of New England). — Spikes slender, 0.4–2 dm. long. (*A. novae-angliae* Scribn.) — Rocky or peaty soils or in thickets, Lab. to B.C., s. to Nfld., N.S., s. N.E., n. Pa., Mich., Wisc., Neb., Colo. and Nev. Late June–Aug. Fig. 177.

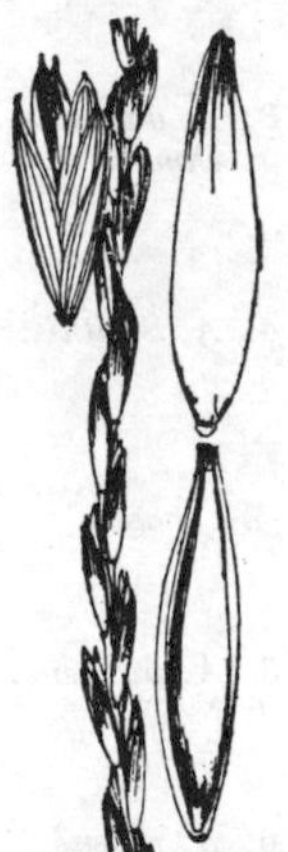

175. A. trachycaulum (typical).

176. A. trachycaulum, v. majus.

177. A. trachycaulum, v. novae-angliae.

178. A. trachycaulum, v. glaucum.

179. A. trachycaulum, v. unilaterale.

Var. **glaùcum** (Pease & Moore) Malte (blue-green). — Plant green or glaucous; spikes slender, 0.5–2 dm. long. (*A. caninum* Am. auth., not Beauv.) — Rocky or gravelly shores or thickets or in bogs, Nfld. to s. B.C., s. to N.S., s. N.E., Pa., Mich., Wisc., Minn., Neb., Colo., Nev. and Calif. June–Aug. Fig. 178.

Var. **unilateràle** (Cassidy) Malte (one-sided). — Spikes stout, 1–2.5 dm. long. (*A. Richardsoni* Schrad.; *A. subsecundum* (Link) Hitchc.) — Rich thickets and shores, Que.; s. Ont. and nw. N.Y. to B.C., s. to Ill., Ia., S.D., Colo., Nev. and Oreg. June–Aug. Fig. 179.

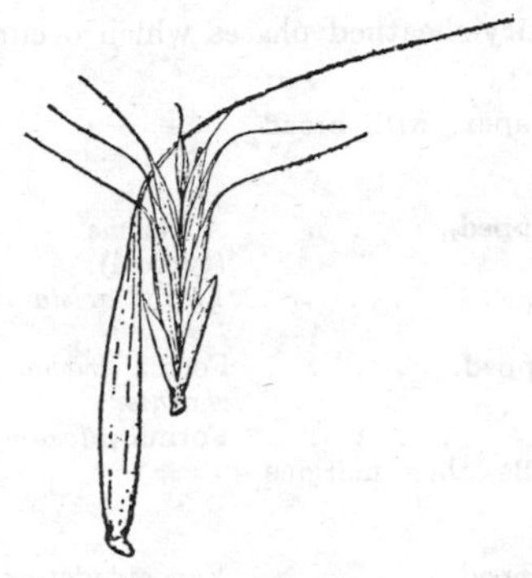
180. A. spicatum.

2. **A. spicàtum** (Pursh) Scribn. & Sm. (spiked). — Tufted, 3–10 dm. high, with short rather stiff and mostly *involute pale leaves;* spike 1–2 dm. long, with *distant strongly compressed spikelets;* glumes 6–10 mm. long, acute or short-awned; *lemmas with strongly arching to divergent awns* 1–2 *cm. long;* rachilla smooth or scabrous; *anthers* 3 *mm. long.* — Dry soils, Keweenaw Co., Mich. (*Farwell*); Yuk. to N.M. and Calif. June, July. Fig. 180.

A. cristàtum (L.) Gaertn. (crested), forming dense tussocks with short and rigid promptly inrolled leaves and short oblong comb-like (distichous) spike of divergent 3–8-flowered subulate-tipped spikelets, is adv. or natzd. about Montreal and on sand and shingle of Fisher's I., N.Y.; Minn.; introd. in the West to hold drifting soils. (Adv. or introd. from Eurasia)

§ Holopỳron Holmb.

3. **A. dasystáchyum** (Hook.) Scribn. (hairy-spiked). — Loosely tufted, with slender wiry rootstocks, and marcescent old leaves about the bases; culms 5–10 dm. high; *leaves pale, firm,* scabrous, *often involute;* spike 0.7–2 dm. long, with the loosely appressed 5–9-flowered *spikelets not strongly flattened; lemmas densely villous to lanate,* scarious-margined; anthers 4–6 mm. long. — Dunes and dry open soil, w. Ont. and Mich. to n. Alta. and B.C., s. to n. Ill., Neb., Colo. and Calif. June, July. Fig. 181.

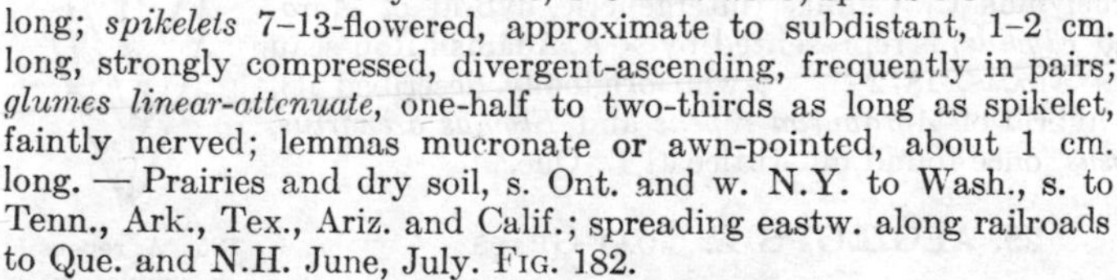
181. A. dasystachyum.

4. **A. Smíthii** Rydb. (in honor of Jared Gage Smith, 1866–). — Glaucous, with elongate *tawny or drab rootstocks* and bases; *culms* rigid, 3–15 dm. high, *their upper nodes with cartilaginous bands shorter than thick; leaves firm or rigid,* scabrous, mostly *involute,* 4–6 mm. wide; spikes 0.8–1.5 dm. long; *spikelets* 7–13-flowered, approximate to subdistant, 1–2 cm. long, strongly compressed, divergent-ascending, frequently in pairs; *glumes linear-attenuate,* one-half to two-thirds as long as spikelet, faintly nerved; lemmas mucronate or awn-pointed, about 1 cm. long. — Prairies and dry soil, s. Ont. and w. N.Y. to Wash., s. to Tenn., Ark., Tex., Ariz. and Calif.; spreading eastw. along railroads to Que. and N.H. June, July. Fig. 182.

182. A. Smithii.

5. **A. púngens** (Pers.) R. & S. (pungent). — *Very glaucous, from wiry widely creeping slender rootstocks; culms* few, rigid, 0.5–1 m. high, *solid at summit; leaves* hard, *with remote coarse ribs,* flat or involute, *tapering to a rigid involute point; spike nearly square in section,* 5–15 cm. long; *spikelets* strongly compressed, *along each side of the thick usually 4-angled rachis, overlapping, mostly alternately divergent,* broadly oblong, 1.3–2 cm. long (excluding awns), 4–9 mm. broad, 7–11-flowered; *glumes coriaceous,* gradually narrowed from above the middle to a mucronate or short-aristate tip, *the ribs and* scabrous *keel broadly rounded;* lemmas acute, mucronate or awned. — Sandy coast, C.B. to Cape Cod, Mass. July, Aug. (Eu.) Fig. 183.

Var. **acadiénse** (F. T. Hubbard) Fern. (of Acadia). — Smaller throughout; *spikelets* 1–1.4 *cm. long,* 3–5 *mm. broad,* 3–5-*flowered.* — Beaches of Bras d'Or Lakes, C.B. July, Aug.

183. A. pungens.

6. **A. rèpens** (L.) Beauv. (creeping), Witch-Grass, Couch-, Quitch- or Quick-Grass, Chiendent (Que.). — Green or glaucous, with *whitish or yellowish* extensively creeping slender *rootstocks; culms* 3–12 dm. high, *hollow at tip; cartilaginous bands of upper nodes longer than thick;* sheaths glabrous or the lower pilose; *blades flat* (tardily inrolled), comparatively soft, scabrous or sparsely pilose above, *finely many-nerved; rachis thin, its internodes rounded on the back,* 2-*edged;* spikes dense or lax, 0.5–2.5 dm. long; spikelets compressed, 2–9-flowered, 0.6–2.2 cm. long; glumes herbaceous, oblong to lanceolate, narrowed from above the middle, strongly and slenderly 5–7-nerved; lemmas obtuse, acute or awned; anthers 4–7 mm. long. — A semicosmop. weed, seemingly indigenous with us on gravelly and sandy shores.

Highly variable. Our chief forms (except glaucous or hairy-sheathed phases which occur in practically all) are indicated below:

a. Glumes oblong, rounded or rather abruptly narrowed at apex, with broad scarious margin. . . *b.*
b. Rachis glabrous except for ciliate edges.
Glumes and lemmas blunt, acute or merely subulate-tipped. . . . . *A. repens* (typical).
Glumes or especially the lemmas definitely awned. . . . . Forma *aristatum.*
b. Rachis pilose or hirsute.
Glumes and lemmas blunt, acute or merely subulate-tipped. . . . . Forma *trichorrhachis.*
Glumes and lemmas definitely awned. . . . . Forma *pilosum.*
a. Glumes lanceolate, gradually tapering from near the middle, their margins narrow or involute. . . *c.*
c. Rachis glabrous except for ciliate edges.
Glumes and lemmas blunt, acute or merely subulate-tipped. . . . . Var. *subulatum.*
Glumes and lemmas definitely awned. . . . . Forma *Vaillantianum.*
c. Rachis pilose or hirsute.
Glumes and lemmas blunt, acute or merely subulate-tipped. . . . . Forma *heberhachis.*
Glumes and lemmas definitely awned. . . . . Forma *setiferum.*

A. rèpens, typical. — Gravelly coast, Nfld. and e. Que. to Me.; and as an aggressive weed (of European origin) s. to s. N.E., N.C. and westw. June–Aug. (Eu.) Fig. 184. — Forma **aristàtum** (Schum.) Holmb. (awned), of similar range. Forma **trichórrhachis** Rohlena (with hairy rachis), Nfld. and e. Que. to Ct. and w. N.Y. Forma **pilòsum** (Scribn.) Fern. (hairy), e. Que. to Ct. and Mich.

Var. **subulàtum** (Schreb.) Reichenb. (awl-shaped). — Coast, Nfld. and e. Que. to Ct.; and as a weed to N.J., w. to Minn., Ia. and Mo. (Eu.) Fig. 185. — Forma **Vaillantiànum** (Wulf. & Schreb.) Fern. (for Sébastien Vaillant, 1669–1722), of similar range. Forma **héberhachis** Fern. (with hairy rachis), Nfld. to R.I. Forma **setíferum** Fern. (bearing bristles), Nfld. to s. and w. N.Y.

× **Agroélymus** C. Camus (intergeneric hybrid of *Agropyron* and *Elymus*) is represented by × **A. Adámsii** Rousseau (for John Adams, 1872– , who originally described it), a sterile hybrid of *Agropyron repens* and *Elymus arenarius,* var. *villosus,* once found on Anticosti I., Que.

184. A. repens (typical).

185. A. repens, v. subulatum.

29. AÈGILOPS L. Goat-Grass

Spikelets 2–5-flowered, plump or cylindric, placed with side against the articulated rachis and closely appressed to it, the rachis-joints upwardly enlarged; glumes flat or rounded on back, obscurely keeled or keelless; otherwise much as in *Agropyron.* — Annuals, nat. of s. Eu. and w. Asia, the mature spike disarticulating at the lowest joint and finally between the spikelets. (Name used by Theophrastus for a kind of wild oat.)

1. A. cylíndrica Host (cylindric). — Tufted annual, 2–6 dm. high; spike 5–8 cm. long; spikelets few; keel of glumes lateral, prolonged into an awn; lemmas of upper spikelets with harsh awns 4–5 cm. long, those of lower spikelets shorter. — Waste places, railroads and fields, Colo. and N.M., eastw. to Ind., Mo. and Okla.; locally about N.Y.; becoming a troublesome weed. (Natzd. from Eu.)

30. LÓLIUM L. Darnel. Ivraie (Que.)

Spikelets several-flowered, solitary in alternate notches of the continuous rachis, one edge of each of the axillary ones resting in a concavity of the rachis, the glume on that side wanting; outer glume firm, 3–many-nerved, exceeding the lowest floret; terminal spikelet with both glumes; rachilla flattened; lemmas convex, 5–7-nerved, with nerves converging above, awned or awnless; grain adherent to palea. — Annuals or perennials, nat. of Eurasia and n. Afr., with simple culms, flat leaves and terminal spikes. (Ancient Latin name.)

Glumes (excluding awns) not reaching the tips of the upper lemmas.
Unexpanded leaves folded; rachis usually smooth except on angles; glumes of upper spikelets exceeding the contiguous floret. . . . . 1. *L. perenne.*

Unexpanded leaves inrolled from each margin; rachis usually roughened; glumes of upper spikelets not exceeding the contiguous floret. . . . . . 2. *L. multiflorum.*
Glumes (excluding awns) usually overtopping the upper lemmas. . . . . . 3. *L. temulentum*

1. L. perénne L. (perennial), Common D., Perennial Ray- or Rye-grass. — Short-lived *perennial;* culms 3–6 dm. high, glabrous; the axis of the inflorescence glabrous except the angles; *leaves* usually not over 4 mm. wide, *folded in the bud; glume shorter than the* 6–10-*flowered spikelet; lemma about* 5–6 *mm. long, awnless.* — Fields and roadsides, chiefly eastw. May–Aug. — This and the following species are cult. as meadow grasses. (Natzd. from Eu.) Fig. 186.

186. L. perenne.

187. L. multiflorum.

Var. cristàtum Pers. (crested). — Inflorescence ovoid, with crowded divergent spikelets. — Local, Del. and D.C. (Adv. from Eu.)

2. L. multiflòrum Lam. (many-flowered), Italian Rye-Grass. — *Annual;* upper portion of culm and convex side of *rachis roughened; leaves convolute in bud; spikelets* 10–20-*flowered; glumes of lowest spikelets often exceeding the contiguous floret;* lemmas 7–8 mm. long, often at least the upper awned, or almost or quite awnless in forma submùticum (Mutel) Hayek (somewhat curtailed). — Fields and roadsides. June–Aug. (Natzd. from Eu.) Fig. 187.

Var. diminùtum Mutel (reduced). — *Perennial,* smaller; *spikelets* 5–9-*flowered; glumes of all the spikelets shorter than the contiguous floret.* — Fields and roadsides. June–Aug. (Natzd. from Eu.)

3. L. temuléntum L. (drunken; from narcotic grains), Bearded or Poison D. — *Annual,* 6–9 dm. high; *glumes fully equaling to overtopping the upper florets;* spikelets 5–7-flowered; lemmas commonly awned. — Grain-fields and waste places, rare, Que. and N.E. to Minn., Mo. and Kans., s. to the Gulf; Pacific States. June–Aug. (Adv. from Eu.) Fig. 188.

31. HÓRDEUM L. Barley. Orge (Que.)

Spikelets 1 (rarely 2)-flowered, 3 together in our species at each joint of the flattened articulate rachis, the middle one sessile and perfect, the lateral pair usually pedicelled, often reduced to awns and together with the glumes of the perfect spikelet simulating a bristly involucre at each joint of the rachis; rachilla of the central spikelet prolonged behind the palea as an awn, sometimes with a rudimentary floret; glumes equal, rigid, narrow-lanceolate, subulate or setaceous, borne at the sides of the dorsally compressed floret which is turned with the back of the palea against the rachis of the spike; lemma obscurely 5-nerved, tapering into an awn; palea slightly shorter, the 2 strong nerves near the margin; grain hairy at the summit, usually adherent to the palea at maturity. — Cespitose annuals or perennials of the Americas, Eurasia and n. Afr., with terminal spikes which disarticulate at maturity, the joints falling with the spikelets attached (rachis continuous in cult. species). (The ancient Latin name.)

188. L. temulentum.

*a.*Lateral spikelets distinctly pedicelled, usually imperfect or rudimentary. . . *b.*
*b.*Glumes of perfect spikelet 3–6 cm. long, bristleform. . . . . 1. *H. jubatum.*
*b.*Glumes 0.8–2 cm. long.
Glumes bristleform; plants perennial. . . . . 2. *H. brachyantherum.*
Glumes (or some of them) lanceolate; plants annual. . . . . 3. *H. pusillum.*
*a.*Lateral spikelets sessile or nearly so, perfect; perennial. . . . . 4. *H. montanense.*

1. H. jubàtum L. (with a mane), Squirrel-tail Grass, Queue d'écureuil (Que.). — Annual or biennial, 3–7 dm. high, erect or geniculate at base; leaves 5 mm. wide or less, scabrous; *spike nodding,* 5–12 *cm. long, about as wide,* pale green to purple; lateral pair of spikelets each reduced to 1–3 spreading awns; *glumes* of perfect spikelets awn-like, 3–6 *cm. long,* spreading; lemmas 6–8 mm. long, with an awn as long as the glumes; all the awns very slender, scabrous. — Coast and open ground, Lab. to Alaska, s. to Nfld., N.S., N.E., L.I., Md., W. Va., Ky., Ill., Mo., Tex. and Mex. June–Aug. Fig. 189. — Often a troublesome weed. (S.Am.;

Siberia; adv. in Eu.) — A hybrid of this and *Agropyron trachycaulum* is thought by some to be the so-called *"Elymus Macounii"*.

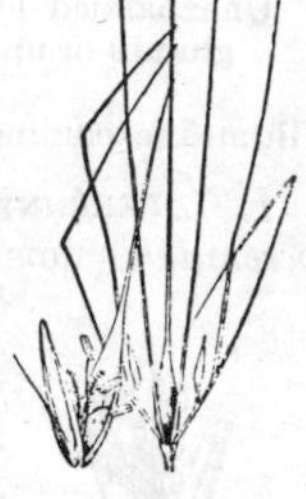

189. H. jubatum.

2. H. brachyantherum Nevski (with short anthers). — Tufted perennial, 2–8 dm. high; leaves flat, up to 1 cm. broad; *spike slender*, 2–8 cm. long, 0.5–1.5 *cm. thick*, erect or nearly so, bronze or greenish; central spikelet perfect, lanceolate, without the awn about 7 mm. long; lateral one imperfect or rudimentary, pedicelled; *glumes* 0.8–2 *cm. long*, scabrous, *all setaceous*, shorter than the lemma; anthers 1–1.5 m m. long. (*H. boreale* Scribn. & Sm., not Gandoger) — Meadows, shores and bottoms, nw. Nfld. and e. Saguenay Co., Que.; Alaska to n. Calif. and Ida. July–Sept. (Ne. Asia) Fig. 190.

190. H. brachyantherum.

3. H. pusillum Nutt. (very small), Little B. — *Annual*, 1–4 dm. high; *leaves* 6 cm. or less long, *erect*, scabrous; *spikes erect*, 2–7 *cm. long*, 1–1.5 *cm. wide;* lateral pair of spikelets abortive; *first glume of each and both glumes of fertile spikelet dilated above the base*, attenuate into a slender awn 8–15 mm. long, equaling the awned lemma. — Plains, roadsides and borders of marshes, often in alkaline soil, Del. to Wash., s. to Fla., La., Tex. and N.M.; occasionally adv. northeastw. May, June.

4. H. montanense Scribn. (of Montana). — *Perennial*, erect or geniculate at base, 6–10 dm. high; leaves 1.2–2 dm. long, 5–8 mm. wide, long-acuminate, scabrous; spikes nodding, 8–17 cm. long, 2–3 cm. wide; *the lateral pair of spikelets nearly sessile, perfect; the middle spikelet 2-flowered or often with the rudiment of a third floret; glumes* 2.3–3.5 *cm. long, subulate-attenuate into slender awns.* (*H. Pammeli* Scribn. & Ball) — Prairies, Ill., Ia., S.D., Mont. and Wyo.; adv. in Que. June–Aug. — Intermediate between *Hordeum* and *Elymus;* closely related to cult. barley.

Cultivated Barley, H. vulgàre L. (common), an annual with continuous rachis and large and mostly perfect lateral spikelets, is spontaneous in waste land; as are Wheat, Tríticum (the classical name) aestìvum L. (of summer), annual, with continuous rachis and plump spikelets solitary and sessile, the glumes ovate and 3-nerved; and Rye, Secàle (ancient name) cereàle L. (cereal), annual, with solitary and sessile spikelets and continuous rachis, but subulate 1-nerved glumes. (All introd. from Eurasia)

32. SITÀNION Raf. Squirrel-tail

Spikelets 2–few-flowered, with the uppermost floret aborted or reduced, 2 at each node of the disintegrating rachis, the bases of the rachis-joints continued as a stipe-like process below the persistent spikelets just above; glumes bristle-like, prolonged into long and short awns, 1–3-nerved; lemmas obscurely 5-nerved, subterete, firm, 2-toothed at apex, the midrib and sometimes the lateral nerves extended into awns; palea firm, about equaling the body of the lemma, with 2 serrulate keels. — Tufted perennials of w. N.Am., with excessively bristly spikes. (Name from the Greek *sitanias*, an adjective applied to a kind of wheat.)

1. S. Hýstrix (Nutt.) J. G. Sm. (hedgehog). — Culms stiffish, 2–5 dm. high, with crowded basal sheaths and long spreading upper leaves; partially included loose long-awned disarticulating spikes about 1 dm. long; the glumes divided to the base into 2 long divergent awns (6–8 cm. long). (*S. longifolium* J. G. Sm.) — Dry open woods, plains and slopes, S.D. to B.C., s. to w. Mo., Tex. and Mex. June–Aug. Fig. 191.

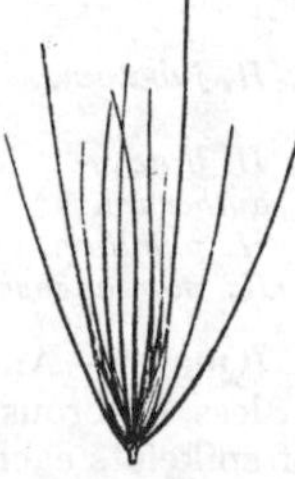

191. S. Hystrix.

33. ÉLYMUS L. Wild Rye, Lyme-Grass

Spikelets 2 (rarely 1)–6-flowered (uppermost florets imperfect), in pairs (sometimes solitary below, rarely in 3's or 4's), sessile at the alternate notches of the continuous rachis; rachilla articulated above the glumes and between the florets; glumes equal, rigid, narrow, 1–3-nerved, acute or awn-pointed, placed edge to edge in front or toward the sides of the florets (which are dorsiventral to the rachis of the spike), simulating an involucre at each joint of the rachis; lemmas convex, obscurely 5-nerved, obtuse, acute or awned from the apex; paleas a little shorter than their lemmas; grain hairy at the summit, adherent to the lemma and palea. — Erect tufted perennials of cool and temp. N. and S.Am., Eurasia and n.Afr., with flat leaves and closely flowered terminal spikes. (Greek name for a kind of grain, from *elyo, rolled up*, from the grain being tightly embraced by lemma and palea.)*

*a.*Extensively creeping, with cord-like rootstock; coarse culms arising from among hard marcescent foliage of preceding year; glumes and lemmas subherbaceous to coriaceous, usually awnless; glumes broadly lanceolate. § PSAMMELYMUS. . . . . 1. *E. arenarius.*

*a.*Tufted, without extensively creeping rootstock; culms slender; glumes and lemmas firm to rigid, at least the lemmas mostly awned; glumes narrowly lanceolate to bristleform. § CLINELYMUS. . *b.*

*b.*Awns straight at maturity (except for arching base); paleas 5–8 (rarely –9) mm. long. . . *c.*

*c.*Glumes flat, at least above the base, 0.8–2 mm. wide. . . *d.*

*d.*Glumes strongly indurated and more or less arched at base, 1–4 cm. long. . . . . 2. *E. virginicus.*

*d.*Glumes not indurated at base, straight, 0.8–1.5 cm. long. . . . . . 3. *E. glaucus.*

*c.*Glumes setiform or linear-setiform, 0.1–0.8 mm. thick. . *e.*

*e.*Lemmas awnless or merely mucronate; stout western species 1–3 m. high, with dense erect spike 1.5–3 dm. long. . . . . . 4. *E. condensatus.*

*e.*Lemmas with long straight awns; slender plants rarely 1 m. high, with spikes 0.5–1.5 dm. long (if plant taller the spike open, with long internodes). . *f.*

*f.*Spike 1–2.5 dm. long; internodes of rachis 3–8 mm. long; spikelets 2–4-flowered; palea 7.5–8 mm. long. . . . . . 5. *E. riparius.*

*f.*Spike 0.5–1.5 dm. long; internodes 1–4 mm. long; spikelets 1–2-flowered; palea 5–6.7 mm. long.

Spikelets and awns obliquely ascending; summit of leaf-sheath with a broad horizontal flange. . . . . 6. *E. villosus.*

Spikelets and awns appressed; summit of sheath narrow. . . . 7. *E. Macounii.*

*b.*Awns outwardly arching at maturity; paleas 8–15 mm. long. . . *g.*

*g.*Glumes setiform, 2–15 mm. long; spikelets 2–3-flowered. . . . . 8. *E. interruptus.*

*g.*Glumes linear-setiform to lanceolate, 1.5–3.5 cm. long; spikelets (2–) 3–7-flowered.

Leaves thin and flat, usually villous above; those of the flowering culms 10–18; palea 9–15 mm. long. . . . . 9. *E. Wiegandii.*

Leaves thick and hard, becoming involute at tip, glabrous or merely scabrous above; those of the flowering culms 4–9; palea 8.5–11 mm. long. . . . . 10. *E. canadensis.*

§ PSAMMÉLYMUS Hack.

1. E. arenàrius L. (of sand), SEA LYME-GRASS, STRAND-WHEAT, SEIGLE DE MER (Que.). — *Coarse* and *very glaucous, from extensively creeping stout rootstocks; culms arising from among crowded marcescent old leaves,* 0.3–1.5 m. high; leaves firm, the lower elongate, the upper shorter, flat or involute, 0.5–1.5 cm. wide; *spike* stiff, dense, 1–3 *dm. long,* 1–3 *cm. in diameter; spikelets usually in pairs,* 3–7-flowered, 2–3 cm. long; *glumes lanceolate,* usually ciliate on the keel or pubescent on the back, acuminate or mucronate, *nearly equaling to exceeding the florets; lemmas awnless,* villous; anthers 4.5–6.5 mm. long. — Typical *E. arenarius,* with culm glabrous at summit, rachis glabrous except for ciliate angles, and firm glumes 1–3-nerved, is Eurasian. Ours is

Var. **villòsus** Mey. (softly hairy). — *Summit of culm and rachis densely pubescent; glumes pliant, mostly pilose on the back,* 3–5-*nerved.* (*E. mollis* Trin.) — Beaches and sands, Greenl. and Arct. N.Am., s. to coast of Mass., shores of L. Michigan and L. Superior, and coast of Calif. June, July. (E.Asia) FIG. 192.

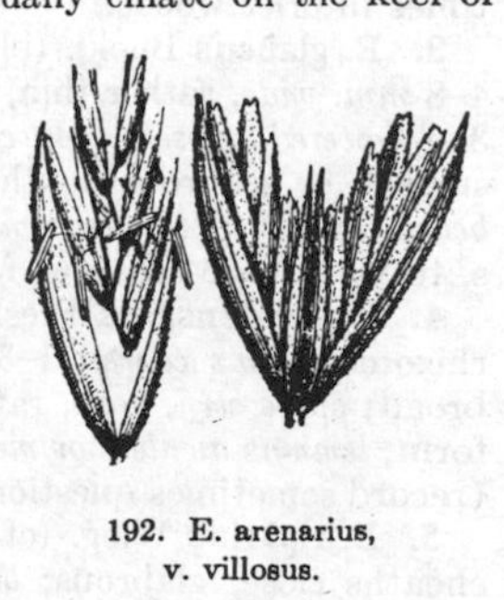

192. E. arenarius, v. villosus.

§ CLINÉLYMUS Hack.

2. E. virgínicus L. (Virginian), TERRELL GRASS. — Green or glaucous; culms few to many in tussocks, 0.3–1.5 m. high; sheath smooth or hairy; blades 3–13 mm. wide, scabrous; *spike stiff, dense,* 0.3–2 dm. long, 1–4 cm. thick, the internodes of the rachis mostly 3–6 mm. long; *spikelets* 2–4-*flowered* (sometimes 1-flowered), *appressed or slightly spreading; glumes strongly indurated and usually unstriated at*

---

* Treatment based largely on that of WIEGAND in Rhodora, xx. 81–90 (1918) and one by FERNALD in Rhodora, xxxv. 187–198 (1933).

the subterete yellowish arching base, flat, green, striate and 0.8–2 *mm. broad above,* 1–4 *cm. long,* pointed to long-awned; lemmas 1–4.5 cm. long (including awns), pointed or with straight awns. — Highly variable; the chief varieties and forms (often treated as species) follow:

*a.*Glumes 1–2.7 cm. long; lemmas 1–3 cm. long. . . *b.*
*b.*Glumes and lemmas awned. . . *c.*
*c.*Base of spike (0.4-2 dm. long) included in or barely exserted from the inflated sheath; glumes mostly 1.5–2 mm. wide; leaves flat, 5-13 mm. wide.
Lemmas and glumes glabrous or merely scabrous-ciliate. . . . . . *E. virginicus* (typical).
Lemmas and glumes villous-hirsute. . . . . Forma *hirsutiglumis.*
*c.*Spike 3–12 cm. long, well exserted (mostly 3–15 cm.) above the less inflated sheath; glumes mostly 0.8–1.6 mm. wide; leaves flat to involute. . . *d.*
*d.*Leaves (and nodes) of flowering culm 6–8; blades mostly flat, up to 13 mm. wide; culms 0.7–1.3 m. high; spike clear green; indurated bases of glumes green or stramineous. . . . . Var. *jejunus.*
*d.*Leaves (and nodes) of flowering culm 4–6; blades firm, often involute, 3–8 mm. wide; culms 3–8 dm. high; spikes whitish-green; bases of glumes white or very pale.
Lemmas and glumes glabrous or merely scabrous-ciliate. . . . . Var. *halophilus.*
Lemmas and glumes hirsute. . . . . Forma *lasiolepis.*
*b.*Glumes and lemmas muticous or merely subulate-tipped. . . . . . . Var. *submuticus.*
*a.*Glumes 2.7–4 cm. long; lemmas mostly 3.5–4.5 cm. long; spikes exserted, 0.4–2 dm. long.
Glumes and lemmas glabrous or merely ciliate-scabrous. . . . . . . Var. *glabriflorus.*
Glumes and lemmas hirsute. . . . . Forma *australis.*

E. virgínicus (typical). — Rich thickets and shores, w. Nfld. to Alta., s. to N.C., W.Va., La. and Tex. Late June–Aug. FIG. 193.—Forma **hirsutiglùmis** (Scribn.) Fern. (with hairy glumes), N.S. to Neb., s. to Va., W.Va. and Tex.

Var. **jejùnus** (Ramaley) Bush (meagre). — Rich thickets, s. Que. and Me. to N.D., s. to Va., La. and Tex. July, Aug.

Var. **halóphilus** (Bickn.) Wieg. (loving salt). *E. halophilus* Bickn. — Sea-coast, N.S. to Va.; reported as local in alkaline soil, Minn. Late July–Oct. — Forma **lasiólepis** Fern. (hairy-scaled), N.S.

Var. **submùticus** Hook. (somewhat blunt). — Rich thickets and alluvium, Gaspé Co., Que.; e. Mass. and R.I.; w. Ont. to Wash., s. to Ky., Mo. and Okla. July, Aug.

193. E. virginicus.

Var. **glabriflòrus** (Vasey) Bush (smooth-flowered). — Rich thickets and alluvium, N.E. to Neb., s. to Fla., Tex. and N.M. Late June–Aug. — Forma **austràlis** (Scribn. & Ball) Fern. (southern), *E. australis* Scribn. & Ball, similar range, sometimes in drier woods.

3. E. glaùcus Buckl. (blue-green). — Glabrous; culms 5–10 dm. high; *leaves* 1.5–2 *dm. long,* 4–8 *mm. wide,* rather thin, flat, scabrous; *spikes slender, the internodes* 8–10 *mm. long; spikelets* 3–6-*flowered, closely appressed; glumes* linear-lanceolate, *not indurated at base,* 3–5-nerved, smooth or scabrous on the nerves, *short-awned, shorter than the nearly smooth lemma which bears an awn twice its own length.* — Moist or dry open thickets and shores, Ont. to Alaska, s. to w. N.Y., Mich., sw. Mo., N.M. and Calif. July, Aug.

4. E. condensàtus Presl (condensed), GIANT WILD RYE. — Tufted, with stout and short rhizome; *culms coarse,* 1–3 *m. high;* leaves and sheaths harsh, the firm pale blades 1–2 cm. broad; *spike stiff, erect,* rather compound, *with spikelets crowded,* 1.5–3 *dm. long*; glumes setiform; *lemmas awnless or merely mucronate.* — Dry plains and slopes, B.C. to Calif., e. to Minn. (record sometimes questioned), Colo. and N.M. July.

5. E. ripàrius Wieg. (of river-banks). — Rather *coarse,* green or glaucous, 0.7–1.8 *m. high;* sheaths close, glabrous; *blades* thin, glabrous, elongate, mostly 1–2 *cm. broad; spike* much exserted, slightly arching, 1–2.5 *dm. long,* 2–4.5 cm. thick; *internodes of rachis mostly* 3–8 *mm. long; spikelets* 2–4-*flowered,* somewhat spreading; *glumes bristle-shaped,* very slender (0.4–0.8 *mm. in diameter*), *indurated toward* the terete unstriated *straight base,* 1.8–3 cm. long; lemmas 2.2–4.5 cm. long, hispidulous, with long straight awns; *palea* 7.5–8 *mm. long.* — Rich thickets and borders of streams, centr. Me. and s. Que. to Wisc., s. to n. Fla., Ky. and Ark. July–Sept.

6. E. villòsus Muhl. (softly hairy). — *Slender,* 5–10 dm. high; *sheaths* usually villous, *their summits expanded into a broad chartaceous horizontal flange;* blades thin, 6–10 mm. wide, commonly villous on the upper surfaces; *spike* long-exserted, 0.5–1.5 *dm. long,* 2–3.5 cm. thick, often slightly nodding; *internodes of rachis* 1.5–3 *mm. long; spikelets* 1 (rarely 2)-*flowered, oblique; glumes bristle-form,* hispid or hirsute, 1.4–3 cm. long, *terete at base, straight; lemmas villous,* with the long *straight awns* 2.3–5 cm. long, or lemmas and glumes glabrous or only scabrous in forma **arkansànus** (Scribn. & Ball) Fern. (of Arkansas); *palea* 5.2–6.7 *mm. long.* (*E. striatus* sensu Hitchc., not Willd.) — Rich (often rocky) woods, thickets and shores, sw. Que. to Wyo., s. to Mass., Ct., N.C., Ala., Okla., Tex. and N.M. June–Aug. FIG. 194.

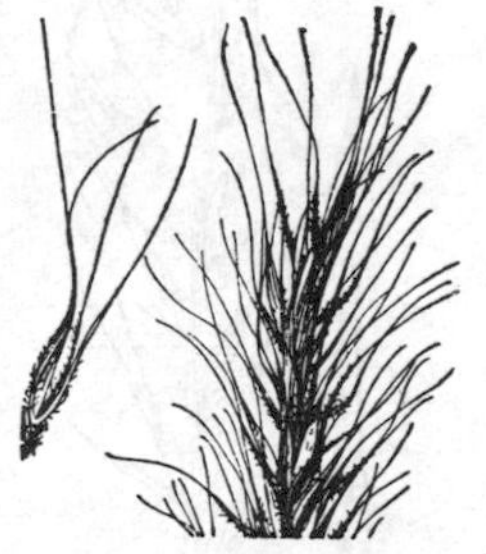

194. E. villosus.

7. E. Macoùnii Vasey (for its discoverer, JOHN MACOUN, 1831–1920). — Slender, 3–10 dm. high; sheaths glabrous or the lower sparsely pilose; *blades* 8–16 *cm. long,* 4 *mm. wide or less,* erect, often involute in drying, scabrous, the lower usually pilose on the upper surface; spikes dense, 4–10 cm. long, 4–8 mm. thick; *spikelets* 1–3-*flowered, the lower solitary and often apparently with* 3 *glumes, the missing spikelet being reduced to a single glume;* glumes linear-setiform, 3-nerved, scabrous, tapering into an awn; **lemmas** 8–10 mm. long, scabrous above, with a slender awn 6–10 mm. long. — Prairies, Minn. to Alaska., s. to Ia., Kans., N.M. and Calif. June–Aug. — Now often considered a hybrid of *Agropyron trachycaulum* and *Hordeum jubatum.*

8. E. interrúptus Buckl. (interrupted). — Culms stout, 9–12 dm. high; *leaves* lax, 1.5–2.5 dm. long, 6–12 mm. wide, scabrous, *somewhat villous above,* setaceous-pointed; spike loose below, 1–2 dm. long; spikelets 2–3-flowered; *glumes subulate, scabrous, varying from a mere point to* 1.5 *cm. long* in the same spike; lemmas hirsute, especially toward the summit, with a divergent awn 2–4 cm. long; palea 8.5–9 mm. long. (*E. diversiglumis* Scribn. & Ball) — Thickets and open woods, Wisc. to N.D. and Wyo., s. to Tenn., Okla., Tex. and Mex. July, Aug. — Sometimes simulated by a hybrid of *E. canadensis* and *Hystrix patula.*

9. E. Wiegándii Fern. (in honor of KARL MCKAY WIEGAND, 1873–1942). — Coarse, 1–2 m. high, more or less glaucous; *cauline leaves* 10–18; the *thin flat blades* 2–3.5 dm. long, 1.3–2.4 *cm. broad, mostly* 60–100-*nerved, usually villous above; spike lax, flexuous,* usually nodding, 1–3.5 dm. long; *rachis often not covered by spikelets,* the middle internodes 5–9 mm. long; spikelets loosely and obliquely ascending, 3–7-flowered; *glumes linear-setiform,* 1.5–3 cm. long; lemmas villous-hirsute, or glabrous or merely scabrous in forma **calvéscens** Fern. (becoming bald), 3.5–4.5 cm. long, the awns outwardly arched in maturity; *palea* 9–15 *mm. long.* — Alluvial soil, Gaspé Co. to L. St. John, Que., s. and sw. to s. N.B., N.E. and Pa. Mid-July–mid-Aug.

10. E. canadénsis L. (of Canada). — Coarse, 0.7–2 m. high, often glaucous; *cauline leaves* 4–9, *thick and hard, becoming involute at tip, glabrous or merely scabrous above, mostly* 20–40 (in very large plants –70)-*nerved,* 5–15 *mm. broad; spikes* usually *dense and stiff,* upright or but slightly nodding, 0.8–2.5 dm. long; *rachis mostly hidden, its internodes* 4–7 *mm. long; spikelets crowded,* appressed to obliquely ascending, 2–5-flowered; *glumes flat* (at least above), 0.5–2 mm. wide, 1.5–3.5 cm. long; lemmas villous-hirsute, or glabrous or merely scabrous in forma **glaucifòlius** (Muhl.) Fern. (with blue-green leaves) = *E. robustus* Scribn. & Sm., 2.8–5 cm. long, the awns outwardly curving when mature; *palea* 8.5–11 *mm. long.* (*E. brachystachys* Scribn. & Ball) — Dry sandy, gravelly or rocky soil, N.B. to s. Alaska, s. to N.C., W.Va., O., Ind., Ill., Mo., Tex., N.M., Ariz. and Calif. July–Oct. FIG. 195.

195. E. canadensis.

34. HÝSTRIX Moench BOTTLE-BRUSH GRASS

Spikelets 2–4-flowered, on very short pedicels, 1–3 together at each joint of the flattened continuous rachis, facing it as in *Elymus,* widely divergent at maturity; glumes reduced to short or minute awns, the first usually obsolete, both often wanting in the upper spikelets; lem-

196. H. patula.

mas convex, rigid, tapering into a long awn; palea strongly 2-keeled; grain pubescent at the summit, free within the lemma and palea. — Perennials of N.Am., Asia and N.Z., with simple culms, flat leaves and loosely flowered spikes. (Name from the Greek *hystrix, a hedgehog,* alluding to the bristly spikes.) ASPERELLA Humb., not Schreb.

1. H. pátula Moench (spreading). — Culms solitary or few, from creeping bases, 0.6–1.5 m. high; leaves divergent, 1–3 dm. long, 0.8–1.5 cm. wide, tapering to both ends, scabrous, or pilose above, sheaths glabrous or sometimes pubescent; spike exserted (rarely partly included), 0.6–2.5 dm. long; *spikelets distant,* at first erect, *soon widely divergent, promptly dropping,* 1–1.5 cm. long excluding the long (1.5–4 cm.) awns; *lemmas glabrous.* (*H. Hystrix* Millsp.; *Asperella Hystrix* Humb.) — Rich or low woods, sw. Me. to N.D., s. to Ga., Tenn., Ark. and Okla. June–Aug. FIG. 196. — Sometimes crosses with *Elymus canadensis.*

Var. **Bigeloviàna** (Fern.) Deam (for JACOB BIGELOW, 1787–1879, who first recorded it). — *Lemmas pubescent.* — More frequent northw., n. N.S. to Wisc., s. to N.E., Pa., O. and Ind.

Tribe IV. AVÈNEAE Nees (see p. 96)

a. Lower floret staminate, with long geniculate awn from the base of lemma; terminal floret perfect, its lemma merely pointed or with short straight awn. . . 35. *Arrhenatherum.*

a. Florets all alike or the uppermost one reduced. . . *b.*

b. Spikelets 2–∞-flowered; rachilla prolonged behind the palea of uppermost floret; perennials, except in *Avena,* the Oat. . . *c.*

c. Falling spikelet carrying the glumes, leaving naked pedicels. . . *d.*

d. Glumes similar, both tapering from base to tip.

Glumes much overtopping florets; uppermost lemma with short claw-like awn; plant velvety. . . . . 36. *Holcus.*

Glumes shorter than column of florets; uppermost lemma with long merely arching awn; plant not velvety. . . . . . . 39. *Trisetum.*

d. Glumes dissimilar, the 2nd upwardly dilated, exceeded by upper floret; lemma awnless or with straightish awn. . . . . . . 37. *Sphenopholis.*

c. Falling spikelet disarticulating above and leaving the glumes as empty husks. . . *e.*

e. Lemmas awnless or short-awned at undivided tip, whitish, lustrous, keeled; rachilla glabrous or nearly so; inflorescence densely spiciform. 38. *Koeleria.*

e. Lemma awned from back or from between apical teeth or, if awnless, the panicle lax and rachilla villous. . . *f.*

f. Awn definitely dorsal, terete; inflorescence an open or spiciform panicle of many spikelets. . . *g.*

g. Awn slender at tip, not jointed. . . *h.*

h. Lemma bidentate at the tapering tip, 5–9-nerved.

Spikelets ascending; glumes rarely equaling column of florets, 1–3-nerved; lemmas keeled, their awns borne from well above the middle; grain free, unfurrowed. . . . . . . 39. *Trisetum.*

Spikelets drooping (rarely ascending); glumes many-nerved, spathiform, longer than and in fruit embracing column of florets; lemma rounded on back, its awn from near or below middle; grain closely adherent to palea and firm lemma, furrowed. . . . . 40. *Avena.*

h. Lemma subtruncate, erose or toothed, 4-nerved, the awn from near or below middle. . . . . 41. *Deschampsia.*

g. Awn blunt and clavate at tip, jointed near the middle, the joint with a circle of stiff hairs. . . . . 42. *Corynephorus.*

f. Awn borne from notch between apical teeth of lemma, flattened; glumes equaling or overtopping densely crowded column of florets; inflorescence a raceme or few-forked panicle of large spikelets. . 44. *Danthonia.*

b. Spikelets 2-flowered; rachilla not prolonged; lemma bidentate, at least the upper dorsally awned; small annuals. . . . . 43. *Aira.*

35. ARRHENATHÈRUM Beauv. OAT-GRASS

Spikelets 2-flowered, the florets approximate, the lower staminate, its lemma bearing a geniculate and twisted awn on the back near the base; upper floret perfect, its lemma short-awned from or near the apex, or awnless; rachilla hairy, prolonged behind the upper palea into a bristle; glumes unequal, acute, thin and scarious; lemmas of firmer texture, 5–7-nerved; palea ciliate on the nerves. — Tall perennials, nat. of Eurasia and n.Afr., with flat leaves and long narrow panicles. (Name from the Greek *arrhen, masculine,* and *ather, awn,* in reference to the awned staminate floret.)

197. A. elatius.

1. A. ELÀTIUS (L.) Mert. & Koch (rather tall), TALL O. — Culms 1 m. or more high, erect; leaves long, linear, 0.5–1 cm. wide, scabrous on both surfaces; panicle purplish, or pale yellowish or greenish in forma FLAVÉSCENS (P. Nielsen) Holmb. (yellowish), and shining, 15–30 cm. long, narrow, the short branches verticillate and usually spikelet-bearing from the base; spikelets 7–8 mm. long;. glumes minutely scabrous, the second about equaling the florets; lemmas scabrous, the awn of the staminate floret about twice the length of its lemma, or in the rare forma BIARISTÀTUM (Peterm.) Holmb. (two-awned) both florets with awns; paleas as long as their lemmas. — Fields and roadsides, Nfld. to B.C., s. to Ga., La., N.M. and Calif.; often cult. June, July. (Natzd. from Eu.) FIG. 197.

Var. BULBÒSUM (Willd.) Spenner (bulbous). — Base bearing moniliform tubers. — Fields and waste places, occasional, e. Mass. to Mich., s. to Ga. and Ala. (Adv. from Eu.)

36. HÓLCUS L.

Spikelets 2-flowered, articulated below the glumes; the lower floret perfect, raised on a curved stipe, awnless; the upper floret staminate (rarely perfect), its lemma bearing a dorsal hooked awn from below the apex; glumes thin, subequal, compressed, navicular, longer than the florets; lemmas somewhat indurated, navicular; paleas thin, nearly as long as the lemmas. — Perennials, nat. of Eu. and Afr., with flat leaves and densely flowered terminal panicles. (A name used by Pliny for a kind of grass.) NOTHOLCUS Nash

198. H. lanatus.

1. H. LANÀTUS L. (woolly), VELVET-GRASS. — *Entire plant grayish, velvety-pubescent, tufted;* culms erect, 3–10 dm. high; leaves 15 cm. long or less, rarely longer, 5–10 mm. wide; panicle purplish, 3–15 cm. long, narrow; spikelets about 4 mm. long, nearly as broad; glumes villous, hirsute on the nerves, the second broader than the first, 3-nerved; lemmas ciliate at the apex; awn of second floret hook-like. — Sterile fields, Nfld. to Ont. and Mich., s. to Ga. and La.; B.C. to Calif. May–July. (Natzd. from Eu.) FIG. 198.

2. H. MÓLLIS L. (soft). — Similar, but *from slender rhizomes,* and with *glabrous culms,* larger spikelets and glabrous glumes. — Locally in Lewis Co., N.Y. and on waste ground elsewhere. (Adv. from Eu.)

37. SPHENÓPHOLIS Scribn.

Spikelets 2–3 (rarely 1)-flowered, the pedicels jointed just below the glumes; rachilla prolonged behind the upper palea as a slender pedicel, articulated between the florets, the glumes and lower floret with joint of pedicel tardily falling together; glumes subequal in length, usually exceeded by the uppermost floret, the first narrow; the second much broader, usually obovate, becoming subcoriaceous in fruit, 3-nerved; lemma chartaceous, its nerves obscure, awnless or rarely with a straight awn below the summit; palea hyaline, narrowed toward the base; grain inclosed within the rigid lemma, free. — Slender N.Am. perennials with narrow terminal panicles. (Name from the Greek *sphen, a wedge,* and *pholis, scale,* referring to the broadly obovate or cuneate second glume.)

*a.*Second glume rounded and somewhat cucullate at summit, with firm chartaceous margins; spikelets 2.5–3 mm. long; inflorescence dense. . . . . 1. *S. obtusata.*

*a.*Second glume rounded at summit or subacute- to acuminate-tipped, not cucullate, with thinner margins; spikelets 3–4.2 mm. long; inflorescences rather lax. . . *b.*

*b.*Lemmas awnless or rarely with awns only 0.2 mm. long. . . *c.*

*c.*Second glume (unfolded) broadly obovate, blunt or abruptly short-tipped; at least the upper lemma with scabrous surface; anthers 1–2 mm. long.

Leaves membranaceous, dark green, flat, often pubescent; 1st glume (folded) linear-lanceolate. . . . . 2. *S. nitida.*

Leaves firm, paler, involute or, if flat, soon inrolling, glabrous; 1st glume slenderly linear. . . . . 3. *S. filiformis.*

*c.*Second glume (unfolded) narrowly obovate to oblanceolate, acute; 1st glume linear-attenuate; lemmas with glabrous surfaces; anthers about 0.5 mm. long. . . . . 4. *S. intermedia.*
*b.*Lemmas (at least the upper) with elongate geniculate awn 1.2–3 mm. long. 5. *S. pallens.*

1. S. obtusàta (Michx.) Scribn. (blunt). — Slender to rather stout, 3–10 dm. high; sheaths pubescent to glabrous; *blades* 4–25 cm. long, *the uppermost one-half to two-thirds as long as its sheath; panicle* rather dense, lance-cylindric, 0.3–2 dm. long, 0.5–2.5 cm. thick, continuous or interrupted, *with spikelets densely crowded on the short appressed branches;* spikelets 2.5–3 mm. long; *glumes* subequal; *the first linear-oblong and rounded at tip; second obovate, rounded-subtruncate and subcucullate at tip, the broad chartaceous margin smooth and shining;* lemmas similar or the second a little scabrous; *anthers* 0.5–0.8 *mm. long.* — Three somewhat definite vars.:

Panicle 0.7–2 dm. long, up to 3 cm. thick, its branches irregularly elongate and not strongly appressed and rounded at summit.
Leaves and their sheaths glabrous or merely scabrous. . . . . . . *S. obtusata* (typical).
Leaves or their sheaths pilose. . . . . Var. *pubescens.*
Panicle 0.3–1(–1.3) dm. long, 0.5–1.5 cm. thick, its tightly appressed branches essentially uniform and strongly rounded; sheaths puberulent. . . . . . Var. *lobata.*

S. obtusàta (typical). — Dry to wet soils, borders of woods, shores, etc., Fla. to Tex., n. to s. Me., s. N.H., N.Y., s. Ont., Mich., Minn. and Neb. May–July. (Bermuda) Fig. 199.

Var. **pubéscens** (Scribn. & Merr.) Scribn. (hairy). — Fla. to La. and Okla., n. to Ct., N.J., s. Ind. and Mo.

Var. **lobàta** (Trin.) Scribn. (lobed). — Fla. to Ariz., n. Mex. and Calif., n. to s. Me., Mass., N.Y., O., Wisc., Minn., N.D., Alta. and B.C. — Spikelets usually pale; those with purple coloring conspicuous are forma **purpuráscens** (Vasey) Waterfall (purplish).

199. S. obtusata.

200. S. nitida.

2. S. nítida (Biehler) Scribn. (shining). — Slender, 3–10 dm. high; sheaths pubescent or glabrous; *leaf-blades dark green,* flat; the upper 1–6 cm. long, one-third to one-fifth as long as sheath; *panicle lax,* 0.5–2 dm. long, with branches loosely spreading in anthesis, later erect, the young flowering panicle up to 5 cm. broad, the mature only 0.5–1 cm. thick; *spikelets* about 3 mm. long, *cuneiform; glumes* subequal, the *first linear-lanceolate* (*folded*), the *second obovate* and *rounded at tip or abruptly pointed; lemmas* oblong, *obtuse, the upper scabrous;* anthers 1–2 mm. long. — Rich woods and rocky slopes, Fla. to Tex., n. to e. Mass., Vt., N.Y., s. Ont., Mich., Ill. and Mo. April–early July. Fig. 200.

3. S. filifórmis (Chapm.) Scribn. (thread-like). — Very slender, 3–8 dm. high; *leaves firm, pale, involute or, if flat, promptly inrolling, glabrous;* panicle much as in no. 2, in anthesis open and lax, in fruit with appressed branches, 0.5–1.5 dm. long; spikelets 3–4 mm. long; *first glume slenderly linear;* upper lemma scabrous. — Dry pine-barrens, woods and clearings, Fla. to Tex., n. to Va. and Tenn. Late April, May.

4. S. intermèdia Rydb. (intermediate). — Slender to rather coarse, 2–10 dm. high; sheaths pubescent or glabrous; *upper blade one-half to quite as long as its sheath,* scabrous on the nerves; panicle 0.3–2 dm. long, rather lax, often nodding, linear-cylindric to lance-oblong; spikelets 3–4.2 mm. long, oblong-lanceolate; *glumes scabrous on the keels,* the *first linear-attenuate;* the *second broadly oblanceolate, acute or subacute; lemmas lanceolate, acute or subacute,* rarely short-awned below the apex, *glabrous* except on the keel near the apex; *anthers* 0.5 *mm. long.* (*S. pallens* of ed. 7, not *Aira pallens* Biehler, basonym) — Meadows, prairies, shores and damp slopes, Nfld. to s. Alaska, s. to Fla., La., e. Kans., Colo. and Ariz. June–Aug. Fig. 201.

201. S. intermedia.

5. S. pállens (Biehler) Scribn. (pale). — Resembling no. 4, 6–9 dm. high; leaves dark green, flat, with glabrous to scabrous sheaths and flaccid smooth blades 2–5 mm. broad; panicle open and lax, flexuous or pendulous, 1.5–2.5 dm. long; *spikelets 3.5–4.5 mm. long; first glume linear-attenuate, with wide scarious margin;* second glume oblong-obovate, broadly scarious-margined; *upper lemma scabrous bearing a geniculate awn* 1.2–3 *mm. long.* — Spring-heads and wet woods, very local, (?) Pa. or (?) O. and se. Va. to S.C. Late May, June.

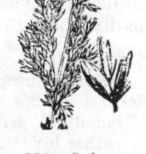

38. KOELÈRIA Pers.

Spikelets 2–4-flowered; rachilla prolonged into a naked pedicel or sterile floret behind the upper palea; glumes unequal, slightly shorter than the florets, membranaceous, acute, the first 1-nerved, the second 3–5-nerved; lemma chartaceous-membranaceous, scarious-margined, faintly 3–5-nerved, acute or mucronate; palea hyaline; grain loosely inclosed within the subrigid lemma, free. — Tufted perennials of temp. reg. with narrow leaves and densely flowered terminal spike-like panicles. (Named for *Georg Ludwig Koeler*, 1765–1807, professor at Mainz and student of grasses.)

1. K. cristàta (L.) Pers. (crested). — Culms erect, 3–6 dm. high, leafy at base; sheaths (at least the lower) retrorsely pubescent; blades flat or becoming involute; panicle cylindrical, 4–15 cm. long, often interrupted at base, pale and shining; spikelets 4–5 mm. long; the glumes and lemmas scabrous. — Dry prairies, sands or open woods, Que. to B.C., s. to Del., O., Ind., Mo., Tex. and Mex.; locally introd. in N.E. (Eurasia) Fig. 202. — Very variable; by Domin much segregated. The American plants need closer study.

202. K. cristata.

39. TRISÈTUM L.

Spikelets 2 or 3 (rarely –12)-flowered; terminal flower rudimentary, often a mere bristle; glumes unequal, awnless, first 1-nerved, second 1–3-nerved and shorter to longer than the spikelet; lemma keeled, 5-nerved, usually chartaceous to scarious, 2-toothed at tapering apex, in ours with a slender dorsal awn or (in the last species) awnless or with a short straight subterminal awn; palea shorter, 2-toothed; stamens 3; grain completely inclosed by but free from lemma and palea. — Tufted perennials of cool or temp. reg., with spike-like or narrow terminal panicles. (Name from *tres, three*, and *seta, a bristle*, from the awned and 2-toothed lemma.)

Panicle spike-like, dense or more or less interrupted. . . . . 1. *T. spicatum.*
Panicle loose and open, with elongate branches.
  Awn of lemma bent (at least when dry), often twisted at base, equaling to exceeding the glumes.
    Upper glume lance-attenuate; lemmas smooth; anthers 2.5–3 mm. long. . 2. *T. flavescens.*
    Upper glume oblanceolate or narrowly obovate; lemmas papillose-scabrous; anthers about 1 mm. long. . . . . 3. *T. pensylvanicum.*
  Awn of lemma wanting or, when rarely present, very short, straight and borne just below the apex. . . . . 4. *T. melicoides.*

1. T. spicàtum (L.) Richter (spiked). — Culms tufted or solitary, erect; sheaths and blades and often the culms soft-pubescent, blades narrow; *panicle* shining, *spike-like*, linear-cylindric to ellipsoid, 1–12 cm. long, dense or more or less interrupted below; 2nd glume broader than the 1st, 3-nerved; lemma with a long (4–5 mm.) divergent awn borne about one-third the way below the acuminate-toothed apex. — A highly variable species of cool reg. Typical *T. spicatum*, with very dense almost capitate bronze panicle 1.5–3 cm. long and 1–2 cm. thick and with lemmas only 3–4 mm. long, of Eurasia, occurs in Arct. Am. Represented with us by

Var. **Maidénii** (Gandoger) Fern. (in honor of Joseph Henry Maiden, 1859–1925). — 1.5–5 dm. high; *panicle* 2–6.5 *cm. long, green, bronze or purple,* very dense except at base; *glumes glabrous* or merely scabrous on the keel, *acute to short-mucronate, the second* 4–5 *mm. long.* — Arct. reg., s. to rocky places, Nfld., mts. and cliffs, e. Que., and B.C. Late June–early Aug. (Eu.) Fig. 203.

203. T. spicatum, v. Maidenii.

Var. **mólle** (Michx.) Beal (soft). — 2–8 dm. high; *panicle* 2.5–12 *cm. long, silvery-green,* becoming whitish-brown, *interrupted; glumes glabrous,* more *attenuate-aristate, the second* 4.5–6.5 *mm. long.* — Ledges and cool shores, Lab. to Yuk. and B.C., s. to Nfld., N.S., N.E., Pa., mts. of N.C., Mich., Wisc., Minn., Colo. and Calif. June–Aug. Fig. 204.

Var. **pilosiglùme** Fern. (with hairy glumes). — 0.3–5 dm. high; *panicle* 1–7 *cm. long, rather dense,* green or bronze; *glumes pilose, the second* 2.5–4.5 *mm. long.* — Exposed rocky places, Lab. to L. Superior, Minn., s. to Nfld., e. Que., C.B., alpine regions of n. N.E., headlands of L. Champlain, Vt. June–Aug.

204. T. spicatum, v. molle.

2. T. FLAVÉSCENS (L.) Beauv. (yellowish), YELLOW OATS. — Tall, up to 8 dm. high; leaves flat, 2–10 mm. broad, sheaths glabrous or somewhat pilose; *panicle loose, lanceolate to ovoid, yellowish,* 0.5–2 dm. long, with obliquely ascending branches; *spikelets* 5–6 mm. long, 2–3-flowered, *in maturity disarticulating below the glumes from the pedicels; glumes* unequal, *acuminate or attenuate, the upper lanceolate; lemmas smooth or slightly scabrous,* with the long bent awn twisted at base; *anthers* 2.5–3 *mm. long.* — Fields and roadsides, occasional, N.E. to Wash., s. to Miss., Mo. and Kans. June, July. (Introd. and natzd. from Eu.)

3. T. pensylvánicum (L.) Beauv. (of Pennsylvania), SWAMP-OATS. — Culms weak, 3–10 dm. high; sheaths and leaves glabrous or lower sheaths pubescent; blades 0.4–1.8 dm. long, 2–8 mm. broad, scabrous; *panicle lax, pale, lanceolate to ovate,* 0.6–2.5 dm. long; *spikelets* 4–7 mm. long, *in maturity not disarticulating from the pedicels; glumes* acute, subequal; *the second* broader, *oblanceolate or narrowly obovate; lemmas* lanceolate, *papillose-scabrous;* the *first* acute or acuminate, *awnless or rarely awned,* the *second with a long* (4–5 *mm.*) *divergent awn* below the acute or 2-toothed apex; *anthers about* 1 *mm. long.* (*Sphenopholis palustris* Scribn.) — Springy meadows and wooded swamps, local, Mass. to O., s. to Fla., Ala. and La. May–July. FIG. 205.

206. T. melicoides.

205. T. pensylvanicum.

4. T. melicoìdes (Michx.) Vasey (resembling *Melica*). — Slender, *glabrous,* with solitary or slightly tufted culms 0.2–1 m. high; leaves long, 0.2–1 cm. wide; *panicle loose,* flexuous, nodding, *silvery-green,* becoming whitish-brown, 0.4–3 dm. long, *the axis and branchlets scabrous; spikelets* 2–4-flowered, 5–9 *mm. long;* glumes unequal, glabrous, the second nearly equaling the florets; *lemmas* smooth or scabrous, *oblong,* obtuse to acute, or slightly 2-toothed at apex, *awnless or rarely with a very short straight rudimentary awn.* — Ledgy or gravelly shores or cool banks, chiefly in calcareous areas, n. and e. Nfld. to n. Mich., s. to s. N.B., centr. and nw. Me., n. Vt. and n. N.Y. Late June–Aug. FIG. 206.

Var. **màjus** (Gray) Hitchc. (larger). — *Some or all the sheaths pilose.* — Similar habitats, usually in different areas, w. Nfld. to w. Ont., s. to n. N.B. and n. Me., n. Vt., n. Mich. and Wisc.

40. AVÈNA L. OAT. AVOINE (Que.)

Spikelets 2–6-flowered; rachilla bearded below the florets; glumes subequal, membranaceous, many-nerved, longer than the lemmas, usually exceeding the uppermost floret; lemmas indurated except toward the summit, 5–9-nerved, bidentate at the apex, bearing a long dorsal twisted awn (the awn straight or wanting in cult. forms); grain pubescent at least at the summit, often adhering to the lemma and palea, furrowed. — Annuals or perennials of temp. reg., with terminal panicles of large spikelets. (The classical Latin name.)

207. A. fatua.

1. A. FÁTUA L. (foolish, useless). — Culms 4–12 dm. high, in small tufts, erect, stout; blades long, 5–8 mm. wide; *panicle loose and open,* the slender branches ascending; *spikelets pendulous,* 2.2–2.5 *cm. long,* excluding the awns; *glumes smooth, striate,* acuminate; *florets* approximate, mostly 3, *promptly falling from the glumes; lemmas* with a ring of hairs at base and more or less appressed-pubescent with *long stiff brownish hairs,* or in forma GLABRÀTA Peterm. (without hair) glabrous; awn inserted about the middle, bent and twisted, 3 cm. long or more. — Fields and waste places, Nfld. to B.C., s. to N.E., Pa., Tenn., Mo., N.M. and Mex. July–Oct. (Natzd. from Eu.) FIG. 207.

2. A. SATÌVA L. (cultivated), the cult. OAT, with loosely subsymmetrical panicle, the 2 florets awnless or with nearly straight awns and tardily falling from the glumes, occurs commonly in waste places but is not long-persistent. (Introd. from Eurasia)

Var. ORIENTÀLIS (Schreb.) Richter (oriental), with very long 1-sided panicle, is likewise casual. (Introd. from Eurasia)

3. A. PUBÉSCENS Huds. (hairy). — Perennial; culms 6–9 cm. high, in small tufts, erect, slender; *sheaths and blades,* at least the lower, *retrorsely pubescent;* panicle rather narrow, the slender flexuous branches erect; *spikelets upright,* 1.2–1.3 *cm. long,* excluding the awns; *glumes 3-nerved, the nerves scabrous;* florets approximate, rachilla-joints clothed with long white hairs; *lemmas scabrous, a tuft of white hairs at the base,* a bent and twisted awn inserted about the middle, 2–2.5 mm. long. — Fields and roadsides, local, Que. to Del. May, June. (Adv. from Eu.)

41. DESCHÁMPSIA Beauv. Hairgrass.

Spikelets 2 (rarely 3)-flowered; rachilla hairy, prolonged behind the upper palea as a hairy bristle; glumes subequal, thin or scarious; lemmas thin, 4-nerved (the midnerve becoming an awn), subtruncate, 2–4-toothed, bearing a slender dorsal awn from or below the middle. — Tufted perennials (our species) of cold and temp. areas, with flat or involute leaves and shining spikelets in loose or narrow panicles. (Named for *Jean Louis Auguste Loiseleur-Deslongchamps*, 1774–1849.)

Glumes shorter than to barely overtopping the upper florets; plants strongly cespitose, with basal tufts of hard setaceous to narrowly linear or folded leaves.
Leaves involute-setaceous; florets approximate (rachilla short); lemmas sharply toothed at apex; awn twisted, much exserted. . . . . . . 1. *D. flexuosa.*
Leaves flat or folded, tardily involute; florets distant; lemmas erose-truncate; awn straight, hardly exserted. . . . . 2. *D. caespitosa.*
Glumes much overtopping the florets; plants with few loosely tufted culms; leaves flat, soft, not markedly tufted. . . . . 3. *D. atropurpurea.*

1. D. flexuòsa (L.) Trin. (zigzag), Common H. — Culms erect, 3–8 dm. high, slender, nearly naked above, *the numerous involute-setaceous basal leaves* 5–20 cm. long; *sheaths scabrous; blades setaceous;* panicle 5–12 cm. long, very loose, rather few-flowered, the subcapillary *flexuous branches* spikelet-bearing near the ends; spikelets bronze or purplish, or pale in forma **flavéscens** Sylvén (yellowish), 4–5 mm. long; *glumes acute; florets approximate; lemmas scabrous, 4-toothed;* awn inserted near the base, 5–7 mm. long, twisted; palea nearly as long as the lemma, scabrous. — Dry open or partially shaded soil, s. Lab. to w. Ont., s. to Nfld., N.S., N.E., Va., w. N.C., Ky., Mich., Wisc. and Minn., and chiefly along the mts. to Ga., Tenn. and Okla. June–Aug. (Eurasia) Fig. 208.

208. D. flexuosa.

Var. **montàna** (L.) Ledeb. (of mountains). — Panicle contracted; spikelets 5–7 mm. long. — Greenl. and Lab. Pen. to Nfld., St. Paul I., N.S., and mts. of e. Que. (Eurasia)

2. D. caespitòsa (L.) Beauv. (bunched), Tufted H. — Culms slender; *basal leaves flat or becoming involute, not setaceous,* 0.3–6 dm. long; *sheaths smooth; blades flat,* scabrous on the upper surface; panicle 0.2–4.5 dm. long, the *scabrous* slender *branches* spikelet-bearing near the ends; spikelets 2–7 mm. long; *glumes acute or blunt; florets distant (rachilla half the length of lower sessile floret); lemmas smooth,* erose-truncate; *awn* from near the base, but little longer than its lemma, *straight, articulated at the base and deciduous;* palea nearly equaling the lemma. — Variable circumboreal species; we have the following vars.:

Culms 0.65–1.7 m. high, 2–6 mm. in diameter at base; leaves flat or only tardily involute; the basal mostly 1.5–6 dm. long, with ligules 5–12 mm. long; lower cauline blades 1.5–4 dm. long; panicles 1.5–4.5 dm. long, diffuse.
Spikelets 3.5–5.5 mm. long. . . . . *D. caespitosa* (typical).
Spikelets 2–3 mm. long. . . . . Var. *parviflora.*
Culms 0.7–7.5 dm. high, 1–2.5 mm. in diameter at base; leaves flat or involute; basal mostly 0.3–3 dm. long, with ligules 2–7 mm. long; lower cauline blades 0.1–0.8 dm. long; panicles 0.2–2.2 dm. long.
Spikelets 3–4.5 mm. long; panicle lax, rather diffuse in anthesis. . . . . Var. *glauca.*
Spikelets 4.5–7 mm. long; panicle commonly contracted in anthesis. . . . Var. *littoralis.*

D. caespitòsa (typical). — Shores and cool banks, Nfld.; Mont. to s. B.C., s. to Colo. and Calif.; locally introd. from Eu. in fields and roadsides, Me. June–Aug. (Eu.)

Var. parviflòra (Thuill.) Richter (small-flowered). — Damp grassland and thickets, local, N.E. (Natzd. from Eu.)

Var. **glaùca** (Hartm.) Lindm. f. (blue-green). — Shores, mts. and damp (often calcareous) soil, Nfld. to Yuk., s. to N.J., Pa., w. Va., W.Va., O., n. Ill., Minn., Ariz. and Calif. Late May–July. (Eu.) Fig. 209.

Var. **littoràlis** (Reut.) Richter (of the seashore). — Damp soils, s. Lab., Nfld. and e. Que. (chiefly alpine); Alaska to mts. of Colo., Utah and Calif. June–Aug. (Eurasia)

209. D. caespitosa, v. glauca.

3. D. atropurpùrea (Wahlenb.) Scheele (dark purple). — Culms erect, 1–5.5 dm. high, slender, leafy; *no tufts of basal leaves;* sheaths smooth; blades flat, soft, 5–10 cm. long, 2–5 mm. wide, nearly glabrous; panicle 4–10 cm. long, with rather few spikelets; the few smooth capillary flexuous branches spreading, sometimes drooping, spikelet-bearing at the ends; spikelets 5–6 mm. long; *glumes acuminate, much overtopping the florets;* lemmas strigose near the summit, erose-truncate and short-ciliate at apex; *awn inserted about the middle, bent,* 3–4 mm. long; palea nearly equaling the lemma. (*Vahlodea* Fries) — Meadows and wet rocks, Lab. to n. Nfld. and mts. of e. Que., n. N.E. and n. N.Y.; Alaska to Colo. and Oreg. July, Aug. (Eurasia) Fig. 210.

210. D. atropurpurea.

42. CORYNÉPHORUS Beauv.

Spikelets 2-flowered, both flowers perfect; rachilla prolonged beyond the florets; glumes awnless, pointed, thin, exceeding the florets; lemmas with awns slightly clavate-thickened at summit; these jointed in the middle and with a circle of stiff hairs at the joint, not exserted. — Tufted perennials nat. of Eu. (Name from the Greek *coryne, club,* and *phoros, bearing;* from the club-shaped awns.)

1. C. canéscens (L.) Beauv. (grayish). — Densely cespitose, 1.5–4 dm. high, glaucous or purplish, with closely tufted convolute-acicular basal leaves; culms slender; panicle dense, narrow, 3–6 cm. long; spikelets 3–4 mm. long. — Sandy fields, roadsides and barrens, Martha's Vineyard I., Mass., Long I., N.Y., and e. N.J. June, July. (Natzd. from Eu.) Fig. 210a.

210a. C. canescens.

43. AÌRA L. Hairgrass

Spikelets 2-flowered, both flowers perfect; glumes thin, somewhat scarious, subequal, acute, awnless, longer than the approximate florets; lemmas bidentate, awned on the rounded back or the lower awnless; palea a little shorter than the lemma; grain included in the slightly indurated lemma and palea and usually adherent to them. — Delicate annuals, nat. of Eurasia. (An ancient Greek name for some grass, perhaps Darnel.) Aspris Adans.

1. A. caryophyllèa L. (like *Caryophyllus* or *Dianthus;* from the tufts of slender leaves). — Culms solitary or few, slender, erect, 8–30 cm. high; blades short, setaceous; panicle open; the silvery shining *spikelets clustered toward the ends of the spreading capillary branches,* 2.5–3 *mm. long,* nearly as broad; *lemma of each floret with a geniculate awn* 3–4 *mm. long from below the middle, the teeth of the apex setaceous.* — Sandy fields and waste places, se. Mass. to Fla. and La.; Lake Co., O.; Pacific coast. May–July. (Natzd. from Eu.) Fig. 211.

2. A. élegans Willd. (elegant). — Similar to the preceding; panicle more diffuse; *spikelets scattered at the ends of the branches,* 1.5–2 *mm. long, mostly shorter than pedicels; lemma of lower floret awnless or with a minute awn just below the apex, the teeth of which are short;* lemma of upper floret bearing a geniculate awn 3 mm. long from below the middle, teeth of apex setaceous. (*A. capillaris* Host, not Savi) — Dry sterile soil, N.J. to Fla. and Tex.; Calif. and Oreg. May, June. (Natzd. from Eu.)

211. A. caryophyllea.

3. A. praècox L. (precocious). — Culms tufted, 0.5–20 cm. high, slender, erect, or lower nodes geniculate; sheaths slightly inflated; blades setaceous; *panicle narrow and dense, the short branches erect,* 1–3 *cm. long; spikelets yellowish,* shining, 3.5–4 *mm. long;* lemmas of both florets bidentate at apex and bearing a geniculate awn 2–4 mm. long from below the middle, the awn of lower floret shorter than that of the upper. — Sandy fields, N.J. and Del. to Va.; B.C. to Calif. May, June. (Natzd. from Eu.)

44. DANTHÒNIA DC. Wild Oat-Grass

Spikelets several-flowered; florets densely crowded into a column, uppermost imperfect or rudimentary; glumes subequal, much longer than the lemmas, usually exceeding the uppermost floret; lemma convex, 2-toothed or bifid at the apex, with a twisted awn between the teeth; awn flat, formed by the extension of the 3 middle nerves of the lemma. — Tufted erect perennials of S. Hemisph., s. Eu. and Asia, and N.Am., with narrow leaves and small terminal panicles or racemes; solitary or few cleistogamous spikelets borne within and at bases of old sheaths. (Named in 1805 for *Etienne Danthoine,* a botanist of Marseilles.)

a. Longer lemmas 2–6.5 mm. long. . . *b.*

b. Culms erect or straight, with stiffly erect panicles, the internodes terete (sometimes compressed or triangular in no. 2); basal crowded leaves much shorter than culm (sometimes elongate in no. 2), the uppermost blade (except sometimes in no. 2) usually becoming remote from the panicle; panicle-branches ascending to erect, tightly appressed in fruit; spiraling base of awn usually dark brown to purplish, strongly contrasting with the paler and straightish summit.

Culms slender, 0.5–1.5 mm. thick (dried) at base, 1–6 dm. high, with terete internodes; panicle remote from upper leaf, mostly with 2–13 spikelets; longer glume 7–11 mm. long, if longer with faint or obscure lateral nerves; base of awn dark-colored. . . . . 1. *D. spicata.*

Culms stout, 1.5–2.5 mm. thick at base, mostly 0.3–1 m. high; the lower internodes terete, or triangular and with one concave side; panicle either remote or closely subtended by upper leaf, dense, usually with 9–20 spikelets; longer glume lance-attenuate, 1.1–2.5 cm. long, prominently 3–7-ribbed; base of awn light to dark brown. . . . . . . 2. *D. Alleni.*

*b.*Culms slightly geniculate at the nodes, the summit usually arching; some of the lower internodes trigonous or compressed and often with the narrower side broadly concave; basal leaves prolonged, one-half as long as to equaling culm, the uppermost nearly reaching to overtopping the panicle; panicle lax and open or with branches finally loosely ascending; the remote lower branches usually strongly divergent, not closely appressed in fruit; spiraling base of awn pale brown or stramineous. . . . . . 3. *D. compressa.*

*a.*Longer lemmas 7–10 mm. long.

Culms 0.5–5 dm. high; panicle dense, 2.5–7.5 cm. long, with appressed-ascending short branches; glumes purple or bronze, spathiform, nearly covering the column of florets; lemmas glabrous except for basal and marginal beard, the terminal short-awned teeth less than one-fourth as long as the blade; boreal species. . . . . 4. *D. intermedia.*

Culms becoming 0.3–1 m. high; panicle lax, 0.4–1.2 dm. long; glumes pale or barely purple-tinged, concealing only the base of the column of florets; terminal long-awned teeth of lemma nearly as long as blade; southeastern species.

Sheaths of leaves pilose; lemma densely silky-villous across back; spiraling base of awn of lemma 2.5–3 mm. long. . . . . 5. *D. sericea.*

Sheaths of leaves glabrous; lemma glabrous except at bearded base and margin; spiraling base of awn of lemma 1–2 mm. long. . . . . . 6. *D. epilis.*

1. D. spicàta (L.) Beauv. (spiked), Poverty-Grass, Junegrass, White Oat-Grass. — Densely cespitose, *brown or drab at base;* the crowded and often curved or twisted glabrous or pilose *basal leaves much shorter than the culms,* involute or flat and slender; *culms terete, erect* or straight, 1–6 dm. high, 0.5–1.5 mm. thick at (dry) base; *cauline leaves* few, the longer and lower ones 3–18 cm. long, *the uppermost* 1–10 cm. long and *becoming remote from the long-stalked panicle;* summit of sheath long-bearded, the truncate *ligule of stiff hairs* 0.4–1.5 *mm. long; panicle stiff, erect, with mostly ascending short branches, these tightly appressed in fruit;* spikelets (1–)2–13; *glumes* slightly unequal or subequal, *the longer* 7–14 *mm. long; lemmas* densely to sparingly pilose or strigose or even glabrous on back, *the larger ones* 2–6.5 *mm. long,* with 2 obliquely triangular-ovate to lance-attenuate awned or awnless teeth, *the twisted base of the central awn dark brown to purplish* and strongly contrasting with the straightish terminal segment. — A polymorphous and very plastic species sadly in need of critical study. The following vars. seem fairly definite, each of them with glabrous or variously pilose foliage:

Column of florets three-fourths to quite as long as the firm glumes; panicle with (1–)2–13(–15) spikelets; lower leaves flat or involute; culms firm, 0.5–1.5 mm. thick at base.

Glumes lance-attenuate, tapering from near base, with 3–5 strong ribs besides the midrib, covering only the base of the column of florets, the sinus at crossing of the glumes one-sixth to one-fourth as high as tip of longest glume; basal marcescent leaves strongly curving and twisting; wide-ranging plant. . . . . *D. spicata* (typical).

Glumes oblong-lanceolate, tapering from near or above middle, with weak or obscure lateral veins, usually covering all but the summit of the column of florets, the sinus at crossing of the glumes usually one-fourth to half as high as tip of longer glume; basal leaves only slightly if at all curved or twisted; northern plant. . . . . Var. *pinetorum.*

Column of florets only one-half to three-fifths as long as the thin and hyaline nerveless or only faintly nerved glumes; spikelets 3–7, scattered; culms delicate, 0.5–1 mm. thick at base; basal leaves filiform-involute, curved; southern. Var. *longipila.*

D. spicàta, typical. — Dry to damp and peaty soil or in thin woodland, Que. to Minn., s. to N.S., N.E., L.I., n. Fla., Ala., Tenn. and Mo. May–July. FIG. 212.

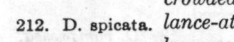

Var. **pinetòrum** Piper (of pine-woods), *D. thermalis* Scribn. — Dry to moist open soil, Nfld. and Côte Nord, Que., to B.C., s. to N.S., N.B., n. N.E., Bruce Pen., Ont., n. Mich., n. Wisc., Minn., Black Hills, S.D., N.M. and Oreg. Late June–Sept.

Var. **longípila** Scribn. & Merr. (long-pilose; a misnomer for some of the plants with glabrous leaves). — Sandy or rocky woods and clearings, N.C. and Ala. to N.M., n. to Ct., Pa., Ky., Mo. and e. Kans. Late May–Sept.

212. D. spicata.

2. D. Álleni Aust. (for its discoverer, TIMOTHY FIELD ALLEN, 1837–1902). — Coarser than nos. 1 and 3; *culms stout,* 1.5–2.5 *mm. thick at base,* mostly 0.3–1 m. high, simple or branching and producing axillary panicles, the lower internodes either terete (*D. Faxoni* Aust.) or trigonous and with 1 side concave; basal leaves stiff and mostly erect, rather broad and often remaining flat, dark green; cauline leaves mostly prolonged, the uppermost either remote from or closely subtending the very dense and compound panicle, the ligule often up to 3 or 4 mm. long; *panicle usually with* 9–20 *spikelets, these densely crowded or glomerate* on the branches, their parts often distorted; *glumes firm, lance-attenuate, strongly* 3–7-*ribbed, the longer* one 1.1–2.5 *cm. long; longer lemmas* 5–6.5 *mm. long,* the twisted base of the central awn pale to dark brown. — Open shores, rocky or arid openings, clearings and recent burns, M.I. to Algoma Distr., Ont., s. to N.S., Me., Mass., Del., mts. to N.C., and O. Late June–Oct. — A perplexing series, probably including hybrids of nos. 1 and 3 and stimulated late-flowering individuals of both, as well as vigorous individuals of disturbed or burned soils; in extreme development very different from nos. 1 and 3 but passing into both; needs close field-observation.

3. D. compréssa Aust. (compressed). — Similar to no. 1; loosely cespitose, *often purplish at base;* the *culms usually geniculate* at base and slightly so *at the nodes* (*the cushions above the nodes often higher on one side*), *usually arching at summit,* slender; *some of the lower and median internodes triangular or compressed, with the narrower side broadly concave; basal leaves mostly flat, prolonged, from one-half to quite as tall as the culm; upper leaf prolonged, usually reaching to exceeding the panicle; panicle lax and open, the* 1 *or* 2 *lower usually forking filiform branches frequently divergent, the branches loosely ascending or spreading in fruit; spiraling base of awn pale brown or stramineous,* not strongly contrasting with the straightish summit. — Woodlands and clearings, s. Que. to O., s. to N.S., N.E., L.I., Va., and upland to Ga. and Tenn. June–Aug.

4. D. intermèdia Vasey (intermediate). — In small tufts, pale drab at base, the erect culms 0.5–5 dm. high; leaves flat or involute, the upper becoming remote from the panicle; *panicle dense,* 2.5–7.5 *cm. long, with appressed-ascending short branches of purple to bronze spikelets; glumes broad, oblong-lanceolate, nearly covering the column of florets,* 1.2–1.5 cm. long, *the sides submembranous,* the midrib prominent, *the lateral ribs weak; larger lemmas* 7–8 *mm. long, glabrous except for bearded base and margins, the summit-teeth broad-based and short-awned, less than a fourth the length of the blade.* — Meadows, peats and gravels, e. Lab. and serpentine or magnesian limestone mts. of w. Nfld. and of Gaspé Pen., Que.; Mackenz. and Alaska, s. to S.D., N.M., Ariz. and n. Calif.; reported from n. Mich. July, Aug.

213. D. sericea.

5. D. serícea Nutt. (silky). — Coarser than our other species; *culms* erect, *becoming* 0.3–1 *m. high;* crowded *basal leaves* elongate, mostly involute, *with pilose sheaths and blades;* cauline leaves either flat or involute, the short uppermost one remote from the long-stalked inflorescence; *panicle* 0.4–1.2 *dm. long, rather lax and open,* with spreading to ascending branches, only slightly closing in fruit; *glumes narrowly lance-attenuate, covering only the base of the column of florets, pale* green, whitish or barely purple-tinged, strongly ribbed, 1.2–1.8 *cm. long; lemmas densely silky-villous, the longer ones* 7–10 *mm. long, their* 2 *slender and long-awned teeth nearly equaling the blade, the long central awn with spiraling base* 2.5–3 *mm. long.* — Dry open woods, clearings and sands, Fla. to La., n. to N.J. and s. Ky. May–early July. FIG. 213.

6. D. épilis Scribn. (hairless). — Similar to no. 5, more slender; *leaves glabrous; lemmas glabrous except for bearded base and margin, the spiraling base of the central awn only* 1–2 *mm. long.* — Sphagnous bogs in pine-barrens, s. N.J.; upland, N.C. to Ga. May, June.

Tribe V. AGROSTÍDEAE Kunth (see p. 96)

a. Lemma membranaceous or scarious, as soft as or not conspicuously harder than glumes, not convolute about the grain. . . *b.*

b. Lemma awnless or dorsally awned, loosely embracing the grain. . . . *c.*

c. Glumes not conspicuously keeled or tightly folded; spikelets in open to spiciform panicles; stigmas plumose or pectinate, exserted from sides of spikelet. *d.*

d. Fruit a utricule, the plump grain loosely surrounded by the easily removed (when moist) thin pellicle. . . . . 45. *Sporobolus.*

d. Fruit a caryopsis, the grain tightly adherent to the pericarp. . . *e.*

e. Floret not stipitate; palea 2- or 3-nerved (or obsolete); stamens mostly 3. . . *f.*

f. Floret disarticulating above and leaving the awnless or merely short-awned glumes as empty husks; perennials, except for some species of no. 49 with loose and open panicles. . . *g.*

g. Palea nearly like the lemma and essentially as large; callus or base of lemma usually with long beard; spikelets 2–15 mm. long.

Awn much shorter than to twice length of blade of lemma; spikelets 2–8 mm. long. . . . . 46. *Calamagrostis.*

Lemmas awnless.

Glumes subequal, slightly exceeding lemma and palea; rachilla prolonged as a bristle beyond the palea; spikelets 8–15 mm. long, in a stiff and dense lance- to linear-cylindric spiciform panicle. . . . . 47. *Ammophila.*

Glumes obviously unequal, shorter than lemma and palea; rachilla not prolonged; spikelets 4–7 mm. long, in an open broad to slender panicle. . . . . 48. *Calamovilfa.*

g. Palea definitely smaller and of more hyaline texture than lemma, or obsolete; floret beardless or very short-bearded at base or, if long-bearded, with awn many times longer than lemma; spikelets 1–4.6 mm. long. . . . . . . 49. *Agrostis.*

f. Floret not promptly disarticulating, spikelet falling intact with the very long-awned pubescent glumes; annual with dense panicle appearing silky from the many slender awns. . . . . 50. *Polypogon.*

e. Floret raised on a short slender stipe; palea 1-nerved or with 2 approximate nerves; stamen 1; spikelet falling intact; panicle large, flexuous. . . . . 51. *Cinna.*

c. Glumes prominently keeled and folded; inflorescence a dense false spike; stigmas with short branches, exserted from summit of floret.

Spiciform panicle partly included in inflated sheath; glumes and 1-nerved lemma awnless, membranaceous, subequal; annual. . . 52. *Heleochloa.*

Spiciform panicle exserted; lemma hyaline, 5-nerved; perennials, rarely annuals.

Glumes awn-tipped; lemma and palea subequal, much shorter than glumes, free to base. . . . . 53. *Phleum.*

Glumes awnless; palea wanting; lemma awned, nearly equaling glumes, its inner margins connate at base. . . . . . . 54. *Alopecurus.*

b. Lemma awn-tipped or mucronate, tightly embracing grain; floret with a hard callus.

Rachilla not prolonged behind the palea as a bristle; lemma awned or awnless; glumes well developed, persistent as husks, if the 1st one obsolete, the plant weak and decumbent and without a rhizome. . . . 55. *Muhlenbergia.*

Rachilla prolonged; lemma long-awned; 1st glume obsolete or nearly so; erect woodland plant with knotty rhizome. . . . . 56. *Brachyelytrum.*

a. Lemma indurated, much harder than the glumes; the fallen floret or spikelet with a definite stipe or callus. . . *h.*

h. Glumes folded or keeled; lemma awned, tightly convolute around the grain; the terete or subterete floret dropping and leaving the empty glumes as husks, with a basal callus.

Awn weak and deciduous; the plump floret with a short callus. . . . 57. *Oryzopsis.*

Awn firm and persistent; the slender floret with callus usually prolonged into an often acute-based slender stipe.

Awn simple. . . . . 58. *Stipa.*

Awn 3-forked. . . . . 59. *Aristida.*

h.Glumes rounded on back, subequal; lemma awnless, its margins inrolled; floret without callus or stipe, falling with the glumes attached, the whole spikelet with a slender pedicel; grain dorsally compressed. . . . . 60. *Milium.*

45. SPORÓBOLUS R. Br. Drop-seed. Rush-Grass

Spikelets 1 (rarely 2)-flowered, awnless, in narrow and spike-like or loose and spreading, often partly included, panicles; lemma as long as or longer than the usually unequal glumes, 1-nerved; palea equaling or exceeding the lemma, often splitting between the strong nerves at maturity; grain readily falling from the spikelet; pellicle loosely inclosing the plump seed, often thin and evanescent, easily removed when moist. — Annuals or perennials of temp. and trop. N. and S.Am., Asia and Afr., with involute or flat leaves. (Name from the Greek *sporos, seed*, and *ballein, to cast forth*, on account of the free rounded grains.)

a.Back of cauline sheaths long-bearded at summit or collar, otherwise glabrous. 1. *S. cryptandrus.*

a.Back of cauline sheaths not long-bearded at summit (except in pilose-leaved species). . . *b*.

b.Panicle contracted or spiciform, linear-cylindric to lanceolate. . . *c*.

c.Culms solitary or tufted; plants without long rhizomes and stolons; leaves not conspicuously distichous. . . *d*.

d.New cauline leaves 4–12; panicles many times shorter than the culms; spikelets 2–7 mm. long. . . *e*.

e.Soft-based annuals, with filiform culms 1–4.5 (rarely –7.5) dm. high; all or nearly all the sheaths spathiform, at least at maturity; terminal panicle usually included or but slightly exserted. . *f*.

f.Spikelets 3.5–6.5 mm. long; grain 1.7–2.2 mm. long.

Lemma pubescent on the sides; leaves glabrous or essentially so. 2. *S. vaginiflorus.*

Lemma glabrous; leaves (especially the lower) papillose-pilose . 3. *S. ozarkanus.*

f.Spikelets 2–3 mm. long; lemmas glabrous on the sides; grain 1–1.5 mm. long. . . . . 4. *S. neglectus.*

e.Hard-based tufted perennials with persistent old leaf-bases; culms 0.2–1 m. or more high; only the upper sheaths spathiform; terminal panicle commonly wholly or partly exserted. . *g*.

g.Sides of lemma pubescent at least toward base.

Spikelets 6–8 mm. long; palea much prolonged beyond the lemma into a subaristate beak. . . . . 5. *S. clandestinus.*

Spikelets 5.5–6 mm. long; palea and lemma subequal, the palea not subaristate. . . . . 6. *S. canovirens.*

g.Sides of lemma glabrous.

Terminal panicle linear-filiform, lax; spikelets 3–5 mm. long; 1st glume linear-lanceolate. . . . . 7. *S. Drummondii.*

Terminal panicle thick-cylindric to lanceolate, dense and stiff; spikelets 5–6.5 mm. long; 1st glume narrowly ovate. . . . 8. *S. asper.*

d.New cauline leaves 2–3; panicle linear-cylindric, one-fourth to half the height of the culm; spikelets 1.5–2 mm. long. . . . . . 9. *S. Poiretii.*

c.Culms terminating rigid branches from long wiry rhizomes and stolons; leaves stiff, strongly distichous. . . . . 10. *S. virginicus.*

b.Panicle open, lax to diffuse. . . *h*.

h.Spikelets 1.5–3.5 mm. long; 1st glume linear-lanceolate to ovate; grain at most 1 mm. in diameter.

Branches of panicle mostly alternate; spikelets 1.5–2.5 mm. long. . . 11. *S. airoides.*

Branches of panicle mostly verticillate or opposite.

Panicle ovoid, 3–12 dm. long; spikelets 1.5 mm. long. . . . . . 12. *S. pyramidatus.*

Panicle lanceolate, 1–2 dm. long; spikelets 3–3.5 mm. long. . . . 13. *S. junceus.*

h.Spikelets 4.5–6 mm. long; 1st glume linear-acicular; grain 2 mm. in diameter. . . . . 14. *S. heterolepis.*

1. S. cryptándrus (Torr.) Gray (with hidden flowers), Sand-Drop-seed. — Tufted, 3–10 dm. high; culms rather stout, erect or somewhat spreading; *sheaths* overlapping, ciliate on the margin and *conspicuously bearded at the summit; blades* 0.6–2 dm. long, 3–5 mm. wide, *flat or* *involute, scabrous on the margin; panicle* lead-colored, usually open, 1–3 dm. long, included at base in the upper sheath, or sometimes contracted and wholly included, *when exserted ovoid to pyramidal* and *with branches naked at base; spikelets* 2–2.5 *mm. long;* first glume about one-third as long as the second; lemma acute, longer than the palea. — Sandy soil, coast, N.H. to L.I.; w. Que. to Wash., s. to N.C., O., Ind., Ill., La., Tex. and Mex.; adv. in s. Me. July–Oct. FIG. 214.

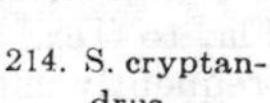

214. S. cryptandrus.

Var. **stríctus** Scribn. (erect). — Leaves more often involute; *panicle* 1–4 dm. long; when exserted *densely spiciform or open-lanceolate, its short branches floriferous nearly to base.* (Var. *involutus* Farw., *S. contractus* Hitchc.) — Mich. to Wash. and southw.; adv. in s. N.Y. and s. Me.

215. S. vaginiflorus.

2. S. vaginiflòrus (Torr.) Wood (with flowers in the sheaths), POVERTY-GRASS. — *Annual;* tufted culms filiform, scabrous, 2–7.5 dm. high, erect to spreading; *leaves* remote, short, about 2 mm. wide, involute at tip, *glabrous* or nearly so; *sheaths* often long-ciliate at orifice, 1.5–5 cm. long, *all or nearly all inclosing panicles; panicles included* or the terminal exserted, 1–5 cm. long; *spikelets* 3.5–6.5 *mm. long;* the acuminate *glumes subequal, about the length of the* acuminate *scabrous appressed-pubescent lemma and the* sharp-pointed *palea; anthers* 1.5–2.2 mm. long, *conspicuous in the exserted panicles; grain* 1.7–2.2 *mm. long.* — Dry, open sterile soil, s. N.B., s. Me. and s. Que. to N.D., s. to Ga. and Tex. Late Aug.–Oct. FIG. 215.

216. S. vaginiflorus, v. inaequalis.

Var. **inaequàlis** Fern. (unequal). — *Palea prolonged far above the glumes and lemma into a slender beak* (0.2–1.8 mm. long) — Centr. Me. to s. Ont., s. to L.I., Pa., Mich., Wisc., Mo., Neb. and Ariz. FIG. 216.

3. S. ozarkànus Fern. (of the Ozarks). — Similar to no. 2; *lower leaves and sheaths papillose-pilose; lemma glabrous,* hardly equaling the glumes. — Chert-glades, sw. Mo. Sept., Oct.

4. S. negléctus Nash (neglected or overlooked). — Similar to no. 2; terminal panicle smaller, commonly purple; *spikelets* 2–3 *mm. long; glumes, lemma and palea* subequal, *thinner or hyaline and shining, glabrous; anthers rarely developed; grain* 1–1.5 *mm. long.* — Dry sterile (often calcareous) soil, w. N.B. to s. Ont. and N.D., s. to Va., Tenn., Mo. and Tex. Aug.–Oct. FIG. 217.

217. S. neglectus.

218. S. clandestinus.

5. S. clandestìnus (Biehler) Hitchc. (concealed). — Tufted perennial; culms slender, 4–12 dm. high; lower *leaves* long, subrigid, *the margins and involute-filiform tips scabrous;* panicle pale, lax, 4–10 cm. long, often partially inclosed in the upper sheath; *spikelets* 6–8 *mm. long; glumes unequal,* acute, the first half the length of the *acute lemma,* the second half that of the *long-acuminate pointed palea; lemma and palea appressed-pubescent toward the base,* the lemma two-thirds the length of the palea. — Dry sandy or rocky soil, Ct. to Fla. and Miss. Late Aug.–Oct. FIG. 218.

6. S. canóvirens Nash (grayish-green). — Similar to no. 5, usually coarser; *leaves hirsute near the base;* panicle stouter, 0.5–2 dm. long; *spikelets* 5.5–6 *mm. long; lemma and palea acute, subequal.* — Dry sandy, gravelly or rocky soil, Ind. and Wisc. to Kans., s. to Miss. and Tex. Sept., Oct.

7. S. Drummóndii (Trin.) Vasey (for its discoverer, THOMAS DRUMMOND, 1780–1835). — Habit of nos. 5 and 6; culms stiff, 5–10 dm. high; leaves firm, 1–3 mm. wide, slender, often involute; *panicle linear-filiform, lax,* 1–2 dm. long; *spikelets* 3–5 *mm. long;* 1st *glume linear-lanceolate,* shorter than the 2nd; *lemma glabrous,* acutish, about equaling the acutish palea. (*S. asper,* var. *Hookeri* (Trin.) Vasey) — Dry prairies and calcareous rocks, Mo. to Miss. and Tex. Sept., Oct.

219. S. asper.

8. S. ásper (Michx.) Kunth (rough). — Perennial; culms stout, 3–10 dm. high; sheaths overlapping; *blades* nearly as long as the culm, the upper exceeding the panicle, *pilose above at the flat base, the long involute-filiform tip scabrous; terminal panicles* 0.5–3 *dm. long, partly included in the large inflated upper sheaths, lateral panicles small* and *usually hidden in the sheaths, or none;* spikelets 5–6.5 mm. long; *glumes unequal, obtuse or subacute,* the first about half as long as the floret; *lemma and palea glabrous, the lemma slightly the longer.* (*S. longifolius* Wood) — Dry open soil, sw. Que., Mass., Vt. and e. N.Y., s. to e. Va.;

O. to N.D., s. to Tenn., La. and Tex.; spreading along roads and railroads elsewhere, often becoming a weed. Aug.-Oct. FIG. 219.

Var. pilòsus (Vasey) Hitchc. (hairy). — Sheaths and blades more generally pilose. — E. Kans. to Tex.

9. S. POIRÉTII (R. & S.) Hitchc. (in honor of JEAN LOUIS MARIE POIRET, 1755–1834), SMUT-GRASS, BLACK-SEED GRASS. — Tufted culms 3–10 dm. high, erect, wiry; *leaves* 10–30 cm. in length, long-attenuate, *those of the culm only* 2 *or* 3; *panicle linear-cylindric, stiff, one-fourth to half the entire length of the plant; spikelets* 1.5–2 *mm. long*, shining, *crowded on the slender erect branches;* glumes obtuse, unequal, the second half as long as the acuminate lemma which is slightly longer than the obtuse palea. (*S. indicus* of ed. 7, not R. Br.) — Dry sandy soil, Fla. to Tex., n. to Va., s. Ky. and se. Mo. May–Oct. (Natzd. from Trop. Am.) — Panicle frequently infected by a black fungus, whence the common names.

10. S. virgínicus (L.) Kunth (of Virginia). — *Extensively creeping*, with *wiry rhizomes and bases*, glabrous; culms ascending, 1.5–6 dm. high; *sheaths crowded*, overlapping; *blades short, firm, involute, conspicuously distichous on the numerous sterile shoots;* panicle exserted, 3–6 cm. long; spikelets 3 mm. long; glumes unequal, the 2nd exceeding the glabrous lemma and palea. — Sandy or muddy shores and marshes, Fla. to Tex. and Mex., n. to Va. (formerly). June–Nov. (Pantrop.)

11. S. airoìdes Torr. (like *Aira*), FINE-TOP SALT-GRASS. — Perennial, forming dense clumps; culms stiff, smooth, 0.3–8 dm. high; leaves smooth, the sheaths pilose at orifice, the hard blades involute; *panicle ovoid to pyramidal*, at first with included base, *later exserted and diffuse*, 1–3 dm. long, *its branches mostly alternate; spikelets* 1.5–2.5 *mm. long; glumes unequal;* the 1st short, narrowly ovate; the 2nd equaling the lemma and palea. — Saline flats, prairies and sands, S.D. to Wash., s. to nw. Mo., e. Kans., Tex. and Mex.; adv. in N.Y. May–Oct.

12. S. PYRAMIDÀTUS (Lam.) Hitchc. (pyramidal). — Tufted; the depressed or decumbent culms 1–4 dm. high; *leaves flat*, short and *broad* (2–4 mm.), *rarely* 1 *dm. long; panicle ovoid*, 3–12 *dm. long, with* spreading-ascending viscid *verticillate or opposite branches* naked below; *spikelets* 1.5 *mm. long;* 1st glume linear-lanceolate, very short. (*S. argutus* (Nees) Kunth) — Roadsides and waste places, Mo., and in se. N.Y., adv. from farther west and south. May–Oct.

13. S. júnceus (Michx.) Kunth (rush-like), WIREGRASS. — Tufted, glabrous, 4–9 dm. high; culms wiry, erect, leafy at base, naked above; the *involute setaceous basal leaves* 1–2.5 *dm. long, spreading; panicle purplish or chestnut, lanceolate*, 1–2 dm. long, *the short verticillate branches spreading; spikelets* 3–3.5 *mm. long;* first glume about one-third the length of the second, which is as long as the glabrous subacute equal lemma and palea. (*S. gracilis* (Trin.) Merr.; *S. ejuncidus* Nash) — Dry sandy soil, Fla. to Tex., n. to se. Va. May–Sept.

14. S. heterólepis Gray (with unequal glumes), NORTHERN DROP-SEED. — Strong-scented, tufted, 0.5–1 m. high; culms rather stout, wiry, erect; basal leaves about half as long as the culm, involute-setaceous; *panicles* becoming *long-exserted*, 0.7–3 dm. long, purple to black, branches ascending; *spikelets* 4.5–6 *mm. long; first glume linear-acicular*, about one-half to two-thirds the length of the floret, the second acuminate, often cuspidate (varying in length in the same panicle), exceeding the glabrous obtuse or subacute equal lemma and palea; *grain* 2 *mm. in diameter; pericarp shining, indurated, splitting the palea.* — Dry trap, limestone or serpentine, local, s. Ct. to se. Pa.; rocks and prairie, w. Que. to Sask., locally s. to n. N.Y., centr. O., n. Ind., Ill., Ark., e. Tex. and Wyo. Aug.-Oct. FIG. 220.

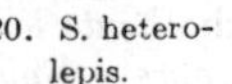
220. S. heterolepis.

46. CALAMAGRÓSTIS Adans. REED-BENTGRASS

Spikelets 1-flowered; rachilla usually prolonged behind the palea into a hairy bristle or pedicel; glumes subequal, usually longer than the floret; lemma awned on the back, usually from below the middle, surrounded at base with often copious long hairs; palea shorter than the lemma, faintly 2-nerved. — Perennials of cool and temp. reg., with mostly simple (sometimes branching) erect culms and many-flowered panicles. (Name compounded of the Greek *calamos, a reed,* and *agrostis, a grass.*)*

*a.*Glumes lanceolate to ovate; lemmas 3–5-nerved below, 4-nerved above the insertion of the awn; callus-hairs shorter than to barely exceeding lemma.
§ DEYEUXIA. . . *b.*

*b.*Awn twisted at base or geniculate, the tip often exserted. . . *c.*

* Treatment largely based on those of STEBBINS in Rhodora, xxxii. 35–57 (1930) and of LOUIS-MARIE, l.c. xlvi. 285–305 (1944).

c.Awn longer than glumes, protruding from tip of spikelet; summits of lemma and palea emarginate-dentate; flowering culms with 1 or 2 (rarely 3) cauline leaves; callus-hairs extending around summit of callus, a band of very short ones between the 2 tufts of longer ones, the latter much shorter than the lemma.
Spikelets 5.5–12 mm. long; awn of lemma 6–7.5 mm. long; caryopsis reddish. . . . . 1. *C. purpurascens.*
Spikelets 4–4.6 mm. long; awn 3–4 mm. long; caryopsis olivaceous. . 2. *C. Lepageana.*
c.Awn shorter than glumes, sometimes projecting at side of spikelet; summits of lemma and palea more or less truncate-emarginate, not dentate; flowering culms with 2–4 cauline leaves; tufts of callus-hairs without shorter connecting series or all elongate and continuous around summit of callus. . . *d.*
d.Summit of leaf-sheath usually glabrous; cauline leaves usually 3. . . *e.*
e.Callus-hairs in 2 lateral tufts, the front of the callus naked; spikelets 2–4.8 mm. long. . . *f.*
f.Longer callus-hairs 1 mm. or less long; spikelets purplish or greenish; plant strongly stoloniferous. . . . . 3. *C. Pickeringii.*
f.Longer callus-hairs 2–3 mm. long.
Loosely stoloniferous, with elongate stolons; culms solitary, scattered; spikelets purplish or purple-tinged; callus-hairs 3 mm. long; awn 3.5–4 mm. long. . . . . 4. *C. lacustris.*
Tufted, with very short basal offshoots; spikelets greenish to pale brown; callus-hairs 2–2.5 mm. long; awn 2 mm. long. . . . 5. *C. Fernaldii.*
e.Callus-hairs in a continuous ring, evenly distributed at summit of callus, 3–4 mm. long, about equaling lemma; spikelets 5–5.2 mm. long. . . . . 6. *C. nubila.*
d.Summit of leaf-sheath pubescent; spikelets greenish; callus-hairs in 2 separated tufts; cauline leaves mostly 4.
Spikelets 3.5–5.5 mm. long; lemma 3–5 mm. long; awn 3 mm. long. 7. *C. Porteri.*
Spikelets 3.3–4.4 mm. long; lemma 3.2–3.6 mm. long; awn 2 mm. long. . . . . 8. *C. perplexa.*
b.Awn straight or barely arched, fine and inconspicuous, included. . . *g.*
g.Panicle during anthesis mostly loose and open, its branches then mostly spreading; lemma membranaceous-translucent at least at summit; callus-hairs equaling or exceeding lemma, of uniform length except for an outer short ring. . . . . 9. *C. canadensis.*
g.Panicle contracted, its rigid branches appressed or ascending during anthesis; lemma firm and opaque; callus-hairs mostly shorter than lemma, of unequal lengths, those on the sides of lemma long and tufted. . . *h.*
h.Prolongation of rachilla bearded only at summit; caryopsis pubescent at summit; spikelets 6–7 mm. long; glumes acuminate-aristate. . . 10. *C. cinnoides.*
h.Prolongation of rachilla bearded throughout; caryopsis glabrous; spikelets 2–5.5 mm. long; glumes merely acute or short-acuminate. . . *i.*
i.Leaves and culms harsh and scabrous, blades mostly flat; ligules erose or lacerate at summit; glumes opaque. . . . . . . 11. *C. inexpansa.*
i.Leaves smooth (sometimes scabrous at tip and margin), mostly becoming involute; ligules entire or barely erose.
Glumes thick and opaque, 1.4–1.8 mm. broad (flattened out); longer branches at most one-fifth the length of panicle. . . 12. *C. labradorica.*
Glumes hyaline and translucent at least at tip, 1–1.2 mm. broad; branches mostly one-fifth to one-third as long as panicle. . . 13. *C. neglecta.*
a.Glumes linear-lanceolate, attenuate; lemmas 3-nerved below, 2-nerved above the insertion of the awn; callus-hairs much exceeding the lemma. § Epigeios. 14. *C. epigejos.*

§ Deyeùxia Hack.

1. C. purpuráscens R. Br. (purplish). — *Cespitose*, the tufts *invested at base with dry whitish marcescent leaves;* culms stiff, 2.5–5 dm. high; *cauline leaves* 1 or 2 (rarely 3), *harsh*, thick, flat, becoming involute; *panicle spike-like, stiff,* slender, lanceolate, 5–12 cm. long, 1–2 cm. thick, *purplish to whitish-brown; spikelets 5.5–12 mm. long; glumes ovate-lanceolate,* exceeding lemma, *thin, scarious, scabrous; lemma 4-cleft and emarginate-dentate at apex,* the coarse *geniculate* dorsal *awn 6–7.5 mm. long, exserted* about 2 *mm.* from summit of spikelet; *callus-*

hairs in 2 tufts shorter than lemma, these tufts separated by a band of much shorter hairs at summit of callus; grain reddish. — Greenl.; calcareous cliffs, Rimouski Co. and L. Mistassini, Que.; Mountain L., Cook Co., Minn.; arct. nw. Am., s. along mts. to S.D., Colo., Nev. and Calif. June–Aug. (Ne. Asia) Fig. 221.

221. C. purpurascens.

2. C. Lepageàna Louis-Marie (for its discoverer, Ernest Lepage, 1905–). — Differing from no. 1 in *weakly stoloniferous base;* culms 3–6.5 dm. high, smooth, with 1 or 2 cauline leaves glabrous on back; panicle slenderly ellipsoid, 6–9 cm. long, 0.6–1 cm. thick; *spikelets* 4–4.6 *mm. long; awn* 3–4 *mm. long; grain greenish.* — Calcareous cliffs at about 800 m. alt., local, Rimouski Co., Que. June, July.

3. C. Pickeríngii Gray (for its discoverer, Charles Pickering, 1805–1878). — *Loosely stoloniferous,* with elongate slender rhizomes; the mostly *solitary culms* 1.5–12 cm. high, *springing from tips of stolons of preceding year, invested at base by loosely shredding marcescent leaves; basal leaves numerous,* flat, *glaucous,* 2–8 mm. wide; *cauline leaves* of the season 2–4, firm, scabrous-margined, *their sheaths glabrous at summit; panicle purplish* to greenish, *glaucous,* slender, *linear-cylindric to lance-ovoid,* with short ascending branches, 3–17 cm. long, 0.5–3 cm. thick; *spikelets 4–5 mm. long;* glumes ovate-lanceolate, strongly keeled, equaling or longer than the obtuse scabrous lemma; *awn geniculate, the tip commonly divergently exserted from side of spikelet; callus-hairs in 2 lateral tufts 1 mm. or less long, only one-fifth to one-fourth as long as lemma, the summit of the callus otherwise without hairs.* — Acid peats or sands, gravels and shores, Nfld. and N.S.; White Mt. reg., N.H.; Green Mts., Vt.; ne. Mass.; Adirondack Mts., N.Y. Late June–early Sept. Fig. 222.

222. C. Pickeringii.

Var. **débilis** (Kearney) Fern. & Wieg. (weak). — Plant more slender, often lower; *spikelets* 2–3.6 *mm. long.* — Nfld. and N.S.; Mt. Katahdin, Me.; White Mt. reg., N.H.; se. N.H. and ne. Mass.; s. N.J.

4. C. lacústris (Kearney) Nash (of lakes or ponds). — Habitally similar to no. 3, 0.5–1 m. high; *leaf-sheaths glabrous or sparsely bearded at summit;* glumes more strongly scabrous; lemma thinner; *callus-hairs* 3 *mm. long, nearly or quite as long as lemma; awn* 3.5–4 *mm. long.* (*C. Pickeringii,* var. Hitchc.) — Shores, damp rocks or gravel or peaty spots, Lab. Pen. to n. Ont., s. locally to sw. Nfld., n.-centr. N.H., Mt. Mansfield, Vt., Adirondack reg., N.Y., n. Mich. and n. Minn. July, Aug.

5. C. Fernáldii Louis-Marie (for its discoverer, 1873–). — *In small clumps with very short basal leafy offshoots, with membranaceous green* narrow scabrous-margined flat *leaves;* culms slender, erect, smooth, 7–9 dm. high, *with 3 very long cauline leaves with glabrous sheaths; panicle pale greenish to pale brown,* narrowly lanceolate, 6–9 cm. long, 0.7–1.5 cm. thick, on very long exserted peduncle; *spikelets* 3–3.7 *mm. long; lemma* 3 mm. long, *its subbasal awn* 2 *mm. long and only slightly twisted; callus-hairs very numerous in 2 separated tufts,* 2–2.5 *mm. long.* (*C. perplexa* in part of ed. 7, not Scribn.) — Wet siliceous cliff on deciduous-wooded mountain-slope, very local, s. Piscataquis Co., Me. July, Aug.

6. C. nùbila Louis-Marie (cloudy; from the type-locality). — *Tufted,* up to 5.5 dm. high; *leaves firm,* scabrous, the basal ones 1–2 dm. long and 5 mm. wide, the summit of the sheath naked; cauline leaves 3, elongate, often overtopping panicle; panicle 1.3 dm. long, dense, with spreading flexuous branches; *spikelets* 5–5.2 *mm. long,* puberulent; lemma 4.2 mm. long, membranaceous, dentate at acuminate apex; awn 3 mm. long, weakly twisted, geniculate; *callus-hairs* 3–4 *mm. long, very abundant, extending continuously around and veiling lemma.* — Lake of the Clouds, Mt. Washington, N.H., little known and but once collected (by *William Boott* in 1862). Aug., Sept.

7. C. Pórteri Gray (for its discoverer, Thomas Conrad Porter, 1822–1901). — Loosely tufted, with slender stolons; culms slender, 0.6–1.2 m. high; *leaves* 1.5–3 dm. long, 4–8 mm. wide, flat, taper-pointed, scabrous, the *summit of the sheath and often the base of the blade bearded; cauline leaves mostly* 4; panicle greenish to bronze, 0.5–2 dm. long, rather loosely flowered; *spikelets* 3.5–5 *mm. long; lemma* 3–5 *mm. long; awn* 3 *mm. long; callus-hairs in* 2 *separated tufts, scanty, a fourth to a third as long as lemma.* — Dry upland woods, very local, Chemung Co., N.Y., to Jackson Co., O., s. to mts. of Pa., w. Va. and W.Va. July, Aug.

8. C. perpléxa Scribn. (confused; apparently an appropriate name). — Similar to no. 7 and presumably a variety of it, slightly glaucous; panicle denser; *spikelets* 3.3–4.4 *mm. long; awn* 2 *mm. long; callus-hairs more abundant, three-fourths length of lemma.* — Rocky woods, very local, Tompkins Co., N.Y. Late July, Aug. — Records from farther north and west belong to no. 4.

9. C. canadénsis (Michx.) Nutt. (Canadian), Blue-joint, Foin bleu (Que.). — In small or large tussocks, 0.5–1.5 m. high, comparatively soft; leaves long, flat, often involute in drying, usually glaucous, those of the often branching culms 4–7; *panicle* purple, lead-colored or greenish, mostly *loose and open during anthesis, with spreading slender branches,* broadly lanceolate to ovoid, 0.6–3 dm. long, 2–10 cm. broad, in fruit more contracted, with ascending branches; spikelets 2–6 mm. long; *glumes* lanceolate or narrowly ovate, barely equaling to slightly exceeding the lemma, *remaining apart in fruit; lemma membranaceous-translucent* at least at summit, erose-truncate; *callus-hairs copious, nearly equaling to exceeding the lemma and the rudiment, of uniform length* except for an outer short ring; *awn delicate, straight or barely arched, included.* — Highly variable.

Spikelets 2–3.8 mm. long; glumes rounded on the back, weakly keeled, acute or acuminate; lemma 1.7–3 mm. long; awn inserted near middle of lemma.
  Panicle loosely flowered; spikelets 2.8–3.8 mm. long; glumes distinctly exceeding the lemma, acute or acuminate. . . . . *C. canadensis* (typical).
  Panicle densely flowered; spikelets 2.2–2.8 mm. long; glumes nearly or quite equalled by lemma, obtuse or nearly acute. . . . . Var. *Macouniana.*
Spikelets 3.8–6 mm. long; glumes narrow, strongly keeled, distinctly acuminate; lemma 3–4.2 mm. long; awn inserted on lower third of lemma.
  Spikelets 3.8–4.5 mm. long; glumes often hyaline on tip and margin, short-scabrous on keel, elsewhere minutely scabrous; lemma 3–3.5 mm. long. . Var. *robusta.*
  Spikelets 4.5–6 mm. long; glumes usually thick and opaque to tip, scabrous or ciliate on keel, elsewhere pilose-hirtellous; lemma 3.5–4.2 mm. long. . . Var. *scabra.*

C. canadénsis (typical). — Meadows, bogs, wet thickets, etc., Nfld. to Mackenz. and B.C., s. to Del., Pa., W.Va., n. O., n. Ind., Ill., Mo., Neb., N.M. and Calif. June–Aug. Fig. 223.

Var. **Macouniàna** (Vasey) Stebbins (for its discoverer, James Melville Macoun, 1862–1920), *C. Macouniana* Vasey. — Local, w. Nfld. and P.E.I.; s. N.H. to N.J. and Pa.; Mich. to Sask., w. to Alta. and Wash., s. to Mo., Neb. and Wyo. June–Aug.

Var. **robústa** Vasey (stout), var. *acuminata* Vasey — Lab. to Alaska, s. to Nfld., e. N.S., coast of Me., mt.-reg. of n. N.E. and n. N.Y., Roan Mt., N.C., L. Superior, Ont. and n. Mich., N.M. and Calif. July–Sept. (Ne. Asia)

223. C. canadensis (typical).

Var. **scàbra** (Presl) Hitchc. (harsh), *C. Langsdorfii* Am. auth., not Trin. — Greenl. and n. Lab. to Alaska, s. to w. Nfld., Gaspé Pen., Que., White Mts., N.H., Smuggler's Notch, Vt., Isle Royale, Mich., Lucile I., Minn., and mts. of Colo. and Calif. July–Sept. (N. Eurasia) Fig. 224.

224. C. canadensis, v. scabra.

10. C. cinnoìdes (Muhl.) Bart. (like *Cinna*). — Glaucous; culms stout, 0.7–1.8 m. high, solitary or few, erect or leaning; leaves very scabrous, sometimes sparingly hirsute, 1.5–3 dm. long, 5–10 mm. wide (those of the innovations shorter, narrow); panicles 0.7–2.3 dm. long, 1–4 cm. thick, spike-like, dense, lanceolate; *spikelets 6–7 mm. long; glumes keeled,* very scabrous, *acuminate-aristate, the tips usually curved outward, exceeding the acuminate lemma* which is awned above the middle; *callus-hairs about half the length of the floret, those of the rudiment copious, confined to the tip,* almost equaling the lemma; *caryopsis pubescent* at least at summit. — Damp sandy or peaty soils, Halifax Co., N.S. (*J. R. Lunt*); s. Me. to s.-centr. N.Y., s. to Ga. and Ala. Mid-July–Oct. Fig. 225.

225. C. cinnoides.

11. C. inexpánsa Gray (unexpanded). — Usually slightly to densely *tufted* and stoloniferous, *glaucous,* the tufts with firm persistent pale sheaths; culms stiff, 0.3–1.2 m. high; *leaves harshly scabrous and hard, flat* and 2–8 *mm. wide,* or involute; new cauline leaves 2–4, elongate, their *erose or lacerate ligules* 2.5–8 *mm. long; panicle pale,* usually rather dense and *spike-like,* lanceolate to subcylindric, 0.5–2 dm. long, *its stiff* appressed or ascending short *branches harsh;* spikelets 3–5.5 mm. long; *glumes mostly opaque,* acute or short-acuminate, their tips connivent in fruit; lemmas slightly shorter, mostly toothed at summit; *callus-hairs* mostly shorter than lemma, *in unequal tufts;* awn straight, included; prolongation of rachilla hairy throughout, 0.6–1.5 mm. long; caryopsis glabrous. Very variable.

Spikelets 4–5.5 mm. long; lemma 3.5–4.5 mm. long; palea 2.7–3.2 mm. long.
  Culms solitary; panicle 1.5–2 dm. long; its longer branches 5–6 cm. long, with 1st internodes 4 cm. long. . . . . *C. inexpansa* (typical).

Culms more or less cespitose; panicle 8–14 cm. long; its longest branches 2–4 cm. long, with 1st internodes 1–2.5 mm. long. . . . . Var. *robusta.*
Spikelets 3–4.5 mm. long; lemma 2.5–3.5 mm. long; palea 1.7–2.6 mm. long.
Compactly flowered, branches of panicle forked and floriferous nearly to base; glumes thick and opaque, usually purple-tinged. . . . . Var. *brevior.*
Loosely flowered, branches forked and floriferous from near or above the middle; glumes thin except near keel, green, rarely purplish. . . . . . Var. *novae-angliae.*

C. inexpánsa (typical). — Rare and local, centr. N.Y. to Alta. — The type seems like an extreme southern development of a species commonly more boreal.

Var. **robústa** (Vasey) Stebbins (stout), *C. hyperborea* Am. auth. in part, not Lange — Rocky, gravelly or peaty soils, Nfld. and e. Que.; Alaska, s. to Colo. and Oreg. Mid-July, Aug.

Var. **brévior** (Vasey) Stebbins (shorter), *C. hyperborea* Am. auth. in part, not Lange — Nfld. to B.C., s. to Gaspé Pen. and Rimouski Co., Que., very locally to White Mts., N.H., n. Vt. and w. N.Y., and more commonly to Ont., Mich., n. Ind., Wisc., n. Ia., Neb., N.M., Ariz. and Calif. July, Aug.

Var. **nòvae-ángliae** Stebbins (of New England). — Damp woods and shaded cliffs, local, e. Me. to n. Vt.

12. C. labradórica Kearney (of Labrador). — Tufted, stiff, 2.5–6 dm. high, with aspect between nos. 11 and 13; *leaves smooth, soft,* 1.5–2.5 *mm. broad, strongly involute, with entire firm ligule,* 1.5–3 *mm. long; panicle* compact, subcylindric, 3–12 cm. long, 0.5–2 cm. thick, *with very short* (*at most one-fifth as long as panicle*) *erect* sparsely scabrous *branches;* spikelets 3.5–4 mm. long; *glumes ovate to ovate-lanceolate,* 1.4–1.8 mm. broad, acute, *chartaceous,* the weak keel hispidulous; lemma slightly shorter, broadly truncate, usually 4-toothed, firm; awn dorsal, straight, included; callus-hairs at most two-thirds as long as lemma. — Marshes and gravelly shores, se. Lab. and e. Saguenay Co. and Anticosti I., Que. July, Aug.

13. C. neglécta (Ehrh.) Gaertn., Mey. & Scherb. (neglected, *i.e.* not recognized previously). — *Culms* solitary or tufted, 1–10 dm. high, *slender; leaves* soft or at least not hard and harsh, *smooth beneath* or merely scabrous at tip and margin, 2–4 *mm. broad, commonly involute; ligule entire or barely erose,* 1.5–3.5 *mm. long; panicle* spike-like, dense or slightly open, lanceolate, 2–15 cm. long, *the appressed-ascending branches one-fifth to one-third its length;* spikelets purplish, bronze or greenish, 2–5 mm. long; *glumes* ovate-lanceolate, *hyaline or translucent* at least at tip, 1–1.2 *mm. broad;* lemma firm, with straight awn and unequally tufted callus-hairs included. — Very variable.

Spikelets 3–5 mm. long; glumes sharply acute or acuminate.
Culms 3–10 dm. high; panicle 5–15 cm. long; callus-hairs half to three-fourths as long as lemma; awn nearly basal. . . . . *C. neglecta* (typical).
Culms 1–4 dm. high; panicle 2–5 cm. long; callus-hairs a quarter to half as long as lemma; awn median. . . . . Var. *borealis.*
Spikelets 2–2.6 mm. long; glumes obtuse or merely acute. . . . . . . Var. *micrantha.*

C. neglécta (typical). — Swales and shores, Greenl. and Lab. to Alaska, s. to Nfld., n.-centr. N.S., s. N.B., n. Me., Mich., Wisc., Colo. and Calif. Late June–Aug. (Eurasia)

Var. **boreàlis** (Laestad.) Kearney (northern). — Greenl., Lab. Pen. and w. Nfld. July, Aug. (N. Eu.)

Var. **micrántha** (Kearney) Stebbins (small-flowered). — Prairies and swamps, Yuk., se. and s. to Wisc., N.D., Colo. and Oreg. June–Aug.

§ Epigeíos W. D. J. Koch

14. C. epigèjos (L.) Roth (upon the ground; from its stoloniferous habit), Feathertop, Bush-Grass. — Extensively creeping, with *slender rigid rhizomes and stolons;* culms firm, 0.6–1.5 m. high; leaves pale, firm, elongate, scabrous, often becoming involute; *panicle slender, crowded,* up to 3.5 dm. long; spikelets often becoming bronze or purplish; *glumes linear-lanceolate, attenuate,* 5–8 mm. long; *lemma* about 3 mm. long, 3-*nerved below,* 2-*nerved above the insertion of the awn;* awn attached above the middle, just below the teeth, projecting forward; *callus-hairs much exceeding the lemma.* — Waste places, local, L.I. (*Beals*). Aug., Sept. (Adv. from Eurasia or n. Afr.)

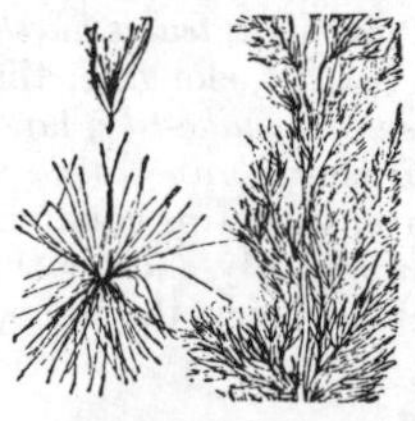
226. C. epigejos, v. georgica.

Var. georgica (K. Koch) Ledeb. (of Georgia, southern Russia). — *Panicle whitish or pale-stramineous; glumes* 4–5 *mm. long; lemma about* 2 *mm. long, with awn often inserted lower down and arched at base.* (*C. arenicola* Fern.) — Sandy woods, thickets and openings, local, e. Mass. to s. Pa. Aug., Sept. (Natzd. from s. Russia) Fig. 226.

47. AMMÓPHILA Host. Sand-Reed. Psamma. Marram

Spikelets 1-flowered, large, awnless, crowded in a long spike-like panicle; rachilla prolonged behind the palea into a hairy bristle; glumes firm, subequal, compressed-keeled, acute; lemma of like texture, surrounded at base by short hairs; palea nearly as long, rather firm, the two nerves close together. — Coarse perennials of Eu. and Atl. N.Am., with creeping rhizomes, rigid culms and involute leaves. (Name from the Greek *ammos, sand*, and *philein, to love.*)

227. A. breviligulata.

1. A. breviligulàta Fern. (with short ligule), Beachgrass. — Culms coarse, stiff, 0.5–1 m. high, from firm widely creeping rhizomes; *leaves* elongate, *the basal arched-recurving*, soon involute, *serrulate-scabrous on the nerves above; ligule chartaceous or coriaceous, rounded,* 1–3 *mm. long; panicle linear-cylindric,* whitish-brown or slightly purple-tinged, 1.3–4 *dm. long; rachis puberulent;* spikelets compressed, 8–12 mm. long; *glumes puberulent,* obtuse to acute; *lemma and palea obtuse;* anthers linear, 6–8 mm. long; caryopsis 3–3.6 mm. long. (*A. arenaria* of e. Am. auth., not Link) — Dunes and sands, coast of Nfld. and s. Lab. Pen. to N.C.; sands of L. St. John, L. Champlain and the Great Lakes. July–Sept. Fig. 227.

A. arenària (L.) Link (of sand), the European Beachgrass, lower, *with elongate thin tapering ligule* 1–3 *cm. long,* puberulent upper leaf-surface, *lanceolate panicle* 1–2 *dm. long,* and longer callus-hairs, spreads from plantings on the coast (as near Provincetown, Mass.). (Introd. from Eu.)

48. CALAMOVÍLFA (Gray) Hack.

Spikelets 1-flowered, awnless; callus densely bearded; glumes chartaceous, unequal, acute; lemma similar, 1-nerved; palea as long as the lemma, broad, deeply furrowed between the strong nerves. — Rather tall rigid N.Am. perennials, with horizontal rhizomes and loosely spreading panicles. (Name from the Greek *calamos, a reed,* and *Vilfa,* a name applied by Adanson to a genus of grasses.)

1. C. brevípilis (Torr.) Scribn. (with short hairs — on the callus). — *Culms* 6–12 dm. high, *from a short horizontal rhizome;* the *basal sheaths indurated and keeled;* blades long, linear, nearly flat or involute; *panicle* purplish, 1.5–3 dm. long, 4–10 (–14) cm. thick, *lanceolate to lance-ovoid,* the slender branches ascending; *pedicels hairy at the summit; spikelets* 4.5–5.5 *mm. long; glumes shorter than the floret, mucronate; callus-hairs less than half the length of the* subequal *scabrous lemma and palea, which are bristly-bearded along the keels.* — Sandy bogs, Pine Barrens of N.J. July, Aug.

Var. **càlvipes** Fern. (with hairless feet). — Leaves more scabrous; *panicle* strongly exserted, *loosely ovoid,* 1.3–2 dm. long, 0.7–1 dm. thick, the branches divergent or loosely ascending; *pedicels naked at apex;* spikelets 4–5 mm. long; lemma and palea more strongly strigose. — Very rare, sphagnous bog, se. Va. — Another rare var. in e. N.C.

228. C. longifolia.

2. C. longifòlia (Hook.) Scribn. (long-leaved). — *Culms* 6–18 dm. high, *from elongate scaly rhizomes,* stout; sheaths usually pubescent, at least on the margins; *leaves* elongated, involute above and *tapering into a long thread-like point; panicle pale,* 1.5–4.5 *dm. long, narrow,* the slender smooth branches erect or ascending; *spikelets* 6–7 *mm. long; glumes acute, the second equal to or exceeding the floret; callus-hairs more than half the length of the smooth lemma and palea.* — Sands, Ont. to Mackenz., s. to n. Ind., n. Ill., Mo., Kans. and Colo. July–Sept. Fig. 228.

49. AGRÓSTIS L. Bentgrass

Spikelets 1 (rarely 2)-flowered; glumes subequal and acute, longer than the broad lemma which is awnless or dorsally awned; palea hyaline, shorter than the lemma, or obsolete; grain loosely inclosed in the lemma. — Annuals or mostly perennials of temp. and cold reg., with usually scabrous leaves, membranaceous ligules and open or contracted panicles. (Old Greek name of grass from *agros, a field,* the place of growth of some species.) *

a. Lemma mostly entire or only shallowly toothed at tip, awnless or with stiff awn at most twice its length; perennials. . . *b.*

* Treatment based largely on that in Rhodora, xxxv. 203–212 (1933).

*b.*Palea 2-nerved, at least half as long as lemma; plants with creeping rhizomes and often with stolons. . . *c.*

*c.*Primary and upper sheaths ventricose; panicle glomerulate; glumes scabrous on back. . . . . 1. *A. verticillata.*

*c.*Sheaths close; panicle more or less open; glumes glabrous (sometimes scabrous on keel).

Ligule of lower and middle leaves as long as or longer than broad, 2–6 mm. long, rounded at tip; panicle with many spikelets on short branches near the rachis, contracted after flowering (or if remaining open, in no. 3, 1.5–3 dm. long with principal leaves 5–9 mm. wide), the spikelets of adjacent verticils not intermingling.

Basal leafy shoots usually numerous, erect and crowded; the base of the plant with widely creeping superficial stolons, not definitely scaly rhizomes; panicle contracting after anthesis, the branches and branchlets becoming erectish and imbricated. . . . . 2. *A. alba.*

Basal leafy shoots mostly decumbent, the plant spreading by scaly subterranean rhizomes; branches of panicle spreading or spreading-ascending in fruit, only the branchlets then appressed to the branches. . . . . 3. *A. gigantea.*

Ligule of lower and middle leaves broader than long, 0.5–1.5 mm. long, truncate; panicle open and diffuse, with branches floriferous chiefly above the middle, the spikelets of adjacent verticils intermingling, the panicle not contracted in fruit; leaves 1–5 mm. wide. . . . 4. *A. tenuis.*

*b.*Palea minute and nerveless or wanting; plants tufted, without stolons (except in no. 10). . . *d.*

*d.*Branches of panicle copiously spinulose-scabrous, spikelet-bearing chiefly toward their tips. . . *e.*

*e.*Spikelets 1.2–2 mm. long, crowded in close spiciform terminal glomerules, short-pedicelled to subsessile; tips of glumes distant in fruit, exposing the grain; anthers roundish, 0.2 mm. long; spring-flowering. 5. *A. hyemalis.*

*e.*Spikelets 2–4 mm. long, mostly distinctly pedicelled; tips of glumes more connivent in age, covering the grain; anthers elongate, 0.4–1 mm. long; summer- or autumn-flowering. . . *f.*

*f.*Mature panicle very diffuse, one-third to two-thirds the entire height of the plant, the widely divergent to reflexed, flexuous branches finally gibbous at base; lemma one-half to two-thirds as long as outer glume. . . . . 6. *A. scabra.*

*f.*Mature panicle scarcely diffuse, lance-ellipsoid to ovoid, one-sixth to (rarely) one-third the entire height of the plant, the stiffish or merely arched branches not at all or but slightly gibbous at base; lemma two-thirds to seven-eighths as long as outer glume.

Mature panicle broadly ovoid, with divergent branches; lemma often long-awned, two-thirds as long as glumes, obscurely nerved; cauline leaves 2–4, with blades 2–10 cm. long, their ligules 1.5–3 mm. long; boreal species. . . . . 7. *A. geminata.*

Mature panicle lanceolate to narrowly ovoid, with strongly ascending branches; lemma awnless or short-awned, prominently nerved, nearly equaling glumes; cauline leaves 5–10, with blades 0.6–2 dm. long, their ligules 4–5 mm. long; southern species. . . . . 8. *A. altissima.*

*d.*Branches of panicle glabrous or only slightly scabrous, forking and floriferous from near or below the middle. . . *g.*

*g.*Flowering panicle open, lanceolate or ellipsoid to ovoid or pyramidal, with elongate divergent or loosely ascending branches (if rarely contracted and short-branched, in var. of no. 11, with spikelets more than 2.5 mm. long). . . *h.*

*h.*Blades of culm-leaves elongate and loosely spreading; panicle 0.5–3.5 dm. long, mostly green, with 5–9 distinct whorls of long branches; lemmas mostly awnless. . . . . 9. *A. perennans.*

*h.*Blades of culm-leaves mostly short and ascending; panicles 1–15 cm. long, mostly bronze or purplish, with 2–4 distinct whorls of branches below the crowded summit; lemmas mostly awned.

Panicle in anthesis lanceolate or slenderly ellipsoid, in fruit closely contracted; spikelets 2–2.5 mm. long; anthers nearly as long as lemma. . . . . 10. *A. canina.*

Panicle in maturity ovoid or pyramidal, not contracting (if rarely slender at maturity with spikelets more than 2.5 mm. long); spikelets 2–4.6 mm. long; anthers a third as long as lemma. . 11. *A. borealis.*

g. Flowering panicle linear-cylindric, dense, 2–10 mm. in diameter, 2–6 cm. long; spikelets about 2 mm. long. . . . . 12. *A. Rossae.*

a. Lemma often sharply 2-toothed at apex, with a slender flexuous awn three to five times its length; annuals. . . . *i.*

i. Palea minute; rachilla not prolonged. . . . . 13. *A. Elliottiana.*

i. Palea nearly as long as lemma; rachilla prolonged behind it as a bristle.

Panicle becoming diffuse, its branches mostly naked at base. . . . . 14. *A. spica-venti.*

Panicle close, linear- or lance-cylindric, its short branches floriferous to base. 15. *A. interrupta.*

1. **A.** VERTICILLÀTA Vill. (whorled), WATER-B. — *Culm strongly decumbent, rooting at the nodes,* often producing stolons; *leaves lanceolate, with* short ligules, the primary and upper with *ventricose sheaths; panicle dense, glomerulate, with crowded sessile or subsessile spikelets* about 2 mm. long; glumes scabrous on the back; lemma half as long, truncate; palea nearly equaling the lemma. — Wet roadsides, ditches, etc., S.C. to Fla., Mex. and Pacific States; locally n. in waste land to Ct. June, July. (Adv. from Eu.) FIG. 229.

229. A. verticillata.

2. **A. álba** L. (white), REDTOP, FOIN FOLLETTE (Que.). — Perennial, *forming turf, with superficial* elongate *stolons and* usually *numerous erect sterile leafy shoots;* culms **0.2–1.3 m. high,** erect or geniculate at base, or in wet places often procumbent and branching at lower nodes; leaves deep green, 3–8 mm. wide, their sheaths close; the *ligule longer than broad, rounded at summit, the larger ones 2–6 mm. long; panicle* ellipsoid to ovoid, purplish to green, comparatively open, 1–3 dm. long, *with spreading to slightly ascending branches floriferous to below the middle and numerous shorter densely floriferous ones,* the longer 3–10 cm. long; *branches and branchlets becoming suberect and appressed in fruit;* spikelets short-pedicelled, lanceolate, 2–3.5 mm. long; glumes subequal, usually scabrous on the keel, awnless, or awned in forma **aristígera** Fern. (bearing awns); lemma two-thirds as long; palea somewhat shorter; anthers 1–1.6 mm. long. (Including *A. stolonifera* L.) — Damp thickets, swales, shores, etc. (indigenous northw.) and fields and roadsides (introd.), Nfld. to Yuk., s. to Ga., La., N.M., Ariz. and Calif. June–Sept. (Eurasia) FIG. 230.

230. A. alba (typical).

Var. **palústris** (Huds.) Pers. (of marshes), CREEPING or CARPET-BENT. — Densely matted, often with abundant superficial repent stolons; culms prostrate, decumbent or erect, slender; *leaves 1–5 mm. wide,* sometimes involute; *panicle in anthesis* stramineous to purple-bronze, *linear- to oblong-cylindric,* 1.5–18 cm. long, 0.3–3 *cm. thick, with closely appressed or ascending very floriferous branches at most 4 cm. long*: panicle in fruit closely contracted. (*A. stolonifera,* var. *compacta* Hartm.; *A. alba,* var. *maritima* Mey.; *A. maritima* Lam.; *A. palustris* Huds.) — Shores, shallow water (fresh to saline) and damp sands, s. Greenl.; se. Lab. to B.C., s. to Va., Ind., Wisc., Minn., N.M. and Calif. June–Sept. (Eu.) FIG. 231.

231. A. alba, v. palustris.

3. **A. gigantèa** Roth (gigantic), BLACK BENT. — Similar to larger extremes of no. 2; *with prolonged subterranean scaly rhizomes; sterile basal shoots decumbent,* often forming superficial stolons; culms up to 1.5 m. high; *leaves 5–9 mm. broad; ligule about as long as broad, 2–4 mm. long; panicle* open, purple to green, 1.5–3 dm. long, *its branches spreading in fruit, only the branchlets then becoming appressed.* (*A. nigra* With.) — Damp shores, e. Que.; and as an introduction w. to Mich., s. to N.J. June–Aug. (Eu.)

4. **A.** TÉNUIS Sibth. (slender), RHODE ISLAND BENT. — Loosely tufted to densely matted, often with subterranean stolons; culms slender, 2–8 dm. high, erect, or decumbent at base; leaves flat, 1–5 mm. wide, the middle and lower with *truncate ligule* 0.5–1.5 *mm. long; panicle ellipsoid to ovoid, not contracted in fruit,* bronze or purplish, rarely green, 0.4–2 dm. long, *the loosely ascending to spreading capillary branches loosely floriferous chiefly from near the middle;* spikelets 1.8–3 mm. long; lemma nearly equaling the glumes, awnless, or in forma ARISTÀTA (Sincl.) Wieg. (awned) awned; *palea about half as long as lemma.* (*A. alba,* var. *vulgaris* Thurb.; *A. vulgaris* With.) — Fields, pastures, roadsides, thickets, etc., Lab. to B.C., s. to N.C., O., Ind., Mo., N.D. and Oreg. June–Sept. (Natzd. from Eu.) FIG. 232. —

232. A. tenuis.

Hybridizes with no. 2. A teratological state (*A. sylvatica* L.) due to nematodes in the spikelets has the glumes, etc., distorted and greatly prolonged.

Var. PÙMILA (L.) Druce (dwarf). — Densely tufted, barely if at all stoloniferous, 0.3–1.5 dm. high; panicle 1–4 cm. long; fruit often blackened by smut. — Sterile open soil, Nfld., St. P. et Miq. and Gaspé Pen., Que. — Probably only a pathological state.

5. A. hyemàlis (Walt.) BSP. (of winter), TICKLEGRASS, HAIRGRASS. — Slightly to densely tufted, 1–8 dm. high; culms very slender, erect, or slightly geniculate at base; leaves erect, involute-filiform or the short cauline with blades sometimes flat and up to 2 mm. wide; *ligules* lacerate, 0.5–4 *mm. long; panicle* purple, rarely green, very loose and open in maturity, slightly exserted, 0.4–3 dm. long, *with* erect or ascending to spreading or finally deflexed *filiform scabrous branches simple or forking only near the tip; spikelets* 1.2–2 *mm. long, crowded in close spiciform terminal glomerules* 2–10 *mm. long, subsessile or short-pedicelled; glumes* subequal, acute, scabrous on the keel, *their tips distinct in fruit, exposing the grain;* lemma 0.5–1 mm. long, blunt; anthers roundish, about 0.2 mm. long. (*A. antecedens* Bickn.) — Dry or damp open sterile soil, bogs or thin woods, Fla. to Tex., n. on and near the Coastal Plain to Mass., s. N.Y., e. Pa., and in the interior n. to Ind., Ill., Minn. and Kans. Mar.–June. FIG. 233.

233. A. hyemalis.

6. A. scàbra Willd. (harsh), HAIRGRASS, FLY-AWAY GRASS, TICKLEGRASS, FOIN FOU (Que.). — Cespitose, 1.5–10 dm. high; culms slender, erect or at base geniculate, fragile; basal leaves involute; cauline leaves flat, up to 5 mm. broad, often inrolling; their *ligules lacerate,* 2.5–7 *mm. long; panicle* purple to green, *very diffuse,* 0.5–4 *dm. long,* finally almost as broad at base, at first included and vaselike at base, with erect branches; *later exserted, with widely divergent to deflexed harshly hispid capillary flexuous branches forking above the middle and* (*in age*) *gibbous at base; spikelets mostly pedicelled,* 2–3 *mm. long; glumes* lance-attenuate, unequal, scabrous on the keel, often discolored at the hyaline margin, becoming *closely connivent in maturity;* lemma 1.3–2 mm. long, awnless, or dorsally awned in forma **Tuckermàni** Fern. (for EDWARD TUCKERMAN, 1817–1886); anthers elongate, 0.4–0.8 mm. long. (*A. hyemalis* in large part, of ed. 7) — Sterile wet, exsiccated or dry open soil, Lab. to Alaska, s. to Md., locally to S.C., O., Ind., Ill., Ia., N.M., Ariz. and Calif. June–Nov. (E. Asia) FIG. 234.

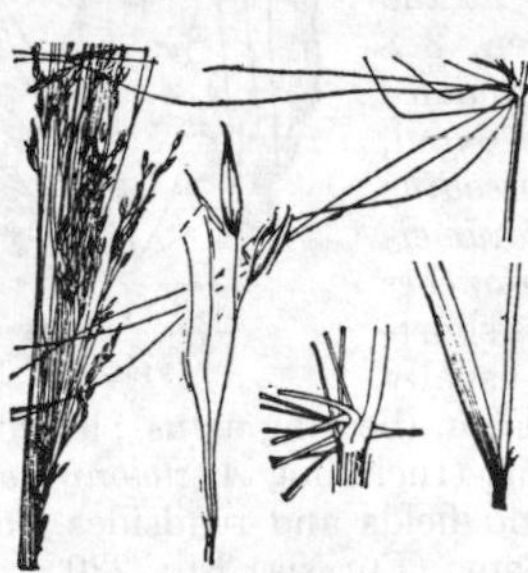
234. A. scabra.

Var. **septentrionàlis** Fern. (northern). — *Spikelets* 3.2–4.3 *mm. long;* glumes subaristate at tip; lemma 2–2.5 mm. long, awnless, or awned in forma **setígera** Fern. (bearing bristles). — Wet sands, peats and barrens, Lab. and e. Saguenay Co., Que., to N.S. Late July–Oct.

7. A. geminàta Trin. (twin; from the branching of the panicle). — Intermediate between nos. 6 and 11, forming small tufts 1–4 (rarely –8) dm. high; basal leaves narrowly linear to involute; *cauline leaves* 2–4, *with flat blades* 2–10 *cm. long* and 1–2 mm. broad, *their ligules* 1.5–4 *mm. long; panicle well exserted, in maturity broadly ovoid, one-sixth to rarely one-third the entire height of the plant,* 0.3–2 dm. long, *very open, with arched-ascending to divergent straightish* (*scarcely flexuous*) *scabrous-hirtellous branches forking mostly above the middle, their bases scarcely gibbous;* spikelets 2–3 mm. long, purple, bronze or green; glumes unequal, ovate-lanceolate; *lemma* about two-thirds as long, obscurely nerved, *dorsally awned,* or awnless in forma **exaristàta** Fern. (without awns). (*A. hyemalis,* var. *geminata* Hitchc.) — Shores, barrens and rocky or peaty habitats, Lab. to Alaska, s. to Nfld., Gaspé Pen., Que., Ont., Minn., Colo. and Calif. July, Aug. FIG. 235. — Apparently hybridizes with nos. 6 and 11.

235. A. geminata.

8. A. altíssima (Walt.) Tuckerm. (tall). — Slightly to densely tufted, with stiff culms 3–10 dm. high, usually *without leafy autumnal offshoots; cauline leaves* 5–10, *with overlapping sheaths and stiff erect* narrowly linear to *involute harsh blades* 0.6–2 *dm. long, their ligules* 4–5 *mm. long; panicle* purple or greenish, partly included to exserted, *lanceolate to narrowly ovoid,* 1–3.3 dm. long, *with few loosely ascending scabrous* capillary *branches,* these *forking into appressed short branchlets above the middle;* spikelets 2.5–3.8 mm. long, short-pedicelled;

236. A. altissima.

glumes lance-attenuate, slightly unequal, *the 2nd nearly equaled by the prominently nerved* awnless or short-awned *lemma* (2.3–3 *mm. long*). (*A. perennans*, var. *elata* Hitchc.; *A. elata* (Pursh) Trin.) — Damp sands, bogs and pine-barrens, nw. Fla. to La. and Miss., n. to N.J. and e. Pa. and, locally, to se. Mass. Aug.–Oct. Fig. 236.

9. A. perénnans (Walt.) Tuckerm. (perennating), Upland Bent. — Erect or geniculate at base, more or less tufted, 3–10 dm. high, the culms often branching below, *commonly with abundant leafy autumnal offshoots at base; leaves flat*, elongate, scabrous; *the cauline* 3–7, *with blades mostly* 1–3 *dm. long*, 2–6 *mm. broad, much longer than their sheaths*, their oblong round-tipped ligules 2.5–5 mm. long; *panicle* green, sometimes bronze-tinged, lance-ellipsoid to narrowly ovoid, 0.5–3.5 dm. long, *with* 5–9 *whorls of distant smooth* or but slightly scabridulous capillary ascending to divergent *branches forking near or below the middle; spikelets* 2–3 mm. long, *with pedicels mostly shorter than or about their length and appressed-ascending;* glumes unequal, acuminate; *lemma* 1.5–2 *mm. long*, shorter than 2nd glume, awnless, or awned in the infrequent forma **chaetóphora** Fern. (bearing bristles); palea very short. — Open woods, thickets, rocky banks and dryish open soil, Que. to Minn., s. to Fla., La. and Tex. Late July–Oct.

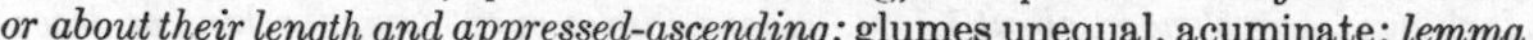

237. A. perennans, v. aestivalis.

Var. **aestivàlis** Vasey (of summer), Thingrass. — *Weak*, erect or decumbent; *ligules* 1–3 *mm. long; panicle very lax*, with divergent branches and mostly divergent branchlets; *many of the spreading pedicels one to three times the length of the spikelets;* lemma awnless, or awned in the rare forma **atheróphora** Fern. (bearing a bristle) = *A. Schweinitzii* Trin. — Woods, thickets and shores, Que. to Wisc., s. to N.C., Tenn., Miss., Mo. and Kans. Late June–Oct. Fig. 237.

10. A. canìna L. (of a dog; this and some other species formerly known as Dog's Grass), Brown or Velvet-Bent. — Forming dense mats, 1–6 dm. high, *often* (*especially late in the season*) *producing long trailing stolons with leaves densely tufted at the nodes; basal leaves involute-setaceous;* the cauline 2–4, flat, 1–2 mm. wide, their blades erect, short, with ligules 2–5 mm. long; *panicle during anthesis lanceolate to slenderly ellipsoid, in fruit strongly contracted*, purplish to yellowish, 2–15 cm. long, *with smooth capillary ascending to slightly spreading branches* forking near the middle; *spikelets* 2–2.5 *mm. long;* lemma shorter than the outer glume, dorsally awned, or awnless in forma mùtica (Gaudin) Döll (blunt); palea minute; *anthers nearly equaling lemma.* — Peaty or siliceous crests, slopes and shores, s. Nfld. to e. Me.; and (introd.) pastures and sterile fields to Minn., s. to Del. and Tenn. June, July. (Eu.) Fig. 238. — Hybridizes with no. 4.

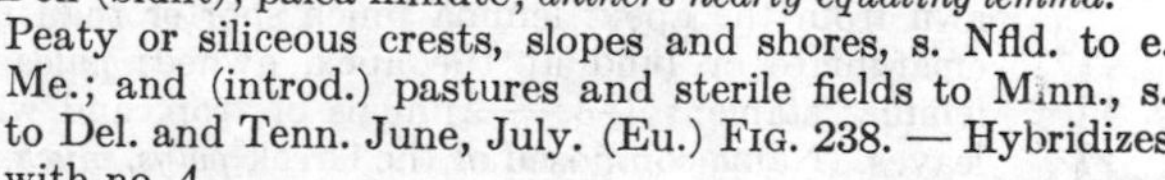

238. A. canina.

239. A. borealis (typical).

11. A. boreàlis Hartm. (northern). — Tufted, 0.2–7 dm. high, *without stolons;* basal *leaves* flat or involute; *the cauline* 2–4, *erect*, 0.2–4 mm. broad, with ligule 1.5–3 mm. long; *panicle* purple, rarely greenish, *in maturity ovoid* (rarely lanceolate), *not contracting*, 1–14 cm. long; *its divergent* (rarely ascending) *branches smooth* or smoothish, forking near the middle; *spikelets* 2–4.6 *mm. long;* glumes narrowly ovate, subequal; lemma nearly as long, usually dorsally awned; palea short; *anthers one-third as long as lemma.* — Variable.

Lemma awned; branches of mature panicle spreading-ascending.
Leaf-blades 0.2–2 mm. broad, often involute; glumes 2–3 mm. long.
Spikelets normal, 2–3 mm. long. . . . . *A. borealis* (typical).
Spikelets with lemmas modified and much prolonged. . . . Forma *macrantha.*
Leaf-blades 1–4 mm. broad, flat; glumes 2.8–4 mm. long. . . . Var. *americana.*
Lemma awnless; branches of mature panicle strongly ascending; spikelets 2.8–4.6 mm. long. . . . . Var. *paludosa.*

A. boreàlis (typical). — Culms 0.4–4.5 dm. high; panicle 1.5–14 cm. long. — Gravelly or rocky open soil, Greenl. and Lab. to Nfld., e. Que. and mts. of n. N.E. and n. N.Y.; Alaska and B.C. July, Aug. (Eurasia) Fig. 239. — Forma **macrántha** (Eames) Fern. (large-flowered), infrequent (perhaps pathological).

Var. **americàna** (Scribn.) Fern. (American). — Culms up to 7.5 dm. high; panicle up to 17 cm. long. — Brooksides and meadows, w. Nfld. and e. Que. to mts. of n. Me. and n. N.H.; Roan Mt., N.C. and Tenn.

240. A. borealis, v. paludosa.

Var. **paludòsa** (Scribn.) Fern. (of marshes). — Dwarf, 0.2–3.5 dm. high; panicle 1–10 cm. long. — Peat or gravel, Lab., w. Nfld. and e. Saguenay Co., Que. Fig. 240.

241. A. Rossae.

12. A. Róssae Vasey (named, in 1892, for its discoverer, Edith Ross). — Densely tufted, 1–2 dm. high; *leaves* about 1 mm. wide, flat or conduplicate, the lower 2–3 cm. long; *the 2–3 cauline confined to the lower half of the culm, their blades 1–2 cm. long* and *exceeded by the sheaths; panicle* long-exserted, *linear-cylindric,* dense, *2–6 cm. long, 2–10 mm. thick; spikelets* purplish or bronze, 2 *mm. long,* on very short smooth branches; glumes ovate, subequal; lemma two-thirds as long; palea minute. — Rocky places, Bonne Bay, Nfld. (*Jansson*); B.C. to Colo. and Calif. July, Aug. Fig. 241.

13. A. Elliottiàna Schultes (for Stephen Elliott, 1771–1830). — *Annual,* delicate, 1–5.5 dm. high; culms forking; *leaves slender, about* 1 *mm. wide; panicles* open, *weak, arching or drooping,* 0.5–2.5 dm. long, purplish to green; glumes about 2 mm. long, scabrous on the keels; *lemma* slightly shorter, with a ring of hairs at base, 5-nerved, *sharply 2-toothed, with a dorsal flexuous very delicate awn nearly* 1 *cm. long; palea minute or wanting.* — Dry or exsiccated sterile soil, Ga. to Tex., n. to se. Va. (locally adv. to Me.), Ky., s. Ind., s. Ill., Mo. and se. Kans. Apr.–early June.

242. A. spica-venti.

14. A. spìca-vénti L. (wind-spike; the panicle teetering in the wind). — *Annual,* 3–8 dm. high, tufted; *leaves flat,* 2–6 *mm. wide; panicle open,* 1–3.5 dm. long, *the slender branches* forking near or below the middle and *naked below;* spikelets 2 mm. long, lustrous, bronze or green; glumes subequal; lemma and *palea nearly equaling the glumes; the scabrous lemma sharply cleft, with a delicate awn 5–7 mm. long; rachilla prolonged back of palea as a naked bristle.* (*Apera* Beauv.) — Roadsides and waste places, local, Me. to Mich., s. to Del., Md., O. and Mo., June–Oct. (Adv. from Eu.) Fig. 242.

15. A. interrúpta L. (interrupted). — Similar to no. 14; *panicle linear- or lance-cylindric, close but interrupted, its short branches floriferous to base; awns* of lemmas 8–12 *mm. long.* — Waste ground, e. Mo.; and on Pacific coast. (Adv. from Eu.)

50. POLYPÒGON Desf. Beardgrass

Spikelets 1-flowered, in a dense spike-like panicle; pedicel disarticulating below the glumes, the ripened spikelet with a short stipe; glumes subequal, entire or 2-lobed, bearing a straight awn from the apex; lemma much shorter than the glumes, broad, emarginate or bifid at the apex, awned; palea smaller than the lemma; stamens 1–3. — Annuals of trop. and warm reg. with flat leaves. (Name composed of the Greek *polys, much,* and *pogon, beard.*)

243. P. monspeliensis.

1. P. monspeliénsis (L.) Desf. (of Montpellier). — Culms 2–6 dm. high, erect from a decumbent base, usually tufted; blades linear, scabrous; *panicle* 3–10 cm. long, *dense, interrupted, pale and soft, silky,* often partly included in the uppermost sheath; spikelets 2.5–3 mm. long. — Waste places and damp soil, e. Va. to Ga., Tex. and Pacific coast; casually n. to Me. and Que. May–Sept. (Natzd. from Eu.) Fig. 243.

51. CÍNNA L. Wood Reedgrass

Spikelets 1-flowered; rachilla articulated below the glumes, forming a short naked stipe below the floret, and prolonged behind the palea into a minute bristle; glumes narrow, hispidulous on the keel; lemma 3–5-nerved, with a short awn from between the minute teeth of the bifid apex; palea 1-nerved, or 2-nerved with the nerves close together; stamen 1. — Tall N.Am. and Eurasian perennials with flat leaves, conspicuous hyaline ligules, and many-flowered nodding panicles. (*Cinna,* a name used by Dioscorides for a kind of grass.)

Spikelets 4.5–6 mm. long; glumes firm, subherbaceous, scabrous-hispid, only the narrow margin hyaline, strongly unequal, the 2nd nearly equaling to exceeding the lemma; anther 1.2–1.5 mm. long. . . . . 1. *C. arundinacea.*

Spikelets 2–4.5 mm. long; at least the 1st glume hyaline except for the midrib.

Panicle contracted, with appressed-ascending branches; glumes unequal, 1st much shorter than lemma, 2nd herbaceous or firm; anther 1.2–1.4 mm. long. 1a. *C. arundinacea,* var. *inexpansa.*

Panicle lax, when well-developed with elongate spreading or drooping branches; glumes subequal, both hyaline, 1st nearly equaling lemma; anther 0.5–0.8 mm. long. . . . . 2. *C. latifolia.*

1. C. arundinàcea L. (reed-like). — Culms erect, 0.5–1.5 m. high, solitary or few, often bulbiform at base; *leaf-blades* 1.5–4 *dm. long,* 0.6–1.8 cm. wide, scabrous; panicle 1–4 dm. long, green or purple-tinged, the branches ascending or spreading, somewhat contracted after flowering; *spikelets* 4.5–6 *mm. long; glumes firm, subherbaceous, scabrous-hispid, strongly unequal, the* 2nd 1–2 *mm. longer than the* 1st, about as long as to exceeding the scabrous minutely awned or awnless lemma; palea 1-nerved; *anther* 1.2–1.5 *mm. long.* — Moist woods and shaded swamps, centr. Me. and sw. Que. to s. Ont. and Minn., s. to Ga., Tenn., Ark. and Tex. Late July–Oct. Fig. 244.

244. C. arundinacea.

Var. **inexpánsa** Fern. & Grisc. (unexpanded). — Panicle closely contracted; *spikelets* 3.7–4.2 *mm. long;* 1st *glume hyaline, glabrous except for scabrous evanescent keel,* 2nd glume shorter than to equaling lemma. — Swamps and wet woods, se. Va. to La. and Okla.

245. C. latifolia.

2. C. latifòlia (Trev.) Griseb. (broad-leaved). — Similar to no. 1, 0.2–1.5 m. high; *blades* 0.6–2.5 *dm. long,* thinner; panicle looser, with more slender branches mostly spreading or drooping; *spikelets* 2–4.5 *mm. long; glumes more delicate and hyaline, subequal, the* 1st *nearly equaling* the short-awned *lemma;* palea 1–2-nerved; *anther* 0.5–0.8 *mm. long.* — Woods, thickets and clearings, se. Lab. Pen. to s. Alaska, s. to Nfld., N.S., s. N.E., L.I., uplands to N.C. and Tenn., Mich., n. Ill., Minn., Colo. and Calif. Early July–Oct. (Eurasia) Fig. 245.

52. HELEÓCHLOA Host

Spikelets 1-flowered, flattened, in dense ellipsoid-ovoid spike-like panicles; glumes awnless, shorter than the 1-nerved lemma which subtends a palea of nearly equal length. — Low cespitose branching annuals nat. of s. Asia and the Mediterr. reg., the numerous spike-like panicles partly included in the inflated sheaths. (Name from the Greek *helos, a marsh,* and *chloa, grass.*)

1. H. schoenoìdes (L.) Host (resembling *Schoenus*). — Usually almost prostrate; leaves rather rigid, tapering to a sharp point; spike 1.5–4 cm. long. — Waste places, local, Mass. to Wisc., s. to Del., Pa. and Mo. July–Sept. (Natzd. from Eu.) Fig. 246.

246. H. schoenoides.

53. PHLÈUM L. Timothy. Phléole or Fléole (Que.)

Spikelets 1-flowered, flattened, in dense cylindrical spike-like panicles; glumes equal, ciliate on the keels, and abruptly awn-pointed, longer than the broad truncate 3–5-nerved hyaline lemma; palea nearly equal, narrow. — Erect simple perennials of cool and temp. reg., with flat leaves and terminal spike-like panicles. (From *phleos,* a Greek name for a kind of reed.)

1. P. praténse L. (of meadows), Common T., Herds' Grass, Mil (Que.). — Culms usually erect, *from* slightly arching *bulbous bases,* solitary or in clumps, 3–10 dm. high, with few or no leafy tufts at flowering time, *minutely scabrous at summit; leaf-sheaths close,* the principal blades 5–10 mm. broad; *panicle slender-cylindric,* green, becoming drab, 0.1–2.2 dm. long, 7–10 mm. thick, stiff and harsh; awns mostly about half as long as bodies of glumes; anthers 1.5–2 mm. long. — Fields, roadsides and clearings, commonly cult. and freely natzd. June–Aug. (Introd. and natzd. from Eu.) Fig. 247. — Spikelets changed to leafy tufts in forma vivíparum (S. F. Gray) Louis-Marie (producing young well-formed); panicle occasionally subtended by a leafy bract. — Passing into

247. P. pratense.

Var. nodòsum (L.) Huds. (knotty; from the bulbous base). — More slender; *culms more decumbent, often with leafy tufts at flowering time,* rarely 7.5 dm. high; *principal leaves* 2–5 *mm. wide; panicle* rarely more than 1 dm. long, 3–6 *mm. thick.* — Fields, roadsides, etc., Nfld. to James Bay, s. to N.E. and N.C. (Natzd. from Eu.)

2. P. alpìnum L. (alpine), Mountain-T. — *Culms from prolonged or creeping bases or*

248. P. alpinum.

cespitose, not bulbous at base, glabrous and smooth at tip, 1–7 dm. high; *upper sheath ventricose; panicle thick-cylindric,* green or drab to purplish, 1–4.5 cm. long, 6–14 mm. thick; awns mostly two-thirds to three-fourths as long as bodies of glumes; anthers 1–1.8 mm. long. — Meadows, damp shores and slopes, Greenl. and Lab. to Alaska, s. to Nfld., Gaspé Pen., Que., alpine reg. of Me. and N.H., n. Mich., N.M. and Calif. July, Aug. (Eurasia) FIG. 248.

GASTRÍDIUM (name from the Greek *gastridion, a small pouch*) VENTRICÒSUM (Govan) Schinz & Thell. (bellied out), differing from *Phleum* in its lustrous slender panicle of spikelets with non-ciliate and prolonged glumes, the very short and villous lemma with a delicate awn much longer than the blade, is sporadic on wool-waste and rubbish. (Adv. from Eu.)

54. ALOPECÙRUS L. FOXTAIL. VULPIN (Que.)

Spikelets 1-flowered, flattened, falling from the axis entire, in slender spike-like panicles; glumes equal, awnless, usually connate at the base, ciliate on the keel; the broad 5-nerved obtuse lemma nearly equaling the glumes, with a slender erect dorsal awn from below the middle, its margins connate near the base; palea none. — Perennials or annuals of temp. and cool reg., with flat leaves and soft dense spike-like panicles. (Name from the Greek *alopex,* fox, and *oura, tail.*)

Spikelets (excluding awns) 4.5–7 mm. long; awn exserted 5–7 mm.; anthers 2.5–3 mm. long.
  Glumes glabrous except for the short-ciliate keels, united to near the middle. 1. *A. myosuroides.*
  Glumes pubescent on the nerves, long-ciliate on the keels, free or barely united at base. . . . . 2. *A. pratensis.*
Spikelets 2–3 mm. long; awn included to exserted about 4 mm.; anthers 0.5–2 mm. long.
  Awn inserted toward base of lemma, geniculate or twisted, exserted 2–4 mm.
    Perennial with geniculate or creeping culms and offshoots; spikelets 2.5–3 mm. long; anthers 1–2 mm. long. . . . . 3. *A. geniculatus.*
    Annual, erect or ascending; spikelets 2–2.6 mm. long; anthers about 0.5 mm. long. 4. *A. carolinianus.*
  Awn inserted near middle of lemma, straight, included, or exserted less than 2 mm. . . . . 5. *A. aequalis.*

1. A. MYOSUROÌDES Huds. (resembling *Myosurus,* Mouse-tail), SLENDER F. — *Annual,* with erect or decumbent culms 2–7 dm. high; leaves scabrous; *panicle slender,* 2–10 cm. long, *4–5 mm. thick; spikelets 6–7 mm. long; glumes glabrous except for the short-ciliate keels, united to near the middle,* slightly shorter than the lemmas; awn exserted 5–7 mm.; anthers 2.5–3 mm. long. (*A. agrestis* L.) — Fields and waste places, local, Mass. to Mich., s. to N.C. and Kans. May–Sept. (Adv. from Eu.) FIG. 249.

249. A. myosuroides.

2. A. PRATÉNSIS L. (of meadows), MEADOW-F. — Perennial, with culms erect or slightly decumbent at base, 3–10 dm. high, from creeping rhizomes; sheaths loose, the upper usually inflated; *panicle* rather stout, 3–12 cm. long, 6–10 *mm. thick;* spikelets about 5 mm. long; *lemmas equaling the acute long-ciliate nearly distinct glumes;* awn usually exserted about 5 mm.; *anthers* 2.5–3 *mm. long.* — Meadows, pastures and damp clearings, Nfld. to s. Ont., s. to Ga., Ind., Mo. and Kans.; Alaska to Id. and Oreg. May–Aug. (Natzd. from Eu.) FIG. 250.

250. A. pratensis.

A. VENTRICÒSUS Pers. (bellied-out), a species resembling no. 2, but more strongly *long-stoloniferous, the glume strongly divergent at summit and much exceeding the lemma, the short awn included,* is appearing in meadows in Nfld. (Natzd. from Eu.)

3. A. GENICULÀTUS L. (bent like the knee-joint), MARSH-F. — Matted *perennial, with decumbent or prolonged and repent shoots,* the ascending tips 1–6 dm. high; sheaths inflated; panicle 1.5–7.5 cm. long, excluding the awns 4–8 mm. thick; *spikelets* 2.5–3 *mm. long; glumes united at base, obtuse, long-ciliate;* lemmas slightly shorter; *awn bent, the exserted portion nearly twice as long as the glumes; anthers* 1–2 *mm. long.* — Ditches, pools and wet clearings, se. Lab. and Nfld. to N.J. and Minn. May–Aug. (Natzd. from Eu.) FIG. 251.

251. A. geniculatus.

4. A. carolinianus Walt. (of Carolina). — Resembling no. 3, but *annual,* tufted, *erect or only slightly geniculate at base,* 1.5–6 dm. high; panicle simple or distinctly ramose; *spikelets* 2–2.6 *mm. long; anthers about* 0.5 *mm. long,* (*A. ramosus* Poir.) — Shores, ditches, fallow fields and low grounds, Fla. to Tex., n. at low alts. to N.J., Pa., O., Ind., Wisc., Minn., S.D., Mont. and Wash. April–July.

5. A. aequàlis Sobol. (equal; referring to glumes and lemma). — Usually glaucous, perennial, the tufted culms erect or decumbent at base, up to 7 dm. high; *sheaths only slightly inflated, the upper* 3.5–10 *cm. long; panicle slender, whitish-drab to mouse-color,* 2.5–8 cm. long, 3–5.5 *mm. thick,* finally long (0.3–2.3 dm.) -exserted; *spikelets* 2–2.5 *mm. long;* glumes obtuse, connate at base, silky, with long-ciliate keels, about equaling the lemmas; *awn attached near middle of lemma, straight and short, included, or exserted* 1–2 *mm.; anthers* 0.6–1 *mm. long.* (*A. fulvus* Sm.; *A. aristulatus* Michx.; *A. geniculatus,* var. *aristulatus* Torr.) — Shallow water, shores, ditches, etc., Nfld. to Alaska, s. to N.S., N.E., Md., O., Ind., Ill., Mo., e. Kans., N.M., Ariz. and Calif. Late May–Sept. (Eurasia)

Var. **nàtans** (Wahlenb.) Fern. (floating). — Lax; *stems creeping or floating; sheaths inflated, the upper* 1–5 *cm. long; panicle* 0.7–3.5 *cm. long,* often purple-tinged, *the base included or finally exserted* 1–5 *cm.* — Shallow pools, nw. Nfld. and adj. e. Que. July–Sept. (Greenl.; Alaska; Eurasia)

252. M. minima.

MÍBORA (meaning unexplained) MÍNIMA (L.) Desv. (tiny), a tiny tufted vernal grass 2–8 cm. high, with filiform mostly naked culms and simple spikes (about 1 cm. long) of tiny purple spikelets without awns, occurs in nurseries in Mass. and N.Y. (Adv. from Eu.) FIG. 252.

55. MUHLENBÉRGIA Schreb.

Spikelets 1 (rarely 2 or 3)-flowered, in contracted to loosely open panicles; a short usually barbate callus below the floret; glumes thin, often aristate; lemma narrow, membranaceous to firm, 3-nerved, awned or awnless, inclosing a thin subequal palea; grain closely enveloped by the lemma. — Genus of N. and S.Am. and e. Asia, our species perennial, often with scaly rhizomes, flat or involute leaves and small spikelets. (Dedicated to *Gotthilf Henry Ernest Muhlenberg,* a distinguished American botanist, 1753–1815, but not one who would have felt honored by the undignified recently made "English" name "MUHLY" for these grasses.)*

a. Panicle contracted, from linear-filiform to lanceolate or subcylindric; spikelets sessile or very short-pedicelled, appressed-ascending, often closely overlapping; lemmas acute or acuminate; culms from tough crowns, scaly rhizomes or decumbent rooting bases. . *b.*

b. Glumes shorter than to barely exceeding lemma, 0.1–3 mm. long; panicle usually arching, its longer branches free from spikelets at base. . . *c.*

c. Culms in dense mats or tussocks, filiform, rigid or subrigid, without elongate scaly rhizomes, at most with slender stolons or, late in the season, bulbiform basal offsets from the hard and knotty crown; leaves stiff, 1–2 mm. wide, quickly involute; panicles slenderly interrupted-spiciform; glumes and lemmas awnless, the latter beardless at base.

Perennial by proliferation from bases of old culms; lemmas glabrous. . 1. *M. Richardsonis.*

Perennial by production of bulbiform basal offshoots; lemmas minutely pubescent. . . . . 2. *M. cuspidata.*

c. Culms 1–few, from hard and elongate scaly rhizomes or, if not rhizomatous, with decumbent bases rooting at the joints; leaves flat or only upon drying becoming inrolled, the principal ones 2–15 mm. broad; panicle slender to thick and dense; lemma (except in no. 11) hairy or bearded at base. . . *d.*

d. Culms weak, without basal rhizomes, reclining and rooting at base; glumes round-tipped or subtruncate, the 1st often minute; panicle soft and flexuous, very slender; lemma awned.

2nd glume 0.2–0.5 mm. long; lemma 1.8–2.2 mm. long, with awn as long as to twice as long (2–5 mm.). . . . . 3. *M. Schreberi.*

2nd glume 1.5–2 mm. long; lemma 2.5–3 mm. long, with awn only half its length. . . . . 4. *M. curtisetosa.*

d. Culms firm, few or solitary from scaly rhizomes; glumes attenuate to pointed or awned tips, both well developed. . . *e.*

e. Glumes ovate, subequal, much shorter than lemma; base of lemma villous or pilose; anthers 0.8–1.5 mm. long; ligule obsolete or shorter than elongate cartilaginous summit of leaf-sheath; median

* In species 5–13 the stolons are frequently much enlarged or distorted as a result of insect-attack.

scales of slender rhizomes closely appressed, oblong to narrowly ovate.

Internodes of culms glabrous or at most minutely scabrous just below the glabrous nodes; spikelets 1.7–2.7 mm. long; lemma awnless or with short awn up to 1.5 mm. long; anthers 0.8–1 mm. long; culms often branching from upper nodes . . . . 5. *M. sobolifera.*

Internodes minutely pilose below and often above the pubescent nodes; spikelets (excluding awns) 3–4 mm. long; lemma tapering into a delicate awn 5–10 mm. long; anthers 1.1–1.5 mm. long; culm simple, only rarely forking. . . . . 6. *M. tenuiflora.*

*e.*Glumes linear-attenuate to narrowly ovate, at least the 2nd nearly to quite as long as the lemma; callus and base of lemma with long straightish beard (except in no. 11); anthers 0.3–0.8 mm. long; ligule usually obvious above the short cartilaginous summit of the sheath; median scales of the rhizomes (except in no. 9) loosely cucullate-arching at base, ovate. . . *f.*

*f.*Callus and base of lemma copiously long-bearded. . . *g.*

*g.*Internodes of culm definitely puberulent below the nodes.

Spikelets loosely disposed on branches of panicle, often slenderly pedicelled; glumes and lemma scarious or hyaline, commonly silvery-green or whitish; glumes unequal; mature grain loosely enveloped and easily freed. . . . . . 7. *M. sylvatica.*

Spikelets closely imbricated on the branches, subsessile or very short-pedicelled; glumes and lemma firm, commonly green or purplish; glumes subequal; mature grain tightly embraced. . . . . 8. *M. mexicana.*

*g.*Internodes of culm glabrous throughout or (in no. 9) barely scabrous at summit; glumes very unequal.

Median scales of rhizome tightly appressed, oblong to narrowly ovate; culm with erect scarcely geniculate simple branches, the internodes glabrous or barely scabrous at summit; panicles linear-cylindric, mostly exserted, 7–15 cm. long; lemma long-awned; anthers 0.8 mm. long. . . 9. *M. brachyphylla.*

Median scales of rhizome loosely cucullate-arching, ovate; culm freely branching, it and the forking branches often geniculate; internodes glabrous; primary panicles lanceolate to slenderly ovoid, little if at all exserted, mostly shorter, sessile or partly included; axillary panicles numerous; lemma awnless or awned; anthers 0.5 mm. long. . . 10. *M. frondosa.*

*f.*Callus and base of lemma glabrous or very sparsely short-setose; culms freely forking, rather bushy-branched; internodes scabrous-puberulent; leaves firm, narrow, erect; panicles mostly included at base; anthers 0.5 mm. long. . . . . . . 11. *M. glabriflora.*

*b.*Glumes linear-lanceolate, subequal, prolonged to long often arching awns, 4.5–8 mm. long, much longer than lemma; panicles spiciform, mostly interrupted, their short and stiff branches densely flowered at base; lemma slenderly villous at base.

Culms usually with several stiffly erect branches from the middle nodes, with internodes glabrous and lustrous (sometimes puberulent at summit); leaf-sheaths keeled; ligule prolonged, conspicuous; anthers 0.5–0.8 mm. long; grain linear-cylindric, 1.8–2.2 mm. long; western. . 12. *M. racemosa.*

Culms simple, rarely with erect basal branches, with internodes puberulent nearly to base; leaf-sheath hardly keeled; ligule short, usually hidden; anthers 1–1.5 mm. long; grain oblong-cylindric, 1.2–1.5 mm. long; northern and eastern. . . . . 13. *M. glomerata.*

*a.*Panicle open and diffuse (contracted only in Nfld. var. of no. 14), slenderly cylindric to broadly ellipsoid; spikelets long-pedicelled, scattered and loosely spreading, purple, bronze or brown; callus not bearded; culms simple or only sparingly branched at base. . . *h.*

*h.*Panicle 0.2–3 dm. long, 0.2–10 cm. thick; spikelets 1.3–2.5 mm. long; glumes and lemma blunt to subacute, awnless.

Culms arising from bases of old culms, the terete bases not rhizomatous; glumes much shorter than glabrous lemma. . . . . 14. *M. uniflora.*

Culms and leafy tufts compressed at base, arising from scaly rhizomes; upper glume and lemma subequal.

Strongly compressed flowering tufts 1–few, from short and thick rhizomes; leaves prolonged, half as long as culm, erect, conduplicate; panicle slenderly cylindric, 1–3 dm. long; lemma puberulent; eastern. 15. *M. Torreyana.*

Slightly compressed flowering tufts numerous and subcespitose, from cord-like slender rhizomatous caudices; leaves short, numerous, spreading or ascending, flat or folded; panicle ovoid, 0.5–2 dm. long; lemma glabrous; western. . . . . 16. *M. asperifolia.*

h. Panicle 2–4.5 dm. long, 0.7–2 dm. thick; spikelets (excluding awns) 3–5 mm. long, with long-tapering glumes and lemma, the latter usually awned; cespitose, without stolons or elongate rhizomes.

Glumes very unequal, bristle-tipped; awn of lemma **5–20 mm.** long. . . 17. *M. capillaris.*

Glumes subequal, merely acutish, not bristle-tipped; lemma awnless or with awn 1–4 mm. long. . . . . 18. *M. expansa.*

1. M. Richardsònis (Trin.) Rydb. (for its discoverer, Sir JOHN RICHARDSON, 1787–1865). — *Matted wiry and very slender perennial;* the ascending fertile *culms springing singly or in tufts from the depressed bases of old culms,* more rarely from slender stolons, 1.5–6 dm. high; *leaves* 1–2 *mm. wide, soon becoming involute,* erect or ascending; *ligule* 2 *mm. long, acute; panicles* exserted, *linear-filiform, interrupted, stiff,* 1–8 cm. long; *spikelets lead-colored, lance-subulate,* about 3 mm. long; *glumes much shorter than the* attenuate slender-pointed tightly rolled *glabrous lemma.* (*Sporobolus Richardsonis* (Trin.) Merr.) — Gravelly shores, calcareous gravels, damp thickets, etc., Anticosti I. and Gaspé Pen., Que.; Restigouche R., N.B.; St. John R. syst., N.B. and Me.; Kennebec R., Me.; Kalamazoo Co., Mich.; Minn. to Alta. s. to Neb., N.M., etc. July–Sept. — Placed by Hitchcock with the more rhizomatous, lower, more decumbent and smaller-flowered western *M. squarrosa* (Trin.) Rydb.

2. M. cuspidàta (Nutt.) Rydb. (abruptly sharp-pointed). — Similar to no. 1, the crowded ***tufts arising from bulbiform offshoots of the dense crowns; ligule* 0.5 *mm. long, erose-truncate;*** panicle 5–10 cm. long, loose; spikelets about 4 mm. long; *glumes two-thirds as long as the minutely pubescent lemma.* (*Sporobolus brevifolius* (Nutt.) Scribn., not *M. brevifolius* Scribn.) — Prairies, gravels, sands and bluffs (often calcareous), n. Mich. to Alta., s. to O., Ky., n. Ill., Mo., Okla. and N.M. Aug.–Oct.

3. M. Schrèberi J. F. Gmel. (for JOHANN DANIEL CHRISTIAN VON SCHREBER, 1739–1810), DROP-SEED, NIMBLE WILL. — *Culms diffuse, flattened,* very slender, *decumbent at base and often rooting at the nodes,* 1.5–6 dm. long, *freely forking into capillary ascending branches;* leaves flat, spreading or loosely ascending, the principal ones 2.5–5 cm. long; ligule very short, truncate and lacerate; panicles filiform to linear-cylindric, the leading ones 0.6–1.8 dm. long, with closely flowered appressed branches, slightly exserted or with bases sheathed; spikelets short-pedicelled; 1*st glume minute* (0.1–0.2 *mm. long*) *or obsolete; the* 2*nd* 0.2–0.5 *mm. long, truncate or blunt;* lemmas green or purple, hairy at base, strongly 3-nerved, excluding the awn 1.8–2.2 mm. long, tapering to a slender awn 2–5 mm. long; palea about equaling blade of lemma; anthers 0.3–0.4 mm. long; grain loosely infolded by the thin lemma, linear-cylindric, reddish-brown, 1–1.4 mm. long. — Woodlands, thickets, dooryards, roadsides, etc., often a troublesome weed, Fla. to e. Tex. and e. Mex., n. to s. N.H., Vt., N.Y., s. Ont., Mich., Wisc., Ia. and Neb. July–Nov. FIG. 253.

253. M. Schreberi.

Var. **palústris** Scribn. (of marshes). — Glumes both well developed, the 2nd up to 1 mm. long. — Swamps, D.C.

4. M. curtisetòsa (Scribn.) Bush (short-bristled). — Somewhat coarser than no. 3, with spikelets slightly larger; 2*nd glume,* 1.5–2 *mm. long; lemma* 2.5–3 *mm. long, with awn only half its length.* — Similar habitats, local, e. Pa.; Ill. and Mo.

5. M. sobolífera (Muhl.) Trin. (bearing sprouts). — Perennial *with hard scaly rhizomes and stolons* 1.5–3 *mm. thick, the scales tightly appressed, the median ones oblong to narrowly ovate;* culms solitary or few, hard, cylindric, erect or strongly ascending, 3–9 dm. high, with few erect branches; *internodes glabrous or at most scabrous just below the glabrous nodes;* leaves flat, pale green, divergent, the primary ones 3–7 mm. wide; *ligule shorter than the elongate cartilaginous summit of the sheath;* panicles filiform to linear-cylindric, long-exserted, the principal ones 0.4–1.7 dm. long; spikelets short-pedicelled, imbricated, 1.7–2.7 mm. long; *glumes* whitish, *ovate,* abruptly cuspidate, *shorter than the* mucronate but *awnless* scabrous *lemma; the latter copiously villous at base;* palea about equaling lemma; anthers 0.8–1 mm. long; grain 1–1.2 mm. long. — Dry rocky (often calcareous) or gravelly woods, shaded ledges, etc., N.H. to Wisc. and Ia., s. to Va., Tenn., Ark. and e. Tex. July–Oct.

6. M. tenuiflòra (Willd.) BSP. (slender-flowered). — Rhizomes much as in no. 5; *culm simple* or rarely a little forking, erect, 0.3–1 m. high, *the internodes short-pilose below and often above the pubescent nodes; leaves flat, dark green, widely divergent,* 0.5–1.5 *cm. wide;* ligule short and hidden; panicle linear-filiform to slenderly cylindric, loose, 0.8–3.3 dm. long, exserted or with base included in sheath; *spikelets* distinctly pedicelled, excluding the awn 3–4 *mm. long;*

glumes ovate, mucronate or acuminate, whitish or purple, *shorter than lemma,* prominently green-keeled, the keel sometimes double or triple; *lemma pruinose-puberulent, tapering to a delicate awn* 5–10 *mm. long,* copiously pilose at base; *anthers* 1.1–1.5 *mm. long; grain* 2–2.3 *mm. long.* — Rocky and gravelly woods and slopes, shaded cliffs, etc., e. Mass. and s. Vt. to Wisc. and Ia., s. to Va., mts. to Ga., Tenn., Ark. and Okla. July–Sept. Fig. 254.

254. M. tenuiflora.

7. M. sylvática Torr. (of woodland). — Perennial with scaly *rhizomes* 2–6 *mm. thick; the scales loosely cucullate-arching at base, ovate, subdistant; culms* solitary or few, rather weak, loosely ascending or arching, *usually very freely forking,* 0.3–1.2 m. tall, *the internodes puberulent below the abruptly enlarged* usually *glabrous nodes and the* commonly *glabrous bases of the sheaths; leaves flat,* spreading or ascending, subflaccid, the larger ones 2–7 mm. wide and 0.8–2 dm. long; *ligule* scarious, subtruncate, *in larger leaves projecting* 0.5–1.5 *mm. above summit of sheath; panicles mostly long-exserted, very slender, flexuous* (*at least some of the branches without flowers at base*), the primary ones 0.8–1.8 dm. long; *spikelets* loosely disposed on the branches, *often slender-pedicelled; glumes very unequal, linear-lanceolate to linear-attenuate,* the 2nd about equaling to slightly longer than blade of lemma, *scarious to hyaline, commonly silvery-green or whitish; lemma* tapering to delicate awn 3–15 mm. long, or awnless in forma **attenuàta** (Scribn.) Palmer & Steyerm. (gradually tapering), with blade 2.3–3 *mm. long, scarious or hyaline and silvery,* bearded at base; *anthers* 0.3–0.6 *mm. long; mature grain nearly or quite free,* 1.4–1.8 mm. long. — Damp thickets, rocky woods, banks of streams, etc., sw. Que. to s. Ont. and Minn., s. to N.E., N.C., Ala., Ark. and ne. Tex. July–Oct.

Var. **robústa** Fern. (stout). — *Culms rigid and stiffly ascending; leaves firm, the broader ones* 5–9 *mm. wide;* panicles fuller; *glumes broadly lanceolate to lance-ovate, shorter than blade of the* long-awned *lemma, the latter* 3–4 *mm. long; anthers* 0.5–0.7 *mm. long; mature grain* more firmly embraced, 1.9–2.1 *mm. long.* — Open woods and thickets, centr. Me. to w. N.Y., s. to ne. Mass., n. R.I., s. Ct. and e. Pa.

8. M. mexicàna (L.) Trin. (Mexican; name given under mistaken idea that the species is Mexican). — In general resembling no. 7, with scaly *rhizomes* 2–6 mm. thick, *their* ovate *mostly approximate scales cucullate-arching at base; culms* 1–several, firm, erect or ascending, 0.2–1.2 m. high, *with stiffly ascending branches, the internodes puberulent* below the abruptly enlarged usually glabrous nodes; *leaves* flat, pale green, *ascending;* the larger ones 0.5–1.8 dm. long, 2–8 mm. wide; *panicles* mostly *exserted,* lanceolate to linear- or oblong-cylindric, *rather stiff,* 0.4–1.8 dm. long, *with the subsessile to short-pedicelled spikelets densely imbricated and often extending down to the bases of the glomerulate and appressed-ascending branches and branchlets; glumes subequal,* lance-attenuate, awnless, or awned in forma **setiglùmis** (S. Wats.) Fern. (with bristle-like glumes), *the blades* shorter than the lemma, *firmly membranaceous, usually green or purplish; lemma firm,* its blade 2.3–3.5 mm. long, awnless, or in forma **ambígua** (Torr.) Fern. (doubtful) with delicate awn up to 1 cm. long; floret stiffly hairy at base; anthers 0.3–0.4 mm. long; *mature grain tightly embraced by lemma,* 1.3–1.6 mm. long. (*M. foliosa* of ed. 7, not (R. & S.) Trin.) — Shores, thickets and damp clearings, Gaspé Pen., Que., to s. B.C., s. to N.B., N.E., Del., upland to w. N.C., O., Ind., Ill., Mo., Kans., N.M., Ariz. and n. Calif. Aug.–Oct.

9. M. brachyphýlla Bush (short-leaved). — Habit somewhat of nos. 6 and 7; *rhizomes slender, with tightly appressed scales; culms* erect or ascending, up to 1.3 m. high, *with numerous elongate* slender simple or forking and *ascending branches* from the middle nodes; *internodes glabrous and lustrous* throughout or merely scabrous at summit; *nodes glabrous;* leaves firm, scabrous, ascending, the larger (primary) ones 0.5–1.5 dm. long and 3–7 mm. broad; the rameal abundant, shorter and narrower; ligule prolonged above summit of sheath; terminal *panicles* usually exserted, the lateral with included bases; the larger ones *lax,* slender, 0.7–1.5 dm. long; *glumes narrowly ovate, very unequal, awned, definitely shorter than lemma;* lemma minutely pubescent and long-bearded at base, with awn 3–8 mm. long. — Low woods, Ind. to Neb., s. to Mo. and e. Tex.; se. Va. Aug.–Oct.

10. M. frondòsa (Poir.) Fern. (leafy). — Perennial with thickish *rhizomes with loosely imbricated scales; culms ascending or geniculate or decumbent and rooting at base, very freely branched* from base or from middle nodes and *often bushy in aspect, the branches often forking; internodes glabrous;* leaves loosely ascending to spreading, rather thin and pliant, the larger ones 0.5–1.5 dm. long and 3–8 mm. broad, *the sheaths loose and laterally compressed; panicles loose and often open, with* spreading-ascending *branches flowering to base,* the terminal panicles exserted or with base included, *smaller axillary ones partly included in many leaf-sheaths;* the terminal lanceolate, slenderly pyramidal or ellipsoid, 0.4–1 dm. long, green to purplish; spikelets soft, 2–3.3 mm.

long; *glumes very unequal, lance-aristate,* the 2nd about equaling the basally bearded acuminate awnless lemma, or in forma **commutàta** (Scribn.) Fern. (changing) with slender awn; anthers 0.5 mm. long; *grain easily loosened,* 1.8–2 mm. long. (*M. mexicana* of ed. 7, not (L.) Trin.) — Damp open woods, thickets, shores, clearings and waste places, N.B. and sw. Que. to s. Ont. and Minn., s. to s. N.E., L.I., n. Ga., Tenn., Mo., and e. Tex. Aug.–Oct.

11. M. glabriflòra Scribn. (smooth-flowered). — Resembling no. 10, very bushy-branched; *culms* more *stiffly ascending or arching, with internodes scabrous-puberulent at summit; leaves firm, erect,* crowded and appressed, mostly 2–4 *mm. broad,* the larger primary ones –7 mm. broad but promptly deciduous; ligule minute; *panicle stiffly linear-cylindric, the terminal* 1.2–6 *cm. long* and hardly to barely exserted, the abundant lateral ones short-exserted or partly included; spikelets 2.5–3 mm. long; *glumes subequal,* acute or acuminate, *awnless; lemma* awnless, *glabrous throughout or with very short basal bearding.* — Dry exsiccated or baked soils, prairies, gravels or rocky slopes, sw. Ind. and Ill. to e. Tex.; local, e. Md., se. Va. and centr. N.C. Aug.–Oct.

12. M. racemòsa (Michx.) BSP. (racemed). — Stiffly ascending perennial with firm slender rhizomes 2–4 mm. thick, their median scales arching from base but mostly appressed and narrowly ovate; *culms* nearly erect, 3–7.5 dm. high, *mostly branching from the middle nodes,* with several erect branches, the *internodes glabrous and lustrous* (rarely puberulent at very summit); *leaves* firm, stiffly ascending, scabrous, 2–6 mm. wide, 8–18 on the main axis, *with the sheath keeled* by the running down of the midrib of the leaf; *ligule prolonged,* 3–5 *mm. long; panicle stiff,* barely if at all flexuous, 3–12 cm. long, 0.4–2 cm. thick, greenish to purplish, *closely lobulate-spiciform* by crowding of the dense, branches, or interrupted at base; *glumes* linear-lanceolate, awned, *subequal,* 4.5–8 *mm. long, much longer than lemma,* scabrous; *anthers* 0.5–0.8 *mm. long; grain* tightly embraced, *linear-cylindric,* 1.8–2.2 *mm. long.* (Var. *ramosa* Vasey) — Dry prairies, rocks and bluffs, Mich. to Sask., s. to Ill., Mo., Kans. and N.M.; adv. along railroads eastw. to n. N.H. Aug.–Oct.

13. M. glomeràta (Willd.) Trin. (in clustered heads). — Upright, from hard slender and elongate scaly rhizomes; *culms simple or with few erect basal branches* (very rarely freely branched from middle), 0.3–1 m. high, stiffly erect; *internodes and nodes puberulent; leaves* erect, firm, scabrous, the larger ones 7–15 cm. long and 2–5 mm. broad; *those of the flowering culm mostly* 7–15, *many of them crowded and overlapping at the middle nodes,* the *sheath scarcely keeled, ligule minute;* panicle long-peduncled, purplish to green, 3–9 cm. long, 0.7–1.6 cm. thick, all but the lowest of *the densely flowered ellipsoid to rounded-obovoid branches closely crowded, the panicle thus appearing densely lobulate-spiciform; spikelets* 4.5–8 *mm. long; glumes* linear-lanceolate, *subequal, once-and-a-half to twice as long as lemma, tapering to long* straight or arching *awns, the keel and the awn copiously hispid* (thus giving the inflorescence a "misty" aspect); *lemma villous below the middle; anthers* 1–1.5 *mm. long; grain* loosely embraced, *oblong-cylindric,* 1.2–1.5 *mm. long.* (*M. racemosa* of ed. 7, not (Michx.) BSP.; *M. setosa* (Biehler) Trin., not (HBK.) Kunth) — Meadows, bogs and wet shores, w. N.S. and s. Me. to s. Ont. and Mich., s. to s. N.E., nw. N.J., Pa., mts. of Va., Ky. and Ind. Aug.–Oct.

Var. **cinnoìdes** (Link) F. J. Herm. (resembling *Cinna*). — *Leaves more scattered,* 5–8(–10), usually not strongly concentrated near middle of culm, up to 8 mm. wide; *panicle* greenish or purplish, usually *more interrupted, the cylindric to oblong-ovoid* often subacute *lower branches* often *subdistant to remote; glumes* slightly broader, *with merely scabrous keel and awn* (the panicle therefore not appearing "misty"). — Bogs, peaty meadows, wet rocks and shores, Nfld. to Alta., s. to N.S., n. Mass., Ct., N.Y., centr. Pa., O., Mich., Wisc., Minn., Wyo., Nev. and Oreg.

14. M. uniflòra (Muhl.) Fern. (one-flowered). — *Loosely matted* delicate species; *the flowering culms arising from axils of old depressed culms,* 0.5–4.5 dm. high, *terete,* the bases without thickened rhizomes; *leaves flat,* 1–2 mm. wide, mostly subbasal; *panicle* long-peduncled, loosely cylindric-ellipsoid, 0.2–2 dm. long, 2.5–6 cm. broad, *very diffuse, with capillary branches and capillary pedicels mostly twice to six times as long as the purple spikelets; spikelets* lance-ellipsoid, 1.3–2 *mm. long; glumes subequal, blunt, much shorter than the glabrous lemma;* occasional spikelets 2-flowered. (*Sporobolus* (Muhl.) Scribn. & Merr.) — Bogs, swales, sandy shores and damp fields, Chicoutimi Co., Que., to Thunder Bay Distr., Ont., s. to N.S., N.E., L.I., N.J., Mich. and Wisc. Late July–Oct.

Var. **térrae-nòvae** Fern. (of Newfoundland). — *Panicle contracted,* 1.2–8 cm. long, 0.2–3 cm. thick, with ascending branches; *many* of the lateral *pedicels only* 1–2 *mm. long;* upper spikelets often 2-flowered, the upper floret pistillate. — Nfld., with transitional forms in N.S.

15. M. Torreyàna (Schultes) Hitchc. (in honor of John Torrey, 1796–1873). — Perennial *with short and stout scaly rhizomes; leafy tufts strongly compressed,* 1–few; the *culms* 3–7 dm. high, *surrounded at base by erect overlapping prolonged conduplicate leaves;* panicle slenderly cylindrical, 1–3 dm. long, loose and open; the long-pedicelled *spikelets* purplish, about 2 mm. long, rarely 2-flowered, *sterile; glumes* blunt, *unequal, the longer about equaling the puberulent lemma.* (*Sporobolus compressus* (Torr.) Kunth) — Peaty or damp sandy pine-barrens, N.J. and Del.; Sumter Co., Ga. Aug.–Nov.

16. M. asperifòlia (Nees & Meyen) Parodi (harsh-leaved), SCRATCHGRASS. — *Culms numerous, decumbent, from elongate slender rhizomes,* 1.5–5 dm. high, slender, terete; *leaves short, firm, harsh, distichous, with crowded and overlapping sheaths; panicle diffuse,* ovoid, 0.5–2 dm. long, *with* capillary divergent branches and *long slender pedicels;* spikelets purplish, 1.5 mm. long; glumes somewhat unequal, shorter than the glabrous lemma and palea. (*Sporobolus asperifolius* Nees & Meyen) — Sandy bottoms, damp sands, etc., Minn. and Sask. to B.C., s. to Ill., Mo., Tex., N.M., Ariz. and Calif. July–Sept. (Mex.; s. S.Am.)

17. M. capillàris (Lam.) Trin. (hair-like), HAIRGRASS. — *In perennial tussocks, with promptly involute* subrigid *leaves* 1–3 dm. long; culms simple, stiff, slender, 0.6–1 m. high; *panicle purple, diffuse,* 2–4.5 dm. long, 0.7–2 dm. thick, *with loosely ascending and freely forking capillary branches; the delicate pedicels much longer than the spikelets; spikelets* (excluding awns) 3–4 *mm. long, slender; glumes unequal, tapering to slender tips,* the 2nd often awned, much shorter than lemma and palea; *lemma* scabrous, *tapering to a delicate awn* 5–15 *mm. long.* — Sandy or rocky woods and clearings, Fla. to e. Tex., n. to Mass., N.J., e. Pa., Ky., s. Ind., s. Ill., Mo. and Kans. Sept., Oct. FIG. 255.

255. M. capillaris.

18. M. expánsa (DC.) Trin. (expanded), HAIRGRASS. — Similar to no. 17, in larger and denser tussocks; *leaves flat,* becoming involute; panicle brown or bronze, more slender; *spikelets* 3.5–5 *mm. long; glumes subequal, not prolonged into slender tips; lemma with awn only* 1–4 *mm. long.* — Bogs, savannas and low pinelands, Fla. to e. Tex., n. to se. Va. Sept., Oct.

56. BRACHYÉLYTRUM Beauv.

Spikelets 1-flowered, few in a narrow panicle; glumes minute, unequal; floret with a short callus, the rachilla prolonged behind the palea into a slender naked bristle; lemma firm, narrow, 5-nerved, terminating in a long straight awn; palea firm, nearly as long as the lemma; grain oblong, inclosed in the lemma and palea. — Perennial, with simple culms from short knotty rhizomes, a single species of N.Am. and e. Asia. (Name composed of the Greek *brachys, short,* and *elytron, husk,* from the minute glumes.)

1. B. eréctum (Schreb.) Beauv. (erect). — Culms erect, 3–10 dm. high; sheaths sparsely retrorse-hispid; blades 3.5–15 cm. long, 0.6–2 cm. wide, lanceolate, very scabrous, pilose on the nerves beneath; panicle narrow, 0.5–2 dm. long; spikelets 1 cm. long (excluding the awns), on capillary pedicels; first glume often obsolete, second sometimes aristate; lemma hispid with hairs 0.2–0.6 mm. long. (*Dilepyrum erectum* (Schreb.) Farw.) — Woods and thickets, w. Mass. to Wisc. and Ia., s. to Ga., Ala., Miss. and La. June–Aug.

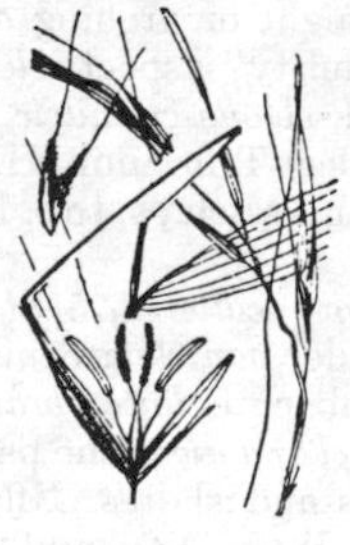

256. B. erectum, v. septentrionale.

Var. **septentrionàle** Babel (northern). — Lemma glabrous, or barely puberulent with hairs less than 0.2 mm. long. — Nfld. to Minn., s. to N.S., N.E., n. N.J., Pa., W.Va. and n. O. FIG. 256.

57. ORYZÓPSIS Michx. MOUNTAIN-RICE

Spikelets 1-flowered, few in narrow to diffuse panicles; glumes rather broad; floret with a short obtuse oblique callus; lemma (not over 1 cm. long) convolute, indurated, including the rather large palea and perfect flower, terminating in a deciduous simple slender awn; grain oblong-ellipsoid, tightly included in the indurated lemma. — Tufted perennials of N. Am. and Eurasia. (Name from *oryza, rice,* and *opsis, appearance,* the grains of some species resembling those of unpolished rice.)

a. Panicle not diffuse; glumes blunt, acute or apiculate, not cuspidate, equaling or but slightly exceeding the smooth or short-pubescent lemma. . . *b.*

b. Leaves flat, 0.4–1.8 cm. wide; spikelets (excluding awns) 6–9 mm. long.
Culms with mostly bladeless sheaths, leaves chiefly basal; panicle simple,

with very short erect branches; glumes apiculate; lodicules three-fourths as long as palea. . . . . 1. *O. asperifolia.*
Culms leafy; the blades elongate; panicle with loosely ascending to spreading branches; glumes acute; lodicules minute. . . . . . . 2. *O. racemosa.*
*b.*Leaves involute, 0.5–2 mm. wide; spikelets (excluding awns) 3–4 mm. long.
Glumes obtuse; awn 0.5–2 mm. long, straight. . . . . 3. *O. pungens.*
Glumes acutish; awn 6–10 mm. long, twisted. . . . . 4. *O. canadensis.*
*a.*Panicle diffuse; glumes long-cuspidate, much exceeding the long-bearded lemma. . . . . 5. *O. hymenoides.*

1. O. asperifòlia Michx. (harsh-leaved). — Culms tufted, 2–7 dm. high, erect or geniculate at the lowest node; *sheaths usually crowded at the base; basal blades evergreen, erect, scabrous,* especially *on the glaucous lower surface,* 4–10 mm. wide, flat, or involute on the margins, attenuate; *culm-leaves tubular sheaths, with blades usually less than* 1 *cm. long; panicle* contracted, 5–12 *cm. long,* the *branches* simple and *erect; spikelets,* excluding awn, 6–8 *mm. long;* glumes subequal, short-ciliate at the apiculate summit; lemma nearly or quite as long as the second glume, sparingly pubescent; *awn* 5–10 *mm. long; lodicules three-fourths the length of the palea; fruit stramineous,* with a densely woolly base. — Woods, thickets and peaty openings, Nfld. to B.C., s. to N.S., N.E., L.I., w.Va., n. Ind., n. Ill., Minn., S.D., and N.M. Late Apr.–July. Fig. 257.

257. O. asperifolia.

2. O. racemòsa (Sm.) Ricker (racemose). — Culms tufted, erect, 3–9 dm. high, *leafy to the summit; leaves* 1–3.5 *dm. long,* 4–18 mm. wide, flat, narrowed toward the base, taper-pointed, *scabrous below, pubescent above;* panicle 0.3–3 dm. long, its *branches* nearly simple and *usually ascending; spikelet,* excluding awn, 7–9 *mm. long;* glumes equal, acute; *lemma* somewhat shorter, pubescent, *becoming black in fruit; awn* 1.5–2.5 *cm. long; lodicules minute.* — Rich, often calcareous and rocky, woods, sw. Me. and sw. Que. to N.D., s. to n. Del., w. Va., W.Va., Ky. and e. Mo. July, Aug. Fig. 258.

258. O. racemosa.

3. O. púngens (Torr.) Hitchc. (sharp-pointed). — *In dense tussocks;* culms 1.5–6 dm. high, erect, slender, simple; sheaths usually crowded at the base, smooth or slightly scabrous; *blades involute-filiform,* the basal ones sometimes as long as the culm, usually half its length, those of the culm short; *panicle* 3–8 *cm. long, branches erect or ascending; spikelets* 3–4 *mm. long; glumes* subequal, *obtuse,* obscurely 5-nerved; lemma usually as long as the glumes, appressed-pubescent; *awn* 0.5–2 *mm. long, straight,* sometimes wanting; palea as long as lemma. — Rocky, sandy or peaty soil, Que. to B.C., s. to N.S., N.E., n. N.J., ne. Pa., nw. Ind., n. Ill., Minn., S.D. and Colo. Late Apr.–June (rarely–Aug.). Fig. 259.

259. O. pungens.

4. O. canadénsis (Poir.) Torr. (of Canada). — Similar to no. 3, 2–9 dm. high; *panicle loose,* 0.4–1.5 *dm. long,* the opposite slender branches loosely ascending; spikelets, excluding awns, about 4 mm. long; *glumes oblong, acutish,* slightly exceeding the pubescent lemma; *awn* 6–10 *mm. long, twisted.* (*Stipa* Poir.) — Siliceous or peaty barrens, thin woods and mountain-slopes, Nfld. to Alta., s. to N.S., Me., N.H., n. N.Y., W.Va., n. Mich., n. Wisc. and n. Minn. June, July. Fig. 260.

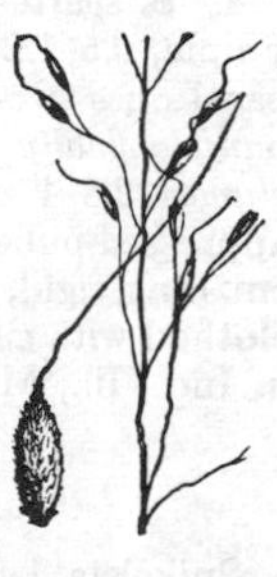

5. O. hymenoìdes (R. & S.) Ricker (membranous; referring to the thin glumes), Silkgrass, Indian Millet. — Tufted, 3–7 dm. high; leaves linear-filiform, elongate; *panicle diffuse,* 0.6–3 dm. long, the slender branches mostly in 2's and divaricate; *spikelets solitary at the tips of long divergent pedicels,* 6–8 mm. long; *glumes* subequal, chartaceous, ovate, *with a long cuspidate tip, nearly twice as long as the lemma; lemma* firm, turgid, *covered with long white beard;* awn 4–6 mm. long, usually dropped from the dark fruit. (*Eriocoma cuspidata* Nutt.) — Sandy prairies and rocky slopes, se. B.C. to se. Calif. and Mex., e. to Man., w. Minn., w. Ia., Kans. and Tex. May–July.

260. O. canadensis.

58. STÌPA L. Feathergrass. Speargrass

Spikelets 1-flowered, in terminal panicles; glumes narrow, acute or bristle-tipped; floret with a bearded usually sharp-pointed callus; lemma convolute, indurated, including the small palea and perfect flower, terminating in a simple strong persistent geniculate twisted awn; grain cylindrical, tightly included in the indurated fruiting lemma. — Rather large tufted perennials of trop. and temp. reg. with often involute leaves. (Name from the Greek *stype, tow*, in allusion to the flaxen appearance of the feathery awns of the original species.)

Glumes 0.5–1.2 cm. long; awn of fruit 2–7.5 cm. long.
  Glumes 5–9 mm. long; lemma appressed-pubescent; awn 2–4 cm. long. . . . 1. *S. viridula.*
  Glumes 9–12 mm. long; lemma glabrous, scabrous at summit; awn 4–7.5 cm. long. . . . . 2. *S. avenacea.*
Glumes 1.5–4 cm. long; awn of fruit 1–2.5 dm. long.
  Glumes 1.5–2.8 cm. long, the upper half aristiform; grain (including callus) 1–1.5 cm. long. . . . . 3. *S. comata.*
  Glumes 2.8–4 cm. long, short-awned at tip; grain 2–3 cm. long. . . . . . 4. *S. spartea.*

1. S. virídula Trin. (greenish), Feather-Bunchgrass. — Culms clustered, 5–10 dm. high, sparingly branched; basal sheaths overlapping, the long usually scabrous involute or subinvolute blades elongated; upper blades shorter, mostly setaceous; *panicle narrow, erect,* 1–2 dm. long; the *branches* mostly in pairs, erect, *rather densely flowered* from near the base; *glumes* 5–9 mm. long, *acuminate-setaceous,* exceeding the *pale appressed-pubescent lemma;* awn 2–4 cm. long; *callus usually rather short.* — Prairies, meadows and thin woods, B.C. and Alta. to N.M., e. to Man., w. Minn., Ia. and Kans.; nw. N.Y. Late May–July.

2. S. avenàcea L. (oat-like), Black Oat-Grass. — Culms tufted, slender, erect or ascending, 3–10 dm. high, leafy at the base; sheaths shorter than the internodes; blades 1–1.5 mm. wide, usually involute, the basal ones one-third to half the length of the culms, those of the culm 4–10 cm. long; *panicle loose,* 1–2 dm. long; the slender *branches* in pairs, *lax, finally spreading; glumes* often purplish, 9–12 mm. long, *acute,* about equaling the *dark brown lemma, which is smooth below, scabrous above and bearing a fringe of short hairs at the summit; awn 4–7.5 cm. long; callus acuminate,* densely covered with brownish hairs. — Dry thin woods and openings, n. Fla. to e. Tex., n. to Mass., N.Y., Mich. and Wisc. April–June (rarely–Sept.). Fig. 261.

261. S. avenacea.

3. S. comàta Trin. & Rupr. (hairy-tufted). — Culms erect, simple, 2–6 (rarely–12) dm. high; *sheaths* mostly crowded at the base, *the upper often loose and inclosing the base of the panicle;* basal blades usually about half the length of the culm, mostly involute-filiform; those of the culm 0.5–1.5 dm. long, 2–4 mm. wide, flat or involute; panicle loose, 1–4 dm. long; branches distant, erect or somewhat spreading, naked below; *glumes* 1.5–2.8 *cm. long, tapering into a slender fragile awn,* much exceeding the sparsely pubescent lemma; *awn* 10–24 cm. long, *pubescent to the geniculation, scabrous and curved beyond;* callus acute. — Dry plains and hills, Yuk. to Lower Calif., e. to Man., Mich., n. Ind., Kans. and Tex.; adv. e. to e. Ont. and w. N.Y.

4. S. spártea Trin. (broom-like), Porcupine-Grass. — Culm rather stout, simple, 0.5–1.2 m. high; sheaths mostly overlapping; blades usually involute, basal ones two-thirds the length of the culm, those of the culm 1–3 dm. long; *panicle finally exserted, narrow,* 1–3 dm. long; branches erect, naked below; *glumes* 2.8–4 *cm. long, attenuate,* exceeding the brownish lemma, which is appressed-pubescent below and nearly or quite glabrous above; *awn* 11–20 cm. long, rigid, scabrous, *minutely pubescent below;* callus acuminate, very sharp-pointed, densely clothed with silky appressed hairs. — Dry prairies and sands, Ont. to B.C., s. to w. Pa., n. O., n. Ind., Ill., Mo., e. Kans. and N.M. May, June. Fig. 262.

262. S. spartea.

59. ARÍSTIDA L. Triple-awned Grass. Needlegrass

Spikelets 1-flowered, in usually narrow panicles; glumes narrow, acute or acuminate; a hard obconical usually hairy callus below the floret; lemma somewhat indurated, convolute, including the thin palea and perfect flower, terminating in a usually trifid awn; grain elongated, tightly included in the lemma. — Tufted annuals or perennials, chiefly of trop. and warm-temp. reg., with narrow leaves. (Name from *arista, a beard* or *awn.*)

a. Tip of lemma with a ring-like thickening (or articulation) at the junction with the column of the awns. § Arthratherum.
  Longer glume 2.5–3 cm. long; grain (below the articulation) 1.3–1.5 cm. long; column twisted, 0.8–1.5 cm. long; awns 3–5 cm. long. . . . . . 1. *A. tuberculosa.*
  Longer glume 1.5–2 cm. long; grain 9–11 mm. long; column straight, 1–2 mm. long; awns 2–3 cm. long. . . . . 2. *A. desmantha.*

*a.*Tip of lemma prolonged without articulation to the bases of the awns. § CHAETARIA. . . *b.*

*b.*Nodes of panicle without woolly tufts; leaf-sheaths not lanate. . . *c.*

*c.*Longer glume 1.2–3 cm. long. . . *d.*

*d.*Lateral awns 2–8 cm. long.

Soft-based annual; 1st glume 3–5-nerved, 2–3 cm. long. . . . . . . 3. *A. oligantha.*

Tough-based perennials; 1st glume 1-nerved, 8–12 mm. long.

Leaves flat; glumes subequal, 10–12 mm. long, overtopping the bases of the awns; awns 1.5–3 cm. long. . . . . 11. *A. purpurascens.*

Leaves involute; glumes very unequal, the 1st barely reaching the bases of the awns, the 2nd 1.5–2.5 cm. long; awns 6–8 cm. long. . . . . 4. *A. longiseta.*

*d.*Lateral awns 0–1.8 cm. long. . . *e.*

*e.*Spikelets remote, mostly solitary along the rachis; glumes 2–5-nerved, the 2nd 2–2.4 cm. long, with awn 3–5 mm. long; lateral awns of lemma 1–6 mm. long, or obsolete. . . . . 5. *A. ramosissima.*

*e.*Spikelets approximate; glumes 1-nerved; the 2nd 0.9–1.5 cm. long, short-awned or merely subulate-tipped; lateral awns 5–15 mm. or more long.

Soft-based annual 2–6 dm. high; small secondary panicles often in basal sheaths; longer glumes 1.2–1.5 cm. long; lateral awns 5–10 mm. long. . . . . 6. *A. basiramea.*

Hard-based perennials 0.3–1.5 m. high; no basal secondary panicles; longer glumes 0.9–1.3 cm. long; lateral awns 1.5–3 cm. long.

Nodes well covered by the herbaceous leaf-sheaths; culms numerous, 0.3–1 m. high; panicles one-third to half the entire height of plant; awns subequal. . . . . 11. *A. purpurascens.*

Nodes exserted from the coriaceous sheaths; culms 1–few, 1–1.5 m. high; panicles one-fourth to one-third the entire height of plant; lateral awns two-thirds to three-fourths as long as central one. . . . . 12. *A. affinis.*

*c.*Longer glume 5–12 mm. long. . . *f.*

*f.*Soft-based annuals. . . *g.*

*g.*Central awn abruptly bent nearly perpendicular to the much shorter (1–15 mm. long) erect lateral ones.

Terminal inflorescence 2–10 cm. long; leaves extending to summit of culm, their sheaths becoming distended; central awn 3–10 mm. long, usually spiraling. . . . . 7. *A. dichotoma.*

Terminal inflorescence 0.8–2.5 dm. long; leaves chiefly below middle of culm, their sheaths rarely distended; central awn 0.5–2.1 cm. long, not spiraling. . . . . 8. *A. longespica.*

*g.*Central and lateral awns about equally ascending, subequal in length, 1–2.2 cm. long.

Awns flat at base; callus obtuse, densely bearded. . . . . . 9. *A. adscensionis.*

Awns terete to base; callus acute, sparsely bearded. . . . . . 10. *A. intermedia.*

*f.*Firm-based perennials (with hard bases or retaining vestiges of old culms). . . *h.*

*h.*Nodes well covered by the herbaceous sheaths; panicles one-third to one-half entire height of plant; 1st glume longer than 2nd; awns subequal, all horizontally divergent or reflexed, 1.5–3 cm. long. . 11. *A. purpurascens.*

*h.*Nodes exserted from the coriaceous sheaths; panicles one-fourth to one-third the entire height of plant; glumes subequal or 2nd longer; middle awn longest, horizontal, perpendicular to the shorter erect or ascending lateral ones.

Glumes 1–1.3 cm. long; lemmas 7–8 mm. long; central (horizontal) awn 1.5–3 cm. long, laterals 1.5–2 cm. long; culms 1–1.5 m. high. 12. *A. affinis.*

Glumes 6–8 mm. long; lemmas 4–5 mm. long; central awn 1.2–2 cm. long, laterals 7–10 mm. long. . . . . 13. *A. virgata.*

*b.*Nodes (at least the lower) of panicle woolly-tufted; young leaf-sheaths commonly lanate; 1st glume often exceeding 2nd; central awn 1.5–3.5 cm. long. . . . . 14. *A. lanosa.*

§ Arthrathèrum (Beauv.) Reichenb.

1. A. tuberculòsa Nutt. (bearing tubercles; from the enlarged bases of branches), Sea-beach-N. — Annual; *culms usually branched below,* 1.5–11 dm. high, *tumid at the joints;* leaves long and involute; panicles rigid, loose, the branches in pairs, one short and about 2-flowered, the other elongated and several-flowered; *glumes* 2–3 *cm. long,* including their slender-awned tips, *not bifid at apex; lemma* 11–15 *mm. long, the twisted bases of the awns of about equal length; callus sharp,* densely bearded, 2–4 *mm. long; awns* divergent, subequal, 3–5 *cm. long.* — Dry sands, coast of Mass. to s.-centr. N.Y., s. to Ga. and Miss.; sw. Mich. and nw. Ind. to Minn. and Ia. Aug.–Oct. (Mex.) Fig. 263.

263. A. tuberculosa.

2. A. desmántha Trin. & Rupr. (with clustered flowers). — Smaller than no. 1; spikelets yellowish-brown; *glumes* subequal, *the upper bifid at apex,* awned from the sinus, 1.5–2 *cm. long; grain* 9–11 *mm. long,* its *column straight,* 1–2 *mm. long;* callus about 2 mm. long; *awns* 2–3 *cm. long.* — Sands, Mason Co., Ill.; Neb. to e. Tex. Aug., Sept.

§ Chaetària Trin.

3. A. oligántha Michx. (few-flowered). — *Annual; culms* tufted, wiry, *branched at base and at all the nodes,* 3–6 dm. high; sheaths loose; blades long, usually involute; panicle or raceme few-flowered, the axis often flexuous and spikelets spreading; *glumes* unequal, *long-awned from a bifid apex,* exceeding the floret, 1*st* 3–5-*nerved* and 2–3 *cm. long,* 2nd strongly 1-nerved; *lemma* 17–20 *mm. long,* scabrous above; awns nearly equal, divergent, 3.5–7 cm. long. — Dry sterile soil, Fla. to Tex., n. to se. N.Y., N.J. and Pa. (adv. to Mass.), W. Va., O., s. Mich., s. Wisc., Ia. and S.D.; Calif. and Oreg. Aug.–Oct. Fig. 264.

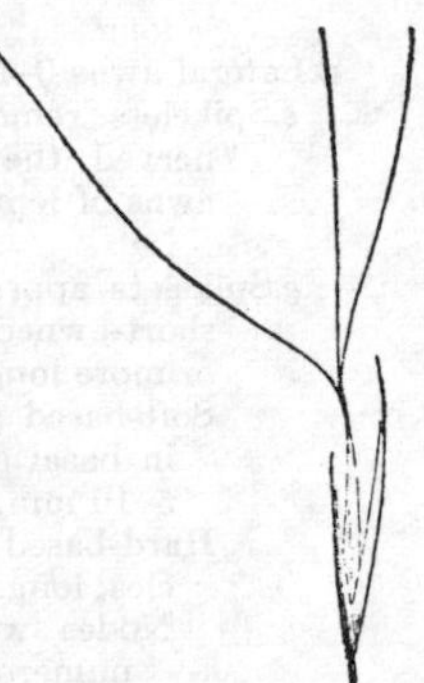

264. A. oligantha.

4. A. longisèta Steud. (long-awned). — *Tufted perennial; culms simple,* mostly 2–3 dm. high; *leaves involute to filiform,* up to 1.5 dm. long; *panicle loose, erect,* of few slender-pedicelled spikelets; *glumes very unequal,* 1-*nerved, the* 1st *barely reaching the bases of the awns, the* 2nd 1.5–2.5 *cm. long;* awns 6–8 cm. long. — West of our range; reaching us as the barely separable

Var. **robústa** Merr. (stout). — Stouter, up to 5 dm. high; leaves longer; panicle stiffer; *awns* 4–6 *cm. long.* (*A. purpurea* of ed. 7, not Nutt.) — Dry plains and foothills, B.C. to n. Mex., e. to sw. Minn., Ia., Kans., Okla. and Tex. June, July.

5. A. ramosíssima Engelm. (very branching). — Annual; culms tufted, wiry, 3–5 dm. high, *repeatedly branching, the branches divergent;* leaves linear to setaceous; panicle loose, the *spikelets mostly solitary along the rachis; glumes* unequal, 3–5-*nerved, awned from the bifid apex;* the 2*nd* 2–2.4 *cm. long, with awn* 3–5 *mm. long,* equaling the lemma; lemma 2–2.3 cm. long; *lateral awns* 1–6 *mm. long or wanting,* erect; *middle awn* 2–3 *cm. long, reflexed by a loose spiral at base.* — Dry sterile soil, s. Ind. to Ia., s. to Tenn., La. and e. Tex. July–Sept.

6. A. basiràmea Engelm. (branching from base). — Annual, 2–6 dm. high, usually *freely branching at the base; culms sparingly branched;* leaves flat, becoming involute at tip; panicles loose, the terminal often partly included in the upper sheaths; *small panicles commonly borne in the basal sheaths; glumes acuminate, unequal,* 1-nerved, 2*nd* 1.2–1.5 *cm. long, the* 1st *about two-thirds as long; lemma about* 1 *cm. long,* excluding the awns; lateral awns 5–10 mm. long, erect or spreading, delicate; the *middle awn stouter,* 1–1.5 *cm. long,* often loosely spiraling when dry. — Dry sandy soil, w. Me. to ne. N.Y.; Mich. to N.D., s. to w. Ill., Mo. and Kans. Aug.–Oct.

7. A. dichótoma Michx. (forked), Poverty-Grass. — Annual; culms usually tufted, wiry, much branched at base and usually forking at the nodes, 0.4–4 (–6) dm. high; *sheaths distended at maturity, borne to the summit of culms;* blades 0.5–2 mm. wide, becoming involute; *panicle simple or subsimple,* linear-lanceolate, *the terminal* 2–10 *cm. long, the lateral often sessile within the sheaths;* glumes subequal, 6–9 mm. long, cuspidate; *lemmas* 5–7 *mm. long,* sparingly appressed-pubescent; *lateral awns* 0.5–2 *mm. long, erect; middle awn* 3–7 *mm. long,* usually *coiled at base* in drying. — Dry sterile soil, Fla. to Tex., n. to N.E., N.Y., O., s. Mich., centr. Ill., Mo. and Kans. Aug.–Oct. Fig. 265. — Passing to

265. A. dichotoma.

Var. **Curtíssii** Gray (for its discoverer, Allen Hiram Curtiss, 1845–1907). — Only slightly branched; panicles with subdistant spikelets;

glumes more unequal, the *2nd 8–12 mm. long,* the 1st two-thirds to three-fourths as long; *lemma 7–10 mm. long, often smoother; central awn 6–10 mm. long, lateral awns* 1.5–4 *mm. long.* (*A. Curtissii* (Gray) Nash) — Fla. to Okla., n. to N.J., Pa., W.Va., Wisc., Minn., Neb. and e. Wyo. Aug.–Oct.

8. A. longespìca Poir. (with long spike). — Annual; culms slender, in small tufts or solitary, branched at base, simple or sparingly branched above, 1.5–6 dm. high; *leaves confined to lower half of culm, their sheaths close;* blades very narrow, usually involute in drying; *terminal raceme or subsimple panicle* 0.8–2.5 *dm. long,* the spikelets often subdistant; *glumes subequal,* 2nd 5–6 *mm. long; lemma* 4–6 *mm. long,* usually mottled; *middle awn horizontal,* 5–13 *mm. long, lateral awns erect,* 1–4 *mm. long.* (*A. gracilis* Ell.) — Sandy soil, Fla. to Tex., n. to se. Ct., and in the interior n. to W.Va., O., se. Mich., Ill., Mo. and Kans. Aug.–Oct.

Var. **geniculàta** (Raf.) Fern. (bent like a knee). — Coarser; panicle simple or freely short-branched; *2nd glume* 5–9 *mm. long; middle awn* 1–2.1 *cm. long, lateral awns* 4–15 *mm. long.* — Extending n. to s. N.H., s. Vt., s. N.Y., n. O., n. Ind. and Mo.

9. A. adscensiònis L. (of Ascension Island). — Annual; culms 1–8 dm. high, tufted, freely branching at base; leaves numerous, with close sheaths and slender elongate often involute blades; panicles terminating the branches, 0.5–2.5 dm. long, rather loose and flexuous; glumes unequal, the 2nd 8–10 mm. long, 1-nerved; lemma about equaling the 2nd glume; *central and lateral awns subequal,* 1–1.5 *cm. long, all ascending, flat at base; callus obtuse, densely bearded.* — Sandy soil, s. Calif. to w. Mo. and Tex.; adv. in N.Y. July–Oct. (Mex.; W.I.; S.Am.; Asia; Afr.)

10. A. intermèdia Scribn. & Ball (intermediate). — Similar to the var. of no. 8; *culms* 3–7 *dm. high, branching at base;* leaves scabrous, 5–15 cm. long, involute; panicle 1–4 dm. long, slender; branches short, appressed; *glumes* attenuate-aristate, subequal or the 2nd longer, 9–10 *mm. long,* scabrous, slightly shorter than the floret; lemma scabrous above the middle, sometimes mottled; *awns terete, subequal and all equally ascending,* 1.5–2.2 *cm. long; callus acute, sparsely bearded.* — Dry soil, Neb. to s. Mich. and Ind., s. to Miss. and Tex. Aug., Sept.

11. A. purpuràscens Poir. (purplish). — Perennial, in tufts, glabrous, or lower sheaths sometimes villous, 0.3–1 m. high; culms erect, simple or sparingly branched; *leaves* 1–2 dm. long, 1–4 mm. wide, *usually involute toward the ends, the herbaceous sheaths well covering the nodes; panicle purplish, one-third to one-half the entire length of the plant,* loosely or rather densely flowered; *glumes* 9–12 *mm. long,* 1-nerved, scabrous; the *1st slightly the longer,* attenuate-aristate, the 2nd aristate from a bidentate apex; lemma 6–7.5 mm. long; *awns subequal,* divergent, not twisted, 1.5–3 cm. long. — Sandy or gravelly soil, Fla. to Tex., n. to Mass., and in the interior n. to O., s. Mich., Wisc., Mo. and e. Kans. Aug.–Oct. (W.I.)

Var. **mìnor** Vasey (smaller). — *Panicle thinner,* often greener; *1st glume* 6.5–9 *mm. long.* — Fla. to Tex., n. to Va. and e. Mo.

12. A. affìnis (Schultes) Kunth (nearly related). — Hard-based, with 1–few stiff erect *culms* 1–1.5 *m. high,* the *nodes exserted from* the sides or summits of the *coriaceous sheaths;* blades becoming involute; *panicles one-fourth to one-third the entire height of plant,* linear-cylindric, virgate; *glumes subequal,* 1–1.3 *cm. long; lemmas* 7–8 *mm. long; central awn thrown horizontally by a semicircular bend,* 1.5–3 *cm. long; lateral awns ascending,* 1.5–2 *cm. long.* — Low pinelands and swamps, Fla. to Tex., n. to N.C. and Ky. Sept., Oct.

13. A. virgàta Trin. (wand-like), Wiregrass. — Similar to no. 12, *lower,* 5–9 *dm. high; glumes* 6–8 *mm. long; lemmas* 4–5 *mm. long; central horizontal awn* 1.2–2 *cm. long,* the shorter ascending *laterals* 7–10 *mm. long.* — Sands, bogs and pinelands, Fla. to Tex., n. to s. N.J. Aug.–Oct.

14. A. lanòsa Muhl. (woolly). — Perennial; culms stout, erect, simple, 0.8–1.5 m. high, in small tufts; *sheaths (at least the lower) woolly; blades flat,* shorter than the culms, 3–6 *dm. long,* 3–6 mm. wide; *panicles* nearly half the length of the entire plant, narrow, rather loosely flowered, nodding, decompound, 3–7 dm. long, with elongate branches; *the chief nodes woolly-tufted; 1st glume* 1.2–2.1 cm. long, acuminate, *notably exceeding 2nd,* the 2nd mucronate from a bidentate apex; lemma spotted, 0.8–1.9 cm. long; lateral awns 10–15 mm. long, the divergent middle awn 1.5–3 cm. long; anthers 5.5–6 mm. long. — Dry sterile soil, Fla. to Tex., n. to s. N.J., W.Va. and se. Mo. Sept., Oct.

Var. **màcera** Fern. & Grisc. (meagre). — *Culms* solitary, *filiform,* 4.5–7 *dm. high; leaves at most* 2 *mm. wide,* equaling or overtopping the inflorescence; *panicle* 1–2.2 *dm. long, with few scattered spikelets* and abbreviated branches: *glumes subequal, the 1st sometimes shorter than the 2nd,* 8–16.2 *mm. long;* middle awn 2.5–3.5 cm. long; *anthers* 3–3.5 *mm. long.* — Dry woods, se. Va. Sept., Oct.

60. MÍLIUM L. MILLET-GRASS

Spikelets 1-flowered, rachilla articulated below the floret; glumes equal, membranaceous, obtuse, with rounded back; lemma slightly shorter, shining, indurated, obtuse, dorsally compressed, the margins inrolled over a similar palea; grain inclosed with the lemma and palea, free. — Our species perennial, with flat leaves and open panicles; small genus of temp. N.Am. and Eurasia, too strongly approaching *Panicum* of the *Paniceae.* (The ancient Latin name of the Millet, the grain of *Milium* somewhat resembling that of Millet, *Panicum miliaceum.*)

266. M. effusum.

1. M. effùsum L. (spread out). — Smooth, often glaucous culms rather slender, simple, 0.7–1.7 m. high; leaves 1–3 dm. long, 0.7–2 cm. wide; panicle loosely elongate-ovoid, 1–2.5 dm. long; the slender branches in remote pairs or fascicles, widely spreading or drooping, spikelet-bearing from about the middle; spikelets 3–3.5 mm. long; glumes minutely scabrous; grain lustrous, equaling the lemma. — Rich (mostly calcareous) woods, thickets and glades, n. Nfld. to w. Ont., s. to N.S., n. and w. N.E., Md., W.Va., centr. O., n. Ind., n. Ill. and Minn. June–Aug. (Eurasia) FIG. 266.

TRÁGUS Hall. (Greek name of some plant mentioned by Dioscorides), of the Tribe ZOYSÌEAE Miq. with slender spiciform racemes of bur-like subglobose 1-flowered spikelets, these covered with longitudinal rows of stout hooked prickles, has two species casual about wool-waste and in rubbish but not long persistent: **T.** RACEMÒSUS (L.) All. (racemed) with distinctly pedicelled spikelets 4–4.5 mm. long; **T.** BERTERONIÀNUS Schultes (for CARLO GIUSEPPE BERTERO, 1789–1831) with nearly sessile spikelets 2–3 mm. long. (Adv. from Old World, often in wool from other regions.)

Tribe VI. CHLORÍDEAE Kunth (see p. 96)

a. Spikelets all alike, each with 1 or more perfect florets. . . *b.*
  b. Spikelets each with 1 perfect flower. . . *c.*
    c. Spikelets with no sterile lemmas or strictly staminate floret. . . *d.*
      d. Rachilla articulated below the glumes, the spikelets falling intact.
        Spikelets and their unequal glumes elongate, the latter not crossing at tip; perennials with tough scaly rhizomes. . . . . . . 61. *Spartina.*
        Spikelets suborbicular, the inflated and navicular subequal glumes with tips crossing; tufted annual. . . . . 62. *Beckmannia.*
      d. Rachilla articulated above the glumes, the falling floret leaving the glumes as empty husks.
        Spikes terminal, digitate, with overlapping spikelets; widely creeping plant. . . . . 63. *Cynodon.*
        Spikes remote in a stiff raceme, with distant spikelets; erect tufted plant. . . . . 64. *Schedonnardus.*
    c. Spikelets with 2–4 empty or staminate lemmas besides the perfect floret, these often small or awn-like. . . . *e.*
      e. Lowest floret perfect; empty lemmas 2; 2nd glume awnless or with terminal awn projected forward; spikes 1–∞.
        Spikelets scattered or remote, appressed. . . . . 65. *Gymnopogon.*
        Spikelets imbricated.
          Spikes approximate or in false whorls; fertile lemma 1-awned or awnless. . . . . 66. *Chloris.*
          Spikes in racemes or solitary; fertile lemma with 3 awns or excurrent nerves. . . . . 67. *Bouteloua.*
      e. Lowest floret neutral, the 3rd perfect; empty lemmas 3 or 4; 2nd glume with divergent dorsal awn; spike solitary. . . . . 68. *Ctenium.*
  b. Spikelets each with 2 or 3 perfect awnless florets; spikes 2 or more. . . *f.*
    f. Inflorescence exserted above the scattered flat leaves, of elongate spikes.
      Spikes digitate; spikelets densely imbricated.
        Rachis of spike prolonged beyond the spikelets; 2nd glume and lowest lemma cuspidate or awned. . . . . 69. *Dactyloctenium.*
        Rachis not prolonged, spikelets extending to tip of spike; glumes and lemmas blunt. . . . . 70. *Eleusine.*
      Spikes in elongate racemose inflorescences. . . . . 71. *Leptochloa.*
    f. Inflorescence subcapitate, of short spikes crowded among groups of sheaths of short pointed leaves terminating branches of a depressed small annual. . . . . 72. *Munroa.*

a.Spikelets dissimilar; plants dioecious or monoecious, the pistillate spikelets in capitate clusters sessile among leaves; staminate spikes elongate, 1–few in exserted racemes, the spikelets imbricated; dwarf perennial, extensively creeping. . . . . 73. *Buchloë*.

61. SPARTÌNA Schreb. CORD- OR MARSH-GRASS

Spikelets 1-flowered, flattened laterally, sessile and closely imbricated in 2 rows along one side of a continuous rachis, forming unilateral spikes which are borne along a common axis; glumes unequal, keeled, acute or bristle-pointed, the second usually exceeding the obtuse thinner 1-nerved lemma; palea equaling or exceeding the lemma. — Coarse perennials of saline to fresh habitats in Eu., Mediterr. reg., N.Am., S.Am. and remote islands of S. Hemisph., with strong creeping rhizomes, rigid simple culms, long tough leaves and often proterogynous flowers. (Name from the Greek *spartine, a cord*, such as was made from the bark of *Spartium* or broom.)

a.Fresh leaves flat (sometimes involute in drying), 0.4–2.5 cm. wide; rhizomes, if hard, 0.4–2 cm. thick; culms 0.5–2.5 cm. thick at base; spikes 5–60 (–100 or more). . . *b*.

b.Margins and tips of leaves and keels of glumes scabrous, harsh to touch; rhizomes and stolons rigid.

Leaves 4–15 mm. wide, orifice of upper sheath plane; spikes 5–30; 2nd glume long-awned. . . . . 1. *S. pectinata*.

Leaves 1–2.5 cm. wide, orifice of upper sheath puckered in drying; spikes 6–100 or more; 2nd glume sharp-pointed, scarcely awned. . . . . . 2. *S. cynosuroides*.

b.Margins and tips of leaves and keels of glumes glabrous, smooth to touch; rhizomes and stolons flaccid. . . . . 3. *S. alterniflora*.

a.Fresh leaves involute, slender, smooth; rhizomes hard, 1–6 mm. thick; culms 1–6 mm. thick at base; spikes 1–9. . . . . 4. *S. patens*.

1. S. pectinàta Link (comb-like), FRESH-WATER CORD-GRASS, SLOUGH-GRASS, HERBE À LIENS or CHAUME (Que.). — *Culms* 0.5–1 *cm. thick at base*, 0.6–2 m. high; rhizomes hard, brownish to purplish, 4–11 mm. thick, closely scaly; *leaves* hard, mostly 3–12 dm. long, 4–15 *mm. wide*, tapering to slender points, keeled, flat, but quickly involute in drying, *scabrous on the margins; orifice of sheath plane; spikes* 5–30, 2–11 *cm. long*, 5–8 *mm. wide including the salient awns*, sessile or short-peduncled, ascending; *glumes serrulate-hispid on the keel*, the 1st acuminate and nearly or quite equaling the floret, the 2*nd tapering to an awn* 3–7 *mm. long; lemma* 7–9 mm. long, *the scabrous midrib abruptly terminating below the* emarginate or 2-toothed *apex*. (*S. Michauxiana* Hitchc.) — Shores, gravels and wet prairies or swamps, Nfld. to Alta. and Wash., s. to N.S., N.E., L.I., w. N.C., W.Va., Ind., Ill., Mo., Tex., N.M. and Oreg. July–Sept. (Eu.) — In forma **variegàta** Vict. (variegated) the margins and midribs of leaves are white.

Var. **Súttiei** (Farw.) Fern. (named in 1920 for its discoverer, GEORGE SUTTIE). — *Spikes mostly* 0.7–1.5 *dm. long*, 3–5 *mm. broad, with appressed awns*, often more peduncled, loosely ascending to spreading. — P.E.I. to Minn., s. to N.S., N.E., N.J., Pa., Ky., Mo. and Okla.

2. S. cynosuroìdes (L.) Roth (for the spikes, like those of *Cynosurus*), SALT REED-GRASS. — *Culms* stout, 1–3 m. high, 1–2 *cm. thick at base;* rhizome hard, deep-seated, 1–2 cm. thick, covered with white scales; *leaves* 1–2.5 *cm. wide*, flat, roughish beneath and on the margins; *orifice of upper sheath puckered in drying; spikes* brownish to purplish, 6–50 *in an open raceme, subdistant or distant and ascending, often definitely peduncled; glumes barely mucronate, the 1st half the length of the lemma, of which the hispid midrib reaches the apex.* — Brackish or fresh tidal marshes, Fla. to Tex., n. along coast to Ct. Aug.–Oct.

Var. **polystáchya** (Michx.) Beal (with many spikes). — *Inflorescence dense, the* 30–100 *or more appressed-ascending or erectish commonly purple spikes only short-peduncled to sessile*. (*S. polystachya* (Michx.) Beauv.) — Saline marshes, n. to Cape Cod, Mass.

3. S. alterniflòra Loisel. (alternate-flowered), SALT-WATER C., HERBE SALÉE (Que.). — Odor strong and rancid; culms 0.1–2.5 m. high, leafy to top; *rhizome flaccid*, covered with white papery scales; *leaves flat* when fresh, becoming involute in drying, 0.4–1.5 cm. wide, *smooth*, succulent, very tough; spikes appressed, 2–15 cm. long, the *rachis* often *projecting as a bristle above the base of the terminal spikelet;* spikelets 9–14 mm. long; *glumes glabrous or only sparingly scabrous below on the keel*, the 1st much shorter than the 2nd; *lemma* 7–10 mm. long, *glabrous or sparingly pilose*. (*S. stricta* Am. auth., not Roth) — Very variable, with 3 pronounced vars.:

Spikelets subremote, the tips at most overlapping the bases of those above; lemma minutely pilose; rachis often continued far beyond the terminal spikelet; plants 0.1–1.4 m. high. . . . . *S. alterniflora* (typical).

Spikelets strongly imbricated, forming close spikes; plants 0.4–2.5 m. high.
Lemma sparingly pilose; rachis often prolonged beyond the upper spikelet. . Var. *pilosa.*
Lemma glabrous; rachis rarely prolonged beyond the upper spikelet. . . . Var. *glabra.*

S. alterniflòra, typical (*S. glabra,* var. *alterniflora* (Loisel.) Merr.). — Saline shores and marshes Nfld. to lower St. Lawrence R., Que., s. to N.J. July–Sept. — Locally natzd. in w. Eu., where it has crossed with the endemic European *S. maritima* (Curtis) Fern. (*S. stricta* (Ait.) Roth), producing the aggressive × *S. Townsendii* H. & J. Groves.

Var. **pilòsa** (Merr.) Fern. (hairy), *S. glabra,* var. *pilosa* Merr. — N.S. to N.C.; also w. N.Y.

Var. **glàbra** (Muhl.) Fern. (glabrous), *S. glabra* Muhl. — Se. N.Y. to Fla. and Tex.

4. S. pàtens (Ait.) Muhl. (spreading), SALT-MEADOW GRASS, HIGH-WATER GRASS, MUSOTTE (Que.). — *Rhizomes wiry, slender,* 1–3 *mm. thick; culms* arising chiefly from matted marcescent shoots of the preceding season, 1–2.5 *mm. thick at base,* 1.5–8 dm. long; *new green cauline leaves of the season usually* 4 (2–5), *filiform or strongly involute, glabrous* on the back, 0.5–2 mm. wide, tapering to slender tips, divergent; *blade of the 2nd from the summit averaging* 1 (0.5–2) *dm. long;* raceme slightly to hardly exserted; spikes 1–4, on a *smooth filiform rachis,* ascending or divergent, 1–6 cm. long, mostly purple; *spikelets* 9–13 mm. long, *loosely imbricated, straightish, ascending, with suberect free tips;* 1st glume linear, about 4 mm. long; *2nd glume* lanceolate, *acuminate,* scabrous on keel and adjacent nerves, glabrous on broad hyaline sides; lemma 5–6 mm. long, obtuse, scarcely equaling the palea. — Saline marshes and brackish shores, sw. Nfld. to lower St. Lawrence R., Que., s. to Va.; inland in w. N.Y. and se. Mich. Late June–Oct.

Var. **monógyna** (M. A. Curtis) Fern. (one-carpelled), WHITE-RUSH. — Mostly coarser; *rhizomes* 2–6 *mm. thick; culms* 1–6 *mm. thick at base,* 0.2–1.5 m. high; *new* green cauline *blades averaging* 6 (5–9), the *2nd from summit averaging* 2 (1–5) *dm. long;* spikes 2–9, purple or stramineous; *spikelets* 7–10 mm. long, *tightly imbricated and strongly arching, with appressed tips; 2nd glume often merely acute or even blunt.* (*S. juncea* (Michx.) Willd.; *S. patens,* var. *juncea* Hitchc.) — Coastal sands and borders of saline or brackish marshes, N.H. to Fla. and Tex. Aug.–Oct. (So. Eu.; n. Afr.)

× **S. caespitòsa** A. A. Eat. (bunched). — A variable series combining characters of nos. 1 and 4; cespitose or subcespitose, 0.6–1 m. high; rhizomes and stolons (when developed) 0.4–1 cm. thick; leaves slender and involute or flat and up to 5 mm. broad, elongate; spikes 2–7, 2–8 cm. long, purple to green; some or all glumes awned, with coarsely scabrous keel. (*S. patens,* var. *caespitosa* (A. A. Eat.) Hitchc.) — Sporadic on borders of saline marshes and strands, Me. to L.I.

62. BECKMÁNNIA Host

Spikelets 1-flowered (in ours), broad, laterally compressed, closely imbricated in 2 rows along one side of a continuous rachis, forming short unilateral spikes; rachilla articulated below the glumes; glumes subequal, inflated, navicular, chartaceous, margin scarious; lemma lanceolate, acuminate, palea nearly as long; grain free within the rigid lemma and palea. — Rather tall erect grasses of cool and temp. N.Am. and Eurasia, with flat leaves and a terminal elongated narrow nearly simple panicle. (Named for *Johann Beckmann,* 1739–1811, professor at Goettingen.)

267. B. syzigachne.

1. B. syzigáchne (Steud.) Fern. (with scissors-like glumes), SLOUGH-GRASS. — Annual, light green; culms solitary or tufted, 5–10 dm. high; sheaths loose, overlapping; blades 1–2.5 dm. long, 5–8 mm. broad, scabrous; panicle 1–2.5 dm. long; the spikes appressed, 1–2 cm. long; spikelets much distended, about 3 mm. long, with 1 perfect and usually 1 imperfect floret, pyriform; glumes rounded-triangular, broadest toward summit, transversely wrinkled; mucronate apex of lemma projecting beyond glumes. (*B. erucaeformis* Am. auth., not Host) — Wet ground, w. Que. to Alaska, s. to Mich., ne. Ill., nw. Ia., Kans., N.M. and Calif.; adv. or natzd. e. to N.Y. and Pa. and s. to Mo. June–Aug. (Asia) FIG. 267.

63. CÝNODON Richard BERMUDA GRASS OR SCUTCH-GRASS

Spikelets 1-flowered, laterally compressed, awnless, alternate in 1 or 2 rows on the 2 margins of a flattened continuous axis, forming unilateral spikes; rachilla prolonged behind the

268. C. dactylon.

palea into a blunt pedicel; glumes unequal, narrow, acute, keeled; lemma broad, navicular, obtuse, ciliate on the keel; palea as long as the lemma, the prominent keels close together, ciliolate; grain free within the lemma and palea. — Low diffusely branched and extensively creeping perennials, ours introd. from Eu., the others Australian, with flat leaves and slender spikes digitate at the apex of the upright branches. (Name from the Greek *cyon, dog,* and *odous, tooth,* from the close rows of tooth-like spikelets.) CAPRIOLA Adans.

1. C. DÁCTYLON (L.) Pers. (with fingers). — Glabrous; culms flattened, wiry; ligule a conspicuous ring of white hairs; spikes 4–6, 2–7 cm. long; spikelets imbricated, 2 mm. long; lemma longer than the glumes. (*Capriola* Ktze.) — Fields, pastures and waste places, abund. southw., locally n. to St. P. et Miq., Mass., N.Y., W.Va., O., Mich., Ill., Ia. and e. Kans. July–Oct. (Natzd. from Eu.) FIG. 268. — Seldom perfects seed.

64. SCHEDONNÁRDUS Steud.

Spikelets 1-flowered, sessile and appressed, alternate and distant along one side of a slender triangular rachis, forming very slender spikes; glumes narrow, unequal, with strong rigid keels, pointed, shorter than the lanceolate acuminate scabrous lemma; palea nearly as long as the lemma; grain free within the subrigid lemma and palea. — A low diffusely branching annual, with short narrow leaves and slender paniculate spikes. (Name from the Greek *schedon, near,* and *Nardus,* from its resemblance to that genus.)

269. S. paniculatus.

1. S. paniculàtus (Nutt.) Trel. (paniculate). — Culms 3–5 dm. high, erect or decumbent at base, leafy below; sheaths and blades smooth; panicle half or more than half the entire height of the plant, its axis usually falcate; spikes solitary and remote, mostly along the convex side, rigid; spikelets 4 mm. long. — Open ground and salt-licks, Man. to Mont., s. to Ill. (formerly), Mo., Tex., N.M. and Ariz. Late May–Oct. (So. S. Am.) FIG. 269. — At maturity the panicle becomes much elongated and decumbent, the axis extending in a broad loose spiral.

65. GYMNOPÒGON Beauv. BEARDGRASS

Spikelets with 1 perfect flower, sometimes 1 or 2 neutral or staminate subsessile florets above the perfect one, remote along one side of a filiform continuous rachis, forming slender unilateral spikes; rachilla prolonged beyond the floret as a slender often awned rudiment; glumes narrow, subequal, rigid, scabrous on the strong keel, equaling or exceeding the florets; lemma thin, bearing a slender straight awn from just below the apex; palea about as long as the lemma. — Perennials of warm reg. of Am. and in Ceylon, with short rather broad rigid leaves and numerous slender spikes, these at first erect, at length widely divaricate or reflexed. (Name composed of the Greek *gymnos, naked,* and *pogon, beard;* alluding to the reduction of the abortive flower to a bare awn.)

1. G. ambíguus (Michx.) BSP. (doubtful). — Culms tufted from a short rhizome, rigid, erect or ascending, 2–5 dm. high; sheaths overlapping; blades often approximate, thick, rigid, spreading, 4–6 cm. long, 0.6–1 cm. or more wide; *spikes* solitary or in 2's along a striate axis, becoming widely divaricate when exserted from the sheath, *spikelet-bearing to the base; awn of floret longer than the glabrous lemma; rudiment long-awned.* — Dry sandy or rocky openings and thin woods, Fla. to Tex., n. to N.J., s. O., s. Ind., Mo. and se. Kans. July–Sept. FIG. 270.

270. G. ambiguus.

2. G. brevifòlius Trin. (short-leaved). — Resembling the preceding; culms more slender, from a decumbent base; leaves 2–4 cm. long, 4–9 mm. wide, involute in drying; *spikes* usually less numerous, more distant, *naked at the base, spikelet-bearing from about the middle; awn shorter than the hairy lemma;* one or two sterile florets sometimes present; rudiment usually awnless. — Sandy or peaty ground, Fla. to La., n. to N.J. and Ark. Aug., Sept.

66. CHLÒRIS Sw. WINDMILL-GRASS

Spikelets with 1 perfect floret, sessile in 2 rows along one side of a continuous rachis, forming unilateral spikes; rachilla prolonged behind the palea and bearing 1 or more rudimentary awned sterile lemmas; glumes unequal, narrow, acute, keeled; lemma often ciliate on the back or margins, 1–3-nerved, the midnerve nearly always prolonged into a slender awn; palea about equaling the lemma; grain free within the lemma and palea. — Usually perennial grasses of N. and S. Am., Austral., Afr. and s. Asia, with flat leaves and digitate spikes. (Named for *Chloris*, mother of Nestor.)

1. C. VERTICILLÀTA Nutt. (whorled). — Culms 1–4 dm. high, erect, or decumbent and rooting at the nodes; sheaths compressed; leaves obtuse, light green; *spikes* several in 1–3 whorls, slender, 7–15 *cm. long*; spikelets 3 mm. long, with awns about 5 mm. long; sterile lemma one. — Prairies, Neb. to Tex. and N.M.; adv. in Mo., Ia., Ill. and Ind. and occasionally eastw. Late May–Oct. FIG. 271. — The ripe inflorescence breaks away and forms a tumbleweed.

271. C. verticillata.

2. C. VIRGÀTA Sw. (wand-like). — Similar, *spikes* 2–8 *cm. long*; *spikelets long-villous or ciliate on margin at summit*, the awn 5–10 mm. long. (*C. elegans* HBK.) — Waste places and old fields, Mo. to La. and westw.; casual in Atlantic States and estab. in e. Pa. (Adv. from the Southwest or the Tropics)

C. GAYÀNA Kunth (for JACQUES GAY, 1786–1864), a strongly *stoloniferous* plant, the inflorescence digitate, the short-awned lemma glabrous except for slight marginal and basal hispidity, is much cult. southw. as RHODES GRASS, and spreads from cult. northw. to se. Mass. (Introd. from Afr.)

TRICHLÒRIS Fourn. (from *tri, three*, and *Chloris*), differing from *Chloris* in having the fertile lemma 3-awned, sometimes occurs with us as **T.** MENDOCÌNA (Phil.) Kurtz (of Mendoza, Argentina), a hard-based perennial, with tall stiff culms, scabrous leaves and a mass of plume-like ascending spikes, found on wool-waste and rubbish but not long persistent. (Adv. from the Southwest, Mex. or S. Am.)

67. BOUTELOÙA Lag. MESQUITE-GRASS. GRAMA-GRASS

Spikelets 1–2-flowered, crowded and sessile in 2 rows along one side of a continuous flattened rachis, which usually projects beyond the spikelets; rachilla prolonged beyond the perfect floret and bearing a sterile (rarely staminate) floret, a second or third rudiment often present; glumes unequal, keeled; lemma broader, 3–5-nerved, 3–5-toothed or -cleft, 3 of the divisions usually awn-pointed; palea about the length of the lemma, bidentate, the 2 keels scabrous; sterile floret sometimes reduced to the awns, rarely obsolete. — Our species perennial, with narrow flat or convolute leaves, and unilateral spikes nearly sessile along a common axis. Large genus of warm reg. of N. and S. Am. (Named for *Claudio Boutelou*, 1774–1842, a Spanish writer upon floriculture and agriculture.)

Spikes 1–6, 2–5 cm. long, curved, of 35–80 densely crowded pectinate spikelets.
§ CHONDROSIUM.
Leaves glabrous or merely scabrous; 2nd glume minutely scabrous; rachis of spike not prolonged. . . . . 1. *B. gracilis.*
Leaves hispid; 2nd glume tuberculate; rachis of spike prolonged beyond the uppermost spikelets. . . . . 2. *B. hirsuta.*
Spikes 15–50, 0.8–2 cm. long, in a slender raceme, of 5–8 spikelets, not pectinate.
§ ATHEROPOGON. . . . . 3. *B. curtipendula.*

§ CHONDRÓSIUM (Desv.) Gray

1. B. grácilis (HBK.) Lag. (slender). — Culms slender, erect, from a short rhizome, leafy at the base, 1.5–5 dm. high; sheaths and *blades glabrous;* the latter about 2 mm. wide, flat or becoming convolute; spikes 1–6, 2–5 cm. long; spikelets 5–6 mm. long; *glumes* narrow, the first about half as long as the *second*, which is *sparsely papillose-pilose on the keel;* fertile lemma pilose, 3-cleft, the divisions awned; *sterile lemma* consisting of 2 truncate lobes and 3 divergent equal awns *with a tuft of long hairs at base;* second rudiment obtuse, awnless. (*B. oligostachya* (Nutt.) Torr.) — Prairies, Wisc., Minn. and Man. to B.C., s. to n. Ill., Mo., Tex. and Mex.; adv. eastw. (S. Am.) July–Sept. FIG. 272.

272. B. gracilis.

2. B. hirsùta Lag. (stiffly hairy). — Culms tufted, erect, 2–5 dm. high, leafy at the base; sheaths smooth; *blades* about 3 mm. wide, flat, *sparsely papillose-hairy, especially on the margins;* spikes 1–4, 1.5–5 cm. long; the *rachis of the spike produced into a prominent point beyond the uppermost spikelets;* spikelets about 5 mm. long; first *glume* setaceous, the *second* equaling the floret and *conspicuously tuberculate-hirsute on the back;* fertile lemma pubescent, 3-cleft, the divisions awn-pointed; *sterile floret* of 2 obtuse lobes and 3 equal awns margined below, *with no tuft of hairs at the base.* — Sandy plains, Wisc. to S.D., and Colo., s. to w. Ill., La., Tex. and Mex.; adv. eastw. July–Sept. FIG. 273.

273. B. hirsuta.

§ Atheropògon (Muhl.) Gray

3. B. curtipéndula (Michx.) Torr. (short-hanging), Tall G. — Culms erect, from short running rhizome, 3–10 dm. high; sheaths pubescent toward the summit; blades 1–3 dm. long, 3–5 mm. wide, flat or involute and setaceous toward the end, scabrous above, sometimes pubescent beneath; *spikes numerous,* 0.8–2 cm. long, spreading or reflexed, in a long mostly 1-sided raceme, *the rachis bifid at the extended apex;* spikelets 7–10 mm. long; first glume less than half the length of the second which is very scabrous on the thickened keel, exceeding the floret; *lemma scabrous,* ending in 3 short slender awns; teeth of palea aristate; *sterile lemma with 2 acute lobes and 3 straight awns, the lateral ones much shorter than the middle awn.* — Dry hills and plains, Ct. to s. Ont., w. to Mont., s. to Ga., Ala., Miss., La., Tex. and Mex. July–Sept. Fig. 274. — The sterile lemma variable, rarely reduced to a single awn.

274. B. curtipendula.

68. CTÉNIUM Panzer Toothache-Grass

Spikelets with 1 perfect flower and 2–5 sterile lemmas, crowded and sessile, pectinate in 1-sided spikes; glumes very unequal, first minute, second nearly as long as the spikelet and bearing a stout horizontally divergent dorsal awn from about the middle; first and second lemmas empty or sometimes with a hyaline palea, awned below the apex, awn erect or ascending; third lemma similar, containing a perfect flower; fourth lemma awnless, with staminate flower or empty; a fifth rudimentary lemma often present. — Rather tall perennials of N. and S. Am., Afr. and Mascarene Ids., with solitary terminal more or less curved spikes. (Name from the Greek *ctenion, a small comb,* from the pectinate appearance of the spike.) Campulosus Desv.

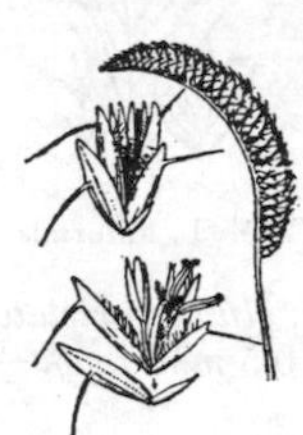

275. C. aromaticum.

1. **C. aromáticum** (Walt.) Wood (aromatic), Lemon- or Orange-Grass. — Culms 1–1.5 m. high, erect, from scaly rhizomes; old sheaths persistent at the base; blades long, flat or involute, stiff; spike 0.5–1.5 dm. long; spikelets 5–7 mm. long; first glume warty-tuberculate on the nerves; florets stiffly ciliate on the margins. — Wet pine-barrens, Fla. to La., n. on Coastal Plain to se. Va. July–Sept. Fig. 275. — Taste very pungent; bruised roots often fragrant.

69. DACTYLOCTÉNIUM Willd. Crowfoot-Grass

Spikelets few-flowered, the uppermost imperfect, sessile and crowded in 2 rows along one side of a continuous rachis which extends beyond the spikelets as a naked point; glumes broad, keeled; lemmas navicular, cuspidate; palea equaling the lemma, acute, deeply folded between the ciliate-winged keels; grain reddish-brown, the loose pericarp transversely wrinkled. — Annual nat. of warm reg. of Old World, with more or less decumbent and creeping base, and 2–6 stout unilateral spikes digitate at the apex of the culm. (Name from the Greek *dactylos, finger,* and *ctenion, a little comb,* alluding to the digitate and pectinate spikes.)

276. D. aegyptium.

1. D. aegýptium (L.) Richter (Egyptian). — Usually glabrous; culms rooting at the lower nodes; spikes 1–5 cm. long; glumes scabrous on the keel, the second cuspidate; the awned tip of lower lemma inflexed, that of the others straight or curved. — Waste and cult. land, abund. southw., locally n. to Mass. and Ill., Sept., Oct. (Natzd. from Old World) Fig. 276.

70. ELEUSÌNE Gaertn. Goose-Grass. Yard-Grass

Spikelets few-flowered, awnless; florets perfect or uppermost staminate, sessile and closely imbricated in 2 rows along one side of a continuous rachis which does not extend beyond the terminal spikelet; glumes unequal, shorter than the floret, scabrous on the keels; lemmas broader, with a thickened 5-ribbed keel; palea shorter, acute, the narrowly winged keel distant; grain black, the loose pericarp marked with comb-like lines, free within the subrigid lemma and palea. — Coarse tufted annuals, nat. of warm areas of the Old World, with stout unilateral spikes digitate or approximate at the apex of the culms. (Name from

277. E. indica.

Eleusis, the town where Ceres, the goddess of harvests, was worshipped.)

1. E. índica (L.) Gaertn. (of India), Wiregrass. — Glabrous; culms flattened, decumbent at base; sheaths loose, overlapping, compressed; spikes 2–10, 2.5–15 cm. long; spikelets appressed, 3–5-flowered, about 5 mm. long. — Yards and waste ground, Que. to Minn., S.D. and southw. July–Oct. (Natzd. from Old World) Fig. 277.

71. LEPTÓCHLOA Beauv. Feathergrass

Spikelets 2–4-flowered, with the uppermost floret usually imperfect or rudimentary, sessile or nearly so, in 2 rows along one side of the slender continuous rachis; glumes and lemmas keeled; the latter 3-nerved, acute, awnless or short-awned, exceeding the palea. — Usually tall annuals of warm reg., with flat leaves and elongated simple panicles composed of the numerous very slender spikes scattered along the main axis. (Name composed of the Greek *leptos, slender*, and *chloa, grass*, from the long attenuated spikes.)

1. L. filifórmis (Lam.) Beauv. (thread-like). — Green or purplish, 0.3–1.2 m. high; sheaths papillose-pilose; panicle 1.5–6 dm. long; its 20–100 or more filiform *stiff spikes* 5–15 cm. long, ascending to spreading; spikelets 1.5–2.5 mm. long; *glumes merely acute, not at all or but rarely overtopping the* 2–4 *florets; lemmas* ovate, notched at summit, *pubescent on midrib and margin* below, *the lower* 1–1.3 *mm. long; grain* 0.7–0.9 *mm. long.* — Sandy, often cult., fields and bottoms, Fla. to Tex. and N.M., n. to Va., and in the interior n. to s. Ind., s. Ill., Mo. and Kans.; adv. ne. to Mass. July–Oct. (Mex.; W.I.; S.Am.) Fig. 278.

278. L. filiformis.

2. L. attenuàta (Nutt.) Steud. (gradually tapering). — Lower, softer; panicles 0.7–3.5 dm. long; their 10–30 *flexuous spikes* 2–11 cm. long, spreading; *glumes aristate, the 2nd overtopping the upper floret; lemmas smoother,* 0.5–1 *mm. long; grain* 0.4–0.5 *mm. long.* — Sandy bottoms, Ill. to La. and Tex. July–Oct.

72. MUNRÒA Torr. False Buffalo-Grass

Spikelets perfect, crowded in 2's or 3's into a subcapitate cluster, the lower 3–4-flowered, the upper 2–3-flowered; glumes of lower spikelets equal, 1-nerved, acute, of upper spikelet unequal, with the first reduced or wanting; lemmas 3-nerved, with excurrent midrib, those of lower spikelet coriaceous, those of upper spikelets membranaceous. — Low tufted grasses, of w. N. and S. Am., with the crowded spikelets clustered among the short and stiff leaves at the ends of wiry branches. (Named for *William Munro*, 1818–1880, distinguished student of grasses.)

279. M. squarrosa.

1. M. squarròsa (Nutt.) Torr. (with recurved tips). — Mats 0.5–5 dm. across; culms scabrous; leaves 1.5–3 cm. long, acerose. — Dry plains, slopes and disturbed soils, Alta. to Ariz., e. to Man., w. Minn., e. Neb., e. Kans., Okla. and Tex. June–Aug. Fig. 279.

73. BÙCHLOË Engelm. Buffalo-Grass

Spikelets unisexual; plants monoecious or dioecious; staminate spikelets 2–3-flowered, sessile in 2 rows along the short 1-sided spikes; glumes unequal, obtuse; lemmas larger, 3-nerved; palea a little shorter than lemma; pistillate spikelets 1-flowered, in nearly capitate 1-sided spikes which are scarcely exserted from the broad sheaths of the upper leaves; glumes indurated, trifid at the apex, united at base and resembling an involucre; lemma narrow, hyaline, inclosing the 2-nerved palea; grain free within the hardened glumes. — A creeping or stoloniferous perennial with narrow flat leaves, and dissimilar staminate and pistillate spikelets borne on the same or on distinct plants. (Name from the Greek *bous, cow* or *ox*, and *chloë, grass.*)

280. B. dactyloides.

1. B. dactyloìdes (Nutt.) Engelm. (like *Dactylis*). — Culms of the staminate inflorescence 1–3 dm. high, the spikes long-exserted; culms of pistillate inflorescence low, much exceeded by the leaves; sheaths overlapping; blades 2 mm. wide or less; staminate spikes 2 or 3, 6–12 mm. long; cluster of pistillate spikelets ovoid, 6 mm. long. (*Bulbilis* Raf.) — Dry plains and prairies, Man., w. Minn. and w. Ia. to Mont., s. to w. La., Tex. and Mex. May–Aug. Fig. 280.

Tribe VII. Phalarídeae Link (see p. 96)

Spikelets with 1 central perfect floret and 2 sterile lemmas; inflorescence spiciform or lobulate-branched, with many closely imbricated spikelets.
Spikelets laterally flattened; glumes subequal; sterile lemmas scale-like, awnless; stamens 3; lodicules evident; fertile lemma indurated in fruit; plant not sweet-scented. . . . . 74. *Phalaris.*
Spikelets subterete; glumes very unequal; sterile lemmas large, awned from back; stamens 2; lodicules 0; fertile lemma remaining membranaceous; plant sweetly fragrant on drying. . . . . 75. *Anthoxanthum.*
Spikelets with 1 perfect central floret with 2 stamens and 2 staminate or empty ones, the staminate with 3 stamens; glumes subequal, broad and scario-membranaceous; inflorescence an open (in fruit contracted) panicle; plant sweetly fragrant. . . . . 76. *Hierochloë.*

74. PHÁLARIS L. Canary-Grass

Spikelets laterally flattened, with 1 perfect terminal flower and 2 sterile lemmas at its base; glumes equal, navicular, much exceeding the florets; sterile lemmas small and narrow, appearing like scales attached to the fertile floret; fertile lemma indurated and shining in fruit, inclosing a faintly 2-nerved palea. — Annuals or perennials of temp. reg., with flat leaves and dense or spike-like panicles. (The ancient Greek name, alluding presumably to the crest-like inflorescence.)

Inflorescence a continuous ovoid spiciform panicle 1.5–4 cm. long; glumes broadly boat-shaped, the keel broadly winged; fruit 2 mm. broad, with 2 entire lanceolate sterile lemmas at base. § Euphalaris. . . . . 1. *P. canariensis.*
Inflorescence a branching or interruptedly spiciform panicle 0.2–2 dm. or more long; glumes lanceolate or narrowly oblong, narrowly or not at all winged; fruit at most 1.5 mm. broad, with sterile lemmas linear and plumose. § Digraphis.
Annual, without rhizomes; panicle closely spiciform; keel of glumes narrowly winged at summit. . . . . 2. *P. caroliniana.*
Perennial, from creeping rhizomes; panicle open or lobate; keel of glumes wingless. . . . . 3. *P. arundinacea.*

§ Euphálaris Godr.

1. P. canariénsis L. (of the Canary Islands), Canary- or Birdseed-Grass, Graines d'oiseaux (Que.). — *Annual,* 3–8 dm. high; *spiciform panicle dense, ovoid,* 1.5–4 cm. long; *spikelets broadly obovate,* 5–6 mm. long, *closely imbricated; glumes broad, whitish, with green veins,* the *keel broadly winged;* fertile lemma brownish; fruit about 5 mm. long, with the 2 basal *entire sterile lemmas lanceolate.* — Waste places and roadsides, rarely persistent. June–Oct. (Adv. from Eu.)

§ Dígraphis (Trin.) Endl.

2. P. caroliniàna Walt. (of Carolina), Maygrass. — *Annual,* 0.2–1 m. high; leaves with narrowly lanceolate blades; *panicle spike-like,* suggesting that of *Phleum,* 1.5–12 cm. long, 7–15 mm. thick; spikelets 5–6 mm. long; *glumes* narrowly oblong; *the keel scabrous, narrowly winged at summit;* fruit 3–4 mm. long, about 1 mm. wide, acuminate; sterile lemmas linear, hairy. — Sandy soil, Fla. to Tex. and Mex., n. to Md., Tenn., Mo., Kans., Colo. and Oreg. May, June.

3. P. arundinàcea L. (reed-like), Reed-C., Roseau (Que.). — *Perennial from creeping rhizomes,* 0.6–2 m. high; leaves flat, elongate, 0.6–2 cm. wide; *panicle 0.5–2 (–3) dm. long, open in anthesis,* tightly contracted in fruit; *spikelets lanceolate,* 4–6 mm. long, pale; fruit 3–4.2 mm. long, 0.7–1.5 mm. broad; *sterile lemmas minute hairy scales.* — Shores, swales and meadows, Nfld. to s. Alaska, s. to e. Md., interior N.C., Ky., Ill., Mo., Okla., N.M., Ariz. and Calif. June–Aug. (Eurasia) — Forma **variegàta** (Parnell) Druce (variegated), with white-striped leaves, both native and introd., is the Ribbon-Grass of gardens.

75. ANTHOXÁNTHUM L. Sweet Vernal Grass. Flouve (Que.). Foin d'odeur (Que.)

Spikelets with 1 perfect central floret and 2 basal sterile lemmas; glumes very unequal; sterile lemmas 2-lobed, hairy, dorsally awned, longer than the fertile floret and falling with it;

fertile lemma truncate, awnless, inclosing a faintly 1-nerved palea and perfect flower; stamens 2. — Fragrant plants with flat leaves and narrow spike-like panicles of proterogynous flowers, nat. of Eurasia and Afr. (Name from the Greek *anthos, flower*, and *xanthos, yellow.*)

281. A. odoratum.

1. A. ODORÀTUM L. (fragrant). — *Perennial* in tufts, without stolons or basal scaly offshoots; culms slender, erect, 2–10 dm. high; leaves rough above, subglabrous to villous, 3–6 mm. broad; panicles continuous or nearly so, 2–7 cm. long, compact, or merely lax at flowering time; *spikelets brownish-green, 8–10 mm. long,* spreading at flowering time; *glumes sparsely long-pilose to hirtellous;* first sterile lemma short-awned below the apex, second bearing a strong bent *scarcely exserted awn* near its base. — Fields, pastures and waste places, Nfld. to s. Ont., s. to Ga., Ala., Miss. and La.; Pacific slope. May–Aug. (Natzd. from Eu.) FIG. 281. — The local forma GIGANTÈUM P. Junge (gigantic), has culms 0.6–1 m. high, panicle 7–14 cm. long and often interrupted, the spikelets 1–1.2 cm. long.

2. A. PUÉLII Lecoq & Lamotte (named in 1847 for TIMOTHÉE PUEL). — Smaller, *annual,* 1–3.5 dm. high; panicles 1–4 cm. long; *spikelets whitish-green,* 5–7 *mm. long; the glabrous glumes* narrower than in no. 1; *the long-exserted awn* blackish at base. (*A. aristatum* sensu Hitchc., not Boiss.) — Dry fields and waste places, N.E. to Ont. and Minn., s. to N.C., W.Va., O. and Miss., local; Pacific slope. June–Sept. (Natzd. from Eu.)

76. HIERÓCHLOË R. Br. HOLY GRASS. FOIN D'ODEUR or HERBE SAINTE (Que.)

Spikelets 3-flowered, the terminal flower perfect, the others staminate or neutral; glumes subequal, shining; sterile lemmas nearly as long as the glumes, navicular, indurated and hairy, each inclosing a 2-nerved hyaline palea and a flower of 3 stamens; fertile lemma similar but smaller, inclosing a 1-nerved palea and perfect flower with 2 stamens. — Fragrant perennials of cool and temp. reg., with flat leaves and terminal panicles. (Name from the Greek *hieros, sacred,* and *chloë, grass;* these sweet-scented grasses being strewn before church-doors on saints' days in the North of Europe.) SAVASTANA Schrank

Lemmas awnless or with minute straight terminal awn less than 1 mm. long; plants stoloniferous.
- Flowering culms soft, soon wilting after anthesis; new cauline blades 2 or 3, short, soft, lanceolate; lemmas awnless. . . . . 1. *H. odorata.*
- Flowering culms firm, persistent; new cauline leaves 3–6, elongate, firm; lemmas awned. . . . . 2. *H. Nashii.*

Lemmas of imperfect florets unequally awned; the longer awn subbasal, bent, 5–8 mm. long; plant cespitose. . . . . 3. *H. alpina.*

1. H. odoràta (L.) Beauv. (fragrant), VANILLA, INDIAN or SWEET GRASS. — *Culms* arising from among dead foliage of preceding year, 3–6(–9) dm. high, *from slender creeping rhizomes, soft, soon shriveling after flowering; cauline leaves 2 or 3, short, lanceolate, with ligules* 1–2 *mm. long;* leaves of sterile shoots prolonged (becoming 2–8 dm. long), scabrous; *panicle pyramidal,* 3–14 *cm. long,* with slender spreading to reflexed lower branches 1–7 cm. long, loose in anthesis, closer in fruit; or culms up to 9 dm. high with *panicle* 2–4 *dm. long and with lower branches* 1–2 *dm. long* in forma **Eàmesii** Fern. (for its discoverer, EDWIN HUBERT EAMES, 1865–1948) of sw. Ct.; spikelets 4–8 mm. long, brownish, bronze or purple; glumes ovate, barely equaling to exceeding the lemmas; *staminate lemmas* hispid-ciliate on the margins and often at summit of keel, *awnless;* fertile lemma hairy at apex. — Meadows, swales and shores (fresh or brackish), Lab. to Alaska, s. to Nfld., N.S., N.E., L.I., n. N.J., Pa., O., Ind., n. Ill., Ia., S.D., N.M. and Ariz.; in the Northeast common on coast and mts. Apr. (southw.)–Aug. (alpine). (Eurasia) — The long leaves of vegetative shoots used for Indian baskets.

2. H. Náshii (Bickn.) Kaczmarek (for GEORGE VALENTINE NASH, 1864–1921). — *Culms hard,* from slender rhizomes, 0.6–1 m. high; *cauline leaves* 3–6, *narrowly linear, firm; the upper* 1–3 *dm. long,* slender-tipped, *with ligules* 5–7 *mm. long; panicle long-exserted,* loose and open, 1.3–4.8 *dm. long, the lower capillary branches* 0.8–2.3 *dm. long;* spikelets 5–8 mm. long; glumes unequal, the 2nd longer than the 1st; *staminate lemmas* 5 mm. long, *with apical straight awns hardly* 1 *mm. long.* — Borders of brackish marshes, s. N.Y. and ne. N.J., local. July, Aug.

3. H. alpìna (Sw.) R. & S. (alpine). — Culms 1–4 dm. high, *tufted;* upper sheaths inflated, *blades* very small, *the lowest and those of the sterile shoots* long and *linear, smooth; panicle contracted,* 2–5 cm. long; spikelets 7–8 mm. long, olivaceous; *staminate lemmas* ciliate on the margins, *the first short-awned below the apex, the second with a longer* (5–8 *mm.*) *bent awn from below the middle;* fertile lemma mucronate. — Greenl. to Alaska, s. on siliceous rock and dry peat to Nfld., and mts. of e. Que., n. N.E., n. N.Y. and B.C. June–Aug. (Eurasia)

Tribe VIII. Orŷzeae Kunth (see p. 96)

Represented with us only by

77. LEÉRSIA Sw. Cutgrass. Whitegrass

Spikelets 1-flowered laterally, perfect, but those in the open panicles usually sterile, those inclosed in the sheaths cleistogamous and fruitful; glumes none; lemma navicular, somewhat indurated, awnless, clasping the palea by a pair of strong marginal nerves; palea of like texture, much narrower, 1-nerved; stamens 1–6. — Perennials of moist ground, in trop. and temp. reg., with mostly rough leaves and short racemes of imbricated spikelets in open panicles. (Named after *Johann Daniel Leers,* a German botanist, 1727–1774.) Homalocenchrus Mieg

*a.*Spikelets oblong, 1–2 mm. wide, loosely imbricated. . . *b.*
*b.*Branches of panicle naked below, floriferous at tips, spreading.
Culms compressed, from clusters of short thick closely scaly rhizomes; leaves scabrous but not bristly-ciliate; spikelets 1–1.3 mm. wide. . 1. *L. virginica.*
Culms terete, from long slender rhizomes; leaves bristly-ciliate; spikelets 1.5–2 mm. wide. . . . . 2. *L. oryzoides.*
*b.*Branches of panicle floriferous to base, closely ascending. . . . 3. *L. hexandra.*
*a.*Spikelets broadly oval to suborbicular, 3–4 mm. wide, closely imbricated. . 4. *L. lenticularis.*

1. L. virgínica Willd. (of Virginia). — Weak; *culms* slender, *compressed,* branched, sprawling or ascending, 0.3–1.2 m. long, *from clustered short thick scaly rhizomes; leaf-blades* lanceolate, *thin,* scabrous but *not ciliate,* 0.3–1.3 cm. wide; *panicle* exserted (rarely included), *simple,* the slender branches stiffly spreading; *spikelets* 2.5–4 *mm. long,* 1–1.3 *mm. broad,* closely appressed; *lemma* smooth or minutely hirtellous, *the keel and margin smooth or minutely ciliolate;* stamens 1 or 2. — Damp woods and thickets, centr. Me. to s. Ont., Minn. and Neb., s. to Ga., Ala., Miss., La. and Tex. July–Oct.

Var. **ovàta** (Poir.) Fern. (ovate). — Ascending or erect, up to 1.6 m. high; *lemma* with margin and often the keel *bristly-ciliate, with cilia up to* 0.6 *mm. long.* — Wet thickets and swales, N.E. to Minn. and Neb., s. to nw. Fla. and Tex. Fig. 282.

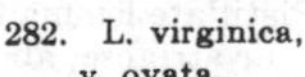
282. L. virginica, v. ovata.

283. L. oryzoides.

2. L. oryzoìdes (L.) Sw. (like *Oryza,* rice), Rice-C. — Ascending or sprawling; *culms terete,* branched, *from slender elongate rhizomes; leaves elongate,* very harsh, *scabrous-hispid on margins* and often on nerves beneath as well as in furrows of sheaths, or leaves smooth and glabrous in forma **glàbra** A. A. Eat. (hairless); *panicle diffusely branched,* lax, exserted or partly included, or wholly included in forma **inclùsa** (Wiesb.) Dörfler (included); *spikelets oblong,* 4–6 mm. long, 1.5–2 *mm. broad;* lemma hispid on the nerves, strongly bristly-ciliate on the keel. — Swamps, ditches and shores, Que. to e. Wash., s. to Fla., Ala., La., Tex., N.M., Ariz. and Calif. (Eu.) Fig. 283. — Forma *inclusa* mostly northw.; forma *glabra* in estuaries or on inundated shores. June–Oct.

3. L. hexándra Sw. (with six stamens). — *Weak, with subterranean stolons;* fertile culms ascending, 0.2–1 m. high; leaf-blades firm, 2–5 mm. wide; *panicle slender, lance-elliptic,* 3–10 cm. long, 0.5–2.5 *cm. thick, the strongly ascending short branches floriferous to base; spikelets* narrowly and *obliquely oblong-lanceolate,* 4–5 mm. long, 1–2 mm. broad, *the acuminate tips curved.* — Shallow water, Fla. to Tex., n. to se. Va. June, July. (Tropics)

4. L. lenticulàris Michx. (lens-shaped), Catchfly Grass. — *Culms* nearly simple, *terete,* decumbent or erect, 0.5–1.3 m. high, with scaly rhizomes; sheaths and blades harsh or smoothish; *blades lanceolate,* 0.6–2 *cm. broad;* panicle nearly simple; *spikelets oval to suborbicular,* 5 *mm. long,* 3–4 *mm. broad, closely imbricated,* strongly ciliate. — Swamps and low woods, w. Fla. to Tex., n. to Md., s. O., Ind., Wisc. and Minn. Aug., Sept.

Tribe IX. Zizanìeae Hitchc. (see p. 96)

Staminate and pistillate spikelets appearing alike, all with subequal membranous lemmas, the staminate below, the pistillate above on the branches; perennial. 78. *Zizaniopsis.*
Staminate and pistillate spikelets dissimilar, the staminate with subequal broad membranous lemmas, and pendulous on lower branches of panicle; the pistillate terete, elongate and erect at summit of panicle; annual. . . . . . . 79. *Zizania.*

78. ZIZANIÓPSIS Döll & Aschers. WATER-MILLET

Spikelets unisexual, the pistillate above, the staminate below on each branch of the panicle, much alike in appearance, laterally compressed; glumes subequal, membranaceous; the first glume of the pistillate spikelet with a short terminal awn, the lemma acute, palea none; glumes and lemma of staminate spikelet acute, nerveless, palea none; stamens 6; grain ovoid, with a chartaceous easily separable pericarp, loosely inclosed in the glumes. — A tall aquatic grass with long leaves and long narrow terminal panicles. (Name from *Zizania* and *opsis*, *appearance*, from likeness to the following genus.)

1. Z. miliàcea (Michx.) Döll & Aschers. (millet-like). — Perennial by a creeping rhizome; culms 1–4 m. high, geniculate at the lower nodes; leaves flat, 3–10 dm. long, 1–3 cm. wide. — Swamps and margins of streams (often tidal), Fla. to Tex., n. to Md., Ky., se. Mo. and Okla. Apr.–June. (Mex.; S.Am.)

79. ZIZÀNIA L. WILD RICE. WATER-OATS. FOLLE AVOINE OR RIZ SAUVAGE (Que.)

Spikelets unisexual, 1-flowered; the pistillate linear, awned, articulated and tardily deciduous on clavate pedicels on the appressed upper branches; the staminate lanceolate, early deciduous, on the expanded lower branches of the same panicle, glumes none in the pistillate spikelet; lemma closely clasping the palea by a pair of strong lateral nerves, a hispid awn from the summit; first glume of staminate spikelet 5-, the second 3-nerved; stamens 6; grain long-cylindrical, blackish, closely enveloped in the inrolled lemma and 3-nerved palea. — Tall aquatic grasses of e. N. Am. and e. Asia, with long leaves and large terminal panicles. (Adapted from the Greek *zizanion*, a weed of wheat-fields.)

1. Z. aquática L. (aquatic). — Annual, with slender to stout simple or basally branching soft culms up to 3 m. high; leaves flat, with long sheaths, and blades 0.3–5 cm. broad; staminate spikelets purplish to stramineous, on capillary pedicels; pistillate spikelets on thick clavate pedicels, promptly disarticulating. — Very variable; the varieties much planted as food for waterfowl, and often naturalized outside the original ranges.

284. Z. aquatica.

Pistillate lemma thin and delicate, slenderly ribbed, opaque, the surface glabrous or strigose; aborted spikelets slender and shriveled, less than 1 mm. thick.
- Plant 1–3 m. tall; leaves 0.8–5 cm. broad; ligules 0.6–2.5 cm. long; body of mature pistillate lemma 1–2 cm. long, usually strigose, its awn 1–7 cm. long. *Z. aquatica* (typical).
- Plant 2.5–10 dm. tall; leaves 3–12 mm. broad; ligules about 3 mm. long; body of pistillate lemma 5–10 mm. long, glabrous or scabrous, its awn 1–8 mm. long. . . . . Var. *brevis*.

Pistillate lemma subcoriaceous, coarsely corrugated, lustrous, the body strigose only in the slender furrows between the broad rounded ridges or at summit; aborted spikelets 1.5–2 mm. thick.
- Leaves 4–15 mm. broad; ligule 3–10 mm. long; lower pistillate branches with 2–6 spikelets. . . . . Var. *angustifolia*.
- Leaves 1–3 cm. broad; ligules 1–1.5 cm. long; lower pistillate branches with 11–29 spikelets. . . . . Var. *interior*.

Z. aquática (typical), *Z. palustris* of ed. 7, not L. — Quiet waters, river-mouths (fresh to brackish), and shores, s. Me. and s. Que. to se. Wisc., s. to w. Fla., n. O., n. Ind. and Ill. June–Sept. FIG. 284.

Var. **brèvis** Fassett (short). — Tidal waters and tributary streams of the St. Lawrence R., s. Que. and e. Ont.

Var. **angustifòlia** Hitchc. (narrow-leaved), *Z. palustris* L.; *Z. aquatica* of ed. 7, not L. — Quiet waters, e. N.B. to Minn., s. to w. N.S., n. N.E., ne. Mass., centr. N.Y., nw. Pa. and n. Ind.

Var. **intèrior** Fassett (inland). — Quiet waters and marshes, Ind. to N.D., s. to Mo. and Tex.

Subfam. II. PANICOÌDEAE (see p. 95)

Spikelets all with the central floret perfect.
- Spikelets perfect, in panicles or in racemes or spikes with continuous rachis; glumes membranaceous, unequal, the lower smaller or obsolete; a lemma of similar texture and appearing like a third glume (the STERILE LEMMA) empty or with a hyaline palea (rarely including a staminate flower) sub-

tending the perfect floret; FERTILE LEMMA and palea indurated, firmly clasped together and covering the free grain, awnless or merely pointed . . Tribe X. PANICEAE (see below).

Spikelets in 2's or 3's at each joint of the often articulated rachis, one usually sessile and perfect, the other or others pedicelled and either (rarely) perfect, staminate or sterile and represented by 2 or 1 empty glumes or mere rudiments or pedicels; fertile lemma hyaline, often awned. . . . . . . Tribe XI. ANDROPOGONEAE. (p. 228).

Spikelets all unisexual, the staminate and pistillate in separate inflorescences or in different parts of the inflorescences; pistillate spikelets indurated, plump, single and with 1 floret pistillate, the other sterile, sunken in hollows of a thick jointed axis and falling included in segments of the disarticulated axis, or, in *Zea*, borne on the surface of a thick axis, awnless; staminate spikelets paired. Tribe XII. MAYDEAE (p. 235).

Tribe X. PANÍCEAE R. Br.

*a.*Spikelets without an involucre of bristles or prickles, in open or dense panicles or 1-sided spikes or racemes. . . *b.*

*b.*Spikelets uniform, all or nearly all fruiting, none subterranean. . . *c.*

*c.*Glumes and sterile lemmas awnless; spikelets in symmetrical panicles or in spikes or spiciform racemes and in 2 (when paired 4) rows along 1 side of the rachis. . . *d.*

*d.*Spikelets in 1-sided spikes or spiciform racemes, in 2 (or 4) rows on 1 side of the rachis. . . *e.*

*e.*Fertile lemma with hyaline margins not inrolled, cartilaginous; spikelets lanceolate to elliptic, pointed; fruit flexible, usually dark-colored. . . . . 80. *Digitaria.*

*e.*Fertile lemma with chartaceous and rigid margins inrolled, hard and resistent; the rigid fruit pale. . . *f.*

*f.*Back of fruit away from rachis of spike; spikelets lanceolate to narrowly oblong, acute or acutish, not paired.

Spikelets pubescent, loosely ascending, with a ring-like callus or blunt-based stipe at base; fertile lemma awned. . . . . . 81. *Eriochloa.*

Spikelets (in ours) glabrous, closely appressed, without basal callus; fertile lemma awnless. . . . . 82. *Axonopus.*

*f.*Back of fruit toward the rachis; spikelets suborbicular to elliptic or ovate, strongly plano-convex, not tightly appressed, borne singly or in pairs (2nd one often obsolete and represented only by its pedicel) along the rachis. . . . . 83. *Paspalum.*

*d.*Spikelets in panicles or, if in close spiciform racemes, the latter not 1-sided.

First glume obsolete or minute; fertile lemma cartilaginous, its delicately hyaline margin not inrolled, acuminate; fruit flexible; culms brittle and easily fractured, the panicles readily broken off. 84. *Leptoloma.*

First glume often present; fertile lemma hard, the rigid margin inrolled, usually blunt; fruit rigid; culms rarely fracturing, the skeleton of the panicle long-persistent.

Spikelets dorsiventrally compressed, in panicles or panicled racemes; 2nd glume and sterile lemma similar, thin and flattish, the 2nd glume 3–9-nerved; fertile lemma sessile. . . . . . 85. *Panicum.*

Spikelets conic-oblong, in dense spiciform panicles; 2nd glume saccate at base, 11-nerved; sterile lemma flat, 3–5-nerved; fertile lemma stipitate. . . . . 86. *Sacciolepis.*

*c.*Glumes or sterile lemma usually awned; inflorescence a panicle of dense 1-sided racemes; coarse annuals. . . . . 87. *Echinochloa.*

*b.*Spikelets very dissimilar, those of terminal panicle with both stamens and pistil but not fruiting; fruit only from basal or subterranean cleistogamous plump spikelets. . . . . 88. *Amphicarpum.*

*a.*Spikelets with an involucre of 1–∞ bristles, or with an involucre of fused prickles forming a bur.

Spikelets subtended by 1–∞ bristles, falling and leaving the bristles persistent; inflorescence a bristly spiciform panicle of innumerable small spikelets. . . . . 89. *Setaria.*

Spikelets (1–5 together) inclosed in an ovoid to globose prickly bur, rather few, the burs dropping intact from rachis of spike. . . . . 90. *Cenchrus.*

80. DIGITÀRIA Heist. FINGER-GRASS. CRAB-GRASS

Spikelets 1-flowered, lanceolate-elliptic, sessile or short-pedicelled, solitary or in 2's or 3's, in two rows on one side of a continuous narrow or winged rachis, forming simple slender racemes which are aggregated toward the summit of the culm; glumes 1–3-nerved, the first sometimes obsolete; sterile lemma 5-nerved; fertile lemma leathery-indurated, with a hyaline margin not inrolled, inclosing a palea of like texture. — Annual, mostly weedy grasses of warm and temp. reg., with branching culms, thin leaves, and digitate or subdigitate inflorescence. (Name from *digitus, a finger.*) SYNTHERISMA Walt.

a. Rachis of racemes with slender-margined angles; first glume usually wanting; culms erect, fruit dark.
  Spikelets minutely pubescent; at least the glume ciliate. . . . . . . 1. *D. filiformis.*
  Spikelets glabrous throughout. . . . . 2. *D. laeviglumis.*
a. Rachis with broad and flat wings; culms often decumbent at base or spreading or creeping. . . *b.*
  b. Spikelets 1.5–2.2 mm. long, on short terete pedicels; 1st glume obsolete.
    Sheaths and blades glabrous; spikelets 1.8–2.2 mm. long; fertile lemma blackish. . . . . 3. *D. Ischaemum.*
    Sheaths and blades pilose; spikelets 1.5–1.7 mm. long; fertile lemma pale. 4. *D. serotina.*
  b. Spikelets 2.5–3.5 mm. long, on short sharply angled pedicels; 1st glume present, small; fruit drab; leaves more or less pubescent. . . . . 5. *D. sanguinalis.*

1. D. filifórmis (L.) Koel. (thread-like), SLENDER C. — Simple or usually branching at the leafy base; culms almost filiform, 0.5–9 dm. high; *lower sheaths hirsute;* blades 0.5–2 dm. long, 1–5 mm. wide, hirsute or glabrous on the lower, scabrous on the upper surface; racemes 1–6, unequal, 1–15 cm. long, very slender; *spikelets* 1.5–2 *mm. long,* mostly in 3's, appressed, the second and third on slender flexuous pedicels; *glume and sterile lemma densely or sparsely villous between the nerves with white gland-tipped hairs;* the glume shorter and narrow, exposing the dark brown acute fertile lemma. — Sterile or sandy soil, n. Fla. to Tex. and Mex., n. to s. N.H., Mass., N.Y., Mich., Ill. and Ia. Aug.–Oct.

Var. **villòsa** (Walt.) Fern. (long-hairy). — Often *coarser and taller* (*up to* 1.4 *m. high*); sheaths often more densely villous; *racemes* sometimes more distant (up to 3 cm. apart), 5–25 *cm. long;* spikelet-clusters usually more distant; *spikelets* 2–2.5 *mm. long.* (*D. villosa* (Walt.) Pers.) — Sandy soil, Fla. to Tex., n. to Va., Tenn., Ill., Mo. and Kans.

2. D. laeviglùmis Fern. (with smooth glumes). — Resembling no. 1, 1.5–7 dm. high; culms filiform, firm, lustrous; lowest sheaths pilose; *leaf-blades glabrous,* 1.5–9 cm. long, 1–3 mm. wide; ligules scarious, subtruncate, erose-dentate, 1 mm. long; racemes 1–3, 0.2–1 dm. long; *spikelets* 1.8–2 mm. long, *strictly glabrous; glume hyaline.* — Dry peaty hollows in granitic ledges, local, Hillsboro Co., N.H. Aug., Sept.

3. D. ISCHAÈMUM (Schreb.) Muhl. (Ancient name, presumably from the Greek *ischaemos, styptic, blood-restraining,* from supposed styptic properties), SMALL C. — *Glabrous;* culms 0.2–4 dm. high, much branched below, prostrate to ascending, often purplish; *leaves* 2–10 *cm. long,* 3–6 mm. wide; *racemes* 1–6, *commonly purplish,* approximate, divergent, *often curved,* 1–9 *cm. long; spikelets* solitary or in 2's, 1.8–2.2 *mm. long; glume and sterile lemma equal, closely short-villous between the nerves, as long as the dark brown fertile lemma.* (*D. humifusa* Pers.) — Cult. and waste ground, P.E.I. to Oreg., s. to Fla., La. and Tex. July–Oct. (Natzd. from Eu.)

Var. **mississippiénsis** (Gattinger) Fern. (of the Mississippi Valley). — Erect or ascending, up to 1 m. high, *usually green; leaves up to* 2 *dm. long* and 6 mm. broad; *racemes stiffer, straight, green, mostly* 6–13 *cm. long.* — Sandy or other dry soil, N.Y. to Minn., s. to e. Va., S.C., Tenn., Mo. and Kans.

4. D. seròtina (Walt.) Michx. (late-fruiting). — Culms tufted or more commonly creeping and forming dense mats; the crowded *sheaths and blades* (up to 8 cm. long and 7 mm. broad) *pilose or villous;* racemes 3–8, terminating ascending branches, 3–10 cm. long; *spikelets* pale, mostly in 2's, 1.5–1.7 *mm. long,* sparsely pubescent between the nerves, on terete pedicels; 1*st glume obsolete, the* 2nd *scarcely half as long as the pale fertile lemma.* — Low sandy ground, Fla. to La., n. to se. Va. June–Sept.

285. D. sanguinalis.

5. D. SANGUINÀLIS (L.) Scop. (stanching blood; from supposed styptic properties). — Culms ascending from a decumbent often creeping base or erect, 3–12 dm. long; nodes and sheaths more or less papillose-hirsute; blades lax, 5–15 cm. long, 4–10 mm. wide, scabrous, often more or less pilose; racemes 3–13, subfasciculate, 0.2–2 dm. long; *spikelets* mostly in pairs, 2.5–3 *mm. long,* glabrous between the smooth or scabrous nerves; 2nd glume about half as long as the *pale or grayish fertile lemma; sterile lemma minutely ciliolate.* — Cult. and waste ground, a troublesome weed, extending n. to s. Can. June–Oct. (Natzd. from Old World) Fig. 285.

Var. ciliàris (Retz.) Parl. (with marginal hairs). — *Spikelets* 3–3.5 *mm. long,* more pubescent; *the ciliation longer.* (Var. *marginata* (Link) Fern.) — Extending n. to Pa., Md., Ill. and Kans. (Natzd. from Old World)

81. ERIÓCHLOA HBK. Cup-Grass

Spikelets with 1 perfect floret, borne singly (rarely paired) in 2 rows on one side of a narrow hairy rachis, with a ring-like disk or callus below the 2nd glume, the back of the fruit turned away from the rachis; 1st glume minute, adnate to the callus; 2nd glume and sterile lemma subequal, acute; the sterile lemma usually inclosing a hyaline palea or a staminate flower; fertile lemma indurated, minutely papillose, mucronate or with a caducous awn. — Annuals or perennials of trop. and warm reg., with terminal panicles of racemosely disposed spike-like racemes. (Name from the Greek *erion, wool,* and *chloë, grass.*)

286. E. contracta.

1. E. contrácta Hitchc. (contracted), Prairie C. — Simple or slightly branched, 0.3–1 m. high, with pubescent flat leaves 3–8 mm. wide; panicle tardily exserted, 1–2 dm. long, with 5–25 ascending spikes 1–2 cm. long; rachis villous; *spikelets pilose,* 3.5–4 mm. long; the *2nd glume and sterile lemma acuminate.* — Low grounds, La. to N.M. and Kans.; adv. in Mo. and Va. July–Oct. Fig. 286.

82. AXÓNOPUS Beauv.

Spikelets 1-flowered, compressed-biconvex, sessile, solitary in two rows on one side of a flattened rachis (which is naked in ours), placed with the back of the fertile lemma turned from the rachis, forming simple spikes; first glume obsolete; lemma and palea indurated but less so than usual in *Paspalum,* margins of the lemma inrolled. — Perennials of warm reg. with 2–several slender spikes digitate or subdigitate at the summit of the culm. (Name from the Greek *axon, axis,* and *pous, foot.*)

1. A. furcàtus (Flügge) Hitchc. (forked). — Tufted, soft, 3–10 dm. high, with long creeping leafy stolons; leaves obtuse, commonly ciliate, 0.5–1 cm. wide; racemes a pair (rarely a 3rd), strongly divergent at tip of the elongate weak culm, 0.5–1 dm. long; spikelets lance-oblong, acute, 4–6 mm. long. — Damp sandy soil and swamps, Fla. to e. Tex., n. to se. Va. and Ark. June–Aug.

BRACHIÀRIA (Trin.) Griseb. (with arms; from the scattered and elongate racemes) has been added to our flora as this goes to press. It differs from no. 82 in having the first glume present and the inflorescence consisting of scattered racemes. Our species is **B. exténsa** Chase (extended). — Annual with decumbent rooting base; rachis of racemes winged; spikelets ovate, glabrous, 4–4.5 mm. long, with broadly ovate blunt first glume. — Low sandy open soil, ditches, etc., Fla. to e. Tex., n. to se. Mo. and Okla. Aug.–Oct. (Cuba)

83. PÁSPALUM L.

Spikelets 1-flowered, mostly plano-convex, nearly sessile, solitary or in pairs, in 2 (or by forking of pedicels in 4) rows on one side of a continuous narrow or dilated rachis, forming simple spike-like racemes; spikelets placed with the back of the fertile lemma (convex side of spikelet) toward the rachis; 1st glume often obsolete, rarely present; 2nd glume and sterile lemma similar, the glume sometimes suppressed; lemma and palea chartaceous-indurated, margins of the lemma inrolled in fruit. — Perennials or annuals of trop. and warm-temp. reg., with 1–several racemes digitate or racemose at the summit of the culm and branches. (Probably from the Greek *paspale, meal.*)

a. Rachis of the racemes herbaceous, spathiform, the margins arching over the spikelets or their bases; soft-stemmed creeping plants with repent shoots.
Racemes of each inflorescence 1–4, 1.5–3 cm. long; rachis abruptly pointed, narrower than and overtopped by the rows of glabrous spikelets. . . . 1. *P. dissectum.*
Racemes 5–50 or more, 2–9 cm. long; rachis acuminate, broader and longer than the rows of pubescent spikelets. . . . . 2. *P. fluitans.*
a. Rachis firm, much narrower than the raceme; culms firm. . . *b.*
b. Spikelets borne singly along the rachis; the short pedicels simple, only very exceptionally forking.
Rhizomes slender, creeping, with numerous elongate repent stolons; spikelets elliptic, acute, 1.3–1.5 mm. wide. . . . . 3. *P. distichum.*

Rhizomes scarcely developed; plant non-stoloniferous, with erect culms; spikelets rounded, obtuse, 2–3.2 mm. broad. . . . . 4. *P. laeve.*

*b.*Spikelets mostly borne in pairs along the rachis, the pedicels forking at base, the 2nd spikelet sometimes rudimentary. . . *c.*

*c.*Spikelets rounded or obtuse at summit, conforming to the fruit, glabrous or only minutely pubescent. . . *d.*

*d.*Spikelets 3–4.2 mm. long. . . *e.*

*e.*Spikelets flattened; 1st glume obsolete; 2nd glume with only the midrib prominent. . . *f.*

*f.*Culms geniculate or decumbent at base, rooting at the lower joints; racemes divergent; spikelets 3–3.2 mm. long. . . . 5. *P. pubiflorum.*

*f.*Culms erect; racemes erect to somewhat divergent.

Spikelets turgidly plano-convex, 3.6–4 mm. long, 2.8–3.1 mm. broad, not conspicuously imbricated, in racemes 6–12 cm. long. . . . . 6. *P. floridanum.*

Spikelets flattened on both faces, 2–3.4 mm. long, 2–2.3 mm. broad, conspicuously imbricated, in racemes mostly 2.5–6 cm. long. . . . . 7. *P. praecox.*

*e.*Spikelets strongly turgid, ellipsoid-ovoid to -obovoid; 1st glume usually present on 1 spikelet of each pair; 2nd glume with 7 strong nerves. . . . . 16. *P. bifidum.*

*d.*Spikelets 1.4–2.8(–3) mm. long. . . *g.*

*g.*Perennials with hard knotty bases; terminal inflorescence with 1–3 (–6) racemes; rachis slender, 0.5–1 mm. broad. . . *h.*

*h.*Spikelets ellipsoid, 1–1.6 mm. broad.

Spikelets glabrous or only sparingly glandular, 1.4–2.1 mm. long; leaves glabrous or pubescent. . . . . . 8. *P. setaceum.*

Spikelets definitely pubescent, 1.8–1.9 mm. long; leaves and sheaths very villous. . . . . 9. *P. debile.*

*h.*Spikelets orbicular or suborbicular, (1.3–)1.5–2.4 mm. broad. . *i.*

*i.*Leaves glabrous, long-villous or very minutely puberulent; nodes glabrous; spikelets glabrous or essentially so.

Culms few to several, loosely ascending or spreading; racemes both terminal and axillary; spikelets plano-convex, crowded but not strongly imbricated. . . . . 10. *P. ciliatifolium.*

Culms usually solitary, erect; racemes only at summit of culm; spikelets strongly flattened, closely imbricated. . . 7. *P. praecox.*

*i.*Leaves very densely short-pilose, subvelutinous; nodes mostly pubescent; spikelets pubescent.

Dense pubescence of sheaths and blades with many long trichomes admixed. . . . . 11. *P. Bushii.*

Dense pubescence uniform, without prolonged trichomes. . 12. *P. psammophilum.*

*g.*Annual with soft base; terminal inflorescence with (2–)4–15 racemes; rachis broadly winged, about 2 mm. wide. . . . . . 13. *P. Boscianum.*

*c.*Spikelets acuminate, the tips prolonged beyond the fruit, long-villous.

Racemes 3–6(–10), loosely ascending; spikelets (excluding hairs) 2.8–4 mm. long, 2–2.5 mm. broad. . . . . 14. *P. dilatatum.*

Racemes (5–)10–20 or more, strictly ascending and overlapping; spikelets 2–3 mm. long, 1.2–1.8 mm. broad. . . . . 15. *P. Urvillei.*

1. P. dissĕctum L. (deeply divided). — Creeping, *forming repent mats; sheaths soft, loose,* glabrous; *blades* 1–6 cm. long, 2–5 *mm. wide; inflorescences* numerous, *each with* 1–4 *racemes* 1.5–3 *cm. long; rachis green, herbaceous, abruptly pointed, narrower than and overtopped by the rows of spikelets; spikelets* ovoid, bluntish, *glabrous,* 2–2.3 *mm. long.* — Wet places or shallow water, Fla. to e. Tex., n. to s. N.J., e. Md., Tenn., s. Ill. and s. Mo. July–Oct. (Cuba)

2. P. flŭitans (Ell.) Kunth (floating). — *Culms soft and spongy,* tufted, *sprawling* or repent, up to 2 m. long; sheaths papillose-hirsute to glabrous; *blades* lanceolate, 0.3–2.5 dm. long, 0.3–2.5 *cm. broad,* smooth or scabrous; *racemes* 5–50 *or more,* 2–9 *cm. long,* finally spreading; *rachis herbaceous, acuminate, broader and longer than the rows of* ellipsoid small (1.2–1.7 *mm. long*) *minutely glandular-pubescent spikelets.* (*P. mucronatum* Muhl.; *P. repens* sensu Chase, not Bergius) — In water, wet places or alluvium, Fla. to e. Tex., n. to se. Va., n. Ky., sw. Ind., Ill., Mo. and ne. Kans. Aug.–Oct.

3. P. dĭstichum L. (two-ranked), Knotgrass. — Creeping and rooting at the nodes, with ascending culms 1–6 dm. high; leaves short, usually crowded, sometimes sparsely hairy on the margins; *racemes* 2 (rarely –4), *strictly terminal*, 1.5–7 cm. long; spikelets singly disposed, 2.5–4 mm. long, 1.3–1.5 mm. wide, ovate, acute, sparsely pubescent; first glume occasionally present. — Ditches and muddy or sandy shores, Fla. to Calif., n. to se. Va., Tenn., Ark., Okla., Utah and Wash. July–Oct. (Mex.; W.I.; S.Am.)

4. P. laève Michx. (smooth). — Slightly tufted, the *slender firm culms erect or ascending*, 0.4–1.3 m. high; leaves rather crowded at base, elongate; the firm blades flat or somewhat folded, 0.3–1 cm. broad, pubescent to glabrous; *inflorescences* terminal and *sometimes a second one short-exserted or included in a lower sheath;* the terminal one usually exserted, with 2–8 distant ascending or spreading racemes 3–12 cm. long; *rachis about* 1 *mm. wide*, with a tuft of hairs at base; *spikelets mostly unpaired, rounded, obtuse, glabrous*, 2.5–3.2 *mm. long*, 2–3.2 *mm. broad*. Variable.

Spikelets slightly longer than broad, 2–2.5 mm. broad.
  Sheaths and blades glabrous or essentially so. . . . . *P. laeve* (typical).
  Sheaths and often the blades strongly pilose. . . . . Var. *pilosum*.
Spikelets orbicular, 2.8–3.2 mm. broad. . . . . Var. *circulare*.

P. laève (typical), *P. angustifolium* LeConte — Damp sandy fields, savannas, thickets and shores, Fla. to Tex., n. to s. N.J., se. Pa., W.Va., s. O. and Ill. July–Oct.

Var. **pilòsum** Scribn. (soft-hairy), *P. plenipilum* Nash — Extending n. to s. N.Y., Pa., Tenn. and s. Mo.

Var. **circulàre** (Nash) Fern. (circular), *P. circulare* Nash — Ga. to e. Tex., n. to s. Ct., se. N.Y., Pa., W.Va., Ind., s. Ill., Mo. and se. Kans.

5. P. PUBIFLÒRUM Rupr. (hairy-flowered). — *Geniculate and decumbent, the lower nodes rooting;* culms stout, 0.5–2 m. long, nodes pubescent; sheaths usually pilose on the scarious margin; blades 1–3 dm. long, 1–2 cm. broad, glabrous or with a few hairs at base; racemes 2–8, 2–10 cm. long; *spikelets* about 3 *mm. long, rounded-obovoid, obtuse, pubescent, mostly in pairs and forming* 4 *rows.* — South of our range; with us as

Var. **glàbrum** Vasey (glabrous). — Smoother; racemes mostly 4–8, 3–13 cm. long; *spikelets glabrous.* (*P. laeviglume* Scribn.) — Open low ground and shallow water, Fla. to e. Tex., n. to e. N.C., w. Ky., s. O., sw. Ind., s. Ill., s. Mo. and se. Kans. June–Aug.

6. P. floridànum Michx. (of Florida). — *Culms erect, robust,* 1–2 *m. high*, from a *stout scaly rhizome; sheaths pubescent; blades* 2–6 *dm. long*, 4–10 mm. wide, *pubescent; racemes* 1–4(–6), stout, *erect or ascending*, 6–12 cm. long; *spikelets paired*, the 2nd sometimes a rudiment, closely crowded, 3.6–4 *mm. long*, 2.8–3.1 *mm. broad*, glabrous. — Damp sand, Fla. to Tex., n. to Va., Tenn., s. Mo. and Okla. Aug., Sept.

Var. **glabràtum** Engelm. (becoming smooth). — *Glabrous or nearly so, glaucous; racemes often longer*, 6–15 *cm. long.* — North to s. N.J., e. Pa., Ark. and se. Kans.

7. P. PRAÈCOX Walt. (early). — Slender, *stiffly erect*, from short scaly rhizomes; culms solitary (rarely 2 or 3), 0.5–1 m. high; *leaves mostly basal*, with 1–3 cauline; *sheaths compressed*, keeled, *the lower purplish*, all or all but the lowest glabrous; *blades firm*, flat or soon folded, *elongate-linear*, 3–7 *mm. broad; racemes* 2–6 (rarely –8), ascending or divergent, *mostly* 2.5–6 *cm. long*, with margined rachis 1.5 mm. broad; *spikelets strongly flattened*, solitary or in pairs, *imbricated* or crowded in 3 or 4 rows, *suborbicular* or elliptic-obovate, *olivaceous or yellow-green, glabrous*, 2–2.8 *mm. long*, 2–2.3 mm. broad. — Swamps and wet pine-barrens, south of our range. Represented with us by

Var. **Curtisiànum** (Steud.) Vasey (for its discoverer, MOSES ASHLEY CURTIS, 1808–1872). — Somewhat stouter, up to 1.5 m. high; *sheaths and blades villous; spikelets* 2.7–3.4 *mm. long.* (*P. lentiferum* Lam.) — Wet pinelands, bogs and swamps, Fla. to Tex., n. to se. Va. Aug.–Oct.

8. P. setàceum Michx. (bristle-like). — *Culms* 1–few, from a short rhizome, ascending or reclining, *filiform*, 2.8–9 dm. long, *with glabrous nodes; leaves mostly near the base*; sheaths pilose to glabrous; blades soft, deep green, 0.5–3 dm. long, 2–15 mm. broad, *ciliate*, the surfaces glabrous or villous; *racemes* slender, the terminal 1–3, straight or arching, 3–9 cm. long, *finally long-peduncled;* axillary racemes often present; rachis slender; *spikelets paired, ellipsoid or ellipsoid-obovoid*, 1.4–2.1 *mm. long*, 1–1.6 *mm. broad, glabrous, or the glume minutely glandular.* — Highly variable, with three fairly pronounced trends:

a. Spikelets 1.4–1.8 mm. long, 1–1.4 mm. broad; leaf-blades 2–8 mm. wide. . . *b.*
b. Leaf-blades densely villous, strongly ascending, 2–6 mm. wide; spikelets 1–1.2 mm. broad. . . . . *P. setaceum* (typical).
b. Leaf-blades glabrous to minutely or sparsely strigose.
  Leaf-blades erect or nearly so, running up the culm, narrowly linear, 1.5–3.5 mm. wide; terminal spike 1; spikelets 1.3–1.4 mm. broad. . . Var. *calvescens*.

Leaf-blades loosely divergent, inclined to be subbasal, broadly linear, 3–8 mm. wide; terminal spikes 1–3; spikelets 1–1.2 mm. broad. . . Var. *longepedunculatum.*

*a.*Spikelets 1.8–2.1 mm. long, 1.3–1.6 mm. broad; leaf-blades 7–15 mm. broad, usually pilose or hirsute; terminal racemes 1–3(–6) . . . . Var. *supinum.*

P. setàceum (typical). — Dry sandy soil, Fla. to Tex., n. to se. Mass., L.I., O. and Ky. June–Oct. (Mex.)

Var. **calvéscens** Fern. (becoming bald). — Bogs, local, se. Va.

Var. **longepedunculàtum** (LeConte) Wood (long-peduncled), *P. longepedunculatum* LeConte — Fla. to Miss., n. to N.J., s. O. and Ky.

Var. **supìnum** (Bosc) Trin. (sprawling), *P. supinum* Bosc — Fla. to La., n. to Del. and Tenn.

9. P. débile Michx. (weak). — Resembling nos. 7 and 8, *stouter,* 0.5–1 m. high; *lower nodes pubescent; sheaths and blades densely villous; terminal racemes mostly* 2–3; *spikelets* 1.8–1.9 *mm. long, distinctly pubescent.* — Sandy soil, Fla. to Tex., n. to L.I. and Okla. July–Sept. (W.I.; Mex.)

10. P. ciliatifòlium Michx. (ciliate-leaved). — Culms few–several, from a knotty base, rather slender, ascending or spreading, 0.3–1 m. long, with *glabrous nodes;* sheaths glabrous or villous; *ligular beard 2–4 mm. long; blades glabrous and ciliate, minutely puberulent or long-pilose,* 0.5–3 dm. long, 0.6–2.5 cm. broad; terminal racemes 1–4, 2–15 cm. long, on a mostly exserted peduncle; axillary racemes 1–3, exserted or included; *spikelets* paired, *orbicular or suborbicular,* 1.9–2.8 *mm. long,* 1.7–2.2 *mm. wide, glabrous* or rarely sparsely and minutely pubescent. Highly variable; of the many proposed segregates the following are best marked:

Surfaces of leaf-blades glabrous throughout or between the long trichomes, not puberulent.

Surfaces quite glabrous. . . . . *P. ciliatifolium* (typical).

Surfaces long-pilose or villous. . . . . Var. *Muhlenbergii.*

Surfaces of leaf-blades minutely puberulent, otherwise smooth or villous . . . Var. *stramineum.*

P. ciliatifòlium (typical). — Dry or moist open places or thin woods, Fla. to Tex., n. to D.C., Tenn. and Mo. June–Oct. (W.I.; Centr.Am.)

Var. **Muhlenbérgii** (Nash) Fern. (in honor of GOTTHILF HENRY ERNEST MUHLENBERG, 1753–1815), *P. pubescens* Muhl. and *P. Muhlenbergii* Nash — Fla. to Tex., n. to s. N.H., s. Vt., N.Y., W.Va., O., s. Mich. and s. Wisc. July–Oct.

Var. **stramíneum** (Nash) Fern. (straw-colored), *P. stramineum* Nash — Sandy soil, Tex. to Ariz. and n. Mex., n. to Ind., Wisc., Minn., Neb. and Colo. June–Sept.

11. P. Búshii Nash (for its discoverer, BENJAMIN FRANKLIN BUSH, 1858–1937). — *Culms erect,* 6–10 dm. high, *with pubescent nodes; lower sheaths densely appressed-pilose, velutinous, as well as sparsely long-villous; blades* firm, 5–20 cm. long, 5–15 mm. wide, softly *pilose and appressed-long-villous on both surfaces;* racemes 2 or 3, 8–12 cm. long; *spikelets* oval to suborbicular, paired, 2–2.2 *mm. long, densely pubescent.* — Dry soil, Ill. to Neb. and Tex. July, Aug.

12. P. psammóphilum Nash (sand-lover). — Culms prostrate or spreading, from a knotty base, 0.25–1 m. long, with mostly *pubescent nodes; sheaths and both surfaces of firm blades velutinous or densely appressed-pilose; ligular beard* 1.5–2 *mm. long;* terminal racemes 1–3, exserted or wholly or partly included; axillary racemes included; *spikelets suborbicular,* 2 *mm. long, densely short-pilose.* — Sands near the coast, se. Mass. to Ga. July–Oct.

13. P. Bosciànum Flügge (for its discoverer, LOUIS AUGUSTIN GUILLAUME BOSC, 1759–1828), BULL-GRASS. — *Soft-based annual,* stout, 0.3–2 m. high, *purplish,* glabrous; leaf-blades 1–4 dm. long, 6–12 mm. wide; *racemes* 2–15, 2.5–9 cm. long, *with a broadly winged rachis about* 2 *mm. wide; spikelets* paired, *appearing* 4*-seriate,* crowded, about 2 mm. long; *glume and* 5*-nerved sterile lemma brownish; fruit dark brown.* — Shores, low woods, etc., Fla. to Tex., n. to Va. and Tenn. June–Oct. (Trop. Am.)

14. P. DILATÀTUM Poir. (dilated), DALLIS-GRASS. — Perennial from a short rhizome, tufted, stoutish, 4.5–17 dm. high, glabrous except the ligules and crowded spikelets; leaves elongated, 4–12 mm. wide; *racemes* 2–10, 5–10 cm. long, *loosely ascending; spikelets ovoid, acuminate,* 2.8–4 *mm. long,* 2–2.5 *mm. broad; glume and sterile lemma silky-villous, overtopping the fruit.* — Meadows, roadsides and borders of ditches, Fla. to Calif., n. to Va., Tenn., Ark., Colo. and Oreg.; occasional on ballast northw. May–Oct. (Introd. and natzd. from the Tropics)

15. P. URVÍLLEI Steud. (for its discoverer, JULES SÉBASTIEN CÉSAR DUMONT D'URVILLE, 1790–1842), VASEY-GRASS. — Stiffly erect, 0.7–2.5 m. high; lower sheaths bearing irritating deciduous bristles; principal leaves glabrous; *racemes* 10–20 *or more, strictly ascending and* *overlapping; spikelets* as in no. 14 but smaller, 2–3 *mm. long,* 1.2–1.8 *mm. broad.* — Dry or moist open soil, roadsides and ditches, Fla. to Mex. and Calif., n. to se. Va. June–Oct. (Introd. and natzd. from the Tropics)

16. P. bífidum (Bertol.) Nash (divided into two parts). — Tall, *erect, slender,* 0.5–1.2 m. high, *from short finger-like simple rhizomes covered with densely pilose overlapping scales; leaves* mostly basal, *strongly glaucous beneath,* broadly linear to linear-oblanceolate, 0.3–1.5 cm. broad, the sheaths villous or glabrous; upper node of culm overtopped by subtending leaf; *uppermost reduced leaf with blade* 2.5–18 *cm. long;* inflorescence subincluded or somewhat elevated, the lowest raceme borne from below to 3 dm. above the upper leaf; *racemes* 2–6, strongly ascending, 0.4–1.6 dm. long, *interrupted,* the slender rachis scabrous-angled, *spikelets in pairs* on puberulent pedicels, *turgid, ellipsoid-ovoid* or *-obovoid,* glabrous, 3.3–4 mm. long, 1 *of each pair with a minute* broadly deltoid or rounded 1*st glume* 0.3–0.6 mm. long; 2*nd glume prominently* 7*-ribbed,* shorter than fruit; *sterile lemma* 5*-nerved.* — South of our range, n. to s. N.C.; represented with us by

Var. **projéctum** Fern. (standing out). — Taller, 1–1.5 m. high; sheaths long-villous; upper node much overtopping (5–16 cm.) leaf; *uppermost blade* 1–3(–7) *cm. long,* the *inflorescence raised* 2.5–4.5 *dm. above it;* racemes 2 or 3, much interrupted, the lower divergent, becoming erect; rachis puberulent; spikelets 4–4.2 mm. long; 1st glume narrowly deltoid, 0.6–1.5 mm. long; fruit included. — Dry pineland, Sussex Co., Va. Aug., Sept. — *P. bifidum* a remarkable and highly localized species, combining characters of *Paspalum* and *Panicum.*

84. LEPTOLÒMA Chase

Spikelets 1-flowered, fusiform, solitary on long capillary 3-angled pedicels; first glume obsolete or very minute, the second 3-nerved and nearly as long as the 5–7-nerved sterile lemma; fertile lemma cartilaginous-indurated, papillose, with a delicate hyaline margin not inrolled, inclosing a palea of like texture; grain free within the lemma and palea. — Tufted perennials, 3 of Australia, 1 in e. N.Am., with flat leaves and very diffuse terminal panicles, which break away at maturity and become tumble-weeds. (Name from the Greek *leptos, slender,* and *loma, border,* in reference to the hyaline margins of the lemma.)

1. L. cognàtum (Schultes) Chase (related; Schultes thinking it close to *Panicum capillare*), Fall Witch-Grass. — Much branched at base, *very brittle,* 3–7 dm. high; lower sheaths pilose; *ligule membranaceous,* 1 mm. long; blades 5–8 cm. long, 4–6 mm. wide, rather rigid, scabrous on the margins; panicle one-third to half the entire height of the plant, short-exserted, crimson; the capillary scabrous subflexuous *branches* at first ascending, soon widely spreading, naked below, *pilose in the axils;* spikelets on scabrous pedicels 1–4 cm. long, acuminate, 2.7–3 mm. long; *glume and sterile lemma with a stripe of appressed silky pubescence between the nerves and on the margins, or the hairs becoming loose and spreading especially on the margins,* very variable in the same panicle; fruit acuminate, chestnut; margins of lemma white. — Dry soil and sand-hills, Fla. to n. Mex. and Ariz., n. to N.H., e. Vt., e. N.Y., O., Mich., Wisc. and Minn. June–Oct.

85. PÁNICUM L. Panic-Grass

Spikelets 1-flowered or rarely with a staminate flower below the terminal perfect one, dorsi-ventrally compressed, in panicles, rarely in racemes; glumes herbaceous, very unequal, the 1st often small, the second subequal to the sterile lemma which often incloses a hyaline palea and rarely a staminate flower; fertile lemma and palea chartaceous-indurated, nerves obsolete, the margins of the lemma inrolled; grain free within the rigid firmly closed lemma and palea. — Annuals or perennials of various habit; a large genus of temp. and trop. reg. (Name from the Latin *panus, an ear of millet.*)*

A. Spikelets subsessile or short-stalked on one side of a rachis; the spiciform racemes forming elongate slender panicles; plants with long creeping rhizomes.
  Spikelets lanceolate, acute, less than 1 mm. thick. . . . . 1. *P. hemitomon.*
  Spikelets ellipsoid, blunt, 2 mm. thick. . . . . 2. *P. obtusum.*
A. Spikelets pedicelled, mostly scattered, in panicles. . . B.
  B. Soft-based annuals or, if perennials, with long linear blades, hard or rhizomatous bases and essentially uniform panicles; the terminal primary panicle 0.1–8 dm. long; basal leaves not forming prostrate winter-rosettes. . . C.
  C. 2nd glume and sterile lemma warty, obscurely nerved; annual. . . . . 3. *P. verrucosum.*

* The key, especially to species nos. 24–72, is strictly artificial, a more natural key here being unpractical.

C.2nd glume and sterile lemma smooth, glabrous, strongly ribbed. . D.

D.Annuals without rhizomes or hardened old bases; culms solitary, often branched from the lowest nodes. . . E.

E.1st glume one-fifth to one-fourth the length of spikelet, broadly rounded to truncate; leaves glabrous, rarely pilose; nodes glabrous. 4. *P. dichotomiflorum.*

E.1st glume one-third to half the length of the spikelet, acute or subacute; sheaths villous or hispid; nodes mostly bearded. . F.

F.Panicles erect; spikelets 1.5–4 mm. long; grain less than 1 mm. broad. . . G.

G.Spikelets acuminate or long-attenuate at tip, lanceolate to lance-ovoid, 2–4 mm. long. . . H.

H.Panicle slender, usually less than half as broad as long; pulvini at bases of panicle-branches glabrous or merely puberulent; spikelets lance-acuminate, 3–3.5 mm. long, mostly at tips of branchlets. . . . . 5. *P. flexile.*

H.Panicle when mature nearly as broad as to broader than long; basal pulvini copiously long-hispid.

Spikelets mostly long-pedicelled, 2–3 mm. long; panicle tardily exserted, its lower ascending branches mostly included during anthesis. . . . . 6. *P. capillare.*

Spikelets borne along the branchlets, mostly short-pedicelled, 2.5–4 mm. long; primary panicles soon exserted and with divergent to reflexed branches. . . . . . . 6a. *P. capillare,* var. *occidentale.*

G.Spikelets short-pointed, ellipsoid, ovoid or obovoid, 1.5–2.5 mm. long; pulvini glabrous or merely short-hispid; terminal panicle (except in obvious dwarfs) rarely half the height of the plant.

Terminal mature panicle long-exserted, the peduncle then becoming one-third as long as to longer than panicle; the pulvini short-hispid or glabrous; the panicle-branches spreading-ascending (not horizontal); spikelets slenderly ellipsoid-ovoid, mostly in 2's at the tips of the branchlets. 7. *P. philadelphicum.*

Terminal mature panicle included at base to short-exserted, the peduncle then becoming rarely one-fourth as long as the panicle; the pulvini glabrous or rarely ciliate; spikelets ovoid.

Panicles broadly ovoid to deltoid, by spreading of the finally widely divergent branches becoming as broad as long; spikelets 1.5–2 mm. long, 0.6–0.7 mm. thick, 2–6 subsessile or short-pedicelled and racemose along the summits of most of the branchlets. . . . . . . 8. *P. Tuckermani.*

Panicles ellipsoid to obovoid, with spreading-ascending branches; spikelets 1.8–2.5 mm. long, 0.9–1.2 mm. thick, more diffusely paniculate. . . . . 9. *P. Gattingeri.*

F.Panicle arching or nodding; spikelets 4.5–5.5 mm. long; grain about 2 mm. broad. . . . . 10. *P. miliaceum.*

D.Perennials with rhizomes or with hardened bases or persistent old stubble; culms simple at base. I.

I.Palea of sterile floret shorter than its glume, not enlarged. . J.

J.Culms terete, hard or rigid; sheaths scarcely compressed; spikelets 2.8–6.5 mm. long, 1.2–3 mm. broad; lower floret staminate; anthers 2–3 mm. long; grains 2–3.5 mm. long.

Green or but slightly glaucous; upper sheaths shorter than internodes; mature panicle exserted; rhizomes with closely imbricated scales 3–12 mm. long. . . . . 11. *P. virgatum.*

Strongly whitened; sheaths all exceeding internodes; base of panicle included or barely exserted; rhizomes with remote elongate scales.

Non-cespitose; culms mostly solitary, 0.3–1 m. high, from long rhizomes; spikelets 5–6.5 mm. long; 1st glume two-thirds to three-fourths length of spikelet. . . . . 12. *P. amarum.*

Densely cespitose, 1–2 m. high; spikelets 4.3–5.5 mm. long; 1st glume half to two-thirds length of spikelet. . . . 13. *P. amarulum.*

J.Culms compressed, firm to soft; sheaths laterally compressed;

spikelets 1.8–4 mm. long, 0.4–1.2 mm. broad; lower floret imperfect; anthers minute; grains 1.3–2.2 mm. long.

Ligules ciliate with hairs 1–3.5 mm. long; leaves 3–8 mm. broad. 14. *P. longifolium.*

Ligules erose or lacerate, not ciliate, 0.5–1 mm. long; leaves 4–18 mm. broad.

Plants with scaly rhizomes and stolons; culms only slightly compressed; blades often pilose above at base; spikelets oblique above 1st glume. . . . . 15. *P. anceps.*

Plants tufted from a short caudex; culms and loose sheaths strongly compressed; blades glabrous above; spikelets straight or essentially so.

Panicles in maturity ovoid, with loosely spreading remote lower branches, or lanceolate with appressed-ascending branches; spikelets 1.7–2.5 mm. long; fruit sessile or barely short-stipitate. . . . . 16. *P. agrostoides.*

Panicles ellipsoid or narrowly ovoid, compact, with stiff ascending branches and stiff divergent branchlets; spikelets 2.4–2.8 mm. long; fruit stipitate. . . . . . 17. *P. stipitatum.*

I. Palea of sterile floret exceeding its glume, thickened and indurated. 18. *P. hians.*

B. Perennials without elongate rhizomes, producing early in the season simple culms with usually terminal panicles 0.1–2.5 dm. long; later, in summer and autumn, branching and bearing axillary fascicles or smaller panicles of mostly cleistogamous and highly fertile flowers; the rameal internodes shortened and the secondary leaves reduced and often crowded; winter-rosettes of short leaves, unlike the cauline, frequent. . . K.

K. Culms tufted, usually densely so, erect or strongly ascending, up to 4.5 dm. high, simple or branching only near base; leaves firm, elongate-linear, 1–4 (rarely –6) mm. broad, erect; autumnal state bearing spikelets, often hidden, in the lower sheaths.

Spikelets acutely beaked, 3–4.5 mm. long, distinctly exceeding the fruit. . . . . 19. *P. depauperatum.*

Spikelets rounded or subacute at tip, not beaked, 2–3.6 mm. long, nearly conforming to the fruit.

Expanded terminal panicle one-sixth to half as broad as long; spikelets 2.7–3.6 mm. long; fruit rounded-ellipsoid or -obovoid, 1.6–2 mm. wide. . . . . 20. *P. perlongum.*

Expanded terminal panicle one-third to three-fourths as broad as long; spikelets 2–2.7 mm. long; fruit narrowly ellipsoid, 1–1.3 mm. wide. . . . . 21. *P. linearifolium.*

K. Culms solitary or commonly tufted, of various habit and stature, in the autumnal state usually forking from middle or upper nodes (if simple or forking only at base, with linear-lanceolate or lanceolate broader and softer blades); spikelets rarely, if ever, borne in the lowest axils. . . L.

L. Spikelets less than 1.5 mm. broad. . . M.

M. Spikelets less than 2 mm. long. . . N.

N. Spikelets at most 0.8 mm. broad. . . O.

O. Ligules obsolete or up to 0.5 mm. long.

Leaves, sheaths, culms and rachis copiously long-villous. . . 23. *P. strigosum.*

Leaves, sheaths, culms and rachis glabrous or rarely sparsely pilose (except for bearded nodes in no. 30).

Culms less than 8 dm. high, with glabrous or appressed-pubescent nodes; blades firm, 1–6 mm. wide; terminal panicle of primary axis 1–7 cm. long; spikelets finely pubescent (or glabrous in no. 57).

Larger basal leaves 2–10 cm. long, 31–51-nerved; lower primary cauline leaves 3.5–8 cm. long, coarsely white-margined; spikelets 1.4–1.7 mm. long.

Lower sheaths and leaves glabrous; spikelets 1.4–1.6 mm. long.

Blades very unequal, the uppermost 1–2.5 cm. long, usually less than half as large as the lower and median cauline ones. . . . . 55. *P. albomarginatum.*

Blades subequal, the uppermost 2–5.5 cm. long. . 56. *P. trifolium.*

Lower sheaths pilose; lower leaf-surfaces minutely pubescent. . . . . 57. *P. tenue.*

Larger basal leaves 1–3.5 cm. long, 11–27-nerved; lower primary cauline leaves 1–4 cm. long, scarcely or only finely white-margined; spikelets 1.3–1.5 mm. long. 58. *P. ensifolium.*

Culms 3–10 dm. high, usually with conspicuously bearded nodes; blades thin, the primary ones 5–15 mm. wide; terminal panicle of primary axis 4.5–12 cm. long; spikelets commonly glabrous (rarely pubescent). 30. *P. microcarpon.*

O. Ligules forming a dense tuft of hairs 1–5 mm. long.

Spikelets at most 1 mm. long; primary blades 1.5–4 cm. long, 3–6 mm. wide. . . . . 42. *P. Wrightianum.*

Spikelets 1.2–2 mm. long.

Primary panicles slenderly ellipsoid to slenderly ovoid, one-fourth to half as thick as long; many or all of the lateral spikelets of the branchlets equaling or exceeding their pedicels. . . . . 40. *P. spretum.*

Primary panicles rhomboid to broadly ovoid, two-thirds to quite as thick as long (sometimes narrower in *P. leucothrix*); most of the spikelets shorter than their pedicels.

Axis of panicle glabrous to minutely puberulent; lower secondary branches of panicle mostly simple; 1st glume subacute, one-fourth to half as long as spikelet; lower and median cauline leaves 2–8 mm. broad.

Tall, 7–9 dm. high; internodes of culm, leaf-sheaths and blades glutinous, with blackish viscid glands intermixed with minute pilosity; lower and median leaves of primary axis 7–8 mm. wide; primary panicles 6–9 cm. long; spikelets 1.7–1.8 mm. long. . . . . 43. *P. glutinoscabrum.*

Lower, 0.5–4.5 dm. high, not viscid or glandular; lower and median leaves of primary axis 2–5 mm. broad; primary panicles 1.5–5 cm. long; spikelets 1.3–1.6 mm. long. . . . . 44. *P. meridionale.*

Axis of panicle pilose (or, if glabrous, with lower secondary branches mostly forking, 1st glume very short and obtuse and leaves broader).

Secondary branches of primary panicles mostly simple; panicles 2–5 cm. long; 1st glume one-third to half as long as spikelet, acutish; larger primary leaves 3–5 mm. wide. . . . . 46. *P. auburne.*

Secondary divisions of lower branches of primary panicles mostly forking; panicles 3–12 cm. long; 1st glume one-sixth to one-third as long as spikelet, rounded, subtruncate or broadly obtuse; larger primary leaves 3–12 mm. broad.

Spikelets 1–1.3 mm. long; sheaths mostly with dense appressed-ascending pubescence. . . . . 41. *P. leucothrix.*

Spikelets 1.3–2.1 mm. long, if only 1.3 mm. long with the sheaths glabrous or divergently pilose. 45. *P. lanuginosum.*

N. Spikelets 0.8–1.2 mm. broad. . . P.

P. Ligule obsolete or not projecting as an obvious brush above the base of the blade. . . Q.

Q. Blades all or partly long-ciliate except at tip, thin and soft; sheaths long-pilose or villous; culms remaining simple or forking only at base; 1st glume about one-third length of spikelet. . . . . 22. *P. laxiflorum.*

Q. Blades non-ciliate or with few cilia only at base, rarely to the middle, firm to coriaceous; culms at maturity branching from middle or upper nodes. . . R.

R. Spikelets ellipsoid to ellipsoid-obovoid, distinctly longer than broad, glabrous or rarely pubescent; primary panicle 3–11 cm. long; cauline leaves merely rounded to subcordate at base, not clasping, the larger 2–10 mm. wide; leaves of winter-rosette 2–9 mm. broad. . . S.

S. Internodes of culms (especially the lower) appressed-pilose or -villous; leaves firm, linear-attenuate, the primary ones 2–5 mm. broad; spikelets obovoid, attenuate at base, pubescent, 1.1–1.4 mm. broad. . . 25. *P. aciculare.*

S. Internodes of culms glabrous or obscurely puberulent; leaves thin, lanceolate or lance-linear, the primary ones 3–10 mm. broad; spikelets ellipsoid, obovoid or lance-ovate, rounded at base, 0.8–1.1 mm. broad. . . . T.

T. Spikelets pubescent, many of the lateral ones (on branchlets of primary panicle) longer than their pedicels; nodes of culm often conspicuously bearded. 31. *P. nitidum.*

T. Spikelets glabrous (rarely pubescent), mostly shorter than their pedicels; nodes smooth or bearded. . . U.

U. Back of fruit highly lustrous, the cellular reticulation obscure; culms firm, ascending and finally bushy and fastigiate-branching and then ascending or often arching or falling over.

1st glume acuminate, ovate, longer than broad, a third length of spikelet; spikelet lance-ovate, acute or acuminate, 2.2–2.4 mm. long. . . . 68. *P. cryptanthum.*

1st glume blunt, rounded, about as broad as long, rarely a third length of spikelet; spikelets ellipsoid to obovoid, blunt or rounded at tip, 1.5–2.2 mm. long.

Cartilaginous annulus of the nodes short, the upper primary ones rarely one-third as high as the breadth of the node.

Spikelets 1.9–2.2 mm. long, all or nearly all shorter than their pedicels; fruits 1.8–2 mm. long, barely half as broad. . . . . 35. *P. dichotomum.*

Spikelets 1.5–1.6 mm. long, the lateral ones on the branchlets often equaling their pedicels; fruits 1.4–1.5 mm. long, more than half as broad. . . . . . 38. *P. caerulescens.*

Cartilaginous annulus of the nodes elongate, the upper primary ones two-thirds to quite as high as the breadth of the node; spikelets ellipsoid-obovoid; fruits 1.4–1.6 mm. long, more than half as broad. . . . . . 37. *P. roanokense.*

U. Back of fruit semiopaque, obviously cellular-reticulate; culms weak, soon reclining, finally sending out strongly divergent weak elongate and prostrate branches. . . . . . 39. *P. lucidum.*

R. Spikelets subglobose or short-pyriform, puberulent; primary panicles 5–25 cm. long; cauline leaves cordate-clasping, the larger 0.7–2.5 cm. wide; larger leaves of winter-rosettes 1–2 cm. broad.

Uppermost primary leaf 2–11 cm. long, 3–15 mm. broad, firm or subcoriaceous; primary panicles 3–13 cm. long, two-thirds to about as broad. . . . . . 53. *P. sphaerocarpon.*

Uppermost primary leaf 0.8–2 dm. long, 1–3 cm. broad, thin, submembranaceous; primary panicles 0.5–2.5 dm. long, one-fourth to half as broad. . . . . 54. *P. polyanthes.*

P. Ligule forming a dense tuft of hairs obviously projecting above base of blade. . . V.

V. Internodes of the culm glabrous.

Primary panicles slenderly ellipsoid to slenderly ovoid, one-fourth to half as thick as long; many or all the lateral spikelets of the branchlets equaling or exceeding their pedicels. . . . . . 40. *P. spretum.*

Primary panicles rhomboid to broadly ovoid, nearly or quite as broad as long; most of the spikelets shorter than their pedicels. . . . . 45. *P. lanuginosum.*

V. Internodes (at least the lowest) of culm pubescent. . . W.

W. 1st glume of young spikelets rounded, subtruncate or with broad-deltoid tip, clearly as short as broad, one-sixth to half length of spikelet.

Axis of panicle (1.5–5 cm. long) glabrous to minutely puberulent; lower secondary branches of panicle mostly simple; 1st glume acutish, one-fourth to half as long as spikelet; lower and median cauline leaves of primary axis 2–5 mm. broad. . . . . 44. *P. meridionale.*

Axis of panicle pilose (or, if glabrous, with lower secondary branches mostly forking; 1st glume very short and obtuse and leaves broader).

Secondary branches of primary panicles mostly simple; panicles 2–5 cm. long; 1st glume one-third to half as long as spikelet, subacute; larger primary leaves 3–5 mm. wide. . . . . 46. *P. auburne.*

Secondary divisions of lower branches of primary panicles mostly forking; panicles 3–14 cm. long; 1st glume one-sixth to one-third as long as spikelet, rounded, subtruncate or broadly obtuse; leaves 3–13 mm. broad.

Elongate primary leaves and nodes of culms mostly 4–7, the blades villous or pilose and 3–10 cm. long; ligule a brush of hairs 3–5 mm. long; spikelets 1.3–2.1 mm. long. . . . . 45. *P. lanuginosum.*

Elongate primary leaves and nodes of culms 6–15, the glabrous blades 6–15 cm. long; ligule up to 1 mm. long; spikelets 1.8–2.2 mm. long. . . . . . 70. *P. mundum.*

W. 1st glume of young spikelets prolonged, acutish, as long as or longer than broad, one-third to half length of spikelet.

Sheaths mostly equaling or exceeding the internodes; only 1–3 internodes of primary axis slightly exserted. 48. *P. subvillosum.*

Sheaths mostly much shorter than the long internodes.

Culms and sheaths villous with widely divergent hairs 3–5 mm. long; spikelets all long-pedicelled (shortest, exceptional, pedicels 3 mm. long). . . . . . . 47. *P. praecocius.*

Culms and sheaths minutely pilose or with longer ascending or appressed pubescence; many spikelets on pedicels only 1–2.5 mm. long. . . . . . . 52. *P. columbianum.*

M. Spikelets 2–4 mm. long. . . X.

X. Blades all or partly long-ciliate except at tip, thin and soft; sheaths long-pilose or villous; culms remaining simple or forking only at base. . . . . 22. *P. laxiflorum.*

X. Blades non-ciliate or with cilia only below the middle, firm; culms at maturity branching at middle or upper nodes. . . Y.

Y. Elongate primary cauline leaves 2–6 (if more, the culm slender), the larger 0.3–1.6 dm. long; culms slender, mostly less than 0.7 (rarely –1) m. high; 1st node above winter-rosette 0.5–3 (rarely –4) mm. thick. . . Z.

Z. Spikelets glabrous; ligules obsolete or minute. . . *a.*

a. Back of fruit highly lustrous, the cellular reticulation obscure; culms firm, ascending, late in season with ascending and finally fastigiate branching. . . *b.*

b. 1st glume rounded to subtruncate at summit; spikelets obovoid, the lateral mostly exceeding their pedicels; primary panicles 3–6 cm. long, slenderly ovoid: culms crisp-puberulent. . . . . 59. *P. lancearium.*

b. 1st glume acute or subacute, with prolonged summit; spikelets ellipsoid to lance-ovoid, mostly long-pedicelled; primary panicles 4–12 cm. long; culms glabrous except sometimes at bearded nodes. . . *c.*

c. Spikelets sharply acute, subulate-tipped; branches of panicle viscid-spotted. . . . . 68. *P. cryptanthum.*

c.Spikelets obtuse to subacute; branches of panicle not viscid-spotted.

Sheaths smooth, not verrucose; spikelets 1.9–2.2 mm. long, obtuse.

Cartilaginous annulus of the nodes short, the upper primary ones rarely one-third as high as the breadth of the node. . . . . 35. *P. dichotomum.*

Cartilaginous annulus elongate, the upper primary ones two-thirds to quite as high as the breadth of node. . . . . 37. *P. roanokense.*

Sheaths (or some of them) with abundant wart-like pale spots; spikelets 2–2.5 mm. long, acutish. . . 36. *P. yadkinense.*

a.Back of fruit semiopaque, obviously cellular-reticulate; culms weak, soon reclining, finally sending out strongly divergent weak elongate and prostrate branches. . . . 39. *P. lucidum.*

Z.Spikelets pubescent; ligules various. . *d.*

d.Spikelets at most 1 mm. broad. . *e.*

e.Culms glabrous except sometimes at the nodes. . . *f.*

f.Lateral spikelets (along the branches) of the primary panicles mostly longer than their pedicels; nodes of culms often heavily bearded.

Leaf-blades glabrous (sometimes ciliate at base); upper sheaths often finely verrucose; spikelets 1.8–2 mm. long. . . . . 31. *P. nitidum.*

At least the lower blades commonly velvety (sometimes glabrous); sheaths not verrucose; spikelets 2–2.7 mm. long.

Leaves all densely soft-pubescent or minutely velvety; spikelets 2–2.4 mm. long. . . . . . 32. *P. annulum.*

Upper and sometimes all leaves glabrous; spikelets 2.2–2.7 mm. long. . . . . 33. *P. mattamuskeetense.*

f.Lateral spikelets of the primary panicle mostly long-pedicelled. . . . . 34. *P. boreale.*

e.Culms minutely ashy-puberulent.

Primary cauline leaves 3–8 mm. broad; spikelets 2–2.1 mm. long, inequilateral, ventrally gibbous. . . . 59. *P. lancearium.*

Primary cauline leaves 5–12 or more mm. broad; spikelets 2.2–3.2 mm. long, equilateral. . . . . 72. *P. commutatum.*

d.Spikelets 1.1–1.4 mm. broad. . . *g.*

g.Spikelets 2–2.5 mm. long. . *h.*

h.Internodes of culms (not sheaths) minutely ashy-puberulent.

Primary cauline leaves 3–8 mm. broad; spikelets 2–2.1 mm. long. . . . . 59. *P. lancearium.*

Primary cauline leaves 5–12 or more mm. broad; spikelets 2.2–3.2 mm. long. . . . . 72. *P. commutatum.*

h.Internodes of culms glabrous, pilose or villous, not closely puberulent. . . *i.*

i.Ligule at most about 1 mm. long.

Culms 5–15 dm. high; leaf-blades lanceolate, divergent or ascending, the primary cauline ones 7–15 mm. broad; primary panicles 5–15 cm. long.

Nodes all or all but the middle and lowest beardless; spikelets 2.2–2.8 mm. long; 1st glume narrowly ovate, acuminate, one-third to two-fifths as long as spikelet.

Basal sheaths velutinous; vernal panicle ellipsoid during anthesis, then rarely more than half as broad as long, the spikelets often as long as to longer than their pedicels and subapproximate along the branchlets. 33. *P. mattamuskeetense.*

Basal sheaths horizontally hirsute to glabrous; vernal panicle ovoid, nearly as broad as long, the scattered spikelets mostly much shorter than their pedicels. . . . . 71. *P. recognitum.*

Nodes during vernal anthesis copiously bearded; spikelets 1.8–2.2 mm. long; 1st glume broadly ovate, blunt, scarcely one-fourth length of spikelet. . . . . 70. *P. mundum.*

Culms 1.5–8 dm. high; leaf-blades lance-linear, firm, ascending or slightly divergent, the primary cauline 2–8 mm. broad; primary panicle 3–8 (rarely –10) cm. long, with scattered spikelets; 1st glume broad-ovate to suborbicular.

Sheaths and lower internodes of culms copiously pilose or villous.

Spikelets pyriform or gradually tapering at base; principal leaves nearly linear, flat or involute.

Blades of autumnal branches strongly involute at tip; primary leaves 2–5 mm. broad; spikelets 1.8–2.4 mm. long, 1.1–1.4 mm. broad. . . . . 25. *P. aciculare.*

Blades of autumnal branches flat; primary leaves 4–8 mm. broad; spikelets 2.3–3.3 mm. long, 1.4–1.6 mm. broad.

Nodes long-bearded; leaves all pubescent. 26. *P. consanguineum.*

Nodes beardless; middle and upper leaves glabrous. . . . . 27. *P. angustifolium.*

Spikelets subtruncate to broadly rounded at base; principal leaves narrowly lanceolate, flat. . . . . 51. *P. Commonsianum.*

Sheaths and internodes glabrous or rarely remotely pilose. . . . . 28. *P. Bicknellii.*

*i.*Ligule with a brush of hairs 2–5 mm. long.

Secondary divisions of lower branches of primary panicles mostly forking; spikelets 1.3–2.1 mm. long. . . . . 45. *P. lanuginosum.*

Secondary branches of primary panicles mostly simple; spikelets 2–2.5 mm. long.

1st glume deltoid-rotund, 0.5–0.8 mm. long, one-fifth to one-fourth length of spikelet; cauline leaves of vernal culms lanceolate, 3.5–6.5 cm. long, glabrous above, their sheaths with coarsely pustular-based hairs; peduncle 0.5–4 cm. long. . . . . 49. *P. Benneri.*

1st glume ovate, one-fourth to two-fifths length of spikelet; cauline leaves lance-linear, 6–11 cm. long, usually villous or pilose above, their sheaths soft-pilose or -villous; peduncles 0.3–1.7 dm. long. . . . . 50. *P. villosissimum.*

*g.*Spikelets 2.6–4 mm. long. . . *j.*

*j.*Lower internodes of culms glabrous, remotely hirsute or only minutely puberulent. . . *k.*

*k.*Principal leaves narrowed to rounded or subcordate, but not definitely cordate, at base, lanceolate to linear.

Leaves thin to firm, lanceolate, divergent to ascending; branches of primary panicle spikelet-bearing nearly to base; spikelets 2.2–2.8 mm. long; 1st glume narrowly ovate, acuminate.

Basal sheaths velutinous; vernal panicle ellipsoid during anthesis, then rarely more than half as broad as long; spikelets often as long as to longer than their pedicels and subapproximate along the branches. . . . . 33. *P. mattamuskeetense.*

Basal sheaths horizontally hirsute to glabrous; vernal panicle ovoid, nearly as broad as long, the scattered spikelets mostly much shorter than their pedicels. . . . . 71. *P. recognitum.*

Leaves thick and firm, linear to linear-lanceolate, mostly ascending; primary panicles with few scattered spikelets, the branches often naked below; spikelets 2.6–3.2 mm. long, 1.1–1.4 mm. broad; 1st glume broad.

Spikelets pyriform or gradually narrowed to base; linear leaves up to 8 mm. broad, the larger 13–31-nerved. . . . . . . 27. *P. angustifolium.*

Spikelets subtruncate to rounded at base; larger primary linear-lanceolate leaves 4–15 mm. wide, 27–75-nerved.

Spikelets 2.3–2.8 mm. long; 1st glume 0.8–1.2 mm. long; primary leaves 4–8 mm. broad. 28. *P. Bicknellii.*

Spikelets 2.9–3.2 mm. long; 1st glume 1.2–2.5 mm. long; primary leaves 8–22 mm. broad.

Branches of primary panicle stiffly ascending to spreading-ascending; midrib prominent nearly to tip of leaf.

Primary cauline leaves 3 or 4, the larger 1.1–1.5 dm. long; nodes not enlarged. 29. *P. calliphyllum.*

Primary cauline leaves 4–8, the larger 1.5–2.2 dm. long; nodes enlarged. . . 67. *P. aculeatum.*

Branches of primary panicle subflexuous, soon divergent; midrib evanescent near middle of blade. . . . . . . 73. *P. mutabile.*

k. Principal leaves cordate to subamplexicaul at base, broadly lanceolate. . . . . 72. *P. commutatum.*

j. Lower internodes copiously pilose to villous.

Spikelets gradually attenuate at base; lower sheaths spreading-villous.

Spikelets rhomboid-ellipsoid, tapering to tip, 3–4 mm. long. . . . . 24. *P. fusiforme.*

Spikelets pyriform, rounded at summit, 2.3–3 mm. long. . . . . 26. *P. consanguineum.*

Spikelets rounded to subtruncate at base; sheaths strigose, appressed-pilose or villous.

Vernal culms 2–6 dm. high; expanded panicle 3–7 cm. broad, usually long-exserted; blades glabrous or glabrate. . . . . 51. *P. Commonsianum.*

Vernal culms 1–3.5 dm. high; expanded panicle 1–2.5 cm. broad, included at base or short-exserted; blades villous. . . . . . . 60. *P. Wilcoxianum.*

Y. Elongate primary cauline leaves 6–12, 1–2.5 dm. long; culms comparatively stout, 0.7–1.5 m. or more high; 1st node above the winter-rosette 3–8 mm. thick.

Sheaths and leaf-blades densely short-villous or velutinous. . 66. *P. scoparium.*

Sheaths and blades glabrous, hispid or hirsute.

Broader primary leaves 0.8–1.5 cm. wide, rounded to slightly narrowed at base; spikelets acute or obtuse.

Spikelets acute.

1st glume ovate, tapering to acute tip, longer than broad, one-third length of spikelet; branches of autumnal phase few, slender, exserted from the primary sheaths.

Spikelets 3 mm. long. . . . . 67. *P. aculeatum.*

Spikelets 2.2–2.4 mm. long. . . . . 68. *P. cryptanthum.*

1st glume suborbicular, abruptly tipped, less than one-fourth length of spikelets; branches of autumnal phase densely leafy, with thickened axes, the bases included in the primary sheaths. . . . . 69. *P. scabriusculum.*

Spikelets obtuse.

Nodes and elongate leaves of vernal culms 6–15, the nodes then long-bearded; rachis of vernal panicle pilose or glabrate; spikelets 1.8–2.2 mm. long; 1st glume triangular-ovate, abruptly tipped, about one-sixth length of spikelet. . . . . 70. *P. mundum.*

Nodes and elongate blades of vernal culms 5–7, the nodes glabrous or glabrate; rachis of panicle glabrous; spikelets 2.2–2.8 mm. long; 1st glume narrowly ovate, tapering to acute tip, one-third length of spikelet. . . . . 71. *P. recognitum.*

Broader primary leaves 1.7–3.5 cm. wide, strongly cordate at base; spikelets obtuse. . . . . 74. *P. clandestinum.*

L. Spikelets 1.5–2.2 mm. broad. . . *l.*

l. Broadest primary cauline leaves 17–35-nerved, 3–8 mm. broad; spikelets 2.5–3.2 mm. long, 1.2–1.6 mm. broad.

Spikelets pyriform or gradually tapering at base, at most 2.8 mm. long; vernal culms 3–6 dm. high.

Nodes long-bearded; leaves all villous. . . . . 26. *P. consanguineum.*

Nodes beardless; middle and upper leaves glabrous. . . . . . 27. *P. angustifolium.*

Spikelets rounded at base, 2.6–3.2 mm. long; vernal culms 1–3.5 dm. high. . . . . 60. *P. Wilcoxianum.*

l. Broadest primary cauline leaves 33–115 (or more) -nerved, 0.5–4 cm. broad; spikelets 2.4–4.5 mm. long, 1.4–2.2 mm. broad. . . *m.*

m. Spikelets 2.4–3.3 mm. long, 1.4–1.7 mm. broad.

Leaf-blades and sheaths soft-pubescent or velvety.

Vernal culms 2.5–7 dm. high; larger primary cauline leaf-blades 7–12 cm. long, 6–12 mm. broad; spikelets 2.9–3 mm. long, the 1st glume one-third to half as long. . . . . . 61. *P. malacophyllum.*

Vernal culms mostly 0.8–1.3 m. high; larger primary cauline leaf-blades 1.2–2 dm. long, 1–1.8 cm. broad; spikelets 2.4–2.6 mm. long, the 1st glume one-fifth to one-fourth as long. 66. *P. scoparium.*

Leaf-blades and sheaths glabrous, pilose or hirsute, not velvety.

Culms 2–8 dm. high; larger primary cauline leaf-blades 5–14 cm. long, 0.5–1.5 cm. broad; spikelets 1.5–2 mm. broad. . 62. *P. oligosanthes.*

Culms 0.6–1.5 m. high; larger primary cauline leaf-blades 1–2.5 dm. long, 1.7–3.5 cm. broad; spikelets 1.4–1.5 mm. broad. . . . . 74. *P. clandestinum.*

m. Spikelets 3.4–4.5 mm. long, 1.7–2.2 mm. broad. . . *n.*

n. Ligule an obvious brush exserted 3–4 mm. above the sheath-orifice; spikelets 4–4.3 mm. long; blades lanceolate, the broader 1–2 cm. broad, rounded at base. . . . . 63. *P. Ravenelii.*

n. Ligule obsolete or up to 1 (rarely –2) mm. long; spikelets 3.4–4 mm. long (if –4.5 mm. long, the ligules obsolete, and blades cordate and ovate-lanceolate, the broader 1.5–3.5 cm. broad). . . *o.*

o. Leaf-blades narrowly lanceolate, rounded at base, the broader primary ones 0.5–1.5 (rarely –2.2) cm. broad, with 33–85 nerves.

Spikelets glabrous or short-pubescent.

Primary panicles ovoid, the branches finally spreading; 1st glume broadly ovate, less than half length of spikelet; leaves spreading or loosely ascending, the upper primary cauline ones 3–10 cm. long. . . . . . 62. *P. oligosanthes.*

Primary panicles slenderly subcylindric to narrowly ellipsoid, the branches erect or ascending; 1st glume narrowly ovate, half length of spikelet; leaves strongly ascending, the upper primary cauline ones 0.7–2 dm. long. . . . . 65. *P. xanthophysum.*

Spikelets long-villous with soft hairs about as long as the diameter of the spikelet; panicle subcylindric to slenderly ovoid or ellipsoid. . . . . 64. *P. Leibergii.*

*o.*Leaf-blades broadly lanceolate or narrowly ovate, cordate at base; the broader primary ones 1.5–4 cm. broad, with 59–115 nerves.

Nodes usually beardless; largest primary cauline leaves four to seven times as long as broad; spikelets 3.4–3.7 mm. long. . . . . 75. *P. latifolium.*

Nodes usually long-bearded; largest primary cauline leaves two-and-a-half to five times as long as broad; spikelets 4–4.5 mm. long. . . . . 76. *P. Boscii.*

Subgen. Eupánicum Godr.

1. P. hemítomon Schultes (halved; from the somewhat 1-sided spikes), Maiden-Cane. — *Culms* stout, 0.5–1.5 m. long, *from elongate rhizomes,* rooting and branching from lower nodes; sheaths loose; blades 1–2 dm. long, about 1 cm. wide, scabrous above; panicle racemiform, 1–3 dm. long, the *remote 1-sided spiciform racemes appressed; spikelets lanceolate, acute,* 2.4–2.8 *mm. long,* barely 1 *mm. broad;* 1*st glume about half as long as 2nd, acute; apex of palea exserted.* — Wet shores, shallow water and borders of swamps, Tex. to Fla., n., locally, to Cape May, N.J. June, July. (S. Am.)

2. P. obtùsum HBK. (blunt). — *Freely stoloniferous;* the flowering branches erect, 2–8 dm. high; sheaths tight; blades firm, 2–7 mm. wide, mostly glabrous; *panicle* slender, 3–12 cm. long, *with few appressed-ascending* 1*-sided spiciform racemes; spikelets ellipsoid,* 3–3.8 *mm. long, about* 2 *mm. thick, obtuse;* 1*st glume blunt, nearly equaling the* 2*nd.* — Damp flats, shores and waste places, Kans. to Tex., n. Mex. and Ariz.; adv. in w. Mo. June–Oct.

3. P. verrucòsum Muhl. (warty). — *Annual, glabrous,* bright green; culms slender, simple or branching below, often geniculate and rooting at lower nodes, 0.2–1.5 m. high; leaves narrowly lanceolate, 2–10 mm. wide, slightly scabrous; *panicles diffuse,* few-flowered; the primary one 0.3–3 dm. long, *its branches* capillary, *spikelet-bearing in the upper half;* spikelets ovoid, 1.7–2 mm. long; 1*st glume one-fourth as long as the obscurely nerved warty* 2*nd glume and sterile lemma;* fruit apiculate. (*P. debile* Ell., not Desf.) — Damp sand or peat, Fla. to e. Tex., n. to Mass. and s. O.; nw. Ind. Aug.–Oct. Fig. 287.

287. P. verrucosum.

4. P. dichotomiflòrum Michx. (with forking inflorescence). — Annual, simple or commonly *divergently branching from base and nodes, geniculate* and decumbent or erect, smooth throughout or foliage sparsely pilose; *culms compressed,* succulent, 0.02–2 m. high or long; lower nodes enlarged; leaves narrowly lanceolate, 0.2–5 dm. long, 0.3–2.5 cm. broad; *panicles diffuse,* exserted or with base included, 1 cm. (in dwarf plants) –4 dm. long; *spikelets oblong-lanceolate to -ovoid,* 1.8–3.6 *mm. long, mostly longer than their scabrous-angled pedicels and secund toward the tips of the branchlets;* 1*st glume rounded-deltoid, one-fifth to one-fourth as long as the* 2*nd* and the sterile lemma. — A variable wide-ranging species; we have the following vars.:

Spikelets oblong-lanceolate, tapering to acuminate tips, (2–)2.6–3.6 mm. long, the 2nd glume and sterile lemma subcoriaceous.

Culms erect or ascending, slightly or scarcely geniculate, with only slightly enlarged nodes; sheaths little if at all inflated; terminal panicles becoming long-exserted, their capillary branches all ascending at maturity; spikelets loosely disposed on the branchlets. . . . . *P. dichotomiflorum* (typical).

Culms loosely ascending or depressed, geniculate, with enlarged lower nodes; lower and primary sheaths inflated; panicles eventually borne at most nodes, their bases included or only slightly exserted, their stiff branches soon horizontally divergent to reflexed; spikelets more crowded. . . . . Var. *geniculatum.*

Spikelets ovoid to slenderly ellipsoid, obtuse or abruptly short-tipped, 1.8–2.2 mm. long, the 2nd glume and sterile lemma submembranaceous; panicles with ascending to spreading (not reflexed) branches, their spikelets often long-pedicelled.

Culms slender, 0.3–6 dm. high or long; leaf-blades 1–8 mm. broad, their surfaces smooth (except sometimes at margin); primary panicles 0.2–2.5 dm. long. Var. *puritanorum.*

Culms coarse, 0.8–2 m. long; primary leaf-blades 0.7–2.5 cm. broad, the upper surface harshly scabrous; primary panicles 2–4 dm. long. . . . . . Var. *imperiorum.*

P. dichotomiflòrum (typical). — Low grounds and waste places, Fla. to Tex. and Mex., n. to s. N.E., s. N.Y., s. Ont., Ind., Ill., Ia., Kans. and Calif. June–Oct. Fig. 288.

Var. **geniculàtum** (Wood) Fern. (bent like the knee-joint). — Similar habitats, often an aggressive weed, Fla. to La., n. to N.S., N.E., N.Y., O., Ill. and Minn.

Var. **puritanòrum** Svenson (of the Puritans). — Damp sands and pond-margins, local, s. N.H. to N.J.; n. Ind. Fig. 289.

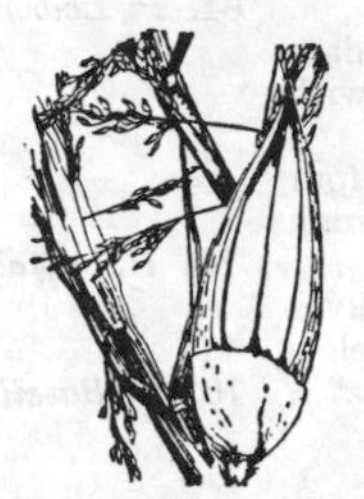
288. P. dichotomiflorum (typical).

289. P. dichotomiflorum, v. puritanorum.

Var. **imperiòrum** Fern. (of the Dominions). — Bottomlands, se. Va.

5. P. fléxile (Gattinger) Scribn. (pliant). — Annual, slender, *erect*, 2–7 dm. high, simple or with erect basal branches; *sheaths papillose-hispid; blades* 1–3 dm. long, 2–7 mm. wide, sometimes glabrous, *erect; panicles* 0.5–3 *dm. long, usually less than half as broad, narrowly ellipsoid with ascending branches and glabrous pulvini; spikelets lance-acuminate,* 3–3.5 *mm. long, solitary at the tips of the branchlets;* 1st glume acute, one-third as long as spikelet; 2nd glume and sterile lemma acuminate, one-third longer than fruit. — Dry or moist, chiefly calcareous ledges, sands and moors, Fla. to e. Tex., n. to w. N.E., sw. Que., s. Ont., s. Mich., Ill., Ia. and S.D. July–Oct. Fig. 290.

6. P. capillàre L. (hair-like), Old-witch Grass, Mousseline (Que.). — Annual, stoutish, simple or with ascending basal branches, 2–8 dm. high; sheaths and usually the blades (0.3–2 cm. wide), *copiously papillose-hispid; panicles* very large and *diffuse,* often purple; *the terminal one often one-third to half the height of the plant, included at base until maturity,* the ascending to finally divergent capillary *branches with hispid basal pulvini,* the ripe panicle brittle and becoming a "tumbleweed"; *spikelets lance-ovoid, short-acuminate,* 2–3 *mm. long, mostly on long pedicels;* 1st glume acute, nearly half as long as the strongly ribbed 2nd glume and sterile lemma. — Open sandy or stony soil or in cult. land, Fla. to Tex., n. to se. Me., sw. Que., s. Ont., s. Man. and Mont. July–Oct. (Bermuda)

290. P. flexile.

291. P. capillare, v. occidentale.

Var. **occidentàle** Rydb. (western). — Mostly lower, 0.2–5(–7) dm. high; *primary panicle commonly exserted,* with more promptly divergent branches, *one-third to two-thirds the height of the plant; spikelets* 2.5–4 *mm. long, long-acuminate, mostly subsessile or short-pedicelled along the ultimate branchlets.* (*P. barbipulvinatum* Nash) — Open sandy or gravelly soil, waste places and cult. fields, se. Que. to s. B.C., s. to N.S., s. N.E., L.I., N.J., W.Va., Ky., Mo., Tex., Ariz. and s. Calif. Fig. 291.

7. P. philadélphicum Bernh. (of Philadelphia). — Annual, slender, ascending, branched at base, 1.5–8 dm. high; sheaths and commonly the erect *narrow* (2–8 *mm. wide*) *blades* hispid; *panicles broadly ellipsoid or obovoid, rather longer than broad, the largest exserted,* diffuse, 0.4–2.5 *dm. long, with loosely spreading-ascending capillary branches; pulvini minutely hispid; spikelets narrowly ovoid,* 1.7–2.2 *mm. long, acute or short-acuminate, mostly short-pedicelled or subsessile in 2's at the tips of the branchlets.* — Rocky or sandy open soil or thin woods, Ga. to e. Tex., n. to s. N.H., sw. Que., N.Y., W.Va., O., s. Mich., s. Wisc. and Ia. June–Oct. Fig. 292.

292. P. philadelphicum.

8. P. Tuckermáni Fern. (for its discoverer, Edward Tuckerman, 1817–1886). — Annual, *usually freely forking and forming depressed or decumbent mats,* rarely (when crowded) simple and erect; culms slender, 0.1–7 dm. long; sheaths and blades (2–10 mm. broad) hispid; *panicles broadly ovoid to rhomboid,* the terminal often slightly exserted, 0.2–2 dm. long, *by spreading of the finally divergent* delicate *branches becoming as*

293. P. Tuckermani.

broad; secondary panicles at many nodes, smaller, rarely completely exserted; *pulvini glabrous; spikelets ovoid, very short-pointed*, 1.5–2 *mm. long*, 2–6 *subsessile or short-pedicelled along the tips of the ultimate branchlets*; 1st glume deltoid-suborbicular, short-acuminate, hardly half the length of the spikelet. — Sandy or gravelly shores or open soils, s. Que. to Minn., s. to N.S., N.E., n. Va., O. and Ind. July–Oct. Fig. 293.

9. P. Gattíngeri Nash (for its discoverer, Augustin Gattinger, 1825–1903). — Annual, resembling no. 8, *decumbent or depressed*, sometimes erect; *culms branching at all the nodes, often repeatedly forking*, up to 1 m. long; *panicles very numerous*, exserted or subincluded, *broadly ellipsoid or obovoid*, the terminal 0.9–2 dm. long, *with spreading-ascending stiff freely forking short branches; pulvini glabrous; spikelets ovoid*, rather turgid, short-pointed, 1.8–2.5 *mm. long*, 0.9–1.2 *mm. broad*. — Sandy shores, fields and roadsides, chiefly calcareous, sw. Que. and w. Mass. to Minn., s. to N.C., Tenn. and Ark. Aug.–Oct. Fig. 294.

294. P. Gattingeri.

10. P. miliàceum L. (millet), Millet, Broom-corn M., Proso. — Coarse hispid-leaved annual up to 1 m. high; *panicle dense*, 1.5–3 dm. long, *arching or nodding in maturity; spikelets* ovoid, 4.5–5.5 *mm. long*, turgid; *grain about* 2 *mm. broad*. — Waste places, N.E., westw. and southw., perhaps not persistent. July–Oct. (Adv. from Old World)

11. P. virgàtum L. (wand-like), Switchgrass. — *Culms terete, hard*, solitary or tufted, erect *from hard closely scaly rhizomes*, 0.3–2 m. high, *green or purplish*, sometimes slightly glaucous; *leaves* firm, elongate-linear, smooth, or *scabrous on the margin* and sometimes sparsely pilose near the base, flat, taper-pointed, 3–15 mm. wide; lower and middle *sheaths exceeding the internodes, the upper shorter; panicles terminal, long-exserted*, 0.5–5 dm. long, with spreading-ascending to strongly appressed branches; *spikelets* ovoid, acuminate, 2.8–6 *mm. long*, strongly nerved; 1st glume pointed, one-third to two-thirds the length of the spikelet; *lower floret staminate; anthers about* 2 *mm. long*, purplish, conspicuous; *grains* 2–3 *mm. long*. Very variable; the best marked vars. with us are the following:

Rhizomes elongate and creeping or, if cespitose, with horizontally divergent branches; culms solitary to many in a loose tussock.
- Spikelets 3.5–6 mm. long; 1st glume two-thirds length of spikelet, acuminate; 2nd glume long-acuminate, conspicuously exceeding fertile lemma and fruit; panicle 2.5–5 dm. long, with loosely ascending branches; culms up to 1 m. high. . . . . *P. virgatum* (typical).
- Spikelets 2.8–3.2 mm. long; 1st glume barely half length of spikelet, blunt; 2nd glume short-acuminate, scarcely to slightly exceeding fertile lemma and fruit; panicle 1–4 dm. long, its branches often appressed; culms 0.5–1 m. high. . . . . Var. *cubense*.

Rhizomes very short, closely interlocking, subascending, forming dense crowns with many closely cespitose culms 0.3–1.1 m. high; spikelets 3.2–4 mm. long; 1st glume two-thirds length of spikelet; 2nd and sterile lemma clearly exceeding fertile lemma and fruit; panicle 0.5–4 dm. long. . . . . Var. *spissum*.

P. virgàtum (typical). — Dry or moist sandy soils, shores, etc., sw. Que. and w. N.H. to Sask., s. to the Gulf States and Centr.Am. July–Sept. (Bermuda; S.Am.)

Var. **cubénse** Griseb. (of Cuba), var. *obtusum* Wood — W.I. and Fla. to Miss., n. on Coastal Plain to Mass. (where local); reported from s. Mich.

Var. **spíssum** Linder (crowded). — Gravelly or sandy fresh to brackish shores and swamps, w. N.S. and centr. Me. to centr. N.Y. and Pa.

12. P. amàrum Ell. (bitter). — *Very glaucous, extensively creeping; rhizomes with remote elongate* (*up to* 1 *dm. long*) *scales; culms mostly solitary*, ascending from the rhizomes, simple or forking at lower nodes, 0.3–1 *m. high; leaves overlapping, longer than* but by involution often exposing *the nodes;* blades firm, glabrous, 1–3 cm. long, 5–12 mm. wide, becoming involute; *panicle slender, elongate*, 0.7–5 dm. long, *the base included or but slightly exserted; spikelets* 5–6.5 *mm. long;* 1st glume three-fourths as long as spikelet, acuminate; 2nd glume exceeding sterile lemma; anthers 2.5–3 mm. long; fruit 3.5 mm. long. (*P. amaroides* Scribn. & Merr.) — Sandy coasts, Tex. to Ga., n. to Ct. Aug.–Nov.

13. P. amárulum Hitchc. & Chase (slightly bitter), Beachgrass. — Similar to no. 12, but *cespitose*, 1–2 *m. high;* leaves longer; *panicle* 4–8 *dm. long; spikelets* 4.3–5.5 *mm. long;* fruit 3–3.5 mm. long. (*P. amarum* of ed. 7, not Ell.) — Sandy coast, Tex. to Fla., n. to s. N.J. July–Sept. (W.I.; Mex.)

14. P. longifòlium Torr. (long-leaved). — Tufted from a short caudex; *tufts compressed at* *base*, 0.2–1 m. high; culms slender, smooth; *leaves firm, erect*, flat; lower sheaths overlapping, 0.5–2.5 dm. long, glabrous, sometimes villous; *basal and lower cauline blades* 1–7 *dm. long*, 3–8 *mm. broad*, glabrous, rarely villous; *ligule ciliate*, 1–3.5 *mm. long; panicles long-exserted*, 0.3–4 dm. long, *with few ascending to horizontally divergent capillary branches* (*the longer* 0.1–1.7 *dm. long*); *spikelets* lanceolate, acuminate, short-pedicelled *along the branchlets*, 2–4 *mm. long;* 1st glume 1–2 mm. long; *2nd glume shorter than to longer than sterile lemma;* grain 0.4–0.9 mm. broad. — Four vars.:

a. Plants 2–9 dm. high; culms 1–3 mm. broad at the 1st exposed node; principal sheaths 0.5–1.5 dm. long, glabrous (or sometimes villous); longer blades 1–3.5 dm. long, 3–6 mm. broad, glabrous, or villous only near base; larger panicles 0.3–2.3 dm. long, 0.1–1.5 dm. broad. . . *b.*

b. Branches and branchlets of panicle ascending, the mature branchlets appressed or subappressed.

Spikelets 2–3 mm. long, the 2nd glume equaling or longer than sterile lemma; longer branches of panicle 3–14 cm. long; grain 0.4–0.7 mm. broad. . . . . *P. longifolium* (typical).

Spikelets 2.6–3.4 mm. long, the 2nd glume usually shorter than sterile lemma; longer branches of panicle 0.1–8 cm. long; grain 0.8–0.9 mm. broad. . . . . Var. *tusketense.*

b. Branches and branchlets strongly (often horizontally) divergent; spikelets 3–4 mm. long, with tips often sharper; 2nd glume equaling or slightly exceeding sterile lemma; longer branches of panicle 5–10 cm. long. . . . Var. *Combsii.*

a. Plants 7–10 dm. high; culms 3–6 mm. broad at the 1st exposed node; principal sheaths 1–2.5 dm. long, usually densely villous; longer blades 3–7 dm. long, often overtopping panicle, 5–8 mm. broad, usually villous to tip; larger panicles 2–4 dm. long, 1–2 dm. broad, in maturity with widely divergent to horizontal branches and branchlets; spikelets 2–2.5 mm. long. . . . . . Var. *pubescens.*

P. longifòlium (typical). — Peaty and sandy swales, bogs and shores, Fla. to Tex., n. to w. N.S., s.-centr. N.H., se. Mass., Ct., e. N.Y., s. O. and Ky. July–Oct. FIG. 295.

295. P. longifolium.

Var. **tusketénse** Fern. (of the Tusket River, N.S.). — Gravelly or peaty shores, Tusket Valley, N.S. July–Sept.

Var. **Còmbsii** (Scribn. & Ball) Fern. (for its discoverer, ROBERT COMBS, 1872–1899). — Wet pine-barrens and low thickets, Fla. to La., n., locally, to se. Va.

Var. **pubéscens** (Vasey) Fern. (hairy). — Fla. to e. Tex., n., locally, to sphagnous swamps of se. Va.

15. P. ánceps Michx. (two-edged). — Plants with *stout scaly* and knotted *rhizomes; culms* stout, few, *only slightly compressed*, 0.3–1.3 m. high, 2.5–6 mm. thick at lowest exposed node; sheaths mostly a little shorter than the internodes, glabrous to papillose-pilose; *ligule a membrane about* 0.5 *mm. long; blades* erect, flat, elongate, 0.4–1.8 cm. wide, *often pilose above near the base; panicles chiefly terminal*, finally exserted, 0.5–4 dm. long, with ascending *remote branches interruptedly floriferous nearly to base*, the appressed branchlets with *arching subsecund ovoid-lanceolate spikelets* 3–4 *mm. long;* lateral panicles, when developed, partly included; 1st glume one-third to half the length of spikelet, 3–5-nerved; *spikelet above* 1st *glume oblique; 2nd glume and sterile lemma* subequal, 5–7-nerved, *prolonged beyond the fruit as beaks;* fruit 1.8–2.2 mm. long. — Swamps, sloughs, low woods, etc., Fla. to Tex., n. to N.J., e. Pa., W.Va., s. O., s. Ind., Ill., Mo. and se. Kans. Late June–Oct. Passing by various transitions to

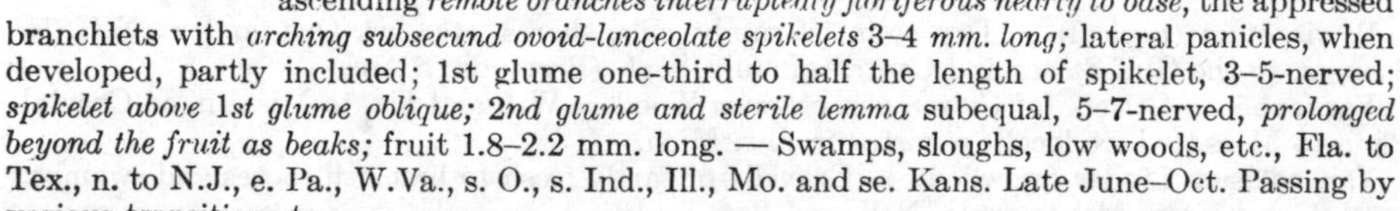

Var. **rhizomàtum** (Hitchc. & Chase) Fern. (with rhizomes). — Often more slender, with 1–3 slender culms up to 1 m. high and 1.5–5 mm. thick at base, the nodes longer-exserted; *rhizomes and stolons more slender and elongate* (*new stolons mostly* 3–12 *cm. long*, 3–7 *mm. thick*); leaves 2–9 mm. broad, firmer; panicles (both terminal and lateral) longer-exserted, 0.5–3 dm. long; *spikelets* 2.4–3 *mm. long.* (*P. rhizomatum* Hitchc. & Chase) — Dry soil of pinelands, etc., n. to e. Va.

16. P. agrostoìdes Spreng. (like *Agrostis*). — *Tufted* from a short caudex, *compressed at the leafy base, glabrous* or pilose near summit of sheath, 0.2–1.8 m. high, often purplish; *sheaths loose, laterally compressed, keeled*, exceeding the internodes; *ligule erose*, 0.5–1 *mm. long; blades* erect, flat, all but the basal elongate, 4–12 *mm. wide*, smooth or scabrous; *panicles at most nodes, the terminal finally exserted, in maturity lanceolate to ovoid*, very *much narrower than to fully as broad as long*, 0.8–3 dm. long, *with the remote lower and middle branches and the lower floriferous branchlets divergent, or all appressed-ascending and compact; spikelets subsecund,* often with 1–8 bristles at base, lanceolate, 1.7–2.5 *mm. long;* 1st glume about half as long, acute, 1–3-nerved; 2nd glume and sterile lemma 5-nerved, acute, the former slightly longer; fruit sessile or on a short (0.2–0.5 mm. long) stipe, 1–1.5 mm. long, 0.5–1 mm. thick. — With three vars.:

Nodes stramineous or pale green, scarcely contrasting in color with the internodes, barely or but slightly constricted upon drying; leaves firm, the larger ones with whitish midrib conspicuous and 0.3–1 mm. broad at base; spikelets ellipsoid to lance-oblong, short-pointed, purplish or bronze (exceptionally green), 0.8–1 mm. in diameter, closely imbricated on the branches or branchlets of the panicle; grain ellipsoid, short-stipitate.

Culms 2–9 dm. high; panicles 0.8–2.5 dm. long, with remote lower and middle branches divergent; spikelets 1.7–2.2 mm. long. . . . . *P. agrostoides* (typical).

Culms up to 1.8 m. high; panicles 0.6–4 dm. long, with ascending to closely appressed branches and branchlets; spikelets 2–2.5 mm. long. . . . . Var. *condensum.*

Nodes darker than and contrasting with the internodes, strongly constricted in drying; leaves submembranaceous; the midrib slender, rarely 0.5 mm. broad; terminal panicles 1.5–4 dm. long, loosely open, the divergent branches and branchlets loosely floriferous; spikelets green or lead-color (rarely purplish), slenderly lance-attentuate, 0.5–0.8 mm. in diameter; grain subcylindric, barely stipitate. . . . . Var. *ramosius.*

P. agrostoìdes (typical). — Sandy or peaty shores and meadows, centr. Me. to w. N.Y., s. to N.C. July–Oct. Fig. 296. — Passing through nondescript colonies into

Var. **condènsum** (Nash) Fern. (condensed), *P. condensum* Nash — Marshes, tidal shores and pond-margins, Fla. to Tex., n. to Mass., s. Ill., Mo. and Kans.

Var. **ramòsius** (Mohr) Fern. (more branched). — Bottomlands, wooded swamps, sloughs, etc., Fla. to e. Tex., n. to Va., s. Ind., s. Ill. and Mo.

296. P. agrostoides.

17. P. stipitàtum Nash (with stipes). — Similar to no. 16, commonly purplish; leaves usually scabrous; *panicles ellipsoid or narrowly ovoid, 1–2 dm. long, stiff, with stiffly ascending branches and short divaricate branchlets densely floriferous to base; spikelets* 2.4–2.8 *mm. long, about* 0.7 *mm. broad; fruit* subcylindric, *with a prominent slender stipe* 0.2–0.4 *mm. long.* — Sandy or peaty swamps, shores and meadows, Ga. to Tex., locally n. to s. Ct., s. N.Y., Pa., W.Va., O., s. Ind. and se. Mo. Aug.–Oct. Fig. 297.

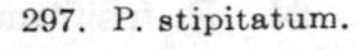
297. P. stipitatum.

18. P. hìans Ell. (gaping). — Tufted from a short caudex, 0.1–1.2 m. high; *culms slender, wiry, erect, or geniculate and rooting at lower nodes;* sheaths keeled, shorter than internodes; blades flat to involute, 2–6 mm. wide, erect; *panicle* terminal, soon exserted, 0.2–3 dm. long, *subsimple or remotely and sparsely long-branched; spikelets* subsessile or short-pedicelled, in short subsecund clusters, ovoid, 1.4–2.4 mm. long, *distended and forced open in maturity by the prolonged, enlarged and indurated sterile palea.* (*Steinchisma hians* (Ell.) Raf.) — Damp shores, roadsides, low woods, etc., Fla. to Tex. and N.M., n. to se. Va., se. Mo. and Okla. June–Oct.

Subgen. Dichanthèlium Hitchc. & Chase

19. P. depauperàtum Muhl. (impoverished). — *Forming dense tussocks* 1.5–4 dm. high; *leaves elongate-linear,* 1–4 (rarely –6) *mm. broad, crowded at the base, firm, erect,* mostly 0.5–1.5 dm. long, *sheaths copiously pilose; culms* filiform, straight or arching, *simple or in the autumnal state forking below,* naked or with 1–3 erect leaves; *vernal panicles lanceolate or narrowly ellipsoid,* 3–10 cm. long, *with few strongly ascending branches* (each with 1–few spikelets); *spikelets* ellipsoid-ovoid, 3–4.5 *mm. long,* glabrous or sparsely pilose, *acutely beaked* by the inrolling of the margins at the tips of 2nd glume and sterile lemma, these strongly 7–9-nerved and *much overtopping the fruit;* fruits 2–2.5 mm. long, 1.3–2 mm. wide; *later panicles reduced to 1–few spikelets in the lowest* and sometimes the other *axils, often hidden by the basal sheaths.* (*P. strictum* Pursh, not R. Br.) — Dry open soil or thin woods, S.C. and Ga. to Tex., n. to s. Que., s. Ont. and Minn., commoner southw. *Vernal panicles* May–Aug.; *autumnal,* July–Oct. Fig. 298.

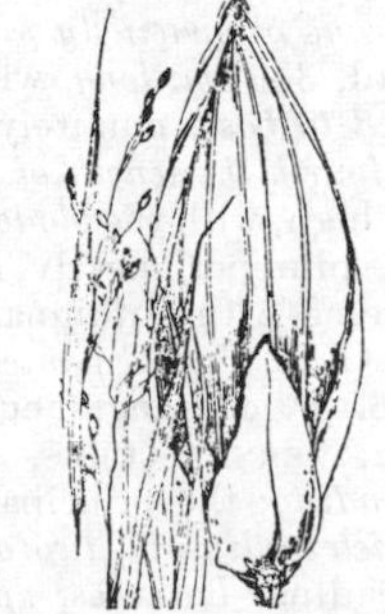
298. P. depauperatum.

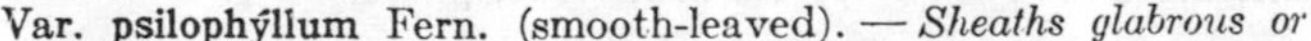
Var. **psilophỳllum** Fern. (smooth-leaved). — *Sheaths glabrous or*

only sparsely pilose. — Common northw., L. St. John, Que., to Wisc., s. to N.S., N.E., Va., upland of N.C. and Tenn., and n. Ind. Forma **cryptóstachys** Fern. (hidden-spiked) has only basal spikelets.

20. P. perlóngum Nash (very elongate). — Resembling no. 19, tussocks looser and smaller; leaf-blades up to 2.5 dm. long; *vernal panicles* 2–8 cm. long, simple or with few erect or slightly spreading-ascending branches, *usually less than one-third as broad as long,* exceptionally broader; *spikelets strongly rounded at summit, pilose; fruit broadly ellipsoid or obovoid,* 2–2.6 mm. long, 1.6–2 *mm. wide.* — Dry prairies, bluffs and thin woods, n. Mich. to s. Man., s. to Ind., Ill., Ark. and Tex. *Vernal panicles* May–Aug.; *autumnal,* July–Oct.

21. P. linearifòlium Scribn. (linear-leaved). — Similar, up to 4.5 dm. high; *sheaths copiously pilose;* blades mostly 1–3.5 dm. long; *terminal panicle* 2–10 cm. long, *when well developed with spreading-ascending branches and one-half to three-fourths* (sometimes only one-third) *as broad as long; spikelets rounded to subacute at tip, 2–2.7 mm. long,* pilose; *fruit narrowly ellipsoid,* 1–1.3 *mm. wide.* — Ga. to Tex., n. to s. Que., s. Ont., n. Mich., Wisc. and Minn., commoner southw. *Vernal panicles* May–July; *autumnal,* July–Oct. Fig. 299.

299. P. linearifolium.

Var. **Wérneri** (Scribn.) Fern. (for its discoverer, William C. Werner, 1851–1935). — *Sheaths glabrous or essentially so.* — Commoner northw., s. Que. to Minn., s. to Va., Ky., Ark. and Tex.

22. P. laxiflòrum Lam. (loosely flowered). — Pale green, forming small *soft tufts* 1–8 dm. high, *without pronounced winter-rosettes;* culms slender, glabrous, with long-bearded nodes, simple or *in the autumnal phase branching from base only; sheaths long-pilose, some* or all of the *hairs* becoming *retrorse;* ligule nearly obsolete; principal *blades elongate-lanceolate, thin* and soft, 0.5–2 dm. long, 3–12 mm. wide, pilose to glabrate or sometimes glabrous, often ciliate or papillose-margined; *vernal panicles* subincluded, finally exserted, usually longer than broad, 3.5–12 cm. long, *lax, the few and remote* spreading-ascending to divergent often sparsely pilose *flexuous branches spikelet-bearing at tip; spikelets ellipsoid-obovoid,* blunt, pilose, 1.8–2.6 *mm. long,* half as broad; 1*st glume broadly deltoid, glabrous, one-fourth to two-fifths the length of the spikelet.* (Including *P. xalapense* HBK.) — Open woods and clearings, Fla. to Tex. and Mex., n. to Md., s. Ind., s. Ill., Mo. and Okla. *Vernal panicles* Apr.–early July; *autumnal,* late June–Sept. (W.I.)

23. P. strigòsum Muhl. (with straight, appressed hairs). — Resembling no. 22, up to 5 dm. high; *culms villous; sheaths and leaf-surfaces long-pilose; blades* chiefly crowded near the base, *lanceolate,* 2.5–8 *cm. long; vernal panicles long-exserted, ellipsoid,* 3–10 cm. long, the *axis and numerous spreading-ascending* capillary *branches long-pilose; spikelets numerous, ellipsoid, rounded at tip, glabrous,* 1.3–1.5 *mm. long, about* 0.7 *mm. broad.* — Sandy woods and bogs, Fla. to La., n. to se. Va. and Tenn. *Vernal panicles* May–July; *autumnal,* July–Sept.

24. P. fusifórme Hitchc. (spindle-shaped). — Resembling no. 25; *primary blades* 2–3 mm. wide, strongly ascending, the principal ones 5–12 *cm. long; spikelets rhomboid, tapering to tip,* 3–4 *mm. long; autumnal blades flat* or barely inrolled at margin. — Dry sandy pinelands and clearings, Fla. to La., n. to se. Va. *Vernal panicles* June–Aug.; *autumnal,* Aug.–Oct. (W.I.; Centr.Am.)

25. P. aciculàre Desv. (bristle-like). — Strongly tufted, *grayish-green,* 2–7.5 dm. high, with winter-rosettes of firm short ovate-lanceolate leaves; vernal culms nearly erect from decumbent bases, *firm,* slender, *appressed-pilose below, glabrous above; lower sheaths pilose, upper glabrous;* ligule less than 1 mm. long; *blades* spreading or ascending, *stiff, taper-pointed, somewhat involute, linear-lanceolate;* the principal ones 4–6 *cm. long,* 2–5 *mm. wide,* 17–23-nerved, *glabrous or sparingly pilose,* with scabrous margin: *vernal panicles* exserted, ovoid, longer than broad, 3–9 *cm. long,* with spreading branches; *spikelets* obovoid, *rounded above, gradually narrowed to base,* minutely pilose, 1.8–2.4 *mm. long,* 1.1–1.4 *mm. broad;* 1st *glume round-deltoid, one-fourth the length of spikelet;* autumnal phase *densely bushy-branched and matted,* up to 5 dm. high, with the *blades strongly involute,* often curved and 1–3 *cm. long,* the autumnal spikelets few, plumper, nearly hidden. (Including *P. chrysopsidifolium* Nash, the taller and looser extreme with autumnal leaves less involute) — Dry sands, Fla. to Tex., n. to s. N.J. (rare) and Okla. *Vernal panicles* May, June; *autumnal,* June–Oct. (W.I.)

26. P. consanguíneum Kunth (closely related). — Similar to no. 25, *coarser,* up to 8 dm. high; vernal culms, *sheaths and blades mostly densely villous; nodes bearded; blades strongly ascending;* the principal ones 7–12 cm. long, 4–8 mm. broad, *the largest* 19–35*-nerved; vernal panicles ellipsoid, two or three times as long as broad,* 2–9 *cm. long, with ascending* (or finally spreading) *branches; spikelets* 2.3–3 *mm. long.* 1.4–1.6 *mm. broad,* pilose; 1*st glume one-third* *length of spikelet;* AUTUMNAL PHASE taller and *less matted* than in no. 25, *the fascicled and fastigiate branches borne from the middle and upper axils, with flat leaves* mostly 3–4 cm. long and 2–3 mm. broad, with few almost hidden spikelets. — Sandy Coastal Plain, n. Fla. to e. Tex., n. to se. Va., and to Tenn. and Ark. *Vernal panicles* May, June; *autumnal,* June–Oct.

27. P. angustifòlium Ell. (narrow-leaved). — Similar to nos. 25 and 26; VERNAL CULMS erect or nearly so, 3–9 *dm. high, the lower internodes puberulent or appressed-pilose, the upper glabrous; nodes not bearded; lower sheaths appressed-pilose, upper glabrous; blades* mostly 5–15 cm. long, the larger 4–8 mm. wide and with 13–31 nerves, stiffly ascending, *glabrous* (or ciliate at base); *vernal panicles* exserted, *ovoid, one-half to four-fifths as broad as long,* 4–10 *cm. long,* open; the few branches subascending, finally divergent, with few spikelets; *spikelets* ellipsoid-obovoid, rounded at tip, attenuate at base, pilose, 2.4–3.3 *mm. long,* 1.4–1.6 *mm. broad;* 1st *glume subacute, one-third length of spikelet;* AUTUMNAL PHASE lower, *stiffly ascending* (sometimes topheavy and leaning), *with stiffly appressed* small *fastigiate flat leaves* hiding the few axillary spikelets. — Sandy pinelands and clearings, Fla. to e. Tex., n. to s. N.J., Tenn. and Ark. *Vernal panicles* May–July; *autumnal,* July–Oct.

28. P. Bicknéllii Nash (for its discoverer, EUGENE PINTARD BICKNELL, 1859–1925). — In small tufts, bluish-green; VERNAL CULMS erect or ascending, slender, 1.5–5 dm. high, *glabrous or lowest internodes puberulent; sheaths glabrous* or lowest sparsely pilose; *ligule almost obsolete; primary cauline blades* 2–4, *ascending, linear-lanceolate, firm, taper-pointed, chiefly* 0.6–1.5 *dm. long and* 4–8 *mm. wide, the largest* 19–29*-nerved, with midrib prominent nearly to tip, glabrous* or basally ciliate; *vernal panicles* broadly ellipsoid or ovoid, 4–10 *cm. long, with* few ascending branches and *scattered long-pedicelled spikelets; spikelets* ellipsoid-obovoid, rounded at tip and at base, pilose or glabrous, 2.3–2.8 *mm. long,* 1.1–1.2 *mm. wide;* 1st *glume* 0.8–1.2 *mm. long;* AUTUMNAL PHASE branching from median nodes, forming bushy crowns with only slightly reduced ascending leaves subtending few-flowered panicles. — Dry thickets and openings, local, sw. Que. to Mich., s. to Ga. and Mo. *Vernal panicles* May–July; *autumnal,* July–Sept. FIG. 300. — Possibly a hybrid of nos. 21 and 35 or related species.

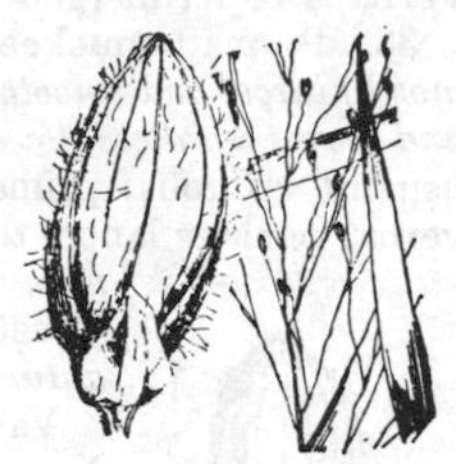
300. P. Bicknellii.

301. P. calliphyllum.

29. P. calliphýllum Ashe (beautiful-leaved). — Similar to no. 28, *coarser,* bright green; vernal culms 2–7 dm. high, glabrous, or the lowest internodes appressed-pilose; *larger primary cauline blades* elongate-lanceolate, thinnish, 1.1–1.5 dm. long, 9–15 *mm. wide,* 47–83*-nerved; vernal panicles* 5–15 *cm. long, one-half to three-fourths as wide,* with stiffly ascending branches; *spikelets* 2.9–3.2 *mm. long;* 1st *glume* 1.2–2.5 *mm. long.* — Dry open woods and clearings, very local, w. Me. to s. Ont., s. to Mass., Ct., Pa., O., Mich. and Mo. *Vernal panicles* June, July; *autumnal,* July–Sept. FIG. 301. — Possibly a hybrid of no. 75 and a narrow-leaved species.

30. P. microcárpon Muhl. (tiny-fruited). — Tufted; VERNAL CULMS slender, 3–10 dm. high, usually with retrorsely *long-bearded* lower (and often upper) *nodes,* otherwise glabrous (rarely pubescent); sheaths (except sometimes the lowest) usually glabrous; *ligule obsolete; blades thin, glabrous* (rarely pubescent), *lanceolate, divergent or somewhat reflexed, the primary ones* 5–15 cm. long and 5–15 *mm. wide; vernal panicles* long-exserted, ovoid, 4.5–12 cm. long, *with many freely forking* ascending to slightly divergent capillary *branches bearing numerous long-pedicelled spikelets; spikelets* ellipsoid-obovoid, green or purple, *glabrous* (very rarely puberulent), 1.5–1.8 *mm. long,* 0.7 *mm. broad;* 1st *glume about one-fourth as long as spikelet;* AUTUMNAL PHASE freely and *repeatedly forking, forming ultimately reclining masses of divergent branchlets and reduced thin blades* (mostly 2–4 cm. long) *with prominently ciliate sheaths and bases* and with small loose axillary and terminal panicles. — Dry to wet woodlands, thickets, openings and swamps, Fla. to Tex., n. to Mass., s. N.Y., W.Va., O., s. Mich., Ill., Mo. and Okla. *Vernal panicles* late May–Sept.; *autumnal,* July–Oct. FIG. 302.

302. P. microcarpon.

31. P. nítidum Lam. (shining; from the grains). — Resembling no. 30 in vernal habit, foliage and bearded nodes; *primary cauline leaves* 5–10 *mm. wide; spikelets pubescent,* 1.8–2 *mm. long,* 1 *mm. broad;* AUTUMNAL PHASE *erect or ascending with fastigiate branchlets bearing* reduced very narrow and *soon involute blades* 1–3 cm. long. — Sandy or peaty soil, Fla. to Tex., locally n. to Va. and se. Mo. *Vernal panicles* mid-May–July; *autumnal,* July–Oct. (W.I.)

32. P. ánnulum Ashe (with a ring — of hairs at nodes). — CULMS solitary or few in small tufts, THE VERNAL erect, 3–9 dm. high, *firm, glabrous,* but *with retrosely bearded nodes;* sheaths pubescent or glabrous, often bearded at summit; *ligule a tuft of hairs less than* 1 *mm. long; blades minutely velutinous or soft-pubescent, the primary cauline ones lanceolate, subcordate at base, divergent,* 5–12 cm. long, 7–13 *mm. wide; vernal panicles* long-exserted, ellipsoid to broadly ovoid, 4–9 cm. long, the flexuous filiform ascending to finally divergent branches *with mostly short-pedicelled to subsessile spikelets; spikelets ellipsoid,* blunt, *subtruncate at base, pubescent,* 2–2.4 *mm. long,* 0.9–1 *mm. wide;* 1st glume one-fourth to two-fifths length of spikelet; AUTUMNAL PHASE *erect, with few short ascending simple or slightly forking axillary branches* with slightly reduced blades and terminal panicles often produced before the falling of the vernal spikelets. — Sandy or rocky sterile open woods and borders of thickets, Ga. to Miss., n., locally, to se. Mass., N.J., e. Pa., Tenn. and se. Mo. *Vernal panicles* June–Aug.; *autumnal,* July–Oct. — Perhaps better merged with the next.

33. P. mattamuskeeténse Ashe (of Mattamuskeet Lake, N.C.). — Resembling no. 32, *mostly larger and smoother;* VERNAL CULMS stoutish, 0.5–1.2 m. high; *nodes or all but the median and lowest beardless; lower and median sheaths and blades velutinous,* the upper usually glabrous; primary cauline blades mostly longer (6–16 cm. long); vernal panicles lance- to ovoid-ellipsoid, up to 12 cm. long; *spikelets* 2.2–2.7 *mm. long.* — Damp sandy or peaty soil, chiefly of the Coastal Plain, e. Mass. to S.C. *Vernal panicles* June–Aug.; *autumnal,* July–Oct. FIG. 303. — Passing freely into

303. P. mattamuskeetense.

Var. **Clùtei** (Nash) Fern. (for its discoverer, WILLARD NELSON CLUTE, 1869–). — Smooth, except sometimes the lowermost nodes, sheaths and blades. (*P. Clutei* Nash) — Similar range.

34. P. boreàle Nash (northern). — CULMS few to many in a stool, THE VERNAL slender, erect or ascending, *glabrous* (or sparsely pilose at base and at nodes), 0.5–7 dm. high; *sheaths glabrous* (rarely sparsely pilose); ligule very short to obsolete; *primary cauline blades glabrous* (rarely sparsely pilose), *lanceolate, subcordate at base, thinnish, ascending* or slightly spreading, 4–14 *cm. long,* 5–16 *mm. broad,* bright green to purplish; *vernal panicles* exserted, ellipsoid or ovoid, 4–12 *cm. long, the loosely ascending flexuous capillary branches with rather scattered long-pedicelled spikelets; spikelets* ellipsoid-obovoid, subacute, *pubescent,* 2–2.2 *mm. long, about* 1 *mm. broad;* 1st glume about one-third length of spikelet; AUTUMNAL PHASE *with few erect simple or subsimple short axillary branches* producing terminal and axillary panicles. — Shores, meadows, fields and thickets, Nfld. to L. St.John, w. to s. Ont. and Minn., s. to N.S., N.E., n. N.J., Pa., n. O. and n. Ind. *Vernal panicles* June–Aug.; *autumnal,* July–Sept. FIG. 304.

304. P. boreale.

35. P. dichótomum L. (forking). — In medium-sized to small tussocks, green to purplish; VERNAL CULMS *very slender,* erect, *glabrous,* 2–7 dm. high, *nodes* usually *beardless; the cartilaginous annulus short, rarely one-third as high as broad; sheaths much shorter than internodes, glabrous* (or with ciliate margins), smooth, or the lowest rarely a little pubescent; ligule nearly obsolete; *primary cauline blades* glabrous, thinnish, narrowly lanceolate, narrowed to base, 5–11 cm. long, 4–10 *mm. broad;* vernal panicles long-exserted, ovoid, lax, 4–11 cm. long, the few flexuous branches spikelet-bearing at tips; *spikelets glabrous, ellipsoid* or narrowly obovoid, *obtuse,* 1.9–2 *mm. long,* 0.9–1 *mm. broad, all or nearly all shorter than their pedicels;* 1st glume one-fourth to one-third length of spikelet; *2nd glume and sterile lemma* faintly nerved, *in maturity often overtopped by the fruit; fruit lustrous, about* 1.8 *mm. long, barely half as broad;* AUTUMNAL PHASE freely branched near the middle, becoming erect to leaning tree-like masses of reduced flat to involute leaves and reduced panicles. — Dry thin woods, thickets and openings, "N.B."; sw. Me. to s.

305. P. dichotomum.

Ont., Mich., Ill. and Mo., s. to Fla. and the Gulf States. *Vernal panicles* May–July; *autumnal*, late June–Nov. Fig. 305. — Passing by numerous transitions to

Var. **barbulàtum** (Michx.) Wood (bearded). — Often coarser; at least *the lower nodes bearded; spikelets* slightly larger (*up to* 2.2 *mm. long*); *fruit mostly covered at maturity.* (*P. barbulatum* Michx.) — Common southward, becoming local northw. to Mass., Ct., s. N.Y., W.Va., O., Mich., Ill. and Mo.

36. P. yadkinénse Ashe (seemingly from Yadkin R., N.C., but cited originally as from Raleigh in the Neuse valley!). — Resembling no. 35, coarser throughout; vernal culms 0.5–1 m. high; *sheaths all or mostly with abundant wart-like spots;* ligule 0.5 mm. long; *primary cauline blades* 6–14 cm. long, 7–15 *mm. wide;* vernal panicles 7–12 cm. long; *spikelets* 2–2.5 *mm. long, acutish,* glabrous or barely puberulent. — Rich or damp woods, thickets, bottomlands and swamps, Ga. to La., n. to N.J., se. Pa., W.Va., Ky., s. Ind. and s. Ill. *Vernal panicles* May–July; *autumnal*, late June–Oct. (Mex.)

37. P. roanokénse Ashe (of Roanoke Island, N.C.). — Resembling nos. 35 and 36, often slightly glaucous; vernal culms erect, 0.5–1 m. high, glabrous; *nodes* smooth; the *cartilaginous annulus elongate, the upper primary ones two-thirds as long as to as long as broad;* sheaths glabrous, not verrucose; *primary cauline blades erect or ascending,* glabrous (or ciliate at base), 6–11 cm. long, attenuate, 3–8 *mm. wide;* vernal panicles 4–10 cm. long, lax, with subascending to divergent branches; *spikelets* ellipsoid-obovoid, glabrous, *about* 2 *mm. long, obtuse,* resembling those of no. 35; *fruit ellipsoid,* 1.4–1.6 *mm. long, more than half as broad.* — Woods or meadows, Fla. to Tex., locally n. to N.Y. and Ct. *Vernal panicles* June, July; *autumnal*, late July–Oct. (W.I.)

38. P. caeruléscens Hack. (bluish). — Tufted, often *glaucous;* VERNAL CULMS slender, 2–8 dm. high, glabrous; *cartilaginous annulus of nodes short,* the upper primary ones rarely one-third as long as broad; sheaths much shorter than internodes, smooth; ligule nearly obsolete; *primary cauline blades ascending or spreading,* often bluish beneath, *linear-lanceolate,* 3–8 cm. long, 3–7 *mm. wide; vernal panicles* short-exserted, *ellipsoid,* 3–7 cm. long, *barely half as wide, with* ascending branches and *short-pedicelled* ellipsoid-obovoid glabrous *spikelets only* 1.5–1.6 *mm. long;* 1st glume one-third length of spikelet; *fruits* 1.4–1.5 *mm. long, more than half as broad,* usually projecting beyond the 2nd glume and sterile lemma; AUTUMNAL PHASE with fascicled branches at the middle and upper nodes, and reduced often involute appressed leaves. — Swamps, pinelands, damp sands and wet woods, Fla. to Miss., locally n. to N.J. *Vernal panicles* May–July; *autumnal*, July–Nov. (W.I.)

39. P. lùcidum Ashe (shining). — Very slender and *weak,* glabrous; VERNAL CULMS at first ascending, *later decumbent or reclining and elongating,* 0.2–1.2 m. long; sheaths short, glabrous; primary cauline *blades thin and flaccid, bright green, lustrous,* glabrous, ascending or widely divergent, 3–7 *cm. long,* 3–6 *mm. broad;* vernal panicles ellipsoid to ovoid, 2.5–7 cm. long, rarely as broad, lax; *spikelets ellipsoid, blunt,* 1.9–2.1 *mm. long, about* 1 *mm. broad, long-pedicelled,* glabrous; 1st glume blunt, two-fifths length of spikelet; 2nd glume and sterile lemma rather prominently nerved; *fruit* usually exserted at tip, *its back semiopaque and obviously cellular-reticulate;* AUTUMNAL PHASE *prostrate, with abundant weak and slender divergent branches,* abundant lustrous small leaves and greatly reduced panicles. — Swamps and wet woods, Fla. to Tex., locally n. to L.I., nw. Ind. and s. Mich. *Vernal panicles* June–Sept.; *autumnal*, July–Nov. (W.I.) Fig. 306.

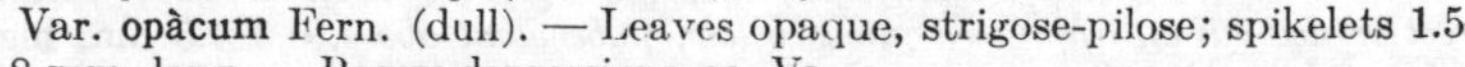

306. P. lucidum.

Var. **opàcum** Fern. (dull). — Leaves opaque, strigose-pilose; spikelets 1.5–1.8 mm. long. — Boggy depressions, se. Va.

40. P. sprètum Schultes (spurned; because Muhlenberg had described it without a name). — VERNAL CULMS solitary or in tufts, from a conspicuous basal rosette, *glabrous* (sometimes pubescent at base), *erect,* rarely a little decumbent, 0.2–1.2 m. high, often purplish, with enlarged glabrous nodes; *sheaths loose,* shorter than internodes, glabrous (or lowest pilose); *ligule a brush of hairs* 2–3 *mm. long; blades of primary cauline leaves* ascending to finally divergent, *subcoriaceous,* lanceolate, *rounded to subcordate at base, glabrous* except for the often ciliate base and the rarely pubescent lower surface, attenuate to tip, 3–14 cm. long, 3–9 *mm. broad; vernal panicles slenderly ellipsoid to slenderly ovoid,* 6–12 *cm. long, one-fourth to half as broad, many or all the numerous lateral spikelets of the ascending glabrous branchlets equaling or exceeding their pedicels; spikelets* ellipsoid, subacute, *pubescent* (rarely glabrous), 1.3–1.6 *mm. long,* 0.7–0.9 *mm. broad,* often purplish; 1st glume one-fourth to one-third length of spikelet; AUTUMNAL PHASE with ascending axillary branches and fascicled branchlets often with

307. P. spretum.

terminal panicles. (Including *P. octonodum* J. G. Sm., with glabrous spikelets, and *P. paucipilum* Nash, with lower leaf-surfaces often minutely pubescent) — Wet peats and sands, Fla. to Tex., n. to N.S., N.E., N.Y., Mich. and Mo. *Vernal panicles* June–Aug., rarely –Oct.; *autumnal*, July–Oct. FIG. 307.

41. P. leucóthrix Nash (white-haired). — Tufted, pale olive-green, often purplish; VERNAL CULMS *slender*, erect, 1–6.5 dm. high, *appressed-pilose*, especially below; *sheaths* shorter than internodes, mostly *with dense appressed-ascending pubescence;* ligule a brush of hairs about 3 mm. long; *blades of primàry cauline leaves firm*, ascending to spreading, narrowly lanceolate, rounded at base, 3–7 *cm. long*, 3–7 *mm. wide, closely short-pubescent or subvelutinous beneath*, glabrous or sparsely villous above; *vernal panicles* long-exserted, 3–8 cm. long, *one-half to three-fourths as broad*, with appressed-pubescent axis and forking branches, *the secondary divisions of the lower branches mostly forking; spikelets* ellipsoid, pubescent, obtuse, 1–1.3 *mm. long*, 0.7 *mm. broad;* 1st glume rounded-deltoid, one-fourth length of spikelet; AUTUMNAL PHASE with long branches from lower and middle nodes, later with appressed fascicled branchlets and only slightly reduced narrow or involute leaves. — Damp sandy pine-barrens, Fla. to e. Tex., locally n. to e. N.C.; s. N.J. *Vernal panicles* May–July; *autumnal*, July–Oct. (W.I.)

42. P. Wrightiànum Scribn. (for its discoverer, CHARLES WRIGHT, 1811–1885). — In small dense tussocks, pale green; VERNAL CULMS *filiform*, erect or ascending, often decumbent at base, 0.5–6 dm. high, *puberulent* at least below the middle, nodes enlarged; sheaths shorter than internodes, glabrous or puberulent; ligule a tuft of hairs 2–3 mm. long; *blades of primary cauline leaves* ascending or spreading, narrowly lanceolate, 1.5–4 *cm. long*, 3–6 *mm. broad*, puberulent or glabrous beneath, pilose above; *vernal panicles* ellipsoid to ovoid, 1.5–6 cm. long, *one-third to two-thirds as broad, with puberulent axis*, freely forking ascending branches *and minute long-pedicelled spikelets; spikelets* plump, ellipsoid, obtuse to subacute, 0.9–1 *mm. long*, 0.5 *mm. broad*, minutely pubescent; 1st glume one-fourth length of spikelet; AUTUMNAL PHASE spreading to decumbent, with numerous ascending fastigiately forking branches from lower and middle axils and terminal as well as reduced axillary panicles. (*P. minutulum* Desv., not Gaudin) — Damp sand or peat, Fla. to Miss., locally n. to L.I. and se. Mass. *Vernal panicles* June–Sept.; *autumnal*, July–Nov. (W.I.) FIG. 308.

308. P. Wrightianum.

43. P. glutinoscàbrum Fern. (glutinous-scurfy). — In *erect* tussocks 7–9 *dm. high;* VERNAL CULMS slender, 1.5–2 mm. thick at base, *with* 5 *elongate internodes covered with mixed cinereous minute puberulence and black wart-like viscid glands;* nodes villous; *primary cauline leaves* lanceolate, up to 7 cm. long and 7–8 *mm. broad*, attenuate, *covered with minute short pilosity mixed with black warts*, the *sheaths* similarly *glutinous and scabrous;* ligule 4–5 mm. long; *vernal panicles* long-exserted, 6–9 *cm. long*, 6–7 cm. thick, *with minutely puberulent rachis and simply forking branches;* AUTUMNAL PHASE *erect*, with short axillary leaf-fascicles and reduced panicles at most 2 cm. long; *spikelets* ellipsoid, subacute, 1.7–1.8 *mm. long*, 0.7–0.8 *mm. thick, short-hispid; lower glume* deltoid-ovate, *subacute*, 0.6–0.7 *mm. long;* upper glume and sterile lemma about equaling lustrous grain. — Boggy spots, local, se. Va. *Vernal panicles* June, July; *autumnal*, July–Sept.

44. P. meridionàle Ashe (southern). — In tussocks, pale green to olivaceous; VERNAL CULMS *filiform, erect or ascending, appressed-pilose below, puberulent above*, 0.5–4.5 dm. high, most leafy below the middle; sheaths appressed-pilose, or the upper puberulent or glabrous; ligule a tuft of hairs usually 1–4 mm. long; *lower and middle blades of primary cauline leaves* 1.5–4 *cm. long*, 2–5 *mm. broad, long-pilose or glabrous* and merely ciliate *on upper surface, glabrous, puberulent or sparsely pilose beneath; vernal panicles very long-exserted, rhomboid*, 1.5–5 *cm. long, with puberulent or glabrous axis and ascending simple or once-forking branches; spikelets* obovoid, rounded at summit, pilose, 1.3–1.6 *mm. long*, 0.8–0.9 *mm. broad*, long-pedicelled; 1*st glume rounded-deltoid, one-fourth to one-third length of spikelet*; fruit broadly ellipsoid; AUTUMNAL PHASE with ascending or erect culms bearing short fastigiate axillary branches of but slightly reduced leaves and reduced axillary panicles. (Including *P. columbianum*, var. *thinium* Hitchc. & Chase) — Dry open habitats and thin woods, sw. N.S.; Mass. to Minn., s. to e. N.C., n. Ga. and n. Ala. *Vernal panicles* June–early Aug.; *autumnal*, July–Nov. FIG. 309. — Passing freely to

309. P. meridionale.

Var. albemarlénse (Ashe) Fern. (of Albemarle Sound, N.C.). — *Coarser; culms* soon spreading and *finally forming depressed mats; pubescence denser; leaves* up to 7 cm. long and 6 mm. broad, *densely short-pubescent above* as well as long-pilose.

(*P. albemarlense* Ashe) — Sandy or argillaceous soil, se. Mass. to N.Y. and W.Va., s. to N.C. and Tenn.; Mich. and nw. Ind. to Minn.

45. P. lanuginòsum Ell. (woolly). — Tufted; VERNAL CULMS mostly slender, erect or ascending, 1.5–10 dm. high. *pubescent at least below*, pubescent to glabrous above; *sheaths shorter than internodes, pilose or villous to glabrous; ligules 3–5 mm. long; larger primary blades* ascending or spreading, lanceolate, rounded and often ciliate at base, acuminate, 3–10 cm. long, 3–12 *mm. broad*, villous or pilose to glabrous; *vernal panicles pyramidal to broad-ovoid*, 3–12 *cm. long*, about as broad, *with flexuous freely forking* and often implexed *ultimately spreading branches and numerous long-pedicelled spikelets; lower branches* often drooping, *their secondary divisions mostly forking; spikelets* obovoid, turgid, *obtuse, pilose*, 1.3–2.1 mm. long, 0.8–1 mm. broad; *1st glume rounded, subtruncate or broadly obtuse, one-sixth to one-third length of spikelet;* AUTUMNAL PHASE fastigiate-branching from middle (and often lower or upper) nodes, the culms often divergent or becoming geniculate or decumbent. — Our most variable species; the following intergrading vars. may be recognized:

*a.*Culms and sheaths velutinous-pilose with subappressed and implexed dense pubescence; blades velutinous, the upper surfaces with close short pubescence and some long trichomes intermixed; spikelets 1.6–1.9 mm. long. . *P. lanuginosum* (typical).

*a.*Culms and sheaths with spreading pilosity or villosity or glabrous; upper surfaces of blades simply pilose or glabrous. . . *b.*

*b.*Axis of panicle spreading-pilose, at least on the lowest internodes; leaf-blades pilose to glabrous above, commonly pubescent beneath; upper sheaths mostly pilose.

Spikelets 1.6–2.1 mm. long; leaf-blades closely short-pilose, sparsely long-pilose or glabrous above. . . . . Var. *fasciculatum.*

Spikelets mostly 1.3–1.5 mm. long; leaf-blades long-pilose above, with hairs mostly 3–6 mm. long. . . . . Var. *implicatum.*

*b.*Axis of panicle glabrous or at most with few appressed hairs; leaf-blades glabrous or sparsely pilose and glabrate above, glabrous or only minutely pubescent beneath; upper sheaths glabrous or somewhat pilose.

Spikelets 1.3–1.6 mm. long. . . . . Var. *Lindheimeri.*

Spikelets mostly 1.6–2 mm. long. . . . . Var. *septentrionale.*

P. lanuginòsum (typical). — Sandy open soil, thin woods, etc., Fla. to Tex., n., rather locally, to s. R.I. *Vernal panicles* late May–July; *autumnal*, July–Nov. FIG. 310.

Var. **fasciculàtum** (Torr.) Fern. (bearing clusters), *P. huachucae* Ashe and *P. huachucae* var. *silvicola* Hitchc. & Chase; *P. tennesseense* Ashe; *P. languidum* Hitchc. & Chase — Dry or moist sterile open soil, thickets and thin woodlands, Fla. to s. Calif., n. to Nfld., s. Que., s. Ont., Minn., S.D., Mont., Ida. and s. B.C. *Vernal panicles* late May–Sept.; *autumnal*, late June–Nov. FIG. 311.

310. P. lanuginosum (typical).

311. P. lanuginosum, v. fasciculatum.

Var. **implicàtum** (Scribn.) Fern. (tangled), *P. implicatum* Scribn. — Similar habitats, Nfld. to s. Ont. and Minn., s. to N.S., s. N.E., s. N.Y., O., Ind., Ill. and Mo. *Vernal panicles* June–Sept.; *autumnal*, July–Nov. FIG. 312.

Var. **Lindheìmeri** (Nash) Fern. (for its discoverer, FERDINAND JACOB LINDHEIMER, 1801–1879), *P. Lindheimeri* Nash — Similar habitats, Fla. to s. Calif., n. to N.S., N.E., N.Y., s. Ont., Mich., Wisc. and Minn. *Vernal panicles* late May–Sept.; *autumnal*, late June–Oct. FIG. 313.

312. P. lanuginosum, v. implicatum.

313. P. lanuginosum, v. Lindheimeri.

Var. **septentrionàle** Fern. (northern). — Similar habitats, N.S. and N.B. to Man., s. to Va., Ky., Mo. and Kans. *Vernal panicles* late May–Aug.; *autumnal*, July–Oct.

46. P. aubúrne Ashe (of Auburn, Ala.). — Resembling the velvety type of no. 45; VERNAL CULMS *grayish-velutinous*, erect or geniculate, *soon becoming depressed*, 2–6 dm. high; sheaths silky-velutinous; *blades of vernal*

314. P. auburne.

culms 3–7 cm. long, 3–5 *mm. broad, ascending, velutinous and sericeous; vernal panicles* short-exserted, ovoid, slightly longer than broad, 2–5 *cm. long,* with *velutinous and sericeous axis and* flexuous ascending to spreading branches *mostly with unforked branchlets;* spikelets long-pedicelled, obovoid, pubescent, 1.3–1.4 mm. long, 0.8–0.9 mm. broad; 1*st glume one-third to half length* of *spikelet, acute;* AUTUMNAL PHASE prostrate or reclining, diffusely branched, forming depressed mats with up-curving branches. — Open sands and sandy open woods, Fla. to La. and Ark., locally n. on Coastal Plain to L.I.; nw. Ind. *Vernal panicles* late May–July; *autumnal,* late June–Nov. FIG. 314.

47. P. praecòcius Hitchc. & Chase (early maturing). — Tufted; VERNAL CULMS slender, ascending or spreading, *commonly geniculate,* 1–5 dm. high, *at least the lower internodes softly villous with long and weak more or less divergent hairs; sheaths much shorter than internodes, spreading-villous;* ligules 3–4 mm. long; *primary cauline blades* firm, *linear-lanceolate, strongly ascending,* 5–9 cm. long, 4–6 mm. broad, *villous with soft hairs* 4–5 *mm. long; vernal panicles* broadly ovoid, 4–6 cm. long, *with* pilose axis and *loosely ascending to spreading* flexuous branches with very *long-pedicelled spikelets; spikelets* obovoid, obtuse, pubescent, 1.8–1.9 *mm. long, about* 1 *mm. broad;* 1st glume one-third to half length of spikelet, elongate-deltoid; AUTUMNAL STATE with ascending sparsely leafy branches of scarcely reduced blades, the terminal panicles often deciduous. — Dry prairies, clearings and open woods, s. Ont. and Mich. to Minn. and Neb., s. to Ill., Mo. and Tex. *Vernal panicles* late May–July; *autumnal,* June–Sept.

48. P. subvillòsum Ashe (somewhat hairy). — Rather densely tufted; VERNAL CULMS erect or ascending, slender, 0.5–5 dm. high, *appressed-pilose, densely so below, with short lower internodes; sheaths mostly equaling or exceeding the internodes,* only the upper ones shorter, appressed-pilose; ligules about 3 mm. long; *primary cauline blades firm, erect or ascending,* narrowly lance-attenuate, 2–7 cm. long, 3–7 mm. wide, *sparsely long-pilose or glabrous above,* short-pilose to glabrous beneath, *the hairs of the upper surface appressed* and 3–5 *mm. long; vernal panicles* long-exserted, *ellipsoid to rhomboid-pyramidal,* 1.5–5.5 *cm. long, one-half to two-thirds as broad, with appressed-pilose* to glabrate *axis and ascending* to barely divergent forking *branches; spikelets* obovoid, obtuse, pubescent, 1.8–2 *mm. long,* 0.9 *mm. broad;* 1*st glume* of young spikelet prolonged, acutish, *as long as or longer than broad, one-third to half length of spikelet;* AUTUMNAL STATE with short appressed branches with scarcely reduced foliage from the lower axils. — Dry sandy soil, Anticosti I. to n. Sask., s. to N.S., s. N.E., L.I., ne. Pa., nw. Ind. and Mo. *Vernal panicles* June–Aug.; *autumnal,* July–Nov. FIG. 315.

315. P. subvillosum.

49. P. Bénneri Fern. (for its discoverer, WALTER MACKINNETT BENNER, 1888–). — Habitally resembling no. 48; *vernal culms* erect, 1.7–3.5 *dm. high,* the *internodes* up to 7.5 cm. long and *long-pilose with ascending hairs; leaves lanceolate,* firm, strongly ascending, *glabrous* on both faces or sparsely and minutely pilose beneath, 3.5–6.5 *cm. long* and 5–8 mm. wide, their *sheaths spreading-hirsute with strongly pustular-based hairs* 1–1.5 *mm. long; vernal panicles* short-exserted *on peduncles only* 0.5–4 *cm. long,* ellipsoid-ovoid, 2.5–6 cm. long, 1.5–4 cm. broad, the *rachis spreading-hirtellous below and with scattered long divergent villi,* the spreading-ascending branches with subsimple branchlets; *pedicels glabrous,* 2–6 mm. long; pubescent *spikelets* as in nos. 48 and 50, 2.2–2.6 *mm. long,* 1.2–1.4 mm. broad; 1*st glume* very short, *deltoid-rotund* and *only* 0.5–0.8 *mm. long, one-fifth to one-fourth length of spikelet.* — Old fields, Hunterdon Co., N.J. *Vernal panicles* late May, June.

50. P. villosíssimum Nash (very long-hairy). — Pale green, tufted; VERNAL CULMS rather slender, 2–7.5 dm. high, erect or ascending; the internodes papillose-villous or pilose, rarely glabrous; nodes long-villous; *sheaths shorter than internodes,* villous, pilose or rarely glabrous; ligule a brush of hairs 2–5 mm. long; *blades* of vernal culms *firm,* ascending or spreading, 6–11 cm. long, 5–10 *mm. wide,* more or less pubescent; *vernal panicles exserted or with base included, ovoid,* 4–10 cm. long, two-thirds to quite as broad, *lax,* with ascending to spreading subsimple to slightly forking stiffish branches *with mostly long-pedicelled spikelets, the secondary branches mostly simple; spikelets* ellipsoid or ellipsoid-obovoid, obtuse, pubescent, 2–2.5 *mm. long,* 1.1–1.2 *mm. broad;* 1st glume one-fourth to two-fifths length of spikelet; AUTUMNAL PHASE ascending to prostrate, with ascending short branches of reduced leaves and mostly hidden panicles. Variable; three vars. have been designated:

Culms and sheaths with abundant horizontally spreading long pubescence. . . *P. villosissimum* (typical).

Culms and sheaths with abundant appressed long pubescence or sparsely short-pilose to glabrous.

Pubescence of culms and sheaths copious, elongate, appressed. . . . . . Var. *pseudopubescens.*

Pubescence of culms very short and sparse or wanting. . . . . Var. *scoparioides.*

P. villosíssimum, typical. — Dry thin woods, thickets and sandy openings, Fla. to Tex., n., somewhat locally, to Mass., N.Y., s. Ont., Mich., Wisc. and Minn. *Vernal panicles* May–July; *autumnal,* June–Oct. (Centr.Am.) Fig. 316.

Var. **pseudopubéscens** (Nash) Fern. (mistaken for *P. pubescens*), *P. pseudopubescens* Nash; *P. ovale* of ed. 7, not Ell. — N. to Ct., L.I., O., Mich., Wisc. and Minn. (Mex.)

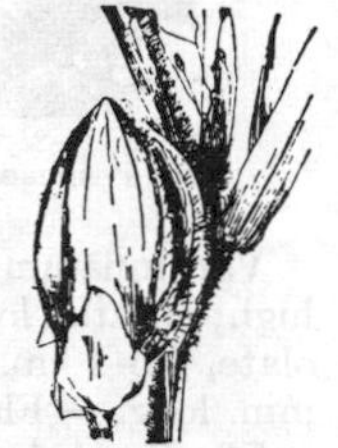

316. P. villosissimum.

Var. **scoparioìdes** (Ashe) Fern. (resembling *P. scoparium*), *P. scoparioides* Ashe — Very local, Mass. and Vt. to Minn., s. to Del., Ind. and Ia. — Possibly a sporadic hybrid of a var. of no. 62 and a small and smoother species, such as a smooth var. of no. 45.

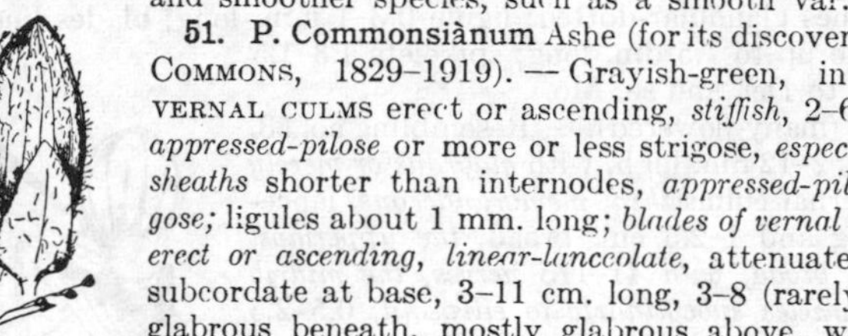

317. P. Commonsianum.

51. P. Commonsiànum Ashe (for its discoverer, Albert Commons, 1829–1919). — Grayish-green, in tussocks; vernal culms erect or ascending, *stiffish,* 2–6 dm. high, *appressed-pilose* or more or less strigose, *especially below; sheaths* shorter than internodes, *appressed-pilose to strigose;* ligules about 1 mm. long; *blades of vernal culms firm, erect or ascending, linear-lanceolate,* attenuate, rounded-subcordate at base, 3–11 cm. long, 3–8 (rarely –11) *mm. broad,* strigose or glabrous beneath, mostly glabrous above, with serrulate mostly inrolled cartilaginous margin; *vernal panicles* long-exserted, *broadly ovoid,* 3–9 *cm. long,* about as broad, *with stiffly spreading branches, mostly simple branchlets and few long-pedicelled spikelets; spikelets* ellipsoid-obovoid, rounded-subtruncate at base, pubescent, 2.5–2.8 *mm. long,* 1.2 *mm. broad;* 1st *glume ovate, two- to three-fifths length of spikelet;* autumnal phase becoming depressed, with stiffly ascending short branches of slightly reduced leaves and reduced panicles. — Sands on or near Coastal Plain, se. Mass. to Fla. — *Vernal panicles* May–July; *autumnal,* July–Nov. Fig. 317.

Var. **Addisònii** (Nash) Fern. (in honor of Addison Brown, 1830–1913). — Slightly smaller; *spikelets* 2–2.5 *mm. long,* 1.1 *mm. broad.* (*P. Addisonii* Nash) — Sands, se. Mass. to S.C., locally inland to centr. N.Y.; nw. Ind.

52. P. columbiànum Scribn. (of the District of Columbia). — Grayish- or bluish-green or purplish, in tussocks; vernal culms *slender, somewhat wiry,* erect, ascending or spreading, 1–9 dm. high, *crisp-puberulent or minutely pilose, the lower* and sometimes all the *internodes also appressed-pilose; sheaths* mostly *shorter than internodes, puberulent,* sometimes also appressed-pilose; *ligule* a tuft of hairs 0.7–1.5 *mm. long; primary cauline blades firm,* with slender white cartilaginous margin, ascending, rounded at base, 3–9 cm. long, 3–8 mm. broad, *glabrous above, appressed-puberulent to glabrous beneath; vernal panicles* finally long-exserted, ovoid, 2–8 cm. long, one-half to seven-eighths as broad, *with ascending to spreading branches and primary axis puberulent to glabrous; spikelets* numerous, *the lateral mostly on pedicels only* 1–2.5 *mm. long,* ellipsoid-obovoid, often purplish, pubescent, 1.2– barely 2 mm. long, about 1 mm. broad; 1*st glume ovate, subacute, one-third to half length of spikelet;* autumnal phase becoming decumbent or loosely spreading, with ascending branches of appressed branchlets with leaves but slightly reduced. (Including *P. tsugetorum* Nash; *P. heterophyllum* Bosc, not Spreng.) — Dry or sandy open ground or thin woods, s. Me. to s. Ont. and Wisc., s. to Ga., Tenn. and Ill. *Vernal panicles* late May–July, rarely–Sept.; *autumnal,* July–Nov. Fig. 318.

318. P. columbianum.

Var. **orícola** (Hitchc. & Chase) Fern. (dweller on the shore). — Culms more densely tufted, the *internodes and sheaths more densely appressed-pilose; blades* small, mostly 2–5 *cm. long,* 2–5 *mm. broad,* mostly loosely *long-pilose above, commonly with mixed short and long pubescence beneath; vernal panicles short-exserted,* 1–3.5 *cm. long; spikelets about* 1.5 *mm. long.* (*P. oricola* Hitchc. & Chase) — Open sandy soil, se. Mass. to e. Va. *Vernal panicles* June, July; *autumnal,* July–Nov.

53. P. sphaerocárpon Ell. (spherical-fruited). — In mostly small tufts, pale green; vernal culms radiate, spreading or ascending, *stiffish,* 1.5–6 dm. high, *glabrous except at the appressed-pilose nodes; sheaths exceeding to nearly equaling the shortish internodes, glabrous* except for ciliate margin, *loose at summit; ligule obsolete* or nearly so; *blades* of vernal culms

319. P. sphaerocarpon.

firm, lanceolate, *cordate-clasping*, 3–12 *cm. long*, 0.7–1.5 *cm. broad*, *with cartilaginous harsh margin* and *obscure midrib*, *the uppermost blade with* 25–55 (rarely –67) *nerves; vernal panicles* long-exserted, 3–13 *cm. long*, *from two-thirds to about as broad*, the axis and ascending branches viscid-spotted; *spikelets short-pyriform to subglobose*, *puberulent*, 1.5–1.8 mm. long, 1–1.3 mm. wide; 1*st glume* broad, obtuse, *one-fourth length of spikelet*; AUTUMNAL PHASE depressed, with few short mostly simple branches and winter-rosettes of broad (1–2 cm.) ovate cartilaginous-margined leaves. — Dry open soil and thin woods, e. Mass. and s. Vt. to s. Ont., Mich., s. Ill., Mo. and se. Kans., s. to the Gulf States, Mex. and n. S. Am. *Vernal panicles* late May–Sept; *autumnal*, July–Nov. FIG. 319. Passing to

Var. **inflàtum** (Scribn. & Sm.) Hitchc. (inflated). — More ascending and slender, up to 1 m. high; sheaths looser, sometimes glandular-dotted; ligule 0.3–1 mm. long; blades linear-lanceolate, 0.5–1 cm. wide; panicle up to 1.5 dm. long; spikelets 1.3–1.5 mm. long. — Fla. to Tex., n. to Del. and se. Mo.

54. P. polyánthes Schultes (many-flowered). — Resembling no. 53, coarser; VERNAL CULMS erect, 2–12 dm. high, with *glabrous or merely puberulent nodes; blades* of vernal culms *thin*, *membranaceous*, lanceolate, mostly 1–2.3 dm. long and 1–2.5 cm. broad, *the uppermost* 0.8–2 *dm. long and* 1–3 *cm. broad, with* 41–115 *nerves, the midrib prominent beneath; vernal panicles subcylindric to ellipsoid*, 0.5–2.5 *dm. long*, *when dry only one-fourth to half as broad*, with numerous spikelets along the often fascicled ascending branches; spikelets 1.5–1.6 mm. long; 1*st glume one-third to two-fifths length of spikelet*; AUTUMNAL PHASE erect, simple, or with simple branches from lower or middle nodes. (*P. multiflorum* Ell., not Poir.) — Woods, thickets and damp siliceous openings, Fla. to e. Tex., n., locally, to se. Mass., s. Ct., Pa., W.Va., s.-centr. O., s. Mich., s. Ind., Ill., Mo. and Okla. *Vernal panicles* June–Aug.; *autumnal*, July–Nov. FIG. 320.

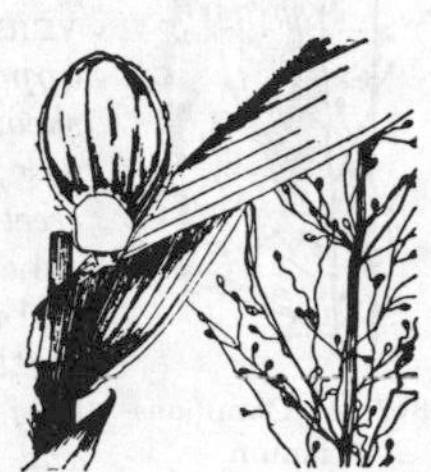
320. P. polyanthes.

55. P. albomarginàtum Nash (white-margined). — Grayish-green to olive, in small tufts; VERNAL CULMS filiform, firm, ascending, 2–7.5 dm. high, *glabrous;* nodes glabrous; *sheaths* mostly much shorter than internodes, *glabrous; ligules less than* 0.5 *mm. long, included; blades* of vernal culms firm, the lower 2.5–5.5 cm. long and 3–6 mm. wide, *with a coarse whitish wire-like margin, glabrous; the uppermost blade greatly reduced, only* 1–2.5 *cm. long and usually less than half as large as the middle and lower blades; vernal panicles* finally long-exserted, pyramidal or broadly ellipsoid, 2.5–7 *cm. long, nearly as broad;* the flexuous branches subascending or spreading, with rather numerous spikelets; *spikelets ellipsoid*, blunt, pubescent, 1.4–1.6 *mm. long*, 0.7–0.8 *mm. thick;* 1*st glume* broad and short, *one-fifth to one-fourth length of spikelet;* AUTUMNAL PHASE with short leafy fascicles of branches from lower axils. — Moist to dry sandy or peaty soil, Fla. to La., n. to Va. and Tenn. *Vernal panicles* late May–July; *autumnal*, July–Oct. (W.I.; Centr. Am.)

56. P. trifòlium Nash (three-leaved; from the few cauline blades). — Vernal phase similar to no. 55, the culms 2–5 dm. high; *leaves nearly uniform*, the blades 2–5.5 cm. long; *autumnal phase with fewer* and small *tufts from middle and upper axils*. — Peaty or sandy soil, Fla. to La., n. to N.J. *Vernal panicles* late May–July; *autumnal*, July–Oct.

57. P. tênue Muhl. (slender). — Closely resembling no. 55; *lower sheaths pilose; lower leaf-surfaces finely pubescent;* panicles 3–5 cm. long; spikelets 1.6–1.7 mm. long, the 1st glume abruptly pointed; autumnal phase slightly branching from middle nodes. — Damp sands, Fla. to e. Va. *Vernal panicles* June, July; *autumnal*, July–Oct. — Might well include nos. 55 and 56 as confluent vars.

58. P. ensifòlium Baldw. (with sword-shaped leaves). — More delicate than no. 55; VERNAL CULMS capillary, erect or leaning, 1–5.5 dm. high, with *prolonged internodes; blades of vernal culms distant, divergent*, 1–4 *cm. long*, 1.5–4 *mm. broad, scarcely or only finely white-margined; vernal panicles* long-exserted, 1.5–4 *cm. long*, nearly as broad, with flexuous branches divergent or the lower deflexed; *spikelets* glabrous or pubescent, 1.3–1.5 *mm. long;* AUTUMNAL PHASE spreading to reclining, with simple elongate branches and terminal as well as axillary panicles; *larger leaves of basal rosette* 1–3.5 *cm. long*, 11–27-*nerved*. — Sphagnous bogs, wet woods and sands, Fla. to La., n. on Coastal Plain to N.J. *Vernal panicles* late May–July; *autumnal*, July–Oct.

59. P. lanceàrium Trin. (a lancer). — Tufted; VERNAL CULMS *stiff, hard*, erect or ascending, often purple-tinged, *minutely ashy-puberulent,* 1.5–5 dm. high; *sheaths* much shorter than internodes, *usually puberulent; ligules nearly obsolete; blades of vernal culms* firm, 2–6 cm. long, 3–8 *mm. broad, glabrous above,* glabrous to puberulent beneath, strongly ciliate at base; *vernal panicles* finally exserted, ellipsoid to pyramidal, 3–6 *cm. long,* two-thirds as broad to nearly as broad, with few spikelets on the flexuous spreading branches; *spikelets ventrally gibbous, pyriform, glabrous or puberulent,* 2–2.1 *mm. long,* 1–1.3 *mm. broad;* 1*st glume broadly obtuse to subtruncate, one-third length of spikelet;* AUTUMNAL PHASE geniculate, with long axillary branches of finally numerous fascicles of reduced flat or involute leaves and reduced panicles. — Pine-barrens and open sands, se. Va. to Fla. and e. Tex. *Vernal panicles* May, June; *autumnal,* June–Nov. (W.I.; Centr. Am.)

Var. **pátulum** (Scribn. & Merr.) Fern. (spreading). — *Weaker;* vernal culms spreading to reclining, *with minutely soft-pilose lower internodes, sheaths and leaf-surfaces and more pubescent spikelets.* (*P. patulum* (Scribn. & Merr.) Hitchc.) — Swamps and low woods, of similar range.

60. P. Wilcoxiànum Vasey (for its discoverer, TIMOTHY ERASTUS WILCOX, 1840–1932). — Tufted, in the vernal state suggesting nos. 19 and 20; VERNAL CULMS *erect, stiffish,* 1–3.5 *dm. high,* appressed- to spreading-*pilose; sheaths mostly overlapping,* appressed-pilose to villous; ligule a dense tuft of hairs about 1 mm. long; *blades of vernal culms subcoriaceous, erect or strongly ascending, linear or narrowly linear-lanceolate; the larger* 0.5–1.5 dm. long, 3–6 *mm. wide, with* 17–27 *nerves; vernal panicles subincluded to short-exserted, ellipsoid-ovoid,* 2–10 cm. long, *about half as wide, with ascending branches; spikelets* few, ellipsoid, pilose, obtuse or rounded at both ends, 2.6–3.2 *mm. long,* 1.2–1.5 *mm. broad;* 1*st glume deltoid-ovate,* subacute, *one-third length of spikelet;* AUTUMNAL PHASE bushy-branched, with overlapping erect reduced leaves and mostly hidden panicles. (Including *P. Deamii* Hitchc. & Chase) — Dry prairies, open places and thin woods, Man. to Colo. and N.M., e. to Wisc., nw. Ind. and ne. Kans. *Vernal panicles* May–July; *autumnal,* June–Sept.

61. P. malacophýllum Nash (soft-leaved). — VERNAL CULMS solitary to few in a tuft, *spreading-pilose,* 2.5–7 *dm. high, with long-bearded nodes; sheaths much shorter than internodes,* less pubescent than culms; ligule 1–1.5 mm. long; *blades of vernal culms* spreading or ascending, *lanceolate,* narrowed to the rounded base, *velvety, the larger* 7–12 *cm. long* and 6–12 *mm. broad; vernal panicles* exserted, *broadly lanceolate to ovoid,* 3–8 cm. long, *one- to three-fourths as broad, with ascending* (finally spreading) *branches with few short-pedicelled lateral spikelets and with shorter basal axillary branches with crowded spikelets; axis and branches of panicle spreading-pilose or villous; spikelets* broadly ellipsoid, *long-pilose,* 2.9–3 *mm. long,* 1.5–1.7 *mm. broad;* 1*st glume ovate, one-third to half length of spikelet;* AUTUMNAL PHASE forming bushy clumps with reduced blades and small panicles. — Sandy or rocky thin woods and openings, Tenn., Mo. and e. Kans. to e. Tex. *Vernal panicles* June, July; *autumnal,* July–Sept.

62. P. oligosánthes Schultes (few-flowered). — Tufted; VERNAL CULMS erect or ascending, 2–8 *dm. high,* glabrous, or pubescent especially below; *primary sheaths* mostly shorter than the internodes, *pilose, hirsute, papillose or glabrous, loose at summit; ligule* a tuft of hairs 1–1.5 *mm. long; blades of vernal panicle firm,* lanceolate, *glabrous or sparsely pilose above,* glabrous, puberulent or finely pubescent beneath, *the larger* 5–14 *cm. long,* 0.5–1.5 *cm. broad and* 33–65-*nerved; primary panicles* finally exserted, *lance-ellipsoid to broadly rhomboid-ovoid,* 4–12 cm. long, *one-half to about as broad,* with loosely ascending to slightly spreading branches; *spikelets ellipsoid-obovoid, sparsely hirsute to glabrous,* 2.9–4 *mm. long,* 1.5–2 *mm. broad;* 1st glume one-third to half length of spikelet; AUTUMNAL PHASE branching, especially from middle and upper nodes, and becoming topheavy or reclining. Highly variable, with the following poorly defined vars.:

Longer branches of vernal panicles with 1–6 remote spikelets 3.5–4 mm. long, the lateral spikelets (along the branches) often on pedicels 0.5–1.5 cm. long; primary blades 5–10 mm. broad, their sheaths appressed-pubescent. . . . *P. oligosanthes* (typical).

Longer branches of vernal panicles with 3–12 spikelets 2.9–3.6 mm. long, the lateral spikelets mostly on pedicels less than 5 mm. long; larger primary leaves 6–15 mm. broad, their sheaths glabrous or spreading-hirsute.
- Spikelets 3.2–3.6 mm. long. . . . . Var. *Scribnerianum.*
- Spikelets 2.9–3 (rarely –3.2) mm. long. . . . . Var. *Helleri.*

P. oligosánthes, typical. — Sandy woods and openings, Fla. to Tex., n. to N.J., Ind., Ill., Mo. and Okla. *Vernal panicles* late May–July; *autumnal,* June–Nov.

Var. **Scribneriànum** (Nash) Fern. (in honor of FRANK LAMSON-SCRIBNER, 1851–1938), *P. Scribnerianum* Nash; *P. macrocarpon* Torr., not Le Conte — Dry thin woods and openings, dry prairies, etc., sw. Me. to s. B.C., s. to Va., mts. of Ga., Tenn., Ark., Tex., Ariz. and Calif. *Vernal panicles* mid-May–July; *autumnal*, July–Nov.

Var. **Hélleri** (Nash) Fern. (for its discoverer, AMOS ARTHUR HELLER, 1867–1944), *P. Helleri* Nash — Open woods and prairies, w. Mo. to La., Tex. and N.M. *Vernal panicles* May, June; *autumnal*, June–Oct.

63. P. Ravenélii Scribn. & Merr. (in honor of HENRY WILLIAM RAVENEL, 1814–1887). — Loosely tufted; VERNAL CULMS 2–7 dm. high, stoutish, erect, *with long appressed pubescence; sheaths* much shorter, lower often equaling or exceeding upper internodes, *pubescent like the culm; ligule an obvious brush exserted 3–4 mm. above the sheath-orifice; blades of vernal culms lanceolate, rounded at base, the larger* 0.7–1.5 *dm. long,* 1–2 *cm. broad,* 51–87*-nerved, densely short-velvety beneath; vernal panicles* subincluded or short-exserted, ellipsoid, *becoming pyramidal* by the final divergence of the branches, 7–12 *cm. long,* becoming as broad or broader, *with few* flexuous *remotely flowered branches; spikelets* ellipsoid, finely and sparsely pubescent, 4–4.3 *mm. long,* 2–2.2 *mm. broad;* 1st *glume ovate, acute, one-third to two-fifths length of spikelet;* AUTUMNAL STATE with short ascending bushy branches of reduced leaves and small panicles from middle and upper axils. — Sandy or rocky thin woods, openings and bottoms, n. Fla. to e. Tex., n. to Del., Md., s. Ky., Mo. and e. Okla. *Vernal panicles* May–July; *autumnal*, June–Oct.

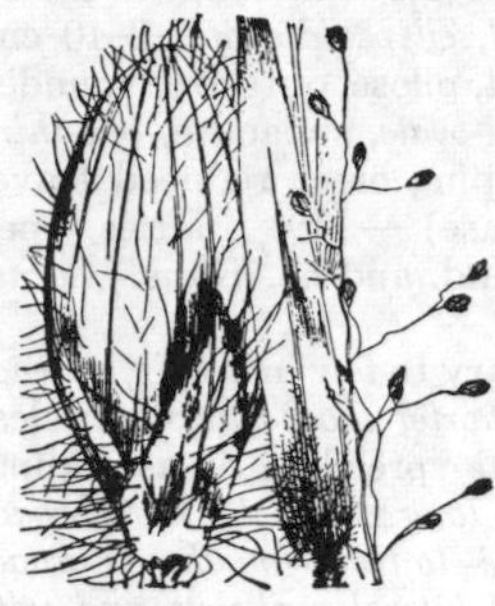
321. P. Leibergii.

64. P. Leibérgii (Vasey) Scribn. (for its discoverer, JOHN BERNHARD LEIBERG, 1853–1913). — Loosely tufted; VERNAL CULMS erect or ascending, *stiff*, geniculate below, 2–9 dm. high, *pilose or scabrous; sheaths* loose, shorter than internodes, *divergently papillose-hispid; ligule nearly obsolete; blades of vernal culms* ascending, *narrowly lanceolate, rounded at base, hispid* especially *beneath,* ciliate toward the base, *the larger* 0.7–1.5 dm. long, 0.7–1.5 *cm. broad; vernal panicles subcylindric to slenderly ovoid or ellipsoid,* 0.5–1.5 dm. long, *less than half as broad, with few* remotely flowered *ascending to erect branches; spikelets* ellipsoid-obovoid, *long-villous with soft hairs,* 3.7–4 *mm. long,* 1.8–2 *mm. broad;* 1st *glume lance-ovate, acutish, half length of spikelet;* AUTUMNAL STATE with few mostly erect branches from middle and lower nodes. — Prairies, meadows and open woods, w. N.Y. to Sask., s. to centr. Pa., O., Ind., Ill., Mo. and e. Kans. *Vernal panicles* mid-May–early Aug.; *autumnal*, June–Sept. FIG. 321.

65. P. xanthophȳsum Gray (with yellow bladders; referring to the plump fruits). — Loosely tufted; VERNAL CULMS radiate-divergent to erect, *firm*, 1.5–6.5 dm. high, *smooth or scabrous; sheaths loose, nearly equaling to slightly exceeding the internodes,* glabrous or papillose-pilose; ligule obsolete or up to 1 mm. long; *blades of vernal culms strongly ascending,* firm, lanceolate, with rounded base; *the upper* 0.7–2 *dm. long,* 0.5–2.2 *cm. broad, glabrous; vernal panicles* becoming long-exserted, *subcylindric to slenderly ellipsoid,* 0.3–1.5 dm. long, *with few stiff erect or strongly ascending branches; spikelets* few, *obovoid, minutely pubescent or glabrous,* 3.7–4 *mm. long,* 2–2.2 *mm. broad;* 1st *glume lance-ovate, acute, half length of spikelet;* AUTUMNAL PHASE with few mostly simple erect branches from lower or middle nodes, often with well developed terminal panicles. — Dry sandy or rocky open soil or thin woods, s. Que. to Sask., s. to N.E., Pa., W.Va., Mich., Wisc. and Minn. *Vernal panicles* June–early Aug.; *autumnal*, July–Sept. FIG. 322.

322. P. xanthophysum.

66. P. scopàrium Lam. (broom-like). — VERNAL CULMS 1–few from a knotted crown, stout, ascending, geniculate below, 0.8–1.3 *m. high, velvety, with a glabrous viscid band below the bearded nodes; sheaths* numerous, much shorter than the internodes, loose, *velvety except at the often viscid summit;* ligule about 1 mm. long; *blades* of vernal culm lanceolate, ascending, spreading or finally sometimes reflexed, *velvety; the larger* 1.2–2 *dm. long,* 1–1.8 *cm. broad;* vernal panicles finally long-exserted, pyramidal or rhomboid, 8–15 cm. long, nearly as broad, with ascending to spreading branches with scattered long-pedicelled spikelets; *spikelets* obovoid, *mucronate at tip,* pubescent, 2.4–2.6 *mm. long,* 1.4–1.5 *mm. broad;* 1st glume ovate, one-fifth to one-fourth length of spikelet; AUTUMNAL PHASE with long repeatedly forking often recurving branches with fascicled branchlets of reduced leaves and small panicles. — Damp thickets, swales and shores, Fla. to Tex., n. to se. Mass., s. N.Y., Ky., s. Mo. and Okla. *Vernal panicles* June–Aug.; *autumnal*, July–Nov. (W.I.)

67. P. aculeàtum Hitchc. & Chase (prickly, from the stiff hairs). — Forming dense tussocks;

VERNAL CULMS 0.5–1 m. high, *scabrous; sheaths* numerous; *the lower stiffly papillose-hispid, with a puberulent ring at summit,* upper glabrous; ligule minute; *blades* of vernal culms *stiff, narrowly lance-attenuate,* rounded at base, *scabrous,* especially above, 1–2 *dm. long,* 9–13 *mm. broad; vernal panicles* ovoid, 8–12 cm. long, about as wide, *with* ascending or finally spreading *scabrous mostly subsimple branches; spikelets* few, chiefly terminal, ellipsoid, *acute,* minutely pubescent, 3 *mm. long;* 1*st glume ovate, acute, one-third length of spikelet;* AUTUMNAL PHASE *with few long exserted* ascending *subsimple or slightly forking branches.* — Swampy woods, se. and e. N.Y. to N.C., rare and little known. *Vernal panicles* June, July; *autumnal,* July–Oct.

68. P. cryptánthum Ashe (with hidden flowers). — Tufted; VERNAL CULMS *very slender,* erect, 7–10 dm. high, *glabrous,* usually with bearded nodes; *sheaths enlarged,* much shorter than the long internodes, *glabrous, with pilose summit,* or the lowest hirsute; ligule minute; *blades* of vernal culms *stiff,* narrowly lance-acuminate, *glabrous,* 1–1.5 *dm. long,* 7–9 *mm. broad; vernal panicles* short-exserted, ovoid, 0.5–1 *dm. long,* about as broad, *with viscid spots on axis and ascending branches; spikelets* broadly *lanceolate, acute, glabrous* or nearly so, 2.2–2.4 *mm. long,* 1 *mm. broad;* 1st glume ovate, acutish, one-fourth to one-third length of spikelet; AUTUMNAL PHASE with few short ascending simple or forking branches of reduced leaves and hidden spikelets. — Savannas and swamps, very local, Tex. to Fla., n. to s. N.J. *Vernal panicles* June, July; *autumnal,* July–Oct.

69. P. scabriúsculum Ell. (somewhat harsh). — VERNAL CULMS erect, 1–1.5 m. high, *scabrous at least near the nodes;* sheaths numerous, usually enlarged below, glabrous or hispid; ligule very short; *blades* of vernal culms *stiff, narrowly lance-attenuate,* glabrous or scabrous, or often pubescent beneath, 1.5–2.5 *dm. long,* 9–15 *mm. broad; vernal panicles* ellipsoid-ovoid, 1–2 dm. long, about two-thirds as broad, *with freely forking* ascending *branches spikelet-bearing nearly to base; spikelets ovoid, acute, glabrous or puberulent,* 2.3–2.6 *mm. long,* 1.1–1.3 *mm. broad;* 1*st glume reniform or suborbicular, one-sixth length of spikelet;* AUTUMNAL PHASE *with short erect densely bushy-forking partially included branches of reduced leaves and often hirsute sheaths.* — Swamps, wet woods and damp sands of the Coastal Plain, Fla. to Tex., n. to N.J. *Vernal panicles* May–July; *autumnal,* July–Nov.

70. P. múndum Fern. (handsome). — In tussocks 0.5–1.4 dm. high; VERNAL CULMS firm, 0.7–3 mm. thick at base, *with* 6–15 *elongate internodes, the lowest internodes cinereous-villous,* the middle and upper ones minutely pilose or puberulent to glabrate; *nodes when young* copiously *long-bearded;* basal rosette with firm glabrous 45–60-nerved broadly lanceolate blades 2–4 cm. long and 8–15 mm. broad; *primary leaves of vernal culm* 6–15, *narrowly lanceolate,* glabrous, 6–15 *cm. long,* 8–13 mm. broad, ciliate at rounded base, attenuate to tip, the sheath glabrous to papillate, the dense *ligule up to* 1 *mm. long; leading panicles* finally exserted, *ellipsoid-ovoid,* 7–12 *cm. long, nearly as broad, the axis spreading-pilose* to glabrate, with freely forking subascending branches; *spikelets* subglobose-*obovoid or ellipsoid, obtuse,* pubescent, 1.8–2.2 *mm. long,* 1–1.2 mm. thick, *long-pedicelled;* the 1*st glume broadly ovate, abruptly tipped, less than quarter length of spikelet;* AUTUMNAL STATE with few ascending branches often terminated by well-formed panicles. — Peaty thickets and bogs, se. Va. and N.C. *Vernal panicles* June, July; *autumnal,* July–Oct.

71. P. recógnitum Fern. (recognized anew). — Differing from no. 70 in its *few* (6 *or* 7) *glabrous internodes* and mostly *glabrous nodes; cauline leaves cordate at base;* the *sheaths horizontally hirsute,* becoming glabrate, with bullate bases of hairs; the *ligule obsolete; rachis of panicle glabrous,* the *branches and pedicels* minutely *barbellate; spikelets ellipsoid, obtuse,* 2.2–2.8 *mm. long,* 1.2 mm. thick; the 1*st glume triangular-ovate and acute, one-third length of spikelet.* — Wet thickets and swamps, R.I. to N.J. and e. Pa. *Vernal panicle* June, July; *autumnal,* July–Oct.

72. P. commutàtum Schultes (changeable). — Tufted; VERNAL CULMS erect (or in swampy woods weak and leaning), stoutish, 2–8 dm. high, *glabrous or sometimes puberulent, with puberulent nodes* and approximate upper internodes; *sheaths* much shorter than the lower internodes, *glabrous, rarely puberulent, with a densely puberulent ring at summit;* ligule nearly obsolete; *blades* of vernal culms *broadly lanceolate, cordate-amplexicaul,* firm to submembranaceous, spreading or ascending, *glabrous or occasionally puberulent,* often ciliate at base; *the larger ones* 5–12 *cm. long,* 1–2.5 *cm. wide,* 53–99-*nerved; vernal panicles* becoming long-exserted, ovoid, 6–12 *cm. long,* about as wide, *with* flexuous *subascending to spreading branches; spikelets ellipsoid, obtuse,* pubescent, 2.2–3.2 *mm. long,* 1.2–1.3 *mm. broad;* 1*st glume ovate,* obtuse, *one-fourth to two-fifths length of spikelet;* AUTUMNAL PHASE with primary panicle often deciduous, the bushy branches borne from the middle and upper nodes. — Thin woods, clearings and open sandy or rocky places, rarely in swamps, common from Fla. to Tex., n., locally, to Mass., N.Y., W.Va., O., Mich., s. Ill. and Mo. *Vernal panicles* May–Aug.; *autumnal,* June–Nov. (W.I.; Mex.)

Var. Joòrii (Vasey) Fern. (for its discoverer, JOSEPH F. JOOR, 1848–1892). — Weaker and more slender, the reclining culms 2–6 dm. long; leaf-blades thin, often curving, 0.7–2 cm. broad; spikelets about 3 mm. long. (*P. Joorii* Vasey) — Low woods, Fla. to Tex., n. to e. Va. and Ark.

Var. Áshei (Pearson) Fern. (for its discoverer, WILLIAM WILLARD ASHE, 1872–1932). — VERNAL CULMS *nearly filiform*, wiry, *puberulent*, 2–5 dm. high; *blades firm, the larger* 4–11 cm. long, 5–11 *mm. broad*, 31–61-*nerved; vernal panicles* 3–8 *cm. long; spikelets* 2.1–2.7 *mm. long;* AUTUMNAL PHASE more bushy-branched. (*P. Ashei* Pearson; *P. umbrosum* LeConte, not Retz.) — Dry thin woods and openings, Mass. to s. Mich., s. to Fla., Ala., Miss., Ark. and Okla. *Vernal panicles* May–July; *autumnal*, July–Nov. — Seems frequently to cross with no. 35, plants of this apparent origin being found from Mass. to S.C.

73. P. mutábile Scribn. & Sm. (changeable). — VERNAL CULMS solitary or in clumps, *grayish-* or *bluish-green*, stiffly erect, 0.4–1.2 m. high, rather slender, *glabrous*, rarely minutely pubescent; *sheaths* much shorter than the lower and middle internodes, *glabrous;* ligule nearly obsolete; *blades* of vernal culms firm, widely spreading to ascending, lanceolate, *narrowed to the* cordate ciliate *base, glabrous, with midrib evanescent near the middle*, 6–17 cm. long, 0.6–1.5 *cm. wide; vernal panicles* ovoid, 0.5–1.5 dm. long, two-thirds to quite as broad, *with* few ascending to spreading branches bearing *scattered long-pedicelled pubescent ellipsoid blunt spikelets* 2.9–3.2 *mm. long* and 1.2–1.3 *mm. broad; 1st glume ovate, subacute, one-third length of spikelet;* AUTUMNAL PHASE with remote erect branches of crowded short branchlets and reduced leaves. — Sandy pinelands and hammocks, Fla. to Miss., n. to e. Va. *Vernal panicles* May–July; *autumnal*, July–Nov.

74. P. clandestìnum L. (hidden). — VERNAL CULMS solitary or in clumps, ascending, 0.6–1.5 *m. high, stout* (*1st node above basal rosette* 3–8 *mm. thick*), *scabrous or harshly pubescent* at least *below the nodes; sheaths* mostly *equaling or exceeding the internodes*, loose, hispid to glabrous, *with a puberulent ring at summit;* ligule 0.5–1 mm. long; *blades of vernal culms* spreading, often reflexed in age, lanceolate, *cordate-amplexicaul; the broader ones* 1–2.5 *dm. long*, 1.7–3.5 *cm. broad*, 51–103-*nerved;* vernal panicles becoming long-exserted, ovoid, 0.7–1.5 dm. long, three-fourths to quite as broad, the forking flexuous fascicled branches ascending to spreading, with frequent shorter basal spikelet-bearing branchlets; *spikelets* ellipsoid-obovoid, pubescent, 2.5–3 *mm. long*, 1.4–1.5 *mm. broad;* 1st glume ovate, one-fourth length of spikelet; AUTUMNAL PHASE with few erect elongate branches with shortened internodes and *overlapping inflated sheaths mostly containing hidden secondary panicles.* — Moist or dry thickets, shores, borders of alluvial woods, etc., w. N.S., centr. Me. and s. Que. to Ia. and e. Kans., s. to n. Fla. and Tex. *Vernal panicles* May–Sept.; *autumnal*, July–Nov. FIG. 323.

323. P. clandestinum.

75. P. latifòlium L. (broad-leaved). — VERNAL CULMS few to several from a knotted crown, erect, 0.2–1 m. high, *glabrous*, or sparsely pubescent at base, *with glabrous* (rarely slightly bearded) *nodes; sheaths* loose, much shorter than the internodes, *mostly glabrous*, with a pilose ring at summit; *leaves of vernal culms broadly lanceolate or lance-ovate*, cordate, *glabrous*, or sparsely pubescent, ciliate at base; *the larger* 0.8–1.8 *dm. long*, 1.5–4 *cm. broad* (*four to seven times as long as broad*), 59–115-*nerved; vernal panicles* loosely rhomboid, 0.5–1.5 dm. long, one-half to three-fourths as broad, *with the few stiffly ascending simple or slightly forking branches remotely spikelet-bearing; spikelets* ellipsoid, short-pedicelled, pubescent, 3.4–3.7 *mm. long*, 1.8–2 *mm. broad;* 1st glume ovate, acutish, one-third to half length of spikelet; AUTUMNAL PHASE with remote elongate branches with shortened upper internodes and partly or wholly included small panicles. — Dry open rich woods, centr. Me. and s. Que. to Minn., s. to N.C., Tenn., Mo. and e. Kans. *Vernal panicles* late May–Aug.; *autumnal*, July–Oct. FIG. 324.

324. P. latifolium.

325. P. Boscii.

76. P. Bóscii Poir. (for its discoverer, LOUIS AUGUSTIN GUILLAUME BOSC, 1759–1828). — Resembling no. 75; differing in culms either glabrous or puberulent; *nodes usually long-bearded; leaf-blades more ovate; the larger* 7–16 *cm. long*, 1.8–3.3 *cm. broad* (*two-*

and-one-half to five times as long as broad); panicle as broad as long, *with puberulent axis and branches; spikelets* 4–4.5 *mm. long,* 2–2.2 mm. broad. — Dry rich woods and thickets, Mass. to Wisc., s. to Fla., Ala., Miss., La. and e. Tex. *Vernal panicles* May–July; *autumnal,* July–Oct. Fig. 325.

Var. **mólle** (Vasey) Hitchc. & Chase (soft to touch). — *Culms downy; sheaths somewhat villous; blades velvety beneath,* more or less pubescent above; axis and branches of panicle somewhat pilose; spikelets more pubescent. — Similar habitats, Ct. to s. Ill. and s. Mo., and southw.

86. SACCIÓLEPIS Nash

Second glume gibbous at the base, 11-nerved, equal to the 3–5-nerved sterile lemma (which incloses a large palea and often a staminate flower), about twice as long as the slightly stipitate fruit; lemma thinner at the apex, the palea free at the tip; spikelets otherwise as in *Panicum.* — Semiaquatic perennials of trop. and warm reg., with narrow spike-like panicles. (Name from the Greek, *saccion, small bag,* and *lepis, scale,* alluding to the saccate second glume.)

1. S. striàta (L.) Nash (with longitudinal lines). — Perennial, stoloniferous; culms fragile, erect from a creeping base, 0.3–2 m. high, branching; sheaths glabrous, or in forma **gíbba** (Ell.) Fern. (humpbacked) the lower and middle and sometimes the upper hirsute; blades 0.4–2 dm. long, about 1 cm. broad, flat, glabrous; panicle lance-cylindric, dense, 0.5–3 dm. long; spikelets 3.5–4 mm. long, lanceolate, acute. — Shallow water, ditches and swamps, Fla. to e. Tex., n. to s. N.J., Tenn. and Okla. Aug.–Oct. (W.I.)

87. ECHINÓCHLOA Beauv.

Spikelets 1-flowered, sometimes a staminate flower below the perfect terminal one, nearly sessile in 1-sided racemes; glumes unequal, often hispid, mucronate; sterile lemma similar and awned from the apex (sometimes mucronate only), inclosing a hyaline palea; fertile lemma and palea chartaceous, blunt to acuminate; margins of the glume inrolled except at the summit, where the palea is not included. — Coarse annuals of trop. and temp. reg., with compressed sheaths, long leaves and terminal panicles of stout racemes. (Name from the Greek *echinos, sea-urchin,* and *chloa, grass,* in allusion to the bristling awns.) *

*a.*Second glume awnless (rarely short-awned in no. 4); spikelets ovoid; fruit ovoid to oval; anthers 0.3–1 mm. long; leaf-sheaths glabrous. . . *b.*

*b.*Coriaceous lemma obtuse to subacute, the tip soft and wilting; spikelets with mostly appressed slender trichomes or smoothish, pustular-based trichomes if present few and marginal. . . . *c.*

*c.*Panicle pale green, linear- to slenderly lance-cylindric, with distant slender and mostly simple branches 1–4 cm. long; bristles and elongate hairs of rachis and nodes of panicle few or none; spikelets 2–2.9 mm. long, glabrous to minutely appressed-pilose. . . . . 1. *E. colonum.*

*c.*Panicle darker green to purple, oblong-cylindric to ovoid or rhomboid, with the longer branches more compound (of crowded glomerules); elongate hairs and bristles often frequent at nodes and along rachis of panicles; spikelets 2.8–3.7 mm. long, usually appressed-pubescent.

Panicle very dense, oblong-cylindric, chocolate-purple, the closely overlapping and ascending branches with incurved tips; spikelets very turgid, awnless; sterile lemmas soft-tipped, with minute and sparse pubescence; lower palea purple. . . . . 2. *E. frumentacea.*

Panicle more open, lance- to broad-ovoid or rhomboid, green to purple, the loosely ascending to spreading branches straight; spikelets scarcely turgid; sterile lemma apiculate and awned, firm-tipped; glumes and sterile lemma more or less strigose-setulose; lower palea whitish. 3. *E. crusgalli.*

*b.*Coriaceous lemma acuminate or subacuminate, with firm tip; spikelets usually echinate, often conspicuously so, the trichomes often pustular-based; setae of nodes and rachis few or wanting (except in var. with spikelets 3.3–4.5 mm. long). . . . . 4. *E. pungens.*

*a.*Second glume awned; spikelets ellipsoid, with slender appressed trichomes not pustular-based; fruit slenderly lanceolate to lance-ellipsoid, acute or acuminate, whitish; anthers 0.6–1.2 mm. long; leaf-sheath hispid, scabrous or smooth. . . . . 5. *E. Walteri.*

1. E. colònum (L.) Link (perhaps a contraction of *colonorum,* of husbandmen). — Culms simple or branching at base, slender, 0.1–1 m. high; leaves with smooth *blades* 3–6 *mm. wide;*

---

* Treatment based chiefly on those of K. M. Wiegand in Rhodora, xxiii. 49–65 (1921) and of N. C. Fassett, ibid. li. 1–3 (1949).

panicle pale green, linear-cylindric to slenderly lanceolate, with distant slender and simple branches 1–3(–4) *cm. long; nodes and rachis of panicle with few or no bristles; spikelets* ovoid, awnless, 2–2.9 *mm. long,* glabrous or sparsely and minutely appressed-pubescent; coriaceous lemma obtuse; anthers 0.7–0.9 mm. long. — Waste places, cult. fields and ditches, Fla. to Mex., n., locally, to N.E., Pa., O., Ill., etc. July–Oct. Forma ZONÀLIS (Guss.) Wieg. (zoned), with leaves cross-banded with purple, is casual. (Adv. and natzd. from Old World)

2. E. FRUMENTÀCEA (Roxb.) Link (grain-bearing), JAPANESE MILLET, BILLION-DOLLAR-GRASS. — Rather coarse, 0.3–1.2 m. high, with smooth lance-attenuate *leaves mostly* 1–2.5 *cm. broad,* the pale midrib broad; *panicle oblong-cylindric, dense, chocolate-purple,* 0.5–1.5 dm. long, 1–4 cm. thick, *of crowded or overlapping* arched-ascending *dense obtuse branches incurved at tip; spikelets turgid, obtuse or soft-tipped, awnless,* glabrous or minutely appressed-pubescent; *lower palea usually purple.* (*E. crusgalli,* var. Wight) — Waste places, roadsides and fields, originally spread from cult. Aug.–Oct. (Introd. from e. Asia)

3. E. CRUSGÁLLI (L.) Beauv. (cock-spur), BARNYARD-GRASS, PIED-DE-COQ (Que.). — Coarse, with culms simple or branching at base, 0.1–1 m. or more high, often depressed or decumbent at base; leaves with smooth or slightly scabrous lanceolate or lance-linear soft blades mostly 0.5–1.5(–2.3) cm. broad; *panicle lanceolate to ovoid or rhomboid,* 0.3–2.5 dm. long, *green to purple-tinged, with* spreading or loosely ascending *straight branches; nodes and rachis of panicle bearing slender bristles; spikelets* ovoid, all or nearly all awnless, or in forma LONGISÈTA (Trin.) Farw. (long-bristled) mostly with long awns 2.8–3.7 *mm. long,* 1.5–2.3 *mm. thick,* either *subglabrous or with appressed slender setiform hairs on the surfaces,* the marginal hairs either slender or pustular-based; sterile lemma apiculate, firm-tipped; *lower palea whitish; tip of coriaceous lemma obtuse, dark, dull, wrinkled and sharply differentiated from the smooth lustrous body, the summit of the body bearing just below the junction a ring of minute* (*microscopic*) *setae;* anthers 0.6–0.85 mm. long. — Waste or cult. ground, etc., a nearly cosmop. weed. June–Nov. (Natzd. from Old World)

4. E. púngens (Poir.) Rydb. (pungent). — Habitally resembling no. 3; leaf-blades scabrous above, 0.5–3 cm. broad; *panicle* 0.5–4.5 dm. long, with widely divergent to ascending branches; *rachis and nodes with few or no bristles* (except in var. *coarctata*); *spikelets* ovoid, 2.5–4.5 *mm. long,* 1.4–2.2 mm. thick, *with glumes and sterile lemma often echinate;* some or all of the trichomes usually with pustular bases; *coriaceous lemma acuminate or subacuminate, with firm tip merging gradually with the lustrous body and without a transitional ring of setae.* (*E. muricata* (Michx.) Fern., not *Panicum muricatum* Retz.) — Highly variable indigenous species; the following vars. with us:

a. Spikelets (excluding long awns) 3.3–4.5 mm. long, 1.8–2.2 mm. thick, often awned; anthers 0.7–0.9 mm. long. . . *b.*
- *b.* Panicle rather open, dark purple to greenish, with lower branches spreading to reflexed; bristles of rachis and nodes sparse or wanting; nerves and margins of 2nd glume and sterile lemma with divergent coarse pustular-based trichomes. . . . . *E. pungens* (typical).
- *b.* Panicle contracted, green, with ascending to appressed branches; pustular-based trichomes wanting or only marginal, the surfaces of glumes and sterile lemma minutely puberulent.
  - Nodes and rachis of panicle with no or few long bristles; spikelets awnless or short-awned; chiefly southwestern. . . . . Var. *ludoviciana.*
  - Nodes and rachis of panicle copiously bearded; spikelets long-awned; southeastern. . . . . Var. *coarctata.*

a. Spikelets 2.5–3.4 mm. long, 1.4–1.8 mm. thick, awnless or very short-awned; anthers 0.3–0.7 mm. long; panicle dense or with ascending and overlapping branches. . . *c.*
- *c.* Panicle green or purple-tinged; 2nd glume and sterile lemma minutely pubescent; pustular-based hairs small and few to wanting. . . . . Var. *Wiegandii.*
- *c.* Panicle commonly deep violet-purple; 2nd glume and sterile lemma with coarse pustular-based divergent hairs on ribs and margin.
  - Spikelets apiculate or short-acuminate; anthers 0.3–0.5 mm. long; nearly transcontinental. . . . . Var. *microstachya.*
  - Spikelets long-acuminate; anthers 0.6–0.7 mm. long; western. . . . . Var. *multiflora.*

E. púngens (typical). — Low grounds, Me. to Minn., s. to N.C., Tenn., Ark. and Okla. Late July–Sept.

Var. **ludoviciàna** (Wieg.) Fern. & Grisc. (of Louisiana), *E. muricata,* var. Wieg. — Miss. to N.M., ne. and n. to e. Va., sw. Pa., W.Va., Tenn. and Mo.

Var. **coarctàta** Fern. & Grisc. (crowded together). — Fla. to se. Va.

Var. **Wiegándii** Fassett (for Karl McKay Wiegand, 1873–1942), *E. muricata*, var. *occidentalis* Wieg.; *E. occidentalis* (Wieg.) Rydb.; *E. pungens*, var. *occidentalis* (Wieg.) Fern. & Grisc.; all as to descr., not as to type. — Me. and N.H. to R.I.; Wisc. to Wash., s. to Ill., Mo., Kans., Tex., N.M., Ariz. and Calif.

Var. **microstáchya** (Wieg.) Fern. & Grisc. (small-spiked), *E. muricata*, var. Wieg.; *E. microstachya* (Wieg.) Rydb. — Sw. Que. and Ont. to Wyo., s. to N.E., Pa., s. Mich., Ill., Mo., Tex. and n. Mex. (W.I.)

Var. **multiflòra** (Wieg.) Fern. & Grisc. (many-flowered), *E. muricata*, var. Wieg. — Minn. to Okla., Tex. and Mex.

5. E. Wálteri (Pursh) Nash (for the early Carolinian botanist, Thomas Walter, ?1740–1789). — Coarse, up to 2.3 m. high; *lower leaf-sheaths hispid or hirsute with pustular-based hairs, or glabrous* in forma **laevigàta** Wieg. (smooth) *or merely scabrous* in forma **brevisèta** Fern. & Grisc. (short-bristled) with green or greenish panicles and awns only 3.5–5 (rarely –20) mm. long; *leaf-blades hard and firm, harshly scabrous*, elongate, 0.5–3 cm. wide; *panicle dense*, lanceolate to lance-ovoid, 0.8–4.5 dm. long, 1.5–10 cm. thick, *arching, purple*, rarely green, *the densely flowered crowded branches appressed-ascending and strongly overlapping; nodes and rachis copiously bearded; spikelets ellipsoid*, 3.5–4.5 mm. long, *strigose with slender nonpustular trichomes; 2nd glume awned;* sterile lemma with awn 1–2.5 cm. long; *fruit whitish, lance-ellipsoid*, acute; *anthers* 0.6–1.2 *mm. long.* — Basic to alkaline marshes, swamps and shallow water, Fla. to Tex., n. near coast to s. N.H., and inland n. to sw. Que., centr. and w. N.Y., s. Ont., Mich., Wisc. and Minn. Aug.–Oct. (W.I.)

88. AMPHICÁRPUM Kunth

Spikelets 1-flowered, of 2 kinds, one kind in a terminal panicle, perfect but not fruitful; the other subterranean, cleistogamous, on slender leafless stems at the base of the culm; the 1st glume of the aërial spikelets variable in size or obsolete; the 2nd and the sterile lemma subequal; lemma and palea indurated, margins of lemma neither hyaline nor inrolled; cleistogamous spikelets much larger, their glumes many-nerved; sterile lemma subrigid; fertile lemma and palea much indurated, acuminate, margins of lemma neither hyaline nor inrolled. — Erect e. Am. annuals or perennials with flat leaves. (Name from the Greek *amphicarpos, doubly fruit-bearing.*)

326. A. Purshii.

1. A. Púrshii Kunth (for its discoverer, Frederick Traugott Pursh, 1774–1820). — Annual; culms erect, branching, 3–8 dm. high; sheaths and blades coarsely hirsute; the latter 1–1.5 dm. long, 0.5–1.5 cm. wide; terminal panicle contracted, up to 2 dm. long; spikelets 4–5 mm. long, ellipsoid; fertile spikelets solitary at tips of subterranean branches, 6–8 mm. long, acuminate. — Damp sandy pinelands, N.J. to e. Md.; e. N.C. and e. Ga. Aug., Sept. Fig. 326.

89. SETÀRIA Beauv. Bristly Foxtail

Spikelets as in *Panicum* but surrounded by few or many persistent awn-like branches which spring from the rachis below the articulation of the spikelets. — Annuals or perennials of trop. and temp. areas, with linear or lanceolate flat leaves and cylindrical spike-like panicles (Name from *seta, a bristle.*) Chaetochloa Scribn.

a. Perennial with firm creeping rhizomes; each spikelet subtended by a fascicle of 8–12 bristles. . . . . 1. *S. geniculata.*
a. Annuals without rhizomes. . . *b.*
  b. Each spikelet subtended by 5–20 bristles. . . . . 2. *S. glauca.*
  b. Each spikelet subtended by but 1–3 bristles (excepting aborted spikelets reduced to bristles). . *c.*
    c. Summit of peduncle retrorse-scabrous; leaf-blades loosely spreading; panicle strongly lobulate, green, 0.5–1.2 cm. thick; bristles subtending spikelets either retrorsely or antrorsely scabrous. . . . . . . 3. *S. verticillata.*
    c. Summit of peduncle upwardly appressed-pubescent; leaf-blades mostly ascending; bristles subtending spikelets antrorsely scabrous. . . *d.*
      d. Mature grains embraced within the falling spikelet, green.
        Panicle only obscurely if at all lobulate, 1–17 cm. long, 0.5–3 cm. thick; grain rugose or cross-wrinkled; leaves 2–17 mm. wide; weeds of dry or disturbed soils.
          Panicle erect or straightish, 0.5–2.3 cm. thick; spikelets 1.6–2.5 mm. long; grain slightly rugose; leaves 2–15 mm. wide, glabrous. 4. *S. viridis.*
          Panicle flexuous or arching, 2–3 cm. thick; spikelets 3 mm. long; grain distinctly cross-wrinkled; leaves 8–17 mm. wide, usually minutely pubescent beneath, strigose above. . . . . . 5. *S. Faberii.*

Panicle conspicuously lobulate, arching or drooping, 0.8–6 dm. long, 2.3–7 cm. thick; grain smooth; leaves 1–4 cm. broad, glabrous; lowland species. . . . . 6. *S. magna.*

d. Mature grains falling free from the spikelet, yellow to red or blackish; panicle lobulate, yellow to purple. . . . . 7. *S. italica.*

1. S. geniculàta (Lam.) Beauv. (bent, like the knee-joint). — *Perennial with short wiry rhizomes or stolons;* culms 1–several, often geniculate at base firm, slender, compressed, 0.4–1 m. high; *leaves* flat, strongly ascending, *mostly overlapping below middle of culm,* long-attenuate, mostly 3–7 cm. wide; peduncle very long; panicle spiciform, dense, yellowish to purplish, 1.5–10 cm. long; *spikelets* 2–3 mm. long, *each subtended by a fascicle of* 8–12 *upwardly scabrous horizontally spreading long bristles; 1st glume about one-third length of spikelet, 2nd longer and with excurrent midrib; sterile lemma equaling the* elliptic-ovate acute striate *transversely rugose fertile lemma.* (Incl. *S. imberbis* R. & S. and many other possibly separable proposed species) — Damp to dry open soil, borders of saline marshes, etc., Fla. to Tex. and Mex., n. to coast of se. Mass. and in interior to Pa., W.Va., Ill., Mo., Kans., N.M. and Calif. July–Oct. (Trop. Am.; temp. S.Am.)

2. S. glaùca (L.) Beauv. (glaucous), Foxtail, Pigeon-Grass, Foin sauvage (Que.). — Annual, with compressed culms often tufted and geniculate at base or erect, 0.1–1.2 m. high; leaves with keeled sheaths, the more or less *loosely spiraling ascending blades* 3–10 mm. wide; *panicle dense, spiciform, cylindric, yellowish,* 1.5–12 cm. long, 0.9–1.4 cm. thick; *bristles* 5–20, 3–8 mm. long; *spikelets* 3 mm. long, *with undulate-rugose fertile lemma.* (Based on *Panicum glaucum* L. (1753) emend. L. (1758); *S. pumila* (Poir.) R. & S.; *S. lutescens* sensu F. T. Hubbard, resting on the invalid *P. lutescens* Weigel) — Weed of cult. or waste places, roadsides, etc., through most of our range and beyond. June–Sept. (Natzd. from Eurasia) Fig. 327.

327. S. glauca.

3. S. verticillàta (L.) Beauv. (whorled). — Annual with weak stems, geniculate and often forking at base, 2–7.5 dm. high; *leaves blue-green,* the *widely divergent* narrowly lanceolate *blades thin; summit of peduncle retrorse-scabrous; panicle interrupted-spiciform,* slender, 2–12 cm. long, 5–12 mm. thick, *green, lobulate with many rounded glomerulate branches, the axis pilose; spikelets about* 2 *mm. long,* with fertile lemma obscurely transverse-rugulose, *each subtended by* 1 relatively short *retrorsely scabrous bristle.* — Waste places, roadsides, etc., s. N.H. to N.D., s., rather locally, to Va., Ala., La. and Tex. July–Oct. (Natzd. from Eurasia)

Var. ambígua (Guss.) Parl. (doubtful). — Axis of panicle merely scabrous; bristles upwardly scabrous. — Less frequent, N.Y. to Ala. (Natzd. from Eurasia)

4. S. víridis (L.) Beauv. (green), Green Foxtail, Bottle-Grass. — Simple or tufted annual, with habit of no. 2; summit of peduncle appressed-pilose; *leaves smooth or glabrous,* 5–15 *mm. wide; panicle spiciform, cylindric, erect,* 1.5–15 *cm. long,* 1–2.3 *cm. thick,* dense, or somewhat open and with branches more evident in forma ramuliflòra D. N. Christiansen (branching-flowered); *spikelets* green, 1.8–2.5 *mm. long,* with *slightly rugose fruit; the green to purplish bristles upwardly scabrous,* spreading-ascending and three or four times as long. — Weed of cult. and other disturbed soil, throughout our range and beyond. June–Oct. (Natzd. from Eurasia)

Var. Weinmánni (R. & S.) Brand (for its discoverer, Johann Anton Weinmann, 1782–1858). — Plant low (0.5–4 dm. high) and slender, often depressed, with narrow (down to 2 mm.) leaves and slender panicles only 1–4 cm. long and 5–10 mm. thick; bristles more ascending and shorter; spikelets 1.6–2.3 mm. long. — Becoming frequent, Nfld. to Ont. and Ia., s. to N.S., N.E., Va., etc. (Natzd. from Eu.)

5. S. Fabèrii Herrm. (named in 1910 for its discoverer, Ernst Faber). — Coarser than no. 4, up to 1.8 m. high; *leaves* 8–17 mm. broad, usually *minutely pubescent beneath and strigose on the upper surface; panicle flexuous or slightly arching,* 1–1.7 dm. long, 2–3 *cm. thick; spikelets* 3 *mm. long; grain transversely and abundantly cross-wrinkled.* — Fields, waste places, roadsides and disturbed soils, e. Mass. to Neb., s. to N.C., Tenn. and Mo. Late July–Oct. (Natzd. from e. Asia)

6. S. mágna Griseb. (large), Giant Foxtail. — Coarse annual *up to* 4 *m. high;* leaves with *blades* mostly 3–7 dm. long and 1–4 *cm. broad;* peduncle appressed-pilose at summit; *panicle conspicuously lobulate with crowded elongate branches, arching or nodding,* 0.8–6 *dm. long,* 2.3–7 *cm. thick,* greenish, becoming stramineous; *grain smooth and lustrous;* bristles very long, ascending. — Swamps, bottomlands and brackish swales, often in disturbed soil, Fla. to Tex., n. to s. N.J. and Md. July–Oct. (W.I.)

7. S. itálica (L.) Beauv. (Italian), Foxtail, German or Hungarian Millet, Millet des

OISEAUX (Que.). — Erect annual, with habit of no. 4, but usually coarser, and with leaves 0.5–3 cm. broad; *panicle lobulate-spiciform*, dense, erect to arching, 1–3 dm. long, 0.7–3.5 *cm. thick; spikelets* 2–3 mm. long, *with green, brown or purplish bristles; grain dropping free from the glumes, yellowish, reddish or blackish*, rugulose to smooth. — A polymorphous species, cult. for many centuries and with a vast number of variations in color of grains and bristles and in size and density of panicle (for a classification of these see F. T. HUBBARD in Am. Journ. Bot. ii. 187, 188 (1915). — Waste places, roadsides, etc., spread from cult. (Introd. from Eurasia)

90. CÉNCHRUS L. SANDBUR. BURGRASS

Spikelets 1-flowered, acuminate, 2–6 together, subtended by a short-pedicelled ovoid or globular involucre of rigid connate spines which is deciduous with the included grains at maturity; glumes shorter than the lemmas; sterile lemma with a hyaline palea; fertile lemma and palea less indurated than in *Panicum*, falcate-acuminate, the lemma not inrolled at the margins. — Annuals or perennials of trop. and temp. reg., our species with simple racemes of spiny burs terminating the culm and branches. (Modification of the old Greek name, *cenchros*, of *Setaria italica*.)

Culms and leafy shoots decumbent, branching from base, annual; larger (upper) leaves with blades (when unfolded) 5–10 mm. broad; spikes included at base or short-exserted, 1.3–2 cm. thick; burs hirsute or villous, with subglobose body and minutely pubescent base.

Leaf-blades membranaceous, flat, smoothish on back; bur hirsute, including the coarser divergent spines 1–1.5 cm. thick; spines terete-subulate, glabrous, or hispid only at base; wide-ranging continental species. . . . . 1. *C. longispinus.*

Leaf-blades coriaceous, promptly infolded, scabrous on back; burs lanate-villous, including the coarser divergent spines 1.3–2 cm. thick; spines flattened, heavily lanate-villous half their length; southern seashore species. 2. *C. tribuloides.*

Culms and leafy shoots in dense ascending or erect clumps from a finally perennial decumbent base; leaves 2–5 mm. broad; terminal (and often lateral) spikes long-exserted, 1–1.5 cm. thick; burs minutely pilose, with ovoid bodies, including the slightly flattened almost subulate glabrous divergent spines 7–10 mm. thick, the prolonged base glabrous. . . . . 3. *C. incertus.*

1. C. longispinus (Hack.) Fern. (long-spined). — *Culms* flattened, branching from base, *decumbent or loosely ascending or spreading*, 1–9 dm. long; *leaves* with slightly inflated sheaths constricted at summit; *the flat* (often becoming involute) *membranous upper blades 4–8 mm. broad*, scabrous above, *nearly smooth beneath*, the hairy ligule 0.5 mm. long; *spikes with included bases or short-exserted*, 1.5–8 cm. long, 1.3–2 *cm. thick; mature burs* stramineous, bronze or purplish, *hirsute, with subglobose body and minutely pubescent base*, including the divergent spines 1–1.5 *cm. thick; spines terete-subulate*, these and the reflexed basal bristles *glabrous, or hispid only at base.* (*C. carolinianus* sensu Hitchc. in ed. 7; *C. pauciflorus* largely of Chase and of Hitchc., not Benth.) — Dry sands of beaches, riverbanks and openings, N.H. to Oreg., s. to N.C., Ky., Mo., Kans., N.M., etc. July–Oct. FIG. 328.

328. C. longispinus.

C. PAUCIFLÒRUS Benth. (few-flowered), native of Mexico and Texas, is more upright and tufted, with leaves only 1.5–4 mm. broad, the summit of the sheath with a chartaceous flange, the spikes 1–5 cm. long and 1–1.5 cm. thick, the stramineous burs 8–12 mm. thick, the larger spines flattened and deltoid at base. It occurs locally about woolen-mills in s. Me.; doubtless elsewhere. (Adv. from the Southwest)

2. C. tribuloides L. (like *Tribulus*, Caltrop), SAND-SPUR. — More robust than no. 1; sheaths loose, more inflated; *blades promptly and strongly involute*, up to 1 cm. broad, *coriaceous, scabrous on back*, often strongly overlapping; *burs* 1.3–2 *cm. thick, lanate-villous, as are the lower halves of the strongly flattened spines.* — Coastal sands, Fla. to La., n. to Staten I., N.Y. July–Oct. (Trop. Am.)

3. C. incértus M. A. Curtis (uncertain). — Culms and leafy shoots *forming dense upright clumps* 0.2–1 m. high from annual and *finally perennial* decumbent bases; *leaves* soft, 2–5 *mm. broad;* terminal (and often lateral) *spikes long-exserted*, 3–10 cm. long and 1–1.5 *cm. thick; burs stramineous, with ovoid body minutely pilose, the prolonged base glabrous, including the slightly flattened lance-subulate smooth divergent spines 7–10 mm. thick;* basal spines and bristles 0–many. — Open sandy soil of Coastal Plain, Fla. to Tex., n. to se. Va., Ark. and Okla. July–Oct.

Tribe XI. ANDROPOGÒNEAE Presl (see p. 189)

a. Rachis slender or merely slenderly clavate, not strongly thickened and cylindric; spikelets not imbedded in it. . *b.*

b. Spikelets uniform, all perfect, in pairs along the rachis.

Rachis continuous, not disarticulating; inflorescence a flabellate panicle of closely overlapping long and slender plumose racemes; both spikelets pedicelled; tall cespitose perennial with narrowly linear attenuate leaves. . . . . 91. *Miscanthus.*

Rachis articulated, the falling segments carrying the ripened spikelets; 1 spikelet of each pair sessile, the other pedicelled.

Racemes very numerous, imbricated in large dense panicles; tall and reed-like erect perennials with elongate linear sessile leaf-blades. . 92. *Erianthus.*

Racemes 1–few, short, greenish, digitate or approximate; low weak and decumbent loosely branched annual with subpetiolate lanceolate leaf-blades. . . . . 93. *Eulalia.*

b. Spikelets dissimilar, one of each group sessile and perfect; each of the other 1 or 2 imperfect, rudimentary or reduced to a pedicel; rachis promptly or tardily disarticulating.

Weak and decumbent annual, the membranous ovate leaf-blade cordate; fertile lemma with dorsal awn. . . . 94. *Arthraxon.*

Stiff and upright, mostly perennial, with linear elongate sessile leaf-blades; fertile lemma terminally awned or awnless.

Racemes solitary, digitately clustered or densely imbricated; rachis several–many-jointed, promptly disarticulating. . . . . 95. *Andropogon.*

Racemes abbreviated, terminating the branches of open panicles; rachis with only 2 or 3 tardily separating joints.

Pedicelled spikelet well developed, often staminate. . . . 96. *Sorgum.*

Pedicelled spikelet reduced to a short hairy pedicel. . . . 97. *Sorghastrum.*

a. Rachis thick, cylindric, articulated, glabrous, with deep excavations embracing the sunken spikelets. . . . . 98. *Manisuris.*

91. MISCÁNTHUS Anderss.

Spikelets 1-flowered, paired on each slender joint of the continuous elongate rachis, uniform, pedicelled. Glumes subequal, membranous, narrow, long-hairy; sterile lemma shorter than the glumes; fertile lemma still smaller, bidentate, bearing a flexuous awn from between the teeth. Tall perennials, nat. of Asia, with long leaves and terminal flabellate panicles of elongate and slender spiciform racemes. (Name from the Greek *mischos, pedicel*, and *anthos, flower.)*

1. M. SINÉNSIS Anderss. (Chinese), "EULALIA". — In dense bushy clumps up to 3 m. high; leaves linear, elongate, crowded on lower half of plant, spreading, with recurving slender tips, sharply serrate; panicle 1–2 dm. long, of very abundant loosely ascending finally whitish plumose racemes 1–2 dm. long. — Roadsides, thickets and old fields, esc. from cult. and natzd. from Fla. to Miss., n. to Mass., N.Y. and O. Sept.–Nov. (Natzd. from e. Asia) — Forms with variegated leaves, var. VARIEGÀTUS Beal (variegated), with leaves longitudinally white-striped, and var. ZEBRÌNUS Beal (like a zebra), with them cross-barred with white, spread to dumps and waste places but apparently do not become natzd.

92. ERIÁNTHUS Michx. WOOLLY BEARDGRASS

Spikelets in pairs, one sessile, the other pedicelled, along the articulate and readily disjointing rachis, both alike, perfect; glumes subequal, firm-membranaceous; the first dorsally flattened, more or less bicarinate, the second keeled above; sterile lemma empty, hyaline, awnless; fertile lemma with an awn usually 1–2.5 cm. long; palea minute, nerveless. — Tall and stout reed-like perennials of warm reg., with elongate flat leaves, racemes crowded in a panicle and commonly clothed with long silky hairs, especially in a tuft around the base of each spikelet (whence the name, from the Greek *erion, wool*, and *anthos, flower*).

a. Awn straight or barely flexuous, directed forward, terete except at the very short base. . . *b.*

b. Panicle linear-lanceolate to lance-oblong, five to twenty times as long as thick; beard at base of spikelets shorter than glumes or wanting, the glumes readily visible in the panicle.

Beard at base of spikelet wanting or of sparse hairs many times shorter than glumes; panicle linear-lanceolate, 1–2 cm. thick, bronze to green, with tightly appressed and erect branches. . . . . 1. *E. strictus.*

Beard 4–6 mm. long, copious; panicle lanceolate to oblong, 3–4 cm. thick.
Cauline leaves and nodes 4–6; the blades 2–10 mm. broad, with 4 or 5 prominent ribs each side of the thick midrib, the uppermost blade greatly reduced and only 4–12 cm. long; panicle lanceolate, bronze to green, 1–1.7 dm. long, soon exserted; glumes strigose-hirtellous, with basal coma 4–5 mm. long; awn 1.6–2 cm. long. . . . 2. *E. coarctatus.*
Cauline leaves and nodes about 10; the blades 1–1.5 cm. broad, with 6–8 prominent ribs each side of midrib, the uppermost sheath including base of oblong stramineous panicle (2–3 dm. long), the uppermost blade 2–2.5 dm. long; glumes glabrous and lustrous, with coma 6 mm. long; awn 8–10 mm. long. . . . . 3. *E. brevibarbis.*
b. Panicle lance-ellipsoid to ovoid or ellipsoid, two to five times as long as thick; beard or at least its longer hairs exceeding the glumes, the latter more hidden in the panicle. . . . . 4. *E. giganteus.*
a. Awn spirally twisted and flat below the middle, the upper half loosely spiraling at base and turned away from the axis of the spikelet.
Beard much longer than and concealing in the panicle the villous blades of the glumes. 5. *E. alopecuroides.*
Beard at most about equaling the blades of the scarcely pubescent glumes. 6. *E. contortus.*

1. E. stríctus Baldw. (erect). — Erect from knotty base; culms glabrous, 1–2 m. high, the 4–6 nodes glabrous or with short deciduous beard; cauline leaves 4–6, stiff, scabrous, lance-linear, with tapering base, 0.5–1 cm. broad, with broad white midrib prominent beneath; *panicle linear-lanceolate,* finally exserted, 1.5–3.5 *dm. long,* 1–2 *cm. thick,* greenish to bronze, *with tightly appressed erect racemes; beard at base of spikelet wanting or of few hairs many times shorter than glume; awn terete except at base, projected straight forward.* — Swampy or damp sandy or peaty woods, thickets and clearings, Fla. to e. Tex., n. to se. Va., Tenn. and se. Mo. July–Oct.

2. E. coarctàtus Fern. (pressed together). — Resembling no. 1; culms 0.7–1.5 m. high, the 4–6 nodes with appressed but deciduous beard; *leaf-blades* 2–10 mm. wide, *with* 3–5 *ribs prominent each side of the coarse midrib; panicle lanceolate,* 1–1.7 *dm. long,* 3–4 *cm. thick,* either long-exserted or *subtended by a slender leaf-blade only* 4–12 *cm. long;* sessile spikelets 6.5–8 mm. long; the *glumes strigose-hirtellous and with a basal coma* 4–5 *mm. long; awn* 1.6–2 *cm. long.* (*E. brevibarbis* sensu Ell. and recent auth., not Michx.) — Damp pinelands, low woods, thickets and swales, Del. to se. Va. Late Sept.–Nov. — A var. farther south.

3. E. brevibárbis Michx. (short-bearded). — Much coarser than no. 2; *culms* up to 2 m. high, *with* 10 *nodes and cauline leaves; leaves* 1–1.5 *cm. broad, with* 6–8 *prominent ribs each side of the midrib; the uppermost leaf prolonged and* 2–2.5 *dm. long,* its sheath *embracing the base of the oblong stramineous* relatively loose *panicle; the latter* 2–3 *dm. long* and 3–4 cm. thick; *glumes of sessile spikelets glabrous* and *lustrous,* the abundant *coma* 6 *mm. long,* the *awn only* 8–10 *mm. long.* — Little known, locally from sw. Ill. to Ark. Sept., early Oct.

4. E. gigantèus (Walt.) Muhl. (gigantic). — Coarse, 1.5–3 *m. high,* with appressed-barbate nodes; summits of sheaths and bases of leaf-blades pilose to glabrate; *blades* 0.75–2.5 *cm. broad; panicle lance-ellipsoid to ovoid or broadly ellipsoid,* 2–6 *dm. long,* 6–15 *cm. thick,* pinkish-brown or purple to bronze or greenish, its peduncle and axis silky-pilose; *beard and longer hairs twice to thrice* the *length of the blades of the glumes; awn terete, projected forward,* 2–2.5 *cm. long.* (*E. saccharoides* Michx.) — Swamps, low woods and swales, Fla. to e. Tex., n. to e. Md. Late Aug.–Oct. FIG. 329. — Passing southw. into

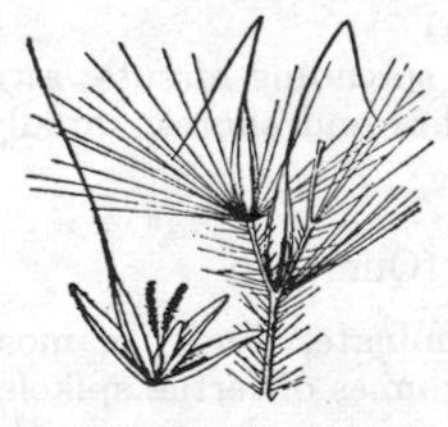

329. E. giganteus.

Var. **compáctus** (Nash) Fern. (compact). — Culms 1.5–2.5 dm. high; leaf-blades 5–10 mm. broad; *panicle* 1–2 *dm. long,* 3.5–8 *cm. thick; beard from slightly longer than to barely twice length of blades of glumes.* (*E. compactus* Nash) — N. C. and Ala., n. to se. N.Y., N.J., e. Pa., D.C., n. Va and Ky.

5. E. alopecuroìdes (L.) Ell. (resembling *Alopecurus*). — Strongly resembling no. 4; *culms glabrous* except at summit; sheaths and blades glabrous, the blades 1.5–2.5 cm. broad; *panicle silvery-whitish to buff or cream-colored,* loose, 1.5–3.5 dm. long; *beard abundant, overtopping the villous blades of glumes; awn flattened and strongly twisted below, the straighter summit divergent.* (*E. divaricatus* sensu Hitchc., not *Andropogon divaricatum* L., basonym) — Sandy or rocky open woods, clearings and thickets, Fla. to Tex., n. to inland Va., very rarely to N.J., and to s. O., s. Ind., s. Ill., Mo. and Okla. Sept., Oct.

6. E. contórtus Ell. (complicated). — Resembling large extremes of no. 2; culms up to 3 m.

high, glabrous except at puberulent or sparsely hairy summit; leaf-blades glabrous, 1–2 cm. broad, scabrous; *panicle* lanceolate or lance-oblong, *brown or purplish*, 1.2–5.5 dm. long; *basal beard shorter than to about equaling the scarcely pubescent blades of the glumes; awns flattened* and twisted below the middle, the summit divergent. — Sandy or peaty dry or moist woods, thickets and clearings, Fla. to e. Tex., n. to Del., Md., Tenn., Ark. and Okla. (July–) Sept.–Nov.

E. Ravénnae (L.) Beauv. (of Ravenna), Ravenna-Grass, Plumegrass, differing from our native species in having 3 (instead of 2) stamens, the smooth culm up to 3 m. high, with narrow (about 1 cm.) leaves and compact erect panicle 2–6 dm. long, the sessile spikelets 4–6 mm. long, the awn only 3–6 mm. long, is cult. and tends to become natzd. from Md. southw. (Introd. from s. Eu.)

93. EULÀLIA Kunth

Spikelets paired, one sessile, the other pedicelled, otherwise alike, on a fragile and disarticulating rachis forming 1–few, when more than 1 digitate, spiciform racemes. 1st glume dorsally flattened, with inflexed margin, 2nd laterally compressed and keeled; sterile lemma hyaline, usually ciliate; fertile lemma awned from tip or awnless, the awn twisted; palea nearly or quite obsolete; grain free. — Annual or perennial grasses nat. of trop. and warm reg. of Old World. (Named in 1829 for *Eulalia Delile*, botanical artist who illustrated Kunth's great volumes on the grasses.) Pollinia Trin.

1. E. vimínea (Trin.) Ktze. (fit for use in basket-work). — Weak reclining or decumbent loosely branched annual with stiff aerial rootlike reflexed processes from the lower nodes, the finally ascending branches up to 1 m. long; leaf-blades lanceolate, 3–8 cm. long, constricted to a subpetiolar base; racemes exserted, 1 or 2(–5), 4–6 cm. long, slender; internodes of common rachis thick, 6–10 mm. long; spikelets green, 4.5–6 mm. long, lanceolate; keel of 2nd glume scabrous or ciliate; fertile lemma awnless. — Waste places, roadsides and shores, local, se. Va. to Ky. and Ala. Oct. (Natzd. from Asia)

Var. variábilis Ktze. (variable). — Fertile lemma with a delicate twisted awn up to 9 mm. long. — Damp roadsides and borders of low woods, Del. and e. Pa. to se. Va. Oct. (Natzd. from Asia)

94. ARTHRÁXON Beauv.

Spikelets 1-flowered, with or without an accompanying pedicelled rudimentary one, sessile in digitate spikes, the rachis fragile and disarticulating. 1st glume blunt, 2nd keeled and blunt or mucronate; sterile lemma hyaline; fertile lemma hyaline or firm, bidentate, bearing a slender awn from the back; grain slenderly linear-terete. — Weak or slender grasses with broad and deeply cordate leaves, nat. of warm reg. of Old World. (Name from the Greek *arthron, joint*, and *axon, axis*, from the articulated rachis.)

1. A. híspidus (Thunb.) Makino (stiffly short-hairy). — Slender and matted, the branching base decumbent, the flowering branches ascending 1–4 dm.; culms filiform, glabrous; sheaths ciliate, smooth or short-hispid; leaf-blade ovate or ovate-lanceolate, 2–6 cm. long, smooth; spikes 2–10 at summit of exserted peduncle or fewer on filiform axillary peduncles, 2–3 cm. long; spikelets lanceolate, green to purple, 4–4.5 mm. long; awn 5–9 mm. long, long-exserted. — Roadsides and damp shores, local, Mo. Sept., Oct. (Adv. from e. Asia)

Var. cryptátherus (Hack.) Houda (with hidden awn). — Sheaths spreading-hirsute; awn minute, not at all or but slightly exserted. — Damp roadsides, ditches and shores, rapidly spreading, Fla. to Ark., n. to N.Y. Sept., Oct. (Natzd. from e. Asia)

95. ANDROPÒGON L. Beardgrass. Barbon (Que.)

Spikelets in pairs (one sessile and perfect; the other pedicelled, staminate, sterile, or most often rudimentary or obsolete) at each joint of the articulate rachis; glumes of fertile spikelet subequal, indurated; the first dorsally flattened, with a strong nerve near each margin, the midnerve faint; second glume keeled above; first lemma empty, hyaline; fertile lemma membranaceous or hyaline, usually awned; palea hyaline, sometimes obsolete. — Tall tufted perennials of trop. and temp. areas; spikes lateral and terminal, the rachis and usually the pedicels long-villous with silky hairs (whence the name, composed of the Greek *aner* (*andr*), *man*, and *pogon, beard*).

a. Racemes solitary on each peduncle of the elongate somewhat virgate inflorescence; the spongy internodes of the rachis and the pedicels enlarged upward, somewhat clavate, with a cup-like depression at summit. § Schizachyrium.

Racemes 3–7 cm. long; pedicels at most nodes paired and truncate or terminated by a sterile rudiment; flowering in autumn. . . . . 1. *A. scoparius.*

Racemes 1–3 cm. long; only the terminal spikelet accompanied by 2 pedicels; the lateral pedicelled spikelet 1, usually fully developed and staminate; flowering from early to late summer. . . . . 2. *A. praematurus.*

a. Racemes 2 or more from a sheath (only exceptionally solitary); internodes and pedicels not club-shaped. . . *b.*

b. Racemes digitately clustered (rarely solitary); joints of rachis filiform. § ARTHOLOPHIS. . . *c.*

c. Pedicelled spikelet staminate (rarely perfect), with both glumes well developed.

Scarcely glaucous; rhizome short and crown-like; awn of sessile spikelet 7–15 mm. long, obviously exserted, geniculate. . . . . 3. *A. Gerardi.*

Glaucous; rhizome elongate; awn of sessile spikelet wanting or up to 6 mm. long, porrect. . . . . 4. *A. Hallii.*

c. Pedicelled spikelet reduced to 1 or 2 glumes, much smaller than the sessile perfect one, or obsolete. . . *d.*

d. Lower leaf-sheaths laterally compressed, keeled and equitant; racemes lax or flexuous, sessile or short-stalked, if long-stalked with peduncles exserted from a fascicle of overlapping and inflated sheaths; stamen 1; sessile spikelet 3–5 mm. long; awn straight and untwisted or, if spiraling at base, with the spiral-column many times shorter than the straight summit. . . *e.*

e. Fascicles of racemes finally exserted from single, either scattered or crowded, spathes.

Peduncles of ultimate spathes rarely 2 cm. long, the spathes not inflated; racemes in 2's (rarely 3's or 4's), with white or silvery hairs; sessile spikelets 3–4 mm. long, with awn 1–2 cm. long. . . 5. *A. virginicus.*

Peduncles 1–6 cm. long, the spathes subinflated; racemes in fascicles of 4–6, with tawny or sordid hairs; sessile spikelets 4.5–5 mm. long, with awn 2–2.5 cm. long. . . . . 6. *A. Mohrii.*

e. Fascicles of racemes included within or slightly exserted from a group of strongly overlapping and inflated sheaths. . . . . . 7. *A. Elliottii.*

d. Lower leaf-sheaths not strongly keeled or equitant; racemes stiff and straight, mostly in pairs terminating long erect peduncles; stamens 3; sessile spikelets 5–6 mm. long, deeply hidden in long silvery-white hairs; awn spiraling to about the middle. . . . . 8. *A. ternarius.*

b. Racemes borne in elongate panicle; joints of rachis strongly flattened. § AMPHILOPHIS. . . . . 9. *A. saccharoides.*

§ SCHIZACHÝRIUM (Nees) Trin.

1. A. scopàrius Michx. (broom-like), BROOM-BEARDGRASS, BROOM, WIREGRASS, BLUESTEM, BUNCHGRASS, PRAIRIE-BEARDGRASS. — In dense or loose *clumps* 0.5–1.5 *m. high,* with stiff and hard culms forming more or less vase-like clusters, the upper half of the culm more or less branching, the nodes often bluish or purplish; *leaves* 3–6 *mm. wide,* often folded, with glabrous or villous sheaths, either green or glaucous; *inflorescence racemose or racemose-paniculate;* the individual *racemes solitary at tips of peduncles or branches,* mostly exserted from the subtending slender and short sheaths; *racemes of spikelets* 3–7 *cm. long,* flexuous or straight; *internodes of rachis and pedicels* of imperfect spikelets pilose, *clavate, with cup-like depression at summit; sessile floret* perfect, 4.5–10 mm. long, *accompanied by* 2 *pedicels, these either truncate or terminated by a sterile rudiment.* — Highly plastic and usually common species, with the following vars. in our area:

a. Joints of rachis beardless for the basal third; the bearding relatively sparse and short, grayish-white; tall mostly southern plants with inflorescence often paniculate; glumes of fertile spikelet 4.5–7 mm. long.

Sheaths copiously villous to glabrous; inflorescence relatively simple, with few appressed-ascending branches. . . . . *A. scoparius* (typical).

Sheaths glabrous or more or less pubescent; inflorescence with very forking fan-like branches, the lateral branches often horizontally divergent. . . Var. *polycladus.*

a. Joints of rachis bearded nearly or quite to base, the bearding often longer and whiter; lower or more slender northern or inland plants. . . *b.*

b. Glumes of fertile spikelet 4.5–6 mm. long; sterile rudiment, including awn, 2.5–5.5 mm. long; bearding comparatively sparse and short. . . . . Var. *frequens.*

b. Glumes of fertile spikelet 6–11 mm. long; sterile rudiment, including awn, 3–10.5 (usually 6 or more) mm. long; bearding abundant and long. . . *c.*

c. Racemes with 5–10 fertile spikelets, usually very flexuous; glumes 7–10 mm. long; rudiment 6.5–10.5 mm. long. . . . . Var. *septentrionalis.*

*c.*Racemes with (8–)11–19 fertile spikelets, rarely flexuous. . . *d.*
*d.*Inflorescence elongate, racemiform, simple or subsimple, its branches little if at all fastigiate; rudiment 3–4.5 mm. long; inland plant. . . Var. *neo-mexicanus.*
*d.*Inflorescence shorter, with abundant fastigiate branching; rudiment 5–8.5 mm. long; plants of Atlantic coast.
Sheaths only slightly compressed, often green; lower cauline blades barely longer than the sheaths; glumes 6–8 mm. long. . . . Var. *ducis.*
Sheaths strongly compressed, usually glaucous; lower cauline blades much longer than sheaths; glumes 8.5–10 mm. long. . . . . . Var. *littoralis.*

A. scopàrius (typical). — Sheaths villous. (Var. *villosissimus* Kearney; *Schizachyrium scoparium* (Michx.) Nash) — Open woods, pinelands and dry clearings, Fla. to Tex., n. on or near Coastal Plain to se. Mass. and inland n. to Ky. and Mo. Sept., Oct. Forma **calvéscens** Fern. (becoming bald) has glabrous sheaths.

Var. **polýcladus** Scribn. & Ball (many-branched). — Dry open woods and clearings, Fla. to Tex. and Mex., n. to N.J., e. Pa. and e. Mo.

Var. **frèquens** F. T. Hubbard (frequent). — Dry sterile soils, w. N.H. and e. Mass. to Minn., s. to Fla., Ala. and Miss. Mid-July–Oct. Fig. 330.

Var. **septentrionàlis** Fern. & Grisc. (northern). — W. N.B. to Ont., s. to n. N.E., ne. Mass., w. Ct., n. N.Y. and n. Mich.

Var. **nèo-mexicànus** (Nash) Hitchc. (New Mexican). — Mont. to Ariz., e. to s. Ont., nw. Pa., O., Ky., Ark. and Tex.; Ottawa R., Que.; Androscoggin R., Me.

330. A. scoparius, v. frequens.

Var. **dùcis** Fern. & Grisc. (of the Duke; for Dukes Co., Mass.). — Sandy soil, Dukes, Nantucket and Barnstable Cos., Mass. July–Oct. — Passing into var. *frequens* and strongly approaching the next.

Var. **littoràlis** (Nash) Hitchc. (of the seashore), *A. littoralis* Nash — Upper borders of sea-beaches and on dunes, s. Ct. to se. Va. Mid-Aug.–Oct. — In extreme development often seeming like a distinct species, but too closely approaching the preceding var., which passes insensibly into var. *frequens.*

2. A. praematùrus Fern. (precocious). — Resembling small extremes of no. 1, glaucous or green to bronze or purplish, 3–6 (–9) dm. high; leaves 2–4 mm. wide, with glabrous sheaths, or in forma **hirtivaginàtus** Fern. (with hirsute sheaths) the lower sheaths hirsute or pilose; *racemes* 1–3 *cm. long, with* 3–7 *internodes* and long-exserted filiform peduncles; *internodes* of flexuous rachis 3–5 *mm. long,* bearded at summit; sessile spikelets 5–7 mm. long; the *pedicelled spikelets solitary* at each node, with well developed glumes 3.5–7 mm. long, *staminate.* — Dry fields and open pineland, S.C. to La., n. to s. N.J. June–Aug.

§ Arthólophis Trin.

3. A. Gerárdi Vitman (for Louis Gérard, French botanist, 1733–1819, who described the species from plants cult. in Provence). — In *coarse* tufts 0.7–1.5 m. high, *from a stout leafy rhizome;* basal leaves firm, scabrous, prolonged, the sheaths glabrous to pilose; *culms* with nodes often bluish, forking above, *with stiffly ascending branches; peduncles* finally long-exserted, stiff, *terminated by* 2–6(–12) *straight ascending spiciform bronze to purplish or green racemes* 3–15 cm. long; *joints of rachis and pedicels* often *ciliate with white hairs* 1–2 *mm. long; sessile spikelet* perfect, usually *scabrous,* lance-subulate, *with geniculate awn* 7–15 *mm. long; pedicelled spikelet* similar but *staminate* (rarely pistillate or perfect), the somewhat flattened and linear-spatulate pedicel often tipped by a white beard 1–3 mm. long (sometimes beardless); the purple anthers conspicuous. (*A. provincialis* Lam., not Retz.; *A. furcatus* Muhl.) — Prairies, swales, shores and dry open ground, Fla. to Tex. and n. Mex., n. to centr. Me., s. Que., s. Ont., Mich., Minn., Man., Sask. and Wyo. June–Sept. Fig. 331. — Passing northw. and westw. into

331. A. Gerardi.

Var. **chrysócomus** (Nash) Fern. (with yellow beard). — Plant *paler green; racemes paler; cilia of rachis-joints and pedicels creamy to yellowish-white,* 2–4 *mm. long; beard at summit of pedicel mostly longer and softer;* the *glumes smoother.* (*A. chrysocomus* Nash) — Sw. Que. to Mont., s. to Ct., n. N.Y., Ind., Ark., Okla., Tex. and N.M.

4. A. Hállii Hack. (for its discoverer, Elihu Hall, 1822–1882). — Resembling var. of no. 3, *with slender stoloniform rhizome; leaves thinner and smoother,* glaucous or pale; racemes 2 or 3 (–5), only 3–8 cm. long; *joints of rachis and pedicels long-bearded,* the hairs white to yellowish; sessile spikelet 8–12 mm. long, the glumes glabrous outside the hispidulous keel; *awn obsolete* *or up to* 6 *mm. long, pointed straight forward.* — Sandhills, draws and dry plains, Mont. to Utah and Ariz., e. to N.D., S.D., e. Ia., Kans., Okla. and Tex. July–Sept.

5. A. virgínicus L. (Virginian), BROOM-SEDGE. — *Leaves* green or glaucous, *with strongly compressed, keeled and equitant sheaths;* culms from slender to stoutish, *with* simple or subsimple and loosely elongate to more densely paniculate or corymbiform inflorescences of *subsessile to short-stalked spathes* 2–6 *cm. long,* these subtending the included or finally exserted racemes; *racemes mostly paired, flexuous,* 1–4.5 cm. long, *with white or silvery hairs; sessile spikelets* 3–4 *mm. long,* the rachis-joints filiform; *awn* straight or spiraling only at base, 1–2 *cm. long;* pedicelled spikelet represented by a subulate scale, or obsolete, the divergent beard longer than the axis of the pedicel. — Highly variable; by some treated as several species:

a. Inflorescence simple or subsimple, prolonged; the scattered spathes exceeding the racemes, mostly 3–5 cm. long; slender plant of dry soils, with flattish leaf-blades 2–5 mm. broad. . . . . *A. virginicus* (typical).

a. Inflorescence forking and branching, from loosely to densely racemose-paniculate to corymbiform. . . . b.

b. Branches of panicle loose, not glomerulate; culms slender; plants mostly of dry soil.

Racemes only 1.5–2 cm. long. . . . . Var. *glaucus.*

Racemes 2.5–4 cm. long. . . . . Var. *tetrastachyus.*

b. Branches of panicle from slightly to strongly glomerulate or corymbiform; culms stout; plants mostly of wet pinelands or swamps. . . c.

c. Inflorescence elongate, not corymbiform; upper leaves shorter than to overtopping culm; spathes and bracteal leaves smooth or but slightly scabrous.

Leaves, especially of basal tufts, heavily white-pruinose; inflorescence rather lax, its upper half only 4–6 cm. in diameter. . . . . . Var. *glaucopsis.*

Leaves green or only slightly glaucous; inflorescences usually dense, their upper halves 0.6–2 dm. in diameter.

Sheaths copiously villous. . . . . Var. *hirsutior.*

Sheaths glabrous or nearly so. . . . . Var. *hirsutior,* forma *tenuispatheus.*

c. Inflorescence strongly corymbiform or subturbinate, often strongly overtopped by the upper leaves; spathes and bracts strongly scabrous. . Var. *abbreviatus.*

A. virgínicus (typical). — Dry open soil, thin woods, etc., Fla. to Tex. and Mex., n. to Mass., N.Y., O., Ind., s. Ill., Mo. and Kans. Late Aug.–Nov. FIG. 332.

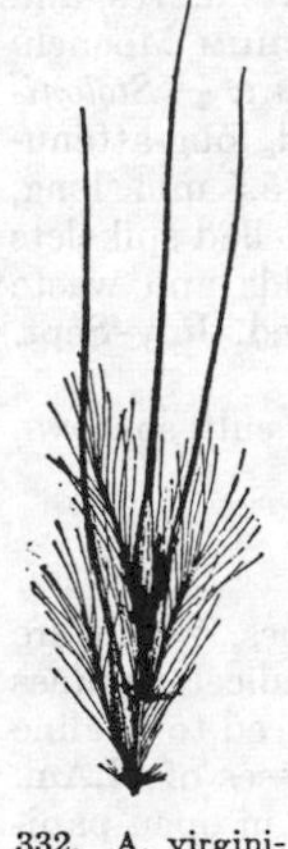

332. A. virginicus.

Var. **glaúcus** Hack. (blue-green; name not always appropriate), *A. capillipes* Nash — Dry sandy pine-barrens, Fla. to Miss., n. to se. Va. Late Aug.–Oct.

Var. **tetrastáchyus** (Ell.) Hack. (four-spiked). — Dry sands, rocks, and pinelands, Fla. to Tex. and Mex., n. to Va., Ky., s. Ill., Mo. and Okla.

Var. **glaucópsis** (Ell.) Hitchc. (with a gray appearance). — Savannas, wet pineland and swamps, Fla. to e. Va.

Var. **hirsùtior** (Hack.) Hitchc. (more hirsute), or forma **tenuispátheus** (Nash) Fern. (thin-spathed), incl. *A. tenuispatheus* Nash — River-swamps, savannas and marshes, Fla. to Mex., n. to s. N.J., Ky., Ark., Okla., s. Nev. and s. Calif. (Trop. Am.)

Var. **abbreviàtus** (Hack.) Fern. & Grisc. (shortened), *A. glomeratus* (Walt.) BSP. — Peaty, boggy or other wet soils, N.C. to se. Mass., s. R.I., L.I., N.J., e. Pa. and D.C. Late Aug.–Oct. — Northward, where only var. *abbreviatus* and typical *A. virginicus* abound, they are wholly distinct. South of our range the former passes by gradual transition through smooth-spathed but habitally identical var. *corymbosus* (Chapm.) Fern. & Grisc. into the smoother form of var. *hirsutior,* thence into var. *tetrastachyus* and through nondescript material to typical *A. virginicus.*

6. A. Mőhrii Hack. (for its discoverer, CHARLES THEODORE MOHR, 1824–1901). — *Leaves* equitant, *with soft-villous sheaths* and scabrous-pubescent firm blades 4–7 mm. broad; culm usually solitary, stout, 0.8–1.5 dm. high, slightly branching above into an open racemose inflorescence; *peduncles* well developed, *up to* 6 *cm. long; spathes subinflated,* 3–6 cm. long; *racemes in fascicles of* 4–6, barely exserted from spathes, *their long hairs sordid or tawny; sessile spikelets* 4.5–5 *mm. long, with awn* 2–2.5 *cm. long;* pedicels terminated by a minute scale or with rudiment wanting. — Wet pine-barrens and bogs, n. Fla. to La.; very locally, se. Va. Sept., Oct.

7. A. Ellióttii Chapm. (for its discoverer, STEPHEN ELLIOTT, 1771–1830). — *Culms tufted,* 0.4–1 m. high, 2.5–4 mm. thick at base; lower sheaths and blades appressed-pubescent to glabrate, the blades 2.5–7 mm. wide, the *upper sheaths greatly enlarged and closely aggregated into an imbricated fascicle,* smooth; the longer of these sheaths 6–12 cm. long, with or without narrow blades and 4–7 mm. thick; *racemes* in 2's (rarely 3's), *included within or exserted from the large sheaths on erect peduncles;* sessile spikelet 3.5–5 mm. long, with geniculate awn 1.5–2.5 cm. long; pedicelled spikelet rudimentary; hairs white. — Dry sterile woods, barrens and fields, Fla. to Tex., n. to N.J., e. Pa., ne. Ky., s. O., s. Ind., s. Ill. and Mo. Sept.–Nov.

Var. **gracílior** Hack. (more slender). — Mostly lower; culms only 1–2.5 mm. thick at base; leaf-blades 1–2.5 mm. broad; larger sheaths 5–8 cm. long, and 2–4 mm. thick. — Fla. to Miss., n. to s. N.J.

8. A. ternàrius Michx. (in threes). — Mostly taller than no. 7; *lower sheaths not strongly keeled nor obviously equitant; fascicles of* 2 (*or* 3) *racemes terminating finally exserted and prolonged erect peduncles;* the densely white-villous *racemes straight,* 3–5 cm. long; *stamens* 3; *sessile spikelets* 5–6 *mm. long, deeply hidden by dense hairs;* first glume scabrous on the sides; *awn spiraling to about the middle,* 1.5–2.5 cm. long. — Dry sandy woods, barrens and openings, Fla. to e. Tex., n. to s. N.J., Del., e. Md., Ky. and s. Mo. Late July–Oct.

Var. **glaucéscens** (Scribn.) Fern. & Grisc. (becoming glaucous). — Glaucous; more slender and smaller; racemes more slender, 4–7 cm. long; 1st glume of sessile spikelet with glabrous sides. (*A. Scribnerianus* Nash) — Sandy pineland, Fla. to se. Va.

§ AMPHÍLOPHIS Trin.

9. A. saccharoìdes Sw. (resembling *Saccharum*). — Culms tufted, simple or branching near base, slender, up to 1.3 m. high; leaves somewhat glaucous, with glabrous blades 4–8 mm. wide; *panicle lanceolate, long-exserted,* 0.5–1.2 dm. long, *composed of appressed-ascending* white-villous *racemes; joints of rachis flattish;* sessile spikelet 3–4 mm. long, usually with *abruptly geniculate twisted awn* 1–1.5 *cm. long; stamens* 3; pedicelled spikelet reduced to a single scale. — Dry prairies, bluffs and draws, chiefly calcareous, Ala. to Mex., n. to s. Mo., Kans., Colo., Ariz. and s. Calif. July–Oct. (Trop. Am.)

96. SÓRGUM Adans. SORGHUM

Similar to *Sorghastrum;* spikelets in pairs, one of them sessile and fertile, the other pedicelled and staminate or sterile; the terminal spikelets in groups of 3, 2 of them pedicelled and empty or staminate; lemma firm. — Tall grasses, nat. of Old World, with elongate leaves, and panicles of finally disarticulating short racemes. (The old oriental name.) SORGHUM Moench.

1. S. HALEPÉNSE (L.) Pers. (of Aleppo), JOHNSON-GRASS or EGYPTIAN MILLET. — *Stoloniferous perennial* with scaly rhizome; culms up to 3 m. high; leaves 1–2 cm. broad, long-attenuate; panicle open, spreading, up to 5 dm. long; sessile spikelet ovoid, silky, 4.5–5.5 mm. long, with an abruptly bent and basally spiralled caducous awn 1–1.5 cm. long; pedicelled spikelets lanceolate. — Cult. southw. and persistent as well as spontaneous in old fields and waste places, locally n. to s. N.E., N.Y., W.Va., O., Ind., Ill. and Ia.; often a bad weed. July–Sept. (Introd. and natzd. from Eurasia)

S. VULGÀRE Pers. (common), SORGHUM or BROOM-CORN, a coarse *annual* much cult. southw., is a casual weed of waste places but hardly persistent. (Introd. from Eurasia)

97. SORGHÀSTRUM Nash

Spikelets sessile at each joint of the slender rachis of the peduncled racemes, which are reduced to 2 or 3 joints, the sterile spikelets reduced (in our species) to hairy pedicels; glumes indurated as in *Andropogon;* sterile lemma thinly hyaline, the fertile lemma reduced to hyaline appendages to the strong awn; palea obsolete. — Perennial grasses of N.Am. (one annual in Afr.), with tall stout culms, the racemes grouped in open panicles. (Named from its resemblance to *Sorgum.*)

333. S. nutans.

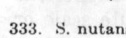

1. S. nùtans (L.) Nash (nodding), INDIAN GRASS, WOOD-GRASS. — *Rhizomatous, the rhizomes scaly;* culms simple, 0.8–2.6 m. high; leaves glaucous, scabrous, elongate, 0.5–1 cm. wide, the sheaths smooth; panicle slenderly ellipsoid, at first loosely expanded, 1–3 (–6) dm. long, contracting and becoming denser and darker in age; *spikelets* lanceolate, 6–8 mm. long, at length drooping, *yellowish- or reddish-brown* and shining, clothed, especially toward base, with fawn-colored hairs; *awn once geniculate,* 1–1.5 *cm. long.* — Dry slopes, prairies and borders of woods, centr. Fla. to Tex. and Mex., n. to centr. Me., s. Que., s. Ont., s. Man., N.D. and Wyo. Aug., Sept. Fig. 333.

2. S. Ellióttii (Mohr) Nash (in honor of Stephen Elliott, scholarly Carolinian botanist, 1771–1830). — More slender, *without scaly rhizomes; panicle looser, chestnut-brown; awn twice geniculate, 2.5–3.5 cm. long.* — Dry sandy fields and borders of woods, Fla. to Tex., n. to e. Md., Tenn. and Ark. Sept., Oct.

98. MANISÙRIS L.

Spikelets in pairs in the excavations at the nodes of a cylindrical articulated axis; one sessile and perfect; the other pedicelled, sterile, with its pedicel adnate to the rachis; glumes of the perfect spikelet awnless, the first coriaceous and covering the excavation in the rachis, the second thinner and navicular; sterile lemma empty or with a rudimentary flower and, like the lemma and palea, hyaline; glumes of sterile spikelet membranaceous. — Perennials with flat narrow leaves and single cartilaginous spikes which disarticulate at maturity, terminating the stem and branches; chiefly trop. (Name said to be from the Greek *manos, necklace,* and *oura, tail,* from the elongate jointed spikes.) Rottboellia of ed. 7, not of L.f.

334. M. rugosa.

1. M. rugòsa (Nutt.) Ktze. (wrinkled). — *Culms tufted, compressed,* 6–12 dm. high; sheaths flattened; leaves 5–10 mm. wide; *spikes 2–7 cm. long,* the *lateral ones on short clustered branches in the axils,* often partly included in inflated sheaths; *first glume of fertile spikelet transversely rugose.* (*Rottboellia* Nutt.; *Coelorachis* Nash) — Low pinelands and pine-barrens, Fla. to Tex., n. very locally, to s. N.J. Aug., Sept., Fig. 334.

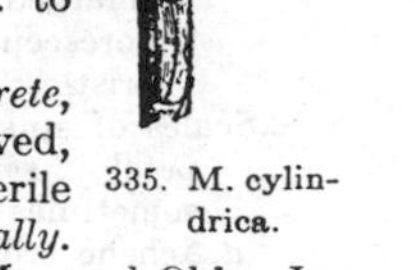

335. M. cylindrica.

2. M. cylíndrica (Michx.) Ktze. (cylindric). — *Culms* 0.3–1 m. high, *terete, from a short rhizome;* leaves 2–3 mm. wide; *spikes* slender, usually curved, 5–15 *cm. long,* terminating the culm and *on elongated axillary peduncles;* sterile spikelet rudimentary; *first glume of fertile spikelet obscurely pitted longitudinally.* (*Rottboellia* Torr.) — Prairies and pine-woods, Fla. to Tex., n. to S.C., Mo. and Okla. June–Aug. Fig. 335.

Tribe XII. Maỳdeae Mathieu (see p. 189)

Represented with us only by

99. TRÍPSACUM L. Gama-Grass. Sesame-Grass

Spikelets unisexual, the staminate ones in pairs at the joints of the terminal portion of the continuous rachis; the pistillate spikelets solitary, embedded in each oblong joint of the cartilaginous thickened articulate rachis below in the same inflorescence, which terminates the culm or its branches; glumes of the staminate spikelet subcoriaceous to submembranaceous, the first dorsally flattened, the second navicular; the first lemma often empty, membranaceous, with a hyaline palea, like the second which incloses a staminate flower; first glume of pistillate spikelet ovate, at length cartilaginous and closing the recess in the rachis, second navicular and coriaceous; florets 2; the lemmas and paleas hyaline, the lower sterile, the upper pistillate. — Tall stout Am. perennials from very thick creeping rhizomes, with broad flat leaves and terminal and axillary spikes separating into joints at maturity. (Name said to come from the Greek *tribein, to rub,* perhaps in allusion to the polished spike.)

1. T. dactyloìdes L. (with fingers, like *Dactylon,* an ancient name of some grass). — Culms 1–2.5 m. high; leaves 3 dm. or more long, 1.5–3.5 cm. wide; spikes 2–3 together at the summit, when their contiguous sides are more or less flattened, or solitary and terete; axillary spikes solitary; glumes of staminate spikelet subcoriaceous, blunt or rounded at tip. — Swales, moist fields, borders of woods, and shores, Fla. to e. Tex. and Mex., n. to s. N.E., N.Y., W.Va., s. Mich., Ill., Ia. and Neb. June–Sept. Fig. 336.

Var. **occidentàle** Cutler & Anders. (western). — Spikes 1 or 2; glumes of staminate spikelet submembranaceous, narrow and elongate, tapering or acuminate at tip. — S. Va. and Tenn.; w. Tex.

336. T. dactyloides.

Còix (from the Greek *coix, a palm*) Lácryma-Jòbi L., Job's-tears, a broad-leaved annual, with inflorescences consisting of often long-stalked globose to ovoid whitish bead-like pistillate spikelets and tufts of scaly staminate spikelets, is occasional on garden-refuse but not long persistent. (Introd. from Asia)

ZÈA (old Greek name of some grass) MAŸS L. (from an aboriginal name), MAIZE OR INDIAN CORN, is casual on garden-refuse and dumps but not persistent.

FAM. 23. CYPERÀCEAE (SEDGE FAMILY)

Grass-like or rush-like herbs with fibrous roots, mostly solid stems (culms), *closed sheaths, and spiked chiefly 3-androus flowers, one in the axil of each of the glume-like imbricated bracts* (scales, glumes), *destitute of any perianth, or with hypogynous bristles or scales in its place; the 1-locular ovary with a single erect anatropous ovule, in fruit forming an achene.* Style 2-cleft with the fruit flattened or lenticular, or 3-cleft and fruit 3-angular. Embryo minute at the base of the somewhat floury albumen. Cauline leaves when present 3-ranked. — A large, widely dispersed family.

a. Flowers all perfect, only rarely some of them (most of them in no. 4) with stamens or pistil abortive; spikelets essentially uniform. . . *b.*

b. Spikelets mostly many-flowered (if only 1-flowered, the spikelets in densely glomerulate heads with 2-ranked scales), with only 1 (more in no. 9) of the lower scales empty. Tribe I. SCIRPEAE. . . . *c.*

c. Scales of the spikelet strictly 2-ranked, conduplicate and keeled, neither white nor petaloid; spikelets clustered.

Inflorescences terminal simple or compound umbels or glomerules; flowers without perianth-bristles; achene without a tubercle or enlarged style-base. . . . . 1. *Cyperus.*

Inflorescences axillary, from the leaf-sheaths; flowers with a perianth of bristles; achene capped by a long tubercle. . . . . 2. *Dulichium.*

c. Scales of spikelet spirally arranged (subdistichous in no. 4, with white petaloid scales and the spikelets crowded above a leafy involucre, and sometimes in no. 3, with single terminal spikelets). . . *d.*

d. Achene crowned by the bulbous or modified persistent style or style-base (or tubercle); flowers without inner subtending scales or bractlets. . . *e.*

e. Culms naked, the basal colored sheaths usually bladeless; spikelet solitary and terminal; perianth-bristles often present. . . . . 3. *Eleocharis.*

e. Culms leafy or leafy-based; spikelets 2–many, in terminal (or axillary) inflorescences subtended by leaves or leafy involucres; perianth wanting.

Spikelets white or whitish, densely crowded in a head above an involucre of several long and broad leaves; most flowers sterile or aborted. . . . . 4. *Dichromena.*

Spikelets dark, in umbels or cymes, with few narrow subtending leaves; flowers mostly perfect.

Leaves flat; spikelets in terminal and axillary cymes; style almost wholly persistent. . . . . 5. *Psilocarya.*

Leaves capillary; spikelets in terminal umbels, some often crowded among basal leaves; style-base only persistent. . . 6. *Bulbostylis.*

d. Achene not crowned by the persistent bulbous base of the style. . . *f.*

f. Flowers without inner subtending scales or dilated sepals.

Style dilated or bulbous at base, deciduous below the enlargement; perianth wanting. . . . . 7. *Fimbristylis.*

Style terete, slender, not dilated at base; perianth of bristles present or wanting.

Perianth of 1–8 bristles or none. . . . . 8. *Scirpus.*

Perianth of very numerous elongate silky bristles. . . . . 9. *Eriophorum.*

f. Flowers with one or more inner subtending scales (bracteoles) or with dilated sepal-like bristles.

Perianth of 3 slender (or suppressed) bristles alternating with 3 broadly dilated ones; basal bracteoles 0. . . . . . . 10. *Fuirena.*

Perianth 0; flower subtended by 1 or 2 basal blunt bracteoles.

Bracteole 1. . . . . 11. *Hemicarpha.*

Bracteoles 2. . . . . 12. *Lipocarpha.*

b. Spikelets mostly 1–2-flowered, with 2–many empty basal scales. Tribe II. RHYNCHOSPOREAE (No. 4 might be sought here).

Achene crowned by the enlarged and modified style or style-base (tubercle); perianth-bristles often present. . . . . 13. *Rhynchospora.*

Achene without tubercle; perianth 0. . . . . 14. *Cladium.*

a. Flowers unisexual, the staminate and pistillate in the same or in different spikes (the latter spikelet-like in appearance).
Achene naked, bony or crustaceous, supported on a disk; spikes few-flowered; lower scales empty; perianth-bristles 0. Tribe III. SCLERIEAE. . . . 15. *Scleria.*
Achene inclosed in a sac (*perigynium*) or spathe. Tribe IV. CARICEAE.
Achene surrounded by a spathe open on one side above the middle; spikes with a single pistillate flower and 1 staminate, or 1-flowered with the sexes separate. . . . . 16. *Kobresia.*
Achene at the base of a closed loose or adnate pouch (*perigynium*) open only at tip or bidentate at summit (if perigynium rarely ruptured or split down one side, the flowers numerous).
Leaf sheathless, the tongue-shaped convolute blade without ligule or midrib, the cartilaginous margin closely crenate-undulate; perigynia and their subtending scales white and petaloid; scapes bractless. . . 17. *Cymophyllus.*
Leaf with sheath, ligule and midrib, entire or, if undulate, not cartilaginous-margined; perigynia firm to membranaceous, colored, their subtending scales firm to herbaceous or membranaceous; culms usually with a bract at least beneath inflorescence. . . . . . 18. *Carex.*

Tribe I. SCIRPEAE Kunth

1. CYPÈRUS L. GALINGALE. UMBRELLA-SEDGE. SOUCHET (Que.)

Spikelets many–few-flowered, mostly flat, variously arranged, mostly in clusters or heads which are commonly disposed in a simple or compound terminal umbel. Scales 2-ranked (their decurrent base often forming margins or wings to the hollow of the joint of the axis next below), deciduous when old. Stamens 1–3. Style 2–3-cleft, deciduous. Achene lenticular or triangular, naked at the apex. — Culms mostly triangular, simple, leafy at base, and with one or more leaves at the summit, forming an involucre to the umbel or head. Peduncles or rays unequal, sheathed at base. Large genus of trop. and temp. reg., ours all flowering in late summer or autumn. (*Cypeiros*, the ancient Greek name.)*

a. Inflorescence a simple or compound umbel or head-like spike; flowers in the axils of all but the lowest scales. . . *b.*
b. Fertile flowers and achenes (1–) 2–many, each achene subtended by a single scale; spikes peduncled or sessile, if sessile with scales not spinulose-serrate. . . *c.*
c. Annuals or short-lived perennials with soft bases and tufted fibrous roots, without stolons or hardened rhizomes or tubers. . . *d.*
d. Styles 2-cleft; achenes lenticular or biconvex. Subgen. PYCREUS. . . *e.*
e. Achenes suborbicular or broadly obovate, nearly or quite as broad as long, black, transversely wrinkled; scales 2 mm. long, yellowish-green. § FLAVESCENTES. . . . . . 1. *C. flavescens*, var. *poaeformis.*
e. Achenes narrowly obovate to oblong, much longer than broad, not transversely wrinkled; scales often tinged with purplish- or golden-brown. . . *f.*
f. Achenes black, nearly equaling the subtending scales; scales fulvous, brown, fuscous or yellowish, 1.4–1.8 mm. long, white-margined, spreading-ascending. § ALBOMARGINATI. . . . . . . 2. *C. albomarginatus.*
f. Achenes drab or brownish, much shorter than the subtending scales; scales 1.5–3.5 mm. long, scarcely white-margined. . . *g.*
g. Scales blunt, their tips closely appressed, commonly tinged with purple or dark brown. § SULCATI.
Styles 2-cleft nearly to base, persistent in fruit and long-exserted beyond the membranous usually lustreless scales; achenes elliptical or narrowly ovate. . . . . . 3. *C. diandrus.*
Styles 2-cleft to about the middle, usually deciduous; scales firm, often lustrous; achenes obovate. . . . . . 4. *C. rivularis.*
g. Scales with the green midrib prolonged into a short subulus, commonly yellow or golden-brown, their tips slightly projecting, giving the spikelet a serrated margin. § POLYSTACHYI.

---

* Highly technical genus, the subgenera distinguished chiefly on recondite characters of the fully mature and disintegrating spikelets. Only material well exhibiting the bases and the mature inflorescences is readily identifiable. The key is quite artificial.

Spikelets linear-lanceolate, 1.5–3 mm. broad; scales 2–3.5 mm. long, oblong-lanceolate, subcoriaceous, lustrous, acute, prominently mucronate; achenes 1.2–1.4 mm. long. . . . 5. *C. filicinus.*

Spikelets linear, 1.2–2 mm. broad; scales 1.5–2 mm. long, narrowly elliptic-ovate, membranaceous, dull, obtuse or only subacute, barely mucronulate; achenes 0.8–1 mm. long. . . . . 6. *C. polystachyos*, var. *texensis.*

d. Style 3-cleft; achenes trigonous. Subgenera EUCYPERUS, MARISCUS, TORULINIUM. . . *h.*

h. Tips of loosely spreading scales very slender and recurved; dwarf annual, rarely up to 1.6 dm. high, when bruised exhaling odor of *Melilotus* or of Slippery Elm (*Ulmus rubra*). § AMABILES. . . . 7. *C. inflexus.*

h. Tips of scales not strongly recurved; if slightly so the plants coarser, without odor of *Melilotus* and the scales closely appressed except at tip. . . *i.*

i. Spikes spherical, subglobose or with digitately radiating spikelets. . . *j.*

j. Culms leafy at base; spikes spherical, subglobose or lobate. . . *k.*

k. Scales 2.5–3 mm. long, the broad green midrib prolonged into a cusp; achenes 1.2–1.5 mm. long. § COMPRESSI. . . . 8. *C. compressus.*

k. Scales 1–2 mm. long; achenes at most 1 mm. long.

Spikelets ovate or broadly oblong, densely crowded into globose or lobate spikes, pale green to pale brown; scales oblong-lanceolate, chartaceous, their tips recurving, about 2 mm. long. § LUZULOIDEI.

Perennial 0.3–7.5 dm. high; achenes linear. . . . 9. *C. virens.*

Annual 0.3–3.5 dm. high; achenes oblong. . . . 10. *C. acuminatus.*

Spikelets linear, fuscous, dark brown or reddish, loosely aggregated into head-like spikes; scales ovate to orbicular, membranaceous, not excurved at tip, about 1 mm. long. § FUSCI.

Spikes loose; scales ovate, the midrib prolonged into a short mucro. . . . . 11. *C. fuscus.*

Spikes dense, often lobate; scales obovate-orbicular, round-tipped, not mucronate. . . . . 12. *C. difformis.*

j. Culms naked at base, the lower sheaths all or nearly all bladeless; spikelets digitately radiating from a very short axis. § HASPANI. . . . . 23. *C. Haspan*, var. *americanus.*

i. Spikes subcylindric, ellipsoid, lanceolate or ovoid. . . *l.*

l. Scales 1–1.5 mm. long; rachilla continuous, not disarticulating at maturity, wingless or very narrowly winged.

Spikelets about 1 mm. wide, crowded, divergent, in cylindric spikes; scales lanceolate, closely imbricated; rachilla with promptly deciduous chaffy wings; achenes 0.8–1 mm. long, short-ellipsoid, pearly-white. § FASTIGIATI. . . . . . 13. *C. erythrorhizos.*

Spikelets about 1.5 mm. wide, loosely fascicled in ovoid or lanceolate spikes; scales obovate to suborbicular, subremote; rachilla wingless or with slender persistent wings; achenes 1–1.5 mm. long, blackish. § IRIAE.

Scales rounded-emarginate, barely mucronulate, yellowish or fulvous; rachilla wingless. . . . . 14. *C. Iria.*

Scales truncate, mucronate, fuscous to golden-brown; rachilla narrowly winged. . . . . 15. *C. microiria.*

l. Scales 1.8–4.5 mm. long; rachilla winged, jointed and disarticulating at base or breaking into segments.

Rachilla not breaking into short segments, the narrow somewhat confluent wings not embracing the achenes; scales 3–4.5 mm. long, yellow or yellow-tinged. § STRIGOSI. . . 25. *C. strigosus.*

Rachilla breaking into short segments with the achenes embraced by broad clasping wings; scales 1.8–3 mm. long. § FERACES.

Tips of scales definitely overlapping the bases of the scales next above on same side of spikelet.

Scales coriaceous or firm, tending to be lustrous, 2–3.5 mm. long, drab to brownish; achenes ellipsoid or slenderly obovoid, 1.5–2 mm. long, becoming gray or blackish. . . . . 16. *C. odoratus.*

Scales membranaceous, dull, 1.5–2.3 mm. long, reddish-brown; achenes oblong, 1–1.5 mm. long, ferruginous or golden-brown. . . . . 17. *C. ferruginescens.*

Tips of thin yellowish-brown to reddish scales scarcely reaching bases of scales next above on same side of spikelet. . . . . . 18. *C. Engelmanni.*

*c.*Perennials with ligneous rhizomes or tubers or producing tuber-bearing stolons. . . *m.*

*m.*Bases not forming hardened rhizomes or sessile tubers, mostly stoloniferous, the stolons often tipped by tubers or new rhizomes; rachillas continuous; the scales of the spikelet gradually deciduous, as they ripen, from base to apex of spikelet. . . *n.*

*n.*Achenes dorsally compressed, with inner face toward the rachilla; stigmas 2; spikelets loosely subracemose. § SEROTINI. . . . . 19. *C. serotinus.*

*n.*Achenes trigonous; stigmas 3. . . *o.*

*o.*Spikelets borne on an elongate rachis, the spikes ovoid to obovoid; scales membranaceous; rachilla winged.

Stolons ligneous; scales keeled, their sides only obscurely nerved or nerveless, fuscous to reddish-purple. § ROTUNDI.

Slender, 1–6 dm. high; stolons filiform, bearing ellipsoid tubers; involucral leaves rarely much exceeding rays of umbel; scales ovate, 2–3.5 mm. long, bluntish. . . . . 20. *C. rotundus.*

Stouter, 0.6–1.2 m. high; stolons thick, forming new rhizomes at their tips; involucral leaves much exceeding rays of umbel; scales elliptic, about 4 mm. long, short-mucronate. . 21. *C. setigerus.*

Stolons weak; scales concave, scarcely keeled, their yellowish to golden-brown sides obviously ribbed. § ESCULENTI. . . . 22. *C. esculentus.*

*o.*Spikelets radiating from a very short axis, forming loosely subglobose head-like spikes (or spikelets altered to bulblets); scales keeled, mucronate; rachilla wingless. § HASPANI.

Rhizome short, without stolons; culms often naked, the basal sheaths commonly bladeless; spikelets linear, 1 mm. wide; scales red or purple, membranaceous, 1–1.5 mm. long. . . 23. *C. Haspan*, var. *americanus.*

Rhizome short and thick, producing slender tuber-bearing stolons; basal leaves well developed; spikelets oblong, about 2.5 mm. wide (or replaced by bulblets); scales pale green to brown, firm, 2–3 mm. long. . . . . 24. *C. dentatus.*

*m.*Bases hard knotty rhizomes or series of tubers; rachillas articulated at base, in some species disarticulating into short articles; styles 3; achenes 3-angled. Subgen. MARISCUS. . . *p.*

*p.*Spikelets flat or strongly compressed, yellowish to golden-brown, strongly divergent, 4–25-flowered; scales thin, linear- to oblong-lanceolate, with green midrib and scarious margin, 7-nerved. § STRIGOSI. . . . . 25. *C. strigosus.*

*p.*Spikelets subulate, subterete or 4-angled, 1–16-flowered; scales firm, lanceolate, oblong or ovate, 9–11-ribbed. §§ LAXIGLUMI and UMBELLATI. . . *q.*

*q.*Spikelets (mature) strongly reflexed, in thick-cylindric to obovoid heads; scales lanceolate. . . *r.*

*r.*Spikelets linear-subterete, soft, 0.8–3 cm. long, 3–10-flowered.

Spikelets greenish-drab, remote in a very open spike, at first ascending, soon reflexed, 1–3 cm. long; achenes linear, 2.5–3 mm. long; denuded rachis with remote scars. . . . 26. *C. refractus.*

Spikelets fulvous to yellowish, crowded in a dense spike, 0.8–1.5 cm. long; achenes linear-oblong, 2–2.5 mm. long; denuded rachis with closely approximate scars. . . . 27. *C. lancastriensis.*

*r.*Spikelets subulate, firm to rigid, 3–11 mm. long, 1–3-flowered.

Culms, leaves and involucres smooth and glabrous; heads subcylindric to slightly obovoid. . . . . 28. *C. retrofractus.*

Culms (at least above), leaves and involucres scabrous.

Heads cylindric to subcylindric or but slightly obovoid; spikelets not pungent. . . . . 29. *C. dipsaciformis.*

Heads obconic, strongly narrowed at base; spikelets pungent. . . . . 30. *C. Plukenetii.*

*q.*Spikelets (except basal) spreading to ascending, not strongly reflexed; scales elliptic-oblong to ovate. . . *s.*

*s.*Scales all acuminate, prominently awned, ascending, 3.5–4.5 mm. long; achenes 2.5–3.5 mm. long; spikelets 6–16-flowered. . . 31. *C. Schweinitzii.*

*s.*Scales blunt or rounded at tip or barely mucronate (terminal empty one in no. 34 subulate), shorter; achenes smaller; spikelets (1–)2–16-flowered. . . *t.*

*t.*Heads loosely hemispherical, with ascending spikelets; scales suborbicular; achenes short-ellipsoid, two-thirds as broad as long. . . . . 32. *C. Houghtonii.*

*t.*Heads spherical to thick-cylindric, with most spikelets horizontally radiating; scales narrower; achene much longer than broad. . *u.*

*u.*Scales with free tips, not strongly overlapping, giving the spikelet a dentate margin; heads nearly spherical, with loosely divergent 2–16-flowered spikelets.

Leaves of involucre smooth; rachilla broadly winged (evident on removal of scales); scales only narrowly margined, the green midrib rarely or not at all excurrent; style 3-cleft to middle. . . . . . . 33. *C. Grayii.*

Leaves of involucre scabrous-margined; rachilla wingless or only narrowly winged; scales with broad hyaline margins, the green midrib of at least the upper ones excurrent as a sharp point; style 3-cleft to base. . . 34. *C. filiculmis.*

*u.*Scales closely overlapping, their tips appressed; spikelets 2–6-flowered. . *v.*

*v.*Heads mostly on distinct rays; scales oblong; achenes linear-oblong, 0.5 mm. wide.

Spikelets 20–40, loosely radiating in subglobose to thick-cylindric heads, 3–6-flowered, their slenderly pointed tips remote. . . . . 35. *C. globulosus.*

Spikelets many more, densely imbricated, 1–3(–5)-flowered, blunt.

Heads globose or globose-ellipsoid, becoming fulvous. 36. *C. ovularis.*

Heads cylindric, green to drab. . . . . . . 37. *C. retrorsus.*

*v.*Heads cylindric, in a dense glomerule, yellowish to pale brown; scales ovate; achenes ellipsoid or ovoid, 1 mm. broad. . . . . 38. *C. cayennensis.*

*b.*Fertile flower 1 (rarely 2); stigmas 2; achene lenticular, with the edge next the rachilla; scales with midrib often spinulose, awn-tipped; spikelets densely crowded into a subspherical head up to 8 mm. long, or 2 or 3 heads crowded in a glomerule. Subgen. KYLLINGA.

Tufted annual; spikelets 1.5–2 mm. long. . . . . 39. *C. tenuifolius.*

Repent rhizomatous perennial; spikelets 3–3.5 mm. long. . . . . 40. *C. brevifolius.*

*a.*Inflorescence elongate, the brown fascicles of many-scaled sterile spikelets from axils of the leafy stem. . . . . 41. × *C. Weatherbianus.*

Subgen. PYCRÈUS (Beauv.) C. B. Clarke

§ FLAVESCÉNTES Kükenth.

1. C. flavéscens L. (yellowish), var. **poaefórmis** (Pursh) Fern. (like a *Poa*). — Annual; culms tufted or solitary, 0.5–4 dm. high; involucral leaves 3, wide-spreading, the longer much exceeding the umbel; umbel small, condensed or with 2–4 short rays up to 4.5 cm. long; *spikelets yellow-green*, spreading or reflexed, linear, obtuse, very flat, 0.3–1.5 cm. long; *scales* ovate, obtuse, thin except for the firm green keel, *about 2 mm. long;* stamens 3, anthers 0.5 mm. long; style deeply 2-cleft; *achenes* flattened, *black, suborbicular to broadly obovate, nearly or quite as broad as long, transversely wrinkled*, when ripe with white incrustation. — Wet sandy soil, Fla. to Tex., n. to se. N.Y., Pa., O., Mich., Ill., Mo. and e. Kans. July–Oct. (W.I.) FIG. 337. — N.Am. representative of a pantropical species.

337. C. flavescens, v. poaeformis.

§ Albomarginàti Kükenth.

2. C. albomarginàtus Mart. & Schrad. (white-margined). — Coarse annual; culms 1–several, smooth, 3–9 dm. high; leaves flat; those of the involucre 3–6, very long; umbel compact or with elongating rays up to 1 dm. long; *spikes* ovoid, becoming lax, 1–3.5 cm. in diameter, *fuscous, fulvous or brown* (rarely yellowish); spikelets linear or linear-lanceolate, subdistant, divergent, 0.7–2 cm. long; *scales ovate to obovate, obtuse,* suberose, *broadly white-margined,* 1.4–1.8 *mm. long, spreading-ascending, only slightly longer than the black round-obovate biconvex achenes.* (*C. flavicomus* ed. 7., not Michx.; *C. sabulosus* of N.Am. auth., not Mart. & Schrad.) — Damp clearings, fields and roadsides, Fla. to Tex., n. to e. Va. and, locally, to L.I. July–Oct. (Trop. Am.; Afr.; Austr.) Fig. 338.

338. C. albomarginatus.

§ Sulcàti Kükenth.

3. C. diándrus Torr. (with two stamens). — Annual; culms slender, tufted or solitary, erect to depressed, 0.2–4.5 dm. long, soft; involucral leaves 3, wide-spreading, very unequal, the longest exceeding the umbel; umbel condensed or with 2–5 unequal rays up to 6 cm. long; spikelets reddish-brown to greenish, spreading or reflexed, lance-oblong, subacute, very flat, 6–32-flowered, 0.4–1.8 cm. long, or 40–50-flowered and 2–2.5 cm. long in the local forma **elongàtus** (Britt.) Fern. (elongate); *scales loosely imbricated,* with appressed tips, ovate, obtuse, *thin and dull,* with a purple-brown margin, 2–2.7 mm. long; stamens 2 (or 3 in upper flowers), anthers 0.3–0.4 mm. long; *style 2-cleft nearly to base, persistent in fruit and long-exserted beyond scales;* achenes drab, elliptical to narrowly obovate. — Wet sandy, gravelly, muddy or peaty soils, S.C. to N.M., n. to sw. N.B., centr. Me., sw. Que., s. Ont., Wisc., Minn. and N.D. June–Oct. Fig. 339.

339. C. diandrus.

4. C. rivulàris Kunth (of streams). — Similar to no. 3, firmer; rays of umbel up to 1.2 dm. long; *spikelets purplish-brown or strongly suffused with purplish,* or greenish or merely stramineous in forma **elùtus** (C. B. Clarke) Kükenth. (washed out), blunt, 0.5–2.3 cm. long; *scales closely imbricated, rather firm and usually lustrous; styles 2-cleft to about the middle, usually deciduous.* — Wet sandy, gravelly, muddy or peaty places, Ga. to Mex., n. to centr. Me., sw. Que., s. Ont., Wisc., Minn., Neb. and Calif. July–Oct. Fig. 340.

340. C. rivularis.

§ Polystáchyi C. B. Clarke

5. C. filicìnus Vahl (fern-like). — Annual; culms slender, soft, densely tufted to solitary, erect or depressed, 0.1–3.5 dm. high; involucral leaves wide-spreading, 1 or 2 of them very long; umbel condensed and sessile or with 1–4 rays up to 5 cm. long, or in forma **Cleavèrii** (Torr.) Kükenth. (named in 1836 for its discoverer, Isaac Cleaver) reduced to a single spikelet; *spikelets linear-lanceolate,* 1.5–3 *mm. broad,* 0.6–2.7 cm. long, *yellowish to golden-brown,* very flat, acute; *scales oblong-lanceolate,* 2–3.5 *mm. long, subcoriaceous, lustrous, acute, the green midrib projecting as a subulus and giving the spikelet a serrate appearance;* stamens 2; style deeply 2-cleft, deciduous; *achenes* narrowly obovate, subtruncate at summit, drab or brownish, 1.2–1.4 *mm. long.* (*C. Nuttallii* Eddy) — Borders of siliceous and saline marshes, brackish to fresh sands, dune-hollows, and rarely fresh pond-shores, Fla. to La., n. to s. Me. Aug.–Oct. (W.I.) Fig. 341.

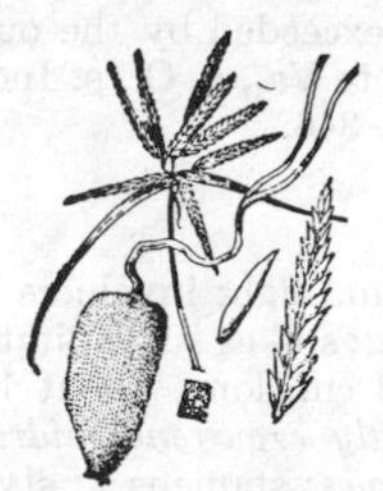
341. C. filicinus.

6. C. polystáchyos Rottb. (many-spiked), var. **texénsis** (Torr.) Fern. (of Texas). — Similar to no. 5; culms solitary or tufted, erect or depressed and matted, with us 0.1–2 (farther south –5) dm. high, slender; umbels sessile or with rays up to 4 cm. long; spikes lax; *spikelets* stramineous, yellowish or ferruginous, linear, 1.2–2 *mm. broad,* 0.5–2 (–3) cm. long; *scales* 1.5–2 mm. long, *narrowly elliptic-ovate, membranaceous, lustreless, obtuse to subacute, barely mucronulate; achenes* 0.8–1 *mm. long.* (Var. *leptostachyus* Boeckl.; *C. Gatesii* and *C. microdontus* Torr.) — Damp sands, peats, shores and clearings, Fla. to Tex. and Mex., n. on Coastal Plain to Cape Cod, Mass., and inland n. to se. Mo. Late-July–Oct. (Trop. Am.; Philippine Ids.) Fig. 342. — Chiefly American variety of a polymorphic pantropical species.

342. C. polystachyos, v. texensis.

§ Amábiles C. B. Clarke

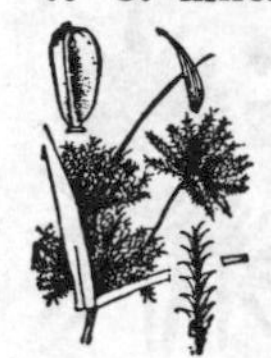
343. C. inflexus.

7. C. infléxus Muhl. (incurved). — Soft-based annual, *with strong odor* (when bruised) *of Melilotus* or of Slippery Elm (*Ulmus rubra*); culms mostly tufted, 0.1–1.6 dm. high, very slender; involucral leaves 2–4, the longer much exceeding the umbel; umbel condensed or with 1–5 rays up to 3 cm. long; *spikelets greenish*, finally pale brown, oblong to linear, flat, 4–24-flowered, 2–9 mm. long, aggregated into dense subcylindric to subglobose spikes 4–15 mm. thick; *scales* 1.5–3 mm. long, oblong, *narrowed to long recurved slender tips;* stamen 1; style 3-parted, deciduous; rachillas persistent after fall of scales; achene slenderly trigonous-obovoid, pale brown, minutely pebbled, 0.8–1 mm. long, 0.3–0.4 mm. broad. (*C. aristatus* of ed. 7, not Rottb.) — Damp sands, silts and alluvium, Fla. to Tex. and Mex., n. to w. N.B., centr. Me., sw. Que., s. Ont., Mich., Minn., s. Man., s. Sask. and s. B.C. July–Oct. Fig. 343.

§ Compréssi Kunth

344. C. compressus.

8. C. compréssus L. (flattened). — Soft-based annual, tufted, ascending or depressed, 0.5–3 dm. high; involucral leaves 3–5, the longest about twice as long as rays of umbel; *umbel* condensed, *loosely hemispherical to rhomboid-obovoid* or with 1–4 rays up to 9 dm. long; *spikelets pale green*, flattened, lanceolate, 10–40-flowered, 0.8–2.5 cm. long, loosely ascending; *scales striped with yellow and white, veiny, the broad green midrib prolonged into a sharp cusp, 2.5–3 mm. long;* stamens 3; style 3-cleft about to middle, deciduous; *achene* pale, *broadly obovoid,* 1.2–1.5 *mm. long,* with concave sides and prominent angles. — Sandy fields, roadsides, ditches, etc., Fla. to Tex., n. to se. N.Y., s. Ind. and Minn. Aug.–Oct. (Natzd. from the South or the Tropics) Fig. 344.

§ Luzuloìdei Kunth

345. C. virens.

9. C. vìrens Michx. (greenish). — Slender, 3–7.5 dm. high, perennial; culm obtusely triangular; leaves and involucre very long, keeled, blades 2–5 mm. wide; umbel compound, many-rayed; *spikelets ovate* (3–6 *mm. long*), *in numerous small greenish heads; achenes pale, linear,* on a slender stipe; scales narrow, acutish, obscurely 3-nerved, chartaceous, the attenuate tips excurved. (*C. pseudovegetus* Steud.) — Wet places, Fla. to Tex., n. to s. N.J., s. Ind., s. Ill., Mo. and se. Kans. July–Oct. (Trop. Am.) Fig. 345.

346. C. acuminatus.

10. C. acuminàtus Torr. & Hook. (tapering at tip). — Slender (0.5–3.5 dm. high); involucre 2–3-leaved; *spikelets ovate, becoming oblong,* 16–30-flowered, *pale,* in globular heads; *scales obscurely 3-nerved, short-tipped;* stamen 1; achene oblong, pointed at both ends, much exceeded by the outwardly curving scale. — Wet places, n. Fla. to Tex., n. to Va., s. O., s. Ind., Ill., Minn. and N.D.; Wash. and Oreg. Aug.–Oct. Fig. 346.

§ Fúsci Kunth

347. C. fuscus.

11. C. fúscus L. (dusky). — Tufted annual with soft culms 0.2–5 dm. high; involucre of 2–4 divergent leaves; umbel condensed or with 1–6 rays up to 4 cm. long; spikes subcapitate; *the purple-brown loosely radiating linear spikelets* 10–40-flowered, 0.3–1.2 cm. long, about 1.5 *mm. broad; scales purple-brown, thin, ovate, with slightly excurrent midrib, nearly veinless, about* 1 *mm. long, little exceeding the achenes;* stamens 2; style 3-cleft, deciduous; *achene trigonous, whitish, ellipsoid, acuminate, about* 1 *mm. long.* — Damp sandy or springy places, local, Mass. to w. N.Y. and Va. July–Sept. (Adv. from Eu.) Fig. 347.

12. C. diffórmis L. (of two forms). — Densely tufted annual, with soft smooth culms 2–7 dm. high, sparingly leafy at base; involucre of 2 or 3 strongly divergent prolonged bracts; *umbel compact,* with 3–8 short rays; *the spikes capitate-globose or lobate, very dense,* 5–15 mm. in diameter; *spikelets crowded,* linear, 4–8 mm. long, *about* 1 *mm. broad,* 10–40-flowered; *scales closely imbricated, appressed, rounded-obovate, scarcely keeled,* castaneous to reddish, with white hyaline margin; *achene about equaling scale,* ellipsoid-obovoid, blunt, greenish to yellowish. — Ditches and wet fields, se. Va., Calif. and Mex. July–Oct. (Adv. from Asia)

§ Fastigiàti Kükenth.

13. C. erythrorhizos Muhl. (red-rooted). — *Annual,* either with coarse obtusely angled erect culms 1.5–9 dm. high or dwarfed and tufted with culms down to 1 cm. long; *roots red;* basal leaves elongate, 2–10 mm. broad, scabrous-margined, with the pale subcoriaceous sheaths purple at base; *involucral leaves* 3–7, *the longest* 2–4 *times as long as the breadth of the umbel,* pale beneath and with broadened subcoriaceous bases; umbel compound, with several sessile rays and 3–9 longer ascending ones up to 1.5 dm. long; *spikes cylindric,* rounded to subtruncate at summit, *rather dense,* up to 4 cm. long, 0.6–3.5 cm. in diameter, *crowded at tips of rays; spikelets* rather *crowded,* divergent, linear, *reddish-brown,* 0.3–2 cm. long, *about* 1 *mm. wide; scales* lanceolate to oblong-ovate, mucronate, *reddish-brown, green-ribbed,* nerveless, 1.2–1.5 *mm. long, closely imbricated; rachillas persistent, bearing freely deciduous chaff-like wings;* stamens 2 or 3; style 3-cleft nearly to middle; *achene* trigonous, *inequilateral,* short-ellipsoid, *about* 0.8 *mm. long, whitish or pearly.* (*C. Halei* Torr.) — Alluvial or damp sandy soil, Fla. to s. Calif., n. to Mass., N.Y., s. Ont., Mich., Wisc., Minn., N.D., Wyo. and Wash. Aug.–Oct. Fig. 348.

348. C. erythrorhizos.

§ Iriae Kunth

14. C. Ìria L. (an old generic name). — Annual; culms 1–several, erect, 1–6 dm. high; involucral leaves 3–5, the longer exceeding the rays; *umbel simple or with* 5–8 *ascending rays* up to 1 dm. long; *spikelets loosely fasciculate and ascending in racemose or racemose-paniculate elongate clusters;* the individual *spikelets* oblong, obtuse, 2.5–10 mm. long, 2–20-flowered, *about* 1.5 *mm. broad; scales subdistant,* spreading-ascending, *broadly obovate,* 1–1.5 *mm. long, rounded-emarginate and barely mucronulate, yellowish-drab or fulvous; rachilla* slender, flexuous, *wingless; achene ellipsoid,* 3-angled, 1 *mm. long,* brown to blackish, densely puncticulate, stipitate. — Clearings, ditches and roadsides, Fla. to Tex., n. to Va. July–Oct. (Natzd. from Old World) Fig. 349.

349. C. Iria.

15. C. microìria Steud. (small *C. Iria*). — Similar; *spikelets more divergent,* in thicker spikes; *scales* suborbicular or broadly obovate, *truncate, mucronate, fuscous or golden-brown; rachilla narrowly winged.* — Fallow fields and waste places, L.I. to e. Pa. Aug.–Oct. (Adv. from Asia)

§ Feràces Kükenth.

16. C. odoràtus L. (fragrant). — Rather coarse annual with stoutish base, erect, 0.1–8 dm. high (in dwarf extremes often forming small tufts); leaves flat, 2–12 mm. broad; *involucral leaves prolonged,* often many times *exceeding* the rather *crowded rays of* the simple or compound *umbel; spikelets* 5–20-flowered, *drab to golden-brown; scales coriaceous or firm, sublustrous,* 2–3.5 *mm. long, their tips definitely overlapping bases of scales next above; rachilla breaking into short segments, with the achenes embraced by broad clasping wings;* achenes ellipsoid or slenderly obovoid, 1.5–2 mm. long, becoming gray or blackish. (*C. ferax* Richard) — Low grounds, often brackish or saline, Trop. Am., n. to Mass., N.Y., se. Mo., Tex. and Calif. Aug.–Oct. (Semicosmop.) Fig. 350.

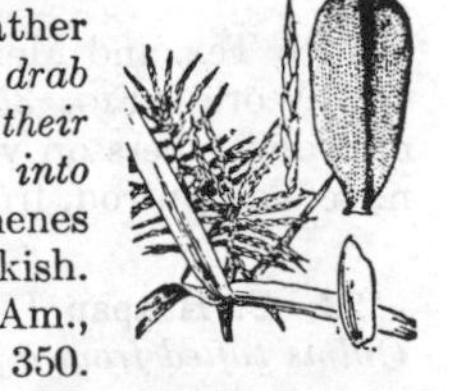

350. C. odoratus.

17. C. ferruginéscens Boeckl. (becoming rusty). — Similar to no. 16, mostly more slender; *spikelets reddish-brown; scales membranaceous, dull,* 1.5–2.3 *mm. long; achenes oblong,* 1–1.5 *mm. long, ferruginous or golden-brown.* — Alluvial and other damp soils, Ct., N.Y. and s. Ont. to Oreg., s. to Va., Ala., Miss., La., Tex., N.M., Ariz. and Calif. Aug.–Oct. — Often confused with no. 13.

18. C. Engelmánni Steud. (for its discoverer, George Engelmann, 1809–1884). — Similar to nos. 16 and 17, but the *spikelets* more slender and terete, *remotely flowered; the tips of the thin yellowish-brown to reddish scales scarcely reaching the bases of those above, thus exposing a part of the axis of the zigzag joints of the narrowly winged rachis;* achenes linear-oblong. — Low grounds, Mass. to s. Ont., Minn. and Neb., s. to se. Va., Ill. and Mo. Aug.–Oct. Fig. 351.

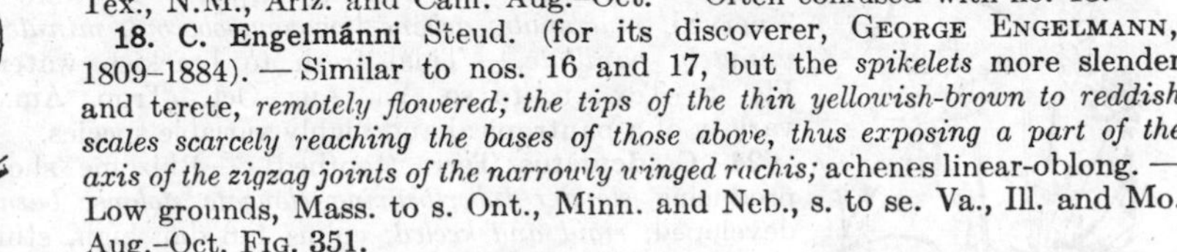

351. C. Engelmanni.

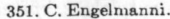

§ Seròtini Kükenth.

19. C. seròtinus Rottb. (late). — *Perennial by soft creeping stems and long stolons; culms* up to 1 m. high, *leafy only at base,* compressed-triangular, *scapiform;* leaves about equaling culm,

6–10 mm. broad, keeled below; involucre mostly longer than rays of the open umbel; *spikes* cylindric, *open, subracemose,* the divergent remote spikelets castaneous and 10–30-flowered; *scales loosely ascending,* slenderly nerved; *achene planoconvex, dorsally compressed, with inner face toward rachilla; styles* 2. — Tidal marshes of Del. R., s. N.J. Aug., Sept. (Adv. from Eurasia)

§ Rotúndi C. B. Clarke

20. C. rotúndus L. (spherical), Nut-grass, Coco-grass. — Perennial by tuber-bearing stolons; culm slender (1–6 dm. high), longer than the leaves; umbel simple or slightly compound, about equaling the involucre; the few rays each bearing 4–9 *dark chestnut-purple* 12–40-flowered *acute spikelets* (0.8–2.5 cm. long); *scales ovate, closely appressed, nerveless* except on the keel, 2–3.5 *mm. long, bluntish;* achenes linear-oblong. — Sandy fields, roadsides and cult. ground, often a troublesome weed, Fla. to Tex. and Mex., n. to Va. and locally to s. N.Y. Aug.–Oct. (Natzd. from Eurasia) Fig. 352.

352. C. rotundus.

21. C. setígerus Torr. & Hook. (bearing bristles). — Similar to no. 20. *stouter,* 0.6–1.2 *m. high,* the *thick stolons forming new rhizomes at their tips;* leaves 0.5–1 cm. broad, nearly as long as culms; *leaves of involucre very much exceeding rays of umbel; scales* stramineous to castaneous, *elliptic, about* 4 *mm. long, short-mucronate;* achenes oblong. (*C. Hallii* Britt.) — Bottomlands and sandy swamps, Mo. and Kans. to Tex. July–Sept.

§ Esculénti Kükenth.

22. C. esculéntus L. (eatable), Yellow Nut-grass, Amande de terre (Que.). — Perennial, *bearing weak filiform stolons terminated by hard tubers;* culms acutely angled, 2–9 dm. high; leaves pale green, 4–9 mm. wide; *involucral leaves* 3–9, *the longest much exceeding* the simple to compound yellowish to golden-brown *umbel;* the latter with several erect short rays and 2–9 strongly ascending longer ones; *spikelets* strongly flattened, mostly 4-ranked along the wing-angled rachis, obtuse, strongly *flattened,* 0.5–1.5 cm. long, or 1.5–3 cm. long and 1.5–2 mm. broad and attenuate to acute tip in forma **angustispicàtus** (Britt.) Fern. (narrow-spiked), or 2–3 cm. long and 2–3 mm. broad and linear and round-tipped in forma **macrostáchyus** (Boeckl.) Fern. (large-spiked); *scales* thin, *yellowish- or golden-brown,* oblong, obtuse or a little mucronate, distinctly nerved, scarious at tip, 2.3–3 mm. long; rachilla with adnate narrow hyaline scales; style 3-cleft; achenes (often not well-developed) lustrous, trigonous, ellipsoid or narrowly obovoid, rounded at summit, 1.2–1.5 mm. long. — Damp sandy soil, often troublesome in cult. ground, Fla. to Tex. and Mex., n. to N.S., s. Que., s. Ont., s. Man. and Wash. Aug.–Oct. (Trop. Am.; Old World) Fig. 353. — Var. satìvus Boeckl. (planted), Chufa, densely cespitose, with crowded tubers on very short stolons, the plant rarely flowering, is cult. for its edible tubers n. to Va. (Introd. from Afr.)

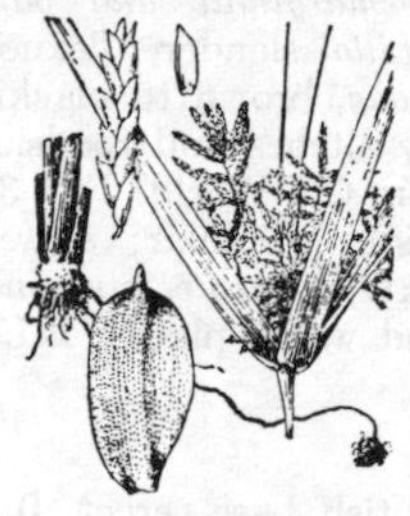
353. C. esculentus.

§ Haspàni Kunth

23. C. Háspan L. (the native name in Ceylon), var. **americànus** Boeckl. (American). — *Culms tufted from a short rhizome, soft,* sharply triangular (mostly flattened in drying), erect, 0.2–1 m. high, *with bladeless* or sometimes blade-bearing *lower sheaths;* involucral leaves mostly 2 and commonly shorter than the filiform elongate longer rays of the open very compound umbel; *spikelets in glomerules, linear,* 0.4–1 *cm. long, about* 1 *mm. wide; scales membranaceous,* tinged with red or purple, 1–1.5 mm. long, oblong, 3-nerved, *mucronate; achene trigonous-obovoid, minute, pearly-white, minutely papillate.* — Tidal fresh to brackish waters and pools, Fla. to Tex., n. to se. Va. Aug.–Oct. (Trop. Am.) — American variety of a pantropical and highly variable species.

354. C. dentatus.

24. C. dentàtus Torr. (toothed). — Rhizome short and thick, producing *slender tuber-bearing elongate stolons; basal leaves* well-developed, *rigid and keeled;* culms 1–6 dm. high, stiff; *umbel erect, shorter than the* 3–4-*leaved involucre; spikelets in subglobose spikes* (or altered into bulblets or leafy tufts), *oblong, about* 2.5 *mm. wide,* 4–9 mm. long, 5–13-flowered; *scales* reddish-brown to pale green, *firm,* 2–3 mm. long, acute, 7-nerved, with prominent mucronate tips, or scales more appressed and 15–40-flowered spikelets 1–1.5 cm. long in forma **ctenóstachys** Fern. (comb-spiked). — Sandy or gravelly shores and damp sands, s. Que. to n. Ind., s. to N.B., N.E., L.I., Del. and Md. July–Oct. Fig. 354.

Subgen. Maríscus (Gaertn.) C. B. Clarke

§ Strigòsi Kükenth.

25. C. strigòsus L. (lean). — Perennial (sometimes flowering the first year), the *base* becoming *a hard corm-like rhizome; culms* 1–several, firm, *smooth,* 0.1–1 m. or more (in very dwarf plants down to 1 mm.!) high, often tufted; leaves soft, flat, 0.1–1.2 cm. wide; involucral leaves 2–7, the longer much exceeding the umbel; umbel simple to very compound (condensed in some dwarf states), usually with 2–15 rays, the longer rays 0.3–4 dm. long; *spikelets yellowish to golden, compressed or flattened, strongly divergent in somewhat open spikes,* 4–14-flowered, 0.5–1.8 cm. long, *deciduous in age; scales golden, with green midrib and paler scarious margin,* strongly conduplicate, oblong-lanceolate, approximate, loosely imbricated, 7-nerved, 3–4.5 mm. long; rachilla bearing narrow whitish somewhat confluent wings; style 3-cleft about one-fourth its length; achene trigonous, linear, acute, less than half length of scale. (Incl. vars. *compositus* and *gracilis* Britt. and var. *compactus* Boeckl.) — Meadows, swales, damp thickets and shores, Fla. to Tex. and N.M., n. to N.E., sw. Que., s. Ont., Mich., Wisc., Minn. and Neb.; Wash. to Calif. Aug.–Oct. Fig. 355.

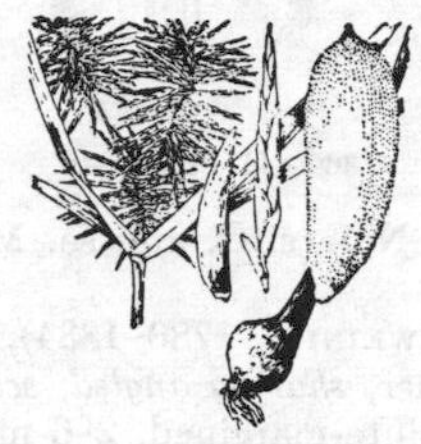
355. C. strigosus.

Var. **robústior** Britt. (stouter). — Plant with 1–few culms; spikelets 12–25-flowered, 2–3 cm. long. — Bogs, marshes, sloughs and wet shores, Fla. to Miss., n., locally, to se. Mass., Conn., L.I., N.J., W.Va., Ind. and Mo.

§§ Laxiglùmi and Umbellàti C. B. Clarke

26. C. refráctus Engelm. (abruptly bent back). — Rhizome short and thick, with 1–few corm-like enlargements; *culm smooth,* 3–9 dm. high; leaves soft and flat, 4–8 mm. broad, slightly scabrous; *rays* usually more or less elongated, *smooth; spikelets* very slender, acuminate, *subterete, in rather loose heads, divaricate or* more or less *reflexed,* 2–6-*flowered,* 1–3 *cm. long;* scales appressed, several-nerved, the lower empty and often persistent after the fall of the rest; joints of rachilla winged, inclosing the *linear achene* (2.5–3 *mm. long*); rachilla disarticulating at base; *denuded rachis with remote scars.* — Dry woods, thickets and rocky banks, N.J. to Ky., s. to Ga., Tenn. and Mo. Aug., Sept. Fig. 356.

356. C. refractus.

27. C. lancastriénsis Porter (of Lancaster, Pa.). — Rhizome short, with 1–few corm-like enlargements; culm stoutish, firm, triangular, smooth, 0.3–1.1 m. high; leaves 3–10 mm. broad; umbel of 6–9 mostly elongate rays; *spikelets fulvous or yellowish, linear-subterete, soft,* 0.8–1.5 *cm. long, very numerous in short-cylindric to obovoid close heads, soon reflexed,* of 3–6 narrow scales, deciduous in age; *achenes linear-oblong,* 2–2.5 *mm. long; denuded rachis with closely approximate scars.* — Sandy or loamy woods, thickets, meadows and clearings, Ga. to Mo., n. to N.J., Pa. and O. Aug.–Oct. Fig. 357.

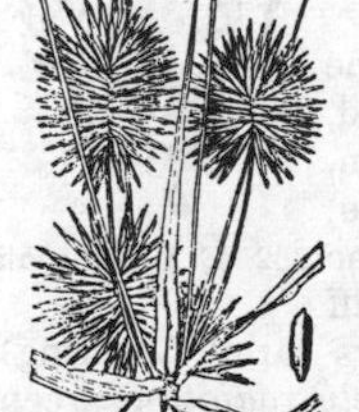
357. C. lancastriensis.

28. C. retrofráctus (L.) Torr. (turned back). — Rhizome elongate, bearing several hard tuber-like enlargements; *culms* slender, *firm to rigid, smooth,* 2–7.5 dm. high, much exceeding the *stiff narrow* (2–5 *mm. broad*) *smooth leaves;* umbel of 3–13 simple *smooth rays,* mostly shorter than the smooth involucre; *spikelets subulate,* firm to rigid, 3–8 *mm. long,* 1–2-*flowered,* densely crowded *in cylindric or slightly obovoid heads* (1–2.5 cm. long), strongly reflexed, golden-brown or drab at maturity; scales closely appressed, the fertile strongly nerved, the terminal involute-subulate; *achenes* linear, 2–2.5 *mm. long;* rachilla disarticulating at base. (*C. hystricinus* Fern.) — Dry sandy soil, N.J. to Ind., s. to Ga., La. and e. Tex. Aug.–Oct. Fig. 358.

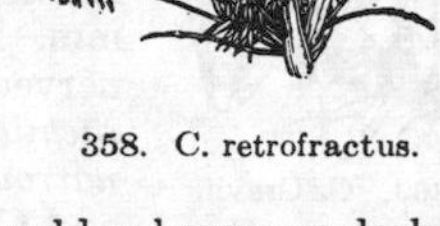

358. C. retrofractus.

29. C. dipsacifórmis Fern. (formed like a teasel). — Habit of no. 28, usually coarser; *culm*

359. C. dipsaciformis.

scabrous, at least above, 2.5–8 dm. high; *leaves* shorter than the culm, *scabrous-hispid above*, 4–9 *mm. wide;* umbel 4–12-rayed, some of the smooth rays equaling the involucre; *spikelets* 1–3-flowered, subulate, rigid, 6–11 *mm. long*, crowded *in cylindric or subcylindric heads* (1.5–4 cm. long), strongly deflexed, yellow-brown at maturity; fertile scales with green midribs; *achene* 3 *mm. long*. — Sandy barrens and dry woods, Ga. to se. Mo., n. to N.J., Md., D.C. and Ky. July–Sept. Fig. 359.

30. C. Plukenḗtii Fern. (for Leonard Plukenet, 1642–1706, one of the original describers and illustrators of American plants). — Habit of nos. 28 and 29; *culms harsh-puberulent*, 0.3–1 m. high; leaves harsh, firm; *involucral leaves* 3–7, mostly shorter than rays, *harsh;* umbel with 4–12 strongly ascending stiff scabrous simple rays up to 2.5 dm. long; *spikelets subulate, pungent*, 1–2-flowered, 3–8 mm. long, greenish, at first spreading but *soon tightly reflexed, very numerous* (100 *or more*) *in turbinate-obovoid dense bur-like heads;* scales 4 or 5, striate, the terminal involute and firm; achenes 2.5–3 mm. long. (*C. retrofractus* of ed. 7, not (L.) Torr.) — Dry or moist sands and rocks, Fla. to Tex., n. to (?) L.I., N.J., s. O. and se. Mo. July–Sept. Fig. 360.

360. C. Plukenetii.

31. C. Schweiníťzii Torr. (for its discoverer, Lewis David de Schweinitz, 1780–1834). — Rhizome short, hard, bearing corm-like branches; *culms* firm, slender, *sharply angled, scabrous*, 1–8 dm. high; leaves firm, flat or revolute-margined, 2–6 mm. wide, short; *involucral leaves* 3–6, the longer exceeding the umbel, *scabrous-margined;* umbel condensed or with 2–10 very unequal smooth slender erect rays up to 1.3 dm. long; *spikes* greenish or drab, *loose, ellipsoid or ovoid, with* irregularly inserted *ascending loosely* 6–16-*flowered spikelets* 1–2.5 cm. long; *scales* greenish-brown, firm, concave, *ovate, acuminate*, 3.5–4.5 *mm. long*, prominently nerved between the *much prolonged green midrib* and the broad hyaline nerveless margin; rachilla with narrow white wings; *achene light brown*, trigonous-ellipsoid, 2.5–3.5 *mm. long*. — Sands, beaches and barrens, sw. Que. to Sask., s. to N.Y., nw. Pa. (adv. to se. Pa.), n. O., n. Ind., Ill., Mo., Okla., Tex. and N.M. July–Sept. Fig. 361. — Apparently crosses with the next.

361. C. Schweinitzii.

32. C. Houghtônii Torr. (for its discoverer, Douglas Houghton, 1809–1845). — Resembling no. 31; *culms obtusely angled, smooth*, 1.5–7 dm. high; *leaves* narrower, *smooth; involucral leaves* 2–5, *smooth; heads loosely hemispherical, with ascending spikelets* 0.5–2 cm. long; *scales roundish*, firm, mucronate, 2–2.8 *mm. long;* rachilla with very narrow wings; *achene dark brown, short-ellipsoid*, rounded at each end, 1–1.5 *mm. long*. — Light, usually dry sandy soil, sw. Que. to Man., locally s. to Mass., L.I., centr. Pa., w. Va., nw. Ind., Ill., and Ia. July–Oct. Fig. 362.

362. C. Houghtonii.

33. C. Graŷii Torr. (for its discoverer, Asa Gray, 1810–1888). — Rhizome hard, with corm-like branches; *culms wiry, filiform, smooth*, obtusely angled, usually tufted, 0.5–3 dm. high; *leaves* gray-green, firm, *nearly filiform, smooth, usually conduplicate*, 0.5–3 mm. wide; *involucral leaves* similar, 3–6, *smooth*, almost filiform, the longer little exceeding the umbel; umbel with 3–14 stiff capillary smooth ascending rays up to 6 cm. long (rays rarely wanting); *heads nearly spherical*, 0.6–2 cm. in diameter, *with loosely radiating* greenish to drab-brown *loosely* 3–9-*flowered* slightly compressed *spikelets* 2.5–9 mm. long; *scales subremote*, ovate, blunt, firm, 2–2.5 mm. long, strongly nerved, *narrow-margined, their green midrib* rarely or *not* at all *excurrent; rachilla with broad hyaline wings; style 3-cleft to middle; achenes* trigonous, *narrowly ellipsoid*, gray or brown, about 2 *mm. long*. — Dry sands of coast and Coastal Plain, Mass. to Fla. July–Oct. Fig. 363.

363. C. Grayii.

34. C. filicúlmis Vahl (with thread-like culms). — Rhizome hard, bearing tough corm-like branches; culms 1–several, wiry, smooth, 0.5–9 dm. high; *leaves* gray-green, firm, narrowly linear to conduplicate, *scabrous on the margins; involucral leaves* similar, 3–6, *with scabrous margins;* umbel reduced to a single sessile dense head or with 1–15 stiff capillary rays up to

9 cm. long; *heads nearly spherical*, greenish to dull brown, *with numerous finally radiating* 2–16-flowered *spikelets; scales* blunt, or the uppermost acute, subcoriaceous or coriaceous, *with broad hyaline margin, the uppermost with slender prolonged tips; rachilla wingless or only narrowly winged; style 3-cleft nearly to base;* achene trigonous, narrowly oblong. — Three vars.:

Scales subcoriaceous or thinnish, yellowish-green or somewhat stramineous to warm brown, 2.8–3.5 mm. long, strongly overlapping; spikelets 6–16-flowered, 0.7–2 cm. long; achenes 1.8–2.3 mm. long; umbel reduced to a sessile glomerule 1.5–3.5 cm. in diameter or with 1–4 rays; culms 1.5–7.5 dm. high. . . . *C. filiculmis* (typical).

Scales coriaceous, greener or in maturity duller brown, 1.8–2.8 mm. long, less imbricated; spikelets 2–8-flowered, 0.3–1.5 cm. long; achenes 1.5–1.8 mm. long; central or sessile glomerule 0.7–2 cm. in diameter.

Umbel a sessile glomerule or with 1–4 unequal rays; involucre of 2–5 bracts, the longest bract 3–15 cm. long; spikes loosely subspherical; culms 0.5–6 dm. high; widely distributed. . . . . Var. *macilentus*.

Umbel with (3–)5–15 elongate rays; involucre of (3–)5–7 bracts, the longest bract (0.5–)1–2.7 dm. long; spikes compactly almost spherical; culms 2–9 dm. high; southeastern. . . . . Var. *oblitus*.

364. C. filiculmis (typical).

365. C. filiculmis, v. macilentus.

C. **filicúlmis** (typical). — Dry rocky, gravelly or sandy ground, Fla. to Tex., n. to Mass., N.Y., O., s. Mich., Ill., Ia. and Neb. Aug.–Oct. (Including *C. Bushii* Britt.) Fig. 364.

Var. **maciléntus** Fern. (meagre). — Centr. Me. and sw. Que. to Minn., s. to Va., O., Ind., Ill. and Mo. Fig. 365.

Var. **oblìtus** Fern. & Grisc. (overlooked). — Sands, Fla. to N.J.

35. C. globulòsus Aubl. (globular). — *Bases without much development of hard bladeless tubers;* culms smooth, tufted, 1–7.5 dm. high; *leaves soft, flat, smooth,* 2–6 mm. broad, often elongate and forming turf; involucral leaves similar, loosely ascending or spreading, prolonged; *umbel with* 3–10 smooth *rays up to 7 cm. long; heads subglobose to thick-cylindric, of* 20–40 *loosely radiating stramineous to green* 3–6*-flowered* lance-cylindric slightly compressed *slender-pointed spikelets; scales closely overlapping, with appressed tips, membranaceous,* ovate-lanceolate or oblong; rachilla broadly winged; achenes linear-oblong, 1.5–2 mm. long, about 0.5 mm. thick. (*C. echinatus* (Ell.) Wood) — Open woods, prairies, roadsides and waste places, Fla. to Tex., n. to Va. (casually to N.J. and Pa.), Mo. and Okla. Aug.–Oct. (Trop. Am.) Fig. 366.

366. C. globulosus.

36. C. ovulàris (Michx.) Torr. (egg-shaped). — Rhizome short, with tuberous enlargements; culms firm, smooth, acutely angled, 0.2–1 m. high; leaves flat or revolute, mostly shorter than the culm, 3–10 mm. wide; *involucre* similar, *of 4–6 widely divergent flat leaves, the longest* 1.2–4.5 *dm. long;* umbel with 1–8(–11) simple ascending smooth rays up to 9 cm. long, rarely a single sessile glomerule; *spikes fulvous, globose-ellipsoid,* definitely longer than thick, *dense,* in maturity 1–2.3 cm. long, 0.8–1.8 cm. thick; *spikelets* linear-lanceolate, *quadrangular,* 2–3-flowered, *radiating; scales* oblong, *closely appressed and imbricated;* achene trigonous-oblong, mucronate, drab or pale brown, about 2 mm. long. (Incl. var. *robustus* Boeckl.) — Sandy swamps, ditches, open woods and barrens, Fla. to Tex., n. to se. N.Y., N.J., Pa., O., Ind., Ill., Mo. and Kans. July–Sept. Fig. 367.

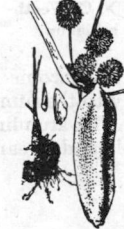

Var. **sphaèricus** Boeckl. (spherical). — Smaller and often more slender; basal and involucral leaves firmer, 1.5–5(–7) mm. wide; longest leaf of involucre 0.5–1.5(–3) dm. long; rays of umbel 1–4(–9); spike globular, 7–12(–15) mm. in diameter. — Ga. to Tex., n. to Va., W.Va., s. O., s. Ind., Mo. and Okla.

367. C. ovularis.

37. C. retrórsus Chapm. (turned backward). — Similar to no. 36; culm firm, more slender, acutely angled, 1.5–9 dm. high; basal and involucral leaves pale green, firm, 1–4 (–7) mm. broad; longest bracts 0.5–2.5 (–5) dm. long; *spikelets cylindric, pale green to drab,* few to very numerous, 2–6 *mm. long, in dense cylindric spikes* 0.5–2 (–3) cm. long and 4–13 mm.

thick, *the crowded spikelets wide-spreading or the basal strongly reflexed;* rays of umbel 0.5–15 cm. long. (*C. cylindricus* (Ell.) Leggett, not Boeckl.; *C. Torreyi* Britt.) — Dry sandy soils, Fla. to Tex., n. to se. N.Y., N.J., Pa., Md., D.C. and se. Ky. Late July–Oct. Fig. 368. — In burns and recent clearings often greatly stimulated, the enlarged spikes then becoming compound, simulating those of the more southern *C. Deeringianus* Britt. & Small.

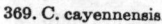

368. C. retrorsus.

Var. **Náshii** (Britt.) Fern. & Grisc. (for its discoverer, George Valentine Nash, 1864–1921). — Culms very slender, 1–6 dm. high; basal and involucral leaves 1.5–3.5 mm. broad; longest bracts 0.5–1.8 dm. long; spikelets 1.5–3.5 mm. long, in loose short-ovoid spikes only 4–9 mm. long and 4–7 mm. thick. (*C. Nashii* Britt.) — Dry sands, pine-barrens, etc., Fla. to Tex., n. to N.J. and D.C.

38. C. cayennénsis (Lam.) Britt. (of Cayenne). — Culms sharply angled, smooth and wiry (2–5 dm. high), much exceeding the smooth flat leaves; *heads* 3–6, cylindric (1–1.7 cm. long), *sessile in a glomerule;* involucral bracts divergent or reflexed; spikelets crowded, 2.5–5 mm. long, dull, pale brown; *scales thin and veiny, the lowest often persistent.* (*C. flavus* of ed. 7, not J. & C. Presl) — Waste ground about Philadelphia. (Adv. from the Tropics) Fig. 369.

369. C. cayennensis.

Subgen. Kyllínga (Rottb.) Suringar

39. C. tenuifòlius (Steud.) Dandy (thin-leaved). — Dwarf usually *tufted annual,* 0.2–3 dm. high, with soft flat leaves; *spikelets* 1.5–2 mm. long, of 3 or 4 2-ranked scales, 1-*flowered, aggregated into a tiny subglobose pale head or 2 or 3 sessile heads confluent* and 4–8 *mm. broad; upper scales of spikelet* ovate, pointed, *rough on keel;* stamens and style 2. (*C. densicaespitosus* Mattf. & Kükenth.; *Kyllinga pumila* Michx.) — Damp soil, Fla. to Tex. and Mex., n. to L.I., Pa., Md., W.Va., O., Ind., Ill., Mo. and se. Kans. Late July–Oct. (Trop. Am.; Afr.) Fig. 370.

370. C. tenuifolius.

40. C. brevifòlius (Rottb.) Hassk. (short-leaved). — *Perennial with long creeping slender rhizomes;* leaves and inflorescence much as in no. 39; *spikelets* 3–3.5 *mm. long.* — Fresh tidal marshes and shores, Del. R., Pa.; Chickahominy R., Va.; Fla. to Tex.; locally adv., sw. Ct. Aug.–Oct. (Trop. and warm-temp. Am.; Afr.; Asia; Austr.; Polynesia)

41. × C. Weatherbiànus Fern. (for its discoverer, Charles Alfred Weatherby, 1875–1949) is a sterile hybrid between no. 24 and *Rhynchospora capitellata,* with fascicles of castaneous axillary spikes, the spikelets with many distichous scales, once found on Cape Cod. Fig. 371.

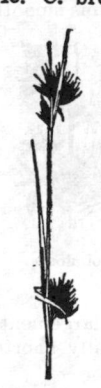

371. × C. Weatherbianus.

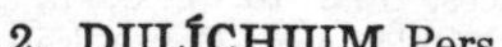

2. DULÍCHIUM Pers.

Spikelets linear, flattened, sessile in 2 ranks on peduncles emerging from the sheaths of the leaves; scales lanceolate, decurrent, forming flat wing-like margins on the joint below. Perianth of 6–9 downwardly barbed bristles. Stamens 3. Style 2-cleft above. Achene flattened, linear-oblong, beaked with the long persistent style. — A perennial herb, with horizontal rhizome and terete simple hollow culms (2–10 dm. high) jointed and leafy to the summit; leaves short and flat, linear, 3-ranked. (Name of uncertain origin.)

372. D. arundinaceum.

1. D. arundinàceum (L.) Britt. (reed-like), Three-way Sedge. — Swamps and margins of pools and streams, Nfld. to B.C., s. to Fla., Tex. and Calif. July–Oct. Fig. 372.

3. ELEÓCHARIS R. Br. Spike-Rush

Spikelet few–many-flowered. Scales imbricated in many (rarely in 2 or 3) ranks. Perianth of 3–12 (commonly 6) bristles, usually rough or barbed downward, rarely obsolete. Style 2–3-cleft, its bulbous base persistent as a tubercle jointed upon the apex of the lenticular or triangular achene. — Leafless (rarely with basal capillary leaves), chiefly perennial, with tufted culms sheathed at the base, from matted or creeping rhizomes; flowering in summer.

Genus nearly cosmop., chiefly of trop. and warm-temp. reg. (Name from the Greek, *elos*, a *marsh*, and *charis*, *grace;* many species being marsh plants.)* Sometimes spelled HELEOCHARIS.

a.Spikelet hardly if at all thicker than the culm; scales firmly persistent. . . *b*.

b.Spikelet cylindric, many-flowered, 1.5–6 cm. long; scales coriaceous, keelless.

Culms terete, septate. . . . . 1. *E. equisetoides*.

Culms quadrangular, nonseptate. . . . . 2. *E. quadrangulata*.

b.Spikelet subulate, 0.7–2.5 cm. long, few-flowered; scales herbaceous, keeled; culms slender, sharply 3-angled. . . . . 3. *E. Robbinsii*.

a.Spikelet thicker or broader than culm; scales deciduous. . . *c*.

c.Tubercle nearly confluent with summit of achene, seemingly a continuation of it but of slightly different texture. . . . *d*.

d.Culms soft, capillary, less than 1 mm. thick, all ascending; stolons terminated by tubers; spikelets 2–7 mm. long, 2–9-flowered, flattened; scales thin.

Culms 1–7 cm. high, in dense mats; spikelets 2–4 mm. long; scales green or pale brown, 1.5–2.5 mm. long; achenes 0.9–1.5 mm. long. . 4. *E. parvula*.

Culms 0.3–4 dm. high, few, from creeping rhizomes; spikelets 4–7 mm. long; scales purple- or brown-tinged, 3–8 mm. long; achenes 1.8–2.5 mm. long. . . . . 5. *E. pauciflora*.

d.Culms firm, flattened, 1–2 mm. broad, 0.1–1.7 m. long, the longer ones arching and often rooting at tip; caudex thick and short, without stolons and tubers; spikelets 0.6–2 cm. long, 12–20-flowered, plump; scales coriaceous. . . . . 6. *E. rostellata*.

c.Tubercle sharply differentiated, usually articulated with achene. . . *e*.

e.Achenes with prominent straight longitudinal ridges separated by numerous slender trabeculae. . . *f*.

f.Culms not spongy; fertile scales scarious or membranaceous, with slender midrib and brown to purple sides; anthers 1 mm. long; bristles wanting or not overtopping tubercle.

Culms capillary, usually angular and sulcate, less than 0.5 mm. thick; spikelets flattened; scales 2–3-ranked, uniform, membranaceous. . . . . 7. *E. acicularis*.

Culms flat, ancipital or with inrolled edges, 0.6–1 mm. broad; spikelets terete; scales many-ranked; empty lowest ones subcoriaceous, larger than the fertile. . . . . 8. *E. Wolfii*.

f.Culms spongy, 0.5–8 cm. high; scales herbaceous, with broad longitudinally sulcate green median band; anthers 0.3–0.4 mm. long; bristles overtopping tubercle. . . . . 9. *E. radicans*.

e.Achenes without prominent longitudinal ridges on the faces, with smooth, rugulose or variously cancellate or reticulate surfaces. . . *g*.

g.Achenes without coarse and deep nearly equilateral honeycomb-reticulation (or if so, in no. 35, the plant with purple rhizomes and stolons, membranaceous scales, and minute tubercle); fertile scales membranaceous to subherbaceous (if coriaceous, the achene much longer than the tubercle). . . *h*.

h.Plants tufted, cespitose (or with soft thread-like rhizomes in nos. 10, 11 and 21), without firm elongate rhizomes and stolons. . . *i*.

i.Achenes biconvex; style 2-cleft. . . *j*.

j.Upper sheaths (at base of culm) loose, with white scarious tips; plant delicately stoloniferous or tufted; achenes olivaceous to black, with acute conical to subulate tubercle.

Outer scales of spikelet ovate-oblong, with prominent green midrib and brown, purplish or pale sides, round-tipped; mature achenes olive to dark brown. . . . . . . 10. *E. olivacea*.

Outer scales oblong-lanceolate, without prominent midrib, pale, obtuse to subacute; mature achene reddish, becoming purple-black. . . . . 11. *E. flavescens*.

j.Upper sheaths close, with herbaceous to coriaceous orifice; plants tufted; achenes whitish, yellowish, greenish or brown (blackish in nos. 12 and 13 with depressed tubercle). . . *k*.

k.Achenes blackish, much broader than the depressed or saucer-shaped tubercle; orifice of upper sheath oblique, with elongate-deltoid tooth.

---

* All measurements of achenes exclude the tubercle. Treatment greatly aided by the papers of H. K. SVENSON in Rhodora, xxxi (1929)–xli (1939).

Spikelet subglobose to ovoid, 2–3 mm. thick; scales firm; achenes 1 mm. long. . . . . 12. *E. geniculata.*

Spikelet lance-ovoid to subcylindric, 1.5–2 mm. thick; scales membranaceous; achenes 0.5 mm. long. . . . . . 13. *E. atropurpurea.*

k. Achenes whitish or greenish to brown, with deltoid to conic-subulate tubercle; orifice of upper sheath subtruncate, with short broad-deltoid tooth (oblique only in no. 23 which has a conic-subulate tubercle). . . *l.*

l. Sheath-orifice oblique, with elongate tooth; culms arching and reclining, many shortened and basal; spikelets lanceolate, loosely flowered; achenes greenish; tubercle subulate, with flange-like base. . . . . 23. *E. intermedia.*

l. Sheath-orifice subtruncate, short-toothed; spikelet lanceolate or cylindric to ovoid or subglobose, usually closely flowered; achenes whitish, yellowish or brown; tubercle deltoid or turban-shaped, thin above. . . *m.*

m. Tubercle deltoid-ovate or bulbiform, strongly constricted at base. . . . . 22. *E. Macounii.*

m. Tubercle deltoid to turban-shaped, not strongly constricted at base. . . *n.*

n. Base of tubercle nearly or quite covering summit of achene. . . *o.*

o. Spikelets broadly ovoid to cylindric, obtuse to subacute; scales obtuse.

Bristles rarely reaching tubercle or much shorter or wanting; tubercle strongly depressed, closely sessile, less than one-fourth the height of achene. 14. *E. Engelmanni.*

Bristles (very rarely wanting) much overtopping tubercle; tubercle broadly deltoid to turban-shaped, one-third to one-half height of achene, with slight constriction at base. . . . . 15. *E. obtusa.*

o. Spikelets lance-acuminate; scales acute. . . . . 16. *E. lanceolata.*

n. Base of tubercle covering about two-thirds of summit of achene.

Tubercle higher than broad, overtopped by bristles; scales reddish or purple. . . . . 17. *E. ovata.*

Tubercle broader than high; bristles rudimentary or wanting; scales greenish or drab. . . . . . . 18. *E. diandra.*

i. Achenes trigonous; style 3-cleft. . . *p.*

p. Culms capillary to bristle-like or angled, not flattened; achenes ellipsoid to pyriform-obovoid, white to brown or olive, much broader than base of tubercle. . *q.*

q. Mature achenes whitish, at most 1 mm. long and 0.5 mm. broad, with minute tubercle; scales 1–2 mm. long.

Scales acute or attenuate to the free or spreading tips; bristles present. . . . . 19. *E. microcarpa.*

Scales broadly rounded above, closely appressed; bristles obsolete. . . . . 20. *E. Brittonii.*

q. Mature achenes yellowish, brown or olive, larger, with prominent tubercle; scales longer. . *r.*

r. Culms stiff, erect, all elongate, in small tufts or solitary, from delicate filiform rhizomes; cylindric-ovoid spikelet and coriaceous appressed whitish-brown scales rounded at summit; achenes 1 mm. long. . . . . . . 21. *E. albida.*

r. Culms weak, arching or reclining, many short ones crowded at base of the dense non-stoloniferous tufts; spikelets lanceolate to lance-ovoid, acutish; scales membranaceous, green, drab or purple; achenes 1.2–1.5 mm. long.

Basal sheaths with truncate orifice; scales reddish, closely appressed; achenes obovoid; tubercle ovate-triangular, two-thirds the breadth of achene. . . . . . . 22. *E. Macounii.*

Basal sheaths with oblique orifice; scales green, with drab or pale brown sides, loosely ascending; achenes slenderly pyriform; tubercle subulate, with basal flange one-fourth the breadth of achene. . . . . . 23. *E. intermedia*

p. Culms strongly flattened, wiry; achenes turbinate-obovoid, black, their summits completely covered by the pale sessile depressed tubercle. . . . . 24. *E. melanocarpa.*

h. Plants with firm reddish, purple or black strong rhizomes or stolons; achenes yellow to tawny or olivaceous; tubercle spongy, bulbiform to depressed, usually separated from achene by a definite constriction. . . *s.*

s. Achenes biconvex, plump, smooth to delicately cancellate or reticulate, mostly covered by subpersistent scales; bristles often present. . . *t.*

t. Upper sheath with oblique herbaceous orifice; achenes smooth or merely obscurely cancellate. . . *u.*

u. Basal scales of spikelet usually 2 or 3 below the thinner fertile scales; culms 0.5–5 mm. thick (in dried material) at summit of upper sheath; perianth-bristles commonly elongate. . *v.*

v. Tubercle elongate, much longer than broad; achenes 1.2–2.1 mm. long, slenderly obovoid or pyriform; culms subterete, rather firm. . . . . 25. *E. palustris.*

v. Tubercle depressed-deltoid, umbonate or broad-ovate, as broad as or broader than long; achenes 1.2–1.6 mm. long, roundish or broad-ovoid.

Culms firm or wiry, subterete; fertile scales loosely ascending, narrowly ovate to lanceolate, mostly acute or attenuate. . . . . 26. *E. Smallii.*

Culms soft, flat or compressed; fertile scales appressed, ovate, obtuse or subacute. . . . . . . 27. *E. macrostachya.*

u. Basal scale solitary, suborbicular, spathiform, usually completely encircling base of spikelet; culms filiform, 0.3–2 (rarely –3) mm. thick at summit of upper sheath; bristles usually wanting. . . *w.*

w. Spikelet closely many-flowered, linear-lanceolate to slenderly ovoid; fertile scales membranaceous, opaque, closely appressed, lower and median ones 1.8–3 mm. long; achenes 0.7–1 mm. broad. . . . . 28. *E. calva.*

w. Spikelet loosely few (5–30)-flowered, lanceolate to ovoid; fertile scales firm, lustrous, loosely ascending, lower and median 3–5 mm. long; achenes 1–1.4 mm. broad.

Achenes broadly obovoid or pyriform; tubercle bulbiform, slenderly conical to lanceolate, commonly higher than broad, 0.2–0.5 mm. broad, covering one-sixth to rarely half the breadth of achene. . . . . 29. *E. halophila.*

Achene ellipsoid to narrowly obovoid; tubercle depressed-deltoid to low-conical, about as broad as high, 0.6–1 mm. broad, covering one-half to three-fourths breadth of achene. . . . . 30. *E. uniglumis.*

t. Upper sheath with subtruncate coriaceous summit; achenes distinctly fine-reticulate. . . . . 31. *E. ambigens.*

s. Achenes trigonous, their surfaces granular, papillate, wrinkled or reticulate, often persistent after falling of scales; bristles mostly wanting or fugacious (persistent in no. 32). . . *x.*

x. Angles of the often persistent achenes not keel-like. . . *y.*

y. Surface of achene minutely roughened by edges or elevated angles of cell-walls, not deeply and coarsely honeycomb-reticulated. . . *z.*

z. Culms subterete, angled or corrugated; scales obtuse or acutish; rhizomes flexuous, cord-like, 0.5–2 mm. thick.

Culms 3–7.5 dm. high; spikelet 7–10 mm. long; fertile scales 2.5–3 mm. long, very numerous; achenes 1.2–1.4 mm. long, obtusely angled; tubercle bulbiform, 0.4–0.5 mm. long; bristles coarse, persistent. . . . . 32. *E. fallax.*

Culms 2–10 cm. high, delicately capillary; spikelet 1.5–4.5 mm. long; fertile scales 1 mm. long, few; achenes 0.7–0.8 mm. long, sharply angled; tubercle a depressed saucer, with central apiculation 0.1 mm. high; bristles wanting. . . . . 33. *E. nitida.*

z. Culms flat, wiry; scales acuminate or attenuate; rhizome subligneous, 2–4 mm. thick. . . . . 34. *E. compressa.*

y. Surface of achene shallowly to deeply and coarsely honeycomb-reticulate, iridescent; rhizome cord-like, 1–3 mm. thick.
Culms 4- or 5-angled; mature achene olivaceous; the reticulations usually deep, without prominent transverse bands. 35. *E. tenuis.*
Culms 6–8-angled; mature achenes yellow to orange, with shallow reticulation, the many transverse bands prominent. . . . . 36. *E. elliptica.*
x. Angles of promptly deciduous achenes keel-like; rhizome stiff, subligneous, 3–6 mm. thick. . . . . 37. *E. tricostata.*
g. Achenes with coarse deep and regular honeycomb-reticulation, iridescent; scales coriaceous or cartilaginous; culms filiform, in dense non-stoloniferous drab-based tussocks.
Spikelets linear-cylindric or slenderly lanceolate, 1.5–2 mm. thick, often proliferous; scales 2 mm. long; achene 1 mm. long. . . . 38. *E. vivipara.*
Spikelets ellipsoid or ovoid, much thicker, not proliferous; scales 2–3.3 mm. long; achene 1.3–2 mm. long.
Culms sharply 3-angled; achenes with 3 thickened and ridge-like angles; tubercle conic-subulate, firm, its base half the breadth of achene. . . . . 39. *E. tortilis.*
Culms terete; achenes obtusely angled; tubercle cap-like, spongy, rounded at summit, as broad as achene. . . . . 40. *E. tuberculosa.*

1. E. equisetoìdes (Ell.) Torr. (like *Equisetum*, horsetail). — *Culms* stout, 0.5–1 m. high, *terete, septate by cross-partitions* 1–5 *cm. apart;* basal sheaths often leaf-bearing; spikelets 2–4 cm. long, cylindric; scales firm, persistent, pale, with scarious margins; *achenes* 2.3–2.6 mm. long, *with transversely linear-rectangular reticulations* and a conical-beaked sessile tubercle; bristles wanting or shorter than achene. (*E. interstincta* of ed. 7, not (Vahl) R. & S.) — Shallow water, n. Fla. to e. Tex., n., very locally, to se. N.C.; e. Md. to e. Mass.; w. N.Y. to Wisc., n. Ill. and Mo. July–Oct. Fig. 373.

373. E. equisetoides.

2. E. quadrangulàta (Michx.) R. & S. (four-angled). — Similar; *culms continuous and sharply* 4*-angled,* 3–9 dm. high, 1.5–4 mm. thick; *upper basal sheath with free tip* 0.5–1 *cm. long;* spikelet 1.5–4 cm. long; *achenes finely reticulated,* 2–2.5 *mm. long,* 1.4–1.8 mm. broad, *with a conical flattened tubercle.* — Pools and creeks (often tidal), chiefly of Coastal Plain, S.C. and Tenn. to Cape May, N.J. June–Oct.

374. E. quadrangulata, v. crassior.

Var. **crássior** Fern. (thicker). — Culms 6–12 dm. high, 3–5.5 mm. thick; *free tip of upper sheath* 1.5–8 *cm. long;* spikelet up to 6 cm. long; *achenes* 2.2–3 *mm. long,* 1.6–2 mm. broad. — Pools and pond-margins, n. Fla. to Tex., locally n. to Mass., Ct., N.Y., s. Ont., O., Mich., Wisc., Mo. and Okla. Fig. 374.

3. E. Robbínsii Oakes (for James Watson Robbins, 1801–1879). — *Fertile culms slender, soft, sharply* 3*-angled* (pressing flat), 1.5–9 dm. long, when submersed *accompanied by numerous capillary much elongated floating sterile culms;* sheath obliquely truncate, with dilated orifice; *spikelets subulate-lanceolate,* 0.7–2.5 cm. long; *scales* 3–9, *herbaceous, keeled, lanceolate, convolute, clasping the long flattened joints of the rachis,* their margins scarious; achene narrowly obovoid, 2–2.5 mm. long, triangular, minutely reticulated, about one-half length of bristles, tipped by the flattened-subulate tubercle. — Shallow ponds, dead-waters, peaty pools, etc., N.S. to Temagami Forest, Ont., and n. Wisc., s. to Del., centr. N.Y. and n. Ind.; ne. N.C.; centr. Ga. to n. Fla. Aug.–Oct. Fig. 375.

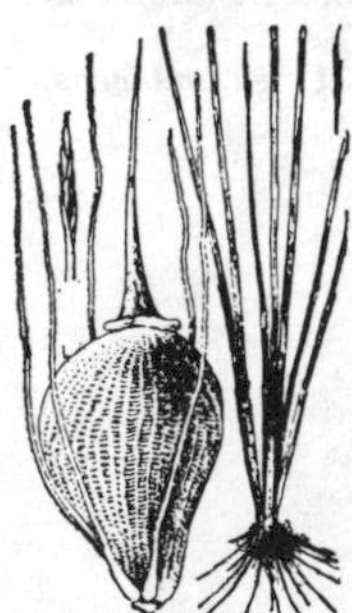

375. E. Robbinsii.

4. E. párvula (R. & S.) Link (very small). — *Densely tufted,* often with filiform stolons and fusiform brown or purple tubers 3–6 mm. long; *culms spongy,* capillary, 1–7 *cm. high; spikelets* ovate, *flattened,* 2–9-flowered, 2–4 *mm. long; scales green to pale brown,* 1.5–2.5 *mm. long,* ovate; *achene* prominently 3-angled, *pale,* 0.9–1.5 *mm. long,* tipped by a minute triangular tubercle; bristles equaling or exceeding achene or wanting. (*Scirpus nanus* Spreng.) — Wet saline or brackish shores, Nfld. to La.; inland, locally, to w. N.Y., Mich. and Minn.; also B.C. to Calif. July–Oct. (W.I.; S. Am.; Eu.; n. Afr.) Fig. 376.

376. E. parvula.

Var. **anachaèta** (Torr.) Svenson (without bristles). — 3–12 cm. high; *spikelets 4–23-flowered, 3–6 mm. long;* scales darker; *achene minutely verrucose; bristles wanting.* — Alkaline to fresh mud, Minn. and w. Ia. to Oreg., s. to La., Tex. and Mex. (W.I.; S.Am.)

5. E. PAUCIFLÒRA (Lightf.) Link (few-flowered). — *Rhizomes and stolons elongate,* often bearing loosely scaly tubers 0.5–1 cm. long; *culms in small tufts,* capillary, furrowed, *firm,* 0.3–4 *dm. high; spikelets* lanceolate to elliptic, flattened, 4–8 *mm. long; scales* 2–7, ovate, acutish, *lustrous, drab to castaneous,* 3–8 *mm. long,* the 2 lowest enlarged; *achene* trigonous-obovoid to -fusiform or planoconvex, 1.8–2.5 *mm. long,* with rectangular reticulation, *tipped by the minute barely differentiated darker tubercle;* bristles longer or shorter than achene. (*Scirpus pauciflorus* Lightf.) — Variable species of Eurasia and N. and S.Am. Typical Eurasian plant has indurated caudices, spikelets 5–8 mm. long, achene equilaterally triangular, lustrous, 1.1–1.3 mm. broad, and bristles regularly and firmly retrorse-barbed. Ours is

377. E. pauciflora, v. Fernaldii.

Var. **Fernàldii** Svenson (for the present author, 1873–). — Caudex softer; spikelets 3–6 mm. long; achene inequilateral, usually opaque, 0.9–1.1 mm. broad; bristles with weak and irregular barbing. — Damp calcareous shores, ledges and swamps, Nfld. to w. Ont., s. to C.B., N.B., n. N.E., nw. N.J., centr. N.Y., nw. Pa., O., Ind., Ill. and Ia. July–Sept. FIG. 377.

6. E. rostellàta Torr. (with small beak). — *In firm tussocks from a thick vertical caudex; culms flattened, wiry,* the fertile erect and 1–6 dm. high; *sterile arching or reclining, up to* 1.7 *m. long, proliferous and rooting at tip;* sheaths coriaceous, drab to brown; *spikelets fusiform to ovoid,* 0.6–2 cm. long; *scales leathery,* obtuse, drab or brown, 3.5–5 *mm. long; achene* trigonous-obovoid, *narrowed into the confluent pyramidal tubercle,* the whole 2–3 mm. long, about equalled by the *rigid bristles.* — Saline, brackish or limy marshes, Fla. to s. Me. and N.S.; centr. N.Y. to B.C., s. to O., Ind., Ill., Tex. and Mex. July–Oct. (W.I.) FIG. 378.

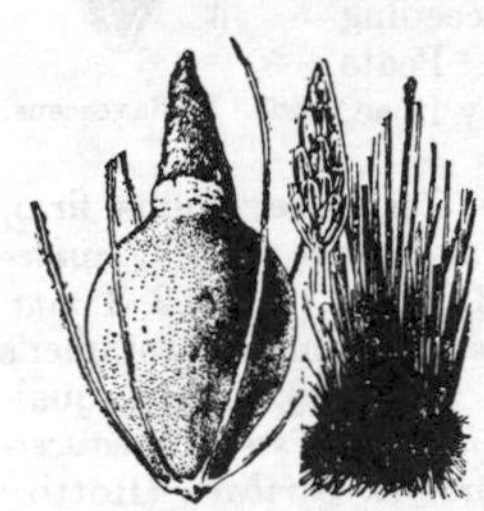
378. E. rostellata.

7. E. aciculàris (L.) R. & S. (needle-shaped). — *Tufted,* forming carpets, *with capillary rhizomes and stolons; culms capillary,* commonly angular and furrowed, 0.2–3 dm. long (often much elongated in deep water); sheaths loose; *spikelets flattened, linear to narrowly ovate,* 2–7 mm. long, 3–15-flowered; *scales membranaceous, with slender greenish midrib,* usually with reddish-brown sides, scarious-margined, acute, 1.3–2.5 *mm. long; anthers about* 1 *mm. long; achenes* (often not developed) *slenderly obovoid or ellipsoid, whitish or pearly,* 0.7–1.2 *mm. long,* obscurely 3-angled, *with distinct straight longitudinal ridges and many transverse cross-ridges;* tubercle minute, conic-triangular to conic-subulate, with flange-like base; bristles 3 or 4, very delicate, equaling achene or wanting. — Damp shores and low grounds, Lab. to B.C., s. to n. Fla., Tenn., Mo., Okla., n. Mex. and s. Calif. July–Oct. (Eurasia) FIG. 379. — Submersed and commonly sterile states, with much elongated delicate culms and rhizomes, have been separated as forma **longicaúlis** (Desmaz.) Hegi (long-stemmed).

379. E. acicularis.

Var. **submérsa** (Hj. Nilss.) Svenson (submersed). — Dwarf, usually sterile; *culms firm, rather fleshy, without longitudinal furrows,* transparent when dry. — Greenl. and Lab. to Alaska, s. to w. Nfld. (Arctic Eu.)

8. E. Wólfii Gray (for its discoverer, JOHN WOLF, 1821–1897). — Tufted, from slender elongate rhizomes; *culms 2-edged, flat or inrolled,* 0.6–1 *mm. broad,* 1.5–6 dm. high; sheaths scarious-margined; *spikelets slender-ovoid, terete,* 5–9 mm. long; *scales* ovate-oblong, obtuse, *the lowest empty ones subcoriaceous and elongate,* the fertile smaller, scarious-margined, purple-striate; *achene* pyriform, lustrous, pale, 1 mm. long, *with* 9 *nearly equidistant obtuse ribs and numerous transverse trabeculae; tubercle depressed-truncate, with apiculate center;* bristles none. — Wet prairies, shores and flats, s. Ind. to s. Sask., s., very locally, to Tenn., La. and Kans.; L.I., N.Y. May–July. FIG. 380.

380. E. Wolfii.

9. E. radìcans (Poir.) Kunth (rooting). — Delicately rhizomatous and stoloniferous, forming close turf 0.5–8 cm. high; *culms spongy,* pale green; sheaths close, membranous, fugacious; *spikelets flattened, elliptic-ovate,* acute, 2–4 *mm. long,* few-flowered; *scales herbaceous,* ovate-lanceolate, *with broad striate median green body* and narrow scarious margins, the lower 1.5–2.5 mm. long; *anthers* 0.3–0.4 *mm. long; achenes* slenderly obovoid, whitish or yellowish, 0.7–1 *mm. long,* 0.2–0.3 mm. thick, *with elevated longitudinal ridges and many close trabeculae;* tubercle conic-subulate, 0.2 mm. long; *bristles* retrorsely barbed, *overtopping achene.* (*E. Lindheimeri* (C. B. Clarke) Svenson) — Wet shores and flats, local, se. Va.; Tex. and n. Mex. to s. Calif., ne., very locally, to Mich. July–Oct. (W.I.; temp. S.Am.) Fig. 381.

381. E. radicans.

10. E. olivàcea Torr. (olive-brown). — *Tufted or with* culms scattered on soft *filiform stolons; culms* compressed-filiform, rather *soft,* often spongy, ascending, diffuse or arching, light green, 0.5–4 dm. long; *sheath loose, with whitish scarious or membranaceous oblique summit;* spikelets ellipsoid to slenderly ovoid, acutish, 4–9 mm. long; *scales* ovate to ovate-oblong, *the outer round-tipped,* loosely appressed-ascending, *with prominent green midrib* and brown to reddish (when submersed green) sides, the lowest 1.5–2.5 mm. long; *achenes olive to dark brown,* lustrous, obovoid, biconvex, 0.8–1 *mm. long,* puncticulate; *tubercle green, saucer-shaped with conic-subulate center* 0.2–0.3 *mm. high;* bristles most often overtopping achene. — Wet sands and peats, Fla. to N.S.; locally inland from Me. to s. Ont. and Minn., thence s. to w. Pa., O. and Mich. June–Oct. Fig. 382.

382. E. olivacea.

11. E. flavéscens (Poir.) Urban (yellowish). — Similar to no. 10; spikelets pale, 2–6 mm. long; *scales pale and scarious, without prominent midrib, the outer oblong to lanceolate,* blunt to acute; *mature achenes red-brown to reddish-black,* 0.6–0.9 mm. long; tubercle with less prominent rim, the conic center only 0.1–0.2 mm. high; bristles shorter than to barely exceeding achene. (*E. flaccida* (Reichenb.) Urban; *E. ochreata* (Nees) Steud.) — Peats and sands of Coastal Plain, Fla. to e. Tex., n. to S.C.; and very locally in se. Va., Del. and s. N.J. June–Oct. (W.I.; Mex.; e. S.Am.) Fig. 383.

383. E. flavescens.

12. E. geniculàta (L.) R. & S. (bending like the knee-joint). — Cespitose; culms firm, subterete, 0.3–4 dm. high; *sheaths tight, with firm oblique orifice; spikelets subglobose to ovoid, obtuse,* closely flowered, 3–5 mm. long, 2–3 *mm. thick; scales firm, brown to purplish,* with paler margin, rounded; stamens 2 (or 3); *achene* obovoid, *lustrous-black* to purple, 1 *mm. long,* nearly equaling the 6–8 coarse bristles; *tubercle* spongy, *whitish, depressed or saucer-shaped.* (*E. capitata* of Am. auth., not (L.) R. Br.; *E. caribaea* (Rottb.) S. F. Blake and var. *dispar* (E. J. Hill) S. F. Blake) — Damp sand and gravel, Fla. to Tex. and s. Calif., n. to N.C.; sands of Great Lakes, s. Ont. to Mich. and nw. Ind. Aug., Sept., Fig. 384.

384. E. geniculata.

13. E. atropurpùrea (Retz.) J. & C. Presl (dark purple). — Dwarf tufted annual; culms erect or arcuate, capillary, 3–12 cm. high; *spikelet lance-ovoid to subcylindric,* 2–8 mm. long, 1.5–2 *mm. thick; scales membranaceous,* ovate, blunt, *with* broad green midrib and *dark brown sides; achenes* strongly flattened, lenticular-obovoid, 0.5 *mm. long, lustrous-black;* tubercle capping one-fourth of top of achene, saucer-shaped, exceeding the short (or 0) slender bristles. — Wet sand, Fla. to Tex. and Mex., locally n. to Ga., Ia., Neb. and Colo. July–Sept. (Trop. regions) Fig. 385.

385. E. atropurpurea.

14. E. Engelmánni Steud. (for its discoverer, George Engelmann, 1809–1884). — Tufted annual, usually with many ascending culms, 1–4 dm. high; sheath-orifice subtruncate, short-toothed; *spikelets slender-cylindric,* usually subacute, 0.4–2 cm. long, 2–3.5 mm. thick; *scales appressed, drab or brown,* with whitish scarious margin, elliptic-obovate, obtuse; *achene* cuneate-obovate, biconvex, *subtruncate at summit,* whitish, becoming brown and lustrous; *tubercle closely sessile, as broad as achene, very depressed-triangular, only* 0.2–0.4 *mm. high,* less than one-fourth the height of achene; *bristles rarely reaching tubercle,* or these wanting or mere rudiments in forma **detónsa** (Gray) Svenson (shaved off.) — Wet sand, peat or mud, s. Me. to Sask. and Wash., s. to e. Va., nw. Ga., Ark., Tex., Ariz., and s. Calif. May–Oct. Fig. 386.

386. E. Engelmanni.

15. E. obtùsa (Willd.) Schultes (blunt). — Similar to no. 14; culms 0.3–7 dm. long, from capillary to 1.5 mm. thick, soft; *spikelets globose-ovoid to ovoid-cylindric,* obtuse, 2–13 mm. long, 2–5 *mm. thick,* closely to loosely many-flowered; *scales* membranaceous, with scarious margins, ovate-oblong to suborbicular, *brown, closely appressed, their arching bases nearly horizontal;* style 3- or 2-cleft; *achene turbinate-obovoid,* somewhat compressed, narrowed at base, pale to deep brown, smooth, lustrous, 1–1.5 mm. long, *much overtopped by the bristles; tubercle broadly deltoid to turban-shaped, acute, one-third to one-half the height*

387. E. obtusa (typical).

of achene and but slightly narrower, with a slight constriction separating them. — Muddy or wet places, C.B. and N.B. to Minn., s. to nw. Fla., Ala., Miss., La. and e. Tex.; also B.C. to n. Cal. May–Oct. (Hawaii) FIG. 387.

Var. **jejùna** Fern. (starved). — *Culms capillary,* 1–10 (rarely –20) *cm. long,* often depressed; *spikelets few-flowered,* 2–5 *mm. long; scales spreading,* drab or purplish; *achene smaller, with tubercle nearly half its height.* — N.S. and N.E. to se. Va.

Var. **Peàsei** Svenson (for its discoverer, ARTHUR STANLEY PEASE, 1881–). — *Bristles none; upper sheaths sometimes with blades up to* 1 *cm. long.* — Wet shores, Que., w. Me. and N.H.

388. E. obtusa, v. ellipsoidalis.

Var. **ellipsoidàlis** Fern. (ellipsoid). — *Spikelets ellipsoid; the scales strongly appressed, their bases obliquely ascending.* — Wet peat and sand, e. Mass. to se. Va. FIG. 388.

16. E. lanceolàta Fern. (lanceolate). — Similar to no. 15; culms capillary, about 2 dm. high; *spikelets lance-acuminate,* 5–8 mm. long; *scales acute,* light brown, scarious; achenes broadly compressed-obovoid, 1 mm. long, exceeded by the coarse bristles; *tubercle* as broad as summit of achene, *elongate-deltoid,* half as high as achene. — Damp places, s. Mo., Ark. and e. Tex. June–Sept. FIG. 389.

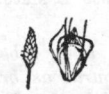

389. E. lanceolata.

17. E. ovàta (Roth) R. & S. (ovate or egg-shaped). — Resembling no. 15; *culms ascending,* 0.3–5 dm. long; spikelets globose-ovoid to ovoid-cylindric, 2–11 mm. long, 2–4 mm. thick, dense; *scales oblong-ovate,* obtuse, appressed, *purplish-brown,* with white margin and pale midrib; achene obpyriform or obovoid, light brown, 1 mm. long; *tubercle deltoid-conic to conic-subulate, usually much longer than broad, about half as broad as achene,* overtopped by the bristles. — Wet open places, local, Nfld. to Minn., s. to N.S., Me., ne. Mass., Ct., centr. N.Y. and n. Ind.; also Wash. Aug.–Oct. (Hawaii; Eurasia) FIG. 390.

390. E. ovata.

Var. **Heùseri** Uechtritz (named in 1866 for its discoverer, PAUL HEUSER). — *Culms arching and depressed, recurving, densely crowded,* of very different lengths; spikelets very dark, the scales more spreading. — Que. to Minn., s. to N.S. and N.E. (Eu.)

18. E. diándra C. Wright (with two stamens). — Similar to no. 17, ascending or diffuse; culms very unequal, capillary, 0.1–5 dm. long; spikelets ovoid, 2–7 mm. long, 1.5–3.5 mm. thick, dense; *scales oblong-ovate, subacute or blunt,* 1–2 mm. long, *at first appressed,* later spreading, *whitish, pale brown or reddish,* with conspicuous green midrib; *achene obpyriform,* pale, 0.8–1 mm. long; *tubercle depressed turban-shape, broader than high, one-third to two-thirds as broad as achene; bristles wanting or rudimentary.* — Silts of rivers and lakes, local, sw. Me. to centr. N.Y. and e. Pa. Late July–Oct. FIG. 391.

391. E. diandra.

19. E. microcárpa Torr. (tiny-fruited). — *Culms tufted, quadrate-filiform, frequently proliferous* from the spikelet as leafy tufts, 0.5–4 dm. long, *weak,* becoming spongy when drowned; *spikelets lanceolate or oblong,* 2–7 *mm. long; scales* ovate, *membranaceous, acute,* 1–1.5 *mm. long, pale* except for green midrib and occasionally 2 brown bands, *with free tips; achene whitish,* smooth, *sharply 3-angled,* obovoid to pyriform, 0.5–*barely* 1 *mm. long, with a minute depressed-conic tubercle, wider than the subtending scales;* bristles shorter than to about equaling tubercle. — Damp sands, swamps and shallow water of Coastal Plain, Fla. to La., n. to se. Va. June–Sept. (W.I.)

Var. **filicúlmis** Torr. (with thread-like culms). — Stiffer; *scales chestnut-brown* except for pale margin, firmer. (*E. Torreyana* Boeckl.) — Similar range, n., locally, to e. Ct. and Tenn.; nw. Ind. FIG. 392.

392. E. microcarpa, v. filiculmis.

20. E. Brittònii Svenson (for NATHANIEL LORD BRITTON, 1859–1934). — Resembling no. 19; culms stiffish; *scales broader than and completely covering the achenes, with rounded appressed tips.* (*E. microcarpa,* var. Svenson) — Wet pine-barrens, pond-shores and bogs, Fla. to Tex., n. to se. N.C.; Cape May, N.J. Aug.–Oct. FIG. 393.

393. E. Brittonii.

21. E. álbida Torr. (whitish). — Tufted, *from a slender creeping base;* culms slender, wiry, striate, 1–4 dm. high, the *basal sheaths with very oblique tips;* spikelet cylindric-ovoid, blunt, 4–9 mm. long; *scales* obtuse, *whitish to light brown, with narrow scarious margin; achenes* smooth, *not glossy,* trigonous-pyriform, 1 mm. long, *contracted below the conic-deltoid pale tubercle* and usually *exceeded by the reddish bristles.* — Damp, chiefly brackish, soil, Fla. to Tex., n. to Md. July–Oct. (W.I.; Mex.) FIG. 394.

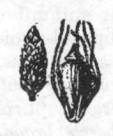

394. E. albida.

22. E. Macoùnii Fern. (for its discoverer, James Melville Macoun, 1862–1920). — Tufted annual, with *weak* subterete *unequal culms* up to 2.5 dm. long, *the shorter crowded at the base and divergent; basal sheaths close, with truncate orifice; spikelets lanceolate,* 7–10 mm. long; *scales oblong, reddish-brown,* obtuse, *closely appressed,* 2–3 mm. long; *achene obovoid, obtusely 3-angled,* olive, lustrous, 1.3 *mm. long; tubercle ovate-triangular,* 0.6 mm. long, *two-thirds breadth of achene* and separated from it by a strong constriction, much overtopped by the bristles. — Borders of marshes, Gatineau Co., Que. Aug., Sept. Fig. 395.

395. E. Macounii.

23. E. intermèdia (Muhl.) Schultes (intermediate). — *Tufted* annual, *with very unequal arching or reclining weak* capillary *culms* up to 4 dm. long, *the shortest crowded at the base; sheaths pale, with oblique orifice; spikelets* lanceolate or ovoid, *acute,* loosely flowered, 2.5–8 *mm. long,* 2–3 mm. thick; *scales oblong,* obtuse to acutish, *drab or brown each side of the green midrib,* 1.7–2.5 *mm. long, loosely ascending; achene* olivaceous, *trigonous-pyriform,* 1.2–1.5 *mm. long; tubercle awl-shaped or subulate-conic, its flange-like base about one-fourth as broad as achene;* longer bristles overtopping tubercle, in forma **Haberèri** Fern. (for its discoverer, Joseph Valentine Haberer, 1855–1925) wanting or rudimentary. (*E. reclinata* Kunth) — Wet calcareous soil, Gaspé Co., Que., to Minn., s. to n. and w. N.E., n. N.J., Md., Tenn., Ill. and Ia. (Forma *Habereri* about Oneida L., N.Y.). Late July–Oct. Fig. 396.

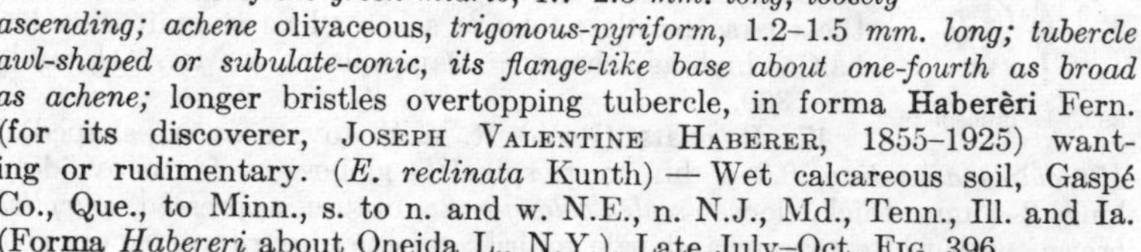

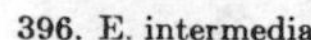

396. E. intermedia.

24. E. melanocárpa Torr. (black-fruited). — Tufted, from a stout caudex; *culms wiry, flattened,* 1–7 dm. long, *some occasionally arching over to ground and rooting at tips or with proliferating spikelets;* upper sheath truncate; spikelets ellipsoid to subcylindric, obtuse, 0.6–1.4 cm. long; *scales ovate, rounded,* buff to reddish-brown, *with broad scarious margin, appressed,* 3–4 *mm. long; achene turbinate-obovoid,* obtusely trigonous, *black, lustrous,* smooth, 1–1.4 *mm. long, truncate at summit; tubercle pale, sessile, depressed, as broad as achene,* with pointed summit. — Sandy or peaty shores and pine-barrens, n. Fla. and Ga.; se. Va.; Del. to se. Mass.; s. Mich. and dunes of n. Ind.; e. Tex. Late June–Oct. Fig. 397.

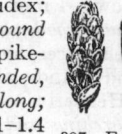

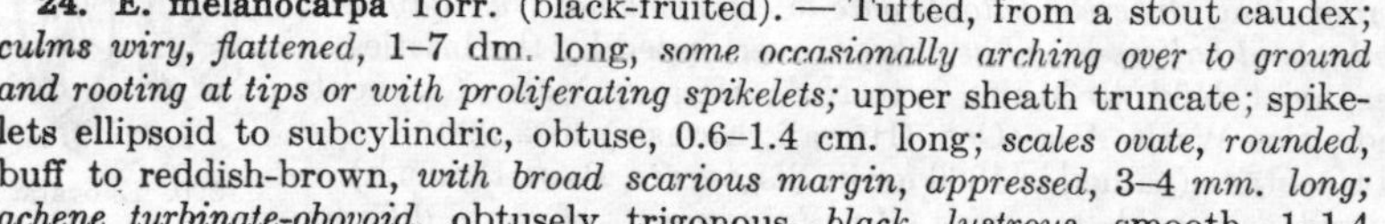

397. E. melanocarpa.

25. E. palústris (L.) R. & S. (of marshes). — Loosely stoloniferous to subcespitose, commonly *with widely creeping reddish slender rhizome; culms* subterete, *firm but not wiry,* 0.1–4 dm. high, 0.5–2 mm. in diameter near base; *sheaths* red or brown, 0.2–2 dm. long, *loose, the upper with oblique herbaceous summit;* spikelets broad-lanceolate to ovoid, 0.5–2 cm. long, 2.5–6 mm. thick; 2 *or* 3 *basal scales firm,* ovate to oblong, obtuse; *fertile scales* numerous, *subappressed,* oblong-ovate, *obtuse to subacute,* thin, *reddish-brown to castaneous,* opaque, with pale scarious margin, the lower and median 3–4 mm. long; *anthers* 1.7–2 *mm. long; achenes narrowly obovoid or pyriform,* yellowish or brown, 1.2–1.7 *mm. long; tubercle spongy, lanceolate to conic-ovoid or slenderly bulbiform, much higher than broad,* separated from achene by a strong constriction; bristles 4, reaching tubercle, or short or wanting. — Margins of ponds and streams or other shallow waters and marshes, Lab. to Alaska, s. to Nfld., N.S., n. N.E., n. Mich., N.D., Wyo., Id. and Oreg. Late June–Sept. (Eurasia).

Var. **màjor** Sonder (larger). — Coarser; *culms* 0.5–2 *m. high,* 1.5–5 *mm. in diameter;* sheaths looser, up to 3 dm. long; spikelets 0.7–2.6 cm. long; *lower and median fertile scales* 3.2–5.5 *mm. long; anthers* 1.7–3 *mm. long; achenes* 1.4–2.1 *mm. long.* (Var. *vigens* Bailey) — Shallow to deep waters or wet places, s. Lab. to B.C., s. to N.S., s. N.E., L.I., N.J., Pa., s. Mich., Ill., Ia., S.D., Wyo. and n. Calif. June–Sept. (Eurasia) Fig. 398.

398. E. palustris, v. major.

26. E. Smállii Britt. (for its discoverer, John Kunkel Small, 1869–1938). — Similar to no. 25; *culms firm and wiry,* slender to stoutish, 0.5–2.5 mm. in diameter at summit of upper sheath, 2.5–9 dm. high; *spikelets* slenderly lance-attenuate to narrowly ellipsoid-ovoid, acute to obtusish, 0.5–2 cm. long, *loosely flowered;* basal firm scales 2 or 3; *fertile scales loosely ascending, with spreading-ascending tips;* the *lower and median lanceolate to narrowly ovate, acute, often slender-tipped,* 3–5 mm. long, *very thin and scarious,* with 2 convergent purple bands; anthers 1–1.8 mm. long; *achenes rounded-obovoid,* 1.2–1.6 *mm. long; tubercle broadly*

399. E. Smallii.

ovate, as broad as or broader than long, often depressed; perianth of 4 delicate bristles or wanting. — Peaty or wet sandy swamps and shores, se. Nfld. to Ont., s. to Md., Pa., w. Va., W.Va., s. Ind., s. Ill. and Mo. Late June–Sept. FIG. 399.

27. E. macrostáchya Britt. (large-spiked). — Resembling no. 25; *culms soft, flattened,* 0.2–1.2 m. high; *sheaths drab or pale brown* except at base; *spikelets* slenderly subcylindric to lanceolate, *usually tapering above or acuminate,* 1–3 cm. long; sterile basal firm scales 2 or 3; *fertile scales very numerous,* membranaceous, *pale brown to purplish,* narrowly ovate, appressed, the lower and median 2–4 mm. long; anthers 1.3–2 mm. long; *achene yellowish or pale brown, obovoid, 1.2–1.6 mm. long; tubercle depressed-deltoid or umbonate, as broad as high,* sessile or essentially so, often overtopped by the 5–8 delicate bristles or bristles wanting. (*E. mamillata* sensu Fern. & Brack., not Lindb. f.) — Marshes, ditches and shores, Minn. to Sask. and B.C., s. to Ill., La., Tex., Mex., and s. Calif. Late May–Aug. FIG. 400.

400. E. macrostachya.

28. E. cálva Torr. (bald). — Loosely stoloniferous to subcespitose, with slender reddish rhizomes and stolons; *culms nearly filiform,* 1–6.5 dm. high, 0.5–1.5 mm. thick at summit of sheath, *terete* or corrugated, rarely subcompressed; *sheaths red or castaneous, very close, with oblique herbaceous summit,* the upper 0.2–1 dm. long; *spikelets* linear-lanceolate or slenderly ovoid, 0.9–1.7 cm. long, 1.5–4 mm. thick, *closely many-flowered; basal firm scale solitary, orbicular or rounded, spathiform, completely encircling bases of lower fertile scales; fertile scales* oblong or ovate, *obtuse, thin and membranaceous, reddish to pale brown, opaque, closely appressed,* the lower and median 1.8–3 *mm. long;* anthers 1.3–1.7 mm. long; *achene* pyriform to narrowly obovoid, yellowish to brown, 1–1.4 *mm. long,* 0.7–1 mm. wide; *tubercle conical,* 0.2–0.45 *mm. broad at base;* bristles 0 or 1–4 and delicate. (*E. palustris,* var. *glaucescens* in part of Am. auth., not *Scirpus glaucescens* Willd., upon which it rests) — Wet shores, marshes and springy spots, Gaspé Pen., Que., to James Bay, w. to Man., s. to N.S., N.E., L.I., Va., Tenn., Ark., Okla. and N.M. June–Sept. (Hawaii; E. Asia) FIG. 401.

401. E. calva.

29. E. halóphila Fern. & Brack. (lover of salt). — Similar to no. 28; culms stiffish, slender, 0.3–6 dm. high, subterete or compressed; *spikelets* lanceolate to slenderly ovoid, 0.3–1.7 cm. long, 2–6 *mm. thick, loosely 5–30-flowered; basal* ovate or rounded *scale spathiform,* completely clasping base of spikelet; *fertile scales subcoriaceous to firm-membranaceous,* reddish to castaneous, with pale margin, *lustrous, the lower and median 3–5 mm. long;* anthers 1.5–2.2 mm. long; *achenes broadly obovoid or pyriform,* yellowish to dark brown or olive, 1.2–1.8 *mm. long,* 1–1.4 *mm. broad; tubercle bulbiform, slenderly conical to lanceolate, higher than broad,* 0.2–0.5 *mm. broad at base, covering one-sixth to rarely half the breadth of achene.* (*E. palustris,* var. *glaucescens* in part of ed. 7, not *Scirpus glaucescens* Willd.) — Saline or brackish shores, Nfld. and s. shore of lower St. Lawrence R., Que., to Va.; James Bay; Cayuga Co., N.Y. Late June–Sept. FIG. 402.

402. E. halophila.

30. E. uniglùmis (Link) Schultes (with one glume). — Similar to no. 29; *achenes ellipsoid to narrowly obovoid; tubercle depressed-deltoid to low-conical, often as broad as high,* 0.6–1 *mm. broad* at base, *covering one-half to three-fourths breadth of achene.* — Wet places, Lab. to B.C., s. to mts. of w. Nfld., n. shore of lower St. Lawrence R., Que., N.D., Wyo. and Oreg. June–Sept. (Eurasia) FIG. 403.

403. E. uniglumis.

31. E. ámbigens Fern. (ambiguous). — Loosely stoloniferous, *with firm rhizomes* purplish-brown and 1.5–2 mm. thick; *sheaths tight,* purplish, *with dark cartilaginous obliquely truncate summit; spikelets lanceolate or slenderly ovoid,* 4–9 mm. long, 2–3.5 mm. thick, loosely few-flowered; *basal scale* ovate, *spathiform,* often with prolonged base, firm; *fertile scales lance-ovate, acutish, pale, with broad white hyaline margin,* the lower and median 2–3 *mm. long;* bristles short or 0; *achene ellipsoid-obovoid,* yellowish to brown, *minutely but distinctly reticulate,* biconvex, 1.2–1.7 mm. long, 1 mm. broad; *tubercle depressed-deltoid,* apiculate, 0.2–0.5 *mm. high,* 0.5–0.7 *mm. broad.* — Pond-margins and marshes, Fla. to La., n. to Elizabeth Ids., Mass. June–early Aug. FIG. 404.

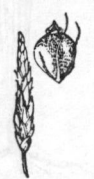

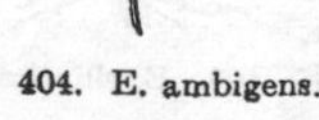
404. E. ambigens.

32. E. fállax Weath. (deceitful). — Stoloniferous, *with reddish slender flexuous rhizomes;* culms in small tufts, subterete or slightly compressed, very slender (0.5–1.1 mm. in diameter), 3–7.5 dm. high; *upper sheaths* reddish, *herbaceous, with entire subobliquely truncate summit; spikelets ovoid to lanceolate, acute,* 7–10 mm. long; *fertile scales* ovate- or obovate-oblong, obtuse, 2.5–3 *mm. long,* castaneous to dark red, with green midrib and narrow pale margin; style 3-fid; *achene trigonous-obovoid, obtusely* 3-*angled,* 1.2–1.4 *mm. long,* about 1 *mm. broad,* slightly reticulate-roughened, persistent after fall of scales; *tubercle bulbiform,* acute, 0.4–0.5 *mm. high;* bristles 3–5, shorter than achene. — Fresh to brackish springy pond-shore, local, n. Cape Cod, Mass. July, Aug. Fig. 405. — Probable local (now extinct) hybrid of no. 26 or 29 and no. 36.

405. E. fallax.

33. E. nítida Fern. (neat). — *Rhizomes* cord-like, purplish, 0.5–1 *mm. thick; culms* tufted or scattered, *delicately capillary,* 4-angled, 2–10 *cm. high,* 0.2–0.4 *mm. thick; basal sheaths* close, greenish or yellowish, *scarcely darkened at summit; spikelets* ovoid, acutish, 1.5–4.5 *mm. long,* 1.5–3 *mm. broad,* loosely flowered; *scales elliptic-oblong, with rounded tips,* purplish-brown, *with very narrow scarious margin,* 1 *mm. long, few,* thin and quickly deciduous; *anthers* 0.5–0.7 *mm. long; achenes pale yellow,* becoming orange, *sharply* 3-*angled,* very minutely wrinkled, 0.7–0.8 *mm. long,* persistent; *tubercle a depressed saucer with central apiculation,* 0.1 *mm. high;* bristles 0. — Damp peaty, sandy or rocky places, Nfld. to s. Alaska, s., very locally, to N.S., n. N.H. and ne. Minn. June–Aug. Fig. 406.

406. E. nitida.

34. E. compréssa Sulliv. (flattened). — *Rhizomes firm or subligneous,* purplish or castaneous, 2–4 *mm. thick; culms flat, wiry,* 1–1.5 mm. broad, 1.5–5 dm. high; sheath truncate, close; spikelets ovoid to ellipsoid, 5–13 mm. long; *scales* ovate to oblong, brown or purplish, *at least the upper with acuminate scarious whitish* often bifid *tips; achene* pyriform-obovoid, golden to brown, *bluntly* 3-*angled to subterete, granular-roughened* or wrinkled, persistent after fall of scales; tubercle depressed- to rounded-conic, with central apiculation; bristles fugacious or wanting. (*E. acuminata* auth., not *Scirpus acuminatus* Muhl.) — Calcareous gravels, sands and peats, w. Que. and n. N.Y. to Sask., s. to ne. Va., Ga., n. Ala., Mo., ne. Tex. and Colo. June, July. Fig. 407.

Var. **atràta** Svenson (blackish). — Spikelets large, with blackish scales. — Anticosti I., Que.; shores of Great Lakes, w. N.Y. to Wisc.

407. E. compressa.

35. E. ténuis (Willd.) Schultes (slender). — *Rhizomes* slender, flexuous and *cord-like,* purplish, 1–3 *mm. thick; culms* in small tufts or scattered, *slender,* 4- or 5-*angled or corrugated,* 0.5–9 dm. long; *sheaths close,* reddish below, *with obliquely truncate usually dark-girdled orifice; spikelets ellipsoid to ovoid,* loosely flowered, 3–10 mm. long; scales ovate, obtuse to acute, membranaceous, reddish to castaneous, with broad pale margin, 2.3–3.5 mm. long, deciduous; anthers 1.5–2 mm. long; *achene* trigonous-obovoid, obtuse-angled, 0.8–1.1 mm. long, olive or drab, *iridescent, with coarse and deep honeycomb-reticulation;* tubercle depressed or pyramidal, one-eighth to one-half height of achene; bristles fugacious or wanting. Very variable; the following confluent varieties are proposed by Svenson:

Culms delicately capillary, rarely more than 3 dm. high, their 4 or 5 angles not wing-like; achenes 0.8–1 mm. long.
  Tubercle acutely pyramidal, often one-fifth as high as achene; walls of reticulations not verrucose-thickened. . . . . *E. tenuis* (typical).
  Tubercle depressed and flattened, about one-eighth height of achene; some or all reticulations with verrucose-thickened walls. . . . . Var. *verrucosa.*
Culms firm, 0.4–0.8 mm. thick, 3–9 dm. high, with 1 or more of the 4 angles wing-like: achenes 0.9–1.1 mm. long; tubercle depressed. . . . . Var. *pseudoptera.*

E. ténuis (typical). — Wet, damp or even dryish sands, gravels and peats, mostly at low altitudes, N.S. and N.E. to O., s. to S.C. and Tenn. May–Sept. Fig. 408.

Var. **verrucòsa** Svenson (covered with warts). — Ky., Ind., Ill. and Ia. to La. and e. Tex.; se. Va.

Var. **pseudóptera** (Weath.) Svenson (falsely winged). — Rich woods and meadows, se. N.Y. and n. N.J. to N.C. and e. Tenn.

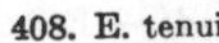

408. E. tenuis.

36. E. ellíptica Kunth (elliptic). — Similar to no. 35; *culms* very slender, 6–8-*angled,* 0.5–7 dm. high; *mature achenes bright yellow to orange,* 0.9–1.1 mm. long, *with shallow reticulation, the many transverse bands prominent;* tubercle depressed-deltoid, with promi-

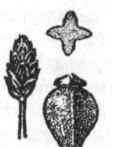

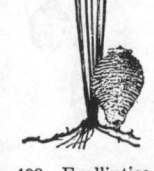

409. E. elliptica.

nent central point. (*E. capitata*, var. *borealis* Svenson) — Damp to dry sand, gravel, shores and turf, Nfld. to Man., s. to N.S., N.E., L.I., n. N.J., Pa., O., s. Ind., s. Ill. and Ia.; s. Alta. and nw. Mont. to sw. B.C. May–Sept. FIG. 409.

410. E. tricostata.

37. E. tricostàta Torr. (three-ribbed). — *Rhizome stiff, subligneous*, 3–6 mm. thick; culms in small tufts, compressed-filiform, stiffish, 2–6 dm. high; sheaths oblique at summit, short-toothed; spikelets densely many-flowered, ellipsoid, becoming cylindric, 0.7–2 cm. long, 2.5–3.5 mm. thick, obtuse; *scales* ovate or oblong, obtuse, rounded at the pale scarious tip, reddish-brown, *about* 2 *mm. long; achenes* golden to brown, trigonous-obovoid, *about* 0.6 *mm. long, with minutely roughened-reticulate sides and prominent keel-like angles; tubercle minute, depressed-conical, acute;* bristles 0. — Sandy or peaty soil of the Coastal Plain, local, n. Fla. to e. S.C.; se. Va.; Del. to Washington Co., R.I., and Nantucket I., Mass. Aug.–Oct. FIG. 410.

38. E. vivípara Link (viviparous). — Tufted from a firm brown vertical rhizome; *culms filiform*, erect, or *usually proliferous at tip* and reclining (often with 2–6 peduncled spikelets at the tip of the culm), 1–3 dm. long; *spikelets linear-cylindric to slenderly lanceolate*, 3–8 mm. long, 1.5–2 mm. thick, *usually proliferous and sterile; scales* appressed, *oblong*, obtuse, 2 *mm. long*, chestnut-sided, with whitish hyaline margin; *achene* 3-angled, obovate, 1 *mm. long*, dark gray, *honeycomb-reticulate;* tubercle pyramidal, narrower than achene; bristles reddish, nearly equaling achene, retrorsely toothed. — Peaty shores and pond-margins, Fla. to se. Va. July–Oct. FIG. 411.

411. E. vivipara.

412. E. tortilis.

39. E. tórtilis (Link) Schultes (twisted). — *In dense tussocks, drab at base; culms sharply triangular*, capillary, lax or reclining, *often spirally twisted*, 2.5–8 dm. long; *sheaths* tight, the upper greenish to drab, *with a very oblique blunt cartilaginous summit; spikelets* ellipsoid or ovoid, acutish, *few-flowered*, 3–9 mm. long; *scales* ovate, obtuse, appressed-ascending, *firm*, with drab or brownish sides, broad pale midrib and scarious margin, 2–2.5 mm. long; *achene* obovoid, turgid, 1.5–2 *mm. long, drab or olivaceous, iridescent, with* 3 *thickened and ridge-like angles and several rows of honeycomb-reticulations; tubercle conic-subulate, firm*, 0.5–1 mm. long, *its nearly free base about half the breadth of achene.* (*E. simplex* of Am. auth., not A. Dietr.) — Springy swamps, wet woods and thickets of the Coastal Plain, Fla. to e. Tex., n. to L.I. June–Sept. FIG. 412.

40. E. tuberculòsa (Michx.) R. & S. (with a tubercle). — Similar to no. 39, stiffer; *the culms obtusely angled or subterete*, 1–7 dm. long; spikelets ovoid to ellipsoid, 0.5–1.3 cm. long; *scales* ovate to suborbicular, rounded at summit, cartilaginous, pale greenish or drab-brown, *appressed to and clasping the achene*, 2.3–3.3 *mm. long; achenes obtusely angled*, obovoid, 1.3–1.6 *mm. long*, drab or olive, iridescent, with numerous longitudinal rows of coarse reticulations; *tubercle cap-like*, whitish, *spongy, rounded at summit, as broad as* and closely crowning, but almost free from the often smaller *achene;* bristles coarse, divergently barbed, or retrorsely barbed in forma **retrórsa** Svenson (reversed), scarcely reaching summit of tubercle. — Wet sandy and peaty shores and swamps, Fla. to Tex., n. to sw. N.S., e.-centr. N.H., ne. Mass., R.I., s. Ct., se. N.Y., Tenn. and Ark. June–Sept. FIG. 413.

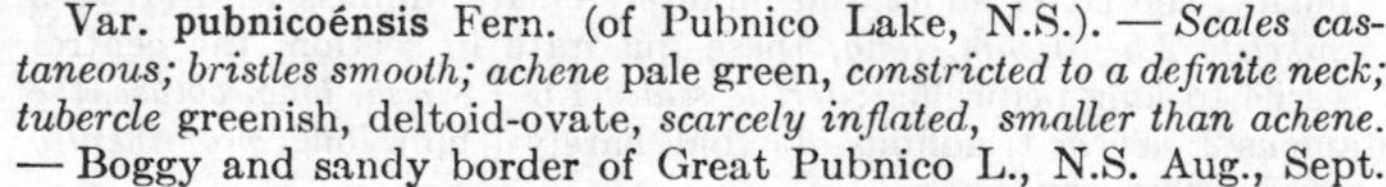

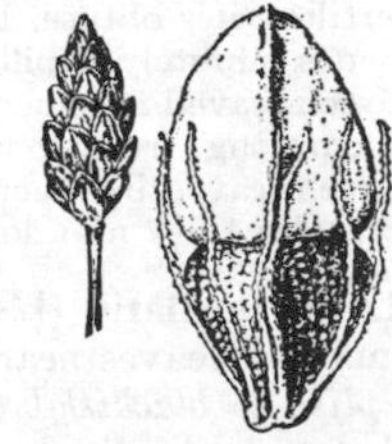
413. E. tuberculosa.

Var. **pubnicoénsis** Fern. (of Pubnico Lake, N.S.). — *Scales castaneous; bristles smooth; achene* pale green, *constricted to a definite neck; tubercle* greenish, deltoid-ovate, *scarcely inflated, smaller than achene.* — Boggy and sandy border of Great Pubnico L., N.S. Aug., Sept.

4. DICHRÒMENA Michx.

Spikelets few-flowered, all but 3 or 4 of the flowers usually imperfect or abortive. Scales somewhat imbricated in 2 ranks, more or less conduplicate or navicular, keeled, white or whitish. Stamens 3. Style 2-cleft. Perianth, bristles, etc., none. Achene lenticular, transversely wrinkled, crowned with the persistent and broad tubercled base of the style. — Culms leafy, from creeping perennial rhizomes; the leaves of the involucre mostly white at the base (whence the name, from the Greek *dis, double*, and *chroma, color*). — A small genus of trop. and warm-temp. Am., sometimes (perhaps rightly) united with *Rhynchospora*.

414. D. colorata.

1. D. coloràta (L.) Hitchc. (colored). — *Culm triangular* (0.25–1 m. high); leaves narrow; those of the involucre 4–7, linear; achene truncate, not margined.— Swamps, chiefly calcareous or brackish, and shallow water, Fla. to Tex., n. to se. Va. July–Sept. Fig. 414.

5. PSILOCÁRYA Torr. Bald Rush

Spikelets ovoid, terete, the numerous scales all alike and regularly imbricated, each with a perfect flower. Stamens mostly 2. Style 2-cleft, its base enlarging and hardening to form the beak of the lenticular or tumid more or less wrinkled achene. — Annuals of Austral., S.Am., W.I. and e. U.S., with leafy culms, the spikelets in terminal and axillary cymes. (Name from the Greek *psilos, naked,* and *caryon, nut.*)

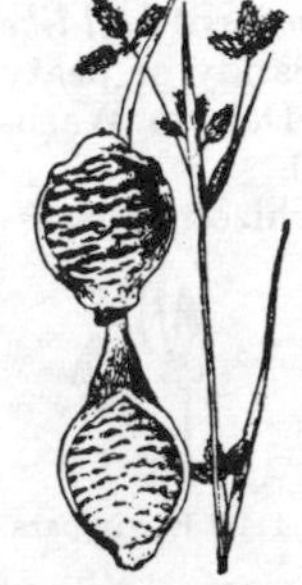

415. P. scirpoides (inflorescence and lower achene); P. nitens (upper achene).

1. P. scirpoìdes Torr. (like *Scirpus*). — Annual, 0.2–4.5 dm. high, leafy; leaves flat; *spikelets ellipsoid-ovoid, blunt, nearly sessile,* 20–30-flowered; *scales oblong-ovate, acute,* reddish-brown; *bracteoles herbaceous, strongly veined,* with prominent green midrib; *achene finely roughened,* somewhat margined, *beaked with a long sword-shaped almost wholly persistent style.* — Peaty and sandy shores and swamps, se. Mass. to e. Md.; s. Mich. and n. Ind. July–Oct. Fig. 415 (inflorescence and lower achene).

Var. **Grìmesii** Fern. & Grisc. (for its discoverer, Earl Jerome Grimes, 1893–1921). — *Spikelets lance-cylindric, the lateral long-pedicelled; scales broadly lance-acuminate,* reddish-brown to castaneous; *bracteoles chartaceous, mostly nerveless.* — Shallow water, wet shores and springy spots, se. Va. and e. N.C. Aug.–Oct.

2. P. nìtens (Vahl) Wood (shining). — Similar; often becoming 5–7 dm. high; *faces of the achene with strong transverse ribs; tubercle depressed, broader than high.* — Wet sands and peats, Fla. to e. Tex., n. to se. N.C.; and locally in Sussex Co., Va., Del., Cape May, N.J., Suffolk Co., L.I., Plymouth Co., Mass., and nw. Ind. Aug.–Oct. Fig. 415 (upper achene).

6. BULBOSTỲLIS (Kunth) C. B. Clarke

Spikelets as in *Fimbristylis*, the comparatively large scales in few ranks. Stamens 2 or 3. Style 2–3-cleft, filiform, glabrous, its base swollen and forming a persistent colored tubercle. Otherwise as in *Fimbristylis;* standing in the same relation to that genus as *Eleocharis* to *Scirpus.* — Leaves primarily basal, narrowly linear or filiform, the sheaths hairy or ciliate. Plants chiefly of trop. and warm-temp. reg. (Name from *bulbus, bulb,* and *stylus, style.*) Stenophyllus Raf.

Achenes prominently and transversely undulate-corrugate or cross-puckered;
  fertile scales obtuse, 1.5–1.8 mm. long. . . . . 1. *B. capillaris.*
Achenes minutely papillate; scales pointed.
  Green-leaved involucre 0 or very short, the longer of the 1 or 2 leaves 0.2–1.5
    cm. long; fruiting scales 0.7–1.3 mm. long. . . . . 2. *B. ciliatifolia.*
  Green-leaved involucre of 3 or 4 bracts, the longest one 1–6 cm. long; fruiting
    scales 1.5–2 mm. long. . . . . 3. *B. coarctata.*

1. B. capillàris (L.) C. B. Clarke (hair-like). — Densely tufted annual 0.3–4 dm. high; culms and leaves nearly capillary, the latter short and minutely ciliate; umbels with (1–)2–9 *purplish to blackish-brown spikelets* 2.5–10 *mm. long,* these quadrate in section, the central spikelet sessile, the lateral sessile to long-pedicelled; *fertile scales* 1.5–1.8 *mm. long, obtuse, the pale midrib not excurrent;* stamens 2; *achene* trigonous-obovoid, barely 1 mm. long, *prominently and transversely undulate-corrugate.* Three vars.:

Plant bearing crowded sessile basal spikelets; spikelets of terminal umbel 3–10
  mm. long, crowded and sessile or the lateral on pedicels shorter than spikelet. *B. capillaris* (typical).
Plant without sessile basal spikelets; spikelets of terminal umbel 2.5–6(–9) mm.
  long.
  Lateral spikelets 2.5–6 mm. long, short-pedicelled, the very unequal pedicels
    0.1–1 cm. long. . . . . Var. *crebra.*
  Lateral spikelets 2.5–9 mm. long, their subequal pedicels 0.6–1 cm. long. . . Var. *isopoda.*

B. capillàris, typical (*Stenophyllus capillaris* (L.) Britt., as to type, and var. *cryptostachys* Fern.) — Dry open soil, s. Me. and sw. Que. to Minn., s. to Va. and Mo. Aug.–Oct.

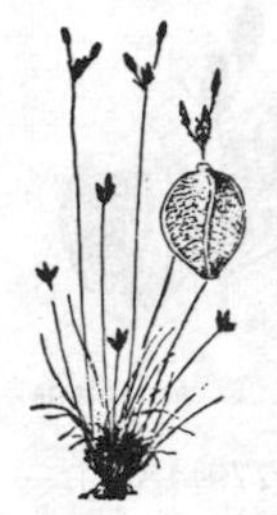

416. B. capillaris, v. crebra.

Var. **crèbra** Fern. (common). — Similar habitats, centr. Me. to Neb., s. to Ga., Tenn., Okla. and Tex. July–Oct. Fig. 416.

Var. **isópoda** Fern. (equal-stalked). — Similar habitats, Md. to s. Ill., s. to Ga., Ala., Ark. and Tex. Late June–Sept.

2. B. ciliatifòlia (Ell.) Fern. (ciliate-leaved; a misnomer). — More delicate; tufts without sessile basal spikelets; umbels simple or compound, the 1–7 rays simple or forking; *green involucre* 0 *or of* 1 *or* 2 *bracts, the longer one* 0.2–1.5 *cm. long; spikelets* castaneous, *acute or acutish,* 1.5–5 *mm. long; fertile scales* 0.7–1.3 *mm. long, the midrib projecting as a point; achenes* smaller, *minutely papillate-roughened,* only obscurely transverse-corrugated. — Sandy pinelands and openings, Fla. to Tex. and Mex., n. to se. Va., Ark. and Okla. July–Sept.

3. B. coarctàta (Ell.) Fern. (drawn-together). — Similar to no. 2, usually taller, up to 6 dm. high; *green involucre of* 3 *or* 4 *leaves, the longer one* 1–6 *cm. long;* cyme open or contracted and dense; *fertile scales* 1.5–2 *mm. long.* — Dry sands and pine-barrens, Fla. to Tex., n. to se. Va. Late Aug.–Oct. This and no. 2 needing critical study.

7. FIMBRISTỲLIS Vahl

Spikelets several–many-flowered, terete; scales all floriferous, regularly imbricated in several ranks. Stamens 1–3. Style 2–3-cleft, often with a dilated or tumid base, which is deciduous from the apex of the naked lenticular or triangular achene. Otherwise as in *Scirpus.* Spikelets in our species umbelled, and the involucre 2–3-leaved. — Large genus of warm reg. (Name compounded of *fimbria, a fringe,* and *stylus, style,* the latter being fringed with hairs in the genuine species.)

*a.*Style 2-cleft; achene lenticular. . . *b.*
*b.*Perennials with firm or hardened bases or with stolons.
Densely cespitose, non-stoloniferous, with coriaceous dark sheaths; scales of spikelets coriaceous, lustrous, glabrous; plant of saline or brackish coast. 1. *F. castanea.*
Cespitose or in small tufts; scales of spikelets submembranaceous, puberulent when young, the outer ones usually persistently so.
In large and dense or small tussocks, non-stoloniferous; bases of tufts bulbous-thickened; plant of acid or subacid habitats. . . . . . 2. *F. Drummondii.*
In small non-bulbous tufts, spreading by slender scaly stolons; plant of brackish to saline coastal area. . . . . 3. *F. caroliniana.*
*b.*Annuals with soft bases and no stolons.
Inflorescence lax, the lateral spikelets on elongate rays; scales broadly ovate; style ciliate. . . . . 4. *F. Baldwiniana.*
Inflorescence a dense glomerule; scales narrow, acuminate; style glabrous. 5. *F. Vahlii.*
*a.*Style 3-cleft; achene 3-angled; soft-based annuals with lance-ovate to linear-cylindric mostly stalked spikelets. . . . . 6. *F. autumnalis.*

1. F. castànea (Michx.) Vahl (chestnut-colored). — *Perennial, rigid, in dense tussocks;* the thickened *base covered with firm dark sheaths;* culms wiry, 0.3–1 m. high, nearly naked; *leaves* pale and firm, *involute;* umbel 3–10-rayed, the rays very unequal, some simple, others forking; *spikelets ovoid to short-cylindric,* 0.7–1.7 *cm. long, the firm somewhat lustrous dark scales all glabrous; stamens* 2 *or* 3; achene broadly obovate, lustrous, minutely striate and reticulated. (*F. spadicea* of ed. 7, not (L.) Vahl) — Sands and brackish shores, Fla. to Tex., n. to L.I. July–Oct. (W.I.) Fig. 417.

417. F. castanea.

2. F. Drummóndii Boeckl. (for its discoverer, Thomas Drummond, 1780–1835). — Densely or openly *cespitose from a knotty rhizome, in large tussocks or smaller tufts, the individual tufts thickened or somewhat bulbous at base,* with pale and thin sheaths, *non-stoloniferous;* leaves becoming involute, scabrous-margined; culms slender, 1.5–7 dm. high; umbel with 1–5 slender rays, each ray with 1–4 brown ellipsoid spikelets; *outer scales puberulent;* achene obovoid, white, becoming darker. (*F. castanea* in part of ed. 7, not (Michx.) Vahl) — Dry or moist pine- or oak-barrens, sterile meadows, prairies, etc., Fla. to e. Tex., n. to L.I., N.J., s. Ont., s. Mich., Ill. and Mo. May–July.

418. F. caroliniana.

3. F. carolinià̀na (Lam.) Fern. (of Carolina). — Similar to no. 2, *in small tufts, spreading by scaly slender stolons;* basal sheaths pale and soft; leaves flat, tardily involute; umbel with the (1–)3–35 ovoid-ellipsoid *spikelets* 0.5–1 *cm. long,* these becoming cylindrical and tapering at summit, or obovoid and rounded above in forma **eucy̆cla** Fern. (well-rounded), mostly on long rays, or all sessile and glomerulate in forma **pycnostáchya** Fern. (with crowded spikes), chestnut-color; scales thin, only the young and the outer puberulent. (*F. castanea* in part of ed. 7; *F. puberula* (Michx.) Vahl) — Brackish or saline sands, dune-hollows and flats, coast of Fla. to Tex., n. to N.J. July–Oct. Fig. 418.

4. F. Baldwinià̀na (Schultes) Torr. (for its discoverer, William Baldwin, 1779–1819). — *Culms* slender (0.5–7 dm. high) *from an annual root, weak,* grooved and flattish; *leaves linear, flat, ciliate-denticulate, glaucous,* sometimes hairy; spikelets ovoid, acute (0.4–1 cm. long); stamen 1; *achene conspicuously* 6–8-*ribbed on each side, and with finer cross-lines,* 1–1.2 mm. long, *with irregular tuberculate surface.* (*F. laxa* of ed. 7, not Vahl; *F. Darlingtoniana* Pennell) — Wet or dryish sterile soils, Fla. to Tex. and Ariz., n. to se. Pa., s. Ill., Mo. and Okla. July–Oct. (W.I.) Fig. 419.

5. F. Váhlii (Lam.) Link (for Martin Vahl, 1749–1804). — *Dwarf tufted annual* (0.3–2 dm. high); the culms, leaves and *very elongated upright bracts filiform;* glomerule 0.3–1 cm. in diameter; spikelets 3–8, subcylindric, greenish or pale brown; the narrow scales acuminate; *achene* minute, *transversely reticulate.* — Damp sands, etc., Fla. to Tex., n. to N.C. and s. Mo.; adv. near Phila. July–Oct. Fig. 420.

6. F. autumnàlis (L.) R. & S. (autumnal). — Soft-based tufted annual, 0.2–2.5 dm. high; culms flat, slender, diffuse or erect; umbel simple, compound or decompound, with 1–several rays, or rays suppressed in forma **brachyáctis** (Fern.) S. F. Blake (short-rayed), and 1–50 *ovate-lanceolate or lance-elliptic* brown or castaneous *spikelets* 3–8 *mm. long; scales* ovate-lanceolate, *with free slender tips; achenes* 0.5–0.75 *mm. long, smooth or minutely roughened.* (*F. Frankii* Steud.; *F. geminata* Kunth) — Sandy or peaty shores and low ground, Ga. to La., n. to centr. Me., sw. Que. and s. Ont. Aug.–Oct. Fig. 421.

Var. **mucronulàta** (Michx.) Fern. (with small points). — Culms up to 4(–6) dm. tall; inflorescence usually compound or decompound, with 5–100 or more *linear-cylindric to slenderly fusiform spikelets* 4–10 *mm. long;* scale-tips more appressed. (*F. autumnalis* of ed. 7; *F. mucronulata* (Michx.) S. F. Blake) — Fla. to Tex. and Mex., n. to se. Mass., Ind., s. Wisc., Mo. and Kans. June–Sept. (Trop. Am.) Fig. 422.

419. F. Baldwiniana.

420. F. Vahlii.

421. F. autumnalis (typical).

422. F. autumnalis, v. mucronulata.

8. SCÍRPUS L. Bulrush

Spikelets few–many-flowered, solitary, or few to very many in a terminal inflorescence subtended by a 1–several-leaved involucre (this when single often appearing like a continuation of the culm); the scales in several ranks. Flowers to all the scales, or to all but one or two of the lowest, perfect. Perianth of 1–6 (or 8) bristles, or sometimes wanting. Stamens 2 or 3. Style 2–3-cleft, simple below, deciduous, or sometimes leaving a tip or point to the lenticular or triangular achene. — Culms sheathed at base. Large and highly variable world-wide genus. (The Latin name of the bulrush.)

a. Inflorescence not subtended by a true persistent involucre, merely with the outermost deciduous scale of the terminal spike or spikelet somewhat longer or larger than the other scales. . . *b.*

b. Spikelet solitary, 3–7 mm. long (excluding bristles), with few stramineous to rufescent scales; achenes sessile or nearly so, beakless or short-beaked, 1.2–2 mm. long. § BAEOTHRYON. . . *c.*

c. Upper basal sheaths membranaceous, the uppermost bearing flat blades 2 cm. or more long; culms 3-angled, scabrous on the angles at summit; perianth-bristles terete, setulose, about equaling or shorter than the beakless achene; upper scales of spikelet with clasping prominently arrow-shaped bases; denuded rachilla flexuous, with flattened or concave thin-margined depressions.

Leaves much shorter than culms; only the outer scale of the spikelet awn-tipped. . . . . 1. *S. Clintonii.*

Leaves about equaling culms; all the scales awn-tipped. . . . 2. *S. verecundus.*

c. Upper basal sheaths coriaceous or firm, the uppermost bearing a firm terete or involute blade 0.3–1.5 cm. long; perianth-bristles flat and smooth or crinkled, exceeding the achene, or wanting; upper scales of spikelets not prominently arrow-shaped at base; denuded rachilla straight, with coriaceous-margined obliquely cup-shaped depressions. . . *d.*

d. Culms more or less wiry, terete, smooth; scales of spikelets 5 or 6; perianth-bristles in maturity 2.5–5 mm. long, about twice length of achene, or wanting; denuded rachilla 1–2 mm. long, with 4 or 5 depressions.

In dense tussocks, non-stoloniferous, 0.5–7.5 dm. high; spikelet 3.5–6 mm. long, the 2 or 3 lower scales with firm awn-tips; anthers 1.3–2.5 mm. long, the connective projecting as a slender purple tip; perianth-bristles about twice length of achene; achene 2 mm. long, beaked. . . . . 3. *S. cespitosus.*

In small tufts, loosely stoloniferous, 0.5–1.7 dm. high; spikelet 3–4 mm. long, the scales awnless; anthers 1.5 mm. long, blunt; perianth-bristles 0; achene 1.2–1.5 mm. long, beakless. . . . . 4. *S. Rollandii.*

d. Culms soft, 3-angled, with concave sides, scabrous on the angles; scales of spikelet 10–20; perianth-bristles finally lengthening to 1–3 cm., white, crinkled; denuded rachilla 2–4 mm. long, with 9–19 depressions. . . . . 5. *S. hudsonianus.*

b. Spikelets crowded into a terminal compound spike 1–2 cm. long; scales castaneous throughout; achenes stipitate and long-beaked, 4.5–5.5 mm. long; perianth wanting or obsolescent. § BLYSMUS. . . . . . 11. *S. rufus*, var. *neogaeus.*

a. Inflorescence subtended by a persistent foliaceous involucre of 1–∞ (sometimes not conspicuous) bracts. . . *e.*

e. Involucre a single terete, triangular or compressed bract simulating the culm and looking like a continuation of it (rarely a 2nd very short and inconspicuous bract); culms naked or leafy only toward base. . . *f.*

f. Inflorescence commonly without elongate branches, either a spike, a spikelet or a glomerule of sessile or subsessile spikelets; culms 3-angled, or if terete slender (less than 3 mm. thick at base) and low. . . *g.*

g. Annuals with tufted slender culms, without rhizomes; anthers less than 1 mm. long; achenes 1.5–2 mm. long, when mature often black or blackish. . . *h.*

h. Spikelets 1–3, 2.5–6 mm. long, 2–3 mm. thick, with 3–10 subdistant scales; these deeply boat-shaped and prominently keeled; achene triangular, subequilateral, the surfaces densely and minutely papillate; culms 2–18 cm. high, setaceous. § ISOLEPIS. . . . 6. *S. koilolepis.*

h. Spikelets 1–12, 4–15 mm. long, 2.5–5 mm. thick, with many closely appressed and imbricated merely concave and but slightly keeled scales; achene biconvex, plano-convex or, if strongly trigonous, conspicuously cross-wrinkled; culms stouter, mostly 0.1–1 m. high. § ACTAEOGETON. . . *i.*

i. Achene with conspicuous transverse corrugations, black when ripe; scales all awn-tipped or cuspidate-acuminate.

Style 2-cleft; achene strongly flattened, plano-convex. . . . 7. *S. Hallii.*

Style 3-cleft; achene 3-angled, subequilateral. . . . . 8. *S. saximontanus.*

i. Achene smooth or only obscurely rugulose or pitted, olivaceous to black; scales blunt or barely mucronulate.

Achene plano-convex (flat on one face, gently rounded to subumbonate on the other), somewhat cuneate-obovate, smooth; bristles, when present, delicate and slender. . . . . . 9. *S. Smithii.*

Achene unequally biconvex, bulged on both faces, more rounded-obovate, more or less pitted; bristles stout (rarely wanting). . 10. *S. Purshianus.*

*g.*Perennials with elongate rhizomes; culms mostly solitary or scattered; anthers 2–3.5 mm. long; mature achenes whitish, drab, olivaceous or brown. . . *j.*

*j.*Spikelets crowded into a terminal castaneous or rufescent compound spike 1–2 cm. long; achene lance-fusiform, long-beaked and -stipitate, 4.5–5.5 mm. long; perianth wanting. § Blysmus. . . 11. *S. rufus*, var. *neogaeus.*

*j.*Spikelets solitary or in glomerules; achenes short-beaked, not stipitate, 2.5–4 mm. long; perianth present. § Schoenoplectus. . . *k.*

*k.*Rhizome soft and weak; culms slender, 0.5–2 mm. thick at summit; upper leaf-sheaths open or readily splitting, with thin friable band and orifice; spikelets oblong-lanceolate to ellipsoid, stramineous or greenish; scales ovate-lanceolate, entire; achenes distinctly trigonous, 2.5–4 mm. long.

Culms terete, nearly filiform; leaves very numerous, capillary and flaccid in submersed plants, or often wanting or few and rigid in emersed colonies; spikelet solitary, 6–13 mm. long; involucre 0.5–6.5 cm. long nearly filiform, with a straight callous tip. . . . . 12. *S. subterminalis.*

Culms triangular, stoutish, with concave sides; leaves 2 or 3, firm, flat, with an obliquely rounded tip. . . . . . . 13. *S. Torreyi.*

*k.*Rhizome firm and hard; culms 1–10 mm. thick; upper leaf-sheaths closed, with firm band and orifice; spikelets ovoid (rarely cylindrical), reddish-brown to castaneous; scales ovate to orbicular, with erose or ciliolate margins; achenes plano-convex or only obscurely triangular in cross-section, 2.5–3 mm. long.

Culms relatively slender, 1–6 mm. thick at the upper sheath, 0.03–1.5 m. high; upper sheath concave but not notched at orifice, bearing an elongate linear sharp-pointed blade 2–9 mm. broad at base; involucre linear, 2–15 cm. long, acute. . 14. *S. americanus.*

Culms stout, 4–10 mm. thick at upper sheath, 0.5–3 m. high; upper sheath with V-shaped notch at orifice, bearing a short lanceolate blunt-tipped blade 8–15 cm. long; involucre lance-triangular, 1–3.5 cm. long, blunt. . . . . 15. *S. Olneyi.*

*f.*Inflorescence with usually elongate branches; culms terete (triangular in no. 20), stout (0.3–2 cm. in diameter at base) and tall, naked above base. § Pterolepis. *l.*

*l.*Culms terete or nearly so; involucre terete; body of achene 1.7–2.8 mm. long, 1.5–3 mm. broad, the slender beak 0.3–1 mm. long. . . *m.*

*m.*Scales fulvous to deep brown or reddish, scario-membranaceous; spikelets ovoid to cylindric; achenes plano-convex, with low rounded back; style 2-cleft.

Scales glabrous or nearly so, minutely if at all spotted, the green midrib prominent and excurrent as an awn; spikelets ovoid to linear-cylindric; culm soft, easily compressed; basal sheath membranaceous, with scarious margin becoming lacerate; bractlets without glandular atoms.

Many spikelets usually in glomerules or close fascicles of 2 or more; scales nearly equaled or exceeded by the mature achenes; anthers with the connective excurrent as a slender tip; perianth-bristles (4–)6, about equaling or slightly overtopping achene; persistent old filaments slender, barely wider than bristles; transcontinental species. . . . . 16. *S. validus*, var. *creber.*

All spikelets slender-pedicelled; scales much overtopping achenes; anther tipped by sessile triangular appendage; perianth-bristles 0 or 1 or 2, very slender and short; persistent old filaments broad, appearing ribbon-like; local plant of e. Me. . . . . 17. *S. Steinmetzii.*

Scales viscid-pubescent, and with the style-branches, bractlets, etc. copiously flecked with red or brown gummy atoms; midrib of scale not prominent, projecting as a short mucro; spikelets slenderly ovoid to linear-cylindric; culm firm; basal sheath firm or subcoriaceous, its margin fibrillose. . . . . 18. *S. acutus.*

*m.*Scales pale brown to drab or whitish-green, firm or subcoriaceous; spikelets lance-acuminate to slenderly ellipsoid, acute to subacuminate, 0.75–2.3 cm. long; achene 3-angled, the back with a definite broad angle; style-branches 3. . . . . 19. *S. heterochaetus.*

*l.*Culms 3-angled, sharply so above; basal sheath with a long triangular-channeled blade; involucre 3-angled; body of achene 4 mm. long, 3-angled, the beak 1–2 mm. long. . . . . 20. *S. etuberculatus.*

*e.*Involucre of 2 or more flat leaves; culms leafy. . . *n.*

*n.*Culms sharply trigonous, solitary or scattered, from corm-like enlargements of elongate moniliform rhizomes; spikelets 1–5 cm. long, 5–11 mm. thick; midribs of scales prolonged into long awns; anthers 2.5–5 mm. long, with a colored hairy subulate tip; achenes 3–4.5 mm. long. § PHYLLANTHELI. . . *o.*

*o.*Achenes equilaterally 3-angled, with angles acute, 4–5 mm. long, pale, embraced by usually 6 strong and persistent bristles; anthers 2.5–4.5 mm. long; plant of fresh waters and alluvium. . . . . 21. *S. fluviatilis.*

*o.*Achenes dorsiventrally compressed and planoconvex to definitely but obtusely trigonous, 2.8–4.5 mm. long, stramineous or olivaceous to deep brown; bristles 2–6 or 0, weak, soon deciduous, rarely present in mature fruit; plants of saline or brackish soils. . . *p.*

*p.*Spikelets fulvous or rufescent, blunt or rounded at apex, ellipsoid-ovoid to thick-cylindric; leaves extending high on culm; summit of sheath strongly ribbed below the semicircular or convex scarious ligule; southern coastal species. . . . . 22. *S. robustus.*

*p.*Spikelets fuscous or castaneous to pale brown or drab or barely fulvous, blunt or tapering to acute or subacute summit; uninjured ligule truncate or with concave summit, if slightly convex strongly V-shaped and subtended by slender nerves; northern and inland species. . . *q.*

*q.*Cauline leaves 3–9, ascending high on the culm; culms 0.5–1.5 m. high, 0.5–2 cm. thick at base; larger leaves 4–12 mm. broad; anthers 3–5 mm. long.

Deep green; involucre of 2–4 bracts, the 1st 1.5–2.5 dm. long, the 2nd 0.8–2 dm. long; spikelets castaneous or fuscous; anthers 3–3.5 mm. long; achenes gradually rounded to base; eastern. 23. *S. maritimus.*

Pale green; involucre of 1 or 2 (rarely 3) bracts, the 1st 0.5–2 dm. long, the 2nd 0 or 1–5 (rarely –10) cm. long; spikelets whitish, pale brown or drab; anthers 3.5–5 mm. long; achenes cuneate at base; western. . . . . 24. *S. paludosus.*

*q.*Cauline leaves 2–4, borne chiefly or wholly below middle of culm; culms 1.5–7.5 dm. high, 2–8 mm. thick at base; larger leaves 1.5–9 mm. broad; involucre of 1 or 2 bracts, 1st bract 0.5–1.7 dm. long, 2nd 0 or 1–9 cm. long; scales fuscous or castaneous to whitish; anthers 2–4 mm. long. . . . . 24. *S. paludosus,* var. *atlanticus.*

*n.*Culms obtusely angled (or sometimes sharply so at summit), from close leafy crowns; spikelets 2–15 mm. long, 1–3 mm. thick, very numerous (35–100 or more) in decompound umbelliform panicles. . . *r.*

*r.*Bristles retrorsely barbed; spikelets in glomerules; culms solitary–few; the short caudex bearing thick scaly stolons. § TAPHROGETON. . . *s.*

*s.*Lower sheaths (at least) red-tinged; bristles barbed nearly to base.

Achenes lenticular; stigmas 2; bristles 4; scales (except terminal) blunt or barely mucronulate; most sheaths red-tinged at base, smooth or barely nodulose when dry. . . . . 25. *S. rubrotinctus.*

Achenes trigonous; stigmas 3; bristles 3 or 6; scales with subulate-acuminate tips; only the lower sheaths red-tinged, prominently septate-nodulose when dry. . . . . 26. *S. expansus.*

*s.*Sheaths uniformly greenish, not red; bristles barbed only above the middle.

New cauline leaves of the season 4–9, the middle and upper distant, not distichous, 0.5–2 cm. broad; internodes prolonged; summit of sheath with thin quickly friable ligule; bristles shorter than to but slightly exceeding achene, or wanting; spikelets pale to dark brown to lead-color.

Scales orbicular-ovate, 1–2 mm. long, merely mucronulate; wide-ranging eastern species. . . . . 27. *S. atrovirens.*

Scales elliptic-ovate, 2–3 mm. long, tapering to a long setulose awn; western. . . . . 28. *S. pallidus.*

New cauline leaves 10–20 or more, distichous, rather crowded, 3–10 mm. wide, the short sheaths with firm ligule; internodes short; bristles about twice length of achene; spikelets rufescent. . 29. *S. polyphyllus.*

*r.*Bristles smooth or with few scattered or ascending hairs (not regularly retrorse-barbed), bent or curled; non-stoloniferous plants in tufts or stools. . . *t.*

*t.*Bristles at maturity scarcely exceeding the scales. § ANDROCOMA.

Cauline leaves 10–20, those near base and middle of culm approximate, distichous, with short and nearly overlapping sheaths and short internodes; umbel terminal, with lower branches widely divaricate and many times longer than central axis; scales of spikelet cucullate-incurved, blunt, the green midrib as broad as the pale margin. . . . . 30. *S. divaricatus.*

Cauline leaves 5–10, becoming remote, not strongly distichous, the internodes prolonged; umbel with more ascending branches; scales flat or flattish, with free tips, the midrib slender.

Spikelets pale brown or rufescent, many of them solitary and on arching to pendulous elongate pedicels; scales with the conspicuous keeled green midrib excurrent as subulate tip.

Principal cauline leaves 3–8 (–10) mm. wide; ultimate raylets and pedicels smooth; curling bristles mostly much longer than achene; wide-ranging species. . . . . . . 31. *S. lineatus.*

Principal cauline leaves (5–) 8–15 mm. wide; ultimate raylets and pedicels scabrous; curling bristles rarely exceeding achenes; southern. . . . . 32. *S. fontinalis.*

Spikelets purple-brown to fuscous, most of them sessile or very short-pedicelled; scales with weak and scarcely excurrent midrib; northeastern species. . . . . 33. *S. Peckii.*

*t.*Bristles at maturity greatly exceeding and often hiding scales, strongly crimped or curled. § TRICHOPHORUM. *u.*

*u.*Spikelets all or nearly all sessile in glomerules of 3–15. . . . 34. *S. cyperinus.*

*u.*Lateral spikelets of each ultimate group usually pedicelled (pedicels abbreviated in exceptional congested inflorescences). . . *v.*

*v.*Involucels and spikelets pale brown or drab to reddish or terra-cotta.

Involucels and scales red-brown to terra-cotta; longer rays of umbel scabrous except at base; southern species, fruiting northward in Sept. and Oct. . . . . 35. *S. rubricosus.*

Involucels and scales whitish-brown (like manila paper) to dull brown or drab; longer rays of umbel smooth except at tip; northern species, fruiting southward from late June–Aug. . . . . 36. *S. pedicellatus.*

*v.*Involucels and spikelets black or blackish.

Base of involucre not glutinous; scales of spikelet 1.5–2 mm. long; anthers 0.3–0.5 mm. long; stigmas 0.5–0.75 mm. long; achenes whitish, cream-colored or drab; northern and upland wide-ranging species. . . . . 37. *S. atrocinctus.*

Base of involucre viscid or glutinous; scales of spikelet 2–3 mm. long; anthers 1–2.5 mm. long; stigmas 1–1.5 mm. long; achenes reddish-brown to castaneous; lowland coastwise species. . . . . 38. *S. Longii.*

§ BAEÒTHRYON (A. Dietr.) Endl.

1. S. Clintònii Gray (for its discoverer, GEORGE WILLIAM CLINTON, 1807–1885). — Densely cespitose, the *bases of the crowded culms* invested *with drab or dull brown membranous shreds* of old culms *and* leaves and with drab *basal sheaths; leaves flat,* 0.5–1 mm. wide, *much shorter than culms,* with blunt callous tip; *culms capillary, acutely trigonous, upwardly scabrous on angles,* 1–3 dm. high; *spikelet solitary,* terminal, *lance-ovoid, acute,* pale brown to rufescent, 3–6 mm. long; *scales* few, appressed-ascending, *the outer one with the green midrib prolonged into a blunt callous-tipped awn* 1–7 mm. long; *the others* thin, ovate, *blunt,* conduplicate, clasping the achene, *sagittate at base, the broad green midrib scarcely reaching the tip;* anthers 1–1.5 mm.

423. S. Clintonii.

long, blunt or barely mucronate; *achene* trigonous-ellipsoid, brown, 1.5–2 mm. long, *rounded to base and to* the scarcely beaked *summit; perianth-bristles terete, setulose* with short spreading or ascending hairs, about equaling or shorter than achene; *denuded rachilla flexuous, with flattened or concave elongate thin-walled depressions.* — Dry or springy argillaceous or slaty ledges, gravel or open woods and turfy shores, local, w. Bonaventure Co., Que., to s. N.B. and centr. Me.; w. N.Y.; Mich. to Minn.; Alta. *Fr.* late May–early July. Fig. 423.

2. S. verecúndus Fern. (bashful). — Similar to no. 1; *basal sheaths* and weathered shreds *light brown or rufescent; leaves* flat, 1–2 mm. wide, *about equaling and finally exceeding culms;* culms 1–4 dm. high; *outer scale* of spikelet *with* an obtuse to subacute *awn 1–6 mm. long; upper scales with similar but shorter awns.* (*S. planifolius* Muhl., not Grimm) — Dry woods and clearings, often about trap and other basic ledges, sw. Me. to O., s. to s. N.E., Del., Md., D.C., upland of Va., W.Va. and Ky., and Mo. *Fr.* mid-May, June. Fig. 424.

424. S. verecundus.

3. S. cespitòsus L. (in tussocks), var. **callòsus** Bigel. (callous). — *Very densely cespitose,* forming hard resistant hassocks; *bases invested with numerous* stramineous to drab *coriaceous imbricated sheaths; culms wiry,* filiform, *terete, smooth,* 0.5–7.5 dm. high; *upper basal sheaths coriaceous, bearing a firm* terete or involute bluntly callous-tipped *blade up to 1.5 cm. long; spikelet* lanceolate to ovoid, 3.5–6 *mm. long; scales* 5 *or* 6, stramineous to brown, *the* 2 *or* 3 *lowest awn-tipped, the lowest awn about equaling or exceeding the uppermost blunt scales; anthers* 1.3–2.5 *mm. long, the connective excurrent as a slender purple* (when fresh) *tip; achene* trigonous-obovoid, somewhat plano-convex, *abruptly pointed,* about 2 *mm. long; perianth-bristles* slender, *flat, barbless, about twice length of achene; denuded rachilla* 1–2 mm. long, *straight, with coriaceous-margined obliquely cup-shaped depressions.* (Subsp. *austriacus* (Palla) Aschers. & Graebn.) — Forming extensive turf on tundra, acid bogs and peat, Arctic regions, s. to Nfld., N.S., Me., mts. of n. N.E., n. and w. N.Y., higher mts. of N.C., Ga. and Tenn., Ont., Mich., n. Ill., Minn., Mont. and Utah. *Fr.* June–Aug. (N. Eurasia) Fig. 425. — Typical *S. cespitosus* of lowland Europe has more scarious margin of orifice of upper leaf-sheath, purple spikelets 6–8 mm. long and more flowers, the perianth-bristles usually upwardly barbellate.

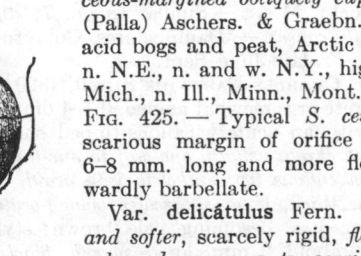

Var. **delicátulus** Fern. (very delicate). — *Culms more slender and softer,* scarcely rigid, *flexuous; basal sheaths black to dark gray, submembranaceous to scarious.* — Calcareous gravels, shores and cliffs, Nfld.; Gaspé Pen., Que.; St. John Valley, n. Me.; Keweenaw Co., Mich.

425. S. cespitosus, v. callosus.

4. S. Rollándii Fern. (for its discoverer, Louis Roland, Frère Rolland-Germain, 1881–). — Much smaller than no. 3, *stoloniferous,* with slender scaly stolons and filiform rhizomes; *culms in small tufts,* 0.5–1.7 dm. high, slenderly filiform; *spikelet ellipsoid-ovoid,* 3–4 *mm. long; scales* ovate, obtuse to subacute, brown or rufescent, the *midrib not excurrent; anthers* 1.5 *mm. long, blunt; perianth* 0; *achene blackish, broadly elliptic-oblong,* 1.2–1.5 *mm. long,* scarcely beaked, plano-convex. (*S. pumilus* sensu Am. auth., not Vahl; *S. alpinus* sensu Am. auth., not Schleicher) — Calcareous ledges, gravels, shores and bogs, Mingan Ids. and Anticosti, Que.; Sask. to s. Alta., s., very locally, to mts. of Colo. *Fr.* July, Aug.

5. S. hudsoniànus (Michx.) Fern. (of Hudson Bay). — In relatively small tussocks or tufts, *with horizontally creeping rhizomes; culms* 1–4.5 dm. high, rather weak, *with scabrous acute angles and concave sides; spikelet* (excluding exserted bristles) 4–7 *mm. long; scales* 10–20, *oblong-lanceolate,* appressed-ascending, thin, *the lowest one with a callous-tipped awn* rarely reaching the tip of spikelet, *the other scales green-keeled but awnless;* achene slenderly trigonous-obovoid, about 1.5 mm. long; *the* 6 *slender flat barbless finally crisped white bristles exserted during anthesis and finally elongating to* 1–3 *cm.;* denuded rachilla as in no. 3 but longer (2–4 mm. long) and with 9–19 depressions. (*Eriophorum alpinum* L.) — Springy or boggy meadows, marly swamps, wet gravels, peat-bogs, etc., often in calcareous districts, se. Lab. to Alaska, s. to Nfld., N.S., N.E., N.Y., Mich., Wisc., Minn. and Mont. *Fr.* June, July. (N. Eurasia) Fig. 426.

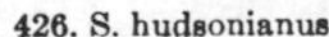

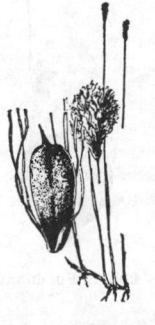

426. S. hudsonianus.

§ Isólepis (R. Br.) Griseb.

6. S. koilólepis (Steud.) Gleason (with hollow scales). — *Tufted annual with sulcate-setaceous culms* 0.2–1.8 dm. high; leaves basal, channeled, setaceous, shorter than culms; inflorescence of 1 sessile spikelet or 2 or 3 in a glomerule, subtended by a slender erect involucre 0.5–2.5 cm. long; *spikelets* ovoid, 2.5–6 *mm. long,* 2–3 *mm. thick, of* 3–10 *rather distant* greenish *deeply boat-shaped sharply keeled and abruptly acuminate scales; achene sharply trigonous, with concave* obovate *closely and minutely papillose faces,* 1.5 mm. long. (*S. carinatus* (Hook. & Arn.) Gray, not Sm.) — Shores, low fields, damp woodroads, etc., Ala. to Mex., n. to Tenn., s. Mo., Okla. and Calif. *Fr.* May, June.

§ Actaeogèton Reichenb.

7. S. Hállii Gray (for its discoverer, Elihu Hall, 1822–1882). — Tufted *annual; culms* soft, slender, *terete,* simple (very exceptionally forking), very unequal, the longer 1–4 dm. long, the shorter down to 1 cm. in length and somewhat crowded at base; basal sheaths bladeless or with a short blade; elongate (up to 1.5 dm.) cauline blade occasional; involucre erect, usually one-half to one-fourth as long as height of culm, tapering to an acutish callous tip; umbel sessile, very rarely with 1 or 2 elongate erect branches up to 2 cm. long; *spikelets* 1–7, slender-ovoid, *becoming cylindric, acute,* 0.5–1.5 cm. long, 2.5–3.5 *mm. thick,* greenish-brown; *scales* ovate, *all cuspidate-acuminate, with green keel* and brown sides, 2.5–3 mm. long, *the outermost commonly much prolonged as a secondary involucre;* stamens 2 or 3, with anthers 0.5–0.9 mm. long; style 2-cleft; *achene* obovate-orbicular, 1.5–2 mm. long, *plano-convex, with very prominent transverse corrugations,* at first whitish, *becoming black;* perianth-bristles wanting. — Peaty and sandy shores, very local, Ga. and Fla. to e. Tex.; Winter P., Middlesex Co., Mass.; Athens, Menard Co., Ill.; s. Mo. and St. Louis, Mo. (formerly). *Fr.* Aug.–early Oct. Fig. 427.

427. S. Hallii.

8. S. saximontànus Fern. (of the Rocky Mts.). — Similar to no. 7; *styles 3-cleft; achene strongly 3-angled, the subequal faces slightly convex.* — Damp shores, Colo. to Mex., e. to S.D., Neb., Kans. and Tex.; Pickaway Co., O. *Fr.* late June–Sept.

9. S. Smíthii Gray (for its discoverer, Charles Eastwick Smith, 1820–1900). — In habit similar to no. 7; culms deep green, subterete or somewhat angled, 0.1–4 dm. long; basal sheaths bladeless or with an acutely callous-tipped short blade; *involucre erect or barely arching, usually one-half to one-third as long as culm, with an acutish callous tip; spikelets* 1–9, *ovoid, acutish,* 0.4–1 cm. long, 3–5 mm. thick, *greenish; scales oblong-ovate, blunt or barely mucronulate, greenish,* becoming pale brown; style 2-cleft; *achene cuneate-obovate,* about 2 mm. long, *smooth, black, lustrous, flat on one face, gently low-convex on the other;* perianth-bristles wanting or merely 1 or 2 minute smooth rudiments, or 4 or 5 very slender and delicate retrorsely setulose ones equaling to exceeding the achene in forma **setòsus** Fern. (bristly), or delicate, elongate and smooth in forma **levisètus** (Fassett) Fern. (with smooth bristles). — Sandy, peaty or muddy (often tidal eastw.) shores, Que. to Minn., s. to s. N.E., N.J., Del., ne. Va., nw. Pa., n. O., n. Ind. and Ill. *Fr.* late July–early Oct. Fig. 428.

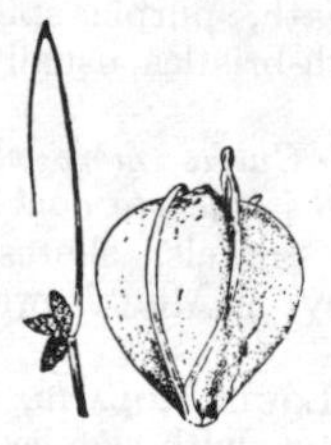

428. S. Smithii.

10. S. Purshiànus Fern. (for its discoverer, Frederick Traugott Pursh, 1774–1820). — Similar to no. 9, usually coarser, with the longer *3-angled paler green culms up to 7* (*rarely* –10) *dm. long; involucre often deflexed at maturity, one-fourth to one-eighth as long as culm, with a rounded callous tip; spikelets* 1–12, ovoid to cylindric, *blunt,* 0.5–1 cm. long, 3.5–5 mm. thick, *brownish or stramineous at maturity; scales orbicular or round-ovate,* with tawny margins; stamens 3; style-branches 2 or 3; *achene round-obovoid, compressed, unequally biconvex, bulged on both faces, more or less pitted; bristles stout, persistent,* 2 *or* 3 *of them overtopping the achene,* or bristles wanting in the local forma **Williámsii** Fern. (for one of its discoverers, Emile Francis Williams, 1858–1929) of e. Mass. and s. Mich. (*S. debilis* Pursh, not Lam.) — Peaty, mossy, muddy or sandy shores, bogs and open acid swamps, Ga. (Fla.?) and Ala., n. to sw. Me., s.-centr. N.H., s. Vt., se. N.Y., e. Pa., W.Va., s. Mich., Wisc. and Minn. *Fr.* Aug.–Oct. Fig. 429.

429. S. Purshianus.

§ Blýsmus (Panzer) Endl.

11. S. rùfus (Huds.) Schrad. (reddish), var. **neogaèus** Fern. (of the New World). — *Extensively creeping*, the *freely forking slender rhizomes* 1–3 mm. thick, *forming dense turf; culms* leafy only toward base, *subterete, compressed, smooth*, 0.5–6 dm. high, *much overtopping the subterete* bluntly callous-tipped *erect leaves; involucre a single bluntly callous-tipped erect bract* 0.5–5 cm. long, *or wanting; spike compound, distichous*, 1–2 cm. long, 0.5–1 cm. broad, *composed of 2 rows of spreading-ascending castaneous to fulvous* 2–5-flowered *spikelets; scales lustrous, conduplicate*, acuminate; *anthers* 2.5–3 *mm. long*, awn-tipped; style-branches 2; *achene* stramineous or olivaceous, smooth, *lance-fusiform, plano-convex, tapering to a stipitate base and a long* often darkened *beak*, 4.5–5.5 *mm. long*, 1–1.7 *mm. broad;* perianth wanting or obsolescent. — Saline or brackish marshes, rarely in fresh peat, Nfld. to lower St. Lawrence R., Que., s. along coast to sw. N.S. and sw. N.B.; shores of Hudson B.; Man. *Fr.* July–Sept. Fig. 430. — Typical Eurasian *S. rufus* with more elliptic and shorter and broader achenes only 3–4.5 mm. long.

430. S. rufus, v. neogaeus.

§ Schoenopléctus Reichenb.

12. S. subterminàlis Torr. (nearly terminal), Swaying Rush. — *Rhizome slender and soft*, the capillary stolons bearing (rarely collected) small moniliform tubers; *leaves of submersed plant capillary and very elongate, numerous, flaccid*, or of the emersed forma **terréstris** (Paine) Fern. (terrestrial) firm and upright and few and short or even wanting; *sheaths whitish, friable; culm terete, capillary*, 0.2–1.4 m. long (responding to depth of water); *involucre a single erect nearly filiform leaf* 0.5–6.5 *cm. long, with a slender straight callous tip; spikelet solitary*, lanceolate to ovoid, 6–13 mm. long, in anthesis 1.5–3, in fruit 4–7 mm. thick, 6–13 mm. long, *green, becoming whitish-brown; scales* ovate-lanceolate, ascending, membranaceous, *becoming pale brown, with excurrent green midrib*, 4–6 *mm. long;* stamens 3, the very slender anthers 3 mm. long and with reddish mucro; *style 3-cleft* to middle, stigmas smooth; *achene trigonous-obovoid*, lustrous, becoming olive-brown, 2.7–3.5 *mm. long, abruptly short-beaked;* bristles downwardly setulose, rarely exceeding beak. — Ponds, deadwaters, quaking bogs and peaty shores, Nfld. to Ont., s. to S.C., e. Ga., Mich., n. Ind., n. Ill. and Mo.; s. Alaska to nw. Mont., Id. and Oreg. *Fr.* late July–Oct. Fig. 431.

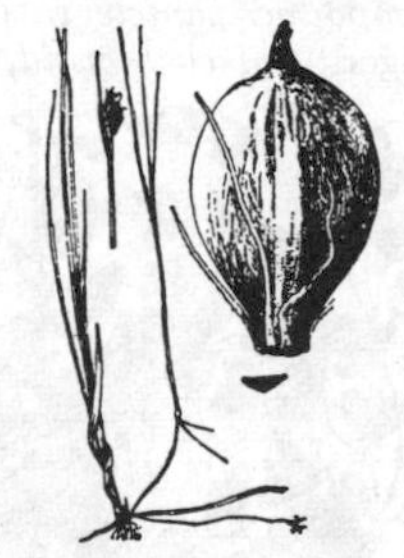

431. S. subterminalis.

13. S. Tórreyi Olney (in honor of John Torrey, 1796–1873). — *Rhizome flaccid*, brownish; *leaves* 2 *or* 3, firm (or in deeply submersed states subflaccid), elongate, triangular-channeled, *with obliquely rounded tip, the inner band of sheath thin and friable; culm* usually solitary, *3-angled, with concave sides*, 0.4–1.5 m. high; *involucre* erect, 3–15 cm. long, *with rounded, usually oblique tip;* spikelets 1–4, cylindric to slenderly ovoid or fusiform, subacute, 1–1.8 cm. long; *scales* ovate, smooth, barely mucronate, *stramineous or pale brown;* style 3-cleft; bristles longer than the *unequally triangular very smooth long-pointed achene* (3.2–4 *mm. long*). — Borders of fresh to brackish ponds and deadwaters, N.B. to Man., s. to s. N.E., L.I., Pa., Ky., Ill., Mo. and S.D.; often ascending to mountain-ponds. *Fr.* July–Sept. Fig. 432.

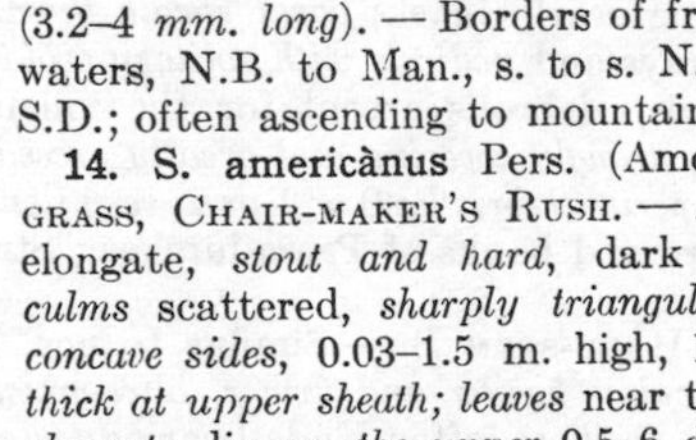

432. S. Torreyi.

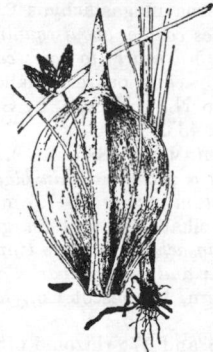

14. S. americànus Pers. (American), Three-square, Sword-grass, Chair-maker's Rush. — *Rhizome* elongate, *stout and hard*, dark brown; *culms* scattered, *sharply triangular, with concave sides*, 0.03–1.5 m. high, 1–6 *mm. thick at upper sheath; leaves* near the base, *elongate*, linear, *the upper* 0.5–6 *dm. long*, 2–9 *mm. broad, sharp-pointed, the firm inner band of sheath with concave summit; involucre erect, slender, pointed*, 2–15 cm. long; spikelets 1–8, glomerulate, ovoid to subcylindric, fulvous, 0.5–2 cm. long; *scales ovate, sparingly ciliate, 2-cleft at apex; anthers tipped with a subulate minutely fringed appendage; style* 2 (rarely 3) *-cleft;* bristles short; *achene* smooth, drab, olivaceous or brown, *plano-*

433. S. americanus.

convex obovoid, short-tipped, 2.5–3 *mm. long.* — Fresh, brackish or saline shores and marshes, Fla. to Tex., n. to Nfld., Que., s. Ont., Mich., Wisc., Minn. and Neb. *Fr.* June–Sept. (Eurasia, and in numerous vars. w. to the Pacific and s. to S. Am.) FIG. 433.

15. S. Òlneyi Gray (for its discoverer, STEPHEN THAYER OLNEY, 1812–1878). — Often coarser than no. 14; *culms with 3 wing-angles and deeply excavated sides, hollow,* 4–10 *mm. thick at upper sheath,* 0.5–3 *m. high; upper leaf-sheath with U- or V-shaped notch at orifice, bearing a lanceolate blunt-tipped blade* 8–15 *cm. long; involucre lance-triangular,* 1–3.5 *cm. long, blunt* to subacute; spikelets 6–12, closely crowded, ovoid, obtuse; *scales suborbicular to orbicular,* the inconspicuous mucronate point shorter than the erose scarious apex; *anthers with very short and blunt minutely bearded tip;* style 2-cleft; achene suborbicular to broadly obovate, mucronate, plano-convex, 2–2.4 mm. long, 1.6–2.2 mm. broad. — Saline to brackish marshes, Fla. to Mex., n. to s. N.H. and w. N.S., Mo., Id. and Oreg.; also n. O. and s. Mich. *Fr.* June–Sept. (W.I. and S.Am.) FIG. 434. — Confused by some with the very different *S. chilensis* Nees & Meyen of temp. S.Am.

434. S. Olneyi.

§ PTERÓLEPIS (Schrad.) Endl.

16. S. VÁLIDUS Vahl (stout), GREAT or SOFT-STEM B. — Rhizome reddish, stout, scaly, horizontal; *culm* pale green, 0.5–3 m. high, 0.3–2 cm. thick at base, *soft and easily compressed,* terete; the *basal sheath membranous, with soon lacerate hyaline margin,* usually bladeless; *decompound panicle* rather stiffly branched, the branches and pedicels minutely pilose, thin-edged; *spikelets ovoid, fulvous or castaneous, many of them in glomerules of 2 or more;* scales ovate, with prominent green keel slightly excurrent, pilose on back to glabrate, with fimbriate-ciliate margins, much longer and broader than the subtended achene; style 2-cleft; perianth-bristles 4–6, very slender, remotely retrorse-setulose above the usually naked basal third, overtopping the plano-convex obovate finally olivaceous to brown achene; anther tipped by a triangular-ovate appendage. — Trop. Am., extending locally n. to se. S.C. Represented with us by

Var. **crèber** Fern. (abundant). — *Panicle usually lax, its rays* and less pubescent *pedicels loosely spreading to pendulous; bractlets brownish, pubescent at tip;* spikelets ovoid, 5–9 mm. long, or linear-cylindric and 9–15 mm. long in forma **megastáchyus** Fern. (with large spikes), some or all in glomerules or some solitary at tips of pedicels; *scales barely to hardly covering the mature achenes, glabrous except for the slightly pubescent emarginate tip and excurrent midrib;* achenes 1.7–2.5 mm. long, 1.3–1.5 mm. broad, or in forma **megastáchyus** 2.3–2.8 mm. long and 1.4–1.8 mm. broad; *bristles coarse, about equaling to but slightly exceeding achene, copiously barbed nearly to base; connective of anther extended as a slender tip.* — Brackish or fresh shallow water and marshes, Nfld. to s. Alaska, s. to N.S., N.E., L.I., Ga., Tenn., Mo., Okla., Tex., N.M., n. Mex. and Calif. *Fr.* June–Sept. FIG. 435.

435. S. validus, v. creber.

17. S. Steinmétzii Fern. (for its discoverer, FERDINAND HENRY STEINMETZ, 1886–). — Similar to var. of no. 16; rhizome nearly black; *basal sheath bearing a prolonged thin blade;* panicle very lax, the pendulous rays and *pedicels with solitary ovoid blunt spikelets* 5–7.5 mm. long; *scales* scario-membranaceous, glabrous except for the villous-ciliate tip and margin, *acuminate,* with very short awn, *strongly covering and greatly exceeding achene; anthers tipped by a triangular sessile appendage; perianth-bristles* 0 *or* 1 *or* 2, *very slender and remotely barbellate, much shorter than achene.* — Shore and banks of Passadumkeag Stream, Penobscot Co., Me. *Fr.* late July–Sept.

18. S. acùtus Muhl. (acute), HARD-STEM B. — Similar to nos. 16 and 17; rhizome often drab or brown; *culms harder and firmer, olive-green; basal sheaths firmer, the margins becoming fibrillose,* often blade-bearing; *panicle stiffer,* with relatively short ascending or divergent (rarely pendulous) rays, or rays suppressed and the inflorescence a dense glomerule in forma **congéstus** (Farw.) Fern. (crowded) FIG. 436a; *bractlets red-spotted, viscid, lacerate-fimbriate at tip; spikelets* mostly in glomerules, *slenderly ovoid to linear-cylindric, acutish,* reddish to fuscous, 1–2 *cm. long; scales* oblong-ovate, *red-dotted, viscid-villous above, hiding the achenes, the similarly colored or paler midrib relatively inconspicuous* and ex-

436a. S. acutus, f. congestus.

tended as a short mucro; achenes becoming black and lustrous. (*S. occidentalis* (S. Wats.) Chase) — Marshes, shores and pond-margins, Nfld. to B.C., s. to N.S., N.E., n. N.J., Pa., O., Ind., Ill., Mo., Okla., Tex., N.M., Ariz. and Calif. *Fr.* Aug., Sept. FIG. 436.

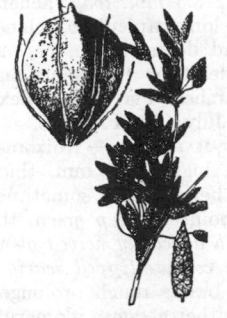

436. S. acutus (typical).

19. S. heterochaètus Chase (with diverse bristles), SLENDER B. — Similar; culms slender, rarely 1 cm. thick, pale green, firm; *panicle with ascending to spreading very slender smooth to barely scabrous rays; bractlets whitish-brown, glabrous; spikelets mostly solitary, pale brown to drab or whitish-green, lance-acuminate to slenderly ellipsoid, acute to subacuminate,* 0.75–2.3 *cm. long; scales firm or subcoriaceous,* deeply emarginate, often slightly red-dotted, *glabrous; bristles* 2–4 (*mostly* 2), *fragile,* unequal, *shorter than achene,* barbellate or smooth; filaments broad; *style 3-cleft; achene trigonous but twice as broad as thick.* — Calcareous or other basic deadwaters, shores and swamps, e. Mass.; sw. Que., w. Vt. and n. N.Y.; Wisc. to N.D., s. to centr. Ky., Ill., Mo. and Okla.; nw. Id., Wash. and Oreg. *Fr.* July–Sept. FIG. 437.

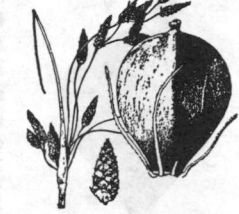

437. S. heterochaetus.

20. S. etuberculàtus (Steud.) Ktze. (without tubercle; Steudel thinking it an aberrant *Rhynchospora*). — *Culm* 1–2 m. high, 3-*angled,* usually *sharply so above,* obtusely so below, the *sheath at base extended into a long slender triangular and channeled leaf;* involucral leaf similar (1–2.5 dm. long), continuing the culm; *spikelets* cylindric (1–2 cm. long), single or sometimes proliferously 2 or 3 together, *nodding on the apices of the* 5–9 *long filiform and flattened peduncles* or rays of the *dichotomous umbel-like corymb,* or the central one nearly sessile; *scales loosely imbricated,* oblong-ovate, acute, *pale,* thin and scarious, with a *greenish nerved back; bristles* 6, *firm, furnished above with spreading hairs rather than barbs,* equaling the slender abrupt beak of the *obovoid-triangular shining achene* (4 mm. long). — Pond-margins and fresh to brackish marshes, very local, Fla. to La., n. to Del. and Mo. *Fr.* June–Aug. FIG. 438. — Often with a 2nd involucral bract, in this character and in its achene and bristles showing alliance with the next species.

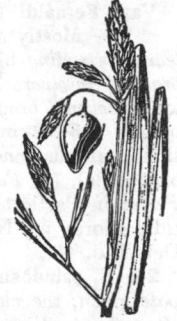

438. S. etuberculatus.

§ PHYLLANTHÈLI Beurl.

21. S. fluviátilis (Torr.) Gray (of rivers), RIVER-B. — Rhizome elongate, moniliform, with thick corm-like enlargements; *culms* stout, sharply angled, *leafy nearly to summit,* 0.7–2 m. high; leaves pale green, flat, 0.7–2 cm. broad, with coriaceous sheaths covering the nodes, the uppermost long-tapering blades and those of the very long involucre much exceeding the *compound umbel; rays* 5–12, mostly *elongate, recurved-spreading,* up to 12 cm. long, smooth, terminated by 1–5 ovoid to subcylindric brown spikelets 1.5–4 cm. long; scales slightly lacerate, the awns much exceeding the cleft tip; *achene obovoid, sharply and exactly triangular, conspicuously pointed, opaque, whitish- to pale brownish-gray,* 4–5 *mm. long, nearly or quite equaled by the* 6 *firm and persistent retrorsely barbed bristles;* anthers 2.5–4.5 mm. long. — Borders of lakes and large streams (usually calcareous), w. N.B. to Sask. and Wash., s. to N.E., Del., Md., ne. Va., w. Pa., O., Ind., Ill., Mo., Kans., N.M. and Calif. *Fr.* July–Sept. FIG. 439.

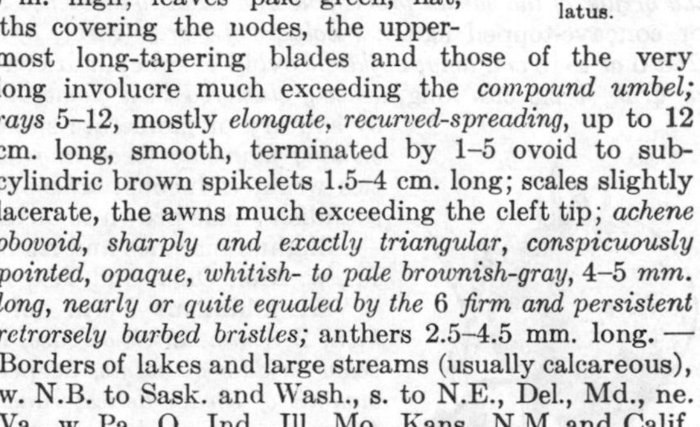

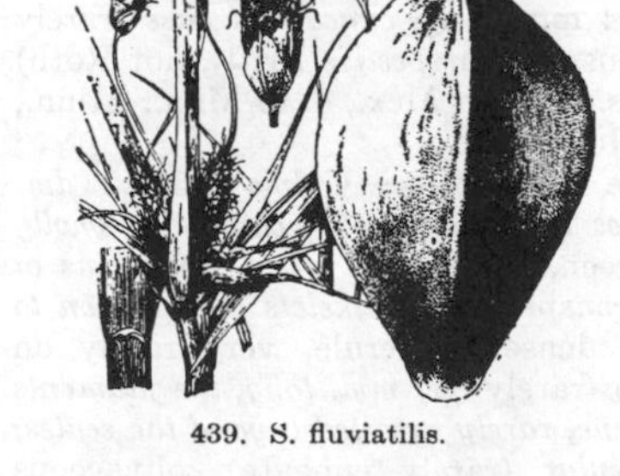

439. S. fluviatilis.

22. S. robústus Pursh (stout). — Rhizome much as in no. 21; culms 0.7–1.5 m. high, leafy nearly to summit; the flat *deep green leaves* 4–10 mm. broad, the upper often overtopping the inflorescence; those of the involucre 3 or 4, very unequal, the longest 2–4 dm. long; *summit of leaf-sheath strongly ribbed below the semicircular or convex dark scarious ligule; spikelets* 1–5 or more, *rufescent, ellipsoid- or conic-ovoid to cylindric, blunt or rounded at tip,* 1.3–3 cm. long, or in forma **protrùsus** Fern. (thrust forward) becoming 3–5 cm. long, 0.8–

440. S. robustus.

1.5 cm. thick, densely glomerulate or 1–3 at tips of slender rays up to 5 cm. long; *scales* all pubescent, *the awn* soon recurving and *many times exceeding the cleft tip; anthers 2–3.5 mm. long;* achene broadly to narrowly obovoid, 3–3.5 mm. long, apiculate, planoconvex, lenticular or with a low and broad dorsal angle, becoming olive to blackish-brown, lustrous; bristles 0 or 1–6, short and caducous. — Brackish or saline coastal marshes, Fla. to e. Mex., n. to Mass. and Calif. (W.I.; e. S.Am.) *Fr.* July–Oct. FIG. 440.

23. S. MARÍTIMUS L. (of the sea), SALT-MARSH-B. — Rhizomes much as in the preceding; culms 0.3–1 m. high, 3–10 mm. thick at base, sharply 3-angled, leafy mostly below (but sometimes above) the middle; *leaves* elongate, taper-pointed, *deep green,* the broader 3–7 mm. broad, *with orifice of sheath delicately nerved about the V-shaped slightly convex- to truncate- or concave-tipped scarious ligule;* involucre of 2–4 bracts, the longest bracts much prolonged and resembling foliage-leaves; *inflorescence* either a dense glomerule or *with* 1–*several glomerulate spikelets terminating the more or less elongate rays; spikelets* ovoid to lance-cylindric, *acute or acuminate, castaneous to black-fuscous,* 1–2 cm. long; *scales with a short awn* projecting from between the two teeth; stamens 3, the anthers 3–3.5 mm. long; bristles mostly 6 (rarely 0), slender and weak, deciduous; style 3 (rarely 2) -cleft; *achene* obovate, 2.8–4 mm. long, 2–2.7 mm. broad, flat on inner, *with a broad low angle on dorsal face,* becoming dark brown, lustrous, *cuneate to the narrow base.* — Casual on waste about ports. (Adv. from Eu.)

Var. **Fernáldi** (Bickn.) Beetle (in honor of the present author, 1873–). — Mostly *coarser; culms up to* 1.5 *m. high* and 1.5 cm. thick at base; *leaves extending* higher, *well above middle of culm, up to* 15 *mm. broad;* 1*st bract of involucre* 1.5–2.5 *dm. long,* 2*nd* 0.8–2 *dm. long;* spikelets 1.2–4 cm. long; *achenes broad- to rounded-obovate, gradually rounded to broadish base,* mostly 2.5–3.2 mm. broad, *with low dorsal angle,* or planoconvex or lenticular and flattened and with 2 style-branches in forma **agònus** Fern. (without angle). (*S. Fernaldi* Bickn.; *S. novae-angliae* Britt.; *S. campestris,* var. *Fernaldi* Bartlett) — Saline to brackish marshes and brackish to fresh tidal shores, ne. N.B., P.E.I., M.I. and C.B. to Va.; w. N.Y. *Fr.* mid-July–Oct. FIG. 441.

441. S. maritimus, var. Fernaldi.

24. S. paludòsus Nels. (of marshes), BAYONET-GRASS, PRAIRIE-B. — *Very pale green;* the *culms 0.3–1.5 m. high, 0.5–2 cm. thick at base; leaves* 3–6, *oftenest extending well above middle of culm, their blades* mostly 0.5–1.5 *cm. broad,* the *veins near the orifice of the sheath prominent and usually thickened* around the V-shaped scarious truncate- or concave-topped ligule; *involucre of* 1 *or* 2 (rarely 3) *bracts,* the 1st bract 0.5–2 dm. long, *the 2nd* 0 *or* 1–10 *cm. long; spikelets* ovoid to subcylindric, acute or acuminate, *whitish to pale brown or drab,* 1–2.5 cm. long, *densely clustered in a glomerule or with* 1–4 elongate *rays* terminated by solitary or glomerate spikelets; *anthers* 3.5–5 *mm. long,* exserted *on long filaments* (*becoming two or three times length of achene*); bristles weak, caducous; *achene* 2.8–4 mm. long, *cuneate at base* (rarely roundish), pale brown to olivaceous. (*S. campestris* Britt., not Roth) — Alkaline marshes and shores, s. B.C. to Mex., e. to Man., Minn., Ia., w. Mo., Okla. and Tex. *Fr.* July–Sept.

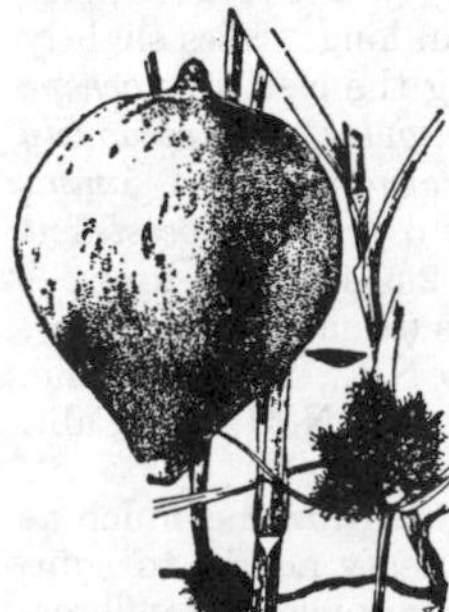

Var. **atlánticus** Fern. (of the Atlantic). — *Culms* 1.5–7.5 *dm. high,* 2–8 *mm. thick at base; leaves* 2–4 (rarely –5), *chiefly or wholly below middle of culms,* deeper green, 1.5–9 *mm. broad,* the *veins at summit of sheath delicate and inconspicuous; spikelets dark brown to fuscous or blackish,* in a single dense glomerule, very rarely on 1–5 divergent rays; *anthers* 2–3.5 (rarely –4) *mm. long, the filaments shorter than to twice length of achene, rarely exserted beyond the scales; achene round-obovate to suborbicular* (rarely cuneate), olivaceous to dark brown. (*S. campestris,* var. *paludosus* Fern., chiefly, not *S. paludosus* Nels., basonym) — Saline marshes and shores, Anticosti I. and Côte Nord, to lower St. Lawrence R., Que., s. to N.J.; centr. and w. N.Y.; James Bay. *Fr.* July–Sept. FIG. 442.

442. S. paludosus, v. atlanticus.

§ Taphrogèton Reichenb.

25. S. rubrotínctus Fern. (red-tinged). — *Culm* rather stout, from a thick stoloniferous caudex, *smooth*, 3–9 dm. high; *leaves* broadly linear, firm, *smooth*, *with most of the* scarcely nodulose *sheaths red-tinged at base*, the *blades* 4–15 *mm. wide;* involucre mostly of 3 leaves, the lowest leaf equaling or exceeding the inflorescence; inflorescence a stiffish decompound umbelliform panicle 0.5–2 dm. high, 3–5 rays longer than the others and stiffly ascending, the shorter ones more numerous and divergent; spikelets 3–7 mm. long, ovoid to lanceolate, few–many in distinct glomerules, or many or all glomerules compacted into dense clusters in forma **confértus** (Fern.) Weath. (crowded), or linear-cylindric and 8–13 mm. long in large distinct loosely radiating glomerules in forma **radiòsus** Fern. (radiate); *scales* ovate, suffused with green and black, all but the terminal *blunt or barely mucronulate; stamens* 2; *bristles* 4; *stigmas* 2; *achenes* whitish, *thin and lenticular.* — Damp open soil, marshes, low thickets, meadows, etc., s. Lab. to Sask., s. to Nfld., N.S., N.E., N.Y., W.Va., Mich., n. Ill., Minn. and Neb. *Fr.* July, early Aug. (–early Sept. at n. limit). Fig. 443.

443. S. rubrotinctus.

26. S. expánsus Fern. (spread-out). — Coarser than no. 25; *culm* 0.6–1.5 m. high, up to 1.5 cm. thick at base, *scabrous* for 1–5 cm. *at summit; leaves with strongly septate-nodulose leathery sheaths* (*only the lower sheaths red-tinged*) *when dry;* the hard and thick *blades scabrous beneath*, paler, 1–2.5 cm. wide; involucre of 3–8 leaves, the lowest leaf usually overtopping the inflorescence; *panicle* (1–)2–3 (–4.5) *dm. high*, with many ascending to loosely divaricate compound rays; spikelets 3–5 mm. long, 3–12 in glomerules 5–10 mm. in diameter, rarely solitary, or short spikelets 20–60 in very dense glomerules in the rare forma **globulòsus** Fern. (having globules); *scales strongly keeled, the keel projecting as a subulate-acuminate tip; stamens* 3; *bristles* 3 *or* 6; *achene with a low dorsal angle.* (*S. sylvaticus* sensu ed. 7, not L.) — Spring-heads, meadows, borders of rills and low thickets, sw. Me. to Mich., s. to Ga. *Fr.* Aug., early Sept. Fig. 444. — Forma **Bissélii** Fern. (for its discoverer, Charles Humphrey Bissell, 1857–1925), *S. sylvaticus*, var. *Bissellii* Fern., has large radiating glomerules 1–2 cm. in diameter, the many linear-cylindric spikelets 6–14 mm. long. Fruiting somewhat earlier.

444. S. expansus.

27. S. atróvirens Willd. (dark green; from the fuscous spikelets). — Culms 1–few from a leafy crown, 0.3–1.8 m. high; *new leaves of season* 4–9, 0.5–2 *cm. wide*, the *summit of the greenish sheath with* V-shaped orifice and a *very thin and promptly friable translucent ligule;* panicle umbelliform, with umbellately forking rays, 0.3–2 dm. high, late in the season or when lopping into water often producing leafy tufts; spikelets greenish-brown or fuscous, ovoid to subcylindric, 2–8 (rarely –10) mm. long, in glomerules; *scales* 1–2 *mm. long, rounded-ovate, mucronulate; bristles* slender, *sparsely and strongly barbed toward summit, shorter than to slightly exceeding the* pointed trigonous *pale achene, or wanting.* — Variable wide-ranging species, with the following vars. and forms:

Pale green; the lower leaves and their sheaths becoming strongly nodulose-septate on drying; blades 0.7–2 cm. wide, with ribs 0.25–0.3 mm. apart; spikelets 3.5–8 mm. long; achene slightly overtopped by perianth-bristles. . . . *S. atrovirens* (typical).

Deeper or warmer green; sheaths and leaves smooth, not superficially nodulose; blades 0.5–1.5 cm. broad, with ribs 0.15–0.2 mm. apart; spikelets 2–4 mm. long.

- Culms several in an ascending clump; leaves firm, opaque; panicle stiff, 0.3–1.5 dm. high, with strongly ascending longer rays 1–12 cm. long; perianth-bristles much shorter than achene or wanting. . . . . Var. *georgianus.*
- Culms solitary, reclining or leaning; leaves membranaceous, translucent; panicle lax, up to 4.5 dm. long, with loosely ascending to divergent slender rays up to 2 or 3 dm. long; bristles exceeding achene. . . . . . . Var. *flaccidifolius.*

S. atróvirens (typical). — Meadows, bogs and low thickets, s. Que. to Sask., s. to Mass., Ct., Ga., Tenn. and Mo. *Fr.* late June–Aug. Fig. 445. — Forma **sychnocéphalus** (S. N. Cowles) S. F. Blake (with crowded heads) has abbreviated rays, the glomerules crowded into 1 or more lobulate masses (Var. *pycnocephalus* Fern.).

Var. **georgiànus** (Harper) Fern. (of Georgia), *S. georgianus* Harper — Nfld. to w. Ont. and Minn., s. to N.S., N.E., L.I., Ga., Tenn. and Ark. *Fr.* June (southw.)–early Sept.

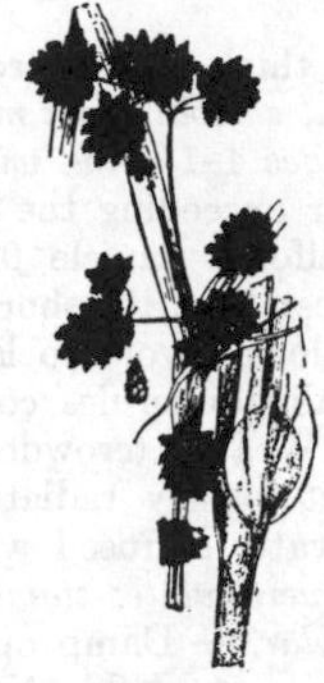
445. S. atrovirens (typical).

(Nfld.) Fig. 446. — The rare forma **angustispicàtus** Fern. (slender-spiked) has slenderly cylindric spikelets up to 1 cm. long. Forma **cephalánthus** Fern. (with flowers in a head) has the rays suppressed, the inflorescence a dense head 1.3–5 cm. in diameter.

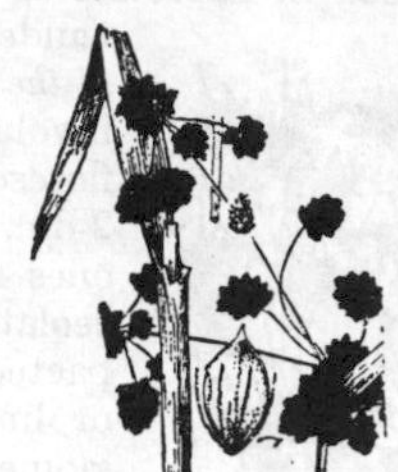
446. S. atrovirens, v. georgianus.

Var. **flaccidifòlius** Fern. (with flaccid leaves). — Bottomlands, se. Va. *Fr.* June. — Rhizome with elongate branches.

28. S. pállidus (Britt.) Fern. (pale). — Similar to no. 27, pale green; spikelets paler brown or drab, mostly in crowded glomerules; *scales* 2–3 *mm. long, with the conspicuous pale midribs prolonged into setulose awns.* (Perhaps better treated as *S. atrovirens*, var. *pallidus* Britt.) — Low grounds, Man. to Wash., s. to Mo., Okla., Tex., N.M. and Ariz. *Fr.* July.

447. S. polyphyllus.

29. S. polyphýllus Vahl (many-leaved). — In small clumps; culms rather slender, 0.5–1.5 m. high, smooth to summit; *leaves* 10–20 *or more, rather crowded near middle of culm, distichous,* 3–10 *mm. wide,* green, thin, long-attenuate, *the short close sheaths with firm ligule; the internodes short;* umbelliform decompound panicle 0.3–1.5 dm. high, with slender divergent and ascending rays; *spikelets rufescent,* 2.5–4 mm. long, in dense glomerules terminating slender pedicels; scales rounded, mucronate, 1–1.5 mm. long; *bristles* 6, *usually twice bent, about twice length* of tiny pale *achene;* panicle often proliferous late in the season. — Low woods, swamps and shores, w. N.E. to s. Ill., s. to Ga. and Tenn. *Fr.* July–Sept. Fig. 447. — Forma **macrostáchys** (Boeckl.) Fern. (with large spikes) has the cylindric spikelets 5–8 mm. long.

§ Andrócoma (Nees) Benth.

30. S. divaricàtus Ell. (divergent). — Slender, *with weak arching to reclining culms* 0.3–1 m. high; *cauline leaves* 10–20, *distichous, rather crowded and with short often overlapping sheaths below the middle,* more scattered above, *the lower and middle internodes short; umbel* very loose and flexuous, *with lower very slender rays* 1–3 *dm. long and strongly divergent, the axis above them very short* and with abbreviated rays; *spikelets* linear-cylindric, 3–6 mm. long (the rachilla finally elongated to 8–15 mm.), 1.5–2 *mm. thick, borne singly, the lateral ones of each group long-pedicelled; scales cucullate-ovate, with incurved blunt tip, the broad green midrib as wide as the pale sides; achene sharply* 3*-angled,* ellipsoid; the bristles short; panicles and leaf-axils frequently proliferating. — Bottomland-woods and -swamps, Fla. to La., n. to e. Va. and se. Mo. *Fr.* June–Aug. (rarely –Oct.). Fig. 448.

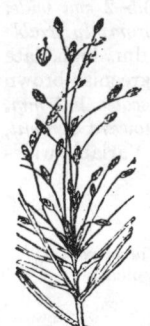

449. S. lineatus.

31. S. lineàtus Michx. (marked with lines; referring to the green line-like keels of the scales). — Culms strongly ascending, firm, remotely 5–10-leaved, with long internodes; leaves 3–8 (–10) mm. wide, pale green, firm; *involucre and involucels pale brown at base; umbels* terminal and sometimes axillary, loose, 0.5–2 dm. high, *subsecund,* the terminal with a 1–3-leaved involucre much shorter than the *long slender nodding-tipped rays; spikelets oblong,* becoming cylindrical, 0.5–1 cm. long, or 1.2–1.8 cm. long in forma **elongàtus** Eames (elongate), 2–3.5 *mm. thick, the lateral ones of each group on smooth pedicels; scales pale brown to rufescent,* ovate, *sharply and slenderly green-keeled, the sharp tips ascending; achene obscurely* 3*-angled, narrowly ellipsoid to fusiform,* long-beaked, papillate; *bristles* curling, *mostly longer than achene.* — Meadows, swales and low thickets, centr. Me. to Ia., s. to Ct., Va., Ala., Miss., Ark., Okla. and e. Tex.; Oreg. *Fr.* June–Aug. Fig. 449.

448. S. divaricatus.

32. S. fontinàlis Harper (of springs). — Resembling no. 31; *leaves* usually *darker green,* thinner, *mostly* 8–15 *mm. wide;* inflorescence looser and with more spreading to divaricate rays;

ultimate raylets and pedicels scabrous; spikelets lance-ovoid; bristles rarely overtopping achenes.— Calcareous spring-heads, bottoms and wet slopes, Fla. to La., n. to e. Va. and Okla. *Fr.* May, June. — Panicles late in season often proliferating.

33. S. Péckii Britt. (for its discoverer, CHARLES HORTON PECK, 1833–1917). — Culms slender, 0.8–1.7 m. high; *leaves pale green,* 5–9 mm. broad, the *margins scabrous; involucre and involucels blackish at base;* inflorescence 0.5–2 dm. high, the 2–5 longest *stiff rays ascending,* the others shorter and ascending or divergent, *the tips scarcely drooping; spikelets* oblong-cylindric, 5–9 mm. long, *mostly sessile or subsessile in glomerules of* 2–7; *scales* oblong-ovate, acutish or obtuse, *blackish-ferruginous* above the pale base; *achene soft, whitish, oblong.* — Meadows and bogs, w. Me. to s. Que. and n. N.Y., s. to w. Ct. and se. N.Y. *Fr.* July, Aug. FIG. 450.

450. S. Peckii.

451. S. cyperinus (typical).

§ TRICHÓPHORUM (Pers.) Darl. (WOOL-GRASSES)

34. S. cyperìnus (L.) Kunth (like *Cyperus*). — In dense tussocks with many curving basal leaves; culms nearly terete (1–1.5 m. high); *leaves* narrowly linear, long, rigid; *those of the involucre* 3–5, *longer than the loose umbel* (1.5–3 dm. long); the tips of the rays at length drooping; *involucels reddish-brown; spikelets* exceedingly *numerous, ovoid, clustered,* woolly at maturity, 3–6 mm. long, or in forma **Andrèwsii** (Fern.) Carpenter (for its discoverer, LUMAN ANDREWS, 1839–1921) linear-cylindric and 7–10 mm. long; *the rust-colored* elongating and crisped *bristles much longer than the soft and pointless reddish-brown scales;* achenes whitish to cream-color or drab, pointed. — Wet meadows and swamps, N.C. to Okla., n. to N.E., N.Y., O., Ind., Ill. and Ia. *Fr.* Aug.–Oct. FIG. 451. — Passing insensibly **northward** to the more general northern

Var. **pélius** Fern. (livid or blackish). — *Involucels drab to blackish,* without red tinge; *scales with little or no red; bristles drab or smoky;* spikelets in forma **condensàtus** (Fern.) S. F. Blake (condensed) aggregated into 1 or more dense heads (FIG. 452). — Nfld. to Minn., s. to N.S., N.E., N.J., Pa., n. Md., W.Va., n. O., n. Ind., n. Ill. and ne. Ia.

452. S. cyperinus, v. pelius, f. condensatus.

35. S. rubricòsus Fern. (like red-ochre). — Coarse and tall (usually 0.7–2 m. high); the culms 2.5–6 mm. thick below the usually ample (1.5–3.5 dm. high, 1.5–2.5 dm. broad) inflorescence; leaves pale green, firm, 5–11 mm. wide; *rays of inflorescence* very elongate, mostly ascending, drooping at tip, *the longer ones scabrous except toward base; involucels red-brown to terra-cotta; spikelets* ovoid, *reddish-brown,* 3–6 mm. long, or in the small (inflorescence often only 1–1.5 dm. long, 1 dm. broad) forma **praelóngus** Fern. (very long) linear-cylindric and 8–12 mm. long, the *ultimate fascicles with a sessile central spikelet, the* (1–) 2 (–4) *lateral spikelets on long pedicels;* curling bristles slightly paler than scales; achenes pale, beaked. (*S. Eriophorum* Michx., illegitimate name). — Swamps, marshes and low woods, Fla. to e. Tex. and Mex., n. to se. Mass., L.I., N.J., se. Pa., Md., W.Va., Ind. and Ill. *Fr.* late Aug.–Oct. FIG. 453.

453. S. rubricosus.

36. S. pedicellàtus Fern. (borne on pedicels). — Similar; culms rather stout, 2–5 mm. thick below inflorescence; *leaves pale green, firm,* 3–10 *mm. broad;* inflorescence ample, 1–3 dm. high; the numerous *ascending subequal rays* slender, with nodding tips, *smooth except at summit; involucels brown to straw-color or drab;* lateral spikelets of each group pedicelled; *scales pale brown to dull brown or drab; wool whitish-brown,* not rufescent. — Alluvial thickets, swales and rich shores, Gaspé Pen., Que., to Minn., s. to N.S., Me., ne. Mass., Ct., n. N.J., N.Y., n. O., n. Ind. and Ia. *Fr.* July, Aug. — Nondescript (perhaps partly hybrid) specimens with darker involucels and spikelets approach the next; they (var. **púllus** Fern., dusky) are much coarser than no. 37, with larger spikelets, smoother rays and later flowering-habit.

37. S. atrocínctus Fern. (black-girdled). — Slender, 0.3–1.8 m. high; culms 1–4 mm. thick below the inflorescence; *leaves bright green,* relatively soft, 2–5 *mm. wide;* inflorescence 0.5–(rarely) 4 dm. high, the very *slender harshly scabrous rays very unequal,* or rays suppressed with the inflorescence forming 1 or few dense subglobose heads in forma **brachýpodus** (Fern.) S. F. Blake (short-stalked), Fig. 455; *involucels and base of involucre black;* spikelets 2.5–6 mm. long, or in the very large forma **grándis** (Fern.) Carpenter (large) 7–10 mm. long in very large inflorescences, the lateral spikelets of each cluster slender-pedicelled; *scales greenish-black,* 1.5–2 *mm. long; anthers* 0.3–0.5 *mm. long; stigmas* 0.5–0.75 *mm. long; achenes whitish to buff; wool drab or olivaceous.* — Meadows and swamps, Nfld. and Côte Nord, Que., to n. Alta., s. to N.S., N.E., n. N.J., Pa., ne. W.Va., Mich., Wisc., ne. Ia., s. Man., s. Sask. and Wash. *Fr.* June, July (southw.)–Sept. (northw.). Fig. 454.

454. S. atrocinctus (typical).

455. S. atrocinctus, f. brachypodus.

38. S. Lóngii Fern. (for its discoverer, Bayard Long, 1885–). — Coarser than no. 37; leaves 3–8 mm. broad; *base of involucre viscid or glutinous;* inflorescence 0.6–2 (–3) dm. high; spikelets ellipsoid, 4–10 mm. long, the lateral very long-pedicelled; *scales* 2–3 *mm. long; bristles whitish,* very long and curled; the 3 *stigmas* 1–1.5 *mm. long; anthers* 1–2.5 *mm. long; achenes ellipsoid-obovoid,* 3-*angled, chestnut- or reddish-brown.* — Meadows, swamps and fresh marshes, locally abundant, e. N.C.; s. N.J. to e. Mass.; w. N.S. *Fr.* June, early July (–mid-Aug. in N.S.) Fig. 456.

456. S. Longii.

9. ERIÓPHORUM L. Cotton-grass. Bog-cotton. Linaigrette (Que.)

Bristles naked, very numerous, silky and becoming greatly elongated. Otherwise as in *Scirpus.* — Spikelets single or clustered or umbellate, when involucrate with leaf-like bracts, upon a leafy or naked stem; scales membranaceous, 1–5-nerved, some of the lowest usually empty. Style very slender and elongated, 3-cleft. Achene acutely triangular. Small, chiefly boreal, genus. (Name composed of the Greek *erion, wool* or *cotton,* and *phoros, bearing.*)*

a. Spikelet solitary, erect; leafy involucre none, the lowest scale enlarged and thickened as a spathe; culm-leaves reduced to mostly bladeless sheaths. . . *b.*

b. Stoloniferous; culms mostly solitary; empty scales at base of spikelet mostly 7 or fewer.

Flowering spikelet broadly obovoid to subglobose, 0.8–1.2 cm. long; scales attenuate; fruiting spikelet depressed-globose, 2–2.5 cm. high; bristles bright white. . . . . 1. *E. Scheuchzeri.*

Flowering spikelet oblong-cylindric, 1.5–2 cm. long; scales bluntish; fruiting spikelet obovoid, 2.5–4 cm. long; bristles cinnamon-color (rarely white). . . . . 2. *E. Chamissonis.*

b. Cespitose, not stoloniferous, forming tufts or tussocks; empty basal scales mostly 10–15. . . . *c.*

c. Spathes and scales without conspicuous pale margin, appressed-ascending; fruiting spikelet obovoid, 1.5–2.5 cm. long; achenes 2–2.3 mm. long, 0.5–1.2 mm. broad.

Culms slender, 3–6.75 dm. high; sheaths scattered; the upper usually high on the culm, scarcely inflated, bladeless; spathe lanceolate to lance-ovate, with broad ribless margin; bristles sordid. . . . . 3. *E. opacum.*

Culms stout and stiff, 0.6–2.2 dm. high; sheaths borne mostly below middle of culm, the upper ampliate-inflated and often with a short blade; spathe ovate, ribbed nearly to margin; bristles bright white. 4. *E. callitrix.*

c. Spathes and scales white-margined, finally divergent or reflexed; fruiting spikelet depressed-globose to broadly obovoid, 2.5–5 cm. in diameter; achenes 2.5–3.5 mm. long, 1.5–2 mm. broad. . . . . . . 5. *E. spissum.*

a. Spikelets 2–several, spreading or drooping on peduncles, or densely clustered; involucre of 1–several leafy bracts. . . *d.*

* Treatment of species 1–5 based on that in Rhodora, xxvii. 203–210 (1925).

*d.*Leaves very slender, 1–1.5 mm. broad, triangular-channeled throughout; involucre a single erect short bract.
Upper cauline leaf with sheath longer than blade. . . . . 6. *E. gracile.*
Upper cauline leaf with sheath shorter than blade. . . . . 7. *E. tenellum.*
*d.*Leaves broader, flat at least below the middle; involucral bracts 2 or more. . . *e.*
*e.*Scales of spikelets greenish, lead-color or blackish, with 1 prominent rib; stamens 3; spikelets loosely peduncled (rarely crowded).
Midrib of scale prominent only below the membranous tip; upper sheaths dark-girdled at summit; anthers 2.5–5 mm. long. . . . . 8. *E. angustifolium.*
Midrib of scale prominent to tip; sheaths not dark-girdled at summit; anthers 1–1.3 mm. long. . . . . 9. *E. viridi-carinatum.*
*e.*Scales of spikelet greenish or straw-colored, with several strong ribs and nerveless reddish margin; stamen 1; spikelets mostly crowded. . . . 10. *E. virginicum.*

1. E. Scheùchzeri Hoppe (for Johann Scheuchzer, 1684–1738). — *Loosely stoloniferous; culms solitary,* soft, 0.5–3.5 dm. high, slightly leafy at base, above usually with 1 bladeless loose membranous-margined black-tipped sheath 2–7 cm. long; leaves of basal shoots 3–12 cm. long, soft, channeled or strongly involute; *flowering spikelet broadly obovoid or subglobose,* 0.8–1.2 *cm. long, in fruit depressed-globose* and 2–2.5 *cm. high; scales* lead-color to blackish, *with slightly paler narrow margins,* the 1–3 outer scales ovate, the others ovate-lanceolate to lance-attenuate; *anthers* 1 *mm. long;* bristles bright white. — Arctic reg., s. in wet peat to n. Nfld., James Bay and Alta. *Fr.* July. (Eurasia) Pl. II, fig. 1.

2. E. Chamissònis C. A. Mey. (for its discoverer, Adalbert von Chamisso, 1781–1838). — Similar; culms firmer, 1–5 dm. high, slender, rarely 1.5 mm. in diameter at base, with reddish-brown blade-bearing sheaths and short slender blades, above with a usually bladeless close sheath 3.5–10 cm. long; *flowering spikelet oblong-cylindric,* 1.5–2 *cm. long, in fruit obovoid* and 2.5–4 *cm. long; scales* brownish-drab to blackish, *with distinct whitish margins,* ovate to ovate-lanceolate, *bluntish; anthers* 1.5–3 *mm. long; bristles reddish or cinnamon-brown,* or white in the local forma **álbidum** (F. Nylander) Fern. (white). (*E. russeolum* Fries) — Wet peat, Lab. to Alaska, s. to Nfld., N.S., N.B., Ont., Minn., Wyo., Id. and Wash. *Fr.* July, early Aug. (Eurasia) Pl. II, fig. 2.

Var. **aquâtile** (Norman) Fern. (aquatic). — Coarse; culms 4–6 dm. high, 2–4 mm. thick at base; basal leaves coarse and elongate, often equaling culms; stolons with bladeless sheaths; bristles paler. — Shallow pools, n. Nfld. (N. Eu.)

3. E. opàcum (Bjornstr.) Fern. (dark; probably for the lead-colored scales). — *Loosely cespitose,* forming small tussocks of 3–17 very slender *terete culms* 3–6.75 *dm. high;* basal leaves nearly filiform, smooth; *sheaths remote, bladeless;* the *uppermost near top of culm, close, barely inflated* toward the dark tip; *flowering spikelet* globose-obovoid, *barely* 1 *cm. long, in fruit* globose and 2–3.5 *cm. broad; spathe lanceolate to lance-ovate, with broad ribless margin; scales lead-color, lance-attenuate, appressed-ascending; anthers* 1–2 *mm. long; pits of denuded receptacle opening obliquely upward,* 75–200, mostly 0.2–0.3 mm. long; achenes cuneate-obovoid, 2–2.3 mm. long, 0.5–1.2 mm. broad; bristles cream-color. (*E. brachyantherum* sensu Hultén, probably not Trautv.) — Calcareous bogs, Nfld. to Alaska, s. to Ont., Sask., s. Alta. and s. B.C. *Fr.* June, July. (Eurasia) Pl. II, fig. 3.

4. E. cállitrix Cham. (beautiful hair). — Similar; lower, stiffer, the 1–6 stoutish culms 0.6–2.2 dm. high; leaves coarser and stiffer; cauline *sheaths mostly toward base of culm, the upper ampliate-inflated and usually with a short blade; spathe ovate, ribbed nearly to margin; anthers* 0.7–1 *mm. long;* achenes ellipsoid-obovoid; bristles bright white. — Greenl. and Baffin I., s. in calcareous peat to n. Nfld. *Fr.* June, July. (Alaska; Asia) Pl. II, fig. 4.

5. E. spíssum Fern. (crowded), Hare's-tail. — *Densely cespitose,* forming broad tussocks *with many stiff trigonous culms scabrous at tip,* 1.5–7 dm. high; basal sheaths brown, long-persistent, fibrillose; *basal leaves* slender, trigonous, *commonly scabrous in lines; upper sheaths* of culm bladeless, *conspicuously inflated above,* veiny-reticulate, with dark membranous tip; flowering *spikelet* 1, erect, obovoid to subglobose, 0.8–1.5 cm. high, *in fruit depressed-globose* and 2.5–5 *cm. broad; scales obovate to ovate-lanceolate,* long-acuminate, lead-color or blackish, *whitish-margined, finally divergent or reflexed;* anthers 1–2 mm. long; *pits of mature denuded receptacle* 25–60, *mostly* 0.4–0.6 *mm. long, opening almost horizontally outward; achenes* 2.5–3.5 *mm. long,* 1.5–2 *mm. broad;* bristles bright white. (*E. callitrix* of ed. 7, not Cham.; *E. vaginatum* of e. Am. auth., not L.) — Acid bogs and peat, Baffin I. to Alaska, s. to Nfld., N.S., N.E., n. N.J., ne. Pa., n. Ind., Wisc. and Minn. *Fr* late Apr.–mid-July. Pl. II, fig. 5.

PLATE II

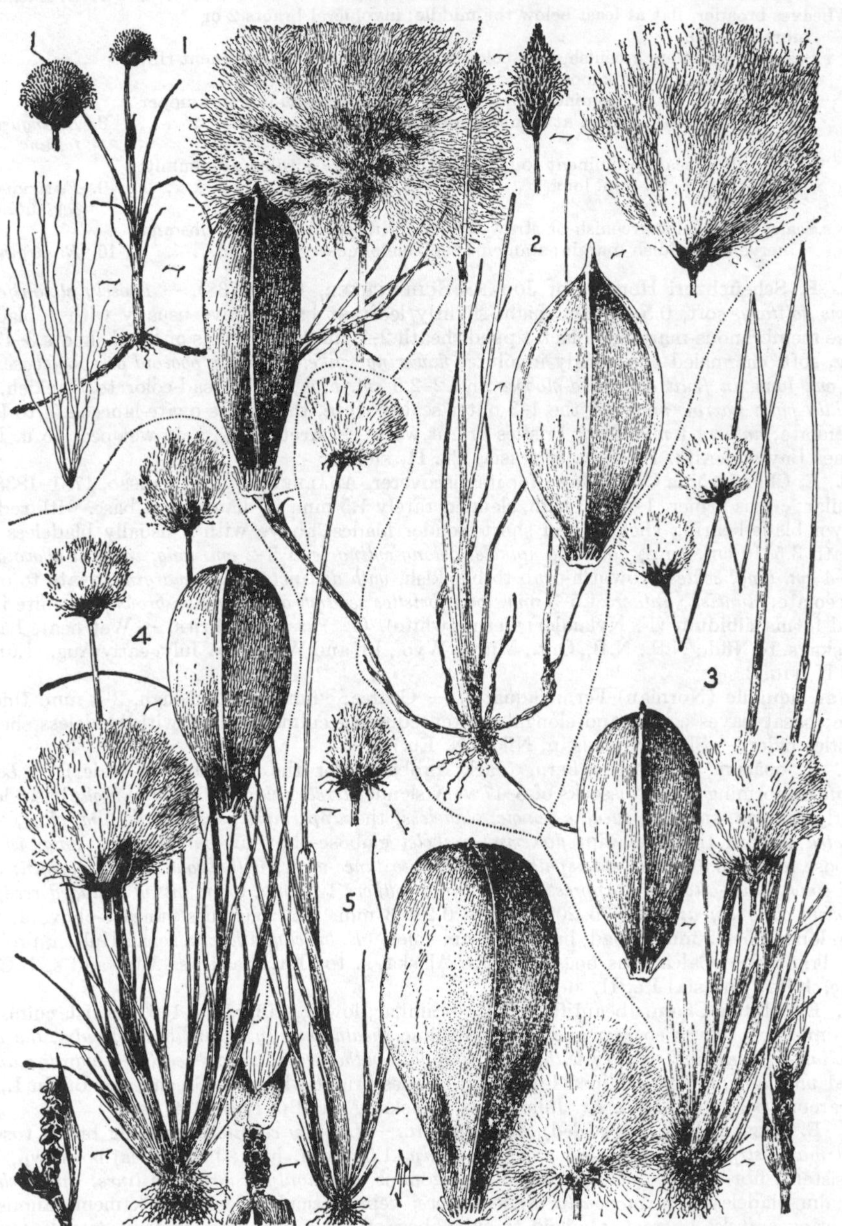

FIG. 1, ERIOPHORUM SCHEUCHZERI; FIG. 2, E. CHAMISSONIS; FIG. 3, E. OPACUM; FIG. 4, E. CALLITRIX; FIG. 5, E. SPISSUM.

Var. **erubéscens** Fern. (becoming red). — Bristles reddish or cinnamon-brown. — Acid peats, Nfld. and adj. Lab.

6. E. grácile W. D. J. Koch (slender). — Weak, very slender; *culms subterete, glabrous,* 2–6 dm. high, commonly *with no young basal leaves at flowering time; sheath of upper culm-leaf* 3.5–5.5 *cm. long, the round-tipped triangular-channeled blade much shorter* (1–4 cm. long) and 1–1.5 mm. broad; *involucral bract* 1, similar, upright, 1–2 cm. long, usually *lead-color or blackish at base; spikelets* 2–5, the lateral on slender minutely pubescent spreading or finally nodding peduncles 0.5–3 cm. long, in anthesis narrowly ovoid and 7–10 mm. long, *in fruit* 1.5–2 *cm. long; scales lead-color or blackish,* all but the uppermost blunt; *anthers* 1–2 *mm. long;* achenes 1.5–2 mm. long; bristles bright white. — Wet peat and inundated shores, Lab. to s. Alaska, s. to Nfld., N.S., N.E., Del., Pa., Mich., n. Ill., Minn., Neb. and Calif. *Fr.* late Apr.–July. (Eurasia)

7. E. tenéllum Nutt. (delicate). — Similar, coarser; *culms obtusely angled, scabrous above,* 2–9 dm. high; *basal green leaves* long, slender, *triangular-channeled, sharp-pointed,* often scabrous; *sheath of upper culm-leaf* 2.5–8 cm. long, *much shorter than the scabrous sharp-pointed blade* (0.3–1.8 dm. long); *involucral bract* 1.5–6 cm. long, *greenish or reddish-brown at base; spikelets* 2–7, mostly on short scabrous peduncles, *in fruit* 2–3 *cm. long; scales greenish to reddish-brown;* achenes 2.5–3 mm. long; bristles milk-white. — Peaty soil, Nfld. to Sask., s. to N.S., N.E., L.I., N.J., Pa., Mich. and Ill. *Fr.* late June–Sept.

Var. **montícola** Fern. (dwelling in mountains). — Culms 1–2 dm. high; upper sheath 1–2 cm. long, blade 3–5 cm. long; spikelet 1. — Alpine bogs, Tabletop Mt., Que.

8. E. angustifòlium Honckeny (narrow-leaved). — Culms soft, obtusely angled, rather slender, 2–6 dm. high; *leaves* flat below the middle, *usually triangular-channeled or conduplicate above the middle,* scabrous-margined, the cauline 1.5–4 mm. broad; *involucral leaves* similar, 2 or 3, *dark purple at least at base; spikelets* 2–10, the lateral *on* divergent or drooping *mostly glabrous stoutish peduncles* 0.5–6 cm. long, *in anthesis ovoid* and 1–2 *cm. long,* in fruit 2–5 cm. long; *scales* lead-color to castaneous, ovate to lanceolate, acutish, *with the slender midrib evanescent below the pale membranous tip; anthers* 2.5–5 *mm. long;* achenes 2.5–3.5 mm. long; bristles bright white to creamy. — Arct. reg., s. in wet bogs and on shores to Nfld., N.S., Me., n. N.Y., n. Ill., N.M. and Oreg. *Fr.* June–Aug. (Eurasia)

Var. **màjus** Schultz (larger). — Stouter, 3–9 dm. high; cauline leaves 4–8 mm. broad. — S. Greenl. and Lab. to Alaska, s. to Nfld., N.S., Mass., Ind., n. Ill., Ia., Neb., Colo. and Oreg. (Eurasia)

9. E. víridi-carinàtum (Engelm.) Fern. (green-keeled). — Culms in small tufts, slender, trigonous, 2–9 dm. high; *leaves flat except at very tip;* basal numerous, elongate, cauline remote and 2–6 mm. wide; *involucral bracts* similar, 2–4, *green throughout or merely brownish at base; spikelets* 3–30, mostly *on* slender simple or forked *minutely hairy peduncles* 1–6 cm. long, or spikelets all sessile or subsessile in a dense glomerule in forma **Fellòwsii** Fern. (for its discoverer, Dana Willis Fellows, 1847–1928), *in anthesis oblong-ovoid* and 0.6–1 *cm. long,* in fruit 1.5–3 cm. long; *scales* greenish-drab to lead-color, *the prominent midrib extending quite to the tip,* in the outermost scale sometimes prolonged; *anthers* 1–1.3 *mm. long;* bristles creamy to buff. — Peats, wet meadows and swampy woods, se. Lab. to s. Alaska, s. to Nfld., N.S., N.E., L.I., N.Y., O., Ind., n. Ill., Ia. and s. B.C. *Fr.* late May–Aug.

10. E. virgínicum L. (Virginian), Tawny C. — Culms slender, wiry, smooth, 0.4–1.2 m. high; *leaves flat except at the very tip, firm, the uppermost with tight sheaths,* 1.5–4 mm. broad; involucral bracts 2–5, divergent, the longest 0.4–1.2 dm. long; *spikelets crowded in a dense glomerule* 1.5–6 cm. in diameter, in anthesis ellipsoid and 6–10 mm. long, *in fruit* 1–2 *cm. long; scales* ovate-oblong, *the strongly striate-ribbed body greenish to straw-color, the thin nerveless margin reddish-brown;* stamen 1; anther 1.5 mm. long; achenes 3–4 mm. long; bristles tawny to copper-color at least at base, or white in forma **álbum** (Gray) Wieg. (white). — Bogs and peaty meadows, se. Lab. to Ont., s. to Nfld., N.S., N.E., L.I., Ga., Tenn., Wisc. and Minn. *Fr.* Aug.–Oct.

10. FUIRÈNA Rottb. Umbrella-grass

Spikelets many-flowered, terete, clustered or solitary, axillary and terminal. Scales imbricated in many ranks, awned below the apex, all floriferous. Perianth of 3 ovate or cordate petaloid scales, these mostly on claws and usually with as many alternating small bristles. Stamens 3. Style 3-cleft. Achene triangular, pointed by the persistent base of the style. — Culms from a usually perennial root, obtusely triangular. Small genus of trop. and warm-temp. reg. (Named for *Georg Fuiren,* a Danish botanist, 1581–1628.)

Awns of the dilated perianth-scales terminal; achenes yellow or brown.
  Annual, tufted; scales of spikelet delicately nerved; dilated perianth-scales oblong or ovate; slender alternating bristles mostly exceeding achene, retrorsely barbed. . . . . 1. *F. pumila.*

Perennial, often with thick rhizome; scales of spikelet coarsely corrugated; dilated perianth-scales rhombic to deltoid-ovate; alternating bristles much shorter than achene or wanting.
Leaf-blades and sheaths copiously spreading-hirsute; bristles of perianth retrorsely barbed. . . . . 2. *F. squarrosa.*
Leaf-blades (except sometimes at base) and middle and upper sheaths glabrous; bristles wanting or short and minutely upwardly barbellate. . 3. *F. breviseta.*
Awns of dilated perianth-scales arising from below tip, retrorsely barbed; achenes whitish; scales of spikelet coarsely corrugated; perennial with thick rhizome. 4. *F. simplex.*

1. F. pùmila Torr. (dwarf). — *Annual,* tufted, the smooth often curving culms 0.5–4.5 dm. high; leaf-sheaths hispid; spikelets in 1–3 glomerules of 2–8, crowded; their *olivaceous scales only delicately nerved, with strongly curving awns; perianth-scales* narrowly to broadly *oblong or ovate,* long-stipitate, *attenuate to a long retrorsely barbed awn;* barbed *bristles* usually *exceeding the yellow-brown achene,* which is *equaled by the persistent style.* (*F. squarrosa* of ed. 7, not Michx.) — Bogs and wet peaty or sandy shores, Fla. to Mass.; s. Mich., nw. Ind. and ne. Ill. Late July–Oct. FIG. 457.

2. F. squarròsa Michx. (with recurved tips). — Similar; *perennial, with thick rhizome;* culms 0.25–1 m. high, erect; sheaths densely pubescent; *scales of spikelet with coarsely corrugated center and straight or but slightly curving awns; perianth-scales rhombic or deltoid-ovate,* with a short thick *smooth terminal awn or point,* the interposed mostly *retrorsely barbed bristles shorter than the yellow achene,* which is twice as long as the persistent style. (*F. hispida* Ell.) — Wet sandy places, Fla. to Tex., n. to N.J., Ky. and Okla. July–Oct. (W.I.) FIG. 458.

458. F. squarrosa.

457. F. pumila.

3. F. brevisèta Coville (with short bristles). — Resembling no. 2; rhizome less developed; *leaves and sheaths glabrous* or bases of blades merely ciliate; *bristles of perianth wanting or very short and upwardly and minutely barbellate.* — Wet sands and peats, Fla. to Tex., n. to se. Va. (rare). Aug.–Oct.

4. F. símplex Vahl (unbranched). — Similar to no. 2, perennial, paler green; leaf-sheaths sparsely pubescent to glabrous; *perianth-scales* ovate-oblong, *the retrorsely barbed awns arising from below the tip,* the alternating *bristles equaling or exceeding the whitish achene.* — Springy places or swamps, Mex. to Kans. and Mo. July–Oct. FIG. 459.

459. F. simplex.

11. HEMICÁRPHA Nees

Spikelet, flowers, etc., as in *Scirpus,* except that there is a minute translucent scale (readily overlooked) between the flower and the axis of the spikelet. Stamen only 1. Style 2-cleft. Bristles or other perianth none. Small genus of trop. and warm-temp. reg. (Name from the Greek *hemi, half,* and *carphos, chaff,* in allusion to the single inner scalelet.)

Scales of spikelet oblong to obovate, short-tipped.
Tips of oblong to narrowly obovate scales spreading or recurving. . . . . 1. *H. micrantha.*
Tips of broadly obovate to rhombic scales appressed. . . . . 2. *H. Drummondi.*
Scales rhombic to narrowly ovate, contracted to long-recurving awns about as long as body. . . . . 3. *H. aristulata.*

1. H. micrántha (Vahl) Pax (tiny-flowered). — Dwarf arching annual (0.2–15 cm. high); involucre 1-leaved, as if a continuation of the bristle-like culm, and usually with another minute leaf; *spikelets* 1–3, *short-cylindric or ovoid* (2–4 mm. long); *scales oblong or narrowly obovate,* brown, *tipped with a short recurved point; achenes cylindric, brown, slightly reticulated, with many close rows of crowded low papillae.* — Sandy borders of ponds and streams, Fla. to Tex. and Mex., n. to s. Me., centr. N.H., centr. N.Y., O., s. Ont., Mich., Wisc., Minn. and Wash. Aug.–Oct. (Trop. Am.) FIG. 460.

460. H. micrantha.

461. H. Drummondi.

2. H. Drummóndi Nees (for its discoverer, THOMAS DRUMMOND, 1780–1835). — Similar; *scales broadly obovate or rhombic, the broad green midrib barely projecting as a blunt appressed tip;* achenes narrowly obovoid, ashy, scarcely reticulated, the papillae fewer and somewhat remote. — Damp sand, etc., Mex. and Tex. to S.D., e. to s. Ont. July–Oct. FIG. 461.

3. H. aristulàta (Coville) Smyth (having small awns). — Spikelets with *ovate or rhombic scales contracted to long recurving awns as long as the body.* (*H. occidentalis* in part of ed. 7, not Gray) — Damp sands, Pacific states, locally e. to Minn., Ia. and Mo. July–Oct.

12. LIPOCÁRPHA R. Br.

462. L. maculata.

Spikelets terete, many-flowered, in a terminal close cluster involucrate by leafy bracts. Scales spatulate, regularly imbricated in many ranks, awnless, deciduous, a few of the lowest empty. Inner scales (bractlets) 2 to each flower, thin, one between the scale of the spikelet and the flower, one between the latter and the axis of the spikelet. Stamens 1 or 2. Style 2–3-cleft. Achene flattish or triangular, naked at the tip. — Culms leafy at base. Small genus of trop. and warm-temp. reg. (Name formed of the Greek *lipos, fat,* and *carphos, chaff,* from the thickness of the inner scales of some species.)

1. L. maculàta (Michx.) Torr. (mottled). — Annual; culm (0.5–3.5 dm. high) much longer than the linear concave leaves; spikelets (3–7 mm. long) green and dark-spotted; inner scales delicate; stamen 1; achene oblong, with a contracted base. — Springy and miry places, Fla. to Ala., n. to Va.; adv. near Phila. July–Oct. (Trop. Am.) FIG. 462.

Tribe II. RHYNCHOSPÒREAE Nees

13. RHYNCHÓSPORA Vahl BEAK-RUSH

Spikelets panicled or variously clustered, ovoid, globular or fusiform, terete, or sometimes flattish; but the scales open or barely concave (not navicular or keeled), the lower commonly loosely imbricated and empty, the uppermost often subtending imperfect flowers. Perianth of bristles. Stamens mostly 3. Achene lenticular, globular, or flattened, crowned by a conspicuous tubercle or beak consisting of the persistent indurated base or even of the greater part of the style. — Chiefly perennials, with more or less triangular and leafy culms; the spikelets in terminal and axillary clusters; flowering in summer; genus nearly cosmop. (Name composed of the Greek *rhynchos, a snout,* and *spora, a seed,* from the beaked achene.) *

*a.*Stout; leaves 0.3–2 cm. broad, their sheaths veiny, cartilaginous at summit; spikelets in flower 1–1.5 cm. long, in fruit flattened and 1.5–3 cm. long; style simple or barely cleft at apex; achenes flattened on the sides, 4.4–6 mm. long; tubercle subulate, 1.5–2.3 cm. long, much exceeding scales. § CALYPTROSTYLIS. . . *b.*

*b.*Bristles much exceeding achenes, 4–6, the longest 4.4–14 mm. long.

Plant non-stoloniferous; inflorescence fastigiate, with stiff ascending branches; achenes 5–5.8 mm. long. . . . . 1. *R. macrostachya.*

Plant loosely stoloniferous; inflorescence diffuse; achenes 4.2–4.8 mm. long. 2. *R. inundata.*

*b.*Bristles much shorter than achene, all but 1–3 of them minute to obsolete, the longest 2–4 mm. long; plant non-stoloniferous; inflorescence diffuse. 3. *R. corniculata.*

*a.*Slender; leaves filiform to flat, narrow or slender, at most 8 mm. wide, their sheaths with a membranous band; spikelets 2–7 (–10) mm. long; style deeply 2-parted, the summit deciduous, leaving the base as a tubercle; achenes lenticular to very plump, 0.9–3(–4) mm. long; tubercle 0.2–2.6 mm. long, usually hidden by scales. § EURHYNCHOSPORA. . . *c.*

*c.*Bristles retrorsely barbed or, if upwardly barbed or smooth, stout and strap-like, pale and exceeding achene. Ser. GLOMERATAE. . . *d.*

*d.*Achene with conspicuous pale wire-like margin, smooth, castaneous, usually lustrous, umbonate, the base prolonged. . . *e.*

*e.*Spikelets 1-fruited, the solitary achene terminating the axis. . . *f.*

*f.*Inflorescence of 2–6 dense globose glomerules (only in poorly developed plants hemispherical and looser), the crowded spikelets ascending to reflexed.

Achenes 1.4–1.6 mm. long, 0.9–1.1 mm. broad. . . . . . . 4. *R. microcephala.*

* The treatment of § *Eurhynchospora* derived largely from the monograph of SHIRLEY GALE, Rhodora, xlvi (April–July, 1944). The figures show silhouettes of inflorescences, nearly × 1, and enlarged drawings of achenes.

Achenes 2–2.4 mm. long, 1.4–1.6 mm. broad. . . . . 5. *R. cephalantha.*

f. Inflorescence of 3–7 turbinate to loosely hemispherical fascicles, the loosely aggregated spikelets ascending to merely divergent. . . . 6. *R. chalarocephala.*

e. Spikelets usually 2- or more-fruited or, if only 1-fruited, with a rudimentary floret borne above the achene.

Achene 1.2–1.4 mm. wide, prominently umbonate, with a pale disk, depressed sides and prominent wire-like margin; inflorescence usually of 4–6 long-peduncled cymes of fascicles. . . . . . . 7. *R. glomerata.*

Achene 0.8–1.2 mm. wide, gradually rounded from the very narrow margin, the surface a uniform brown; inflorescence of 2–6 compact irregularly lobed fascicles, more rarely cymose-fasciculate, all but lowest fascicles short-stalked. . . . . 8. *R. capitellata.*

d. Achenes inconspicuously margined, finely granulate to slightly rugulose, with short dark broken lines running from the brownish margin between the roughenings toward the disk. . . *g.*

g. Spikelets white, becoming whitish-brown; stamens mostly 2; bristles 10–12, villous at base; achenes 0.9–1.2 mm. wide. . . . . . . 9. *R. alba.*

g. Spikelets deep brown to castaneous; stamens 3; bristles 6, serrulate, not villous below; achenes 0.6–1 mm. wide.

Fascicles of spikelets 3 or 4, from lower as well as upper axils; spikelets 2–2.8 mm. long; achene 1.2–1.3 mm. long; endemic in pine-barrens of N.J. and Del. . . . . 10. *R. Knieskernii.*

Fascicles 1 or 2, near summit of culm; spikelets 5–7 mm. long; achene 1.8–2.6 mm. long; wide-ranging in limy soils. . . . . . 11. *R. capillacea.*

c. Bristles upwardly serrulate or, if rarely smooth, capillary and fragile, rarely wanting. . . *h.*

h. Bristles heavily plumose below with dense long hairs; plant tufted, with capillary leaves; spikelets only 1–5, 4–7 mm. long; the lower on a divergent short branch; achenes 2–2.6 mm. long and 1.6–2 mm. wide, horizontally rugulose. Ser. Plumosae. . . . . 12. *R. oligantha.*

h. Bristles upwardly serrulate, with at most a few basal hairs, or smooth. . . *i.*

i. Surface of achene smooth. . . *j.*

j. Culms with bulbous bases and without lateral fascicles; spikelets 1-flowered, whitish-brown to tawny; bristles wanting or 1–3 and very short rudiments at base of achene; achene 1.4–1.7 mm. long; the tubercle apiculate, very short. Ser. Chapmaniae. . . . . 13. *R. pallida.*

j. Culms not bulbous-based, usually with lateral fascicles; spikelets 2–6-flowered, dark brown; bristles 5 or 6, well developed (or in nos. 17 and 18 sometimes rudimentary); achenes 0.6–1.5 mm. wide (up to 1.7 mm. wide in no. 16 with bristles as long as or longer than achene). . . *k.*

k. Achenes compressed-pyriform, prolonged at base, 0.6–1.1 mm. wide. Ser. Fuscae.

Loosely stoloniferous, with culms solitary or few; spikelets 2- or 3-flowered, 4.5–7 mm. long; achenes 1.3–1.7 mm. long; tubercle long-attenuate, 1–1.3 mm. long; northern. . . . 14. *R. fusca.*

Densely cespitose; spikelets 3–6-flowered, 3–4 mm. long; achenes 1–1.3 mm. long; tubercle compressed-deltoid, 0.4–0.6 mm. long; southeastern. . . . . 15. *R. filifolia.*

k. Achenes broadly ellipsoid to suborbicular, not prolonged at base, 1.1–1.7 mm. wide. Ser. Fasciculares. . . . *l.*

l. Tubercle prolonged, 1–2.6 mm. long; bristles nearly equaling to exceeding tubercle. . . . . 16. *R. gracilenta.*

l. Tubercle deltoid or deltoid-subulate, 0.4–0.7 mm. long; bristles much shorter than tubercle or obsolete.

Erect, stiff, 0.4–1.5 m. high; leaves firm, 1–4 mm. broad; terminal corymb of fascicles 1–5.5 cm. broad; spikelets 3–5 mm. long, with awned or mucronate scales exceeding achenes; tubercle 0.4–0.7 mm. long. . . . . 17. *R. fascicularis.*

Reclining or diffuse; culms filiform, 1–7 dm. long; leaves flaccid, at most 1 mm. broad; terminal corymb 0.5–1.1 cm. broad; spikelets 2.5–3.2 mm. long; scales blunt, shorter than achenes; tubercle 0.3–0.4 mm. long. . . . . 18. *R. debilis.*

i. Surface of achene transversely rugulose to ridged or with alveoli forming a honeycomb-pattern. . . . *m.*

m. Lax and reclining, the filiform to narrowly linear leaves at most 1 mm. wide; lax culms filiform, 2.5–6 dm. long; small terminal cyme with few mostly pedicelled spikelets 2–4 mm. long; scales roundish, obtuse. Ser. RARIFLORAE. . . . . 19. *R. rariflora.*

m. Erect or strongly ascending; leaves flat, 1–8 mm. wide; culms 0.1–1.5 m. high; cyme compound or decompound; scales often mucronate or awned. . . *n.*

n. Tubercle conic-apiculate, subterete below, buttressed and embraced at base by the narrowed neck of the honeycomb-alveolate achene; spikelets 1-fruited. Ser. HARVEYAE.

Spikelets 4–5.5 mm. long; achene 2–2.4 mm. long, 1.8–2.2 mm. wide. . . . . 20. *R. Grayii.*

Spikelets 2.5–3 mm. long; achene 1.5–1.8 mm. long, 1.3–1.6 mm. wide. . . . . 21. *R. Harveyi.*

n. Tubercle deltoid to elongate, compressed, closely capping and not embraced by summit of transversely rugulose or ridged achene; spikelets 1–10-fruited. . . . *o.*

o. Cyme stiff, erect but with divergent branches, consisting of glomerules of spikelets. Ser. GLOBULARES. . . . . . . 22. *R. globularis.*

o. Cyme usually flexuous or arching, consisting of ascending fastigiate fascicles of spikelets (or in the loosely stoloniferous no. 23 cyme very open, with divaricate or reflexed branches). Ser. CADUCAE. . . *p.*

p. Bristles overtopping tubercle. . . *q.*

q. Leaves 1.5–4 mm. broad; spikelets fusiform, 4.5–7 mm. long; achene slenderly oblong-ellipsoid, 2–2.5 mm. long; tubercle triangular-subulate, 0.8–1.2 mm. long. . . 23. *R. inexpansa.*

q. Leaves 4–8 mm. broad; spikelets ovoid, 3–5 mm. long; achene ellipsoid-obovoid to suborbicular, 1–1.8 mm. long; tubercle deltoid, 0.2–0.8 mm. long.

Long-stoloniferous; cyme open, with remote divaricate to reflexed long branches; spikelets 3–4 mm. long, 4–12-flowered; achene 1–1.3 mm. long, with smooth tubercle 0.2–0.4 mm. long. . . . . 24. *R. miliacea.*

Short-rhizomatous or cespitose; cyme corymbose-paniculate, with many ascending branchlets and fascicles of 3–6-flowered spikelets 4–5 mm. long; achene 1.4–1.8 mm. long, with setose tubercle 5–8 mm. long. . 25. *R. caduca.*

p. Bristles much shorter than achene or wanting; depressed tubercle 0.2–0.3 mm. long.

Spikelets slenderly ovoid, 3–5 mm. long; scales pale-margined, awned; achenes ellipsoid to narrowly obovoid, 1.3–2 mm. long. . . . . 26. *R. Torreyana.*

Spikelets subglobose-ovoid, 2–2.5 mm. long; scales blunt or merely mucronate, dark brown to fulvous; achene broadly obovoid to suborbicular, 1–1.3 mm. long. . . . 27. *R. perplexa.*

§ CALYPTROSTŸLIS (Nees) Pax

1. R. macrostáchya Torr. (large-spiked), HORNED-RUSH. — *Bulbous-thickened at base, with erect autumnal shoots usually crowded about the fruiting culm; leaves* firm, 0.35–1.2 *cm. broad;* culms stout, smooth, 0.15–11 dm. high; *inflorescence* up to 1 m. long, the *peduncles strongly ascending or fastigiate; terminal glomerules* sessile or short-stalked, *with* (1–)10–50 *spikelets; spikelets* reddish-brown, in maturity 1.5–3 *cm. long; achenes flattened on the faces,* narrowly obovate, thick-stipitate, 5–5.4 *mm. long,* 2.6–3.1 mm. broad; *tubercle* pale, subulate, serrulate, 1.7–2.2 *cm. long, its base* 1–1.8 *mm. broad; bristles* 6, upwardly barbellate, *the longest* 1.1–1.4 *cm. long.* — Wet sand and peat, Fla. to Tex., n. to sw. Me., Mass., Oswego Co., N.Y., Mo. and se. Kans. *Fr.* late July–Oct. FIG. 463.

463. R. macrostachya.

Var. **colpóphila** Fern. & Gale (lover of estuaries). — Plant 0.8–1.75 m. high; *leaves and bracts greatly prolonged, subflaccid,* 0.9–1.5 *cm. broad; achenes* 5–5.8 mm. long, 3–3.8 *mm. broad; tubercle* 1.8–2.3 cm. long, its *base* 1.8–2.4 *mm. broad.* — Fresh tidal marshes, Chesapeake drainage, Md. to rivers of se. Va. entering Albemarle Sd.

2. R. inundàta (Oakes) Fern. (inundated), HORNED-RUSH. — Smaller, *loosely stoloniferous; leaves* 3–7 *mm. broad; culms* rather slender, 2–6 *dm. high; inflorescences* 1–2.5 *dm. long, diffuse, the glomerules with* 1–6 *spikelets;*

achenes 4.2–4.8 *mm. long,* 2.2–2.6 *mm. broad; tubercle* 1.5–1.7 *cm. long; longest bristles* 0.9–1.1 *cm. long.* (*R. macrostachya,* var. *inundata* (Oakes) Fern.) — Inundated pond-margins and wet peat, very local, Fla. (acc. to *Small*); Del. and s. N.J.; s. and e. L.I.; Washington Co., R.I.; w. Barnstable and se. Plymouth Cos., Mass. *Fr.* late July–Sept.

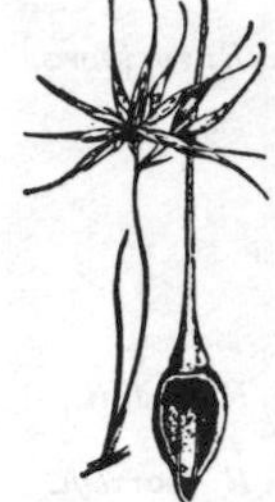

464. R. corniculata.

3. R. corniculàta (Lam.) Gray (horned), Horned-rush. — Similar to no. 2, *not stoloniferous; leaves* 0.6–2 *cm. wide; culms* 0.5–2 *m. high; inflorescences* diffuse, *up to* 1 *m. long; achenes* 5–6 *mm. long,* 2.8–3.3 *mm. broad,* twice as broad as base of tubercle (1.2–2 cm. long); *bristles much shorter than achene, the longest* 2–4 *mm. long, others nearly or quite obsolete.* — Swamps and ditches, Fla. to La., n. to Del., Ky. and Mo. *Fr.* late June–Sept. (W.I.) Fig. 464.

Var. **intèrior** Fern. (inland). — *Achenes* 4.4–5.3 *mm. long,* 2.4–2.8 *mm. broad,* the summit scarcely broader than the base of tubercle. — Ala. to Tex., Ark. and s. Ind.

§ Eurhynchóspora Griseb.

Ser. Glomeràtae Small

4. R. microcéphala Britt. (tiny-headed). — In tussocks 3–8.5 dm. high; leaves linear, 1–3 mm. wide, often involute; culms slender, ascending, subterete; *glomerules dense, globose to hemispherical,* 1 *terminal and* 2–5 *lateral ones* on included axillary peduncles, 1–1.8 cm. broad, *the inflorescence occupying one-half to two-thirds the height of the culm; spikelets* brown to castaneous, slenderly lance-attenuate, 3–4 *mm. long,* 1-*fruited, ascending, spreading and reflexed, the* 1 *floret terminal;* scales ovate to lanceolate, acute, tightly inrolled at tip; *bristles* 6, *strap-like, retrorsely barbellate,* their tips convergent around and slightly exceeding tubercle; *achene* 1.4–1.6 *mm. long,* 0.9–1.1 *mm. broad,* its smooth body suborbicular, *with a short narrow gynophore and a pale wire-like margin,* lustrous, brown; tubercle subulate-attenuate, 0.7–1.1 mm. long, its base narrower than summit of achene. (*R. axillaris,* var. *microcephala* Britt.) — Swamps, savannas and wet pinelands, Fla. to Miss., n. on Coastal Plain to N.J. *Fr.* late July–Oct. (Cuba) Fig. 465.

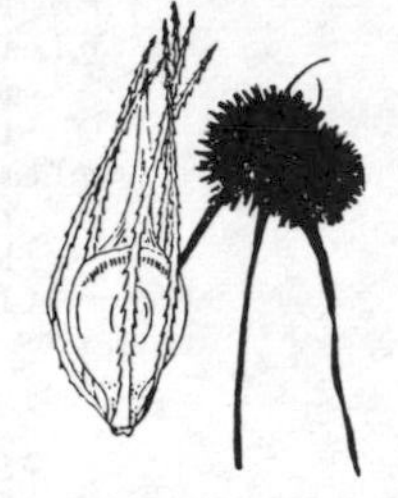

465. R. microcephala.

5. R. cephalántha Gray (with flowers in heads). — Coarser than no. 4; *leaves* 1.5–3 *mm. wide,* often inrolled in drying; culms 0.4–1 m. high; *inflorescence* 1–2.8 dm. long (rarely longer), *occupying the upper fifth of the culm, with* 1 terminal and solitary or 2 or 3 (rarely 4) *subglobose to hemispherical glomerules* 1.3–2.2 *cm. in diameter,* or terminal one lobulate and larger; bracts conspicuous; *spikelets ovoid,* castaneous, 4–6 *mm. long,* crowded, 1-flowered; scales tightly inrolled at apex; *bristles* 6, *strap-like, their margins and upper surfaces retrorsely barbed,* or in forma **antrórsa** Gale (forward) upwardly hispid; *achene* lustrous, smooth, 2–2.4 *mm. long,* 1.4–1.6 *mm. wide;* the lenticular suborbicular body usually with definite shoulders, a broad wire-like edge, depressed sides, and tapering to slender gynophore; *tubercle* 1.4–2.4 *mm. long.* (*R. axillaris* sensu Britt., not *Schoenus axillaris* Lam., basonym) — Sphagnous bogs and savannas, Ga. to La., n. on Coastal Plain to N.J. *Fr.* July–Oct. Fig. 466.

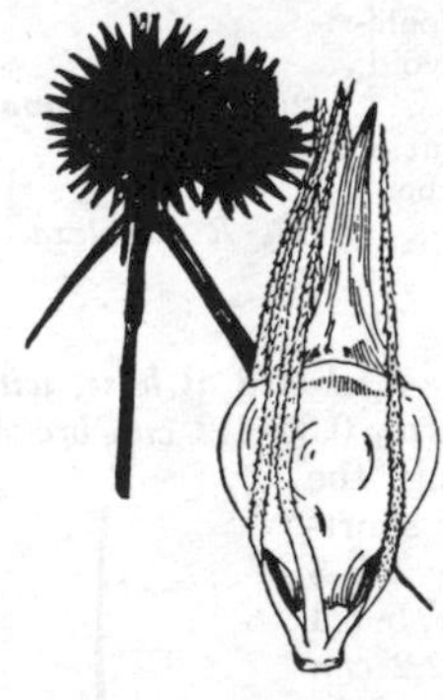

466. R. cephalantha.

Var. **pleiocéphala** Fern. & Gale (many-headed). — Usually *coarser* and taller (0.6–1.1 m. high); *leaves* flat, 2.5–4.5 *mm. wide; inflorescence* 1.4–5 dm. long, *occupying the upper fourth to half of culm; glomerules* 4–7, 1.8–2 *cm. thick,* 1–3 *terminal or subterminal, the others remote;* scales fulvous, less often castaneous. — Flat pinelands, savannas and shallow pools of Coastal Plain, Fla. to La., n. to Va.

6. R. chalarocéphala Fern. & Gale (loose-headed). — Similar to nos. 4 and 5, 2–8 dm. high; leaves flat, 1–2 mm. wide; *inflorescence of* 3–7 remote *turbinate* (or the terminal loosely hemispherical) mostly 2–5-*lobed fascicles,* the terminal one 0.9–1.8 cm. broad; *spikelets loosely aggregated, lanceolate, all ascending or merely spreading;* the fertile floret 1, abruptly terminating the axis; *scales cinnamon-brown,*

467. R. chalarocephala.

lanceolate, closely involute around achene and tubercle; *achene* lenticular, obovoid, 1.4–1.7 mm. long, 0.9–1 *mm. wide, the gynophore prolonged and slender;* subulate-attenuate tubercle 1–1.6 mm. long. — Pond-margins, swamps, wet pinelands and ditches, Coastal Plain, N.J. to Fla. *Fr.* late July–Oct. FIG. 467.

7. R. glomeràta (L.) Vahl (in compact clusters). — Cespitose, 0.5–2 *m. high; leaves* firm, *flat,* 2.5–7 *mm. broad; inflorescence occupying the upper third to half of the culm,* 0.3–1 *m. long, its axillary peduncles freely forked; the narrow cymes* arching, *with fascicles or loose glomerules of spikelets* metallic-gray beneath; *spikelets* ovoid, 4.5–6.5 *mm. long,* deep brown or castaneous, subsessile, ascending to spreading, 2- *or* 3-*fruited or, if* 1-*fruited, with a sterile floret above the fertile one;* scales loosely imbricated, apiculate but soon erose or torn; margins of the 6 strap-like bristles retrorsely and heavily echinate, the tips connivent about tips of tubercles; *achene* 1.5–1.8 mm. long, 1.2–1.4 *mm. wide,* its suborbicular smooth *umbonate* lustrous *body with* definite shoulders, and *prominent wire-like margin, with short and thick gynophore;* tubercle compressed-subulate, 1.3–1.8 mm. long, its base nearly covering summit of achene. (Var. *paniculata* Chapm.) — Wet peaty or sandy soil, Fla. to e. Tex., n. on Coastal Plain to N.J. and inland n. to interior N.C., Tenn., Ark. and Okla. *Fr.* July–Oct. FIG. 468.

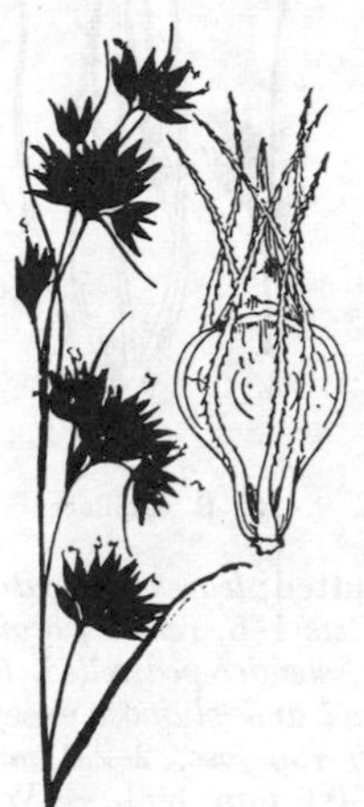

468. R. glomerata.

469. R. capitellata.

8. R. capitellàta (Michx.) Vahl (with small heads). — Similar to no. 7, smaller, 0.1–1.5 m. high; leaves 0.5–3.5 mm. broad; culms slender; *inflorescence* 0.2–6 *dm. long,* brown throughout, *the terminal cyme composed of* 1–*several ultimate turbinate to subglobose fascicles terminating short branchlets; lateral fascicles* 1–5 *on barely exserted or included peduncles; spikelets* 3–5 *mm. long,* (1–)2–5-*fruited; scales* short-mucronulate, *caducous,* often forced off by maturing achene; *bristles weaker than in no.* 7, *the* retrorse *barbules diminishing downward,* or in forma **controvérsa** (S. F. Blake) Gale (reversed) bristles upwardly serrulate, or in forma **discùtiens** (C. B. Clarke) Gale (shaken off) the bristles smooth; *achene pyriform,* 1.3–1.8 mm. long, 0.8–1.2 *mm. broad, not prominently umbonate, narrow-margined;* tubercle 0.9–1.6 mm. long. (*R. glomerata* of ed. 7, not (L.) Vahl) — Low grounds, damp shores, etc., n. Fla. to e. Tex., n. to n. N.B., n. Me., sw. Que., s. Ont. and Wisc. *Fr.* July–Oct. FIG. 469. — Rarely crosses with *Cyperus dentatus,* producing × **C. Weatherbiànus** Fern. (p. 248).

9. R. álba (L.) Vahl (white). — *Tufted,* often forming dense tussocks; leaves setaceous to linear, 0.5–2.5 mm. wide, the upper short and inconspicuous; culms 1–8 dm. high; *fascicles* turbinate, *at first milk-white, in maturity whitish-brown,* the terminal rather dense and 0.4–1.6 cm. broad, the axillary smaller and on short erect peduncles; *spikelets* lance-ovoid, sessile, 2 (rarely 3)-*flowered,* if only 1-fruited with a terminal rudiment, 3.7–5 mm. long; *stamens* 1–3, *commonly* 2; *bristles* 10–12, *retrorsely barbed,* or in forma **laevisèta** Gale (smooth-bristled) barbless, *often pilose at base,* the longer reaching about to tip of tubercle; *achenes* narrowly to broadly obovoid, *long-stipitate,* lustrous, 1.5–2 *mm. long,* tapering above to the *sessile* flattened *triangular-lanceolate tubercle* (0.6–1.2 *mm. long*). — Bogs or wet peats and sands, Nfld. to s. Alaska, s. to N.S., N.E., Md., rarely e. Va. and w. N.C., W.Va., O., n. Ind., n. Ill., Minn., Id. and Calif. *Fr.* July–Sept. (W.I.; Eurasia) FIG. 470. — Hybridizes with no. 8.

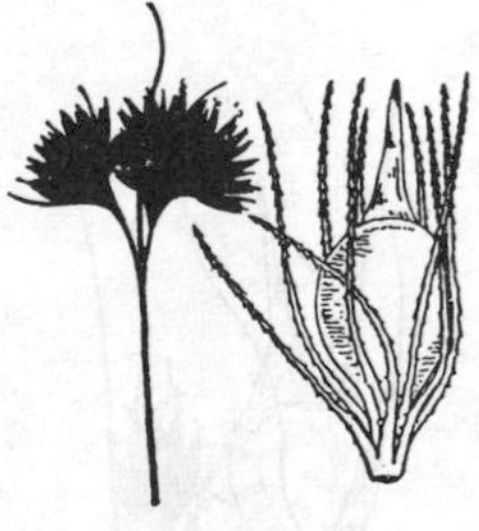

470. R. alba.

471. R. Knieskernii.

10. R. Knieskérnii Carey (for its discoverer, PETER D. KNIESKERN, 1800–1871). — Tufted; leaves narrowly linear, becoming involute, up to 1.8 mm. wide, short; *culms filiform,* 1–6 dm. high, *bearing cymes of spikelets in nearly all the axils; spikelets crowded in close cymes* 3–10 *mm. broad,* brown, ellipsoid-ovoid, 2–2.8 *mm. long,* 2- or 3-fruited; scales caducous; *bristles retrorsely barbed, about equaling body of* *achene; achene* smooth, obovoid, *prolonged at base,* 1.2–1.3 *mm. long,* 0.6–0.8 *mm. broad; tubercle triangular, about* 0.5 *mm. long.* — On iron-ore in pine-barrens, s. N.J. and Sussex Co., Del. *Fr.* late July–Sept. Fig. 471.

11. R. capillàcea Torr. (hair-like). — Forming tussocks 0.5–4.5 dm. high; *leaves setaceous or narrowly linear,* the uppermost commonly overtopping the inflorescence; culms capillary; *fascicles small, ellipsoid or ovoid, with* 1–10 *spikelets,* the terminal 2–8 mm. broad, the 1 axillary fascicle subsessile or short-peduncled; *spikelets* lanceolate, sessile or subsessile, brown, 5–7 *mm. long,* 1–5-*fruited; scales pale-margined;* stamens 3; *bristles* 6, rarely 12, retrorsely barbed, or smooth in forma **levisèta** (E. J. Hill) Fern. (smooth-bristled), *the longer often overtopping the tubercle; achenes* narrowly oblong-obovoid, 1.8–2.6 *mm. long, the long gynophore half as long as the body; tubercle* lanceolate, 0.8–1.6 *mm. long.* — Damp calcareous ledges, bogs and sands, w. Nfld. to Sask., s. to N.B., Me., n. Vt., nw. Ct., n. N.J., se. Pa., mts. of sw. Va. and e. Tenn., O., Ind., n. Ill., Mo. and S.D. *Fr.* July–Sept. Fig. 472.

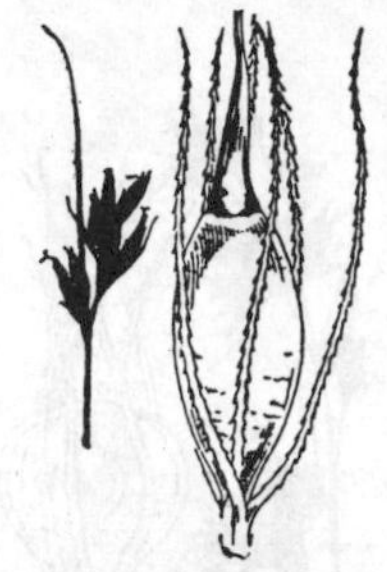
472. R. capillacea.

473. R. oligantha.

Ser. Plumòsae (C. B. Clarke) Small

12. R. oligántha Gray (few-flowered). — Tufted; *leaves and culms filiform;* culms 1.5–5 dm. high, leafless; *spikelets* 1–5, *remote, ovoid-fusiform,* olivaceous or brown, 4–7 *mm. long, slender-pedicelled, the lower pedicel divergent; bristles* of perianth *broad and plumose* except at tip; *achenes* ellipsoid-obovoid, *transversely rugulose,* 2–2.6 *mm. long,* 1.6–2 *mm. broad; tubercle a depressed ring with a conical center,* 0.3–0.6 mm. high. — Wet pine-barrens, sands and peats, Fla. to e. Tex., locally n. to N.C.; Del. and s. N.J. *Fr.* Aug., Sept. (Centr. Am.) Fig. 473.

Ser. Chapmániae Gale

13. R. pállida M. A. Curtis (pale). — Tufted; leaves narrowly linear to involute-filiform, stiffish; *culms bulbous at base,* 4.4–9.5 dm. high, harshly angled above; *cymes depressed-corymbose,* hemispherical to turbinate, *strictly terminal, whitish-brown* to tawny, dense, 0.5–2.5 cm. broad; spikelets ovoid-attenuate, 4–6 mm. long; *bristles wanting or* 1–3 *rudimentary; achenes ellipsoid-obovoid,* 1.4–1.7 *mm. long; tubercle apiculate,* 0.1–0.4 *mm. long, its base covering barely half the top of the achene.* — Wet pine-barrens and bogs of the Coastal Plain, local, N.C. and se. Va.; e. Md. to L.I. *Fr.* July–Sept. Fig. 474.

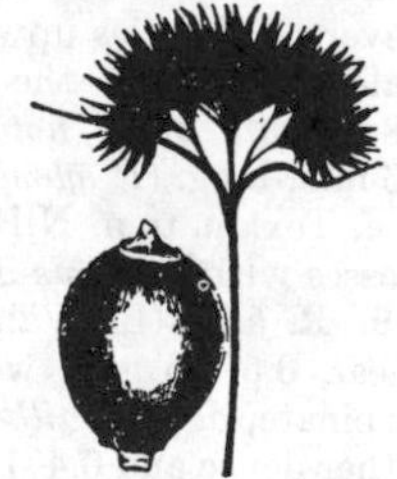
474. R. pallida.

Ser. Fúscae (C. B. Clarke) Gale

14. R. fúsca (L.) Ait. f. (sooty). — *Loosely stoloniferous; leaves setaceous or involute,* the basal short; *culms* filiform, *solitary or few* from tips of stolons, 0.5–5 dm. high; *fascicles* terminal and from upper axils; *the terminal* simple or compound, 0.3–2 *cm. broad,* the 1 or 2 axillary smaller and on exserted peduncles; *spikelets lanceolate to lance-ovoid,* acute, sessile or subsessile, 2- or 3-fruited, 4.5–7 *mm. long,* deep brown or castaneous; *bristles* 5 or 6, upwardly barbed, *the longer equaling or overtopping the tubercle; achene* pyriform or obovoid, smooth, plump, 1.3–1.7 *mm. long,* 1–1.1 mm. wide; *tubercle* greenish, *long-attenuate,* 1–1.3 *mm. long, serrulate-ciliate.* — Bogs and wet peat or sand, Nfld. and N.S. to Ont., s. to N.E., L.I., Del. and s. Mich. *Fr.* June–Oct. (Eu.) Fig. 475.

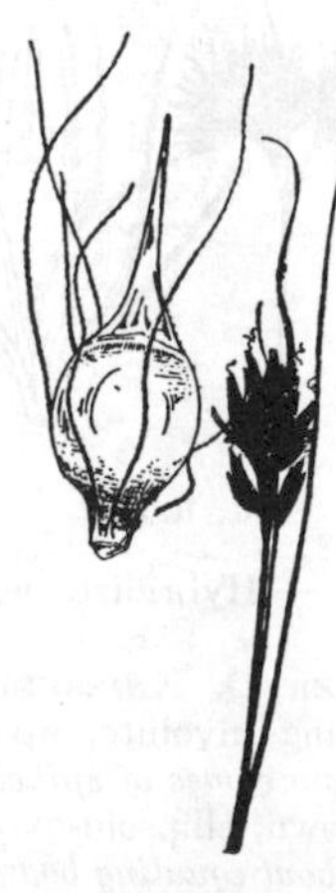
475. R. fusca.

15. R. filifòlia Gray (thread-leaved). — Forming dense tussocks; *leaves and culms filiform* or the cauline leaves often flat and 2 mm. wide; culms 1–7.5 dm. high, bearing *terminal close corymbiform cymes* 5–15 mm. broad *and usually smaller ones from the upper* 1 *or* 2 *axils;* spikelets lanceolate, 3–4 mm. long, 3–6-flowered; scales ferruginous, loose, caducous; *bristles* (or some of them) *overtopping the*

476. R. filifolia.

achene, upwardly barbed; *achene* slenderly obovoid, 1–1.3 *mm. long*, smooth; *tubercle triangular*, 0.4–0.6 *mm. long, its margin hispid.* — Wet pine-barrens, pond-margins and swamps, Fla. to e. Tex., n., locally, on Coastal Plain to Va.; Sussex Co., Del. and Cape May, N.J. *Fr.* July–Sept. Fig. 476.

Ser. Fasciculàres Gale

16. R. gracilénta Gray (slender). — Tufted, very slender; *leaves of basal tufts capillary, the cauline linear-setaceous or flat and* 1–2.5 *mm. wide*, the uppermost exceeding the inflorescence; culms capillary, lax, 2–10 dm. high; terminal cymes corymbiform, rather dense, 0.3–1.5 cm. broad; axillary fascicles (when present) few-flowered, 1–3, short-peduncled; spikelets lance-ovoid, acute, brown, subsessile or short-pedicelled, 1- or 2-fruited, 3–5.5 mm. long; scales awned; *bristles* commonly 6, delicate, *upwardly hispid*, nearly *equaling to slightly overtopping the tubercle; achenes* ellipsoid-ovoid to suborbicular, *plump, evenly rounded on the edges*, 1.1–2 *mm. long; tubercle compressed, triangular-subulate, with prolonged flattish tip*, 1–2.6 *mm. long.* — Wet pine-barrens, swales and peats, chiefly of the Coastal Plain, n. Fla. to e. Tex., n. to N.J., e. Md., n. Va., Tenn. and Ark. *Fr.* July–Sept. Fig. 477.

477. R. gracilenta.

17. R. fasciculàris (Michx.) Vahl (bundled or clustered). — Tufted, erect; *leaves linear*, 1–4 *mm. wide;* culms subterete or obtusely angled, stoutish, 0.4–1.5 m. high; *cymes* compact, *turbinate to hemispherical*, brown, 5–10 *mm. broad, the terminal* 3–7 *in broad irregular corymbs* 1–5.5 *cm. broad;* axillary corymbs or cymes 1–3, on stiffish erect short peduncles; *spikelets* ascending, ovoid-lanceolate, 3.5–5 *mm. long*, 1–3-fruited, *with a terminal rudiment; scales* brown to fuscous, lance-ovate, *with curved awn*, caducous; *bristles* 5 or 6, usually *much shorter than achene or nearly obsolete, upwardly serrulate; achene ellipsoid or ovoid*, 1.5–2 *mm. long*, 1.2–1.5 *mm. broad; tubercle obtusely triangular*, 0.4–0.7 *mm. high, capping the achene.* — Wet pinelands, savannas, pond-margins and ditches, Fla. to e. Tex., n. to se. Va. *Fr.* late July–Sept. (W.I.; Centr. Am.) Fig. 478.

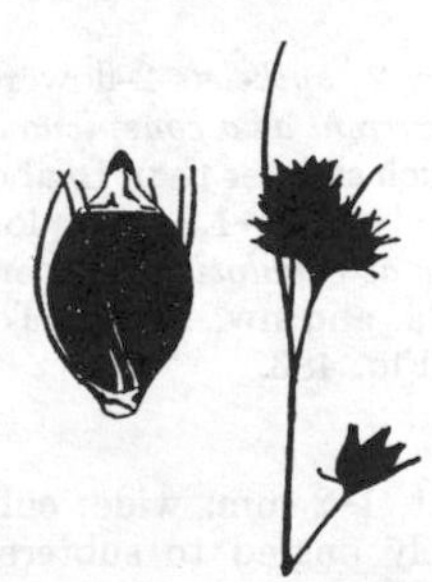

479. R. fascicularis, v. distans.

478. R. fascicularis (typical).

Var. **dístans** (Michx.) Chapm. (distant; from remote fascicles). — More slender, 4–7.5 dm. high; leaves rarely 2 mm. wide; *terminal corymb* 0.5–2.5 *cm. broad; spikelets* 3–4 *mm. long; scales mucronate; bristles exceeding achene; achene* 1.3–1.5 *mm. long; tubercle narrower-based.* (*R. distans* Michx.) — Damp pinelands, Fla. to Miss., n. to se. Va. (Bermuda) Fig. 479.

18. R. débilis Gale (weak). — In tussocks; *leaves filiform*, weak and *loosely spreading or* firmer, *flat and up to* 1 *mm. broad*, much shorter than culms; *culms filiform*, subterete, *loosely ascending to reclining*, 1–7 *dm. long; fascicle* 1, *terminal*, rarely a smaller one below, the terminal crowded, corymbiform, 0.5–1.1 *cm. broad; spikelets* thick-ovoid, 2.5–3.2 *mm. long*, 1- or 2-fruited, with a rudiment above, with the *mature achenes evident; scales shorter than achenes*, friable, caducous, fuscous or castaneous; bristles rarely half as long as achene; *achene broadly ovoid or orbicular*, biconvex, 1.3–1.5 mm. long, 1.4–1.6 *mm. broad*, smooth, castaneous; *broad-based tubercle compressed-deltoid, apiculate*, 0.3–0.4 *mm. long.* — Damp sandy or peaty soil, n. Fla. and Ala., n. on Coastal Plain to se. Va. *Fr.* June–Sept. Fig. 480.

480. R. debilis.

Ser. Rariflòrae Gale

19. R. rariflòra (Michx.) Ell. (few-flowered). — Tufted, *lax and reclining; culms and leaves filiform* or the latter involute, the former 2–6 dm. long; *spikelets few, on elongate filiform pedicels in lax* slightly compound *cymes*, 2.5–4 mm. long, 1–3-fruited; scales round-ovate,

481. R. rariflora.

obtuse, castaneous, often forced off by maturing fruit; *bristles much shorter than achene*, upwardly serrulate, or rudimentary, their bases ciliate; *achenes* ellipsoid or obovoid, 1.3–1.5 *mm. long, with sharp transverse wrinkling and linear longitudinal cancellation; tubercle obtusely deltoid*, 0.3–0.6 *mm. high.* — Damp peats and sands, chiefly on the Coastal Plain, Fla. to Tex., n. to s. N.J. and Tenn. *Fr.* late June–Aug. (W.I.; Centr. Am.) FIG. 481.

Ser. HÁRVEYAE Gale

20. R. Grayii Kunth (for ASA GRAY, 1810–1888). — In stout tussocks, erect; leaves 2–4 mm. wide, flat, firm; culms obtusely angled, smooth, stiff, 0.3–1 m. high, with long internodes; terminal cyme 1–1.5 cm. broad, of 1–3 glomerules of few spikelets; lateral glomerules 1–4, smaller, on exserted peduncles; *spikelets* ovoid, 2–3-flowered, 1-*fruited*, 4–5.5 *mm. long;* scales mucronate, sandy-brown to castaneous, tightly imbricated, scarcely friable; stamens 3 or 6; *bristles brittle*, upwardly hispidulous, shorter than achene to overtopping tubercle; *achene tumid above*, suborbicular, *the base compressed*, 2–2.4 *mm. long*, 1.8–2.2 *mm. wide, honeycombed with minute shallow pits; tubercle conic-apiculate*, 0.4–0.6 mm. long, *buttressed and partly clasped at base by the narrowed summit of the achene.* — Dry sandy pine-barrens, Fla. to e. Tex., n. on Coastal Plain to se. Va. *Fr.* June–Aug. (Cuba) FIG. 482.

482. R. Grayii.

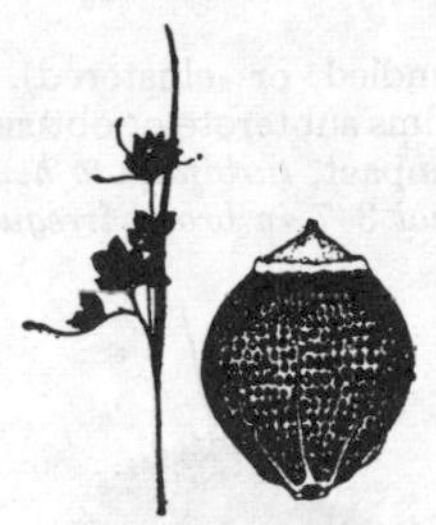

483. R. Harveyi.

21. R. Hárveyi W. Boott (for its discoverer, FRANCIS LEROY HARVEY, 1850–1900). — Similar to no. 20, *in softer-based* and *smaller tussocks;* leaves 1.5–3 mm. wide; *terminal cyme* 0.8–2.5 *cm. broad*, of 1–4 small glomerules; lateral glomerules 1 or 2; *spikelets* 2-flowered, 1-fruited, 2.5–3 *mm. long; midrib of scales excurrent as a conspicuously recurving awn;* stamens 3; *bristles delicate*, much shorter than to about equaling achene; *achene* ovoid to suborbicular, 1.5–1.8 mm. long, 1.3–1.6 *mm. wide, the surface-pits isodiametric or obscured and appearing as rugulosities;* tubercle 0.4–0.5 mm. long. — Wet to dryish peats, pinelands and openings, Ga. and nw. Fla. to Tex., n. to se. Va., n.-centr. N.C., n. Miss., Ark. and Okla. *Fr.* June, July. FIG. 483.

Ser. GLOBULÀRES Gale

22. R. globulàris (Chapm.) Small (globular). — Tufted; leaves flat, 1–2 mm. wide; culms slender, smooth, stiff, erect or ascending, 0.5–7.5 dm. high, obtusely angled to subterete; *terminal cyme* 0.5–3.5 *cm. broad, with* 0-few *stiffly ascending unequal branches terminated by lobulate glomerules of* 3–8 *spikelets, the subtending bracts short and inconspicuous*, the *glomerules* 3–8 *mm. in diameter;* lateral cymes 1–6, smaller, on erect mostly exserted peduncles; *spikelets broadly ovoid to subrotund*, 1.5–3 *mm. long*, 1- *or 2-fruited; scales* papery, closely involute, suborbicular, *blunt or mucronate, soon splitting;* bristles shorter than achene; *achene broadly obovoid to subglobose*, tumid above, compressed at base, 1–13 *mm. long, finely cancellate, transversely ridged to rugulose; tubercle* deltoid-conical, 0.3 *mm. long, its basal annulus with rounded edges.*— Peaty, sandy or argillaceous depressions, chiefly of Coastal Plain, Fla. to e. Tex., n. to Del. and Md.; n. Calif. *Fr.* June, July (–Oct.) FIG. 484.

484. R. globularis (typical).

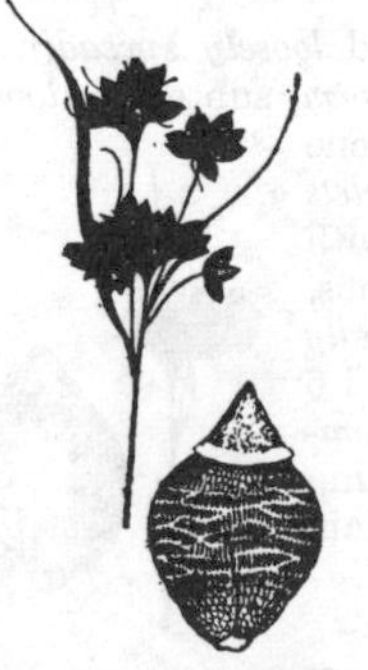

485. R. globularis, v. recognita.

Var. **recógnita** Gale (rediscovered). — *Coarser; leaves* 2–5 *mm. broad;* culms 0.15–1 m. high, the *bracts prolonged;* terminal cymes larger, with more numerous *glomerules* 5–10 *mm. in diameter, with many crowded spikelets; spikelets* 2.5–4 *mm. long; achenes* 1.3–1.6 *mm. long*, 1.2–1.5 *mm. wide; tubercle* 0.3–0.6 *mm. high.* (*R. cymosa* sensu Torr. and later auth., not Ell.) — Low grounds, Fla. to e. Tex., n. to N.J., se. Pa., e. Md., interior N.C., Tenn., Mo. and Okla.; n. O. to ne. Ill. *Fr.* July–early Sept. (W.I.; Centr. Am.) FIG. 485.

Ser. CADÙCAE Gale

23. R. inexpánsa (Michx.) Vahl (unexpanded). — In tussocks; *leaves narrowly linear or involute, 1.5–4 mm. wide;* culms 0.3–1.2 m. high, flexuous above; *cymes loosely paniculate, flexuous or nodding,* the terminal 0.3–1.3 dm. long, the 2–6 axillary ones nodding; *spikelets fusiform,* chestnut-brown, strongly ascending, 4.5–7 mm. long, *mostly pedicelled,* 1–4-fruited; scales acute or awned, caducous; *bristles* delicate, upwardly serrulate, *much overtopping the tubercle; achenes slenderly oblong-ellipsoid,* transversely wrinkled, 2–2.5 *mm. long,* pale, *with setose margins;* tubercle triangular-subulate, 0.8–1.2 mm. long. — Borders of low woods and thickets, swamps and ditches, Coastal Plain, Fla. to e. Tex., n. to se. Va. and Ark. *Fr.* July–Sept. FIG. 486. — Spikelets frequently changed to globular echinate galls.

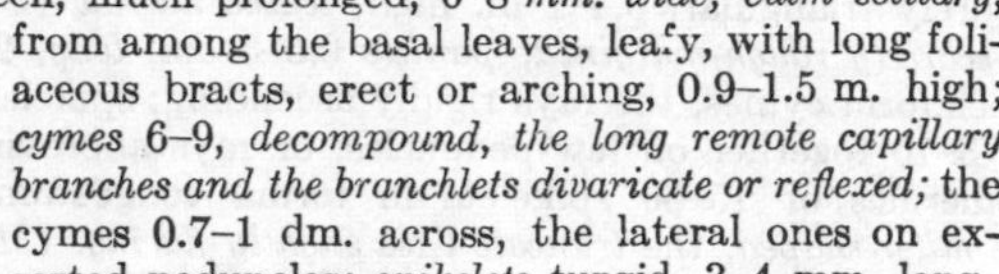

486. R. inexpansa.

24. R. milià cea (Lam.) Gray (Millet-like). — *Stoloniferous,* the tufts scattered; *leaves* pale green, much prolonged, 6–8 *mm. wide; culm solitary,* from among the basal leaves, leafy, with long foliaceous bracts, erect or arching, 0.9–1.5 m. high; *cymes 6–9, decompound, the long remote capillary branches and the branchlets divaricate or reflexed;* the cymes 0.7–1 dm. across, the lateral ones on exserted peduncles; *spikelets* turgid, 3–4 mm. long, distant, *on slender pedicels, 4–12-flowered, 3–10-fruited;* scales aristulate, caducous, exposing spikelets of fruit; bristles delicate, spreading, exceeding tubercle; *achenes* broadly obovoid, transversely ridged, longitudinally striate, pale, 1–1.3 *mm. long,* 0.9–1.1 mm. wide; *tubercle* depressed-conic, 0.2–0.4 *mm. long.* — Wet woods, mediacid to calcareous swamps and margins of pools, Coastal Plain, Fla. to La., n., locally, to se. Va. *Fr.* June, July. (W.I.) FIG. 487.

487. R. miliacea.

25. R. cadùca Ell. (quickly falling). — *Culms* solitary or few from a stoutish rhizome, 0.7–1.5 m. high, *acutely 3-angled; leaves flat,* pale, 4–8 *mm. wide; cymes corymbose-paniculate, loosely and freely forking, with ascending capillary branches; the terminal* 0.4–1.5 *dm. high,* 2–7 *cm. broad,* with numerous small fascicles, *the* 3–5 *axillary* smaller cymes *flexuous on filiform peduncles; spikelets* deep-brown, ovoid, 4–5 *mm. long;* scales thin, acute or awned, caducous; *bristles* delicate, upwardly serrulate, *overtopping the tubercle;* achenes ellipsoid-obovoid to suborbicular, transversely wrinkled, 1.4–1.8 mm. long; *tubercle* broadly deltoid, *setulose,* 5–8 mm. long. — Pools, swamps and borders of wet woods, Fla. to e. Tex., n. to se. Va., nw. Ga., n. Ala. and nw. Ark. *Fr.* late June–Sept. FIG. 488.

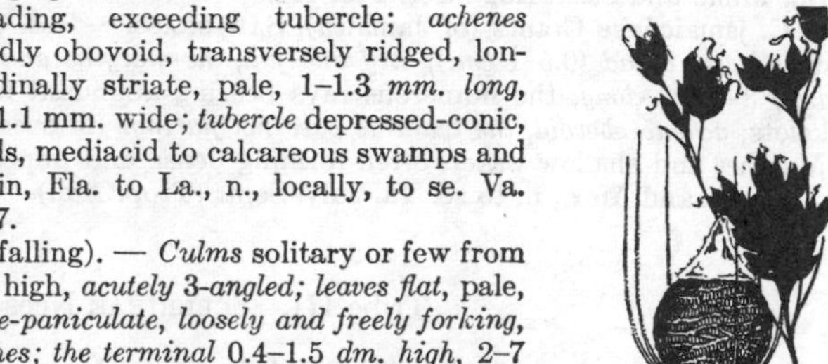

488. R. caduca.

26. R. Torreyàna Gray (for its discoverer, JOHN TORREY, 1796–1873). — Tufted; *leaves* slenderly linear, *involute in drying,* 2–3 mm. wide; culms slender, subterete, 0.2–1 m. high; *cymes paniculate, lax,* often flexuous, the terminal 1–8 cm. long, 0.5–4 cm. broad, the axillary ones smaller and often nodding on filiform peduncles; *spikelets* drab to brown, ellipsoid, 3–5 mm. long, with pale-margined awned scales, *ascending, mostly pedicelled; bristles wanting or very short,* closely appressed and upwardly serrulate; *achenes strongly compressed, ellipsoid to narrowly obovoid, transversely rugose* with undulating vertically lineolate ridges, 1.3–2 *mm. long,* 0.8–1.2 *mm. broad; tubercle sessile, completely capping achene,* compressed-conical, 0.2–0.5 *mm. high.* — Damp to dryish sands and peats of the Coastal Plain, Ga. to se. Mass. *Fr.* mid-July–Sept. FIG. 489.

27. R. PERPLÉXA Britt. (ambiguous). — Similar to no. 26; *lateral cymes on short ascending peduncles; spikelets subglobose-ovoid,* 2–2.5 *mm. long,* scattered or rather crowded; *scales ferruginous or fuscous, broadly ovate, blunt or mucronulate;* bristles none or 1–3 rudiments; *achene broad-obovoid to suborbicular,* flattened, 1–1.3 *mm. long,* 0.9–1.2 mm. broad, with few prominent transverse ridges and numerous longitudinal striae; *tubercle broadly deltoid with straight sides meeting in a terminal angle,* pale, compressed, 0.2–0.3 mm. long. — Wet peat, shallow pools, and flat pineland, chiefly on Coastal Plain, Fla. to e. Tex., n. to se. N.C. and centr. Tenn.; represented with us by

489. R. Torreyana.

Var. **virginiàna** Fern. (Virginian). — Often coarser; *spikelets* 2.5–3 *mm.*

long, crowded and sessile or subsessile; *tubercle depressed and broadly rounded.* — Similar habitats, se. Va. *Fr.* mid-June–Sept.

14. CLÀDIUM P. Br. Twig-rush

Spikelets ovoid or oblong, of several loosely imbricated scales; the lower scales empty, one or two above bearing a staminate or imperfect flower; the terminal flower perfect and fertile. Perianth none. Stamens 2. Style 2–3-cleft, deciduous. Achene ovoid or globular, somewhat corky at the summit, or pointed, without any tubercle, in which it differs from *Rhynchospora.* — Plants of trop. and temp. reg. (Name from the Greek *cladion, a branchlet,* from the repeatedly branched cyme of the original species.)

490. C. mariscoides.

1. C. mariscoìdes (Muhl.) Torr. (like *Mariscus*). — Perennial, stoloniferous; culm obscurely triangular, 0.4–1 m. high; *leaves narrow* (1–3 *mm. wide*), *channeled, scarcely rough-margined; panicle* 0.5–3 *dm. long,* 2–5 *cm. broad,* of 2–4 umbelliform cymes, the rays rigidly ascending; spikelets clustered in heads of 3–10 together on few peduncles, or rays abbreviated or obsolete and glomerules of 15–30 spikelets in forma **congéstum** Fern. (crowded); *achene miter-shaped,* the *truncate base slightly flaring.* (*Mariscus mariscoides* (Muhl.) Ktze.) — Swamps, marshes and sandy shores or pond-margins, either fresh or brackish, Fla. and Ala., n. to sw. Nfld., N.S., s. Que., n. Ill., Minn. and Sask. Aug.–Oct. Fig. 490.

2. C. jamaicénse Crantz (of Jamaica), Saw-grass. — Tall (1–3 m.) and coarse; *leaves broad* (0.5–1 *cm.*), *stiff and flat, the margins and midrib beneath harshly serrate; panicle* 3–9 *dm. long,* the numerous rays bearing abundant fascicled small chestnut-colored spikelets; *achene obovoid, the truncate base not flaring.* (*Mariscus jamaicensis* (Crantz) Britt.) — Marshes and shallow water, often forming dense and impenetrable, very scratchy thicket, Fla. to Tex. and Mex., n. to se. Va. July–Sept. (Trop. Am.)

Tribe III. Sclerìeae Nees

15. SCLÈRIA Bergius Nut-rush

Flowers monoecious; the pistillate spikelets 1-flowered, usually intermixed with clusters of few-flowered staminate spikelets. Scales loosely imbricated, the lower empty. Stamens 1–3. Style 3-cleft. Achene globular to ovoid or ellipsoid, stony, bony, or enamel-like in texture. — Perennials (sometimes annuals) of trop. and warm-temp. reg., with triangular leafy culms, mostly from creeping rhizomes, or soft-based annuals. Inflorescence, in our species, of terminal and axillary clusters, the lower clusters often peduncled. (Name from the Greek *scleria, hardness,* from the indurated fruit.)

a. Achenes smooth, papillose, granulose or warty, not reticulated or wrinkled; hypogynium (disk beneath achene) not lobed, forming a 3-cornered cushion; perennials with elongate rhizomes. . . *b.*

b. Achenes smooth and lustrous. . . *c.*

c. Hypogynium at base of achene without tubercles, its surface with minute (microscopic) pebbling or processes. . . *d.*

d. Membranous band on inner side of leaf-sheath glabrous or nearly so below the puberulent ligule; achenes 1.2–2.5 mm. high.

Achene subglobose to oblate, strongly rounded at summit, nearly or quite as broad as long, 2–2.5 mm. high, 2–2.7 mm. broad; culms 2.5–6 mm. thick at base; leaves 5–9 mm. broad; plant of dry to moist soils. . . . . 1. *S. triglomerata.*

Achene subglobose to ovoid, only slightly rounded to summit, 1–1.8 mm. high, 1.2–1.8 mm. broad; culms 1–2 mm. thick at base; leaves 1–2.5 mm. broad; plant of wet peat. . . . . 2. *S. minor.*

d. Membranous band puberulent to tomentulose; achenes ovoid to ovoid-subglobose, longer than thick, 2.2–3.3 mm. high.

Culms 1–few, from ligneous elongate or loosely forking rhizomes, flowering only at summit; surface of hypogynium tuberculate with low rounded pebbling; plant of dry soil. . . . . 3. *S. nitida.*

Culms numerous, cespitose, from a relatively soft and densely branched base, with capillary lateral flowering branches as well as terminal inflorescences; surface of hypogynium covered with prolonged lance-acuminate scale-like laminae (seen under magnification); plant of swamps. . . . . 4. *S. flaccida.*

c. Hypogynium bearing 8 or 9 small subglobose hard tubercles between its summit and the base of the achene, its surface smooth; achene ovoid, 2.5–3.5 mm. long; culms tufted, from relatively soft closely branching rhizome. . . . . 5. *S. oligantha.*

b. Achenes papillose, granulose or warty; hard rounded tubercles borne at summit of hypogynium.

Tubercles 6 or 9; achene 1.5–2 mm. in diameter. . . . . 6. *S. pauciflora.*

Tubercles 3, entire or lobed; achene 2–3 mm. in diameter. . . . . . 7. *S. ciliata.*

a. Achenes reticulate or wrinkled; hypogynium 3-lobed or wanting; annuals or short-lived perennials with soft bases and no elongate rhizomes.

Inflorescences terminal and axillary panicles, at least the terminal ones with elongate leafy bracts; hypogynium with 3 broad lobes; achene 1.8–2.5 mm. high, reticulate or pitted.

Lateral panicles on filiform and elongating drooping peduncles; achene pubescent, its reticulations irregular, with spirally arranged ridges. . 8. *S. Muhlenbergii.*

Lateral panicles subsessile or on short ascending peduncles; achene glabrous, regularly pitted or reticulated in vertical rows. . . . . . . 9. *S. reticularis.*

Inflorescence an interrupted spike of sessile minutely bracted glomerules; hypogynium obsolete; achene about 1.5 mm. in diameter, rough-wrinkled, thick-stipitate. . . . . 10. *S. verticillata.*

1. S. triglomeràta Michx. (with three clusters). — Coarse, *with large knotty ligneous forking rhizome; culms* clustered, coarse, 2.5–6 *mm. thick at base,* 3-angled, 0.5–1 m. high; *leaves* pale to yellowish-green, 5–9 mm. broad, linear and *scarcely narrowed up to the short tip,* hard, scabrous, glabrous, or the sheaths and midrib beneath pilose; *the membranous inner band of the ventral side of the sheath glabrous or nearly so below the sharply differentiated glabrous to pubescent ligule;* inflorescences terminal and often lateral erect-peduncled fascicles or panicles, the terminal fascicles often in 3's; *achenes subglobose to oblate, lustrous, smooth,* white to buff, *broadly rounded at summit,* nearly or *quite as broad as long,* 2–2.5 *mm. high,* 2–2.7 *mm. broad;* hypogynium 3-angled, crustaceous, minutely pebbled. — Dry to moist thin woods and openings, Fla. to Tex., n. to Mass., (?) Vt., N.Y., s. Ont., O., s. Mich., s. Wisc., s. Minn. and Kans. June–Sept. Fig. 491.

491. S. triglomerata.

2. S. mìnor (Britt.) Stone (smaller). — Bluer-green and more slender; *culms* 1–2 *mm. thick at base,* 3–7.5 dm. high, *filiform; leaves* 1–2.5 *mm. broad, attenuate to prolonged slender tips,* glabrous or merely scabrous; terminal inflorescences mostly of 2 fascicles; *achenes* smooth, lustrous, *subglobose to ovoid,* 1–1.8 *mm. high,* 1.2–1.8 *mm. broad.* (*S. triglomerata,* vars. *gracilis* and *minor* Britt.) — Peaty or boggy depressions, S.C. to s. N.J. June, July.

3. S. nítida Willd. (shining). — *Rhizomes* ligneous, *elongate* and simple or, if freely forking, with long horizontal branches; culms 1–few, stiffly erect, 4.5–9 dm. high, more slender than in no. 1, coarser than in no. 2; *leaves blue-green,* hard, scabrous, *long-attenuate,* 2–8 mm. broad; *inner band of leaf-sheath puberulent to tomentulose and not strongly differentiated from the puberulent to tomentulose ligule; inflorescences terminal,* of 1 or 2 approximate fascicles; *achenes* smooth, lustrous, *ovoid to ovoid-globose,* gradually rounded to tip, *longer than thick,* 2.8–3.3 *mm. high,* 2–2.8 *mm. thick; hypogynium* (under high magnification) tuberculate *with low rounded pebbling.* — Dry (rarely damp) sands, pinelands and barrens, Ga. to Ky. and N.J. June–Sept.

4. S. fláccida Steud. (flaccid). — More *cespitose* than no. 3, *from* relatively *dense and soft rhizome with ascending branches; culms* softer, arching and *flexuous; leaves lax, thin,* glaucous, 4–9 mm. broad, *scarcely narrowed up to the short rounded tip, inner band of sheath densely pubescent; inflorescences* terminal and *on pendulous capillary axillary peduncles,* the terminal one of 2 or 3 approximate fascicles; *achene* smooth, lustrous, ovoid, *longer than thick,* 2.5–3.2 mm. long, 2.2–2.8 mm. thick; *hypogynium* (under high magnification) *densely covered with prolonged lance-acuminate scale-like laminae.* — Swamps and damp clearings, Fla. to La., n., locally, to se. Va. June, July.

5. S. oligántha Michx. (few-flowered). — Rhizome nodulose, relatively soft, with crowded or *cespitose soft sharply 3-angled culms* 3–7.5 dm. high; *leaves thin,* 2–6 mm. wide, *glabrous* *except for the scabrous to hispid margins and midribs beneath; inflorescences terminal and on flexuous axillary peduncles*, the terminal mostly of 2 fascicles and much overtopped by the bracts; achene ovoid, smooth, lustrous, longer than thick, 2.5–3.5 mm. long; *hypogynium* minutely pebbled, 3-cornered, *bearing at summit* 8 *or* 9 *hard rounded tubercles beneath the achene.* — Rich or moist woods and clearings, Fla. to Tex. and e. Mex., n. to Va., D.C., Ky., s. Ind. and sw. Mo. May–early July. Fig. 492.

492. S. oligantha.

6. S. paucifl̀ora Muhl. (few-flowered). — Rhizome subligneous, elongate, forking; *culms* few–several, *very slender,* triquetrous, scabrous-angled, glabrous or barely pubescent, 1.5–7.5 dm. high; *leaves narrowly linear,* 1–2.5 *mm. broad, glabrous or only sparingly hirtellous;* inflorescences of 1–3 glomerulate terminal small fascicles and sometimes capillary-peduncled smaller axillary ones; bracts narrow, ciliate, the longer much exceeding the terminal inflorescence; *achene globose,* white, *transversely verrucose-papillose,* 1.5–2 *mm. in diameter;* the *hypogynium* or disk *bearing* 3 *pairs of minute rounded tubercles.* — Dry to moist pinelands, barrens and peats, Fla. to Tex., n. to N.J., e. Pa., Ky., s. Mo. and se. Kans. June–Sept. (W.I.) Fig. 493.

493. S. pauciflora (typical).

Var. **caroliniàna** (Willd.) Wood (of Carolina). — Similar, but *culms and leaves copiously pilose.* — N. Fla. to Tenn. and Mo., n. to sw. N.H., centr. Mass., se. N.Y., D.C., n. O., s. Mich. and s. Ill.

Var. **kansàna** Fern. (of Kansas). — Copiously pilose; *the pairs of large tubercles separated by a smaller intermediate one.* — Sandy soil, Cherokee Co., Kans. Fig. 494.

494. S. pauciflora, v. kansana.

7. S. ciliàta Michx. (with marginal hairs). — Similar to no. 6; *culms* 2–6 dm. high, slender, *glabrous,* or slightly pubescent below; *leaves* firm, 1–2.5 *mm. wide, soon becoming revolute;* inflorescences 0.7–2.5 cm. long, of 1–3 (mostly 1 or 2) fascicles; *bracts very short-ciliate; scales smooth or smoothish; achenes* globose, 2–3 *mm. in diameter,* white, *with verrucose-tuberculate surfaces; tubercles* 3, *broad,* entire or notched. — Sandhills, pinelands and barrens, Fla. to Tex., n. to se. Va. and s. Mo. July–Oct. Fig. 495.

495. S. ciliata (typical).

Var. **Ellióttii** (Chapm.) Fern. (in honor of Stephen Elliott, 1771–1830). — Coarser; *culms* stout, *pubescent; leaves flat,* 3–6 *mm. wide;* inflorescences 1.5–3.5 cm. long, fuller and denser; *bracts with fimbriate-ciliate bases; scales pubescent;* tubercles more often lobed. (*S. Elliottii* Chapm.) — Sandy or argillaceous open woods and barrens and peaty meadows and bogs, Fla. to Tex., n. to Va., Mo. and Okla. June–Aug. Fig. 496.

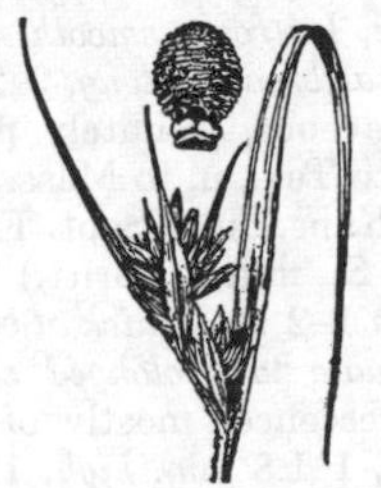

496. S. ciliata, v. Elliottii.

8. S. Muhlenbérgii Steud. (for Gotthilf Henry Ernest Muhlenberg, 1753–1815). — Annual or short-lived perennial with soft base; culms soft and weak, tufted, 0.2–1 m. high, loosely ascending to diffuse, trigonous or compressed, glabrous; leaves flaccid, glabrous, 2–7 mm. wide; *inflorescences terminal and pendulous axillary panicles, the latter on recurving capillary peduncles,* the former 1.5–4 cm. long, loosely forking, the ascending branches with scattered small fascicles; *achene* subglobose, 2–2.5 *mm. in diameter,* white to buff, *irregularly pitted-reticulated or pitted-rugose with the more or less pubescent ridges often somewhat spirally arranged,* much broader than the 3-*lobed* small *hypogynium.* (*S. reticularis,* var. *pubescens* Britt.; *S. setacea* sensu Core, not Poir.) — Bogs, sphagnous swales and wet pinelands, Fla. to Tex., n. to L.I., N.J., e. Pa., D.C., Tenn. and s. Mo.; nw. Ind. Aug.–Oct. (Trop. Am.) Fig. 497.

497. S. Muhlenbergii.

9. S. reticulàris Michx. (reticulate). — Smaller than no. 8, annual; culms 1–few, erect, 1–6.5 dm. high; *leaves* 1–4 *mm. wide; lateral inflorescences subsessile or on very short erect peduncles;* terminal panicle close, 0.5–3 cm. long; *achene globose-ovoid or -obovoid,* 1.8–2.2 *mm. high,* drab to olive-brown, *glabrous, with vertical rows of regular shallow pits or reticulations.* — Damp sandy shores and depressions, locally abundant, Fla. to se. S.C.; Md. and Del. to e. Mass.; nw. Ind. Aug.–Oct. (Mex.) Fig. 498.

498. S. reticularis.

10. S. verticillàta Muhl. (in whorls). — Slender annual, fragrant in drying; culms 1–5 dm. high, filiform, trigonous; leaves slenderly linear, 0.5–2 mm. wide, long-attenuate, soft, the margins soon revolute, the sheaths either glabrous or retrorse-pilose; *fascicles* (1–)2–6, few-flowered, *sessile in a greatly interrupted spike* 1–8 cm. long, the bracts short and setaceous, the scales brown and pointed; *achenes globose-obovoid,* white or whitish, *about* 1.5 *mm. long, apiculate, with transverse low laminate ridges; hypogynium wanting or represented only by a short stipe-like disk.* — Calcareous bogs, shores and wet rocks, local, nw. Ct. to s. Ont. and Minn., s. to L.I., nw. N.J., Pa., O., Ind., Ill., Mo. and Tex.; also se. N.C. and nw. Fla., w. to La., etc. July–Sept. (Close allies in Trop. Am.) FIG. 499.

499. S. verticillata

Tribe IV. CARÍCEAE Kunth

16. KOBRÈSIA Willd.

Spikelets unisexual and one-flowered or with two flowers (one pistillate, one staminate) in short spikes aggregated in elongate heads or panicles; the pistillate flower consisting of a spathiform glume (homologous with the perigynium of *Carex*) wrapping about the base of the achene and subtended by the scale of the spikelet. — Perennial herbs of northern reg., resembling the first subgenus (*Vigneae*) of *Carex*, but with the perigynium replaced by the open glume which has its margins connate at base. (Named for *Von Kobres*, a nobleman of Augsburg and patron of botany in Willdenow's time.)

1. K. simpliciúscula (Wahlenb.) Mackenz. (slightly simple). — In dense tussocks from short and dense cespitose horizontal rhizomes; culms firm, obtusely angled, strict, 0.3–5.3 dm. high; leaves in crowded brown-based tufts, filiform or promptly involute, firm, marcescent, mostly shorter than culms; head linear-lanceolate to slenderly ellipsoid, 1–4 cm. long, of approximate appressed-ascending brown spikes 3–8 mm. long, the terminal spike staminate, the others androgynous or pistillate and few-flowered; scales castaneous, white-margined; glumes slenderly ellipsoid, 2.5–2.8 mm. long, castaneous, lustrous, chartaceous, the margins free nearly to base; achene linear-oblong, slightly exserted. (*K. caricina* Willd.) — Greenl. to Alaska, s. to damp calcareous peats, gravels and rocky slopes of Nfld., Côte Nord and Anticosti I., Que., shores of Hudson Bay and high mts. of Colo. July, Aug. (Eurasia)

17. CYMOPHÝLLUS Mackenz.

Flowers unisexual; the staminate borne in a brush above the capitate-clustered pistillate ones, of three stamens in the axil of a scale; pistillate flowers much as in *Carex*, with three or four stigmas, the stipitate achene surrounded by an inflated and persistent perigynium and often accompanied by a short setiform rachilla; pistillate scales white and petaloid. — A monopodial perennial with the foliage of each branch of the rhizome consisting of 4 or 5 membranaceous closed basal sheaths and a single broad and tongue-shaped convolute blade without sheath, ligule or midrib and quite lacking in the epidermis the bulliform cells characteristic of leaves of most *Cyperaceae*, the cartilaginous margins minutely undulate and thus appearing serrulate; the leafless and bractless culms flat. (Name from the Greek *cyma, a wave,* and *phyllon, leaf,* from the minutely undulate leaf-margin.) — A very primitive, relict genus.

1. C. Fràseri (Andr.) Mackenz. (for its discoverer, JOHN FRASER, 1750–1811), FRASER'S SEDGE. — Culms 2–5 dm. high, smooth and stiff; leaves pale, closely many-ribbed, very thick and persistent, developed after the flowering, becoming 1–7 dm. long and 2–4.5 cm. broad; pistillate half of spike with 20–30 milk-white inflated obscurely nerved perigynia 5–6 mm. long and exceeding the rounded white scales; stamens conspicuous, with slender white anthers. (*Carex Fraseri* Andr.) — Rich upland woods and banks of streams, sw. Va. and e. W.Va. to w. S.C. and e. Tenn., local. May–July. FIG. 500.

500. C. Fraseri.

18. CÀREX L. SEDGE. LAÎCHE (Que.)

Flowers unisexual, destitute of floral envelopes, disposed in spikes; the staminate consisting of three stamens in the axil of a bract or SCALE; the pistillate comprising a single pistil with a bifid or trifid style, forming in fruit a hard achene which is inclosed in a sac (PERIGYNIUM) borne in the axil of a bract or SCALE. Staminate and pistillate flowers borne in different parts of the spike (spike *androgynous*), or in separate spikes on the same culm, or rarely the plant dioecious. — Perennial grass-like herbs with mostly triangular culms, 3-ranked leaves with sheath, ligule and midrib, and spikes in the axils of leafy or scale-like bracts, often aggregated into heads. An exceedingly critical world-wide genus, the study of which should be attempted only with complete and fully mature specimens. (The classical Latin name, of obscure signification; derived by some from the Greek *keirein, to cut*, on account of the sharp leaves — as indicated in the English name *Shear-grass.*)*

Subgen. I. VIGNEA

Spikes mostly uniform and sessile, bearing the staminate flowers at base or apex or sometimes scattered (spikes unisexual in a few species); stigmas 2 and achenes lenticular (except in no. 14, with strong creeping stems and with firm dentate-margined perigynia 1–1.4 cm. long); cladoprophylla (bracteoles at base of spike or branch of inflorescence) rarely present. For subgenus II see p. 295.

*A.*Spike 1. . . *B.*

*B.*Spike androgynous (*i.e.* staminate at apex); pistillate scales round-tipped; achene with a slender bristle (rudimentary rachilla) at base, within the perigynium; plants tufted, with filiform leaves; arctic-alpine species.

Spike ovoid; perigynia tapering to a slender stipe, appressed-ascending, 3–3.5 mm. long. . . . . § 1. NARDINAE, p. 300.

Spike subglobose, with slender staminate tip; perigynia sessile, rounded at base, soon spreading, 2–3 mm. long. . . . . § 2. CAPITATAE, p. 301.

*B.*Spike unisexual, androgynous or gynecandrous (staminate at base, pistillate above); pistillate scales acute or acuminate; no rachilla present in perigynium; lowland, north-temperate species.

Culms filiform, solitary, from filiform rhizomes and stolons; perigynia convex on upper face. . . . . § 3. DIOICAE, p. 301.

Culms densely cespitose, wiry; perigynia plane on upper face. 57. *C. exilis* of. § 15. STELLULATAE, p. 315.

*A.*Spikes 2 or more; no rachilla present in perigynium. . . *C.*

*C.*Some or all spikes androgynous (with terminal staminate flowers or their remnants), not gynecandrous (with upper flowers pistillate, lower staminate); or in no. 14 inflorescences dioecious, perigynia dentate, firm, 1–1.4 cm. long and stigmas 3. . . *D.*

*D.*Rhizomes slender or cord-like, extensively creeping or elongating; culms mostly solitary, from stolons or offshoots of the rhizome. . . *E.*

*E.*Monoecious; anthers less than 5 mm. long; perigynia 2–6 mm. long; styles 2; achenes lenticular. . . *F.*

*F.*Perigynia wingless or essentially so, dorsally cleft, if bidentate tardily so.

Culms simple, arising from subterranean rhizomes.

Spikes crowded into an apparently continuous head 4–14 mm. long, without awned or blade-bearing bracts; pistillate scales obtuse; perigynia membranaceous, usually subinflated, loosely covering the small achene; leaves strongly involute; culms obtusely angled, 0.1–10(–25) cm. high; arctic-alpine species. . . . . § 4. FOETIDAE, p. 301.

Spikes distinct, forming a definitely interrupted head, at least the lower with awned or short-leafy bracts; pistillate scales acute; perigynia coriaceous, plano-convex, thin-edged, closely enveloping the large achene; leaves flat, canaliculate or tardily inrolled; culms mostly taller; plants of temp. areas. . . . § 5. DIVISAE, p. 302.

Culms arising from axils of dried leaves of preceding year along a wiry elongate, forking and leaning or reclining superficial culm-like stem; spikes few in an irregular ovoid to ellipsoid head. . . . . § 6. CHORDORRHIZEAE, p. 302.

---

* In constructing the key the treatments of KÜKENTHAL in *Das Pflanzenreich* iv[20] (1909), and of MACKENZIE in the *North American Flora*, xviii (1931–35) have been much relied upon. The dates are for mature but not over-ripe fruit. The perigynial characters are here based on study of mature plants. In general the perigynia at the tip of the spike are less characteristic than those nearer the middle; and, if possible, the latter alone should be used in critical comparisons.

F. Perigynia with distinct thin wing or very thin margin, often sharply bidentate at apex; spikes often dissimilar, some androgynous, others staminate, others pistillate or variously mixed. . . . . § 7. ARENARIAE, p. 302.

E. Dioecious; anthers 4.5–6 mm. long; perigynia 1–1.4 cm. long; styles 3; achenes trigonous. . . . . § 8. MACROCEPHALAE, p. 303.

D. Rhizome short or, if sometimes elongate (in no. 47), not freely stoloniferous and with leafy tufts or culms approximate; leafy and fertile shoots mostly cespitose or subcespitose; mature perigynia often spreading. . . *G.*

G. Perigynia strongly compressed and plano-convex or unequally biconvex, usually more than 4 to each spike. . . *H.*

H. Spikes 2–12(–15), in mostly simple interrupted or close heads; perigynia coriaceous or firm, plano-convex, not inflated, though sometimes spongy-based; culms slender, mostly firm, not wing-angled. . . . . § 9. BRACTEOSAE, p. 304.

H. Spikes numerous, in paniculate spiciform heads, usually 2–several on each lateral branch. . . *I.*

I. Leaf-sheaths close; blades firm, 1–5(–8) mm. wide; culms slender and firm; perigynia firm, flat or merely convex on inner face, 2–3.5 mm. long.

Inner band of leaf-sheath usually cross-puckered, at least in age; bracts (at least the lower) setaceous, often overtopping the branches and spikes; scales awned; perigynia flat on inner face, stramineous, yellow or green. . . . . . § 10. MULTIFLORAE, p. 308.

Inner nerveless band of leaf-sheath not cross-puckered; bracts mostly short and inconspicuous or wanting; scales blunt, acute or merely mucronate; perigynia more or less biconvex, brown or drab to purplish or blackish. . . . . . § 11. PANICULATAE, p. 309.

I. Leaf-sheath loose; blades soft to firm, 2.5–17 mm. broad, the ventral band of the sheath usually either cross-puckered or dotted; culms soft, flattened under pressure; perigynia spongy or corky at base, thin and soft, yellowish, stramineous (finally brown) or greenish, more or less inflated . . . . . . § 12. VULPINAE, p. 310.

G. Perigynia ellipsoid-ovoid, nearly terete, 2–3 mm. long, strongly rounded at summit, only 1–3 (rarely 4) in the scattered spikes; plant delicate. 47. *C. disperma* of § 13. HELEONASTES, p. 312.

C. Some (especially the terminal) or all spikes gynecandrous, with the staminate flowers at base or scattered, not at apex (except in no. 47 and in dioecious individuals). . . *J.*

J. Perigynia with rounded to very narrow borders, without well defined winged margins, thickened or corky at base. . . *K.*

K. Perigynia with rounded margins, ascending or merely spreading-ascending, densely and minutely puncticulate under magnification (except in nos. 55 and 56), of soft or membranaceous texture (firm in no. 47).

Perigynia 1.5–4 mm. long, densely puncticulate under magnification, nearly beakless or with short subentire beak. . . . . . § 13. HELEONASTES, p. 311.

Perigynia 4–5.5 mm. long, not puncticulate, with long slender bidentate beak, closely appressed-ascending. . . . . . . § 14. DEWEYANAE, p. 314.

K. Perigynia with thin but scarcely winged margins, ascending to horizontally divergent or reflexed in maturity, firm, not puncticulate, very spongy at base, with bidentate or obliquely cut beak. . . . § 15. STELLULATAE, p. 315.

J. Perigynia with thin or winged margins, mostly with concave inner faces, not spongy-thickened, ascending or with spreading tips. . . . . . § 16. OVALES, p. 318.

Subgen. II. EUCAREX

Some of the spikes strictly pistillate, the staminate flowers in distinct or mixed spikes; stigmas 3 and achenes trigonous, or if stigmas 2 and achenes lenticular some or all of the spikes peduncled; cladoprophylla at bases of peduncles. For subgenus I see p. 294.

*L.*Style jointed at or near base, soon disarticulating from mature achene and shriveling, the achenes then beakless or short-apiculate; perigynia 1.5–10 mm. long. . . *M.*

*M.*Staminate scales tightly clasping the rachis, their margins united at base.

Spike solitary, terminal, linear-oblong, without leafy bracts, the staminate tip sessile; rachis straight, filiform; scales shorter than the numerous blunt membranaceous appressed perigynia. . . . . § 17. POLYTRICHOIDEAE, p. 329.

Spikes 2–4, axillary; the lower pistillate scales, at least of some spikes, prolonged as foliaceous bracts; staminate spikes stalked; rachis zigzag, winged; perigynia few, plump, firm, long-beaked. . . . . § 18. PHYLLOSTACHYAE, p. 329.

*M.*Staminate scales loosely ascending, with free margins. . . *N.*

*N.*Spike solitary, terminal. . . *O.*

*O.*Monoecious, the spikes staminate at summit; scales whitish, brown or, if purple, with pale hyaline margins; perigynia glabrous, or barely puberulent at summit, firm, obtusely angled.

Spike continuous, the staminate portion not stalked, the pistillate not strongly interrupted; pistillate flowers several–many, approximate; perigynia 3–4 mm. long, 1.5-2 mm. thick.

Densely cespitose, with filiform or involute-acicular rigid leaves fibrillose at base; perigynia dull, minutely puberulent at summit. § 19. FILIFOLIAE, p. 330.

Loosely repent and stoloniferous, with tough cord-like rhizomes; leaves flat, not fibrillose at base; perigynia glabrous.

Spikes ovoid to lanceolate; pistillate scales acuminate, pale, caducous; perigynia brilliantly lustrous, castaneous to blackish. . . . . § 20. OBTUSATAE, p. 330.

Spikes linear-cylindric; pistillate scales blunt, purple to brown, with pale margins, persistent; perigynia opaque or only sublustrous, stramineous to olivaceous or brownish. . . . § 21. RUPESTRES, p. 330.

Spike interrupted, the staminate portion stalked and prolonged, the pistillate of 1–3 remote flowers; perigynia 5–6.5 mm. long, 2.5–3 mm. thick. . . . . § 22. FIRMICULMES, p. 330.

*O.*Dioecious, the spikes of staminate plants soon shriveling; scales purple to castaneous; perigynia pubescent (at least at summit), thin, with thin or sharp angles.

Leaves mostly shorter than the bractless stiff culms; scales blunt or merely acute, narrower and shorter than perigynia; these ovoid to ellipsoid, 2.5–3 mm. long, hairy nearly to the short-stipitate base; boreal species. . . . . § 23. SCIRPINAE, p. 331.

Leaves much exceeding the filiform loosely bracted culms; middle and upper scales prolonged into cusps, concealing the perigynia; these slenderly oblanceolate, 4–5 mm. long, slenderly long-stipitate, pubescent at tip; southern species. . . . . . . § 24. PICTAE, p. 331.

*N.*Spikes 2 or more, the lower pistillate or partly so, the terminal staminate or partly so. . . *P.*

*P.*Achenes only obscurely 3-angled, with rounded or convex sides, closely filling body of perigynium; perigynia 2–4.5 mm. long, 0.7–2.5 mm. thick, in plump spikes 2–15(–18) mm. long; foliaceous bract at base of inflorescence (excluding inflorescences at base of plant) sheathless, barely short-sheathed or wanting. . . *Q.*

*Q.*Perigynia coriaceous, glabrous, lustrous; spikes all terminating erect culms, not basal; culms filiform, 0.2–1.8 dm. high; leaves mostly canaliculate or plicate, 0.5–1.5 mm. wide.

Stoloniferous; culms sharply angled, glabrous; pistillate spikes subglobose, with acute or acuminate fulvous lance-ovate scales; perigynia divergent-ascending, 2.5–3 mm. long. . . . . . § 25. LAMPROCHLAENAE, p. 331.

Densely cespitose; culms subterete, smooth; pistillate spikes ellipsoid, with orbicular to broadly ovate round-tipped purple-black scales; perigynia appressed, 2 mm. long. 133. *C. terrae-novae* of § 28. DIGITATAE, p. 335.

*Q.*Perigynia membranaceous, pubescent, if glabrous throughout borne largely in spikes crowded among the leaf-bases.
Pistillate scales rough-cuspidate; summit of achene capped by a dark annulus with projecting central apiculation. . . . . . . § 26. PRAECOCES, p. 331.
Pistillate scales smooth; summit of achene merely short-apiculate, without dark annulus. . . . . § 27. MONTANAE, p. 332.

*P.*Achenes definitely 3-angled and with flat or concave sides (at least above) or lenticular and 2-sided. . . *R.*

*R.*Perigynia tightly filled to tip by the achenes, 2–5 mm. long.
Leaves and culms glabrous; sheaths below the non-basal pistillate spikes tubular or spathiform, without flat green blades; perigynia 1.5–3.5 mm. long (if up to 4.5 mm. long stipitate and with some filiform recurving peduncles much longer than the spikes), minutely beaked (if beak up to 0.5 mm. long, the lustrous and glabrous perigynium less than 1 mm. thick).
Staminate spike overtopping the racemosely disposed pistillate ones; scales purple to dark brown; perigynia strongly tapering to stipitate at base, stramineous to purple or brown, 2–4.5 mm. long; style not bulbous-based; leaves flat or plicate. . . . § 28. DIGITATAE, p. 335.
Staminate spike overtopped by the subcorymbose pistillate ones; scales whitish to pale brown; perigynia gradually tapering but not stipitate at base, olivaceous to blackish, glabrous, 1.5–2 mm. long; base of style bulbous-thickened; leaves setaceous. . . . . § 29. ALBAE, p. 337.
Leaves and culms pubescent; bract below lowest pistillate spike nearly or quite sheathless, with flat blade; perigynia 3.5–5 mm. long, about 1.5 mm. thick, pubescent, with slender beak about 0.75 mm. long. . . . . § 30. TRIQUETRAE, p. 337.

*R.*Perigynia not tightly filled by achene, at least the summit usually empty (except for style), 1.5–10 mm. long. . . *S.*

*S.*Styles 2; achenes lenticular. . . *T.*

*T.*Culms slender to nearly filiform, 0.3–7 dm. high; leaves only 1–4 mm. wide; bracts sheathing; staminate spike 1 (sometimes pistillate above); pistillate spikes thick-cylindric, ellipsoid, ovoid or obovoid, 0.5–3 cm. long.
Leaves soft, green, flat and straight; pistillate scales contrasted with the plump subterete white-pulverulent or orange and smooth perigynia; the latter 1.7–3 mm. long; plants of damp habitats. . . . . § 31. BICOLORES, p. 337.
Leaves firm, dry, whitish-green, conduplicate and curving; pistillate scales brown, conforming to the flat, thin and lustrous oblong-lanceolate perigynia; these 5–6 mm. long; arctic-alpine xerophyte. 180. *C. misandroides* cf. § 41. FERRUGINEAE, p. 352.

*T.*Culms mostly coarser, 0.5–10 dm. or more high; leaves 1–12 mm. or more broad; bracts nearly or quite sheathless; staminate spikes 1–3 or more; pistillate spikes slenderly cylindric to ellipsoid, 0.5–10 cm. long; perigynia compressed, lenticular, planoconvex or biconvex.
Scales aristate, subulate-tipped or muticous, mostly much exceeding the perigynia; pistillate spikes nearly all peduncled; achenes often constricted on one side near the middle. § 32. CRYPTOCARPAE, p. 338.
Scales obtuse to acute, not aristate, if subulate-tipped with the upper spikes mostly sessile, shorter than to exceeding perigynia; achenes not constricted. . . . . § 33. ACUTAE, p. 340.

*S.*Styles 3 (only exceptionally 2 or 4); achenes trigonous (only exceptionally lenticular or 4-angled). . . *U.*

*U.*Bract at base of inflorescence (excluding rare basal spikes) sheathless or barely sheathing (if sheath prolonged, in a few species of § *Hirtae*, the perigynia very plump and nearly round in cross-section, often pubescent, and with strongly bidentate beak). . . *V.*

V.Leaves and perigynia glabrous (perigynia barely scabrous-hispid in some species). . . *W*.

W.Perigynia beakless or with minute beak with entire, emarginate or barely bidentate orifice. . . *X*.

X.Perigynia compressed, strongly appressed-ascending, 1–2 mm. broad.

Roots with glabrous surfaces (except for rootlets); terminal spike pistillate except at base (staminate in no. 152 and exceptionally in others); short bracts with dark auricles. . . . . § 34. ATRATAE, p. 343.

Roots covered with dense felt-like pubescence; terminal spike typically staminate. . . . . § 35. LIMOSAE, p. 345.

X.Perigynia plump, squarrose to spreading-ascending, 1.7–3.5 mm. thick.

Terminal spike staminate throughout; perigynia ellipsoid, rhomboid, ovoid or obovoid, longer than broad, not transversely rugose, 2.8–5 mm. long. . . . . § 36. PENDULINAE, p. 347.

Terminal spike pistillate except at base; perigynia compressed-globose-obovoid, as broad as long, transversely rugose, nerveless, 2.5–3 mm. long. . . . § 37. SHORTIANAE, p. 347.

W.Perigynia with prolonged beak.

Leaves membranaceous, ribbon-like, 0.6–1.8 cm. broad; staminate spike 1, its base overtopped by the uppermost of the 3–8 pistillate spikes; perigynia green, membranaceous, loose, scabrous, prominently 3-angled, 2.5–4 mm. long, the curved hyaline beak nearly equaling the body. . . . . § 38. ANOMALAE, p. 348.

Leaves firm, 1.5–5 mm. broad; staminate spikes 1–3, the long peduncle much overtopping the 1 or 2 pistillate ones; perigynia drab-brown, coriaceous, close, roundish in cross-section, 4–6 mm. long, tapering to a firm bidentate beak much shorter than the body. 166. *C. Walteriana*, var. *brevis* of § 39. HIRTAE, p. 348.

V.Leaves (or sheaths) or perigynia or both pubescent.

Plants with long horizontal stolons and rhizomes; the leaves and culms in small tufts; leaves glabrous (except in no. 171, with perigynia 5–9 mm. long and with sharply bidentate beak); terminal spike or spikes ordinarily staminate throughout; perigynium 2.5–9 mm. long, 1.7–3 mm. thick, with sharply bidentate beak (except in no. 170 with long hyaline and weakly toothed beak). . . § 39. HIRTAE, p. 348.

Plants cespitose, without elongate stolons; leaves or sheaths or both pubescent; terminal spike staminate only at base or throughout; perigynia 2–3.5 mm. long, 1.5–2 mm. broad, with entire or barely notched orifice. . . . . § 40. VIRESCENTES, p. 350.

U.Bract at base of inflorescence with a prolonged closed and tubular sheath; perigynia glabrous. . . *Y*.

Y.Perigynia ascending or not strongly divergent, the orifice entire, oblique or but slightly notched (if orifice deeply notched, the perigynia not strongly ribbed and the bracts ascending). . . *Z*.

Z.Terminal spike regularly pistillate except at base.

Loosely stoloniferous; culms capillary, 0.8–1.5 dm. high, leafless except at base; leaves strongly glaucous, 1–2 mm. wide, less than 1 dm. long; spikes thick-cylindric to obovoid, mostly less than 1 cm. long; pistillate scales blackish-purple, blunt, with slender midrib not excurrent; perigynia beakless, glaucous. 205. *C. livida*, var. *rufinaeformis* of § 48. PANICEAE, p. 360.

Cespitose; culms coarser and taller, leafy with broader green leaves and longer bracts; spikes linear- to oblong-cylindric, mostly longer; pistillate scales pale or with excurrent midrib; perigynia green to brown.

Pistillate scales ciliate at tip; perigynia 5–7 mm. long, tapering to short straight beak. 192. *C. venusta* of § 43. SYLVATICAE, p. 354.

Pistillate scales non-ciliate; perigynia 2.5–6 mm. long, beakless, very short-beaked or with long curved beak. § 42. GRACILLIMAE, p. 352.

Z. Terminal spike regularly staminate throughout, only in rare exceptions with a few pistillate flowers. . . *a.*

a. Culms usually arising from centers of leafy tufts or from among leaves, slender and firm, not winged or easily compressed, long-persistent after falling of fruit; basal leaves linear, elongate and with prolonged narrow tips, 1–10(–15) mm. broad. . . *b.*

b. Cespitose, either densely or loosely, without horizontal slender stolons (except in no. 197); perigynia green to brown, not glaucous; scales pale brown to green or white. . . *c.*

c. Pistillate spikes linear- to oblong-cylindric, the lower peduncled and usually drooping or loosely spreading; perigynia lanceolate to fusiform or ovoid, obscurely nerved or nerveless, the pale orifice entire or barely bidentate. . . *d.*

d. Leaves flaccid, not overwintering; perigynia ovoid, thin-angled, opaque, produced into a long curving slender entire beak. 181. *C. prasina* of § 42. GRACILLIMAE, p. 352.

d. Leaves firmer, the overwintering ones subcoriaceous; perigynia not thin-angled (if long-beaked with beak straight), glabrous and lustrous, or puberulent. . . *e.*

e. Bases purple or purplish; lower sheaths at base of culm without green blades; pistillate spike linear- to oblong-cylindric; perigynia gradually tapering to a conical beak, with oblique finally bidentate orifice. . . . . . . § 43. SYLVATICAE, p. 354.

e. Bases drab or brown; lowest sheaths at bases of culms with elongate green blades; pistillate spikes oblong-cylindric.

Pistillate spikes 2–4 mm. thick; their scales blunt or short-tipped, promptly deciduous; perigynia lance-ovoid, 2–4 mm. long, less than 1 mm. thick, definitely 3-angled, tapering to a conic tip, the orifice truncate and entire. . . . . . . § 44. CAPILLARES, p. 356.

Pistillate spikes 5–10 mm. thick; their scales long-acuminate, subpersistent; perigynia ovoid or ellipsoid, 5–6 mm. long, 2.5–3 mm. thick, obscurely angled or strongly rounded in cross-section, the long beak soon cleft. § 45. LONGIROSTRES, p. 356.

c. Pistillate spikes oblong-cylindric, erect or ascending, only rarely with elongate peduncles; perigynia ellipsoid, oblong-ovoid, obovoid or subglobose, opaque (sublustrous in nos. 198 and 199, with impressed nerves), beakless or very short-beaked, with entire or barely emarginate orifice.

Perigynia with elevated ribs, 1.5–4 mm. long. . . § 46. GRANULARES, p. 356.

Perigynia with impressed nerves, 3–6 mm. long. . § 47. OLIGOCARPAE, p. 357.

b. Loosely stoloniferous, with small tufts of leaves and culms and horizontal slender rhizomes and stolons; perigynia glaucous; scales dark purple to greenish. § 48. PANICEAE, p. 360.

a. Culms from lateral buds or sometimes central, relatively weak, sometimes wing-angled, readily compressed or flattened, soon shriveling after maturity of fruit; leaves linear to oblong and short-pointed, up to 4 cm. broad; sheaths of bracts often loose and subinflated; perigynia fusiform to fusiform-obovoid, tapering to the conic tip and the gradually narrowed spongy base. . . § 49. LAXIFLORAE, p. 362.

Y. Perigynia usually soon divergent to reflexed, rarely remaining ascending, subinflated to barely obtuse-trigonous, many-ribbed, the bidentate beak with erect teeth; pistillate spikes subglobose to thick-cylindric, at least the upper usually sessile, rarely up to 2.5 cm. long. . . . . § 50. EXTENSAE, p. 368.

L. Style continuous, not jointed at base, firm and persistent at summit of achene; perigynia frequently inflated, 2.5–20 mm. long, often with emarginate or bidentate beak. . . *f*.

f. Spike solitary, bractless, pistillate below, staminate above; the 1–12 soon reflexed and promptly falling slenderly subulate perigynia 0.5–1.5 mm. thick; pistillate scales caducous. . . . . § 51. ORTHOCERATES, p. 370.

f. Spikes 1–several, at least the lowest leafy-bracted; perigynia more than 1.5 mm. thick; pistillate scales more persistent. . . *g*.

g. Perigynia obconic or broadly obovoid, inflated, truncate or abruptly rounded above to a long subulate beak. . . . . . . § 52. SQUARROSAE, p. 370.

g. Perigynia subulate to ovoid or subglobose, tapering or only gradually curving to beak. . . *h*.

h. Perigynia firm or of thick texture, scarcely inflated, green or olivaceous to brown, pubescent to glabrous, ascending to spreading, in linear- to oblong-cylindric spikes; culms 1–few from stoloniferous bases. . § 53. PALUDOSAE, p. 371.

h. Perigynia thin or papery, green to stramineous, brown or purplish, glabrous (if hirtellous strongly inflated and in subglobose spikes); culms in dense tussocks (if few or solitary from prolonged stolons and with turgid perigynia). . . *i*.

i. Pistillate scales with scabrous awns equaling or longer than the blades; pistillate spikes elongate, densely flowered, the (20–) 30–200 or more perigynia angled to subterete, 3–9 mm. long and 1–4 mm. thick, strongly ribbed, often stipitate, ovate to lanceolate, and with beak one-third as long as to longer than body, with firm teeth. . § 54. PSEUDO-CYPEREAE, p. 373.

i. Pistillate scales blunt to cuspidate, the awn, if present, less than half as long as blade and smooth; pistillate spikes cylindric to subglobose, 1–many-flowered; perigynia only obscurely angled or with roundish cross-sections, stipitate to sessile, often short-beaked. . . *j*.

j. Perigynia 8–20 mm. long, if less than 10 mm. long in subglobose spikes; staminate spike solitary (often more in § 57).

Perigynia subulate to slenderly lanceolate, 1–3 mm. thick, delicately nerved, barely inflated, soon deciduous.

Style-branches ending below tip of beak; teeth of beak abruptly recurved. . . . . § 55. COLLINSIANAE, p. 375.

Style-branches exceeding beak; teeth of beak erect or ascending. . . . . § 56. FOLLICULATAE, p. 375.

Perigynia lanceolate to ovoid or flask-shaped, 3–8 mm. thick, strongly ribbed, usually much inflated, persistent. . . . . § 57. LUPULINAE, p. 376.

j. Perigynia 2.5–10 mm. long, in spikes with elongate axes; staminate spikes 1–several (usually 2 or more in the species with perigynia 7–10 mm. long). . . . . § 58. VESICARIAE, p. 378.

Subgen. I. VÍGNEA (P. Beauv.) Kükenth. (see p. 294)

§ 1. NÁRDINAE Tuckerm. (see p. 294) — A single species with us:

1. C. nárdina Fries (smelling like nard). — *Densely tufted*, with persistent brown sheaths, 3–9 cm. high; *leaves* bristleform, curving or straight, *often overtopping the* obtusely angled

smooth short *culms; spike* solitary, bractless, *ellipsoid or ovoid*, 4–10 mm. long, the short tip staminate; *pistillate scales* castaneous, ovate, *rounded at summit; perigynia appressed-ascending*, 3–3.5 *mm. long, ellipsoid*, plano-convex, slightly exceeding scales, *narrowed below to a conspicuous stipe*, tapering to a short beak with a dorsal cleft; stigmas 2; *achenes* lenticular, *exceeding the bristleform rachilla.* — Greenl. to Alaska, s. to calcareous schists of Shickshock Mts., L. Mistassini and Hudson B. reg., Que. July, Aug. (Eurasia) Fig. 501.

501. C. nardina.

§ 2. Capitàtae Christ (see p. 294) — A single species with us:

2. C. capitàta L. (head-like). — *Densely tufted*, brown and *purple at base*, wiry, 0.7–5 dm. high; *leaves* filiform, *shorter than the slender culms; spike* solitary, bractless, *subglobose, with a slender staminate tip, the pistillate body* 3–6 *mm. in diameter;* pistillate scales broadly ovate, rounded above, brown, with pale hyaline margin; *perigynia* at first ascending, *soon divergent*, 2–3 *mm. long*, ovate, plano-convex, *sessile, rounded at base*, short-beaked; stigmas 2; *rachilla nearly equaling the* lenticular *achene.* (Incl. *C. arctogena* H. Sm.) — Greenl. to Alaska, s. in peats and gravels to nw. Nfld., higher mts. of Que. and N.H., James Bay reg., and locally along the Cordillera to Mex. and Calif. Late June–Aug. (Eurasia; s. S.Am.) Fig. 502.

502. C. capitata.

§ 3. Dioìcae Tuckerm. (see p. 294) — A single species with us:

3. C. gynócrates Wormsk. (dominant female; from the stout pistillate spike). — *Culms filiform, solitary or few, from filiform creeping stolons and rhizomes*, 0.2–3 dm. high, mostly overtopping the setaceous leaves; *spike* 0.5–2 cm. long, either *staminate* and linear-cylindric, *pistillate* and thick-cylindric, *or staminate above and pistillate below; pistillate scales* oblong-ovate, *acute or acuminate*, brown or reddish, staminate scales paler and blunter; *perigynia* at first ascending, *finally divergent to reflexed, plump, coriaceous, slenderly ovoid, thin-edged, dorsally nerved, conic-beaked, spongy-based*, yellowish, olive or brown, lustrous, 2.5–4 mm. long; stigmas 2. — Peaty soils, Greenl. and Baffin I. to Alaska, s. to Nfld., C.B., N.B., n. Me., N.Y., w. Pa., n. Mich., Wisc., Minn., Man., mts. of Colo. and B.C. June–Aug. (N. Eurasia) Fig. 503.

503. C. gynocrates.

§ 4. Foètidae Tuckerm. (see p. 294)

Heads subglobose or globose-ovoid; scales and subinflated perigynia soon divergent; leaves smooth or barely scabrous near tip. . . . . 4. *C. maritima.*
Heads ellipsoid to thick-cylindric; scales and flattish perigynia appressed-ascending; leaves spinulose-scabrous toward tip. . . . . 5. *C. Langeana.*

4. C. marítima Gunn. (of the sea). — Extensively creeping, with slender cord-like rhizomes and stolons; *leaves* 3–15 cm. long, in ascending tufts, *strongly involute, smooth* or nearly so, often curving; *culms* 0.1–2.5 dm. high, *arching*, often shorter than leaves, *obtusely angled, smooth; spikes* 3–5, *closely crowded in an apparently continuous head* 8–14 mm. in diameter, androgynous, with staminate tips insignificant; *bracts subtending lowest spikes papery*, lustrous, brown, *obtuse or mucronate; anthers* 1.7–2.5 *mm. long;* pistillate scales scarious, brown, usually with broad pale margins, ovate, obtuse; *perigynia* finally *divergent* and slightly exceeding the scales, *membranaceous, subinflated*, 3.5–5 mm. long, *nearly nerveless*, ovate, rounded at base to a short stipe, the short beak with a dorsal cleft; styles 2. (*C. incurva* Lightf.) — Greenl. to Alaska, s. to turfy open calcareous slopes and shores of nw. Nfld., adj. Que. and James Bay reg. July, Aug. (Eurasia) Fig. 504.

504. C. maritima.

Var. **setìna** (Christ) Fern. (of bristles; from the slender leaves). — Very dwarf; *culms* 1 *mm.*–2 *cm. long; leaves* bristleform, 1–3.5 *cm. long; fruiting heads* 4–7 *mm. in diameter; perigynia* 3–3.5 *mm. long.* — Calcareous clay and gravel, Ingornachoix Bay, Nfld.; Arct. reg.

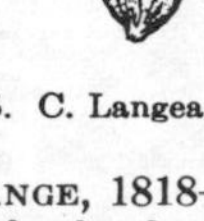

505. C. Langeana.

5. C. Langeàna Fern. (for its discoverer, Johann Martin Christian Lange, 1818–1898). — Similar to no. 4; *leaves spinulose-scabrous* toward tip; culms 1–9 cm. high; *heads ellipsoid to thick-cylindric*, 3–5 *mm. thick*, fulvous; *scales and flattish perigynia appressed-ascending*, the scales barely pale-margined; *anthers* 1 *mm. long.* — Peaty limestone-barrens, Ingornachoix Bay, Nfld.; Greenl. July, Aug. Fig. 505. — Possibly a hybrid of nos. 3 and 4.

§ 5. Divìsae Christ (see p. 294)

Rhizomes filiform; leaves canaliculate, 1–2 mm. broad; culms obtusely angled, smooth, 0.5–4 dm. high; perigynia 2.5–3 mm. long. . . . . 6. *C. stenophylla*, var. *enervis*.

Rhizomes stoutish, ligneous, 2–6 mm. thick; leaves flat, 2–5 mm. broad; culms acutely angled, scabrous toward summit, 2–7.5 dm. high; perigynia 3–4 mm. long.

Spikes 5–15, in an interrupted cylindric head 1–5 cm. long; beak of perigynium one-half as long as body; western species. . . . . 7. *C. praegracilis*.

Spikes 2–7, in an irregularly oblanceolate to ovoid head 1–3 cm. long; body of perigynium three to five times as long as beak; Atlantic species. . . . . 8. *C. divisa*.

6. C. stenophýlla Wahlenb. (slender-leaved). — *Rhizomes and stolons filiform;* leafy shoots tufted, low; the *canaliculate subrigid leaves* 1–2 *mm. broad,* curved; *culms* strict, slender, *obtusely angled, smooth,* 0.5–2.5 dm. high, overtopping the leaves; spikes 5–6, androgynous, densely crowded in a subcontinuous ellipsoid to ovoid brown or ferruginous head 0.7–1.5 cm. long; lowest bracts with awn-tips; *perigynia ascending, coriaceous,* plano-convex, *thin-edged,* mostly covered by the ovate acute scales, *closely enveloping the* large lenticular *achene,* ovate, 3–3.5 mm. long, definitely nerved, with a short scabrous-margined dorsally cleft beak. — Eurasia — Represented with us by

Var. **enérvis** (C. A. Mey.) Kükenth. (nerveless). — Up to 4 dm. high; *spikes distinct, in a definitely interrupted head; perigynia* 2.5–3 *mm. long,* often exceeding scales, *nerveless* or nearly so. (*C. stenophylla* of Am. auth., not Wahlenb.; *C. Eleocharis* Bailey) — Man. to Yuk., s. on dry plains and bluffs to Ia., Kans., N.M., Utah and e. Oreg. Late May–July. (Asia) Fig. 506.

506. C. stenophylla, v. enervis.

7. C. praegrácilis W. Boott (very slender). — *Rhizomes subligneous,* elongate, forking, dark-purple to blackish, 2–6 *mm. thick, sending up scattered* erect *tufts of flat* scabrous-margined *leaves* 2–3 *mm. broad and slender* arched-ascending *acutely angled scabrous culms* 2–7.5 dm. high; *spikes* 5–15, androgynous, *forming a linear- to lance-cylindric interrupted* brown *head* 1–5 *cm. long;* lowest bracts awned or with short blade; pistillate scales ovate, acute, concealing the perigynia; *perigynia coriaceous,* plano-convex, ovate, 3–4 *mm. long,* appressed-ascending, slightly nerved on the back, the inner face nerveless, *with* a serrulate *beak one-half as long as body;* styles 2. (*C. marcida* Boott, not J. F. Gmel.) — Low open ground and prairies, Yuk. to Mex., e. to Man., n. Mich., Ia., Mo. and Okla. June, July. (S.Am.) Fig. 507.

507. C. praegracilis.

508. C. divisa.

8. C. divìsa Huds. (separated). — Resembling no. 7, paler green; leaves 2–5 mm. broad; culms 1–5 dm. high, firm; *spikes* 3–7 *in an irregularly oblanceolate to ovoid head* 1–3 *cm. long; perigynia ribbed on both faces, the beak several times shorter than the body.* — Coastal sands, local, Calvert Co., Md. May, June. (Natzd. from Eu.) Fig. 508.

§ 6. Chordorrhìzeae Fries (see p. 294) — A single species:

9. C. chordorrhiza L. f. (with cord-like roots). — *Stems cord-like or wiry, prolonged, leaning or reclining, bearing few tufts of narrow leaves; culms terminal and lateral, arising from axils of shriveled leaves of preceding year,* 1–4.5 dm. high, smooth, obtusely angled; *spikes* 3–5, androgynous, *in* an ovoid, ellipsoid or irregularly deltoid *head* 0.5–1.5 *cm. long;* pistillate scales rounded, deep brown, acuminate; *perigynia compressed-ovoid to subglobose,* coriaceous, 2–3.5 mm. long, nerved on both faces, with a short beak dorsally cleft. — Quagmires and inundated bogs, s. Baffin I. to Alaska, s. to Nfld., e. Que., centr. Me., sw. Vt., centr. N.Y., n. Ind., n. Ill., n. Ia. and Sask. Late May–Aug. (Eurasia) Fig. 509.

509. C. chordorrhiza.

§ 7. Arenàriae Kunth (see p. 295)

a. Thin border of perigynium narrow, extending nearly to the base; inflorescences chiefly without leafy bracts; northern and continental species. . . *b*.

b. Inner band of leaf-sheath green-nerved nearly to summit; spikes 12–25 in a slender head 3–7 cm. long; perigynia 2.5–5 mm. long, the beak one-fourth to one-half as long as body.

Nodes of culm (at least the upper) exserted from leaf-sheaths; principal spikes nearly uniform, pale, subglobose to short-ovoid, 6–9 mm. long; scales obtuse to cuspidate; perigynia 2.5–4.5 mm. long. . . . . . 10. *C. Sartwellii*.

Nodes covered by upper sheaths; inflorescence more continuous above;

spikes rufescent, ovoid or ellipsoid, 8–15 mm. long; scales acute; perigynia 4–5 mm. long. . . . . 11. *C. disticha.*
*b.*Inner band of leaf-sheath nerveless; spikes 2–8 in a head 1.5–3.5 cm. long; perigynia 5–6 mm. long, the beak two-thirds as long as body. . . . . 12. *C. foenea.*
*a.*Thin border of perigynium dilated from below the middle into a broad greenish wing; lower spikes often subtended by leafy bracts; species of Atlantic sands. 13. *C. arenaria.*

10. C. Sartwéllii Dew. (for its discoverer, Henry Parker Sartwell, 1792–1867). — Rhizome ligneous, covered with fibrillose black scales, sending up tufts of elongate attenuate scabrous leaves and scattered stiff *slender sharp-angled scabrous culms* 0.3–1.2 m. high; *upper nodes of culm exserted from leaf-sheath; inner band of sheath green-nerved nearly to the prolonged summit; spikes* 12–25, *the pistillate subglobose or short-ovoid,* 6–9 *mm. long,* pale fulvous, some spikes (especially the crowded upper ones) staminate, the lowest with spathiform bracts, all *in an interrupted linear- to ellipsoid-cylindric head* 3–7 *cm. long; pistillate scales obtuse* or mucronate; *perigynia* membranaceous, planoconvex, broadly ovate, 2.5–4.5 *mm. long,* slenderly nerved on both faces, *contracted to* a serrulate *beak one-fourth to one-third as long as body;* styles 2. — Calcareous bogs, marshes and swales, sw. Que. to n. B.C., s. to centr. and w. N.Y., O., Ind., Ill., Mo., Neb. and Colo. June, July. Fig. 510.

510. C. Sartwellii.

11. C. dísticha Huds. (two-ranked). — Similar to no. 10; *nodes all covered;* summit of leaf-sheath scarcely prolonged beyond base of blade; *lower pistillate spikes ovoid or ellipsoid,* 8–15 *mm. long,* darker brown; *pistillate scales acute; perigynia* 4–5 *mm. long, with beak one-half as long as body.* (*C. intermedia* Good.) — Bogs, Hastings Co., Ont. (*Macoun*); natzd. (?) from Eu. along the St. Lawrence R., near Montreal, Que. June, July. (Eu.)

12. C. foènea Willd. (hay-like). — Rhizome cord-like or subligneous, extensively creeping, with fibrillous brown scales; *leaves* stiff, 1–3 mm. wide; *the inner band of the sheath hyaline, nerveless;* culms slender, scattered, overtopping the leaves, sharply angled, 1–9 dm. high; *spikes* 2–8, *in a head* 1.5–3.5 *cm. long,* some gynecandrous, others sometimes all staminate, others often androgynous, *the terminal usually with prolonged staminate base; scales* lustrous, brown, with whitish margin, lance-ovate, *acute; perigynia* appressed-ascending, planoconvex, narrowly ovate- or oblong-lanceolate, 5–6 *mm. long,* many-nerved, *with narrow greenish margin, tapering to a beak two-thirds as long as body.* (*C. siccata* Dew.) — Dry open soil, s. Me. and sw. Que. to Mackenz., s. to N.J., N.Y., O., Ind., Ill., Neb., N.M. and Ariz. May–July. Fig. 511.

511. C. foenea.

13. C. arenària L. (of sand). — Resembling no. 12, coarser; *leaves about equaling the curving culms* (0.7–5 dm. high); *spikes* 3–15 in a lanceolate interrupted head 3–7 dm. long, *the lower usually subtended by bracts with blades* 1–3 *cm. long;* scales lance-acuminate, ferruginous; *perigynia* coriaceous, lance-ovate, 4–5 mm. long, *the margin from near the middle dilated into a broad thin wing* extending to the short bidentate beak. — Coastal sands and dunes, Md. and Va., appearing like a native; long supposed to be adventive. May–July. (Eu.) Fig. 512.

512. C. arenaria.

§ 8. Macrocéphalae Kükenth. (see p. 295) — One species with us:

14. C. Kobomùgi Ohwi (native Japanese name). — *Dioecious,* extensively creeping, the stout rhizomes and stolons and bases of flowering culms covered with long shredding brown scales; *leaves* coriaceous, 4–8 *mm. broad, pungent, papillate-serrulate;* culms arising from among persistent old leaves, coarse, subcylindric, smooth, 1–3 dm. high; *spikes* very numerous, *crowded into* an ovoid or ellipsoid *head* 3–6 cm. long and 2–4 *cm. thick;* bracts and *scales herbaceous,* green, becoming drab, *serrate* near base; *anthers* 4.5–6 *mm. long; perigynia* soon divergent, coriaceous, lance-ovate, 10–14 *mm. long,* erose near base; *styles* 3; *achenes trigonous.* — Spreading on dune-sands, Ocean Co., N.J., to e. Va. May–July. (Natzd. from Asia) Fig. 513.

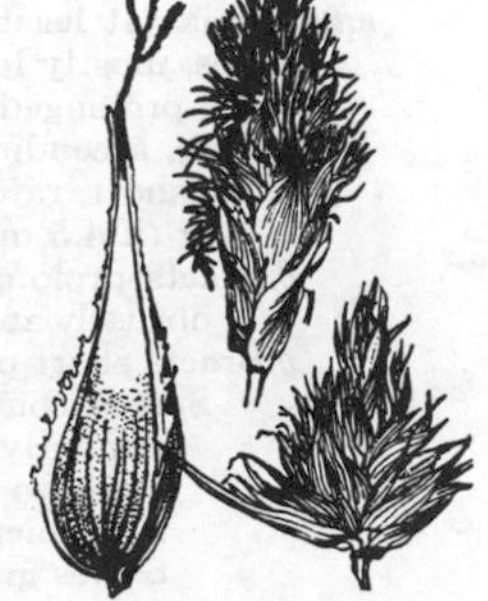
513. C. Kobomugi.

§ 9. BRACTEÒSAE Kunth (see p. 295)

*a.*Leaf-sheaths close or tight, usually not prominently reticulate or septate on the back; blades 0.7–4.5 mm. broad. . . *b.*

*b.*Perigynia conspicuously spongy-thickened below the middle, the nerve-like margin inflexed. . . *c.*

*c.*Margins of perigynia smooth; scales brown, acuminate, soon deciduous; spikes mostly approximate.

Perigynia biconvex, ovoid, deep- to brownish-green, the lower half to two-thirds spongy and corrugated. . . . . . . 15. *C. retroflexa.*

Perigynia planoconvex, lanceolate to lance-ovate, whitish-green or straw-colored, the flat inner face nerveless, only the basal fourth or fifth distended. . . . . 16. *C. texensis.*

*c.*Margins of perigynia minutely serrulate above; scales whitish or brown-tinged, obtuse to subacute, persistent; spikes mostly remote.

Broader leaves 1.8–3 mm. wide, deep green, the lowest sheaths 1.5–2.5 mm. in diameter; culms stiffish, suberect; perigynia deep green, 3–4.5 mm. long, broadly ellipsoid-ovoid, rather abruptly short-beaked, those of lower spikes (6–)9–20, all soon divergent; stigmas strongly coiled. . . . . 17. *C. convoluta.*

Broader leaves 0.8–2 mm. wide, the lowest sheaths 0.7–1.7 mm. in diameter; culms weak, lax; perigynia 2–3.5 mm. long, narrowly ellipsoid to broadly oblanceolate, those of lower spikes 2–8 (–12), loosely ascending, tardily divergent.

Culms loosely ascending; broadest leaves 1.3–2 mm. wide, pale green; perigynia pale, 2.8–3.5 mm. long, gradually tapering to the obscurely bidentate tip, those of lowest spikes 5–12; stigmas elongate, flexuous or recurved. . . . . 18. *C. rosea.*

Culms reclining or diffusely spreading, capillary; broadest leaves 0.8–1.2 mm. wide, deep green; perigynia deep green, 2–3 mm. long, abruptly contracted to clearly bidentate beak, those of lowest spikes 2–6; stigmas mostly tightly coiled. . . . . . . 19. *C. radiata.*

*b.*Perigynia of nearly uniform thickness and texture, not spongy-thickened at base (except in no. 24), their margins little, if at all, incurved. . . *d.*

*d.*Inflorescence densely capitate, lobate-globose to subglobose-ovoid, 0.7–2 cm. long; perigynia 2–3.5 mm. long.

Sheaths thickened at summit; scales of the spikes long-awned; perigynia ovate, broadest near middle, tapering to base.

Perigynia 2–3 mm. long, about 1.5 mm. broad. . . . . 20. *C. cephalophora.*

Perigynia 3–3.5 mm. long, 2–2.5 mm. broad. . . . . 21. *C. mesochorea.*

Sheaths not thickened at summit; scales merely acute or short-cuspidate; perigynia cordate-deltoid, broadest at the subtruncate base. . 22. *C. Leavenworthii.*

*d.*Inflorescence elongate, linear- to ellipsoid-cylindric or slenderly ovoid, interrupted, 1.5–5(–8) cm. long. . . *e.*

*e.*Perigynia completely covered by the brown or fulvous scales, 2.7–3.5 mm. long, about 1.25 mm. broad; northwestern species. . . . . 23. *C. Hookerana.*

*e.*Perigynia (at least the margins or beaks of the upper) exceeding the scales, mostly longer and broader than in no. 23. . . *f.*

*f.*Ligule prolonged, many times longer than broad; perigynia 4–6 mm. long, ascending in approximate spikes; scales fulvous. . . . . 24. *C. spicata.*

*f.*Ligule short, rarely (in no. 29) longer than broad; perigynia 3–4 mm. long (if 4.5 mm. long, in moniliform inflorescences). . . *g.*

*g.*Bracts prolonged, many times exceeding the inflorescence; culms obtusely angled, smooth; southwestern species. . . . . . . 25. *C. arkansana.*

*g.*Bracts short or wanting; culms acutely angled, scabrous above.

Spikes contiguous (lower rarely remote) in a dense oblong to slenderly ovoid inflorescence 1.5–4 cm. long; perigynia broadly ovate to suborbicular, 3–4 mm. long, 2–3 mm. broad; achenes suborbicular; ligules as short as broad.

Scales greenish, whitish or pale brown, long-awned, nearly or quite equaling the coriaceous spreading-ascending perigynia; indigenous species.

Bracts of inflorescence slender-based; perigynia longer and broader than scales, 3–3.5 mm. long, 2–2.5 mm. broad; achene about 2 mm. long. . . . . 26. *C. Muhlenbergii.*

Bracts broad-based; perigynia mostly covered by the broad

scales, 3.5–4 mm. long, 2.5–3 mm. broad; achene 2.2–2.5 mm. long. . . . . 27. *C. austrina.*

Scales ferruginous, merely mucronate, much shorter than the membranous finally squarrose perigynia; adventive. . . 28. *C. Pairaei.*

Spikes all or all but the upper remote, in a moniliform inflorescence 5–8 cm. long; perigynia membranaceous, much exceeding the pale acuminate scales, narrowly ovate, 3.5–4.5 mm. long, appressed; achenes oval; ligules slightly longer than broad; adventive. . . . . 29. *C. divulsa.*

a. Leaf-sheaths loose and usually prominently reticulate or septate-nodulose on the back, the ventral band usually weak and easily fractured; blades 3–10 mm. broad.

Scales long-acuminate to awn-tipped, nearly or quite as long and three-fourths as broad as perigynia; the latter firm, 2–3 mm. broad; spikes crowded. . . . . 30. *C. gravida.*

Scales obtuse to acute, much shorter and narrower than perigynia.

Membranous inner band of leaf-sheath firm, its summit concave; blades 3–6 mm. broad; spikes contiguous in heads 2.5–5 cm. long; perigynia 3.5–5.5 mm. long, 2–3.2 mm. broad. . . . . 31. *C. aggregata.*

Membranous inner band of leaf-sheath friable, its summit truncate; blades mostly 5–10 mm. broad; perigynia 3–4.5 mm. long, 1.5–2.5 mm. broad.

Spikes crowded into a head 1.5–4 cm. long; scales obtuse to acutish, not acuminate; perigynia membranaceous, flat on inner face, wing-margined only above middle; ligule longer than broad. . . . 32. *C. cephaloidea.*

Spikes (except the approximate uppermost) remote, in a moniliform head 3–15 cm. long; scales acuminate; perigynia subcoriaceous, concave on inner face, narrowly wing-margined to base; ligule as broad as long. . . . . 33. *C. sparganioides.*

15. C. retrofléxa Muhl. (bent backward). — In dense clumps 1–7.5 dm. high; leaves flat, 1–3 mm. wide, with close sheaths; culms stiff, erect, about equaling or exceeding leaves, slender; *spikes* 3–9, *closely approximate* or the lowest subdistant, *in a* cylindric *head* 1–4 *cm. long,* without leafy bracts or the lowest bracted; *perigynia corky-gibbous and corrugated below the middle,* green to brown, strongly rounded at base, compressed-ovoid, 2.5–3 mm. long, *contracted to a smooth* bidentate *beak, only slightly exceeding the brownish acute and soon deciduous scales.* — Dry rocky or sandy woods and thickets, Fla. to Tex., n., locally, to s. N.H., Vt., N.Y., O., s. Mich., Ill., Mo. and e. Kans. May, June. FIG. 514.

514. C. retroflexa.

16. C. texénsis (Torr.) Bailey (of Texas). — More slender than no. 15, 1–3 dm. high; leaves 0.7–1.5 mm. wide; spikes 2–8 in a slender head 0.6–3 cm. long; *perigynia planoconvex, lanceolate to lance-ovate, whitish- or pale green, the flat inner face nerveless, only the basal fourth distended.* — Rocky or sandy woods and fields, Ga. to Tex., n., locally, to s. N.J. (adv.), e. Md., s. O., Ky. and se. Mo. (*C. retroflexa,* var. Fern.) Apr.–early June.

17. C. convolùta Mackenz. (rolled up). — In dense tussocks, *full green; sheaths* close, *the lowest* 1.5–2.5 *mm. in diameter; leaf-blades* elongate, *the broader* 1.8–3 *mm. wide; culms stiffish* but slender, *erect or suberect,* 3–9 dm. high; *spikes* 4–7, *distant* or the upper approximate in a moniliform head 2.5–6 cm. long; *the lower* (fuller) *spikes* (6–)9–20-*flowered,* often subtended by a linear-setaceous elongate bract; *perigynia deep green,* 3–4.5 *mm. long, broadly ellipsoid-ovoid, spongy-gibbous below, mostly nerveless,* rather *abruptly contracted to a short* bidentate *serrulate-margined beak,* all soon divergent, much exceeding the whitish rounded scales; *stigmas tightly coiled.* (*C. rosea* of ed. 7, not Schkuhr) — Dry woods and thickets, N.B. to s. Man., s. to S.C., Ala., Ark. and e. Kans. Mid-May (southw.)–mid-July (northw.) FIG. 515.

515. C. convoluta.

18. C. ròsea Schkuhr (rose-like; from rosettes of perigynia). — Similar to no. 17, in looser tufts, *paler; lowest sheaths* 0.7–1.7 *mm. in diameter; broader blades* 0.8–2 *mm. wide; culms weaker,* loosely ascending, filiform, 2–7.5 dm. long; *lower spikes with the* 5–12 *loosely ascending finally divergent pale green narrowly ellipsoid perigynia* 2.8–3.5 *mm. long, gradually tapering to the obscurely bidentate tip; stigmas elongate, flexuous or recurving,* rarely if ever tightly coiled. (Var. *minor* Boott) — Rich woods and thickets, Gaspé Pen., Que., to N.D., s. to Ga., Ala. and La. May–early July.

19. C. radiàta (Wahlenb.) Dew. (radiate). — Looser than no. 18, deep green; *culms diffusely spreading or reclining*, capillary, 2–6 dm. long; *broadest leaves* 0.8–1.2 *mm. wide*, soft; *lowest spikes* frequently subtended by elongate bracts, 2–6-*flowered; perigynia* narrowly ellipsoid or oblanceolate, *deep green*, 2–3 *mm. long, abruptly contracted to clearly bidentate beak; stigmas tightly coiled*. (*C. rosea*, var. *radiata* (Wahlenb.) Dew.) — Open woods and thickets, centr. Me. and s. Que. to O., s. to Del., Md., W.Va., and in mts. to N.C. and Tenn. Late May–mid-Aug.

516. C. cephalophora.

20. C. cephalóphora Muhl. (bearing heads). — Strict but soft, densely cespitose, 2–8 dm. high; leaves 2–4.5 mm. wide, long-attenuate, the sheath thickened at summit; *head lobate-subglobose or ovoid*, 0.7–1.8 *cm. long, the spikes crowded*, the lower 1 or 2 spikes usually with a setaceous bract; *perigynia elliptic-ovate, tapering to base*, 2–3 *mm. long, about* 1.5 *mm. broad*, slightly longer than the acute or rough-cusped scale. — Dry woods and openings, Fla. to Tex., n. to centr. Me., sw. Que., s. Ont., s. Mich. and se. Man. May–July. FIG. 516.

21. C. mesochòrea Mackenz. (midland). — Coarser than no. 20, up to 1 m. high; leaves firmer and harsher; *perigynia* 3–3.5 *mm. long*, 2–2.5 *mm. broad*, ovate. — Dry open soil, Mass. to Wisc., s. to L.I., Va., Tenn., Mo. and Tex. May–July.

517. C. Leavenworthii.

22. C. Leavenwórthii Dew. (for its discoverer, MELINES CONKLIN LEAVENWORTH, 1796–1862). — Resembling no. 20, usually more lax, 1–6.5 dm. high; leaves 1–3 mm. wide; head 7–20 mm. long; *perigynia cordate-deltoid, subtruncate at the broad base*, exceeding the acutish rarely cuspidate scales.— Open woods and (often adventive) grasslands, Fla. to Tex., n. to s. N.J., Pa., s. Ont. s. Mich., Ill., Ia. and Kans. May, June. FIG. 517.

518. C. Hookerana.

23. C. Hookeràna Dew. (in honor of Sir WILLIAM JACKSON HOOKER, 1785–1865). — Cespitose, with dark bases, 1.5–4.5 dm. high, very slender; leaves pale green, flat, thin, 1–2.5 mm. wide, the dark-margined ligule broader than long; culms very slender, acutely angled, rough above; *inflorescence linear-cylindric*, 2–5 cm. long, with 5–10 subdistant to approximate appressed slenderly ovoid spikes; *bracts with brown broad scarious bases; scales pale brown, prominently awned, completely covering the perigynia; perigynia lance-ovate, membranaceous*, 2.7–3.5 *mm. long, about* 1.25 *mm. wide*, obscurely nerved on back. — Plains, prairies and dry banks, Thunder Bay Distr., Ont., to N.D. and Alta. June, July. FIG. 518.

519. C. spicata.

24. C. SPICÀTA Huds. (spiked). — Culms tufted, 1.5–8 dm. high, rough, longer than the narrow (1.5–3 mm. wide) leaves; *ligule prolonged, much longer than broad; spikes* 4–10, approximate or slightly distant, *forming a somewhat interrupted cylindric head* 1.5–4.5 *cm. long; scales green, fulvous or tawny*, acuminate, nearly as broad but scarcely as long as the perigynia; *perigynia heavy, narrowly ovate, lustrous*, 4–6 *mm. long, nerveless*, ascending to spreading. (*C. muricata* of ed. 7 and many auth., hardly L.; *C. contigua* Hoppe) — Dry fields, roadsides and groves, N.S. and s. Me. to s. Ont., s. to e. Va. and n. O. Mid-May–July. (Natzd. from Eurasia) FIG. 519.

520. C. arkansana.

25. C. arkansàna Bailey (of Arkansas). — Densely cespitose, *with smooth obtusely angled slender culms* 1.5–6 dm. high; leaves pale, 1–2 mm. wide, smooth, or scabrous-margined, with ligule shorter than broad; *spikes* 3–6, *subglobose, squarrose, greenish*, in a leafy-bracted interrupted inflorescence 1.5–3 cm. long; *bracts much prolonged, the lowest and longest* 0.5–2.5 *dm. long, much overtopping inflorescence;* scales triangular-ovate, short-acuminate, pale, narrower than perigynia; perigynia round-ovate to suborbicular, 3.5–4 mm. long, 2.25–3 mm. wide, pale, membranaceous, nerved on back, abruptly serrate-beaked. — Damp prairies and low woods, ne. Tex. to w. Ark. and sw. Mo. May, June. FIG. 520.

521. C. Muhlenbergii.

26. C. Muhlenbérgii Schkuhr (for GOTTHILF HENRY ERNEST MUHLENBERG, 1753–1815). — *Stiff* throughout; culms tufted at the tips of slightly prolonged stout rhizomes covered with drab sheaths of past years, 0.15–1.3 m. high; *leaves* firm, pale, often falcate, *often folded*, harsh, 2–4 mm. broad; culms harsh above, sharply angled; *inflorescence rather dense*, 1.5–4 *cm. long, thick-oblong to linear-cylindric, of* 3–10 *distinct pale green to finally drab globose-ovoid spikes; bracts* setaceous, *slender-based*, often evident; *scales pale*, ovate, *rough-awned*, nearly or quite equaling but *narrower than perigynia; perigynia suborbicular* or broadly ovate, *firm*, plano-convex, 3–3.5 *mm. long*, 2–2.5 *mm. broad*, very strongly nerved on both faces; *achene about* 2 *mm. long*.— Dry woods, openings and fields, Fla. to Tex., n. to s. Me., s. N.H., sw. Que., s. Ont., O., Mich., Wisc. and Minn. June, July. FIG. 521. — Passing insensibly into the relatively unimportant

Var. **enérvis** Boott (nerveless). — Perigynia nerveless or with only short basal nerves. (*C. plana* Mackenz.) — Fla. to Tex., n. to N.E., s. Ont., Mich., Ill., Ia. and Neb.; said by Mackenzie to prefer calcareous soils.

27. C. austrìna (Small) Mackenz. (southern). — In habit similar to no. 26; leaves 2.5–4.5 mm. wide, commonly flat; the whitish-green spikes more squarrose; *bracts with dilated bases; scales broader, nearly or quite covering the perigynia; perigynia* 3.5–4 *mm. long*, 2.5–3 *mm. broad*, strongly nerved on back, nearly or quite nerveless on inner face; *achene* 2.2–2.5 *mm. long*. — Calcareous glades and rocky prairies, Tex. to Ark., n. to Kans., Mo. and Ind. May, June. FIG. 522.

522. C. austrina.

523. C. Pairaei.

28. C. PAIRAÈI F. W. Schultz (named in 1868 for MICHEL PAIRA). — Resembling no. 24; more slender; *ligule wider than long; inflorescence* more open, *with the lower* of the 3–8 *spikes remote*, forming a somewhat open inflorescence 2–3 cm. long; *scales ferruginous, merely mucronate*, small; *perigynia much exceeding scales*, becoming somewhat squarrose, membranaceous, *broadly ovate*, plano-convex, 3–3.5 *mm. long, about* 2 *mm. broad*, nerveless or obscurely nerved at base, attenuate to short beak. (*C. echinata* sensu Kükenth., not most auth.) — Fields and roadsides, local, N.B., Pa. and N.D. June, July. (Adv. from Eu.) FIG. 523.

29. C. DIVÚLSA Stokes (separated). — Culms many from summit of stout ligneous rhizome, 0.3–1 m. high, scabrous at triangular summit; leaves flat, 2–3 mm. broad, the *ligule slightly longer than broad; spikes* 6–8, subglobose, all but the upper *remote, forming a moniliform inflorescence* 5–8 *cm. long*; scales whitish, ovate, acuminate; *perigynia* membranaceous, greenish, *narrowly ovate, appressed, much exceeding scales*, 3.5–4.5 *mm. long, rarely* 2 *mm. broad; achene oval*. (*C. virens* sensu Mackenz., probably not Lam.) — Grassland and rocky slopes, local, Pa. and D.C. June. (Adv. from Eurasia) FIG. 524.

524. C. divulsa.

30. C. grávida Bailey (heavy; with fruit). — *Leaf-sheaths loose*, nodulose-septate on back, *with nerveless inner band truncate at summit*; the lax blades 3.5–8 mm. broad; *culms* thin and *sharply angled*, scabrous, 0.3–1 m. high; *spikes* greenish to drab, subglobose, *approximate* in a dense to slightly open ellipsoid head 1–3 cm. long, or in forma **laxifòlia** (Bailey) Kükenth. (loose-leaved) 3–5 cm. long; *scales long-acuminate to awn-tipped, nearly or quite as long and three-fourths as broad as perigynium; perigynia firm*, drab or pale brown, narrowly to broadly ovate, 3.5–5.5 mm. long, 2–3 mm. broad, *gradually tapering from near the middle into a beak with sharp teeth about* 1 *mm. long*, nerveless on inner, faintly nerved or nerveless on outer face. — Swales, shores, prairies and plains, sw. Ont. to N.D. and Wyo., s. to O., w. Ind., Ill., Mo. and Kans. Late May, June. FIG. 525. — Passing gradually into

525. C. gravida.

Var. **Lunelliàna** (Mackenz.) F. J. Herm. (for JOEL LUNELL, 1851–1920). — *Perigynia* with broadly ovate to suborbicular bodies, *abruptly tapering from well above the middle into short beaks about* 0.5 *mm. long*, often more strongly ribbed on back. (*C. Lunelliana* Mackenz.) — W. Ind., s. Mich. and Ill., sw. to Ark., Okla., Tex. and N.M. May.

31. C. aggregàta Mackenz. (crowded in a mass). — Habitally similar to no. 30; *leaves* 3–6 *mm. broad*, the *summit of the firm inner band of the sheath concave*, the ligule shorter than broad; *head linear-cylindric*, slightly interrupted, 2.5–5 *cm. long; scales narrowly ovate, acuminate, definitely shorter and narrower than perigynia; perigynia* membranaceous, ovate, 3.5–5.5 *mm. long*, 2–3.2 *mm. broad, gradually tapering to beak*, plane and nerveless on inner, slightly nerved on outer face. — Rich woods, thickets and meadows, N.Y. to Ia., s. to N.J., D.C., Ky., Mo. and Okla. May, early June. FIG. 526.

526. C. aggregata.

32. C. cephaloídea Dew. (head-like). — Lax, very green, 0.3–1.2 *m. high; leaves soft, rather*

527. C. cephaloidea.

flaccid, mostly 5–8 *mm. broad; the veinless inner band of the sheath friable, its summit truncate,* the *ligule much longer than broad;* culms soft and weak; *head* 1.5–4 *cm. long,* rather *dense,* green, becoming pale brown; *scales ovate, obtuse to acute, much shorter and narrower than perigynia; perigynia* spreading, narrowly ovate, pale green, *membranaceous,* 3–4.5 mm. long, 1.5–2.5 mm. broad, tapering gradually to beak, nerveless, *flat on inner face, narrowly wing-margined only above middle.* — Rich woods, thickets, bottomlands and swales, s. Que., s. Ont. and Minn., s. to w. N.B., N.E., nw. N.J., Pa., O., Ind., Ill. and ne. Ia. Late May–early July. Fig. 527.

33. C. sparganioìdes Muhl. (like *Sparganium*). — Rather lax, 0.4–1 m. high; *leaves* soft, mostly 5–10 *mm. broad; thin sheaths friable, with truncate summit and broad ligule; the lower leaves conspicuously clothing the base of the culm; spikes* 6–15, the upper approximate, *the lower remote* (lowest 0.5–6 cm. apart) *in a moniliform green head* 3–15 *cm. long,* lowest spikes often compound; *scales acuminate, much shorter and narrower than perigynia; perigynia subcoriaceous,* greenish, becoming yellow or golden, *concave on inner face, narrowly wing-margined to base,* 3–4 mm. long, 1.5–2.3 mm. broad, the beak sharp-toothed, nerves inconspicuous or absent. — Rich woods, sw. Que. and w. N.H. to Owen Sound, Ont., Minn. and S.D., s. to Va., Tenn., Mo. and Kans. June, July. Fig. 528.

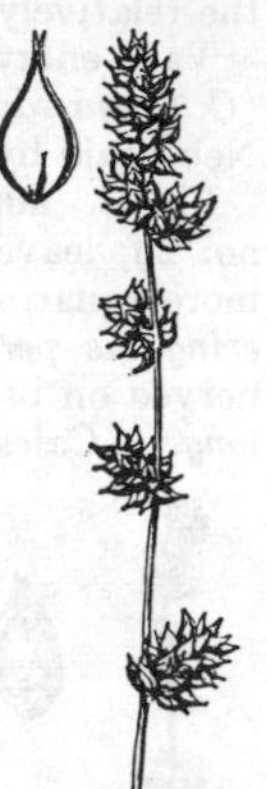

528. C. sparganioides.

§ 10. Multiflòrae Kunth (see p. 295)

Leaves mostly equaling or exceeding height of culm; perigynia narrowly ovate, tapering gradually to a beak one-half to three-fourths as long as body, green to stramineous, 1–1.8 mm. broad. . . . . 34. *C. vulpinoidea.*

Leaves mostly shorter than height of culm; perigynia broader, abruptly short-beaked, stramineous to golden-brown, 1.5–3 mm. broad.

Summit of membranous inner band of sheath of upper leaves truncate or only slightly rounded; perigynia appressed-ascending, broadly ovate, 1.6–2.4 mm. broad, their subtending scales evident. . . . . 35. *C. annectens.*

Summit of membranous band of sheath of upper leaves prolonged as a broad tongue; perigynia widely spreading, reniform to suborbicular, 2.5–3 mm. broad, hiding the scales; southwestern. . . . . 36. *C. triangularis.*

34. C. vulpinoídea Michx. (resembling *C. vulpina,* with inflorescence like a fox's tail). — Cespitose, mostly stiffish, 0.2–1 m. high; culms firm, very scabrous above; *leaves* mostly *flat, longer than the culms,* 2–6 mm. wide, scabrous, *long-attenuate, their close sheaths with cross-puckered membranous band; heads linear- to lance-cylindric,* 2–15 cm. long, dense to somewhat open, *stiff and straight* (rarely lax and arching), the clusters of spikes crowded, or in forma **segregàta** (Farw.) Raymond (separated) distant in a moniliform inflorescence, *greenish to yellowish or dull brown, with* 1–several *setaceous bracts; perigynia ovate to ovate-lanceolate, mostly ascending,* crowded, 1.7–3 mm. long, 1–1.8 *mm. broad, tapering gradually to a beak one-half to two-thirds as long as the flat-faced body,* greenish to stramineous; scales long-awned. — Low grounds, Nfld. to s. B.C. and south beyond our range. June–Aug. Fig. 529.

C. setàcea Dew. (bristle-like), an obscure plant with very slender perigynia, the leaves either longer or shorter than the culms, seems like a sterile hybrid of nos. 34 and 35. Fig. 530.

35. C. annéctens Bickn. (connecting). — Resembling no. 34; *mature culms definitely taller than the firm leaves; summit of inner band of leaf-sheath truncate or only slightly convex;* head stiff, 2–10 cm. long, often less compound and with fewer and shorter bracts than in no. 34, becoming yellowish to warm brown or drab; *scales shorter-awned; perigynia broadly ovate,* 2.5–3.5 *mm. long,* 1.6–2.4 *mm. broad, with prominent* notched *beak much shorter than body,* commonly nerved on back. (*C. setacea,* var. *ambigua* (Barratt) Fern.) — Dry or moist sterile, often sandy, soils, s. Me. to Wisc., s. to Va., Tenn. and Ind. (reported by Mackenzie from farther south and southwest). Late May–July. Fig. 531.

Var. **xanthocárpa** (Bickn.) Wieg. (yellow-fruited). — Head compact, 2–5(–7) cm. long; perigynia mostly 2.2–2.7 mm. long and 1.5–1.8 mm. broad, with shorter obscurely notched beak. (*C. brachyglossa* Mackenz.) — More local eastw.; commoner westw.; centr. Me. and sw. Que. to Wisc. and Ia., s. to Va., O., Ind., Ill., Mo. and Kans.

36. C. triangulàris Boeckl. (triangular). — Resembling no. 35; *summit of inner band of leaf-sheaths prolonged into a broad tongue;* head dense, 3–5 cm. long; *perigynia horizontally* *spreading, thus hiding the scales, reniform to suborbicular,* 2.5–3 *mm. broad,* short-beaked. — Swales, meadows and low prairies, Miss. to Tex., n. to se. Mo. and Okla. May, June. Fig. 532.

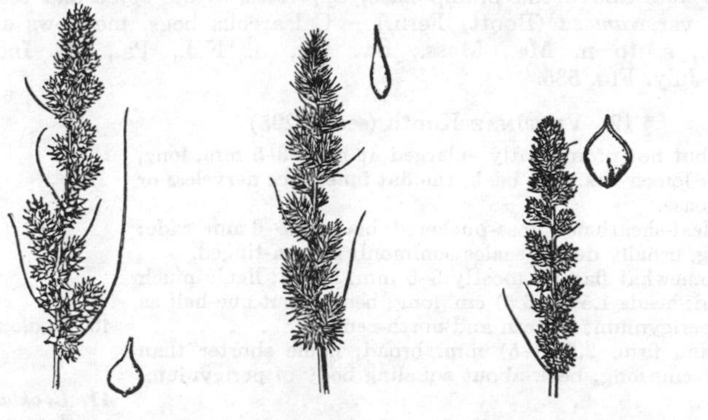
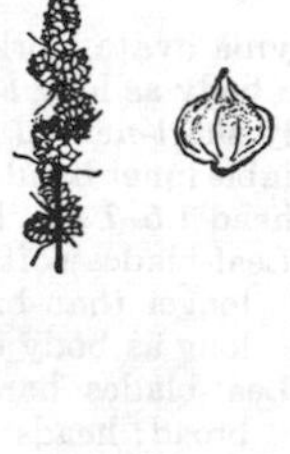

529. C. vulpinoidea. 530. C. setacea. 531. C. annectens. 532. C. triangularis.

§ 11. Paniculàtae Kunth (see p. 295)

Leaves 2.5–8 mm. broad; panicle lax, 0.7–1.8 dm. long; perigynia gibbous, broadly obovoid, abruptly beaked. . . . . 37. *C. decomposita.*

Leaves 1–3 mm. broad; panicle or head dense and stiff or open, 1–10 cm. long; perigynia slightly biconvex, ovate to lanceolate, tapering to beak.

Summit of veinless band of leaf-sheath pale; head of spikes straight, dense or slightly open, 1–5(–8) cm. long, dark brown; perigynia olivaceous to blackish, lustrous, with convex inner face, soon wide-spreading, not wholly covered by the scales. . . . . 38. *C. diandra.*

Summit of veinless band of leaf-sheath discolored, yellow-brown to bronze; head lax and open, often moniliform or loosely branched, 3–10 cm. long, pale brown; perigynia stramineous to brown, with flat inner face, appressed and covered by the scales. . . . . 39. *C. prairea.*

37. C. decompósita Muhl. (decompound), Cypress-knee Sedge. — In loose stools, the *obtuse-angled or almost terete* loosely spreading and arching *culms* 0.5–1.5 *m. high; leaves* long-sheathed, *the firm blades* 2.5–8 *mm. broad,* channeled below, mostly *overtopping the culm; panicle open, flexuous,* 0.7–1.8 *dm. long,* olive-brown or darker, the lower branches 1–4 cm. long; spikes crowded; the crowded *firm* shining *perigynia* spreading, *gibbous, broadly compressed-obovoid or obpyramidal,* olivaceous to blackish, *tapering to stipitate base, very abruptly beaked,* 2–2.5 mm. long, nearly or quite as broad. — Wooded swamps, often on floating logs and bases of trees (most often *Taxodium*), local, Fla. to La., n. to e. Va., w. N.Y., O., s. Mich. and se. Mo. June–Aug. Fig. 533.

533. C. decomposita.

534. C. diandra.

38. C. diándra Schrank (two-stamened). — In loose tussocks from a short rhizome, 0.2–1.1 m. high, the slender obtuse-angled harsh-tipped *culms mostly overtopping the narrow* (1–3 *mm. broad) plicate leaves; summit of veinless pale band of leaf-sheath not conspicuously discolored,* often dotted; *head stiff,* rarely a little flexuous, *dense or but slightly open,* 1–5(–8) *cm. long,* 5–12 *mm. thick, dark brown; perigynia soon widely divergent, narrower and mostly longer than scales, olivaceous to blackish, shining, firm, strongly convex on back, slightly so on inner face, ovate, rounded to subtruncate at base,* stipitate, *tapering to long beak,* with prominent nerves along center of back, 2–3.5 mm. long. — Peaty swamps, bogs, etc., oftenest calcareous, s. Lab. to Yuk. and se. Alaska, s. to Nfld., N.S., N.E., N.J., Pa., O., Ind., Ill., Ia., Neb., Colo. and Calif. Late May–Aug. (Eurasia) Fig. 534.

535. C. prairea.

39. C. praìrea Dew. (of prairie). — Differing from no. 38 in the *darkened (yellow-brown to bronze) summit of veinless band of leaf-sheath;* more *sharply* *angled culms; head flexuous and open, moniliform to loosely paniculate, pale brown,* 3–10 *cm. long; perigynia stramineous to brown,* but slightly lustrous, ovate to lanceolate, 2.5–4 mm. long, *flat on inner face* above the plump base, *appressed in the spike and covered by the scales.* (*C. diandra*, var. *ramosa* (Boott) Fern.) — Calcareous bogs, meadows and wet thickets, Que. to Alta., s. to n. Me., Mass., nw. Ct., n. N.J., Pa., O., Ind., Ill., Ia. and Neb. Late May–July. Fig. 535.

§ 12. Vulpìnae Kunth (see p. 295)

*a.*Perigynia ovate, corky but not prominently enlarged at base, 3–5 mm. long, the body as long as or longer than the beak, the flat inner face nerveless or only short-nerved at base.
Friable inner band of leaf-sheath not cross-puckered; blades 2.5–6 mm. wide; head 1.5–7 cm. long, usually dense; scales commonly brown-tinged.
Leaf-blades soft, somewhat flaccid, mostly 3–6 mm. broad; ligule much longer than broad; heads 1.5–5(–5.5) cm. long; beak about one-half as long as body of perigynium; eastern and north-central. . . . . . . 40. *C. alopecoidea.*
Leaf-blades hard and firm, 2.5–4(–5) mm. broad; ligule shorter than broad; heads 4–7 cm. long; beak about equaling body of perigynium; southwestern. . . . . 41. *C. oklahomensis.*
Friable inner band closely cross-puckered; blades mostly 5–10 mm. wide; head (2–)3–7.5 cm. long, usually interrupted; scales whitish; beak nearly as long as body of perigynium. . . . . 42. *C. conjuncta.*
*a.*Perigynia subulate-lanceolate above the corky or spongy base, 4–9 mm. long, the beak longer than the body. . . *b.*
*b.*Perigynia 4–7 mm. long, gradually tapering from base to apex; panicle or head 1.5–10 cm. long, dense.
Membranous band of leaf-sheath friable, commonly cross-puckered, the unthickened summit prolonged; heads paniculate-compound, often with elongating branches, 1.5–10 cm. long, thick-cylindric, lanceolate or ovoid, 1–4 cm. in diameter, soon becoming yellowish or brown. . 43. *C. stipata.*
Membranous band firm, with hard cartilaginous concave or obliquely subtruncate summit-border, rarely cross-puckered; heads less paniculate, linear-cylindric to lanceolate, 1.5–5(–6) cm. long, 1–2 cm. thick, green, becoming drab. . . . . 44. *C. laevivaginata.*
*b.*Perigynia 6–9 mm. long, abruptly enlarged below into a disk-like base; panicle dense or interrupted and strongly flexuous, 6.5–25 cm. long.
Membranous band of leaf-sheath thin and friable, without firm summit, strongly speckled with purple; panicle yellow-green to yellow-brown; scales with brown or green sides; outer face of perigynium prominently nerved over the bulbous base; plant of Miss. Basin. . . . . . 45. *C. crus-corvi.*
Membranous band firm, with thickened orifice, with few or no purple dots; panicle whitish- or bluish-green to -drab; scales with white sides; the whitish strongly thickened bulbous base of perigynium scarcely nerved; southeastern. . . . . 46. *C. Bayardi.*

40. C. alopecoídea Tuckerm. (like a fox). — Rather stout but soft, 0.4–1 m. high, the *weak culms wing-angled, with concave sides; leaves thin and soft,* often longer than culms, *mostly* 3–6 *mm. broad,* the nodulose-septate *sheath with friable unwrinkled dotted inner band, the ligule much longer than broad; heads dense, stramineous or tawny,* thick-cylindric to lanceolate, 1.5–5.5 *cm. long,* 0.8–1.5 cm. thick; *scales brownish in age; perigynia* ovate, 3–4 *mm. long,* 1.5–2 mm. broad, prominently stipitate, *corky at base, flat on the inner* nerveless *face,* brown-nerved on convex outer face, *tapering into* serrulate *beak about half as long as body,* ascending, about equaling or little exceeding scale. — Calcareous meadows, swales and low thickets, rather local, s. Que., s. Ont. and Minn., s. to n. and w. N.E., n. N.J., Ind., Ill. and n. Ia. June, July. Fig. 536.

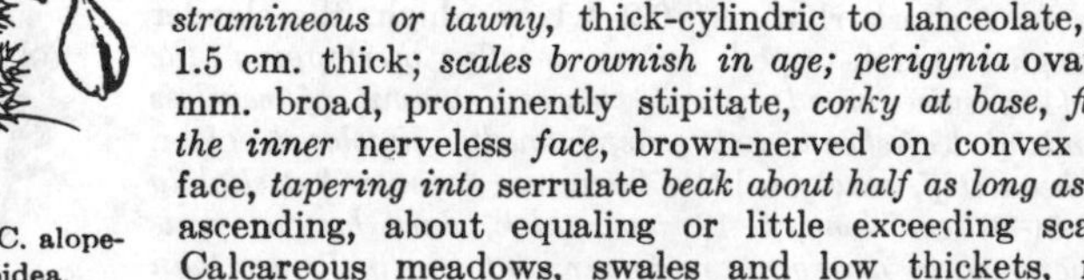
536. C. alopecoidea.

41. C. oklahoménsis Mackenz. (of Oklahoma). — Resembling no. 40; *leaves firm and hard,* 2.5–5 mm. wide, mostly shorter than culm, the *ligule shorter than broad; head* coarser, 4–7 *cm. long; perigynia* 4–5 mm. long, *the beak about equaling body.* — Calcareous meadows, Ark. to e. Tex., n. to sw. Mo. and Okla. May, June. Fig. 537.

537. C. oklahomensis.

42. C. conjúncta Boott (joined together). — Resembling no. 40, 0.5–

538. C. conjuncta.

1.2 m. high, the wing-angled culms soft; *leaves* soft, flaccid, 5–10 *mm. broad, the inner band closely cross-puckered; head usually interrupted, green,* 2–7.5 *cm. long; perigynium* lance-ovate, 3.5–5 mm. long, pale green, *whitish and thickened below, the beak nearly as long as the body and almost equaling or a little exceeding the cuspidate whitish scale.* — Calcareous bottoms, glades and swales, local, N.Y. to S.D., s. to Va., Tenn., Mo. and e. Kans. June. Fig. 538.

43. C. stipàta Muhl. (crowded). — Cespitose, *yellowish-green,* with fibrillose bases, stout, 0.4–1 m. high; *the thick spongy and soft culms sharply angled, with concave sides;* leaves flat and soft to flaccid, 4–8 mm. broad; the *sheath* coarsely reticulate, *its membranous band prolonged and very thin and friable and commonly cross-puckered* at least between the nerves; *head paniculate-spiciform, dense, yellowish-green,* becoming brown, 1.5–10 *cm. long, thick-cylindric to lanceolate or ovoid,* 1–4 *cm. thick; perigynia thin, spongy at base, subulate-lanceolate to narrowly ovate, tapering from base to long slender beak,* 4–5 mm. long, *brown-nerved,* the roughish *beak* about *twice the length of the body* and much exceeding the pale scales. — Low grounds, s. Lab. to s. Alaska, s. to Nfld., N.S., N.E., L.I., N.C., Tenn., Mo., Kans., N.M. and Calif. May–Aug. (E. Asia) Fig. 539. — Passing southw. into

539. C. stipata.

Var. **máxima** Chapm. (largest). — Mostly taller; leaves 0.8–1.7 cm. broad; perigynia 5–6 mm. long. (*C. uberior* (Mohr) Mackenz.) — Fla. to Tex., n. to N.J., Pa., s. Ind. and Mo.

540. C. laevivaginata.

44. C. laevivaginàta (Kükenth.) Mackenz. (smooth-sheathed). — Resembling no. 43, greener and more slender; the *leaf-sheaths* closer and less reticulate; *the veinless inner band firm, usually without cross-puckering, its concave to obliquely subtruncate summit cartilaginous-thickened;* the leaf-blades 3–6 mm. wide; *head greener* when young, *linear-cylindric to lance-ovoid,* 1.5–6 *cm. long,* 1–2 *cm. thick,* less paniculate than in no. 43; *perigynia* widely spreading, *green, becoming drab,* with pale nerves, 4.5–7 *mm. long.* — Boggy or swampy woods and meadows, Me. to Minn., s. to Fla., Ala. and Mo. May–July. Fig. 540.

45. C. crús-córvi Shuttlw. (crow-spur). — Stout, glaucous, 0.5–1 m. high; culms rough at least above, sharply triangular with concave sides, mostly shorter than the leaves; leaves flat, much prolonged, rough-serrulate, 5–12 mm. wide, the septate-nodulose *sheath with inner veinless band thin and friable, strongly purple-speckled; panicle much branched and compound,* dense or *more often open and flexuous,* 6.5–25 *cm. long,* 2–6 cm. thick, *greenish to yellow-brown; scales brown or green,* with 3-nerved green center, narrow; *perigynia* 6–9 *mm. long, nearly subulate above the abruptly enlarged long-stipitate bulbous base; the outer convex face strongly nerved to base, the beak thrice the length of the body,* deeply and sharply cleft. —Wooded swamps, bottoms and sloughs, nw. Fla. to e. Tex., n. to O., s. Mich., Wisc., s.Minn. and e. Neb. June, July. Fig. 541.

541. C. crus-corvi.

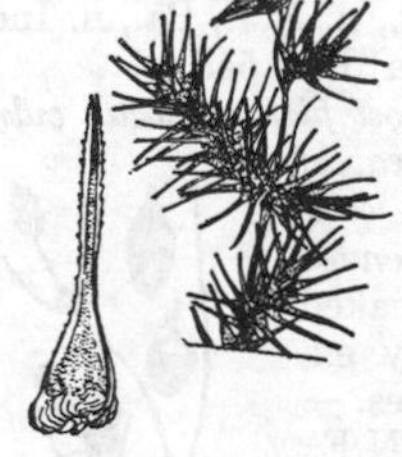
542. C. Bayardi.

46. C. Bayárdi Fern. (for one of its discoverers, Bayard Long, 1885–). — Resembling no. 45, *heavily white-glaucous; leaves* firmer; *the inner band firm, with few or no purple dots, the orifice thickened; panicle whitish- or bluish*-green to -drab; scales white; *the white and strongly bulbous base of the short-stipitate perigynium nerveless or scarcely nerved;* teeth of beak shorter. (*C. crus-corvi,* var. *virginiana* Fern.; *C. virginiana* Fern., not Woods) — Calcareous bottomlands, se. Va. to ne. S.C. June. Fig. 542.

§ 13. Heleonástes Kunth (see p. 295)

a. Perigynia ellipsoid-ovoid, nearly terete, unequally biconvex, rounded to summit, subcoriaceous, brown and lustrous when ripe, 1–few in each remote spike; staminate flowers terminal; loosely cespitose to substoloniferous delicate plant. . . . . *47. C. disperma.*

*a.*Perigynia compressed, planoconvex, membranaceous; staminate flowers basal or scattered. . . *b.*

*b.*Inflorescence flexuous; the 1 or 2 uppermost spikes divaricate-pedunculate, raised high above the lower one, the latter subtended by a prolonged bract many times its length; spikes loosely 1–5-flowered; perigynia 2.5–4 mm. long, appressed-ascending; plant delicate, loosely subcespitose. . . 48. *C. trisperma.*

*b.*Inflorescence straightish, continuous, with prolonged basal bract only occasional in no. 54; spikes (3–)6–30-flowered; perigynia 1.5–3.5 mm. long. . . *c.*

*c.*Head of 2–4 subglobose closely approximate silvery loosely-flowered spikes, subglobose to ovoid or ellipsoid, 0.5–2 cm. long; perigynia beakless, 3–3.5 mm. long; plant loosely cespitose. . . . . 49. *C. tenuiflora.*

*c.*Head elongate, slender to thick-cylindric, of 2–many spikes, 1–15 cm. long; perigynia definitely (often short-) beaked, 1.5–3.5 mm. long; plants more densely cespitose. . . *d.*

*d.*Perigynia broadest near the middle, gradually tapering to base; spikes 2–8(–11), if more than 4 not strongly imbricated in the head.

Spikes 2–4, approximate; scales castaneous or rufescent. . . . . 50. *C. bipartita.*

Spikes 3–11, all or all but the uppermost segregated.

Culms smooth; leaves yellowish-green; terminal spike with prolonged staminate base; perigynia closely covered by the obtuse fulvous scales; plant halophytic. . . . . 51. *C. Mackenziei.*

Culms scabrous above; leaves glaucous to deep green; terminal spike not greatly prolonged at base; perigynia broader and longer than the pale scales; not halophytic.

Glaucous; leaves 2–4 mm. broad; spikes elongate (rarely subglobose), 4–12 mm. long, with 10–30 appressed perigynia; these pale green to whitish-brown, smooth or at most sparsely serrulate at base of the inconspicuous beak; plant chiefly paludal. . . . . 52. *C. canescens.*

Green; leaves 1–2.5 mm. broad; spikes ellipsoid to subglobose, 3–7 mm. long, with 5–10 soon loosely spreading perigynia; these green, becoming brown, with well-developed serrulate beak; mostly in drier habitats. . . . . 53. *C. brunnescens.*

*d.*Perigynia broadest near base, ovate, rounded to subcordate at base, long-beaked; spikes mostly 5–15 or more, usually overlapping or crowded in a dense oblong head. . . . . 54. *C. arcta.*

47. C. dispérma Dew. (two-seeded). — *Exceedingly slender*, 0.5–6 dm. high, loosely tufted or slightly stoloniferous; leaves flat, soft and weak, mostly shorter than the culm; *spikes with* 1–3 (*or the terminal with* 3–6) *perigynia and* minute *staminate flowers at summit, scattered* at the summit of the subcapillary culm, the bracts obsolete or the lowest very short; *perigynia plump, ellipsoid-ovoid, nearly terete,* 2–2.8 mm. long, *rounded at summit to minute entire beak,* submembranaceous, becoming brown and lustrous, finely nerved, longer than the white scale, usually at length splitting and exposing the dark achene. (*C. tenella* Schkuhr, not Thuill.) — Mossy or damp woods and clearings, s. Lab. to Alaska, s. to Nfld., N.S., N.E., n. N.J., Pa., n. Ind., Wisc., Minn., S.D., N.M. and Calif. May–Aug. (Eurasia) Fig. 543.

543. C. disperma.

48. C. trispérma Dew. (three-seeded). — Loosely cespitose; the *almost filiform weak culms* 2–7 dm. long, usually *much overtopping the soft narrow* (1–2 *mm. broad*) *leaves;* the 2 or 3 *spikes with* 2–5 *perigynia,* staminate at base, *the terminal* 1 *or* 2 *spikes on a divergent peduncle-like summit of the culm, the lower one subtended by a very prolonged bract many times its length;* the finely many-nerved beaked elliptic-oblong *perigynia* 3.3–4 *mm. long,* 1.6–1.8 mm. broad, slightly exceeding the ovate-oblong pale obtuse to mucronate-acuminate scales. — Mossy woods, clearings and bogs, s. Lab. to Sask., s. to Nfld., N.S., N.E., n. N.J., Md., mts. to w. N.C. and e. Tenn., W.Va., O., n. Ind., Ill. and Minn. June–Aug. Fig. 544.

544. C. trisperma.

Var. **Billíngsii** Knight (named in 1906 for Francis M. Billings). — Leaves nearly setaceous, 0.2–0.5 mm. wide; the 1 or 2 spikes with 1–3 perigynia 2.5–3.3 mm. long. — In sphagnum, Nfld. to Ont., s. to N.S., N.E., N.J. and Pa.

× **C. tríchina** Fern. (hair-like), an evident hybrid of nos. 48 and 49, is occasional from s. Lab. to n. Me. and Mich.

49. C. tenuiflòra Wahlenb. (thin-flowered). — Loosely cespitose; the firm subcapillary

545. C. tenuiflora.

culms 1–6 dm. high, mostly exceeding the very narrow (0.5–2 mm. broad) pale green leaves; *spikes* 2–4, *subglobose, loosely* 3–10-*flowered, silvery, closely approximate in an ovoid, ellipsoid or subglobose head* 0.5–2 *cm. long;* perigynia plump, planoconvex, elliptic-oval or -oblong, 3–3.5 mm. long, 1.5–1.7 mm. broad, pale and finely puncticulate, almost beakless, about equaled by the ovate or ovate-oblong whitish scale. — Bogs and mossy woods or pond-margins, Nfld. and e. Saguenay Co., Que., to Alaska, s. to N.B., centr. Me., w. Mass., N.Y., Mich., Wisc., Minn., Man., Sask. and s. Alta. June–Aug. (Eurasia) Fig. 545.

50. C. bipartìta Bellardi (two-parted; in reference to the style). — Cespitose; *culms* smooth below, scabrous at summit, 0.5–3 *dm. high*, capillary; leaves mostly much shorter than culms, flat to involute; *spikes* 2–4, *castaneous or fulvous, approximate or imbricated in a head* 0.7–2 *cm. long*, ellipsoid to rounded, *the lower spikes* 4–10 *mm. long and* 2.5–5 *mm. thick*, the terminal, including the slender staminate base, longer; *perigynia pale brown to reddish*, ellipsoid to ovoid or fusiform, 1.3–3.8 mm. long, *beaked*, faintly nerved to nerveless; *the scales ferruginous to rufescent, with pale margins.* — Very variable arct. and boreal species, represented with us by 3 vars.:

Plant relatively stiff and upright, with flat leaves 1–3 mm. broad; culms obtusely angled below; perigynia ellipsoid-lanceolate to broadly obovoid, reddish-brown, abruptly beaked, 2.5–3.8 mm. long; scales obtuse; plant of fresh habitats. . . . . *C. bipartita* (typical).

Plant relatively weak, the involute or narrow and flaccid leaves 0.5–1.5 mm. broad; culms more acutely angled, curving or loosely reclining; perigynia olivaceous or pale brown, 1.3–3 mm. long; scales obtuse or acutish; halophytic.
  Perigynia broadly ellipsoid, ovoid or obovoid, 1.3–1.9 mm. long, abruptly beaked. . . . . Var. *amphigena.*
  Perigynia fusiform, 2.5–3 mm. long, tapering to slender beak. . . . . . Var. *glareosa.*

C. bipartìta (typical), *C. bipartita* of Allioni's descr., not of most auth.; *C. Lachenalii* Schkuhr; *C. lagopina* Wahlenb. — Arct. reg., s. to turfy slopes and alpine ravines of n. Nfld., Shickshock Mts., Gaspé Pen., Que., and Mont. Late July–early Sept. (Eurasia) Fig. 546.

546. C. bipartita (typical).

Var. **amphígena** (Fern.) Polunin (of both hemispheres), *C. glareosa*, var. *amphigena* Fern.; *C. marina* sensu Mackenz., not Dew. — Greenl. and e. Arct. Am., s. on saline or brackish shores to nw. Nfld., e. Que., ne. N.B., St. Lawrence R. w. to Temiscouata Co., Que., and James Bay; Alaska. Late June–early Aug. (Eurasia) Fig. 547.

547. C. bipartita, v. amphigena.

Var. glareòsa (Wahlenb.) Polunin (gravelly), *C. glareosa* Wahlenb. — Brackish shores, Baffin I., s. to n. Nfld., lower St. Lawrence R., Que., and James B.; Alaska. (Eu.)

51. C. Mackénziei Krecz. (in honor of Kenneth Kent Mackenzie, 1877–1934). — *Glaucous- to yellow-green, loosely cespitose and stoloniferous;* the *smooth and soft* nearly filiform *culms* 1–4.5 dm. high, mostly overtopping the soft and flat narrow (1–2.5 mm. broad) leaves; *head* 1.5–5.5 *cm. long, of* 2–6 *scattered ovoid to thick-cylindric fulvous spikes*, the lower spikes 5–12 mm. long and rounded to the base, *the terminal one with prolonged staminate base; perigynia* fusiform-ovoid, faintly nerved, 2.5–3.3 mm. long, 1.6–2 mm. broad, *conic-rostrate*, usually abruptly contracted to substipitate base, *closely covered by the obtuse fulvous scales.* (*C. norvegica* Willd., not Retz.) — Saline or brackish marshes and shores, sw. Greenl. and Lab. to Me.; Hudson Bay; Alaska. June–Aug. (Eurasia) Fig. 548.

548. C. Mackenziei.

× **C. pseudohélvola** Kihlm. (false *C. helvola*, yellowish), *C. helvola* of ed. 7, not Blytt, is a sterile hybrid of nos. 51 and 52, occasional from se. Lab. Pen. to Me. (Eu.)

52. C. canéscens L. (canescent). — *Glaucous*, cespitose, 1.5–8 dm. high, *the soft culms scabrous above* and mostly *overtopping the soft and flat pale leaves; head* 2–15 *cm. long, with at least the lower* and sometimes all the 4–8 *whitish to silvery-brown appressed-ascending subcylindric, ellipsoid or ovoid spikes distant or apart; these* 4–12 *mm. long,* 10–30-*flowered, the terminal one with staminate base not prolonged; perigynia* ovoid-oblong, pale, 1.8–3 mm. long, *smooth or at most sparsely serrulate at base of the inconspicuous beak*, more or less nerved on both faces, somewhat exceeding the ovate pointed scale. — Highly variable circumboreal species; 3 confluent vars. with us:

Spikes oblong-ovoid to cylindric, 6–12 mm. long; perigynia 2.3–3 mm. long, 1.3–1.7 mm. broad, often serrulate near summit.
Head 2–7 cm. long; all but the lowest spikes approximate or but slightly remote, the latter 0.5–2.5 cm. apart. . . . . *C. canescens* (typical).
Head 6–15 cm. long, all but the uppermost spikes distant, the two lower 2–4 cm. apart. Var. *disjuncta*.
Spikes short-ovoid to subglobose, 4–7 mm. long; perigynia scarcely or barely 2 mm. long, often quite smooth; head 2–5 cm. long. . . . . Var. *subloliacea*.

C. canéscens (typical). — Swamps, shallow pools and swales, Greenl. and Lab. to Alaska, s. to Nfld., N.S., N.E., n. N.J., N.Y., W.Va., O., Ind., Wisc., Minn., Ariz. and Calif. May–Aug. (Eurasia) Fig. 549.

Var. **disjúncta** Fern. (separated). — Nfld. to Minn., s. to N.S., N.E., L.I., Va., O. and Ind. Fig. 550.

Var. **subloliàcea** Laestad. (somewhat like *C. loliacea*). — Lab. to n. Alta., s. to Nfld., N.S., N.E., N.J., Ind., Wisc., Minn., Wyo., Id. and Wash. (Eurasia) Fig. 551.

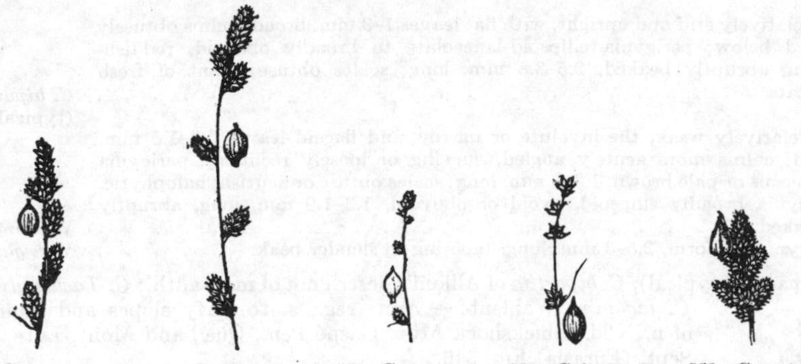

549. C. canescens (typical). 550. C. canescens, v. disjuncta. 551. C. canescens, v. subloliacea. 552. C. brunnescens, v. sphaerostachya. 553. C. arcta.

53. C. brunnéscens (Pers.) Poir. (brownish). — More slender and *greener* than no. 52; *leaves* firm, erect, 1–2.5 *mm. broad; culms* nearly filiform, *rather stiff*, 1–3(–5) dm. high, scabrous above; *head straight and stiffish*, 1.5–3.5 *cm. long, with* approximate to slightly distant short-oblong, *ellipsoid or subglobose greenish to brown loosely* 5–10-*flowered spikes* 3–7 *mm. long, the lower spikes only* 1–10 *mm. apart; perigynia loosely spreading when mature, green to pale brown*, 2–2.7 mm. long, 1–1.5 mm. broad, *serrulate at base of the distinct beak*, exceeding the acute ovate brownish to silvery scales. — Greenl. and Lab. to Alaska, s. on rocky or turfy summits and slopes to Nfld., Que., mts. of n. N.E. and n. N.Y., Wyo., Utah and Wash. Late June–Aug. (Eurasia)

Var. **sphaerostáchya** (Tuckerm.) Kükenth. (round-spiked). — *Lax*, with soft or flaccid prolonged leaves; the filiform *culms loosely ascending to arching*, 0.2–1.2 *m. high; spike flexuous*, 2–7 cm. long, *with the subglobose* greenish *spikes nearly all distant, the lowest* 1–2.5 *cm. apart.* — Dry or moist woods, clearings and rocky slopes, s. Lab. to B.C., s. to Nfld., N.S., N.E., N.J., Pa., upland to N.C. and Tenn., O., Mich., Wisc., Minn., Colo. and Oreg. Late May–Aug. (Eurasia) Fig. 552.

54. C. árcta Boott (contracted). — *Pale green* or somewhat glaucous; *culms very soft, in loose stools*, 1.5–6 dm. high, often overtopped by the *soft flat leaves* (2.5–4 mm. broad); head dense, oblong, with 5–15 ovoid or subcylindric spikes (6–11 mm. long); *perigynia cordate-ovate*, with a rather definite beak, strongly nerved on the outer, faintly on the inner face, 2–3 mm. long, 1.2–1.5 mm. broad, somewhat exceeding the acute often brown-tinged scales. — Wet woods, alluvial thickets, shores and swales, Gaspé Pen., Que., to B.C., s. to N.B., n. N.E., N.Y., s. Ont., n. Mich., Wisc., Minn., Mont., Id. and n. Calif. June–Aug. Fig. 553.

§ 14. Deweyànae Tuckerm. (see p. 295)

Leaves 1–2.5 mm. wide; spikes lance-cylindric; perigynia 1–1.3 mm. broad, strongly nerved; scales oblong. . . . . 55. *C. bromoides*.
Leaves 2–5 mm. wide; spikes ovoid or ovoid-cylindric; perigynia 1.6–1.9 mm. broad, faintly nerved or nerveless; scales ovate. . . . . 56. *C. Deweyana*.

55. C. bromoìdes Schkuhr (like *Bromus*). — *Very slender* and lax, *green*, scarcely glaucous; the culms 3–8 dm. long, mostly exceeding the soft flat leaves; *inflorescence loosely subcylindric*, 2–5.5 cm. long, of 2–6 *approximate or slightly scattered lance-cylindric spikes* (0.5–2 *cm. long*); beak of the *perigynium* one-half to two-thirds as long as the *strongly nerved* body, slightly *exceeding* the *oblong pointed scale*. — Rich low woods and swamps, Fla. to La., n. to w. N.B., n. N.E., s. Que., s. Ont. and Wisc.; also Hidalgo, Mexico. May–July. Fig. 554.

554. C. bromoides.

56. C. Deweyàna Schwein. (for Chester Dewey, 1784–1867). — Very lax, *glaucous;* the culms 2–12 dm. long, much exceeding the soft flat leaves; *inflorescence flexuous*, 2–6 cm. long; the 2–7 *spikes* 3–12-*flowered* (5–12 *mm. long*), the upper subapproximate or scattered; the *lowest very remote* (1–3 *cm. apart*), usually subtended by an elongate slender bract; beak about half as long as the body of the perigynium, somewhat exceeding the *ovate acuminate or short-cuspidate pale scale*. — Rich open woods and banks, s. Lab. to B.C., s. to Nfld., N.S., N.E., Pa., O., Mich., Wisc., n. Ia., S.D., Colo. and Id. May–Aug. Fig. 555.

555. C. Deweyana.

Var. **collectànea** Fern. (gathered together). — Head ellipsoid, only 1–3 cm. long, the 2–4 spikes contiguous, the lowest only 2–10 mm. apart. — Gaspé Pen., Que. — Closely simulating *C. leptopoda* Mackenz. of w. N.Am.

§ 15. Stellulàtae Kunth (see p. 295)

*a.*Spike 1, terminal, pistillate, staminate or pistillate above and with a linear-clavate prolonged staminate base, only exceptionally with 1–several small lower spikes; densely cespitose, with rigid involute leaves and stiff filiform culms, the soon divergent plump perigynia about 3 mm. long. . . . . 57. *C. exilis.*

*a.*Spikes normally 2 or more; leaves not rigid. . . *b.*

*b.*Perigynia broadest near base, with serrulate beaks (if broadest at middle, in no. 59, the spikes contiguous, the ascending perigynia linear-oblong, the fulvous scales obtuse). . . *c.*

*c.*Beak of perigynium (2.2–3.3 mm. long) only minutely notched, not conspicuously bidentate at tip, at most one-third as long as body; culms slender, often capillary; leaves 0.25–3 mm. wide.

Perigynia stramineous, appressed-ascending in maturity, linear- to lance-oval or slenderly fusiform, scarcely 1 mm. broad; achene linear or narrowly oblong; local plants of n. Me.

Scales acute or acuminate, nearly covering the weakly nerved perigynia; perigynia quickly rupturing and exposing the achene. . . 58. *C. elachycarpa.*

Scales rounded, much shorter than the strongly nerved firm and unrupturing perigynia. . . . . 59. *C. Josselynii.*

Perigynia green, becoming brown, widely spreading or recurving in maturity, oblong-ovate to deltoid, 1–2 mm. broad; achene ovate to suborbicular; wide-ranging.

Leaves flat or convolute, 1–3 mm. broad, firm, mostly shorter than the erect or ascending firm culms; scales brown or fulvous, with pale margin; inner face of perigynium nerveless or weakly nerved. . . . . 60. *C. interior.*

Leaves soon convolute, 0.5–1 mm. broad, flaccid, mostly equaling or exceeding the loosely ascending to reclining culms; scales whitish; inner face of perigynium strongly nerved. . . . . 61. *C. Howei.*

*c.*Beak of perigynium (2.5–4 mm. long) sharply bidentate at tip. . . *d.*

*d.*Midribs of scales evanescent below the flat hyaline tip.

Culms filiform, 0.5–1.5 mm. thick (in dry specimens) at base, 1.5–6 dm. high; leaves 1–2.5 mm. broad; perigynia narrowly ovate, stramineous or brownish, 0.8–1.8 mm. broad, the slender beak one-third to two-thirds as long as body.

Spikes variously staminate, mixed or pistillate; scales castaneous; the middle and upper ones acuminate, nearly as long as perigynia; perigynia 2.5–3 mm. long; calcicolous species. . . . . 62. *C. sterilis.*

Spikes pistillate, with staminate flowers at base; scales pale, blunt, about one-half as long as perigynia; perigynia 2.5–3.5 mm. long; oxylophytic species. . . . . 63. *C. echinata.*

Culms coarser, 1.5–3.5 mm. thick at base, mostly becoming 4–12 dm. high; leaves 1.5–5 mm. wide; scales pale brown to whitish; perigynia ovate to broadly cordate-deltoid, green or greenish, yellowish only in full maturity, mostly 1.2–2.5 mm. broad, tapering to a thick beak one-fourth to one-half as long as body; oxylophytic.

Spikes with 8–40 perigynia; scales about one-half as long as perigynia; perigynia ovate to broadly cordate-deltoid, mostly 2–2.5 mm. broad, the inner face prominently nerved. . . . . . 64. *C. atlantica.*

Spikes with 6–15 perigynia; scales about two-thirds as long as perigynia; perigynia ovate, 1.2–2 mm. broad, the inner face nerveless or nearly so. . . . . 65. *C. Wiegandii.*

d. Midribs of scales prominent to (or essentially to) the tip.

Well developed perigynia plump and firm, green, becoming dull brown, 2.5–3 mm. long, 2–2.5 mm. broad, the reniform to suborbicular body nearly or quite as broad as long, short-beaked. 66. *C. incomperta.*

Well developed perigynia thinner, submembranaceous, becoming stramineous to yellow-brown, 2.5–4 mm. long, at most 2 mm. wide, the ovate to lanceolate body definitely longer than broad, long-beaked.

Leaves 2.5–4 mm. broad; scales much shorter than body of perigynium; plant of southern mts. . . . . 67. *C. Ruthii.*

Leaves 0.75–2.5 mm. broad; scales equaling or exceeding body of perigynium; transcontinental mostly lowland plants.

Spikes 3–7, approximate or remote in a head 2–7.5 cm. long; perigynia ovate, 2.7–4 mm. long, up to 2 mm. broad, nerved on inner face; ligule longer than broad. . . . . . . 68. *C. cephalantha.*

Spikes 2–5, approximate (rarely remote) in a head 1–3 cm. long; perigynia lanceolate to lance-ovate, 2.5–3.5 mm. long, 1–1.4 mm. broad, nerveless on inner face; ligule as broad as long. . 69. *C. angustior.*

b. Perigynia broadest near middle, elliptic-ovate, thin, conspicuously nerved, tapering to a narrow substipitate base and a very short smooth beak; weak plant with soft culms and mostly remote green spikes. . . . . 70. *C. seorsa.*

57. C. exilis Dew. (meagre). — *In dense* brown-based *tussocks, with wiry involute leaves and* mostly overtopping *subrigid culms* 1.5–7 dm. high; *spike solitary* (very rarely 1–several smaller ones), 1–3 cm. long, either *wholly staminate* and slender, *wholly pistillate* and thick-cylindric, *or gynecandrous;* pistillate scales broadly ovate, acutish; *perigynia* soon divergent, *ovate-lanceolate,* about 3 mm. long, *with serrulate thin margins, strongly convex on the outer, flattish and few-nerved on the inner face.* — Peaty bogs and swales, locally abundant, se. Lab. to James Bay, s. to Nfld., N.S., Me., Mass., R.I., Del., N.Y., s. Ont. and n. Mich. May–Aug. Fig. 556. — Hybridizes with no. 62.

556. C. exilis.

58. C. elachycárpa Fern. (small-fruited). — Densely cespitose; the *wiry compressed-capillary culms* 2–5.5 dm. high, *strongly scabrous above,* much taller than the flat (1–2 mm. wide) leaves; head slenderly cylindric, 1–2.5 cm. long, of 2–7 *appressed-ascending yellow-green and brown* staminate (clavate), mixed or pistillate (ovoid) small *spikes* 3–5 *mm. thick; scales* dull brown, ovate, *acuminate or acute, nearly covering the appressed-ascending perigynia; perigynia lanceolate or lance-oblong,* 2.5–3 *mm. long, less than* 1 *mm. broad, weakly nerved,* with very short barely notched beak, *soon rupturing and exposing the pale oblong achene.* (*Kobresia* Fern.) — Springy calcareous shores, Aroostook R., Me. June, July. Fig. 557.

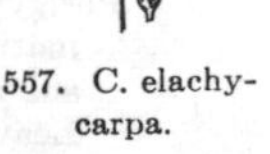

557. C. elachycarpa.

59. C. Josselýnii (Fern.) Mackenz. (in honor of John Josselyn, first Maine botanist, author of *New England's Rarities* in 1672; and of the Josselyn Botanical Society of Maine). — Habitally like no. 58; *culms subterete, smooth except at tip;* leaves longer, spikes 2–5, pistillate except at base; *scales rounded, narrower and much shorter than perigynia; perigynia fusiform to slenderly ovate, plump, firm, not rupturing,* green, becoming stramineous, 2.5–3.3 mm. long, barely 1 mm. broad, *strongly nerved.* — Meadows and damp shores, St. John R., n. Me. Late June, July. Fig. 558.

558. C. Josselynii.

60. C. intèrior Bailey (inland). — Cespitose, the *erect or ascending firm slender culms,* 1.5–5 dm. high, sharply angled, smooth except at scabrous summit, mostly overtopping the *firm ascending flat or convolute leaves* (1–3 *mm. broad*)*;* head 1–2(–3) cm. long,

559. C. interior.

of 2–4(–6) distant or approximate subglobose green (finally brown) spikes about 4 mm. in diameter, the terminal spike often with prolonged clavate staminate base; *scales brown or fulvous, with broad scarious rounded summit, about half as long as body of perigynium; perigynia olive-green, becoming brown, plump and firm, oblong-ovate to deltoid, 2.25–3.25 mm. long, 1–2 mm. broad,* the broad *short beak merely notched,* the margin thickened, *nerveless* or essentially so, or definitely nerved in forma **keweenawénsis** (F. J. Herm.) Fern. (of Keweenaw Pen., Mich.), *on inner face,* finally wide-spreading or recurved; achene round-ovate. (*C. scirpoides* Schkuhr, not *C. scirpoidea* Michx.) — Damp or wet, often calcareous, soils, se. Lab. to B.C., s. to Nfld., N.S., N.E., Del., Pa., W.Va., O., Ind., Ill., Mo., Kans., N.M., n. Mex. and n. Calif. May–Aug. Fig. 559.

61. C. Hówei Mackenz. (in honor of Eliot Calvin Howe, 1828–1899). — More lax and slender than no. 60, the *flaccid arching or reclining culms* up to 7.5 dm. long and *often equaled or exceeded by the weak slender* (0.5–1 *mm. wide*) *soon convolute leaves;* head 1–2.5 cm. long, of 2–4 (or 5) mostly separated spikes; *scales whitish; perigynia strongly nerved on the inner face.* (*C. scirpoides,* var. *capillacea* (Bailey) Fern.) — Sphagnous or mossy swamps, thickets and woods, Fla. to La., n. to N.S., s. Me., s. N.H., (? Vt.), N.Y., O. and s. Mich. Late April–early Aug. Fig. 560.

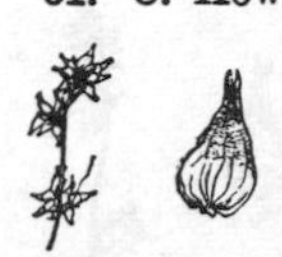

560. C. Howei.

62. C. stérilis Willd. (sterile). — Habit of no. 60; the firm filiform culms 0.5–1.5 mm. thick at base, 1.5–6 dm. high; leaves shorter, canaliculate to involute, firm, 1–2.5 mm. broad; *spikes sometimes staminate and slender, or mixed or even wholly pistillate,* 3–6, approximate or scattered in slender heads 2–3 cm. long; *pistillate scales castaneous, subcoriaceous, the middle and upper ones acuminate and nearly as long as the perigynia, the midrib obscure,* the margins hyaline; *perigynia* stramineous, narrowly ovate, 2.5–3 *mm. long,* about 1.5 mm. wide, firm, with slightly elevated margins, slightly nerved on both faces, the slender sharply bidentate beak shorter than the body. — Wet calcareous soils, Nfld. to s. Ont. and Minn., s. to M.I., w. Ct., n. N.J., Pa., O., Ind. and Ill. Late May–Aug. Fig. 561.

561. C. sterilis.

63. C. echinàta Murr. (prickly). — Similar to no. 60; spikes mostly pistillate, more echinate from spreading of perigynia; *scales paler, blunt, about half as long as the perigynia; the latter 2.5–3.5 mm. long, nerveless or nearly so on the inner face.* (*C. Leersii* Willd.; *C. stellulata* Good.) — Greenl. and Lab., s. in peaty soils to Nfld., e. Que. and n. Ont. Late July–Sept. (Eurasia) Fig. 562.

562. C. echinata.

64. C. atlántica Bailey (Atlantic). — Rather *coarse,* the *stiff culms* 1.5–3.5 *mm. thick at base,* at first overtopped by the leaves, soon lengthening to 4–12 *dm. high,* harsh above; *leaves* pale green, firm, much prolonged, 1.5–4 *mm. wide,* scabrous above; *heads* 2–6 cm. long, *of* 3–6 somewhat separated to approximate echinate *hard 8–40-flowered olivaceous spikes mostly* 7–12 *mm. long and* 6–8 *mm. thick; scales pale brown to whitish, about one-half as long as the perigynia, obtuse to subacute,* the midnerve not reaching the tip; *perigynia soon strongly spreading, greenish,* becoming yellowish in full maturity, *plump, strongly spongy at base, ovate to broadly cordate-deltoid,* 2.5–3.3 mm. long, 2–2.5 *mm. broad,* tapering to a short thick bidentate beak, *strongly nerved on both faces.* (*C. sterilis* in large part of ed. 7, not Willd.) — Mossy or peaty soils, Fla. to Tex., n. to N.S., e. N.B., se. and s. Me., s. N.H., Mass., e. N.Y., N.J. and D.C. May–early Aug. Fig. 563.

563. C. atlantica.

65. C. Wiegándii Mackenz. (in honor of Karl McKay Wiegand, 1873–1942). — Similar to no. 64; *culms smoother and softer, pressing flat; leaves thin,* 2–5 mm. broad, *smooth* except at tip and sometimes the margins; *spikes only 6–15-flowered,* rather smaller; *scales about two-thirds length of perigynia; the latter* ovate, thinner, 1.2–2 *mm. wide, the inner face nerveless or nearly so.* (*C. sterilis* in part of ed. 7, not Willd.) — Boggy or peaty soils, Nfld. to Ont., s. to N.S., Me., ne. Mass. and n. N.Y. June–Aug. Fig. 564.

564. C. Wiegandii.

66. C. incompérta Bickn. (undiscovered). — Resembling slender states of no. 64, weaker, with very *slender culms smooth* except at summit, 0.2–1 m. high; *leaves* shorter than to exceeding culms, *smooth except on margin and at tip,* 1.5–2.5 mm. wide; *heads* 1–5 cm. long, *usually moniliform, with* 2–4(–5) subglobose *green* (finally brown) *6–15-flowered spikes* 4–5 mm. thick; *scales whitish to pale brown,* ovate, *acute, barely as long as body of perigynium,*

565. C. incomperta.

the keel-like green midrib prominent to the tip; perigynia soon widely divergent to recurving, *green* (becoming dull brown), very plump at base, *firm,* 2.5–3 *mm. long,* 2–2.5 *mm. broad, the reniform to suborbicular body nearly or quite as broad as long, short-beaked, strongly nerved on both faces,* the beak sharply bidentate. — Swamps, wet woods and damp peats, Fla. to Tex., n. to sw. Me., s. N.H., Mass., N.Y., ne. O. and s. Mich. May–early Aug. Fig. 565.

67. C. Rùthii Mackenz. (for its discoverer, Albert Ruth, 1844–1932). — Resembling no. 68, stouter; *culms about* 3 *mm. thick at base,* rather soft (flattened in drying), 3–6 dm. high; *leaves* flat, thin, pale green, 2.5–4 *mm. wide,* scabrous above, the firm inner band of the sheath with cartilaginous summit; head moniliform, 3.5–7.5 cm. long, of 4–7 ellipsoid-globose echinate 5–15-flowered pale green spikes 5–7 mm. thick; *scales* pale brown, membranous, narrow, *much shorter than body of the perigynium,* the prominent green midrib extending to or beyond the tip; *perigynia* narrowly ovate, 2.8–3.8 mm. long, the yellowish body *nerveless on the inner face,* the slender green bidentate beaks wide-spreading or recurving. — Brooksides and woods in the mts., sw. Va. to Ga. and e. Tenn. May–July. Fig. 566.

566. C. Ruthii.

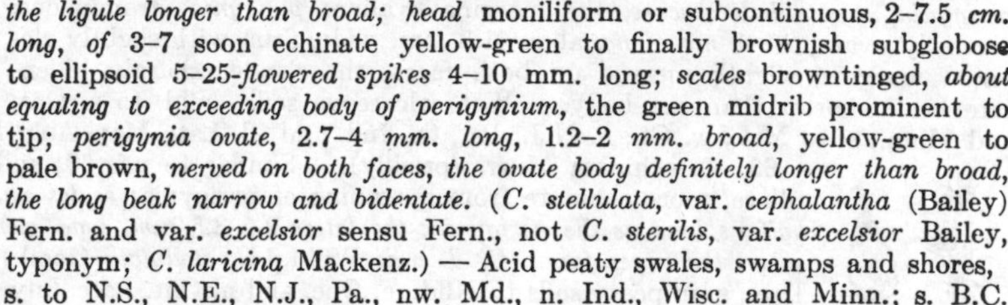

567. C. cephalantha.

68. C. cephalántha (Bailey) Bickn. (with flowers in a head). — Densely cespitose; the *firm slender* sharply angled and somewhat scabrous *culms* 0.2–1.4 m. high, finally *exceeding the firm narrow* (1.5–2.5 *mm. wide*) *scabrous* flat to canaliculate *leaves;* sheaths not strongly thickened at orfícei, *the ligule longer than broad; head* moniliform or subcontinuous, 2–7.5 *cm. long, of* 3–7 soon echinate yellow-green to finally brownish subglobose to ellipsoid 5–25-*flowered spikes* 4–10 mm. long; *scales* browntinged, *about equaling to exceeding body of perigynium,* the green midrib prominent to tip; *perigynia ovate,* 2.7–4 *mm. long,* 1.2–2 *mm. broad,* yellow-green to pale brown, *nerved on both faces, the ovate body definitely longer than broad, the long beak narrow and bidentate.* (*C. stellulata,* var. *cephalantha* (Bailey) Fern. and var. *excelsior* sensu Fern., not *C. sterilis,* var. *excelsior* Bailey, typonym; *C. laricina* Mackenz.) — Acid peaty swales, swamps and shores, Nfld. to Ont., s. to N.S., N.E., N.J., Pa., nw. Md., n. Ind., Wisc. and Minn.; s. B.C. and Wash. June–Aug. Fig. 567.

69. C. angústior Mackenz. (more slender). — Smaller than no. 68; the very slender culms 1–3 (rarely –6) dm. high; leaves 0.75–2 mm. broad, *the ligule as broad as long; head* 1–2(–3) *cm. long,* of 2–5 approximate to remote finally echinate subglobose to short-ellipsoid *spikes* 4–6 *mm. long; perigynia lanceolate to lance-ovate,* 2.5–3.5 *mm. long,* 1–1.4 *mm. broad, nerveless on the inner face.* (*C. stellulata,* var. *angustata* Carey) — Swales, swampy thickets and shores, Lab. to n. Ont., s. to Nfld., N.S., N.E., Del., Md., D.C., mts. to N.C. and Tenn., n. Mich., Wisc. and Minn.; s. B.C. to Colo., Nev. and Calif. June–early Aug. Fig. 568. — Individuals with spikes separated in the head have been called var. *gracilenta* Clausen & Wahl.

568. C. angustior.

569. C. seorsa.

70. C. seórsa Howe (separated). — Forming loose stools; the *soft and weak* (*flattened in drying*) *culms* 2–7.5 dm. high, *equaling or exceeding the thin pale leaves* (2–4 mm. wide); *head moniliform,* 2.5–7 cm. long, *of* 2–7 *green* rounded to ellipsoid 5–20-*flowered spikes* 3.5–6 *mm. thick,* the *terminal* one usually with long clavate staminate base; *scales pale,* rounded, *with green or brownish midrib keel-like to tip; perigynia green, broadest near middle, elliptic-ovate, thin,* wide-spreading, 2–3 mm. long, *conspicuously nerved, tapering to a narrow substipitate base and a very short smooth beak.* — Wet woods and swamps, Mass. to Ga.; also interior N.Y., n. O., n. Ind. and Mich. April–early July. Fig. 569.

§ 16. Ovàles Kunth (see p. 295)

a. Bracts wanting or setaceous, if broad at most twice as long as the head of distinct spikes; perigynia one-fifth to four-fifths as broad as long, with lanceolate to suborbicular bodies. . . *b.*

b. Tips of perigynia clearly exceeding their subtending scales, the margins of the perigynia exposed at least above. . . *c.*

c. Spikes 1.5–2.5 cm. long, long-cylindric, pointed; perigynia 7–10 mm. long, narrowly oblong-lanceolate, thin and scale-like. . . . . 71. *C. muskingumensis.*

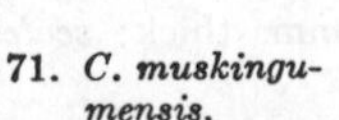

c.Spikes 3–15(–20) mm. long (if 1.5 cm. or more, plumper and with shorter and broader perigynia); perigynia 3–7.7 mm. long (if 7 mm. or more broadly ovate). . . *d*.

d.Perigynia at most 2 mm. broad. . . . *e*.

e.Perigynia thin and scale-like, barely distended over the achene.

Sheaths close; leaf-blades 1–3 mm. wide; those of the sterile culms few, ascending and near summit of culm; spikes stramineous or fulvous, lustrous; perigynia 4–7 mm. long, 1.2–2(–2.6) mm. broad, the wing extending continuously to base. . . . 72. *C. scoparia*.

Sheaths loose and loosely ribbed; leaf-blades 2.5–8 mm. wide; those of the numerous sterile culms abundant, spreading and somewhat distichous; spikes green or dull brown, opaque; perigynia 3–5 mm. long, 1–1.5 mm. broad, the wing abruptly narrowed above the base.

Inner band of the leaf-sheath with firm summit; head continuous or nearly so, not strongly flexuous; tips of perigynia closely appressed or strongly ascending. . . . . 73. *C. tribuloides*.

Inner band of the sheath with weak or friable summit; head commonly moniliform, at least below, flexuous; tips of perigynia loosely ascending to recurved. . . . . 74. *C. projecta*.

e.Perigynia thicker, firm, obviously distended over the achene. . . *f*.

f.Tips of mature perigynia rosulate-spreading, concealing the scales; spikes globose or subglobose; wing of perigynium abruptly contracted below the middle; leaves 3–7 mm. broad, with loose sheaths. . . . . 75. *C. cristatella*.

f.Tips of mature perigynia ascending to but slightly spreading, not concealing the scales; spike ovoid, obovoid or ellipsoid; wing of perigynium continuous; leaves 1–6 mm. broad, mostly with tight sheaths. . . *g*.

g.Perigynia narrowly lanceolate, 0.4–1 mm. broad, gradually tapering to long slender acute or subacute base and to prolonged slender beak. . . . . 76. *C. Crawfordii*.

g.Perigynia ovate or broadly lanceolate to obovate, 1–2 mm. (or more) broad, with rounded to broadly cuneate bases. . . *h*.

h.Perigynia ovate to broadly lanceolate, broadest below the middle or toward the base. . . *i*.

i.Heads of primary inflorescences compact, of crowded to approximate or merely separated spikes, not moniliform and flexuous (late culms in species under 2nd "*i*" may bear crowded spikes), 1–4 cm. long.

Perigynia nerveless or only faintly nerved at base on inner face, 2–4 mm. long, 1–2 mm. broad. . . . . 77. *C. Bebbii*.

Perigynia distinctly nerved on inner face (if nerveless in no. 87 the perigynium broader), 3–5 mm. long, the well developed ones 1.5–2 mm. broad (or broader).

Hyaline inner band of close leaf-sheath scarcely prolonged at summit beyond base of blade, the blade 2–4 mm. wide; scales exceeding body of loosely ascending perigynia; perigynia 3.5–5 mm. long.

Leaves of fertile culms 3 or 4; sterile leafy shoots none or exceptional; scales deep brown or fulvous, nearly as long as beak of perigynium; perigynia 1.7–2 mm. broad, clearly 5–7-nerved on inner face; northern species. . . . . 78. *C. tincta*.

Leaves of fertile culms 4–7; sterile shoots with approximate terminal leaves frequent; scales pale, barely reaching beak of perigynium; perigynia 2–3 mm. broad, nerveless or nearly so on inner face; more southern. . . . . 87. *C. molesta*.

Hyaline inner band of loose sheath prolonged at summit beyond base of leaf-blade; the latter mostly 3.5–6.5 mm. broad, 4–7 on the fertile culms; sterile culms with leaves subapproximate at summit frequent; scales pale, shorter than body of perigynium; perigynia greenish, 3–4 mm. long, 1.5–2 mm. wide, the beaks soon spreading or slightly recurving. . . . . 79. *C. normalis*.

i.Heads of primary inflorescences lax or moniliform (those of delayed or autumnal culms sometimes compact), 2.5–7 cm. long.

Well-developed leaves of fertile culms 4–7, mostly 3.5–6.5 mm. broad, the sheaths loose and relatively thick; sterile culms with leaves approximate at summit frequent. . . . . 79. *C. normalis.*

Well-developed leaves of fertile culms 3–5, 0.5–3.5 mm. broad, the slender sheaths tight; sterile leafy culms wanting.

Inner side of leaf-sheath with prolonged hyaline veinless band; scales merely acute; perigynia greenish, becoming stramineous, firm, well distended over the achene, 3–4.5 mm. long, 1.5–2 mm. broad, faintly 3–7-nerved or nerveless on inner face.

Moniliform head commonly nodding from above lowest spike; spikes (except terminal) rounded to tapering but not strongly clavate at base; perigynia 3–4.5 mm. long. . . . . 80. *C. tenera.*

Moniliform head nearly straight to arching, not strongly nodding; spikes with prolonged clavate staminate bases; perigynia 3–3.5 mm. long. . . 81. *C. festucacea.*

Inner side of leaf-sheath veiny nearly to summit; scales long-acuminate to awn-tipped; perigynia stramineous to brown; thin and almost scale-like, 4.8–6 mm. long, 2–3 mm. broad, clearly about 10-nerved on inner face. . . . . 93. *C. hormathodes.*

*h.*Perigynia obovate to rounded-oblong or elliptical, broadest near or above the middle.

Leaves rigid, strongly glaucous; the chartaceous summit of the inner band of the sheath much prolonged beyond the 2-auricled base of the blade; head moniliform, usually flexuous or nodding; spikes highly lustrous, silvery to pale brown; scales nearly covering perigynia; plant of dunes and maritime sands northeastw. . . . . . . 91. *C. silicea.*

Leaves softer and greener; hyaline summit of inner band of sheath little if at all prolonged above auricle-less base of blade; head continuous or but slightly interrupted, erect or arching; spikes opaque or but slightly lustrous, pale green, becoming pale brown; scales definitely exceeded by perigynia; wide-ranging, not of dune-sands.

Spikes opaque, mostly segregated in an interruptedly cylindric arching head 2–6 cm. long; scales acuminate, with pointed tips; body of perigynium obovate to suborbicular. . . . . 82. *C. albolutescens.*

Spikes sublustrous, crowded to subapproximate in an ovoid, rhomboid or subcylindric stiffly erect head 1–4 cm. long; scales obtuse to subacute; body of perigynium nearly elliptical. . . . . 83. *C. Longii.*

*d.*Perigynia more than 2 mm. wide. . . *j.*

*j.*Scales acute to blunt, without awn-tips. . . *k.*

*k.*Bases of leaf-blades without evident auricles, the blades not conspicuously glaucous; plants of various habitats, not primarily of sand-dunes. . . *l.*

*l.*Perigynia thin and scale-like, barely distended over the achene, 1.2–2.6 mm. broad, lanceolate to narrowly ovate. . . . . 72. *C. scoparia.*

*l.*Perigynia thicker and firmer, usually well distended over the achene (if thin and scale-like wider and broadly ovate to obovate). . . *m.*

*m.*Beak of perigynium short and broad (triangular), gradually tapering into the firm chartaceous green to pale brown broadly elliptic to rhombic or roundish body of the perigynium (3–4.5 mm. long, 1.8–3.4 mm. broad).

Leaf-sheaths tight, blades mostly 2–3 mm. broad; spikes ellipsoid to obovoid, gradually rounded or tapering at base; perigynia 1.8–2.5 mm. broad, nerved on inner face. 83. *C. Longii.*

Leaf-sheaths loose, blades mostly 3–6 mm. broad; spikes conic-ovoid, broadly rounded to subtruncate at base; perigynia 2–3.4 mm. broad, nerveless on inner face. . . 84. *C. cumulata.*

*m.*Beak of perigynium elongate and narrow above, more abruptly differentiated from the ovate, obovate or suborbicular body of the perigynium, the latter 3.5–7.7 mm. long and 1.5–5 mm. broad. . . *n.*

n. Perigynium pale green to dull brown, 3.5–4.5 mm. long, 1.5–2.5 mm. broad, its obovate to obovate-suborbicular body broadest above the middle. . . . . 82. ***C. albolutescens.***

n. Perigynium ferruginous to stramineous or greenish, 4–7.7 mm. long, 2–5 mm. broad, its ovate to orbicular or reniform body broadest at base or below the middle. . . *o.*

o. Perigynia firm, well distended over the achene, mostly opaque (except the wings) to transmitted light, 4–5.5 mm. long. . . *p.*

p. Well-formed perigynia 2–3.5 mm. broad, the transparent or subtranslucent nerveless or nerved wing about one-fourth as wide as either side of the opaque body; transcontinental or inland species. . . *q.*

q. Spikes tapering to subacute summit, 2–6, approximate or crowded, with appressed perigynia; scales acuminate; perigynia gradually tapering from the ovate body to a subacute base. . . 85. ***C. suberecta.***

q. Spikes rounded or obtuse at summit, 3–10, the tips of the perigynia spreading-ascending; scales blunt to acuminate; perigynia broadly rounded to subtruncate at base.

Spikes conical to slightly rounded at summit, narrowed below, often with clavate or turbinate base, 3–8 in an open often slightly moniliform head 2–4.5 cm. long; scales acuminate, nearly equaling beak of perigynium; body of firm pale brown perigynium broadly ovate to suborbicular, its wing rarely nerved. . . . . . . 86. ***C. brevior.***

Spikes broadly rounded at summit and base, the lateral ones without prolonged bases; scales blunter, reaching only to base of beak.

Head compact, 1–3 cm. long, of 2–5 (–"8") crowded or approximate whitish-green to pale drab spikes; perigynia firm, chartaceous, ovate, 2–3 mm. broad, pale green, becoming drab, their translucent narrow wings nerveless or rarely nerved; leaves mostly 2–3.5 mm. broad. . . . . 87. ***C. molesta.***

Head usually open and submoniliform, (2–)3–7.5 cm. long, of (3–)4–9 mostly separated yellow-green, stramineous or ferruginous spikes; perigynia thin, membranaceous, pale-stramineous, with broadly ovate to suborbicular body 2.5–3.5 mm. broad, the translucent wings often 1–2-nerved; leaves 3–4.5 mm. wide. . . . . 88. ***C. Merritt-Fernaldii.***

p. Well-formed perigynia 3.5–5 mm. broad, with reniform to orbicular bodies, the subtranslucent 1–2-nerved wing about one-third as wide as either side of the opaque body; species of southern Coastal Plain. . 89. ***C. reniformis.***

o. Perigynia membranaceous and thin, translucent, ovate, 5.5–7.7 mm. long, 2.7–4.8 mm. broad, pale stramineous. 90. ***C. Bicknellii.***

k. Bases of rigid glaucous leaf-blades with a pair of rounded auricles; silvery-lustrous to pale brown spikes in arching moniliform heads; the scales nearly covering the perigynia; plant of northeastern coastal sands. . . . . 91. ***C. silicea.***

j. Scales awn-tipped.

Leaf-blades harsh, mostly 2.5–5.5 mm. wide; spikes pale green to drab or pale brown, approximate; body of perigynium obovate or obovate-rotund, broadest near summit. . . . . . 92. ***C. alata.***

Leaf-blades smooth except toward tip, mostly 1–2.5 mm. broad; spikes ferruginous, warm-brown or stramineous, mostly scattered in flexuous heads; body of perigynium lanceolate to suborbicular, broadest near base.

Spikes rhomboid-ellipsoid, tapering gradually to summit and

base, the terminal tapering to short staminate base; perigynia lance-ovate, tapering gradually to appressed-ascending beaks. 93. *C. hormathodes.*

Spikes more rounded, the terminal broadly rounded to the prolonged staminate base; perigynia round-ovate to suborbicular, abruptly narrowed to loosely spreading-ascending beaks. . . 94. *C. straminea.*

b. Tips of perigynia equaled or exceeded by their subtending scales, the margins nearly or completely covered. . . *r.*

r. Bases of stiff very glaucous leaves with 2 rounded auricles; head arching or nodding, of silvery-brown or whitish clavate-based conic-ovoid spikes; plant of northeastern coastal sands. . . . . . . 91. *C. silicea.*

r. Bases of leaves not auricled, the blades neither very stiff nor strongly glaucous; spikes whitish-green or brownish to ferruginous or darker; plants rarely on coastal sands. . . *s.*

s. Head of approximate to crowded spikes, erect, 1–4 cm. long. . . *t.*

t. Perigynia lance-subulate, 0.7–1.5 mm. broad, tapering to subacute prolonged base and narrow beak, scarcely winged below middle; species of centr. Me. . . . . 95. *C. oronensis.*

t. Perigynia ovate, 1.7–3 mm. broad, winged to base.

Perigynia thin and scale-like, except over the achene, membranaceous, concave to flat on inner face, 1.7–2.5 mm. broad, with broad thin wing; achene 1–1.7 mm. broad.

Head irregularly globose-ovoid, 1–2 cm. long; scales castaneous or purple-black, with silvery scarious margins, obtuse; arctic-alpine species. . . . . 96. *C. macloviana.*

Head interruptedly subcylindric to ellipsoid or ovoid, 1.8–5 cm. long; scales pale brown or stramineous, acute or acuminate; lowland or Great Plains species.

Well-developed cauline leaves of fertile culms 4–7; sterile leafy tufts numerous; spikes fulvous, obtuse; perigynia greenish, becoming brown; achene about 1 mm. broad; eastern. . . . . 97. *C. leporina.*

Well developed cauline leaves of culms 2 or 3; sterile tufts few or none; spikes whitish-drab, acute; perigynia drab, yellow at base; achene 1.5–1.7 mm. broad; northwestern. . . . 98. *C. xerantica.*

Perigynia plump and heavy, coriaceous, flat to convex on the inner face, 2–3 mm. broad, the wing very narrow and firm; achene 1.8–2.1 mm. broad. . . . . 99. *C. adusta.*

s. Head moniliform and flexuous or at least with the lower spikes remote, 1.5–7 cm. long.

Spikes brown to ferruginous; perigynia nerveless or only short-nerved on inner face, 4–6.5 mm. long.

Inflorescence with olive-brown to bronze spikes; perigynia 4–5 mm. long, 1.9–2.7 mm. broad. . . . . 100. *C. aenea.*

Inflorescence with silvery-brown or pale ferruginous spikes; perigynia 4.5–6.5 mm. long, 1.5–2 mm. broad. . . . . . . 101. *C. praticola.*

Spikes silvery-green to whitish-brown; perigynia usually with strong nerves the entire length of the inner face, 3–4.5 mm. long. . . 102. *C. argyrantha.*

a. Bracts leaf-like and much prolonged, forming a conspicuous involucre, the lowest 1–2 dm. long; spikes crowded in a dense head; perigynia lance-subulate, about 5 mm. long, barely 1 mm. broad. . . . . . . 103. *C. sychnocephala.*

570. C. muskingumensis.

71. C. muskingumênsis Schwein. (of Muskingum River, Ohio). — Cespitose, 3.5–10 dm. high, *with numerous sterile shoots* and few fertile culms; *leaves of sterile culms very numerous, somewhat crowded and almost distichous, of fertile culms fewer* and more ascending, *light green, firm,* 3–7 *mm. wide, subcordate at junction with the loose sheath; head* 4–8.5 cm. long, *of* 5–13 *lance-cylindric pointed* appressed greenish *spikes* 1.5–2.5 *cm. long;* scales brownish, about half as long as perigynia; *perigynia flat and scale-like,* narrowly lanceolate, 7–10 *mm. long,* about 2.5 mm. broad, becoming stramineous. — Low woods, bottomlands and swamps, s. Ont. to e. Man., s. to Ky., Mo. and e. Kans. July, Aug. Fig. 570.

72. C. scopària Schkuhr (broom-like). — Cespitose, 0.2–1 m. high; culms slender, mostly erect, harsh near summit on the angles; *leaves* firm, yellow-green, 1–3 *mm. wide, those of the few sterile culms subterminal and ascending, the slender sheath close; head* compact to moniliform, *of* 3–12 *stramineous to fulvous lustrous* ascending ovoid to obovoid *spikes* 0.5–1.6

cm. long; scales brownish, with paler margins, shorter than the perigynia; *perigynia thin and scale-like, only slightly distended over the achene, stramineous to brown*, lanceolate to narrowly ovate, 4–7 *mm. long*, 1.2–2.6 *mm. broad, with thin wing continuous to base*, nerved on both faces. — Highly variable wide-ranging species, the following vars. and forms recognized:

a. Scales stramineous or pale brown, scarcely contrasted with the perigynia; perigynia lanceolate to lance-ovate, 4–7 mm. long, 1.2–2.6 mm. broad. . . *b.*

b. Spikes tapering to acute or subacute tips. . . *c.*

c. Tips of spikes merely acute or subacute.

Spikes contiguous to imbricated in the erect or but slightly arching head.

Head 1.5–5 cm. long, 0.7–1.5 cm. thick, the spikes all ascending. . *C. scoparia* (typical).

Head 1.5–3 cm. long, 1.3–2.5 cm. thick, some of the crowded spikes divergent. . . . . Forma *condensa.*

Spikes all or at least the lower ones remote in a flexuous loosely moniliform head 3.5–6 cm. long. . . . . Forma *moniliformis.*

c. Tips of approximate spikes prolonged, slender-acuminate to caudate. . Forma *peracuta.*

b. Spikes turbinate, broadly truncate at summit, in loose to moniliform heads 3–5 cm. long. . . . . Forma *subturbinata.*

a. Scales castaneous to blackish, strongly contrasted with perigynia; perigynia 4–5 mm. long, 2 mm. wide, the body elliptic-ovate; spikes crowded in a head 1.3–2.5 cm. long. . . . . Var. *tessellata.*

C. scopària (typical). — Moist to dry open grounds, thickets and open woods, Nfld. to B.C., s. to N.S., N.E., S.C., Tenn., Ark., N.M. and Oreg. May–Aug. Fig. 571. — Hybridizes with nos. 74, 77, 82 and 84. — Forma **condénsa** (Fern.) Kükenth. (condensed), Nfld. to Ont. and Minn., s. to N.S., N.E., N.C., Mich., and Neb. Fig. 572. — Forma **monilifórmis** (Tuckerm.) Kükenth. (like a chain of beads), Nfld. to B.C., s. to N.S., N.E., N.Y. and Va. — Forma **peracûta** Fern. (very acute), N.S. to Ont. — Forma **subturbinàta** (Fern. & Wieg.) Fern. (somewhat top-shaped), Nfld. to Mich., s. to N.S., N.E. and Tenn.

571. C. scoparia (typical).

572. C. scoparia, f. condensa.

Var. **tessellàta** Fern. & Wieg. (checkered). — N. S. to e. L.I.

73. C. tribuloìdes Wahlenb. (like *Tribulus*, the Caltrop). — Loosely cespitose; *culms* angled and harsh at summit, 0.3–1.2 m. high, *slightly exceeding to overtopped by* the slender pointed *leaves; the latter* with firm to weak *deep to pale green* blades 2.5–8 *mm. broad; the sheaths loose, prominently veiny, the veiny inner band with firm veinless summit; sterile culms numerous, with many spreading and somewhat distichous leaves;* fertile culms with 4–10 leaves; *head* continuous or interrupted at base, straight or curving, *pale green to drab, dull,* 2–6 cm. long, 0.7–2 cm. thick, *of* 6–15 ascending slenderly ellipsoid to obovoid *blunt spikes* 6–12 mm. long; scales pale, narrower and shorter than perigynia; *perigynia thin and scale-like*, lanceolate, *pale green to drab*, 3–5 *mm. long*, 1–1.5 *mm. broad, the narrow wing abruptly tapering below the middle.* — Bottomlands, swales and low woods, Fla. to La. and Okla., n. to N.B., s. Que., s. Ont., Mich., Wisc. Minn. and Neb. June–Sept. Fig. 573. — Hybridizes with the next.

573. C. tribuloides.

74. C. projécta Mackenz. (projecting). — Similar to no. 73, of laxer habit; the *inner band of the sheath of the flaccid leaves weaker, friable;* leaves of fertile culms 4–6; *head* moniliform and *flexuous* to dense, 3–8 cm. long, 0.5–1.5 cm. thick, *of* 6–15 ellipsoid-obovoid *green to stramineous or brownish loose spikes* 4–10 mm. long; *perigynia with loosely ascending, spreading or recurving tips.* (*C. tribuloides*, var. *reducta* Bailey) — Swales, thickets and damp woods, Nfld. to Man., s. to N.S., N.E., N.J., D.C., W. Va., O., n. Ind., n. Ill. and Mo.; s. B.C. June–Aug. Fig. 574. — Hybridizes with nos. 72 and 73.

574. C. projecta.

75. C. cristatélla Britt. (with small crests). — Habitally similar to nos. 73 and 74; *leaves*

575. C. cristatella.

soft and flat, 3–7 *mm. broad,* often equaling the culms, *with loose sheaths green-striate on inner band; head* slenderly cylindric to ellipsoid, 2–4 cm. long, 0.8–1.6 cm. thick, *of* 6–15 *crowded or approximate subglobose pale green to pale brown densely flowered rosulate spikes* 5–10 mm. in diameter, or spikes all distant in a flexuous moniliform head in forma **catellifórmis** (Farw.) Fern. (chain-like); scales pale, much narrower and shorter than perigynia; *perigynia* narrowly ovate, *firm and rather distended over the achene,* 3–4 *mm. long,* 1.2–1.5 *mm. broad, the narrow wing abruptly contracted above the base, in maturity with the recurving pale tips forming dense rosettes and hiding the scales.* (*C. cristata* Schwein., not Clairv.) — Swales, damp woods and bottoms, sw. Que. to N.D., s. to N.H., Del., Md., D.C., W.Va., Ky., Mo. and Neb. June–Aug. Fig. 575.

76. C. Crawfórdii Fern. (in honor of Ethan Allan Crawford, early settler in the White Mts.). — Cespitose, 1–8.25 dm. high, the slender erect sharply angled culms mostly exceeding the leaves; leaves 1–4 mm. broad, yellowish-green, the inner band of the close sheath hyaline; spikes brown to stramineous, slenderly ellipsoid, 3–11 mm. long, 3–15, approximate in a slenderly cylindric to thick-oblong or ellipsoid head 1–3 cm. long and 0.5–2 cm. thick; scales pale brown, dull, nearly as broad as perigynia but much exceeded by the prominent slender beaks; *perigynia narrowly lanceolate, very narrowly winged, much distended over the achene,* 3–4 mm. long, 0.4–1 *mm. broad, gradually tapering to acute or subacute base and long narrow beak.* — Damp to dry open ground, rarely in woods, Nfld. to B.C., s. to N.S., N.E., L.I., nw. N.J., high mts. of Tenn., Mich., Wisc., Minn., Man., Sask., Ida. and Wash. June–Sept. Fig. 576.

576. C. Crawfordii.

77. C. Bébbii Olney (for Michael Schuck Bebb, 1833–1895). — Densely cespitose, 2–9 dm. high, the erect sharply angled slender culms smooth except at summit and mostly overtopping the leaves; fertile culms with 3 or 4 well developed ascending firm blades 1.5–4.5 mm. broad, the *tight sheath with* white hyaline *inner band concave at summit;* sterile culms frequent, their leaves subapproximate at summit; *head compact,* ovoid to ellipsoid, *brownish* (or soon becoming so), 1–2.5 *cm. long, of* (3–)5–12 *globose-ovoid to ellipsoid ascending spikes* 5–9 *mm. long, their crowded perigynia ascending; scales brown, their bases nearly as broad as the perigynia, their narrowed tips slightly exceeding body of perigynia; perigynia* firm, narrowly ovate, 2–4 *mm. long,* 1–2 *mm. broad, the inner face nerveless or only obscurely nerved at base.* — Swales, meadows, etc., Nfld. to e. Alaska, s. to N.S., Me., Mass., w. Ct., L.I., n. N. J., w. N.Y., n. O., n. Ind., Ill., n. Ia., Neb., Colo., Ida. and Wash. June–Aug. Fig. 577. — Hybridizes with no. 72.

577. C. Bebbii.

78. C. tíncta Fern. (tinged). — Mostly taller than no. 77, 3–9 dm. high, the arched-ascending slender *culms much overtopping the leaves; fertile culms with* 3–4 *well-developed leaves; the close sheaths with hyaline inner band not prolonged at summit,* often cross-rugulose, the *blades* 2–4 *mm. wide; sterile leafy culms wanting or few; heads* ellipsoid-ovoid to cylindric, 1.5–4 *cm. long, of* 4–8 crowded or approximate ellipsoid-ovoid ascending *brown spikes* 6–10 mm. long, their perigynia loosely ascending; *scales deep brown or fulvous, nearly equaling beak of perigynium; perigynia* narrowly ovate, 3.5–5 *mm. long,* 1.7–2 *mm. broad,* stramineous, becoming brownish, *clearly* 5–7-*nerved on the inner face.* — Rich woods, thickets and fields, w. N.B. and s. Que. to Mich., s. to n. and w. N.E. and N.Y.; Alta. and Wash. Mid-June, July. Fig. 578.

578. C. tincta.

79. C. normàlis Mackenz. (at right angles). — Coarser than nos. 77 and 78, up to 1.5 m. tall; well-developed *leaves of fertile culms* 4–7, *the loose somewhat mottled sheaths with inner hyaline band prolonged at summit,* the *blades* mostly 3.5–6.5 *mm. broad; sterile culms frequent,* with subapproximate leaves at summit; heads straight, subcylindric to lance-oblong, 1.5–4 cm. long, of 3–10 subapproximate to crowded *pale green spikes,* or the head flexuous and moniliform and 3–7 cm. long with all or all but the terminal spikes remote in forma **perlónga** Fern. (prolonged) fig. 580; *scales pale, shorter than body of perigynium; perigynia pale green* (becoming stramineous), *narrowly ovate,* 3–4 *mm. long,* 1.5–2 mm. broad, *about* 7-*nerved on inner face,*

579. C. normalis (typical).

580. C. normalis f. perlonga.

the beaks soon spreading or slightly recurving. (*C. mirabilis* Dew., not Host) — Rich woods, thickets and swales, Me. to Man., s. to N.C., e. Tenn., Mo. and Okla. Late May–Aug. FIG. 579.

80. C. ténera Dew. (slender). — Cespitose; the *very slender culms* 3–7.5 dm. high, smooth except at summit, and *much exceeding the leaves;* culms with 3–5 well-developed flat *blades* 0.5–2.5 *mm. broad, the tight slender sheaths with prolonged* smooth *hyaline band on inner side; inflorescence* moniliform, *stramineous* (rarely green), *flexuous or nodding,* 2.5–5 cm. long, *of* 3–8 *round-based* ovoid or ellipsoid *spikes* 4–8 *mm. long; scales* ovate, *acute,* greenish to brownish, narrower and shorter than perigynia; *perigynia stramineous* or becoming so, narrowly *ovate, firm,* clearly distended over the achene, 3–4 *mm. long,* 1.5–1.8 *mm. broad,* their tips not conspicuously spreading, *faintly* 3–7-*nerved or nerveless on the inner face.* (*C. straminea* of ed. 7, not Willd.) — Meadows, woodlands and moist to dry openings, Gaspé Co., Que., to Alta., s. to N.B., N.E., N.C., O., Ind., Ill., Mo., S.D. and Mont. Late May–Aug. FIG. 581.

581. C. tenera (typical).

Var. **echinòdes** (Fern.) Wieg. (bristly). — Perigynia prolonged, 4–4.5 mm. long, their tips very prominently spreading. — S. Que. to Man., s. to w. N.E., N.Y., Mich., Ia. and S.D. FIG. 582.

582. C. tenera, v. echinodes.

81. C. festucàcea Schkuhr (fescue-like). — Similar to no. 80; blades up to 3.5 mm. broad; the *inflorescence straight or arching, not nodding; spikes all long-clavate at base,* 0.6–1 *cm. long, grayish-green; perigynia* 3–3.5 *mm. long, greenish.* — Swales, low woods, etc., Ga. to La. and Okla., n. to Mass., N.Y., O., s. Mich., Ill. and Ia. Late May–early July. FIG. 583.

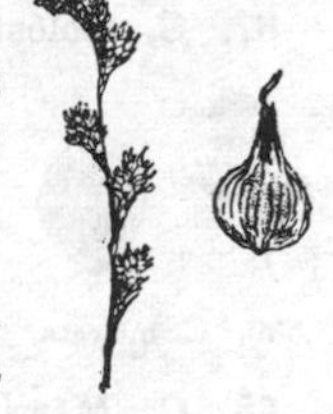

583. C. festucacea.

82. C. albolutéscens Schwein. (whitish-yellow). — Cespitose; the slender culms 0.3–1.2 m. high, smooth except at tips, well overtopping the pale green leaves; *leaves* of fertile culms 3–5, *their tight slender sheaths veined on the inner band with* the short hyaline *summit slightly prolonged* beyond the sloping bases of the blades; the latter pale green, 2–3.5 mm. wide; *head slightly open and moniliform, erect or arching,* 2–6 *cm. long, of* 3–10 *opaque greenish to pale brown ellipsoid spikes* 6–10 mm. long; *scales* pale, exceeding body of perigynium, *acuminate, slender-tipped; perigynia* appressed, *greenish to pale brown,* firm but thin, 3.5–4.5 mm. long, 1.5–2.5 mm. wide, veined on both faces, *the body obovate to suborbicular, broadest near summit,* abruptly short-beaked. (*C. straminea* sensu Mackenz., not Willd.) — Wet woods, thickets, peats, etc., Fla. to Tex., n. on or near the Coastal Plain to N.S., s. N.H., Mass., and N.Y., and in the interior n. to centr. Pa., s. Mich., Ill. and Mo. Late May–Aug. FIG. 584. — Hybridizes with no. 72.

584. C. albolutescens.

83. C. Lóngii Mackenz. (in honor of BAYARD LONG, 1885–). — Somewhat stiffer and stouter than no. 82; the *culms* 0.15–1.2 m. high, *harsher toward summit; leaf-blades* 2–3 mm. broad, firmer and *harsher, orifice of sheath concave; head stiffly erect, dense to but slightly open,* 1–4 cm. long, *ovoid, rhomboid or subcylindric,* of 2–10 *slightly lustrous* green to silvery-brown ellipsoid to obovoid *spikes; scales obtuse or subacute,* lustrous, much narrower than perigynia; *perigynia* green to pale brown, 3–4.5 mm. long, 1.8–2.5 mm. broad, firm, *elliptic to rhombic, broadest near the middle,* with firm short beak. (*C. albolutescens* of ed. 7, in part, not Schwein.) — Wet or damp sandy, argillaceous or peaty soils, Fla. to Tex. and Mex., n. to sw. Me., Mass. and N.Y., s. Ont., s. Mich. and nw. Ind. May–Sept. (Bermuda; Guatemala) FIG. 585.

585. C. Longii.

84. C. cumulàta (Bailey) Mackenz. (piled up). —Coarser than no. 83; culms 1.5–9 dm. high; *leaf-sheaths loose,* the thickish and firm *blades mostly* 3–6 *mm. broad* (except in obvious dwarfs); *head compact* or slightly open, *stiffly erect, of* 3–30 *crowded* (sometimes compound) *or approximate spikes,* thick-cylindric to ellipsoid or ovoid, 1–4.5 cm. long, or the spikes mostly 0.7–2 cm. apart in moniliform heads up to 1 dm. long in forma **solùta** Fern. (separated); *spikes conic-ovoid, dense, green and brown, broadly rounded to truncate at base; perigynia* planoconvex, rhombic-ovate to suborbicular, with short appressed firm deltoid tips, 2–3.4 *mm. broad, nerveless on inner face.* — Dry or moist acid soils, P.E.I. and e. N.B. to Sask., s. to N.S., N.E., N.J., Pa., O., Mich. and se. Man. June–Sept. FIG. 586. — Hybridizes with no. 72.

586. C. cumulata.

85. C. suberécta (Olney) Britt. (suberect). — In tussocks; the slender culms 0.3–1 m. high; leaves mostly 2–3 mm. wide, with long close sheaths, the fresh blades of the year 3–5 on fertile culms; *heads* becoming *tawny or ferruginous*, subcylindric to irregularly ovoid, 1–3 cm. long, *of 2–6 approximate or crowded spikes; spikes tapering to subacute at summit*, with appressed perigynia; *scales acuminate; perigynia* 4–5 mm. long, 2.2–2.8 mm. broad, *gradually tapering from the ovate body to the subacute base*, nerveless or nearly so on inner face. — Meadows, shores and prairies, s. Ont. to Minn., s. to w. Va., O., Ind., Ill. and Mo. Late May–July. Fig. 587.

587. C. suberecta.

86. C. brèvior (Dew.) Mackenz. (shorter). — Stiffish, in clumps; culms 0.3–1 m. high; new leaves of the year on fertile culms 3–6, 1–4 mm. wide, with close sheaths; *head open to slightly moniliform, 2–4.5 cm. long; the* 3–8 brownish to greenish *spikes conical to rounded above, somewhat narrowed and often clavate or turbinate at base; scales acuminate, nearly equaling the spreading-ascending beaks of perigynia; perigynia broadly ovate to suborbicular*, usually 4–5.5 mm. long, 2.5–3.5 mm. broad, *rounded to subtruncate at base*, nerveless or faintly few-nerved on inner face, *the wing rarely nerved.* (*C. festucacea*, var. *brevior* (Dew.) Fern.) — Dry open soil, Me. to s. B.C., s. to Del., D.C., Tenn., Ark., Tex., N.M. and Oreg. May–July. Fig. 588.

588. C. brevior.

87. C. molésta Mackenz. (troublesome). — Like no. 86 in having leaves mostly 2–3.5 mm. broad; *compact heads* 1–3 *cm. long*, usually *of 2–5 crowded or approximate whitish-green to pale drab spikes broadly rounded to summit and base* (the lateral not clavate); *scales* blunter, *reaching only to base of beak; perigynia firm, chartaceous, ovate,* 2–3 *mm. broad, pale green, becoming drab,* their translucent narrow wings nerveless or only rarely nerved. — Dry or slightly moist open grounds, borders of woods, etc., Mass. and Vt. to Sask., s. to Del., D.C., Tenn., Ark., Kans. and Colo. May–July. Fig. 589.

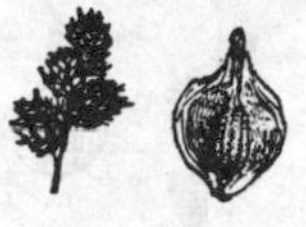
589. C. molesta.

88. C. Mérritt-Fernáldii Mackenz. (for the discoverer, 1873–). — Similar to nos. 86 and 87, mostly taller; *leaves 3–4.5 mm. wide; head usually open and submoniliform,* (2–)3–7.5 *cm. long, of* (3–)4–9 *mostly separated yellow-green, ferruginous or stramineous round-based and round-tipped spikes; scales blunt or short-cuspidate, barely reaching base of beak of perigynium; perigynia thin, membranaceous, pale stramineous, with broadly ovate to suborbicular body* 2.5–3.5 *mm. broad,* the translucent wings often 1–2-nerved. (*C. festucacea* of ed. 7, not Schkuhr) — Dry gravelly or rocky banks, dryish meadows and borders of woods, Que. to s. B.C., s. to N.E., N.Y., Mich., Wisc., Minn., Kans. Mont., Ida. and n. Calif. June–early Aug. Fig. 590.

590. C. Merritt-Fernaldii.

89. C. renifórmis (Bailey) Small (kidney-shaped). — Habit of the preceding; sheaths close and slender; *head flexuous or arching,* 2–5 cm. long, of 3–6 distinct to remote silvery-brown to green ellipsoid to obovoid round-tipped spikes with loosely ascending beaks; scales silvery-brown, very narrow; *perigynia* thin, 4–5 mm. long, 3.5–5 *mm. broad, with reniform to suborbicular bodies and conspicuous beaks, the subtranslucent wings often 1–2-nerved and about one-third as wide as either side of the opaque body.* — Cypress-swamps, wet woods and sloughs, Fla. to Tex., n. to se. Va., Ark. and Okla. May, June. Fig. 591.

591. C. reniformis.

90. C. Bicknéllii Britt. (in honor of Eugene Pintard Bicknèll, 1859–1925). — Habit of no. 85; leaves 2–4.5 mm. wide; *head* 2–6 cm. long, *of 3–7 silvery-brown to greenish ovoid, obovoid or subglobose usually turbinate-based spikes with prominent loosely ascending beaks; perigynia thin and membranaceous, translucent, ovate, 5.5–7.7 mm. long, 2.7–4.8 mm. broad,* pale stramineous, broadly winged, about 10-nerved on each face. — Dry slopes, thickets and fields, local eastw., centr. Me. to Sask., s. to Del., Pa., O., Ind., Ill., Ark., Okla. and N.M. May–July. Fig. 592.

592. C. Bicknellii.

91. C. silícea Olney (of sand). — In *stiff* clumps, *usually glaucous or whitened;* culms slender, wiry, 1.5–8 dm. high; *leaves* rigid, 2–4.5 mm. broad, often involute, *with* long sheaths, *bases of blades with a pair of rounded auricles; head arching, usually moniliform* (sometimes paniculate-branching), 2.5–10 cm. long, *of* 3–7(–12) mostly remote *ellipsoid to slenderly obovoid slender-based whitish to silvery-brown spikes; scales broad,* hyaline, *slightly shorter than to covering the closely appressed perigynia; perigynia*

593. C. silicea.

firm and opaque, 3.5–5 mm. long, 2–3 mm. broad, *with short deltoid beak, broadly winged; the body 3-5-nerved on the inner, 6–12-nerved on the outer face.* — Maritime sands and rocks, sw. Nfld., M.I. and Gaspé Pen., Que., to Md. June–Aug. Fig. 593.

92. C. alàta T. & G. (winged). — Relatively coarse, in tussocks; culms sharply angled 0.4–1.4 m. high; *leaves* firm, *scabrous-margined, mostly* 2.5–5.5 *mm. wide,* 3–7 to each fertile culm, the sheaths green and strongly nerved nearly or quite to the narrow subchartaceous auricle; *head* cylindric to ovoid, 2–6.5 cm. long, *of* 3–11(–17) *crowded, approximate or subdistant ovoid to ellipsoid silvery-brown to greenish spikes,* with strongly ascending perigynia; *scales* acuminate, *usually rough-awned,* much narrower than perigynia; *perigynia* firm, *pale brown to greenish, obovate or obovate-suborbicular,* with prominent beak, broad-winged, 4–5 mm. long, 2.5–4 *mm. broad, very faintly nerved or nerveless.* — Marshes and low woods, Fla. to Tex., n. to Mass., N.Y., Pa. and s. Mo.; and from N.Y. inland to s. Mich. and n. Ind. Late May–July. Fig. 594.

594. C. alata.

93. C. hormathòdes Fern. (necklace-like). — Densely tufted, with slender sharply angled loosely ascending to spreading culms smooth except at summit, 0.2–9.5 dm. high; *leaves smooth except at tip,* 1–2.5 *mm. wide,* strongly ascending, the close sheaths veined nearly to orifice; *heads moniliform and flexuous or arching to erect and crowded* (on late culms often densely congested and with compound spikes), 3–6.5 cm. long, frequently with subtending elongate bracts; *spikes ferruginous, warm-brown or stramineous,* 3–9 (on late culms up to 15), *rhomboid-ellipsoid, tapering gradually to summit and base, the terminal one tapering to short staminate base,* the lateral 8–15 mm. long and 5–9 mm. thick, or spikes only 5–8 mm. long and 3–6 mm. thick with the perigynia only 4–5 mm. long and culms only 1.5–6 dm. high in forma **invìsa** (W. Boott) Fern. (invisible) fig. 596; *scales lance-attenuate to aristate-tipped; perigynia fulvous, lance-ovate, tapering gradually to appressed ascending beaks,* 4.8–6 mm. long, distinctly about 10-nerved on either face. — Brackish to fresh marshes, sands and rocks, near the coast, St. P. et Miq. and w. Nfld. and lower St. Lawrence R., Que., to Va.; also Tippecanoe Co., Ind. (*Deam*). Late May–Aug. Fig. 595.

595. C. hormathodes (typical).

596. C. hormathodes, f. invisa.

94. C. stramínea Willd. (straw-colored). — Differing from no. 93 in usually taller culms, 3.5–10 dm. high; head moniliform and flexuous, 3.5–8 cm. long; *spikes browner,* 3–8, *more rounded, the terminal broadly rounded to the prolonged staminate base;* scales brownish; *perigynia round-ovate to suborbicular, abruptly contracted to loosely spreading-ascending beaks.* (*C. hormathodes*, var. *Richii* Fern.; *C. Richii* Mackenz.) — Fresh swamps and swales, Mass. to s. Mich., s. to Del., Md., D.C. and s. Ind. Mid-May–early July. Fig. 597.

597. C. straminea.

95. C. oronénsis Fern. (of Orono, Me.). — In loose stools; the sharply angled culms 0.5–1 m. high, harsh above; leaves smooth, 2.5–4 mm. broad, much shorter than the culms; *head erect, thick-cylindric to ellipsoid or lanceolate,* 2–3 *cm. long, of* 3–9 *crowded to subapproximate dark brown spikes; spikes* ascending, rhomboid-ovoid, pointed, 0.5–1 *cm. long; scales dark,* with pale hyaline margins, *covering the perigynia; perigynia lance-subulate,* about 4 mm. long, 0.7–1.5 *mm. broad, tapering to subacute prolonged base and narrow beak, scarcely winged below the middle.* — Fields, meadows and clearings, Penobscot Valley, Me. June, July. Fig. 598.

598. C. oronensis.

96. C. maclovìana D'Urv. (for the Falkland Islands, early named for Maclovius or St. Malo). — In dense tussocks 1–5 dm. high; the stiff culms obtusely angled and smooth below, sharply angled and scabrous at summit; leaves short and crowded, firm, mostly 2.5–4.5 mm. wide; *heads irregularly globose-ovoid,* 1–2 *cm. long, very dark, with* 3–8 *crowded and mostly divergent spikes;* spikes ovoid to ellipsoid, rounded below, blunt; *scales castaneous or purple-black with silvery scarious margins, obtuse, equaling or exceeding the perigynia; perigynia ovate, flattish, membranaceous,* about 4 mm. long and 2 mm.

599. C. macloviana.

broad, *the slender beak with terete tip.* — Greenl. and Lab. to alpine meadows of Gaspé Co., Que.; nw. Am., where represented by several doubtfully distinct "species." July, Aug. (Arct.-alpine Eu.; s. S. Am.) Fig. 599.

97. C. leporìna L. (of a hare). — Tufted, with spreading or ascending stiff culms 2–8 dm. high; leaves mostly short and firm, 1.5–4 mm. broad, the cauline 4–7; sterile leafy tufts numerous; *head interruptedly subcylindric to ellipsoid or ovoid,* 1.8–3.5 *cm. long, of* 3–6 *obovoid or ellipsoid approximate or subapproximate brown to ferruginous ascending spikes* (0.8–1.4 mm. long); *scales pale brown, acuminate, covering the perigynia; perigynia* green, becoming pale brown, *thin and scale-like, except over the achene, membranaceous, ascending,* 3.8–4.5 mm. long, 1.8–2.3 mm. broad; *achene* about 1 *mm. broad.* — Swales, damp thickets, roadsides, old fields and pastures, Nfld. to N.Y. and e. Pa.; summit of Roan Mt., N.C. and Tenn. June–Aug. (Natzd. from Eu.) Fig. 600.

600. C. leporina.

98. C. xerántica Bailey (dry). — *Stiff,* the wiry acute-angled culms harsh above, 3–6 dm. high; *leaves* short, mostly near the base; the *cauline* 2 *or* 3, *hard, pale,* 1–3 mm. broad; sterile tufts few or none; *head linear-cylindric,* 2–5 *cm. long, of* 3–6 *distinct ascending rhomboid acute brownish-white spikes* (8–13 mm. long); scales pale, acute, covering perigynia; *perigynia* appressed, 4–4.8 mm. long, 2–2.5 mm. broad, drab, *the inner face nerveless or barely nerved at the golden-yellow base;* achene 1.5–1.7 *mm. broad.* — Dry plains and hills, Man. to Alta., s. to Minn., S.D. and N.M. June, July. Fig. 601.

601. C. xerantica.

99. C. adústa Boott (swarthy). — *Culms stiffly erect,* smooth, 2–8 dm. high; leaves usually shorter, 2–5 mm. wide, firm; *head erect, stiff,* linear-cylindric to lance-ovoid or ellipsoid, 1–4 cm. long, often with a stiff erect basal bract, *of* (1–)2–15 simple or compound *full and rounded approximate brownish to bronze-green or olivaceous spikes* (6–12 mm. long); *scales concealing the perigynia; perigynia plump and heavy, coriaceous, flat to convex on inner face,* 4–5 mm. long, 2–3 *mm. broad, the wing very narrow and firm; achene* 1.8–2.1 *mm. broad.* — Dry open woods, gravels, rocks and clearings, in acid soils, Nfld. to Mackenz., s. to N.S., N.B., Me., n. N.Y., Mich., Wisc., Minn. and Sask.; adv. in s. B.C. June–Sept. Fig. 602.

602. C. adusta.

100. C. aènea Fern. (bronzy). — *Culms* smooth and wiry but more or less *flexuous at tip,* 0.2–1.2 m. high; leaves much shorter, rather soft and flat, 2–4 mm. broad; *head loosely cylindric or moniliform,* with at least the lower spikes remote (rarely crowded or even paniculate, especially in late-flowering specimens), commonly nodding from above lowest spike, 1.5–7 cm. long, *of* (1–)3–7(–18) obovoid *mostly clavate-based bronze, olivaceous or ferruginous spikes* (0.8–2.5 cm. long); *scales covering the perigynia; perigynia hard and thickish,* 4–5 *mm. long,* 1.9–2.7 *mm. broad, nerveless or only short-nerved on inner face; achene* 1.3–1.7 *mm. broad.* — Open woods, dry slopes, rocks and gravel, Lab. to Yuk. and e. Alaska, s. to Nfld., N.S., N.E., N.Y., centr. Pa., Mich., Wisc., Minn., S.D. and Mont. May–Aug. Fig. 603.

603. C. aenea.

101. C. pratícola Rydb. (prairie-dweller). — Differing from no. 100 in very slender culms only 2–7 dm. high; leaves 1–3.5 mm. broad; delicately flexuous moniliform to barely crowded head *with silvery-green to whitish-brown or pale ferruginous slender spikes; perigynia flatter and thinner,* lance-ovate, 4.5–6.5 *mm. long,* 1.5–2 *mm. broad.* (*C. pratensis* Drej., not Host) — Open woods, meadows, prairies and clearings, Greenl. and Lab. to Alaska, s. (very rarely in the East) to Nfld., Gaspé Pen., Que., n. Me., Parry Sd., Ont., n. Mich., N.D., Colo. and Calif. June–Aug. Fig. 604.

604. C. praticola.

102. C. argyrántha Tuckerm. (silvery-flowered). — Similar to no. 100, up to 1 m. high; *leaves* loose, *pale green* or glaucous, 2–4.5 mm. wide; *heads arching,* or *nodding* from near base, *more or less moniliform* to linear-cylindric and interrupted, *with* 4–15 *silvery-green to whitish-brown* clavate-based ovoid to subglobose *spikes* (6–17 mm. long), some of the latter often crowded at summit of head; *scales pale, covering perigynia; perigynia pale green to drab,* 3–4.5 *mm. long, usually with strong nerves on inner face.* (*C. foenea*

605. C. argyrantha.

of ed. 7, not Willd.) — Dry woods, thickets and clearings, N.B. to Ont., s. to N.S., N.E., Md., N.C., W.Va., n. O., Mich. and Minn. June–Aug. Fig. 605.

103. C. sychnocéphala Carey (many-headed). — Tufted; culms smooth, 2–6 dm. high; *leaves soft*, ascending, 2–4 mm. broad; *bracts leaf-like and much prolonged, unequal, forming a conspicuous involucre; the lowest longest, 1–2 dm. long; spikes* 4–10, ellipsoid, 8–15 mm. long, *forming a dense ovoid to ellipsoid head; perigynia lance-subulate, about* 5 *mm. long, barely* 1 *mm. broad.* — Meadows, open woods and clearings, sw. Que. to Alta., s. to N.Y., Mich., Wisc., Ia., S.D. and Mont. July, Aug. Fig. 606.

606. C. sychnocephala.

Subgen. II. Eucàrex Coss. & Germ. (see p. 295)

§ 17. Polytrichoìdeae Tuckerm. (see p. 296) — A single species:

104. C. leptàlea Wahlenb. (delicate). — Densely *cespitose* to substoloniferous; *the* very *lax* and *soft leaves* mostly shorter than the culms, 0.5–1.3 *mm. wide; culms capillary*, obtusely angled, *flaccid*, 1–7 dm. long; *spike* 1, *terminal, linear-oblong*, 0.4–1.6 cm. long, the terminal staminate portion short; *staminate scales tightly cucullate, their margins united nearly up to the middle;* pistillate scales pale brown, with prominent green midrib; *perigynia* subalternate, *closely appressed, oblong to narrowly ellipsoid, blunt, beakless*, 2.5–3.5 mm. long, *longer than the caducous scales.* — Mossy or wet woods, swales and clearings, Lab. to Alaska, s. to Nfld., N.S., N.E., upland of N.C. and Tenn., Mo., N.D., Colo. and n. Calif. June–Aug. Fig. 607.

607. C. leptalea (typical).

608. C. leptalea, v. Harperi.

Var. **Hárperi** (Fern.) Stone (for its discoverer, Roland Macmillan Harper, 1878–). — Perigynia strongly overlapping, more slender, 3.5–5 mm. long; scales whitish. (*C. Harperi* Fern.) — Wet mossy woods and bogs, Fla. to Tex., n. to s. N.J., se. Pa. and Ark. May, June. Fig. 608.

§ 18. Phyllostáchyae Tuckerm. (see p. 296)

Scales all (or all but the uppermost) broadly foliaceous, green throughout, embracing and mostly hiding the perigynia, the lowest foliaceous bract 3–6 mm. broad; staminate scales about 3.
- Perigynia ellipsoid-ovoid, empty at summit, 4.5–6 mm. long, tapering to conical beak 2–3 mm. long. . . . . 105. *C. Backii.*
- Perigynia obovoid, tightly filled by achene, 4 mm. long, with beak 0.5–1 mm. long. . . . . 106. *C. saximontana.*

Scales scarious-margined, only the lowest 1 or 2 prolonged and foliaceous; these 1–2 mm. broad, narrower than and hardly embracing the perigynia; staminate flowers 6–20.
- Perigynia ellipsoid-fusiform, gradually tapering into a stout triangular serrate beak; staminate scales 6–12, tapering at summit. . . . . . . 107. *C. Willdenowii.*
- Perigynia subglobose, abruptly contracted to a slender beak; staminate scales 8–20, truncate. . . . . 108. *C. Jamesii.*

105. C. Báckii Boott (in honor of Sir George Back, 1796–1878). — Forming dense mats; leaves dark green, 3–6 mm. broad, stiffish, very abundant and overtopping the very unequal culms; spikes solitary, terminating short and long capillary culms (0.1–3 dm. long); *staminate scales about* 3, *somewhat tubular, tightly inclosing the rachis, their margins united below; pistillate flowers* 2–5, *on a flattened zigzag rachis; scales very broad and leaf-like, green throughout, many times overtopping and at base embracing and hiding the perigynia, the lowest* 3–6 *mm. broad; perigynia ellipsoid-ovoid, 2-edged, empty at summit,* 4.5–6 *mm. long, tapering to conical beak* 2–3 *mm. long.* (*C. durifolia* Bailey) — Dry rocky or sandy woods and bluffs, Gaspé Pen., Que., to Sask., s., rather locally, to n. and w. N.E., n. N.J., n. Pa., Mich., Wisc., Minn. and Neb. May–July. Fig. 609.

609. C. Backii.

610. C. saximontana.

106. C. saximontàna Mackenz. (of the Rocky Mountains). — Similar to no. 105, paler; *perigynia obovoid, tightly filled by achene,* 4 *mm. long, with* *beak* 0.5–1 *mm. long.* — Dry woods and thickets, Man. to B.C., s. to w. Minn., Neb., Colo., Utah and e. Oreg. June, July. Fig. 610.

611. C. Willdenowii.

107. C. Willdenòwii Schkuhr (in honor of Karl Ludwig Willdenow, 1765–1812). — Similar to the two preceding; leaves soft and pale, 1.5–4 mm. broad; spike more compact, with 3–9 pistillate flowers; *scales* of pistillate flowers *scarious-margined, only the lowest* 1 *or* 2 *prolonged and foliaceous;* these 1–2 *mm. broad, narrower than and hardly embracing the perigynia; staminate scales* 6–12, *tapering at summit; perigynia ellipsoid-fusiform, gradually tapering into a stout triangular serrate beak.* — Rocky woods, nw. Fla. to Tex., n., locally, to e. Mass., s. Vt., N.Y., s. Ont., s. Mich. and Minn. May–July. Fig. 611.

612. C. Jamesii.

108. C. Jàmesii Schwein. (for its discoverer, Edwin James, 1797–1861). — Very similar to no. 107; pistillate flowers 1–3, scattered; *staminate scales* 8–20, *truncate; perigynia subglobose, abruptly contracted to a slender beak.* — Rich, mostly calcareous, woods, N.Y. and s. Ont. to Mich. and Ia., s. to Va., Tenn., Mo. and Kans. May, June. Fig. 612.

§ 19. Filifòliae Tuckerm. (see p. 296) — A single species with us:

109. C. filifòlia Nutt. (thread-leaved). — *Forming dense tussocks* 0.5–3 dm. high; *leaves filiform or involute-acicular*, crowded, with persistent old brown sheaths, *fibrillose at base;* culms similar, obtusely angled; *spike solitary, terminal*, 1–3 cm. long, *the long staminate upper portion linear-cylindric, the basal pistillate part plump; scales* obtuse, with broad white scarious margins, *the pistillate completely concealing the young perigynia; perigynia obovoid to subglobose*, about 3 mm. long, *opaque and usually minutely puberulent* below the minute hyaline truncate beak; *rudimentary rachilla present within the perigynium.* — Dry plains and hills, Yuk. to N.M., e. to Man., w. Minn., S.D., Neb. and Tex.; ne. Lab. May, June. Fig. 613.

613. C. filifolia.

§ 20. Obtusàtae Tuckerm. (see p. 296) — A single species with us:

110. C. obtusàta Lilj. (obtuse). — *Creeping by slender cord-like dark rhizomes;* the stiff channelled pale leaves (1–1.5 mm. wide) and the sharply angled slender culms (0.5–2 dm. high) arising in small scattered tufts; *spike solitary*, terminal, lanceolate to ovoid, staminate at summit, pistillate at base, 0.5–1.4 cm. long; *scales pale, the caducous pistillate ones acuminate; perigynia* ellipsoid-obovoid, round-angled, 3–3.5 mm. long, coriaceous, *highly lustrous, castaneous to blackish*, finely sulcate, with short hyaline-tipped beak. — Dry plains, bluffs and rocky slopes, Yuk. and Alaska to s. B.C. and N.M., eastw. to Man., w. Minn. and S.D.; old (unverified) report from Nfld. May–July. (N. Eurasia) Fig. 614.

614. C. obtusata.

§ 21. Rupéstres Tuckerm. (see p. 296) — A single species with us:

111. C. rupéstris Bellardi (of rocks). — Habitally similar to no. 110, the rhizomes and old sheaths paler; leaves coarser, 1–3 mm. broad, usually more densely tufted and arching; culms 0.3–1.8 dm. high; *spike linear-* to *lance-cylindric*, 0.6–2.3 cm. long, mostly staminate, *the* 1*-few basal pistillate flowers approximate to scattered and appressed-ascending; scales obtuse, the pistillate ones broad and rounded at the pale hyaline summit*, with brown to purple centers; *perigynia* ellipsoid-obovoid, 3–4 mm. long, *stramineous to olivaceous or brownish, barely lustrous*, with a short terete beak. — Greenl. and Ellesmerel., s. to rocks, gravels or alpine barrens of Nfld., Anticosti I., Shickshock Mts., Gaspé Pen., Que., Ung. and ne. Man.; nw. Am. s. to n. Mont. Late June–Aug. (Eurasia) Fig. 615.

615. C. rupestris.

§ 22. Firmicúlmes Kükenth. (see p. 296) — A single species with us:

112. C. Geỳeri Boott (for its discoverer, Carl Andreas Geyer, 1809–1853). — Rhizome more or less elongate, firm; leaves and culms in scattered to approximate tufts; culms sharply angled, firm, slender, 1–3.5 dm. high; *leaves* elongating, becoming erect, 2–3.5 mm. broad, firm, dark green, *with scabrous margins; spike solitary, terminal, interrupted, the* pale and elongate (0.5–2.5 cm. long) *staminate portion stalked, the pistillate base with* 1–3 *remote large flowers; pistillate scales* stramineous, with paler hyaline border, usually *much longer and broader than and somewhat sheathing the perigynia; perigynia* ellipsoid-obovoid, 5–6.5 *mm. long,* 2.5–3 *mm. thick,* firm, trigonous, stramineous or brownish, the abrupt short truncate beak often denticulate at base. — Woods and thickets, Alta. and B.C., s. to Colo., Utah and n. Calif.; calcareous rocky woods, local, Centre Co., Pa. May, June. Fig. 616.

616. C. Geyeri.

§ 23. Scirpìnae Tuckerm. (see p. 296) — A single species with us:

113. C. scirpoìdea Michx. (like *Scirpus*). — *Dioecious,* with subcespitose to creeping purplish base; pistillate plants mostly stiff, 1–7 dm. high; staminate plants smaller; *leaves flat, shorter than culms,* 1.5–4 mm. broad; *spike solitary* (rarely with small rudimentary basal spike), *densely cylindrical,* 1–4 cm. long; *scales* blunt or merely acute, *narrower than perigynia,* oblong-obovate, *ciliate,* purplish-brown or darker; *perigynia* ovoid, short-pointed, *sharply angled, very hairy,* 2–3 mm. long, appressed-ascending. — Greenl. to Alaska, s. to turfy, peaty or rocky situations and alpine areas of Nfld., C.B., n. N.E., ne. N.Y., s. Ont., n. Mich., Man., Colo. and e. B.C. June–Aug. (Eurasia) Fig. 617.

617. C. scirpoidea.

Var. **scirpifórmis** (Mackenz.) O'Neill & Duman (with the form of *Scirpus*). — Scales with broad white hyaline margins. (*C. scirpiformis* Mackenz.) — Lab. to n. Man., s. to dry crests of w. Nfld., N.D. and Mont.

Var. **convolùta** Kükenth. (convolute or rolled up). — *Leaves convolute or subfiliform,* 0.5–1.5 *mm. broad.* — Calcareous ledges, gravels and open woods near L. Huron in Ont. and Mich.

§ 24. Píctae Kükenth. (see p. 296) — A single species with us:

114. C. pícta Steud. (painted). — *Dioecious,* from stout slightly creeping radiately branching rhizomes, the old centers up to 6 dm. across; *leaves* flat and firm, persisting through the winter, *at least twice longer than the culms; culms filiform, axillary, with loose bladeless sheathing bracts; spike* bractless, *linear-cylindric to linear-clavate,* 2–6 *cm. long,* the staminate smaller than the pistillate; *pistillate scales* oblong-obovate, short-ciliate, purple, with paler midrib, *the middle and upper ones prolonged into cusps and concealing the perigynia; perigynia slenderly oblanceolate,* 4–5 *mm. long,* thin, sharply angled, *slenderly long-stipitate, pubescent at tip.* — Wooded slopes and bluffs, local, Ind. to Minn., s. to Ga., Ala. and La. Fig. 618.

618. C. picta.

§ 25. Lamprochlaènae Drej. (see p. 296) — A single species with us:

115. C. supìna Willd. (lying back). — *Stoloniferous,* with slender cord-like brown stolons and dense tufts of leaves and culms; culms erect, slender, sharply 3-angled, scabrous above, 0.3–1.8 dm. high, mostly overtopping the crowded leaves; *leaves flat or channeled,* 0.5–1.5 *mm. wide, harsh along the long-attenuated tip;* inflorescence of 2 or 3 approximate or subapproximate spikes, the terminal staminate one pale, the pistillate ones subglobose and about 5 mm. in diameter; *pistillate scales lance-ovate, acuminate, fulvous; perigynia spreading-ascending, coriaceous, glabrous, lustrous,* pale brown, becoming fuscous, *plump-obovoid,* 2.5–3 mm. long; the short cylindric beak with oblique hyaline orifice. — Dry rocks and sands, Greenl. and Baffin I.; Keewatin to Mackenz. and Alaska., s. to n. Minn. and Sask. (Eurasia) Fig. 619.

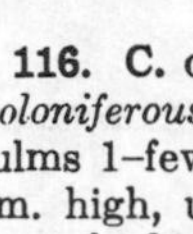

619. C. supina.

§ 26. Praecòces Christ (see p. 297) — A single species with us:

116. C. caryophyllèa Lat. (its leaf-tufts suggesting *Caryophyllus* or *Dianthus*). — Slightly *stoloniferous, stiff, scabrous,* with fibrillose bases and scattered leafy tufts; culms 1–few from each tuft, slender, firm, erect or arching, smoothish, 0.5–4 dm. high, usually much taller than the leaves; leaves short, flat or plicate, curved, firm, 2–3 mm. broad; *staminate spike* clavate, usually *sessile; pistillate spikes* 2 or 3, *contiguous,* sessile, or the lowest short-peduncled and subtended by a bract scarcely as long as itself, all ellipsoid to short-cylindric, the lowest 0.7–1.5 cm. long; *pistillate scales rough-cuspidate; perigynia* trigonous-obovoid, about 2.5 mm. long, *thinly hispid-hirsute,* the very short beak erose or entire; ***achene capped by a dark annulus with a projecting central apicu-***

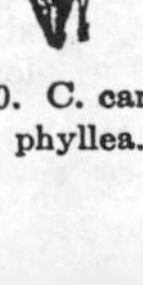

620. C. caryophyllea.

lation. — Dry roadsides and grasslands, s. Me. to e. N.Y. and D.C. May, June. (Natzd. from Eu.) Fig. 620.

§ 27. Montànae Fries (see p. 297)

a. Culms all elongate; the inflorescences all normally with terminal spike staminate, none of them hidden and merely pistillate. . . *b.*

b. Body of perigynium subglobose to short-ovoid, about as long as thick, only moderately close over the achene.

Plant with long slender leafless scaly-bracted stolons; bases of leafy shoots coarsely fibrillose. . . . . 117. *C. pensylvanica.*

Plant without leafless horizontal stolons; bases not at all or only slightly fibrillose. . . . . 118. *C. communis.*

b. Body of perigynium slenderly ellipsoid to fusiform-obovoid, definitely longer than thick, tightly investing achene. . . *c.*

c. Perigynia much exceeding scales; staminate spike 2–6 mm. long, sessile.

Pale green; culms ascending, much exceeding leaves; scales with broad white margins; perigynia slenderly ellipsoid to fusiform, 3–4 mm. long, gradually narrowed to thick spongy base, copiously hirsute. . 119. *C. Peckii.*

Deep green; culms setaceous, the longer ones arching or recurving, mostly shorter than leaves; scales with purple margins; perigynia 2.5–3 mm. long, the ellipsoid-obovoid body abruptly narrowed to a slender stipe, minutely puberulent. . . . . 124. *C. deflexa.*

c. Perigynia equaled, exceeded by or barely exceeding scales, mostly strongly contracted to slender stipe, 2–4 mm. long, appressed-pubescent to puberulent; staminate spike often longer. . . *d.*

d. Principal leaves, when mature, 3–7 mm. broad, their ligules longer than broad; inflorescence 1–8 cm. long; perigynia 2.5–4 mm. long, about 1.5 mm. thick. . . . . 118. *C. communis.*

d. Principal leaves 0.5–2.5 mm. broad, their ligules broader than long; inflorescence 0.5–5 (rarely –7) cm. long; perigynia 2–3.5 mm. long, 0.5–1.5 mm. thick. . . *e.*

e. Staminate (usually) and pistillate spikes sessile; perigynia 2.5–3.5 mm. long, 1–1.5 mm. thick; plant densely cespitose (loosely so in no. 122).

Plant densely cespitose, without horizontal elongate stolons; staminate spike usually sessile; perigynia green to olivaceous above the stipitate base.

Culms weak, loosely spreading, arching, recurving or reclining; head dense; all or all but the lowermost spikes aggregated into an irregular subglobose, obovoid, ovoid or ellipsoid glomerule 1–2 cm. long. . . . . 120. *C. Emmonsii.*

Culms firmer, erect or strongly ascending; the distinct to distant spikes in an interruptedly linear-cylindric head 1–4.5 (–8) cm. long. . . . . 121. *C. artitecta.*

Plant loosely cespitose, with prolonged horizontal fibrillose cord-like stolons; staminate spike usually short-stalked; perigynia whitish-green. . . . . 122. *C. physorhyncha.*

e. Staminate and lowest pistillate spike short-peduncled; perigynia pale, thin, 2–2.5 mm. long, 1–1.25 mm. thick; plant weak, loosely cespitose, pale. . . . . 123. *C. novae-angliae.*

a. Culms of various lengths, the longer ones with staminate or both staminate and pistillate spikes; others but slightly exserted or included within the basal leaf-sheaths, with pistillate spikes or with them both staminate and pistillate. . . *f.*

f. Remnants of old leaves soft, only slightly if at all shredded; scales blunt or acute, much shorter than mature perigynia.

Leaves soft, loosely ascending or spreading; longer culms capillary, smooth except at tip, flexuous, arching or recurving; staminate spike 2–5 mm. long; perigynia 2.5–3 mm. long, with thick beak about 0.5 mm. long. 124. *C. deflexa.*

Leaves firm, stiff, erect or stiffly ascending; longer culms coarser, harsh above, stiffly ascending; staminate spike up to 1.5 cm. long; perigynia 3–4.5 mm. long, with lanceolate beak 0.7–1.5 mm. long. 125. *C. Rossii.*

f. Remnants of old leaves persisting as stiff tufted shreds; scales acuminate or sharp-pointed, nearly equaling to exceeding perigynia.

Body of perigynium ellipsoid-oblong, one-half to two-thirds as broad as long; southern species. . . . . 126. *C. nigromarginata.*

Body of perigynium broadly ellipsoid or ellipsoid-obovoid to subglobose, four-fifths to quite as broad as long; northern and eastern species.
  Leaves relatively soft, 1.5–2.5 (–3) mm. broad; perigynia pubescent; achenes blackish.
    Pistillate scales lance-ovate, tapering to long or acuminate tips; perigynia 3.2–4.7 mm. long, with ellipsoid or ellipsoid-obovoid bodies, the beak 0.9–1.7 mm. long. . . . . 127. *C. umbellata.*
    Pistillate scales broadly ovate or ovate-oblong, with short acute tips; perigynia 2.2–3.3 mm. long, with globose-ovoid bodies, the beak 0.5–1 mm. long. . . . . 128. *C. abdita.*
  Leaves hard, firm, very scabrous, the broader ones 2.5–5 mm. broad; scales lance-ovate, long-tapering; perigynia as in no. 127 but usually glabrous, the beak 1–2.5 mm. long. . . . . . . 129. *C. tonsa.*

117. C. pensylvánica Lam. (of Pennsylvania). — *Strongly stoloniferous,* the small or dense leafy tufts with reddish bases and usually *with persistent brush-like tufts of fibers, the cord-like horizontal stolons fibrillose;* leaves 1–3 mm. wide, soft or firm, ascending, the over-wintering ones loosely reclining; culms slender, erect, sharply rough-angled, 0.5–4 dm. high; terminal staminate spike clavate, sessile to short-stalked, 0.8–2 cm. long, reddish to whitish-brown; *pistillate spikes* 1–4, *approximate or slightly separated,* globose or ovoid, 3–12 mm. long, the lowest often leafy-bracted; *pistillate scales* ovate to lanceolate, subobtuse to acute or acuminate, *nearly equaling to exceeding perigynia, reddish-purple or brown* or rarely paler; *perigynia 2.5–4 mm. long, pubescent; the subglobose to thick-ovoid body about as long as thick,* moderately tight over the achene. — Very variable. Three vars. recognizable:

Leaves relatively soft, smoothish or only slightly scabrous; perigynium 1.3–1.8 mm. in diameter, obtusely 3-angled.
  Perigynium 2–3 mm. long, the short beak one-fourth to one-fifth as long as the body.
    Leaves at flowering time much shorter than to nearly equaling culms, 1.5–3 mm. broad. . . . . *C. pensylvanica* (typical).
    Leaves very narrow, 1–1.5 mm. wide, equaling or commonly overtopping the culms. . . . . Forma *gracilifolia.*
  Perigynia 3–4 mm. long, the slender beak two-thirds as long as to nearly equaling the body. . . . . Var. *distans.*
Leaves stiffer and firmer, strongly scabrous; perigynia 3–3.5 mm. long, the sub-globose body scarcely angled and 1.5–2.2 mm. thick. . . . . . Var. *digyna.*

C. pensylvánica (typical). — Open dry soil or open woods, s. Que. to s. Ont. and N.D., s. to Me., Mass., L.I., S.C., Tenn. and Ia. — Forma **gracilifòlia** (Peck) Kükenth. (slender-leaved), mostly in thickets and woods. April–June. Fig. 621. — Hybridizes with nos. 118 and 127.

621. C. pensylvanica (typical).

Var. **dístans** Peck (distant), var. *lucorum* (Willd.) Fern., *C. lucorum* Willd. — Woods, thickets and openings, s. Que. and s. Ont., s. to N.S., N.E., L.I., N.J., upland to N.C. and Tenn., Mich. and Wisc. May–July. Fig. 622.

622. C. pensylvanica, v. distans.

Var. **dígyna** Boeckl. (2-styled), *C. heliophila* Mackenz. — Plains, prairies and openings, Ont. to Alta., s. to w. N.Y., Ind., Ill., Mo., Kans. and N.M. May–July.

118. C. commùnis Bailey (growing in colonies). — *Forming close tufts* or tussocks, *without elongate stolons, the bases rarely fibrillose;* culms 1–6 dm. high, sharply angled, rather soft, mostly well overtopping the leaves; *leaves* flat, *the overwintering or mature ones* 3–7 *mm. broad, with ligules longer than broad,* the bases purplish; heads interruptedly linear-cylindric, 1–8 cm. long; the 1–5 *pistillate spikes mostly distinct,* often remote, 0.5–1 cm. long, the lowest often leafy-bracted; staminate spike linear-clavate, sessile or stalked, green to reddish-purple, 0.3–2 cm. long, about 1.5 mm. thick; *pistillate scales exceeding or slightly exceeded by perigynia,* usually with purplish sides, sometimes pale throughout; *perigynia* 2.5–4 mm. long, pubescent, *with subglobose to thick-ellipsoid body,* the elongate base spongy, the short beak broad. — Woods and clearings, Gaspé Pen., Que., to Ont. and Minn., s. to N.S., N.E., Md., upland to Ga., W.Va., Ky. and Ark. May–July. Fig. 623.

623. C. communis.

119. C. Péckii Howe (for Charles Horton Peck, 1833–1917). — *Pale green,* loosely cespitose; *culms* 1–6.5 dm. high, ascending, *much exceeding*

624. C. Peckii.

leaves, straightish; leaves soft, pale and narrow (1.5–3 *mm. wide*); *pistillate spikes* 1–3, globose or thick-ovoid, all *approximate* or the lowest slightly remote, naked or subtended by a narrow bract, the head 0.8–2 cm. long; *staminate spike sessile, often hidden in the head; scales* pale brown or reddish, *with broad white hyaline margins, much shorter than perigynia; perigynia copiously short-hirsute, slenderly ellipsoid to fusiform, 3–4 mm. long, gradually narrowed to thick spongy base,* the slender beak with bidentate hyaline orifice. (*C. albicans* of ed. 7, not Willd.) — Dry calcareous rocky slopes, rich open woods and ravines, Gaspé Pen., Que., to Yuk., s. to w. N.B., n. N.E., n. N.J., N.Y., s. Ont., Mich., Wisc., Minn., S.D., s. Alta. and s. B.C. May–early Aug. Fig. 624.

120. C. Emmónsii Dew. (in honor of Ebenezer Emmons, 1798–1863). — *In dense tussocks,* the soft and persistent overwintering *leaves* (0.5–1.5 *mm. broad*) and those of the season loosely spreading, the *old leaves mostly much longer than the weak arched, recurving or reclining filiform culms* (0.3–4.5 dm. long), the bases of the tufts purple; *head glomerulate,* 1–2 *cm. long, subglobose, obovoid, ovoid or ellipsoid*; the 2 or 3 (rarely 4) globose to ovoid *spikes closely aggregated,* or the lowermost remote, 3–8 mm. long; staminate spike 3–8 mm. long, sessile, often nearly hidden; *pistillate scales* oblong-obovate; their sharp-pointed summits *narrower than and about equaling the perigynia, loosely spreading-ascending; perigynia slenderly ellipsoid, olive or deep green* and minutely pubescent above, 2.5–3.3 *mm. long,* 1–1.3 *mm. broad,* with short spongy base and slender beak. (*C. varia* of ed. 7, not Muhl.; *C. albicans* sensu Mackenz., not Willd.)— Dry woods, thickets and clearings, Fla. and Ala., n. to P.E.I., N.S., se. Me., s. N.H., s. Vt., N.Y., O., Mich. and Wisc. April–July. Fig. 625.

625. C. Emmonsii.

626. C. artitecta.

121. C. artitécta Mackenz. (closely covered). — Stiffer and more erect than no. 120, the *firmer culms erect or ascending; head interruptedly linear-cylindric,* 1–3.5 *cm. long;* staminate spike 0.35–1.5 cm. long; pistillate spikes 1–4, scattered, the lower 3–13 mm. apart; *scales more appressed* to perigynia; beak of perigynium about 1 mm. long. (*C. varia* Muhl., not Lamnitzer; *C. varia,* var. *colorata* Bailey) — Dry woods and clearings, S.C. to e. Tex., n. to sw. Me., s. N.H., sw. Que., s. Ont., Mich., Ill., Ia. and Kans. April–June. Fig. 626.

Var. **subtiliróstris** F. J. Herm. (slender-beaked). — Heads 3–8 cm. long, the lower spikes 1–4 cm. apart; scales prolonged, less appressed; beak of perigynium 1–2 mm. long. — Local, Mass. to Ind. and Tenn. — Anomalous plant, about as well treated as an extreme of no. 120.

122. C. physorhýncha Liebm. (bellows-beaked). — Superficially resembling slender states of no. 117, with *loose tufts* of leaves and culms *fibrillose at base* and *bearing elongate horizontal fibrous-scaly leafless rhizomes;* culms slender, somewhat to hardly exceeding the leaves, 1.5–5 dm. high; leaves pale green, flat, scabrous, 1.7–3 mm. broad; head interrupted, 1.5–4 cm. long; *staminate spike* clavate, *usually peduncled; pistillate spikes* 1–4, *remote,* sessile or short-stalked; pistillate scales brown, about equaling perigynia; *perigynia whitish-green, slenderly ellipsoid-obovoid,* slightly pubescent, 2–3 mm. long, *about* 1 *mm. thick,* the beak barely half the length of the body, the stipe slender. — Sandy or rocky woods, Fla. to Tex. and Mex., n. to se. Va., se. Mo. and Okla. May, June. Fig. 627.

627. C. physorhyncha.

628. C. novae-angliae.

123. C. nòvae-ángliae Schwein. (of New England). — Very slender and soft, *loosely cespitose; culms weak,* filiform-trigonous, 1–4 dm. long, *little or not at all longer than the soft pale green leaves* (0.7–1.5 *mm. wide*); *head interruptedly linear-cylindric,* 1.5–6 cm. long, with weak axis; *staminate spike usually short-peduncled, very slender,* 0.5–1 *cm. long,* 0.5–1 *mm. thick; pistillate spikes* 1–3, *remote, the lower short-peduncled* and usually leafy-bracted, loosely 3–10-flowered, ellipsoid to subglobose; scales pale, hyaline, ovate, often tinged with brown or red; *perigynia thin, pale,* 2–2.5 *mm. long,* 1–1.25 *mm. thick,* minutely pubescent, the yellowish base spongy, the beak short. — Woodlands and damp slopes, Nfld. to Ont., s. to N.S., Me., Mass., nw. Ct., n. Pa. and Wisc. June–Aug. Fig. 628.

124. C. defléxa Hornem. (bent down). — Loosely to densely *tufted; leaves soft, loosely ascending or spreading,* 1–3 mm. wide; *culms* 0.2–4 *dm. long, setaceous, smooth except at tip, flexuous, curved or spreading,* little exceeding to shorter than leaves, some of them often crowded among the leaf-bases; *staminate spike* 2–5 *mm. long,* sometimes hidden in the head; *pistillate spikes* 2 or 3, 2–8-flowered, green or green and reddish-brown, *all aggregated into a head; or the lowest one slightly remote, short-peduncled* and subtended by a leafy bract; *perigynia* finely pubescent, green, becoming brownish, 2.5–3 *mm. long, with plump body, the thick beak about* 0.5 *mm. long, longer than the reddish-brown scales.* — Woods, clearings and turfy slopes, Nfld. to s. Yuk., s. to N.S., Me., Mass., N.Y., s. Ont., Mich., Wisc. and Minn. May–Aug. (Greenl.) Fig. 629.

629. C. deflexa.

125. C. Róssii Boott (for Sir John Ross, 1777–1856). Similar to no. 124, *stiffer,* more densely cespitose; *leaves firm, erect or stiffly ascending; culms* coarser, *harsh above, stiffly ascending; staminate spike up to* 1.5 *cm. long;* scales usually with long cuspidate tips; *perigynia* 3–4.5 *mm. long, with lanceolate beak* 0.7–1.5 *mm. long.* — Rocky slopes, bluffs and dry openings, Thunder Bay Distr., Ont., to Yuk., s. to n. Mich., Minn., S.D., Colo., Utah and Calif. Late May, June. Fig. 630.

630. C. Rossii.

126. C. nigromarginàta Schwein. (black-margined). — Densely or loosely cespitose, *forming large mats, fibrillose at base; overwintering leaves firm, scabrous, prolonged to* 2–5 *dm., much longer than the culms,* 1–4 mm. broad; *culms of various lengths,* very slender, rough-angled, flexuous or arching, the longer ones 0.3–2.5 cm. long, *the shorter frequently crowded among leaf-bases;* heads irregularly ovoid or ellipsoid, the 2 or 3 ellipsoid or ovoid pistillate spikes approximate to the single staminate spike, the latter 5–15 mm. long and purplish-brown to pale green; *scales of pistillate spikes ordinarily purplish* (sometimes pale); *perigynia* barely pubescent, *fusiform or with slenderly ellipsoid body,* 3–4 mm. long, 1.3–1.6 mm. thick, equaling or shorter than the scales. — Dry woods, thickets and clearings, Fla. to La., n. to s. Ct., se. N.Y., Pa., s. Ind. and se. Mo. April–June. Fig. 631.

631. C. nigro-marginata.

Var. **floridàna** (Schwein.) Kükenth. (of Florida). — Less cespitose, the *stolons horizontally elongate;* old leaf-bases less fibrillose; scales uniformly pale or light brown; achenes less definitely triangular. (*C. floridana* Schwein.) — Locally n. to se. Va.

127. C. umbellàta Schkuhr (bearing umbels). — *Low* and conspicuously *cespitose, forming dense mats,* the closely ascending tufts *bristly-fibrillose at base; leaves* membranaceous to firm, scabrous, 1.5–3 *mm. broad; culms* of various lengths, *sometimes all crowded among the leaf-bases,* or *sometimes a few of them elongate to* 0.3–2 *dm. and stiffly ascending* as forma **vicìna** (Dew.) Wieg. (near), sharply angled, scabrous, bearing either staminate or pistillate spikes or both; pistillate spikes 0.5–1 cm. long, mostly sessile; staminate spikes 8–12 mm. long; *pistillate scales lance-ovate, gradually tapering to long or acuminate tips,* green, with brownish tinge; *perigynia* 3.2–4.7 *mm. long, finely pubescent, with ellipsoid-obovoid bodies* 1.25–2.2 mm. thick, *the beak* 0.9–1.7 *mm. long;* achene blackish. (*C. rugosperma* Mackenz.) — Dry sandy, argillaceous or rocky soil, P.E.I. to s. Ont. and Minn., s. to N.S., N.E., Md., upland Va., Ind., Ill. and Mo. April–July. Fig. 632.

632. C. umbellata.

128. C. ábdita Bickn. (hidden). — Very similar to no. 127; *pistillate scales broadly ovate or ovate-oblong, with short acute tips,* not long-tapering and gradually acuminate; *perigynia* 2.2–3.3 *mm. long, with globose-ovoid bodies* 1–1.8 mm. thick, *the beak* 0.5–1 *mm. long.* (*C. umbellata,* var. *brevirostris* Boott; *C. umbellata* sensu Mackenz., not Schkuhr) — Open woods, clearings and fields, Nfld. and s. Lab. Pen. to Sask., s. to N.B., N.E., L.I., Va., Tenn., Ill. and Minn.; Vancouver I. April–July. Fig. 633.

633. C. abdita.

129. C. tónsa (Fern.) Bickn. (shaved). — Coarser than no. 127; the very scabrous *hard and firm leaves becoming* 2.5–5 *mm. broad;* culms mostly short, very rarely 1 dm. high; *pistillate scales lance-ovate, long-tapering, about equaling to much longer than perigynia; perigynia* as in no. 127, *glabrous* or essentially so. (*C. umbellata,* var. *tonsa* Fern.) — Dry sands, rocks and openings, Gaspé Pen., Que., to n. Alta., s. to N.S., N.E., Va., Ind., Wisc. and Minn. April–July. Fig. 634.

634. C. tonsa.

§ 28. Digitàtae Fries (see p. 297)

Terminal spike pistillate at base, staminate above; capillary peduncles of lower (and often basal) spikes much longer than the spike; pistillate scales abruptly awned; perigynium 4–4.5 mm. long, its slender spongy stipe about equaling the glabrous or barely pubescent body. . . . . 130. *C. pedunculata.*

Terminal spike staminate to base; no basal or prolonged peduncles; pistillate scales blunt; perigynia 2–3.5 mm. long, not long-stipitate (except in no. 133).

Leaves flat, becoming 1.5–4 mm. broad; perigynia 2.5–3.5 mm. long, pubescent, nearly beakless.

Pistillate spikes cylindric, 1–2 cm. long; scales exceeding perigynia; staminate spike 1.5–2.5 cm. long, often stalked. . . . . 131. *C. Richardsonii.*

Pistillate spikes subglobose, 4–7 mm. long; scales much shorter than perigynia; staminate spike 3–6 mm. long, often sessile. . . . . 132. *C. concinna.*

Leaves soon plicate, short and curving, 1–1.5 mm. wide; perigynia 2–2.5 mm. long, glabrous and lustrous, slender-beaked. . . . . 133. *C. terrae-novae.*

130. C. pedunculàta Muhl. (peduncled). — Low and diffuse, 0.5–3 dm. high, forming purple-based lax mats; *leaves* abundant, dark green, overwintering, flat, becoming firm and 2–5 mm. broad, *mostly longer than the weak culms; staminate spike small, usually pistillate at base; pistillate spikes* 2–4 on each culm, *scattered and long-peduncled from green spathe-like sheaths*, erect or spreading, many *other spikes nearly or quite radical* and very long-stalked, *all 3–8-flowered; scales* green to purple, *truncate and cuspidate or aristate*, mostly a little longer than perigynia; *perigynia* smooth or slightly pubescent above, 4–4.5 *mm. long*, sharply angled, slenderly obovoid, *with slender spongy stipe* and minute beak. — Rich woods and slopes, Nfld. to Sask., s. to N.S., N.E., Del., upland to nw. Ga., O., Mich., nw. Ill., ne. Ia. and S.D. April, May (–July northw.). — Fig. 635.

635. C. pedunculata.

636. C. Richardsonii, f. exserta.

131. C. Richardsònii R. Br. (for its discoverer, Sir John Richardson, 1787–1865). — *Stoloniferous*; the small tufts of leaves and slender harsh ascending culms (1–3 dm. high) with brownish fibrillose bases; leaves firm, 2–4 mm. wide; culms bearing near base purple or brown spathiform sheaths with or without short green blades, and at summit purple or brown *sheathing spathes subtending the spikes; staminate spike* stout and mostly short-peduncled, 1.5–2.5 *cm. long; pistillate spikes* 1–3, *appressed-ascending*, linear- to lance-cylindric, their short peduncles included or barely exserted from the sheaths, or twice or thrice length of sheath in forma **exsérta** Fern. (exserted), fig. 636, 1–2 cm. long; *scales exceeding perigynia, brown, with a conspicuous white-hyaline margin*, pointless or obtuse; *perigynia ellipsoid-obovoid*, 2.5–3.5 mm. long, *firm, hairy*, the very short beak entire or erose. — Calcareous rocks, barrens, sands, open woods and prairies, sw. Vt. to n. Alta., s. to w. N.Y., n. O., n. Ind., n. Ill., Minn. and S.D.; very rare eastw. May, June.

132. C. concínna R. Br. (beautiful). — Loosely cespitose, with frequently prolonged caudices; *culms smooth* except at summit, *subfiliform, curved, flexuous or loosely ascending*, 0.3–2 dm. long, *with* thin colored spathiform basal bracts and *short terminal tubular bracts;* leaves dark green, soft, often curving, 1–3 mm. wide; *staminate spike 3–6 mm. long, often sessile; pistillate spikes 2 or 3, subglobose, 4–7 mm. long*, approximate, or the lower remote; *perigynia hairy, slenderly trigonous-ovoid*, blunt, 2.5–3 mm. long, *much exceeding the dark* pale-margined *roundish scales.* — Calcareous woods, cool banks and mossy knolls, Nfld. and Saguenay Co., Que., to Alaska, s. to n. N.B., Ont., n. Mich., ne. Wisc., S.D. and Colo. June, July. Fig. 637.

637. C. concinna.

133. C. térrae-nòvae Fern. (of Newfoundland). — *In small dense tussocks* with drab or pale brown bases; *culms* mostly overtopping leaves, *stiffly erect, obtusely angled*, 1–12 cm. high; *leaves plicate*, attenuate, very crowded, *curving*, long persistent, 1–1.5 *mm. wide;* staminate spike 2–6 mm. long, sessile, very slender, with few hyaline-margined scales; *pistillate spikes* 1–3, sessile or nearly so, *closely appressed-ascending, ellipsoid or thick-cylindric*, 2–7 mm. long, the lowest subtended by a short colored spathiform scarious bract; *scales purple or darker, shorter than the imbricated perigynia, caducous; perigynia slenderly ovoid-ellipsoid, fusiform, slenderly substipitate, 2–2.5 mm. long, glabrous and lustrous*, firm, tightly filled by achene, *with* hyaline-tipped *subtruncate cylindric beak* 0.3–0.5 *mm. long.* — Calcareous barrens of n. and w. Nfld.; L. Mistassini, Que. July, early Aug. Fig. 638. — Local representative of the arct.-alpine *C. glacialis* Mackenz.

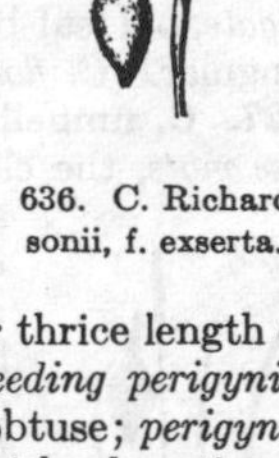

638. C. terrae-novae.

§ 29. Álbae Aschers. & Graebn. (see p. 297) — A single species with us:

134. C. ebúrnea Boott (like ivory). — Tufted from a *rigid pale brown stoloniferous base; culms capillary, wiry,* smooth, 0.7–4 dm. high, *naked at summit except for pale tubular truncate sheathing bracts; leaves involute-filiform,* mostly shorter than culms; *staminate spike* 4–8 mm. long, sessile or nearly so, *overtopped by the two or more upper spikes, the inflorescence corymbiform; pistillate spikes* 2–4, *on* slender *erect peduncles,* lanceolate, ovoid or ellipsoid, 2–6 mm. long; pistillate *scales whitish or pale brown,* thin, shorter than perigynia, caducous; *perigynia olivaceous to blackish, glabrous,* becoming lustrous, 1.5–2 *mm. long; base of style bulbous-thickened.* — Calcareous ledges, gravels or sands, Nfld. to se. Alaska, s. to N.S., w. N.E., Va., Ala., Ark. and Tex. May–Aug. Fig. 639.

639. C. eburnea.

§ 30. Triquètrae Carey (see p. 297) — A single species with us:

135. C. hirtifòlia Mackenz. (hairy-leaved). — Loosely cespitose, *pubescent throughout;* culms ascending, straight or arching, 0.2–8 dm. high, sharply angled, soft, pilose, harsh above; *leaves pale green, flat and soft,* 0.5–1 *cm. wide,* mostly shorter than culms; *bract* at base of inflorescence *nearly or quite sheathless, flat,* green; staminate spike pale, slenderly clavate, 0.8–2 cm. long, sessile or nearly so; pistillate spikes 2–4, appressed-ascending, approximate or subdistant, the lower short-peduncled, short-cylindric, 0.7–2.3 cm. long, loosely flowered; *perigynia very hairy, sharply 3-angled, firm,* 3.5–5 *mm. long, about* 1.5 *mm. thick, with slender* minutely toothed *beak about* 0.75 *mm. long,* about equaling the truncate and rough-cuspidate thin scales. (*C. pubescens* Muhl., not Poir.) — Rich (often calcareous) woods and meadows, w. N.B. to s. Ont. and Minn., s. to N.S., N.E., Md., Ky., Mo. and e. Kans. May–June. Fig. 640. — Hybridizes with no. 182, producing × **C. Sullivántii** Boott.

640. C. hirtifolia.

§ 31. Bicolòres Tuckerm. (see p. 297)

Bract shorter than to barely equaling inflorescence, with broad basal auricles; spikes (except sometimes the lowest) crowded into a subturbinate corymbiform glomerule, the terminal one strongly pistillate above the staminate base; scales purple-black, much shorter than the white-granulose perigynia; culms weak, reclining, arching or flexuous. . . . . 136. *C. bicolor.*

Bract foliaceous, usually exceeding inflorescence, the base without auricles; spikes distant to approximate; terminal one staminate throughout or pistillate at summit; culms ascending to erect.

  Terminal spike pistillate above; scales brown to purplish, their rounded summits much shorter than the crowded white-pulverulent dry perigynia. . . . 137. *C. Garberi.*

  Terminal spike staminate throughout (rarely slightly pistillate); scales whitish to fulvous, short-pointed; perigynia less crowded, fleshy, orange (drying brownish), minutely puncticulate. . . . . 138. *C. aurea.*

136. C. bícolor Bellardi (two-colored). — Loosely cespitose or substoloniferous, forming small pale mats; *culms subcapillary, curving, flexuous or arching,* 1–15 *cm. long; leaves* flat or channeled, *long-attenuate,* often curving, 1–2.5 mm. broad; *bract auricled at base, the short blade rarely overtopping the inflorescence; spikes crowded,* or the lower apart, *in a fastigiate glomerule; terminal spikes pistillate above; the others* (1–4) all pistillate, thick-ellipsoid, *closely flowered,* 5–10 mm. long; *scales* purple-black, obtuse or short-mucronate, *with evanescent* green *mid-rib,* slightly *shorter than the white-granulose* biconvex beakless stipitate membranaceous ellipsoid or obovoid *perigynium* (1.7–3 mm. long); *stigmas* 2; *achene suborbicular, biconvex, about half as long as perigynium.* — Damp calcareous gravels and peats, nw. Nfld.; also Greenl., Baffin I., Hudson Bay reg. and s. Alaska. July, early Aug. (Eurasia) Fig. 641.

641. C. bicolor.

137. C. Gárberi Fern. (for its discoverer, Abram Paschal Garber, 1838–1881). — Coarser than no. 136, more stoloniferous, the leafy tufts often fibrillose at base; *culms stiffly erect,* 0.5–4 dm. high; *leaves stiffish, flat, erect,* 1.5–5 *mm. broad,* shorter than to much overtopping culms; *bracts not auricled, the lower* (often others) *much exceeding inflorescence; spikes* 3–7, *densely thick-cylindric, the upper crowded,* the lower distant (sometimes basal ones on long peduncles), the *terminal staminate only at base,* 0.8–3 cm. long; *scales* membranaceous, broadly oblong to obovate, *brown to purplish* (rarely green),

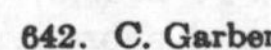
642. C. Garberi.

rounded at tip or barely mucronulate, shorter than perigynia; *perigynia* ellipsoid to obovoid, plump and biconvex, *dry and white-papillate,* 2–2.5 mm. long, with rounded beakless summit, short-stipitate. (*C. bicolor*, in part, of ed. 7, not All.; *C. Hassei* as to northern and eastern plants of Mackenz., not Bailey) — Calcareous sands, gravels and ledges, especially near the Great Lakes, L. Mistassini, Que., to nw. Ont., s. to n. and w. N.Y., O., Mich. and nw. Ind. June, July. Fig. 642.

Var. **bifària** Fern. (in two parts). — Weaker and more slender; culms flexuous or arching, slender, up to 6 dm. high; leaves thinner and softer, 1–2.5 mm. broad; spikes less crowded; perigynia 2.5–3 mm. long, with long stipes. — Calcareous shores and shaded ledges, e. Que., n. N.B. and n. Me.; Alaska to s. Alta. and s. B.C. Late June–Aug.

643. C. aurea.

138. C. aùrea Nutt. (golden). — Resembling no. 136, slender, pale green; culms 0.3–5.5 dm. high; leaves 1–3 mm. broad; 2 *or* 3 *bracts much prolonged; terminal spike most frequently staminate throughout,* sometimes pistillate at summit; pistillate spikes less approximate, more loosely flowered; *scales white to fulvous,* short-pointed; *perigynia fleshy at maturity, plump, orange, drying brown, minutely puncticulate.* — Meadows, springy banks and damp shores (chiefly calcareous), Nfld. to Alaska, s. to N.S., N.E., n. Pa., n. O., n. Ind., n. Ill., Minn., Neb., N.M. and Calif. June, July. Fig. 643.

§ 32. Cryptocárpae Tuckerm. (see p. 297)

a. Plant with elongate horizontal leafless stolons; culms 1–few from each crown; achene narrowly invaginated near the middle on one side; halophytic or of tidal shores.
- Pistillate spikes pendulous (rarely erect); scales with awns longer than the blades. . . . . 139. *C. paleacea.*
- Pistillate spikes erect or ascending (rarely pendulous); scales acuminate or with awns much shorter than the bodies.
  - Scales of pistillate spikes stramineous to rufescent or dark purple; perigynia plano-convex; pistillate spikes mostly erect. . . . . . 140. *C. salina.*
  - Scales blackish or dark purple; perigynia biconvex; pistillate spikes loosely spreading to pendulous. . . . . . 141. *C. Lyngbyei.*

a. Plants densely to loosely cespitose, without horizontal leafless stolons; culms several–many; at least the lower pistillate scales with long rough awns; plants of fresh habitats.
- Perigynia smooth or nearly so; the sides nerveless or sometimes with a single median nerve reaching the apex; achenes oblong to obovate, variously bent or contorted, often with a deep invagination on one or both margins near the middle. . . . . . 142. *C. crinita.*
- Perigynia manifestly but minutely granular-papillate, with 2–4 distinct nerves on each face extending nearly or quite to tip, lenticular, scarcely inflated; achene broadly ovate to suborbicular, not at all bent or contorted. . . . . . 143. *C. Mitchelliana.*

139. C. paleàcea Wahlenb. (chaffy). — Forming small tussocks *with prolonged rope-like* scaly but *leafless horizontal stolons; sterile erect leafy tufts* soon greatly *overtopping the culms,* 0.2–1.7 m. high, *the pale flat smooth* strongly ribbed *blades* 3.5–12 *mm. broad;* the *sharp-angled* relatively slender *culms* 0.1–1 m. high, *leafy,* their blades elongate, the upper bracteal ones prolonged; *pistillate spikes* 2–6, scattered, 2–8 cm. long, 0.8–2 *cm. thick,* often staminate at tip, *pendulous or loosely spreading, the lowest on filiform peduncles* 1.5–7.5 *cm. long,* or spikes only 1.2–2.5 cm. long and erect or strongly ascending with lower peduncles only 0.3–2 cm. long in forma **erectiúscula** Fern. (somewhat erect); staminate spikes 2–4, unequal, the long-stalked terminal one 2–6 cm. long; *pistillate scales stramineous to brown, the ascending awns longer than the blades and greatly exceeding the perigynia; perigynia* rounded-ellipsoid, *plano-convex,* 2.5–3 *mm. long,* pale, few-nerved or nerveless, with nearly entire beak 0.3–0.5 mm. long; stigmas 2; achene usually invaginated on one side near the middle. (*C. maritima* O. F. Muell., not Gunn.) — Saline or brackish marshes and shores,

644. C. paleacea.

Lab. to lower St. Lawrence R., Que., s., along coast to Mass.; Hudson Bay, Que. and Ont. June–Aug. (Eu.) Fig. 644. — Hybridizes with forms of no. 140.

140. C. salìna Wahlenb. (of salt). — Stoloniferous; differing from no. 139 in its leaves and culms sometimes smaller; *pistillate spikes short-peduncled to subsessile and erect, if long-stalked and pendulous only* 3–6 *mm. thick and with short-pointed scales; perigynia elliptic,* somewhat granular, 3–3.5 mm. long; *scales* purple or fuscous, lanceolate to ovate, *blunt to acuminate or cuspidate, the short awn, when present, much shorter than the blade.* — Polymorphic boreal halophyte. We have the following vars.:

Culms 3–18 cm. high, slender, often curving, obtusely angled; leaves 1–2.5 mm. wide, the margins involute toward apex; pistillate spikes 1–3, ellipsoid, 0.5–1.5 cm. long, 3–4 mm. thick; bases of lower foliaceous bracts dilated and partly sheathing bases of spikes; pistillate scales ovate, blackish (rarely pale), blunt, with midrib scarcely excurrent. . . . . Var. *subspathacea.*

Culms 1–9 dm. high, stiffish, erect; leaves 2–9 mm. wide, flat or with revolute margins; pistillate spikes 1–4, more cylindric, 1–8 cm. long, 3–10 mm. thick; bases of foliaceous bracts not dilated nor sheathing; pistillate scales ovate to lanceolate, acuminate or short-awned, with excurrent midrib.

Spikes all (or all but lowest) erect, short-peduncled to subsessile, the lowest peduncle 0.2–2 cm. long.

Culms 1–3 dm. high, obtusely angled; leaves 2–4 mm. wide, with revolute margins; pistillate spikes 1–3 cm. long, 3–4 mm. thick. . . . . . *C. salina* (typical).

Culms 1.5–9 dm. high, subacutely angled; leaves 2–9 mm. wide, flat; pistillate spikes 2–8 cm. long, 4–10 mm. thick. . . . . Var. *kattegatensis.*

Spikes (at least the pistillate) all filiform-peduncled and pendulous, the lowest peduncles 2–6 cm. long; pistillate spikes linear-cylindric, 2–6 cm. long, 3–6 mm. thick. . . . . Var. *pseudofilipendula.*

C. salìna (typical), *C. lanceata* Dew. — Saline or brackish shores, Greenl. and Lab. to n. Nfld. and Côte Nord, Gaspé Pen. and lower St. Lawrence R., Que.; Hudson Bay; Alaska. July, Aug. (N. Eu.) Fig. 645.

Var. **subspathàcea** (Wormskj.)Tuckerm. (somewhat sheathed), *C. subspathacea* Wormskj. — Saline shores, Greenl. and Baffin I. to Saguenay Co., Que.; Hudson Bay; Alaska. July, Aug. (N. Eurasia) Fig. 646.

Var. **kattegaténsis** (Fries) Almq. (of the Kattegat Strait), *C. recta* Boott. — Saline or brackish shores, marshes and swales, Lab. and shores of Hudson Bay, s. to Nfld., N.S., Me. and Mass. July, Aug. (N. Eu.) — Hybridizes with no. 139 and with no. 148.

Var. **pseudofilipéndula** Kükenth. (simulating *C. filipendula;* drooping on threads). — Local, s. Lab., Nfld. and Anticosti I., Que. (N. Eu.)

646. C. salina, v. subspathacea.

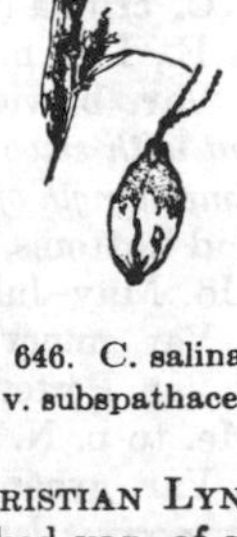

645. C. salina (typical).

141. C. Lýngbyei Hornem. (for its discoverer, Hansen Christian Lyngbye, 1782–1837). — Resembling no. 139 or the pendulous-spiked var. of no. 140; forming dense clumps *with hard leaves and subligneous stout dark stolons;* culms erect, 2–9 dm. high; *leaves firm,* revolute-margined; *pistillate spikes blackish or dark purple, pendulous* on capillary elongate peduncles, cylindric, 1–5 cm. long, 6–10 *mm. thick; scales dark,* rarely awned; *perigynia coriaceous,* 2.5–3.5 mm. long, elliptic to obovate, *biconvex.* — Greenl. and Lab., s. on seashores and coastal bluffs to Côte Nord, Anticosti I. and Gaspé Pen., Que.; Alaska to Calif. July, Aug. (Eurasia) Fig. 647.

647. C. Lyngbyei.

142. C. crinìta Lam. (long-haired). — *In* large or small *stools, without elongate leafless stolons,* usually *with several to many sharp-angled* often rough *culms* 0.3–1.6 m. high; *leaves* flat, more or less rough on nerves and margins, 4–12 mm. broad; the lowest shorter and *at the base of the culm reduced to fibrillose sheaths; pistillate spikes* 2–6, cylindric, variously peduncled, arching, pendulous or erect, 1–10 cm. long, *the lower subtended by foliaceous prolonged bracts; scales* greenish-brown, *rough-awned,* spreading or ascending, *equaling to twice to four times the length of the perigynia; perigynia often* more or less *inflated,* ovate-lanceolate or elliptic to suborbicular, 2–4 mm. long, *smooth or nearly so, the sides nerveless or sometimes with a single median nerve reaching the apex; achenes oblong to obovate,* lenticular, *variously bent or contorted, often with a deep invagi-*

nation on one or both margins near the middle; style bent. — Highly variable; the following vars. best marked:

a.Sheaths smooth and glabrous. . . b.

b.Pistillate spikes densely flowered; the somewhat spreading and crowded perigynia inflated, thick-ovoid to obovoid, loosely investing and obviously longer than achene; scales (at least the lower) long-awned. . . c.

c.Culms 0.3–1.6 m. high; leaves 4–12 mm. broad; pistillate spikes 3.5–10 cm. long, loosely spreading or drooping; perigynia 2–4 mm. long; scales (at least the lower) slightly exceeding to four times as long as perigynia.

Pistillate spikes rarely, if at all, staminate at apex; awns of lower scales two to four times as long as perigynia, upper scales definitely longer than perigynia; the latter 2–3 (–3.5) mm. long and 1–2 mm. thick; broadly ranging. . . . . *C. crinita* (typical).

Pistillate spikes often staminate at apex; awns of lower scales about equaling to twice length of perigynia, the upper scales barely as long as to but slightly longer than perigynia; the latter strongly inflated, 3–4 mm. long and 2–3 mm. thick; plant of Coastal Plain or of the South. Var. *brevicrinis.*

c.Culms 3–6 dm. high; leaves 4–5 mm. broad; pistillate spikes 1–3.5 cm. long, erect or strongly ascending; perigynia 2 mm. long; scales about twice as long as perigynia. . . . . Var. *minor.*

b.Pistillate spikes more loosely flowered; the ascending perigynia not inflated, closely investing and barely longer than achene, ellipsoid or ovoid; scales lance-attenuate. . . . . Var. *Porteri.*

a.Sheaths of lower leaves rough-hispidulous with short stiff ascending setae; perigynia ascending, moderately inflated, loosely investing the achene, chiefly ovoid; staminate spikes often pistillate at tip.

Culms 0.5–1.6 m. high; leaves 4–12 mm. broad; staminate spikes usually pistillate at tip; pistillate spikes drooping, 2.5–10 cm. long; perigynia 3–4 mm. long. Var. *gynandra.*

Culms 3–8 dm. high; leaves 4–6 mm. broad; staminate spikes often pistillate except at base; pistillate spikes erect or ascending, 1–3.5 cm. long; perigynia 3 mm. long. . . . . Var. *simulans.*

C. crinìta (typical). — Swales, damp thickets and low woods, Nfld. to n. Man., s. to N.S., N.E., L.I., n. Ga., Tenn. and Mo. Late May–Aug. — Hybridizes with nos. 144, 151 and 165.

Var. **brevicrinis** Fern. (with short hair). — *As coarse as* var. *gynandra* and *simulating it, but with smooth lower sheaths; lower foliaceous bract* 2.5–4 (*av.* 3.6) *dm. long, mostly three or more times length of axis of inflorescence; perigynia strongly inflated and crumpled.* — Wooded swamps and bottoms, N.C. to e. Tex., n. on or near Coastal Plain to s. N.E., and inland n. to Ky. and Mo. May–July.

Var. **mìnor** Boott (smaller). — Low woods, N.S., N.E. and N.Y. to Minn.

Var. **Pórteri** (Olney) Fern. (for its discoverer, Thomas Conrad Porter, 1822–1901). — Local, Me. to n. N.Y.

Var. **gynándra** (Schwein.) Schwein. & Torr. (with stamens and pistils). — Resembling var. *brevicrinis, but with sheaths of lower leaves scabrous, lower foliaceous bract* 1.2–4 (*av.* 2.5) *dm. long, averaging twice length of axis of inflorescence; perigynia ovoid, only slightly inflated, not crumpled.* (*C. gynandra* Schwein.) — Nfld. to Ont. and Wisc., s. to N.S., N.E., Md., D.C. and upland of N.C. and Tenn.; commoner northw. — Hybridizes with no. 165.

Var. **símulans** Fern. (imitating). — Nfld. to n. Ont., s. to N.S., Me., ne. Mass. and Vt., often at high altitudes.

143. C. Mitchelliàna M. A. Curtis (in honor of Elisha Mitchell, 1793–1857). — Resembling var. *gynandra* of no. 142; lower sheaths slightly hispidulous; leaf-blades thinner, only 2.5–9 mm. wide; terminal spikes usually staminate throughout; pistillate spikes often more slender; scales relatively short; *perigynia* 2.5–3.5 mm. long, *manifestly granular with numerous minute papillae, distinctly 2–4-nerved to apex or near it on both faces, lenticular, scarcely inflated,* distinctly longer than achene; *achene* broadly ovate to suborbicular, *with regular outline, not at all contorted or invaginated;* style straight. — Swales, swamps and wet woods, Fla. to e. Tex., n. on or near Coastal Plain to Mass., inland n. to Tenn. Late May–early Aug.

§ 33. Acùtae Fries (see p. 297)

a.Pistillate spikes erect or stiffly ascending; perigynia with tips plane or not twisted. . . b.

b.Culms solitary or few from small crowns, with lower sheaths only slightly if at all fibrillose; or, if culms numerous in dense tussocks, the plant with

pale soft and smooth leaves much overtopping the culms, their bases not fibrillose; perigynia (1.8–)2–3.5 mm. long.
Perigynia nerveless or only obscurely nerved near base; leaves 2–8 mm. broad.
Culms leafy and tall, 0.2–1.5 m. high, obtusely angled (or acute only above); leafy basal offshoots approximate to the fertile shoots, erect; leaves glaucous, scabrous on veins and margins; lowest bract usually equaling or overtopping inflorescence; terminal staminate spike 2–5 cm. long. . . . . **144.** *C. aquatilis.*
Culms leafy only near base, 0.2–4.5 dm. high, acutely angled; leafy basal offshoots mostly from long horizontal stolons; leaves dark green, smooth (rarely a little scabrous); lowest bract rarely equaling inflorescence; terminal staminate spike 0.5–2.5 cm. long. . . . . **145.** *C. Bigelowii.*
Perigynia nerved; leaves 1–3 mm. wide.
Stoloniferous, with elongate horizontal stolons; leaves mostly shorter than culms; green midrib of the purple to blackish scales very slender; coastwise species. . . . . **146.** *C. nigra.*
Stolons wanting, plant forming dense tussocks; leaves mostly much overtopping culms; green central portion of scales about as broad as the darker margins; transcontinental plant of fresh shores and swales. . . . . **147.** *C. lenticularis.*
b. **Culms numerous in stools, mostly exceeding leaves; lower sheaths strongly fibrillose or not; perigynia 2–2.75 mm. long.**
Pistillate spikes often clavate or tapering at base, often with staminate tips, the lower frequently peduncled; scales appressed-ascending; perigynia elliptic or ovate, not inflated.
Lower sheaths fibrillose; ligule longer than broad; lower pistillate spikes oftenest remote; perigynia granular-papillate above middle. . . . **148.** *C. stricta.*
Lower sheaths not fibrillose; ligule as broad as long; pistillate spikes commonly overlapping; perigynia granulose only at tip. . . . **149.** *C. Emoryi.*
Pistillate spikes cylindric, rarely attenuate at base, rarely staminate at tip, 1–4 cm. long, mostly sessile; scales soon divergent; perigynia obovoid to subglobose, inflated; lower sheaths rarely fibrillose. . . . . **150.** *C. Haydenii.*
a. **Pistillate spikes (at least the lower) curving, arching or drooping; perigynia with elongate finally bent or twisted beaks; cespitose, from stout freely branching rhizomes; leaves soft, dark, 3–5 mm. broad, the bracteal ones short.** . . . . **151.** *C. torta.*

144. C. aquátilis Wahlenb. (aquatic). — *Glaucous,* in small tufts, *with erect leafy tufts* and with cord-like horizontal stolons; *culms* erect, *obtusely angled,* smooth, except sometimes at the sharply angled summit, 0.2–9 dm. high; *leaves very glaucous,* firm, *equaling or exceeding the culms,* flat or channeled, 2–5 mm. broad, *slightly scabrous* on keel and margin and at long-attenuate tip, *their bases not fibrillose; bracts prolonged, the lower equaling to overtopping the spikes;* staminate spikes 1 or 2, the peduncled upper one 2–2.5 cm. long; pistillate spikes 3–5, erect, cylindric, dense, with rounded summit and tapering base, approximate or subdistant, sessile, or the lower short-peduncled, usually staminate at tip, 2–6 cm. long, 3–5 mm. thick; scales purplish (rarely paler), blunt to cuspidate, shorter than to exceeding perigynia; *perigynia flat,* elliptic or elliptic-obovate, 2.5–3 mm. long, 1.25–1.75 mm. broad, *nerveless or nearly so,* with minute entire beak; stigmas 2; achene lenticular. — Shallow pools, pond- and river-margins and swales, often alpine with us, Arct. Am., s. to Nfld., C.B., M.I., L. Mistassini, Que., Hudson Bay reg., N.M. and Calif. July, Aug. (Circumpolar) — Passing southw. into

Var. **áltior** (Rydb.) Fern. (taller). — Tall, 0.3–1.5 m. high; *culms more regularly acute-angled and scabrous at summit;* leaves 2.5–8 mm. broad; staminate spikes 2–5, the terminal one 3–5 cm. long; pistillate spikes 3–10 cm. long, 3–7 mm. thick, scattered; *perigynia more rounded-elliptic or -obovate, 2.3–3.3 mm. long,* 1.5–2.3 *mm. broad,* slightly nerved. (Var. *substricta* Kükenth.; *C. substricta* (Kükenth.) Mackenz.) — In sweet or calcareous waters and marshes, Nfld., and Côte Nord, Que., to B.C., s. to N.S., n. and w. N.E., n. N.J., N.Y., s. Ont., O., Mich., Ind., Wisc., Mo., Neb., Colo. and Oreg. Late May–Aug. — Hybridizes with no. 142.

145. C. Bigelòwii Torr. (for its discoverer, JACOB BIGELOW, 1787–1879). — *Stoloniferous, with cord-like horizontal leafy-tipped stolons; culms* 1–few from the leafy tufts, *smooth* or nearly so, *slender,* 0.2–4.5 *dm. high, acutely angled; leaves dark green,* smooth (rarely scabrous), firm, *mostly subbasal,* 3–7 mm. broad, *becoming revolute in drying; lowest bract rarely equaling inflorescence;* staminate spike usually peduncled, 0.3–2.5 cm. long, frequently with some pistillate flowers; *pistillate spikes* 1–6, approximate or distant, *erect,* sessile (or lowest peduncled), *linear-cylindric to clavate,* (0.5–)2.5 *cm. long,* 3–6 *mm. thick* (when attacked by fungus often short and thick, with greatly distended or elongate infertile perigynia); scales oblong to obovate, obtuse to mucronate, dark purple, with pale midrib, shorter than to exceeding perigynia; *perigynia* broadly elliptic to obovate, or narrowly elliptic to linear-lanceolate and in very slender-peduncled spikes in forma **anguillàta** (Drej.) Fern. (from *anguilla*, an *eel*, from the slender spikes), *flattish*, greenish to purple, *nerveless*, 2.5–3.5 mm. long; stigmas 2; achene lenticular. (*C. rigida* of ed. 7, not Good. nor Schrank; *C. concolor* sensu Mackenz., not R. Br.) — Greenl. and Baffin I. to Alaska, s. to bleak treeless barrens of Nfld., alpine areas of e. Que. and n. N.E. and n. N.Y., and reg. of Hudson Bay. July–Sept. (N. Eu.)

146. C. nìgra (L.) Reichard (black). — Loose to slightly cespitose, 0.5–6 dm. high, *strongly stoloniferous, the leafy and fertile tufts mostly terminating horizontal stolons;* culms sharply angled, smooth, or roughish at summit; *leaves* 1–3 *mm. wide*, soft, mostly shorter than culms, *bluish or glaucous, their margins involute in drying;* staminate spike solitary, more or less peduncled, or with smaller basal spikes; pistillate spikes 2–3(–4), erect, subdistant to approximate, usually the tips of the lower reaching those above, cylindric, 0.8–4.5 cm. long, 4–6 mm. thick, densely flowered or sometimes loosely so below, often staminate at tip, the lowest usually subtended by a bract 2–10 cm. long; *scales purple to blackish, with slender green midrib*, ovate, obtuse, much *narrower and* usually *shorter than perigynia; perigynia appressed, oval to round-ovate, mostly fine-striate toward base*, 2.5–3 mm. long, bright green to *tawny*, the short beak nearly entire; stigmas 2; achene lenticular. (*C. Goodenowii* J. Gay; *C. acuta* sensu Mackenz., not L.) — Open turf, swales, gravels, rocks, etc., s. Greenl. and se. Lab., s. to Nfld., lower St. Lawrence R., Que., N.S., Me., Mass. and R.I., mostly within 50 miles of the sea. May–Sept. (Eurasia)

Var. **strictifórmis** (Bailey) Fern. (with the form of *C. stricta*). — *Densely cespitose, forming erect clumps* 0.2–1 m. high, with leaves often equaling or exceeding culms; staminate spike usually long-peduncled; pistillate spikes more distant, 1.5–7 cm. long; stolons fewer. — Nfld. and se. Lab. to lower St. Lawrence R., Que., s. to N.S., Me., Mass. and R.I. — Simulating no. 148.

147. C. lenticulàris Michx. (lens-shaped). — *In dense tussocks, without elongate horizontal stolons, pale throughout*, 1–6 dm. high; culms slender, erect, sharply angled, usually harsh at summit; *leaves very narrow* (1–3 *mm. broad*), numerous, usually *much overtopping the culms;* spikes 3–8, more or less aggregated, or the lowest remote, the terminal one either staminate or androgynous, mostly sessile, erect, 1–4.5 cm. long, 2.5–4 mm. thick; *scales* obtuse, about half as long as to equaling perigynia, *the broad median green band about as broad as the brown or purplish margins; perigynia* lance-ovate to suborbicular, minutely granular, *brown-nerved*, 1.8–3.5 mm. long, short-stipitate, the tip empty and entire; stigmas 2; achene lenticular. — We have the following vars.:

Pistillate scales oblong or elliptic, 2–3.5 mm. long; perigynia lance-ovate to elliptic or rhombic, acute or acutish at both ends, 2.2–3.5 mm. long.
- Terminal spike wholly or mostly staminate. . . . . *C. lenticularis* (typical).
- Terminal spike mostly pistillate, staminate only at base.
  - Perigynia elliptic or oval, 2.2–3 mm. long, with very short stipe and tip. . . Var. *Blakei.*
  - Perigynia lance-ovate to subrhombic, 2.7–3.5 mm. long, with prolonged stipe and long tapering empty tip. . . . . Var. *albi-montana.*

Pistillate scales very short-oblong to suborbicular, 1.5–2 mm. long; perigynia broadly oval to suborbicular, rounded at both ends. . . . . Var. *eucycla.*

C. lenticulàris (typical). — Gravelly shores, meadows and swales, Lab. to Mackenz., s. to Nfld., N.S., Me., N.H., w. Mass., N.Y., s. Ont., Mich., Minn., Man., Sask., Id. and s. B.C. June–Sept. Fig. 648.

Var. **Blàkei** Dew. (for its discoverer, Joseph Blake, 1814–1888). — S. Lab. to Ont., s. to Nfld., N.S., Me., N.H., w. Mass., n. N.Y., n. Mich. and n. Minn.

Var. **àlbi-montàna** Dew. (of the White Mountains). — N. Lab.; and chiefly on borders of alpine brooks, Nfld., Gaspé Pen., Que., and White Mts., N.H.

Var. **eucỳcla** Fern. (well rounded). — Upper Humber and Exploits Rivers, Nfld.

148. C. strícta Lam. (erect). — *In dense and broad* (often high) *stools*, with crowded erect leafy and fertile shoots, horizontal stolons wanting or scarce; *some or all lowest sheaths strongly fibrillose* on the inner side; *culms* erect, *slender and firm, sharply angled*, with concave sides, rough above, 0.5–1.3 m. high, *overtopping the leaves;* leaves prolonged, long-attenuate, scabrous-margined, rather stiff, channelled and keeled below, 1.5–5.5 mm. broad; ligule longer than broad; 1 or 2 lowest bracts leafy and equaling the inflorescence; staminate spike 1, peduncled, or with 1 or 2 sessile ones at base; *pistillate spikes* 1–4, erect, sessile or the lowest short-peduncled, scattered, cylindric to clavate, densely flowered, or loosely so at base, *often staminate-tipped*, 2–11 cm. long, 3–6 mm. thick, or only 1–2 cm. long and densely flowered at base and usually without staminate apex in forma **brévior** House (shorter), or linear-cylindric and only 2–3 mm. thick and 2–7 cm. long and long-attenuate at base with the lower peduncled in forma **xerocárpa** (S. H. Wright) Kükenth. (dry-fruited); *scales* oblong-obovate to lanceolate, *appressed-ascending*, obtuse to acute, reddish-brown, with pale midrib, shorter than to exceeding perigynia; *perigynia ovate to elliptic*, unequally biconvex, *granular-papillate* (under a lens) *above the middle, appressed-ascending*, 2.2–2.75 mm. long, becoming tawny, *lightly few-nerved*, beakless, with entire tip; stigmas 2; achenes lenticular. — In acid or subacid swamps, swales and low woods, forming "nigger-heads", N.B. to Ont., s. to N.S., N.E., L.I., N.C., O., Ind., Ill. and Minn. May–Aug. — Hybridizes with var. of no. 148. — Passing insensibly to

648. C. lenticularis.

Var. stríctior (Dew.) Carey (straighter). — Often more glaucous, of looser habit, forming freely stoloniferous colonies of smaller tussocks; basal sheaths only sparingly fibrillose; pistillate spikes mostly 2.5–7.5 cm. long and frequently staminate at tip, or only 0.7–2 cm. long and without staminate tips in forma **curtíssima** (Peck) Kükenth. (very short). (*C. strictior* Dew.) — Similar habitats, often in calcareous swamps, Que. to Ont. and Minn., s. to N.S., N.E., N.C., Tenn. and Ia.

149. C. Emòryi Dew. (in honor of William Hemsley Emory, 1811–1887). — Resembling no. 148; *loosely cespitose*, forming colonies or open beds, *freely stoloniferous; lower sheaths not fibrillose*, the basal ones deeper purple or red; culms stouter than in no. 148; *leaf-sheaths* more *septate-nodose* (when dry); *ligule as broad as long; lower bract* prolonged, *equaling to much exceeding inflorescence;* pistillate spikes 3–6, strongly overlapping (rarely distant), 2–10 cm. long; *perigynia* ovate or obovate, becoming stramineous, abruptly short-beaked, *slightly granular-papillate only at apex*. — Swamps, river-margins and shores, oftenest in basic or calcareous waters, Fla. to Tex. and N.M., n. to nw. N.J., N.Y., O., Ind., Wisc., Minn., Man. and Colo. May, June.

150. C. Haydénii Dew. (for its discoverer, Ferdinand Vandeveer Hayden, 1829–1887). — Resembling no. 148, *in looser clumps*, but *without prolonged stolons;* basal sheaths rarely fibrillose; leaves stiffish; ligule as long as or longer than broad; *pistillate spikes* 2 or 3, *dense, rarely staminate at tip or long-attenuate at base*, sessile or nearly so, 1–4 cm. long, 4–7 *mm. thick; scales very sharp and wide-spreading, much longer than perigynia; perigynia obovoid to subglobose, inflated.* (*C. stricta*, var. *decora* Bailey) — Meadows, swales and thickets, mostly in rich soils, St. P. et Miq. to L. Mistassini, Que., w. to Ont., s. to N.B., N.E., N.J., Pa., O., Ind., Ill., Mo. and Neb. May–Aug.

649. C. torta.

151. C. tórta Boott (twisted). — Erect, *in open clumps from stout forking rhizomes* and cord-like roots; culms stout at base, sharply angled, smooth or roughish above, 2–9 dm. high; *leaves soft*, dark, 3–5 mm. broad, *flat, those of the culms very short; bracts short; pistillate spikes* 2–6 (rarely compound), mostly somewhat approximate, or the lower remote, the upper sessile and ascending, but the others often *spreading or drooping, curved or arching, slender*, 1.5–9 cm. long, 3–6 mm. thick; staminate spike 1 (rarely 2), peduncled, 1.5–4 cm. long, occasionally with pistillate flowers; pistillate scales purple-margined, obtuse, shorter than perigynia; *perigynia lance-ovate*, 2.5–3 mm. long, green, *the slim upper half empty and more or less tortuous*, the beak entire or erose. — By streams, rarely in swamps, Gaspé Pen., Que., to Minn., s. to N.S., N.E., Del., Md., upland to Ga. and Tenn., and Ark. May–July. Fig. 649. — Hybridizes with no. 142.

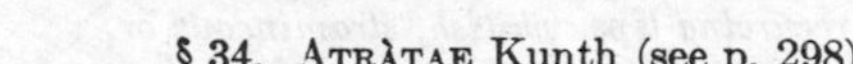

§ 34. Atràtae Kunth (see p. 298)

a. Cespitose, without long horizontal stolons; leaves 5–15 at the base of each culm, the basal sheaths not becoming fibrillose; scales blunt or acute, not awn-tipped, 1.5–4 mm. long; upland or boreal plants. . . *b*.

b. Pistillate scales blunt to acute, 1.5–3 mm. long, much shorter than perigynia; inflorescence erect (if flexuous, with terminal spike wholly staminate); pistillate spikes glomerulate or, if scattered, mostly erect.

Terminal spike staminate; pistillate spikes subdistant, the lowest peduncled; scales with a slender pale midrib. . . . . 152. *C. stylosa.*

Terminal spike mostly pistillate; all spikes crowded into an irregular dense glomerule, sessile or nearly so; scales dark throughout.

Densely cespitose; culms stiffly erect, 0.5–3 dm. long; perigynia trigonous-obovoid, abruptly short-beaked, 2–2.5 mm. long, becoming reddish or purple, strongly granular-papillose, the tips not recurving; alpine. . . . . 153. *C. norvegica.*

Loosely cespitose; culms weak, loosely ascending, 1.5–6 dm. high; perigynia ellipsoid or ellipsoid-obovoid, prolonged, 2.5–3.5 mm. long, tapering to beak, in age with widely spreading or recurving tips, whitish, stramineous or pale brown, sparingly granulose; woodland species. . . . . 154. *C. media.*

*b.***Pistillate scales acuminate (rarely blunt), 3–4 mm. long, purple-brown, about equaling perigynia; inflorescence arching or flexuous, the lateral spikes slender-peduncled, the terminal staminate only at base.** . . . . . 155. *C. atratiformis.*

*a.***Scarcely cespitose, culms 1–few from small tufts terminating long horizontal stolons; leaves 2–4 at base of each culm; basal sheaths finally becoming fibrillose; scales (or some of them) long-awned, 3–10 mm. long; paludal or marsh plant.** . . . . 156. *C. Buxbaumii.*

152. C. stylòsa C. A. Mey. (with a style). — *Cespitose*, forming dense tussocks with strong roots; *culms* 1–5 dm. high, slender, straight or curving, mostly divergent, scabrous except at base, *sharply angled; leaves* crowded, subrigid, flat, pale green, becoming revolute, shorter than culms, 1.5–3 *mm. wide, the inner band of the sheath soon fracturing;* bract narrow, rarely reaching staminate spike; *terminal spike staminate* (exceptionally pistillate at summit); *pistillate spikes* 1–3(–4), *thick-cylindric, subdistant, at least the lowest peduncled*, erect, 0.5–1.8 cm. long, 5–7 mm. thick; *scales* blunt to acutish, *nearly equaling or exceeding perigynia*, dark purple, *with slender pale midrib* and pale hyaline margin; *perigynia slenderly trigonous-ellipsoid*, inflated, 3–3.5 *mm. long*, long-stipitate, the surface pebbled, brown or purplish, nerveless or nearly so, with minute entire beak, the often persistent style-base prominent; *achene ellipsoid-obovoid, gradually rounded at summit*, with attenuate base. — Alaska. — Represented with us by

Var. nigritélla (Drej.) Fern. (diminutive, meaning small black). — Leaves less rigid, usually less scabrous, or smooth, *the inner band of the sheath remaining unbroken; culms* more slender, *commonly smooth except at tip;* pistillate spikes 4–6 mm. thick; *scales much shorter than perigynia; perigynia more broadly obovoid*, 2–2.5 *mm. long*, less stipitate; *achene more rounded-oblong, its summit broadly rounded to subtruncate.* (*C. nigritella* Drej.) — Peaty, turfy or gravelly soils, Greenl. and Lab. to Nfld., e. Saguenay Co., Que. and Ung. Late June–Aug. (N. Eurasia) Fig. 650.

650. C. stylosa, v. nigritella.

153. C. norvégica Retz. (Norwegian). — Habit of no. 152; culms stiff, 0.3–3 dm. high; leaves soft, 1.5–3 mm. wide; *spikes all pistillate or the terminal one staminate at base only, closely aggregated in a lobate glomerule, sessile* or nearly so; bract very short; *scales dark throughout; perigynia trigonous-obovoid, abruptly short-beaked*, 2–2.5 *mm. long, becoming reddish or purple, strongly granular-papillate, the tips erect.* (*C. alpina* Sw., not Schrank; *C. Halleri* of ed. 7, not Gunn., as to type; *C. Vahlii* Schkuhr) — Greenl. and Baffin I., s. to turfy slopes and crests of n. Nfld., Shickshock Mts., Gaspé Pen., Que., and Hudson Bay, Ung. and Man. July, Aug. (N. Eurasia) Fig. 651.

651. C. norvegica.

154. C. mèdia R. Br. (intermediate). — More *loosely cespitose* than no. 153, the weaker and arched-ascending culms 1.5–6 dm. high; leaves soft, loosely spreading or arching; *perigynia ellipsoid to ellipsoid-obovoid, more prolonged*, 2.5–3.5 *mm. long, tapering to beak, in age with widely spreading to recurving tips, whitish, stramineous or pale brown, less conspicuously granulose.* (*C. Vahlii*, var. *inferalpina* sensu Fern., not *C. inferalpina* Wahlenb., basonym; *C. angarae* Steud.) — Mossy, often calcareous, woods, thickets and shores, s. Lab. to Alaska, s. to Gaspé Pen., Que., n. N.B., n. Mich., n. Wisc., n. Minn., s. Man., s. Alta., Id. and s. B.C. Late June–early Aug.

155. C. atratifórmis Britt. (with form of *C. atrata*, blackish). — Loosely cespitose, with purple bases; *culms* slender, erect, sharply angled, roughish above, 0.2–1.1 m. high, *flexuous at tip;* leaves soft, warm

652. C. atratiformis.

green, flat, shorter than culms, 2–5 mm. broad; *spikes* 3–6, reddish-brown to purple-black, the terminal staminate only at base, *the* others *pistillate* and *on slender arching or recurving peduncles,* thick-cylindric or ellipsoid, 1–2.5 *cm. long; scales acuminate or blunt, purple-brown throughout,* 3–4 *mm. long, about equaling the perigynia;* perigynia broadly ellipsoid or ovoid, thin and puncticulate, very dark, 2.5–3 mm. long, the very short beak with barely notched orifice. (*C. atrata,* var. *ovata* (Rudge) Boott) — Brooksides, ravines and damp slopes, oftenest calcareous or circumneutral, Lab. to Yuk., s. to Nfld., Gaspé Pen., Que., n. N.B., n. Me., mts. of n. N.E., n. Mich., Sask. and s. Alta. Mid-June–early Sept. Fig. 652.

× **C. quirponénsis** Fern. (of Quirpon, Newfoundland) is an infertile hybrid of nos. 153 and 155, in n. Nfld.

156. C. Buxbaùmii Wahlenb. (for Johann Christian Buxbaum, 1693–1730). — Loosely tufted, the *small clusters of leaves and culms terminating long horizontal stolons;* culms 0.2–1 m. high, sharply angled, rough at summit; *leaves* narrow, *pale green and glaucous, sharply keeled,* 1.5–4 mm. broad, *only* 2–4 *at the bases of the culms, the old sheaths becoming fibrillose; spikes* 2–5(–7), *sessile* and approximate or the lowest short-stalked, from globular to thick-cylindric, 0.7–2 cm. long, or 2–5 cm. in forma **macrostáchya** (Hartm.) Kükenth. (large-spiked), the terminal spike turbinate and pistillate above, or staminate throughout in forma **heterostáchya** Anderss. (variable-spiked); *staminate scales very long-lanceolate; the pistillate lance-ovate and often long-awned, equaling or exceeding perigynia,* 3–10 *mm. long,* purple-black, fuscous or brown, with green midrib, or whitish or pale brown in forma **dilùtior** Kükenth. (weaker-colored); *perigynia* elliptic and beakless, *whitish and granular,* nearly nerveless, 2.5–4 mm. long, the orifice entire. (*C. polygama* Schkuhr, not J. F. Gmel.) — Wet shores, swamps and bogs, Nfld. to s. Alaska, s. to N.S., N.E., e. Va., upland of N.C., Ky., Ark., Colo. and Calif. Mid-May–early Aug. (Eurasia) Fig. 653.

653. C. Buxbaumii.

§ 35. Limòsae Tuckerm. (see p. 298)

a. Plant with loosely forking rhizomes or with horizontal stolons; culms 1–few from each tuft; pistillate scales obtuse to acute, not long-acuminate.
- Pistillate spikes oblong-cylindric, rarely staminate at tip, 2–30-flowered, 0.5–2(–2.5) cm. long; scales as broad as or broader than perigynia; rhizomes loosely forking, 1–3 mm. thick; leaves 0.5–3 mm. broad; boreal species.
  - Culms obtusely angled; leaves green, flat; peduncle of uppermost pistillate spike 1–10 mm. long; pistillate scales purple-brown to blackish, spathiform, embracing bases of young perigynia, promptly deciduous, leaving persistent old perigynia. . . . . 157. *C. rariflora.*
  - Culms sharply angled; leaves glaucous, involute or corrugated; peduncle of uppermost pistillate spike 0.5–3 cm. long; pistillate scales often paler, hardly spathiform, persistent. . . . . 158. *C. limosa.*
- Pistillate spikes linear-cylindric, often staminate at tip, the longest 2–5.5 cm. long, many-flowered; scales much narrower than perigynia; plant spreading by simple horizontal scaly stolons 4–6 mm. thick; leaves 2.5–6 mm. broad; se. species. . . . . 159. *C. Barrattii.*

a. Plant densely to loosely cespitose, forming clumps with few–many sharp-angled culms; leaves flat, scarcely glaucous; pistillate spikes thick-cylindric to subglobose, mostly pendulous; scales long-acuminate, much narrower but usually longer than perigynia. . . . . 160. *C. paupercula.*

157. C. rariflòra (Wahlenb.) Sm. (scantily flowered). — *In small tufts from loosely forking rhizomes* 1–3 *mm. thick,* the roots felted, the individual erect *tufts with solitary erect obtusely angled smooth culms* 0.5–3.5 dm. high; *leaves green, flat,* mostly short, 1–2 mm. wide; *bracts with dark auricled sheaths and short slender blades;* terminal spike staminate, peduncled, 0.5–2 cm. long; *pistillate spikes* 1–3, *oblong-cylindric,* 0.5–1.5 *cm. long,* ascending or pendulous, the lowest long-peduncled, *the uppermost on peduncles* 1–10 *mm. long; scales purple-brown to blackish, spathiform, embracing the bases and covering the perigynia, promptly deciduous, leaving persistent perigynia;* perigynia ellipsoid to ovoid, compressed-trigonous, 3–3.5 mm. long, glaucous, beakless. — Greenl. and Baffin I. to Alaska, s. on peaty barrens, bogs and pond-margins of Nfld., St. P. et Miq., M.I., Gaspé Pen., Que., and James Bay; Mt. Katahdin, Me. July, Aug. (N. Eurasia) Fig. 654.

654. C. rariflora.

158. C. limòsa L. (growing in mud). — *Spreading by prolonged slender* *loosely forking rootstocks, with long felt-covered roots; culms rising singly from along old rootstocks,* slender, *sharply angled, rough above,* 1.5–6 dm. high; principal *leaves* in separate tufts, *glaucous,* long-attenuate, *involute or corrugated;* terminal spike staminate, long-peduncled, 1–3 cm. long; *pistillate spikes* 1–3, *oblong-cylindric,* subdistant at summit of culm, ascending to pendulous, 1–2.5 cm. long, the lowest long-peduncled (very rarely the lowest basal and on exaggeratedly prolonged peduncle), *the uppermost on a peduncle* 0.5–3 *cm. long; scales stramineous to brown,* covering the perigynia, *hardly spathiform, persistent;* perigynia much as in no. 157, 2.5–4 mm. long. — Wet bogs, peaty quagmires, pond-margins, etc., Lab. to s. Alaska, s. to Nfld., N.S., N.E., Del., Pa., n. O., n. Ind., n. Ill., ne. Ia., Sask., Mont. and Calif. May–Aug. (Eurasia) Fig. 655. — Hybridizes with nos. 157 and 160.

655. C. limosa.

159. C. Barráttii Schwein. & Torr. (in honor of Joseph Barratt, 1796–1882). — *Spreading by simple horizontal scaly stolons* 4–6 *mm. thick;* the small leafy tufts with 1–few *strict concave-sided* sharply angled smoothish *culms* 3–9 *dm. high; leaves firm,* prolonged, pale green or glaucous, 2.5–6 *mm. broad,* flat, becoming revolute, *the basal sheaths becoming fibrillose; staminate spike* peduncled, *dark purple,* 2.5–5.5 *cm. long,* often with 1 or 2 small basal secondary spikes; *pistillate spikes linear-cylindric, often staminate at tip, the longest* 2–5.5 *cm. long,* 5–7 *mm. thick,* loosely spreading to pendulous, on thread-like peduncles; bracts short and narrow, purple-auricled; *scales much narrower* and mostly shorter *than perigynia,* obtuse, purple; perigynia narrowly ovoid, trigonous, 2.5–3 mm. long, granular and puncticulate, stramineous or purple-tinged, with very short entire beak. (*C. littoralis* Schwein., not Krocker) — Peaty swamps, pinelands and wet woods, on or near Coastal Plain, N.C. to N.J., L.I. and centr. Ct.; mts. of Ala. and Tenn. April–July. Fig. 656.

656. C. Barrattii.

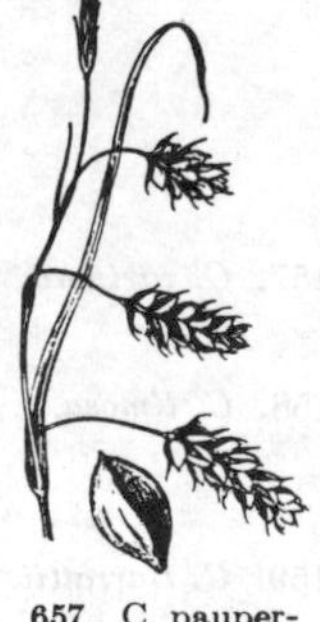

657. C. paupercula, v. irrigua.

160. C. paupércula Michx. (stunted). — Loosely to densely cespitose, *forming clumps* with yellowish-tomentulose roots; *culms acutely angled at summit,* 1–8 dm. high; leaves flat, pale green but scarcely glaucous, lax, shorter than culms; *lowest bract* as wide as leaves and *exceeding inflorescence;* staminate spike short-peduncled; *pistillate spikes* 2 or 3, subglobose to thick-cylindric, *mostly pendulous* on capillary peduncles; *scales* castaneous to stramineous, *long-acuminate, much narrower* but (except in one var.) *usually longer than perigynia, soon deciduous, thus exposing the subpersistent perigynia;* perigynia broadly ovate to suborbicular, nerved in the middle, pale, papillate, 3–4 mm. long, 1.5–2.5 mm. broad. (*C. magellanica* of N.Am. and Eu. auth., not Lam.) — Variable boreal species:

a. Pistillate scales 5–8 mm. long, much exceeding the perigynia. . . *b.*

b. Culms smooth except rarely at very base of inflorescence.

Pistillate spikes subglobose-ovoid, 4–8 mm. long; the 3–12 perigynia 3–3.5 mm. long and 1.5–2 mm. broad; staminate spike 7–11 mm. long; culms 1–2.5 dm. high; alpine. . . . . *C. paupercula* (typical).

Pistillate spikes thick-cylindric, 1–1.9 cm. long; scales castaneous to stramineous; the more numerous perigynia 3.2–4 mm. long and 1.7–2.5 mm. broad; staminate spikes 0.7–1.8 cm. long; culms 1–8 dm. high. . Var. *irrigua.*

b. Culms scabrous-serrulate on angles for 2–12 cm. below the inflorescence, more slender, 3–8 dm. high; spikes cylindric, 1–1.8 cm. long; scales pale brown to stramineous; perigynia as in preceding. . . . . Var. *pallens.*

a. Pistillate scales 3–4 mm. long, about equaling perigynia. . . . . . . Var. *brevisquama.*

C. paupércula (typical). — Lab. and Ung., s. to alpine peats of Shickshock Mts., Gaspé Pen., Que., and of White Mts., N.H. July, Aug. — Passing insensibly to

Var. irrígua (Wahlenb.) Fern. (well-watered). — Peaty or sphagnous bogs, meadows and woods, Lab. to Alaska, s. to Nfld., N.S., Me., Mass., N.Y., Mich., Wisc., Minn., Sask., Colo., Utah and Wash. June–Aug. (Eurasia) FIG. 657. — Through pale-scaled extremes passing into the strictly North American

Var. **pállens** Fern. (pale). — Similar habitats, Nfld. to Ont., s. to N.S., Me., nw. Ct., ne. Pa., N.Y., n. O., n. Ind., Wisc. and Minn.

Var. **brevisquàma** Fern. (short-scaled). — Local, e. Que. — Perhaps a hybrid of no. 160 with no. 158.

§ 36. PENDULÌNAE Fries (see p. 298)

Loosely cespitose, with long horizontal stolons, 1–6 dm. high; pistillate scales awnless; perigynia granulose-roughened, 3 mm. long; style abruptly bent; adventive in dry fields northward. . . . . 161. *C. flacca.*
Cespitose, without long stolons, (0.3–)5–12 dm. high; pistillate scales long-awned; perigynia minutely papillate or smooth, 2.8–5 mm. long; style straight; southern paludal natives.
Pistillate scales gradually tapering or rounded to the awn; perigynia rhomboid-ovoid, inflated, green, becoming brown, 4–5 mm. long, strongly ribbed. 162. *C. Joori.*
Pistillate scales retuse and notched below the awn; perigynia ovoid to obovoid, not inflated, glaucous, 2.8–3.5 mm. long, essentially nerveless. . . . . . 163. *C. glaucescens.*

161. C. FLÁCCA Schreb. (flabby). — *Very stoloniferous and glaucous,* with horizontal stolons; culms stiff, slender, 1–6 dm. high; *leaves firm,* short, 3–6 mm. broad, *with* revolute *scabrous margins;* bract equaling or exceeding inflorescence, sheathless; staminate spikes 1 or 2, peduncled, the terminal 2–3.5 cm. long; pistillate spikes 1–3, slenderly cylindric, 1.5–7 cm. long, 4–6 mm. thick, remote, mostly peduncled, erect or ascending; scales blunt or mucronate, purplish, with pale midrib, slightly shorter than perigynia; *perigynia subglobose to ellipsoid, plump,* spreading-ascending, 3 *mm. long, granulose-roughened,* stramineous to purplish, with minute entire beak; *style abruptly bent.* (*C. glauca* Scop.) — Dry fields and roadsides, N.S., Que., Ont. and Mich. June, July. (Natzd. from Eu.) FIG. 658.

658. C. flacca.

162. C. Joòri Bailey (for its discoverer, JOSEPH F. JOOR, 1848–1892). — *In small clumps* from stout horizontal rhizomes; culms 0.4–1.3 *m. high, firm,* acutely angled, scabrous above; basal sheaths coriaceous, castaneous; *leaves pale green or glaucous,* stiff, *much prolonged to attenuate tips,* 5–10 *mm. broad,* with scabrous revolute margins; lower bract barely sheathing, the upper ones setaceous; staminate spike 1.5–7 cm. long, on a long scabrous peduncle (rarely 1 or 2 small secondary spikes); *pistillate spikes* 3–6, cylindric, 1.5–7 *cm. long,* 6–9 *mm. thick,* sometimes staminate at tip, all or all but uppermost filiform-peduncled, erect to pendulous; *the scales and perigynia squarrose; pistillate scales* reddish-brown with green prolonged midrib, *gradually tapering or rounded to base of the rough awn; perigynia rhomboid-ovoid, inflated, green* and often glaucous *to brown,* 4–5 *mm. long, strongly ribbed,* wth a flattish-conical entire-orificed beak about 1 mm. long. (*C. macrokolea* of ed. 7, not Steud.) — Bottomlands, swampy woods and shores, Fla. to Tex., n. to e. Md., w. Tenn. and se. Mo. Aug.–Oct. FIG. 659. — Hybridizes with the next.

659. C. Joori.

163. C. glaucéscens Ell. (rather glaucous). — Resembling no. 162, usually strongly glaucous; culms up to 1.5 m. high; *leaves* 3–7 *mm. wide,* with red-dotted sheaths; *staminate spike* 1, thick-clavate; *pistillate scales retuse or notched below the awn; perigynia more glaucous, ovoid or rhomboid, not inflated,* 2.8–3.5 *mm. long, essentially nerveless.* (*C. verrucosa* of ed. 7, not Muhl.) — Swales, shallow water and wooded swamps, Fla. to La., n. to e. Md. July, Aug. FIG. 660. — The large clavate staminate spikes with long anthers very conspicuous in early summer; hybridizes with no. 162.

660. C. glaucescens.

§ 37. SHORTIÀNAE Bailey (see p. 298) — A single species:

164. C. Shortiàna Dew. (for its discoverer, CHARLES WILKINS

Short, 1794–1863). — In clumps from a thick rhizome; culms 2–9 dm. high, sharply angled, firm, smooth except at summit, usually shorter than leaves; leaves flat, rather thin, scabrous, 0.4–1 cm. broad; *spikes* 3–6, somewhat approximate near top of culm, the lowest 2 or 3 short-peduncled, ascending, 1–3.5 *cm. long*, 4–6 *mm. thick, evenly cylindric,* densely flowered, *the terminal one clavate and staminate below the long pistillate summit;* scales thin and blunt, divergent, about as long as perigynia; *perigynia squarrose,* somewhat *globose-obovoid, as broad as long,* olive to brown, sharp-edged but *nerveless, transversely rugose,* 2.5–3 *mm. long,* with entire orifice. — Rich woods, bottomlands and meadows, chiefly calcareous, s. Ont. to Ia., s. to w. Va., Tenn., Mo. and Okla. May, June. Fig. 661.

661. C. Shortiana.

× **C. Deàmii** F. J. Herm. (for Charles Clemon Deam, 1865–) is a hybrid of nos. 164 and 237.

§ 38. Anómalae Carey (see p. 298)
— A single species with us:

662. C. scabrata.

165. C. scabràta Schwein. (rough). — In small tufts from widely creeping horizontal stolons; *culms* 2–8 dm. high, *sharp-angled and scabrous; leaves membranaceous, ribbon-like,* 0.6–1.8 *cm. broad,* dark green, *scabrous;* bracts essentially sheathless; *spikes* 4–9, scattered, the *upper* 1 or 2 *sessile and overlapping the terminal staminate one;* the remainder often long-peduncled, 1–6 cm. long, cylindric, compact; *scales acute and rough-tipped,* green-nerved, about as long as body of perigynium; *perigynia green, membranaceous, loose,* ovoid-trigonous, *prominently angled,* 2.5–4 *mm. long, scabrous,* prominently few-nerved, *the curved hyaline* slightly toothed *beak nearly as long as the body.* — Wet woods, glades and meadows, Gaspé Pen., Que., to Ont., s. to N.S., N.E., n. Del., n. Md., upland to S.C. and Tenn., n. O. and Mo. June–Aug. Fig. 662. — Hybridizes with no. 142.

§ 39. Hírtae Tuckerm. (see p. 298)

a. Leaf-sheaths and blades glabrous; backs of scales glabrous; beak of perigynium with teeth less than 1 mm. long. . . *b.*

b. Perigynium glabrous, coriaceous, 4–6 mm. long, 2–3 mm. thick, strongly impressed-nerved; species of Coastal Plain bogs. . . . . 166. *C. Walteriana,* var. *brevis.*

b. Perigynium pubescent, smaller (except in no. 167, a northern species of dry soils with prominently ribbed perigynium). . . *c.*

c. Beak of perigynium firm, with acerose teeth; lower bract elongate, nearly reaching to much overtopping the terminal staminate spike; staminate spikes 1–3, elevated on a long peduncle usually well above the upper pistillate spike.

Perigynia 4–7 mm. long, 2–3 mm. thick, conspicuously ribbed, short-hirtellous; plant of dry habitats. . . . . 167. *C. Houghtonii.*

Perigynia 2.5–5 mm. long, 1.2–3 mm. thick, the obscure ribs mostly hidden under dense pilosity; plants of damp to paludal or aquatic habitats.

Leaves filiform-convolute except at base, smooth and wiry, 0.5–2 mm. thick; culms obtusely angled and smooth except sometimes at tip; teeth of perigynia 0.2–0.5 mm. long; achene ellipsoid. . . 168. *C. lasiocarpa,* var. *americana.*

Leaves flat, with revolute margins, scabrous, 2–5 mm. wide; upper 2 or more dm. of culm acutely angled and scabrous; teeth of perigynia 0.3–0.8 mm. long; achene broadly obovoid to obovoid-ellipsoid. . . . . 169. *C. lanuginosa.*

c. Beak of perigynium hyaline, whitish and only obscurely cleft at tip; lower bract rarely reaching the often solitary (rarely with small secondary spikes) subsessile to very short-peduncled staminate spike. . . . . 170. *C. vestita.*

a. Leaf sheaths and backs of staminate and pistillate scales pilose; beak of perigynium with pungent teeth 1–2 mm. long. . . . . 171. *C. hirta.*

166. C. WALTERIÀNA Bailey (for THOMAS WALTER, 1740–1789). — Extensively creeping by long cord-like scaly horizontal stolons; tufts of leaves and culms scattered; culms sharply angled, smooth or slightly rough, mostly shorter than the leaves, 0.3–1.2 m. high; *leaves firm*, pale green, septate-nodulose, strongly channeled at base, *flat but becoming involute* above, 2.5–9 *mm. broad*, with long-attenuate harsh tips; lowest bract prolonged; staminate spikes 1 or 2, usually on a peduncle; *pistillate spikes* 1 *or* 2, erect, mostly sessile, distant, cylindric, 1–6 cm. long, 0.5–1 cm. thick; ascending *perigynia* about twice as long as the thin ovate acute or short-cuspidate reddish-purple, green-centered and hyaline-margined scales, *ovoid, plump, coriaceous, pilose, impressed-nerved*, 4–6 *mm. long*, 2–3 *mm. thick*, with strong short beak purplish between the firm short (0.3–0.5 mm. long) teeth. (*C. striata* Michx., not Gilib.) — Pine-barren swamps and pond-margins, Fla. to S.C. — Represented with us by

663. C. Walteriana, v. brevis.

Var. **brévis** Bailey (short). — *Perigynia glabrous.* (*C. striata*, var. Bailey) — Similar habitats, S.C. to N.J., L.I. and se. Mass. May–Aug. FIG. 663.

167. C. Houghtònii Torr. (for its discoverer, DOUGLAS HOUGHTON, 1809–1845). — Extensively creeping, *with slender prolonged stolons* and small scattered tufts of leaves and 1–few *stiff sharply angled scabrous culms* 1.5–9 dm. high; *leaves flat*, often with revolute margins, 2–8 mm. broad, strongly ascending; *lower bract prolonged*, sheathless or with very short sheath; staminate spike usually 1 (or with small basal secondary ones) on a scabrous peduncle; *pistillate spikes* 1–3, sessile or the lowest short-stalked, erect, cylindric to subglobose, 1–4.5 *cm. long*, 7–12 *mm. thick*, compact; scales thin-margined, acute or awned, shorter than perigynia; *perigynia* ovoid, 4–7 *mm. long*, 2–3 *mm. thick*, *short-hirtellous, conspicuously ribbed*, the beak terminated by pungent teeth 0.5–0.8 mm. long. — Dry acid sands, gravels and clearings, Nfld. to Sask., s. to N.S., n. N.E., N.Y., Mich., Wisc. and Minn. June–Aug. FIG. 664.

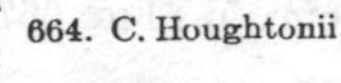
664. C. Houghtonii.

168. C. LASIOCÁRPA Ehrh. (hairy-fruited). — In tufts from long horizontal rhizomes and stolons; *culms* quill-like at base, slender, *obtusely angled and smooth*, 0.3–1.2 m. high; *leaves* light green, *filiform-convolute* except at base, *very slender* (0.5–2 *mm. thick*), *smooth* up to the acicular sometimes scabrous tips; lower bract prolonged, sheathless or short-sheathing; staminate spikes 1–3, elevated on a scabrous peduncle, the terminal much the longest (2–7 cm. long) and linear-cylindric; pistillate spikes 1–3, distant, sessile or nearly so, erect, cylindric to ellipsoid, 2–6 cm. long; scales acuminate-mucronate, reddish- or purplish-brown, with broad green center, about equaling perigynia; *perigynia oblong-ovoid*, 4–6 *mm. long*, 1.5–2 *mm. thick*, *the obscure nerves hidden by griseous pilosity, beaked, with sharp teeth* 0.7–1 *mm. long; achene obovoid.* (*C. filiformis* of auth., not L.) — Eurasia. — Represented with us by

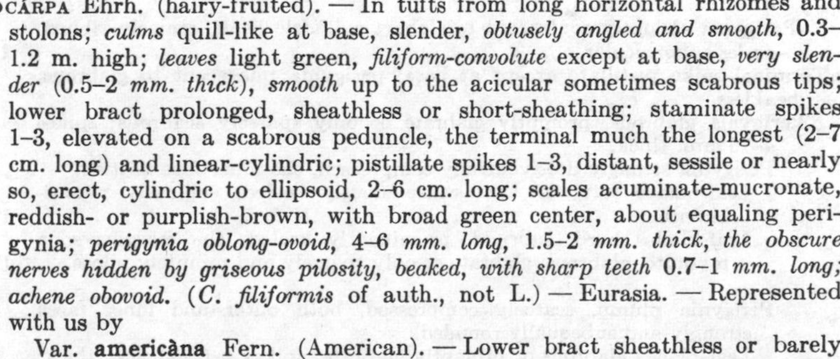

665. C. lasiocarpa, v. americana.

Var. **americàna** Fern. (American). — Lower bract sheathless or barely short-sheathed; pistillate spikes 0.8–3 (very rarely –6) cm. long; scales frequently awn-tipped or cuspidate; *perigynia ovoid-ellipsoid*, 3–4.5 *mm. long, with very short beak, the teeth only* 0.2–0.5 *mm. long; achene trigonous-ellipsoid or ellipsoid-obovoid.* (*C. filiformis* of ed. 7, not L.) — Bogs, peaty swales, inundated shores and shallow water, Nfld. to B.C., s. to N.S., N.E., L.I., n. N.J., Pa., n. O., n. Ind., n. Ill., ne. Ia., Man., Sask., Id. and Wash. May–Aug. (Ne. Asia) FIG. 665.

169. C. lanuginòsa Michx. (woolly). — Differing from no. 168 in its *flat* pale *scabrous leaves* 2–5 *mm. broad, their margins revolute;* upper third or fourth of *culm acutely angled and scabrous; perigynia* rather plumply ovoid, 2.5–5 mm. long, 1.2–3 *mm. thick*, often with well-developed neck, the *teeth* 0.3–0.8 *mm. long; achene trigonous-obovoid.* — Rich meadows, swales and shores, Anticosti I., Que., to B.C., s. to N.B., N.E., Va., Tenn., Ark., Okla., Tex., N.M., Ariz. and s. Calif. May–Aug.

170. C. vestita Willd. (clothed). — *Stiff, with* extensively creeping *hard* scaly *rhizomes and stolons; culms sharply angled,* 0.3–1 m. high; leaves firm, harsh, 2–5 mm. wide, with revolute margins; *bracts short, rarely reaching staminate spike; staminate spike clavate,* usually 2–5 cm. long (rarely with smaller basal spikes), *subsessile to short-peduncled,* or only about 1 cm. long and hidden among pistillate spikes in forma **Kénnedyi** (Fern.) Kükenth. (for its discoverer, George Golding Kennedy, 1841–1918); pistillate spikes 1–3, subapproximate or the lower remote (very rarely subradical), often staminate at tip, ellipsoid or thick-cylindric, 0.8–3 cm. long, compact; scales shorter than perigynia; *perigynia* ovoid to obovoid, nerved, *stiffly hairy,* 3–4 mm. long, short-beaked, *the beak often purple and white-hyaline at orifice which may become split in age.* — Dry sandy woods and clearings northw., often in damper woods southw.; s. Me. to e. N.Y., s. to Va. May–Aug. Fig. 666. — The large reddish-brown staminate spike with scales white- and hyaline-bordered conspicuous in anthesis.

666. C. vestita.

171. C. hírta L. (rough). — Widely creeping, forming loose open carpets; *culms* slender, firm, 2–6 dm. high, *obtusely angled; leaves* soft and *flat, generally sparsely hairy,* 2–4 mm. broad, *their sheaths densely hirsute;* staminate spikes 1–3 on a long peduncle; pistillate spikes 2 or 3, distant, more or less short-peduncled, erect or nearly so, 1.5–4 cm. long, rather loose; bracts elongate; *scales (both staminate and pistillate) pilose on back; perigynia subcoriaceous, slenderly conic-ovoid,* nerved, soft-hairy, 5–9 *mm. long, their long beaks deeply cleft, the* 2 firm spreading-ascending *teeth* 1–2 *mm. long.* — Dry fields, roadsides, etc., local, P.E.I. to Mich., s. to N.S., Mass., N.J., Pa. and D.C. June–Aug. (Natzd. from Eu.) Fig. 667.

667. C. hirta.

§ 40. Virescéntes Kunth (see p. 298)

a. Terminal spike staminate throughout (rarely 1–few scattered perigynia); perigynia glabrous.
- Perigynia rounded or narrowed to beakless tip, nerveless or only faintly nerved; scales oblong-ovate. . . . . 172. *C. pallescens,* var. *neogaea.*
- Perigynia abruptly contracted to a short cylindric beak, coarsely ribbed; scales suborbicular. . . . . 173. *C. Torreyi.*

a. Terminal spike pistillate except at base; perigynia pubescent to glabrous, beakless. *b.*

b. Perigynia glabrous, promptly glabrate or only sparsely scabrous; spikes 4–10 mm. thick.
- Perigynia strongly dorsiventrally compressed, the inner face flattish.
  - Leaf-blades soft, flat, not conspicuously keeled, copiously villous-hirsute. . . . . 174. *C. hirsutella.*
  - Leaf-blades hard, subrigid, prominently keeled, the margins soon revolute, glabrous, glabrate or only sparsely and minutely pilose. . 175. *C. complanata.*
- Perigynia plump, scarcely compressed, both outer and inner faces strongly and subequally rounded.
  - Leaf-blades glabrous or promptly glabrate; spikes 4–6(–7) mm. thick; anthers 1–2 mm. long; pistillate scales ovate, 2–3 mm. long; perigynia 2–3 mm. long. . . . . 176. *C. caroliniana.*
  - Leaf-blades soft-pilose; spikes 6–10 mm. thick (excluding tips of scales); anthers 3–4 mm. long; pistillate scales lance-ovate, long-attenuate, 3–6 mm. long; perigynia 3–4 mm. long. . . . . . . 177. *C. Bushii.*

b. Perigynia copiously soft-pilose; spikes 2–5 mm. thick; leaves pubescent.
- Spikes linear-cylindric, 1.5–4 cm. long, 2–4 mm. thick, subacute, usually tapering at base; anthers 1.5–2.5 mm. long; perigynia ellipsoid-ovoid, strongly ribbed on back. . . . . 178. *C. virescens.*
- Spikes subglobose to thick-cylindric, 0.5–1.8 cm. long, 3–5 mm. thick, obtuse, rounded to subtruncate at base; anthers 0.7–1.6 mm. long; perigynia broadly ovoid, more finely nerved. . . . . . . 179. *C. Swanii.*

172. C. palléscens L. (rather pale), var. **neogaêa** Fern. (of the New World). — Cespitose;

culms **slender, erect,** acutely angled, *pubescent,* 1.5–7.5 dm. high; *leaves* pale green, membranaceous, flat, *pilose* (*especially* the lower and *the lower brown sheaths*), 2–5 mm. wide; *lower bract* scarcely sheathed, prolonged, *often with undulate-margined base; terminal spike staminate,* linear-clavate, subsessile to short-peduncled; pistillate spikes 2–4, thick-cylindric to subglobose, 0.5–2.2 cm. long, 4.5–7 mm. thick, densely flowered, short (very rarely long) -peduncled to subsessile, erect or spreading, subapproximate; *pistillate scales oblong-ovate,* cuspidate to acute, about equaling perigynia; *perigynia oblong-ellipsoid, glabrous, strongly rounded to the beakless entire orifice, nerveless or faintly nerved,* 2.5–3 mm. long. — Meadows, grasslands, thickets and glades, Nfld. to Ont., s. to N.S., N.E., N.J., Pa., O. and Mich. May–Aug. FIG. 668. — Typical Eurasian *C. pallescens* has the perigynium more gradually rounded or tapering to a definite but very short beak.

668. C. pallescens, v. neogaea.

173. C. Tórreyi Tuckerm. (in honor of JOHN TORREY, 1796–1873). — In habit like no. 172; culms 1.5–5 dm. high; leaves 1.5–3 mm. wide; lowest bract rather short; pistillate spikes 1–3, 0.5–1.5 cm. long; *pistillate scales suborbicular; perigynia obovoid, strongly ribbed, abruptly short-beaked,* the beak cylindric. (*C. abbreviata* Prescott) — Meadows, copses and wooded slopes, n. Alta. to Minn., S.D. and Colo. June, July. FIG. 669.

669. C. Torreyi.

174. C. hirsutélla Mackenz. (slightly hirsute). — Loosely cespitose; *culms* slender, wiry (in deep shade weak), *pubescent,* acutely angled, 2–9 dm. high; *leaves soft, flat, pilose-hirsute,* prolonged, 1.5–4 mm. wide, their sheaths copiously pilose; *spikes* 2–5, all pistillate, *the terminal one with clavate staminate base* and usually longer than the subapproximate lower ones, *thick-cylindric to subglobose,* mostly sessile, 4–7 *mm. thick;* staminate scales pale, acuminate; pistillate scales ovate, blunt or short-cuspidate, shorter than perigynia; *perigynia glabrous, rounded-obovoid, short-pointed,* prominently nerved on back, less so on inner face, obtusely angled, *dorsiventrally compressed, with flattish inner face,* 2.2–3 mm. long. (*C. triceps,* var. *hirsuta* (Willd.) Bailey) — Open woods, clearings, fields and meadows, se. Me. and sw. Que. to s. Ont., Mich. and Ia., s. to Ct., e. Va., upland to S.C. and Ala., Miss., Ark., Okla. and e. Tex. May–July.

175. C. complanàta Torr. & Hook. (evened out or flattened). — Smoother than no. 173, the *glabrous or glabrate culms* soon overtopping the leaves and lengthening to 0.3–1.2 m. high; *leaves firm, subrigid, glabrous or glabrate, strongly keeled and with soon revolute margins; perigynia olivaceous, narrowly ellipsoid,* 2.5–3.5 *mm. long.* (*C. triceps* Michx., not Schrank) — Dry to moist sandy or argillaceous woods, clearings and fields, Fla. to Tex., n., mostly on Coastal Plain, to N.J., and inland n. to Tenn. and Ark. Mid-May–early July (rarely –Aug.) FIG. 670.

670. C. complanata.

176. C. caroliniàna Schwein. (of Carolina). — Habit of the preceding; *culms glabrous or glabrate,* becoming 0.3–1.2 m. high; *leaves with glabrous or promptly glabrate flat soft prolonged blades* 2–5 mm. wide; *spikes* 4–6(–7) *mm. thick; anthers* 1–2 *mm. long; pistillate scales ovate,* 2–3 *mm. long; perigynia plump, subterete and turgid, scarcely flattened,* olivaceous, 2–3 *mm. long.* (*C. triceps,* var. *Smithii* Porter) — Low woods, bottomlands, meadows, etc., Ga. to Tex., n. to N.J., Pa., W.Va., O., Ind., Mo. and Okla. May, June.

177. C. Búshii Mackenz. (for its discoverer, BENJAMIN FRANKLIN BUSH, 1858–1937). — Habit of nos. 174–176; the stiff culms sparsely pubescent; *leaves with soft-pilose blades* 2.5–6 mm. broad; spikes 2 or 3, 6–10 mm. thick (excluding scale-tips), the very large terminal one 1.2–2.5 cm. long; *anthers* 3–4 *mm. long; pistillate scales lance-ovate, long-attenuate,* 3–6 *mm. long,* mostly exceeding the perigynia; *perigynia very plump, rounded in section,* 3–4 *mm. long.* — Meadows (chiefly calcareous), fields, prairies and open woods, w. Mass. and R.I. to Mich. and Ia., s. to Md., W.Va., Miss., Ark. and Tex. Mid-May–early July.

178. C. viréscens Muhl. (greenish). — Densely cespitose, the *leaves and culms pubescent* and erect or strongly ascending; culms harsh above, 0.3–1.5 m. high; *leaves* pale green, *mostly shorter than culms,* 2–4 mm. broad; *spikes* 2–5, the terminal staminate at base and pistillate above, the others pis-

671. C. virescens.

tillate, 1.5–4 *cm. long*, 2–4 *mm. thick*, *linear-cylindric*, *subacute*, *usually tapering at base*, subdistant, the lower often peduncled; scales thin, whitish, acute, shorter than perigynia; *anthers* 1.5–2.5 *mm. long; perigynia slightly pilose, ellipsoid-ovoid*, compressed, 3-angled, substipitate, with entire orifice, *strongly ribbed on back*, 2–2.5 mm. long. — Dry woods, thickets and clearings, sw. Que. to Ind., s. to L.I., Ga., Tenn. and Mo. Late May–July. FIG. 671. — Hybridizes with nos. 182, 189 and var. of 190.

672. C. Swanii.

179. C. **Swánii** (Fern.) Mackenz. (in honor of CHARLES WALTER SWAN, 1838–1921). — Smaller than no. 178, *more copiously pilose; culms* 1.5–6(–8) dm. high, *often overtopped by the soft-pilose leaves;* the *thick-cylindric to subglobose spikes* 0.5–1.8 *cm. long*, 3–5 *mm. thick*, *broadly rounded at summit, broadly rounded to subtruncate at base; anthers* 0.7–1.6 *mm. long; perigynia broadly ovoid, finely nerved, copiously pilose.* (*C. virescens*, var. *Swanii* Fern.) — Woods, thickets and clearings, N.S. to s. Ont. and Wisc., s. to s. N.E., L.I., N.C., Tenn. and Ark. Late May–July. FIG. 672.

§ 41. FERRUGÍNEAE Tuckerm. (see p. 297) — A single species with us:

180. C. **misandroìdes** Fern. (resembling *C. misandra;* the latter name meaning man-hater, from the suppressed staminate flowers). — *In* loose tough and *dry clumps, with* hard scaly rhizomes and *mats of marcescent whitish- or gray-green conduplicate and curving wiry leaves* from brown bases; culms obtusely angled, smooth, 0.5–4.5 dm. high; lowest bract long-sheathing, slender; *spikes borne very irregularly*, 1–7, either approximate at summit of culm, all or partly scattered or even basal, some or all on capillary arching peduncles or the upper subsessile, the terminal 1 usually staminate, *the* others *pistillate* throughout or staminate at tip, *lance-ellipsoid to ovoid* and 0.5–3 *cm. long; scales brown to dark purple*, very thin, *conforming to the perigynia; perigynia thin, flattish*, rarely 3-angled, membranaceous, *oblong-lanceolate*, lustrous, 5–6 *mm. long, much longer than the lenticular stipitate achene; style-branches* 2 (rarely 3). — Dry calcareous barrens, ledges and rock-slides, local, sw. Nfld.; Gaspé Co. and L. Mistassini, Que. July, Aug. FIG. 673. — Anomalous representative of an Asiatic and arctic-alpine section, the other species regularly with 3 style-branches and trigonous achenes.

673. C. misandroides.

§ 42. GRACÍLLIMAE Carey (see p. 299)

*a.*Sheaths and blades glabrous or the former with only few scattered hairs.
- Spikes 3–5 mm. thick; pistillate scales acute or awned; perigynia ovoid, sharp-angled, tapering to a long curved beak, 3–4 mm. long. . . . . 181. *C. prasina.*
- Spikes 1–3 mm. thick; pistillate scales blunt or short-cuspidate; perigynia slenderly ellipsoid, obtusely angled, beakless, 2–3.5 mm. long. . . . 182. *C. gracillima.*

*a.*Sheaths and often the blades pilose. . . *b.*
- *b.*Leaves 1.5–3 mm. wide; spikes 2–3 mm. thick; perigynia beakless, 2.5–3 mm. long, 0.8–1.3 mm. thick; scales obtuse or short-cuspidate. . . . . . 183. *C. aestivalis.*
- *b.*Leaves 3–10 mm. wide; spikes 3–6 mm. thick; perigynia short-beaked, 3.5–6 mm. long, 1.5–2.5 mm. thick.
  - Pistillate scales much shorter than perigynia, if cuspidate with cusps much shorter than blade; perigynia 3.5–5 mm. long.
    - Peduncles mostly shorter than spikes; lateral spikes linear-cylindric, 1.5–4.5 cm. long, 3–5 mm. thick, floriferous to base; scales acuminate to cuspidate; southern species. . . . . 184. *C. oxylepis.*
    - Peduncles mostly longer than spikes; lateral spikes 1–2.5(–3) cm. long, 3.5–6 mm. thick, usually with small empty basal scales; scales obtuse, acute or merely short-cuspidate; northern species. . . . 185. *C. formosa.*
  - Pistillate scales mostly long-attenuate or awned, the slender tip about equaling the body and about equaling to exceeding perigynia; spikes 4–6 mm. thick; perigynia 4.5–6 mm. long. . . . . 186. *C. Davisii.*

181. C. **prásina** Wahlenb. (leek-green). — Loosely cespitose from tough crowns, brown at base; culms sharply angled, with concave sides, smooth except at summit, 2.5–8 dm. high, exceeded by or slightly exceeding leaves; *leaves* pale green, *thin and weak*, 2.5–6 *mm. wide; bracts* prolonged, usually overtopping inflorescence, *the pale sheath short;* terminal spike short-peduncled, pale, pistillate at summit or wholly staminate; *lateral spikes* 2–4, subapproximate, *pale green*, spreading to pendulous, mostly peduncled, linear-cylindric, 1.5–6 cm. long, 3–5 *mm. thick*, loosely flowered; *pistillate scales nearly colorless*, very thin, *acute or* *awned*, shorter than perigynia; *perigynia pale green*, thin, *nearly nerveless*, *ovoid*, *sharp-angled*, 3–4 mm. long, *produced into a long curving slender entire* or merely toothed *beak*. — Rich low woods and glades, Me. and s. Que. to s. Ont., and Mich., s. to Ct., e. Va., w. S.C. and Tenn. May, June. FIG. 674.

674. C. prasina.

182. C. gracíllima Schwein. (very slender). — In loose clumps, purplish at base; culms slender, diffuse to erect, 0.2–1 m. high, smooth; *leaves* membranaceous, rarely reaching inflorescences, deep green, those of the sterile tufts 5–9 mm. broad, *glabrous and with glabrous sheaths;* lower bract erect, prolonged, with tubular sheath; *terminal spike pistillate at summit* (very rarely wholly staminate); *lateral spikes* (2–)3 or 4, the upper approximate to the staminate one, the lowest remote, mostly slender-peduncled, drooping, spreading or loosely ascending, *linear-cylindric*, 1–6 cm. long, 2–3 *mm. thick; pistillate scales whitish, obtuse or short-cuspidate*, about half length of perigynia; *perigynia appressed-ascending, green*, becoming drab, *membranaceous, slenderly ellipsoid, obtusely angled*, obtuse to subacute, *beakless*, 2–3.5 *mm. long*, 1–1.7 mm. thick, with entire orifice. — Woodlands, thickets and meadows, Nfld. to se. Man., s. to N.S., N.E., Va., upland of N.C. and Tenn., and Mo. May–July (–Aug.) FIG. 675. — Hybridizes with nos. 135, 178 and 183.

675. C. gracillima.

Var. **macérrima** Fern. & Wieg. (very lean). — Low, only 3–5 dm. high; *leaves* 3–5 *mm. broad;* lateral *spikes* 1.5–3 cm. long, 1.5–2 *mm. thick; perigynia subdistant, subcoriaceous, brown, acutely angled*, 2–2.8 mm. long, 1–1.2 mm. broad. — Meadows, Bay of Islands, Nfld.

× **C. Sullivántii** Boott (for WILLIAM STARLING SULLIVANT, 1803–1873) is a very striking hybrid of no. 182 with no. 135.

183. C. aestivàlis M. A. Curtis (of summer). — Resembling no. 182, smaller and more slender; *culms* 2.5–9 dm. high, *sparsely pubescent; leaves* 1–3 *mm. wide, with pubescent sheaths; spikes loosely ascending* (rarely drooping), *loosely flowered*, 2–3 mm. thick; scales obtuse or short-cuspidate; perigynia slenderly ellipsoid-lanceolate, only 0.8–1.3 mm. thick. — Wooded or rocky slopes, e. Mass. and sw. N.H. to w. N.Y., s. to mts. of Ga. and Tenn. June–Aug. FIG. 676.

× **C. aestivalifórmis** Mackenz. (like *C. aestivalis*) is a local hybrid of no. 183 with no. 182, with which its author has erroneously placed several hybrids from quite outside the range of no. 183.

676. C. aestivalis.

184. C. oxýlepis Torr. & Hook. (sharp-scaled). — Similar to nos. 182 and 183; culms often slightly pubescent, 0.3–1.2 m. high; *leaves firmer*, often pubescent, 3–8 mm. broad, *with pilose sheaths; spikes linear-cylindric*, 1.5–4.5 *cm. long*, 3–5 *mm. thick, short-peduncled; scales acuminate to cuspidate, slightly shorter than perigynia; perigynia* lance-ellipsoid, *prominently nerved*, 3.5–5 *mm. long, prominently angled*, green, becoming pale brown, *short-beaked*. — Rich hardwoods, Fla. to Tex., n. to e. Va., Tenn. and se. Mo. Late April–early June. FIG. 677.

677. C. oxylepis.

185. C. formòsa Dew. (handsome). — Coarser than no. 184; *leaves membranaceous; lateral spikes oblong-cylindric*, 1–2.5 (–3) *cm. long*, 3.5–6 *mm. thick, mostly much shorter than the recurving filiform peduncles and with scattered empty scales along the tips of the peduncles; scales obtuse, acute or merely short-cuspidate; perigynia inflated, obscurely nerved*. — Calcareous woods, thickets and meadows, sw. Que. to s. Ont. and Minn., s. to w. Ct., N.Y., s. Mich., Wisc. and ne. Ia. May, June. FIG. 678.

678. C. formosa.

186. C. Davísii Schwein. & Torr. (for EMERSON DAVIS, 1798–1866, amateur student of *Carex*). — Resembling no. 185; *lateral spikes fertile to base*, 1.5–4.5 *cm. long*, 4–6 mm. thick, *erect or ascending, subsessile or only short-peduncled; scales mostly long-attenuate to long-awned, the slender tips about equaling to exceeding perigynia; perigynia inflated, oblong-ovoid, several-nerved, with* short slightly *bidentate beak*. — Rich calcareous woods, meadows and shores, w. N.E. to Minn., s. to Md., Tenn., Mo., Okla. and Tex. May, June. FIG. 679.

679. C. Davisii.

§ 43. Sylváticae Boott (see p. 299)

a. Leaves pilose; pistillate spikes oblong-cylindric, 0.8–2.5 cm. long, 4–8 mm. thick; beak of perigynium half as long as body, bidentate. . . . . . . 187. *C. castanea.*

a. Leaves glabrous or merely scabrous-hirtellous; pistillate spikes linear-cylindric, 1.5–8 cm. long, 2–6 mm. thick. . . *b.*

b. Pistillate spikes remotely 1–6(–8)-flowered; perigynia hispid, the slender beak nearly one-half as long as the lance-subulate body; northwestern species. . . . . 188. *C. assiniboinensis.*

b. Pistillate spikes mostly 10–50 (or more)-flowered; perigynia glabrous and lustrous, if puberulent with beak less than one-fourth length of body. . . *c.*

c. Perigynia nerveless or with few basal nerves.

Nerves at base of perigynium definite but short, the beak very much shorter than the body; inner band of lower bract-sheath hyaline and glabrous.

Basal leaves becoming 6–10 mm. broad; pistillate scales awned or cuspidate; perigynia tight, 3–5 mm. long, definitely 3-angled, strongly stipitate; achene sessile, longer than the style. . . . . 189. *C. arctata.*

Basal leaves 2–7 mm. broad; pistillate scales rarely awned; perigynia loose, subinflated, 4–10 mm. long, usually obscurely 3-angled, fusiform-lanceolate; achene stipitate, shorter than the style. . . . 190. *C. debilis.*

Nerves wanting; beak about as long as the definitely 3-angled ellipsoid body; scales long-awned; inner band of lower bract-sheath herbaceous, with firm chartaceous summit. . . . . 191. *C. sylvatica.*

c. Perigynia with about 10 strong uniform nerves from base to orifice, stipitate; inner band of lower bract-sheath coriaceous, scabrous at summit; Coastal Plain species. . . . . 192. *C. venusta.*

187. C. castànea Wahlenb. (chestnut-colored). — *Loosely cespitose,* with tufts of leaves and culms from purplish to castaneous crowns; *culms* slender, *often lateral on the caudex, pilose,* 0.2–1.1 m. high; *leaves flat,* 2.5–7.5 mm. broad, *pilose,* much shorter than culm; staminate spike 0.7–2 cm. long, short-peduncled; *pistillate spikes* 2–5, approximate, widely spreading or drooping on filiform peduncles, *oblong-cylindric,* 0.8–2.5 *cm. long,* 4–8 *mm. thick,* densely flowered, *tawny; scales brown, acute, thin; perigynia slenderly conic,* 3.5–5 mm. long, membranaceous, nerved, glabrous, *the* hyaline-tipped *bidentate beak about half as long as body.* — Calcareous woods, thickets, shores and meadows, Nfld. to L. Mistassini, Que., w. to Thunder Bay Distr., Ont., s. to C.B., N.B., n. and w. N.E., N.Y., Mich., Wisc. and Minn. Late May–July. Fig. 680. — Hybridizes with no. 189.

680. C. castanea.

188. C. assiniboinénsis W. Boott (of Assiniboine R., Can.). — Loosely cespitose, the small crowns with reddish-purple bases; culms 3–7.5 dm. high, very slender, angled, smooth or slightly scabrous; *leaves* glabrous, 2–3 *mm. wide;* bracts short; staminate spike long-peduncled, 2–3 cm. long; *pistillate spikes* 2 or 3, *very remote, alternately and remotely* 1–6 (–8)*-flowered,* 1.5–3 cm. long; scales green and stramineous, lance-acuminate or short-awned, about equaling perigynia; *perigynia lance-subulate,* 5–6.5 *mm. long, hispid, with slender beak about half as long as body.* — Rich woods, thickets and shores, s. Man. to Wisc., n. Ia. and S.D. June, early July. Fig. 681.

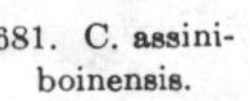

681. C. assiniboinensis.

189. C. arctàta Boott (contracted). — Cespitose, *with purple bases and* usually abundant *firm dark green overwintering basal leaves* 6–10 *mm. broad;* culms 0.2–1 m. high, obtusely angled, firm, smooth, ascending or arching; lower bract broad and short, long-sheathing, the inner band veinless and hyaline; staminate spike short-peduncled or subsessile among the upper pistillate ones; *pistillate spikes* 3–5, the upper approximate, the lowest distant and loosely ascending to flexuous on a filiform peduncle, *linear-cylindric,* 1.5–8 cm. long, 3–5 mm. thick, loosely flowered; *scales awned or cuspidate,* pale green or whitish, with darker midrib; *perigynia spreading-ascending, tight, lustrous, with short prominent basal nerves,* 3–5 *mm. long, definitely* 3*-angled, abruptly and conspicuously stipitate, abruptly contracted to a short beak; achene sessile, longer than style.* — Woods, thickets and clearings, Nfld. to w. Ont., s. to N.S., Me., Mass., w. Ct., Pa., n. O., Mich., Wisc. and Minn. June–Aug. — Hybridizes with no. 178 and more frequently with no. 187, then producing

× **C. Knieskérnii** Dew. (for its discoverer, Peter D. Knieskern, 1800–1871), *C. arctata* ×

castanea; C. castanea, var. *Knieskernii* (Dew.) Mackenz. — Occurring more widely than most hybrids.

190. C. débilis Michx. (weak). — Resembling no. 189; *more slender and lax, with basal leaves only 2–7 mm. broad;* spikes often overtopped by the leafy bracts; *scales rarely awned,* white-margined to tawny; *perigynia mostly subinflated, soft and thin,* 4–10 *mm. long, usually only obscurely 3-angled, fusiform-lanceolate, faintly nerved at base; achene stipitate, shorter than style.* A very variable species, with well-marked but confluent vars.:

a. Perigynia glabrous and lustrous; scales rarely cuspidate. . . *b.*
b. Perigynia mostly overlapping, obscurely angled; pistillate spikes 1.5–6 cm. long; midrib of pistillate scales usually evanescent at tip.
Scales whitish; perigynia 6–10 mm. long. . . . . *C. debilis* (typical).
Scales stramineous to greenish-brown; perigynia 4.5–7 mm. long.
Basal leaves 2–4 mm. broad, thin; spikes loosely spreading to pendulous; perigynia stramineous to rusty, twice as long as scales. . Var. *Rudgei.*
Basal leaves 4–7 mm. broad, firmer; spikes stiffer, simply spreading to erect; perigynia greener, barely one-third longer than scales. . . Var. *strictior.*
b. Perigynia mostly remotely alternate, not overlapping, firmer and more definitely trigonous; spikes 4–8 cm. long; midribs of pistillate scales sometimes excurrent; basal leaves subcoriaceous, 4–7 mm. wide. . . . Var. *interjecta.*
a. Perigynia minutely puberulent, often more nerved, 5–9 mm. long.
Pistillate scales with midrib evanescent below the tip. . . . . Var. *intercursa.*
Pistillate scales with excurrent midrib. . . . . Var. *pubera.*

C. débilis (typical). — Low woods, thickets, swamps and clearings, Fla. to Tex., n. to se. Mass., L.I., N.J., Pa., Ky., s. Ind. and Ark. May, June. Fig. 682.

682. C. debilis (typical).

Var. **Rúdgei** Bailey (in honor of Edward Rudge, 1763–1846), *C. flexuosa* Muhl.; *C. tenuis* Rudge, not Gmel. — Open woods, thickets and meadows, Nfld. to Ont., Wisc. and Minn., s. to N.S., N.E., L.I., N.C., Tenn. and Mo. May–Aug. Fig. 683. — Hybridizes with no. 179.

Var. **stríctior** Bailey (straighter). — Woods and clearings, centr. Me. to Vt. and Mass., mostly at high altitudes (up to 1600 m.). July, Aug.

Var. **interjécta** Bailey (thrown between). — Woods and clearings, n.-centr. N.H. to s. Mich., s. to Ct., n. N.J., Pa., and mts. of w. Va. and e. Tenn. June, July. — Very distinct in its extreme development.

Var. **intercúrsa** Fern. (run between). — Swampy woods and clearings, se. Va. and e. N.C.

Var. **pùbera** Gray (puberulent), *C. allegheniensis* Mackenz. — Low woods and meadows, Pa. to Ga., e. of the Alleghenies. May, June.

683. C. debilis, v. Rudgei.

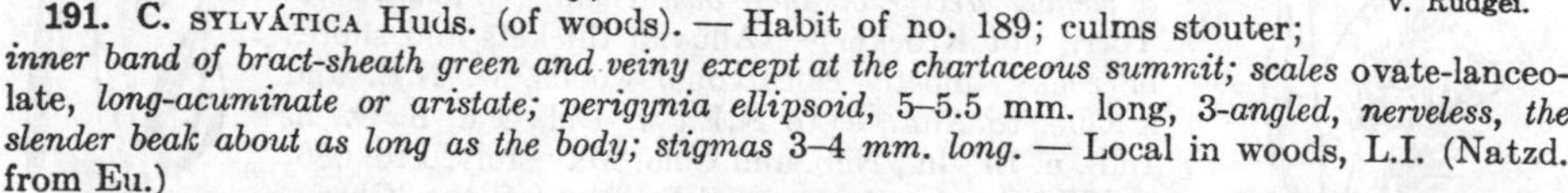

191. C. sylvática Huds. (of woods). — Habit of no. 189; culms stouter; *inner band of bract-sheath green and veiny except at the chartaceous summit; scales* ovate-lanceolate, *long-acuminate or aristate; perigynia ellipsoid,* 5–5.5 mm. long, 3-*angled, nerveless, the slender beak about as long as the body; stigmas* 3–4 *mm. long.* — Local in woods, L.I. (Natzd. from Eu.)

192. C. venústa Dew. (pleasing, graceful). — Cespitose, with purplish bases; culms slender, firm, sharply angled, 0.4–1.1 m. high; basal leaves overwintering and becoming subcoriaceous, 4.5–8 mm. broad; *inner band of lower bract-sheath coriaceous, scabrous at summit;* staminate spike (sometimes with a few perigynia) 2–5 cm. long, short-peduncled, usually only partly overtopping the uppermost pistillate one; *pistillate spikes* 3 or 4, the lowest remote, short-peduncled and erect to loosely spreading or drooping, *linear-cylindric,* 2.5–5.5 cm. long, 4–6 *mm. thick;* scales brownish, shorter than perigynia, promptly deciduous; *perigynia* ascending, imbricated, *green,* becoming pale brown, *lance-ellipsoid,* 5.5–8 *mm. long, strongly about* 10-*nerved from base to orifice, short-pubescent,* red-dotted, *stipitate.* — Sphagnous bogs, wet mossy woods and pinelands of Coastal Plain, Fla. to se. Va. May, June.

684. C. venusta, v. minor.

Var. **mìnor** Boeckl. (smaller). — Perigynia glabrous. (*C. oblita* Steud.) — Similar habitats, Ga. to La., n. to N.J. and L.I. May–early July. Fig. 684.

C. **brúnnea** Thunb. (brown) of § **Grácilles**, with spikes all pistillate, usually with staminate tips, 1–2 cm. long, often 2 or 3 together at a node, stiff and harsh narrow leaves, the stipitate and long-beaked brown strongly costate perigynia hispid, has been found on waste at Salem, Mass., *fr.* in Sept. (Adv. from Asia, the Malayan reg. or Australia)

§ 44. Capillàres Aschers. & Graebn. (see p. 299) — A single species with us:

193. C. capillàris L. (hair-like). — *In dense tussocks; culms* filiform, erect, 0.2–2 *dm. high,* smooth; *leaves flat* or finally involute, 0.5–2 *mm. wide,* the lower 0.2–1 dm. long, *with drab or brown bases; spikes* 2–4, *approximate* (or the lower up to 2 cm. distant) *in a racemose-corymbiform inflorescence* 1–4(–6) cm. long; staminate spike pale, sometimes with a few pistillate flowers, filiform-peduncled, often overtopped by the upper pistillate one; *pistillate spikes* 1–3, slender-peduncled, *oblong-cylindric, 6–8-flowered,* 4–10 *mm. long,* 2–4 *mm. thick; scales pale, blunt or short-tipped, promptly deciduous; perigynia olivaceous,* firm, *lustrous, lance-ovoid,* subfusiform, 2–3 *mm. long, less than* 1 *mm. broad, definitely 3-angled, tapering to a conic beak with truncate or entire orifice.* — Arct. Am., s. to exposed mostly calcareous habitats of Nfld., M.I., Gaspé Pen., Que., alpine reg. of Mt. Washington, N.H., and high mts. of Colo. and Utah. Late June–Aug. (Eurasia) Fig. 685.

685. C. capillaris.

Var. **màjor** Blytt (larger). — Looser, mostly in smaller tufts, 2–6 *dm. high;* basal *leaves* 1.5–4 *mm. broad,* the longer 0.6–3 dm. long; *inflorescence elongate,* loosely racemose, (4–)6–20 *cm. long; pistillate spikes up to 20-flowered,* 0.7–1.7 *cm. long; perigynia* 2.5–4 *mm. long.* (Var. *elongata* Olney) — Damp, springy or mossy calcareous woods, thickets, shores and wooded swamps, s. to Nfld., Que., s. N.B., n.-centr. Me., n. Vt., w. N.Y., Mich., Wisc., Minn., Man., Sask., etc. June, July. (Eurasia) — Although by some considered merely a luxuriant form, var. *major* is widely dispersed with us at low altitudes, typical *C. capillaris* arct.-alpine.

§ 45. Longiróstres Kükenth. (see p. 299)

Rhizomes and bases of leafy tufts coarsely fibrillose; perigynia subglobose-ovoid, barely inflated, nerveless (except sometimes at base), abruptly contracted to slender beak as long as body; northern. . . . . 194. *C. Sprengelii.*

Rhizomes and bases not fibrillose; perigynia conic-ovoid, inflated, prominently ribbed, tapering to beak much shorter than body; southern. . . . . 195. *C. cherokeensis.*

194. C. Sprengélii Dew. (in honor of Kurt Sprengel, 1766–1833). — Cespitose, *with coarse strongly fibrillose dull brown or drab bases;* culms 0.3–1 m. high, slender, sharply angled, scabrous at summit; leaves flat, loose, 3–4 mm. wide; staminate spikes 1–4, peduncled; *pistillate spikes* 2–5, oblong-cylindric, 1–5 cm. long, 8–10 mm. thick, rather loosely flowered, *slender-peduncled, drooping or loosely ascending; scales awned or acute,* pale, about equaling perigynia; *perigynia with* green to straw-colored *lustrous globose-ovoid ribless bodies abruptly contracted to slender erect beaks their own length.* (*C. longirostris* Torr., not Krocker) — Alluvial thickets and shores or rocky (mostly calcareous) woods, w. N.B. and s. Que. to Alta., s. to N.E., w. Del., Pa., n. O., n. Ind., n. Ill., Ia., Neb. and Colo. May–July. Fig. 686.

686. C. Sprengelii.

195. C. cherokeénsis Schwein. (of the Cherokee country). — Resembling no. 194, *with darker non-fibrillose bases; culms* 2–7 dm. high, *obtusely angled, smooth;* leaves 3–6 mm. broad; staminate spikes whitish; *pistillate spikes* 2–10, *remote, often in 2's or 3's,* 1.5–5 cm. long, 5–9 mm. thick; *scales acuminate,* pale; *perigynia conic-ovoid, inflated, strongly ribbed,* 5–6 mm. long, *tapering into beak much shorter than body.* — Low calcareous woods, swamps and prairies, Fla. to Tex., n. to Ga., Tenn., se. Mo. and Okla. April, May. Fig. 687.

687. C. cherokeensis.

§ 46. Granulàres O. F. Lang (see p. 299)

Plant cespitose, with several culms from a crown; leaves flat, those of the season flaccid, 3–12 mm. wide; staminate spike sessile or short-peduncled; lowest pistillate spike rarely basal. . . . . 196. *C. granularis.*

Plant loosely stoloniferous, with culms solitary; leaves often folded, firm, glaucous, 1.5–4 mm. wide; staminate spike long-peduncled; lowest pistillate spike often basal. . . . . 197. *C. Crawei.*

196. C. granulàris Muhl. (granular). — *In small or large tussocks, with overwintering leaves becoming firm* or subcoriaceous, often glaucous, *5–12 mm. broad;* culms slender, loosely spreading to ascending, 1–6(–9) dm. high; *leaves of the season flat, flaccid,* warm green to glaucous; bracts broad and elongate; *staminate spike* 0.5–2.5 cm. long, *sessile among the upper pistillate ones or short-peduncled; pistillate spikes* 2–4, scattered or the upper ones approximate, all but the upper peduncled, erect or ascending, *compact,* short-ellipsoid to oblong-cylindric, 0.5–3.5 cm. long, 4–6 *mm. thick;* scales pale brown, acuminate to awned, less than half length of perigynia; *perigynia olive-green to -brown,* puncticulate, *inflated-ovoid to subglobose, usually strongly nerved,* 2.5–4 mm. long and 1.5–2.5 *mm. thick,* with a short entire or emarginate straight or slightly bent beak. — Calcareous or rich woods, meadows and bottomlands, Vt. to s. Ont. and Minn., s. to Fla., Ala., Miss., La. and e. Kans. May–early July. Fig. 688. — Through a nondescript var. *recta* Dew. (*C. rectior* Mackenz.), with relatively small perigynia more faintly and slenderly nerved, passing into

688. C. granularis (typical).

Var. **Haleàna** (Olney) Porter (named in 1871 for its discoverer, Thomas J. Hale). — *Pistillate spikes* 3–5 *mm. thick; perigynia less inflated, oblong-ellipsoid or -ovoid,* 2.3–3 mm. long, 1–1.5 *mm. thick,* slightly more pointed, often with finer nerves. (*C. Haleana* Olney; *C. Shriveri* Britt.) — Calcareous shores, meadows and woods, w. Gaspé Pen., Que., to Sask., s. to n. and w. N.E., Va., Tenn., Mo. and Kans. May–July. Fig. 689.

689. C. granularis, v. Haleana.

197. C. Cráwei Dew. (for its discoverer, Ithamar Bingham Crawe, 1792–1847). — *Loosely stoloniferous, with small tufts of stiff glaucous often folded leaves* 1–4 *mm. broad; culms mostly solitary,* 0.2–4 dm. high; *staminate spike* 1–3 cm. long, *long-peduncled; pistillate spikes* 2–4 (rarely with secondary ones), *distant, the lowest often basal,* short-peduncled, or the upper sessile, erect, compact, 1–3 cm. long, 4–6 mm. thick; scales obtuse or short-pointed; *perigynia ellipsoid to oblong-ovoid,* often resinous-dotted, nearly nerveless or few-nerved, 3–3.5 mm. long, *pale green to pale brown, with minute hyaline-tipped beak.* — Calcareous shores, gravels, meadows and glades, Mingan Ids., Que., to Alta., s. to n. Me., nw. Ct., n. N.J., N.Y., n. Ala., Mo., Kans., Wyo. and Wash. June, July. Fig. 690.

C. microdónta Torr. & Hook. (small-toothed), differing from no. 197 in *leaves 3–6 mm. wide; pistillate spikes 7.5 mm. thick; perigynia definitely ribbed, 3–4.5 mm. long,* the *beak strongly bidentate.* — Prairie and limestone glades, Miss. to Tex., n. to Mo. and Okla.

690. C. Crawei.

§ 47. Oligocárpae Carey (incl. § *Griseae* Bailey) (see p. 299)

a. Perigynia loose to close, obscurely angled, much longer than achene, empty above, round to subtruncate at base, scarcely or barely beaked, (1–)6–45 imbricated in rather dense spikes; leaves 1.5–15 mm. thick.

Axis of inflorescence and peduncles of spikes scabrous; perigynia 3–4 mm. long, 1–2 mm. thick, oblong- or ellipsoid-conic, stramineous, sublustrous; leaves 1.5–5 mm. broad.

Upper pistillate spikes closely aggregated about the sessile to but short-peduncled staminate spike (5–10 mm. long), many times overtopped by the very long bracts. . . . . 198. *C. katahdinensis.*

Upper pistillate spikes well separated; staminate spike usually long-peduncled, 1–2.5 cm. long, overtopping or barely exceeded by the longer bracts. . . . . 199. *C. conoidea.*

Axis of inflorescence and peduncles of spikes smooth; perigynia 3–6 mm. long, 1.5–3 mm. thick, green, becoming pale brown to fulvous, opaque; leaves 1.5–15 mm. broad.

Leaves thin and flaccid to firm, green, rarely glaucous, 1.5–10 mm. wide;

bracts elongate-linear, with cylindric tight sheaths; pistillate spikes 3–20-flowered.
Perigynium distended or close and usually not cross-wrinkled, 1.5–2.5 mm. thick; achene gradually rounded to the style-base, ellipsoid or ellipsoid-obovoid. . . . . 200. *C. amphibola.*
Perigynium conspicuously puckered and wrinkled toward summit, 2–3 mm. thick; achene truncate or subtruncate at summit, broadly cuneate-obovoid. . . . . 201. *C. corrugata.*
Leaves firm to coriaceous, usually glaucous, 5–15 mm. wide; bracts linear-lanceolate, with sheaths broadened upward to summit; pistillate spikes 7–60-flowered. . . . . 202. *C. flaccosperma.*

*a.*Perigynia tight, definitely angled, nearly filled by achene, narrowed to base, rather definitely beaked, 1–9, loosely disposed or remote on a flexuous axis; leaves 2–7 mm. broad.
Sheaths glabrous, leaf-blades 2–5 mm. wide; perigynia 3.5–4 mm. long. . 203. *C. oligocarpa.*
Sheaths pubescent, blades 3–7 mm. wide; perigynia 4–5 mm. long. . . . 204. *C. Hitchcockiana.*

198. C. katahdinénsis Fern. (of Mt. Katahdin, Me.). — Densely cespitose; culms 0.1–1.8 dm. high, erect, slender, sharply angled; leaves 3–4 mm. broad, flat; *bracts foliaceous, many* (2–6) *times overtopping the spikes; upper pistillate spikes closely aggregated about the sessile to but short-peduncled staminate spike* (5–10 *mm. long*), *oblong-cylindric,* 5–15 mm. long, 5–15-flowered, the lowest often remote and sometimes subbasal, the *axis and peduncles scabrous;* scales whitish, with green awns, mostly shorter than perigynia; perigynia ellipsoid-oblong, 3–4 mm. long, sublustrous, with many impressed nerves, beakless. — Gravelly and rocky siliceous shores, local, centr. Nfld.; L. St. John, Que.; Mt. Katahdin, Me. July, Aug. Fig. 691.

691. C. katahdinensis.

199. C. conoídea Schkuhr (cone-shaped). — In small to large dense tussocks; culms slender, erect or ascending, 1–7 dm. high; leaves flat, warm green, 1.5–5 mm. wide; *bracts* similar, *the uppermost rarely if at all overtopping the usually peduncled staminate spike* (1–2.5 *cm. long*); *pistillate spikes well-separated on a scabrous axis,* oblong-cylindric, 0.7–2.5 cm. long, closely flowered, erect, the lower mostly peduncled, the uppermost often sessile; scales loosely spreading and rough-awned, equaling or exceeding perigynia; *perigynia oblong-conic,* 3–4 *mm. long, sublustrous, stramineous,* impressed-nerved, gradually tapering to a point, the orifice entire. — Moist grassy places, Nfld. to sw. Ont. and Minn., s. to N.S., N.E., Del., upland to N.C., O., Ind., Ill. and Ia. May–Aug. Fig. 692.

200. C. amphíbola Steud. (ambiguous). — In loose or dense clumps; culms slender, firm, central or somewhat lateral from the tufts of leaves, smooth except for scabrous summit; *leaves warm green to but slightly glaucous,* 1.5–10 mm. broad, *flat,* slightly to strongly scabrous on margins and veins beneath; *axis of inflorescence smooth; pistillate spikes oblong-cylindric,* scattered, or the upper sometimes approximate, 0.7–3 *cm. long,* 3–8 *mm. thick,* sessile or *on* short ascending *smooth peduncles, with few–20 imbricated perigynia; bracts* foliaceous and *elongate-linear, their close sheaths cylindric;* scales blunt or cuspidate; *perigynia with many impressed nerves, thick-cylindric or obovoid and turgid, with rounded base and summit, to tighter, more slenderly oblong-cylindric and tapering to base and entire orifice,* 4–5.5 *mm. long,* 1.5–2.5 *mm. thick; achenes rounded to style-base, ellipsoid to ellipsoid-obovoid.* — Highly variable, with four somewhat localized but freely confluent vars.:

692. C. conoidea.

Perigynia relatively tight or scarcely inflated, obtusely angled, somewhat tapering to base and apex, 4–4.7 mm. long, 1.4–2.2 mm. thick; staminate spike usually standing well above the scattered pistillate ones; pistillate spikes 3–6 mm. thick.
Bases and slender branches of the rhizome purple; leaves thin, flaccid, only slightly scabrous, 1.5–4 mm. broad; culms 1.5–5 dm. high; pistillate spikes 0.7–2 cm. long, 3–10-flowered. . . . . *C. amphibola* (typical).

Bases and more crowded and coarser branches of rhizome mostly brown (rarely slightly purple-brown); leaves firmer, stiffish, more scabrous, mostly 3–7.5 mm. broad; culms 1.5–8 dm. high; pistillate spikes 1–2.5 cm. long, 5–20-flowered. . . . . Var. *rigida.*

Perigynia inflated when mature, only obscurely round-angled, rounded to base and orifice, 4–5.5 mm. long, 1.7–2.5 mm. thick; staminate spike frequently partly hidden among the upper pistillate ones; pistillate spikes remote or the uppermost approximate, 5–8 mm. thick.

Principal leaves 4–10 mm. broad; culms 2–8 dm. high; longer pistillate spikes 1–3 cm. long, oblong-cylindric, mostly 7–20-flowered; perigynia oblong-cylindric, 4–5.5 mm. long. . . . . Var. *turgida.*

Principal leaves 1.5–5 mm. broad; culms 1–4 dm. high; pistillate spike short-cylindric to subglobose, 0.7–2 cm. long, 3–7-flowered; perigynia short-cylindric to inflated-obovoid, 4–4.5 mm. long; southwestern. . . . . . Var. *globosa.*

C. amphíbola (typical).— Calcareous hardwoods and rich slopes, Fla. to e. Tex., n. to Del., e. Pa., Tenn. and Ark. May, early June. FIG. 693.

693. C. amphibola (typical).

Var. rígida (Bailey) Fern. (stiff). — Rich woods, bottomlands and meadows, Fla. to Tex., n. to w. Mass., N.Y., Ind., Ill. and Mo. Late April–June. FIG. 694.

Var. túrgida Fern. (inflated), *C. grisea* of most auth., not Wahlenb. — Similar habitats, St. John R., N.B., to s. Ont. and Minn., s. to Mass., Ct., Ga., Ala., La. and e. Tex. May–July.

Var. globòsa Bailey (globose), *C. bulbostylis* Mackenz. — La. and Tex., n. to Mo. Late April, May.

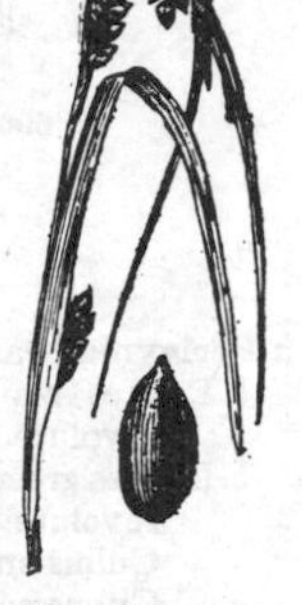

694. C. amphibola, v. rigida.

201. C. corrugàta Fern. (wrinkled). — Resembling var. *turgida* of no. 200, warm green; *perigynia with empty summits cross-puckered and wrinkled; achenes truncate, cuneate-obovoid.* — Calcareous bottomlands, Nottoway R., se. Va.; Tenn. Val., n. Ala. Late April, May.

202. C. flaccospérma Dew. (flaccid-fruited). — In small loose tussocks, usually *more or less glaucous,* sometimes light green; *culms* central or lateral from the crowns, spreading to ascending, slender, firm, *obtusely angled,* leafy, 1–5 dm. high; *basal leaves* overwintering, *submembranaceous to firm,* 5–15 *mm. broad; bracts linear-lanceolate, their loose sheaths enlarged upward; pistillate spikes* 2–4, distant, the lower often nearly basal, oblong-cylindric, erect, short-peduncled, 1–3.5 cm. long, 5–8 mm. thick, *with* 7–60 imbricated *stramineous to yellow-brown perigynia;* staminate spike 1–2.5 cm. long, usually peduncled and elevated above the upper pistillate one; *scales brown* or red-tinged, hyaline, *one-third to half as long as perigynia; perigynia firm, not inflated,* oblong-subcylindric, obscurely trigonous, striate-nerved, *mostly* 4–6 *mm. long,* with base rounded to subtruncate, the summit slightly tapering to the orifice. — Rich, often calcareous, woods, bottomlands and swamps, Fla. to Tex., n. to se. Va., Tenn. and se. Mo. Late April–early June. FIG. 695. — Southw. passing into

695. C. flaccosperma (typical).

Var. glaucòdea (Tuckerm.) Kükenth. (gray-green). — Stiffer and *more glaucous; leaves coriaceous, thick,* 0.4–1 cm. broad; *staminate spike sessile or short-peduncled, its base often overlapped by the upper pistillate spikes; perigynia* 3–5 *mm. long.* (*C. glaucodea* Tuckerm.) — Calcareous woods and meadows, Ala. to La., n. to Mass., N.Y., s. Ont., O., Ind., Ill. and Mo. May–early July. FIG. 696.

203. C. oligocárpa Schkuhr (few-fruited). — In small clumps; culms sharply triangular, slender, 0.7–5 dm. high; *leaves flat,* 2–5 *mm. broad, with glabrous sheaths;* bracts elongate, spreading-ascending; staminate spike sessile or stalked; *pistillate spikes* 2–4, scattered, stalked, or the uppermost sessile, *loosely* (1–)2–8-*flowered,* erect, 0.5–1.5 cm. long, with rachis evident; *scales* loosely spreading, *rough-awned, longer than perigynia; perigynia* 3.5–4 *mm. long, tight, definitely trigonous, impressed-nerved,* obovoid, *tapering to base, abruptly contracted to a conspicuous* mostly oblique *beak* with entire orifice. — Calcareous woods and copses, nw. Fla. to Tex., n. to Vt., sw. Que., s. Ont., O., s. Mich., Ill., Ia. and Kans. May–July. FIG. 697.

204. C. Hitchcockiàna Dew. (in honor of EDWARD HITCHCOCK, 1793–1864). — Similar to no. 203, usually coarse; culms up to 7 dm. high; *leaf-sheaths pubescent, the blades mostly* 3–7 *mm. wide;* staminate *spike* often longer (1–3 cm. long); *pistillate* 1–2.5 *cm. long; perigynia* 4–5 *mm. long.* — Calcareous or rich woods, sw. Que. and Vt. to s. Ont. and Wisc., s. to Va., Tenn. and Mo. May–July. FIG. 698.

696. C. flacco-sperma, v. glaucodea.

697. C. oligocarpa.

698. C. Hitchcockiana.

§ 48. Paníceae Tuckerm. (see p. 299)

a. Perigynia beakless or with a very short oblique tip. . . *b.*

b. Leaves white-glaucous, 0.5–3.5 mm. wide, quickly becoming plicate or involute. . . . . 205. *C. livida.*

b. Leaves green or only slightly glaucous, 1.5–7 mm. broad, flat, in age becoming revolute (rarely slightly involute at base only).

Culms smooth throughout. . . . . 206. *C. panicea.*

Culms scabrous at summit.

Lower sheaths mostly blade-bearing; principal blades 1.5–7 mm. broad; pistillate spikes dense, except sometimes at base; the crowded and imbricated perigynia 1.5–2.5 mm. thick.

Leaves grayish and stiffish, 2.5–7 mm. wide; lower pistillate spikes 5–10 mm. thick; perigynia abruptly contracted at apex, becoming turgid. . . . . 207. *C. Meadii.*

Leaves green, submembranaceous, 1.5–4.5 mm. wide; pistillate spikes 3–5 mm. thick; perigynia tapering to tip, not turgid. . . . 208. *C. tetanica.*

Lower (purple) sheaths largely bladeless; principal blades thin, green, 1.5–4 mm. wide; pistillate spikes loosely alternate-flowered; perigynia about 1.5 mm. thick, not turgid, tapering to short outwardly curved beak. . . . . 209. *C. Woodii.*

a. Perigynia with a prolonged straight beak.

Rhizomes and stolons hard and stout; lower sheaths of leafy tufts bladeless; culms stiff, harsh above; spikes stiffly erect, densely many-flowered. . . . 210. *C. polymorpha.*

Rhizomes and stolons soft and slender; lower sheaths blade-bearing; culms flexuous, smooth; spikes loosely ascending, spreading or drooping, loosely few-flowered. 211. *C. vaginata.*

205. C. lívida (Wahlenb.) Willd. (pale lead-colored). — *White-glaucous* and loosely *stoloniferous;* the slender smooth culms 0.5–4.5 dm. high, solitary or few from stolons bearing marcescent basal leaves; *leaves* 0.5–3.5 *mm. wide, quickly becoming plicate or involute;* sterile leafy tufts frequent; terminal spike staminate (rarely pistillate), 0.7–2.5 cm. long; pistillate spikes 1–3, subapproximate and subsessile, or the lowest rarely on a prolonged scabrous basal peduncle, cylindric, 0.7–2.5 cm. long; lowest bract usually blade-bearing; *scales purple or brown, with broad green center and pale margins,* obtuse to acutish, closely appressed to the 5–15 ascending perigynia; perigynia glaucous, granular, rhomboid-ellipsoid to fusiform or oblong-ovoid, nerved, 2.2–4.6 mm. long, beakless, the very short point straightish, with entire orifice. — Three vars.:

Culms 0.5–3 dm. high, arching; pistillate spikes 0.7–1.5 cm. long, the lower one frequently remote, subradical and long-peduncled; perigynia 2.2–3.2 mm. long, rounded or obtuse at summit.

Culms 0.5–3 dm. high; terminal spike staminate throughout, 0.7–1.5 cm. long. . . . . *C. livida* (typical).

Culms 1–1.5 dm. high; terminal spike pistillate except at base. . . . . Var. *rufinaeformis.*

Culms 1.5–4.5 dm. high, erect, rarely with basal spikes; pistillate spikes 0.7–2.5 cm. long; staminate spike 1.5–2.5 cm. long; perigynia 3.2–4.6 mm. long, tapering to acute or slenderly conical summit. . . . . Var. *Grayana.*

C. lívida (typical). — Wet calcareous soil, local, n. Nfld. and s. Alta. July, Aug. (N. Eu.) **Var. rufinaefórmis** Fern. (like *C. rufina*, reddish). — Wet calcareous soil, n. Nfld.

699. C. livida, v. Grayana.

Var. **Grayàna** (Dew.) Fern. (for its discoverer, Asa Gray, 1810–1888), *C. livida* of ed. 7, not (Wahlenb.) Willd. — Calcareous meadows, bogs and depressions, Lab. to Alaska, s. to Nfld., M.I., n. Me., ne. Mass., w. Ct., s. N.J., centr. N.Y., Mich., Wisc., Minn., Sask., Ida. and n. Calif. May–July. (N. Eu.) Fig. 699.

206. C. paníceа L. (like millet, from *panus*, an ear of millet, with which this species was early confused). — In loose brown-based tufts from horizontal stolons and rhizomes, stiffish, strict, glaucous-green or bluish; *culms smooth*, angled, 1–6 dm. high; leaves in numerous sterile tufts, those of the culms 3–8, mostly basal, 2–6 cm. broad, firm, flat, finally revolute; *bracts* broad and short, *the* long-sheathed *lowest one with blade rarely one-fourth as long as inflorescence; staminate spike* 1.5–3 cm. long, *smooth-peduncled* to subsessile; *pistillate spikes* 1–3, scattered, mostly *on smooth peduncles*, erect, rather compact, or loose below, 1–3 cm. long, 5–7 mm. thick; scales purple-margined, mostly shorter than perigynia; *perigynia* ovoid, stramineous or purple, somewhat *turgid*, shining, puncticulate, scarcely nerved, 2.5–5 mm. long, 1.7–2.5 mm. thick, with a short purplish truncate tip. — Meadows and grasslands, locally abundant, se. Nfld. to Ct.; Minn. May–July. (Natzd. from Eu.) Fig. 700.

700. C. panicea.

701. C. Meadii.

207. C. Meàdii Dew. (for its discoverer, Samuel Barnum Mead, 1799–1880). — Resembling no. 206; *culms more sharply angled, scabrous above*, the bases drab; *leaves grayish-green, stiff*, 2.5–7 *mm. broad, their margins often involute at base; lowest bract one-half to quite as long as inflorescence; staminate and lower* (often exserted) *pistillate spikes scabrous-peduncled, the latter thick-cylindric* and *dense*, 1–3.5 cm. long, 5–10 *mm. thick;* scales purplish-brown, with hyaline margin; *perigynia* puncticulate, *obovoid, becoming turgid, abruptly contracted to a short entire curved beak, strongly ribbed*, 3–5 mm. long, 1.7–2.5 *mm. thick*. (*C. tetanica*, var. *Meadii* (Dew.) Bailey) — Calcareous meadows, prairies and depressions, Ga. to Tex., n. to w. N.J., Pa., s. Ont., O., Mich., Wisc., Minn., Man. and Sask. May, June. Fig. 701.

208. C. tetánica Schkuhr (rigid). — More *slender* than no. 207, *the whitish filiform* subterranean *stolons numerous;* culms in small brown- to purple-based tufts, slender, 1.2–7 dm. high; *leaves green, submembranaceous*, 1.5–4.5 *mm. broad*, in age becoming revolute; bracts prolonged; *pistillate spikes* 0.7–4 cm. long, 3–5 *mm. thick, more loosely flowered at base; perigynia strongly tapering* to base and *to outwardly curved tip, not turgid*, 2.5–4 *mm. long*, 1.5–2.3 *mm. thick.* — Calcareous bogs, meadows, swales and low woods, e. Mass. to Man., s. to Ct., Va., O., Ind., Ill., Ia. and S.D. May–July. Fig. 702.

209. C. Woòdii Dew. (named in 1846 for one of its discoverers, William A. Wood). — Much like no. 208; *bases of leafy tufts with several bladeless purple sheaths; leaves thin, green*, 1.5–4 *mm. broad*, mostly overtopping the weak and very slender culms; *pistillate spikes loosely alternate-flowered*, 1.5–3.5 cm. long, 3–5 mm. thick; scales purplish-brown or paler, often acuminate; *perigynia* 3.5–4 mm. long, *about* 1.5 *mm. thick*, not turgid, tapering to a short outwardly curved tip. (*C. tetanica*, var. *Woodii* (Dew.) Bailey) — Rich calcareous, usually dry, woods, w. Ct. and N.Y. to s. Man., s. to D.C., W.Va., O., n. Ind., n. Ill. and Mo. May, June.

702. C. tetanica.

703. C. polymorpha.

210. C. polymórpha Muhl. (of many forms). — Stout, *from stout and firm cord-like widely creeping rhizomes; culms* 3–6 dm. high, *stiffly erect*, harsh above, *solitary from old leafy tips; new leafy tufts with bladeless basal sheaths;* foliage-leaves firm, flat, becoming revolute, 3.5–6 mm. broad; *pistillate spikes* 1 or 2, *stiffly erect, densely many-flowered,* 1.5–4 (rarely–6) cm. long, 6–10 mm. thick; staminate spike (or spikes) peduncled, purplish; lower bract short, erect; *pistillate scales* reddish-brown, *obtuse; perigynia* ovoid, obscurely nerved, 4–5.5 mm. long, obtusely angled, *the very long and straight beak oblique or lipped at the orifice.* — Dry sandy open woods and clearings, very local, s. Me. and centr. Mass. to Md. June–Aug. FIG. 703.

211. C. vaginàta Tausch (sheathed). — *Very slender* and more or less *diffuse, with* wide-creeping *slender soft rhizomes and stolons; culms smooth,* flexuous, 1–5 (–8) dm. high; leaves 1.5–5 mm. wide, soft, mostly shorter than culms; staminate spike long-peduncled; *lower bract spathiform,* with very short blade; *pistillate spikes* 1–3, *all peduncled and loosely ascending to spreading, loosely* 3–20-*flowered; scales* loose, *acute,* shorter than perigynia; *perigynia* thin, nerveless, 3–5 mm. long, slenderly ovoid, *with slender cylindric bidentate hyaline-orificed beak.* (*C. sparsiflora* (Wahlenb.) Steud.; *C. saltuensis* Bailey; *C. altocaulis* (Dew.) Britt.) — Mossy woodlands and calcareous swamps and bogs, Baffin I. to Alaska, s. to Nfld., M.I., N.B., n. Me., w. Vt., n. N.Y., s. Ont., n. Mich., n. Wisc., n. Minn., Sask., s. Alta. and s. B.C. June–Aug. (N. Eurasia) FIG. 704. — Although the Am. plant is frequently separated from the Eurasian, no stable character is evident in the series and most variations are readily matched on the two continents.

704. C. vaginata.

§ 49. LAXIFLÒRAE Kunth (see p. 300)

a. Culms without green leaves, the numerous long tubular-spathiform concave-tipped sheaths purplish and numerous from base to remote upper spike; staminate spike purple, peduncled; leaves of basal rosette firm, evergreen, the new ones appearing after ripening of fruit, lanceolate, 1.5–3 cm. broad, the basal ones reduced and purple. . . . . 212. *C. plantaginea.*

a. Culm-leaves and lower bracts with flat green blades; staminate spike pale to deep brown or brownish-purple; leaves of basal rosettes or tufts 0.15–4 cm. broad. . . *b.*

b. Blades of upper cauline leaf and lowest bract shorter than to only twice or thrice the length of their sheaths; perigynia sharply 3-angled.

Basal leaves bright green, the tufts purple-based, the evergreen blades mostly 2–3.7 dm. long and 0.8–1.7 cm. broad; culms ascending, 2.5–8 dm. high; perigynia 5–6.5 mm. long. . . . . 213. *C. Careyana.*

Basal leaves glaucous, the tufts brown- or drab-based, the less evergreen blades mostly 1.1–3 dm. long and 1.2–3 cm. broad; culms weak, loosely spreading, 1–3(–4) dm. long; perigynia 2.5–4.5 mm. long. . . . . 214. *C. platyphylla.*

b. Blades of upper cauline leaf and lowest bract many times longer than their sheaths. . . *c.*

c. Perigynia acutely or subacutely angled, with plane faces (obtusely angled in rare var. of no. 216, with leaves only 1.5–3.5 mm. wide); pistillate spikes 0.7–3 cm. long; leaves 1.5–12 mm. broad.

Lowest bract greatly overtopping inflorescence; upper bract spathe-like; staminate spike crowded among the upper pistillate ones, 4–12 mm. long; pistillate scales blunt. . . . . 215. *C. abscondita.*

Lowest bract only slightly if at all overtopping inflorescence; upper ones not spathiform; staminate spike peduncled or elevated above upper pistillate one, 0.8–2.5 cm. long; pistillate scales acuminate or awned.

Leaves green, 1.5–5 mm. broad; lateral spikes pistillate to base, interruptedly linear-cylindric, 1–3 cm. long, 3–4 mm. thick, ascending or but slightly spreading (loosely spreading on long-exserted peduncles only in a var. with perigynia 2.5–3 mm. long). 216. *C. digitalis.*

Leaves glaucous or pale green, mostly 6–12 mm. broad; many lateral spikes with 1–3 staminate flowers or empty scales at base, thick-cylindric, 0.5–2 cm. long, 4–6 mm. thick, the lower and sometimes all loosely spreading to pendulous on long filiform peduncles; perigynia 2.8–4 mm. long. . . . . 217. *C. laxiculmis.*

*c.*Perigynia with rounded angles at least below the middle, or barely angled, long-stipitate; leaves 0.25–4 cm. broad. (In this subsection care should be taken to guard against blasted and distorted perigynia, which are frequent.). . *d.*

*d.*Sterile leafy basal tufts at fruiting time without prolonged culms, forming rosettes with well developed leaves 0.7–4 cm. broad, the broadest blades 23–67-nerved; pistillate spikes interruptedly linear- or oblong-cylindric, remotely or loosely alternate-flowered; perigynia strongly 24–45-nerved (no. 224 with new basal sprouts often subsessile, has the broadest leaves 15–35-nerved, the pistillate spikes dense, the perigynia nerveless or very delicately 7–21-nerved).

Culms wing-angled; basal leaves 1–4 cm. broad; broader cauline leaves and lowest bract 0.8–2 cm. broad; bracts frondose, the upper much overtopping the inflorescence; longest pistillate spikes 1–2.5 cm. long, linear-cylindric, remotely flowered, overtopping or only slightly exceeded by staminate spike; midribs of pistillate scales barely or not at all excurrent; anthers 1.4–2.2 mm. long. . 218. *C. albursina.*

Culms wingless; basal leaves 0.7–2.5 cm. broad; broader cauline leaves and lowest bract 2–10 mm. broad, the bracts exceeded by or only slightly overtopping inflorescence; longest pistillate spikes 1.5–5 cm. long; midribs of at least the lower pistillate scales excurrent as awns; staminate spike prominent; anthers 2–4.5 mm. long.

Leaves membranaceous, smooth or barely scabridulous, their ribs not prominent; pistillate spikes distantly flowered, 2–3 mm. thick; perigynia broadly fusiform-obovoid, 3–4.5 mm. long, half as thick, the spongy stipe 0.5–1 mm. long; stigmas 2 mm. long; staminate spike sessile or short-peduncled, its scarious pale brown to whitish scales with usually excurrent midrib. . 219. *C. laxiflora.*

Leaves subcoriaceous, scabrous, their ribs prominent; pistillate spikes 4–8 mm. thick, with subapproximate outward-curving perigynia; perigynia slenderly fusiform-obovoid, 4–5.5 mm. long, less than half as thick, the spongy stipe 1.3–2 mm. long; stigmas 3–5 mm. long; staminate spike short- or long-peduncled, its chartaceous broad white scales with green midrib not excurrent. . . . . 220. *C. striatula.*

*d.*Sterile leafy tufts at fruiting time with culms elongating to 0.5–1.5 dm. high (often subsessile in no. 224), their terminal blades 2–15 mm. broad, the broadest ones 13–35-nerved; pistillate spikes linear- to oblong-cylindric, interrupted or with all or all but the lowest perigynia usually approximate or imbricated, the spikes 0.5–4 cm. long. . . *e.*

*e.*Mature normal perigynia strongly asymmetrical, broadly rounded on one side above, with the tip becoming nearly perpendicular to the long axis.

Upper pistillate spikes usually approximate and close to the subsessile pale staminate spike; bracts usually much overtopping inflorescence; rachis of pistillate spike with thin wing-like angles; perigynia 3–4.5 mm. long, the short and broad tip not prolonged. 221. *C. blanda.*

Upper pistillate spike usually solitary and remote; staminate spike elevated on a peduncle, brown or purplish (rarely white); bracts rarely overtopping inflorescence; rachis not wing-angled; perigynia 2.5–3.5 mm. long, the strongly curved beak often proboscis-like. . . . . 222. *C. gracilescens.*

*e.*Mature normal perigynia nearly symmetrical, straightish or barely curved, with the beak projected forward or but slightly arched.

Perigynia broadly obovoid, less than twice as long as broad, 2.5–3.5 mm. long, strongly ribbed, broadly rounded above to an abrupt short-conic tip much shorter than the stipe; pistillate spikes linear-cylindric, much interrupted, the longer 1–4 cm. long, scattered. . . . . 223. *C. ormostachya.*

Perigynia fusiform-obovoid, twice to thrice as long as broad, 2.5–5.5 mm. long, nearly nerveless to slenderly nerved, tapering to a prolonged beak as long as or longer than the stipe; pistillate spikes closely flowered, except sometimes at base, the longer 0 5–3 cm. long.

Pistillate spikes linear-cylindric, 3–4 mm. thick; perigynia 1–1.5 mm. thick, nerveless or faintly few-nerved; sterile leafy new basal tufts at fruiting time either subsessile or with elongating culms; northern and upland species. . . . . . 224. *C. leptonervia.*

Pistillate spikes oblong-cylindric, 4.5–10 mm. thick; perigynia 2–3 mm. thick, many-nerved; new basal leafy tufts on elongating culms; southern or Coastal Plain species.

Bracts overtopping the crowded upper pistillate and subsessile approximate pale staminate spike; perigynia erect, whitish-green to brown. . . . . 225. *C. crebriflora.*

Bracts much shorter than inflorescence; pistillate spikes scattered and remote; staminate spike long-peduncled (or with small pistillate one at base), conspicuous; perigynia outwardly arching, warm green to brown. . . . . . 226. *C. styloflexa.*

212. C. plantagínea Lam. (plantain-like). — *In purple-based tufts, with broadly lanceolate firm* somewhat bullate dark green basal *evergreen leaves* 1.5–3 *cm. broad; culms* arising laterally from old bases, slender, *covered from base to summit by tubular-spathiform purplish loose sheaths with short concave colored tips,* 2–6 dm. high, shriveling before development of new basal leaves; staminate spike purple, raised on a usually long peduncle, the long anthers conspicuous; *pistillate spikes remote,* 2–4, their bases included or but slightly exserted, linear-cylindric, 1–2.5 cm. long, *alternately flowered;* scales triangular-ovate, pale-hyaline with darker center, pointed; *perigynia* oblong-ovoid, sharply 3-angled, 3–5 mm. long, scabrous, *with erect or curving conical beak.* — Rich hardwoods, N.B. to se. Man., s. to Me., n. Mass., Ct., Pa., upland to Ala. and Tenn., Wisc. and Minn. April–early June. Fig. 705.

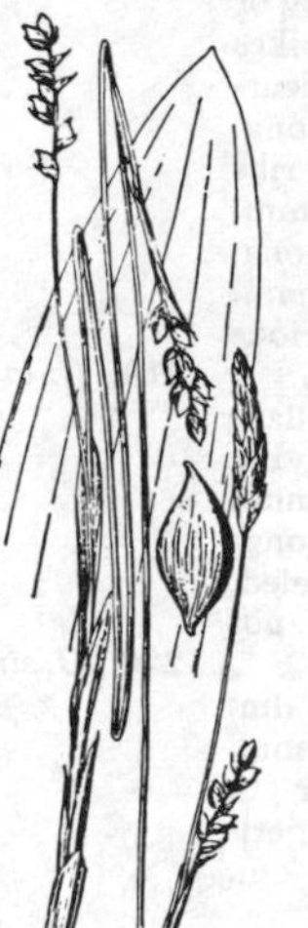

705. C. plantaginea.

213. C. Careyàna Torr. (in honor of John Carey, 1797–1880). — *In purple-based tufts, with lanceolate* firm *evergreen bright green basal leaves mostly* 2–3.7 *dm. long and* 0.8–1.7 *cm. broad; culms* mostly lateral from the crowns, slender, *ascending,* 2.5–8 *dm. high, with few blunt and short lower leaves, the upper cauline leaf and the lowest bract with flat blade shorter than to somewhat longer than their long sheaths;* staminate spike thick, deep brown, short-peduncled; pistillate spikes 2 (rarely 3), the upper short-peduncled or subsessile at base of staminate peduncle, the lower remote and longer-peduncled, thick-cylindric, 0.7–2 cm. long, 3–8-flowered; scales white and purplish, with sharp green midrib, shorter than perigynia, awned or acute; *perigynia* sharply 3-angled, ovoid, *olivaceous,* finely many-nerved, 5–6.5 *mm. long,* tapering to base and to conical entire beak. — Rich hardwoods, local, N.Y. to s. Ont. and Mich., s. to upland Va., O., Ind., Ill. and Mo. May, June. Fig. 706.

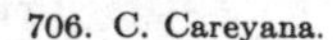

706. C. Careyana.

214. C. platyphýlla Carey (broad-leaved). — *In pale-based tufts, with broadly lanceolate glaucous* hardly evergreen *leaves mostly* 1.1–3 *dm. long and* 1.2–3 *cm. broad; culms subcapillary,* numerous, *flexuous or loosely spreading,* 1–3 (rarely –4) *dm. long, weak and soon shriveling; cauline leaves and bracts with acute lanceolate blades mostly twice or thrice the length of their sheaths;* staminate spike pale; *pistillate spikes* 2–4, scattered, mostly short-peduncled, *interruptedly linear-cylindric, remotely* 2–10-*flowered;* scales pale, cuspidate or acute; *perigynia* green, ellipsoid-ovoid, many-striate, 2.5–4.5 *mm. long,* tapering to substipitate base and short conic beak. — Rich deciduous woods and rocky slopes, s. Me. to sw. Que., w. to s. Ont., s. to s. N.E., N.C. and Tenn. Late April–June. Fig. 707.

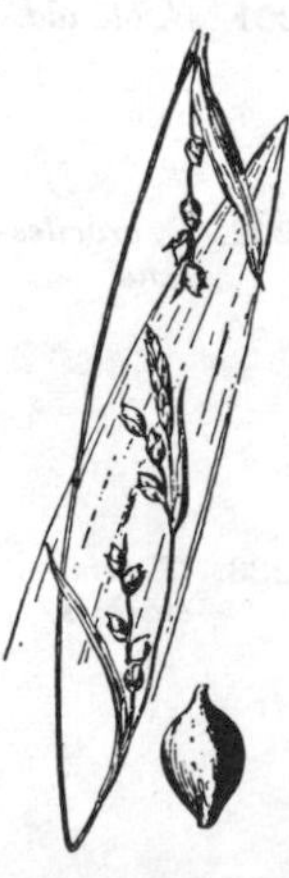

707. C. platyphylla.

215. C. abscóndita Mackenz. (concealed). — In dense low clumps, with glaucous or pale green thin *leaves soon spreading or recurving* and 1–3 *dm. long* and 4–9 *mm. broad; culms short,* 0.5–2 *dm. long, more or less hidden among the leaf-bases, with very prolonged bracts; lowest bract nearly basal and*

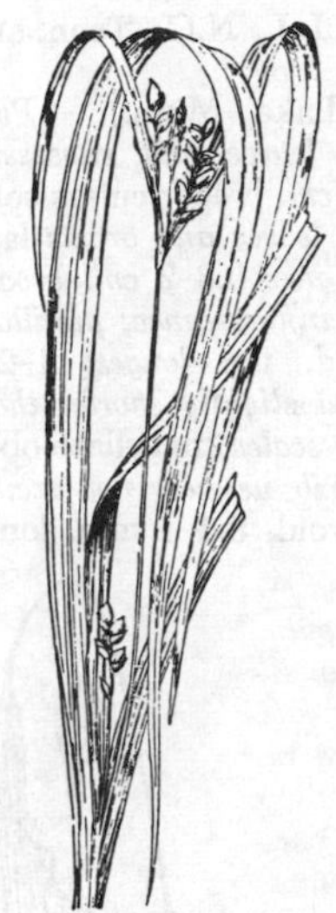
708. C. abscondita.

much overtopping the inflorescence; upper bracts subspathiform, exceeding inflorescence; lowest *pistillate spike* often remote or subbasal and on prolonged peduncle; *upper ones approximate with the small* (4–12 *mm. long*) *staminate one* at base of long upper bract, oblong-cylindric, 0.5–1.5 cm. long; perigynia tawny, slenderly trigonous-ovoid, twice longer than the thin *blunt scales,* membranaceous, 2.5–3.5 mm. long, essentially beakless. (*C. ptychocarpa* Steud., not Link) — Rich hardwoods, Fla. to La., n. to se. Mass., sw. R.I., se. Ct., L.I., N.J., e. Pa., D.C. and s. Ind. May–July. Fig. 708.

Var. **rostellàta** Fern. (with small beak). — Leaves firmer, very glaucous; perigynia 3–4.5 mm. long, with definite beak; pistillate scales 2–3 mm. long, half as long as perigynia. — Wet peaty pine-barrens and pinelands, se. Va. to Fla.

× **C. absconditifórmis** Fern. (like *C. abscondita*) is a local, sterile hybrid, exactly combining the characteristics of nos. 215 and 217.

216. C. digitàlis Willd. (of a finger; from the slender spikes). — In dense clumps, *with erect or ascending green* flat *leaves* 1.5–5 *mm. broad;* culms slender, ascending, 0.5–6 dm. high; *upper bracts only slightly exceeding or shorter than inflorescence; staminate spike subsessile to long-peduncled, elevated above the upper pistillate one,* 0.8–2.5 *cm. long; pistillate spikes scattered,* the upper subsessile or stalked and ascending; the lower mostly peduncled, *linear-cylindric, alternately flowered,* 1–3 *cm. long,* 3–4 *mm. thick, with pistillate flowers borne to the base;* scales pale, acute; perigynia rhomboid-ovoid to lance-fusiform, angled, 2.5–4 mm. long, barely short-beaked, much longer than scale, finely many-nerved. — Three vars.:

Perigynia rhomboid-ovoid, definitely angled, flat-faced, nearly symmetrical or only slightly oblique at the short and scarcely beaked summit, mostly 2.5–3 mm. long.
  Culms 0.5–5 dm. high; staminate spike approximate to uppermost pistillate one or elevated on a peduncle only 0.2–3 (–4.5) cm. above its base; upper pistillate spikes subsessile to short-peduncled; rhizome not coralline. . . *C. digitalis* (typical).
  Culms 3–6 dm. high; staminate spike on peduncle 0.5–2 dm. long; pistillate spikes all long-peduncled, the lower commonly pendulous; rhizome coralline-knotty, the crowns bulbiform. . . . . Var. *macropoda.*
Perigynia lance- to slenderly ovoid-fusiform, asymmetrically curved to a prolonged tip, obscurely angled, 3–4 mm. long; culms 1.2–4 dm. high. . . . . Var. *asymmetrica.*

C. digitàlis (typical). — Dry or dryish hardwoods and glades, centr. Me. to s. Ont. and Wisc., s. to Ga., nw. Fla., Ala., Miss. and Mo. May–July. Fig. 709.

Var. **macrópoda** Fern. (long-stalked). — Rich woods, Ala. and La., n. to e. Va., Md. and s. Ind. May, June.

Var. **asymmétrica** Fern. (unsymmetrical). — Calcareous hardwoods, Fla. and Ala., n. to se. Va. and centr. Pa.

× **C. copulàta** (Bailey) Mackenz. (coupled), *C. digitalis,* var. *copulata* Bailey and *C. laxiculmis,* var. *copulata* (Bailey) Fern., seems to be a hybrid of nos. 216 and 217.

217. C. laxicúlmis Schwein. (loose-culmed). — In low dense clumps with *leaves* of basal rosettes *glaucous* or pale green, loosely curving or spreading, 1.5–4.5 dm. long and mostly 6–12 *mm. broad; culms* numerous, *loosely ascending or spreading, weak,* very slender, 1.5–6 dm. long; *staminate spike peduncled, raised well above the upper pistillate one,* 1–2.5 *cm. long;* lower bracts prolonged, the upper shorter and rarely overtopping the inflorescence; *pistillate spikes often with* 1–3 *staminate or empty basal scales, thick-cylindric,* 0.5–2 *cm. long,* 4–6 *mm. thick, the lower* and sometimes all *loosely spreading to pendulous on long filiform peduncles;* scales hyaline, sharply keeled, the lower cuspidate, the upper acute, shorter than to about equaling perigynia; *perigynia* spreading-ascending, *sharply trigonous,* ovoid, barely short-beaked, many-nerved, 2.8–

709. C. digitalis.

710. C. laxiculmis.

4 *mm. long.* — Rich woods and glades, s. Me. to s. Ont. and Wisc., s. to L.I., N.C., Tenn. and Mo. May–early July. FIG. 710. Hybridizes with nos. 215 and 216.

218. C. albursìna Sheldon (from its abundance near White Bear Lake, Minn.). — Pale green, with overwintering *lanceolate or lance-oblong leaves and subsessile new basal tufts of similar 23–67-nerved leaves* (1–)1.5–4 *cm. broad;* culms soft, *wing-angled,* 1.7–3.5 *mm. thick,* 2–6 dm. high; *cauline leaves and bracts* lanceolate or lance-linear, soft and smooth, *the broader ones* 0.8–2 *cm. broad,* the *upper bracts* frondose and *greatly overtopping the inflorescence; pistillate spikes interruptedly linear-cylindric, remotely flowered,* the *longest* 1–2.5 *cm. long, the upper approximate and overtopping or but slightly shorter than the pale alternate-flowered staminate spike; pistillate scales* flabellate-obovate, truncate or emarginate, whitish, *the green midrib usually not excurrent;* rachis thin-angled; perigynia long-stipitate, obovoid, 3.5–4 mm. long, strongly 27–35-nerved, with short conical ascending or curving entire-orificed tip. (*C. laxiflora,* var. *latifolia* Boott) — Calcareous hardwoods and ravines, sw. Que. and Vt. to Minn., s. to Ct., n. N.J., n. Md., Pa., upland to Va., Tenn. and Ark. Late April–June. FIG. 711.

711. C. albursina.

219. C. laxiflòra Lam. (loosely flowered). — More slender than no. 218; *basal leaves* linear- to oblong-lanceolate, 0.7–2.5 *cm. broad, membranaceous, smooth or barely scabridulous,* the broadest 25–51-nerved; *culms wingless,* 0.7–2 *mm. thick,* 2–6 dm. high; *broader cauline leaves and lower bracts* linear-lanceolate, 3.5–10 *mm. wide,* the upper bracts often equaling or overtopping inflorescence, their sheaths with entire or merely erose angles; *spikes scattered,* or the upper approximate; *the pistillate ones interruptedly linear-cylindric, distantly flowered,* 1.5–5 *cm. long,* 2–3 *mm. thick, the uppermost approximate to the sessile or but short-peduncled pale staminate one; staminate scales scarious, pale brown to whitish, usually with excurrent midrib;* pistillate scales pale, oblong-obovate, mostly acute, mucronate or cuspidate; *perigynia broadly fusiform-obovoid,* 3–4.5 *mm. long, half as thick,* strongly 24–36-nerved, *with spongy stipe* 0.5–1 *mm. long,* and *with a straight or slightly oblique slenderly conical tip; stigmas* 2 *mm. long.* (Var. *patulifolia* (Dew.) Carey; *C. anceps* Muhl.) — Rich hardwoods and loamy slopes, nw. N.S., and centr. Me. to Ont. and Wisc., s. to s. N.E., L.I., Ga. and Tenn. Late April–June. FIG. 712.

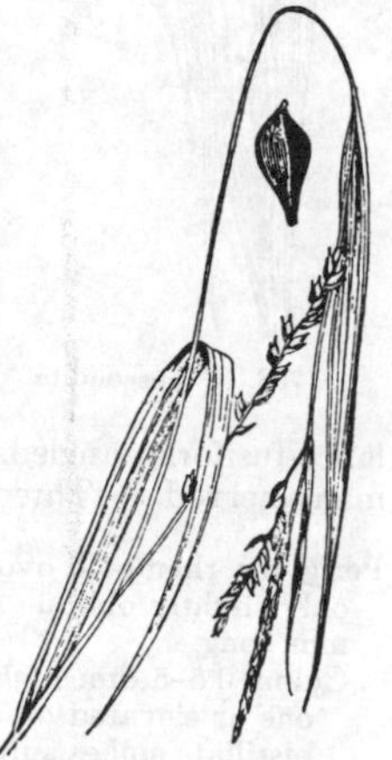

712. C. laxiflora.

Var. **serrulàta** F. J. Herm. (finely saw-toothed). — Scabrous, *bracts* linear-lanceolate, *their sheaths with serrulate angles.* — Local, N.Y. and Pa. to Mich., Ind. and Tenn.

220. C. striátula Michx. (with fine longitudinal lines). — Resembling no. 219; *basal leaves linear, subcoriaceous, scabrous, prominently ribbed,* 6–15 mm. broad; culms scabrous above; *cauline leaves* and bracts linear, *scabrous,* the *broader ones* 2–7 *mm. wide;* upper bracts usually shorter than inflorescence; *pistillate spikes* remote, linear- to oblong-cylindric, 4–8 *mm. thick, with subdistant to imbricated perigynia; staminate spike* conspicuous, 2–3.5 cm. long, usually peduncled, *its chartaceous white round-tipped scales with midrib usually not excurrent; perigynia outward-curving, slenderly fusiform-obovoid,* 4–5.5 *mm. long, less than half as thick, the stipe* 1.3–2 *mm. long; stigmas* 3–5 *mm. long.* (*C. laxiflora,* var. *Michauxii* Bailey) — Rich hardwoods, Fla. to Tex., n. to Ct., se. N.Y., N.J., Pa. and s. Ind. May, June. FIG. 713.

713. C. striatula.

221. C. blánda Dew. (charming). — In tussocks, green or slightly glaucous; *basal* leafy tufts with slightly prolonging culms, the *blades* 4–12 *mm. wide;* fertile culms mostly 1–3 mm. thick at base, with slightly scabrous-erose angles; cauline leaves with loose sheaths, the larger blades 2.5–9 mm. broad; *bracts usually much overtopping inflorescence; upper pistillate spikes usually approximate and close to the subsessile whitish staminate one,* 0.5–3 cm. long, *with smooth wing-angled rachis,*

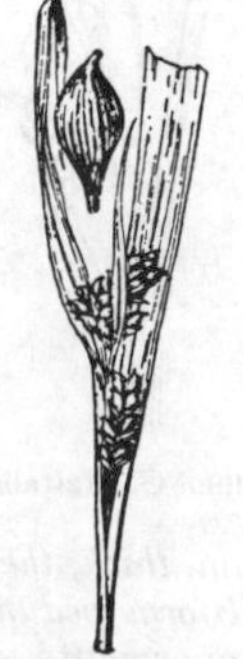

714. C. blanda.

the lower spikes remote and often peduncled; anthers 2–3.5 mm. long; *perigynia approximate or crowded*, spreading-ascending, ellipsoid-obovoid, 3–4.5 *mm. long*, 23–30-nerved, *strongly asymmetrical, with abruptly bent short and broad beak.* (*C. laxiflora*, var. Boott) — Woods, bottomlands, thickets and meadows, sw. Que. to s. Ont. and N.D., s. to e. Mass., R.I., Ct., Ga., Ala., Miss., La. and e. Tex. April–June. Fig. 714.

222. C. graciléscens Steud. (rather slender). — Similar to no. 221, usually more slender, with culms only 0.5–1.5 mm. thick at base; lowest sheaths more often purplish when fresh; basal blades 3–8 mm. broad; larger cauline leaves 1.5–5 mm. broad, with closer sheaths; *bracts rarely overtopping inflorescence; staminate spike elevated on a peduncle*, often purplish; *pistillate spikes mostly scattered*, often slender-peduncled, 0.5–2.5 cm. long, *the rachis not winged; perigynia* 2.5–3.5 *mm. long, the strongly curved beak usually proboscis-like.* (*C. laxiflora* sensu Wieg., not Lam.; *C. laxiflora*, var. *gracillima* Boott) — Rich woods, thickets and bottoms, Vt. and sw. Que. to s. Ont. and Wisc., s. to Mass., Ct., N.C., Ala., La. and e. Tex. April–June. Fig. 715.

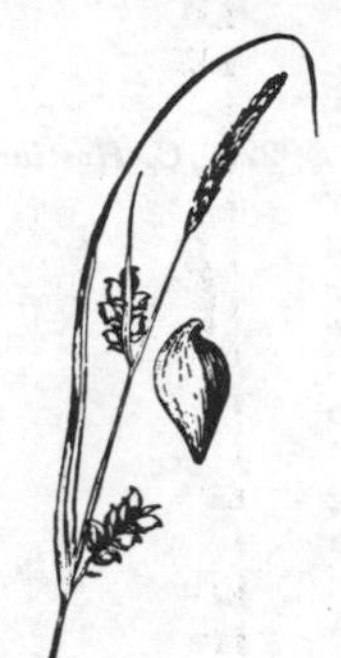
715. C. gracilescens.

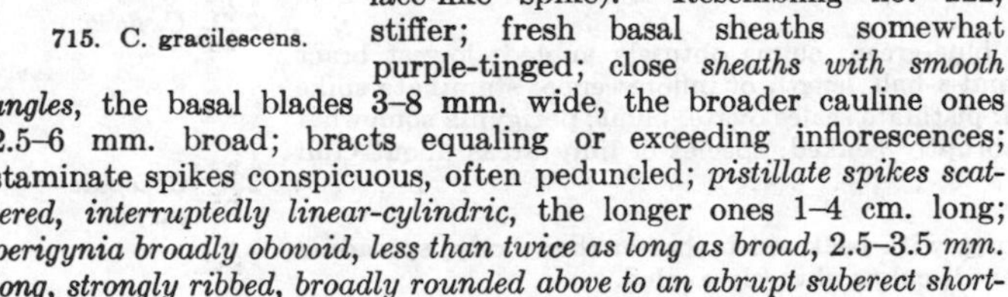

223. C. ormostáchya Wieg. (with necklace-like spike). — Resembling no. 222, stiffer; fresh basal sheaths somewhat purple-tinged; close *sheaths with smooth angles*, the basal blades 3–8 mm. wide, the broader cauline ones 2.5–6 mm. broad; bracts equaling or exceeding inflorescences; staminate spikes conspicuous, often peduncled; *pistillate spikes scattered, interruptedly linear-cylindric*, the longer ones 1–4 cm. long; *perigynia broadly obovoid, less than twice as long as broad*, 2.5–3.5 *mm. long, strongly ribbed, broadly rounded above to an abrupt suberect short-conic tip much shorter than the stipe.* — Sandy or rocky woods and clearings, Saguenay Co., Que., to L. Superior reg., Ont., s. to N.S., Me., Mass., n. Pa., N.Y., n. O., Mich. and Wisc. May–early July. Fig. 716.

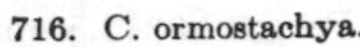
716. C. ormostachya.

224. C. leptonérvia Fern. (finely nerved). — Green, relatively slender; new basal rosettes either subsessile or slightly elevated on culms, their blades 3–10 mm. broad; cauline sheaths erose-scabrous on angles, their larger blades 2.5–7 mm. broad; culms often retrorse-scabrous; bracts mostly overtopping inflorescence; *staminate spike usually approximate to or partly hidden by pistillate spikes, upper pistillate spikes approximate*, lower often remote, *linear-cylindric*, subdensely to alternately flowered, *the longer* 1–3 *cm. long and* 3–4 *mm. thick*, the rachis smooth; *perigynia erect, fusiform-obovoid*, 2.5–4 *mm. long*, 1–1.5 *mm. thick, nerveless or faintly nerved, tapering to prolonged straightish beak as long as the slender stipe.* (*C. laxiflora*, var. Fern.) — Low woods, clearings and thickets, Nfld. and s. Lab. Pen. to n. Minn., s. to N.S., N.E., n. N.J. and Pa., upland to N.C. and Tenn., n. O., Mich. and Wisc. May–July. Fig. 717.

717. C. leptonervia.

225. C. crebriflòra Wieg. (closely flowered). — Similar to the preceding, brown-based; basal leaves 3–8 mm. broad, scabrous-margined; cauline leaves 3–5 mm. broad, their sheaths somewhat scabrous on the angles; culms scabrous; *bracts overtopping inflorescence; pistillate spikes oblong-cylindric, the longer ones* 0.5–2 *cm. long and* 4.5–10 *mm. thick, the upper approximate and nearly hiding the pale staminate one; perigynia fusiform-obovoid*, 3.5–4.5 *mm. long*, 2–3 *mm. thick*, many-nerved, *erect*, whitish-green or -brown, *tapering to erect beak as long as the stipe.* — Bottomlands and low woods, Fla. to Tex., n. to se. Va. Late April–early June.

226. C. styloflé xa Buckl. (with curved style). — More lax and slender than no. 225; *bracts mostly shorter than inflorescences; staminate spike long-peduncled*, or bearing a small pistillate one at base; *pistillate spikes scattered*, the upper sessile or subsessile, *the lowest often nearly basal and filiform-peduncled*, rather dense, 1–2 cm. long; *perigynia outwardly arching*, fusiform to fusiform-ovoid, warm green to brown, 3.5–4.5 mm. long, tapering to tips, *long-stipitate.* (*C. laxiflora*, var. Boott) — Low woods, wet moss, peaty spring-heads, etc., Fla. to Tex., n. to Ct., se. N.Y., e. Pa., D.C. and Ky. April–June. Fig. 718.

718. C. styloflexa.

§ 50. EXTÉNSAE Fries (see p. 300)

a. Leaves and bracts involute, rigid; pistillate spikes thick-cylindric; perigynia obliquely ascending, firm, olivaceous, short-beaked; achene rhomboid-ellipsoid, tapering to both ends; plant of coastal sands. . . . . 227. *C. extensa.*

a. Leaves and bracts flat, pliable; pistillate spikes thick-cylindric to subglobose; perigynia loosely ascending to divergent or reflexed, membranaceous or inflated, pale green, stramineous or purplish; achenes broadly obovoid, rounded to subtruncate at summit. . . *b.*

b. Loosely tufted, with short decumbent leafy-tipped stolons; all bracts long-sheathed, with erect blades; perigynia ascending; achenes 2 mm. long, 1.4–1.8 mm. broad; plant of Nfld. and region of Gulf of St. Lawrence. . 228. *C. Hostiana.*

b. Densely cespitose, with erect shoots; all but lowest bract sheathless or nearly so, with divergent to reflexed blades; perigynia soon squarrose-spreading to deflexed; achenes 1.2–1.8 mm. long, 1–1.4 mm. broad (a freely hybridizing series, treated by many authors as one plastic and quite unstable species). . . *c.*

c. Perigynia 3–6 mm. long, mostly inflated, with beak half as long as to equaling body, the outer ones finally reflexed.

Leaves yellowish- or warm-green; culms acutely angled at summit; lowest bract twice to four times length of inflorescence; staminate spike sessile to short-stalked, or sometimes partly or wholly pistillate; pistillate scales lanceolate to lance-ovate and acuminate; perigynia lance-ovate to subulate, gradually tapering to beak; wide-ranging species. . . . . 229. *C. flava.*

Leaves glaucous or blue-green; culms obtusely angled; lowest bract one-half to once-and-a-half length of inflorescence; staminate spike usually peduncled; pistillate scales ovate, blunt; perigynia somewhat obovoid, rather abruptly beaked; species of limy areas about Gulf of St. Lawrence. . . . . 230. *C. lepidocarpa.*

c. Perigynia 2–3.5 mm. long, with beak much shorter than body, spreading or only slightly reflexed; culms obtusely angled.

Summits of sheaths of upper leaves and lowest bract prolonged and convex; perigynia definitely inflated, 2–3.5 mm. long, the beak nearly half as long as body; plant of Nfld. and e. Canada. . . . 231. *C. demissa.*

Summits of sheaths of upper leaves and lowest bract concave or truncate; perigynia barely or scarcely inflated, 2–3 mm. long, the beak one-third as long as body; transcontinental species. . . . . . . 232. *C. viridula.*

227. C. EXTÉNSA Good. (stretched out). — In *strict* tussocks 1.5–8 dm. high, *with wiry involute leaves* shorter than the firm slender culms; *bracts similar, divergent, all but the lowest sheathless; pistillate spikes* 2–4, the lowest remote and short-peduncled, the remainder subapproximate and sessile, *thick-cylindric*, 0.5–2.5 cm. long, compact; scales blunt, brown-edged, shorter than *perigynia;* these *obliquely ascending, olivaceous, firm*, ovoid, narrowed at base, strongly nerved, 3–4 mm. long, the short beak sharply 2-toothed; *achene rhomboid*, 2–2.5 *mm. long, tapering to both ends.* — Coastal sands, local, L.I. and Coney I., N.Y.; Norfolk Co., Va. June–Aug. (Natzd. from Eu.) FIG. 719.

719. C. extensa.

228. C. Hostiàna DC. (for NIKOLAUS THOMAS HOST, 1761–1834). — Loosely cespitose, *with short decumbent stolons* bearing erect tufts of bright green firm leaves 2–4 mm. broad; culms few, 3–4.5 dm. high, obscurely trigonous; *pistillate spikes* 2–4, *slenderly ovoid to cylindric*, 1–1.5 cm. long, *densely flowered, remote*, sessile or slightly exserted *from the long sheath of the erect bracts; perigynia obliquely ascending*, submembranaceous, ellipsoid-ovoid, yellowish-green, 3 mm. long, short-beaked, longer than the castaneous white-margined subacute ovate scales; staminate spike long-peduncled or with reduced pistillate spikes at base, 1–2 cm. long; achene closely invested, broadly obovoid, nearly 2 mm. long and 1.4–1.8 mm. broad. (*C. Hornschuchiana* Hoppe) — Marshes, Miquelon, St. P. et Miq. July, Aug. (Eu.)

720. C. Hostiana, v. laurentiana.

Var. **laurentiàna** Fern. & Wieg. (of the Gulf of St. Lawrence). — Coarser; culms up to 6 dm. high; leaves 2.5–4.5 mm. broad; pistillate spikes up to 2 cm. long; perigynia 3.5–5 mm. long; staminate spike 1.5–3 cm. long. (*C. fulvescens* Mackenz.) — Calcareous marshes, glades and shores, w. Nfld., St. P. et Miq., and Anticosti I., Que. Mid-July–Sept. FIG. 720.

× **C. pseùdo-fúlva** Fern. (simulating *C. fulva*) is a hybrid of var. *laurentiana* with no. **230.**

229. C. flàva L. (yellowish). — In *yellow-green* clumps 1–8 dm. high; leaves flat, soft but stiffish, 1.5–6 mm. wide; culms slender, firm, 1–8 dm. high, sharply angled at summit; *bracts* leaf-like, 1.5–6 mm. wide, *divergent to reflexed, only the lowest sheathed, these twice to four times length of inflorescence;* pistillate spikes 2–6, aggregated, or the lower remote, subglobose to ellipsoid-cylindric, 0.7–2 cm. long, rather loosely flowered, 6–14 mm. thick; staminate spike sessile or short-stalked, rarely long-stalked, 0.5–2 cm. long, sometimes partly pistillate above or wholly pistillate; *pistillate scales* brown to greenish, *lanceolate to lance-ovate, acuminate; perigynia slenderly ovoid to subulate,* yellowish or light green, often becoming warm brown, *usually inflated,* loosely spreading, *the lower reflexed and obliquely curving,* strongly nerved, 3.2–6 mm. long, *tapering to* curved prominently bidentate *beak half as long to as long as the body; achene broadly obovoid, broadly rounded at summit,* 1.4–1.8 mm. long. — Highly variable:

Pistillate scales brown, conspicuous until reflexing of lower perigynia; pistillate spikes ellipsoid to subglobose, 1–1.4 cm. thick; perigynia yellowish, 4–6 mm. long, with the serrulate or smooth beak 1.5–2.7 mm. long; leaves 2–6 mm. wide.
  Perigynia lance-ovoid, 1.3–2 mm. thick. . . . . *C. flava* (typical).
  Perigynia subulate, 1–1.2 mm. thick. . . . . Var. *gaspensis.*
Pistillate scales pale and relatively inconspicuous, promptly hidden by the recurving perigynia; pistillate spikes mostly ellipsoid to subcylindric, 6–10 mm. thick; perigynia greenish, 3.2–4.2 mm. long, with smooth beak 1.2–1.5 mm. long; leaves 1.5–4 mm. broad. . . . . Var. *fertilis.*

C. flàva (typical), incl. *C. laxior* (Kükenth.) Mackenz., at least as to type. — Meadows, swales and shores, Nfld. and se. Lab. Pen. to se. Alaska, s. to N.S., N.E., n. N.J., Pa., n. O., n. Ind., Wisc., Minn. and Mont. June–Aug. (Eu.) FIG. 721.

721. C. flava (typical).

Var. **gaspénsis** Fern. (of Gaspé Pen., Que.). — Calcareous shores, Nfld. and Anticosti I. to L. Mistassini, Que., s. to ne. Me. and n. Vt.

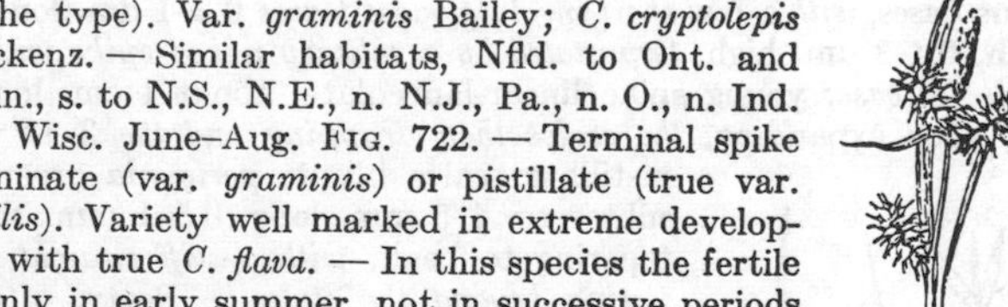

Var. **fértilis** Peck (fertile; all the spikes pistillate in the type). Var. *graminis* Bailey; *C. cryptolepis* Mackenz. — Similar habitats, Nfld. to Ont. and Minn., s. to N.S., N.E., n. N.J., Pa., n. O., n. Ind. and Wisc. June–Aug. FIG. 722. — Terminal spike staminate (var. *graminis*) or pistillate (true var. *fertilis*). Variety well marked in extreme development but too freely confluent with true *C. flava.* — In this species the fertile culms are usually produced only in early summer, not in successive periods up to autumn.

722. C. flava, v. fertilis.

× **C. xánthina** Fern. (yellowish) is a hybrid of typical *C. flava* and *C. Hostiana,* var. *laurentiana.*

× **C. Pieperiàna** Junge (named in 1904 for a local German botanist, G. R. PIEPER) is a hybrid of typical *C. flava* and no. 230.

× **C. subviridula** (Kükenth.) Fern. (somewhat green) is a hybrid of typical *C. flava* and no. 232.

723. C. lepidocarpa.

230. C. lepidocárpa Tausch (scaly-fruited). — Similar to no. 229; *glaucous,* in small tufts, with the 1–few *culms obtusely angled; lowest bract only one-half to once-and-a-half as long as inflorescence,* 1–2.5 mm. broad; *staminate spike* 1–3 cm. long, *peduncled* (rarely subsessile); *pistillate spikes* 1–3, *remote,* the lowest 1–20 cm. apart, ellipsoid-ovoid, 5–10 mm. thick, densely flowered, *with* squarrose to recurving inflated *greenish to whitish-brown perigynia* 2.5–6 *mm. long, the obovoid body rather abruptly rounded to the short beak; scales ovate, blunt.* — Calcareous bogs, swamps and gravels, w. Nfld.; Mingan Ids., Anticosti I., Gaspé Pen. and L. Mistassini, Que.; M.I. July, Aug. (Eu.) FIG. 723. Hybridizes with var. of no. 228 and with no. 229.

231. C. demissa Hornem. (low). — Resembling no. 229; *culms* 0.5–3 dm. high, loosely ascending or arching, *obtusely angled;* leaves deep green, flat, usually curving, exceeding to shorter than culms, 2–3 mm. wide; *summits of sheaths of upper leaves and of the lower widely divergent* to reflexed and prolonged

724. C. demissa.

bracts elongated and convex; staminate spike sessile or short-stalked, 0.5–2 cm. long; pistillate spikes 2–4, the upper crowded, globose-ovoid, 6–8 mm. thick; *scales* ovate, *blunt,* brown; *perigynia inflated,* green to stramineous, 2–3.5 mm. long, *with beak nearly half as long as body.* (*C. Oederi* of most authors, not Retz.) — Boggy, peaty and turfy acid soils, Nfld., e. Que., M.I. and N.S.; rarely on ballast southw. June–Sept. (Eu.) FIG. 724. — In this and no. 232 new fruiting culms continue to develop through summer and autumn.

232. C. virídula Michx. (greenish). — More slender than no. 231, densely matted; culms erect, numerous and continuously produced through the summer, 0.2–4.5 dm. high, with mostly stiffly ascending pale green leaves 1–3.5 mm. wide; *summit of sheaths of upper leaves and lowest* divergent *bract concave or subtruncate;* staminate spike peduncled to sessile, 0.5–2 cm. long, frequently becoming partly or wholly pistillate; *pistillate spikes* 2–20, the upper usually contiguous, sometimes compound, ellipsoid to subglobose, pale green to drab, 4–7 *mm. thick,* densely flowered; scales obtuse, brownish; *perigynia* horizontally spreading to slightly reflexed, *scarcely inflated,* greenish to stramineous, 2–3 *mm. long, with beak one-third the length of the body.* (*C. Oederi,* var. *pumila* sensu ed. 7, not Retz.; *C. chlorophylla* Mackenz., the individuals with numerous pistillate spikes, the terminal spike largely pistillate; *C. Oederi,* var. *prolifera* H. B. Lord) — Damp, often calcareous, gravels, shores, muddy spots and springy places, Nfld. to se. Alaska, s. to N.S., N.E., n. N.J., Pa., n. O., n. Ind., Ill., Wisc., Minn., N.D., N.M. and Calif. June–Sept. FIG. 725. Hybridizes with no. 229.

725. C. viridula.

§ 51. ORTHOCERÀTES Koch (see p. 300)

Lowest sheaths blade-bearing; perigynia with a slender process (the rachilla) projecting from the orifice and exceeding the styles. . . . . 233. *C. microglochin.*
Lowest sheaths bladeless; perigynia without rachilla. . . . . 234. *C. pauciflora.*

233. C. microglòchin Wahlenb. (with minute bristle). — Loosely tufted from short-stoloniferous bases, *with* a few *short involute basal leaves* 0.2–1 dm. long; *culms* 1–few, filiform, stiff, smooth, 0.4–3 dm. high, *terminated by a solitary naked spike with staminate flowers at tip, the pistillate at base;* young spike linear-lanceolate, about 1 cm. long, with appressed-ascending scales, soon expanding, *the scales then dropping and the* 3–12 *perigynia* becoming *reflexed;* pistillate scales blunt; *perigynia* greenish to drab, lance-subulate, subterete, 4–5 mm. long, 0.5–1 mm. thick, spongy-based, slenderly tapering to beak, *with a stiff straight rachilla* (rare in *Carex*) *projecting beyond the styles.* — Peaty, often calcareous, soils, Greenl. and Lab. to Alaska, s. to nw. Nfld., Côte Nord and Anticosti, Que.; shores of James Bay, and Rocky Mts. to Colo. July, Aug. (Eurasia; s. S. Am.) — A very primitive and ancient species, transitional to the subantarctic genus *Uncinia.*

726. C. pauciflora.

234. C. paucifl̀ora Lightf. (few-flowered). — Coarser than no. 233; *lowermost sheaths bladeless;* culms up to 6 dm. high; *perigynia* 1–6, yellowish or straw-colored, 6–8 *mm. long,* 0.7–1.5 *mm. thick, without a rachilla.* — Acid peat and mossy bogs, Lab. to s. Alaska, s. to Nfld., N.S., n. and w. N.E., n. Pa., n. Ind., Wisc., Minn. and Wash. June, July. (Eurasia) FIG. 726. — The prolonged and persistent old culms of the preceding year often much overtop the new fruiting culms of the season.

§ 52. SQUARRÒSAE Carey (see p. 300)

Culms obtusely angled; bracts greatly prolonged, about as long as height of culms; terminal spike often wholly staminate; pistillate spikes 0.8–1.2 cm. thick, the rough-awned scales much exceeding the wide-spreading perigynia. 235. *C. Frankii.*
Culms acutely angled, many times higher than length of bracts; terminal spike pistillate except at base, 1–2.2 cm. thick, the scales much shorter than the perigynia.
  Pistillate scales sharp-tipped or awned; beaks of perigynia widely divergent or the lowest reflexed. . . . . 236. *C. squarrosa.*
  Pistillate scales blunt; beaks of perigynia subappressed-ascending. . . . . 237. *C. typhina.*

235. C. Fránkii Kunth (for its discoverer, JOSEPH C. FRANK, 1782–1835). — In dense

727. C. Frankii.

but small tussocks; *culms obtusely angled*, 1–8 dm. high, *greatly overtopped by the broad ribbon-like bracts;* leaves and lower bracts 4–9 mm. broad, very prolonged, rough on nerves, deep green; *terminal spike staminate throughout or pistillate at summit;* fully *pistillate spikes* 2–7, *cylindric*, green, becoming brownish, 1–4 cm. long, 8–12 *mm. thick*, the upper overlapping, the lower more remote on a zigzag axis; *perigynia strongly squarrose, obconic,* inflated, *abruptly truncate below the prolonged subulate firm-toothed beak*, 4–5 mm. long, *exceeded by the long rough awns of the scales.* — Calcareous meadows, bottoms and low, rich woods, Ga. to Tex., n. to e. Pa., centr. and w. N.Y., O., s. Mich., Ill., Mo. and Kans. June–Sept. Fig. 727.

728. C. squarrosa.

236. C. squarrŏsa L. (with wide-spreading parts). — In dense or loose stools; *culms acutely angled*, rough above, 3–9 dm. high; leaves warm green, 2.5–6 mm. broad, weak, roughish, exceeding culms; *bracts slender*, somewhat overtopping inflorescence; *spikes solitary, less often 2–4, the terminal staminate only at base, all subglobose to thick-cylindric or ellipsoid,* 1–3 cm. long, 1–2.2 *cm. thick; pistillate scales sharp-pointed or awned, shorter than the* obconic inflated *perigynia; the latter* subtruncate at summit and *with* rough central subulate toothed *beaks horizontally spreading* to *reflexed.* — Calcareous bottomlands, meadows and low woods, Ct. to s. Ont., Minn. and e. Neb., s. to N.C., Tenn. and Ark. June–Sept. Fig. 728. — Hybridizes with no. 249.

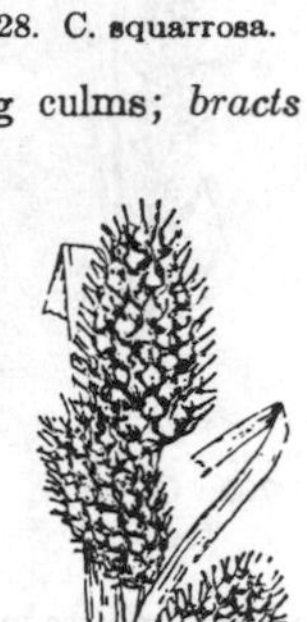
729. C. typhina.

237. C. typhina Michx. (resembling *Typha*, cat-tail). — Similar, coarser; *leaves gray-green,* 0.5–1 *cm. broad; spikes* 1–6, subcylindric to lance-ellipsoid, 1.5–5 *cm. long,* 1–1.7 *cm. thick; pistillate scales blunt,* mostly hidden; *beaks of perigynia turned upward,* sometimes subappressed. (*C. typhinoides* Schwein.) — Calcareous meadows and wooded bottomlands, sw. Que. to Wisc. and e. Ia., s. to sw. Me., Ct., Ga., Ala. and La. June–Sept. Fig. 729. — Hybridizes with no. 164.

§ 53. Paludŏsae Fries (see p. 300)

a. Teeth of perigynia 0.1–1 mm. long, thick, erect or but slightly curved; perigynia glabrous; leaf-blades and sheaths glabrous. . . *b.*

b. Perigynia compressed, flattened-trigonous, papillate, 3.5–4 mm. long, in slenderly cylindric spikes 6–7 mm. thick; local adventive near Boston. 238. *C. acutiformis.*

b. Perigynia subterete or only obscurely angled, smooth, in oblong to thick-cylindric spikes. . . *c.*

c. Spikes 1–2 cm. long and 5–7 mm. thick; pistillate scales deep purple-red, nearly equaling perigynia; perigynia spongy, 4–5 mm. long; leaves rigid, 2–3 mm. broad; adventive on St. Lawrence R., near Montreal. 239. *C. nutans.*

c. Spikes (1–)2–10 cm. long and 1–1.5 cm. thick; pistillate scales paler, half as long as to equaling perigynia; perigynia firm, 5.5–8 mm. long; leaves firm, not rigid, 0.4–1.5 cm. broad; wide-ranging natives.

Lower sheaths of fertile culms purple, bladeless, soon becoming fibrillose on inner side; mature perigynia prominently ribbed. . . . 240. *C. lacustris.*

Lower sheaths whitish or brownish, blade-bearing, not usually fibrillose; mature perigynia nerveless or very delicately impressed-nerved. 241. *C. hyalinolepis.*

a. Teeth of perigynia 1–3 mm. long, slender and pungent (only 0.5–1 mm. long in no. 245 with densely pubescent perigynia); perigynia (5–12 mm. long), leaf-blades and sheaths glabrous or pubescent. . . *d.*

d. Perigynia glabrous.

Leaf-blades chiefly 0.4–1.2 cm. broad, pubescent beneath; summit of inner band of pilose middle and upper sheaths not evidently nerved; perigynia lance-ovate to lanceolate, 7–12 mm. long, with mostly curving teeth 1.6–3 mm. long. . . . . 242. *C. atherodes.*

Leaf-blades 2–8 mm. broad, glabrous beneath; summit of inner band of glabrous middle and upper sheaths pinnately veined; perigynia ovoid, 5–7 mm. long, with erect to barely curving teeth 1–1.8 mm. long. . . 243. *C. laeviconica.*

d. Perigynia pubescent.

Ribs of mature perigynia strong and evident; teeth 0.7–2.7 mm. long. 244. *C. trichocarpa.*

Ribs obscure, mostly hidden in dense pubescence; teeth 0.5–1 mm. long. . . . . 245. *C. subimpressa.*

238. C. acutifórmis Ehrh. (in form like *C. acuta*). — Stout, 0.4–1.2 m. high, with cord-like stolons; culm thick and sharply angled; leaves broad, flat and *glaucous*, much prolonged; *pistillate spikes* 2–5, all but the uppermost peduncled, spreading or drooping, *slenderly cylindrical*, 2–5.5 cm. long, 6–7 *mm. thick*, loosely flowered below; *perigynia compressed, flattened-trigonous*, ovoid, *papillate and opaque*, 3.5–4 *mm. long, the short beak slightly toothed; scales* rough-awned, purplish, *exceeding perigynia.* — Boggy meadows and borders of saline marshes, local, e. Mass. June, July. (Natzd. from Eu.) Fig. 730.

730. C. acutiformis.

239. C. nùtans Host (nodding). — Creeping by hard stolons; culms 3–6 dm. high, slender, nodding during anthesis; *leaves* overtopping culms, *rigid*, harsh, 2–3 *mm. wide; pistillate spikes* 2–4, *oblong or oblong-cylindric*, 1–2 *cm. long*, 5–7 mm. thick; *scales red-purple*, lance-acuminate, *about equaling perigynia; perigynia spongy-coriaceous*, ovoid-conical, 4–5 *mm. long*, olivaceous, *glabrous*, opaque, impressed-nerved, the short beak deeply cleft into broad-based teeth. — Low grounds, local, St. Lawrence R., near Montreal, Que. July. (Adv. from Eu.)

240. C. lacústris Willd. (of lake-margins). — Large and stout from creeping rhizomes, stoloniferous, 0.6–1.3 m. high; culms sharply angled and harsh at summit; *basal sheaths purplish, bladeless, these and blade-bearing sheaths soon fibrillose on inner side;* blades glaucous, often conspicuously septate-nodulose from margin to midrib, mostly 0.6–1.5 cm. broad, the evident *ligule elongate; pistillate spikes* short-oblong to elongate, 2–10 *cm. long*, 1–1.5 *cm. thick*, erect or the lower sometimes drooping, densely flowered except at base; *perigynia olivaceous, coriaceous, glabrous, lance-ovoid*, lightly *many-nerved*, 5.5–7 *mm. long*, tapering to thick beak, *with broad-based* erect to arching *teeth* 0.3–1 *mm. long;* scales blunt to awned, mostly about half as long as perigynia. (*C. riparia* of ed. 7, not Curtis; *C. riparia*, var. *lacustris* (Willd.) Kükenth.) — Calcareous or circumneutral swamps and shallow water, Anticosti I., Que., to s. Man., s. to N.S., N.E., Va., O., Ind., Ill., Ia. and S.D.; Id. Late May–Aug. Fig. 731.

731. C. lacustris.

241. C. hyalinólepis Steud. (with hyaline scales). — Differing from no. 240 in its *whitish to pale brown blade-bearing lower sheaths, the latter rarely fibrillose*, the blades slightly or scarcely septate near the midrib, the *ligule short and broad; perigynia smooth and nerveless or very delicately impressed-nerved*, often more ovoid, 6–8 mm. long, often equaled by awns of the scales. (*C. riparia*, var. *impressa* S. H. Wright; *C. impressa* Mackenz.) — Calcareous or brackish swamps, swales and shores, Fla. to Tex., n. to s. N.J., se. Pa., s. Ont., O., s. Mich., s. Wisc., Ia. and Neb. May–July.

242. C. atheròdes Spreng. (like an ear of wheat). — Resembling no. 240, usually not so coarse; the purple basal sheaths bladeless and becoming fibrillose; culms smooth, sharply angled; *principal leaves* 0.4–1.2 *cm. broad, pubescent beneath; middle and upper sheaths pilose, the inner band with a plane firm summit;* pistillate spikes 2–4, cylindric, 2–12 cm. long, mostly ascending; *perigynia glabrous, strongly ribbed, lance-ovoid to lanceolate*, 7–12 *mm. long*, tapering to slender beaks *with slender firm often outwardly curving teeth* 1.6–3 *mm. long;* scales aristate, longer than to shorter than perigynia. (*C. aristata* R. Br., not Honck.; *C. trichocarpa*, var. *aristata* Bailey) — Calcareous meadows, swales and shores, Ont. to Mackenz. and Yuk., s. to N.Y., O., Ind., Wisc., Mo., Neb., Colo., Utah and Oreg.; Knox Co., Me. June–Aug. (Eurasia) Fig. 732.

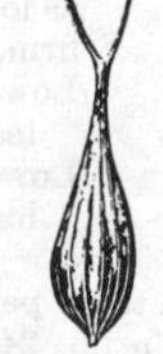

732. C. atherodes.

243. C. laevicónica Dew. (with smooth cone). — Similar to no. 242, smaller; culms only 3–7.5 dm. high; *leaves* 2–8 *mm. broad, glabrous, the inner band of the glabrous sheath strongly pinnate-veined at summit;* pistillate spikes 1.5–7.5 cm. long; *perigynia ovoid*, 5–7 *mm. long, with erect to barely curving teeth* 1–1.8 *mm. long*, the nerves less prominent or obscure.

733. C. laevi-conica.

(*C. trichocarpa*, var. *Deweyi* Bailey) — Calcareous marshes, bottomlands and wet prairies, Man. and Sask., s. to Ill., Mo., Kans. and Mont. June, July. Fig. 733.

734. C. trichocarpa.

244. C. trichocárpa Muhl. (hairy-fruited). — Resembling no. 243, slender; culms up to 1 m. or more tall; leaves scabrous, otherwise glabrous, 3–8 mm. wide, the *summit of the inner band of the sheath purple-tinged;* pistillate spikes cylindric to ellipsoid, 1.5–9 cm. long; *perigynia* ovoid to ovoid-lanceolate, *strongly ribbed, short-hairy,* 5–10 mm. long, longer than the acute scales, *the teeth of the beak* 0.7–2 *mm. long.* (Incl. var. *turbinata* Dew.) — Calcareous swales, marshes and bottomlands, sw. Que. and Vt. to Ont. and Minn., s. to Ct., Del., Pa., O., Ind., Ill. and n. Ia. June–Aug. Fig. 734.

245. C. subimpréssa Clokey (somewhat impressed). — Similar to no. 244; culms 2–6 dm. high; *perigynia* ovoid, 5–7 mm. long, *the impressed nerves nearly hidden by the close soft pubescence.* — Alluvium and low woods, Ind., Ill. and Mo. June.

§ 54. Pseùdo-Cypèreae Tuckerm. (see p. 300)

a. Perigynia closely investing base of achene, not inflated, greenish, becoming pale brown, compressed and 2-edged, all or at least the lower ones soon reflexed; plants in dense clumps, with firm strongly septate-nodose blades and prolonged ligules.
  Perigynia 3–5 mm. long; their teeth nearly parallel, erect and 0.5–1 mm. long. . . . . 246. *C. Pseudo-Cyperus.*
  Perigynia 5–7 mm. long; their teeth arched-divergent and 1.2–2 mm. long. 247. *C. comosa.*
a. Perigynia loosely investing achene, inflated, subterete, straw-colored or yellow-green, becoming warm brown, ascending or spreading, not reflexed; leaves less conspicuously septate-nodose, the ligules about as broad as long. . . b.
  b. Culms in clumps, firm, sharply trigonous; pistillate spikes subtruncate at base; perigynia spreading or spreading-ascending; staminate scales rough-awned.
    Perigynia lance- or slender-ovoid, 15–20-nerved, only slightly inflated, 1.5–2 mm. thick; achene cuneate-obovoid. . . . . 248. *C. hystricina.*
    Perigynia with ovoid to subglobose strongly inflated bodies 2–4 mm. thick, 8–10-nerved; achene ellipsoid.
      Pistillate spikes thick-cylindric to subglobose-cylindric, 1.4–2 cm. thick; perigynia ovoid, tapering to beak shorter than to about equaling body; leaves and bracts mostly 3–7 mm. broad. . . . 249. *C. lurida.*
      Pistillate spikes slender-cylindric, 0.8–1.3 cm. thick; perigynia with subglobose bodies abruptly contracted to beak equaling or longer than body; leaves 2–4 mm. wide. . . . . 250. *C. Baileyi.*
  b. Culms solitary at tips of elongate stolons, weak, easily compressed, smooth; pistillate spikes slenderly cylindric, mostly approximate, narrowed to base; perigynia ascending, about 8-nerved; staminate scales smooth and awnless. . . . . 251. *C. Schweinitzii.*

246. C. Pseùdo-Cypèrus L. (old generic name, false *Cyperus*). — Tall and rather stout, 0.5–1 m. high, in clumps; culms thick and very sharply triangular, rough throughout; leaves very long, rough-margined, 0.5–1 cm. broad, firm, prominently septate-nodulose; *pistillate spikes* 3–5, slenderly peduncled and more or less drooping, somewhat contiguous or scattered, 2.5–7.5 cm. long, *slenderly cylindric,* 8–11 *mm. thick, very closely flowered; perigynia soon strongly reflexed, more or less 2-edged, many-costate,* 3–5 *mm. long, greenish to drab; the beak shorter than the body, with erect to slightly divergent sharp teeth* 0.5–1 *mm. long;* scales with very long rough awns about equaling perigynia. — Shallow water, swamps and bogs, Nfld. to Sask., s. to N.S., Me., Mass., n. Ct., n. N.J., n. Pa., Mich., Wisc., Minn. and N.D. June–Aug. (Eurasia and n. Afr.) Fig. 735.

247. C. comòsa Boott (bearded). — Resembling no. 246, mostly coarser, 0.5–1.5 m. high; leaves 0.6–1.6 cm. broad; *pistillate spikes thicker,* 1.3–1.7 *cm. thick,* less densely flowered; *perigynia* 5–7 *mm. long,* less tightly reflexed; *beak about as long as body, its arched-outcurving teeth* 1.2–2 *mm. long.* — Swamps and shallow water, Fla. to La., n. to w. N.S., centr.

735. C. Pseudo-Cyperus.

Me., sw. Que., s. Ont., Mich., Wisc., Minn. and Neb.; Id. and Wash. to Calif. June–Aug. Fig. 736. —Crosses with no. 258.

248. C. hystricina Muhl. (porcupine-like).— *In* small to broad *clumps;* culms slender, erect, *firm, sharply trigonous, scabrous,* 0.15–1 m. high; *leaves firm, scabrous,* green, 3–10 mm. broad; *pistillate spikes* 2–5, subapproximate to scattered, the lower peduncled and spreading to drooping, *subtruncate at base,* thick-cylindric, 1–4 cm. long, 1–1.5 cm. thick, or prolonged to 6 cm. in forma **Dúdleyi** (Bailey) Wieg. (for William Russell Dudley, 1849–1911); *perigynia* greenish to stramineous, inflated, ovoid, about 6 mm. long, prominently 15–20-*nerved, spreading,* the slender beak sharply toothed; *staminate scales awned.* —Swales, swamps and shores, chiefly calcareous, Gaspé Co., Que., to Alta. and Wash., s. to P.E.I., N.B., N.E., N.J., D.C., upland to Tenn., Mo., Okla., Tex., N.M., Ariz. and Calif. Fig. 737. — Crosses with no. 249. — Specific name originally *hystericina,* presumably an error, since the German equivalent was given as *"Stachelschweinartige Segge"* or sedge like a porcupine (Hystrix).

736. C. comosa.

249. C. lùrida Wahlenb. (sallow). —Very variable in size, 0.2–1.05 m. high, *in dense clumps; culms* angled and smooth, *firm; leaves* long and loose, warm- or yellowish-green, *mostly* 3–7 *mm. broad,* rough, the similar bracts much elongate; *pistillate spikes* (1–)2–4, thick-cylindric to subglobose, 1–7.5 cm. long, 1.4–2 *cm. thick, truncate or broadly rounded at base,* variously disposed, the 1 or 2 upper sessile and nearly erect to drooping, the others more or less peduncled, approximate to remote, very densely flowered; *perigynia membranaceous,* lustrous, becoming yellow-brown, spreading-ascending, *about* 10-*nerved,* ovoid, 6–9 mm. long, 2.5–4 *mm. thick, strongly inflated, the ovoid body barely equaling the slender long conical beak; staminate spike single, its scales with long rough awns; achene cuneate-obovoid.* —Swales, swamps and wet woods, one of the commonest species, Fla. to Tex. and e. Mex., n. to N.S., N.B., Me., s. Que., Ont. and Minn. June–Oct. Fig. 738. — Hybridizes with nos. 236, 248, 258, 261 and 262.

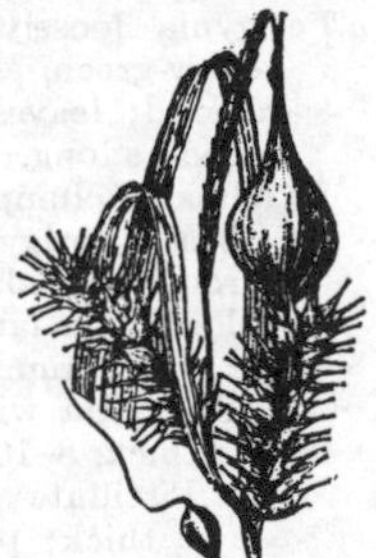

738. C. lurida.

737. C. hystricina.

250. C. Baileyi Britt. (in honor of Liberty Hyde Bailey, 1858–). — More slender than no. 249; culms 2–7 dm. high, more sharply angled; *leaves mostly* 2–4 *mm. wide; pistillate spikes* more slender, 1–4 cm. long, 0.8–1.3 *cm. thick; perigynia* 5–7 mm. long, 2–2.5 *mm. thick, abruptly contracted to a beak as long as or longer than the subglobose body.* (*C. lurida,* var. *gracilis* (Boott) Bailey) — Swampy woods and meadows, sw. Que. and n. N.H. to Mich., s. to uplands of Va. and Tenn. June–Sept. Fig. 739.

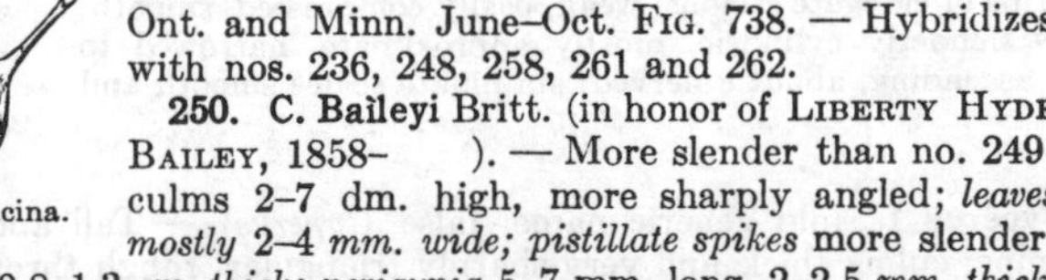

251. C. Schweinítzii Dew. (in honor of Lewis David de Schweinitz, 1780–1834). — Soft but erect, 2–7 dm. high, *yellow-green,* becoming stramineous in drying; *culms solitary or few from widely creeping rhizomes, soft, easily flattened, smooth; leaves thin and soft,* 0.5–1 cm. broad, the radical longer than the culms, the others mostly shorter; *pistillate spikes* 3–5, slenderly cylindrical, *tapering at base, mostly approximate and ascending,* 2.5–7.5 cm. long, 8–13 mm. thick; *perigynia* papery, inflated, lance-ovoid, 5–7

740. C. Schweinitzii.

739. C. Baileyi.

mm. long, *about* 8-*nerved*, long-beaked, *ascending*, slightly exceeding the rough-awned scales; *staminate scales awnless*. — Calcareous swamps, meadows and low woods, sw. Vt. to s. Ont. and Mich., s. to w. Ct., nw. N.J., Pa., upland to N.C. and Tenn., and Mo., local; probably much exterminated in the past by clearing of rich lands. May–July. FIG. 740.

§ 55. COLLÍNSIAE Mackenz. (see p. 300) — A single species:

252. C. Collínsii Nutt. (in honor of ZACCHEUS COLLINS, 1764–1831). — In dark tussocks with *very slender weak or arching obtusely angled smooth culms* 0.1–1.1 m. long; leaves soft and thin, much shorter than culms, 1.5–5 mm. wide; lower bracts elongate, sheathing, the upper reduced; *pistillate spikes* 2–5, remote, sessile to short-peduncled, *loosely subglobose to thick-cylindric*, often slightly staminate at tip, *with* 1–8 *widely divergent or reflexed subulate* green slender-beaked *promptly deciduous perigynia* 0.8–1.5 *cm. long, the style-branches borne well below the very slender tip, the two rigid teeth soon abruptly reflexed;* staminate spike solitary, whitish, rarely 1 cm. long; pistillate scales lanceolate, persistent after fall of perigynia. (*C. subulata* Michx., not J. F. Gmel.) — Sphagnum of springy wooded swamps, local, s. R.I. to e. Pa., s. to S.C. and Ga. June–Sept. FIG. 741.

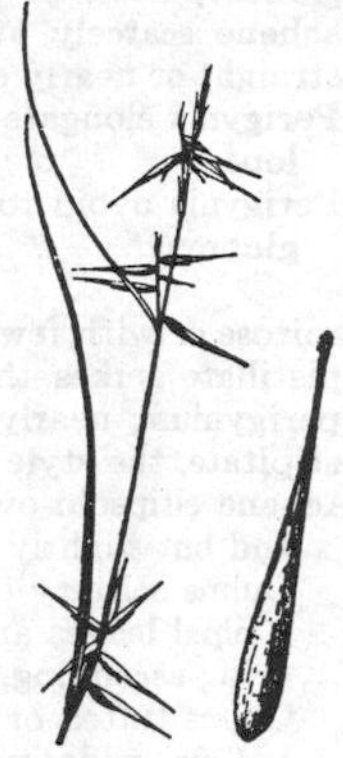
741. C. Collinsii.

§ 56. FOLLICULÀTAE Mackenz. (see p. 300)

Leaves 1.5–3.5 (–4) mm. broad; summits of sheaths of bracts concave; staminate spike 0.6–1.5 cm. long; scales of pistillate spike awnless, less than half the length of perigynia. . . . . 253. *C. Michauxiana.*

Leaves 3.5–17 mm. broad; summits of sheaths of bracts prolonged; staminate spike 1.2–4 cm. long; scales of pistillate spike mostly cuspidate or awned, one-half as long to nearly equaling perigynia. . . . . 254. *C. folliculata.*

253. C. Michauxiàna Boeckl. (in honor of ANDRÉ MICHAUX, 1746–1802). — Densely cespitose, yellow-green; the slender smooth erect culms 2–6 dm. high; *basal leaves* crowded, firm, flat, shorter than culms, 1.5–3.5 (–4) *mm. broad; bracts* prolonged, overtopping inflorescence; *the lowest* 1–3 *mm. broad, its sheath with hyaline band concave at summit; pistillate spikes* 2–4, loosely subglobose or ellipsoid, the lowest remote and peduncled, *the uppermost* crowded and *often hiding the pale sessile staminate spike* (6–15 *mm. long*); *pistillate scales obtuse to acute, less than half length of perigynia; perigynia* at first appressed-ascending, soon loosely divergent, green to stramineous, *lance-subulate*, scarcely inflated, 8–13 mm. long, soon falling and leaving empty scales. (*C. abacta* Bailey) — Acid peats and wet sands, Nfld. to Ont., s. to N.S., n. N.E., w. Mass., n. N.Y. and n. Mich., often ascending to alpine areas. June–Aug. (E. Asia) FIG. 742.

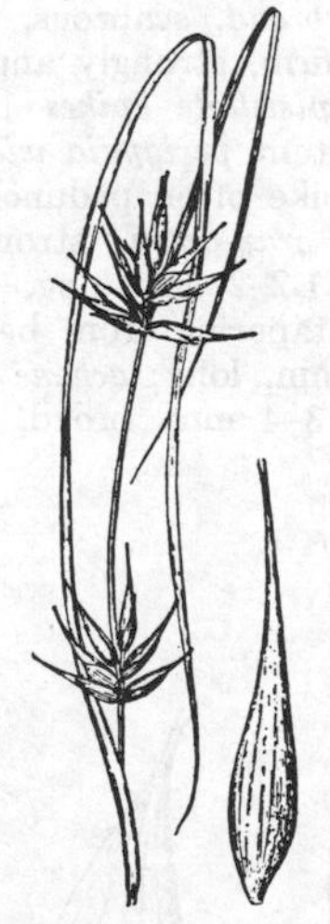
742. C. Michauxiana.

254. C. folliculàta L. (bearing follicles). — Much coarser; 0.3–1.2 m. high; *basal leaves mostly* 0.6–1.7 *cm. broad; bracts* greatly prolonged; *the lowest one* 5–10 *mm. wide, the inner band of the sheath prolonged at summit;* pistillate spikes remote or the upper approximate, the lower peduncled and often spreading or drooping, mostly pistillate to tip; *staminate spike* sessile or short-peduncled, 1.2–2.5 *cm. long; pistillate scales cuspidate or awned, three-fourths as long as to exceeding perigynia; perigynia lance-conic, slightly inflated,* 1–1.5 cm. long, 2.5–3.5 *mm. thick at base.* — Peaty thickets, swampy woods and swales, Nfld. to Wisc., s. to N.S., N.E., N.C. and Tenn. June–Aug. FIG. 743. — Southeastw. passing into

743. C. folliculata.

Var. **austràlis** Bailey (southern). — Greener, more slender and lower (up to 8 dm. high); *basal leaves* 3.5–10 *mm. wide; lowest bracts* 3–7 *mm. wide; pistillate spikes often staminate at tip;* staminate spike 2–4 cm. long; *pistillate scales awnless or barely awned, about half as long as the lance-subulate scarcely inflated perigynia* (2–2.5 *mm. thick at base*). (*C. lonchocarpa* Willd.;

C. Smalliana Mackenz.) —Wet woods, bottomlands and swamps of the Coastal Plain, Fla. to La., n. to s. N.J. May–July.

§ 57. LUPULÌNAE Tuckerm. (see p. 300)

a. Densely cespitose, without elongate stolons; pistillate spikes globose or subglobose; beak of perigynium much shorter than body; ligule circular; achene scarcely stipitate, the persistent style (within the perigynium) straight or nearly so.
   Perigynia elongate-rhomboid, cuneate at base, firm, opaque, often hispidulous. . . . . 255. *C. Grayii.*
   Perigynia ovoid to lanceolate, rounded at base, membranaceous, lustrous, glabrous. . . . . 256. *C. intumescens.*

a. Cespitose or with few or solitary culms, bearing horizontal prolonged stolons; pistillate spikes thick-cylindric, ellipsoid or subglobose-ellipsoid; beak of perigynium nearly equaling to longer than body; ligule ovate; achene stipitate, the style usually spiraling or contorted. . . *b.*
 b. Achene ellipsoid-ovoid, definitely longer than broad, with uniform angles and but slightly depressed sides.
   Culms solitary or nearly so, from slender wide-creeping rhizomes; principal leaves and lower bracts 3–6 mm. wide; perigynia loosely spreading-ascending, 1–1.4 cm. long. . . . . 257. *C. louisianica.*
   Culms tufted or cespitose; principal leaves and lower bracts mostly 0.6–1 cm. wide; perigynia appressed-ascending, 1.3–2 cm. long. . . . . 258. *C. lupulina.*
 b. Achenes rhomboid to broadly obconic, nearly as broad as to broader than long, with nipple-tipped angles and depressed sides.
   Perigynia spreading-ascending, pale green, becoming yellowish-brown, conic-ovoid, gradually tapering to beak about as long as body; achene rhomboid, subequally tapering to base and apex; staminate spikes 1 or 2. 259. *C. lupuliformis.*
   Perigynia soon almost horizontally divergent, deep green, becoming paler green, abruptly narrowed to beak two to four times length of distended body; achene depressed, broader than high, subtruncate at summit; staminate spikes (1–)2–5. . . . . 260. *C. gigantea.*

255. C. Graỳii Carey (for ASA GRAY, 1810–1888). — In clumps 0.3–1 m. high, very leafy, *pale or gray-green; leaves firm,* 0.45–1.4 *cm. broad,* scabrous, the cauline with prolonged loose sheaths; culms firm, strongly angled above; upper leaves and bracts prolonged; *pistillate spikes* 1 or 2, *globose,* 3–4 cm. in diameter, of 6–30 persistent *perigynia widely radiating from a common center,* the lowest spike often peduncled; staminate spike 1; *perigynia firm, opaque, gray-green,* strongly ribbed, inflated, glabrous, *elongate-rhomboid,* 1.2–2 cm. long, 5–8 mm. thick, *cuneate to* sessile *base,* conically tapering from below middle to short beak with sharp teeth 1–2 mm. long; *achene* obtusely angled, *plump-obovoid,* 5–6 mm. long, 3–4 mm. broad, the sides convex at summit; scales ovate, much shorter than perigynia. — Calcareous meadows and alluvial woods, sw. Que. to Ia., s. to w. N.E., L.I., n. N.J., Pa., O., Ind., Ill. and n. Ark. June–Oct. FIG. 744. — Passing into the often more southern

744. C. Grayii.

Var. **hispídula** Gray (short-hispid). — Perigynia, at least when young, hispidulous. — Ga. to Miss. and Mo., n. to centr. Ct., e. N.Y., O., s. Mich., Ind. and Ill., apparently tbe only var. from se. Va. to Miss.

256. C. intumèscens Rudge (swelled-up). — More *slender* than no. 255, *dark green; leaves and bracts* 2.5–8 *mm. wide, soft and lax; pistillate spikes* 1–3, often aggregated, subglobose, 1.5–3 *cm. thick, the* (1–)5–15 *perigynia spreading and loosely ascending* (*not strongly reflexed*); *perigynia full green,* glabrous, *lustrous, thin* and inflated below, conic-ovoid, 5–8 *mm. thick at the rounded base,* 1–1.7 cm. long; *achene trigonous-ellipsoid, broadest near*

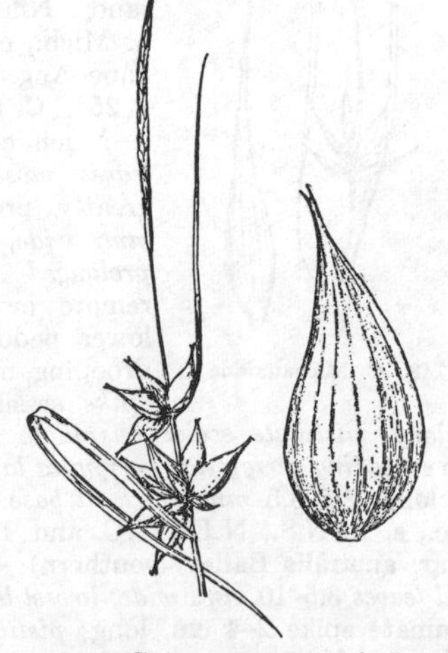

745. C. intumescens (typical).

middle, acuminate, 4–5.5 mm. long, *with flat to slightly concave sides*, 3 mm. broad. — Alluvial woods, meadows and swales, one of the commonest species, Fla. to Tex., n. to N.S., centr. Me., sw. Que., N.Y., s. Ont., Mich. and Wisc. Late May–Sept. Fig. 745. — Hybridizes with nos. 257 and 261. Northw. passing into

Var. **Fernáldii** Bailey (for its discoverer, 1873–). — *Perigynia lanceolate or lance-ovoid*, barely inflated, 3–5 *mm. thick* at base; *achene trigonous-obovoid, broadest near rounded summit*, 3–4 mm. broad. — Nfld. to Man., s. to n. and w. N.E., N.Y., n. Pa., mts. of N.C., Mich., Wisc. and Minn. June–Sept. Fig. 746. — Forma **ventriòsa** Fern. (bellied-out), with similar range and achenes, has the perigynia strongly inflated, conic-ovoid and 5–8 mm. thick.

746. C. intumescens, v. Fernaldii.

257. C. louisiánica Bailey (of Louisiana). — *Culms solitary* or mostly so *from* leafy *tips of slender elongate stolons and rhizomes*, slender, smooth, 2–7.5 dm. high; *leaves* and bracts soft, roughish, 3–6 *mm. broad*, often overtopping the inflorescence; *staminate spike* usually *long-peduncled* (sometimes with a small secondary one); *pistillate spikes* 2–4, *scattered*, sessile, or the lower short-peduncled, thick-cylindric to subglobose, 2–4.5 cm. long, 2–3 cm. thick; the rather *few* green to stramineous ascending *perigynia* conic-ovoid, thin, bladdery, 1–1.4 cm. long, with a rather abrupt slender-conic sharply 2-toothed beak, twice as long as the firm acuminate scales; *achene ellipsoid-ovoid, trigonous*, olivaceous, 2.5–3.5 *mm. long*, 1.7–2 *mm. thick*. (*C. Halei* Carey, not Dew.) — Wooded bottomlands and swamps, Fla. to Tex., n. to s. N.J., D.C., Ky., s. Ind. and Mo. June–Oct. Fig. 747. — Hybridizes with no. 256.

747. C. louisianica.

258. C. lupulìna Muhl. (hop-like). — Very stout and leafy, *in* small or large *tussocks* with creeping rhizomes, 0.3–1.3 m. high; *leaves* 0.6–1 *cm. broad*, loose; bracts broad and elongate; *pistillate spikes* 2–6, *mostly approximate and sessile or nearly so at top of culm*, or the lower short-peduncled and distant, *subglobose to thick-cylindric*, very heavy and *densely flowered with* pale green to drab-brown *appressed-ascending perigynia*, *mostly* 2.5–6 *cm. long and* 2–3.5 *cm. thick; perigynia* much inflated, firm but rather soft, 1.3–2 *cm. long;* pistillate scales firm, lance-ovate, mostly much shorter than perigynia; staminate spike sessile or subsessile; *achenes rhomboid-ovoid, much longer than broad*, with nearly uniform angles, about 4 mm. long, 2–3 mm. wide. — Swamps and wet woods, Fla. to Tex., n. to N.S., Me., s. Que., N.Y., s. Ont., Mich., Wisc. and Minn. June–Oct. Fig. 748. — Hybridizes with nos. 247, 249, 261, 263 and 265. — Northw. passing into

748. C. lupulina (typical).

Var. **pedunculàta** Gray (having peduncles). — Staminate spike sessile to long-peduncled; *pistillate spikes* mostly *subdistant to remote* (pistillate inflorescence 1–3 dm. long), *mostly peduncled;* the lower on peduncles 1–12 cm. long, *chiefly* 3–10 cm. long and 2–2.5 *cm. thick*. — Less common, s. Que. and Ont. to Ga. and Ill. — Resembling no. 259.

259. C. lupulifórmis Sartwell (with form of *C. lupulina*). — Resembling the var. of no. 258, stout and tall, up to 1.2 m. high; differing in its broader leaves (0.7–1.5 cm. wide); *staminate spike often peduncled; pistillate spikes* 3–5, crowded toward summit or scattered, *sessile to short-peduncled*, 3–8 cm. long and 2–3.5 cm. thick; *perigynia stramineous to fulvous*, thin, *lance- to slenderly ovoid-conic*, 1.3–2 cm. long; *achene broadly trigonous-rhomboid, about as broad as long, cuneate below, gradually tapering above, the summits of the angles nipple-like*. — Calcareous swamps, meadows and prairies, Vt. to Minn., s. to Ct., Va., Ky., La. and e. Tex. July–Oct. Fig. 749.

749. C. lupuliformis.

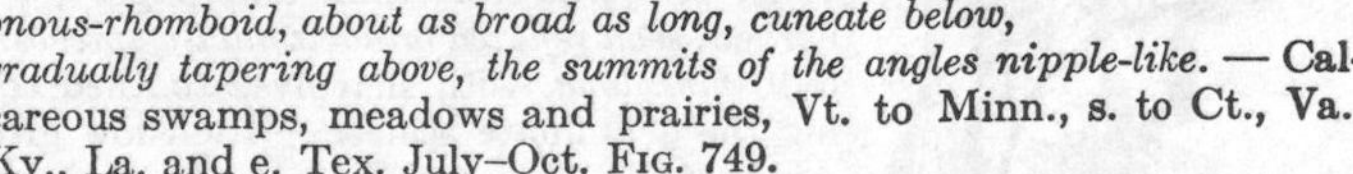

× C. **Macoùnii** Dew. (for John Macoun, 1831–1920) is an apparent hybrid with no. 261.

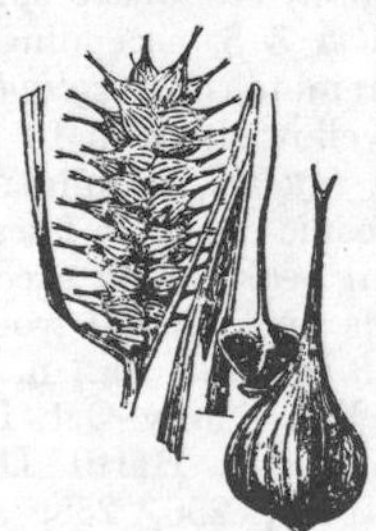

750. C. gigantea.

260. C. gigantèa Rudge (gigantic). — *Loosely tufted* or with 1–few stoutish culms 0.4–1.2 m. high, *from wide-creeping rhizomes and stolons; leaves* ribbon-like, *full green*, prolonged, 0.6–1.6 cm. wide; bracts greatly prolonged; staminate spikes (1–)2–5; *pistillate spikes green*, 2–4(–5), *scattered,* the lowest peduncled, 3–8 cm. long, 2–3 cm. thick; *perigynia swollen below, abruptly contracted to a slender beak two to four times length of body, spreading at right angles* or nearly so, *never becoming yellow; achene depressed, broader than high,* with nipple-like high angles, the base obconic, *the summit subtruncate or broadly rounded.* — Bottomlands and swamps, Fla. to Tex., n. to Del., Va., s. Ind., Ky. and Mo. Late May–Sept. Fig. 750.

§58. Vesicàriae Tuckerm. (see p. 300)

*a.*Stigmas 3; achene trigonous; beak sharply 2-toothed; pistillate scales blunt, acute, acuminate or short-awned; pistillate spikes 0.5–2 cm. thick. . . *b.*

*b.*Perigynia soon reflexed or horizontally spreading, inflated, in thick-cylindric mostly approximate spikes; sheaths loose, the brown membranous inner band truncate to prolonged at summit, the lax blades 0.4–1 cm. broad; lower bracts many times the length of inflorescence. . . . . 261. *C. retrorsa.*

*b.*Perigynia ascending to merely spreading (basal ones rarely reflexed in no. 262 with firm leaves); sheaths close, the firm inner band concave at summit; blades 1–12 mm. broad, firm; lower bract short or prolonged. . *c.*

*c.*Culms mostly thick and spongy at base, 1–few from long horizontal rhizomes and stolons, generally smooth and obtusely angled (except at very summit); leaves when dry prominently septate-nodulose, pale green or glaucous, 2–15 mm. broad. . . . . 262. *C. rostrata.*

*c.*Culms slender, rarely spongy-based, more acutely angled, often harsh above; leaves not conspicuously septate-nodulose (except in no. 264 with filiform-involute blades), not glaucous, 1–7 mm. broad. . . *d.*

*d.*Achene asymmetrical, invaginated on one side; perigynia thin and papery, strongly inflated, 5–6.5 mm. thick; lowest bract many times longer than inflorescence. . . . . 263. *C. Tuckermani.*

*d.*Achene symmetrical, not laterally invaginated; perigynia membranaceous to firm, lanceolate to ovoid, 2–5 mm. thick; lowest bracts shorter than to thrice length of inflorescence. . . *e.*

*e.*Leaves filiform-involute, wiry; culms 1–few in tufts from hard scaly elongate stolons and rhizomes; staminate spike 1; pistillate spikes 3–15-flowered, 0.7–2 cm. long; scales blunt or bluntish, broadly ovate, conspicuous; perigynia subcoriaceous, very short-beaked; achene round-obovoid, nearly as broad as long. . . . . . 264. *C. oligosperma.*

*e.*Leaves flat, flexible; culms in dense tussocks or few; staminate spikes 1–4; pistillate spikes 1–7.5 cm. long, 20–100- or more-flowered; scales lanceolate or lance-ovate, acute or acuminate, more hidden; perigynia prominently beaked; achene ellipsoid to slenderly obovoid, much longer than broad.

In loose tussocks or small tufts from elongate horizontal stolons and rhizomes; perigynia subcoriaceous, 3–5 mm. thick, the beak minutely roughened; species of Atlantic States and Provinces. 265. *C. bullata.*

In dense clumps without prolonged horizontal stolons; perigynia membranous, 2–4 mm. thick, with smooth beak; transcontinental species. . . . . 266. *C. vesicaria.*

*a.*Stigmas 2 (only exceptionally 3); achene lenticular; beak entire or barely emarginate; pistillate scales blunt; pistillate spikes 4–8 mm. thick. . . . 267. *C. saxatilis.*

261. C. retrórsa Schwein. (turned backward). — In large but loose tussocks 0.3–1 m. high; culms coarse, obtusely angled and smooth or nearly so; *leaves and greatly prolonged bracts lax,* dark green, 4–10 *mm. broad,* soft, septate-nodose, mostly longer than culms, *the loose sheaths with the brown inner band truncate to prolonged at summit;* staminate spikes 1–4, sessile or short-peduncled; *pistillate spikes* 3–8, ascending to spreading, *closely crowded,* or the lower ones remote, *thick-cylindric,* 1.5–8 cm. long, 1.2–2 cm. thick, becoming yellowish-green to fulvous; *perigynia* very thin and papery, *much inflated,* prominently nerved, *soon reflexed or horizontally spreading,* conic-ovoid or -falcate, 5–10 *mm. long,* with sharply 2-toothed conical beak, much exceeding the acuminate scales. — Rich low grounds and alluvial woods, Anticosti I., Que., to B.C., s. to N.S., N.E., n. N.J., Pa., n. O., n. Ind., n. Ill., n. Ia., S.D., Colo., Utah and Wash. July–Oct. Fig. 751. — Hybridizes with nos. 249 (producing × **C. Hártii** Dew. (named for Samuel Hart Wright, 1825–1905), 256, 258, and 259 (producing × C. **Macoùnii** Dew.).

751. C. retrorsa.

262. C. rostràta Stokes (beaked). — *Culms thickened and spongy at base,* there 2–15 mm. thick, 1*-few from horizontal rhizomes and* *stolons, smooth and obtusely angled* (except at summit), coarse to slender, 1–10 mm. thick above upper sheath, 0.3–1.2 m. high; *leaves* when dry *prominently septate-nodulose, pale green to glaucous,* 2–12 mm. broad, firm, nearly equaling to exceeding culms, *the inner band firm and concave at summit;* bracts strongly ascending; staminate inflorescence peduncled, of 2–4 distinct spikes; *pistillate spikes* cylindric to slenderly ovoid, (1–)2–4, sessile, or the lower peduncled, 1–15 cm. long, mostly *dense,* becoming yellowish to tawny; *perigynia* ovoid, flask-shaped or slenderly conic, *ascending* (basal sometimes reflexed), *mostly inflated,* 3–10 mm. long. — Variable boreal species, represented with us by the following vars.:

*a.*Pistillate scales oblong to ovate, blunt or merely acute, not prolonged at tip, shorter than perigynia (3–6 mm. long).
Beak of perigynium prominently toothed, the teeth 0.3–0.7 mm. long; leaves canaliculate to flat, 2–8 mm. broad; pistillate spikes 2–4, cylindric, 1.5–7 cm. long; perigynia prominently inflated-ovoid, membranaceous, 3–5(–6) mm. long, mostly abruptly beaked. . . . . *C. rostrata* (typical).
Beak emarginate, the teeth 0.1–0.2 mm. long; leaves canaliculate to involute, 4 mm. broad; pistillate spike 1, 2.7 cm. long; perigynia barely inflated, oblong-conic, 5.5–6 mm. long, attenuate to beak. . . . . Var. *anticostensis.*

*a.*Pistillate scales narrowly ovate to linear-lanceolate, tapering to acuminate tip or awned, often nearly equaling or the lower often exceeding the perigynia (4–10 mm. long); leaves mostly flat.
Culms coarse, 0.2–1 cm. thick above upper sheath, 0.3–1.2 m. high; leaves 0.1–1.2 cm. broad; pistillate spikes cylindric, 2–15 cm. long, 0.9–2 cm. thick; perigynia crowded, 4–10 mm. long. . . . . . . Var. *utriculata.*
Culms very slender, barely 1 mm. in diameter, 3–5 dm. high; leaves 2–5 mm. broad; pistillate spikes ovoid to short-oblong, 1–2.5 cm. long; perigynia few, 4–6 mm. long. . . . . Var. *ambigens.*

C. rostràta (typical), *C. inflata* sensu Rendle & Britten and others, not Huds. — Shallow water, wet shores and swamps, s. Greenl. and Lab. to Alaska, s. to Nfld., ne. N.S., n. N.B., n. Vt., n. Mich., Sask., mts. to Colo. and s. Calif. July–Sept. (Eurasia) Fig. 752.

Var. **anticosténsis** Fern. (of Anticosti). — Quiet water, Anticosti I., Que. — Needs further study.

Var. **utriculàta** (Boott) Bailey (bottle-shaped), *C. utriculata* Boott; *C. inflata* var. *utriculata* (Boott) Druce — Shallow water and swamps, s. Greenl. and Lab. to B.C., s. to Nfld., N.S., N.E., L.I., Del., D.C., W.Va., upland of e. Tenn., O., Ind., Wisc., Minn., S.D., N.M. and Calif. June–Oct. (Eurasia) Fig. 753. Hybridizes with no. 265.

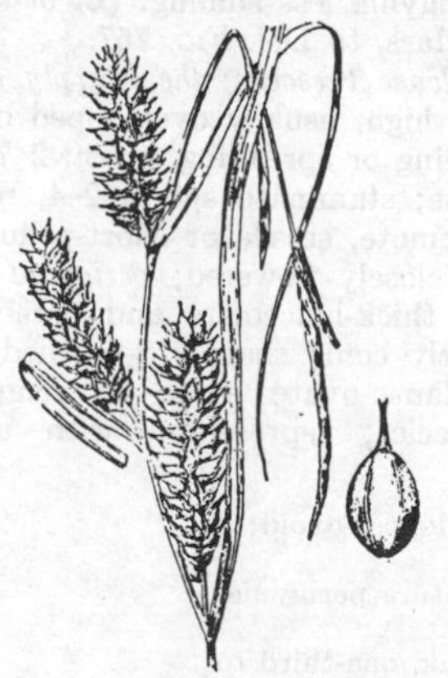
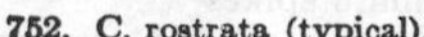
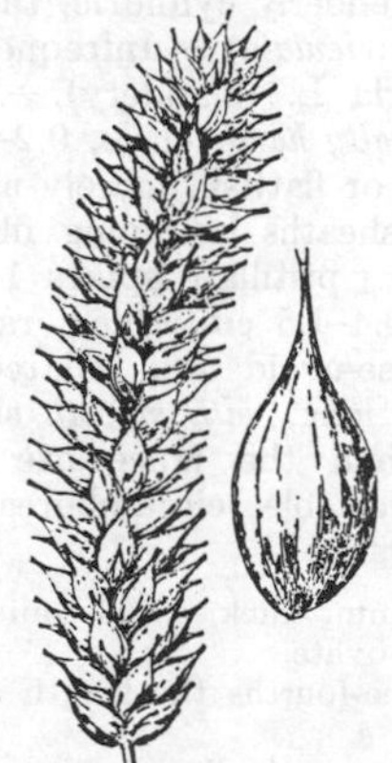

752. C. rostrata (typical). 753. C. rostrata, v. utriculata. 754. C. Tuckermani.

Var. **ámbigens** Fern. (doubtful). — Pools and swamps, Gaspé Pen., Que., n. N.B. and n. Me.; reported from Rupert R. July–Sept.

263. C. Tuckermáni Boott (in honor of Edward Tuckerman, 1817–1886). — In loose clumps, with slender erect to arching culms up to 1 m. high; leaves 3–5 mm. wide, green; *bracts* leaf-like, *greatly prolonged and lax;* staminate spikes 2 or 3, slender-peduncled or the upper sessile; *pistillate spikes* 2 or 3, slender-peduncled or the upper sessile, *thick-cylindric,* 2–6 cm. long, 1.2–1.8 cm. thick, *loosely flowered; perigynia* lustrous, *extremely membranaceous and bladder-like,* yellowish-green, becoming light brown, strongly nerved, 7–10 mm. long, 5–6.5 *mm. thick,* the slender conic beak sharply 2-toothed; scales acute to acuminate, much shorter than perigynia; *achene* ellipsoid-trigonous, deeply *invaginated on one side,* 3–4 mm. long, with long sigmoid persistent style. — Rich or calcareous meadows, swales and low woods, Que. to s. Ont. and Minn., s. to N.B., n. and w. N.E., n. N.J., Pa., O., Ind., Ill. and ne. Ia. June–Aug. Fig. 754. Hybridizes with no. 258.

264. C. oligospérma Michx. (few-seeded). — *Culms* 1–few *from hard scaly and elongate rhizomes and stolons, filiform, stiff,* 0.2–1 m. high; leaves and *bracts filiform-involute, wiry; staminate spike* 1, peduncled; *pistillate spikes* 1 or 2, rarely 3, sessile or the lowest short-peduncled, *globular or thick-ellipsoid,* 0.7–2 *cm. long,* 3–15-*flowered; perigynia* shining, turgid, *subcoriaceous,* ovoid, *gradually contracted to the very short beak; scales conspicuous, broadly ovate, blunt or bluntish; achene round-obovoid, nearly as broad as long.* — Peat-bogs, acid swamps and shallow water, Lab. to Mackenz., s. to Nfld., N.S., N.E., Pa., O., Ind., Wisc., Minn. and Sask. June–Aug. Fig. 755.

755. C. oligosperma.

756. C. bullata.

265. C. bullàta Schkuhr (inflated). — *Culms* slender, few to numerous *from creeping rhizomes and stolons,* sharply angled, 0.3–1 m. high; leaves firm, 2–5 mm. broad, flat or channeled; their sheaths often becoming fibrillose, concave at orifice; staminate spikes 1–3; *pistillate spikes* 1 *or* 2, *remote,* thick-cylindric to subglobose, erect, sessile or short-peduncled, 1–5 cm. long, 1–2 cm. thick; *perigynia* ascending to slightly spreading, turgid, *firm,* lustrous, becoming yellowish-green to brown, 6–10 mm. long, *the roughish or serrulate conic-cylindric beak* sharply 2-toothed at tip, much exceeding the lanceolate acute to acuminate scales. (Var. *Greenii* (Boeckl.) Fern.) — Acid swales, meadows and bogs, w. N.S. and sw. Me. to e. N.Y., s. to Ga. and Tenn. June–Oct. Fig. 756.

× **C. Òlneyi** Boott (for Stephen Thayer Olney, 1812–1878), *C. bullata* sensu Man. ed. 7, not Schkuhr, is a coarser plant, with leaves 4–6 mm. wide; pistillate spikes mostly 2, more slenderly cylindric, the perigynia less shining. (*C. bullata* × *C. rostrata,* var. *utriculata*) — Infrequent, Mass. to Del. Fig. 757.

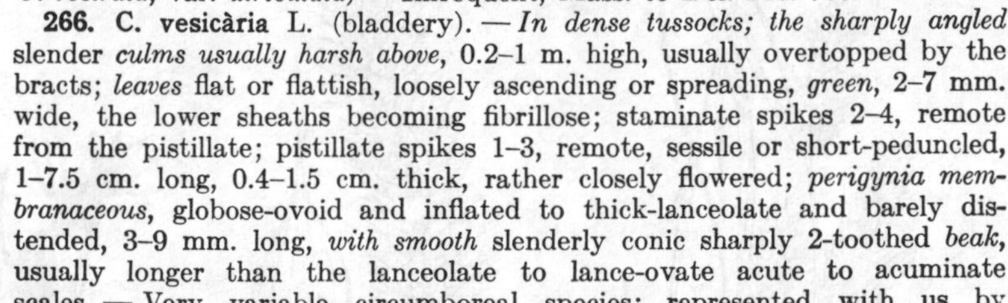
757. × C. Olneyi.

266. C. vesicària L. (bladdery). — *In dense tussocks; the sharply angled* slender *culms usually harsh above,* 0.2–1 m. high, usually overtopped by the bracts; *leaves* flat or flattish, loosely ascending or spreading, *green,* 2–7 mm. wide, the lower sheaths becoming fibrillose; staminate spikes 2–4, remote from the pistillate; pistillate spikes 1–3, remote, sessile or short-peduncled, 1–7.5 cm. long, 0.4–1.5 cm. thick, rather closely flowered; *perigynia membranaceous,* globose-ovoid and inflated to thick-lanceolate and barely distended, 3–9 mm. long, *with smooth* slenderly conic sharply 2-toothed *beak,* usually longer than the lanceolate to lance-ovate acute to acuminate scales. — Very variable circumboreal species; represented with us by the following freely intergrading vars.:

a. Perigynia strongly inflated, 2–3.5 mm. thick, ovoid-conic to globose-ovoid; scales lance-acuminate to lance-ovate. . . *b.*

b. Pistillate scales one-half to three-fourths the length of mature perigynia, pale, rarely castaneous. . . *c.*

c. Perigynia slenderly ovoid-conic, gradually tapering to beak, one-third to one-half as thick as long, in maturity 7–9 mm. long; pistillate spikes 2–7.5 cm. long, 1–1.5 cm. thick; leaves 4–7 mm. wide. . . . . . *C. vesicaria* (typical).

c. Perigynia with subglobose to globose-ovoid bodies, one-half to two-thirds as thick as long, rather abruptly beaked, 3–8 mm. long; leaves 2–5 mm. wide. . . *d.*

d. Pistillate spikes 1–1.5 cm. thick; perigynia 5–8 mm. long; scales long-acuminate.

Pistillate spikes 2 or 3, cylindric, mostly 2–7.5 cm. long. . . . . Var. *monile.*

Pistillate spikes 1 or 2, ellipsoid to cylindric-ovoid, 1–2 (–2.5) cm. long. . . . . Var. *distenta*.
d. Pistillate spikes 5–8 mm. thick; perigynia 3–5 mm. long. . . . . Var. *jejuna*.
b. Pistillate scales nearly equaling or exceeding the abruptly beaked plump perigynia, purple. . . . . Var. *laurentiana*.
a. Perigynia scarcely to but slightly inflated, lanceolate to slenderly ovoid, slenderly tapering, 4–6 mm. long, 1.5–2 mm. thick; scales ovate; pistillate spikes mostly 2, 1–4 cm. long, 4–8 mm. thick. . . . . Var. *Raeana*.

C. **vesicària** (typical). — Meadows, swales and shores, Nfld. to B.C., s. to N.B., N.E., Pa., Mich., Wisc., Minn., Colo. and Calif. June–Aug. (Eurasia) Fig. 758. — Less common with us than the next.

Var. **monìle** (Tuckerm.) Fern. (necklace). — Similar habitats, more common, Que. and Ont., s. to N.S., N.E., Del., Pa., O., Ind., Ill. and Mo. Fig. 759.

Var. **disténta** Fries (swollen). — Local, Nfld. to Wisc., s. to N.S., n. N.E. and n. N.Y. (Eu.)

Var. **jejùna** Fern. (meagre). — S. Lab. to Thunder Bay Distr., Ont., s. to Nfld., N.S., N.E., N.Y., Mich. and Wisc.

Var. **laurentiàna** Fern. (of region of Gulf of St. Lawrence). — Local, n. Nfld. and e. Saguenay Co., Que., s. to St. P. et Miq., M.I., and sw. N.B.

759. C. vesicaria, v. monile.

Var. **Raeàna** (Boott) Fern. (in honor of John Rae, 1813–1893), *C. Raeana* Boott — Nfld. to Sask., s., locally, to Me. and White Mts., N.H. Fig. 760.

758. C. vesicaria (typical).

267. C. saxátilis L. (found among stones). — Low, 2–4 dm. high, the 1–few culms and the tufts of leaves arising from purplish rhizomes; *leaves* flat, *becoming involute*, 2–5 mm. wide, nearly or quite equaling the culms; staminate spike 1 (rarely 2), clavate, short-peduncled; *pistillate spikes* 1–3, *purple-black, subglobose to thick-cylindric,* 0.5–2 cm. long, 5–8 *mm. thick; perigynia ovoid, usually nerveless,* 3–4 *mm. long, with short entire beak, slightly exceeding the blunt purple scales; stigmas* 2; *achene lenticular, only partly filling the inflated perigynium.* — Arct. reg., s. to Lab., etc. (Eurasia) — Represented with us by

760. C. vesicaria, v. Raeana.

Var. **rhomàlea** Fern. (stout). — Rhizomes and bases purple-brown to drab; *leafy tufts and bases of culms* 3–7 *mm. thick;* culms 2–8 dm. high, smooth; leaves 2–4 mm. wide; *staminate spikes* (1–) 2 *or* 3, *linear-cylindric; anthers* 2–4 *mm. long; pistillate spikes* stramineous to purple, in maturity 1–3 cm. long, 5–8 *mm. thick; perigynia* 3–5 mm. long, *tightly filled by the achene;* scales obtuse to subacute. — Sandy, gravelly or peaty shores and margins of pools, Lab., Nfld. and Que. to n.-centr. Me. July–Sept. Fig. 761.

761. C. saxatilis, v. rhomalea.

Var. **miliàris** (Michx.) Bailey (like millet). — More slender; *leafy tufts and bases of culms* 2–4 *mm. thick; leaves* 1–2.5 *mm. wide, involute in drying; staminate spike* 1, rarely 2; *anthers* 1–2.5 *mm. long; pistillate spikes* stramineous to purple, 0.5–3 cm. long, 4–6(–7) *mm. thick; perigynia* 2.5–3.5 *mm. long, tightly filled by achene.* (*C. miliaris* Michx.) — Peaty or gravelly damp soils, e. Lab. to Hudson Bay, s. to Nfld., N.B. and n.-centr. Me. Fig. 762.

762. C. saxatilis, v. miliaris.

× **C. mainénsis** Porter (of Maine), *C. saxatilis*, vars. *rhomalea* and *miliaris* × *C. vesicaria*, is a group of sterile plants frequent where the two species are together, including *C. Grahami* sensu Man. ed. 7, not Boott. Similar hybrids of *C. saxatilis*, var. *miliaris* and *C. rostrata* occur.

FAM. 24. ARÀCEAE (Arum Family)

Plants with acrid or pungent juice, simple or compound often veiny leaves, and flowers crowded on a spadix, which is usually subtended by a spathe. — Floral envelopes none, or of 4–6 sepals. Fruit usually a berry. Seeds with fleshy albumen or none, but filled by the large fleshy embryo. A large family, chiefly tropical. Herbage abounding in slender rhaphides. — The genuine *Araceae* usually have no floral envelopes and are almost all monoecious or dioecious; but our last two genera, with more highly developed flowers, are not to be separated from them.

*a.*Spathe well developed, fleshy or petaloid. . . *b.*
*b.*Spadix slender and elongate; flowers unisexual, without a calyx.
Flowers covering only the base of the spadix; leaves (in ours) compound. 1. *Arisaema.*
Flowers almost completely covering spadix; leaves simple. . . . . . 2. *Peltandra.*
*b.*Spadix short-cylindric to globose; flowers (or the lower ones) perfect.
Spadix short-cylindric, subtended by a flat open petaloid spathe; flowers without a perianth. . . . . 3. *Calla.*
Spadix globose or ellipsoid, enveloped in a fleshy ovoid spathe; sepals 4, hooded. . . . . 4. *Symplocarpus.*
*a.*Spathe obscure or like the foliage-leaves: flowers perfect, with perianths of 6 sepals.
Spadix naked, terminating the terete scape; leaf-blade oblong. . . . . 5. *Orontium.*
Spadix much overtopped by the sword-like spathe resembling the foliage-leaves. . . . . 6. *Acorus.*

1. ARISAÈMA Mart. Indian-turnip. Dragon-Arum

Spathe convolute below and mostly arched above. Flowers monoecious or by abortion dioecious. Staminate flowers above the pistillate, each of a cluster of almost sessile 2–4-locular anthers opening by pores or chinks at the top. Pistillate flowers each of a 1-locular ovary containing 5 or 6 erect orthotropous ovules; in fruit becoming a 1–few-seeded scarlet berry. — Low Am., Asiatic and Afr. perennial herbs; the tuberous rhizome, tuber or corm sending up a simple scape sheathed with the petioles of the simple or compound veiny leaves. (Name from the Greek *aris*, a kind of *arum*, and *haima*, *blood*, from the spotted leaves of some species.)

*a.*Leaves palmately divided into 3 (rarely 5) subequal leaflets; the ovate usually arching summit (hood) of the spathe flat; spadix club-shaped or cylindric, blunt, included; fruiting head ovoid or subglobose. . . *b.*
*b.*Inrolled tubular half of spathe obscurely or at most shallowly furrowed; summit of tube with a rounded-horizontal curve or pronounced flange.
Lateral leaflets strongly oblique at base, conspicuously rounded on lower side; tube of spathe 3.5–7 cm. long, the summit-flange 2–8 mm. broad, the broadly oblong-ovate abruptly acuminate hood 3–6 cm. broad; fruiting head 3–6 cm. long; seeds depressed-globose or reniform, deeply invaginated at base. . . . . 1. *A. atrorubens.*
Lateral leaflets with sides subequally narrowed; tube of spathe 2.5–4.5 cm. long, the summit-flange 0.5–2 mm. broad; the narrowly oblong or lance-ovate attenuate hood 2–3 cm. broad; fruiting head 1.5–2 cm. long; seeds obovoid, narrowed to a broad stipe. . . . . . . 2. *A. triphyllum.*
*b.*Inrolled tubular half of spathe sharply and deeply corrugated (when fresh), the ridges white; summit of tube passing by a gradual curve into the attenuate narrowly ovate hood (2–6 cm. broad); fruiting head 2.5–3.5 cm. long; seeds obovoid, narrowed to a thick stipe. . . . . . . 3. *A. Stewardsonii.*
*a.*Leaf pedately divided into 5–15 very unequal leaflets; the oblong summit of the spathe with inrolled margins; spadix slender, tapering, long-exserted; fruiting head slenderly conical. . . . . 4. *A. Dracontium.*

1. A. atrórubens (Ait.) Blume (dark red), Jack-in-the-pulpit, Indian-turnip, Oignon sauvage or Petit Prêcheur (Que.). — *Tuber* subglobose, up to 6 cm. thick, *its basal offsets sessile; leaves* 1–3, the 3 (rarely 5) ovate-acuminate leaflets *often pale or glaucous beneath; the lateral leaflets strongly oblique, the lower half* much broader than the acute-based upper side and *rounded at base; tube of spathe only slightly if at all corrugated*, 3.5–7 *cm. long, its summit curved horizontally and with a flange or flap* 2–8 *mm. wide; hood* full purple, or purple to bronze and with pale longitudinal stripes in forma **zebrìnum** (Sims) Fern. (striped), or green throughout in forma **víride** (Engler) Fern. (green), *arching* over the spadix, *broadly ovate-oblong, rather abruptly acuminate,* 3–6 *cm. broad; spadix* cylindric or club-shaped, *its rounded summit* (when dry) 2.5–8 *mm. thick; fruiting heads* ovoid to subglobose, 3–6 *cm. long; seeds* 1–3, *depressed-globose or reniform, deeply invaginated at base.* — Rich woods and thickets, N.B. and adj. Que. to se. Man., s. to N.E., S.C., Tenn., Mo. and e. Kans. Late April–June. — Probably hybridizes with nos. 2 and 3.

2. A. triphýllum (L.) Schott (three-leaved), Small Jack-in-the-Pulpit. — Usually (but not always) smaller than no. 1; *tuber bearing elongate scaly stolons; leaves green on both sides, lateral leaflets acute at base; tube of spathe not corrugated,* 2.5–4.5 *cm. long, its summit horizontally* *curved* and *bordered by a very narrow flange* 0.5–2 *mm. wide;* hood clear green, or solid dark purple in forma **pusillum** (Peck) Fern. (very small), without pale stripes above the base, *ascending or only slightly arching, narrowly oblong- to lance-ovate, attenuate,* 3–5 cm. long, 2–3 cm. broad, or 5.5–9 cm. long and 2–4 cm. broad in the larger southern forma **attenuàtum** (Small) Engler (attenuate) of Va. and W.Va. to Fla. and Ala.; *spadix* slender, its summit (when dry) 1.5–3 *mm. thick; fruiting heads* 1.5–2 *cm. long; seeds obovoid,* narrowed to a broad stipe, sometimes shrunken but not clearly invaginated at base. (*A. triphyllum,* var. *pusillum* Peck) — Wet woods, swamps and peat-bogs of Coastal Plain and Piedmont, Ga. to Ky., n. to se. N.Y., Ct. and se. Mass. Mid-May–early July.

3. A. Stewardsònii Britt. (for one of its discoverers, Stewardson Brown, 1867–1921), Jack-in-the-pulpit, Indian-turnip, Bog-onion. — Similar to nos. 1 and 2; tuber frequently setting off stolons or stalked plantlets; *leaflets green and lustrous beneath, the lateral with subequilateral acute bases; tube of spathe sharply and deeply corrugated* when fresh, green or purple, *with white ridges,* 3–6.5 cm. long, *the summit tapering gradually into the narrowly ovate* (2–6 cm. broad) *attenuate* green to purple-striped arching to ascending *hood;* summit of spadix (dry) 2–6 mm. thick; *fruiting head* 2.5–3.5 *cm. long;* seeds as in no. 2. (*A. triphyllum,* var. *Stewardsonii* (Britt.) Stevens) — Wet or swampy woods and thickets, N.S. and P.E.I. to Minn., s. to n. N.J. and Pa., and along the mts. to N.C. Mid-May–mid-July.

4. A. Dracóntium (L.) Schott (Greek name for a kind of arum), Green Dragon, Dragon-root. — *Leaf usually solitary, pedately divided* into 5–15 oblong-lanceolate pointed leaflets; *spadix often androgynous, tapering to a long and slender point* beyond the oblong and convolute-pointed greenish spathe. (*Muricauda Dracontium* (L.) Small) — Rich or alluvial woods, thickets and swales, Fla. to Tex., n., locally, to s. N.H., Vt. and sw. Que., s. Ont., Mich. and Wisc. May, June.

Pinéllia (named in 1830 for an Italian botanist, *Pinelli*) ternàta (Thunb.) Breitenbach (in threes), differing from *Arisaema* in being monoecious, the slender and long-exserted spadix (suggesting that of *A. Dracontium*) covered with staminate flowers, these separated from the pistillate base by a constriction in the tube of the slender spathe, the ternate leaves often bearing bulbets on the petiole, the rhizome elongate and brittle, is spreading in gardens, plantings and lawns of se. N.Y. and n. N.J., fl. in summer. (Adv. from e. Asia)

2. PELTÁNDRA Raf. Arrow-Arum

Spathe elongated, convolute throughout or with a dilated blade above. Flowers thickly covering the long and tapering spadix throughout (or only its apex naked). Anther-masses sessile, naked, covering the upper part of the spadix, each of 4–6 pairs of locules embedded in the margin of a thick and shield-shaped connective, opening by terminal pores. Ovaries at the base of the spadix, each surrounded by 4–5 distinct scale-like white staminodia, 1-locular, bearing 1–few amphitropous ovules at the base. Berries in an ovoid fleshy head enveloped by the base of the leathery spathe. — Stemless Atl. N.Am. herbs, with mostly sagittate- or hastate-based palmately 3-nerved and pinnately veined leaves, and simple scapes from a thick fibrous or subtuberous root. (Name from the Greek *pelte, a small shield,* and *aner, stamen,* from the shape of the latter.)

1. P. virgínica (L.) Schott & Endl. (Virginian), Tuckahoe. — Scape 2–3.5 dm. high, in anthesis about equaling the leaves; blades sagittate, hastate or usually with basal lobes (see below); *spathe tightly convolute throughout,* with wavy margin, *green or with the tightly appressed margin pale, not constricted at junction of limb and tube; staminate portion of the white or whitish spadix several times longer than the pistillate, the flowers covering it essentially to apex, the upper three-fourths of spathe rotting away; the fruiting* scape recurving; the *persistent base of the fruiting spathe* 3–6 cm. long, *with a beak-like summit;* ovules several; *berry* green or amber, *when dried* 0.6–1.2 *cm. long,* when long-submersed becoming distended; seeds 1(–3). — Protean as to leaf-outline; the following forms have been named:

a. Leaf-blade (above tip of petiole) broad (up to 1.8 dm. wide), with well-developed basal lobes 2–13 cm. long. . . . *b.*

b. Blade 1–2.2 dm. long, 0.8–1.8 dm. broad; basal lobes 3–7.5 cm. broad, shorter than breadth of blade, deltoid to rounded. . . . *c.*

c. Leaf almost equilaterally triangular, 1.8–3 dm. across between tips of the obtuse to subacute basal lobes; these three-fourths as long as blade; sinus open. . . . . Forma *latifolia.*

c. Leaf elliptic or elliptic-oblong, not equilateral, 1–2 dm. across between tips of lobes; these one-fourth to half as long as blade.

Leaf broadly rounded, about as wide as long, scarcely curving to the

broadly rounded basal lobes; sinus nearly or quite closed. . . . Forma *rotundata.*
Leaf oblong, with somewhat divergent basal lobes narrowed to tip; sinus open. . . . . *P. virginica* (typical).

b. Blade 1–2.5 dm. long, 3–12 cm. broad, acute or acuminate; basal lobes 1–3.5 cm. broad, lanceolate to linear.
Blade 6–12 cm. broad; basal lobes widely divaricate, once to twice as long as breadth of blade. . . . . Forma *hastifolia.*
Blade 3–5.5 cm. broad; basal lobes scarcely divaricate, short. . . . . Forma *brachyota.*

a. Leaf-blade narrow (2.5–5 cm. wide), generally acuminate, with rounded, oblique or cordate base; basal lobes (ears), when present, very short (1–2 cm. long) and unequal.
Leaves oblong to oblong-lanceolate, 4.5–5 cm. broad, with or without unequal basal ears. . . . . Forma *heterophylla.*
Leaves narrowly lanceolate, 2.5–3.5 cm. broad, without basal ears. . . Forma *angustifolia.*

P. virgínica (typical) or forma **latifòlia** (Raf.) Blake (broad-leaved), forma **rotundàta** Blake (rounded), forma **hastifòlia** Blake (halberd-leaved), forma **brachyòta** Blake (short-eared), forma **heterophýlla** (Raf.) Blake (variable-leaved) or forma **angustifòlia** (Raf.) Blake (narrow-leaved). — Swamps and borders of ponds and slow streams, Fla. to Tex., n. to s. Me., N.H., Vt., sw. Que., n. N.Y. and s. Ont. April (southw.)–July (northw.).

2. P. luteospàdix Fern. (with yellow spadix). — *Limb of spathe with open or widely spreading white border* 1–1.7 *cm. broad* each side of the middle, *constricted at base, soon deliquescing and by circumscission leaving a truncated fruiting spathe* 5.5–8 *cm. long; spadix orange-yellow, the apical* 1–2 *cm. without flowers; dried berries* 1–1.5 *cm. long.* — Bottomlands and wet woods, se. Va. to Fla. June.

3. CÁLLA L. Water-Arum

Spathe ovate (abruptly pointed, the upper surface white), persistent. Lower flowers perfect and 6-androus; the upper often of stamens only. Filaments slender; anthers 2-locular, opening lengthwise. Ovary 1-locular, with 5–9 erect anatropous ovules. Berries (red) distinct, few-seeded. — A low perennial boreal herb, growing in cold bogs or water, with a long creeping rhizome bearing cordate long-petioled leaves and solitary scapes. (A name used by Pliny, of unknown meaning.)

1. C. palústris L. (of marshes), Wild Calla. — Quagmires, pond-margins and wet bogs, Nfld. to Alaska, s. to N.S., N.E., N.J., Pa., Ind., Wisc. and Minn. April–Aug. (Eurasia) — An occasional plant, forma **polyspathàcea** Vict. & Rousseau (with many spathes) has 2 or 3 spathes.

4. SYMPLOCÁRPUS Salisb. Skunk-cabbage. Tabac du diable or Chou puant (Que.)

Stamens 4, opposite the sepals, with at length rather slender filaments; anthers extrorse, 2-locular, opening lengthwise. Style 4-angled and subulate; stigma small. Ovule solitary, suspended, anatropous. Fruit a globular or ovoid mass, composed of the enlarged and spongy spadix, inclosing the spherical seeds just beneath the surface, which is roughened with the persistent fleshy sepals and pyramidal styles. — Perennial herb, with a strong odor like that of the skunk, and also somewhat alliaceous, a stout erect rhizome, and a cluster of very large and broad entire veiny leaves, preceded in earliest spring by the nearly sessile spathes, which barely rise out of the ground. (Name from the Greek *symploce, connection,* and *carpos, fruit,* in allusion to the coalescence of the ovaries into a compound fruit.)

1. S. foètidus (L.) Nutt. (foul-smelling). — Leaves ovate, cordate, becoming 3–6 dm. long, short-petioled; spathe spotted and striped with purple and yellowish-green, ovoid, with inrolled margins. (*Spathyema foetida* (L.) Raf.) — Wet meadows or swampy woods and thickets, Que. to se. Man., s. to w. N.S., N.E., L.I., Va., upland to Ga. and Tenn., W.Va., O., Ind., centr. Ill. and Ia. Feb.–May. (A var. in e. Asia)

5. ORÓNTIUM L. Golden Club

Spathe incomplete and distant, merely a leaf-sheath investing the lower part of the slender scape and bearing a small and imperfect bract-like blade. Lower flowers with 6 concave sepals and 6 stamens; the upper ones with 4. Filaments flattened; anthers 2-locular, opening obliquely lengthwise. Ovary 1-locular, with an anatropous ovule. Fruit a blue-green utricle. — An aquatic perennial, with a deep rhizome, and long-petioled entire oblong and nerved leaves. (Name said to have belonged to some plant growing in the Syrian river *Orontes.*)

1. O. aquáticum L. (aquatic). — Sandy, muddy or peaty shores and shallow water, Fla. to La., n. to Mass., centr. N.Y., W.Va. and Ky. Apr.–June.

6. ÁCORUS L. Sweetflag. Flagroot. Calamus. Belle-Angélique or Redote (Que.)

Sepals 6, concave. Stamens 6; filaments linear; anthers reniform, 1-locular, transversely opening. Ovary 2–3-locular, with several pendulous orthotropous ovules in each locule. Fruit at length dry, gelatinous inside, 1–few-seeded. — Aromatic, especially the thick creeping rhizome (*calamus* of the shops); small genus of N. Hemisph. Leaves ensiform; the upper and more foliaceous prolongation of the scape may be considered as a kind of open spathe. (Latin name of an aromatic plant.)

1. A. Cálamus L. (old name for a reed). — Leaves 0.5–1 m. or more long, 0.7–2 cm. broad; the leaf-like spathe similar, prolonged 3–8 dm. beyond the spadix; spadix 4–9 cm. long, in fruit becoming 1.2–2 cm. thick. — Wet places and borders of quiet water, P.E.I. to Mont. and Oreg., s. to N.S., N.E., Fla. and Tex. May–Aug. (Early introd. and natzd., but usually sterile, in Eu.)

FAM. 25. LEMNÀCEAE (Duckweed Family)

Minute stemless plants, floating free on or in the water, destitute of distinct stem and foliage, being merely a frond or fronds, producing one or few monoecious flowers from the edge or upper surface, and often descending roots from underneath; ovules rising from the base of the locule. Fruit a 1–7-seeded utricle. Seed large. Embryo straight. — The simplest, and some of them the smallest, of flowering plants, propagating by the proliferous growth of a new individual from a cleft in the edge or base of the parent frond, also by autumnal fronds in the form of minute bulblets, which sink to the bottom of the water, but rise and vegetate in spring; the flowers (in summer) and fruit scarce, in some species hardly ever seen. — These plants may be regarded as very simplified *Araceae*.

Frond with 1 or more rootlets and 2 reproductive pouches.
- Rootlets 2–several; frond 4–15-nerved. . . . . 1. *Spirodela.*
- Rootlet 1; frond 1–5-nerved. . . . . 2. *Lemna.*

Frond without rootlets, with 1 reproductive pouch.
- Frond thick, ovoid or ellipsoid to globose. . . . . 3. *Wolffia.*
- Fronds thin, slenderly ligulate. . . . . 4. *Wolffiella.*

1. SPIRODÈLA Schleid.

Anther-locules bilocellate by a vertical partition and longitudinally dehiscent. Ovules 2. Rootlets several, with axile vascular tissue. Otherwise as *Lemna;* small semicosmop. genus. (From the Greek *speira, a cord,* and *delos, evident.*)

763. S. polyrhiza.

1. S. polyrhìza (L.) Schleid. (many-rooted), Water-flaxseed, Lentille d'eau (Que.). — Fronds round-obovate (3–8 mm. long), thick, purple and rather convex beneath, dark green above, palmately 5–11-nerved, with 6–18 roots. — Pools and pond- and stream-margins, Fla. to Tex. and Mex., n. to P.E.I., N.E., se. Que., s. Ont., Minn., Neb., Nev. and s. B.C. (Temp. and trop. reg.) Fig. 763.

2. S. oligorrhìza (Kurtz) Hegelm. (few-rooted). — Fronds oblong or narrowly obovate, 2–4 mm. long, 1.5–2.5 mm. broad, with only 4–6 nerves and 2–3 (rarely –6) roots. — Quiet waters, local, Mo. and doubtless elsewhere. (Warm and temp. reg.)

2. LÉMNA L. Duckweed. Duck's-meat

Flowers produced from a cleft in the margin of the frond, usually three together surrounded by a spathe; two of them staminate, consisting of a stamen only; the other pistillate, of a simple pistil; the whole therefore imitating a single diandrous flower. *Stam. Fl.* Filament slender; anther 2-locular, didymous; the locules dehiscent transversely. *Pist. Fl.* Ovary 1-locular; style and truncate or funnel-shaped stigma simple. Ovules and seeds 1–7. — Fronds 1–5-nerved, producing a single rootlet beneath (which is destitute of vascular tissue), proliferous from a cleft in the margin toward the base. Plants of temp. and trop. reg. (Name of a water-plant mentioned by Theophrastus.)

*a.*Fronds 6–10 mm. long, long-stalked, forming large tangled and connected mats. 1. *L. trisulca.*
*a.*Fronds 2–5 mm. long, sessile or nearly so, forming tiny rosettes or freely separating. . . *b.*
*b.*Fronds linear- to elliptic-oblong, twice to thrice as long as broad, sharply pointed at base, 0.5–1.5 mm. broad, nerveless or obscurely 1-nerved. . 2. *L. valdiviana.*
*b.*Fronds broadly oblong, obovate or suborbicular, one-half to three-fourths as broad as long, rounded or blunt at base, 1.2–3 mm. broad, obscurely to clearly 3-nerved.
Seed orthotropous; root-tip pointed. . . . . 3. *L. perpusilla.*
Seed amphitropous; root-tip rounded. . . . . 4. *L. minor.*

1. L. trisúlca L. (three-furrowed), Star-D. — *Fronds oblong to oblong-lanceolate* (6–10 *mm. long*), *attenuate at base into a slender stalk,* denticulate at the tip, very obscurely 3-nerved, often without rootlets, usually several series of offshoots remaining connected; *spathe sac-like;* seeds ovate, amphitropous, with small round operculum. — Ponds and springy places, M.I. to s. B.C., s. to N.S., N.E., Ala., Ark., Tex. and Mex. (Old World) Fig. 764.

764. L. trisulca.

2. L. valdiviàna Phil. (of Valdivia). — *Fronds linear-elliptic to narrowly oblong,* small (0.5–4 mm. long), 0.5–1.5 *mm. wide,* rather thick, usually somewhat falcate, *opaque, obscurely 1-nerved; spathe broad-reniform;* utricle long-ovate, pointed by the long style; *seed orthotropous,* oblong, *with a prominent acute operculum.* (*L. cyclostasa* C. H. Thompson, ascribed to Chev., who had no such species) — Quiet waters, Fla. to Tex. and Mex., n. to Mass., N.Y., O., s. Mich., Ill., Mo., Wyo. and Oreg. (Trop. Am.) Fig. 765.

765. L. valdiviana.

Var. **abbreviàta** Hegelm. (shortened). — *Fronds broadly elliptic or oval,* 1.5–2.3 *mm. broad, translucent, veinless.* — Fla. to Mex. and s. Calif., n. to se. Va. (Trop. Am.)

3. L. perpusílla Torr. (very tiny). — *Fronds obovate or roundish-obovate,* oblique (2–3 mm. long), *obscurely 3-nerved;* utricle ovate; style rather long; *seed orthotropous,* ovate or oval, obtuse, *with scarcely apiculate operculum.* (Incl. var. *trinervis* Aust. or *L. trinervis* (Aust.) Small, individuals with thinner and translucent fronds) — Fla. to Okla., n., locally, to Mass., N.Y., O., Ill. and Neb.

4. L. minor L. (smaller), Lentille d'eau or Merde de grenouille (Que.). — *Fronds round- to elliptic-obovate* (2–5 mm. in diameter), rather thick, *very obscurely 3-nerved; spathe sac-like;* utricle short-urceolate, tipped with a short style; seed oblong-obovate, *amphitropous, with prominent rounded operculum.* — Quiet waters, Fla. to s. Calif., n. to St. P. et Miq., s. Lab. Pen., Ont. and B.C. (Old World) Fig. 766.

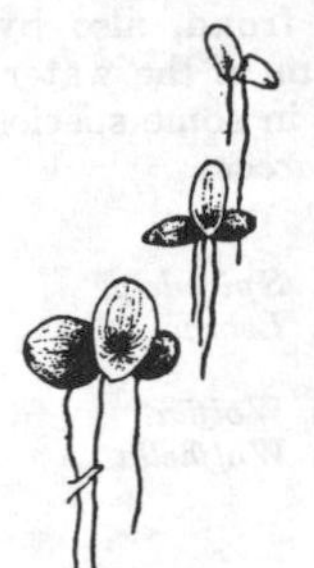
766. L. minor.

3. WÓLFFIA Horkel Water-meal

Flowers central, bursting through the upper surface of the globular (or in some species flat) and loosely cellular frond, only 2: one consisting of a single stamen with a 1-locular 2-valved anther; the other of a globular ovary, tipped with a very short style and a depressed stigma. Ovule orthotropous, rather oblique in the locule. Utricle spherical. Albumen thin. — Fronds rootless, proliferous from a cleft or funnel-shaped opening at the base, the offspring soon detached; no rhaphides. — The simplest and smallest of flowering plants, floating as little grains in or on water, in warm reg. (Named for *Johann Friedrich Wolff,* 1788–1806, who wrote on *Lemna* in 1801.) Bruniera Franch. (mispelled Bruneria by Small)

Not dotted; upper surface strongly convex. . . . . 1. *W. columbiana.*
Dotted.
Upper surface flattish. . . . . 2. *W. punctata.*
Upper surface low-conical. . . . . 3. *W. papulifera.*

1. W. columbiàna Karst. (of Colombia). — *Globose to ellipsoid,* 0.7–1.5 mm. long, very loosely cellular, light green all over, *not dotted;* stomata 1–6; the opening at the base circular and with a thin border. (*Bruniera columbiana* (Karst.) Nieuwl.) — Floating rather beneath the surface of quiet waters, Fla. to La., n. to Mass., sw. Que., N.Y., s. Ont., Mich., Wisc. and Minn. (Trop. Am.) Fig. 767.

767. W. columbiana.

2. W. punctàta Griseb. (dotted). — *Oblong*, smaller and more densely cellular, *flattish* and deep green with many stomata *above*, tumid and pale below, *brown-dotted all over;* anterior edge sharp; opening at base, circular. (*Bruniera punctata* (Griseb.) Nieuwl.) — Floating on the surface of quiet water, Fla. to Tex., n., very locally, to e. Md., nw. N.Y., s. Ont., Mich., Ill. and s. Minn. (W.I.)

3. W. papulífera C. H. Thompson (bearing a papilla). — *Lower surface hemispherical, the upper* flattish at the margin, *rising at the center to a single low papilla.* — Local, Va. to Ill., Ky., Mo. and Kans. (Mex.) — Perhaps not separable from no. 2.

4. WOLFFIÉLLA Hegelm.

Flowers and fruit unknown. Frond (in ours) linear-attenuate or flagellate, falcate or sigmoid, many times longer than wide, punctate, solitary or cohering at the base and radiating in a stellate manner. Pouch single, triangular, basal. — Small genus of imperfectly known Am. plants. (Name a diminutive of *Wolffia.*)

1. W. floridàna (J. D. Sm.) C. H. Thompson (of Florida). — Fronds hollow, gradually attenuate from base to flagelliform apex, 6–8 mm. long. — Quiet waters, Fla. to Tex., n., locally, to Mass. and se. Mo.

FAM. 26. XYRIDÀCEAE (Yellow-eyed Grass Family)

Rush-like herbs with narrow leaves sheathing the base of a naked scape, which is terminated by a head of perfect 3-androus flowers, with extrorse anthers, glumaceous calyx, and a regular colored corolla; the 3-valved mostly 1-locular capsule containing several or many orthotropous seeds with a minute embryo at the apex of fleshy albumen.

1. XỲRIS L. Yellow-eyed Grass

Flowers single in the axils of coriaceous scale-like bracts, which are densely imbricated in a head. Sepals 3; the 2 lateral navicular and persistent; the anterior one larger, enwrapping the corolla in the bud and deciduous with it. Petals 3, yellow (rarely white), with claws which more or less cohere, fugacious. Fertile stamens 3, inserted on the claws of the petals, alternating with 3 sterile filaments which are cleft and in our species plumose or bearded at the apex. Style 3-cleft. Capsule ellipsoid, free, 1-locular, with 3 parietal more or less projecting placentae, 3-valved, many-seeded. — Ours nearly all (except no. 1), perennials. Genus of S. Hemisph., Atl. N. Am. and e. and s. Asia. (*Xyris*, Greek name of some plant with 2-edged leaves, from *xyron, a razor.*)

a. Annual or biennial, in small fibrous-rooted flattened tufts; leaves membranaceous, whitish-green, 1.5–4 cm. long; scapes 1–3 dm. high, setiform, disarticulating at base; spike 5–6 mm. long, 3–4.2 mm. thick, its few bracts with free tips; upper third of keel of included lateral sepals very minutely denticulate; southeastern species. . . . . 1. *X. Bayardi.*

a. Perennials with crowns or rhizomes; leaves firm, rarely membranaceous, full green, blue-green or fulvous, mostly more than 4 cm. long; scapes 0.3–12 or more dm. high, mostly not disarticulating at base (disarticulating in the short-leaved and slender- and low-scaped no. 7, with blue-green leaves up to 13 cm. long, compact spike 4–9 mm. long and 3–6 mm. thick, its bracts tightly appressed, and crest of lateral sepals definitely toothed nearly to base; and in no. 8, a northern densely cespitose and slenderly rhizomatous plant, with metallically lustrous scapes, spikes 4–10 mm. long, with tightly appressed bracts, and tips of lateral sepals exserted); spikes 0.4–3.3 cm. long, 3–17 mm. thick. . . *b.*

b. Tips of lateral sepals completely covered by subtending bracts. . . . *c.*

c. Lateral sepals with ciliate or ciliolate keel.

Plant strongly bulbous, with stiff erect, gray- or blue-green (often spiraling) leaves 0.5–4 mm. broad; keel much narrower than wings of lateral sepals, usually with terminal tuft of hairs. . . . . . 2. *X. torta.*

Plant scarcely bulbous, with fleshy or leathery yellow- to fulvous-green brown-based equitant leaves 4–10 mm. broad; keel and wings of lateral sepals subequal, without terminal hairs. . . . . . . 3. *X. ambigua.*

c. Lateral sepals with toothed or erose, not ciliate, keel. . . *d.*

d. Keel with toothed crest only above the middle; longer leaves (except in obviously young or dwarfed individuals) 0.8–7 dm. long, not papillate when dry; scapes not easily disarticulating at base; mature spikes 0.5–3 cm. long. . *e.*

*e.*Base bulbous-thickened; lowest leaves represented by short firm bulb-scales; principal leaves rufescent, at least below middle, coriaceous, the longer 2–10 dm. long and 6–13 mm. broad; mature spikes 1.5–3 cm. long. . . . . 4. *X. platylepis.*

*e.*Base not bulbous; leaves not strongly reddened (except sometimes at very base); mature spikes 0.4–1.5 cm. long.

Leaves (when dry) translucent; the larger 20–40-nerved and 6–14 mm. broad, half as long as to longer than scapes; seeds about 25-ribbed. . . . . 5. *X. difformis.*

Leaves opaque or subopaque to transmitted light, except in submersed form; the larger 3–19 (–23)-nerved and 0.5–6 mm. broad, mostly a third to half as long as mature scapes; seeds about 13-ribbed. . . . . 6. *X. caroliniana.*

*d.*Keel with toothed crest nearly its entire length; longer leaves 3–13 cm. long, membranaceous, blue-green, minutely papillate toward base when dry; scapes readily disarticulating at base, bristleform; spike compact, 4–9 mm. long. . . . . 7. *X. Curtissii.*

*b.*Tips of lateral sepals projecting definitely beyond tips or sides of their subtending bracts. . . *f.*

*f.*Scapes bristleform, 0.3–4.5 dm. high, readily disarticulating at base; leaves 0.2–1.5 dm. long, 0.5–2.5 mm. wide; mature spikes 4–10 mm. long, 2–7 mm. thick; keels of lateral sepals entire, or finely toothed at tip. . . . . 8. *X. montana.*

*f.*Scapes coarser, 3–12 dm. high, persistent; larger leaves much longer and (except in the bulbous no. 12) broader; mature spikes 1–3 cm. long, 0.6–2 cm. thick; keels more strongly fringed. . . *g.*

*g.*Keel of lateral sepals erose to slightly lacerate.

Mature spike ovoid to obovoid or subglobose, 1–1.7 cm. long; northeastern. . . . . 9. *X. Congdoni.*

Mature spike ellipsoid-subcylindric, 1.5–3.3 cm. long; southeastern. 10. *X. Smalliana.*

*g.*Keel with a conspicuous fringe much longer than breadth of blade.

Non-bulbous; leaves ligulate, subcoriaceous, the larger 6–10 mm. broad; scapes 6–12 dm. high, scabrous; spikes ovoid. . . . . 11. *X. fimbriata.*

Bulbous; leaves linear, rigid, often spiraling, 1–4 mm. broad; scapes 3–7 dm. high, smooth; spike slenderly ellipsoid to slenderly cylindric. . . . . 12. *X. flexuosa.*

1. X. Bayárdi Fern. (for one of its discoverers, BAYARD LONG, 1885–). — Tiny *annual or biennial, with* 1–few *flabelliform tufts of whitish-green membranous* lance- or linear-ensiform minutely denticulate *white-margined* curving *leaves 1.5–4 cm. long* and 2–4 mm. broad; *scapes* mostly 1, *filiform, 1–3 dm. high,* glabrous, *disarticulating at base; spike few-flowered,* ellipsoid-ovoid, *acute,* 5–6 *mm. long* and 3–4.2 *mm. thick; bracts* elliptic-oval, *with free tips,* brown, with white-hyaline margin and distinct green back; *lateral sepals* obliquely linear-lanceolate, 3–3.5 mm. long, 0.3–0.5 mm. wide, *the upper third of the very slender dorsal keel minutely denticulate;* seeds 0.45–0.5 mm. long. — Sandy and peaty pond-shore, very local, se. Va.

2. X. tórta Sm. (twisted). — *Bulbous; the hard bulbs pale brown,* 4–10 *mm. thick,* 5–12 *mm. high, commonly clustered* and forming tufts of few–50; *leaves narrowly linear, stiff,* often spiraling, 0.5–4 mm. broad, *gray- or blue-green; scapes* 1–∞, 1.5–8 dm. high, *slender,* stiff, often spiraling; *spike* ovoid, 5–18 mm. long, 4–10 *mm. in diameter, rounded at summit;* bracts brown throughout or with green center not sharply defined, completely covering calyx; *lateral sepals with minutely ciliate keel much narrower than wings, with terminal tuft of hairs.* (*X. flexuosa* of ed. 7, not Muhl.) — Damp to dryish siliceous or argillaceous peaty depressions and shores, Ga. to Ark. and Tex., n. to s.-centr. N.H., Mass., N.Y., and upland of W.Va. and O. to N.C. and Tenn.; also n. O. to Minn. and Ia. July–Sept. FIG. 768.

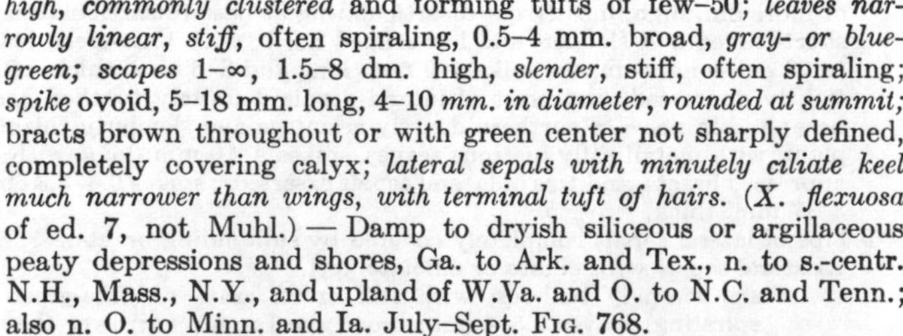

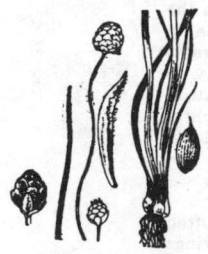

768. X. torta.

Var. **macrópoda** Fern. (large-based). — *Bulbs castaneous,* 1–2 *cm. thick,* 1.5–2 *cm. high, solitary or few;* scapes up to 9 dm. high; *spike* slenderly ellipsoid-ovoid, *subacute; lateral sepals scarcely or barely barbellate at tip.* — Peaty and boggy depressions, Coastal Plain, se. Va.

3. X. ambígua Beyrich (doubtful). — *Cespitose, with* numerous *flattened leafy shoots* from a short central rhizome, the base often bearing persistent bristly shreds of old leaves; *leaves* ligulate to broadly linear, *subcoriaceous or coriaceous, yellow- to fulvous-green, brown-based* and in age equitant, the larger 4–10 *mm. broad; scapes stout,* winged above, 0.5–1 m. high; spike ovoid or ellipsoid, 1–2.5(–3) cm. long, 6–15 mm. in diameter, with tightly appressed pale brown green-centered broadly rounded bracts; *keel of lateral sepals ciliolate, about as broad as wings, without apical tuft;* sepals large. — Wet peaty habitats, Fla. to e. Tex., n. on Coastal Plain to se. Va. July–Sept. (W.I.)

4. X. platýlepis Chapm. (broad-scaled). — Coarse; *base bulbous-thickened, the lowest leaves represented by short firm bulb-scales; elongate leaves rufescent below the middle, coriaceous, the longer ones* 2–10 *dm. long and* 6–13 *mm. broad;* scapes stout, 6–12 dm. high; *spike slenderly ellipsoid,* 1.5–3 *cm. long,* 0.8–1.7 cm. in diameter; bracts very firm, brown, broadly rounded; lateral sepals with keel lacerate-toothed above the middle. — Wet pinelands and swamps, Fla. to La., n. on Coastal Plain to e. Va. July–Sept.

5. X. diffórmis Chapm. (of two forms). — Resembling no. 6, coarser; *leaves tapering from middle to base and apex, translucent when dry, the larger* 20–40*-nerved and* 6–14 *mm. broad, half as long as to longer than scapes;* scapes rather slender, 3.5–8 dm. high, strongly flattened at summit; spike subglobose-ovoid, 1–2 cm. long; seeds about 25-ribbed. — Damp peats and sands, Fla. to Okla. and e. Tex., n. on Coastal Plain to Del. July–Oct.

6. X. caroliniàna Walt. (Carolinian). — *Soft-based,* tufted or cespitose, *the bases flattened; leaves* linear or linear-lanceolate, *opaque or subopaque, the longer* 0.5–6 *mm. wide and* 3–19 (–23)*-nerved, mostly one-third to one-half as long as mature scapes,* or leaves flaccid and much elongate (5–8 mm. wide) and often translucent in forma **fláccida** Fern. (flaccid) of estuaries, N.J. and Pa. to Va.; scapes 1–several, 0.5–9 dm. high, slender; spike ovoid, in maturity 0.5–1.4 cm. long, 0.4–1.1 cm. in diameter; *bracts* yellowish or pale brown, *with conspicuous broad green center,* completely covering calyx; *lateral sepals with erose or jagged keel;* seeds about 13-ribbed. — Wet peaty or sandy soil, Fla. to La., n. to N.S., se. and centr. Me., n.-centr. N.H., se. Vt., e. N.Y., thence w. to Parry Sd., Ont., Wisc. and n. Ind. July–Sept. Fig. 769. — Forma **phyllólepis** Fern. (leaf-scaled) is an aberrant plant with bracts and floral parts changed to leafy tufts.

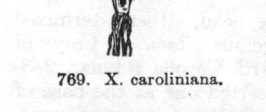

7. X. Curtíssii Malme (for its discoverer, Allen Hiram Curtiss, 1845–1907). — Dwarf, cespitose; *leaves soft and thin, blue-green, minutely papillate at base when dry, the longer ones* 3–13 *cm. long,* 2–4 mm. wide; *scapes setiform, readily disarticulating at base,* 1.5–3.5 dm. high; *spike compact,* ellipsoid-ovoid, 4–9 *mm. long,* 3–6 mm. in diameter; bracts few, drab and pale brown; *keel of lateral sepals with the crest toothed nearly its entire length.* (*X. neglecta* Small, not A. Nilss.) — Wet pine-barrens and bogs, Fla. to Miss., n. on Coastal Plain to se. Va. July–Sept.

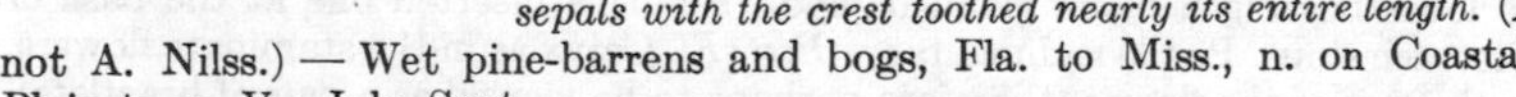

769. X. caroliniana.

8. X. montàna Ries (of mountains). — Tufted or *densely matted,* with slightly forking rhizomes, the bases soft and flattened; *leaves* 0.2–1.5 dm. long, 0.5–2.5 *mm. broad,* soft; *scapes filiform,* 0.3–4.5 dm. high, wiry, straight or spiraling, when fresh often with metallic lustre, *readily disarticulating at base; spike broadly lanceolate to slenderly ovoid,* when mature 4–10 *mm. long and* 2–7 *mm. in diameter;* bracts few, appressed, brown throughout or with inconspicuous small green center, often erose-ciliolate, in forma **bracteòsa** Haberer (with conspicuous bracts) the lowest bracts elongate to 5–15 mm.; *lateral sepals with slightly exserted* brown or purple *tips, the keel entire or but slightly toothed at summit.* — Wet peat and sand, Nfld. to Ont., s. to N.S. and, locally, to N.E., n. N.J., ne. Pa. and n. Mich. July–early Sept. Fig. 770.

770. X. montana.

9. X. Cóngdoni Small (for its discoverer, Joseph Whipple Congdon, 1834–1910). — *Coarse,* tufted; bases soft, much compressed; *leaves linear-attenuate,* 1.5–6 dm. long, 3–10 *mm. broad,* brown or reddish at base; scapes 0.3–9 dm. high; *spikes ovoid to obovoid or subglobose,* 1–1.7 *cm. long,* 0.9–1.6 *cm. thick; bracts loosely imbricated, brown with a broad green center; lateral sepals conspicuously exserted, the keel jagged or erose-lacerate.* (*X. Smalliana* of ed. 7, not Nash; *X. Smalliana,* var. *Olneyi* Gleason) — Quagmires and peaty shores, sw. Me. to N.J. Late July–Sept. Fig. 771.

771. X. Congdoni.

10. X. Smallìana Nash (for its discoverer, John Kunkel Small, 1869–1938). — Similar to no. 9; leaves 3–7 mm. broad; *spike ellips-*

oid-subcylindric, 1.5–3.3 *cm. long*, 0.7–1.2 *cm. thick*. (*X. elata* in part of ed. 7, not Chapm.) — Bogs and pond-margins, Fla. to La., n. on Coastal Plain to Cape May, N.J. July–Sept.

11. X. fimbriàta Ell. (fringed). — *Coarse, soft-based; leaves strap-shaped, subcoriaceous*, the larger 6–10 *mm. wide*, fulvous; *scapes* 6–12 *dm. high, scabrous; spike ovoid*, 1.1–2.3 cm. long, 0.8–1.3 cm. thick; bracts loose; *lateral sepals nearly twice as long as the bracts*, conspicuously *long-fringed* above the middle. — Swamps and pond-margins, Fla. to Miss., n. on Coastal Plain, locally, to N.J. July–Sept. Fig. 772.

772. X. fimbriata.

12. X. flexuòsa Muhl. (flexuous). — *The coarse lustrous castaneous bulbs clustered*, often closely spiraling, their broad-based scales firm; *leaves rigid, often spiraling*, 1–4 *mm. broad; scapes* slender, *smooth*, wiry, 3–7 dm. high; *spike slenderly ellipsoid to slenderly cylindric*, 1–2.5 cm. long, *acute; lateral sepals* conspicuously *exserted, long-fringed above the middle*. (*X. arenicola* Small) — Sandy and peaty pine-barrens or bogs, Fla. to Ark. and e. Tex., n. on Coastal Plain to N.J. July–Sept. (W.I.) Fig. 773.

773. X. flexuosa.

FAM. 27. ERIOCAULÀCEAE (Pipewort Family)

Aquatic or marsh herbs, stemless or short-stemmed, with a tuft of fibrous roots, a cluster of narrow and often loosely cellular grass-like leaves, and naked scapes sheathed at the base, bearing dense heads of monoecious or rarely dioecious small 2–3-merous flowers, each in the axil of a scarious bract; the perianth double or rarely simple, chaffy; anthers introrse; the fruit a 2–3-*locular* 2–3-*seeded capsule;* seeds pendulous, orthotropous; embryo at the apex of mealy albumen. — Chiefly trop. plants, a few in northern temp. reg.

Perianth of 2 or 3 sepals and 2 or 3 petals, these free in the pistillate flowers; stamens twice as many as corolla-lobes; anthers 2-locular. . . . . . . 1. *Eriocaulon.*
Perianth of 3 sepals and 0 petals; stamens 3; anthers 1-locular. . . . . . 2. *Lachnocaulon.*

1. ERIOCAÙLON L. Pipewort

Flowers monoecious and androgynous, *i.e.* both kinds in the same head, either intermixed or the central ones staminate and the exterior pistillate, rarely dioecious. *Stam. Fl.* Calyx of 2 or 3 keeled or navicular sepals, usually spatulate or dilated upward. Corolla tubular, 2–3-lobed, each of the lobes bearing a black gland or spot. Stamens inserted one at the base of each lobe and one in each sinus. Pistils rudimentary. *Pist. Fl.* Calyx as in the staminate flowers, often remote from the rest of the flower (therefore perhaps to be viewed as a pair of bractlets). Corolla of 2 or 3 separate narrow petals. Stamens none. Ovary often stalked, 2–3-lobed; style 1; stigmas 2 or 3, slender. Capsule membranaceous, loculicidal.— Leaves mostly smooth, loosely cellular and pellucid, flat or concave above. Flowers, also the tips of the bracts, etc., usually white-bearded or woolly. Our species are all stemless, wholly glabrous excepting at the base and the flowers, with a depressed head and dimerous flowers. Plants of temp. and trop. reg. of e. N. Am., S. Am., s. and e. Asia, Afr. and Austral., only no. 3 in w. Eu. (Name compounded of the Greek *erion, wool*, and *caulos, a stalk*, from the wool at the base of the scape in the original species.)

a. Leaves obtuse; bracts of involucre acute, pale. . . . . 1. *E. decangulare.*
a. Leaves attenuate to slender tips; bracts obtuse, mostly dark. . . *b.*
b. Scapes 10–12-ribbed; heads 0.7–1.5 cm. in diameter; chaff obtuse. . . . 2. *E. compressum.*
b. Scapes 4–7-striate; heads 2.5–10 mm. in diameter; chaff acutish.
Mature heads depressed-globose, the abundant and crowded marginal flowers tending to push back and hide the involucre; chaff and flowers fringed by abundant elongate white clavate trichomes; seeds subglobose, rarely globose-ellipsoid. . . . . 3. *E. septangulare.*
Mature heads depressed-hemispherical, loosely fewer-flowered; involucre not reflexed nor hidden; chaff and flowers glabrous or nearly so; seeds ellipsoid, rarely subglobose. . . . . 4. *E. Parkeri.*

1. E. decangulàre L. (ten-angled). — *Leaves obtuse*, varying from lanceolate to linear-subulate, rather rigid, 6–40 cm. long; scapes 10–12-ribbed (3–9 dm. high); head hemispherical, becoming globose (6–14 mm. in diameter); scales of the involucre acutish, straw-color or light brown; *chaff* (bracts among the flowers) *pointed.* — Swamps, pools and wet pine-barrens, Fla. to Tex., n. on Coastal Plain to e. Pa. and N.J. July–Sept.

2. E. compréssum Lam. (flattened). — *Leaves spreading* (5–12 cm. long), *linear-subulate,* rigid, or when submersed thin and pellucid, *tapering* gradually *to a sharp point,* mostly shorter than the sheath of the 10–12-*ribbed scape;* scales of the involucre very obtuse, turning lead-color; *chaff obtuse.* — Swamps, pools and wet pine-barrens, Fla. to e. Tex., n. on Coastal Plain to se. N.C.; e. Md. to s. N.J. May–July.

3. E. septangulàre With. (seven-angled), WHITE-BUTTONS, DUCKGRASS. — *Leaves* subulate, *fragile, pellucid,* 0.1–2.4 dm. long, with 3–7 rows of large reticulations; *scapes* stiff or soft and fragile, 0.2–6 (or in deep water -20) dm. long, 4–7-*striate; heads* at first campanulate to hemispherical, *by development of the crowded flowers* and fruits *the margins becoming recurved and the mature head depressed-globose,* 3–10 mm. in diameter; *involucre blackish to drab or greenish, its narrowly obovate bracts* at first appressed-ascending, then spreading, *finally recurved; the acutish chaff and* lead-colored *flowers bearded, at least at top, with hard white club-shaped trichomes; seeds subglobose to short-ellipsoid,* stramineous, with darker tips, delicately reticulate, 0.5–0.7 mm. long. (*E. articulatum* (Huds.) Morong) — Shallow pools or streams and on sandy, gravelly or peaty shores, Nfld. to w. Ont., s. to N.S., N.E., L.I., N.J. and Del., centr. Va., n. O., n. Ind., Wisc. and Minn. July–Sept. (W. Ireland and w. Scotland) FIG. 774.

4. E. Párkeri Robins. (for its discoverer, CHARLES F. PARKER, 1820–1883). — Similar to small states of no. 3, plant green to purple; leaves 1–6 cm. long; scapes 0.1–2 dm. high; *mature heads rather loosely flowered, drab, depressed-hemispherical,* 2.5–7 mm. broad; *involucre pale; the bracts appressed, ascending to merely spreading, not hidden in fruit; chaff and flowers glabrous or merely ciliolate with minute trichomes; seeds* mostly *ellipsoid,* rarely subglobose. — Tidal mud and estuaries, Va. to Me.; estuary of St. Lawrence, Que. July–Oct. FIG. 775.

775. E. Parkeri.

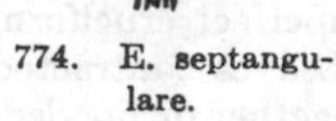

774. E. septangulare.

2. LACHNOCAÙLON Kunth HAIRY PIPEWORT. BOG-BUTTONS

Flowers monoecious, etc., as in *Eriocaulon.* Calyx of 3 sepals. Corolla none. *Stam. Fl.* Stamens 3; filaments below coalescent into a clavate tube around the rudiment of a pistil, above separate and elongated; anthers 1-locular. *Pist. Fl.* Ovary 3-locular, surrounded by 3 tufts of hairs (in place of a corolla). Stigmas 3, two-cleft. — Scape slender, bearing a single head, 2–3-angled, hairy. Small genus of se. U.S. (Name from the Greek *lachnos, wool,* and *caulos, stalk,* from the hairy scapes.)

1. L. ánceps (Walt.) Morong (two-edged). — Leaves linear-subulate, tufted, villous. — Low pine-barrens and bogs, Fla. to e. Tex., n. to se. Va. June–Aug.

FAM. 28. BROMELIÀCEAE (PINEAPPLE FAMILY)

Herbs (or scarcely woody plants, nearly all tropical), the greater part epiphytes, with persistent dry or fleshy and channelled crowded leaves sheathing at the base, usually covered with scurf. — Flowers hermaphrodite, mostly regular, with 3 sepals and 3 petals. Stamens 3 or 6. Ovary inferior or superior, 3-locular, the style slender and elongate; stigmas 3; the anatropous ovules on axile placentae. Fruit a capsule crowned by the persistent calyx, or a berry; seeds (in ours) with a crown of hairs. Embryo in a depression in the mealy endosperm, near the hilum. — Epiphytes, mostly tropical.

1. TILLÁNDSIA L. LONG MOSS

Perianth plainly double, 6-parted; the 3 outer divisions (sepals) membranaceous; the 3 inner (petals) colored; all connivent below into a tube, spreading above, lanceolate. Stamens 6, hypogynous, or the alternate ones adhering to the base of the petals; anthers introrse. Ovary free; style filiform; stigmas 3. Capsule cartilaginous, 3-locular. Seeds several or many in each locule, clavate, pointed, raised on a long hairy-tufted stalk, like a coma. Scurfy-leaved epiphytes of trop. and warm-temp. Am. (Named for *Elias Tillands,* 1640–1693, professor at Abo, who, as a student crossing directly from Stockholm, was so seasick that he returned to Stockholm by walking more than 1000 miles around the head of the Gulf of Bothnia and

hence assumed his surname (*by land*); the genus erroneously supposed by Linnaeus to dislike water.)

1. T. usneoìdes L. (like *Usnea*), Spanish or Black Moss. — Stems filiform, branching, pendulous; leaves filiform; peduncle short, 1-flowered; flower yellow. (*Dendropogon usneoides* (L.) Raf.) — Forming long hanging tufts, on branches of trees, chiefly in swamps, Fla. to Tex. and Mex., n. on Coastal Plain to e. Va. Apr.–Nov. (Trop. Am.)

FAM. 29. COMMELINÀCEAE (Spiderwort Family)

Herbs with fibrous or sometimes thickened roots, jointed and often branching leafy stems, and chiefly perfect and 6-androus often irregular flowers, with the perianth free from the 2–3-locular ovary, and having a distinct calyx and corolla, viz., 3 persistent commonly herbaceous sepals and 3 petals, the latter ephemeral or decaying or deciduous. Stamens hypogynous, often some of them sterile; anthers with 2 separated locules. Style 1; stigma undivided. Capsule 2–3-locular, 2–3-valved, loculicidal, 3–several-seeded. Seeds orthotropous. Leaves entire, parallel-veined, sheathing at base, the uppermost dissimilar and forming a kind of spathe. — Chiefly tropical.

Inflorescences borne in folded spathiform bracts; fertile stamens 3; two petals usually larger than the 3rd. 1. *Commelina.*
Inflorescence not borne in folded spathes; petals nearly or quite equal.
  Inflorescence racemose, paniculate, fascicled or with solitary flowers; fertile stamens 3 or 2. 2. *Aneilema.*
  Inflorescence umbellate; fertile stamens 6 or 5. . . . . 3. *Tradescantia.*

1. COMMELÌNA L. Dayflower

Flowers irregular. Sepals somewhat colored, unequal; the 2 lateral partly united. Two lateral petals rounded or reniform, on long claws, the odd one smaller. Stamens unequal, 3 of them fertile, one of which is bent inward; 3 of them sterile and smaller, with imperfect cruciform anthers; filaments naked. — Often procumbent and rooting at the nodes. Leaves contracted at base into sheathing petioles; the floral one cordate and clasping, folded together or hooded, forming a spathe inclosing the flowers, which expand for a single morning and are recurved on their pedicel in bud and after wilting. Petals mostly blue. Flowering all summer. Ours all with perennial roots, or propagating by striking root from the joints. Genus of trop. and warm-temp. reg. (Dedicated to the early Dutch botanists, *Commelin*, on account of the 2 showy petals and 1 less conspicuous petal, Linnaeus referring to the three botanists of that name, two of whom, *Jan*, 1629–1692, and *Kaspar*, 1667–1731, were conspicuous botanists, while the third "died before accomplishing anything in Botany".)

Annuals, decumbent at base or creeping, with slender root-fibers; margins of spathes free to base; seeds pitted or reticulate.
  Leaf-sheaths rarely if ever long-ciliate at summit; spathes on peduncles 1–5 cm. long; the 2 larger petals with ovate blades 0.8–1.5 cm. long; seeds 3.5–4 mm. long, rugose. 1. *C. communis.*
  Leaf-sheaths long-ciliate at summit; spathes on filiform peduncles 0.5–2 cm. long; the 2 larger petals with reniform blades 3–5 mm. long; anthers 5; seeds 2–2.5 mm. long, reticulate. . . . . 2. *C. diffusa.*
Perennials with thick roots; margins of spathes united below; seeds smooth.
  Creeping by elongate rhizomes; leaf-sheaths without flanges at summit, ciliate with long appressed reddish bristles; petals all blue, subequal. . . . . . 3. *C. virginica.*
  Roots a fascicle of fleshy fibers; leaf-sheaths with spreading herbaceous flanges ciliate with pale hairs; 2 blue petals much larger than the 3rd, white one. 4. *C. erecta.*

1. C. commùnis L. (growing in colonies). — Annual with slender roots and erect or decumbent and rooting bases; the stems and branches ascending or depressed, smoothish; *leaves fleshy,* with lanceolate or lance-ovate blades 4–12 cm. long, *the sheaths glabrous on margin and summit; spathes* 1.5–3 cm. long, *on peduncles* 1–5 *cm. long,* round-cordate when opened out, *the margins free;* posterior flower of the inflorescence exserted from spathe on long stalk; *larger petals* clawed, ovate, *with pale violet-blue blades,* 1–1.5 cm. long, 3rd petal paler and smaller; lobed sterile anthers yellow; *seeds rugose,* 3.5–4 *mm. long.* — Dooryards, ditches, roadsides and groves, Mass. to Wisc., s. to N.C., Ala., Ark. and e. Kans. June–Oct. (Natzd. from Asia)

Var. lùdens (Miquel) C. B. Clarke (deceiving). — Leaf-sheath more often ciliolate; posterior flower often reduced to a rudiment, less exserted; larger petals intense violet-blue, 0.8–1 cm. long; *lobed sterile anthers with madder centers.* — Similar habitats, N.J. and Pa. to Va. and Ky. (Natzd. from Asia)

2. C. diffùsa Burm. f. (diffuse). — Stems filiform, creeping, rooting at nodes, forming mats; *leaf-sheaths with long-ciliate summits;* blades membranaceous, ovate-lanceolate to oblong, acuminate, 2–8 cm. long; spathes 1–2.5 cm. long, acuminate; racemes 1 or 2, the lower of 1–3 flowers and rarely bearing fruit; the upper maturing capsules, often long-exserted; *petals all sky-blue or paler, the 2 upper with reniform blades* 3–5 *mm. long,* the lower smaller and saccate; anthers 5; *seeds* black, 2–2.5 *mm. long, reticulate.* (*C. nudiflora* sensu many auth., not L.) — Wooded bottomlands, marshes and waste places, trop. Am., n. to e. Va., s. O., s. Ind., s. Ill., Mo. and e. Kans., and as a casual weed to e. Mass. July–Oct. (Pantrop. and warm-temp. reg.)

3. C. virgínica L. (of Virginia). — Coarse *perennial with strong creeping and forking rhizome;* stems smooth, simple or forking, 0.4–1.2 m. high, erect (rarely reclining); *leaf-sheaths* long-cylindric, *the erect summits fringed with long reddish bristles; blades* broadly lanceolate, *scabrous above,* mostly 1–2 dm. long and 2.5–5 cm. broad; *spathes crowded* in terminal clusters, their margins united at base; *petals* blue, *subequal, the* 2 *larger* clawed, *with round-ovate to reniform blades* 1 *cm. or more long;* seeds reddish-fuscous, about 6 mm. long, smooth or puberulent. (*C. hirtella* Vahl) — Rich low woods, thickets and clearings, Fla. to Tex., n. to s. N.J., s. Pa., Md., W.Va., Ky., s. Ill., Mo. and Kans. July–Oct.

4. C. erécta L. (erect). — *Perennial with fascicled fleshy roots;* stem slender, erect or with arching bases, then ascending, simple or forking; *leaf-sheaths* hairy or smooth, *with horizontal or spreading foliaceous pale-ciliate flange;* blades lanceolate to linear; spathes with margins united above the base, often scattered; upper petals blue (rarely white), showy; lower one white or whitish; seeds brown, smooth, about 3 mm. long. — Our most variable species, the vars. freely confluent:

Stems (0.45–) 0.6–1.2 m. high; larger leaves of primary axis lanceolate to lance-ovate, (0.9–)1–1.5 dm. long, (1.5–)2–4 cm. broad; mature spathes (2.2–)2.5–3.6 cm. long. . . . . *C. erecta* (typical).

Stems 1–4.5(–7) dm. high; larger leaves linear to linear-lanceolate, 4–20 mm. broad; mature spathes 1–3 cm. long.
Longer leaves 4–10 cm. long; mature spathes 1–2(–2.5) cm. long. . . . . Var. *angustifolia.*
Longer leaves 7–15 cm. long; mature spathes mostly 2.5–3 cm. long. . . . . Var. *Deamiana.*

C. erécta (typical).—Spathe glabrescent to finely hirtellous, or long-villous with white hairs at base in forma **intercúrsa** Fern. (intermediate). (*C. virginica* of ed. 7, not L.) — Loamy or sandy soil or rocky slopes, in woods or openings, n. Fla. to Tex., n. to se. N.Y., Pa., W.Va., s. Ill., Mo. and Kans. June–Oct.

Var. **angustifòlia** (Michx.) Fern. (narrow-leaved). — Spathes finely hirtellous, or villous at base with long white hairs in forma **críspa** (Wooton) Fern. (crisp) = *C. crispa* Wooton — Dry sands, Fla. to Tex. and n. Mex., n. to Del., W.Va., s. Ill., Mo. and Neb. — Forma **albìna** Fern. (whitish) has all petals white.

Var. **Deamiàna** Fern. (for its discoverer, Charles Clemon Deam, 1865–). — Dry sands, s. Mich. and n. Ind. to Kans. and Tex.

2. ANEILÈMA R. Br.

Flowers regular. Sepals membranaceous, persistent. Petals all alike, ovate to obovate, sessile. Stamens 2 or 3, with bearded or naked filaments, one anther of different size from the others; staminodia 2–4. Capsule 2–3-locular, the locules with 1–several seeds. Seeds with hard rugose testa. — Plants resembling *Commelina,* with spatheless flowers axillary or in terminal racemes or panicles; chiefly of trop. and warm reg. of the Old World, a few American. (Name from the Greek *aneu, without,* and *eilema, spathe.*)

1. A. Keìsak Hassk. (native Japanese name). — Extensively creeping, prostrate or reclining, rooting at lower nodes, the forking stems 0.3–1.8 m. long; leaves narrowly lanceolate, the often folded blade and sheath confluent; flowers from the upper axils, eventually forming elongate leafy-bracted racemes; petals lilac to bluish-purple, 7–9 mm. long; capsule ellipsoid-ovoid, exserted. — Fresh tidal marshes and shores, se. Va. Sept., Oct. (E. Asia)

3. TRADESCÁNTIA L. Spiderwort

Flowers regular. Sepals herbaceous. Petals all alike, ovate, sessile. Stamens all fertile; filaments bearded. Capsule 2–3-locular, each locule 1–2-seeded. — Perennials. Stems mucilaginous, mostly upright, simple or forking, leafy. Leaves keeled. Flowers ephemeral, in umbelled cymes, axillary and terminal, often produced through the summer; involucral bracts mostly like the foliage-leaves. Plants of temp. and trop. Am. (Named for *John*, the elder *Tradescant*, ? –1637, gardener to Charles the First of England.)*

a. Cymes sessile or only short-stalked, much surpassed by the leaf-like bracts; funicular scar on seed linear. . . b.
b. Upper leaf-blades much broader than the sheaths. . . c.
c. Sepals 4–10 mm. long, not inflated or conspicuously foliaceous.
Pedicels 1–1.7 cm. long, pilose to glabrate; sepals 4–10 mm. long; capsule 4–6 mm. long. . . . . 1. *T. subaspera.*
Pedicels 2–3.2 cm. long, glandular; sepals 9–10 mm. long; capsule 6–8 mm. long. . . . . 2. *T. ozarkana.*
c. Sepals 1–1.4 cm. long, more or less inflated, conspicuously foliaceous; pedicels 2–3.2 cm. long, pilose; capsule 5–7 cm. long. . . . 3. *T. Ernestiana.*
b. Upper leaf-blades narrower than to about as broad as the sheaths. . d.
d. Sepals glandless. . . e.
e. Sepals pubescent their entire length; leaves not glaucous; pedicels pilose.
Stem glabrous, or merely puberulent above, with 2–5 nodes; leaf-blades glabrous; pedicels 1.5–3.5 cm. long; sepals herbaceous, inflated; petals 1.2–1.8 cm. long. . . . . 4. *T. virginiana.*
Stem pilose, with 1 or 2 nodes; leaf-blades pilose; pedicels 4–6 cm. long; sepals somewhat petaloid, not inflated; petals 1.8–2.2 cm. long. . . . . 5. *T. Tharpii.*
e. Sepals glabrous or merely bearded at tip; leaves glaucous; pedicels glabrous, 0.7–2.5 cm. long. . . . . 6. *T. ohiensis.*
d. Sepals more or less glandular, rarely glabrate. f.
f. Stem glabrous, or merely puberulent above, with 2–6 nodes; leaves glabrous; pedicels 1–3.3 cm. long.
Green, not glaucous; pedicels 1.8–3.3 cm. long, heavily viscid-pubescent; sepals 1–1.3 cm. long; petals 1.8–2 cm. long. . . . 7. *T. bracteata.*
Glaucous; pedicels 1–2 cm. long, sparsely pubescent to glabrous; sepals 4–10 mm. long; petals 0.7–1.6 cm. long. . . . . . . 8. *T. occidentalis.*
f. Stem villous, with 1 or 2 nodes; leaves pilose; pedicels 4–6 mm. long, villous; sepals 8–10 mm. long; petals 1–1.5 cm. long. . . . 9. *T. longipes.*
a. Cymes both sessile and long-peduncled; bracts very short, 1–7 mm. long; funicular scar roundish. . . . . 10. *T. rosea.*

1. T. subáspera Ker (somewhat harsh). — Stems stoutish, more or less flexuous above, 4–8 dm. high, with 6–10 nodes, pilose to glabrate; leaves dark green, firm, with *elliptic-lanceolate blades* 1–5 cm. broad, *constricted into a petiolar base*, glabrous or slightly puberulent, ciliolate; umbellate cymes terminal and usually from upper nodes, the upper ones sessile; bracts wide-spreading, up to 1.5 dm. long; *pedicels* 1–1.7 *cm. long, pilose or glabrate; sepals* 4–10 *mm. long, not inflated*, pubescent with glandular or eglandular hairs or glabrate; petals 1–1.5 cm. long, blue, rarely white; capsule 4–6 mm. long; seeds slightly elongate, 2–3 mm. long, with linear funicular scar. (*T. pilosa* Lehm.) — Woods, thickets and clearings, sometimes ruderal, W.Va. to Ill., s. to Tenn. and Mo.; introd. and natzd. in e. Pa. June, July.

Var. **montàna** (Shuttlew.) Anders. & Woodson (of mountains). — Stem little if at all flexuous above; lateral cymes mostly peduncled. (*T. comata* Small) — Similar habitats, Va. and W.Va. to nw. Fla. and Ala.

2. T. ozarkàna Anders. & Woodson (of the Ozark region). — Similar to no. 1; stem 1–5 dm. high, with 4–7 nodes; leaves succulent, pale, slightly glaucous; *pedicels* 2–3.2 *cm. long, glandular; sepals* 9–10 *mm. long*, glandular; petals pale roseate to white; *capsule* 6–8 *mm. long;* seeds 3–4 mm. long. — Wooded slopes and ravines or roadsides, Ozark reg. of sw. Mo., w. Ark. and e. Okla. April–June.

3. T. Ernestiàna Anders. & Woodson (in honor of Ernest Jesse Palmer, 1875–). — Similar to no. 2; stem scarcely flexuous, 2–4 dm. high, glabrous or sparsely pilose, with 2–5 nodes; *leaves* membranaceous, dull green, with broad *blade contracted directly into the sheath;* cyme terminal, solitary; *pedicels* 2–3.2 *cm. long, pilose; sepals* 1–1.4 *cm. long, somewhat inflated, strongly foliaceous*, uniformly pilose; petals deep blue, rarely roseate, about 1.5 cm. long; capsule 5–7 mm. long; seeds 2–3 mm. long. — Rocky woods and bluffs, sw. Mo., w. Ark. and e. Okla. April, May.

4. T. virginiàna L. (of Virginia). — *Stems* erect or ascending, 0.5–3.5 dm. high, *with* 2–5 *nodes, glabrous, or merely puberulent above; leaves* thin, dull green, the linear-lanceolate long-

* Treatment derived largely from that of Anderson & Woodson, Contrib. Arn. Arb. no. ix. (1935).

acuminate blades *glabrous*, 1–3.5 dm. long, 0.5–2.5 cm. broad; cymes solitary and terminal, sometimes peduncled at uppermost node; *pedicels* 1.5–3.5 *cm. long*, pilose; *sepals uniformly pilose, inflated*, 0.9–1.6 cm. long, foliaceous; *petals* 1.2–1.8 *cm. long*, variously blue, purple or roseate, or white in forma **albiflòra** Britt. (white-flowered); capsule 4–7 mm. long; seed 2–3 mm. long. — Woods, thickets, meadows and roadsides, Ct. to Wisc., s. to nw. Ga., Tenn. and Mo.; cult. and esc. ne. to Me. Late April–July.

5. T. Thárpii Anders. & Woodson (in honor of BENJAMIN CARROLL THARP, 1885–). — Low; *stems* during anthesis only 2–6 cm. high (or acaulescent), later lengthening to 1–2 dm., *villous, with* 1 *or* 2 *nodes; leaves* firm, green, often with reddish hyaline margin, the linear-lanceolate blade 1.5–3 dm. long, 0.9–2.5 cm. broad, loosely *pilose;* cymes many-flowered, solitary, terminal; bracts wide-spreading; *pedicels* 4–6 *cm. long*, pilose; *sepals* 1.2–1.6 cm. long, *somewhat petaloid*, purplish, rarely greenish, not inflated, uniformly pilose; *petals* 1.8–2.2 *cm. long*, purple or blue. — Rocky prairies, sw. Mo. and Kans. to Tex. April, May.

6. T. ohiénsis Raf. (of Ohio). — *Glaucous and glabrous;* stems 2–6.5 dm. high, with 3–8 nodes; leaves firm, the linear-lanceolate blade up to 4.5 dm. long, 0.5–4.5 cm. broad, glabrous, except for the frequently pilose sheath; cyme terminal, solitary, or a peduncled one at the upper node; bracts divaricate or reflexed; *pedicels* 0.7–2.5 *cm. long, glabrous; sepals* 0.5–1.5 cm. long, glaucous, *glabrous, or bearded only at tip;* petals 0.8–2 cm. long, blue, purple or rose, rarely white. (*T. canaliculata, barbata* and *reflexa* Raf.) — Thickets, meadows, woodlands, railroad-gravels, etc., Mass. to Minn. and Neb., s. to Fla., Ala., Miss., La. and Tex.; often cult. and esc. April–June.

7. T. bracteàta Small (bracted). — Similar to the preceding, *green, not glaucous; stems* 0.5–4.5 dm. high, *glabrous, or minutely puberulent above*, with 2–4 nodes; leaves bright green, stiff, glabrous; cyme solitary and terminal, rarely a peduncled one at upper node; *pedicels* 1.8–3.3 *cm. long, heavily viscid-pubescent; sepals* 1–1.3 *cm. long, viscid-pubescent; petals about* 2 *cm. long*, roseate, rarely blue. — Prairies, roadsides, railroad-gravels, etc., Mich. to Mont., s. to s. Ind., Ill., Mo. and Kans. May–July.

8. T. occidentàlis (Britt.) Smyth (western). — *Stems* rather stout, 0.5–6 dm. high, *glabrous and glaucous*, with 3–6 nodes; *leaf-blades* mostly 1–6 dm. long, 0.2–2 *cm. broad, glaucous and glabrous;* cymes solitary and terminal or with 1 or 2 peduncled ones at the upper nodes; bracts sharply reflexed or divergent, like the foliage-leaves; *pedicels* 1–2 *cm. long, sparsely pubescent to glabrous; sepals* 4–10 *mm. long*, glaucous or purplish, uniformly glandular-pubescent, rarely glabrate; *petals* 0.7–1.6 *cm. long*, blue, purple or roseate; seed 2–4 mm. long. — Sandy plains and prairies, w. Wisc. and Man. to Mont., s. to La., Tex. and Mex. May–July.

9. T. lóngipes Anders. & Woodson (long-stalked). — *Acaulescent* or nearly so, the *villous stem* lengthening to 2–10 cm., *with* 1 *or* 2 *nodes; leaves* firm, deep green, loosely *pilose*, with linear-lanceolate blade 1.5–2.2 dm. long and 0.6–1.2 cm. broad; cyme terminal; *bracts* like foliage-leaves or longer, *loosely ascending; pedicels* 4–6 *mm. long*, purplish, *villous; sepals* 8–10 *mm. long*, somewhat *petaloid, purplish, viscid-pilose; petals* 1–1.5 *cm. long*, pink, purple or blue. (*T. brevicaulis*, in part, of ed. 7, not Raf.) — Rocky or sandy woods and hills, Mo. May.

10. T. RÒSEA Vent. (rose-colored). — *Roots slender and fibrous; stems* slender, *freely forking*, 2.5–4 dm. high, glabrous or nearly so, with 2–4 nodes; leaves pale green, with linear-lanceolate blade as broad as or broader than sheath, 1.2–2 cm. long, 4–14 mm. broad; *cymes terminal and on long peduncles; bracts greatly reduced*, 1–3 *mm. long, scarious*, ascending; pedicels 8–12 mm. long, glabrous; sepals 4–6 mm. long, glabrous; *petals pink or roseate*, 7–9 *mm. long;* capsule trigonous-obovoid, 2–2.5 mm. long; *seeds* 1.5–2 mm. long, *with roundish funicular scar*. (*Cuthbertia rosea* (Vent.) Small) — Sandy barrens and woods, ne. Fla., n. to N.C. — Represented with us by

Var. **gramínea** (Small) Anders. & Woodson (grass-like). — *Densely cespitose*, 1–3 dm. high, often pilose below, the nodes 3–6; *leaf-blades narrower than sheaths*, 1–5 *mm. broad; bracts more foliaceous*, 2–7 *mm. long;* pedicels 8–15 mm. long; *capsule more subglobose*, 2–3 mm. long. (*Cuthbertia graminea* Small) — Sandy pinelands and oak-scrub, s. Fla. to se. Va. June–Sept.

FAM. 30. PONTEDERIÀCEAE (PICKERELWEED FAMILY)

Aquatic or paludal herbs with perfect more or less irregular flowers from a spathe; the petal-like 6-merous perianth free from the 3-locular ovary; the 3 *or* 6 *mostly unequal or dissimilar stamens inserted in throat of perianth.* — Perianth with the 6 divisions colored alike, *imbricated* in 2 rows in the bud, the whole together sometimes revolute-coiled after flowering, then withering away, or the base thickened-persistent and inclosing the fruit. Anthers introrse. Ovules anatropous. Style 1; stigma 3-lobed or 6-toothed. Fruit a perfectly or incompletely 3-locular many-seeded capsule or a 1-locular 1-seeded utricle. Embryo slender, in floury albumen.

Stamens 6; perianth funnelform, the limb bilabiate.
Petioles inflated, blades reniform or elliptic; perianth 4–6 cm. long; capsule 3-locular, many-seeded. . . . . 1. *Eichornia.*
Petioles not inflated, blades ovate to linear-lanceolate; perianths much shorter; fruit a 1-seeded utricle. . . . . 2. *Pontederia.*
Stamens 3; perianth salverform, with slender tube; capsule many-seeded. . . . 3. *Heteranthera.*

1. EICHHORNIA Kunth WATER-HYACINTH

Perianth funnelform, slightly 2-lipped; the limb spreading, somewhat oblique, its broad and thin lobes slightly unequal. Stamens 6, the 3 upper all included, the 3 lower more exserted; anthers oblong, basifixed. Ovary 3-locular, with numerous ovules. Capsule surrounded by the marcescent perianth, membranaceous, many-seeded. — Aquatic herbs nat. of S.Am. and Afr., the rhizomes and stems floating, rooting at the nodes, the long-petioled leaves rounded or broad; the showy flowers spicate-racemose or panicled, subtended by a bladeless or small-bladed sheath. (Named for *Johann Albrecht Friedrich Eichhorn* of Berlin, 1779–1856.) PIAROPUS Raf.

1. E. CRÁSSIPES (Mart.) Solms (with thick footstalk). — Base of petiole spongy-inflated; blade suborbicular, reniform or rounded-elliptic, 3–10 cm. long, usually broader; flowers several, bluish-purple, 4–6 cm. long and broad. — Ponds, quiet streams and ditches, Fla. to Tex., n. to Va. and Mo., a troublesome plant farther south, there clogging waterways. (Introd. from S.Am.)

2. PONTEDÈRIA L. PICKERELWEED

Perianth funnelform, 2-lipped; the 3 upper divisions united to form the 3-lobed upper lip; the 3 lower spreading and their claws, which form the lower part of the curving tube, more or less separate or separable to the base; tube after flowering revolute-coiled. Stamens 6; the 3 anterior long-exserted; the 3 posterior (often sterile or imperfect) with very short filaments, unequally inserted lower down; anthers versatile, oval, blue. Ovary 3-locular; two of the locules empty, the other with a single suspended ovule. Utricle 1-locular. — Stout Am. herbs, with thick creeping rhizome, producing erect long-petioled leaves, and a 1-leaved stem bearing a spike of violet-blue ephemeral flowers. Basal leaves with a sheathing stipule within the petiole. (Dedicated to *Guilio Pontedera*, 1688–1756, professor at Padua.)

1. P. cordàta L. (heart-shaped). — *Leaves soft, the cauline with petioles averaging* 4.5 *cm. long;* blades cordate at base, narrowly deltoid-ovate, tapering with straight sides from base to apex, or broadly ovate and gradually curved from the broad cordate base to blunt summit in forma **latifòlia** (Raf.) House (broad-leaved), or truncate to tapering at base and narrowly deltoid to linear-lanceolate in forma **angustifòlia** (Pursh) Solms (narrow-leaved), or blade wanting and submersed leaves linear and translucent or blade up to 5 mm. broad in forma **taènia** Fassett (ribbon-like); *spike dense,* with a spathe-like bract; *young and commonly the mature flowers white-villous;* upper lobe of perianth marked with a pair of yellow spots (rarely all white); *mature fruits* crested with 6 toothed ridges, 6–10 *mm. long; seeds* 3.5–4.5 *mm. long,* 2–2.5 *mm. in diameter.* — Muddy shores or in shallow water, P.E.I. to s. Ont., s. to N.S., N.E., L.I., n. Fla., Mo. and Okla. June–Nov.

2. P. lanceolàta Nutt. (lance-leaved). — *Leaves* firmer or *harder, the cauline with petioles averaging* 2.7 *cm. long;* blades lance-oblong to lance-linear, narrowed to base, or narrowly deltoid-ovate and tapering with straight sides from the truncate to shallowly cordate base to apex in forma **trullifòlia** Fern. (trowel-leaved), or ovate and gradually curving from the broad deeply cordate base in forma **brasiliénsis** (Solms) Fern. (of Brazil); *spike rather loose; young flowers glandular and sometimes hirtellous, glandular or glabrate in age; mature fruits* 5–6 *mm. long; seeds* 2.7–3.2 *mm. long,* 1.8–2 *mm. in diameter.* — Similar habitats, Fla. to Tex., n., locally, to Del. May–Sept. (W.I.; S.Am.)

3. HETERANTHÈRA R. & P. MUD-PLANTAIN

Perianth with a slender tube; the limb somewhat equally 6-parted, ephemeral. Stamens in the throat, usually unequal; anthers erect. Capsule 1-locular or incompletely 3-locular by intrusion of the placentae. — Low Am. and Afr. herbs, in mud or shallow water, with a 1–few-flowered spathe bursting from the sheathing side or base of a petiole. (Name from the Greek *hetera, different,* and *anthera, anther,* from the dissimilar anthers of the original species.)

Stamens unequal, 2 with ovate yellow anthers, 1 with larger greenish anther; leaves long-petioled, with broad opaque blades.
Blades round-reniform to cordate-ovate, acute. . . . . 1. *H. reniformis.*

Blades oblong, obtuse. . . . . 2. *H. limosa.*
Stamens alike, all 3 with sagittate anthers; leaves sessile, linear, translucent. . 3. *H. dubia.*

1. H. renifórmis R. & P. (kidney-shaped). — *Leaves round-reniform to cordate and acute;* spathe 3–10-flowered; flowers white or pale blue; 2 *posterior filaments with ovate yellow anthers, the 3rd with larger greenish anther;* capsule incompletely 3-locular. (*H. peduncularis* Benth.) — Creeping in mud or floating in shallow water, Fla. to Tex. and Mex., n. to Ct., e. N.Y., Ky., s. Ill., Mo. and Neb. Aug.–Oct. (Trop. Am.)

2. H. limòsa (Sw.) Willd. (of mud). — Similar; *leaves oblong or lance-oblong, obtuse at both ends;* spathe 1-flowered; flowers larger, blue, or white in forma **albiflòra** Benke (white-flowered). — Sloughs, pond-margins and mud, Fla. to N.M. and Mex., n. to Ky., s. Ill., Minn., Neb. and Colo. July–Sept. (Trop. Am.)

3. H. dùbia (Jacq.) MacM. (doubtful), Water-Stargrass. — Submersed, or stranded as forma **terréstris** (Farw.) Vict. (of land), *branching grass-like plant with linear* translucent *sessile leaves;* the 1-flowered terminal spathes becoming lateral by prolongation of the stem; *flowers* small, *pale yellow,* with long thread-like tube, often cleistogamous; *anthers all sagittate;* capsule 1-locular, with 3 parietal placentae. (*Zosterella dubia* (Jacq.) Small) — Streams and quiet waters or their argillaceous or calcareous shores, Fla. to Tex. and Mex., n. to sw. Que., s. Ont., Minn., Id. and Oreg. June–Sept. (Trop. Am.)

FAM. 31. JUNCÀCEAE (Rush Family)

Grass-like or rush-like herbs with small regular and hypogynous persistent flowers, 3 *glumaceous sepals, and* 3 *similar petals,* 6 *or* 3 *stamens with* 2*-locular anthers, a single short style,* 3 *filiform hairy stigmas, and an ovary either* 3*-locular or* 1*-locular with* 3 *parietal placentae, forming a loculicidal* 3*-valved capsule.* Seeds anatropous, with a minute embryo inclosed at the base of the fleshy albumen. — Flowers liliaceous in structure, but somewhat sedge-like in aspect and texture.*

Capsule perfectly or imperfectly 3-locular, with many seeds; plants never hairy. 1. *Juncus.*
Capsule 1-locular, 3-seeded; leaves and young stems frequently hairy. . . . 2. *Luzula.*

1. JÚNCUS L. Rush. Bog-Rush. Jonc (Que.)

Capsule 3-locular, or 1-locular by the placentae not reaching the axis. Stamens 3 or 6, when 3 opposite the 3 sepals. Seeds numerous. — Chiefly perennials, and in wet or dry open soil or water, with pithy or hollow and simple (rarely branching) stems, glabrous narrow to filiform leaves and cymose or clustered small (greenish, stramineous or brownish) flowers, chiefly in summer. Cosmop. genus. (The classical name.)

A. Individual flowers prophyllate, *i.e.* each subtended by a pair of bracteoles besides the bractlet at base of pedicel, scattered or solitary on the branches of the cyme or slightly clustered at their tips, very rarely in glomerules; leaves never nodulose-septate. . . *B.*

B. Leaves with very slender and elongate flat, involute, channeled or setaceous blades (the basal reduced or wanting in one var. of no. 2); inflorescence subtended by 1 or more slender bracteal leaves, terminal (pseudolateral only in no. 15); culms very slender, rarely 2 mm. in diameter at base. § Poiophylli. . . *C.*

C. Leaf-sheaths gradually tapering at summit, not auricled; inflorescence one-fourth to nine-tenths entire height of plant; soft-based annual, little if at all tufted (though often branching at base and simulating a crown of stems). . . . . 1. *J. bufonius.*

C. Leaf-sheaths auricled or prolonged at summit; inflorescence rarely one-fourth the entire height of plant; perennials with elongate rhizomes or with densely tufted leaves and culms, with dried remnants of culms of preceding year often showing. . . *D.*

D. Auricles of leaf-sheaths deeply cleft, borne opposite the blade; basal sheaths all or nearly all without prolonged blades; inflorescence 1–4-flowered, many times overtopped by the upper leaves; capsule very firm, tapering to a long beak; seeds few, irregularly semirhombic; arctic-alpine. . . . . 2. *J. trifidus.*

D. Auricles entire, at the base of the blade; lower sheaths mostly blade-

---

* In the *Juncaceae* the figures vary: showing in some the habit, base, basal sheaths or portion of culm; in all or nearly all the inflorescence, enlarged mature flower and the enlarged seed.

bearing; inflorescences several–many-flowered, exceeding, equaling or overtopped by upper leaf; capsules rather thin-walled, rounded to retuse at summit; seeds very numerous, obliquely oblong, ovoid or obovoid to spindle-shaped. . . *E.*

E. Rhizome horizontal, becoming slender and elongate; leaf-sheaths extending about half-way up the culm; sepals obtuse, with incurved tips, often of two colors.

Anthers about thrice the length of the filaments; capsule ellipsoid-ovoid, equaling or but slightly exceeding the perianth. . . . 3. *J. Gerardi.*

Anthers scarcely longer than filaments; capsule globose-obovoid, distinctly exserted. . . . . 4. *J. compressus.*

E. Rhizome short and erect, mostly hidden in the tussock of crowded crowns; leaf-sheaths confined to base or lower third of plant; sepals acute or sharp-pointed, uniformly colored (except for scarious margins), their tips ascending to spreading-ascending. . . *F.*

F. Leaves numerous, flat or involute, if terete very slender, unlike the culms; inflorescence terminal, its longer involucral bract many times shorter than culm; perianth closely ascending or spreading-ascending; capsule with thinnish walls; seeds slender, apiculate or slender-tailed, not coarsely reticulate. . . *G.*

G. Sepals spreading-ascending, their tips not closely appressed to capsule; seeds 0.25–0.6 mm. long, without prolonged white tails. . . *H.*

H. Capsule definitely 3-locular.

Inflorescence greatly overtopping to barely shorter than the involucre; perianth 2.5–3.5 mm. long; eastern to midwestern. . . . . 5. *J. secundus.*

Inflorescence overtopped by involucre; perianth 4–6 mm. long; southwestern. . . . . 6. *J. kansanus.*

H. Capsule 1-locular, or only imperfectly 3-locular with the partial partitions extending only part-way to axis of capsule. . . *I.*

I. Auricles of uninjured leaf-sheath prolonged or tongue-like.

Leaf-sheaths whitish to drab, their margins broadly scarious and friable, the whitish scarious auricles lance-triangular or -oblong, two to four times as long as broad; partitions of capsule narrow, extending less than half-way to the axis; transcontinental species. 7. *J. tenuis.*

Leaf-sheaths (at least the inner) purplish, with firmer non-friable margins; the truncate or round-tipped auricles firmer, drab to fuscous, about as broad as long; partitions extending half-way to axis of capsule; species of Atlantic States. . . . . . . 8. *J. platyphyllus.*

I. Auricles of uninjured leaf-sheath gradually rounded, following curve of sheath-summit, scarcely prolonged. . . *J.*

J. Leaves flat or merely involute in drying; sheaths of basal leaves stramineous, brown or drab (rarely purplish); continental or inland species.

Auricles submembranaceous or firm, pale brown to drab, sheaths membranaceous-margined; perianth strongly ascending or erect, 3–4 mm. long, about equaling capsule. . . . . 9. *J. interior.*

Auricles coriaceous, yellowish or amber, sheaths firm; perianth 4–6 mm. long, spreading-ascending, overtopping capsule. . . . . 10. *J. Dudleyi.*

J. Leaves terete, or merely slender-channeled on upper side; inner sheaths purplish when fresh, outer brown; auricles firm; perianth 3.5–5 mm. long, exceeding the brown and highly lustrous capsule; coastwise species. 11. *J. dichotomus.*

G. Sepals erect, closely appressed to capsule; seeds 0.5–1.4 mm. long, with white tails. . . *K.*

K. Petals attenuate to subulate at tip, with very narrow scarious margins; capsule shorter than perianth; inflorescence pale stramineous. . . . . 12. *J. oronensis.*

K. Petals blunt or merely acutish, with broad scarious margins; capsules exserted.

Inflorescence green or greenish-stramineous, subtended by

a stiffly erect bract 1–8 cm. long; capsule greenish; seeds pale brown, 1–1.4 mm. long, with white tails half as long as body. . . . . 13. *J. Vaseyi.*

Inflorescence brown to castaneous when mature, subtended by 1 or more flexuous bracts up to 0.5–2.5 dm. long; capsule olive-brown, reddish or castaneous; seeds reddish, 0.5–0.6 mm. long, with minute white tails. . . . 14. *J. Greenei.*

*F.*Leaves few, terete and resembling the culms; inflorescence falsely lateral, overtopped by a terete bract looking like a continuation of and a third as long as to almost equaling the culm; perianth stiffly spreading-ascending; capsule hard-walled; seeds irregularly obconic, beaked, strongly stipitate, 0.6–0.8 mm. long, the castaneous body coarsely ribbed and cross-ribbed. . . . 15. *J. coriaceus.*

*B.*Leaves and culms terete or slightly compressed, uniform, 0.5–5 mm. thick at base, with long tubular or open chartaceous to subcoriaceous often mucronate but otherwise bladeless basal sheaths; inflorescence appearing as if lateral (pseudolateral), subtended by an erect or divergent bract appearing like a continuation of the culm. § Genuini. . . *L.*

*L.*Culms with coarse to very fine longitudinal furrows or striae, in tussocks (or, if from an elongate and slender rhizome, with green or pale stramineous flowers about 3 mm. long), the flowers 1.5–4.2 mm. long; anthers shorter than to but slightly longer than filaments; seeds 0.5–1 mm. long. . . *M.*

*M.*Rhizome very slender and cord-like, extensively creeping; culms mostly scattered, capillary and delicate, 1–6 dm. high; cyme few-flowered, with mostly simple branches; flowers about 3 mm. long, greenish; stamens 6, anthers much shorter than filaments; capsule broadly rounded at summit; boreal species. . . . . 16. *J. filiformis.*

*M.*Rhizome short or stout, the culms cespitose or approximate in rows along rhizome, 1–5 mm. thick at summit of sheath, 0.3–2 m. high; cyme (except sometimes in no. 18) compound and many-flowered; the flowers 1.5–4.2 mm. long, green to brown; stamens 6 or 3, the anthers about equaling filaments; summit of capsule conical, rounded or emarginate. . . *N.*

*N.*Capsule mucronate or pointed, with coriaceous or subcoriaceous walls; local species, chiefly of Atlantic States.

Stamens 6; flowers 2–3 mm. long, brown; castaneous ovoid capsule conically tapering to beak.

Plant glaucous; cyme with stiffly ascending branches; perianth about equaling capsule, 3 mm. long; petals and sepals similar, lance-attenuate; seeds 0.5 mm. long. . . . . 17. *J. inflexus.*

Plant green; cyme with divergent branches; perianth 2–2.7 mm. long, much shorter than capsule; petals ovate or oblong, obtuse, unlike the lance-attenuate sepals; seeds 0.7–1 mm. long. . . . . 18. *J. gymnocarpus.*

Stamens 3; plant deep green; cyme lax and open, with loosely spreading or recurving branches; flowers 3–3.6 mm. long, greenish, the lance-attenuate sepals and petals about equaling the trigonous-oblong olivaceous subtruncate but mucronate capsule; seeds 0.6 mm. long. . . . . 19. *J. Griscomi.*

*N.*Capsule thin-walled, emarginate, depressed or subtruncate at summit, beakless or barely mucronulate; perianth 1.5–4.3 mm. long; semi-cosmop. species. . . . . 20. *J. effusus.*

*L.*Culms not furrowed nor striate, when dry at most irregularly wrinkled, rising from firm extensively creeping and forking rhizome; sepals with deep brown to castaneous sides; anthers two to four times length of filaments; capsule brown to blackish, with subconical summit; seeds about 1 mm. long. . . . . 21. *J. balticus.*

*A.*Individual flowers eprophyllate, *i.e.* each subtended only by the bractlet at base of the very short pedicel. . . *O.*

*O.*Leaves not septate. . . *P.*

*P.*Culms and pungent terete rigid leaves similar, up to 1.5 m. high, arising singly or scattered from coarse horizontal rhizome; basal sheaths coriaceous, elongate and bladeless; inflorescence pseudolateral, the single terete pungent involucral leaf appearing like a continuation of the culm; flowers in glomerules. § Thalassii. . . . . . . 22. *J. Roemerianus.*

P.Culms and basal leaves clearly differentiated, the latter (if terete) not pungent; inflorescence terminal; flowers solitary, few or glomerulate. . . *Q*.

Q.Leaves filiform, terete or deeply channeled or involute; inflorescence of 1–few heads; flowers with rather conspicuous white, whitish, yellowish or intensely colored perianths; seeds 1.5–4 mm. long, including the long pale appendages; boreal plants. § Stygii.

Non-stoloniferous, forming large or small tussocks; perianths whitish to pale pinkish or stramineous; capsule pale.

Involucre erect, with lanceolate narrow-margined base; capsule 6–9 mm. long, strongly exserted; seeds 3–4 mm. long. . . . 23. *J. stygius*, var. *americanus*.

Involucre somewhat divergent, with ovate broadly clasping base; capsule 3–5 mm. long, included or barely exserted; seeds 1.5 mm. long. . . . . 24. *J. albescens*.

Stoloniferous, the culms rising singly from tips of stolons; perianths and capsules castaneous; seeds 2.5–3.7 mm. long. . . . . . . 25. *J. castaneus*.

Q.Leaves flat, at most folded when dry; inflorescence of few–many heads; flowers with greenish to brown perianths; seeds 0.3–0.7 mm. long, merely apiculate; plants of temp. or warm reg. § Graminifolii. . . *R*.

R.Culms weak, from tufts of flaccid leaves and fibrous roots or from trailing or floating elongate repent stems; glomerules 1–few; perianth herbaceous, green, 5–10 mm. long; stamens 3; capsule subulate. . . . . 26. *J. repens*.

R.Culms firm, erect, from knotty rhizomes or firm stolons; glomerules few–very many; perianth scarious or subcoriaceous, 2–7 mm. long; stamens 3 or 6; capsule plump. . . *S*.

S.Perianth 5–7 mm. long; anthers 6, yellow; style elongate; capsule abruptly long-beaked; northern and northwestern. . . . . 27. *J. longistylis*.

S.Perianth 2–3.5 mm. long; anthers 3, red to dark purple; capsule beakless; plants of southern, eastern and central states. . . *T*.

T.Rhizome thick and knotty or with short and thick finger-like basal offshoots; culms numerous in tussocks, or few or solitary.

Slender culms 1.5–6(–8) dm. high, 1–2(–3) mm. thick at base, mostly several in tussocks; leaves soft, the basal ones 1–5 mm. broad; cymes 1–10 cm. long, with 2–30(–50) glomerules; stamens shorter than sepals, quickly shriveling. . 28. *J. marginatus*.

Stouter; culms 0.6–1.3 m. high, 3–5 mm. thick at base, solitary or few; leaves hard and firm, the basal 4–7 mm. broad; cymes 0.3–2 dm. long, with 30–200 glomerules; stamens equaling or exceeding sepals, persistent and often conspicuous in fruit. . . . . 29. *J. biflorus*.

T.Rhizome slender, flexuous, cord-like, the scaly stolons up to 2 dm. long and 1.5–3 mm. thick; culms 1–3, mostly solitary, 0.4–1 m. high. . . . . 30. *J. Longii*.

O.Leaves septate or nodulose, *i.e.* with septa at regular intervals, terete, compressed or filiform. § Septati. . . *U*.

U.Flowers in dense spherical heads; sepals and petals linear- or narrowly lance-acicular; seeds merely pointed to blunt. . . *V*.

V.Plants with filiform extensively creeping and branching rhizomes and subterranean stolons; involucre or uppermost leaf usually overtopping inflorescence; stamens 6; capsule subulate or subulate-lanceolate, exserted.

Flowers 3–4 mm. long, reddish-brown; petals equaling or exceeding sepals. . . . . 31. *J. nodosus*.

Flowers 4–5 mm. long, greenish to dull brown; petals much shorter than sepals. . . . . 32. *J. Torreyi*.

V.Plants without filiform branching rhizomes, either in small tufts from fibrous roots or with short and thick finger-like often clustered rhizomes; involucre much shorter than inflorescence; stamens 3. . . *W*.

W.Tufted, with many fibrous roots; leaves compressed; inflorescence 1.2–3 dm. long, with the longer naked internodes of the branches 6–16 cm. long; capsule subulate, exserted.

Leaves gladiate, 3–8 mm. broad; fruiting heads 8–12 mm. in diameter; valves of capsule after dehiscence remaining united at summit, thus forming a solid beak. . . . . 33. *J. polycephalus*.

Leaves straight, not gladiate, 2–4 mm. broad; fruiting heads 1.2–1.5 cm. thick; valves of capsule usually separating to tip or, if remaining united after dehiscence, not forming a solid beak. . . 34. *J. validus.*

W. Tufted or with solitary culms from thick and short finger-like whitish rhizome or clustered rhizomes; leaves terete or subterete, slender; inflorescences 1–15 cm. long, with the longer naked internodes of the branches 0–8 cm. long. . . *X.*

X. Capsules subulate, exserted, their valves, after dehiscence, united above as a beak.

Blade of uppermost leaf equaling to longer than its sheath; sepals hard and firm, pale green to drab, becoming dull brown; anthers exserted. . . . . 35. *J. scirpoides.*

Blade of uppermost leaf much shorter than its sheath; sepals soft and flexible, deep reddish-brown, becoming darker; anthers included. . . . . 36. *J. megacephalus.*

X. Capsule ovoid-conic, only one-half to two-thirds as long as perianth; heads pale green to dull brown. . . . . . 37. *J. brachycarpus.*

U. Flowers in hemispherical heads, or in narrower heads, or only 1–3, if heads sometimes subspherical then the petals lanceolate to ovate. . . *Y.*

Y. Seeds with definite caudate tips. . . *Z.*

Z. Culm and leaves papillose-scabrous; stamens 6; seeds 2–3 mm. long. 38. *J. caesariensis.*

Z. Culms and leaves smooth; stamens mostly 3; seeds 0.7–1.8 mm. long. . . *a.*

a. Flowers, with mature fruit, 2–5 mm. long; sepals and acute petals firm, often subrigid; seeds spindle-shaped, with conspicuous tails. . . *b.*

b. Cyme (when well developed) ovoid or broader, one-fourth as to nearly as broad as long; glomerules few–very many-flowered, turbinate to subglobose; capsule shorter than to slightly exceeding perianth.

Stiffly ascending; cyme one-fourth to half as broad as high, with stiffly ascending stoutish branches; mature fruit and perianth 3.8–5 mm. long; capsule abruptly short-pointed, rarely tapering; seeds 1.3–1.8 mm. long, their tails two-thirds to quite as long as the body. . . . . . 39. *J. canadensis.*

Loosely ascending; cyme one-third to nearly as broad as long, with filiform (often flexuous) loosely ascending to horizontal branches; mature fruit and perianth 2–4 mm. long; capsule more tapering; seeds 0.7–0.9 mm. long, their tails only one-third as long as body. . . . . 40. *J. subcaudatus.*

b. Cyme elongate, strict and narrow, three to six times longer than broad; glomerules turbinate, 2–few-flowered; capsule much longer than perianth, gradually tapering; seeds about 1 mm. long, their tails half as long as body. . . . . 41. *J. brevicaudatus.*

a. Flowers, with mature fruit, 2.5–3.5 mm. long; cyme open and diffuse; sepals and blunt petals soft and scarious-margined; glomerules mostly 2–5-flowered; seeds ellipsoid, their short tails about one-tenth as long as body. . . . . 42. *J. brachycephalus.*

Y. Seeds merely dark-pointed or blunt, not caudate. . . *c.*

c. Stamens 3. . . *d.*

d. Basal leaves quill-like, firm, few and erect, clearly septate; culms erect or ascending, firm, 0.2–1 m. or more high; cyme well developed, with many heads; heads rarely leafy-tufted; petals acute; capsule short-beaked or attenuate. . . *e.*

e. Capsule about equaling to one-third longer than perianth, the mature flower (including capsule) 2–3.5 mm. long. . . *f.*

f. Heads relatively few, 3–50 (rarely more); perianth 2–3.5 mm. long; capsule lanceolate to lance-ovoid or -oblong, gradually arching to short beak or apiculation.

Perianth 2.5–3.5 mm. long; capsule lance-ovoid or -ellipsoid, about equaling perianth. . . . . **43. *J. acuminatus.***

Perianth 2–2.5 mm. long; capsule slenderly lanceolate, definitely exserted. 44. *J. debilis.*

f. Heads usually 50–500; perianth 2–2.5 mm. long; capsule ovoid to broadly ellipsoid, broadly rounded to nearly beakless summit.

Leaves weakly septate, the septa not very evident in dry material; capsule deep brown; fibrous roots ending in fleshy fusiform tuberous enlargements; plant of southern Coastal Plain. 45. *J. Elliottii.*

Leaves prominently nodose-septate; capsule greenish or stramineous; roots without enlargements; plant of Mississippi and western Gulf drainage. . . . . . . 46. *J. nodatus.*

e. Capsule nearly or quite twice as long as perianth, lance-linear, gradually tapering to tip, the mature flower 3.5–6 mm. long. 47. *J. diffusissimus.*

d. Basal leaves capillary, flaccid, crowded, barely septate; culms weak, ascending, arching or declined, 0.5–4 dm. long, capillary, often submersed or floating; heads 1–few, remote, often accompanied by leafy tufts; petals blunt; capsule with truncate-rounded beakless summit. 48. *J. bulbosus.*

c. Stamens 6. . . *g.*

g. Culms stiffly erect, 0.3–1.3 m. high, arising singly or in rows from the stout and hard horizontal rhizome; leaves erect.

Lower elongate tubular chartaceous dark-colored sheaths bladeless; cauline blade (rarely 2), median, terete, equaling or overtopping the stiffly branched cyme.

Heads hemispherical; perianth 3–4 mm. long; sepals and petals deep brown, narrowly lance-attenuate, firm; beak of capsule not exserted. 49. *J. militaris.*

Heads subglobose; perianth 2 mm. long; sepals and petals grayish to whitish, oblong, obtuse, with membranous margins; beak of capsule exserted. . . . . 50. *J. pervetus.*

Lower sheaths short, open, with short blade; cauline leaves several, compressed, much shorter than culm; cyme strongly decompound, the ultimate branchlets spreading; perianth 3 mm. long; sepals and petals lance-subaristate; beak of capsule exserted. 51. *J. acutiflorus.*

g. Culms softer, slender, erect to depressed or creeping, 0.1–7.5 dm. high or long; basal leaves with green blades; inflorescence usually overtopping upper leaves. *h.*

h. Flowers few–many in definite heads, rarely if ever replaced by fascicles of leaves; basal and cauline leaves quill-like, with a single central tube; style short; capsule brown to blackish.

Sepals acuminate; branches of cyme divergent. . . . . . 52. *J. articulatus.*

Sepals blunt; branches of cyme erect or ascending. . . . . 53. *J. alpinus.*

h. Flowers solitary or in 2's or 3's, often replaced by stiff fascicles of reduced leaves; basal and cauline leaves with 2 or more tubes, filiform or very slender; style elongate; capsule pale. . . *i.*

i. Culms erect or ascending from a horizontal rhizome; flowers secund and mostly sessile along branchlets of an open dichotomous cyme; fascicles of reduced leaves confined to cyme; anthers much longer than filaments.

Rhizome soft, slender, about 1 mm. thick; sepals obtuse; northern species. 54. *J. pelocarpus.*

Rhizome subligneous, stout, 2.5–6 mm. thick; sepals acute; southeastern. 55. *J. abortivus.*

i. Stems repent or floating, bearing scattered fascicles of reduced leaves; roots tufted; flowers 1–3 on 1–3 remote axillary to terminal filiform peduncles; anthers about equaled by filaments. 56. *J. subtilis.*

§ POIOPHÝLLI (Buchenau) Vierh.

1. J. bufônius L. (of toads; from its occurrence in damp places), TOAD-R. — *Annual;* culm low and slender, 0.3–3.5 dm. high, simple or branching at base; few flat *leaves with sheaths gradually tapering to blade; cyme open, often one-fourth to nine-tenths entire height of plant; flowers remote,* 3–7 mm. long, *scattered along the branches,* whitish, greenish or pale brown, rarely viviparous or converted to leafy tufts, or in forma **congéstus** (Schousboe) Wahlb.

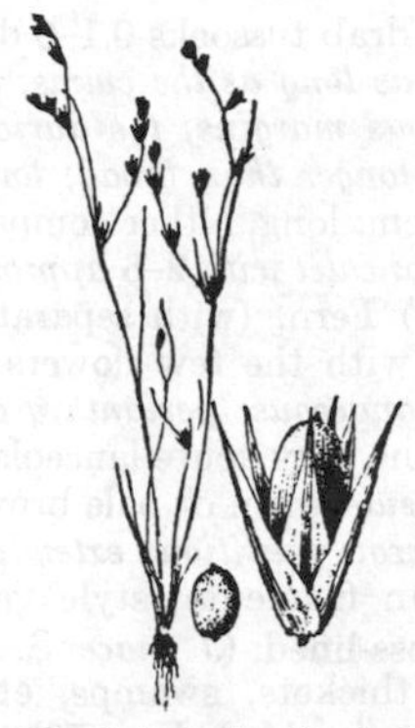
776. J. bufonius (typical).

(congested) with 2–4 flowers crowded in glomerules; sepals and petals linear-lanceolate, subulate-tipped; filaments slightly shorter than anthers; *seeds slenderly ovoid to ellipsoid*, 0.3–0.5 mm. long. — Damp or exsiccated open ground, roadsides, etc., throughout our range and beyond. June–Nov. (Semicosmop.) Fig. 776.

Var. **halóphilus** Buchenau & Fern. (lover of salinity). — More fleshy throughout; *flowers often in 2's or 3's; whitish petals obtuse; seeds short-cylindric, abruptly truncated at the ends.* — Brackish or saline soils, s. Lab. to lower St. Lawrence, Que., s. to coast of Mass.; w. N.Y.; James Bay; Sask. to Neb. and Colo. (Eu.) Fig. 777.

777. J. bufonius, v. halophilus.

2. J. trífidus L. (three-forked). — Stems hard, *densely tufted from matted creeping rhizomes*, erect (1–4 dm. high), sheathed and *mostly leafless at base, 2–3-leaved at the summit;* flowers brown (3–4 mm. long); sepals ovate-lanceolate, acute, equaling or rather shorter than the ovoid *beak-pointed deep brown capsule;* anthers much longer than the filaments; seeds few, oblong, angled (1.5–2 mm. long), short-tailed. — Arct. reg., s. to acid or sterile rocky or sandy barrens and mts. of Nfld., and bare mts. of Que., n. N.E. and n. N.Y. June–Aug. (Eurasia) Fig. 778.

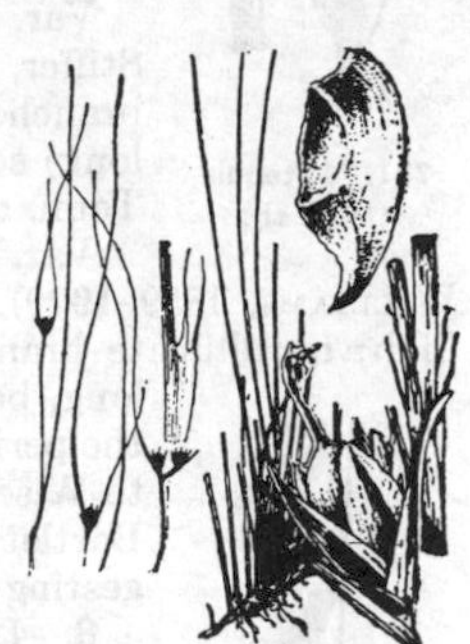
778. J. trifidus.

Var. **monánthos** (Jacq.) Bluff & Fingerhuth (one-flowered). — Looser and more slender, up to 6 dm. high, the *numerous basal leaves often equaling the culm;* flowers 1 (rarely 2). — Local, dry cliffs and crests, centr. N.H. to e. N.Y., s. to mts. of N.C. (Eu.)

3. J. Gerárdi Loisel. (for Louis Gérard, 1733–1819, Provençal botanist), Black Grass. — *Rhizomes and slender stolons dark, horizontally spreading;* culms in small tufts, scarcely flattened, stiffly ascending, 1.5–8 dm. high; *leaf-sheaths extending about half-way up the culm,* the *blades* soft and *green;* cyme contracted, 0.1–1(–1.5) dm. long, *usually exceeding the bracteal leaf; the brown and green flowers 2–3.5 mm. long,* mostly sessile or very short-pedicelled; the oval-oblong *sepals with incurved tips and scarcely equaling the ellipsoid-ovoid* obtuse mucronate *capsule; anthers about thrice length of filaments;* style as long as ovary; seeds obovoid, 0.4–0.5 mm. long, delicately ribbed and cross-lined. — Saline marshes and saline spots, Nfld. and Gaspé Pen., Que., s. along coast to Fla., and locally inland (sometimes as railroad-weed) to Ind. and Minn. June–Sept. (Eurasia; n. Afr.) Fig. 779. — Perhaps better treated as a var. of no. 4.

779. J. Gerardi.

Var. **pedicellàtus** Fern. (having pedicels). — Cyme 1–2 dm. long; flowers mostly on pedicels 3–10 mm. long, the perianths 3.3–5 mm. long. — Local on coast, Me. to R.I.

4. J. compréssus Jacq. (compressed; from the flattened culms). — Similar to no. 3, *glaucous; culms flattened;* margins and auricles of leaf-sheaths more delicate; *cyme usually overtopped by an elongate bract;* perianth usually paler; *capsule globose-obovoid, distinctly exserted,* castaneous; *filaments nearly equaling anthers.* — Fresh to brackish open soil, Nfld., N.S. and P.E.I. to e. Ont. July–Sept. (Eurasia)

5. J. secúndus Beauv. (one-sided). — Tufted from short crowns, *strict*, 0.6–4.5 dm. high; the *short* flat *but very narrow leaves crowded, rarely half as long as culm*, their sheaths with rounded membranous auricles; *cyme overtopping* (rarely exceeded by) *the longer involucral leaf*, 2–14 cm. high; *its branches mostly incurved-ascending, with flowers chiefly along the inner side; perianth* 2.5–3.5 *mm. long*, ascending, *barely exceeding the distinctly* 3-*locular capsule;* anthers exceeding filaments. — Dry open sterile soil, rock-pockets, clearings, etc., s. Me. to s. Ind., s. to N.C., Tenn. and Mo. June–Oct. Fig. 780.

6. J. kansânus F. J. Herm. (Kansan). — Much *stouter* than no. 5, 2.5–5 dm. high; *leaves very thick* but flat, 0.7–1.3 mm. broad, their *auricles firm; cyme* denser, *overtopped by involucre; flowers* somewhat ferruginous, 4–6 *mm. long, the rigid sepals longer than the petals;* the 6 stamens with filaments half as long as anthers; *capsule about as long as petals, clearly 3-locular;* seeds 0.5–0.6 mm. long, short-tailed. — Dry prairie and rocky slopes, Mo. and Kans. June–Aug.

780. J. secundus.

7. J. ténuis Willd. (slender). — In more or less dense pale green to drab tussocks 0.1–6 dm. high; *leaves flat*, in drying often involute, soft, *often one-half to quite as long as the culms; the green or drab sheaths with whitish friable broad scarious margins; the auricles* of uninjured leaves *lance-triangular to -oblong, much longer than broad; lower bract or bracts prolonged* above the cyme; *cyme* 1.5–15 cm. long, either compact or with unequal ascending branches, *each branch or branchlet with 2–6 approximate flowers*, or in forma **discretiflòrus** (F. J. Herm.) Fern. (with separated flowers) the very loose cyme up to 2 dm. long and with the few flowers of each branch or branchlet scattered; *prophylla membranaceous; perianth green or stramineous*, 3–4.5 mm. long, spreading-ascending; the very acute lanceolate *sepals* spreading in fruit, *much longer than capsule; capsule* green or pale brown when ripe, ovoid, *retuse, falsely 3-locular, with the narrow partitions extending less than half-way to axis;* anthers much shorter than filaments; style very short; seeds 0.3–0.4 mm. long, delicately ribbed and cross-lined. (*J. macer* S. F. Gray) — Damp, wet or dry open soils, roadsides, thickets, swamps, etc., throughout our range and beyond. June–Sept. (Eu.; n. Afr.) Fig. 781. — Passing into the two following:

781. J. tenuis (typical).

Var. **anthelàtus** Wieg. (with lateral branches exceeding central axis). — Stiffer, 4–9 dm. high; cyme 0.5–2 dm. long, its strongly ascending elongate branches and branchlets with flowers (mostly 2.5–3.5 mm. long) scattered along the inner side. — Me. to Ind., s. to Ga., Tenn. and Mo. Fig. 782.

Var. **Williámsii** Fern. (for its discoverer, Emile Francis Williams, 1859–1929). — Low (1–5 dm high) and slender; cyme 2–9 cm. high; its ultimate branches widely divergent to arched-recurving, 0.5–2 cm. long, bearing flowers from base to apex along the upper side; the perianth usually smaller than in the typical var. — E. Que. to Wisc., s. to N.S., N.E. and N.Y. Fig. 783. — *J. monostichus* Bartlett proves to be a fungus-infected state somewhat suggesting this var.

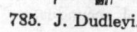

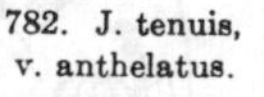

782. J. tenuis, v. anthelatus.

783. J. tenuis, v. Williamsii.

8. J. platyphýllus (Wieg.) Fern. (flat-leaved; an undistinctive name, the plant originally separated as a var. of the terete-leaved no. 11). — In dense tussocks; the *inner basal sheaths purple-tinged, with firm unfriable margins, the truncate or round-tipped firm-membranaceous drab or fuscous auricles about as broad as long;* leaf-blades at first flat, in drying becoming involute; culms firm, erectish, 0.3–1 m. high, leafy only toward base; cyme 0.3–2.3 dm. high, overtopped by the involucre, the ascending crowded to open branches floriferous along the upper side or toward tips; *flowers* much as in no. 7, pale brown, *with firm bracteoles and prophylla; partitions extending half-way to axis of capsule.* (*J. dichotomus*, var. Wieg.) — Dry or moist acid soils, e. Me. to w. N.Y., s. to Fla. June–Sept.

9. J. intèrior Wieg. (inland). — In habit like nos. 7 and 8; leaves flat to involute, the *pale brown to drab sheaths membranaceous-margined; the submembranaceous to firm pale brown to drab auricles short, gradually rounded, following the curve of the sheath-summit;* cyme overtopped by longer bracteal leaf, much as in typical no. 7 or more compact, 0.2–1 (–1.5) dm. high; *perianth mostly stramineous, strongly erect, 3–4 mm. long, about equaling capsule.* — Prairies, sand-plains, open ground, roadsides, etc., O. and Mich. to Wash., s. to Ind., Ill., Mo., Kans. and Tex. May–July. Fig. 784.

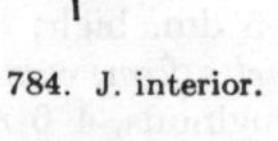

784. J. interior.

10. J. Dúdleyi Wieg. (for its discoverer, William Russell Dudley, 1849–1911). — In habit resembling nos. 7 and 9; *basal sheaths firm, stramineous to drab* (rarely purple-tinged); *auricles coriaceous, yellowish or amber, short and rounded;* the flat to involute blades mostly less than half as long as the culm; cyme 1–7 cm. long, overtopped by 1 or more bracteal leaves, the flowers chiefly clustered at tips of erect branches; *perianth* greenish-yellow (rarely purple-tinged), 4–6 *mm. long, spreading-ascending, longer than capsule;* partitions extending half-way to axis of capsule; filaments slightly exceeding anthers; seeds 0.35–0.45 mm. long. — Damp to dryish calcareous or sweet soil, Nfld. to B.C., s. to N.S., N.E., N.J., Pa., D.C., mts. to Va., Tenn., Mo., Okla., N.M. and Ariz. June–Sept. Fig. 785.

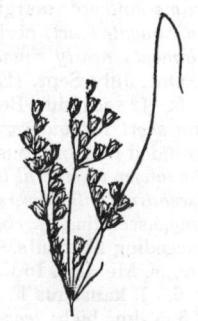

785. J. Dudleyi.

11. J. dichótomus Ell. (2-parted). — In hard olive or dark green tussocks with the *inner firm sheaths purple-tinged, the outer brown; leaf-blades filiform,* merely slenderly channeled on upper side, stiffish, *two-thirds as long as culms, the short rounded auricles firm and cartilaginous;* culms subrigid, 0.4–1 m. high; cyme loose to dense, 2–8 cm. long, often with 1-sided forked branches, *mostly longer than single prolonged involucral leaf; perianth greenish-brown,* 3.5–5 *mm. long;* the *lustrous* lanceolate sharp-pointed *sepals spreading in fruit,* about *as long as the highly lustrous ovoid* short-beaked *brown or castaneous* obscurely 1-*locular capsule;* anthers nearly equaling filaments. — Damp sands and peats on or near Coastal Plain, Fla. to Tex. and Mex., n. to se. Mass. June–Oct. (Trop. Am.) FIG. 786.

786. J. dichotomus.

12. J. oronénsis Fern. (for type-locality, Orono, Me.) — In dense tussocks with firm pale brown to pink sheaths; leaves subterete or promptly involute, the basal auricles membranous and rounded; culms stiffly erect, 2.5–6 dm. high; *cyme* elongate, 2.5–9 dm long, *overtopped by the erect bracts,* subdichotomous, *pale straw-color,* the flowers mostly secund along the suberect branches; *perianth* 4–5 *mm. long, pale stramineous, the* subequal *lance-subulate sepals and petals narrowly scarious-margined, tightly appressed to the shorter* trigonous truncate-emarginate *capsule; seeds* 1 *mm. long, sigmoid-fusiform, with white tails one-fourth length of brown body.* — Damp thickets and swamps, local, Me.; Alta. July, Aug. FIG. 787.

787. J. oronensis.

13. J. Vàseyi Engelm. (for its discoverer, GEORGE VASEY, 1822–1893). — Densely tufted, the stiffly erect culms 1.2–8 dm. high; leaves nearly terete, very slightly channeled; *cyme green or greenish,* 1–4 cm. high, *subtended by a stiff erect bract* 1–8 *cm. long; sepals lanceolate,* acute, *appressed; petals bluntish, with broad scarious margins;* anthers equaling filaments; *capsule greenish,* 4.5–6 *mm. long, exserted; seeds* slender, 1–1.4 *mm. long, with white tails half as long as pale brown body.* — Damp thickets, shores, etc., Côte Nord, Que., to n. Alta., s. to N.B., Me., n. N.Y., Mich., Ill., Ia., Colo. and Utah. July, Aug. FIG. 788.

788. J. Vaseyi.

14. J. Greènei Oakes & Tuckerm. (for its discoverer, BENJAMIN DANIEL GREENE, 1793–1862). — Differing from no. 13 in its deeply channeled leaves; *cyme* 1–6 cm. long, *brown or fulvous, becoming castaneous, subtended by* 1 *or more flexuous bracts up to* 0.5–2.5 *dm. long; perianth* 4–5 *mm. long; capsule olive-brown, reddish or castaneous; seeds ovoid, reddish,* 0.5–0.6 *mm. long, with minute white tails.* — Dry to moist siliceous or argillaceous soil, sand-dunes, mountain-slopes, etc., sw. N.S. to St. Maurice Co., Que., w. to Algoma Distr., Ont., s. to s. N.E., L.I., N.J., n. O., n. Ind., n. Ill. and Minn. June–Sept. FIG. 789.

15. J. coriàceus Mackenz. (leathery; from the firm perianth). — In tough tussocks 0.3–1 m. high; *at least the inner sheaths bearing terete* or at most compressed *long culm-like leaves; cyme falsely lateral, overtopped by a terete bract looking like a continuation of and a third as long as to about equaling the culm,* the flowers crowded or borne at the tips of more or less elongate branches or branchlets; *perianth with firm spreading-ascending lustrous sepals* and petals; *capsule subglobose, shining, firm-walled, about equaled by the spreading sepals; seeds irregularly obconic, beaked, strongly stipitate,* 0.6–0.8 *mm. long, the castaneous body coarsely ribbed and cross-ribbed.* (*J. setaceus* of ed. 7, not Rostk.) — Swamps, marshes, low pinelands and wet woods, Fla. to e. Tex., n. to Cape May, N.J., Ky., Ark. and Okla. June–Sept. FIG. 790.

789. J. Greenei.

790. J. coriaceus.

§ GENUÌNI (Buchenau) Vierh.

16. J. filifórmis L. (thread-like). — *Rhizome slender, cord-like,* creeping and forking; *culms filiform, subdistant or scattered along the rhizome* or in small tufts, 1–6 dm. high; *cyme few-flowered, with mostly simple branches* 0.2–2 *cm. long, or nearly capitate; flowers* about 3 *mm. long, greenish; sepals* lanceolate (petals a little shorter and less pointed), mostly *longer than the broadly ovoid round-tipped capsule; stamens* 6, *anthers*

791. J. filiformis.

much shorter than filaments; seeds about 0.5 mm. long, indistinctly reticulated, short-pointed at each end. — Wet shores, bogs, etc., s. Greenl. and Lab. to Alaska, s. to Nfld., N.S., Me., ne. and w. Mass., N.Y., uplands of Pa. and W.Va., Mich., Minn., Wyo., Utah and Oreg. June–Aug. (Eurasia) Fig. 791.

17. J. inflèxus L. (incurved). — *Glaucous;* the culms densely tufted from a short stout rhizome, 0.3–1.5 m. high, 1–3 mm. thick near base, the brown basal sheaths paler at summit; *cyme many-flowered, with strongly ascending branches* up to 8 cm. long; *perianth brownish, about* 3 *mm. long, about equaling capsule; sepals and petals firm, lance-attenuate; stamens* 6, anthers slightly longer than filaments; *capsule ovoid, conically tapering to beak, castaneous,* lustrous, *firm-walled; seeds about* 0.5 *mm. long,* reticulate. — Meadows, swales and damp roadsides, local, N.Y. to Mich. July, Aug. (Natzd. from Eu.) Fig. 792.

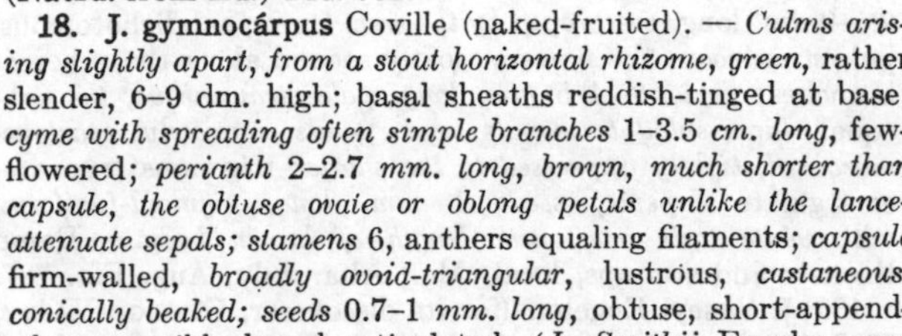

18. J. gymnocárpus Coville (naked-fruited). — *Culms arising slightly apart, from a stout horizontal rhizome, green,* rather slender, 6–9 dm. high; basal sheaths reddish-tinged at base; *cyme with spreading often simple branches* 1–3.5 *cm. long,* few-flowered; *perianth* 2–2.7 *mm. long, brown, much shorter than capsule, the obtuse ovate or oblong petals unlike the lance-attenuate sepals; stamens* 6, anthers equaling filaments; *capsule* firm-walled, *broadly ovoid-triangular,* lustrous, *castaneous, conically beaked; seeds* 0.7–1 *mm. long,* obtuse, short-appendaged at each end, many-ribbed and reticulated. (*J. Smithii* Engelm., not Kunth) — Mossy swamps, very local, mts. of e. Pa. to mts. of N.C. and Tenn.; also Walton Co., Fla. Aug. Fig. 793.

793. J. gymnocarpus.

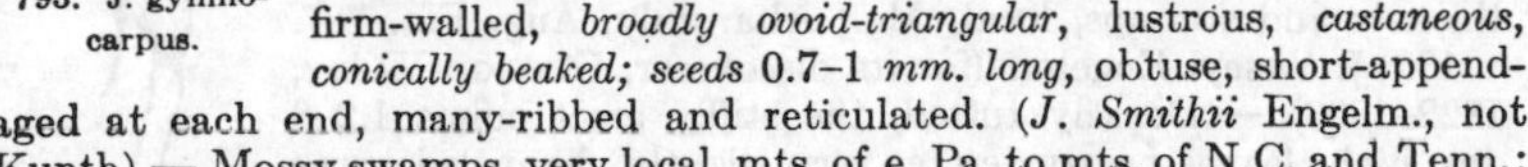

792. J. inflexus.

19. J. Gríscomi Fern. (for one of its discoverers, Ludlow Griscom, 1890–). — Resembling no. 20; *culms* soft, *bright green,* about 1 m. high, 2–3 mm. thick at base, distinctly furrowed; *cyme very lax and open, with loosely spreading or recurving filiform forking branches up to* 4–7 *cm. long; flowers* 3–3.6 *mm. long,* greenish, *many of them on pedicels up to* 1 *cm. long;* lance-attenuate sepals and petals about equaling the *trigonous-oblong olivaceous subtruncate but mucronate (mucro firm,* 0.3–0.5 *mm. long) capsule;* seeds 0.6 mm. long, golden-brown, unequally ellipsoid, with short purplish apiculations, obscurely transverse-reticulate. — Springy woodland, local, Princess Anne and James City Cos., Va. June.

20. J. effùsus L. (loosely spreading), Soft R. — *In arched-ascending dense tussocks from a stout mostly hidden rhizome; culms* soft to firm, 0.4–2 *m. high,* very finely many-striate or with 10 or more coarser ridges and furrows, *green, not glaucous;* basal sheaths chartaceous, strictly erect, more or less clasping bases of culms, mucronate; *cyme many-flowered, with forking branches, open and obviously anthelate to densely compacted;* involucral bract erect or ascending; *flowers* 1.5–4.3 *mm. long, stramineous or greenish to brown;* sepals lance-attenuate, narrowly margined; *stamens* 3 (rarely –6); *style minute; capsule thin-walled,* pale to dark brown, *depressed or emarginate at summit,* or in one var. rounded, *beakless or barely murconulate; seeds about* 0.5 *mm. long, short-apiculate, transversely reticulate.* — A highly variable species of all temp. reg., with well marked but morphologically confluent geographic varieties. We have the following:

a. Perianth usually somewhat spreading from base; the sepals soft and pliable (inclined to wrinkle in drying), linear-lanceolate, 1.5–3 mm. long. . . *b.*

b. Culms not obviously sulcate, very finely many-striate; base of involucral leaf not prominently dilated; capsule emarginate to obtuse and beakless. . . *c.*

c. Culms relatively stout, 1.5–5 mm. thick at summit of upper sheath, 0.4–1.5 m. high; uppermost sheath 0.6–1.7 dm. long, pale (rarely dark) brown to stramineous above; capsule emarginate at summit.

Culms relatively soft, easily compressed, pale green; inflorescence irregular, somewhat open, with ascending and spreading evident branches up to 3–5.5 cm. long; perianth white-stramineous; capsule pale brown; plant of Nfld. and Maritime Provinces. . . . . . *J. effusus* (typical).

Culms firm, resistant to pressure, deep green; inflorescence dense, subglobose or lobulate, 1–4.5 cm. in diameter; perianth stramineous or greenish, often brown-tinged; capsule deeper brown; wide-ranging northw. . . . . Var. *compactus.*

c. Culms slender, 1–2 (rarely –3) mm. thick at summit of upper sheath, 0.3–1 m. high, almost filiform above; uppermost sheath slender, 3–12 cm. long, dark brown or green above; inflorescence lax or compact, 1.5–4.5

cm. in diameter; perianth stramineous or greenish, often suffused with reddish-brown; capsule olive-brown, obtuse, scarcely or rarely emarginate. . . . . Var. *decipiens.*

*b.*Culms coarsely 12–15-sulcate near inflorescence, soft and easily compressed, pale green, 1.5–3 mm. thick at summit of upper sheath, 5–10 dm. high; base of involucral leaf dilated; inflorescence tightly congested, subglobose or lobulate, dark brown, 1–2.5 cm. in diameter; capsule obtuse or retuse, blunt or apiculate. . . . . Var. *conglomeratus.*

*a.*Perianth more ascending to appressed; the sepals firm to subrigid (not wrinkling), 2.2–4.3 mm. long, lance-attenuate. . . *d.*

*d.*Culms costate-angulate or sulcate, firm, deep green, 1–4 mm. thick at summit of upper sheath, 0.3–1.2 m. high; basal sheaths purplish at least below, the uppermost 0.4–2 dm. long; inflorescence loose or slightly compact, 1.5–8 cm. in diameter; sepals ascending, not appressed.

Perianth 3–4.3 mm. long; sepals definitely exceeding petals and capsule. Var. *Pylaei.*

Perianth 2.2–3 mm. long; sepals, petals and capsule subequal. . . . . Var. *costulatus.*

*d.*Culms finely many-striate (when dry), usually easily compressed, 2–5 mm. thick at summit of upper sheath, 0.2–2 m. high; basal sheaths drab or brownish, or stramineous above, the uppermost 0.9–3 dm. long; inflorescence open, often diffuse, 4–14 cm. in diameter; sepals and subequal petals with appressed tips, about equaling to shorter than the pale brown to greenish capsule. . . . . Var. *solutus.*

794. J. effusus (typical).

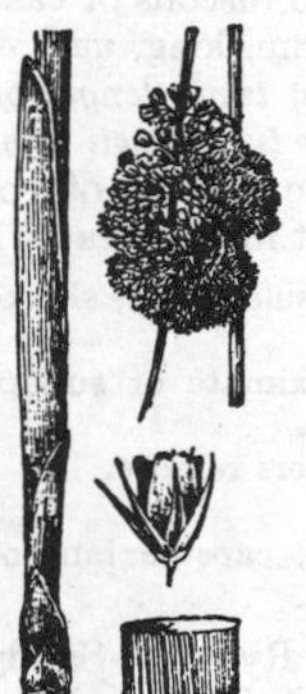

795. J. effusus, v. compactus.

796. J. effusus, v. decipiens.

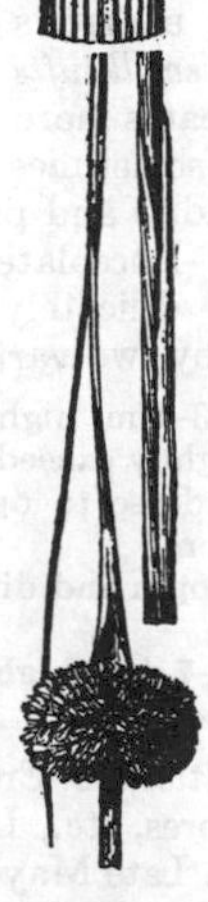

797. J. effusus, v. conglomeratus.

J. effùsus (typical). — Peaty swamps and thickets, margins of pools, etc., s. Nfld. and P.E.I. July–Sept. (Eurasia, etc.) Fig. 794.

Var. **compáctus** Lej. & Court. (compact). — Swales, wet thickets, bogs, etc., Nfld. to Ont., s. to N.S., N.E., N.Y., W.Va., Mich. and Wisc. July–Sept. (Eu.) Fig. 795.

Var. **decipiens** Buchenau (deceitful). — Swales and damp open ground, Nfld. to Ont., s. to N.E., N.Y. and Mich. Late June–early Sept. (E. Asia) Fig. 796.

Var. **conglomeràtus** (L.) Engelm. (massed together), *J. conglomeratus* L. in part; *J. Leersii* Marsson — Boggy, peaty or swampy places, Nfld., St.P. et Miq. and N.S.; Franklin, Ct. (*Woodward*). July–Sept. (Eu.) Fig. 797. — In Europe generally maintained as a species; with us the differential characters not so clear.

Var. **Pylaèi** (Laharpe) Fern. & Wieg. (for its discoverer, Auguste Jean Marie Bachelot de la Pylaie, 1786–1856). — Swampy thickets, peats and wet clearings, Nfld. to Algoma Distr., Ont., s. to N.S., N.E., Pa., w. Va., W.Va., n. Ind., Wisc. and Minn. Late June–Sept. Fig. 798.

Var. **costulàtus** Fern. (with small ribs). — Peaty or damp sandy soil, C.B. to s. Que., s., especially toward the coast or on the Coastal Plain, to s. N.E., L.I. and S.C. June–Sept. Fig. 799.

Var. **solùtus** Fern. & Wieg. (loosened). — Swampy or marshy ground, Fla. to e. Tex., n. to Nfld., Que. and Ont. June–Sept. Fig. 800. — Usually the commonest var.

798. J. effusus, v. Pylaei.

799. J. effusus, v. costulatus.

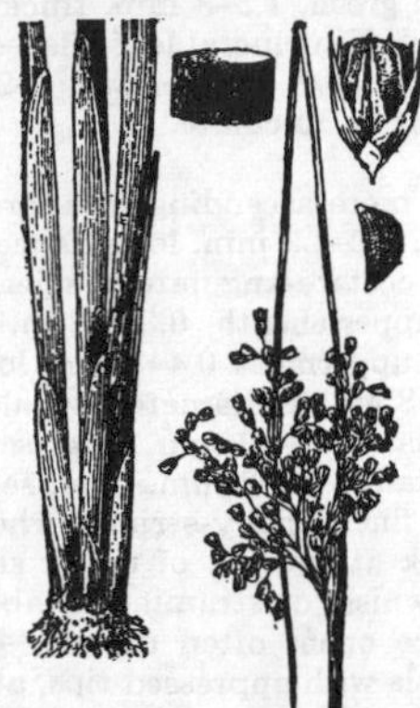

800. J. effusus, v. solutus.

21. J. bálticus Willd. (of the Baltic). — *Rhizome firm, extensively creeping and forking; culms in small tufts or scattered, slender, terete,* erect or arching, *with smooth unfurrowed surface;* basal sheaths more or less lustrous, stramineous to fuscous or castaneous; cyme usually loosely forking (sometimes subglomerate); *sepals* 3.5–6 mm. long, *with deep brown to castaneous sides,* green midrib and pale margins; *anthers two to four times length of filaments; capsule* trigonous-ovoid to -lanceolate, *pyramidal-conic to mucronate tip, brown to blackish; seeds* 0.8–1 *mm. long,* bluntish, delicately reticulate. — Highly variable species of cooler temp. reg.; represented with us by two varieties, both with anthers about four times as long as filaments:

Culms 0.3–1 m. high, 1–2.5 mm. thick at base; capsule ovoid, shorter than to but slightly exceeding perianth.
  Cyme dense to open, 1–9 cm. long, with approximate or subapproximate flowers. . . . . Var. *littoralis.*
  Cyme open and diffuse, 0.4–1.5 dm. long, the flowers remote. . . . . . Forma *dissitiflorus.*
Culms 3–7 dm. high, about 1 mm. thick at base; capsule lanceolate, long-exserted. . . . . Var. *stenocarpus.*

Var. **littoràlis** Engelm. (of shores), *J. litorum* Rydb. — Sandy brackish to fresh shores, etc., Lab. to B.C., s. to Nfld., N.S., N.E., L.I., Pa., O., Ind., Ill., Mo., etc. Late May–Sept. Fig. 801. — Forma **dissitiflòrus** Engelm. (with scattered flowers), more local.

Var. **stenocárpus** Buchenau & Fern. (slender-fruited). — Alluvial shores, meadows and swales, Gaspé Pen. and Rimouski Co., Que.; James Bay reg., Ung. July–Sept.

§ Thalássii (Buchenau) Vierh.

22. J. Roemeriànus Scheele (for its discoverer, Karl Ferdinand von Roemer, 1818–1891). — *Culms* stout and *rigid* (0.5–1.5 m. high), *the apex, as well as the similar hard terete leaves, pungent; cyme* compound, open and spreading, *pseudolateral, brown,* 3–several greenish or light brown flowers (3–3.5 mm. long) in a cluster; sepals lanceolate, sharp-pointed, longer than the obtusish petals; anthers much longer than the broad filaments; styles shorter than the ovary; *seeds* 0.7 *mm. long,* very delicately ribbed. — Brackish marshes, Fla. to Tex., n. to Md. May–Oct.

J. marítimus Lam. (maritime), with ellipsoid capsule exceeding the calyx, the seeds with long caudate tips, a widely dispersed species with interrupted range, once occurring on marshes of Staten I., N.Y., has been exterminated there.

801. J. balticus, v. littoralis.

§ Stýgii Fries

23. J. stýgius L. (from its boggy habitat; Linnaeus, in his *Lachesis Lapponica,* writing of an area which "consists chiefly of marshes, here called *stygx.*

A divine could never describe a place of future punishment more horrible than this country, nor could the Styx of the poets exceed it"). — *Culms* 1–3 dm. high, filiform, *in* small or large *tufts*, 1–3-leaved below, naked above, *the leaves filiform; heads* 1–4, *of* 1–4 *flowers about the length of the sheathing scarious awl-pointed bract; flowers* pale, *white and reddish*, 3–4 mm. long; sepals lanceolate, acute; petals obtusish, three-fourths the length of the trigonous-ovoid acute or acuminate pale capsule (5–6 mm. long), as long as the slender stamens; filaments many times longer than the oblong anthers; recurved stigmas shorter than the style; seeds oblong, with a very loose coat prolonged at both ends, 2–2.5 mm. long. — Eurasia. With us as

Var. **americànus** Buchenau (American). — Often taller, 1–4.5 dm. high; heads 1 or 2; *flowers* larger, 4.5–5.5 *mm. long; the distinctly mucronate-tipped capsule longer*, 6–9 *mm. long; seeds* 3–4 *mm. long.* — Wet moss, bogs and bog-pools, s. Lab. to Sask., s. to Nfld., C.B., M.I., N.B., Me., N.Y., Mich. and Minn. July, Aug. (Centr. Eu.) Fig. 802.

802. J. stygius, v. americanus.

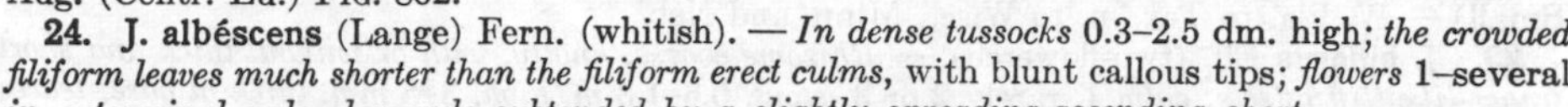

24. J. albéscens (Lange) Fern. (whitish). — *In dense tussocks* 0.3–2.5 dm. high; *the crowded filiform leaves much shorter than the filiform erect culms*, with blunt callous tips; *flowers* 1–several *in a terminal pale glomerule subtended by a slightly spreading-ascending short spathiform bract; perianth whitish or pink-tinged*, 3–5 *mm. long;* stamens 6, about equaling perianth, the ovate anthers much shorter than filaments; style nearly obsolete, the stigmas elongate; *capsule about equaling perianth*, trigonous-cylindric, *rounded at summit; seeds* spindle-shaped, 1.5 *mm. long, the broad-based white tails shorter than the darker body.* — Greenl. to Alaska, s. to damp peaty or boggy calcareous areas of n. and w. Nfld., Côte Nord and Mingan Ids., Que., Ung., Keewatin and mts. to Colo. and Utah. Late June–Sept. (N. Eurasia) Fig. 803.

25. J. castàneus Sm. (chestnut-colored). — *Strongly stoloniferous; the solitary stiff culms erect from tips of stolons*, 1–4 dm. high, leafy; *leaves* erect, *canaliculate;* heads solitary and terminal or 1 or 2 others on erect rays, 1–several-flowered; *lower involucral leaf overtopping inflorescence; flowers deep brown, in maturity* 6–10 *mm. long;* sepals linear-lanceolate, acute, longer than or about equaling the narrow obtuse petals; *capsule* castaneous to purple-black, *conspicuously exserted; seeds very slenderly fusiform*, 2.5–3.7 *mm. long, the tails twice to thrice the length of the body.* — Arct. reg., s. to se. Lab., calcareous alpine meadows and brooksides, Shickshock Mts., Gaspé Pen., Que., Ung., n. Man. and mts. to N.M. July, Aug. (Eurasia) Fig. 804.

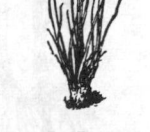

804. J. castaneus.

803. J. albescens.

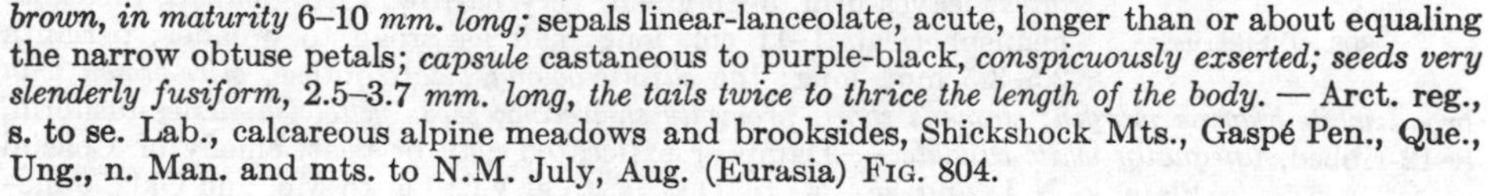

§ Graminifòlii Buchenau

26. J. rèpens Michx. (creeping and rooting at the nodes). — *Stems* ascending, 0.5–2 dm. high, *from fibrous annual roots, at length creeping or floating and greatly elongate; leaves* short, linear, *flaccid, those of the culm nearly opposite and fascicled;* heads few in a loose leafy cyme, 3–12-flowered; *flowers green*, 0.5–1 *cm. long; sepals and petals rigid, lance-subulate, sepals as long as the linear-triangular obtuse capsule*, the petals much longer; stamens as long as the sepals; filaments much longer than the oblong anthers; seeds obovoid, slightly pointed, very delicately ribbed and cross-lined. — Miry shores, pools, ditches, etc., Fla. to Tex., n. to Del., Tenn., Ark. and Okla. June–Oct. Fig. 805. — Often forming extensive mats; submersed states simulate *Heteranthera dubia.*

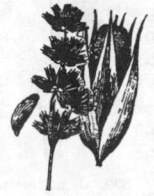

805. J. repens.

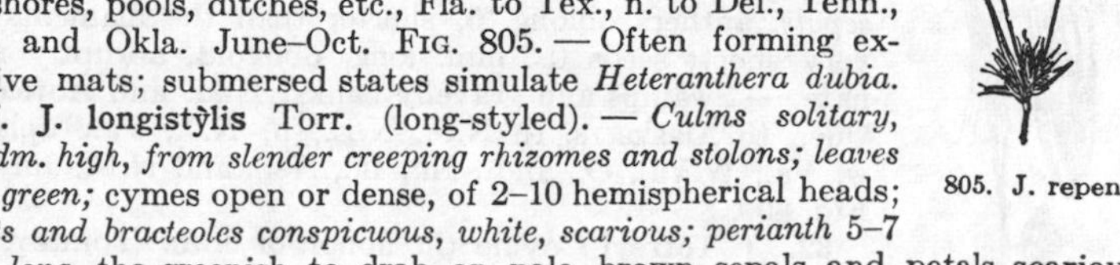

27. J. longistỳlis Torr. (long-styled). — *Culms solitary*, 2–7 *dm. high, from slender creeping rhizomes and stolons; leaves pale green;* cymes open or dense, of 2–10 hemispherical heads; *bracts and bracteoles conspicuous, white, scarious; perianth* 5–7 *mm. long*, the greenish to drab or pale brown sepals and petals scarious-margined; stamens yellow; *style elongate; capsule abruptly long-beaked;* seeds ellipsoid, 0.5 mm. long. — Damp gravelly or sandy shores and prairies, B.C. to Man., s. to Calif., Ariz., N.M. and Neb.; local, n. Mich.; Detroit R., Ont.; and sw. Nfld. June–Aug. Fig. 806.

806. J. longistylis.

807. J. marginatus.

28. J. marginàtus Rostk. (margined; from the hyaline margins of the petals). — *Cespitose, in* small or dense *tussocks from a short thick and often knotty rhizome; culms slender,* 1.5–6(–8) *dm. high,* 1–2(–3) *mm. thick at base; leaves* green, flat, *soft, the basal ones* 0.4–2 *dm. long* and 1–4(–5) *mm. broad,* the uppermost blade 0.2–1.5 dm. below the inflorescence; *cymes* 1–10 *cm. long,* open or somewhat compact, *with* 2–30 (rarely –50) *glomerules* of 2–12 flowers each; *perianth* 3–3.5 *mm. long, the reddish-brown sepals sharply acute, the ovate to oblong and blunt* to mucronate *petals with green center clearly separated from the hyaline margin by a brown band; stamens* 3, *shorter than sepals,* with reddish anthers, *quickly shriveling; capsule rounded, thin-walled, beakless; seeds brown,* 0.5 mm. long, plump, many-ribbed, apiculate at ends. — Moist siliceous, argillaceous or peaty soil, Fla. to e. Tex., n. to w. N.S., centr. Me., centr. N.H., Vt., N.Y., s. Ont., O., Mich., Mo. and Kans. June–Sept. Fig. 807. — Passing insensibly to

Var. **setòsus** Coville (bristly). — Petals lance-attenuate, subulate-tipped. (*J. setosus* (Coville) Small) — W. Fla. to Tex., n. to Wisc., Minn. and Neb.

29. J. biflòrus Ell. (two-flowered). — *Rhizome coarse, knotty, with or without thick and short finger-like stiff offsets; culm* 1, rarely 2 or 3, *stout,* 0.6–1.3 *m. high,* 3–5 *mm. thick at base; leaves hard and firm, the basal ones* 4–7 *mm. wide* and 1.5–3 dm. long, the uppermost blade 1–2.5 dm. below inflorescence; *cyme* 0.5–2 *dm. long,* the ascending *branches and branchlets bearing* 30–200 *remote* 2–several-flowered distant or scattered *glomerules,* or the compact cyme 3–10 cm. long and with crowded or approximate many-flowered glomerules in forma **ádinus** Fern. & Grisc. (close); perianth much as in no. 28, often smaller; *stamens equaling sepals, becoming firm and persistent, the dark purple anthers often showing in fruit; seeds reddish-castaneous,* 10–16-ribbed, ellipsoid-fusiform, dark at bases of short appendages. (*J. aristulatus* of ed. 7, not Michx.) — Peaty or wet sandy bogs, shores, low fields, etc., Fla. to Tex., n. to Cape Cod and Nantucket I., Mass., se. N.Y., N.J., e. Pa., n. O., s. Mich., Ill., Mo. and Okla. Late May–Sept. Fig. 808.

808. J. biflorus.

30. J. Lóngii Fern. (for one of its discoverers, Bayard Long, 1885–). — In habit similar to no. 29; the *usually solitary* (rarely –3) *culm* more slender, 0.4–1 m. high, *arising from the tips of or from among flexuous cord-like scaly stolons up to* 2 *dm. long and* 1.5–3 *mm. thick;* leaves firm, deep green, very narrow; *cyme compact,* oblate to hemispherical, 1–11 cm. long, half as broad to broader; perianth 2.5–3.5 mm. long; the *elliptic-oblong petals* obtuse, *olive-brown with broad white hyaline margin; stamens short, promptly shriveling; seeds yellowish,* lance-fusiform, 8–12-ribbed, *unequally white-caudate.* — Damp or exsiccated clay or peat, chiefly of Coastal Plain, s. N.J. and se. Pa. to D.C., s. to e. Va.; La. to Mo. and Okla. June–Aug.

§ Septàti (Buchenau) Vierh.*

31. J. nodòsus L. (knotty; from the septate leaves). — *Stem* erect, 1.5–6 dm. high, *slender, from a creeping thread-like and tuber-bearing rhizome,* mostly with 2 or 3 slender leaves; *heads spherical, reddish-brown,* few or several, rarely single, 8–20-flowered, 7–11 mm. in diameter, *overtopped by the involucral leaf; flowers* 3–4 *mm. long;* sepals nearly as long as the slender triangular taper-pointed 1-locular capsule; *petals equaling or exceeding sepals;* anthers oblong, 6, shorter than the filaments; style very short; seeds 0.5 mm. long, obovoid, abruptly mucronate. — Swamps and gravelly banks, Nfld. and Anticosti I., Que., to Alaska, s. to N.S., N.E., n. N.J., Pa., upland to w. Va., W.Va., O., Ind., Ill., Ia., Neb. and N.M. July, Aug. Fig. 809.

809. J. nodosus.

32. J. Tórreyi Coville (in honor of John Torrey, 1796–1873). — Similar to the last; stem stouter, 0.4–1 m. high, with thick leaves; *heads* few and large, 1–1.5 *cm. in diameter,* 30–80-flowered, *greenish or dull brown; flowers* 4–5 *mm. long;* petals somewhat shorter than sepals; anthers linear, shorter

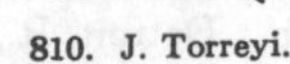
810. J. Torreyi.

* In members of this section the inflorescence is often changed to masses of horn-like galls.

than the filaments.—Low, often sandy soil, N.Y. to Sask. and Wash., s. to Md., D.C., upland of w. Va., Ala., Miss., Mo., Tex., n. Mex. and Calif.; locally adv. along railroads, roadsides, etc., e. to N.E. and N.J. July–Oct. FIG. 810.

33. J. polycéphalus Michx. (many-headed). — Plant *robust*, about 1 m. high, tufted, *with fibrous roots; leaves laterally compressed, flattened under pressure, gladiate,* 3–8 *mm. wide, with incomplete septa,* or narrower and with septa complete; *cyme* 1.2–3 *dm. long, with divergently ascending long primary branches, its longer naked internodes* 6–16 *cm. long; heads globose, in fruit* 8–12 *mm. in diameter;* sepals and petals linear-setaceous; stamens shorter than petals, the anthers included; *subulate capsule exserted, after dehiscence the valves remaining united above and forming a solid beak.* — Sandy pond-margins, ditches, savannas, etc., Fla. to La., n. to N.C., very doubtfully to se. Md., and to Okla. Late June, July. FIG. 811.

811. J. polycephalus.

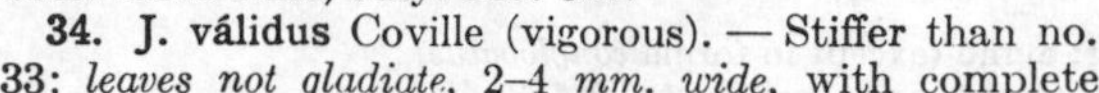

34. J. válidus Coville (vigorous). — Stiffer than no. 33; *leaves not gladiate,* 2–4 *mm. wide,* with complete septa; *fruiting heads* 1.2–1.5 *cm. in diameter; capsule, at dehiscence, splitting into its flat valves or with the valves barely united at summit, not forming a definite beak.* (*J. polycephalus,* in part, of ed. 7) — Sandy swales and damp prairies, n. Fla. to Tex., n. to Ga., sw. Mo. and Okla. July.

35. J. scirpoìdes Lam. (resembling *Scirpus*). — *Stem* erect, 2.5–9 dm. high, rather slender, *from a thick horizontal whitish rhizome,* bearing about 2 terete leaves with wide and open sheaths, and a *cyme,* 5–15 cm. long, *of* few or many *densely flowered pale green to dull brown spherical heads, much longer than the involucral leaf,* its branches erect and often elongated; heads 6–13 mm. in diameter, 15–40-flowered; flowers 3–4 mm. long; *sepals and petals rigid,* subulate and (especially the sepals) *bristly-pointed, at length pungent,* as long as the stamens and nearly equaling the oblong-triangular taper-pointed 1-locular capsule; *anthers* very small, *exserted;* style elongated or very short; seeds ovoid, abruptly pointed at each end, 0.5 mm. long. — Damp sandy soil, shallow pools, wet pinelands, etc., Fla. to Tex., n. to L.I. and Staten I., N.Y., N.J., e. Pa., Ind., sw. Mich., Ill., Mo. and Okla. Mid-June–Oct. FIG. 812.

Var. **meridionàlis** Buchenau (southern). — Heads lobulate, up to 2 cm. in diameter. — Fla. to e. Va. and locally to L.I.

812. J. scirpoides.

36. J. megacéphalus M. A. Curtis (large-headed). — Stouter than no. 35, 0.45–1.1 m. high, *from a less defined rhizome; uppermost leaf with blade greatly reduced;* cyme 1–9 cm. long; the *deep reddish-brown or darker* spherical *heads* 8–12 *mm. thick; sepals and petals soft and flexible,* about equaling capsule; *anthers included,* longer than in no. 35; seeds more slender. — Brackish to fresh marshes, swales, dune-hollows, etc., Fla. to La., n. to e. Va. and doubtfully to se. Md. Late June–Aug.

37. J. brachycárpus Engelm. (short-fruited). — In habit suggesting no. 35; stem erect, 4–9 dm. high, from a thick white horizontal rhizome, bearing about 2 leaves and 2–10 densely flowered spherical heads, 7–11 mm. in diameter, in a slightly spreading, crowded to open cyme much exceeding the involucral leaf; *flowers pale green to dull brown,* 4 mm. long; *capsule ovoid-conic, only one-half to two-thirds as long as perianth;* anthers much shorter than filaments; style very short; *seeds* 0.3 *mm. long,* abruptly apiculate. — Damp siliceous, argillaceous or peaty soil, S.C. to Tex., n. to e. Va. and locally to e. Mass., ne. O., s. Ont., s. Mich., n. Ind., Ill., Mo. and Okla. June–Sept. FIG. 813.

813. J. brachycarpus.

38. J. caesariénsis Coville (of New Jersey). — *Culms* 1–few, tufted from a short crown-like (rarely elongate) rhizome, 4–9 dm. high, rigid and, *with firm* terete septate *leaves, scabrous; cyme* 0.3–1.7 dm. long, with strongly ascending branches, *greatly overtopping the short involucral leaf;* heads 2–10-flowered, turbinate to hemispherical, scattered or approximate; *flowers in maturity* 5–6 *mm. long,* subtended by a firm brown bracteole; *sepals firm, rigid, green,* strongly nerved, with pale margin, *lanceolate,*

814. J. caesariensis.

subulate-tipped, nearly equaling similar petals; *stamens* 6; *capsule narrowly trigonous-ovoid, tapering to acute tip,* lustrous-brown, incompletely 3-locular, the style (in anthesis) conspicuous; *seeds 2–3 mm. long,* subcylindric-fusiform, the long tails white to reddish. (*J. asper* Engelm., not Sauzé & Maillard) — Sphagnous swamps and spring-fed mossy woods, local, s. N.J. and Anne Arundel Co., Md., to se. Va. July–Oct. Fig. 814.

39. J. canadénsis J. Gay (Canadian). — *Culms tufted,* stiffly erect, slender to stoutish, 0.2–1.2 m. high, *smooth;* leaves terete, clearly septate, slender; cyme open and loosely anthelate to dense and congested, usually much overtopping the short erect involucral leaf, 0.1–3.3 dm. long; heads turbinate, hemispherical or subglobose, 2–many-flowered; *mature fruit and perianth* 3.8–5 *mm. long, the* 3 *green lanceolate subrigid subulate-tipped sepals and* the similar *petals about equaling to shorter than the brown to stramineous commonly round-tipped and short-beaked capsule; seeds fusiform,* 1.3–1.8 *mm. long,* with 30–40 weak longitudinal ribs and weak, if any, cross-bars, *their white tails two-thirds to quite as long as the body.* — A polymorphous species:

a. Capsule plump, gradually rounded at summit to the rather abrupt short beak. . . *b.*
b. Perianth 2.5–3.3 (rarely –3.5) mm. long; cyme (except in forma *conglobatus*) with spreading-ascending branches (rays) and branchlets, 0.4–3 dm. high. . . *c.*
c. Heads chiefly or wholly scattered in anthelate fashion along the branches of the open cyme; cyme 0.4–3 dm. high, with some elongate branches.
Heads densely 8–20-flowered, hemispherical to subglobose. . . . . *J. canadensis* (typical).
Heads turbinate to subhemispherical, 2–7-flowered. . . . . . . Forma *apertus.*
c. Heads all or many densely crowded into irregular glomerules or masses, globose, many-flowered, the glomerules sessile or on short rays up to 1–3 cm. long. . . . . Forma *conglobatus.*
b. Perianth 3.5–4.6 mm. long; cyme with stiffly erect branches and branchlets, 0.3–1.5 (–2) dm. high. . . . . Var. *sparsiflorus.*
a. Capsule slender, gradually attenuate to acute tip, long-exserted; cyme open, 1–3.3 dm. high; glomerules hemispherical to subglobose, many-flowered; perianth 3–4 mm. long. . . . . Var. *euroauster.*

J. canadénsis (typical). — Marshy places, usually in subacid or acid soils, s. Que., s. Ont. and Minn., s. to N.S., N.E., L.I., Ga., Tenn. and La. Fig. 815. — Forma **apértus** Fern. (open), more local, N.S. to Ga. — Forma **conglobàtus** Fern. (crowded together), especially near the coast, often at borders of brackish or saline marshes but locally inland from Me. to Minn. and s. to Ga. and n. Ill. Fig. 816. July–Oct.

Var. **sparsiflòrus** Fern. (sparsely flowered; referring ineptly to the few heads). — Nfld. to Laurentide Mts., Que., s. to N.S., e. Me. and, locally, se. Mass.

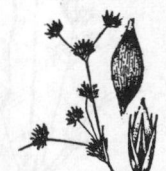

Var. **euroaúster** Fern. (southeastern). — Local, se. Va. to Ga.

815. J. canadensis (typical).

816. J. canadensis. f. conglobatus.

40. J. subcaudàtus (Engelm.) Coville & Blake (somewhat tailed). — Densely tufted, *loosely ascending to sprawling,* the very slender culms 2–9 dm. long; leaves slenderly terete; *cyme loose and open, with few spreading-ascending to horizontally divergent filiform branches,* 0.3–2.5 *dm. long, one-third to nearly as broad; the remote hemispherical to subglobose heads* 3–35, *many of them terminating slender branches* 1–7 *cm. long;* mature fruit and perianth 2–4 mm. long; *sepals and petals greenish, lance-linear, conspicuously ribbed or corrugated;* capsule brownish, slightly exserted, tapering at summit; *seed* 0.7–0.9 *mm. long,* fusiform, the plump body *with* 21–25 *strong longitudinal ribs with distinct cross-bars, the white tails one-third as long as body.* (*J. canadensis,* var. Engelm.) — Bogs, mossy woods, shaded spring-heads, etc., Ga. to se. Mo., n. to se. Mass., s. Ct., se. N.Y., n. N.J., Pa. and W.Va. July–Oct. Fig. 817.

817. J. subcaudatus.

Var. planisépalus Fern. (with flat sepals). — Perianths 2–3 mm. long; the flat scarcely ribbed sepals and petals greener on the back; the mature capsule strongly exserted. — Savannas, bogs and spruce-swamps, w. N.S.

41. J. brevicaudàtus (Engelm.) Fern. (short-tailed). — Culms slender, 1.5–7 dm. high, bearing few deep brown 3–7-*flowered heads in a* somewhat *erect contracted cyme* 2.5–15 cm. long and *three to six times longer than broad;* perianth 2.5 mm. long; sepals acute, the petals rarely obtusish, much shorter than the prismatic gradually pointed usually deep brown capsule; *seeds about* 1 *mm. long, their tails half as long as the body.* — Muddy or wet places, Lab. to n. Alta., s. to Nfld., N.S., N.E., L.I., Pa., upland to mts. of N.C., s. Ont., Mich., n. Ill., Minn. and Man. June–Sept. Fig. 818.

818. J. brevicaudatus.

42. J. brachycéphalus (Engelm.) Buchenau (short-headed). — Culms slender, 2.5–7 dm. high, bearing numerous small 3–5-flowered heads in a large (0.5–2.5 dm. long) spreading, open or diffuse cyme; perianth greenish or light brown; *sepals rather soft and scarious-margined, shorter than the blunt petals* and the brown abruptly short-beaked capsule; stamens usually 3, but in forma **hexándrus** Martin (with six stamens) 6; *seeds ellipsoid,* barely 1 mm. long, *their short white tails about one-tenth as long as the body.* — Calcareous shores, marshes, meadows, etc., C.B. to Ont., s. to n. and w. N.E., n. N.J., Pa., O., Ind. and Ill. July–Sept. Fig. 819.

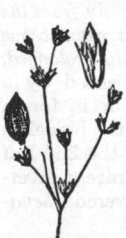

819. J. brachycephalus.

43. J. acuminàtus Michx. (tapering to tip; from the sepals). — Culms tufted, erect, slender, 3–10 dm. high, bearing about 2 leaves and an open cyme; *heads rather few* and large, 0.5–1 cm. broad, 5–many-flowered, greenish, at length straw-colored or darker; *perianth* 2.5–3.5 *mm. long; sepals and petals* lance-subulate, sharp-pointed, equal, *as long as the ovoid-prismatic short-pointed* 1-locular straw-colored or light brown *capsule;* anthers a little shorter than the filaments; style almost none; seeds small, 0.3–0.4 mm. long, acute at both ends, ribbed and reticulated. — Damp soil, Ga. to Tex. and Mex., n. to w. N.S., centr. Me., s. N.H., Mass., N.Y., s. Ont., Mich., Wisc. and Minn. Late May–Aug. Fig. 820. — The heads in autumn, especially when lopping into water, proliferating. Forma **obtusàtus** F. J. Herm. (obtuse) has the capsules rounded-subtruncate at apex.

820. J. acuminatus.

821. J. debilis.

44. J. débilis Gray (weak). — Similar to no. 43, weaker (the culms and leaves even elongating and flaccid when permanently submerged); cyme 0.2–2 dm. long, open, with 3–35 scattered 2–9-flowered heads (in a sterile, perhaps pathological, state the non-fruiting flowers greatly increased); greenish *perianth* 2–2.5 *mm. long; capsule more narrowly lanceolate, a third longer than the perianth.* — Wet places, shores, etc., Ga., nw. Fla., Ala., Tenn. and Mo., n. to R.I., Ct., N.Y. and Ky. June–Oct. Fig. 821.

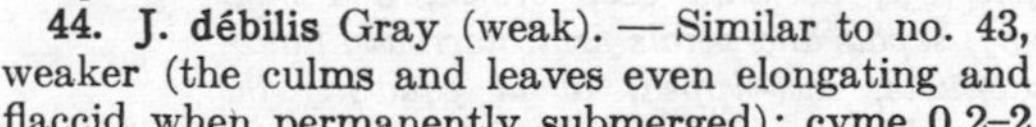

45. J. Ellióttii Chapm. (for its discoverer, Stephen Elliott, 1771–1830). — Fibrous *roots* mostly *ending in fleshy thick-fusiform tubers* 1–2 *cm. long; leaves firm, erect, terete, septate but the septa not prominent;* culms 1–several, erect or arching, slender, 3–9 dm. high; *cyme with erect or suberect slender branches* and branchlets, 0.3–1.8 dm. high, mostly 3–7 cm. broad, *all but the smallest with* 50–200 *small* turbinate to hemispherical 3–8-*flowered heads* (*mostly 4–7 mm. broad*); *perianth* 2–2.5 *mm. long,* the drab to pale brown sepals and petals lance-acuminate; stamens 3; *capsule ovoid, rounded to minute beak, castaneous to blackish, about equaling perianth; seeds reddish,* ellipsoid, minutely pointed at each end, *clearly reticulate.* — Wet swales, bogs, pond-margins, etc., Fla. to e. Tex., n. on Coastal Plain to Del. Mid-May–Aug.

46. J. nodàtus Coville (with nodes, *i.e.* septa). — Coarser than no. 45; *roots* (apparently) *not tuberiferous; leaves* (dry) *strongly nodose;* culms stoutish, 0.6–1.2 m. high; *cyme more diffuse,* 1–3 *dm. long and mostly* 0.8–2.5 *dm. broad, with loosely ascending to divergent main branches; heads mostly* 200–500; *capsule greenish or stramineous; seeds pale brown, scarcely reticulate.* (*J. robustus* (Engelm.) Coville, not S. Wats.) — Sloughs, shores, wet woods, etc., La. and Tex., n. to Ind., Ill., Mo. and Kans. June–Aug. Fig. 822.

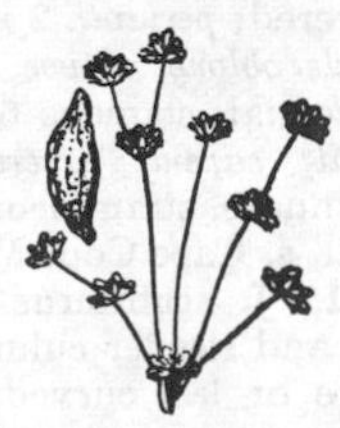
822. J. nodatus.

47. J. diffusíssimus Buckl. (most diffuse). — As slender as no. 44;

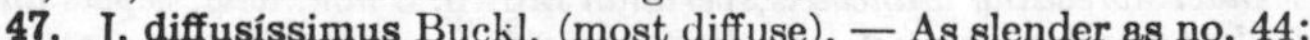

823. J. diffusissimus.

erect or ascending, 2–7.5 dm. high; *cyme very diffuse, loosely dichotomous, mostly* 1–3 *dm. long and nearly as broad;* turbinate to hemispherical few-flowered *heads in maturity appearing subechinate* from the divergent capsules; *mature flower* (*including fruit*) 3.5–6 *mm. long; the greenish to pale brown lance-linear prismatic tapering capsule nearly or quite twice as long as perianth;* greenish or pale brown sepals and petals linear-subulate. — Wet argillaceous, siliceous or boggy ground, Ga. to Tex., n. to Va., Ky., s. Ind., Mo. and Kans. June, July. FIG. 823.

48. J. bulbòsus L. (bulbous; from the hard and thickened but hidden bases). — Densely matted, *forming carpets of crowded filiform faintly septate flaccid leaves* 2–10 *cm. long,* from slightly knotty bases and fibrous roots; *culms filiform, weak, decumbent, prolonged and floating or ascending,* 0.5–4 *dm. or more long,* when permanently immersed much longer and with prolonged leaves; *cyme lax, simple or but slightly forking, with* 1–8 few-flowered turbinate to hemispherical *often viviparous or leafy-tufted heads;* perianth 3.5–4 mm. long; *petals oblong, obtuse,* about equaling oblong acute sepals; stamens 3 (very rarely 6); *capsule* barely exserted, cylindric-ovoid, *rounded to retuse at summit,* barely mucronulate; seeds 0.5 mm. long, obovoid, apiculate, greenish-brown or reddish, with rectangular reticulation. (*J. supinus* Moench) — Margins and siliceous or peaty shores of pools and streams, often floating, se. Nfld.; St. P. et Miq.; Sable I., N.S. Late July–Sept. (Eu.; n. Afr.; Canary Ids.; Azores) FIG. 824.

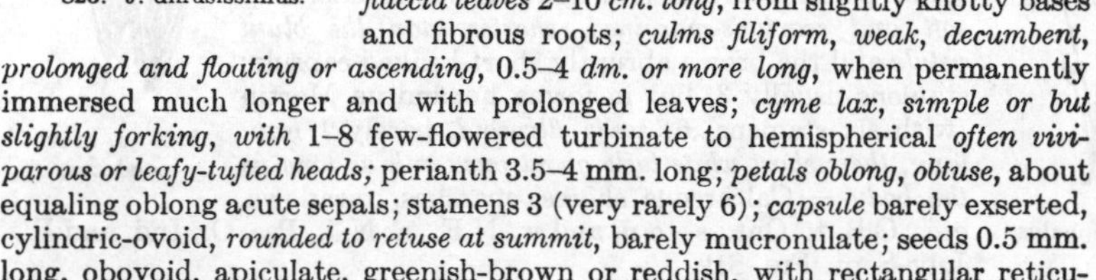

824. J. bulbosus.

49. J. militàris Bigel. (soldierly; the extensive companies of erect plants with rigid pointed leaves, rising high above cymes, suggesting soldiers on parade). — *Rhizome stout, subligneous, horizontal; culms stout, firm* (but hollow), *erect, rising singly from the rhizome,* 0.3–1.2 *m. high; lower elongate tubular colored chartaceous sheaths bladeless; cauline leaf* usually 1, *submedian, greatly prolonged, erect, overtopping the cyme,* the uppermost leaf represented by a colored bladeless or nearly bladeless scarious sheath, or in forma **subnùdus** Fern. (almost naked) the submedian leaf and the bladeless sheath wanting, or in forma **bifróns** Fern. (two-leaved), the 2nd leaf with well developed blade; *cyme with stiffly ascending* (rarely divergent) *branches,* 0.5–1.8 dm. long; *heads* 5–12(–25)-flowered, *hemispherical; perianth deep brown,* 3–4 *mm. long; sepals and petals firm, narrowly lance-attenuate, awl-tipped, about equaling the narrowly ovoid-trigonous* lustrous taper-beaked 1-locular *capsule;* stamens 6, anthers longer than filaments; seeds 0.6 mm. long, obovoid, abruptly short-pointed. — Sandy, gravelly or peaty margins of lakes and ponds, usually standing in water, e. Md., Del., N.J. and ne. Pa., n. and ne. to Nfld., N.B., n. Me., s.-centr. N.H., se. Vt., n. N.Y., Temagami Distr., Ont., and Mich.; old collection of *Drummond* from Ala. Late June–Sept. FIG. 825. — Sometimes producing, in deep water, numberless long capillary submersed leaves from the rhizome.

825. J. militaris.

50. J. pérvetus Fern. (old veteran; the plant perhaps a last relict). — More slender than no. 49; basal sheaths shorter; *culms rigid,* erect or spiraling; *firm* terete *leaves* more slender, *scarcely overtopping cyme; cyme* 2.5–10 cm. long, *its branchlets divergent; heads subglobose,* 10–30-flowered; *perianth* 2 *mm. long, drab, pale brown or whitish; sepals and petals oblong, obtuse, with membranous margins;* some flowers often unisexual; stamens 6, sometimes wanting; *filament and anther subequal; capsule exserted,* 3 mm. long, ovoid-prismatic, subulate-attenuate, stramineous or rufescent; seeds 0.4–0.6 mm. long. — Fresh to brackish swale, very local, s. Cape Cod, Mass. Sept., Oct. FIG. 826.

826. J. pervetus.

51. J. acutiflòrus Ehrh. (acute-flowered). — *Rhizome hard, stout, horizontal,* bearing the tall and slender culms in rows; culms 0.3–1.3 m. high, compressed at base; *leaves compressed,* more or less curved, *numerous,* suberect, greatly elongate, prominently nodose; *cyme very decompound, with anthelate slender spreading branchlets; perianth* brown, 3 *mm. long; sepals and somewhat longer petals narrowly lanceolate, almost awned;* stamens 6; *capsule* trigonous, ovoid, *tapering to long exserted beak*; seeds about 0.5 mm. long, slender, merely pointed. — Damp open woods, local, s. Nfld. and St.P. et Miq. Aug., Sept. (Eu.)

52. J. articulàtus L. (jointed; from the septate leaves). — *Culms loosely tufted* from a short creeping horizontal often coralline-tuberiferous rhizome, arched-ascending to erect, sometimes branching below or becoming stoloniform, 1–6 dm. high, with 1–3 slender leaves; *cyme* short, 0.2–1.5 dm. long, *spreading;* the hemispherical *heads 3–many-flowered; flowers brown,* 2.5–3 mm. long; *sepals acuminate; petals* a little longer than the sepals, *shorter than the slender-conic incompletely 3-locular deep chestnut-brown shining capsule; anthers as long as the filaments;* ovary attenuate into a *short style;* seeds 0.5 mm. long, obovoid, attenuate below, abruptly pointed above. — Wet ground, Nfld. to B.C., s. to N.S., N.E., L.I., N.J., N.Y., W.Va., n. O., n. Ind., n. Ill., Minn. and Oreg. July, Aug. (Eurasia) Fig. 827. — Hybridizes with nos. 31, 39, and 53, and more frequently with no. 41, then producing × **J. fulvéscens** Fern. (reddish-brown).

827. J. articulatus.

Var. **obtusàtus** Engelm. (bluntish). — Inflorescence paler; the greenish flowers smaller; the abruptly mucronate pale brown capsule shorter and duller. — Nfld. to Mich., s. to N.J., often in brackish soil.

53. J. alpìnus Vill. (alpine). — Culms erect or slightly decumbent, 0.3–5.3 dm. high, from a creeping rhizome, with 1 or 2 slender erect quill-like leaves; *cyme meagre,* 1–14 cm. long, *with* 1–few *erect branches, bearing* 1–10(–15) *dense rounded-hemispherical dark brown heads; flowers sessile or equally short-pedicelled,* 2–2.5 mm. long; *sepals oblong, obtuse,* mucronate, usually longer than obtuse petals, *as long as or shorter than the obtuse short-pointed* incompletely 3-locular castaneous *capsule;* anthers as long as filaments; style short; seeds fusiform, 0.5 mm. long. — Wet shores, marshes, etc., usually calcareous, Nfld. to B.C., s. to Gaspé Pen., Que., ne. N.B., Minn., Rocky Mts., etc. July–Aug. (Eurasia) Fig. 828.

828. J. alpinus (typical).

Var. **fuscéscens** Fern. (becoming swarthy). — *Cyme with more spreading-ascending branches,* 0.3–3 *dm. long, the* 3–80 *hemispherical heads greenish or pale brown.* — Calcareous shores and damp soils, sw. Que. to Sask., s. to w. Vt., N.Y., n. O., n. Ind., Ill. and Mo.

Var. **rariflòrus** Hartm. (sparsely flowered). — *Heads ellipsoid or elongate-hemispherical, 2–10-flowered,* dark or light *brown, some of the central flowers elevated well above the others on long pedicels.* (Var. *insignis* Fries; *J. nodulosus* Wahlenb.; *J. Richardsonianus* Schultes) — Greenl. and Lab. to Alaska, s. to Nfld., N.B., centr. Me., N.Y., s. Ont., n. Ind., Ill., Minn., Neb., s. Alta. and Wash. Fig. 829. — In forma **ùniceps** (Hartm.) Krok & Lagerstedt (one-headed) the tiny plant is only 3–15 cm. high, with 1 cauline leaf and only 1 or 2 heads (in nw. Nfld.).

829. J. alpinus, v. rariflorus.

× **J. alpinifórmis** Fern. (with form of *J. alpinus*) is a hybrid (in Nfld.) of typical *J. alpinus* with no. 52.

× **J. nodosifórmis** Fern. (with form of *J. nodosus*) is a hybrid (in Nfld. and in Gaspé Co., Que.) of typical *J. alpinus* with no. 31.

54. J. pelocárpus Mey. (fruit of the mud). — *Rhizome whitish, filiform, firm but hardly ligneous, about* 1 *mm. thick;* culms very slender, erect or ascending, 0.6–5 dm. high, with few *almost filiform* nodulose *leaves with* 2 *or more slender tubes; cyme* compound, *loosely branching,* one-third to two-thirds height of plant, *with* 1–3 *flowers at forks and along the capillary branches,* the *flowers often changed to promptly deciduous slender bulblets consisting of reduced firm leaves;* flowers 2.5 mm. long, greenish, often tinged with red; *sepals and petals oblong, obtuse,* the petals longer; *anthers much longer than filaments; capsule* 1-locular, *slender-beaked,* pale brown, 3–4 mm. long, 1–1.5 mm. thick; style slender; seeds obovoid, short-pointed, about 0.5 mm. long; in forma **submérsus** Fassett (submersed) the plant consisting chiefly of tufts of flaccid basal leaves without culms. — Damp shores, pools and wet sand, s. Lab. to Algoma Distr., Ont., s. to Nfld. and St.P. et Miq., N.S., N.E., L.I., Del., Pa., n. Ind., Wisc. and Minn. Mid-July–

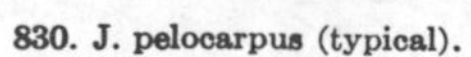

830. J. pelocarpus (typical).

Sept. FIG. 830. — The proliferous plants usually sterile and much larger than the fertile and with larger and more diffuse cymes.

Var. **sabulonénsis** St. John (of Sable Island). — Prostrate or depressed, greatly reduced in all parts, 2–6 cm. high, the culms curving; septa of leaves obscure; flowers approximate; capsule 2.5–3.5 mm. long. — Wet peat, sands and pools, sw. Nfld. and Sable I., N.S. July, Aug. FIG. 831.

831. J. pelocarpus, v. sabulonensis.

55. J. abortìvus Chapm. (aborted; from the change of flowers to bulblets). — Similar to no. 54; *rhizome subligneous*, 2.5–6 *mm. thick; culms* fewer and more scattered, *stiffly erect*, 3–8.5 dm. high; *leaves coarser, about* 1 *mm. thick;* cyme one-sixth to one-half height of plant; *sepals acute.* — Moist pine-barrens and pond-margins, Fla. to se. Va. Mid-July–Sept.

832. J. subtilis.

56. J. subtìlis Mey. (slender). — *Forming dense creeping or floating mats,* reddish (when fresh), in water becoming 4 or 5 dm. long and with elongate capillary leaves, on shore forming rosettes 0.5–2 dm. broad with a tuft of primary leaves 2 or 3 cm. long and *repent branches bearing small fascicles of small leaves and axillary or terminal flowers* either sessile or short-pedicelled; flowers and capsule much as in no. 54, but filaments longer. — Margins and shores of ponds and streams, Nfld. and Côte Nord to L. St. John, Que., s. to n.-centr. Me. and s. Que. Late July–Sept. (Greenl.) FIG. 832.

2. LÚZULA DC. WOODRUSH

Capsule 1-locular, 3-seeded, 1 often carunculate seed to each parietal placenta. — Perennials of cold and temp. reg., often hairy, frequently in dry ground, with flat and soft usually hairy leaves, and spiked, crowded, or umbelled flowers. (From *Gramen Luzulae*, or *Luxulae*, diminutive of *lux, light*, — a name given to one of the species from its shining with dew.) JUNCOIDES Adans.; JUNCODES Ktze.

a. Flowers solitary at the tips of the ultimate branches of the inflorescence.
- Inflorescence an umbel, the filiform peduncles 1–few-flowered; flowers 2.5–4.5 mm. long; seeds with a long curved appendage (caruncle). . . . . 1. *L. acuminata.*
- Inflorescence a loose decompound cyme; flowers 2–3 mm. long; seeds not appendaged. . . . . 2. *L. parviflora.*

a. Flowers crowded in spikes or glomerules. . . . *b.*
- *b.* Flowers white or whitish, the glomerules in a diffuse corymbiform inflorescence 0.3–2 dm. high. . . . . 3. *L. luzuloides.*
- *b.* Flowers pale to deep brown or brownish-purple, when paler greenish; inflorescences mostly shorter. . . *c.*
  - *c.* Leaves 1–4 mm. wide, with subulate (often involute) tips; bracts at bases of flowers and prophylla ciliate-fimbriate; arctic-alpine species.
    - Inflorescence a dense arching or nodding prolonged panicle or spike. . 4. *L. spicata.*
    - Inflorescence a dense subcapitate corymb of glomerules, either with or without ascending to divergent rays. . . . . 5. *L. confusa.*
  - *c.* Leaves often broader, flat, with blunt callous tips; bracts at bases of flowers and prophylla entire or merely lacerate. . . *d.*
    - *d.* Plants with small decumbent crowns connected by short horizontal stolons; the usually solitary flowering stem 0.3–2 dm. high; spikes 2–6, subglobose, all but the central one on horizontal to recurving rays; anthers two to five times as long as filaments. . . . . . 6. *L. campestris.*
    - *d.* Plants densely cespitose, erect; the 2–many erect flowering stems 0.5–6 (–9) dm. high; spikes ovoid, cylindric or subglobose, few–20 or more; anthers shorter than to but slightly longer than filaments. . . *e.*
      - *e.* Caruncle 0.1–0.2 mm. long, forming a mere pale tip only one-sixth to one-tenth full length of seed; spikes subglobose to ovoid; perianth deep brown to fuscous-blackish, 2–3 mm. long; capsule pointed, nearly black; arctic-alpine species. . . . . 7. *L. sudetica.*
      - *e.* Caruncle of fully developed seed 0.3–0.7 mm. long, a fifth to a third the full length of seed (if only 0.2 mm. long, in no. 9, the pale spikes cylindric). . . *f.*
        - *f.* Rays (when developed) of umbel erect to spreading-ascending, not horizontally divergent or reflexed and elongate; spikes ovoid to cylindric. . . *g.*

g. Mature spikes 6–9 mm. thick, 0–10 (rarely more) of them borne on elongate rays or their branches; perianth 2.5–4.5 mm. long; mature seeds 1.5–2 mm. long; plants without white coralline tubers. . . . . 8. *L. multiflora.*

g. Mature spikes 4–5 mm. thick (or, if 6–7 mm. thick, in no. 10, the plant producing whitish coralline tubers in the crowns), 2–20 of them borne on elongate rays or their branches; perianth 1.8–3 mm. long; seeds 1–1.5 mm. long.

Base of plant not producing tubers; spikes 4–5 mm. thick; perianth 1.8–2.3 mm. long; caruncle broadly rounded at tip; northern species. . . . . 9. *L. pallescens.*

Base of plant bearing firm whitish coralline tubers; spikes 5–7 mm. thick; perianth 2–3 mm. long; caruncle conically tapering to tip; southern species. . . . . 10. *L. bulbosa.*

f. Rays of umbel very slender, elongate, horizontally divergent to reflexed; spikes subglobose; perianth 2.5–4 mm. long; seeds 1.2–1.6 mm. long; southern species. . . . . . 11. *L. echinata.*

1. L. acuminàta Raf. (tapering to point). — *Plant loosely cespitose,* often somewhat stoloniferous, 1–3.7 dm. high; *leaves lance-linear,* hairy, the basal 0.5–1.2 cm. wide; *umbel mostly simple, with loosely ascending, spreading or drooping filiform 1- or 2-flowered rays 1–4 cm. long;* flowers 2.5–4.5 (in fruit) mm. long; sepals and petals broadly lanceolate, pale brown to stramineous, with hyaline margins, shorter than the conic-ovoid lustrous capsule; *seeds subglobose, 2.5 mm. long, with a large curved caruncle.* (*L. saltuensis* Fern.) — Woods, clearings and bluffs, s. Nfld. to Sask., s. to N.S., N.E., N.C., upland to n. Ga. and n. Ala., Ind., Ill., Minn. and S.D. April, May. Fig. 833.

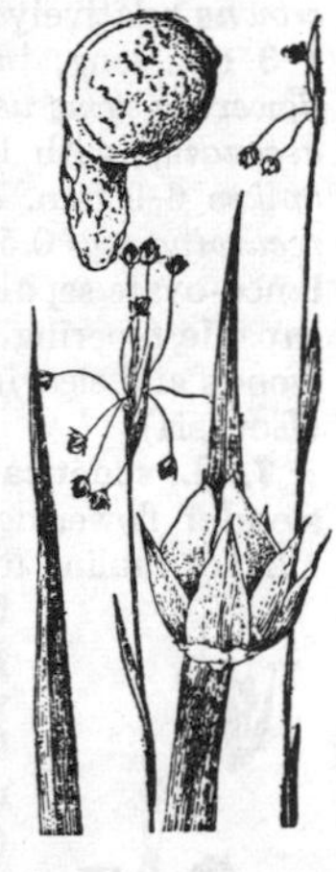

Var. **carolìnae** (S. Wats.) Fern. (of Carolina). — Some or all *rays of umbel forking, 2–5-flowered;* perianth often dark-castaneous; basal leaves up to 1.5 cm. wide. (*L. carolinae* S. Wats.) — Calcareous wooded slopes and bluffs, Pa. and O., s. to Ga. and Ala. Late March–May.

2. L. parviflòra (Ehrh.) Desv. (small-flowered). — Forming small clumps, with 1–several *flowering stems mostly 2–6 dm. high; basal leaves flat, glabrous* or nearly so, 0.5–1.5 *cm. broad, pale green,* the 5–7 cauline erect leaves much smaller; *cyme decompound, 4–12 cm. high, with long slender ascending or arching* and shorter divergent *rays, many-flowered;* prophylla (at bases of flowers) entire; *flowers about* 2 (in fruit 3) *mm. long;* the pointed sepals and petals similar, *castaneous or brownish;* stamens 6; capsule trigonous-ovoid, blunt, exceeding perianth; *seed* 1.3 mm. long, obliquely ovoid, *without caruncle,* castaneous, lustrous. — Arct. reg., s. to damp woods, thickets, slopes and summits of Nfld., Shickshock Mts., Que., and mts. to Ariz. and Calif. July, Aug. (Eurasia)

Var. **melanocárpa** (Michx.) Buchenau (black-fruited). — Usually taller and laxer; *flowering stems* 0.4–1 *m. high; cymes* 0.5–2 *dm. high, very lax,* the shorter rays loosely spreading to reflexed; *perianth pale brown to stramineous; capsule brown to black.* — Lab. to Alaska, s. to damp thickets and open woods of Nfld., C.B., n. N.E., and mts. of N.E. and N.Y., n. Mich., Wyo. and Calif. June–Aug. (Eurasia) Fig. 834.

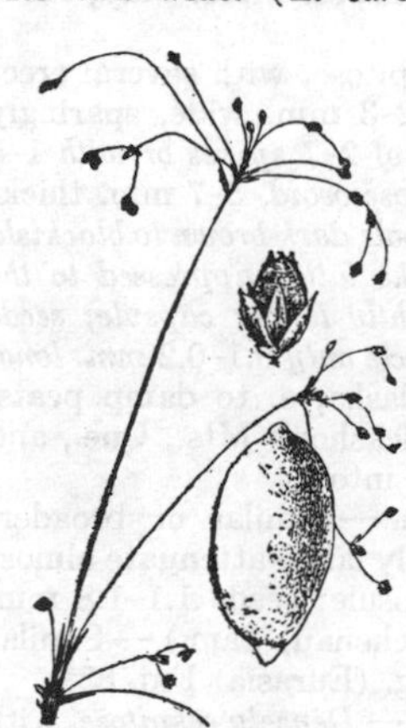
834. L. parviflora, v. melanocarpa.

833. L. acuminata.

3. L. luzuloìdes (Lam.) Dandy & Wilmott (resembling *Luzula;* Lamarck having placed it in *Juncus*). — Loosely cespitose, 3–8 dm. high; *leaves* long, *linear, erect,* more or less hairy, *the basal* 3–8 *mm. wide; inflorescence diffusely corymbiform,* 0.3–2 *dm. long,* the ultimate *branches terminated by glomerules of* 3–8 *whitish flowers; sepals and petals* lanceolate, *acute,* the *sepals distinctly shorter;* capsule castaneous, about equaling perianth; seed obliquely ovoid, about 1.2 mm. long. (*L. nemorosa* Mey., not Baumg.) — Open groves, shaded roadsides, lawns, etc., local, N.S. to Ont. and Minn., s. to s. N.E., s. N.Y. and e. Pa. June. (Natzd. from Eu.) Fig. 835.

835. L. luzuloides.

4. L. spicàta (L.) DC. (spiked). — Densely cespitose, 1–4.5 dm.

836. L. spicata.

high; *leaves channelled,* 1–4 mm. wide, narrowly linear, *with slender inrolled tips;* sessile *glomerules* of flowers *crowded in* interrupted or dense brown *arching or nodding spiked panicles* 1–4 cm. long; *floral bracts and prophylla long-ciliate;* sepals bristle-pointed, scarcely as long as the abruptly short-pointed capsule; seeds merely with a roundish projection. — Arct. reg., s. to gravels, talus, peaty openings, etc., Nfld. and alpine areas of Que., n. N.E., ne. N.Y., Colo., Ariz. and Calif. June–Aug. (Eurasia) Fig. 836.

5. **L. confùsa** Lindeberg (confused; originally with other species). — Cespitose, 0.5–4 dm. high, stiffly ascending, with purplish basal sheaths; *leaves channelled, subulate-tipped; spikes crowded in a close* subglobose to ovoid *glomerule or borne* singly or clustered *at the tips of* 1–4 *ascending to arching filiform peduncles* 1–7 cm. long; *bractlets and prophylla fimbriate-ciliate;* flowers 2–2.5 mm. long, the fuscous to castaneous lanceolate acute petals and sepals subequal; *capsule* about equaling perianth, *trigonous-spherical, obtuse;* seed 1.2 mm. long, castaneous. — Arct. reg., s. to alpine areas of Shickshock Mts., Que., Mt. Katahdin, Me., Mt. Washington, N.H., Ung., Keewatin and n. Alta. July, Aug. (Eurasia) Fig. 837.

837. L. confusa.

6. **L. campéstris** (L.) DC. (of low fields). — *Low and tufted, the tufts or crowns* relatively small and *scattered, connected by stolons and slender rhizomes* 1–3 cm. long; basal leaves 2–3 mm. wide, their margins long-hairy below; *flowering stems usually* 1 *to each crown,* 0.3–2 dm. high, *decumbent or spreading-ascending,* with 1–3 small cauline leaves; *umbel of* 2–6 *subcapitate* dark brown *spikes* 6–9 mm. in diameter, *all but the central sessile one on horizontally divergent to arched-recurving rays* 0.5–2 *cm. long;* perianth about 3 mm. long, the dark brown (but paler margined) lance-ovate sepals and the similar petals subequal; *anthers two to five times as long as filaments;* capsule tapering at tip, about equaling perianth; seed olivaceous, with long caruncle. — Open woods and clearings, Avalon Pen., Nfld.; locally adv. from Eu. in lawns, se. Mass. May, June. (Eurasia)

7. **L. sudética** (Willd.) DC. (of the Sudetic Mts.). — Densely cespitose, with several erect slender flowering stems 0.5–4.5 dm. high; *basal leaves deep green,* 2–3 mm. wide, sparingly ciliate; cauline leaves narrower and shorter; *umbel a dense glomerule of* 2–7 *spikes or with* 1–5 filiform ascending *rays* 0.3–3 *cm. long; spikes subglobose-ovoid,* 5–7 mm. thick; *perianth* 2–2.5 *mm. long, the broadly lanceolate acute sepals dark brown to blackish,* with paler margin, *slightly longer than the petals; the latter appressed to the blackish or deep brown short-pointed subequal or slightly longer capsule; seeds* reddish-brown, 1–1.6 *mm. long, tipped by a short caruncle only* 0.1–0.2 *mm. long.* (*L. campestris* var. *alpina* Gaudin) — Greenl. to Alaska, s. to damp peats, meadows and slopes of Nfld., alpine meadows of Shickshock Mts., Que., and Hudson Bay reg. July, Aug. (Eurasia) Passing freely into

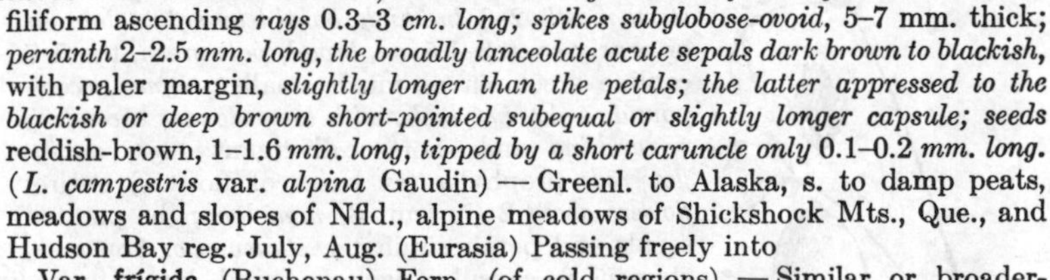

838. L. sudetica, v. frigida.

Var. **frígida** (Buchenau) Fern. (of cold regions). — Similar or broader-leaved; perianth 2.2–3 mm. long, the subequal narrowly lance-attenuate almost bristle-tipped sepals and petals overtopping the capsule; seeds 1.1–1.8 mm. long. (*L. campestris*, var. Buchenau; *L. frigida* (Buchenau) Sam.) — Similar range, s. to bogs, peaty barrens, etc., Nfld. July, Aug. (Eurasia) Fig. 838.

8. **L. multiflòra** (Retz.) Lejeune (many-flowered). — *Densely cespitose,* with numerous erect flowering stems up to 9 dm. high; basal leaves abundant, copiously ciliate, pale green, 1–7 mm. broad; *umbel open or congested, with* 1–10 *ascending to erect rays* (rays usually wanting in one var.); *spikes ovoid to cylindric, in maturity* 6–9 *mm. thick,* 4–11 mm. long; *perianth* 2.5–4.5 *mm. long;* sepals lanceolate, acuminate or acute, about equaling petals; capsule broadly rounded to short tip; *seeds* 1.5–2 *mm. long, the more or less bulbiform caruncle* (0.3-0.7 mm. long) *occupying nearly one-third its length.* — Highly variable circumboreal species. We have the following vars.:

Umbels more or less open, with usually 1 sessile or subsessile central spike, the other spikes on ascending rays of varying lengths (0.5–6 cm. long); perianth 2.5–3.3 mm. long, about equaled or overtopped by the capsule.

Tussocks relatively full; basal leaves 2–7 mm. broad; flowering stems mostly 1.5–6.7 dm. high; sepals pale brown in center; capsules medium brown to fulvous. . . . . *L. multiflora* (typical).

Tussocks relatively small; basal leaves 1–4 mm. broad; flowering stems 1 4.3 dm. high; sepals deep brown to fuscous, with pale margins; capsules deep chestnut to blackish. . . . . Var. *fusconigra.*

Umbel condensed, many or all spikes sessile or subsessile, without or with some elongate rays; perianth 3–4.5 mm. long, pale, overtopping capsules.
Flowering stems 0.5–4.25 dm. high; umbel usually with some stiff rays 0.5–2.5(-3.5) cm. long; spikes 4–8 mm. long, 6–7 mm. thick; perianth 3–4 mm. long; seeds 1.5–1.7 mm. long, the caruncle slenderly c. . . . . Var. *acadiensis.*
Flowering stems (with us) 2–9 dm. high; umbel usually capitate, rarely with 1–few slender rays 2–9 cm. long; spikes 6–11 mm. long, 7–9 mm. thick; perianth 3.5–4.5 mm. long; seeds 1.6–2 mm. long, the large caruncle rounded at base. . . . . Var. *congesta.*

L. multiflòra (typical), *L. campestris*, var. *multiflora* (Retz.) Čelak. — Fields, meadows and open woods, usually common, Nfld. and Côte Nord, Que., to B.C., s. to N.S., N.E., L.I., n. Del., ne. Md., Pa., upland to Va., W.Va., Ky., Ind., Ill., Minn., etc. Fig. 839. — Farther west becoming involved with the series known as *L. comosa* Mey. April–July.

Var. **fusconìgra** Čelak. (tawny-black), *L. campestris*, var. *frigida* sensu ed. 7, in large part, not Buchenau — Peaty barrens, turfy slopes, clearings and fields, Nfld. and Que., s. to N.S., centr. N.H. and n. N.Y., and locally along coast to Nantucket I., Mass. June–Aug. (Eurasia) Fig. 840.

839. L. multiflora (typical).

840. L. multiflora, v. fusconigra.

841. L. multiflora, v. acadiensis.

842. L. multiflora, v. congesta.

Var. **acadiénsis** Fern. (of Acadia). — Open woods, meadows, peaty barrens, etc., Nfld. and Gaspé Pen., Que., s. to N.S., s. N.B. and se. Me. June–Aug. Fig. 841.

Var. **congésta** (Thuill.) Koch (crowded). — Damp thickets, peaty barrens, shores, etc., Nfld. and St. P. et Miq. July, Aug. (Eurasia) Fig. 842.

9. L. palléscens (Wahlenb.) Bess. (becoming pale). — With habit of no. 8; *basal leaves* pale, 1–4 *mm. broad;* flowering stems very slender, 1–5.25 dm. high; *umbel* compact or open, *with* 6–20 *or more slenderly cylindric* pale to brown *spikes, many of them* solitary or clustered *on* ascending to slightly spreading *rays* 0.5–3.5 *cm. long; spikes* 4–5 *mm. thick,* 3–8 mm. long; *perianth* 1.8–2.3 *mm. long,* the acuminate sepals and petals about equaling capsule; *seeds slender,* 1–1.5 *mm. long, the rounded-oblong caruncle* 0.2–0.4 *mm. long.* — Meadows, open woods, clearings and rocky slopes, Nfld.; Gaspé Pen., Que.; and, locally, centr. Vt. and n. N.Y. June–early Aug. (Eurasia) Fig. 843.

843. L. pallescens.

10. L. bulbòsa (Wood) Rydb. (bulb-bearing). — In small pale tufts *with numerous whitish coralline or knotty bulb-like tubers borne at base* (these often hidden by the drab leaf-bases); *basal leaves* 2–7 *mm. wide;* flowering stems 1–4.5 dm. high; *umbel* much as in no. 9, *with* 6–20 *or more cylindric* pale brown *spikes,* the ascending rays 0.5–7 cm. long; *spikes* 5–7 *mm. thick,* 5–12 mm. long; *perianth* 2–3 *mm. long; seeds* 1.2–1.5 *mm. long, the conically tapering caruncle* 0.4–0.6 *mm. long.* (*L. campestris*, var. Wood) — Siliceous or argillaceous open woods, clearings and fields, Fla. to e. Tex., n. to se. Mass., s. Ct., N.J., Pa., nw. Ind., Ill., Mo. and e. Kans. Late March–June. Fig. 844.

844. L. bulbosa.

845. L. echinata.

11. L. echinàta (Small) F. J. Herm. (hedgehog-like). — Plant with general aspect of no. 8, the tussocks mostly smaller; flowering stems slender, 1–4.5 dm. high; *umbel lax, with* 2–10 (–16) *spikes, all but the central sessile spike on filiform* simple or forking *rays, some of the rays horizontally divergent or reflexed* and 1–5 cm. long, ascending rays up to 7 cm. long; *spikes subglobose or ovoid,* 6–9 *mm. thick; perianth* 3–4 *mm. long,* the dark brown *sepals much overtopping capsules; anthers nearly twice length of filaments; seeds* 1.2–1.6 *mm. long, the conic-tapering caruncle* 0.5–0.6 *mm. long.* — Woods, thickets and clearings, Ga. and Ala., n. to se. Mass., se. N.Y., N.J., Pa., W.Va., O., s. Ind. and s. Ill. April–June. FIG. 845.

Var. **mesochòrea** F. J. Herm. (midland). — More delicate; *spikes* 5–7 *mm. thick; perianth about* 2.5 *mm. long,* often paler, *barely equaling or shorter than capsule;* anthers less than twice length of filaments. — Woods and thickets, w. N.C. to Miss. and e. Tex., n. to W.Va., O., Ind., Ill., Ark. and Okla. April–June.

FAM. 32. LILIÀCEAE (LILY FAMILY)

Herbs, or rarely woody plants, with regular and symmetrical almost always 6-*androus flowers; the perianth not glumaceous, free from or adnate to base of the chiefly* 3-*locular ovary; the stamens* 1 *before each of the sepals, petals or lobes* (6 *altogether, exceptionally* 4), *with* 2-*locular anthers; fruit a few–many-seeded capsule or berry; the small embryo inclosed in copious albumen.* Seeds anatropous or amphitropous (orthotropous in *Smilax*). Flowers not from a spathe, except in *Allium* and *Nothoscordum,* the outer and inner ranks of the perianth colored alike (or nearly so) and generally similar, except in *Trillium.* — A complex cosmop. family of many technically distinguished tribes (by some treated as separate families). For wholly satisfactory understanding flowers, fruits and roots are essential.

*a.*Flowers perfect or polygamomonoecious (if dioecious, spicate-racemose or (in *Asparagus*) the true leaves reduced to scales), not in axillary umbels; petioles without tendrils. . . *b.*

*b.*Sepals and petals distinct, except sometimes at base (if united into a tubular gamophyllous perianth, the flowers drooping and without scurfy surface). . . *c.*

*c.*Leaves basal or scattered (or, if in whorls on the stem, numerous and the flowers large and campanulate with long styles and the fruit a capsule). . . *d.*

*d.*Fruit a capsule. . . *e.*

*e.*Leaves not both evergreen and rigidly sword-shaped; stem herbaceous; style or styles (except in no. 1, with small spicate-racemose flowers) definite. . . *f.*

*f.*Style none; stigma sessile; filaments woolly. . . . . . . 1. *Narthecium.*

*f.*Style or styles definite; filaments glabrous. . . . *g.*

*g.*Styles 3, distinct. . . . *h.*

*h.*Stigmas linear, forming a line along the inner side of each style; capsule loculicidal (opening into each locule through the dorsal suture). . . *i.*

*i.*Leaves needle- or bristle-like; raceme with bristle-like bracts; seeds 2 in each locule. . . . . . 2. *Xerophyllum.*

*i.*Leaves flat; racemes bractless or nearly so; seeds numerous.

Flowers perfect; sepals and petals lilac, several-nerved; capsule 3-lobed at summit, each valve divergently 2-lobed; appendages of seeds tapering. . . . 3. *Helonias.*

Flowers dioecious; sepals and petals white, 1-nerved; capsule unlobed; appendages of seeds wing-like. . . 4. *Chamaelirium.*

*h.*Stigmas terminal; capsule septicidal (dehiscent through the partitions between the locules). . . *j.*

*j.*Anthers 2-locular, oblong or ovoid; leaves 2-ranked, equitant. 5. *Tofieldia.*

*j.*Anthers confluently 1-locular, peltate or cordate; leaves not 2-ranked and equitant. . . *k.*

*k.*Stems and inflorescences glabrous; flowers perfect or polygamous (rarely monoecious); seeds ovoid, ellipsoid or linear-cylindric, barely winged or wingless. . . *l.*

*l.*Sepals and petals without basal glands.

Stem scapose; inflorescence a simple raceme; sepals and petals free from ovary, rounded at summit, about equaled by stamens. . . . . 6. *Amianthium.*

Stem leafy; inflorescence a panicle; sepals and petals adnate to base of ovary, usually attenuate, much longer than stamens. . . . . 7. *Stenanthium.*

l. Sepals and petals bearing basal glands. . . . 8. *Zigadenus.*

k. Stems (above) and axis of inflorescence pubescent; flowers polygamomonoecious; seeds flat, broadly winged.

Sepals and petals slender-clawed and with a pair of glands at base of blade. . . . . . . 9. *Melanthium.*

Sepals and petals (at least of fertile flowers) without claws, not gland-bearing. . . . . . . 10. *Veratrum.*

g. Style 1, either cleft or uncleft; capsule loculicidal. . *m.*

m. Style 3-cleft; flowers solitary, at first terminal, but by prolongation of the slender axis later appearing axillary, the perianth slenderly campanulate. . . . . 11. *Uvularia.*

m. Style uncleft or stigma only 3-cleft; leafy axis not prolonging after flowering. . . *n.*

n. Flowers in terminal umbels subtended by spathes (or flowers replaced by bulblets); bulb tunicated.

Seeds 1 or 2 in each locule; plant with odor of onion. 12. *Allium.*

Seeds several in each locule; plant without onion-odor. . 13. *Nothoscordum.*

n. Flowers variously disposed, not in terminal umbels with spathes; plants without onion-odor. . . *o.*

o. Fibrous-rooted, not bulbous; leaves basal; scape bearing large lily-like flowers, with sepals and petals united at base into a funnelform tube.

Leaves linear or ensiform, keeled; flowers ascending, yellow to orange. . . . . 14. *Hemerocallis.*

Leaves dilated, flat or fluted; flowers spreading, bluish, lilac or white. . . . . 15. *Hosta.*

o. Bulbous; leaves and flowers various. . . *p.*

p. Bulbs not tunicated, solid or of numerous tooth-like scales; stems leafy or with a pair of broad flat leaves; flowers large, funnelform or campanulate with spreading or recurving segments; anthers extrorse.

Stem leafy; bulb consisting of tooth-like scales; seeds flattened. . . . . 16. *Lilium.*

Stem with a pair of leaves near the ground; bulb solid; seeds plump. . . . . 17. *Erythronium.*

p. Bulbs tunicated; leaves basal, slender; scapes terminated by racemes or corymbs of small to medium-sized flowers; anthers introrse. . . *q.*

q. Perianth with distinct spreading sepals and petals.

Pedicels jointed; filaments filiform; style thread-like. . . . . 18. *Camassia.*

Pedicels not jointed; filaments flattened; style 3-sided. . . . . 19. *Ornithogalum.*

q. Perianth gamophyllous, a globular to ovoid or urceolate tube, 6-lobed at tip. . . . . 20. *Muscari.*

e. Leaves evergreen, rigid or leathery, sword-like; stem woody; racemes or panicles tall, with showy white to greenish flowers; the 3 stigmas sessile. . . . . 21. *Yucca.*

d. Fruit a berry. . . *r.*

r. Leaves reduced to scarious scales, the apparent (phyllodial) leaves filiform; stem repeatedly branched. . . . . 22. *Asparagus.*

r. Leaves broad, flat and herbaceous; stem simple or few-forked. . . *s.*

s. Sepals and petals distinct; inflorescences terminal and not 1-sided (or on supra-axillary twisted or contorted peduncles in no. 27). . . *t.*

t. Scapose, with oblong or oval finely ciliolate basal leaves and terminal umbels or racemose corymbs. . . . . 23. *Clintonia.*

t. Leafy-stemmed. . . *u.*

u. Inflorescences terminal. . . *v.*

v. Flowers in elongate racemes or panicles, white, 2–5 mm. long, terminating the simple stem.

Sepals 3; petals 3; ovary 3-locular, leaves narrowed at base. . . . . 24. *Smilacina.*

Sepals 2; petals 2; ovary 2-locular; leaves cordate at base. 25. *Maianthemum.*

v. Flowers 1–few, yellowish, greenish, tawny or mottled, with sepals and petals 1.5–2.5 cm. long, at the tips of stem and branches. . . . . 26. *Disporum.*

*u.*Inflorescences 1–2 (–3)-flowered, from the axils or supra-axillary, the simple or forking capillary peduncles usually abruptly bent or contorted. . . . . 27. *Streptopus.*

*s.*Sepals and petals united into a tubular to campanulate drooping perianth, free only at summit.

Stem very leafy; peduncles axillary, 1–several-flowered; perianth cylindric, greenish, stramineous or whitish. . . . . 28. *Polygonatum.*

Leaves few, sheathing base of scape; raceme many-flowered, 1-sided; perianth bell-shaped, white. . . . . . . 29. *Convallaria.*

*c.*Leaves in 1 or 2 whorls at summit (or middle and summit) of stem; common style short or wanting, the 3 branches stigmatic down the inner side; fruit a berry.

Leaves in 2 whorls, the lower whorl with more than 3 leaves; flowers umbellate from the upper whorl; sepals and petals alike, yellowish; style-branches elongate, capillary, exserted. . . . . . 30. *Medeola.*

Leaves in a terminal whorl of 3; flower solitary, terminal; sepals green, herbaceous; petals colored, petaloid; style-branches shorter than petals. . . . . 31. *Trillium.*

*b.*Sepals and petals united into a cylindric scurfy-roughened tube; flowers ascending in a terminal raceme; ovary partly inferior; fruit capsular; leaves chiefly rosulate. . . . . 32. *Aletris.*

*a.*Flowers dioecious, yellowish or greenish, in axillary or supra-axillary umbels; fruit a berry; leaves netted-veined; petioles often bearing tendrils. . . . 33. *Smilax.*

Artificial Key to Genera

a. Flowers or inflorescence terminal. . . *b.*

*b.*Flowers solitary. . . *c.*

*c.*Stem with the numerous leaves alternate or in several whorls. . . *d.*

*d.*Leaves elongate, scattered or in whorls on the erect and simple stem from a scaly bulb; flowers (lilies) large and showy, campanulate or broadly funnelform; anthers versatile. . . . . 16. *Lilium.*

*d.*Leaves oblong to ovate, alternate; stem slender, usually forking, from fleshy fibrous roots or slender rhizomes; flowers medium-sized; anthers basifixed.

Flowers stramineous to orange, slenderly campanulate to subcylindric; capsule 3-angled, seemingly axillary by prolongation of stem and branches; leaves equilateral. . . . . 11. *Uvularia.*

Flowers greenish to white or mottled, open-campanulate; fruit a berry terminating stem or branches; leaves inequilateral. . . . 26. *Disporum.*

*c.*Stem 2-leaved or with a terminal whorl of 3 (exceptionally more) leaves.

Leaves 2, elongate, parallel-veined, near surface of ground; stem from deep-seated bulb; flower campanulate, with sepals like the petals; fruit a 3-angled capsule. . . . . 17. *Erythronium.*

Leaves 3, broad and netted-veined, in a whorl at summit of stem; base a short rhizome; sepals green and herbaceous, unlike the colored petals; fruit a berry. . . . . 31. *Trillium.*

*b.*Flowers in clusters. . . *e.*

*e.*Flowers very large and showy, campanulate or funnelform (lilies).

Flowering stem nearly or quite leafless, from fibrous roots; leaves basal; sepals and petals united below into a funnelform tube.

Leaves broadly linear, keeled; flowers ascending, yellow to orange. 14. *Hemerocallis.*

Leaves broad, several-ribbed; flowers blue, lilac or white, ascending, spreading or drooping. . . . . 15. *Hosta.*

Flowering stem with numerous linear to oval leaves, from a scaly bulb; sepals and petals free. . . . . 16. *Lilium.*

*e.*Flowers medium or small (not true lilies). . . *f.*

*f.*Flowers in umbels. . . *g.*

*g.*Leaves basal; umbel terminating a scape. . . *h.*

*h.*Bulbous; leaves not ciliolate; umbel subtended by 2 or 3 scarious bracts; fruit a capsule.

Plant with onion-odor; perianth not reflexed in fruit. . . . 12. *Allium.*

Plant without onion-odor; perianth reflexed in fruit. . . . . 13. *Nothoscordum.*

*h.*Not bulbous; rhizome slender and elongate; leaves ciliolate; umbel spatheless; fruit a berry. . . . . 23. *Clintonia.*

*g.*Leaves borne on flowering stem.

Stem simple, with 2 whorls of leaves, 1 near middle, 1 at summit. 30. *Medeola.*

Stem forking, with alternate leaves. . . . . 26. *Disporum.*

f. Flowers in racemes, corymbs or panicles. . . *i.*
i. Sepals and petals distinct or slightly united only at base, the perianth-lobes then much longer than tube. . . *j.*
j. Woody evergreens, with rigid or leathery sword-like leaves with marginal shreddy fibres, and panicles of large white campanulate flowers. 21. *Yucca.*
j. Herbs; leaves mostly not evergreen, without shreddy marginal fibres; flowers medium to small. . . *k.*
k. Filaments conspicuously bearded; stigma sessile; linear-leaved, with greenish-yellow small flowers in racemes. . . . . . 1. *Narthecium.*
k. Filaments glabrous; style or styles definite. . . *l.*
l. Styles 3. . . *m.*
m. Anthers oblong or ovate, 2-locular. . . *n.*
n. Leaves 2-ranked and equitant, linear; inflorescence often glutinous. . . . . 5. *Tofieldia.*
n. Leaves not 2-ranked and equitant. . . *o.*
o. Leaves abundant, linear-acicular; pedicels much longer than the white flowers. . . . . 2. *Xerophyllum.*
o. Leaves few, broad, the lower oblanceolate or spatulate; pedicels short.
Flowers lilac, perfect, in a dense raceme at summit of hollow scape; capsule depressed, obcordate, 3-lobed, the valves 2-lobed. . . . . 3. *Helonias.*
Flowers white, dioecious, in an elongate spiciform raceme at summit of leafy stem; capsule ellipsoid or slenderly obovoid, not deeply lobed. . . . . 4. *Chamaelirium.*
m. Anthers peltate or reniform, with confluent locules. . . *p.*
p. Axis of inflorescence glabrous; seeds obovoid to oblanceolate or linear, narrowly winged or wingless. . . *q.*
q. Sepals and petals without shining colored glands at base.
Sepals and petals rounded at summit, about equaling stamens, free from base of ovary. . . . . 6. *Amianthium.*
Sepals and petals acuminate, much longer than stamens, adnate to base of ovary. . . . . 7. *Stenanthium.*
q. Sepals and petals with 1 or 2 shining or colored glands at base. . . . . 8. *Zigadenus.*
p. Axis of inflorescence pubescent; seeds flat, broad-winged.
Sepals and petals with narrow claws and conspicuous glands (nectaries) at base of blade, free from ovary. 9. *Melanthium.*
Sepals and petals clawless and glandless, usually adnate to base of ovary. . . . . 10. *Veratrum.*
l. Style single, sometimes cleft at tip. . . *r.*
r. Bulbous, with linear basal leaves; fruit a capsule.
Flowers blue; filaments thread-like. . . . . 18. *Camassia.*
Flowers greenish-white; filaments broad. . . . . . 19. *Ornithogalum.*
r. Not bulbous, arising from elongate rhizomes; leaves oblong to ovate; fruit a berry. . . *s.*
s. Leaves basal, ciliolate; corymb or umbel with few long-pedicelled flowers 0.8–2 cm. long. . . . . 23. *Clintonia.*
s. Leaves alternate on stem, not ciliolate; white flowers 2–5 mm. long, sessile or pedicelled, in elongate racemes or panicles.
Leaves narrowed at base; flowers 3-merous. . . . . 24. *Smilacina.*
Leaves cordate at base; flowers 2-merous. . . . . 25. *Maianthemum.*
i. Sepals and petals united except at summit into a tubular or campanulate perianth. . . *t.*
t. Bulbous, with slenderly linear leaves; flowers blue or bluish, in a dense terminal raceme. . . . . 20. *Muscari.*
t. Not bulbous; leaves broader; flowers white to yellow.
Flowers white, recurving, in a 1-sided raceme; perianth not scurfy; fruit a berry. . . . . 29. *Convallaria.*
Flowers white or yellow, ascending in a spiral spike-like raceme, scurfy-roughened; fruit a capsule. . . . . 32. *Aletris.*
a. Flowers or inflorescences in the axils of alternate stem-leaves. . . *u.*
u. Perianth tubular, the sepals and petals free only at tip; flowers 1–few, drooping from the axils. . . . . 28. *Polygonatum.*
u. Perianth of distinct sepals and petals. . . *v.*

*v.*Flowers 1–2 (rarely –3), pendulous; herbaceous plants with parallel-veined leaves and no tendrils. . *w.*

*w.*Plant freely bushy-branched; true leaves represented by small scales, the apparent leaves filiform phyllodia. . . . . 22. *Asparagus.*

*w.*Plants simple or few-forked, with dilated leaves.

Flowers stramineous or yellow, 1.5–4.5 cm. long, borne at tips of stem and branches but by prolongation of axis the ascending capsules seemingly axillary. . . . . 11. *Uvularia.*

Flowers purple to cream-color or greenish, smaller, hanging on usually contorted peduncles from leaf-axils; fruit a berry. . . . 27. *Streptopus.*

*v.*Flowers numerous in ascending umbels; woody or herbaceous plants with netted-veined leaves and often with tendrils; fruit a berry. . . . . 33. *Smilax.*

1. NARTHÈCIUM Huds. Bog-Asphodel

Perianth-segments 6, linear-lanceolate, yellowish, persistent. Anthers linear, introrse. Seeds ascending, appendaged at each end with a long bristleform tail. — Rhizome creeping, bearing linear equitant leaves and a simple stem or scape terminated by a simple dense bracteate raceme; pedicels bearing a linear bractlet. Remarkably localized plants in e. N. Am., Eu. and Japan. (Name from the Greek *narthecion, a chest* or *box* for holding ointments, ironically given to the European *N. ossifragum*, which was formerly supposed to break the bones of sheep feeding on it.)

1. N. americànum Ker (American), Yellow Asphodel. — Stem 2.5–4 dm. high; leaves 0.7–1.5 mm. wide, 7–9-nerved; raceme dense (2–5 cm. long); perianth-segments narrowly linear (4–5 mm. long), scarcely exceeding the stamens. (*Abama americana* (Ker) Morong) — Boggy pine-barrens and savannas, N.J. and Del. June, July.

2. XEROPHÝLLUM Michx.

Perianth widely spreading; segments (white) oval, distinct, without glands or claws, 5–7-nerved, at length withering, about the length of the subulate filaments. Anthers 2-locular, short, retrorse. Styles filiform, stigmatic down the inner side, persistent. Capsule globular, 3-lobed, obtuse (small). Seeds collateral, 3-angled, not margined. — Two N. Am. herbs with the stem simple, from a thick tuberous rhizome, bearing a simple dense bracteate raceme of showy flowers, and thickly beset with needle-shaped leaves, the upper ones reduced to bristle-like bracts; those from the base in a dense tuft, reclined, rough on the margin, dry and rigid. (Name from the Greek *xeros, dry,* and *phyllon, leaf.*)

1. X. asphodeloìdes (L.) Nutt. (like Asphodel), Turkeybeard. — Stem 0.4–1.5 m. high. — Sandy pinelands, N.J., Del. and se. N.C.; mountain-woods, Va. to Tenn. and Ga. May–July.

3. HELÒNIAS L.

Perianth of 6 spatulate-oblong lilac segments, persistent, several-nerved, glandless, shorter than the filiform filaments. Anthers 2-locular, roundish-oval, blue, extrorse. Styles revolute, stigmatic down the inner side, deciduous. Capsule obcordately 3-lobed, loculicidally 3-valved; the valves divergently 2-lobed. — A smooth perennial, with many oblong-spatulate or oblanceolate evergreen flat leaves, from a tuberous rhizome, producing in early spring a stout hollow sparsely bracteate scape 3–6 dm. high and sheathed with broad bracts at the base and terminated by a simple and short dense raceme. Bracts of inflorescence obsolete; pedicels shorter than the flowers. (Name probably from the Greek *helos, a swamp,* the place of growth.)

1. H. bullàta L. (swollen), Swamp-pink. — Swamps and bogs, Staten I., N.Y., and N.J. to e. Va.; mts. of Pa. to nw. Ga. April, May.

4. CHAMAELÍRIUM Willd. Devil's-bit

Perianth of 6 spatulate-linear spreading 1-nerved segments, white (drying yellowish) in staminate, greenish in pistillate flowers, withering-persistent. Filaments and (white) anthers as in *Helonias;* pistillate flowers with rudimentary stamens. Styles linear-clavate, stigmatic along the inner side. Capsule ellipsoid, not lobed, of a thin texture, loculicidally 3-valved from the apex. Seeds linear-oblong. — Smooth herb, with an erect stem from a (bitter) thick and abrupt tuberous rhizome, terminated by a virgate spiciform raceme 1–3 dm. long of small bractless flowers; pistillate plant more leafy than the staminate. Leaves flat, lanceolate, the lowest spatulate, tapering into a petiole. (Name formed of the Greek *chamai, on the ground,* and *leirion, lily,* the genus having been founded on a dwarf undeveloped specimen.)

1. C. lùteum (L.) Gray (yellow), Blazing-star, Rattlesnake-root, Fairy-wand. — Stem 2–12 cm. high; staminate flowers divergent, short- or long-pedicelled; pistillate flowers and capsules 7–14 mm. long, ascending. (Including *C. obovale* Small) — Meadows, thickets and rich woods, Fla. to Ark., n. to w. Mass., N.Y., s. Ont., O., Mich. and Ill. May–July. — Variable in size of flowers and fruits; needing further study.

5. TOFIÈLDIA Huds. False Asphodel

Perianth more or less spreading, persistent; the sepals (white or greenish) concave, oblong or obovate, without claws, 3-nerved. Filaments subulate; anthers short, innate or somewhat introrse, 2-locular. Styles subulate; stigmas terminal. Seeds oblong, horizontal. — Slender perennials of N. Hemisph. and the Andes, mostly tufted, with short or creeping rhizomes, and simple stems leafy only at the base, bearing small flowers in a close raceme or spike. Leaves 2-ranked, equitant, linear, grass-like. (Named for *Thomas Tofield*, 1730–1779, an English botanist.)

Glabrous; fruiting raceme 5–8 mm. thick, its pedicels solitary; seeds not appendaged. . . . . 1. *T. pusilla.*
Pubescent or glandular above; raceme 1–2 cm. thick, its pedicels mostly in 3's or 4's; seeds tailed.
 Perianth not becoming rigid; capsule thin-walled; seeds with a contorted tail at each end. . . . . 2. *T. glutinosa.*
 Perianth rigid about the firm capsule; appendages of seed short. . . . 3. *T. racemosa.*

1. T. pusílla (Michx.) Pers. (very small). — *Glabrous;* scape filiform, leafless or rarely 1-bracted, 0.2–3 dm. high; leaves strongly equitant, in numerous basal flabelliform short tufts; *raceme* subglobose to cylindric, *becoming* 0.5–4.5 cm. long and 5–8 *mm. thick;* flowers whitish or greenish; *capsules* stramineous, 2–3 *mm. long; seeds unappendaged.* (*T. minima* (Hill) Druce, based on illegitimate name; *T. palustris* sensu Am. auth., not Huds.; *T. borealis* Wahlenb.) — Greenl. to Alaska, s. in calcareous marshes or on rocks to Nfld., Anticosti I. and Gaspé Pen. and Ungava Distr., Que., n. Mich., Minn., Mont. and s. B.C. June, July. (Eurasia)

2. T. glutinòsa (Michx.) Pers. (sticky). — *Stem* bracted or bractless, 0.5–5 dm. high, *with the mostly fascicled pedicels heavily glutinous with dark glands;* leaves broadly linear, one-half to two-thirds length of scape; raceme elongating to 1–8 cm., becoming 1–2 cm. thick, *flowers fascicled; sepals and petals whitish, petaloid; capsule* stramineous or red, *thin-walled,* 4–8 mm. long; *seeds with a contorted tail at each end.* (*Triantha glutinosa* (Michx.) Baker) — Calcareous marshes, damp ledges and shores, Nfld. to Man., s. to N.S., n. N.E., centr. N.Y., W.Va., mts. to Ga., n. O., n. Ind., n. Ill. and Minn. June–Aug.

3. T. racemòsa (Walt.) BSP. (racemed). — Similar to no. 2; stem (above) and *pedicels minutely glandular;* raceme up to 19 cm. long; *perianth-segments becoming firm around the* smaller *firm capsule; seed short-appendaged.* (*Triantha racemosa* (Walt.) Small) — Wet sand, clay or peat of Coastal Plain, Fla. to Tex., n. to N.J. June–Aug.

6. AMIÁNTHIUM Gray Fly-poison

Perianth widely spreading; the free sepals and petals oval or obovate, without claws or glands, persistent. Filaments capillary. Anthers, capsules, etc. nearly as in *Melanthium.* Styles filiform. Seeds 1–4 in each locule. — A glabrous herb, the simple stems from a bulbous base or coated bulb, scape-like, few-leaved, terminated by a simple dense raceme of initially white flowers, the white persistent perianth turning green or purplish in age. A single poisonous species. (Name from the Greek *amiantos, unspotted,* and *anthos, flower;* the name formed with more regard to euphony than to good construction, alluding to the lack of glands on the perianth.) Chrosperma Raf.

1. A. Muscaetóxicum (Walt.) Gray (fly-poison). — *Leaves broadly linear,* elongated, obtuse, 0.4–2.7 cm. wide; *raceme simple; capsule abruptly 3-horned;* seeds ellipsoid, with a fleshy red coat. — Low sandy grounds, bogs and open woods, Fla. to s. Mo. and Okla., n. along the mts. to W.Va. and Pa. and on Coastal Plain to L.I. May–early July. — Bulb very poisonous.

7. STENÁNTHIUM Gray

Perianth spreading; sepals and petals narrowly lanceolate, tapering from base to an attenuate to bluntish tip, much longer than the stamens, coherent at base to ovary. Capsule 3-locular, 3-beaked at summit. Seeds elongate, wingless or nearly so. — Smooth, slightly bulbous perennials of N.Am. and e. Asia, with virgate stem leafy toward base, long and grass-like conduplicate and keeled leaves, and numerous small flowers in compound racemes, forming a long terminal panicle; flowering in summer. (Name from the Greek *stenos*, *narrow*, and *anthos*, *flower*, from the narrow sepals and slender panicles.)

1. S. gramíneum (Ker) Morong (grass-like), Featherbells. — Stem leafy below or with reduced leaves up to the panicle, 0.25–1.9 m. high, from slightly bulbous base; leaves ascending, 0.4–3 cm. broad; panicle elongated, 0.7–9 dm. long, lanceolate to lance-ovoid, the racemiform branches bearing wholly staminate or rarely some perfect sessile or short-pedicelled whitish, greenish or bronze-purple flowers, the terminal unbranched spiciform axis with subsessile to pedicelled perfect flowers; perianths 3–10 mm. long; the sepals and petals linear-lanceolate, attenuate (rarely bluntish); capsules oblong-subcylindric to ovoid-urceolate, 6–10 mm. long, with 3 short spreading beaks; seeds obliquely lanceolate, 5–8 mm. long. — With 3 confluent vars.:

a. Stem (dry) 4–10 mm. thick at lowest exposed internode; leaves rather crowded below, rapidly diminishing below panicle, firm to coriaceous, mostly opaque, the larger ones 4–15 mm. wide, their prominently raised ribs producing a corrugated surface; panicle lax, the branches distant, the flowers mostly subremote along the often flexuous branches; perianth 3–8(–10) mm. long; capsules ovoid-urceolate, 6–9 mm. long, on spreading to reflexed pedicels; seeds 5–5.5 mm. long.
 Stem 0.5–1.9 dm. high, 4–10 mm. thick at base; perianth 5–10 mm. long. *S. gramineum* (typical).
 Stem 0.25–1 m. high, 1.5–5 mm. thick at base; perianth 3–4.5 (–5) mm. long. Var. *micranthum*.
a. Stem (dry) 7–15 mm. thick at lowest exposed internode, up to 1.8 m. high; leaves crowded and numerous nearly up to panicle, thin and membranaceous, translucent, the larger ones 1–3 cm. wide, their ribs mostly immersed in the tissue; panicle usually dense, with flowers crowded along the stiffly ascending branches; perianth 5–10 mm. long; capsules oblong-subcylindric, 9–10 mm. long, crowded and ascending to horizontally spreading; seeds 5–8 mm. long. . . . . Var. *robustum*.

S. gramíneum (typical). — Rich woods, thickets and borders of bottoms, nw. Pa. to Ill. and Mo., s. to se. Va., upland to nw. Fla., Ala., La. and e. Tex. Mid-June–Sept.

Var. **micránthum** Fern. (tiny-flowered). — Upland woods, w. Va. and e. Tenn. to nw. Fla. and n. Ala. July–Sept.

Var. **robústum** (S. Wats.) Fern. (stout), *S. robustum* S. Wats. — Se. Pa., Md. and D.C. to Ind. July–Sept.

8. ZIGÁDENUS Michx.

Flowers perfect or polygamous. Perianth withering-persistent, spreading; the petals and petal-like sepals oblong or oval, 1–2-glandular near the more or less narrowed but rarely definitely unguiculate base. Stamens free from sepals and petals and about as long. Anthers, styles and capsules nearly as in *Melanthium*. Seeds angled, rarely margined. — Smooth and often glaucous N.Am. and Asiatic perennials with rhizomes or bulbs, and with rather large panicled or racemed white to yellow, greenish or bronze flowers in summer. (Name composed of the Greek *zygos*, *a yoke*, and *aden*, *a gland*, the glands being sometimes in pairs.)

a. Stem from creeping rhizome; glands 2, distinct, above the narrowed base of each sepal and petal. . . . . 1. *Z. glaberrimus*.
a. Stem from bulbous-thickened base; gland solitary, or 2-lobed at summit. . . *b.*
 b. Perianth coherent with base of ovary, 0.7–1.5 cm. long, equaling to about half the length of capsule; glands conspicuous, obcordate or 2-lobed, reaching nearly to middle of petal and sepal.
  Middle and upper bracts of inflorescence herbaceous, tapering to a firm subulate tip; sepals and petals strongly suffused on the back with green, bronze or purple; capsule ovoid-conic, 5–8 mm. in diameter, barely exceeding perianth. . . . . 2. *Z. glaucus*.
  Middle and upper bracts with scarious margins and summits; sepals and petals pale, without or with only small darkened spot outside at base; capsule lance-conic, 4–6 mm. in diameter, twice as long as perianth. 3. *Z. elegans*.
 b. Perianth free from ovary, 4–8 mm. long, only one-third to one-fourth length of capsule; glands smaller, basal. . . *c.*
  c. Outer coats of bulbous base papery, not fibrous; lower and middle bracts of panicle or raceme scarious; mature pedicels 1–2.5 cm. long.
   Lower bracts prolonged, caudate, deciduous; sepals and petals 6–8 mm. long; capsules ellipsoid. . . . . 4. *Z. Nuttallii*.

Lower bracts short, persistent; sepals and petals 4–5 mm. long; capsules slenderly conical. . . . . 5. *Z. densus.*
c.*Outer coats fibrous; lower and middle bracts herbaceous; mature pedicels 0.4–1.5 cm. long. . . . . 6. *Z. leimanthoides.

1. Z. glabérrimus Michx. (very smooth). — *Rhizome subligneous,* blackish, horizontal, *elongate;* stem 6–12 dm. high, leafy; basal leaves elongate, linear, attenuate, firm; panicle loosely pyramidal, 1–2.5 dm. long; flowers perfect; *petals and sepals* 1–1.5 cm. long, white, lance-ovate, *acute, with 2 distinct glands above the short but definite claw; capsule lance-conic, barely equaling the connivent old perianth.* — Savannas and wet pinelands, Fla. to La., n. to se. Va. July–Sept.

2. Z. glaûcus Nutt. (blue-green), White Camass. — *Bulbous,* the outer bulb-coats fibrous, very glaucous; stems 2.5–9 dm. high, few-leaved; *leaves* chiefly basal, *coriaceous,* linear, channelled, elongate, obtuse to subacute, 0.5–2 cm. broad; cauline reduced, becoming bracts; inflorescence a few-forked *elongate-lanceolate to open-ovoid panicle* (rarely a simple raceme) 0.5–4.5 dm. long; *its bracts* like the upper leaves, reduced, *herbaceous, tapering to firm subulate tips; base of perianth coherent with ovary,* the elliptic-oblong to narrowly ovate or obovate *sepals and petals* 7–15 mm. long, creamy-white, *strongly suffused on the back with green* (*or bronze or purplish*), with *the conspicuous basal* greenish or bronze *glands obcordate or 2-lobed at summit and about half the length of blade; capsule ovoid-conic,* with recurving beaks, 1–1.4 *cm. long,* 5–8 *mm. in diameter, barely exceeding the finally connivent perianth;* seeds oblanceolate, 4–5 mm. long. (*Z. chloranthus* of ed. 7, probably not Richards.; *Anticlea glauca* (Nutt.) Kunth) — Calcareous gravel, cliffs, shores and bogs, e. Saguenay Co., Que., to Minn., s. to ne. N.B., w. N.Y., n. O., n. Ind. and Ill.; mts. of Va. and N.C. Mid-July–Sept.

3. Z. élegans Pursh (elegant), White Camass, Alkali-grass. — Similar; *leaves more attenuate at tip, thinner; inflorescence* commonly *a slender* loose cylindric *raceme,* rarely a panicle; *middle and upper bracts with scarious margins and summits,* blunt to mucronate; pedicels usually more slender; *perianth paler, without or with only small darkened spot outside at base; capsule lance-conic,* 1.3–2.2 *cm. long,* 4–6 *mm. in diameter, twice as long as perianth;* seeds 5–6 mm. long. (*Anticlea elegans* (Pursh) Rydb.) — Prairies, meadows and calcareous rocks, Alaska to Ariz. and N.M., e. to Man., Minn., Ia. and Mo. June, July.

4. Z. Nuttállii Gray (for its discoverer, Thomas Nuttall, 1786–1859), Death-Camass, Poison Camass. — *Outer bulb-coats papery, not fibrous;* stem stout, 3–7.5 dm. high, leafy-bracted; leaves mostly basal, coriaceous, falcate; *raceme* (rarely paniculate-branched) thick-cylindric, 1–2 dm. long, *its scarious bracts caudate,* deciduous; *perianth* yellowish-white, *free; sepals and petals* 6–8 *mm. long,* oval, each *with an obovate basal gland; capsules* erect, *ellipsoid, thin-walled, three or four times length of perianth, on filiform pedicels* 1.5–2.5 *cm. long.* (*Toxicoscordion Nuttallii* (Gray) Rydb.) — Prairies and calcareous rocks, Tenn. to Kans. and Tex. May.

5. Z. dénsus (Desr.) Fern. (dense), Black Snakeroot, Crow-poison. — Bulbous, the barely thickened *bulbs with smooth coats;* leaves linear, narrow; stem slender, 0.4–1.5 m. high, remotely bracted; raceme simple (rarely branching below), subcylindric, 0.5–2 (rarely –4) dm. long, 3–5 cm. thick; *bracts small, firm, brownish, persistent;* perianth creamy-white to pink, nearly or wholly free; *sepals and petals* 4–5 *mm. long,* each usually with a very small gland at base when fresh; *capsules slenderly conical, on pedicels* 1–2 *cm. long.* (*Z. angustifolius* (Michx.) S. Wats.; *Amianthium ang.* Gray; *Tracyanthus ang.* Small) — Damp pinelands and bogs, Fla. to e. Tex., n. to se. Va.; mts. of w. N.C. and Tenn. May, June. — Strongly resembling *Amianthium Muscaetoxicum* and similarly poisonous; but often with the perianth-glands (when fresh) of *Zigadenus.*

6. Z. leimanthoides Gray (like *Leimanthium*). — Resembling no. 4; *outer bulb-coats fibrous;* stem slender, 0.5–2.5 m. high; leaves elongate-linear; *flowers more crowded in panicled racemes* (central axis 1.2–3 dm. long); *lower and middle bracts of panicle herbaceous; sepals and petals* creamy or yellow, each *with a deeper yellowish spot on the contracted base; capsules slender-conic, on pedicels only* 0.4–1.5 *cm. long.* (*Oceanorus* Small) — Sandy pinelands and bogs of the Coastal Plain and Piedmont, Ga. to s. La.; mts., Ala. to Va. and W.Va.; Coastal Plain, Del., N.J. and L.I. June–Aug.

9. MELÁNTHIUM L.

Perianth of 6 separate and free widely spreading somewhat cordate or oblong and hastate or oblanceolate sepals raised on slender claws, at first cream-colored or greenish. Filaments shorter than the divisions of the perianth, adhering to their claws, often to near the summit, persistent. Anthers cordate or reniform, confluently 1-locular, peltate after opening, extrorse. Capsule ovoid-conical, 3-lobed, of 3 inflated membranaceous several-seeded carpels; seeds flat, broadly winged. — Stems tall and leafy, from a thick rootstock, roughish-downy above, as well as on the open and ample pyramidal panicle (composed chiefly of simple racemes), the terminal part of the panicle mostly fertile. Leaves linear to oblanceolate or oval, not plaited; small N.Am. genus. (Name composed of the Greek *melas, black,* and *anthos, flower,* from the darker color which the persistent perianth assumes after expanding.)

Sepals and petals each with a conspicuous double gland at base; the oblong-ovate to suborbicular blade often auricled at base.
 Leaves broadly linear; blades of sepals and petals oblong to ovate, flat, blunt, much longer than claw; stamens borne at or above middle of claw. . . . 1. *M. virginicum.*
 Leaves oblanceolate; blades of sepals and petals suborbicular, undulate-crisped, abruptly pointed, little longer than claw; stamens borne at base of claw. 2. *M. hybridum.*
Sepals and petals without glands at base, oblanceolate, tapering to claw; stamens borne near base of claw. . . . . 3. *M. parviflorum.*

1. M. virgínicum L. (Virginian), Bunchflower — Stem 0.8–1.7 m. high, scurfy above; *leaves firm, broadly linear, attenuate,* 1.5–3 *cm. broad;* panicle 1.5–4.5 dm. long, with ascending to spreading lateral branches; flowers creamy, changing to green or purplish, scurfy outside; *blades of sepals and petals broadly oblong to ovate, round-based, cordate or hastate, flat, blunt,* 5–8 *mm. long, two or three times length of claw,* with 2 dark glands at base; *stamens borne at or above middle of claw;* capsule erect, ovoid, with furrows between the round-backed carpels, 1.3–1.8 cm. high, 3-beaked; seeds whitish, narrowly obovate, about 10 in each locule, 5–7 mm. long. — Meadows, swales, savannas and low thickets, n. Fla. to Tex., n. to s. N.Y., O., Ind., Ill. and Ia. Mid-June, July.

2. M. hýbridum Walt. (hybrid; from the mixture of fertile and infertile flowers). — More slender, up to 1.2 m. high; *leaves* thinner, *oblanceolate, acuminate,* 3–6 *cm. wide; blades of sepals and petals suborbicular, abruptly pointed, undulate-crisped,* 2.5–5.5 *mm. long, about equaling slender claw,* with 2 dark glands at base; *stamens borne at base of claw;* capsules subtruncate at summit; seeds 4–8 in each locule, about 8 mm. long. (*M. latifolium* Desr.) — Open woods and slopes, local, Piedmont and upland reg., Ga. to s. N.Y. and s. Ct. Late June, July.

3. M. parviflòrum (Michx.) S. Wats. (small-flowered). — Slender, 0.6–1.6 m. high; *leaves submembranous, elliptic to narrowly obovate, petioled,* short-tipped, 5–10 *cm. wide;* panicle loose and open, 0.4–1 m. long; *sepals and petals oblanceolate, tapering to claw, without glands,* 4–6 mm. long; *stamens borne near base of claw,* shorter than in nos. 1 and 2; seeds 4–6 in each locule, about 5 mm. long. (*Veratrum* Michx.) — Rich woods in the mts., w. Va. and W.Va. to Ga. and e. Tenn. July–early Sept.

10. VERÀTRUM L. False Hellebore. Varaire (Que.)

Perianth of 6 spreading obovate-oblong (greenish or brownish to blackish) divisions, more or less contracted at the base (but not clawed), nearly free from the ovary, not gland-bearing. Filaments free from and shorter than the sepals, recurving. Anthers, pistils, fruit, etc. nearly as in *Melanthium.* — Somewhat pubescent perennials of N. Hemisph., with simple stems from a thickened base producing coarse fibrous roots (very poisonous), 3-ranked plaited and strongly veined leaves, and racemed-panicled dull or dingy flowers; in summer. (Latin name of the Hellebore.)

1. V. víride Ait. (green), White Hellebore, Itchweed, Indian Poke, Tabac du diable (Que.). — *Stem stout, very leafy* to the top, 6–20 dm. high; *leaves broadly oval,* pointed, *sheath-clasping; panicle pyramidal,* the *dense spike-like racemes* spreading; *perianth yellowish-green,* moderately spreading, *the segments ciliate-serrulate; ovary glabrous;* capsule many-seeded. — Swamps and low grounds, Saguenay Co., Que., to Minn., s. to N.B., N.E., Md., and upland to Ga. and Tenn. May–July.

2. V. Woòdii Robbins (in honor of Alphonso Wood, 1810–1881). — *Stem slender, sparingly leafy,* 8–14 dm. high; *leaves oblanceolate,* only the lowest sheathing; *panicle very narrow; perianth greenish-purple to purple-black, with entire segments; ovary tomentose,* soon glabrate; capsule few-seeded. — Woods, hillsides and bluffs, O. to Ia., s. to Mo. and Okla. July–Sept.

11. UVULÀRIA L. Bellwort. Wild-oats. Merry-bells

Perianth slenderly campanulate, lily-like; the narrow sepals and petals stramineous to orange-yellow, nearly parallel, gibbous at base; stamens much shorter, barely adherent to base of petals. Style deeply cleft. Capsule coriaceous to submembranaceous, 3-angled, either truncate and 3-lobed at summit or ellipsoid and pointed, loculicidal, tardily dehiscent. Seeds few in each locule, globose or obovoid. — Low e. N.Am. herbs with a short or elongate rhizome, simple or usually forking above, bearing lance-oblong to oval perfoliate or sessile divergent leaves, the 1–few pendulous flowers at first terminal, but in fruit seemingly subaxillary through prolongation of the stem and branches. (Name "from the flowers hanging like the uvula, or palate".) Including Oakesia Wats. or Oakesiella Small

Leaves perfoliate; rhizome short, with fleshy fibers; capsule truncate at summit.
 Leaves usually glabrous; sepals and petals granular-pubescent within. . . . 1. *U. perfoliata.*
 Leaves usually pubescent beneath; sepals and petals smooth within. . . . 2. *U. grandiflora.*
Leaves sessile, not perfoliate; rhizome short or elongate; capsule ellipsoid.
 Rhizome very short, bearing several approximate fleshy fibers; leaves bright green both sides; ovary and capsule sessile at summit of receptacle. . . . 3. *U. pudica.*
 Rhizome elongate, with scattered fibers; leaves pale beneath; ovary and capsule raised on a stipe. . . . . 4. *U. sessilifolia.*

1. U. perfoliàta L. (perfoliate). — Rhizome very short, bearing a cluster of long fleshy fibers; stem slender, 2–6 dm. high; *upper bladeless sheath of stem* 3–5.5 *cm. long; leaves glaucous,* oblong to ovate-lanceolate, perfoliate at base, usually *glabrous,* 1–4 *below the fork,* the mature leaves 3.5–11 cm. long; the *sterile branch* (when present) finally elongating, *with* 1–4 *leaves; perianth* 2–3.5 cm. long, the *segments glandular-roughened within;* stamens shorter than style, tip of connective acuminate; locules of *truncate capsule* with 2 dorsal ridges and 2-beaked at apex. — Thin woods, thickets and clearings in acid to circumneutral soils, Mass. and s. Vt. to s. Ont., s. to n. Fla., Ala., Miss. and n. La. Late Apr.–early June.

2. U. grandiflòra Sm. (large-flowered). — Larger than no. 1, up to 7.5 dm. high in maturity, *not glaucous; upper bladeless sheath* 5–12 *cm. long; leaves* rather larger, 4.5–13 cm. long, usually *whitish-pubescent beneath,* 1 *or* 2 *below the fork;* the *sterile branch* (when developed) *with* 4–8 *leaves; perianth* deeper orange-yellow, 2.5–5 cm. long, *its segments smooth within; stamens exceeding style, obtusely tipped;* capsule obtusely lobed. — Rich woods and thickets, chiefly calcareous, sw. Que. to N.D., s. to Ga., Ala., Ark. and Okla. Apr.–early June.

3. U. pudìca (Walt.) Fern. (modest). — *Rhizome* much as in nos. 1 and 2, *very short, with several approximate fleshy fibers; stem* 1–4.5 dm. high, *with minutely puberulent lines; leaves bright green and shining on both sides,* oval, rounded to the *sessile* base, minutely serrulate, in maturity with cross-reticulations prominent beneath; perianth straw-yellow, 1.5–3 cm. long; common style shorter than to exceeding the *acute anthers; ovary sessile at summit of receptacle; capsule* membranaceous, *trigonous-ellipsoid, not stalked* above the receptacle, 1.5–3.3 cm. long. (*U. puberula* Michx.; *Oakesia puberula* Wats.; *Oakesiella puberula* Small) — Woods, Va. and W.Va. to Ga. and Ala. Apr., May.

Var. **nítida** (Britt.) Fern. (shining). — Stem glabrous; leaves thinner, the reticulations not prominent beneath, the margins less scabrous; capsule usually short. (*U. nitida* (Britt.) Mackenz.) — Sandy pine-barrens, L.I. and N.J.; se. Va. May, early June.

4. U. sessilifòlia L. (sessile-leaved), Wild-oats. — Resembling no. 3 but *rhizome elongate, with scattered roots;* stem glabrous; *leaves glaucous beneath,* the reticulations not conspicuous in maturity; *anthers obtuse; ovary and capsule distinctly stalked above the receptacle.* (*Oakesia* S. Wats.; *Oakesiella* Small) — Woods, thickets and clearings, N.B. to N.D., s. to N.S., N.E., Ga., Ala. and Mo. Apr.–mid-June.

Cólchicum L. (from ancient *Colchis* on the Black Sea), a genus of Eurasia, with deep-seated corm and long-tubed crocus-like flowers but with superior ovary, is much cult. C. autumnàle L. (autumnal), with large lanceolate vernal leaves and late autumnal purplish flowers, spreads locally to meadows and fields. (Introd. from Eu.)

12. ÁLLIUM L. Onion. Garlic. Leek. Ail (Que.)

Perianth of 6 entirely colored sepals and petals which are distinct, or united at the very base, 1-nerved, often becoming dry and scarious and more or less persistent; the 6 filaments subulate or dilated at base. Style persistent, filiform; stigma simple or only slightly 3-lobed. Capsule lobed, loculicidal, 3-valved, with 1–2 or more ovoid-reniform amphitropous or campylotropous black seeds in each locule. — Strong-scented and pungent herbs of N. Hemisph.; the leaves and usually scapose stem from a coated bulb; flowers in a simple umbel, some or all of them frequently replaced by bulblets; spathe 1–3-valved. (The ancient Latin name of the Garlic.)

a. Umbel bearing or changed to small bulbs. . . *b.*

b. Scape naked above the subbasal soft and smooth flattish leaves; spathe usually of 3 broad whitish short-beaked valves; outer bulb-coats coarsely fibrous; perianths, when present, equaling or exceeding stamens. . . . . 1. *A. canadense.*

b. Stem leafy nearly to middle; leaves minutely toothed or scabrous on margins, keels or ribs; spathe 1-or 2-valved; outer bulb-coats membranous or if striate, only tardily a little fibrous. . . *c.*

c. Leaves flattened at least toward summit; spathes herbaceous.

Leaf-blades flat, not twisted, 0.5–1.5 cm. wide, acute; spathe of 1 deciduous long-beaked valve; bulbs white, with membranous coats; perianth, when present, whitish or greenish, its lance-acuminate segments exceeding the stamens. . . . . 2. *A. sativum.*

Leaf-blades thick and channeled at base, flattened above, round-tipped, narrowly linear, firm, twisting before anthesis; spathe of 2 united but finally separate prolonged slender unequal valves; bulbs with fuscous to pale finely striate outer coats; perianth yellow-brown, drying purplish, with truncate-rounded segments slightly shorter than stamens. . . . . 3. *A. oleraceum.*

c. Leaves subterete, hollow, striate, easily crushed, acute; spathe of 1 caducous scarious bract. . . . . 4. *A. vineale.*

a. Umbel without bulbs. . . *d.*

d. Leaves narrow or slender, linear or terete, present at flowering time; capsule slightly lobed, each locule with 2–several ovules or seeds. . . *e.*

e. Pedicels equaling or exceeding the flowers. . . *f.*

f. Leaves extending one-third to half-way up the stiff stem (0.3–1.8 m. high); 3 interior filaments with dilated bases, 3-cuspidate near apex.

Leaves terete or subterete, slender; stem slender; umbel lax, hemispherical to subspherical, 2–5 cm. broad. . . . . 4. *A. vineale.*

Leaves flat or folded, 1–3 cm. broad; stem stout; umbel globose, very dense, 6.5–8 cm. in diameter. . . . . 5. *A. Ampeloprasum.*

f. Leaves basal or nearly so, flat or channelled, narrowly linear; scape slender, 1–6 dm. high; umbel hemispherical to loosely subspherical, 1.5–6 cm. broad; filaments all plane or the inner ones at most 1-toothed on each margin. . . *g.*

g. Sepals and petals shorter than the finally exserted stamens; ovary and capsule with 6 high crests below the summit; outer bulb-coats not strongly fibrous.

Scape strongly arched at summit, the umbel nodding; pedicels loosely arching or flexuous; leaves thin, soft and flat. . . . 6. *A. cernuum.*

Scape erect, not arched at summit; pedicels stiffly ascending or spreading; leaves thick, hard and rounded on the back, channelled. . . . . 7. *A. stellatum.*

g. Sepals and petals longer than the stamens; outer bulb-coats coarsely fibrous or ropy. . . *h.*

h. Summit of ovary or capsule without crests.

Sepals and petals lanceolate, thin, loosely spreading or recurving in fruit. . . . . 8. *A. mutabile.*

Sepals and petals ovate, thick and firm, closely appressed to fruit. . . . . 9. *A. Drummondi.*

h. Summit of ovary or capsule with 6 crests. . . . . 10. *A. textile.*

e. Pedicels much shorter than pink or roseate flowers, the umbel subcapitate; leaves terete. . . . . 11. *A. Schoenoprasum.*

d. Leaves elliptic-lanceolate, fleshy, 2–6 cm. broad, shriveling before anthesis; bulbs clustered on a rhizome; ovary and fruit deeply 3-lobed, with 1 ovule or seed in each locule. . . . . 12. *A. tricoccum.*

1. A. canadénse L. (Canadian), Wild Garlic. — *Bulb* ovoid to subglobose, up to 2.5 cm. thick, *the outer brown or drab coat fibrous*, the inner coats white and lustrous; *leaves* subbasal, narrowly linear, *soft and flaccid, flattish*, slightly keeled, *smooth;* scape erect, 2–6 dm. high; *spathe* usually *of 3 ovate* short-beaked *whitish valves*, subpersistent; *umbel chiefly a capitate cluster of whitish* to purplish *bulbs;* flowers few or none, on filiform loosely ascending pedicels; *perianth whitish or pink-tinged, the narrowly lanceolate sepals and petals equaling or exceeding the stamens.* — Low woods, thickets and meadows, Fla. to Tex., n. to w. N.B., s. Que., s. Ont., Wisc. and Minn. May–early July.

2. A. sativum L. (sown), Garlic. — Coarser; *bulb* up to 3 cm. or more thick, *its coats* *membranous; leaves extending nearly to middle of stem, firm, flat,* 0.5–1.5 *cm. wide,* scabrous on margin and keel, *acute;* stem up to 1.2 m. high; *spathe a single green and herbaceous long-beaked deciduous valve;* umbel bearing bulbs; *perianth, when present,* whitish or greenish, *exceeding the stamens; inner filaments dilated at base, 3-cleft, the 2 lateral segments prolonged into flexuous awns,* the short central one bearing the anther. — Roadsides, pastures, fields, etc., N.Y. to Ind., s. to Ky., Tenn. and Mo., esc. from cult. June, July. (Introd. from Old World)

3. A. OLERÀCEUM L. (like a vegetable), WILD GARLIC. — Somewhat resembling no. 4; *outer coats of bulb minutely ribbed,* not fibrous; stem rather weak, often flexuous, 2–6 dm. high, leafy nearly to middle; *leaves* narrowly linear, *twisting or spiraling before anthesis, channelled at base, becoming flattened and firm toward the rounded apex,* minutely scabrous on margins and keel; *spathe of 2 unequal herbaceous slender persistent valves with free beaks but at first united below, later separating downward, the longer up to 2 dm. long;* umbel with bulbs and flowers; *perianths yellow-brown suffused with green,* often purplish in drying, *about equaling the stamens, on* slender *spreading to deflexed unequal pedicels;* segments obtuse to acutish; *filaments slender, without appendages.* (*A. carinatum* sensu Am. auth., not L.) — Borders of woods, thickets, roadside-banks, etc., the plants not colonial, e. Pa. to e. Va. Late July, Aug. (Natzd. from Eu.)

4. A. VINEÀLE L. (of vineyards), FIELD-GARLIC. — *Bulbs often crowded,* the older and secondary basal ones making dense mats, *their outer coats membranous but friable; stem becoming stiff,* erect, *leafy to near middle,* 0.3–1.2 m. high; *leaves at first terete,* later becoming channelled, *hollow, striate, the young ones easily compressed, attenuate; spathe usually* 1, *scarious, short,* beaked, *its edges united above; umbel projecting through base of the deciduous spathe,* subcapitate, 2–5 cm. in diameter, *with numerous* (sometimes proliferating) *bulbs* and few–many flowers on ascending to spreading pedicels, or umbel wholly of bulbs in forma COMPÁCTUM (Thuill.) Aschers. (compact), or nearly or quite *without bulbs* and with abundant flowers in forma CAPSULÍFERUM (W. D. J. Koch) Aschers. & Graebn. (bearing capsules); *perianth* greenish to purple, *its* lanceolate to elliptic *segments obtuse to acutish* and about equaling the stamens; *filaments dilated at base, the inner ones 3-cuspidate at summit.* — Grassland, fallow fields, etc., too abundant, Ga. to Ark., n. to Mass., N.Y., O., Mich. and Ill. Late May–July. (Natzd. from Eu.)

5. A. AMPELÓPRASUM L. (leek of the vineyard), WILD LEEK. — *Bulb* large, *with papery coats, splitting at flowering time* by pressure of the very numerous stalked acuminate bulblets borne from near the base within; *stem coarse, up to 1.8 m. high, scabridulous, with long-sheathing leaves nearly up to the middle; blades flat, thick, somewhat folded, 1–3 cm. broad,* soon shriveling, *scabrous on margin and keel,* the tip hooded; *spathe* 1-valved, globose, scarious, the firm herbaceous beak about equaling the blade, *deciduous by circumscission from the base; umbel globose* or hemispherical, *with innumerable flowers on radiating pedicels; perianth with connivent gibbous* oblong-lanceolate *segments* verrucose-scabrous on the back, greenish or white, purple-tinged at tip; *filaments* exserted, 3-*cuspidate at tip, the 2 lateral sterile cusps prolonged into flexuous awns thrice the length of the central fertile one.* — Òld World species, represented with us by

Var. ATROVIOLÀCEUM (Boiss.) Regel (dark violet). — Umbel globose, 6.5–8 cm. in diameter; flowers deep purple, smooth on back. — Grasslands, York Co., Va. June. (Natzd. from se. Eu. and the Orient)

A. PÓRRUM L. (Latin name of leek), LEEK, differing from no. 5 in slender bulb, *softer and smooth green* but solid *stem, soft smooth leaves* more nearly basal, *their sheaths membranaceous, spathe with soft beak, flowers pink or whitish* and shorter-pedicelled, the *filaments with short lateral cusps,* persists after cult. or about refuse. (Ancient cult. plant, source unknown). **A.** CÈPA L. (Latin name of onion), the common ONION, differing in having *broadly ovoid bulb, hollow terete leaves and stem,* umbel with or without bulbs, *pedicels much exceeding the white or greenish flowers,* filaments slender or the alternate ones with broad and short-toothed base, and **A.** FISTULÒSUM L. (hollow-stemmed), the WELSH ONION, *with slender bulbs* fascicled, the leaves and stem inflated, the *pedicels about equaling* the *flowers,* similarly persist (the former of unknown origin, the latter introd. from Asia).

6. A. cérnuum Roth (nodding), WILD ONION. — *Bulbs* 1–several, slender, *crowning a barely persistent rhizome,* their outer coats membranous; leaves flat and soft, 2–8 mm. wide; scape angulate-margined, 2–7.5 dm. high, *arching at summit;* spathe 2-valved, whitish, scarious, deciduous; *umbel nodding* or bent to one side; *pedicels* many or few, *loosely arching or flexuous; perianth* pink to deep roseate or white, its segments obtuse or rounded at summit and *much shorter than stamens; ovary and capsule bearing 2 high crests just below summit of each valve.* (Incl. *A. allegheniense* Small, form with deep-colored flowers, and *A. oxyphilum* Wherry, sparsely flowered form with whitish flowers) — Ledges, gravels, rocky or wooded slopes and crests, ascending to high alts., N.Y. to B.C., s. to Ga., Tenn., Mo., Tex. etc. July, Aug.

7. A. stellàtum Fraser (starry), WILD ONION. — Differing from no. 6 in having only 1 or 2 bulbs borne on a rhizome, their outer coats finely lineolate-reticulate; *leaves thick, hard and rounded on back, channelled; scape stiffly erect, not arching at summit,* terete except at summit; spathes firmer and more persistent; *umbel erect, with stiffly ascending to spreading pedicels; outer perianth-segments acute or acutish.* — Rocky prairies, slopes, shores and ridges, O. to Sask., s. to Ill., Mo., Kans. and Tex. July, Aug.

8. A. mutábile Michx. (changeable; Michaux having confused two different species), WILD ONION. — *Bulb with coarsely fibrous outer coats; leaves* narrowly linear, *attenuate, firm, channelled;* scape 2–6 dm. high, terete; spathe of 2 or 3 ovate abruptly acuminate persistent valves; umbel many-flowered, with *filiform* ascending to spreading *pedicels twice or thrice length* of lavender to white *perianth; sepals and petals thin, lance-acuminate, loosely spreading or recurving in fruit, shorter than the stamens; ovary and capsule without crests.* — Prairies, calcareous barrens, bluffs, etc., n. Fla. to Tex., n. to N.C., Mo. and Neb. May, June.

9. A. Drummóndi Regel (for its discoverer, THOMAS DRUMMOND, 1780–1835), WILD ONION. — Usually smaller than no. 8; the scapes 1–4.5 dm. high; pedicels subtended by longer bracteoles; *sepals and petals ovate, firm and thick, closely appressed to capsule.* (*A. Nuttallii* S. Wats.) — Dry prairies and calcareous hills, sw. Mo. and Kans. to Tex. May.

10. A. téxtile Nels. & Macbr. (woven), WILD ONION. — *Bulb with intricately fibrous outer coats;* leaves 2–4 mm. wide, firm, channelled; scape 1–3 dm. high, firm; spathes much as in nos. 8 and 9; umbel resembling that of no. 9; *sepals and petals thin, loosely enveloping capsule, ovate to lanceolate, acuminate; ovary and capsule with 2 crests at summit of each valve.* (*A. reticulatum* Don, not Presl) — Dry prairies, open woods and calcareous rock, Man. to Alta., s. to Ia., Neb. and N.M. May, June.

11. A. SCHOENÓPRASUM L. (Greek name, rush-leek), CHIVES, CIBOULETTE or BRÛLOTTE (Que.). — *Bulbs crowded and persistent* from a rhizomatous base, *making close sod,* the oblong-ovoid bulbs with membranous coats; *leaves* slender, *terete or semiterete, hollow,* smooth, nearly equaling to overtopping the scape; scapes 1.5–5 dm. high; slender terete spathe pink or whitish, splitting into 2 or 3 ovate acute or short-acuminate valves shorter than the umbel; *umbel hemispherical to subpyramidal,* sometimes becoming subglobose, 2.5–4 cm. broad; *pedicels shorter than* or lowest about equaling *flowers;* perianth pink or purple, 1–1.2 cm. long, the sepals oblong- or ovate-lanceolate and acute. — Much cult., often esc. and persisting northw. June–Aug. (Introd. from Eurasia)

Var. **laurentiànum** Fern. (of region of Gulf of St. Lawrence). — *Bulbs solitary, rarely* 2; *leaves* coarser, *rarely reaching inflorescence;* scape coarser, 2–6 dm. high; *umbel* 2.3–3.3 *cm. in diameter;* pedicels much shorter than flowers; *perianth* 8–10 *mm. long,* deeply colored, the *sepals ovate to ovate-lanceolate.* — Calcareous rocks, gravels and shores, w. Nfld. to Algoma Distr., Ont., s. to N.B., n. N.Y. and s. Ont.; Mont., Wash. and Oreg. June–Aug. (E. Asia)

Var. **sibíricum** (L.) Hartm. (Siberian). — Similar to preceding var.; *bulbs* 1 *or* 2 (rarely –6); *leaves coarse, rarely reaching umbel; scape coarse,* 2–7.5 *dm. high; umbel* 3.5–5 *cm. in diameter; perianths* 1–1.4 *cm. long,* pale to deep pink; *sepals lanceolate to lance-linear, attenuate.* (*A. sibiricum* L.) — Calcareous or basic rock, gravels and shores, Gaspé Pen., Que., to Alaska, s. to N.S., n. N.E., Mich., Minn., Colo. and Wash. June–Aug. (Eurasia)

12. A. tricóccum Ait. (3-locular), WILD LEEK, RAMP, AIL DES BOIS (Que.). — *Bulbs* slenderly ovoid, 2–5 cm. long, *forming a crown on a rhizomatous base, with fleshy coats; leaves elliptic-lanceolate,* fleshy, *flat,* 1–3 dm. long, 2–6 *cm. wide,* appearing in early spring and *usually shriveled before flowering time;* scape slender, firm, 1.5–4 dm. high; umbel hemispherical, with straight ascending to divergent pedicels; perianth white or whitish, its oblong sepals and petals equaling the nearly distinct filiform-subulate filaments; *capsule deeply 3-lobed, each locule 1-seeded.* (*Validallium* Small) — Rich woods and bottoms, w. N.B. and s. Que. to Minn., s. to N.S., N.E., Del., Md., upland to Ga. and Tenn., Ill. and Ia. June, July.

13. NOTHOSCÓRDUM Kunth FALSE GARLIC

Flowers greenish or yellowish-white. Capsule obovoid, somewhat lobed, obtuse, with the style obscurely jointed on the summit; locules several-ovuled and -seeded. Filaments filiform, distinct, adnate at base. — Bulb tunicated, not alliaceous. Otherwise as in *Allium.* Small genus of N. and S.Am. and China. (Name from the Greek *nothos, false,* and *scordon, garlic.*)

1. N. biválve (L.) Britt. (two-valved). — Scape 1.5–3.5 dm. high; bulb small, often bulbiferous at base; leaves narrowly linear; flowers few, on slender pedicels; the segments narrowly oblong, about 1 cm. long; ovules 4–7 in each locule. — Low sandy woods, grassland, prairies and disturbed soil, Fla. to Tex. and Mex., n. to Va., s. O., Ind., s. Ill. and Neb. April, May, rarely Oct., Nov. (W.I.)

14. HEMEROCÁLLIS L. DAY-LILY. LIS D'UN JOUR (Que.)

Perianth funnelform, lily-like; the short tube inclosing the ovary, the spreading limb 6-parted, the 6 stamens inserted on its throat. Anthers as in *Lilium*, but introrse. Filaments and style long and filiform, declined and ascending; stigma simple. Capsule (at first rather fleshy) 3-angled, loculicidally 3-valved, with several black spherical seeds in each locule. — Showy perennials, nat. of Eurasia, with fleshy-fibrous roots or tubers; the long and linear keeled leaves 2-ranked at the base of the tall scapes, which bear at the summit several bracted and large flowers; these collapsing and decaying after expanding for a single day (whence the name, from the Greek *hemera, a day*, and *callos, beauty*).

1. H. FÚLVA L. (reddish-yellow), COMMON ORANGE D. — Leaves 1–2 cm. broad; scape 0.5–2 m. high, 3–15-flowered; flowers 1–1.3 dm. long, tawny-orange, deeper-colored toward center, the 3 inner perianth-segments (petals) wavy-margined and obtuse. — Roadsides, borders of fields and thickets, through our range. May–July. (Introd. and natzd. from Eurasia) — Originally derived from gardens, rapidly spreading by branching rhizomes and tuberous roots; rarely perfecting seeds with us.

Var. KWÁNSO Regel (native Japanese name). — Flowers double, continuing to expand over a longer period. — Cult., and locally natzd. in e. Va. (Introd. and natzd. from Asia)

2. H. FLÀVA L. (yellow), YELLOW D. — Smaller; with fragrant yellow flowers. — Locally spread from cult., N.E. to Mich. and Pa. May, June. (Introd. from Asia)

15. HÓSTA Tratt. PLANTAIN-LILY or FUNKIA

Perianth much as in no. 14, but with enlarged throat; the bluish, lavender or whitish flowers loosely ascending, divergent or drooping. The 6 stamens inserted on the tube or hypogynous. — Ornamental rhizomatous perennials, nat. of e. Asia, with fibrous roots, broad petioled several-ribbed basal leaves, the scapes often with bracts. (For *Nicolaus Thomas Host*, 1761–1834.) FUNKIA Spreng.*

1. H. PLANTAGÍNEA (Lam.) Aschers. (Plantain-leaved), FRAGRANT P. — Rhizome stout, branching, producing large circular mats; *leaf-blade cordate- or subcordate-ovate, up to* 2.5 *dm. long and* 1.5 *dm. broad, prominently* 7–9*-ribbed;* a large leaf-like bract below each flower; *flowers crowded, ascending or spreading, white, very fragrant,* 1–1.3 *dm. long,* 6–7 cm. broad at summit; *capsules long-cylindric, up to* 7 *cm. long,* 6–7 mm. thick. — Cult. and frequently spreading to roadside-thickets and waste places. Aug.–Oct. (Introd. from e. Asia)

2. H. VENTRICÒSA (Salisb.) Stearn (bellied out), BLUE P. — Leaves cordate or narrowed to base, up to 2 dm. wide; *scape* up to 1 m. high, *naked or with* 2 *or* 3 *remote small bracts; raceme elongate, with many spreading or finally drooping lavender-purple non-fragrant flowers about* 5 *cm. long; perianth blue* white-streaked within, *slender below the abruptly dilated throat; capsules* 3–4 *cm. long,* about 5 mm. thick. (*H. caerulea* Tratt., not Jacq.) — Cult.; natzd. and spreading in rich woods along streams, N.J., Pa., Del. and Md.; doubtless elsewhere. July–Sept. (Natzd. from e. Asia)

3. H. JAPÓNICA (Thunb.) Voss (Japanese), NARROW-LEAVED P. — *Leaves flat;* the blade broadly *lanceolate, up to* 1 *dm. long and* 4 *cm. broad; scapes bracted below the slender and lax raceme; flowers spreading, finally drooping, pale lilac, about* 4 *cm. long,* the *tube gradually expanding to the throat; capsules clavate,* about 2 *cm. long.* (*H. lancifolia* Tratt.) — Cult. and spreading to thickets, roadsides and waste places. July–Sept. (Introd. from e. Asia)

16. LÍLIUM L. LILY. LIS (Que.)

Perianth funnelform or campanulate, colored, of 6 divisions, spreading or recurved above, deciduous. Anthers linear to oblong, extrorsely attached near the middle to the tapering apex of the long filament, at first usually included but at length versatile, the locules dehiscent by a lateral or slightly introrse line. Style elongate or very short (dimorphic); stigma 3-lobed. Capsule subcylindric to ovoid; seeds densely packed in 2 rows in each locule. — Bulbs scaly producing simple stems with numerous alternate-scattered or whorled narrow sessile leaves, and from one to several large and showy flowers in summer. Plants of temp. N. Hemisph. (The classical Latin name.)

a. Flowers erect; sepals and petals narrowed below into claws; bulbs not rhizomatous. . . *b.*

---

* Treatment derived from that of L. H. BAILEY in Gentes Herbarum, ii, fasc. 3 (1930).

b. Sepals and petals loosely ascending, with suberect to slightly spreading blunt or merely acutish tips, gradually tapering into the short claws.

Petals and sepals glabrous, blotched near base with purple-brown oblong to roundish spots, their veins (seen by transmitted light) without pustules; leaves whorled or scattered. . . . . 1. *L. philadelphicum.*

Petals and sepals pubescent at base within, without large spots, their veins (seen by transmitted light) bearing semi-lanceolate to semi-ovate or -obovate pale to dark pustules; leaves scattered. . . . 2. *L. bulbiferum.*

b. Sepals and petals more loosely ascending, with outwardly curved attenuate tips, rounded to the slender elongate claws. . . . . 3. *L. Catesbaei.*

a. Flowers nodding, the buds erect, their peduncles arching before anthesis, then becoming erect in fruit; sepals and petals sessile; bulbs rhizomatous. . *c.*

c. Stems smooth and glabrous; upper leaves without bulblets; sepals and petals glabrous or nearly so within at base. . *d.*

d. Sepals and petals with ascending to spreading but not strongly recurved tips, the tips when outwardly arched not reaching the base of the perianth-tube; margins and nerves (beneath) of leaves bearing minute (often microscopic) spicules.

Flower nearly horizontal, with ascending sepals and petals 4–6 cm. long, these spotted inside nearly to apex; leaves 3–8 cm. long, in 3–6 whorls, the fuller whorls with 3–9 leaves. . . . . . 4. *L. Grayi.*

Flower nodding, with arching and outwardly curving sepals and petals 5–9 cm. long, these rarely spotted more than two-thirds their length; leaves mostly (5–) 6–18 cm. long in 6–11 whorls (besides scattered blades), the fuller whorls with 7–15 leaves. . . . . . . 5. *L. canadense.*

d. Sepals and petals strongly recurving, their tips reaching or surpassing the base of the perianth-tube. . . . *e.*

e. Leaves lanceolate, attenuate to base and apex, green, submembranaceous or membranaceous, with lateral veins evident beneath, the longer blades 5–18 cm. long.

Margins and often the veins of lower leaf-surfaces bearing minute spicules; perianth-segments with obscurely delimited green basal zone 5–12 mm. long; midrib of back of sepal low and rounded, obscure, of back of petal elevated and with rounded edges; anthers oblong; stigma broadly and shallowly 3-lobed; plant of the interior, flowering from late June through July. . . . . 6. *L. michiganense.*

Margins and veins of lower leaf-surfaces glabrous or with rounded papillae; basal green zone of perianth-segments clearly defined, 1–1.5 cm. long; midrib of back of sepal elevated and with 2 sharp ridges, of back of petal flat-topped and acute-margined; anthers linear; stigma deeply 3-lobed; southern and eastern plant, flowering from mid-July to early Sept. 7. *L. superbum.*

e. Leaves obovate to oblanceolate, blunt to abruptly short-tipped, glaucous (when fresh), fleshy to coriaceous, their lateral veins obscure, the longer blades 4–10 cm. long. . . . . 8. *L. Michauxii.*

c. Stems scabrous, more or less arachnoid above; the upper leaves subtending bulblets; sepals and petals copiously pubescent within at base of midrib. 9. *L. tigrinum.*

1. L. philadélphicum L. (of Philadelphia), Wild Orange-red L., Wood-L. — *Stem* 0.2–1.05 m. high, *glabrous, with* 2–6 *whorls of* lanceolate to lance-linear (rarely elliptic-oblong) *leaves* and often a few scattered blades; *flowers* 1–5, *erect, open-campanulate,* 5.5–8.5 cm. high, from faded orange to deep orange-red, or yellow in forma **flaviflòrum** E. F. Williams (yellow-flowered), *with oblong to roundish purple spots within;* the lanceolate or lance-oblong *sepals and petals glabrous, blunt or merely short-acuminate, tapering below into claws;* capsule 3–5.5 cm. long, broadly round-subtruncate at summit, slightly tapering at base. — Dry thickets, open woods and clearings, e. and centr. Me. to s. Que. and w. to s. Ont., s. to s. N.E., Del., Md., W.Va. and uplands to N.C. and Ky. Mid-June–mid-Aug. (–Sept.) — Northward and westw. passing insensibly to

Var. **andìnum** (Nutt.) Ker (of the Andes; Nuttall having extended the term to cover the Rocky Mts.). — *Leaves mostly scattered,* sometimes whorled at the upper 1–3 nodes, linear to lanceolate; flower intense red to scarlet. (*L. umbellatum* Pursh) — Dry or moist woods, glades, prairies, swamps and openings, w. Que. to B.C., s. to Ky., Ill., Ia., Neb. and N.M.

2. L. bulbíferum L. (bearing bulbs), Orange L. — Resembling var. of no. 1, with scattered leaves, the *uppermost leaves sometimes subtending axillary bulblets; summit of stem and peduncle sparsely pubescent; sepals and petals* of firmer texture, *pubescent within at base; their* *veins* (when seen by transmitted light) *showing semilanceolate to semiovate or semiobovate pale to dark pustules.* (*L. croceum* Chaix) — Roadsides and fields, Bellechasse Co., Que., locally to e. N.Y., esc. from cult. July. (Introd. from Eu.)

3. L. Catesbaèi Walt. (for its discoverer, Mark Catesby, (?)1679–1749), Southern Red L., Leopard- or Pine-L. — Habitally similar to no. 1; *bulb-scales slender, usually terminated by erect elongate linear blades; cauline leaves all scattered, appressed-ascending, greatly decreasing toward summit of stem,* linear to narrowly linear-lanceolate, *attenuate, the broader ones 2–12 mm. wide;* flower solitary, erect, red; *sepals and petals rounded to long slender claws, the long-tapering blades recurving from above the middle and about thrice the length of the claw,* the broadest blades 4.5–9.5 (av. 7) cm. long and 1–2.6 (av. 1.8) cm. broad; capsule gradually narrowed to a beak. — Damp pinelands, bogs and swamps, Fla. to La., n. to N.C. and (?) s. Ill.

Var. **Lôngii** Fern. (for one of its discoverers, Bayard Long, 1885–), Long's Red L. — *Bulb-scales plump, without leafy tips; lower and middle cauline leaves blunt and oblong,* 6–13 mm. wide; *sepals and petals loosely ascending, scarcely recurving at tip, the larger blades about twice length of claw,* 4.5–6.5 *(av.* 5.5) *cm. long and* 1.8–3.4 *(av.* 2.6) *cm. broad; capsule broadly rounded at summit.* — Wet pinelands, savannas and bogs, Ala. and Ga., n. to se. Va. Late July–Aug.

4. L. Graỳi S. Wats. (for its discoverer, Asa Gray, 1810–1888), Gray's L., Orange-bell L. — *Stem* 0.5–1 m. high, *with* 3–6 *whorls of* 3–9 *leaves each; the firm* lanceolate to lance-ovate or -oblong *blunt to acutish leaves* 3–8 *cm. long,* minutely scabrous on margin and nerves beneath; *flowers* 1–4, long-peduncled, *nearly horizontal* to slightly nodding, *open-campanulate,* 4–6 *cm. long,* deep reddish-orange, dark-spotted nearly to summit; *sepals and petals* broad-oblong, *with barely spreading short-pointed tips.* — Meadows, swamps and low woods among the mts., Va., N.C. and Tenn. Mid-June, July.

5. L. canadénse L. (Canadian), Wild Yellow or Canada L. — Stem 0.5–2 m. high, glabrous, terminated by 1-*few nodding flowers,* or flowers up *to* 22 *in* 1–3 *or* –4 *umbels; leaves* narrowly to broadly lanceolate, attenuate, *often scabrous along the margins and veins beneath with minute spicules, in* 6–11 *whorls,* besides scattered blades; *those of the fuller whorls* 7–15, *attenuate,* 0.5–1.8 *dm. long, one-tenth to one-fifth as broad; nodding flower yellow to orange,* or red in forma **rûbrum** Britt. (red), spotted within for half or two-thirds its length, *with thick tube, the sepals and petals outwardly arching from near or below the middle; the petals* 1.2–2 *cm. broad; filaments suberect;* anthers oblong; stigma deeply 3-lobed; capsules erect. — Meadows, low thickets and wet woods, Gaspé Pen., Que., to O., s. to N.S., N.E., e. Md., Pa. and upland to Va. and Ky. Mid-June–early Aug., *av.* of *fl. per.* July 6. — Passing into

Var. **editòrum** Fern. (of uplands). — Median cauline *leaves elliptic to oblong,* acute or subacute, *scarcely attenuate,* one-fifth to half as broad as long; *flowers red, with slender tube,* the sepals and petals arching from near or above the middle; *petals* 0.8–1.3 *cm. wide.* — Mountains and upland dry woods, Pa. to s. Ind., s. to Ala.

6. L. michiganénse Farw. (of Michigan), Michigan L. — Somewhat intermediate between nos. 5 and 7; *leaves* lanceolate, attenuate to base and apex, *their margins and nerves beneath* (except in shaded habitats) *with minute spicules; flowers* 1–several, *if* 2 *or more* chiefly *in* 1–3 *umbels,* nodding, orange or orange-red, *with sepals and petals strongly recurving, their tips reaching or surpassing the base of tube; midrib of sepals usually low, rounded and inconspicuous* on the back, of petals slightly ridged and rounded; basal green zone inside 5–12 mm. long; *filaments strongly outward-arched; anthers oblong,* 8–12 (–17) *mm. long; stigma broadly and shallowly* 3-*lobed;* bulbs yellowish. — Bogs, meadows, low thickets or woods, wet prairies, etc., s. Ont., to Man., s. to Tenn. and Ark. Late June, July, *av.* of *fl. per.* July 6.

7. L. supérbum L. (superb), Turk's-cap-L. — Resembling no. 6; up to 2.5 m. high; *margins and nerves* (beneath) *of leaves smooth or with low rounded papillae; flowers* if 2 or more *alternately disposed,* less often umbellate; basal green zone of sepals and petals 1–1.5 cm. long; *midrib at back of sepal elevated and sharply ridged, of petals acute-margined; anthers linear-arcuate,* 17–25 *mm. long; stigma deeply* 3-*lobed;* bulbs whitish. — Peaty meadows, swales, wet sand and swampy woods, Ga. and Ala., n. to se. N.H., Mass., N.Y., and (?) s. Ind. Mid-July–early Sept., *av.* of *fl. per.* Aug. 1.

8. L. Michaùxii Poir. (for its discoverer, André Michaux, 1746–1802), Turk's-cap- or Carolina L. — Stem 0.3–1.2 m. high; *leaves coriaceous or fleshy, glaucous when fresh, obovate or oblanceolate, blunt to abruptly short-tipped, their lateral veins obscure beneath, the larger blades* 4–10 *cm. long, scattered or in* 1–4 *whorls;* flowers 1–3, nodding, resembling those of no. 7; stigma with low broad lobes. (*L. carolinianum* Michx., not Bosc) — Pine- and oak-woods, n. Fla. to s. La., n. to Va., ascending in the mts. to open slopes and summits. Aug.

9. L. tigrìnum L. (spotted like a tiger), Tiger-L., Martagon or Patagon (Que.). —

Stem tall, purplish, *scabrous, cobwebby above; leaves scattered, with broad bases,* the lower prolonged, *the upper* shorter and *subtending axillary bulbs;* flowers large, showy, nodding, orange- or pinkish-red, mottled within with purple spots; the *sepals and petals* soon recurving, *their midribs within copiously pubescent at base;* filaments outwardly arching; fruit rarely, if ever, formed with us. — Roadsides and dry thickets, N.E. to N.D., s. to Va., spread from cult. Aug., early Sept. (Introd. and natzd. from e. Asia)

Tùlipa (old name) sylvéstris L. (of woodland), a wild tulip of Europe, readily recognized by its solitary subscapose large yellow flowers, 6-divided perianth and thickish subsessile stigma, is estab. in meadows, e. Pa. (Adv. from Eu.)

17. ERYTHRÒNIUM L. Dog's-tooth-Violet. Fawn-Lily

Perianth lily-like, of 6 lanceolate recurved or spreading divisions, deciduous, the 3 inner usually with a callous tooth on each side of the base and a groove in the middle. Filaments 6, subulate; anthers oblong-linear, often dimorphic. Style elongated. Capsule obovoid, contracted at base, 3-valved, loculicidal. Seeds rather numerous. — Nearly stemless herbs of temp. N. Hemisph., with two smooth and shining flat leaves tapering into petioles and sheathing the base of the commonly (with us) one-flowered scape, rising from a deep solid scaly bulb. Flowers rather large, nodding, in spring. (The Greek name for the purplish-flowered European species, from *erythros, red.*)

Plants vegetatively propagated by subterranean offshoots or small buds borne from the bulb; perianth 1.8–4 cm. long.
  Flowers wholly or partly yellow (rarely brown); several lateral veins of sepals and petals with outward-arching forking tips; stigmas erect. . . . . . 1. *E. americanum.*
  Flowers pinkish- or bluish-white; outer veins of sepals and petals projected forward or only slightly arching; stigmas spreading to recurving. . . . 2. *E. albidum.*
Plants vegetatively propagated by prolonged offshoots from near middle of stem; perianth 1–1.4 cm. long, rose-colored. . . . . 3. *E. propullans.*

1. E. americànum Ker (American), Yellow Adder's-tongue, Trout-Lily, Ail doux (Que.). — Bulb deep-seated, sending out elongate propagating shoots; young plants with elliptic solitary usually mottled leaves; *mature plants with* the pair of elongate-elliptic or elliptic-lanceolate fleshy *leaves mottled or plain,* the larger of the pair 2–5 cm. broad; peduncle 0.5–2 dm. long, arching at summit; *flower* 1.8–4 cm. long, *pale yellow within,* often spotted near base, or chestnut-brown in forma **castàneum** L. B. Sm. (chestnut); *sepals and petals with rounded subcordate bases, their lateral veins strongly forking and outward-arched;* style upwardly clavate, *stigmas erect.* — Rich woods, bottomlands and meadows, strongly colonial, N.B. to Ont., and Minn., s. to N.S., N.E., L.I., Ga., Tenn., Ark. and Okla. Late March–June. — Highly variable, needing more study.

2. E. álbidum Nutt. (whitish), White Dog's-tooth-Violet. — Similar to no. 1, vegetatively propagating by long offsets from the bulbs; *leaves rarely mottled,* the larger of the flat paired ones 1.3–4 cm. broad; *flower pinkish- to bluish-white; sepals and petals without subcordate bases,* arched-spreading or recurving, *their rather prominent veins running parallel* to the tip or only the outermost slightly arched *and forking;* style more slender below the clavate summit; *stigmas spreading or recurving.* — Woods and thickets, s. Ont. to Minn., s. to Ga., Ky., Mo. and Okla. April–June.

Var. **mesochòreum** (Knerr) Rickett (midland). — Bulb producing shorter mostly included offsets; *leaves* firmer and narrower, *the broader one of each pair* 0.5–2 *cm. broad,* often folded; *sepals and petals loosely ascending, not strongly recurving, their veins more obscure* and the outer ones often arching and forking. (*E. mesochoreum* Knerr) — Woods and meadows, Ia. and Neb., s. to Mo., Okla. and Tex.

3. E. propúllans Gray (sprouting forth). — *Offshoot arising from about midway on the stem, prolonging and forming a terminal bud;* paired leaves relatively small, acuminate, the larger one 1–2 cm. broad; *peduncle* filiform, *shorter than to about equaling leaves; perianth pink or roseate,* 1–1.4 *cm. long,* the sepals and petals with spreading or recurving tips; anthers very short; style slender, the stigmas united. — Rich woods of flat bottomlands, local, Goodhue and Rice Cos., Minn. April, May.

18. CAMÁSSIA Lindl.

Perianth slightly irregular, of 6 blue or purple spreading 3–7-nerved divisions; filaments filiform. Style filiform, the base persistent. Capsule short and thick, 3-angled, loculicidal, 3-valved, with several black roundish seeds in each locule. — Scape and linear leaves from a coated bulb; the flowers in a simple raceme, mostly bracted; 2 N.Am. species. (From the native American Indian name, *quamash* or *camass.*) QUAMASIA Raf.

1. C. scilloìdes (Raf.) Cory (like *Scilla*), EASTERN CAMASS, WILD HYACINTH. — Scape 1.5–7 dm. high; leaves keeled, elongate, 0.4–1.5 cm. broad, or in forma **Petersénii** Steyerm. (named in 1938 for its discoverer, OSCAR PETERSEN) 1.5–4 cm. broad, green, or in forma **variegàta** Steyerm. (variegated) with whitish markings; raceme elongated; bracts longer than the pedicels; divisions of the perianth pale blue, 3-nerved, 10–14 mm. long; capsule acutely triangular-globose. (*C. esculenta* Robins., not Lindl.; *Quamasia esculenta* Coville; *C. hyacinthina* Britt.) — Low fields, meadows and open woods, Ala. to Tex., ne. and n. to sw. Pa., s. Ont., se. Mich., s. Wisc., e. and s. Ia. and e. Kans. May, June.

SCÍLLA L. (old Latin name), SQUILL, an Old World genus of bulbous plants with narrow radical leaves and racemes of blue to white flowers differing from those of *Camassia* in having 1-nerved perianth-segments, has many cult. species: **S.** NONSCRÍPTA (L.) Hoffmgg. & Link (without writing; old name given because the petals are not veined like writing as they were thought to be in the *Hyacinth*), the ENGLISH BLUEBELL or HAREBELL, with linear leaves 1 cm. or more wide, scape up to 5 dm. high, and several divergent lavender flowers about 2 cm. long and subtended by long bracts, and **S.** SIBÍRICA Andr. (Siberian), with oblanceolate leaves and few deep blue flowers minutely bracted and on scapes 1–2 dm. high; both spreading from cult. to roadsides, vacant lots and old fields. (Introd. from Eurasia)

19. ORNITHÓGALUM L. STAR-OF-BETHLEHEM

Perianth of 6 (mostly white) spreading 3–7-nerved divisions. Filaments 6, flattened-subulate. Style 3-sided; stigma 3-angled. Capsule roundish-angular, with few dark and roundish seeds in each locule, loculicidal. — Scape and linear channelled leaves from a coated bulb. Flowers bracted; pedicels not jointed. Genus nat. of Old World. (A whimsical name from the Greek *ornis, a bird,* and *gala, milk.*)

1. O. UMBELLÀTUM L. (umbelled), NAP-AT-NOON. — *Scape* 1–2.5 *dm. high; flowers* 5–6, *on long and spreading-ascending pedicels; perianth* 1–2 *cm. long,* its oblong or oblong-lanceolate segments green along back. — Roadsides, grasslands, thickets, etc., Nfld. to Ont. and Neb., s. to N.S., N.E., N.C., Miss., Mo. and Kans. Late April–June. (Introd. and natzd. from Eu.)

2. O. NÙTANS L. (nodding). — Coarser; *scapes* 3–6 *dm. high;* leaves 0.5–1 cm. broad; inflorescence a *raceme with ascending to finally nodding short-stalked flowers; perianth* 2.5–3.5 *cm. long,* with oblong segments. — Low meadows and fields, N.Y. to Md. and D.C. April, May. (Introd. from Eu.)

20. MUSCÀRI Mill. GRAPE-HYACINTH

Perianth globular or ovoid, minutely 6-toothed (blue, rarely pink or white). Stamens 6, included; anthers short, introrse. Style short. Capsule loculicidal, with 2 black angular seeds in each locule. — Leaves and scape (in early spring) from a coated bulb; the small flowers in a dense raceme, sometimes musk-scented (whence the name). Small genus nat. of Mediterr. reg. and sw. Asia.

1. M. BOTRYOÌDES (L.) Mill. (like a cluster of grapes). — *Leaves flat, broadly channelled, linear-oblanceolate, broadened upward from the attenuate base to the broader* (3–8 *mm. broad*) *summit;* raceme conic-cylindric or -ovoid, 2–6 cm. long; *perianths* blue-violet, *subglobose,* later ellipsoid, 3–5 mm. long, longer than the divergent pedicels, with odor suggesting English violets; capsule subglobose-angulate; seeds subglobose-obovoid, black, rugulose. — Pastures, fields, roadsides and fence-rows, spread from cult., N.E. to Minn., s. to Va., Mo. and Kans., locally abund. April, May. (Introd. and natzd. from Eu.)

2. M. RACEMÒSUM (L.) Mill. (racemose). — *Leaves slenderly linear-subterete, furrowed or slenderly channelled, only* 1.5–3 *mm. broad, lax;* raceme dense, ovoid; perianth ellipsoid, with pale teeth, with heavily musky (some say plum-like) odor. — Fields, roadsides, lawns, fallow ground, etc., often an obnoxious weed southw., Mass. to Mich., s. to Ga. and Miss. April, May. (Natzd. from Eu.)

M. COMÒSUM Mill. (with much hair), a coarser plant, with long firm linear-lanceolate attenuate leaves, elongate raceme (1–2 dm.), perianth about 1 cm. long and on longer scattered pedicels, persists or spreads locally from cult. (Introd. from Eu.)

21. YÚCCA L. BEARGRASS. SPANISH-BAYONET

Perianth of 6 large white or greenish oval, oblong or lanceolate flat withering-persistent segments, the 3 petals broader, longer than the 6 stamens. Stigmas 3. Capsule subcylindric to ovoid, somewhat 6-sided, 3-locular, or imperfectly 6-locular by a partition from the back, fleshy, at length loculicidally 3-valved from the apex. Seeds very many in each locule, flattened. — Stems woody, short, in ours bearing persistent narrow leaves and an ample panicle or raceme of showy flowers. Genus of trop. and warm-temp. N.Am. (A native Haitian name.)

Inflorescence a much-branched panicle 1.5–3 m. high, raised far above the leaves on a tall peduncle; leaves 2–10 cm. broad; eastern.
  Leaves oblong-lanceolate, oblanceolate or subspatulate; the mature ones very thick and, when dry, scabrous, concave, bluntish or with short cucullate tip, 3.5–10 cm. broad; axis and branches of panicle glabrous; flowers 5–7 cm. long; petals 2–3 cm. broad, rounded to short tips; style about 1 cm. long. 1. *Y. filamentosa.*
  Leaves linear-lanceolate, long-attenuate to a slender point, flat, 2–4 cm. broad, smooth or scabridulous; axis and branches of panicle pruinose-pilose; flowers 3–5 cm. long; petals 1–2 cm. broad, gradually acuminate; style obsolete or up to 5 mm. long. . . . . 2. *Y. Smalliana.*
Inflorescence a simple racemose panicle or with few low branches, 0.3–1 m. long, nearly or quite reached by the leaf-tips; leaves 0.5–1.5 cm. wide; western. 3. *Y. glauca.*

1. Y. filamentôsa L. (bearing slender threads), SILKGRASS, SPOONLEAF-YUCCA. — Caudex short; *with* many spreading-ascending *thick concave rigid (when dry harshly scabrous) oblong-lanceolate, oblanceolate or subspatulate bluntish or short-tipped leaves, usually hooded below the spine,* the outer short and sometimes spoon-shaped, *the inner* 3–8 *dm. long and* 3.5–10 *cm. broad,* their margins with numerous loosening and twisting long threads; *peduncle and panicle tall,* together up to 3 m. high; *the open freely branching panicle* 1.5–2 *m. long, its axis and branches glabrous; flowers* whitish, 5–7 *cm. long;* the *petals rounded to abrupt short tips,* 2–3 *cm. broad;* filaments irregularly spiculate-papillose; *style about* 1 *cm. long;* capsule thick-cylindric to short-ovoid, often constricted at or near middle, 1.5–4.5 cm. long; seeds semiorbicular, 6–7 mm. long, 3–5.5 mm. broad, lustrous-black. (Var. *concava* (Haw.) Baker; Y. *concava* Haw.) — Dry sands of beaches, dunes, old fields and pinelands, Coastal Plain, Ga. to s. N.J. Often cult. and estab. northw. June–Sept.

2. Y. SMALLIÀNA Fern. (in honor of JOHN KUNKEL SMALL, 1869–1938), ADAM'S-NEEDLE. — Habitally similar to no. 1, up to 4 m. high; *leaves thinner, flat, smooth or merely scabridulous, linear-lanceolate, long-attenuate from middle to the prolonged narrow point,* up to 1 m. long, 2–4 cm. broad; *axis and branches of panicle pruinose-pilose; flowers* 3.5–5 *cm. long; petals gradually acuminate,* 1–2 *cm. broad;* filaments pruinose-pilose, especially below; *style nearly obsolete or up to* 5 *mm. long;* capsule more slenderly cylindric, 4–6 cm. long; seeds slightly larger. (Y. *filamentosa* sensu most auth., not L.) — Sands, old fields and bluffs, Fla. to La., n. to N.C. and Tenn.; often esc. from cult. northw. June–Sept.

3. Y. glaùca Nutt. (blue-green), SOAPWEED. — *Stem decumbent,* with 1 or more erect crowns; *leaves hard and rigid, linear-attenuate, rounded on back, the margins inrolled,* pale, 5–12 *mm. broad,* tapering to long pungent tips; *peduncle little overtopping the leaves; panicle racemiform,* with no or only basal slightly elongate branches, 0.3–1 *m. long;* flowers 3.5–6 cm. long; petals short-pointed; *style green, tumid;* capsule 6-sided, about 7 cm. long; seeds 1–1.3 cm. long. — Dry plains and sandhills, Ia. and N.D. to Mont., s. to w. Mo., Tex., N.M. and Ariz. Late May–July.

Var. **mŏllis** Engelm. (soft). — Leaves flatter and flexible, up to 1.5 cm. wide; seeds 8–10 mm. long. (Y. *arkansana* Trel.) — Sw. Mo. and Ark. to Tex.

22. ASPÁRAGUS L. ASPARAGUS. ASPERGE (Que.)

Perennials, with much branched stems from thick and matted rootstocks, and small greenish-yellow axillary dioecious flowers on jointed pedicels. The narrow, commonly filiform, so-called leaves being really branchlets, functioning as leaves, clustered in the axils of little scales which are the true leaves. — Nat. of Old World. (The ancient Greek name.)

1. A. OFFÍCINALIS L. (of the shops), GARDEN A. — Sandy fields and roadsides n. to s. Que. and s. Ont. June. (Introd. from Eu.)

23. CLINTÒNIA Raf.

Perianth of 6 divisions, lily-like, deciduous. Filaments long and filiform; anthers extrorsely fixed at a point above the base. Ovary ovoid-subcylindric, 2–3-locular; style long. — Short-stemmed N. Am. and Asiatic perennials, with slender creeping rhizomes, bearing a naked peduncle sheathed at the base by the bases of 2–4 large oblong or oval ciliate leaves; flowers umbelled, rarely single. (Dedicated to *De Witt Clinton*, 1769–1828, prominent statesman, several times governor of New York.)

Flowers 1–8, greenish-yellow, 1.2–2 cm. long; berry blue, with several seeds (or at least ovules) in a locule. . . . . 1. *C. borealis*.
Flowers 5–30, white, speckled with green and purple, 5–8 mm. long; berry black, with 2 ovules (or seeds) in a locule. . . . . 2. *C. umbellulata*.

1. C. boreàlis (Ait.) Raf. (northern), Corn-Lily, Bluebead-Lily. — Leaves 1–3 dm. long, coriaceous; scapes (sometimes bracted) usually taller, with *loose umbels of 2–8 flowers and sometimes secondary lower smaller umbels* (or umbel reduced to 1 flower); *perianth campanulate, greenish-yellow,* downy outside, especially when young, 1.2–2 *cm. long,* the segments with spreading tips; *ovary with several ovules in a locule; berry blue,* or white in forma **albicárpa** Killip (white-fruited), *short-ovoid.* — Woods and thickets (extending to subalpine meadows), Lab. to Man., s. to Nfld., N.S., N.E., mts. of Ga. and Tenn., Mich., Wisc. and Minn. May–Aug. (northw.).

2. C. umbellulàta (Michx.) Morong (with small umbel), Speckled Wood-Lily. — *Umbel* more regular and compact, 5–30-*flowered; perianth-segments* more spreading, blunter, *white, speckled with green and purple,* 5–8 mm. long; *ovules* 2 in a locule; berries black, globular to ellipsoid. (*Xeniatrum* Small) — Rich woods, w. N.Y. and e. O., s. along the upland to Ga. and Tenn. May–early July. — *C. alleghaniensis* Harned, with the fruit with a slight blue tinge may be a hybrid of nos. 1 and 2.

24. SMILACÌNA Desf. False Solomon's-seal

Perianth 6-parted, spreading, withering-persistent. Filaments 6, slender; anthers short, introrse. Ovary 3-locular, with 2 ovules in each locule; style short and thick; stigma obscurely 3-lobed. Berry globular, 1–2-seeded, at first greenish or yellowish-white, often speckled with madder-brown, at length a dull subtranslucent ruby-red. — Perennial herbs of boreal and n.-temp. reg., with simple stems from creeping or thickish rhizome, alternate nerved mostly sessile leaves, and white, sometimes fragrant, flowers. (Name a diminutive of *Smilax*, name used by Tournefort (1700) for these plants; not at all from the *Smilax* of L. and later botanists.) Vagnera Adans.

Flowers on very short pedicels in a terminal racemose panicle; sepals and petals 1.5–3 mm. long, exceeded by the stamens; rhizome thick, fleshy, continuous or barely forking. . . . . 1. *S. racemosa*.
Flowers long-pedicelled in a simple raceme; sepals and petals 4–6 mm. long, exceeding the stamens; rhizome slender, forking, stoloniform.
  Upright, with the 7–12 lanceolate or oblong to oval sessile or somewhat clasping leaves minutely downy beneath; raceme nearly sessile. . . . . . 2. *S. stellata*.
  Weak, leafy and fertile shoots intermixed from different branches of the rhizome; the 2–4 oblanceolate to elliptic glabrous leaves narrowed to subpetiolar bases; raceme peduncled. . . . . 3. *S. trifolia*.

1. S. racemòsa (L.) Desf. (racemed), False Spikenard, Solomon's zigzag. — *Rhizome fleshy,* brownish, *knotty,* elongated, *often* 1 *cm. or more thick; stem* 0.5–1 *m. high,* stiffly *arched-ascending,* slightly zigzag, *bearing* 5–12 or more *firm spreading* oblong to oval-lanceolate taper-pointed *abruptly short-petioled leaves; the larger blades* 1–2.5 *dm. long and* 3.5–9.5 *cm. broad; panicle* sessile or on a peduncle usually less than half its length, *ovoid to pyramidal,* 0.7–1.7 dm. long, 3–10 cm. in diameter, three-eighths to three-fourths as broad as long, *its longer racemes* 2–6 *cm. long* and with 8–24 flowers, *the axis and rachises pilose-hirtellous; perianth* 1.5–3 *mm. long, shorter than the stamens.* (*Vagnera* Morong) — Woods, clearings and bluffs, Que. to B.C., s. to N.S., N.E., Va., upland of N.C. and Tenn., Mo., etc. Mid-May–July. Passing southward into

Var. **cylindràta** Fern. (cylindrical). — Mostly smaller; rarely 7.5 dm. high; larger leaves of mature plants 0.85–1.7 dm. long and 3.5–6 cm. broad; *peduncle half as long as to longer than the nearly cylindric panicle; the latter* 4.5–8.5 (rarely –13) *cm. long,* 1.5–3 *cm. in diameter, one-fourth to three-eighths as broad as long,* its longest racemes 1–2.5 cm. long and 6–10-flowered. — Ga. to Ariz., n. to s. N.H., Mass., s. N.Y., s. Ont., O., Mich., s. Wisc., Kans. and Colo.

2. S. stellàta (L.) Desf. (starry). — *Stoloniferous, extensively creeping,* the rather slender pale rhizomes freely forking; *stem erect to arching,* 2–6.5 *dm. high, very leafy; leaves* lanceolate to oblong, tapering at tip, *subapproximate, sessile or slightly clasping, minutely pilose beneath,* somewhat glaucous, submembranaceous to firm, the larger ones 4–15 cm. long and 1–4 cm. broad; *raceme nearly sessile,* simple, 2–4.5 cm. long, with scattered pedicels about as long as the flowers; perianth 4–6 mm. long, exceeding the stamens; rachis glabrous to minutely hirtellous.

(*Vagnera* Morong) — Gravelly or alluvial shores, bluffs, thickets, open meadows, etc., Nfld. to B.C., s. to N.S., N.E., N.J., Pa., upland to w. Va., W.Va., O., Ind., Ill., Mo., Kans., N.M., Ariz. and s. Calif. May–early Aug. (northw.) — Passing along the coast into

Var. **crássa** Vict. (thick). — Stiff, 1–4 dm. high; *leaves closely crowded, cordate or abruptly rounded at base, oval to ovate-oblong, obtuse to subacute, very thick and heavy,* the larger ones 4–10 cm. long and 1.5–5 cm. broad. — Gravelly or sandy sea-strands, headlands, bluffs and sea-cliffs, se. Lab., Nfld. and lower St. Lawrence R., Que., s. along coast to Ct.; river-gravels, n. Me.

3. S. trifòlia (L.) Desf. (three-leaved). — *Extensively creeping; the slender rhizomes and stolons whitish,* soft, *sending up sterile and flowering shoots; the latter* weak, 0.5–2 dm. high, *bearing* 2–4 *glabrous oblanceolate to elliptic leaves narrowed to subpetiolar bases; raceme* slender, loose and open, *peduncled,* 2–6 cm. long, the long-pedicelled flowers with perianth exceeding the stamens. (*Vagnera* Morong) — Bogs, mossy woods and peaty shores, Lab. to Mackenz., s. to Nfld., N.S., Me., Mass., Ct., n. N.J., Pa., n. O., Mich., n. Ill., Minn., Man., Sask., and s. Alta., ascending to alpine areas. May–Aug. (Siberia)

25. MAIÁNTHEMUM Weber

Perianth 4-parted and stamens 4. Ovary 2-locular; stigma 2-lobed. Otherwise as in *Smilacina.* — Flowers solitary or fascicled, in a simple raceme upon a low 1–3-leaved stem. Leaves ovate- to lanceolate-cordate. Small genus of N. Hemisph. (Name from the Latin *Maius, May,* and the Greek *anthemon, a flower.*)

1. M. canadénse Desf. (Canadian), False or Wild Lily-of-the-valley, Two-leaved Solomon's-seal, Muguet (Que.). — Extensively creeping, the freely forking filiform rhizomes bearing stalked tuberous enlargements; a single cordate leaf accompanying the flowering stem; the latter erect, glabrous, 0.5–2.5 dm. high, bearing (1–) 2 or 3 *glabrous* leaves; lower cauline leaf 2–10 cm. long by 1.5–5.5 cm. broad; raceme loosely subcylindric, 1.5–5 cm. long; flowers sweetly fragrant, the perianth-segments about 2 mm. long. (*Unifolium* Greene) — Woods and recent clearings, often ascending to subalpine areas, Lab. to Man., s. to Nfld., N.S., N.E., L.I., Del., Pa., upland to Ga. and Tenn., and Ia. May, June, rarely–Aug. (northw.).

Var. **intèrius** Fern. (interior). — More or less *pubescent;* leaves often larger and bluer-green, the largest 4–13 cm. long by 2–9 cm. broad. — No. B.C. to s. Alta., e. to Ont., O. and n. and w. N.Y.

26. DÍSPORUM Salisb.

Perianth campanulate, the 6 lanceolate or linear segments deciduous and carrying the basally attached stamens. Filaments filiform, much longer than the linear-oblong blunt anthers. Ovary with 2 ovules (in our species) suspended from the summit of each locule; style one; stigmas 3, short, recurved-spreading, or sometimes united into one! Berry ovoid or subcylindric, pointed, 3–6-seeded, red. — More or less pubescent low herbs of N.Am. and e. and s. Asia, with creeping rhizomes, erect stems sparingly branched above, closely sessile ovate to oblong leaves, and greenish, yellow or purplish drooping flowers on slender terminal peduncles and solitary or few in an umbel. (Name from the Greek *dis, double,* and *spora, seed,* in allusion to the 2 ovules in each locule.)

1. D. lanuginòsum (Michx.) Nicholson (woolly), Yellow Mandarin. — Up to 7.5 dm. high; young branchlets, lower surfaces of leaves and, more permanently, *veins beneath minutely pilose-tomentulose; leaves* taper-pointed, rounded or slightly cordate at base, *the larger mature ones* 6.5–11.5 *cm. long and* 2.5–5 *cm. broad,* the 2 terminal ones approximate; *perianth* 1.3–2 cm. long, *its greenish unspotted narrowly lanceolate and broad-clawed segments* only slightly recurving, *much longer than the stamens;* anthers 3–4 mm. long; *fruit glabrous, smooth.* — Rich woods, w. N.Y. and s. Ont., s. to Ga., Ala. and Tenn. May, early June.

2. D. maculàtum (Buckl.) Britt. (spotted), Nodding Mandarin. — *Young branchlets and veins of lower leaf-surfaces hispid with stiff spreading hairs;* terminal *leaves* more approximate, *the larger mature ones* 1.5–1.8 *dm. long and* 6–9 *cm. broad; perianth* 1.5–2.5 cm. long, *yellow, spotted with dark purple; the broadly lanceolate slender-clawed segments* spreading, *shorter than the stamens; fruit villous and rugose.* — Rich woods, O. and s. Mich., s. to Ga., Ala. and Tenn. Late April, May.

27. STRÉPTOPUS Michx. Twisted-stalk

Perianth recurved-spreading from a campanulate base, deciduous; the 6 divisions lanceolate, acute, the inner (petals) keeled. Anthers sagittate, extrorse, fixed near the base to the short flattened filaments, tapering above to a slender entire or 2-cleft point. Berry roundish to ovoid, many-seeded. — Herbs of cooler half of N. Hemisph. with rather stout stems from a short or creeping rhizome, ordinarily forking and divergent branches, ovate and taper-pointed sessile or clasping membranaceous leaves, and small extra-axillary flowers, either solitary or in pairs, on slender filiform peduncles which are abruptly bent or contorted near the middle (whence the name, from the Greek *streptos, twisted,* and *pous, foot* or *stalk*).*

Perianth-segments wide-spreading or recurving from near the middle; anthers lance-subulate, entire, many times exceeding filaments; stigma subentire or merely lobed. . . . . 1. *S. amplexifolius.*

Perianth-segments with only the tips recurved in age; anthers ovate, 2-horned, shorter than to about equaling filaments; stigma 3-cleft at tip. . . . . . 2. *S. roseus.*

1. S. amplexifòlius (L.) DC. (clasping-leaved), White Mandarin. — Rhizome usually short and thick; stem forking or sometimes simple, 2–9 dm. high; *leaves amplexicaul;* peduncles simple or forked, in anthesis 1–3 cm., in fruit 1.5–8 cm. long; *perianth* greenish-white to deep purple, *its lance-attenuate strongly falcate or recurving segments* 7–13 *mm. long; anthers lance-subulate, entire, much longer than filaments; stigma subentire to merely lobed.* — Wide-ranging species, in its typical form southern European. We have three vars.:

Stem (except at base) and lower leaf-surfaces glaucous and glabrous; the stem usually forking, 3–9 dm. high; perianth whitish-green; berries thick-ellipsoid to globose, 1–2 cm. long, scarlet.
Leaf-margin entire or essentially so. . . . . Var. *americanus.*
Leaf-margin minutely denticulate (10–25 teeth per cm.). . . . . Var. *denticulatus.*
Stem and lower leaf-surfaces greener; the stem hispid and simple or forked, 2–6 dm. high; leaves definitely ciliate-denticulate, smooth or hirtellous beneath; perianth deep purple to roseate; berries subglobose, about 1 cm. in diameter, deep red. Var. *oreopolus.*

Var. **americànus** Schultes (American), Liverberry, Scootberry. — Moist woods and thickets, Greenl. and Lab. to Man., s. to Nfld., N.S., N.E., N.Y. and n. Mich., and along mts. to N.C.; Alaska to N.M. and Ariz. May–July. (E. Asia)

Var. **denticulàtus** Fassett (with fine teeth). — About the upper Great Lakes, Ont., Mich., Wisc. and Minn.; s. Alaska to n. Calif. (E. Asia)

Var. **oreópolus** (Fern.) Fassett (mountain-dweller), *S. oreopolus* Fern. — Peaty thickets and openings, n. Nfld., Mingan Ids. and subalpine areas of Shickshock Mts., Que., Mt. Katahdin, Me., and Mt. Washington, N.H. July, Aug.

2. S. ròseus Michx. (rose-colored). Rose Mandarin, Rognons de coq (Que.). — Stem forked or simple, 2.5–6 dm. high, often hispidulous above with multicellular hairs, not glaucous; *leaves sessile, not clasping, strongly corrugated,* the margins *ciliate-denticulate; peduncles* 1–2.5 cm. long, usually *hispid; perianth pink or rose-purple; its lanceolate segments* 6–12 mm. long, *with only the tips recurved in age; anthers ovate, 2-horned, shorter than filaments; stigma 3-cleft; berries subglobose or obscurely 3-lobed, about 1 cm. in diameter, cherry-red.* Three vars.:

Rhizome matted, the short internodes mostly hidden by roots; leaves copiously ciliate (22–60 cilia per cm.); perianth-segments glabrous within, rarely with low papillae; sepals 7–11-nerved.
Pedicels glabrous. . . . . *S. roseus* (typical).
Pedicels ciliate with multicellular hairs. . . . . Var. *perspectus.*
Rhizome slender, rooting at the distant nodes; leaves with 20–30 cilia per cm.; perianth-segments minutely papillate within; sepals 3–5-nerved. . . . . Var. *longipes.*

S. ròseus (typical). — Mountain-woods, Pa. to N.C. and Tenn. May. — Passing into

Var. **perspéctus** Fassett (well-known). — Woods and moist thickets, s. Lab. to Ont., s. to Nfld., N.S., N.E., n. N.J., Pa. and Mich., and along the mts. to Ga. and Ky. Apr.–July.

Var. **lóngipes** (Fern.) Fassett (long-stalked), *S. longipes* Fern. — Woods, s. Ont. and s. Man., s. to nw. Pa., n. Mich., Wisc. and Minn.

28. POLYGÓNATUM Mill. Solomon's-seal. Sceau-de-Salomon (Que.)

Perianth cylindrical, 6-lobed at the summit; the 6 stamens inserted on or above the middle of the tube, included, anthers introrse. Ovary 3-locular, with 2–6 ovules in each locule; style

* Treatment following that of N. C. Fassett in Rhodora, xxxvii. 88–113 (1935).

slender, deciduous by a joint; stigma obtuse or capitate, obscurely 3-lobed. Berry globular, black or blue; the locules 1–2-seeded. — Perennial herbs of N. Hemisph., with usually simple stems from a creeping knotted rhizome, naked below, above bearing nearly sessile or half-clasping nerved leaves, and axillary nodding greenish flowers; pedicels jointed near the flower. (Name from the Greek *polys, many,* and *gonu, knee;* alluding to the numerous joints of the rhizome.)

Leaves pilose or hirtellous on nerves beneath.
  Stem, peduncles and pedicels glabrous; native species. . . . . 1. *P. pubescens.*
  Stem, peduncles and pedicels hirsute; introduced species. . . . . 2. *P. latifolium.*
Leaves glabrous on both surfaces.
  Stem slender, 2–9 dm. high; leaf-blades flat, nearly or quite sessile; the larger ones (seen by transmitted light) 46–120-nerved, 5.5–15 cm. long and 1.2–6 cm. broad; the terminal small blade 20–66-nerved; peduncles 1–3(–5)-flowered; perianth 1–1.7 cm. long, its lobes 3–4 mm. long; enlarged terminal joint of fruiting pedicel 0.7–1.5 mm. long, often as broad. . . . . . . 3. *P. biflorum.*
  Stem stout, 0.6–2 m. high; leaf-blades somewhat puckered, not quite flat, narrowed to subpetiolar clasping or sheathing bases; the larger ones 110–220-nerved, 0.9–2.5 dm. long and 3.5–13 cm. broad; the terminal small blade 58–112-nerved; peduncles 2–10-flowered; perianth 1.7–2 cm. long, its lobes 5–6.5 mm. long; enlarged terminal joint of fruiting pedicel 1–3 mm. long, often longer than broad. . . . . 4. *P. canaliculatum.*

1. P. pubéscens (Willd.) Pursh (hairy). — *Rhizome* whitish, 1–1.8 cm. thick, *superficial; stem* slender, *glabrous,* 0.2–1.1 m. high; *leaves pilose or hirtellous beneath along the nerves, sessile or very short-petioled,* lance-oblong to elliptic-oval, 4–14 cm. long, 1.5–7.5 cm. broad; *peduncles* 1 or 2 (–4)-flowered, *glabrous, the lowest usually in the* 1st *or* 2nd (rarely 3rd or 4th) *axil; perianth* slender, 7–13 *mm. long;* stamens inserted high on the perianth-tube. (*P. biflorum,* in part, of ed. 7, not Ell.) — Woods, wooded bluffs and about boulders in woods, s. Que. to s. Man., s. to N.S., N.E., ne. Md., Pa., upland to S.C., W.Va., Ky., Ind., Wisc. and Ia. May, June. — Forma **fúltius** Fern. & Harris (supported) is a sterile form with flowers subsessile or very short-pedicelled from axils of leafy branches, local, e. Mass.

2. P. LATIFÒLIUM (Jacq.) Desf. (broad-leaved). — 3–4.5 dm. high, the arching *stem hirsute above; leaves* few, elliptic, abruptly acuminate, *definitely petioled,* membranaceous, *hirtellous on veins beneath; peduncles* 2–3-flowered, *with the pedicels hirsute;* perianth thick-cylindric, whitish, 1.2–1.7 cm. long. — Roadside-thickets, local, e. Mass. June. (Adv. or introd. from Eu.)

3. P. biflòrum (Walt.) Ell. (two-flowered). — *Rhizome* 0.6–1.5 *cm. thick, buried in soil,* with less constricted joints than in no. 1; *stem slender,* 1.5–5 *mm. thick* below lowest leaf, 2–9 *dm. high; leaves flat, glabrous on both sides, sessile or nearly so,* narrowly lanceolate to broadly ovate, green or glaucous beneath; *the largest ones* (of each plant) *with* 46–120 *nerves and* 5.5–15 *cm. long by* 1.2–6 *cm. broad; the terminal small ones* 20–66-*nerved; peduncles* 1–4 *cm. long,* 1–3 (–5)-flowered, *the lowest usually borne from the* 3rd (1st–5th) *axil; pedicels becoming* 0.5–2 *cm. long; perianth slenderly cylindric,* 1–1.7 *cm. long, its* oblong *lobes* 3–4 *mm. long; filaments* commonly *papillate or granulose,* slender, inserted near middle of perianth-tube; *enlarged terminal joint of fruiting pedicel cup-shaped or campanulate,* with the rim flaring, 0.7–1.5 *mm. long, often as broad;* seeds subglobose, pale brown, 2.7–3.5 mm. long. — Dry to moist, sandy, loamy or rocky woods and thickets, Fla. to Tex., n. to Ct., N.Y., s. Ont., s. Mich., Ill., Ia. and Neb. May–late June. — Forma **ramòsum** (McGivney) Fern. (branching), a rare form in s. Mich. and n. Ind., has its normally elongate peduncles replaced by leafy branches bearing very short axillary peduncles.

4. P. canaliculàtum (Muhl.) Pursh (channelled). — Coarser than no. 3; *rhizome* 1.5–3 *cm. thick,* not much constricted, deeply buried; *stem stout,* 0.5–1.3 *cm. thick* at lowest leaves, 0.6–2 m. high; *leaves* more or less *corrugated and with puckered margin,* not really flat, mostly *narrowed to* broad *subpetiolar clasping or sheathing bases; the larger ones with* 110–220 *nerves and* 0.9–2.5 *dm. long by* 3.5–13 *cm. broad,* the smallest terminal ones 58–112-nerved; *peduncles becoming* 1.5–9 *cm. long,* 2–10-flowered, *pedicels becoming* 1–4 *cm. long,* the *lowest peduncle commonly borne from the 4th or 5th* (3rd–8th) *axil; perianth thick-cylindric,* 1.7–2 *cm. long, its lobes* 5–6.5 *mm. long;* filaments broad, smooth or merely granulose; *enlarged terminal joint of fruiting pedicel subcylindric to slenderly campanulate* (except at flaring summit), 1–3 *mm. long, usually longer than broad;* seeds 3–4.5 mm. in diameter. (*P. commutatum* (R. & S.) Dietr.) — Rich woods, alluvial thickets, river-silts, etc., Ct. valley, N.H., to s. Man., s. to S.C., Tenn., Mo. and Okla. May, June.

29. CONVALLÀRIA L. Lily-of-the-valley. Muguet (Que.)

Perianth campanulate, white, with 6 short recurved lobes. Stamens 6, included, inserted on the base of the perianth; anthers introrse. Ovary 3-locular, tapering into a stout style; stigma triangular. Ovules 4–6 in each locule. Berry few-seeded, red. — Perennial herbs of Eurasia and Alleghanian Am., glabrous, stemless, with slender running rhizomes, 2 or 3 oblong leaves, and an angled scape bearing a one-sided raceme of sweet-scented nodding flowers. (From *convallis, a valley.*)

1. C. majàlis L. (blooming in May). — *Spreading by stolons, colonial, forming dense carpets;* scape elongate, so that the *flowers* are *borne opposite the middles or upper halves of the leaves,* 1–2 dm. high; *leafy axis* up to base of upper leaf 5–12 *cm. high,* 2- or 3-leaved, the *larger leaves* 1–2 *dm. long and* 3–7.5 *cm. broad, the veins and cross-partitions* (as seen by transmitted light) *relatively faint; bracts* lanceolate, 4–10 *mm. long, much shorter than the longer pedicels; seeds nearly globose.* — Roadsides, thickets and open woods, abundantly spread from cult. Late April–June. (Introd. and natzd. from Eu.)

2. C. montàna Raf. (of mountains). — *Plants scattered, not colonial;* scape and *raceme shorter than leafy axis or barely reaching lower half of lowest leaf; leafy axis* 1.5–2 *dm. high; larger leaves* 1.5–3 *dm. long and* 4–12 *cm. broad,* strongly nerved, the *nerves and cross-partitions sharply visible by transmitted light; longer bracts linear,* 0.8–2 *cm. long, equaling or exceeding the longer pedicels; seeds lenticular or oblate, broader than high.* (*C. majuscula* Greene) — Rocky or sandy woods and slopes, mts. of Va. and W.Va. to n. Ga. and e. Tenn. May, June.

30. MEDÈOLA L. Indian Cucumber-root. Concombre sauvage or Jarnotte (Que.)

Perianth recurved; the 3 sepals and 3 petals oblong and alike, pale greenish-yellow, deciduous. Stamens 6; anthers shorter than the slender filaments, oblong. Styles stigmatic along the upper side, recurved-diverging from the globose ovary, long and filiform, deciduous. Berry globose, dark purple, 3-locular, few-seeded. — A perennial herb, with a simple slender stem rising from a horizontal white tuber (with taste of cucumber), bearing near the middle a whorl of 5–9 obovate-lanceolate leaves; also another of 3 (rarely 4 or 5) much smaller ovate ones at the top subtending a sessile umbel of small recurved flowers. (Named after the sorceress *Medea,* for its imagined great medicinal virtues.)

1. M. virginiàna L. (of Virginia). — Stem 2–9 dm. high, clothed with flocculent and deciduous wool. — Rich woods, Que., Ont. and Minn., s. to N.S., N.E., Fla., Ala. and La. May, June.

31. TRÍLLIUM L. Trillium. Wakerobin. Birthroot. Trille (Que.)

Sepals 3 (exceptionally 2 or 4–8), lanceolate or oblong, spreading or reflexed, herbaceous, persistent. Petals 3 (or 4–8), withering in age. Stamens 6; anthers linear, on short filaments, adnate. Styles subulate or slender, spreading or recurved above, persistent, stigmatic along the inner side. Seeds ovoid, horizontal, several in each locule. — Low perennial herbs of temp. N. Am. and e. Asia, with a stout and simple stem rising from a short and praemorse tuber-like rhizome, bearing at the summit a whorl of 3 (exceptionally 0, 2 or 4–8) dilated more or less ribbed but netted-veined leaves, and a terminal large flower; the young plants with a single leaf-blade terminating the stem-like petiole; flowering in spring. (Name from *tres, three;* all the parts being in threes.) * — Aberrations are not rare, with the calyx and sometimes petals changed to leaves, or the leaves and parts of the flower reduced or increased in number.

*a.*Flower sessile, erect or with ascending petals. . *b.*

*b.*Leaves sessile or essentially so; sepals ascending, not reflexed; petals clawless or at most with dilated claw-like bases. . . *c.*

*c.*Leaves broadly oval to subrhombic, 2.5–12 cm. broad; petals 2–11.5 cm. long; filaments broad, much shorter than anthers; anthers 1–2 cm. long, the connective prolonged at tip; common style obsolete, the 3 stigmatic branches borne directly on ovary; rhizome heavy, brown or fuscous, 1–2.5 cm. thick. . . *d.*

*d.*Petals broad-based, not claw-like below; summit of stem and leaves mostly glabrous. . . *e.*

*e.*Leaves 3.5–10 cm. long, 1.5–7.5 cm. broad, usually mottled; sepals

* Treatment derived in part from those of W. A. Anderson in Rhodora, xxxvi. 119–128 (1934) and of Wiegand & Eames, Fl. Cayuga Lake Basin (1926).

1.5–4 cm. long; petals 2–4 cm. long; stamens half as long as petals; anther-connective broad, flat, prolonged 1–3 mm. . . . . . 1. *T. sessile.*

*e.*Leaves 7–17 cm. long, 6.5–12.5 cm. broad; sepals 3–7.5 cm. long; petals 4–11.5 cm. long; stamens about one-third length of petals; anther-connective scarcely prolonged, or forming a narrow arching beak less than 1 mm. long.

Leaves obscurely or scarcely mottled; sepals 3.5–7.5 cm. long; petals 4–11.5 cm. long, dark purple to greenish. . . . . . 2. *T. cuneatum.*

Leaves strongly mottled; sepals 3–4 cm. long; petals 4–6 cm. long, greenish- to buttercup-yellow. . . . . 3. *T. luteum.*

*d.*Petals gradually narrowed to broad claws; stamens nearly half as long as petals; anther-connective prolonged; summit of stem and bases of leaves beneath usually hirtellous. . . . . . . 4. *T. viride.*

*c.*Leaves narrowly oblong or oblong-lanceolate, 1–2.5 cm. broad; petals 1.2–2 cm. long; filaments slender, nearly equaling to longer than anthers; anthers 3–8 mm. long, connective not prolonged; common style slender, prolonged; rhizome whitish, 3–10 mm. thick. . . . . 10. *T. pusillum,* var. *virginianum.*

*b.*Leaves definitely petioled; sepals abruptly reflexed; petals definitely clawed. 5. *T. recurvatum.*

*a.*Flower peduncled. . *f.*

*f.*Leaves sessile or nearly so, if short-petioled broadly rhombic in outline and with acutely 6-angled ovary and fruit. . . *g.*

*g.*Leaves ovate to rhombic, acuminate, 0.35–2 dm. broad; common style obsolete, the 3 stigmatic branches borne directly on summit of ovary; rhizome coarse, brown or fuscous, 1.5–3 cm. thick. . . *h.*

*h.*Stigmas short and stout, tapering from base to recurved tip, about half as long as the ovary; petals 1.5–5.5 cm. long, 0.4–3.5 cm. broad, of essentially constant color; leaves broadly rhombic, the margin from the broad middle to the base nearly straight to slightly concave. . . *i.*

*i.*Ovary whitish or pale; petals with ascending bases, spreading from near the middle, white, creamy or pinkish; anthers shorter than to about equaling stigmas; peduncles arching under the leaves or strongly declined.

Leaves with short obscure petioles; peduncles 0.5–4 cm. long, recurved; petals 1.5–2.5 cm. long, 0.5–1.7 cm. broad; anthers 2.5–6.5 mm. long, pink, about twice as long as filaments. . . 6. *T. cernuum.*

Leaves sessile; peduncles 3–12 cm. long, straight, erect, horizontal or reflexed; petals 2–5 cm. long, 1–3.5 cm. broad; anthers 6–15 mm. long, creamy-white, at least twice as long as filaments. . 7. *T. flexipes.*

*i.*Ovary purple (pale in obvious albinos); petals spreading from base, dark purple, purple-brown, yellow-green or white, 1.5–5.5 cm. long, 1–3 cm. broad; peduncle straight, erect, divergent or (rarely) deflexed, 0.5–10 cm. long. . . . . 8. *T. erectum.*

*h.*Stigmas slender, of uniform diameter, erect or spreading; mostly more than half as long as ovary; petals 4–8.5 cm. long, 1.7–4 cm. broad, white to pink, their bases erect, their tips spreading; leaves only slightly rhombic, the margin from middle to base slightly convex; peduncle erect, 2–12 cm. long. . . . . 9. *T. grandiflorum.*

*g.*Leaves narrowly oblong or oblong-lanceolate, blunt, 1–2.5 cm. broad; filaments slender; common style definite; ovary and fruit obtusely 3-angled; rhizome whitish, 3–10 mm. thick. . . . . 10. *T. pusillum.*

*f.*Leaves definitely petioled; filaments slender; ovary and fruit obtusely 3-angled, not winged.

Stem 3–15 cm. high; leaves oval or ovate, obtuse, 1.5–5 cm. long; petals white, or pink-striped at base, spreading from erect base, oblong, obtuse; common style slender, elongate; fruiting peduncle recurved; berry depressed-globose, 6–8 mm. long . . . . 11. *T. nivale.*

Stem up to 5 dm. high; leaves ovate, acuminate, 0.5–1.7 dm. long; petals white with purple or pink marking at base, acute; common style obsolete; fruiting peduncle ascending; berry ovoid or ellipsoid, 1.4–1.8 cm. long. . . . . 12. *T. undulatum.*

1. T. séssile L. (sessile), Toadshade. — Stem 1–3 dm. high, from a stout brown rhizome 1–2 cm. thick; *leaves* (rarely 2, 4 or 5) *oval,* obtuse or abruptly short-tipped, rounded to sessile base, often mottled, 3.5–10 *cm. long,* 1.5–7.5 *cm. broad;* flower sessile, erect; *sepals spreading,* 1.5–4 *cm. long; petals erect-spreading, narrowly oblong to oblanceolate,* 2–4 *cm. long, blunt,* maroon, or yellow-green in forma **viridiflòrum** Beyer (green-flowered); *stamens half as long as petals, the broad flat anther-connective prolonged* 1–3 *mm.*; berry globose, 6-angled, about 1.2 cm. in diameter. — Rich woods, Ga. to Miss. and Ark., n. to se. Va., w. N.Y., O., Ind., Ill. and centr. Mo. April–early June. — Flower aromatic.

2. T. cuneàtum Raf. (wedge-shaped; from base of petal). — Coarser than no. 1; *leaves* round-ovate, abruptly short-acuminate, *obscurely to scarcely mottled,* 7–17 *cm. long,* 4–11.5 *cm. broad; sepals* 3.5–7.5 *cm. long; petals* 4–11.5 *cm. long,* 1–2.5 *cm. broad, deep purple or maroon to greenish,* acutish; *stamens about one-third as long as petals,* the filament very short and broad, the anther 1–2 cm. long, the *anther-connective prolonged into an arched triangular-ovate beak* 0.2–0.7 *mm. long.* (*T. Hugeri* Small) — Rich woods, nw. Fla. to Miss., n. to upland of N.C. and Ky. April, May. — Flower slightly ill-scented.

3. T. lùteum (Muhl.) Harbison (yellow). — Similar to no. 2 (which may be a var. of it), 2–6 dm. high; leaves more strongly mottled, 7–14 cm. long; *sepals* 3–4 *cm. long; petals* 4–6 *cm. long, greenish- to buttercup-yellow,* bluntish; *anthers* 1–1.3 *cm. long;* berry broadly ovoid, 2–2.5 cm. long. — Rich woods, Ala. to Ark., n. to Ky. and Mo. April, May. Flowers lemon-scented.

4. T. víride Beck (green). — Resembling nos. 2 and 3, more inclined to be *hirtellous at summit of stem and at base of lower leaf-surfaces; leaves more narrowly oval or ovate and more tapering to acuminate tip,* 3.5–8.5 cm. broad; sepals 4–5.5 cm. long; *petals* linear-lanceolate or -oblanceolate, greenish, *narrowed below to broad claw-like base,* 3.6–8 cm. long, 4–12 *mm. wide; stamens nearly half as long as petals, anthers* 1.2–2.3 cm. long, *the connective prolonged as an arching beak* 1 *mm. or more long.* (*T. viridescens* Nutt.) — Rich woods, Ala. to La. and Okla., n. to N.C., sw. Va., Ill., Mo. and e. Kans. April, May.

5. T. recurvàtum Beck (recurved). — Stems 1.5–4.5 dm. high, *from a pale horizontal rhizome* 0.6–1.2 *cm. thick; leaves definitely petioled,* the petiole 0.8–2.5 cm. long; blade broadly elliptic or oval to obovate or lance-ovate, 4.5–11 cm. long; flower sessile; *sepals strongly reflexed,* lance-acuminate, 1.5–3 cm. long; *petals* erect or ascending, *clawed,* oblong-lanceolate to -ovate, 2–4.5 cm. long, acute, red-brown, maroon or purplish, or greenish-yellow with red claws and stamens in forma **lùteum** Clute (yellow), or yellow to yellow-green with yellow stamens in forma **Shaỳi** Palmer & Steyerm. (named in 1935 for its discoverer, W. F. SHAY); filaments 3–4 mm. long, anthers 5–10 mm. long, their connectives projecting as short arching beaks; stigmas erect, with recurved tips. — Rich woods, O. to Ia., s. to Ala., Miss. and Ark. April, May. — Aberrant individuals may have some or all sepals, petals and stamens turned to broad veiny leaves.

6. T. cérnuum L. (nodding), NODDING T. — Stems 1.5–6 dm. high, 1–several, *from a thick* brown or fuscous *ascending rhizome up to* 3 *cm. thick; leaves* rhombic-ovate to -subrotund, acuminate, *constricted at base into a short petiole,* the blade 4–15 cm. long and broad; *peduncle* 0.5–2.5 (–3.5) *cm. long, recurved, the flower nodding and nearly hidden under the leaves;* sepals lance-acuminate; *petals* whitish to creamy or pale pink, or in forma **Tángerae** Wherry (for its discoverer, LOUISE FORNEY ARNOLD TANGER, 1889–) deep roseate, *with recurving tips,* oblong-lanceolate, 1.5–2.5 cm. long, 5–9 mm. broad; *anthers pink,* 2.5–4.5 *mm. long,* slightly longer than to about equaling filaments; ovary white to pink, stigmas stout; berry broadly ovoid, dark red. — Damp or peaty woods and thickets, oftenest in acid soils, Nfld. and Gaspé Pen., Que., to Wisc., s. to N.S., N.E., Del., Md., upland to Ga., and W.Va. Late April (southw.)–early July (northw.) — Flower sweet-scented.

Var. **macránthum** Wieg. (large-flowered). — Peduncle 1–4 cm. long; petals oblong-oval to -obovate, 1–1.7 cm. broad; anthers 4–6.5 mm. long. — Richer, often calcareous, mucky soil or alluvium, Vt. to Mackenz., s. to sw. N.E., Pa., e. Tenn., O., Ind., Ill. and Ia.

7. T. fléxipes Raf. (with bent foot-stalk). — Similar to no. 6, often coarser; leaves sessile, 0.75–2 dm. long and broad; *peduncle* 3–12 *cm. long, straight, erect, divergent or reflexed; petals* 2–5 *cm. long,* 1–3.5 *cm. broad,* white, *not recurved,* or petals maroon or purple with the filaments and ovary sometimes so in forma **Walpòlei** (Farw.) Fern. (named for one of its discoverers, BRONSON ALVA WALPOLE, 1890–); *anthers* 6–15 *mm. long, creamy-white,* at least twice as long as filaments; *ovary white or pale.* (*T. declinatum* (Gray) Gleason, not Raf.; *T. Gleasoni* Fern.) — Bottomlands and damp rich woods, centr. N.Y. to s. Minn., s. to Md., Tenn. and Mo. April–June.

8. T. eréctum L. (erect), STINKING BENJAMIN, PURPLE T., SQUAWROOT. — Rhizome stout, brown, up to 3 cm. thick; stems 1–several, stoutish, 1.5–6 dm. high; leaves broadly rhombic-ovate to -rotund, acuminate, sessile, 4–19 cm. long and broad; peduncle straight, erect, divergent or declined, 0.5–10 cm. long; sepals lanceolate to lance-oblong, acuminate, about equaling petals; *petals spreading from base,* 1.5–5.5 *cm. long,* 1–3 *cm.* broad, lance- to narrowly ovate-oblong, acute or acuminate, crimson to dark purple or purple-brown, or in forma **Cáhnae**

(Farw.) Louis-Marie (named in 1925 for its discoverer, "Mrs. Cahn") purple at base and whitish above, or in forma **viridiflòrum** (Hook.) Peattie (green-flowered) greenish, or in forma **lùteum** Louis-Marie (yellow) clear yellow, or in forma **albiflòrum** R. Hoffm. (white-flowered) white (= var. *album* (Michx.) Pursh); *stamens exceeding the stout spreading or recurved stigmas; ovary purple* (except in pale-flowered forms); berry dark red (except in albinos), depressed-ovoid, 6-angled. — Rich woods, w. Gaspé Pen., Que., to Ont. and Mich., s. to N.S., N.E., n. Del., Pa., W.Va. and upland to Ga. and Tenn. April–early June. — Flower ill-scented. The unusual forma **polýmerum** Vict. (with several members; of the cycles) has leaves, sepals, petals, stamens or styles or more than one of these cycles increased to 4 or more. Many trivial so-called forms based on size and intensity of coloration of flower have been proposed.

Var. **blándum** Jennison (charming). — Petals ovate to elliptic, rounded or merely acute at tip, creamy-white; ovary pale or dark; flower not ill-scented. (Var. *album* sensu many auth., not (Michx.) Pursh) — Chiefly in the mts. of Tenn. and Ky., locally n. to N.E., N.Y. and O.

9. T. grandiflòrum (Michx.) Salisb. (large-flowered). — Stem 1.5–4.5 dm. high; ordinarily with 3 sessile or subsessile round-ovate to subrhombic-oval *leaves gradually rounded to subcuneate base, acuminate,* 0.5–3 dm. long, or leaves lance-elliptic and petals narrowly lanceolate in forma **elongàtum** Louis-Marie (elongate), or leaves wanting in forma **Chándleri** (Farw.) Vict. (named for its discoverer, Benjamin F. Chandler, 1870–1918), or leaves 3 and definitely petioled in forma **lirioìdes** (Raf.) Vict. (like *Lirium*, an old name for *Lilium*), or leaves and 1 or more cycles of flower in 4's or 5's in forma **polýmerum** Vict. (with many members; of cycles), or leaves, sepals, petals and carpels 2 each in forma **dímerum** Louis-Marie (in twos); *peduncle erect or ascending,* 2–12 cm. long; sepals acuminate, 2.5–5 cm. long; *petals oblong, narrowly obovate or oblanceolate to suborbicular, 4–8.5 cm. long, 1.7–4 cm. broad, their bases erect, their tips spreading,* usually white or white fading to pale pink, or striped green and white in forma **striàtum** Louis-Marie (striped), or green throughout in forma **víride** Farw. (green), or with carpels and stamens changed to white petals in forma **petalòsum** Louis-Marie (having petals); *filaments* stout, *shorter than the long slender pale anthers, these persistent about the fruit; stigmas slender, of uniform diameter,* erect or spreading, *mostly more than half as long as the pale ovary;* berry ovoid, 6-angled. — Rich woods and thickets, Ga. to Ark., n. to w. Me. (local), w. N.H. (local), s. Que., s. Ont., Mich., Wisc. and Minn. April–June. — Our handsomest, most fickle and sporting species, with many scores of aberrant forms described (as well as the above), these belonging more to the field of teratology than to taxonomy!

10. T. pusíllum Michx. (very small). — *Rhizome whitish, 3–10 mm. thick;* stem slender, 1–3 dm. high; *leaves* nearly or quite *sessile, narrowly oblong or oblong-lanceolate, blunt,* 2–8.5 cm. long, 1–2.5 *cm. broad,* 3- or 5-veined; peduncle 0.7–3 cm. long, ascending; sepals spreading, oblong, blunt, 1.3–4 cm. long, 0.4–1.5 cm. broad; *petals white, changing to pink and then to purple,* lanceolate, 1.6–3.5 *cm. long,* 4–18 mm. broad, nearly equaling to slightly longer than sepals, spreading-ascending; *filaments slender, anthers* slender and 3–9 *mm. long; common style slender,* the *ovary and berry obtusely* 3-angled, *the latter barely* 1 *cm. long.* (Incl. *T. ozarkanum* Palmer & Steyerm.) — Damp or dryish acid woods, se. S.C.; s. Mo. and nw. Ark. April.

Var. **virginiànum** Fern. (Virginian). — Flower sessile or subsessile; petals 1.2–2 cm. long, 3–5 mm. wide, mostly shorter than sepals. — Damp woodlands, se. Va. Early May.

11. T. nivàle Riddell (snowy), Dwarf White or Snow-T. — Dwarf, 3–15 cm. high; *leaves oval or ovate, obtuse,* 1.5–5 *cm. long, definitely petioled; peduncle* short, erect, ascending or, at least *in fruit, recurving; petals white, or pink-striped at base, oblong, obtuse,* 1.2–3 cm. long, spreading from an erect base; *filaments slender; common style slender, elongate;* berry depressed-globose, 6–8 mm. long. — Rich woods, clearings, shaded ledges, etc., w. Pa. to Minn., s. to Ky. and Mo. Late March–May.

12. T. undulàtum Willd. (wavy), Painted T. — Rhizome thickish, brown; stems 1–several, 1–5 dm. high; *leaves definitely petioled, ovate, acuminate,* 0.5–1.7 *dm. long,* 3–13 cm. broad, or leaves and sometimes cycles of flowers in 4's–8's in forma **polýmerum** Vict. (with many members; of cycles), or sepals enlarged and nearly like the foliage-leaves with the latter in 1–3 whorls and the petals 3–6 in forma **Cleavelándicum** (Wood) Fern. (in honor of Parker Cleaveland, 1780–1858); peduncle erect or ascending; sepals lance-acuminate, spreading, green, or in forma **striàtum** Louis-Marie (striped) white-striped; *petals* lanceolate to oblong-ovate, *wavy-margined, acute, white with conspicuous red to purplish blotch and stripes at base;* common style obsolete; *berry ovoid to ellipsoid,* 1.4–1.8 cm. long, scarlet or red, obscurely 3-angled. — Acid or subacid woods and swamps, Gaspé Pen., Que., to e. Man., s. to N.S., N.E., n. N.J., Pa., W.Va., upland to Ga. and Tenn., Mich. and Wisc. April–June. — Exceptional plants have but 2 leaves, with the flowers dimerous.

32. ÁLETRIS L. COLICROOT. STARGRASS

Perianth cylindrical, wrinkled and roughened outside by thickly set points, the tube adhering to the base of the ovary, 6-cleft at the summit. Stamens 6, inserted at the base of the lobes; filaments and anthers short, included. Style subulate, 3-cleft at the apex; stigmas minutely 2-lobed. Capsule ovoid, beaked, enclosed in the roughened perianth; seeds numerous, minute, costate. — E. N. Am. and e. Asiatic perennial and smooth stemless herbs, very bitter, with short and thick rhizome and a spreading rosette of thin and flat lanceolate leaves; the small flowers in a spike-like raceme terminating a naked slender scape (3–10 dm. high). (*Aletris*, a female slave who ground corn; in allusion to the apparent mealiness of the perianths.)

1. A. farinôsa L. (mealy; from the granuliferous perianth), UNICORN-ROOT. — Leaves firm, 0.5–2 dm. long; scapes 3–9 dm. high, with remote small bracts; raceme 0.8–3.5 dm. long, densely to subremotely flowered; *perianth tubular, whitish*, with granulate surface, *its lobes lance-oblong*, marcescent, *shrinking* in maturity *and thus often exposing the long abrupt beaks, these about as long as the plump body of the capsule.* — Dry or moist peats, sands and gravels, Fla. to Tex., n. to sw. Me., s. N.H., centr. Mass., se. N.Y., s. Ont., Mich. and Wisc. Late May–Aug.

2. A. lùtea Small (yellow). — Resembling no. 1 in its *long tubular perianth up to 1 cm. long*, but this less scurfy, and *yellow, saffron-color or dull orange; capsule included, gradually tapering to short beaks about half as long as the body.* — Pinelands and bogs, Fla. to La., n. to Ga. and, very locally, to se. Va. Late May, June. Presumably a hybrid of nos. 1 and 3.

3. A. aùrea Walt. (golden). — Leaves membranaceous, 3–12 cm. long; *raceme remotely flowered; perianth broadly campanulate, orange-yellow*, smoother, *its lobes short-ovate; beaks of capsules included, about as long as plump body.* — Damp pine-barrens and bogs, Fla. to Tex., n. to se. Va. and Md. (reported n. to N.J.). June, July (rarely Sept.)

33. SMÌLAX L. GREENBRIER. CATBRIER

Flowers dioecious, in umbels on axillary peduncles, small, greenish, yellowish or bronze, regular; the perianth-segments distinct, deciduous. Filaments linear, inserted on the very base of the perianth; the introrse anthers linear or oblong, fixed by the base, apparently 1-locular. Ovary of fertile flowers 3-locular (1-locular with single stigma in *S. laurifolia*); stigmas thick and spreading, almost sessile; ovules 1 or 2 in each locule, pendulous, orthotropous. Fruit a small berry. — Shrubby or herbaceous plants of the Tropics, N.Am., the Mediterr. reg. and e. Asia, usually climbing or supported by a pair of tendrils on the petiole of the broad ribbed and netted-veined simple leaves. (An ancient Greek name of an evergreen oak.)

*a.*Stems annual, herbaceous, without prickles; leaves membranaceous to subcoriaceous in age; ovules 2 (rarely 1) in each locule. § NEMEXIA. . . *b.*

*b.*Plant climbing by tendrils borne from axils of most of the middle and upper leaves, the simple or branching stem thus greatly elongating; peduncles borne chiefly from axils of dilated leaves. . . *c.*

*c.*Leaves green or barely paler beneath; peduncle shorter than to once-and-a-half length of subtending leaf; umbels 10–35-flowered; plants chiefly of East and Southeast.

Blade of leaf more or less deltoid-ovate to subhastate, becoming firm, glabrous, hardly lustrous, with 3 prominent ribs running into the often bent or twisted tip; umbel becoming loosely spherical; pedicels 3–10(–15) mm. long; perianth 1.5–2.5 mm. long; filaments shorter than to about equaling anthers. . . . . 1. *S. Pseudo-China.*

Blade of leaf broadly to narrowly ovate, membranaceous, lustrous and minutely pilose or hirtellous (rarely glabrous) on nerves beneath, with 5 delicate veins extending into the abrupt flat tip; umbel hemispherical; pedicels (5–)7–22 mm. long; perianth 4–6 mm. long; filaments longer than anthers. . . . . 2. *S. pulverulenta.*

*c.*Leaves glaucous or pale beneath, becoming firm, oblong-ovate to cordate-rotund; umbels becoming globose, with 20–100 or more flowers; pedicels 5–20 mm. long; perianth 3–5 mm. long; filaments longer than anthers; wide-ranging species.

Petioles of principal leaves 1–4.5 cm. long, the lower leaf-surface glabrous; bladeless bracts below the dilated leaves appressed-ascending; berries blue; chiefly eastern. . . . . 3. *S. herbacea.*

Petioles of principal leaves 2.5–9 cm. long; the lower leaf-surface minutely pubescent; bladeless lower bracts spreading-ascending; berries black; plant of the interior. . . . . 4. *S. lasioneura.*

b. Plant erect or nearly so, up to 1 m. high, without or with only weak upper tendrils; some or all peduncles borne from axils of bladeless bracts below the leaves. . . . . 5. *S. ecirrhata.*

a. Stem woody, often prickly, at least at base; ovules solitary; leaves glabrous, deciduous or evergreen. § EUSMILAX. . . *d.*

d. Leaves ovate, lanceolate, panduriform or rounded, mostly rounded to cordate at base, 5–9-ribbed, with the 3 middle ribs prominent beneath (at least at base); many petioles bearing tendrils; stigmas 2; ovule and berry 3-locular. . . *e.*

e. Leaves green or nearly so beneath. . . *f.*

f. Peduncles mostly shorter than subtending petioles.

Stems prickly at base, their slender and lithe branches and branchlets prickleless or rarely with small subterete prickles; leaves ovate to ovate-oblong, submembranaceous, smooth; berries red. . . . 6. *S. Walteri.*

Strong stems and the subrigid branches bearing strong flattened prickles; leaves from narrowly ovate to suborbicular or reniform, rounded to cordate at base, becoming coriaceous, often muriculate at base beneath; berries blue-black, with bloom. . . . . . 7. *S. rotundifolia.*

f. Peduncles mostly longer than subtending petioles. . . *g.*

g. Prickles hard and rigid, broad-based, pale or merely dark-tipped; peduncles ascending to divergent, only rarely recurving, 0.3–3 cm. long; pedicels 2–7 mm. long; stem 4-angled or terete.

Stems either terete or 4-angled, glabrous; leaves submembranaceous to thin-coriaceous, narrowly to broadly ovate to suborbicular or reniform, margins not thickened, veins and veinlets slender; peduncles 0.3–1.5 cm. long; berries blue with bloom, usually 2-seeded; wide-ranging. . . . . 7. *S. rotundifolia.*

Stems 4-angled, their bases stellate-scurfy; leaves thick-coriaceous and hard, with thickened margins, deltoid to panduriform, the coarse veins and veinlets prominent; peduncles 0.7–3 cm. long; berries black, mostly 1-seeded; southern. . . . . 8. *S. Bona-nox.*

g. Prickles weak and pliant, terete, more or less bristle-like, mostly blackish; stems terete; leaves thin-coriaceous to membranaceous, with thin margins and delicate veins, elliptic- to broadly ovate or shallowly lobulate or fiddle-shaped; peduncles divergent to drooping, 1.5–6.5 cm. long; pedicels 5–12 mm. long; berry black, with 1 large seed. . . . . 9. *S. tamnoides.*

e. Leaves strongly whitened beneath, narrowly to broadly ovate or rounded, submembranaceous; slender terete stem with subulate prickles. . . 10. *S. glauca.*

d. Leaves oblong to oblong-linear or -lanceolate, narrowed to very short petioles, heavy-coriaceous, the thick midrib much more prominent beneath than the 2–4 lateral ones; many leaves without tendrils; stigma and locule of ovary and fruit 1. . . . . 11. *S. laurifolia.*

§ NEMÉXIA (Raf.) A. DC.

1. S. Pseùdo-Chìna L. (false China-root). — Stem herbaceous, climbing by means of tendrils or in young plants erect; *upper bracts at base of stem gradually changing to foliage-leaves, the lowest leaves distinctly smaller and narrower than those above; fully developed leaves green both sides, more or less deltoid-ovate or panduriform,* rarely only cordate-ovate, becoming firm, slender-petioled, *the blade with* 3 *prominent nerves running into the abruptly bent or twisted tip,* the midrib often excurrent; peduncles slender, spreading-ascending to divergent, slightly shorter than to once and a half length of subtending leaf; *umbel becoming loosely spherical,* 10–35-flowered; *pedicels* 3–10 (–15) *mm. long; perianth* greenish, 1.5–2.5 *mm. long,* faintly scented; *filaments shorter than to about equaling anthers; styles* (best seen immediately after fall of sepals and petals) *slenderly clavate,* recurving, caducous; *berries blue to black, with sweet and date-like pulp.* (*S. tamnifolia* Michx.; *Nemexia tamnifolia* Small) — Sphagnous swales, bogs, borders of low woods or damp sands, chiefly on Coastal Plain, L.I. to D.C., s. to S.C. and e. Ga. June, early July.

2. S. pulverulénta Michx. (powdery; from the fine pubescence). — Stem herbaceous, usually high-climbing by tendrils; *basal bracts abruptly changing to fully developed long-petioled* (3–9 *cm.*) narrowly to broadly *ovate leaves, the membranaceous blade lustrous green and often with minute whitish pubescence,* especially on nerves, *beneath, usually with* 5 *delicate nerves running* *into the abruptly acuminate plane tip;* long-peduncled *umbel hemispherical,* with 10–35 greenish flowers on *pedicels 5–22 mm. long; perianth 4–6 mm. long; filaments longer than anthers; styles lingulate;* berry black. (*S. herbacea,* var. Gray; *Nemexia* Small)—Rich, mostly calcareous, woods and thickets, n. Ga., Tenn. and Mo., n. to R.I., L.I., n. N.J., Pa., W.Va., O., s. Ind. and s. Ill. Late April, May.

3. S. herbàcea L. (herbaceous), CARRION-FLOWER, JACOB'S-LADDER, RAISIN DE COULEUVRE (Que.). — In habit similar to no. 2; *bladeless bracts at base of stem appressed-ascending; leaves glaucous or pale and glabrous beneath, on petioles* 1–4.5 *cm. long,* from narrowly oblong-ovate to nearly round, acute or broadly obtuse, tapering, truncate or cordate at base, *becoming firm;* peduncles up to twice length of subtending leaves, becoming strongly divergent in fruit; *umbels becoming globose, with* 20–100 *or more flowers; pedicels* 5–20 *mm. long; flowers carrion-scented,* 3–4 *mm. long;* filaments longer than anthers; styles lingulate; *berry with bluish bloom.* (*Nemexia* Small) — Rich or alluvial thickets, meadows and low woods, w. N.B. and s. Que. to Man. (acc. to *Rydberg*), s. to N.E., L.I., Va., upland to Ga. and Ala., Tenn. and Mo. May, June. — Foliage and habit very variable, not clearly resolved into definite vars.

4. S. lasioneùra Hook. (hairy-nerved). — Habitally similar to no. 3; *bladeless bracts at base of stem spreading-ascending; leaves thinner, minutely pubescent* and pale *beneath, on petioles* 2.5–9 *cm. long;* globular umbels crowded; pedicels 5–15 mm. long; perianth 3.5–5 mm. long; berries black. (*S. herbacea,* var. A. DC.; *Nemexia* Rydb.) — Alluvial or rich thickets, borders of woods, etc., s. Ont. and O. to Sask. and Mont., s. to w. Ga., Ala., Ark., Okla. and Colo. May, June.

5. S. ecirrhàta (Engelm.) S. Wats. (without tendrils). — *Erect or leaning, up to* 1 *m. high, without tendrils or with only 1–few weak terminal ones; lower third or half of stem with bladeless loosely ascending bracts, these subtending the lower or sometimes all the filiform elongate peduncles; leaves* very long-petioled, thinnish, broadly elliptic-ovate to subrotund, cordate or rounded at base, minutely *pilose and pale beneath* when young; umbels hemispherical, becoming subglobose, rather few-flowered; pedicels 5–13 mm. long; perianth 4–5 mm. long; filaments and anthers subequal; *berry 3-seeded.* (*Nemexia* Small) — Open woods, s. Ont. to Minn. and S.D., s. to Tenn. and Mo. May, early June.

§ EUSMÌLAX A. DC.

6. S. Wálteri Pursh (for THOMAS WALTER, (?)1740–1789), WALTER'S or RED-BERRIED G., RED-BERRIED BAMBOO. — *Slender and lithe* woody vine, with widely creeping slender rhizomes, clambering over bushes; *the lower half of the stem with scattered subulate prickles, the terete branches nearly or quite without prickles;* tendrils numerous; *leaves submembranaceous, smooth,* green both sides, *ovate to ovate-oblong, with rounded bases,* mostly 0.6–1 dm. long and 3–6 cm. broad; *peduncles mostly shorter than petioles;* flowers greenish to bronze; *berries bright red.* — Swampy or boggy thickets, low pinelands, etc., Fla. to La., n. to N.J. Late April–early June. — Persistent fruit handsome over winter.

7. S. rotundifòlia L. (round-leaved), COMMON G. OR C., BULLBRIER, HORSEBRIER. — *Tough woody vine,* from long slender rhizomes, *with strong greenish subrigid* terete to quadrangular stems and *branches bearing stout flattened prickles;* tendrils numerous; *leaves* narrowly *ovate to suborbicular or reniform, with rounded to cordate bases,* bright green both sides, lustrous, *becoming subcoriaceous,* mostly 4.5–10 cm. long, *often muriculate on back near base and on short petiole; peduncles* 0.3–1.5 *cm. long,* ascending to divergent, pedicels 2–7 mm. long; flowers greenish to bronze; *berries blue-black, with bloom,* mostly 2-seeded. (Incl. var. *quadrangularis* (Muhl.) Wood) — Moist to dryish thickets and woods, an obnoxious pest, Fla. to e. Tex., n. to N.S., s. Me., s. N.H., N.Y., s. Ont., O., Ind., s. Ill., se. Mo. and Okla. Late April–June (rarely –Aug.).

8. S. Bòna-nòx L. (good-night; the Spanish *buenas noches* for West Indian species recorded by Clusius). — Straggling to climbing, from ligneous thickened and knotty rhizomes and slender subterranean stolons, *with quadrate stems* and branches, the stem and main branches *stiffly prickly,* the branchlets with or without prickles; *leaves* very variable, *from strongly panduriform to broadly ovate,* with or without broadened bases, or (in var. *exauriculata*) lanceolate and without dilated base, *coarsely ciliate or eciliate,* becoming *stiffly coriaceous, strongly reticulate, with thickened margins; peduncles* 0.7–3 *cm. long, ascending, divergent or slightly recurving,* longer than subtending petioles; pedicels 3–6 mm. long, glaucous; berries black, with a bloom; *seed* usually *solitary,* ellipsoid-obovoid to subglobose, 4–4.8 *mm. long,* 3.5–4.2 *mm. broad.* — Four vars. with us:

a. Leaves bristly-ciliate, often mottled with white. . . *b.*

b. Blades deltoid to panduriform; shrubs low, with intricately forking slender stems, rarely climbing high.

Leaf-blade deltoid-ovate to panduriform, the broad rounded basal lobes (when developed) gradually tapering to the narrower deltoid to oblong summit, the latter 1.5–4 cm. broad at base. . . . . *S. Bona-nox* (typical).

Leaf-blade with horizontally divergent oblong basal lobes perpendicular to the narrowly oblong to linear-lanceolate summit, the latter 0.5–2 cm. broad. . . . . Var. *hastata*.

b. Blades narrowly lanceolate, cordate at base, not panduriform; stems stouter, apparently climbing. . . . . Var. *exauriculata*.

a. Leaves of fertile branches eciliate or with few caducous weak cilia, green, rarely mottled (leaves of new offsets frequently mottled and ciliate), from broadly cordate-ovate or deltoid-ovate to shallowly panduriform, the broad basal lobes not prolonged, the blades 3–8 (on sprouts –12) cm. broad; stouter, often high-climbing. . . . . Var. *hederaefolia*.

S. Bòna-nòx (typical), CHINA-BRIER, BULLBRIER, TRAMP'S-TROUBLE. — Dry to moist sand of dunes, clearings, fields and thickets, Fla. to Tex. and e. Mex., n. to Md., Ky., s. Ind., s. Ill., Mo. and Kans. May.

Var. **hastàta** (Willd.) A. DC. (halberd-shaped). — Similar habitats, local, Fla. to Tex., n. to e. N.C. and, less characteristically, se. Va.

Var. **exauriculàta** Fern. (without ear-lobes). — Norfolk Co., Va., little known.

Var. **hederaefòlia** (Beyrich) Fern. (ivy-leaved). — Rich or damp woods and wet thickets, Fla. to Tex. n. to Del. (isolated on Nantucket I., Mass.), Tenn., s. Ill., Mo. and Kans. May, June.

9. S. tamnoìdes L. (like *Tamnus*, Bryony), CHINA-ROOT, HELLFETTER. — *Stout-based* high-climbing shrub from short knotty rhizomes, *with* lithe *terete glabrous stems and* branches, *very prickly below and sparsely* so *above with* blackish subulate relatively flexible *unequal prickles; leaves submembranaceous*, lustrous green, *some or all panduriform*, of narrowly to broadly ovate outline, rounded to cordate base, *the veins slender, the reticulations delicate*; principal blades 5–11 cm. long and 3–8.5 cm. broad; *peduncles divergent to drooping*, slender, 1.5–6.5 *cm. long*, much longer than subtending petioles; *pedicels* very slender, 5–12 *mm. long;* perianth bronze to greenish; fruit black, mostly 1-seeded, *the lustrous red-brown subglobose seed* 5–6.5 *mm. in diameter*. (*S. Pseudo-China* in part of ed. 7, not L.) — Low woods and thickets, chiefly on Coastal Plain, Fla. to e. Tex., n. to e. Va. and Kans. May.

Var. **híspida** (Muhl.) Fern. (with stiff hairs), BRISTLY G. — Often more bristly; *leaves thinner, regularly ovate to elliptic or rounded*, the principal ones 6–12 cm. long and 4–11 cm. broad (on vigorous very bristly sprouts up to 2 dm. long and 1.7 dm. broad); fresh ripe berry bitter; seeds 1 or 2, when solitary 4–6 mm. in diameter, when 2 smaller and compressed. (*S. hispida* Muhl.) — Rich, often calcareous, woods, thickets and bottoms, N.Y. and s. Ont. to Minn. and S.D., s. to sw. Ct., Va., upland to Ga. and Ala., Ark. and e. Tex. Mid-May–early July.

10. S. glaùca Walt. (whitened with bloom), SAWBRIER, WILD SARSAPARILLA. — Lithe freely climbing and entangling shrub with thick and knotty rhizomes and with slender terete often glaucous stems with relatively scattered stiff prickles; *leaves glaucous or whitened beneath* (the bloom melted off by extreme heat in drying) and sometimes above, elliptic or ovate to reniform, with rounded or subcordate bases, *the lower surface densely papillose or hirtellous-pulverulent;* peduncles slender, arching to drooping, longer than subtending petioles; berries blue with bloom, sometimes blackish. — Dry to moist sandy thickets, open woods, fields, etc., Fla. to e. Tex., n. to N.J., W.Va., s. O., s. Ind., s. Ill. and se. Mo.

Var. **leurophýlla** Blake (smooth-leaved). — Leaves glabrous or smooth beneath. — Much commoner, at least with us, farther south less distinctive; Fla. to Okla., n. to s. N.E., se. N.Y., e. Pa., inland Va., W.Va. and se. Mo. Late May, June.

11. S. laurifòlia L. (laurel-leaved), LAUREL-LEAVED G., "BAMBOO", BLASPHEME-VINE. — Evergreen high-climbing vine, with knotty thickened subligneous rhizomes and with strong terete stems armed, especially below and on vigorous sprouts, with rigid terete prickles; *tendrils intermittent, few or wanting on flowering branchlets; leaves heavily coriaceous*, short-petioled, with *the thick midrib much more prominent beneath than the 2–4 lateral ones, the oblong to oblong-linear or -lanceolate blades* 0.6–2 *dm. long* and 1–7.5 cm. broad; umbels short-stalked, often crowded and subpaniculate along the branchlets; *stigma solitary, ovary 1-locular;* berries becoming black. (*S. lanceolata* L., not recent auth.) — Swamps and low grounds, Fla. to Tex., n. to N.J. and Tenn. *Fl. late summer and autumn of 1st season, lasting over winter.* (W.I.)

S. LANCEOLÀTA of recent auth., not L., the southern shrub with reddish berries, **S. SMÁLLII** Morong, has been erroneously recorded from Va. True *S. lanceolata* L. is a narrow-leaved extreme of no. 11.

FAM. 33. HAEMODORÀCEAE (Bloodwort Family)

Perennial stoloniferous herbs with fibrous roots, equitant leaves, and perfect 3-androus regular woolly flowers; the tube of the 6-lobed perianth coherent with the whole surface, or with the lower part of the 3-locular ovary. — Anthers introrse. Capsule crowned or inclosed by the persistent perianth, 3-locular, loculicidal, 3–many-seeded. A small family, chiefly of the S. Hemisph., closely simulating the tribe *Conostylideae* of the *Amaryllidaceae.* Ours with dense compound cymes of dingy yellow flowers.

1. LACHNÁNTHES Ell. Redroot

Perianth 6-parted down to the adherent ovary. Stamens opposite the 3 larger or inner divisions (petals); filaments long, exserted; anthers soon curved or coiled, attached near the base. Style filiform, exserted, declined. Capsule globular. Seeds few on each fleshy placenta, flat and rounded, fixed by the middle. — Leaves clustered at the base and scattered on the stem, which is hairy at the top and terminated by a dense compound cyme of dingy yellow and loosely woolly flowers (whence the name, from the Greek *lachne, wool,* and *anthos, flower*). — A single species:

1. L. tinctòria (Walt.) Ell. (used in dyeing). — Rhizomes and long stolons slender, with red juice; flowering stem 2–9 dm. high; the corymb 3–15 cm. broad. (*Gyrotheca* Morong) — Sandy and peaty shores and swamps, Fla. to La., n. to se. Va.; Del. to se. Mass.; Queens Co., N.S.; becoming weedily aggressive in artificial cranberry-bogs, there sometimes adv. outside the natural range. July–Sept. (W.I.)

FAM. 34. DIOSCOREÀCEAE (Yam Family)

Plants usually with twining stems from large tuberous roots or knotted to smooth rhizomes and ribbed and netted-veined petioled leaves, small dioecious 3- or 6-androus and regular flowers, with the 6-cleft calyx-like perianth adherent in the pistillate plant to the 3-locular ovary. Styles 3, distinct. — Ovules 1 or 2 in each locule, anatropous. Fruit usually a membranaceous 3-angled or -winged capsule.

1. DIOSCORÈA L. Yam

Flowers very small, in axillary panicles or racemes. Capsule loculicidally 3-valved by splitting through the winged angles. Seeds flat, with a membranaceous wing. — Large genus of trop. and warm reg. (Dedicated to the Greek naturalist of the 1st century, A.D., *Dioscorides.*)

a. Leaves regularly cordate; perianth-segments oblong-lanceolate to elliptic; fruits about as long as broad; plants without axillary tubers, with rhizomes; indigenous. . . *b.*

b. Stem erect, or twining only at summit; 1 or more of the lower nodes with leaves in whorls of 4–7, others with opposite leaves; upper leaves alternate; larger leaf-blades 7–15 cm. long; capsules 2.5–3 cm. long; seeds 1.8 cm. broad, with thin hyaline wing broader than the orbicular embryo. . . 1. *D. quaternata.*

b. Stem twining; leaves all alternate or only the lowermost subapproximate or in whorls of 3; larger leaf-blades 5–11 cm. long; capsules 1.2–2.5 cm. long; seeds 0.7–1.2 cm. broad.

Internodes of stem glabrous; leaves sparsely pilose to glabrous beneath; staminate inflorescence becoming lax, with definite internodes conspicuous between the glomerules of flowers; pistillate flowers 5–18; seeds with thin pale hyaline wing broader than the dark oval embryo (3–5 mm. broad). . . . . 2. *D. villosa.*

Internodes more or less hirtellous when young; leaves closely velvety-pilose beneath; staminate inflorescence with subapproximate glomerules of flowers; pistillate flowers 1–4; seeds uniformly dark brown, firm and corky except for the narrow (1 mm. broad) thin wing. . . . . . . 3. *D. hirticaulis.*

a. Leaves more or less panduriform; perianth-segments round-ovate; fruits much broader than long; plants, late in the season bearing axillary tubers, with large tuberous fleshy root; introduced. . . . . 4. *D. Batatas.*

1. D. quaternàta (Walt.) J. F. Gmel. (in fours). — *Stems* 1–3 m. long, *erect below,* leaning or twining above; 1 *or more of the lower nodes with leaves in whorls of* 4–7, *others with opposite leaves,* the upper leaves alternate; longer petioles 5.5–10 cm. long; *blades* cordate-ovate, long-acuminate, green on both sides or paler beneath, glabrous to minutely pilose beneath, *the larger ones* 7–15 *cm. long; staminate* inflorescence axillary, loosely paniculate, the *glomerules of flowers becoming* 3–10 *mm.* apart; pistillate flowers 3–10; *capsules* 2.5–3 *cm. long,* ellipsoid or obovoid, as long as or longer than broad; *seeds* 1.8 *cm. broad;* embryo orbicular, 5 mm. in diameter, wing broader. (Including *D. glauca* Muhl.) — Rich woods and limy slopes, ne. Pa. to s. Ill. and Mo., s. to nw. Fla., La. and Okla. May.

2. D. villòsa L. (soft-hairy). — *Stem twining, glabrous; leaves all alternate* or only the lower 3 whorled, the larger ones on petioles 2–7 cm. long, *blades short-pilose on the nerves beneath,* or glabrous in forma **glabrifòlia** (Bartlett) Fern. (smooth-leaved), *the larger ones* 5–11 *cm. long; staminate inflorescences* loosely paniculate, *in maturity with the small glomerules separated by* 2–4 *mm.; pistillate flowers* 5–18; capsules suborbicular to short-obovoid, as long as broad or slightly shorter, 1.5–2.5 cm. long; *seeds* 7–11 *mm. broad, the broad pale hyaline wing conspicuously contrasting with the dark embryo* (3–5 mm. broad). (*D. paniculata* Michx.) — Wet woods, thickets and swamps, s. N.E. to Minn., s. to se. Va., O., Tenn., Ark. and e. Tex. May–Aug.

3. D. hirticaùlis Bartlett (with short-bristly stems). — Similar to no. 2; *some internodes of stem sparsely hirtellous; larger leaf-blades* 5–10 cm. long, *gray or whitish beneath with close pubescence; staminate inflorescences remaining close, with subapproximate glomerules; pistillate flowers* 1–4; capsules reniform or oblate-obovoid, barely as long as broad, 1.2–1.8 cm. long; *seeds firm or subcoriaceous, uniformly dark brown,* 1–1.2 cm. broad, *the thin wing only* 1 *mm. broad.* — Bogs and peaty depressions, s. N.J. to Ga. June. — The staminate inflorescences often modified by galls.

4. D. Batátas Dcne. (from *Batatas,* the sweet potato), Chinese Yam. — *Root a large farinaceous tuber; leaves* mostly *panduriform;* staminate inflorescences with rather crowded flowers; *perianth-segments round-ovate; capsules* (rare with us) *much shorter than broad,* with 3 flat wing-like lobes nearly 1 cm. broad; *small potato-like* and promptly deciduous *tubers borne* late in the season *in the leaf-axils.* — Thickets, old house-sites and waste places, n. to N.J. (Introd. and natzd. from Asia)

FAM. 35. AMARYLLIDÀCEAE (Amaryllis Family)

Perennials with bulbs, corms or rhizomes and scapose or leafy terminally flowering stems, with linear to oblong radical leaves and regular (*or nearly so*) *perfect* 6-*androus flowers, the tube of the corolline* 6-*parted perianth partly or wholly coherent with the* 3-*locular ovary, the lobes imbricated in bud.* — Anthers introrse. Style single. Capsule 3-locular, 6–many-seeded. Seeds anatropous or nearly so, with a straight embryo in the axis of fleshy albumen.

a. Perianths and fruit glabrous; base a tunicated bulb or thick crown. . . . *b.*
- *b.* Flowers terminating leafless scapes; perianth with spreading or funnelform showy limb, deciduous; leaves thin and elongate, from tunicated bulbs. . . *c.*
  - *c.* Flowers loosely spreading or nodding.
    - Perianth with a slender tube and with a campanulate or cup-like crown arising from summit of tube, perianth wide-spreading. . . . . . . 1. *Narcissus.*
    - Perianth without slender tube or crown, campanulate, the segments green-tipped. 2. *Leucojum.*
  - *c.* Flowers erect or ascending.
    - Scapes terminated by 2–several bracts and an umbel of sessile (rarely 1) flowers; tube of perianth filiform, much longer than the slender-funnelform throat; sepals and petals linear, divergent; perianth surmounted by a large flaring crown connecting the bases of the elongate filaments. . . . . 3. *Hymenocallis.*
    - Scapes with 1 terminal bract and a single flower; sepals and petals broad; crown wanting.
      - Tube of perianth much shorter than the broadly funnelform throat and limb; anthers versatile, on elongate filaments. . . . . . 4. *Zephyranthes.*
      - Tube much longer than the throat and spreading limb; anthers dorsifixed near base, on short filaments. . . . . 5. *Cooperia.*
- *b.* Flowers spicate on a leafy-bracted stem; perianth tubular-funnelform, subpersistent; leaves thick, oblong; base a thick crown at summit of thick rhizome. . . . . 6. *Agave.*

a. Perianths pubescent on outside; base a corm, rhizome or tuber.
- Cormose, with 1–few-flowered scapes; perianth completely covering ovary and often with connivent segments forming a beak above the capsule; sepals and petals without basal beard. . . . . 7. *Hypoxis.*

Rhizomatous, stoloniferous, the leafy stem terminated by a many-flowered corymb; summit of ovary free from perianth; woolly tufts at bases of sepals and petals. . . . . 8. *Lophiola*.

1. NARCÍSSUS L. NARCISSUS

Perianth-tube elongate, with a cup-like or campanulate crown at junction of tube and spreading limb. Stamens inserted in tube, included, the anthers basifixed. — Scapes and linear or ensiform leaves arising from a tunicated bulb; flowers solitary or umbelled, from a scarious sheath, loosely spreading to nodding, the perianth yellow or white. Genus nat. of Eurasia and n. Afr. (Named for mythological *Narcissus*.)

1. N. PSEÙDO-NARCÍSSUS L. (false Narcissus), DAFFODIL. — *Flower* solitary, *yellow*, the perianth-tube broadened upward, *the campanulate* slightly 6-lobed or undulate-margined *crown often longer than the* spreading *perianth-segments.* — Fields and open groves, locally natzd. April, May. (Introd. from Eu.)

2. N. POÉTICUS L. (of the poets), POETS' N. — *Flower* solitary, *white*, the tube slender, *the shallow* cup-like *red-margined crown much shorter than the* rotate *perianth-segments.* — Fields, meadows and borders of groves, sometimes natzd. April–June. (Introd. from Eu.)

Other species, **N.** INCOMPARÁBILIS Mill., **N.** JONQUÍLLA L. (JONQUIL) and some others, sometimes persist or spread from gardens.

2. LEUCÒJUM L. SNOWFLAKE

Elongate tube and crown wanting; the white green-tipped subequal segments ascending, forming an ovoid or campanulate perianth. Stamens erect, inserted at the base of the perianth, the anthers much longer than the filaments and opening by longitudinal slits. — Scapes and linear-ensiform leaves from tunicated more or less fascicled bulbs, the nodding flowers exserted from the papery sheath on long pedicels. Nat. of Eu. and Mediterr. reg. (Name from the Greek *leucos, white,* and *ion, a violet.*)

1. L. AESTÌVUM L. (of summer), SUMMER-S. — Leaves 1–1.5 cm. wide; scape 3–8 dm. high, arched-ascending, in fruit reclining; flowers 2–∞; capsule obovoid. — Locally abundant in meadows and low woods, N.S. to Va. April–early June. (Introd. and natzd. from Eu.)

GALÁNTHUS (from *gala, milk,* and *anthos, flower*) NIVÀLIS L. (snowy), SNOWDROP, a small plant differing from *Leucojum* in having the inner perianth-segments much shorter than the outer, the anthers opening at tip, wanders very slightly from cult. (Introd. from Eu.)

3. HYMENOCÁLLIS Salisb. SPIDER-LILY

Perianth-tube very slender, elongate, the limb with linear spreading segments; crown showy, forming a large cup connecting the bases of elongate filaments; anthers versatile; capsule firm, few-seeded. — Trop. and warm-temp. Am. herbs with scapes and leaves from a tunicated bulb. Flowers white (in ours) to purplish, showy, sessile in a terminal umbel subtended by 2 or more bracts. (Name composed of the Greek *hymen, a membrane,* and *callos, beauty.*) Sometimes merged with PANCRATIUM L.

1. H. occidentàlis (Le Conte) Kunth (western). — Leaves liguliform, glaucous, 3–5 dm. long, 18–36 mm. broad; scape 3–6-flowered; bracts narrow, 5 cm. long; perianth-tube about 8–10 cm. long, the linear segments scarcely shorter; the crown 2.5–3 cm. long, tubular below, broadly funnelform above, the margin deltoid and entire, or 2-toothed and erose, between the white filaments, which are twice longer; anthers yellow; style green. — Marshy banks of streams, Ga. to Ala., n. to Ky., s. Ind., s. Ill. and se. Mo. July–Sept.

4. ZEPHYRÁNTHES Herb. ZEPHYR-LILY

Perianth funnelform, from a tubular base; the 6 divisions petal-like and similar, spreading above; the 6 stamens inserted in the naked throat. Capsule membranaceous, 3-lobed. Trop. and warm-temp. Am. bulbous herbs. (From the Greek *zephyros, west,* and *anthos, flower.*)

1. Z. Atamásco (L.) Herb. (aboriginal name), ATAMASCO-LILY. — Leaves bright green and shining, very narrow, channelled, the margins acute; scape 2–3.5 dm. high; peduncle short; spathe 2-cleft at the apex; perianth white and pink, 6–9 cm. long; stamens and style declined.

(*Atamosco* Greene) — Rich woods and damp clearings, Fla. to Miss., n. to Va. Late April–early June.

5. COOPÈRIA Herb. RAIN-LILY

Perianth-tube very long and slender, the limb widely spreading and 6-parted, the short stamens borne on the throat. Spathe single, membranaceous. Capsule depressed-globose; seeds numerous. — Small N. Am. genus, the leaves grass-like from a tunicate bulb. (Named in honor of *Daniel Cooper*, ?1817–1842, an English botanist.)

1. C. Drummóndii Herb. (for its discoverer, THOMAS DRUMMOND, 1780–1835). — Scape slender, 2–5 dm. high; perianth white or rose-tinged, the stalk-like tube often 1 dm. in length. — Prairies and limestone-hills, se. Kans., Okla. and Tex. July–Sept.

6. AGÀVE L. AMERICAN ALOE

Perianth tubular-funnelform, persistent, 6-parted; the divisions nearly equal, narrow. Stamens 6; anthers linear, versatile. Capsule coriaceous, many-seeded; seeds flattened. — Am. plants; the leaves thick and fleshy, often with cartilaginous or spiny teeth, clustered at the base of the many-flowered scape, from a thick fibrous-rooted crown. (Name from the Greek *agaue*, *noble*, — not inappropriate as applied to *A. americana*, the Century-plant.)

1. A. virgínica L. (Virginian) FALSE ALOE; RATTLESNAKE-MASTER. — Herbaceous; leaves entire or denticulate, green, or purple-mottled in forma **tigrìna** (Engelm.) Palmer & Steyerm. (like a tiger); scape 1–2 m. high; flowers scattered in a loose virgate spike, greenish-yellow, fragrant; perianth 18–24 mm. long, its narrow tube twice longer than the erect lobes, flowers fragrant at night. (*Manfreda* Salisb.) — Dry woods, thickets and open slopes, Fla. to Tex., n. to e. S.C., interior Va., W.Va., s. O., s. Ind., s. Ill. and Mo. Late June–Aug.

7. HYPÓXIS L. STARGRASS

Perianth pilose without, its tube completely coherent with the ovary, the sepals and petals yellow to whitish within, usually greenish on the back, connivent at least after anthesis, commonly forming a beak-like crown to the fruit (deciduous in a few species). Seeds globular to ellipsoid, with pebbled or sculptured testa. — Stemless small herbs with grassy and usually hairy linear leaves and slender 1–several-flowered scapes from a corm-like short vertical rhizome. Genus wide-spread in S. Hemisph., N. Am. and s. Asia. (Old name taken over by Linnaeus, *hypoxys, somewhat acid.*)*

Scapes or most of them long-exserted from the sheaths, chiefly 0.5–5.5 dm. long, (1–)2–7-flowered: anthers versatile; capsule indehiscent, permanently retaining the beak of connivent sepals and petals; seeds sharply muricate, the rostrate hilum subterminal. Subgen. EUHYPOXIS.
- Scapes mostly 2–7-flowered; petals oblong or narrowly elliptic, bluntish, 2.5–5 mm. broad. 1. *H. hirsuta.*
- Scapes 1–3 (rarely –4)-flowered; petals lance-acuminate, 1–2.5 mm. broad.
  - Leaves and scapes flaccid, loosely divergent; leaves membranaceous, glabrous or nearly so, 0.45–1.4 cm. broad. . . . . 2. *H. leptocarpa.*
  - Leaves and scapes erect or nearly so, firm; leaves pilose 1–4 mm. broad. 3. *H. micrantha.*

Scapes included or but slightly exserted from the sheaths, finally up to 1–8 cm. long, 1-flowered; anthers basifixed; capsule circumscissile at summit, throwing off the perianth-beak; seeds with reticulate or low-pebbled surface, the rostrate hilum lateral. Subgen. IANTHE.
- Perianth expanding, 1–2 cm. broad; sepals and petals 5–9 mm. long, 2–4 mm. broad; the former only sparsely pilose on back; capsule 3–4 mm. long, tardily dehiscent, the beak (before circumscission) 6–10 mm. long. . . . . . 4. *H. sessilis.*
- Perianth not expanding, when laid open 5 mm. broad; sepals and petals 2–3 mm. long, 0.5 mm. broad, the former long-bearded; capsule 8–10 mm. long, promptly circumscissile and dehiscent, beak (before circumscission) 4–5.5 mm. long. 5. *H. Longii.*

Subgen. EUHYPÓXIS Baker

1. H. hirsùta (L.) Coville (stiffly hairy). — *Corm subglobose* to ellipsoid, 0.5–2 cm. thick, covered with membranaceous pale to brown sheaths; leaves linear, 1–6 dm. long, 1–8 mm.

---

* Treatment based largely on that of AMELIA BRACKETT in Rhodora, xxv. 120–147 (1923).

846. H. hirsuta.

broad; *scapes* filiform, ascending to reclining, 0.4–3.5 *dm. long, mostly* 2–7-*flowered;* pedicels sparsely pilose, elongate; ovary short-pilose, the mature capsule sparsely so; sepals and petals yellow above, greenish, short-pilose and soon glabrate on back, lanceolate to oval, 0.5–1.5 cm. long; the *petals round-tipped,* 2.5–5 *mm. broad; capsule* ellipsoid, sparsely pilose, or scapes, pedicels, perianth and fruit permanently white-villous with crowded hairs 3–4 mm. long in the southern forma **villosíssima** Fern. (most hairy), 2–6 mm. long, *indehiscent, permanently crowned by the beak; seeds black,* lustrous, *sharply muricate* (under magnification), 0.8–1.3 *mm. in diameter, the* beak and *rostrate hilum subterminal.* — Open woods and meadows, Fla. to Tex., n. to sw. Me., s. N.H., Mass., N.Y., O., Ind., Wisc., Man. and N.D. Late April–Sept. Fig. 846.

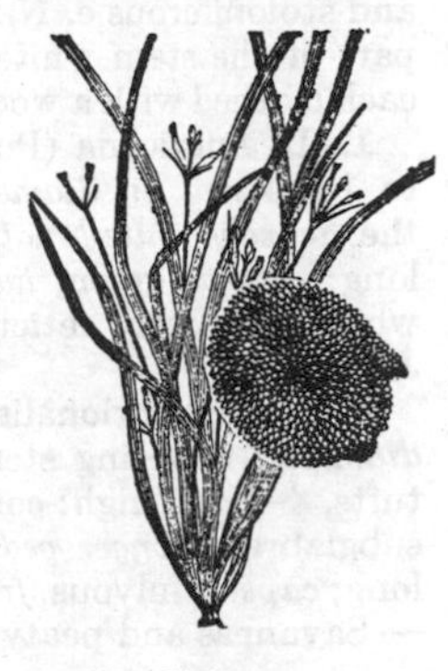
847. H. leptocarpa.

2. H. leptocárpa Engelm. (slender-fruited). — *Flaccid, the leaves and scapes loosely spreading to prostrate; leaves membranaceous, glabrous* or nearly so, 0.45–1.4 *cm. broad,* 1.5–7.5 dm. long; *scapes* capillary, 1–3 (*rarely*–4)-*flowered; sepals and petals lance-attenuate,* 1–2 *mm. broad;* capsules 4–10 mm. long, glabrous or glabrate, crowned by persistent perianth-beak; seeds as in no. 1, with lower and blunter murication. (*H. hirsuta,* var. *leptocarpa* (Engelm.) Brackett) — Hammocks, low woods and bottoms, Fla. to Tex., n. to se. Va., e. Ky. and se. Mo. May–July. Fig. 847.

848. H. micrantha.

3. H. micrántha Pollard (tiny-flowered). — Resembling slenderest no. 1; corm 4–12 mm. thick, blackish-sheathed; *leaves* 1–4 (very rarely –6) mm. broad, *firm, pilose; scapes* 0.3–1.8 dm. long, 1–3-*flowered; sepals and petals lance-acuminate, the petals* 1–2.5 *mm. broad; seeds dark-brown,* more finely muriculate. — Pinelands, bogs and sandy clearings, Fla. to Tex., locally n. to se. Va. April–Sept. (Isle of Pines) Fig. 848.

Subgen. Iánthe (Salisb.) Baker

850. H. Longii.

4. H. séssilis L. (sessile; referring to the flowers). — *Corm cylindric to ellipsoid,* 0.5–1 *cm. thick, covered with dark brown sheaths;* leaves firm, 0.7–3 dm. long, 1–5 mm. broad, sparsely pilose or glabrate; *scapes essentially wanting or up to* 8 *cm. long,* erect, 1-*flowered; sepals and petals yellow above, wide-spreading,* 5–9 *mm. long,* 2–4 *mm. broad, the sepals only sparsely pilose on the back; anthers basifixed; capsule* pyriform, 3–4 *mm. long, tardily dehiscent* and circumscissile; *the tardily deciduous beak before circumscission* 6–10 *mm. long,* fully twice as long as the capsule; *seeds ellipsoid,* 1.5–2 *mm. long,* the friable testa *iridescent and golden-brown or olive with very fine reticulations, the rostrate hilum lateral.* — Pinelands, very local, Fla. to Tex., n. to se. Va. Late April, May. Fig. 849.

849. H. sessilis.

5. H. Lóngii Fern. (for one of its discoverers, Bayard Long, 1885–). — Similar to no. 3; basal sheaths paler; *flowers included in the sheaths, not expanding* (cleistogamous?), their scapes later becoming exserted and up to 6 cm. long; *sepals* 3 *mm. long,* 0.5 *mm. wide, densely bearded and overtopped by long white hairs; petals white,* 2 *mm. long,* 0.5 *mm. wide; capsule* thick-clavate, 0.8–1 *cm. long, promptly circumscissile and dehiscent* into 3 membranous white valves; *beak before circumscission* about half as long as capsule, 4–5.5 *mm. long; seeds olive-black,* iridescent, the lateral hilum prominent. — Damp peaty and sandy open soil, very local, se. Va. June–Aug. Fig. 850 (the perianth mechanically opened).

8. LOPHÌOLA Ker

Divisions of the perianth nearly equal, spreading, longer than the 6 stamens, which are inserted at their base. Anthers fixed by the base. Capsule ovoid, free from the perianth except below the middle, pointed by the subulate style, which finally splits into 3 divisions, one terminating each valve. Seeds numerous, oblong, ribbed, anatropous. — Slender rhizomatous and stoloniferous e. N.Am. herbs with linear and nearly smooth leaves; inflorescence and upper part of the stem whitened with soft matted wool. Perianth-lobes naked only toward the tip, each clothed with a woolly tuft near the base (whence the name, from the Greek *lophia, a mane*).

1. L. americàna (Pursh) Wood (American), GOLDEN-CREST. — *Rhizome* slender, thickening to 1.5–4 *mm. in diameter;* flowering stem solitary, 3–5.3 dm. high; corymb 3–10 cm. broad, the *densely white-lanate pedicels becoming* 1–6 *mm. long;* mature (closed) perianth 5–7 mm. long; *capsule* green, *adnate at least half its length to the perianth; seeds rounded at both ends,* whitish, coarsely reticulate. — Bogs in the Pine Barrens, N.J., and n. Del. (formerly). Late June, July.

2. L. septentrionàlis Fern. (northern). — *Rhizome* more thickened, *becoming* 5–8 *mm. in diameter;* flowering stems solitary or often subcespitose, with many approximate leafy erect tufts, 4–7 dm. high; corymb 4–15 cm. broad, more open, less heavily lanate, in fruit becoming subglabrate; *longer pedicels becoming* 0.7–2 (–2.5) *cm. long;* mature (closed) perianth 6–9 mm. long; *capsule* fulvous, *free almost to base; seeds caudate-tipped at base*, drab, more finely reticulate. — Savannas and peaty shores, local, w. N.S. Aug., Sept.

FAM. 36. IRIDÀCEAE (IRIS FAMILY)

Herbs with equitant 2-ranked leaves and regular or irregular perfect flowers; the 3 petals and 3 petal-like sepals convolute in the bud, the tube adnate to the 3-locular ovary; the 3 distinct or monadelphous stamens alternate with the petals, with extrorse anthers. — Flowers from a spathe of 2 or more leaves or bracts, usually showy. Style single, usually 3-cleft; stigmas 3, opposite the locules of the ovary, or 6 by the parting of the style-branches. Capsule 3-locular, loculicidal, many-seeded. Seeds anatropous; embryo straight, in fleshy albumen. Rhizomes, tubers, or corms mostly acrid.

a. Flower regular or nearly so; sepals and petals similar and essentially equal, spreading. . . *b.*
- *b.* Bulbous; style-branches opposite the anthers, slender, 2-parted, divergent below middle of anthers. . . . . 1. *Nemastylis.*
- *b.* Not bulbous; style-branches alternate, simple, erect, short.
  - Stem from long rhizome; perianth mottled; style-branches clavate; filaments free except at base; capsule promptly and completely dehiscent, the seeds long-persistent on the axis. . . . . 2. *Belamcanda.*
  - Stem 2-edged, from fibrous roots and very abbreviated rhizome; style surrounded by tube of filaments, the stigmas filiform; capsule tardily dehiscent or dehiscent only at summit. . . . . 3. *Sisyrinchium.*

a. Flowers with dissimilar sepals and petals, the former spreading or recurved, the latter spreading or erect; style-branches petal-like and arching over the anthers; stigma a thin plate beneath apex of style-branch; rhizomatous or tuberous. . . . . 4. *Iris.*

1. NEMASTỲLIS Nutt. CELESTIAL LILY

Sepals and petals similar and nearly equal, spreading. Style short, opposite the anthers, its slender 2-parted branches clasping and exserted between the anthers; stigmas minute, terminal. Capsule obovoid, truncate, dehiscent at the summit. Seeds globose or angled. — Stem terete, with few plicate leaves and few fugacious flowers from 2-bracted spathes. Small Am. genus. (Name from the Greek *nema, a thread*, and *stylis, style*, for the slender style-branches.)

1. N. geminiflòra Nutt. (twin-flowered), PRAIRIE-IRIS. — Stem 1–6 dm. high; spathes 2-flowered; flowers pale blue-purple, 4–7 cm. broad, the divisions oblong-obovate; capsule 1–1.3 cm. long. (*N. acuta* Herb.) — Prairies, calcareous glades and rich woods, La. and Tex., n. to w. Tenn., Mo. and se. Kans. April–June.

2. BELAMCÁNDA Adans. BLACKBERRY-LILY

Sepals and petals widely and equally spreading, all nearly alike, oblong, with a narrowed base, naked. Stamens monadelphous only at base; anthers oblong. Style clavate, 3-cleft. Capsule pyriform; the valves at length falling away, leaving the central column covered with the globose black and fleshy-coated seeds, imitating a blackberry (whence the popular name). — Perennial, nat. of Asia, with rhizomes, foliage, etc., of an *Iris;* the branching stems (0.5–1 m. high) loosely many-flowered; the orange-yellow flower mottled with crimson-purple spots. (An East Indian name for the species.)

1. B. CHINÉNSIS (L.) DC. (of China). — Roadside-thickets, open woods, etc., near settlements, Ct. to Neb. and s. beyond our limits. June, July. (Introd. and natzd. from Asia)

3. SISYRÍNCHIUM L. BLUE-EYED GRASS

Sepals and petals alike, spreading, commonly apiculate or mucronate. Capsules globular, obtusely 3-angled. Seeds globular. — Low slender perennials, chiefly of Am. (rarely in Austral., Mauritius and Eu.), with fibrous roots, linear to lanceolate leaves, 2-edged or winged stems, and fugacious subumbelled small flowers from a usually 2-leaved spathe. (Name used by Theophrastus for some plant, later transferred to the present genus.)*

*a.*Spathes sessile and terminating simple stems, mostly subtended by an erect leaf-like bract. . . *b.*

*b.*Spathes mostly 2 (sometimes 1 or 3) together. . . *c.*

*c.*Spathes with firm green outer scales resembling the foliaceous bracts but subaristate; filaments free above; anthers 4.5 mm. long; local plant of se. Mich. . . . . 1. *S. hastile.*

*c.*Spathes with more scarious scales; filaments united to summit; anthers at most 2.5 mm. long.

Stem flattened, distinctly wing-margined; leaves flat, 1–3 mm. wide, the old ones marcescent, not bristle-like; perianths white or pale violet; plant of interior. . . . . 2. *S. albidum.*

Stem filiform; leaves slenderly capillary, up to 0.5 mm. thick, the old remnants disintegrating into bristles; perianths violet; plant of se. Coastal Plain. . . . . 3. *S. capillare.*

*b.*Spathe solitary. . . *d.*

*d.*Outer elongate bract with margins free to base; perianths white or pale blue (rarely yellow); chiefly western and southwestern. . . . . . . 4. *S. campestre.*

*d.*Outer elongate bract with margins united above base; perianth violet (exceptionally white). . . *e.*

*e.*Pedicels soon strongly spreading to recurving, much exceeding the inner bract; margins of outer bract united only slightly above base; stems very slender, barely margined; leaves 1–2 mm. broad; capsules 2–4 mm. high. . . . . 5. *S. mucronatum.*

*e.*Pedicels erect or ascending, barely or but slightly overtopping inner bract; margins of outer bract united 2–6 mm.; stems winged, 1–3 mm. wide; capsules 3–6 mm. high. . . . . 6. *S. montanum.*

*a.*Spathes all or nearly all peduncled from the axil of the leaf-like bract, the individual spathes usually with subequal bracts. . . *f.*

*f.*Old leaf-bases persisting as tufts of erect bristle-like fibers; capsules 3–5 mm. high. . . . . 7. *S. arenicola.*

*f.*Old leaf-bases soon deciduous, or persisting as loose irregular shreds.

Stems much flattened and broadly winged, except in dwarf individuals 2–6 mm. wide; peduncles flat and winged; inner bracts of spathe 1.5–2 cm. long; capsules 4–6 mm. high; plant deep green, drying blackish. . 8. *S. angustifolium.*

Stems slender, narrowly margined, 1–3 mm. wide; peduncles filiform; inner bract of spathe 1–1.5 cm. long; capsules 3–4.5 mm. high; plant pale green or glaucous, hardly blackened in drying. . . . . 9. *S. atlanticum.*

1. S. hastile Bickn. (shaft of a spear; from the long-pointed firm scales). — Stiff and erect, dull green, about 4 dm. high, the *stem* (1–1.5 mm. wide) *narrowly margined but not winged; leaves* firm and *stiff,* slender and *conduplicate,* barely 1 mm. broad, except at the flattened base; the 2 *spathes* closely sessile, each 4-bracted; the *lance-attenuate* strongly nerved *inner bracts* 1.5–2.5 cm. long, much exceeded by the linear outer bract; *filaments free above; anthers* 4.5 *mm. long.* — Sandy shores, Belle Isle, Detroit River, Mich. May, June. — Very little known; seeming unique among our species. Needs careful collecting.

2. S. álbidum Raf. (whitish). — Erect, pale green or glaucous, 1.5–4.5 dm. high; *stems* 1–3 *mm. wide, flattened and slightly winged,* usually twice exceeding *flat leaves; spathes* 2 (rarely 1 or 3), *with lance-acuminate* pale or purple-tinged *inner bracts* 1.3–2.3 cm. long, usually twice

* The fugacious perianths render flowering material difficult to identify. The most stable characters are in fruiting material.

exceeded by the erect outer foliaceous bract; *pedicels* ascending to outwardly arching, *with exserted tips;* flowers about 1 cm. long, whitish or pale violet; *filaments united to summit; anthers at most* 2.5 *mm. long.* — Prairies and dry, often sandy, open soil or thin woodland, Ga. and Ala. to La., n. to N.C., s. Ont., O., s. Mich., s. Wisc. and Mo.; casual northeastw. May, June.

3. S. capillàre Bickn. (hair-like). — Excessively slender, erect; *old leaves persisting as stiff basal fibers; stems filiform,* 1.5–5.25 dm. high, iridescent when fresh and with tight prolonged spiraling *leaves capillary,* glaucous or iridescent, *at most* 0.5 *mm. thick; spathes* 2 (rarely 1 or 3), closely subtended by an erect setaceous bract 2.5–8 cm. long; the spathes 1–1.3 *cm. long,* with hyaline-margined bracts; pedicels exserted, erect or with arching tips; perianth blue-violet, 6–8 mm. long; *capsules pale,* 2–3 *mm. high.* — Flat pinelands, bogs and wet pine-barrens, local, Fla. to se. Va. Late Apr., May.

4. S. campéstre Bickn. (of low fields). — Cespitose, glaucous, slender, 1–5 dm. high; *stems* erect, *flat,* 1–3 *mm. broad,* winged, usually exceeding the flat and slightly broader leaves; *outer elongate foliaceous bract* 2.5–4.5 cm. long, *with margins free to base; spathe* solitary, sessile, *gibbous at base,* green or pink-tinged, half as long as outer bract; pedicels slightly exserted, ascending; *perianth white to pale blue,* or in the local forma **flaviflòrum** (Bickn.) Steyerm. (yellow-flowered) yellow (*S. flaviflorum* Bickn.); *capsules* 2–4 *mm. high.* — Prairies and dry open soil, Wisc. to Man., s. to Ill., La. and Tex. Apr.–June. Fig. 851.

851. S. campestre.

852. S. mucronatum.

5. S. mucronàtum Michx. (with a short straight point; from the tips of the sepals and petals). — Tufted, green, 1.5–4.5 dm. high, erect; *stems* flat, 0.5–1.5 *mm. broad, barely margined; leaves* 1–2 *mm. broad; outer foliaceous bract with margins united only slightly above base;* spathe usually purple or purple-tinged, 1–2 cm. long; *pedicels soon strongly spreading to recurving, much exceeding inner bract;* perianth violet (rarely white); *capsules* straw-color to yellow-green, 2–4 *mm. high.* — Meadows, fields and open woods, ne. Me. to Wisc., s. to N.C. May, June. Fig. 852.

6. S. montànum Greene (montane; name inappropriate with us). — Tufted, *erect, pale green, not often blackening in drying; leaves* 1–3 *mm. wide,* erect or ascending; *stem simple* (rarely slightly forked), *pale,* 1–6 dm. high, *distinctly flattened and wing-margined; spathes pale green or stramineous, the outer bract with margins united* 2–6 *mm. above the base,* 2–8 cm. long, the inner 1.5–3.5 cm. long; perianth blue-violet; *fruiting pedicels erect or strongly ascending, shorter than to slightly overtopping inner bract; capsules whitish-green to straw-color or pale brown,* 3–6 *mm. high.* — Shores, meadows and damp open soil, w. Nfld. and Anticosti I., Que. to Mackenz. and n. B.C., s. to s. Que., s. Ont., w. N.Y., n. Ind., n. Ill., n. Ia., Neb., Colo., etc. June, July. Fig. 853.

853. S. montanum.

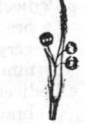

Var. **crèbrum** Fern. (frequent). — *Greener, mostly darkened in drying; spathes deeper-green, often purple-tinged,* the inner bract 2–6.5 cm. long; *capsules green, dull brown, drab or purple-tinged, when ripe dark to blackish.* (*S. angustifolium* of ed. 7 and of Bickn., not Mill.) — Dry to moist open soil, Nfld. to Ont., s. to N.S., N.E., Pa. and mts. to w. Va. May–July.

7. S. arenícola Bickn. (growing in sand). — Tufted from a fibrous base, *the old leaf-bases persisting as crowded erect bristles; leaves and stem* green or but slightly glaucous, *inclined to darken in drying; leaves* erect, *firm,* 1–3.5 mm. wide; *stems* erect, mostly *with* 2–8 *forks or peduncles,* 1.5–5.2 dm. high, the main axis flattened and winged, 1.5–3.5 cm. wide; peduncles 2–11 cm. long, slender, arched-ascending, simple or forking; spathes 1.3–2 cm. long, pale brown; perianths blue-violet; *fruiting pedicels erect or ascending,* exserted; capsules pale to deep brown, drying dark, 3–5 mm. high. (*S. carolinianum* Bickn., not Klotzsch; *S. fibrosum* Bickn.) — Dry, mostly sandy soil on or near Coastal Plain, Fla. and Ala., n. to se. Mass. and w. N.S.; also s. Mich. as the scarcely separable *S. Farwellii* Bickn. May–early July.

8. S. angustifòlium Mill. (narrow-leaved). — *Loosely tufted,* ascending to geniculate and spreading, the *leaves and stems mostly deep green* (rarely glaucescent) *and drying blackish; leaves submembranaceous,* 1.5–6 mm. wide, shorter than to exceeding flowering stems; *stems broadly winged* (except in young or crowded individuals), *flexuous* or even geniculate, 1–5 dm. high, *mostly forking; peduncles* 2–5, *loosely ascending, winged,* 2–15 cm. long; *lowest foliaceous bract slightly shorter than to overtopping the flowering spathes, the latter* 1.5–2 *cm. long, with mostly subequal bracts; perianth pale blue, changing to violet; fruiting pedicels long, slender, outwardly arching or recurving, much overtopping their spathes, few* (*mostly* 1–5); *capsules dark or blackish in drying,* 4–6 *mm. long.* (*S. gramineum* Curtis and of Lam. in part; *S. grami-*

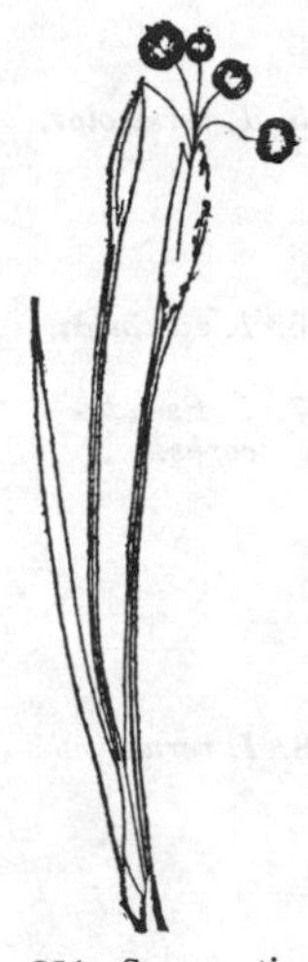

noides Bickn.; also including *S. intermedium* Bickn., a nondescript series, perhaps hybrid, intermediate between nos. 5 or 6 and 8) — Meadows, low woods and thickets or damp shores, Fla. to e. Tex., n. to se. Nfld., s. Que., s. Ont., O., Ind., Ill., Mo. and e. Kans. May–July. FIG. 854. — Simple-stemmed plants are distinguished from var. of no. 6 by softer texture and few loosely spreading to recurving long pedicels.

854. S. angustifolium.

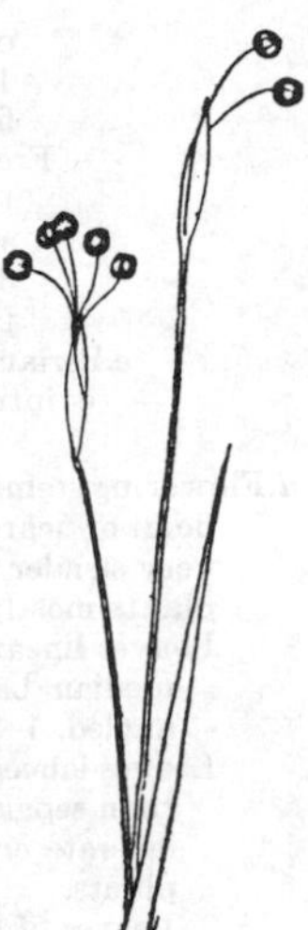

9. S. atlánticum Bickn. (Atlantic). — *More slender* than no. 8, *pale green or glaucous, hardly blackened in drying,* ascending to erect; *leaves firm, very pale,* 1–3 *mm. wide, mostly much overtopped by the flowering stems; stems wiry, slender, narrowly margined,* 1–3 *mm. wide,* 2–7 dm. high, usually *forking into* 2–4 often geniculate *filiform peduncles; spathes pale green or purple-tinged,* 1–1.5 *cm. long;* perianth blue-violet; *fruiting pedicels ascending,* exserted; *the mostly* 3–10 brown or finally dark *capsules* 3–4.5 *mm. high.* (Incl. *S. apiculatum* Bickn.) — Damp to dry meadows, swales, marshes and low woods, Fla. to Miss., n. to w. N.S., s. Me., s.-centr. N.H., Vt. and se. N.Y., and in the interior n. to O. and s. Mich. May–July. FIG. 855.

855. S. atlanticum.

4. ÌRIS L. IRIS. FLEUR-DE-LIS

Tube of the flower more or less prolonged beyond the ovary. Stamens distinct; the oblong or linear anthers sheltered under the overarching petal-like branches of the style, which bear the stigma in the form of a thin lip or plate under the apex; most of the style connate with the sepals and petals into a tube. Capsule 3–6-angled, usually coriaceous. Seeds depressed-flattened or plump, usually in 2 rows in each locule. — Perennials of N. Hemisph., with ensiform, grassy or lanceolate leaves and large showy flowers; ours with creeping and more or less tuberous rhizomes. (*Iris, the rainbow.*)*

a. Flowering stem (0.5–)2–10 dm. or more tall, simple or often forking, with elongate basal leaves, arising from a usually stout and chiefly subterranean rhizome (rhizome slender and superficial in no. 1); perianth-tube much shorter than sepals, the latter mostly much larger than the petals; capsules 1.5–10 cm. long; plants often paludal. . . *b.*

b. Rhizome superficial or nearly so, producing rope-like stolons 2–5 mm. thick; leaves 2–9 mm. broad; ovary and capsule acutely 3-angled. . . . 1. ***I. prismatica.***

b. Rhizome deep-seated or barely superficial, without conspicuous superficial stolons, 1–3 cm. in diameter; leaves becoming 0.5–3 cm. or more broad; ovary and capsule obtusely 3-angled or with 6 angles. . . *c.*

c. Petals bristle-tipped, involute or tubular, 1–2 cm. long; capsule thin-walled (easily impressed), beakless, with compressed-pyriform seeds; boreal species of maritime bluffs, headlands, cliff-crests and upper beaches or dunes. . . . . 2. ***I. Hookeri.***

c. Petals not bristle-tipped, oblanceolate, cuneate or obovate, longer; capsule hard-walled, usually beaked; seeds not pyriform; more southern or wide-ranging often paludal plants. . . *d.*

d. Ovary and capsule 6-angled; lower or axillary spathes sessile to very short-peduncled; leaves flaccid; inland or southern species.

Flowering stem ascending, up to 1 m. high; leaves up to 1.5 cm. broad; lateral spathes, when present, from upper axils; perianth copper-colored or reddish. . . . . 3. ***I. fulva.***

Flowering stem weak or reclining, 1.5–5.3 dm. long; leaves 1.5–3 cm. broad; lateral spathes from all but lowest axils; perianth blue or blue-purple. . . . . 4. ***I. brevicaulis.***

d. Ovary and capsule obtusely 3-angled; lower axillary spathes, when present, terminating long branches; leaves firm; wide-ranging plants. . *e.*

e. Perianth bluish, violet or purple, its tube constricted above the ovary; indigenous.

Fresh leaf-tufts purplish at base; spathe-bracts mostly papery or scarious, 3–6 cm. long; papillae of greenish-yellow basal blotch

---

* Some statements of characters derived from the synopsis by R. C. FOSTER, Contrib. Gray Herb., no. cxix (1937).

of sepal shorter than thickness of sepal; capsule symmetrical, 1.5–6 cm. long, its inner surface lustrous; seed lustrous, with finely and regularly pitted firm coat; northern and northeastern. . . . 5. *I. versicolor.*

Fresh leaf-tufts buff or pale brown at base; spathe-bracts firm, up to 1.4 dm. long; pubescence of bright yellow basal blotch of sepal as long as thickness of sepal; capsule often asymmetrical, 3–10 cm. long, its inner surface dull; seed with irregularly deep-pitted brittle corky coat; southern and inland species. . . . . 6. *I. virginica.*

e. Perianth yellow or yellowish, its tube not constricted above the ovary; introd. and spread from cult. . . . . 7. *I. Pseudacorus.*

a. Flowering stem 0.5–1.5 dm. high, simple, without long basal leaves, from superficial or nearly superficial rhizomes or stolons 2–8 mm. thick; perianth-tube very slender, nearly as long as or longer than sepals; capsule 1–3.2 cm. long; plants mostly not paludal.

Leaves linear-attenuate, firm; sepals and petals subequal; yellow or orange median basal band of sepal not fimbriate-margined; capsule obtusely angled, 1–3.2 cm. long; southeastern. . . . . 8. *I. verna.*

Leaves lanceolate or lance-ensiform, herbaceous or succulent; petals smaller than sepals; white and purple patch at base of orange-white fimbriate or lacerate crest of sepal; capsule sharply 3-angled, 1–1.7 cm. long; inland plants.

Leaves of fans (leafy tufts) 0.5–2.5 cm. broad; perianth-tube thread-like, 4.5–7 cm. long; sepals obovate or spatulate; capsule about 1 cm. long; woodland species southward. . . . . 9. *I. cristata.*

Leaves 0.5–1.5 cm. broad; perianth-tube 1.3–1.8 cm. long; sepals cuneate; capsules 1–1.7 cm. long; plant of upper Great Lakes. . . . . . . 10. *I. lacustris.*

1. I. prismática Pursh (like a prism; from the sharply angled ovary and capsule), SLENDER BLUE FLAG. — *Rhizome and numerous superficial or but slightly buried* scaly *stolons only* 2–5 *mm. thick,* these bearing terminal reddish-based *tufts of firm erect faintly ribbed leaves only* 2–5 *mm. wide* and becoming 3–6 dm. long; *flowering stem* single, *arising from tip of last-year's stolon* and surrounded at base by fibrous remnants of old leaves, *terete, solid, very slender,* 2–7.5 dm. high, with 1–4 remote erect leaves, simple or with 1 or 2 ascending 1- or 2-flowered branches mostly borne well below the terminal 1–3-flowered one; *spathe-valves pale brown, wholly or partly scarious-membranaceous, narrowly lanceolate,* 2–4 *cm. long,* rarely with an outer foliaceous bract up to 8 cm. long; *pedicels* flattened on inner side, *mostly exserted, the longer ones of the terminal cluster becoming* 3–7 *cm. long, articulated below the ovary; flower pale bluish or blue-violet,* 5–7 *cm. broad; perianth-tube about* 3 *mm. long; sepals* veined below with deep violet, *the ovate or obovate to roundish blade* 1.3–2 *cm. broad and rather abruptly narrowed to the broadish claw,* yellowish at base; *petals* oblanceolate to oblong-obovate, 3.5–4.5 cm. long, 0.7–1.5 *cm. broad;* style-branches 2–3 cm. long, the serrate crests nearly quadrate; filaments equaling or longer than the bluish anthers (0.8–1.2 cm. long); *capsule sharply 3-angled, almost winged,* 1.5–4.5 cm. long, *the oblong flattish or concave faces* 6–14 (av. 9.6) *mm. broad;* seeds in 1 row in each locule, compressed but with convex sides, obovoid to D-shaped or nearly square, 3–4 mm. long. — Brackish or saline to fresh marshes, sands, shores or meadows, along coast, ne. Md. and Del. to sw. Me.; e. C.B.; rarely inland 30 mi. from tidal water. May (southw.)–mid July (C.B.).

Var. **austrìna** Fern. (southern). — Coarser; *leaves of sprouts* 5–9 *mm. broad, prominently sulcate-ribbed; spathe-valves firmer, subcoriaceous,* either brown or green and foliaceous; *flower* 8–10 *cm. broad; perianth-tube* 5–6 *mm. long; sepals* spatulate-obovate 1.2–1.7 cm. broad, *the blade gradually tapering into the* slender-based *claw; faces of capsule* 10–14 *(av.* 12.5) *mm. broad; seeds* 4–6 *mm. long.* — Acid swamps, wet barrens, and shallow pools, mts. of Tenn. and N.C. to Ga., e. to Piedmont and Coastal Plain of se. Va. and N.C. May, June.

2. I. Hoòkeri Penny (for Sir WILLIAM JACKSON HOOKER, 1785–1865), BEACHHEAD-IRIS or -FLAG. — *Densely cespitose, with many crowded upright crowns* with marcescent old leaves, *the crowded fans of new leaves with erect or strongly ascending blades* (0.5–)1–5.2 dm. long and (0.3–)0.5–1.4 cm. broad; leaves with cross-bands of white in forma **zonàlis** (Eames) Fern. (zoned); *flowering stems several to many* from a single clump, simple or with 1 or 2 erect branches, 0.5–6 *dm. high,* with leaves mostly subbasal, *either naked above or with a single remote bracteiform lanceolate blade* 0.5–1.5 *dm. long borne nearly midway on the stem; outer bracts of spathe firm,* herbaceous, *lanceolate to oblong-ovate, arching to sharp point,* 2.5–6(–9) cm. long, the inner ones similar or scarious-tipped, longer or shorter; *flower* 7–11 cm. broad, *deep to pale blue or blue-violet,* or white or whitish in forma **pallidiflòra** Fern. (pale-flowered); *sepals with broadly rounded* *blade* 2.5–4 cm. broad and strongly contracted to the broad and short claw, *a diffuse white patch at base of blade; petals reduced to small involute or tubular bristle-tipped insignificant rudiments only* 1–2 *cm. long;* style-branches with bifid toothed summit; *capsule thin-walled, blunt, thick-ellipsoid, with rounded angles,* 2–4 cm. long; *the dark brown lustrous seed compressed-pyriform,* 4–6 mm. long, *with prominent raphe,* soon loosening and rattling in the capsule. (*I. setosa* Pall., var. *canadensis* M. Foster; *I. canadensis* Small) — Turfy crests of headlands, rocky slopes, upper borders of beaches, dunes, etc., within reach of ocean-spray, coast of Lab., Nfld. and maritime Que. up the St. Lawrence to Charlevoix and Montmagny Cos., s. along coast to Knox Co., Me. June–Aug.

3. I. fúlva Ker (reddish-yellow), Red Iris. — *Rhizome* branching, *about* 1.5 *cm. thick;* leaves linear-ensiform, bright green, flaccid, arched-reflexing above, 6–10 dm. long, up to 1.5 cm. broad; *flowering stem* up to 1 m. high, *simple or sessile-branched above;* cauline leaves prolonged; short branches subtended by long tapering erect leaves; *spathes very unequal, the herbaceous green outer one* 1 *dm. or more long,* the partly scarious inner ones shorter; *flower copper-colored or reddish-brown,* variegated with green or blue lines, about 1 dm. broad; *sepals* obovate or oblong-elliptic, with short claw *reflexed; petals* similar but slightly smaller, *reflexed;* style-branches 2 cm. long, with rounded crest; *capsule hexagonal,* ovoid or ellipsoid, up to 5 cm. long; *large seeds flattened, with corky or spongy outer coat.* — Sloughs, ditches, muddy shores and swampy woods, Ala. to La., n. to sw. Ill. and se. Mo. May, June.

4. I. brevicaùlis Raf. (short-stemmed), Lamance I. — Rhizome rather slender, 1–2.5 cm. in diameter; *stem flexuous, loosely ascending to depressed,* compressed, 1.5–5.3 dm. long; *basal leaves lax,* 3–6 or more dm. long, 1.5–3 cm. broad; *spathes terminal and subsessile or short-peduncled from all but lowest axils, subtended by broad and very prolonged leaves* 2–6 *dm. or more long;* spathe-bracts subequal, up to 5 cm. long, the outer pair green, the inner scarious-margined; *flower deep blue or blue-purple; ovary prominently* 6*-angled;* sepals 7–9.5 cm. long, 2.5–3 cm. broad, the ovate blade slightly longer than the greenish-yellow dark-striped claw; the latter with a yellowish-white summit; oblanceolate petals slightly shorter than sepals; style-branches greenish, with entire or toothed subquadrate to semi-ovate crests; *capsule hexagonal,* ovoid or ellipsoid, 3–5 cm. long; seeds irregularly circular, with thick coat. (*I. foliosa* Mackenz. & Bush) — Swamps, bottoms, borders of rich woods, etc., Ala. to e. Tex., n. to O., Ind., Ill., Mo. and Kans. Late May, June.

5. I. versícolor L. (variously colored), Blue Flag, Poison Flag, Clajeux (Que.). — Rhizome stout, deeply buried to superficial; *leaves firm, ascending, pale green to grayish,* the basal ones (0.1–)0.2–1 m. long, 0.5–3 cm. broad, when fresh commonly purplish at base; flowering stem simple or with 1 or 2 branches above, the cauline ascending leaves prolonged; *spathe-bracts papery or scarious or the outer subherbaceous,* 3–6 cm. long; *pedicels slender,* 1 *or more of them elongating to equal or exceed spathe; ovary* 3*-angled;* perianth-tube constricted above the ovary, infundibuliform, up to 1.2 cm. long; *sepals* 4–7 cm. long, 2–3.5 cm. broad, *blue-violet,* or perianth white in forma **Murrayàna** Fern. (named in 1936 for its discoverer, Andrew Murray), with darker veins, *often with greenish-yellow base and claw,* the basal patch *closely papillose-pubescent with papillae no longer than thickness of sepal;* petals about half to two-thirds as long as sepals, 0.5–2 cm. wide; style-branches up to 3.5 cm. long, with entire or toothed crest; *capsule* thick-cylindric to ellipsoid, *firm,* with definite beak, *obtusely* 3*-angled,* 1.5–6 cm. long, *tardily opening, often persistent over winter, the inner surfaces as if varnished; seeds D-shaped,* 5–8 mm. long, *with sublustrous finely and regularly pebbled firm brown coat.* — Marshes, meadows, ditches or turfy shores, s. Lab. to Man., s. to Nfld., N.S., N.E., w. Va., w. Pa., n. O., Mich., Wisc. and Minn. Mid-May (southw.)–mid-Aug. (northw.) — Hybridizes with nos. 2 and 6.

6. I. virgínica L. (Virginian), Southern Blue Flag. — Differing from no. 5 in *flaccid* and greener *leaves; those of the basal tufts buff* or pale brown at base, soon *arched-recurving or falling to ground; flowering stem weak, soon low-arching and maturing fruit on the ground or in water; spathe-bracts firmer, usually subherbaceous,* up to 14 cm. long; *sepals* with obovate or obovate-oval blade 3–4 cm. broad, *with* prominent *yellow midrib expanding to a broad bright yellow pubescent patch at base; the hairs elongate, as long as thickness of blade;* petals obovate to obovate-spatulate, two-thirds to four-fifths as large as sepals; *capsule ovoid, ellipsoid or thick-cylindric,* 3–7 *cm. long,* 1.3–2.5 *cm. thick, often asymmetrical, brittle-walled, dull or scarcely lustrous on inner surface, early disintegrating; seed roundish to irregularly D-shaped,* 3–6 mm. thick at back, 5–8 mm. broad, *with an irregularly deep-pitted brittle corky coat.* (*I. caroliniana* S. Wats.) — Marshes, bottomlands, wet savannas and shallow water, Fla. to e. Tex., n. on or near Coastal Plain to e. Va. May.

Var. **Shrèvei** (Small) E. Anders. (named in 1927 for Ralph Shreve). — Leaves firmer, more ascending to suberect; flowering stem firmer and less inclined to fall to ground; capsule often long-cylindric, 5–10 cm. long, 1.4–2 cm. thick, prominently beaked; seeds often thinner. (*I. Shrevei* Small) — Marshes, ditches, wet shores, shallow water, etc., sw. Que. to Minn., s. to upland of N.C., n. Ala., Tenn., Ark. and e. Kans. Mid-May (southw.)–Mid-July (northw.)

7. I. Pseudácorus L. (false *Acorus*), Yellow I. of Europe. — Habitally similar to nos. 3, 6 and 7; *rhizome pink-fleshed; spathes* much as in no. 6, *the outer sharply keeled; perianth of a general yellow color, its green tube not constricted above the ovary;* sepals recurving, with broad blade abruptly contracted at base and usually with 2 ridges; petals erect, somewhat spoon-shaped. — Spread from cult. to marshland, meadows, brooksides, etc., Nfld. to Minn. and southw. June–Aug. (Introd. from Eu.)

Other Old World species or their hybrids, **I.** germanica L., **I.** pallida Lam., etc., occasionally wander from cult. These should be identified through works on cult. plants.

8. I. vérna L. (vernal), Dwarf or Violet I. — *Loosely and superficially creeping, with slender or cord-like whitish rhizomes and rootless straight stolons* 2–4 *mm. thick; these lengthening to* 2.5–10 *cm. between the thick rooting crowns and bearing remote dark scales; flowering scape and* 1 or 2 *flowers* 0.5–1.5 *dm. high,* with mostly 5-bracted *spathe of scarious or subscarious pale or brownish bracts* extending nearly to summit; the upper lance-attenuate bracts longest; *perianth with long nearly filiform tube,* the blue-purple to violet narrowly obovate *sepals and petals subequal;* the *sepals with a deep yellow or orange* oblong or lanceolate *minutely pubescent median basal band; mature leaves linear-attenuate, firm,* marcescent, up to 3.5 dm. long, 3–8 *mm. broad,* with 9–19 ribs; *capsule obtusely angled, sessile or very short-stalked,* 1–2 *cm. long.* — Dry to damp acid sands or sandy peats and sandy pine-barrens of Coastal Plain and outer Piedmont, Md. and D.C. to e. S.C. and (?) Ga. Late March–early May.

Var. **Smalliàna** Fern. (for John Kunkel Small, 1869–1938, who clearly defined it), Upland Violet I. — *Rhizome stout, dark, torulose,* 4–8 *mm. thick, with very short contractions between the heavy rooting imbricate-scaly elongate internodes, with short thick branches forming compact clumps; spathe-bracts* 5–8, *firmer, often greener; mature leaves* 5–13 *mm. broad,* 15–29 (probably more)*-ribbed; capsule* 2–3.2 *cm. long, its stalk up to* 2.5 *cm. long.* — Dry woods, openings and barrens, Piedmont and mts., Pa. and W.Va., s. to nw. Fla. and Ala. Mid-April–mid-May.

9. I. cristàta Ait. (crested), Crested Dwarf I. — *Rhizome and stolons slender, their pale papery scales fewer than in no.* 8; *leaves lanceolate or broadly lance-ensiform,* green, *herbaceous,* at first 0.5–2.5, in maturity up to 4 dm. long, 0.5–2.5 *cm. broad; flowering stem* up to 4.5 cm. high, *with sheathing green leaves, the larger leaves like the blades of vegetative shoots;* thread-like tube of perianth 4.5–7 cm. long; *sepals* obovate to spatulate, *lilac-purple, with a whitish patch at end of the* 3*-ridged toothed orange-white crest,* the white patch streaked with dark purple; *petals smaller; capsule sharply* 3*-angled, about* 1 *cm. long,* hidden in the spathe. (*Neubeckia* Alef.) — Rich woods, wooded bottoms and ravines or bluffs, D.C. to Ind. and Mo., s. to N.C., Ala., Miss., Ark. and e. Okla. Late April, May.

10. I. lacústris Nutt. (of lakes). — Smaller and *with more slender rhizomes than no.* 9; *leaves* of fans more spreading and *narrower* (0.5–1.5 cm. broad); *spathe-bracts more scarious; perianth-tube* 1.3–1.8 *cm. long, thickened upward; sepals cuneate,* emarginate; capsule 1–1.7 cm. long. — Beaches, sandy woods and bogs near the Great Lakes, L. Huron, Ont., to L. Superior, Wisc. Late May–early July.

Cánna (Latin name for a reed) fláccida Salisb. (flaccid), a yellow-flowered member of the Cannáceae, nat. from Fla. to S.C., persists on garden-refuse in Va.

FAM. 37. MARANTÀCEAE (Arrowroot Family)

Herbs with distichous pinnately veined commonly asymmetrical leaves, irregular perfect flowers, and strongly reduced asymmetrical androecium, only one-half of one anther polleniferous, the other half as well as the anthers of the remaining stamens sterile and petaloid. — Ovary inferior; locules 3 or by abortion fewer, 1-ovuled. Style single, more or less unilateral or declined. Seeds arillate; embryo curved, in copious albumen.

1. THÀLIA L.

Erect scapose aquatic herbs with ovate-lanceolate long-petioled leaves, colored caducous bracts, and open panicles of showy usually purple flowers. Sepals 3, equal or nearly so, usually much shorter than the 3 nearly or quite distinct petals. Staminodia somewhat connate, petaloid, one of them enlarged, deflexed and lip-like. Small genus of trop. and warm-temp. Am. (Named for *Johann Thal,* a German physican and naturalist who died in 1583.)

1. T. dealbàta Roscoe (whitewashed). — White-powdery; scapes 1–2 m. high; leaf-blades ovate-lanceolate, acute at apex, rounded or subcordate at base; corolla and bracts pale blue, the staminodia purple or violet. — Pond-margins and swamps, Fla. to Tex., n. to S.C. and se. Mo. July–Oct.

FAM. 38. BURMANNIÀCEAE (BURMANNIA FAMILY)

Small annual herbs, often with minute and scale-like leaves, or those at the root grass-like, or perennial saprophytes; the flowers perfect, with a 6-cleft corolla-like perianth, the tube of which adheres to the 1-locular or 3-locular ovary; stamens 3 and distinct, opposite the inner divisions of the perianth; capsule many-seeded, the seeds very minute. — A small, chiefly trop. family.

1. BURMÁNNIA L.

Ovary 3-locular, with the thick placentae in the axis. Filaments 3, very short. Style slender; stigma capitate, 3-lobed. Capsule often 3-winged. Small mostly trop. herbs. (Named for *Johannes Burmann*, 1706–1779, a Dutch botanist.)

1. B. biflòra L. (two-flowered). — Slender (7–16 cm. high), 1–several-flowered; perianth 5 mm. long, bright blue, 3-winged. — Peaty bogs on Coastal Plain, Tex. to Fla., n. to se. Va., rare. Mid-July–Oct.

2. THÍSMIA Griffith

Ovary 1-locular, with 3 stalked (soon breaking) parietal placentae. Stamens 6 (rarely only 3), with nearly obsolete filaments, the 6 anthers sagittate or quadrangular. Style short and thick; stigmas 3, simple or 2-lobed. Fruit fleshy, cupuliform. — Fleshy saprophytic mostly trop. herbs with erect campanulate or urceolate 6-lobed perianth, the leaves reduced to scales. (Name an anagram of that of *Thomas Smith*, English plant-anatomist who died about 1825.)

1. T. americàna N. E. Pfeiff. (American). — White, hyaline, with erect or curved stem 0.3–1 cm. high; flower 0.8–1.5 cm. long, bluish-green, the tube with 6 conspicuous and 6 minor nerves; petals connate at apex; stamens borne on ring at throat of perianth, united into a tube curving down into perianth-tube. — Among tall herbs, low prairies, Chicago, Ill., only once collected. Aug., Sept. — A most remarkable species, to be sought again, constituting, with the single species, *T. Rodwàyi* F. v. Muell. of Tasmania and New Zealand, the § RODWÀYA Schlechter.

FAM. 39. ORCHIDÀCEAE (ORCHIS FAMILY) *

Herbs, distinguished by perfect zygomorphic gynandrous flowers, with 6-merous (sometimes apparently 5-merous) *perianth adnate to the 1-locular ovary, with innumerable ovules on 3 parietal placentae, and with either 1 or 2 fertile stamens, the pollen usually cohering in masses.* Perianth usually of 6 divisions; the 3 outer (sepals) mostly of the same texture as the 3 inner (petals). Of the inner series, one, termed the *lip*, differs from the rest in shape, and is sometimes prolonged at the base into a *spur*. The *lip* is really the posterior petal, but by a twist of the pedicel or ovary of half a turn it is more commonly directed downward and becomes apparently anterior. At the base of the lip, in the axis of the flower, is the *column*, composed of a single fertile stamen, or, in *Cypripedium*, of two stamens and the rudiment of a third, variously coalescent with the style. Anther 2-locular, each locule containing one or more masses of pollen (*pollinia*), or the pollen granular. Stigma viscid or (in *Cypripedium*) rough. Fruit a 1-locular 3-valved capsule. Flowers solitary or in racemes or spikes, often showy, each flower usually subtended by a bract. Leaves parallel-nerved, solitary, or several and alternate, sometimes apparently opposite or whorled. Perennials, often with corms or with tuberoid-roots; sometimes rootless saprophytes. — A cosmop. family comprising more than 10,000 species; largely dependent on insects for pollination.

a. Fertile anthers 2, lateral, the sterile one changed to a dilated fleshy staminodium borne above the terminal stigma; pollen granular, not in masses. Tribe I. CYPRIPEDIEAE.
Stem leafy at least at base; perianth spreading; lip an inflated sac. . . . . 1. *Cypripedium.*
a. Fertile anther solitary. . . *b.*
b. Anther persistent; pollinia prolonged at base of anther into filaments

---

* Treatment based partly on that of OAKES AMES in ed. 7, with much aid from ALBERT M. FULLER'S Orchidaceae of Wisconsin: Bull. Public Mus. Milwaukee, xiv. no. 1 (1933).

(caudicles) which are attached to viscid disks or glands. Tribe II. Ophrydeae.
Viscid disks (glands) contained in a pouch (or bursicule) of the rostellum. 2. *Orchis*.
Viscid disks naked, not in a pouch. . . . . 3. *Habenaria*.
b.Anthers caducous or readily detachable. . . *c*.
c.Pollen granular or powdery. Tribe III. Neottieae. . . *d*.
d.Anther terminal. . . *e*.
e.Lip not saccate at base. . . *f*.
f.Lip free; column wingless or winged only at summit. . . *g*.
g.**Column wingless at summit; flower resupinate.** . . *h*.
h.Column toothed at apex; anther decumbent, its locules facing downward; pollen-grains smooth.
Pollen-grains simple. . . . . 4. *Pogonia*.
Pollen-grains in tetrads.
Leaves solitary or alternate. . . . . 5. *Cleistes*.
Leaves 5 or 6(–10), whorled. . . . . 6. *Isotria*.
h.Column entire or merely lobed at apex; anther erect, with outward-facing locules; pollen-grains in tetrads, reticulate or pitted. . . . . 7. *Triphora*.
g.**Column prominantly winged at apex; flower not resupinate.** 8. *Calopogon*.
f.Lip adherent to the winged column. . . . . 9. *Arethusa*.
e.Lip saccate at base, beardless. . . . . 10. *Epipactis*.
d.Anther dorsal. . . *i*.
i.Lip concave or trough-like, not deeply cleft; leaves basal or, if cauline, alternate and narrow. . . *j*.
j.Lip descending or porrect; flowers spicate.
Lip not saccate, appendaged at base. . . . . . . 11. *Spiranthes*.
Lip saccate, unappendaged. . . . . 12. *Goodyera*.
j.Lip ascending; flowers loosely racemose. . . . . . 13. *Ponthieva*.
i.Lip flat, notched or cleft at summit; leaves opposite, cauline, broad. 14. *Listera*.
c.Pollen waxy or soft, not granular, smooth. Tribe IV. Epidendreae. . . *k*.
k.Pollen-masses 4. . . *l*.
l.Pollen masses unappendaged. . . *m*.
m.Plant leafless, with yellowish-green or yellow to brown, fuscous or purple sheathed stems and coralline-branching rhizomes. 15. *Corallorhiza*.
m.Plant with 1 or more leaves. . . *n*.
n.Flowers in racemes; perianth 1–10 mm. long; lip not saccate. . . *o*.
o.Petals filiform or linear; lip entire, undulate or 2-cleft at apex; leaves present during anthesis, green, herbaceous.
Lip ovate or ovate-oblong, tapering to tip; leaves alternate at base of stem or solitary on scape. . . . . . . 16. *Malaxis*.
Lip obovate or quadrate-oblong, broad at summit; 2 leaves basal. . . . . 17. *Liparis*.
o.Petals oblong; lip deeply 3-lobed; solitary firm corrugated blue-green and striate overwintering basal leaf often shriveled at flowering time. . . . . 18. *Aplectrum*.
n.Flower solitary, terminal, large and showy, the lip trough-like or suggesting a sugar-scoop, 2–2.5 cm. long; leaf basal, overwintering. . . . . 19. *Calypso*.
l.Each of the 4 pollen-masses attached by a very short filament to the viscid disk or gland; lip 3-lobed. . . . . . . 20. *Tipularia*.
k.Pollen-masses 8, united into a single fascicle; leafless, the purple-sheathed stem from coralline rhizome; lip 3-lobed. . . . . . 21. *Hexalectris*.

Artificial Key to Genera

a.Fertile anthers 2, one on each side of column, a sterile one dorsal on the column and modified to a petaloid staminodium; lip a showy closed or merely cleft inflated pouch 1.5-7 cm. long. . . . . 1. *Cypripedium*.
a.Fertile anther 1, terminal or dorsal near summit of column; lip various, if slightly pouch-like much smaller than above. . . *b*.
b.Flower 1 (rarely 2 or 3), not definitely spicate or racemose. . . *c*.
c.Leaves in a whorl of 5 or 6(–10) at summit of scape; sepals 1.3–3.5 cm. long; lip 3-lobed. . . . . 6. *Isotria*.
c.Leaves solitary or scattered. . . *d*.
d.Leaf basal, round-ovate to oval, petioled, overwintering; scape naked; lip shaped like a broad trough or sugar-scoop; plant of mossy coniferous woods northw. . . . . 19. *Calypso*.

*d.*Leaf or leaves basal or cauline, linear to oblong, elliptic or narrowly obovate; stem leafy or with sheathing bracts. . . *e.*

*e.*Lip erect, borne above sepals and petals; plants of peats and bogs. 8. *Calopogon.*

*e.*Lip extended forward parallel with or drooping below sepals and petals. . . *f.*

*f.*Leaves hidden in sheathing bracts until after anthesis, linear; scape from solid bulb; sepals and petals united at base; lip erect at base and there united with column, the dilated upper half abruptly bent downward; plant of bogs. . . . . 9. *Arethusa.*

*f.*Leaves developed on flowering stem before flowering, broader; the stem from slender roots or elongate tuber; sepals and petals free. . . *g.*

*g.*Stem arising from slender elongate roots, bearing 1 (or 2) submedian narrowly lanceolate to elliptic or narrowly obovate leaf; the usually solitary terminal flower ascending; anther articulated at base; plants of peats and bogs.

Petals spreading; lip flattish, its dilated apex lacerated, with the upper surface crested or bearded. . . . . . 4. *Pogonia.*

Petals, lip and column connivent; lip trough-like, with narrowed apical lobe, merely crested above. . . . 5. *Cleistes.*

*g.*Stem arising from a tuber, soon stoloniferous, with scattered small ovate leaves; flowers nodding from upper axils; anther rigidly attached to column; fragile woodland saprophyte. . 7. *Triphora.*

*b.*Flowers 2–very many, in racemes or spikes. . *h.*

*h.*Flowers with a distinct slender and elongate spur. . . *i.*

*i.*Leaves green, herbaceous, developed in the spring, present at flowering time; roots slender to tuberous, the tuberoids not connected as a chain-like series along the rhizome; anther persistent; capsules ascending to divergent.

Caudicles of pollinia convergent, contained in a special pouch or bursicule. . . . . 2. *Orchis.*

Caudicles divergent, not contained in a bursicule. . . . . . 3. *Habenaria.*

*i.*Leaf solitary, basal, purple beneath, coriaceous, prominently nerved and plaited, developed in autumn and overwintering, disappearing before or soon after flowering time; rhizome with a row of superficial plump tubers; anther quickly shriveling; capsules reflexed; s. and se. woodland plant. . . . . 20. *Tipularia.*

*h.*Flowers spurless or merely with a small basal pouch. . . *j.*

*j.*Leaves wanting; the yellow to brown or purple fleshy stem with colored sheaths, arising from a coralline rhizome.

Lip with a callus on each side of the base of the midnerve; pollen-masses 4, free. . . . . 15. *Corallorhiza.*

Lip with 5 or 6 longitudinal ridges; pollen-masses 8, fascicled; southern woodland plant. . . . . 21. *Hexalectris.*

*j.*Leaf or leaves present; plant usually without coralline rhizome. . . *k.*

*k.*Base of plant a horizontal moniliform rhizome with 2 or more large corm-like tubers separated by slender strands; single basal leaf developed in summer or autumn and overwintering, firm, blue-green, corrugate-striate, shrivelling by or soon after flowering time; capsules reflexed; woodland plant. . . . . 18. *Aplectrum.*

*k.*Base of plant not a moniliform rhizome; leaves fleshy to membranaceous, developed before flowering time, if overwintering fleshy and soft. . . *l.*

*l.*Leaves chiefly in a basal rosette or paired at base. . . *m.*

*m.*Leaves spreading, in a basal rosette; roots slender, elongate and fleshy, or elongate tuberoids. . . *n.*

*n.*Flowers sessile or essentially so in a spike or spiciform raceme, white, creamy or yellowish-green; lip projected forward or descending; pollen-masses 2; wide-ranging.

Leaves membranaceous, soon shriveling, green; lip with a horn-like callosity on each side near base, rarely saccate. 11. *Spiranthes.*

Leaves fleshy, evergreen, mostly with white or colored reticulations or veins; lip without basal callosities, usually saccate. . . . . 12. *Goodyera.*

*n.*Flowers pedicelled in a lax raceme, whitish with green veins; lip erect or ascending; pollen-masses 4; southern woodland plant. . . . . 13. *Ponthieva.*

*m.*Leaves 2, ascending, from large bulb or corm; lip obovate or quadrate-oblong. . . . . 17. *Liparis.*

*l.*Leaves cauline, or if basal solitary, opposite or scattered. . . *o.*
*o.*Leaves opposite or subopposite, paired, cordate-ovate to elliptic-oblong, borne from below to above middle of slender stem; flowers loosely racemose; perianth greenish, smoky or purplish.
Stem arising from fibrous roots; leaves sessile; lip 2-lobed or 2-cleft at summit; anther borne on back of the column. 14. *Listera.*
Stem arising from a solid tuber; lower leaf of pair petioled; lip entire; anther borne between terminal teeth of column. 16. *Malaxis.*
*o.*Leaves scattered, alternate or solitary. . . *p.*
*p.*Perianth 1.5–4.5 cm. broad. . . *q.*
*q.*Flowers pink (rarely white), lilac or roseate, 2–4.5 cm. broad, resupinate, the slender-stalked but upwardly dilated and bearded lip erect; floral bracts tiny. . 8. *Calopogon.*
*q.*Flowers in axils of leafy bracts, 1.5–2.5 cm. broad, with descending lip.
Weak woodland saprophyte, with small oval leaves; flowers few, nodding from upper axils, pink or white; sepals and petals lanceolate; lip with 2 marginal lobes, not saccate. 7. *Triphora.*
Strong and tall (up to 9 dm.) with veiny lanceolate to ovate leaves; flowers numerous in elongate raceme, green and purplish; sepals and petals ovate; lip inflated or saccate at base, with broadly cordate apical half. 10. *Epipactis.*
*p.*Perianth smaller, white or creamy to greenish.
Flowers sessile in a more or less spirally twisted spike, white, creamy or stramineous; stem arising from elongate fleshy roots or slender tuberoids; cauline (mostly subbasal) leaves much longer than broad. . . . . . 11. *Spiranthes.*
Flowers slender-pedicelled, in a raceme; perianth tiny, greenish; stem arising from solid corms; solitary cauline leaf or basal leaves broad and short. . . . . 16. *Malaxis.*

Tribe I. Cypripedìeae Lindl.

1. CYPRIPÈDIUM L. Lady's-slipper. Moccasin-flower, Sabot de la Vierge (Que.)

Sepals spreading, all three distinct or in most cases two of them united into one under the inflated sac-like lip. Petals mostly spreading, linear or oblong. Column declined, on each side a fertile stamen with its short filament bearing a 2-locular anther; pollen loose and pulpy or powdery-granular, the face of the anther converted into a viscid film; on the upper side of the column a dilated petaloid but thickish staminodium or infertile stamen; stigma terminal, obscurely 3-lobed, moist and roughish. — Roots coarsely fibrous. Leaves many-nerved and plaited, sheathing at the base. Stems pubescent. Flowers solitary or few, mostly large and showy; genus of N. Hemisph. (Name incorrectly Latinized from *Cypris, Venus,* and *pedilon, shoe,* therefore by some purists spelled *Cypripedilum.*)

*a.*Flowering stem leafy nearly to the summit, the alternate leaves becoming distant above; lip with a rounded opening near base in front. . . *b.*
*b.*The 3 sepals all separate; lip reticulate, prolonged at apex into a conical point. . . . . 1. *C. arietinum.*
*b.*The 2 lower sepals often united; lip not strongly reticulate, broadly rounded at apex. . . *c.*
*c.*Sepals and petals acuminate, attenuate-tipped or acute; upper sepal lanceolate to narrowly ovate; petals mostly longer than lip, often spirally twisted, linear-attenuate to lance-acuminate.
Lip yellow, 2–5 cm. long; upper sepal 2–8 cm. long; petals 2.5–9 cm. long. . . . . 2. *C. Calceolus.*
Lip white, purple-striped within, 1.8–2.3 cm. long; upper sepal 2–3 cm. long; petals 2.5–3.5 cm. long. . . . . 3. *C. candidum.*
*c.*Sepals and petals blunt, white; upper sepal suborbicular to elliptic-oval; petals shorter than lip, flat; lip 3–5 cm. long, white or suffused with pink or pale purple. . . . . 4. *C. reginae.*
*a.*Flowering stem a naked scape, with 2 basal leaves; lip with a deep fissure extending down the front, drooping, bag-like. . . . . 5. *C. acaule.*

1. C. arietinum R. Br. (like a ram's head), Ram's-head L. — Stem *slender,* 1.5–3.6 dm. high, *with* 3 *or* 4 *elliptic-lanceolate to oblong bluish-green* firm smooth *leaves;* flower solitary; *sepals* *madder-purple*, streaked with green, *all 3 distinct, about 2 cm. long*, the upper lanceolate or lance-ovate and acuminate, the others linear-attenuate and more or less undulate or twisted; *lip about 2 cm. long, irregularly triangular, strongly reticulate with purplish veins, tapering to the* greenish or yellowish *conical apex*. (*Criosanthes arietina* House) — Damp or mossy woods or bogs, sw. Que. to Man., s., rarely, to n. N.E., centr. and w. Mass., N.Y., Mich., Wisc. and Minn. Late May, June. (E. Asia) The rare forma **albiflòrum** House (white-flowered) has the flower white.

2. C. Calcèolus L. (a small shoe). — Stem with ovate to lanceolate pointed leaves; flowers 1 or 2, terminal; *sepals* purplish, the upper broadly lanceolate or lance-ovate one opposite the lip, *the 2 lateral ones united under the lip;* petals wide-spreading, lance-linear, flat or slightly spiraling, slightly longer than lip, purple-brown; lip yellow, often purple-spotted near orifice, inflated, rounded at apex, 2.5–4 cm. long, slightly dorsiventrally compressed; staminodium stalked, ovate, with rounded to subcordate base. — Eurasia; represented with us by three slightly confluent vars.:

Upper sepal 2.5–7 cm. long, long-acuminate, abruptly narrowed to subcuneate at base; petals linear or linear-lanceolate, loosely descending or drooping, spirally twisted, yellowish (often with purple stripes) to purplish, 3.5–9 cm. long; lip 2–5 cm. long; staminodium triangular, with tapering, truncate or subcordate base; stem 1.5–7 dm. high, with 3–6 oval to ovate-lanceolate leaves, the larger blades 0.7–2 dm. long and 2–12 cm. broad.

Stem 1.5–5.5 dm. high, 3 or 4 (–5)-leaved; largest leaves 2–9 cm. broad; sepals usually madder-purple, the upper one 2.5–5 cm. long, the others united; petals 3.5–5 cm. long; lip 2–4 cm. long; staminodium truncate or tapering at base; flower strongly fragrant; chiefly of bogs, swamps or cool shores and wet rocky places. . . . . Var. *parviflorum*.

Stem 2–7 dm. high, 4–6-leaved; largest leaf 4–12 cm. broad; sepals and petals usually greenish-yellow, often streaked with purple lines; upper sepal 4–7 cm. long, the others united below or distinct; petals 5–9 cm. long; lip 3–5 cm. long; staminodium often more rounded to subcordate at base, its stalk longer than in the preceding; flower less fragrant than in the preceding; chiefly of mesophytic forest. . . . . Var. *pubescens*.

Upper sepal 2–4 cm. long, merely acute, usually rounded at base, the lateral ones smaller and united; petals oblong-lanceolate, flat or merely undulate, 2.5–4 cm. long, purple to stramineous; lip 2–4 cm. long; staminodium cordate at base; stem 0.6–2.5 dm. high, 2–4-leaved; largest leaves 3–10 cm. long and 1.5–4.5 cm. broad. . . . . Var. *planipetalum*.

Var. parviflòrum (Salisb.) Fern. (small-flowered), Small Yellow L., Small M., Small Golden Slipper (*C. parviflorum* Salisb.). — Bogs (chiefly calcareous), mossy swamps and woods, wet shores, damp rocks, etc., Nfld. to n. B.C., s. to N.S., N.E., N.J., Pa., mts. of Ga. and Tenn., O., Ind., Ill., Mo., Tex., N.M., Utah and Wash. May–July. — In its largest development passing through an intermediate (possibly hybrid) *C. flavescens* DC. to

Var. pubéscens (Willd.) Correll (pubescent), Large Yellow L. or Moccasin-flower, Golden Slipper, Whip-poor-will Shoe (*C. pubescens* Willd.). — Dry to moist mesophytic (usually rich) woodland, N.S. and centr. Me. to Minn., s. to s. N.E., Ga., Ala., Tenn. and Mo. Mid-April–mid-June.

Var. planipétalum (Fern.) Vict. & Rousseau (with flat petals). — Limestone-barrens and calcareous talus, thickets and shores, n. and w. Nfld.; and Mingan Ids., Anticosti and e. Gaspé Pen., Que. July, early Aug. — Our nearest approach to Eurasian *C. Calceolus*.

× **C. Andrewsii** A. M. Fuller (named in 1932 for its discoverer, Edward Palmer Andrews) is a hybrid of *C. Calceolus*, var. *parviflorum* with no. 3, with ovate-lanceolate sepals 2.5–3.7 cm. long, lanceolate petals greenish suffused with purple and 3–4 cm. long, the creamy-white lip 2–2.5 cm. long, the staminodium very narrowly triangular.

× **C. Favilliànum** J. T. Curtis (named in 1932 for its discoverer, Stoughton Willis Faville) is a hybrid of *C. Calceolus*, var. *pubescens* and no. 3, with sepals and petals yellowish-green striped with brown, the upper sepal ovate and about 3.5 cm. long, the petals (8 mm. wide) 4 cm. long, the lip (3 cm. long) yellow on expanding but soon becoming white, the staminodium oblong to ovate.

3. C. cándidum Muhl. (white), Small White L. — Stem 1.5–4 dm. high, with mostly approximate erect lanceolate or narrowly elliptic leaves 1–4 cm. broad; flower 1 (rarely 2), slightly fragrant; *sepals and petals greenish-yellow*, often with purple lines; *upper sepal lanceolate, 2–3 cm. long*, the *lateral lanceolate pair united* nearly or quite to tip; *petals linear-lanceolate, long-attenuate*, 2.5–3.5 *cm. long; lip white*, purple-striped within, obliquely obovoid, 1.8–2.3 cm.

long; *staminodium oblong to narrowly ovate.* — Calcareous meadows, prairie, mossy glades, etc., w.-centr. N.Y. to N.D., s. to n. N.J., e. Pa., Ky. and Mo., becoming very rare. May, June.

4. C. reginae Walt. (of the queen), SHOWY L. — Coarser than our other species, densely hirsute, 3–8.3 dm. high, with strongly corrugated and suberect overlapping large ovate leaves; flowers 1 or 2 (rarely 3 or 4), very handsome; *sepals and petals shorter than lip, blunt, white; upper sepal suborbicular to elliptic-oval,* at first erect, soon concave and arching over the lip, 3–4.5 cm. long; lower pair united and narrower; petals ovate-lanceolate, flat, wide-spreading; *lip* subglobose or inflated-obovoid, *white, strongly suffused with pink to roseate broad bands on the front,* or in forma **albolàbium** Fern. & Schub. (white-lipped) wholly white, 3–5 cm. long. (*C. hirsutum* of ed. 7, not Mill.; *C. spectabile* Salisb.) — Mossy (chiefly calcareous) swamps, bogs or woodland-glades or (northw.) damp calcareous slopes or shores, Nfld. to Man., s. to N.S., n. and w. N.E., nw. N.J., centr. and w. Pa., mts. to Ga. and Tenn., Mo. and N.D. Mid-May (southw.)–mid-Aug. (northw.) — The hairs of the fresh foliage sometimes cause serious eczema. Plant liable to extinction through raids by nurserymen and would-be cultivators.

5. C. acaùle Ait. (stemless), STEMLESS, COMMON or TWO-LEAVED L. or NERVE-ROOT. — *Leaves* 2, oblong- to obovate-elliptic, *basal; flowering stem a naked scape* 1–5.5 dm. high; flower (rarely 2) terminal; sepals lanceolate to lance-ovate, yellow-green shaded with purple; petals lanceolate, greenish-brown, flat or slightly twisted; *lip drooping, inflated-saccate, obovoid, very veiny,* pink or roseate, or lip white and sepals and petals pale in forma **albiflòrum** Rand & Redfield (white-flowered), 3.5–7 cm. long, *with a slender fissure extending down the* front; staminodium rhombic. (*Fissipes* Small) — Dry acid soil of woodlands, or northward often in bogs, moss or wet woods, Nfld. to n. Alta., s. to N.S., N.E., L.I., Ga., Ala., Tenn., Minn., Man. and Sask. April (southw.)–July (northw.)

Tribe II. OPHRÝDEAE Lindl.

2. ÓRCHIS L. ORCHIS

Flowers ringent. Sepals and petals nearly equal. Lip turned downward, coalescing with the base of the column, spurred below. Anther-locules contiguous and parallel. Pollen cohering in numerous coarse waxy grains, which are collected on a cobwebby elastic tissue into two large masses (one filling each anther-locule) borne on slender stalks, the bases of which are attached to the glands or viscid disks of the stigma; the two glands contained in a common little pouch, or bursicule, placed just above the orifice of the spur. — Acaulescent, with 1 or 2 (rarely 3) basal leaves; flowers (in ours) pink or purple (rarely white), in a loose raceme; rhizome short, with fleshy roots. Large genus of N.Hemisph., chiefly of Eurasia. (Greek *orchis, testicle,* the ancient name, from the roundish tuberoids of some species.)

1. O. rotundifòlia Banks (round-leaved), SMALL ROUND-LEAVED O. — *Leaf solitary,* orbicular to elliptic, 3–8 *cm. long;* scape naked, 0.4–2.5 dm. high, with 2–9 *flowers* 1.2–1.5 *cm. long;* sepals and upper petals roseate, ovate-oblong, the *lateral sepals spreading; lip* white, spotted with purple, 6–8 mm. long, 3-*lobed,* the lateral lobes oblong, the larger median one dilated and notched at apex; spur slender, depending. — Mossy calcareous swamps and woods, Nfld. to Yuk., locally s. to N.B., n. (and formerly w.) N.E., N.Y. (formerly), Wisc., Minn. and Mont. June, July. (Greenl.)

2. O. spectábilis L. (showy), SHOWY O. — *Leaves* 2 (rarely 3), oblong-obovate, lustrous, 0.7–2 *dm. long;* scape 4–5-angled, 0.3–1.2 dm. high, with 3–12 *flowers* mostly exceeded by the foliaceous bracts and 2–3 *cm. long; sepals and petals contiguous, forming a* lilac to roseate, or white in forma **Gordinièrii** (House) Weath. (named in 1923 for its discoverer, H. C. GORDINIER), *vaulted galea behind the column; lip* white, rarely pink, ovate, *undivided;* spur blunt. (*Galeorchis* Rydb.) — Rich, mostly calcareous, woods, Que. and Ont., s. to N.B., N.E., Ga., Ala., Tenn., Mo. and ne. Kans. Apr.–early June.

3. HABENÀRIA Willd. REIN-ORCHIS. FRINGED ORCHIS

Flowers usually small, in loose or dense racemes or spikes. Sepals spreading, mostly similar; petals erect, connivent with the upper sepal. Lip entire, toothed or fringed laterally, or tripartite, the divisions cuneate to flabelliform and variously toothed or fimbriate. Spur shorter or longer than the lip. Glands or viscid disks (to which the pollen-masses are attached) naked and exposed, separate, sometimes widely so. In some of our species the stigma has two or three appendages. — Glabrous plants with one or more leaves. Tuberoids elongated, fusiform or somewhat palmate, or roots sometimes slender. An amphigean genus often separated by local authors into numerous genera. (Name from *habena, a thong* or *rein,* in allusion to the shape of the lip or spur of some species.)

*a.*Lip simple, not fringed, either entire, crenulate, lobed or toothed. . . *b.*

*b.*Lip oblong, cuneate, obovate or ovate, toothed at apex or along margin. . . *c.*

*c.*Teeth or lobes of lip terminal. . . *d.*

*d.*Cauline leaves 3–several; roots fleshy and tuberous-thickened; spike (excluding bracts) becoming much longer than thick; at least the lower bracts exceeding the ascending ovaries; sepals lanceolate or lance-ovate, tapering; spur much shorter than ovary.

Spike dense, 1–2 cm. thick (including bracts); flowers strongly vanilla-scented; sepals and petals similar, lance-ovate, stramineous or creamy-white, 2.7–4 mm. long; summit of lip cleft into 3 subequal elongate porrect lobes; spur slenderly cylindric; plant of calcareous barrens near Straits of Belle Isle. . . . . 1. *H. straminea.*

Spike lax and open, including lower bracts 1.5–8 cm. thick; flowers green; lanceolate petals much narrower than ovate sepals, about 7 mm. long; apex of lip with 2 or 3 short teeth, the central tooth short or obsolete; spur rounded and saccate; wide-ranging plant of woods and thickets. . . . . 2. *H. viridis.*

*d.*Cauline leaf 1 (well developed), rarely 2; roots slender, not much thickened; spike up to twice as long as thick; bracts shorter than soon divergent ovaries; sepals and petals yellowish or yellowish-green, ovate, blunt, similar; lip with 3 low rounded terminal teeth; spur slenderly clavate, equaling or exceeding ovary. . . . . . . 3. *H. clavellata.*

*c.*Teeth of lip marginal (sometimes wanting in no. 5).

Spike slender, excluding longer bracts 1–2 cm. thick; flowers greenish or greenish-yellow; lip oblong, with hastate base, and with a basal median thick elongate tubercle on upper surface. . . . . . . 4. *H. flava.*

Spike stoutish, 2–3.5 cm. thick; flowers golden-yellow; lip ovate, entire, crenulate or with short marginal teeth, without tubercle at base. . . . . 5. *H. integra.*

*b.*Lip linear-lanceolate to oblong or ovate, entire. . . *e.*

*e.*Perianth orange- or golden-yellow; lip oval to oblong; rare plant of s. N.J. 5. *H. integra.*

*e.*Perianth greenish to white. . . *f.*

*f.*Principal leaves 2 or more; plant (0.1–) 0.2–1 m. high. . . *g.*

*g.*Leaves several, cauline, extending up the stem, at least the upper tapering to acute tips; spike densely flowered or, if more open, 1.5–8 cm. thick. . . *h.*

*h.*Stems slender; leaves linear or linear-lanceolate, firm, keeled, often longitudinally folded, the 2 or 3 lowest prolonged and 5–10 mm. broad; spike with ultimately subdistant flowers with divergent lance-subulate straight ovaries; perianth white, lip ascending. 6. *H. nivea.*

*h.*Stem usually stoutish; leaves soft and herbaceous, flat, the lower mostly broader; spike rather dense, with ascending ovaries; lip drooping or merely upward-curving.

Flowers greenish, the sepals and petals herbaceous.

Lip lanceolate to oblong-oval, not greatly broadened at base. 7. *H. hyperborea.*

Lip with dilated rhombic or ovate base. . . . . . 8. ×*H. media.*

Flowers bright white, the sepals and petals soft and petaloid.

Spike slenderly cylindric, 1–2 cm. thick, 0.4–3 dm. long, spicily fragrant; ovaries plump, the flowers sessile; lip with dilated ovate base; spur 4–5 mm. long; northern. . . . 9. *H. dilatata.*

Spike open, racemiform, 6–9 cm. thick, 5–10 cm. long, not strongly fragrant; ovaries slenderly subulate, resembling pedicels; lip linear-oblong, narrowed at base; spur 2.5–3.5 cm. long; southern. . . . . 17. Var. of *H. blephariglottis.*

*g.*Principal leaves 2 or 3, basal. . . *i.*

*i.*Basal leaves 2 or 3, linear-spatulate to oblanceolate, 1–3 cm. broad, membranaceous and translucent, soon wilting; scape slender, naked; spike interruptedly linear-filiform, remotely flowered, in anthesis 1–1.5 cm. thick; perianth 2–4 mm. long; sepals 1-nerved. . . . . 10. *H. unalascensis.*

*i.*Basal leaves 2 (exceptionally 3), orbicular, broadly elliptic or broad-oblong, thickish, persistent, flat on the ground, 0.2–2 dm. broad; spike or raceme 2–8 cm. thick; perianths much larger than in no. 10; sepals 3–5-nerved. . . *j.*

*j.*Scape naked (exceptionally 1-bracted), 0.7–2.5 dm. high, fleshy; flowers (ovaries) sessile in a spike 2–4 cm. thick; sepals lanceolate or lance-ovate; spur 0.9–2.5 cm. long. . . . . . . 11. *H. Hookeri.*

*j.*Scape bracted nearly to summit, firm, 0.7–5.2 dm. high; flowers pedicelled in a loose raceme 2–8 cm. in diameter; upper sepal orbicular, the 2 lateral ovate; spur 0.8–4.5 cm. long.
Spur 0.8–2.7 cm. long. . . . . 12. *H. orbiculata.*
Spur 3–4.5 cm. long. . . . . 13. *H. macrophylla.*
*f.*Leaf solitary, basal, obovate to spatulate-oblanceolate, rounded above or blunt; plant 0.5–4 dm. high, with filiform scape; spike racemiform, 1–15 cm. long. . . . . 14. *H. obtusata.*
*a.*Lip simple and fringed or 3-parted and either fringed or toothed. . . *k.*
*k.*Lip simple. . . *l.*
*l.*Flowers orange or orange-yellow; southern species.
Spike dense, 2–3.5 cm. thick; style (slender brightly colored summit of ovary) 3–5 mm. long; fully grown unexpanded perianth-buds and lateral sepals and petals 3.5–4 mm. long; spur 5–9 mm. long, about half length of ovary. . . . . 15. *H. cristata.*
Spike lax and open to slightly dense, 4–8 cm. thick; style (slender colored summit of ovary) 7.5–12 cm. long; fully grown perianth-buds 6–8 mm. long; spur 1.5–2.5 cm. long, nearly equaling to longer than ovary. . . . . 16. *H. ciliaris.*
*l.*Flowers white; wide-ranging boreal to austral species. . . . . . 17. *H. blephariglottis.*
*k.*Lip 3-parted. . . *m.*
*m.*Flowers yellowish-green, greenish-white or cream-colored, with or without bronze or rose tints; middle lobe of lip cuneate to slender claw; glands of anther linear to oval; larger lower leaves 1–4, the largest 0.5–3.5 cm. broad.
Perianth 4–6 mm. long; lip 7–15 mm. long; spur 1–1.7 cm. long; flower yellow-green, dirty-yellow or tinged with bronze or rose. . . . . 18. *H. lacera.*
Perianth 8–10 mm. long; lip 1.5–2 cm. long; spur 2–4 cm. long; flower creamy-white to whitish-green. . . . . 19. *H. leucophaea.*
*m.*Flower lilac-pink to deep rose-purple or violet (white only in albinos); middle lobe of lip fan-shaped, as broad as to much broader than long, subtruncate to broadly tapering to short base; glands of anther suborbicular; larger lower leaves 2–5, the largest 2–9 cm. broad. *n.*
*n.*Lip deeply lacerate-fringed (except in rare forms).
Inflorescence 2.5–4.5 cm. in diameter; perianth 4–7 mm. long; lip 6–16 mm. broad. . . . . 20. *H. psycodes.*
Inflorescence 5–9 cm. in diameter; perianth 9–12 mm. long; lip 1.8–3 cm. broad. . . . . 21. *H. fimbriata.*
*n.*Lip shallowly erose, its short teeth triangular or blunt, the terminal lobe deeply emarginate. . . . . 22. *H. peramoena.*

1. **H. stramínea** Fern. (straw-colored). — *Stem* 1–3.5 dm. high, from tuberous-thickened roots, *leafy; the* 2–4 *lower leaves oblong-obovate,* obtuse or subacute, 2–7 cm. long and 0.8–3.5 cm. broad, upper leaves gradually reduced; *spike cylindric,* rather dense, 3–10 cm. long, 1–2 *cm. thick;* lance-acuminate bracts twice length of fusiform ovary; flowers strongly vanilla-scented; *sepals straw-colored, translucent,* narrowly ovate, attenuate, *obviously* 3*-veined,* 2.7–4 *mm. long; lateral petals similar,* about equaling sepals, *strongly* 3*- or* 5*-veined; lip thin,* 3–5 *mm. long, broadly cuneate and* 3*-lobed at apex; lobes lance-deltoid, subequal, clearly* 2*- or* 3*-nerved; spur cylindric-clavate,* obtuse, 2–3 *mm. long.* (*H. albida* (L.) R. Br., var. F. Morris) — Limestone-barrens, n. and nw. Nfld. July, Aug. (Greenl.; Iceland; Faroe Ids.)

2. **H. víridis** (L.) R. Br. (green), var. **interjécta** Fern. (thrust between), Frog-Orchis. — *Tuberoids* thick and fleshy, *often palmate; stem leafy,* 0.5–3.5 dm. high; the *lower* of the 2–5(–7) *leaves oblong to oblong-obovate and blunt,* the upper narrower and smaller ones acute; *spike* at first subcapitate, *becoming cylindric,* 0.2–2 dm. long *and rather lax, with prominent subascending foliaceous bracts; lower bracts lanceolate or narrowly ovate, once-and-a-half to thrice as long as subtended flowers, the middle and upper bracts shorter than to barely exceeding the flowers; flowers green* or tinged with bronze; *narrowly lanceolate petals much narrower than ovate sepals, about* 7 *mm. long; lip oblong to oblong-cuneate, the truncate apex with* 2 *short oblong or deltoid teeth,* a median tooth rarely developed. — Rich (often calcareous) turfy shores, meadows, thickets and woods, nw. Nfld.; Ung. to Alaska, s. to n. Mich., n. Wisc., n. Ia., S.D., Colo. and s. B.C. Mid-May (southw.)–Aug. (Ne. Asia) — Exactly intermediate between typical Eurasian *H. viridis* and the following.

Var. **bracteàta** (Muhl.) Gray (bracted), Bracted or Long-bracted Green Orchis. — 1.3–5.5 dm. high; *bracts linear-lanceolate, soon widely divergent, the lower and median two to six* *times, the upper up to twice as long as flowers; lip narrower.* (*H. bracteata* R. Br.; *Coeloglossum bracteatum* Parl.) — Rich woods, thickets and meadows, Nfld. to s. Alta., s. to N.S., N.E., N.J., Pa., upland to S.C. and Tenn., O., Ind., Ill. and Ia. May (southw.)–Aug. (northw.). (E. Asia)

3. H. clavellàta (Michx.) Spreng. (like a little club; from the spur), Green Woodland Orchis. — Very slender, 0.7–4.5 dm. high, from slender only slightly fleshy roots, *with 1 well developed lower leaf and few greatly reduced upper ones; lower leaf narrowly oblanceolate, spatulate or narrowly oblong,* blunt, *tapering to subpetiolar base,* 0.5–2.3 *dm. long and* 0.7–2.7 *cm. broad (twelve to five times as long as broad); spike oblong-cylindric,* 2–6 cm. long, 1.5–3 cm. thick, *with few subdistant finally divergent flowers* often turned so that the lip projects from the side; *bracts* lanceolate or lance-ovate, acuminate, *shorter than ovary; perianth greenish or greenish yellow or -white, 2–4 mm. long; sepals and petals oblong-ovate, obtuse, subequal and equaling the lip,* the lateral sepals spreading; *lip cuneate-oblong, with 3 rounded short apical teeth; spur slenderly clavate, closely appressed to and equaling or exceeding ovary.* (*Gymnadeniopsis* and *Denslovia* Rydb.) — Mossy or wet sandy woods, thickets, spring-heads and shores, Fla. to Tex., n. to Mass. (very locally to N.B.), N.Y., O., Ind., s. Wisc. and s. Minn. July, Aug. — Passing northward into

Var. **ophioglossoìdes** Fern. (resembling *Ophioglossum*). — *Large lower leaf oval, oblong or broadly oblanceolate, rounded to tapering to essentially sessile base,* 3–17 cm. long by 1–4 *cm. broad (twice to six times as long as broad).* — Nfld. and Côte Nord, Que., to w. Ont., s. to N.S., N.E., N.J., Mich., Wisc. and Minn. July–Sept.

4. H. flàva (L.) R. Br. (yellow; a misnomer), Pale Green Orchis. — Plant with elongate fleshy tuberoids, 1–7 dm. high, *slender, with* (1–)2(–3) *large distant leaves borne from slightly below to slightly above the middle* and 1–4 much smaller bracteal ones above; *larger leaves lanceolate to lance-oblong or -elliptic, tapering to apex,* 0.6–2 dm. long, 1–5 cm. broad; *bracted peduncle (above uppermost large leaf)* 5–15 *cm. long; spike loosely cylindric, with distant flowers,* 0.4–2 dm. long, 1–2 cm. thick; *floral bracts narrowly lance-attenuate, only the lowest slightly exceeding or merely equaling the greenish flowers, the others mostly shorter; sepals oblong- to rhombic-ovate or suborbicular,* 2–5.5 *mm. long; petals similar,* broader, like the sepals *often crenulate above; lip ovate to suborbicular, usually hastate at base, the margin with few teeth or crenations, with a prominent elongate median tubercle below the middle; spur slenderly clavate, equaling or exceeding ovary.* (*Perularia* Farw.; *P. scutellata* (Nutt.) Small) — Swampy woods, bottomlands, swales and wet shores, Fla. to e. Tex., n. to Md., Ky., s. Ind., s. Ill. and e. Mo.; Yarmouth and Queens Cos., N.S. June (southw.)–Sept. (northw.). — Flowers sweetly fragrant.

Var. **herbìola** (R. Br.) Ames & Correll (grass-green). — *Stouter; larger leaves* 2–5, broader and less attenuate, *extending toward summit of stem, gradually decreasing in size;* peduncle shorter; *spike compact, its floral bracts longer, the lower often much exceeding flowers; lip quadrate-oblong.* (Var. *virescens* sensu Fern., not *Orchis virescens* Muhl., basonym) — Swales, meadows, shores and woods, N.B. to Ont., s. to N.S., N.E., Del., Md., upland to N.C. and Tenn., W.Va., O., Ind., Ill., Wisc. and Mo.

5. H. íntegra (Nutt.) Spreng. (entire; from the lip), Southern Yellow Orchis. — Plant stiffly erect, 3–6 dm. high, *from a vertical slenderly napiform tuberoid;* leaves several, the 1 or 2 lower ones elongate-lanceolate and acuminate, 1–3 cm. broad, the upper ones rapidly reduced to slender bracts; *spike dense, oblong-cylindric,* 2.5–9 *cm. long,* 2–3.5 *cm. thick; bracts lance-subulate, divergent, about equaling the slender subulate-tipped ovaries; flowers horizontally divergent, golden- or orange-yellow; perianth* 3–4.5 *mm. long;* lateral sepals ovate or obovate; petals oblong; *lip oval to oblong, its margin entire, crenulate or toothed;* spur slender, about half as long as ovary. (*Gymnadeniopsis* Rydb.) — Savannas, sphagnous bogs and wet pine-barrens, Fla. to e. Tex., n. to e. N.C. and Tenn.; very rare and nearly extinct, s. N.J. Aug., Sept.

6. H. nívea (Nutt.) Spreng. (snowy), Snowy Orchis, Bog-torch. — Slender, *stiffly erect;* the stem ridged, 2–9 dm. high, *from a plump ellipsoid or fusiform vertical tuberoid; principal leaves 2 or 3 from near base of stem, linear to linear-lanceolate, firm, keeled,* often folded, *prolonged,* attenuate, 5–10 *mm. broad,* upper leaves reduced to bracts; *spike* slenderly conical, becoming thick-cylindric, 4–15 cm. long and 1.5–2.5 *cm. thick; bracts and ovary much as in no. 5; flowers soon divergent; perianth white,* 4–6 mm. long; *lip* slightly longer, *linear or linear-lanceolate, entire, erect.* (*Gymnadeniopsis* Rydb.) — Savannas, argillaceous meadows and bogs, Fla. to Ark. and e. Tex., n. to e. N.C.; Del. and s. N.J. Aug., Sept.

7. H. hyperbòrea (L.) R. Br. (far-northern), Northern or Leafy Northern Green Orchis. — Plant 1–4 dm. high, with thickish soft stems, from tuberous-thickened roots; *leaves extending up the stem, the larger lower ones narrowly oblong to oblong-lanceolate,* the larger ones 2–10 cm. long and 0.5–3 cm. broad, *soft and herbaceous, flat;* spike cylindric, open to rather dense, 1.5–9 cm. long; floral bracts herbaceous, the lower longer than or about equaling ascending *greenish or greenish-yellow flowers;* perianth about 4 mm. long; upper sepal ovate, subconnivent with lanceolate petals as a hood, lateral sepals oblong-lanceolate; *lip oblong-oval or elliptic, entire, blunt, descending;* spur slender, shorter than plump ovary; flowers with strong fragrance suggestive of *Convallaria.* (*Platanthera* Lindl.; *Limnorchis* Rydb.) — Greenl. to Alaska, s. to peaty thickets and limestone-barrens of n. Nfld. and Hudson Bay, Ung. July, Aug. (Iceland; ne. Asia) — Passing into

Var. **huronénsis** (Nutt.) Farw. (of Lake Huron). — Mostly coarser, 0.15–1 m. high; lower leaves up to 2 dm. long and 5 cm. broad; spike dense to rarely open, 0.4–3 dm. long; flowers somewhat larger; *lip lanceolate, tapering gradually to tip;* flowers only faintly odorous. (*H. hyperborea* of ed. 7; *H. huronensis* (Nutt.) Spreng.; *Platanthera huronensis* (Nutt.) Lindl.; *Limnorchis huronensis* Rydb.) — Peaty bogs, thickets and woods, or even in dryish woods, s. Lab. to Alaska, s. to Nfld., N.S., N.E., Pa., O., Ind., Ill., Minn., Neb., N.M. and Oreg. Late May (southw.)–July (northw.). — Hybridizes with no. 9, producing no. 8. Stout and tall plants with dense spikes often occur with slender or low individuals with narrow attenuate leaves (down to 5 mm. wide) and with slender interrupted spikes.

8. × **H. mèdia** (Rydb.) Niles (intermediate). — As in no. 7; perianth herbaceous, greenish-yellow or stramineous, but the *lip more or less dilated at base.* (*H. dilatata,* var. Ames; *Limnorchis* Rydb.) — Growing with *H. hyperborea,* var. *huronensis,* and *H. dilatata* and evidently a hybrid of them.

9. H. dilatàta (Pursh) Hook. (dilated; from the base of the lip), LEAFY WHITE ORCHIS, BOG-CANDLE, SCENT-BOTTLE. — Habitally like no. 7, 0.15–1 m. high; leaves often narrow, 0.5–2, rarely –4 cm. wide; *flowers milk-white, strongly spicy-fragrant, with suggestion of clove; sepals and petals of soft petaloid texture; lip rhombic and dilated at base,* narrowed above. (*Platanthera* Lindl.; *Limnorchis* Rydb.; *L. fragrans* Rydb.; *H. fragrans* Niles) — Swales, meadows, bogs and wet woods, Lab. to Alaska, s. to Nfld., N.S., N.E., n. N.J., Pa., Mich., Wisc., Minn., S.D., N.M. and Calif. Mid-June (southw.)–Aug. (Iceland)

10. H. unalascénsis (Spreng.) S. Wats. (of Unalaska), ALASKAN ORCHIS. — *Slenderly scapose,* 1.5–6 dm. high, *from* 1 *or* 2 *ovoid or ellipsoid tuberoids; leaves* 2 or 3 (rarely 4), *basal, linear-spatulate to oblanceolate,* 1–3 *cm. broad, membranaceous, translucent, soon wilting; scape nearly naked* or with very short and narrow bracts; *spike interruptedly linear-filiform, remotely flowered, in anthesis* 1–1.5 *cm. thick,* elongating to 1–4.5 dm.; *flowers green,* with disagreeable (often sickening) odor; *perianth* 2–4 *mm. long; sepals* 1-*nerved,* lance-ovate, *the* 2 *lateral ones adnate below to margin of lip;* narrower petals and upper sepal connivent as a hood; *lip* lance-ovate, *with elevated base;* spur slender. (*Piperia* Rydb.) — Dry to moist calcareous evergreen woods or open limestone-flats, Anticosti; shores of Lakes Huron and Superior, Ont.; Alaska to Wyo., Utah and Calif. Mid-June–early Aug.

11. H. Hoòkeri Torr. (for Sir WILLIAM JACKSON HOOKER, 1785–1865), HOOKER'S ORCHIS. — Plant with very fleshy elongate tuberoids; *leaves* 2 (exceptionally 3), *orbicular, broadly elliptic or oblong-obovate, flat on the ground, thick,* deep green above, paler and lustrous beneath, 0.5–1.6 *dm. long,* 2.5–12 *cm. broad; scape fleshy, naked* (rarely with 1 bract), 0.7–2.5 *dm. high; flowers greenish or greenish-yellow, sessile in a strict spike* 0.5–2.5 dm. long and 2–4 *cm. thick,* rather remote, fragrant; *upper lance-acuminate sepal* 7–11 *mm. long, lateral sepals* 8.5–11.5 *mm. long; petals* 6–9 *mm. long; lip* lanceolate, upcurving at tip, 9–13 *mm. long; spur* 1.4–2.5 *cm. long;* capsule 1.5–2.2 cm. long. (*Platanthera* Lindl.; *Lysias* Rydb.) — Dryish woods, Que. and Ont., s. to N.S., N.E., Pa., O., nw. Ind., ne. Ill. and ne. Ia. Late May–early Aug.

Var. **abbreviàta** Fern. (shortened). — *Leaves* 2.5–7 *cm. long* and 2–5 cm. broad; *scape* 5–9 *cm. high; spike* 2.5–9 *cm. long; flowers yellow or yellowish-white; upper sepals ovate, obtuse,* 5–7 *mm. long, lateral sepals* 6–8.5 *mm. long; petals* 5–7 *mm. long; lip* 6–10 *mm. long, deltoid-lanceolate, with blunt* and scarcely prolonged *tip; spur* 9–13 *mm. long; capsules* 9–14 *mm. long.* — In humus of *Empetrum, Juniperus, Vaccinium,* etc., on limestone-barrens, nw. Nfld. Mid-July–early Aug.

12. H. orbiculàta (Pursh) Torr. (rounded; from the leaves), ROUND-LEAVED ORCHIS. — Resembling no. 11, but with thinner *basal leaves* 0.6–2 *dm. long and broad; scape* drier and usually *definitely bracted,* 1.5–5.2 *dm. high,* with 1–5 bracts; *raceme lax and open,* 0.7–2.5 *dm. long and* 3.5–8 *cm. thick, with lower internodes* 0.8–2.5 *cm. long; flowers greenish-white; pedicels filiform,* 6–13 *mm. long; perianth* 8–12 *mm. long; upper sepal roundish, the* 2 *lower ovate and slightly drooping or reflexed;* petals lanceolate, ascending; *lip narrowly lanceolate to linear-oblong,* blunt, abruptly deflexed, 1–1.5 *cm. long; spur* slenderly clavate, reflexed, often arching at tip, 1.6–2.7 *cm. long;* mature capsule 1.5–2 cm. long, 4–6 mm. thick. (*Platanthera* Lindl.; *Lysias* Rydb.) — Dry to moist woods, Nfld. to Thunder Bay Distr., Ont., and Minn., s. to N.S., N.E., n. N.J., Pa., W.Va., upland to S.C. and Tenn., O., n. Ind. and n. Ill.; with a var. from s. Alaska to Mont., Ida. and Wash. Late June–early Aug. — Passing into

Var. **Lehórsii** Fern. (for its discoverer, Mathurin Le Hors, 1886–). — *Leaves* 0.6–9 cm. long, 4–5.5 *cm. broad; scape* 7–11 *cm. high*, with 0–2 (rarely 3) bracts; *raceme dense*, 2–5.5 *cm. long*, 2–4 *cm. in diameter; lower internodes* 1–8 *mm. long; pedicels stout*, 2–4 *mm. long; spur* 8–15 *mm. long*. — Open heathy barrens, local, St. P. et Miq., with transitional forms on bare mountainous areas of w. Nfld.

13. H. macrophýlla Goldie (large-leaved). — Resembling no. 12 but larger throughout; leaves very thin, up to 2.5 dm. long and broad; raceme 0.8–3.5 dm. long, open; *flowers whiter, twice as large as in no.* 12; *upper sepal broadly ovate, the* 2 *lateral ones ascending; lip linear*, 1.5–2.2 *cm. long; spur* 3.2–4.5 *cm. long;* mature capsules 1.8–2.5 cm. long, 4–6 mm. thick. (*Lysias* House) — Rich old woods, se. Nfld.; C.B. to n. Wisc., s. to n. N.E., ne. and centr. Mass., nw. Ct. and N.Y. Late June–early Aug.

14. H. obtusàta (Pursh) Richards. (bluntish), Blunt-leaf Orchis, One-leaf Rein-orchis. — Plant *with filiform or very slender naked* (rarely 1-bracted) *scape* 0.5–4 *dm. high and much overtopping leaf*, from elongate slender fleshy-tuberous roots; *leaf* 1, *basal, obovate to spatulate-oblanceolate, rounded above or blunt;* the blade 2.5–12 cm. long and 1–5.5 cm. broad, tapering to petiolar base; *raceme slender, spiciform*, 2–15 *cm. long, remotely flowered; flowers greenish-white; perianth* 5–6 *mm. long;* upper sepal roundish, arching as a hood; 2 lateral sepals lance-oblong, reflexed; narrow arching petals ascending; lip linear-lanceolate, 5–7 mm. long, deflexed; spur slender, about equaling lip. (*Platanthera* Lindl.; *Lysiella* Rydb.) — Mossy, chiefly coniferous or mixed woods, s. Lab. to Alaska, s. to N.S., n. N.E., w. Mass., N.Y., Mich., n. Wisc., n. Minn. and Colo. Late June–Sept.

Var. **collectànea** Fern. (drawn together). — *Scape* 5–12 cm. high, *nearly equaled by leaf; raceme ellipsoid or oblong-cylindric, densely flowered*, 1–4.5 *cm. long*. — Mossy thickets, damp slopes, open peaty barrens, etc., Lab. to Alaska, s. to n. and w. Nfld., Côte Nord, Que., Ung. and n. Man. July, Aug.

15. H. cristàta (Michx.) R. Br. (crested), Crested Yellow Orchis. — Slender, 2–7.5 dm. high, from fleshy elongate roots; leaves several, rapidly reduced upward, the 1 or 2 larger lower ones narrowly lanceolate to lance-oblong, firm; *spike dense* (sometimes open), slenderly conic, *becoming thick-cylindric*, 2–13 cm. long, 2–3.5 *cm. thick; flowers yellow to orange; ovary lanceolate, its prolonged colored style* 3–5 *mm. long*, about equaled by bract; *unexpanded perianth and expanded concave sepals and petals* 3.5–4 *mm. long;* upper sepal roundish, lateral ones ovate; petals cuneate, fringed at summit; *lip broadly ovate, deeply fringed; spur* 5–9 mm. long, *about half length of ovary*. (*Blephariglottis* Raf.) — Dry to moist open soil, thickets, woods and bogs, Fla. to Tex., Ark. and Tenn., n. to N.J.; formerly se. Mass., very rare. July–early Sept. — Hybridizes with the next, producing × **H. Chapmáni** (Small) Ames (for Alvan Wentworth Chapman, 1809–1899), and with no. 17, producing × **H. Cánbyi** Ames (for its discoverer, William Marriot Canby, 1831–1904).

16. H. ciliàris (L.) R. Br. (fringed), Yellow Fringed Orchis, Orange-plume. — Larger throughout than no. 15, with 1 or 2 finger-like tuberoids, 3–9 dm. high; *spike racemiform* because of slender pedicel-like ovaries, looser, 4–8 *cm. thick; ovary linear-subulate, its colored style* 7.5–12 *mm. long; unexpanded orange perianth and orbicular lateral sepals* 6–8 *mm. long; petals linear-lanceolate or oblong*, commonly hidden by the lateral sepals; *lip oblong or oblong-ovate*, long-fringed; *spur* 1.5–2.5 *cm. long, nearly equaling to exceeding ovary*. (*Blephariglottis* Rydb.) — Bogs, peaty or sandy woods or thickets or dryish swales and slopes, Fla. to e. Tex., n. to s. N.E. (now rare), N.Y., and rarely s. Ont., s. Mich., s. Wisc. and Mo. July–early Sept.

17. H. blephariglóttis (Willd.) Hook. (eyelid-tongued; from the fringed lip), White Fringed Orchis. — Habitally like no. 16, with very long fleshy tuberous-thickened roots and often 1 or 2 thicker and shorter tuberoids, 1–9 dm. high; racemiform spike thick-cylindric, 0.25–1.8 dm. long, (2.5–)3.5–7 *cm. thick*, soon subtruncate across the summit; bracts lanceolate; *flowers bright white to creamy; ovary* very slender, 1.2–2.5 *cm. long, in fruit becoming plump-lanceolate; petals spatulate, slightly toothed at apex or entire; lip ovate- to lance-oblong, with irregular capillary fringe mostly shorter than the breadth of the median disk; spur* 1–2.5 *cm. long*. (*Blephariglottis* Rydb.; *B. holopetala* (Lindl.) Niles) — Wet boggy or peaty soil, S.C. to Nfld., P.E.I., N.B., n. N.E. and s. Que., thence inland to Muskoka Distr., Ont., and Mich. Late June (southw.)–Sept. (northw.).

Var. **conspícua** (Nash) Ames (conspicuous). — 0.2–1.2 m. high; *inflorescence* lax, 6–9 *cm. thick*, larger-flowered; *ovaries* 2.2–3 *cm. long; spur* 3–5 *cm. long*. (*Blephariglottis conspicua* Nash) — Bogs, swamps and damp pine-barrens, Fla. and Ala. to w. N.C. and se. Va. Aug.

Var. **integrilàbia** Correll (entire-lipped). — Like preceding var. but *lip entire or merely undulate; spur* 2.5–3.5 *cm. long*. — Western N.C. and Cumberland Plateau of Ky. to Coastal Plain of Ala. and Miss. Late July–early Sept.

18. H. lácera (Michx.) Lodd. (torn or lacerated), RAGGED ORCHIS. — Plant 2–8 dm. high, with 4–9 leaves; *lower leaves narrowly oblong to oblanceolate, the larger ones* 1–4, *the largest* 1–3.5 *cm. broad* and 0.6–2 dm. long, the upper bracteiform; *raceme* slender and spiciform to thick-cylindric, rather loosely flowered, 0.5–3 dm. long and 2–6 *cm. thick; flowers* short-pedicelled to subsessile, fragrant, *yellowish-green or -white to sordid-yellow or bronzy; perianth* 5–6 *mm. long; lip cleft* nearly to base *into* 3 *clawed divisions* 1–1.5 *cm. long, the* inequilateral *lower divisions widely divergent, the terminal one cuneate into a very slender claw, all deeply dissected into long capillary segments; glands of anther oblong-linear,* as long as stalk of pollen-masses. (*Blephariglottis* Farw.) — Dry to wet meadows, old fields, clearings, thickets, alluvial or wet woods and savannas, M.I. to Ont. and Minn., s. to N.S., N.E., L.I., Fla., Ala., Miss. and Tex. June (southw.)–early Sept. (northw.). — Hybridizes with no. 20, producing the local × **H. Andrewsii** M. White (for ALBERT LEROY ANDREWS, 1878–); also said to hybridize with no. 21.

Var. **térrae-nòvae** Fern. (of Newfoundland). — 1–4.5 dm. high, with 2–4 (rarely –6) leaves; larger leaves 4–10 cm. long; raceme 3–10 cm. long; *flowers variously colored, with mixtures of yellow-white, bronze, pink and purple,* strongly pungent-scented; *perianth* 4–5 *mm. long; lip* 7–10 *mm. long, with relatively short fringe.* — Bog-barrens, peaty tundra and alpine meadows, locally very abundant, Nfld., St. P. et Miq., and C.B. and Sable I., N.S. July–Sept.

19. H. leucophaèa (Nutt.) Gray (ashy), PRAIRIE ORCHIS or PRAIRIE WHITE FRINGED ORCHIS. — Larger throughout than no. 18, 0.2–1.2 m. high; larger lower leaves 1.5–3.5 cm. broad; raceme lax, few–many-flowered, 0.3–2 dm. long; flowers creamy-white to whitish-green, deliciously fragrant; *perianth* 8–10 *mm. long; lip* 1.5–2 *cm. long; spur* 2–4 *cm. long; glands transversely oval.* (*Blephariglottis* Farw.) — Wet prairie or open (often tamarack-) swamps, or bogs and shores, now rare, local and sporadically appearing, s. Ont. to N.D., s. to centr. N.Y., O., n. Ind., Ill., La. and Kans.; open tamarack-swamp, s. Aroostook Co., Me. Mid-June–Aug.

20. H. psycòdes (L.) Spreng. (butterfly-like), SMALL PURPLE FRINGED ORCHIS, SOLDIER'S-PLUME. — Plant 2–9 dm. high; larger cauline leaves 2–5, oblanceolate-elliptic, the largest ones 2–7 cm. broad; *raceme* 2.5–4.5 *cm. thick,* 0.5–2.5 dm. long; *flowers lilac-pink to deep rose-purple,* or white in forma **albiflòra** (Bigel.) R. Hoffm. (white-flowered), fragrant; *perianth* 4–7 *mm. long;* lower sepals oval; petals narrowly flabelliform or spatulate, denticulate or entire toward tip; *lip* 6–16 *mm. broad, usually cleft nearly to base into* 3 *broad and usually lacerate divisions,* the two lateral ones divergent, *the very broad terminal one subtruncate or broadly tapering to short base; spur nearly equaling to exceeding ovary.* (*Blephariglottis* Rydb.) — Meadows, damp thickets, alluvial or springy shores, low woods, etc., Nfld. and Côte Nord, Que., to nw. Ont., s. to N.S., N.E., L.I., N.J., Pa., upland to Ga. and Tenn., O., Ind., Ill. and Ia. Late June–Aug. — Local departures occur, such as forma **vàrians** (Bryan) Fern. (varying) with terminal division of lip obsolete, and forma **ecalcaràta** (Bryan) Dole (without spur) with spur lacking and entire narrowly oblong lip resembling the sepals and petals.

21. H. fimbriàta (Ait.) R. Br. (fringed), PURPLE or LARGE PURPLE FRINGED ORCHIS. — Similar to no. 20, larger; larger lower leaves 2.5–9 cm. broad; *raceme* 5–9 *cm. in diameter;* flowers pale lilac to roseate, or white in forma **albiflòra** Rand & Redfield (white-flowered); *perianth* 9–12 *mm. long; lip* 1.8–3 *cm. broad.* (*Blephariglottis grandiflora* (Bigel.) Rydb.) — Rich woods, thickets and meadows, s. Nfld. and N.B. to se. Ont. and w.-centr. N.Y., s. to N.S., N.E., N.J., Del. (formerly), Pa., W.Va. and mts. to N.C. and Tenn. Early June (southw.)–mid-Aug. (northw.), in adjacent areas flowering about two weeks earlier than no. 20. — Hybridizes with no. 20 and, perhaps, with no. 18. Several aberrant forms occur, among them forma **mentotónsa** Fern. (with shaved chin) in which the cuneate terminal division of the lip is naked or barely erose-denticulate.

22. H. peramoèna Gray (very beautiful), PURPLE FRINGELESS ORCHIS, PRIDE-OF-THE-PEAK. — Resembling no. 21, but flowers rose-purple to purple-violet; *divisions of lip merely erose-denticulate, the lateral ones truncate, the terminal emarginate.* (*Blephariglottis* Rydb.) — Meadows, bogs, alluvial thickets and low woods, w. N.J. (rare) to Ill. and se. Mo., s. to Del., Md., and upland to Ala. and Tenn. June–Aug.

Tribe III. NEOTTÌEAE Lindl.

4. POGÒNIA Juss. POGONIA or BEARD-FLOWER

Flowers 1 or 2, terminal, each with a reduced foliaceous bract; sepals and petals free. Lip spatulate-cuneate, unlobed or 3-lobed, with 2 basal calli, the dilated summit lacerate, its upper surface densely crested or bearded. Column with a deep dentate-margined excavation or clinandrium at summit. Anther articulated at base, freely moving, arching, its locules facing the base of the clinandrium. Pollen-grains distinct, not in tetrads. Capsule subtruncate, the perianth deciduous from it. — Roots slender, cord-like, elongate, budding and setting off new long-petioled basal leaves and upright flowering stems with 1 (rarely 2) submedian sessile or subsessile leaf. Small genus of e. N.Am. and e. Asia. (Name from the Greek *pogonias, bearded.*)

1. P. ophioglossoìdes (L.) Ker (like *Ophioglossum*). — Roots horizontal, up to 8 dm. long; basal petioled leaves (rarely collected) lanceolate to oblong or narrowly obovate; flowering stem slender, erect, 0.5–6 dm. high; cauline leaf (rarely wanting) narrowly lanceolate to narrowly obovate or elliptic, 1–15 cm. long, ascending; perianth pale to deep pink, or white in forma **albiflòra** Rand & Redfield (white-flowered), 1.5–3 cm. long, the blunt oblong to narrowly elliptic *sepals and the broader petals spreading; beard or crest of entire lip of* 3 *rows of elongate yellow- or brown-tipped processes;* flower fragrant. — Sphagnous bogs, peaty swales, savannas and wet mossy shores, Nfld. and M.I., w. to n. Ont., s. on Coastal Plain and outer Piedmont to N.S., N.E., L.I., Fla., thence to e. Tex., and inland to Pa., Tenn., Ind., n. Ill. and Minn. May (southw.)–Aug. (northw.). — In rare teratological plants 2 or 3 lips may occur and the flower become otherwise modified by the column becoming petaloid, etc.

Var. **brachypògon** Fern. (short-bearded). — Internodes of roots short, the basal leaves and flowering stems numerous; *sepals and petals porrect, not wide-spreading; beard of lip almost wanting or of very short knobs.* — Gravelly beaches of ponds, w. N.S. — Simulating the eastern Asiatic *P. japonica* Reichenb. f.

5. CLEÌSTES Richard Spreading Pogonia

Similar to *Pogonia* in habit, foliage and flower, but the petals, lip and column connivent; the lip trough-like, oblong-cuneate, with erose margins, abruptly contracted to a prolonged narrowly ovate to triangular apical lobe, a prominent ridge extending down the middle and on the apical lobe, dividing into 3 slender lobulate crests. Pollen-grains in tetrads. — Small genus of trop. Am. and e.-temp. N.Am., at least ours only sparingly spreading by root-shoots. (Name from the Greek *cleistos, closed,* from the connivent petals, lip and column.)

1. C. divaricàta (L.) Ames (divergent; from the wide-spreading sepals). — Stem erect, 2.2–7.2 (av. 4.5) dm. high; basal leaf (not often collected) long-petioled; median cauline leaf firm, narrowly oblong, 6.5–15 (av. 10) cm. long; peduncle (from base of median leaf to base of floral bract) 0.9–2 (av. 1.5) dm. long; flower 1 (rarely 2); *sepals linear or linear-lanceolate,* brownish, with attenuate tips, the upper slightly shorter than *the spreading lateral ones,* these 4.5–7 *cm. long; petals* pink to white, narrowly oblong to oblanceolate, connivent over the column and base of lip, 3.5–5 *cm. long,* 8–14 *mm. broad;* lip about equaling or slightly exceeding petals; ovary and stipe during anthesis 2.5–4.5 (av. 3.25) cm. long. (*Pogonia* R. Br.) — Damp pine-barrens, peaty thickets, etc., on Coastal Plain, n. Fla. to N.C., very rarely to s. N.J. June, early July.

Var. **bifària** Fern. (two-fold; from the two areas of occurrence). — 1.5–5(–6.5) dm. high; median leaf 3.5–13 (av. 7.6) cm. long; peduncle above median leaf 0.3–1.6 (av. 1) dm. long; *longer sepals* 3–4.5 *cm. long;* petals 2–3 cm. long, 5–10 mm. wide; ovary and stipe during anthesis 1.2–3.5 (av. 2.6) cm. long. — Sandy upland woods, mountain-crests and slopes, Cumberland Plateau of Ky. and Blue Ridge of w. N.C., coming out to peats and sands of Coastal Plain, se. N.C. to Fla. and La.

6. ISÓTRIA Raf. Whorled Pogonia

Flower much as in *Pogonia* but terminating the naked peduncle which is subtended by a whorl of 5 or 6 (rarely –10) involucral leaves. Lip 3-lobed at summit, sessile, with a median crest. Pollen-grains in tetrads. — Small eastern N. Am. genus, with propagation by root-sprouts, a short vertical rhizome, and erect scaly-based but otherwise naked scape and 1 (rarely 2) terminal flower, with perianth deciduous after pollination of stigma. (Name from the Greek *isos, equal,* and *treis, three,* from the equal sepals.)

1. I. verticillàta (Willd.) Raf. (whorled). — Horizontal cord-like roots elongate, frequently sending up new basal leaves or flowering scapes; scapes (up to whorl of involucral leaves) 1–3.7 dm. high; involucre at first small and embracing erect young flower, the *leaves* soon spreading in a whorl and *finally enlarging to acutish obovate to rhombic-elliptic* subcuneate *green blades* 5–10 *cm. long;* peduncle obliquely ascending or erect, shorter than mature leaves; *sepals linear,* yellow-green below, *madder-purple above,* porrect, subequal, 3.5–6 *cm. long;* petals porrect, lance-oblong, yellow-green, 1.5–2.5 cm. long; *lip* narrowly oblong-obovate; its body green, with a slightly papillose linear median crest, 3-lobed, *the undulate-subtruncate terminal white lobe and smaller lateral lobes striped purple and white; capsule* erect, 2.5–3.5 *cm. long, its peduncle*

2.5–6 *cm. long.* (*Pogonia* Nutt.) — Acid or mediacid woodlands, locally abundant, Fla. to e. Tex., n. to sw. Me., centr. N.H., Vt., N.Y., O., s. Mich. and Mo. May, June. — Sterile plants strongly resembling *Medeola.*

2. I. medeoloìdes (Pursh) Raf. (resembling *Medeola*), Small Whorled P. — Roots shorter, only rarely producing sprouts; *scape and leaves gray with bloom; whorl soon reflexing,* its narrow rhombic-elliptic *leaves becoming* 3.5–6 *cm. long; flowers* 1 or 2, of a general *greenish-yellow* color; *sepals* arching, 1.5–2.5 *cm. long; petals* 1.2–1.5 *cm. long; lip pale green,* narrowly obovate, its 3 lobes subequal; *capsule* 1.3–2.5 *cm. long, its peduncle* 0.5–2 *cm. long.* (*I. affinis* Rydb.; *Pogonia affinis* Aust.) — Dry woodland, very rare, local and in small colonies, N.H. and Vt., s. to N.C.; se. Mo. May–early July.

7. TRÍPHORA Nutt.

Flowers nodding from upper leaf-axils, quickly fading; sepals and petals free. Lip 3-lobed, the terminal lobe largest, without calli at base. Column semiterete at base, laterally dilated near middle and at the entire apex. Anther rigidly attached to top of column, erect, with the locules facing outward; pollen-grains in tetrads. — Small woodland plants of trop. Am. and e. temp. N.Am. with an elongate tuber and vegetatively propagating by short slender stolons tipped by tubers; leaves alternate, clasping, the upper subtending racemose flowers; marcescent perianth often persisting in fruit. (Name from the Greek *treis, three* and *phoros, bearing,* from the frequently three flowers.)

1. T. trianthóphora (Sw.) Rydb. (bearing three flowers), Nodding Pogonia. — Stem fragile, 0.3–3 dm. long, often hidden among fallen leaves, from a subcylindric tuber and soon bearing horizontal stolons at base; leaves small, ovate; flower in bud nodding, in anthesis ascending, in fruit pendulous (or remaining erect) on slender pedicels, pale pink or whitish, 1–1.5 cm. long; sepals lanceolate, exceeding lip, the upper one arching over the lip, the lower ones spreading; petals similar, shorter; lip with 3 green median longitudinal lines; anther subcylindric, with red-purple flanges; pollen purple. (*Pogonia* BSP.; *P. pendula* Lindl.) — Humus of hardwood-forests, usually developing only at remote periods, Fla. to e. Tex., very locally n. to sw. Me., N.H., Vt., N.Y., O., s. Mich., s. Wisc. and Ia. — Colonies with capsules all erect are the inconstant var. *Schaffneri* Camp.

8. CALOPÒGON R. Br. Grass-pink. Swamp-pink

Flowers in a loose spiciform raceme, resupinate. Sepals and petals spreading, distinct. Lip linear-oblong at base, dilated and bearded above with numerous clavate hairs, papillose at the apex. Column free, slender, winged at the summit; anther terminal, operculate; pollen-masses 4 (2 in each anther-locule); pollen-grains connected by filaments. — Scape from a solid bulb-like tuber, sheathed below by the base of the commonly solitary elongate leaf, naked above. Small genus of Atl. N. Am. (Name composed of the Greek *calos, beautiful,* and *pogon, beard.*) Limodorum L. in part.

1. C. pulchéllus (Salisb.) R. Br. (pretty). — Plant 0.7 (northw.)–9.75 (southw.) dm. high, with 1 (rarely 2) linear or linear-lanceolate to narrowly oblong *leaf* 0.2–5 cm. *broad and many times as long;* scape (rarely a 2nd axillary one) much overtopping leaf; raceme with (1–) 2 (northw.) –10 (southward exceptionally –20) *magenta to rose-pink or lilac,* or in forma **albiflòrus** (Britt.) Fern. (white-flowered) white, showy *flowers* 2.5–4.5 *cm. broad,* these sessile in axils of tiny acute bracts along a slightly geniculate rachis; *sepals concave;* the lateral ones ovate-lanceolate and falcate, broader than the upper; petals lanceolate, obtuse, constricted near middle; *lip* as if hinged, with projecting flanges at base, *its white beard with yellow and crimson tips; terminal blade flabelliform, subtruncate at summit and* 8–15 *mm. broad,* cuneate to the long stalk; *wings of column rhombic.* (*Limodorum tuberosum* BSP., not L.) — Sphagnous bogs or peaty meadows, savannas and wet shores, Fla. to Tex., n. to Nfld., Que., Ont. and Minn. Mid-May (southw.) –Aug. (northw.).

Var. **latifòlius** (St. John) Fern. (broad-leaved). — *Leaves broadly lanceolate, oblanceolate or narrowly oblong-ovate, often paired, barely equaling to much exceeding the short scape; tuber very large* (*up to* 2 *cm. thick*) and dark-coated; plant 1.5–2 dm. high. — M.I., and Sable I., N.S. Late July, Aug.

2. C. pállidus Chapm. (pale). — *Smaller throughout,* 2–6 dm. high, the *scape filiform;* leaf narrowly linear, 2–5 mm. broad; *flowers* 1–10, scattered, 2–3 *cm. broad, lilac to whitish; lip cuneate-obovate,* 5–7 *mm. broad at summit, abruptly short-pointed, with* 2 *basal auricles; wings of column broadly triangular.* (*Limodorum* Mohr) — Damp pine-barrens, Fla. to La., n. to se. Va. June–early Aug.

9. ARETHÙSA L. Arethusa

Flowers ringent. Sepals and petals nearly alike, erect, united at base, arching over the column. Lip partly erect, the apical half abruptly recurved. Column adherent to the lip, dilated above, petal-like; anther lid-like, attached by a well defined membrane, 2-locular; pollen-masses 2 in each locule of the anther, powdery, granular. — Scape smooth, from a solid white or greenish bulb. Leaf solitary, linear, nerved, hidden in the sheaths of the scape, protruding after the flower opens. Two species: ours and one in Japan. (Named for the nymph, *Arethusa.*)

1. A. bulbòsa L. (bulbous), Swamp-pink. — Plant 0.6–3.5 dm. high, from an ovoid bulb; scape terminated by a solitary flower 2.5–5 cm. long, rarely 2-flowered; sepals and petals magenta-pink, or in forma **subcaerùlea** Rand & Redfield (bluish) bluish-lilac, or in forma **albiflòra** Rand & Redfield (white-flowered) white; the sepals oblong, acute or obtuse, the lateral ones falcate; the petals oblong, obtuse or obscurely pointed; lip oblong, narrowed toward the base, with 3–5 fringed yellow or white crests; margin of lip fimbrillate, spotted and striated with magenta-crimson or plain; column denticulate or toothed at the dilated apex; stigma protuberant, turned down; capsule erect, 2.5–4 cm. long, long-beaked, capped by marcescent perianth. — Sphagnous bogs and peaty meadows, Nfld. to Ont. and Minn., s. to N.S., N.E., L.I., Del., Md., mts. to S.C., O., n. Ind. and Wisc., and rare in La.; rapidly becoming extinct (the bulbs only loosely attached in the moss) s. of Nfld. and Canada. Late May (southw.)–Aug.

10. EPIPÁCTIS Sw. Helleborine

Flowers in a loose or somewhat dense bracteose raceme. Sepals ovate-lanceolate, strongly keeled. Petals shorter, ovate, acute. Lip strongly saccate at base; the apical part broadly cordate, acute, with a raised callus in the middle and two inconspicuous nipple-like protuberances on each side near the point of union with the sac. Column broad at the top, the basal part narrower; anther sessile, behind the broad truncate stigma, on a slender-jointed base; pollen farinaceous, becoming attached to the gland capping the small rounded beak of the stigma. — Stem leafy. A small genus of N. Hemisph. (Ancient name for Hellebore.) Serapias L. in part. Helleborine Mill. Amesia Nels. & Macbr.

1. E. Helleborìne (L.) Crantz (Greek name for Hellebore). — Plant 0.3–9 dm. high; leaves clasping stem, ovate to lanceolate, acute, conspicuously nerved; raceme 0.5–2.5 dm. long, often 1-sided; perianth about 8mm. long, green, suffused with madder-purple; lip similarly colored, darker within, the apical portion as if jointed with the sac, with 2 tubercles at base. (*E. latifolia* (L.) All.; *Serapias* L.) — Woods, thickets, ravines and alluvium near settled areas, appearing as if spontaneous and soon spreading, sw. Que. and Ont., s. locally to N.H. and w. N.E., N.J., Pa., D.C. and Mo. Late June–Sept. (Introd. and natzd. from Eu.)

11. SPIRÁNTHES Richard Ladies'-tresses. Pearl-twist

Perianth usually somewhat ringent. Lateral sepals narrow, the upper sepal united with the petals. Lip short-stalked, with a callus protuberant within on each side of the base, the somewhat dilated summit spreading or recurved, crisped, wavy, or rarely toothed or lobed. Column short, bearing the ovate stigma on the front and the sessile or short-stalked (mostly acute or pointed) 2-locular erect anther on the back; pollen-masses 2 (1 in each locule), narrowly obovoid, each 2-cleft and split into thin and tender plates of granular pollen united by elastic filaments, coherent to the narrow viscid gland, which is set in the slender or tapering thin beak terminating the column. After the removal of the gland, the beak is left as a 2-toothed or forked tip. — Small semicosmop. genus, with roots commonly clustered. Stem bracted above, leaf-bearing below or at the base. Flowers small, white, yellowish- or greenish-white, in a more or less spirally twisted spike (whence the name, from the Greek *speira*, a *coil* or *spiral*, and *anthos*, *flower*). Gyrostachys Pers. Ibidium Salisb.

a. Flowers in a single rank, usually secund. . . *b.*

b. Leaves basal, ovate or elliptic, broad-petioled, 1–4 cm. long, soon wilted or quite absent at flowering time; rachis, bracts and ovaries glabrous; perianth 2–5 mm. long. . . *c.*

c. Root a solitary (rarely 2) vertical finger-like tuber; leaves wanting at flowering time; perianth 2–3 mm. long; lip thin, loosely cellular, white or whitish. . . . . 1. *S. tuberosa.*

c. Roots 2–several, divergent, tuberous-thickened but slender; leaves present or wilted or absent at flowering time; perianth 4–5 mm. long; lip thick, with green central band.

Leaves commonly present at flowering time, thin and semitranslucent, the veins and simple or subsimple veinlets evident; spike secund or with few remote spirals, the series of flowers elongate; perianth forming a slender tube 1–1.75 (–2.2) mm. thick; dilated summit of lip with a broad white border. . . . . 2. *S. lacera.*

Leaves commonly absent at flowering time, thick and opaque, the hidden (except by transmitted light) veinlets more branched and forming a fine mesh; spike strongly spiraling, with several short series of flowers; perianth ringent, its tube 1.5–2.5 mm. thick; white margin of summit of lip very narrow. . . . . 3. *S. gracilis.*

*b.*Leaves basal and extending slightly up stem, linear or elongate-lanceolate or -oblanceolate, persistent or subpersistent; rachis, bracts and ovaries pubescent; perianth 4–11 mm. long. . . *d.*

*d.*Lip pubescent on the outside, its basal callosities curved, the veins obscure; perianth 6–11 mm. long.

Lip ovate or ovate-lanceolate, broadest between the callosities and the middle, tapering to a merely undulate apex. . . . . 4. *S. vernalis.*

Lip oblong, broadest across the basal callosities, the apex dentate-lacerate. . . . . 5. × *S. laciniata.*

*d.*Lip glabrous on the outside, oblong, merely undulate at top, its basal callosities straight, its veins prominent; perianth 4–6 mm. long. . . . 6. *S. praecox.*

*a.*Flowers in 2 or more ranks, more or less crowded in the spike. . . *e.*

*e.*Lateral sepals free from and not connivent with the hood; lip not strongly constricted below the summit. . . *f.*

*f.*Lip white or whitish, pubescent, ovate or ovate-oblong, the 2 basal callosities prominent; rachis of spike villous; flowering in late summer and autumn.

Plant 1–10.5 dm. high; larger leaves 0.5–4 dm. long; spike 2–18 cm. long, 1.5–3 cm. thick; perianth 7–12 mm. long, somewhat ringent; plants of open sterile dry to moist or swampy places.

Roots fleshy and tuberous-thickened, descending, up to 8 cm. long, not stoloniferous; flowering stem 1–6 dm. high, 1–3 mm. thick at base; basal leaves thick and firm, pale green, the larger 2–15 mm. broad, rapidly reduced to cauline bracts; lateral sepals equaling hood; lip white, not strongly constricted above base. . . . . 7. *S. cernua.*

Roots slender and cord-like, horizontal, finally ending in new plants (stoloniferous), up to 2 dm. long; flowering stem up to 1.05 m. high, its base 3–7 mm. thick; leaves membranaceous, dark green, the larger 1–3.5 cm. wide, the cauline only gradually reduced; lateral sepals much shorter than hood; lip constricted above broad base, suffused with yellow-green. . . . . 8. *S. odorata.*

Plant very slender, 1.5–4 dm. high; leaves submembranaceous, the longer ones 4–15 cm. long; spike 1.5–6 cm. long, 1–1.3 cm. thick; perianth 4–5 mm. long; plant of rich woods. . . . . 9. *S. ovalis.*

*f.*Lip bright yellow, quadrate-oblong, glabrous, the basal callosities obscure and deeply imbedded; rachis of spike merely puberulent; perianth 5–7 mm. long; leaves membranaceous, lanceolate, oblanceolate or oblong; flowering in early summer. . . . . 10. *S. lucida.*

*e.*Lateral sepals united at base, connivent with upper sepal and petals as an upward-arching hood; lip panduriform, strongly contracted below the broadly deltoid to roundish summit. . . . . 11. *S. Romanzoffiana.*

1. S. tuberŏsa Raf. (tuberous), Little L. — Very slender, 1.5–5.25 dm. high, the scape with few insignificant sheaths below middle; *root tuberous-thickened,* 1 (sometimes 2 or 3), *vertical; leaves basal, ovate,* blunt, 1–2 cm. long, *usually decayed or wanting at flowering-time; spike very slender* 2–11.5 cm. long, *wholly 1-sided or with few spiral twists in the glabrous rachis; flowers* slightly *distant,* not overlapping; *perianth slenderly tubular, 2–3 mm. long, its white sepals and petals porrect; lip white,* obovate-oblong, *of thin loosely cellular texture,* its apex crisped, the 2 basal callosities slender. (*S. Beckii* of ed. 7 and most auth., not Lindl.; *Ibidium Beckii* sensu House) — Dry, chiefly sandy or argillaceous fields and borders of woods, Ga. to Miss., n. to N.J., O. and s. Ind. July–Oct.

Var. **Graўi** (Ames) Fern. (for its discoverer, Asa Gray, 1810–1888). — 0.7–4.5 dm. high; *spike* 1.5–8 cm. long, *of numerous mostly approximate spirals of crowded or imbricated flowers.* (*S. Grayi* Ames) — Similar habitats, e. Mass. to e. Va. and, less characteristically, to N.C. Late July–Sept.

2. S. lắcera Raf. (irregularly cleft), Northern Slender L. — *Roots fleshy-thickened, several,* *divergent; leaves basal, present or becoming wilted at flowering time, ovate or ovate-oblong, thinnish and membranaceous,* 1–4 cm. long, *obviously veined, the veinlets simple or but slightly forking;* slender scape (with spike) 1–5.25 dm. high; *spike very slender,* 2–16 cm. long, *wholly* 1*-sided or with* 1*-few remote spiral twists in the glabrous rachis and with several-flowered long series of slightly distant flowers; perianth white,* 4–5 *mm. long, scarcely ringent, the porrect sepals and petals forming a slender tube* 1–1.75 (rarely –2.2) *mm. thick; lip trough-like* below, *the green coloring of the trough extending only slightly down the base of the flaring white* crisp-margined *summit of the lip.* (*S. gracilis* in part of ed. 7) — Dry to moist siliceous, argillaceous or peaty meadows, dune-hollows, barren fields and thickets, Que. to Mackenz., s. to N.S., N.E., L.I., locally to Va., w. N.C. (up to 1000 m. alt.) and Tenn., s. Ont., n. Ind., n. Ill., Wisc., Minn. and s. Man. Mid-June–Sept. (av. Aug. 5).

3. S. grácilis (Bigel.) Beck (slender), SOUTHERN SLENDER L. — Coarser than no. 2; *leaves rarely present at flowering time, thick and opaque, the finer mesh of more forking veinlets seen only by transmitted light; spike open-cylindric, strongly spiraling, with many short series of few secund approximate flowers; perianth gaping or ringent, the bases of the broader sepals and petals forming a tube* 1.5–2.5 *mm. thick; dilated summit of lip green except for a narrow white margin.* (*Ibidium* House) — Dry or moist sterile open soil, thickets or open woods, Fla. to e. Tex., n. to sw. Me., se. N.H., s.-centr. Vt., se., centr. and w. N.Y., O., Ind., s. Wisc., Mo. and Okla. Late July–Oct. (av. Sept. 2).

4. S. vernàlis Engelm. & Gray (of spring; the original material from near the southern limit, there vernal). — Plant 0.15–1 m. high, with elongate fleshy roots; *leaves basal and often extending slightly up the scape, linear or linear-lanceolate, firm, erect, tapering to both ends,* the longer ones 0.6–2 dm. long; *spike* 0.6–2 dm. long, *secund or loosely spiraling,* the *rachis, bracts and ovaries downy-pubescent;* bracts ovate- to lance-acuminate, exceeding ovaries, with broadly hyaline margins; *perianth* yellowish or creamy, 6–11 *mm. long;* upper sepal and oblong obtuse petals adherent at base and forming a hood; lateral sepals lanceolate; *lip pubescent beneath, ovate or ovate-lanceolate, broadest between the* 2 *curved callosities and the middle,* trough-like toward base, *with broad yellow or yellow-green elongate cushions each side of midrib,* these hiding the veins, *the summit crisped or short-dentate.* (Incl. × *S. intermedia* Ames; *Ibidium vernalis* House) — Dry to moist siliceous, argillaceous or gravelly fields, clearings and swales, Fla. to e. Tex., n. to s. N.E., se. N.Y., N.J., Md., s. O., Tenn., Mo. and e. Kans. Late May–Sept.

5. × S. laciniàta (Small) Ames (slashed). — Differing from no. 4 in the greater development of *capitate pubescence about the inflorescence;* cauline leaves more developed; *flowers* more *remote* in the usually 1-sided spike; *lip oblong, broadest across the callosities, the apex jagged-toothed or lacerate.* (*Ibidium* House; *S. praecox* of ed. 7, not S. Wats.) — Bogs, marshes, shallow ponds, etc., Fla. to e. Tex., n. to N.J. July–Sept. — Apparently a hybrid of nos. 4 and 6.

6. S. praècox (Walt.) S. Wats. (precocious; in the South, whence originally described). — Similar to nos. 4 and 5; *perianth* white, 4–6 *mm. long; lip glabrous beneath, oblong, merely undulate at summit, with straight basal callosities, the upper surface usually green-veined.* (*Ibidium* House) — Meadows, swales and open woods, Fla. to Tex., n. to L.I. June–Sept.

7. S. cérnua (L.) Richard (nodding; from the downward-arching perianths), COMMON or NODDING L., SCREW-AUGER. — Plant 1–6 dm. high, *from tuberous-thickened fleshy descending roots* 2–8 *cm. long; leaves mostly radical,* linear-lanceolate to narrowly oblanceolate or oblong, *firm, pale green,* the *larger ones* 0.4–3 dm. long and 0.2–1.5 *cm. broad; the cauline leaves changing to nearly bladeless sheathing smaller bracts; scape slender,* 1–3 *mm. thick at base* (dried), pubescent at summit; *spike* 2–15 cm. long, 1.5–3 *cm. thick,* rather *dense, of* 2 *or* 3 *irregular series of* slightly vanilla-scented *white or creamy* divergent to slightly recurving *flowers;* the *rachis pubescent; perianth* 7–10 *mm. long,* exceeding the acuminate bract; *upper lanceolate sepal and* 2 *lanceolate petals somewhat connivent at base and forming a hood;* 2 *lateral sepals* similar but more attenuate and with inrolled margins, *somewhat spreading, equaling the hood; lip oblong or narrowly oblong-ovate, entire below, crisped and undulate-margined toward apex, white, with* 2 *submedian pubescent curved basal auricles; seeds with* 2 *or more embryos.* (*Ibidium* House) — Wet to dryish open soil or in bogs, low thickets or on shores, Fla. to e. Tex., n. to N.S., n. N.E., sw. Que., s. Ont., Mich., Wisc., Minn. and S.D. Mid-Aug. (northw.)–early Nov. (southw.).

Var. **ochroleùca** (Rydb.) Ames (yellowish-white). — Floral bracts longer; perianth whitish to stramineous or greenish; lip narrowly ovate, its callosities longer; *seeds with a single embryo;* fragrance strong (often pungent and disagreeable). — Dry barrens, rocky slopes, sterile fields and open woods, locally throughout.

8. S. odoràta (Nutt.) Lindl. (fragrant), MARSH L. — Coarser than no. 7, *up to* 1.05 *m. high; roots slender, tough, cord-like, horizontal,* 0.8–2 *dm. long, finally tipped with vegetative buds or new plants (stoloniferous); leaves membranaceous, dark or deep green, the larger ones* 2–4 *dm. long* *and* 1–3.5 *cm. wide; cauline leaves only gradually reduced upward; scape* 3–7 *mm. thick at base* (dried); *spike* 0.6–1.8 dm. long, *with regular spirals of* dull yellowish-white *flowers; perianth* 9–12 *mm. long, the lateral sepals shorter than the hood; lip constricted between the subrhombic base and the* obtuse *summit, veined and suffused with yellowish-green;* flowers strongly vanilla-scented; embryo single. (*S. cernua*, var. *odorata* Correll) — Fresh tidal or inundated marshes and inundated shores, Fla. to e. Tex., n. to Md. and Tenn. Sept., Oct.

9. S. ovàlis Lindl. (oval; from the lip). — Very slender, 1.5–4 dm. high, with fleshy tuberous-thickened descending roots; *leaves basal and cauline, submembranaceous, deep green, lustrous,* oblanceolate, tapering to long petiolar base, *the larger ones* 4–15 *cm. long* and 0.5–1.3 *cm. broad,* the lower cauline ones well developed; *spike* 1.5–6 *cm. long,* 1–1.3 *cm. thick,* compact; *perianth white,* 4–5 *mm. long, slender and nearly tubular, the lip and lance-linear lateral sepals porrect; lip ovate, membranaceous, white,* slightly veined, with rhombic base and scarcely flaring rounded summit. (Perhaps *S. montana* Raf.) — Rich (often calcareous) woods, e. Va. and W.Va. to s. Ind., Ky. and Mo., s. to n. Fla., Ala., Miss. and e. Tex. Sept., Oct.

10. S. lùcida (H. H. Eat.) Ames (shining). — Strongly resembling no. 9, 0.5–3 dm. high; *leaves mostly radical, oblong to oblanceolate,* narrowed to short base, *dark green, lustrous, membranaceous, the larger ones* 3–15 cm. long and 0.7–2.3 *cm. broad; scape glabrous at summit or* (*like the rachis*) *merely puberulent;* spike 1–11 cm. long, slenderly cylindric, of 3 or 4 spiraling ranks; *perianth white, slenderly subcylindric,* 5–7 *mm. long; lip quadrate-oblong, yellow,* with about 5 slender veins, obtuse, *the basal callosities obscure and imbedded.* (*S. plantaginea* (Raf.) Torr.) — Alluvial or damp rocky (often calcareous) shores and slopes, rich damp thickets and meadows, ne. N.B. to s. Ont. and Minn., s. to N.S., n. and w. N.E., N.J., Pa., upland to N.C. and Tenn., O., Mich., Ind. and Ill. May–July.

11. S. Romanzoffiàna Cham. (in honor of Chamisso's patron, Nikolai Rumiantzev, Count Romanzoff, 1754–1826), Hooded or Romanzoff's L. — Plant with fleshy thickish tuberous roots, 0.3–5 dm. high, habitally *resembling no. 7 but differing in having bracts often longer* than flowers; creamy-white to straw-colored *flowers* (6–12 *mm. long*) *with the* 3 *sepals and* 2 *petals somewhat united and forming an upward-arching hood;* the *lip* (laid open) *panduriform, strongly constricted below the broadly deltoid to rounded entire summit, its callosities minute and globular;* flowers strongly fragrant (suggesting almonds). (*Ibidium* House; *I. strictum* House; *S. stricta* Nels.) — Meadows, swamps, clearings and thickets, Lab. to Alaska, s. to Nfld., N.S., n. and w. N.E., N.Y., Pa., O., Mich., Wisc., ne. Ia., S.D., Colo., Utah and Calif. Mid-July–early Sept. (Scotl. and Ireland)

12. GOODYÈRA R. Br. Rattlesnake-plantain. Lattice-leaf

Lip more or less saccate, with a straight or recurved tip, sessile, entire, without callosities at base. Upper sepal and the petals united into a hood over the lip. Anther borne on the back of the short column; pollen-masses 2, the narrow gland to which they are attached held by the forked or 2-toothed beak which terminates the column. — Root of thick fibres from a somewhat fleshy creeping rhizome. Leaves all basal, dark green, or reticulate-veined with white. Scape, spiciform raceme, and the whitish flowers glandular-downy. Rosulate evergreen plants of N. Hemisph. (Named for *John Goodyer*, English botanist, 1592–1664.) Epipactis Boehm., not Sw.

a. Raceme loosely flowered or one-sided; lip with elongate tip, its margin flaring or recurved. . . *b.*
b. Leaves 1–7 cm. long, mostly reticulate-veined; perianth 4–5 mm. long; lip saccate, with recurving tip.
Leaves 1–3 cm. long, 5-nerved; raceme 1-sided; perianth 4 mm. long; anther short, blunt or blunt-tipped; beak shorter than body of stigma; lip strongly saccate, with recurved tip. . . . . 1. *G. repens.*
Leaves 2–7 cm. long, 5–9-nerved; raceme loosely spiraling; perianth 5 mm. long; anther acuminate; beak as long as or longer than body of stigma; lip less saccate, its tip less recurved. . . . . 2. *G. tesselata.*
b. Leaves 4–10 cm. long, without evident fine reticulate venation; raceme 1-sided; perianth 8–9 mm. long; lip scarcely saccate, elongated, with incurved margin. . . . . 3. *G. oblongifolia.*
a. Raceme densely oblong-cylindric; leaves reticulate with slender white veins; lip globose-ventricose, with short blunt tip, the margin neither recurved nor flaring. . . . . 4. *G. pubescens.*

1. G. rèpens (L.) R. Br. (repent), Dwarf R. — Slightly creeping, with slender stolons; *leaves* ovate to oblong-lanceolate, 1–3 *cm. long,* 5-nerved, *with subhorizontal dark veins;* flowering stem (including raceme) 1–2.5 dm. high; *raceme one-sided*, rather lax, in anthesis, 2.5–6 cm. long; *perianth 4 mm. long; lip strongly saccate-inflated, with recurved tip; anther short, blunt or blunt-tipped; beak shorter than body of stigma.* (*Epipactis* Crantz; *Peramium* Salisb.) — Damp mossy coniferous or mixed woods, Anticosti I. to n. B.C., s. very locally to Thunder Bay Distr., Ont., and s. Alta. July, Aug. (Eurasia) — More generally with us as

Var. ophioìdes Fern. (resembling a snake), HERBE ÉCARTANTE (Que.). — *Veins of leaves bordered by conspicuous broad white pencilings* (when fresh). — (*Epipactis repens*, var. A. A. Eat.; *Peramium ophioides* Rydb.) — Damp mossy woods, s. Lab. Pen. to Alaska, s. to Nfld., N.S., n. and w. N.E., n. N.J., N.Y., mts. to N.C. and Tenn., Mich., Wisc., Minn., S.D., etc. July, Aug.

2. G. tesselàta Lodd. (like a mosaic). — Coarser than no. 1, 1–3.5 dm. high; *leaves 2–7 cm. long, 5–9-nerved*, fleshy, green above, *with subhorizontal to oblique slightly interlacing veins; the veins bordered by irregular generally pale green penciling*, the whole blade often irregularly mottled with dark and light green (or rarely unmottled); *raceme loosely spiraling*, in anthesis 3–16 cm. long; *perianth* 5 *mm. long; anther acuminate; beak as long as or longer than body of stigma; lip slightly saccate, its tip but slightly recurved.* (*Epipactis* A. A. Eat.; *Peramium* Heller) — Dry to moist woodland, Nfld. to e. Man., s. to N.S., N.E., N.Y., Mich., Wisc. and Minn. July, Aug.

3. G. oblongifòlia Raf. (oblong-leaved), GIANT OR MENZIES' R. — Usually coarser than other species, 2–4.5 dm. high; *leaves firm*, 4–10 *cm. long*, oblong-lanceolate to -oval, *green above, often with broad white midrib or center* or sometimes mottled pale and dark green; *raceme one-sided, during anthesis* 6–18 *cm. long; perianth* 8–9 *mm. long; lip scarcely saccate, elongated, with incurved margin; anther* ovate, *long-attenuate;* slender beak longer than body of stigma. (*G. decipiens* (Hook.) F. T. Hubbard and *G. Menziesii* Lindl.; *Epipactis decipiens* Ames; *Peramium decipiens* Piper and *P. Menziesii* Morong) — Dry coniferous or mixed woods, n. C.B.; Gaspé Pen. to n. N.B., n. Me. and Montmorency Co., Que.; Bruce Pen. to Algoma Distr., Ont., s. to n. Mich. and n. Wisc.; s. Colo. and se. Utah to N.M. and Ariz.; nw. Mont. and s. B.C., s. along mts. to Calif. July, Aug.

4. G. pubéscens (Willd.) R. Br. (pubescent), DOWNY R. — Plant 1.5–5 dm. high, often colonial and somewhat repent; *leaves* ovate to ovate-lanceolate, 3–6.5 cm. long, *dark green above, with* 5 or 7 white nerves (midnerve broad) and *many fine white reticulating veins; raceme dense, oblong-cylindric,* in anthesis 3–11 cm. long; perianth 4–5.5 mm. long; *lip globose-ventricose, with short blunt tip; anther blunt; stigma with* 2 *very short teeth.* (*Peramium* MacM.; *Epipactis* A. A. Eat.) — Dry or moist woods, Me. and sw. Que. to w. Ont., s. to Fla., Ala., Tenn. and se. Mo. July–early Sept. — The colonial plants very handsome throughout the winter.

13. PONTHIÈVA R. Br. PONTHIEVA. PONTHIEU'S ORCHID

Flowers loosely racemose, often glandular. Sepals free, spreading. Petals triangular, asymmetrical. Lip uppermost, raised on the column, ascending, its abruptly dilated blade concave or folded. Column short, nearly terete, dilated above, the erect rostellum concave on back; anther erect, with contiguous locules; pollen-masses in 2 pairs, pendulous from a gland of the rostellum. Fibrous-rooted herbs of trop. and warm-temp. Am., with reticulate rosulate membranaceous leaves, and simple scapes with few sheathing bracts. (Named for *Henri de Ponthieu*, who collected plants of the Caribbean reg. for Sir Joseph Banks in 1778.)

1. P. racemòsa (Walt.) Mohr (racemose), SHADOW-WITCH. — Leaves narrowly oblong-elliptic to narrowly obovate, lustrous-green, broadly short-petioled, the larger 5–13 cm. long; scape slender, 2–5.2 dm. high, with loose raceme 0.6–2 dm. long; pedicels spreading-ascending; sepals, petals and lip whitish, with green veins; median sepal oblong to elliptic-lanceolate, 6–7 mm. long, the lateral ones ovate and slightly shorter; petals about equaling lateral sepals; lip slender-clawed from near middle of column, erect, reniform, slender-tipped, 4–5 mm. long. — Damp rich woods, often in shell-marl, Fla. to Tex., n. to se. Va. Sept. (W.I.; Centr. Am.; S.Am.)

14. LÍSTERA R. Br. TWAYBLADE

Sepals and petals nearly alike, spreading or reflexed; lip mostly drooping, longer than the sepals, 2-lobed or 2-cleft at the summit. Column wingless. Stigma with a rounded beak; anther borne on the back of the column at the summit, erect, ovate; pollen powdery, in two masses, joined to a minute gland. — Roots fibrous. Stem bearing in the middle a pair of nearly opposite sessile leaves. The small flowers greenish or madder-purple, in a terminal raceme. Small genus of N. Hemisph. (Dedicated to *Martin Lister*, 1638–1711, a celebrated English naturalist.)

*a.*Column very short, at most 0.5 mm. long; lip not dilated above, deeply cleft, with two linear or narrower segments.
Lip with a tooth on each side at base, not auricled, its lobes linear; rachis and pedicels glabrous. 1. *L. cordata.*
Lip without basal teeth, auricled, its lobes setaceous; rachis and pedicels glandular. . . . . 2. *L. australis.*
*a.*Column 2–3 mm. long; lip with broad or dilated summit, shallowly notched or cleft at most one-third its length. . . *b.*
*b.*Lip auricled at base, oblong, ciliate; pedicels and ovary glabrous. . . . *c.*
*c.*Lip cleft one-fourth to one-third its length, with the sinus narrow and with short incurved-clasping basal auricles. . . . . 3. *L. auriculata.*
*c.*Lip merely emarginate with a broad open sinus at tip, the basal auricles divergent and 1.5 mm. long and nearly as broad. . . . . . . 4. *L. borealis.*
*b.*Lip not auricled at base, dilated above.
Leaves oval to round-ovate, 2–6.5 cm. long, mostly longer than peduncle of raceme; pedicels and ovary glandular; lip cuneate, ciliate. . . . 5. *L. convallarioides.*
Leaves reniform-ovate, 1.5–3 cm. long, much shorter than peduncle of raceme; pedicels and ovary glabrous; lip broadly obovate, eciliate. . . 6. *L. Smallii.*

1. L. cordàta (L.) R. Br. (heart-shaped), Heartleaf T. — Very slender, *glabrous except near the leaves,* 0.5–3 dm. high; *leaves* about midway on stem, *cordate- to rounded-ovate,* 1–3 *cm. long, much shorter than peduncle* of *raceme; the latter* 1–7 cm. long, loose and open, *glabrous;* pedicels 2–3 mm. long, glabrous, ascending to spreading, much longer than subtending bract; *flower about* 2 *mm. across,* purplish to greenish or stramineous; *sepals ovate, the upper arched above the very short lip;* petals ovate-lanceolate; *lip lance-linear,* 3–5 *mm. long, deeply divided into* 2 *narrowly linear spreading lobes, the base with a pair of horn-like teeth;* column minute. — Mossy knolls and (southw.) mossy woods, Greenl. and Lab. to Alaska, s. to Nfld., N.S., s. N.E., n. N.J., mts. of Md. and W.Va. to N.C., Mich., Wisc., Minn., N.M. and Oreg. Late May (southw.)–Aug. (northw.). (Eurasia)

2. L. austràlis Lindl. (southern), Southern T. — Habitally similar to no. 1, *stiffer and stouter; leaves firmer and thicker, narrowly ovate,* 1–4 cm. long; *raceme* open, 1.5–10 cm. long, *usually longer than erect peduncle; pedicels glandular, longer* (*up to* 8 *mm.*); flower reddish- or greenish-purple; sepals ovate, wide-spreading; *short petals oblong, recurving; lip* 10 *mm. long, linear, deeply cleft into linear-setaceous porrect lobes, with a small tooth at the sinus* and an inversely T-shaped fold at base; column 1.5 mm. long. (*Ophrys* House) — Damp woods or, northward, in sphagnous thickets and bogs, Fla. to La., n. to s. Va. and Tenn. and very locally to sw. Que., se. Ont. and w. N.Y. Late Apr. (southw.)–early July (northw.).

3. L. auriculàta Wieg. (eared), Auricled T. — Plant erect, 0.4–2.5 dm. high; *leaves oblong-ovate to broadly elliptic or ovate, sometimes suborbicular,* rounded at base, rounded to subacute at apex, 1.5–6.5 cm. long; *peduncle glandular-pubescent, very short; raceme* 2–10 cm. long, *its rachis pubescent; bracts and stoutish short ascending pedicels glabrous;* flowers pale green or greenish-white; sepals lance-ovate; petals narrowly oblong, bluntish, spreading, half as long as lip; *lip broadly oblong, slightly constricted near middle,* 6–8 *mm. long,* watery- or greenish-white, *ciliate, not narrowed at base, there with* 2 *short incurved-clasping auricles, cleft at tip, with slender sinus and porrect blunt lobes;* column about 2.5 mm. long. — Alluvial banks, calcareous silts or crevices, alder-thickets and arbor-vitae swamps, Nfld. to Ung. and n. Ont., s. to Gaspé Pen., Que., n. N.E. and n. N.Y., Mich., Wisc. and Minn. Late June–early Aug.

4. L. boreàlis Morong (northern), Northern T. — Differing from no. 3 in its narrower oblong to narrowly elliptic-oval firmer leaves 1.5–4 cm. long; bracts smaller; the lanceolate sepals and almost linear wide-spreading petals three-fourths length of lip; *lip emarginate at apex, with a large tooth in the broad open sinus, its base bearing a pair of oblong divergent auricles* 1.5 *mm. long and* 1.25 *mm. wide.* — Dryish humus, chiefly beneath dwarf spruce and other shrubs, on limestone, local, nw. Nfld.; Mingan Ids. and Anticosti, Que.; Hudson Bay reg.; Mackenz. to Alaska, s. to mts. of Colo., Ida., and s. B.C. June, July.

5. L. convallarioìdes (Sw.) Nutt. (resembling *Convallaria*), Broad-lipped T. — Relatively stout, 0.7–3 dm. high; *stem densely glandular-pubescent above* the leaves; *leaves broadly oval to subrotund,* pale green, 2–6.5 *cm. long; raceme short-peduncled,* open, 2–13 cm. long, *its rachis, slender pedicels and bracts glandular;* flower rather large; *oblong-lanceolate sepals and narrowly linear petals* 4.5–5 *mm. long, reflexed; lip about* 9 *mm. long,* watery- or whitish-green, *ciliate, cuneate to slender stipe-like base, with retuse apex and round-tipped broad lobes,* the basal fourth bearing on each margin a short triangular tooth; *column slender,* 3 *mm. long.* — Damp peaty or mossy glades, woods, thickets, swamps and shores, Nfld. to Alaska, s. to N.S., n. N.E., mts. to N.C. and Tenn., Mich., Wisc., Minn., Utah, Nev. and Calif. June–Aug.

6. L. Smállii Wieg. (for its discoverer, John Kunkel Small, 1869–1938), Small's T. — Habitally suggesting nos. 1 and 2, *very slender*, 1–3 dm. high; *stem glandular above; leaves firm*, deep green, *reniform-ovate*, 1.5–3 *cm. long, many times shorter than the slender peduncle; raceme* very loose and open, 2–10 cm. long, *its rachis glandular*, the *short bracts and elongate* (*up to* 1 *cm. long*) *filiform pedicels glabrous;* sepals lanceolate; petals lance-linear, a third as long as lip, spreading or reflexed; *lip broadly obovate, eciliate,* 6–9 *mm. long,* the dilated *summit with an open sinus, there bearing a broad oblong to obovate tooth* 1 *mm. long, base of lip sessile,* with a tooth on each side; column thick, 1.5 mm. long. (*L. reniformis* Small, not D. Don; *Ophrys Smallii* (Wieg.) House) — Humus of damp woods and thickets, or in bogs of mountain-reg., Pa. and W.Va., s. to Ga. and e. Tenn. July, Aug.

Tribe IV. Epidéndreae Lindl.

15. CORALLORHÌZA Chatelain Coral-root

Perianth somewhat ringent, gibbous or obscurely spurred at base. Sepals and petals narrow, nearly alike, 1–3-nerved; lateral sepals ascending, forming with the lip the gibbosity or short spur which is mostly adnate to the ovary. Lip slightly adherent to the base of the compressed column. Anther terminal; pollen-masses 4, soft-waxy, free. — Brownish, purplish or yellowish herbs destitute of green foliage, with branched and toothed coral-like underground rhizomes sending up a simple scape which has sheaths in place of leaves and a raceme of colored flowers. Fruit reflexed. Small genus of N. Hemisph. (Name from the Greek *corallion, coral,* and *rhiza, root.*) Corallorrhiza R. Br.

*a.*Lip 3-lobed or with a prominent lateral tooth on each margin.
Plant greenish, greenish-yellow or rarely brownish; flowers yellowish-green to slightly brown-tinged; lip notched on each side toward base, with low subtruncate basal lobes, white and unspotted or rarely dotted with red or purple; capsules greenish; chiefly flowering in spring. . . . . . 1. *C. trifida.*
Plant madder-purple, brown or warm-yellowish; flowers usually spotted with purple or red; lip auricled and with prolonged basal lobes; capsules brown or fulvous; flowering in summer. . . . . 2. *C. maculata.*
*a.*Lip unlobed, entire or merely denticulate. . . *b.*
*b.*Perianth 3–7 mm. long, not obviously striped; lip broadly oval to roundish; capsule 5–12 mm. long. . . *c.*
*c.*Perianth about 7 mm. long; sepals and petals somewhat spreading; lip about 5 mm. long; flowering in spring. . . . . 3. *C. Wisteriana.*
*c.*Perianth 3–4 mm. long; sepals and petals nearly connivent as a hood; lip 2.5–3 mm. long; autumnal. . . . . 4. *C. odontorhiza.*
*b.*Perianth 0.8–2 cm. long; sepals and petals conspicuously purple-veined; lip tongue-shaped; capsule 1.5–2 cm. long. . . . . 5. *C. striata.*

1. C. trífida Chatelain (3-parted; from the 3-lobed lip), Early or Pale C. — *Rhizome whitish,* with somewhat intricate branching, forming flattened platforms; *scape* erect, 0.8–3 dm. high, with 3 or 4 close sheaths toward base, 2–4 *mm. thick at base;* raceme 2–10-flowered, up to 8 cm. long; pedicels very short, at first ascending, later reflexed; *sepals* narrowly oblong to ligulate, bluntish, 4.5–6 *mm. long, greenish-yellow, often brown-tinged,* the lateral oblique; *petals* similar, *often red-dotted; lip* nearly equaling petals, *oblong, subtruncate or broadly rounded and notched at tip, with low basal lobes, whitish, with many red or purplish dots.* — Dryish tundra, thickets, bogs and woods, Greenl. and Lab. to Alaska, s. to n. Nfld., Côte Nord, Que., Ung., n. Ont., and mts. to Wyo. and Oreg. (Eurasia) — With us mostly as

Var. **vérna** (Nutt.) Fern. (vernal). — *Plant pale yellow or yellowish-green; sepals yellow-green,* at most brownish-tipped, *these and the similar petals linear-lanceolate, unspotted; lip white, undotted or with very few basal reddish dots,* 3–4 *mm. long, with an abruptly narrowed recurved tip; capsules greenish, becoming drab,* 6–10 *mm. long.* (*C. trifida* of ed. 7) — Damp woods, thickets and swamps, Nfld. to B.C., s. to N.S., N.E., N.J., Pa., mts. to Ga. and Tenn., O., n. Ind., Wisc., Mo., S.D., Colo. and Oreg. May–July. — Scapes solitary to densely clustered.

2. C. maculàta Raf. (mottled or spotted), Spotted C. — Usually *coarser* than no. 1; *rhizome brownish; scape* (with raceme) *up to* 6 dm. high and 1 *cm. thick at base, yellow to warm brown, fuscous or purple-brown,* the similarly colored sheaths extending well above the middle; *raceme* few–35- or more-flowered, 0.4–2 *dm. long; perianth* whitish, *usually spotted with red or purple,* 5–10 *mm. long; lip with* 2 *basal auricles and* 2 *prolonged basal lobes; capsules brown or fulvous,* 1–2 *cm. long.* — Dry woods, Nfld. to B.C., s. to N.S., N.E., L.I., Va., upland of N.C. and Tenn., O., Ind., Wisc., Minn., S.D., Colo. and Calif. Late June–Aug. — Fresh plants occur in the following color-forms (until dry): typical plant with scapes, sheaths and perianth yellow, the lip spotted; forma **flávida** (Peck) Farw. (yellowish) similar but lip unspotted; forma **intermèdia** Farw. (intermediate) with scape and sheaths dull brown or fuscous, perianth yellow-brown, lip spotted; forma **punícea** (Bartlett) Weath. & Adams (reddish-purple) the scape, sheaths and perianths reddish-purple, lip spotted.

3. C. Wisteriàna Conrad (for its discoverer, CHARLES JONES WISTER, 1782–1865), WISTER'S C. — Coarser than no. 1, 1–4.5 dm. high; *scape* somewhat bulbous at base, from a small coralloid rhizome, *tinged with red or purple,* commonly flexuous toward summit, with paler striate sheaths below; *raceme lax and open,* 2–13 cm. long; *flowers soon nearly horizontal,* finally somewhat reflexed; *perianth* 6–8 *mm. long, the reddish- or purplish-backed* linear *sepals somewhat spreading, their paler inner faces flecked with purple lines;* petals pale, slightly shorter and broader, purple-flecked; *lip* 5 *mm. long, deflexed, roundish, with basal claw, unlobed, white, with purplish dots, emarginate; capsule* 8–12 *mm. long.* — Rich, chiefly deciduous woodland, Fla. to e. Tex., n. (becoming very local) to N.J., Pa., W.Va., O., s. Ind., s. Ill., Mo. and S.D. Late March–May.

4. C. odontorhìza (Willd.) Nutt. (tooth-rooted), LATE or AUTUMN C. — Similar to no. 3; smaller throughout, 1–3 dm. high; scape often greener, slender; *raceme* 3–20-flowered, 1.5–7 *cm. long; perianth* 3–4 *mm. long; sepals and petals approximate as a hood, subequal; lip* white, 2.5–3 *mm. long, crinkly- or erose-margined,* purple-rimmed and -dotted; *capsules* slender-pedicelled, pendulous, 5–8 *mm. long.* — Dry woodlands, sw. Me. to Minn., s. to Ga., Ala., Miss. and Mo. Aug.–early Oct. — Forma **flávida** Wherry (yellowish) has scape, sheaths and perianth yellow (without the usual purple tones) and the lip unspotted.

5. C. striàta Lindl. (striped), STRIPED C. — As coarse as no. 2, 1.5–5 dm. high; stoutish scape and sheaths madder-purple, or in forma **fúlva** Fern. (yellowish-brown) yellowish to orange-brown as also the perianths; *raceme* 0.3–2 *dm. long;* flowers subapproximate, often very numerous; *perianth usually purplish,* 0.8–2 *cm. long, the lance-ovate translucent conspicuously purple-veined sepals and petals loosely approximate and forming a broad arching hood; lip tongue-shaped, usually madder-purple* (yellow-brown in forma *fulva*), striped only at base, *about as long as other petals,* abruptly drooping; *capsules* strongly reflexed, 1.5–2 *cm. long.* — Calcareous or rich woods, Gaspé Pen., Que.; sw. Que. to w. Ont., s. to nw. N.Y., s. Ont., Mich., ne. and n. Wisc. and ne. Minn.; s. Alta. and s. B.C., s. to Wyo., Ida., and Calif. Late May–Aug.

16. MALÁXIS Sw. MALAXIS. ADDER'S-MOUTH

Sepals lanceolate, oblong or ovate, spreading. Petals lanceolate, filiform or linear, spreading. Lip auricled at base, narrowing toward the summit, entire or cleft. Column very small, terete, with 2 teeth or auricles at the summit and the erect anther between them; pollen-masses 4, in one row (2 in each anther-locule), cohering in pairs, waxy, without stalks, filaments, or gland. — Low herbs of Am. and Eurasia from solid tubers producing simple stems which bear 1 or few leaves and a raceme of tiny mostly greenish flowers. (*Malacos, weak* or *delicate,* from the frail character of no. 1.) Including MICROSTYLIS (Nutt.) Eat.

*a.*Leaves 2–5, alternate, basal, subtending axillary new tubers; sepals narrowly ovate, 2 of them erect, the 3rd drooping; petals half as long, lance-ovate, blunt; lip ovate, conspicuously green-nerved, erect, much shorter than lateral sepals; capsule crowned by the persistent porrect perianth. . . . 1. *M. paludosa.*

*a.*Leaves solitary or 2(–3), subbasal or high on stem, without axillary tubers; sepals lanceolate to ovate; petals linear to thread-like; lip not conspicuously veined, with entire tip or 2-cleft. . . *b.*

*b.*Leaf 1 (rarely 2); flower greenish to yellowish. . . *c.*

*c.*Lip drooping, entire, abruptly long-pointed. . . . . 2. *M. brachypoda.*

*c.*Lip finally ascending, 2-lobed at summit, with a median tooth.

Raceme oblong-cylindric, in anthesis 1.3–2.5 cm. thick; pedicels 4–10 mm. long; lip oblong-oval, shallowly cordate at base, the 2 lateral apical lanceolate lobes much prolonged, the central tooth minute. . 3. *M. unifolia.*

Raceme slenderly cylindric, in anthesis 5–10 mm. thick; pedicels 2–4.5 mm. long; lip broadly cordate-deltoid, the 2 lateral apical lobes deltoid and shorter than the broad basal ones, the median tooth triangular. . . . . 4. *M. Bayardi.*

*b.*Leaves 2 (rarely 3); lip orange to vermilion, erect, entire, cordate-ovate, the basal auricles broadly rounded, the tip not prolonged. . . . . 5. *M. floridana.*

1. M. paludòsa (L.) Sw. (boggy), BOG-M. or A. — Plant very slender, 0.4–2.5 dm. high; *slender rhizome surmounted by* 2–5 *alternate* oblanceolate or narrowly obovate *leaves* 0.5–3 cm. long, *these eventually subtending* small *tubers;* scape filiform; *raceme very slender,* 1–13 cm. long, *with* short-pedicelled ascending *yellowish-green tiny flowers; sepals narrowly ovate,* 2 *of them erect, the 3rd drooping; petals lance-ovate, blunt, half as long as sepals; lip ovate, erect, much shorter than lateral sepals, conspicuously green-nerved; capsule crowned by the porrect persistent perianth.* — In wet sphagnum, n. Minn. and adj. Ont.; Alaska. July, Aug. (Eurasia)

2. M. brachýpoda (Gray) Fern. (short-pedicelled). — Plant slender, with a plump ovoid corm, 0.5–2.5 dm. high; leaf solitary, subbasal to definitely cauline, oblong to broadly elliptic-oval, 1.7–9 cm. long, or in forma **bifòlia** (Mousley) Fern. (two-leaved) a smaller second leaf below the usual one; *raceme slender,* 2–12 cm. long, *in anthesis* 5–8 *mm. thick; flower-bud just before expansion ovate,* 1.5–2 *mm. long; pedicel and ovary during anthesis* 1.5–3 *mm. long; expanded perianth yellowish-green,* 3–5 *mm. broad,* its segments strongly divergent, *perpendicular to axis of ovary, finally reflexed and appressed to capsule; lip drooping, entire,* ovate, *with slender pointed tip.* (*M. monophyllos* sensu most Am. auth., not (L.) Sw.; *Microstylis monophyllos* of ed. 7, not (L.) Lindl.) — Damp calcareous gravels, talus, peats, swales and bogs, s. Lab. Pen. to Man., s. to Nfld., M.I., N.B., n. and w. N.E., n. N.J., Pa., mts. to Tenn., n. Ind., Wisc., Minn., and mts. to Tex., Colo. and Calif. June–early Aug.

3. M. unifòlia Michx. (one-leaved), Green A. — Habitally similar to no. 2, 0.6–3.5 dm. high; *darker green leaf* usually about midway on scape, or in forma **bifòlia** Mousley (two-leaved) a secondary leaf above the usual one; *raceme oblong-cylindric,* 1–10 cm. long, *in full anthesis* 1.3–2.5 *cm. thick,* the crowded and unexpanded summit broad and truncate, *the mature divergent pedicels* 4–10 *mm. long;* flowers green or greenish; *lower sepals ovate, spreading, upper one oblong-linear, suberect;* petals thread-like, reflexed; *lip* at first drooping but by torsion of the pedicel *becoming erect, oblong-oval, shallowly cordate at base, the* 2 *lateral apical lobes lanceolate and prolonged, thrice as long as broad, with a minute central tooth.* (*Microstylis* BSP.) — Dry or moist woods, borders of swamps or bogs or on gravelly slopes, Nfld. to Sask., s. to N.S., N.E., Fla., Ga., Ala., La. and Tex. Late May (southw.) – Aug. (northw.). Fig. 856.

856. M. unifolia.

4. M. Bayárdi Fern. (for its discoverer, Bayard Long, 1885–). — Similar to no. 3, smaller throughout, 1.2–2.7 dm. high; *raceme linear-cylindric, in anthesis* 1.5–12 cm. long and 5–10 *mm. thick; pedicels in anthesis ascending,* 2–4.5 *mm. long;* flowers greenish, 2.5–3.5 mm. long, petals linear; *lip broadly cordate-deltoid,* 2.5 *mm. long, nearly as broad, soon erect, its small basal lobes subdivergent and* 1–1.4 *mm. long, the* 2 *lateral and terminal deltoid blunt lobes only* 0.4–0.6 *mm. long, with the triangular median tooth* 0.2–0.3 *mm. long.* — Dry sandy woods and clearings, local, Pa. to N.C. July–Sept. Fig. 857.

5. M. floridàna (Chapm.) Ktze. (of Florida), Florida M. or A. — *Two* (rarely 1 or 3)*-leaved from near the base to below middle of scape,* 1–3.5 dm. high; *leaves* subopposite or remote, unequal, elliptic-oblong to broadly oval; *the lower* (usually larger) *one narrowed to a petiolar base,* 2.5–8.5 cm. long; scale prolonged; raceme lax, 3–15 cm. long; pedicels slender, elongate; the 2 lateral oblong and the median lanceolate somewhat porrect *sepals greenish;* petals filiform, green, reflexed; *lip broadly cordate, the subacute entire tip not prolonged, the narrow basal auricles broadly rounded, the shield orange to vermilion,* the margin paler and hyaline. (Often confused with the larger green-lipped West Indian *M. spicata* Sw., with basal auricles of lip prolonged and subtruncate) — Rich, calcareous or marly woods, slopes and shores, Fla. to S.C.; se. Va. Aug., early Sept.

857. M. Bayardi.

17. LÍPARIS Richard Twayblade

Sepals oblong-lanceolate. Petals linear or filiform. Lip entire. Column 2–3 mm. long, curved, stout at base, with narrow wings above; anther terminal, operculate; pollen-masses 4 (2 in each anther-locule), slightly united in pairs, without stalks, filaments, or gland. — Low herbs of trop. and temp. reg., with solid bulbs or tubers, a pair of basal leaves, a low scape and a few-flowered raceme. (Name from the Greek *liparos, fat* or *shining,* from the smooth and lustrous leaves.)

1. L. lilifòlia (L.) Richard (with leaves of *Lilia,* an old group including *Erythronium* and *Convallaria*), Lilia-leaved T. — Plant 0.6–3 dm. high; *leaves* elliptical or narrowly ovate, *green,* lustrous; scape angled; raceme of 5–40 *pale mauve-purple and greenish,* or in forma **viridiflòra** Wadmond (green-flowered) pale green throughout, *flowers on filiform widely divergent* (in fruit ascending) *pedicels as long as the flowers; the flower about* 1 *cm. across;* sepals oblong-lanceolate but usually tightly involute or convolute, thus appearing thread-like, similar, recurved; petals pendent, usually madder-purple; *lip broadly cuneate-obovate, translucent, usually madder-purple,* 7–10 *mm. long,* with clasping auricles at base; *capsules* clavate, erect *on long pedicels.* (*L. liliifolia* (Sw.) Lindl.) — Loamy or sandy woods and clearings, s. N.H. to s. Minn., s. to s. N.E., Ga., Ala., Tenn. and Mo. May–early July.

2. L. Loesélii (L.) Richard (for Johann Loesel, 1607–1655), Loesel's, Bog- or Yellow T. — Smaller than no. 1, with lanceolate to lance-ovate *yellow-green* strongly keeled *leaves;* flowering stem (and raceme) 0.5–3 dm. high; *flowers* 2–25, *about* 5 *mm. broad, yellowish-green, on short ascending pedicels; lip concave, oblong- to obovate-spatulate, about* 5 *mm. long,* arching; *capsules longer than their pedicels.* — Bogs, peaty meadows and damp thickets, Gaspé Pen., Que. to Sask., s. to N.S., N.E., N.J., Pa., Md., upland to Ala., O., n. Ind., n. Ill., e. Mo. and N.D. June, July. (Eu.)

18. APLÉCTRUM (Nutt.) Torr. Putty-root. Adam-and-Eve

Perianth neither gibbous nor with any trace of a spur or sac at base. Lip free, 3-lobed, with three longitudinal crests. Column compressed; pollen-masses 4. — Scape from near the summit of a globular corm-like tuber. Leaf solitary; petiole distinct. The slender naked rootstock producing each year a globular solid tuber or corm, often 2.5 cm. in diameter (filled with exceedingly glutinous matter), which sends up late in summer a large oval many-nerved plaited leaf lasting through the winter; early in the succeeding summer the scape appears, terminated by a loose raceme of lurid flowers. A single species. (Name from *a-privative, without,* and the Greek *plectron, spur.*)

1. A. hyemàle (Muhl.) Torr. (of winter; from the conspicuous evergreen leaf). — Rootstock like a coarse chain of 2–4 large biennial corm-like tubers separated by intermediate slender strands; leaf firm, blue-green, corrugate-striate, the blade 0.7–2 dm. long; scape stout, 3–4.5 dm. high, with remote sheathing bracts, the leaf often shriveled at flowering time; flowers 8–20, about 1 cm. long, divergent, purplish-green or madder-purple, or in forma **pállidum** House (pale) pale yellowish throughout; sepals long, greenish or yellowish, tinged with madder-purple; petals shorter, arching over the column, oblong, obtuse, yellowish, tinged with madder-purple above; lip white or nearly so, sparingly marked with magenta; plump capsules reflexed. — Rich woods, now becoming rare, sw. Que. and Vt. to Sask., s. to s. N.E., Ga., Tenn. and Ark. May, June.

19. CALÝPSO Salisb.

Sepals and petals similar, ascending, spreading, oblong-lanceolate, acute, magenta-crimson, rarely white. Lip larger than the rest of the flower, saccate, with three longitudinal rows of yellow (or white) glass-like hairs in front and with a translucent apron-like appendage (formed by the overlapping of the lip) spotted with madder-purple, the sac (bearing two conspicuous horns at its base) whitish, with irregular usually purple-madder markings. Column winged, having the operculate anther just below the apex; pollen-masses waxy, 2, each 2-parted, all sessile on a square gland. — A single circumboreal species with solitary basal leaf and a 1-flowered scape; sometimes with a coralline rhizome below the tuber. (Named for the goddess *Calypso.*)

1. C. bulbòsa (L.) Oakes (with a bulb). — Plant 0.5–2 dm. high; tuber superficial; leaf oval to round-ovate, veiny, bluish-green, with undulate margin, 2–6 cm. long, developing in autumn, overwintering, and shriveling soon after flowering season, the slender petiole 3-angled; scape smooth, with membranous sheathing bracts, it and the leaf from separate buds; pedicel of flower subtended by a petaloid bract; lip resembling a sugar-scoop, 2–2.5 cm. long, marked with purple-madder, or white in forma **cándida** Hylander (white); erect capsule tipped by the marcescent perianth. (*C. borealis* Salisb.; *Cytherea bulbosa* House) — Cool, mossy woods, chiefly calcareous, often about bases of *Thuja,* Lab. to Alaska, s. (now very locally) to w. Nfld., N.S., n. N.E., n. N.Y., Mich., Wisc., Minn., Ariz. and Calif. Mid-May–early July. (Eurasia)

20. TIPULÀRIA Nutt. Cranefly Orchis

Flowers greenish, tinged with madder-purple, numerous in an elongated loose bractless raceme. Sepals oblong-oval, obtuse, upper sepal narrower. Petals oblong, obtuse. Lip with a slender spur, 3-lobed; lateral lobes obtuse, obscurely toothed; apical lobes broad at base, margin deflexed at the middle, apex expanded. Column wingless; anther operculate, terminal; pollen-masses 2, waxy, each 2-parted, connected by a linear stalk with the transverse small gland. — Tubers connected in a horizontal series, producing in autumn a single ovate slender-petioled nerved and plaited leaf, purplish beneath, and in summer a long slender scape. Two species: ours, and another Himalayan. (Name from fancied resemblance of flowers to insects of the genus *Tipula*.)

1. T. díscolor (Pursh) Nutt. (two-colored, from the contrast of upper and lower leaf-surfaces). — Leaf 5–13 cm. long, disappearing before anthesis; filiform scape 2–4 dm. high; raceme 1–3 cm. long; spur about 2 cm. long, twice as long as ovary; capsules slender, reflexed. — Hardwood-forest, Fla. to e. Tex., n. to Va., and locally to L.I. and se. Mass., w. N.Y., O. and s. Ind. July, Aug.

21. HEXALÉCTRIS Raf.

Sepals and petals nearly equal, free, somewhat spreading, several-nerved; perianth not gibbous nor spurred at base. Lip obovate, 3-lobed, with 5 or 6 prominent ridges down the middle, the middle lobe somewhat concave. Pollen-masses 8, united into a single fascicle. — Leafless plants with stout or somewhat coralline annulated rhizomes. A single species. (Name probably from the Greek *hex, six,* and *alectryon, cock,* from the crest of the lip.)

1. H. spicàta (Walt.) Barnh. (spicate), Crested Coral-root. — Plant 2–9 dm. high, with stout sheathing purplish scales; flowers in a raceme up to 3 dm. long, bractless, madder-purple, about 2 cm. long; sepals narrowly oval, obtuse; petals similar, shorter. (*H. aphylla* (Nutt.) Raf.) — Dry woods, Fla. to Ariz. and n. Mex., n., locally, to Md., n. Va., W.Va., s. O., s. Ind. and Mo. June–Aug.

Class II. DICOTYLEDÒNEAE, Dicotyledons

Subclass I. ARCHICHLAMÝDEAE (Choripétalae, Apétalae and Polypétalae)

FAM. 40. SAURURÀCEAE (Lizard's-tail Family)

Flowers perfect or dioecious, in dense spikes, without perianth, with 3–4 *indehiscent* 1-*seeded carpels free or united at base; ovule free, orthotropous; fruit dry; stamens* 2–6 *or more, hypogynous.* — Succulent herbs with jointed stems, alternate entire leaves and spikes of flowers naked or subtended by an involucre; N. Am. and Asiatic.

1. SAURÙRUS L. Lizard's-tail

Stamens mostly 6 or 7, hypogynous, with distinct filaments. Fruit somewhat fleshy, wrinkled, of 3–4 indehiscent carpels united at base. Stigmas recurved. Seeds usually solitary, ascending.— Perennial paludal herbs, one in e. N. Am., one in e. Asia, with cordate converging-ribbed petioled leaves, without distinct stipules; flowers (each with a small bract adnate to or borne on the pedicel) crowded in a slender virgate and naked-peduncled terminal spike or raceme (its appearance giving rise to the name from the Greek *sauros, a lizard,* and *oura, tail*).

1. S. cérnuus L. (nodding), Water-dragon, Swamp-lily. — Extensively creeping, stoloniferous, the rhizome aromatic; flowering stems 4–9 dm. high, naked below, simple or forking; spike 1–3 dm. long, peduncled, nodding at tip; flowers white; filaments long and capillary. — Swamps and shallow water, Fla. to Tex., n. to R.I., Ct., sw. Que., N.Y., s. Ont., s. Mich., Ill., Mo. and se. Kans. June–Sept.

FAM. 41. SALICÀCEAE (Willow Family)

Dioecious (or by exception monoecious) *trees or shrubs with both kinds of flowers in aments, one to each bract* (scale), *without perianth; the fruit a* 1-*locular and* 2–4-*valved capsule, with* 2–4 *parietal or basal placentae, bearing numerous seeds furnished with long silky down.* — Stigmas 2–4, often 2-lobed. Seeds ascending, anatropous, without albumen. Cotyledons flattened. Leaves alternate, undivided, with scale-like and deciduous or else leaf-like and persistent stipules. Wood soft and light; bark bitter.

Aments ascending or divergent, rarely drooping; bracts entire or merely toothed; each flower with 1 to 4 basal glands; stamens 1–12; style 1 (or none) with 2 simple or bifid stigmas; buds with 1 scale. . . . . 1. *Salix.*

Aments soon arching or pendulous; bracts mostly lacerate; each flower with a symmetrical or oblique basal cup-like disk; stamens 8–30 or more; stigmas dilated or prolonged, 2–4; buds of several imbricated scales. . . . . . . 2. *Populus.*

1. SÀLIX L. Willow. Osier. Saule (Que.)

Staminate flowers of (1–)2 or 3–12 distinct or united stamens, with 1 or 2(–4) glands at base of pedicel. Pistillate flowers with 1 to 4 basal glands; style 1 or none, the 2 stigmas simple or 2-cleft. — Trees or shrubs of cold to warm-temp. reg., with mostly terete branches. Leaves from narrowly linear to orbicular. Buds covered by a single scale, with an inner usually adherent membrane. Aments expanding before (*precocious*), with (*coaetaneous*) or after (*serotinous*) the leaves. (The classical Latin name.)* — Species largely insect-pollinated and very freely hybridizing.

Synoptic Key to Sections and Species

A. Bracts of ament yellowish to whitish, falling before maturing of capsule; aments borne on short lateral leafy branchlets, expanding with or after the leaves; stamens 2–12, their filaments free and hairy toward base; style short or obsolete, the stigmas emarginate or bifid. . . *B.*

B. Stamens 3–12. . *C.*

C. Petioles glandless at summit; flowers in more or less definite whorls, the whorls of the staminate aments remote, of the pistillate distant to approximate; torus of staminate flowers with 2 glands, of pistillate mostly with 1; filaments mostly arching. . *D.*

D. Leaves green both sides; glands of staminate flowers free; branchlets brittle at base. § 1. Nigrae. . . . . 1. *S. nigra.*

D. Leaves whitened or glaucous beneath.

Branchlets red- or purple-tinged, brittle at base; petioles 3–7 mm. long; stipules large, roundish, persistent; glands of staminate flower lobulate or forming a false disk, the gland of pistillate flower clasping base of pedicel; capsule granulose-roughened. § 2. Bonplandianae. . . . . 2. *S. caroliniana.*

Branchlets yellowish, tenacious and flexible; petioles mostly 1–3 cm. long; stipules minute, promptly deciduous; glands of staminate and pistillate flowers elongate, free; capsule smooth. § 3. Triandrae. . . . . 3. *S. amygdaloides.*

C. Petiole bearing heavy glands or gland-tipped teeth at summit or at junction with blade; flowers spirally arranged in dense to lax aments; 2 glands of staminate flower more or less united and forming a lobed false disk, 2 glands of pistillate flower free; filaments straight. § 4. Pentandrae.

Leaves green and lustrous on both surfaces, acuminate or with prolonged subcaudate tip; staminate aments 2–6 cm. long, pistillate becoming 2–5 cm. long; capsule 5–6.5 mm. long, the firm valves opening in late spring or early summer.

Capsule conic-subulate, its pedicel twice as long as upper gland; mature leaves short-acuminate; introd. . . . . . 4. *S. pentandra.*

Capsule conic-ovoid, its pedicel about four times as long as upper gland; mature leaves with long-attenuate tail-like tip; native. 5. *S. lucida.*

Leaves pale or whitened beneath, merely acuminate; staminate aments 1–1.5 cm. long, pistillate 2–3.5 cm. long; capsule conic-subulate, 7–10 mm. long, its indurated valves often not opening until late summer or autumn; pedicel twice length of upper gland. . . . 6. *S. serissima.*

B. Stamens 2. . . *E.*

E. Leaves petioled, lanceolate to narrowly ovate, long-acuminate, serrate; aments terminating scattered short leafy shoots; large trees; introd. . . *F.*

F. Young petioles more or less glandular or viscid at summit; pistillate flowers with 2 glands; capsules subulate. § 5. Fragiles.

Leaves green both sides, glabrous from the first, in maturity 2.5–4 cm. broad, coarsely undulate-serrate; capsules about 5 mm. long. 7. *S. fragilis.*

Leaves pale beneath, silky when young, glabrate, 0.5–1.5 cm. broad, finely repand-serrate; capsules 1–1.5 mm. long. . . . . . 8. *S. babylonica.*

F. Young petioles glandless; pistillate flowers with 1 gland; capsule conic-ovoid, 3–5 mm. long. § 6. Albae. . . . . 9. *S. alba.*

E. Leaves linear to oblong-lanceolate, short-acuminate to merely subacute,

---

*In constructing keys much help has been gained from the critical studies of Camillo Schneider summarized in Journ. Arn. Arb. iii. 61–125 (1921), and by Hugh M. Raup in Sargentia, iv. 82–127 (1943).

sessile or nearly so, remotely denticulate with projecting teeth; aments often clustered at ends of crowded branchlets; stoloniferous shrubs or small trees. § 7. LONGIFOLIAE. . . . . 10. *S. interior*.

A. Bracts of ament yellow or brown to fuscous, blackish or black-tipped, persistent, present (marcescent) until shriveling of ament; stamens 2 (or 1). . . *G*.

G. Stamens with distinct filaments. . . *H*.

H. Aments falsely terminal, peduncled, borne just below tips of branches on side of stem opposite terminal leaf; glands of both staminate and pistillate flowers 2–4, often forming a lobed false disk; ovary sessile; leaves entire or very finely toothed, commonly with revolute margins, coriaceous, the round-tipped orbicular to elliptic or obovate blade reticulate, with nerves impressed into dark green upper surface, whitish beneath; boreal species, calcicolous. § 8. RETICULATAE.

Bracts and ovaries pubescent.

Leaves glabrous or promptly glabrescent, conspicuously reticulate beneath; buds glabrous and glutinous; anthers yellow to purplish; prostrate in mats, the branches often rooting.

Petioles twice to several times longer than buds; fruiting aments 1–3 cm. long, long-peduncled; bracts yellowish-brown to reddish; circumpolar species. . . . . 11. *S. reticulata*.

Petioles shorter than to but slightly longer than buds; fruiting aments 0.6–1.8 cm. long, short-peduncled; bracts fuscous to dark purple; local, along Straits of Belle Isle. . . . . . 12. *S. jejuna*.

Leaves appressed-villous beneath with dense lustrous white pubescence, reticulations thus hidden; petioles many times shorter than heavy blades; buds closely pubescent; anthers yellow; stout-branched ascending (rarely prostrate) low shrub. . . . . . 13. *S. vestita*.

Bracts, ovaries and lustrous buds glabrous; leaves glabrescent; depressed small shrub with rooting trunks and branches; local, w. Nfld. 14. *S. leiolepis*.

H. Aments terminal or subterminal or from axils of preceding year. . . *I*.

I. Staminate flowers with 2 basal glands; pistillate with 2 glands or only 1. . . *J*.

J. Trunks and main branches subterranean, stoloniferous and rooting at nodes, filiform; the very short and slender upper branchlets above ground, bearing 2–4 glabrous reticulate rounded slender-petioled leaves and subterminal 2–8-flowered tiny aments. § 9. HERBACEAE. . . . . 15. *S. herbacea*.

J. Trunks and main branches above ground, stouter, ligneous; branchlets with several leaves; aments several–many-flowered, terminating branches or subtended by leafy bracts. . *K*.

K. Stamen 1; ovary glabrous; bracts heavily bearded; depressed shrub with leaves lustrous above, pale beneath. § 10. UVA-URSI. 16. *S. Uva-ursi*.

K. Stamens 2; ovary pubescent, at least at base, or glabrous. . . *L*.

L. Bracts fuscous or blackish or with blackish tips, long-bearded toward apex; ovary sessile or subsessile; glands short, rarely twice as long as thick; branches prostrate. § 11. OVALIFOLIAE.

Branchlets brittle on drying; leaves dark green and brilliantly lustrous above; aments erect, perpendicular to the prostrate axis of branchlet, the pistillate becoming 3–10 cm. long; capsules lance-conic, 7–10 mm. long. . . . . 17. *S. arctophila*.

Branches flexible; leaves pale green, opaque or barely lustrous above; aments merely ascending, 1–5.5 cm. long; capsules ovoid-conic, 5–8 mm. long (in our vars.). . . . . . . 18. *S. arctica*.

L. Bracts uniformly colored, yellowish to fulvous, neither black-tipped nor fuscous. . . *M*.

M. Ovary sessile or very short-pedicelled, villous, tomentose or glabrous, the pedicel much shorter than glands, these many times longer than thick; expanding leaves villous, tomentose or glabrous, not silvery-silky. § 12. GLAUCAE.

Bracts and ovary pubescent, the bracts yellowish or pale brown.

Leaves oblong or barely oblong-obovate, ascending to suberect and overlapping, 1–3.5 cm. long, 0.5–1.5 cm. broad, on thick petioles shorter than well developed axillary buds; aments in anthesis subglobose to short-ellipsoid, lengthening to 0.5–2.7 cm. long; bracts pilose but usually not long-bearded; anthers ellipsoid-globose. . . . . 19. *S. brachycarpa*.

Leaves elliptic to ovate, obovate or suborbicular, divergent or loosely ascending, rarely strongly imbricated; those of fertile branches 1–9 cm. long and 0.5–5 cm. broad, on slender petioles longer than axillary buds; aments cylindric, 1–7 cm. long; bracts long-villous, usually long-bearded; anthers oblong-ellipsoid. . . . 20. *S. cordifolia.*

Bracts and ovary glabrous, the bracts olivaceous; glabrous or promptly glabrate leaves and aments shaped like those of no. 19. . . . . 21. *S. chlorolepis.*

M. Ovary long-pedicelled, with pedicel three or four times length of short glands, minutely silvery-silky; expanding leaves silvery-silky beneath. § 13. ARGYROCARPAE. . . . . 22. *S. argyrocarpa.*

I. Staminate and pistillate flowers each with 1 ventral gland at base. . . *N.*

N. Aments coetaneous (flowering as leaves begin to unfold) or serotinous (flowering after leaves are partly grown), usually subtended by leafy bracts or foliage-leaves. . . *O.*

O. Foliage and bark of young growth (when dry) with strong and long-persistent balsamic-resiniferous fragrance; leaves slender-petioled, short-oval to oblong-lanceolate, membranaceous, subpellucid, prominently reticulate beneath; stipules 0 or minute and caducous; bracts whitish, thin and elongate; pedicel becoming six to eight times length of broad gland. § 14. BALSAMIFERAE. 23. *S. pyrifolia.*

O. Foliage not strongly balsamic. . . *P.*

P. Vigorous vegetative shoots often bearing persistent toothed stipules; leaves mostly toothed (except in the thick-leaved § *Chrysanthae*); tall or, if low, stout-branched shrubs, not obviously stoloniferous. . . *Q.*

Q. Aments few–many along the slender and elongate branches of preceding year; leaves mostly with numerous teeth; overwintering buds on new shoots axillary; styles 0–1.5 mm. long. . . *R.*

R. Aments usually flowering from apex to base, the pistillate ones remaining compact; their bracts brown to blackish, strongly bearded or villous; fruiting pedicels very short to thrice length of subtending bracts; ovary and capsule glabrous (tomentose when young in no. 30, with dark brown long-villous bracts), the style usually definite. § 15. CORDATAE. . . *S.*

S. Leaves sharply acute or tapering to acuminate tip. . . *T.*

T. Ovaries and capsules glabrous. . . *U.*

U. Flowering (2nd-year) branches with brown or darker bark, or, if occasionally yellow, the species northeastern. . . *V.*

V. Leaves gradually tapering to acuminate tip, lanceolate to oblong-ovate (rarely oblanceolate or obovate), each margin with 25–90 forward-arching to depressed or low and rounded callous-tipped or -margined (or when young gland-tipped) simple teeth; petioles (2–) 5–35 mm. long; bracts of ament oblong to obovate, deep brown or reddish to blackish; style 0.2–1.2 mm. long. . . *W.*

W. Leaves merely paler- or gray-green, not glaucous or whitened beneath; styles 0.2–0.5 mm. long; capsules 4–7 mm. long.

Leaves broadly oblong-lanceolate to ovate, mostly strongly cordate, the mature ones one-fifth to two-thirds as broad as long; pistillate aments dense, in anthesis with appressed-ascending ovaries, the crowded capsules on very short pedicels shorter than to barely exceeding bracts; boreal species. 24. *S. cordata.*

Leaves oblong-lanceolate, subcordate, rounded or tapering at base, mature blades one-eighth to one-third as broad as long; pistillate aments in anthesis with widely (often horizontally) divergent ovaries, the widely divergent capsules on pedicels as long as to much longer than bracts; wide-ranging in temperate areas. . . . . 25. *S. rigida.*

W. Leaves strongly whitened or glaucous beneath; styles 0.6–1.2 mm. long; capsules 6–11 mm. long.

Leading shoots glabrous or rarely pubescent; leaves oblong to ovate, sublustrous above, glabrous or glabrate, subacuminate, not slender-tipped; fruiting aments 2–10 cm. long; northeastern boreal species. . . 26. *S. glaucophylloides.*

Leading shoots densely and permanently puberulent to tomentose; leaves narrowly lanceolate to ovate-oblong, dull green above, permanently pilose (at least on nerves) beneath, the tip caudate-attenuate; fruiting aments 6–10 cm. long; western and southwestern. . . . . 27. *S. eriocephala.*

V. Leaves acute or abruptly short-acuminate from well above the middle, oblong-ovate, each margin with 81–137 horizontally or subhorizontally divergent and permanently gland-tipped prolonged often compound teeth; mature petiole 2–10 (–15) mm. long; bracts of aments oblong, pale brown, long-bearded; style 0.7–1.5 mm. long. . . . . 28. *S. syrticola.*

U. Flowering branches with yellow or yellowish bark, becoming gray; petioles yellowish; leaves yellow-green, glaucous beneath, acuminate; fruiting aments 2–5 cm. long, dense; capsules 4–5 mm. long; style 0.2–0.5 mm. long; northwestern. . 29. *S. lutea.*

T. Ovaries and at least bases of capsules tomentose; new shoots and young leaves densely tomentulose; leaves (except on sprouts) not cordate; fruiting aments 4–10 cm. long; northeastern, boreal. . . . 30. *S. laurentiana.*

S. Leaves blunt or rounded at tip, those of 2nd-year branches 2–6 cm. long.

Leaves dark green and lustrous, mostly darkened in drying, tapering to blunt tips, mostly without stipules; fruiting aments 2–3 cm. long; boreal transcontinental species. . . . 31. *S. myrtillifolia.*

Leaves pale green, opaque, scarcely darkening, broadly rounded above, mostly subtended by persistent stipules; fruiting aments 0.5–2 cm. long; local shrub of Que. . . . . 32. *S. obtusata.*

R. Aments flowering from base to apex, the fruiting ones becoming very lax and open; bracts yellowish-brown, sparsely pilose; fruiting pedicel three or four times length of subtending bract; ovary silky; style obsolete or nearly so. § 16. Fulvae. . . . . 33. *S. Bebbiana.*

Q. Aments 1–4, when more than 1 approximate at summits of stout branches of preceding year; leaves entire or nearly so; overwintering large buds on new shoots terminal or subterminal; styles 1.5–2.5 mm. long; with us only in Nfld. and Gaspé Pen. § 17. Chrysantheae.

Leaves oblong-ovate to suborbicular, glabrous or promptly glabrate; ovaries and capsules glabrous. . . . . 34. *S. calcicola.*

Leaves narrowly to broadly oblong, flocculent-tomentose; ovaries and capsules tomentulose. . . . . . 35. *S. Wiegandii.*

P. Vigorous vegetative shoots without stipules; leaves entire, with revolute margins, elliptic, oblong or oblong-obovate to oblanceolate, those of flowering branches only 1–5 cm. long; aments terminating leafy branchlets, short and thick, mostly fulvous; style very short or obsolete; ovaries and capsules glabrous or nearly so; low slender basally creeping and stoloniferous shrubs of peaty habitats. § 18. Roseae.

Fruiting pedicels filiform, glabrous, twice as long as the greenish-yellow nearly glabrous bracts; ovary and capsule glabrous; stigmas essentially sessile; wide-ranging. . . . 36. *S. pedicellaris.*

Fruiting pedicels thick, pilose, shorter than to slightly exceeding the dark pubescent bracts; capsule slightly pilose or glabrescent; style 0.4–0.7 mm. long; endemic of Que. . . 37. *S. hebecarpa.*

N. Aments precocious (flowering long before expansion of leaves). . . *X.*

X. Fruiting pedicel three to six times length of gland; pubescence of sprouts, leaves and ovaries not flocculent-tomentose. . . *Y.*

Y. Leaves glabrous or mostly soon glabrate, the young with deciduous fulvous hairs; flowering aments 2.5–7 cm. long, fruiting ones 2.5–12 cm. long and 2–3 cm. thick, without expanded basal bracts; capsules 7–12 mm. long; style definite; stigmas slender and elongate. §19. Discolores. . . . . 38. *S. discolor.*

Y. Leaves permanently pubescent to glabrate beneath; flowering aments 0.5–6 cm. long, the fruiting ones 1–9 cm. long and 0.5–2 cm. thick; capsules 3–9 (in some introd. species 7–10) mm. long; style definite or obsolete; stigmas short and thick. . . *Z.*

Z. Capsules 3–9 mm. long; leaves linear- to oblong-lanceolate to -oblanceolate or obovate; stipules semiovate to -lanceolate or wanting; indigenous. § 20. Griseae. . . *a.*

a. Leaves dull green or dull grayish, thickish, entire or undulate, acute to blunt (not acuminate); fruiting aments scarcely or not at all leafy-bracted at base; capsules with prolonged slender beak about as long as plump base, dull-pubescent; style nearly obsolete. . . . . 39. *S. humilis.*

a. Leaves bright green above, lustrous (at least when young) beneath with silky or velvety pubescence, membranaceous, acuminate, serrulate to serrate-dentate, inclined to blacken in drying; mature fruiting aments on short bracted peduncles; capsules 3–6 mm. long, gradually tapering or rounded to tip, with glistening pubescence. . . *b.*

b. Branchlets flexible and elastic; stipules 0 or minute and promptly deciduous; leaves entire or nearly so at base, usually glabrate; bracts of ament pale brown to yellowish; fruiting aments loose, 1.5–4 cm. long; capsules 5–9 mm. long, gradually tapering to blunt tip; style obsolete. . . . . 40. *S. gracilis.*

b. Branchlets brittle at base; stipules of vigorous sprouts abundant; leaves toothed to base, permanently pubescent beneath; bracts of ament dark brown to black; fruiting aments rather dense, 2–6 cm. long; capsules 3–5 mm. long.

Pubescence of lower leaf-surface minute, silky; leaf-margin finely and regularly serrulate; stipules of sprouts lanceolate, deciduous; capsules 2–5 mm. long, broadly rounded above; style obsolete. . . . 41. *S. sericea.*

Pubescence of lower leaf-surface plush-like or velvety; leaf-margin coarsely and unequally serrate-dentate, with longer teeth of mature leaf broad-based, gland-tipped, horizontally divergent and 1–2 mm. long; stipules broadly lanceolate to ovate, subpersistent; capsules 5 mm. long, slenderly conical; style definite. 42. *S. coactilis.*

Z. Capsules 7–10 mm. long; leaves elliptic or oval to obovate, their margins coarsely undulate; stipules semicordate or semireniform; introd. species, spread from cult. § 21. Capreae.

Branches glabrous or puberulent and glabrate; staminate aments flowering from base to apex; fruiting aments lax and open, the pedicels five or six times length of gland. 43. *S. Caprea.*

Branches velvety-tomentose; staminate aments flowering from summit to base; fruiting aments dense, the pedicels three to five times length of gland. . . . . . . 44. *S. cinerea.*

X. Fruiting pedicel at most twice length of gland (or in no. 46, with flocculent tomentum on branches and leaves, up to four times as long). . . *c.*

c. Pubescence of young branchlets, leaves and ovaries a dull whitish flocculent tomentum; stipules lanceolate or semiovate, subpersistent; bracts of ament pale brown. § 22. CANDIDAE.

Leaves oblong to linear-lanceolate, plane, the revolute margin entire or undulate; pedicels at most twice length of gland; style slender, elongate; wide-ranging northw. . . . 45. *S. candida.*

Leaves oblong-lanceolate, rugulose, the revolute margin with obscure gland-tipped crenations; pedicels two to four times length of gland; style very short to obsolete; endemic of Nfld. and of Mingan Ids., Que. . . . . . . 46. *S. cryptodonta.*

c. Pubescence of branchlets or leaves not flocculent, frequently lustrous or wanting; style definite. . . *d.*

d. Leaves linear-lanceolate or oblanceolate to elliptic or oblong, glabrous, pilose or silky-velutinous beneath; capsules definitely pedicelled; indigenous northern shrubs or small trees. § 23. PHYLICIFOLIAE. . . *e.*

e. Leaves green or merely grayish beneath, lanceolate to elliptic or oblong, acute or blunt, entire or obscurely toothed, glabrous or sparsely pilose, membranaceous; branchlets not pruinose (except in no. 47 with glabrous leaves and sessile aments); capsules 5–8 mm. long; arctic-alpine or confined to Nfld. or e. Que. . . *f.*

f. Aments essentially sessile, their basal bracts not definitely foliaceous; leaves glabrous; branchlets glabrous or pruinose; capsules 5–6 mm. long; arctic-alpine species. 47. *S. planifolia.*

f. Aments on leafy peduncles; leaves pilose to glabrous; branchlets pubescent or glabrous; capsules 6–8 mm. long. . *g.*

g. Branchlets densely cinereous-pilose or -velutinous; young foliage pilose, mature glabrescent, crenate- or undulate-dentate; fruiting aments 2–6 cm. long, on leafy peduncles 2–10 mm. long; style 0.6–1.1 mm. long. . . . . 48. *S. paraleuca.*

g. Branchlets glabrous; leaves glabrous or promptly glabrate; fruiting aments 5–10 cm. long, on leafy peduncles 0.5–3 cm. long; style 1–1.5 mm. long; endemics of Nfld.

Stigmas strongly reflexed, 1.5–2 mm. long; pedicels scarcely 1 mm. long; young leaves minutely pilose, glandular-dentate. . . . . 49. *S. ancorifera.*

Stigmas ascending, 0.5 mm. long; pedicels 1–2 mm. long; young leaves glabrous, entire or undulate-crenate.

Bracts of ament blackish, 3.5–4.5 mm. long; fruiting aments 5–10 cm. long. . . . . 50. *S. pedunculata.*

Bracts yellowish, 2–3 mm. long; fruiting aments 3–7 cm. long. . . . . 51. *S. amoena.*

e. Leaves strongly whitened and either silky-velutinous or glabrescent beneath, linear-lanceolate to broadly lanceolate or oblanceolate, attenuate to tip, essentially entire, thick and firm; branchlets pruinose; fruiting aments 2–5 cm. long; capsules 4–5 mm. long; wide-ranging boreal species. . . . . 52. *S. pellita.*

d. Leaves elongate-linear to lanceolate, prolonged to attenuate slender tips, minutely silvery-satiny beneath; aments sessile or subsessile; capsule nearly sessile, minutely puberulent, 6–8 mm. long; tree spreading from cult. § 24. VIMINALES. 53. *S. viminalis.*

G. Stamens with the 2 hairy filaments united to summit but with 2 often united red anthers; leaves lingulate or oblanceolate, glaucescent, subopposite; aments dense, precocious; capsules obtuse, sessile, 2–3 mm. long. § 25. HELIX. . . . . 54. *S. purpurea.*

ARTIFICIAL KEY TO STAMINATE MATERIAL *

a. Flowers each with 2 glands at base. . . *b.*

b. Stamens 3 or more. . . *c.*

---

* Staminate material of the following species unknown: nos. 12, 14, 22, 24, 30, 32, 37, 46, 48, 49, 50, 51 and 52.

c. Flowers tufted and more or less whorled along rachis of ament; summit of petiole or its junction with leaf-blade glandless; glands at base of flower distinct; trees of temp. and warm reg.

Branchlets red- or purple-tinged, brittle at base; stipules conspicuous, rather persistent.

Expanding leaves green both sides; wide-ranging, N.B. westw. and southw. . . . . 1. *S. nigra.*

Expanding leaves glaucous or whitened beneath; southern, n. to Md., D.C., Ky., s. Ind., s. Ill. and Mo. . . . . 2. *S. caroliniana.*

Branchlets yellowish, tenacious and flexible; leaves whitened beneath; stipules minute or wanting; from sw. Que. and Vt. to Wash., s. to w. Mass., N.Y., O., Ind., Ill., Mo., etc. . . . . 3. *S. amygdaloides.*

c. Flowers spirally arranged; summit of petiole or its junction with leaf-blade bearing coarse glands; glands more or less united as an irregular cup at base of flower; northern shrubs or small trees.

Leaves green and lustrous on both surfaces, acuminate; staminate aments 2–6 cm. long.

Leaves (mature ones) short-acuminate; introd. . . . . . 4. *S. pentandra.*

Leaves (mature) with long-attenuate tail-like tips; native. . . 5. *S. lucida.*

Leaves whitened beneath, merely short-acuminate; staminate aments 1–1.5 cm. long. . . . . 6. *S. serissima.*

b. Stamens 1 or 2. . . *d.*

d. Stamen 1 or of 2 completely united filaments.

Stamen 1; depressed arctic-alpine shrub; alternate glandular-toothed elliptic to obovate leaves 0.5–2.5 cm. long, lustrous-green above, pale beneath; aments terminating leafy branchlets. . . . . . 16. *S. Uva-ursi.*

Stamen of 2 wholly united filaments; erect shrub; the leaves often opposite or subopposite, narrowly oblanceolate, glaucescent, purplish, denticulate to entire, 5–10 cm. long; aments sessile on old branches, before expansion of leaves. . . . . 54. *S. purpurea.*

d. Stamens 2, nearly or quite distinct. . . *e.*

e. Filaments densely long-villous below middle; bracts of aments yellow; aments inclined to crowd or cluster at ends of crowded branchlets; leaves linear to oblong-lanceolate, short-acuminate to subacute, sessile or nearly so, remotely denticulate with projecting teeth; freely stoloniferous, forming colonies on open shores. . . . . . 10. *S. interior.*

e. Filaments sparsely villous or pilose at base to glabrous; aments 1–few or scattered; leaves broader, if linear not divergently denticulate. . . *f.*

f. Filaments pilose at base. . . *g.*

g. Leaves lanceolate to ovate-lanceolate, acuminate or slender-tipped, becoming 5–16 cm. long; large introd. trees.

Summit of young petiole viscid or glandular.

Young leaves green and glabrous, the mature ones 1–1.5 dm. long and 2.5–4 cm. broad, coarsely undulate-serrate; staminate aments becoming 3–5 cm. long and about 1 cm. thick; branches not "weeping". . . . . 7. *S. fragilis.*

Young leaves silky, pale beneath, the mature ones 2–12 cm. long and 0.5–2 cm. broad, finely serrulate; aments mostly shorter and more slender; branches and branchlets "weeping". . . . . 8. *S. babylonica.*

Summit of young petiole not viscid; leaves whitened beneath, silky when young, finely serrate, in maturity 5–12 cm. long and 1–3 cm. broad. . . . . 9. *S. alba.*

g. Leaves broader, shorter, not acuminate; small native shrubs of Nfld., Anticosti or Gaspé Pen.

Aments falsely terminal, on naked peduncles borne just below tip of branchlet from side of stem opposite terminal round to obovate leaf.

Leaves glabrous, conspicuously reticulate beneath; buds glabrous and glutinous; prostrate, matted, with slender branches often rooting. . . . . 11. *S. reticulata.*

Leaves with appressed silky villosity beneath with glistening white hairs, reticulate above, buds pubescent; stout-branched ascending or depressed shrub. . . . . . 13. *S. vestita.*

Aments borne at tips of scattered axillary branchlets.

Leaves suberect, overlapping, oblong or barely oblong-obovate, on thick petioles shorter than well-developed axillary buds; aments subglobose to short-ellipsoid, 0.5–2.7 cm. long; bracts not long-bearded; anthers ellipsoid-globose. . . . . 19. *S. brachycarpa.*

Leaves loosely spreading or ascending, elliptic to ovate, obovate or suborbicular, on slender petioles longer than axillary buds; aments cylindric, elongate; bracts long-villous; anthers elongate. . . . . 20. *S. cordifolia.*

f. Filaments glabrous; low shrubs of Nfld., Anticosti or Gaspé Pen., or alpine southw. . . *h.*

h. Trunks and main branches subterranean, stoloniferous and rooting at nodes, filiform; flowering branchlets 2–4-leaved; leaves glabrous, reticulate, roundish, slender-petioled; subterminal tiny aments 2–8-flowered. . . . . 15. *S. herbacea.*

h. Trunks and main branches above ground, stouter, ligneous; branches several-leaved; aments many-flowered. . . . . . § *Ovalifoliae*, § *Glaucae*, § *Argyrocarpae* (see pp. 508–510).

a. Flowers each with 1 basal gland. . . *i.*

i. Filaments pilose at base. . . *j.*

j. Aments expanding with the grayish-pilose leaves (coaetaneous); bracts yellowish-brown, sparsely pilose; wide-ranging shrub or small tree. 33. *S. Bebbiana.*

j. Aments expanding before the leaves (precocious); bracts dark purple or black at least at tip. . . *k.*

k. Aments 0.5–1.3 cm. thick, 1–2.5 cm. long; first flowers expanding at base or near middle of ament, thence progressively to apex.

Branches slender, yellow or greenish to reddish-purple, flexible and elastic; aments usually beginning to flower near middle, 1–1.3 cm. thick; bracts pale brown, with black tip. . . . . 40. *S. gracilis.*

Branches coarser, dark or fuscous, brittle at base; aments flowering from base to apex, 5–8 mm. thick; bracts blackish. . . . . . 41. *S. sericea* and 42. *S. coactilis.*

k. Aments ellipsoid to thick-cylindric, 1–5 cm. long, 1–2.5 cm. thick; bracts dark purple to blackish.

Aments 1.5–2.5 cm. thick, flowering from apex to base; native coarse shrub or small tree. . . . . 38. *S. discolor.*

Aments 1–2 cm. thick; spread from cult.

Branchlets by 2nd year glabrous, brown and shining; wood under the bark smooth; aments 1.5–2 cm. thick, flowering from base upward; anthers yellow. . . . . 43. *S. Caprea.*

Branchlets of 2nd year tomentose, blackish, opaque; wood under the bark striate; aments 1–1.5 cm. thick, flowering from apex downward; anthers red or orange. . . . . 44. *S. cinerea.*

i. Filaments glabrous. . . *l.*

l. Aments expanding as the leaves expand. . . *m.*

m. Slender and low basally creeping and obviously stoloniferous shrub of peaty or boggy habitats, with entire exstipulate fulvous elliptic to oblong-obovate or oblanceolate leaves developed at flowering time; aments with nearly glabrous greenish-yellow bracts. . . . . . 36. *S. pedicellaris.*

m. Stouter, not obviously stoloniferous, or if so with toothed leaves and bracts heavily pubescent. . . *n.*

n. Young growth with strong balsamic-resiniferous fragrance; bracts yellowish or pale brown. . . . . 23. *S. pyrifolia.*

n. Young growth without evident balsamic fragrance.

Flowering branchlets borne from many axils of the elongate branches; leaves toothed, those of flowering branchlets without conspicuous stipules; internodes of branches glabrous or closely pubescent. . . . . § *Cordatae* (see p. 511).

Flowering branchlets 1–4, crowded near summits of coarse stems, with abbreviated long-villous internodes; new leaves entire, subtended by conspicuous firm stipules. . . . . § *Chrysantheae* (see p. 514).

l. Aments expanding before opening of leaf-buds. . . . *o.*

o. Branchlets floccose with dull white tomentum. . . . . 45. *S. candida.*

o. Branches and branchlets glabrous or merely cinereous-puberulent or -tomentulose, not flocculent.

Branches cinereous-puberulent or -tomentulose; aments soon divergent or recurving, 0.5–3 cm. long; young anthers purplish or violet. 39. *S. humilis.*

Branches glabrous or at most pruinose; aments ascending to divergent, 2–3.5 cm. long; anthers yellow.

Branches short, brittle-based; aments distant on the branches, their bases 1–2.5 cm. apart; native, boreal and arctic-alpine. 47. *S. planifolia.*

Branches long, flexible and elastic; aments subapproximate, their bases rarely 1 cm. apart; introd. . . . . 53. *S. viminalis.*

Artificial Key to Pistillate or Fruiting Material

a. Aments supra-axillary and falsely terminal, borne on naked peduncles from side of stem opposite the terminal leaf. . . *b.*

b. Bracts and ovaries pubescent. . . *c.*

c. Leaves glabrous or promptly glabrescent, conspicuously reticulate beneath; prostrate in mats, the slender branches often rooting; shrubs of Nfld.

Petioles twice to several times longer than axillary buds; fruiting aments 1–3 cm. long, long-peduncled. . . . . 11. *S. reticulata.*

Petioles shorter than to slightly longer than buds; fruiting aments 0.6–1.8 cm. long, short-peduncled. . . . . 12. *S. jejuna.*

c. Leaves beneath with dense appressed silky villosity; petioles short; coarse-branched mostly ascending shrub of Nfld., Gaspé, etc. . . 13. *S. vestita.*

b. Bracts and ovaries glabrous; leaves glabrescent; depressed shrub of Nfld. 14. *S. leiolepis.*

a. Aments borne along the branches from axils of preceding year or terminating bracted or leafy peduncles or shoots. . . *d.*

d. Leaves of vigorous shoots chiefly subopposite or opposite, firm, glaucescent, oblanceolate, 1–2 cm. broad; aments often paired at the nodes, expanding before leaves, sessile, in maturity only 5–6 mm. thick; bracts gray-bearded; capsule plump-ovoid, white-pubescent, 2–3 mm. long, sessile, with very short style and divergent stigmas. . . . . . . 54. *S. purpurea.*

d. Leaves alternate (or subopposite only in no. 15); aments borne singly along branches or at tips of branchlets. . . *e.*

e. Trunk and principal branches subterranean, rooting at nodes, filiform, not strongly woody; short slender upper branchlets above ground, bearing 2–4 glabrous rounded reticulate slender-petioled leaves and tiny 2–8-flowered terminal aments; arctic-alpine. . . . . . . 15. *S. herbacea.*

e. Trunks and main branches above ground, stouter, woody; branchlets several–many-leaved; aments several–many-flowered. . . *f.*

f. Flowers and capsules tufted in whorls along the rachis of ament; leaves linear- to ovate-lanceolate, serrate; bracts of ament pale, soon deciduous; indigenous trees.

Leaves green both sides; branchlets brittle at base. . . . . 1. *S. nigra.*

Leaves whitish or glaucous beneath.

Branchlets reddish or purplish, brittle at base; petioles 3–7 mm. long; stipules large, roundish, persistent; gland of pistillate flower clasping pedicel; capsule granulose. . . . . . 2. *S. caroliniana.*

Branchlets yellowish, tenacious and pliable; petioles mostly 1–3 cm. long; stipules small, caducous; gland free; capsule smooth. 3. *S. amygdaloides.*

f. Flowers and capsules spirally arranged. . . *g.*

g. Bracts of ament yellowish, whitish or pale brown, deciduous before ripening of capsule. . . *h.*

h. Petiole bearing heavy glands or gland-tipped teeth at summit or at junction with the serrate blade; pistillate flowers with 2 free glands at base. . . *i.*

i. Leaves green and lustrous on both surfaces; pistillate aments 2–5 cm. long; capsule 5–6.5 mm. long, its valves opening in late spring or early summer.

Capsule conic-subulate, its pedicel about twice length of upper gland; mature leaves short-acuminate; introd. . . . . 4. *S. pentandra.*

Capsule conic-ovoid, its pedicel about four times length of upper gland; mature leaves with long-attenuate tail-like tip; indigenous. . . . . 5. *S. lucida.*

i. Leaves pale or whitened beneath, merely acuminate; pistillate aments 2–3.5 cm. long; capsule conic-subulate, 7–10 mm. long, its indurated valves often not opening until late summer or autumn; pedicel twice length of upper gland. . . . . 6. *S. serissima.*

h. Petioles glandless, or only weakly glandular or viscid in no. 7 (with coarsely undulate-margined leaves). . . *j.*

*j.*Leaves petioled, lanceolate to narrowly ovate, long-acuminate, serrate or coarsely crenate; aments terminating scattered leafy peduncles; introd. trees. . . *k.*

*k.*Young petioles viscid at summit; pistillate flowers with 2 glands; capsules subulate.

Leaves green both sides, glabrous from the first, in maturity 2.5–4 cm. broad, coarsely undulate-serrate; capsule about 5 mm. long. . . . . 7. *S. fragilis.*

Leaves pale beneath, silky when young, 0.5–2 cm. broad, finely repand-serrate; capsules 1–1.5 mm. long. . . . 8. *S. babylonica.*

*k.*Young petioles not viscid; pistillate flowers with 1 gland; capsule conic-ovoid, 3–5 mm. long. . . . . . 9. *S. alba.*

*j.*Leaves sessile or nearly so, linear to oblong-lanceolate, short-acuminate to merely subacute, remotely denticulate; aments often crowded near ends of branches; native stoloniferous shrub or tree. . . . . 10. *S. interior.*

*g.*Bracts of ament pale to blackish, persistent until after opening of capsules. . . *l.*

*l.*Ovary, pedicel and capsule glabrous; aments flowering as the leaves unfold (coaetaneous) or after they are fairly developed (serotinous). . . *m.*

*m.*Bracts of ament glabrous, olivaceous; aments in anthesis subglobose to short-ellipsoid, in fruit 0.5–2 cm. long; leaves oblong, elliptic or oblanceolate, glabrescent, 1–2.5 cm. long; dwarf shrub of Mt. Albert, Gaspé Co., Que. . . . . . . 21. *S. chlorolepis.*

*m.*Bracts bearded, pilose or villous. . . *n.*

*n.*Leaves entire.

Slender basally creeping and stoloniferous shrub with glabrous flexible branches; leaves elliptic, oblong or oblong-obovate to oblanceolate, fulvous when young; stipules 0; aments lax in fruit, terminating leafy branchlets; bracts of ament pale, barely pubescent; style obsolete; wide-ranging in peaty habitats. . . . . . . 36. *S. pedicellaris.*

Stout woody-branched and gnarled shrub, the new short internodes densely villous; leaves oblong-ovate to suborbicular, green; stipules large, subpersistent; aments dense, subterminal or crowded at summits of branches; bracts of ament dark, heavily bearded; style 1.5–2.5 mm. long; low shrub of calcareous rock, Nfld., Gaspé, etc. . 34. *S. calcicola.*

*n.*Leaves serrate or dentate. . . *o.*

*o.*Dwarf prostrate arctic-alpine shrub with elliptic to obovate glandular-serrate leaves 0.5–2.5 cm. long. . . . . 16. *S. Uva-ursi.*

*o.*Taller upright shrubs with larger leaves; chiefly of temperate areas. . . *p.*

*p.*Stipules 0 or minute and caducous; new growth and foliage with strong persistent balsamic or resiniferous fragrance; leaves membranaceous, slender-petioled, short-oval to oblong-lanceolate; bracts of ament whitish. . . . . 23. *S. pyrifolia.*

*p.*Stipules often or usually borne on vigorous shoots; foliage not strongly balsamic; leaf-blades firm to subcoriaceous in maturity, with thickish petioles; bracts of aments brown to blackish. . . . . . . § *Cordatae* (see p. 511).

*l.*Ovary, pedicel and at least base of capsule pubescent. . . *q.*

*q.*Aments expanding as leaves unroll (coaetaneous) or after they are well grown (serotinous). . . *r.*

*r.*Slender shrub, creeping in moss and stoloniferous, with slender glabrous flexible branches; stipules 0; leaves entire, fulvous, revolute-margined, oblong to oblong-obovate, those of flowering branches 1–3 cm. long; aments terminating leafy branchlets, fulvous; capsules glabrescent; style 0.4–0.7 mm. long; in peats of Shickshock Mts., Que. . . . . 37. *S. hebecarpa.*

*r.*Coarser, with stoutish branches, not obviously with creeping and stoloniferous subterranean stems. . . *s.*

*s.*Expanding leaves, ovaries and capsules minutely silvery-silky with glistening hairs; leaves narrow, repand-crenate, tapering to both ends; arctic-alpine. . . . . 22. *S. argyrocarpa.*

s. Expanding leaves, ovaries and capsules variously pubescent to glabrescent, not glistening-silvery; leaves broad or narrow. . *t.*

t. Young shoots and leaves flocculent-tomentose; oval to oblong entire revolute-margined coriaceous leaves prominently reticulate beneath, 1.5–4.5 cm. long; cordate-ovate firm stipules glandular-serrate, persistent on fruiting branches; fruiting aments thick, with black sordid-villous bracts; style about 2 mm. long; low and tough shrub of Nfld. . . . . 35. *S. Wiegandii.*

t. Young growth not flocculent; stipules usually deciduous from fruiting branches or wanting. . . *u.*

u. Bracts of ament dark brown to black or black-tipped. . *v.*

v. Young branches densely tomentose; leaves oblong to oblong-obovate, short-acuminate, 6–18 cm. long; aments 4–10 cm. long; ovaries and bases of capsule tomentose; pedicel 1–2 mm. long; tall shrub or small tree along lower R. and Gulf of St. Lawrence, Que. . . . . 30. *S. laurentiana.*

v. Young branches glabrous; leaves oblong or elliptic or obovate, not acuminate, 1–4.5 cm. long; capsules minutely pubescent to glabrescent, sessile or subsessile; prostrate shrubs of Nfld., Anticosti or alpine areas.

Branchlets brittle on drying; leaves dark green, brilliantly lustrous above; aments erect, perpendicular to prostrate axis, the pistillate becoming 3–10 cm. long; capsule lance-conic, 7–10 mm. long. . . . . 17. *S. arctophila.*

Branchlets flexible; leaves pale green, opaque or barely lustrous above; aments merely ascending, 1–5.5 cm. long; capsules ovoid-conic, 5–8 mm. long. . . . . 18. *S. arctica.*

u. Bracts of ament yellowish to fulvous, not darkened at tip.

Ovary and capsule sessile or nearly so; the pedicel much shorter than the 2 glands; aments dense: low to prostrate shrubs of Nfld., Anticosti or mts. of Gaspé.

Leaves overlapping, suberect, 0.5–1.5 cm. broad, the thick petioles shorter than well developed axillary buds; aments in anthesis subglobose to short-ellipsoid, lengthening to 0.5–2.7 cm. long; bracts merely pilose. . . . . 19. *S. brachycarpa.*

Leaves scarcely overlapping, divergent to loosely ascending, 0.5–5 cm. broad, their slender petioles longer than axillary buds; aments cylindric, 1–7 cm. long; bracts long-villous, often long-bearded. . . . . 20. *S. cordifolia.*

Ovary and capsule pedicelled, the mature pedicels three or four times length of subtending bract; gland 1; upright shrub or small tree of temp. reg. 33. *S. Bebbiana.*

q. Aments expanding before unrolling of leaf-buds (precocious). . . *w.*

w. Young branchlets, leaves and capsules with dull whitish flocculent tomentum.

Leaves oblong to linear-lanceolate, plane, the revolute margin entire to undulate; pedicels at most twice length of gland; style slender, elongate; wide-ranging in low grounds. . . . . 45. *S. candida.*

Leaves oblong-lanceolate, rugulose, the revolute margin with obscure gland-tipped teeth; pedicel two to four times length of gland; style obsolete or very short; endemic of Nfld. and of Mingan Ids., Que. . . . . . 46. *S. cryptodonta.*

w. Leaves and branchlets not floccose-tomentose. . . *x.*

x. Fruiting pedicels three to six times length of gland; wide-ranging in temp. reg. . . *y.*

y.Leaves glabrous or mostly soon glabrate, the very young ones with deciduous fulvous hairs; flowering aments 2.5–7 cm. long, the fruiting ones 2.5–12 cm. long and 2–3 cm. thick; style definite; stigmas slender and elongate. 38. *S. discolor.*

y.Leaves permanently pubescent to glabrate beneath, if glabrate with fruiting aments only 1–5 cm. long and capsules 3–6 mm. long.

Capsules 3–6 mm. long; leaves linear- to oblong-lanceolate to -oblanceolate or obovate, entire to serrate; stipules semiovate to -lanceolate or wanting; indig. . . . . § *Griseae* (see p. 515).

Capsules 7–10 mm. long; leaves elliptic or oval to obovate, their margins coarsely undulate; stipules semicordate or semireniform; introd. . . . . . § *Capreae* (see p. 517).

x.Fruiting pedicels at most twice length of gland.

Leaves linear-lanceolate or oblanceolate to elliptic or oblong, glabrous, pilose or silky-velutinous beneath; capsule definitely short-pedicelled; indig. northw. . . § *Phylicifoliae* (see p. 518).

Leaves elongate-linear or -lanceolate, prolonged to attenuate slender tips, minutely silvery-satiny beneath; aments sessile or nearly so; capsule nearly sessile; introd. . . . . 53. *S. viminalis.*

Artificial Key to Material with mature Foliage (including Sprouts)

a.Prostrate or with creeping or repent stems; leaves merely acutish to rounded at tip, not acuminate, 0.5–9 cm. long; shrubs of Nfld., Anticosti, Gaspé Pen. or alpine areas southw. . . *b*.

b.Branches slender, rooting at nodes or the main branches subterranean and rooting; leaves glabrous or glabrate, prominently reticulate beneath.

Leaves whitish or silvery-gray beneath, entire or low-crenate, with revolute margins; of calcareous rock or gravel, Nfld.

Petiole two-fifths as long as to about equaling the suborbicular to oblong blade, the latter 1.5–5.4 cm. long. . . . . 11. *S. reticulata.*

Petiole many times shorter than blade, the latter narrowly to broadly elliptic, oblong or oblong-oval, 0.5–2 cm. long.

Leaves entire, those of sprouts at first villous-ciliolate and with bearded slender stipules; of barrens along Straits of Belle Isle. 12. *S. jejuna.*

Leaves crenate, those of sprouts glabrous and exstipulate; local, sw. Nfld. . . . . 14. *S. leiolepis.*

Leaves green both sides, crenate-serrate; of peaty or mossy alpine areas on acid rock. . . . . 15. *S. herbacea.*

b.Branches superficial, ligneous, not rooting at nodes; leaves not conspicuously reticulate beneath. *c*.

c.Leaves densely clothed beneath with glistening white appressed long villosity, dark green and deeply reticulate-rugose above, with rounded summits; internodes enlarged upward, prominently angled or keeled. . 13. *S. vestita.*

c.Leaves without long white lustrous appressed villosity. . . *d*.

d.Leaves of fertile branches as well as of sprouts accompanied by large persistent stipules; the new shoots heavily long-villous with spreading hairs; tips of short branchlets terminated by large subglobose or oblate buds for next season's flowering.

Leaves oblong-ovate to suborbicular, glabrous or promptly glabrate. 34. *S. calcicola.*

Leaves narrowly to broadly oblong, flocculent-tomentose. . . . 35. *S. Wiegandii.*

d.Leaves of fertile branches without stipules; overwintering buds small and elongate, chiefly axillary. . *e*.

e.Young and often mature leaves and branchlets grayish-pilose or villous. . . . . 20. *S. cordifolia.*

e.Young leaves and branchlets glabrous or promptly glabrate.

Leaves entire or merely undulate, 1–4.5 cm. long.

Branchlets brittle on drying; leaves dark green, brilliantly lustrous above. . . . . 17. *S. arctophila.*

Branchlets flexible; leaves pale green and opaque or barely lustrous above. . . . . 18. *S. arctica.*

Leaves serrate or serrate-dentate.

Closely depressed in flat mats; leaves crowded, finely glandular-serrate, 0.5–2.5 cm. long; of sterile or acid rock or alpine areas. . . . . 16. *S. Uva-ursi.*

Merely somewhat and exceptionally depressed; leaves not crowded, coarsely serrate-dentate, 2–7 cm. long; of damp calcareous soils, Nfld. . . . . 31. *S. myrtillifolia*, var.

*a.*Erect or ascending, depressed only in extreme conditions; leaves blunt to acute or acuminate. . *f.*

*f.*Summit of petiole or its junction with the blade bearing coarse glands; mature leaves coriaceous, lustrous above, acuminate, tapering to petiole, one-half to one-sixth as broad as long; branchlets usually lustrous.

Leaves green on both surfaces.

Leaves merely short-acuminate, the mature ones 3.5–10 cm. long and 1.5–3 cm. broad; spread from cult. . . . . 4. *S. pentandra.*

Leaves with long-attenuate tail-like tips; mature blades 5–17 cm. long and 0.5–5 cm. broad; indig. . . . . 5. *S. lucida.*

Leaves whitish beneath, short-acuminate, mostly 4–8 cm. long and 1–3 cm. broad. . . . . 6. *S. serissima.*

*f.*Summit of petiole glandless or only viscid; if glandular, teeth at base of blade the latter rounded to cordate at base and not highly lustrous. . . *g.*

*g.*Leaves glabrous. . *h.*

*h.*Branchlets whitish-pruinose.

Leaves lance-linear to narrowly oblong or oblanceolate, mostly acuminate, subcoriaceous, strongly whitened beneath, 4–12 cm. long by 0.5–2 cm. broad (on sprouts often broader), one-tenth to one-fourth as broad as long, the lateral veins prominent beneath; of stream-banks etc., wide-ranging northw. . . . . . . 52. *S. pellita.*

Leaves elliptic-lanceolate to oblong or broadly elliptic, blunt or merely acute, membranaceous, merely paler beneath, 2.5–7 cm. long by 0.7–3.5 cm. broad, one-fourth to three-fifths as broad as long, the lateral veins not very prominent; Nfld. and Lab. westw., southward subalpine. . . . . 47. *S. planifolia.*

*h.*Branchlets not pruinose. . . *i.*

*i.*Leaves often opposite or subopposite, narrowly oblanceolate, glaucous and purple-tinged; introd. . . . . . 54. *S. purpurea.*

*i.*Leaves alternate. . . *j.*

*j.*Leaves elongate and acuminate or acute, mostly 3–15 cm. or more long and one-tenth to two-thirds as broad as long; shrubs or trees of low altitudes of temp. reg. . . *k.*

*k.*Leaves subsessile or with petioles rarely 3 mm. long, linear to oblong-lanceolate, short-acuminate to merely subacute; remotely denticulate with projecting teeth; stoloniferous and loosely colonial shrubs or small trees. . . . . 10. *S. interior.*

*k.*Leaves distinctly petioled, the petioles 0.3–3 cm. long, the blades narrowly lanceolate to ovate. . . *l.*

*l.*Vigorous vegetative sprouts with conspicuous subpersistent stipules. . . *m.*

*m.*Leaves green or merely paler, not whitened, beneath.

Leaves narrowly lanceolate, long-attenuate, often falcate, five to ten times as long as broad, finely serrulate; branchlets very brittle at base; tree of broad range. . 1. *S. nigra.*

Leaves mostly broader, oblong-lanceolate to ovate, gradually tapering to acuminate (but not long-attenuate) tip; coarse shrubs, rarely small trees. . . . 24. *S. cordata* and 25. *S. rigida.*

*m.*Leaves glaucous or whitened beneath.

Leaves lanceolate to lance-ovate, often falcate, long-attenuate at tip, gradually rounded to tapering at base; branchlets reddish or purple-tinged, brittle; southern tree. . . . . 2. *S. caroliniana.*

Leaves lanceolate to oblong, elliptic, ovate or obovate, merely acute or acuminate, not long-attenuate nor falcate; shrubs, rarely small trees.

Leaves cordate or broadly rounded at base, lustrous above, glabrous from the first, closely serrulate; northern shrubs.

New shoots with brown or dark bark; leaves dark

green above; boreal, northeastern. . . . . . . 26. *S. glaucophylloides.*

New shoots with yellow bark; leaves pale green above; northwestern. . . . . . 29. *S. lutea.*

Leaves tapering to base, irregularly crenate-serrate, the unfolding ones with caducous fulvous hairs. . . . 38. *S. discolor.*

l. Vigorous vegetative sprouts without stipules or these very small and caducous.

Branchlets brittle at base, easily broken off; introd. trees.

Leaves green both sides, coarsely undulate-serrate, the mature ones 1–1.5 dm. long and 2.5–4 cm. broad; branchlets not "weeping". . . . . 7. *S. fragilis.*

Leaves whitened or pale beneath, finely serrulate, the mature ones 2–12 cm. long and 0.5–2 cm. broad; branchlets "weeping". . . . . 8. *S. babylonica.*

Branchlets tenacious and flexible, not easily broken off.

Indig. shrubs or small trees.

Leaves lanceolate to lance-ovate, coriaceous, tapering to caudate-attenuate tip, the larger blades 1–5 cm. broad; the petioles mostly 1–3 cm. long; large shrub or small tree. . . . . 3. *S. amygdaloides.*

Leaves narrowly lanceolate, membranaceous, acuminate but not caudate-tipped, the larger blades 0.3–2 cm. broad; petioles 5–12 mm. long; small shrub. 40. *S. gracilis.*

Introd.; spread from cult. . . . . . 9. *S. alba.*

j. Leaves shorter and blunt to merely acute, not long-acuminate, rarely (except in the balsamic-resiniferous thin-leaved no. 23) more than 8 cm. long. . . *n.*

n. Foliage and buds strongly balsamic-resiniferous; leaves membranaceous, slender-petioled, short-oval to oblong-lanceolate, without stipules, or stipules minute and caducous. . . 23. *S. pyrifolia.*

n. Foliage not obviously balsamic-resiniferous; leaves firmer. . . *o.*

o. Slender basally creeping and stoloniferous shrubs with flexible branches and no stipules; leaves entire, with revolute margin, elliptic, oblong, oblong-obovate or oblanceolate, fulvous when young, those of fruiting branches only 1–5 cm. long; species of peaty or boggy habitats.

Branches erect, 0.3–1 m. high; shrub of acid bogs and wet meadows at low alts. . . . . 36. *S. pedicellaris.*

Branches loosely ascending or spreading, 1–3 dm. long; shrub of serpentine alpine area, Shickshock Mts., Gaspé Pen. . . . . 37. *S. hebecarpa.*

o. Coarser, not obviously creeping or stoloniferous; boreal or alpine.

Internodes very short, heavily spreading-villous; stipules coarse and persistent, often to 2nd year; overwintering buds terminal, large, subglobose or oblate; leaves oblong-ovate to suborbicular, heavily coriaceous, on very short and thick petioles; of calcareous barrens and alpine areas, Nfld. and Gaspé. . . . . . . 34. *S. calcicola.*

Internodes elongate or short, glabrous or essentially so; stipules 0 or soon falling (subpersistent in no. 32); leaves relatively thin.

Stipules conspicuous and subpersistent; leaves pale green to fulvous, opaque, elliptic to subrotund, broadly rounded at summit, coarsely serrate-dentate; petioles slender and elongate; along Gaspé rivers. . . . 32. *S. obtusata.*

Stipules 0 or minute (or conspicuous only in no. 31).

Dwarf, 1–2 dm. high, with wide-spreading branches; leaves oblong, elliptic or oblanceolate, 1–2.5 cm. long; petioles 2–3 mm. long; on serpentine-barrens of Mt. Albert, Gaspé Pen. . . . . . . 21. *S. chlorolepis.*

Coarser and taller; leaves 2–8 cm. long.

Leaves rounded or blunt at tip, mostly rounded at base, regularly dentate-serrate, those of fertile branches with thick petioles 2–5 mm. long;

vigorous sprouts often with broad coarsely toothed stipules; calcicolous. . . . . 31. *S. myrtillifolia.*

Leaves acute to bluntish, tapering or gradually rounded to base, entire, undulate or low-crenulate, with slender petioles 4–12 mm. long; vigorous sprouts exstipulate or with tiny caducous stipules; of acid to somewhat calcareous soils.

Wide-ranging from Lab. and Nfld. westw. and northwestw., s. to subalpine or alpine areas of Gaspé and n. N.E. . . . . 47. *S. planifolia.*

Known only from calcareous n. and w. Nfld. . . 49. *S. ancorifera.* 50. *S. pedunculata* and 51. *S. amoena.*

*g.*Leaves pubescent at least beneath. . . *p.*

*p.*Pubescence of young shoots and expanding leaves a dull white flocculent tomentum; stipules often persistent.

Leaves oblong to linear-lanceolate, 4–12 cm. long, plane, erect or suberect and inclined to overlap in tufts on the stout branches, the revolute margin entire or undulate; wide-ranging, in calcareous swamps. . . . . 45. *S. candida.*

Leaves oblong-lanceolate or oblanceolate to oblong-ovate or oval, 2–6 cm. long, rugulose, spreading or loosely ascending; branches slender; endemics of Nfld. and Que.

Upright shrub 2–4 m. high, the elongate branches ascending; overwintering buds axillary, elongate; leaves firm but flexible, their revolute margins with appressed-crenate gland-tipped teeth; by streams or ponds or in wet glades, Nfld. and Que. . . . 46. *S. cryptodonta.*

Depressed and gnarled, 1–3 dm. high, with divergent branching; overwintering buds 1–few, terminal or subterminal, subglobose; leaves subrigidly coriaceous, the revolute margin entire; of calcareous barrens, nw. Nfld. . . . . . . 35. *S. Wiegandii.*

*p.*Pubescence not flocculent. . . *q.*

*q.*Lower surfaces of all or at least young leaves with lustrous white or silvery pubescence. . . *r.*

*r.*Leaves suborbicular or broadly elliptic to obovate, with broadly rounded summit, 1–8 cm. long, coriaceous, dark green and reticulate-rugose above, silvery beneath with long appressed satiny hairs (rarely glabrate); short internodes broadened upward, angled or keeled; low shrub of limestones, Nfld. and Côte Nord, Anticosti and Gaspé Pen., Que. . . . . 13. *S. vestita.*

*r.*Leaves linear-lanceolate to cordate-ovate, pointed or acuminate. . . *s.*

*s.*Branchlets whitish-pruinose; leaves lance-linear to narrowly oblong or oblanceolate, subcoriaceous, 4–12 cm. long and 0.5–2 cm. broad (on sprouts sometimes larger); tall shrub or small tree of stream-banks, etc., wide-ranging northw. . 52. *S. pellita.*

*s.*Branchlets not pruinose. . . *t.*

*t.*Leaves broadly cordate-lanceolate to -ovate, the margin with many coarse forward-arching to depressed gland- or callus-tipped teeth; sprouts heavily pubescent, their large stipules toothed. . . . . 24. *S. cordata.*

*t.*Leaves linear or lanceolate to narrowly elliptic or narrowly oblong, narrowed to base. . . *u.*

*u.*Leaves entire, repand-undulate or with depressed undulate-crenate margins.

Low arctic-alpine shrub with narrowly lanceolate or oblanceolate to elliptic often undulate-crenate blunt or merely acute leaves 2–6 cm. long. . . . . . . 22. *S. argyrocarpa.*

Coarse shrub or small tree with narrowly lance-linear entire long-attenuate leaves up to 2.5 dm. long; spread from cult. . . . . 53. *S. viminalis.*

*u.*Leaves definitely serrate or dentate. . . *v.*

*v.*Leaves subsessile or with petioles rarely 3 mm. long, linear to oblong-lanceolate, short-acuminate to merely subacute, remotely denticulate with projecting teeth. 10. *S. interior.*

v. Leaves longer-petioled, lanceolate, acute to acuminate, with many serrations.
Branchlets tenacious at base, the longer ones and sprouts flexible; stipules 0 or narrow and caducous.
Low shrub of swales and damp habitats; leaves membranaceous, often entire toward base, commonly blackening in drying; indig. northw. . . . 40. *S. gracilis.*
Tree; leaves firm, serrulate to base, not usually blackening in drying; spread from cult. . . . 9. *S. alba.*
Branchlets brittle at base; stipules of vigorous sprouts large and often persistent.
Lower leaf-surface with minute silky pubescence; leaf-margin finely and regularly serrulate; stipules of sprouts lanceolate, deciduous. . . . . . 41. *S. sericea.*
Lower leaf-surface with plush-like or velvety pubescence; leaf-margin coarsely and unequally serrate-dentate; the larger teeth of mature blades broad-based, gland-tipped, horizontally divergent, 1–2 mm. long; stipules broadly lanceolate to ovate, subpersistent. . . . . . 42. *S. coactilis.*

q. Lower surfaces of all or at least younger leaves with opaque cinereous pubescence or sometimes more fuscous or fulvous lustrous hairs. . . *w.*

w. Leaves entire or merely undulate.
Young branchlets, 2nd year branches and lower leaf-surfaces densely cinereous-tomentulose or -pilose; wide-ranging. . . 39. *S. humilis.*
Young branchlets glabrous or sparingly pilose; lower leaf-surfaces with looser or sparse pilosity or villosity.
Leaves oblong or barely oblong-obovate, ascending to suberect and overlapping, 1–3.5 cm. long, 0.5–1.5 cm. broad, on thick petioles shorter than well developed axillary buds; dwarf shrub 1–4 dm. high, on serpentine tableland of Mt. Albert, Gaspé Pen., Que. . . . . 19. *S. brachycarpa.*
Leaves mostly broader and larger, on slender petioles longer than axillary buds.
Low arctic-alpine shrub with elliptic to ovate, obovate or suborbicular obtuse to acute leaves not prominently reticulate beneath. . . . . 20. *S. cordifolia.*
Medium to tall shrub or small tree, wider-ranging across much of continent in temp. areas, the mostly acute leaves becoming reticulate beneath. . . . . 33. *S. Bebbiana.*

w. Leaves definitely toothed. . . *x.*

x. Leaves cordate or strongly rounded to base; sprouts bearing large rounded stipules. . . . . § 15. *Cordatae* (see p. 511).

x. Leaves narrowed or rounded at base; sprouts with only small stipules or with stipules wanting. . . . *y.*

y. Indigenous. . . *z.*

z. Young branchlets, 2nd-year branches and at least lower leaf-surfaces densely cinereous-tomentulose or -pilose. 39. *S. humilis.*

z. Young branchlets glabrous or pilose; lower leaf-surfaces with looser or sparse pilosity or villosity.
Leaves bright green above, the young with soon deciduous fulvous hairs; coarse shrub or small tree of damp or wet habitats; wide-ranging. . . . . . 38. *S. discolor.*
Leaves dull or opaque above.
Branches (except for pubescence on the youngest) lustrous; leaves firm, only finely reticulate beneath; local shrub of e. Que. . . . . 48. *S. paraleuca.*
Branches dull, not shining; leaves membranaceous, strongly reticulate beneath; wide-ranging shrub or small tree. . . . . 33. *S. Bebbiana.*

y. Introduced; spread slightly from cult.
Branches glabrous or puberulent and glabrate; leaves one-half to three-fourths as broad as long. . . . . 43. *S. Caprea.*
Branches velvety-tomentose; leaves one-third to half as broad as long. . . . . 44. *S. cinerea.*

§ 1. Nìgrae Loud.

1. S. nìgra Marsh. (black), Black W. — *Shrub or,* when fully developed, *tree* 3–20 m. high, *with flaky dark brown to blackish bark; branchlets brittle at base* but tough and flexible above, *the youngest greenish* and somewhat pilose, soon glabrate and darker, with angles running down from leaf-bases; axillary buds becoming 2–3 mm. long; *leaves* beginning to expand at flowering time, the first ones lanceolate and pilose, *later lengthening to elongate-lanceolate, often falcate, and attenuate to slender tip,* becoming glabrous, *deep green both sides,* but midrib often minutely pilose, *glandular-serrulate,* with pilose petiole, *the fully developed leaf* 0.5–1.5 *dm. long and* 5–15 *mm. wide,* those of sprouts often larger; the *sprouts with auricled acute glandular-serrate stipules; aments terminating leafy shoots,* 2–8 cm. long, *slender-cylindric, with the flowers clustered in apparent whorls* along the rachis; the yellowish *bracts* crisp-villous on the inner surface, *soon falling; staminate flowers with* 2 *basal glands,* the 3–5 free *arching filaments hairy below; pistillate flowers with* 1 *basal gland,* the style nearly obsolete; fruiting ament 3–7 cm. long, becoming dense; *capsules* ovoid-conical, fulvous, *glabrous,* 3–5 mm. long, *the wide-spreading pedicels much longer than the free gland.* — Banks of streams, shores and rich low woods, N.B. to N.D., s. to s. N.E., L.I., N.C., locally to Ala., Tenn., Ark., etc. *Fl.* April–June. Fig. 858. — Hybridizes with nos. 2, 3, 5 and 9.

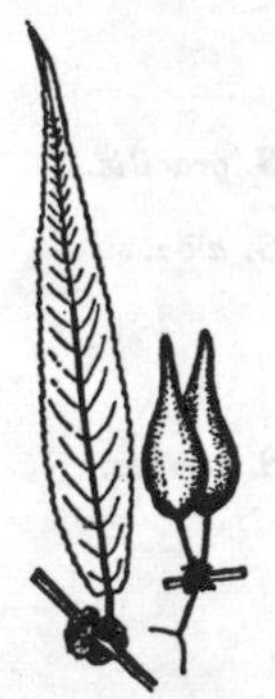

858. S. nigra.

§ 2. Bonplandiànae Schneid.

2. S. caroliniàna Michx. (Carolinian), Ward's W. — Resembling no. 1, rarely 10 m. high; *bark gray,* deeply checkered; *brittle-based branchlets* yellowish, *commonly becoming reddish or purplish, more or less pubescent; leaves whitened or glaucous beneath, firm, lanceolate to lance-ovate,* 1–3 cm. wide; *stipules* on sprouts conspicuous, *obtuse; aments* up to 1 dm. long, *the fruiting ones lax and rather open;* bracts villous outside toward base; *glands of staminate flower lobulate or forming a false disk,* the *stamens* 4–8; *gland of pistillate flower clasping base of pedicel; capsules* usually *granulose-roughened,* 4–6 mm. long, long-pedicelled. (*S. longipes* Shuttlew.; *S. Wardi* Bebb) — River-banks, shores and low woods, Fla. to Tex., n. to Md., D.C., W.Va., s. Ind., s. Ill., Mo., Kans., etc. *Fl.* May, early June. (Cuba) Fig. 859. — Hybridizes with no. 1.

859. S. caroliniana.

§ 3. Triándrae Dumort.

3. S. amygdaloìdes Anderss. (resembling *Amygdalus,* the peach), Peach-leaved W. — Coarse shrub or small tree up to 12 m. high, with reddish-brown or darker bark; *branchlets tenacious, flexible, somewhat drooping, yellowish or darker; leaves lanceolate to lance-ovate,* long-attenuate, *pale green above, whitened beneath,* closely serrulate, 0.5–1.5 dm. long and 1–5 *cm. broad, on petioles* 0.5–3 *cm. long; sprouts with minute stipules or none;* aments on short leafy branchlets, resembling those of no. 1; 2 *glands of staminate and of pistillate flowers elongate, free; capsule lanceolate,* 4–5 *mm. long, smooth.* — Shores, low woods and swamps, sw. Que. and Vt. to Wash., s. to w. Mass., N.Y., O. Ind., Ill., Mo., etc. *Fl.* late April–early June. Fig. 860. — Flowers occasionally perfect. — Hybridizes with nos. 1 and 25.

860. S. amygdaloides.

× **S. Glatfélteri** Schneid. (for its discoverer, Noah Miller Glatfelter, 1837–1911), a hybrid of nos. 1 and 3, is abundant near St. Louis and sometimes elsewhere, when the two parents commingle.

§ 4. Pentándrae Dumort.

4. S. pentándra L. (with 5 stamens), Bay-leaved W. — Shrub or small tree up to 7 m. high, *resinous-fragrant;* branchlets lustrous; overwintering buds blackish, slender, lustrous; *stipules of sprouts small, caducous; leaves gummy on expanding, shining, green both sides* or barely paler beneath, *broadly lanceolate to ovate-oblong, acute or short-acuminate,* 3.5–10 *cm. long,* 1.5–3 *cm. broad,* becoming coriaceous; *petiole* up to 1 cm. long, *with large glands toward summit; aments expanding with leaves,* on leafy peduncles, the bracts and flowers

861. S. pentandra.

spirally crowded; *bracts greenish-yellow*, oblong, hairy at base; *staminate aments* deep yellow, 2–6 cm. long, 1–1.5 *cm. thick; 2 glands of staminate flower often united at base, the 5 filaments straight* and hairy below middle; *pistillate ament* yellow-green, 2–5 cm. long, *dense*, 1 *cm. thick;* 2 *glands of pistillate flower free;* capsule conic-subulate, 5–6 mm. long, glabrous, its pedicel twice as long as upper gland. — Spread from cult., N.S. to Ont., s. to N.E., Pa., Md., O., Ill. and Ia. *Fl.* May, early June. (Introd. from Eu.) Fig. 861.

5. S. lùcida Muhl. (shining), Shining W. — Resembling no. 4; large shrub, or small tree up to 9 m. high, with trunks 1.2 dm. through; branchlets glabrous; *leaves* lanceolate to lance-ovate, *green both sides or barely paler beneath;* the unfolding ones often blunt, crisp with rufescent or sordid caducous hairs; *the later and mature ones long-attenuate into a tail-like tip, glabrous,* finely serrate, the fully *mature blades* 0.5–1.7 *dm. long* and 2–5 *cm. broad; summit of petiole or its junction with blade with coarse glands or glandular teeth;* stipules of sprouts oblong to semicircular; staminate aments orange-yellow, 2–5 cm. long, thick-cylindric or slenderly ovoid; *pistillate aments becoming 2–5 cm. long; capsule opaque,* greenish to pale brown or stramineous, *conic-ovoid,* 5–6.5 *mm. long, its pedicel about four times length of upper gland, its valves opening in late spring or early summer.* — Low grounds, shores, swamps, etc., se. Lab. to n. Man., s. to Nfld., N.S., N.E., L.I., Del., Md., O., Ind., Ill., Ia. and S.D. *Fl.* late April–June. Fig. 862. — Hybridizes with nos. 1, 6 and 9.

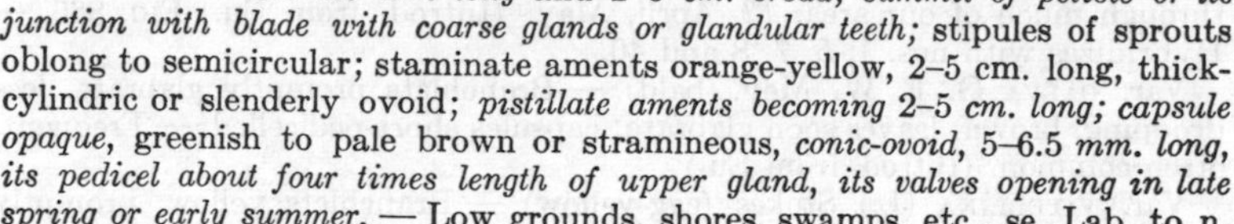

862. S. lucida.

Var. **angustifòlia** Anderss. (narrow-leaved). — Smaller throughout, often down to 0.2–1 m. high; mature narrowly lanceolate leaves 3–8 cm. long, 0.5–2 cm. broad. — Lab. to Man., s. to Nfld., N.S., n. N.E., n. N.Y., Bruce Pen., Ont., Wisc. and Minn.

Var. **intónsa** Fern. (unshaved). — Coarse shrub or small tree; new *branchlets and lower surfaces of large leaves permanently pubescent with sordid or rufous hairs.* — Nfld. and Côte Nord, Que., to Ont., s. to N.B., centr. Me., n. N.Y. and n. Ind. — Hybridizes with no. 1.

6. S. seríssima (Bailey) Fern. (very late), Autumn-W. — Shrub, differing from nos. 4 and 5 in its *short-acuminate* very firm *elliptic-lanceolate to oblong-lanceolate leaves whitened beneath,* 4–10 cm. long and 1–3 (on sprouts –4) cm. broad; *staminate aments 1–1.5 cm. long; pistillate aments becoming in maturity 2–3.5 cm. long,* with loosely divergent capsules; *olivaceous to brown-tinged finally lustrous indurated capsule conic-subulate, 7–10 mm. long, opening in late summer or autumn;* the pedicel twice exceeding upper gland. — Calcareous marshes and bogs, Nfld. and Anticosti, Que., to Alta., s. to M.I., w. Ct., n. N.J., Pa., n. O., n. Ind., Wisc., Minn., N.D. and Colo. *Fl.* May–July. Fig. 863. — Hybridizes with no. 5.

863. S. serissima.

§ 5. Frágiles W. D. J. Koch

7. S. fragilis L. (brittle), Crack-W. — Tree up to 30 m. high, with thick rough bark, *branches divergent;* the young *branchlets glabrous* or nearly so, lustrous, *very brittle at base; buds viscid; leaves lanceolate, glabrous* or promptly glabrate, *green both sides,* in maturity 1–1.5 *dm. long and 2.5–4 cm. broad* (on sprouts to oblong-ovate and to 2 dm. long by 7 cm. wide), *coarsely undulate-serrate (with 4–7 large teeth per cm.); petioles* 1–1.5 cm. long, *viscid toward summit when young;* aments terminating leafy shoots, slender, the staminate 3–5 cm. long, the pistillate becoming 5–7 cm. long; bracts oblong, pale, ciliate, caducous; *stamens usually 2, with 2 glands at base; capsule subulate-conical,* about 5 mm. long, with subsessile stigmas, *the pedicel about twice length of the 2 glands.* — Spread from cult. to roadsides, borders of woods, etc., Nfld. to Ont., s. to N.S., N.E., Va., Ky., Ill. and Ia. *Fl.* April–early June. (Introd. from Eu.) Fig. 864.

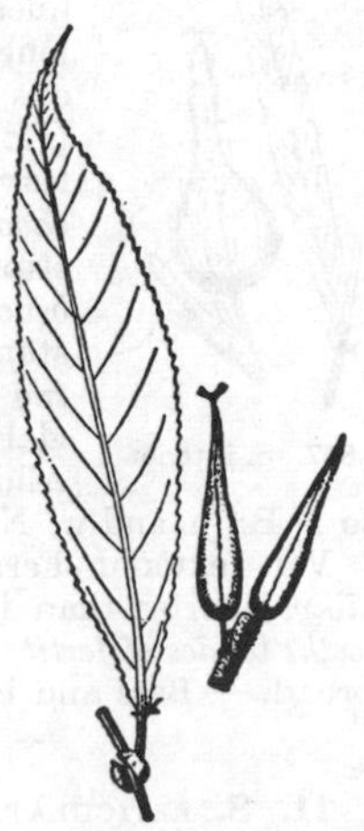

864. S. fragilis.

×S. RÙBENS Schrank (reddish), a common hybrid of nos. 7 and 9, with leaves glaucous beneath, silky when young, the pistillate flowers often with only 1 gland, is frequent. (Introd. from Eu.)

8. S. BABYLÓNICA L. (of Babylon), WEEPING W. — Differing from no. 7 in its *pendulous ultimate branches and branchlets; leaves whitened beneath, silky when young,* then glabrate, narrower, *in maturity* 2–12 *cm. long and* 0.5–2 *cm. wide, with finer serration; aments* 1.5–2 *cm. long; capsules plumper,* 1–1.5 *mm. long, nearly sessile.* — Much cult., locally spread to river-banks and shores, Que. and Ont., southw. *Fl.* April, May. (Introd. from Eurasia) FIG. 865. — Hybridizes with nos. 7 and 9.

865. S. babylonica.

§ 6. ÁLBAE Borrer

9. S. ÁLBA L. (white), WHITE W. — Differing from no. 7 in more diffuse branching, with often pendent *branchlets silky,* olivaceous to brown, *flexible and not brittle at base; leaves white-silky, finely serrulate, in maturity* 5–12 *cm. long and* 1–3 *cm. broad, the petiole not viscid; pistillate flowers with* 1 *gland; capsule conic-ovoid,* 3–5 *mm. long, nearly sessile.* — Much planted; natzd. through much of our area. *Fl.* April, May. (Introd. from Eu.) FIG. 866. — Hybridizes with nos. 1, 5, 7, 8 and 40.

866. S. alba.

Var. CÁLVA G. F. W. Mey. (bald). — Branchlets promptly glabrate, less drooping, brown; leaves soon glabrate; capsules short-pedicelled. — Frequent, often common. (Introd. from Eu.)

Var. VITELLÌNA (L.) Stokes (egg-yellow). — Branchlets yellow, promptly glabrate; leaves soon nearly glabrous. — The commonest var. with us. (Introd. and natzd. from Eu.)

×**S. Hankensònii** Dode (for its discoverer, EDWARD L. HANKENSON, 1845–1910), a hybrid of *S. alba,* var. *vitellina* and *S. nigra,* with green branchlets of the latter, leaves and stipules of the latter but white beneath, and aments much as in the former, occurs as large trees by a brook in Wayne Co., N.Y.

×**S. Jésupi** Fern. (for its discoverer, HENRY GRISWOLD JESUP, 1826–1903), a frequent and characteristic hybrid of nos. 5 and 9, differs from *S. lucida* by its narrow leaves not lustrous above but pale beneath, its narrow stipules, its slender aments and its few (3–5) stamens; from *S. alba* by its long slender leaf-tips, ferruginous pubescence of young leaves, glandular summit of petiole, more tapering and longer-pedicelled capsule and the bracts dentate as in *S. lucida.* (*S. pameachiana* sensu Anderss., not Barratt)

§ 7. LONGIFÒLIAE Anderss.

10. S. intèrior Rowlee (inland), SANDBAR-W. — *Extensively colonial* stoloniferous shrub or small tree up to 5.5 m. (usually only 1–4 m.) high, with many grayish stems; branchlets glabrous or promptly glabrate, brown; *leaves* glabrous or promptly glabrate, *firm, veiny, linear to narrowly lanceolate,* or on sprouts linear- or lance-oblong, *acuminate, remotely dentate with sharp* (often prolonged) *teeth;* the *blades in maturity mostly* 5–14 *cm. long and* 5–10 (on vigorous sprouts –15) *mm. broad,* tapering to short petioles, or in forma **Wheèleri** (Rowlee) Rouleau (for its discoverer, CHARLES FAY WHEELER, 1842–1910) leaves often broader and more or less permanently silvery-silky; *leafy lateral shoots abundant, often continuing for some time to bear lower secondary aments;* stipules minute or wanting; *aments slender, with pale deciduous yellowish bracts;* staminate aments 2–4 cm. long, the secondary later ones smaller; stamens 2, the filaments crisp-pubescent at base; *pistillate aments lengthening to* 8 *cm.; capsules slender, scattered,* 7–9 mm. long, thinly silky, often glabrate, with nearly obsolete styles. (*S. longifolia* Muhl., not Lam.) — Alluvial soils, often pioneering on bars and beaches, e. Que. to Alaska, s. to N.B., n. and w. N.E., Del., Md., O., Ind., Ill., Ark., Okla., etc. *Fl.* May, June. FIG. 867.

867. S. interior.

Var. **extèrior** Fern. (outer; from its marginal range). — *Leaves permanently* somewhat *silky,* shorter than in forma *Wheeleri, short-tipped, oblong-lanceolate, with nearly suppressed teeth; blades of fertile branchlets* 2–5 *cm. long,* of vegetative shoots 3–7 cm. long and 1–1.5 cm. broad. — Bars and beaches of Aroostook River, Me., and of Susquehanna R., se. Pa.

§ 8. RETICULÀTAE Fries

11. S. RETICULÀTA L. (netted). — *Depressed,* with prostrate or barely ascending slender branches 0.5–2 dm. long; *buds ovoid, glabrous, glutinous;* stipules glandular, caducous; *leaves* *few, on long slender* reddish *petioles, suborbicular to obovate or oval, broadly rounded at summit, firm,* dark green and rugose above, paler and *conspicuously reticulate beneath,* entire or nearly so, 1.5–4.5 cm. long, the young blades silky, the older often somewhat so beneath; *ament on long naked usually villous peduncle, borne subterminally but from the side of the branchlet opposite the terminal leaf,* very slender, 1–3 cm. long; *glands 2–4, more or less united into a dissected cup surrounding base of the* 2 hairy-based *stamens or the base of the ovary;* bracts greenish, with dark red or fuscous tips, villous; anthers dark red or purple; *ovary sessile; the ovoid capsule* densely white-pubescent, 3–4 *mm. long;* stigmas short and thick, reddish. — Arctic-alpine reg. of Eurasia and N. Am. — Represented with us by

Var. **semicálva** Fern. (partly bald). — Branchlets and leaves glabrous from the first; peduncle glabrous or only sparsely pubescent; bracts of staminate ament glabrous or promptly glabrate, of pistillate ament yellowish to pale reddish; anthers yellow; ovary and capsule thinly puberulent. — Lab. Pen., s. to calcareous barrens of n. and nw. Nfld. *Fl.* late June, July. — The members of § *Reticulatae* often considered the most primitive living members of the genus, on account of the subterminal naked peduncles, 2–4 often united glands, etc.

12. S. jejūna Fern. (insignificant). — In small mats, from creeping subterranean stolons; differing from no. 11 in its very *short petioles only* 5 *mm. long* (*shorter than to barely longer than buds*); *leaf-blades short-oblong to broadly elliptic,* 0.5–2.5 cm. long, entire, those of sprouts at first villous-ciliolate; *stipules* (on sprouts) *slender, long-bearded; fruiting aments* 0.6–1.8 *cm. long,* 7–9 *mm. thick, terminating more or less pilose branchlets, the peduncle very short; bracts fuscous to dark purple, pilose;* capsule villous-tomentose or glabrate; *style distinct,* 0.3 *mm. long.* — Calcareous barrens and cliffs, se. Lab. and n. Nfld., along Straits of Belle Isle. *Fl.* late June, July. Fig. 868.

868. S. jejuna.

13. S. vestìta Pursh (clothed). — *Stout-branched mostly ascending shrub* with dark gray somewhat angled branches, up to 1 m. high; *branches with short internodes enlarged upward and with prominent keels or angles; leaves very thick,* orbicular to obovate, elliptical or oblong, *very short-petioled,* 1–8 cm. long, dark green and usually deeply rugose above, *whitened beneath and usually with dense long appressed white silky villosity; aments* much as in no. 11, slender, *on shorter hairy peduncles; bracts silky;* ovary and capsule pubescent, the latter 5–7 mm. long. — Variable, the following vars. with us:

a. Leaves permanently clothed beneath with dense pubescence; overwintering buds pubescent; fruiting aments dense. . . . *b.*
- b. Leaves orbicular to broadly obovate, broadly rounded to emarginate at summit, the upper surface strongly rugose-reticulate; staminate aments 0.6–1.5 cm. long; fruiting ones 0.5–3 cm. long.
  - Erect or ascending; leaves 1.5–7 cm. long; fruiting aments 1–3 cm. long. . . . *S. vestita* (typical).
  - Depressed, forming prostrate mats; leaves 1–1.6 cm. long; fruiting aments 5–6 mm. long. . . . . Forma *mensalis.*
- b. Leaves elliptical, oblong or narrowly oblong-obovate, less rounded to subacute at apex, plane above; staminate aments 1–3 cm. long, fruiting ones 2–5 cm. long. . . . . Var. *erecta.*

a. Leaves only thinly silky, glabrescent, obovate, 3–8 cm. long; staminate aments 1.7–3.5 cm. long; pistillate aments remotely flowered at base. . . . Var. *psilophylla.*

S. vestìta (typical). — Calcareous rocky soil, Lab. and Ung., s. to n. and w. Nfld., and Mingan Ids., Anticosti and Gaspé Pen., Que., often subalpine. *Fl.* late June–Aug. Fig. 869. — Forma **mensàlis** Fern. (of a table-land; from its type-locality), bleak windswept alpine barren of Table Mt., Port-au-Port Bay, Nfld.

Var. **erécta** Anderss. (erect). — Local, sw. Nfld.; Mingan Ids., and e. Gaspé Co., Que.; Alta., B.C., Mont. and Oreg.

Var. **psilophýlla** Fern. & St. John (bare-leaved). — Calcareous and schistose cliffs, Mingan Ids., Shickshock Mts., Gaspé Pen., and L. Mistassini, Que.

869. S. vestita (typical).

14. S. leiólepis Fern. (smooth-scaled). — Suggesting no. 12 or small-leaved no. 11, the repent branches coarser; *glabrous* or nearly so *throughout;* the elliptic round-tipped *leaves* 0.7–2 cm. long, *conspicuously reticulate beneath, crenate, those of sprouts glabrous and estipulate, petioles only* 1–4 *mm. long; fruiting aments ellipsoid,* 5–11 *mm. long,* the *glabrous peduncle only* 1–2.5 *mm. long; bracts olivaceous or reddish-yellow, glabrous; capsules* sessile, conic-ovoid, obtuse, *glabrous, purplish,* 3.5–

870. S. leiolepis.

4.5 mm. long, the style nearly obsolete. — Mossy knolls on calcareous tableland, Table Mt., Port-au-Port Bay, Nfld. *Fl.* late June, July. Fig. 870.

§ 9. Herbàceae Borrer

15. S. herbàcea L. (herbaceous). — *Tiny shrub with trunks and main branches subterranean, stoloniferous and rooting at nodes, filiform; the very short and slender upper branchlets above ground, bearing* 2–4(–6) *glabrous* lustrous membranaceous *reticulate* rounded slender-petioled *leaves* 1–3 cm. long; *aments subterminal,* ovoid, 2–8 (rarely–10)-*flowered,* 5–8 mm. long; bracts obtuse, concave, glabrous or nearly so, yellow-green with darker margin; capsule subsessile, conic-subulate, with slender style. — Arct. reg., s. to mossy alpine areas on granitic, siliceous or schistose mts. of Nfld. and Gaspé Pen., Que.; Mt. Katahdin, Me.; White Mts., N.H.; Adirondack Mts., N.Y. *Fl.* late June–Aug. (Eurasia) Fig. 871.

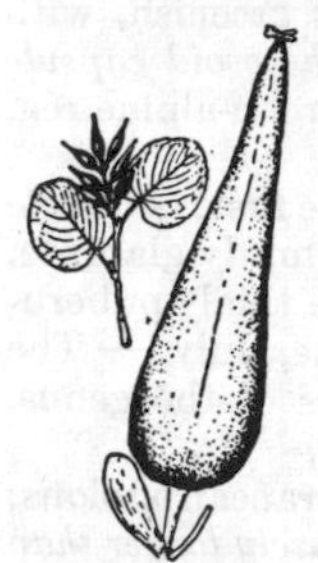

871. S. herbacea.

× **S. Peàsei** Fern. (for its discoverer, Arthur Stanley Pease, 1881–), combining characters of nos. 15 and 16, but much coarser than no. 15, with trailing ligneous branches up to 5 dm. long and 3–6 mm. thick, the slender branchlets flexuous, the prominently reticulate-veiny crenate leaves oblong to narrowly obovate and tapering to base, the fruiting aments densely many-flowered and 1.5–3 cm. long, occurs on mossy banks, King's Ravine, Mt. Adams, N.H.; also about Hudson Bay.

§ 10. Ùva-úrsi Fern.

16. S. Ùva-úrsi Pursh (bearberry). — *Prostrate, with stout base,* the *ligneous freely forking branches trailing and forming dense mats* 2–9 dm. across; *leaves elliptical to obovate,* 0.5–2.5 *cm. long, firm* (often marcescent), *lustrous-green above,* pale beneath, acute or obtuse, *finely and remotely glandular-serrate,* glabrous, or in forma **lasiophýlla** Fern. (hairy-leaved) pilose above; *aments* terminating leafy lateral shoots, *thick-cylindric, with* obovate *long-silky bracts rose-red at tip; stamen normally only* 1 (rarely 2); *fruiting aments lengthening to* 1.5–3 *cm.* and becoming slenderly cylindric, densely flowered above, often sparsely so at base; *capsule* conic-ovoid, brownish, pedicelled, *with short style and slender basal gland.* — Greenl. and e. Arct. Am., s. to barrens of Nfld. and St. Paul I., N.S., and alpine areas of Que., n. N.E. and n. N.Y. *Fl.* June, July. Fig. 872. — The aberrant forma **phyllólepis** Fern. (with leafy scales) has the bracts of the pistillate aments changed to leaves and the ovaries abortive.

872. S. Uva-ursi.

§ 11. Ovalifòliae Rydb.

17. S. arctóphila Cockerell (Arctic-lover). — Depressed, *with trailing* dark-barked *branches and brittle-based leafy branchlets; leaves dark green and lustrous above,* elliptic to obovate, acutish to rounded at tip, gradually rounded to cuneate to slender petiole, 1–4.5 cm. long, entire or essentially so; *aments erect, perpendicular to the prostrate axis; bracts fuscous, with blackish tip, long-bearded at apex;* stamens 2, the filaments pilose at base; *pistillate aments elongating in fruit to* 3–10 *cm.*; ovary and capsule pubescent or glabrescent, or in forma **lejocárpa** (Anderss.) Fern. (smooth-carpelled) glabrous, subsessile, with short and broad gland; *capsule lance-conic,* 7–10 *mm. long,* the style long and slender. — Greenl. and e. Arct. Am., s. to barrens, meadows or alpine reg. of Nfld., Anticosti and Shickshock Mts., Gaspé Pen., Que., and Mt. Katahdin, Me., and barrens about Hudson Bay, Ung. *Fl.* mid-June, July. Fig. 873. — Hybridizes with no. 16.

873. S. arctophila.

18. S. Árctica Pallas (Arctic). — Similar to no. 17, the trunk more deeply in the ground, the horizontal branches stouter, these and the *branchlets flexible; leaves paler green, dull or opaque, cobwebby-villous when young,* soon glabrate, obovate, often retuse, cuneate or tapering at base, 2.5–7.5 *cm. long; aments not strongly divergent, nearly parallel with axis of branchlet;* bracts blackish, obtuse, hairy; pistillate aments becoming 2.5–9 cm. long; *gland elongate, much exceeding short pedicel.* — Siberia and Arct. nw. Am. — Represented with us by three varieties with much more slender and flexible branches and smaller leaves (1.5–5 cm. long) and fruiting aments 1–5.5 cm. long:

Young leaves villous above; the mature ones thick and chartaceous, broadly elliptic-ovate to suborbicular, broadly cuneate to subcordate at base, rounded to barely subacute at apex, 1.5–3.5 cm. long, 1.2–3 cm. broad; fruiting aments dense, up to 3.5 cm. long; capsules about 7 mm. long. . . . . Var. *kophophylla.*

Young leaves glabrous; fruiting ament lax at base.
Leaves thin-papery, oval, elliptic or broadly obovate, rounded above, rounded to broadly cuneate at base, in maturity 1.5–5 cm. long and 1–3 cm. broad, the lateral nerves joining midrib at a broad angle; fruiting aments 3–5.5 cm. long and 1.8 cm. thick; capsules 7–8 mm. long. . . . . Var. *araioclada.*
Leaves firm and thick, narrowly oval to narrowly oblong-oblanceolate or -obovate, mostly acute at both ends, the tip often plicate-acuminate, in maturity 1.5–3 cm. long and 0.8–1.8 cm. broad, the lateral nerves meeting midrib at a sharp angle; fruiting aments 1–2 cm. long, about 1.2 cm. thick; capsules 6 mm. long. Var. *antiplasta.*

Var. **kophophýlla** (Schneid.) Polunin (blunt-leaved). — Canadian e. Arct., s. to serpentine slopes and tablelands of w. Nfld. and tableland of Mt. Albert, Gaspé Pen., Que. *Fl.* late June–Aug. Fig. 874.

874. S. arctica, v. kophophylla.

Var. **araióclada** (Schneid.) Raup (slender-branched). — Torngat Mts., n. Lab.; serpentine barrens, Mt. Albert, Que.; Rocky Mts., Alta.; Selkirk Mts., B.C.

Var. **antiplásta** (Schneid.) Fern. (similar). — Serpentine slopes and tableland, Mt. Albert, Que.

× **S. Waghórnei** Rydb. (for its discoverer, Arthur Charles Waghorne, 1851–1900) seems to be a hybrid of *S. arctica*, var. *kophophylla* and a form of no. 20.

§ 12. Glaùcae Fries

19. S. brachycárpa Nutt. (short-carpelled). — *Dwarf ascending shrub* 1–4 *dm. high,* with fuscous or blackish bark, the short internodes upwardly enlarged; *new branchlets very short; the oblong to barely oblong-obovate thick white-tomentulose* entire or subentire *blunt leaves ascending and strongly overlapping,* 1–3.5 *cm. long by* 0.5–1.5 *cm. broad, on thick petioles shorter than the well developed pilose axillary buds; aments in anthesis subglobose to short-ellipsoid, the pistillate in maturity lengthening to* 0.5–2.7 *cm. long; bracts yellowish to fulvous, pilose; anthers ellipsoid-globose; ovary and capsule subsessile, with* 2 *basal glands,* white-tomentulose; capsule 4–5 mm. long. — Calcareous or magnesian rock or gravel, Anticosti and Mt. Albert, Gaspé Co., Que.; Ung. to n. B.C., s. to Man., Sask., Colo., Utah and Oreg. *Fl.* late June–Aug. Fig. 875.

875. S. brachycarpa.

Var. **antimìma** (Schneid.) Raup (closely resembling). — Leaves soon green and glabrescent; branchlets smoother; capsules merely cinerous-villous. — Serpentine slopes and tableland, Mt. Albert, Que.; Ung. to Alaska.

× **S. gaspeénsis** Schneid. (of the Gaspé Peninsula) is an evident hybrid, on Mt. Albert, of no. 19 with no. 21.

20. S. cordifòlia Pursh (heart-leaved). — Depressed or ascending shrub *up to* 2 *m. high* (in sheltered situations), *with flexible branchlets; leaves* mostly loosely divergent or ascending, *on slender petioles longer than axillary buds*; the elliptic to ovate, obovate or orbicular *blades* 1–9 cm. long and 0.5–5 cm. broad, rather thin, *mostly villous or pilose when young; aments slenderly cylindric,* terminating short leafy branchlets, *the pistillate ones becoming* 1–7 *cm. long; bracts yellowish to fulvous,* villous; stamens 2, with glabrous filaments, *anthers elongate-ellipsoid;* ovary and capsule pubescent, style elongate. — Highly variable mostly calcicolous species; the following varieties with us:

a. New branchlets and young leaves more or less densely villous or silky. . . *b.*
b. Leaves of fruiting branches oval to ovate or obovate, cordate to broadly rounded at base, mostly 3–7 cm. long and 2–5 cm. broad, those of sprouts larger and definitely cordate. . . . . *S. cordifolia* (typical).
b. Leaves of fruiting branches not cordate, if rounded at base mostly smaller and narrower. . . *c.*
c. Mature leaves mostly 2.5–6 cm. long, oblanceolate, oblong, elliptic or narrowly obovate.
Leaves glabrate in maturity or only sparsely silky-villous along nerves beneath. . . . . Var. *callicarpaea.*
Leaves permanently and rather densely villous. . . . . Var. *intonsa.*
c. Mature leaves mostly only 1–2.5 cm. long, orbicular to short-oblong or narrowly obovate.
Leaves elliptic, short-oblong or narrowly obovate, acute or subacute, 0.5–1.5 cm. broad. . . . . Var. *Macounii.*
Leaves orbicular to short-oblong, rounded at summit, 0.7–2 cm. broad. Var. *eucycla.*

a. New branchlets and young leaves glabrous or essentially so, obovate, 2.5–9 cm. long, 2–5 cm. broad. . . . . Var. *tonsa.*

S. cordifòlia (typical). — Lab. Pen., s. to calcareous mts., barrens and gravels of nw. Nfld. and Mingan Ids., Que., very local. Fig. 876.

876. S. cordifolia (typical).

Var. **callicarpaèa** (Trautv.) Fern. (with beautiful fruit). — Greenl. and Arct. e. Canada, s. to limestones and schists of n. and w. Nfld., St. Paul I., N.S., and Shickshock Mts., Gaspé Pen., Que.; the commonest var. *Fl.* late June, July. Fig. 877.

Var. **intónsa** Fern. (unsheared). — Lab. Pen., s. to n. Nfld. and Shickshock Mts., Que.

Var. **Macoùnii** (Rydb.) Schneid. (for its discoverer, John Macoun, 1831–1920). — Lab. Pen., s. to Notre Dame Bay and Bonne Bay, Nfld., and Shickshock Mts., Que.

Var. **eucỳcla** Fern. (well rounded). — Prostrate on turfy or boggy calcareous or magnesian barrens, n. and w. Nfld.; near Straits of Belle Isle, Saguenay Co., Que.

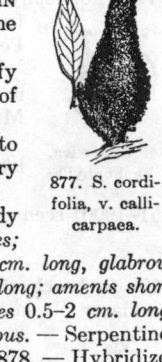

877. S. cordifolia, v. callicarpaea.

Var. **tónsa** Fern. (sheared). — Canadian e. Arct., s. to n. and nw. Nfld. and Shickshock Mts., Que. — Leaves very thin; aments large, up to 9 cm. long.

21. S. chlorólepis Fern. (green-scaled).—Dwarf, with woody trunks 1–2 *dm. high, with wide-spreading glabrous branches; leaves crowded, oblong, elliptic or oblanceolate,* firm, 1–2.5 *cm. long, glabrous or glabrescent,* essentially entire, the *thick petioles* 2–3 *mm. long; aments short, at first subglobose to ellipsoid* as in no. 19, the *fruiting ones* 0.5–2 *cm. long; anthers subglobose; bracts olivaceous, glabrous; capsules glabrous.* — Serpentine-tableland, Mt. Albert, Gaspé Co., Que. *Fl.* July, Aug. Fig. 878. — Hybridizes with no. 19.

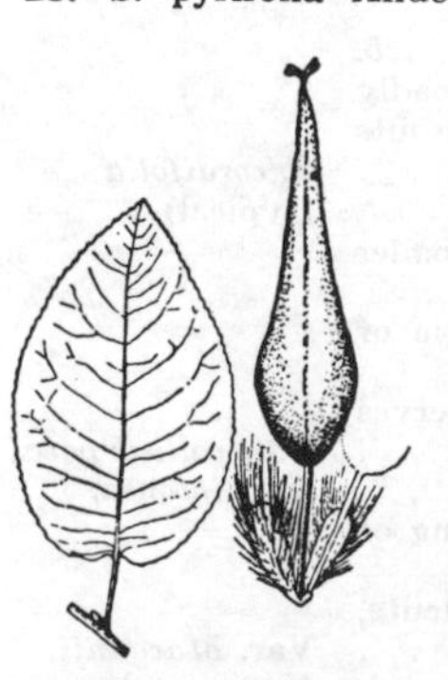
878. S. chlorolepis.

§ 13. Argyrocárpae Fern.

22. S. argyrocárpa Anderss. (silvery-carpelled). — Ascending shrub 0.2–1.7 m. high; *earliest leaves obovate and obtuse; the later ones narrowly lanceolate, oblanceolate or narrowly elliptic, undulate-crenate,* blunt or merely acute, *tapering to both ends,* 2–6 *cm. long, silvery-silky beneath with minute glistening hairs; stipules minute, fugacious; fruiting aments lax,* 1.5–2.5 *cm. long;* bracts pubescent, yellowish; *glistening-silvery capsule* slenderly conic, tapering to long style, *its slender pedicel three or four times length of the usually* 2 *glands;* filaments glabrous, the basal glands 2. — Lab. Pen., s. to alpine or subalpine meadows or wet rock of Shickshock Mts., Gaspé Pen., Que., and White Mts., N.H. *Fl.* June–Aug. Fig. 879.

879. S. argyrocarpa.

× **S. Graỳi** Schneid. (for its discoverer, Asa Gray, 1810–1888), a hybrid of nos. 22 and 47, differs from the former in its coarser and taller habit, pruinose branches, larger and broader leaves less silky, and pistillate aments larger; from the latter in its uniformly glaucous hue, the young leaves not purplish, the styles red-purple instead of yellow.

§ 14. Balsamíferae Schneid.

23. S. pyrifòlia Anderss. (with leaves of the pear), Balsam-W. — Shrub, rarely small tree up to 7 m. high, *with strong balsamic-resiniferous fragrance;* branches with shining reddish-brown or olivaceous bark, the branchlets brittle at base; *stipules* 0 *or minute and caducous; leaves very thin and pellucid when young,* then purple-tinged; the *mature ones* firmer, *short-oval to ovate or oblong-lanceolate,* pale and *prominently reticulate beneath, those* of flowering shoots obtuse or rounded at summit, *of later and leading shoots acute or acutish,* cordate to tapering at base, *on long slender petioles;* mature blades 3–10 cm. long, *slightly glandular-serrulate; aments terminating leafy-bracted branchlets; bracts whitish, very thin, elongate;* staminate aments 2–4 cm. long, 1–1.5 cm. thick, the 2 long filaments glabrous; *pistillate aments becoming* 3–9 *cm. long* and 1.5–2 cm. thick; *the loosely divergent slender pedicels finally six to eight times as long as the broad gland;* capsule glabrous, lance-conic, with short style. (*S. balsamifera* Barratt) — Low thickets and borders of woods, s. Lab. to Mackenz. and n. B.C., s. to Nfld., N.S., n. N.E., n. N.Y., Mich., Wisc.,

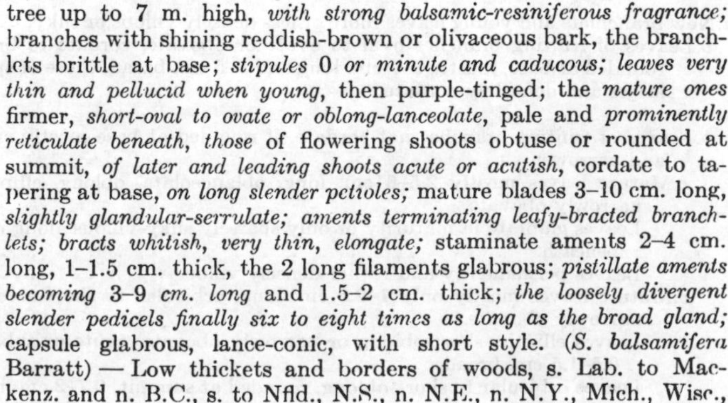
880. S. pyrifolia.

Minn., Man., Sask., etc., ascending to subalpine areas. *Fl.* Mid-May–Aug. Fig. 880. — **Hybridizes with no. 38.**

§ 15. Cordàtae Barratt

24. S. cordàta Michx. (heart-shaped). — Coarse shrub 0.3–1.5 m. high, the new and often the older *branchlets densely velutinous with gray villosity; stipules* large, *obliquely semicordate-ovate,* usually *with* 10–22 (*av.* 14) *gland-tipped or soon glandless teeth on the longer margin; leaves broadly lance-oblong to oblong-ovate, long-acuminate, gradually tapering from below or near the middle,* the young and often the mature *densely gray-villous to subsericeous* on both surfaces or *especially beneath,* the *mature blades* 3–13 cm. long and (1.5–) 2–6 *cm. broad, each margin with* 25–90 (*av.* 55) *forward-arching at first gland-tipped but soon glandless mostly simple teeth;* mature *petiole* (2–)5–35 (av. 13) mm. long, *villous;* staminate aments scarcely 2 cm. long, subtended by hardly foliaceous bracts, subsessile; *filaments barely exserted* beyond the bright white beard of the obovate blackish to brown bracts; *anthers roundish;* pistillate aments in maturity 2–6 cm. long; their bracts narrowly ovate, fuscous or brown, the bright white beard only slightly longer; *ovaries* glabrous, *in anthesis appressed-ascending, in fruit more spreading, on pedicels shorter than bracts;* style 0.2–0.5 mm. long. (*S. adenophylla* Hook., not sensu Schneid.; not *S. cordata* Muhl.) — Gravelly or sandy shores, beaches and dunes, se. Lab. Pen. to n. Ont., s. to Nfld., N.S., n. Me., e. Cape Cod, Mass., n. N.Y., Simcoe and Bruce Cos., Ont., and n. Mich. *Fl.* late May, June. Fig. 881. — Probably hybridizes with nos. 25 and 28.

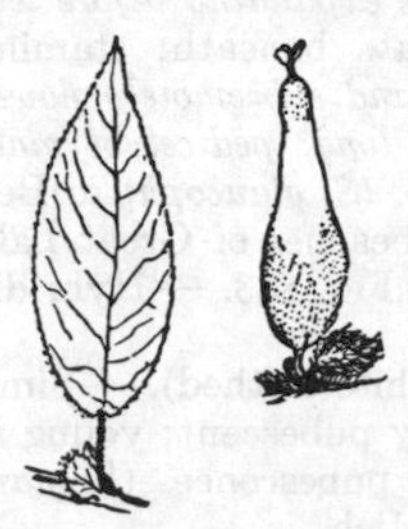

881. S. cordata.

Var. **abràsa** Fern. (rubbed off.) — Branchlets, petioles and leaf-blades glabrous or promptly glabrate except for pubescent midrib. — Nfld. to L. St. John, Que., s. to N.S., Gaspé Pen., Que., n. N.B. and n. Me.

25. S. rígida Muhl. (stiff). — Shrub 0.3–3 m. or more high, resembling no. 24, *with conspicuously red to claret-purple expanding leafy tips; branchlets glabrous or glabrate;* stipules with fewer broad-based rarely gland-tipped dentations; *leaves oblong-lanceolate, subcordate to broadly rounded at base, glabrous* or promptly glabrate, or in forma **móllis** (Palmer & Steyerm.) Fern. (soft) midribs as well as petioles and branchlets all soft-pubescent, the *mature blades one-sixth to one-third as broad as long,* the larger blades 1.5–4.5 (av. 2.75) *cm. broad; stamens long-exserted,* the *anthers elongate; pistillate aments in anthesis with widely* (*often horizontally*) *divergent ovaries, the widely divergent capsules on pedicels as long as to much longer than bracts.* (*S. cordata* Muhl., not Michx.) — Banks of streams, pond-shores and low thickets, s. Nfld. to Ont., s. to N.S., N.E., N.C., Miss., Ark. and Kans. *Fl.* April–June, Fig. 882. — Hybridizes with nos. 3, 10, 24, 26, 28, 33, 36, 38, 40, 41 (× *S. myricoides,* see below), 42, 45, and with some introd. and cult. spp. — unfortunate behavior for a species called *S. rigida.*

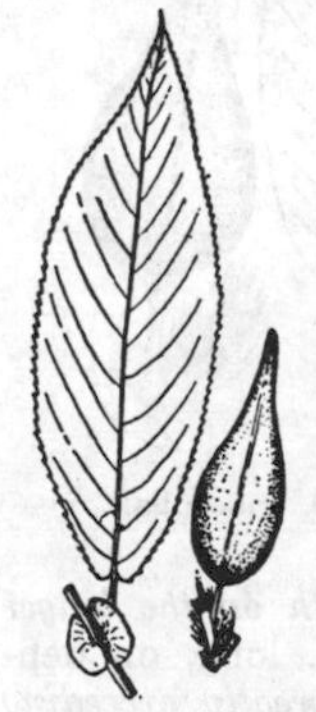

882. S. rigida.

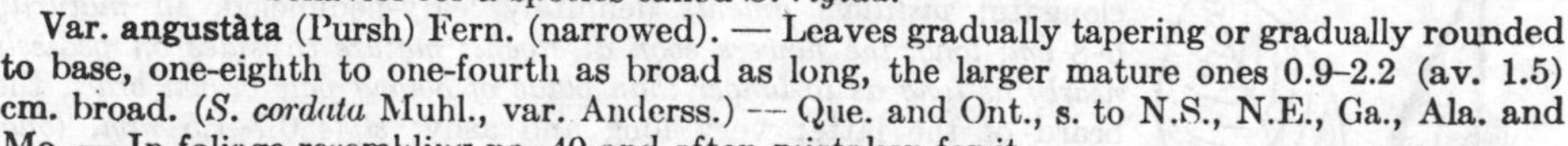

Var. **angustàta** (Pursh) Fern. (narrowed). — Leaves gradually tapering or gradually rounded to base, one-eighth to one-fourth as broad as long, the larger mature ones 0.9–2.2 (av. 1.5) cm. broad. (*S. cordata* Muhl., var. Anderss.) — Que. and Ont., s. to N.S., N.E., Ga., Ala. and Mo. — In foliage resembling no. 40 and often mistaken for it.

× **S. myricoides** (Muhl.) Carey (resembling *Myrica*), a very common hybrid of no. 25 with no. 41, has cinereous often canescent-pubescent shoots, branchlets brittle at base; leaves very long-tapering to apex, narrowed to base, glaucous and often sericeous beneath; stipules small; ovaries or capsules often slightly silky, the pedicels very short. — Found throughout the coincident ranges.

26. S. glaucophylloìdes Fern. (resembling *S. glaucophylla*). — *Coarse shrub or small tree* up to 5 m. high; branchlets brown or castaneous (rarely yellowish), glabrous or promptly glabrate and lustrous, or in forma **lasióclada** Fern. (with hairy sprouts) permanently grayish-velvety; *leaves oblong or lanceolate to narrowly ovate, lustrous-green above, white-glaucous beneath, partially grown at anthesis,* blackening in drying; the mature blades subcoriaceous, 3–12 cm. long and 1.5–6 cm. broad, *acute or subacute to short-acuminate,* rounded to cordate or tapering at base, closely crenate or crenate-serrate with gland-tipped teeth, the petioles 3–12 mm. long; stipules semicordate, coarsely toothed, subpersistent; *aments expanding with the leaves,* subtended by 3–5 small leaves; *bracts dark brown to blackish,* oblong to obovate, obtuse, 1.5–2 mm. long, white-bearded; *staminate aments* 2–4 cm. long, the prolonged filaments glabrous, *the anthers oblong; pistillate aments close*, in maturity 2–6 cm. long; *ovary and capsule glabrous;* the *capsule* conic-subulate, 7–10 *mm. long, its pedicel* 1–1.5 *mm. long.* — Gravelly shores and rich thickets, chiefly in calcareous soil, Nfld. to n. Ont., s. to Gaspé Pen., Que., n. N.B. and n. Me. *Fl.* late May, early June.

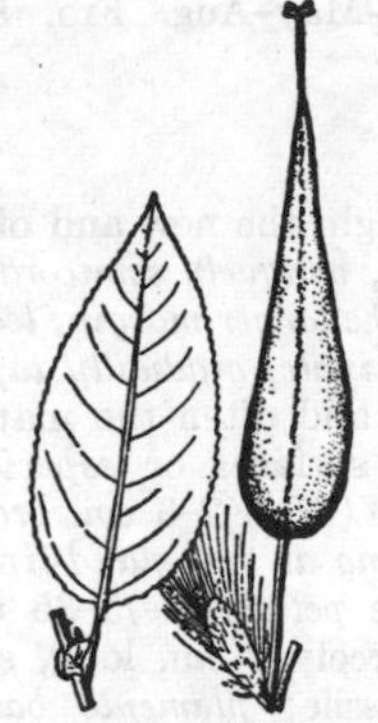

883. S. glaucophylloides, v. glaucophylla.

Var. **glaucophýlla** (Bebb) Schneid. (with blue-green leaves). — *Spreading shrub* 1–2.5 m. high, glabrous or promptly glabrescent; *leaves less developed at flowering time; aments expanding before leaves well unfolded;* mature leaves more reticulate beneath; staminate aments 3–5 cm. long; *pistillate aments lax and subremotely flowered in anthesis, elongating in fruit to* 6–10 *cm. long; pedicel of mature capsule* 2–4 *mm. long, often overtopping bract.* (*S. glaucophylla* Bebb, not Bess. nor Anderss.) — Sand-dunes and beaches of Great Lakes, Ont. and O. to Wisc. and Ill. *Fl.* early May. Fig. 883. — Hybridizes with no. 25.

Var. **albovestìta** (C. R. Ball) Fern. (white-clothed). — Similar to var. *glaucophylla* but branchlets densely pubescent; young and often older leaves clothed with dense white pubescence. (*S. glaucophylla*, var. C. R. Ball) — Dunes of Great Lakes, N.Y. and Ont. to Mich.

27. S. eriocéphala Michx. (cottony-headed). — Coarse shrub or tree up to 16 m. high, with black bark; *leading shoots permanently and densely puberulent to tomentose; leaves narrowly lanceolate to ovate-oblong, dull green above, glaucous and permanently pilose* (at least on nerves) *beneath, attenuate to tail-like tip;* the mature blades 8–15 cm. long and 1–4 cm. broad, minutely serrulate; *stipules subreniform,* much smaller than in nos. 24–26; *aments almost precocious,* expanding before leaves are much developed, resembling those of no. 26, the fruiting ones 6–10 cm. long. (*S. missouriensis* Bebb) — Bottomlands and shores, sw. Ind. to Minn. and S.D., s. to Ky., Mo. and Neb. *Fl.* late March, April. Fig. 884.

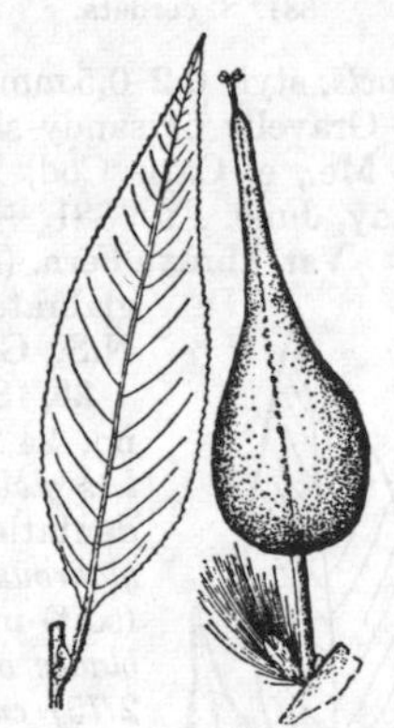

884. S. eriocephala.

28. S. syrtícola Fern. (sand-dweller), Sand-dune-W. — *Low colonial shrub* 1–3 m. high, *with* numerous ascending *stout gray-tomentulose to -puberulent branches; leaves oblong-ovate, acute or abruptly short-acuminate from well above the middle, densely lustrous-pubescent,* the mature ones 3.5–9.5 cm. long and 2–6 cm. broad; *each margin with* 81–137 (*av.* 105) *horizontally or subhorizontally divergent and permanently gland-tipped prolonged often compound teeth;* mature petiole 2–10 (–15) (av. 6.3) mm. long, thick; *mature and larger stipules with* 24–40 (*av.* 32) *mostly gland-tipped and straight teeth on the longer margin;* staminate aments in full anthesis 2–4 cm. long, on definitely leafy-bracted peduncles; elongate *filaments greatly exceeding* the long *ashy-white beard of the oblong pale brown bracts,* anthers elongate; *pistillate aments* definitely leafy-peduncled, *in maturity* 6–8 *cm. long, the flowers soon divergent; mature capsules on pedicels nearly as long as to longer than blade of oblong pale brown bract,* the beard of the latter very long and ashy; *style* 0.7–1.5 *mm. long.* (*S. adenophylla* sensu Bebb, of Schneid. and of C. R. Ball, in large part, not Hook.) — Dunes and beaches of the Great Lakes, w. Ont., Mich., Ind., ne. Ill. and e. Wisc. *Fl.* May, early June. Fig. 885. — Hybridizes with no. 25 and perhaps 24.

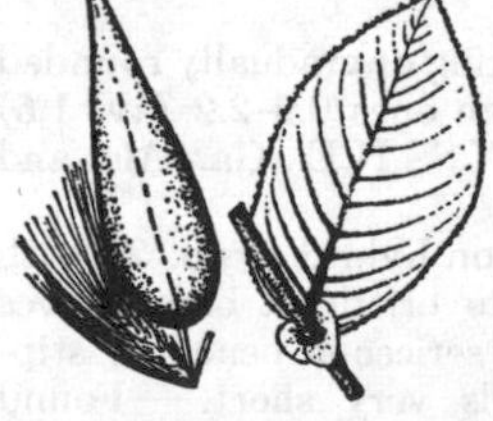

885. S. syrticola.

29. S. lùtea Nutt. (yellow). — *Gray-barked* shrub or small tree; the *branches and branchlets yellow, becoming gray; leaves yellow-green above, glaucous beneath,* resembling those of nos. 25 and 26, *firm,* acuminate, with rounded to cordate or attenuate bases; stipules rounded; *aments on short leafy-bracted peduncles, the dense fruiting ones* 2–5 *cm. long;* bracts blackish-tipped; *capsule glabrous,* 4–5 *mm. long,* long-pedicelled; *style* 0.2–0.5 *mm. long.* — Stream-banks, n. Ont. to Alta., s. to nw. Ia., Neb., Colo., Utah and Calif. *Fl.* late May, early June. Fig. 886.

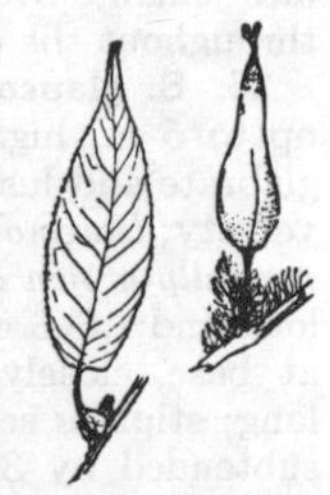

886. S. lutea.

30. S. laurentiàna Fern. (of the River and Gulf of St. Lawrence). — *Coarse shrub or small* widely branching *tree* up to 5 m. high; *branchlets stout, densely canescent-tomentose,* the tomentum sublustrous; *leaves of* *fertile branches oblong to oblong-ovate, at first silky-tomentose on both surfaces; later glabrescent and lustrous, dark green above* and glaucous beneath, 6–18 *cm. long and* 3–4.5 *cm. broad* (on sprouts up to 2 dm. long and 9.5 cm. broad), subentire or shallowly crenate, acute, short-acuminate or even cuspidate, *subcordate* (on sprouts often cordate) *or rounded at base* to pubescent petiole; *pistillate aments on short leafy branches,* dense, on canescent-tomentulose peduncles, *in fruit* 4–10 *cm. long;* bracts oblong, obtuse, dark brown, long-bearded; *ovaries and capsules* (at least at base) *tomentulose,* the subulate-conic capsules 5–7 mm. long; pedicels about 1 mm. long, one-half to twice exceeding gland; style about 0.5 mm. long. — Bluffs and slopes near lower St. Lawrence R. and Gulf, Straits of Belle I. to Mingan Ids. and n. Gaspé Co., w. to Matane Co., Que.; L. Mistassini, Que. to James Bay, Ung. and n. Ont. *Fl.* June. Fig. 887.

887. S. laurentiana.

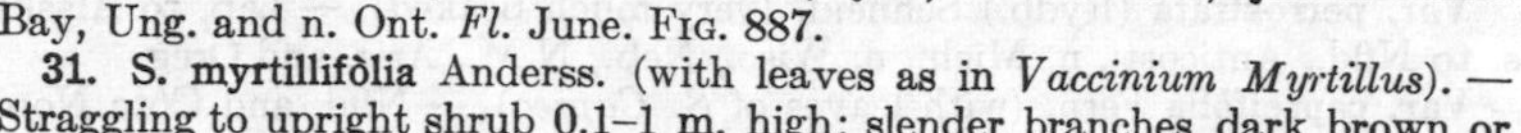

31. S. myrtillifòlia Anderss. (with leaves as in *Vaccinium Myrtillus*). — Straggling to upright shrub 0.1–1 m. high; slender branches dark brown or fuscous, smooth; *leaves dark green and lustrous on both sides or only slightly paler beneath, oblong to narrowly oblong-obovate, tapering to blunt or bluntish tips,* rounded to subcordate at base, *regularly dentate-serrate, the young ones darkening in drying, those of fertile branches mostly without stipules;* the mature blades 2–8 cm. long and 1–3 cm. broad, *their thick petioles only* 2–5 *mm. long;* vigorous sprouts often with broad coarsely toothed stipules and larger leaves; *aments expanding with or after the leaves,* on short small-bracted peduncles, the staminate 1.5–2.5 cm. long; their bracts pale brown to fuscous, obovate, rounded above, with grayish beard; pistillate aments in maturity 2–3 cm. long; *capsule glabrous,* conic-subulate, *distinctly pedicelled.* — Calcareous swamps, shores and alpine meadows, se. Lab. to Alaska, s. to Nfld., Anticosti I. and Gaspé Pen., Que., James Bay reg. of Ung. and n. Ont., Man., Sask., s. Alta. and s. B.C. *Fl.* late June, July.

Var. **brachýpoda** Fern. (short-stalked). — Prostrate or very low and depressed; *leaves glaucous beneath;* bracts of ament black; *capsules sessile or subsessile.* — Calcareous barrens and summits, n. and w. Nfld.; L. Mistassini, Que. — Resinous-balsamic in drying.

32. S. obtusàta Fern. (blunt). — Low slender shrub 0.5–1 m. high, with slender lustrous brown glabrous branches; *leaves pale green, opaque, oblong or the terminal ones suborbicular, broadly rounded at summit,* 2–5 cm. long, 1–3 cm. broad, glabrous or when young puberulent or pilose, the midrib beneath retaining pubescence, purplish when young, becoming subcoriaceous, broadly dentate-serrate; slender petioles 5–12 mm. long; *stipules cordate, persistent,* obscurely glandular-dentate, up to 5 mm. or more long; *pistillate aments sessile,* 0.5–2 *cm. long; bracts* oblong-ovate, obtuse, somewhat *fuscous, loosely villous; capsules* glabrous, conic-subulate, *reddish or yellowish,* 2–3 *mm. long;* style very short; short pedicel exceeding the gland. — Gravels along R. St. Anne des Monts, Gaspé Co., and by L. Mistassini, Que. *Fl.* June.

§16. Fúlvae Barratt

33. S. Bebbiàna Sarg. (in honor of Michael Schuck Bebb, 1833–1895, American student of the genus), Long-beaked W., Chaton or Petit Minou (Que.). — Shrub or small tree with furrowed gray or brown bark; *new and often older branchlets gray with pilosity or tomentum; leaves oblong to ovate or oblong-oblanceolate, toothed or subentire,* acute or acutish, *at first gray- or whitish-pilose to -tomentose,* either permanently so or glabrescent, *often rugose-veiny beneath, opaque above; aments expanding with the young leaves, flowering from base to apex;* flowers each with 1 basal gland; *bracts yellowish-brown, sparingly pilose;* filaments pilose at base; fruiting aments very lax and open; *fruiting pedicels three or four times length of subtending bract; ovary silky; capsule blunt, somewhat pubescent; style obsolete or nearly so.* — One of the widest-ranging and most variable species. The following varieties are recognized:

a. Capsules 5–9 mm. long, their pedicels 2–5 mm. long. . . *b.*

b. Fruiting aments 2.5–5 cm. long; mature pedicels 3–5 mm. long; leaves subentire to dentate or dentate-serrate. . . *c.*

c. Leaves at first gray-pubescent, soon only slightly so to glabrescent, ovate-oblong to oblanceolate, those of fertile branches becoming 3–8 cm. long and 1–3 cm. broad; slender branchlets at first pubescent, often losing pubescence by 2nd year; chiefly shrubs.

Mature leaves reticulate-rugose beneath. . . . . *S. Bebbiana* (typical).

Mature leaves plane and scarcely reticulate beneath. . . . . . Var. *perrostrata.*

c. Leaves densely and permanently pilose-tomentose beneath and somewhat so above, oblong to ovate, elliptic or obovate, in maturity 2–4 (on sprouts to 6) cm. broad; coarse branchlets densely pubescent to 2nd year; coarse shrub or tree to 4 m. high. . . . . Var. *capreifolia.*

b. Fruiting aments 6–8 cm. long; pedicels 2–3.5 mm. long; young leaves glandular-toothed, heavily tomentose; coarse shrub or small tree, the stout branchlets densely and permanently pubescent. . . . . . Var. *projecta.*

a. Capsules 9–12 mm. long, on pedicels 5–8.5 mm. long; fruiting aments 4–8 cm. long and 2.5–3 cm. thick; leaves scarcely rugose, sparsely pilose, soon glabrate, mostly 5–10 cm. long; branchlets densely pubescent 1st year; coarse shrub or tree up to 6 m. high, with trunk up to 1.5 dm. in diameter. . . Var. *luxurians.*

S. Bebbiàna (typical), *S. rostrata* Richards., not L. — Moist to dry thickets, Nfld. to Alta., s. to N.S., N.E., L.I., Md., Pa., O., Ind., Ill. and Ia. *Fl.* late April–early June. Fig. 888. — Hybridizes with nos. 25, 26, 38, 39, 40 and 45.

Var. **perrostràta** (Rydb.) Schneid. (very much beaked). — Lab. to Alaska, s. to Nfld., Anticosti, n. Mich., n. Wisc., Neb., N.M., Ariz. and Oreg.

Var. **capreifòlia** Fern. (with leaves of *S. Caprea*). — Nfld. and Côte Nord, Que., s. to N.S.

Var. **projécta** (Fern.) Schneid. (thrust forward). — Coast near Bay of Islands, Nfld. — Little known; probably a distinct species.

Var. **luxùrians** Fern. (luxuriant). — Near lower St. Lawrence R., Rimouski Co. to Gaspé Co., Que.

888. S. Bebbiana.

§ 17. Chrysántheae W. D. J. Koch

34. S. calcícola Fern. & Wieg. (growing on lime). — *Stout* woody-branched and *gnarled shrub*, either prostrate or ascending and up to 5 dm. high, divergently branched; *new very short internodes densely villous, older ones glabrate; leaves oblong-ovate to suborbicular, green, glabrous or glabrate, entire or nearly so, subtended by persistent or subpersistent large coriaceous dentate stipules*, the short- and thick-petioled *coriaceous leaf-blades* mostly 1.5–5.5 *cm. long* and 1–4.5 *cm. broad; aments thick, sessile, borne from near or at tip of branchlets from large subglobose* pubescent *overwintering buds*, expanding as new leaves begin to show; bracts dark, heavily bearded; *staminate aments subglobose to thick-ellipsoid*, 1.5–3.5 cm. long and 1.5–2 cm. thick; filaments glabrous; *pistillate aments* in maturity 3–7.5 cm. long, *dense, erect or nearly so; capsule glabrous; styles* 1.5–2.5 *mm. long.* — Calcareous rocks and barrens, Baffin I. to nw. coast of Hudson Bay, s. to n. and nw. Nfld., alpine bluffs of Shickshock Mts., Gaspé Pen., Que., and James Bay, Ung.; mts., Alta. *Fl.* June. Fig. 889.

35. S. Wiegándii Fern. (for its discoverer, Karl McKay Wiegand, 1873–1942). — Depressed or ascending as in no. 34, 1–3 dm. high; *young branches and buds whitish-tomentose; leaves oval or oblong, becoming* 1.5–4.5 *cm. long and* 0.4–2.8 *cm. broad*, stiffly coriaceous, sparsely *flocculent-tomentose on both faces*, the margins revolute, the *veins of lower surface prominent and pitted-reticulate; stipules glandular-serrate; pistillate aments on short leafy peduncles, in maturity* 2.7–6 cm. long and 1–1.5 *cm. thick;* the peduncle and rachis villous or tomentose; bracts sordidly crisp-villous; *capsule sparsely villous-tomentose or glabrate;* staminate aments 1–1.5 cm. long; style about 2 mm. long. — Limestone-barrens near St. John and Ingornachoix Bays, nw. Nfld. *Fl.* late June, early July.

889. S. calcicola.

§ 18. Róseae Anderss.

36. S. pedicellàris Pursh (having pedicels). — *Slender creeping and stoloniferous subsimple to loosely branching shrubs* 0.3–1 m. high; *the flexible glabrous branches erect, without stipules; expanding leaves purplish; mature blades entire, with revolute margin*, elliptic, oblong, oblong-obovate or oblanceolate, *those of fertile branches only* 1–5 *cm. long; aments terminating leafy branchlets; bracts yellow, weakly pilose to glabrous;* staminate aments 1–3 cm. long, slender, with glabrous filaments; *pistillate aments fulvous or yellowish, thick, in maturity* 1–3 *cm. long, loosely few-flowered;* capsules glabrous, 5–10 mm. long, the *mature loosely divergent filiform glabrous pedicels twice as long as bracts; stigmas essentially sessile.* — Three vars.:

Leaves obovate-oblong to broadly oblanceolate, obtuse or subacute, in maturity 1–2.5 cm. broad; fruiting aments lax but not conspicuously open; capsule ovoid at base, tapering to blunt-tipped beak, 5–8 mm. long.

Leaves green, not glaucous, beneath. . . . . *S. pedicellaris* (typical).

Leaves whitened and glaucous beneath. . . . . Var. *hypoglauca.*

Leaves oblanceolate to linear-oblong, acute at each end, 6–10 mm. wide, glaucous beneath; fruiting ament very lax, with distant pedicels; capsule subulate, 7–10 mm. long. . . . . Var. *tenuescens*.

S. pedicellàris (typical). — Acid bogs and sphagnous shores, local, sometimes subalpine, St. P. et Miq. to B.C., s. to Gaspé Pen., Que., n. N.E., N.Y., s. Ont., n. Mich., Wisc. and Wash. *Fl.* May–July. Fig. 890.

Var. **hypoglaùca** Fern. (glaucous beneath). — Usually more common, Lab. to B.C., s. to Nfld., N.S., N.E., n. N.J., Pa., O., Ind., n. Ill., n. Ia., Man. and Oreg. *Fl.* late April–early June. — Said to hybridize with no. 25.

Var. **tenuéscens** Fern. (becoming thin). — Locally abundant, centr. Me. to n. N.H., s. to ne. Mass.

37. S. hebecárpa Fern. (with hairy carpels; only when young). — Resembling no. 36; the *branches loosely spreading or ascending*, only 1–3 dm. long; mature leaves of flowering branches 1–3 cm. long, glaucous beneath; *bracts of ament dark, pilose; ovary and capsule (at least at base) pilose, glabrate; mature capsules on thick pilose pedicels shorter than or barely equaling bracts; style* 0.4–0.7 *mm. long.* — Boggy or mossy depressions, serpentine and schistose tableland, Mt. Albert, Gaspé Co., Que. *Fl.* June.

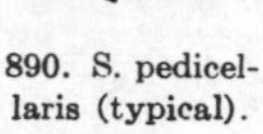

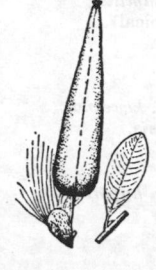

890. S. pedicellaris (typical).

§ 19. Discolòres Barratt

38. S. díscolor Muhl. (parti-colored; from the leaves), Large Pussy-W., Petit minou or Chaton (Que.). — Large shrub or small gray-barked tree up to 6 m. high; *branchlets glabrous or*, if at first pilose, *soon glabrate and lustrous; aments expanding long before the leaves (precocious), sessile, naked or nearly so at base*, thick-cylindric; the white-bearded bracts dark red or brown to blackish; each flower with 1 basal gland; staminate aments 1.5–5 cm. long; *pistillate aments* 2.5–7 cm. long, *in fruit becoming* 2.5–12 *cm. long and* 2–3 *cm. thick; capsule* 7–12 *mm. long, minutely pubescent*, long-beaked, *with definite style*, the pedicel much longer than gland; *leaves expanding after ripening of fruit, often with caducous fulvous hairs on expanding, promptly glabrate, bright green above, whitened beneath*, lanceolate to narrowly obovate or elliptic, petioled, irregularly crenate-serrate especially near the middle, the mature blades of fertile branches mostly 3–10 cm. long and 1–3 cm. broad; vigorous sprouts with larger leaves and stipules. — Damp thickets or shores, often in swamps, Lab. to Alta., s. to Nfld., N.S., N.E., L.I., Del., Md., Ky., Mo., S.D. and Mont. *Fl.* late Feb.–May (northw.) Fig. 891. — Hybridizes with nos. 23, 25, 26, 29, 33, 39, 40 and 45.

Var. **latifòlia** Anderss. (broad-leaved). — *Branchlets with dull* (not lustrous) *bark, often puberulent until 2nd year; leaves often holding rusty pubescence on lower surface*, broadly lanceolate to narrowly ovate or obovate, becoming 2.5–5 cm. broad. (Var. *eriocephala* of ed. 7, not *S. eriocephala* Michx.) — Local through the range; perhaps with admixture of no. 39.

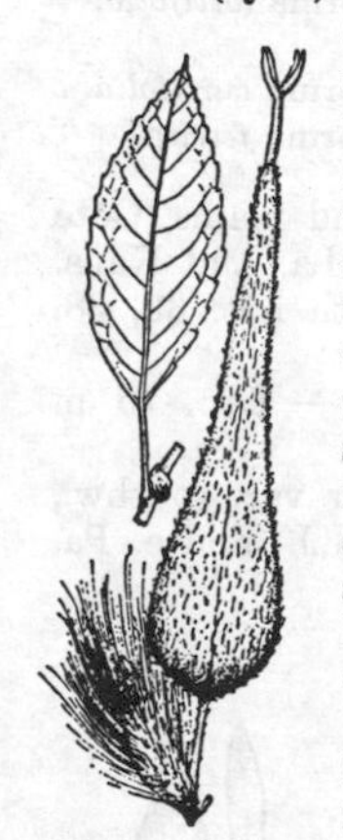

891. S. discolor.

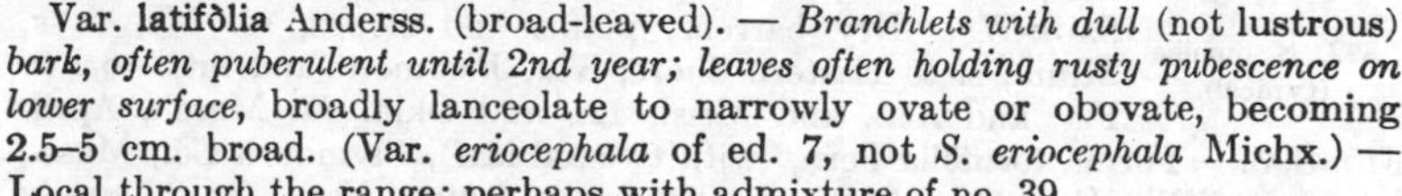

Var. **Òveri** C. R. Ball (for its discoverer, William Henry Over, 1866–). — *Branchlets glabrous but dull; leaves glabrous from the first or promptly glabrate, oblong-ovate to narrowly obovate, mostly* 3–6 *cm. broad.* — Nfld. and Côte Nord, Que., to Man., s. to P.E.I., N.B., Me., Ia., S.D. and Mont.

§ 20. Gríseae Borrer

39. S. hùmilis Marsh. (low), Small Pussy-W. — Shrub 0.25–3 m. high, *with wand-like flowering branches* more or less *cinereous-pubescent or, if glabrate, dull* (not lustrous); *aments precocious, sessile*, often in long series, *globular or ovoid on expanding*, soon lengthening, *the young ones often recurving; staminate aments in anthesis* 0.5–3 *cm. long*, 0.7–2.3 cm. thick; *the unopened anthers usually red or purplish;* pistillate aments lengthening in maturity to 1–8 cm.; *capsules cinereous*, very *long-beaked*, 7–9 mm. long, *their pedicels* 1.5–2 *mm. long; style very short; stigmas short and thick; leaves expanding after maturity of fruit*, narrowly to broadly oblanceolate to narrowly obovate or subelliptic, *broadest near or above middle, opaque* (rarely lustrous) *and gray-green above*, cinereous-pubescent to glabrescent and grayish beneath, *obtuse or acute (not acuminate), entire or undulate-dentate;* stipules of sprouts narrow or wanting. — Highly variable, with two strongly defined but intergrading series often treated as species:

a. Shrub 1–3 m. high; flowering branches 2–5 mm. thick; staminate aments 1–3 cm. long, 1–2.3 cm. thick; fruiting aments 2–5 (–8) cm. long; leaves mostly 4–10 cm. long, short-petioled. . . . *b*.

b. Leaves permanently soft-pubescent beneath, the pubescence mostly or completely hiding at least the veinlets.
Mature leaves oblanceolate, sublanceolate or subelliptic, three or four times as long as broad, mostly 1–2 (–3) cm. broad, blunt or acutish, white-tomentose beneath when young, in maturity grayish-pilose or puberulent, the pubescence mostly hiding the veinlets. . . . . . *S. humilis* (typical).

Mature leaves narrowly to broadly obovate or elliptic or oblanceolate, acute to round-tipped, up to half as broad as long, mostly 1.5–5 cm. broad, the dense whitish plush-like tomentum obscuring or hiding the lateral veins. . . . . Var. *keweenawensis.*

b. Leaves promptly glabrate or barely puberulent beneath, the prominent veinlets becoming conspicuous beneath; mature blades 0.7–2(–3) cm. broad. . . . . Var. *hyporhysa.*

a. Shrub 0.25–1 m. high; flowering branches rarely more than 2 mm. thick; staminate aments 5–12 mm. long, 7–8 mm. thick; fruiting aments 1–2.5 cm. long; leaves mostly 1.5–7 cm. long and 0.5–1.5 (–1.7) cm. broad, short-petioled to subsessile. . *c.*

c. Leaves dull or opaque and more or less grayish above, whitened or gray beneath. . . *d.*

d. Leaves narrowly oblanceolate to linear-lanceolate, 1.5–7 cm. long, 0.5–1.5 cm. broad, acute or subacute.
Leaves flat. . . . . Var. *microphylla.*
Leaves spirally twisted. . . . . Forma *tortifolia.*

d. Leaves narrowly cuneate-obovate, 1.3–3 cm. long, 0.6–1.7 cm. broad, rounded at summit. . . . . Forma *curtifolia.*

c. Leaves bright lustrous green above, otherwise as in var. *microphylla.* . . Forma *festiva.*

S. hùmilis (typical), PRAIRIE-W. or GRAY W. — Dry thickets, openings and plains, Côte Nord, Que., to Minn., s. to N.S., N.E., L.I., N.C., Ky., n. La. and Kans. *Fl.* March–early June (northw.). FIG. 892. — Hybridizes with nos. 33, 38, 40 and 41.

Var. **keweenawénsis** Farw. (of Keweenaw Pen., Mich.). — Lab. to n. Ont., s. to Nfld., N.S., n. N.E., n. N.Y., n. Mich. and n. Wisc.

Var. **hyporhỳsa** Fern. (wrinkled beneath). — The commoner var. southw., Fla. to e. Tex., n. on or near Coastal Plain to e. Ct., L.I., N.J., and e. Pa. and inland n. to W.Va., O., s. Mich., s. Wisc., Ia. and Okla.

Var. **microphỳlla** (Anderss.) Fern. (small-leaved), *S. tristis* Ait.; *S. humilis*, var. *tristis* (Ait.) Griggs, DWARF GRAY W., SAGE-W. — Dry barrens, plains and slopes, se. Me. to s. Minn., s. to Eastern Shore, Va., Piedmont and mts. to nw. Fla. and Ala., and Miss., La. and Okla. *Fl.* March, April. FIG. 893. — Forma **tortifòlia** Fern. (with twisted leaf), Plymouth Co., Mass.; forma **curtifòlia** Fern. (short-leaved) and forma **festìva** Fern. (gay), locally abund., w. Cape Cod, Mass.

892. S. humilis (typical).

893. S. humilis, v. microphylla.

40. S. grácilis Anderss. (slender). — Shrub *with slender ascending or erect green or olive-brown* (sometimes glaucous-pruinose) *tough and elastic* glabrous or glabrate *branches* 1–3 m. high; *aments appearing with the leaves* on leafy peduncles, *flowering from near middle to base and apex; bracts linear-lanceolate to narrowly oblanceolate or spatulate, yellowish or brown,* only slightly pubescent; staminate aments ellipsoid-obovoid, 1–2 cm. long, 1–1.3 cm. thick; *pistillate aments* at first small (1–2 cm. long), *in fruit becoming* 1.5–4 *cm. long and on leafy peduncles up to 2 cm. long, finally lax, with* divergent *long-beaked but blunt finely silvery-silky capsules* 5–7 mm. long, these *on slender* pubescent *pedicels* 2.5–4 *mm. long; style very short or obsolete; leaves often silvery- to fulvous-silky on expanding,* either permanently so or glabrate, *linear or lanceolate, acuminate, lustrous green above,* whitish beneath, *membranaceous,* the young ones soon darkened in drying, *entire or only obscurely and minutely denticulate,* mostly overlapping and ascending on the branchlets, *in maturity* 2.5–7 cm. long and 3–11 *mm. broad; stipules* 0 or minute and caducous. (*S. petiolaris*, var. *rosmarinoides* (Anderss.) Schneid. and var. *angustifolia* Anderss.) — Meadows and swales, Que. to Alta., s. to n. Mass., w. Ct., n. N.Y., n. Mich., n. Wisc. and Minn. *Fl.* May, June.

894. S. gracilis, v. textoris.

Var. **textòris** Fern. (of the basket-maker). — Usually *coarser; capsules up to 9 mm. long; mature leaves glabrate or rarely silky, 4–10 cm. long and up to 2 cm. broad, strongly serrate-dentate,* except at base, *with gland-tipped teeth.* (*S. petiolaris* sensu Am. auth., not Sm.) — S. Que. to Man., s. to N.B., N.E., n. N.J., ne. and centr. Pa., O., Ind., Ill., n. Ia. and Neb. Fig. 894. — Hybridizes with nos. 9, 25, 33, 38, 39, 41 and 45.

× **S. subserícea** (Anderss.) Schneid. (somewhat silky) is a hybrid of the var. of no. 40 with no. 41, coarser than no. 40, the young leaves more loosely sericeous, the larger glabrate leaves puberulent on midrib above; bracts oblong, with rounded black tips.

41. S. serícea Marsh. (silky), Silky W. — Shrub up to 3 or 4 m. high, *with* brown glabrous or glabrate *brittle-based branchlets; aments precocious,* their *bracts dark brown to blackish; staminate aments flowering from base to apex,* 1–2.5 cm. long; *fruiting aments rather dense,* 2–5 cm. long; *capsules oblong-ovoid, round-tipped, 3–5 mm. long, glistening-silky; style obsolete;* pedicel slender, 1–2 mm. long; *leaves* expanding at maturity of fruit, *lustrous beneath with very minute appressed silky pubescence, dark green above,* narrowly lanceolate, tapering to petiole and to *acuminate tip,* 0.4–1 dm. long and 1–2.5 cm. broad, *finely and regularly serrulate to base,* blackening in drying; *stipules of sprouts* lanceolate, *deciduous.* — Low thickets and banks of streams, s. Que. to Wisc. and Ia., s. to N.S., N.E., L.I., S.C., Tenn. and Mo. *Fl.* late March–early May. Fig. 895. — Hybridizes with nos. 25, 39, 40, 42 and 45.

895. S. sericea.

42. S. coáctilis Fern. (felted). — Coarser than no. 41, with blackish branchlets; *aments expanding with young leaves, leafy-bracted at base;* pistillate aments in anthesis 2–3.5 cm. long, in fruit up to 6 cm. long; *capsules* 5 mm. long, *conic-subulate, white-villous, the style definite; leaves heavily felted beneath with at first somewhat fulvous then white plush-like dense pubescence; mature blades* oblong- to narrowly ovate-lanceolate, tapering or rounded to base, 0.5–1.5 *dm. long and* 1.5–4 *cm. broad, very coarsely serrate-dentate with horizontally divergent unequal gland-tipped teeth, the longer broad-based teeth* 1–2 *mm. long;* stipules broadly lanceolate to ovate, subpersistent. — Rich swamps and river-banks, s. Que., w. N.B. and n. and e.-centr. Me. *Fl.* May. Fig. 896. — Hybridizes with nos. 25 and 41.

896. S. coactilis.

§ 21. Cápreae Bluff & Fingerhuth

43. S. Cáprea L. (she-goat; because leaves eaten by her — as if she distinguished as to species or genus), Goat-W. — Coarse shrub or small gray-barked tree, the stout *branches* at first finely pubescent or glabrous, but *wholly glabrate or glabrous, brown and lustrous by 2nd year, the wood smooth under the bark; buds becoming smooth; aments precocious,* strongly divergent, sessile; the *staminate ovoid,* 1.5–4 cm. long by 1.5–2 cm. thick, *beginning to flower from base upward;* bracts elliptic to narrowly obovate; filaments smooth or slightly hairy at base, *anthers yellow;* fruiting aments 5–10 cm. long, becoming lax; capsules flask-shaped, gray-pubescent, beaked, 6–7 mm. long, style very short; pedicel elongate, many times exceeding gland; *leaves broadly ovate to elliptic,* broadest near middle, subcordate, rounded or narrowed at base, dentate to subentire, *half to three-quarters as broad as long,* 4–10 cm. long, on slender petioles up to 2 cm. long, soft-pubescent on expanding, *the upper surface soon glabrous and shining;* stipules of sprouts subpersistent. — Cult. and spread to thickets, locally, N.E. to O. and Md. *Fl.* March, April. (Introd. from Eu.) Fig. 897.

897. S. Caprea.

44. S. cinèrea L. (ashy), Gray W. — Differing from no. 43 in its *branchlets blacker and more permanently (to 2nd year) and heavily tomentose; wood striate under the bark; buds permanently pubescent; staminate aments expanding from apex downward,* 2–3 cm. long, more slenderly ovoid-cylindric, fruiting ones more slender; filaments more hairy at base; *anthers reddish,* becoming orange; *leaves obovate to elliptic-lanceolate, permanently pubescent on both faces, one-third to half as broad as long.* — Spread from cult., N.S. and Mass., rarely southw. *Fl.* April, early May. (Introd. from Eu.) Fig. 898.

898. S. cinerea.

§ 22. Cándidae Schneid.

45. S. cándida Flügge (white), Hoary W. — Shrub 0.5–2 m. high; *branchlets* stout, when young (up to 2nd year) *whitish-tomentose with flocculent dull wool*, the older branches reddish; *aments* precocious, cylindric, *densely flowered*, the fruiting ones 3–5 cm. long; *bracts pale brown; young anthers red; capsule densely white-woolly; fresh style dark red, elongate; pedicel short, at most twice length of elongate dark gland; leaves* oblong to linear-lanceolate, dull white-tomentose when young, retaining much pubescence, or in forma **denudàta** (Anderss.) Rouleau (denuded) glabrescent, 4–12 *cm. long, plane, erect or suberect and inclined to overlap in tufts on the stout branches, the revolute margin entire or undulate.* — Calcareous bogs and thickets, Lab. to B.C., s. to Nfld., M.I., N.B., n. and w. N.E., n. N.J., Pa., n. O., n. Ind., n. Ill., n. Ia., S.D. and Colo. *Fl.* April–June. Fig. 899. — Hybridizes with nos. 33, 36, 38, 41 and some others.

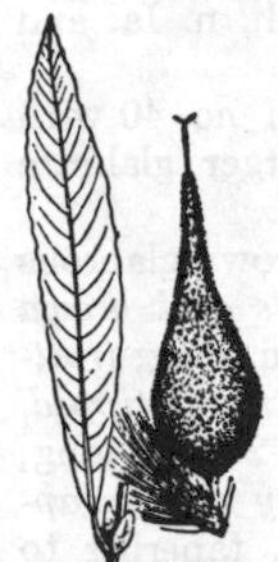
899. S. candida.

× **S. rubélla** Bebb (reddish) is an obvious hybrid of nos. 45 and 25; × **S. Clárkei** Bebb (named in 1878 for its discoverer, Daniel Clarke) is an equally obvious hybrid of nos. 45 and 40.

46. S. cryptodónta Fern. (with hidden teeth). — Shrub 2–4 m. high, *differing from no.* 45 *in its more slender and lustrous older branches; leafy-peduncled fruiting aments only* 2–3 *cm. long; style obsolete or very short; pedicel two to four times length of gland; leaves oblong-lanceolate, rugulose, the revolute margin with obscure gland-tipped teeth.* — Shores and thickets, Nfld., and Mingan Ids., Que. *Fl.* May–June. Fig. 900.

900. S. cryptodonta.

§ 23. Phylicifòliae Dumort.

47. S. planifòlia Pursh (flat-leaved). — Divaricately much branched shrub 0.2–3 m. high; *the glabrous branchlets purplish or sometimes pruinose and glaucous; aments precocious, sessile,* with few rudimentary leaves at base, *distant along the branches;* bracts dark, silky-villous; filaments glabrous; *fruiting aments becoming* 3.5–4.5 *cm. long,* dense; *capsules whitish-silky,* 5–6 *mm. long,* conic-rostrate from an ovoid base, short-pedicelled, the definite style yellow (drying black); *leaves glabrous,* elliptic-lanceolate to oblong, subequally acute to obtuse at both ends, remotely and minutely repand-toothed, 2.5–7 cm. long, *dark green and shining above, glaucous beneath,* becoming subcoriaceous; *stipules obsolete.* (*S. phylicifolia* of ed. 7, not L.) — Lab. to Alta., s. to damp thickets and stream-banks, chiefly in subacid areas of Nfld., Côte Nord, Que., higher mts. of Gaspé Pen., Que., and n. N.E. *Fl.* June–early Aug. Fig. 901. — Hybridizes with no. 22, producing × **S. Gray̆i** Schneid.

901. S. planifolia.

48. S. paraleùca Fern. (partly white). — Resembling no. 47, more upright, up to 4 m. high; *new branchlets densely cinereous-pilose or velutinous,* the older ones becoming glabrate and lustrous; *leaves* oblanceolate to oblong or oblong-ovate, *at first strongly pilose,* finally glabrate, 2.5–10 cm. long, 1.3–3 cm. wide, *crenate with gland-tipped teeth; fruiting aments* 2–6 cm. long, *on leafy peduncles* 2–10 *mm. long;* bracts brown to fuscous; *style* 0.6–1.1 *mm. long.* — Shores and rich thickets, Côte Nord and Gaspé Pen., Que. *Fl.* June. Fig. 902.

902. S. paraleuca.

49. S. ancorífera Fern. (bearing anchors; from the long reflexed stigmas). — Shrub with glabrous and lustrous fuscous branchlets; *young leaves* narrowly obovate, *finely rufescent-pilose,* glabrate, *remotely glandular-dentate; pistillate aments* in anthesis 3.5–4.5 cm. long, *on leafy peduncles up to* 8 *mm. long; style* 1 *mm. long, the stigmas reflexed and* 1.5–2 *mm. long.* — Little known shrub, Bay of Islands, Nfld. *Fl.* May, early June. Fig. 903.

903. S. ancorifera.

50. S. pedunculàta Fern. (peduncled). — Shrub 0.5–1.5 m. high, *with divergent to decumbent glabrous* and fuscous finally shining *branchlets; leaves glabrous,* oblong-lanceolate to narrowly obovate, sublustrous above, glaucous beneath, *entire or undulate; stipules lanceolate,* glandular-dentate, deciduous; *pistillate aments on leafy peduncles* 1–2.5 *cm. long, in fruit* 5–10

904. S. pedunculata.

cm. long and 1.5–2 *cm. thick; bracts fuscous or blackish,* ovate-oblong, 3.5–4.5 *mm. long; capsules* lance-ovoid, long-beaked, 6–8 *mm. long, pilose with short glistening hairs; style* 1.2–1.5 *mm. long, the erect oblong stigmas much shorter;* pedicel 1–2 mm. long, twice as long as gland. — Damp calcareous areas, n. and nw. Nfld. *Fl.* June. FIG. 904.

905. S. amoena.

51. S. amoèna Fern. (charming). — Differing from no. 50 in more erect branching; *new branchlets* sparsely pilose and glabrate, *castaneous; fruiting aments* 3–7 *cm. long, lax toward base, flexuous to somewhat nodding; bracts yellowish,* 2–3 *mm. long; style* 1 *mm. long; pedicel but slightly exceeding gland.* — Calcareous thickets along Straits of Belle Isle, Nfld. *Fl.* June, early July. FIG. 905.

906. S. pellita.

52. S. pellita Anderss. (clad in skins). — Large shrub or small tree; the *olivaceous to reddish branchlets whitened by a pruinose coat; aments* precocious, *subsessile,* with small leaves at base, the *fruiting ones* 2–5 *cm. long; bracts dark brown; capsules* conic-ovoid, 4–5 *mm. long, densely white-pubescent; style slender, elongate;* pedicel slightly exceeding elongate gland; *leaves linear-lanceolate to broadly lanceolate or oblanceolate, attenuate to tip, essentially entire, thick and firm,* mostly 4–12 cm. long, *green and glabrous above, whitened and silky-velvety beneath,* or glabrescent but white beneath in forma **psila** Schneid. (smooth); stipules none or minute and caducous. — Streambanks and rich thickets, s. Lab. to n. Ont., s. to Nfld., N.S., n. N.E. and n. Mich. *Fl.* May, June. FIG. 906.

§ 24. VIMINÀLES Bluff & Fingerhuth

53. S. VIMINÀLIS L. (bearing withes), OSIER. — Shrub or small tree up to 8 m. high; *branches long and flexible,* minutely silky when young, soon glabrous and lustrous; *leaves elongate-linear or -lanceolate, elongated to slender tips,* entire, undulate, *mostly* 1.5–2.5 *dm. long and about* 1 *cm. broad, minutely silvery-satiny beneath;* stipules slender, caducous; *aments sessile,* dense, *rather crowded along the branches,* precocious; bracts blackish-tipped; staminate aments slenderly cylindric, 2–3 cm. long; pistillate aments becoming 4–6 cm. long in fruit; *capsules sessile or nearly so,* minutely puberulent, 6–8 mm. long; *style elongate.* — Spread from cult., Nfld. and Que., s. to N.S. and N.E. *Fl.* late April–June. (Introd. from Eu.) FIG. 907.

907. S. viminalis.

× S. SMITHIÀNA Willd. (named for Sir JAMES EDWARD SMITH, 1760–1840), a hybrid of nos. 53 and 43, with aments more distant and thicker than in no. 53; the lanceolate or lance-oblong leaves merely acute, undulate-dentate, 1–3 cm. broad, silvery-pilose beneath, has spread from cult. from P.E.I. and N.S. to N.E. and Wisc. (Introd. from Eu.)

S. INCÀNA Schrank (hoary), a low shrub with linear-lanceolate revolute-margined leaves shorter than in no. 53 and merely acute, the fruiting aments short-peduncled and only 1–2 cm. long, the smoothish capsules pedicelled, has spread from cult. in Que., N.S., and N.E. (Introd. from Eu.)

§ 25. HÉLIX Dumort.

54. S. PURPÙREA L. (purple), PURPLE OSIER, BASKET-W. — Shrub up to 6 m. high, the *slender often purple or red flexible elongate* ascending *branches* mostly glabrous; *leaves lingulate or oblanceolate, glaucescent and purple-tinged, subopposite,* denticulate or entire, 5–10 cm. long, 1–2 cm. broad; *aments often paired,* precocious, slenderly cylindric, sessile or subsessile; the *staminate ones* 2–3 cm. long, 5–6 *mm. thick, fuscous; the 2 hairy filaments and often* (not always) *the 2 red anthers united as* 1; bracts obovate, dark-tipped; *capsules plump-ovoid, white-pubescent,* 2–3 *mm. long, sessile,* with very short style and divergent tiny stigmas. — Low grounds, Nfld. to Ont. and Wisc., s. to N.S., N.E., Va., W.Va., O., Ill. and Ia. *Fl.* April, early May. (Originally introd. from Eu. for basket-making.) FIG. 908.

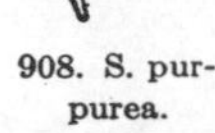

908. S. purpurea.

2. PÓPULUS L. POPLAR. ASPEN. PEUPLIER (Que.)

Flowers from a cyathiform disk which is symmetrical or obliquely lengthened in front. Stamens 6–60 or more, with distinct filaments. Stigmas 2–4. Capsules 2–4-valved. — Trees of N. Hemisph., with dilated and often broadly ovate, rounded or cordate toothed blades and terete, channelled or laterally flattened petioles. Buds scaly, frequently covered with resinous varnish. Aments elongate and commonly drooping during anthesis or in fruit, appearing before the leaves, their bracts (scales) mostly deeply dissected. (Classical Latin name.)

Key to Staminate Flowering Material*

a. Overwintering terminal buds not heavily glutinous, 6–10 mm. long; bracts of ament with 3–7 linear to lance-attenuate bearded long segments; stamens 6–12, with blunt anthers; trunk (well above base) and branches smoothish.
- Buds glabrous and lustrous; bracts of ament 3–5-cleft. . . . . 1. *P. tremuloides.*
- Buds canescent-pubescent, dull; bracts 5–7 cleft. . . . . 2. *P. grandidentata.*

a. Overwintering terminal buds 1–2.5 cm. long, often viscid or glutinous; bracts of ament fimbriate on margin or summit with abundant flexuous filiform unbearded segments; stamens 12–60 or more; trunk furrowed, with narrow plates or (in no. 7) smoothish. . . *b.*

b. Overwintering terminal buds 1.5–2.5 cm. long, heavily glutinous; staminate aments drooping; stamens 18–60 or more, their anthers blunt or emarginate.
- Vigorous sprouts often 4-angled or winged; overwintering buds not strongly balsamic-fragrant, the terminal one with 6 or 7 scales, the outer scale puberulent at base; bark of mature tree furrowed; wide-ranging from Gulf States to southernmost Canada. . . . . . . 5. *P. deltoides.*
- Vigorous sprouts and branches terete; overwintering buds strongly balsamic-fragrant, glabrous, the terminal one with 5 scales; bark of upper trunk and branches smoothish. . . . . 7. *P. balsamifera.*

b. Overwintering terminal buds 1–1.5 cm. long, downy at base, only slightly viscid, not balsamic; staminate aments erect on expanding, later arching, elongating and drooping; stamens 12–20; anthers apiculate; tree of inundated swamps southw. . . . . 9. *P. heterophylla.*

Key to Pistillate Flowering and Fruiting Material

a. Bracts of ament with 3–several linear-lanceolate segments or short deltoid teeth bearded at summit; ovary borne in an oblique unsymmetrical unlobed or merely undulate- or entire-margined disk; stigmas with slender terete branches; capsules slenderly conic-lanceolate or -oblong; overwintering terminal buds 3–10 mm. long. . . *b.*

b. Bracts 3–7-cleft; stigmas spreading-ascending, each with 2 subulate-attenuate branches; capsules 5–9 mm. long; nat.
- Overwintering buds glabrous, lustrous; bracts of ament 3–5-cleft; stigmas with strongly dilated and recurved basal lobes; expanding leaves thin and glabrous or promptly glabrate. . . . . 1. *P. tremuloides.*
- Overwintering buds canescent-pubescent, dull; bracts of ament 5–7-cleft; stigmas without recurved basal lobes; expanding leaves thick and white-felted. . . . . 2. *P. grandidentata.*

b. Bracts 7–9-cleft or with margin merely short-dentate; stigmas horizontally divergent, with cylindric blunt branches; capsules 3–5 mm. long, tomentose; overwintering buds pubescent; expanding leaves whitish-tomentose; introd.
- Bracts of ament with many short deltoid teeth; stigmas 2-branched. . . 3. *P. alba.*
- Bracts of ament deeply 7–9-lacerate; stigmas 4-branched. . . . . . 4. *P. canescens.*

a. Bracts of ament fringed by 9–very numerous filiform beardless segments; ovary borne in symmetrical or but slightly oblique cup- or saucer-like disk; stigmas dilated or with dilated lobes; capsule ovoid; overwintering terminal buds 1–2.5 cm. long. . . *c.*

c. Bracts fringed by very many slender segments; stigmas with broad lobulate or crenate spreading lobes. . . . *d.*

d. Overwintering terminal buds 1.5–2.5 cm. long, heavily glutinous, glabrous or merely glutinous-puberulent, not downy; ovary in a shallowly cup- or saucer-shaped disk with entire or shallowly undulate margin; style very short or obsolete; capsule surmounting persistent disk, slenderly ovoid, subsessile or very short-pedicelled; expanding leaves not white-felted.
- Overwintering buds not strongly balsamic, with 6 or 7 scales, outer scales puberulent at base; disk obliquely unsymmetrical; vigorous

* Staminate trees of species nos. 3, 4 and 8 apparently not found wild with us.

sprouts often 4-angled or winged; bark of mature tree furrowed; wide-ranging from Gulf States to extreme s. Canada. . . . . . 5. *P. deltoides.*

Overwintering buds strongly balsamic, with 5 glabrous scales; disk symmetrically saucer-shaped; sprouts and branchlets terete; bark of upper trunk and branches smoothish.

Branchlets glabrous; young leaves glabrous or nearly so, cuneate to rounded or barely subcordate at base; nat. northward. . . . 7. *P. balsamifera.*

Branchlets pubescent; young leaves pubescent, with cordate base; spread from cult. . . . . 8. × *P. gileadensis.*

*d.*Overwintering terminal bud 1–1.5 cm. long, pale-pubescent at base, only slightly viscid; disk a cup with deeply and regularly lobed margin, promptly deciduous; ovary globose-ovoid, only twice as high as disk, with long slender style; capsule naked-based, plump-ovoid, 0.8–1.2 cm. long, on slender elongate pedicel; expanding leaves white-felted; tree of inundated swamps southward. . . . . 9. *P. heterophylla.*

*c.*Bracts with 9–11 slender segments; stigmas appressed-reflexed against the ovary, with entire ovate lobes. . . . . 6. *P. nigra.*

KEY TO WELL-GROWN FOLIAGE-MATERIAL *

*a.*Petioles flattened or compressed toward summit. . . *b.*

*b.*Leaves ovate to suborbicular or elliptic.

Each margin finely crenate-serrate with 20–40 or more teeth. . . . 1. *P. tremuloides.*

Each margin with 5–15 coarse and remote unequal deltoid teeth. . . 2. *P. grandidentata.*

*b.*Leaves deltoid or rhombic, copiously dentate.

Mature blades of short branches (not sprouts) 6–12 cm. broad, with basal glands; nat. . . . . 5. *P. deltoides.*

Mature blades 3–8 cm. broad, without basal glands; introd. . . . 6. *P. nigra.*

*a.*Petioles terete. . . *c.*

*c.*Blade cuneate to rounded at base, without a sinus or the sinus open; introd. species or no. 7 nat. and boreal. . . *d.*

*d.*Leaves whitish- or canescent-felted or -pilose beneath, angulate-lobed or with coarse blunt teeth, not acuminate nor with balsamic frageance; introd.

Leaves (especially of sprouts or strong shoots) angulate-lobed, heavily white-felted beneath. . . . . 3. *P. alba.*

Leaves not lobed, merely dentate, more sparsely pubescent. . . . 4. *P. canescens.*

*d.*Leaves metallic-lustrous beneath, with strong balsamic fragrance, lanceolate to cordate-ovate, acuminate, closely fine-toothed; boreal.

Leaves cuneate to rounded or subcordate at base, lanceolate to narrowly ovate, glabrous or but sparsely pubescent on midrib beneath. . . 7. *P. balsamifera.*

Leaves broadly cordate-ovate, pubescent beneath, strongly so on nerves and midrib. . . . . 8. × *P. gileadensis.*

*c.*Blade broadly rounded at base, blunt or merely acute at apex, the deep sinus usually closed; young foliage and new shoots white-downy; tree of inundated swamps southw. . . . . 9. *P. heterophylla*

§ LEÙCE Duby (WHITE POPLARS OR ASPENS)

1. **P. tremuloìdes** Michx. (like Eurasian *P. tremula*), QUAKING ASPEN, TREMBLING ASP, QUIVER-LEAF, TREMBLE (Que.). — Tree up to 20 m. high, *with smooth or smoothish greenish-gray to whitish bark; branchlets slender, glabrous, flexible, reddish-brown,* spreading or ascending, or in forma **péndula** Jaeger & Beissner (pendulous) drooping, *with remote lustrous glabrous axillary buds* (*the terminal* 8–10 *mm. long*)*; leaves of short branches ovate to rhombic or suborbicular, glabrous or promptly glabrate, short-acuminate,* subcuneate, broadly rounded or subcordate at base, *as long as or longer than broad,* in maturity 2–8 cm. long and 1.8–7 cm. broad, or in forma **renifórmis** Tidestr. (kidney-shaped) reniform or oblate and abruptly short-tipped (as broad as or broader than long), *each margin closely crenate-serrate with* 20–40 *or more short and regular teeth,* membranaceous; sprout-leaves larger; *petiole* elongate, *almost filiform,* only 0.5–1 *mm. thick at laterally flattened summit;* aments soon pendulous; *bracts deeply divided into* 3–5 *attenuate long-bearded segments; stamens* 6–12 *in an obliquely prolonged entire-margined disk; ovary glabrous,* with short stout style; *stigmas with strongly dilated and recurved basal lobes,* each with 2 subulate-attenuate spreading-ascending branches; *capsule slenderly oblong- or*

* Only typical and normal foliage included; wholly atypical foliage, coming out late in the season after stripping in early summer by caterpillars, too inconstant for inclusion.

lance-conic, 5–9 mm. long. (*P. graeca* of some auth., not Ait.; *P. atheniensis* of some auth., not K. Koch) — Dry open woods and recent burns, Lab. to Alaska, s. to Nfld., N.S., N.E., L.I., Va., upland Tenn., n. Mo., Alta., etc.; westw. beyond our range replaced by other vars. or species.

Var. **magnífica** Vict. (magnificent). — *Branchlets strongly lignified and conspicuously torulose with approximate gray nodes 6–12 mm. thick, with approximate leaf-scars, the branchlets often drooping and brittle at base; mature leaf-blade heavily coriaceous,* with principal nerves coarse and prominent beneath, *reniform to round-ovate, abruptly tipped, as broad as to broader than long* (3.5–9 *cm. long,* 4–10 *cm. broad*)*; petiole stoutish, its broad summit* 2–4 *mm. thick.* — River-banks and -terraces and woodland, Gaspé Pen., Que., to Thunder Bay Distr., Ont., s. to n. N.E., sw. Que., and Bruce Pen., Ont.

2. P. grandidentàta Michx. (large-toothed), LARGE-TOOTHED ASPEN. — Tree with bark similar to that of no. 1; *branchlets* stiff, *the young ones tomentulose; buds canescent-pubescent; expanding leaves heavily white-felted; mature blades* glabrate, coriaceous, narrowly to broadly ovate, short-acuminate, 4–12 cm. long, cuneate to rounded at base, *each margin with* 5–15 *coarse remote unequal deltoid teeth;* aments coarser than in no. 1, their *bracts 5–7-cleft; stigmas without dilated and reflexed bases; ovary and capsule puberulent, on a pubescent pedicel.* — Dry woods, slopes and recent burns, s. Que. to s. Ont. and Minn., s. to N.S., N.E., N.C., Tenn. and ne. Mo. Rarely crosses with no. 1.

3. P. ÁLBA L. (white), WHITE or SILVER-LEAVED P., ABELE. — Rapidly growing tree with grayish-white smooth (or basally cracked) bark and spreading branches; *young branchlets and overwintering buds canescent; expanding leaves white-felted; mature leaves* (especially of sprouts or strong shoots) *palmately angulate-lobulate with* 3–5 *coarse blunt lobes and few teeth, white- or gray-tomentose beneath,* on tomentose petioles; *bracts of aments with many short deltoid teeth; stigmas with* 2 *terete horizontally divergent branches; ovary and capsule* 3–5 *mm. long, tomentose.* — Cult. and spreading by suckers (especially after destruction of parent trunk) on roadsides and borders of fields, often obnoxious. (Introd. and natzd. from Eu.)

4. P. CANÉSCENS (Ait.) Sm. (gray), GRAY P. — Resembling no. 3; *leaves scarcely lobed, merely dentate, more sparsely pubescent,* gray beneath; *bracts of ament deeply 7–9-lacerate; stigmas with* 4 *branches.* — Cult. and spreading like the last, N.E. to Minn. and southw. (Introd. and natzd. from Eu.)

§ AEGÌRUS S. F. Gray (COTTONWOODS)

5. P. deltoìdes Marsh. (deltoid; from outline of leaf), COTTONWOOD, NECKLACE-P., LIARD (Que.). — Tree up to 30 m. or more high, with *old bark gray, ridged and scaly,* younger bark yellow-green and smoother; branches ascending to spreading; *overwintering buds* 1.5–2.5 *cm. long, heavily glutinous, with* 6 *or* 7 *scales, the outer scales puberulent at base; leaves of short branches triangular-ovate,* 6–12 *cm. long and broad, subtruncate to subcordate* at base, *acuminate, coarsely crenate-dentate,* ciliolate, *bearing* 2 *or* 3 *basal glands; petiole flattened at summit; vigorous shoots and sprouts angled,* the sprout-leaves very large; *bracts of ament marginally fimbriate with many flexuous thread-like beardless segments; stamens* very many (*–60 or more*) *in an obliquely prolonged persistent entire or shallowly undulate disk,* the anthers blunt; ovary and capsule glabrous, the *stigmas wide-spreading, with broadly dilated lobes toothed; capsule ovoid.* (*P. balsamifera* sensu Sargent, not L.) — River-banks, bottomlands and rich woods, sw. Que. to Man., s. to w. N.E., and southw. beyond our range; westward passing into other vars. or closely related species.

Var. **missouriénsis** (Henry) Rehd. (of Missouri). — *Leaves definitely longer than broad, up to* 1.6 *dm. long,* with 3 or 4 basal glands. (Var. *angulata* (Michx. f.) Sarg., not Ait.) — Common southw., less so northw.

6. P. NÌGRA L. (black), BLACK P. of Eu. — Differing from no. 5 in more furrowed bark, *terete branches and branchlets; leaves rhombic,* only 3–8 cm. broad, *with no basal glands; bracts of ament with* 9–11 *slender segments; stigmas appressed-deflexed against the ovary, with entire ovate lobes.* — Locally spread from cult. (Introd. from Eu.) — The so-called var. ITÁLICA Muenchh. (Italian), LOMBARDY P., with strongly ascending and brittle branches forming a columnar tree, spreads from cult., an infertile and often diseased derivative of a single freakish individual, propagated by sprouts. (Introd. from Eu.)

§ TACAMAHÁCA Spach (BALSAM-POPLARS)

7. P. balsamífera L. (balsamic), BALSAM-P., HACKMATACK, TACCAMAHAC, LIARD (Que.). — Small or large trees up to 30 m. high, with trunk up to 2 m. in diameter, the upper part smooth; *overwintering buds large, heavily coated or saturated with yellow gummy strongly balsamic-fragrant resin, the terminal ones with* 5 *scales; branchlets lustrous, terete; leaves glabrous, with metallic lustre on pale lower surface, broadly lanceolate to ovate,* acuminate, *subcuneate to rounded or* *subcordate at base, on terete petioles,* the blades on short branches, 5–12 cm. long; *bracts of ament fringed* at broad summit *by very many flexuous bristle-like segments;* stamens many on an oblique entire disk; ovary and glabrous capsule subtended by a symmetrical persistent saucer-shaped disk; stigmas sessile, with coarsely toothed rounded lobes; capsule thick-walled, ovoid. (*P. Tacamahacca* Mill.) — River-banks or -gravels (northw. as a pioneer), etc., Lab. to Alaska, s. to Nfld., N.S., n. N.E., N.Y., Ont., Mich., Wisc., Minn., nw. Neb. and Colo.

Var. **subcordàta** Hylander (slightly heart-shaped). — Leaves cordate- or subcordate-ovate, often strongly asymmetrical at base, slightly pubescent on veins beneath, the petiole slightly pubescent. (*P. Michauxi* Dode as to plant, in part, not the plant of Michx. f. cited as the basis; *P. balsamifera*, var. *Michauxi* Henry in part only) — Ung. to Thunder Bay Distr., Ont., s. to Gaspé Pen., Que., n. N.E., n. N.Y. and n. Mich.

× **P. Jáckii** Sarg. (for JOHN GEORGE JACK, 1861–1949) is a local hybrid of nos. 7 and 5.

8. × **P. gileadénsis** Rouleau (of Gilead; from the colloquial name), BALM-OF-GILEAD. — Large tree, differing from no. 7 in broadly cordate-ovate leaves pubescent especially beneath; staminate aments wanting. (*P. candicans* of most auth., not Ait.) — Spread from cult. by sprouts and cuttings, Nfld. to Ont., s. into N. States. (Origin unknown; probably an infertile hybrid of nos. 5 and 7.)

§ LEUCOÌDES Spach

9. P. heterophýlla L. SWAMP- or BLACK COTTONWOOD, DOWNY P. — Tree up to 30 m. high, with *old bark in narrow plates; young branchlets whitish-tomentose,* becoming glabrate and lustrous; *overwintering buds* 1–1.5 *cm. long, canescent-tomentose toward base; leaves cottony upon expanding,* soon glabrescent, *the broadly ovate round-tipped to merely acute blades cordate and usually with deep and closed sinus; staminate aments erect on expanding,* later arching and drooping; bracts long-fringed; *stamens* 12–20 in a nearly symmetrical shallow disk, the *anthers apiculate; disk of pistillate flowers a symmetrical deeply lobed cup* half as high as ovary, *deciduous;* ovary globose-ovoid, with long slender style, stigmas with deeply lobed broad branches; *capsule naked at base,* plump-ovoid, 0.8–1.2 cm. long, *on slender elongate pedicel.* — Inundated swamps and bottomlands, n. Fla. to La., n. on or near Coastal Plain to centr. Ct., se. N.Y., se. Pa. and inland n. to O., s. Mich., Ill. and Mo.

FAM. 42. MYRICÀCEAE (WAX-MYRTLE FAMILY)

Monoecious or dioecious shrubs or trees with both kinds of flowers in short scaly aments, and resinous-dotted often fragrant leaves. — Differing from the Birches chiefly in the 1-locular ovary with a single erect orthotropus ovule and the drupe-like nut. Involucre and perianth none.

Leaves entire, dentate, serrate, or at most incised, without stipules; ovary and fruit naked at base or with 2–4 entire basal short bracts. . . . . 1. *Myrica.*
Leaves pinnatifid, with semicordate stipules; ovary and fruit overtopped by 8 linear-subulate persistent bracts. . . . . 2. *Comptonia.*

1. MYRÌCA L.

Mostly dioecious; flowers solitary under a scale-like bract and with 2–4 short entire basal bracts not overtopping the fruit (or bracts wanting); pistillate flowers in ovoid or cylindric aments, staminate aments ellipsoid or thick-cylindric, these from axillary scaly buds; stamens 2–20; filaments somewhat united below; anthers 2-locular. Fruit globose or ovoid, in ours with waxy coat or resinous atoms. — Shrubs or trees chiefly of trop. and temp. reg. with entire or toothed often evergreen leaves without stipules. (*Myrike*, the Greek name of the Tamarisk or some other fragrant shrub, perhaps from the Greek *myrizein, to perfume;* the name later transferred to this aromatic genus.)

a. Aments borne at the summits of the preceding year's branchlets, the crowded staminate ones with flowers overtopped by broad lustrous bracts; ovary and fruit ovoid, with lustrous resin-atoms, bearing 2 wing-like adnate bracts. Subgen. GALE. 1. *M. Gale.*
a. Aments borne on the old wood, chiefly below the leafy tips, scattered, the staminate flowers exceeding the bracts; ovary and fruit globose, in maturity white or drab with heavy wax. Subgen. MORELLA. . . *b.*
b. Leaves of flowering branches oblong to narrowly obovate, mostly 1–3 cm. broad, their upper surfaces with waxy atoms remote or wanting; new branchlets villous (rarely glabrous), mostly without abundant wax-coating; fruits 3–4.5 mm. in diameter.

Bark of mature branches whitish-gray or drab; leaves dull above, membranaceous, deciduous (subpersistent southw); inflorescences all borne below the leafy tips; young fruit densely pubescent, ripe fruit 3.5–4.5 mm. in diameter. . . . . 2. *M. pensylvanica.*

Bark of mature branches blackish; leaves lustrous above, coriaceous, evergreen; inflorescences below or in the axils of the old leaves; young fruit glabrous, ripe fruit 3–3.5 mm. in diameter. . . . . 3. *M. heterophylla.*

b. Leaves of flowering branches narrowly cuneate-oblanceolate to narrowly cuneate-obovate, coriaceous, evergreen, mostly 0.5–2 cm. broad, with abundant wax-atoms on the lustrous upper surface; new branchlets waxy, glabrous or sparsely pilose; young fruits glabrous, mature ones 2–4 mm. in diameter.

Coarse shrub or small tree, non-colonial; leaves of fertile branches narrowly oblanceolate, tapering to acute base and apex, mostly 4–9 cm. long; fruits 2–3 mm. in diameter. . . . . 4. *M. cerifera.*

Dwarf freely stoloniferous colonial shrub; leaves oblanceolate to narrowly obovate, obtuse, mostly 1.5–4 cm. long; fruits 3–4 mm. in diameter. . 5. *M. pusilla.*

Subgen. GÀLE Endl.

1. M. Gàle L. (old generic name), SWEET GALE, "MEADOW-FERN", PIMENT ROYAL or BOIS-SENT-BON (Que.). — Shrub 0.3–2 m. high, usually with strongly ascending brown branches; *leaves cuneate-oblanceolate, grayish,* more or less pubescent beneath, slightly serrate toward apex, *deciduous;* flowers early in spring; the *staminate aments crowded, with large lustrous brown bracts, pistillate aments becoming cone-like, borne at the summit of last year's wood; nuts ovoid,* imbricated, resin-dotted, 2-*winged by the two thick ovate scales which coalesce with its base.* (*Gale palustris* (Lam.) Chev.) — Shallow water and swamps, Lab. to Alaska, s. to Nfld., N.S., N.E., L.I., N.Y., s. Ont., Mich., Wisc., Minn., and Oreg.; mts. to N.C. and Tenn. Apr.–June. (Eurasia)

Var. **subglàbra** (Chev.) Fern. (almost glabrous). — Leaves glabrous or nearly so. — S. Lab. and Nfld. to Ont., s. to N.S., N.E., n. N.J., e. Pa. and Mich.

Subgen. MORÉLLA Benth. (WAX-MYRTLES)

2. M. pensylvánica Loisel. (of Pennsylvania), BAYBERRY or CANDLEBERRY. — Stout stiffly branched shrub 0.3–2 m. high (rarely a tree up to 4.5 m. high, with trunk 1.2 dm. in diam.); *branches mostly whitish-gray or drab, the young ones* villous, pilose or glabrate, mostly *without sessile glands; leaves elliptic, oblanceolate or obovate,* 1–3 *cm. broad, membranaceous, deciduous* or subpersistent, *opaque* and often pubescent above, pale beneath and with scattered wax-atoms, commonly pilose on the nerves; *inflorescences all borne below the leafy tips;* staminate aments with flowers about equaling or exceeding bracts; pistillate aments maturing into dense clusters of bony globular nuts covered with white or gray wax, *the young fruits densely pubescent, the mature* 3.5–4.5 *mm. in diameter.* (*M. carolinensis* of ed. 7, not Mill.) — Dry or wet sterile soil near the coast or on the Coastal Plain, N.C. to e. N.B., M.I. and s. Nfld., inland locally to n. O. May–July — Crosses with no. 4, producing × **M. Macfarlànei** Youngken (for JOHN MUIRHEAD MACFARLANE, 1855–1943).

3. M. heterophýlla Raf. (variable-leaved), WAX-MYRTLE. — Coarse shrub or small tree with *blackish branches,* the *leafy branchlets black-pilose; leaves* resembling those of no. 2, but mostly acute or acutish, *coriaceous, evergreen; inflorescences either below or in the axils of the leaves; young fruit glabrous, mature fruit* 3–3.5 *mm. in diameter.* (*M. Curtissi,* var. *media* Chev.) — Dry or moist thickets and woods, La. to Fla., n. to inner Coastal Plain of Va. and s. N.J. Apr.–June.

Var. **Curtíssi** (Chev.) Fern. (for its discoverer, ALLEN HIRAM CURTISS, 1845–1907). — Branchlets glabrous. (*M. Curtissi* Chev.) — Similar range, rare with us.

4. M. cerífera L. (bearing wax), WAX-MYRTLE or CANDLEBERRY. — Shrub or tree up to 12 m. high (trunk then up to 2 dm. in diam.); *young branchlets waxy,* glabrous or but sparsely pilose; *leaves* of fertile branches *narrowly oblanceolate, tapering to acute base and apex, mostly* 4–9 *cm. long* and 0.5–2 cm. broad, *fulvous or yellow-green, coriaceous, evergreen, heavily coated on both surfaces with waxy granules; fruits* glabrous, when mature 2–3 *mm. in diameter.* (*Cerothamnus* Small) — Thickets, woods and swamps, Fla. to Tex., n. to s. N.J. and Ark. Apr.–June. — Crosses with nos. 2 and 5.

5. M. pusílla Raf. (very small), DWARF WAX-MYRTLE. — Low stoloniferous *shrub forming* *colonies* 0.2–2 *m. high;* branchlets waxy, glabrous or nearly so; *leaves* coriaceous, like those of no. 4 but *oblanceolate to obovate and obtuse, mostly* 1.5–4 *cm. long; fruits* 3–4 *mm. in diameter.* (*M. pumila* (Michx.) Small; *Cerothamnus pumilus* Small) — Dry or moist pine-barrens and woods, Fla. to Tex., n. to s. Del. and Ark. Apr.–June.

2. COMPTÒNIA L'Hér. Sweet-fern

Monoecious or dioecious; the staminate flowers in flexuous cylindric aments, with reniform-cordate pointed scaly bracts and 3–6 stamens; pistillate flowers in globular bur-like aments; ovary surrounded by 8 long linear-acicular bracts persistent around the conical or barrel-shaped hard smooth nut; leaves pinnatifid, membranaceous, deciduous, with semicordate stipules. — Low pubescent shrub, with fragrant foliage, fruit and twigs, the round-lobed leaves appearing later than the flowers. (Named for *Henry Compton,* 1632–1713, Bishop of London, a cultivator and patron of botany.) — A single species.

1. C. peregrìna (L.) Coult. (foreign — to the original European author). — Shrub 0.3–1.5 m. high; *new branchlets villous or pilose; leaves* linear-lanceolate, dark green above, pale beneath, *pilose on one or both surfaces; staminate aments subapproximate; fruiting aments* 1.5–2.5 *cm. in diameter;* bracts of pistillate flowers ciliate at least below; *nuts* 4–5 *mm. long.* (*Myrica* (L.) Kuntze; *M. asplenifolia* of ed. 7, not L.) — Open sterile woodlands, clearings and pastures, C.B. to Man., s. to Va., W.Va., O., nw. Ind., ne. Ill. and Minn., and in the upland to n. Ga. and Tenn. Apr.–June.

Var. **asplenifòlia** (L.) Fern. (with leaves like *Asplenium*). — Lower; *new branchlets minutely puberulent; leaves only sparsely short-puberulent or glabrous; staminate aments remote; fruiting aments* 0.8–1.5 *cm. in diameter; nuts* 3–4 *mm. long.* (*Myrica asplenifolia* L., as to type.) — Dry sandy pinelands and barrens, L.I. to Va. March–May.

FAM. 43. LEITNERIÀCEAE (Corkwood Family)

Dioecious shrub or small tree with both kinds of flowers in aments opening before the leaves; the staminate aments many-, the pistillate few-flowered; calyx and corolla none; stamens 3–12, *whorled, with short distinct filaments, hypogynous; ovary 1-locular with solitary ascending amphitropous ovule and thickish terminal style with lateral groove.* Leaves simple, entire, alternate; stipules obsolete or none. Flowers solitary in the axils of ovate pubescent scales, sessile. Fruit an obovoid somewhat compressed leathery drupe.

1. LEITNÈRIA Chapm.

Characters of the family. The only genus, with a single species. (Named in memory of Dr. *E. F. Leitner,* a German naturalist who traveled and was killed in Florida during the Seminole War.)

1. L. floridàna Chapm. (of Florida), Corkwood. — Stout arborescent shrub 1–7 m. high; leaves oblong or obovate, somewhat canescent-tomentose on the lower surface; staminate catkins about 3 cm. long, the pistillate half as long; drupe 1–2 cm. long. — Swamps, n. Fla. to Tex., n. to Ga. and s. Mo. March. — Wood lighter than cork.

FAM. 44. JUGLANDÀCEAE (Walnut Family) *

Trees with alternate pinnate leaves and no stipules, the leaflets glandular-dotted beneath; flowers monoecious, the staminate in aments with or without a calyx adnate to the bract and the two bractlets; the pistillate solitary or in a small cluster or spike, with a bract, two or three bractlets and a regular 4-lobed calyx (or this absent) adherent to the incompletely 2–4-locular but only 1-ovulate ovary. Ovule orthotropous, erect, at the apex of the incomplete primary partition. Fruit similar to a dry drupe, the fibrous-fleshy or woody husk or exocarp (ripened bract and bractlets or involucre) fused until maturity with the crustaceous or bony nutshell or endocarp (ripened carpels), containing a 2–4-lobed seed. Albumen none. Cotyledons fleshy and oily, sinuous or corrugated, 4-lobed; radicle short, superior. — A small family of important trees, consisting chiefly of the two following genera:

Pith of branchlets separating into thin plates; staminate aments separate, sessile, on last year's branchlets, in the axils of the fallen leaves of the previous season; stamens 8–40; staminate and pistillate flowers with 4 small sepals; style-branches (stigmas) elongate; nut with indehiscent husk and irregularly furrowed shell. . . . . 1. *Juglans.*

* Treatment prepared with very important aid of Wayne E. Manning.

Pith continuous; staminate aments in fascicles of 3, the fascicles subsessile to long-stalked in the axils of bud-scales on new growth; stamens 3–8; staminate and pistillate flowers usually without sepals; stigmas short; husk of fruit splitting or partly splitting into valves, the nutshell smooth or merely reticulate. . . . . 2. *Carya.*

1. JÙGLANS L. WALNUT. NOYER (Que.)

Staminate aments sessile, separate, though often superposed, near the apex of the preceding year's growth; stamens 8–40, the floral receptacle adnate to the bract, 2 bractlets and usually 4 sepals, 1–3 of the sepals sometimes reduced to minute teeth; filaments free, very short. Pistillate flowers solitary or several together in a cluster or short spike on a peduncle at the end of the branch, with a bract, 2 often irregularly toothed bractlets, and 4 small sepals. Style short, with 2 (rarely 3) elongate style-branches, the inner surfaces of which are deeply fringed and stigmatic; style-branches carinal (above the center of the carpel). Fruit with a fibrous-fleshy indehiscent (in our species) husk (exocarp) and a mostly rough irregularly furrowed nutshell or endocarp. Am. and Eurasian trees, with pinnate leaves of many serrate (in our species) leaflets. Pith chambered (in plates). (Name contracted from *Jovis glans*, the *nut of Jupiter.*)

1. J. cinèrea L. (ashy), BUTTERNUT, WHITE W. — *Leaflets* 7–17, oblong-lanceolate, pointed, rounded at base, *downy with fascicled hairs* especially *beneath*, the *petioles and branchlets downy with clammy hairs; leaf-scars on branchlets with a hairy fringe along upper margin;* staminate flowers subsessile, attached at the end, with the floral receptacle elongated; sepals often much reduced; *fruit ellipsoid, clammy,* pointed; the nut deeply sculptured and rough with ragged ridges, 2-locular at the base. — Rich woods and river-terraces, w. N.B. to N.D., s. to Ga. and Ark. Apr.–June. — Trunk up to 30 m. high, with gray bark, widely spreading branches and light brown wood.

2. J. nìgra L. (black), BLACK W. — *Leaflets* 11–17 (–"23"), ovate-lanceolate, taper-pointed, somewhat cordate or unequal at base, smooth above, *the lower surface and* the *petioles minutely downy, the hairs solitary or in pairs; leaf-scars on branchlets notched, without hairy fringe;* staminate flowers clearly stalked from beneath, with the floral receptacle round; *fruit spherical,* or ellipsoid in forma **oblónga** (Marsh.) Fern. (oblong), roughly dotted; the nut corrugated, 4-locular at top and bottom. — Rich woods, w. Mass. to Minn., s. to Fla. and Tex. Apr.–June. — Tree up to 50 m. high, with dark bark and valuable brown wood.

2. CÀRYA Nutt. HICKORY. HICORIER OR CARYER (Que.)

Staminate aments in fascicles of 3 (rarely 4 or 5 on individual trees) in the axils of bud-scales; stamens 3–8 (–"10"?), adnate to the bract and 2 bractlets (1 or 2 sepals rarely present in individual flowers); filaments short or none, free. Pistillate flowers 2–10 in a cluster or short spike on a peduncle terminating the shoot of the season; bract and typically 3 bractlets sepal-like in flower, a true calyx absent. Stigmas sessile, 2, these sometimes divided, with a stigmatic disk at their base, papillose, commissural (above the lines connecting the carpels), usually persistent. Fruit with a 4-valved firm and at length dry husk (exocarp or involucre), this often falling away from the smooth and crustaceous or bony nutshell or endocarp which is incompletely 2-locular and at the base mostly 4-locular. — Fine timber-trees of e. N.Am. and e. Asia, with hard and very tough wood, and scaly buds from which in spring are put forth usually both kinds of flowers, the staminate below and the pistillate above the leaves. Nuts ripening and falling in October. (*Carya*, an ancient Greek name of the Walnut.) SCORIA Raf. (1808); HICORIUS Raf. (1817); HICORIA Raf. (1836)

a. Scales of overwintering buds paired, valvate, 4 or 6; scars of bud-scale more or less broad and separate, not forming a true ring; leaflets 5–17, lanceolate or nearly so, often falcate; some or all of staminate aments from separate lateral scaly buds near the summit of shoots of the preceding season; husk (exocarp) prominently keeled or winged at the sutures; shell of nut or outer ends of septa, as seen in cross-section, usually with powder-filled cavities. § APOCARYA. . . *b.*

b. Fascicles of staminate aments sessile or nearly so; stamens 3–6; leaflets 9–17; scales of overwintering buds with clusters of bright yellow hairs; fruit subterete, elongate; nut subcylindric, slenderly ellipsoid or ovoid, blunt to acute, essentially 2-locular at base; kernel sweet. . . . . . . 1. *C. illinoensis.*

b. Fascicles of staminate aments peduncled; leaflets 5–13(–15); scales of overwintering buds without clusters of bright yellow hairs; fruit compressed, ovoid, obovoid or subglobose; nut flattened, nearly or quite as broad as high, usually beaked by the two persistent indurated stigmas, 2- or 4-locular at base; kernel bitter.

Overwintering buds reddish-brown, with caducous yellow glands; leaflets 7–13(–15), narrowly to broadly lance-falcate; stamens 3–6; husk winged to base, the mature valves completely separating; nut essentially 2-locular at base, the shell shallowly furrowed and wrinkled, red-brown; tree of river-swamps southw. . . . . 2. *C. aquatica.*

Overwintering buds permanently yellow-scurfy; leaflets 5–9(–11), barely or scarcely falcate; stamens 4; husk wingless near base, in maturity splitting only to below middle; nut 4-locular in its lower half, smoothish, gray; wide-ranging northw. to s. Canada. . . . . 3. *C. cordiformis.*

*a.*Scales of overwintering buds imbricate, 6–12; scars of bud-scale very narrow, confluent, forming a definite usually ciliate ring; leaflets 3–9, scarcely falcate; fascicles of staminate catkins all in the axils of bud-scales at the base of the leafy shoot of the season; husk scarcely or barely keeled on sutures; nut-shell without cavities. § Eucarya. . *c.*

*c.*Overwintering terminal bud 1.3–3.5 cm. long, of 10–12 scales; innermost scales accrescent (enlarged) after expansion, becoming colored, fleshy and 2.5–8 cm. long; husk mostly 3–12 mm. thick; nut commonly with prominent angles; kernel sweet; branchlets usually stout; young branchlets usually more or less pubescent or tomentose, especially early in season; bark exfoliating in loose plates (except in no. 6 with tomentose new growth). . . *d.*

*d.*Young branchlets, petioles and lower leaf-surfaces glabrous or merely scurfy or pilose, not tomentose; outer narrowly tipped dark brown scales of terminal bud persistent; accrescent inner scales of terminal buds becoming 5–8 cm. long and 2.5–4 cm. broad; husk splitting to base into separate valves; bark of mature tree shaggily exfoliating in plates.

First-year branchlets reddish or olive-brown, becoming gray; larger accrescent bud-scales puberulent above; leaflets 5(–7); the serrations strongly ciliate when young, some or all of them with a persistent dense tuft of hairs on one or both sides of apex of tooth; nut whitish or whitish-brown, its shell 1–2 mm. thick. . . . . 4. *C. ovata.*

First-year branchlets orange or light tan, becoming gray; larger accrescent bud-scales glabrous above; leaflets 7 or 9, usually ciliate when young, but without dense subterminal tufts of hairs on serrations; nut yellowish- to reddish-brown, its shell 3–6 mm. thick. . . . 5. *C. laciniosa.*

*d.*Young branchlets, petioles, rachises and lower leaf-surfaces tomentose with curly fascicled hairs; outer dark bud-scales of terminal bud deciduous in the autumn, exposing silky inner bud-scales; accrescent inner scales of terminal bud 2.5–4 cm. long and about 1.3 cm. broad, pubescent on both surfaces; leaflets (5–)7–9; husk splitting to middle or below middle; nut reddish-brown, its very tough shell 5–6 mm. thick; bark deeply furrowed. . . . . 6. *C. tomentosa.*

*c.*Overwintering terminal buds 6–12 mm. long, of 6–10 scales; outermost dark scales mostly deciduous in the autumn; innermost scales slightly accrescent, at most 6 cm. long; husk 1.5–4(–5) mm. thick; nut with rounded to angled surface; branchlets usually slender; young branchlets glabrous (except in no. 10 and rarely in no. 9); bark usually close, furrowed or merely flaky or with small plates; leaflets without subterminal dense tufts of hairs on serrations. . . *e.*

*e.*Young branchlets and overwintering buds glabrous or with pale pubescence; staminate aments not rusty-pubescent; nut pale brown, the shell not conspicuously reticulated; bark gray. . *f.*

*f.*Overwintering buds 8–12 mm. long, not notably lepidote; leaflets 5 or 7, at first often yellow-glandular, but not silvery-lepidote; terminal leaflet 3.5–12 cm. broad; nut scarcely ridged or ridged only toward summit.

Bark close, becoming furrowed and ridged; young leaflets not scurfy; staminate aments 5–8 cm. long, with villous bracts; pistillate flowers and their bracts hoary-tomentose; husk in late autumn tardily splitting or remaining closed, dark brown, shining; kernel sweetish or acrid. . . . . 7. *C. glabra.*

Bark soon breaking into small plates or scales; young leaflets yellowish-scurfy; staminate aments 0.8–1.7 dm. long, puberulent; pistillate flowers and their bracts yellow-lepidote; husk in late autumn promptly splitting to base, light brown, dull, warty; kernel sweet. . . . . 8. *C. ovalis.*

f. Overwintering terminal buds about 6 mm. long, with only 6–9 scales, silvery-lepidote; staminate aments silvery-lepidote and pubescent, only 5–12 cm. long; pistillate flowers covered with yellow scales; expanding leaves silvery-scaly beneath; leaflets 7 or 9, the terminal one 2–6 cm. broad; fruit covered with yellow scales, the husk tardily splitting to base; nutshell ridged nearly to base. . . . . . . 9. *C. pallida.*

e. Young branchlets, overwintering buds, young leaves, staminate aments, etc. reddish- or rusty-pubescent; nutshell dark reddish-brown, coarsely reticulate with pale ridges; bark dark; western and southwestern tree. 10. *C. texana.*

§ Apocàrya DC.

1. C. illinoénsis (Wang.) K. Koch (of Illinois), Pecan. — Tree up to 50 m. high, with buttressed base and very thick furrowed and ridged bark; *overwintering buds* flattened, *with paired and valvate narrow scales covered with jointed hairs; fascicles of staminate aments sessile or nearly so, from separate lateral buds near summit of shoots of preceding year;* stamens 3–6; *leaflets* 9–17, *oblong-lanceolate, those of lower pairs with falcate and slightly recurving tips,* terminal leaflet only slightly broader than upper ones; *fruit elongate,* subcylindric, with *keels along the sutures extending to base; ripe husk (exocarp) splitting to below middle; nut reddish-brown, elongate, subcylindric,* slenderly ellipsoid or ovoid, *2-locular at base,* shell rather thin, *kernel sweet.* (*C. Pecan* (Marsh.) Engl. & Graebn. and *Hicoria Pecan* Britt., based on the inadequately described *Juglans Pecan* Marsh.) — Bottomlands, Ind. to Ia., s. to Ala., Miss., La. and e. Tex. — Crosses with no. 6, producing × **C. Schnéckii** Sarg. (for its discoverer, Jacob Schneck, 1843–1906); with no. 5, producing × **C. Nussbaumèrii** Sarg. (named in 1918 for its discoverer, J. J. Nussbaumer); and with no. 3 (see below).

2. C. aquática (Michx. f.) Nutt. (aquatic), Water-H., Bitter Pecan. — Differing from no. 1 in having the *reddish-brown overwintering buds covered with caducous yellow glands; fascicled staminate aments peduncled, the individual aments pedicelled; leaflets* 7–13(–15), *lance-falcate, glabrous;* stamens 3–6, their subtending bract longer than broad; *fruit* subglobose, ellipsoid or obovoid, *winged to base, the ripe husk with valves separating to base; nut laterally compressed, reniform or subglobose,* firm-beaked, 2-locular to base, the *reddish-brown shell furrowed and wrinkled; kernel bitter.* (*Hicoria* Britt.) — Bottomlands and swamps, the bases often inundated, Coastal Plain, Fla. to Tex., n. to Va., sw. Ill. and se. Mo.

3. C. cordifórmis (Wang.) K. Koch (heart-form; from the nut), Pignut, Bitternut, Swamp-H., Noyer dur (Que.). — Differing from no. 2 in the *lanceolate valvate scales of overwintering buds sulfur-yellow with persistent scurf; leaflets* 5–9(–11), *lanceolate to lance-ovate, hardly falcate, pubescent beneath; stamens* 4; *fruit cylindric to compressed,* the *husk wingless at base, splitting only to below middle; nut 4-locular in lower half, the shell smoothish, gray.* — Wet to dry woods, stream-banks and swamps, Fla. to Tex., n. to N.H., sw. Que., s. Ont., Mich., Wisc., Minn. and se. Neb. — Crosses with no. 1, producing × **C. Brownii** Sarg. (named in 1913 for its discoverer, George M. Brown); and with no. 4, producing × **C. Làneyi** Sarg. (named in 1913 for C. C. Laney).

§ Eucàrya DC.

4. C. ovàta (Mill.) K. Koch (ovate; from shape of fruit), Shagbark- or Shellbark-H., Noyer blanc (Que.). — Tree with gray *bark soon splitting into long and loosening plates; branchlets* minutely scurfy-pubescent, *soon glabrate, reddish-brown and lustrous; overwintering terminal buds ovoid,* 1.3–2.5 *cm. long, about* 1 *cm. thick, of* 10–12 *imbricated scales; outermost scales* dark, with narrow loose elongate tips, *persistent through winter;* the *innermost scales accrescent, becoming fleshy or petaloid, yellowish or purple-tinged and* 6–8 *cm. long and* 2.5–4 *cm. broad,* narrowly obovate, *pubescent on both faces; leaves with glabrous or glabrate petioles and* 5(–7) *broadly lanceolate, ovate or obovate soon glabrous or glabrate leaflets, the* 3 *upper leaflets* much *longer than the lower;* the *terminal* petiolulate *one* commonly largest, *becoming* 1–2 *dm. long and* 5–11 *cm. wide; most of serrations with a dense tuft of persistent hairs on one or both sides near apex of tooth;* staminate aments stalked, in a peduncled fascicle, usually glandular-hirsute, the subtending bracts of the flowers greatly prolonged; stamens 4–6; *fruit subglobose to broadly obovoid,* with depressed summit, 3.5–6 *cm. long; husk* 3–12 *mm. thick, splitting to base into separate valves; nut* subglobose-reniform to obovoid or ellipsoid, *with* 4 *angles, the whitish-brown shell* 1–2 *mm. thick; kernel sweet.* (*Hicoria* Britt.) — Rich woods, bottoms and slopes, Fla. to e. Tex., n. to s. Me., sw. Que., s. Ont., centr. Mich., centr. Wisc., s. Minn. and se. Neb.

Var. pubéscens Sarg. (pubescent). — Similar but branchlets, petioles and rachises densely and more permanently pubescent, many of the hairs fascicled; *leaflets permanently soft-downy beneath.* — Bottoms, Ga. to Miss., n. to sw. N.H. and Tenn.

Var. **Nuttállii** Sarg. (for THOMAS NUTTALL, 1786–1859), SMALL-FRUITED SHAGBARK. — *Fruit much smaller,* the subglobose *nut* strongly compressed laterally and *only* 1.5–2 *cm. long.* — Mass. to Pa. and Mo.

Var. **fraxinifòlia** Sarg. (ash-leaved). — *Leaflets* lanceolate to narrowly oblanceolate, *the larger ones only* 2.5–5 *cm. wide; fruit obovoid, only* 3.5–4 *cm. long.* — S. Ont. to Ia., s. to w. N.Y., O., Ind. and Okla.

5. C. laciniòsa (Michx.) Loud. (full of flaps or folds; from the loosening plates of bark), BIG SHELLBARK, KINGNUT. — Differing from no. 4 in stouter and less pointed overwintering buds; first-year *branchlets orange-brown or light tan,* often puberulent; *larger accrescent bud-scales glabrous above; leaflets* 7 or 9, *oblong-lanceolate or oblong-obovate;* the *upper ones usually larger* than in no. 4 (1.2–2.3 *dm. long* and 7–12 *cm. broad*), *acuminate, soft-pubescent beneath; serrations without subterminal dense tufts of hairs;* fruit 4–6.5 cm. long; *nut* 4–6-angled, *larger* (3–6 *cm. long*) *with yellowish- to reddish-brown shell.* — Rich woods and bottoms, N.Y. to Ia. and Neb., s. to Va., e. Tenn., Ala., La. and e. Kans.

6. C. tomentòsa Nutt. (tomentose), MOCKERNUT (in early New York Dutch Moker-noot, heavy-hammer nut), WHITE-HEART H. — *Bark* close and *deeply furrowed, not exfoliating; overwintering terminal buds* 1.5–2 cm. long, the *outer dark scales caducous in autumn; the larger accrescent inner scales* 2.5–4 *cm. long and about* 1.3 *cm. broad, pubescent on both faces; young branchlets* usually *tomentose; leaves* fragrant from abundant glands; the *petiole, rachis and lower surfaces permanently tomentose;* leaflets 5–9, essentially equilateral, lance-oblong to -obovate, acuminate, pale to orange-brown beneath; the upper ones largest, 1–2 dm. long and 5–12 cm. broad; anthers with oblong round-tipped locules; fruit globose to obovoid or thick-ellipsoid, 3.5–5 cm. long; the *husk* 3 *or* 4 *mm. thick, splitting to middle or slightly beyond; nut* globose to ellipsoid, only slightly compressed, 4-angled, *the reddish-brown very hard and tough shell* 5–6 *mm. thick;* kernel sweet. (*C. alba* K. Koch, not Nutt.) — Dry to moist woodland, Fla. to e. Tex., n. to Mass., Vt., N.Y., s. Ont., s. Mich., Ill., Ia. and Neb.

7. C. glàbra (Mill.) Sweet (smooth), PIGNUT. — *Bark close, becoming furrowed and ridged,* light gray; *overwintering terminal buds acute,* pale brown, 9–12 *mm. long, glabrous at first, becoming silky-pubescent in autumn* after the loss of the outer bud-scales; young branchlets reddish-brown, glabrous; *petioles, rachis and leaflets glabrous,* or the latter merely pilose on nerves beneath; *leaflets* 5 (rarely 7), oblong or oblong-lanceolate, *the terminal one* 8–17 cm. long and 3.5–6(–12) *cm. broad,* often with conspicuous tufts of hairs beneath; *staminate aments* 5–8 *cm. long, with* long-acuminate *villous bracts;* stamens 4–6; pistillate flowers and their bracts hoary-tomentose; *fruit* compressed, *obovoid, dark brown, smooth, shining,* 1.5–3.5 cm. long by 2–3 cm. thick; *husk* 1.5–2.5 mm. thick, *tardily opening by* 1 *or* 2 *sutures, or indehiscent;* nut pale brown, scarcely ridged, the shell rather thick; *kernel sweetish or acrid.* — Dry woods and slopes, e. Mass. and Vt. to s. Ont. and s. Ill., s. to e. Va. and upland to Ga., Ala. and Miss. — Many trees appear to have fruit intermediate between this and the next species.

Var. **megacárpa** Sarg. (large-fruited). — Leaflets larger, the terminal one 2–2.5 dm. long; stamens 4 or 6; fruit 2.5–5 cm. long; husk 3.5 mm. thick. — N.Y. to s. Ill., s. to Fla., Ala. and La.

8. C. ovàlis (Wang.) Sarg. (oval), SWEET PIGNUT, FALSE SHAGBARK. — Resembling no. 7 but *old bark often shaggy or breaking into small plates or scales; overwintering buds blunter and usually stouter; young branchlets scurfy; expanded leaves yellowish-scurfy;* leaflets more commonly 7, the *terminal mature leaflet* 4–8(–12) *cm. broad; staminate aments* 0.8–1.7 *dm. long,* puberulent; *pistillate flowers and their bracts yellowish-lepidote; fruit light brown, dull, minutely warty; husk promptly splitting to base; nut pale to deeper brown;* kernel sweet. — Several vars. recognized, some of these perhaps better considered as mere forms:

a. Petioles, rachises and leaflets glabrous or soon glabrate. . . *b.*
  b. Leaflets not conspicuously glandular beneath; exocarp wingless or only very narrowly keeled at sutures. . . *c.*
    c. Fruit ellipsoid to subglobose.
      Nut scarcely or not at all angled, ovoid to ellipsoid, acute at apex. . *C. ovalis* (typical).
      Nut angled, broadly ellipsoid, cordate or subcordate at apex. . . . Var. *obcordata.*
    c. Fruit obovoid. . . . . Var. *obovalis.*
  b. Leaflets glandular beneath, viscid; fruit with winged sutures, small (1.3–1.6 cm. thick); inner surface of fresh husk with a resinous odor; nut angled. Var. *odorata.*
a. Petioles, rachises and lower leaflet-surfaces permanently pubescent; fruit usually obovoid. . . . . Var. *hirsuta.*

C. ovàlis (typical), *C. microcarpa* Nutt. — Rich woodlands and bluffs, Mass. to Wisc. and Ia., s. to Ga., Ala. and Miss.

Var. **obcordàta** (Muhl. & Willd.) Sarg. (inversely heart-shaped). — Similar range, southwestw. into Mo.

Var. **obovàlis** Sarg. (obovate). — So. N.E. to Mo., s. to Ga., Ala. and Ark.

Var. **odoràta** (Marsh.) Sarg. (fragrant). — Mass. to s. Ont., s. to Ga., Miss. and Mo.

Var. **hirsùta** (Ashe) Sarg. (stiffly hairy). — Sw. N.H. to N.C.

9. C. pállida (Ashe) Engl. & Graebn. (pale), Pale H. — Bark very pale to dark gray, on upland trees deeply furrowed; branchlets pubescent to glabrous; *overwintering buds silvery-lepidote, the terminal one about* 6 *mm. long and of* 6–9 *scales; staminate aments silvery-lepidote,* 5–12 *cm. long; pistillate flowers yellow-lepidote; expanding leaves silvery-lepidote beneath; leaflets* 7 *or* 9, lanceolate, *long-attenuate, resiniferous and aromatic, pale beneath, the terminal one* 1–1.5 dm. long and 2–5 *cm. broad; rachises* and under sides of midribs of leaflets *at first with fascicles of curly hairs,* becoming glabrate; *fruit yellow-lepidote,* 1.5–4 cm. long, the *husk* 3–4 mm. thick and *tardily splitting to base; nut with ridges running nearly to base,* thin-shelled (2–3 mm. thick); kernel sweet. — Woods, Fla. to La., n. to upland N.C. and Tenn. and on or near Coastal Plain to s. N.J.

10. C. texàna Buckl. (Texan), Buckley's or Black H. — Bark dark brown to blackish, deeply furrowed; *young branchlets and overwintering buds* (terminal one 8–12 mm. long), *rusty-pubescent, the bud-scales hairy-tufted at tip; leaflets* 5 or 7, broadly oblong-lanceolate, when unfolding heavily *red-pubescent* and pale-scaly *beneath, the terminal one* 1–1.5 dm. long and 5–6 *cm. broad; staminate aments* 5–8 *cm. long, rusty-pubescent; fruit* subglobose to ellipsoid or ellipsoid-obovoid, aromatic, 3–4 cm. long, *yellow-scurfy; husk* 2–4 mm. thick, *splitting to middle or nearly to base; nut brown, with surface coarsely reticulate with pale ridges;* shell 2 mm. or more thick; kernel sweet. (Incl. var. *arkansana* Sarg. and *C. Buckleyi* Sarg.) — Dry sandy woods or rocky slopes, La. and Tex., n. to s. Ind., s. Ill., Mo. and e. Okla.

Var. **villòsa** (Sarg.) Little (villous). — *Leaflets more permanently and densely pubescent, oblong-elliptic to oblanceolate;* husk tardily splitting or indehiscent. — E. Mo.

FAM. 45. CORYLÀCEAE (Hazel Family)

Monoecious (rarely dioecious) trees or shrubs with alternate simple straight-veined leaves and deciduous stipules; the staminate flowers in aments; the pistillate clustered, spiked, or in a scaly ament, the 1-locular and 1-seeded nut with or without a foliaceous involucre. Ovary 2-locular, with 2 pendulous anatropous ovules in each locule; fruit seemingly 1-locular and 1-ovuled; styles 2. Seed with no albumen, filled with the embryo, and with 1 integument. Betulaceae.

a. Pistillate flowers in a short ament or head, 2 to each bract and each with 1 or more bractlets which enlarge into a foliaceous involucre to the nut; staminate flowers in pendulous aments, without a calyx; stamens 3 or more to each bract, adnate to it, filaments often forked. Tribe I. Coryleae. . . *b.*

b. Bract of staminate flower with 2 bractlets inside; pistillate flowers few in heads; involucre leafy, inclosing the acorn-like nut. . . . . 1. *Corylus.*

b. Bract of staminate flowers simple; pistillate flowers in aments; nut small or achene-like.

Each ovary and nut in a bladdery closed involucre. . . . . 2. *Ostrya.*

Each small nut subtended by an enlarged flat leafy bract. . . . . . . 3. *Carpinus.*

a. Pistillate and staminate flowers both in scaly aments, 2 or 3 to each bract, the pistillate flowers without a calyx or involucre, the small compressed nut often winged; staminate aments pendulous, the flowers with 2–4-parted calyx and 2 or 4 stamens. Tribe II. Betuleae.

Bracts of pistillate ament thin, 3-lobed or unlobed, deciduous with or soon after the nuts; stamens 2, bifid. . . . . 4. *Betula.*

Bracts of pistillate ament woody, 3–5-lobed at tip, long-persistent; stamens 4. . . . . 5. *Alnus.*

Tribe I. Corýleae Meisn.

1. CÓRYLUS L. Hazel. Hazelnut. Filbert. Noisetier (Que.). Coudrier (Que.)

Staminate flowers consisting of 8 (half-) stamens with 1-locular anthers, their short filaments and pair of scaly bractlets cohering more or less with the inner face of the scale of the ament. Pistillate flowers several from a scaly bud; ovary tipped by the short limb of the adherent calyx, one of the ovules sterile; style short; stigmas 2, red, elongated and slender. Nut ovoid or subglobose, inclosed in a leafy or partly coriaceous cup or involucre consisting of the two bractlets enlarged and often grown together and lacerated at the border. Cotyledons very thick (raised to the surface in germination), sweet and edible; the short radicle included. — Shrubs or small trees of the N. Hemisph. with thinnish doubly toothed leaves (folded lengthwise in the bud), flowering in early spring; staminate aments single or fascicled from scaly buds in the axils of the preceding year, the pistillate terminating early leafy shoots. (The classical name, probably from the Greek *corys, a helmet,* from the involucre.)

1. **C. americàna** Walt. (American), AMERICAN H. — Shrub up to 3 m. high, with *buds rounded at summit; twigs, petioles and involucres more or less glandular-bristly,* or essentially glandless in forma **missouriénsis** (A. DC.) Fern. (of Missouri); *staminate aments* mostly *peduncled, the hairs at the tip* of the scales *barely exceeding the long reddish point* and bracteoles extending beyond the margins of the scales; *leaves roundish-cordate; fruiting involucre broadly open* down to the globose bony-shelled nut, *of 2 broad foliaceous* cut-toothed almost distinct *bracts, downy.* — Thickets, centr. Me. to Sask., s. to Ga., Mo. and Okla. Apr., May.

Var. **indehíscens** Palmer & Steyerm. (unopening). — Involucre with bracts united on one side (sometimes partly or wholly so on the other) to summit. — N.C. to Okla., n. to Mo.

2. **C. cornùta** Marsh. (horned), BEAKED H. — *Twigs, petioles, etc., without glands,* except near nodes; *buds acute; staminate aments sessile or subsessile, the tuft of hairs at the tip of the scales much exceeding the short tip* and bracteoles, barely projecting beyond margins of scale; *leaves ovate to ovate-oblong; fruiting involucre of united bracts, much prolonged as a slender foliaceous beak* beyond the thin-shelled nut, *densely bristly,* or smooth in the local forma **inérmis** Fern. (unarmed). (*C. rostrata* Ait.) — Rich thickets, clearings and borders of woods, Nfld. to s. B.C., s. to Ga., Tenn., Mo., e. Kans. and Colo. Apr., May.

2. ÓSTRYA Scop. HOP-HORNBEAM. IRONWOOD

Staminate flowers consisting of several stamens in the axil of each bract; filaments short, often forked, bearing 1-locular (half-) anthers, their tips hairy. Pistillate flowers a pair to each deciduous bract, each of an incompletely 2-locular 2-ovuled ovary, crowned with the short-bearded border of the adherent calyx, tipped with 2 long-linear stigmas, and inclosed in a tubular bractlet, which in fruit becomes a closed bladdery ellipsoid involucre very much larger than the small smooth nut; these inflated involucres loosely imbricated to form a sort of strobile in appearance like that of the Hop. — Slender trees of the N. Hemisph., with very hard wood, brownish furrowed bark, and foliage resembling that of Birch; leaves open and concave in the bud, more or less plaited on the straight veins. Flowers appearing with the leaves; the staminate aments 1–3 together from scaly buds at the tips of the branches of the preceding year; the pistillate single, terminating short leafy shoots of the season. (The Greek name for a tree with very hard wood.)

1. **O. virginiàna** (Mill.) K. Koch (Virginian), AMERICAN HOP-H., LEVERWOOD, BOIS DE FER (Que.). — Leaves oblong-ovate, taper-pointed, very sharply double-serrate, downy beneath, with 11–15 principal veins; buds acute; involucral sacs bristly-hairy at the base; new branchlets glabrous or sparsely pilose and glabrate, or stipitate-glandular in forma **glandulòsa** (Spach) Macbr. (glandular). — Rich woods, N.S. to Man., s. to Va., uplands of Ga. and Tenn., Mo. and Okla. Apr., May.

Var. **lásia** Fern. (hairy). — Branchlets densely and subpersistently villous. — Coastal Plain, Fla. to Tex., n. to Va. and less characteristically to se. Mass., inland from Gulf coast to Tenn., s. Ill., Ia. and S.D.

3. CARPÌNUS L. HORNBEAM. IRONWOOD

Staminate flowers similar to those of *Ostrya.* Pistillate flowers several, spiked in a sort of loose terminal ament, with small deciduous bracts, each subtending a pair of flowers; the single involucre-like bract open, enlarged in fruit and foliaceous, merely subtending the small ovoid several-nerved nut. — Trees or tall shrubs of the N. Hemisph., with close gray bark, in this and in the slender buds and straight-veined leaves resembling the Beech; leaf-buds as in *Ostrya.* (The early Latin name.)

1. **C. caroliniàna** Walt. (of Carolina), AMERICAN H., BLUE OR WATER-BEECH, CHARME (Que.). — Leaves oblong to narrowly oblong-ovate, 1–3.5(–4) cm. broad, 2.5–8 cm. long, acute, low-serrate (teeth rarely 1 mm. high); bracts of the fruiting aments blunt or pointed, entire or with 1 or 2 blunt teeth; calyx-lobes when mature prolonged. — Rich woods and swamps, Fla. to e. Tex., n. to Md., Tenn. and s. Ill. Apr. — Bark ashy-gray.

Var. **virginiàna** (Marsh.) Fern. (Virginian). — Bark blue-gray; leaves oval or narrowly ovate, abruptly caudate-tipped, 5–12 cm. long, 2.5–6 cm. broad, with sharp and slender teeth (the larger 1–3 mm. high); fruiting bracts acute, usually with 1–5 sharp teeth; mature calyx-lobes short. — N.E. to Ont., and Minn., s. to uplands of N.C., Tenn., and Ark. Apr., May.

Tribe II. Betuleae Spach

4. BÉTULA L. Birch. Bouleau (Que.)

Staminate flowers 3 (the bractlets 2) to each peltate scale or bract of the aments, consisting each of a calyx of one scale bearing 4 short filaments with 1-locular anthers (or strictly of two 2-parted filaments, each division bearing an anther-locule). Pistillate flowers 2 or 3 to each usually 3-lobed (unlobed in no. 14 and rarely in no. 9) bract, without bractlets or calyx, each a naked ovary, becoming a winged (wingless in no. 14) and scale-like nutlet (or small samara) crowned by the two spreading stigmas. — Outer bark often separable in sheets, that of the branchlets dotted. Buds sessile, scaly. Staminate aments terminal and lateral, sessile, formed in summer, remaining naked through winter and expanding in early spring, with or preceding the leaves; pistillate aments (strobiles) ovoid to cylindrical, usually terminating very short 2-leaved early lateral branches of the season (spurs). Important genus of N. Hemisph. (The ancient Latin name.)

a. Leaves of fruiting branches with 8 or more pairs of nerves impressed above; fruiting aments 1 cm. or more thick, cylindric to ovoid, their bracts rather persistent; wing of fruit not broader than the seed-bearing body; trees with brown to reddish or yellowish bark. Ser. Costatae. . . *b.*

b. Bark and twigs sweet-aromatic; leaves membranaceous, ovate or ovate-oblong, with rounded or cordate bases, regularly serrate, green both sides; pistillate aments sessile, erect or not pendulous, ovoid to thick-cylindric.

Bark dark brown, close, in age merely furrowed and ashy-brown; bracts of pistillate ament glabrous. . . . . 1. *B. lenta.*

Bark yellowish- or silvery-gray, loosely exfoliating in irregular films, or occasionally dark brown and close; bracts of pistillate ament pubescent or with ciliate lobes. . . . . 2. *B. lutea.*

b. Bark not aromatic, pinkish-white to terra-cotta or pale brown, shaggily exfoliating; leaves firm, rhombic-ovate, cuneate to subtruncate at base, irregularly dentate-serrate, white and when young downy beneath; staminate aments cylindric, peduncled, with soft-downy bracts. . . . 3. *B. nigra.*

a. Leaves of fruiting branches with 7 or fewer pairs of prominent veins; bracts of mature pistillate aments readily deciduous. . . *c.*

c. Wing as broad as or broader than body of fruit; leaves definitely petioled, deltoid to oblong or ovate, acute to blunt; fruiting aments slenderly cylindric, 1–6 cm. long, definitely peduncled; trees or coarse shrubs with white, whitish or dark papery bark. Ser. Albae. . . *d.*

d. Bark opaque, chalky- or ashy-white, close, the layers not readily exfoliating; staminate ament usually 1 and, before expanding, pointing stiffly forward; leaves glabrous (or merely glutinous) on both sides, abruptly ending in prolonged tail-like tips (caudate); fruiting aments 1–2.5 cm. long; the mature bracts nearly horizontally divergent, crowded, 3–4.5 mm. long, uniformly ashy-puberulent on back. . . . . . . 4. *B. populifolia.*

d. Bark lustrous, creamy- or pinkish-white to warm brown, in maturity often exfoliating or separating into layers; staminate aments 1–few; leaves not prominently caudate-tipped; fruiting aments (except in dwarf shrubs) mostly larger; their bracts (except in the introduced no. 5) ascending, glabrous or pilose on back, the lobes ciliate. . . *e.*

e. Samaras 2.5–6.5 mm. broad, the wings broader than the body; trees or coarse shrubs. . . *f.*

f. Leaves glabrous or soon glabrate on both sides; young shoots glabrous or glabrate or merely with resinous warts. . . *g.*

g. Trees with whitish bark; leaves deltoid-ovate, acuminate from broad base, those of fertile branches 3–10 cm. long; staminate aments 4–10 cm. long; lateral lobes of pistillate bracts divergent, larger than terminal lobe; species of low alt.

Leaves of fertile branches 3–7 cm. long, their cuneate to truncate bases entire; fruiting aments 2–4 cm. long, their bracts divergent; introd. species. . . . . 5. *B. pendula.*

Leaves of fertile branches 6–10 cm. long, their rounded bases toothed except near petiole; fruiting aments 2.5–5 cm. long, their bracts ascending; indig. species. . . . . . . 6. *B. caerulea-grandis.*

*g.*Shrub with dark close bark; leaves ovate, merely acute to blunt, those of fruiting branches 1.5–4.5 cm. long; staminate aments 1.5–3.5 cm. long; lateral lobes of pistillate bracts ascending, scarcely broader than terminal lobe; subarctic-alpine species. . 7. *B. minor.*

*f.*Leaves pubescent beneath, at least when young, or on veins or in their axils; young vegetative shoots pubescent or puberulent.

Buds lustrous with resin; leaves merely acute, those of fertile branches 3–5 cm. long; mature fertile aments 1.5–3 cm. long; introd. species. . . . . 8. *B. alba.*

Buds scarcely resinous; leaves mostly acuminate, those of fertile branches 2.5–10.5 cm. long; mature fertile aments 1.5–6.5 cm. long; indig. species. . . . . 9. *B. papyrifera.*

*e.*Samaras 2–3.5 mm. broad, the wings scarcely to barely as broad as the nutlet; new sprouts pubescent; leaves elliptic, rhombic-oval or ovate; shrub with close dark bark. . . . . 10. *B. borealis.*

*c.*Wings narrower than to rarely as broad as body of fruit, or wanting in no. 14; shrubs, or no. 11 a small tree, with dark scarcely exfoliating bark, small obovate to reniform leaves with 2–5 pairs of veins and erect sessile or short-peduncled fruiting aments only 0.5–3 cm. long. Ser. Humiles. . . *h.*

*h.*Bracts of pistillate aments 3-lobed; nutlets definitely winged. . . *i.*

*i.*Alleghenian tree 6–8 m. high; branchlets and leaf-blades glabrous, eglandular; blades of fertile branches suborbicular and subdigitately veined, their slender petioles villous; middle lobe of pistillate bracts broad and low. . . . . 11. *B. uber.*

*i.*Northern depressed to upright shrubs up to 3 m. high; branchlets pubescent, puberulent or glandular, rarely quite glabrous; blades of fertile branches obovate to orbicular or reniform, pinnately veined, with short thick petioles; middle lobe of pistillate bracts prolonged.

Young branchlets and vigorous shoots villous to puberulent or glabrous, glandless or only remotely glandular; leaves submembranaceous, whitish or gray beneath, pubescent to glabrous, glandless or sparingly glandular; shrub of N. States and Can. 12. *B. pumila.*

Young branchlets and vigorous shoots covered with large resinous wart-like glands; leaves glutinous, green both sides, glabrous or nearly so; arctic-alpine. . . . . 13. *B. glandulosa.*

*h.*Bracts of pistillate aments unlobed, entire, oblong; nutlets wingless; dwarf shrub with cinereous branchlets and tiny flabelliform imbricated leaves, of Nfld., s. Lab. Pen. and N.S. . . . . 14. *B. Michauxii.*

Series Costàtae Regel

1. B. lénta L. (tough or flexible; from the twigs), Cherry-, Sweet or Black B., Mountain-mahogany, Merisier rouge (Que.). — *Bark* of trunk *dark brown, close,* in age becoming ashy-brown and furrowed, very sweet-aromatic; leaves ovate or ovate-oblong from a more or less cordate base, acuminate, sharply and finely double-serrate, when mature bright green above and glabrous except on the veins beneath; *fruiting catkins short-cylindric* (1.5–2.5 cm. long); the *scales firm and smooth, with* short and *divergent lobes.* — Rich woods, sw. Me. to s. Que. and e. Ont., s. to s. N.E., L.I., Md., and upland Ga. and Tenn. *Fl.* mid-April, May. — Forma **laciniàta** Rehd. (slashed) is a rare form with deeply lacerate leaves.

2. B. lùtea Michx. f. (yellow), Yellow or Gray B., Merisier (Que.). — *Bark* of trunk *yellowish- or silvery-gray, detaching in very thin filmy layers,* less aromatic; leaves slightly or not at all cordate and often narrowed toward the base, duller green above and usually more downy on the veins beneath; *fruiting aments narrow-ovoid to subglobose, their firm and subcoriaceous bracts* more or less *pubescent,* 5–8 mm. long, *with ciliate lobes,* and *cuneate basal portion* 1–2.5 *mm. long.* (*B. alleghaniensis* Britt.) — Rich woods, Gaspé Co., Que., to Ont., s. to N.S., N.E., L.I., Del., Md., upland to N.C. and Tenn., W.Va., n. Ind., n. Ill., and ne. Ia. *Fl.* May, June. — In forma **fállax** Fassett (deceitful) the dark bark often close, thus simulating that of no. 1.

Var. **macrólepis** Fern. (large-scaled), "Witch-Hazel" of Nfld. — *Bracts of pistillate aments foliaceous, in maturity* 8–13 *mm. long, the elongate basal portion* 2.5–6 *mm. long.* — Similar range, n. to Nfld. and s. Lab. Pen., Que., generally commoner northw.

× **B. Sandbérgi** Britt. (for its discoverer, John Herman Sandberg, 1848–1917), a hybrid of no. 9 and *B. pumila,* var. *glandulifera,* Sandberg's B. — Shrub or bushy tree 2–10 m. high, with dark brown close outer bark, *inner bark without wintergreen-flavor;* young branchlets rusty-pubescent, late in season glabrescent and more or less glandular; *leaves* ovate, rhombic, elliptic or obovate, *rounded to cuneate at base, with* 4 *or* 5 *pairs of veins,* 2.5–5.5 cm. long, acute to blunt, serrate to serrate-crenate, firm, dull green and reticulate above, *pale and pubescent to glabrate beneath, resinous-dotted; fruiting aments cylindric,* 2–2.5 cm. long, 5–7 *mm. thick, their peduncles about* 1 *cm. long;* bracts puberulent, ciliate; nutlets narrower than wings. — Tamarack-swamps, local, Wisc. to Sask. and Mont. (Descr. after Rosendahl & Butters).

× **B. Purpùsii** Schneid. (for Joseph Anton Purpus, 1860–1933), a hybrid of nos. 2 and 12. — Differing from preceding in generally lower habit, *grayer bark, the inner wintergreen-flavored;* spur-shoots numerous, with crowded leaves or scars; *leaves with* 5–7 *pairs of veins, broadly cuneate to rounded or subcordate at base; fruiting aments* 1–1.2 *cm. thick, ovoid-cylindric, on peduncles* 3–8 *mm. long;* bracts larger. — Local in tamarack-swamps, Bruce Pen., Ont., Mich., n. Ind. and n. Ill. to Minn. (Descr. after Rosendahl & Butters).

3. B. nìgra L. (black), River- or Red B. — Tree with *shaggy and freely exfoliating greenish-brown, pinkish or terra-cotta bark* and reddish twigs; *leaves firm, rhombic-ovate, cuneate or subtruncate at base, irregularly dentate-serrate, whitish beneath* and when young downy; *petioles, peduncles and thick-cylindric pistillate aments tomentose;* bracts of latter with nearly equal oblong-linear lobes. — Bottomlands and borders of streams at low alts., Fla. to e. Tex., n. to e. N.Y. and sw. Ct. (locally in ne. Mass. and se. N.H.), Pa., W.Va., O., s. Mich., s. Wisc., s. Minn. and e. Kans. *Fl.* April, May.

Series Álbae Regel

4. B. populifòlia Marsh. (poplar-leaved) White, Gray, Fire- or Oldfield-B., Bouleau rouge (Que.). — *Bushy tree up to* 10 *m. high, the main trunk or trunks usually sprouting from base; bark close, not exfoliating, chalky-white,* with dark elongate markings; *branchlets* slender, wiry, *glabrous,* often warty; *leaves somewhat lustrous,* often gummy, *glabrous, triangular, tapering from nearly or quite truncate base to long tail-like tip,* tremulous on short petioles; *staminate ament* 1 (rarely 2), *projecting stiffly forward* before opening; *fruiting aments* slender-stalked, ascending, 1–2.5 *cm. long;* mature *bracts nearly horizontally divergent,* crowded, 3–4.5 mm. long, uniformly *puberulent on back;* fruits with thin wings broader than nutlet. — Sterile dry to wet soil, P.E.I. to Laurentide reg., Que., w. to sw. Ont., s. to N.S., N.E., L.I., Del., Pa., upland of Va., n. O. and n. Ind. *Fl.* Mid-April, May. — Hybridizes with no. 9. — Forma **incisifòlia** Fern. (with incised leaves) with leaves deeply dissected, rare.

× **B. caerùlea** Blanch. (blue), a hybrid of nos. 4 and 6, with aments much as in no. 4, bark and bluish leaves more like those of no. 6, is frequent with the two parents.

5. B. péndula Roth (drooping), European B. — Small tree with whitish exfoliating bark; *branchlets in mature trees pendulous, glabrous; leaves deltoid-ovate, acuminate from* broad *entire truncate to cuneate base,* the *blades of fertile branches* 3–7 *cm. long;* staminate aments 1–3, 4–9 cm. long; *fruiting aments* 2–4 *cm. long,* their puberulent to glabrous *bracts with divergent or arched-recurving lateral lobes larger than terminal lobe.* — Cult., occasionally spread to thickets, open woods, etc., N.S. to Ont., s. to N.E., Pa., Mich., Wisc. and Ia. — Forma dalecárlica (L.f.) Schneid. (of Dalecarlia), Cut-leaved B., with lacerately dissected leaves, cult. and occasionally running wild. (Introd. from Eu.)

6. B. caerùlea-grándis Blanch. (large blue), Blue B. — *Large tree with creamy- to pinkish-white freely exfoliating bark;* branchlets and *twigs* loosely *spreading,* scarcely pendulous, *glabrous* or merely gummy; *leaves bluish, lustrous, broadly deltoid-ovate, acuminate,* the gradually *rounded base toothed except near petiole,* the *blades on fertile branches* 6–10 *cm. long; fruiting aments* 2.5–5 *cm. long, with appressed-ascending bracts,* the latter *with the divergent lateral lobes larger than the terminal one.* — Dry woods, Gaspé Pen. to Montmorency Co., Que., s. to N.S., n. N.E. and e. N.Y. *Fl.* May, early June.

7. B. mìnor (Tuckerm.) Fern. (smaller), Dwarf White B. — Depressed to upright *coarse shrub* 0.3–2 m. or more high, the *tough branches with dark close bark; new branchlets glabrous, often bearing glandular warts; leaves ovate, acute to blunt, glabrous,* often glutinous, *those of the fruiting branches* 1.5–4.5 *cm. long; staminate aments* 1.5–3.5 *cm. long; fruiting aments* cylindric, peduncled, 1.5–3.5 *cm. long;* fertile *bracts* 3–6 mm. long; *their lateral lobes ascending, scarcely broader than the terminal lobe; samaras* 2.5–5 *mm. broad,* their wings broader than achene. (*B. papyrifera,* var. Tuckerm.; *B. alba,* var. Fern.) — Acidic rocky barrens, peats and alpine summits, Lab. Pen., s. to Nfld., Shickshock Mts., Gaspé Pen. and Laurentide Mts., Que., and higher mts. of n. N.E. and ne. N.Y. *Fl.* June, July — Simulating *B. odorata,* var. *tortuosa* of arct. Eurasia and Greenl.

8. B. álba L. (white), White B. of Europe. — Tree with whitish exfoliating bark; branches and *branchlets spreading,* the latter *densely pubescent, without glands; winter-buds glutinous and shining; leaves* ovate to subrhombic, *merely acute,* pubescent on expanding, usually *hairy in axils of veins beneath, those of fruiting branches* 3–5 *cm. long; fruiting aments* 1.5–3 *cm. long,* peduncled; *fertile bracts appressed-ascending, puberulent.* (*B. pubescens* Ehrh.) — Cult. and spreading to roadsides, thickets, etc., Nfld. to N.E., w. to Mich. (Introd. from Eu.)

9. B. papyrífera Marsh. (bearing paper), PAPER-, CANOE-, or WHITE B., BOULEAU BLANC (Que.). — Tree or coarse shrub with *bark of young trunks* (rarely of old trees) *warm brown, this commonly soon exfoliating and exposing* the *creamy- to pinkish-white thick and finally exfoliating bark of most mature trees;* branches and branchlets ascending, spreading or rarely drooping, the *vigorous new shoots pubescent and often glandular; winter-buds not conspicuously glutinous; leaves* slender-petioled, firm-membranaceous to subcoriaceous, *pubescent beneath when expanding, usually retaining tufts of hairs in the axils of* and minute pilosity along the *veins beneath,* the elliptic-oblong to ovate or cordate serrate or serrate-dentate *blades* mostly *acuminate; those of fertile branches* 2.5–10.5 *cm. long;* staminate aments (1–)2–3(–4), 5–10 cm. long; *fruiting aments* cylindric, peduncled, (1.5–)2.5–6.5 *cm. long; bracts* pubescent to glabrous, *with rhombic* divergent to ascending *lateral lobes* (wanting in one var.) much *broader than middle lobe; samaras* 3.5–6 *mm. broad,* the *wings as broad as or broader than* usually *hispidulous achene.* — Polymorphous transcontinental species. We recognize the following vars. and forms:

*a.*Leaves merely rounded to tapering at base. . . *b.*
*b.*Bracts of pistillate aments 3-lobed; peduncle usually shorter than fruiting ament; the latter 2.5–6.5 cm. long. . . *c.*
*c.*Mature fertile bracts 3.5–7 mm. long, with divergent lateral lobes; samaras 3.5–5 mm. broad. . . *d.*
*d.*Branchlets spreading or ascending, not strongly drooping; leaves of fertile branches broadly ovate, mostly rounded at base; pistillate aments mostly solitary on the spurs. . . *e.*
*e.*Bark of trunks of fruiting trees (or shrubs) creamy- to pinkish-white, very soon exfoliating.
Leaves membranaceous to firm, hardly lustrous. . . . . . . *B. papyrifera* (typical).
Leaves thick and leathery, lustrous above. . . . . . . Forma *coriacea.*
*e.*Bark of fruiting trunks warm brown and smooth, only on oldest bases with smooth outer brown layer exfoliating. . . . . . . Var. *commutata.*
*d.*Branchlets pendulous; leaves of fertile branches narrowly ovate to ovate-lanceolate, only slightly rounded to gradually tapering to petiole; pistillate aments often in fascicles of 2–4 on the spurs. . . . Var. *pensilis.*
*c.*Mature fertile bracts 7–10 mm. long, with ascending lateral lobes; samaras 6–8 mm. broad; leaves ovate, with rounded bases; fruiting aments solitary or paired.
Peduncles of fruiting aments 0.5–1.5 cm. long, many times shorter than ament. . . . . Var. *macrostachya.*
Peduncles 2–3 cm. long, one-half to essentially as long as pendulous ament. . . . . Forma *longipes.*
*b.*Bracts unlobed or with merely rudimentary lateral lobes, elliptic-oblong; pistillate aments 1.5–2 cm. long, about equaled by arched-recurving peduncle; leaves rhombic-oval, dentate. . . . . . . Var. *elobata.*
*a.*Leaves definitely cordate at base; bracts of mature pistillate aments 5–10 mm. long, mostly with ascending lobes; bark of mature trunks warm brown to creamy- or pinkish-white. . . . . Var. *cordifolia.*

B. papyrífera (typical). — Woods, especially on slopes, Lab. to Alaska, s. to Nfld., N.S., N.E., N.Y., upland of Pa. and W.Va., n. O., n. Ind., n. Ill., n. Ia., S.D., etc. *Fl.* late April–June. — Sometimes crosses with no. 4. — Forma **coriàcea** Fern. & Wieg. (leathery), dunes of L. Ont., N.Y.

Var. **commutàta** (Regel) Fern. (changeable). — Woods, Lab. to sw. Que., s. to Me. and ne. Mass.; nw. N.Am.

Var. **pénsilis** Fern. (pendent). — Nfld. to w. Que., s., locally, to N.S., Me., ne. Mass. and n. N.Y.

Var. **macrostáchya** Fern. (large-spiked). — N. Nfld. to Rimouski Co., Que., s. to N.S. and n. Me. — Forma **lóngipes** Fern. (long-stalked), Gaspé Pen., Que.

Var. **elobàta** (Fern.) Sarg. (without lobes). — By alpine brooks, upper serpentine-slopes, Mt. Albert, Gaspé Co., Que. — A remarkable extreme, known only from young material; mature fruit needed.

Var. **cordifòlia** (Regel) Fern. (heart-leaved). — Lab. to Algoma Distr., Ont., s. to Nfld., N.S., N.E. (rare southw.), n. N.Y., high Blue Ridge, N.C., Mich., Wisc. and n. Ia.

10. B. boreàlis Spach (northern). — Upright to depressed *shrub* or shrubby small tree *with close dark bark; new branchlets and sprouts densely villous-tomentose to pilose; leaves broadly ovate, subrhombic, elliptic or* (rarely) *obovate, pubescent when young, more or less so in maturity, those of fertile branches* 1.5–5.5 cm. long, *short-acute to obtuse,* doubly dentate-serrate, *short-petioled;* fruiting aments 1–3 cm. long; their *bracts with ascending lobes; samaras* 2–3.5 mm. broad, the *wings from scarcely as broad to about as broad as the nutlet.* — Calcareous or magnesian (more rarely acidic) soils, Lab. and Ung., s. to Nfld., Anticosti I. and Gaspé Pen., Que., C.B., Mt. Katahdin, Me., and n. Vt., often alpine. *Fl.* June, July.

Series Hùmiles D. J. Koch

11. B. ùber (Ashe) Fern. (fruitful). — *Small* dark-barked *tree 6–8 m. high; branchlets glabrous; leaves suborbicular,* broadly rounded at both ends or subcordate at base, *glabrous; those of fertile branches about 2 cm. long, their principal veins subbasal, the margins coarsely dentate; petioles slender, elongate, villous; pistillate aments* compact, *ellipsoid-subcylindric, sessile, erect,* 1–1.5 *cm. long; fertile bracts* coriaceous, *strongly ribbed,* glabrous, *the broad and low middle lobe and the broad lateral lobes subequal; samara broadly cuneate, the wings broadened upward* and narrower than to nearly as broad as nutlet. — Mts. of Smyth Co., Va., little known.

12. B. pùmila L. (dwarf), Low or Swamp-B. — Erect, 0.5–3 m. high, or prostrate and matted; young branchlets pubescent to glabrescent or rarely quite glabrous, glandless or sparingly glandular-dotted; *leaves submembranaceous, pale green to whitish beneath,* obovate, orbicular or reniform, coarsely dentate or dentate-serrate, 0.8–7 cm. long, both surfaces finely reticulate; fruiting aments 0.7–3 cm. long, 5–9 mm. thick, sessile or nearly so, erect; *middle lobe of pistillate bracts narrow and much longer than lateral lobes;* samara suborbicular, the wings barely as wide as the nutlet. — Very variable:

*a.*Leaves of fertile branches cuneate-obovate, definitely longer than broad, conspicuously paler beneath, 1–4 cm. long; those of vegetative sprouts rounder and broader and up to 7 cm. long; upright (rarely depressed) shrubs 0.5–3 m. high. . . *b.*
- *b.*Branchlets and leading shoots soft-downy or puberulent.
  - Branchlets and lower surfaces of leaves without glands; leaves mostly pubescent beneath. . . . *B. pumila* (typical).
  - Branchlets and lower leaf-surfaces more or less glandular-warty; leaves sparingly pubescent or glabrescent. . . . Var. *glandulifera.*
- *b.*Branchlets and leaves glabrous or promptly glabrate, without glands. . . Var. *glabra.*

*a.*Leaves of fertile branches broadly obovate and more rounded to base to orbicular or reniform, nearly or quite as broad as long, less whitened beneath, 0.8–3 cm. long, those of vegetative sprouts roundish to reniform and up to 5 cm. long; branchlets and at least younger lower leaf-surfaces pubescent, glandless; prostrate or depressed to divergently branched and up to 0.5 m. high. . . . . . Var. *renifolia.*

B. pùmila (typical). — Bogs and wooded swamps (often calcareous), Nfld. to Ont., s. to C.B., n.-centr. Me., nw. Ct., n. N.J., N.Y., n. O., n. Ind., and Wisc. *Fl.* May, June.

Var. **glandulífera** Regel (bearing glands), *B. glandulifera* (Regel) Butler. — Similar habitats, w. Que. to B.C., s. to n. N.Y., n. Ind., n. Ill., n. Ia., N.D. and Mont. Hybridizes with no. 2.

Var. **glàbra** Regel (hairless). — Local, Mich. and n. Ind.

Var. **renifòlia** Fern. (with kidney-shaped leaves). — Peats, tundra, rocky slopes, crests and alpine barrens, Lab. Pen., s. to Nfld., St. Paul I., N.S., M.I., P.E.I. and ne. N.B. *Fl.* June–Aug.

13. B. glandulòsa Michx. (glandular), Dwarf B. — Depressed and matted or ascending and up to 2 m. high*; young branchlets* glabrous or at most minutely puberulent and *conspicuously dotted with resinous wart-like glands; leaves* obovate to suborbicular, 0.5–3 cm. long, *green on both sides, coriaceous, glutinous,* glabrous or minutely glandular-puberulent, slightly reticulated; fruiting aments 0.5–2.5 cm. long, 3–7 mm. thick. (Including var. *sibirica* (Ledeb.) Blake = var. *rotundifolia* (Spach) Regel, too difficult to distinguish in our material). — Arct. Am., s. on acidic rocky barrens, crests and summits to Nfld., alpine areas of Gaspé Pen., Que., n. N.B., Mt. Katahdin, Me., White Mts., N.H., and Adirondack Mts., N.Y., n. Ont., Black Hills, S.D., etc. *Fl.* June–Aug.

14. B. Michaùxii Spach (for its discoverer André Michaux, 1746–1802), Michaux's or Newfoundland Dwarf B. — *Tiny shrub* with slender creeping subterranean bases; *the erect or ascending* simple or branching *stems* 1–6 dm. high, *minutely cinereous-tomentulose; leaves broadly flabelliform-obovate,* 5–10 *mm. long,* about as broad, *glabrous, lustrous, coria-*

ceous, coarsely reticulate, incised around the rounded to subtruncate summit, closely imbricate-ascending along the branches; *staminate and pistillate aments sessile among the imbricated leaves of the very short ascending spurs; fruiting aments* 5–10 *mm. long; bracts simple, entire or nearly so, oblong to oblong-ovate, narrower than the thick-rimmed wingless* ovate to roundish **nutlets,** *tightly appressed,* with often spreading tips. (*B. terrae-novae* Fern.) — Bogs, heaths and acid peaty barrens, se. Lab. to Ung.; Nfld.; local, N.S. *Fl.* June, July. — Our representative of the arct. and subarct. *B. nana* L., which has 3-lobed bracts, thin-winged samaras, longer aments, much longer styles, etc.

5. ÁLNUS B. Ehrh. Alder. Aulne or Aune (Que.)

Staminate aments with 4 or 5 bractlets and 3 (rarely 6) flowers upon each short-stalked peltate bract; each flower usually with a 3–5-parted calyx and as many stamens; filaments short and simple; anthers 2-locular. Pistillate aments ovoid or ellipsoid; the fleshy bracts each subtending 2 flowers and a group of 4 tiny scalelets adherent to the bracts of the ament, which are woody in fruit, cuneate-obovate, truncate or 3–5-lobed. — Shrubs or trees of N. Hemisph. and w. S. Am., with few-scaled leaf-buds and solitary to often racemose- or cymose-clustered aments. (The ancient Latin name.)

a. Flowers developed with the expanding leaves; pistillate aments from scale-covered buds, loosely racemose; nutlets with conspicuous thin wing; overwintering buds sessile, of 3–6 unequal imbricated scales. Subgen. Alnaster. 1. *A. crispa.*

a. Flowers developed long before (in earliest spring) or long after (in autumn) the expanding leaves; aments all from naked buds; nutlets essentially wingless; overwintering buds stipitate, of 2 or 3 subequal scales. Subgen. Alnus. . . *b.*

b. Vernal-flowering, fruiting the same season; young inflorescences terminal; leaves plicate in bud, their lateral veins straight; fruiting cones 3–12 or more, in irregular inflorescences, 0.5–1.5 cm. thick. . . *c.*

c. Shrubs, either low or up to 8 m. high and branching from base or barely tree-like; leaves simply or doubly serrate or serrulate; fruiting cones short- and thick-stalked to sessile, the terminal lobes of their bracts prolonged and ascending.

Cortex of trunks with linear transverse whitish lenticels; axis of young or flowering inflorescence arching, without right angles, the pistillate branch or branches (in monoecious specimens) then drooping and appearing to be below the staminate ones; leaves broadest at or below middle, with rounded to subcordate bases, ovate, oval, subelliptic or roundish, often repand or doubly serrate, scarcely glutinous, in maturity with cross-veins beneath prominent and forming ladder-like reticulation. . . . . 2. *A. rugosa.*

Cortex with shorter and fewer darker lenticels or these obscure; axis of young inflorescence with abruptly geniculate or right-angled bends, the pistillate branches erect or strongly divergent, thus appearing above the staminate ones; leaves obovate to obovate-elliptic, broadest at or above middle, cuneate to but slightly rounded at base, mostly simply serrulate, glutinous when young, in maturity with lower surface with only weak cross-veins. . . . . 3. *A. serrulata.*

c. Round-headed tree with erect trunk; leaves dentate or lobulate and denticulate; fruiting cones all on long slender stalks, the terminal lobes of their bracts low and depressed. . . . . . . 4. *A. glutinosa.*

b. Late summer- or early autumn-flowering, fruiting the 2nd year; inflorescences mostly axillary; leaves not plicate in bud, their lateral veins arching; fruiting cones 2 or 3, racemose, all slender-stalked, 1.5–2 cm. thick, the terminal lobes of their bracts low and broad. . . . . . . 5. *A. maritima.*

Subgen. Alnáster Endl.

1. A. críspa (Ait.) Pursh (crisped; from the leaf-margin), Green or Mountain-A., Aulne vert (Que.). — Ascending and bushy shrub up to 3 m. high, the *young branches and peduncles glabrous or but sparsely pubescent and glabrate; leaves* round-oval, ovate or slightly cordate, *glutinous and glabrous or with veins beneath slightly pubescent,* irregularly *serrulate or biserrulate* with very fine and sharp crowded teeth, with the margins often puckered, in maturity 3–8 cm. long, or in the dwarf and prostrate alpine small-coned (8–10 mm. long) forma **strágula** Fern. (forming carpets) only 2–4.5 cm. long; *pistillate aments slender-stalked,* 2–10 *in simple* or corymbosely clustered *racemes,* in maturity 1–2 cm. long. (*A. viridis* sensu Am. auth., not DC.;

A. Alnobetula sensu Am. auth., in part, not K. Koch) — Lab. to Alaska, s. on rocky shores, slopes and mts., to Nfld., Que., higher mts. of n. N.E., n. N.Y. and w. N.C., Ont., Mich., Wisc., Minn., Man., Sask., and Alta. *Fl.* June–Aug.

Var. **móllis** Fern. (soft). — Usually coarser; *young branches and peduncles permanently soft-pubescent; leaves permanently covered beneath with dense soft pubescence,* the mature blades 3–12 (on vigorous sprouts –20) cm. long. (*A. mollis* Fern.) — Chiefly at low alts., s. Lab. to Algoma Distr., Ont., s. to Nfld., N.S., n. N.E., centr. and w. Mass. and N.Y. *Fl.* May, June.

Subgen. Álnus Endl.

2. A. rugòsa (Du Roi) Spreng. (wrinkled), Speckled A., Verne or Aulne blanchâtre (Que.). — Spreading or loosely ascending shrub, rarely slightly tree-like, up to 5 (rarely to 8) m. high, the main trunks or basal branches 0.1–1.5 dm. in diameter, covered with warm brown to blackish-gray *cortex marked with whitish linear lenticels up to 7 mm. or more long;* inflorescences developed before unfolding of leaves; *axes of young or flowering inflorescences arching,* without right angles, *the pistillate branch or branches* (in monoecious shrubs) *then drooping and appearing to be below the staminate aments; leaves ovate, oval, subelliptic or rounded, broadest below or near the middle, with rounded to subcordate bases, barely if at all glutinous, oftenest doubly serrate or serrate-dentate, often repand, undulate; the mature blades with cross-veins prominent beneath and forming scalariform reticulation; fruiting cones on short thick stalks to sessile, in irregular open or close cymes;* terminal lobes of cone-bracts prolonged, erect or arching. (*A. incana* sensu Tuckerm. and later Am. auth., not (L.) Moench) — Variable; the following vars. and forms recognized:

a. Leaves green or fulvous, not glaucous, beneath.
  Lower surfaces of leaves glabrous or promptly glabrate, only the principal veins or their axils sometimes permanently pilose. . . . . *A. rugosa* (typical).
  Lower surfaces of leaves permanently soft-pilose or subvelutinous to touch. Forma *Emersoniana.*
a. Leaves glaucous or whitened beneath. . . *b.*
b. Lower surfaces of leaves glabrous or promptly glabrate.
  Leaves ovate or oval to round-elliptic, with low toothing. . . . . . Var. *americana.*
  Leaves narrowly elliptic to ovate-lanceolate, lacerate or jagged-toothed. Forma *tomophylla.*
b. Lower surfaces of leaves densely soft-pilose or subvelutinous to touch. . . Forma *hypomalaca.*

A. rugòsa (typical), *A. incana,* var. *hypochlora* sensu Fern., not Callier — Low grounds, swamps, margins of streams, etc., w. N.S. to n. Mich., s. to s. N.E. and locally to n. and e. Pa. and n. Ind. *Fl.* March–May. — Forma **Emersoniàna** Fern. (for George Barrell Emerson, 1797–1881, distinguished student of trees and shrubs), scattered, often abund.

Var. **americàna** (Regel) Fern. (American), *A. glauca* Michx. f.; *A. americana* (Regel) K. Koch — Lab. to Hudson Bay and Sask., s. to Nfld., N.S., Me., Mass., mts. of Pa., Md. and W.Va., n. O., n. Ind., Wisc., and ne. Ia. *Fl.* April–early June. — Forma **tomophýlla** Fern. (with lacerate leaves) a rare cut-leaved form in Nfld. and Me. Forma **hypomálaca** Fern. (soft beneath) throughout the range, often commoner.

3. A. serrulàta (Ait.) Willd. (finely saw-toothed), Common A. — Differing from no. 2 in having cortex with smaller or obscure pale dots; *axis of young inflorescence with* 1 *or more abruptly geniculate or right-angled bends, the pistillate branches erect or strongly divergent and often appearing above the staminate aments;* the latter usually more slender than in no. 1; *leaves obovate or obovate-elliptic, broadest at or above middle, cuneate to but slightly rounded at base, simply serrulate,* only exceptionally strongly undulate, *the expanding ones glutinous and often aromatic;* the mature blades *with lower surface delicately reticulate or with only weak cross-veins.* (*A. rugosa* sensu K. Koch and most Am. auth. subsequently, not Spreng.) — The following vars. and forms recognized:

a. Principal leaves definitely obovate, cuneate or subcuneate to subacute base, those of vigorous 1st year's shoots obtuse to acute, of fertile branches of 2nd year one-third to two-thirds as broad as long.
  Lower surfaces of mature leaves glabrous or strongly glabrescent. . . . *A. serrulata* (typical).
  Lower surfaces of mature leaves permanently and densely pilose-tomentulose, plush-like to touch. . . . . Forma *noveboracensis.*
a. Principal leaves broadly elliptic-obovate to broadly oblong-elliptic or sub-

rotund (though broadest at or above the middle), gradually rounded at base, those of fertile branches of 2nd year mostly three-fifths to nine-tenths as broad as long. . . *b.*

*b.*Lower surfaces of mature leaves glabrous or strongly glabrescent.

Leaves gradually rounded to subacute (more rarely acute) at apex, mostly 6–15 cm. long; staminate aments 3–7 cm. long. . . . . . Var. *subelliptica.*

Leaves broadly retuse or emarginate at summit, mostly 2–5 cm. long; staminate aments 2 cm. long. . . . . Forma *emarginata.*

*b.*Lower surfaces of mature leaves permanently and densely pilose-tomentulose, plush-like to touch.

Large shrub or small tree; principal leaves 6–12 cm. long; staminate aments 3–7 cm. long; ripe cones 1–2 cm. long. . . . . Forma *mollescens.*

Dwarf shrub 0.5–1 m. high; principal leaves 2–5 cm. long; staminate aments 1.3–1.8 cm. long; ripe cones 6–12 mm. long. . . . . . Forma *nanella.*

A. serrulàta (typical). — Swamps, wet woods, stream-margins, etc., n. Fla. to La., n. to sw. N.S. (local), centr. and s. Me., s. N.H., centr. Vt., N.Y., W.Va., O., Ind., Ill., Mo. and se. Okla. *Fl.* Feb.–May. — Forma **noveboracénsis** (Britt.) Fern. (of New York), *A. noveboracensis* Britt., local, often abund., throughout the broad range.

Var. **subellíptica** Fern. (nearly elliptic). — Ga. to s. N.H., Mass. and N.Y. — Forma **emarginàta** Fern. (notched at summit), local in bogs of Hartford Co., Ct.; forma **molléscens** Fern. (becoming soft), scattered through the range; forma **nanélla** Fern. (very small), local in sphagnous bogs and thickets, se. Va.

4. A. GLUTINÒSA (L.) Gaertn. (gummy), BLACK A. of Europe. — *Tree with full rounded crown,* up to 35 m. high, *with erect* dark-barked *trunk, the young glabrous branchlets and expanding leaves very gummy; leaves dark green, flabellate-obovate to suborbicular, the dentate or denticulate margin often lobulate; fruiting cones all on long slender stalks, their bracts with depressed terminal lobes.* (*A. vulgaris* Hill) — Spread from cult. and locally natzd., Nfld. to Ill., s. to Del. and Pa. *Fl.* March–May. (Introd. from Eu.)

A. INCÀNA (L.) Moench (grayish), WHITE A. of Eu., with which no. 2 has long been needlessly confused, a tree differing from no. 4 in whitish-gray bark, more elliptic to ovate doubly toothed and serrate leaves whitened and often heavily pubescent beneath, and shorter-stalked cones, is planted and may sometimes escape from cult. (Introd. from Eu.)

5. A. marítima (Marsh.) Nutt. (of sea-shore), SEASIDE A. — Shrub or small tree up to 10 m. high, with bark finally reddish-brown; branchlets at first pubescent, soon glabrate; *leaves not plicate in bud,* oblong, ovate or obovate, *cuneate to long petiole, dark green above, remotely serrulate, with slender arching veins* and delicately reticulate veinlets; *aments expanding in late summer, the inflorescence or the individual aments often axillary; pistillate aments 2 or 3 in a raceme, slender-stalked; fruit maturing 2nd year; ripe cones* 1.5–2 *cm. thick, their bracts with depressed and broad terminal lobes.* — Pond-shores and stream-banks near coast of Del. and Md. *Fl.* late Aug., Sept.

FAM. 46. FAGÀCEAE (BEECH FAMILY)

Monoecious trees or shrubs with alternate simple straight-veined leaves, deciduous stipules, the staminate flowers in aments or capitate clusters, the pistillate solitary or slightly clustered, the 1-locular and 1-seeded nut inclosed (or partly inclosed) in a cupule consisting of more or less consolidated bracts, which become indurated. Ovary 3–7-locular; ovules 1 or 2 in each locule (usually only 1 ripening); styles 3. Seed with no albumen, filled by the embryo and with 2 integuments.

Involucre of pistillate flower and fruit 2–4-valved, prickly; nut compressed or triangular; leaves unlobed.

Staminate flowers in a rounded head, this pendulous on a slender peduncle; nuts sharply triangular; bark smooth. . . . . 1. *Fagus.*

Staminate flowers in a stiff moniliform ament; nuts lenticular or compressed, not triangular; bark very rough. . . . . 2. *Castanea.*

Involucre a saucer-like, cup-shaped or turbinate cupule, made up of prickleless imbricated scales; nut circular in cross-section; leaves lobed or unlobed. . . 3. *Quercus.*

1. FÀGUS L. BEECH. HÊTRE (Que.)

Staminate flowers with deciduous scale-like bracts; calyx campanulate, 5–7-cleft; stamens 8–16; filaments slender; anthers 2-locular. Pistillate flowers usually in pairs at the apex of a short peduncle, invested by numerous subulate bractlets, the inner bractlets coherent at base to form the 4-lobed involucre; calyx-lobes 6, subulate; ovary 3-locular with two ovules in each locule; styles filiform, stigmatic along the inner side. Nuts usually 2 in each urceolate and soft-prickly coriaceous involucre, which divides to below the middle into 4 valves. Cotyledons thick, folded and somewhat united, but rising above ground and expanding in germination. — Trees of the N. Hemisph., with a close and smooth ash-gray bark, a light horizontal spray, and undivided strongly straight-veined leaves. Flowers appearing with the leaves, the yellowish staminate ones in pendulous heads from the lower, the pistillate from the upper axils of the leaves of the season. (The classical Latin name, from the Greek *phagein, to eat,* in allusion to the esculent nuts.)

1. F. grandifòlia Ehrh. (large-leaved). — Large tree; leaves pale- or yellowish-green, oblong-ovate, mostly cuneate at base, distinctly and often coarsely serrate, the veins beneath silky, the leaf otherwise glabrous or glabrate, or the under side short-pubescent and with veins villous in forma **pubéscens** Fern. & Rehd. (pubescent); prickles of the mature grayish to yellowish fruiting involucre subulate-filiform, recurving or spreading, mostly 4–7 mm. long. — Rich uplands, rarely in swamps, P.E.I. to w. Ont., s. to Va., W.Va. and the Great Lakes states. Apr., May.

Var. **caroliniàna** (Loud.) Fern. & Rehd. (of Carolina). — Leaves darker, thicker, often rounded or subcordate at base, the teeth smaller, glabrous, or leaf soft-pubescent beneath in forma **móllis** Fern. & Rehd. (soft); mature fruiting involucre usually rufous-tomentose, its fewer prickles with free tips mostly less than 3 mm. long. — Fla. to Tex., n. to se. Mass., O., Ind., s. Ill. and Mo.

2. CASTÀNEA Mill. Chestnut. Châtaignier (Que.)

Staminate flowers interruptedly clustered in long and naked cylindrical aments; calyx mostly 6-parted; stamens 8–20; filaments slender; anthers 2-locular. Pistillate flowers usually 3 together in an ovoid scaly prickly involucre; calyx with a 6-lobed border crowning the 3–7-locular 6–14-ovuled ovary; abortive stamens 5–12; styles linear, exserted, as many as the locules of the ovary; stigmas small. Nuts coriaceous, inclosed, usually 2–3 together or solitary in the involucre. Cotyledons very thick, somewhat plaited, cohering, remaining under ground in germination. — Leaves strongly straight-veined, undivided. Flowers later than the leaves, cream-color; the aments axillary near the ends of the branches, wholly staminate or the upper androgynous, with the pistillate flowers at the base. Trees or shrubs of the N. Hemisph. (The ancient Latin name, from the Greek *castana, chestnut.*)

Leaves glabrous on both sides, long-acuminate; involucres of fruit with glabrous spines; the 2 or 3 (rarely –9) nuts flattened on at least 1 side. . . . . . 1. *C. dentata.*
Leaves pubescent beneath; involucres of fruit with pubescent spines; the solitary nut not flattened.
  Branchlets glabrous; leaves of fertile branches broadly lanceolate to oblong, acuminate, minutely pubescent to glabrate beneath, their coarse marginal teeth 3–8 mm. long; fruiting involucres 2.3–3 cm. in diameter, their spines 1–1.3 cm. long. . . . . 2. *C. ozarkensis.*
  Branchlets pubescent; leaves of fertile branches oblong to obovate, merely acute or obtuse, velvety beneath, their small marginal teeth 1–3 mm. long; fruiting involucres 2–2.5 cm. in diameter, their spines 3–7 mm. long. . . . 3. *C. pumila.*

1. C. dentàta (Marsh.) Borkh. (toothed), Chestnut. — A large rough-barked tree up to 30 m. high; *leaves oblong-lanceolate, pointed,* serrate with coarse pointed teeth, acute at base, when mature *smooth and green both sides; nuts* 2 or 3 (rarely 7–9) in each involucre, *flattened* on one or both sides, very sweet. — Dry gravelly or rocky, mostly acid soil, Ga. and nw. Fla. to Miss., n. to s. Me., N.H., Vt., N.Y., s. Ont., s. Mich. and s. Minn. June–early Aug. — In recent years extensively destroyed by the Chestnut-bark disease.

2. C. ozarkénsis Ashe (of the Ozarks). — Tree up to 20 m. high; *twigs* gray, *glabrous; leaves* of fertile shoots *broadly lanceolate to oblong, acuminate, minutely pubescent* to glabrate *beneath, coarsely toothed, teeth* 3–8 *mm. long; fruiting involucres* 2.3–3 *cm. in diameter, their spines pubescent* and 1–1.3 *cm. long; nut solitary, not flattened.* — Woods and rocky slopes, Miss. and La. to s. Mo. and Okla. June.

3. C. pùmila (L.) Mill. (dwarf), Chinquapin. — Spreading shrub or small tree; *leaves oblong, acute, serrate* with pointed teeth, *whitish-downy beneath,* those of fertile branches mostly 0.7–1.5 dm. long; *involucres small, often in spikes; mature bur about* 2.5 *cm. in diameter, closely covered by* scales bearing *short erect crowded bristles, with* small *ovoid* pointed very sweet *plump (not flattened) nut.* — Dry woods and thickets, Fla. to Tex., n. to N.J. (locally in Norfolk Co., Mass.), e. Pa., Tenn. and Ark.

Var. **Áshei** Sudw. (for William Willard Ashe, 1872–1932). — *Scales of the bur remote, with horizontally divergent bristles, the surface of the bur with broad open areas.* — Ga. to e. Tex., n. to se. Va. and Ark.

C. neglécta Dode (overlooked), with thin leaves broader and more glabrate beneath than in no. 3, is sometimes considered a hybrid of nos. 1 and 3 but may be merely a shade-state of no. 3.

3. QUÉRCUS L. Oak. Chêne (Que.)

Staminate flowers in naked aments; bracts caducous; calyx 2–8-parted or -lobed; stamens 3–12; anthers 2-locular. Pistillate flowers scattered or somewhat clustered, consisting of a nearly 3-locular and 6-ovuled ovary, with a 3-lobed stigma, inclosed by a scaly bud-like involucre which becomes an indurated cup (*cupule*) around the base of the rounded nut or acorn. Cotyledons remaining underground in germination; radicle very short, included. — Flowers greenish, yellowish or reddish. Staminate aments single or often several from the same lateral scaly bud, filiform or moniliform and hanging in all our species. Hard-wood trees or shrubs of temp. and warm N. Hemisph. All the species inclined to hybridize freely; these hybrids not here defined. (Classical Latin name.)

a. Bark pale, often scaly; leaves or their lobes or teeth with included (not excurrent) veins; stamens 6–8; scales of cup more or less corky, woody or knobby at base; stigmas sessile or nearly so; aborted ovules at the base of the perfect seed; inner surface of shell of nut glabrous; fruit maturing the first year, the kernel often sweetish. Subgen. Lepidobalanus, White Oaks. . . *b.*

b. Leaves deciduous, sinuate-toothed or -lobed. . . *c.*

c. Leaves lyrate or sinuate-pinnatifid. . . *d.*

d. Mature leaves glabrous beneath. . . . . 1. *Q. alba.*

d. Mature leaves finely pubescent beneath. . . *e.*

e. Scales of cup naked-tipped, not awned.

Fruit nearly sessile; the fine-scaled saucer-shaped cup one-third to half as high as the ovoid acorn. . . . . 2. *Q. stellata.*

Fruit short-peduncled; the coarse-scaled cup nearly covering the depressed-globose acorn. . . . . 3. *Q. lyrata.*

e. Upper scales of cup long-awned. . . . . 4. *Q. macrocarpa.*

c. Leaves coarsely sinuate-toothed but not lobed (except slightly in no. 5). Chestnut-Oaks. . . *f.*

f. Fruiting peduncle 2.5–6 cm. long, much exceeding the petioles. . . . 5. *Q. bicolor.*

f. Fruit sessile or very short-peduncled. . . *g.*

g. Cup 2.5–3 cm. broad, its scales free to the base. . . . . 6. *Q. Michauxii.*

g. Cup at most 2.5 cm. broad, only the small tips of the scales distinct. . . *h.*

h. Leaves with acute or pointed teeth.

Leaves with 8–13 teeth on each margin; tree. . . . . . 7. *Q. Muehlenbergii.*

Leaves with 3–7 teeth on each margin; low shrub. . . . . 8. *Q. prinoides.*

h. Leaves with somewhat rounded teeth. . . . . 9. *Q. Prinus.*

b. Leaves coriaceous, evergreen, entire or rarely spiny-toothed. . . . . 10. *Q. virginiana.*

a. Bark often dark, furrowed but rarely flaky; leaves mostly deciduous, if lobed with acute lobes or teeth, the midrib in unlobed blades, the principal veins in lobed blades usually excurrent as bristles (at least when young); stamens mostly 4–6; scales of cup membranaceous; styles long and spreading; abortive ovules near top of perfect seed; inner surface of shell tomentose; fruit maturing the second year. Subgen. Erythrobalanus, Red and Black Oaks. . . *i.*

i. Leaves pinnatifid or pinnately lobed, slender-petioled, not coriaceous, the lobes or teeth conspicuously bristle-tipped. . . *j.*

j. Mature leaves green or greenish on both sides, glabrous beneath except for tufts of hair in axils of veins in some species. . . *k.*

k. Longest lobes of leaf shorter than to about equaling (never twice as long as) the breadth of the broadish median portion of the blade.

Winter-buds 5–7 mm. long, not strongly angled, chestnut-brown, soon glabrate; cups saucer-shaped to cupuliform, with tightly appressed brown merely puberulent scales. . . . . 11. *Q. rubra.*

Winter-buds 6–12 mm. long, strongly angled or grooved, gray-tomentose; cups turbinate or cupuliform, with thin loosely imbricated pale pubescent scales, the free tips of the upper scales forming a loose fringe. . . . . 17. *Q. velutina.*

k. Longest lobes of leaf two to six times as long as breadth of narrower part of median portion of blade. . . *l*.

l. Upper scales of cup tightly appressed, not forming a definite fringe. . . *m*.

m. Expanded saucer-shaped portion of cups 3–5 mm. high, 1–1.5 cm. broad. 12. *Q. palustris.*

m. Cups higher or broader. . . *n*.

n. Cups brown or castaneous, the scales finally glabrate and lustrous. 13. *Q. coccinea.*

n. Cups ashy with persistent dull pubescence.

Mature leaves of fertile branches 1.5–2 dm. long, on petioles 4–6 cm. long; cups saucer-shaped to turbinate, 2–3 cm. broad. . . . . 14. *Q. Shumardii.*

Mature leaves 7–16 cm. long, on petioles 2–5 cm. long; cups turbinate or goblet-like, 1–1.8 cm. broad.

Cup gradually sloping to base; acorn 1.2–2 cm. long; northwestern. . . . . 15. *Q. ellipsoidalis.*

Cup abruptly contracted to a thick stipe-like base 2–7 mm. long; acorn 2–2.8 cm. long; southwestern. . . . . . 16. *Q. Nuttallii.*

l. Upper scales of pubescent cup loosely imbricated or forming a loose fringe, thin and scarious-tipped.

Bark dark brown to blackish; leaves dark green above, with broadly rounded, subtruncate or rarely cuneate bases; petioles mostly 3–6 cm. long; scales of cup lanceolate or lance-ovate, tapering to tip, the free tips of the upper loosely ascending or spreading when dry; wide-ranging large tree. . . . . 17. *Q. velutina.*

Bark bluish-gray; leaves gray-green, cuneate to slightly rounded at base; petioles 0.5–1.5 cm. long; scales of cup oblong, with broadly rounded tips, the free tips of the upper ones incurved; small tree of s. Coastal Plain. . . . . 18. *Q. laevis.*

j. Mature leaves whitish, grayish or fulvous and pubescent beneath.

Leaves of fertile branches 1–2.5 dm. long, fulvous-pubescent beneath; cup turbinate, about half covering acorn, 2–2.5 cm. broad, its attenuate large scales loosely ascending; acorn 1.5–2.5 cm. high. . . 17. *Q. velutina.*

Leaves whitish- or grayish-tomentose beneath; cup saucer-shaped, shallow, 1–1.8 cm. broad, with closely appressed scales; acorn 1–1.5 cm. high.

Leaves of fertile branches 1–3 dm. long, with long tapering often falcate lobes; petioles 2–6 cm. long; scales of cup obtuse or truncate; large southern tree. . . . . 19. *Q. falcata.*

Leaves of fertile branches 5–12 cm. long, with short triangular or ovate lobes; petioles 0.7–3.5 cm. long; scales of cup tapering to tip; low bushy northern or upland tree. . . . . 20. *Q. ilicifolia.*

i. Leaves entire or with few teeth (or somewhat 3–5-lobed at summit), commonly bristle-pointed; acorns globular or nearly so, small (rarely 2 cm. long). . . *o*.

o. Leaves widening or dilated upward, often undulate, sinuate or lobulate above. . . *p*.

p. Leaves broadly obovate or obovate-oblong, tapering to rounded or cordate broad base, mostly 1–2.5 dm. long, the dilated often shallowly lobulate summit nearly as broad, scurfy-pubescent beneath; cup turbinate, or turbinate-hemispheric, 1.5–2 cm. broad. . . . 21. *Q. marilandica.*

p. Leaves oblanceolate to narrowly obovate or subrhombic, cuneate at base, definitely longer than broad, mostly 0.5–1.5 dm. long, the lower surface glabrous or soon glabrate; cup saucer-shaped, 1–1.5 cm. broad.

Leaves cuneate-obovate, -spatulate or -oblanceolate, broadest near summit; scales of cup acutish, tightly appressed. . . . . 22. *Q. nigra.*

Leaves rhombic-lanceolate, -oblong or -obovate, broadest near middle; scales of cup obtuse, rather loose, the upper ones forming a scarious fringe. . . . . 23. *Q. laurifolia.*

o. Leaves not dilated upward, generally entire (sometimes lobulate at summit). . . *q*.

q. Leaves densely tomentose beneath. . . *r*.

r. Branchlets glabrous or promptly glabrate; leaves membranaceous, dark green above.

Leaves permanently stellate-pubescent beneath, 1.2–6 cm. wide; cups 1.5–2 cm. broad, cupuliform or turbinate. . . . . . . 24. *Q. imbricaria.*

Leaves silky-tomentose beneath, 0.7–3 cm. wide; cups 9–12 mm. broad, shallowly saucer-shaped. . . . . 26. *Q. Phellos.*

*r.*Branchlets heavily tomentose, becoming puberulent; leaves coriaceous, pale green above; cups saucer-shaped, 1–1.5 cm. broad. . 25. *Q. incana.*

*q.*Leaves glabrous or promptly glabrate (rarely silky) beneath. . . *s.*

*s.*Leaves thinnish or membranaceous, deciduous in autumn.

Blades on fertile branches slightly rhombic to lanceolate or oblanceolate, 2–5 cm. wide, the midrib rarely excurrent; cup deeply saucer-shaped to turbinate, 1.5–2 cm. broad, its scales with free tips. . . . . 23. *Q. laurifolia.*

Blades linear to narrowly oblong, with strongly excurrent midrib, 0.7–3 cm. wide; cup shallowly saucer-shaped, 9–12 mm. broad, its scales tightly appressed. . . . . 26. *Q. Phellos.*

*s.*Leaves thick and coriaceous, evergreen or overwintering; cup shallowly saucer-shaped, 1–1.5 cm. broad, its scales with free tips. . 27. *Q. hemisphaerica.*

Subgen. LEPIDOBÁLANUS Endl. WHITE OAKS

1. Q. álba L. (white), WHITE O. — *Large tree with flaky light-colored bark; branchlets glabrous or glabrate,* becoming reddish-brown; young leaves white-lanate beneath; *mature leaves glabrous and whitened,* or green in forma **víridis** Trel. (green), *beneath,* obovate to obovate-oblong, mostly 1–2.5 dm. long, *cuneate to petioles* 0.8–2 *cm. long,* the *blades with obliquely descending sinuses* one-eleventh to one-half as deep as the breadth of the central portion of the blade, the narrower region of the latter 0.5–2.5 cm. broad, the longer of the 4–10 ascending narrowly oblong entire to slightly lobulate or apically cut lobes 2.5–8 cm. long and 0.8–2(–3) cm. broad, or in forma **latíloba** (Sarg.) Palmer & Steyerm. (broad-lobed) the blades usually cleft less than half-way to midrib and with the round-tipped broadly oblong lobes 1.5–3(–4) cm. broad, or in forma **repánda** (Michx.) Trel. (with slightly uneven margin) the sinuses shallow with the round-tipped lobes as broad as long; peduncle rather short; cup shallowly bowl-shaped, drab, rough with tuberculate puberulent scales, about one-third as long as the ovoid to ellipsoid acorn (2–3 cm. long); branches wide-spread, forming a broad crown; leaves reddish to violaceous in autumn; sometimes producing basal young plants by budding from the roots. — Dry woods, Fla. to Tex., n. to centr. Me., s. Que., s. Ont. and Minn. FIG. 909. — Hybridizes with no. 2, producing × **Q. Fernowi** Trel. (for BERNHARD EDUARD FERNOW, 1851–1928); with no. 3; with no. 4, producing × **Q. Bebbiàna** Schneid. (for MICHAEL SCHUCK BEBB, 1833–1895); with no. 5, producing × **Q. Jackiàna** Schneid. (for JOHN GEORGE JACK, 1861–1949); with no. 6, producing × **Q. Beàdlei** Trel. (for CHAUNCEY DELOS BEADLE, 1866–   ); with no. 7, producing × **Q. Deàmi** Trel. (for CHARLES CLEMON DEAM, 1865–   ); with no. 8, producing × **Q. Fáxoni** Trel. (for CHARLES EDWARD FAXON, 1864–1918); with no. 9, producing × **Q. Saùlii** Schneid. (for JOHN SAUL, 1819–1897); and with *Q. Robur,* producing × **Q. bimundòrum** E. J. Palmer (of two worlds).

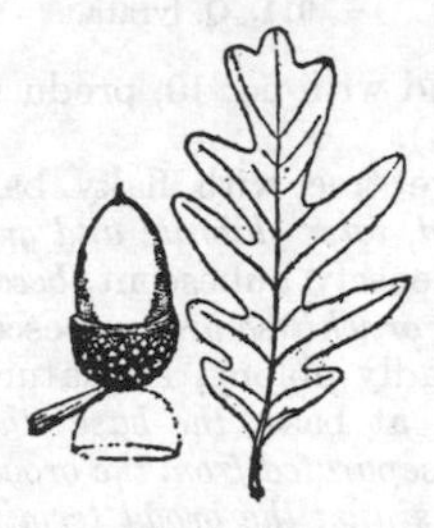
909. Q. alba.

Q. RÒBUR L. (ancient Latin name), ENGLISH O., differing from no. 1 in darker and furrowed bark, smaller *cordate- to truncate-based short-petioled to subsessile leaves, elongate* (3–8 *cm. long*) *peduncles,* and deeper cup, is much planted eastw. and spreads locally to roadsides and borders of woods. (Introd. from Eu.)

2. Q. stellàta Wang. (with star-shaped hairs), POST-O. — *Small stiffly branched tree* rarely 20 m. high, or coarse shrub, with gray-brown fissured and flaky bark; *the branchlets tomentose; leaves grayish- or brownish-downy underneath, with stellate hairs intermixed, very thick, hard and harsh above, short-petioled,* the obovate *blades* very variable, mostly 1–2 dm. long, *usually with horizontally divergent broad* and upwardly dilated *blunt truncate or emarginate large lobes* (*suggesting a Swiss cross*), *the middle pair of lobes much longer than the short and entire basal ones and separated from them by a broad open sinus* (on sprouts often unlobed and suggesting the outline of leaves of no. 21), *rounded to subcuneate at base; fruit sessile or nearly so, the* deeply saucer-shaped fine-scaled *cup one-third to half as high as the ovoid acorn.* — Dry sterile soil, Fla. to Tex., n. to se. Mass., s. R.I., s. Ct., se. N.Y., e. Pa., W.Va., centr. O., s. Ind., Ill., Ia. and Kans. FIG. 910. — Foliage most variable. — Hy-

910. Q. stellata.

bridizes with nos. 1, 3, 4, 5, 8 and 9. Of the many named variations the best-marked with us is

Var. **Margarétta** (Ashe) Sarg. (chivalrously named in 1903 for MARGARET HENRY WILCOX, who two years later became Mrs. Ashe). — Usually shrubby, rarely arborescent; leaves, smoother, even glabrescent, 0.4–1 dm. long, with short subuniform spreading or ascending entire lobes. (*Q. Margaretta* Ashe; *Q. stellata*, var. *Boyntoni* Sarg.) — Similar habitats, Fla. to Tex., n. to e. Va. and locally to se. Mass., sw. Mo. and Okla.

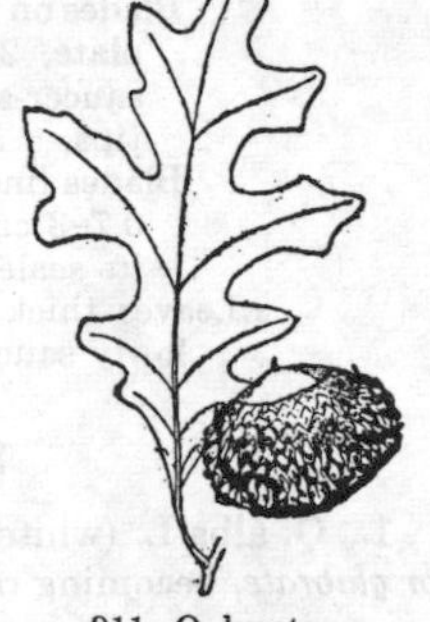

911. Q. lyrata.

3. Q. lyràta Walt. (lyre-shaped), OVER-CUP O., SWAMP-POST-O. — Gray-barked tree up to 30 m. high; *young branchlets pubescent, becoming glabrate; leaves* crowded at tips of branchlets, obovate-oblong, acute at base, more or less deeply 6–8-*lobed, the young white-tomentose beneath,* the mature glabrescent or whitened beneath, or in forma **víridis** Trel. (green) green; *the basal lobes much shorter, more triangular and separated from the upper* oblong to triangular *ones by a broad sinus; upper surface dark green and smooth;* peduncles short; *cup depressed to subglobose, pubescent, nearly covering the depressed acorn, the lower scales blunt and rugged, the upper thin and acutish and forming an incurving fringe.* — Bottomlands and wet woods, n. Fla. to e. Tex., n. along Coastal Plain to s. N.J. and inland n. to sw. Ind., s. Ill., se. Ia. and e.-centr. Mo. FIG. 911. — Hybridizes with no. 1; and with no. 10, producing × **Q. Cómptonae** Sarg. (for its original discoverer, Miss C. C. COMPTON).

912. Q. macrocarpa.

4. Q. macrocárpa Michx. (large-fruited), MOSSY-CUP O. — Large tree with flaky bark; *branchlets at first pubescent and orange-tinged, later glabrate and gray,* sometimes with cork-wings; *leaves* at first densely pubescent, *becoming glabrate and lustrous-green above and gray or whitish* and pubescent to glabrate *beneath,* obovate-oblong to broadly oblong, in maturity 1–3 dm. long, cuneate to slightly rounded at base; *the basal third with* 1–3 *pairs of short triangular lobes, these separated from the broader upper half or two-thirds by a well developed sinus; the broad terminal portion bluntly oblong- or ovate-lobulate,* or in forma **olivaefórmis** (Michx. f.) Trel. (olive-shaped; from fruit of the original collection) the leaf deeply cleft nearly to midrib into often narrower and more elongate lobes; *peduncle short or obsolete; cup deep,* 2–5 cm. across, *thick and woody, its upper scales tapering to fringe-like tips, one-third to half (or more) covering the acorn.* — Bottomlands, rich woods and fertile slopes, w. N.B. and centr. Me. (local) and s. Que. to Man., s. to w. N.E., Del. (formerly), Pa., W.Va., upland to N.C., Tenn., Ark., Okla. and Tex. — Very variable in foliage and shape and size of fruit. FIG. 912. — Hybridizes with nos. 1 and 5; and with no. 7, producing × **Q. fállax** E. J. Palmer (deceitful).

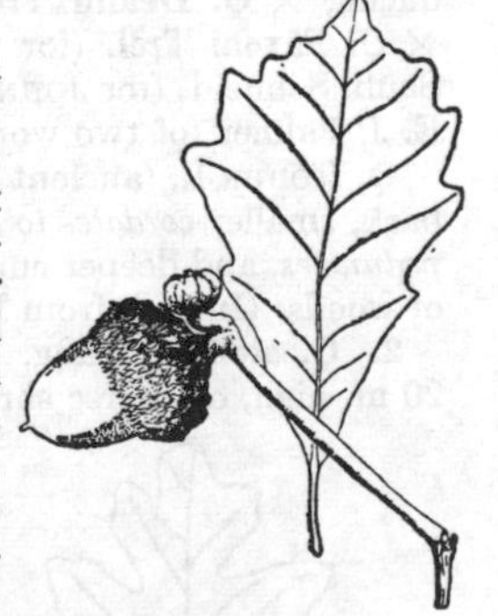

913. Q. bicolor.

5. Q. bícolor Willd. (two-colored), SWAMP-WHITE O. — Tree with flaky gray bark; *leaves obovate or oblong-obovate,* cuneate at base, *coarsely sinuate-crenate* and *often* rather *pinnatifid* than toothed, *usually soft-downy and white-hoary beneath,* the primary veins lax and little prominent; *fruiting peduncle 2.5–6 cm. long, much exceeding petioles; cup* one-third to half as long as acorn, *woody, the upper scales awn-pointed* and sometimes forming a moss-like fringed margin; acorn 2–3 cm. long. — Bottomlands, stream-margins and swamps, s. Me. and s. Que. to s. Minn. and Neb., s. to s. N.E., L.I., Del., Md., n. W.Va., upland to Ga. and Ky., Ark. and Okla. FIG. 913. — Hybridizes with no. 1; with no. 2, producing × **Q. substellàta** Trel. (somewhat stellate); with no. 3, producing × **Q. humidícola** E. J. Palmer (growing in moisture), and with no. 4, producing × **Q. Híllii** Trel. (for ELLSWORTH JEROME HILL, 1833–1917).

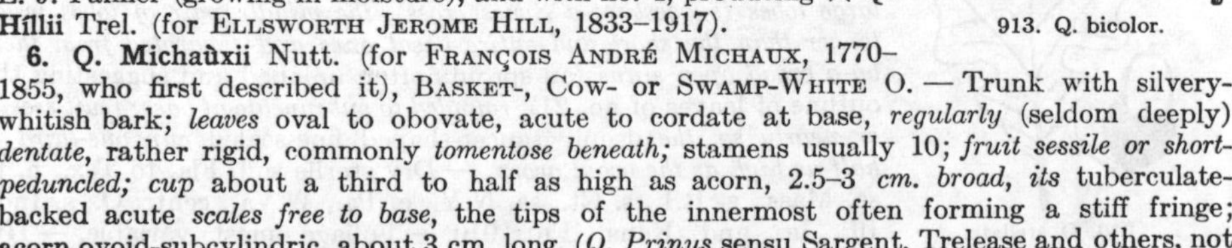

6. Q. Michaùxii Nutt. (for FRANÇOIS ANDRÉ MICHAUX, 1770–1855, who first described it), BASKET-, COW- or SWAMP-WHITE O. — Trunk with silvery-whitish bark; *leaves* oval to obovate, acute to cordate at base, *regularly* (seldom deeply) *dentate,* rather rigid, commonly *tomentose beneath;* stamens usually 10; *fruit sessile or short-peduncled; cup* about a third to half as high as acorn, 2.5–3 *cm. broad, its* tuberculate-backed acute *scales free to base,* the tips of the innermost often forming a stiff fringe; acorn ovoid-subcylindric, about 3 cm. long. (*Q. Prinus* sensu Sargent, Trelease and others, not

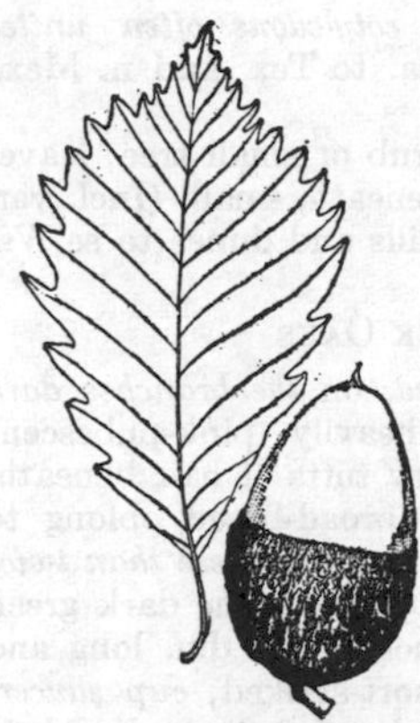
914. Q. Michauxii.

L.) — Inundated bottoms, stream-borders and swamps, Fla. to Tex., n. to N.J., s. Ind., s. Ill. and se. Mo. FIG. 914. — Hybridizes with no. 1.

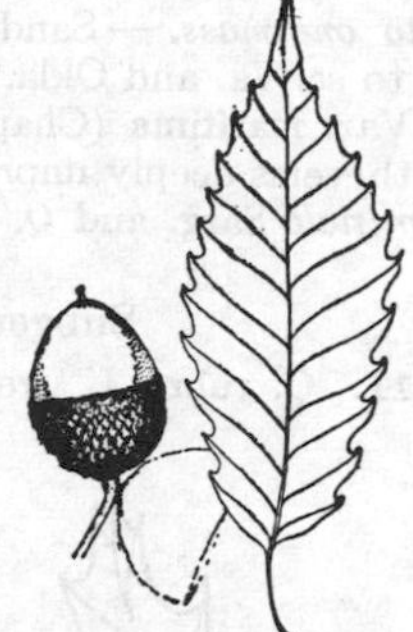
915. Q. Muehlenbergii.

7. Q. Muehlenbérgii Engelm. (for GOTTHILF HENRY ERNEST MUHLENBERG, 1753–1815), YELLOW O., CHESTNUT-O. — Bark thin, eventually flaky; *leaves* 1–2 dm. long, *slender-petioled,* oblong or even lanceolate, or in forma **Alexánderi** (Britt.) Trel. (for its discoverer, SAMUEL ALEXANDER, 1841–1917) obovate to obovate-oblong, usually acute or acuminate, most often obtuse to rounded at base, *almost equally and rather sharply toothed,* with 8–13 *teeth on each margin;* peduncle short to obsolete; *cup* a third to half as long as acorn, *thin, of small appressed scales;* acorn globose to ovoid, 1.5–2 cm. long. — Dry calcareous slopes and ridges or on rich bottoms, n. Fla. to e. Tex., n. to nw. Vt., n. N.Y., s. Ont., s. Mich., s. Wisc., s. Minn. and e. Neb. FIG. 915. — Hybridizes with nos. 1 and 4.

8. Q. prinoìdes Willd. (resembling *Q. Prinus*), CHINQUAPIN-OAK, DWARF CHESTNUT-O. — Slender *shrub* usually 1–3 m. high, rarely arborescent; branchlets glabrate or glabrous; *leaves oblong or oblong-ovate or -obovate,* 5–12 *cm. long, undulate, with* 3–7 *bluntish teeth on each margin,* whitish-tomentose beneath, bright green above, the *petioles* 0.4–1 *cm. long; cup* as in no. 7 but smaller (1–2 *cm. broad*), *with fine tuberculate scales.* — Dry plains, rocks, thickets, and borders of woods, sw. Me. to Minn. and Neb., s. to s. N.E., L.I., Md., se. Va. (rare), upland to Ala., Tenn., Ark., Okla. and Tex. FIG. 916. — Hybridizes with no. 1; also with no. 2, producing × **Q. stelloides** E. J. Palmer (resembling a star).

Var. **ruféscens** Rehd. (reddish). — Young branchlets reddish-pubescent; lower surfaces of leaves with reddish hairs mixed with the whitish ones or reddening the veins. — Se. Mass., locally to N.C.

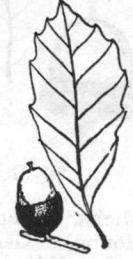

916. Q. prinoides.

9. Q. Prìnus L. (Greek name of the European Oak), CHESTNUT-O., ROCK-CHESTNUT-O. — Tree with furrowed bark; branchlets glabrous or pilose and soon glabrate; *leaves* heavy and coriaceous, obovate to oblong or lanceolate, often abruptly acuminate, with acute to obtuse base, yellow-green and lustrous above, *undulately crenate with broad rounded teeth, pale and minutely downy or glabrescent beneath;* primary ribs 10–16 pairs, straight, prominent beneath; fruiting peduncle shorter than petioles or almost obsolete; *cup* mostly tuberculate *with* hard and stout *scales united at base;* acorn ovoid or ellipsoid, broadly rounded above, 2–3 cm. long, up to 2 cm. thick. (*Q. montana* Willd.) — Dry or rocky woods, bluffs and crests (mostly siliceous), sw. Me. to s. Ont. and centr. Ind., s. to e. Va., upland (up to 1400 m. alt.) to Ga., n. Ala. and n. Miss. FIG. 917. — Hybridizes with no. 1; with the introd. *Q. Robur,* producing × **Q. Sargéntii** Rehd. (for CHARLES SPRAGUE SARGENT, 1841–1927) and with no. 2, producing × **Q. bernardiénsis** W. Wolf (of St. Bernard College, Ala.).

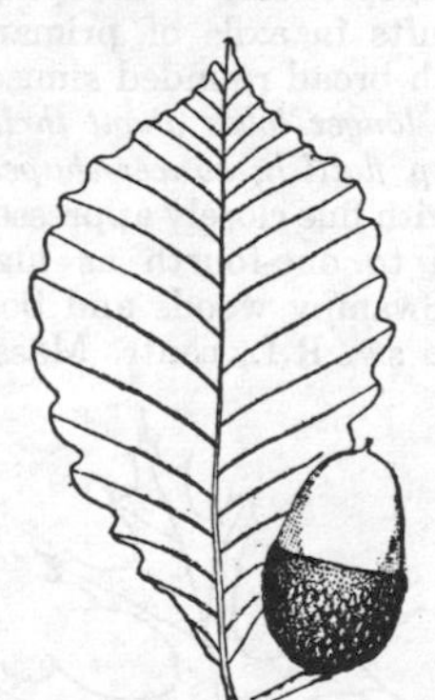
917. Q. Prinus.

10. Q. virginiàna Mill. (Virginian), LIVE O. — Tree with broad crown, the thick dark bark furrowed and finally flaky; *branchlets stiffly divergent, at first gray-pubescent,* finally glabrescent; *leaves coriaceous, evergreen, narrowly oblong or elliptic to narrowly obovate, entire, or* the later ones *sharply toothed* above, often revolute-margined, lustrous-green above, grayish to whitened and pubescent beneath, or in forma **viréscens** (Sarg.) Trel. (greenish) green and glabrescent beneath, the veins inconspicuous, the blades at most 6 cm. long and 2 cm. broad, or in forma **macrophýlla** (Sarg.) Trel. (large-leaved) the blades 7–10 cm. long and 3–6 cm. broad; peduncle usually well-developed, puberulent; cup canescent, turbinate, about 1.5 cm. broad, with thinnish acute

918. Q. virginiana.

dorsally keeled small scales, about one fourth height of acorn; *cotyledons often united into one mass.* — Sandy, dry to wet soil, chiefly near the coast, Fla. to Tex. and n. Mex., n. to se. Va. and Okla. (Cuba, Mex. and Centr. Am.) Fig. 918.

Var. **marítima** (Chapm.) Sarg. (maritime), Dwarf Live O. — Shrub or small tree; leaves with veins deeply impressed above, prominent and often reticulate beneath, small. (Incl. var. *geminata* Sarg. and *Q. geminata* Small) — Fla. to Miss., n. on sandhills and dunes to se. Va.

Subgen. Erythrobálanus Spach Red and Black Oaks

11. Q. rùbra L. (red), Red O. — *Bark* hard and *deeply furrowed,* on *the branches dark gray or brown; leaves* slender-petioled, heavily pink-pubescent when young, soon *glabrate* except for axillary tufts of hair beneath, *coarsely pinnatifid, with the longer* oblique broad-based oblong to narrowly ovate sharply cleft *lobes about as long as to less than twice as long as the breadth of the broad median portion of the* dark green (beneath yellow-green) *blades,* the latter mostly 1–2 dm. long and 0.5–1.3 dm. broad; fruit sessile or very short-stalked; *cup saucer-shaped or flattish,* with a narrow raised border, 1.8–3 *cm. broad,* of rather fine closely appressed scales, *very much shorter than the* ovoid or ellipsoid *acorn* (2–3 cm. long). (*Q. maxima* (Marsh.) Ashe; *Q. borealis,* var. *maxima* Ashe) — Dry or upland woods, Ga. to Okla., n. to P.E.I., N.B., s. Que., s. Ont., s. Mich., Wisc., s. Minn. and Neb. Fig. 919. — Hybridizes with nos. 12, 13, 24 and 26; also with no. 17, producing × **Q. Hawkínsiae** Sudw. (for its discoverer, Mrs. Eugene Hawkins, the name originally published in the masculine as "*hawkinsi*").

919. Q. rubra.

Var. **boreàlis** (Michx. f.) Farw. (northern), Northern Red O., Gray O. — *Bark of upper trunk and branches paler gray and smoother* (suggesting *Fagus*); *cup deeper and turbinate to hemispherical, inclosing about a third of the acorn,* 1.5–2.5 cm. broad. (*Q. ambigua* Michx. f., not Humb. & Bonpl.; *Q. borealis* Michx. f.; *Q. rubra,* var. *ambigua* (Michx. f.) Houba) — Generally more northern, Que. to Algoma Distr., Ont., s. to N.S., n. N.E., n. and w. N.Y., nw. Pa., mts. to N.C. (at about 1250 m. alt.), n. Mich., Wisc. and Ia. — Hybridizes with no. 20.

12. Q. palústris Muenchh. (of marshes), Pin-O., Spanish O. — Bark gray-brown, smooth except in age; *branchlets very slender,* soon glabrate, spreading or *drooping; leaves thin,* lustrous, glabrate, except for large tufts in axils of primary veins, *mostly* 0.5–1.5 *dm. long, deeply pinnatifid,* with broad rounded sinuses and oblong to narrowly ovate toothed lobes, the *longer lobes about thrice as long as breadth of median portion of blade; cup flattish, saucer-shaped,* 1–1.5 *cm. broad, its expanded portion* 3–5 *mm. high,* with fine closely appressed ovate sparsely puberulent scales, only one-third to one-fourth as high as the *nearly spherical acorn* (1–1.5 *cm. long*). — Swampy woods and bottoms at low alts., N.C. to n. La. and Okla., n. to sw. R.I., centr. Mass., se. N.Y., e. Pa., s. Ont., s. Mich., Ill. and Ia. Fig. 920. — Hybridizes with var. of no. 14, producing × **Q. mutábilis** Palmer & Steyerm. (variable); with no. 17, producing × **Q. vàga** Palmer & Steyerm. (unfixed); and with no. 26, producing × **Q. Schochiàna** Dieck (described in 1892).

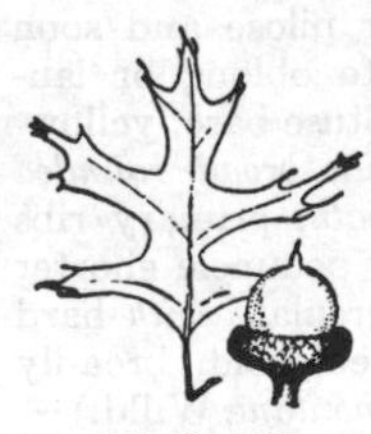
920. Q. palustris.

13. Q. coccínea Muenchh. (scarlet), Scarlet O. — *Bark* gray or light brown, *the inner reddish;* branchlets soon glabrous; *leaves,* at least in full-grown trees, bright green, shining above, *glabrous beneath,* turning bright red in autumn, elliptic- or oblong-ovate, 0.7–1.5 dm. long, truncate or broadly subcuneate to slender petioles 3–6 cm. long, deeply sinuate-pinnatifid with broad rounded sinuses and long narrowly oblong pinnatifid-toothed lobes; buds small, sparingly pilose above; *cup turbinate, or hemispherical with a conical base,* 1.5–2.2 *cm. broad, brown or castaneous, with tightly appressed finally glabrate and lustrous scales,* a third to half covering the ovoid to subglobose acorn (1.3–2 cm. long). — Dry light soil, s.

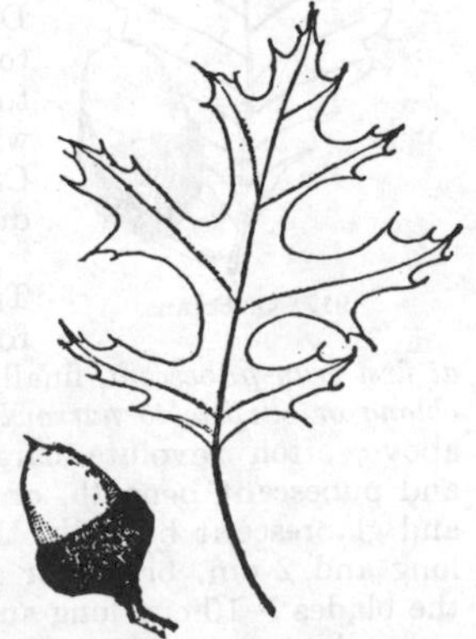
921. Q. coccinea.

Me. to Ont., s. to s. N.E., L.I., Va., upland (up to 1600 m.) to Ga. and Ala., n. Miss., Ark., and Okla. Fig. 921. — Hybridizes with no. 11, producing × **Q. Bénderi** Baenitz (named in 1903 for Georg Bender); and with no. 20, producing × **Q. Robbínsii** Trel. (for James Watson Robbins, 1801–1879).

14. Q. Shumárdii Buckl. (for Benjamin Franklin Shumard, 1820–1869, State Geologist of Texas in 1860), Shumard's Red O. — Resembling no. 13; the base often buttressed; bark deeply and broadly furrowed or broken into plates; *winter-buds glabrous; mature leaves of fertile branches* 1.5–2 *dm. long, on petioles* 4–6 *cm. long, glabrate* and lustrous *except for large brown tufts in axils beneath; cup flattish, saucer-shaped,* with incurved margin, 2–3 *cm. broad, grayish- or ashy-puberulent, about one-fourth as high as acorn, its scales closely appressed; acorn* oblong-ovoid, 1.8–3.7 *cm. high,* 1.3–2.5 cm. thick. (*Q. texana* of ed. 7, not Buckl.) — Rich woods, bottoms or calcareous slopes, Fla. to e. Tex., n. to Md., Wayne Co., W.Va., s. O., s. Ind., s. Ill., s. Ia. and se. Kans. Fig. 922.

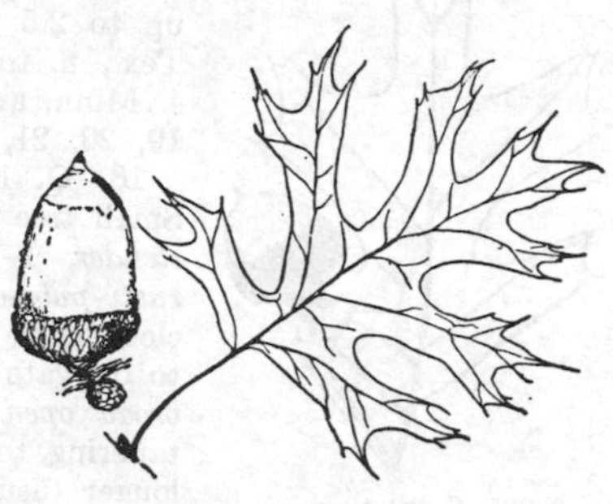

922. Q. Shumardii.

Var. **Schnéckii** (Britt.) Sarg. (for Jacob Schneck, 1843–1906), Schneck's Red O. — Differing in its *cups deeper, strongly rounded below, about a third as high as the* often smaller *acorns.* — Similar range. — Hybridizes with no. 12.

15. Q. ellipsoidàlis E. J. Hill (ellipsoid), Jack-O. — Small to medium-sized tree resembling nos. 12 and 13, with smooth or only shallowly furrowed gray bark, the *inner bark yellow; leaves* much as in no. 12, 7.5–12.5 *cm. long, on petioles* 3–5 *cm. long; cups ashy-gray, pubescent, turbinate to goblet-shaped, gradually sloping to base,* 1–1.8 *cm. broad,* with obtuse appressed scales, *one-half to one-third as high as the ellipsoid-cylindric acorn* (1.5–2 *cm. high*), *or in* forma **depréssa** (Vasey) Trel. (depressed) *the acorn thicker and oblong-ovoid to subglobose and* 1.2–1.6 *cm. high.* — Dry or moist siliceous to argillaceous woods, s. Mich. to s. Man., s. to n. O., n. Ind., n. Ill. and n. Ia. Fig. 923. — Hybridizes with no. 17, producing × **Q. palaeolithícola** Trel. (lover of Palaeozoic rock).

923. Q. ellipsoidalis.

16. Q. Nuttállii E. J. Palmer (for Thomas Nuttall, 1786–1859), Nuttall's O. — Slender, often pyramidal tree suggesting no. 12, with gray or slate-colored finally darker smooth or slightly fissured bark; *leaves* small, as in no. 15, but with the *longer blades more tapering at tip,* 8–16 cm. long, on slender petioles 2–5 cm. long; *cups deeply turbinate* to cupuliform, gray-puberulent, 1.6–1.8 cm. broad, *abruptly contracted below to a thick stipe-like base* 2–7 *mm. long; acorns* thick-cylindric to oblong-ovoid, broadly rounded at base, 2–2.8 *cm. high,* 1.2–2 cm. thick; winter-buds pubescent or ciliate at tip. — Bottoms and low grounds, Ala. to Tex., n. to se. Mo. and Ark. Fig. 924.

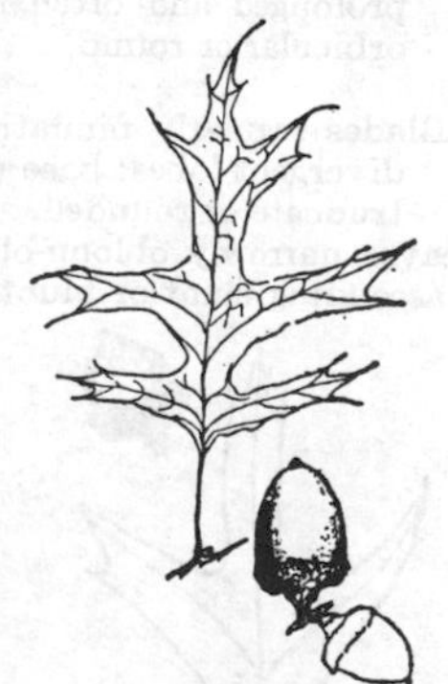

924. Q. Nuttallii.

17. Q. velùtina Lam. (velvety; from the young foliage), Black or Yellow-barked O., Quercitron. — Tree with dark bark, the *inner bark yellow or orange;* young branchlets tomentose, soon glabrescent, or in forma **missouriénsis** (Sarg.) Trel. (of Missouri) permanently pubescent as are the lower surfaces of the leaves; *winter-buds* 6–12 *mm. long, strongly angled or grooved, gray-tomentose;* leaves of several different types on different adult trees, dark green and glabrous above, more or less fulvous and tardily glabrescent beneath, 1–2.5 dm. long, of an ovate-oblong to obovate outline, in the typical and relatively rare form (as described by Lamarck) obovate and only very shallowly lobulate (chiefly a juvenile state), or in forma **macrophýlla** (Dippel) Trel. (large-leaved) cut less than half-way to midrib and with the broadly oblong toothed lobes many times broader than the slender sinuses, or in the usually common forma **dilaniàta** Trel. (torn to pieces) resembling the leaf of no. 18 in having round-based sinuses extending half to three-fourths to midrib and the oblong terminally toothed lobes often narrower than the broader sinuses, or in the local forma **pagodaefórmis** Trel. (shaped like a pagoda) with the usually large blade with deep and very broad round-based sinuses and the long mostly acuminate lobes subentire or only slightly toothed; *cup turbinate or cupuliform, with thin loosely imbricated pale pubescent lanceolate scales, the free tips of the upper scales forming a loose fringe,* the *cup about half as long as the large* (1.2–1.9 cm. broad, up to 2.5 cm. high) ovoid *acorn.* — Dry woods, n. Fla. to e. Tex., n. to s. Me., centr. N.H., Vt., N.Y., s. Ont., Mich., Ill., s. Minn. and se. Neb. Fig. 925. — Hybridizes with nos. 11, 15, 19, 20, 21, 23, 24, 25 and 26.

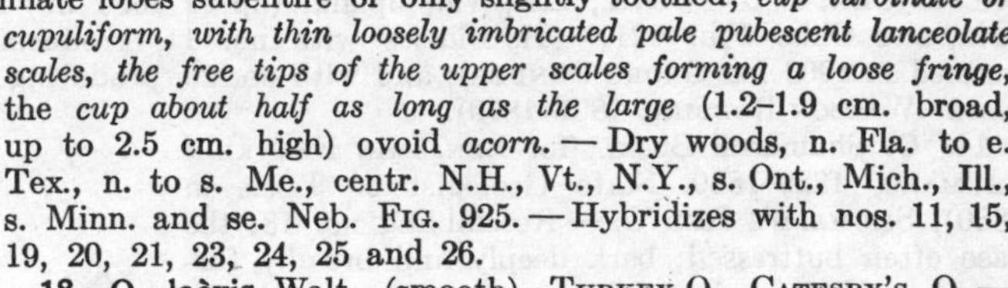

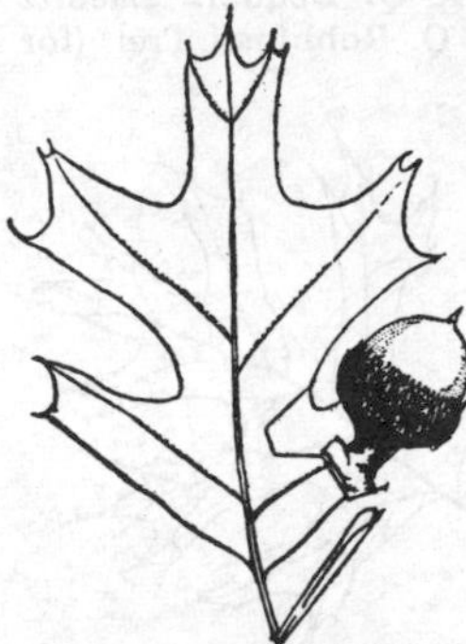

925. Q. velutina, f. macrophylla.

18. Q. laèvis Walt. (smooth), Turkey-O., Catesby's O. — Small tree with *bluish-gray* furrowed or flecked *bark; winter-buds slender,* 1–1.3 cm. long, *their thin- and erose-margined scales rusty-pubescent at tip;* branchlets at first pubescent with fascicled hairs, later glabrate; *leaves gray-green,* broadly triangular to obovate or oval, *with cuneate or tapering bases, very deep and broad open sinuses* and 1 or 2 (–3) unequal pairs of divergent tapering to oblong entire or terminally cleft lateral lobes, the longer (usually lowest) lobes 3–10 cm. long; *petioles* 0.5–1.5 *cm. long; cups turbinate,* 1.8–2.5 cm. broad, *with gray-pubescent oblong- to round-tipped scales, the free tips of the upper series arching over into the inner summit of the cup;* acorn white-tomentose at summit, 1–2 cm. high. (*Q. Catesbaei* Michx.) — Dry sandy woods and barrens on the Coastal Plain, Fla. to La., n. to se. Va. Fig. 926. — Hybridizes with nos. 19 and 25.

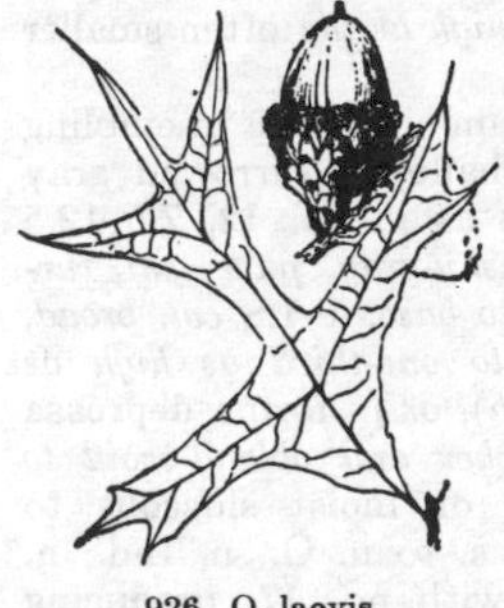

926. Q. laevis.

19. Q. falcàta Michx. (sickle-shaped), Spanish O. — Tree with gray smooth or finally shallowly furrowed and flaky bark; young branchlets at first gray-tomentose, becoming glabrate and reddish the second year; the ovoid acute buds puberulent; *leaves* very variable, those *of fertile branches* dark green above, *grayish-tomentulose beneath,* 1–3 *dm. long, on petioles* 2–6 *cm. long; cups saucer-shaped,* 1.2–1.6 cm. broad, *with appressed obtuse or truncate canescent scales,* one-fourth to one-third as high as the subglobose to round-ellipsoid warm brown nut. — Extremely variable in foliage; the following are the most clearly defined of many described variations:

Leaves ovate to obovate, with well-developed spreading to spreading-ascending lateral lobes, these attenuate and strongly bristle-tipped.
  Blades with 1 or 2 (–3) very unequal pairs of lateral lobes, one pair greatly prolonged and often falcate; base of blade usually prolonged and semi-orbicular or round. . . . . *Q. falcata* (typical).
  Blades regularly pinnatifid with (2–) 3–5 pairs of scarcely falcate subequal divergent lobes; base (below lowest short lobes) short and cuneate to sub-truncate or rounded. . . . . Var. *pagodaefolia.*
Leaves narrowly oblong-obovate, uncleft except toward summit, with 3 short ascending blunt or bluntish terminal lobes; base cuneate to rounded. . . . Var. *triloba.*

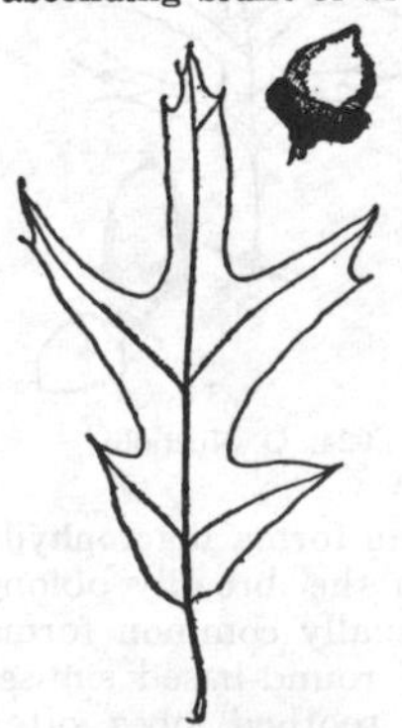

927. Q. falcata (typical).

Q. falcàta (typical), *Q. rubra* sensu Sargent and others, not L.; *Q. digitata* (Marsh.) Sudw. — Moist to dry woods, Fla. to e. Tex., n. to s. N.J., e. Pa., W.Va., s. O., s. Ind., s. Ill., s. Mo. and e. Okla. Fig. 927. — Hybridizes with no. 24, producing × **Q. ánceps** E. J. Palmer (two-edged) and with no. 26, producing × **Q. subfalcàta** Trel. (somewhat sickle-shaped).

Var. **pagodaefòlia** Ell. (with leaves like a pagoda), *Q. pagodaefolia* Ashe; *Q. Pagoda* Raf. — Chiefly on bottomlands or near streams, n. Fla. to La., n. to s. N.J., D.C., Ky., s. Ind., s. Ill. and se. Mo.

Var. **tríloba** (Michx.) Nutt. (three-lobed), *Q. triloba* Michx. — Rather local, Fla. to e. Tex., n. to L.I., N.J., se. Pa., W.Va., s. Ind. and se. Mo. — Hybridizes with no. 26, producing × **Q. ludoviciàna** Sarg. (of Louisiana).

20. Q. ilicifòlia Wang. (holly-leaved), Bear-O., Scrub-O. — Tough straggling or intricately branched *shrub or small dark-barked tree* up to 6 m. high; branchlets at first hoary, becoming dark and glabrate the second year; winter-buds blunt, with loosely imbri-

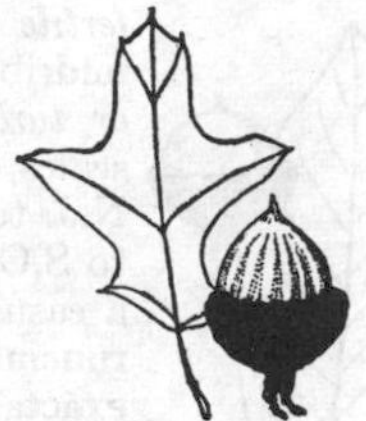
928. Q. ilicifolia.

cated scales; *leaves* of fertile branches dark green above, *whitish-felted beneath,* 5–12 *cm. long, angulate-pinnatifid with short triangular or ovate pointed lobes,* the *petioles* 0.7–3.5 *cm. long; cup saucer-shaped to cupuliform, frequently abruptly enlarged above base, its closely appressed scales tapering to tip;* acorn small, about 1 cm. high. — Dry rocky, gravelly or sandy barrens, se. Me. to N.Y. and W.Va., s. to s. N.E., L.I., Md. and upland to w. N.C. Fig. 928. — Hybridizes with no. 11, producing × **Q. Fernáldii** Trel. (for its unwitting discoverer, 1873–1950); with no. 13; with no. 17, producing × **Q. Rèhderi** Trel. (for Alfred Rehder, 1863–1949); with no. 21, producing × **Q. Bríttoni** W. T. Davis (for Nathaniel Lord Britton, 1859–1934); and with no. 26, producing × **Q. Gíffordi** Trel. (for one of its discoverers, J. C. Gifford).

21. Q. marilándica Muenchh. (of Maryland), Black Jack, Jack-O. — Small tree or coarse shrub with blackish bark soon divided into rectangular scaly blocks; branchlets at first scurfy, finally glabrescent; winter-buds rusty-pubescent; *leaves broadly obovate or obovate-oblong, tapering to rounded or cordate broad base, mostly* 1–2.5 *dm. long, the dilated often shallowly lobulate summit nearly as broad, scurfy-pubescent beneath; cup turbinate or turbinate-hemispheric,* 1.5–2 *cm. broad,* with loosely imbricated tapering scales. — Dry siliceous or argillaceous barrens and sterile woods, Fla. to Tex., n. to Md. and more locally to L.I., N.J., e. and s. Pa., W.Va., s. O., s. Ind., Ill., Mo. and se. Neb. Fig. 929. — Hybridizes with no. 17, producing × **Q. Búshii** Sarg. (for Benjamin Franklin Bush, 1858–1937); with no. 20; with no. 24; with no. 25; and with no. 26, producing × **Q. Rúdkini** Britt. (named in 1882 for its discoverer, William H. Rudkin).

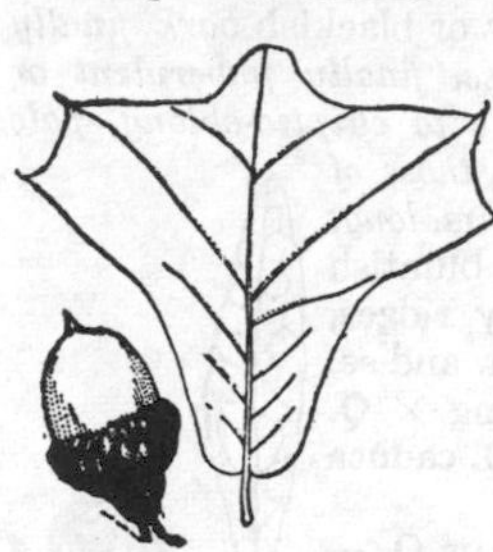
929. Q. marilandica.

22. Q. nìgra L. (black), Water-O., Possum-O. — Tree, or in exposed habitats depressed and shrubby, with smoothish brown bark; branchlets slender, glabrous, grayish in maturity; winter-buds ovoid, with loosely imbricated puberulent scales; *leaves cuneate-obovate, -spatulate, or -oblanceolate, broadest near the often dilated summit, entire or apically shallowly undulate or with* 3 *short rounded lobes,* or deeply cleft into 3 narrow oblong elongate lobes in forma **tridentífera** (Sarg.) Trel. (three-toothed), the *mature blades* glabrous (or merely hairy-tufted beneath in axils of veins), rather thin, blue-green above, paler green beneath, those *of fertile shoots* mostly 4–10 *cm. long and* 2–5 *cm. broad, the long cuneate base tapering to a short petiole;* small *cups saucer-shaped,* their *tightly appressed scales acutish,* much shorter than the small acorn. — Dry woods or borders of streams or bottomlands, Fla. to Tex., n. on Coastal Plain to Del. and inland n. to Ky., se. Mo. and Okla. Fig. 930. — Hybridizes with nos. 18, 21 and 25, and with no. 26, producing × **Q. Capèsii** W. Wolf (named in 1945 for its discoverer, A. Capesius).

930. Q. nigra.

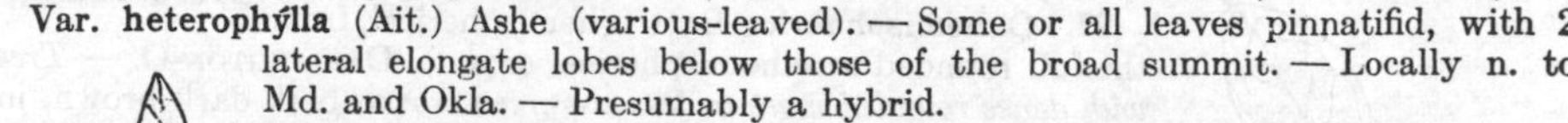

Var. **heterophýlla** (Ait.) Ashe (various-leaved). — Some or all leaves pinnatifid, with 2 lateral elongate lobes below those of the broad summit. — Locally n. to Md. and Okla. — Presumably a hybrid.

23. Q. laurifòlia Michx. (laurel-leaved), Laurel-leaved O. — Tree with gray scaly bark; branchlets glabrous; *leaves thin and membranaceous, slightly rhombic to lanceolate or oblanceolate, broadest near middle,* cuneate to short petioles, *with reticulate venation evident by transmitted light, glabrate and beneath often retaining tufts of axillary hairs, those of fertile branches* 6–12.5 *cm. long and* 2–4 (–5) *cm. broad, their midribs rarely excurrent, promptly deciduous; cup* deeply saucer-shaped, cupuliform or *turbinate,* 1.5–2 *cm. broad,* 7–10 *mm. high; its* ovate tapering *scales* with scarious margins and deciduous scarious tips, *the innermost forming a loose fringe;* acorn pubescent, thick-ovoid, rounded above, 1–1.3 cm. high. (*Q. obtusa* (Willd.) Ashe; *Q. rhombica* Sarg.) — Low woods, bottoms and borders of swamps, Fla. to e. Tex., n. on Coastal Plain to Cape May, N.J. Fig. 931.

931. Q. laurifolia.

24. Q. imbricària Michx. (overlapping; from the use of the wood by early settlers of Illinois, in Michaux's time, for shingles), Shingle-O., Laurel-O. — Tree with *bark* brown, *eventually with shallow fissures and low ridges; branchlets glabrate, lustrous;* winter-buds with shining scales; *leaves lanceolate, lance-elliptic or oblong,* rather *thin, lustrous-green above, permanently stellate-pubescent beneath, entire,* blunt to acutish, with a bristle-tip; *those of*

932. Q. imbricaria.

fertile branches mostly 8–15 *cm. long* and 1.2–6 cm. wide, the yellowish midrib prominent beneath; *petioles* 0.5–2 *cm. long; cups shallowly cupuliform or turbinate,* 1.5–2 *cm. broad,* with thin ovate obtuse to acute pubescent scales, one-third to half as high as the acorn. — Rich woods and bottoms, N.J. to s. Wisc., Ia., and se. Neb., s. to n. Del., D.C., upland and Piedmont to S.C., Tenn., Ark. and e. Kans.; very local, Norfolk Co., Mass. (probably a casual introduction). Fig. 932. — Hybridizes with no. 11, producing × **Q. runcinàta** (A.DC.) Engelm. (saw-toothed); with no. 12, producing × **Q. exácta** Trel. (with accurately spread teeth); with no. 14, producing × **Q. Egglestòni** Trel. (for Willard Webster Eggleston, 1863–1935); with no. 17, producing the relatively frequent × **Q. Leàna** Nutt., Lea's O. (for Thomas Gibson Lea, 1785–1844); and with no. 21, producing × **Q. tridentàta** Engelm. (three-toothed).

25. Q. incàna Bartr. (hoary), Turkey-O., High-ground Willow-O., Blue Jack, Sand-Jack. — Small trees with the gray or blackish *bark finally in squarish blocks; young branchlets heavily tomentose, finally puberulent or glabrescent; leaves coriaceous, lanceolate or oblanceolate to elliptic-oblong, pale green above, white- or grayish-tomentose beneath, entire* or nearly so, *those of fertile branches mostly* 4–9 *cm. long and* 1.5–2.5 *cm. broad, petioles* 3–8 *mm. long; cup saucer-shaped,* 1–1.5 *cm. broad,* with pale-pubescent blunt or bluntish scales; acorn about 1.5 cm. high. (*Q. cinerea* Michx.) — Dry sandy ridges, pine-barrens or dunes of the Coastal Plain, Fla. to Tex., n. to se. Va. and se. Okla. Fig. 933. — Hybridizes with no. 19; with no. 21, producing × **Q. cravenénsis** Little (of Craven Co., N.C.); with no. 22, producing × **Q. cadùca** Trel. (promptly falling).

933. Q. incana.

26. Q. Phéllos L. (Greek for the cork-oak, *Q. Suber*), Willow-O. — Tree with bark finally fissured and split into irregular plates; branchlets glabrous; winter-buds acute; *leaves linear or narrowly oblong, acute, bristle-tipped,* 4–11 cm. long, 0.7–3 *cm. broad* (when first expanding very slender and involute), *arching or rounded at base,* often falcate, *submembranaceous, deciduous in autumn,* green and *glabrous above,* paler and *glabrous or promptly glabrate beneath,* or in forma **intónsa** Fern. (unshaved) permanently silky-tomentose beneath with whitish hairs; *cup shallowly saucer-shaped,* 9–12 *mm. broad,* the gray-pubescent *closely appressed* scales with rounded tips; acorn subspherical, gray-pubescent, about 1 cm. high. — Swamps, bottoms and other low ground, or sandy uplands, n. Fla. to e. Tex., n., chiefly on Coastal Plain, to L.I., N.J., se. Pa., and inland n. to sw. Ill., e. Mo. and e. Okla. Fig. 934. — Hybridizes with no. 11, producing the frequent × **Q. heterophýlla** Michx. f. (various-leaved); with no. 14, producing × **Q. moultonénsis** Ashe (for the Moulton Valley, Tenn.), with no. 17, producing × **Q. filiàlis** Little (for the filial generation "of the cross") and with nos. 19, 20, and 21.

934. Q. Phellos.

27. Q. hemisphaèrica Bartr. (hemispherical, from the full rounded semihemispherical crown), Darlington-O. — *Tree with dense rounded more or less evergreen crown;* bark dark brown, in age blackish, deeply fissured and broken into ridges, very thick; branchlets glabrous, often drooping; *leaves coriaceous, thick and dense, opaque to transmitted light, lustrous-green and glabrous on both surfaces, entire,* the margins often revolute; *those of fertile branches oblong-lanceolate or -oblanceolate to narrowly elliptic or oblong,* 3–10 *cm. long and* 1–3 *cm. broad,* acute to blunt, *cuneate to very short petiole;* leaves of vegetative sprouts sinuate-serrate or -dentate; *cup shallowly saucer-shaped,* 1–1.5 *cm. broad and* 2–5 *mm. high, flattish at base,* the scales with deciduous scarious free tips. (*Q. laurifolia* sensu Ell. and later auth., not Michx.) — Dry to damp woods, sandhills and barrens, Fla. to e. Tex., n. on outer Coastal Plain to se. Va. Fig. 935.

935. Q. hemisphaerica.

FAM. 47. ULMÀCEAE (Elm Family)

Trees or shrubs with alternate mostly distichous and inequilateral pinnately veined simple leaves and fugacious stipules; flowers polygamous or polygamodioecious; calyx with 3–7 *lobes or sepals; stamens of the same number, erect in bud; ovary 1-locular, with 2 styles or stigmas; fruit a samara, nut or drupe; seed without endosperm.* — Widely dispersed in temp. and trop. reg.

Buds along the twigs somewhat spreading; flowers (or some of them) borne on the branchlets before the leaves or after their fall; calyx campanulate or turbinate, lobed; anthers extrorse; fruit a samara or nut, embryo straight; leaves with several prominent parallel veins. (Tribe I. ULMEAE).
Flowers perfect, preceding or following the leaves; fruit flat, winged; leaves mostly double-serrate. . . . . 1. *Ulmus.*
Flowers partly unisexual, appearing with the leaves, the staminate along the branchlets, the 1–3 perfect or pistillate in the leaf-axils; fruit an ellipsoid nut, wingless; leaves mostly simple-serrate. . . . . 2. *Planera.*
Buds appressed; flowers all on new leafy shoots of the year; sepals distinct; anthers introrse; fruit a rounded drupe, embryo curved; leaves prominently 3-veined at base. (Tribe II. CELTIDEAE). . . . . 3. *Celtis.*

Tribe I. ÚLMEAE Agardh

1. ÚLMUS L. ELM. ORME (Que.)

Calyx campanulate, 4–9 cleft. Stamens 3–9, with long and slender filaments. Ovary 1–2-locular, with a single anatropous ovule suspended from the summit of each locule; styles 2, short, diverging, stigmatic along the inner edge. Fruit a 1-locular and 1-seeded membranaceous samara. Albumen none; cotyledons large. — Flowers purplish or yellowish, in lateral clusters. Leaves strongly straight-veined, short-petioled and oblique or unequally somewhat cordate at base. Stipules small, caducous. Trees and shrubs of e. N.Am. and Eurasia. (The classical Latin name.) *

*a.*Flowers nearly sessile or very short-pedicelled, not pendulous; samaras suborbicular to broadly elliptic, their margins not ciliate.
Stamens 5–9; stigmas pink; samaras 1–2 cm. broad, pubescent at center of each side; leaves 1–2 dm. long; branches not corky. . . . . . . 1. *U. rubra.*
Stamens 3–5; stigmas white; samaras 1–1.5 cm. broad, glabrous; leaves 2–8 cm. long.
Branches often corky-winged; leaves strongly asymmetrical at base, doubly serrate, scabrous above; samaras mostly longer than broad. 2. *U. procera.*
Branches all slender; leaves nearly symmetrical, mostly simply serrate, smooth; samaras as broad as or broader than long. . . . . . 3. *U. pumila.*
*a.*Flowers long-pedicelled, soon pendulous; samaras elliptical, ovate or oblong, densely ciliolate-fringed on each margin. . . *b.*
*b.*Flowers vernal; calyx with short lobes; filaments longer than anthers.
Flowers in fascicles, the axis only slightly elongating; samaras glabrous over the seed; branches not corky-thickened. . . . . 4. *U. americana.*
Flowers in elongating racemes; samaras usually pubescent on the sides; branches often corky-thickened.
Samaras broadly elliptic, 0.9–1.5 cm. broad inside the short marginal fringe; leaves with petioles 3–10 mm. long; mature blades 8–15 cm. long, 3–9 cm. broad; cork confined to older branchlets. . . . . 5. *U. Thomasi.*
Samaras lance-ovate, 3–5 mm. broad inside the very long fringe; leaves subsessile; the mature blades 3–9 cm. long, 1–4 cm. broad; young branchlets often corky. . . . . 6. *U. alata.*
*b.*Flowers autumnal; calyx cleft nearly to base; filaments shorter than anthers. 7. *U. serotina.*

1. U. rùbra Muhl. (red), SLIPPERY or RED E. — Winter-buds downy with rusty hairs; *leaves ciliate,* ovate-oblong to obovate, taper-pointed, 1–2 *dm. long, fragrant in drying, soft-downy beneath* at least when young, *very harsh above; branchlets and pedicels downy; flowers nearly sessile in subglobose glomerules;* calyx-lobes and *stamens* 5–9; *stigmas pink; samaras roundish, glabrous except for the pubescent center of each side,* 1–2 *cm. broad.* (*U. fulva* Michx.) — Rich soil (often calcareous), w. Fla. to Tex., n. to N.E., sw. Que., s. Ont., Minn. and N.D. Mar.–early May FIG. 936. — Small or middle-sized tree with very mucilaginous inner bark.

936. U. rubra.

2. U. PROCÈRA Salisb. (tall), ENGLISH E. — Winter-buds minutely pale-pubescent; *branchlets often corky-winged; leaves* elliptic or ovate, short-acuminate, 5–8 cm. long, slightly scabrous above, *with tufts of hair in the axils beneath; flowers short-pedicelled,* in small glomerules; calyx-lobes and *stamens* 3–5;

* The figures show fruiting inflorescences slightly reduced.

stigmas white; samaras roundish, *glabrous* throughout, 1–1.2 *cm. broad.* (*U. campestris* in large part of ed. 7) — Roadsides, thickets and borders of woods, s. N.E. and N.Y. to Va., and doubtless elsewhere, spreading from cult., often by suckers. Apr. (Introd. from Eu.) Fig. 937.

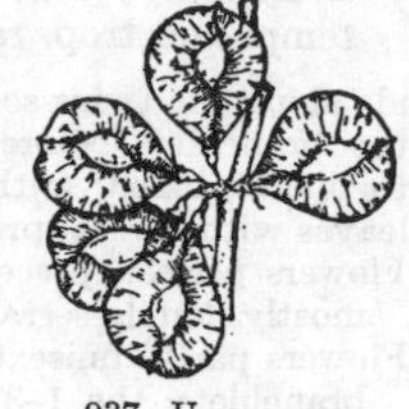

937. U. procera.

U. glàbra Huds. (smooth), *U. campestris* in part of L., Wych E., with branches not corky, without basal suckers, with samaras as large as in no. 1 but glabrous throughout and on glabrous pedicels, and with non-ciliate abruptly pointed leaves, tends to spread from cult. (Introd. from Eu.)

3. U. pùmila L. (dwarf), Dwarf E. — *Branchlets very slender*, glabrous or glabrate; *leaves elliptic to lanceolate, smooth*, mostly *simply serrate*, 2–7 cm. long, *the bases subequal;* stamens 4 or 5; *samaras* round-obovate, *mostly broader than long*, 1–1.5 *cm. broad, glabrous;* a small tree or shrub. — Often planted; natzd. from Minn. to Kans. (Introd. from Asia)

4. U. americàna L. (American), American or White E. — Buds glabrous or pale-pubescent; *branchlets* glabrous or sparingly pilose, *not corky-thickened;* leaves ovate-oblong to oval, abruptly pointed, 5–15 cm. long, glabrous or scabrous above, glabrous or soft-pubescent and glabrate beneath; *flowers in close fascicles, the pedicels elongating and becoming pendulous;* calyx with 5–9 roundish lobes; *samaras* elliptic, 1–1.2 cm. long, *glabrous* except for the finely ciliolate margin. — Rich soil, especially along streams or in lowlands, Gaspé Pen., Que., to Sask., s. to N.S., N.E., n. Fla., La. and Tex. Mar.–early May. Fig. 938. — A large and well-known ornamental tree, variable in habit, usually with gradually spreading branches and pendulous branchlets. — Four forms, found more or less throughout the range, are recognized: forma **péndula** (Ait.) Fern. (hanging down) with leaves smooth or smoothish above and the young branchlets pubescent; forma **laèvior** Fern. (smoother) similar, with young branchlets glabrous; forma **álba** (Ait.) Fern. (white) with leaves harshly scabrous above and the young branchlets pubescent; forma **intercèdens** Fern. (standing between) similar, with young branchlets glabrous.

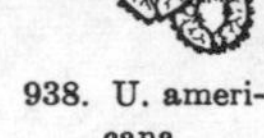

938. U. americana.

5. U. Thómasi Sarg. (for its discoverer, David Thomas, 1776–1859), Rock- or Cork-E. — Buds downy-ciliate; branchlets pubescent, *the older branches often developing cork-ridges; leaves* ovate-oblong or oval, short-acuminate, 8–15 *cm. long,* 3–9 *cm. broad, on petioles* 3–10 *mm. long,* glabrous and very smooth above, the veins very straight; *flowers in elongating loose pendulous racemes up to* 5 *cm. long; samaras* broadly elliptic, *pubescent,* 0.9–1.5 *cm. broad within the very narrow ciliate margin.* (*U. racemosa* Thomas, not Borkh.) — Rich woods and calcareous uplands, w. Que. and w. N.E. to Minn. and S.D., s. to Tenn., Mo. and e. Kans. Apr., early May. Fig. 939.— Round-topped trees.

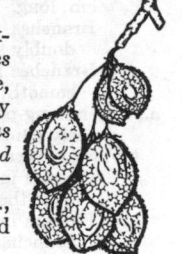

939. U. Thomasi.

6. U. alàta Michx. (winged), Wahoo, Winged or Small-leaved E. — *Bud-scales and branchlets nearly glabrous; branches,* at least some of them, *corky-winged;* leaves downy beneath, ovate-oblong and oblong-lanceolate, acute, thickish, 3–9 *cm. long,* 1–4 *cm. broad, subsessile;* racemes short; *samaras lance-ovate,* 3–5 *mm. broad between the long-ciliate margins,* pubescent or glabrate. — Low, rarely upland, woodlands and thickets, Fla. to Tex., n. to Va., Ky., s. Ind., s. Ill. and Mo. March. Fig. 940. — Small round-topped tree, sometimes up to 20 m. high.

940. U. alata.

7. U. seròtina Sarg. (late), September-E. — Tree of moderate size; leaves narrowly obovate, acuminate, doubly serrate, paler and soft-pubescent beneath; flowers racemose; *calyx cleft nearly to the base,* its divisions very narrow; *filaments shorter than anthers; samaras elliptic, copiously long-ciliate,* 2-horned, 1–1.5 cm. long. — Limestone-hills and bottoms, s. Ky. and s. Ill. to Ga., Ala. and Ark. Sept., Oct. Fig. 941.

2. PLÀNERA J. F. Gmel. Planer-tree. Water-Elm

Flowers monoeciously polygamous. Calyx 4–5-cleft. Stamens 4–5. Ovary ovoid, 1-locular, 1-ovuled, with 2 spreading styles which are stigmatose down the inner side, in fruit becoming coriaceous. — N.Am. tree with small leaves like those of Elms, the flowers appearing with them in small axillary clusters. (Named for *Johann Jakob Planer*, 1743–1789, a German botanist and professor at Erfurt.)

941. U. serotina.

1. P. aquática (Walt.) J. F. Gmel. (aquatic), WATER-E. — Nearly glabrous; leaves ovate-oblong, small; fruit stalked in the calyx, beset with irregular rough projections. — Swamps, Coastal Plain, n. Fla. to Tex., n. to N.C., Ky., s. Ill. and se. Mo. Apr. — A rather small tree.

Tribe II. CELTÍDEAE Dumort.

3. CÉLTIS L. NETTLE-TREE. HACKBERRY. SUGARBERRY. BOIS INCONNU (Que.)

Calyx 5–6-parted, persistent. Stamens 5–6. Ovary 1-locular, with a single suspended ovule; stigmas 2, long and pointed, recurved. Cotyledons folded and crumpled. — Flowers greenish, axillary, the pistillate solitary or in pairs, pedicelled, appearing with the leaves; the lower flowers usually staminate only, fascicled or racemose along the base of the branchlets of the season; trunks often with cork-ridges. Trees and shrubs with thin-fleshed sweet drupes, chiefly in N. Hemisph. (Name of Pliny's for the Lotus with sweet berries described by Herodotus, Dioscorides, Theophrastus and Homer and thought by some to be of this genus with sweet-pulped drupes.) — The e. Am. species and vars. too often seeming confluent. Fully mature fruit important for identification.

Mature drupes ellipsoid-ovoid, -obovoid or subspherical, 8–11 mm. long, commonly short-beaked, evidently puckered in drying, purple-black to fuscous, brown or orange-red; stone 7–9 mm. long; leaves subtending fruits definitely 10–40-toothed on the broader margin, broadly to narrowly ovate and rather abruptly long-acuminate, yellowish-green beneath. . . . . 1. *C. occidentalis.*

Mature drupes 5–8 mm. long, beakless or nearly so, not much puckered in drying, orange or brownish to cherry-red; stone 5–7 mm. long; leaves subtending fruits entire or sparingly few-toothed (if many-toothed uniformly green on both faces and only gradually long-acuminate).

Leaves of fertile branchlets (not always of sterile leaders) entire or nearly so, ovate, one-half to three-fourths as broad as long, 2–8 cm. long, gray-green on both faces or darker above, blunt, acute or only short-acuminate; shrub or small tree of dry or exposed habitats. . . . . 2. *C. tenuifolia.*

Leaves of fertile branchlets entire or copiously serrate, narrowly lanceolate to lance-oblong or lance-ovate, less than half as broad as long, 4–10 cm. long, uniformly pale green on both surfaces, tapering gradually to long-attenuate or -acuminate often falcate tip; tree (often tall) mostly of bottom-land and rich low ground. . . . . 3. *C. laevigata.*

1. C. occidentàlis L. (western; as contrasted with the Old World). — Large or small tree or low shrub greatly varying in response to habitat; old bark deeply furrowed and checkered or warty; *leaves of fruiting branchlets (those subtending flowers and fruits)* broadly to narrowly and *obliquely ovate, definitely* 10–40-*serrate-toothed on at least the broader* (often on both) *margin, mostly two-fifths to four-fifths as broad as long, abruptly long-acuminate, mostly* 3–12 (*av.* 8) *cm. long and* 1.2–9 (*av.* 4.5) *cm. broad;* blades of leading shoots longer; *drupe purple-black to fuscous, buff or orange-red, ellipsoid-ovoid, -obovoid or subspherical, 8–11 mm. long, commonly with a short thick beak; stone 7–9 mm. long, 5–8 mm. thick.* — Excessively variable, passing freely from one var. to another and suspected of hybridizing with nos. 2 and 3. The best marked vars. are as follows:

Leaves coriaceous, scabrous, gradually acuminate; drupes spherical or nearly so, orange-red to fuscous, on pedicels 0.3–1.5 cm. long. . . . . *C. occidentalis* (typical).

Leaves submembranaceous or membranaceous, smooth; drupes ellipsoid-ovoid to -obovoid, finally blackish, black-purple or dark brown, on pedicels mostly 0.8–3.5 cm. long.

Principal leaves conspicuously inequilateral, one side very much broader and more strongly rounded to cordate at base, averaging more than half as broad as long, the larger ones subtending fruits 3–9 cm. broad at base. . Var. *pumila.*

Principal leaves more nearly equilateral, the slightly broader side tapering gradually to cuneate or slightly rounded base; the blades ovate-lanceolate, averaging less than half as broad as long, the larger ones on fruiting branchlets 1.5–5 cm. broad at base. . . . . Var. *canina.*

C. occidentàlis (typical), MICOCOULIER (Que.).— Tree up to 30 m. high, or coarse shrub. (Var. *crassifolia* (Lam.) Gray; *C. crassifolia* Lam.) — Dry to moist and rich woods, river-banks, rocky barrens, sands, etc., Mass. to Ida., s. to n. Fla., Tenn., Ark. and Okla. *Fr.* Oct., Nov.

Var. **pùmila** (Pursh) Gray (dwarf; the type being the very low extreme). — Tree up to 20 m. high or low or shrubby. (*C. occidentalis* of most Am. auth., not L.) — Damp to dry woodlands, bluffs, shores and slopes, sw. Que. to N.D., s. to s. N.E., Ga., Ala., Ark. and Okla.

Var. **canina** (Raf.) Sarg. (of a dog, from the colloquial name, "DOGBERRY"). — Tree up to 30 m. high or low or shrubby. — Bottoms, rich banks or rocky slopes, sw. Que. to Utah, s. to s. N.E., Ga., Tenn., Mo. and Okla.

2. C. tenuifòlia Nutt. (thin-leaved). — Shrub or small tree up to 8 m. high; *leaves of fertile branchlets entire or very sparingly toothed* (on sprouts mostly toothed), *ovate, thin and smooth, one-half to three-fourths as broad as long, 2–8 cm. long,* 1–4 cm. broad, *blunt, acute or merely short-acuminate, gray-green on both surfaces or darker green above; drupes spherical, beakless,* 5–8 *mm. in diameter, glaucous, orange to brown or cherry-red,* not much puckered in drying, on pedicels 3–10 (–13) mm. long; *stone 5–7 mm. long, 4.8–6 mm. broad.* (*C. pumila* of most Am. auth., not Pursh) — Dry crests, bluffs and slopes, n. Fla. to La., n. to Pa., n. Va., Ind. and Mo. *Fr.* Sept., Oct.

Var. **georgiàna** (Small) Fern. & Schub. (of Georgia). — *Pubescent; leaves coriaceous, scabrous above.* (*C. georgiana* Small; *C. pumila,* var. *georgiana* Sarg.) — Dry crests and slopes, Ga. to La. and Okla., n. to D.C., Va., Ky., Ind., Mo., and e. Kans.

3. C. laevigàta Willd. (smooth). — Tree up to 30 m. high; *leaves* membranaceous or submembranaceous (thick and firm in one var.), *uniformly pale green on both faces,* with conspicuous veins, *lanceolate or lance-oblong* (exceptionally lance-ovate), *entire or serrate, the tail-like and often curving tip much prolonged; leaves of fruiting branchlets less than half as broad as long, 4–10 cm. long and* 1.5–3.5 (–4.5) *cm. broad; drupes subspherical, 5–8 mm. in diameter, beakless, orange, brown or red,* on pedicels 6–15 mm. long; *stone 4.5–7 mm. long, 5–6 mm. broad.* — Three vars. with us:

Leaves thin and membranaceous or submembranaceous, smooth or only sparingly hirtellous above.
- Leaves entire. *C. laevigata* (typical).
- Leaves copiously serrate. . . . . Var. *Smallii.*

Leaves coriaceous, scabrous above, entire. . . . . Var. *texana.*

C. laevigàta Willd. (typical). — Tall tree. (*C. mississippiensis* Bosc) — Bottomlands and low woods, Fla. to Tex. and Mex., n. to Va., Ky., s. Ind., s. Ill., Mo. and Okla. *Fr.* Oct., Nov. (W.I.)

Var. **Smállii** (Beadle) Sarg. (for JOHN KUNKEL SMALL, 1869–1938). — Similar habitats, Fla. to La., n. to se. Va., Ky., s. Ill. and Mo.

Var. **texàna** Sarg. (Texan). — *Leaves* tending more to lance-ovate, *thick, firm, scabrous,* entire; shrub or bushy tree. — Bluffs, rocky slopes, dry woods, etc., sw. Ill., Mo. and Kans. to Tex. and N.M.

FAM. 48. MORÀCEAE (MULBERRY FAMILY)

Trees or shrubs (rarely herbs) with milky juice and mostly alternate simple or palmately lobed leaves, the fugacious stipules leaving a scar; flowers unisexual, commonly in heads, aments, or hollow receptacles or on disks, the calyx often becoming fleshy in fruit; filaments inflexed in the bud; anthers 2-locular, opening longitudinally; ovary of 2 carpels (or one carpel aborted), usually 1-locular, with a pendulous ovule and 2 filiform styles; fruit an achene or drupe, with usually curved embryo and fleshy endosperm or none. — An economically important family, chiefly trop.

Leaves toothed, often lobed; branches without spines; staminate flowers in elongate aments; syncarps (fruiting spikes or heads of mature flowers) 1–2 cm. thick, fleshy.
- Buds with 3–6 scales; syncarp cylindric or ellipsoid, juicy, the fruits not protruding. . . . . 1. *Morus.*
- Buds with 2 or 3 scales; syncarp globose, with protruding fruits. . . . . 2. *Broussonetia.*

Leaves entire; branches with axillary spines; staminate flowers in rounded racemes; syncarp globose, 0.7–1.5 dm. in diameter, covered with a dry rind. 3. *Maclura.*

1. MÒRUS L. MULBERRY

Winter-buds with 3–6 scales. Flowers monoecious or dioecious. Calyx 4-parted; lobes ovate. Stamens 4; filaments elastically expanding. Ovary 2-locular, one of the locules smaller and disappearing; styles stigmatic down the inside. Achene ovate, compressed, covered by the succulent berry-like calyx, the whole spike thus becoming a thickened ellipsoid and juicy (edible) multiple fruit. — Trees of warm reg. of the N. Hemisph. (The classical Latin name.)

1. M. rùbra L. (red), RED M. — *Leaves* cordate-ovate, serrate, *rough above, downy beneath,* pointed (on young shoots often lobed); flowers frequently dioecious; *fruit dark purple or red,* 3–6 cm. long. — Rich woods, Fla. to Tex., n. to sw. Vt., N.Y., s. Ont., Minn. and S.D. Apr.–early June. — Large or small tree, ripening its blackberry-like fruit in late June and July.

2. M. álba L. (white), White M. — *Leaves glabrous or glabrate above and merely hairy-tufted in axils beneath,* coarsely toothed, *blunt or short-pointed* and oblique-based; syncarps shorter, paler. — Spread from cult., especially from N.Y. southw. and westw. (Introd. and natzd. from Asia)

2. BROUSSONÈTIA L'Hér.

Winter-buds with 2 or 3 scales. Flowers dioecious; the staminate in flexuous aments; calyx 4-parted; stamens 4; filaments inflexed in bud. Pistillate flowers in dense globular tomentose heads. Leaves alternate, ovate, toothed, often irregularly lobed, pubescent and more or less scabrous. Small genus nat. of e. Asia. (Named for *Auguste Broussonet*, 1761–1807, of Montpellier, physician and naturalist.)

1. B. papyrífera (L.) Vent. (bearing paper), Paper-Mulberry. — Syncarp about 2 cm. in diameter, orange, with red protruding fruits. (*Papyrius* Ktze.) — Spreading from cult. to roadsides and thickets, s. N.E. to Mo. and southw. Apr., May. (Introd. and natzd. from Asia)

Fìcus (ancient name) Càrica L. (from Caria in Asia Minor), the Fig, a large shrub or small tree with connate stipules inclosing the terminal buds, the scabrous leaves palmately 3–5-lobed and the minute flowers borne inside a closed pyriform receptacle, has spread from cult. to clearings and waste places southw. (Introd. from Asia)

3. MACLÙRA Nutt. Osage Orange. Bois-d'Arc

Flowers dioecious; the staminate in loose short racemes, with 4-parted calyx, and 4 stamens inflexed in the bud; the pistillate in a dense globose head, with a 4-cleft calyx inclosing the ovary. Style filiform, long-exserted; ovule pendulous. Fruit an achene buried in the greatly enlarged fleshy calyx. Albumen none. Embryo recurved. — Tree with entire pinnately veined leaves, axillary peduncles and stout axillary spines; a monotypic N.Am. genus. (Named for the early American geologist, *William Maclure*, 1763–1840.)

1. M. pomífera (Raf.) Schneid. (pome-bearing). — Tree 10–20 m. high or large shrub; leaves ovate to oblong-lanceolate, pointed, mostly rounded at base, lustrous; mature syncarp roughly orange-like, 0.7–1.5 dm. in diameter, disappointingly dry and hard. (*Toxylon* Raf.) — Spread from cult. to roadsides and clearings, n. to s. N.E., N.Y., O., Ind., Ill. and Ia. Late May, June. (Introd. and natzd. from Ark. and Tex.)

FAM. 49. CANNABINÀCEAE (Hemp Family)

Harsh aromatic watery-juiced herbs with palmately nerved and usually lobed or divided leaves and persistent stipules; flowers dioecious, the staminate loosely racemed or panicled, with 5 calyx-segments and 5 stamens; pistillate flowers densely clustered, the cupuliform entire calyx closely investing the 1-locular ovary; style 2-cleft to base; fruit a usually glandular achene with curved or coiled embryo and fleshy endosperm. — Small family of economically important plants of the N. Hemisph.

Stem erect; leaves with 3–7 elongate leaflets; pistillate flowers scattered in spikes along the normally leafy branches. . . . . 1. *Cannabis.*
Stem twining; leaves 3–7-lobed or uncleft; pistillate flowers in close cone-like axillary spikes, in fruit covered by the undivided bracts. . . . . . . 2. *Humulus.*

1. CÁNNABIS L. Hemp. Marijuana. Chanvre (Que.)

Flowers green, the staminate in axillary compound racemes or panicles, with 5 sepals and 5 drooping stamens. Achene crustaceous. Embryo simply curved. — A tall roughish annual nat. of Asia, with digitate leaves of 3–7 linear-lanceolate coarsely toothed leaflets, the upper alternate; the inner bark of very tough fibers. (The ancient Greek name, thought by some to come from the Persian name, *Kanab.*)

1. C. satìva L. (sown). — Stem simple up to the erect-branched inflorescence, scabrous, 0.2–2 m. or more high. — Waste grounds and roadsides, chiefly sporadic and derived from foreign packing, Que. to B.C. and southw. June–Oct. (Adv. from Asia)

2. HÙMULUS L. Hop. Houblon (Que.)

Flowers dioecious; the staminate in loose axillary panicles, with 5 sepals and 5 erect stamens. Pistillate flowers in short axillary and solitary spikes or aments; bracts foliaceous, imbricated, each 2-flowered, in fruit forming a sort of membranaceous strobile. Achene invested with the enlarged scale-like calyx. Embryo coiled in a flat spiral. — Twining harshly scabrous perennials of N. Hemisph. with stems almost prickly downward, and mostly opposite cordate and palmately 3–7-lobed leaves. (A late Latin name, of Teutonic origin.)

Sinuses between upper lobes of leaves mostly broad and open; lower leaf-surfaces bearing yellow resinous granules; fruiting spikes much enlarged, their broad membranous bracts covering the fruits, not bristly. . . . . 1. *H. Lupulus.*
Sinuses mostly closed or very narrow; leaves without granules; fruiting spikes barely enlarged, their long-attenuate herbaceous bracts much narrower than fruits, bristly-hispid. 2. *H. japonicus.*

1. H. Lùpulus L. (early generic name), Common H. — **Leaves mostly 3–5(–7)-lobed or those of branches sometimes uncleft, cordate, rounded to ovate, acuminate or with acuminate lobes, the** ***sinuses between the upper lobes mostly rounded and open; the lower surface bearing*** **yellow,** ***resinous*** bitter and aromatic ***granules, the petiole rarely equaling the blade;*** **staminate panicles mostly 5–12 cm. long;** ***fruiting ament*** **loose, subglobose or ovoid,** ***its membranous blunt or short-pointed bracts roundish and completely covering achenes.*** — **Nat. in alluvial thickets, N.B. to Mont., s. to n. and w. N.E., n. Pa., W.Va., e. Ky., O., Ind., Ill., Mo., Kans. and N.M.; also introd. from Eu., long cult. and estab. in waste places, fence-rows, old house-sites, etc. throughout our range.** July, Aug. — The native plant sometimes called *H. americanus* Nutt., *H. Lupulus* var. *neomexicanus* Nels. & Cockerell and *H. neomexicanus* (Nels. & Cockerell) Rydb.; its characters evasive.

2. H. japónicus Sieb. & Zucc. (Japanese), Japanese H. — **Leaves mostly 5–7-lobed,** ***the deeper and narrower sinuses often closed, the scabrous surfaces without waxy granules; petioles often longer than blade;*** **staminate panicles up to 4 dm. long; fruiting ament compact, its** ***attenuate herbaceous bracts narrower than the achenes and bristly-hispid.*** (Possibly *H. scandens* (Lour.) Merr., but that name based upon a plant described as *fruticose* and with *glabrous leaves.*)— **Roadsides, waste places and fence-rows, N.E. to Mich., s. to Va. and Mo. July–Oct. (Introd. and natzd. from Asia)**

FAM. 50. URTICÀCEAE (Nettle Family)

Herbs, shrubs or small trees, often with stinging hairs, simple leaves with stipules, cymose or spicate flowers unisexual, calyx with 2–5 imbricate or valvate lobes, filaments inflexed in the bud, anthers 2-locular and opening longitudinally, ovary 1-locular with an erect ovule and a simple style or stigma, fruit an achene with straight embryo and usually with endosperm. — **Ours all herbs with watery juice and greenish flowers.**

*a.*Calyx of pistillate flower of 2–5 separate or nearly separate sepals. . . *b.*
*b.*Plant beset with stinging bristles.
Leaves opposite; sepals 4 in both pistillate and staminate flowers; stigma capitate; achene straight, erect, inclosed by the 2 upper sepals. . . . 1. *Urtica.*
Leaves alternate; sepals of staminate flowers 5; stigma elongate-subulate; achene deflexed, nearly naked. . . . . 2. *Laportea.*
*b.*Plant essentially smooth and shining, without bristles, opposite-leaved; sepals 3 or 4, those of pistillate flower unequal; achene partly naked, erect. 3. *Pilea.*
*a.*Calyx of pistillate flower tubular or cupuliform, inclosing the achene; plants without stinging bristles.
Leaves opposite, serrate; flower-clusters spicate, not involucrate; style long and filiform. . . . . 4. *Boehmeria.*
Leaves alternate, entire or essentially so; flowers in involucrate-bracted clusters; stigma tufted. . . . . 5. *Parietaria.*

1. URTÌCA L. Nettle. Ortie (Que.)

Flowers monoecious or rarely dioecious, clustered, the clusters mostly in racemes, spikes or loose heads. ***Stam. Fl.*** **Sepals 4. Stamens 4, inserted around the cupuliform rudiment of a pistil.** ***Pist. Fl.*** **Sepals 4, in pairs; the 2 outer smaller and spreading; the 2 inner flat or concave, in fruit membranaceous and inclosing achene. — Stipules in our species distinct. Flowers greenish. Genus semicosmop.** (The classical Latin name, from *uro, to burn.*)

a. Tough-stemmed perennials, with staminate and pistillate flowers chiefly in separate inflorescences many times longer than petioles. . . *b.*
b. Leaf-surfaces and upper internodes without (or essentially without) long bristles; inflorescences monoecious (rarely dioecious). . . *c.*
c. Petioles of principal cauline leaves 1.5–8.5 cm. long, one-fifth to three-fourths length of blade, many times longer than stipules.
Stems glabrous or setulose above, only sparingly pilose; leaves glabrous or only sparingly pilose beneath; those subtending lowest inflorescences with 13–23 pairs of teeth. . . . . 1. *U. gracilis.*
Stems cinereous-pilose or -puberulent above; leaves cinereous beneath; those subtending lowest inflorescences with 19–38 pairs of teeth. . 2. *U. procera.*
c. Petioles of principal cauline leaves 0.7–2.5 cm. long, one-sixth to one-fourth length of the glabrous or pilose-hirsute coarsely toothed blade; stipules half as long as to longer than petioles. . . . . 3. *U. viridis.*
b. Leaf-surfaces or internodes (upper as well as lower) bearing numerous bristles 0.75–2 mm. long; inflorescences dioecious. . . . . 4. *U. dioica.*
a. Annuals with staminate and pistillate flowers mixed in inflorescences shorter than petioles.
Flower-clusters lax and elongating. . . . . 5. *U. urens.*
Flower-clusters compact, spherical or ellipsoid. . . . . 6. *U. chamaedryoides.*

1. U. grácilis Ait. (slender). — Slender, 0.3–1 m. high, mostly *monoecious* (rarely dioecious); *stem glabrous above or somewhat setulose and sparingly pilose; leaves lanceolate to ovate,* rounded to cordate at base, slender-petioled, *glabrous* on both surfaces *or sparingly pilose beneath; those subtending the lowest inflorescences* 5–15 cm. long, *with* 13–23 *pairs of coarse teeth* and *petioles* 1.5–5.5 cm. long (*one-fifth to three-fifths as long as blade*); *stipules* thin, greenish or stramineous, *glabrous to pilose, very much shorter than petioles; inflorescences* usually forking, *slender,* mostly *moniliform* or *interrupted.* (*U. Lyallii* of ed. 7, in part, not S. Wats.) — Thickets and rich damp soil, Nfld. to Alaska, s. to N.S., n. and w. N.E., centr. N.Y., W.Va., Minn., N.M. and Oreg. July–Sept. Fig. 942.

942. U. gracilis.

943. U. procera.

2. U. procèra Muhl. (tall). — Similar, often taller (up to 3.2 m. high); *stem* only slightly if at all bristly but *strongly cinereous-pilose or -puberulent* above; *leaves oblong-lanceolate to narrowly lance-ovate,* usually *cinereous-puberulent beneath; those subtending lowest inflorescences* 6.5–18 cm. long, *with* 19–38 *pairs of finer teeth; stipules* submembranaceous, *cinereous-puberulent.* (*U. gracilis* of ed. 7, not Ait.) — Thickets and roadsides, C.B. and P.E.I. to w. Ont. and N.D., s. to N.C. and La. July–Oct. Fig. 943.

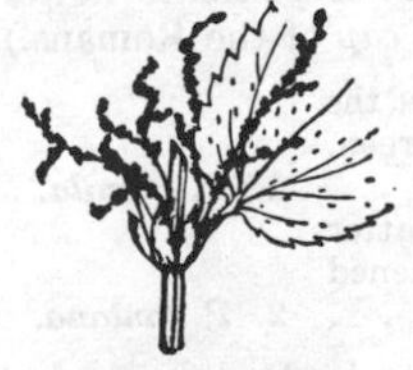
944. U. viridis.

3. U. víridis Rydb. (green). — Stouter, 2–8 dm. high, monoecious; *stem glabrous or pilose above;* primary *leaves lanceolate to narrowly ovate,* firm, rounded at base, *glabrous* on both surfaces *or pilose-hirsute beneath; those subtending lowest inflorescences* 3.5–10 cm. long, *with* 11–25 *pairs of coarse teeth and petioles* 0.7–2.5 *cm. long* (*one-sixth to one-fourth length of blade*); *stipules* lanceolate, with glabrous scarious margins, *half as long as to longer than petioles; inflorescences dense and thick; the staminate* 3–6 *mm. thick.* — Thickets and gravelly shores, s. Lab. to coast of Me.; Alta. to S.D. and N.M. Late June–Sept. Fig. 944.

4. U. dioìca L. (dioecious), Stinging N. — Very bristly and stinging *with stiff bristles* 0.75–2 *mm. long* on stem, leaves or both, 3–9 dm. high; *lower and median leaves cordate-ovate,* very coarsely toothed, the surfaces glabrous except for the numerous bristles, or glabrous and essentially without bristles in forma glàbra Hartm. (smooth), or both appressed-pubescent and very bristly in forma holoseríceA Fries (all of silk); the upper leaves ovate to lanceolate, all with blades many times longer than petiole; inflorescences with simple to much-forked branches, *dioecious.* — Waste places,

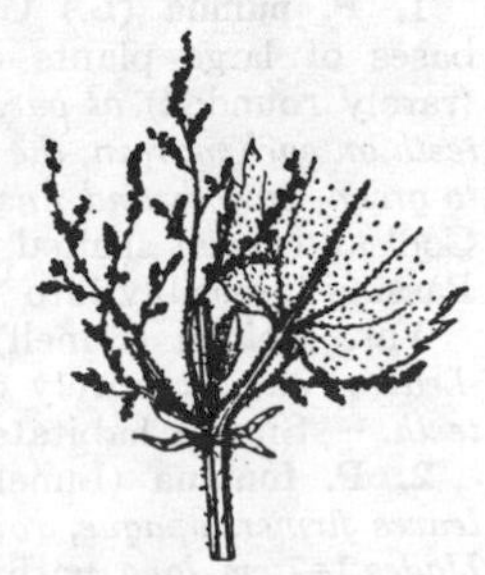
945. U. dioica.

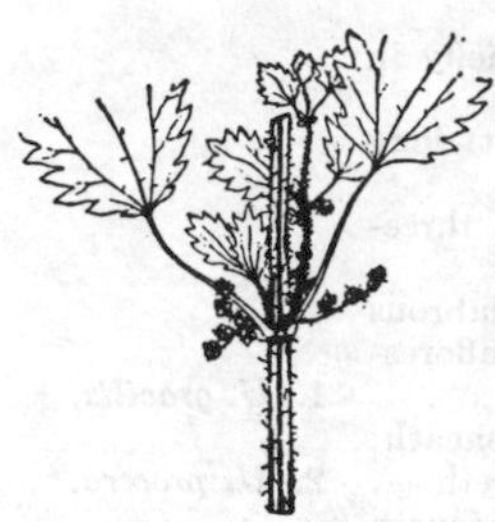

946. U. urens.

roadsides, etc., Nfld. to Man. (acc. to Rydberg), s. to N.S., N.E., Va. and Ill. Late June–Sept. (Natzd. from Eurasia) FIG. 945. — By F. J. Herm. considered to include nos. 1, 2, and 3 as a single var., *U. dioica*, var. *procera* (Muhl.) Wedd.

5. U. ùrens L. (burning or stinging), BURNING N., DOG-N. — Soft-stemmed annual 1–7 dm. high; stem with abundant stinging bristles (whence the name); *leaves* long-petioled, *elliptic or oval*, with long spreading teeth; *flower-clusters* 2 *in each axil, loose and slightly elongating but shorter than petioles.* — Waste ground and about houses, Nfld. to B.C., s. to N.S., N.E., N.Y., Pa., W.Va., Ill. Mo. and Calif. June–Sept. (Natzd. from Eurasia) FIG. 946.

947. U. chamaedryoides.

6. U. chamaedryoìdes Pursh (resembling *Chamaedrys*).— Slender annual 2–7 dm. high, with stinging bristles; *leaves ovate and mostly cordate;* the upper ovate-lanceolate, coarsely serrate-toothed; *flower-clusters globular,* 1–2 in each axil, and spicate at the summit. — Bottomlands, rich woods and waste places, Fla. to Tex. and Mex., n. to W.Va., Ky., se. Mo. and Okla.; adv. northw. to Mass. Late March–May (–Oct.). FIG. 947.

2. LAPÓRTEA Gaud. WOOD-NETTLE

Flowers monoecious or dioecious, clustered, in loose cymes; the upper widely spreading and chiefly or entirely fertile; the lower mostly sterile. *Stam. Fl.* Sepals and stamens 5, with a rudiment of an ovary. *Pist. Fl.* Calyx of 4 sepals, the 2 outer or one of them usually minute, and the 2 inner much larger. Stigma hairy down one side, persistent. Achene ovate, flat, reflexed on the winged or margined pedicel, nearly naked. — Perennial herb (with us; arborescent in the Tropics) with large serrate leaves and axillary stipules. (Named for *François L. de Laporte*, Count of Castelnau, entomologist of the 19th century.)

1. L. canadénsis (L.) Wedd. (of Canada). — Stem 0.3–1 m. high; leaves ovate, coarsely toothed, slender-tipped, strongly feather-veined, 0.7–2 dm. long, long-petioled; fruiting cymes divergent; stipule single, 2-cleft. (*Urticastrum divaricatum* Ktze.) — Low woods and banks of streams, St. P. et Miq. and Que. to Man., s. to N.S., N.E., Fla., Ala., Miss., Mo. and Okla. July–Sept.

3. PÍLEA Lindl. RICHWEED. CLEARWEED. COOLWORT

Flowers monoecious or dioecious. *Stam. Fl.* Sepals and stamens 3–4. *Pist. Fl.* Calyx 3-parted, each of the unequal segments subtending a concave scale-like staminodium; the straight ovary with a sessile penicillate stigma. Achene compressed, partly exserted. Stingless, mostly glabrous and low N. Am. and trop. herbs, with united stipules; the staminate flowers often mixed with the pistillate. (Named from shape of larger sepal of pistillate flower in original species, which partly covers the achene, like the *pileus* or *felt cap* of the Romans.)

Petioles of the larger leaves one-third as long as to about equaling blade; the latter up to 1.5 dm. long, commonly cuneate; fruit stramineous or green, smooth and unspotted or with purple markings, the margins not paler. . . . 1. *P. pumila.*
Petioles of larger leaves one-fifth to one-half the length of blade; the latter usually rounded at base, 1–7 cm. long; fruit black or dark olive, roughened by roundish bosses, the margin colorless. . . . . 2. *P. fontana.*

1. P. pùmila (L.) Gray (dwarf). — Low, 1–7 dm. high, simple to bushy-branched, the bases of large plants decumbent; *leaves lustrous, translucent,* ovate, long-pointed, *cuneate* (rarely rounded) *at base; the largest on each plant* 0.2–1.5 dm. long, *with* 3–11 *coarse rounded teeth on each margin, the petiole one-third to as long as the blade; fruit* narrowly ovate, *stramineous to green, smooth and unspotted or with purple markings, the margins not paler.* (*Adicea* Raf.) — Cool and moist shaded places, P.E.I. to Ont., s. to N.E., Va., Tenn., Ia. and S.D. July–Oct. Passing insensibly into

Var. **Deàmii** (Lunell) Fern. (for its discoverer, CHARLES CLEMON DEAM, 1865–). — *Leaves rounded* (rarely cuneate) *at base, the largest on each plant with* 11–17 *often less rounded teeth.* — Similar habitats, w. N.Y. to Kans., s. to Fla., La. and Tex. Aug.–Oct.

2. P. fontàna (Lunell) Rydb. (of springs). — Smaller, simple or but slightly branched; *leaves firmer, opaque, rounded at base, on petioles one-fifth to half the length of the blade; largest blades* 1–7 *cm. long,* with 3–9 low rounded teeth on each side; *fruit broadly ovate, black or blackish,* *roughened by rounded bosses, the margin colorless.* — Swamps, wet shores and springheads, w. N.Y. to N.D., s. to n. Fla., Ind., Ill. and Neb. Aug.–Oct.

4. BOEHMÈRIA Jacq. False Nettle

Flowers monoecious or dioecious, clustered; the staminate much as in *Urtica;* the pistillate with a tubular or urceolate entire or 2–4-toothed calyx inclosing the ovary. Style elongate-subulate, stigmatic and papillose down one side. Achene elliptical, closely invested by the dry and persistent compressed calyx. — No stinging hairs. Am., Asiatic and tropical (where woody). (Named for *Georg Rudolph Boehmer*, 1723–1803, professor at Wittenberg.)

1. B. cylíndrica (L.) Sw. (cylindric), Bog-Hemp. — Perennial, smoothish or somewhat pubescent; stem (3–9 dm. high) simple; leaves chiefly opposite (rarely all alternate), ovate to ovate- or oblong-lanceolate, thin or thinnish, smooth or but slightly scabrous, mostly long-petioled, pointed, serrate, 3-nerved; stipules distinct; flowers dioecious, or the two kinds intermixed, the small clusters densely aggregated in simple and elongated axillary spikes, the staminate spikes interrupted, the pistillate often continuous, frequently leaf-bearing at the apex. — Moist or shady ground, Fla. to Tex., n. to Me., s. Que., s. Ont. and Minn. Late July–Oct. (W.I.)

Var. **Drummondiàna** Wedd. (for its discoverer, Thomas Drummond, 1780–1835). — Leaves oblong-lanceolate, less sharply pointed, harshly scabrous above, usually pubescent beneath, short-petioled. (Var. *scabra* Porter; *B. scabra* Small; *B. Drummondiana* Wedd.) — In more open habitats, Fla. to Tex., n. to s. N.E., N.Y., Mich., Ill. and Neb.

5. PARIETÀRIA L. Pellitory

Flowers monoeciously polygamous; the staminate, pistillate and perfect intermixed in the same cymose axillary clusters; the staminate much as in the last; the pistillate with a tubular or campanulate 4-lobed and -nerved calyx inclosing the ovary and the ovoid achene. — Widely distributed diffuse or tufted herbs, not stinging, with alternate entire 3-ribbed leaves, and no stipules. (The ancient Latin name, from *paries, wall,* from the habitat of the original species.)

Blades of middle and upper leaves (those subtending inflorescences) in all but obviously starved individuals lance-attenuate, 2–7 cm. long; bracts in well developed plants twice to thrice length of mature calyces. . . . . . . 1. *P. pensylvanica.*

Blades of middle and upper leaves oblong, ovate or rhombic, 1–4 cm. long; bracts shorter than to once-and-a-half length of mature calyces.
- Bracts linear or linear-lanceolate, mostly exceeding mature calyces. . . 2. *P. floridana.*
- Bracts oblong or narrowly elliptic, shorter than to but slightly overtopping calyces. . . . . 3. *P. obtusa.*

1. P. pensylvánica Muhl. (of Pennsylvania). — Low, annual, simple or sparingly branched, with ascending branches, minutely downy; *blades of middle and upper leaves oblong-lanceolate, long-tapering,* 2–7 cm. long, *several times longer than petiole,* thin, the upper surface *roughened with opaque dots; flowers 2 or 3 times shorter than the linear or linear-lanceolate involucral bracts;* stigma sessile. — Rocky or gravelly shaded ground or waste places, Fla. to Tex. and Mex., n. to s. and w. N.E., sw. Que., s. Ont., Minn., N.D., Mont. and s. B.C.; gravelly beach, Cranberry I., Me. May–Sept. Fig. 948.

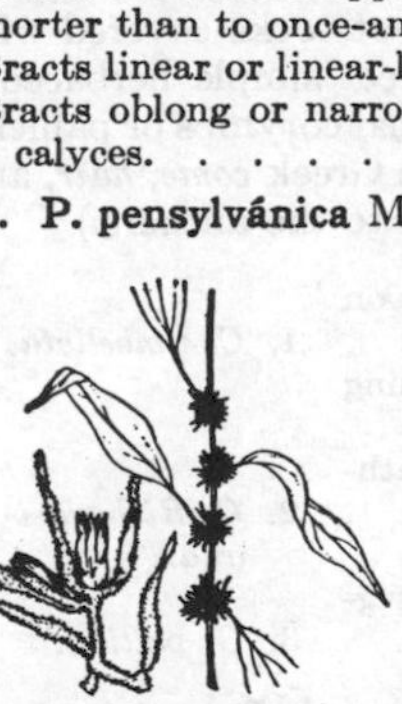

948. P. pensylvanica.

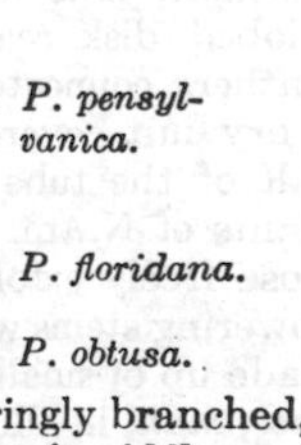

949. P. floridana.

2. P. floridàna Nutt. (of Florida). — Loosely, often diffusely, branched; *leaf-blades ovate to rhombic, short-acuminate,* 1–4 *cm. long;* involucres of *linear or linear-lanceolate bracts* mostly about *one-and-a-half times length of mature calyces.* — Woodlands, thickets and bases of cliffs, Fla. to Tex., n. to N.C. and Ky.; Pawtuckaway Mt., Rockingham Co., N.H. (*A. A. Eaton*). June–Sept. Fig. 949.

3. P. obtùsa Rydb. (blunt). — Similar to no. 2, often cinereous-pubescent, at least on young parts; *leaves broad-ovate to ovate-oblong,* hardly acuminate, *round-tipped; involucral bracts oblong to narrowly elliptic, shorter than to but slightly overtopping calyces.* — Shaded places, Rocky Mt. reg. into n. Mex., e. to s. Mo. and Tex. Fig. 950.

950. P. obtusa.

P. nummulària Small (of money-changers or coin-like; from the roundish leaves), a prostrate or assurgent small plant with reniform to rhombic-ovate or suborbicular obtuse long-petioled leaves 0.5–1 cm. broad, the achenes scarcely 1 mm. long, a wanderer from Florida, has appeared about streets of Wilmington, Del.

FAM. 51. SANTALÀCEAE (Sandalwood Family)

Herbs, shrubs or trees with entire leaves; the 4–5-cleft calyx valvate in the bud, its tube coherent with the 1-locular ovary; ovules 2–4, suspended from the apex of a stalk-like free central placenta which rises from the base of the locule, but the (indehiscent) fruit always 1-seeded. — Seed destitute of any proper seed-coat. Stamens equal in number to the lobes of the calyx, and inserted opposite them into the edge of a fleshy disk. Style 1. A small family, chiefly trop.

Herbs with extensive creeping mostly subterranean stems and simple erect flowering stems.
- Flowers in small cymules borne in terminal corymbs or panicles, whitish, perfect; fruit a dry nut surmounted by the free summit of the calyx. . . . 1. *Comandra.*
- Flowers 2–4 in peduncled cymules from the axils of the middle leaves, bronze, purplish or green, the central one perfect, the others staminate; fruit a red juicy false drupe completely covered by the adherent fleshy calyx. . . . 2. *Geocaulon.*

Shrubs.
- Leaves alternate; flowers in spikes or racemes; fruit pyriform, 2–2.5 cm. long. 3. *Pyrularia.*
- Leaves opposite; staminate flowers in umbels, pistillate solitary; fruits 1–1.5 cm. long.
  - Peduncle of staminate umbel elongate; the calyx-tube elongate, turbinate; pistillate flower without bracts below calyx-limb; fruit globose. . . . 4. *Nestronia.*
  - Peduncle of staminate umbel very short-stalked; the calyx rotate; pistillate flower bearing 4 sepal-like bracts at summit of ovary; fruit ellipsoid. . . 5. *Buckleya.*

1. COMÁNDRA Nutt. Bastard-Toadflax

Flowers perfect. Calyx campanulate or urceolate, free from the summit of the ovary, the turbinate limb with petaloid ascending whitish lobes, the tube bearing an elongate shallowly 5-lobed disk reaching its summit, its short free lobes much exceeded by the filaments. Anthers connected by a tuft of hairs with the calyx-lobes. Style filiform, prolonged. Fruit a dry nut covered by the lower half or two-thirds of the coriaceous calyx-tube, the upper half of the tube forming a free neck below the calyx-lobes. — Smooth sometimes parasitic plants of N.Am. and se. Eu., with creeping or sprawling subligneous rootstocks covered with loose freely exfoliating corky or papery whitish-brown cortex, with erect simple herbaceous flowering stems with alternate sessile or subsessile leaves, and with terminal corymbs or panicles made up of small umbels or cymules of whitish flowers. (Name from the Greek *come, hair,* and *aner, man,* in allusion to the hairs of the calyx-lobes which are attached to the anthers.)

- Fully developed inflorescence an ellipsoid panicle 3–12 cm. high, its cymules on strongly divergent branches. . . . . 1. *C. umbellata.*
- Fully developed inflorescence a corymb, the cymules on strongly ascending branches.
  - Green or greenish; leaves lanceolate to ovate, submembranaceous; perianth-segments 2–3 mm. long; fruit subglobose, 3–5 mm. in diameter. . . . . 2. *C. Richardsiana.*
  - Glaucous; leaves linear to narrowly lanceolate, acute, firm; perianth-segments 3–5 mm. long; fruit ovoid, 6–10 mm. long. . . . . 3. *C. pallida.*

1. C. umbellàta (L.) Nutt. (bearing umbels). — Rootstock under ground; flowering stems 1.5–4 dm. high, branched, very leafy; *leaves* oblong, *thin, pale beneath,* 1–3.5 cm. long, *the pale midrib prominent beneath; inflorescence an ellipsoid panicle with* many cymules of small flowers on *divergent branches;* calyx-tube conspicuously continued as a neck to the dry *globular-urceolate fruit; the calyx-segments oblong,* 2–3 *mm. long;* fruit green or drab, its body 4–6 mm. in diameter. — Dry sterile or acid ground, centr. Me. to Mich., s. to Ga. and n. Ala. Late April–June.

2. C. Richardsiàna Fern. (for its discoverer, George Henry Richards, 1838–1922). — *Rootstock superficial or nearly so,* very elongate and freely branching; flowering stems 0.5–3 dm. high, often tufted, very leafy; *the lanceolate to ovate green leaves submembranaceous, not obviously paler beneath, obscurely veiny; inflorescence a corymb* 1–4 cm. broad, *of* 1–6 *dense cymules on strongly ascending branches; calyx-segments* 2–3 *mm. long; fruit subglobose,* 3–5 *mm. in diam-*

eter. — Calcareous gravelly, sandy or marly soils, Nfld. and s. Lab. to Man., s. to C.B., M.I., P.E.I., e. N.B., s. Que., w. Vt., n. and w. N.Y., Ky., Ind., Mo. and e. Kans. May–Aug.

3. C. pállida A. DC. (pale). — *Leaves narrower, more glaucous and acute, linear to narrowly lanceolate* (or those upon the main stem oblong), all acute or somewhat cuspidate, *firm; calyx-segments* 3–5 *mm. long; fruit ovoid,* 6–10 *mm. long.* — Dry hills and plains, Man. to B.C., s. to Minn., Tex., N.M. and Ariz. May, June.

2. GEOCAÙLON Fern. NORTHERN COMANDRA

Central flower of each 2–4-flowered cymule hermaphrodite, others staminate. Calyx herbaceous, campanulate in the central, turbinate in the other flowers; the limb rotate, with ovate acute bronze or greenish lobes; the tube completely adnate to the ovary. Disk salverform, borne from base of the throat, its elongate lobes about equaling the filaments. Anthers connected by a tuft of hairs with the bases of the calyx-lobes. Style conical and short. Fruit a juicy ovoid-globose false drupe, the succulent calyx-tube completely surrounding the nut and crowned by the sessile vestiges of the crown and lobes. — Smooth N.Am. plant with filiform brown or reddish smooth creeping rootstock, with erect herbaceous simple stems, with entire alternate leaves and with solitary slender-peduncled 2–4-flowered cymules borne in the middle axils, the staminate flowers caducous. (Name from the Greek *ge, earth,* and *caulos, stalk,* from the long slightly subterranean but scarcely modified stems.)

1. G. lívidum (Richards.) Fern. (lead-colored). — Flowering stems 0.7–3 cm. high; leaves membranous or flaccid, livid or purplish, elliptic to narrowly obovate; peduncles 1–3, in fruit 1–2 cm. long; calyx about 4 mm. broad; drupe solitary (rarely 2), scarlet, 6–10 mm. in diameter. (*Comandra* Richards.) — In moss or damp humus, Lab. to Alaska, s. to Nfld., N.S., e. Me., mts. of n. N.E. and n. N.Y., ne. O., n. Mich., n. Minn., Sask., s. Alta. and s. B.C. Late May–early Aug.

3. PYRULÀRIA Michx. OILNUT. BUFFALO-NUT

Calyx 4–5-cleft, the lobes recurved, hairy-tufted at base in the staminate flowers. Stamens 4 or 5, on very short filaments, alternate with as many rounded glands. Pistillate flowers with a pyriform ovary invested by the adherent tube of the calyx, naked at the flat summit; style short and thick. Fruit fleshy, pyriform. — Shrubs or trees, the following of e. N.Am., two others Himalayan, with alternate short-petioled deciduous leaves and small greenish flowers in short and simple spikes or racemes. (Name a diminutive of *Pyrus, pear,* from the shape of the fruit.)

1. P. pùbera Michx. (minutely pubescent). — Shrubby, straggling (1–4 m. high), minutely downy when young; leaves obovate-oblong, acute or pointed at both ends, soft, very veiny, minutely pellucid-punctate; spike few-flowered, terminal; calyx 5-cleft; fruit 2–2.5 cm. long. — Rich woods, thickets and alluvium, mts. of Pa. and W.Va. to Ga. and Ala. May, June. — Parasitic on roots of deciduous trees and shrubs; whole plant, especially fruit, with an acrid poisonous oil.

4. NESTRÒNIA Raf.

Calyx 4–5-lobed. Staminate flowers in 3–8-flowered slender-peduncled umbels, their tubes turbinate, the anthers connected with the lobes by a tuft of hair; the pistillate solitary, jointed upon short peduncles springing from opposite axils. Leaves oval, thin, deciduous, short-petioled. — A single species. (Name said by its author to be derived from a Greek word for *Daphne.*) DARBYA Gray

1. N. umbéllula Raf. (with small umbels). — Low shrub, 3–8.5 dm. high; leaves 3–6 cm. long, mostly acute; flowers small, greenish; drupes at length globose, 1–1.3 cm. in diameter. — Parasitic on roots of deciduous trees, upland Va. to Ga. and Ala. May, June.

5. BÚCKLEYA Torr.

Similar to *Nestronia.* Staminate umbel very short-peduncled, the calyx rotate, its lobes not connected with the anthers. Pistillate flower with 4 sepal-like bracts at summit of ovary; fruit retaining or eventually dropping the bracts. — Large freely branching shrubs with opposite lanceolate subsessile leaves, parasitic on roots of *Tsuga,* one species of e. N.Am. and three e. Asiatic. (Named for *Samuel Botsford Buckley,* American botanist, 1809–1884.)

1. B. distichophýlla Torr. (with two-ranked leaves). — Shrub up to 4 m. high; leaves distichous, 2–6 cm. long; flowers greenish; fruit ellipsoid, 1–1.5 cm. long. — Very local, with *Tsuga,* w. Va. and N.C. and e. Tenn. *Fl.* May.

FAM. 52. LORANTHÀCEAE (Mistletoe Family)

Shrubs, chiefly parasitic on trees, with calyx or calyx-tube adnate to the inferior ovary, which forms a drupe. Stamens the same number as the calyx-lobes (in staminate flowers) or merely rudimentary (in the pistillate). — Leaves coriaceous, greenish, yellowish or brown, or reduced to scales, entire, opposite or whorled. Seed solitary, without testa, with copious endosperm and large embryo. Chiefly trop.

Leaves broad, thick, greenish; drupe pulpy, globose. . . . . 1. *Phoradendron.*
Leaves reduced to tiny connate thin brown scales; drupe dry, compressed. . . 2. *Arceuthobium.*

1. PHORADÈNDRON Nutt. False Mistletoe

Flowers small, dioecious, in short ament-like jointed spikes, usually several to each short fleshy bract or scale, and sunk in the joint. Calyx globular, 3 (rarely 2–4)-lobed; in the staminate flowers a sessile anther borne on the base of each lobe; in the pistillate flowers the calyx-tube adhering to the ovary; stigma sessile, obtuse. Drupe 1-seeded, pulpy. — Yellowish-green woody parasites on the trunks or branches of trees, with jointed much-branched stems and thick firm persistent leaves, chiefly of trop. Am. (Name composed of the Greek *phor, a thief,* and *dendron, tree,* from the parasitic habit.)

1. P. flavéscens (Pursh) Nutt. (yellowish), American Mistletoe. — Leaves obovate or oblanceolate, glabrous, 2–5 cm. long; spikes 1–2 cm. long; drupes white or creamy. — On various deciduous trees, Fla. to e. Tex., n. to N.J., e. Pa., W.Va., s. O., s. Ind., s. Ill., s. Mo. and se. Kans. Sept., Oct.

2. ARCEUTHÒBIUM Bieb.

Calyx mostly compressed; the staminate usually 3-parted, the pistillate 2-toothed. Anthers a single orbicular locule, opening by a circular slit. Drupe compressed, on a short recurved pedicel. — Parasitic on *Pinaceae,* glabrous, with rectangular branches and connate scale-like leaves; species scattered over N. Hemisph. (From the Greek *arceuthos,* the *juniper,* and *bios, life,* the plants being parasites on *Juniperus* and related trees.)

1. A. pusíllum Peck (tiny), Dwarf Mistletoe, Petit Gui (Que.). — Very dwarf, the scattered or clustered flowering stems (springing from rhizomatous stems in the cambium of the host) 0.6–2 cm. high, simple or branched, olive to castaneous or purplish; scales obtuse; flowers solitary in most axils; fruit slenderly ellipsoid, 2–3.5 mm. long, olivaceous or brown. (*Razoumofskya* Ktze.) — On branches of *Picea* and rarely *Larix* and *Pinus Strobus,* Nfld. to Ont., s. to N.S., N.E., n. N.J., n. Pa., Mich., Wisc. and Minn. Apr.–June. — Often causing "witches' brooms".

FAM. 53. ARISTOLOCHIÀCEAE (Birthwort Family)

Twining shrubs or low herbs with perfect flowers, the conspicuous lurid to purple calyx valvate in bud and coherent (at least at base) to the 6-locular ovary, which forms a many-seeded 6-locular capsule or berry in fruit. Stamens 5–12, more or less united with the style; anthers adnate, extrorse. — Leaves petioled, mostly cordate and entire. Seeds anatropous, with a large fleshy raphe and a minute embryo in fleshy albumen. A small family of bitter-tonic or stimulant, sometimes aromatic, plants of trop. and temp. reg.

Stemless herbs, with creeping rhizomes, the flowers borne adjacent to or between the rounded to cordate or deltoid basal leaves; stamens 12, with more or less distinct filaments. . . . . 1. *Asarum.*
Caulescent herbs or twining shrubs with alternate leaves and lateral flowers; stamens 6, the sessile anthers adnate to the stigma. . . . . 2. *Aristolochia.*

1. ÁSARUM L. Asarabacca. Wild Ginger. Asarette (Que.)

Calyx regular; the limb 3-cleft or -parted. Petals 0–3, when present rudimentary, subulate, alternate with the calyx-lobes. Tips of the filaments usually continued beyond the anther into a point. Capsule rather fleshy, globular, bursting irregularly or loculicidal. Seeds large, thick. — Stemless perennial herbs of N. Hemisph., with aromatic-pungent rhizomes bearing 2 or 3 scales, then one or two reniform, cordate or hastate leaves on long petioles, and short-peduncled flowers close to the ground in the axils or (in no. 5) terminal, in spring. (An ancient Greek name, *asaron,* of obscure derivation.)

*a.*Calyx-lobes wholly adnate to the ovary, the tips inflexed in bud; filaments slender, much longer than the short anthers; style barely 6-lobed at summit, with 6 radiating thick stigmas; leaves (in ours) membranaceous, not evergreen, a single pair terminating branches of the superficial rhizome. § EUASARUM. . . . 1. *A. canadense.*

*a.*Calyx-tube inflated, campanulate or flask-shaped, its base adnate to the lower half of the ovary, its 3 lobes short; anthers sessile or nearly so, oblong-linear; styles 6, fleshy, diverging, 2-cleft, bearing a thick extrorse stigma below the cleft; leaves coriaceous, evergreen, often mottled, usually a single one produced each year; rhizomes clustered and ascending, or creeping and subterranean. § CERATASARUM. (Clearly distinct in our region but passing into § *Euasarum* in Eurasia.). . . *b.*

*b.*Sinus of principal leaves shallow, 0.5–3 cm. deep, one-fifth to one-third length of the ovate to reniform blade; calyx broadly campanulate or urceolate, not strongly constricted at throat, 0.8–2.5 cm. in diameter at throat; anther-connective blunt. . . . *c.*

*c.*Rhizome short and stout, bearing many thick and elongate roots and 1–several erect or ascending short leafy crowns. . . . *d.*

*d.*Mature calyx 1–2.5 cm. long, 0.8–1.5 cm. in diameter below the lobes; anther-locules 1.7–2.5 mm. long.

Calyx 1–2.5 cm. long, its lobes 5–10 mm. long, longer than broad; anther-locules exceeded by connective. . . . . 2. *A. virginicum.*

Calyx 1–1.5 cm. long, its lobes 2–3 mm. long, shorter than broad; anther-locules equaled by connective. . . . . 3. *A. Memmingeri.*

*d.*Mature calyx 2.5–5 cm. long, 1.5–2.5 cm. in diameter below the lobes; anther-locules 2.5–3.5 mm. long. . . . . 4. *A. Shuttleworthii.*

*c.*Rhizome slender, horizontally and widely creeping and branching, with mostly solitary leaves scattered or terminating slender subterranean stolons. . . . . 5. *A. Lewisii.*

*b.*Sinus of principal leaves 1.5–5 cm. deep, mostly broad and open and one-third to two-fifths length of the deltoid, subhastate or triangular-cordate blade; calyx slenderly ovoid to thick-fusiform, strongly constricted at summit of tube or narrowed to tip, there 0.4–1 cm. in diameter; anther-connective prolonged to a point.

Mature calyx elongate, thick-fusiform or flask-shaped, 1.5–3.5 cm. long, strongly constricted below the large lobes. . . . . 6. *A. arifolium.*

Mature calyx slenderly ovoid, 1–2 cm. long, gradually narrowed to the thick beak formed by the short lobes. . . . . 7. *A. Ruthii.*

§ EUÁSARUM A. Br. WILD GINGER. GINGEMBRE SAUVAGE (Que.)

1. A. canadénse L. (of Canada). — *Rhizome creeping, elongate; plant soft-pubescent* especially *on the petioles and calyces; leaves membranaceous, broadly reniform* with broad open sinus, often abruptly short-acuminate, or suborbicular or round-ovate with closed sinus and broadly round-tipped in the local forma **Phélpsiae** Fern. of n. N.Y., 0.6–1.8 dm. broad when mature; *calyx* campanulate, brown-purple inside, the *lobes deltoid-ovate to lance-attenuate.* Highly variable; the following freely intergrading varieties the best marked:

Calyx-lobes with caudate tips 0.5–2 cm. long.

Lobes deltoid- to oblong-ovate, 1–2.5 cm. long, narrowed to a slender tip 0.5–1.5 cm. long. . . . . *A. canadense* (typical).

Lobes narrowly to broadly lance-attenuate, 1.5–3.5 cm. long, tapering gradually from base to slender tip. . . . . Var. *acuminatum.*

Calyx-lobes scarcely caudate-tipped or merely mucronate.

Lobes deltoid, 5–12 mm. long, abruptly reflexed. . . . . . . Var. *reflexum.*

Lobes attenuate, 1.2–2 cm. long, ascending or only tardily reflexing. . . . Var. *ambiguum.*

A. canadénse (typical). — Rich woods and about shaded calcareous ledges, Gaspé Pen., Que., to Minn., s. to N.C., Ky. and Ill. April, May. — Forma **Phélpsiae** Fern. (for ORRA ALMIRA PARKER PHELPS, 1867– , its discoverer), known only about limestone-ledges, local, St. Lawrence Co., N.Y.

Var. **acuminàtum** Ashe (tapering at tip), *A. acuminatum* (Ashe) Bickn. — Vt. to Minn., s. to Va., Tenn. and Mo.

Var. **refléxum** (Bickn.) Robins. (bent backward), *A. reflexum* Bickn. — Ct. to se. Man., s. to N.C., Ky. and Mo.

Var. **ambíguum** (Bickn.) Farw. (obscure), *A. ambiguum* (Bickn.) Daniels — Nw. N.J. to Mich., s. to Va., O., Ind., Ill., Mo. and e. Kans.

§ CERATÁSARUM A. Br. (HEXASTYLIS Raf.) HEARTLEAF

2. A. virgínicum L. (Virginian). — *Rhizome ascending, short and stout, with long cord-like roots, forking into* 1–several *short* and *thick leafy crowns; leaves* long-petioled, *coriaceous, evergreen,* often mottled, *reniform to cordate-ovate;* the blades 2.5–9.5 cm. long and 2–8.5 cm. broad, *with narrow sinus* 0.5–3 *cm. deep and one-third to two-fifths length of blade; calyx* basal, slender-peduncled, often hidden by fallen leaves, *broadly campanulate-urceolate,* fleshy or leathery, accrescent, dark purple to fuscous (when buried paler), 1–2.5 *cm. long,* 0.8–1.5 *cm. in diameter at the slight constriction below the spreading-ascending lobes; these longer than broad,* 5–10 *mm. long; anther-locules* 1.7–2.5 mm. long, *slightly exceeded by the blunt connective.* (Incl. *A. minus* and *heterophyllum* Ashe and *Hexastylis virginica* and *heterophylla* Small) — Acid soils of sandy, peaty or rocky woods, e. Va. to W.Va., s. to S.C. and Tenn. March–May.

3. A. Memmíngeri Ashe (named in 1897 for EDWARD READ MEMMINGER). — Very similar to no. 2 but the leaf more rounded and the small *calyx only* 1–1.5 *cm. long, with lobes shorter than broad and only* 2–3 *mm. long; anther-connective not longer than locules.* (*Hexastylis* Small) — Acid woodlands, e. W.Va. and sw. Va., along mts. to Ga. May, June. Perhaps only a local form of no. 2.

4. A. Shuttlewórthii Britten & Baker (for ROBERT JAMES SHUTTLEWORTH, 1810–1874, English collector who designated it as new). — Like no. 2 but *with thinner* and often larger *leaves;* mature *calyx* 2.5–5 *cm. long and* 1.5–2.5 *cm. in diameter below the lobes, the latter* 0.8–1.5 *cm. long and about* 1 *cm. broad; anther-locules* 2.5–3.5 *mm. long.* (*Hexastylis* Small; *A. grandiflorum* Small, not Klotzsch; *A. macranthum* (Shuttlew.) Small, not Hook. f.) — Upland woods and ravines, w. Va. and W.Va., s. to nw. Ga. and Ala. May–July.

5. A. Lewísii Fern. (for its discoverer, JOHN BARZILLAI LEWIS, 1868–). — *Rhizome subterranean, slender, cord-like,* whitish, *horizontal, greatly elongate and branching; leaves solitary and scattered or single at tips of slender stolons,* coriaceous, more or less mottled, *deltoid, deltoid-ovate, ovate or reniform,* 2–8 cm. long; flower solitary, somewhat nodding, terminating rhizome or stolons; *calyx campanulate,* grayish-brown or paler and glabrous outside, dark purple and *villous within,* 2–3 *cm. long,* 1.3–2 *cm. in diameter,* the broad lobes ascending to slightly spreading. — Moist to dry acid woodland of outer Piedmont area, se. Va. and N.C. (there also on Coastal Plain). April, May.

6. A. arifòlium Michx. (Arum-leaved). — Habitally similar to no. 2, usually with only 1–few crowns; *leaf-blade deltoid, subhastate or triangular-cordate,* 0.3–1.6 *dm. long,* 0.4–1.5 *dm. broad,* the *broad sinus* 1.5–5 *cm. deep and one-third to two-fifths the length of the blade; calyx elongate, thick-fusiform or flask-shaped,* 1.5–3.5 *cm. long, strongly constricted to a neck below the large spreading-ascending lobes; locules of anthers exceeded by the slender tip of the connective.* (*Hexastylis* Small) — Woodlands, Fla. to La., n. on Coastal Plain to se. Va. and inland n. to sw. Va. and Tenn. Late April–June.

7. A. Rùthii Ashe (for its discoverer, ALBERT RUTH, 1844–1932). — Differing from no. 6 in its *slenderly ovoid calyx* 1–2 *cm. long, gradually tapering from base to the scarcely constricted summit; the erect lobes only* 1–3 *mm. long and forming a short narrowed beak.* (*Hexastylis* Small) — Upland woods, sw. Va. and Ky., s. to Ala. May, June.

2. ARISTOLÒCHIA L. BIRTHWORT

Calyx tubular; the tube variously inflated above the ovary, mostly contracted; sessile anthers wholly adnate to the short and fleshy 3–6-lobed or -angled style. Capsule naked, septicidally 6-valved. Seeds very flat. — Twining, climbing or sometimes upright perennial herbs or shrubs of trop. and warm reg., with alternate leaves and lateral or axillary greenish or lurid-purple flowers. (Name from the Greek *aristos, best,* and *lochia, delivery,* from supposed value in aiding childbirth; the curved flower, with summit and base together, suggesting, by the doctrine of signatures, the human foetus in the womb.)

a. Calyx-tube strongly curved. . . . *b.*
b. Erect slender herb 1–4.5 dm. high; flowers chiefly on short basal branches. 1. *A. Serpentaria.*
b. High twining shrubs; flowers borne along the stems.
Nearly glabrous; leaves abruptly pointed; peduncle with a leafy bract; calyx purple, with abrupt flat border, glabrous. . . . . 2. *A. durior.*
Soft-pubescent; leaves blunt; peduncle bractless; calyx yellowish, with reflexed limb, pubescent. . . . . 3. *A. tomentosa.*
a. Calyx-tube straight; erect herb; flowers in axillary fascicles. . . . . . . 4. *A. Clematitis.*

1. A. Serpentària L. (old name for this and other reputed cures for snake-bite), VIRGINIA SNAKEROOT. — *Stems slender, erect,* from a knotty rhizome, 1–4.5 dm. high; *leaves* few, *ovate* *to ovate-oblong, cordate at base,* acuminate, slender-petioled; *the larger membranaceous blades* 4–14 cm. long, 2–7 *cm. broad; flowers* (very rarely at summit of stem) *on short slender-bracted* (often slightly subterranean) *basal branches;* the deep purple to chocolate-colored *calyx strongly curved,* 1–1.5 *cm. long,* the broad flaring limb obtusely 3-lobed; capsule ellipsoid to subglobose, 0.8–1.3 cm. long; after dehiscence radiate-stelliform, resembling a 6-lobed *Geaster,* 3.5–4 cm. across; seeds obovoid, 4–5 mm. long, resembling grape-stones with pebbled yellowish white back and sides. — Rich, often calcareous, woods, Fla. to Tex., n. to sw. Ct., se. N.Y., O., Ind., s. Ill., centr. Mo. and se. Kans. May–July.

Var. **hastàta** (Nutt.) Duchartre (halberd-shaped). — *Leaf-blades linear- to broad-lanceolate or lance-deltoid, long-attenuate,* 0.5–2.5 *cm. broad,* the rounded auriculate *basal lobes divergent.* (*A. hastata* Nutt.) — Fla. to Tex., n. to Va. and se. Mo.

2. A. dùrior Hill (tougher), Dutchman's-pipe, Pipe-vine. — High-twining shrubby vine, *nearly glabrous; leaves cordate-ovate to cordate-reniform, membranaceous, abruptly pointed,* the blades on fertile branches 0.9–2 dm. long and broad, on vegetative shoots up to 4 dm. across; *peduncles* axillary, *with a clasping leaf-like bract,* borne along the branches; *calyx about 3 cm. long, strongly curved like a Dutch pipe, brown-purple,* contracted at mouth, *with abrupt flat* obscurely 3-lobed *limb;* capsule cylindric-ovoid, 5.5–8 cm. long, 2.5 cm. in diameter, firm, the valves separating only in age. (*A. macrophylla* Lam.) — Rich woods and banks of streams, sw. Pa. and W.Va., s. in the uplands to Ga. and Ala.; much cult. and locally natzd. eastw. to N.E. and N.J. May, June.

3. A. tomentòsa Sims (tomentose). — Similar to no. 2; but *young branchlets, lower surfaces of leaves and flowers densely white-pubescent; leaves* ovate to round-reniform, *blunt,* the blades of fertile branches 0.5–1.5 cm. long; downy *peduncles bractless; calyx yellowish, with an oblique dark purple orifice and a rugose reflexed limb;* capsules about 3 cm. in diameter. — Rich, chiefly alluvial or calcareous, woods, Fla. to e. Tex., n. to N.C., sw. Ind., s. Ill., Mo. and se. Kans.; locally natzd. in w. N.Y. May, June.

4. A. Clematìtis L. (name used by Dioscorides, some think for this plant). — *Erect glabrous herb* with corrugated stem 0.7–1.5 m. high; leaves cordate-deltoid to reniform; *flowers in axillary cymose fascicles; calyx-tube straight,* slender, open, with ampliate 6-lobed limb, the lobes appendaged; fruit subglobose, 2.5–3 cm. long. — Waste places and roadsides, local, N.Y., O., and Md. June–Aug. (Introd. from Eu.)

FAM. 54. POLYGONÀCEAE (Buckwheat Family)

Herbs or shrubs with alternate simple leaves, and stipules in the form of sheaths (ocreae, these sometimes obsolete) *above the swollen joints of the stem; the flowers mostly perfect, with a more or less persistent calyx, a 1-locular ovary bearing 2 or 3 styles or stigmas, and a single usually erect orthotropous seed.* Fruit usually an achene, compressed or 3–4-ángled or -winged. Stamens 4–12, inserted on the base of the 3–6-cleft calyx.

*a.*Flowers subtended by an involucre, several–many; stamens 9; stipules 0. . . 1. *Eriogonum.*
*a.*Flowers without involucre; stamens 4–8. . . . *b.*
*b.*Plants without tendrils (if climbing the herbaceous stem twining); stipular sheaths manifest, at least when young; calyx not laterally winged; ovules erect from base of locule. . . . *c.*
*c.*Outer sepals wide-spreading or reflexed, not enlarged, the inner ones erect and enlarged in fruit.
Sepals 4; stigmas 2; achene lenticular, flat, with orbicular wing; leaves mostly basal, reniform. . . . . 2. *Oxyria.*
Sepals 6; stigmas 3; achene 3-angled; leaves mostly elongate, extending up stem. . . . . 3. *Rumex.*
*c.*Sepals all ascending or erect (or 2 outer tardily reflexed in no. 7), equal or subequal, often petal-like or with petaloid margins; achene triangular or lenticular. . . . *d.*
*d.*Style 2-cleft to base, persistent as 2 rigid deflexed and hooked beaks of achene; flowers remote on very elongate slender axes, 1–3 in each fascicle, soon deflexed, the greenish calyx not enlarged in fruit. . . 4. *Tovara.*
*d.*Styles 2 or 3, deciduous, not hooked; flowers solitary or in fascicles in axils of leaves or bracts, in spiciform panicles or in panicled or corymbed racemes; calyx green to whitish or roseate. . . *e.*
*e.*Flowers fascicled in axils of leaves or foliaceous bracts, in spiciform panicles or in panicles or corymbs of racemes; embryo curved.
Achene lenticular or 3-angled, included in closely embracing and enlarging appressed calyx (or, if exserted, borne in axillary

fascicles and the leaves narrow); embryo marginal, curved around one side of albumen; plants of various habit, not heath-like. . . . . 5. *Polygonum.*

Achene 3-angled, exserted, or only loosely embraced by the unchanging but soon shriveling calyx; plants smooth annuals with hastate, cordate or deltoid leaves and corymbose racemes; embryo in center of albumen, its broad cotyledons spiraling around the ascending radicle. . . . . 6. *Fagopyrum.*

*e.*Flowers single and pedicelled in axils of scales (ocreolae), the pedicels recurving in fruit and jointed toward base, borne in panicled racemes; calyx petaloid, loosely ascending, its 2 outer sepals sometimes reflexed in fruit; embryo straight, marginal, slender; leaves firm, narrow, soon disarticulating; heath-like plants. . . . . 7. *Polygonella.*

*b.*Plants shrubby, climbing by tendrils borne at ends of branches; stipules obsolete; mature sepals with wing decurrent down pedicel; ovule hanging from apex of a slender stalk. . . . . 8. *Brunnichia.*

1. ERIÓGONUM Michx. UMBRELLA-PLANT

Flowers perfect, involucrate; involucre 4–8-toothed or -lobed, usually many-flowered; the more or less exserted pedicels intermixed with narrow scarious bracts. Calyx 6-parted or -cleft, colored, persistent about the achene. Stamens 9, upon the base of the calyx. Styles 3; stigmas capitate. Achene triangular. Embryo straight and axial, with foliaceous cotyledons. — Leaves entire, without stipules. N. Am. (mostly western). (Name from the Greek *erion, wool,* and *gonu, knee,* from the woolly stems and leaves and swollen joints of the plants.)

1. E. longifòlium Nutt. (long-leaved). — Perennial, erect; *leaves* oblanceolate, acute or acutish, canescent beneath, *the lower cuneate at base; sepals* linear, caudate-attenuate, *villous-canescent.* — Rocky or sandy open woods and glades, s. Mo. and Kans. to Tex. June–Sept.

2. E. Álleni S. Wats. (for its discoverer, TIMOTHY FIELD ALLEN, 1837–1902). — Perennial, erect; *leaves* oblong, canescent-tomentose beneath, flocculent, or glabrate above, *the lower rather abrupt at base;* inflorescence leafy; *sepals* elliptical, yellow, *nearly glabrous.* — Shaly slopes and barrens, local, W.Va. and Va. July–Sept.

2. OXÝRIA Hill MOUNTAIN-SORREL

Outer sepals smaller and spreading, the inner broader and erect (but unchanged) in fruit. Stamens 6. Stigmas 2, sessile, tufted. Achene lenticular, thin, flat, much larger than the calyx, surrounded by a broad veiny wing. Embryo straight, in center of the albumen, slender. — Low arctic-alpine perennial, with round-reniform and long-petioled leaves chiefly from the rootstock, obliquely truncate sheaths, and small greenish to crimson flowers clustered in panicled racemes on a stoutish 1–2-leaved stem. (Name from the Greek *oxys, sour,* from the acid leaves.)

1. O. dígyna (L.) Hill (with two carpels). — Arct. reg., s. to cool rocky slopes, Nfld., and alpine areas of Gaspé Pen., Que., and N.H.; mts. of w. N. Am., s. to N. M., Ariz. and s. Calif. June–Aug. (N. Eurasia)

3. RÙMEX L. DOCK. SORREL

Calyx of 6 sepals; the outer 3 sepals herbaceous, sometimes united at base, often spreading in fruit; the 3 inner larger, somewhat colored (in fruit called *valves*) and convergent over the 3-angled achene, veiny, often bearing a grain-like tubercle lying along one side of the albumen. — Herbs with small and numerous (chiefly greenish) flowers which are mostly crowded and commonly whorled in panicled racemes; the petioles somewhat sheathing at base. (The ancient Latin name.)* — Genus essentially cosmop.

*a.*Flowers perfect or monoeciously polygamous; leaves not sagittate or hastate, slightly or not at all acid to taste. Subgen. LAPATHUM (DOCKS). . . *b.*

*b.*Stem ascending to depressed, producing axillary branches or leaf-tufts; leaves entire or nearly so, fleshy, mostly pale. § AXILLARES. . . *c.*

*c.*Mature fruiting valves 2–3.5 cm. broad, all without grains, or grains merely rudimentary; stipules dilated at summit; western sp. with long creeping rootstock. . . . . 1. *R. venosus.*

*c.*Mature fruiting valves 2–6 mm. broad, 1 or more of them with obvious grain; stipular sheaths close, cylindric; tap-root vertical. . . *d.*

*d.*Fruiting pedicels clavate, strongly deflexed, straightish and stiff, the

---

* Some details taken from RECHINGER fil. in Field Mus. Pub. Bot. xvii. no. 1 (1937). Only fully mature fruiting material of most species can be easily identified.

longer ones two to five times as long as the subacuminate valves; all 3 valves with lanceolate grains; aquatic or paludal species. . . . 2. *R. verticillatus.*

d. Fruiting pedicels filiform, curved, shorter than to hardly twice length of blunt valves; grains fusiform-ovoid to ovoid-ellipsoid; plants of dry to moist but scarcely subaquatic habitats.

Grains narrowly fusiform-ovoid, much narrower than wings of valve; stem ascending or erect; branches of panicle ascending; plants not primarily maritime.

Leaves narrowly lanceolate, 1.5–3 cm. broad, long-attenuate; mature valves deltoid-ovate, all 3 grain-bearing. . . . . . 3. *R. mexicanus.*

Leaves ovate- to oblong-lanceolate, the principal ones 2.5–5.5 cm. broad; mature valves broadly ovate, only 1 of them with a plump grain. . . . . 4. *R. altissimus.*

Grains 3, ovoid or ellipsoid, much broader than narrow wings of valves; lower branches of panicle nearly horizontally divergent; prostrate or procumbent; chiefly maritime species. . . . . . 5. *R. pallidus.*

b. Stem ascending to spreading, usually without axillary branches (at most with exceptionally small axillary leaves); leaves green to fulvous, often crenulate, crisped or undulate. § Simplices. . . *e.*

e. Valves entire or merely dentate or denticulate. . *f.*

f. Enlarged and prominent grain on none or only 1 valve. . . *g.*

g. Stem arising from deep tap-root and from last year's basal rosette of leaves; leaves twice to many times longer than broad. . . *h.*

h. Mature valves reniform, broadly rounded at summit, definitely broader than long, 4–7 mm. long; all midribs without grains or 1 with small rudimentary grain; weedy species. . . . . . 6. *R. domesticus.*

h. Mature valves broadly ovate to ovate-orbicular, tapering to ovate tip, as long as or longer than broad. . *i.*

i. Leaves plane or nearly so; those of basal rosette and base of stem 2–18 cm. broad, mostly with broadly rounded to cordate bases; mature valves 6–9 mm. long; achenes 2–3.5 mm. long.

All 3 valves without grains; achene 2–3 mm. long; lowest leaves 0.5–2 dm. long, 2–10 cm. broad; boreal native. . . 7. *R. fenestratus.*

One valve with large and plump grain; achene 2.8–3.5 mm. long; lowest leaves 2–4.5 dm. long, 0.7–1.8 dm. wide; adv. species. . . . . 8. *R. Patientia.*

i. Leaves with crumpled and crisped margins; those of basal rosette and base of stem 1.5–8 cm. broad, narrowed to base; mature valves 4–6 mm. long; achenes 1.5–2.5 mm. long; introd. weed. 11. *R. crispus.*

g. Stem arising from creeping rhizome, without basal rosette; leaves round-ovate to reniform-suborbicular, the lower ones 1–3 dm. broad; all 3 valves without grains. . . . . 9. *R. alpinus.*

f. Enlarged and prominent grain (or grains) on all 3 valves. . *j.*

j. Fruiting racemes upright, forming a compact panicle; mature valves 4–7.5 mm. broad, broadly ovate to suborbicular, their wings much broader than grains.

Leaves flat, entire or with finely crenulate margin; indig. aquatic or paludal species. . . . . 10. *R. orbiculatus.*

Leaves with crumpled and puckered margin; natzd. weed of drier habitats. . . . . 11. *R. crispus.*

j. Fruiting racemes very slender, loosely spreading-ascending; mature valves oblong, about 1.5 mm. broad, only slightly broader than grain; introd. weed. . . . . 12. *R. conglomeratus.*

e. Valves with long sharp salient teeth or marginal bristles. . *k.*

k. Perennials, with firm stems; valves bearing marginal teeth; weeds of fresh soil.

Slender, 2–7 dm. high, with slender, much interrupted divaricate racemes forming an open panicle one-half to three-fourths height of plant; basal leaves firm, pale green, merely undulate, 4–15 cm. long, 1.5–6 cm. broad. . . . . 13. *R. pulcher.*

Coarser, 0.6–1.5 m. high, with erect racemes forming a panicle rarely half height of plant; basal leaves membranaceous, warm green, with red veins, crenulate, 1–3 dm. long, 5–15 cm. broad. . . . 14. *R. obtusifolius.*

k. Annuals or biennials with soft hollow stems; valves with marginal bristles; natives of saline or brackish soil.

Grain turgid, ellipsoid-ovoid; marginal bristles about as long as breadth of valve; species of Atlantic area. . . . . 15. *R. persicarioides.*

Grain linear-lanceolate; marginal bristles longer than breadth of valve; transcontinental species. . . . . 16. *R. maritimus.*

a. Flowers dioecious (rarely polygamous); at least the basal leaves usually sagittate or hastate, acid. . . . *l.*

l. Plant spreading by long slender rootstocks, or cespitose in no. 18; sepals of fruiting calyx not evidently enlarged in fruit. Subgen. ACETOSELLA (SORRELS).

Spreading by long slender rootstocks; basal leaves hastate or, if unlobed, lanceolate or lance-oblong; achene exserted from the deciduous calyx; wide-ranging weed. . . . . 17. *R. Acetosella.*

Cespitose; basal leaves narrowly linear, erect, simple or rarely lobed; achene covered by persistent calyx. . . . . 18. *R. graminifolius.*

l. Plants without slender rootstocks; valves of fruiting calyx enlarged, thin and veiny. Subgen. ACETOSA (SORRELS).

Leaves linear or lanceolate, some of them hastate, with narrow divaricate basal lobes; fruiting valves 2.5–4 mm. broad; southern native. . 19. *R. hastatulus.*

Leaves oblong or broadly lanceolate, sagittate; fruiting valves 4–5 mm. broad; northern introd. . . . . 20. *R. Acetosa.*

Subgen. LÁPATHUM (Campd.) Rech. f. (DOCKS)

§ AXILLÀRES Rech. f.

1. R. venòsus Pursh (veiny; from the conspicuous valves), "WILD BEGONIA", "WILD HYDRANGEA", SOUR GREENS. — *Stems arising from a creeping rootstock,* 2–6 dm. high, with ascending branches; *stipules dilated upward; leaves* slender-petioled, ovate to oblong or lanceolate, pale, flat, entire, *mostly* 4–12 *cm. long;* panicle nearly sessile, very dense in fruit; *valves rose-color,* entire, *without grains, cordate-reniform,* very veiny, 1.3–2 *cm. long,* 2–3.5 *cm. broad;* achene 5–7 mm. long, nearly as broad. — Sandy soil, local, Wisc. *Fr.* May–July. (Adv. from farther west) FIG. 951.

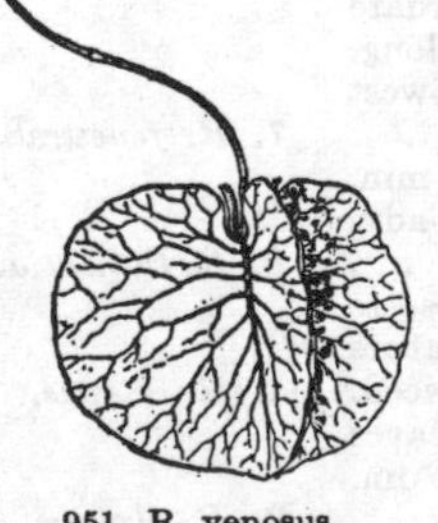

951. R. venosus.

2. R. verticillàtus L. (whorled), SWAMP- or WATER-D. — *Erect,* from deep tap-root, smooth and *pale,* 0.5–1.5 m. high, *simple except for short axillary fascicles;* principal *leaves* plane, entire or nearly so, *narrowly to broadly lanceolate to lance-oblong,* petioled, thickish when fresh, *almost papery when dried; panicles with ascending racemes elongate* and nearly leafless, the whorls of flowers distant to subapproximate; *fruiting pedicels strongly reflexed, slenderly clavate, straightish and stiff, two to five times length of fruiting calyx; valves* rhombic- or deltoid-cordate, slightly longer to shorter than broad, with a narrowed tip, coarsely rugose-reticulate, *each with a long lanceolate grain.* (Incl. *R. floridanus* Meisn.) — Wet or inundated swamps, margins of streams and pools or in standing water, Fla. to e. Tex., n. to sw. Que., s. Ont., Mich., Wisc. and Minn. *Fr.* June–Sept. FIG. 952.

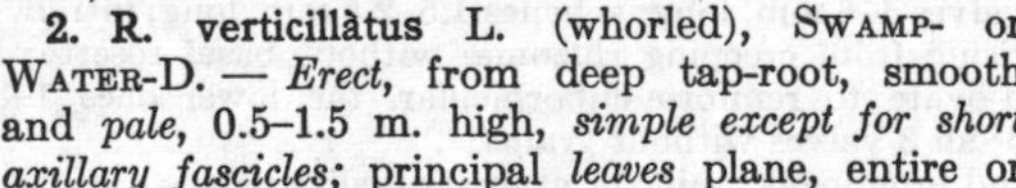

952. R. verticillatus.

3. R. mexicànus Meisn. (Mexican). — Upright, with numerous short to elongate branches; leaves linear-lanceolate to narrowly oblong, pale green or glaucous; *panicle* very dense, its *branches strict or strongly ascending;* pedicels shorter than or sometimes exceeding the *olive- to ruddy-brown* deltoid-ovoid *calyx; valves* 3.5–6 mm. long, *the tips much exceeding the* narrowly ellipsoid to subulate *brown grains;* achenes 1.7–2.3 mm. long. (Incl. *R. trianguli-valvis* (Danser) Rech. f.) — Rich soil (sometimes brackish), Nfld. and Côte Nord, Que., to B.C., s. to M.I., e. N.B. (as an adventive to N.E. and s. N.Y.), O., Ind., Ill., Mo., Tex., N.M., Mex. and Calif. *Fr.* July–Sept. FIG. 953.

953. R. mexicanus.

4. R. altíssimus Wood (tallest), PALE D. — Rather tall (1–2 m. high); *leaves ovate- or oblong-lanceolate,* acute, pale, thickish, obscurely veiny; the cauline 7–15 cm. long, contracted at base into a short petiole; racemes spike-like and panicled, nearly leafless; whorls crowded; *pedicels nodding, shorter than the fruiting calyx; valves broadly ovate* or obscurely

954. R. altissimus.

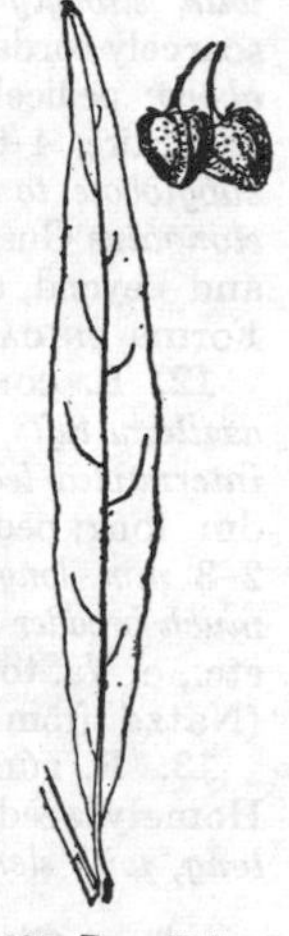
955. R. pallidus.

cordate, obtuse or acutish, entire, loosely reticulated, one with a conspicuous grain, the others with a thickened midrib or naked. — Alluvial or other rich soil, Ala. to e. Tex., n. to w. N.Y., O., Mich., Wisc., Minn., Neb. and Colo.; adv. e. to N.E., N.J., e. Va., etc. *Fr.* late May–Aug. FIG. 954. — Exceptional plants have 3 large grains.

5. R. pállidus Bigel. (pale), WHITE D., SEABEACH-D. — Depressed or ascending; root white; *leaves glaucous, narrowly lanceolate,* or the lowest oblong; the *lowest branches of the dense panicle spreading at nearly right angles;* pedicels much shorter than the *whitish-brown fruiting calyx; valves* deltoid-ovate, 3–4 mm. long, the *tips but slightly exceeding the conspicuous whitish* ovoid or lance-ellipsoid large *grains;* achenes 2–3 mm. long. — Saline or brackish marshes, beaches and rocks, Nfld. to lower St. Lawrence, Que., s. along coast to L.I. *Fr.* mid-June–Sept. FIG. 955.

§ SÍMPLICES Rech. f.

6. R. DOMÉSTICUS Hartm. (occurring about the house). — Erect, 0.2–1.5 m. high; rosette- and lower cauline leaves narrowly oblong or oblong-lanceolate, flat or slightly undulate; fruiting racemes erect, crowded into a dense elongate leafy panicle; *fruiting valves reniform-ovate, definitely as broad as long, with broadly rounded summit, 4–7 mm. long,* the wings coarsely veiny, the *midribs without grains or 1 of them rarely with a tiny grain;* achenes 2.3–3 mm. long. (*R. Patientia* in large part of ed. 7, not L.) — Fields, roadsides, waste places, etc., Nfld. to N.E., and less often to Wisc. *Fr.* late June–Oct. (Natzd. from Eu.) FIG. 956.

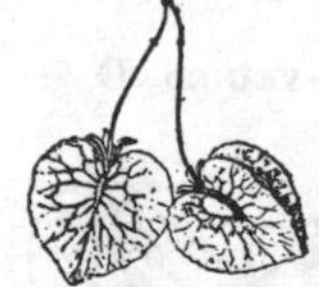
956. R. domesticus.

957. R. fenestratus.

7. R. fenestràtus Greene (with windows; from the reticulate valves). — Tall and upright, up to 2 m. high, commonly reddish-tinged; *rosette- and basal cauline leaves* flattish, lanceolate to lance-ovate, *tapering from rounded to cordate base to narrow tip;* the *blades mostly 0.5–2 dm. long* and 2–10 (rarely to 15) *cm. broad;* panicle dense, with erect racemes, in fruit red or pale pinkish-brown; *mature valves scario-membranaceous, broadly ovate to ovate-rotund,* entire or nearly so, reticulate-veiny, 6–9 mm. long, without grains or rarely with 1 poorly developed grain; achene 2–3 mm. long. (*R. occidentalis* of ed. 7, not S. Wats.) — Lab. Pen. and n. Man. to Alaska, s. in rich swales or on shores to Nfld., N.S., e. Me., James Bay, Sask., Mont. and n. Calif. *Fr.* Aug., Sept. FIG. 957.

8. R. PATIÉNTIA L. (the old colloquial name), PATIENCE or PATIENCE-D. — Resembling nos. 6 and 7; *rosette- and basal leaves broadly cordate-lanceolate or -oblong, 2–4.5 dm. long and* 0.7–1.8 *dm. broad,* plane or barely undulate; mature *valves broadly ovate,* 6–8.5 *mm. long, one of them with a plump grain,* the others without or with poorly developed grains; *achenes* 2.8–3.5 *mm. long.* — Waste places and rich thickets and roadsides, N.E. to Minn. and westw., s. to e. Pa., w. N.Y., Mo., Okla., etc., local. *Fr.* mid-June, July. (Natzd. from Eurasia)

9. R. ALPÌNUS L. (alpine). — *Stem arising from creeping rootstock,* without overwintering rosette; *leaves suborbicular, reniform or rounded-ovate,* with broadly rounded summit, *the larger ones 1–3 dm. broad;* panicle with crowded ascending racemes; *fruiting valves firm, cordate-ovate, acutish, 4–6 mm. long; grains wanting.* — Fields and meadows, local, w. N.S. and s. Me. *Fr.* June–Aug. (Adv. from Eu.)

10. R. orbiculàtus Gray (disk-shaped; from the valves), WATER-D. — Tall and stout (1–2 m. high); leaves oblong-lanceolate, rather acute at both ends, transversely veined and with obscurely erose-crenulate margins; the lowest, including the petiole, 3–6 dm. long, the middle rarely truncate or obscurely cordate at base; racemes upright in a large compound panicle, nearly leafless; whorls crowded; *pedicels obscurely jointed; valves orbicular or round-ovate,* very obtuse, obscurely cordate at base, *finely reticulated,* entire or repand-denticulate, *all grain-bearing.* (*R. Britannica* sensu ed. 7, not L.) — Wet meadows, swamps and shores, Nfld. to N.D., s. to N.S., N.E., N.J., Pa., O., Ind., Ill., Ia. and Neb. *Fr.* late June–Sept. FIG. 958.

958. R. orbiculatus.

11. R. CRÍSPUS L. (curled), YELLOW D. — Smooth, 0.9–1.6 m. high; *leaves* *with strongly wavy-curled margins, lanceolate,* acute, the lower truncate or scarcely cordate at base; *whorls crowded in prolonged wand-like racemes, leafless above;* pedicels with tumid joints; *valves round-cordate, obscurely denticulate* or entire, 4–6 mm. broad, mostly all grain-bearing; *the grains very plump, subglobose to ellipsoid, with rounded ends or sometimes narrower.* (Incl. *R. elongatus* Guss.) — Cult. and waste soil, old fields, etc., throughout our range and beyond, a common weed. *Fr.* June–Sept. (Natzd. from Eu.) Fig. 959. — Forma unicallòsus Peterm. (with one grain) has only 1 valve with a grain.

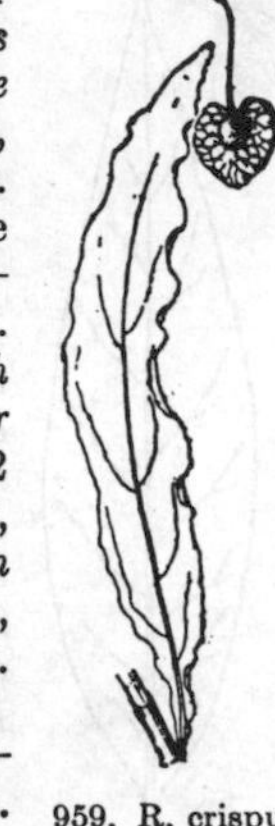
959. R. crispus.

12. R. conglomeràtus Murr. (clustered). — *Stems* 1–several, *often with axillary tufts, branched from toward base into many loosely ascending, slender interrupted leafy racemes; larger leaves oblong to narrowly oblong-ovate,* 0.6–2 dm. long; pedicels short; *outer sepals appressed-ascending; valves oblong, entire, 2–3 mm. long,* about 1.5 *mm. broad, each with a relatively large plump grain much broader than the wings.* — Roadsides, waste ground, ditches, open shores, etc., e. Va. to e. S.C., local; O., Ind. and Mich.; Pacific states. *Fr.* June, July. (Natzd. from Eu.)

13. R. púlcher L. (beautiful; name scarcely appropriate), Fiddle-D. — Homely weed similar to no. 12; the *slender stem diffuse or ascending, 2–7 dm. long, with slender widely divergent much interrupted racemes forming an open panicle up to three-fourths height of plant; leaves firm, pale green,* oblong, sometimes constricted and panduriform at base; *the larger ones 4–15 cm. long and* 1.5–6 *cm. broad, with entire or undulate margin;* flowers crowded, green, on *thick short pedicels articulated near middle; mature valves ovate, coriaceous, 4–6 mm. long,* 2.5–4.5 mm. wide, reticulate, *with long marginal teeth,* or in forma anodóntus Haussk. (without teeth) the teeth very short or obsolete; each valve with a plump tubercle. — Waste places, roadsides, etc., Fla. to Mex. and s. Calif., n. to Md., locally to L.I., Ark., Okla., etc. *Fr.* late May, June. (Natzd. from Eu.) Fig. 960.

960. R. pulcher.

14. R. obtusifòlius L. (blunt-leaved), Bitter, Blunt-leaved or Red-veined D. — Stem roughish, *erect,* 0.6–1.5 *m. high, with terminal panicle of upright racemes* with slightly remote whorls of flowers; *basal leaves broadly to narrowly ovate,* long-petioled, *membranaceous, cordate to subcordate at base, rounded or blunt at apex, usually red-veined; the larger 1–3 dm. long and 5–15 cm. broad, crenulate;* upper leaves oblong to lanceolate, acute; *valves* triangular-ovate, *submembranaceous,* 3–5 mm. long, *with prominent teeth,* only 1 valve (rarely all) with a plump grain. — Waste places, roadsides, shaded byways, etc., throughout our range and beyond. *Fr.* June–Sept. (Natzd. from Eu.) Fig. 961.

961. R. obtusifolius.

15. R. persicarioides L. (resembling *Persicaria*). — *Soft- and hollow-stemmed annual or biennial, with* 1–several depressed or ascending simple or divergently branching *stems 0.2–6 dm. high; basal leaves linear- to oblong-lanceolate, cordate or subcordate; larger cauline leaves similar,* with subcordate to cuneate bases, *flat,* somewhat undulate, short-petioled; glomerules many-flowered, the lower remote, the upper more approximate or contiguous, subtended by narrow leaves; *pedicels* filiform, *articulated near base,* slightly longer than mature calyx; *mature valves about 2 mm. long, 1 mm. wide, about as wide as length of marginal bristles; the plump ellipsoid-ovoid grains rounded at both ends, nearly covering breadth of valve,* whitish to creamy-yellow when fresh; achene about 1 mm. long. — Sandy beaches and borders of coastal marshes, very local, lower St. Lawrence, Que.; M.I.; P.E.I.; ne. N.B.; s. N.S.; ne. Mass.; thought by Linnaeus to have reached him from Va. *Fr.* Aug., Sept. Fig. 962.

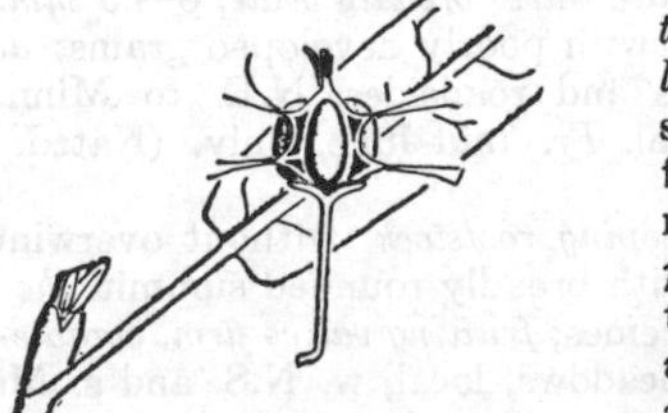

962. R. persicarioides.

16. R. marítimus L. (maritime). — Very similar to no. 15, often taller; *basal and lower cauline leaves flat, narrowly lanceolate, tapering gradually to base; tubercle linear-lanceolate, brownish or reddish;*

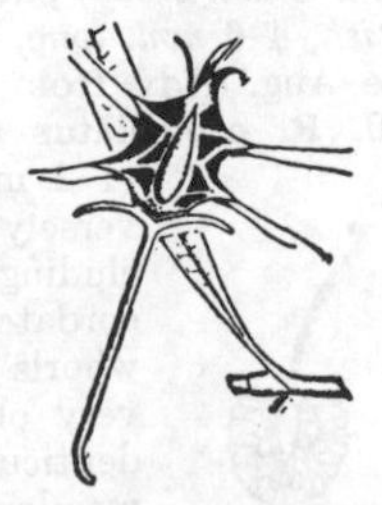
963. R. maritimus (typical).

marginal bristles once-and-a-half to twice as long as breadth of mature valve; achene 1.3–1.4 mm. long. — Waste places and cult. ground, infrequent, Me. to Pa. *Fr.* July, Aug. (Adv. from Eu.) Fig. 963.

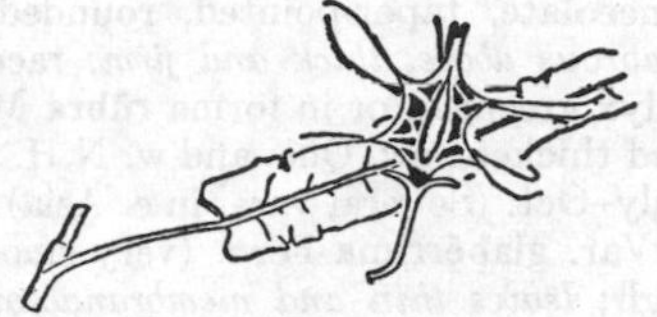

964. R. maritimus, v. fueginus.

Var. **fuegìnus** (Phil.) Dusén (of Fuegia), Golden D. — *Basal and lower cauline leaves broadly to narrowly lanceolate, with more or less crisped margin and cordate to truncate base;* fruiting racemes golden-brown. (*R. fueginus* Phil.; *R. persicarioides*, mostly, of ed. 7, not L.) — Saline, brackish or alkaline marshes and shores, and as a weed in disturbed soils, Anticosti I., Que. to B.C., s., mostly along coast, to L.I., and inland s. to w. N.Y., s. Ont., Ill., Ark., N.M., Ariz. and Calif. *Fr.* July–Oct. (S. and Andean S. Am.) Fig. 964.

Subgen. Acetosélla (Meisn.) Rech. f. (Sorrel)

17. R. Acetosélla L. (old generic name, meaning Little Sorrel), Sheep-S., Common S., Oseille (Que.), Surette (Que.). — *Spreading by slender running rootstocks;* the leafy tufts *with more or less hastate leaves*, with the 2 basal lobes somewhat divergent; *the terminal lobe lanceolate or oblanceolate and 2–12 mm. broad, subacute*, or in forma integrifòlius (Wallr.) G. Beck (entire-leaved) all or most of the basal leaves unlobed; *flowering stem 0.5–5 dm. high*, the slender reddish to yellowish racemes erect in panicles; flowers nodding on short jointed pedicels; *pistillate flowers with achene exserted from and soon divested of the calyx.* — Ubiquitous weed of worn-out fields and sour soils, difficult to eradicate (except by sweetening soil), throughout our range and beyond. *Fr.* June–Oct. (Natzd. from Eu.) Fig. 965.

Var. pyrenaèus (Pourret) Timbal-Lagrave (Pyrenaean). — *Coarser*, up to 7 dm. high; *basal hastate leaves with terminal lobe obovate or elliptic*, usually *blunt or rounded at apex*, 1.2–3 *cm. broad.* — Local, Que. to Va. and Kans. (Natzd. from Eu.)

965. R. Acetosella.

18. R. graminifòlius L. (grass-leaved). — *Tufted, without creeping rootstocks*, 0.5–2 dm. high; *leaves linear, erect, simple or rarely slenderly hastate; calyx twice as long as achene, persistent.* — Arct. reg., s., very locally, to coastal rocks of sw. Nfld. *Fr.* July–Oct.

Subgen. Acetòsa (Campd.) Rech. f. (Sorrel)

19. R. hastátulus Baldw. (slightly halberd-shaped). — *Stems* 1–several, *from tap-root*, erect, 0.15–1.3 m. high; *basal* long-petioled *leaves* lanceolate, oblanceolate or oblong, blunt, *with* 2 (or 4) *widely divergent basal lobes, or merely undulate;* upright panicle with many ascending racemes, leafless; *flowers dioecious, on slender recurving pedicels, about* 2.5 *mm. long;* the *valves broadly ovate*, without dorsal grains; *in fruit enlarged* and stramineous to pink, *reticulate* and 2.5–4 *mm. broad.* — Sandy soil, Fla. to Tex., n. near coast to N.C.; very local, s. N.J., e. L.I. and ne. Mass.; inland n. to s. Ill., Mo. and Kans. *Fr.* May–Aug. Fig. 966.

966. R. hastatulus.

20. R. Acetòsa L. (old generic name), Garden-S., Oseille (Que.). — Coarser, *tufted from tap-root; leaves oblong or broadly lanceolate, sagittate;* stem succulent, ribbed; *fruiting valves* 4–5 *mm. broad.* — Fields, meadows and roadsides, Greenl. and Lab. to Alaska and B.C., s. to Nfld., N.S., n. and w. N.E. and n. Pa. *Fr.* June–Sept. (Introd. and natzd. from Eu.) Fig. 967. — Old-fashioned (but still excellent) salad-plant.

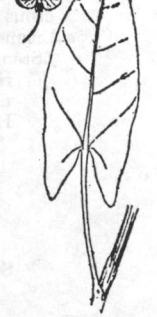

967. R. Acetosa.

4. TOVÀRA Adans. Jumpseed

Calyx unequally 4-parted, mostly herbaceous and greenish, persistent. Stamens 5. Style 2-parted to base, deflexed and hooked, persistent, becoming rigid. Achene lenticular, included in the close calyx; embryo marginal, curved around one side of the albumen. — Tall perennials of e. N. Am. and e. Asia, with knotty rhizome, the flowers remote in elongate and slender stiffish spiciform racemes, 2 or 3 in each fascicle, these soon deflexed; the greenish calyx not enlarged in fruit. (Named presumably for *Simon a Tovar*, Spanish physician of the 16th century.)

1. T. virginiàna (L.) Raf. (Virginian). — Rhizome thick, knotty, subligneous; stem terete, upright, 0.6–1.5 m. high; sheaths cylindric, hairy, ciliate-fringed; *leaves* ovate to broadly lanceolate, taper-pointed, rounded to base, usually pubescent beneath, *strigose and often scabrous above, thick and firm;* racemes 1–several, the leading one becoming 2–5 dm. long; calyx greenish, or in forma **rùbra** Moldenke (red) reddish. (*Polygonum* L.) — Rich woodlands and thickets, sw. Que. and w. N.H. to s. Ont. and Minn., s. to Fla., Ala., Miss., La. and e. Tex. July–Oct. (Several vars. in e. Asia)

Var. **glabérrima** Fern. (very smooth). — Slender, *from elongate cord-like rhizome,* 0.45–1 m. high; *leaves thin and membranaceous, glabrous or promptly glabrate.* — Rich low woods and wooded bottoms, se. Va. to Fla.

5. POLÝGONUM L. KNOTWEED. SMARTWEED. RENOUÉE (Que.)

Calyx 4–6 (mostly 5)-parted, the divisions often petaloid and brightly colored, enlarging, persistent and embracing the lenticular or 3-angled achene. Stamens 3–9. Styles or stigmas 2 or 3, deciduous. Embryo placed in a groove on the outside of the albumen and curved halfway around it, the radicle and usually the cotyledons slender. — Nearly cosmop. and highly variable genus of herbs (usually with us) or shrubs, belonging in several natural sections; flowers solitary, in fascicles in axils of leaves or bracts, or in spiciform panicles (here called *spikes*), or in panicles or corymbs of racemes. (Name from *poly, many,* and *gonu, knee* or *joint;* from the thickened joints of the stem.)* Including BISTORTA (L.) Adans., PERSICARIA Mill., TRACAULON Raf., TINIARIA Webb & Moq. and PLEUROPTERUS Turcz.

a. Flowers in axillary fascicles, or spicate, with foliaceous bracts; leaves and bracts jointed upon a very short petiole adnate to the short sheath (ocrea) of the finally 2-lobed or lacerate scarious stipules; calyx usually more or less herbaceous, mostly 5–6-parted; stamens 3–8, the 3 inner filaments broad at base; styles 3, achenes 3-angled; cotyledons incumbent; albumen horny; glabrous annuals (with us). § 1. AVICULARIA. . *b.*

b. Stem terete or nearly so, prostrate or ascending; flowers or fascicles of flowers subtended by and greatly overtopped by leaves, not forming partially naked spikes. . . *c.*

c. Achenes highly lustrous. . . *d.*

d. Foliage whitish-gray; sepals loosely ascending or spreading, oval or elliptic to obovate, their margins broadly petaloid; achenes trigonous-ovoid, dark, 3–5.3 mm. long; depressed (except when crowded) on maritime sands and beaches.

Fresh foliage strongly whitened; ocreae of lower nodes 7–10 mm. long; sepals obovate, pink-margined; achene blackish, 3–4 mm. long, its faces 1.6–2.2 mm. broad; species of Atl. coast southward. 1. *P. glaucum.*

Fresh foliage only slightly whitish; ocreae of lower nodes 4–8 mm. long; sepals oval, white-margined; achene dark olivaceous, 4.5–5.3 mm. long, its faces 3–3.5 mm. broad; plant of e. Can. 2. *P. Raii.*

d. Foliage bluish- to yellowish-green or fulvous; sepals appressed to achene, their margins slightly to scarcely petaloid; achenes greenish, olivaceous or brown. . . *e.*

e. Achene strongly exserted from calyx. . . *f.*

f. Stem and branches prostrate or loosely spreading, fulvous, scarcely ridged when fresh, flexible; leaves 1–4.5 cm. long; ocreae only tardily dissected; northeastern maritime plants.

Hardly succulent, with slender simple or subsimple remotely leafy branches; ocreae 1–1.3 cm. long, prominently nerved, their margins appressed; leaves linear-oblanceolate, acutish, 2–4.5 cm. long, submembranaceous, the lateral veins evident beneath; achene ovate-lanceolate, greenish, 6 mm. long, gradually tapering above. . . . . 3. *P. oxyspermum.*

Succulent, the fleshy stem with simple to much divided branches, their tips with approximate (often subopposite) leaves; ocreae 3–6 mm. long, flaring above, almost veinless; leaves fleshy, elliptic or narrowly elliptic-obovate, rounded at tip, scarcely veiny, 1–3 cm. long; achene broadly ovoid, olive-brown, about 4 mm. long, abruptly slender-tipped. . . . . . 4. *P. Fowleri.*

f. Stem and branches decumbent-ascending to erect, mostly 0.2–1 m. or more high, greenish, firm, strongly ridged; leaves green, with lateral nerves evident beneath, mostly 1–6 cm. long; ocreae soon lacerate into veiny shreds.

* Many characters, especially of sections, taken from A. N. STEWARD'S *Polygoneae of Eastern Asia,* Contrib. from Gray Herb., no. lxxxviii (1930); others from the studies of E. E. STANFORD in Rhodora, xxvii–xxix (1925–1927). Mature achenes are essential for identification.

Decumbent, with spreading, subascending or prostrate branches; internodes mostly shorter than leaves; leaves oval, oblong, narrowly elliptic or elliptic-obovate, blunt, the larger blades 0.5–2 cm. broad; mature calyx subrhombic, the sepals with conspicuous petaloid margins and prominent midribs; achene olivaceous, inequilateral, broadly ovoid, 3.5–4.8 mm. long and 2.5–3 mm. wide; northeastern maritime species. . . . 5. *P. allocarpum.*

Erect or strongly ascending, with upright branches; lower internodes mostly longer than leaves; leaves narrowly lanceolate or oblanceolate to nearly linear, the longer blades 2–10 mm. broad; mature calyx ovoid, the sepals often with inconspicuous pale margins, not prominently ribbed; achene equilateral; wide-ranging coastal and inland plants.

Achene lanceolate, pale green, 4–6.5 mm. long. . . . . . 6. *P. exsertum.*

Achene dark brown, the normal ones ovoid, included, only 2.2–3.5 mm. long; late proliferating ones elongate to 4 mm. and exserted. . . . . 8. *P. ramosissimum.*

e. Achene included within the calyx or its tip barely exserted.

Pedicels included within the sheathing ocreae.

Stem decumbent, with spreading, subascending or prostrate branches; leaves oval, oblong, narrowly elliptic or elliptic-obovate, the larger blades 0.5–2 cm. broad; achene 3.5–4.8 mm. long. . . . . 5. *P. allocarpum.*

Stem and branches ascending; leaves linear-oblong, the larger blades 1–5 mm. broad; achene 2 mm. long. . . . . . 7. *P. prolificum.*

Pedicels when mature exserted above the sheathing ocreae; stem and branches erect, up to 2 m. high; leaves narrowly lanceolate to oblanceolate or nearly linear, acutish; achene 2.2–4 mm. long. 8. *P. ramosissimum.*

c. Achene dull or barely sublustrous.

Stem scarcely ridged, flexible; leaves fulvous or warm green, long-petioled, oblong-obovate to spatulate, their petioles much exceeding subtending nerveless ocreae; upper crowded spatulate divergent leaves appearing subopposite; achene 3.5–4 mm. long; maritime boreal native species. . . . . 9. *P. boreale.*

Stem firm, ridged or corrugated; leaves rarely fulvous, sessile or very short-petioled; achene 2–3 mm. long; weedy widespread plants.

Leaves and margins of sepals yellowish-green; leaves elliptic, blunt; mature pedicels exserted from the fibrous ocreae. . . . . . 10. *P. erectum.*

Leaves blue-green; margins of sepals purplish, pink or white; pedicels all or mostly included.

Erect or decumbent-ascending; leaves crowded, elliptic to obovate, broadly rounded above; fruiting calyx 3.4–4 mm. long, strongly constricted below the subrostrate tip; inner sepals white-margined; achene olivaceous. . . . . 11. *P. achoreum.*

Depressed, loosely spreading or erect (especially when young or crowded); leaves scattered to approximate, linear to oblong or elliptic, acute or blunt; fruiting calyx 2–3 mm. long, not constricted and rostrate; sepals purple or pink; achene dark brown. 12. *P. aviculare.*

b. Stem and branches sharply angled, erect; flowers or fascicles of flowers mostly subtended by short bracts, thus forming slender spiciform terminal inflorescences.

Leaves strongly plicate; flowers erect. . . . . 13. *P. tenue.*

Leaves flat or their margins merely revolute; flowers nodding on recurving pedicels. . . . . 14. *P. Douglasii.*

a. Flowers fascicled in terminal dense to loose spiciform panicles or racemes without foliaceous bracts or in subcapitate or paniculate inflorescences; ocreae rarely hyaline or deeply lacerate; leaves not jointed to petiole. . . *g.*

g. Inflorescence elongate, spiciform; stems not prickly. . . *h.*

h. Stem simple (in ours), arising from a thickened more or less bulbiform contorted caudex, stiffly erect, with basal (as well as cauline) leaves and a single terminal spike. § 2. Bistorta.

Spike usually dimorphic; terminal flowers showy with petaloid calyx but sterile, lower half of spike bearing only ovoid bulbils. . . . . 15. *P. viviparum.*

Spike uniform, with showy fertile flowers.

Radical leaves attenuate at base to a long wingless petiole; achene rhombic-obovoid, cuneate to base. . . . . 16. *P. bistortoides.*

Radical leaves broadly rounded to cordate at base, the summit of the petiole broadly winged; achene broadly ovoid, rounded at base. . 17. *P. Bistorta.*

*h.*Stem (except in starved individuals) branching, from fibrous roots or slender stolons and rhizomes, leafy above the base; spikes terminating main axis and branches. § 3. PERSICARIA. . . *i.*

*i.*Ocreae (sheaths) nearly or quite without marginal cilia. . . *j.*

*j.*Perennial from elongate slender extensively creeping rhizomes and stolons, aquatic, amphibious, paludal or riparian. . . *k.*

*k.*Spikes 1–3 (or 4) straight, erect on terminal (or finally by elongation of leafy axis subterminal) peduncles; upper leaves scarcely reduced; wide-ranging northern and inland plants; aquatic or stranded or in inundated swamps.

Aquatic state: stems, branches and leaves floating, glabrous or glabrescent.

Peduncle glabrous; leading spike thick-cylindric, -ellipsoid or ovoid, 1–4 cm. long, 1–2 cm. thick when mature; ocreolae glabrous; floating leaves oblong or narrowly oval to oblong-lanceolate, glabrous on both surfaces, rounded to barely subcordate at base. . . . . 18. *P. amphibium.*

Peduncle pubescent; leading spike slenderly cylindric, 4–18 cm. long, 7–15 mm. thick when mature; ocreolae pubescent; floating leaves lanceolate to lance-ovate, glabrous or glabrescent, strongly rounded to cordate at base. . . . . . 19. *P. coccineum.*

Terrestrial state: stems or branches ascending; leaves pubescent or glabrous.

Ocreae with a broad horizontal herbaceous flange at summit. 18. *P. amphibium.*

Ocreae cylindric, without horizontal flange at summit.

Spike thick-cylindric, -ellipsoid or ovoid, 1–4 cm. long, 1–2 cm. thick; leaf-surfaces harshly scabrous with stiff strigose hispidity less than 1 mm. long; local adventive. . 18. *P. amphibium.*

Spike slenderly cylindric, the leading one 4–18 cm. long, 7–15 mm. thick; leaf-surfaces glabrous to pubescent with softer and longer trichomes; indigenous. . . . . . . 19. *P. coccineum.*

*k.*Spikes several–many in a forking terminal panicle, loosely ascending, linear-cylindric, 5–10 mm. thick; upper leaves much smaller than lower; southern or southeastern tall branching plant of swamps and borders of pools. . . . . 20. *P. densiflorum.*

*j.*Annual (or no. 21 sometimes perennial) without elongate rootstocks and stolons, not aquatic; spikes (in well developed plants) paniculately or subcorymbosely disposed, the bracteal leaves much smaller than the primary foliage; achene lenticular.

Style long-exserted and stamens included in some plants, style included and stamens exserted in others; achene convex on both faces; southwestern. . . . . 21. *P. longistylum.*

Style and stamens usually included, flowers uniform or, if dimorphic, occurring together; one or both faces of achene concave; wide-ranging or northern.

Peduncles and axis of inflorescence with obvious divergent stalked glands or strigae (glabrous in a var. in n. Ohio); spikes thick, erect, mostly 1–5 cm. long, pink to purplish (or white); achene 2.2–3.5 mm. broad. . . . . 22. *P. pensylvanicum.*

Peduncles and axis of inflorescence glabrous or with inconspicuous sessile capitate glands; spikes erect to drooping, 1–8 cm. long, purplish, pink or green; achene 1.5–2.8 mm. broad.

Spikes pink or purplish, paniculate, erect, arching or pendulous, on definite glabrous to sparsely glandular peduncles, 1–8 cm. long; mature calyx ovoid to rhomboid, constricted toward summit to form a thick beak overtopping achene; achene 1.8–2.2 mm. long, 1.5–2 mm. broad. . . . 23. *P. lapathifolium.*

Spikes green (exceptionally purplish), subcorymbosely paniculate or sessile on spiciform erect branches, erect or nearly so, 1–5 cm. long; axis copiously glandular; mature calyx round to broadly ovoid, gradually rounded at summit, about equaling or slightly shorter than achene; achene 2.5–3.5 mm. long, 2.3–2.8 mm. broad. . . . . 24. *P. scabrum.*

i. Ocreae (sheaths) with summit bearing well-developed cilia or bristles. . . *l.*

l. Summit of ocreae expanded into a horizontally divergent herbaceous flange.

Perennial by long tough rootstocks; leaves lanceolate, short-petioled; spikes 1–4 on erect glabrous peduncles, 1–4 cm. long; native of marshy habitats. . . . . 18. *P. amphibium*, var. *stipulaceum.*

Annual with fibrous roots; leaves ovate to broadly oblong, long-petioled; spikes few–many, paniculate, slenderly cylindric, arching to drooping, 2–10 cm. long, on long closely pubescent peduncles; garden-escape. . . . . 25. *P. orientale.*

l. Summit of ocrea without spreading herbaceous flange. . . *m.*

m. Annual with fibrous roots and without elongate subligneous rootstock or autumnal basal offshoots. . . *n.*

n. Summit of stem and peduncles bearing abundant gland-tipped hairs; leaves scabrous-hispid; spikes slenderly cylindric, arching to pendulous. . . . . 26. *P. Careyi.*

n. Summit of stem and peduncles not obviously glandular; leaves mostly smooth or nearly so. . . . *o.*

o. Mature (at least dry) calyx punctate or dotted with glands (under magnification); spikes slender, arching, usually much interrupted at base, the lowest flowers commonly extending down to upper leaf-axils.

Calyx greenish or with purple tips; achene dull or opaque. 27. *P. Hydropiper.*

Calyx white; achene lustrous. . . . . 33. *P. punctatum.*

o. Mature calyx not dotted; spikes lax or dense (at least above), at least the longer ones slender-peduncled. . . . *p.*

p. Spikes slender, 3–5 mm. thick, sparsely flowered; leaves without dark mottling.

Cilia of ocreolae much shorter than sheath; achene lenticular or trigonous, 2–3 mm. long.

Leaves membranaceous, oblong or oblanceolate, cuneate at base; spikes interrupted at base, the upper leaves often subtending remote lower groups of flowers; calyx whitish, greenish or pinkish, in fruit 2.5–3.5 mm. long; achene 2.5–3 mm. long. . . 28. *P. dubium.*

Leaves firm, lance-linear, attenuate to base and apex; spikes subcontinuous; calyx claret-purple, in fruit 2–2.5 mm. long; achene 2–2.4 mm. long. . . . . 29. *P. minus*, var. *subcontinuum.*

Cilia of ocreolae bristleform, equaling or longer than sheath; spike subcontinuous; calyx roseate or purplish; achene trigonous, about 2 mm. long. . . . . . 30. *P. cespitosum*, var. *longisetum.*

p. Spikes dense, thick-cylindric or ellipsoid, 4–11 mm. thick; calyx pink or purple (rarely greenish or white); leaves firm, often with a dark blotch on upper surface.

Mature primary spikes 7–11 mm. thick; calyx in anthesis 2–3 mm. long, when mature the sepals with prominently reticulate bases; achene 2.5–3 mm. long; leaves often purple-blotched; ubiquitous weed. . . . . 31. *P. Persicaria.*

Mature primary spikes 4–6.5 mm. thick; calyx in anthesis 1.8–2 mm. long, when mature scarcely or barely reticulate; achene 2 mm. long; leaves mostly green above; plant of sandy or gravelly shores, N.S. and s. N.E. . . . . 32. *P. puritanorum.*

m. Perennials with subligneous forking and elongate rootstocks, often producing perennating autumnal basal offshoots. . . . *q.*

q. Mature (at least dry) calyx punctate or with glandular dots.

Spikes interrupted at base, 2–many ocreolae and their fascicles distant from those above; ocreolae with few cilia much shorter than sheath or eciliate; sepals white; fruiting calyx 3.5–5 mm. long, copiously dark-punctate or glandular; achene included, trigonous or lenticular, 3–4 mm. long; leaves lanceolate to lance-oblong or oblong, 0.6–4.5 cm. broad.

Rootstocks and rooting branches 2–5 mm. thick; ocreae glabrous, fragile; leaves 0.6–3.5 cm. broad; spike loose and open except at summit; ocreolae mostly distant or subdistant, ciliate; fruiting calyx 3.5–4 mm. long; achene 3–3.5 mm. long, about 2 mm. broad, lenticular or trigonous. . . . . 33. *P. punctatum.*

Rootstocks and rooting branches 3–8 mm. thick; ocreae strigose, firm; leaves mostly 2–4.5 cm. broad; spike continuous except at base; ocreolae contiguous, the truncate summit eciliate; fruiting calyx 4–5 mm. long; achene 3.5–4 mm. long, 2–3 mm. broad, trigonous. . . . . 34. *P. robustius.*

Spikes continuous, rather dense, except for occasionally remote lower fascicles; bristly cilia of ocreolae numerous, one-half to two-thirds length of sheath; sepals dull purple or greenish; fruiting calyx 1.8–3 mm. long, with whitish glands; achene with exserted tip, 1.5–2.5 mm. long, 1.2–2 mm. broad, trigonous; leaves narrowly lanceolate to almost linear-attenuate, mostly 0.5–2 cm. wide. . . . . . 36. *P. opelousanum.*

*q.*Mature calyx without glandular dots or punctation. . *r.*

*r.*Aquatic or of wet shores and swales, with decumbent slender rooting bases and loosely ascending mostly simple slender fertile branches, smooth or rarely scabridulous, 1–4 mm. thick above rooting base; leaves glabrous or merely scabrous, linear to lanceolate, 0.5–2.5 cm. wide; ocreae short-strigose, ciliate with bristles 1–7 mm. long or eciliate; calyx pink, purplish or greenish, rarely white, in maturity 1.8–4 mm. long.

Spike bright pink, roseate or rarely white; fruiting calyx elongate-ovoid, covering tip of achene; achene 2–3 mm. long; leaves narrowly to broadly lanceolate. . . . . 35. *P. hydropiperoides.*

Spike dull purple to green; fruiting calyx thick-rhomboid to subglobose-ovoid, scarcely covering achene; achene slightly exserted, 1.5–2.5 mm. long; leaves narrowly lanceolate to linear. . . . . 36. *P. opelousanum.*

*r.*Mesophytic or paludal, chiefly of low woods, thickets, swamps and clearings; stem erect from horizontal rhizome, 2.5–6 mm. thick toward base, often branching from median nodes; leaves lanceolate, commonly strigose (smooth in one var.), the primary ones 1.2–3.5 cm. wide; ocreae long-strigose or spreading-hirsute, fringed with bristles 5–15 mm. long; calyx white, in maturity 3–3.5 mm. long. . . . . . . 37. *P. setaceum.*

*g.*Inflorescence of more or less open panicles, the stems not prickly and the leaves broad and mostly cordate to truncate at base; or inflorescence capitate to racemose, the weak stems reflexed-prickly, the leaves sagittate or hastate. . . *s.*

*s.*Prickly-stemmed weak annuals with sagittate or hastate leaves; flowers capitate to short-racemose. § 4. Echinocaulon.

Leaves sagittate, the basal lobes directed back; flowers in subspherical heads; common style elongate, 3-parted; achene sharply 3-angled. . 38. *P. sagittatum.*

Leaves hastate, the basal lobes horizontally spreading; flowers few in short racemes; common style nearly wanting, 2-parted; achene lenticular. . . . . 39. *P. arifolium.*

*s.*Prickleless; plants either slender and (when well developed) twining or coarse, erect and somewhat shrubby; leaves rounded, ovate or broadly lanceolate, with cordate to truncate or cuneate bases; flowers paniculate. . . *t.*

*t.*Leaves roundish to ovate, cordate to truncate or cuneate at base; 3 sepals usually enlarged in fruit and keeled or expanded into wings; plants slender and twining or upright and bushy-branched. § 5. Tiniaria. . . *u.*

*u.*Slender, usually twining, primary leaves cordate to rounded at base. . . *v.*

*v.*Bases of sheaths (ocreae) mostly retrorsely barbed (rarely beardless); inflorescences consisting of axillary and terminal panicles of racemes; calyx white, the sepals obscurely keeled, not winged in fruit, on long slender pedicels; achene shining and smooth. . 40. *P. cilinode.*

*v.*Bases of sheaths beardless; inflorescences of axillary and terminal clusters and (or) simple racemes of greenish to yellowish flowers and green, yellowish or roseate fruiting calyx. . . *w.*

*w.*Outer sepals keeled, barely or hardly winged in fruit; flowers short-pedicelled, in fruit 3.5–4.5 mm. long, greenish; achene granular, opaque, 3–4 mm. long; weedy annual. . . . . 41. *P. Convolvulus.*

*w.*Outer sepals winged in fruit; flowers long-pedicelled, in fruit becoming 4–10 mm. long; achene smooth and shining; native perennials.

Mature fruiting calyx 4–6 mm. long, with herbaceous greenish wings 0.25– barely 1 mm. broad; achene 2–3.5 mm. long and 1.5–2 mm. broad; fruiting racemes 1–1.5 cm. thick. . 42. *P. cristatum.*

Mature fruiting calyx 8–10 mm. long, with thin scarious yellowish-brown to roseate wings 1.5–3 mm. broad at summit; achene 3.5–6 mm. long, 2.5–4 mm. broad; fruiting racemes 2–3 cm. thick. . . . . 43. *P. scandens.*

*u.*Erect coarse and tall bushy-branched perennials; primary leaves cordate or truncate to cuneate at base; spread from cult.

Leaves truncate to cuneate at base, abruptly pointed, the larger ones 0.5–1.5 dm. long. . . . . 44. *P. cuspidatum.*

Leaves mostly cordate, gradually tapering or arching to tip, the larger ones 1.5–3 dm. long. . . . . 45. *P. sachalinense.*

*t.*Leaves oblong-ovate to broadly lanceolate, with short rounded auricles at base; coarse, shrubby, with subterminal panicles of white flowers; sepals not enlarged in fruit. § 6. Aconogonon; spread from cult. . 46. *P. polystachyum.*

§ 1. Aviculària Meisn. Knotweed. Knotgrass

1. P. glaùcum Nutt. (blue-green), Seabeach-K. — Depressed or prostrate (except when young or crowded) annual with firm stems and branches; *ocreae* silvery, those *of lower nodes* 7–10 *mm. long; leaves whitish-gray or -green, linear-oblong, firm,* crowded along the branches; *sepals broadly obovate, petaloid, pink, loosely ascending or slightly spreading; achene exserted, lustrous, becoming blackish,* 3–4 *mm. long,* the *faces* 1.6–2.2 *mm. broad.* (*P. maritimum* sensu Ell.. and later Am. botanists, not L.) — Sandy seabeaches, saline pond-shores and dune-hollows, local, Mass. to Ga. Late July–Nov. Fig. 968.

968. P. glaucum.

2. P. Raìi Bab. (for John Ray, 1627–1705, pioneer English botanist). — Looser and *less whitened than no.* 1; *longer ocreae only* 4–8 *mm. long; leaves lance-oblong to lance-linear,* flat, *not* specially *crowded; sepals oval, with white petaloid spreading summits; achenes* ovoid, *lustrous, exserted, olivaceous,* 4.5–5.3 *mm. long, the faces* 3–3.5 *mm. broad.* — Sandy or gravelly seabeaches, strands or dune-hollows, Nfld., M.I. and ne. N.B., s. along coast to Sable I. and sw. N.S. Late July–Sept. (Nw. Eu.) Fig. 969.

969. P. Raii.

970. P. oxyspermum.

3. P. oxyspérmum Mey. & Bunge (sharp-fruited). — *Loosely divergent-branched* and prostrate to subascending, *with slender flexible scarcely furrowed stem and simple or subsimple elongate branches, the internodes elongate; longer ocreae* 1–1.3 *cm. long, appressed, prominently nerved; leaves membranaceous, linear-oblanceolate, acute, green,* scarcely glaucous, the larger ones 2–4.5 cm. long and 2–7 mm. wide, petioled; *flowers* 1–3 *in remote axils,* about 4 mm. long, turbinate-campanulate; *the oblong obtuse sepals erect, appressed, roseate- or whitish-margined; achenes shining, exserted, olivaceous, ovoid-lanceolate,* about 6 *mm. long and* 2–3 *mm. broad.* (*P. acadiense* Fern.) — Sandy and gravelly strands, Bras-d'Or Lakes, C.B. Late July–Sept. (Baltic reg. of Eu.) Fig. 970.

4. P. Fowleri Robins. (for its discoverer James Fowler, 1829–1923). —

Loosely ascending to prostrate, finally much branched, *succulent, fulvous or warm green; stems scarcely furrowed, flexible; ocreae* scarious, *flaring at summit, the larger ones 3–6 mm. long, almost veinless; leaves petioled, fleshy, elliptic to narrowly elliptic-obovate, broadly rounded above, scarcely veiny,* 1–3 cm. long and 3–15 mm. broad; *the upper bracteal ones crowded, divergent and often appearing subopposite; sepals oblong,* round-tipped, *with roseate margins; achenes broadly ovoid, lustrous, exserted, olive-brown, about 4 mm. long, abruptly slender-tipped.* — Sandy or gravelly seashores, se. Lab. and Côte Nord to lower St. Lawrence, Que., s. to Nfld. and outer coast of N.S. Mid-July–Sept. Fig. 971.

971. P. Fowleri.

5. P. allocárpum Blake (with different fruit). — *Decumbent, with* spreading, subascending or prostrate *firm, corrugated branches,* with long lower internodes; *ocreae soon lacerate into scarcely veiny shreds up to 1 cm. long; leaves thin, blue-green, oval, oblong, narrowly elliptic or elliptic-obovate, blunt to acute,* short-petioled, *the larger ones* 1.5–6 cm. long and 0.5–2 *cm. broad;* pedicels of the 1–4 axillary flowers included; *calyx green, in maturity* 3.7–4.5 *mm. long, subrhombic,* with the *green oblong appressed round-tipped sepals* conspicuously *roseate- or white-margined,* their backs *with strong midribs; achenes broadly ovoid, shining,* exserted or included, *inequilateral, olivaceous,* 3.5–4.8 *mm. long and* 2.5–3 *mm. broad.* — Sandy, gravelly or rocky seacoast, Nfld. and Anticosti I. to lower St. Lawrence, Que., s. to coast of Me.; shores of James Bay. Late July–Oct. Fig. 972.

972. P. allocarpum.

6. P. exsértum Small (exserted or thrust out). — *Erect, with strongly ascending branches, up to* 1.5 *m. high,* the strong stem terete, *the lower internodes elongate; ocreae soon becoming brown, deeply cleft and lacerate into veiny shreds; leaves lanceolate to lance-linear or the earlier often oblanceolate,* pale or yellowish-green, *the longer primary ones* 3–6 *cm. long,* petioled; flowers 1–4, their *pedicels rarely exserted* beyond the ocreae; *calyx slenderly campanulate,* 3.5–4 mm. long, *the green herbaceous oblong sepals closely appressed to the base of the long-exserted achene; achene lustrous, lanceolate, pale green,* 4–6.5 *mm. long.* — Saline, brackish or other alkaline soils, along the coast from P.E.I. and N.B. to sw. Que., s. along coast, locally to N.J.; prairies and shores, Ind. to Sask. and Alta., s. to Ill., Mo. and Neb. Late July–Oct. Fig. 973.

973. P. exsertum.

7. P. prolíficum (Small) Robins. (prolific). — Erect or ascending, *with numerous upcurving branches* 1.5–8 dm. high; *leaves bluish-green, linear-oblong, firm,* blunt to acutish, *the larger ones* 1–5 *mm. broad; pedicels included within the* soon lacerate *sheathing ocreae; sepals* 1.3–1.8 *mm. long, oblong,* green, *with narrow white or roseate margins,* closely appressed to the *included achene; achene lustrous,* ovoid, *about* 2 *mm. long.* — Saline or brackish marshes or shores, s. Me. to Va.; prairies, waste places, etc., Minn. to Sask. and Wash., s. to Ark., Okla. and e. Tex. July–Nov. Fig. 974.

974. P. prolificum.

8. P. ramosíssimum Michx. (much branched), Bushy Knotweed. — *Erect* and usually with many simple or bushily forking firm upright corrugated branches, *up to 2 m. high, the whole plant yellowish-green;* leaves narrowly lanceolate to oblanceolate or linear, acute; flowers 1-several in the upper axils; *pedicels soon elongating and exserted* out of the sheathing ocreae; calyx green, campanulate, about 3 mm. long, *the oblong sepals yellowish-green with yellow margins; achene trigonous-ovoid, lustrous, dark brown, included,* 2.2–3.5 *mm. long, or by proliferation late in the season prolonged to 4 mm. and exserted.* — Sandy and light soils, shores, roadsides, etc., N.E. and sw. Que. to Wash., rare eastw., common inland, s. to Del., Pa., O., Ind., Ill., Mo., Okla., Tex. and N.M. July–Oct. Fig. 975. — Eastw. more generally represented by forma **atlánticum** Robins. (Atlantic), with sepals roseate-margined, not drying yellowish (*P. atlanticum* (Robins.) Bickn.), chiefly coastal, s. Me. to Del., inland to Minn. and Ia.

975. P. ramosissimum.

9. P. boreàle (Lange) Small (northern). — Slightly *fleshy, fulvous or warm green; stems slender, flexible, scarcely corrugated, loosely ascending to prostrate,* they or the divergent branches 0.5–4 dm. long; *ocreae nearly nerveless, loosely flaring upward,* tardily lacerate; *leaves long-petioled, oblong-obovate to spatulate,* the lower blades 1–3 cm. long and 3–18 mm. broad; *the bracteal* ones smaller, *widely divergent and by crowding of flowering nodes appearing subopposite;* calyx about 4 mm. long, the *broad sepals with white to pink petaloid margins; achenes ovoid, opaque,* brown, *3.5–4 mm. long.* (*P. islandicum* Meisn. ex Small) — Greenl. and Lab. to sandy, gravelly or rocky coast of n. Nfld. and Côte Nord, Que.; shores of Hudson Bay; Alaska. Aug.–Oct. (N. Eu.) Fig. 976.

976. P. boreale.

10. P. eréctum L. (erect). — Rather *stout, erect or ascending,* usually bushy-branched, 0.2–1 m. high, *the hard stem corrugated; leaves yellowish-green, firm, elliptic, blunt, the larger primary ones* 2.5–7 cm. long and 1–2.5 *cm. broad,* the rameal and bracteal ones smaller and more approximate; *ocreae soon fibrous-lacerate, deciduous; mature pedicels soon exserted; mature calyx yellow-green,* herbaceous, *about 3 mm. long,* rhomboid; its oblong blunt sepals appressed, with yellow-green to whitish margin; *achene broadly ovoid, dull,* brown, *included, about 3 mm. long.* — Disturbed soil, waste places, etc., often weedy, sw. Que. and n. Vt. to Wisc. and Ia., s. to s. Me., s. N.E., L.I., Ga., Tenn., Mo., and Kans. Aug.–Oct. Fig. 977.

977. P. erectum.

11. P. achòreum Blake (without a native land). — Resembling in habit no. 10, *blue-green,* more compact, 1–3.5 dm. high; *ocreae* bright white, *nonfibrous, persistent; leaves broadly rounded above,* the larger ones up to 3 cm. long and 1.5 cm. wide, the rameal approximate; *pedicels included* in the sheathing ocreae; *mature calyx 3.5–4 mm. long, 2–2.5 mm. wide, bluish-green, ovoid-fusiform* or *subrhombic, strongly contracted below the subrostrate tip; sepals unequal, the 3 outer with* or without narrow pale margins and with *cucullate tips, hiding the 2 short white-margined inner ones; achenes inequilaterally trigonous, dull,* included or barely exserted, *olivaceous, about* 2.5 *mm. long and* 1.8 *mm. wide.* — Disturbed soil, roadsides, saline marshes, etc., n. Nfld. to Alaska, s. to N.S., Vt., N.Y., Ont., Mich., Wisc., Mo., S.D., Mont. and Ida. July–Sept. Fig. 978.

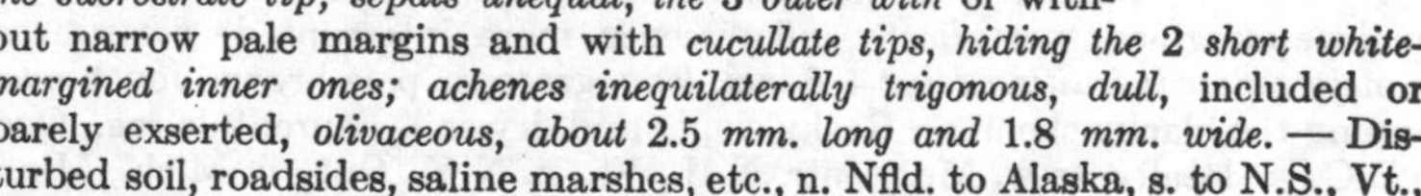

978. P. achoreum.

12. P. AVICULÀRE L. (pertaining to birds, which eat the young leaves and the achenes),* KNOTWEED, TRAÎNASSE (Que.), HERBE À COCHONS (Que.). — Prostrate or loosely ascending to upright weedy annual, with the main stem corrugated; *leaves blue-green,* linear to oblong or elliptic, scattered to approximate; *pedicels soon exserted beyond the hyaline and flaccid soon torn ocreae; fruiting calyx 2–3 mm. long, with roseate to purple sepals scarcely 2 mm. long; achenes dull or barely sublustrous, dark brown, 2–2.5 mm. long, included* or the tip barely exserted. — Heteromorphous and semicosmop. weed, with a multitude of variations. Three vars. recognized:

Leaves relatively thin, the larger ones of the primary axis and branches 1–4 cm. long, linear to lanceolate, oblanceolate or narrowly elliptic; achenes dull, minutely granular-striate.

Depressed or loosely spreading to ascending; leaves linear to narrowly lanceolate, the larger ones 1.5–5 mm. broad, their margins flat. . . . . . . *P. aviculare* (typical).

Loosely ascending to erect, with relatively few ascending branches; larger leaves oblong-lanceolate to narrowly lance-elliptic, 5–15 mm. broad, their margins often crinkled. . . . . Var. *vegetum.*

Leaves subcoriaceous or thick and firm, oblong to oblong-lanceolate, the larger ones 1–3 cm. long, obtuse; stem and branches prostrate or spreading; achenes dull to sublustrous, minutely puncticulate. . . . . Var. *littorale.*

---

* P. AVICULARE is the historically earliest name for a vast series of apparently apomictic variations: some of them treated in Europe as true species, again as subspecies, varieties, races or forms. Many of these aggressive and rather nondescript plants wander to our waste lots and dumps, thence to spread. Their proper elucidation is impossible without time-consuming and most cautious study and reference to the Old World types and treatments.

P. AVICULÀRE (typical), incl. var. *angustissimum* Meisn., *P. neglectum* Bess. and *P. heterophyllum* Lindm. f. — Disturbed soil, roadsides, gardens, etc., ubiquitous throughout our area and much beyond. June–Nov. (Natzd. from Eu.) FIG. 979 (branch).

979. P. aviculare (typical) and leaf of v. vegetum.

Var. VÉGETUM Ledeb. (vigorous), incl. var. *monspeliense* (Pers.) Aschers. and *P. aequale* Lindm. f. — Similar habitats, throughout. FIG. 979 (leaf).

Var. **littoràle** (Link) W. D. J. Koch (of the shore), incl. *P. littorale* Link and *P. buxiforme* Small — Seashores, salt-marshes and alkaline habitats inland, also in rich non-saline soils, Nfld. to Alaska, s., especially along coast, to S.C., and inland s. to interior N.Y., Tenn., Ark., Okla., etc. Late July–Nov. FIG. 980.

980. P. aviculare, v. littorale.

13. P. ténue Michx. (slender). — *Stem* slender, erect or ascending, *angled*, with strongly ascending branches, 0.5–4.5 dm. high; *leaves* narrowly lanceolate to linear, firm, nearly erect, *plicate, furrowed on each side of the midrib*, acute at each end; *flowers 1 or 2 in axils of greatly reduced bracteal leaves, thus forming slender interrupted terminal spiciform inflorescences, erect, subsessile or short-pedicelled; fruiting calyx* 3–3.5 *mm. long, with subequal ovate tapering* pale to roseate *sepals tightly appressed over the inclosed achene;* achene dark brown to dull black, 2.5 mm. long. — Dry open soil, s. Me. to s. Minn., s. to Ga., Ala., Ark. and Okla. Late June–Oct. FIG. 981.

Var. **protrùsum** Fern. (protruding). — *Fruiting calyx only* 1.5–2 *mm. long, the round-ovate outer sepals exceeding the inner ones; black achene strongly exserted.* — Dry sands, local, se. Va.

981. P. tenue.

14. P. Douglásii Greene (for its discoverer, DAVID DOUGLAS, 1798–1834). — Similar to no. 13, more loosely branched, 1–7.5 dm. high; *leaves flat* or merely revolute-margined, *not plicate;* inflorescence more interrupted; *flowers soon strongly reflexed; fruiting calyx* 4–6 *mm. long*, green or pale brown, much overtopping the black achene. — Rocky slopes and dry soil, Laurentide reg., Que., to B.C., s., locally, to w. Me., centr. N.H., Vt., n. N.Y., Ont., n. Mich., Minn., Okla., N.M. and Calif. July–Sept. FIG. 982.

982. P. Douglasii.

§ 2. BISTÓRTA L. BISTORT

15. P. vivíparum L. (with young well developed on the parent-plant), ALPINE B. — *Caudex or rhizome thick*, somewhat farinaceous, *short* and usually contorted; *radical leaves* (except the often broad lowermost) slender-petioled, *lance-oblong, lanceolate or lance-linear, acute or subacute, attenuate or narrowed to base*, firm, lustrous above, pale and glabrous or minutely pilose beneath, 2–10 cm. long, 0.2–2.5 cm. broad, the margin above strongly lineolate; lower cauline leaves similar, shorter-petioled to sessile, upper much reduced; flowering stem glabrous, erect, 0.3–4.5 dm. high; *ocreae brown, scarious, open down one side; spike* 2–10 cm. long *with the lower flowers replaced by ovoid bulbils* (or young leafy plantlets), the *upper flowers sterile and with pink or white petaloid calyces*, or in forma **florígerum** G. Beck (bearing flowers) all axils floriferous, or in forma **bulbígerum** G. Beck (bearing bulbs) the whole spike bulbiferous. (*Bistorta* S. F. Gray) — Arct. reg., s. on cool or damp slopes, gravels or wet rock to Nfld. and St. P. et M., e. Que., Thunder Bay Distr., Ont., n. Mich. and Minn.; and subalpine and alpine areas to Que., n. N.E. and N.M. Late June–Aug. (Eurasia)

Var. **alpìnum** Wahlenb. (alpine). — *Lower leaves oblong to oblong-ovate, blunt, rounded or cordate at base*, 1.5–7 cm. long, 1–3 cm. broad. — Côte Nord and Gaspé Pen., Que.; Alaska to Wyo. (Eurasia)

16. P. bistortoìdes Pursh (resembling *P. Bistorta*). — Coarser than no. 15; *basal leaves lanceolate, oblanceolate or lance-oblong, attenuate to the slender petiole*, 1–2.5 *dm. long*, 2–5 cm. broad, *with very broad midrib; stem* 0.25–1 *m. high; spike floriferous throughout, fertile, thick-cylindric*, 1–6 cm. long, 1–1.5 *cm. thick; calyx* pink or white, 5–6 *mm. long; achene pale brown, lustrous, rhombic-obovoid, cuneate at base*, 3.5–4 mm. long. — Along rills in subalpine meadows, sw. Nfld. (*Pease*); Alaska to N.M., Ariz. and Calif. July–Sept.

17. P. BISTÓRTA L. (old generic name, double-twist, from the often sigmoid rhizome)

BISTORT of Europe. — As coarse as no. 16 or coarser; *radical leaves narrowly ovate, subtruncate to cordate at base,* 0.6–2 dm. long, the long *petiole broadly winged at summit;* spike much as in no. 16, pink; *achene dark brown, trigonous-ovoid, rounded at base.* — Local weed of fields, meadows and cult. ground, N.S. and e. Mass. May, June. (Adv. from Eu.)

§ 3. PERSICÀRIA L. SMARTWEED

18. P. AMPHÍBIUM L. (amphibious), WATER-S. — *Perennial by slender and tough forking rhizomes, stolons and rooting stems;* branches elongate, simple, subuniformly leafy to summit; the *leaves elliptic-oval to lanceolate,* petioled; ocreae cylindric, with or without spreading and ciliate summit; *spikes* 1–4, *dense, straight, erect on erect* or *terminal* (or by prolongation of axis subterminal) *glabrous peduncles, thick-cylindric, -ellipsoid or ovoid, the leading one 1–4 cm. long and* in maturity 1–2 *cm. thick;* calyx pink or roseate, in anthesis with spreading blunt sepals, in fruit closed; stamens and styles dimorphic, some flowers with the 2-parted styles 3–4 mm. long and stamens 1–2 mm. long, thin, anthers mostly shrunken and included, others with stamens 4–6.5 mm. long, with well developed exserted anthers and style about 3 mm. long; achene lenticular, broadly ovoid to orbicular, 2.5–3 mm. long. — Highly variable and plastic circumboreal and S. Afr. species. The following varieties and ecological forms recognized with us:

a. Aquatic forms; leaves floating, mostly on long slender petioles, the blade elliptic-oblong to oblong-lanceolate, smooth on both sides; ocreae without horizontal herbaceous flanges.
- Leaves oblong-lanceolate, tapering gradually from near or below middle to apex, broadly rounded to cordate at base; mature spikes 1–1.5 cm. thick; ocreolae deltoid or rounded-rhombic. . . . . *P. amphibium* (typical).
- Leaves elliptic to elliptic-oval, gradually tapering or rounded at base, rarely subcordate; mature spike 1.5–2 cm. thick; ocreolae elongate-triangular, acutish. . . . . Var. *stipulaceum,* f. *fluitans.*

a. Terrestrial or stranded forms; leaves lanceolate, spreading to ascending, short-petioled; the young stems, ocreae and leaves glabrous to strongly pubescent. . . *b.*

b. Margins of new ocreae bearing an herbaceous horizontally divergent often bristly-ciliate flange.
- New growth of stem and surfaces of leaves glabrous or nearly so. . . . Var. *stipulaceum.*
- New growth of stem shaggily villous or hirsute; leaves villous. . . . Var. *stipulaceum,* f. *hirtuosum.*

b. Margins of ocreae ascending, without horizontal flange (except in transitional plants).
- Leaves smooth or with slender appressed pilosity; spike 1.5–2 cm. thick; ocreolae narrowly triangular. . . . . Var. *stipulaceum,* f. *simile.*
- Leaves harshly scabrous on both surfaces with firm nearly subulate hairs less than 1 mm. long; spike 1–1.5 cm. thick; ocreolae broadly rounded-triangular. . . . . *P. amphibium,* f. *terrestre.*

P. AMPHÍBIUM (typical). — Local about the harbor, Yarmouth, N.S., chiefly or wholly as forma TERRÉSTRE (Leers) Blake (Adv. from Eu.)

Var. **stipulàceum** Coleman (having stipules), *P. Hartwrightii* Gray; *Persicaria Hartwrightii* Greene — Meadows, open or wooded swamps, shores and ditches, s. Lab. to Alaska, s. to Nfld., N.S., N.E., N.J., Pa., O., Ind., Ill., Mo., Neb., Colo., etc. — In less inundated habitats, or on gravelly shores, slopes, prairies, etc., becoming the densely villous forma **hirtuòsum** (Farw.) Fern. (shaggy); at the margins of lakes or quiet waters passing through forma **símile** Fern. (resembling; in this case forma *terrestre*) to the fully aquatic extreme with oblong or elliptic long-petioled glabrous leaves, forma **fluitáns** (Eat.) Fern. (floating), *P. natans* (Michx.) Eat., not *P. amphib.*, var. *natans* Moench, this, in turn, when stranded readily changing to typical var. *stipulaceum* or to forma *simile.* June–Sept.

19. P. coccíneum Muhl. (scarlet; hardly descriptive). — Usually coarser than no. 18, from stout creeping rootstocks; the *terrestrial forms* decumbent to suberect and *more or less scabrous, their lanceolate to lance-ovate leaves long-attenuate and* 1–2 *dm. long; ocreae close,* mostly *strigose; aquatic forms* smoother, *their* longer-petioled *leaves* thinner and *with rounded to cordate bases; peduncles pubescent* and often glandular; *spikes slenderly cylindric, the leading ones 4–18 cm.*

long and 7–15 mm. thick; ocreolae pubescent. (*P. Muhlenbergii* (Meisn.) S. Wats.; *Persicaria* (Meisn.) Small) — Highly variable; the following vars. and forms recognized:

Terrestrial forms with ascending stems; upper leaves merely rounded at base, glabrous or pubescent.
  Spikes mostly 4–8 cm. long; petioles attached near base of ocreae; leaves and upper internodes of stem glabrous or finely strigose. . . . . *P. coccineum* (typical).
  Spikes 4–18 cm. long; petioles mostly attached midway on the ocreae; leaves and upper internodes densely canescent-pubescent. . . . . Var. *pratincola.*
Aquatic, with glabrous floating leaves and branches.
  Leaves broadly lanceolate to ovate, scarcely acuminate, the upper ones cordate and 4–8 cm. broad. . . . . Forma *natans.*
  Leaves lance-acuminate, the upper ones tapering to merely rounded at base and 2–5 cm. broad. . . . . Var. *rigidulum.*

P. coccíneum (typical). — Shores and margins of ponds or streams, wet prairie, etc., Que. to Wash., s. to N.S., N.E., N.C., Ky., Ark., Okla., Tex., and Calif. July–Sept. Forma **nàtans** (Wieg.) Stanford (swimming), chiefly submersed, with the leaves floating.

Var. **pratíncola** (Greene) Stanford (dwelling on the prairie). — Prairie and shores, Ind. to N.D., s. to Ark., Okla., Tex. and Mex. — Possibly better treated as a more pubescent form.

Var. **rigídulum** (Sheld.) Stanford (stiffish). — In ponds, lakes and quiet streams, Ont. to Wash., s. to n. N.Y., Ia. and Oreg. — Leaves floating or ascending; possibly better considered a glabrous aquatic state of narrow-leaved typical *P. coccineum.*

20. P. densiflòrum Meisn. (densely flowered). — *Perennial by strong forking rootstocks;* the firm decumbent and rooting *stems soon ascending,* subsimple or with erect branches, 0.6–2 m. high, glabrous; *ocreae* cylindric, *the lowermost often with very short caducous cilia, the others eciliate,* united at base to the short petiole; *leaves lanceolate, acuminate to both ends,* the principal ones 0.7–3 dm. long, glabrous, with midrib prominent beneath; *spikes several in a narrow open panicle, erect or slightly arching, slenderly cylindric,* 2–11 cm. long; *ocreolae turbinate, much exceeded by the mature pedicels;* calyx white or whitish, about 3 mm. long; *styles* 2; achene biconvex, blackish-brown, lustrous, broad-ovoid to suborbicular, about 1.5 mm. broad. (*Persicaria portoricensis* (Bertero) Small) — Wet swampy woods, thickets and margins of shallow pools, W.I. and Fla. to Tex., n. on or near Coastal Plain to s. N.J. and to se. Mo. Aug.–Oct. (Trop. S.Am.)

21. P. longistỳlum Small (long-styled). — Annual (or possibly short-lived perennial), superficially resembling no. 22, branching, up to 1 m. high, the *slender stem glabrous below, minutely glandular and pilose above;* nodes conspicuously swollen; ocreae thin-membranaceous, upwardly broadened, entire or very short-ciliate, friable; leaves lanceolate to lance-ovate, glabrous, usually glandular-dotted beneath; *spikes* panicled, long-peduncled, erect, 2–8 cm. long, *about 1 cm. thick; pedicels exserted;* calyx 2–3.5 mm. long, pink; *some plants bearing flowers with long exserted style and included stamens, other plants* (more fertile) *with flowers having short style and exserted stamens; achene* ovate to orbicular, *biconvex,* 2.5–3 mm. long, 2–2.5 *mm. wide,* dull or slightly shining. (*Persicaria* Small) — Low grounds, La. to N.M., n. to s. Ill., Mo. and Kans. July–Oct.

22. P. pensylvánicum L. (Pennsylvanian), Pinkweed. — Annual, ascending to erect (in one var. depressed), up to 1.2 m. high; *summit of stem and peduncles with abundant divergent stipitate glands, or* in one var. only slightly glandular but *copiously strigose-hispid;* ocreae thin-scarious, cylindric, soon friable, eciliate, the upper often glandular or strigose; leaves lanceolate to elliptic or oval; *spikes dense, erect, pink to purplish* (rarely white), subglobose-ovoid to thick-cylindric, 1–6 cm. long, 1–1.5 cm. thick, the longer ones peduncled; *flowers all alike or if occasionally dimorphic, borne in the same inflorescence,* with *stamens and style usually included, mostly cleistogamous,* with much imperfect pollen; *pedicels soon exserted; achenes lenticular,* round-ovoid to orbicular, 2.2–3.5 *mm. broad, flattened on one face, concave on other,* somewhat shining. (*Persicaria* Small) — Highly variable; the following vars. recognized:

a. Leaves evidently strigose, lanceolate. . . *b.*
  b. Peduncles covered with spreading gland-tipped hairs; stamens (6–)8.
    Calyx 3–4 mm. long; achene 3–3.5 mm. long, 2.2–2.8 mm. broad. . . *P. pensylvanicum* (typical).
    Calyx 4–6 mm. long; achene 3.5–4 mm. long, 3–3.2 mm. broad. . . . Var. *rosaeflorum.*
  b. Peduncles strigose-hispid, without or with very few glands; stamens 6; achenes 2.2–2.8 mm. broad. . . . . Var. *durum.*
a. Leaves smooth and glabrous or promptly glabrescent; achenes 2.5–3.5 mm. broad. . . *c.*

c.Peduncles and axis of inflorescence copiously stipitate-glandular.
Stem erect, ascending or depressed; leaves lanceolate to narrowly lance-ovate, acuminate; the principal ones 0.5–2 dm. long; spikes mostly peduncled, the larger ones cylindric, 2–5 cm. long; stamens 7–8. . Var. *laevigatum*.
Stem depressed or subascending; leaves lance-elliptic to oval, blunt or bluntish, purple-blotched, mostly 2–5 cm. long; spikes short-peduncled to sessile, subglobose to short-ellipsoid, 1–2 cm. long; stamens 5 or 6. Var. *nesophilum*.
c.Peduncles and axis of inflorescence glabrous. . . . . . Var. *eglandulosum*.

P. pensylvánicum (typical). — Depressed to erect, up to 1.2 m. high; flowers bright pink or roseate, or in forma **albìnum** Fern. (albino) white. — Damp shores, thickets, clearings and disturbed or cult. soil, Fla. to Tex., n. to w. N.S., Mass., N.Y., s. Ont., s. Mich., s. Minn. and Okla. Late May–Oct.

Var. **rosaeflòrum** J. B. S. Norton (rose-flowered). — Flowers deep-roseate. — Brackish to fresh clearings, fields and openings, Del., Md. and Va. Aug.–Nov. — Often too close to the typical var.

Var. **dùrum** Stanford (rough). — Fla. to Tex., n. on or near Coastal Plain to se. Va. July–Oct.

Var. **laevigàtum** Fern. (smooth). — Glands deep colored; flowers roseate or pink, or glands pale or yellowish and leaves and flowers pale in forma **albíneum** Farw. (white). — The common plant northw., Côte Nord, Que., to Minn., S.D. and Colo., s. to N.S., N.E., L.I., Va., upland N.C., Tenn., Mo., and Okla. July–Oct.

Var. **nesóphilum** Fern. (island-lover). — Sandy beaches of ponds near sea, Nantucket and Martha's Vineyard Ids., Mass., and Block I., R.I., to s. L.I. Aug.–Oct.

Var. **eglandulòsum** J. C. Myers (glandless). — Habit of var. *laevigatum*, strictly glabrous throughout; flowers pale. — Calcareous shores, Ottawa and Essex Cos., Ont.

23. P. lapathifòlium L. (dock-leaved). — Annual, usually branching, erect to depressed, up to 2.5 m. high, *glabrous or with sessile glands on lower surfaces of leaves and somewhat on peduncles;* leaves narrowly to broadly lanceolate to ovate or elliptic, often glandular-punctate beneath or sometimes incanous; *ocreae cylindric, eciliate* or the lowest with very short cilia; *inflorescence in well developed plants paniculate; spikes slenderly cylindric, dense,* 1–8 cm. long, 5–9 *mm. thick, erect, arching or somewhat pendulous;* ocreolae with entire or very short-ciliolate margins; *calyx pink to purplish* above the often greenish base, *in maturity ovoid to rhomboid, constricted toward summit to form a thick beak of sepals overtopping the achene;* stamens and style included; *achene* lustrous, ovate or oval, *flattened, with one side concave,* 1.8–2.2 *mm. long,* 1.6–2 *mm. broad, included.* (*Persicaria* S. F. Gray) — Variable species of temp. reg.; our vars. are:

a.Leaves lanceolate, broadest near base, attenuate to tip. . b.
b.Leaves green on both surfaces except for sessile glands beneath, up to 2.5 dm. long and 5 cm. broad; longer spikes arching to pendulous, 2–8 cm. long. . . . . *P. lapathifolium* (typical).
b.Leaves white-pubescent beneath, narrowly lanceolate, the longer 3–10 cm. long and 3–10 mm. wide; spikes 1–few, erect, 1–3.5 cm. long. . . . Var. *salicifolium*.
a.Leaves elliptic, oblong-ovate or subrhombic, broadest well above base or near middle, not long-attenuate. . . c.
c.Erect, 1–2 m. high; leaves oblong-ovate, gradually rounded to petiole, the principal ones 1–2 dm. long and 4–7 cm. broad, thin and membranaceous; spikes mostly 2.5–6 cm. long. . . . . Var. *ovatum*.
c.Prostrate or depressed, the trailing and much-forked branches 1–4.5 dm. long; leaves subrhombic, cuneate at base, firm, often with large purple blotch above, the principal ones 2.5–7 cm. long and 1.5–3.5 cm. broad; spikes 1–4 cm. long. . . . . Var. *prostratum*.

P. lapathifòlium (typical). — **Plant up to 2.5 m. high, usually with many arching to pendulous slender pink spikes.** (Incl. *P. nodosum* Pers. and *P. incarnatum* Ell.) — Swampy thickets, shores, damp clearings and cult. fields, Nfld. to B.C., s. and sw. beyond our limits. July–Nov. (Eu.)

Var. **salicifòlium** Sibth. (willow-leaved). — Plant small, the simple to slightly branched very slender stems 0.5–6 dm. high. (Var. *incanum* (Willd.) W. D. J. Koch; *P. tomentosum*, var. *incanum* sensu ed. 7, not (Schmidt) Gurke) — Chiefly on gravelly shores, Nfld. to Alta., s. to N.S., N.E., N.J., e. Pa., n. Mich., Minn., S.D., Ida. and Calif. July–Oct. (Eu.)

Var. **ovàtum** A. Br. (ovate). — Peaty pond-shores, M.I. Aug.–Oct. (Eu.)

Var. PROSTRÀTUM Wimm. (prostrate). — Shores, railroad-yards and waste places, local, N.S. to Alta. and Wash., s. to N.J., e. Pa. and Calif. Aug.–Oct. (Natzd. from Eu.)

24. P. scàbrum Moench (harsh). — Annual, up to 1 m. high, the simple to forking stem glabrous except for abundant sessile glands toward summit; *leaves* lanceolate to lance-oblong, pale green, acute to barely acuminate, *at least the lower retaining more or less flocculent white tomentum on the lower surface;* axis of inflorescence with abundant sessile glands; *spikes* thick, *green* (rarely purplish), *erect, the lateral mostly sessile or short-stalked,* 1–5 cm. long and 0.8–1.5 *cm. thick; mature calyx round or broadly ovoid, the sepals not forming a beak, about equaling to shorter than the often exserted style; achene orbicular to broadly ovate, with concave sides,* 2.5–3.5 *mm. long* and 2.3–2.8 *mm. broad.* (*P. Persicaria,* ssp. *tomentosum* Schrank) — Cult. ground, waste places, shores, etc., Nfld. to Alta., s. to N.S., n. N.E., N.Y., Mich., Minn. and Oreg. July–Oct. (Natzd. from Eu.)

25. P. orientàle L. (oriental), Kiss-me-over-the-garden-gate, Princess-feather, Bâton de Saint-Jean (Que.), Monte-au-ciel (Que.). — Tall branching *soft-hairy annual; leaves ovate to broadly oblong,* sharply acuminate, *the larger broadly rounded or cordate at base, long-petioled;* inflorescence an *open panicle of long-peduncled dense cylindric ascending to drooping bright rose-colored spikes; ocreae* or some of them *expanded at summit into a broad horizontal herbaceous* and ciliate *flange; spikes* 2–10 *cm. long;* achene orbicular to reniform, biconvex, dull. (*Persicaria* Spach) — Waste places, roadsides, etc., spread from cult. June–Oct. (Introd. from Eurasia)

26. P. Càreyi Olney (for its discoverer, John Carey, 1797–1880). — Erect or ascending simple or branching annual, 0.15–1.5 m. high; *stem and peduncles glandular-bristly; leaves* narrowly lanceolate, attenuate at both ends, *hispid and scabrous;* ocreae ciliate, sometimes slightly margined; *spikes slenderly cylindric, arching to recurving, rather loosely flowered,* up to 1 dm. long; *pedicels exserted* from the ciliate ocreolae; *calyx purple or reddish; stamens* 5 (rarely 8); achene lenticular, lustrous. (*Persicaria* Greene) — Low thickets, swamps, recent burns, clearings and cult. ground, Me. to Ont., s. to s. N.E., Del., e. Pa., n. Ind., Wisc. and Minn. July–Oct.

27. P. Hydrópiper L. (old name, Water-pepper), Common Smartweed. — Depressed, assurgent or erect annual up to 6 dm. high, green or more or less reddened, *intensely acrid or peppery; leaves* lanceolate, up to 9 cm. long, *sessile or decurrent into short petioles,* glabrous or glabrescent, *with wavy margin;* ocreae scarious, brown, truncate, with cilia 1–2 mm. long; *spikes arching at tips of branches, slender and lax, with distant cleistogamous fascicles extending down to and often inclosed in upper sheaths;* ocreolae turbinate, greenish or red-tipped, eciliate or short-ciliate; pedicels included or exserted; *calyx greenish or red-tipped,* in fruit 2–4.5 mm. long, *covered with dark sessile glands,* mostly 4-parted; *achene dull, minutely punctate-striate,* lenticular or trigonous, 2–3.5 mm. long, dark brown to black, *its tip often slightly exserted.* (Including var. *projectum* Stanford; *Persicaria* Opiz) — Damp soils, either nat. or adv., throughout our range and beyond. June–Nov. (Eurasia)

28. P. dùbium Stein (doubtful). — Annual, *somewhat intermediate between nos. 27 and 31; leaves membranaceous, green, oblong to oblanceolate, cuneate or tapering to* subsessile or short-petioled *base;* ocreae ciliate; *spike slender, more or less interrupted,* especially at base, 3–5 *mm. thick, remote fascicles extending down to sheaths of upper leaves;* ocreolae short-ciliate; *calyx whitish, greenish or with pink sepals, in maturity,* 2.5–3.5 *mm. long; achene lenticular or trigonous,* 2.5–3 *mm. long,* somewhat shining. (*P. laxiflorum* Weihe; *P. mite* sensu recent European auth., not Pers.) — Ditches, roadsides and waste places, local, Que. to Va. and Ind. Late June–Oct. (Natzd. from Eu.) — Probably, as often considered by European botanists, a series of hybrids of nos. 27 and 31.

29. P. mìnus Huds. (lesser), var. **subcontínuum** (Meisn.) Fern. (almost continuous; from the spike which in typical *P. minus* is interrupted). — Weak annual suggesting small forms of no. 31; *leaves dull green, linear-lanceolate, attenuate to base and apex,* mostly 3–10 mm. wide, *divergent;* ocreae with few short cilia; *spikes slender-peduncled,* 1–5 cm. long, *linear-cylindric, loosely* but almost continuously *flowered,* 3–5 *mm. thick; ocreolae very short-ciliate; calyx claret-purple, in fruit* 2–2.5 *mm. long; achene* biconvex or trigonous, lustrous, 2–2.4 *mm. long.* — Roadsides, swamps, ditches, etc., local, Mass. to Ind. and se. Pa. Aug.–Oct. (Natzd. from Eu.)

30. P. cespitòsum Blume (tufted), var. **longisètum** (DeBruyn) Stewart (with long bristles). — Slender simple or *loosely branching* prostrate or finally ascending annual, *often with the filiform branches longer than the indefinite main axis;* ocreae ciliate; *leaves* elliptic, ovate, lanceolate or subrhombic, *tapering about equally to apex and subsessile base,* glabrous above, ciliate on margins and veins beneath; *spikes* terminal, 1–few, slender-peduncled, *linear-cylindric,* 1–4 cm. long, 3–5 *mm. thick, loosely flowered,* often with distant lower flowers; *ocreolae with bristle-form cilia equaling or longer than the sheath; calyx purple or roseate,* 2–2.5 mm. long, on exserted pedicels; *achene trigonous, lustrous,* 2–2.4 mm. long. — Roadsides, waste places, shores and other damp spots, locally abund., Mass. to Ill., s. to Del., Md. and Ky. June–Oct. (Natzd. from se. Asia)

31. P. Persicària L. (old generic name, said to come from the leaves resembling those of *Persica*, the peach), Lady's-thumb, Heart's-ease. — *Nearly smooth and glabrous* annual with simple or branching erect, ascending or decumbent stem 0.2–1 m. high; ocreae thin, ciliate; *leaves* narrowly to broadly *lanceolate, often purplish-blotched above,* tapering to base and apex, firm, or in the inundated forma submérsum Erikson (submersed) very thin and pellucid, the *primary ones* 3–15 *cm. long and* 0.5–3 *cm. broad; spikes* 1–several and paniculate, *glabrous-peduncled* or the secondary ones short-peduncled to subsessile, *oblong- or thick-cylindric, dense,* 7–11 *mm. thick,* the leading ones 1.5–4.5 cm. long; ocreolae usually ciliolate; *calyx pink or purplish to pink and green,* or white in the rare forma albiflorum Millsp. (white-flowered), 2–3 (in fruit –4) *mm. long,* the *mature sepals with prominently reticulate bases; achene* lenticular or trigonous, lustrous, 2.5–3 *mm. long.* — Ubiquitous weed of damp clearings, cult. ground, roadsides, shores, etc., throughout our range and beyond. June–Oct. (Natzd. from Eu.)

Var. angustifòlium Beckh. (narrow-leaved). — Leaves linear-lanceolate, 3–5 mm. wide, strigose on both faces; bristle-like cilia of ocreae half as long as strigose sheath. — Local, waste places and railroad-gravel, N.Y. to Va. (Adv. from Eu.)

Var. ruderàle (Salisb.) Meisn. (of rubbish). — Prostrate or depressed in freely forking mats; leaves oblong- or narrowly rhombic-lanceolate, blunt to acute, scarcely acuminate, often scabrous, the primary ones only 2–5 cm. long; ocreae short-ciliate; spikes subglobose to short-cylindric, 0.5–1.5 cm. long, short-peduncled to sessile. — Locally abundant, St. P. et Miq. and Que. to Oreg., s. to N.S., N.E., Va., W.Va. and n. Ind. (Natzd. from Eu.)

32. P. puritanòrum Fern. (of the Puritans). — Smaller in most parts than no. 31, the slender prostrate to erect stem finally divergently much branched, 0.3–5.5 dm. long; *leaves mostly green above,* lanceolate or narrowly rhombic-lanceolate, tapering from middle to subacute to acute apex and to cuneate short-petioled or subsessile base, more or less strigose beneath, the larger ones 3–10 cm. long and 0.5–1.6 cm. broad; *spikes* dense, *slenderly cylindric,* the leading ones 1–3.5 cm. long and 4–6.5 *mm. thick; calyx* rose-pink or purple, 1.8–2, *in fruit* 2.4–2.6 *mm. long, smooth or only faintly reticulate; achenes* black, lustrous, 2 *mm. long.* — Sandy or gravelly pond-shores, w. N.S.; sw. Me.; se. Mass. and s. R.I., locally inland to s.-centr. Mass. July–Oct.

33. P. punctàtum Ell. (dotted), Water-Smartweed. — *Annual, or perennial with elongate tough branching rootstock* 2–5 *mm. thick,* glabrous or essentially so, loosely spreading or ascending, 0.03–1 m. high; perennial or annual; *ocreae scarious, glabrous* or nearly so, soon rupturing, ciliate; leaves narrowly to broadly lanceolate to lance-oblong, usually punctate, the *short petiole gradually broadened at base; spikes* 1–several and paniculate, *linear-cylindric to moniliform,* 3–8 *mm. thick,* up to 15 cm. long; *ocreolae slightly distant to very remote, ciliate; calyx* greenish below, *with white* (rarely roseate) *sepals, glandular-punctate, in fruit* 3.5–4 *mm. long; achene lenticular or trigonous, lustrous,* 3–3.6 *mm. long and about* 2 *mm. wide.* — Wide-ranging species of N. and S.Am. and Asia, represented with us by 3 vars.:

Plant up to 1 m. high; leaves not fleshy, strongly punctate, acuminate, up to 1.5 dm. long; spike 2–15 cm. long; calyx heavily punctate.

- Perennial by tough rootstocks, exceptionally annual; leaves deep green, firm, rather hard when dry, the primary ones 6–15 cm. long and 1–3.5 cm. broad; principal spikes clearly peduncled, without remote fascicles of flowers extending down to upper leaf-axils, subcontinuous except at base, with most of the ocreolae equally subdistant, the flowering axis 2–9 cm. long. . *P. punctatum* (typical).
- Annual, exceptionally perennial; leaves pale or yellowish-green, thin, the primary ones 2–10 cm. long and 0.3–2 cm. broad; principal spikes loosely moniliform except at tip, the ocreolae and fascicles of flowers increasingly distant, becoming remote downward often into the upper leaf-axils, the flowering axis thus prolonged to 3–15 cm. in length. . . . . . . Var. *leptostachyum.*

Plant 0.3–2 dm. high, divergently branched below; leaves fleshy, sparsely punctate, 2–6 cm. long, round-tipped to subacute; flowers usually extending well down into leaf-axils; free flowering axis 1–5 cm. long; calyx sparsely punctate; estuarine. . . . . Var. *parvum.*

P. punctàtum (typical), *P. acre* HBK. — Wet places, shores, swamps, etc., Fla. to Mex. and s. Calif., n. to s. N.H., Mass., N.Y., O., Mich., Ill., Neb., etc. July–Oct. (Trop. Am.; s. Asia)

Var. leptostáchyum (Meisn.) Small (slender-spiked). — Similar habitats, Que. to B.C., s. to N.S., N.E., L.I., Va., Tenn., La., Tex. and Mex.

Var. **párvum** Vict. & Rousseau (small). — Fresh to brackish tidal shores, St. Lawrence R., Que., to lower Delaware R., N.J. and Pa.

34. P. robústius (Small) Fern. (more robust). — *Coarser* than no. 33, *perennial with stout subligneous forking rootstocks and rooting decumbent bases of stems 3–8 mm. thick, sending off strong leafy basal shoots in autumn;* ascending *flowering stems* 0.3–2 *m. or more high, stout; ocreae cylindric, close, firm, strigose,* truncate, *with stiff bristleform cilia 3–8 mm. long; leaves firm,* dark green, lanceolate to oblong or elliptic, the *primary ones* 0.4–1.8 dm. long and 2–4.5 *cm. broad;* the short *petiole abruptly dilated at base; spikes* long-peduncled, paniculate, *continuous* except sometimes at base; *ocreolae truncate, eciliate,* all but the lowest *contiguous; mature calyx 4–5 mm. long; achene trigonous,* 3.5–4 *mm. long and* 2–3 *mm. broad.* (*P. punctatum*, var. *robustius* Small) — Margins of ponds or quiet streams or in inundated places, Fla. to Mex., n. to w. N.S., s. N.H., Mass., N.Y. and Mo. Late July–early Nov. — In aspect suggesting no. 37.

35. P. hydropiperoìdes Michx. (resembling *P. Hydropiper;* at least to the original author), Mild Water-pepper. — Perennial *with slender tough branching and extensively creeping rootstocks and decumbent finally ascending slender simple or subsimple flowering branches* rooting at base, *smooth or rarely scabridulous, 1–4 mm. thick at base; ocreae short-strigose, ciliate with bristles 1–7 mm. long or eciliate, leaves linear-lanceolate to lance-oblong, ascending, mostly smooth* or smoothish, 0.5–2.5 *cm. wide; spikes erect, cylindric to almost filiform,* 2–8 cm. long, often interrupted at base; ocreolae ciliate to eciliate; *calyx roseate or purplish,* rarely greenish or white, *not dotted, some with stamens only* and only 1.5–2 mm. long, *others with stamens and carpel and becoming 2.5–4 mm. long, these in fruit elongate-ovoid and completely covering achene; achene* lustrous, 2–3 *mm. long,* about 2 mm. wide, sharply trigonous, with concave sides (in two vars. outside our range both trigonous and lenticular). — Wide-ranging N.Am. species with many well-differentiated geographic vars.; ours are as follows:

*a.*Spikes mostly borne singly at tips of or along branches of inflorescence; flowering stems rarely up to 1 m. high, with leaves rarely 1.5 dm. long; fruiting calyx 2.5–3.5 mm. long; achenes 2–3 mm. long; sterile flowers not forming bulbils. . *b.*
*b.*Cilia of ocreae 2–7 mm. long, of ocreolae 0.5–2 mm. long; achene 2.5–3 mm. long. . . *c.*
*c.*Cilia of ocreae 2–4 mm. long, of ocreolae 0.5–1 mm. long; leaves lanceolate, 1–2.5 cm. wide; spikes linear-cylindric to slenderly moniliform, 3–6 mm. thick; wide-spread. . . . . *P. hydropiperoides* (typical).
*c.*Cilia of ocreae 4.5–7 mm. long, of ocreolae 1–2 mm. long; southern.
Leaves lanceolate, the larger 1.5–2 cm. wide, not long-attenuate; spike lax, 3–5 mm. thick; pedicels long-exserted; calyx roseate; southwestern. . Var. *Bushianum.*
Leaves linear or linear-lanceolate, 4–14 mm. broad, long-attenuate; spike dense, 5–8 mm. thick; pedicels scarcely exserted; calyx purplish and green; southeastern. . . . . Var. *euronotorum.*
*b.*Cilia of ocreae wanting or at most 1.2 mm. long, of ocreolae wanting or up to 0.4 mm. long; leaves lanceolate, the larger 0.8–3.2 cm. broad; achene 2–2.5 mm. long.
Spikes lax, the leading ones 4–7 cm. long; pedicels long-exserted from the eciliate or short-ciliate slender puberulent ocreolae; fruiting calyx 2.5–3 mm. long; leaves scabrous beneath; Virginian. . . . . . . Var. *breviciliatum.*
Spikes close, the larger ones 2–3.5 cm. long; pedicels included or barely exserted from the upwardly flaring glabrous mostly eciliate ocreolae; fruiting calyx 3–3.5 mm. long; leaves glabrous; Nova Scotian. . . . Var. *psilostachyum.*
*a.*Spikes or many of them sessile and digitately clustered in groups of 2–4 at the tips of the erect peduncles, dense, 7–10 mm. thick; plant very tall (1–1.5 m.); leaves narrowly lance-attenuate, mostly 1.5–2 dm. long; ocreae short-ciliate; ocreolae eciliate or nearly so; sterile flowers often present, these becoming elongate bulbils about 5 mm. long; perfect flowers with purple calyx becoming 3.5–4 mm. long; achene about 3 mm. long; Nova Scotian. Var. *digitatum.*

P. hydropiperoìdes (typical). — Shallow water or wet shores, swales or inundated bottoms, Fla. to Tex., n. to N.S., N.B., s. Que., s. Ont., Mich., Wisc., Minn. and Neb. June–Nov. Hybridizes with no. 34. — Typically with smooth or smoothish leaves and pink to purplish flowers; flowers white in forma **leucochránthum** A. H. Moore (white-flowered); leaves and stem definitely strigose in forma **strigòsum** (Small) Stanford (strigose).

Var. **Bushiànum** Stanford (for its discoverer, Benjamin Franklin Bush, 1858–1937). — Ky. to Kans. and Okla.

Var. **euronotòrum** Fern. (of southeast winds). — Se. Va. and se. S.C.

Var. **breviciliàtum** Fern. (with short cilia). — Wet swales and ditches, se. Va.

Var. **psilostáchyum** St. John (smooth-spiked). — Sable I. and Shelburne Co., N.S.

Var. **digitàtum** Fern. (having digits). — Peaty and boggy lake-margins, Yarmouth and Shelburne Cos., N.S. July–Oct., beginning to flower nearly a month later than typical *P. hydropiperoides* in the same reg.

36. P. opelousànum Riddell (of Opelousas, La.). — *Resembling narrowest-leaved no.* 35; *stems often more branching and more nearly erect,* up to 1.3 m. high; leaves narrowly lanceolate to lance-linear, 0.5–1.5 cm. wide, commonly minutely glandular beneath; *spikes* mostly peduncled, *continuous* or subcontinuous, *slenderly cylindric,* up to 6 cm. long; ocreolae slenderly funnelform, with spreading cilia 1–2 mm. long; pedicels exserted; *calyx greenish-white or green and purplish,* undotted, *in maturity thick-rhomboid to subglobose-ovoid, not covering tip of achene and* 2.5–3 *mm. long; achene slightly exserted,* 1.5–2.5 *mm. long.* — (*P. hydropiperoides,* var. Stone; *Persicaria* Small) — Shallow water, wet shores, inundated swales and sandy or peaty thickets or bottoms, Fla. to Tex. and e. Mex., n., mostly on or near Coastal Plain, to Mass., se. N.Y. and N.J. and inland n. to Tenn., s. Ill., s. Mo. and Okla. Late July–Oct.

Var. **adenócalyx** Stanford (with glandular calyx). — Calyx bearing numerous large white sessile glands. — More frequent in e. and se. Mass., and s. R.I. and local in s. Ct. — Plants either perennial or annual, possibly a hybrid of *P. opelousanum* and *P. punctatum,* var. *leptostachyum.*

37. P. setàceum Baldw. (bristly). — Stoutish to rather slender firm-stemmed *perennial with elongate horizontal subligneous rhizome; the erect stem* 2.5–6 *mm. thick at base,* simple or *more commonly branching from middle and upper nodes,* 0.4–1.5 m. high; *ocreae* cylindric, *long-strigose or -hirsute or -villous,* truncate, *bearing marginal closely ascending bristles* 5–15 *mm. long; leaves* lanceolate, dark green, *strigose or rarely smooth, the primary ones* 0.7–1.8 dm. long and 1.2–3.5 *cm. broad;* spikes 1–several, paniculate, mostly peduncled, erect, very slender, the longer ones 3–8 cm. long; pedicels exserted; *calyx in maturity* 3–3.5 *mm. long, the sepals white;* achene 2–3 mm. long, trigonous, lustrous. — Variable in pubescence, with 3 vars.:

Leaves strigose above (especially when young), firm.
  Strigosity of upper leaf-surface of mixed short and long hairs, the latter 0.8–1.5 mm. long; hairs of lower surface similar, usually abundant; ocreae copiously strigose-villous. . . . . *P. setaceum* (typical).
  Strigosity of upper leaf-surface wholly or chiefly of stiff hairs only 0.2–0.5 (rarely 0.8) mm. long; lower surface similarly strigose to glabrous; ocreae remotely strigose. . . . . Var. *interjectum.*
Leaves glabrous or essentially so on both surfaces; ocreae sparsely appressed-strigose. . . . . Var. *tonsum.*

P. setàceum (typical). — Low woods, swamps, shores and wet clearings, Fla. to Tex., n. to s. N.J., Ark. and Okla. Mid-July–Oct.

Var. **interjéctum** Fern. (thrown in between). — Similar habitats, se. Mass. to Oswego Co., N.Y., w. to s. Mich., s. to s. R.I., L.I., se. Va., and Ky.

Var. **tónsum** Fern. (shaved). — Wooded swamps, s. N.J. to S.C.; Ky. to Okla.

P. NEPALÉNSE Meisn. (of Nepal), a depressed to ascending annual of § *Cephalophilon,* with the lower truncate-based leaves ovate, the upper ones with broadly winged clasping bases, the flowers in dense heads subtended by clasping leafy bracts, has appeared locally on a river-bank in Ct. (Adv. from Asia)

§ 4. ECHINOCAÙLON Meisn. (TRACAULON Raf.) TEARTHUMB, TEARGRASS, SCRATCHGRASS

38. P. sagittàtum L. (arrow-shaped), ARROW-LEAVED TEARTHUMB, ARROW-VINE, GRATTE-CUL (Que.). — *Weak* more or less geniculate, simple or mostly loosely and divergently branching, mostly leaning on or supported by other vegetation; 4-*angled stems and branches and midribs of leaves beneath retrorse-prickly; upper internodes retrorsely barbed essentially to summit, not greatly elongate; leaves oblong- to narrowly ovate-lanceolate, sagittate, the primary ones* 3–10 cm. long and 0.7–2.8 cm. broad, *averaging two-fifths as broad as long, the upper ones well developed;* peduncles glabrous; *flowers capitate,* pink to white; *stamens mostly* 8; *common style elongate,* 3-*parted; achene sharply trigonous,* 3–3.5 mm. long. (*Tracaulon* Small) — Low grounds, Fla. to Tex., n. to Nfld., Que., Ont. and Sask. June–Oct. (Asia) — Forma **chloránthum** Fern. (green-flowered), with ciliation of leaf-margin and bristles of midrib wanting or reduced, and green calyces, occurs on tidal estuaries from Me. to N.Y.

Var. **graciléntum** Fern. (slender). — Stems greatly prolonged and slender, the *upper internodes very much longer than the lower and median ones, smooth except at base; leaves narrowly* *lance-sagittate, the primary ones* 3.5–6.5 cm. long and 5.5–10.5 mm. wide, *averaging one-sixth as broad as long, upper leaves reduced to tiny bracteiform blades.* — Fresh to brackish tidal marshes, se. Va.

39. P. arifòlium L. (with the leaf of *Arum;* i.e., the European *A. maculatum*), HALBERD-LEAVED TEARTHUMB. — Coarser than no. 38; *lower and median leaves hastate with horizontally divergent basal lobes,* the upper often unlobed or sagittate, the *surfaces more or less pubescent:* the lower slender-petioled, with *terminal lobe attenuate* and 4–15 cm. long by 1–7 cm. broad; *peduncles glandular-bristly; flowers few in short raceme,* pink or purplish to white; *stamens* 6; *common style nearly obsolete, 2-parted; achene biconvex,* 4–4.2 *mm. broad,* 3–3.2 *mm. thick, its sides often umbonate.* (*Tracaulon* Raf.) — Wet places, oftenest tidal marshes, Del., Md. and D.C. to Fla. July–Oct.

Var. **pubéscens** (Keller) Fern. (pubescent; a character not confined to the var.). — *Achene* smaller, lenticular, 3–3.5 *mm. broad,* 2.2–2.6 *mm. thick, the gently rounded sides not umbonate.* (Var. *lentiforme* Fern. & Grisc.) — Wet places, P.E.I. to s. Ont. and Minn., s. to N.J., Pa., O. and Ind. July–Sept.

P. PERFOLIÀTUM L. (with leaf surrounding stem), a species of e. Asia, is becoming estab. in nurseries, etc. in Pa. and may become a troublesome weed. Its ocreae are expanded into leafy perfoliate blades, the true leaves deltoid and basally peltate. (Adv. from Asia)

§ 5. TINIÀRIA Meisn. (BILDERDYKIA Dumort; TINIARIA Webb & Moq.) CLIMBING BUCKWHEAT, NIMBLE-WILL, CORNBIND

40. P. cilinòde Michx. (with ciliate nodes). — Perennial; the puberulent or *short-pilose stem soon twining,* freely branching and high-climbing, or erect and ending in panicles without elongating and twining and only 1.5–6 dm. high in forma **eréctum** (Peck) Fern. (erect); *leaves* cordate-ovate, *acuminate, pilose beneath,* the larger ones slender-petioled and 5–12 cm. long by 2.5–10 cm. broad; *ocreae bearing a ring of reflexed bristles at base; inflorescences* consisting *of* (usually many) axillary and terminal *panicles* (sometimes moniliform axillary spiciform racemes); *flowering calyx white* or pink-tinged, *in fruit wingless and only obscurely keeled and* 4–5 *mm. long; achene* lustrous, blackish, *trigonous-ellipsoid,* 3–4 *mm. long.* (*Bilderdykia* Greene; *Tiniaria* Small) — Dry thickets, rocky slopes and borders of woods, Nfld. to Sask., s. to N.S., N.E., n. N.J., Pa., W.Va., upland to N.C. and Tenn., Mich., Wisc. and Minn. June–Aug.

Var. **laevigàtum** Fern. (smooth). — *Stems glabrous; leaves glabrous beneath* except for puberulent ribs, *the nodes at base often beardless; fruiting calyx* 3.5–4 *mm. long; achene* broadly *trigonous-obovoid to -subglobose, scarcely* 3 *mm. long.* — Rocky summits of Spruce Knob, W.Va., and of Peaks of Otter, Va.

41. P. CONVÓLVULUS L. (old generic name for any twiner, early botanists, like too many modern ones, being more impressed by habit than by floral morphology), BLACK BINDWEED, CHEVRIER (Que.). — Slightly *roughish annual;* stem soon twining, or sometimes merely procumbent; leaves cordate, ovate, dull green; *flowers in short axillary clusters or axillary and terminal interrupted or moniliform spike-like racemes; pedicels* mostly *shorter than puberulent* green or purplish-tipped *calyx; fruiting calyx* subrhomboid-ovoid, 4–5 *mm. long,* trigonous, *with scarcely developed keels; achene minutely roughened, dull.* — Weed of cult. and waste ground, throughout our areas and beyond. May–Nov. (Natzd. from Eu.)

Var. SUBALÀTUM Lej. & Court. (somewhat winged). — *Fruiting calyx* 5–8 *mm. long, the* 3 *keels with pale scarious wings* 0.4–1 *mm. broad at summit.* — N.E. to e. Pa. and Mich.; common northw. and eastw. (Natzd. from Eu.) — Plant often stouter and climbing higher, suggesting no. 42.

42. P. cristàtum Engelm. & Gray (crested). — Perennial twiner (rarely fruiting 1st year); *leaves* cordate- or deltoid-ovate, *usually with broad and shallow sinus; the larger blades* 2–9 *cm. long,* their petioles 1–2 cm. long, upper leaves reduced and short petioled to sessile; inflorescences of axillary clusters and racemes; *racemes* slender, interrupted, *usually leafy-bracted below, in fruit* 2–15 cm. long and 1–1.5 *cm. thick;* pedicels equaling or exceeding fruit, reflexed; *mature fruiting calyx* 4–6 *mm. long, with herbaceous greenish* entire, toothed or crimped *wings* 0.25 *to barely* 1 *mm. broad; achene lustrous,* trigonous, 2–3.5 *mm. long,* 1.5–2 *mm. broad.* (*P. dumetorum* sensu ed. 7 and Am. auth., not L.) — Dry to moist rocky to argillaceous slopes, open woods, thickets, sands and clearings, Fla. to Tex., n. to e. Mass., Vt., N.Y., Ind., Ill., Minn. and Kans. July–Oct.

43. P. scándens L. (climbing), CLIMBING FALSE BUCKWHEAT. — More extensively twining and with longer stems than no. 42; *leaves* cordate-ovate to slightly hastate, acuminate, *the larger ones* 5–13 *cm. long; mature racemes mostly* 0.5–2 dm. long, 2–3 *cm. thick; fruiting calyx* 8–10 *mm. long, with thin scarious pale brown, stramineous or roseate wings* 1.5–3 *mm. broad at* *summit; achene* 3.5–6 *mm. long,* 2.5–4 *mm. broad.* — Damp thickets, low woods, bottoms, shores and openings, Fla. to Tex., n. to N.S., N.B., n. N.E., s. Que., Ont. and Man. Late Aug.–Nov.

P. Aubérti L. Henry (named in 1907 for its discoverer, P. Georges Aubert), a slightly woody twiner with wavy-margined truncate to cordate bluntish leaves and slender glandular-pubescent racemes of winged fruits, spreads from cult. to waste places. (Introd. from China)

44. P. cuspidàtum Sieb. & Zucc. (abruptly pointed), Japanese Knotweed. — Rapidly *spreading by stout subterranean rhizomes and offshoots; stem erect, glaucous,* often *mottled, widely bushy-branched,* 1–2.5 *m. high;* ocreae membranous, tubular; *leaves* petioled, *round-ovate, truncate to slightly cuneate at base, abruptly cuspidate,* becoming firm and 0.5–1.5 *dm. long; flowers* greenish-white, *dioecious, in forking axillary panicles;* fruiting calyx wing-angled, 8–9 mm. long; achene shining, trigonous, about 4 mm. long. (*P. Sieboldi* of some auth., not Meisn.) — Waste places and neglected gardens, rapidly spreading and becoming obnoxious, Nfld. to Ont. and Minn., s. to Md. Aug., Sept. (Introd. from e. Asia)

45. P. sachalinénse F. Schmidt (of Sachalin Island), Sachaline, Giant Knotweed. — Similar to no. 44, coarser and taller (*up to* 3.5 *m. high*); stem angular-striate; *leaves cordate, up to* 3 *dm. long, gradually tapering to tip;* flowers greenish. — Similarly spreading and monopolizing neglected gardens and waste ground, Mass. and N.Y. to Md. Aug., Sept. (Introd. from e. Asia)

§ 6. Aconogónon Meisn.

46. P. polystáchyum Wall. (many-spiked). — Somewhat *shrubby,* 1–2 m. high, with many branches; *leaves short-petioled to subsessile, broadly lanceolate to lance-ovate, with cordate to truncate base and* 2 *rounded basal lobes,* 0.5–1.5 dm. long; *inflorescence a terminal or subterminal panicle* of racemes; *calyx about* 4 *mm. long, white, not enlarged in fruit.* — Waste places and neglected yards, rapidly spreading, as yet local, N.S. and N.E. Aug., Sept. (Introd. from e. Asia)

6. FAGOPỲRUM Mill. Buckwheat. Sarrasin (Que.)

Calyx petal-like, equally 5-parted. Stamens 8. Styles 3; stigmas capitate. Achene 3-sided. Embryo large, in the center of the albumen, which it divides into 2 parts, with very broad and foliaceous plaited and twisted cotyledons. — Annuals, nat. of Eurasia, with triangular-cordate or hastate leaves, semicylindrical sheaths, and corymbose racemes of white, greenish, or rose-colored flowers. (Latin *fagus,* the *beech,* and Greek *pyros, wheat,* from the resemblance of the grain to the beech-nut; hence the English name *Buckwheat,* from the German *Buche,* beech.)

1. F. sagittàtum Gilib. (arrow-shaped), Buckwheat. — *Flowers white,* with honey-bearing yellow glands interspersed between the stamens; *achene* acute, *smooth and shining.* (*F. esculentum* Moench; *Polygonum Fagopyrum* L.) — Waste places, old fields, roadsides, etc., spread from or persistent after cult. June–Sept. (Introd. from Asia)

2. F. tatáricum (L.) Gaertn. (Tartarian), India-wheat. — *Flowers greener,* smaller, shorter-pedicelled; *achene very dull and roughish,* the sides sulcate. — Similarly spread from or persistent after cult., especially in e. Canada and N.E., local elsewhere. (Introd. from Asia)

7. POLYGONÉLLA Michx. Jointweed

Calyx 5-parted, petaloid, loosely persistent about the achene, the 3 inner divisions often enlarging in fruit, in which case the outer are usually spreading. Stamens 8. Styles 3, and achene 3-angular. Embryo slender, straight or nearly so, toward one side of the albumen. — Slender glabrous N.Am. annuals or perennials, with alternate narrow leaves jointed at the base, and rather rigid truncate or oblique naked sheaths or bracts. Flowers on solitary pedicels (nodding in fruit) jointed near the base, borne in slender panicled racemes. (Diminutive of *Polygonum.*)

Annual; leaves linear-filiform, promptly disarticulating; sepals nearly uniform,
  the outer not recurved in fruit. . . . . 1. *P. articulata.*
Perennial, shrubby or suffruticose; leaves linear to oblanceolate, mostly persistent; inner sepals much larger than the small reflexing outer ones.
  Suffruticose, 2.5–5 dm. high; leaves oblanceolate or spatulate; fruiting racemes
    4–5 mm. thick; mature calyx 2–2.5 mm. broad. . . . . 2. *P. polygama.*
  Shrubby, 3.5–12 dm. high; leaves linear-filiform; fruiting racemes 12–16 mm.
    thick; mature calyx 3.5–6 mm. broad. . . . . 3. *P. americana.*

1. P. articulàta (L.) Meisn. (jointed). — *Annual,* erect, branching, glaucous, 1–6 dm. high; leaves linear-filiform, deciduous; flowers rose-color or white, or dark purple in forma **atròrubens**

Fern. (dark red), nodding, in very slender racemes: *sepals subequal, all erect and connivent about the fruit*; *ripe calyx about* 2 *mm. broad;* achene slightly exserted, smooth. (*Delopyrum* Small) — Dry sands, sw. Me. to s. Ont. and Minn., s. to ne. N.C., n. Ind., Wisc. and Ia. Late Aug.–Oct. Fig. 983. — Erroneously called in N.E. "HEATHER"; an anomalous species with calyx of *Polygonum.*

983. P. articulata.

2. P. polýgama (Vent.) Engelm. & Gray (polygamous), October-flower. — *Suffruticose,* 2.5–5 dm. high, mostly bushy-branched, making a rounded panicle of ascending filiform branches and delicate racemes; *leaves oblanceolate or spatulate,* obtuse, the *larger ones* 2–3 *mm. broad; racemes* slender, *in maturity* 4–5 *mm. thick; sepals* lemon-yellow to pinkish or pale-brown, *the outer narrow and reflexed in fruit, the others enlarging and broad; fruiting calyx* 2–2.5 *mm. broad;* achene included. — Dry sands and pine-barrens, Ga. to se. Va. (very local). Sept., Oct.

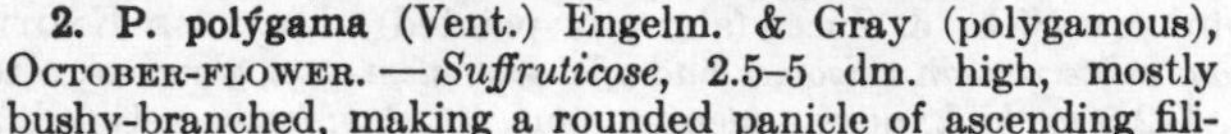
984. P. americana.

3. P. americàna (Fisch. & Mey.) Small (American). — *Shrubby, up to* 12 *dm. high,* loosely branched at summit; *leaves linear-filiform; mature racemes* 12–16 *mm. thick; calyx* as in no. 2, larger, *when ripe* 3.5–6 *mm. broad.* — Dry sands, Ga. to se. Mo., Ark. and Tex. Late Aug.–Oct. Fig. 984.

8. BRUNNÍCHIA Banks

Calyx-divisions somewhat petal-like, oblong, connivent and coriaceous in fruit, the base and almost the whole length of the pedicel winged on one side. Stamens 8; filaments capillary. Styles 3, slender; stigmas depressed-capitate. Ovule pendulous on a slender erect funiculus; seed erect, 6-grooved. Achene obtusely triangular, partly 3-locular, inclosed in the indurated calyx. — Somewhat shrubby, with grooved stems, climbing by tendrils from the ends of the branches. (Named for *M. T. Brünnich,* a Norwegian naturalist of the 18th century.)

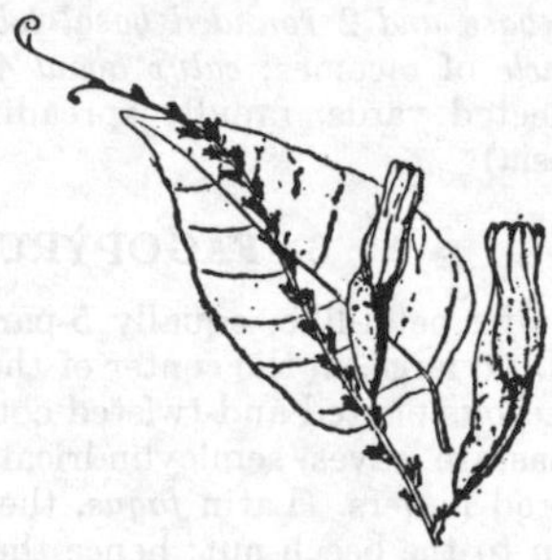
985. B. cirrhosa.

1. B. cirrhòsa Gaertn. (having tendrils), Ladies'-eardrops. — Terete woody stems up to 2 cm. in diameter at base; leaves ovate, pointed, entire; petioles dilated at base, but with no distinct sheath or stipules; flowers greenish, 2–5 in a fascicle, crowded in axillary and terminal racemes; fruiting calyx with the wing 2.4–2.7 cm. long. — River-banks, swamps, low woods and thickets, chiefly on Coastal Plain, Fla. to Tex., n. to S.C., w. Ky., s. Ill., se. Mo. and Okla. *Fr.* Aug.–Oct. Fig. 985.

FAM. 55. CHENOPODIÀCEAE (Goosefoot Family)

Chiefly herbs with inconspicuous flowers, more or less succulent, with mostly alternate leaves and no stipules or scarious bracts; minute usually greenish flowers, with the free calyx imbricated in the bud; the stamens as many as its lobes or occasionally fewer and inserted opposite them or on their bases; the 1-locular ovary becoming a 1-seeded thin utricle or rarely an achene. Embryo coiled into a ring around the mealy albumen (*when there is any*) *or else conduplicate or spiral.* Calyx persistent, mostly inclosing the fruit. Styles or stigmas 2, rarely 3–5. — Mostly inert or non-poisonous, often weedy plants; several are potherbs or vegetables, such as Spinach and Beet.

a. Embryo coiled into a ring about the usually copious central albumen; leaves flat; stem not jointed. . . *b.*
b. Calyx of all or of staminate flowers 3–5-parted or toothed. . . *c.*
c. Flowers perfect (stamens rarely wanting) or with pistillate ones intermixed, mostly clustered or panicled; calyx persistent in fruit. . . *d.*
d. Fruiting calyx bearing a wing or spines. . . *e.*
e. Fruiting calyx with a horizontal wing.
Leaves sinuate-dentate; wing continuous; endosperm of seed abundant. . . . . 1. *Cycloloma.*
Leaves entire; wing 5-lobed; endosperm almost none. . . . . 2. *Kochia.*
e. Fruiting calyx with 1 or more lobes ending in a spine; endosperm scanty; leaves entire. . . . . 3. *Bassia.*

d.Fruiting calyx without wing or spines.
Calyx 3–5-toothed, becoming saccate and reticulated; leaves pinnatifid. . . . . 4. *Roubieva.*
Calyx 5 (rarely 4)-parted, not changing in fruit or merely becoming succulent; leaves various. . . . . 5. *Chenopodium.*
c.Flowers unisexual (monoecious or dioecious).
Pistillate flowers mostly without a calyx, embraced between 2 bracts; pericarp of utricle usually free from seed. . . . . . . 6. *Atriplex.*
Pistillate flowers with a 3–4-parted perianth, not covered by bracts; pericarp of utricle adherent to seed. . . . . . . 7. *Axyris.*
b.Calyx of 1 sepal.
Larger leaves hastate; flowers glomerulate in upper axils; utricle ovoid, somewhat compressed, with wingless margin. . . . . 8. *Monolepis.*
Larger leaves linear; flowers mostly solitary in axils of reduced upper bracts; utricle plano-convex, with thin or winged margin. . . . 9. *Corispermum.*
a.Embryo horseshoe-shaped, conduplicate or spiral; endosperm nearly or quite wanting. . *f.*
f.Stem fleshy, jointed; leaves reduced to appressed opposite scales or teeth, with the axillary flowers sunk into the fleshy stem; calyx utricle-like; embryo horseshoe-shaped or conduplicate. . . . . 10. *Salicornia.*
f.Stem scarcely fleshy, not jointed; leaves slender and elongate, alternate; flowers axillary, not sunken in stem; embryo spiral.
Calyx wingless; leaves fleshy, not spinescent; embryo flattish-spiral. . 11. *Suaeda.*
Calyx horizontally winged in fruit; leaves spiny-tipped; embryo conic-spiral. . . . . 12. *Salsola.*

1. CYCLOLÒMA Moq. Winged Pigweed

Flowers perfect or pistillate, bractless. Calyx with the concave lobes strongly keeled, at length appendaged by a broad and continuous horizontal scarious wing. Stamens 5. Styles 3 (rarely 2). — A much branched coarse annual, with alternate sinuate-toothed petioled leaves, and very small scattered sessile flowers in open panicles. (Name composed of the Greek *cyclos, a circle*, and *loma, a border*, from the encircling wing of the calyx.)

1. **C. atriplicifòlium** (Spreng.) Coult. (with leaves of *Atriplex*). — Diffuse, 1–8 dm. high, more or less arachnoid-pubescent or glabrate, light green or often deep purple; leaves oblong, coarsely sinuate-dentate, often deciduous before maturing of fruit; calyx 3–4 mm. broad. — Sandy soil, Ind. to Man., s. to Ill., Ark., Tex., n. Mex. and Ariz.; spread eastw. to Que., N.E. and N.J. Late June–Oct. Fig. 986.

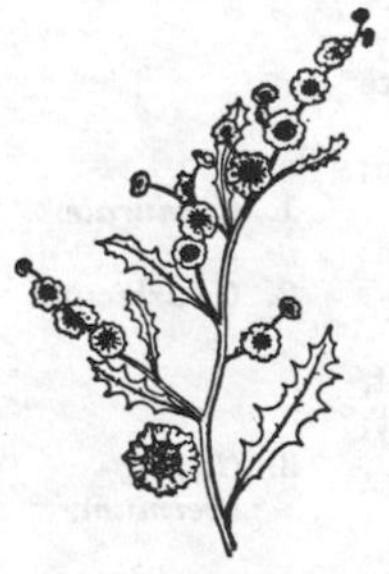

986. C. atriplicifolium.

2. KÒCHIA Roth

Differing from *Cycloloma* in its narrowly linear entire leaves, the lobed wing of fruiting calyx, the membranaceous seed-coat and the lack of endosperm. — Nat. of Old World and w. N.Am. (Named for *Wilhelm Daniel Joseph Koch*, great German botanist, 1771–1849.)

1. **K. Scopària** (L.) Roth (broom-like), Summer-cypress, Belvedere, Petit soldats (Que.). — Erect pyramidal or ovoid-topped upright branching and closely leafy annual up to 1.5 m. high; leaves narrowly lance- or linear-attenuate, often ciliate at base, strongly reddened in maturity; flowers in small axillary clusters, sessile; each sepal at length developing a narrow thickish dorsal wing or appendage. — Spread from cult. from N.E. w. and s. Aug.–Oct. (Introd. from Eurasia)

3. BÁSSIA All.

Differing from *Cycloloma* and *Kochia* in having the 5 erect or incurved calyx-lobes finally developing elongate spinescent tips or conical tubercles. — Herbs or shrubs with narrow to terete leaves, nat. of the Old World. (Named for *Ferdinand Bassi*, Curator of the Botanic Garden of Bologna in the 18th century.) Echinopsilon Moq.

1. **B. hirsùta** (L.) Aschers. (hirsute). — Hirsute annual with very divergent or arcuate bushy-branched stems up to 4 dm. high; *leaves narrowly linear, blunt, subterete,* fleshy; *flowers solitary,* axillary; *fruiting calyx with* the triangular *lobes coalescent and each bearing a dorsal conical tubercle.* (*Echinopsilon* Moq.) — Saline or brackish soil, sometimes on waste ground

inland, Mass. to Md., often appearing as if indig. but probably originally adv. Aug.–Oct. (Natzd. from Eurasia)

2. B. HYSSOPIFÒLIA (Pall.) Ktze. (hyssop-leaved). — Coarser, taller and more upright, with stiff whitish stems; *leaves flat, oblong-linear, acute,* the floral ones abbreviated and oblong; *flowers in glomerules,* forming interrupted spikes; *each mature calyx-lobe bearing a subulate prickle.* — Waste places, Mass. and se. N.Y.; sw. U.S. July–Sept. (Adv. from Eurasia)

4. ROUBIÈVA Moq.

Flowers minute, perfect or pistillate, solitary or 2–3 together in the axils. Calyx urceolate, 3–5-toothed, contracted at the apex and inclosing the fruit. Stamens 5, included; styles 3, exserted. Fruit membranaceous, compressed, glandular-dotted. Seed vertical. Embryo annular. — Perennial glandular herb, with alternate pinnatifid leaves. (Dedicated to *G. J. Roubieu,* professor at Montpellier early in last century.)

1. R. MULTÍFIDA (L.) Moq. (much divided). — Prostrate or ascending, branching and leafy; leaves lanceolate to linear (1.2–1.8 cm. long), deeply pinnatifid with narrow lobes; fruiting calyx obovate. (Perhaps better as *Chenopodium* L.) — Waste ground, local, Mass. to Ind., s. to N.C. and Ala. July–Oct. (Adv. from S.Am.)

5. CHENOPÒDIUM L. GOOSEFOOT. PIGWEED. ANSÉRINE (Que.)

Flowers all bractless. Calyx 5 (rarely 4)-parted or -lobed, more or less enveloping the fruit. Stamens mostly 5; filaments filiform. Styles 2, rarely 3. Seed lenticular, horizontal (*i.e.,* with its greatest diameter at right angles to the floral axis) or vertical; embryo coiled partly or fully around the mealy albumen. — Weeds, usually with a white mealiness or glandular. Flowers mostly sessile in small clusters collected in panicles of spikes. (Name from the Greek *chen, a goose,* and *pous, foot,* in allusion to the shape of the leaves.)* — Our species are mostly annuals, flowering through late summer and autumn. Genus essentially cosmop.

a. Seeds horizontal. . . *b.*
- *b.* Embryo completely encircling the endosperm; leaves and inflorescence not glandular, often farinose. . . *c.*
  - *c.* Upper surface of rhombic leaves lustrous, the leaf-margin with acuminate teeth.
    - Seeds dull, finely dotted, 1.2–1.5 mm. broad, with acute margin; inflorescences short, spreading, loose. . . . . 1. *C. murale.*
    - Seeds lustrous, smooth, about 1 mm. broad, with obtuse margin; inflorescence suberect, moniliform. . . . . 2. *C. urbicum.*
  - *c.* Upper surface of leaves dull or opaque. . . *d.*
    - *d.* Plant glabrous and green throughout; leaves oblong to oval, blunt, entire; cymes or spikes of flowers borne from base to summit of plant; seed lustrous, about 1 mm. broad. . . . . 3. *C. polyspermum.*
    - *d.* Plant more or less farinose, at least about inflorescence (if green throughout in no. 4, the leaves with acuminate tips and teeth); inflorescences borne chiefly toward summit of plant or its branches. . . *e.*
      - *e.* Larger leaves rounded or cordate at base, thin, acuminate and with large acuminate teeth, green throughout; pericarp easily removed from lustrous seed; seed 1.5–3 mm. broad. . . . . . . 4. *C. hybridum.*
      - *e.* Larger leaves attenuate to subtruncate at base (if rounded thick and bluntish or merely acutish); seed 0.5–1.5 mm. broad (up to 2 mm. in no. 12 with thick leaves and with pericarp closely adherent to seed). . . *f.*
        - *f.* Pericarp free from or easily removed from seed but see no. 9. . . *g.*
          - *g.* Principal leaves linear, narrowly oblong or narrowly lanceolate, entire or merely hastate-based, short-petioled.
            - Pericarp free from seed. . . . . 5. *C. leptophyllum.*
            - Pericarp adherent to but easily removed from seed. . . 6. *C. pallescens.*
          - *g.* Principal leaves ovate-lanceolate, broadly oblong, oval or rhombic, on long petioles. . . *h.*
            - *h.* Leaves thin, oblong to lanceolate, green and scarcely farinose, the larger ones 3–6 cm. long; seeds 1.3 mm. broad; plant erect, very slender, up to 1 m. high. . . . . . . 7. *C. Boscianum.*

---

* Treatment largely derived from that of PAUL C. STANDLEY in N. Am. Fl. xxi. 9–21 (1916).

h. Leaves thick, rhombic to rounded- or deltoid-ovate, about as broad as long, farinose, 1–3 cm. long; seed 0.8–1 mm. broad; plants low and spreading.

Plant not ill-scented; calyx-lobes keeled on back; pericarp free from the lustrous seed. . . . . 8. *C. incanum.*

Plant ill-scented; calyx-lobes rounded on back; pericarp adherent to dull seed. . . . . 9. *C. Vulvaria.*

f. Pericarp closely adherent to and with difficulty removed from seed. . . *i.*

i. Plant not ill-scented; leaves thickish, rhombic, ovate or lanceolate; flowers mostly in thick glomerules, forming heavy interrupted spikes or cymes; seed 1–2 mm. broad. . . *j.*

j. Principal leaves lanceolate to elliptic, narrowly ovate or oblong, entire or nearly so, green, glabrous or nearly so; glomerules scattered in slender and interrupted spikes; seed 1–1.2 mm. broad. . . . . 10. *C. lanceolatum.*

j. Principal leaves rhombic to broadly ovate, more or less farinose; glomerules crowded on dense branches; seed 1.3–2 mm. broad.

Young stem and foliage farinose; leaves glaucous, strongly farinose beneath; fresh calyx-lobes scarcely keeled; seed 1.3–1.5 mm. broad. . . . . 11. *C. album.*

Young stem often purple or red; leaves dark green or yellowish, only sparingly farinose; fresh calyx-lobes sharply keeled; seed 1.5–2 mm. broad. . . . . 12. *C. paganum.*

i. Plant usually ill-scented; leaves thinnish, oblong, ovate or rhombic, sinuate; flowers in small glomerules, the branches of the inflorescence slender; calyx-lobes sharply keeled; seeds 0.8–1 mm. broad. . . . . 13. *C. Berlandieri.*

b. Embryo only partly encircling the endosperm; leaves and inflorescences glandular. . . *k.*

k. Pericarp without glandular dots; inflorescence loosely forking, the flowers solitary, some of them pedicelled; slender annuals.

Lobes of leaf not angled, acute; calyx-segments horned; flowers puberulent-glandular. . . . . 14. *C. graveolens.*

Lobes of leaf angled, obtuse; calyx-segments without horns; flowers pubescent. . . . . 15. *C. Botrys.*

k. Pericarp glandular-dotted; inflorescence paniculate or spicate, the flowers sessile in glomerules; coarse annual or perennial. . . . . 16. *C. ambrosioides.*

a. Seeds vertical or mostly vertical, with a few horizontal. . . *l.*

l. Plant glandular-villous; seeds 0.5 mm. broad. . . . . 17. *C. carinatum.*

l. Plant glabrous or merely pruinose with short inflated hairs. . . *m.*

m. Flowers in small clusters; calyx herbaceous to slightly fleshy, not becoming succulent and bright red in fruit. . . *n.*

n. Seeds both vertical and horizontal on same plant, 0.5–1 mm. broad; style-branches very short; small annuals. . . *o.*

o. Plant farinose with inflated white hairs; calyx herbaceous.

Seed 0.6 mm. broad, smooth, with acutish margin. . . . . 18. *C. glaucum.*

Seed 0.8–1 mm. broad, tuberculate, with obtuse margin. . . . 19. *C. salinum.*

o. Plant glabrous, green to fulvous.

Flowers in leafy spikes; calyx fleshy; seed 0.8–1 mm. broad. . 20. *C. rubrum.*

Flowers in axillary glomerules; calyx herbaceous; seed 0.5–0.6 mm. broad. . . . . 21. *C. humile.*

n. Seeds all vertical, 1.5 mm. broad; style-branches long and slender; coarse perennial. . . . . 22. *C. Bonus-Henricus.*

m. Flowers in large spicate glomerules 0.5–1.5 cm. thick, in maturity succulent, brilliant red and berry-like; seed ovoid to oblong, erect.

Inflorescence naked above; seed with acutish margin. . . . . 23. *C. capitatum.*

Inflorescence leafy to summit; seed obtuse-margined. . . . . . 24. *C. foliosum.*

1. C. MURÀLE L. (of walls). — Annual, 1–6 dm. high, mostly with divergent branches from base to summit; *leaves bright green, lustrous above,* long-petioled, *triangular to rhombic-ovate, acuminate,* 3–8 cm. long, *with long narrowly triangular unequal teeth; racemes or spikes divergent, loosely somewhat corymbed;* calyx-lobes scarcely keeled; *seed* horizontal, *dull, finely dotted,* 1.2–

1.5 *mm. broad, the margin acute.* — Waste places, Que. to B.C., s. beyond our limits. July–Nov. (Natzd. from Eu.)

2. C. ÚRBICUM L. (of cities). — Similar to no. 1, *more upright* and taller, *with erect or ascending elongate moniliform inflorescences of sessile glomerules; seed* 0.7–1 *mm. broad, smooth, lustrous,* the *margin obtuse.* — Waste places, local, N.B. and N.S. to Ont., s. to s. N.E., Md., O., Ind., Ill. and Mo. July–Oct. (Natzd. from Eu.)

3. C. POLYSPÉRMUM L. (many-seeded). — *Green throughout,* up to 9 dm. high, the lower branches divergent or prostrate; *leaves oblong to oval, blunt, opaque, entire,* 3–8 cm. long, petioled; *loose to dense cymes or spikes of flowers borne from base to summit of plant;* calyx-lobes thin, spreading in age; horizontal *seed about* 1 *mm. broad, black, lustrous,* the margin rounded. — Local in waste places or cult. ground, Que. and Ont., s. to N.E., N.J., Md., Tenn. and Ia. Aug., Sept. (Adv. from Eu.)

4. C. hýbridum L. (hybrid; a misnomer), var. **gigantospérmum** (Aellen) Rouleau (gigantic-seeded), MAPLE-LEAVED G. — Upright glabrous annual, *green throughout,* 0.2–1.5 m. high, widely much branched, the branches acutely angled; *leaves thin, bright green,* long-petioled, ovate, *acuminate, rounded to subcordate at base,* 3–17 cm. long, *with* 1–few *triangular-acuminate long lobes; panicle loose, terminal,* with leafless branches; *calyx with thin* lanceolate to ovate *segments; pericarp* adherent to seed, *thin, easily rubbed off; seed* horizontal, lenticular, black, *lustrous, with low ridges radiating from center,* 1.5–2 *mm. in diameter.* — Rocky woods, thickets, clearings and waste places, Que. to B.C., s. to N.S., N.E., L.I., Del., interior Va., Ky., Mo., Okla., N.M. and Calif. July–Oct. — Typical *C. hybridum* of Eurasia has smaller seeds.

Var. **Standleyànum** (Aellen) Fern. (for PAUL CARPENTER STANDLEY, 1884–). — Seed 2–3 mm. broad. — Minn. and S.D. to Mo. and N.M.

5. C. leptophýllum Nutt. (slender-leaved). — Erect annual, 1–8 dm. high, simple or branched below, commonly *with erect branches above, copiously whitish-farinose above,* becoming greenish; *leaves short-petioled, linear, linear-lanceolate or linear-oblong, entire,* or the lowest subhastate, *thick and firm, farinose;* glomerules of flowers dense, in interrupted panicled spikes; *calyx* farinose, *completely inclosing fruit, its lobes keeled; pericarp free from seed;* seed horizontal, 1 mm. broad, lustrous, the margin obtuse. — Sandy coast, Me. to Va.; dry soil, Ont. to Alta., s. to O., Ind., Ill., Mo., Okla., Tex. and Mex., often adv. elsewhere. Aug.–Oct.

6. C. palléscens Standl. (rather pale). — Differing from no. 5 in its *divergently and loosely branching habit;* more nearly naked open inflorescence; *pericarp loosely adherent to seed; seed* 1.5 *mm. broad.* — Dry sandy or stony hills, Mo. to Tex. and N.M.

7. C. Bosciànum Moq. (for its discoverer, LOUIS AUGUSTIN GUILLAUME BOSC, 1759–1828). — Slender erect loosely and divergently branched annual 3–10 dm. high, green and nearly glabrous; *thin triangular-oblong to lanceolate leaves attenuate into slender petioles,* the lower sinuate-dentate or all entire, *the larger* ones 3–6 *cm. long;* flowers solitary or few in small glomerules along the very slender branches; calyx barely farinose; *calyx-lobes rounded, barely keeled; pericarp free; seed* horizontal, lustrous, 1.3 *mm. broad.* — Open woods, thickets, roadsides and waste places, Ga. to Tex., n. to N.E., sw. Que., s. Ont., Minn. and N.D. July–Oct.

8. C. INCÀNUM (S. Wats.) Heller (hoary). — Much branched and strongly farinose, 1–5 dm. high; *leaves thick, rhombic to rounded- or deltoid-ovate,* 1–3 *cm. long,* about as wide, the larger ones sinuate; flowers glomerulate in flexuous panicled spikes; *calyx* farinose, its oval or oblong *lobes keeled; pericarp free* from lustrous seed; *seed* 0.8–1 *mm. broad,* its margin obtuse. — Waste places, local, Me. and Mo. July–Oct. (Adv. from farther west)

9. C. VULVÀRIA L. (old generic name), STINKING G. — Low-branching *ill-scented annual with lower branches spreading* and 1.5–4 dm. long, copiously farinose; *leaves rhombic- to rounded-ovate,* 1–3.5 cm. long, nearly as wide, *entire,* thick, farinose beneath; *inflorescence leafy, of small panicles* of spikes; *calyx* densely farinose, *its lobes rounded on the back; pericarp adherent to dull seed* (about 1 *mm. broad*). — Waste places, local, Que., Ont. and Wisc., s. to Md. and Ind. July–Oct. (Adv. from Eu.)

10. C. LANCEOLÀTUM Muhl. (lanceolate), PIGWEED. — Erect annual up to 1.5 m. high, with strongly ascending branches, nearly or quite glabrous; *leaves* long-petioled, *lanceolate to elliptic, narrowly ovate or oblong, thickish, entire,* or the lower with low rounded teeth, *green,* or sparsely farinose beneath; *glomerules of flowers in slender interrupted upright spikes or cymes; calyx-lobes obtusely keeled; pericarp closely adherent to seed; seed* horizontal, lustrous, 1–1.2 *mm. broad.* — Waste or cult. ground, Nfld. to B.C., s. beyond our limits. July–Oct. (Natzd. from Eu.)

11. C. ÁLBUM L. (white), PIGWEED, LAMB'S-QUARTERS, POULETTE GRASSE or CHOU GRAS (Que.). — Similar, 0.2–2 m. high, often more branching, *more farinose; leaves glaucous, strongly farinose beneath,* at least when young; *the blades of the principal ones rhombic or rhombic-ovate, more or less sinuate-dentate or -serrate; glomerules farinose,* rather *crowded* in paniculate spikes or cymes; *seed* 1.3–1.5 *mm. broad.* — Weed of cult. and waste ground, throughout our area and beyond. June–Oct. (Natzd. from Eu.)

12. C. paganum Reichenb. (of the country), Pigweed. — Coarser than no. 11, the stem often purplish or red-tinged; *leaves rhombic to ovate, dark green to yellowish-green, only sparingly farinose; calyx-lobes sharply keeled when fresh; seed* 1.5–2 *mm. broad,* less shining. — Cult. and waste ground, Que. to Alaska, s. beyond our limits. July–Oct. (Natzd. from Eu.)

13. C. Berlandièri Moq. (for its discoverer, Jean Louis Berlandier, 1805–1851). — *Ill-scented* slender annual *with very slender arched-ascending branches,* up to 1.5 m. high; principal *leaves* long-petioled, *thin, oblong, ovate or rhombic, sinuate or dentate,* 1.2–4 cm. long; branches of inflorescence slender, moniliform; calyx slightly farinose, its *lobes sharply keeled; pericarp closely adherent; seed* horizontal, 0.8–1 *mm. broad,* often puncticulate. — Open soil, Fla. to Tex. and Mex., n. to N.C., Mo. and N.D., and adv. to Pa., Ind. and Ill. July–Nov.

14. C. gravèolens Lag. & Rodr. (strong-smelling). — *Strong-smelling* erect annual 1–6 dm. high, *slender, often paniculately branched; leaves* slender-petioled, deltoid-ovate to oblong, *laciniately or almost pectinately divided,* glabrous or viscid above, *glandular beneath; slenderly forking dichotomous branches of inflorescence with a sessile perfect flower at each fork and usually pedicelled sterile flowers, with spinulose pedicels; calyx-lobes corniculate-appendaged, glandular-puberulent;* pericarp adherent; *seed* 0.5–0.8 *mm. broad, the embryo only partly encircling the endosperm.* (*C. incisum* Poir.) — Waste or cult. ground, local, s. Me. and Mass. Aug.–Oct. (Adv. from sw. U.S., Mex. or S.Am.)

15. C. Bòtrys L. (old generic name, meaning a bunch of grapes, from the inflorescence), Jerusalem-oak, Feather-geranium. — *Aromatic glandular villous and viscid* mostly erect-branched annual 1–6 dm. high; *leaves sinuate-pinnatifid,* with *obtuse angled lobes; cymes spreading, loose, leafless, in elongate virgate panicle; flowers subsessile in axils or some of them pedicelled,* about 1 mm. long, pubescent; calyx-segments oblong or ovate; *seed* 0.6 *mm. broad, opaque.* — Waste places, roadsides and disturbed soil, often cult., Que. to Wash., s. to N.S., N.E., Va., Ky., Mo., etc. July–Oct. (Introd. from Eurasia)

16. C. ambrosioìdes L. (resembling *Ambrosia*), Mexican-tea. — *Coarse aromatic smoothish* annual or perennial *with* stout root and *stout mostly glabrous or merely waxy-pruinose* ascending simple or branching *stem* up to 1.5 m. high; leaves short-petioled, oblong or lanceolate, repand-toothed or nearly entire, the upper tapering at both ends; *spikes* densely flowered, *leafy or intermixed with leaves,* forming a slenderly pyramidal panicle; *calyx* about 1 mm. long, *glandular-dotted,* inclosing fruit; *pericarp* thin, *glandular-dotted; seed* horizontal or sometimes vertical, 0.6–0.8 *mm. broad,* embryo only partly encircling endosperm. — Waste places, cult. ground, etc., common southw., n. to N.E., N.Y., s. Ont., Wisc., Ia., etc. Aug.–Nov. (Natzd. from trop. Am.)

Var. anthelmínticum (L.) Gray (destroyer of intestinal worms), Wormseed. — Leaves (especially the lower) often more laciniate-pinnatifid; spikes very slender and elongate, leafless or with greatly reduced leaves. — Often more common. (Natzd. from trop. Am.)

Var. chilénse (Schrad.) Spegaz. (of Chile). — *Stems and branches loosely white-villous or hirsute;* inflorescence much as in last. (*C. vagans* Standl.) — Weed of waste or cult. soil, n. to N.J., e. Pa., Va. and W.Va. (Natzd. from S.Am.)

17. C. carinàtum R. Br. (keeled; from the strongly hooded calyx-segments). — *Depressed* or ascending *glandular-villous* annual, *with lower branches elongate;* slender-petioled *leaves* oblong or oblong-ovate, 1–4 cm. long, *coarsely and obtusely sinuate,* the upper reduced, *both surfaces glandular-villous; flowers in short axillary clusters; calyx about* 0.6 *mm. long, densely glandular, its segments keeled or hooded;* pericarp thin; *seed vertical,* 0.5 *mm. broad.* — Waste places, roadsides, etc., local, Mass., N.Y., N.J. and e. Pa.; Mo. to Tex. and Calif. Aug.–Oct. (Natzd. from Austral.)

18. C. glaùcum L. (blue-green), Oak-leaved G. — Low, spreading, rather densely *farinose with inflated white hairs;* slender-petioled *leaves* oblong to oblong-ovate, 1–5 cm. long, sinuate, *green and glabrous above, densely farinose beneath; flowers densely clustered in small axillary and terminal spikes, glabrous,* calyx only partially inclosing fruit; *style-branches short;* pericarp free; *seeds vertical and horizontal,* 0.6 *mm. broad, smooth, with acutish margin.* — Waste places, cult. land, roadsides, etc., Que. to Sask., s. to P.E.I., N.B., N.E., Va., O., Ind., Ill., Mo., Neb., etc. July–Oct. (Natzd. from Old World)

19. C. salìnum Standl. (of saline soil). — Similar to no. 18; leaves often more deltoid-ovate; *seeds* 0.8–1 *mm. broad, finely tuberculate, with rounded margin.* — Waste ground, Mo. July–Oct. (Adv. from farther west) — Perhaps an extreme of no. 18.

20. C. rùbrum L. (red), Coast-Blite. — Stem angled, simple to much branched, *erect,* up to 8 dm. high, *glabrous,* rather *fleshy, green to fulvous; principal leaves deltoid- or rhombic-ovate,* *up to* 1.5 *dm. long, tapering to cuneate base and acuminate tip,* hastate or coarsely toothed, the uppermost narrowly lanceolate to linear and often entire; *flowers crowded in axillary spikes,* forming a virgate leafy panicle; *calyx fleshy,* becoming reddish, its obtuse lobes 2–5; *style-branches short; seed usually vertical,* 0.8–1 *mm. broad,* its margin rounded. — Salt-marshes or saline soils, Nfld. to Wash., s. to N.B., N.E., N.J., w. N.Y., Mo., Neb., N.M., Ariz. and s. Calif. Aug.–Nov. (Eu.)

21. C. hùmile Hook. (low-growing). — *Low* annual, *widely branched at base,* 0.3–2.5 dm. high; *stem whitish,* slender; *leaves rhombic-ovate, obovate or suborbicular,* or the upper narrower, 0.7–3 cm. long, *entire or shallowly sinuate, green and glabrous; flowers in axillary glomerules, green; calyx herbaceous;* style branches short; *seed* vertical, 0.5–0.6 *mm. broad, sharp-margined.* — Saline or brackish soil, N.S., s. N.B. and s. Me.; Man. to B.C., s. to Neb., Colo. and Calif. Aug.–Oct.

22. C. Bònus-Henrìcus L. (old name), Good King Henry. — *Coarse perennial,* 0.2–3 dm. high, *with simple or subsimple* sparingly farinose to *glabrate stout stems; principal leaves broadly triangular-hastate,* 0.5–1.5 *dm. long, entire or subsinuate,* fleshy, glabrous; flowers in dense glomerules in slenderly panicled short spikes; *calyx herbaceous; style-branches long and slender; seed vertical, compressed-globose,* 1.5 *mm. thick.* — Waste places, roadsides, etc., local, Que. to Ia., s. to N.S., N.E., L.I. and Pa. Late May–Sept. (Natzd. from Eu.)

23. C. capitàtum (L.) Aschers. (with heads), Strawberry-Blite, Indian-paint. — Glabrous erect simple or branching annual 1–6 dm. high; long-petioled *leaves triangular, somewhat hastate,* sinuate, bright green; *flowers in globose distinct to confluent glomerules in the axils, in fruit becoming succulent, bright red and* 0.5–1.5 *cm. thick, forming interrupted spikes free from leaves at summit; seed ovoid to oblong, erect, with acutish margin.* (*Blitum* L.) — Light soil, clearings, burns, etc., Gaspé Pen., Que., to Alaska, s. to N.B., n. and w. N.E., n. N.J., Pa., Mich., Wisc., Minn., etc. Late May–Aug. (Eurasia)

24. C. foliòsum (Moench) Aschers. (leafy). — Similar to no. 23, more branched, smaller throughout; *leaves lacerate-serrate or -dentate; smaller leaves extending to summit of stem and branches, greatly exceeding the small* (5–10 *mm. thick*) globular *glomerules; seed with obtuse margin.* (*Blitum virgatum* L., not *C. virgatum* Thunb.) — Waste places, local, Me., Mass. and N.Y. June–Aug. (Adv. from Eu.)

6. ÁTRIPLEX L. Orach. Arroche (Que.)

Flowers monoecious or dioecious; the staminate like the flowers of *Chenopodium,* but sterile by the abortion of the pistil; the pistillate consisting simply of a naked pistil inclosed between a pair of appressed foliaceous bracts, which are enlarged in fruit and sometimes united. Seed vertical. Embryo coiled into a ring around the albumen. In one section, including the Garden Orach, there are some pistillate flowers with a calyx like that of the staminate, but without stamens, and with horizontal seeds as in *Chenopodium.* — Herbs (ours annuals), usually mealy or scurfy with bran-like scales and with spicate-clustered flowers in summer and autumn. Genus essentially cosmop. (The ancient Latin name.)

a. Margins of fruiting bracts free except at base. . . *b.*

b. Pistillate flowers of two kinds, some with regular 3–5-lobed calyx, others without; fruiting bracts suborbicular, 1–1.5 cm. broad; foliage green and glabrous. . . . . 1. *A. hortensis.*

b. Pistillate flowers uniform, all without calyx; fruiting bracts mostly smaller. . . *c.*

c. Lower leaves and branches opposite or subopposite; leaves green or soon becoming green, entire, slightly dentate or hastate; calyx of staminate flowers 4-cleft.

Spiciform branches of inflorescence naked except at base; freely tuberculate fruiting bracts 1–5 mm. long; seed 1–2 mm. broad. . . 2. *A. patula.*

Spiciform inflorescences leafy-bracted; fruiting bracts less tuberculate, 5–12 mm. long; seed 2–4 mm. broad. . . . . . 3. *A. glabriuscula.*

c. Lower as well as upper leaves and branches alternate (the lowest rarely subopposite); young stems and leaves grayish-farinose or lepidote, leaves mostly dentate; calyx of staminate flowers 5-cleft.

Erect, up to 1 m. high, with ascending branches; fruiting bracts dentate from base to apex; seed 1.5–2 mm. broad. . . . . 4. *A. rosea.*

Prostrate or depressed, the horizontal lower branches or stems 1–6 dm. long; fruiting bracts entire above the 2 median marginal teeth; seed 3–4 mm. broad. . . . . 5. *A. sabulosa.*

a. Margins of fruiting bracts united to or above middle; leaves silvery-scurfy or -mealy.

Leaves deltoid or subrhombic to hastate, petioled; fruiting bracts scarcely compressed, their margins lacerate; seed 1.5 mm. long; western inland species. . . . . 6. *A. argentea.*

Leaves oblong, oval or obovate, mostly subsessile; fruiting bracts compressed, their margins dentate; seed 2 mm. long; coastal species. . . . 7. *A. arenaria.*

1. A. HORTÉNSIS L. (of the garden), GARDEN O. — *Coarse* annual up to 1.5 m. high, *green* to yellowish or reddish, *glabrous; lower leaves* somewhat *triangular to ovate and subcordate,* 0.5–2 *dm. long,* blunt, *entire or merely denticulate;* inflorescence paniculate, of terminal and axillary spikes; *pistillate flowers of two sorts, some with* 3–5-*lobed calyx and no bracts, others with* 2 *subrotund entire bracts* (*finally* 1–1.5 *cm. broad*) *and no calyx; seeds* 2–4 *mm. broad.* — Waste places, roadsides, etc., spread from cult., N.E. and s. Que. to Mont., s. to L.I., Ky., Mo., etc. Aug.–Oct. (Introd. from Asia)

2. A. pátula L. (spreading). — Simple to much branching *slender-stemmed glabrous* (or when young farinose) annual, *with lower and sometimes nearly all primary branches opposite or subopposite* and widely divergent, up to 1.5 m. high; *leaves green,* fleshy, *lanceolate or linear to ovate- or triangular-hastate, petioled; inflorescences interrupted-spiciform,* terminating stems and branches, *only the basal glomerules with foliaceous bracts;* pistillate flowers uniform, without calyx, all inclosed by a pair of bracts; *fruiting bracts ovate-triangular or rhombic-hastate, green, herbaceous,* entire or toothed, somewhat muricate on back, *the margins free to below middle,* mostly 1–5 *mm. long; staminate calyx 4-cleft to base; seed* 1–2 *mm. broad.* — Highly variable; three pronounced but confluent vars. with us:

Lower primary leaves hastate, often dentate.

Principal leaves lance-hastate, entire or slightly dentate; upper leaves narrower and not hastate. . . . . *A. patula* (typical).

Principal leaves broadly triangular- to oval-hastate, often dentate. . . . Var. *hastata.*

Lower primary and rameal leaves linear, rarely subhastate or toothed. . . . Var. *littoralis.*

A. pátula (typical). — Saline, brackish or rich soils, both coastal and inland, Nfld. to B.C., s. to N.S., L.I., N.C., Ind., Ill., Mo., etc., both nat. and adv. July–Nov. (Eurasia) — The very similar var. **bracteàta** Westlund (with bracts), but with fruiting bracts 1–1.5 cm. long, occurs on marshes of C.B. (Eu.)

Var. **hastàta** (L.) Gray (halberd-shaped), *A. hastata* L. — Similar habitats, often commoner, Nfld. to B.C., s. to N.S., N.E., L.I., S.C., O., Ind., Ill., Mo., etc. June–Nov. (Eurasia)

Var. **littoràlis** (L.) Gray (of seashore), *A. littoralis* L. — Coast, P.E.I., N.B. and N.S. to N.J. and Pa.; L. Champlain, Vt., to s. Ont., Ind. and Wisc. (Eu.)

3. A. glabriúscula Edmondston (smoothish). — Differing from no. 2 in stouter habit; very fleshy leaves triangular to ovate; *spiciform inflorescences leafy-bracted to tip; fruiting bracts less tuberculate on back,* 5–12 *mm. long; seed* 2–4 *mm. broad.* — Seacoast, Greenl., Lab. Pen. and n. Man., s. to Nfld. and lower St. Lawrence, Que., thence to s. N.E. Aug.–Nov. (N. Eu.)

4. A. RÒSEA L. (roseate). — *Erect rather stiff* annual *up to* 1 *m. high, the branches arched-ascending, hoary-mealy; leaves alternate,* short-petioled to subsessile, the principal ones rhombic-ovate or oval, *sinuate, mealy,* the reduced upper ones entire; flowers in axillary glomerules or interrupted terminal spikes; *staminate calyx 5-cleft; fruiting bracts dentate, united to middle, firm,* often warty, 4–10 mm. long; *seed* 1.5–2 *mm. broad.* — Waste places and roadsides, N.Y. to Wisc. and southw. Aug.–Oct. (Adv. from Eurasia)

5. A. sabulòsa Rouy (full of sand). — *Depressed or prostrate, with lower alternate or subopposite weak branches* 1–6 *dm. long, lepidote-farinose; leaves* rhombic-ovate, blunt, *irregularly crenate-dentate* above cuneate base, *permanently white-lepidote or farinose,* 1.5–3 *cm. long,* short-petioled; 1–6 *flowers glomerulate in most axils; fruiting bracts* rhombic, *with* 1–3 *median low teeth near middle,* the whitish base coriaceous, the white-lepidote subherbaceous limb 6–9 mm. long; *seed* 3–4 *mm. broad.* (*A. maritima* E. Hallier, not Crantz) — Sands bordering Gulf of St. Lawrence, M.I., P.E.I. and ne. N.B. Aug.–Oct. (Nw. Eu.)

6. A. ARGÉNTEA Nutt. (silvery), SALTBUSH. — Much branched *gray-scurfy* annual, stiffly divaricate-branching, 0.2–1 m. high; *principal leaves deltoid to subrhombic or hastate, petioled; fruiting bracts with margins united to middle, lacerate; seed* 1.5 *mm. long.* — Alkaline flats, Man. to Mont., s. to Kans., Okla., N.M. and Calif.; adv. eastw. to Mich., O. and Mo. July–Oct.

7. A. arenària Nutt. (of sand), SEABEACH-O. — Widely branched *gray-scurfy* annual, ascending to depressed; *leaves oblong, oval or obovate, short-petioled to subsessile, entire, undulate* or sparsely toothed; pistillate flowers in axillary glomerules; *fruiting bracts* roundish, *compressed, their margins united to middle, dentate,* the backs muricate and reticulate; *seed* 2 *mm. long.* — Sandy coastal beaches and borders of salt-marshes, W.I. and Fla. to Tex., n. to s. N.H. Aug.–Oct.

SPINÀCIA L. (the old name) OLERÀCEA L. (fit for a garden vegetable), SPINACH, **a familiar fleshy annual with bracts of pistillate flowers united to summit, the stigmas 4 or 5, occurs as a casual waif on dumps, etc. (Introd. from Eurasia)**

7. ÁXYRIS L.

Flowers monoecious, without bracts or bractlets, pilose or villous; the staminate with hyaline 3–5-parted perianth and 2–5 stamens; the pistillate solitary in the axils, with perianth of 3 or 4 unequal segments. Ovary compressed, orbicular, the short style with 2 elongate stigmatic branches. Utricle inclosed in perianth, obovoid to spherical, winged or crested at summit. Pericarp closely adherent to seed, the latter erect, obovoid, compressed. Embryo slender, horseshoe-shaped, embracing abundant endosperm. — Alternate-leaved mostly pubescent annuals nat. of Asia and Russia. (Name probably from the Greek *axyros, unshorn.*)

1. A. AMARANTHOÌDES L. (like *Amaranthus*), RUSSIAN PIGWEED. — Upright, 0.2–1.2 m. high, usually with divergently ascending filiform branches naked below and leafy or floriferous at summit and forming an elongate raceme or panicle; leaves lanceolate to elliptic-ovate, stellate-pubescent beneath; leaves of flowering branches bracteiform; staminate spike slender, naked, terminal; fruit smooth, 2-winged at summit, the mature bracts and calyx becoming whitish and conspicuous; seed 2.5 mm. long, silvery. — Waste places, roadsides, cult. fields, etc., e. Que. to Alta., s. to N.S. and the N. States, rapidly spreading. June–Oct. (Natzd. from Siberia and Russia)

8. MONÓLEPIS Schrad.

Flowers small, glomerate in the upper axils. Sepal 1, green, entire, bract-like, fleshy, obtuse. Utricle moderately flattened. Seed vertical, much compressed. Embryo annular about copious albumen. — Small genus of Am. annuals. (Name from the Greek *monos, solitary,* and *lepis, scale.*)

1. M. Nuttalliàna (R. & S.) Greene (for its original describer, THOMAS NUTTALL, 1786–1859), POVERTYWEED. — Branched from the base, 0.7–3 dm. high, somewhat fleshy, rather pale green, scarcely or not at all mealy; leaves narrow, slender-petioled, hastate, passing gradually into foliaceous bracts. — Saline and alkaline soil, from the Great Plains westw., extending e. to Man., N. and S.D., Neb., Mo., Okla. and Tex.; casual as a weed e. to N.E. June–Oct.

9. CORISPÉRMUM L. BUGSEED

Calyx of a single delicate sepal on the inner side. Stamens 1 or 2, rarely 5. Styles 2. Fruit oval, flat, with the outer face rather convex and the inner concave, sharp-margined; seed vertical. Embryo slender, coiled around a central albumen. — Low branching annuals of N. Hemisph., with narrow linear alternate 1-nerved leaves. (Name from the Greek *coris, a bedbug,* and *sperma, seed.*)

Fruit with definite marginal wing.
  Spikes dense; fruit 3.5–4.5 mm. long. . . . . 1. *C. hyssopifolium.*
  Spike very slender and open; fruit 2–3 mm. long. . . . . 2. *C. nitidum.*
Fruit wingless or essentially so, 2.5–3 mm. long; spike subcontinuous. . . . 3. *C. orientale,* var. *emarginatum.*

1. C. hyssopifòlium L. (hyssop-leaved). — Stiffish, much branched, up to 6 dm. high, more or less stellate-villous when young, soon glabrate, often becoming reddish; leaves narrowly linear, pointed, 2–7 cm. long; *spikes* dense to slightly open, 4–8 *mm. thick; imbricated bracts* lanceolate to ovate, scarious-margined, pointed, *the lower* 6–10 *mm. long, the median and upper mostly broader than fruit; fruit strongly winged,* 3.5–4.5 *mm. long, the wing about* 0.5 *mm. wide.* — Sandy beaches, dunes and openings, w. Que. to Wash., s. to n. Ind., Ill., Mo., Okla. and Mex.; adv. e. to e. N.Y. Aug.–Oct. (Eurasia) FIG. 987.

987. C. hyssopifolium.

2. C. nítidum Kit. (neat). — More slender than no. 1; *spikes very slender and open; the distant bracts mostly narrower; fruit* 2–3 *mm. long,* with narrower wing. — Sands, nw. Ind. to Id., s. to Mo., Kans., Tex. and Ariz. Aug.–Oct. (Eu.)

3. C. ORIENTÀLE Lam. (eastern), var. EMARGINÀTUM (Rydb.) Macbr. (with shallow terminal notch). — Resembling no. 1; *spikes rather dense, 5–6 mm. thick; bracts broadly ovate to elliptic,* mostly 4–7 mm. long, *with very narrow scarious margin; fruit 2.5–3 mm. long,* essentially *wingless.* (*C. emarginatum* Rydb.) — Sands, w. Mo. Aug.–Oct. (Adv. from farther west)

10. SALICÓRNIA L. Glasswort. Samphire

Flowers perfect, 3 together immersed in each hollow of the thickened upper joints, forming a spike; the two lateral sometimes sterile. Stamens 1 or 2. Styles 2, united at base. Seed vertical, without albumen. Embryo thick, the cotyledons incumbent upon the radicle. — Low plants of saline soil, semicosmop., with succulent leafless jointed stems, and opposite branches; the flower-bearing branchlets forming the spikes. (Name composed of *sal, salt,* and *cornu, a horn;* saline plants with horn-like branches.)

Stem annual; middle flowers higher than the lateral ones.
  Scales mucronate-pointed and conspicuous, especially when dry. . . . . 1. *S. Bigelovii.*
  Scales blunt or bluntish, inconspicuous.
    Joints of spike much longer than thick, conspicuously exceeding the middle flower; seed 1.3–2 mm. long. . . . . 2. *S. europaea.*
    Joints about as thick as long, scarcely exceeding middle flower; seed 1–1.2 mm. long. . . . . 3. *S. rubra.*
Perennial with prolonged subligneous depressed stems and rhizomes; flowers nearly equal in height; seed 0.7–1 mm. long. . . . . 4. *S. virginica.*

1. S. Bigelòvii Torr. (for its discoverer, Jacob Bigelow, 1787–1879), Dwarf Saltwort. — *Simple-stemmed annual or with strongly ascending branches,* 0.3–3 dm. high, the cylindric-clavate green to red *spikes* 1–10 cm. long and *4–6 mm. thick, obtuse; the joints thicker than long; the pointed margins of the scales divergent, conspicuous as cusps when mature; middle flower half higher than lateral ones, occupying nearly the whole length of the joint;* seed pubescent, 1–1.5 mm. long. (*S. mucronata* Bigel., not Lag.) — Salt-marshes, s. Me. to S.C. Aug.–Nov.

2. S. europaèa L. (European), "Samphire", Chickenclaws, Pigeonfoot, Corail (Que.). — Annual, *erect,* 0.5–4 dm. high, simple or simply or very much branched, green, in autumn yellow, orange or red; *spikes slender,* 1–10 cm. long, *1.5–3 mm. thick, tapering at tip; their joints longer than thick; scales blunt or bluntish, sometimes notched, inconspicuous;* seeds 1.3–2 mm. long. (*S. herbacea* L., including *S. ramosissima* Woods and some other habital variants often recognized in Eu.) — Salt-marshes along coast, Ga., n. to N.S., M.I., P.E.I. and e. N.B.; inland in salt-licks and -marshes to Mich., Wisc. and Ill.; Pacific coast. Aug.–Nov. (Eurasia; Afr.)

Var. **símplex** (Pursh) Fern. (simple). — Simple or sparsely branched, weakly ascending or reclining, the long tapering simple spikes 3–10 cm. long and 3–5 mm. thick. (*Var. pachystachya* (W. D. J. Koch) Fern.) — Salt-marshes and sea-strands, Nfld. and P.E.I. to s. N.H.; inland in N.Y. and Mich. — Mostly later than typical form of sp. (Eu.)

Var. **prostràta** (Pall.) Fern. (prostrate). — Stem ascending to depressed, 0.3–2 dm. long, in better-developed plants prostrate or with lower branches prostrate; spikes few to very many, loosely spreading to drooping, 0.5–3 cm. long, 2–3 mm. thick. (*S. prostrata* Pall., including the simpler-stemmed *S. pusilla* Woods and *S. gracillima* Moss) — Salt-marshes and sea-strands, Nfld. to lower St. Lawrence, Que., s. to Knox Co., Me. (Eurasia)

3. S. rùbra Nels. (red). — *Bushy-branched* (0.5–2 dm. high), the abundant simple or forking branches ascending, turning red in autumn; scales broadly triangular, blunt or subacute; spikes slender-cylindric (2–3.5 mm. thick), blunt, rather closely jointed; *flowers crowded, the middle one* higher than the others and *usually reaching the tips of the joints;* fruit pubescent; *seed* 1–1.2 *mm. long.* — Low alkaline soil, Man. to B.C., s. to w. Minn., S.D., Wyo., Nev. and Wash. Aug.–Oct.

4. S. virgínica L. (Virginian), Perennial Saltwort, Leadgrass, Woody Glasswort. — *Forming extensive perennial mats, with subligneous rhizomes freely forking in sand, the long branching superficial stems trailing,* greenish, turning lead-color or pale brown; *spikes mostly solitary at tips of ascending branches or peduncled* along some axes, 1.5–6 cm. long, loosening in age; *the mature scales with prominent horizontally divergent firm margins;* joints of spike thicker than high, the *flowers subequal in height; seed* 0.7–1 *mm. long.* (*S. perennis* sensu Standl. in part, not Mill.; *S. ambigua* Michx.) — Sandy sea-strands and borders of salt-marshes, s. N.H. to S.C., perhaps beyond. Aug.–Oct.

11. SUAÈDA Forsk. Sea-Blite

Flowers sessile in the axils of leaves or leafy bracts. Calyx 5-parted, fleshy, inclosing the fruit (utricle) and often carinate or crested. Stamens 5. Stigmas 2 or 3. Seed vertical or horizontal, with a flat-spiral embryo, dividing the scanty albumen (when there is any) into two portions. — Fleshy halophytic plants, with alternate nearly terete linear leaves, nearly cosmop. (An Arabic name.) Dondia Adans.

Leaves linear or slender-cylindric, not broadened at base; plants of Atlantic coast.
  Seed 2 mm. broad. . . . . 1. *S. maritima.*
  Seed 1.2–1.5 mm. broad.
    Sepals rounded (not carinate) on the back. . . . . 2. *S. Richii.*
    Sepals (or some of them) carinate on the back.
      1 or 2 sepals much more cucullate-carinate than the others. . . . . 3. *S. americana.*
      Sepals subequally carinate. . . . . 4. *S. linearis.*
Leaves strongly dilated at base; plants of western plains. . . . . 5. *S. depressa.*

1. S. marítima (L.) Dumort. (maritime). — Comparatively low, 0.5–4 (rarely 5 or 6) dm. high, ascending or depressed, subsimple or with spreading-ascending or decumbent subsimple branches, or even forming depressed mats; *leaves usually glaucous, acutish, semicylindric* (*flat above, convex beneath*), 5 cm. or less long; those of the flowering branches shorter, and much exceeding the 1–4 axillary flowers; sepals pale green, rounded or obscurely keeled on the back; seed red-brown or black. (*Dondia* Druce) — Salt-marshes and sea-strands, e. Que. to Va. July–Oct. (Eurasia)

2. S. Ríchii Fern. (for its discoverer, William Penn Rich, 1849–1930). — Stems *procumbent*, forming mats 5 dm. or less across (sometimes fruiting when 1 cm. long); *leaves dark green, not glaucous, subcylindric, dorsally compressed, obtuse*, the lower 1.5 cm. or less in length; those of the flowering branches broader and shorter (4–5 mm. long); seed black. (*Dondia* Heller) — Salt-marshes and sea-strands, se. Nfld. to ne. Mass. Aug.–Oct.

3. S. americàna (Pers.) Fern. (American). — Stems *procumbent*, the branches 2 or 3 dm. long, only the abundant *densely flowered spiciform ultimate branches* ascending; *leaves linear, thickish* (*flat above*), *acute*, the lower about 2 cm. long, those subtending the crowded flowers broader and shorter; *sepals very irregular*, 1 or 2 strongly keeled. (*Dondia* Britt.) — Salt-marshes and sea-strands, Côte Nord, Que., s. to s. Me. Aug.–Oct. — Ripe plant deep red or purple.

4. S. lineàris (Ell.) Moq. (linear). — *Erect or ascending*, 2–9 dm. high, profusely branched; the *slender branches ascending or wide-spread, not procumbent;* leaves narrowly linear, acute, deep green, not glaucous, the lower 4 cm. or less long, the upper similar but shorter; sepals subequally carinate. (*Dondia* Heller) — Sandy coast, W.I. and Fla. to Tex., n. to s. Me. Aug.–Oct.

5. S. depréssa (Pursh) S. Wats. (depressed). — *Decumbent* or erect, branching from the base; *leaves broadest at base*, the cauline 1–4 cm. long, the floral lanceolate to ovate; one or more of the calyx-lobes *very strongly carinate or crested.* (*Dondia* Britt.) — Saline soil, Man. and w. Minn. to Neb., and westw. across the plains, s. to Tex., etc.; adv. in Mo. Aug.–Oct.

12. SÁLSOLA L. Saltwort. Soude (Que.)

Flowers perfect, with 2 bractlets. Calyx 5-parted, its divisions at length horizontally winged on the back, the wings forming a broad scarious border. Stamens mostly 5. Styles 2. Seed horizontal, without albumen. — Herbs or slightly shrubby branching plants with fleshy and rather terete or subulate leaves and sessile axillary flowers; the genus widely dispersed. (*Salty*, from the Latin *salsus.*)

1. S. Kàli L. (old generic name from the Persian, a large carpet), Common S., Barilla-plant. — Annual, diffusely branching, bushy, *pubescent* (rarely glabrous); *leaves* all alternate, *awl-shaped, stiffish, prickly-pointed;* flowers single; calyx with converging lobes forming a sort of beak over the fruit, the yellowish to lead-colored wings nearly orbicular and spreading. — Sandy seashores, Nfld., s. Lab. Pen. and e. Que., s. to Ga. July–Oct. (Eurasia)

Var. **caroliniàna** (Walt.) Nutt. (Carolinian). — *Plant glabrous* or but sparsely pubescent; *wings larger*, roseate. — Sw. Nfld.; Mass. to Fla.

Var. tenuifòlia Tausch (slender-leaved), Russian Thistle, Chardon de Russie (Que.). — Erect or ascending, very bushy; *leaves* especially on the young and vegetative stems *longer* (3–7 cm. in length), more slender, *filiform;* flowers somewhat variable but apparently showing no constant difference from those of the typical form. (*S. Tragus* Reichenb., not L.; *S. pestifer* Nels.) — Sandy shores, cult. fields, roadsides and wastes, Que. to B.C., s. to N.S., N.E., N.Y., O., Ind., Ill., Mo., etc. July–Oct. (Natzd. from Eurasia) — A weed of rapid spread in the last half-century, particularly across the plains.

FAM. 56. AMARANTHÀCEAE (AMARANTH FAMILY)

*Weedy herbs with nearly the characters of the preceding family, but the flowers mostly imbricated with dry and scarious persistent bracts; these often colored, commonly 3 in number.** — The greater part of the family tropical.

Leaves alternate; anthers 2-locular.
Filaments united at base; ovary with 2–8 ovules; flowers perfect. . . . . . 1. *Celosia.*
Filaments separate and distinct; ovary with 1 ovule.
Flowers monoecious, dioecious or polygamous, all with a calyx of 5 or sometimes 3 distinct sepals, these not falling off with the fruit. . . . 2. *Amaranthus.*
Flowers dioecious; pistillate flowers without a calyx. . . . . 3. *Acnida.*
Leaves opposite; anthers with 1 locule.
Calyx of 5 sepals; filaments united at base into a cup; flowers paniculate. 4. *Iresine.*
Calyx 5-cleft, the sepals united below; filaments united into a tube; flowers spicate. . . . . 5. *Froelichia.*

1. CELÒSIA L. CELOSIA

Flowers subtended by a bract and two bractlets. Calyx scarious, in fruit erect and (in our species) concealing the utricle. Stamens 5. Fruit a thin membranaceous utricle, circumscissile or dehiscing irregularly, ovoid or subglobose. Small trop. herbs. (Name from the Greek *keleos, burning*, on account of the seared appearance of the flowers.)

1. C. ARGÉNTEA L. (silvery). — Erect glabrous herb, 3–12 dm. tall; leaves lanceolate, short-petioled, acute; inflorescence a simple dense cylindrical spike; sepals white, pink or red, or inflorescence fasciated and forming a broad and flattened somewhat amorphous structure in var. CRISTÀTA (L.) Ktze. (crested), the COCKSCOMB; sepals much longer than bracts; style conspicuous, exserted. — Much cult. and frequently found in refuse and waste places. (Introd. from the Tropics)

2. AMARÁNTHUS L. AMARANTH

Flowers 3-bracted. Calyx glabrous. Stamens 5, 2 or 3, separate; anthers 2-locular. Stigmas 2 or 3. Fruit an ovoid 1-seeded utricle, 2–3-beaked at the apex, mostly longer than the calyx, opening transversely or sometimes bursting irregularly or indehiscent. Embryo coiled into a ring around the albumen. — Coarse chiefly annual plants, with alternate and entire petioled leaves, and small green or purplish flowers in axillary or terminal spiked clusters in late summer and autumn; genus semicosmop. (except in cold reg.). (Greek *amarantos, unfading*, because the dry calyx and bracts do not wither.)

a. Utricle dehiscent, thin. . . *b.*
b. Utricle regularly circumscissile, the top falling off as a lid; no spines in leaf-axils. . . *c.*
c. Erect or strongly ascending, relatively coarse; principal leaves with blades mostly 0.5–3 dm. long; flowers in terminal and axillary simple or panicled spikes; sepals and stamens 5 (3 in no. 6). . . . *d.*
d. Sepals of pistillate flowers narrow-clawed, spatulate; flowers dioecious; terminal spike prolonged, rarely with very elongate secondary spikes at base.
Principal leaves ovate, lanceolate or rhombic, tapering to tip; bracts twice to thrice length of flowers. . . . . 1. *A. Palmeri.*
Principal leaves oblong-lanceolate to oblanceolate, rounded at tip; bracts mostly shorter than flowers. . . . . 2. *A. Torreyi.*
d. Sepals of pistillate flowers scarcely clawed, oblong to ovate-lanceolate; flowers monoecious; inflorescences paniculate or of simple spikes.
Calyx of pistillate flowers 1.5–2 mm. long, the bracts slightly longer to twice as long.
Lateral spikes of terminal panicle divergent; sepals of pistillate flowers obtuse or rounded at apex; bracts red or purple (rarely green); utricle longer than calyx. . . . . 3. *A. cruentus.*
Lateral spikes of terminal panicle spreading-ascending to erect; sepals (at least the outer) of pistillate flowers acute; bracts green or merely red-tinged; utricle shorter than to about equaling calyx. . . . . 4. *A. hybridus.*

* Many characters derived from treatment of PAUL C. STANDLEY in N. Am. Fl. xxi[2] (1917).

Calyx of pistillate flowers about 3 mm. long, the bracts 2–3 times as long, green; lateral spikes of terminal panicle strongly ascending.

Terminal panicle usually lobulate, with crowded lateral spikes 1–5 cm. long; sepals of pistillate flowers round-tipped, truncate or emarginate; stamens 5. . . . . 5. *A. retroflexus.*

Terminal panicle with stiff much-prolonged central spike; the few lateral spikes also prolonged, erect and often 4–12 cm. long; sepals of pistillate flowers acute; stamens usually 3. . . . 6. *A. Powellii.*

*c.*Depressed or ascending, relatively slender; principal leaves with blades 0.5–7 cm. long; flowers in small axillary clusters; stamens 2 or 3.

Erect or ascending, diffusely branched; bracts subulate, pungent, much exceeding calyx; seeds about 0.8 mm. broad. . . . . 7. *A. albus.*

Prostrate; bracts ovate-oblong, short-acuminate, scarcely or only slightly exceeding calyx; seeds 1.3–1.5 mm. broad. . . . . 8. *A. graecizans.*

*b.*Utricle irregularly circumscissile or bursting; a pair of sharp spines in axils of many leaves. . . . . 9. *A. spinosus.*

*a.*Utricle indehiscent or irregularly bursting; spineless plants; bracts inconspicuous. . . *e.*

*e.*Stems scarcely or but slightly fleshy; leaves rhombic, ovate or lanceolate; flowers 1–2 mm. long; seeds 0.8–1.1 mm. broad. . . *f.*

*f.*Utricles rugose.

Flowers chiefly in terminal panicled slender and elongate spikes; leaves plane, the principal ones 2–5 cm. broad. . . . . 10. *A. viridis.*

Flowers in small rounded axillary glomerules; leaves crisped, 0.5–1 cm. broad. . . . . 11. *A. crispus.*

*f.*Utricles smooth. . . *g.*

*g.*Leaves acute or obtuse, not emarginate.

Principal leaf-blades 1.5–3 cm. long, 1–1.5 cm. broad, acute; flowers monoecious; utricle conic-ovoid, much exceeding calyx. . . . . 12. *A. deflexus.*

Principal leaf-blades 4.5–7 cm. long, 1–2.5 cm. broad, obtuse; flowers dioecious; utricle subglobose, about equaling calyx. . . . 13. *A. ambigens.*

*g.*Leaves emarginate or retuse.

Flowers mostly in terminal or axillary prolonged spikes 2–8 cm. long; bracts at most half as long as calyx; seeds 1 mm. broad. . 14. *A. lividus.*

Flowers only in short axillary glomerules, if terminal the spike rarely 1 cm. long; bracts about equaling calyx; seeds 0.8 mm. broad. . . . . 15. *A. ascendens.*

*e.*Stems very fleshy; leaves orbicular to round-obovate; flowers 3–4 mm. long; seeds 2–2.5 mm. broad. . . . . 16. *A. pumilus.*

1. A. Pálmeri S. Wats. (for its discoverer, Edward Palmer, 1831–1911). — Tall, up to 1.5 m. high, simple or with short erect branches; *principal leaves* long-petioled, *with ovate, lanceolate or rhombic tapering blades* 0.5–1.7 cm. long; *inflorescences consisting of prolonged erect nearly continuous* (interrupted at base) *terminal spikes* 1–3.5 *dm. long,* with or without basal smaller and axillary spikes, *dioecious; bracts* pungent, *twice to thrice length of flowers;* sepals of staminate flowers oblong, acute, 2–3 mm. long, green or red-tinged; of pistillate flowers spatulate with slender claw, obtuse, or the outer acute; stamens 5; utricle circumscissile; seeds 1.3 mm. across. — Waste places, wool-waste, railroads, etc., local, Mass. to Pa., w. to Mo. and Kans. Sept., Oct. (Adv. from farther southwest)

2. A. Tórreyi (Gray) Benth. (for John Torrey, 1796–1873). — Similar to no. 1; *leaf-blades oblong-lanceolate to oblanceolate, rounded at tip,* the principal ones 5–8 cm. long; inflorescence with few–many prolonged and interrupted spikes; *bracts mostly shorter than flowers;* seeds about 1 mm. across. — Dry open soil, Ia. to Colo., s. to Tex. and N.M.; adv. in Mich., Pa. and Va. July–Oct.

3. A. cruéntus L. (stained with blood), Purple A. — Coarse, 0.3–2 m. high; leaves long-petioled, the primary ones with ovate-lanceolate to elliptic or rhombic blades up to 3 dm. long and 1 dm. broad; *terminal inflorescence a large red or purple* (rarely green) *panicle with prolonged slender terminal spike and many horizontally divergent lateral ones;* flowers monoecious; *bracts about equaling to but slightly exceeding flowers; sepals of pistillate flowers* oblong, *about* 1.5 *mm. long,* at least the inner *round-tipped;* stamens 5; *utricle distincly exceeding calyx,* circumscissile; seed about 1 mm. broad. (*A. paniculatus* L.) — Frequently cult.; spread as a weed or waif to roadsides and waste places, s. Que. southw. and westw. Aug.–Oct. (Introd. and adv. from Asia)

A. caudàtus L. (tailed), with the thick and prolonged terminal spike arching or drooping from near its base; the broad sepals of the pistillate flowers with overlapping margins, rarely spreads from cult. (Introd. from trop. Am.)

4. A. HÝBRIDUS L. (hybrid), GREEN A., PIGWEED, WILD BEET. — Similar to no. 3, greener, often with ascending branches, the bases and roots often reddish; leaf-blades mostly 0.5–1.5 dm. long; *spikes* usually in panicles, *green, the lateral ones spreading-ascending to suberect; bracts nearly twice as long as sepals; sepals* (at least the outer) *of pistillate flowers acute; utricle shorter than to about equaling calyx.* — Waste places, cult. fields, etc., N.E., westw. and southw. beyond our limits. Aug.–Oct. (Semicosmop. weed)

5. A. RETROFLÉXUS L. (bent backward), GREEN A., PIGWEED, WILD BEET, CANNE (Que.). — Coarser than no. 4; *terminal panicles mostly lobulate;* the *crowded ascending round-tipped lateral spikes* 1–5 *cm. long,* forming a leafy-bracted elongate panicle; bracts about twice as long as calyx; *sepals of pistillate flowers about* 3 *mm. long,* linear-oblong, *rounded, truncate or emarginate at tip;* stamens 5; seeds about 1 mm. broad. — Waste or cult. ground, P.E.I., westw. and southw. beyond our limits. Aug.–Oct. (Semicosmop. weed)

6. A. POWÉLLII S. Wats. (for its discoverer, JOHN WESLEY POWELL, 1834–1902). — Similar to no. 5; *terminal panicle of few erect stiff spikes, the central one* 1–2.5 *cm. long, the lateral ones* 4–12 *cm. long;* bracts 2–3 times length of calyx; *sepals of pistillate flowers acute; stamens usually* 3. — Waste places, Me. to Pa. Aug., Sept. (Adv. from w. U.S. and Mex.)

7. A. álbus L. (white), TUMBLEWEED. — Pale green, slender, with whitish *erect or ascending* slender stems *diffusely branched,* 0.1–1 m. high; leaves obovate to spatulate-oblong, obtuse or retuse, with veiny blades 1–7 cm. long; *flowers* monoecious, *in small axillary clusters,* greenish; *bracts subulate, pungent, much exceeding calyx;* sepals 3; stamens 3; utricle rugose, longer than calyx; *seeds about* 0.8 *mm. broad.* (*A. graecizans* of ed. 7, not L.) — Disturbed or waste ground, s. Que. westw. and southw. beyond our limits. July–Oct. (Semicosmop. weed, originally described from Virginia) FIG. 988.

Var. **pubéscens** (Uline & Bray) Fern. (hairy). — Plant viscid-puberulent. — Waste ground and open plains, Colo., Nev. and N.M., e. to Mo.

988. A. albus.

8. A. graecìzans L. (simulating *A. graecus*), TUMBLEWEED. — Similar to no. 7, *prostrate or decumbent,* the less rigid stem often purplish; inflorescences more compact; *bracts ovate-oblong, short-acuminate, scarcely or only slightly exceeding calyx;* utricle scarcely rugose; *seeds* 1.3–1.5 *mm. broad.* (*A. blitoides* S. Wats.) — Disturbed or waste ground, s. Que., westw. and southw. beyond our range, mostly adv. with us from farther west. July–Oct. FIG. 989.

989. A. graecizans.

9. A. SPINÒSUS L. (spiny), THORNY A. — Smooth, bushy-branched; stem reddish; *leaves* rhombic-ovate or ovate-lanceolate, dull green, a pair of *spines in their axils;* upper clusters sterile, forming long and slender spikes; the fertile globular and mostly in the axils; flowers yellowish-green, small; *utricle* thinnish, *bursting or irregularly circumscissile.* — Disturbed and waste ground, N.E. to Minn., s. to the Tropics. July–Oct. (Semicosmop. weed of warm countries)

10. A. VÍRIDIS L. (green). — Slender, ascending, usually branching from base, up to 9 dm. high, *glabrous; leaves* long-petioled, *the rhombic or ovate thin glabrous blades plane, the principal ones* 4–8 cm. long and 2–5 *cm. broad, obtuse and emarginate; flowers mostly in terminal panicles of slender elongate spikes* (4–12 *cm. long,* 4–8 *mm. thick*); bracts ovate or lanceolate, acute, much shorter than flowers; sepals oblong, cuspidate, 1–1.5 mm. long; stamens 3; *utricle thin-walled, rugose,* equaling or exceeding calyx; seeds about 1 mm. broad. (*A. gracilis* Desf.) — Waste ground, local, Mass. and southw. July–Oct. (Adv. from Old World) FIG. 990.

990. A. viridis.

11. A. CRÍSPUS (Lesp. & Thév.) A. Br. (crinkled). — Smaller than no. 10, slender, procumbent, freely branching, *pubescent; leaf-blades* rhombic-ovate to -lanceolate, acute, *crisped,* 0.5–2.5 *cm. long,* 0.5–1 *cm. broad; flowers in small rounded axillary glomerules; utricle rugose; seeds about* 0.8 *mm. broad.* — Waste and disturbed ground, local, N.Y. and southw. Aug.–Oct. (Adv.; nat. reg. unknown)

12. A. DEFLÉXUS L. (turned back). — Similar to the two preceding, low and spreading or ascending, glabrous, or pubescent at tips; *principal leaf-blades rhombic-ovate to lanceolate,* 1.5–3 *cm. long,* 1–1.5 *cm. broad, acute; flowers* monoecious, *mostly in dense lobulate terminal spikes or short-branched* slender *panicles up to* 1 *dm. long;* bracts ovate, cuspidate, about 2 mm. long; *utricle conic-ovoid, smooth,* with fleshy walls, *much exceeding calyx.* — Waste ground, local, Mass. and southw. July–Oct. (Semicosmop. weed)

13. A. ámbigens Standl. (doubtful). — Rather stout, erect, little branched, glabrous;

principal leaf-blades 4.5–7 *cm. long,* 1–2.5 *cm. broad, obtuse; flowers dioecious* or polygamo-dioecious; the staminate in terminal paniculate spikes up to 5 cm. long and in dense axillary spikes; the pistillate glomerules in simple or paniculate drooping spikes up to 1.5 dm. long; bracts deltoid-lanceolate, half the length of the calyx; *utricle subglobose, smooth, about equaling calyx;* seeds 1.1 mm. broad. — Damp soil, n. Ill. and se. Minn.

14. A. LÍVIDUS L. (lead-colored). — Rather coarse, procumbent or ascending, the *stems* up to 1 m. long or high, *often red; leaf-blades* firm, rhombic-ovate, *with rounded retuse or emarginate tip, the principal ones* 2–8 *cm. long,* long-petioled; *flowers in* dense axillary glomerules and in *terminal thick and dense spikes* 2–8 *cm. long; bracts* ovate to oblong, *less than half as long as calyx;* sepals oblong, obtuse or acutish; *utricle conic-ovoid, overtopping calyx,* smooth; *seeds* 1 *mm. broad.* — Waste ground, local, Mass. and N.Y. to Md. Aug.–Oct. (Adv. from the Tropics) FIG. 991.

991. A. lividus.

15. A. ASCÉNDENS Loisel. (ascending). — Similar to no. 14, more slender and smaller, paler green, the leaves more scattered; *leaf-blades* 1–3 *cm. long,* deeply retuse; *flowers mostly in small axillary glomerules, if terminal in spikes rarely* 1 *cm. long; bracts about equaling calyx; utricle globose-ovoid; seeds* 0.8 *mm. broad.* (*A. viridis* sensu Standl., not L.) — Waste ground, local, Mass., N.Y. and southw. Aug.–Oct. (Pantrop. weed)

16. A. pùmilus Raf. (dwarf), SEABEACH A. — Prostrate to ascending, *with very fleshy stems* forming mats; *leaves* crowded near tip of branches, *fleshy, the orbicular or round-obovate* emarginate *blades* mostly 1–2 *cm. long, decurrent into broad petioles;* flowers in axillary glomerules; bracts half as long as calyx; *sepals* of pistillate flowers 5, oblong, obtuse, 3–4 *mm. long;* stamens 5; *utricle* fleshy, indehiscent, *ovoid,* 5 *mm. long,* rugulose; *seeds* 2–2.5 *mm. broad.* — Seabeaches, very local, Nantucket and Martha's Vineyard Ids., Mass., to N.C. July–Oct.

3. ACNÌDA L. WATER-HEMP

Habit of *Amaranthus.* Bracts 1–3, unequal. Staminate calyx of 5 thin oblong mucronate-tipped sepals, longer than the bracts; stamens 5, the anther-locules united only at the middle. Stigmas 2–5, often long and plumose-hispid. Fruit somewhat coriaceous and indehiscent, or a thin membranous utricle dehiscing irregularly (rarely circumscissile), usually 3–5-angled. — Small N. Am. genus. (Name from the Greek *a, not,* and *cnide, nettle,* from the general resemblance to species of nettle, but without stinging hairs.)

Fruit indehiscent, with firm and close ribbed or angled pericarp; bracts of pistillate flowers much shorter than utricle, blunt; seed obovate, 2–3.5 mm. long; succulent plants of tidal shores and marshes. . . . . 1. *A. cannabina.*
Fruit dehiscent or indehiscent, the thin and loose pericarp not angled; bracts of pistillate flowers nearly equaling to longer than utricle, pungent; seeds 0.6–1 mm. long; scarcely succulent plants of fresh or inland habitats.
 Utricle irregularly dehiscent or indehiscent. . . . . 2. *A. altissima.*
 Utricle regularly circumscissile. . . . . 3. *A. tamariscina.*

1. A. cannábina L. (resembling *Cannabis,* hemp). — Erect, slender to very stout, fleshy, glabrous, 0.3–2 m. high or more; leaves lanceolate to lance-linear, long-petioled, acuminate to base and apex, the longer blades 0.2–2 dm. long; flowers in elongate simple or paniculate slender often leafy-bracted subcontinuous to interrupted spikes, the lower and axillary and sometimes all in dense glomerules; *bracts lax, much shorter than fruit, blunt;* staminate flowers with ovate-oblong obtuse to acute thin sepals; *fruit obovoid,* 2.5–4 *mm. long, indehiscent, with firm and close ribbed or angled pericarp; seeds obovate,* 2–3.5 *mm. long.* — Salt-marshes and tidal shores, s. Me. to Fla. July–Oct.

2. A. altíssima Riddell (very tall). — Similar to no. 1, erect; leaf-blades narrowly ovate to lanceolate, 2.5–15 cm. long; pistillate panicles of erect or spreading densely flowered spikes; *bracts acuminate, pungent, exceeding flowers; utricle thin, not angled, ovoid to globose, about* 1.5 *mm. long,* smooth or roughened, *irregularly dehiscent or indehiscent; seeds roundish,* 0.6–0.8 *mm. broad.* (*A. tuberculata* Moq.) — River-banks, low grounds and disturbed soil, Ont. to Colo., s. to Ky., Ind., Ill. and Mo.; adv. eastw. to N.E. Aug.–Oct. Passing into the two following:

Var. **subnùda** (S. Wats.) Fern. (almost naked). — Stem prostrate or ascending, 1–4 dm. long, more branched; leaves with lanceolate to spatulate or narrowly obovate obtuse blades only 0.7–7 cm. long; pistillate flowers in very dense globular distinct glomerules 0.7–1.5 cm. in diameter; bracts shorter than in typical var. (*A. tuberculata,* var. Robins.) — Mud, sands and shores, sometimes in waste places, sw. Que. to Wisc. s. to Ky., Ill. and Mo.

Var. **prostràta** (Uline & Bray) Fern. (prostrate). — Similar to last var., but leaves acute or obtuse; glomerules scattered, loose and few-flowered, 4–6 mm. in diameter. (*A. tuberculata,* var. Robins.) — Similar habitats, sw. Que. to S.D., s. to Tenn. and Mo.; adv. in N.E. and N.Y.

3. A. tamariscina (Nutt.) Wood (resembling *Tamarix*). — Resembling typical var. of no. 2, stout, erect, up to 1.5 m. high; leaves with oblong to lanceolate blades up to 1 dm. long; *spikes* mostly naked, *very slender, stiff and straight*, single or panicled; *utricle definitely circumscissile.* — River-gravels and sands, alluvial fields and waste places, Mich. to S.D., s. to Ky., Ark., Tex. and N.M.; adv. in N.J. and Del. Aug.–Oct.

4. IRESÌNE P. Br. Bloodleaf

Flowers mostly polygamous or dioecious, 3-bracted. Calyx of 5 sepals. Stamens mostly 5. Fruit a globular utricle, not opening. — Herbs or half-shrubs of trop. and warm parts of Am. and Afr., with opposite petioled leaves, and minute scarious white flowers crowded into clusters or spiked and branching panicles; the calyx, etc., often bearing long wool (whence the name, from the Greek *eiresione, a wreath or staff entwined with fillets of wool.*)

1. I. rhizomatòsa Standl. (having rhizomes). — Stoloniferous perennial with slender horizontal rhizomes; stems mostly solitary, erect, 0.3–1.5 m. high, sparsely pubescent or glabrous, mostly simple up to inflorescence; leaves ovate, acute or long-acuminate, decurrent into slender petioles; pistillate panicle up to 3 dm. long, open; bracts and calyx silvery-white, the sepals faintly 1-nerved. (*I. paniculata* of ed. 7, not Poir.) — Low woods, Md. to s. Ill. and Kans., s. to e. Va., Ala., La. and Tex. Aug.–Oct.

5. FROELÍCHIA Moench Cottonweed

Flowers perfect, 3-bracted. Calyx tubular, 5-cleft at the summit, below 2–5-crested lengthwise or tubercled and indurated in fruit, inclosing the indehiscent thin utricle. Filaments united into a tube, bearing 5 oblong 1-locular anthers and as many sterile ligulate appendages. — Hairy or woolly Am. herbs, with opposite sessile leaves and spiked scarious-bracted flowers. (Named for *Joseph Aloys Froelich*, a German botanist, 1766–1841.)

1. F. floridàna (Nutt.) Moq. (of Florida). — *Stout* annual, *the stem simple or with few well-developed erect branches, up to* 2 m. high and 7 *mm. thick*, tomentulose, especially below, often viscid above; *leaves elliptic-lanceolate*, tapering from near or below middle to apex and base, *the larger blades* 0.5–1.2 *dm. long* and 0.5–2.5 *cm. broad*, papillose-puberulent above, sericeous-tomentose beneath; lowest internode of rachis of terminal inflorescence commonly 1–2 dm. long; *spikes* dense, 1–1.3 *cm. thick*, the terminal up to 1 dm. long, whitish, with pale to fuscous bracts; *mature fruiting calyx bearing* 2 *longitudinal wing-like toothed crests.* — Sandy soil, Fla. to Miss., n. to Del. July–Sept.

Var. **campéstris** (Small) Fern. (of low fields). — Principal leaves more oblanceolate or subspatulate, broadest above middle, rounded to apex, often silky above; lowest internode of terminal inflorescence usually only 2–10 cm. long. (*F. campestris* Small) — Sandy soil, Ind. to Minn. and Neb., s. to Mo., Okla. and Tex.

2. F. grácilis (Hook.) Moq. (slender). — Slender annual, the *stem commonly with divergent basal branches;* the larger plants 1.5–4.5 (–6) dm. high, with *stems* 1–2 *mm. in diameter; leaves* linear to narrowly lanceolate, *the larger* 3–9 *mm. broad, the upper much reduced*, silky on both surfaces; *spikes* 7–8 *mm. thick; mature fruiting calyx with two vertical rows of nearly distinct slender rigid processes.* — Sandy soil, Ia. to Colo., s. to Ark., Tex. and Mex.; adv. eastw. to N.Y., N.J. and Va. July–Sept.

Gomphrèna (modification of Pliny's *Gromphaena*, a species of Amaranth) globòsa L. (globose), Globe-Amaranth or Immortelle, a low annual with subglobose solitary heads of crimson to white crowded flowers, the filaments united into an elongate tube, spreads from cult. but rarely long persists. (Introd. from trop. Asia)

FAM. 57. NYCTAGINÀCEAE (Four-o'clock Family)

Herbs (or in the tropics often shrubs or trees) with mostly opposite and entire leaves, stems tumid at the joints, a delicate tubular or funnelform calyx which is colored like a corolla, its persistent base constricted above the 1-locular 1-seeded ovary and indurated into a sort of nut-like pericarp; stamens few, slender and hypogynous; embryo coiled around the outside of mealy albumen, with broad foliaceous cotyledons (in *Abronia* monocotyledonous by abortion). — Represented in our gardens by the Four-o'clock or Marvel-of-Peru (Mirábilis Jalápa), in which the calyx is commonly mistaken for a corolla, the cup-like involucre of each flower exactly imitating a calyx.

1. MIRÁBILIS L. Four-o'clock. Umbrella-wort

Flowers 1–5 in each 5-lobed involucre. Calyx with short or long tube and a campanulate to salverform colored limb. Stamens 3–5, hypogynous. Style filiform; stigma capitate. Fruit achene-like, ribbed or angled — Perennial herbs with opposite leaves and clustered flowers or involucres, wide-ranging in warm or temp. and trop. N. and S. Am., rarely Asia (*mirabilis, wonderful*, name originally assigned to our no. 1). Including Oxybaphus L'Hér. and Allionia Loefl.

a. Involucre herbaceous, cup-like, not enlarging, 1-flowered; calyx salverform, with prolonged tube and variously colored corolla-like limb. True Mirabilis. 1. *M. Jalapa.*
a. Involucre membranaceous, broad and open, enlarging and veiny in fruit, 3–5-flowered; calyx with very short tube and campanulate pink or purple deciduous limb. Oxybaphus. . . *b.*
b. Leaves ovate, the larger ones definitely petioled and with rounded to cordate bases. . . . . 2. *M. nyctaginea.*
b. Leaves oblong, lanceolate or linear, attenuate to sessile or short-petioled bases.
Stem densely hirsute or pilose at least below; leaves oblong to lanceolate; fruit with broad smooth angles. . . . . 3. *M. hirsuta.*
Stem glabrous or merely with puberulent lines toward base.
Leaves oblong or lanceolate; fruits with tuberculate angles. . . . . 4. *M. albida.*
Leaves linear to linear-lanceolate; fruits with smooth angles. . . . 5. *M. linearis.*

1. M. Jalápa L. (old generic name from Jalapa in Mex.), Four-o'clock, Marvel-of-Peru. — Stout-rooted, with ascending branched slender stem up to 1 m. high; leaves ovate, the lower primary ones petioled, acuminate, bright green, often viscid; involucres clustered at leafy-bracted tips of stiffly ascending branches; *involucres campanulate, herbaceous, with erect teeth; calyx salverform,* 3–6 *cm. long, with long slender tube and* reddish-purple (in cult. often yellow, white or variegated) *corolla-like limb;* fruits ovoid. — Spread from cult. to waste places, roadsides, etc., freq. southw., rarely persistent northw. June–Oct. (Introd. from trop. Am.)

2. M. nyctagínea (Michx.) MacM. (resembling *Nyctago* or *Mirabilis*). — Nearly smooth; stem becoming repeatedly forked, 0.3–1.5 m. high; *leaves broadly ovate, cordate;* inflorescence but slightly pubescent; *pedicels* slender, *becoming* 1 *cm. in length,* the lower axillary and solitary, the upper crowded upon short floral axes; involucres at length very large, 2 cm. in diameter; fruit cylindric-obovoid, 4 mm. in length, rather acutely angled. (*Oxybaphus* Sweet; *Allionia* Michx.) — Prairies and rich soil, Wisc. and Man. to Mont., s. to Ala., La. and Tex.; adv. and often weedy eastw. to Atl. States. June–Oct. — The leaves vary to oblong or ovate-lanceolate and abrupt or even cuneate at the base.

3. M. hirsùta (Pursh) MacM. (hirsute). — More or less *glandular-hirsute,* especially about the nodes and the usually contracted inflorescence, 3–9 dm. high; *leaves lanceolate to linear-lanceolate, sessile* and cuneate at base or narrowed to a short petiole; stamens often 5; fruit with thickened obtuse angles. (*Allionia* Pursh; *Oxybaphus* Sweet) — Dry open soil, Wisc. to Sask. and Wyo., s. to Mo., Okla., Tex. and N.M.; adv. eastw. to N.E. June–Oct.

4. M. álbida (Walt.) Heimerl (whitish). — Similar but *smoother;* stem whitish; *leaves oblong,* elongated, obtuse; flowers in weak individuals few, axillary, in stronger ones numerous in a terminal panicle; *angles of fruit tuberculate.* (*Allionia* Walt.; *Oxybaphus* Sweet) — Meadows and openings, Ga. to Tex., n. to S.C., Tenn., Ia. and Kans.; rarely adv. northeastw. July–Oct.

5. M. lineàris (Pursh) Heimerl (linear). — Often tall, *glabrous* except the more or less hirsute peduncles and involucres; *leaves linear,* thick and glaucous, often elongated, 5–15 cm. long; *angles of fruit smooth.* (*Allionia* Pursh; *Oxybaphus* Robins.) — Dry open soil, Minn. to Mont., s. to w. Mo., Okla., Mex. and Ariz.; rarely adv. to N.E. July–Oct.

FAM. 58. PHYTOLACCÀCEAE (Pokeweed Family)

Plants with alternate entire leaves and perfect flowers, having the general characters of Chenopodiaceae, *but usually a several-locular ovary composed of as many carpels united in a ring, and forming a berry in fruit.*

1. PHYTOLÁCCA L. Pokeweed

Calyx of 5 rounded and petal-like sepals. Stamens 5–30. Ovary of 5–12 carpels united in a ring, with as many short separate styles, in fruit forming a depressed-globose 5–12-locular berry, with a single vertical seed in each locule. Embryo curved in a ring around the albumen.

— Tall and stout perennial herbs of trop. and warm reg., with large petioled leaves, and terminal racemes which become lateral and opposite the leaves. (Name compounded of the Greek *phyton, plant,* and the middle Latin *lacca, crimson-lake;* in allusion to the crimson coloring matter which the berries yield.)

1. P. americàna L. (American), Poke, Scoke, Garget, Pigeonberry. — A smooth plant, with a rather unpleasant odor, and a very large poisonous root (often 1–1.5 dm. in diameter) sending up stout stalks at length 2–3 m. high; calyx white or pinkish; stamens and styles 10; ovary green; berries in long racemes, dark purple, ripe in autumn. (*P. decandra* L.) — Rich low ground, recent clearings and roadsides, Fla. to Tex., n. to N.E., s. Que., N.Y. and s. Ont. July–Oct. — Rind of mature stem rich purple; young leafy sprouts used as greens.

FAM. 59. AIZOÀCEAE

A miscellaneous group, *chiefly of fleshy or succulent plants, with mostly opposite leaves and usually no stipules.* Differing from *Caryophyllaceae* and *Portulacaceae* by having the ovary and capsule 1–several-locular, and the stamens and petals sometimes numerous, as in *Cactaceae* (but the petals wanting in most of the genera). Seeds with the slender embryo curved about mealy albumen. Our genera apetalous and with the calyx free from the ovary.

a. Leaves opposite, mostly unequal; calyx with definite tube. . . *b.*
b. Succulent; calyx-lobes colored within like petals; capsule circumscissile.
Stipules none; ovary 3–5-locular; seeds numerous. . . . . 1. *Sesuvium.*
Stipules present; ovary 1–2-locular; seeds 1–5. . . . . 2. *Trianthema.*
b. Non-succulent; calyx-lobes not petaloid; capsule 3-valved. . . . . . . 3. *Geocarpon.*
a. Leaves whorled; capsule 3–5-valved.
Glabrous; flowers pedicelled; sepals distinct to base. . . . . 4. *Mollugo.*
Tomentulose; flowers sessile; calyx campanulate, cleft only to middle. . . 5. *Glinus.*

1. SESÙVIUM L. Sea-Purslane

Calyx 5-parted, purplish inside, persistent, free. Petals none. Stamens 5–60, inserted on the calyx. Styles 3–5, separate. Capsule 3–5-locular, many-seeded, circumscissile, the upper part falling off as a lid. — Usually prostrate maritime herbs of trop. and warm-temp. reg., with succulent stems, opposite leaves and axillary or terminal flowers. (An unexplained name.)

1. S. marítimum (Walt.) BSP. (of the sea). — Annual, procumbent or sometimes erect; leaves oblong- to obovate-spatulate, obtuse; flowers sessile; stamens 5. — Damp coastal sands, Fla. to Tex., n., very locally, to se. N.Y. July–Oct. (W.I.)

2. TRIÁNTHEMA L. Sea-Purslane

Calyx-lobes 5, concave, colored within, with dorsal horn-like appendage from beneath the apex. Stamens 5–10, when 5 alternating with calyx-lobes. Ovary truncate, 1–2-locular; styles or stigmas 2, central, but in ours 1 and becoming excentric. Capsule short-cylindric or turbinate, 1–5-seeded, tardily circumscissile, the upper portion thickened, coriaceous or suberose, with mostly 2 rounded marginal crests. — Prostrate herbs or undershrubs, trop. and sub-trop. (Name from the Greek *treis, three,* and *anthemon, flower.*)

1. T. Portulacástrum L. (with flowers of *Portulaca,* purslane). — Diffuse dichotomously branched herb; leaves obovate or suborbicular, 1–2 cm. long; flowers purplish within, closely crowded in axils of branches. — Alkaline soils, Fla. to Tex. and Mex., n. on waste ground, locally, to N.J. and Mo. Aug.–Oct. (Pantrop.)

3. GEOCÁRPON Mackenz.

Calyx turbinate below, the 5 3-nerved green lobes erect. Petals none. Style none, stigmas 3. Capsule 1-locular, 3-valved, the sharp tips alternating with and exceeding the stigmas, many-seeded. — Small winter-annual with opposite connate leaves and axillary flowers. (Name from the Greek *ge, earth,* and *carpos, fruit.*)

1. G. mínimum Mackenz. (smallest). — Branching from base, 1–4 cm. high; leaves cucullate, ovate to narrowly elliptic, 3–4 mm. long; calyx 4–5 mm. long, slightly exceeding the capsule. — Sandy barrens, rare, sw. Mo. April, May.

4. MOLLÙGO L. Indian Chickweed. Mollugine (Que.)

Sepals 5, white inside. Stamens hypogynous, 5 and alternate with the sepals, or 3 and alternate with the 3 locules of the ovary. Stigmas 3. Capsule 3-locular, 3-valved, loculicidal, the partitions breaking away from the many-seeded axis. Seeds estrophiolate. — Low rather inconspicuous annuals of temp. and trop. reg., much branched; the stipules obsolete. (An old name for *Galium Mollugo*, transferred to this genus, probably from the similarly whorled leaves.)

1. M. VERTICILLÀTA L. (whorled). CARPETWEED. — Prostrate, forming mats; leaves spatulate, clustered in whorls at the joints, where the 1-flowered pedicels form a sort of sessile umbel; stamens usually 3. — Sandy river-banks, roadsides and cult. grounds, Fla. to Tex. and Mex., n. to N.S., N.E., s. Que., s. Ont., etc., rapidly migrating northw. June–Nov. (Weedy immigrant from Trop. Am.)

5. GLÌNUS L.

Calyx campanulate, cleft to middle. Stamens 3–5. Stigmas 3–5. Capsule 3–5-valved, loculicidal. Seeds strophiolate, with slender funiculus. — Tomentulose annuals of warm reg., with falsely verticillate leaves. (Name from the Greek *glinos, sweet juice;* application not evident.)

1. G. LOTOÌDES L. (like *Lotus*). — Depressed or ascending stellate-pubescent annual; larger leaves broadly obovate to orbicular, 0.5–2 cm. broad, petioled, accompanied by smaller blades. — Low grounds, La. to s. Calif., n. to w. Mo. and Okla. July–Sept. (Natzd. from s. Eu.)

TETRAGÒNIA (four-angled) TETRAGONIOÌDES (Pall.) Ktze. (like *Tetragonia*) or *T. expansa* Murr., NEW ZEALAND SPINACH, freely branching succulent herb with deltoid-ovate leaves abruptly narrowed at base and axillary obconic apetalous green flowers and indehiscent fruits, sometimes escapes from cult. (Introd. from e. Asia) MESEMBRYÁNTHEMUM (midday-flower) CRYSTALLÌNUM L. (crystalline), ICE-PLANT, fleshy annual, with vesicular-celled flat fleshy leaves and numerous petals and stamens, is commonly cult. and sometimes occurs in rubbish. (Introd. from Afr.)

FAM. 60. PORTULACÀCEAE (PURSLANE FAMILY)

Herbs with succulent leaves and essentially regular but unsymmetrical flowers, **viz.,** *sepals fewer than the petals; the stamens opposite the petals when of the same number, but often indefinite; otherwise nearly as* Chickweeds. — Sepals 2. Petals 5, or sometimes none. Stamens mostly 5–20. Styles 2–8, united below or distinct, stigmatic along the inside. Capsule 1-locular (somewhat 3-locular in *Talinum*), with few or many campylotropous seeds rising on stalks from the base. Embryo curved around mealy albumen. — Insipid and non-poisonous herbs, with entire leaves. Corolla opening only in sunshine, mostly ephemeral, then shriveling.

a. Calyx-tube adnate to the lower half of ovary and capsule; calyx-lobes coming off with the summit of capsule; mostly depressed branching plants. . . . 1. *Portulaca.*
a. Calyx and ovary free. . . *b.*
  b. Calyx persistent in fruit; opposite-leaved.
    Petals 3–5, somewhat unequal, often connate at base; stamens 3–5; ovules 2 or 3; leaves 1–several pairs; roots fibrous. . . . . . . 2. *Montia.*
    Petals 5, showy, equal, distinct; stamens 5; ovules usually 6; leaves a single pair; root cormatose or thickened. . . . . 3. *Claytonia.*
  b. Calyx deciduous; scapose or subscapose, with rosette of slender basal leaves. 4. *Talinum.*

1. PORTULÁCA L. PURSLANE. POURPIER (Que.)

Calyx 2-cleft; the tube cohering with the ovary below. Petals 5, rarely 6, inserted on the calyx with the 7–20 stamens, fugacious. Style mostly 3–8-parted. Capsule 1-locular, globular, many-seeded, opening transversely, the upper part (with the upper part of the calyx) separating as a lid. — Fleshy herbs of trop. and warm reg., with mostly scattered leaves. (An old Latin name, of unknown meaning.)

a. Leaves flat, naked or nearly so in the axils; calyx-lobes carinate; petals yellow. . *b.*
  b. Leaves broadly obtuse to truncate, rarely retuse; calyx-lobes pointed in bud by the projecting keel; seeds obtusely granulate.
    Prostrate; the principal leaves 1–3 cm. long; stamens 7–12. . . . . . 1. *P. oleracea.*
    Ascending; leaves 2.5–5 cm. long; stamens 12–18. . . . . 2. *P. neglecta.*
  b. Leaves often retuse; calyx-lobes obtuse in bud; seeds sharply tuberculate. 3. *P. retusa.*
a. Leaves nearly terete, linear-subulate, hairy in the axils; calyx-lobes not keeled; petals copper-color or red. . . . . 4. *P. parvula.*

1. P. OLERÀCEA L. (like a garden-vegetable), COMMON P., "PUSLEY". — *Prostrate,* very smooth; *leaves obovate* or cuneate; flowers sessile (opening only in sunny mornings); sepals keeled; petals pale yellow; *stamens* 7–12; *style deeply* 5–6-*parted;* flower-bud flat and acute. — Cult. and waste ground, too common southw., n. to s. Can. June–Nov. (Natzd. from Eu.)

2. P. neglécta Mackenz. & Bush (overlooked). — Similar to no. 1, *ascending,* forming broad clumps; *leaves* 2.5–5 *cm. long,* more spatulate; *stamens* 12–18. — Rich bottomlands, Ark. and Mo. July–Sept.

3. P. retùsa Engelm. (notched). — *Leaves often retuse;* calyx-lobes obtuse in the bud; petals small or minute; *style* shorter, 3–4-*cleft; seeds* larger, *sharply tuberculate;* otherwise similar to no. 1. — Rocky glades, sands, etc., sw. Mo. to Utah, s. to Ark., Tex. and Ariz. July–Sept.

4. P. pârvula Gray (small). — Ascending or spreading, copiously *hairy in the axils; leaves linear-subulate, nearly terete,* 6–12 mm. long; *petals reddish or copper-color* and yellow. (*P. pilosa* of ed. 7, not L.) — Barrens and sands, w. Mo. to Colo., s. to Okla., Tex., N.M. and Mex. June–Sept.

P. GRANDIFLÒRA Hook. (large-flowered), PORTULACA of the flower-gardens, persists and spreads casually after cult. (Introd. from S. Am.)

2. MÓNTIA L.

Sepals 2, persistent. Petals 3–5, usually a little unequal and often connate at the base. Stamens as many, adhering to the base of the petals. Style-branches 3. Ovules few. Seeds 2–3. — Annuals or fibrous-rooted sometimes rhizomatose or stoloniferous perennials, ours with opposite leaves; genus widely dispersed. (Named for *Giuseppe Monti,* 1682–1760, professor of botany at Bologna.)

Dwarf, with leaves linear-oblong to spatulate-oblanceolate, 0.5–7 mm. broad; mature calyx 1–1.5 mm. long; petals white, unequal, barely exceeding calyx, united at base into a gamopetalous corolla split down one side; boreal and maritime plants.
  Flowers in 2–8-flowered lateral cymes; branches usually terminated by a tuft of leaves above flowering portion; seeds about 1 mm. long, plump, slightly lustrous. . . . . . 1. *M. rivularis.*
  Flowers 1 or 2 from the axils and tips of stem and branches; seeds 1–1.6 mm. long, flattened-lenticular, highly lustrous. . . . . . 2. *M. lamprosperma.*
Coarser; leaves 0.5–1.3 cm. broad; calyx 3 mm. long; petals distinct, roseate, much longer than sepals; perennial, propagating by filiform stolons. . . . 3. *M. Chamissoi.*

1. M. rivulàris Gmel. (of rills), WATER-BLINKS. — *Deep green perennial with trailing and forking branches* 1–3 dm. long, *rooting at the nodes;* leaves oblanceolate to spatulate-obovate, 2–7 mm. wide; *flowers chiefly in* 2–8-*flowered lateral cymes; branches usually ending in a leafy tuft above the flowers* and fruits; *calyx* 1–1.5 *mm. long;* petals whitish, somewhat unequal, spreading, barely longer than calyx, united at base into a gamopetalous corolla split down one side; *seeds about* 1 *mm. long, plump, slightly lustrous* and reticulate. (*M. fontana,* in part, of ed. 7, probably not of L.) — Rills, pools and ditches, local, se. Nfld.; e. N.B. June–Sept. (Eu.) FIG. 992.

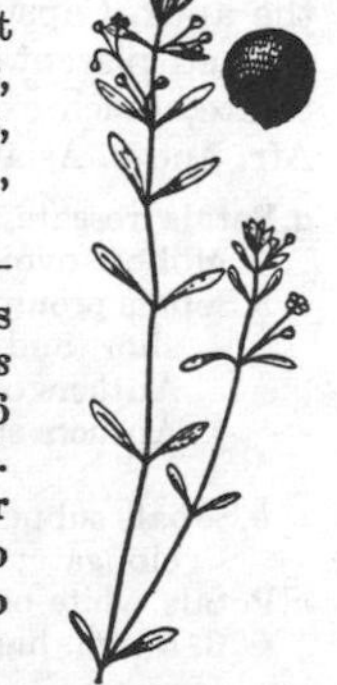
992. M. rivularis.

993. M. lamprosperma.

2. M. lamprospérma Cham. (glossy-seeded), BLINKS. — Smaller and paler, *annual;* the simple or forking weak stems not prolonged and repent at base; *flowers solitary* or *in pairs from the axils and tips,* without leafy tuft above; *seeds* 1–1.6 *mm. long,* strongly flattened, *lenticular, highly lustrous.* (*M. fontana,* in part, of ed. 7, perhaps not of L.) — Springy or seeping slopes, wet shores and brackish spots, Arct. reg., s. to mts. of nw. Nfld. and Laurentides, Que., and along coasts to Me. and s. Alaska. June–Sept. (Eurasia) FIG. 993.

3. M. Chamissòi (Ledeb.) Durand & Jackson (for its discoverer, ADALBERT VON CHAMISSO, 1781–1838). — Procumbent or ascending, *propagating by slender bulblet-bearing runners;* leaves several pairs, oblong-spatulate, 2.5–5 cm. long; *petals* 5, *pale rose-color, much exceeding the calyx;* seeds densely muriculate. (*Crunocallis Chamissonis* Rydb.) — Dripping sandstone-ledges, Winona Co., Minn., local; Alaska to N.M., Ariz. and Calif. June, July.

M. PERFOLIÀTA (Donn) Howell (perfoliate), lax with rounded long-petioled basal leaves and a pair of broad connate leaves below the clustered white flowers, and **M. SIBIRÍCA** (L.) Howell (Siberian), similar but with cauline leaves distinct and petals pink, are casually adv. about yards and shaded roadsides, probably esc. from cult. (Both adv. from w. N.Am.)

3. CLAYTÒNIA L. Spring-beauty

Sepals 2, ovate, free, persistent. Stamens 5, adhering to the short claws of the petals. Style 3-cleft at the apex. Capsule 1-locular, 3-valved, 3–6-seeded. — Perennials of N. Am., Asia, Austral. and N.Z., our two species sending up simple stems in early spring from a small deep-seated tuber, bearing a pair of opposite leaves and a loose raceme of showy flowers. Corolla rose-color with deeper veins, opening for more than one day! (Named in honor of *John Clayton*, ?–1773, one of our earliest botanists, who contributed to Gronovius the materials for the Flora Virginica.)

Leaves broadly oblanceolate or spatulate-oblong to oval; bract at base of lowest pedicel scarious, 1.5–3 mm. long, 1.5–3 mm. broad; longitudinally folded older sepals often subacute to acuminate in profile. . . . . 1. *C. caroliniana.*

Leaves linear, linear-oblanceolate or, rarely, lance-attenuate; lowest bract firmly herbaceous, 2–10 mm. long, 2–6 mm. broad; folded older sepals usually blunt. 2. *C. virginica.*

1. C. caroliniàna Michx. (of Carolina). — Corm 0.5–2.7 cm. thick, bearing 1–8 fragile succulent flowering stems 0.3–3 dm. long; *leaves broadly oblanceolate or spatulate-oblong to oval or narrowly ovate*, petioled or rarely sessile; the blade 1.3–10 cm. long, 0.5–3.3 cm. broad; raceme (1–) 2–11-flowered; *bract subtending lowest pedicel scarious*, 1.5–3 *mm. long and broad;* pedicels at first erect, finally spreading or recurving; calyx in flower 4.8–8 mm., in fruit 5–9 mm. long; the *sepals* obtuse to acute, *when folded often appearing subacute to acuminate;* petals 0.9–1.5 cm. long, white or pink with deeper stripes. — Rich open woods, alluvial thickets, upland slopes, etc., w. Nfld. to Sask., s. to N.S., N.E., mts. of N.C. and Tenn., n. Ill. and Minn. Mar.–July (northw.)

2. C. virgínica L. (of Virginia). — Similar; corm 1–5 cm. thick, bearing 2–40 flowering stems; *leaves linear or linear-oblanceolate, tapering to each end*, 4–15 cm. long, 0.2–1 cm. wide, or lanceolate and 1–2.5 cm. wide in forma **robústa** (Somes) Palmer & Steyerm. (robust); raceme 5–19-flowered; *lowest bract firm, herbaceous*, 2–10 *mm. long*, 2–6 *mm. broad; sepals blunt or bluntish;* petals broadly oblong to obovate, one and one-half to three times length of sepals, or petals narrowly oblong and only 3–5 mm. long and exceeded by sepals and stamens in forma **micropétala** Fern. (with tiny petals). (*C. media* Small) — Rich woods, thickets and clearings, sw. Que. and w. and s. N.E. to s. Ont. and Minn., s. to Ga., Ala., Miss., La. and Tex. Mar.–May.

4. TALÌNUM Adans. Fameflower

Sepals 2, distinct and free, deciduous. Petals 5, ephemeral. Stamens 5–∞. Style 3-lobed at the apex. Capsule 3-locular at the base when young, 3-valved, with many seeds on a globular stalked placenta. — Ours perennials, subscapose from a thickish rhizome. Leaves linear, subterete, much exceeded by the peduncles. Flowers white or rose-colored, cymose. Small Am., Afr. and s. Asiatic genus. (Derivation of name obscure.)*

*a.*Petals roseate, 8–13 mm. long; stamens 8–30 or more; capsules globose or globose-ovoid. . . *b.*

*b.*Sepals promptly deciduous; petals 5, about 8 mm. long; stamens 8–25; style short and stout; capsule 3–5 mm. high.

Anthers oblong; stigmas very short; seeds smooth. . . . . 1. *T. teretifolium.*

Anthers spherical; stigmas elongate; seeds rough. . . . . 2. *T. rugospermum.*

*b.*Sepals subpersistent; petals 8–10, 1–1.3 cm. long; stamens 30 or more; style elongate; capsule 6–7 mm. high. . . . . 3. *T. calycinum.*

*a.*Petals white or pale pink, 3–7 mm. long; stamens mostly 5; capsule ellipsoid, 3–5 mm. high. . . . . 4. *T. parviflorum.*

1. T. teretifòlium Pursh (with leaves round in cross-section). — Rhizome fleshy, thick, ascending, bearing 1–∞ short or elongate caudices; the subcylindric *leaves* 2–5 cm. long and 1–2 *mm. thick, blunt or mucronate;* scapes 1–2 dm. high; cyme two to four times forked; the *bracts all subtending well-developing branches or pedicels; sepals promptly deciduous; petals* 5, *about* 8 *mm. long, roseate;* stamens 15–20, *anthers oblong;* stigmas very short; capsule globose-ovoid, 3–5 mm. high; *seeds smooth* and lustrous. — Dry rocks (often serpentine) and sands, Ga. and Ala., n. to barrens of se. Pa., W.Va. and Tenn.; formerly in sw. Ct. June–Sept. — Flowers expanding from noon to mid-afternoon.

* Treatment based largely on that of N. C. Fassett in Rhodora, xxx. 205, 206 (1928).

2. T. rugospérmum Holzinger (roughened-seeded). — Similar to no. 1; rhizome more slender; *leaves* usually more slender, *with a curved mucronate tip; many* of the *bracts subtending aborted flowers;* stamens 12–25, *anthers spherical; style cleft one-third its length; seeds* finely *rugose.* — Dry sands and sandstone, nw. Ind. and Ill. to e. Minn. and ne. Ia. June, July. — Flowers expanding in late afternoon.

3. T. calycinum Engelm. (with large calyx). — Coarser than no. 1; leaves blunt; *sepals* somewhat *persistent* about the fruit; *petals* bright red, 8–10, 1–1.3 *cm. long; stamens* 30 *or more; style elongate;* anthers elongate; *capsule* 6–7 *mm. high;* seeds smooth. — Cliffs, dry barrens and sands, Ill. to Neb., s. to Ark. and Tex. June–Aug.

4. T. parviflòrum Nutt. (small-flowered). — Smaller throughout; root elongate and branching; scapes 3–15 cm. high; sepals promptly deciduous; *petals white or pale pink,* 3–7 *mm. long; stamens mostly* 5; *capsule ellipsoid,* 3–5 mm. high; seeds smooth. — Sterile rock and sands, s. Minn. to Colo., s. to Ark., Tex. and Mex. May–Aug.

FAM. 61. CARYOPHYLLÀCEAE (Pink Family)

Herbs with mostly opposite entire leaves and symmetrical 4–5-merous flowers with or without petals; the distinct stamens no more than twice the number of the sepals, either hypogynous or perigynous; styles 2–5 (or rarely united into one); seeds several or usually many, attached to the base or to the central column of the 1-locular (rarely 3–5-locular) capsule, or 1 in a utricle, with a slender embryo coiled or curved around the outside of mealy albumen (in *Dianthus* nearly straight). — Bland herbs; the stems usually swollen at the joints; uppermost leaves rarely alternate. Leaves often united at the base. Calyx persistent. Styles stigmatic along the inside. Seeds amphitropous or campylotropous. Including Illecebraceae.

a. Fruit a 1-seeded utricle (indehiscent).
  Stipules and petals 0; calyx an indurated urceolate cup, its lobes much overtopping the utricle; flowers green, sessile in upper axils. Tribe I. Scleranthеae. . . . . 1. *Scleranthus.*
  Stipules well developed; sepals or calyx-lobes nearly distinct; flowers pedicelled in the axils or in cymes. Tribe II. Paronychieae.
    Leaves opposite; petals 0; style-branches or stigmas 2; utricle membranaceous. . . . . 2. *Paronychia.*
    Leaves alternate; petals 5; stigmas 3, sessile; utricle crustaceous. . . . 3. *Corrigiola.*
a. Fruit a several–many-seeded capsule; petals oftenest present. . . *b.*
  b. Sepals distinct or essentially so; petals (when present) and stamens borne at base of sessile ovary or on a basal disk; petals without claws. Tribe III. Alsineae. . . *c.*
    c. Leaves with scarious stipules; petals uncleft at tip.
      Leaves opposite; styles and valves of capsule 3. . . . . 4. *Spergularia.*
      Leaves whorled; styles and valves of capsule 5. . . . . 5. *Spergula.*
    c. Leaves without stipules. . . *d.*
      d. Capsule ovoid or ellipsoid, dehiscent by as many or twice as many valves or teeth as there are carpels.
        Styles alternating with sepals, 4 or 5.
          Petals entire or none; leaves linear-filiform or -subulate. . . . 6. *Sagina.*
          Petals deeply notched; leaves cordate-ovate. . . . . . . 7. *Myosoton.*
        Styles opposite sepals, usually 3.
          Petals entire or barely emarginate; valves of capsule entire, bifid or 2-parted; leaves linear-filiform to dilated. . . . . 8. *Arenaria.*
          Petals deeply notched or bifid (rarely wanting); valves of capsule bifid or 2-parted; leaves linear to ovate. . . . . 9. *Stellaria.*
      d. Capsule cylindrical, often bent near summit or curved, dehiscent by twice as many equal teeth as there are carpels.
        Petals notched or 2-cleft; styles 5 (3 in one species with flowers solitary in axils); seeds laterally somewhat compressed, attached edgewise; flowers not umbellate. . . . . 10. *Cerastium.*
        Petals denticulate, not 2-cleft; styles 3; seeds dorsally flattened, attached by the inner face; flowers in peduncled umbels. . . . . 11. *Holosteum.*
  b. Sepals united into a tube or cup; petals with basal claws, borne with the stamens on the stipe of the stalked ovary. Tribe IV. Sileneae. . . *e.*
    e. Calyx naked at base, without bracteoles; seeds globular or reniform, laterally attached; embryo curved or coiled. . *f.*
      f. Styles 5; capsule opening by 5 or 10 teeth.
        Calyx-lobes leaf-like, 2–3 cm. long, much overtopping the unappendaged petals and the capsule; styles opposite petals. . . . . . 12. *Agrostemma.*

Calyx-lobes not prolonged; petals with or without appendages at base of blade; styles alternate with petals. . . . . . . 13. *Lychnis.*

*f.*Styles 2 or 3; capsule with 3, 4 or 6 valves.

Calyx 10-nerved; styles 3; capsule 3- or 6-valved. . . . . . . 14. *Silene.*

Calyx 5-nerved or angled or terete and only obscurely nerved; styles 2; capsule 4-valved.

Calyx subcylindric and obscurely nerved or ovoid and 5-angled; capsule dehiscent by short teeth. . . . . 15. *Saponaria.*

Calyx top-shaped or campanulate; capsule deeply cleft. . . . 16. *Gypsophila.*

*e.*Calyx with scarious or foliaceous bracts at base, cylindric; seeds compressed, attached by their flat or concave face; embryo nearly straight. 17. *Dianthus.*

Tribe I. SCLERÁNTHEAE DC.

1. SCLERÁNTHUS L. KNAWEL

Sepals 5, united below into an indurated cup, inclosing the utricle. Stamens 10 or 5. Styles 2, distinct. — Insignificant little weeds, nat. of Old World, with subulate leaves, obscure greenish clustered flowers and no stipules. (Name from the Greek *scleros, hard,* and *anthos, flower;* from the hardened calyx-tube.)

1. S. ÁNNUUS L. (annual). — Much branched, spreading (7–12 cm. high); flowers sessile in the forks; calyx-lobes scarcely margined. — Waste places and roadsides, P.E.I. to Minn., s. to Fla. and Miss. Mar.–Oct. (Natzd. from Eu.)

Tribe II. PARONYCHÌEAE DC.

2. PARONÝCHIA Mill. WHITLOW-WORT

Sepals 5, distinct or essentially so. Petals 0 or minute teeth or setiform staminodia. Stamens 2–5. Style 2-cleft at apex or stigmas 2 and sessile. Utricle inclosed in or slightly exceeding the câlyx. Radicle ascending or turned downward. — Dry herbs of temp. and warm reg., with stipules, the insignificant flowers clustered in dense or loose cymes or in the forks of the stem. (Greek name for a *whitlow* or *felon,* a disease of the nails, and for plants with whitish scaly parts, once supposed to cure it.) Including ANYCHIA Michx. and ANYCHIASTRUM Small.*

*a.*Perennials with elongate stipules and linear to acicular leaves; flowers sessile in terminal corymbs. § ACONYCHIA.

Flowers hidden by the conspicuous silvery bracts, hairy; leaves flat, silky. 1. *P. argyrocoma.*

Flowers much exceeding the short silvery bracts, glabrous; leaves linear-bristleform, glabrous. . . . . 2. *P. virginica.*

*a.*Annuals or short-lived perennials with short stipules; leaves oblong, elliptic or spatulate; flowers short-pedicelled in the forks or in cymules on the branches. § ANYCHIA. . . *b.*

*b.*Perennial, forming depressed mats, freely branching from the crown, the hard branches becoming 3–6 dm. long. . . . . 3. *P. riparia.*

*b.*Annuals with slender erect stems branching well above the base, or if depressed and basally branching with longer branches only 0.5–2 dm. long.

Sepals flat and plane, much shorter than the mature subglobose utricle, their tips not evidently mucronulate; stems glabrous. . . 4. *P. canadensis.*

Sepals (at least in age) longitudinally ribbed, rarely much exceeded by the slenderly obovoid utricle, usually mucronulate to subulate-tipped; branches puberulent. . . . . 5. *P. fastigiata.*

§ ACONÝCHIA Fenzl

1. P. argyrócoma (Michx.) Nutt. (with silvery locks), SILVERLING. — Forming broad tufts, freely branched, *few of the branches fertile; leaves* linear, *flat, permanently silky;* inflorescence densely cymose, surrounded by conspicuous large silvery bracts; *calyx* hairy, 4–5 *mm. long,* the sepals with flattish usually hairy awns; petals mere teeth between the stamens. — On or among acid rocks, mts. of Ga. and e. Tenn. to Va. and W.Va. July–Sept.

Var. **albimontàna** Fern. (of the White Mountains). — Branches often more floriferous; *leaves glabrate, the margins involute;* cyme mostly lax; *calyx* 3.3–4 *mm. long,* its awns less flattened and less pubescent. — Bare granitic slopes, mts. (or sandy river-banks) of N.H. and w. Me.; ledges near mouth of Merrimac R., Mass., local. Late June–Sept.

2. P. virgínica Spreng. (of Virginia). — Smooth, tufted; *stems* (1–4 dm. high) *ascending*

* Treatment of § *Anychia* largely based on that in Rhodora, xxxviii. 416-421 (1936).

from a rather woody base; *leaves* (1.2–3.6 cm. long) *and bracts narrowly subulate; cymes open, repeatedly forked;* sepals short-pointed; minute bristles in place of petals. (*P. dichotoma* Nutt., not DC.) — Rocky places, w. Md. and W.Va. to mts. of N.C. and e. Tenn.; Ark., Okla. and Tex. July–Oct.

§ Anýchia (Michx.) Fenzl

3. P. ripària Chapm. (of river-banks). — *Depressed perennial* (sometimes annual farther southw.); *the matted branches trailing or spreading* 3–6 *dm. from the subligneous crown,* much forking; leaves oblong, acute, the larger ones 1–2 cm. long, with stipules 3–6 mm. long; branches diffuse, glabrous or puberulent, with many terminal and axillary cymules; calyces 1–1.5 mm. long, the corrugated glabrous sepals mucronulate. — Sandy shores, borders of woods and sandhills, n. Fla. and Ala., n. to se. Va. Late June–Sept.

4. P. canadénsis (L.) Wood (Canadian), Forked Chickweed. — *Glabrous annual,* very slender and erect, many times forked above and bushy-branched, 1–4 dm. high, *with elongate internodes; leaves thin, elliptic;* inflorescence diffuse, the minute pedicelled flowers in the axils of leaves and leafy bracts; stipules and stipular bracts short; *sepals flat, blunt, scarcely corrugated; styles distinct* or nearly so; *utricle exceeding calyx, subglobose.* (*Anychia* Ell.) — Thin or dry woods and rocky or sandy openings, s. N.H. to s. Ont. and Minn., s. to Ga., Tenn., Ark. and e. Kans. Late June–Oct.

5. P. fastigiàta (Raf.) Fern. (with crowded erect branches), Forked Chickweed. — Similar to no. 4, *stiffer, with shorter to crowded internodes; stem pilose or puberulent; leaves firm,* oblanceolate to narrowly obovate or narrowly elliptic; stipules and stipular bracts longer, attenuate; *sepals usually corrugated, mucronulate to subulate-tipped; styles united below; capsule* obovoid, *included or barely exserted.* — Very variable, with four fairly defined vars.:

a. Stem stiffly ascending, the freely branching summit flabelliform in outline; leaves grayish, those of the primary axis 1–2 cm. long; stipular bracts lance-attenuate; sepals definitely corrugated; united styles much shorter than ovary. . . *b.*
b. Sepals with minutely mucronulate tips.
Stipular bracts (subtending the flowers) shorter than the calyces. . . *P. fastigiata* (typical).
Stipular bracts equaling or exceeding the calyces. . . . . Var. *paleacea.*
b. Sepals with subulate awns 0.2 mm. long. . . . . Var. *Nuttalli.*
a. Stem diffusely to horizontally branched, forming intricate mats; leaves greener, the larger (primary) only 0.7–1.2 cm. long; stipular bracts ovate-lanceolate; sepals less corrugated to plane, blunt, very minutely mucronulate; united styles nearly or quite equaling ovary. . . . . Var. *pumila.*

P. fastigiàta (typical), *Anychia polygonoides* Raf. — Dry woods or rocky or sandy openings, Mass. to Minn., s. to Fla. and Tex. July–Sept. Fig. 994.

Var. **paleàcea** Fern. (chaffy). — Del. and Pa. to Ill., s. to Va. and Tenn.

Var. **Nuttálli** (Small) Fern. (named in 1925 for Lawrence W. Nuttall, its discoverer). — Mts. of Pa.

Var. **pùmila** (Wood) Fern. (dwarf), *Anychiastrum montanum* Small. — Pa. to Ga. and Ala.

Herniària L. (from its reputed medicinal virtues), Herniary, an Old World genus of matted herbs with intricately branched slender stems, opposite oblong crowded leaves, minute scarious stipules, tiny green granular flowers crowded in axillary cymes, the calyx deeply 5-parted and inclosing the indehiscent 1-seeded capsule, is casually adv. on waste ground: H. glàbra L. (smooth), glabrous except for hairs on leaf-margins, local, Me. to N.Y.; H. cinèrea DC. (ashy), the plant gray with short hispidity, sands near Green Bay, Wisc. (Adv. from Eu.)

994. P. fastigiata (typical).

3. CORRIGÌOLA L. Strapwort

Calyx with 5 concave lobes. Petals 5, pale, about equaling calyx-lobes. Stamens 5. Stigmas sessile. Utricle crustaceous, included. — Mostly alternate-leaved herbs with well-developed stipules, nat. of Eurasia, Afr. and S. Am. (Name from *corrigia, a shoe-string,* perhaps from the slender depressed stems.)

1. C. littoràlis L. (of the seashore). — Glaucous depressed slender annual, with stems or branches 1–4 dm. long; leaves lance-oblong, with semisagittate stipules; flowers crowded in terminal cymes or axillary; petals white, oblong to narrowly obovate; utricle 3-nerved. — Waste places and grassland, local, Mass. to Pa. and e. Md. June–Sept. (Adv. from Eu.)

Tribe III. ALSÍNEAE DC.

4. SPERGULÀRIA J. & C. Presl SAND-SPURREY

Sepals 5. Petals 5, entire. Stamens 2–10. Styles and valves of the many-seeded capsule 3, very rarely 5 (when the valves alternate with the sepals!). Embryo not coiled into a complete ring. — Low herbs, ours mostly annuals or biennials, chiefly on or near saline areas, with filiform or linear opposite leaves, and sometimes smaller ones clustered in the axils; stipules scaly-membranaceous; flowering all summer. (Name a derivative of *Spergula*.)* TISSA and BUDA Adans.

a. Leaves fascicled, subterete, strongly mucronate; stipules lanceolate or narrowly triangular-acuminate, much longer than broad; capsules about equaling glandular-pilose calyx; seeds rounded or angular, dark brown, deeply sculptured, 0.4–0.6 mm. long. . . . . 1. *S. rubra.*

a. Leaves scattered or but slightly fascicled, fleshy, blunt or only short-mucronate; stipules broadly triangular to ovate, nearly or quite as broad as long; capsule usually exceeding glabrous or glandular-pilose calyx; seeds brown, not deeply sculptured, 0.6–1.4 mm. long (if deeply sculptured and only 0.4–0.5 mm. long, seeds black and pyriform). . . *b.*

b. Seeds brown, 0.5–1.4 mm. long, not deeply sculptured; leaves 0.6–2 mm. wide; mature capsule ovoid, 3.6–8 mm. long, usually much exceeding calyx.

Stipules slightly longer than broad, acuminate or subequilaterally triangular, 2–6 mm. long; leaves short-mucronate; sepals pilose or glabrous; seeds 0.5–1.1 mm. long.

Leaves mostly not fascicled, 0.6–1.5 mm. wide; stipules triangular, at most 4 mm. long; longest fruiting pedicels 1–10 mm. long; stamens 2–5; mature capsule 3.5–6 mm. long; seeds wingless, rarely with thin friable and erose wing up to 0.4 mm. wide. . . . . 2. *S. marina.*

Leaves mostly fascicled, up to 2 mm. wide; stipules ovate, up to 6 mm. long; longest fruiting pedicels mostly 1–2 cm. long; stamens 9 or 10; capsules 5.5–9 mm. long; seeds mostly with a firm wing. . . . . 3. *S. media.*

Stipules mostly broader than long, truncate or apiculate, 1–2.8 mm. long; leaves blunt; sepals glabrous; stamens 2–5; seeds 0.8–1.4 mm. long, often with erose or friable wing. . . . . 4. *S. canadensis.*

b. Seeds black, with silvery tinge, 0.4–0.5 mm. long, pyriform, deeply sculptured; leaves 0.5–1 mm. wide; stipules usually broader than long, 1–2 mm. long; capsules globose, 2.5–4 mm. long, about equaling calyx. . . 5. *S. diandra.*

1. S. rùbra (L.) J. & C. Presl (red). — Annual or short-lived perennial, simple or diffusely matted or prostrate, with branching stems up to 3.5 dm. long, glabrous or sparsely glandular-pilose; *leaves fascicled, strongly mucronate, subterete, not fleshy,* linear-filiform, up to 2.5 cm. long, 0.4–1.2 *mm. wide; stipules lance- or narrowly triangular-attenuate,* lustrous white, 2.5–5 *mm. long;* cyme leafy, many-flowered, usually glandular, the lower bracts like the foliage-leaves; pedicels filiform, glandular-pubescent, the lower becoming 3.5–13 mm. long, spreading or reflexed; *sepals lanceolate,* usually *pubescent,* 3.5–5 mm. long, exceeding the pink petals; *stamens usually* 10, or fewer and with some aborted ones; *capsule about equaling calyx; seeds semiobovate to gibbous-cuneate or angled, wingless,* 0.4–0.6 *mm. long, deeply sculptured* with closely interwoven vermiform pattern with minute dark papillae over the surface particularly near the swollen rim. — Dry sandy, gravelly or sterile soils, Nfld. to Minn., s. to Va. and very rarely to Ala.; w. N.Am.; apparently indig. about Gulf of St. Lawrence, elsewhere natzd. from Eu. Late May–Oct. FIG. 995.

995. S. rubra.

2. S. marìna (L.) Griseb. (marine). — *Fleshy* annual, erect or more often diffuse, with branches up to 3.5 dm. long; *leaves fleshy, rarely fascicled, bluntly mucronate,* up to 4 cm. long, 0.6–1.5 mm. broad, glandular-pubescent or glabrous; *stipules triangular,* about as long as broad or slightly longer, 2–4 mm. long; cyme usually lax, the lower and sometimes the upper bracts foliaceous and elongate or the upper much abbreviated; sepals ovate, blunt, glandular or glabrous, 2.4–5 mm. long, much exceeding the white or pink petals; *stamens* 2–5; *capsule ovoid,* equaling or exceeding calyx, 3.5–6.5 *mm. long; the lower fruiting pedicels* 1–10 *mm. long; seeds* pale brown or reddish, opaque, *not sculptured,* glandular-papillate, or without papillae in the minor var. **leiospérma** (Kindb.) Gürke (smooth-seeded), 0.5–

996. S. marina.

* Treatment derived from that of R. P. ROSSBACH in Rhodora, xlii (1940).

0.9 mm. long, *wingless or with thin friable wing.* (*S. salina* J. & C. Presl) — Saline or brackish soils, Que. to B.C., s. along coast to Fla. and locally inland to centr. N.Y., Ill., Tex. and N.M. June–Oct. (Eurasia; Baja Calif.; S.Am.) Fig. 996.

3. S. mèdia (L.) C. Presl (intermediate). — Coarser than no. 2, often biennial or perennial; *leaves fascicled* in the axils, 1–5 cm. long, 0.8–2 *mm. broad,* usually glabrous; *stipules triangular-ovate,* as long as or longer than broad, 2.5-6 *mm. long; lower fruiting pedicels mostly* 1–2 *cm. long;* sepals narrowly ovate, mostly 3–6 mm. long, glabrous, rarely pubescent, longer than the white petals; *stamens* 9 *or* 10; *mature capsules* 5.5–9 *mm. long,* much exceeding calyx; *seeds* dark brown, smooth or with only delicate markings, 0.6–1.1 mm. long, *mostly with a firm wing.* (*S. marginata* (DC.) Kittel; *S. alata* Wieg.) — Saline soils, Shelter I. and near Onondaga and Cayuga Lakes, N.Y.; also Pacific coast and S.Am. June–Oct. (Natzd. from Eu.) Fig. 997.

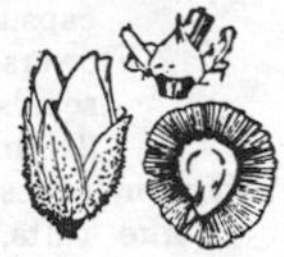

997. S. media.

4. S. canadénsis (Pers.) Don (Canadian). — Annual, *glabrous nearly throughout;* the ascending to prostrate stems up to 2.5 dm. long; *leaves* fleshy, *blunt,* 0.6–2 mm. broad; *stipules mostly broader than long, truncate or apiculate,* 1–2.8 *mm. long;* cyme leafy; sepals ovate, blunt, 2.2–3.2 mm. long, mostly exceeding the white or pink petals; *stamens* 2–5; fruiting pedicels glabrous (rarely sparsely glandular), 4–15 mm. long; *seeds* brown, rounded above, 0.8–1.4 *mm. long, wingless or* entirely or partially *surrounded by a white erose wing, surface nearly smooth* or minutely reticulate. — Saline or brackish soils, Lab. and lower St. Lawrence, Que., to L.I.; Alaska and B.C. June–Oct. Fig. 998.

998. S. canadensis.

5. S. diándra (Guss.) Boiss. (with two stamens). — Small annual, prostrate or diffuse, with stems up to 1.5 dm. long; leaves glandular, short-mucronate, 1–2.5 cm. long, 0.5–1 mm. wide; *stipules* about as broad as long, 1–2 *mm. long;* sepals ovate-lanceolate, 2.5–3.5 mm. long, glandular, longer than white or pink petals; stamens 4–7; *capsules almost globose,* 2.5–4 *mm. long, equaling or slightly exceeding calyx; seeds black, with silvery tinge, pyriform,* 0.4–0.5 *mm. long, with interwoven vermiform pattern* or sometimes smooth or with scattered papillae, wingless. (*S. atheniensis* Heldr. & Sart.) — Gravelly soil, local, near shore of Buzzards Bay, Mass.; Ga.; Pacific coast. June–Oct. (Adv. from Eu.) Fig. 999.

999. S. diandra.

5. SPÉRGULA L. Spurrey

Stamens 5 or 10. Styles 5. The 5 valves of the capsule opposite the sepals. Embryo spirally annular. Leaves in whorls. Otherwise as *Spergularia;* herbs nat. of Old World. (Name from *spargere, to scatter;* from the sowing of the seeds to produce quick-growing early forage.)

1. S. arvénsis L. (of cultivated fields), Corn-S., Grippe (Que.). — Annual, *bright green,* scarcely or not at all viscid; leaves numerous, in whorls, filiform, *channeled at base* (2–5 cm. long); stipules minute; petals white; *seeds plump, roughened with minute whitish papillae, with very narrow wing-margin.* — Weed of cult. grounds, Nfld. to Alaska, s. to Va., Mo. and Calif. Mar.–Oct. (Natzd. from Eu.)

Var. satìva (Boenn.) Reichenb. (sown — formerly in Europe as forage for cattle and sheep). — *Dull green* and very viscid or essentially glabrous; *seeds* obscurely reticulate, *without papillae.* — Fields and waste places, infrequent, w. N.E. to O. and Va. (Natzd. from Eu.)

2. S. pentándra L. (with five stamens). — Smaller, rarely 2 dm. high; *leaves* filiform, *scarcely channelled at base;* flowers few; petals lanceolate, acute; *seeds flattish, smooth, with wing as broad as the body.* — Locally as a weed, s. N.J. April. (Adv. from Eu.)

6. SAGÌNA L. Pearlwort

Sepals 4 or 5. Petals 4 or 5, undivided, or often none. Stamens as many as the sepals, rarely twice as many. Styles as many as the sepals and alternate with them. Capsule many-seeded, 4–5-valved to the base; valves opposite the sepals. — Little matted herbs of cool or temp. reg. with filiform or subulate leaves, no stipules, and small flowers terminating the stems or branches; in spring and summer. (Name from *sagina, fattening;* previously applied to *Spergula,* which is still planted in Eu. as early forage.)

a. Upper leaves slender, not bearing dense bulb-like tufts in their axils; petals inconspicuous, shorter than to barely exceeding sepals or wanting; seeds brown. . . *b.*

b. Annual, usually without a persistent basal rosette, erect or ascending; pedicels not hooked at summit; capsules 1–2 mm. thick; seeds delicately marked with slender ridges. . . . . 1. *S. decumbens.*

*b.*Perennial, with a central rosette of leaves (except in densely crowded individuals), forming depressed mats; pedicels hooked at summit after flowering, later becoming straight; capsules 1.7–2.5 mm. thick; seeds with pebbled or closely reticulated surfaces.
Flowers terminal and lateral, 4 (rarely 5)-merous; pedicels 0.3–2 cm. long; sepals 1–2.4 mm. long, oval to suborbicular, spreading in fruit; capsule 2–3.2 mm. long. . . . . 2. *S. procumbens.*
Flowers mostly terminal, 5 (rarely 6)-merous; pedicels 1–3 cm. long; sepals 2–3 mm. long, oval to oblong, appressed to capsule; the latter 3–5 mm. long. . . . . 3. *S. saginoides.*
*a.*Upper leaves of flowering branches scale-like, subtending dense axillary bulb-like tufts, the branches thus becoming moniliform; petals conspicuous, about twice as long as sepals; seeds black, strongly pebbled. . . . . . 4. *S. nodosa.*

1. S. decúmbens (Ell.) T. & G. (depressed but with ascending tips). — *Annual,* usually without a basal rosette; *stem* capillary, *erect, ascending* or rarely decumbent, simple or slightly forking, 0.2–1.5 dm. high; leaves linear-subulate, slenderly mucronate; *pedicels* filiform, *straight, not hooked after anthesis,* mostly 0.5–2.7 cm. long; *sepals* 5 (rarely 4), oblong or elliptic, obtuse, 1–2.5 mm. long, *closely appressed-ascending;* petals 0 or 1–5 and rudimentary or sometimes equaling or slightly exceeding sepals; stamens 3–10; *capsule* slenderly ovoid, 2–3.5 mm. long, 1–2 *mm. thick, its* 5 (rarely 4) *valves nearly erect after dehiscence; seeds* reddish-brown, *delicately marked with slender ridges,* 0.2–0.3 mm. long. (Incl. var. *Smithii* (Gray) S. Wats.) — Moist or dryish sandy fields, paths and open spots in woods, Fla. to Tex., n. to e. Mass., s. Vt., s. N.Y., Ky., Ill., Mo., and e. Kans. Mar.–Aug.

2. S. procúmbens L. (lying on the ground), Birdseye. — *Matted perennial with a* dense *central rosette* and often with similar rosettes borne on the branches (in dwarfed crowded plants rosettes undeveloped); *rosette-leaves bristle-tipped,* 0.5–2 cm. long; the few to many simple or forking and often proliferous usually *procumbent leafy stems* 0.15–1.5 dm. long; *flowers terminal and lateral,* 4 (rarely 5)-merous; *pedicels hooked at summit after flowering,* becoming straight in maturity, 0.3–2 *cm. long; sepals oval to suborbicular,* rounded at summit, 1.5–2.4 *mm. long, spreading in fruit;* petals much shorter than sepals, watery-white or quite wanting; stamens 4 (rarely 5); *capsule* ovoid, 2–3.2 *mm. long, its valves spreading-ascending after dehiscence;* seeds pebbled, 0.3–0.4 mm. long. — Damp open soil, shores, etc., Nfld. to Minn., s. to N.S., N.E., Del., Md., W.Va. and O. Apr.–Nov. (Eurasia)

Var. **compácta** Lange (compact). — Very dwarf; leaves mostly 2–6 mm. long; sepals 1–1.5 mm. long; petals wanting; capsule barely 2 mm. long. — Arct. reg., s. to Nfld. and lower St. Lawrence, Que. (Eu.)

3. S. saginoides (L.) Karst. (resembling *Sagina;* placed by Linnaeus in *Spergula*). — Resembling no. 2; leaf-tips shorter; *branches decumbent or ascending,* forming mats 0.1–1.5 dm. broad; *flowers mostly terminal,* 5 (rarely 6)*-merous; pedicels* 1–3 *cm. long,* erect, hooked at summit after anthesis, becoming straight; *sepals* 2–3 *mm. long, oval to oblong, appressed-ascending;* stamens 10 (rarely 5); *capsule* conic-ovoid, 3–5 *mm. long, its valves erect* after dehiscence. (*S. Linnaei* Presl) — Wet alpine rocks and river-silts, Gaspé Co., Que.; Ung. to Alaska, s. to N.M. and Calif. June–Aug. (Greenl.; Eurasia)

4. S. nodòsa (L.) Fenzl (knotty). — Tufted perennial; *stems* simple to much forked, capillary, 0.2–2 dm. long, *moniliform from the bulb-like fascicles of axillary reduced leaves,* glabrous; *petals* 5, *white, about twice as long as sepals; seeds black, pebbled.* — Damp rocky, gravelly or peaty soil, e. Lab. to Hudson Bay, s. to coast of e. Me. and of L. Superior, n. Mich. and Minn. July–Sept. (Eu.)

Var. **pubéscens** Mert. & Koch (pubescent). — Peduncles, etc., more or less glandular-puberulent. (Var. *glandulosa* Aschers.) — E. Nfld. to coast of ne. Mass. (Eu.)

7. MYOSÒTON Moench Giant Chickweed

Sepals 5. Petals 5, deeply notched. Stamens 10. Styles 5, alternate with the sepals. Capsule broadly ovoid, 1-locular, opening by 5 bifid valves, many-seeded. — Perennial, nat. of Old World, with angled stems, cordate-ovate leaves and scattered axillary flowers. (Name from *mys, mouse,* and *ous, ear,* from the soft leaves.) Malachia Fries, altered by many authors to Malachium.

1. M. aquáticum (L.) Moench (aquatic). — Stems prolonged, minutely glandular-pilose, loosely branched, cymose at summit; leaves of the main stems sessile, more or less clasping or short-petioled, mostly 3–8 cm. long, acuminate; the rameal leaves smaller, often petioled; flowers from upper axils, on filiform viscid pedicels deflexed in fruit; sepals lance-acuminate, 6–9 mm. long; petals white, exceeding sepals; seeds dark, tubercular-roughened. (*Stellaria*

Scop.; *Alsine* Britt.) — Meadows, alluvial thickets and shores, local, e. Ont. to Minn., s. to s. N.E., Del., Pa., W.Va., O. and Ill. June–Oct. (Natzd. from Eu.)

8. ARENÀRIA L. Sandwort. Sabline (Que.)

Sepals 5. Petals 5, entire, sometimes barely notched, rarely wanting. Stamens 10. Styles 3, rarely more or fewer, opposite as many sepals. Capsule short, splitting into as many or twice as many valves as there are styles, few–many-seeded. — Low usually tufted herbs, with sessile exstipulate leaves and small usually white flowers; genus nearly world-wide. (Name from *arena, sand*, in which many of the species grow.) — Many Europeans, and Rydberg and Small in this country, have broken the genus into several; the differential characters relied upon are too confluent in some Am. species.

a. Leaves (in ours) dilated, lanceolate, linear-oblong or oblanceolate to ovate. . . *b.*

b. Plant not fleshy; stamineal disk inconspicuous or wanting; capsule ovoid or conic to cylindric, its valves 2-cleft; seeds reniform, with marginal hilum. . . *c.*

c. Ovules and young seeds with a pale spongy appendage (strophiole) at the hilum; stems scattered, the terminal peduncles often becoming lateral by elongation of the stems or branches; seeds 1–1.6 mm. long. § 1. Moehringia (Moehringia).

Leaves oval to narrowly oblong, obtuse or merely subacute, puberulent; bractlets of peduncle blunt or merely acute; sepals obtuse, 2–3 mm. long; capsule slenderly conic to ovoid. . . . . . . 1. *A. lateriflora.*

Leaves lance-acuminate, glabrous or nearly so; bractlets linear-aristate; sepals acuminate, 2.5–6 mm. long; capsule globose-ovoid. . . . 2. *A. macrophylla.*

c. Ovules and seeds without a strophiole; stems solitary or somewhat tufted, not prolonged beyond the inflorescences; seeds (in ours) 0.4–0.8 mm. long. § 2. Arenaria (in the restricted sense). . . *d.*

d. Annual with rough-pubescent finally diffusely branched stem and ciliate ovate to lanceolate scabrous sessile leaves; sepals scabrous, strongly 3–5-nerved; seeds strongly rugose, opaque, 0.4–0.6 mm. long. . . 3. *A. serpyllifolia.*

d. Perennials with filiform creeping (mostly subterranean) stems; flowering stems smooth or soft-pubescent, not diffusely branched; leaves elliptic to lanceolate or oblanceolate, non-ciliate or softly ciliolate only at base; sepals smooth, only the midrib sometimes prominent; seeds smooth and lustrous or only obscurely pebbled, 0.6–0.8 mm. long.

Closely matted, with flowering stems 0.1–1.5 dm. high, 1–4-flowered; leaves glabrous, thickish, not punctate; arctic-alpine species. . . 4. *A. humifusa.*

Loosely reclining or leaning, with lax stems prolonged (up to 1.3 m.), and axillary flowers; leaves thin, ciliate at base, punctate; southern species. . . . . 5. *A. lanuginosa.*

b. Plant fleshy and succulent (brittle in drying); petals and stamens inserted on a conspicuous 10-lobed disk; capsule globose or ovoid, 6–12 mm. in diameter, with entire valves; seeds pyriform, with nearly basal hilum, 3–4.5 mm. long. § 3. Ammodenia (Ammodenia; Honkenya; Adenarium). 6. *A. peploides.*

a. Leaves mostly linear, subulate or bristle-like; valves of capsule entire, merely emarginate or only tardily cleft. § 4. Alsine (Minuartia; Alsine; Subulina; Alsinopsis). . . *e.*

e. Sepals nerveless or obscurely nerved.

Root slender; stem capillary, herbaceous; leaves soft, not crowded on the glandless flowering stems; sepals oblong to oval; petals usually notched at summit; capsule thin, barely exceeding calyx. . . . . . 7. *A. groenlandica.*

Root stout, perpendicular; stem stoutish, subligneous; leaves rigid, crowded at base of flowering stems; the latter glandular at summit; sepals roundish-oval; petals entire; capsule firm, twice the length of calyx. . . . . 8. *A. caroliniana.*

e. Sepals prominently ribbed. . . *f.*

f. Sepals firm, with subulate tips; valves of capsule with thick wire-like margins, entire at tip. . . *g.*

g. Leaves nearly nerveless or with midrib conspicuously stronger and usually longer than the lateral nerves; seeds 0.5–1 mm. long. . . *h.*

h. Leaves firm or rigid, the midrib prominent to the tip; pedicels and calyx glabrous; petals entire; seeds 0.6–1 mm. long.

Branches conspicuously marcescent, bearing numerous dense fascicles of rigid leaves; petals about twice length of sepals. . 9. *A. stricta.*

Branches not strongly marcescent; fascicles few or none; petals shorter than to barely equaling sepals. . . . . 10. *A. dawsonensis.*

h. Leaves soft, the midrib evanescent; pedicels and base of calyx often glandular; petals notched; seeds 0.5–0.6 mm. long. . . . . 11. *A. patula.*

g. Leaves with 3 essentially equal strong ribs; seeds 0.4–0.6 mm. long; arctic-alpine species, glabrous or glandular-pubescent. . . . 12. *A. rubella.*

f. Sepals herbaceous, with soft or scarious tips; valves of capsule thin-margined, emarginate; arctic-alpine species.

Basal shoots few, short, herbaceous, bearing obscurely keeled leaves; flowering stem 1–3-flowered; petals 0.5–1 mm. broad, shorter than to barely exceeding sepals; capsule 4–6 mm. long. . . . . . 13. *A. sajanensis.*

Basal branches crowded, trailing, freely forking, subligneous, densely clothed with highly marcescent thick-ribbed leaves; flowering stems 1-flowered; petals 2–2.5 mm. broad, much exceeding sepals; capsule 0.6–1 cm. long. . . . . 14. *A. marcescens.*

§ 1. MOEHRÍNGIA (L.) Fries

1. A. lateriflòra L. (flowering on the side), GROVE-S. — Rhizome filiform, widely creeping; flowering stems capillary, retrorsely puberulent, simple to loosely branched, 0.5–4 dm. long; *leaves oval to narrowly oblong, obtuse or at most subacute, spreading,* in remote pairs, *puberulent,* pellucid-punctate; the larger ones 0.7–3.5 cm. long, 0.2–1.5 cm. broad; *peduncles* filiform, terminal and lateral, the former often *becoming lateral by elongation of the upper axillary branch,* 1–4 cm. long, with 1–6 flowers on naked or 2-bracteolate pedicels; *bractlets blunt or merely acute; sepals* spreading or ascending, *oblong or oval, obtuse, 2–3 mm. long; petals* white (rarely pink), *oblong-obovate, obtuse, 3.5–8 mm. long, long petals associated with long filaments, short with short; capsule slenderly conic to ovoid, pale,* 4–7 mm. long, the deeply 2-parted *valves slightly divergent after dehiscence; seeds* black, lustrous, 1–1.3 *mm. long,* commonly *with a conspicuous pale strophiole at the hilum.* (*Moehringia* Fenzl) — Gravelly or turfy shores, meadows, borders of woods, etc., Lab. to Alaska, s. to Nfld., N.S., N.E., Md., O., Ind., ne. Ill., Mo., S.D., and N.M. May–Aug. (Eurasia) FIG. 1000.

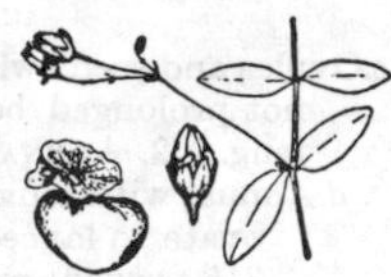
1000. A. lateriflora.

1001. A. macrophylla.

2. A. macrophýlla Hook. (large-leaved). — Similar to no. 1; stems 0.5–1.5 dm. high, simple or but slightly branched; *leaves ascending, lance-acuminate,* 1–7 cm. long, *glabrous or nearly so;* the 1–few peduncles terminal or becoming lateral, 0.5–2.5 cm. long; *bractlets linear-aristate; sepals lanceolate to lance-ovate, acuminate,* 2.5–6 *mm. long;* petals 3–6 mm. long; *capsule globose-ovoid, its firm greenish* 2-cleft *valves strongly divergent after dehiscence; seeds* 1.3–1.6 *mm. long,* many without a strophiole. (*Moehringia* Torr.) — Local on sandy or rocky wooded slopes or shores, chiefly in magnesian or basic areas, Lab. and Ung., s. to Shickshock Mts., Que., serpentines and traps of w. N.E., L. Sup., Ont., and Wisc.; s. B.C. to Colo., N.M. and s. Calif. May–Aug. (Asia) FIG. 1001.

§ 2. ARENÀRIA (in restricted sense)

3. A. SERPYLLIFÒLIA L. (with leaves of *Serpyllum,* Thyme), THYME-LEAVED S. — *Annual,* simple to intricately forking, 0.2–2 dm. high; *branches cinerous-puberulent; leaves ovate, sessile, acuminate, ciliate, scabrous, 2–7 mm. long; inflorescence a leafy nearly regular panicle of dichotomous cymes;* pedicels straight, 3–10 mm. long; *fruiting calyx* 3–4 *mm. long,* 2–3 *mm. broad at base, of* lanceolate to lance-ovate acuminate 3-nerved *scabrous sepals; petals* oblong, *shorter than sepals; capsule ovoid or flask-shaped,* its 2-cleft olive valves hard and resistent; *seeds globose-reniform, opaque, strongly rugose,* 0.6 mm. long. — Dry sterile fields, roadsides, etc., Que. to B.C., s. to N.S., N.E., L.I., Fla., Tenn., Mo., etc. Apr.–Aug. (Natzd. from Eu.) FIG. 1002.

1002. A. serpyllifolia.

Var. TENÙIOR Mert. & Koch (more slender). — Diffuse, often sprawling; *leaves lance-ovate or lanceolate; panicle* irregular, elongate, *of slender racemiform branches; fruiting calyx 2–3 mm. long, 1.2–2 mm. broad at base; capsule elongate-ovoid to subcylindric, its pale membranaceous valves fragile;* seeds 0.4–0.5 mm. long. (*A. leptoclados* Guss.) — Similar habitats, Que. to Mich., s. to N.S., N.E. and Ga. (Natzd. from Eu.)

4. A. humifùsa Wahlenb. (spreading on the ground). — Densely or loosely *matted perennial, the filiform creeping* (mostly subterranean) *stems forming close carpets* 0.2–2 dm. broad; *leaves* imbricated or remote, *lanceolate, oblanceolate or oblong, glabrous and fleshy,* when dry *obscurely 1-nerved, 2–10 mm. long; flowering stems* decumbent at base, 0.1–1.5 dm. high, remotely leafy, 1–4-*flowered,* the puberulent pedicels 0.2–3 cm. long; *calyx cylindric, rounded at base,* 3–5 mm. long, in fruit 2–2.5 mm. thick; *sepals oblong,* obtuse or acutish, *nerveless or only obscurely 1-nerved, closely appressed; petals* narrowly oblong, 5–7-nerved, *barely equaling sepals;* anthers flesh-pink; *capsule cylindric,* exceeding calyx, 4.5–5.5 *mm. long,* olive or castaneous, *its valves deeply split; seeds* 0.6–0.7 mm. long, brown, *shining,* obscurely rugulose. (*A. cylindrocarpa* Fern.) — Greenl., Ellesmerel., n. Lab. and Ung.; calcareous or magnesian gravels and alpine ledges of w. Nfld. and of Shickshock Mts., Que.; Yuk. and Alaska to s. Alta. and s. B.C. July, Aug. (N. Eu.) Fig. 1003.

1003. A. humifusa.

5. A. lanuginòsa (Michx.) Rohrb. (wooly; from the fine lines of pubescence). — Perennial, with filiform creeping subterranean stems, *the flowering stems weak and reclining or leaning, up to* 1.3 *m. long,* with lines of minute pubescence; *leaves thin, narrowly elliptic,* 1–2.5 cm. long, *acute, narrowed to ciliate bases,* often pseudoverticillate, *punctate; flowers axillary, on slender pedicels;* sepals ovate, acute, puncticulate, petals commonly smaller or wanting; seeds smooth and shining. — Rich or damp woods, shaded ditches, etc., Fla. to Tex. and Mex., n. to se. Va. May–Aug. (W.I.; S.Am.) Fig. 1004.

1004. A. lanuginosa.

§ 3. Ammodènia (Patrin) B. & H.

6. A. peploìdes L. (resembling *Peplis Portula*), Seabeach-S., Sea-Chickweed, Sea-Purslane. — *Fleshy, rooting and much branched deep in sand;* the leafy branches of the season procumbent, forking; *leaves fleshy, oblong, elliptic, lanceolate or ovate,* 0.5–4.5 *cm. long;* flowers in terminal leafy cymes or scattered in the upper axils; sepals ovate to lanceolate, 4–7 mm. long; petals 6, spatulate-obovate, about equaling sepals; stamens 8 or 10, with the petals inserted on a *conspicuous 10-lobed disk; capsule globose or ovoid, leathery,* 6–12 *mm. in diameter,* with 3–5 entire valves; *seeds* few, *pyriform, with hilum nearly basal,* 3–4.5 *mm. long.* — A variable circumpolar species of maritime sands. We have four vars.:

Flowers few to many in terminal leafy broad cymes; branches flaccid, 1–2 mm. thick.
- Branches freely forking, 0.5–2 dm. high; leaves ovate to elliptic, 0.5–2 cm. long; sepals 4–5 mm. long; capsule 6–8 mm. in diameter, thick-walled, opaque; seeds dark brown, rugulose. . . . *A. peploides* (typical).
- Branches simple or subsimple, 1.2–3.5 dm. high; leaves broadly lanceolate to elliptic-ovate, the middle ones 2–4.5 cm. long; sepals 5–7 mm. long; capsule 7–9 mm. in diameter, thin-walled, translucent when dry; seeds light brown, smooth. . . . . Var. *maxima.*

Flowers few, terminal and axillary, not in broad cymes.
- Branches flaccid, 0.3–2 dm. long, 1–2 mm. thick, often purple-tinged; leaves ovate to elliptic, narrowed at base, slightly fleshy, 0.5–1.5 cm. long; pedicels slender; capsule 5–8 mm. in diameter; seeds lustrous, only slightly rugulose. Var. *diffusa.*
- Branches stiff, very fleshy or coriaceous, 1.5–5 dm. long, 2.5–6 mm. thick; leaves oblong or oblong-ovate, scarcely narrowed at base, coriaceous, the middle ones 1–3 cm. long; pedicels thick; capsule 8–12 mm. in diameter; seeds papillose, only slightly lustrous. . . . . Var. *robusta.*

A. peploìdes (typical), *Ammodenia* Rupr.; *Honkenya* Ehrh. — Arct. coasts, s. to n. Nfld., M.I., James Bay and s. Alaska. June–Aug. (Eurasia) Fig. 1005.

Var. **máxima** Fern. (greatest). — Sands, w. Nfld.; Alaska. (Ne. Asia)

Var. **diffùsa** Hornem. (loosely spreading). — Arct. coasts, s. to Straits of Belle I., Nfld. (Eurasia)

Var. **robústa** Fern. (stout), *Ammodenia maritima* (Raf.) Bickn.; *Am. pepl.,* var. *maritima* (Raf.) Stone — Coastal sands from Straits of Belle I. to lower St. Lawrence, Que., s. to Md. (reported from se. Va.). May–Sept.

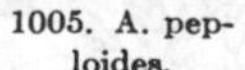

1005. A. peploides.

§ 4. Alsìne (Scop.) B. & H.

7. A. groenlándica (Retz.) Spreng. (of Greenland), Mountain-S., "Mountain-Daisy". — *Tufted, forming dense mats of short leafy shoots* 1–13 cm. across; *stems glabrous,* filiform, depressed, decumbent or suberect, simple to freely forking, 0.2–1.5 dm. high, 1–30-flowered; *leaves linear, obtuse, soft,* often flaccid, or the basal narrowly oblanceolate and 3–15 mm. long, uppermost cauline (below 1st fork) 2–9 mm. long; pedicels erect or spreading, filiform, becoming 0.6–2.3 cm. long; *calyx campanulate,* 3–5 mm. long, *the essentially nerveless oblong to oval scarious-margined sepals obtuse; petals* obovate, usually *retuse,* white, 6–10 *mm. long* (rarely wanting); *capsule* globose-ovoid to slender-conical, *slightly exserted, with entire thin valves;* seeds reddish-brown, rugose, 0.7–0.8 mm. long. (*Minuartia* Ostenf., *Sabulina* Small) — Greenl. and Lab. Pen.; granitic ledges and gravel, s. Nfld., higher mts. of Gaspé Pen., Que., and of n. N.E. and n. N.Y.; coast, N.S. to e. Me. June–Aug. Fig. 1006. — Passing freely to

1006. A. groenlandica.

Var. **glàbra** (Michx.) Fern. (glabrous). — Less tufted, usually *with few if any leafy basal shoots;* stems erect or strongly ascending, 0.7–2.7 dm. high, 1–50-flowered; uppermost cauline leaves 0.8–3 cm. long; pedicels becoming 1.2–4.5 cm. long; *petals* 4–8 *mm. long.* (*A. glabra* Michx.; *Minuartia glabra* (Michx.) Mattf.) — Mts. of Ga., Tenn. and N.C., locally n. on granitic or siliceous rock of Shawangunk and Catskill Mts., N.Y., wooded ledges of Ct. and R.I., hills and lesser mts. of N.H. and Me. and coast of Me. May–Sept.

8. A. caroliniàna Walt. (of Carolina), Pine-barren-S., Longroot. — *Densely matted, with stout vertical root; the stiff forking stems* prostrate or decumbent, *bearing stiff* ascending *leafy sterile branches* covered *with rigid linear-subulate* triangular-channeled *leaves* 4–7 mm. long; *flowering stems erect,* 0.5–1.5 dm. high, remotely leafy, *glandular* especially *at summit,* 1–13-flowered; calyx campanulate, 3–4 mm. long, the *obscurely veined sepals* appressed-ascending, *roundish-oval; petals* spatulate-obovate, *entire,* 7–9 mm. long, *green at base;* capsule ovoid, firm and lustrous, with entire blunt valves 7–9 mm. long; seeds brown, rugulose, 1 mm. long. (*Minuartia* Mattf., *Sabulina* Small) — Dry sands, nw. Fla. and Ga. to se. Va. (very rare); Del. and N.J.; very local, Staten I., L.I. and s. R.I. Late May–July (rarely Sept.). Fig. 1007.

1007. A. caroliniana.

9. A. stricta Michx. (upright), Rock-S. — Loosely tufted, becoming diffuse; *stems filiform, wiry, branching from the marcescent leafy base; leaves* subulate-setaceous or -linear, *strongly ribbed, rigid,* the principal ones 0.6–1.7 cm. long, usually *subtending conspicuous dense fascicles* of shorter leaves; *flowering branches* 0.5–3 dm. high, *glabrous, with numerous fascicles of leaves,* loosely cymose above, 3–30-flowered; *sepals firm,* ovate, *acuminate, strongly* 3–5-*ribbed,* 4–5 mm. long; *petals* oblong-obovate, about *twice as long as sepals; capsule* slenderly ovoid, about equaling or slightly overtopping calyx, *valves entire;* seeds nearly black, rugose, 1 mm. long. (*Sabulina* Small; *A. Michauxii* Hook. f.) — Dry calcareous or magnesian (rarely siliceous) ledges and gravel, n. N.H. and sw. Que. to w. Ont., s. to S.C., Ky. and Ark. May–July. Fig. 1008.

1008. A. stricta.

Var. **texàna** Robins. (of Texas). — Stiffer, 1–2 dm. high; *leaves* 5–10 *mm. long, connate at the enlarged nodes;* inflorescence denser; sepals lanceolate, almost cartilaginous, the edges inrolled. (*A. texana* Britt.; *Sabulina texana* Rydb.) — Rocky hills, O. to Neb., s. to Ark. and Tex. May, June.

10. A. dawsonénsis Britt. (of Dawson, Yukon). — Resembling no. 9, without marcescent basal shoots, the wiry stems *leafy only at base,* 0.2–2.3 dm. high; *leaves firm, with* 1 *nerve* or angle *more prominent than the other* 2, often proliferous in the axils, the principal ones 0.3–1.6 cm. long; inflorescences 1–17-flowered, the rigid *pedicels ascending or slightly divergent;* calyx 3–5.5 mm. long; *sepals narrowly ovate, subulate-tipped,* glabrous, *granular-roughened,* with broad hyaline margin, the midrib prominent; *petals* narrowly oblong, *shorter than or barely equaling the sepals; capsule ovoid, slightly exserted,* valves entire; seeds 0.6–0.9 mm. long. (*A. litorea* Fern.; *Sabulina litorea* and *dawsonensis* Rydb.) — Calcareous ledges and gravel, s. Lab. to Yuk., s. to Nfld., e. Que., Ung., w. Ont., Minn., S.D., and Alta. June–Aug. Fig. 1009.

1009. A. dawsonensis.

11. A. pàtula Michx. (spreading). — *Diffusely branched annual,* with capillary stem 0.5–3 dm. high; *leaves* slightly fleshy, linear-filiform, 0.5–1.5 mm. broad, the *midrib evanescent,* or leaves mostly 2–2.3 mm. broad and up to 3.5–5 cm. long in forma **robústa** Steyerm. (stout); stems and leaves densely glandular, or glabrous and with pedicels, leaves and sepals glabrous in forma **Pítcheri** (Nutt.) Steyerm. (for Zina Pitcher, 1797–1872), or stem and leaves glabrous but pedicels and sepals glandular in forma **mèdia** Steyerm. (intermediate); *pedicels divergent,* 0.5–4.5 cm. long, *often glandular at summit as is the base of calyx;* sepals lance-attenuate, strongly 3–5-ribbed, 3.5–6 mm. long; *petals obcordate,* about

1010. A. patula.

twice the length of sepals; capsule about equaling calyx, its blunt valves entire; *seeds* 0.5–0.6 *mm. long.* (*Minuartia* Mattf.; *Sabulina* Small) — Limestone-barrens, cliffs and open woods, s. O. to e. Kans., s. to Ala., Ark. and Tex. Late Apr.–June. Fig. 1010.

1011. A. rubella.

12. A. rubélla (Wahlenb.) Sm. (reddish). — Densely or loosely tufted, forming mats 0.1–1.5 dm. across, more or less glandular-pilose, or in forma **èpilis** (Fern.) Polunin (without hair) glabrous; basal shoots bearing crowded *linear-subulate* equally *3-ribbed leaves* 2.5–14 mm. long; flowering stems decumbent at base, filiform, rather frail, 0.2–2 dm. high, 1–7-flowered; *pedicels* 0.2–2.5 *cm. long; calyx in fruit slender-campanulate,* 2.5–5 mm. long; sepals lanceolate to lance-ovate, acuminate, with 3 uniform parallel ribs; *petals narrowly oblong or spatulate, shorter than sepals; capsule cylindric to slenderly ovoid,* 3–5-valved, equaling or exceeding calyx; *seeds* reddish-brown, rugose, 0.4–0.6 *mm. long.* (*A. verna* L., vars. *rubella* S. Wats., *propinqua* Fern. and *pubescens* Fern.; *A. propinqua* Richards.) — Arct. reg., s. on calcareous or magnesian rocks and gravel to Nfld., Que., n. Vt., N.M., Ariz. and Calif. June–Aug. (Eurasia) Fig. 1011. — Polymorphous species with many named forms.

13. A. sajanénsis Willd. (of the Saiansk Mts., southwest of Lake Baikal, Siberia). — *Stems prostrate or decumbent, forming soft mats* 0.3–1.5 *dm. broad, with herbaceous or subherbaceous moss-like basal shoots* with obscurely keeled linear leaves 0.5–1 cm. long; *flowering stems* 1–3-*flowered,* minutely glandular above, 0.1–1.3 dm. high; calyx 3.5–6 mm. long; *sepals* 4–7, often unequal, strongly ascending, *oblong or linear, 3-nerved, glabrous or minutely pilose, blunt or attenuate, with* entire or erose *purplish scarious tips; petals linear-spatulate,* 0.5–1 *mm. broad, about equaling to slightly exceeding sepals; anthers* whitish, 0.2–0.3 *mm. long; capsule* slender-conical, 4–6 *mm. long;* the *membranaceous pale valves* lance-oblong, blunt, emarginate; *seeds* olivaceous, smooth, 0.7–1 *mm. long.* — Arct. reg., s. to magnesian rocks and gravel, Mt. Albert, Que., Ung. and high Rocky Mts. to Ariz. July–Sept. (Eurasia) Fig. 1012.

1012. A. sajanensis.

14. A. marcéscens Fern. (marcescent, retaining old dry leaves). — Densely matted, *the crowded and trailing* freely forking *subligneous branches* 0.5–2.5 *dm. long, densely clothed with highly marcescent thick-ribbed round-tipped* glabrous *coriaceous leaves* 4–8 mm. long and 0.3–0.5 mm. wide; *flowering stems* ascending from tips of leafy shoots, 2–5 *cm. long, 1-flowered,* glandular-pilose, with 2–4 pairs of remote reduced leaves; peduncle 0.6–1.5 cm. long; *calyx purplish or fuscous, turbinate-campanulate,* 3.8–5 *mm. long,* pilose at base; *sepals oblong, obtuse,* keeled; *petals spatulate to narrowly obovate,* white or lilac, yellow-based, 6–8 *mm. long,* 2–2.5 *mm. broad; anthers* 0.5–1 *mm. long; capsule subcylindric,* 0.6–1 *cm. long,* the linear-oblong firm stramineous valves emarginate; *seeds smooth,* 1–1.2 *mm. long, with* the micropyle prolonged into *a beak.* — Magnesian ledges and gravels, w. Nfld.; Mt. Albert, Que. July, Aug. Fig. 1013.

1013. A. marcescens.

9. STELLÀRIA L. Chickweed. Starwort

Sepals 4–5. Petals (white) 4–5, deeply 2-cleft, sometimes none. Stamens 8, 10 or fewer. Styles 3, rarely 4, opposite as many sepals. Capsule ovoid to globose, 1-locular, opening by twice as many valves as there are styles, several–many-seeded. Seeds naked. — Flowers solitary or cymose, terminal or appearing lateral by the prolongation of the stem from the upper axils. Genus nearly cosmop. (Name from *stella, a star,* in allusion to the star-shaped flowers.) Alsine L. in part, not Wahlenb.

*a.*Leaves ovate, the lower and middle distinctly petioled; weak stems and branches filiform, not angled; calyx often pilose. . . . . 1. *S. media.*

*a.*Leaves linear, lanceolate or elliptic (rarely ovate), mostly sessile or essentially so; stems and branches 4-angled; calyx glabrous (except sometimes in no. 2). . . *b.*

*b.*Leaves elliptic or elliptic-oblong, in maturity 1–4 cm. broad; cymes with large leafy bracts; capsule globose, much shorter than calyx; vernal perennial with ligneous base. . . . . 2. *S. pubera.*

*b.*Leaves linear, lanceolate or narrowly elliptic, 1–7 (very rarely –12) mm. broad; cymes (if developed) with very small bracts; capsule ovoid (or in no. 3 globose and as long as calyx). . . *c.*

*c.*Stems pubescent in lines; axis of inflorescence and pedicels puberulent; flower 1.5–2 cm. broad; capsule inflated, globose. . . . . . 3. *S. Holostea.*

c.Stems and pedicels glabrous; flowers mostly smaller; capsule ovoid. . . *d*.

d.Flowers in well-developed cymes with small scarious bracts. . . *e*.

e.Cymes terminal or terminal and axillary, long-peduncled; sepals firm, ribbed, in maturity 3.5–8 mm. long; petals conspicuous.

Inflorescence terminal, long-peduncled, loosely dichotomous, with many flowers, the fruiting pedicels divergent; capsules pale. . 4. *S. graminea.*

Inflorescences terminal and axillary, few-flowered, the fruiting pedicels ascending.

Sepals lance-acuminate, sharp-pointed, 5–8 mm. long; petals 8–12 mm. long; capsule pale. . . . . 5. *S. palustris.*

Sepals elliptic to lance-ovate, blunt or merely subacute, 3–7 mm. long; petals 4–8 mm. long; capsule blackish. . . . . . 6. *S. longipes.*

e.Cymes all (or nearly all) axillary; sepals almost nerveless, in maturity 2.5–4 mm. long.

Leaves linear, long-attenuate; cymes long-peduncled, open, with strongly divergent branches; flowers 5–7 mm. broad, with evident petals; capsules 4–6 mm. long; seeds smooth. . . . 8. *S. longifolia.*

Leaves oblong or elliptic-lanceolate; cymes sessile or short-peduncled, with ascending branches; flowers smaller, with inconspicuous or no petals; capsules 3–3.5 mm. long; seeds pebbled. 9. *S. Alsine.*

d.Flowers pedicelled, in the axils of foliage-leaves or, if cymose, with leaf-like bracts. . . *f*.

f.Plants fleshy, matted, with oval, elliptic or narrowly oblong flat blunt leaves 0.2–1.5 cm. long, many of the axils bearing sterile tufts or branchlets.

Very fleshy; leaves oval or elliptic, mostly 0.2–1 cm. long; seeds smooth. . . . . 10. *S. humifusa.*

Slightly fleshy; leaves linear-oblong, the larger 0.8–1.5 cm. long; seeds rugose-roughened. . . . . 11. *S. crassifolia.*

f.Plants not fleshy, with mostly ascending stems; leaves linear to lanceolate, narrowly ovate or narrowly spatulate, up to 4(–6) cm. long; axillary sterile fascicles few or none.

Leaves firm, carinate, sharp-tipped; petals 4–8 mm. long, exceeding sepals; capsules blackish. . . . . 7. *S. monantha.*

Leaves soft or thin, scarcely keeled, blunt or merely acute; petals 0 or much shorter than sepals; capsules pale brown.

Leaves linear-spatulate, with obscure midrib, glabrous throughout; capsule subglobose-ovoid, about equaling calyx. . . . 12. *S. fontinalis.*

Leaves elliptic or narrowly ovate to linear, with definite midrib, ciliolate at base; capsule conic-ovoid, much exceeding calyx. 13. *S. calycantha.*

1. S. mèdia (L.) Cyrillo (intermediate), Common C., Mouron des oiseaux (Que.). — *Weak* annual or perennial *with* trailing, matted or loosely ascending *terete stems* (varying greatly with habitat) *pubescent in lines*, 0.1–8 dm. long; *lower and median leaves ovate, with rounded bases and often ciliate petioles*, the blade 0.5–4 cm. long; upper leaves sessile; calyx usually pilose or villous, especially at base; petals shorter than calyx, 2-parted or wanting; stamens 3–10; capsule ovoid, equaling or exceeding calyx; seeds tuberculate. (*Alsine* L.) — Dooryards, cult. ground, waste places and damp woods and thickets, throughout our range and beyond, now a cosmop. weed. Feb.–Dec. (sometimes all winter). (Natzd. from Eurasia) — Excessively variable in size, development of petals, number of stamens, etc., and much subdivided by European authors. The most marked variety with us is

Var. glabérrima G. Beck (very smooth). — Relatively small, densely matted, *glabrous throughout;* fruiting calyx promptly disarticulating. — Lawns and roadsides e. Va. to S.C. (Natzd. from Eu.)

2. S. pùbera Michx. (puberulent), Great or Star-C. — *Tough-based perennial*, often with dimorphic ascending stems *puberulent above*, 4-angled; *leaves elliptic to elliptic-oblong*, those of the later sterile or sometimes few-flowered shoots larger than in the early chiefly fertile stems, varying from 2–10 cm. long and 1–4 *cm. broad; primary inflorescences broad leafy cymes*, the *puberulent pedicels* at first ascending, later divergent; sepals ovate to oblong-lanceolate, 4–6 mm. long, obtuse or acutish, shorter than petals, only inconspicuously if at all ciliate; petals sometimes cleft half their length, sometimes nearly to base; stamens 10; capsule globose, shorter than calyx. (*Alsine* Britt.) — Rich woods and shaded rocky slopes, N.J. to Ill., s. to n. Fla. and Ala. Late March–May.

Var. **silvática** (Béguinot) Weath. (of the woods). — Median leaves of sterile shoots with petioles 1–2 cm. long, oval or ovate; sepals 7.5–11 mm. long, acute or acuminate, equaling or exceeding petals, at least the outer ciliate toward base. (*Alsine tennesseensis* Small, in part) — Ct. (introd.) and N.Y. to Ind., s. to Tenn.

3. S. Holóstea L. (for *Holosteum*, with which it was early associated), Greater Stitchwort. — Slender perennial with scabrous-angled slender ascending flowering stems 1.5–3 dm. high; *leaves firm, scabrous* on margin or midrib beneath, *lance-attenuate*, sessile; *axis of cyme and* long divergent *pedicels puberulent;* bracts herbaceous; *flowers* 1.5–2 *cm. broad;* petals exceeding the lance-acuminate sepals, notched to middle; *capsule inflated, globose, about equaling calyx;* seeds wrinkled on sides, tuberculate on back. — Spread from cult. to roadsides and rocky woods, local, N.E. to W.Va. and N.C. May, June. (Introd. from Eu.)

4. S. gramínea L. (grass-like), Common Stitchwort. — Stem weak, reclining or ascending, 2–5 dm. high, rhombic in section; *leaves lanceolate*, firm; *broadest a little above the ciliolate base* (leaves of vegetative shoots shorter and broader, ovate- to oblong-lanceolate, the lower subpetiolate); *cyme terminal, long-peduncled*, loosely dichotomous and *diffuse*, many-flowered, *fruiting pedicels divergent; bracts small, scarious;* sepals firm, lanceolate, acute, 3-nerved, 3.5–5 mm. long; petals spreading, slightly longer than sepals; capsule pale brown; seeds finely roughened. — Grasslands, Nfld. to Ont. and Minn., s. to N.S., N.E., L.I., S.C., W.Va., O., Ind., and Mo. May–Oct. (Natzd. from Eu.)

5. S. palústris Retz. (of marshes). — Similar to no. 4, slender, erect, 3–5 dm. tall, pale green to glaucous; *leaves linear-lanceolate, glabrous throughout; cymes terminal and axillary*, long-peduncled, *few-flowered, with long erect pedicels; sepals lance-acuminate, sharp-pointed,* 5–8 *mm. long; petals* 8–12 *mm. long;* capsule pale. (*S. Dilleniana* Moench; *S. glauca* With.) — Low grasslands and shores, local, Que. and Mich. June–Sept. (Natzd. from Eu.)

6. S. lóngipes Goldie (long-stalked). — Erect or decumbent or in exposed situations densely tufted, essentially glabrous, green or glaucous, 0.5–3 dm. high; *leaves stiff*, linear to lanceolate or lance-oblong, mostly attenuate from near base, spreading to stiffly ascending, 1–4.5 cm. long; *inflorescences terminal and axillary, either few-flowered cymes or solitary flowers; pedicels elongate, erect* or strongly ascending; *bracts small, scarious; sepals* firm, *elliptic to lance-ovate, blunt or merely subacute,* 3–7 *mm. long; petals* 4–8 *mm. long; capsule slenderly ovoid, exserted, shining, deep brown to blackish;* seeds with low pebbling. — Gravelly, turfy and open wooded situations, Greenl. to Alaska, s. to Nfld., M.I., ne. N.B., Megantic Co., Que., n. and w. N.Y., n. Ind., Minn., N.M., Ariz. and Calif. June–Aug. — Polymorphic, green or glaucous, lax or dense, broad-leaved or narrow-leaved colonies in different habitats of single small areas.

7. S. monántha Hultén (one-flowered). — *Differing from no.* 6 *in less rigid habit, softer and usually broader leaves, and slightly larger flowers from the axils of non-scarious leaves.* — Greenl. and Lab. to Alaska and ne. Asia, s. to swales and meadows of n. Nfld., alpine slopes of Gaspé Pen., Que., se. B.C. and ne. Wash.

8. S. longifòlia Muhl. (long-leaved). — Similar to no. 5; stems weak, reclining or ascending, often with rough angles, 1.5–5 dm. high; *leaves linear*, glabrous, acutish at both ends, commonly *attenuate*, spreading or loosely ascending; *cymes scaly-bracted, soon becoming lateral* by prolongation of leafy tip, *long-peduncled, open, with strongly divergent branches; flowers* 5–7 *mm. broad*, on spreading or deflexed pedicels; *sepals* 2.5–4 *mm. long, almost nerveless;* petals about equaling sepals; *capsule stramineous*, 4–6 mm. long; *seeds smooth.* — Damp thickets, meadows and shores, s. Nfld. to Alaska, s. to N.S., N.E., Va., Ky., Mo., Neb. and N.M. May–July. (Eurasia)

9. S. Alsìne Grimm (for *Alsine*, inclusive early name for several genera). — Trailing or diffuse annual or biennial with weak prolonged stems, leaving the *naked and sessile or short-stalked few-flowered cymes lateral; leaves oblong or elliptic-lanceolate*, rather fleshy, mostly 0.7–2.5 cm. long; pedicels ascending; bracts small, scarious; *sepals* lanceolate, herbaceous, 2–3 *mm. long; petals wanting or inconspicuous; capsule* pale, slender, 3–3.5 *mm. long;* seeds pebbled. (*S. uliginosa* Murr.) — Springheads, borders of rills and wet places, Nfld. to Del., Md., Pa. and w. N.Y. May–Oct. (Eu.)

10. S. humifùsa Rottb. (spreading on the ground). — *Very fleshy* and matted, with ascending or trailing stems; *leaves* crowded or becoming scattered, *oval or elliptic, flat, blunt,* 0.2–1 *cm. long, many of the axils bearing sterile tufts or branchlets;* flowers solitary in the axils or few in leafy cymes; petals 5–7 mm. long, a little longer than the fleshy-herbaceous blunt sepals; capsule rounded, usually shorter than sepals; *seeds smooth.* — Greenl. to Alaska, s. on saline or brackish shores and borders of marshes to Nfld., N.S. and Hancock Co., Me.; also (?) upper St. John R., Me.; Pacific coast s. to Oreg. June–Aug. (Eurasia)

11. S. crassifòlia Ehrh. (fleshy-leaved). — Less fleshy than no. 10, the flaccid freely branching slender stems matted, depressed or diffuse, 0.5–4.5 dm. long; *leaves linear-oblong, the larger* 0.8–1.5 *cm. long*, acutish; flowers solitary or few in leafy cymes, on filiform ascending or loosely spreading pedicels; *sepals* lanceolate, 2.5–4 mm. long, *slightly shorter than* the narrow *petals;* capsule slenderly conical, pale, barely exserted; *seeds rugose.* — Springheads, borders of rills and wet depressions, Lab. to Alaska, s. along coast to Nfld., M.I., P.E.I. and e. N.B., and inland s. to Mich., Ill., Minn., N.D. and Colo. July, Aug. (Eurasia)

12. S. fontinàlis (Short & Peter) Robins. (of springs). — *Flaccid,* with slender dichotomously branching stems; *leaves linear-spatulate,* obtuse or subacute, up to 2.5 cm. long, *glabrous throughout, the midrib obscure;* flowers solitary in the forks and at tips of branches, on slender ascending stalks; *sepals oblong, obtuse,* 2.5–3.5 *mm. long;* petals 0; *capsule subglobose-ovoid,* about equaling calyx. — Springs, rills and wet rocks, upland of Ky. and Tenn. April–June.

13. S. calycántha (Ledeb.) Bong. (calyx-flowered). — Weak perennial with slender rootstocks and stolons; stems glabrous or only a little scabrous, 0.3–5 dm. long, simple or loosely branched; *leaves soft but with definite midrib, narrowly ovate or elliptic to lanceolate or lance-linear, acute or acutish, minutely ciliolate at base;* flowers solitary in forks of stem and terminal or in open leafy cymes; *sepals broadly lanceolate to oblong,* acute or blunt, 2–5.5 *mm. long;* petals 0 or shorter than sepals and watery-white; *capsule conic-ovoid,* thin-walled, *pale brown, much exceeding calyx.* — Polymorphous boreal species. We have the following vars.:

a. Mature calyx 2–3.5 (–4) mm. long; mature (but unopened) capsule 3–5 mm. long. . . *b.*

b. Leaves ovate, ovate-lanceolate or elliptic-lanceolate, the primary ones 0.7–2.5 cm. long. . . . . *S. calycantha* (typical).

b. Leaves lanceolate to lance-linear, the primary ones 2.5–7 cm. long.

Upper bracteal leaves long and but slightly reduced, herbaceous throughout; flowers few, axillary and terminal. . . . . Var. *isophylla.*

Upper leaves much reduced to short slightly scarious-margined bracts; flowers numerous, in terminal cymes. . . . . Var. *floribunda.*

a. Mature calyx 4–5 mm. long; mature capsule 5–6 mm. long; leaves nearly uniform, lanceolate, the primary ones 3–7 cm. long. . . . . Var. *laurentiana.*

S. calycántha (typical), *S. borealis* Bigel. — Wet places, Greenl. to Alaska, s. to Nfld., P.E.I., Que., n. N.E. (chiefly alpine), n. N.Y., Colo. and Oreg. May–Sept. (Eurasia)

Var. **isophýlla** Fern. (equal-leaved), *S. borealis,* var. Fern. — Similar habitats, Nfld. to Alaska, s. to N.S., N.E., Pa., W.Va., Mich., Minn., Man., Colo., Utah and se. Calif. — Passing insensibly to the two following:

Var. **floribúnda** Fern. (full of flowers), *S. borealis,* var. Fern. — Nfld. to B.C., s. to N.S., N.E., Pa., Mich., Wisc., Minn., Ariz. and Oreg.

Var. **laurentiàna** Fern. (of the Gulf or River St. Lawrence). — Wet or shaded places, Anticosti I. and lower St. Lawrence R., Que.

10. CERÁSTIUM L. Mouse-ear Chickweed

Sepals 5, rarely 4. Petals as many, 2-lobed or -cleft, rarely entire, often wanting in some of the flowers. Stamens 10 or fewer. Styles mostly 5, rarely 4 or 3, opposite the sepals. Capsule 1-locular, usually elongated, often curved, membranaceous, opening at the summit by twice as many teeth as there were styles, many-seeded. Seeds rough. Herbs, widely dispersed in cool and temp. reg. (Name from the Greek *cerastes, horned,* alluding to the shape of the slender and often curved capsule.)*

a. Styles 5; capsules with 10 finally circinate-revolute teeth and 10 nerves; median internodes of flowering stems usually pubescent all around. . . *b.*

b. Perennials, with elongate prostrate or creeping matted basal branches or offshoots. . . *c.*

c. Bracts of inflorescence herbaceous or only the uppermost and smallest slightly scarious-margined; petals broad, with contiguous margins, ascending, longer than sepals; seeds 0.6–1.7 mm. in diameter. . . *d.*

d. Plant green or whitened with pubescence; seeds 0.6–1 (rarely –1.3) mm. in diameter, with close testa.

Pubescence, at least of basal tufts, long and more or less entangled, especially at tips of leaves, or even lanate; cyme not strongly dichotomous; pedicels usually long and ascending; sepals in fruit 6–10 mm. long; capsule 0.95–2 cm. long; seeds 1–1.3 mm. in diameter. . . . . 1. *C. alpinum.*

---

* The figures show a fertile branch (reduced) and a seed (enlarged).

Pubescence short, pilose to hirsute, often scanty; cyme (when well-developed) dichotomous (as in *C. vulgatum*); sepals in fruit 4.2–8 mm. long; capsule 8.5–12 mm. long; seeds 0.6–1 mm. in diameter. . . 2. *C. beeringianum.*

d. Plant purplish or fulvous; seeds 1.3–1.7 mm. in diameter, with loose or inflated testa. . . . . 3. *C. terrae-novae.*

c. Bracts of inflorescence broadly scarious-margined or only the lowermost wholly herbaceous; petals narrower, without contiguous margins and only slightly exceeding sepals, or, if broad and conspicuously longer than sepals, widely spreading in anthesis; seeds 0.35–0.7 mm. in diameter.

Basal leafy branches or offshoots herbaceous, with few or no axillary tufts, their hirsute leaves oblong; petals about equaling or only slightly exceeding sepals, narrow and cleft to middle, the claw ciliate. 4. *C. vulgatum.*

Basal branches and offshoots becoming dry and marcescent, bearing conspicuous axillary fascicles or leafy tufts, their mostly linear to oblong leaves not hirsute; petals twice to thrice length of sepals, their broad lobes spreading during anthesis, the claw glabrous. . 5. *C. arvense.*

b. Annuals or winter-annuals, without persistent decumbent or creeping sterile leafy offshoots. *e.*

e. Stem weak; leaves thin, narrowly oblong to oblanceolate, the larger 1–8 cm. long; capsules twice to thrice as long as the oblong blunt or merely acutish sepals; indig. species.

Pedicels, at least the lower ones, 1–5.5 cm. long, hooked at tip, much longer than the capsules. . . . . 6. *C. nutans.*

Pedicels 2–10 mm. long, not hooked at tip, shorter than to but slightly longer than capsules. . . . . 7. *C. brachypodum.*

e. Stem firm; leaves firm, broadly elliptic to ovate, obovate or spatulate, the longer 0.3–3 cm. long; capsules about equaling to twice length of lance-attenuate sepals; introd. weeds of fields and roadsides. . . *f.*

f. Bracts of inflorescence herbaceous. . . *g.*

g. Cyme at first glomerulate, in maturity with the terminal clusters dense; fruiting pedicels shorter than to about equaling calyx; capsule curving. . . . . 8. *C. viscosum.*

g. Cyme open and lax; fruiting pedicels two to five times length of calyx; capsule nearly straight.

Plant viscid; bracts not long-tufted at tip; petal-claws and filaments glabrous. . . . . 9. *C. tetrandrum.*

Plant pilose, not viscid (except sometimes at summit); bracts terminating in long tufts of hairs; petal-claws and filaments ciliate. . . . . 10. *C. brachypetalum.*

f. Bracts with scarious-membranaceous summit or margin.

Bracts with upper half scarious; sepals with broad scarious margins and tips; petals with simple veins, shallowly notched. . . . 11. *C. semidecandrum.*

Bracts and sepals with very narrow scarious margins and tips; petals with branching veins, deeply bifid. . . . . 12. *C. pumilum.*

a. Styles 3 or 4; the 6 or 8 teeth of the capsule not circinate-revolute; median internodes of stem glabrous or merely with lines of short hairs below the nodes. . . . . 13. *C. cerastoides.*

1. C. alpinum L. (alpine). — Densely or loosely matted perennial; *basal offshoots with the leaves bearing entangling pubescence at tip;* leaves of the season oval, oblong, narrowly ovate or lanceolate, 0.5–2 cm. long; *flowering stems weak,* 0.1–3 dm. long; bracts similar to foliage-leaves or only slightly scarious-margined; *flowers* 1–6; *pedicels straight,* strongly ascending, finally spreading, *much longer than the calyx, in fruit becoming* 1.5–5 *cm. long; sepals* ovate-lanceolate, in anthesis 5.5–9, *in fruit* 6–10 *mm. long; petals* cuneate-obovate, *about twice length of calyx,* the 2 oblong lobes *ascending;* capsule nearly straight, 0.95–2 *cm. long; seeds* tuberculate, 1–1.3 *mm. in diameter.* — Very variable circumpolar species. We have 3 vars.:

Pubescence of stems, pedicels and calyx short, straight and spreading.

Summit of stem, pedicels, etc. hirsute, not glandular. . . . . *C. alpinum* (typical).

Summit of stem, pedicels, etc. glandular. . . . . Var. *glanduliferum.*

Pubescence of stem, etc., a long entangling pale, often flocculent, glandless tomentum. . . . . Var. *lanatum.*

C. **alpìnum** (typical). — Arct. reg., s. to rocky shores of Straits of Belle I., Nfld., w. Ung. and Keewatin. Late June–Sept. (Eurasia) Fig. 1014.

Var. **glandulíferum** Koch (bearing glands). — Similar range, s. to Pistolet Bay, Nfld. (Eurasia)

Var. **lanàtum** (Lam.) Hegetschw. (woolly). — South to n. Nfld. and Mingan Ids., Que. (Eurasia)

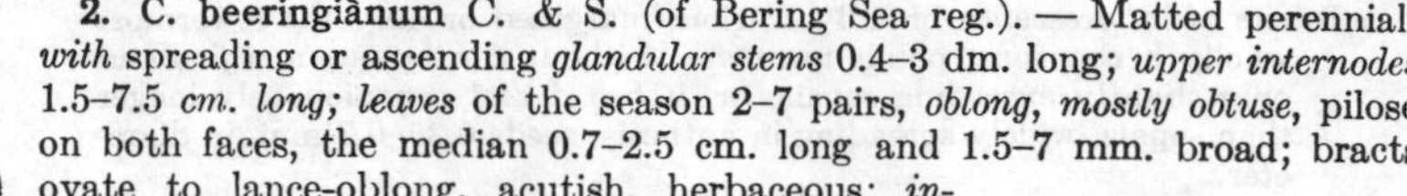

1014. C. alpinum.

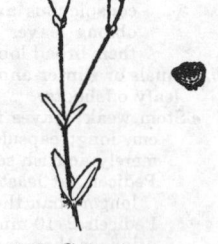

2. C. beeringiànum C. & S. (of Bering Sea reg.). — Matted perennial, *with* spreading or ascending *glandular stems* 0.4–3 dm. long; *upper internodes* 1.5–7.5 *cm. long; leaves* of the season 2–7 pairs, *oblong, mostly obtuse*, pilose on both faces, the median 0.7–2.5 cm. long and 1.5–7 mm. broad; bracts ovate to lance-oblong, acutish, herbaceous; *inflorescence* simple to *dichotomous*, 1–14-*flowered; pedicels mostly ascending* (rarely nodding at tip) in maturity, 0.5–3 cm. long; *sepals* in anthesis 3.5–7, in fruit 4.2–8 mm. long, *broadly lanceolate to oblong-ovate, obtuse*, the inner conspicuously scarious-margined; *petals* bluntly 2-lobed, 6–8 (rarely –9) *mm. long*, ascending, only slightly exceeding calyx, marcescent; capsule 8.5–12 mm. long; fruiting calyx campanulate, 4–6.5 mm. broad at summit; seeds 0.6–1 mm. in diameter, bluntly papillose, not angled. — Calcareous ledges and gravels, Lab. to w. Nfld. and Gaspé Pen. (ascending to 1000 m. alt.) and Rimouski Co., Que.; Alaska and Yuk. to Ariz. June–Aug. (Ne. Asia) Fig. 1015.

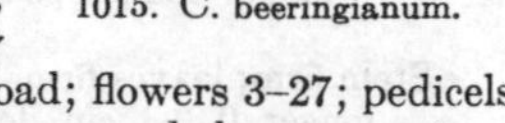

1015. C. beeringianum.

Var. **grandiflòrum** (Fenzl) Hultén (large-flowered). — Coarser; stems up to 4 dm. long; *upper internodes becoming* 0.4–1.2 *dm. long; leaves broadest near base*, lanceolate to lance-oblong or narrowly ovate, *mostly acutish*, the median 1–4.2 cm. long and 0.3–1.6 cm. broad; flowers 3–27; pedicels after anthesis nodding at tip or strongly divergent, up to 4 cm. long; *sepals* longer, *acute or acuminate; petals* 0.9–1.2 *cm. long*. (*C. Fischerianum* sensu Fern. & Wieg., not Ser.) — Damp open slopes or calcareous cliffs and gravel, n. Nfld.; Gaspé Pen., Que.; Alaska. (Ne. Asia)

C. Regélii Ostenf. (for Edouard August von Regel, 1815–1892), a little-known Arct. species, smaller than no. 2, with nearly glabrous stems, short elliptic leaves and rounded sepals 4.5–6 mm. long, may reach n. Nfld. Fuller material is necessary for satisfactory identification.

3. C. térrae-nòvae Fern. & Wieg. (of Newfoundland). — Tufted perennial, *fuscous or suffused with purple;* stems loosely ascending to suberect, densely glandular-hirsute, or hirsute and glandless in forma **Waghórnei** Fern. & Wieg. (for Arthur Charles Waghorne, 1851–1900), very leafy; internodes short, the median 0.5–2.5 cm. long; new *leaves* purplish, 5–10 *pairs, elliptic-oblong, obtuse*, densely glandular-hirsute (except in forma *Waghornei*), 0.5–1.4 cm. long and 1.5–3.5 mm. broad; bracts ovate-lanceolate, scarcely scarious; *flowers* 1–3; the slender *mostly erect pedicels* becoming 1.5–2.5 cm. long; sepals ovate-oblong, broadly scarious-margined, 5.5–6.5, becoming 6–7 mm. long; petals obtusely 2-lobed, ascending, twice as long as calyx; *capsule* straight, 0.9–1.3 *cm. long; seeds* 1.3–1.7 *mm. in diameter, angulate; the vesicular or loose testa with some faces prominently papillate, others with rows of small transverse ridges.* — Serpentine-gravel, sands and rocky tablelands, w. Nfld. July, Aug. Fig. 1016.

1016. C. terrae-novae.

4. C. vulgàtum L. (common), Common M. — Short-lived *matted perennial* with depressed basal leafy offshoots; flowering stems 1–6.5 dm. high, hirsute, or glandular in forma **glandulòsum** (Boenn.) Druce (glandular), the median internodes becoming 2–12 cm. long; *leaves of the season* 3–7 pairs, *oblong to narrowly oval, hirsute on both surfaces*, the median 0.5–4 cm. long and 1.5–15 mm. broad; *bracts* similar but smaller, *broadly scarious at summit and margin; inflorescences* 3–60-flowered, forming terminal *ultimately very dichotomous cymes*, at first rather compact, *in fruit with the lower pedicels divergent or reflexed* and two to four times length of calyx; sepals 4–7 mm. long, ovate-lanceolate, acute, hirsute, scari-

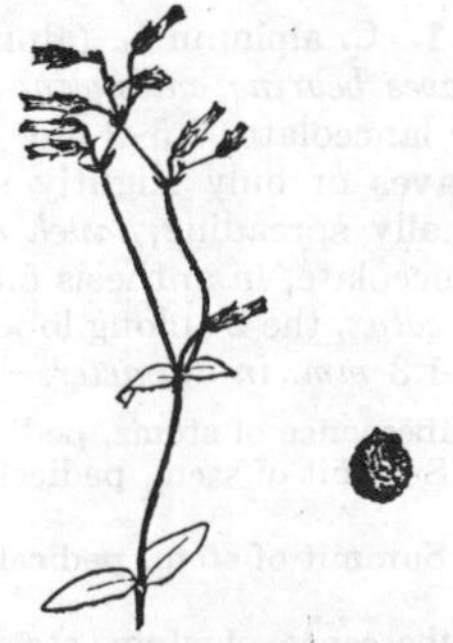

1017. C. vulgatum.

ous-margined; *petals* 4–8 mm. long, *narrow, about equaling or very slightly exceeding sepals,* cleft to middle, *with ciliate claw;* capsule curved, 7–11 mm. long; seeds 0.5–0.7 mm. in diameter, reddish, tuberculate. (*C. triviale* Link) — Roadsides, fields and cult. grounds, in all inhabited reg., throughout our area and beyond. Early spring–late autumn (sometimes through winter). (Natzd. from Eurasia) Fig. 1017.

Var. HOLOSTEOÌDES Fries (resembling *Holosteum*). — Plant glabrescent or with lines of minute hairs on the stems; leaves elliptic or oblong, dark green, round-tipped, ciliate. — Waste places, local, ne. N.S. and e. Va. (Adv. from Eu.)

5. C. arvénse L. (of cultivated ground), Field-C. — Matted or tufted *perennial with depressed* or trailing *tough basal branches bearing marcescent firm leaves and abundant and conspicuous axillary fascicles or leafy tufts;* flowering branches ascending, simple to freely branched, 0.2–4 dm. high; leaves linear-subulate to narrowly ovate, 0.5–6.5 cm. long, 0.5–13 mm. broad, mostly confined to lower two-thirds of branch; *bracts scarious-margined;* sepals 4.5–8.5 mm. long; *petals twice or thrice length of sepals, the broad lobes spreading in anthesis, the claw glabrous;* capsule cylindric, equaling to much exceeding calyx; seeds reddish, 0.35–0.7 mm. in diameter, the testa close and tuberculate. — A heteromorphous species of cold and temp. reg. of N. and S. Hemisph. Ours are tentatively placed as follows:

a. Internodes of flowering stem villous or pilose with reflexed non-glandular hairs. . . *b.*

b. Leaves linear to oblong or lanceolate; those of the flowering stems mostly 0.5–3.5 cm. long and 0.5–4 (–5) mm. broad, tapering to base. . . . . *C. arvense* (typical).

b. Leaves lanceolate to ovate, those of the flowering stems often more rounded at base, mostly (2–) 3–6.5 cm. long and (3–) 5–13 mm. broad.

Stems and leaves more or less pilose or the leaves glabrous above, their blades mostly lanceolate, up to 6.5 cm. long and 3–10 mm. broad. . Var. *villosum.*

Stems and leaves very densely velvety- or tomentose-villous with long white pubescence; leaves oblong-lanceolate to lance-ovate, 2–4 cm. long and 0.8–1.3 cm. broad. . . . . Var. *villosissimum.*

a. Internodes all glandular-hispid, the gland-tipped short hairs often intermixed with glandless ones. . . . . Var. *viscidulum.*

C. arvénse (typical). — Plant compact or lax; leaves densely pilose or the upper surface glabrate. (Incl. *C. campestre* and several other proposed spp. of Greene) — Gravelly, turfy or rocky basic soils, often a weed in grasslands, Lab. to Alaska, s. to Nfld., P.E.I., N.B., N.E., Del., Md., Wisc., S.D., N.M. and Calif. Late April–Aug. (alpine). (Eurasia) Fig. 1018. — A complex series, needing close study.

Var. **villòsum** (Muhl.) Hollick & Britt. (long-hairy). — Mostly tall, 1–4.5 dm. high, with long internodes and peduncles; leaves gray with close pilosity, or green and glabrous or promptly glabrate above in forma **oblongifòlium** (Torr.) Pennell (oblong-leaved). (Incl. *C. velutinum* Raf.) — Thin rocky soil and cliffs, s. Ont. to Ida., s. to Va., Tenn. and Mo. April–June.

Var. **villosíssimum** Pennell (very long-hairy). — Depressed, forming widely spreading mats, very densely velutinous- or tomentose-villous. — Serpentine-barrens, Chester Co., Pa.

1018. C. arvense.

Var. **viscídulum** Gremli (sticky). — Compact or lax, with narrowly linear to lanceolate or oblong leaves; whole plant glandular. (Incl. *C. confertum, C. occidentale* and *C. oreophilum* Greene) — Cliffs and gravel, St. Paul I., N.S.; Alaska, s. to N.D., Colo., Utah and Calif. (Eu.)

C. tomentòsum L. (tomentose), Snow-in-Summer, a depressed and matted perennial with stems, leaves and calyx densely white-woolly or tomentose, is cult. in rockeries and borders, and sometimes spreads to wild habitats. (Introd. from Eurasia)

6. C. nùtans Raf. (nodding). — *Weak annual,* the simple or loosely rather *flaccid* viscid-pilose stem 0.5–

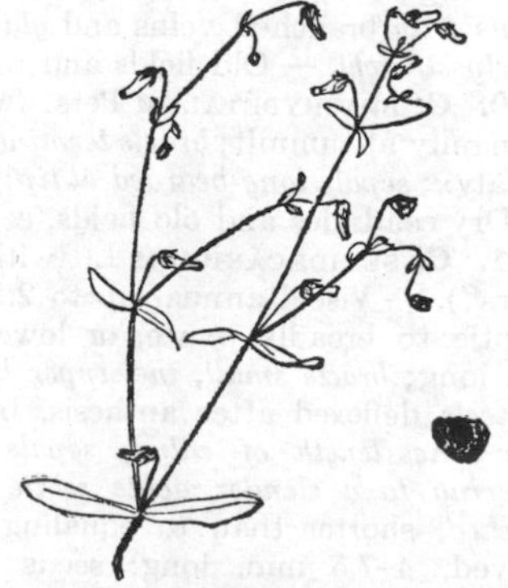

1019. C. nutans.

6 dm. high, the *median internodes* 0.2–1 *dm. long; leaves oblong-lanceolate to narrowly obovate, acute or acutish, thin,* the median 0.5–8 cm. long and 0.2–1.5 cm. broad; *bracts* similar but smaller, *herbaceous;* inflorescence loose, simple to dichotomous, 1–41-flowered; *pedicels* filiform, *ascending or spreading-ascending, with hooked tips, in fruit* 1–5.5 *cm. long; sepals* 2–5.5 mm. long, *oblong-lanceolate, thin, blunt,* pilose; petals narrowly obovate (wanting in cleistogamous flowers), with glabrous claw, cleft nearly to middle, exceeding calyx; *capsule curved,* 0.4–1.3 cm. long; seeds about 0.5 mm. in diameter, reddish-brown, bluntly papillate. (*C. longepedunculatum* of auth., possibly of Muhl.) — Alluvium, rich wooded slopes, calcareous rocks, etc., sw. Que. and w. N.E. to B.C., s. to n. Fla., Ala., Ark., Tex., N.M. and Ariz. March–June. Fig. 1019.

7. C. brachýpodum (Engelm.) Robins. (short-stalked). — Similar to no. 6, smaller; *leaves obtuse,* rarely 3 cm. long; *pedicels not hooked at tip, shorter than to but slightly longer than capsules.* — Open woods, slopes and meadows, Ga. to Ariz. and Mex., n. to se. Va., Tenn., Ill., N.D., Alta. and Wash. Late March–June. Fig. 1020.

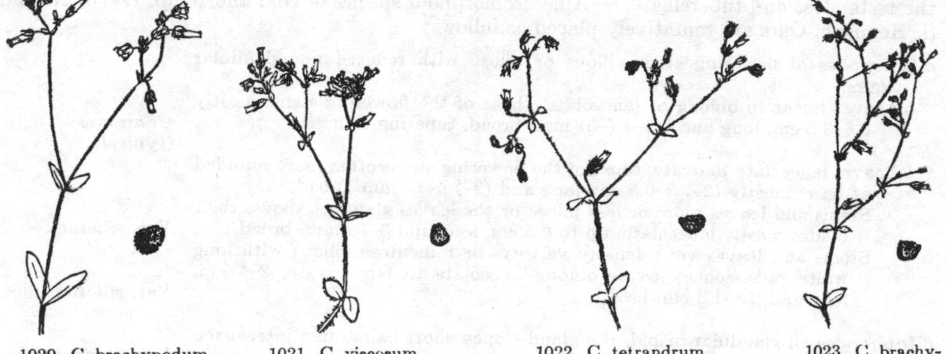

1020. C. brachypodum. 1021. C. viscosum. 1022. C. tetrandrum. 1023. C. brachypetalum.

8. C. viscòsum L. (somewhat sticky). — Erect or ascending *viscid annual,* simple to freely branched below, 0.6–3 dm. high; *leaves elliptic to narrowly oblong-obovate, rounded or obtuse at tip,* hairy on both sides, the median 0.5–3 cm. long; *bracts* very small, *herbaceous; inflorescence at first glomerulate, in maturity* looser but *with the terminal clusters dense; pedicels in anthesis shorter than, in fruit shorter than to but slightly longer than calyx; sepals* ovate-lanceolate, *sharply acute,* firm, *with scarious margins,* 3–5 mm. long; *petals oblong, deeply notched,* narrow, their margins not meeting, slightly exceeding sepals, or in forma apétalum (Dumort.) Mert. & Koch (without petals) minute or wanting; stamens 10, filaments glabrous; capsules slender, slightly curved, 5–9 mm. long; seeds minute, pale brown, finely muricate. — Waste places, fields and roadsides, s. Nfld.; Fla. to Tex. and Calif., n. to se. Mass., N.Y., O., Ill., S.D. and B.C. March–July. (Natzd. from Eu.) Fig. 1021.

9. C. tetrándrum Curtis (with four stamens). — *Viscid annual* up to 2 dm. high, resembling no. 8, but the *open and lax cyme dichotomous; pedicels two to five times length of calyx; petals with* branched veins and *glabrous claws; stamens 4 or 5, with glabrous filaments; capsule nearly straight.* — Old fields and roadsides, se. Va. March, April. (Adv. from Eu.) Fig. 1022.

10. C. brachypétalum Pers. (with short petals). — Similar to no. 9, *pilose,* rarely viscid, then only at summit; *bracts terminating in a tuft of hairs;* fruiting pedicels once to thrice length of calyx; *sepals long-bearded at tip; claws of petals ciliate; stamens* 10, *their filaments long-ciliate.* — Dry roadsides and old fields, e. Va. April, May. (Adv. from Eu.) Fig. 1023.

11. C. semidecándrum L. (with half of ten stamens, "filaments 5 sterile, 5 bearing anthers"). — Viscid annual up to 2.5 dm. high, simple or usually branching from base; leaves elliptic to broadly ovate, or lowest spatulate, thick, obtuse, hispidulous, the median 0.3–2 cm. long; *bracts small, the upper half scarious; cyme becoming loose and open,* 2–50-flowered; *pedicels* deflexed after anthesis, becoming erect or ascending in fruit and *one and a half to four times length of calyx; sepals* 3–4 mm. long, *lanceolate,* firm and *glandular on back, tapering to a slender acute white scarious tip; petals* narrow, *with simple veins, shallowly notched,* shorter than to equaling sepals; *stamens mostly* 5 (sometimes 10); capsule barely curved, 4–7.5 mm. long; seeds 0.4–0.5 mm. in diameter, pale brown, rugose. — Sandy fields, cult. land and roadsides, se. Mass. to N.C. April–June. (Natzd. from Eu.) Fig. 1024.

1024. C. semidecandrum.

12. C. pùmilum Curtis (dwarf). — Similar to no. 11, but leaves spatulate to oblong; *bracts and sepals with very narrow scarious margins and tips; petals with branching veins, deeply 2-cleft.* — Sandy fields, s. N.J. April, May. (Adv. from Eu.) Fig. 1025.

1025. C. pumilum.

13. C. cerastoìdes (L.) Britt. (like *Cerastium*, Linnaeus having placed it in *Stellaria*). — *Matted perennial, with numerous prostrate or trailing* very leafy slender *glabrous or lineolate-pilose* simple or forking *branches* up to 1.5 dm. long; *leaves fleshy, glabrous,* linear-lanceolate to narrowly elliptic, 0.7–1.5 cm. long, 1–4 mm. broad, obtuse, often *subtending axillary fascicles* or elongate sterile shoots; cymes 1–7-flowered; pedicels ascending or slightly divergent, 1–3 cm. long; *sepals glabrous, thin,* narrowly oblong, *obtuse,* scarious-margined, 5–7 mm. long; petals broad, deeply notched, much exceeding calyx; *styles* 3 *or* 4; *capsule thick-cylindric, nearly straight,* 7–10 mm. long, *the* 6 *or* 8 *teeth not circinate-revolute;* seeds about 1 mm. in diameter, pale brown or olivaceous, bluntly muricate. (*C. trigynum* Vill.) — Arct. reg., s. to wet rocks and brook-beds, mts. of nw. Nfld. and Gaspé Pen., Que. July, Aug. (Eurasia) Fig. 1026. — A noteworthy species, forming a transition to *Stellaria.*

1026. C. cerastoides.

11. HOLÓSTEUM L. Jagged Chickweed

Sepals 5. Petals 5, usually jagged or denticulate at the point. Stamens 3–5, rarely 10. Styles mostly 3. Capsule ovoid, 1-locular, many-seeded, opening at the top by 6 teeth. Seeds rough, flattened on the back, attached by the inner face. Annuals or biennials, nat. of Old World, with several (white) flowers in an umbel borne on a long terminal peduncle. (Name used by Dioscorides for some unidentified plant, from the Greek *holos, all,* and *osteon, bone.*)

1. H. umbellàtum L. (with umbels). — Glaucous or very pale; leaves oblong; peduncle and upper part of stem glandular; pedicels reflexed after flowering. — Fields, roadsides and cult. ground, e. Mass. to Ga., Ky. and O. Mar.–May. (Natzd. from Eu.)

Tribe IV. Silèneae DC.

12. AGROSTÉMMA L. Corn-Cockle

Calyx ovoid, with 10 strong ribs; the elongated teeth (in ours 2–3 cm. long) exceeding the 5 large unappendaged petals. Stamens 10. Capsule 1-locular. Leaves linear. — Tall silky annual or biennial nat. of Old World. (Name from the Greek *agros, field,* and *stemma, crown.*)

1. A. Githàgo L. (old generic name), Purple Cockle, Nielle des blés (Que.). — Flowers 2.5–4 cm. in diameter; petals purplish-red, paler toward the claw and spotted with black. — Grainfields and, less frequently, by roadsides. June–Sept. (Natzd. from Eu.)

13. LÝCHNIS L. Campion

Styles 5, rarely 4, capsules opening by as many or twice as many teeth; otherwise nearly as in *Silene.* Herbs of temp. or cold reg. (Ancient Greek name for a scarlet or flame-colored species, from *lychnos, a flame.*)

a. Plant densely white-woolly; flowers few, on stiff peduncles; calyx-teeth filiform, twisted; teeth of capsule same number as styles. . . . . . . 1. *L. Coronaria.*
a. Plant green (sometimes lanate in no. 8); flowers several to many; calyx-teeth not twisted; valves of capsule either same number as styles or bifid. . . *b.*
b. Loosely branching leafy biennials or short-lived perennials; some or all flowers dioecious or variously polygamo-dioecious, with more or less inflated fruiting calyx; valves of capsule distinctly 2-toothed.
Calyx-teeth triangular-lanceolate, merely acute; capsule globose, with wide mouth, its teeth recurved. . . . . 2. *L. dioica.*
Calyx-teeth lance-linear, attenuate; capsule conic-ovoid, with narrow mouth, its teeth erect or only slightly spreading. . . . . 3. *L. alba.*
b. Erect perennials with simple or only sparsely branched stems; flowers perfect

or mostly so; calyx scarcely inflated; valves of capsule entire or barely notched. . . *c.*

*c.*Stem loosely long-villous; leaves ovate, cordate, very numerous; inflorescence dense, hemispherical; corolla scarlet. . . . . 4. *L. chalcedonica.*

*c.*Stem glabrous or only short-pubescent above base; cauline leaves linear, lanceolate or oblanceolate, narrowed at base; inflorescence open or, if dense, with elongate axis; corolla not scarlet. . . *d.*

*d.*Stem glabrous or merely viscid-pilose; cymes broad, many-flowered; petals showy, pink or roseate; valves of capsule entire. . . *e.*

*e.*Stem sparsely pilose; cyme open-paniculate; petals deeply 4-cleft; ovary and capsule sessile within the calyx, unilocular at base, septicidally dehiscent. . . . . 5. *L. Flos-cuculi.*

*e.*Stem glabrous except at the rarely viscid nodes; cyme forming a rather dense thyrsiform or corymbiform panicle; petals 2-cleft or notched; ovary and capsule stipitate within the calyx, 5-locular when young, loculicidally dehiscent.

Stem 0.5–4 dm. high, non-glutinous; calyx campanulate, with short broadly deltoid teeth; petals deeply 2-cleft. . . . . 6. *L. alpina.*

Stem up to 8 dm. high, glutinous above and at the nodes; calyx slenderly cylindric, becoming clavate, with narrow acute teeth; petals merely emarginate. . . . . 7. *L. Viscaria.*

*d.*Stem densely viscid-puberulent (rarely lanate) throughout; inflorescence long and virgate, slender, the flowers on appressed-erect pedicels; petals included or but slightly exserted; valves of capsule cleft. . . 8. *L. Drummondii.*

1. L. Coronària (L.) Desr. (old generic name, meaning like a crown or used in making garlands), Rose-C., Mullein-Pink. — Perennial, *covered with dense white wool throughout,* stems up to 8 dm. high; leaves oblong or oval; *flowers few on stiff ascending peduncles; calyx* ovoid-ellipsoid, with alternating prominent ribs and *with linear-subulate twisted teeth; petals* red-purple (rarely white), much exceeding calyx, *with large rounded limb, with 2 rigid acute lanceolate basal scales;* capsule sessile; seeds black. — Open rocky woods, roadsides and waste places, Me. to s. Ont., s. to Conn., Del., Pa., O. and Ind. June–Aug. (Introd. and natzd. from s. Eu.)

2. L. dioìca L. (dioecious), Red C. — Biennial or short-lived perennial, with *loosely forking simply pilose stems* 0.3–1 m. high; leaves oval to lance-oblong, pubescent, the lower 4–12 cm. long; *cyme loosely dichotomous* with prolonging leafy-bracted branches; *flowers dioecious,* almost odorless, *diurnal, those at the forks short-pedicelled; calyx* hairy, *inflated, with triangular-lanceolate* acute *teeth;* petals showy, roseate or pink (rarely white), with obovate 2-lobed limb; *capsule globose, with wide mouth, its teeth recurved after dehiscence.* (*Melandrium* Coss. & Germ.) — Waste places, roadsides, etc., Nfld. to Ont., s. to N.S., N.E., Del., Pa., O., Ill. and Mo. Late May–Sept. (Natzd. from Eurasia)

3. L. Álba Mill. (white), White Cockle or Campion, Evening-Lychnis. — Similar to no. 2 and hybridizing with it; plant *glandular-pubescent above; flowers* white to pink, *opening in the evening,* fragrant, *those at the forks longer-pedicelled; calyces* of staminate flowers ellipsoid, of the pistillate ovoid and inflated at maturity; *the teeth elongate, lance-linear, attenuate; capsule conic-ovoid, with narrow opening, the teeth erect or only slightly spreading.* (*Melandrium* Garcke) — Roadsides, borders of fields and waste places, Que. to B.C., s. to N.S., N.E., N.C., O., Ind., Ill., Mo., Kans., Colo., Utah and Calif. Late May–Sept. (Natzd. from Eurasia) — Often confused with the annual *Silene noctiflora,* which has smaller and perfect flowers and 3 styles.

4. L. chalcedónica L. (of Chalcedon on the Bosporus), Scarlet Lychnis, Maltese-cross, London-pride. — *Villous* perennial with tall *very leafy stems; leaves* membranaceous, *ovate, cordate; inflorescence dense, hemispherical,* 4–12 cm. broad; the *calyx becoming strongly clavate at base; petals scarlet* (rarely white), bifid. — Spreading from cult. to thickets, open woods and roadsides, P.E.I. to s. N.E., w. to Minn. June–Aug. (Introd. from Asia)

5. L. Flòs-cùculi L. (cuckoo-flower), Ragged-Robin, Cuckoo-flower. — Perennial with spatulate basal leaves; *stems* slender, erect, *sparingly pilose, viscid above;* cauline leaves remote, narrowly lanceolate; *cyme open-paniculate;* calyx campanulate, becoming subglobose, glabrous; *petals* deep rose, or white in forma albiflòra Sylvén (white-flowered), *the spreading limb deeply 4-cleft into unequal linear lobes; ovary and capsule sessile within the calyx, unilocular at base, septicidally dehiscent.* — Fields, meadows and swales, Que. to N.Y., s. to N.S., N.E. and Pa. Late May–July. (Natzd. from Eu.)

6. L. alpìna L. (alpine). — Perennial *with dense rosettes* of spatulate to linear basal leaves 1–3.5 cm. long and 2–5 mm. broad; *flowering stems* solitary or few, *stiffly erect,* 0.5–3 dm. high, *glabrous*, with 2–4 pairs of erect leaves up to 3 cm. long; *cyme densely corymbiform*, with or without remote lower branches; *calyx campanulate, with short broadly deltoid teeth*, in anthesis 3–5 mm. long; *petals* roseate, including claw, 6–8 mm. long, *deeply 2-cleft, the crown reduced to 2 short scales; ovary and capsule stipitate, 5-locular when young, loculicidally dehiscent, the valves entire.* (*Viscaria* G. Don) — Eu. Our representative is

Var. **americàna** Fern. (American). — *Coarser*, up to 4 dm. tall; *radical leaves* subcoriaceous, *up to* 6.5 *cm. long; cauline leaves 3–7 pairs, the larger up to 5.5 cm. long* and 2.5–10 mm. broad; *flowering calyx* 5–7 *mm. long; petals*, including claw, 8.5–14.5 *mm. long*, 3–6 mm. broad, roseate, or white in the rare forma **albiflòra** (Lange) Fern. (white-flowered). — Greenl. and Lab. s. to magnesian barrens and gravels of Nfld. and Gaspé Co., Que. July, Aug.

7. L. Viscària L. (an old generic name, from the viscid stem), German Catchfly. — Coarser than no. 6, up to 8 dm. high, the *stem glutinous* above and at the nodes; leaves elongate-linear or narrowly oblanceolate, *often 2 dm. long; calyx about 1 cm. long, slenderly cylindric, becoming clavate, its narrow teeth acute; petals merely emarginate*, roseate. — Cult. and esc. locally to roadsides and thickets, N.E. to O. May–July. (Introd. from Eurasia)

8. L. Drummóndii (Hook.) S. Wats. (for its discoverer, Thomas Drummond, 1780–1835). — Cinereously *viscid-puberulent* (rarely lanate) *throughout; caudex stout with 1–∞ virgate flowering stems;* leaves linear to narrowly oblanceolate; the virgate and elongate inflorescence with *appressed-erect pedicels;* calyx ovoid-cylindric, *petals included or barely exserted; valves of capsule cleft.* (*Wahlenbergella* Rydb.) — Dry plains and hillsides, B.C. to Ariz., e. locally to Man., Minn., sw. Mich. and Neb. June–Aug.

14. SILÈNE L. Catchfly. Campion

Calyx 5-toothed, 10–many-nerved, naked at the base. Stamens 10. Styles 3, rarely 4. Capsule 1-locular, sometimes 3-locular at least at the base, opening by 3 or 6 teeth at the apex. — Flowers solitary or in cymes. Petals mostly crowned by a scale at the base of the blade. — Extensive genus of temp. and cold reg. (Name adopted by Linnaeus from earlier authors; said to have come from mythological *Silenus*, referring to the viscid excretions of many species, the intoxicated foster-father of Bacchus being described as covered with foam.)

a. Calyx glabrous or only minutely puberulent at base. . . *b.*

b. Densely matted arctic-alpine species, with aspect of a coarse moss, the approximate marcescent leaves linear; flowers solitary and mostly sessile at tips of the crowded stems. . . . . 1. *S. acaulis.*

b. Stems elongating, with the pairs of leaves distant, broader; flowers several. . . *c.*

c. Calyx more or less inflated, becoming ovoid to campanulate or subglobose, the nerves not prominent; petals white or pale pink; stem not glutinous; perennials (no. 3 biennial). . . *d.*

d. Leaves all opposite; petals 2-lobed.

Glaucous, with thick merely acute leaves; inflorescence a long-peduncled small-bracted cymose panicle; petals without crowns.

Calyx ovoid, much inflated, not appressed to the stipitate included capsule, the 20 nerves connected their entire length by anastomosing veinlets. . . . . 2. *S. Cucubalus.*

Calyx cylindric-ellipsoid, only slightly inflated, the summit appressed to the subsessile exserted capsule, 20-nerved at base, 10-nerved above, the veins hardly anastomosing. . . . 3. *S. Cserei.*

Green, with thin attenuate-tipped leaves; flowers scattered, solitary and divergent in the upper leaf-axils; petals with a small crown. . 4. *S. nivea.*

d. Leaves mostly in whorls of 4; inflorescence an elongate open panicle; petals fringed. . . . . 5. *S. stellata.*

c. Calyx tight over the capsule, the 5–10 prominent straight ribs not evidently anastomosing; petals roseate, purple or whitish; stem usually with glutinous areas; annual or biennial.

Flowers loosely panicled, on filiform pedicels; fruiting calyx ovoid; petals transient or wanting, rarely conspicuous; capsule subsessile within calyx. . . . . 6. *S. antirrhina.*

Flowers in close terminal corymbiform cymes, short-stalked; calyx clavate; petals showy, pink or roseate (rarely white), of long duration; capsule long-stipitate. . . . . 7. *S. Armeria.*

a. Calyx pilose. . . *e.*

e. Annual or biennial, with soft bases and weedy habit; petals with limb usually shorter than calyx. . . *f.*

f. Calyx about 30-ribbed, conic, tapering from broad base to the long-acuminate teeth. . . . . 8. *S. conica.*

f. Calyx 5–10-nerved. . . *g.*

g. Inflorescence simple or dichotomously racemose, with 1-sided racemes.

Inflorescence dichotomously forking; flowers nodding in anthesis, the fruits appressed-ascending; calyx-ribs pectinate-hirsute; limb of petals cleft to below middle; coarse, 0.4–1.5 m. high. . . . . 9. *S. dichotoma.*

Inflorescence a simple raceme; flowers ascending; calyx villous or glandular-pilose; limb of petals notched only above middle; slender, 1–4.5 dm. high.

Fruiting calyx 7–10 mm. long, ovoid-campanulate, rounded at base, erect or ascending, with long lance-attenuate teeth; petals white (rarely purple), the small limb entire or merely crenate or emarginate. . . . . 10. *S. gallica.*

Fruiting calyx 1.2–1.8 cm. long, bladdery-ellipsoid, with cylindric base, on divergent or nodding pedicels, the teeth obtuse; petals roseate, the broad limb notched. . . . . 11. *S. pendula.*

g. Inflorescence an open cymose panicle; coarse weedy annual or biennial, viscid-villous; fruiting calyx inflated-ovoid, 1.5–2.3 cm. long, with prominent green ribs. . . . . 12. *S. noctiflora.*

e. Perennials with firm caudex or with tufts of basal leaves; limb of petals usually (except in no. 18) as long as or longer than calyx. . *h.*

h. Inflorescence corymbiform or the flowers subtended by broad foliaceous bracts; fruiting calyx 1–3 cm. long; petals pink to crimson or scarlet, the limb 0.8–2.5 cm. long. . . *i.*

i. Inflorescence a corymb of small cymes; fruiting calyx 1–2.2 cm. long; petals pink, the limb 0.8–1.5 cm. long; plant cespitose, the stems 0.4–3 dm. high. . . . . 13. *S. caroliniana.*

i. Inflorescence elongate or the pedicelled flowers in axils of broad leaves; fruiting calyx 1.5–3 cm. long; petals scarlet or red, the limb 1–2.5 cm. long; stems 1–few, elongate, 0.2–1.5 m. tall.

Stem stiffly erect, 0.7–1.5 m. high, closely puberulent-pubescent, with 15–30 pairs of round-based sessile firm leaves; petals scarlet, uncleft. . . . . 14. *S. regia.*

Stem weak, shorter, villous at least above, with 2–8 pairs of narrow-based or short-petioled thin leaves; petals mostly cleft.

Cauline leaves oblanceolate; calyx 2–2.5 cm. long; petals deep red to crimson, the limb 2–2.5 cm. long. . . . . . . 15. *S. virginica.*

Cauline leaves all elliptic to round-ovate; calyx 2.5–3 cm. long; limb of scarlet petals 1–2 cm. long. . . . . 16. *S. rotundifolia.*

h. Inflorescence a terminal panicle with small bracts; fruiting calyx 8–12 mm. long; petals white or pink, the limb 5–15 mm. long.

Stem 0.5–1.2 m. high; the 5–10 pairs of large ovate leaves rounded to sessile bases; petals cleft into a fine fringe. . . . . . 17. *S. ovata.*

Stem 1.5–5 dm. high; the 2–5 pairs of narrowly lanceolate to oblanceolate leaves tapering to narrow bases; petals with 2 linear lobes. . 18. *S. nutans.*

1. S. ACAÙLIS L. (stemless), MOSS-CAMPION. — Dwarf, *forming dense hassocks or tussocks of crowded marcescent branches densely clothed with small linear leaves; flower solitary, terminal; calyx* tubular, slender, *glabrous,* 7–11 mm. long, about equaling the peduncle; petals lilac or purple, notched or entire; capsule twice as long as calyx. — Eurasia. Ours is

Var. exscàpa (All.) DC. (without a scape). — *Calyx campanulate, 4–6 mm. long,* often purple-tinged, subsessile or short-stalked, or the peduncle equaling or longer than calyx in forma **caulèscens** (Vaccari) Fiori (with a stem); petals pink or lilac-tinged, or white in forma **albiflòra** Hartz (white-flowered); *capsule ovoid, barely exserted.* — Arct. reg., s. to gravelly, rocky or turfy barrens and cliffs of Nfld. and of St. Paul I., N.S., and alpine areas of Que. and N.H.; w. N.Am. s. to Mont. and Wash. June–Aug. (Eurasia)

2. S. CUCÙBALUS Wibel (old generic name), BLADDER-CAMPION, MAIDEN'S-TEARS, PÉTARDS (Que.). — *Perennial, glaucous,* glabrous (rarely hispid), *with* many *depressed-ascending or decumbent stems* from a branching or substoloniferous caudex; *leaves lanceolate, lance-oblong or lance-ovate,* at least the upper rounded at base, *fleshy; cyme long-peduncled, open-paniculate,* with stiff branches, few–many-flowered, the upper bracts small; *calyx ovoid, membranaceous, conspicuously bladdery-inflated, glabrous, with fine anastomosing veinlets connecting the* 20 *obscure nerves;* petals white (rarely pinkish), 2-lobed, crownless; *capsule* stipitate, *included.* (*S. latifolia* (Mill.) Britten & Rendle; *S. vulgaris* (Moench) Garcke; *S. inflata* Sm.) — Roadsides, borders of fields, gravelly shores, etc., Nfld. to B.C., s. to N.S., N.E., Va., Tenn., Mo., Kans., Colo. and Oreg. Apr.–Aug. (Natzd. from Eurasia)

Var. LATIFÒLIA (Reichenb.) G. Beck (broad-leaved). — Very tall, up to 1 m. or more; cauline leaves thin, 0.8–1 dm. long, 3–5.5 cm. broad. (*S. bosniaca* (G. Beck) Hand.-Maz.) — Local, Albany Co., N.Y. (Adv. from Eu.)

3. S. CSÈREI Baumg. (dedicated in 1816 to Baumgarten's patron, WOLFGANG VON CSEREI). — Similar to no. 2, but *biennial; stems 1–few, erect; leaves* very thick and glaucous, *the cauline oblong or oblong-ovate, clasping; panicle* simple and *elongate* or with 2 erect slender branches; *calyx* glabrous, *cylindric-ellipsoid, the summit appressed to the subsessile exserted capsule,* 20-*nerved at base,* 10-*nerved above,* the veins hardly anastomosing. (*S. fabaria* sensu Rydb., not Sibth. & Sm.) — Railroads, roadsides and waste places, N.Y. to Mont., s. to s. N.J., Pa., O., Ind. and Mo., rapidly spreading. Late May–Sept. (Natzd. from se. Eu.)

4. S. nívea (Nutt.) Otth (snowy), SNOWY CAMPION. — Perennial with loosely spreading bases; stems slender, ascending or leaning, glabrous, 0.4–1.5 dm. long, simple or forking above; *leaves thin, green,* lanceolate, *long-attenuate at tip; flowers few, on finally divergent slender pedicels in the upper leaf-axils;* calyx glabrous, membranaceous, scarcely veiny, subcylindric, in fruit broader and with obconic base; petals white, with cuneate obovate slightly notched limb and a small crown. (*S. alba* Muhl., *nomen subnudum*) — Rich woods and alluvium, local, Pa. to Minn. and S.D., s. to Md., Va., Tenn. and Mo.; cult. and locally natzd. ne. to Me. June–Aug.

5. S. stellàta (L.) Ait. f. (star-like), STARRY CAMPION, WIDOW'S-FRILL. — Perennial with strong branching crown; stems stiff, ascending, 0.4–1 m. high; all except lowest and uppermost *leaves in whorls of* 4, ovate or lanceolate, long-acuminate, *minutely puberulent beneath,* smooth to scabrous above; *panicle* simple and *elongate* or with a few elongate ascending branches, with small foliaceous bracts; calyx campanulate, inflated, not strongly veiny, glabrous or merely puberulent; corolla 2 cm. broad; *petals* white, crownless, *with margin cut into a fringe.* — Woods and clearings, Mass. to Minn., s. to Ga., Ala., Ark., Okla. and e. Tex. July–Sept.

6. S. antirrhìna L. (with leaves as in *Antirrhinum*), SLEEPY CATCHFLY. — Erect or ascending *annual or biennial,* 0.5–8 dm. high, *paniculately branched, the slender branches and branchlets* of the panicle ascending (in weak or shaded plants loosely spreading), *often with usually* dark *glutinous areas,* or nonglutinous in forma **Deaneàna** Fern. (named for WALTER DEANE, 1848–1930); cauline leaves lanceolate or oblanceolate; *flowers slender-pedicelled;* flowering *calyx* fusiform, *in fruit becoming campanulate-ovoid, glabrous,* with straight nerves, the short teeth often purple; *petals* small, *transient,* pink or purplish, or white above in forma **bícolor** Farw. (two-colored), more or less emarginate, or petals wanting in forma **apétala** Farw. (without petals) *i.e.* var. *divaricata* Robins.; *capsule subsessile* within the closely embracing calyx. — Dry or rocky open woods, fields, waste places, etc., sw. Que. to B.C., s. beyond our range. May–Sept. (Mex.; S.Am.; adv. in Eu.)

7. S. ARMÈRIA L. (early confused with *Armeria,* Thrift), GARDEN- or SWEET-WILLIAM-CATCHFLY, NONE-SO-PRETTY. — Glaucous annual with smooth erect stem and branches; leaves elliptic or ovate-lanceolate; *cymes terminal, close, corymbiform;* flowers short-pedicelled; *calyx* slender, *clavate, prominently ribbed; petals pink or roseate,* or white in forma ALBIFLÒRA Sylvén, subentire or minutely toothed, crowned with subulate scales; *capsule long-stipitate.* — Cult. and esc. to roadsides and waste places, Que. to Minn., s. to N.B., N.E., Va., O., Ind., Ill. and Ia. June–Oct. (Introd. from Eurasia)

8. S. CÓNICA L. (conical). — Annual or biennial, *puberulent,* 1.5–5 dm. high; leaves linear-lanceolate, acute; flowers slender-stalked and erect, in a stiffish cyme; *calyx strongly about* 30-*ribbed, minutely hispid, conic, tapering from the thick base to the long-acuminate teeth* (*half length of calyx*); petals roseate, small; stamens pilose; capsule sessile within calyx. — Sandy fields and roadsides, local, Mass. to Mich., s. to Del. and O. May–July. (Adv. from Eu.)

9. S. DICHÓTOMA Ehrh. (forking), FORKING CATCHFLY. — Biennial, 0.4–1.5 *m. high,* hirsute; leaves lanceolate or oblanceolate; *upper half of stem divided into long simple or forking erect branches with secund flowers borne in long spiciform racemes; flowers* subsessile or short-stalked, *nodding in anthesis, the fruits appressed-ascending; calyx* 10-ribbed, *the ribs pectinate-hirsute,* in fruit slenderly ellipsoid; *limb of white* (rarely pink) *petals cleft to below middle,* the crowns small and obtuse. — Fields and waste places, Que. to Mont., s. to beyond our limits. June–Sept. (Natzd. from Eu.)

10. S. GÁLLICA L. (French). — Annual or biennial, slender, 1–4.5 dm. high; stem hirsute or hispid; lower *leaves* spatulate, obtuse, *the upper lanceolate; racemes simple,* 1-sided, up to 1.5 dm. long, *the short-pedicelled flowers in the axils of bracteal leaves; fruiting calyx* 7–10 *mm. long, rounded at base, erect or ascending,* villous-hirsute, *with long lance-attenuate teeth;* petals very small, the white, or in forma QUINQUEVÚLNERA (L.) Mert. & Koch (five-notched) purple or maroon, *limb entire or merely crenate or emarginate;* seed-faces excavated. (*S. anglica* Am.

auth., not L.) — Waste places, roadsides, etc., local, N.S. to Mich., s. to N.J., Ill. and Mo.; Pacific states. Late June–Sept. (Adv. from Eu.)

11. S. péndula L. (pendulous), NODDING CATCHFLY. — Coarser than no. 10, crisp-pubescent; lower leaves spatulate-obovate, upper ovate-lanceolate; *petals roseate, the notched limb about* 1 *cm. long; fruiting calyx* 1.2–1.8 cm. long, *bladdery-ellipsoid, with cylindric base, on divergent or nodding pedicels, the teeth obtuse; seeds with plain faces.* — Cult. and locally spreading to roadsides and fields. July–Sept. (Introd. from Eu.)

12. S. noctiflòra L. (flowering at night), NIGHT-FLOWERING CATCHFLY, STICKY COCKLE. — *Annual* or winter-annual up to 1 m. high, *viscid-villous;* leaves lanceolate or ovate-lanceolate, the lower spatulate; *flowers* fragrant, opening at night, erect, *rather few, in an open cyme; calyx* cylindric, *in fruit becoming inflated-ovoid,* 1.5–2.3 *cm. long,* with prominent green ribs tending to anastomose, and subulate teeth; petals pinkish at base, the deeply cleft limb creamy-white. — Cult. ground (often a troublesome weed) and waste places, throughout our range and beyond. Late May–Sept. (Natzd. from Eu.) — Sometimes confused with *Lychnis alba.*

13. S. caroliniàna Walt. (of Carolina). — *Usually cespitose, perennial;* ascending *stems* 0.4–2 *dm. high,* puberulent to hirsute, glandular toward summit; *radical leaves oblong-spatulate to oblanceolate, obtuse,* up to 1.2 dm. long, 0.5–3 *cm. broad, densely short-pubescent on both surfaces;* cauline leaves linear-oblong, acute, up to 5 cm. long and 0.8 cm. broad; *cymes in a rather dense corymb; flowers short-stalked; calyx slenderly cylindric,* 1.5–1.7 cm. long, *densely glandular;* limb of pink petals slightly notched or entire, 0.8–1.3 cm. long, the claws slightly exceeding calyx. — N.C. and S.C., and in forms transitional to the next in Va. and Md.

Var. **pensylvánica** (Michx.) Fern. (of Pennsylvania). — Stems 0.4–3 dm. high; *rosette-leaves often acute,* 1–2 *cm. broad, glabrous on both surfaces;* cauline leaves ovate or elliptical, acute to obtuse; *calyx* 1–1.8 *cm. long.* (*S. pensylvanica* Michx.) — Dry sandy, gravelly or rocky woods and openings, s. N.H. to e. O., s. to N.C. and e. Tenn. April–early June.

Var. **Whérryi** (Small) Fern. (for EDGAR THEODORE WHERRY, 1885–). — Similar to the last, but *pubescence of summit of stem and thicker calyx without glands;* cauline leaves lanceolate; *calyx* 1.5–2.2 *cm. long;* claws of petals little if at all exceeding calyx, limbs 1–1.4 cm. long. (*S. Wherryi* Small) — Local, O. to Mo. and Ala. April, May.

14. S. règia Sims (royal), ROYAL CATCHFLY. — Perennial; *stems erect,* 0.7–1.5 *m. high, closely pulverulent-pubescent, with* 15–30 *pairs of* lance-ovate *round-based* puberulent *firm sessile leaves;* panicle elongate, ellipsoid, leafy-bracted, the branches and pedicels strongly ascending; *calyx* 2–2.5 *cm. long,* cylindrical, becoming fusiform in fruit, glandular-pilose; *petals with scarlet subentire limb* 1.5–2 *cm. long.* — Dry woods, barrens and prairies, Ga. to Okla., n. to O., Ind., s. Ill. and Mo. June–Aug.

15. S. virgínica L. (Virginian), FIRE-PINK. — Perennial; *stems slender, weak, villous* at least above, 2–6 *dm. long; cauline leaves* 2–6 *pairs, thin, oblanceolate, glaucous* or merely ciliate, the basal spatulate, all *tapering at base; flowers* few, *loosely cymose,* long-pedicelled, the inflorescence leafy-bracted; *fruiting calyx* 2–2.5 *cm. long,* with round-tipped teeth; *limb of deep red to crimson petals* 2–2.5 *cm. long,* usually 2-cleft. — Open woods, clearings and slopes, w. N.Y. (formerly) and s. Ont. to Minn., s. to Ga., Ala., Ark. and Okla. April–June.

16. S. rotundifòlia Nutt. (round-leaved), ROUND-LEAVED CATCHFLY. — Weak, the villous to merely glandular-puberulent branching stems with 4–8 *pairs of thin elliptic to round-ovate* abruptly pointed *leaves; flowers* few, long-pedicelled, *in the axils of foliage-leaves; calyx* 2.5–3 *cm. long; limb of scarlet petals* 1–2 *cm. long,* 2-cleft and cut-toothed. — Crevices and talus of dry siliceous (rarely calcareous) cliffs and ledges, W.Va. and s. O. to Ga. and Ala. June, July.

17. S. ovàta Pursh (ovate). — Perennial; stems firm, 0.5–1.2 m. high, erect, smooth; *cauline leaves* 5–10 *pairs,* large, *ovate, long-acuminate, sessile by round bases; panicle terminal,* slender, *with small scarious bracts; calyx* pale, minutely pubescent, slenderly tubular, 8–10 *mm. long,* with deltoid-acuminate teeth; *petals white, the deeply and dichotomously cleft limb* 1–1.5 *cm. long.* — Rich woods, local, w. N.C. and se. Ky. to uplands of Ga., Ala. and Ark. Aug.

18. S. nùtans L. (nodding). — *Subcespitose* perennial, with slender stems *up to* 5 *dm. high;* basal leaves tufted, oblanceolate, long-petioled; *cauline leaves* 2–5 *pairs, narrowly lanceolate to oblanceolate, tapering to narrow bases;* panicle terminal, slender, with small bracts; *flowers at first nodding,* the fruits erect; *calyx* slender, viscid, green-nerved, 1–1.2 *cm. long,* in fruit ovoid, with slender base; *petals* whitish to pink, 5–10 *mm. long, with* 2 *linear lobes.* — Cult. and locally spread to roadsides, etc., Me. to s. N.Y. and O. May–July. (Introd. from Eurasia)

15. SAPONÀRIA L.

Calyx slenderly ovoid or subcylindric, 5-toothed, obscurely nerved, naked. Stamens 10. Styles 2. Capsule 1-locular, or incompletely 2–4-locular at base, 4-toothed at the apex. —

Coarse annuals or perennials, nat. of Old World, with large flowers. (Name from *sapo, soap;* the mucilaginous juice of no. 1 forming a lather with water.)

1. S. officinàlis L. (of the shops), Soapwort, Bouncing-Bet, Herbe à savon (Que.). — Flowers in corymbed clusters; *calyx terete;* petals crowned with an appendage at the top of the claw; leaves oval-lanceolate. — Roadsides, etc., throughout. July–Sept. — A stout perennial, with large rose-colored flowers, these commonly double in northern part of range. (Natzd. from Eu.)

2. S. Vaccària L. (old generic name, from vacca, cow), Cowherb, Cow-Cockle. — Annual, glabrous; flowers in corymbed cymes; *calyx 5-angled,* enlarged and wing-angled in fruit; petals pale red, not crowned; leaves ovate-lanceolate. (*Vaccaria vulgaris* Host; *V. Vaccaria* Britt.) — Waste places and cult. ground, occasional. June–Sept. (Adv. from Eu.)

16. GYPSÓPHILA L.

Calyx slenderly turbinate or campanulate, 5-nerved, 5-toothed, naked at base. Petals not crowned. Stamens 10. Styles 2. Capsule 1-locular, 4-valved at the apex, sessile. — Slender often glaucous annuals or perennials, nat. of Old World, with numerous small flowers. (Name from the Greek *gypsos, gypsum,* and *philein, to love,* from the habitat of some species.)

a. Stoutish glaucous plants with leaves 0.5–1.5 cm. broad; capsules spherical. . . b.
b. Hard-based perennials 0.6–2 m. high; longer pedicels 0.5–1.2 cm. long; calyx 1.5–3 mm. long; blade of petal about equaling calyx.
Cauline leaves narrowed to base; calyx-teeth suborbicular, the green center much narrower than the scarious margin. . . . . 1. *G. paniculata.*
Cauline leaves subcordate at base; calyx-teeth ovate, the green center and scarious margin subequal in breadth. . . . . 2. *G. perfoliata.*
b. Slender-based annual up to 6 dm. high; longer pedicels 1.2–3.5 cm. long; calyx 3–5 mm. long; blade of petal 2–3 times as long as calyx. . . . 3. *G. elegans.*
a. Frail green annual with capillary stems and branches; leaves 0.5–1 mm. broad; capsules ellipsoid-ovoid. . . . . 4. *G. muralis.*

1. G. paniculàta L. (paniculate), Baby's-breath. — *Strong-based glaucous perennial* up to 1 m. high, naked at base; *leaves* lance-acuminate, *attenuate at base,* 1-nerved; inflorescence a large panicle, with corymbiform branches; *pedicels 5–12 mm. long; calyx 1.5–3 mm. long, the teeth suborbicular, their green centers much narrower than the white-scarious margin; blade of petal about equaling calyx,* white, spatulate; *capsule globose,* about equaling calyx; seeds tuberculate in rows. — Cult. and locally spread to sandy or rocky roadsides and waste places, N.E. to Man., s. to Ind. and Neb. June–Aug. (Introd. from Eurasia)

2. G. perfoliàta L. (perfoliate). — Similar to no. 1, up to 2 m. high; *leaves subcordate* at base, 3–5-nerved; *calyx-teeth ovate, the green center and scarious margin subequal in breadth;* petals purplish; seeds sparsely rugose. — Cult. and locally spread to seabeaches and roadsides, Ct. and doubtless elsewhere. June–Oct. (Introd. from Eurasia)

3. G. élegans Bieb. (elegant). — Glaucous *annual or short-lived perennial, up to 6 dm. high;* lower leaves spatulate, upper lance-acuminate and short-connate at base; panicle open, corymbiform, the *longer pedicels 1.2–3.5 cm. long; calyx 3–5 mm. long,* hemispherical, *divided to base into ovate* obtuse, *white-margined lobes; petals* white or roseate, *twice or thrice exceeding calyx;* seeds rugose-tuberculate. — Cult. and spreading to roadsides and waste places. May, June. (Introd. from Eurasia)

4. G. muràlis L. (of walls). — Freely branched *low annual up to 2 dm. high,* with capillary stems and branches; *leaves linear-filiform,* 1–2 cm. long, 0.5–1 *mm. broad;* flowers on slender pedicels, in the forks; calyx turbinate, 3–4 mm. long, whitish, with broad green ribs, the teeth blunt; petals purplish, twice as long as calyx; *capsule ellipsoid-ovoid.* — Roadsides, fields, etc., local, Me. to Ont., s. to N.J., Ind. and Minn. July–Oct. (Natzd. from Eu.)

17. DIÁNTHUS L. Pink. Carnation. Oeillet (Que.)

Calyx cylindrical, nerved or striate, 5-toothed, subtended by 2 or more imbricated bractlets. Stamens 10. Styles 2. Capsule 1-locular, 4-valved at the apex. Seeds flattish on the back; embryo scarcely curved. — Ornamental chiefly Old World plants, of well known aspect and value in cult. (Name from the Greek *Dios, Jupiter,* and *anthos, flower, i.e.* Jove's own flower.)

a. Flowers sessile or closely crowded in glomerules or fascicles. . b.
b. Glomerules of flowers almost or completely surrounded by an involucre of broad scarious bracts. . . . . 1. *D. prolifer.*
b. Glomerules and inflorescences subtended by slender foliaceous bracts.

Branching annual, pubescent; flowers few in small obconic clusters at tips of branches. . . . . 2. *D. Armeria.*

Cespitose perennial with simple erect stems, glabrous or merely scabrous; fascicles of flowers crowded into a broad and dense corymbiform cyme. 3. *D. barbatus.*

a. Flowers all or nearly all solitary and long-stalked, 1–many on a stem.

Leaves linear-lanceolate, those of the depressed basal shoots 0.5–2 cm. long; calyx 1.5–2 cm. long, the long-tipped basal bracts half as long; petals with dentate margins. . . . . 4. *D. deltoides.*

Leaves linear, those of the erect basal tufts 3–10 cm. long; calyx 1.8–3 cm. long, the abruptly short-pointed basal bracts only one-fourth to one-third length of calyx; petals lacerate above middle. . . . . . . 5. *D. plumarius.*

1. D. pròlifer L. (proliferous, bearing young on offshoots), Childing P. — Smooth slender *annual,* 2–5 dm. high, *with* few linear leaves and 1-*few long peduncles bearing an involucre of* 2 *or* 3 *pairs of oval obtuse scarious bracts* which completely hide the sessile calyces within; flowers expanding singly; petals red, entire or crenulate, only slightly exserted; seeds concave, elliptic. — Sandy fields and roadsides, local, s. N.Y. to Del., Va., Ky. and O. May–Oct. (Natzd. from Eu.)

2. D. Armèria L. (early placed with *Armeria,* Thrift), Deptford P. — *Stiff* simple or usually *slenderly branched annual* or biennial up to 8 dm. high; the *stem pilose;* leaves linear or linear-lanceolate; *flowers few in obconic clusters at tips of branches, the clusters subtended by* narrow strongly ribbed *pubescent bracts;* calyx pubescent; its basal bracteoles elongate, about equaling the tube; petals roseate, dotted with white, the margin toothed. — Dry fields and roadsides, s. Que. and s. Ont., s. to N.S., N.E., L.I., Ga., Ky. and Mo. May–July. (Natzd. from Eu.)

3. D. barbàtus L. (bearded), Sweet William. — *Densely tufted perennial with simple erect glabrous stems;* leaves lanceolate; *inflorescence a corymbiform cyme* of crowded fascicles, with glabrous slender bracts and calyx; petals roseate to white. — Cult. and spreading to roadsides, groves, etc., Que. to N.D., s. to N.E., Del., etc. June–Aug. (Introd. from Eu.)

4. D. deltoìdes L. (triangular), Maiden-P. — Low perennial with *elongate trailing basal offshoots bearing linear-lanceolate green ciliate-margined and -keeled leaves* 0.5–2 *cm. long;* flowering stems ascending, slender, 1–4 dm. high, with 5–8 pairs of narrow erect leaves; *flowers* 1-few, 1.5–2 *cm. broad,* mostly long-stalked; *calyx* slenderly cylindric, 1.5–2 *cm. long,* with lanceolate teeth, *the basal bractlets half as long and prolonged into slender tips; petals* roseate, with dark circle at bases of limbs, *dentate.* — Roadsides, dry fields and open woods, N.E. to N.J., Ill. and Mich. Late May–Aug. (Natzd. from Eu.)

5. D. plumàrius L. (feathery), Grass- or Garden-P. — *Glaucous* perennial *with erect basal tufts of whitish linear-attenuate firm smooth leaves up to* 1 *dm. long;* flowering stems decumbent or erect, stiff, with 3–5 pairs of erect leaves; *flowers* larger than in no. 4, 3–4 *cm. broad, with fragrance of clove; calyx* thicker, 1.8–3 *cm. long, its basal scales one-fourth to one-third as long, abruptly pointed; petals* pale pink or white, *lacerate* above. — Spread from cult. to sandy fields and roadsides, local, N.E. to Mich. May–Aug. (Introd. from Eu.)

FAM. 62. CERATOPHYLLÀCEAE (Hornwort Family)

Aquatic herbs with whorled finely dissected leaves and minute axillary and sessile monoecious flowers without floral envelopes but with an 8–12*-cleft involucre in place of a calyx, the pistillate a simple* 1*-locular ovary with a suspended orthotropous ovule; seed filled by a highly developed embryo with a very short radicle, thick oval cotyledons, and a plumule consisting of several nodes and leaves.* — Consists only of the genus

1. CERATOPHÝLLUM L. Hornwort. Cornifle (Que.)

Staminate flowers of 10–20 stamens with large sessile anthers. Fruit an achene, beaked by the slender persistent style. — Herbs growing under water; the sessile leaves cut into thrice-forked narrow divisions (whence the name from the Greek *ceras, a horn,* and *phyllon, leaf*). Small genus of wide dispersal.

1. C. demérsum L. (submerged). — Lowest leaves of young seedling simple; stems prolonged, branching, brittle and stiffly branched or more cord-like and flexuous; *leaves* finely *dissected into capillary to linear and flattened serrate divisions; fruit* (rarely collected) compressed, ellipsoid, *wingless, smoothish,* 4.5–5.5 *mm. long, with* 2 *basal spines* 2–5 mm. long; style 4.5–6 mm. long. — Quiet waters, Que. to n. B.C., s. beyond our range. July–Sept. — Very variable in length, breadth and toothing of leaf-divisions, brittleness or flaccidity of stems, etc.; needs careful collecting and study.

2. C. echinàtum Gray (bearing prickles). — Lowest leaves of seedling cleft; *leaves* of prolonged stem and branches *with entire capillary divisions*, the uncleft base of the leaf often more expanded than in no. 1; *fruit narrowly winged by confluent bases of the 2–5 lateral spines, roughened on surface, 5–7 mm. long;* style 5–10 mm. long. (*C. demersum*, var. Gray) — Quiet waters, Fla. to Tex. and Mex., n. to sw. N.B., s. Me., sw. Que., N.Y., O., Mich., Ill. and Minn.; less often collected than no. 1.

FAM. 63. NYMPHAEÀCEAE (Water-lily Family)

Aquatic perennial herbs with horizontal (rarely erect) *rhizomes and peltate or cordate leaves floating or emersed; the ovules borne on the sides or back* (*or when solitary hanging from the summit*) *of the locules, not on the ventral suture; the embryo inclosed in a little pouch* at the end of the albumen next the hilum, except in *Nelumbo*, which has no albumen. Cotyledons thick and fleshy, inclosing a well-developed plumule. — Flowers axillary, solitary. Vernation involute. Rhizomes apparently endogenous.

Leaves with basal sinus, not peltate; petals numerous, often persistent, hypogynous or variously adnate to surface of the compound 8–30-locular ovary (formed by union of many carpels); ovules numerous, inserted over inner face of the locules; stigmas radiate (as in the Poppy); fruit many-seeded, berry-like, with firm rind; petioles and peduncles arising from rhizomes. Subfam. I. Nymphaeoideae.
- Flower subglobose, with 5, 6 or more concave yellow or red- or green-tinged sepals; petals small, thick, stamen-like, inserted with the stamens under the ovary, mostly persistent; stigmas radiate and sessile on a disk; fruit ovoid or urceolate, naked (except sometimes at base). . . . . . 1. *Nuphar.*
- Flower widely expanding, then closing, with 4 sepals green or purplish outside; petals fleshy, showy, variously colored, adnate to ovary, the outer passing gradually into sepals, the inner into stamens; stigma radiate from around a central globular projection at summit of ovary, extended into slender incurved projections; fruit depressed-globose, often surrounded by bases of decayed petals. 2. *Nymphaea.*

Leaves (at least floating or emersed ones) centrally peltate; sepals, petals and stamens hypogynous.
- Sepals and petals passing gradually into each other, with the indefinitely numerous stamens promptly deciduous; receptacle enlarged, broadly turbinate, with the numerous 1-seeded pistils as immersed pits in the broad summit, the imbedded but soon free nut-like fruits globular; petioles and tall peduncles arising from the rhizomes. Subfam. II. Nelumbonoideae. . . . 3. *Nelumbo.*
- Sepals and petals each 3 or 4, persistent; stamens definite in number; carpels free and distinct, coriaceous, indehiscent, 1–3-seeded, on the dorsal suture; stems leafy and prolonged, coated with mucilage, the small flowers axillary. Subfam. III. Cabomboideae.
  - Leaves all peltate and undivided, all alternate; petals dull purple; stamens 12–18. 4. *Brasenia.*
  - Leaves peltate at surface of water, the submerged ones capillary-dissected and opposite or whorled; petals white, yellow-spotted at base; stamens 3–6. 5. *Cabomba.*

Subfam. I. NYMPHAEOÌDEAE

1. NÙPHAR Sm. Yellow Pond-lily. Cow-lily. Spatter-dock.

Water- or Marsh-collard. Lis d'eau jaune (Que.)

Sepals 5, 6, or sometimes more, roundish, concave. Petals numerous, small and thickish, stamen-like or scale-like, inserted with the very numerous short stamens on the receptacle under the ovary, not surpassing the disk-like 6–25-rayed sessile stigma, mostly persistent and at length recurved. Fruit ovoid, naked, usually ripening above the water. Aril none. — Rhizome creeping, cylindrical. Leaves with a deep sinus at the base. Flowers yellow or sometimes tinged with purple, produced all summer. Genus widely dispersed in N. Hemisph. (Name said to be of Arabic origin.) Nymphaea L., in part. Nymphozanthus L. C. Richard

a. Floating leaves 3.5–10 cm. long, on compressed-filiform petioles; sepals 1–1.8 cm. long; petals and stamens promptly deciduous; fruit naked at base, smooth, green or red-tinged, constricted to a slender neck only 2.5–6 mm. in diameter; stigmatic disk with 6–10 deep crenations or lobes and 6–10 rays. 1. *N. microphyllum.*

a. Floating or emersed leaves mostly larger, on thicker petioles; sepals and capsules mostly longer; petals and stamens persistent at base of the fruit; the fruit 1.5–5 cm. high, gradually narrowed to thick neck; mature stigmatic disk 0.8–2.5 cm. in diameter, entire to crenate, with 7–25 rays. . . *b*.

b. Petioles strongly flattened from above, often margined; leaf-blade usually floating, the sinus closed or narrow.

Floating blades 0.5–2 dm. long, 0.45–1.4 dm. broad; flowers 2–2.5 cm. high, when laid open 4–6 cm. broad; outer petals 4–6 mm. long, 2–4 mm. broad; anthers 3–5 mm. long; disk carmine, deeply undulate or lobed, with 8–15 stigmatic rays; mature fruit smooth or barely ridged, 1.5–2.5 cm. thick, the neck 5–10 mm. thick. . . . . 2. × *N. rubrodiscum.*

Floating blades 0.7–3.5 dm. long, 0.5–2.5 dm. broad; flowers 2.5–4.5 cm. high, when laid open 5–11 cm. broad; outer petals 7–9 mm. long, 4–8 mm. broad; anthers 3.5–10 mm. long; disk green or yellowish, entire or only slightly undulate, with 7–25 stigmatic rays; mature fruit ridged, 2–3.5 cm. thick, the neck 0.8–1.8 cm. thick. . . . . 3. *N. variegatum.*

b. Petioles nearly circular or broadly elliptic in cross-section, not margined. . . *c*.

c. Blades of long-petioled floating or emersed firm leaves ovate or rounded-oblong to orbicular, less than twice as long as broad; short-petioled basal filmy submersed leaves roundish, often not developed. . . *d*.

d. Floating leaves glabrous beneath; petioles and peduncles glabrous.

Leaf-blades floating (when emersed reclining or rarely erect), orbicular to rounded-oblong, the narrow sinus only 3–6 cm. wide between the broadly rounded basal lobes; flowers 2–3 cm. high, when spread open 5–7 cm. broad; band of stamens short-cylindric, of 3–5 circles; stigmatic rays 8–15 (mostly 12); fruit 1.5–3 cm. high, the mature disk 0.8–1.8 cm. across.

Bases of sepals, fruit and disk green; anthers 3–5.5 mm. long; southeastern. . . . . 4. *N. fluviatile.*

Bases of sepals (inside) and disk red; anthers 7–8 mm. long; southwestern. . . . . 5. *N. ozarkanum.*

Leaf-blades erect (in permanently deep water sometimes floating), ovate to suborbicular, the sinus broadly V- or U-shaped and 4–15 cm. wide between the usually bluntly triangular basal lobes; flowers 3–4 cm. high, when spread open 6–10 cm. broad, yellow and green; band of stamens truncate-conic, of 5–8 circles; stigmatic rays 9–23 (mostly 15–18); fruit 2–5 cm. high, the mature disk pale green to yellowish and 1.3–2.5 cm. across. . . . . 6. *N. advena.*

d. Floating leaves minutely pubescent beneath, their petioles often so at summit.

Petiole glabrous or nearly so; fruit with the summit-rim 2.8–3.5 cm. across, entire; stigmatic disk with entire margin, 2–2.4 cm. broad; stigmatic rays ending 2–4 mm. from margin of disk; seeds lustrous, glabrous; southwestern. . . . . 7. *N. ovatum.*

Petiole usually densely pubescent; fruit with summit-rim deeply undulate, 1.3–1.5 cm. across; stigmatic disk with rays extending nearly to margin; seeds dull, densely puberulent on one side; eastern. . . . . 8. *N. puteorum.*

c. Blades of long-petioled floating firm leaves ovate-oblong or narrowly ovate to narrowly oblong-lanceolate, two to five times as long as broad; short-petioled filmy submersed blades of similar outline, usually abundant; flowers rather small, yellow and green.

Blades of floating leaves narrowly ovate or ovate-oblong, two to three times as long as broad, 0.7–1.8 cm. wide, the sinus 5.5–10 cm. deep and 3–7 cm. wide. . . . . 9. × *N. interfluitans.*

Blades of floating leaves narrowly oblong to narrowly oblong-lanceolate, three to five times as long as broad, 5–12 cm. wide, the sinus only 2–5 cm. deep and less than 3 cm. wide. . . . . 10. *N. sagittifolium.*

1. N. microphyllum (Pers.) Fern. (small-leaved). — Very slender, with rhizomes only 1–2 cm. thick, their leaf-scars 5–9 mm. broad; *petioles compressed-filiform;* submersed leaves filmy, round-reniform; *floating blades* broadly elliptical, 3.5–10 *cm. long,* with a narrowly V-shaped sinus; peduncles cord-like, slender; *flowers* 1.5–3 *cm. broad;* the 5 elliptic to obovate *sepals standing apart* in full anthesis and fruit, 1–1.8 *cm. long,* yellow; *petals and stamens promptly deciduous; fruit naked,* globose-ovoid or flask-shaped, *smooth,* 1–2 *cm. high, abruptly constricted to a slender neck only* 2–5 *mm. thick; mature stigmatic disk* 2.5–6 *mm. broad,* 6–10-*rayed, red,* often bordered with yellow; seeds about 3 mm. long, ellipsoid. (*Nymphaea* Pers.; *Nymphozanthus* (Pers.) Fern.) — Pond-margins and deadwaters, Nfld. to Man., s. to N.S., N.E., n. N.J., Pa., Mich., Wisc. and Minn. June–Oct.

2. × **N. rubrodíscum** Morong (with red disk). — Coarser than no. 1, rhizome 1–2.5 cm. thick; floating leaves on dorsally compressed thick petioles; the *blades* 0.5–2 *dm. long,* 0.45–1.4 *dm. broad,* with closed to narrowly V-shaped sinus; *flowers* 2–2.5 *cm. high, when laid open* 4–6 *cm. broad;* sepals broadly elliptic to rounded-obovate, overlapping, the petaloid parts deep yellow, often red-tinged within at base; *outer petals* 4–6 *mm. long and* 2–4 *mm. broad; anthers* shorter than filaments, 3–5 *mm. long; disk carmine, with* 8–15 *rays; mature fruit* purplish, 1.5–2.5 *cm. high and thick, the neck* 5–10 *mm. thick; seeds* 2.5–3.5 *mm. thick.* (*Nymphaea* (Morong) Greene; *Nymphozanthus* (Morong) Fern.) — Pond-margins and deadwaters, Nfld. to Ont., s. to N.S., N.E., N.J., e. Pa., N.Y., Mich., Wisc. and Minn. June–Sept. — Exactly combining the characters of nos. 1 and 3 and evidently a fertile hybrid of them.

3. N. variegàtum Engelm. (variegated), BULLHEAD-LILY, PIED-DE-CHEVAL (Que.). — Relatively coarse; rhizome 2.5–7 cm. in diameter, with leaf-scars 1–2.5 cm. broad; *petioles flattened above, with a median ridge running down from the midrib; floating blades* 0.7–3.5 *dm. long,* 0.5–2.5 *dm. broad,* broadly ovate, *strongly rounded at summit and with broad rounded basal lobes and closed or narrow sinus; flowers* 2.5–4.5 *cm. high, when laid open* 5–11 *cm. broad; sepals* leathery, the outer broadly ovate or elliptic, *the* 3 *inner* suborbicular or round-obovate, deep yellow, *suffused inside at base with purple; outer petals* 7–9 mm. long, 4–8 *mm. wide; anthers* 3.5–10 *mm. long; disk green or yellowish, with* 7–25 *stigmatic rays; mature fruit strongly ridged,* often purplish, 2–4.3 *cm. high,* 2–3.5 *cm. thick, with neck* 0.8–1.8 *cm. thick;* seeds 2.5–4 mm. thick. (*Nymphaea* (Engelm.) G. S. Mill., *N. advena,* var. *variegata* Fern. and *N. americana* Mill. & Standl., not *Nuphar americanum* Provancher; *Nymphozanthus variegatus* (Engelm.) Fern.) — Pond-margins, slow streams and pools, Lab. to Alaska, s. to Nfld., N.S., N.E., Del., Md., O., n. Ind., n. Ill., Ia., Neb., S.D. and Mont. May–Sept.

4. N. fluviátile (Harper) Standl. (of a river). — *Leaf-blades floating* or reclining on mud, *orbicular to round-oblong,* 1.2–2.2 dm. long, *the narrow sinus* 3–6 *cm. wide between the bases of the rounded lobes,* the petioles subterete; *flowers* 2–3 *cm. high, when spread open* 5–7 *cm. broad; outer sepals green, inner ones green and yellow; stamens in* 3–5 *circles, the anthers* 3–5.5 *mm. long; fruit green,* 1.5–3 cm. high, *the mature green disk* 0.8–1.8 cm. across, *with* 8–15 *short stigmatic rays.* (*Nymphaea* Harper; *Nymphozanthus* (Harper) Fern.) — Creeks, rivers, ponds and swamps, Ga. to se. Va. June–Oct.

5. N. ozarkànum (Mill. & Standl.) Standl. (of the Ozark reg.). — Similar to no. 5; leaf-blades often more elliptic, with deeper sinus; *sepals red or red-tinged within; anthers* 7–8 *mm. long; stigmatic disk orange-red.* (*Nymphaea* Mill. & Standl.) — Ponds, quiet streams and sloughs, s. Mo., Ark. and Okla. May–Oct.

6. N. ádvena (Ait.) Ait. f. (immigrant; when grown in Europe). — *Leaves erect* (floating only in exceptionally deep water); the *subterete petiole usually rising above water-level; blades erect,* ovate to rounded-oblong, 1–4 dm. long, *with broad open V- or U-shaped sinus* 4–15 *cm. wide between the often subtriangular basal lobes; flowers* usually on erect peduncles, raised above the water, 3–4 *cm. high, when spread open* 6–10 *cm. broad;* outer sepals green outside, inner ones yellow, with greenish tips, only rarely tinged with red; *stamens in* 5–8 *circles; disk pale green to yellowish, with* 9–23 *stigmatic rays; fruit* erect or lopping into mud, furrowed, *green,* 2–5 *cm. high,* with very short and thick neck, the mature disk 1.3–2.5 cm. across. (*Nymphaea* Ait.; *Nymphozanthus* Fern.) — Tidal waters, pond-margins and swamps, Fla. to Tex. and e. Mex., n. to coast of N.E., centr. N.Y., nw. Pa., O., s. Mich., Wisc. and Neb. May–Oct. (W.I.)

7. N. ovàtum (Mill. & Standl.) Standl. (ovate; from outline of leaf). — *Floating leaves subelliptic-ovate, broadly rounded above,* 2–3.5 cm. long, the blade *minutely silky-pubescent beneath,* with narrow open sinus, the petiole glabrous or sparsely pubescent; *flower green and yellow,* 1–2.5 cm. high, 3–4 cm. broad, when spread open 6.5–8 cm. across; *stamens in* 5 *or* 6 *circles; fruit globose-ovoid,* 3–4 *cm. high,* 3–4 *cm. broad, the high rim of the green to yellowish disk entire, the smooth-centered depressed disk* 2.2–3 *cm. across; stigmatic rays ending* 2–4 *mm. from margin of disk; seeds ovoid to obovoid, lustrous and glabrous,* 3.5–5 mm. long. — E. Tex., n. to se. Kans. June–Oct.

8. N. puteòrum Fern. (of pits; from the type-station). — Differing from no. 7 in the usually *dense puberulence of petiole; leaf-blade more tapering to apex; fruit* somewhat smaller, *with deeply scalloped summit of the rim, this only* 1.3–1.5 *cm. across; center of disk usually bearing papillae* *resembling stigmatic rays; rays extending nearly to margin of disk; seeds quadrate-obovate, dull and puberulent at least on the back,* often lustrous ventrally. — Local, Sussex Co., Va.

9. × N. interflùitans Fern. (floating between). — Intermediate between nos. 6 and 8; *filmy basal leaves abundant, the broadly oblong or oblong-ovate blade* about as long as the petiole, 2–3.5 dm. long, 0.7–1.8 *dm. wide, with* puckered margin, rounded tip and *broad rounded sinus; blade of floating leaf narrowly ovate or ovate-oblong,* 2–3.5 dm. long, 0.7–1.8 *dm. wide,* broadly rounded at tip, *with acute sinus* 5.5–10 *cm. deep and* 3–7 *cm. wide;* flowers and poorly developed fruits as in no. 8. — Forming extensive colonies in upper fresh tidal reaches of Chickahominy R., Va., there occurring between the mid-current band of no. 8 and the tidal-marsh area of no. 6. May–Oct.

10. N. sagittifòlium (Walt.) Pursh (arrow-leaved). — *Submersed filmy leaves* abundant, *narrowly oblong to oblong-lanceolate,* 1.5–4.5 dm. long, 8–12 *cm. broad,* round-tipped, with V-shaped sinus, the margin puckered; *floating firm blades narrowly oblong to narrowly oblong-lanceolate,* 1.5–4 (in emersed plants ascending and down to 1) dm. long, 5–12 *cm. wide, the sinus only* 2–5 *cm. deep and less than* 3 *cm. wide; flowers* 2–3 *cm. high, when spread wide open* 4–6 *cm. broad,* yellow and green; *stamens* forming a subcylindric truncated band *of* 3–6 *circles;* stigmatic rays 10–14 (mostly 12); capsule 2.5–3.5 cm. high, *the mature green disk* 1.3–1.5 *cm. across.* — Forming a band at mid-stream, upper fresh tidal reaches of Chickahominy R., Va.; coastal streams, se. N.C. and ne. S.C. May–Oct.

2. NYMPHAÈA L. Water-lily. Water-nymph. Lis d'eau (Que.)

Sepals 4, nearly free. Petals numerous, in many series, the innermost gradually passing into stamens, imbricately inserted all over the ovary. Stamens indefinite, inserted on the ovary, the outer with dilated filaments. Ovary 12–35-locular, the concave summit tipped by a globular projection at the center, around which are the radiate stigmas; these projecting at the margin and extended into linear and incurved sterile appendages. Fruit depressed–globular, usually covered by the bases of the decayed petals, maturing under water. Seeds enveloped by a sac-like aril. — Flowers white, pink, yellow, or blue, very showy. Genus widely dispersed in trop. and temp. reg. (Ancient name from the Water-nymphs.)* Castalia Salisb.

Leaves orbicular to round-reniform, not mottled, 0.5–6 dm. across; flowers 0.5–3 dm. across; sepals 2.8–10 cm. long; receptacle at line of insertion of sepals circular; petals 17–32; stamens 35–106; fruit closely covered by the perianth; rhizome horizontal, forking.
- Flower strongly fragrant; sepals purple or green outside; petals mostly with tapering ovate tip; inner filaments narrower than anthers; seeds 1.5–2.3 mm. long; branches of rhizome not strongly constricted at base. . . . . . 1. *N. odorata.*
- Flower odorless or with faint odor; sepals green outside; petals broadly rounded at summit; filaments broader than anthers; seeds 2.8–4.4 mm. long; branches of rhizome constricted at base, tuber-like, readily detachable. 2. *N. tuberosa.*

Leaves ovate to obovate, mottled when young, 4–9 cm. broad; flowers odorless, 3.3–8 cm. broad; sepals 1.5–4 cm. long; receptacle at line of insertion of sepals obtusely 4-angled; petals 8–17; stamens 12–40; fruit naked, the sepals erect, not meeting over its summit; rhizome unbranched, erect. . . . . . . 3. *N. tetragona.*

1. N. odoràta Ait. (fragrant), Fragrant Water-lily, Pond-lily. — *Rhizome* horizontal, elongate and continuous, *forking, the branches not strongly constricted at base nor readily detached;* leaves arising along the rhizome; *petioles purplish-green to red;* the nearly *orbicular blades* floating, depressed into mud or (in shallow water or marshes) ascending, flat, green above, *usually purple beneath,* 0.5–2.5 dm. across, with narrow sinus; *flowers very fragrant,* expanded on 3 or 4 days from early morning to about noon, 5–15 cm. broad; *sepals often purplish on back,* ovate to ovate-lanceolate, 2.8–8 cm. long, 1–2.5 cm. broad, with rounded tips; *petals* 17–32, 1–2.2 cm. broad, white, or the outer often all roseate in forma **rùbra** Guillon (red), *gradually tapering above to ovate rounded tips;* stamens 36–100, *the inner filaments narrower than their anthers; seeds* ellipsoid, 1.5–2.3 *mm. long,* exceeded by the aril. (Incl. var. *minor* Sims and var. *rosea* Pursh which are the typical form; *Castalia* (Ait.) Woodville & Wood) — Ponds, deadwaters, bog-pools (where often dwarf), etc., throughout most of our range and beyond. June–Sept.

Var. stenopètala Fern. (narrow-petaled). — Leaves as in typical var.; *peduncle stiff and erect, raising flower above the water; sepals during anthesis reflexed, drab- or fuscous-green on back and, like the petals, lance-acuminate and only* 0.8–1.5 *cm. broad.* — Shallow water, Great Dismal Swamp, Va.

* Key-characters taken partly from Wiegand & Eames, Flora of the Cayuga Lake Basin (1926).

Var. gigantèa Tricker (very large). — Coarser in general; *leaf-blades* thicker, very prominently ribbed and *dull purple to green beneath*, 1.5–6 dm. across, *the edges (when growing) strongly upturned;* flowers 1–2 dm. broad; *sepals* 5–10 cm. long, 2.4–3.6 *cm. broad*, green to purplish on back. (*Castalia* (Tricker) Fern.; *C. lekophylla* Small) — Similar habitats, Fla. to Tex. and Mex., n. on or near Coastal Plain to se. Mass., and inland n. to Ill. and Okla. (W.I.; S. Am.)

2. N. tuberòsa Paine (bearing tubers). — Usually as coarse as second var. of no. 1; *rhizome with readily disarticulating branches strongly constricted at base and tuber-like; petioles green, striped above with brown; leaf-blades green beneath* or rarely dull purple, flat and floating or somewhat elevated above the water, 1–4 dm. across; *flower odorless or barely fragrant*, 1–2.5 cm. broad, opening for 3 or 4 days from about 8 *a.m.* to early afternoon; sepals green on the back; *petals broadly rounded at summit; filaments broader than their anthers; seeds* 2.8–4.4 *mm. long.* (*Castalia* (Paine) Greene) — Pond-margins and slow streams, sw. Que. to n. Ont., Minn. and Neb., s. to Md., O., Ind., Ill. and Ark. June–Sept.

N. Álba L. (white), forma **ròsea** Hartm. f. (rose-colored), resembling *N. odorata*, forma *rubra*, but with petioles more approximate on the rhizome, the *flowers faintly scented only on opening*, 0.7–1.5 dm. across, the 12–24 roseate *petals broadly rounded at summit*, spreads rapidly in cult. and tends to become natzd. (Introd. from Eu.)

3. N. tetragòna Georgi (four-angled). — *Rhizome erect, unbranched*, the slender petioles and peduncles arising from its summit; *leaf-blades broadly ovate to obovate*, relatively thin, *mottled when young*, 4–9 *cm. broad*, with deep sinus; *flowers* opening in the afternoon, 3.3–8 *cm. broad; sepals* 1.5–4 *cm. long; receptacle at line of insertion of sepals obtusely 4-angled; petals* 8–17, *white, with crimson lines; stamens* 12–40; *sepals becoming stiffly erect and not arched over the naked fruit;* seeds 3 mm. long. (*N. Leibergi* Morong; *Castalia Leibergi* Morong) — Pond-margins and swamps, Que. and n. and nw. Me. to Keewatin, s., locally, to n. Mich., n. Minn., Ida. and Wash. June–Sept. (Asia and n. Eu.)

Subfam. II. NELUMBONOÌDEAE

3. NELÚMBO Adans. Sacred Bean

The only genus of the subfamily; two species, ours and one found from se. Asia to Austral. (Name Ceylonese.)

1. N. lùtea (Willd.) Pers. (yellow), Yellow Nelumbo, Water-Chinquapin, Pond-nuts, Wonkapin. — Leaves usually raised high out of the water, circular, 3–6 dm. in diameter, with the center depressed or cupped; flower pale yellow, 1.2–2.5 dm. broad; anthers tipped with a slender hooked appendage. (? *N. pentapetala* (Walt.) Fern.) — Ponds, quiet streams and estuaries, Fla. to Tex., n. locally to s. N.E., N.Y., s. Ont., Minn. and Ia. July–Sept. (W.I.) — Tubers farinaceous and edible. Seeds also eatable. Embryo like that of *Nymphaea*, on a large scale; cotyledons thick and fleshy, inclosing a plumule of 1 or 2 well-formed young leaves, inclosed in a delicate stipule-like sheath.

N. nucífera Gaertn. (bearing nuts), the oriental Sacred Lotus, with pink flowers, has become locally estab. through cult. (Introd. from Asia)

4. BRASÈNIA Schreb. Water-shield

Sepals 3 or 4. Petals 3 or 4, linear, sessile. Stamens 12–18; filaments filiform; anthers innate. Pistils 4–18, forming little clavate indehiscent fruits; stigmas linear. Seeds 1–2, pendulous on the dorsal suture! — Rootstock creeping. Leaves alternate, long-petioled, centrally peltate, oval, floating. Flowers axillary, small, dull purple. A single species. (for Christopher Brasen.)

1. B. Schréberi Gmel. (for Johann Christian Daniel Schreber, 1739–1810), Purple Wen-dock. — Leaves entire or shallowly crenate, 2–10 cm. across. — Ponds and slow streams, Fla. to Tex., n. to P.E.I., s. Que., s. Ont. and Minn.; and s. B.C. to Oreg. June–Sept. (Trop. Am.; Asia; Afr.; Austral.) — Stems, petioles and lower surfaces of leaves heavily covered with a viscid jelly.

Subfam. III. CABOMBOÌDEAE

5. CABÓMBA Aubl.

Sepals 3. Petals 3, oval, biauriculate above the very short claw. Stamens 3–6; anthers short, extrorse. Pistils 2–4, with small terminal stigmas. Seeds 3, pendulous. — Slender, mainly submersed, with opposite or verticillate capillary-dissected leaves, a few of them floating, alternate and centrally peltate. Flowers single on long axillary peduncles. Small genus of N. and S. Am. (Probably an aboriginal name.)

1. C. caroliniàna Gray (of Carolina), FANWORT. — Floating leaves linear-oblong or -obovate, often with a basal notch; flower 1.2–1.8 cm. broad, white with yellow spots at base; stamens 6. — Ponds and quiet streams, Fla. to Tex., n. to Va., s. Ill. and e. Mo., and natzd. n. to Mass., N.Y. and O. May–Sept.

FAM. 64. RANUNCULÀCEAE (CROWFOOT FAMILY)

Herbs or sometimes woody plants, with a colorless and usually acrid juice, polypetalous, or apetalous with the calyx often colored like a corolla, hypogynous; the sepals, petals, numerous stamens and many or few (rarely single) pistils all distinct and unconnected. Flowers regular or irregular. Sepals 3–15. Petals 2–15 or wanting. Stamens indefinite, rarely few. Fruits either dry capsules or seed-like (achenes) or berries. Seeds anatropous (when solitary and suspended the raphe dorsal), with hard albumen and a minute embryo. Leaves often dissected, their stalks dilated at the base, sometimes with stipule-like appendages. — A large family, including some acrid-narcotic poisons.

a. Pistils forming achenes or utricles borne in heads or dense spikes; sepals petaloid or petals well developed and plane. . . *b.*
b. Sepals 3–20, imbricated in bud; herbs, never climbing; leaves alternate or radical, only the upper sometimes opposite or whorled. Tribe I. ANEMONEAE. . . *c.*
c. Petals regularly present; sepals mostly 5.
Sepals plane, without basal spurs; receptacle short, the fruit forming a globose head or ellipsoid to short-cylindric spike; seed (within the achene) ascending, with ventral raphe. . . . . 1. *Ranunculus.*
Sepals prolonged at base into a spur; receptacle filiform, the fruiting spike slender and tail-like; seed suspended, with dorsal raphe. . . 2. *Myosurus.*
c. Petals wanting or represented only by modified stamens (staminodia); sepals 3–many. . . *d.*
d. Leaves all alternate; flowers in racemes, panicles or corymbs; achenes often utricular.
Leaves simple, palmately lobed; seed erect in the 4-angled utricle. . 3. *Trautvetteria.*
Leaves ternately compound or decompound; seed suspended in the ridged but terete to compressed achene. . . . . 4. *Thalictrum.*
d. Leaves at summit of stem or peduncle opposite or whorled, forming an involucre; peduncles solitary or umbellate, 1-flowered; achenes not inflated. . . *e.*
e. Stigma strictly terminal, broad and depressed at flowering time, subsessile; achenes strongly 8–10-ribbed; roots fusiform, tuberous. 5. *Anemonella.*
e. Stigma on the inner side of the style; achenes without prominent ribs; roots (in ours) not tuberous.
Leaves simple, the radical 3-lobed; the involucral unlobed, close to the flower, calyx-like; plant acaulescent. . . . 6. *Hepatica.*
Leaves compound or dissected, the involucral remote from the flower; plant caulescent. . . . . 7. *Anemone.*
b. Sepals 4, valvate in bud, showy; petals none or small and transitional to stamens; styles elongate and plumose; opposite-leaved herbs or slightly woody climbers. Tribe II. CLEMATIDEAE. . . . . 8. *Clematis.*
a. Pistils forming follicles (pods opening down 1 side) or berries. Tribe III. HELLEBOREAE. . . *f.*
f. Ovules and seeds more than 2; roots and rootstocks not conspicuously yellow and bitter or, if so, the latter filiform and plant acaulescent; herbs. . . *g.*
g. Sepals showy, petaloid, not caducous. . . *h.*
h. Flowers regular and symmetrical. . . *i.*
i. Petals wanting or very small or represented merely by nectaries. . . *j.*
j. Caulescent, with 1 or more cauline leaves; follicles sessile or essentially so.
Leaves ternately compound, the primary divisions long-petiolulate. . . . . 9. *Isopyrum.*
Leaves simple, unlobed or deeply palmate-lobed.
Petals 0; leaves rounded or reniform, unlobed. . . . . . 10. *Caltha.*
Petals 5–20, fleshy, slender, with a nectariferous pit above the base; leaves palmately lobed or dissected. . . . 11. *Trollius.*

j.Acaulescent, with trifoliolate evergreen leaves; naked scape 1-flowered; follicles long-stalked. . . . . 12. *Coptis.*

i.Petals prolonged at base into a hollow spur; leaves ternately decompound. . . . . 13. *Aquilegia.*

h.Flowers irregular and unsymmetrical, sepals 5, petals 2 or 4.

Upper sepal spurred; petals 4 (rarely 2), the upper pair with long spurs inclosed in the spur of the calyx. . . . . 14. *Delphinium.*

Upper sepal hooded, helmet-like or saccate, covering the 2 long-clawed small petals. . . . . 15. *Aconitum.*

g.Sepals small, caducous; flowers in racemes, regular; petals stamen-like or none; leaves decompound.

Flowers in slender elongate often paniculate racemes; pistils 1–8, developing into follicles. . . . . 16. *Cimicifuga.*

Flowers in short and thick single racemes; pistil 1, forming a berry-like indehiscent fruit. . . . . 17. *Actaea.*

f.Ovules 2, seeds 1 or 2; roots and rhizomes yellow and bitter with berberine.

Herb with palmately lobed leaves and single terminal flower; sepals 3; petals 0; pistils in a head forming a raspberry-like group of 2-seeded berries. . . . . 18. *Hydrastis.*

Shrub with pinnate leaves and loose compound racemes; sepals 5; petals 5, 2-lobed; follicles inflated, 1-seeded. . . . . . . 19. *Xanthorhiza.*

Tribe I. ANEMÒNEAE DC.

1. RANÙNCULUS L. CROWFOOT. BUTTERCUP. RENONCULE (Que.)

Annuals or perennials; cauline leaves alternate. Flowers solitary or somewhat corymbed, yellow, rarely white. Petals plane or concave, often with a nectariferous spot or pit at base, mostly 5. (Sepals and petals rarely only 3, the latter often more than 5). Stamens mostly numerous, occasionally few. Large nearly worldwide group of herbs. — (A Latin name for a little frog; applied by Pliny to these plants, the aquatic species growing where frogs abound.)*

a.Achenes transversely ridged or rugose; petals white, the yellowish base bearing a naked (not scale-covered) nectariferous pit; plants aquatic or amphibious, rooting at the nodes. § BATRACHIUM. . . *b*.

b.Plants with dilated leaves only, these shallowly 3–5-lobed, subtruncate to broadly cordate at base. . . . . 1. *R. hederaceus.*

b.Plants with finely dissected leaves only. . . *c*.

c.Submersed leaves flaccid, flabelliform to rounded, collapsing when withdrawn from water, with a definite petiole raising the blade well above the adnate tapering stipule; achenes beakless or essentially so. . . 2. *R. trichophyllus.*

c.Submersed leaves firmer, nearly orbicular, holding form when withdrawn, the blade essentially sessile at summit of the dilated stipule; achenes definitely beaked.

Achenes numerous (30–80), usually about 40, about 1–1.5 mm. long, the beak 0.2–0.5 mm. long; stipular sheath one-fourth to three-fourths free from the short petiole. . . . . 3. *R. subrigidus.*

Achenes fewer (8–30), usually about 16, about 1.5 mm. long, the beak 1 mm. long; stipular sheath from three-fourths to entirely adnate to petiole. . . . . 4. *R. longirostris.*

a.Achenes with smooth, pitted, pebbled, muricate, longitudinally nerved or rarely pubescent sides, very rarely cross-ribbed (no. 17); petals yellow or yellowish (white only in albinos), often with the nectariferous spot or pit covered by a scale on the claw. . . *d*.

d.Achenes longitudinally nerved; fleshy, chiefly halophytic plant with filiform repent stolons and roundish crenate-dentate leaves. § HALODES. . . 5. *R. Cymbalaria.*

d.Achenes not longitudinally nerved; plants of various habit; leaves simple or divided. . . *e*.

e.Achenes pubescent, obovoid, not compressed, blunt and beakless; roots or many of them tuberous-thickened, enlarged toward apex; leaves all cordate-ovate; sepals 3; petals 8–12. § FICARIA. . . . . . . 6. *R. Ficaria.*

---

* The specific distinctions in *Ranunculus* are so generally in the fruit that mature achenes are important for proper identification. The figures show a characteristic leaf (reduced), a flower or fruiting head (reduced) and usually an enlarged achene.

*e.*Achenes glabrous (though sometimes muricate or pebbled), firm, compressed, usually tipped by the style or persistent style-base; roots fibrous or merely fusiform-thickened, tapering at apex; sepals mostly 5 (3 in a few species); petals (1–3–) mostly 5–10. § EURANUNCULUS. . . *f.*

*f.*Aquatic, amphibious or terrestrial, with elongate creeping stems or stolons; emersed leaves palmately lobed or divided; submersed leaves either capillary-multifid or merely thin and pellucid. . . *g.*

*g.*Stems leafy and rooting at some of the nodes; petals 3–8; achenes obliquely ovate, semiobovate or suborbicular, including the falcate to mammiform beak 1.5–3.5 mm. long. . . *h.*

*h.*Leaves (submersed) with numerous filiform to linear segments; sepals 5, 2.5–8 mm. long; petals golden-yellow, 0.35–1.7 cm. long; achenes 1.5–3.5 mm. long, including the prolonged falcate beak.

Submersed leaves ternately decompound into linear-filiform segments; sepals 5–8 mm. long; petals 0.6–1.7 cm. long; fruiting heads (5–)8–13 mm. long; mature achenes prominently corky-thickened at base and along ventral margin, including beak 2.5–3.5 mm. long. . . . . . . 7. *R. flabellaris.*

Submersed leaves nearly orbicular, with 3–5 cuneate and linear-cleft lobes; sepals 2.5–4 mm. long; petals 3.5–5 mm. long; fruiting heads 4–6 mm. long; mature achenes without much corky thickening, 1.5–2 mm. long. . . . . 8. *R. Gmelini.*

*h.*Leaves all alike, with 3–5 obovate or oblong entire or shallowly notched lobes; sepals 3, 1.5–2 mm. long; petals 3, pale yellow, 1.5–4 mm. long; fruiting heads 2–4 mm. long; achenes 1–1.5 mm. long, scarcely beaked, with sessile depressed or mammiform stigma. . . . . 9. *R. hyperboreus.*

*g.*Stems filiform and stoloniferous, creeping, bearing scattered slender-petioled 3-lobed leaves and erect scapes; petals 6–10; achenes obliquely ovoid-fusiform, 3–4.5 mm. long, nearly equaled by the slender hooked beak. . . . . 10. *R. lapponicus.*

*f.*Terrestrial or amphibious; the flowering stem arising from a definite crown or tuft of leaves, or, if creeping and with habit of the preceding, with simple and unlobed leaves. . . *i.*

*i.*Leaves all simple, entire or merely toothed, linear (or linear-filiform) to ovate. . . *j.*

*j.*Perennials with elongate creeping bases or stems rooting at the lower nodes; sepals 2–7 mm. long (or down to 2 mm. in no. 14 with matted and prostrate filiform stems); achenes 1.3–2.5 mm. long. *k.*

*k.*Middle and upper leaves acuminate, the median blades 0.5–1.3 dm. long; sepals 5–7 mm. long; achenes 2–2.5 mm. long, the thin keel passing gradually to subulate deflexed beak 0.6–1.5 mm. long. . . . . 11. *R. ambigens.*

*k.*Middle and upper leaves not acuminate, the larger 0.5–5 cm. long; sepals 2–4 mm. long; achenes 1.3–2 mm. long, barely tipped or very short-beaked.

Stems quill-like, stoutish, 1.5–5 mm. thick at base, erect, ascending or trailing; flowers forming a loose corymb; sepals 3–4 mm. long; petals broadly obovate or roundish, 4–7 mm. broad, 9–13-nerved; stamens 25–50; carpels 25–50; fruiting head 3.5–5 mm. long. . . . . 12. *R. Flammula.*

Stems filiform, repent and matted or tufted; flowers mostly solitary; sepals 2–2.8 mm. long; petals oblong, 1–3 mm. broad, 3–9-nerved; stamens 10–20; carpels 15–20; fruiting head 1.5–3 mm. long. . . . . 13. *R. reptans.*

*j.*Short-lived annuals with erect or ascending solitary or tufted stems; sepals 1–3 mm. long; achenes 1–1.3 mm. long, essentially beakless.

Inflorescence becoming diffusely paniculate; stem and axis of panicle often proliferating in summer and autumn by leafy shoots; sepals 2–3 mm. long; petals mostly 5, 3–8 mm. long; stamens 12–20. . . . . 14. *R. laxicaulis.*

Inflorescence an interrupted raceme with remote flowers, not proliferating; sepals 1–1.6 mm. long; petals 1–3(–5), 1.5–2.5 mm. long; stamens 3–10. . . . . 15. *R. pusillus.*

*i.*Leaves (or at least some of them) deeply cleft or divided. . . *l.*

*l.*Achenes smooth or minutely pitted or pebbled, rarely (no. 16) slightly cross-ribbed, not harshly muricate; flowers or inflorescences mostly terminal, the peduncles elongating in fruit. . . *m.*

*m.*Achenes not strongly flattened and without indurated or wing-like margin, 1–2.2 mm. long; basal leaves frequently undivided or merely lobed. . . *n.*

*n.*Base and margin of achene corky-thickened; the firm middle of the face smooth, pebbled or obscurely cross-ribbed; fleshy essentially glabrous plant of shallow water or wet habitats; the lower and middle cauline leaves long-petioled. 16. *R. sceleratus.*

*n.*Base and margin of achene not corky, the face plane; plants not fleshy, neither aquatic nor amphibious; cauline leaves short-petioled to sessile. . . *o.*

*o.*Basal leaves all deeply cleft; peduncles scapiform, the longer in fruit one-third to nine-tenths the entire height of the plant.

Radical leaves 0.5–1.5 cm. broad; flowering stems weak and flexuous, 1–13 cm. long; sepals 2–3.5 mm. long, linear to oblong, glabrescent; petals 1.5–3 mm. long; fruiting head 2–4 mm. thick; achenes 1–1.3 mm. long; receptacle glabrous. . . . . 17. *R. pygmaeus.*

Radical leaves 1.3–4 cm. broad; flowering stems erect, 1–3 dm. high; sepals 5–8 mm. long, ovate, villous on back; petals 0.7–1 cm. long; fruiting head 5–7 mm. thick; achenes 1.8–2.2 mm. long; receptacle villous. . 18. *R. pedatifidus.*

*o.*Basal leaves or some of them merely shallowly crenate or dentate. *p.*

*p.*Stamens in 1 or 2 cycles; sepals glabrous, subglabrous or only sparingly pilose; fruiting heads 2.5–5.5 mm. thick; receptacle linear-cylindric or -lanceolate to ellipsoid. . . *q.*

*q.*Achenes 1.5–2 mm. long, with firm subulate beak.

Stems weak, flexuous, villous, 1–4-flowered; sepals spreading, 3–5 mm. long; petals 4–6 mm. long; alpine species. . . . . 19. *R. Allenii.*

Stems firm, erect, glabrous, corymbose-cymose, many-flowered; sepals reflexed, 2.5–3.5 mm. long; petals 1.5–2.3 mm. long; Alleghenian species. . . 20. *R. allegheniensis.*

*q.*Achenes 1.1–1.5 mm. long, the beak obsolete or soft and minute. . . *r.*

*r.*Roots slender and fibrous; the simple radical leaves 1–9.5 cm. broad, with 9–45 marginal teeth; sepals 2.5–5 mm. long, sessile; fruiting heads 3–4.5 mm. thick; achenes lustrous; receptacle hairy. . . 21. *R. abortivus.*

*r.*Roots fusiform-thickened; simple radical leaves 0.8–2.5 cm. broad, with 5–13 marginal teeth; sepals 2–3.5 mm. long, abruptly contracted to a claw; fruiting heads 2.5–3.3 mm. thick; achenes dull; receptacle glabrous.

Uncleft basal leaves cuneate, rounded or barely subcordate (rarely cordate) at base, often mixed with divided leaves; petals 2–3.5 mm. long, about equaling sepals. . . . . 22. *R. micranthus.*

Uncleft basal leaves cordate or subcordate, rarely mixed with divided ones; petals much exceeding sepals, 4–8 mm. long. . . . 23. *R. Harveyi.*

*p.*Stamens in 3–5 cycles; sepals copiously long-villous; fruiting heads 6–10 mm. thick; receptacle obovoid. . 24. *R. rhomboideus.*

*m.*Achenes strongly flattened, usually wing-margined, the body 2–5 mm. long; basal leaves all or nearly all compound. . . *s.*

*s.*Petals 2–6 mm. long, shorter than or barely exceeding sepals; achenes (excluding beaks) 2–3.3 mm. long. . . *t.*

*t.*Erect, without stolons; petals oblong or oblong-obovate, 1–3 mm. broad; achenes 2–2.7 mm. long, with beak 0.5–1 mm. long.

Perennial from short and thick corm-like base; basal leaves 3–5-lobed; cauline similar, few, the lowest usually about midway on the stem; fruiting head subglobose, 4.5–8 mm. long; beak of achene long, recurved and hooked. . . . . 25. *R. recurvatus.*

Annual or perennial with soft base; leaves numerous, uniformly spaced, ternately compound, with petiolulate leaflets; fruiting head ovoid to cylindric, 0.6–1.8 cm. long; beak of achene deltoid, erect or ascending, 0.5–1 mm. long. . . . . . . 26. *R. pensylvanicus.*

t. Ascending, declined or trailing, with leafy stolons; petals broadly obovate or rounded, 2.5–5 mm. broad; achenes 2.5–3.3 mm. long, with a straight flat-subulate beak 1–1.5 mm. long. . . . . 27. *R. Macounii.*

s. Petals 5–17mm. long, definitely exceeding the sepals; achenes (excluding beaks) 2–5 mm. long. *u.*

u. Stigma terminating the long style, 0.2–0.5 mm. long, usually deciduous in fruit; fruiting head 7–15 mm. thick; beak of achene subulate, 1.3–3 mm. long, submarginal. . . *v.*

v. Sepals 3.5–5 mm. long, promptly reflexed; petals oblong; achenes 10–20, the body 3.7–5 mm. long; the marginal wing 0.5–1 mm. wide, separated from the face by a high acute ridge; southern species with repent branches after flowering. . . . . . . 28. *R. carolinianus.*

v. Sepals 5–11 mm. long, spreading or tardily reflexed; petals obovate to oblong; achenes more numerous, the body 2–4.3 mm. long, the wing narrower or separated by a low ridge. . *w.*

w. Plants not developing trailing branches, with mostly tufted slender stems; roots often fusiform-thickened; lowest leaves commonly much smaller and less cleft than the dissected later basal ones; stipules inconspicuous, tapering to the petioles; fruiting heads 7–11 mm. thick; body of achene 2–3.5 mm. long, beak 1.3–2 mm. long. . . *x.*

x. Fleshy roots 0.8–1.5 dm. long; basal leaves palmately 3-parted or 3–5-divided, their segments or leaflets oblanceolate, obovate or rhombic; fruiting head 7–11 mm. thick; achene with broad dorsal keel. . 29. *R. hispidus.*

x. Fleshy roots 1.5–5.5 cm. long, thick; later basal leaves oblong to ovate, pinnately divided into linear or oblong leaflets or linear-segmented leaflets; fruiting head 5–8 mm. long; achenes narrowly margined. . . . . 30. *R. fascicularis.*

w. Plant coarse, soon developing trailing or repent elongate leafy branches; flowering stems 1–few; roots fibrous but coarse; lower leaves all ternately compound, with petiolulate 2–3-cleft and incised ovate to rhombic large leaflets; stipules commonly conspicuous, rounded at summit; fruiting heads 9–15 mm. thick; body of achene 3–4.5 mm. long, beak 1.5–3 mm. long. . . . . 31. *R. septentrionalis.*

u. Stigma nearly covering one side of the short style, 0.3–1 mm. long, persistent in fruit; fruiting head 4.5–10 mm. thick; beak of achene deltoid to deltoid-subulate, 0.4–1.4 mm. long. . *y.*

y. Sepals spreading.

Basal leaves ternately divided into mostly petiolulate leaflets; stems often repent or sending out elongate trailing branches; beak of achene submarginal, deltoid-subulate, 0.7–1.4 mm. long; receptacle villous. . . . . 32. *R. repens.*

Basal leaves pedately 5-parted into sessile divisions; stem erect; beak nearly central, deltoid, 0.4–0.8 mm. long; receptacle glabrous (except sometimes at tip). 33. *R. acris.*

y. Sepals promptly and tightly reflexed; achenes suborbicular, with nearly central beak 0.4–0.8 mm. long.

Perennial with subglobose corm 0.5–3.5 cm. thick; sepals 5–8 mm. long; petals 8–14 mm. long; fruiting head 7.5–10 mm. thick; achenes smooth. . . . . 34. *R. bulbosus.*

Annual with soft base; sepals 2–4 mm. long; petals 5–8 mm. long; fruiting head 4.5–8 mm. thick; achenes (mature) low-tuberculate or smooth. . . 35. *R. Sardous.*

l. Achenes harshly muriculate on the faces; flowers axillary, their peduncles mostly shorter (2–25 mm. long) than the subtending deeply lobed leaves; petals and sepals subequal. . . . . 36. *R. parviflorus.*

§ Batràchium DC.*

1. R. hederàceus L. (resembling *Hedera*, the Ivy; from the lobed leaves). — Stems *closely creeping*, succulent; *only dilated leaves developed*, these slender-petioled, *with reniform obtusely and shallowly 3–5-lobed blades 5–18 mm. broad; flowers 4–6 mm. wide;* arching peduncles 2–20 mm. long; petals white; stamens 6–10; carpels 10–16; mature achenes 1–1.5 mm. long, sharply keeled, glabrous, rugose; receptacle glabrous. (*Batrachium* S. F. Gray) — Shallow pools, rills, and wet sandy depressions, se. Nfld.; se. Pa. and Md. to S.C. April–Aug. (Eu.) Fig. 1027.

1027. R. hederaceus.

2. R. trichophýllus Chaix (with hair-like leaves), White Water-C. — Stems elongate in water or abbreviated on mud, *bearing only finely dissected flaccid foliage which collapses upon removal from water; leaf-blades* flabelliform to rounded, *those of the primary stems* (and usually of the branches) *with a distinct free petiole 2–18 mm. long above the* sheathing and mostly tapering *stipular base,* the expanded blade 1.5–5 cm. long; peduncles 1–6 cm. long; flowers 0.7–1.7 cm. broad; the white petals rarely overlapping at margin, 2–3 times as long as sepals; stamens mostly 10–15; *carpels* 8–35 (*usually* 16–24), *the style promptly deciduous; achenes* rugose, often hispid, 1.25–1.5 mm. long, *the beak obsolete or a mere tooth* near the summit of the ventral margin; receptacle usually hirsute. (*Batrachium* Bosch; *R. aquatilis,* var. *capillaceus* of ed. 7) — Fresh to brackish waters, Lab. to Alaska, s. to Nfld., N.S., n. and w. N.E., n. N.J., Pa., Minn., S.D., N.M., Ariz. and Lower Calif. July–Sept. (Eurasia) Fig. 1028. Passing into

1028. R. trichophyllus.

Var. **eradicàtus** (Laestad.) W. B. Drew (pulled-up by the roots; Laestadius supposing the plant to have been uprooted by ice-action). — Plants small and weak; *expanded leaf-blades* 0.5–2.5 *cm. long;* flowers 0.7–1 cm. broad; *stamens* 5–10; *carpels* 8–20 (*usually* 12–16); *achenes* 1–1.5 mm. long, *usually smooth.* (*R. confervoides* Fries) — Pools and wet shores, Greenl. and Lab. to Alaska, s. to Nfld., Anticosti I. and Gaspé Pen., Que., Ung. and Wyo. (Eu.)

Var. **calvéscens** W. B. Drew (becoming bald). — Resembling the typical form, but with *achenes smoothish and glabrous; receptacle glabrous or nearly so.* — Se. Nfld. and Que. to Mich., s. to N.S., N.E. and Pa.

3. R. subrígidus W. B. Drew (somewhat stiff), White Water-C. — Resembling no. 2; *leaves less flaccid, often firm, tending to hold their rounded form when withdrawn from water; the reniform or suborbicular blades of the primary leaves sessile or subsessile at the summit of the dilated stipular base,* the rameal leaves sometimes petioled; *stipule one-fourth to three-fourths free from the short petiole;* peduncles 1.5–8 cm. long; petals broad, often overlapping at margin; stamens 10–22; *carpels* 30–80 (*av.* 40), *with a* subequal long *persistent style; achenes* 1–1.5 mm. long, *with a firm subulate beak* 0.2–0.5 *mm. long.* (*R. circinatus* in part of ed. 7, not Sibth.) — Calcareous or brackish waters, w. Nfld. and e. Que. to Alta. and Wash., s. to w. Mass., Mich., Minn., S.D., Tex. and Mex. June–Sept. Fig. 1029.

1029. R. subrigidus.

4. R. longiróstris Godr. (long-beaked), White Water-C. — Similar to no. 3; *leaves firmer, often stiff; the stipular base* larger, *from three-fourths to entirely adnate to* the very short *petiole;* flowers 1–2 cm. broad; *carpels* 8–30 (*av.* 16); *achenes* about 1.5 mm. long, *with beak averaging* 1 *mm. long.* (*Batrachium* F. Schultz; *R. circinatus* in part of ed. 7, not Sibth.) — Calcareous waters, sw. Que. to Oreg., s. to Del., Pa., Tenn., Kans., Tex., N.M. and Ariz. May–Sept. Fig. 1030.

1030. R. longirostris.

§ Halòdes Gray

5. R. Cymbalària Pursh (for *Cymbalaria*, having similar leaves), Seaside C. — Tufted, with

* Treatment of this section largely derived from that of W. B. Drew in Rhodora xxxviii (1936).

filiform repent stolons, fleshy, glabrous throughout, or with peduncles or petioles pilose in forma **hebecaùlis** Fern. (hairy-stemmed); *leaves* long-petioled, *roundish or reniform,* 0.2–3.5 cm. long, *crenate or dentate; scapes* up to 2.5 dm. long, 1–10-*flowered;* flowers 6–9 mm. broad, with bright yellow narrow petals; fruiting heads globose-ovoid to cylindric, 3–14 mm. long, 3–6 mm. thick; *achenes thin-walled,* obliquely obovate, with an erect beak, *longitudinally nerved,* 1.6–2.3 mm. long. *Oxygraphis* Prantl; *Halerpestes* Greene) — Saline or brackish shores, rarely in fresh habitats (then chiefly as forma *hebecaulis*), s. Greenl. and Lab. to Alaska, s. to Nfld., N.S., N.E., L.I., N.J., w. N.Y., n. Ill., Ark., Kans., Tex. and Mex. May–Oct. (S.Am.; Eurasia) FIG. 1031.

1031. R. Cymbalaria.

§ FICÀRIA Boiss.

6. R. FICÀRIA L. (old generic name, from *Ficus*, the fig, from the tuberous roots, like fig-warts), LESSER CELANDINE, PILEWORT, FICAIRE (Que.). — *Roots* both fibrous and *clavate and tuberous;* stems succulent, decumbent, 1–3 dm. long, often bearing axillary bulblets (when the flowers are usually sterile); *leaves* petioled, *cordate-ovate,* succulent and *lustrous; sepals* 3, 5–8 mm. long; petals 8–12, lustrous, yellow, fading to whitish, oblong; stamens numerous; fruiting head globose; *achenes pubescent* (commonly abortive with us), *obovoid, beakless,* 3–4 mm. long. — Old garden-plant, occasionally spreading to waste places or open woods and locally estab. Apr.–June. (Introd. from Eu.)

§ EURANÚNCULUS Gray

7. R. flabellàris Raf. (fan-like), YELLOW WATER-C. or -B. — Amphibious aquatic; submersed or floating plant with elongate stout hollow glabrous stem; its *leaves* remote, alternate, 0.3–1.5 dm. long, *ternately decompound into flaccid and pellucid linear-filiform segments; stranded* or shore *plants,* forma **ripàrius** Fern. (of river-banks) = *R. delphinifolius,* var. *terrestris* Farw., not *R. multifidus,* var. *terrestris* Gray, upon which it was based, *with* shortened stem and *leaves reduced in size and thickened;* extreme specimens with clustered long-petioled leaves with 3–5 cuneate-obovate deeply cleft lobes, the petioles and blades either glabrous or densely pubescent; flowers 1–7, on long thick often proliferating peduncles; *sepals* 5, ovate to suborbicular, 5–8 *mm. long; petals* golden-yellow, broadly obovate, 0.6–1.7 *cm. long; fruiting head* subglobose, 5–13 *mm. long; anthers* oblanceolate to oblong, 1–1.5 *mm. long, only slightly broader than the clavate filaments; achenes* obliquely ovate, gibbous, *when mature prominently corky-thickened at base and along the ventral margin, including the flat-subulate beak* 2.5–3.5 *mm. long.* (*R. delphinifolius* Torr.) — Quiet waters and muddy shores, Me. to s. B.C., s. to N.C., La., Kans., Utah and Calif. May, June. FIG. 1032.

1032. R. flabellaris.

8. R. Gmèlini DC. (for the discoverer, JOHANN GEORG GMELIN, 1709–1755), var. **Hoòkeri** (D. Don) Benson (for Sir WILLIAM JACKSON HOOKER, 1785–1865), SMALL YELLOW WATER-C. or -B. — Amphibious, the *elongate stems creeping or floating* in shallow water or sprawling and decumbent on muddy shores or in meadows, *freely rooting at nodes; submersed leaves flaccid,* long-petioled, pellucid, *semiorbicular to orbicular, with* 3–5 *cuneate linear-cleft lobes,* 1.5–8 *cm. broad;* emersed or floating blades thicker (or all leaves thick, reduced in size and either subglabrous or pilose in the wholly emersed form), reniform to orbicular; bracteal ones 0–2, very small; inflorescence 1–4-flowered, often proliferous-prolonged; *sepals* round-obovate, 2.5–4 *mm. long; petals* 5, golden-yellow, obovate, 3.5–5 *mm. long; anthers* ellipsoid, 0.5–1 *mm. long, broader than and sharply separated from the slender filaments; fruiting head* subglobose, 4–6 *mm. long; achenes* firm, *without or with only slight corky thickening* at base and margin, including the slender style 1.5–2 *mm. long.* (*R. Purshii* Richards.; *R. Gmelini,* var. *Purshii* Hara) — Spring-rills, clear cold pools, shores and meadows, Nfld. to Alaska, s. to N.S., e. Me., Mich., Ia., N.D., N.M. and Oreg. July, Aug. FIG. 1033. (Typical *R. Gmelini* chiefly in Asia)

1033. R. Gmelini, v. Hookeri.

Var. **prolíficus** (Fern.) Hara (prolific). — Stems or branches strongly ascending, 1.5–4.5 dm. high, 3–50-flowered; simple or subsimple bracteal leaves numerous, the lower 1–4 cm. long. — Meadows and exsiccated pools, similar range. — No ripe achenes known; plant somewhat suggesting *R. sceleratus.*

9. R. hyperbòreus Rottb. (high-northern). — Very small, fleshy, forming mats; *leaves* 0.5–2 cm. broad, *deeply 3-lobed,* the blunt lobes entire or notched; *peduncles in fruit becoming* 2–15 *mm. long; sepals* 3, 1.5–2 *mm. long; petals* 3, 1.5–4 *mm. long; fruiting heads* 2–4 *mm. long;* achenes 1–1.5 mm. long. — Shallow fresh or brackish water and wet shores, Greenl. to

1034. R. hyperboreus.

Alaska, s. to n. Nfld. and e. Saguenay Co., Que.; Alta. to Mont. June–Aug. (N. Eurasia) Fig. 1034.

10. R. lappónicus L. (of Lapland). — *Stems filiform, extensively creeping bearing* scattered slender simple or rarely forking *scapes* or 1-leaved flowering stems 0.5–2.3 dm. high; *basal leaves* mostly solitary, slender-petioled, reniform, *with* 3 *cuneate-flabelliform* 3–7-*cleft lobes*, 1.5–5 cm. broad; *sepals* 3, ovate, 3–6 mm. long; petals 6–10, oblong, 4–6 mm. long, yellow or whitish; *achenes* 5–15, obliquely ovoid-fusiform, acutely margined, 3–4.5 mm. long, *slightly longer than the slender hooked style.* — Subarct. reg., s., locally, in moss and wet woods to n. Me., Ont., ne. Minn., Alta. and B.C. June, July. (Eurasia) Fig. 1035.

1035. R. lapponicus.

11. R. ámbigens S. Wats. (uncertain), Water-plantain-Spearwort. — *Coarse, the decumbent and stoloniferous stem rooting at the lower nodes,* 0.5–2 *cm. thick at base,* finally ascending or erect, simple or forking, 3–8 dm. high; basal *leaves* and those of stolons oblong-lanceolate; *the middle and upper lance-acuminate,* 0.5–1.3 *dm. long,* 0.8–3 cm. broad, remotely undulate-serrulate, with dilated petioles; corymb (1–)3–23-flowered, the peduncles ascending; *sepals* narrowly ovate, 5–7 *mm. long; petals* lustrous-yellow, *oblong,* 5–8 mm. long, 2–3.2 *mm. broad;* stamens 25–30; fruiting head globose-ovoid, 4–7 mm. long; *achenes* obliquely cuneate-obovate, compressed, *puncticulate-pebbled,* 2–2.5 *mm. long, the thin keel passing gradually into a long* (0.6–1.5 *mm.*) *subulate deflexed beak.* (*R. laxicaulis* of ed. 7, not (T. & G.) Darby; *R. obtusiusculus* of recent auth., not Raf.) — Sloughs, ditches and muddy swamps, local, Me. to s. Ont. and Minn., s. to Ga., Tenn. and n. La. June–Sept. Fig. 1036.

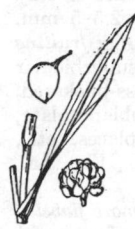

1036. R. ambigens.

12. R. Flámmula L. (old generic name, from *flamma, flame,* from the burning and acrid juice), Spearwort. — Similar to no. 11; repent and decumbent; *base* 1.5–5 *mm. thick,* stem erect or ascending and 2–6 dm. long; *lower leaves ovate-oblong to oblanceolate,* long-petioled, *with blades* 2–5 *cm. long* and 0.6–1.3 cm. broad, undulate or remotely dentate; upper leaves narrower and shorter; corymbose cyme lax, (1–)2–30-flowered; *sepals* 3–4 *mm. long; petals* golden-yellow, *broadly obovate to roundish,* 4–7 *mm. long,* 4–7 *mm. broad,* sessile or nearly so, 9–13-*nerved; stamens* 25–50; *carpels* 25–50; *fruiting head* globose, 3.5–5 *mm. long; achenes* ovate to roundish, plump, 1.6–2 *mm. long, very short-tipped.* — Springy thickets and pond-margins, se. Nfld. and St. P. et Miq.; sw. N.S.; Oreg. July–Sept. (Eurasia) Fig. 1037.

1037. R. Flammula.

Var. **angustifòlius** Wallr. (narrow-leaved). — Slender, decumbent or trailing, sometimes rooting at upper nodes; leaves all lanceolate to linear-lanceolate, the blades 1–3 cm. long and 2.5–6 mm. wide, essentially entire. — Se. Nfld.; w. N.Am. (Eurasia)

13. R. réptans L. (creeping), Creeping Spearwort. — *Stems filiform, repent,* rarely ascending, or in running water sometimes much elongated and trailing as forma **submérsus** Glück (submerged); *basal leaves* tufted, *filiform or linear and bladeless or linear-lanceolate or -oblanceolate;* leaves of the repent or flexuous branches smaller, 1, 2 or 3 at a node; *flowers usually solitary; sepals* 2–2.8 *mm. long; petals narrowly obovate to oblong,* 2.5–5 *mm. long,* 1–3 *mm. broad,* usually with definite claw, 3–6-*nerved; stamens* 10–20; *carpels* 15–20; *fruiting head* hemispherical to spherical, 1.5–3 *mm. long;* achenes plump, 1.3–2 mm. long, with a distinct short beak. (*R. Flammula,* vars. *reptans* Mey. and *filiformis* Hook.) — Damp sandy, gravelly or muddy shores, etc., Greenl. to Alaska, s. to Nfld., N.S., Me., Mass., N.Y., Mich., Wisc., Minn., Colo. and Oreg. June–Aug. (Eurasia) Fig. 1038.

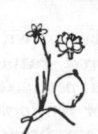

1038. R. reptans.

Var. **ovális** (Bigel.) T. & G. (oval). — Blades well developed, lanceolate to oval, up to 3 cm. long and 6 mm. broad. — Nfld. to B.C., s. to N.S., N.E., n. N.J., Pa., Mich., Ariz. and Calif. June–Aug.

14. R. laxicaùlis (T. & G.) Darby (loose-stemmed). — *Annual,* or perennial only by flabelliform offshoots, very slender, 1.5–5 dm. high, becoming *diffusely paniculate; stem and axis of panicle often proliferating late in season by flabelliform leafy offshoots;* leaves denticulate to serrulate; the lower long-petioled, ovate to oblong (1–4.5 cm. long), the upper oblong to linear; *sepals* 2–3 *mm. long; petals mostly* 5, 3–8 *mm. long; stamens* 12–20; *achenes subglobose,* 1–1.3 *mm. long, essentially beakless,* the minute style deciduous. (*R. oblongifolius* of ed. 7, not Ell.; *R. texensis* Engelm.) — Wet places, Fla. to Tex., n. to s. Ct., s. Ind., Ill., Mo. and Kans. April–June (Aug.–Oct.). Fig. 1039.

1039. R. laxicaulis.

15. R. pusillus Poir. (very small). — *Annual*, resembling no. 14; stem weak, ascending 1–6 dm. high, simple or few-branched, the *branches becoming interruptedly racemose by elongation of the inflorescence, not proliferating;* leaves entire or denticulate, oval or oblong to linear; *sepals* 1–1.6 *mm. long; petals* 1–3 (–5), 1.5–2.5 *mm. long; stamens* 3–10; achenes plump, smooth, or in a more sw. var. minutely papillate. — Swamps, ditches and shallow pools, Fla. to Tex., n. to se. N.Y., O., s. Ind. and e. Mo.; Calif. April–June. Fig. 1040.

1040. R. pusillus.

16. R. sceleràtus L. (cursed, *i.e.* growing in vile places), Cursed C. — Annual or short-lived perennial, *fleshy*, usually *glabrous;* stems 1–3, erect, usually freely branching, 0.8–6 dm. high; *radical and lower leaves succulent, long-petioled*, or petioles elongating in deep water to 0.5 m. long and blades floating in forma **nàtans** Glück (swimming), 3–5-parted, the *cuneate divisions with oblong or roundish lobes or crenate teeth;* upper leaves obovate, with entire or only slightly toothed divisions, or simple; peduncles pilose; sepals 2.5–5 mm. long, pilose; petals pale yellow, oblong to obovate, 2–5 mm. long; *fruiting heads cylindric to globose,* 4–10 *mm. long,* 3–7 *mm. thick; achenes* suborbicular to obovate, *corky-thickened at base*, smooth or obscurely cross-puckered, 1–1.4 *mm. long, barely mucronate;* receptacle obovoid or oblanceolate, glabrous or but slightly hairy. — Sloughs, pools, springy places, etc., in brackish to fresh habitats, Nfld. to Alaska, s. to N.S., N.E., Fla., La., Ark., N.M. and Calif. May–Aug. (Eurasia) Fig. 1041.

1041. R. sceleratus.

Var. **multífidus** Nutt. (much divided). — Divisions of lower *leaves longer and more flabelliform, much narrowed at base, cleft into numerous oblong to lanceolate lobes; median leaves with numerous lanceolate to linear lobes.* — Ne. N.B.; n. Mich.; w. Minn. to Alaska, s. to N.M. and Ariz.

17. R. pygmaèus Wahlenb. (dwarf), Dwarf B. — Caudex simple or forking, bearing 1–10 *slender weak* ascending or reclining 1-*flowered stems* 0.1–1.3 *dm. high; radical leaves* slender-petioled, *flabelliform to reniform, deeply cleft into* 3–5 *lobes*, glabrous, 0.5–1.5 *cm. broad;* cauline leaves 1 or 2; the lower similar to the radical, but deeper-cleft into 3–5 linear-oblanceolate segments; upper leaves similar or a simple oblanceolate blade, sessile; *sepals* 5, 2–3.5 *mm. long; petals* oblong-obovate, *pale yellow,* 1.5–3 *mm. long;* stamens 10–20; *fruiting head ovoid to cylindric*, 3–5 mm. long, 2–3 *mm. thick; achenes* semi-obovate, plump, straightish on the ventral face, gibbous on the dorsal, 1–1.3 *mm. long*, with a slender subulate recurving beak at the ventral margin; *receptacle lance-cylindric, glabrous.* — Greenl. to Alaska, s. to alpine reg. of Shickshock Mts., Gaspé Pen., Que., and mts. of Colo. July–early Sept. (Eu.) Fig. 1042.

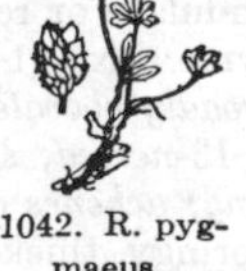
1042. R. pygmaeus.

Var. **Langeànus** Nathorst (for its discoverer, Johan Lange, 1818–1898). — Less tufted; basal leaves mostly divided into 3 deeply lobed petiolulate leaflets; fruiting heads 5–7.5 mm. long; beak straighter. — Greenl.; ne. Lab.; wet amphibolite rocks at 950–1000 m., Mt. Albert, Gaspé Co., Que.; Mont.

18. R. pedatífidus Sm. (divided like a bird's foot), var. **leiocárpus** (Trautv.) Fern. (with smooth carpels). — Caudex simple or forking, bearing 1–10 *erect* rather stout *simple or slightly forking stems* 1–3 dm. high, *with erect naked peduncles becoming* 0.5–2 *dm. long; radical leaves* long-petioled, *cordate-ovate or reniform, pedately cleft nearly to base into* 5–9 *simple or lobed segments,* 1.3–4 cm. broad; cauline leaves sessile or subsessile, the lower cleft nearly to base into long linear or oblanceolate entire segments, the upper 3-cleft or simple; *sepals villous on the back*, oblong, 5–8 *mm. long; petals* bright yellow, *obovate or rounded,* 0.7–1 *cm. long;* stamens numerous; *fruiting head ellipsoid or thick-cylindric,* 0.7–1 cm. long, 5–7 *mm. thick; achenes* plump, glabrous, 1.8–2.2 *mm. long, with a curved subulate* often deciduous *beak about one-fourth their length;* receptacle elongate, villous. (*R. affinis* R. Br.) — Arct. reg., s. to calcareous or basic cliffs or turfy slopes of nw. Nfld., Shickshock Mts., Gaspé Pen., Que., Ung., Keewatin and Alta. Late June–Aug. (N. Eurasia) Fig. 1043. — Typical *R. pedatifidus*, with pubescent achenes, is in w. N.Am. and Asia.

1043. R. pedatifidus, v. leiocarpus.

19. R. Allénii Robins. (for its discoverer, John Alpheus Allen, 1863–1916). — Caudex bearing 1–4 simple or forking slender *slightly villous* 1–4-flowered *flexuous stems* 0.6–2.8 *dm. high; radical leaves* long-petioled, *suborbicular to reniform, deeply* 7–11-*crenate*, sometimes 3-lobed, 1–3.3 cm. broad; cauline leaves 2–4, cleft nearly to base into 3–5 elliptic, lance-oblong or

1044. R. Allenii.

spatulate entire or slightly notched segments; uppermost leaves subsessile, simple and lanceolate or 2–3-parted; peduncles villous at summit; *sepals* 5, ovate, sparingly villous, 3–5 *mm. long; petals* 5, bright yellow, *broadly obovate,* 4–6 *mm. long;* stamens 12–20; *fruiting head ovoid or short-cylindric,* 5–7 mm. long, 4–5 *mm. thick; achenes* semiobovate, slightly compressed, 1.5–2 *mm. long, with a slender subulate beak; receptacle subcylindric, loosely long-villous.* — Wet gravel and flood-plains of alpine brooks, n. Lab. and sw. Ung.; Shickshock Mts., Gaspé Pen., Que. July, Aug. Fig. 1044.

20. R. alleghenіénsis Britt. (of the Alleghenies). — Resembling no. 21, usually *glaucous* when fresh; stems 2–7 dm. high, corymbose-cymose; *radical leaves thickish,* opaque, *on glabrous petioles,* simple or palmately cleft or divided, the simple ones cordate at base, *crenate or dentate,* the divided ones (when developed) with 3 petiolulate rhombic to obovate lobed and cleft leaflets; cauline similar to the divided basal ones, but with narrower entire or toothed leaflets, the lower petioled, the upper sessile; *sepals* oblong to ovate, reflexed, 2.5–3.5 *mm. long; petals* pale yellow, narrowly obovate, 1.5–2.3 *mm. long; fruiting heads globose to globose-ovoid,* 4–5 *mm. thick; achenes* suborbicular, compressed, 1.8–2 *mm. long, with a slender-subulate firm curved or spiraling beak; receptacle* elongate, *hairy.* — Rich woods and rocky calcareous or basic slopes, rarely in meadows, rather local, e. Mass. and Vt. to se. O., s. to upland of S.C. and Tenn. Late Apr.–July. Fig. 1045.

1045. R. allegheniensis.

21. R. abortìvus L. (abortive, *i.e.* with reduced styles and petals), Kidneyleaf-B. — Roots slender and fibrous; *stems* 1–7, *glabrous,* or minutely pilose above, *green,* erect, forking, 1.2–8.5 dm. high, few–many-flowered; *radical leaves somewhat fleshy, glabrous, on glabrous or only sparingly silky petioles,* simple, cleft or divided; *the simple ones reniform to round-ovate or orbicular,* 1–9.5 cm. broad, crenate or dentate *with* 9–45 *teeth along the length of the outer margin;* divided basal leaves (often wanting) with 3 obovate toothed simple or lobed leaflets; cauline leaves very variable, simple or divided, the lower petioled; uppermost divided leaves sessile, with narrower segments; *sepals* 2.5–5 *mm. long, sessile,* reflexed; *petals oblong to oval,* 1.5–4 *mm. long; fruiting head globose or ellipsoid,* 2.5–7 mm. long, 3–4.5 *mm. thick; achenes* suborbicular, *lustrous,* 1.2–1.5 *mm. long, with a very minute straight or curved style; receptacle* lanceolate to ellipsoid, *hairy.* — Four varieties:

a. Basal uncleft leaves merely cordate at base, with open sinus. . . *b.*
b. Peduncles, upper internodes and leaf-surfaces glabrous or essentially so.
Uppermost and often all cauline leaves divided to base into narrow segments; radical leaves simple or some of them divided. . . . . *R. abortivus* (typical).
All cauline as well as radical leaves simple and undivided. . . . . . Var. *indivisus.*
b. Young peduncles (at least toward base), upper internodes and often the surfaces of youngest cauline leaves minutely pilose. . . . . . Var. *acrolasius.*
a. Basal uncleft leaves orbicular, with a narrow or nearly closed sinus. . . . Var. *eucyclus.*

R. abortìvus (typical). — Low woods, thickets, clearings and damp slopes, n. Fla. to Okla., n. to s. Me., s.-centr. N.H., Vt., N.Y., O., Mich., Wisc., Minn. and s. Sask. Late March–June (–Aug.). Fig. 1046. — The rare forma **coptidifòlius** Fern. (with leaves of *Coptis*) has the numerous cauline leaves divided into 3 mostly petiolulate flabelliform to rhombic or oblate leaflets, known only from Lancaster Co., Pa.

Var. **indivìsus** Fern. (undivided). — Alluvial woods, se. Va.

Var. **acrolásius** Fern. (hairy at summit). — Low woods, clearings and damp slopes, s. Lab. to B.C., s. to Nfld., N.S., Me., Mass., n. R.I., n. Ct., w. N.Y., n. Mich., Black Hills, S.D., and Colo. April–Aug.

Var. **eucỳclus** Fern. (well rounded). — Nfld. to w. N.Y. and Ct.

1046. R. abortivus.

22. R. micránthus Nutt. (tiny-flowered). — *Roots fusiform-thickened,* mostly 1.5–3 mm. thick, fascicled; *stems* 1–8, rather slender, usually forking, *villous,* 5–23-flowered, to 1.5 dm. high; *radical leaves loosely spreading, with slender villous petioles,* dark green and lustrous, simple or palmately cleft or divided, firm, *the simple ones* oblate to rhombic-obovate and 1.5–4.5 *cm. broad, with* 11–19 *broad blunt flat-topped teeth above the entire subcordate to cordate base; the other basal leaves* (when present) *with* 3 cuneate- to rhombic-oblanceolate to -obovate lobes or *petiolulate* dentate or cleft *leaflets; lower cauline leaves* short-petioled, *with* 3 simple or deeply cleft oblanceolate to narrowly obovate mostly *petiolulate villous leaflets;* upper similar, subsessile, less divided; *sepals abruptly contracted to a claw,* reflexed, 2–3.2 mm. long; *petals linear to narrowly oblong,* pale yellow, 2–3.5 *mm. long; fruiting head ellipsoid or cylindric,* 3.5–6 mm. long, 2.5–3.5 *mm. thick; achenes* rather plump, *dull,*

1047. R. micranthus, v. delitescens.

1048. R. Harveyi.

minutely pitted, 1.1–1.4 mm. long, with minute soft style; *receptacle linear, glabrous* or only slightly hairy at top. — Rich woods and calcareous banks, Md. to Ill., s. to Ga., Ala. and Ark. March–early May.

Var. **delitéscens** (Greene) Fern. (hiding). — More slender, weak, *pale green, with dull foliage; the simple basal leaves cuneate to rounded or subtruncate at base,* 0.5–2.5 cm. broad, *their* 5–13 *teeth ovate or triangular.* — Thin soil, chiefly on rocky or dryish slopes in partial shade, Mass. to Ill., s. to N.C. and Ark. Fig. 1047.

Var. **cymbalístes** (Greene) Fern. (cymbal-like). — As slender as the last; *simple dull basal leaves deeply cordate,* with few ovate or triangular teeth. — Dry woods, Ind.

23. R. Hárveyi (Gray) Britt. (for its discoverer, Francis Leroy Harvey 1850-1900). — Resembling no. 22, pilose to glabrate; *the uncleft basal leaves cordate or subcordate,* rarely mixed with deeply divided leaves; *petals* 4–8 *mm. long, much exceeding sepals.* — Ledges, bluffs or ravines, w. Ala., Mo. and Ark. (to be expected in e. Okla.). April, early May. Fig. 1048.

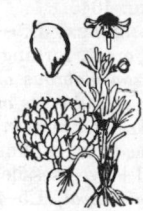

24. R. rhomboìdeus Goldie (rhombic), Prairie-B. or -C. — Roots thickened; *stems* 1–5, forking, loosely villous to glabrous, 0.5–2.5 dm. high, 2–8-*flowered; radical leaves* slender-petioled, *from obovate to rhombic or oblate, cuneate, rounded or rarely subcordate at base, crenate or dentate above the entire base,* 1–6 cm. long, 0.8–4.5 cm. broad; lower cauline similar or flabelliform and deeply cleft and lobed; middle and upper with 3–7 linear or oblanceolate divisions; *sepals* oblong, *copiously long-villous,* reflexed, 3.5–8 mm. long; *petals* 5–7, *narrowly obovate,* deep yellow, 5–9 *mm. long; stamens* in 3–5 *series,* subpersistent; *fruiting heads globose,* 6–10 *mm. in diameter; achenes* very numerous, *obliquely pyriform,* slightly compressed, about 2 mm. long, *with a minute erect slender beak; receptacle obovoid,* glabrous or sparingly hairy. (*R. ovalis* sensu Britt. and others, not Raf.) — Prairies, rich warm slopes, etc., formerly abundant, now rare or local, ne. B.C. to Neb., e. to Ont., Mich. and Ill.; reported (more than a century ago) from near Montreal, Que.; largely exterminated through cult. of land. April, May. Fig. 1049.

1049. R. rhomboideus.

25. R. recurvàtus Poir. (bent backward; referring to the styles). — *Rhizome* short and thick, somewhat *cormlike,* bearing 1–3 upright or ascending *pliant villous-hirsute,* rarely glabrous as forma **laevicaùlis** Harger (smooth-stemmed), 2–16-flowered *stems* 1.5–7.5 dm. high; *radical leaves* long-petioled, *round-ovate,* cordate, 3–5-*cleft, mostly to or below the middle*; the lobes rhombic-ovate, cleft or dentate, *setulose at least above; cauline leaves similar, petioled,* somewhat more deeply cleft, *the lowest usually midway on the stem*; the uppermost smaller, sometimes simple; sepals long-hirsute, strongly reflexed, 3–7 mm. long; *petals* 5, *oblong, pale yellow,* 2–5 *mm. long; fruiting heads globose or globose-ovoid,* 4.5–8 *mm. long; achenes obovate, thin, pitted, ventrally narrow-margined,* 2.3–2.7 *mm. long excluding the long recurved uncinate beak; receptacle clavate, strongly tuberculate-ridged, hirsute.* — Damp or swampy rich woods, brooksides, etc., Nfld. to Man., s. to N.S., N.E., Fla., Ala., Miss., La. and e. Tex. May–July. Fig. 1050.

1050. R. recurvatus.

Var. **adpressípilis** Weath. (with appressed hairs). — Hairs of the stem closely appressed, short and scattered. — Centr. N.Y. to mts. of Va. and Tenn.

26. R. pensylvánicus L. f. (of Pennsylvania), Bristly C. — Annual or perennial with fibrous roots; *stems* erect or ascending, 0.1–1.2 m. high, *soft, hollow, spreading-hirsute or -hispid,* branching, 3–many-flowered; *leaves numerous, nearly uniform* on the stem, *ternately compound, the petiolulate* cuneate to round-based *hirsute leaflets* sparingly to numerously *cleft into cuneate-lanceolate* laciniate to serrate-dentate *segments;* sepals bristly outside, reflexed, 2.5–5.5 mm. long; *petals oblong-obovate,* pale yellow, fading whitish, 2–5 *mm. long,* 1.3–3 *mm. broad; fruiting head ovoid to cylindric,* 0.6–1.8 *cm. long,* 5–9 *mm. thick; achenes* numerous, obliquely obovate, *thin,* with olive or pale brown sides, *thin-mar-*

1051. R. pensylvanicus.

gined, 2–2.5 mm. long excluding the *erect or ascending deltoid-acuminate green beak* (0.5–1 *mm. long*); *receptacle slender-cylindric or lanceolate, thin-ridged, hirsute, 5–16 mm. long.* — Wet meadows, alluvium, ditches, etc., s. Lab. to Alaska, s. to Nfld., N.S., N.E., Del., Pa., W.Va., O., Ind., Ill., Ia., Neb., Colo. and Oreg., more doubtfully to N.M. and Ariz. July–Sept. Fig. 1051.

27. R. Macoùnii Britt. (for its discoverer, John Macoun, 1831–1920). — Perennial with thick fibrous roots, finally *with trailing branches or leafy stolons; stems ascending, declined or trailing,* soft and hollow, hispid or hirsute, 2–24-flowered, 0.2–1.2 m. long; *radical leaves* hispid-petioled, glabrous or sparingly hispid, ternately compound; *the petiolulate leaflets ovate to obovate,* 2–3-*cleft into* lacerate or dentate *acute lobes;* cauline leaves scattered, similar, their leaflets with cuneate-subtruncate entire bases; sepals hispid, reflexed, 3–6 mm. long; *petals broadly obovate or roundish,* deep yellow, 3.5–6 *mm. long,* 2.5–5 *mm. broad; fruiting heads globose or ovoid,* 6–10 *mm. in diameter; achenes* thin, minutely pitted, narrow-margined, 2.5–3.3 *mm. long excluding the ascending or oblique straight flat-subulate beak* (1–1.5 *mm. long*); *receptacle ellipsoid to subcylindric,* ridged, *hirsute,* 4–7 mm. long. — Alluvial thickets, low woods and swales, s. Lab. to Alaska, s. to Nfld., e. Que., n. Mich., Ia., Kans., N.M., Ariz. and Calif. June–Aug. Fig. 1052.

1052. R. Macounii.

28. R. caroliniànus DC. (of Carolina). — Tufted perennial with thick fibrous roots, *after flowering developing long trailing or repent leafy branches; earliest basal leaves small,* ovate, 3-lobed or 3-cleft; *the later larger,* long-petioled, *with* 3 *mostly petiolulate* rhombic cuneate 3-*cleft or* 3-*divided* and sharply toothed *leaflets; flowering stems slender, flexuous,* subglabrous, pilose or spreading-hirsute, 0.5, elongating to 5 dm. long, *finally producing trailing branches;* flowers 1–10; *sepals* 3.5–5 *mm. long, promptly reflexed; petals oblong,* 8–12 mm. long, 2.5–7 mm. broad; *fruiting heads* subglobose, 7–13mm. in diameter, *with only* 10–20 *achenes;* the latter obliquely rounded-ovate, *with body* 3.7–5 *mm. long; the marginal wing* 0.5–1 *mm. wide, separated from the face by a high acute ridge; beak submarginal, erect, lance-subulate,* 1.5–2.5 *mm. long, with short deciduous terminal stigma.* (*R. palmatus* of most auth., not Ell.; *R. septentrionalis* var. *pterocarpus* Benson) — Low woods, thickets and shores, Fla. to Tex., n. to Md., W.Va., s. Ind., s. Ill., Mo. and Neb. Late March–early June. Fig. 1053.

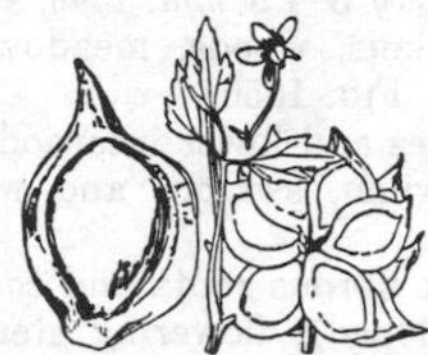
1053. R. carolinianus.

29. R. híspidus Michx. (stiffly hairy). — Tufted slender perennial, with fusiform-thickened roots 0.8–1.5 dm. long; earliest leaves suborbicular, merely dentate to deeply 3-cleft; *later leaves* larger, *palmately* 3-*parted or* 3–5-*divided, with obovate or rhombic coarsely cleft to laciniate segments,* the *terminal leaflet often petiolulate; petioles* spreading-villous to appressed-pilose, *with inconspicuous basal stipules; flowering stems* 1–4, *slender, flexuous,* simple or forking, 1–4.8 dm. high, 2–8-flowered, spreading-villous to appressed-pilose; *sepals* 5.5–7.5 *mm. long, spreading,* only tardily reflexing; *petals obovate to oblong,* 7–14 mm. long, 5–10 *mm. broad; fruiting heads* 7–11 mm. in diameter, *with* 15–40 *achenes;* the latter 3–3.5 *mm. long* (*excluding subulate erect beak* 1.3–2 mm. long), 2.5–3.3 *mm. broad, with dorsal keel* 0.4–0.5 *mm. broad.* — Rich moist woods, Ga. to Ark., n. to s. N.Y., O., Ind., Ill. and Mo. Late March–May. Fig. 1054.

1054. R. hispidus.

Var. **fálsus** Fern. (deceptive). — Foliage much as in typical *R. hispidus; pubescence usually appressed* (sometimes spreading); *achenes* 2–2.5 *mm. long,* 2–2.3 *mm. broad, with dorsal keel about* 0.2 *mm. broad.* (*R. palmatus* Ell.; perhaps *R. marilandicus* Poir. and *R. hispidus,* var. *marilandicus* (Poir.) Benson) — Dry open or rocky woods, Mass. and s. Vt. to Wisc. and Ia., s. to S.C., O., s. Mich. and Ark. Mid-April–June.

Var. **eurýlobus** Benson (wide-lobed). — *Radical leaves simple or* 3-*lobed, the lobes or sessile terminal leaflet broadly rhombic to suborbicular;* pubescence either spreading or appressed; mature achenes not seen. — Rich woods and bottoms, Md. and W.Va. to Ga. and Ala.

30. R. fasciculàris Muhl. (clustered; referring to the thickened roots), Early B. or C. — Similar to no. 29; the *tuberous-thickened*

1055. R. fascicularis.

roots plumper, *only* 1.5–5.5 *cm. long; stems* very weak, *silky-canescent,* 0.2–2.5 dm. long; *later rosette-leaves oblong to ovate, pinnately* 3–5*-divided, the divisions or leaflets with linear or oblong obtuse lobes;* petals linear-oblong to obovate, pale yellow, 6–15 mm. long, 2–7 mm. broad; *fruiting head* 5–8 *mm. long; achenes narrow-margined,* 2.2–3.2 *mm. long,* with subulate straight or curved beak 1.3–2 mm. long. — Thin soil in open woods or exposed hills and ledges (often calcareous), s. N.H. to Ont. and Minn., s. to Ga. and Ia. April, May. Fig. 1055.

Var. **apricus** (Greene) Fern. (open to the sun). — Later rosette-leaves less divided, simple, shallowly lobed or with the 3 somewhat confluent divisions unlobed or but slightly cleft. — Mich. to Ia., and more characteristically s. to Miss., La. and Tex. March, April.

31. R. septentrionalis Poir. (northern), Swamp-B. — Perennial, with thick fibrous roots and *eventually coarse trailing or repent elongate branches; stems* 1*–few, ascending, declined or trailing,* soft and hollow, hirsute with spreading to ascending or appressed pubescence or glabrous; *lower leaves all ternately compound,* long-petioled, subglabrous to strigose-hispid; the *petiolulate leaflets* ovate to rhombic and 2–3-cleft and sharply incised, *cuneate to rounded-truncate at base; the middle leaflet* of the larger blades 3–12 *cm. long, on a petiolule* commonly 1–5 *cm. long; stipules of radical* and lower cauline *leaves coriaceous, brown, rounded at summit; sepals* hirsute to glabrous, *spreading,* tardily reflexed, 6–10 mm. long; petals rounded-obovate, 7–15 mm. long, 4.5–10 mm. broad; *stigma terminating the long style,* 0.2–0.5 mm. long, soon deciduous; *fruiting heads* globose-ovoid, 0.8–1.4 cm. long, 0.9–1.5 *cm. thick; achenes minutely pitted,* thin-margined, *the body* 3–4.5 *mm. long, the straight to falcate flat-subulate beak* 1.5–3 *mm. long.* — Alluvial thickets, woods, meadows, etc., e. Que. to Man., s. to N.E., Md., Ky. and Mo. Late April–July. Fig. 1056.

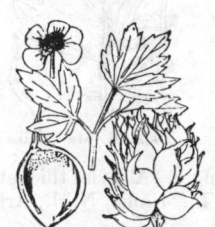

1056. R. septentrionalis.

Var. **caricetorum** (Greene) Fern. (growing among sedges). — Petioles and lower internodes densely retrorse-hispid. (*R. sicaeformis* Mackenz. & Bush) — Alluvium, swamps and wet prairies, w. Md. and O. to Minn., Ia. and Mo. April, May.

32. R. repens L. (creeping), Creeping B. — Perennial with thick fibrous roots and *commonly trailing or repent elongate branches* (these sometimes undeveloped); flowering stems ascending; *leaves dark green,* often mottled with white; *the lower ternately divided into mostly petiolulate* deeply lobed or cleft rhombic *leaflets,* the terminal leaflet often 3-parted or -divided; middle leaflet of the larger leaves 1.5–6 cm. long, subsessile or on a petiolule up to 5.5 cm. long; sepals hirsute, spreading; *petals* 5–9, or very numerous in var. *pleniflorus* (double-flowered), *golden-yellow, rounded-obovate,* 6–17 mm. long, 4.5–15 *mm. broad; stigma nearly covering one side of the short* (0.5–1 mm. long) *style, persistent; fruiting heads globose* or globose-ovoid, 7–10 *mm. thick; achenes* obliquely obovate, flattish or biconvex, minutely pitted, *narrow-margined,* 2–3.5 *mm. long excluding the deltoid-subulate beak* (0.7–1.4 *mm. long*)*; receptacle villous.* — Very variable; the following variations are the most important:

a. Middle leaflet of basal leaves cuneate to subtruncate at base; petals 5–9; stamens numerous. . . *b.*
b. Lobes and teeth of leaves deltoid or ovate to oblong, obtuse or bluntish. . . *c.*
c. Trailing or repent branches or stolons present.
Stems and petioles distinctly pubescent.
Pubescence appressed. . . . . *R. repens* (typical).
Pubescence wide-spreading. . . . . Var. *villosus.*
Stems and petioles glabrous or nearly so. . . . . Var. *glabratus.*
c. Trailing or repent branches wanting. . . . . Var. *erectus.*
b. Lobes and teeth of the much cleft leaves lanceolate to linear, acuminate. . Var. *linearilobus.*
a. Middle leaflet of basal leaves rounded or subcordate at base; petals very numerous. . . . . Var. *pleniflorus.*

R. repens (typical). — Wet open ground, ditches, yards, etc., a rapidly spreading weed, s. Lab. to Ont., s. to Nfld., N.S., N.E., L.I., N.C. and Mo.; w. N.Am. May–Sept. (Natzd. from Eu.) Fig. 1057.

1057. R. repens.

Var. **villosus** Lamotte (soft-hairy). — Wet to dry open soil, s. Lab. to Mich. and s. N.Y. (Natzd. from Eu.)

Var. **glabratus** DC. (becoming smooth). — Nfld. and Que. to N.J.; w. N.Am. (Natzd. from Eu.)

Var. **erectus** DC. (erect). — Local, Nfld. and Que. to N.E. (Natzd. from Eu.)

Var. LINEARÍLOBUS DC. (linear-lobed). — Roadsides, etc., N.E. (Natzd. from Eu.)

Var. PLENIFLÒRUS Fern. (double-flowered). — Old gardens, ditches and roadsides, infrequent, N.E. to O. and N.C.; w. N.Am. (Probably originally from Eu.)

33. R. ÁCRIS L. (acrid), TALL or COMMON B., BOUTON-D'OR (Que.). — Perennial, with a *short thickish praemorse erect rhizome* and a simple or forking crown; *stem erect,* forking, 0.1–1.5 m. high, rather firm, fistulous, glabrous, appressed-pilose or hirsute especially below; *rosette-leaves* long-petioled, angulate-rounded, *pedately* 5–7*-parted, the sessile divisions deeply cleft and parted* into linear or narrowly lanceolate toothed segments; *peduncles not furrowed;* sepals spreading, villous, 3.5–7 mm. long; petals mostly 5 (0–20), obovate, deep yellow, sometimes cream-color, 6–15 mm. long, 6–14 mm. broad; *stigma nearly covering one side of the short curved style, persistent;* fruiting heads subglobose, 5–7.5 mm. in diameter; *achenes flattish, smooth, definitely margined,* 2.5–3.5 *mm. long, with a deltoid-acuminate* erect or curved *beak* 0.4–0.8 *mm. long;* receptacle glabrous (except sometimes at tip). — Fields and clearings, a ubiquitous weed northeastw., s. Lab. and Nfld. to Ont. and Minn., s. to N.S., N.E., L.I., Va., O., Ind., Ill., Mo. and ne. Kans. May–Aug. (Natzd. from Eu.) FIG. 1058.

1058. R. acris.

Var. LATISÉCTUS G. Beck (with broad segments). — Basal leaves with broad obovate or cuneate segments cleft less than half-way to base. (Var. *Steveni* of ed. 7, not Lange) — Nfld. to s. Ung., s. to N.S., N.E., N.Y. and N.J. (Natzd. from Eu.)

34. R. BULBÒSUS L. (bulbous), BULBOUS B. or C. — Base a *subglobose corm* 0.5–3.5 cm. thick; stems erect, forking, 0.5–6.5 dm. high, silky-villous to glabrate; *rosette-leaves ternately compound, the* 3 *roundish divisions with* 3 *notched leaflets,* the terminal leaflet often petiolulate; *peduncles angled or furrowed at summit; sepals reflexed,* 5–8 *mm. long,* villous outside; *petals rounded-obovate,* 0.8–1.4 *cm. long,* nearly as broad; *stigma nearly covering one side of the short curved style, subpersistent; fruiting head* subglobose-ovoid, 7.5–10 *mm. thick; achenes* obliquely obovate to suborbicular, flattened, distinctly margined, 2.5–3 mm. long, *with a deltoid beak* 0.4–0.8 *mm. long; receptacle hairy.* — Pastures, fields, open woods and roadsides, Nfld. to Ont., s. to N.S., N.E., L.I., Ga. and La.; w. N.Am. Late March–June. (Natzd. from Eu.) FIG. 1059.

1059. R. bulbosus.

Var. VALDEPÙBENS (Jord.) Briq. (very hairy). — Copiously hoary-villous. — R.I. and N.Y. to Va. (Natzd. from Eu.)

Var. DISSÉCTUS Babey (dissected). — Leaflets of rosette-leaves much dissected into narrowly oblong or linear segments. — Occasional, N.E. to Mich. and Va. (Natzd. from Eu.)

35. R. SARDÒUS Crantz (old name used by Pliny for some such plant of Sardinia). — Resembling no. 34, *annual; sepals* 2–4 *mm. long; petals* 5–8 *mm. long; fruiting heads* 4.5–8 *mm. thick;* mature achenes low-tuberculate or smooth. (*R. parvulus* L.) — Roadsides and waste places, locally abund., N.B. to Mo. and southw.; w. N.Am. May–Aug. (Natzd. from Eu.) FIG. 1060.

1060. R. Sardous.

1061. R. parviflorus.

36. R. PARVIFLÒRUS L. (small-flowered). — *Annual, hairy, diffuse, slender;* lower leaves roundish-cordate, 3-cleft, coarsely toothed or cut, the upper 3–5-parted; *flowers axillary, their short* (2–25 *mm. long*) *peduncles mostly shorter than the subtending leaves;* petals and sepals subequal; *carpels and achenes minutely hispid and rough,* beaked, narrowly margined. — Roadsides and waste places, Fla. to Tex., n. to N.Y., Ky. and Mo.; w. N.Am. Late March–June. (Natzd. from Eu.) FIG. 1061.

2. MYOSÙRUS L. MOUSETAIL

Sepals 5, spurred at the base. Petals 5, small and narrow, raised on a slender claw, at the summit of which is a nectariferous hollow. Stamens 5–20. Achenes numerous, somewhat 3- or 4-sided, crowded on a very long and slender spike-like receptacle (whence the name, from the Greek *myos, of a mouse,* and *oura, a tail*), the seed suspended. — Little annuals, with tufted narrowly linear-spatulate radical leaves, and naked 1-flowered scapes; genus widely dispersed. Flowers small, greenish.

1. **M. mínimus** L. (very small). — Spur nearly equaling blade of sepal; stamens 10–18; fruiting spike 1.5–5 cm. long; achenes quadrate, blunt, appressed. (Incl. *M. Shortii* Raf.) —

Damp argillaceous or calcareous soils, fallow fields, etc., Fla. to Tex., n. to e. Va., s. Ont., Ill., Minn. and Sask. Late March–June. (Eu.)

3. TRAUTVETTÈRIA Fisch. & Mey. FALSE BUGBANE

Sepals 3–5, usually 4, concave, petal-like, very caducous. Petals none. Achenes numerous, capitate, membranaceous, compressed, somewhat 4-angled and inflated. Seed erect. — A perennial herb, with alternate palmately lobed leaves, and corymbose white flowers. (For *Ernst Rudolph von Trautvetter*, 1809–1889, distinguished Russian botanist.)

1. T. caroliniénsis (Walt.) Vail (of Carolina), TASSEL-RUE. — Stem 0.5–1.5 m. high; basal leaves long-petioled, 1–4 dm. broad, 5–11-lobed; the lobes variously dissected, serrate to lacerate-toothed. — Banks of streams, wooded bluffs and prairies, sw. Pa. to Mo. and nw. Fla. June–Aug. (Japan)

4. THALÍCTRUM L. MEADOW-RUE. PIGAMON (Que.)

Sepals 4–5, petal-like or greenish, usually caducous. Petals none. Achenes 4–15, grooved or ribbed, or else inflated. Stigma unilateral. Seed suspended. — Perennials of the N. Hemisph., with alternate 2–3-ternately compound leaves, the divisions and the leaflets stalked; petioles dilated at base. Flowers in corymbs, panicles or racemes, often polygamous or dioecious. (A name of some plant mentioned by Dioscorides.)*

a. Flowers in a simple lax raceme terminating the usually leafless filiform scape (0.4–3.5 dm. high); leaves in a small basal rosette, coriaceous, lustrous. . . 1. *T. alpinum.*

a. Flowers in panicles or corymbs; plants coarser and usually taller, with cauline leaves. . . *b.*

b. Flowers perfect; filaments expanded above into an elliptic or oblanceolate petal-like blade; anthers 0.4–0.6 mm. long, blunt; fruits flat, scimitar-shaped, on stipes one-half to two-thirds their length, with a minute point-like style. . . . . 2. *T. clavatum.*

b. Flowers unisexual or polygamo-dioecious; filaments mostly slender; anthers longer, apiculate- or subulate-tipped; fruits plump, less curved, sessile or stalked, beaked by the subpersistent subulate long-stigmatose style. . . *c.*

c. Key to pistillate and fruiting material. . . *d.*

d. Carpels and fruits long-stipitate; flowers dioecious, with filiform filaments; Alleghenian plants.

Stem from short crown; sepals lanceolate; young style 1.5–2.5 mm. long; body of mature achene ovoid, 3–4 mm. long. . . . . . 3. *T. coriaceum.*

Stem from long cord-like horizontal rhizome; sepals ovate; young style 3–4 mm. long; body of mature achene obliquely lanceolate, 4–7 mm. long. . . . . 4. *T. Steeleanum.*

d. Carpels and fruits sessile or nearly so (or, if long-stipitate, the flowers chiefly polygamous, the filaments upwardly dilated and the plants northern). . . *e.*

e. Middle and upper cauline leaves long-petioled, barely expanded at flowering time, their semiovate to lunate stipules green and herbaceous; flowers in spring; fruit dropping in May or June. . 5. *T. dioicum.*

e. Middle and upper cauline leaves sessile to only short-petioled, well developed at flowering time, their ovate to suborbicular stipules brown and subcoriaceous; flowers in summer; fruit in summer and autumn. . . *f.*

f. Leaflets glabrous or, if pilose beneath, without capitate glands or waxy atoms. . . *g.*

g. Stigmas 1–4.5 mm. long; sepals longer than broad, 2–5.5 mm. long; leaflets mostly toothed or lobed; northern, northeastern or western species. . . *h.*

h. Stem from a cord-like horizontal rhizome; leaflets glabrous, membranaceous (rarely coriaceous); flowers dioecious, slenderly or virgately racemose-thyrsoid; stigma lanceolate.

Cauline leaves 4–6; leaflets thin, mostly 1.5–5 cm. broad, their veins not prominent, the veinlets (seen by transmitted light) forming a coarse open reticulum; stigmas 2.5–4.5 mm. long; ripe carpels 6–11 mm. long. . . . 6. *T. confine.*

Cauline leaves 1–3; leaflets coriaceous, mostly 0.8–2 cm. broad, their veins prominent beneath, veinlets forming small closed reticulations; stigmas 1–2.5 mm. long; ripe carpels 3–4 mm. long. . . . . 7. *T. venulosum.*

---

* Treatment based in part on that of BERNARD BOIVIN in Rhodora, xlvi. (1944).

*h.*Stem from a short thick crown; leaflets firm when mature, with veins prominent beneath; flowers dioecious or polygamous, corymbose-paniculate; stigma linear, 1.3–4.5 mm. long.

Sepals lanceolate to narrowly ovate, acuminate, mostly slender-tipped; stigma 2–4.5 mm. long, about equaling body of carpel; filaments, when present, filiform; flowers chiefly dioecious. . . . . 8. *T. dasycarpum.*

Sepals broadly oblong, elliptic, obovate or lanceolate, rounded at summit (rarely acutish); style and stigma 1.3–3.5 mm. long, shorter than carpel-body; filaments, when present, clavate-thickened above the middle; stamens and carpels often in same flower. . . . . . 10. *T. polygamum.*

*g.*Stigmas 0.7–1.5 mm. long; sepals suborbicular, 1 mm. long; leaflets mostly entire; southeastern species. . . . . . . 11. *T. macrostylum.*

*f.*Leaflets bearing sessile or short-stalked rounded waxy atoms on lower surface. . . . . 9. *T. revolutum.*

*c.*Key to staminate material. . . *i.*

*i.*Leaflets glabrous or pilose beneath, without waxy atoms. . . *j.*

*j.*Filaments delicately filiform, soon drooping and frequently entangling in age; flowers dioecious. . . *k.*

*k.*Middle and upper leaves long-petioled, barely expanded at flowering time; flowers in early spring; anthers slenderly linear, 2–3.5 mm. long, with minute apiculate tips. . . . . 5. *T. dioicum.*

*k.*Middle and upper cauline leaves sessile or only short-petioled, well developed at flowering time; flowers in late spring or summer. . . *l.*

*l.*Anthers 2–5 mm. long, with subulate tip 0.4–1 mm. long; leaflets with veins scarcely prominent beneath, glabrous, membranaceous.

Stem from short caudex; sepals 3.5–4 mm. long; anthers 2–3.5 mm. long, with subulate tip about 0.4 mm. long. . . 3. *T. coriaceum.*

Stem arising from long cord-like horizontal rhizome; sepals 3–5 mm. long; anthers 2.5–5 mm. long, with subulate tip 0.4–1 mm. long.

Alleghenian species; leaflets (by transmitted light) with a fine closed reticulum of veinlets; carpels stipitate. . . 4. *T. Steeleanum.*

Boreal species; leaflets with coarse open reticulum; carpels sessile. . . . . 6. *T. confine.*

*l.*Anthers 1.5–3.5 mm. long; leaflets prominently veiny beneath, coriaceous to membranaceous, their veinlets forming a closed reticulum.

Stem arising from a cord-like slender rhizome, 2–7.5 dm. high, with 1–3 petioled cauline leaves; leaflets glabrous; panicle thyrsoid, with suberect branches, leafy-bracted only at base; sepals oblong, round-tipped; anthers 2–3.5 mm. long; northwestern species. . . . . 7. *T. venulosum.*

Stem arising from a stout caudex, 0.6–2 m. or more tall, with 3–7 sessile cauline leaves; leaflets pubescent or glabrous beneath; inflorescence open, paniculate-corymbose, the branches leafy-bracted at base; sepals lance- to ovate-acuminate; anthers 1.5–3.2 mm. long; wide-ranging inland. 8. *T. dasycarpum.*

*j.*Filaments dilated upward, ascending, pendulous only in age, rarely entangling.

Anthers 0.7–2 mm. long; mature filaments (after dropping of sepals) 3.5–6.5 mm. long, linear-clavate, dilated chiefly above the middle; sepals longer than broad, 2–5.5 mm. long; northern and northeastern species. . . . . 10. *T. polygamum.*

Anthers 0.5–1 mm. long; mature filaments 2–4 mm. long, oblanceolate, dilated upward from near the base; sepals suborbicular, 1.5–2 mm. long; southeastern species. . . . . 11. *T. macrostylum.*

*i.*Leaflets bearing sessile or short-stalked rounded waxy atoms on lower surfaces. . . . . 9. *T. revolutum.*

§ HOMOTHALÍCTRUM (Fries) Boivin

1. T. alpìnum L. (alpine), ALPINE M. — Rhizome slender, often stoloniferous; *leaves chiefly in a basal rosette, dark glossy green above,* whitened and prominently veined beneath, 0.1–1.2 *dm. long,* often forming broad mats, the terminal petiolulate division with 5–7 cuneate-obovate to rounded slightly lobed to dentate *leaflets* 1.5–9 *mm. long; stem capillary, simple* (rarely forked), *scapose or with* 1 *leaf,* 0.4–3.5 *dm. high, terminated by a simple* (rarely forking) *loose raceme* 0.2–1.5 dm. long; pedicels arching; flowers perfect but appearing dioecious through the early development and dropping of the stamens and the tardy enlargement of the carpels; sepals smoky or pearl-color; the 8–12 purple filiform filaments pendulous, the yellow to brown linear apiculate anthers 1.4–2 mm. long; stigmas purple and deltoid when young, elongating, deciduous; achenes prominently ribbed, 2.2–3.8 mm. long. — Greenl. and Lab. to Alaska, s. on peaty or turfy swales, glades or bogs or on wet calcareous ledges or gravel to Nfld., Anticosti and the Gaspé Pen., Que., Colo., Utah, Nev. and Calif. Mid-June–early Aug. (Eurasia)

§ PHYSOCÁRPUM DC.

2. T. clavàtum DC. (club-shaped; from the filaments), LADY-RUE. — *Very slender, from a tuft of fleshy fusiform roots,* 1.5–6 dm. high, glabrous, 1–3-leaved; radical leaves biternate, with thin suborbicular coarsely crenate leaflets; inflorescence lax; the few *flowers* long-stalked, *with white obovate petaloid sepals* 2.5–3.5 mm. long, *the filiform-based* erect *filaments expanded above into a white elliptic or oblanceolate petal-like blade, with blunt anthers only* 0.4–0.6 *mm. long; the flat scimitar-shaped minutely beaked fruits on slender stipes one-half to two-thirds their length.* — Brooksides, mts. of Va., W.Va. and Ky. to Ga. and Ala. May–July.

§ HETEROGÀMIA Boivin

3. T. coriàceum (Britt.) Small (leathery). — *Caudex stout,* with the *root-fibers bright yellow;* stem 0.9–1.5 m. high, with a large and *loose pyramidal panicle; common petioles of cauline leaves usually well developed;* leaflets broadly obovate to suborbicular, 3–9-toothed or -lobed, firm-membranaceous, pale and glabrous beneath; *flowers dioecious; carpels and fruits long-stipitate, the young style* 1–2 *mm. long; the body of the mature fruit obliquely ovoid,* 3–4 *mm. long;* filaments filiform, shorter than the long-linear subulate-tipped anthers. (*T. caulophylloides* Small) — Upland woods, Pa. and W.Va., s. to Ga. and Tenn. May, June.

4. T. Steeleànum Boivin (for its discoverer, EDWARD STRIEBY STEELE, 1850–1942). — Similar to no. 3; *caudex long and cord-like, roots not yellow;* flowering stem 0.5–2 m. high; *leaflets* submembranaceous, 2.5–5 cm. broad, *with a fine and mostly closed reticulum* (seen by transmitted light); *young style* 3–4 *mm. long; body of mature long-stalked fruit obliquely lanceolate,* 4–7 *mm. long;* sepals of staminate flowers scarious, oblong, blunt, 4–5 mm. long; filaments filiform; anthers long-linear, 4–5 mm. long, with subulate tip 0.8 mm. long. — Rich woods, thickets and alluvium, sw. Pa. and Md. to mts. of Va. and W.Va. May, June.

5. T. dioìcum L. (dioecious), EARLY M., QUICKSILVER-WEED. — *Caudex erect;* stem 2–7.5 dm. high, with 1–3 *distinctly petioled* glaucous or pale *leaves* below the inflorescence, *the upper with semiovate to lunate green stipules; leaflets delicate, mostly not fully expanded at flowering time,* later firmer and spreading, reniform to obovate, with obtuse lobes or rounded broad teeth; *panicles dioecious, terminal and axillary;* sepals of staminate plant thin, stramineous to purple, oblong or oval, 2.5–4.3 mm. long; of the pistillate firmer and greener or purple, smaller; stamens pendulous, with filiform filaments and yellow *linear mucronate-acuminate anthers* 2–3.5 *mm. long; fresh stigma* purple, *filiform,* 1.5–3 *mm. long,* completely *covering the deciduous style; achenes subsessile, ellipsoid, short-pointed,* 2.6–4.5 *mm. long,* regularly ridged and furrowed. — Rich rocky woods, ravines and alluvial terraces, Ga. and Ala. to Mo., n. to St. P. et Miq. (?), centr. Me., sw. Que., s. Ont., Minn. and N.D. Apr., May (–July).

6. T. confìne Fern. (bordering; originally recognized near the international boundary). — *Caudex horizontal, cord-like;* stem 0.2–1 m. high, 4–6-leaved below the *loosely racemose panicle;* upper *leaves* short-petioled or subsessile, *pale or glaucous, the scarious-margined stipules rounded-ovate to suborbicular; leaflets thinnish,* reniform to obovate, slightly 3-lobed or with broad obtuse teeth, mostly 1.5–5 cm. broad, *with veins not prominent beneath,* the *veinlets* (by transmitted light) *forming a coarse open reticulum; flowers* usually *strictly dioecious; sepals of staminate plant scarious,* oblong, *yellowish-green to purplish,* 3–5 mm. long, oblong; *anthers* 2.5–5 *mm. long, linear, with terminal subulus* 0.5–1 *mm. long,* on pendulous filiform filaments; sepals of pistillate plant thick, ribbed, sometimes foliaceous, 1.5–3 mm. long; carpels 4–13, only 1–9 ripening; *fresh stigma* purple, *lanceolate,* 2.5–4.5 *mm. long, persistent; achenes sessile,* fusiform to ovoid, *long-beaked,* 6–11 *mm. long,* regularly and equally ridged and furrowed. (*T. venulosum* sensu Britt., as to eastern plant, not Trel.) — Alluvial or shingly calcareous shores and talus,

Mingan and Anticosti Ids., Que., to James Bay and L. Winnipeg, s. to Gaspé Pen., Que., n. and w. N.B., n. Me., Champlain Valley of Vt. and N.Y., n. Mich., n. Wisc. and ne. Minn. June, July.

7. T. venulòsum Trel. (with evident veinlets). — Smaller than no. 6, 2–7.5 dm. high; *cauline leaves* 1–3, *their coriaceous leaflets strongly veiny beneath* and *mostly* 0.8–2(–3) *cm. broad, their veinlets* (seen by transmitted light) *forming small closed reticulations;* sepals of staminate flowers 3–4 mm. long; *anthers* 2–3.5 *mm. long, the subulus* 0.1–0.2 *mm. long; stigmas* 1–2.5 *mm. long; achenes* 3–4 *mm. long.* — Prairies, thickets, open woods and shores, sw. Ung. to n. B.C., s. to n. Wisc., n. Minn., S.D. and Wyo. Late May–July.

§ Leucocòma (Greene) Boivin

8. T. dasycárpum Fisch. & Lall. (hairy-carpelled), Purple M. — *Caudex short and thick, erect;* stem 0.6–2 m. high, often purple; upper leaves sessile or subsessile; their ovate to suborbicular stipules brown; *leaflets firm, with veins prominent beneath, usually covered beneath with fine non-glandular pubescence; inflorescences* corymbose-paniculate, mostly *dioecious; sepals lanceolate to narrowly ovate, acuminate,* commonly *slender-tipped,* 3–5 mm. long; *filaments filiform,* about 4 mm. long, *soon drooping and entangling; anthers* oblong-linear, 1.5–2.5 *mm. long, with subulate tip only* 0.1–0.2 (–0.4) *mm. long; stigma* 2–4.5 *mm. long,* about equaling ovoid body of carpel. — Meadows, swamps and damp thickets, Ont. to Alta., s. to O., Ind., Ill., Mo., Kans., N.M. and Ariz. Late May–July.

Var. **hypoglaùcum** (Rydb.) Boivin (glaucous beneath). — Plant *glabrous throughout; leaflets membranaceous; filaments* 4–7 *mm. long; anthers* 2.2–3.2 *mm. long;* stigma 2.5–5 mm. long; *body of mature carpel lanceolate.* — Minn. and S.D. to s. B.C., s. to La., Tex. and Ariz.

9. T. revolùtum DC. (with edge rolled back), Purple, Skunk- or Wax-leaved M. — Similar to no. 7, *leaflets glandular beneath with sessile or short-stalked rounded waxy atoms,* or strictly glabrous in the local forma **glàbrum** Pennell; *sepals* ovate, *oblong* or oblanceolate, erose, 3–4 mm. long, or shorter in pistillate plants; filaments weak, soon drooping, 4.5–5.5 mm. long; anthers 1.2–3 mm. long; *carpels glandular or waxy-puberulent; stigma* linear-filiform, 1.8–3.5 *mm. long,* usually deciduous; achenes short-stipitate, tipped by base of style, puberulent, 3.8–6 mm. long. — Dry open woods, thickets, prairies and barrens, occasionally in meadows, Mass. to s. Ont., s. to n. Fla., Ala., Tenn. and Mo. May–July (–Sept.).

10. T. polýgamum Muhl. (polygamous), Tall M., Muskrat-weed, King-of-the-meadow. — Glabrous or pubescent but not glandular, 0.5–2.6 m. high; cauline leaves sessile; leaflets rather firm, roundish to oblong, commonly with mucronate lobes or tips, sometimes puberulent beneath; panicles very compound, with broad rounded to flattish top; flowers white (rarely purplish), the pistillate ones usually with some stamens; sepals 2–3.5 mm. long, oblong or obovate, blunt; *anthers not drooping, oblong, blunt,* 0.7–2 *mm. long, on white upwardly dilated (clavate)* stiff *filaments* 3.5–5 *mm. long;* carpels and fruit glabrous or pubescent; *style and linear stigma* 1.3–2.5 *mm. long;* body of fruit 2.5–5 mm. long, subsessile or short-stalked. — Meadows, low thickets and swamps, Nfld. to Ont., s. to N.S., N.E., L.I., Ga. and Tenn. June–Aug.

Var. **hebecárpum** Fern. (hairy-carpelled). — Larger throughout, pubescent; *inflorescence strongly corymbose; sepals* 3–5.5 *mm. long, elongate; filaments* 5–6.5 *mm. long; anthers* 1.2–2 *mm. long; style and stigma* 2–3.5 *mm. long;* body of fruit 4–5.5 mm. long, pubescent or glabrous. — S. Lab. to nw. Que., s. to Nfld., N.S., n. N.E. and n. N.Y., ascending to subalpine meadows.

Var. **intermèdium** Boivin (intermediate). — *Leaflets coriaceous, with revolute margins; inflorescence an elongate panicle; anthers* 1.5–2 *mm. long.* — Local, s. N.E. — Possibly a hybrid of no. 10 with no. 9.

11. T. macrostýlum (Shuttlw.) Small & Heller (long-styled). — Smaller than no. 9, up to 1.5 m. high; *leaflets* thick, *often entire* or some of them shallowly 3-lobed; *flowers smaller, scattered on long pedicels; sepals suborbicular, in staminate flowers* 1.5–2, *in pistillate flowers about* 1 *mm. long; filaments dilated upward from near the base,* 2–4 *mm. long; anthers* 0.5–1 *mm. long; stigmas only* 0.7–1.5 *mm. long; fruits* sessile or subsessile, *their bodies* 3 *mm. long.* — Wooded shores, bluffs and meadows, se. Va. and N.C. June–Aug.

5. ANEMONÉLLA Spach

Involucre compound, at the base of an umbel of flowers. Sepals 5–10, usually white or pink and conspicuous. Petals none. Achenes 4–15, ovoid, terete, strongly 8–10-ribbed, sessile. Stigma terminal, broad and depressed. — A low glabrous perennial; leaves all radical, compound. (Name a diminutive of *Anemone,* to which this plant has sometimes been referred.)

1. A. thalictroìdes (L.) Spach (like *Thalictrum*), Rue-Anemone. — Stem and slender petiole of radical leaf (1–3 dm. high) rising from a cluster of thickened tuberous roots; leaves 2–3-ternately compound; leaflets roundish, somewhat 3-lobed at the end, cordate at the base, long-petiolulate, those of the 2–3-leaved 1–2-ternate involucre similar; flowers (1–)2–several in an umbel (rarely with secondary umbels); sepals petal-like, oval, 5–15 mm. long, pink or white, or green and leaf-like in forma **chlorántha** Fassett (green-flowered). (*Syndesmon* Hoffmgg.) — Open woods, sw. Me. to Minn., s. to nw. Fla., Ala., Miss., Ark. and Okla. Early April–June. — Sepals, stamens and involucre sometimes variously modified: in forma **Favilliàna** Bergseng (for its discoverer, in 1926, STOUGHTON FAVILLE) the stamens and pistils are all changed to sepals.

6. HEPÁTICA Mill. LIVERLEAF. HEPATICA. NOBLE LIVERWORT. TRINITAIRE (Que.)

Leaves cordate and 3-lobed, thickish and persistent through the winter, the new ones appearing later than the flowers, which are single on hairy scapes. Small genus of N. Hemisph., sometimes united with *Anemone*. (Feminine of the Latin *hepaticus, pertaining to the liver*, from the shape of the leaves.)

1. H. americàna (DC.) Ker (American). — Leaves on long pilose petioles, the mature *blade oblate-reniform, the lobes rounded at summit and usually broader than long; involucral bracts broadly elliptic to broadly oval, rounded at summit;* sepals oblong to short-oval, 6–15 mm. long, bluish-lavender, or deep-purple in forma **purpùrea** Farw. (purple), or rosy-pink in forma **rhodántha** Fern. (rose-flowered), or white in forma **cándida** Fern. (shining white); *mature achenes fusiform to lance-subulate,* 1–1.4 *mm. thick,* tipped by a slender (deciduous) style. (*H. triloba* of ed. 7, not Gilib.; *H. Hepatica* Am. auth., not (L.) Karst.) — Dry woods, N.S. to Man., s. to n. Fla., Ala. and Mo. (Jan.–) March–early June. Forma **Cáhnae** Farw. (named in 1924 for its discoverer, Mrs. CAHN) has the blue sepals very numerous.

2. H. acutíloba DC. (acute-lobed). — Leaves less oblate, the *lobes acute or acutish,* usually longer than broad, frequently notched (with several small lobes); *involucral bracts lanceolate, oblong or narrowly oval, acute to obtusish, not rounded;* sepals bluish, or white in forma **albiflòra** R. Hoffm. (white-flowered), or pink in forma **ròsea** R. Hoffm. (rosy); *mature achenes lance-ovoid,* 1.4–1.8 *mm. thick,* usually with less definite style. — Rich, often calcareous, woods, w. Me. to Minn., s. to Ga., Ala. and Mo. March–June (rarely Sept.–March). — Forma **plèna** Fern. (full) has the white sepals very numerous, the stamens changed to sepals. — Species usually in separate areas from no. 1, hybridizing with it when they mingle.

7. ANEMÒNE L. ANEMONE. ANÉMONE (Que.)

Sepals few or many, petal-like. Petals none, or in no. 11 resembling abortive stamens. Achenes pointed or tailed, flattened, not ribbed. Seed suspended. — Perennial herbs of temp. and cool reg., with radical leaves; those of the stem 2 to 9 together, opposite or whorled and forming an involucre remote from the flower; peduncles 1-flowered, solitary or umbellate. (The ancient Greek and Latin name, a corruption of *Na 'mān,* the Semitic name for *Adonis,* from whose blood the crimson-flowered *Anemone* of the Orient is said to have sprung.)

*a.*Styles short, not plumose; sepals 0.5–2.5 cm. long; staminodia none. ANEMONE proper. . . *b.*
- *b.*Carpels and achenes densely long-pubescent, in maturity long-woolly and forming dense woolly heads. . . *c.*
  - *c.*Tuberous-rooted; sepals 10–20, linear; flowering stem solitary, 1-flowered. 1. *A. caroliniana.*
  - *c.*Non-tuberous; sepals commonly 4–8, broader. . . *d.*
    - *d.*Plant with slender creeping rhizome and mostly 1-flowered stem; bases of stems and petioles glabrous or silky; involucral leaves sessile or subsessile, their lobes and divisions blunt. . . . . 2. *A. parviflora.*
    - *d.*Plants with stout branching caudex, mostly 2–many-flowered; bases of stems and petioles villous; involucral leaves petioled, their lobes and divisions acute or acutish. . . *e.*
      - *e.*Leaves dissected into numerous linear or narrowly lanceolate lobes; styles filiform, often deciduous in fruit. . . . . 3. *A. multifida.*
      - *e.*Leaves with 3–5 oblanceolate or obovate divisions, these with broad lobes; styles subulate, persistent in fruit. . . *f.*
        - *f.*Involucre mostly 5–9 (rarely only 3)-leaved; peduncles mostly naked; plant cinereously silky-pubescent; styles crimson. . . 4. *A. cylindrica.*
        - *f.*Primary involucre 2–3 (rarely –5)-leaved; some of the peduncles bearing involucels; plants greener, loosely pubescent or glabrate; styles pale or merely crimson-tipped.

Divisions of leaves mostly cuneate at base (with straight sides); mature anthers 0.7–1.2 mm. long; fruiting heads 7–11 mm. thick, with ascending or subascending styles. . . . 5. *A. riparia.*

Divisions of leaves mostly convex at base; mature anthers 1.2–1.6 mm. long; fruiting heads 1.2–1.5 cm. thick, with divergent styles. . . . . 6. *A. virginiana.*

*b.*Carpels and achenes glabrous or pubescent, usually not woolly, in maturity forming loose globular heads. . . *g.*

*g.*Involucre sessile or subsessile; stems mostly proliferous-branching, 1–6-flowered; sepals white; achenes flattened, cuneate-obovate to reniform, broadly wing-margined, strigose, becoming glabrate, 4–6 mm. broad. 7. *A. canadensis.*

*g.*Involucre with petioled leaves; stems simple, 1-flowered; sepals pink- or crimson-tinged to white; achenes fusiform to slenderly ellipsoid, less than 2 mm. thick, short-hirsute to villous. . . *h.*

*h.*Terminal leaflet of involucral leaf not at all or but rarely incised or deeply cleft, the lateral leaflets merely serrate or dentate; veins and veinlets of sepals strongly anastomosing below the usually free tips.

Rhizome 2–5 mm. thick; involucral leaves glabrous or nearly so, with leaflets 2–8.7 cm. long; sepals 1.3–2 cm. long; achenes 3.5–3.8 mm. long, minutely hirsute. . . . . 8. *A. lancifolia.*

Rhizome 1–2 mm. thick; involucral leaves pilose on both sides, with leaflets 1–2 cm. long; sepals about 8 mm. long; achenes 2.5–3 mm. long, soft-villous to lanate. . . . . 9. *A. minima.*

*h.*Terminal leaflet of involucral leaf incised, the lateral often cleft; veins of sepals simple or subsimple, slightly forking above middle, the branches free to the tip or but slightly anastomosing. . . . . 10. *A. quinquefolia.*

*a.*Styles long and hairy, in fruit forming feathery tails; sepals 1.5–4 cm. long; stamens accompanied by gland-like staminodia. § PULSATILLA. . . . 11. *A. patens,* var. *Wolfgangiana.*

ANEMÒNE proper

1. A. caroliniàna Walt. (of Carolina). — *Stem slender, simple,* 0.5–3 dm. high, *from a tuberous rhizome; radical leaves once or twice* 3-*parted or -cleft;* involucre low on the stem, 3-parted, its cuneate divisions 3-cleft; *sepals* 10–20, *oblong-linear,* white to roseate, deep violet in forma **violàcea** Clute (violet), 1–2 cm. long; *fruiting head ellipsoid,* 1–1.5 cm. long, with woolly entangled achenes. — Calcareous sands, gravels and prairies, Fla. to Tex., n. to N.C., Ind., Wisc., Minn., and S.D. April, May.

2. A. parviflòra Michx. (small-flowered). — *Rhizome slender, freely forking, stoloniferous,* its branches terminated by rosettes of *lustrous dark green leaves and solitary flowering stems* 0.5–3.5 dm. high; *basal leaves* petioled, *with* 3 *cuneate-obovate divisions* blunt-lobed at summit; involucre sessile, with 2 or 3 narrow-lobed leaves; *sepals* 4–7, white above, *silky and bluish at base beneath,* 0.7–1.5 *cm. long,* 4–9 *mm. broad; styles filiform, erect;* fruiting head woolly, globose-ovoid, 6–11 mm. long. — N. Lab. to Alaska, s. in wet or dry calcareous soil to w. Nfld., Gaspé Pen. and L. Mistassini, Que., Thunder Bay Distr., Ont., N.D., Colo., Ida. and Oreg. June–Aug. (E. Asia)

3. A. multífida Poir. (much cleft). — *Loosely cespitose,* the larger plants *with a multicipital caudex,* each crown with slender *commonly proliferous silky-villous stems* 0.5–3 dm. high; *basal leaves* long-petioled, 2–3 *times ternately divided into linear or narrowly lanceolate acuminate lobes;* the blades 1.5–8 cm. long; involucral leaves 3, similar, petioled; leading peduncle usually naked, others mostly with involucels; sepals 5–10 mm. long, oblong to oval, yellowish-white within, yellowish, greenish or purplish and silky outside, usually 5, or 5 and bright red in forma **sanguínea** (Pursh) Fern. (red), or 14–16 and red in forma **polysépala** Fern. (with many sepals); *anthers* 0.7–1 *mm. long; styles filiform, erect; fruiting head subglobose, short-ovoid or short-cylindric,* 0.8–1.8 cm. long, 0.8–1.1 cm. thick, woolly. (Var. *hudsoniana* DC.; *A. hudsoniana* Richards.) — Dryish slaty or calcareous gravel and ledges, local, Nfld. to Alaska, s. to n. and w. N.B., n. Me., n. Vt., n. N.Y., n. Mich., S.D., N.M. and Calif. Mid-May–early Aug. (S.Am.)

Var. **Richardsiàna** Fern. (for its discoverer, GEORGE HENRY RICHARDS, 1838–1922). — Stems 1.5–7 dm. high, several-flowered; sepals 1.1–1.7 cm. long, bright red, or milk-white in forma **leucántha** Fern. (white-flowered). — River-gravels, Gaspé Pen., Que.; Hudson Bay reg.; Black R., N.Y.; Rocky Mts.

4. A. cylíndrica Gray (cylindric), THIMBLEWEED, LONG-HEADED A. — *Caudex simple or rarely with* 2 *crowns,* each with 1 strict 1–8-flowered *densely cinereous-villous stem* 3–7 dm. high; *basal leaves* long-petioled, *cinereous beneath, with* 5 *cuneate-obovate* deeply lobed and cleft *divisions,* the middle division 1.5–6 cm. long; *involucre with* 5–9 (very rarely only 3) *petioled leaves,* 3 *of* *them large, the others smaller,* being the crowded involucels at bases of the *commonly naked peduncles;* sepals leathery, densely silky outside, oblong to oval, whitish-drab to creamy or greenish, 0.5–1.5 cm. long, or thin and white and obovate to ovate in forma **álbida** Farw. (whitish); *anthers* 0.8–1.2 *mm. long; fruiting head slender-cylindric,* 2–4 *cm. long,* 0.6–1 *cm. thick,* woolly; the *subulate crimson styles* spreading-ascending, with usually recurved tips. — Dry open soil, prairies and slopes, w. Me. to Alta., s. to N.J., Pa., O., Ind., Ill., Mo., ne. Kans., N.M. and Ariz. May–July.

5. A. ripària Fern. (of the river-bank), Thimbleweed. — Resembling no. 4; caudex stout, bearing 1–4 *sparingly villous to glabrate* 1–10-flowered *stems* 0.3–1.3 m. high; *basal leaves sparingly pubescent, with* 3–5 *oblanceolate to obovate mostly cuneate* deeply lobed and cleft *divisions,* the middle divisions 0.3–1 dm. long; *involucre with* 3 (rarely 2 or 4) *uniform* petioled *leaves* similar to basal; *leading peduncle naked, the others with involucels* and sometimes proliferating; sepals 4–6, thin and petaloid, at least the inner broadly oblong to oval, rounded at summit and 1.3–2 cm. long, 0.8–1.5 cm. broad, milk-white, or bright red in forma **rhodántha** Fern. (rose-flowered), or thick and leathery and greenish or greenish-white and oblong-acuminate and only 0.7–1.3 cm. long and 2.5–5 mm. wide in forma **inconspícua** Fern. (inconspicuous); *anthers* 0.7–1.2 *mm. long; fruiting head* cylindric, 1.2–3 cm. long, 7–11 *mm. thick, whitish-brown, the subulate pale* or brown- to purple-tipped *styles ascending or subascending.* — Calcareous or slaty ledges, gravelly shores or thickets, w. Nfld. to s. B.C., s. to N.S., N.B., n. and w. N.E., N.Y., Mich., Ill. and Minn.; forma *rhodantha,* Grand River, Gaspé Co., Que. Late May–July.

6. A. virginiàna L. (Virginian), Thimbleweed. — Coarser than no. 5, *more loosely villous; leaves more pubescent,* with *the divisions often broader and less cuneate at base;* sepals leathery, greenish or greenish-yellow, or strongly tinged with red above in forma **rubrosépala** House (red-sepaled), very pubescent on back, narrowly oblong, acuminate, 0.7–1.3 cm. long, or in forma **leucosépala** Fern. (white-sepaled) thinner and petaloid and white with the larger ones less pubescent on back and obovate and rounded above and 1.2–1.7 cm. long; *anthers* 1.2–1.6 *mm. long; fruiting head* 1.2–1.5 *cm. thick, with firmer divergent styles.* — Dry or rocky open woods, thickets and slopes, centr. Me. to Minn., s. to Ga., Tenn., Ark. and Kans. Late June–Aug.

7. A. canadénsis L. (Canadian). — *Rhizome slender* and tough, producing a few short crowns, each crown with 1–3 *slender villous usually proliferous-branching stems* 2–7 dm. high and bearing a large primary involucre and 1 or 2 series of involucels; *basal leaves* long-petioled, *the strongly veiny blade pilose on veins beneath,* 0.5–1.5 *dm. broad,* deeply 5–7-parted, the cuneate-based acuminate divisions mostly 3-cleft; *involucre and involucels sessile, the primary involucre of* 3 *cuneate-obovate deeply* 2–3-*parted* coarsely toothed *divisions;* primary peduncle naked, others with involucels of 2 leaves; *sepals* 5, *white,* unequal, oblong to obovate, *the larger* 1–2.5 *cm. long;* fruiting heads globose; *achenes flat, broadly wing-margined, cuneate-obovate to orbicular or reniform,* rounded or truncate at summit, strigose-hirsute or glabrate, 3.5–6 *mm. long and broad, abruptly beaked by the equally long subulate-filiform erect style.* — Damp thickets, meadows and gravelly shores, chiefly calcareous (or alluvial), Anticosti I. and Gaspé Pen., Que., to B.C., s. to N.S., centr. Me., w. N.E., n. N.J., Pa., W.Va., O., Ind., Ill., Mo., Kans. and N.M.; spreading from cult. elsewhere. May–July.

8. A. lancifòlia Pursh (lance-leaved). — *Rhizome* horizontal, 2–5 *mm. thick,* crisp, whitish when fresh, covered with tooth-like scales; flowering stem solitary, glabrous or nearly so, 1.4–3.5 dm. high; radical leaves (chiefly young plants) solitary, long-petioled, with 3 sessile leaflets, the 2 lateral deeply cleft; *involucre* toward summit of stem, *its* 3 *short-petioled leaves with* 3 *rhombic to narrowly ovate dentate leaflets* 2–8.7 cm. long; *sepals* 4–7, whitish, 1.3–2 cm. long; *their veins* numerous, *freely forking,* subparallel *and strongly anastomosing;* stamens very numerous, their filaments shorter than to less than twice as long as carpels; achenes minutely hirsute, fusiform, 3.5–3.8 mm. long, with a straight or arching obliquely thick-subulate beak 1–1.5 mm. long. (*A. trifolia* of ed. 7, not L.) — Damp woods and thickets in the Alleghenies, s. Pa. and W.Va. to Ga. and Ky. April–June.

9. A. mínima DC. (very small). — Similar to no. 8, dwarf; *rhizome* 1–2 *mm. thick; basal leaves* with filiform petioles; their rhombic-ovate leaflets sharply dentate above, entire and cuneate below, *pilose on both sides;* flowering stems filiform, 1–1.5 dm. high; *involucral leaves* slender-petioled, *their pilose rhombic leaflets* 1–2 *cm. long, sharply toothed above the middle; sepals* oblong, *about* 8 *mm. long* and 4 mm. wide; *achenes* few, *ellipsoid,* 2.5–3 *mm. long, softly villous or almost lanate,* the recurving beak about 1 mm. long. — Very little known, mts. of Va. April, May. — Perhaps a dwarf form of no. 8.

10. A. quinquefòlia L. (five-leaved), Wood-A. — Rhizome whitish, crisp, horizontal, toothed, 1–4 mm. thick; radical leaves (young plants) long-petioled, with 3 or by division seemingly 5 leaflets; flowering *stem* solitary, *glabrous* or nearly so, slender, 0.5–3 dm. high, *with* 3 (2–4) *long-petioled glabrous or glabrate involucral leaves* toward summit; *the* 3–5 cuneate-obovate to rhombic or lanceolate *leaflets acuminate, incised,* 1–5 cm. long, *the lateral often deeply cleft; sepals* commonly 5 (4–9), whitish, tinged with pink or crimson outside or throughout, oblong to oval, 0.6–2.5 cm. long; *their veins simple or subsimple,* nearly parallel, *slightly forking above the middle, the branches free to the tip or rarely slightly anastomosing;* achenes densely short-hirsute, fusiform, 3.5–4.5 mm. long, the curved subulate beak 1–2 mm. long. — Open woods, thickets and clearings, Que. to N.C., w. N.Y. and, locally, O. and Ky. April–June.

Var. **intèrior** Fern. (inland). — Stem spreading-villous above. — N. Ont. and e. Man., s. to s. Ont., Ky., Mich., Ill. and Ia.

§ Pulsatílla Pers.

11. A. pàtens L. (spreading), var. **Wolfgangiàna** (Bess.) Koch (for its discoverer, Johann Friedrich Wolfgang, Russian botanist early in 19th century), Pasque-flower, Prairie-smoke, Hartshorn-plant. — *Silky-villous,* 0.5–4 dm. high, from a brown crown; *leaves* ternately divided, the lateral divisions 2-parted, the middle ones stalked and 3-parted, *the segments and those of the sessile cup-like involucre deeply cleft into linear or narrowly lanceolate acute lobes; sepals* 5–7, blue, purple or white, 1.5–4 *cm. long,* spreading in anthesis; *stamens usually accompanied by minute or indistinct gland-like staminodia; carpels numerous in a head, with long hairy styles, in fruit plumose as in* Clematis. (Var. *Nuttalliana* Gray; *Pulsatilla hirsutissima* Britt.; *P. ludoviciana* Heller) — Prairies and exposed slopes, Arct. nw. Am., s. to n. Mich., Ill., Mo., Tex., N.M., Utah, and Wash. April–June. (Siberia)

Tribe II. Clematídeae DC.

8. CLÉMATIS L. Clematis

Sepals (normally 4) valvate in bud, the margins often induplicate. Petals none or small, transitional into stamens. Stamens numerous, with adnate anthers. Carpels numerous in a head, long-styled, in fruit forming achenes. Seeds suspended; raphe dorsal. — Opposite leaved herbs or slightly woody vines, the latter climbing by the bending or clasping of the leaf-stalks. Large genus of temp. reg. (*Clematis,* a name of Dioscorides for a climbing plant with long and lithe branches, from *clema, a shoot.*)

a. Flowers cymose-paniculate, with white or whitish sepals widely spreading, dioecious, the pistillate flowers with sterile stamens; anthers blunt, in ours at most 4 mm. long; half-woody climbers. § Flammula. . . *b.*

b. Leaflets toothed or dissected, membranaceous to subcoriaceous; sepals 6–12 mm. long; anthers 0.6–1.5 mm. long; achenes pilose or villous-hirsute.

Leaves 3 (lower rarely 5)-foliolate. . . . . 1. *C. virginiana.*

Leaves 5–7-foliolate (lower sometimes more divided). . . . . 2. *C. ligustici-folia.*

b. Leaflets entire or merely undulate-crenate or -cleft, becoming coriaceous; sepals 10–17 mm. long; anthers 2–4 mm. long; achenes appressed-silky. 3. *C. dioscorei-folia.*

a. Flowers solitary on long peduncles, purple, bluish or greenish-yellow, with large sepals, hermaphrodite. . . *c.*

c. Sepals thick and leathery, erect and connivent at base or throughout; petals and staminodia wanting; anthers long and linear, pointed; leaves simple or wholly or in part pinnate. § Viorna. *d.*

d. Leaves all or at least the upper ones of the main axis slender-petioled and pinnate or pinnately divided; stems climbing or at least the upper leaves with twining tips. *e.*

e. Sepals connivent throughout or recurved only at tip; tails of fruit densely plumose. . *f.*

f. Stem climbing; leaves pinnate or those of the branches sometimes simple. . . *g.*

g. Stems 6-angled, corrugated in drying; nodes of stems, peduncles, lower surfaces of green leaves and backs of sepals minutely pilose. 4. *C. Viorna.*

g. Stems subterete, only slightly corrugated in drying, these, the peduncles, surfaces of glaucous leaves and backs of sepals glabrous.

Leaflets coriaceous, prominently reticulated; flower blue-lavender with greenish tips. . . . . 5. *C. versicolor.*

Leaflets membranaceous, only faintly nerved; flower reddish-purple. . . . . 6. *C. glauco-phylla.*

f. Stem erect, simple or bushy-branched, not climbing or the upper leaves slightly catching to supports; all but the upper leaves simple and subsessile, the uppermost compound (rarely simple); flower reddish- or bluish-purple, the sepal-tips abruptly recurved. 7. *C. Addisonii.*

e. Sepals with upper half soon recurving, the margin thin; tails of fruit glabrous or appressed-pubescent, not plumose.

Leaflets coriaceous, reticulated; upper half of sepals with narrow and entire margin and with recurved point. . . . . 8. *C. Pitcheri.*

Leaflets thin and membranaceous, not conspicuously reticulated; upper half of sepal with broad wavy margin, spreading. . . . 9. *C. crispa.*

d. Leaves all simple (or merely cleft), none pinnate, all sessile to very short-petioled; stem erect, not climbing. . . *h*.

h. Leaves thinnish to slightly coriaceous; carpels and achenes pilose, not tomentose, the mature achenes 3–4 mm. broad; tails of fruits with straight beard extending nearly to the tips; plants of Atlantic and Alleghenian regions. . . *i*.

i. Carpels and achenes with appressed pubescence, that of the summit pointing forward; lowest beard of fruiting tails ascending or spreading-ascending; fruiting head with tails loosely separated at margin.

Branches (when developed) only exceptionally overtopping main axis; larger leaves of primary stem 6–12 cm. long, silky-pilose to glabrate beneath; flowers 2–3.5 cm. long, the cinereous backs of sepals densely silky-villous; mature fruiting peduncle lengthening to 5–19 cm. long, much overtopping the subtending leaves; mature fruiting head 5–10 cm. in diameter, the tails 3–6 cm. long. 10. *C. ochroleuca.*

Branches (when developed) mostly overtopping main axis; larger leaves of primary axis 5–7.8 cm. long, glabrescent beneath; flowers about 2 cm. long, the greenish backs of the sepals only minutely pilose; mature fruiting peduncles 1–5 cm. long, shorter than subtending leaves; mature fruiting head 4–5.5 cm. in diameter, the tails 2–3 cm. long. . . . . 11. *C. viticaulis.*

i. Carpels and achenes horizontally to retrorsely villous at summit; lowest beard of fruiting tails similarly divergent; fruiting head compact, 4–6 cm. in diameter, the tails strongly arcuate-recurving; fruiting peduncles 3–9 cm. long; central axis often overtopped by branches. . . . . 12. *C. albicoma.*

h. Leaves heavily coriaceous, strongly reticulated beneath, the upper crowded; summit of carpels and achenes and base of otherwise beardless tail densely tomentose; mature achenes 5–6 mm. broad; plant of Ozark region and westward. . . . . 13. *C. Fremontii.*

c. Sepals thin and petaloid, marginless, divergent from base; outer stamens altered into petaloid staminodia; anthers short; leaves 3-foliolate. § ATRAGENE. 14. *C. verticillaris.*

§ FLÁMMULA DC.

1. C. virginiàna L. (Virginian), VIRGIN'S-BOWER, DEVIL'S-DARNING-NEEDLE, HERBE AUX GUEUX (Que.). — *Leaves* simply *3-foliolate; leaflets thin,* ovate, and often subcordate, *incisely few-toothed* and somewhat lobed, glabrous or beneath sparingly pilose and glabrate, or permanently and densely pilose beneath in forma **missouriénsis** (Rydb.) Fern. (of Missouri); panicles corymbiform, with numerous creamy-white flowers; sepals 6–12 mm. long; anthers 0.6–1.5 mm. long; achenes brown or rufescent, pilose or villous-hirsute. — Low grounds, thickets and borders of woods, Gaspé Pen., Que., to Man., s. to N.S., N.E., Ga., Ala., Miss., La. and e. Kans. July–Sept.

2. C. ligusticifòlia Nutt. (with leaves of *Ligusticum*). — Very similar, but the *leaves 5-foliolate or quinate-ternate;* leaflets small, 1.5–4 cm. broad, pale green, thickish, of firm texture; achenes drab or pale brown, densely pilose. — Climbing over bushes, Man. to B.C., s. to w. Mo., Kans., N.M., Ariz. and Calif.; locally adv. in e. Pa. July, Aug.

3. C. DIOSCOREIFÒLIA Lévl. & Vaniot (with leaves of *Dioscorea*). — *Glabrous* high-climber; *leaves coriaceous, the* 5 *round-ovate to suborbicular leaflets* with strongly rounded tips and cordate bases, *their margins undulate-crenate or entire;* panicles corymbiform; *sepals* 1–1.7 *cm. long; anthers* 2–4 *mm. long,* linear; *achenes appressed-silky.* (*C. paniculata,* var. *discoreifolia* Rehd.) — Thickets, roadsides, fencerows, etc., occasionally spread from cult., Mass. to Va. July–Sept. (Introd. from e. Asia)

Var. ROBÚSTA (Carr.) Rehd. (stout). — Leaflets thinner, ovate, acute, round-based or subcordate, entire or undulate. (*C. paniculata* Thunb., not J. F. Gmel.) — Similar habitats, Mass. to Fla. and Tenn. (Introd. from e. Asia)

§ VióRNA Reichenb.

4. C. Vióma L. (old name of the European *C. Vitalba* L., transferred by Linnaeus to this very different species, Gerarde (1597) saying of the European vine "called commonly Viorna quasi vias ornans, of decking and adorning waies and hedges, where people trauell, and thereupon I have named it the Traueilers Ioie"), LEATHER-FLOWER, VASE-VINE. — Habit of nos. 5 and 6; *leaflets* 3–7, *bright green, membranaceous*, ovate to lanceolate or oblong, *acute*, sparsely pubescent beneath; peduncles naked or 2-bracted; *sepals* 1.5–2.5 cm. long, oblong-lanceolate, acuminate, *the slender tip recurving;* achenes ovate, strigose, with thick obtuse margin; style very plumose, usually sordid. — Rich woods and thickets, Ga. to Tex., n. to Pa., O., Ind., s. Ill. and se. Ia. May–Aug.

Var. **fláccida** (Small) Erickson (weak). — Leaflets flaccid, velvety beneath. (*C. flaccida* Small) — Warren Co., Ky.

5. C. versícolor Small (variously colored), LEATHER-FLOWER. — Climbing, glabrous or nearly so; leaves pinnate; *leaflets* oval, *blunt or round-tipped, coriaceous, strongly reticulated,* 3–7 cm. long; peduncles with 2 simple bracts near base; *sepals dull purple or bluish-lavender, with greenish tips*, lanceolate, coriaceous, *glabrous without*, with upper pale margin pilose, 2–3 cm. long, slightly recurved at tip; *achenes* strigose, *with prominent lateral keel;* style very plumose. — Rocky woods, barrens and bluffs, w. Ky. and w. Tenn., s. Mo., sw. Ark. and se. Okla. May, June.

6. C. glaucophylla Small (with blue-green leaves), LEATHER-FLOWER. — Similar to no. 4; *leaflets submembranaceous* or becoming slightly firm, *glaucous beneath, without prominent reticulation*, ovate, acute or obtuse; *peduncles* 2-bracted or naked, *often several in a leafy corymb; sepals reddish-purple, puberulent*, lance-acuminate, *with a long recurving subulate tip; achenes* rhombic-suborbicular, strigose, *the lateral keel obscure;* style very plumose, sordid. — Rich woods and river-banks, nw. Fla. to se. Okla., n. to w. Va., Ky. and Mo. May–July.

7. C. Addisônii Britt. (for its discoverer, ADDISON BROWN, 1830–1913). — *Suberect*, 6–9 dm. high; *leaves* all or many of them *simple, sessile*, broadly ovate, deep green above, glaucous beneath, obtuse; *the later ones pinnate, with prehensile petiolules* and elliptic-ovate leaflets; flowers and fruit much as in no. 4. — Alluvial soil, Blue Ridge, w. Va. May–July.

8. C. Pítcheri T. & G. (for its discoverer, ZINA PITCHER, 1797–1872). — Calyx campanulate; the dull purplish *sepals with narrow and slightly margined recurved points; tails of the fruit filiform and naked or short-villous;* leaflets 3–9, ovate or somewhat cordate, entire or 3-lobed, much reticulated; uppermost leaves often simple. — Thickets and borders of woods, w. Ind. to Ia. and se. Neb., s. to w. Tenn., Ark. and Tex. June–Aug.

9. C. críspa L. (crinkled), BLUE JASMINE. — Calyx cylindrical below, the upper half of the bluish-purple *sepals* (2.5–4.5 cm. long) *dilated* and widely spreading, with *broad and wavy thin margins; tails of the fruit silky* or glabrate; leaflets 5–9, thin, varying from ovate or cordate to lanceolate, entire or 3–5-parted. — Wet woods and swamps, Fla. to Tex., n. to se. Va., s. Ill. and s. Mo. Apr.–Aug.

10. C. ochroleûca Ait. (yellowish-white), CURLY-HEADS. — *Erect*, branching from some or all the nodes or rarely simple, 3–7 dm. high, the *branches rarely overtopping the central axis, the young growth silky-pilose*, glabrescent; *leaves* entire or shallowly lobed, rounded at base, *the mature coriaceous*, reticulate-veiny, *silky-pilose to glabrate beneath*, glabrous above, *the larger primary ones* broadly to narrowly ovate, 6–12 *cm. long;* flowers solitary, 2–3.5 cm. long, on long peduncles; sepals dull yellow to purplish, canescent on the back; *anthers coarsely bearded, the spreading beard* 1–1.5 *mm. long; mature fruiting peduncle lengthening to* 5–19 *cm., much overtopping the subtending leaves; mature fruiting head* 5–10 *cm. in diameter, the curving styles loosely separated at margin; achenes subsymmetrical, appressed-pilose to summit;* the mature fruiting *style* 3–6 *cm. long, with ascending cinnamon-brown to buff or yellowish-white beard.* — Woods, thickets, rocky slopes and clearings, se. N.Y. to e. Pa., s. to Ga. April, May.

11. C. viticaûlis Steele (with stem of *Vitis*). — Resembling no. 10; *stem* usually *with elongate branches overtopping the central axis; larger leaves of primary axis* 5–7.8 *cm. long*, glabrescent beneath; *flowers about* 2 *cm. long, the greenish backs of the sepals only minutely pilose;* mature *fruiting peduncles* 1–5 cm. long, *shorter than subtending leaves; fruiting head* 4–5.5 *cm. in diameter, very loose and open*, with few fruits, the brownish plumose *tails* 2–3 *cm. long.* — Shaly barrens of Bath Co., Va.

12. C. albícoma Wherry (white-locked). — Smaller than no. 10, the *main axis* 2–3 *dm. high and usually much overtopped by the branches*, the stems loosely pilose or glabrescent; *leaves* at first sparingly pilose on veins beneath, otherwise glabrous, soon quite glabrate, *the larger ones* 4–7.5 *cm. long and* 1.5–4.5 *cm. broad;* flowers 1.7–2.8 cm. long, purplish, the backs of the sepals white-villous; *fruiting peduncles* 3–9 *cm. long,* pilose; *fruiting head compact,* 4–6 *cm. in diameter; the tails strongly arcuate-recurving,* drab or whitish-gray; *summits of carpels and achenes and bases of tails with horizontally divergent to slightly reflexed long villi.* — Shaly barrens and slopes, w. Va. and e. W.Va. May, June.

Var. **coáctilis** Fern. (felty). — Main axis 2–4.5 dm. high; *stems and lower leaf-surfaces persistently pilose-tomentose; larger leaves* 6–10 *cm. long and* 4–9 *cm. broad;* sepals tomentulose on back. — Mts. of Va. May, June.

13. C. Fremóntii S. Wats. (for its discoverer, John Charles Fremont, 1813–1890), Fremont's C. — Stems 1.5–4.5 dm. high; *leaves crowded, very thick,* often coarsely toothed, orbicular to rounded-ovate or oblong, obtuse, sparingly villous-tomentose, those of main axis 3–5 paired, 5–13 cm. long and 2–11 cm. broad; *peduncles* 0.5–4 *cm. long, much shorter than the subtending leaves; backs of sepals glabrous; achenes villous-tomentose at summit,* 5–6 *mm. broad; styles glabrous except at tomentose base,* 1.5–3 cm. long. — Calcareous hills, glades and dry prairies, Neb. and Kans. May, June. Represented with us by

Var. **Rièhlii** Erickson (for its original collector, Nicholas Riehl, 1808–1852). — Less compact; stems up to 7 dm. high, with prolonged internodes; leaves thinner, those of main axis 3–6 pairs, elliptic-lanceolate to ovate, obtuse or acute, the larger ones 8–14 cm. long and 4–8.5 cm. broad; peduncles lengthening to 4–6 cm. — Calcareous glades and barrens, e. Mo.; plains of w.-centr. Kans.

§ Atrágene (L.) DC.

14. C. verticillàris DC. (whorled; from the flowers in axils of paired leaves), Purple C. — Woody-stemmed climber, almost glabrous; leaves trifoliolate, with slender common and partial petioles; leaflets ovate or slightly cordate, pointed; flower mauve-purple, 5–7.5 cm. broad, *the thin and marginless* almost translucent broadly lanceolate to lance-oblong *sepals divergent from base,* 3–3.5 cm. long, sparsely pilose or glabrescent; *outer stamens altered into petaloid staminodia;* anthers short; mature plumose styles about 5 cm. long. (*Atragene americana* Sims) — Rocky (often calcareous) slopes and open woods, e. Que. to Man., s. to N.E., Del., ne. Md., W.Va., O., Mich., Wisc. and ne. Ia. May, June.

Var. **cacùminis** Fern. (of peaks or crests). — Sepals oblong-elliptic, rounded above, 2–3.3 cm. long, thicker, cinereous-pilose on back. — High slopes and crests, Blue Ridge, Va. and N.C.

C. orientàlis L. (oriental), a climber with finely dissected foliage and yellow flowers 3–5 cm. across, with the sepals spreading to reflexed, is beginning to spread from cult. (Introd. from Asia)

C. Viticélla L. (little vine) of § Viticélla Link, a woody vine with 3–7-*foliolate pinnate leaves,* the leaflets narrowly ovate, *the* 1–few terminal and axillary *long-peduncled roseate to violet flowers with wide-spreading large sepals, the styles glabrous,* is much cult. and sometimes escapes to thickets. (Introd. from Asia)

Tribe III. Helleboreae DC.

9. ISOPỲRUM L.

Sepals 5, petal-like, deciduous. Stamens 10–40. Pistils 3–6 or more, pointed with the styles. Follicles ovate or oblong, 2–several-seeded. — Slender smooth N.Am. and Asiatic perennial herbs, with 2–3-ternately compound leaves; the leaflets 2–3-lobed. Flowers axillary and terminal, white. (The ancient Greek name of a *Fumaria,* transferred to our plant probably because of similar foliage.)

1. I. biterǹatum (Raf.) T. & G. (biternate). — Petals none; filaments white, clavate; pistils 3–6 (commonly 4), divaricate in fruit, 2–3-seeded; seeds smooth. — Rich or calcareous woods and thickets, s. Ont. to Minn., s. to nw. Fla., Ala., Mo. and Tex. April, May. — Root-fibers sometimes tuberous-thickened.

10. CÁLTHA L. Marsh-Marigold. Populage (Que.)

Sepals 5–9, petal-like. Pistils 5–10, with scarcely any styles. Follicles compressed, spreading, many-seeded. — Glabrous perennials of cool or temp. reg., with round and cordate or reniform large leaves. (Latin name of a strong-smelling yellow flower, probably the common Marigold, from *calathos, a cup*)

1. C. palústris L. (of swamps), "Cowslip", King-cup, May-blob, Souci d'eau (Que.). —

Stem hollow, furrowed, erect or decumbent, elongating in maturity to 1.5–8 dm.; leaves roundish to open-reniform, dentate; *sepals* broadly oval to narrowly obovate, *deep yellow or orange,* 1–2.5 cm. long, 0.5–2 cm. broad; follicles few, 1–1.5 cm. long, with prominent beak, ascending or slightly recurved-spreading. (Including vars. *flabellifolia* and *radicans* of ed. 7) — Swamps, wet meadows and wet woods, Lab. to Alaska, s. to Nfld., N.S., N.E., S.C., Tenn., Ia. and Neb. Early April–June (–Aug. northw.). (Eurasia)

2. C. nàtans Pall. (floating). — Stems commonly floating; leaves ovate-reniform, thin, subentire; flowers small (1–1.2 cm. broad); *sepals white or pinkish;* carpels numerous (3 mm. long), in a globose head. — In ponds or on muddy shores, Alta. to Alaska, se. to n. Minn. June–Sept. (N. Asia)

11. TRÓLLIUS L. Globe-flower

Sepals 5–15, petal-like. Petals small, 1-lipped, the concavity near the base. Stamens and pistils numerous. Follicles 9 or more, many-seeded. — Smooth perennials of N. Hemisph., with palmately parted and cut leaves, suggesting those of *Ranunculus,* and large solitary terminal flowers. (Name a latinization of *Troll, a globe,* from *Trollblume,* the Germanic vernacular designation.)

1. T. láxus Salisb. (loose), Spreading G. — Leaves 5–7-parted; pale greenish-yellow sepals 5–6, spreading; petals 15–25, inconspicuous, much shorter than the stamens. — Rich meadows and swamps, rare or local, w. Ct. to Mich., s. to Pa.; by old records n. to w. N.H. and w. Me. April, May.

12. CÓPTIS Salisb. Goldthread. Tisavoyanne jaune or Savoyane (Que.)

Sepals 5–7, petal-like, deciduous. Petals 5–7, small, fleshy, hollow at the apex. Stamens 15–25. Pistils 3–9, on slender stalks. Follicles divergent, membranaceous, pointed by the style, 4–8-seeded. — Low smooth N.Am. and Asiatic perennials, with ternately divided evergreen radical leaves, and small whitish flowers on scapes. (Name from the Greek *coptein, to cut,* alluding to the divided leaves.)

1. C. groenlándica (Oeder) Fern. (of Greenland), Canker-root. — Rhizomes filiform, bright yellow, bitter; leaves evergreen, lustrous, slender-petioled; leaflets 3, short-petiolulate, cuneate-obovate, sharply toothed, obscurely 3-lobed; scape slender, 7–13 cm. high, 1-flowered; sepals spatulate to elliptic-lanceolate, 1–3 mm. broad, obtuse or subacute, narrowed to sessile base; blade of petal usually rounded-obovate; body of mature follicle 5–9 mm. long, with beak 2.5–4 mm. long; seeds about half filling follicle, rounded in section. — Mossy woods and swamps, s. Greenl. and Lab. to Man., s. to Nfld., N.S., N.E., N.J., mts. of N.C. and Tenn., n. O., n. Ind. and n. Ia. May–July — Formerly confused with the Asiatic and Alaskan *C. trifolia* (L.) Salisb.

Three Eurasian plants of our gardens occasionally spread locally from cult. and the first, at least, is inclined to persist: Eránthis (from *er, spring,* and *anthos, flower*) hyemàlis (L.) Salisb. (of winter), Winter-Aconite, a dwarf perennial herb with palmately multifid radical leaves, the scape bearing a single large yellow cup-shaped flower subtended by a leaf, the 5–8 narrow sepals deciduous, the petals merely small 2-lipped nectaries, the stipitate follicles several-seeded; Helléborus (the Greek name) víridis L. (green), Green Hellebore or Winter-Aconite, a perennial with large pedate leaves and solitary nodding flower in winter and early spring, with 5 broad persistent finally green sepals and small tubular 2-lipped petals, the firm follicles sessile; Nigélla (from *niger, black*) damascèna L. (of Damascus), Fennel-flower or Love-in-a-mist, slender annual with leaves divided into capillary segments, the uppermost leaf subtending and overtopping the bluish flower, the 5 deciduous sepals petaloid, the follicles connate.

13. AQUILÈGIA L. Columbine. Ancolie (Que.)

Sepals 5, regular, colored as are the petals. Petals 5, all alike, with a short spreading lip, produced backward into large hollow spurs, much longer than the calyx. Pistils 5, with slender styles. Follicles erect, many-seeded. — Perennials of N. Hemisph., with 2–3-ternately compound leaves, the leaflets lobed. Flowers large and showy, terminating the branches. (Name by some derived from *aquila, the eagle,* from supposed resemblance of the curved spurs to claws; by others from *aqua, water,* and *legere, to collect,* from the evident fluid at the base of the hollow spur.)*

* Varietal characters derived from the revision of *Aquilegia* by P. A. Munz in Bailey, Gent. Herb. vii. fasc. 1 (1946).

Spurs straight or but slightly arching, with oblique nectariferous tips; flowers scarlet or red (rarely pale) and yellow; beaks (styles) of mature follicles 1.5–2 cm. long. . . . . 1. *A. canadensis.*

Spurs strongly hooked, the nectariferous tips conspicuously turned back; sepals and spurs blue, purple, pink or white; beaks (styles) of mature follicles 2–10 mm. long.

Flowers 2.5–5.5 cm. long, blue, purple, pink or white; beaks of follicle 5–10 mm. long. . . . . 2. *A. vulgaris.*

Flowers 1.5–2.5 cm. long; sepals and spurs blue or purple; blades of petals yellowish-white; beaks of follicles 2–5 mm. long. . . . . . . 3. *A. brevistyla.*

1. A. canadénsis L. (Canadian), WILD C., "MEETINGHOUSES", "HONEYSUCKLE", GANTS DE NOTRE-DAME (Que.). — Slender, 0.15–1 m. high; *flowers* 3–5.3 cm. long, nodding so that the straight or arching spurs turn upward; *sepals reddish- to greenish-yellow; petals scarlet or bright red, yellow within;* or flower salmon-pink in the local forma **Phippénii** (J. Robins.) R. Hoffm. (named in 1880 for its discoverer, GEORGE D. PHIPPEN), yellow in the still rarer forma **flaviflòra** (Tenney) Britt. (yellow-flowered), and white in the very rare forma **albiflòra** House (white-flowered); stamens and style long-exserted; fruit erect, of 5 parallel ascending *follicles* with ultimately outcurving summits *tipped by persistent styles* 1.5–2 cm. long. — Very variable in stature, size and shape of sepals, direction of follicles and of habitat. The following confluent vars. recognized:

*a.*Sepals 1–1.4 cm. long, ovate to ovate-lanceolate. . . *b.*

*b.*Spurs slender; sepals 6–8 mm. wide.

Basal leaves biternate, their leaflets cleft one-third to half their length; lamina of petal 6–8 mm. long; spur 2–2.5 cm. long; follicles soon spreading toward summit; eastern. . . . . *A. canadensis* (typical).

Basal leaves bi- or triternate, their leaflets mostly cleft to below middle; lamina of petal 5–6 mm. long; spur 1.8–2.2 mm. long; southwestern. Var. *latiuscula.*

*b.*Spurs stout; sepals 5–7 mm. wide; follicles erect; northwestern. . . . Var. *hybrida.*

*a.*Sepals 15–20 mm. long, oblong-ovate, 6–7 mm. broad, spurs stout; follicles erect; wide-ranging. . . . . Var. *coccinea.*

A. canadénsis (typical), including the color-forms noted above. — Rocky wooded or open, often springy, slopes, rarely in swamps, Nfld. (?) and Que., s. Ont. and Wisc., s. to N.E., Ga. and Tenn. April–July.

Var. **latiúscula** (Greene) Munz (broadish). — Ia. and Neb. to Ark., Okla. and Tex.

Var. **hýbrida** Hook. (hybrid). — N. Mich. to Man., s. to Ia. and Neb.

Var. **coccínea** (Small) Munz (scarlet). — N.H. to Minn., s., mostly in upland, to S.C., Ala. and Mo.

2. A. VULGÀRIS L. (common), GARDEN- or EUROPEAN C., GANTS DE NOTRE-DAME (Que.). — Stiff, stoutish; *flowers* 2.5–5.5 *cm. long, blue, purple, pink or white; the spurs short and thick, with strongly recurved nectariferous tips; follicles* erect, *tipped by styles* 5–10 *mm. long.* — Roadsides and borders of woods and fields, Nfld. to Ont., s. to N.S., N.E., Pa. and Ia. May–July. (Introd. from Eu.)

3. A. brevistỳla Hook. (short-styled). — Slender, 0.2–1 m. high; *flowers* 1.5–2.5 *cm. long,* ascending or nodding; *lamina of the yellowish-white sepals little shorter than the blue or purple petals* and the bluish hooked spur; follicles tipped by styles only 2–5 mm. long. — Rock-crevices, open woods and meadows, n. Alta. to Alaska, se. and s. to Minn., S.D. and s. Alta. June, July.

14. DELPHÍNIUM L. LARKSPUR

Sepals 5, irregular, petal-like; the upper one prolonged into a spur at the base. Petals 4 (rarely only 2, united into one), irregular, the upper pair continued backward into long spurs which are inclosed in the spur of the calyx, the lower pair with short claws. Pistils 1–5, forming many-seeded follicles. — Leaves palmately divided or cut. Flowers in terminal racemes. Large genus of N. Hemisph. and w. S.Am. (Name from *delphinus, dolphin,* in allusion to the shape of the flower, which is sometimes not unlike the classical figures of the dolphin.)

_a._Petals 2, united into 1 body; carpel and follicle 1; annual with leaves dissected into narrowly linear or filiform segments. § CONSOLIDA. . . . . . . 1. *D. Ajacis.*

_a._Petals 4, distinct; carpels and follicles 3–5; perennials. § DELPHINASTRUM. . . _b._

_b._Key to flowering material. . . _c._

_c._Raceme open, with elongate spreading-ascending pedicels; flowers 2.5–3.7 cm. long.

Stem succulent, 1–9 dm. high; flowers blue or variegated with white, expanding in spring. . . . . 2. ***D. tricorne.***

Stem firm, 0.5–1.2 m. high; flowers bluish-purple; the sepals spotted with yellow or brown; the petals with conspicuous yellow beard, expanding in early summer. . . . . 3. ***D. Treleasei.***

*c.*Raceme virgate, with erect or suberect pedicels; flowers 1.4–2.5 (–3.2) cm. long. . *d.*

*d.*Stem glabrous except in inflorescence, 0.6–2 m. high; leaves with broad cuneate or cuneate-lanceolate divisions with few lanceolate lobes. 4. ***D. exaltatum.***

*d.*Stem closely and minutely pubescent, 0.3–1.5 m. high; ultimate lobes of leaves narrowly linear.

Flowers prevailingly deep blue or violet (rarely white); spur once-and-a-half length of upper sepal; southern, 0.3–1 m. high. 5. ***D. caroliniānum.***

Flowers white or whitish (rarely a little bluish); spur once-and-a-half to twice length of upper sepal; northwestern and western species 0.5–1.5 m. high. . . . . 6. ***D. virescens.***

*b.*Key to fruiting material. . . *e.*

*e.*Follicles strongly divergent in maturity, mature in May and in early June; seeds with tight close smooth coat; stem succulent, glabrous, relatively low. . . . . 2. ***D. tricorne.***

*e.*Follicles erect and parallel (except at tip), later; seeds with a loose cellular coat; stem firm. . . *f.*

*f.*Stem glabrous, 0.5–2 m. high.

Leaves mostly basal, the cauline only 1–3, relatively small, with linear lobes; raceme loose and open with spreading-ascending pedicels; lower pedicels 3–14 cm. long. . . . . 3. ***D. Treleasei.***

Leaves numerous and subuniformly alternate to summit of stem, with broad lanceolate lobes; raceme strict, virgate, with erect pedicels at most 1–4.5 cm. long. . . . . 4. ***D. exaltatum.***

*f.*Stem closely and minutely pubescent, 0.3–1.5 m. high.

Seeds winged; stems 0.3–1 m. high. . . . . 5. ***D. caroliniānum.***

Seeds wingless, stem 0.5–1.5 m. high. . . . . 6. ***D. virescens.***

§ Consólida DC.

1. D. Ajàcis L. (the plant, according to Ovid, springing from the blood of *Ajax*, as shown by the letters A I A on one of the petals), Rocket-L. — Slender *annual* 3–10 dm. high, simple or with ascending branches; *leaves* several, *with very numerous narrowly linear to filiform segments;* terminal racemes elongate, spiciform, loosely many-flowered; flowers blue, violet, purple or pink, or in forma Álbum R. H. Cheney (white) white, with slender horizontal spur; petals 2 united; *follicle* 1, pubescent; seeds covered with broken ridges. — Roadsides, waste places and old fields, partly spread from cult., N.S. to Minn. and s. and sw. beyond our limits. May–Sept. (Adv. and introd. from Eu.)

§ Delphinástrum DC.

2. D. tricórne Michx. (three-horned; from the three divergent follicles), Dwarf L. — *Stem* simple, *succulent,* 1–9 *dm. high,* from a cluster of thick and short tuberous roots; leaves mostly basal, only few and remote above, pedately 5-parted, the divisions cleft and laciniate into a few narrow lobes; *raceme open and loose, with elongate spreading-ascending pedicels,* the rachis hirtellous to glabrous; *flowers* 2.5–3.5 *cm. long, blue or violet or variegated with white,* or white throughout in forma **albiflorum** Millsp. (white-flowered), the long straightish spur ascending; *follicles strongly divergent in maturity; seeds with a tight smooth coat.* — Rich woods and calcareous slopes, Pa. to Minn. and Neb., s. to Ga., Ala., Ark. and Okla. April, May (–early June).

3. D. Treleàsei Bush (in honor of William Trelease, 1857–1945). — Essentially *glabrous* throughout; *stem firm,* slender, 0.5–1.2 m. high; *leaves mostly basal,* their segments deeply cleft, *the long acute lobes linear; cauline leaves only* 1–3, *relatively small and remote; raceme* lax and open, *with spreading-ascending pedicels, the lower pedicels* 3–14 *cm. long; flowers* 2.5–3.7 cm. long, *bluish-purple* (rarely white), *the lamina of each sepal more or less distinctly spotted with yellow and brown; petals with a conspicuous yellow beard;* seeds with a loose wrinkled coat. — Calcareous barrens, slopes and glades, sw. Mo. May, June.

4. D. exaltàtum Ait. (very tall), Tall L. — *Stem* slender, *glabrous* except at summit, 0.6–2 *m. high; leaves numerous, extending subuniformly to summit of stem,* deeply 3–5-cleft; *the divisions* narrowly cuneate or cuneate-lanceolate, *divergent,* 3-*cleft* or the lateral 2-cleft *into lanceolate lobes; raceme elongate, virgate,* commonly panicled at base; *flowers* blue or white, 1.4–2.2 *cm.*

long, canescent-puberulent; seeds with a loose cellular coat. — Rich woods, thickets and rocky slopes (ascending to high mountain-summits), s.-centr. Pa. to O., s. to Ala. July–Sept.

5. D. caroliniànum Walt. (of Carolina). — *Stem* simple or with few erect branches, *sparsely hirtellous below, glandular-hispid above*, 0.3–1 m. high; lower *leaves* several, deeply 3–5-parted, *the divisions twice or thrice cleft into narrowly linear lobes;* cauline leaves scattered, the upper less dissected; raceme slender, virgate, the remote *deep blue or violet* (rarely white) *flowers* 2–3.2 cm. long, *the ascending or horizontal slightly arching spur once-and-a-half length of upper sepal;* seeds winged, rugose. (*D. azureum* Michx.) — Dry open woods, sandhills, barrens and fields, Fla. to Tex., n. to Va., Ky., Mo. and Okla., May, June.

Var. **Nortoniànum** (Mackenz. & Bush) Perry (in honor of John Bitting Smith Norton, 1872–). — *Pubescence longer* and more copious, *tending to be floccose;* seeds more rugose. (*D. Nortonianum* Mackenz. & Bush) — Woods, prairies and glades, s. Mo., Ark. and Okla.

Var. **críspum** Perry (crisped). — *Stem densely puberulent or minutely pilose* (especially above) *with non-glandular incurved hairs.* — Dry woods and barrens, O. to Ill., s. to Ark. and Okla.

6. D. viréscens Nutt. (greenish). — Similar to no. 5, 0.5–1.5 *m. high; stem glandular-pubescent above, glandless and crisp-pubescent at base; flowers* less scattered in the raceme, *white or whitish* (rarely pale blue); *spur once-and-a-half to twice length of upper sepal; seeds wingless,* strongly rugose-squamellate. (*D. albescens* Rydb.) — Prairies, barrens and dry open woods, nw. Wisc. to Man., s. to Mo., Okla., and Tex. Late May–July.

Var. **Penárdi** (Huth) Perry (for its discoverer, Eugène Penard, collector in w. N. Am. late in 19th century). — *Upper part of stem glandless and crisp-pubescent, lower half glandular.* (*D. Penardi* Huth) — Neb. and Colo., s. to w. Mo., Okla. and Tex.

15. ACONÌTUM L. Aconite. Monkshood. Wolfsbane

Sepals 5, petal-like, very irregular; the upper one (helmet) hooded or helmet-shaped, larger than the others. Upper petals 2, consisting of small spur-shaped bodies raised on long claws and concealed under the helmet; other petals 6 or fewer, much reduced or wholly wanting. Pistils 3–5. Follicle several-seeded. Seed-coat usually wrinkled or scaly. — Perennials of N. Hemisph. with palmately cleft or dissected leaves, and showy flowers in racemes or panicles. (The ancient Greek and Latin name, of uncertain origin.)

a. Upper sepal with broad helmet-shaped summit; flowers deep blue, violet or purple (rarely white); rachis of inflorescence glabrous or nearly so or spreading-hirsute; root turnip-like. . . *b.*

b. Rachis of inflorescence glabrous or very sparsely incurved-pilose.

Leaves divided to base into pinnately many-cleft segments, the ultimate small lobes linear; inflorescence a long spiciform raceme; hooded sepal 2–2.7 cm. high, with beak 2–5 mm. long. . . . . 1. *A. Napellus.*

Leaves mostly divided only partly to base, the divisions cuneate or oblong-obovate; inflorescence loosely paniculate, with short clusters of flowers; hooded sepal 1.5–1.8 cm. high, with beak 5–9 mm. long. . . . . 2. *A. uncinatum.*

b. Rachis copiously spreading-hirsute; raceme or open panicle lax, few-flowered; hooded sepal 1.4–1.7 cm. long. . . . . 3. *A. noveboracense.*

a. Upper sepal with elongate-conic summit; flowers white to yellowish; rachis closely pilose with incurved hairs; roots slender. . . . . 4. *A. reclinatum.*

1. A. Napéllus L. (old generic name from *Napus, turnip;* from shape of root), Garden-M. or -W. — Tall, erect, simple or rarely branching above, from a turnip-like root; *leaves* very numerous, *divided to base into pinnately many-cleft segments, the ultimate small lobes linear;* raceme many-flowered, strict, spiciform, *the rachis and pedicels nearly glabrous but with remote incurved hairs;* flowers blue-violet or purplish (rarely white); *hooded sepal 2–2.7 cm. high, broadly rounded to the visor-like beak* (2–5 *mm. long*); follicles 3 or 4. — Roadsides, borders of thickets or fields and old house-sites, Nfld. to Ont. and N.Y., spreading from cult. July–Sept. (Introd. from Eu.)

2. A. uncinàtum L. (hooked; from the tip of the helmet), Wild M. — Glabrous except for the summits of the pedicels, slender, erect or reclining or leaning, from tuberous-thickened root; *leaves firm, deeply* 3–5-*lobed,* the lobes rhombic-ovate and coarsely toothed; *flowers blue, in a loose panicle* of small clusters; *hooded sepal erect, broadly rounded-conical,* 1.5–1.8 *cm. high,* prominently beaked in front (*beak* 5–9 *mm. long*); follicles turgid. — Low woods and damp slopes, Pa. to Ind., s. to Ga. and Ala. Aug.–Oct.

Var. **acùtidens** Fern. (sharp-toothed). — Lobes of principal leaves narrower, cuneate, acuminate, acutely serrate or incised. — Md. to Va. and mts. of N.C. and Tenn.

3. A. noveboracénse Gray (of New York). — Erect to reclining, 0.2–1 m. high, from a tuberous-thickened root, leafy, simple to paniculately branched, *the summit and strict loosely flowered racemes spreading-hirsute;* leaves deeply parted, the broadly cuneate divisions 3-cleft and incised, glabrous, or sparsely hairy near margin; flowers blue; *hooded sepal* 1.4–1.7 *cm. high, gibbous-ovoid, with broad rounded summit, the beak* 3–7 *mm. long;* follicles thick-cylindric. — Rich woods, shaded ravines and damp slopes, local, se. N.Y. to Wisc. and Ia. June, July.

4. A. reclinàtum Gray (reclining), TRAILING W. — Stem trailing or leaning, sometimes ascending, 1–3 m. long, from slender roots; *leaves deeply* 3–7-*cleft,* the lower orbicular in outline; the divisions cuneate, incised, often 2–3-lobed; inflorescence a loose panicle, *the rachis and pedicels closely pilose with incurved hairs; flowers white to yellowish; hooded sepal* 1.5–2.3 *cm. high, soon horizontal, the elongate-conical summit with a straight beak in front.* (Incl. *A. vaccarum* Rydb.) — Woods among the mts., W.Va. and w. Va. to Ga. June–Sept.

16. CIMICÍFUGA L. BUGBANE. RATTLETOP

Sepals 4 or 5, falling off soon after the flower expands. Petals, or rather transformed stamens, 1–9, small, on claws, 2-horned at the apex. Stamens as in *Actaea.* Pistils 1–8, forming dry dehiscent follicles in fruit. — Perennials of N. Hemisph. with 2–3-ternately divided leaves, the leaflets cut-serrate, and white flowers in elongated virgate racemes. (Name from *cimex, a bug,* and *fugere, to drive away.*)

Pistils 3–8, stipitate; stigma minute; seeds chaffy-coated. . . . . 1. *C. americana.*
Pistil 1 (rarely 2 or 3), sessile; stigma broad and flat; seeds smooth. . . 2. *C. racemosa.*

1. C. americàna Michx. (American), AMERICAN or MOUNTAIN-B., SUMMER-COHOSH. — Stem 0.6–2 m. high; leaves 2–3-ternate and then pinnately 3–5-foliolate; the ovate and oblong leaflets incised and dentate or the terminal one 3-cleft, acuminate; inflorescence a lax and very elongate terminal raceme with shorter lateral ones; petals 2-horned, with basal concave nectary; *pistils* 3–8, *shorter than the slender stipes; style subulate, tipped by the minute introrse stigma; follicles* flattened, membranaceous, about 1 cm. long, *long-stipitate; seeds* 6–8, *in a single row, laterally flattened, covered with scarious scales.* — Moist woods, chiefly along the mts., Pa. and W.Va. to Ga. and Tenn. Aug., Sept.

2. C. racemòsa (L.) Nutt. (with racemes), BLACK SNAKEROOT, BLACK COHOSH. — Stem 1–2.6 m. high, from a knotted rhizome; leaves 2–3-ternately and then often quinately compound; leaflets subcuneate to subcordate at base, mostly 3–10 cm. long, or leaves irregularly pinnately decompound with leaflets much smaller, narrower and laciniate or incised in the rare forma **dissécta** (Gray) Fern. (dissected); racemes few, virgate, erect, becoming 3–9 dm. long; petals 1- or 2-horned; *ovary* 1 (rarely 2 or 3), *not stipitate, the short thick style tipped by the depressed broad stigma; follicles ovoid; seeds horizontal in a double row, with smooth close coat.* — Rich woods, w. Mass. to s. Ont., s. to Ga., Tenn. and Mo.; spread from cult. in n. and e. N.E. June–Sept.

Var. **cordifòlia** (Pursh) Gray (cordate-leaved). — Leaflets few (about 9), very large (1–2.5 dm. long), at least the terminal one deeply cordate. (*C. cordifolia* Pursh) — Damp woods, mts. of sw. Va., N.C. and Tenn. — Said to flower later than the typical form.

17. ACTAÈA L. BANEBERRY. NECKLACEWEED. COHOSH

Sepals 4 or 5, falling off when the flower expands. Petals 4–10, small, flat, spatulate, on slender claws. Stamens numerous, with slender white filaments. Pistil single; stigma sessile, depressed, 2-lobed. Seeds smooth, flattened, and packed horizontally in 2 rows. — Perennials of N. Hemisph., with ample 2–3-ternately compound leaves, the ovate leaflets sharply cleft and toothed, and a short and thick terminal raceme of whitish flowers, followed by berry-like indehiscent fruits. (Ancient name of the Elder, transferred by Linnaeus to this genus.)

1. A. rùbra (Ait.) Willd. (red), RED B., SNAKEBERRY, POISON DE COULEUVRE (Que.). — Raceme ovoid to subcylindric, in fruit becoming 3–10 cm. long; *pedicels filiform,* more or less minutely pilose with fulvous hairs; *petals* rhombic- or lance-spatulate, *tapering to summit; stigmas during anthesis slightly elevated above summit of ovary,* in fruit contracted and relatively inconspicuous; *fruits cherry-red,* ovoid-ellipsoid, lustrous; seeds 10–16, 3–4 mm. long. — Woods and thickets, s. Lab. to n. B.C., s. to Nfld., N.S., N.E., L.I., n. N.J., N.Y., W.Va., O., Ind., Ia., S.D., Colo., Utah and Oreg. *Fl.* May–July; *fr.* Aug.–Oct. — Fruit mildly poisonous, disagreeable to taste. — Forma **neglécta** (Gillman) Robins. (overlooked), has fruit ivory-white, on filiform pedicels (*A. eburnea* Rydb.; *A. alba* sensu Mackenz. and Rydb., not Mill.), similar range, often more abundant.

2. A. pachýpoda Ell. (with thick pedicels), WHITE B., WHITE C., DOLL'S-EYES. — Raceme ellipsoid to subcylindric, in fruit becoming 3–17 cm. long; *pedicels stout, in maturity nearly as thick as the peduncle*, red; *petals* slender, *mostly truncate*, seeming like modified stamens; *stigma during anthesis broadly sessile; fruits* globose-ovoid, *white, capped by the red or purple broad sessile stigma*, or fruit red in forma **rubrocárpa** (Killip) Fern. (with red carpels); seeds (3–)5–10, 4–5 mm. long. (*A. alba* sensu Bigel. and later auth., not Mill.) — Rich woods and thickets, P.E.I. to s. Man., s. to N.S., N.E., Ga., Ala., La. and Okla. *Fl.* May, June; *fr.* July–Oct.

18. HYDRÁSTIS Ellis ORANGEROOT. YELLOW PUCCOON

Pistils 12 or more in a head, 2-ovuled; stigma flat, 2-lipped. Ovaries becoming a head of crimson 1–2-seeded berries in fruit. — Low e. N.Am. and e. Asiatic perennial herbs sending up in early spring, from a thick and knotted yellow rhizome, a single radical leaf and a simple hairy stem, which is 2-leaved near the summit and terminated by a single greenish-white flower. (Name suggested by the leaf of *Hydrophyllum canadense*, with which this plant was early confused.)

1. H. canadénsis L. (Canadian), GOLDEN-SEAL, "TUMERIC". — Leaves rounded, cordate at the base, 5–7-lobed, doubly serrate, veiny, when full grown in summer 1–2 dm. wide. — Rich woods, Vt. to Minn. and Neb., s. to Ga., Ala., Ark. and (formerly) e. Kans. April, May. — Much sought for medicine and largely exterminated.

19. XANTHORHÌZA Marsh. SHRUB-YELLOWROOT

Sepals 5, regular, spreading, deciduous. Pistils 5–15, with 2 pendulous ovules. Fruit 1-seeded, oblong, the short style becoming dorsal. — A low shrubby plant; the bark and long roots deep yellow and bitter. Flowers polygamous, brown-purple, in compound drooping racemes, appearing along with the 1–2-pinnate leaves from large terminal buds in early spring. (Name compounded of the Greek *xanthos, yellow*, and *rhiza, root.*) ZANTHORHIZA L'Hér., alternative spelling.

1. X. simplicíssima Marsh. (most simple, *i.e.*, unbranched). — Stems slender, 2–6 dm. high; leaflets cleft and toothed. (*Z. apiifolia* L'Hér.) — Damp woods, thickets and stream-banks, N.Y. to W.Va., s. to Fla. and Ala.; spreading from cult. elsewhere. April, May.

FAM. 65. BERBERIDÀCEAE (BARBERRY FAMILY)

Shrubs or herbs with the sepals and petals both imbricated in the bud, usually in two rows of 3 (*rarely* 2 *or* 4) *each; the hypogynous stamens as many as the petals and opposite them; anthers opening by* 2 *valves or lids hinged at the top.* (*Podophyllum* is an exception in having more numerous stamens, the anthers opening along the sides; *Jeffersonia*, in having the sepals in one row.) *Pistil single.* — Filaments short. Style short or none. Fruit a berry or a capsule. Seeds few or several, anatropous, with albumen. Embryo small, except in *Berberis*. Leaves alternate, with dilated bases or stipulate.

*a.*Herbs. . . *b.*
- *b.*Leaves simple or with 2 large leaflets; petals white, thin, showy; fruit a berry or capsule. . . *c.*
  - *c.*Flowering stem usually with 2 leaves; leaves simple; fruit a berry.
    - Flower solitary, usually in the fork between the 2 leaves; stamens 12–18, anthers opening longitudinally; berry ovoid, yellowish, many-seeded, 2.5–5 cm. long. . . . . 1. *Podophyllum.*
    - Flowers in terminal cyme; stamens 6, anthers with terminal valves; berries globose, blue, 2–4-seeded, about 1 cm. in diameter. . . . . 2. *Diphylleia.*
  - *c.*Flower terminating a scape; leaves basal, divided into 2 large half-ovate leaflets; fruit a pyriform capsule, opening horizontally by a lid. . . 3. *Jeffersonia.*
- *b.*Leaves ternately compound; petals smaller than sepals, thick and gland-like, greenish, yellowish or bronzy; ovary soon bursting, the 2 (or 1) ovules maturing as blue spherical drupe-like seeds. . . . . 4. *Caulophyllum.*

*a.*Shrubs with prickles, yellow wood, yellow flowers and 1–few-seeded red berries. 5. *Berberis.*

1. PODOPHÝLLUM L. MAY-APPLE. MANDRAKE. POMME DE MAI (Que.)

Flower-bud with three green bractlets which early fall away. Sepals 6, fugacious. Petals 6 or 9, obovate. Stamens twice as many as the petals in our species; anthers linear-oblong, not opening by uplifted valves. Ovary ovoid; stigma sessile, large, thick and undulate. Fruit a large fleshy berry. Seeds covering the very large lateral placenta, in many rows, each seed inclosed in a pulpy aril. — Perennial herbs of e. N.Am. and e. Asia, with creeping rhizomes and thick fibrous roots. Stems usually 2-leaved, 1-flowered. (Name from the Greek *pous, podos, a foot,* and *phyllon, a leaf,* probably referring to the stout petioles of the radical leaf.) — Genus sometimes placed in the *Ranunculaceae.*

1. P. peltàtum L. (shield-shaped), WILD JALAP. — Flowerless stems terminated by a large round 5–9-lobed leaf peltate in the middle; flowering stem ordinarily with a pair of terminal deeply 3–7-parted leaves (rarely with the leaves alternate or even 3, or in forma **aphýllum** Plitt (without leaves) leafless) and a solitary flower nodding in the fork (peduncle rarely adnate to petiole or with a leafy bract); petals waxy-white, 2.5–4 cm. long; stamens 12–18; fruit ovoid, 2.5–5 cm. long, very rarely 2–8 carpels or separate fruits in forma **polycárpum** Clute (many-carpeled), yellow when ripe (in midsummer), or red in forma **Deàmii** Raymond (for CHARLES CLEMON DEAM, 1865–), sweet and edible. — Rich woods, thickets and pastures, Fla. to Tex., n. to w. Que., s. Ont. and Minn.; spread from cult. elsewhere. Early April–early June.

2. DIPHYLLEÌA Michx. UMBRELLA-LEAF

Sepals 6, fugacious. Petals 6, oval, flat. Stamens 6. Ovary ellipsoid; stigma depressed, subsessile. Ovules 5 or 6 attached to one side of the locule below the middle. Berry globose, few-seeded. Seeds oblong, with no aril. — Glabrous perennials of e. N.Am. and e. Asia, with thick horizontal rhizomes, sending up each year either a very large centrally peltate and cut-lobed rounded umbrella-like radical leaf on a stout petiole; or a flowering stem bearing two similar (but smaller and more 2-cleft) alternate leaves which are peltate near one margin, and terminated by a cyme of white flowers. (Name from the Greek *dis, double,* and *phyllon, leaf.*)

1. D. cymòsa Michx. (with flowers in a cyme). — Stem coarse, 0.6–1 m. high; root-leaves 3–7 dm. broad, 2-cleft, each division 5–7-lobed; petals narrowly obovate, 1–1.5 cm. long; berries on long ascending pedicels, blue, about 1 cm. in diameter. — By mountain-streams, Va. to Ga. May–Aug.

3. JEFFERSÒNIA Bart. TWINLEAF

Sepals 4, fugacious. Petals 8, oblong, flat. Stamens 8; anthers oblong-linear, on slender filaments. Ovary ovoid, soon gibbous, pointed; stigma 2-lobed. Capsule pyriform, opening half-way round horizontally, the upper part making a lid. Seeds many, in several rows on the lateral placenta, with a fleshy lacerate aril on one side. — Perennial glabrous herbs of e. N.Am. and e. Asia, with matted fibrous roots, long-petioled radical leaves parted into 2 half-ovate leaflets, and simple naked 1-flowered scapes. (Named in honor of *Thomas Jefferson,* 1743–1826, third President of the United States.)

1. J. diphýlla (L.) Pers. (two-leaved). — Low; flower white, 2.5 cm. broad, the parts rarely in threes or fives; the two leaflets entire to shallowly sinuate-dentate, or in forma **lobàta** Clute (lobed) with 2–6 marginal lobes becoming 1–4 cm. long. — Woods, N.Y. and s. Ont. to Wisc. and ne. Ia., s. to Ala. April, May.

4. CAULOPHÝLLUM Michx. BLUE COHOSH

Sepals 6, with 3 or 4 small bractlets at the base, ovate-oblong. Petals 6 thick gland-like somewhat reniform or hooded bodies, with short claws, much smaller than the sepals, one at the base of each of them. Stamens 6. Pistil gibbous; style short; stigma minute and unilateral; ovary bursting soon after flowering by the pressure of the 2 erect enlarging seeds, and withering away; the spherical seeds naked on their thick seed-stalks, looking like drupes, the fleshy integument turning blue; albumen horny. — A perennial glabrous herb, with matted knotty rootstocks, sending up in early spring a simple and naked stem terminated by a small raceme or panicle of yellowish-green flowers, and a little below bearing a large triternately compound sessile leaf (whence the name from the Greek *caulos, stem,* and *phyllon, leaf,* the stem seeming to form a stalk for the greatly expanded leaf).

1. C. thalictroìdes (L.) Michx. (resembling Meadow-rue, *Thalictrum*), PAPOOSE-ROOT. — Whole plant glaucous; stems 3–7.5 dm. high; leaflets obovate-cuneate, 2–3-lobed, a smaller biternate leaf often at the base of the panicle; flowers appearing while the leaf is yet small. — Rich woods, N.B. to se. Man., s. to N.S., N.E., mts. of S.C., Tenn. and Mo. Late April–early June. (E. Asia)

5. BÉRBERIS L. Barberry

Sepals 6, roundish, with 2–6 bractlets outside. Petals 6, obovate, concave, with two glandular spots inside above the short claw. Stamens 6. Stigma circular, depressed. Fruit a 1–few-seeded berry. Seeds erect, with a crustaceous integument. — Shrubs, nat. of all continents but Austral., with yellow wood and inner bark, yellow flowers mostly in umbels or racemes, sour berries, and 1–9-foliolate leaves. Stamens sensitive, springing forward when the base is touched. (Name derived from *Berbêrys*, the Arabic name of the fruit.)

Prickles (or many of them) forking; leaves toothed; flowers in racemes; berries juicy.
  Branches brown; adult leaves remotely toothed; berries ovoid, in subumbelliform racemes. . . . . 1. *B. canadensis.*
  Branches gray; adult leaves numerously spinulose-toothed; berries ellipsoid, in elongate racemes. . . . . 2. *B. vulgaris.*
Prickles simple; leaves entire; flowers solitary or in small umbels; berries dryish. 3. *B. Thunbergii.*

1. B. canadénsis Mill. (Canadian; a misnomer), American or Allegheny-B. — Shrub 3–9 dm. high, with arching *brown branches;* leaves repandly toothed, the teeth less bristly-pointed than in no. 2; *racemes few-flowered;* petals notched at the apex; *berries ovoid;* otherwise as in the next. — Dry woodlands and bluffs, mts. of W.Va. and Va. to Ga., w. to se. Mo.; not in Canada. May.

2. B. vulgàris L. (common), Common B., Épine-vinette (Que.). — Upright shrub, 1–3 m. high, with gray bark; leaves scattered on the fresh shoots of the season, mostly reduced to sharp triple or branched spines, from the axils of which the next season proceed rosettes or fascicles of obovate-oblong closely bristle-toothed leaves (the short petiole jointed!) and drooping *many-flowered racemes;* petals entire; *berries ellipsoid,* scarlet. — Thickets, pastures and fencerows, N.S. to Del. and Pa., w. to Minn., Ia., and Mo., abundantly natzd. and thoroughly wild in e. and s. N.E. May, June. (Natzd. from Eu.) — Forma enùclea West (without seed), with seedless berries, local in Mass.

3. B. Thunbérgii DC. (in honor of Carl Peter Thunberg, 1743–1828), Japanese B. — Compact shrub 0.5–1.5 m. high, with brown branches and *simple spines; leaves entire,* spatulate or narrowly obovate, small; *flowers solitary in the axils or few in an umbel; berries ellipsoid to globose, dry.* — Spreading from cult. to pastures and other open habitats, N.S. to Mich., s. to N.C. and Mo. April, May. (Introd. from Asia)

Akèbia (Japanese name) quinàta Dcne. (in fives), an Asiatic twining shrub of the family Lardizabalàceae, with digitately 5-foliolate leaves, flowers with 6 petaloid purple sepals, 3 or more carpels with sessile stigmas, and large purple berries, tends to spread from cult. (Introd. from Asia)

FAM. 66. MENISPERMÀCEAE (Moonseed Family)

Woody climbers with palmate or peltate alternate leaves, no stipules, the sepals and petals similar, in three or more rows, imbricated in the bud; hypogynous, dioecious, 3–6-gynous; fruit a 1-seeded drupe, with a large or long curved embryo in scanty albumen. — Flowers small. Stamens several. Ovaries nearly straight, with the stigma at the apex, but often incurved in fruiting so that the seed and embryo are bent into a crescent or ring. Chiefly a trop. family.

Petals and sepals both present; anthers 4-locular; seed incurved or crescent-shaped.
  Stamens, petals and sepals each 6. . . . . 1. *Cocculus.*
  Stamens 12–24, slender; petals 6–8. . . . . 2. *Menispermum.*
Petals none; anthers 2-locular; seed saucer-shaped. . . . . 3. *Calycocarpum.*

1. CÓCCULUS DC. Coralbeads

Sepals, petals and stamens 6, alternating in threes, the two latter short. Anthers 4-loculed. Pistils 3–6; style pointed. Drupe and seed as in *Menispermum.* — Flowers in axillary racemes or panicles; small genus of e. N.Am., Hawaii, Asia and Afr. (An old name, a diminutive of *coccus, a berry.*)

1. C. carolìnus (L.) DC. (of Carolina), Red-berried Moonseed, Snailseed. — Minutely pubescent; leaves downy beneath, ovate or cordate, entire or sinuately or hastately lobed, variable in shape; flowers greenish, the petals in the staminate ones auriculate-inflexed below around the filaments; drupe red (as large as a small pea). (*Cebatha* Britt.; *Epibaterium* Britt.) — Rich woods and thickets, Fla. to Tex., n. to se. Va., Ky., s. Ill., Mo. and se. Kans. July, Aug.

2. MENISPÉRMUM L. Moonseed

Sepals 4–8. Petals 6–8, short. Stamens 12–24 in the staminate flowers, as long as the sepals; anthers 4-locular. Pistils 2–4, raised on a short common receptacle; stigma broad and flat. Drupe globular, the mark of the stigmas near the base, the ovary in its growth after flowering being strongly incurved so that the (wrinkled and grooved) laterally flattened stone takes the form of a large crescent or ring, the slender embryo therefore horseshoe-shaped; cotyledons filiform. — Two species: ours and one in e. Asia. Flowers white or whitish, in small and loose axillary panicles. (Name from the Greek *men, moon,* and *sperma, seed.*)

1. M. canadénse L. (Canadian), Yellow Parilla, Raisin de couleuvre (Que.). — Leaves peltate near the edge, 3–7-angled or -lobed. — Rich thickets and stream-banks, w. Que. and w. N.E. to se. Man., s. to Ga., Ala., Ark. and Okla. June, July. — Drupes black, with a bloom, ripe in Sept., looking like frost-grapes.

3. CALYCOCÁRPUM Nutt. Cupseed

Sepals 6, petaloid. Petals none. Stamens 12 in the staminate flowers, short; anthers 2-locular. Pistils 3, fusiform, tipped by a radiate many-cleft stigma. Drupe globular; thin crustaceous putamen hollowed out like a cup on one side. Embryo foliaceous, cordate. — Flowers greenish-white, in long racemose panicles. A single species. (Name from the Greek *calyx, a cup,* and *carpos, fruit.*)

1. C. Lȳoni (Pursh) Gray (for its discoverer, John Lyon, English botanical explorer of N.Am. prior to 1818). — Leaves large, thin, deeply 3–5-lobed, cordate at the base; the lobes acuminate; drupe 2.5 cm. long, black when ripe. — Rich or alluvial soils, Fla. to La., n. to Ky., s. Ill., Mo. and e. Kans. May, June. — Climbing to tops of trees.

FAM. 67. MAGNOLIÀCEAE (Magnolia Family)

Trees or shrubs with the leaf-buds covered by membranous stipules; flowers polypetalous, ***hypogynous, with many stamens and carpels; the calyx and corolla colored alike, in three or*** *more rows of three, imbricated (rarely convolute) in the bud.* — Sepals and petals deciduous. Anthers adnate. Pistils many, mostly crowded together and covering the prolonged receptacle, cohering with each other and in fruit forming a sort of fleshy or dry cone. Seeds 1 or 2 in each carpel, anatropous; albumen fleshy; embryo minute. — Leaves alternate, not toothed, marked with minute transparent dots, feather-veined. Flowers single, large. Bark aromatic and bitter.

Leaves not truncate; petals white to yellowish or greenish; fruits coherent in a cone, coriaceous-baccate, dorsally dehiscent as follicles. . . . . 1. *Magnolia.*
Leaves emarginate-truncate; petals greenish-yellow, marked with orange; fruits samara-like, indehiscent, falling singly. . . . . 2. *Liriodendron.*

1. MAGNÒLIA L. Magnolia

Sepals 3. Petals 6–9. Stamens imbricated, with very short filaments, and long anthers opening inward. Pistils coherent, forming a fleshy and rather woody cone-like red fruit; each carpel at maturity opening on the back from which the 1 or 2 berry-like seeds hang by an extensile thread. — Trees and shrubs of temp. e. and trop. N. Am. and e. and Himalayan Asia. (Named for *Pierre Magnol*, 1638–1715, professor of botany at Montpellier.)

a. Leaves thin-coriaceous, more or less evergreen, obtuse, glaucous beneath, 0.8–1.5 dm. long; flower subglobose, 3–5 cm. long; fruiting cone 3–5 cm. long. . . . . 1. *M. virginiana.*
a. Leaves membranaceous, deciduous, 1.3–9 dm. long; flower slender-campanulate to more open, larger; fruiting cone 5–12 cm. long. . . . *b.*
b. Leaves of flowering branches somewhat scattered, rounded or tapering at base, acuminate, soft-pubescent beneath; petals dull green and glaucous or yellow-tinged, 5–7 cm. long; styles filiform, promptly deciduous; follicles rounded and blunt. . . . . 2. *M. acuminata.*
b. Leaves of flowering branches approximate in umbrella-like cluster; petals white, 0.6–2.3 dm. long; styles thick, persistent; follicles beaked.
Stipules, young flower-buds and follicles densely pubescent; leaves 3–9 dm. long, with broad basal auricles, canescent-tomentose beneath. 3. *M. macrophylla.*

Stipules, young flower-buds and follicles glabrous; leaves 2–7 dm. long, glabrous or glabrate.
Leaves acute at base. . . . . 4. *M. tripetala.*
Leaves auricled at base. 5. *M. Fraseri.*

1. M. virginiàna L. (Virginian), SMALL or LAUREL-M., BAY, SWEET or SWAMP-BAY, BEAVER-TREE, LAUREL. — Large shrub or small tree up to 10 m. high; *leaf-buds silky;* branchlets glabrous or glabrate; *leaves* oval to broadly lanceolate, 0.8–1.5 *dm. long, obtuse, subcoriaceous, evergreen* (or deciduous northw.), *glaucous but glabrous beneath; flower subglobose,* white, changing to pale brown, very fragrant; *petals roundish-obovate,* 3–5 *cm. long; fruiting cone* ellipsoid, 3–5 *cm. long,* the follicles glabrous; seeds 8–10 mm. long. — Swamps and low woods, Fla. to Miss., n. to Pa., N.J. and locally to e. Mass. and Tenn. May–July (–autumn).

Var. **austràlis** Sarg. (southern). — Taller, up to 20 m. or more high; *young branchlets, petioles, veins and often lower surfaces of leaves white-silky.* — Fla. to e. Tex., n. to e. S.C. and locally to se. Va. and Ark.

2. M. acuminàta L. (acuminate), CUCUMBER-TREE. — *Leaves thin, oblong, pointed, green* and a little pubescent beneath, 13–25 cm. long; *flower slenderly campanulate, glaucous-green* tinged with yellow, 8 cm. long; *petals* oblanceolate to cuneate, 5–7 *cm. long;* fruiting cone 5–8 cm. long, *the filiform styles promptly deciduous; follicles rounded and obtuse.* (*Tulipastrum* Small) — Rich woods, w. N.Y. and s. Ont. to s. Ill., s., chiefly in the uplands, to Ga., Ala. and Ark. May, June. — Tree up to 30 m. high; fruit when young resembling a small cucumber.

3. M. macrophýlla Michx. (large-leaved), GREAT-LEAVED M. — *Leaves obovate-oblong, cordate* at the narrowed base, pubescent and *white beneath; flower open-campanulate, white, with a purple spot* at base; petals ovate, 15 cm. long; cone of fruit ovoid or subglobose; follicles pubescent. — Rich woods, w. Fla. to La., n. to W.Va., Ky. and Ark. May, June. — Tree 6–12 m. high; leaves 3–9 dm. long, somewhat clustered on the flowering branches.

4. M. tripétala L. (with three petals), UMBRELLA-TREE, UMBRELLA-M. — *Leaf-buds glabrous; leaves crowded in an umbrella-like circle at summits of flowering branches,* obovate-lanceolate, *pointed at both ends,* soon glabrous, 3–6 dm. long; fruiting cone elongate, 0.8–1.2 dm. long; *follicles glabrous.* — Rich woods, Ga. to Ark., n. to s. Pa., W.Va., O., Ky. and Mo.; locally spread from cult. northw. to centr. Mass. May. — Small tree.

5. M. Fràseri Walt. (in honor of JOHN FRASER, 1750–1811, Walter's publisher), EAR-LEAVED UMBRELLA-TREE. — *Leaves oblong-obovate or spatulate, auriculate at the base,* glabrous, 2–5 dm. long; petals obovate-spatulate, with narrow claw, 1 dm. long. — Swamps and along streams, upland of Va., W.Va. and e. Ky. to Ga. and Ala. May. — Slender tree up to 15 m. high.

2. LIRIODÉNDRON L. TULIP-TREE

Sepals 3, reflexed. Petals 6, in two rows, making a campanulate corolla. Anthers linear, opening outward. Pistils flat and squamelliform, narrow, imbricated and cohering in an elongated cone, dry, falling away whole, like a samara or key, indehiscent, 1–2-seeded in the small cavity at the base. — Two (perhaps one) species, of e. N. Am. and China. (Name from the Greek *lirion, lily* or *tulip,* and *dendron, tree.*)

1. L. Tulipífera L. (the old generic name; tulip-bearing), TULIP-POPLAR. — Leaves very smooth, with 2 lateral lobes near the base, and 2 at the apex, which appears as if cut off abruptly by a broad shallow notch; petals 5 cm. long, greenish yellow marked with orange; cone of fruit 7.5 cm. long. — Rich soil, Worcester Co., Mass., to s. Ont., Wisc., and southw. May, June. — A most beautiful tree, sometimes 40 m. high and with trunk 2–3 m. in diameter in the Western and Southern States, the timber commonly called POPLAR or WHITEWOOD.

FAM. 68. CALYCANTHÀCEAE (CALYCANTHUS FAMILY)

Shrubs with opposite entire leaves, no stipules, the sepals and petals similar and indefinite, the anthers adnate and extrorse, and the cotyledons convolute; the fruit suggesting a rose-hip. Chiefly represented by the genus:

1. CALYCÁNTHUS L. CAROLINA ALLSPICE

Calyx of many sepals united below into a fleshy inversely conical cup (with some leaf-like bractlets growing from it); the lobes lanceolate, mostly colored like the petals, which are similar, in many rows, thickish and inserted on the top of the closed calyx-tube. Stamens numerous, inserted just within the petals, short; some of the inner ones sterile (destitute of anthers). Pistils several or many, inclosed in the calyx-tube, inserted on its base and inner face. — Aromatic shrubs of e. N. Am. and e. Asia, with brownish-maroon flowers terminating leafy shoots. (Name composed of the Greek *calyx, a cup* or *calyx*, and *anthos, flower.*) Butneria Duham.

1. C. fértilis Walt. (fertile). — *Leaves oblong* or ovate, thin, either blunt or taper-pointed, scabrous above, *bright green and glabrous*, or in forma **nãnus** (Loisel.) Schelle (dwarf) glaucous; flowers less fragrant than in no. 2. (*Butneria* Kearney) — Rich woods among the mts., Ga. and Ala., n. to s. Pa. and s. O. Late Apr.–June.

2. C. flóridus L. (flowering). — *Leaves oval, soft-downy underneath;* flowers when crushed yielding strong fragrance suggesting strawberries. (*Butneria* Kearney) — Rich woods, Fla. to Miss., n. to Va. and W.Va. Apr.–Aug. — Cult. northw.

FAM. 69. ANNONÀCEAE (Custard-apple Family)

Trees or shrubs with naked buds and no stipules, a calyx of 3 sepals, and a corolla of 6 thickish petals in two rows, valvate in the bud, hypogynous, polyandrous. — Anthers adnate, extrorse; filaments very short. Pistils several or many, separate or cohering in a mass, fleshy or pulpy in fruit. Seeds anatropous, large, with a minute embryo at the base of the ruminated albumen. — Leaves alternate, entire, feather-veined. Flowers axillary, solitary. Trop. and in temp. e. N. Am.

1. ASÎMINA Adans. North American Pawpaw

Petals 6, enlarging after the bud opens; the outer set larger than the inner. Stamens numerous in a globular mass. Pistils few, ripening 1–4 large thick-cylindric pulpy fruits; seeds several, horizontal, flat, inclosed in a fleshy aril. — Shrubs or small trees with unpleasant odor when bruised; the lurid flowers solitary from the axils of leaves of preceding year. (Name from *Asiminier* of the French colonists, from the Indian name *assimin.*)

1. A. tríloba (L.) Dunal (three-lobed), Pawpaw. — *Trees* 3–12 m. high; young shoots and expanding leaves clothed with rusty down, soon glabrous; *leaves* obovate-oblong to -lanceolate, pointed, *when mature* 1.5–3 *dm. long; flowers* appearing with the leaves, 3–4 *cm. broad, on villous pedicels becoming* 1–3 *cm. long; petals* dull purple, veiny, *round-ovate, very unequal, the outer ones three or four times as long as the calyx; style definite; fruits* (3–)7–13 *cm. long,* green or at length dark brown, the pulp sweet and edible in autumn; *seeds somewhat flattened, mostly* 2–3.3 *cm. long.* — Rich woods and alluvium, Fla. to Tex., n. to N.J., w. N.Y., s. Ont., Mich., Ill., se. Ia. and se. Neb. Apr., May.

2. A. parviflòra (Michx.) Dunal (small-flowered), Dwarf Pawpaw. — *Shrub* 0.5–1.3 m. high; *leaves* firmer, 0.5–1.7 *dm. long; flowers about* 2 *cm. broad, on tomentulose pedicels becoming* 1–8 *mm. long; petals oblong or ovate, subequal; stigma sessile; fruits* 2–3 (–6) *cm. long; seeds plump,* 1–1.5 *cm. long.* — Pinelands or oak-woods, n. Fla. to Miss., n. to se. Va. Apr.

FAM. 70. LAURÀCEAE (Laurel Family)

Aromatic trees or shrubs with alternate simple leaves mostly marked with minute pellucid dots, and flowers with a regular calyx of 4 *or* 6 *colored sepals imbricated in* 2 *rows in the bud, these free from the* 1-*locular and* 1-*ovuled ovary and mostly fewer than the stamens; anthers opening by* 2 *or* 4 *uplifted valves.* — Flowers clustered. Style single. Fruit a 1-seeded berry or drupe. Seed anatropous, suspended, with no albumen, filled by the large almond-like embryo.

Leaves evergreen; flowers perfect, in panicles; stamens 12, 3 of them sterile; fruiting pedicels slender. . . . . 1. *Persea.*
Leaves deciduous; flowers dioecious or polygamo-dioecious; stamens of staminate flowers 9.
  Leaves often lobed; flowers in peduncled corymbiform racemes; anthers 4-locular, 4-valved; pistillate flowers with 6 rudiments of stamens; fruiting pedicels strongly clavate. . . . . 2. *Sassafras.*
  Leaves mostly unlobed; flowers sessile or short-stalked in small clusters; pistillate flowers with 12–18 rudiments.
    Mature leaves coriaceous, narrowly oblong; flowers 2–4 in loose umbels; anthers 4-locular, 4-valved. . . . . 3. *Litsea.*
    Mature leaves membranaceous, oblong-obovate; flowers numerous in dense compound umbels; anthers 2-locular, 2-valved. . . . . 4. *Lindera.*

1. PÉRSEA Mill. Red Bay

Flowers perfect, with a 6-parted calyx persistent at the base of the berry-like fruit. Stamens 12, in four rows, the 3 of the innermost row sterile and gland-like, the rest bearing 4-locular anthers (*i.e.*, with each proper locule divided transversely into two) opening by as many uplifted valves; the anthers of 3 stamens turned outward, the others introrse. — Trees of trop. and warm-temp. Am., one on the Canary Ids., with persistent entire leaves and small panicled flowers. (An ancient name of some oriental tree.)

1. P. Borbònia (L.) Spreng. (old name for *Persea*). — Tree of medium size; branchlets puberulent to glabrate or glabrous; leaves lanceolate to lance-oblong, soon shining above, pale and glabrous or soon glabrate beneath; common peduncle 1–7 cm. long; drupes dark blue or blackish, the fruiting peduncle red. (*Tamala* Raf.) — Woods, wooded swamps and shores, Fla. to Tex., n. to Del. May–July. — Usually less common than forma **pubéscens** (Pursh) Fern. (pubescent), *P. pubescens* (Pursh) Sarg. and *P. palustris* (Raf.) Sarg., with densely tomentose branchlets and lower leaf-surfaces.

2. SÁSSAFRAS Nees Sassafras

Flowers dioecious, with a 6-parted spreading calyx; the staminate with 9 stamens inserted on the base of the calyx in 3 rows, the 3 inner with a pair of stalked glands at the base of each; anthers 4-locular, 4-valved; pistillate flowers with 6 short rudiments of stamens and an ovoid ovary. Drupe ovoid (blue), supported on a clavate and rather fleshy reddish pedicel. — Trees, ours and two in e. Asia, with spicy-aromatic bark, and very mucilaginous twigs and foliage; leaves deciduous, often lobed. Flowers greenish-yellow, naked, in clustered and peduncled corymbed racemes, appearing with the leaves, involucrate with scaly bracts. (The aboriginal name, applied by the early French settlers in Florida.)

1. S. álbidum (Nutt.) Nees (whitish), White S. — Tree up to 40 m. high, with pale green twigs, the young ones glabrous or essentially so, often glaucous; leaves ovate, entire, or some of them 3-lobed, glabrous or essentially so from the first. — Woods and thickets, Va. to Ark., n. to sw. Me., s. N.H., s. Vt., N.Y., O., Mich., and Ill. April–June.

Var. **mólle** (Raf.) Fern. (soft), Red S. — Twigs closely pubescent or puberulent; leaves densely pubescent when young, permanently so beneath. (*S. variifolium* (Salisb.) Ktze.; *S. officinale* Nees & Eberm.) — Fla. to Tex., n. to sw. Me., s. N.H., Mass., N.Y., O., Ind., Ill., se. Ia., Mo. and se. Kans. — Sprouting freely, and weed-like southw.

3. LÍTSEA Lam.

Flowers dioecious, with a 6-parted deciduous calyx; the staminate with 9 stamens in 3 rows; their anthers all introrse, 4-locular, 4-valved; pistillate flowers with 12 or more rudiments of stamens and a globular ovary. Drupe globular. — Shrubs or trees, chiefly of trop. Old World, with entire leaves, and small flowers in axillary clustered umbels. (Name of Chinese origin.)

1. L. aestivàlis (L.) Fern. (of summer), Pond-spice. — Shrub 2 or 3 m. high, with forking and divaricate zigzag branchlets; flowers yellow, about 6 mm. broad, appearing before the leaves from exposed overwintering buds; leaves narrowly oblong, coriaceous, entire, subapproximate, 2–6 cm. long, 0.5–1.5 cm. broad; drupe red. (*L. geniculata* (Walt.) B. & H.; *Glabraria geniculata* Britt.) — Ponds and wet swamps, very rare, Fla. to La., n. to se. Va. (at least formerly) and Tenn. March, Apr.

4. LÍNDERA Thunb. Wild Allspice. Feverbush

Flowers polygamodioecious, with a 6-parted open calyx; the staminate with 9 stamens in 3 rows, the inner filaments 1–2-lobed and gland-bearing at base; anthers 2-locular and 2-valved; pistillate flowers with 15–18 rudiments of stamens in 2 forms, and a globular ovary. Drupe obovoid, red, the stalk not thickened. — Deciduous-leaved shrubs of e. N. Am. and e. and s. Asia, with honey-yellow flowers in almost sessile lateral umbel-like clusters, appearing before the leaves (in our species); the clusters composed of smaller groups or umbels, each of 4–6 flowers and surrounded by an involucre of 4 deciduous scales. Leaf-buds scaly. (Named for *Johann Linder*, 1676–1723, early Swedish botanist.) Benzoin Fabricius

1. L. Benzòin (L.) Blume (old name for some member of the *Lauraceae*), Spicebush, Benjamin-bush. — *Nearly smooth* (2–5 m. high); leaves oblong-obovate, tapering at base, pale and glabrous beneath; drupes red, or yellow in fòrma **xanthocárpa** (G. S. Torr.) Rehd.

(yellow-fruited). (*Benzoin aestivale* sensu Nees, not *Laurus aestivalis* L.) — Damp woods and brooksides, sw. Me. to s. Ont., s. Mich. and Ill., s. to N.C., Ky., Mo. and se. Kans. March–May.

Var. **pubéscens** (Palmer & Steyerm.) Rehd. (pubescent). — Leaves pubescent at least on the nerves beneath. — Fla. to Tex., n. to N.J., e. Pa., Ky., s. Mich., Ill. and Ia.

2. L. melissifolia (Walt.) Blume (with leaves of *Melissa*), Jove's-fruit. — *Shrub* 0.3–1.8 *m. high, with pubescent branches and buds; leaves narrowly elliptic-oval or oblong, rounded to cordate at base,* up to 1.5 dm. long, *minutely pilose on both faces; drupe obovoid,* larger than in no. 1. (*Benzoin* Nees) — Swamps and pond-margins, very local, Fla. to La., n. to N.C. and s. Mo.

FAM. 71. PAPAVERÀCEAE (Poppy Family)

Herbs (rarely shrubby) with milky, colored or watery juice, flowers regular, or seemingly irregular but laterally compressed and with dimerous regularity, with 2 or 4 (rarely 3) sepals, 6–very many stamens, embryo minute, at base of fleshy albumen. — Petals 4–12, distinct or slightly united. Stamens hypogynous. Fruit a capsule with mostly nerve-like parietal placentae, or 1-seeded and indehiscent. Leaves alternate, without stipules. Two usually very distinct subfamilies:

Subfam. PAPAVEROÌDEAE (Poppy Subfamily)

Juice milky or colored; flowers regular; sepals 2 or 3, large, fugacious; petals 4–12 (many more in cult. "double" flowers), spreading, imbricated and often crumpled in the bud, early deciduous; stamens many, distinct; fruit a 1-locular capsule (in *Papaver* imperfectly many-locular, in *Glaucium* 2-locular). Peduncles mostly 1-flowered; juice narcotic and acrid.

a. Acaulescent, from a stout rhizome with orange-red juice; leaves basal, rounded, palmately lobed or undulate; scape naked; petals white, 6–12 (usually 8), not crumpled in bud; capsule 1-locular, 2-valved. . . . . 1. *Sanguinaria.*

a. Caulescent; juice yellow or white; leaves pinnately veined or pinnately divided or lobed; petals 4, crumpled in bud. . . *b.*

b. Capsule 2–4-valved, the ripe valves separating to the base; petals yellow. . . *c.*

c. Ovary and ovoid capsule hirsute, with long style; stigmas and placentae 3–4. . . . . 2. *Stylophorum.*

c. Ovary and linear-cylindric to spatulate-oblanceolate capsule glabrous or merely scabrous, with 2 nearly or quite sessile stigmas and 2 placentae. . . *d.*

d. Petals 4, deep yellow; capsule linear or linear-cylindric; flowers solitary or umbelled, axillary or terminal.

Flowers slender-pedicelled on long axillary peduncles; capsule smooth, 1.5–5 cm. long; seeds crested. . . . . 3. *Chelidonium.*

Flowers solitary on short and thick axillary and terminal peduncles; capsule scabrous, 1.5–2.5 dm. long; seeds crestless. . . . 4. *Glaucium.*

d. Petals 0; the two sepals cream-colored; capsule spatulate-oblanceolate, 1.5–2.5 cm. long; flowers on long peduncles. . . . . . . 5. *Macleaya.*

b. Capsule 4–20-valved, dehiscent only at summit or to the middle.

Stigmas united into a 4–20-rayed sessile crown; capsule opening by pores under the stigma, not prickly; leaves not prickly. . . . . . . 6. *Papaver.*

Stigmas 3–6; capsule opening by as many valves near the summit; leaves prickly. . . . . 7. *Argemone.*

1. SANGUINÀRIA L. Bloodroot. Sanguinaire or Sang-dragon (Que.)

Sepals 2. Petals 8–12, spatulate-oblong. Stamens about 24. Style short; stigma 2-grooved. Capsule ellipsoid or fusiform, turgid, 1-lŏcular, 2-valved. Seeds with a large crest. — Low perennial; its thick prostrate rhizomes (with abundant red-orange acrid juice) sending up in earliest spring a palmate-lobed or undulate leaf and 1-flowered scape. Flower mostly white, handsome, the bud erect, the petals not crumpled. (Name from the color of the juice, from *sanguinarius, bleeding.*)

1. S. canadénsis L. (Canadian), Red Puccoon. — Rhizome simple or forking, elongate; *leaves* at full development of flowers usually nearly reaching them, *pale green* but not strongly glaucous beneath, *membranaceous,* becoming *in maturity* 1–2.8 *dm. broad, the margins of the broad basal lobes and summits of the upper narrower ones coarsely dentate or crenate;* mature petiole 2–4 dm. high, distinctly overtopping the mature capsule; petals white, or pink in forma **Colbyòrum** Benke (named in 1933 for its discoverer, Earl H. Colby), very numerous, the flower "double" in forma **múltiplex** (E. H. Wilson) Weath. (manifold). — Rich woods, e. Que. to Man., s. to N.S., N.E., Pa., Ky., Ill. and e. Kans. Late March–May. Passing insensibly to

Var. **rotundifòlia** (Greene) Fedde (round-leaved). — Rhizome often shorter, less branched or simple; *flowers at full development more generally overtopping the leaves; leaf-blade subcoriaceous,* usually *strongly glaucous beneath, in maturity* 0.7–2 *dm. broad, lobed or unlobed, the margins dentate, barely undulate or entire;* mature petiole 0.7–2 (–3 dm. high), overtopping or even overtopped by the capsule. — Commoner southw., n. Fla. to e. Tex., n. to L.I., N.J., Pa., O., Ind. and Wisc. Late March–May.

2. STYLÓPHORUM Nutt. CELANDINE-POPPY

Sepals 2, hairy, Petals 4. Style distinct, columnar; stigma 2–4-lobed. Capsules bristly, 2–4-valved to the base. Seeds conspicuously crested. — A perennial low herb, with stems naked below and with 2 opposite leaves (or sometimes 1–3-leaved) and umbellately 1–few-flowered at the summit; the flower-buds and the capsules nodding. Leaves pinnately parted or divided. Juice yellow. (From the Greek *stylos, style,* and *phoros, bearing;* the long style being one of the distinctive characters.)

1. S. diphýllum (Michx.) Nutt. (two-leaved), WOOD-POPPY. — Leaves pale beneath, smoothish, deeply pinnatifid into 5 or 7 oblong sinuate-lobed divisions, and the basal leaves often with a pair of small distinct leaflets; peduncles equaling the petioles; flower deep yellow (5 cm. broad); stigmas 3 or 4; capsule ovoid. — Rich woods and bluffs, w. Pa. to Wisc., s. to sw. Va., Tenn. and Mo. Mar.–May.

3. CHELIDÒNIUM L. CELANDINE. SWALLOWWORT

Sepals 2. Petals 4. Stamens 16–24. Style almost none; stigma 2-lobed. Capsule linear-cylindric, smooth, 2-valved, the valves opening from the bottom upward. Seeds crested. — Biennial herbs nat. of Eurasia, with brittle stems, saffron-colored acrid juice, pinnately divided or 2-pinnatifid and toothed or cut leaves, and small yellow flowers in a pedunculate umbel; buds nodding. (Ancient Greek name, from *chelidon,* the *swallow,* because, according to Aristotle and other early scholars, with its saffron juice the mother-swallows bathed the eyes and strengthened the sight of their young.)

1. C. MÀJUS L. (larger), HERBE AUX VERRUES (Que.). — Leaves glaucous and glabrous beneath; pedicels and sepals glabrous, or in forma HYPÓTRICHUM (Aznavour) Hayek (with hairs beneath) the lower leaf-surfaces minutely puberulent on the nerves and the pedicels and sepals villous. — Rich damp soil about towns, Que. and Ont., s. to N.S., N.E., n. Ga., Tenn. and Mo. Mar.–Aug. (Natzd. from Eu.)

4. GLAÙCIUM Mill. HORN-POPPY. SEA-POPPY

Sepals 2. Petals 4. Style none; stigma 2-lobed or 2-horned. Capsule very long and linear, completely 2-locular by a spongy false partition; seeds crestless. — Annuals or biennials with saffron-colored juice, clasping leaves, and solitary yellow flowers. Nat. of Old World. (The Greek name, *Glaucion,* from the glaucous foliage.)

1. G. FLÀVUM Crantz (yellow). — Lower leaves pinnatifid; upper ones sinuate-lobed and toothed, cordate-clasping; capsule rough, 1.5–2.5 dm. long. (*G. Glaucium* (L.) Karst.) — Waste ground and sandy shores, local, se. Mass. and e. R.I. to Mich., s. to Va. and W.Va. June–Aug. (Natzd. from Eu.)

5. MACLEÀYA R.Br. PLUME-POPPY. TREE-CELANDINE

Sepals 2. Petals 0. Style none; stigma 2-lobed. Capsule oblanceolate, 2-locular, splitting to base, 4- or 6-seeded. — Tall perennials, nat. of e. Asia, with large roundish and deeply lobed petioled leaves and panicles of flowers with creamy sepals and very numerous (about 30) stamens with long anthers. (Named in 1826 for *Alexander Macleay,* early Secretary of New South Wales.)

1. M. CORDÀTA (Willd.) R.Br. (heart-shaped). — Stem erect, up to 2.5 m. high; leaves alternate, distant, the larger ones 1–3 dm. long, whitened beneath; panicle elongate; sepals creamy. (*Bocconia* Willd.) — Spread from cult. to waste places, fields and roadsides, Mass. to Ill., s. to Del. and Md. July, Aug. (Introd. from e. Asia)

6. PAPÀVER L. Poppy. Pavot (Que.)

Sepals mostly 2. Petals mostly 4. Stigmas united in a flat 4–20-rayed crown, resting on the summit of the ovary and capsule; the latter short and turgid, with 4–20 many-seeded placentae projecting like imperfect partitions, opening by as many pores or chinks under the edge of the stigma. — Herbs with a white juice; the flower-buds nodding. Genus nat. of all continents but S. Am. (The ancient name.)

1. P. somníferum L. (sleep-bringing), Common P. — *Smooth*, glaucous; leaves clasping, wavy, incised and toothed; *capsule globose;* corolla mostly white or purple. — Near dwellings, by roadsides and in waste, Nfld. to N.D. and southw., an esc. from cult. June–Sept. (Introd. from Eu.)

2. P. Rhoèas L. (an old Greek name), Corn-P., Coquelicot (Que.). — *Bristly;* leaves pinnatifid; *capsules obovoid*, turbinate; corolla bright scarlet, often dark at center. — Rubbish-heaps and, rarely, fields, N.S. to N.D., s. to N.E., Va., Mo. and Kans. Late May–Oct. (Adv. from Eu.)

3. P. dùbium L. (doubtful). — Pinnatifid leaves and the long stalks *bristly; capsules clavate, smooth;* corolla light scarlet. — Cult. fields and waste grounds, Mass. to N.C., Tenn. and Mo. Late May–July. (Natzd. from Eu.)

California Poppy, Eschschòltzia (in honor of *Johann Friedrich Eschscholtz,* 1793–1831), califórnica Cham. (Californian), with leaves ternately dissected into narrow segments, sepals coherent into a slender-tipped hood and pushed off by the expanding petals, and capsule long and slender, is casually persistent after cult. and about dumps. (Introd. from Calif.)

7. ARGEMÒNE L. Prickly Poppy. Devil's-fig

Sepals 2 or 3, often prickly. Petals 4–6. Style almost none; stigmas 3–6, radiate. Capsule ellipsoid, prickly, opening by 3–6 valves at the top. Seeds crested. — Am. annuals or biennials, with prickly bristles and yellow juice. Leaves sessile, sinuate-lobed, and with prickly teeth, often blotched with white. Flower-buds erect, short-peduncled. (Name of an herb mentioned by Pliny.)

1. A. intermèdia Sweet (intermediate). — Stout, very glaucous; *peduncles leafy; corolla white*, 8–10 cm. in diameter. — Sandy and gravelly prairies, shores and waste places, Ill. to Ida., s. to Mo., Okla., Tex. and N.M. Late May–Sept.

A. mexicàna L. (Mexican), with flowers sessile or nearly so and petals orange, yellow or creamy, and **A. álba** Lestib. f. (white), with naked-peduncled white flowers, casually spread from cult. (Introd. from the Southwest)

Subfam. FUMARIOÌDEAE

Delicate herbs with watery juice, compound dissected leaves; flowers seemingly irregular, laterally compressed and with dimerous regularity; sepals small, scale-like; flattened corolla closed, the 4 petals in 2 pairs, the outer with usually spreading tips and one or both spurred or saccate at base, the inner pair narrower and with callous-crested tips united over the stigma; stamens in 2 sets of 3 each, opposite the larger petals, their filaments often united, the middle anther of each set 2-locular, the lateral ones 1-locular. Fruit 1-locular, either 1-seeded and indehiscent or several-seeded with 2 parietal placentae and deciduous valves. Flowers mostly racemose; juice bitter but innocuous. (Fam. *Fumariaceae*)

Corolla bigibbous or 2-spurred, the 2 outer petals alike; capsule several-seeded.
  Petals united into a spongy persistent subcordate corolla; seeds crested; plant high-climbing by leaf-stalks. . . . . 8. *Adlumia.*
  Petals but slightly cohering into a cordiform or 2-spurred usually deciduous corolla; seeds crestless; plant not climbing. . . . . 9. *Dicentra.*
Corolla with only 1 petal spurred at base, deciduous.
  Fruit a few–several-seeded dehiscent slender capsule; seed with crest or aril; ours biennials. . . . . 10. *Corydalis.*
  Fruit 1-seeded, globose, indehiscent; seed without crest; annual. . . . . . 11. *Fumaria.*

8. ADLÙMIA Raf. Climbing Fumitory. Mountain-fringe

Petals all permanently united into a cordate-ovate corolla, becoming spongy-cellular and persistent, inclosing the small few-seeded capsule. Seeds not crested. Stigma 2-crested. Filaments monadelphous below in a tube which is adherent to the corolla, diadelphous at the summit. — A climbing biennial, with thrice-pinnate leaves, cut-lobed delicate leaflets, and ample panicles of drooping white or purplish flowers. (Dedicated to Major *John Adlum*, 1759–1836, amateur botanist.)

1. A. fungòsa (Ait.) Greene (spongy). — Wet or recently burned woods or rocky wooded slopes, local, e. Que. to Ont. and Minn., s. to N.E. and mts. of N.C. and Tenn. June–Oct. — Handsome delicate vine climbing by the slender young leaf-stalks over high bushes; often cult. and frequently escaping.

9. DICÉNTRA Bernh. Bleeding-heart. Coeurs-saignants des bois (Que.)

Petals slightly cohering into a cordiform or 2-spurred corolla, either deciduous or withering-persistent. Stigma 2-crested and sometimes 2-horned. Filaments slightly united into two sets. Capsule 10–20-seeded. Seeds crested. — Low often stemless N.Am. and Asiatic perennials (as to our wild species) with ternately compound and dissected leaves, and racemose nodding flowers. Pedicels 2-bracted. (Name from the Greek *dis, twice*, and *centron, a spur;* — accidentally printed Diclytra in the first instance, which by an erroneous conjecture was changed afterwards into Dielytra.) Bikukulla Adans. Bicuculla Millsp.

Inflorescence a simple raceme; scapes 1–2.5 dm. high below the lowest flower; base of plant with grain-like or tuber-like enlargements or bulbiform clusters.
  Corolla divergently 2-spurred at base, the spurs prolonged and arching; base a crowded mass of grain-like scales or thickened leaf-bases. . . . . 1. *D. Cucullaria.*
  Corolla cordate, the saccate bases short and rounded; fleshy yellow grains (thickened leaf-bases) borne along the filiform rhizome. . . . . 2. *D. canadensis.*
Inflorescence a thyrse, with the several–many cordate-based flowers in lateral clusters; scape 2.5–4 dm. high; base a stout crown. . . . 3. *D. eximia.*

1. D. Cucullària (L.) Bernh. (an old generic name, hood-like), Dutchman's-breeches, Breeches-flower. — Scape and slender-petioled leaves from a sort of *granulate bulb;* lobes of leaves linear; *corolla with* 2 *divergent spurs* longer than the pedicel; *crest of the inner petals minute.* (*Bikukulla* Millsp.) — Rich woods, Gaspé Pen., Que., to N.D., s. to N.S., N.E., Ga., Ala., Mo. and e. Kans.; Wash. and Ore. Apr.–early June. — A very delicate plant, sending up in early spring, from the cluster of grain-like tubers crowded together in the form of a scaly bulb, the finely cut leaves and the slender scape, bearing 4–10 pretty, but odd, white flowers tipped with cream-color. — The local forma **purpuritíncta** E. H. Eames (purple-tinged) has the calyx deep purple and the corolla pink, deep orange at the flexure.

2. D. canadénsis (Goldie) Walp. (Canadian), Squirrel-corn. — Subterranean shoots bearing scattered *grain-like tubers* (resembling peas or grains of Indian corn, yellow); leaves as in no. 1; *corolla merely cordiform*, the spurs very short and rounded; *crest of the inner petals conspicuous, projecting.* (*Bikukulla* Millsp.) — Rich woods, sw. Que. to Minn., s. to N.E., N.C., Tenn., and Mo. Apr., May. — Flowers greenish-white tinged with rose, with the fragrance of hyacinths.

3. D. exímia (Ker) Torr. (choice), Turkey-corn, Staggerweed. — Subterranean shoots scaly; divisions and lobes of the leaves broadly oblong; *scape 2.5–4 dm. high, terminated by a thyrse of many flowers borne in lateral clusters; corolla* flesh-color to purple, cordate at base, *narrowed into a long neck*, the lax tips of the outer petals acuminate, the inner surpassed by the apex of the crest. — Rocky woods and cliffs, Ga. and Tenn. to se. and w. N.Y. and W.Va.; also cult. and sometimes esc. Late Apr.–Sept.

10. CORÝDALIS Medic. Corydalis

Corolla 1-spurred at the base (on the upper side), deciduous. Style persistent. Capsule many-seeded. Seeds crested or arillate. Flowers in racemes. Our species are biennial, leafy-stemmed, and pale or glaucous; large genus of N. Hemisph. (The ancient Greek name of the crested lark.) Capnoides Adans. Capnodes Ktze.

a. Stem erect; flowers roseate, with yellow tips; capsules ascending, linear-cylindric, mostly 3–5 cm. long, 1.5–2 mm. thick. . . . . . . 1. *C. sempervirens.*
a. Stem diffuse or lax; flowers yellow; capsules ascending to pendulous, mostly shorter and thicker. . . *b.*
  b. Outer petals wing-crested on back. . . *c.*
    c. Flowers 6–14 mm. long, with spur one-fifth to one-third length of corolla; capsules smooth and glabrous. . . *d.*
      d. Dorsal crest definitely 3–4-toothed; capsules spreading to drooping on

long filiform pedicels; seeds minutely reticulate, with acute margins; flowers all with spurs and full development. . . . . 2. *C. flavula.*

d. Dorsal crest entire or rarely toothed; capsules on very short erect pedicels, erect to arched-ascending; seeds smooth, with obtuse margins; later flowers smaller, much reduced, spurless, cleistogamous.

Terminal raceme rarely overtopping leafy branches, compact; the fruiting raceme compact above, its longer lower internodes 2–10 (–12) mm. long; capsule erect, thick-cylindric, scarcely torulose, 6–15 mm. long. . . . . 3. *C. micrantha.*

Terminal raceme commonly elevated above foliage, soon prolonged and remotely flowered, the longer internodes becoming 1.5–2.5 cm. long; capsule arched-ascending, slenderly cylindric, torulose, 1.5–2.5 cm. long. . . . . 4. *C. Halei.*

c. Flowers 12–17 mm. long, with spur about as long as body; capsules densely pruinose-puberulent, with the vesicles transparent when fresh, erect. . 5. *C. crystallina.*

b. Outer petals merely keeled on the back, not crested, the spur barely half as long as the body; flowers 1–2 cm. long.

Seeds smooth, with rounded or acute margins; capsules pendulous, recurving or ascending. . . . . 6. *C. aurea.*

Seeds reticulated (under high magnification), acute-margined; capsules erect or strongly ascending. . . . . 7. *C. campestris.*

1. C. sempérvirens (L.) Pers. (evergreen or overwintering), PALE C., ROCK-HARLEQUIN. — Very glaucous annual or biennial, *erect,* 0.1–1.3 m. high; racemes loosely panicled; *flowers pink, with yellow tips,* the spur of corolla very short and rounded; *capsules* erect or ascending, *linear-cylindric, mostly* 3–5 *cm. long,* 1.5–2 *mm. broad.* — Rocky places and recent clearings, Nfld. to Alaska, s. to N.S., N.E., n. Ga., Tenn., Ill., Minn., Mont. and s. B.C. May–Sept.

2. C. flàvula (Raf.) DC. (yellowish), YELLOW-HARLEQUIN, YELLOW FUMEWORT. — Slender, *soon diffuse and low-branching,* 1.3–5 dm. high; lower leaves long-petioled, finely dissected; *flowers pale yellow, conspicuously bracted,* 6–8 *mm. long; dorsal crest of outer petals* 3–4-*toothed; capsules* torulose, *on filiform spreading or pendulous pedicels; seeds* acutely margined, *minutely rugose-reticulate,* arils loose. (*Capnodes* Ktze.) — Rocky or sandy slopes, shores and open woods, Ct. to s. Ont. and Minn., s. to Va., Tenn., La. and e. Kans. Apr., May.

3. C. micrántha (Engelm.) Gray (tiny-flowered), SLENDER FUMEWORT. — Similar to no. 2, rather firmer; bracts narrower and smaller; *earliest flowers* 8–13 *mm. long, in compact* 3–12-*flowered racemes; dorsal crest entire or sparingly toothed; terminal raceme elongating in fruit to* 1.5–6 *cm., rarely overtopping the foliage, the upper flowers or capsules remaining approximate, the lower becoming* 2–10(–12) *mm. apart; later flowers smaller, few in dense clusters, with paler small petals, spurless, cleistogamous; capsules* on erect pedicels 0.5–3 mm. long, *erect or stiffly ascending, thick-cylindric, scarcely torulose,* 6–15 *mm. long;* seeds smooth, with rounded margins. — Sandy or gravelly fields, rocky slopes, etc., Ill. to Minn. and Neb., s. to Tenn., Ark. and Okla. Late March–May; *cleist. fls.* June.

4. C. Hàlei (Small) Fern. & Schub. (for its discoverer, JOSIAH HALE, ?1791–1856). — Differing from no. 3 in its habit; the ascending *stem and branches* generally taller, *up to 4 dm. high and less leafy; terminal raceme commonly elevated above the foliage, soon prolonged and distantly* 4–20-*flowered, the raceme in fruit usually* 5–20 *cm. long,* with the *longer internodes* 1.5–2.5 *cm. long; capsules* on pedicels 2–5 mm. long, *slender, torulose,* 1.5–2.5 *cm. long, ascending or outwardly arching.* — Sandy soil of Coastal Plain, Fla. to e. Tex., n. to ?Va. and inland n. to se. Mo. April, early May; *cleist. fls.* later.

5. C. crystallìna Engelm. (bearing crystals), MEALY C. — Erect, mostly branching at base, 1–5 dm. high; pedicels short, erect; *corolla bright yellow,* 1.2–1.7 *cm. long, the spur nearly as long as the body;* crest very broad, usually toothed; *capsules* terete, erect, densely *covered with transparent vesicles;* seeds acutely margined, tuberculate. (*Capnodes* Ktze.) — Open woods, prairies and barrens, Ia. to Ark. and Tex. Apr.–June.

6. C. aúrea Willd. (golden), GOLDEN C. — Diffusely branched annual or biennial 1–6 dm. high, glaucous; *flowers* 1–2 *cm. long,* golden-yellow, the slightly curved *spur barely half as long as the body,* shorter than the pedicel, the *outer petals merely keeled but not crested on the back; capsules loosely spreading or pendulous,* torulose; *seeds smooth, with rounded margins.* (*Capnodes* Ktze.) — Calcareous rocky slopes, gravelly or sandy shores and open woods, e. Que. to Alaska, s. to Vt., N.Y., W.Va., O., Ill., Mo., Tex., N.M. and Calif. May–July.

Var. **occidentàlis** Engelm. (western). — *Capsules ascending* or only slightly spreading, on shorter pedicels; *seeds with thinner or acute margins.* (*C. montana* Engelm.; *Capnoides* Britt.) — Rocky woods, barrens and prairies, Ill. to Mont., s. to Mo., Okla., Tex. and Mex. Apr., May.

7. C. campéstris (Britt.) Buchholz & Palmer (of the Plains). — Resembling no. 5; the stout capsules ascending; the *seeds* (under magnification) ***with concentric rings of fine reticulations***, acute-margined. — Bluffs, barrens and prairies, Ill. to Neb., s. to Ark. and Tex. Apr.-June.

11. FUMÀRIA L. Fumitory. Earth-smoke

Corolla 1-spurred at the base. Style deciduous. Fruit indehiscent, small, globular, 1-seeded. Seeds crestless. — Branched and leafy-stemmed annuals nat. of Eurasia and Afr., with finely dissected compound leaves, and small flowers in dense racemes or spikes. (Name from *fumus, smoke*, presumably from the nitrous odor of the roots when first pulled from the ground.)

F. officinalis L. (of the shops, from its early repute in medicine), Common F. — Sepals ovate-lanceolate, acute, sharply toothed, narrower and shorter than the corolla (which is flesh-color tipped with crimson); fruit slightly notched. — Cult. and waste ground, somewhat local and casual. May–Aug. (Adv. from Eu., where many microspecies have been proposed)

FAM. 72. CAPPARIDÀCEAE (Caper Family)

Herbs (when in northern regions) ***with cruciform flowers, but the 6 or more stamens not tetradynamous****, a 1-locular capsule with 2 parietal placentae, and reniform seeds.* — Capsule as in *Cruciferae*, but with no partition; seeds similar, but the embryo coiled rather than folded. Leaves alternate, mostly palmate. Often with the acrid or pungent qualities of *Cruciferae* (as in *capers*, the flower-buds of *Capparis spinosa*).

Petals laciniate, very unequal; gland on the receptacle tubular. . . . . . . 1. *Cristatella.*
Petals entire or merely notched at apex; gland solid.
  Petals notched at apex; stamens 8–32; capsule sessile or short-stiped. . . . 2. *Polanisia.*
  Petals entire; stamens 6; capsule long-stiped. . . . . 3. *Cleome.*

1. CRISTATÉLLA Nutt.

Petals laciniate, cuneate-flabelliform, white to yellowish, the 2 anterior smaller, all conspicuously clawed. Stamens 6–14, purplish. Torus depressed, bearing a tubular gland on the upper side behind the declined ovary. Capsule ascending, short-stalked, linear-cylindric. Seeds cochleate-reniform. — Leafy viscid and puberulent N.Am. annuals, with petioled palmately 3-foliolate leaves and small racemose flowers in the upper axils. (Diminutive of *crista, a crest*, probably alluding to the lacerate petals.)

1. C. Jàmesii T. & G. (for its discoverer, Edwin James, 1797–1861). — Stem 1–4 dm. high, branching; leaflets linear or narrowly oblong; petals 3–4 mm. long; stamens 6–9, little longer than the petals; capsule about 2 cm. long, its stipe about equaling the gland. — Sandy and gravelly soils, Ill. to Colo., s. to La. and Tex. June–Aug.

2. POLANÍSIA Raf. Clammyweed

Petals with claws, notched at the apex. Stamens 8–32, unequal. Receptacle not elongated, bearing a gland behind the base of the ovary. Capsule linear or oblong, veiny, turgid, many-seeded. — Fetid annuals of warm reg., with glandular or clammy hairs. Flowers in leafy racemes. (Name from the Greek *polys, many*, and *anisos, unequal*, points in which the genus differs in its stamens from *Cleome*.)

1. P. gravèolens Raf. (strong-smelling). — Leaves with 3 oblong leaflets; *flowers 4–6 mm. long;* petals whitish; calyx and filaments purplish; *stamens about* 11, *scarcely exceeding the petals; style short; capsule slightly stipitate above the calyx.* — Gravelly shores and banks, w. Que. and Vt. to Man., s. to Md., Tenn., Ark. and Okla. June–Sept.

2. P. trachyspérma T. & G. (rough-seeded). — *Flowers* larger (8–10 *mm. long*); the *stamens* (12–16) *long-exserted; style* 4–6 *mm. long; capsule sessile or nearly so* within the calyx; seeds usually rough. — Sandy and gravelly soil, N.D. to B.C., s. to sw. Mo., Okla., Tex., N.M. and Ariz.; casually adv. eastw. June–Sept.

3. CLEÒME L.

Petals entire, with claws. Stamens 6. Receptacle somewhat produced between the petals and stamens and bearing a gland behind the stipitate ovary. Capsule linear to oblong, many-seeded. — Our species annuals with bracteate racemes; genus widespread in trop. and warm reg. (Name of uncertain derivation, early applied to some mustard-like plant.)

1. C. serrulàta Pursh (finely toothed), Stinking-clover. — *Glabrous; leaves 3-foliolate;*

leaflets lance-oblong, mostly entire; petals white or rose-colored, short-clawed; stipe of capsule as long as pedicel. (*Peritoma* DC.) — Prairies, damp sands and waste places, Man. to Wash., s. to Ill., Mo., Kans., N.M. and Ariz.; occasionally adv. eastw. May–Sept.

2. C. spinòsa Jacq. (spiny), Spider-flower. — *Viscid-pubescent; leaflets* 5–7, lanceolate, serrulate; petals white or rose-colored. — Cult. and spreading to waste grounds or alluvium, Fla. to Tex., n. to Mass., N.Y., O., Ind., etc. July–Sept. (Introd. and natzd. from the Tropics)

FAM. 73. CRUCÍFERAE (Mustard Family)*

Herbs with a pungent watery juice and cruciform tetradynamous regular flowers; fruit a silique or silicle. Sepals 4, deciduous. Petals 4, hypogynous, their spreading limbs forming a cross. Stamens 6, two of them inserted lower down and shorter (rarely only 4 or 2). Fruit usually 2-locular by a thin septum stretched between the two marginal placentae, from which, when ripe, the valves separate, either much longer than broad (a *silique*) or short (a *silicle*), sometimes indehiscent and nut-like, or separating across into 1-seeded joints. Seeds campylotropous, without albumen, filled by the large embryo, which is curved or folded in various ways: *i.e.*, the *cotyledons accumbent*, viz., their margins on one side applied to the radicle, so that the cross-section of the seed appears thus O=; or else *incumbent*, viz., the back of one cotyledon applied to the radicle, thus O||. In these cases the cotyledons are plane; but they may be folded upon themselves and around the radicle, as in *Brassica*, where they are *conduplicate*, thus O ». In *Leavenworthia* alone the whole embryo is straight. — Leaves (except in *Lunaria*) alternate; stipules none. Flowers in terminal racemes or corymbs; pedicels rarely bracted. A large and natural family of pungent or acrid, but not poisonous plants. The fruit and seeds give the chief characters of the genera.

Key to Flowering Material

a. Petals yellowish to orange. . . *b.*
  b. Leaves all or nearly all simple, not deeply divided. . . *c.*
    c. Flowering stems arising from a definite rosette or with rosettes borne on basal offshoots. . . *d.*
      d. Rosette-leaves pubescent or scurfy; flowering stem leafy to or above middle.
        Cauline leaves not clasping.
          Leaves loosely pubescent; base of stem and axis of raceme villous; sepals hairy. . . . . 1. *Draba.*
          Leaves, stems, axis of raceme and sepals scurfy. . . . . . 5. *Lesquerella.*
        Cauline leaves (at least the lower) sagittate-clasping. . . . . . 43. *Arabis.*
      d. Rosette-leaves glabrous; flowering stem leafy only toward base. . . 22. *Diplotaxis.*
    c. Flowering stems without definite basal rosette. . . *e.*
      e. Cauline leaves not clasping.
        Plant (especially leaves and young stems) closely pubescent with appressed stellate or bifurcate hairs.
          Pubescence of many-pronged hairs . . . . 4. *Alyssum.*
          Pubescence of 2–3-parted hairs. . . . . 31. *Erysimum.*
        Plant glabrous or with simple hairs.
          Flowering stem leafy only toward base; sepals erect in anthesis; petals slender-clawed; ovary linear-cylindric. . . . . . 22. *Diplotaxis.*
          Flowering stem usually leafy nearly to summit; sepals spreading; petals cuneate at base, scarcely clawed; ovary short. . . . 32. *Rorippa.*
      e. Cauline leaves (at least the lower) clasping.
        Style wanting, stigma sessile; inflorescence a broad corymb of many racemes; pedicels promptly reflexed after anthesis. . . . . 11. *Isatis.*
        Style elongate; inflorescence of simple elongate racemes; pedicels not reflexed.
          Leaves lanceolate to oblong, acutish; petals 2–6 mm. long; style filiform, abruptly terminating the globose to pyriform ovary.
            Stem hispid to summit; flowers 2–3 mm. long; ovary globose, soon becoming reticulate, with 1 or 2 ovules. . . . . . 17. *Neslia.*
            Stem glabrous at least toward summit; flowers 4–6 mm. long; ovary pyriform, smooth, with many ovules. . . . . . 16. *Camelina.*

---

* The tribal distinctions are highly technical and often contradictory; they are here omitted. The keys are wholly artificial. The sequence of genera is that of Watson and Robinson in Gray, Synoptical Flora of North America.

Leaves broadly elliptic, with rounded summit, glabrous and glaucous; petals 7–10 mm. long; ovary slender, tapering into the thick style. . . . . 24. *Conringia.*

*b.*Leaves (at least the lower) deeply pinnatifid, lyrate or pinnate. . . *f.*

*f.*Pubescence of simple hairs or wanting. . . . *g.*

*g.*Racemes without bracts (except rarely the lowest pedicel). . . . *h.*

*h.*Petals 1–2 cm. long.

Ovary and silique with corky transverse partititions separating the ovules and seeds; petals conspicuously veined. . . . . 19. *Raphanus.*

Ovary and silique without corky separating partitions.

Petals 1–1.2 cm. long, usually not conspicuously veined. . . 20. *Brassica.*

Petals 1.5–2 cm. long, with violet veins. . . . . 23. *Eruca.*

*h.*Petals 0.5–10 mm. long.

Ovules in 2 rows in each locule. . . . . 22. *Diplotaxis.*

Ovules in 1 row in each locule.

Cauline leaves clasping the stem by deeply pinnatifid bases; plants perennial (rarely annual) with crowded overwintering basal leaves. . . . . 35. *Barbarea.*

Cauline leaves not clasping, if so with entire or shallowly toothed bases and plant annual, or biennial with overwintering rosette much loosened or wanting during anthesis.

Sepals erect during anthesis; petals 6–10 mm. long; ovary and silique with prominent beak. . . . . 20. *Brassica.*

Sepals loosely spreading during anthesis; petals 0.5–8 mm. long; ovary and silique (or silicle) without prominent beak.

Annuals or biennials of weedy and dry habitats; sepals linear to narrowly oblong; style very short and thick or none; ovary slenderly linear to subulate, with each valve 1–3-nerved. . . . . 26. *Sisymbrium.*

Perennials to annuals chiefly of damp or aquatic habitats; sepals ovate to elliptic; style definite, usually slender; ovary short and broad to short-cylindric, the valves nerveless. . . . . 32. *Rorippa.*

*g.*Racemes with pinnatifid bracts subtending all or all but the uppermost pedicels.

Scabrous; ovary linear-cylindric, tapering gradually to the broad stigma. . . . . 21. *Erucastrum.*

Smooth and glabrous; ovary lance-elliptic, flat, wing-margined, abruptly tipped by a filiform style. . . . . 38. *Selenia.*

*f.*Pubescence stellate; leaves finely dissected; flowers small. . . . 28. *Descurainia.*

*a.*Petals purple, pink or white (sometimes yellow at base). . . . *i.*

*i.*Leaves all or nearly all simple, not deeply divided. . . . *j.*

*j.*Flowering stems arising from a definite rosette or with rosettes borne on basal offshoots. . . . *k.*

*k.*Flowering stem a leafless scape 1–15 cm. high. . . . *l.*

*l.*Leaves flat, firm; ovary flat or elongate, with a usually definite style; plants of dry or rocky habitats.

Petals cleft to middle; annual of fields and roadsides southward. . 1. *Draba.*

Petals rounded or merely emarginate at summit; cespitose perennials of alpine areas or of Nfld. or Gaspé.

Crowns (at summit of stout tap-root) short and stout, dark; leaves fleshy, linear-lanceolate, 1–4 cm. long, without definite midrib; sepals broadly elliptic. . . . . 29. *Braya.*

Crowns dividing into pale filiform branches; leaves dry and thin, linear to obovate, 3–15 mm. long; sepals narrower.

Leaves slender-petioled, without prominent midrib, ovate or oval; style obsolete. . . . . 41. *Cardamine.*

Leaves sessile, with midrib prominent beneath, linear, oblanceolate or obovate; style slender and definite. . . . 1. *Draba.*

*l.*Leaves subulate, prolonged, soft; ovary globose or ovoid, with sessile stigma; aquatic or subaquatic. . . . . 12. *Subularia.*

*k.*Flowering stem with 1 or more leaves. . . . *m.*

*m.*Cauline leaves sessile, with sagittate bases.

Ovary inverted-triangular or cuneate. . . . . 13. *Capsella.*

Ovary slender, elongate. . . . . 43. *Arabis.*

*m.*Cauline leaves not sagittate at base. . . . *n.*

*n.*Fleshy mostly maritime boreal plants with succulent broad leaves; ovary plump, globose to ellipsoid. . . . . 15. *Cochlearia.*

n.Not fleshy, narrow-leaved; ovary flat or slender. . . *o*.

o.Ovary slenderly linear or linear-cylindric.

Annual or winter-annual with slender tap-root and soft base; style obsolete; weed of fields and roadsides southward. . 27. *Arabidopsis*.

Biennial or perennial with strong bases or crowns, if seemingly annual with definite styles; mostly natives of rocky habitats.

Inflorescence capitate in anthesis; petals 3–4 mm. long; cauline leaves mostly only 1–3 mm. wide. . . . . 29. *Braya*.

Inflorescence corymbose or racemose in anthesis; petals longer; cauline leaves mostly broader. . . . . . . 43. *Arabis*.

o.Ovary lanceolate to ovate or elliptic.

Weak annual or winter-annual; flowers 1 mm. long; style none; plant of Nfld. and Lab. . . . . . . 14. *Hutchinsia*.

Obvious strong-based perennials or biennials with larger flowers; if winter-annuals southern and with larger flowers. 43. *Arabis*.

j.Flowering stems without definite basal rosettes. . . *p*.

p.Fleshy maritime plants with succulent leaves; ovary plump.

Biennial, forming circular 1st-year rosettes of ovate to rounded leaves; flowers 2–5 mm. long, with white or pink petals; ovary globose to ellipsoid, not beaked; plants from Gulf of St. Lawrence northward. . . . . 15. *Cochlearia*.

Annual without rosettes; flowers larger, with lilac petals; ovary jointed, the upper segment forming a blunt beak; plants mostly more southern. . . . . 18. *Cakile*.

p.Not especially fleshy and succulent. . . *q*.

q.Cauline leaves clasping the stem.

Upper leaves perfoliate, surrounding the stem. . . . . . . 8. *Lepidium*.

Upper leaves merely sagittate or auricled at base, not surrounding the stem.

Ovary notched at summit, very flat, with sessile stigma or short style barely projecting out of the notch.

Pubescent; petals 1.5 mm. long. . . . . 8. *Lepidium*.

Glabrous; petals 2–4 mm. long. . . . . 6. *Thlaspi*.

Ovary not notched, plump or slender and elongate, tipped by style or stigma.

Pubescent; cauline leaves oblong, remotely dentate, sessile, cordate- or sagittate-clasping; inflorescence a corymb of racemes; petals white, without long claws, 3–4 mm. long; ovary subglobose or cordate, with filiform style; field-weeds. . . . . 9. *Cardaria*.

Glabrous; cauline leaves acuminate to prolonged, with slightly auricled bases, closely (often doubly) sharp-toothed; inflorescences elongate racemes; petals purple, 7–12 mm. long, with exserted claws; ovary slenderly linear, without definite style. . . . . 36. *Iodanthus*.

q.Cauline leaves not clasping. . . *r*.

r.Ovary short, lanceolate to ovate or rounded.

Leaves hoary with stellate or forking hairs.

Hairs stellate; petals 2-parted. . . . . 2. *Berteroa*.

Hairs 2-pointed, attached in the middle; petals entire. . 3. *Lobularia*.

Leaves glabrous or with mostly simple hairs.

Weak annual 0.3–2 dm. high; basal leaves about 1 cm. long; flowers about 1 mm. long; plant of n. Nfld. and Lab. . . 14. *Hutchinsia*.

Coarser; leaves and usually flowers larger; more southern.

Petals 1–2 mm. long (or wanting); ovary flat; annuals or biennials, rarely perennial. . . . . 8. *Lepidium*.

Petals 5–8 mm. long; ovary plump; coarse perennial. . 34. *Armoracia*.

r.Ovary elongate, slenderly linear or cylindric.

Cauline leaves deeply cordate, deltoid-ovate, long-petioled; bruised plant with odor of onion. . . . . 25. *Alliaria*.

Cauline leaves rounded, tapering or short-auricled at base, narrower, sessile or short-petioled; plant without onion-odor.

Blades of petals raised above sepals on very long slender claws.

Glabrous; cauline leaves (or some of them) with auricled bases, long-acuminate to base and apex, copiously (often doubly) sharp-toothed; sepals about 3 mm. long; limb of petal broad, 3–6 mm. long. . . . . 36. *Iodanthus*.

Hispid or hirsute; cauline leaves not auricled; sepals 3.5–9 mm. long.

Cauline leaves numerous, lanceolate to ovate, sharply toothed; flowering pedicels 5–10 mm. long; sepals 5–9 mm. long; limb of petal 1–1.5 cm. long. . . . . . 30. *Hesperis.*

Cauline leaves few, oblong to oblanceolate, shallowly dentate; flowering pedicels 0.5–4 mm. long; sepals 3.5–6 mm. long; limb of petal 3–5 mm. long. . . . . . 37. *Chorispora.*

Blades of petals not raised above sepals; cauline leaves few, blunt and blunt-toothed.

Annual or biennial field-weed; axis of raceme and backs of sepals hispid; stigma broad, obviously 2-lobed. . . . 22. *Diplotaxis.*

Perennials of spring-heads, rills or wet woods; axis of raceme and backs of sepals glabrous; stigma minute, not obviously lobed. . . . . 41. *Cardamine.*

*i.*Leaves (or larger ones) deeply pinnatifid, lyrate, pinnate or palmately divided. . . *s.*

*s.*Cauline leaves 2 or 3, palmately compound or deeply cleft; flowers rather large; rhizomes fleshy, toothed or jointed; early maturing woodland plants. . . . . 40. *Dentaria.*

*s.*Leaves principally pinnatifid, lyrate or pinnate, not palmate. . . *t.*

*t.*Petals 0.5–8 mm. long (or wanting). . . *u.*

*u.*Ovary lance-oblong or ovate to orbicular or obcordate.

Flowering stems scapose or nearly so, naked or with reduced leaves, arising from a basal rosette; flowers minute; ovary emarginate. 7. *Teesdalia.*

Flowering stem leafy, simple or forking.

Petals 0.5–2 mm. long; style 0 or barely exserted from a notch; plants relatively small.

Ovary notched at summit, the stigma borne in the notch.

Erect or ascending with long racemes terminating stem and branches; ovary roundish, flat, smooth. . . . . . . 8. *Lepidium.*

Depressed or in mats, with short racemes terminating and borne along the branches; ovary reniform, plump, roughened. . . . . 10. *Coronopus.*

Ovary not notched, nearly square across summit, with small sessile stigma. . . . . 14. *Hutchinsia.*

Petals 5–8 mm. long; style definite, terminal; plants coarse and terrestrial or with prolonged aquatic stems. . . . . . 34. *Armoracia.*

*u.*Ovary slenderly linear.

Minutely stellate-pubescent and glandular-granular; leaves mostly bi- or tripinnate; petals creamy-white. . . . . 28. *Descurainia.*

Glabrous or with some simple hairs; leaves simply pinnate or pinnatifid; petals white or pinkish.

Succulent perennial, with prolonged stems rooting in water; petals 4–5 mm. long; nectariferous glands large and horseshoe-shaped; pedicels after anthesis promptly divergent or recurving. . . . . 33. *Nasturtium.*

Hardly succulent; the flowering stems growing from basal rosettes (these sometimes disappearing before maturing of fruit); petals 1.5–4 mm. long; nectariferous glands small, not horseshoe-shaped; pedicels spreading-ascending to suberect.

Leaflets of rosette-leaves and of at least the lower cauline leaves distinct and petiolulate (leaf pinnate). . . . 41. *Cardamine.*

Leaflets sessile, often more or less confluent along the rachis (leaf pinnatifid).

Stem relatively weak, usually simple and ascending; lower leaves with 1–6 pairs of membranous obovate to rounded lateral segments, the broad terminal segment much larger; pedicels slender, becoming 2–15 mm. long; silique slenderly linear, distinctly beaked; ovules and seeds wingless; plants of damp or wet natural habitats. . . . . 41. *Cardamine.*

Stem stiff, often several divergent from the rosette; lower leaves with 5–14 pairs of oblong to linear segments, terminal segment similar; pedicels short and stout, rarely lengthening to 5 mm.; silique broadly linear, with scarcely evident beak; ovules and seeds margined; plant of disturbed soil and fallow fields southward. . . . 42. *Sibara.*

*t.*Petals 1–2 cm. long. . . *v.*

*v.*Acaulescent or nearly so, the simple scapes or elongate racemes arising from a basal rosette, the ascending filiform scapes or pedicels 2–10 (or more) cm. long in anthesis, longer in fruit. . . . . 39. *Leavenworthia.*

*v.*Leafy-stemmed, with elongate racemes; pedicels shorter. . . *w.*

*w.*Coarse annuals or biennials, more or less hispid with stiff hairs; lower leaves runcinate, pinnatifid or lyrate; plants of disturbed soils.

Petals 1 cm. long. . . . . 22. *Diplotaxis.*

Petals 1.5–2 cm. long.

Root thickened (radish); flowering pedicels 1–2 cm. long. . . 19. *Raphanus.*

Root not thickened; flowering pedicels 1–3 mm. long. . . . 23. *Eruca.*

*w.*Slender perennial, glabrous, with pinnate leaves; leaflets of lower leaves petiolulate; plants of meadows, grasslands and springy spots. . . . . 41. *Cardamine.*

Key to Fruiting Material

*a.*Leaves simple, pinnatifid or pinnate. . . *b.*

*b.*Fruit at most four times as long as broad (silicle). . . *c.*

*c.*Fruit longitudinally dehiscent; the falling valves exposing the thin septum and seeds. . . *d.*

*d.*Silicle obviously flattened. . . *e.*

*e.*Fruits in regular racemes. . . *f.*

*f.*Silicle flattened parallel to the broad septum. . . *g.*

*g.*Leaves entire or toothed, not strongly pinnatifid; racemes without leafy bracts; fruits sessile or nearly so (not stipitate); pubescence often stellate or forked.

Flowering stems arising from a definite rosette or with rosettes borne on basal offshoots; fruit lanceolate or oblong to broadly ovate; seeds wingless. . . . . 1. *Draba.*

Flowering stems without well developed basal rosettes; fruit orbicular or, if oblong, with winged seeds.

Fruit oblong or elliptic, the valves flat; seeds winged; petals deeply cleft, white; stems tall, up to 1 m. high. . . . 2. *Berteroa.*

Fruit orbicular or nearly so, the valves convex; seeds wingless; petals entire; stems low.

Hairs of stems and leaves 2-pointed, attached at middle; petals white. . . . . 3. *Lobularia.*

Hairs stellate; petals yellow. . . . . 4. *Alyssum.*

*g.*Leaves pinnately dissected, with flowers subtended by pinnate bracts; silicle flat, stipitate; small glabrous annual. . . 38. *Selenia.*

*f.*Silicle flattened contrary to the narrow septum. . . *h.*

*h.*Silicle orbicular or obovate to lanceolate. . . *i.*

*i.*Silicle notched or emarginate at summit; plants glabrous or with simple hairs.

Flowering stems scapiform, leafless, 1–2 dm. high, borne from a basal rosette; petals very unequal. . . . . 7. *Teesdalia.*

Flowering stems leafy, mostly taller, with or without basal rosette; petals equal.

Seeds 2 or more in each locule. . . . . 6. *Thlaspi.*

Seeds 1 in each locule. . . . . 8. *Lepidium.*

*i.*Silicle not notched at summit.

Coarse rhizomatous perennial 1–2 m. high, glabrous or with simple pubescence; basal leaves 2–3 dm. long; racemes crowded in a large corymbiform panicle; seeds 1 in each locule. . . . . 8. *Lepidium.*

Dwarf annual with filiform stems up to 2 dm. high; basal leaves about 1 cm. long, stellate-pubescent; raceme slender; seeds 2 or more in each locule. . . . 14. *Hutchinsia.*

*h.*Silicle inverted-triangular, broad at summit. . . . . 13. *Capsella.*

*e.*Fruits solitary, terminating simple or forked scapes or peduncles 0.5–1.5 cm. long, from among the basal leaves; fruit oblong. . . . . 39. *Leavenworthia.*

*d.*Silicle globose to ovoid, obovoid or pyriform, scarcely if at all flattened.

Leaves subulate, soft and flaccid, glabrous, in basal tufts; plant aquatic or subaquatic, with naked few-fruited scapes 2–15 cm. high. . . . . 12. *Subularia.*

Leaves flat; plants terrestrial, usually coarser; flowering stem leafy.

Stems and leaves more or less covered with scales or appressed stellate hairs; flowering stems arising from basal rosettes of narrow leaves; petals yellow. . . . . 5. *Lesquerella.*

Stems and leaves glabrous or with some or all hairs simple.

Glabrous, succulent, maritime; rosette-leaves long-petioled, ovate to reniform, entire or remotely dentate; petals white to purplish. . . . . 15. *Cochlearia.*

Glabrous or pubescent, not maritime or succulent; rosette-leaves (if present) short-petioled, narrower, variously toothed or lobed.

Cauline leaves sagittate-clasping, unlobed; silicle pyriform, 5–9 mm. long, the firm valves 1-nerved. . . . . . 16. *Camelina.*

Cauline leaves not sagittate; silicle globose, ovoid, ellipsoid or obovoid, smaller, nearly or quite nerveless.

Capsule 2-locular; lower leaves not capillary-dissected.

Perennials or annuals with slender bases and mostly pinnate or pinnatifid leaves and yellow petals. . . . 32. *Rorippa.*

Perennial with coarse tap-root; the long-petioled oblong to ovate basal leaves crenate, 2–4 dm. long; petals white. . . . . 34. *Armoracia.*

Capsule 1-locular; weak aquatic perennial with capillary-dissected submersed leaves; petals white. . . 34. *Armoracia.*

*c.*Fruits or their locules indehiscent or very tardily dehiscent, or separating merely by transverse partitions. . *j.*

*j.*Fruits pendulous, cuneate-oblong, flat, 0.8–1.5 cm. long. . . . . 11. *Isatis.*

*j.*Fruits ascending or divergent, not pendulous, plump. . *k.*

*k.*Fruits subglobose, cordate, reniform or short-obovoid, 1.5–4 mm. long.

Leaves pinnately divided; racemes compact, borne along the depressed stems; silicles reniform, wrinkled, notched at apex, the 2 closed valves separating in maturity. . . . . 10. *Coronopus.*

Leaves undivided; stems ascending; racemes terminal or in corymbs; silicles tipped by persistent styles.

Silicles cordate or subglobose, inflated, smooth; racemes in corymbs; petals white. . . . . 9. *Cardaria.*

Silicles subglobose, firm, reticulated; racemes elongated, loosely disposed; petals yellow. . . . . 17. *Neslia.*

*k.*Fruits elongate, 1.3–7 cm. long, their surfaces neither wrinkled nor reticulate.

Plant glabrous, fleshy, of sea- and lake-beaches; fruit 1.3–2.5 cm. long, its beak formed by the disarticulating upper joint containing an erect seed. . . . . 18. *Cakile.*

Plant hispid, not fleshy, of fields and waste ground; fruit 3–7 cm. long, continuous, the beak not articulated; seeds all pendulous. 19. *Raphanus.*

*b.*Fruit 4–∞ times as long as broad (silique). . . *l.*

*l.*Siliques indehiscent or disarticulating by cross-partitions into seed-bearing joints.

Plant succulent and glabrous; silique 1.5–2 cm. long, of 2 joints, the terminal joint blunt and containing an erect seed; plant of beaches and shores. . . . . 18. *Cakile.*

Plant not succulent, hispid; silique 2.5–7 cm. long, moniliform, of several joints or jointless and divided inside by spongy tissue, the slender beak pointed; seeds all pendulous; plants of fields and waste places.

Siliques 3–5 mm. thick, on slender pedicels 1–2 cm. long, the locules 1-seriate; beak tipped by broad emarginate stigma. . . . 19. *Raphanus.*

Siliques 1.5–2 mm. thick, on stout pedicels 2–5 mm. long, the locules 2-seriate; slender beak tapering to the minute stigma. . . 37. *Chorispora.*

*l.*Silique longitudinally dehiscent, the falling valves exposing the seeds and the septum. . . *m.*

*m.*Flowering stem arising from a definite rosette of basal leaves. . . *n.*

*n.*Leaves simple, undivided or merely sinuate or lyrately lobed. . . *o.*

*o.*Fruits solitary, terminating simple or forking scapes or peduncles 0.5–1.5 dm. long, these borne from among lyrate small leaves. 39. *Leavenworthia.*

*o.*Fruits in definite racemes. . . *p.*

*p.*Hairs of foliage, stem, etc. all or in part forking; ripe valves of silique stiffish, not strongly coiling upon falling.

Siliques flat or flattish.

Siliques linear-oblong to lanceolate, 0.3–2.2 cm. long; seeds definitely in 2 rows in each locule. . . . . 1. *Draba.*

Siliques linear, 1.5–10 cm. long; seeds in 1 row or only obscurely in 2 rows. . . . . 43. *Arabis*.

Siliques terete or quadrate-cylindric.

Rosette-leaves fleshy, linear-oblanceolate, entire; caudex stout; scapes naked; siliques lance-subulate, 4–9 mm. long, 10–16-ovulate; high-northern plants (Nfld. with us). 29. *Braya*.

Rosette-leaves thinner, oblong, oblanceolate or narrowly ovate, often toothed; caudex slender or scarcely developed; flowering stems leafy; siliques slenderly linear-terete or -tetragonal, 1–3 cm. long, many-seeded.

Perennial with slender branching caudices; siliques terete, torulose, 1.2–3 cm. long; cells of septum obliquely or transversely elongate. . . . . . . 29. *Braya*.

Annual without branching caudex; siliques tetragonal-cylindric, with straight margins, 1–1.5 cm. long; cells of septum vertically elongate. . . . . 27. *Arabidopsis*.

p.Hairs all simple or often quite wanting; ripe valves of silique elastic, coiling or rolling into rings upon falling. . . . . 41. *Cardamine*.

n.Leaves deeply pinnatifid or pinnate.

Leaves bi- or tripinnate or pinnatifid; siliques slenderly cylindric to clavate; cotyledons incumbent; petals yellow or yellowish. . 28. *Descurainia*.

Leaves once pinnate or pinnatifid; siliques flattened; cotyledons accumbent; petals white to purple.

Leaves pinnate, at least the lower with distinct leaflets; valves of silique elastic, coiling upon dropping; seeds wingless. . . 41. *Cardamine*.

Leaves stiffly pinnatifid; valves stiff, not coiling; seeds winged. 42. *Sibara*.

m.Flowering stem without definite basal rosette. . . *q*.

q.Raceme with lower flowers subtended by leafy bracts; leaves deeply pinnatifid or bipinnatifid; siliques slender, 4-angled, 2.5–3.5 cm. long, their valves keeled. . . . . 21. *Erucastrum*.

q.Raceme with flowers ebracteate. . . *r*.

r.Siliques flattened.

Leaves and stems glabrous or with simple hairs; siliques 0.5–3 cm. long.

Cauline leaves auricled, sharply and often doubly toothed; valves of silique 1-nerved, not elastic. . . . . . . 36. *Iodanthus*.

Cauline leaves not auricled, if simple merely sinuate or dentate; valves nerveless, elastic and coiling after falling. . . 41. *Cardamine*.

Leaves or stems usually bearing some forked hairs; siliques up to 10 cm. long. . . . . 43. *Arabis*.

r.Siliques terete or 4-angled. . . *s*.

s.Stems bearing closely appressed straight 2-pronged hairs attached near the middle (malpighiaceous hairs). . . . . 31. *Erysimum*.

s.Stems glabrous or with simple or stellate hairs. . . *t*.

t.Cauline leaves sagittate- or cordate-clasping.

Silique tapering to a terete indehiscent beak 0.8–2 cm. long; seeds globose; cotyledons conduplicate; petals deep yellow. 20. *Brassica*.

Silique very slender, tapering to thick short style; seeds oblong or flat; cotyledons not conduplicate; petals creamy to yellow.

Leaves simple, not lyrate; petals pale yellow or creamy.

Cauline leaves elliptic, cordate-clasping; silique angled; seeds plump, in 1 row; cotyledons incumbent. 24. *Conringia*.

Cauline leaves lanceolate to oblong, sagittate-clasping; silique terete; seeds flat, obscurely 2-seriate; cotyledons accumbent. . . . . 43. *Arabis*.

Leaves, or some of them, lyrate-pinnatifid; seeds oblong to quadrate; cotyledons accumbent; petals deep yellow. 35. *Barbarea*.

t.Cauline leaves not clasping. . . *u*.

u.Fruit terminated by a conical to flat or 3-angled indehiscent beak.

Seeds globose, in 1 row in each locule. . . . . . . 20. *Brassica*.

Seeds ovoid or ellipsoid, in 2 rows in each locule.

Siliques terete, the beak slenderly conical; lower pedicels 0.5–3.5 cm. long. . . . . 22. *Diplotaxis*.

Siliques 4-angled, tipped by a flat triangular-lanceolate beak; lower pedicels 0.5–5 mm. long. . . . . . 23. *Eruca*.

u.Fruits dehiscent to tip, without long indehiscent beak. . . *v*.

*v.*Pubescence of leaves and stem stellate or forked; leaves bi- or tripinnate into many segments; siliques slenderly cylindric to clavate. . . . . 28. *Descurainia.*
*v.*Pubescence simple or wanting; siliques not clavate. . . *w.*
*w.*Annuals or biennials without creeping bases; leaves simple, merely toothed or pinnatifid or lyrate; valves of silique nerved.
Leaves lanceolate to rounded-ovate, regularly toothed.
Leaves deltoid-cordate, serrate, the lower long-petioled; siliques angled, 2.5–5 cm. long. . . 25. *Alliaria.*
Leaves lanceolate to narrowly ovate, dentate, short-petioled to sessile; siliques terete, 5–14 cm. long. . . . . 30. *Hesperis.*
Leaves or many of them lyrate or deeply pinnatifid.
Stems terete; leaves pinnatifid; cotyledons incumbent. . . . . 26. *Sisymbrium.*
Stems angulate-corrugate; lower leaves lyrate or pinnatifid; cotyledons accumbent. . . . . . 35. *Barbarea.*
*w.*Perennials with creeping bases; leaves pinnate; valves of siliques nerveless.
Succulent, with stems rooting at nodes; divisions of leaves roundish or oblong, mostly entire; petals white. . . . . 33. *Nasturtium.*
Not succulent, spreading by firm subterranean rhizomes; divisions of leaves lanceolate to linear, toothed or incised; petals yellow. . . . . 32. *Rorippa.*
*a.*Leaves palmately divided or cleft into 3 or more leaflets or segments; rhizome fleshy, toothed or jointed. . . . . 40. *Dentaria.*

1. DRÀBA L.

Silicles or siliques oval, oblong, or even linear, flat; the valves plane or slightly convex; the partition broad. Seeds several or numerous, in 2 rows in each locule, marginless. Cotyledons accumbent. Filaments not toothed. — Low herbs, chiefly of the N. Hemisph. (some, often suffruticose, in S.Am.), with entire or toothed leaves, and white or yellow flowers; pubescence often stellate. (Name from *drabe, acrid*, applied by Dioscorides to some cress.) *

*a.*Petals rounded or emarginate at summit; flowering stems with 1 or more leaves above basal rosette or, if scapose, with perennial bases bearing remnants of old leaves. . . *b.*
*b.*Scapose perennials with persistent shreds of old leaves below the living rosettes; scapes filiform, 1–11 cm. high; rosette-leaves 3–15 mm. long, 1–4.5 mm. broad. . . *c.*
*c.*Leaves with glabrous surfaces, at most slightly ciliate near tip. . . . . 1. *D. Allenii.*
*c.*Leaves copiously pubescent. . . *d.*
*d.*Leaves bristly with simple and variously forking trichomes. . . . . 2. *D. rupestris.*
*d.*Leaves canescent-pannose with minute stellate trichomes; simple hairs wanting or sparse.
Leaves cuneate-obovate to broadly oblanceolate, obtuse; siliques glabrous; seeds 14–18, 0.7–1 mm. long. 3. *D. nivalis.*
Leaves linear or linear-oblanceolate, acute; siliques stellate-hirtellous; seeds 8–10, 1.2–1.8 mm. long. . . . . . . 4. *D. Peasei.*
*b.*Flowering stem if perennial often taller and with 1 or more leaves above the basal rosette, if annual or biennial and low with at least 1 pair of cauline leaves; rosette-leaves mostly larger. . . *e.*
*e.*Perennial, or nos. 5 and 6 often biennial, with basal rosettes; the crowns bearing dead remnants of old leaves below the living rosettes. . . *f.*
*f.*Flowering stem simple or with erect or strongly ascending branches; leaves rarely laciniate; style many times shorter than silicle, at most 1.8 mm. long. . . *g.*
*g.*Petals deep yellow; ovaries and silicles densely villous-hirsute with divergent to reflexed hairs 0.5–1 mm. long. . . . . . 5. *D. mingan-ensis.*

* Treatment based on that in Rhodora, xxxvi. (1934). The figures show in many cases the habit (much reduced); in some a rosette-leaf; in many the trichomes of the upper leaf-surface (much enlarged); in FIGS. 1067 and 1070 the trichomes of the stem; in many the silicle (enlarged); and in FIGS. 1064 and 1065 an enlarged seed.

g. Petals white; ovaries and silicles glabrous to minutely short-pubescent. . . *h.*

h. Biennials (rarely short-lived perennials) with lower leaves of rosette shriveling soon after anthesis; subspherical rosette of 1st year loosening and elongating to form many-leaved flowering stem; axis of raceme and pedicels densely pilose-tomentulose to villous. . . . . 6. *D. incana.*

h. Perennials; branches of caudex usually with fibrous shreds of old leaves at summit below living rosette; new rosettes well developed at flowering time; axis of raceme and pedicels glabrous, sparsely hirtellous or stellate-pubescent. . . *i.*

i. Foliage with all or many of its hairs simple or elongate and irregularly forking, with or without admixed stellate hairs. . *j.*

j. Siliques linear to linear-lanceolate, 2–2.5 mm. broad; sepals 0.4–0.9 mm. broad; rosette-leaves linear-oblanceolate, 1–5 mm. broad; cauline leaves lanceolate, 1.5–6 mm. broad. . 7. *D. clivicola.*

j. Siliques elliptic to oblong-lanceolate, mostly 2.5–4 mm. broad; sepals 1–2.3 mm. broad; rosette-leaves oblanceolate to narrowly obovate, up to 9 mm. broad; cauline leaves oblong to ovate, 3–11 mm. broad.

Rosette-leaves narrowly oblanceolate, hispid with many simple or elongate and furcate hairs; cauline leaves ovate; mature central racemes from one-third to nearly full height of plant; sepals 1–1.5 mm. broad. . . . . . 8. *D. norvegica.*

Rosette-leaves cuneate-oblanceolate, with numerous stellate and several–few simple and elongate furcate hairs; cauline leaves oblong; mature central raceme one-fourth to one-half full height of plant; sepals 1.3–2.3 mm. broad. . . 9. *D. laurentiana.*

i. Foliage with close stellate pubescence forming, at least on expanding leaves, a pannose coat; simple trichomes wanting or only rare (except as cilia) on the rosette-leaves. . . *k.*

k. Silicles plump, glabrous; sepals 1.5–2 mm. long, 1 mm. broad; seeds closely but irregularly imbricated, often turned oblique to septum. . . . . 10. *D. pycnosperma.*

k. Silicles strongly flattened (rather plump in no. 13, with densely tomentulose valves); sepals 2–3.5 mm. long, 1–2.3 mm. broad; seeds not imbricated, lying flat against septum. . . *l.*

l. Silicles glabrous or only sparingly hirtellous or scabrous, strongly flattened; racemes usually bractless. . . *m.*

m. Cauline leaves mostly rounded at base; mature siliques definitely veiny, usually flat; style obsolete or thick and short (up to 0.5 mm. long); fruiting pedicels stoutish, short, the lowest 1–6 mm. long.

Stems hirsute, especially on lower internodes, with abundant simple divergent trichomes overtopping the stellate hairs; cauline leaves 6–25 (average 10). 9. *D. laurentiana.*

Stems closely stellate-pannose, sparsely or not at all hirsute below; cauline leaves 1–8, rarely –14 (av. 4). 11. *D. glabella.*

m. Cauline leaves narrowed or only slightly rounded at base; mature siliques scarcely or only obscurely veiny, often twisted, very thin; style slender, 0.5–1 mm. long; fruiting pedicels slender, lowest 3–15 mm. long. . . . . 12. *D. arabisans.*

l. Silicles densely stellate-tomentulose, only slightly compressed; racemes usually leafy-bracted at base. . . . 13. *D. lanceolata.*

f. Flowering stem with strongly divergent branches; leaves laciniate or subpectinate; style filiform, 1.5–3 mm. long, one-fourth to one-third as long as spirally twisted stellate- pubescent silique. . . . . . 14. *D. ramosissima.*

e. Dwarf annuals or biennials with bractless racemes; flowering stems leafy or at least with one pair of leaves above the base. . . *n.*

n. Silicles 1.7–6 mm. long, 6–16-seeded; petals (when developed) 2–3 mm. long; stems simple, or branching nearly to summit, with the numerous small leaves strigose with variously forking trichomes.

Stems with abbreviated corymbiform branches from the middle and upper axils; silicles linear-ellipsoid, 4–6 mm. long, minutely stellate-puberulent; seeds 1–1.5 mm. long. . . . . 15. *D. aprica.*

Stems mostly with elongate or leafy branches (or simple); silicles oblong-ellipsoid, 1.7–5 mm. long, glabrous; seeds 0.5–0.8 mm. long. 16. *D. brachycarpa.*

*n.*Silicles flattish, 3–18 mm. long, 15–80-seeded; petals (when developed) 2–5 mm. long; stems simple, or forking only below, hispid below as are the leaves. . . *o.*

*o.*Flowers uniform, with yellowish narrowly cuneate petals about 2 mm. long; leaves scattered nearly to the slender elongate raceme. . . 17. *D. nemorosa.*

*o.*Flowers heteromorphic, some with broad white petals 3.5–5 mm. long, others smaller, others apetalous; leaves mostly near base; flowering stems subscapiform, with short and thick flowering racemes.

Leaves obviously dentate, hispid with stipitate and sessile forking hairs; rachis and pedicels pubescent. . . . . . . 18. *D. cuneifolia.*

Leaves entire or nearly so, hirsute-ciliate with simple hairs (stellate on lower surface); rachis and pedicels glabrous. . . . . . 19. *D. reptans.*

*a.*Petals deeply cleft; dwarf annuals or winter-annuals; scapes naked, arising from basal rosettes. . . . . 20. *D. verna.*

1. D. Allénii Fern. (for its discoverer, John Alpheus Allen, 1863–1916). — Forming close mats; the caudices freely branched, with *filiform whitish branches covered with white subulate persistent old midribs; rosettes* lax; *their leaves thin,* oblanceolate, subacute, *glabrous,* the expanding ones ciliate with simple hairs, 0.5–1.5 cm. long, 1–3 *mm. wide, with midrib prominent beneath;* scapes filiform, naked (very rarely with 1 leaf near base), glabrous, 1–8 cm. high; mature raceme with rachis 0.2–2.5 cm. long, 2–8-flowered, the flowers short (up to 5 mm.) -pedicelled; *sepals narrowly oblong,* 1.2–2 mm. long, 0.5–1 *mm. wide,* glabrous; *petals* white, emarginate, 2–3 *mm. long,* 1–2 *mm. wide; silicles* glabrous, oblong-lanceolate, acute, 2.7–7 mm. long, 1–2 *mm. wide,* style definite, *valves reticulate-nervose, septum with a median fold.* — Calcareous schists and limestones, alpine and subalpine areas, Shickshock Mts., Gaspé Pen., Que.; ? Mt. Katahdin, Me. Late June, July. Fig. 1062.

1062. D. Allenii.

2. D. rupéstris R. Br. (growing on rocks). — Similar; the shorter *branches of the caudex retaining marcescent leaves or their shreds; rosette-leaves linear-oblanceolate to oblong, hispid with simple or variously forking hairs,* the midrib delicate and evanescent; *scape* capillary, naked, up to 1.1 dm. high, *hirtellous* below like the leaves; *mature raceme with rachis up to 6 cm. long,* 3–20-flowered; pedicels hirtellous, up to 4 mm. long; *sepals* 1.5 mm. long, *hirtellous;* petals white, 2.5–4 mm. long; *silicles* hirtellous with simple or forking hairs, *suberect or strongly ascending, oblong,* 3–8 mm. long, short-styled, 12–30-seeded. — Greenl., Lab. and Ung. (Eu.), represented with us by

Var. **leiocárpa** O. E. Schulz (smooth-fruited). — Silicles glabrous. — Calcareous gravel, nw. Nfld. July, early Aug. (Eu.) Fig. 1063.

1063. D. rupestris, v. leiocarpa.

3. D. nivàlis Lilj. (growing near snow). — Forming mats 1–10 cm. broad; *branches of caudex* clothed *with marcescent shreds* of dead leaves *below the* small compact *subglobose rosettes; leaves cuneate-obovate to broadly oblanceolate, obtuse, subcoriaceous,* 3–11 mm. long, 1–4.5 *mm. broad, with firm subulate midrib* prominent beneath, *canescent-pannose with minute stellate hairs;* scape filiform, up to 10 cm. high, stellate-tomentulose; mature raceme up to 6 cm. long, 1–13-flowered; sepals stellate-pilose, 1.5–2 mm. long; petals white, cuneate, 2.5–3 mm. long, 1.2–2 mm. broad; *silicles* lanceolate to narrowly oblong, *glabrous,* 4–9 mm. long, short-styled; the 14–28 seeds 0.7–1 mm. long. — Arct. reg., s. on calcareous rocks to n. Nfld., Gaspé Pen., Que., Ung., and n. Man. Late June–Aug. (N. Eurasia) Fig. 1064.

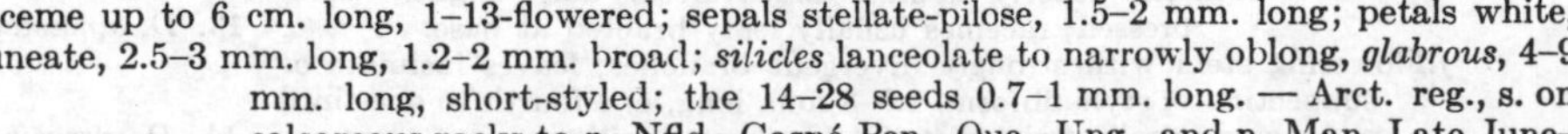

1064. D. nivalis.

4. D. Peàsei Fern. (for its discoverer, Arthur Stanley Pease, 1881–). — Similar; rosettes lax; *leaves linear or linear-oblanceolate, acute,* thick, 3–6.5 mm. long, 1–1.5 *mm. wide,* strongly ribbed, *loosely and coarsely stellate-pannose, finally glabrate,* subcoriaceous and lustrous; scapes filiform, 2 cm. high, stellate-hirsute; flowers unknown; fruiting raceme with rachis 1–2 cm. long, the stellate-hirtellous lower pedicels 4–6 mm. long; *silicles*

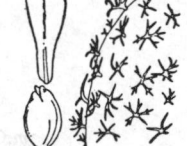

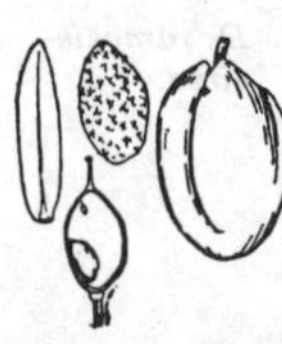

1065. D. Peasei.

3–6, *stellate-hirtellous, reticulate-veiny, elliptic,* 3–5.5 *mm. long,* 2–2.5 *mm. wide, with style about* 1 *mm. long; the* 8–10 *seeds* 1.2–1.8 *mm. long.* — Limestone-talus, Cape Rosier, Gaspé Co., Que. June. FIG. 1065.

5. D. minganénsis (Vict.) Fern. (of the Mingan Ids., Que.). *Biennial or perennial* with simple or multicipital caudex; *rosette-leaves oblanceolate,* entire or subentire, 0.5–9 cm. long, 1.5–17 mm. broad, more or less *canescent with stalked and few-rayed irregularly stellate hairs; flowering stems* 1-many, simple or branched, 0.4–4 dm. high, *pilose-hirsute with simple, forking and stellate hairs,* very leafy; *cauline leaves* 7–25, oblong or narrowly ovate, *rounded to a sessile base,* 0.5–4.2 cm. long, 0.2–2 cm. broad, *stellate-tomentulose;* mature racemes 3–9 cm. long, 1.5–3.5 cm. thick, usually standing well above the leaves, mostly 10–30-flowered; *pedicels spreading to arched-ascending,* pilose; sepals 2.3–3.3 mm. long, oblong or oval, with villous back; *petals yellow,* 3.5–5.5 mm. long, narrowly cuneate-obovate; *silicles lanceolate to linear-oblong, usually twisted,* 1–1.8 *cm. long, densely hirtellous with long divergent to retrorse simple or forked hairs* 0.5–1 *mm. long,* 30–40-seeded; *style* 1–1.5 *mm. long.* — Calcareous ledges and gravel, Mingan Ids. and Rimouski Co., Que., and Ung. Late June–early Aug. FIG. 1066.

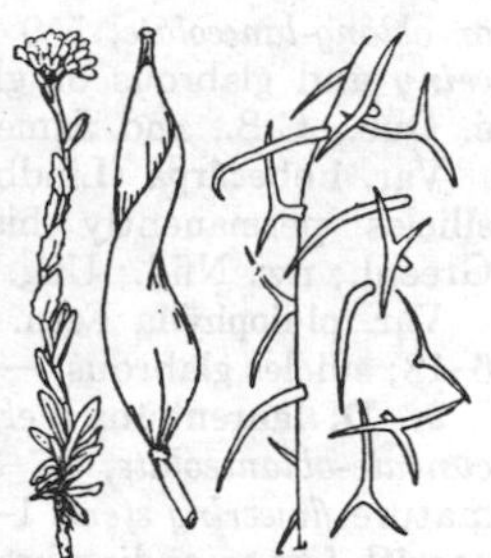
1066. D. minganensis.

6. D. incàna L. (hoary). — Biennial (rarely perennial), with simple to multicipital caudex; *subspherical rosettes of* 1st *year loosening and elongating to form the usually leafy* (*up to* 50 *or more leaves*) *flowering stem; rosette-leaves* lanceolate or oblanceolate, *ciliate with long simple or bifurcate hairs;* lower *cauline leaves* narrow, the upper gradually broader and oblong to ovate, *hirsute; flowering stem usually very leafy, pilose-hirsute to tomentulose or villous especially above and on the axis of raceme with implicated hairs;* mature racemes elongate, 1–15 cm. long, few–100-flowered, the *lower flowers often leafy-bracted;* petals white; *silicles oblong or elliptic to lanceolate,* 4–14 mm. long, *with short or obsolete styles,* glabrous, 16–40-seeded. — Turfy, gravelly or ledgy calcareous areas, Greenl. and Lab. to Nfld., M.I., Gaspé Co., Que., Ung. and islands of L. Sup., Mich. June–Aug. (Eurasia) FIG. 1067.

1067. D. incana.

Var. **confùsa** (Ehrh.) Lilj. (confused). — Silicles pubescent. — S. to. Nfld., M.I., ne. N.B. and shores of James Bay, usually commoner.

7. D. clivícola Fern. (dweller on slopes). — Loosely matted perennial; *rosettes lax; the linear-oblanceolate acute often incised leaves* 1–5 *mm. wide, hirtellous* with simple, forking and stellate hairs; *flowering stems filiform,* flexuous, becoming 0.3–3 dm. high, *hirtellous at base; cauline leaves* 3–13, *lanceolate to narrowly ovate,* 0.5–1.7 cm. long, 1.5–6 *mm. wide, hispid, incised-serrate or entire;* fruiting raceme elongate, lax, 2–13 cm. long, 5–19-flowered; *the lowest pedicels* 4–10 *mm. long, often subtended by a leafy bract; sepals oblong,* 1.6–2 *mm. long,* 0.4–0.9 *mm. wide, glabrous* or glabrate; *petals* white, 1–1.5 *mm. wide; siliques* glabrous, *linear or linear-lanceolate,* 7–12 mm. long, 2–2.5 *mm. wide,* with capitate stigma and reticulate-veiny valves, 16–26-seeded. — On alpine and subalpine calcareous schists, Shickshock Mts., Gaspé Pen., Que. Late June–early Aug. FIG. 1068.

1068. D. clivicola.

8. D. norvégica Gunn. (Norwegian). — Similar to no. 7, coarser; *rosette-leaves* narrowly oblanceolate to narrowly obovate, *hispid* with numerous simple, bifurcate and stellate trichomes; *mature fruiting stems* 0.1–2 dm. high, *hirsute,* especially below, *with simple and variously forking trichomes, with* 1 (rarely 0)–5 *ovate hispid leaves* 3–10 *mm. broad; mature central racemes elongated to* one-third to six-sevenths *full height of plant,* 5–25-flowered; *pedicels short,* stellate-hirtellous, *the lowest* 1–5 *mm. long; sepals* 1.8–2.6 mm. long, 1–1.5 *mm. broad; petals* white, 2–3 *mm. wide; silicles oblong*

1069. D. norvegica.

or oblong-lanceolate, 5–9 mm. long, 2–3.8 *mm. wide, with style* 0.2–0.5 *mm. long, the valves veiny* and glabrous or glabrate; seeds 14–28. — Calcareous ledges, gravel and turf, Nfld., e. Que., C.B., and James Bay, Ung. June–early Aug. (Eu.) Fig. 1069.

Var. **hebecárpa** (Lindbl.) O. E. Schulz (hairy-fruited). — Flowering stem hispid to summit; silicles permanently hispid with mixed simple and forking hairs. — Similar habitats, Greenl.; nw. Nfld.; Ung. (Eu.)

Var. **pleiophýlla** Fern. (many-leaved). — Fruiting stems up to 2.7 dm. high; cauline leaves 6–18; silicles glabrous. — Similar habitats, nw. Nfld. and e. Que.

9. D. laurentiàna Fern. (of the Gulf of St. Lawrence). — Matted perennial; *rosette-leaves cuneate-oblanceolate,* 0.7–3.8 *cm. long,* 4–9 *mm. wide,* thick and firm, *densely stellate-pannose;* mature *flowering stems* 1–3.5 dm. high, stoutish, simple, or with few ascending branches, *pilose-hirsute below* with mixed simple, furcate and stellate hairs; *cauline leaves* (3–)6–25, *oblong,* mostly 1–3 cm. long, 3–11 *mm. wide,* coarsely dentate, *stellate-pannose;* central fruiting raceme 3–15 cm. long, 6–30-flowered, lax, the lower pedicels 1.5–6 mm. long; *sepals* oval, broadly white-margined, 2.3–3 mm. long, 1.3–2.3 *mm. wide; petals* white, broadly obovate, 4.5–5 *mm. long,* 2.5–4 *mm. wide; silicles* glabrous, elliptic to oblong-lanceolate, flat or twisted, 5–14 mm. long, 2–6 *mm. wide, with rugulose or veiny valves* and very short style; *seeds* 20–40, 1–1.3 *mm. long.* — Calcareous turf and gravel near Gulf of St. Lawrence, n. Nfld. to Mingan Ids., Que. June, July. Fig. 1070.

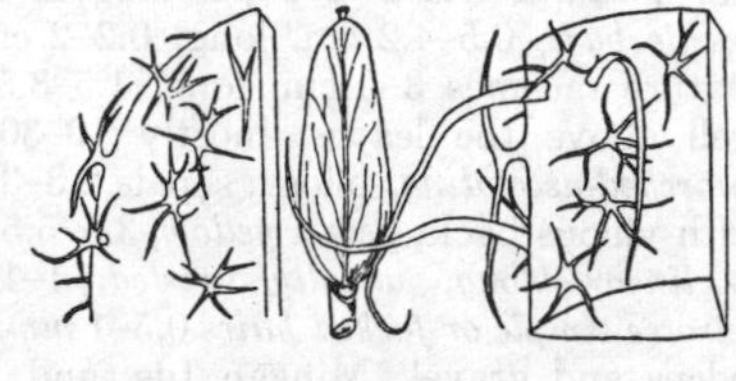
1070. D. laurentiana.

1071. D. pycnosperma.

10. D. pycnospérma Fern. & Knowlt. (with crowded seeds). — Forming mats up to 2.5 dm. broad; the pale slender mostly forking *caudices* retaining shreds of old leaves and *often with persistent bases of stellate trichomes; rosettes flat,* their *leaves cuneate-spatulate to narrowly rhombic,* 0.7–5 cm. long, 2–15 mm. broad, entire or shallowly dentate above middle, *finely and closely stellate-pannose;* flowering *stems* simple or loosely upright-branching, in fruit 0.5–3.3 dm. high, very slender, *loosely to densely stellate-pubescent, often with admixture of forking and simple hairs; cauline leaves* 1–4, *ovate to oblong, sparsely hirsute* with simple, bifurcate and stellate hairs *or glabrate; fruiting racemes very lax and slender,* 0.15–2 dm. long, 0.6–1.8 *cm. thick,* with pedicels mostly 3–12 mm. apart; *sepals oblong,* 1.5–2 *mm. long,* 1 *mm. broad; petals* white, 2 *mm. broad; silicles plump,* compressed-ovoid, or -oblong, 2.5–10 mm. long, 1.8–3.3 mm. broad, glabrous. with very short style; *seeds* 10–32, *closely imbricated and often oblique to septum,* 1–1.4 mm. long, — On limestone near Gulf of St. Lawrence, nw. Nfld. and e. and n. Gaspé Co., Que. June, July. Fig. 1071.

11. D. glabélla Pursh (becoming smooth). — Matted perennial; *rosette-leaves* cuneate-oblanceolate to spatulate, *attenuate to petiolar base,* 0.7–4.5 cm. long, 2–10 mm. broad, *closely stellate-pannose; flowering stems* simple to freely forking, 0.5–4 dm. high, *stellate-pannose,* often glabrate at summit; *cauline leaves* 1 (rarely 0)–5, *ovate to oblong, rounded or subamplexicaul at base,* dentate or entire, 0.8–4 cm. long, 2–12 mm. broad; mature *racemes lax, the longer* 0.3–1.5 dm. long, mostly *with* 5–15(–20) *subdistant silicles; pedicels finally ascending, stoutish, the lowest mostly* 1–6 *mm. long; sepals* 2–3 mm. long, 1–2 *mm. broad,* white-margined; *petals* white, 3–5.5 mm. long, 2–4 *mm. broad; silicles narrowly to broadly lanceolate, acute to subacute,* 6–13 mm. long, 1.5–3 *mm. wide, with obsolete or very short and thick style, flattish, the valves definitely veiny,* glabrous or hirtellous; *seeds* 18–38, 0.7–1 *mm. long.* (*D. hirta* of auth., not L.) — Arct. reg., s., chiefly on calcareous rock, to Nfld., Que., L. Champlain, N.Y., and shores of Hudson Bay. June–Aug. (Eurasia)

1072. D. glabella, v. orthocarpa.

1073. D. glabella, v. megasperma.

Var. **orthocárpa** (Fern. & Knowlt.) Fern. (straight-fruited). — Similar; cauline leaves 5–8; longer racemes mostly 10–35-flowered. (*D. arabisans*, var. *orthocarpa* Fern. & Knowlt.) — Limestone-rocks, Mingan Ids. and Gaspé Co. to Temiscouata Co., Que.; Keewatin. Fig. 1072.

Var. **megaspérma** (Fern. & Knowlt.) Fern. (large-seeded). — Coarser; cauline leaves 5–14; silicles elliptic to oblong-ovate, obtuse, 2.5–5 mm. wide; seeds 1–1.3 mm. long. — S. Lab., Nfld., e. Que. and ne. N.B. Fig. 1073.

12. D. arábisans Michx. (passing to *Arabis*). — Forming loose mats up to 2.5 dm. across; *rosette-leaves* oblanceolate or spatulate, *attenuate to a petiolar base*, 0.7–7 cm. long, 0.2–1.6 cm. broad, *thin, closely and minutely stellate-pannose*, in age sometimes glabrate; *flowering stems* slender, simple or *with* often *flexuous loosely ascending branches*, 0.5–4.5 dm. high, *glabrous or stellate-tomentulose*, often glabrate above; *cauline leaves* 3–12, oblanceolate, oblong or narrowly obovate, *cuneate or but slightly rounded at base*, 0.5–4.5 cm. long, 0.2–1 cm. broad; *racemes rather lax in fruit*, the *central ones* 7–25-flowered, *becoming* 2–12 *cm. long and* 1.3–3 *cm. thick; pedicels slender*, glabrous or nearly so, *divergent or arched-ascending, the lowest* 3–15 (–25) *mm. long;* sepals glabrous or sparsely hirtellous, 1–1.5 mm. wide; petals white, 5–6.5 mm. long, 2.8–3.8 mm. broad; *silicles very thin, glabrous*, narrowly lanceolate to narrowly ovate or elliptic, *commonly acuminate, usually twisted*, sometimes flat, 5–15 mm. long, 1.5–3 mm. broad, *with slender style* 0.5–1 *mm. long* and *smooth and veinless valves; seeds* 12–36, 1.1–1.7 *mm. long.* — Argillaceous or calcareous rocks, Nfld. to Ont., s. to N.B., Me., Vt., N.Y., Mich., Wisc. and Minn. May–July. Fig. 1074.

1074. D. arabisans (typical).

1075. D. arabisans, v. canadensis.

Var. **canadénsis** (Brunet) Fern. & Knowlt. (Canadian). — Low, 0.7–1.5 dm. high; silicles elliptic-ovate, 3–8 mm. long, 2–4 mm. broad. — Scattered in the general range, Nfld. to n. Mich. Fig. 1075.

13. D. lanceolàta Royle (lanceolate). — Forming loose mats up to 1.5 dm. across; *rosette-leaves crowded*, oblanceolate to spatulate, 0.7–4 cm. long, 1–8 mm. broad, *cinereous with dense and soft stellate tomentum; flowering stems* 0.5–3.5 dm. high, *stellate-tomentulose* and short-pilose; cauline leaves 2–10, lanceolate to narrowly ovate, 0.5–3 cm. long; *racemes often leafy-bracted at base, in maturity one-third to four-fifths full height of plant; pedicels strongly ascending in fruit;* sepals pilose; petals white, 3–5 mm. long; *silicles linear-lanceolate to elliptic-ovoid, plump*, 4–14 mm. long, 1.5–2.5 mm. broad, twisted or plane, *the convex valves densely stellate-pannose;* style up to 0.75 mm. long; seeds 20–48. (*D. stylaris* auth., not J. Gay; *D. cana* Rydb.) — Calcareous cliffs and slopes, Gaspé Co., Que., to Yuk., s. to N.B., n. N.E., Bruce Co., Ont., n. Mich., ne. Wisc., Colo. and Utah. May, June. (Asia) Fig. 1076.

1076. D. lanceolata.

14. D. ramosíssima Desv. (much branched). — Forming broad mats (up to 3 dm. broad), with elongate branching caudices; *rosette-leaves* membranaceous, cuneate-oblanceolate, 1–5 cm. long, 2–15 mm. broad, minutely stellate-pilose, *with horizontally divergent narrow teeth* or merely low-serrate; *flowering stems* 0.7–4.5 dm. high, *divergently branched above, forming corymbosely paniculate inflorescences*, pubescent with mixed simple, forking and stellate hairs; cauline leaves 2–24, lanceolate to ovate, coarsely serrate to deeply pectinate; racemes becoming very loose; *petals* white, 5–7 *mm. long; silicles long-pedicelled, linear-lanceolate to elliptic-oblong, usually spirally twisted, stellate-pubescent*, 3–11 mm. long, with filiform *style* 1.5–3 *mm. long; seeds* 7–15, 1.2–1.8 *mm. long.* — Calcareous cliffs and rocks, Va., W.Va., Ky., Tenn. and N.C. Late April–early June. Fig. 1077.

1077. D. ramosissima.

Var. **glabrifòlia** E. L. Br. (smooth-leaved). — Leaves and stem glabrous or essentially so. — Limestone-cliffs of Ky. R., Ky.

15. D. aprìca Beadle (of sunny spots). — Annual or winter-annual, with slender stem 0.5–3.5 dm. high, simple or loosely branched and *stellate-puberulent*, rarely with elongate branches above but usually *with abbreviated or subsessile short corymbs* in the middle and upper axils; lower leaves obovate to rhombic-oval, coarsely 2–4-dentate, thin, petioled, 0.7–2 cm.

1078. D. aprica.

1079. D. brachycarpa.

long, minutely strigose-hirtellous with stellate trichomes; cauline leaves remote, up to 17 in number, narrow, entire or sparsely dentate; *flowers apparently hermaphrodite, some apetalous, others petaliferous;* sepals narrowly oblong, 0.8–1 mm. long; *petals* (when present) white, *spatulate,* 2.5–3 mm. long; *silicles linear-ellipsoid, 4–6 mm. long, minutely stellate-puberulent,* with minute style; *seeds* 6, *1–1.5 mm. long.* — Woods and clearings, local, n. Ga. to Ark. and se. Mo. Apr., May. Fig. 1078.

16. D. brachycárpa Nutt. (short-fruited). — *Annual or biennial* (forming rosettes the 1st year); *stems simple or* more commonly *bushy-branched,* either from base or from upper axils, stellate-hirtellous or -strigose, 0.4–2 dm. high; basal leaves petioled, 0.5–2 cm. long; cauline 3–12, sessile, lower elliptic to obovate, upper narrower; *racemes in fruit lax,* up to 7.5 cm. long, with spreading-ascending short pedicels; *flowers di- or trimorphic,* the smaller apetalous and cleistogamous, others with small narrow petals, others with obovate white petals 2–3 mm. long; *silicles oblong-ellipsoid, glabrous,* 1.7–5 mm. long; seeds 10–16, 0.5–0.8 mm. long. — Dry open soil, grassland and waste ground, Fla. to Tex., n. to Va., Ind., s. Ill., Mo. and Kans.; Oreg. Mar.–May. Fig. 1079.

17. D. nemorȏsa L. (of woods). — *Annual or winter-annual,* 0.5–3 dm. high, *simple or with few ascending branches; stem hispid* with mixed simple, bifurcate and stellate hairs; *rosette-leaves oblong-obovate or elliptic, obtuse, with pubescence as on stem; cauline leaves* 0–7 *on primary axis,* 1–5 *on branches,* oblong to ovate, acute, with 2–6 pairs of teeth; *fruiting raceme lax, elongating to nearly full height of plant; petals pale yellow,* turning whitish, *narrowly cuneate, about* 2 *mm. long;* fruiting pedicels widely spreading to subascending, often arching, 0.7–2.5 cm. long; *silicles narrowly oblong,* 3–13 *mm. long, minutely strigulose-hispid;* seeds 30–50. — Dry open soil, n. B.C. to Nev. and Utah, e. to w. Ont. and Mich. May, June. (Eurasia) Fig. 1080.

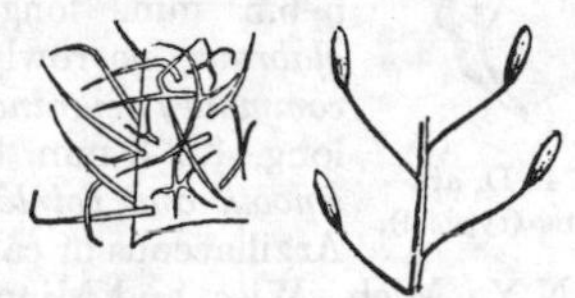

1080. D. nemorosa.

Var. **lejocárpa** Lindbl. (smooth-fruited). — Silicles glabrous. (*D. lutea* Gilib.) — Alask. to Colo., e. to Hudson Bay, L. Sup., Ont. and Minn.; adv. e. to Que. (Eurasia)

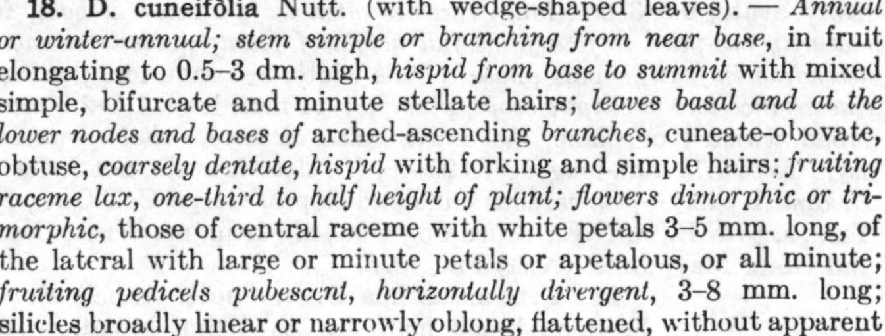

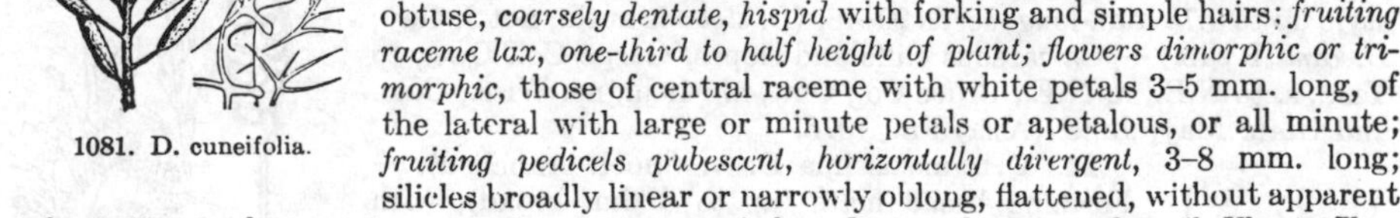

1081. D. cuneifolia.

18. D. cuneifȏlia Nutt. (with wedge-shaped leaves). — *Annual or winter-annual; stem simple or branching from near base,* in fruit elongating to 0.5–3 dm. high, *hispid from base to summit* with mixed simple, bifurcate and minute stellate hairs; *leaves basal and at the lower nodes and bases of* arched-ascending *branches,* cuneate-obovate, obtuse, *coarsely dentate, hispid* with forking and simple hairs; *fruiting raceme lax, one-third to half height of plant; flowers dimorphic or trimorphic,* those of central raceme with white petals 3–5 mm. long, of the lateral with large or minute petals or apetalous, or all minute; *fruiting pedicels pubescent, horizontally divergent,* 3–8 mm. long; silicles broadly linear or narrowly oblong, flattened, without apparent style, 6–16 mm. long, strigose-setulose, 20–48-seeded. — Dry rocky or sandy soil, Fla. to Tex. and n. Mex., n. to Ky., s. Ill., Mo., Kans. and Colo. Mar.–May. Fig. 1081.

19. D. réptans (Lam.) Fern. (creeping). — Similar to no. 17, the larger plants with depressed or diffuse branches, 0.2–1.5 dm. high, *hispid below, glabrous above and on the rachis and pedicels; leaves in a basal rosette and* 1 or 2 pairs or groups of 3 *crowded above it or at the bases of the otherwise scapiform branches,* spatulate-obovate, *entire or nearly so, bristly-ciliate,* the lower surface with stellate and forking hairs; *fruiting raceme short and subumbelliform, its axis* 0.3–4 cm. long, *glabrous;* flowers dimorphic; *fruiting pedicels divergent or arched-ascending;* silicles linear or narrowly oblong, 0.5–2.2 cm. long, glabrous; seeds 15–60. (*D. caroliniana* Walt.) — Dry sands and ledges, Ga. to N.M., n. to e. Mass., R.I., Ct., se. and n. N.Y., e. Pa., s. Ont., Mich., Wisc., Minn., N.D. and Colo.; also Wash. and Oreg. Late March–early June. Fig. 1082.

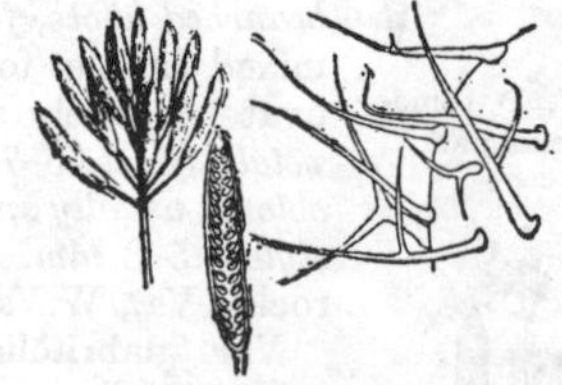

1082. D. reptans.

Var. **micrántha** (Nutt.) Fern. (tiny-flowered). — Silicles minutely hispid. (*D. micrantha* Nutt.; *D. coloradoensis* Rydb.) — Chiefly western, n. to Ill., Minn., S.D., Mont. and Wash.

1083. D. verna.

20. D. VÉRNA L. (of spring), WHITLOW-GRASS. — *Small annual or winter-annual; leaves all in a basal rosette,* oblanceolate or spatulate, more or less pilose on upper surface with mixed hairs, 0.5–2.5 cm. long; *scapes filiform,* unequal, in fruit elongating to 0.3–3 dm. high; *raceme* in fruit *becoming elongate and lax;* sepals 1–2 mm. long; *petals* white, *bifid half their length,* 1.5–2.5 mm. long; silicles narrowly oblong-elliptic, 4–10 mm. long, 1.5–2.3 mm. broad, glabrous, with 40–60 seeds. (*Erophila verna* (L.) Mey.) — Roadsides, fields and open, dry places, Mass. to Ill., Ky., Tenn. and N.C. Mar.–early June. (Natzd. from Eu.) FIG. 1083. — Passing into

Var. BOERHAÀVII Van Hall (in honor of HERMAN BOERHAAVE, 1668–1738). — Silicles broadly elliptic, elliptic-obovate or rounded, 2.5–6 mm. long, 2–4 mm. broad; seeds fewer. (Var. *aestivalis* Lej.; *Erophila Boerhaavii* (Van Hall) Dumort.) — Mass. to s. Ont., s. to Ga. and Tenn. (Natzd. from Eu.)

2. BERTERÒA DC.

Silicle elliptic, with slender style; seeds several, winged. Petals white, 2-parted. Pubescence stellate. — Tall leafy branching plants, separated from *Alyssum* by deeply cleft white petals and margined seeds, nat. of Old World. — (Named for *Carlo Guiseppe Bertero*, Piedmontese botanist, 1789–1831.)

1084. B. incana.

1. B. INCÀNA (L.) DC. (hoary), HOARY ALYSSUM. — Gray-green, 0.3–1 m. high, branched above; leaves entire, lanceolate; *silicles canescent-pubescent, plump,* 2.5–3.5 *mm. thick.* — Fields and waste places, N.S. to Mont., s. to N.E., N.J., Pa., W. Va., O., Ind., Ill., Mo. and Kans. June–Sept. (Natzd. from Eu.) FIG. 1084.

2. B. MUTÁBILIS (Vent.) DC. (changeable). — Similar; *silicles sparingly pubescent or glabrate, flattish,* 4.5–6 *mm. broad.*— Roadsides and cult. ground, Mass. and Kans.; doubtless elsewhere. June–Sept. (Adv. from Eu.)

3. LOBULÀRIA Desv. SWEET ALYSSUM

Silicle as in *Alyssum.* Petals white, entire. Cotyledons accumbent. Hairs of the stem and leaves 2-pointed, appressed, attached in the middle. — Low perennials with narrow leaves, nat. of Old World. (Latin *lobulus*, a little lobe, probably referring to the 2-lobed hairs.)

1. L. MARÍTIMA (L.) Desv. (of the sea). — Slightly hoary; leaves linear; flowers small, honey-scented. (*Alyssum* Lam.; *Koniga* R. Br.) — Often cult., and occasionally spontaneous, N.E. to Mich. and southw. June–Nov. (Introd. from Eu.)

4. ALÝSSUM L.

Silicle small, orbicular, with only one or two wingless seeds in a locule; valves nerveless, somewhat convex, the margin flattened. Flowers yellow or white. Cotyledons accumbent. Plant stellate-pubescent. — Low herbs nat. of Old World (one in Alaska). (Greek name of plant reputed to check hydrophobia, from *a*-privative and *lyssa, rabies.*)

1. A. ALYSSOÌDES L. (old generic name, like *Alyssum*). — Dwarf hoary annual with stems simple or usually several from the base, arched-ascending, 1–3 dm. high, and with linear-spatulate leaves, pale yellow or whitish cuneate petals little exceeding the persistent calyx, and orbicular sharp-margined 4-seeded silicle, the style minute. — Grassland, roadsides and waste places, Que. to B.C., s. to N.J., Pa., W.Va., O., Ind., Ill., Minn., Utah and Calif. April–June (Natzd. from Eu.) FIG. 1085.

1085. A. Alyssoides.

A. SAXÁTILE L. (living on rocks), GOLDEN-TUFT, a profuse matted perennial with brilliant golden-yellow flowers and glabrous suborbicular silicles with long persistent style, is beginning to spread from cult. (Introd. from Eu.)

2. A. PETRAÈUM Ard. (of rock). — Biennial, with 1st year's rosettes of round-tipped oblanceolate dentate gray-green minutely stellate-pubescent leaves; the simple or loosely branching flowering stem 2–3 dm. high, with narrowly oblanceolate leaves; the glabrous silicles elliptic, about 5 mm. long and tipped by a slender style. Cliffs in Oneida Co., N.Y. Late May, June. (Adv. from Eu.)

5. LESQUERÉLLA S. Wats.

Silicle mostly globular or inflated, with a broad orbicular to ovate hyaline partition nerved to the middle, the hemispherical or convex thin valves nerveless. Seeds few or several, in 2 rows, flat. Cotyledons accumbent. Filaments toothless. — Low Am. herbs, hoary with stellate hairs or lepidote. Flowers mostly yellow, the petals spatulate. (Named for *Leo Lesquereux*, distinguished bryologist and paleobotanist, 1805–1889.)

a.Stout-based perennials with long-lived caudices; flowering stems arising from among the basal leaves, the terminal bud of the rosette not developing a flowering stem.

Rosette-leaves spatulate to oblanceolate, 1–5 cm. long; sepals elliptic, round-tipped; petals obovate; silicles 3.7–7 mm. in diameter, glabrous or lepidote, on ascending pedicels. . . . . 1. *L. Purshii.*

Rosette-leaves broadly linear to narrowly oblanceolate, 3–10 cm. long; sepals lanceolate, tapering; petals oblanceolate to spatulate; silicle 3–4 mm. in diameter, densely pilose, on recurving pedicels. . . . . . . 2. *L. ludoviciana.*

a.Slender-rooted annuals, biennials or short-lived perennials, without heavy caudex; central bud of the rosette developing into a flowering stem. . . *b*.

b.Silicles glabrous, 2–4 mm. in diameter, on ascending pedicels.

Silicles 2–3 mm. in diameter, sessile or barely stipitate; ovules 2 in each locule; filaments abruptly dilated at base. . . . . 3. *L. angustifolia.*

Silicles 3–4 mm. in diameter, on slender stipes 1–2 mm. long; ovules 8–10 in each locule; filaments linear throughout. . . . . 4. *L. gracilis*

b.Silicles pilose, 1–2.5 mm. in diameter, on divergent pedicels. . . . . . 5. *L. globosa.*

1. L. Púrshii (S. Wats.) Fern. (for its discoverer, FREDERICK TRAUGOTT PURSH, 1774–1820). — Perennial, with stout simple to multicipital caudex covered with bases of old leaves and stems; flowering stems decumbent, arcuate-ascending or erect, from among the old basal leaves, 0.2–2 dm. long, silvery-stellate; *rosette-leaves spatulate or oblanceolate,* petioled, 1–5 *cm. long,* silvery, stellate-pannose; *cauline leaves* 2–8, *narrowly to broadly oblanceolate,* small; *flowers* 2–8 *on* scurfy *ascending pedicels; sepals elliptic, rounded at summit,* 2.7–5 mm. long; *petals obovate,* 5–8 mm. long; *silicles* globose or subglobose, *lepidote to glabrous,* 3.7–7 *mm. in diameter, on ascending pedicels* 2–16 mm. long, the ovules 4–6 in each locule. — Calcareous cliffs, barrens and gravels, n. and w. Nfld. and Anticosti; n. B.C. July, early Aug.

2. L. ludoviciàna (Nutt.) S. Wats. (of old Louisiana). — Similar to no. 1, but the stellate pubescence coarser; stems 1–4 dm. long; *rosette-leaves broadly linear to narrowly oblanceolate,* 3–10 *cm. long,* very densely stellate-pilose, scabrous; *cauline leaves* 8–15, similar, smaller; *flowers* 15–50 *or more; sepals lanceolate, tapering to tip,* 4–7 mm. long; *petals oblanceolate to spatulate; silicles densely stellate-pilose,* 3–4 *mm. in diameter, on arched-recurving pedicels* 1–2 cm. long. (*L. argentea* of ed. 7, not S. Wats.) — Sands and gravels, Illinois R., Ill.; Minn. to Mont., s. to Kans., Colo. and Ariz. May–July.

3. L. angustifòlia (Nutt.) S. Wats. (narrow-leaved). — Slender annual or biennial with slender tap-root; *flowering stems filiform,* numerous, simple or branching, 1–2.5 dm. high, appressed-canescent; *rosette-leaves* oblanceolate to elliptic, entire to sublyrate, *sparsely pubescent,* slender-petioled, 1–2 cm. long; cauline leaves 5–10, linear or narrowly oblanceolate, 1–2 cm. long, entire; flowers 5–20, on filiform ascending pedicels; sepals lance-oblong, 3–5 mm. long; petals narrowly obovate, 5–7 mm. long; *filaments abruptly dilated at base; silicles* globose, *glabrous, sessile or short-stipitate,* 2–3 *mm. in diameter,* on ascending pedicels up to 1 cm. long, mature style 3–4 mm. long; *ovules* 2 *in each locule.* — Limestone-barrens, Mo. and Ark. to Tex. Apr.–June.

4. L. grácilis (Hook.) S. Wats. (slender). — *Coarser;* stems erect to decumbent, 3–5 dm. high, remotely stellate-hirtellous; rosette-leaves 3–8 cm. long; *cauline leaves* oblanceolate to broadly linear, repand, lyrate or entire, 1–4.5 *cm. long; filaments linear, not dilated at base; silicles* 3–4 *mm. in diameter, on slender stipes* 1–2 *mm. long; ovules* 8–10 *in each locule.* — Prairies, fields and waste places, se. Ia. to Tex. Apr.–June. FIG. 1086.

1086. L. gracilis.

5. L. globòsa (Desv.) S. Wats. (globose). — *Biennial or short-lived perennial,* loosely canescent-stellate; stems several, simple or branched, 2–5 dm. high; radical leaves oblong, entire or repand-dentate, 2.5–5 cm. long; cauline leaves numerous, oblanceolate to linear, entire or repand; petals spatulate, 4–5 mm. long; *silicles stellate-pilose,* 1–2.5 *mm. in diameter, sessile* (nonstipitate), *on* filiform *divergent pedicels;* mature styles 2–3 mm. long; *ovules* 1 *or* 2 *in each locule.* — Calcareous bluffs, se. O., Ky. and Tenn. Apr.–June. FIG. 1087.

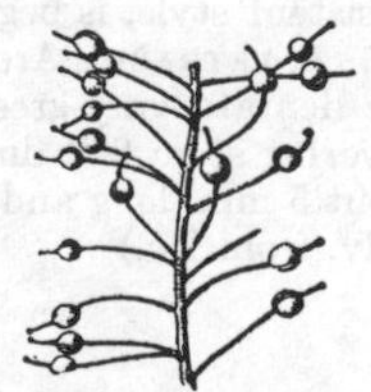

1087. L. globosa.

6. THLÁSPI L. Penny-Cress. Cents (Que.)

Silicle orbicular, obovate, or obcordate, flattened contrary to the narrow partition, the midrib or keel of the navicular valves extended into a wing. Seeds 2–8 in each locule. Cotyledons accumbent. Petals equal. — Low plants nat. of Old World, with undivided radical leaves, the cauline leaves sagittate and clasping, and small white or purplish flowers. (Name from the Greek *thlaein, to crush*, from the flattened silicle.)

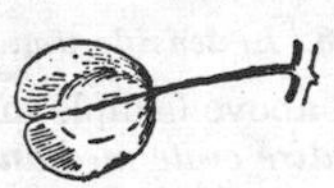
1088. T. arvense.

1. T. arvénse L. (of cultivated land), Field-P., Mithridate-Mustard, Frenchweed, Fanweed, Stinkweed. — Smooth erect annual up to 8 dm. high; lowest leaves petioled, narrowly obovate, soon dropping; *middle and upper membranaceous leaves oblong*, dentate or entire, sessile, with sagittate base; petals white, 3–4 mm. long; *silicles suborbicular to rounded-oblong, 1–1.8 cm. long, broadly winged, deeply emarginate; seeds* compressed, ovoid, *blackish, with concentric ridges, 2–2.3 mm. long.* — Roadsides, waste places and fields, Greenl. and Lab. to Alaska, s. beyond our limits; abund. northw. and often a troublesome weed of grainfields. April–Aug. (Natzd. from Eu.) Fig. 1088.

2. T. perfoliàtum L. (perfoliate). — More slender, 1–3 dm. high; *stem-leaves ovate-sagittate*, nearly entire; petals 2–3 mm. long; *silicles obliquely obovate, 4–6 mm. long*, narrowly winged; *seeds yellow-brown, smooth*, 1–1.5 *mm. long.* —Waste places and fields, s. Ont. to Md., Va., Ky. and Mo. March, April. (Natzd. from Eu.)

Ibèris (ancient Greek name) amàra L. (bitter), Candytuft, a stiff branching annual with toothed or cut narrow leaves, the showy white or crimson flowers with the 2 lower petals much larger than the others, and flat rounded silicles with the long style projecting from the terminal notch, spreads from gardens, but is hardly naturalized. (Introd. from Eu.)

7. TEESDÀLIA R. Br.

Silicle rounded or obovate, emarginate, flattened contrary to the partition, concave, the valves navicular and strongly margined above. Seeds 2 in each locule. Cotyledons accumbent. Petals unequal, 2 larger than the others. Longer filaments with a scale-like appendage at base. — Small annuals or winter-annuals nat. of Eurasia and n. Afr., with pinnately lobed rosulate leaves and scapiform or slightly forking stems, with small white flowers. (Named for *Robert Teesdale*, 17th century English botanist and horticulturist.)

1. T. nudicaùlis (L.) R. Br. (naked-stemmed). — Rosette-leaves oblanceolate, obtuse, 2–3 cm. long, blunt-lobed; stems 1–many, erect or arched-ascending, 1–2 dm. high; flowers about 1 mm. long; silicles 1.5–4 mm. long. — Sandy fields, roadsides and waste places, local, se. Mass. to N.C. March–May. (Natzd. from Eu.)

8. LEPÍDIUM L. Pepperwort. Peppergrass. Tonguegrass.

Silicle roundish, much flattened contrary to the narrow partition, promptly dehiscent; valves navicular. Seeds solitary in each locule, pendulous. Cotyledons incumbent, or in no. 5 accumbent! Flowers small, white or greenish. — Herbs of temp. and warm reg., with slender ascending racemes and uniform slender divaricate pedicels. (Name from the Greek *lepidion, little scale*, alluding to the fruit.)

*a.*Upper leaves perfoliate, rounded. . . . . 1. *L. perfoliatum.*

*a.*Upper leaves tapering or merely sagittate or auriculate-clasping, elongate. . . *b.*

*b.*Silicles rounded-ovate, not evidently notched, tipped by the nearly sessile stigma; middle and upper leaves tapering at base. . . . . 2. *L. latifolium.*

*b.*Silicles notched or emarginate at apex, the style little if at all projecting above the notch. . . *c.*

*c.*Densely villous; cauline leaves sagittate-clasping; silicles 5–6 mm. long. 3. *L. campestre.*

*c.*Glabrous or only minutely pubescent; cauline leaves tapering to base; silicles shorter. . . *d.*

*d.*Stamens 6; silicles 5–6 mm. long, prominently winged throughout, on thick ascending pedicels. . . . . 4. *L. sativum.*

*d.*Stamens 2 (rarely 4); silicles 2–4 mm. long, barely if at all winged (except sometimes at summit), on slender spreading pedicels. . . *e.*

*e.*Petals equaling or exceeding sepals; cotyledons accumbent (the edges against the radicle as shown in cross-section of seed); silicles 2.5–4 mm. long. . . . . 5. *L. virginicum.*

*e.*Petals wanting or shorter than sepals; cotyledons incumbent (with the back of 1 against the radicle), rarely oblique; silicles mostly 2–3 mm. long.

Branches bearing many short corymbiform racemes in the leaf-axils as well as longer (2–4 dm. long) terminal ones (in fruit). . 6. *L. ramosissimum.*

Branches bearing simple naked elongate racemes (in fruit mostly 0.5–1.5 dm. long).

Fetid; lower leaves bipinnatifid; silicles oval or elliptic, gradually narrowed to the apical teeth. . . . . . 7. *L. ruderale.*

Nearly scentless; lower leaves pinnatifid or toothed; silicles rounded-obcordate or round-obovate, broadly curved at summit. . . . . 8. *L. densiflorum.*

1. L. perfoliàtum L. (through the leaf). — Annual, divergently branched above (simple in starved specimens), 1–5 dm. high; lower *leaves* dissected and elongate, *the entire ovate median ones deeply cordate-clasping, the upper more rounded and perfoliate;* petals linear, a little longer than sepals; stamens mostly 6; *silicles* on spreading-ascending pedicels, *rhombic-ovate,* 3.5-4 mm. long, tapering to slightly notched summit. — Roadsides, fields and waste places, local eastw., N.E. to Mich., s. to Pa., Mo., and Kans., becoming frequent westw. across the continent. *Fr.* June, July (Natzd. from Eu.)

2. L. latifòlium L. (broad-leaved). — *Tall perennial,* with subterranean rhizomes, 1–2 m. high, glaucous, nearly or quite glabrous; basal leaves long-petioled, with oblong dentate blades up to 3 dm. long; *cauline leaves* lanceolate, oblong or rhombic, entire or toothed, the upper *narrowed to sessile or short-petioled bases; racemes short and corymbiform, crowded along the branches of the large corymbiform panicle;* petals spatulate, 1.5 mm. long; stamens 6; *silicles rounded-ovate, 2 mm. long, with nearly sessile stigma.* — Beaches, tidal shores and waste ground, local, coast of Mass. to L.I. *Fr.* Aug.–Oct. (Natzd. from Eu.)

1089. L. campestre.

3. L. campéstre (L.) R. Br. (of fields), Cow-Cress. — *Densely short-villous annual or biennial,* simple to freely branching, 1.5–7 dm. high; basal leaves petioled, entire to pinnatifid; *cauline leaves* numerous, overlapping, *sagittate-clasping,* suberect, denticulate; petals slightly exceeding sepals; stamens 6; *silicles oblong-ovate,* 5–6 *mm. long,* slightly emarginate, *concave above,* often papillose. — Waste places, roadsides and fields, freq. throughout our range and westw. *Fr.* May–Sept. (Natzd. from Eu.) Fig. 1089.

4. L. satìvum L. (sown), Garden-Cress. — Glabrous annual 2–8 dm. high, glaucous; *leaves all dissected, the lower bipinnatifid;* racemes strict, elongate; petals 2 mm. long; *stamens* 6; *silicles oblong-ovate,* 5–6 *mm. long, broadly winged* and concave, *on short and thick ascending pedicels.* — Roadsides and waste places, Que. to N.S. and N.E., casual esc. from cult. elsewhere. *Fr.* July–Sept. (Introd. from Eu.) Fig. 1090.

1090. L. sativum.

5. L. virgínicum L. (Virginian), Poor-Man's Pepper. — Annual or biennial, smooth or minutely pubescent, green, 2–9 dm. high; basal leaves incised, pinnatifid or pinnate; cauline leaves ascending, lanceolate or linear, incised to entire; racemes elongate; *petals equaling or exceeding sepals;* stamens 2 (rarely 4); *silicles nearly orbicular,* emarginate at summit, 2.5–4 *mm. long,* with very short style, on slender subascending to divergent pedicels; *cotyledons accumbent.* — Dry open soil, roadsides and waste places, throughout our range (rare at the extreme north) and westw. and southw. *Fr.* June–Nov. (Adv. in Eu.) Fig. 1091.

1091. L. virginicum.

6. L. ramosíssimum Nels. (much branched). — Resembling no. 8, but stems and branches more pulverulent and intricately branched; *lower cauline leaves few-toothed, upper linear and entire; inflorescences with many short corymbiform racemes in the leaf-axils as well as terminal longer racemes becoming 2–4 dm. long;* petals linear, shorter than the oblong sepals; *silicles elliptic, tapering somewhat to summit,* 2.5–3.5 mm. long; cotyledons incumbent. — Dry open soil, Rocky Mts. and Great Plains, spreading eastw. on roadsides and along railroads to e. Que. *Fr.* June–Oct. (Adv. from farther west)

7. L. ruderàle L. (of rubbish). — *Fetid* biennial or annual 1.5–5 dm. high; *basal leaves bipinnatifid,* the lower cauline divided, the upper linear; *racemes mostly simple, elongating* to 1 dm. or more; *sepals linear; petals* 0; *silicles oval or elliptic-ovate, gradually narrowed to the apical teeth,* 2–3 mm.

1092. L. ruderale.

long; cotyledons incumbent. — Roadsides and waste places, Nfld. to Mich., O. and Del. *Fr.* May–Sept. (Natzd. from Eu.) Fig. 1092.

1093. L. densiflorum.

8. L. densiflòrum Schrad. (with crowded flowers). — Resembling nos. 6 and 7, *nearly scentless*, much branched, especially above; *basal leaves* oblanceolate, *coarsely toothed to pinnatifid;* cauline leaves narrower, entire or but slightly toothed; *racemes* numerous, ascending, *up to* 1.5 *dm. long; petals* 0 or rudimentary; *silicles rounded-obcordate or round-obovate, strongly rounded at summit,* about 2.5 mm. long; cotyledons incumbent. (*L. apetalum* of ed. 7, not Willd.; *L. neglectum* Thell.) — Roadsides and waste places, Que. to Mont. and Oreg., southw. through most of our range. *Fr.* June–Sept. Fig. 1093. — A weedy species throughout its range, probably adv. in our area.

9. CARDÀRIA Desv.

Silicle ovoid, subglobose or cordate, more or less inflated, indehiscent or only tardily dehiscent, with thin fenestrate or entire septum, the navicular valves nerveless or 1-nerved. Nectar-glands large and well developed, completely surrounding bases of the individual stamens. Seeds 2–4, pendulous, wingless. Cotyledons incumbent. Filaments free, toothless. — Perennials nat. of the Old World, with cauline leaves dentate, clasping or sessile, and small white flowers in corymbed racemes. (Name from the *cordiform* fruit of the first species.) Hymenophysa C. A. Mey.

1. C. Dràba (L.) Desv. (from early inclusion under *Draba*), Hoary Cress. — *Perennial* by subterranean rhizomes; stems stoutish, erect or decumbent, somewhat hoary, 2–5 dm. high; *leaves* oblong, mostly dentate, the basal 4–10 cm. long and petioled, *the cauline sessile and sagittate-clasping; racemes in* terminal *corymbs;* petals 3–4 mm. long, exceeding sepals; *silicles* on divergent filiform pedicels, *reniform or depressed-cordate,* 3–4 *mm. long, barely if at all notched, tipped by a slender style* 1 *mm. long.* (*Lepidium Draba* L.) — Roadsides, waste places and fields, local, N.S. to D.C. and westw., a troublesome weed farther west. *Fr.* late June–Aug. (Natzd. from Eu.) Fig. 1094.

1094. C. Draba.

2. C. pubéscens (C. A. Mey.) Rollins (pubescent), var. **elongàta** Rollins (elongate), Whitetop. — Minutely pubescent, 1–4 dm. high; leaves sagittate- or cordate-clasping; racemes 4–10 cm. long; pedicels slender, spreading-ascending; *fruits obovoid-subglobose, strongly inflated,* 2.5–3.5 *mm. in diameter, nearly equaled by the slender persistent style.* (*Hymenophysa pubescens* sensu Am. auth., not C. A. Mey.) — Roadsides and fields, Wash. to Calif., e., locally, to Mich. and Pa. June–Sept. (Adv., presumably from Asia)

10. CORÓNOPUS Trew Wart-Cress. Swine-Cress

Silicle flattened contrary to the narrow partition; the two locules indehiscent, strongly wrinkled or tuberculate, 1-seeded. Cotyledons narrow and incumbently folded transversely. — Diffuse or prostrate fetid annuals or biennials nat. of Old World, with minute whitish flowers. Stamens often only 2. (Name from the Greek *korone, crow,* and *pous, foot,* from the deeply cleft leaves.) Carara Medic.; Senebiera Poir.

1. C. dídymus (L.) Sm. (twin, referring to the fruit). — Forming mats up to 8 dm. broad; leaves 1–2-pinnately parted; silicles notched at apex, rough-wrinkled. (*Carara* Britt.; *Senebiera* Pers.) — Waste places, roadsides and cult. fields, locally abund., Fla. to Tex., n. to Nfld. April–Oct. (Natzd. from Eu.) Fig. 1095.

C. procúmbens Gilib. (procumbent), with less divided leaves, the *tuberculate silicles not notched,* frequently appears about ports but is apparently not persistent. (Adv. from Eu.)

1095. C. didymus.

11. ÍSATIS L. Woad

Silicle flat, pendulous, cuneate-oblong to ovate or even orbicular, strongly and simply ribbed on each side, indehiscent, 1-seeded, with sessile stigma. Seed pendulous, marginless. Cotyledons incumbent, rarely accumbent. — Erect annual or biennial herbs nat. of Old World, with simple uncleft leaves and small yellow flowers. (Greek name for some plant producing a dark dye.)

1. I. tinctòria L. (used for dying), Dyer's W., Asp-of-Jerusalem. — Glaucous, glabrous, 5–9 dm. high; radical leaves obovate or oblong, petioled, coarsely toothed, 4–10 cm. long; cauline narrower, auriculate-clasping; inflorescence a broad corymb of many racemes; silicle firm, becoming dark, oblong with tapering base, 0.8–1.5 cm. long. — Roadsides, local, se. Nfld.; w. Va.; doubtless elsewhere. *Fr.* June–Oct. (Introd. from Eu.) — Ancient source of an indigo dye.

12. SUBULÀRIA L. Awlwort

Silicle ovoid or globular, with a broad partition; the turgid valves 1-nerved. Seeds several. Cotyledons long and narrow, incumbently folded transversely, *i.e.*, the cleft extending to the radicular side of the curvature. Style none. — Two dwarf stemless perennials of the N. Hemisph., aquatic or littoral; the tufted leaves awl-shaped (whence the name from *subula*, an *awl*). Scape naked, few-flowered, 2–15 cm. high. Flowers minute, white.

1. S. aquática L. (aquatic). — Our only species. — Sandy or gravelly margins (emersed or immersed) of lakes and slow streams, Nfld. to Alaska, s., somewhat locally, to N.S., n. N.E., n. N.Y., Wyo. and Calif. July–Oct. (Greenl.; Eurasia)

13. CAPSÉLLA Medic. Shepherd's-purse. Shovelweed

Silicle obcordate-triangular, flattened contrary to the narrow partition; the valves navicular, wingless. Seeds numerous. Cotyledons incumbent. — Annuals or winter-annuals with rosulate leaves, nat. of Eurasia; petals small, white or pink; style almost none. (Name a diminutive of *capsa*, a box.)

Silicles 5–10 mm. long.
  Petals 2–4 mm. long, distinctly exceeding sepals; silicles with straight or slightly convex sides. . . . . 1. *C. Bursa-pastoris.*
  Petals 1–1.5 mm. long, shorter than to but slightly exceeding sepals; silicles with sides concave toward summit. . . . . 2. *C. rubella.*
Silicles 2.5–3 mm. long, with convex sides; petals longer than sepals. . . . 3. *C. gracilis.*

1. C. Búrsa-pastòris (L.) Medic. (Shepherd's-pouch, *Bursa* an old generic name), Pick-pocket, Tabouret (Que.). — Rosette-leaves either toothed but undivided or variously cleft or runcinate-pinnatifid; stem 1–6 dm. high, smooth or somewhat hirsute; primary cauline leaves sagittate, sessile; *petals 2–4 mm. long, distinctly exceeding sepals;* fruiting raceme much elongated; *silicles with straight or slightly convex sides, shallowly emarginate to subtruncate,* 5–10 mm. long. (*Bursa* Britt.) —Weed of roadsides, cult. ground, etc., throughout our range and beyond. March–Dec. (sometimes all winter). (Natzd. from Eu.) — Extremely variable in foliage and outline of silicle. Upon these characters Almquist proposed sixty-three forms or elementary species.

Var. bífida Crépin (two-cleft). — *Silicles deeply notched at summit, the notch about* 1 *mm. deep.* — More local, Nfld. to Va. (Adv. from Eu.)

2. C. rubélla Reut. (reddish). — Similar, mostly smaller, often reddish; *petals 1–1.5 mm. long, shorter than to but slightly exceeding sepals; silicles with the sides concave toward summit, deeply emarginate,* 5–8 mm. long. — Similar habitats, locally abundant, Lab. and Nfld., s. in coastwise states to S.C., La. and Tex.; also Pacific N.Am. March–Dec. (Natzd. from Eu.)

3. C. grácilis Gren. (slender). — Plant slender; flowers as in no. 1; *silicles* emarginate, *only 2.5–3 mm. long, with convex sides.* — Weed of disturbed or cult. soils, local, se. Va. April–June. (Adv. from Eu.)

14. HUTCHÍNSIA R. Br.

Silicle elongate-oval to lanceolate or elliptic, entire at apex, flattened contrary to the narrow partition, the valves wingless. Seeds 2–several in each locule. Cotyledons incumbent. Style none. — Annuals or winter-annuals of Eurasia and N.Am., with entire or pinnately lobed leaves; petals small, white. (Named for *Ellen Hutchins*, 1785–1815, Irish botanist.)

1. H. procúmbens (L.) Desv. (prostrate).— Annual with simple or branching filiform stems 0.3–2 dm. high; radical leaves about 1 cm. long, pinnately lobed or entire, stellate-pubescent; cauline leaves scattered, linear or oblanceolate; flowers about 1 mm. long; fruiting raceme slender, open, with divergent pedicels; silicle 2–4 mm. long. (*Hymenolobus* Nutt.) — Damp calcareous or alkaline ledges, gravels and flats, s. Lab. and n. Nfld.; Alta. and B.C. to Colo., Nev. and Calif. June–Sept. (Eurasia)

15. COCHLEÀRIA L. Scurvygrass

Sepals short and broad, rounded at apex. Petals obovate or cuneate, white or pinkish. Stamens straight. Style usually slender. Capsule subglobose to ellipsoid, very turgid, often obcompressed. Seeds 2–several, biseriate in the locules. — Succulent boreal and halophytic biennials, with entire, angulate or toothed thick leaves; most difficult to classify. (Name from *cochlear, a spoon,* from the form of the leaves in some species.)

Silicles ellipsoid-oblong, with nearly sessile stigma. . . . . 1. *C. groenlandica.*
Silicles compressed-subglobose, -reniform, -ovoid or -obovoid, with slender style.
 Longer fruiting pedicels 2–8 mm. long; mature (dried) silicles tapering to acute to obtuse base and apex, 2.5–5 mm. broad, the dried valves prominently reticulate. . . . . 2. *C. tridactylites.*
 Longer fruiting pedicels 0.5–2 cm. long; mature (dried) silicles broadly rounded to emarginate at base and apex, 4–7 mm. broad, the dried valves smooth or only obscurely reticulate. . . . . 3. *C. cyclocarpa.*

1. C. groenlándica L. (of Greenland). — Rosette-leaves ovate to rounded-deltoid or reniform, truncate or shallowly subcordate at base, entire or obscurely dentate, the blades 0.5–2 cm. long; *cauline leaves spatulate-oblanceolate to oblong,* entire or obscurely toothed; stems 1–many, the central often very short, the others 0.4–4 dm. high; *flowers* 2–3 *mm. long,* at first densely corymbose, the raceme soon elongating; basal (and sometimes all) leaves absent in fruiting plants; *ellipsoid-oblong smooth silique* 4–8 mm. long and 2–4 broad, *tipped by the nearly sessile stigma; seeds yellow-brown,* 1–1.4 *mm. long.* — Rich damp soil, mostly near the coast, Arct. Am. s. to n. Nfld. July, Aug. (Eurasia)

2. C. tridactylìtes Banks (with three fingers). — Similar; rosette-leaves entire or 3–5-dentate; *cauline leaves oblanceolate to narrowly obovate, sharply* 2–8-*toothed* near the middle; stems or longer branches 0.2–2.5 dm. high; *flowers* 3–5 *mm. long; longer fruiting pedicels* 2–8 *mm. long,* stiff; *mature (dried) siliques* compressed-subglobose to -ovoid, *tapering to acute or obtuse base and apex,* 3–5 *mm. long,* 2.5–5 *mm. broad, the dried valves prominently reticulate, the slender style* 0.6–0.8 *mm. long; seeds* darker brown, 1.3–1.6 *mm. long.* — Calcareous, brackish or other rich soils, near the coast, Straits of Belle Isle, Lab. Pen., to Anticosti I., Que., s. to Notre Dame and Ingornachoix Bays, Nfld., and M.I. June–Aug.

3. C. cyclocárpa Blake (round-fruited). — Similar to no. 2, usually coarser; rosette-leaves usually cordate, up to 3.5 cm. long; *cauline leaves oblanceolate to rotund,* sessile, entire or *coarsely* 2–4-*toothed;* stems 0.6–4.5 dm. high; flowers 3–5 mm. long; *longer fruiting pedicels* 0.5–2 *cm. long, often arching; mature (dried) silicles compressed-subglobose or -reniform, broadly rounded to subcordate base and emarginate summit,* 3.5–8 *mm. long,* 4–7 *mm. broad, the dried valves smooth or only obscurely reticulated,* the slender style 0.6–0.8 mm. long; *seeds* dark brown, 1.4–2 *mm. long.* — Calcareous damp soil, near the coast, Nfld. and adj. Saguenay Co., Que., to Mingan Ids. and Anticosti I., Que. Late June–Sept.

16. CAMÉLINA Crantz FALSE FLAX

Silicle obovoid or pyriform, pointed, margined, with filiform style; partition broad; valves 1-nerved. Seeds numerous, oblong. Cotyledons incumbent. Flowers small, yellow. — Erect herbs nat. of Old World, with sagittate-clasping leaves. (Name from the Greek *chamai, dwarf,* and *linon, flax.*)

1. C. SATÌVA (L.) Crantz (sown; formerly, for its seeds, containing oil used in soap-making), GOLD-OF-PLEASURE. — Annual, 3–9 dm. high, erect, usually with ascending branches; *stem glabrous or with minute closely appressed stellate hairs;* leaves lanceolate, the lowest tapering to petioles, the others with sagittate sessile bases; racemes elongating; *silicles mostly* 7–9 mm. long, *three to four times as long as style,* 6–7 *mm. thick,* on pedicels 1.2–3 cm. long, capsule-walls soon hardening; *seeds mostly* 1–1.5 *mm. long,* pale yellowish-brown. — Roadsides, waste places and cult. fields, occasional, Que. to B.C., s. to N.S., N.E., S.C., Ill., Mo., Kans., Colo. and Calif. April–Aug. (Adv. from Eu.) FIG. 1096.

1096. C. sativa.

2. C. MICROCÁRPA Andrz. (small-fruited). — Similar, usually more slender; *stem and leaves harsh with elongate simple and branching hairs;* pedicels often shorter; *silicles about* 5 *mm. long* and 4–5 *mm. thick, about twice as long as style,* the walls remaining thin; *seeds* darker, *mostly less than* 1 *mm. long.* — Roadsides, waste places and fields, more common than no. 1, Nfld. to B.C., s. to N.S., N.E., Va., Tenn., Mo., Tex., etc. April–Sept. (Natzd. from Eu.)

17. NÉSLIA Desv. BALL-MUSTARD

Silicle subglobose, compressed, slender-beaked, indehiscent, 1-locular or obscurely 2-locular, the surface reticulated after drying. Seed 1 (rarely 2). Cotyledons incumbent. Style slender.

Flowers small, yellow. — Erect herbs nat. of Old World. (Named in 1813 for *J. A. N. de Nesle* of Poitiers.)

1. N. PANICULÀTA (L.) Desv. (with panicles). — Slender annual or biennial, somewhat stellate-pubescent, simple up to the inflorescence; leaves oblong, sagittate-clasping; racemes elongate; pedicels slender, spreading, 5–9 mm. long; silicle 2–3 mm. in diameter. — Grainfields and waste places, Nfld. to B.C., s. to N.S., N.E., N.J., Pa., O., Ind., Ill. and S.D. June–Sept. (Natzd. from Eu.) FIG. 1097.

1097. N. paniculata.

MYÁGRUM (from *mys, mouse*, and *agra, trap*) PERFOLIÀTUM L. (through a leaf), a stiff glaucous annual, with obtuse auriculate-clasping lanceolate leaves, and firm cuneate-lyrate silicles with 1 seed in the basal half and 2 rounded empty locules below the beak, and EUCLÍDIUM (Greek *eu, well*, and *kleio, closed*) SYRÌACUM (L.) R. Br. (Syrian), similar, but with oblanceolate short-petioled toothed leaves and long-beaked densely hirsute small ovoid 2-locular and 2-seeded silicles, have occurred in waste places but seem not to have long persisted. (Both adv. from Eurasia)

18. CAKÌLE Hill SEA-ROCKET. CAQUILLIER (Que.)

Silicle short, transversely 2-jointed, fleshy, the upper joint separating at maturity; each joint indehiscent, 1-loculed and 1-seeded, or the lower sometimes seedless. Seed erect in the upper, suspended in the lower joint. Cotyledons obliquely accumbent. — Fleshy annuals of N.Am., Eu. and Afr. Flowers purplish. (An old Arabic name.)

1. C. edéntula (Bigel.) Hook. (without teeth). — Stem simple to widely branched, 0.5–8.5 dm. high; leaves obovate or oblanceolate, sinuate-toothed, narrowed to base; *upper joint of silicle ovoid or rarely ovoid-lanceolate, short-beaked, its articulating base without pits or the pits rudimentary; articulating summit of lower joint without processes or these barely developed.* — Sandy or gravelly beaches and seacoast, s. Lab. to S.C.; local about Great Lakes. — July–Sept. (Iceland; Azores) — Fresh young foliage and stems with flavor of horseradish.

Var. **lacústris** Fern. (of lake-shores). — *Upper joint of silicle ovoid-lanceolate, long-beaked, its articulating surface with 2 deep and 4 shallow pits; articulating summit of lower joint with 2 long and 4 short subulate processes.* — Strands of Lakes Ont., Erie, Huron and Mich.

C. MARÍTIMA Scop. (maritime), with leaves linear or dissected to midrib into linear segments, the upper locule of the fruit prolonged and lanceolate or ensiform, the summit of the lower one usually with 2 divergent teeth, is sporadic on coastal rubbish. (Adv. from Eu.)

RAPÍSTRUM (Latin name of the wild rape) RUGÒSUM (L.) All. (wrinkled), a yellow-flowered annual or biennial with transversely 2-jointed appressed fruits on thickened pedicels, the upper 8-ribbed globose joint abruptly slender-beaked and much thicker than the lower joint, is sporadic in waste ground. (Adv. from Eu.)

19. RÁPHANUS L. RADISH. RADIS (Que.)

Silique torulose or cylindric, tapering upward, indehiscent, several-seeded, continuous and spongy within between the seeds, or moniliform by constriction between the seeds, with no proper partition. Style long, tipped by the broad emarginate stigma. Seeds spherical and cotyledons conduplicate. — Annuals or biennials, nat. of Old World. (Name from the Greek *raphanos, quickly appearing*, alluding to the rapid germination.)

1. R. RAPHANÍSTRUM L. (generic name used by Tournefort), WILD R., JOINTED CHARLOCK. — More or less hispid, 2–8 dm. high; leaves lyrate, rough; petals 1.5–2 cm. long, pale yellow, turning whitish, with violet veins, or pale yellow and with yellow veins in forma SULPHÙREUS (Babey) Hayek (sulphur-colored), or white and with violet veins in forma ÁLBUS (Schuebler & Martens) Hayek (white), or white and with pale or greenish veins in forma CÁNDIDUS (Opiz) G. Beck (clear white), or purple-violet throughout in forma PURPÙREUS (Reichenb.) Domin (purple); *silique 2–8-seeded, moniliform, 4–6 mm. thick, slender-beaked, transversely divided when ripe into 1-seeded segments*, the beak glabrous, or in forma HÍSPIDUS (Lange) Thell. (hispid) the beak harsh with stiff appressed hairs. (*Raphanistrum innocuum* Moench) — Weed of grainfields, waste places, etc., Nfld. to Man., s. to N.S., N.E., Va., Ky., Ia. and Kans. April–Nov. (Natzd. from Eu.) FIG. 1098.

1098. R. Raphanistrum.

1099. R. sativus.

2. R. sativus L. (planted), Radish. — Similar; petals pale purple; *silique thick* (6–10 *mm. in diameter*), *continuous, 2–3-seeded, with conical beak.* — Persistent in old fields and waste places. (Introd. from Eu.) Fig. 1099.

20. BRÁSSICA L. Mustard. Turnip. Moutarde (Que.). Chou (Que.)

Silique slender or thickish, nearly terete or 4-sided, with a stout often 1-seeded beak; valves 1–5-nerved. Seeds globose, in 1 row. Cotyledons conduplicate. — Annuals or biennials, nat. of Old World, with yellow petals about 1 cm. long. Lower leaves mostly lyrate, incised or pinnatifid. (The Latin name of the Cabbage.) Including Sinapis L.

a. Beak of silique large, flat or conspicuously angled, usually containing one seed in an indehiscent locule; leaves not clasping. (Sinapis).
- Leaves all pinnatifid; beak ensiform, longer than the densely bristly body of the silique. . . . . . . 1. *B. hirta.*
- Middle and upper leaves rhombic to oblong, merely toothed; beak 4-angled, 2-edged, about half as long as the smooth or sparsely bristly body of the silique. . . . . . . 2. *B. Kaber.*

a. Beak terete or slenderly conical, empty. (True Brassica). . . b.

b. Leaves not clasping.
- Plant glabrous or nearly so, glaucous; pedicels elongate, slender, spreading; silique loosely ascending, 3–5 cm. long. . . . . . . 3. *B. juncea.*
- Plant hirsute, green; pedicels short and thick, appressed; siliques closely appressed, 1–2 cm. long. . . . . . . 4. *B. nigra.*

b. Leaves (at least the middle and upper) clasping.
- Glaucous, remotely hispid at least when young; petals pale yellow, their blades 6–7 mm. long. . . . . . . 5. *B. Rapa.*
- Green, glabrous; petals deeper yellow, smaller. . . . . . . 6. *B. Napus.*

1. B. hírta Moench (rough-hairy), White M. — *Leaves petioled*, deeply *lyrate-pinnatifid*, the lobes sinuate-dentate; *siliques* 1.5–3.5 cm. long, on horizontally divergent pedicels, *white-bristly* (*densely so toward base*), the valves prominently 3-nerved, few-seeded, *terminated by a flattened ensiform decurrent-based beak equaling or longer than the dehiscent body;* seeds pale. (*B. alba* of Am. auth., not Gilib.; *Sinapis alba* L.) — Waste places and roadsides, local, P.E.I. to s. B.C., s. to N.S., N.E., Del., D.C., Ia., etc. *Fr.* June–Aug. (Introd. from Eurasia) — Cult. as a source of mustard and as salad.

2. B. Káber (DC.) L. C. Wheeler (Persian name), Charlock, Moutarde d'été (Que.). — More or less hispid, especially at base; *lower leaves on hispid petioles*, obovate, lyrate-pinnatifid; *middle and upper leaves nearly or quite sessile, oblong to ovate or rhombic, acute*, dentate, sparsely pilose; *pedicels short and thick, ascending*, 2.5–3 mm. long; silique 2–2.5 cm. long, the beak only 0.6–1.2 cm. long. — Eurasia. Represented with us by

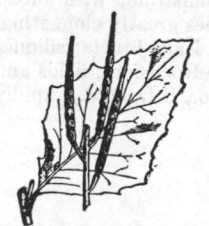

Var. pinnatífida (Stokes) L. C. Wheeler (pinnatifid), Crunch-weed. — *Fruiting pedicels* 3–7 *mm. long; siliques* 2.5–4.5 *cm. long*, 3–4 *mm. thick, cylindric*, glabrous or nearly so, at most *only slightly torulose, terminated by a 4-angled 2-edged* 1- or 2-seeded *beak* 1–1.5 *cm. long.* (*B. arvensis* (L.) Rabenh., not L.; *B. sinapistrum* Boiss.; *Sinapis arvensis* L.) — Common weed of cult. and waste ground, throughout our range and beyond. *Fr.* June–Aug. (Natzd. from Eurasia)

1100. B. Kaber, v. Schkuhriana.

Var. Schkuhriàna (Reichenb.) L. C. Wheeler (for Christian Schkuhr, 1741–1811). — *Siliques slender, strongly torulose*, 3–5.5 cm. long, 1.5–2 *mm. thick, attenuate into* a usually *curving beak.* — Less common, throughout. (Natzd. from Eurasia) Fig. 1100.

3. B. júncea (L.) Coss. (rush-like, probably from its slender habit), Chinese or Leaf-M. — *Nearly glabrous*, somewhat *glaucous;* lower leaves lyrate and petioled; *upper leaves* oblong, entire or dentate, *tapering to short petioles; pedicels slender, spreading; siliques continuous with* the *slenderly conical seedless beak, up to* 3.5 *cm. long.* — Cult. fields, etc., usually common. *Fr.* June–Sept. (Natzd. from Eurasia) Fig. 1101.

1101. B. juncea.

Var. crispifòlia Bailey (curly-leaved), Curled M. — *Leaves* with *crisped or curled* deeply cleft margins. (*B. japonica* of ed. 7, not Thunb.) — Less common, mostly esc. from cult. (Introd. from Eurasia)

4. B. nìgra (L.) Koch (black), Black M. — *Hirsute* with scattered hairs, *green;* leaves slender-petioled, the lower with large terminal lobe and a few small lateral ones; *siliques* short, 1–2 *cm. long,* glabrous, torulose, indistinctly quadrangular, *short-pedicelled, appressed and overlapping at maturity, the empty subulate beak* 1–3 *mm. long;* seeds black. — Waste places and cult. fields, throughout our range and far beyond. *Fr.* June–Oct. (Natzd. from Eurasia) Fig. 1102. — Principal source of table-mustard.

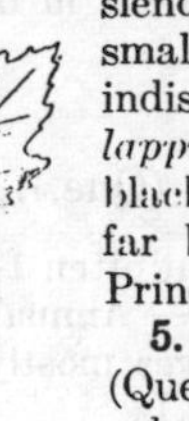

1102. B. nigra.

5. B. Ràpa L. (old generic name), Bird's Rape, Navette (Que.). — Glabrous or when young remotely hispid, *glaucous, succulent;* lower leaves sparingly toothed or pinnatifid, petioled; *upper leaves sessile by a cordate-auriculate base,* subentire, oblong-lanceolate, often constricted above base; *petals pale yellow, their blades* 6–7 *mm. long; pedicels spreading, slender; siliques erect,* 3–7 *cm. long,* gradually narrowed into a slender beak 0.8–2 cm. long; seeds dark brown. (*B. campestris* L.) — Weed of cult. fields, etc., often too common, throughout. *Fr.* June–Oct. (Natzd. from Eurasia) Fig. 1103.

1103. B. Rapa.

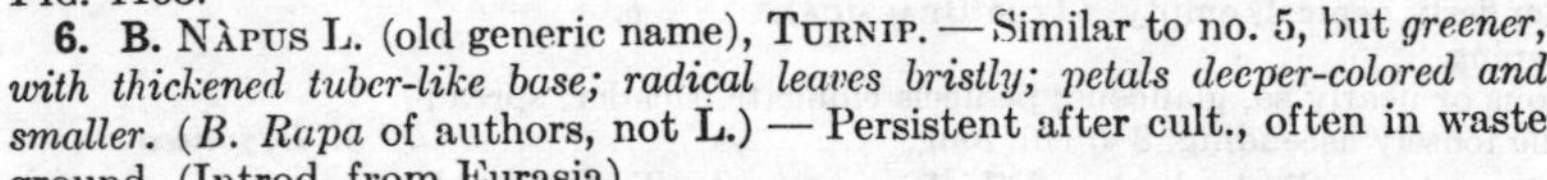

6. B. Nàpus L. (old generic name), Turnip. — Similar to no. 5, but *greener, with thickened tuber-like base; radical leaves bristly; petals deeper-colored and smaller.* (*B. Rapa* of authors, not L.) — Persistent after cult., often in waste ground. (Introd. from Eurasia)

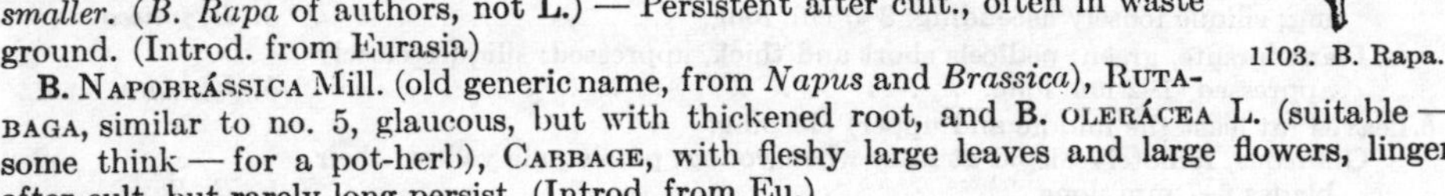

B. Napobrássica Mill. (old generic name, from *Napus* and *Brassica*), Rutabaga, similar to no. 5, glaucous, but with thickened root, and **B. oleràcea** L. (suitable — some think — for a pot-herb), Cabbage, with fleshy large leaves and large flowers, linger after cult. but rarely long persist. (Introd. from Eu.)

21. ERUCÁSTRUM Presl

Silique 4-angled, slender, elongate, tapering to a conic beak, the valves keeled; the lateral nerves delicate, reticulate-anastomosing, more or less torulose. Seeds ovoid or ellipsoid, 1-rowed. Cotyledons conduplicate. — Annual or short-lived herbs nat. of Eu. and the Mediterr. reg., somewhat like *Brassica,* but with lower flowers of the racemes leafy-bracted. (Name from resemblance to *Eruca.*)

1. E. gállicum (Willd.) O. E. Schulz (French). — Annual, 2–8 dm. high; stem retrorsely pubescent with simple hairs; leaves oblong, deeply pinnatifid or bipinnatifid, with broad rounded to truncate sinuses; petals pale yellow, about 5 mm. long; racemes greatly elongating, the lower slender divergent or ascending pedicels subtended by small leafy bracts; siliques 2.5–3.5 cm. long, 1–2 mm. thick. (*E. Pollichii* Schimper & Spenner) — Roadsides, fields and waste places, local, Nfld. and N.S. to B.C., s. to Pa., W.Va., Ind., Mo., S.D., etc., rapidly spreading. June–Oct. (Natzd. from Eu.)

22. DIPLOTÁXIS DC.

Seeds ovoid, in two rows in each locule; other characters as in *Brassica.* — Leaves toothed or pinnatifid; the yellow, white or purplish petals with slender claws. — Genus nat. of Old World, somewhat arbitrarily separated from *Brassica.* (Name from the Greek *diplous, double,* and *taxis, row,* alluding to the biseriate seeds.)

Pedicels and backs of sepals glabrous or essentially so; petals yellow when fresh; lowest fruiting pedicels 1–3.5 cm. long.
  Annual or biennial, the stems herbaceous at base; leaves mostly near base of plant; sepals 3–5 mm. long; lowest mature siliques 2–3.5 cm. long, attenuate to sessile base. . . . . 1. *D. muralis.*
  Perennial, the stems suffruticose at base; leaves extending well up the stems; sepals 5–8 mm. long; lowest mature siliques 3–5 cm. long, stipitate. . . . 2. *D. tenuifolia.*
Pedicels and backs of sepals hispid; petals white, veined with pink or purple; lowest fruiting pedicels 0.5–1 cm. long. . . . . 3. *D. erucoides.*

1. D. muràlis (L.) DC. (growing on walls). — *Annual or biennial,* the smooth or only sparingly hispid stem *leafy only near the* branching *herbaceous base; leaves* oblong, coarsely toothed to deeply incised or pinnatifid, *the basal rosulate;* pedicels and calyx glabrous; *sepals* glabrous 3–5 *mm. long;* petals yellow (sometimes drying purplish) about twice length of sepals; fruiting raceme elongate, open; mature pedicels spreading-ascending, the lowest 1–3 cm. long; *siliques* erect, 2–3.5 cm. long, linear-terete, attenuate at both ends, *sessile.* — Waste places and roadsides, locally abund., especially near the coast, Que. and Ont., s. to N.S., N.E., N.J., Pa., O., Mich., Ill. and Ia. June–Sept. (Natzd. from Eu.)

2. D. TENUIFÒLIA (L.) DC. (slender-leaved). — Similar; *perennial, the stems suffruticose at base, leafy up to the peduncles;* leaves deeply pinnately parted; *sepals* broader, 5–8 *mm. long;* petals deep yellow, twice to thrice length of sepals; lowest fruiting pedicels 2–3.5 cm. long; *siliques* 3–5 *cm. long, stipitate.* — Waste places and roadsides, local, N.B. to Ont., s. to N.S., N.E., Va. and Mich.; doubtless of wider range. July–Oct. (Natzd. from Eu.)

3. D. ERUCOÌDES (L.) DC. (resembling *Eruca*). — Biennial or annual, scabrous; leaves obovate, coarsely dentate or lobed; *racemes leafy-bracted at base; sepals and pedicels hirsute; petals white with pink or purple veins,* twice to thrice length of sepals; *siliques* sessile, 2–3.5 cm. long, *on* thick *pedicels* 0.5–1 *cm. long.* — Waste places, Gaspé Co., Que. July–Sept. (Adv. from Eu.)

23. ERÙCA Mill. GARDEN-ROCKET

Silique thickish, somewhat 4-sided, tipped by a large style persisting as a flattish triangular-lanceolate beak. Seeds ellipsoidal, slightly compressed, borne in 2 rows. — Annuals or biennials, nat. of Mediterr. reg., with pinnatifid leaves and rather large flowers; the petals ochroleucous to yellowish or purplish, with violet veins. (The classical Latin name, appreciatively used by Pliny, from the Greek *ereugomai, belch.*)

1. E. SATÌVA Mill. (sown). — Coarse erect annual; flowers on thick pedicels only 0.5–5 mm. long; petals 1.5–2 cm. long; siliques fusiform, 4-angled, erect on short stout pedicels. — Waste places and cult. grounds, e. Ont. to N.D., s. to N.J., Pa., Ill., Mo., etc. Late May–Oct. (Natzd. from Eu.)

24. CONRÍNGIA Link HARE'S-EAR-MUSTARD

Siliques long, slender, 4-angled, somewhat rigid. Seeds oblong, one row in each locule. Cotyledons incumbent. — Glabrous annuals nat. of Old World, with sessile elliptic entire amplexicaul leaves. (Named for *Hermann Conring,* 1606–1661, professor at Helmstadt.)

1. C. ORIENTÀLIS (L.) Dumort. (eastern). — Simple or slightly branching, 0.2–8 dm. high, somewhat succulent; petals pale yellow, 7–10 mm. long; siliques 4–15 cm. long, tapering into the thick style. — Waste places, new fields, etc., Que. to Man. and westw., s. beyond our range. June–Aug. (Natzd. from Eu.)

25. ALLIÀRIA B. Ehrh. GARLIC-MUSTARD

Silique long, linear, angled; valves keeled, 3-nerved; stigma simple, sessile or nearly so. Oval sepals caducous. Pubescence simple or none. — Ours biennial, nat. of Eurasia, with deltoid-ovate cordate dentate petiolate leaves and small white flowers. (Name from *Allium, onion* or *garlic,* referring to the odor.)

1. A. OFFICINÀLIS Andrz. (of the shops, from former repute in medicine).—Tall; siliques 2.5–5 cm. long, spreading, borne on short thick pedicels. (*A. Alliaria* Britt.) — Roadsides, open woods and near habitations, local, Que. and Ont., s. to Va., Ky. and ne. Kans. April–June. (Introd. and natzd. from Eu.)

26. SISÝMBRIUM L.

Siliques cylindric and prismatic or long-subulate, the valves 1–3-nerved; stigma 2-lobed. Seeds oblong, marginless, in 1 row in each locule. Cotyledons incumbent. Calyx open. — Flowers yellow. Pubescence of simple hairs or none. Leaves pinnatifid to entire. Ours mostly annuals or biennials from the Old World. (Latinized from an ancient Greek name for some plant of this family.)

*a.*Siliques subulate, 1–2 cm. long, on short erect pedicels, closely appressed to rachis. § VELARUM. . . . . 1. *S. officinale.*

*a.*Siliques cylindrical, elongate, 2–10 cm. long, spreading or loosely ascending. § EUSISYMBRIUM. . . *b.*

*b.*Fruiting pedicels about as thick as the long (5–10 cm.) siliques; segments of leaves elongate-linear. . . . . . 2. *S. altissimum.*

*b.*Fruiting pedicels much more slender than the siliques, the latter 2–6 cm. long; segments of leaves triangular-ovate to lanceolate.
Petals much exceeding sepals; siliques 2–3.5 cm. long. . . . . . . 3. *S. Loeselii.*
Petals and sepals subequal; siliques 4–6 cm. long. . . . . 4. *S. Irio.*

§ Velàrum DC

1104. S. officinale, v. leiocarpum.

1. S. officinàle (L.) Scop. (of the shops, from former repute in medicine), Hedge-Mustard. — Stiffly divaricate-branching annual up to 6 dm. high, hirsute at least at base; leaves slender-petioled, runcinate-pinnatifid with toothed segments or the upper hastate to lanceolate and entire; *racemes spiciform;* flowers pale yellow, about 3 mm. long; *siliques subulate, tapering from near base to summit,* 1–2 *cm. long, pubescent* or tomentulose, *close-appressed to rachis, on very short pedicels.* (*Erysimum* L.) — Waste places, throughout, less general than the var. May–Oct. (Natzd. from Eu.)

Var. leiocárpum DC. (smooth-fruited). — Siliques essentially glabrous; plant greener. — Common weed throughout our range and beyond. May–Oct. (Natzd. from Eu.) Fig. 1104.

§ Eusisýmbrium Gren. & Godr.

2. S. altíssimum L. (tallest), Tumble-Mustard. — Tall loosely branching annual up to 1.5 m. high, sparsely spreading-hirsute at base; *leaves* pinnatifid *with long linear segments;* flowers loosely racemose, pale yellow, 6–8 mm. long, the petals exceeding the sepals; *siliques slenderly cylindric, rigid,* spreading-ascending, 5–10 *cm. long, glabrous, hardly thicker than the short thickish pedicels.* (*Norta* Britt.) — Fields, roadsides and waste places, throughout our range and beyond, rapidly spread in the last half-century. May–Aug. (Natzd. from Eu.) Fig. 1105.

1105. S. altissimum.

3. S. Loesélii L. (for Johann Loeselius, 1607–1655, who early recognized the species). — Stiffly branched annual up to 1 m. high, usually retrorse-hispid; *leaves petioled,* runcinate-pinnatifid, *the lobes triangular;* racemes much elongating; *petals* yellow, 5.5–6 *mm. long, about twice length of sepals; siliques* linear, 2–3.5 *cm. long,* about 0.7 mm. broad, *on capillary spreading-ascending pedicels* 0.7–1.5 cm. long; seeds scarcely 1 mm. long. — Waste places and old fields, as yet local, N.E. to Ida. May–Nov. (Adv. from Eu.)

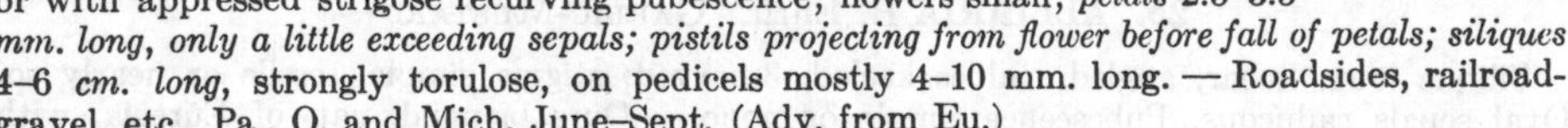

4. S. Ìrio L. (Latin name for a siliquose plant). — Similar to no. 3, mostly lower (up to 6 dm. high), usually much branched from base, glabrous or with appressed strigose recurving pubescence; flowers small; *petals* 2.5–3.5 *mm. long, only a little exceeding sepals; pistils projecting from flower before fall of petals; siliques* 4–6 *cm. long,* strongly torulose, on pedicels mostly 4–10 mm. long. — Roadsides, railroad-gravel, etc., Pa., O. and Mich. June–Sept. (Adv. from Eu.)

Other species, recently appearing as casual weeds, may spread. These include S. orientàle L. (eastern), similar to no. 2, but short-pilose, with lobes of leaves ovate, the siliques sparsely puberulent; and S. polycerÀtium L. (with many horns), with the tiny flowers 1–3 in the axils of repand-dentate hastate leafy bracts, the torulose siliques 1–2.5 cm. long. (Both adv. from Eurasia)

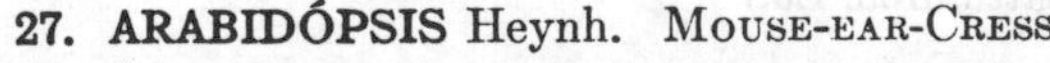

27. ARABIDÓPSIS Heynh. Mouse-ear-Cress

Silique tetragonal-cylindric; midrib of septum so broad and thin as to be wholly obscure; valves with lateral nerves very slender and anastomosing with the midrib by reticulations. Seeds numerous, ovoid, in 1 or 2 ranks in each locule. Cotyledons incumbent. — Old World annual to biennial herbs with basal rosettes of petioled leaves, the cauline ones short-petioled or sessile to sagittate-clasping. Petals white or purplish, sometimes yellow. The pubescence of forking hairs or with these intermixed with simple ones. (Name from *Arabis* and *opsis, aspect,* from resemblance to the genus *Arabis.*)

1. A. Thalìana (L.) Heynh. (for Johann Thal, who described it in the 16th century). — Slender, simple or branched, hairy at base, 0.5–4.5 dm. high; rosette-leaves oblong or narrowly obovate, rounded at summit, entire or repand, scabrous with stipitate-branching hairs; cauline leaves remote, sessile, narrowed at base, acutish; raceme at first dense, then elongating; petals white, spatulate, 3–4 mm. long, twice as long as sepals; siliques on capillary divergent to loosely ascending pedicels 0.5–1.5 cm. long, slender, 1–1.5 cm. long, less than 1 mm. wide; stigma sessile; seeds 40–68, 1-seriate, barely 0.5 mm. long. (*Sisymbrium* J. Gay; *Stenophragma* Čelak.) — Dry fields, roadsides and waste places, Mass. to Mich. and Ill., s. beyond our limits. March–June. (Natzd. from Eu.)

28. DESCURAÌNIA Webb & Berthelot Tansy-Mustard

Silique linear-cylindric to slenderly clavate, the stigma small, entire, the valves 1-nerved. Seeds elliptic or oblong, marginless, in 1 or 2 rows in each locule. Cotyledons incumbent. Calyx spreading-ascending. Flowers small, yellow or whitish. Pubescence forked or stellate, often reduced to minute granules or mixed with simple hairs or glands. — Annual or biennial herbs of Am. and Eu., with divided leaves, the basal rosette usually withered at flowering time. (Named for *François Déscourain*, 1658–1740, French apothecary and botanist.)

Siliques linear-cylindric, about 1 mm. thick; seeds 1-ranked in each locule; plants not glandular.
Fruiting pedicels spreading-ascending, 0.8–1.5 cm. long; siliques 1–3 cm. long, usually arcuate, loosely ascending; seeds 10–20 in each locule. . . . 1. *D. Sophia.*
Fruiting pedicels strongly ascending, 3–6 mm. long; siliques 5–10 mm. long, appressed-ascending and overlapping; seeds 4–8 in each locule. . . . 2. *D. Richardsonii.*
Siliques clavate, 5–12 mm. long, 1–2 mm. thick; seeds 2-ranked in each locule; plant often glandular or glutinous. . . . . 3. *D. pinnata.*

1. D. Sòphia (L.) Webb (old generic name), Sagesse des Chirurgiens (Que.). — Slender branching stellate-pubescent annual up to 7.5 dm. high; *leaves* bi- or tripinnate *with fine linear to oblanceolate segments;* racemes (especially central one) elongate; sepals mostly exceeding the greenish-yellow petals, 2–2.5 mm. long; *fruiting pedicels spreading-ascending,* 0.8–1.5 *cm. long; siliques linear-cylindric,* 1–3 *cm. long,* about 1 mm. thick, usually arcuate, *loosely ascending,* with very short style; septum 2 (rarely 3) -nerved; *seeds* 10–20 *in each locule, strictly 1-ranked,* oblong-ellipsoid, about 0.8 mm. long. (*Sisymbrium* L.; *Sophia multifida* Gilib.) — Roadsides and waste places, locally abund. eastw., common west of our range, Gaspé Co., Que., to Wash., s. to N.S., N.B., N.E., Del., Pa., Ill., Kans., Colo., Utah and Calif. Late May–Aug. (Natzd. from Eu.)

2. D. Richardsônii (Sweet) O. E. Schulz (for its discoverer, Sir John Richardson, 1787–1865). — *Canescent,* 0.2–1.2 m. high; *lower and rosette-leaves* bi- or tripinnate or -pinnatifid, *with obtuse segments;* upper leaves smaller, less dissected, with narrower segments; flowers yellow; calyx 1–1.5 mm. long, equaled or exceeded by the petals; *fruiting pedicels strongly ascending,* 3–6 *mm. long; siliques cylindric to slenderly lanceolate,* 5–10 *mm. long, appressed-ascending and overlapping;* seeds strictly 1-ranked and 4–8 in each locule, 0.75–1 mm. long. (*Sisymbrium canescens,* var. *Hartwegianum* of ed. 7, not *S. Hartwegianum* Fourn.) — Calcareous gravels, prairies and roadsides, Côte Nord, Que., to Yukon, s. to Gaspé Pen., Que., centr. Me., Minn., Kans. and Colo. July–Sept.

3. D. pinnàta (Walt.) Britt. (pinnate). — Similar to no. 2, 1–7 dm. high; the stem, rachis, etc. *densely canescent to somewhat glandular;* flowers whitish; *fruiting pedicels* 0.6–1.5 *cm. long, nearly horizontally divergent; siliques clavate,* 5–9 mm. long; *seeds 2-ranked* in each locule. (*Sisymbrium canescens* Nutt.) — Dry sands and waste places, Fla. to Tex., n. to se. Va. March–May.

Var. **brachycárpa** (Richards.) Fern. (short-fruited). — *Green,* not canescent, more or less pubescent and glandular; flowers slightly larger; *fruiting pedicels* 0.8–1.6 cm. long, *spreading-ascending;* siliques up to 12 mm. long. (*Sisymbrium canescens,* var. S. Wats.) — Dry sandy or rocky soils, Gaspé Pen., Que., to Mackenz., s. to n. N.H., Vt., N.Y., W.Va., Tenn., Ark., Tex. and Colo. May–Aug.

29. BRAỲA Sternb. & Hoppe

Siliques cylindric to lance-subulate, the septum of peculiar and characteristic structure with its cells elongated transversely or obliquely. Flowers white or purplish, capitate in anthesis. — Arct. and boreal perennials, with low or tufted habit and mostly branched hairs, the slender, often fleshy leaves without prominent veins. (Named for *Franz Gabriel, Count de Bray,* 1765–1832, of Rouen.)

Flowering stems leafless, 1–15 cm. high; rosette-leaves fleshy, linear-oblanceolate, entire; siliques lance-subulate, non-torulose, 4–9 mm. long.
Stems glabrous or only sparsely pilose; sepals deciduous, usually green; petals 4–5.5 mm. long, broadly obovate, the limb white; siliques glabrous. . . 1. *B. Longii.*
Stems copiously pilose; sepals subpersistent, purplish; petals 1.5–4 mm. long, spatulate, white, becoming pink; siliques hirtellous. . . . . 2. *B. Fernaldii.*

Flowering stems leafy, 0.2–3.7 dm. high; rosette-leaves thinner, oblanceolate, often sinuate or toothed; siliques linear-terete, torulose, 1.2–3 cm. long. . . . 3. *B. humilis*.

1. B. Lóngii Fern. (for its discoverer, BAYARD LONG, 1885–). — Cespitose, with stout tap-root and multicipital caudex; *leaves all basal*, rosulate, *fleshy*, linear-oblanceolate, *entire*, obtuse, 1–4 cm. long, 1–3 mm. broad, *glabrous; stems scapose*, 1–10 (in fruit rarely –15) *cm. high, glabrous or sparsely pilose;* flowering raceme dense, corymbiform, soon elongating to 1–7 cm. with mostly overlapping fruits; *sepals greenish*, rarely purplish, glabrous, 2–3 mm. long, elliptic or oblong, rounded above; *petals* 4–5.5 *mm. long, broadly obovate, white*, except for the violet base; pistil lanceolate, glabrous, 10–16-ovulate; *siliques lance-subulate, plane*, 4–9 *mm. long*, 1–2 *mm. broad*, the thick style 0.4–0.8 mm. long. — Pockets and crevices of limestone, nw. Nfld. *Fl.* July; *fr.* Aug.

2. B. Fernáldii Abbe (for one of its discoverers, **1873–1950**). — Smaller; leaves 0.5–2 mm. broad; *scapes copiously gray-pilose; sepals purplish*, 1.5–2.5 mm. long; *petals* 1.5–4 *mm. long, spatulate-oblanceolate*, white, *changing to pink; ovary and silique hirtellous*, the slender style 0.5–1 mm. long. — Limestone-barrens, nw. Nfld. July, Aug.

Another species, found on limestone-barrens of n. Nfld., with densely villous and closely crowded linear leaves but with flowers and fruit unknown, has been incorrectly associated with the northwestern *B. Richardsonii* (Rydb.) Fern. Fertile material is much needed.

3. B. hùmilis (C. A. Mey.) Robins. (low). — More delicate; *rosette-leaves thinnish, oblanceolate, often toothed or sinuate*, either hirsute or glabrous; *stems* 0.2–3.7 *dm. high, simple or forking*, more or less pubescent, *with scattered* narrow entire or toothed *leaves;* flowers white or purplish; *siliques linear-terete, torulose*, erect, 1.2–3 *cm. long*, about 1 mm. in diameter, minutely pubescent. (*Sisymbrium* C. A. Mey.; *Torularia* O. E. Schulz) — Calcareous gravels and cliffs, local or rare, Greenl. to Alaska, s. to w. Nfld., Anticosti, James Bay, Alta. and B.C. July, Aug. (Asia)

Var. **leiocárpa** (Trautv.) Fern. (smooth-fruited). — Siliques glabrous. (Var. *novae-angliae* (Rydb.) Fern.; *Pilosella novae-angliae* Rydb.; *Arabidopsis novae-angliae* (Rydb.) Britt.) — Calcareous cliffs and talus, n. Vt.; L. Sup. reg. of Ont. and Mich. June, July. (Siberia)

30. HÉSPERIS L. ROCKET. JULIENNE (Que.)

Silique very slender, nearly cylindrical; stigma lobed, erect. Seeds in 1 row in each locule, oblong, marginless. Cotyledons incumbent. — Biennial or perennial, nat. of Old World, with serrate sessile or petiolate leaves, and large white or purple flowers. (Name from the Greek *hesperos, evening* or *evening-star*, from the evening fragrance of the flowers.)

1. H. MATRONÀLIS L. (matronly, from the old name, MOTHER-OF-THE-EVENING, the flowers then becoming fragrant), DAME'S-VIOLET, JULIENNE DES DAMES (Que.). — Tall; leaves lanceolate, acuminate; petals purple (sometimes albino); siliques 5–14 cm. long. — Roadsides, thickets and open woods, Nfld. to Ont. and westw., s. to N.S., N.E., Ga., O., Ind., Ia. and Kans., spread from cult. May–Aug. (Introd. and natzd. from Eu.)

BÙNIAS L. (name used by Dioscorides) ORIENTÀLIS L. (oriental), a tall biennial with the leaves lyrate-pinnatifid toward base, slender glandular racemes of yellow flowers, oblique and slightly rugulose slenderly ovoid beaked fruits, is beginning to appear in meadows, etc., as yet somewhat casual. (Adv. from Eurasia)

31. ERÝSIMUM L. TREACLE-MUSTARD. VÉLAR (Que.)

Silique linear, 4-sided, the valves keeled by a strong midrib; stigma broadly lobed. Seeds in 1 row in each locule, oblong, marginless. Cotyledons in ours (often obliquely) incumbent. — Chiefly biennials of Am., Eurasia and n. Afr., mostly with yellow flowers; the leaves not clasping. Pubescence of appressed 2–3-parted hairs. (Name from the Greek *eryomai, help* or *save*, from the supposed medicinal properties of some species.)

a. Pedicels filiform, divergent or spreading-ascending; petals 3–5 mm. long; siliques 1–3 cm. long, subterete, green and glabrous or only sparsely pubescent; annual or biennial. . . . . 1. *E. cheiranthoides*.

a. Pedicels thick; petals 0.6–3 cm. long; siliques 2–11 cm. long, terete or 4-angled. . . *b*.

b. Divergently branching annual; pedicel scarcely enlarged at junction with silique; the latter green and glabrous or only remotely strigose, torulose. 2. *E. repandum*.

b. Simple or with mostly ascending branches, perennial or biennial; pedicel with dilation at summit, at junction with the non-torulose canescent-strigose silique.

- Calyx 6–10 mm. long; limb of petals 2.5–6 mm. long, 1–3 mm. broad; siliques 2–6 cm. long, on erect or suberect pedicels.
  - Limb of petals 2.5–3 mm. long, 1.5–2 mm. broad; fruiting raceme elongate (up to 2 or 3 dm. long) and open; siliques about 1 mm. broad; seeds 1–1.3 mm. long. . . . . 3. *E. inconspicuum.*
  - Limb of petals 4.5–6 mm. long, 2–3 mm. broad; fruiting raceme crowded and corymbiform at summit; siliques 1.5–2 mm. broad; seeds 1.6–2 mm. long. . . . . 4. *E. coarctatum.*
- Calyx 1–1.5 cm. long; limb of petals 6–12 mm. long, 4–12 mm. broad; siliques 5–11 cm. long, on loosely ascending to spreading pedicels.
  - Canescent, with thick and firm leaves, the cauline 2–10 mm. broad; blade of petals 4–6 mm. broad. . . . . 5. *E. asperum.*
  - Green, with thin submembranaceous leaves, the cauline 0.5–2 cm. broad; blade of petals 7–12 mm. broad. . . . . 6. *E. arkansanum.*

1. E. cheiranthoìdes L. (resembling *Cheiranthus*), Wormseed-Mustard, Herbe au chantre (Que.). — *Annual* or winter-annual, 0.1–1.3 m. high, simple or with ascending branches, *green or only slightly grayish*, very leafy, minutely roughened; *leaves* lanceolate, *scarcely toothed*, finely pubescent with trifid hairs; *flowers* small, *on divergent to spreading-ascending filiform pedicels; petals* yellow, *3–5 mm. long; siliques subterete*, slender, 1–3 *cm. long, green and glabrous or only sparsely pubescent;* seeds oblong, warm brown, about 1 mm. long. (*Cheirinia* Link) — Waste places, cult. fields and rich meadows, throughout our range and beyond. June–Sept. (Natzd. from Old World) Fig. 1106.

1106. E. cheiranthoides.

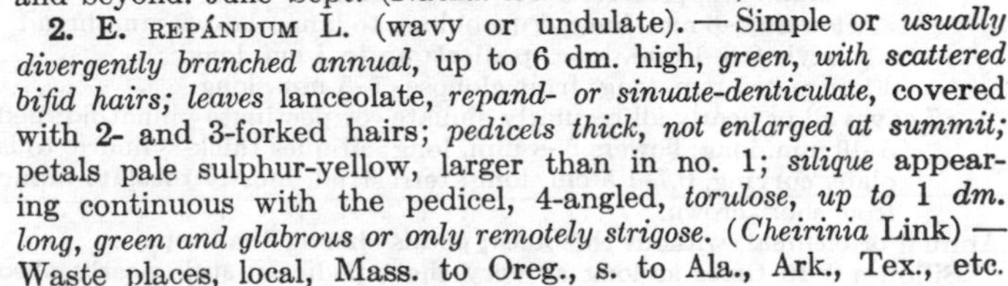

2. E. repándum L. (wavy or undulate). — Simple or *usually divergently branched annual*, up to 6 dm. high, *green, with scattered bifid hairs; leaves* lanceolate, *repand- or sinuate-denticulate*, covered with 2- and 3-forked hairs; *pedicels thick, not enlarged at summit;* petals pale sulphur-yellow, larger than in no. 1; *silique* appearing continuous with the pedicel, 4-angled, *torulose, up to 1 dm. long, green and glabrous or only remotely strigose.* (*Cheirinia* Link) — Waste places, local, Mass. to Oreg., s. to Ala., Ark., Tex., etc. May–July. (Adv. from Eurasia)

3. E. inconspícuum (S. Wats.) MacM. (inconspicuous). — Erect *perennial;* the stems stiff, up to 6 dm. high, *cinereous and scabrous*, with 2- or 3-pointed hairs, *simple or with ascending branches; leaves* linear, *mostly entire*, the crowded radical ones sometimes repand-dentate; *pedicels* thick, *with dilated tip;* calyx 6–8 mm. long, with linear-oblong sepals; *petals* pale yellow, *the limb* 2.5–3 *mm. long and* 1.5–2 *mm. broad; fruiting raceme* elongate (to 2 or 3 dm.), *becoming open and lax; siliques* erect, 4-angled, cinereous, 2–5 *cm. long, about* 1 *mm. broad*, with short thick style and broad 2-lobed stigma; *seeds* 1–1.3 *mm. long.* (*E. parviflorum* Nutt., not Pers.) — Dry open soil, n. B.C. to Nev., e. to Ont., Minn., N.D., and S.D.; and adv. s. to Mo. and Kans. and e. to N.E. and N.S. June, July. Fig. 1107.

1107. E. inconspicuum.

4. E. coarctàtum Fern. (pressed together). — Greener than no. 3, 0.4–7.5 dm. high; radical leaves oblanceolate, entire or obsoletely dentate; cauline leaves narrowly lanceolate to oblanceolate, entire, obtuse to acute; *racemes* finally elongating below but *the summit crowded and corymbiform;* pedicels erect, thick; calyx 6–10 mm. long; *limb of petals* 4.5–6 *mm. long and* 2–3 *mm. wide; siliques* 4.5–6 *cm. long*, 1.5–2 *mm. broad*, 4-angled, cinereous; *seeds* 1.6–2 *mm. long.* — Calcareous cliffs and gravels near Gulf of St. Lawrence, w. Nfld.; Mingan Ids., Anticosti I. and e. Gaspé Pen., Que. June–Aug.

5. E. ásperum (Nutt.) DC. (harsh), Western Wallflower, Prairie-Rocket. — *Stiff cinereous* perennial, with simple or sparsely branching stems 1–4.5 dm. high; *leaves thick and firm*, the numerous radical ones *repand-dentate*, the linear or narrowly lanceolate cauline ones entire or repand-toothed; *pedicels loosely ascending to divergent;* calyx 1–1.3 cm. long; *limb of the yellow petals* 6–10 mm. long, 4–6 *mm. broad; siliques stiffly divergent*, 4-angled, cinereous, 6–11 *cm. long*, with broad 2-lobed stigma. — Dry prairies, bluffs and sands, Mont. to N.M., e. to Man., Minn., Neb., Kans. and Okla. May, June.

6. E. arkansànum Nutt. (of Arkansas), Western Wallflower. — *Greener* than no. 5, biennial or perennial, with slender mostly solitary stems 0.3–1.2 m. high; *leaves green, thin*, lanceolate, entire or remotely denticulate, 0.5–2 *cm. broad;* calyx 1–1.5 cm. long; *limb of deep* *yellow to copper-colored petals* 7–12 *mm. broad; siliques loosely ascending or erect,* on divergent to ascending pedicels. (*E. elatum* Nutt.) — Rocky, gravelly or sandy open woods, bluffs, plains, and shores, O. to Wash., s. to s. Ill., Mo., Tex., N.M. and Calif. May–July.

32. RORÍPPA Scop. Yellow Cress

Fruit a short silique or a silicle, varying from slender to globular, terete or nearly so; valves strongly convex, nerveless. Seeds usually numerous, small, turgid, marginless, in 2 irregular rows in each locule (except in *R. sylvestris*). Cotyledons accumbent. — Aquatic to terrestrial plants of temp. reg., with yellow petals, small nectariferous glands, and commonly pinnate or pinnatifid leaves, usually glabrous. (Name derived from an old Saxon word, *Rorippen*, for these plants, and often unjustifiably altered to Roripa.) Radicula Hill; Nasturtium in large part, of many auth., not R. Br.

a. Perennial, with creeping to ascending rhizomes or tap-roots; petals much longer than sepals. . . *b.*

b. Leaves pinnately parted into toothed or cleft divisions; siliques slenderly linear-cylindric, 1–2.5 cm. long, the style 0.5–1 mm. long; seeds 0.6–0.8 mm. long. . . . . 1. *R. sylvestris.*

b. Leaves merely toothed to pinnatifid; siliques thick-cylindric, lanceolate or ellipsoid to globose, 0.1–1.5 cm. long, the style 1–3 mm. long; seeds about 1 mm. long. . . *c.*

c. Middle and upper leaves dentate, crenate or irregularly pinnatifid; pedicels 7–17 mm. long; siliques 1–7 mm. long, not strongly curved; aquatic or paludal plants with greatly prolonged bases.

Flowers 4–10 mm. long; fruit ellipsoid to lanceolate, 3–7 mm. long.

Flowers 8–10 mm. long; fruit ellipsoid, 1–2 mm. broad, with style 1–2 mm. long; pedicels 1–1.7 cm. long. . . . . 2. *R. amphibia.*

Flowers 4–6 mm. long; fruit oblong to lanceolate, 2 mm. broad, with style 0.8–1 mm. long; pedicels up to 1 cm. long. . . . . . 3. *R. prostrata.*

Flowers 3–5 mm. long; fruit globose, 1–3 mm. long. . . . . . . 4. *R. austriaca.*

c. Leaves all or nearly all regularly sinuate- or pectinate-pinnatifid; pedicels 5–10 mm. long; flowers 6–8 mm. long; siliques thick-cylindric to lanceolate, curving, 0.7–1.5 cm. long; terrestrial, mostly rosulate-branching from short crown. . . . . 5. *R. sinuata.*

a. Annual or biennial, without rhizomes; petals shorter than sepals.

Siliques 6–10 times as long as very short pedicels; style nearly obsolete; petals wanting or up to 0.5 mm. long; stamens 4; ovules about 200. . 6. *R. sessiliflora.*

Siliques shorter than to 4 times as long as pedicels; style definite; petals 1.2–2 mm. long; stamens 6; ovules 44–60.

Low, mostly rosulate-branching from crown, the stems 1–3 dm. long; sepals of well-developed flowers 1.5 mm. long; petals 1.2 mm. long; siliques 2–4 times length of pedicels. . . . . 7. *R. obtusa.*

Mostly taller (up to 1.3 m. high), rarely rosulate-branching; sepals 2 mm. long; petals 1.7–2 mm. long; siliques shorter than to twice length of pedicels. 8. *R. islandica.*

1. R. sylvéstris (L.) Bess. (of woods), Creeping Y. — *Perennial with subterranean creeping shoots;* stems spreading or ascending, 1–6 dm. high, simple to freely branched; *leaves thin, pinnately parted; the divisions toothed or cut,* lanceolate or linear; pedicels filiform, divergent; *petals* bright yellow, *exceeding sepals; siliques slenderly linear-cylindric,* 1–2.5 *cm. long,* the style 0.5–1 mm. long; seeds 0.6–0.8 mm. long. (*Nasturtium* R. Br.; *Radicula* Druce) — Meadows, shores and roadsides, often a troublesome weed, Nfld. to Ont., s. to N.B., N.E., Va., Ky. and Mo., rapidly spreading. May–Sept. (Natzd. from Eu.)

2. R. amphíbia (L.) Bess. (growing equally on land or in water). — *Perennial with long rooting* and branching *bases* and prolonged floating, creeping or loosely ascending fistulous stems; *leaves oblanceolate to narrowly obovate,* entire, dentate, serrate or jagged, or the lower leaves pectinate in forma variifòlia (DC.) Hayek (variable-leaved); inflorescence corymbose-paniculate; *pedicels* 7–15 *mm. long; flowers* 8–10 *mm. long, the petals longer than the sepals; fruit ellipsoid,* 3–7 *mm. long,* with slender style; seeds about 1 mm. long. — Quiet waters and shores, or even on dry roadsides, rapidly spreading, Que., N.E. and N.Y. June, July. (Natzd. from Eu.) — Probably two or more species or vars. to be separated when better material is at hand.

3. R. prostràta (Bergeret) Schinz & Thell. (prostrate). — Similar to no. 2; stem depressed or erect, from vertical tap-root, mostly 3–9 dm. high; leaves mostly cleft; *flowers smaller* (only 4–6 *mm. long; fruit oblong to lanceolate, about* 2 *mm. broad,* about 5 mm. long; *style barely* 1 *mm. long; mature pedicels at most about* 1 *cm. long.* — Waste places, wet meadows, etc., e. Pa. (Adv. from Eu.)

4. R. austrìaca (Crantz) Bess. (Austrian). — Similar to no. 2; *flowers* small, 3–5 *mm.*

long; fruit globose, 1–3 *mm. long.* — Low fields and muddy shores, local, N.Y. to Wisc., Ia. and N.D. June, July. (Adv. from Eu.)

5. R. sinuàta (Nutt.) Hitchc. (strongly wavy). — *Perennial,* with subterranean rhizomes, *mostly rosulate-branching* at summit of caudex, the stems 1–4 dm. long; *leaves* oblong or elliptic-lanceolate, *regularly sinuate- or pectinate-pinnatifid,* with nearly entire linear-oblong lobes; pedicels 5–10 mm. long; flowers 6–8 mm. long, the petals exceeding the sepals; *siliques thick-cylindric to lanceolate, curving,* 0.7–1.5 *cm. long,* tipped by the slender style; seeds about 1 mm. long. (*Radicula* Greene) — Sandy or rocky shores, fields, bottoms and roadsides, w. Ont. to Wash., s. to Mich., Ill., Mo., Okla., Tex., N.M., Ariz. and Calif. Late April–July. Fig. 1108.

1108. R. sinuata.

6. R. sessiliflòra (Nutt.) Hitchc. (sessile-flowered). — Erect or ascending simple to branched annual or biennial 1–4 dm. high; radical leaves often persistent, oblong-obovate, obtusely crenate to lyrate-pinnatifid; *cauline leaves* narrowly obovate or oblong, *shallowly and obtusely toothed; flowers minute, nearly sessile; petals wanting or up to* 0.5 *mm. long; stamens* 4 (rarely 5); *siliques* linear-oblong to short-oval, 5–12 mm. long, *six to ten times as long as the very short pedicels; style nearly obsolete; ovules about* 200. (*Radicula* Greene) — Muddy or sandy bottoms, river-flats and wet places, nw. Fla. to Tex., n. to Ind., Wisc., Minn. and Neb.; Potomac R., D.C.; James R., e. Va. April–Oct.

7. R. obtùsa (Nutt.) Britt. (obtuse). — Similar to no. 6, much branching, *mostly rosulate-branched from crown;* the slender *stems* 1–3 *dm. long; leaves pinnately parted or divided,* the divisions roundish and obtusely toothed or repand; flowers minute, short-pedicelled; *sepals* of well developed flowers 1.5 *mm. long; petals* 1.2 *mm. long; stamens* 6; *siliques* linear-oblong to short-oval, 3–8 mm. long, *two to four times length of pedicels, the style definite; seeds* 44–60. (*Radicula* Greene) — Muddy or sandy shores and wet places, s. B.C. to Calif., e. to O., W.Va., Ill., Mo. and Tex.; said to be adv. in D.C. May–Sept.

Var. **íntegra** (Rydb.) Vict. (entire). — Leaves spatulate or spatulate-rhombic, merely dentate. — St. Lawrence R., Que.; Rocky Mt. reg.

8. R. islándica (Oeder) Borbás (of Iceland). — Annual or biennial, 0.1–1.3 m. high, simple or branched; leaves pinnate, pinnatifid or merely toothed; pedicels filiform, elongate; *sepals about* 2 *mm. long; petals* 1.7–2 *mm. long;* stamens 6; *siliques slenderly ellipsoid to subglobose,* straight or arcuate, 2–10 mm. long, 1–4 mm. thick, *shorter than to twice the length of the pedicels;* style short but usually definite; seeds 44–60, 0.4–0.9 mm. long. — Very variable. We have the following vars.:

Leaves all or nearly all pinnate or deeply pinnatifid; the numerous segments lanceolate and dentate, decurrent along the rachis; siliques slenderly ellipsoid, often curved, 4–10 mm. long, equaling the pedicels; thin-leaved glabrous plant. . . . . *R. islandica* (typical).

Lower leaves merely pinnatifid, runcinate or uncleft, the middle and upper coarsely toothed to subentire; plants relatively coarse, up to 1.3 m. high, the leaves (except in submersed states) firm.

Siliques slenderly ellipsoid or subcylindric, 3–9 mm. long, 1–2.5 mm. in diameter; glabrous throughout or stem hispid below. . . . . . . Var. *Fernaldiana.*

Siliques short-ellipsoid, ovoid or thick-lanceolate to subglobose, 2–5.5 mm. long, 1.7–4 mm. in diameter; base of stem or lower leaves frequently (not always) hispid. . . . . Var. *hispida.*

R. islándica (typical), *R. palustris* (L.) Bess.; *Radicula palustris* (L.) Moench — Wet shores and waste places, partly nat., partly adv. from Eu.; Anticosti I., Que. to Mich., s. to N.S., N.E., N.J. and Pa. May–Sept. (Greenl.; Eurasia)

Var. **Fernaldiàna** Butt. & Abbe (in honor of the present author, 1873–), *Radicula palustris* of ed. 7; *Rorippa isl.,* var. *microcarpa* (Regel) Fern., not forma *microcarpa* (G. Beck) Thell. — Wet shores and damp openings, s. Lab. to B.C., s. to N.S., N.E., Va., Tenn., La., Tex., N.M. and Calif. May–Oct. (E. Asia) Fig. 1109. — Passing insensibly to the next var. — Forma **reptabùnda** Fern. (creeping) of exsiccated pond-margins of n. N.H. and Minn. has long creeping and rooting stems with fascicles of mostly simple leaves.

1109. R. islandica, v. Fernaldiana.

Var. **híspida** (Desv.) Butt. & Abbe (stiffly hairy), *R. hispida* (Desv.) Britt.; *R. islandica,* var. *glabrata* (Lunell) Butt. & Abbe; *Radicula palustris,* var. *hispida* (Desv.) Robins. — Shores, wet openings and waste places, Nfld. to B.C., s. to N.S., N.E., Fla., La., Tex., N.M. and Oreg. (W.I.) — Siliques frequently with 3–4 (sometimes 6) carpels. Fig. 1110.

1110. R. islandica, v. hispida.

33. NASTÚRTIUM R. Br. Watercress. Cresson (Que.)

Silique linear-cylindric, plump, the convex valves nerveless. Nectariferous glands horseshoe-shaped. Seeds 2-seriate in each locule, small, turgid, marginless. Cotyledons accumbent. — Aquatic or paludal herb, nat. of Old World, with succulent smooth stem, often pinnate leaves and white petals twice the length of the calyx. (Name from *nasus tortus, a wry or twisted nose*, alluding to the effect of the pungent qualities of the plant.) — The generic name erroneously applied to members of *Tropaeolum* of the *Tropaeolaceae*, a Mex. and S. Am. genus familiar in cult.

1. N. officinàle R. Br. (of the shops), True W., Cresson de fontaine (Que.). — Perennial, with creeping or floating freely rooting stems; leaflets 3–11, roundish to oblong, somewhat fleshy, nearly entire; siliques (0.8–) 1–2.7 cm. long, curving, ascending, on divergent pedicels. (*Radicula Nasturtium-aquaticum* (L.) Britten & Rendle; *Rorippa Nasturtium-aquaticum* (L.) Schinz & Thell.; *N. Nasturtium-aquaticum* (L.) Karst.) (Introd. from Eu.) — Three vars.:

Silique 1.5–2.5 mm. broad, 0.7–1.8 cm. long, beakless or with thick style 0.1–1 mm. long, on horizontally divergent to ascending pedicels; longest (lowest) pedicels 0.6–1.5 cm. long.
Terminal leaflet roundish to oval; lateral leaflets rounded at base, elliptic, oval or obovate . . . . *N. officinale* (typical).
Terminal and elongate lateral leaflets nearly oblong. . . . . Var. *siifolium*.
Silique 1–1.5 mm. broad, (1–) 1.5–2.7 cm. long, tapering to a slender style 0.5–2 mm. long, on arched-recurving to loosely ascending pedicels; longest pedicels 0.8–2.5 cm. long. . . . . Var. *microphyllum*.

N. officinàle (typical). — Brooks, springheads, rills and cool waters, originally cult., now generally natzd. throughout our area and beyond. April–Oct. Fig. 1111.

Var. siifòlium (Reichenb.) Koch (with leaves of *Sium*). — Local, Mass. to Ia. and westw., s. to N.C., Tex. and Calif.

Var. microphýllum (Boenn.) Thell. (small-leaved). — Leaflets rather smaller, the lateral ones mostly tapering to short petiolules. — Nfld. to Ont., s. to Fla. and Mich.

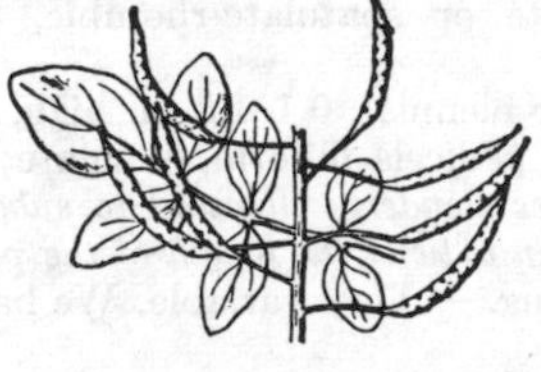

1111. N. officinale.

34. ARMORÀCIA Gaertn., Mey. & Scherb.

Silicle subglobose, obovoid or ellipsoid, the convex valves nerveless; stigma capitate or 2-lobed, broad. Seeds in 2 rows in each locule, turgid and wingless. Cotyledons accumbent. — Perennials of N. Hemisph. with deep roots or rhizomes, leaves undivided or pinnatifid (submersed ones often capillary-dissected), flowers white. (Ancient name of the Horseradish.)

1. A. lapathifòlia Gilib. (dock-leaved), Horseradish, Raifort (Que.). — *Coarse, with thick vertical root; stem upright,* 0.6–1.3 *m. high; basal leaves oblong or oblong-ovate, long-petioled,* with coarse midrib, up to 4 dm. long, *crenate; cauline leaves* lanceolate, smaller, *the lower often pinnatifid,* the upper crenate; racemes elongating, panicled; *fruiting pedicels spreading-ascending; silicle* subglobose or globose-obovoid (often not well developed), 2-*locular, capped by the very short style and depressed-capitate stigma.* (*A. rusticana* (Lam.) Gaertn., Mey. & Scherb.; *Radicula Armoracia* (L.) Robins.; *Rorippa Armoracia* (L.) Hitchc.) — Moist ground, spread from cult., throughout. May–July. — Root the source of the familiar condiment. (Introd. and natzd. from Eu.)

2. A. aquática (Eat.) Wieg. (aquatic), Lake-Cress. — *Weak, from a slender rootstock;* the stems simple and prolonged or branching; *basal leaves* once to thrice *pinnately dissected into numerous capillary divisions;* emersed leaves oblong, entire, serrate or pinnatifid, or dissected like the basal ones in forma **capillifòlia** Vict. & Rousseau (hair-leaved); racemes becoming lax, with *divergent to reflexed long pedicels; silicle ovoid,* 1-*locular, the long style tipped by the* 2-*lobed stigma.* (*Radicula* Robins.; *Rorippa* Britt.; *Neobeckia* Greene) — Lakes and quiet streams, w. Que. and Ont., s. to Fla. and La. June–Aug.

35. BARBARÈA R. Br. Winter-Cress

Silique linear, terete or somewhat 4-sided, the valves being keeled by a midnerve. Seeds in a single row in each locule, marginless. Cotyledons accumbent. — Mostly Eurasian and N. Am.

biennials or perennials with overwintering basal rosettes, the cauline leaves with clasping bases; flowers yellow. (Anciently called the Herb of St. Barbara, the seed of no. 2 being sown in w. Eu. near St. Barbara's day, in mid-December.) CAMPE Dulac

Beak of the silique slender, 1.5–3 mm. long; uppermost leaves incised, coarsely dentate, angulate or lobed, but rarely pinnatifid. . . . . 1. *B. vulgaris.*
Beak thickish, 0.5–1 (rarely –2) mm. long; uppermost leaves usually lyrate-pinnatifid or at least not angulate-dentate.
Basal leaves with 10–20 lateral leaflets; petals deep yellow, 6–8 mm. long; siliques 4–8 cm. long; seeds 1–1.5 mm. broad, grayish-brown, foveolate-reticulate; weed of fields, southw. . . . . 2. *B. verna.*
Basal leaves simple or with 2 or 4 small lateral basal leaflets; petals pale yellow, 2.5–5 mm. long; siliques 2–3.5 cm. long; seeds 0.8–1 mm. broad, warm brown, minutely rugulose; indigenous n. species. . . . . 3. *B. orthoceras.*

1. B. vulgàris R. Br. (common), COMMON W., YELLOW ROCKET, HERBE DE SAINTE-BARBE or CRESSON DE TERRE (Que.). — Smooth (very rarely hirsute at base) biennial or perennial; lower leaves lyrate (rarely simple); the terminal lobe much the largest, elliptic-oblong to suborbicular; the lateral smaller lobes narrower, 1–4 pairs, or wanting; *upper leaves* obovate to rounded, *coarsely dentate, angulate or lobed,* but rarely pinnatifid; petals narrowly obovate, 5.5–8 mm. long, 2–3 mm. broad, bright yellow; flowers somewhat racemose even in anthesis; *siliques* slender-pedicelled, *with a slender beak* (style) 1.5–3 *mm. long; seeds* short-oblong to quadrate, 1–1.5 mm. long, the *lustrous* grayish surface *rugulose.* — Very plastic Eurasian species, natzd. with us. Our vars. are as follows:

Pedicels ascending to appressed or erect; siliques erect or strongly ascending, closely overlapping, the raceme dense.
Siliques (excluding beak) 1.5–3 cm. long. . . . . *B. vulgaris* (typical).
Siliques (excluding beak) 0.8–1.5 cm. long. . . . . Var. *sylvestris.*
Pedicels spreading; siliques arcuate-ascending to horizontally divergent, not imbricated, the raceme lax and open.
Siliques (excluding beak) mostly 2–3 cm. long. . . . . Var. *arcuata.*
Siliques (excluding beak) mostly 0.7–1.5 cm. long. . . . . . . Var. *brachycarpa.*

B. VULGÀRIS (typical), *B. stricta,* in part, of ed. 7, not Andrz. — Meadows, brooksides and damp woods, Nfld. to Ont., s. to N.S., N.E., L.I., Va., W.Va., O., Ind., Ill., Mo. and Kans. April–Aug. (northw.). (Natzd. from Eu.) — A double-flowered, usually sterile form, forma PLÈNA Fern. (double), cult. and slightly natzd. from Que. to Pa.

Var. SYLVÉSTRIS Fries (of woods). — Anticosti I., Que., to e. Pa., local. (Natzd. from Eu.)

Var. ARCUÀTA (Opiz) Fries (bow-like), *B. vulgaris* of ed. 7. — Similar habitats, St. P. et Miq. and Que. to Ill., s. to N.S., N.E., Va. and Ky. (Natzd. from Eu.) — Ordinarily glabrous but in forma HIRSÙTA (Weihe) Fern. (hirsute) with basal leaves villous-hirsute.

Var. BRACHYCÁRPA Rouy & Foucaud (short-fruited). — Local, Anticosti I., Que., to Ct. (Natzd. from Eu.)

2. B. vérna (Mill.) Aschers. (of early spring), EARLY W., BELLE-ISLE CRESS. — Biennial, glabrous; *leaves all pinnatifid; the basal with* rounded-oval to -oblong terminal lobe and 10–20 smaller *lateral lobes; petals* 6–8 *mm. long, bright yellow;* pedicels 3–8 mm. long, nearly as thick as the *long* (4–8 *cm.*) slightly flattened *rigid* ascending *siliques, with thick style* 0.5–1 *mm. long; seeds* 1–1.5 *mm. broad, grayish-brown, foveolate-reticulate.* (*B. praecox* (J. E. Sm.) R. Br.) — Weed of fields, Fla. to Calif., n. to Ct., N.Y., O., Ind. and Mich. Late March–May. (Natzd. from Eu.)

3. B. orthóceras Ledeb. (with straight horns). — Strict, the stem and lower leaves often purple-tinged; basal leaves oblong to elliptic, simple or with 2 or 4 small basal lobes; *lower and middle cauline leaves* more decidedly *lyrate-pinnatifid,* ordinarily with 4–12 small lobes; *uppermost leaves* oblong to narrowly obovate, *lyrate-pinnatifid,* with few basal lobes; racemes in anthesis dense, in fruit elongate and slender; *petals pale yellow,* 2.5–5 *mm. long; siliques* subterete or compressed, not conspicuously angled, 2–3.5 *cm. long,* somewhat *crowded, appressed or strongly ascending,* on thick pedicels; *the style thick,* 0.5–1(–2) *mm. long; seeds* 0.8–1 *mm. broad, warm-brown, minutely rugulose.* (*B. stricta* of many Am. auth., not Andrz.; *B. americana* Rydb.) — Banks of streams, in swamps or on wet rocks, Lab. to Alaska, s. to Nfld., Gaspé Pen., Que., St. John R., Me., alpine reg. of N.H., Bruce Pen., Ont., n. Mich., Wisc., Minn., Man., Ariz. and Calif. June–Sept. (Asia)

36. IODÀNTHUS T. & G. Purple Rocket

Silique long, slender, somewhat flattened; valves 1-nerved; stigma entire but slightly elongated over the placentae. Seeds 1-ranked in each locule, oblong, marginless. Cotyledons essentially accumbent. — Erect Am. perennial with purplish flowers. (Name from the Greek *iodes, violet-colored*, and *anthos, flower.*)

1. I. pinnatífidus (Michx.) Steud. (pinnately cleft). — Glabrous, 3–9 dm. high; root-leaves round or cordate, on slender petioles; cauline leaves auricled, ovate-oblong and ovate-lanceolate, sharply and often doubly toothed, tapering to each end, the lower into a winged petiole, rarely bearing a pair or two of small lateral lobes; petals with long claws exserted, the broad limb 3–6 mm. long; siliques 1.8–3 cm. long, on short diverging pedicels, pointed by a short style. — River-banks and alluvium, Ala. to Tex., n. to w. Pa., W.Va., O., Ind., Ill. and Minn. May, June.

Lunària L. (from *luna, the moon*), Honesty, distinguished from the preceding by its deltoid-ovate leaves and the elliptic or rounded to lanceolate very flat silicles up to 2.5 cm. broad, has two cult. species which wander to roadsides and waste ground but are hardly naturalized: L. ánnua L. (annual), annual or biennial, the silicles broadly elliptic and rounded at both ends; and L. redivìva L. (living again), perennial, the silicles lance-oblong and pointed at each end. (Introd. from Eu.)

37. CHORÍSPORA DC.

Silique curving, elongate, lomentaceous, with attenuate, 2-seriate, indehiscent, 1-seeded locules, the slender beak subulate and tipped by the minute stigma. Seeds solitary in each locule, pendulous, margined or marginless. Cotyledons accumbent. — Branching glandular or pilose herbs nat. of Asia, with purple or yellow flowers and entire or pinnatifid leaves. (Name from *choris, asunder*, and *spora, a seed*, from the alternating locelli.)

1. C. tenélla (Willd.) DC. (very slender). — Sparsely glandular-hirtellous annual, mostly branched, 2–5 dm. high; lower leaves runcinate; middle and upper lanceolate to oblong, petioled, undulate-dentate; sepals 3.5–6 mm. long; petals purple, the claw exserted, the narrow limb 3–5 mm. long; siliques upwardly curving, including the long slenderly subulate beak 3–4 cm. long. — Roadsides and fields, local, Mass. to Ia. and Neb. and w. to Wash. *Fr.* May–July. (Adv. from Asia)

Matthìola (for the early Italian botanist, *Pietro Andrea Matthiola*) incàna (L.) R. Br. (grayish), the common Stock or Gilliflower, an erect much-branched perennial with entire obtuse linear-oblong heavy leaves, large purple or reddish (in cult. variously colored) flowers, compressed narrowly linear siliques with sessile stigma, and winged flattened seeds in 1 row; and var. ánnua (Sweet) Voss (annual), Ten-weeks Stock, an annual garden variety, sometimes spread from cult. but do not long persist. (Introd. from Eu.)

38. SELÈNIA Nutt.

Silicle large, oblong-elliptical, flat; the valves nerveless. Seeds in 2 rows in each locule, rounded, broadly winged; cotyledons accumbent; radicle short. — A low annual with once or twice pinnatifid leaves and leafy-bracteate racemes of yellow flowers. (Name from the Greek *selene, the moon*, with allusion to *Lunaria*, which this genus somewhat resembles in its silicles.)

1. S. aùrea Nutt. (golden). — Simple or branching from base, 0.5–2 dm. high; lobes of the small simply pinnatifid leaves entire or toothed; silicles 1–2 cm. long, on elongate spreading pedicels, beaked by the long slender style. — Cherty barrens or bluffs and sandy soil, sw. Mo., se. Kans., Ark. and e. Okla. April, May.

39. LEAVENWÓRTHIA Torr.

Silique broadly linear or oblong, flat; the valves nerveless but minutely reticulate-veined. Seeds in a single row in each locule, flat, surrounded by a thick wing. Embryo straight or the short radicle only slightly bent in the direction which, if continued, would make the orbicular cotyledons accumbent. — Small N.Am. winter-annuals, glabrous and often stemless, with lyrate leaves and short 1–few-flowered scapelike peduncles. (Named for Dr. *Melines Conklin Leavenworth*, 1796–1862, a southern botanist.)

1. L. uniflòra (Michx.) Britt. (one-flowered). — Scapes 5–15 cm. high; leaf-lobes usually numerous (7–15); petals purplish or nearly white with a yellowish base, obtuse; *siliques not torulose,* oblong to linear (1.2–3 cm. long); style short. — Rocky woods, glades and limestone-barrens, Ala. to Ark., n. to s. O., s. Ind. and Mo. March–early May.

2. L. torulòsa Gray (cylindric, with constrictions). — Similar, but *siliques torulose* even when young, linear; style 2–4 mm. long; seeds acutely margined rather than winged; petals emarginate. — Cedar-glades, Ky. to Ala. March, April.

40. DENTÀRIA L. Toothwort. Pepperroot

Silique lanceolate, flat. Style elongated. Seeds in one row, wingless, the funiculus broad and flat. Cotyledons petioled, thick, very unequal, their margins somewhat infolding each other. — Perennials of damp woodlands of the N. Hemisph., with long fleshy sometimes interrupted scaly or toothed rhizomes of a pleasant pungent taste; stems leafless below, bearing 2 or 3 petioled compound or deeply cleft leaves about or above the middle, and terminated by a corymb or short raceme of large white or purple flowers. (Name from *dens,* a *tooth,* presumably from the toothed rhizomes of some species.)

Rhizome elongate, prominently toothed; larger leaflets or principal segments of cauline leaves 1.3–5.5 cm. broad.
- Rhizome continuous; central leaflet of basal leaves 5–8.5 cm. long and 3.5–6.5 cm. broad; cauline leaves mostly 2, approximate; the central leaflet broadly elliptic to ovate, usually 7–10 cm. long and 2.5–4.5 cm. broad. . . . . . 1. *D. diphylla.*
- Rhizome with slender constrictions; central leaflet of basal leaves 2.5–5(–6) cm. long and 2–3(–5) cm. broad; cauline leaves 2 or 3, approximate or distant; the central leaflet or its middle segment lanceolate to narrowly elliptic, 2.5–6 cm. long and 1.3–3.5 cm. broad. . . . . 2. *D. maxima.*

Rhizome of readily separable fusiform to ellipsoid toothless or only obscurely toothed tubers; larger leaflets or principal segments of cauline leaves 0.1–3.5 cm. broad.
- Rachis of inflorescence more or less hirsute; flowering stem without basal leaf. 3. *D. laciniata.*
- Rachis glabrous.
  - Flowering stem with basal leaf unlike the cauline; cauline leaves with simple or merely incised leaflets 0.5–3.5 cm. broad. . . . . 4. *D. heterophylla.*
  - Flowering stem without basal leaf; leaflets of cauline leaves dissected into 3–7 long linear segments 1–3 mm. broad. . . . . 5. *D. multifida.*

1. D. diphýlla Michx. (two-leaved), Snicroûte (Que.), Carcajou (Que.). — *Rhizome long and continuous,* often branched, *the annual segments slightly or not at all tapering at the ends;* stems in anthesis 1.5–3 dm. high, stoutish; *leaves* 3-foliolate, the basal and *cauline* similar, *the former with* lateral leaflets obliquely ovate, the *central leaflet broadly ovate* and 5–8 *cm. long and* 3.5–6.5 *cm. broad, with coarse rounded teeth; cauline leaves* 2 (rarely 3), *opposite or approximate; the central leaflet broadly elliptic to ovate, usually* 7–10 *cm. long and* 2.5–4.5 *cm. broad, coarsely blunt-toothed;* sepals 5–8 mm. long, about half the length of the white petals; silique rarely maturing. — Rich woods, Gaspé Pen., Que., to s. Ont., s. to N.S., N.E., S.C., Ky. and Mich. Mid-April–early June. Fig. 1112.

1112. D. diphylla.

× **D. anómala** Eames (anomalous) seems to be a hybrid of nos. 1 and 3. A hybrid of nos. 1 and 2 has also been recorded.

2. D. máxima Nutt. (greatest). — *Rhizome* interrupted, *consisting of several elongate strongly toothed segments which are constricted at each end,* the older commonly retaining shreds of old stems; *radical leaves sharply toothed* or sometimes cleft into long segments, *the central leaflet or large segments* 2.5–5(–6) *cm. long and* 2–3(–5) *cm. broad; cauline leaves* 2 or 3, *approximate or distant; the central sharply-toothed leaflet or its middle segment lanceolate to narrowly elliptic,* 2.5–6 *cm. long and* 1.3–3.5 *cm. broad, sharply toothed* or cleft, *more or less ciliolate;* flowers much as in no. 1; siliques rarely maturing. — By woodland-streams or on calcareous wooded slopes, s. Me. to Wisc., s. to s. N.E., Pa., W.Va. and Tenn. April, May.

3. D. laciniàta Muhl. (slashed). — *Tubers* deep-seated, yellow-brown, *fusiform,* 1–4 cm. long, *smooth or only obscurely toothed; stem without basal leaf,* the *cauline leaves* 3 (rarely 2), whorled or slightly distant, each *with 3 variously cleft leaflets;* these sometimes simple and *sharply toothed to incised* or sometimes even entire, or cleft to base into 5–9 sharply toothed to incised segments, the leaflets or long segments varying from linear to oblong or oblanceolate; *rachis of inflorescence more or less hirsute;* sepals 6–9 mm. long; petals 1–2 cm. long, white or purplish; siliques slenderly lance-subulate, including the long beak

1113. D. laciniata.

2.5–5 cm. long. — Rich woods, wooded bottoms and calcareous rocky banks, w. Que. and Vt. to Minn. and Neb., s. to Fla., Ala., La. and e. Kans. March–May. FIG. 1113. — Most variable in the breadth and degree of toothing and incision of the leaflets, several such variations often found together. Of more stability is

Var. **coalèscens** Fern. (coalescing). — Leaves simple, those of the cauline whorl with lanceolate or oblanceolate to obovate partially cleft digitate or pinnatifid blades; tubers ellipsoid, 1–2 cm. long.— Rich deciduous woods, Princess Anne County, Va.

× **D. incisifòlia** Eames (cut-leaved) seems to be a hybrid of nos. 3 and 2.

4. D. heterophýlla Nutt. (various-leaved). — Tubers as in no. 3, often more slender; *basal leaf approximate to or borne at bottom of flowering stem, with* 3 *rhombic-obovate or ovate crenate or dentate* or lobed *leaflets* 1.5–8 cm. long; *cauline leaves* 2 or 3, opposite, whorled or scattered, *with* 3 *narrowly oblong* entire or toothed *leaflets* 1.5–8.5 dm. long, 0.5–3.5 cm. broad; sepals purple-tinged; petals pink or purplish, 1.2–2 cm. long; siliques as in no. 3. — Rich woods, n. N.J. to s. O., s. to Ga. and Tenn. April, May.

5. D. multífida Muhl. (much divided). — Tubers as in nos. 3 and 4; *flowering stem without basal leaves,* very slender; *cauline leaves* 2 or 3, approximate, *with* the 2 or 3 *long-stalked leaflets cut into* 3–7 *long linear entire or sparingly toothed segments* 1–3 *mm. broad; rachis of raceme glabrous;* sepals 4–6 mm. long; petals 2 or 3 times as long as sepals; capsules much as in no. 3.— Alluvial and rich woods, Ga. and Ala., n. to s. O. and s. Ind. April. — Resembling slenderest forms of no. 3 but distinguished by the glabrous axis of the raceme.

41. CARDÁMINE L. BITTER CRESS

Silique linear, usually flattened, opening elastically from the base; the valves nerveless and veinless or nearly so; placentae and partition thick. Seeds in a single row in each locule, wingless; the funiculus slender. Cotyledons accumbent, flattened, equal or nearly so, petiolate. — Mostly glabrous perennials of cold and temp. reg., leafy-stemmed, often growing along watercourses and in wet places. Flowers white or purple. (A Greek name, *kardamon,* used by Dioscorides for some cress.)

a. Dwarf densely tufted alpine perennial, rarely 1 dm. high, with simple ovate leaf-blades 0.2–1.5 cm. long and 1–5 flowers. . . . . 1. *C. bellidifolia.*

a. Taller, not tufted, with mostly larger simple or pinnate leaves and usually more flowers. . *b.*

b. Tuberous-based upright perennial with simple cauline leaves and flowers 0.7–2 cm. long.

Stem glabrous; cauline leaves 4–14, the lower 2–5 petioled; petals white. 2. *C. bulbosa.*

Stem more or less hirsute; cauline leaves 2–5(–6), all sessile or 1 or 2 lowest petioled; petals rose-purple or pink-tinged. . . . . 3. *C. Douglassii.*

b. Fibrous-rooted; flowers mostly smaller or, if large, the plant with prostrate branches or with pinnate leaves. . . *c.*

c. Petals 5–15 mm. long; perennials. . . *d.*

d. Leaves simple or, if pinnate, with terminal leaflet 1.5–5 cm. broad and with only 1 or 2 pairs of lateral leaflets; petals 5–10 mm. long; southern species.

Stem diffuse, reclining or with trailing branches; leaves all simple, rounded, their petioles not appendaged. . . . . 4. *C. rotundifolia.*

Stem erect, simple or nearly so; leaves usually with 1 or 2 pairs of lateral and a large angulate-lobed terminal leaflet, the upper petioles with long arching sagittate basal appendages. . . . . 5. *C. Clematitis.*

d. Leaves pinnate, with numerous leaflets or segments, the terminal leaflet of the cauline leaves much narrower; petals 1–1.5 cm. long; northern species. . . . . 6. *C. pratensis.*

c. Petals 1.5–4 mm. long (wanting in no. 12); plants annuals or biennials with overwintering rosettes, or short-lived perennials. . . *e.*

e. Cauline leaves with prolonged sagittate-auriculate bases, the numerous leaflets acuminate. . . . . 7. *C. impatiens.*

e. Cauline leaves without basal auricles, the leaflets or blades mostly blunt. . . *f.*

f. Petioles of cauline leaves hirsute-ciliate at base.

Stems flexuous, hirsute, with 4–10 membranaceous leaves similar to the radical, their rounded or ovate angulate-dentate leaflets 0.4–2 cm. broad and on slender petiolules 2–15 mm. long; stamens 6; siliques spreading-ascending, with style 0.7–1 mm. long, 20–24-seeded. . . . . 8. *C. flexuosa.*

Stem stiffly ascending, glabrous, with 2–6 firm leaves quite unlike the basal, their narrow leaflets 0.5–4 mm. broad and cuneate-narrowed into the short petiolules; stamens usually 4; siliques stiffly erect, with thick style rarely 0.5 mm. long, 22–36-seeded. 9. *C. hirsuta.*

*f.*Petioles of cauline leaves naked at base. . . *g.*

*g.*Flowers petaliferous, 1.5–4 mm. long, pedicelled; siliques 1–3 cm. long, linear, on slender pedicels 2–15 mm. long; leaves pinnate or pinnatifid.

Stems (except in submersed bases) hispid or hispidulous at base; leaflets with bases decurrent along, or confluent with, rachis; terminal leaflet of cauline leaves usually broader than lateral leaflets; ripe silique with slender style 0.5–2 mm. long; seeds 1–1.5 mm. long. . . . . 10. *C. pensylvanica.*

Stems glabrous throughout; leaflets quite distinct, not decurrent along the rachis; terminal leaflet of cauline leaves scarcely broader than lateral leaflets, linear to oblanceolate; ripe silique scarcely beaked, the mature style 0.2–0.7 mm. long; seeds 0.7–0.9 mm. long. . . . . 11. *C. parviflora.*

*g.*Flowers apetalous, 0.7–1.2 mm. long, subsessile; siliques 5–10 mm. long, lanceolate, on thick pedicels 0.5–1.5 mm. long; leaves roundish, simple or with 2 round lateral segments. . . . . . 12. *C. Longii.*

1. C. bellidifòlia L. (with leaves like English Daisy, *Bellis*). — *Dwarf, tufted* perennial 2–11 *cm. high; leaves ovate, entire,* or sometimes with a blunt lateral tooth, 0.2–1.5 *cm. long,* slender-petioled; flowers 1–5; siliques upright, linear, 1.5–3 cm. long, the style very short and thick. (Incl. var. *laxa* Lange) — By alpine brooks, in cold ravines or on wet mossy rocks, Greenl. to Alaska, s. to n. Lab., Shickshock Mts., Gaspé Pen., Que., Mt. Katahdin, Me., White Mts. N.H., Alta. and Oreg. June–Sept. (No. Eurasia)

2. C. bulbòsa (Schreb.) BSP. (bulbous), Spring-Cress. — *Stem* simple or rarely branched above, erect, *glabrous, from tuberous base, in anthesis* 1.5–5 *dm. high;* radical leaves long-petioled, oblong to cordate-ovate or reniform; *cauline leaves* 4–14, scattered, ovate or rounded to lanceolate, *the* 2–5 *lower petioled;* the upper sessile, entire, undulate or remotely dentate; *sepals greenish,* with white margins; *petals white,* 7–16 mm. long; *siliques* slenderly lanceolate, tapering to a slender style tipped by a conspicuous stigma, *the lower on pedicels* 1.5–2.5 *cm. long;* seeds oval. — About springs, in bottomland-woods or in meadows, n. Fla. to Tex., n. to se. N.H., Vt., sw. Que., s. Ont., Mich., Wisc., Minn. and e. S.D. Late March–June. — Forma **fontinàlis** Palmer & Steyerm. (of springs and rills) resembles no. 4 in its unthickened bases and roots and its cauline leaves broadly ovate or roundish and subcordate. Running water, Mo.

3. C. Douglássii (Torr.) Britt. (for its discoverer, David Bates Douglass, 1790–1849). — Similar to no. 2; *stem* usually *hirsute,* especially above, in anthesis 0.7–3 dm. high; radical leaves mostly suborbicular; *cauline leaves* 2–5(–6), *all sessile or the* 1 *or* 2 *lowest petioled; sepals purple-tinged; petals rose-purple or pink-tinged,* or white in forma **albídula** Farw. (whitish), 0.7–2 cm. long; *lowest siliques on pedicels* 1.5–4 *cm. long.* (*C. bulbosa,* var. *purpurea* (Torr.) BSP.) — Calcareous springy places, rich woods and bottomlands, Ct. to s. Ont. and Wisc., s. to Va., Tenn. and Mo. Mid-March–mid-May, usually some weeks earlier than no. 2.

4. C. rotundifòlia Michx. (round-leaved), Mountain-Watercress. — *Stems branching, weak or decumbent, making long runners; root fibrous;* leaves all much alike, roundish, somewhat angled, often cordate at the base, petioled; petals white, 5–10 mm. long; siliques linear-subulate, equaled or exceeded by the pedicels, slender-beaked. — Springy places and brooksides, mts. of N.C., n. to N.J., Pa., w. N.Y., O. and Ky. May, June. — Fruiting raceme often proliferous at tip.

5. C. Clematìtis Shuttlw. (for *Clematis* — or *Clematitis* —, probably for its lax habit). — Glabrous and lax, with *slender rhizome;* small radical leaves reniform or cordate, with or without a pair of smaller lateral leaflets; *cauline leaves on sagittately appendaged petioles;* terminal leaflet mostly 3-lobed; petals 6–8 mm. long; siliques 2.5–3 cm. long, much exceeding the pedicels. — Springy places on the mts., sw. Va. to Ala. and Tenn. May, June.

6. C. praténsis L. (of meadows), Cuckoo-flower. — Stem ascending from a short rhizome, simple; *leaflets numerous, those* of the lower leaves rounded and stalked, *of the upper leaves oblong or linear,* entire or slightly angle-toothed; *petals* (white or rose-color) *thrice the length of* *the calyx;* siliques (very rare) 2–3 cm. long, 2 mm. broad; style short. — Highly variable boreal species, usually reproducing vegetatively by young plantlets springing from bases of leaflets:

Lateral leaflets of basal leaves with ovate to reniform blades; the terminal obovate to reniform, 0.35–3 cm. long; stem 1.5–5 dm. high.
Terminal leaflet of basal leaves distinctly crenate or dentate, with 3–9 teeth; petals pink or white (often changing to pink in drying).
Petals in a single series; lateral leaflets of middle and upper cauline leaves linear or oblong, sessile or nearly so. . . . . *C. pratensis* (typical).
Petals very numerous; lateral leaflets of middle and upper leaves oblanceolate to oblong-obovate, petiolulate. . . . . Forma *plena.*
Terminal leaflet of basal leaves entire or obscurely toothed; lateral leaflets of middle and upper cauline leaves usually distinctly petiolulate; petals white. Var. *palustris.*
Lateral leaflets of basal leaves with linear, oblong or elliptic blades; the terminal oblong to ovate, 1–6 mm. long; stem 0.5–2.5 dm. high. . . . . Var. *angustifolia.*

C. PRATÉNSIS (typical), LADY'S-SMOCK. — Meadows and lawns, Nfld. to O. and Pa. May–early July. (Natzd. from Eu.) — Forma PLÈNA G. Beck (double-flowered) in wet meadows, Middlesex Co., Mass. May, early June. (Natzd. from Eu.)

Var. **palústris** Wimm. & Grab. (of marshes). — Calcareous shallow water, bogs, springs and swampy woods, Lab. to Mackenz. and n. B.C., s. to Nfld., Gaspé Pen., Que., w. N.E., n. N.J., N.Y., w. Va., W.Va., n. O., n. Ind., n. Ill. and Minn. May–early July. (Eu.)

Var. **angustifòlia** Hook. (narrow-leaved). — Greenl. to Alaska, s. to n. Lab.; calcareous swales and peaty barrens, n. Nfld. July. (Eurasia)

7. C. IMPÀTIENS L. (impatient; for the explosive siliques). — Biennial 1.5–8 dm. high, erect, glabrous, simple or branched below; *leaves very numerous* (6–20), *membranaceous,* the basal rosulate, with 2–4 pairs, *the lower and middle cauline with* 6–9 *pairs of* narrowly ovate or lanceolate *acuminate* sharply toothed or lacerate *leaflets, the bases of the principal leaves sagittate-auriculate;* petals wanting or white and up to 2.5 mm. long; siliques on spreading-ascending to erect pedicels, very slender, 1.5–2 cm. long, with long attenuate style and 10–24 seeds. — Shaded grassland, etc., s. N.H. and e. Pa., local. June, July. (Adv. from Eu.)

8. C. flexuòsa With. (flexuous). — Dull green annual, biennial or short-lived perennial with *flexuous* ascending sparsely *hirsute* soft *stems* 1–5 *dm. high; leaves* pinnate, *membranaceous,* the rosulate basal ones with 3–6 pairs of slenderly petiolulate crenate-dentate suborbicular to obovate leaflets; *cauline leaves* similar, *not greatly reduced in size, with hirsute-ciliate petiole, the principal rounded-ovate to oblong or reniform leaflets* 0.4–2 *cm. broad and on petiolules* 0.2–1.5 *cm. long;* petals white, 2–4 mm. long; stamens 6; *siliques on loosely spreading-ascending pedicels,* slender, 1.2–2.5 cm. long, *with style* 0.7–1 *mm. long,* 20–24-*seeded.* — Springy calcareous banks or in shade, Nfld. to L. St. John, Que. June, July. (Eurasia)

9. C. HIRSÙTA L. (with stiff hairs). — Annual or biennial *with stiffly ascending glabrous stems* 0.5–3 dm. high; *leaves mostly on lower half of plant,* crowded at base, 2–6 on the stem; *radical leaves with hirsute-ciliate petioles* and 1–3 *pairs of thickish* orbicular to ovate petiolulate *leaflets; cauline leaves much smaller, their petioles* or bases *ciliate, their* 2–3 *pairs of narrow firm leaflets* 0.5–4 *mm. broad and cuneate-narrowed into short petiolules,* often *strigose on upper surface;* petals white, 1.5–2 mm. long; *stamens* 4; *siliques stiffly erect,* on thick short pedicels, *with thick style* 0.3–0.5 mm. long, 22–36-*seeded;* valves on dropping rolling into a tight ring. — Roadsides, lawns, old fields, etc., se. N.Y. to Ill., s. to Ga. and Ky. March, April. (Natzd. from Eu.)

10. C. pensylvánica Muhl. (of Pennsylvania). — Biennial or short-lived perennial 0.5–7.5 dm. high; *stem hispid at base,* otherwise mostly glabrous, erect or, when wholly or partly submersed, decumbent or trailing and then glabrous throughout, simple or much branched; rosette-leaves with 1–6 pairs of elliptic, obovate or rounded glabrous leaflets with bases decurrent along or confluent with the rachis, dentate or undulate, the terminal leaflet largest; *cauline leaves with membranaceous leaflets linear-oblanceolate to obovate, their bases oblique and confluent with the rachis, the terminal leaflet usually much broader than the lateral;* petals white, 1.5–4 mm. long; stamens 6; *siliques* narrowly linear, 1–3 *cm. long,* on slender pedicels 2–15 mm. long, *beaked by a tapering style* 0.5–2 *mm. long; seeds* 1–1.5 *mm. long.* — Springs, rills, wet clearings, etc., our commonest species, s. Lab. to B.C., s. to Nfld., N.S., N.E., L.I., Fla., Ala., Ark., Tex. and Oreg. March–Aug. (–Dec.) — Excessively variable: in dryish habitats stiffly erect and with narrow rather firm leaflets; in shade with membranaceous elongate leaflets; when submerged or in summer and autumn often producing broad (even suborbicular) membranous leaflets on proliferating old stems. An excellent substitute for Watercress.

Var. **Brittoniàna** Farw. (for NATHANIEL LORD BRITTON, 1859–1934). — Upper and often all cauline leaves simple and uncleft or pinnatifid, the lower with or without strongly confluent leaflets. — Less common, Nfld. to Minn., s. to Fla. and Tex.

11. C. parviflòra L. (small-flowered). — Annual or biennial 1–2(–4) dm. high, erect, simple to much branched from below; *stem glabrous throughout;* basal leaves with about 5 pairs of oblong mostly entire distinct leaflets; *cauline leaves with* 5–8 pairs of distinct (not decurrent) *linear entire leaflets, the terminal leaflet scarcely broader;* petals white, 2–2.5 mm. long; *siliques* erect, 1–2 cm. long, less than 1 mm. broad, on subhorizontal pedicels 7–10 mm. long, 22–36-seeded, *beakless or with style up to* 0.7 *mm. long; seeds* 0.7–0.9 *mm. long.* Eurasia. — Represented with us by

Var. **arenícola** (Britt.) O. E. Schulz (growing in sand). — Often stouter; *leaflets of rosette-leaves* commonly *obovate to suborbicular and usually with* 1 *or* 2 *pairs of teeth; cauline leaves with* 2–6 *pairs of* linear to oblanceolate *leaflets; petals* 2.5–3.5 *mm. long;* fruiting pedicels 4–8(–10) mm. long; siliques 1.5–3 cm. long; *seeds* 26–46. (*C. arenicola* Britt.) — Dry woods, shaded or exposed ledges and sandy soils, n. Fla. to Tex., n. to s. N.S., sw. N.B., centr. Me., sw. Que., Thunder Bay Dist., Ont., Wisc. and Minn.; Oreg. to s. B.C. April–Aug.

12. C. Lóngii Fern. (for one of its discoverers, Bayard Long, 1885–). — Differing from no. 10 in erect to prostrate stem glabrous at base, flexuous; *leaves simple and reniform to suborbicular with cordate base, or rarely with* 1 *pair of rounded leaflets; flowers apetalous,* 0.7–1.2 *mm. long, subsessile; siliques lanceolate,* 5–10 *mm. long, on thick pedicels* 0.5–1.5 *mm. long.* — Tidal estuary of Cathance R., Me.; estab. by the late *F. F. Forbes* on the Charles R., Mass.; estuaries confluent with Chesapeake Bay, Md. and Va. June–Sept.

42. SIBÀRA Greene

Silique linear, compressed, the stiff valves faintly 1-nerved at base or nerveless. Seeds in a single row in each locule, winged or wingless. Cotyledons accumbent. Sepals erect. Petals white or purple, clawed. — Annuals or short-lived biennials with simple or loosely branching stems, pinnatifid or pinnate leaves and stiffly racemose flowers and fruits; a genus of warm reg. of Am., intermediate between *Cardamine* and *Arabis.* (Name an anagram of *Arabis.*)

1. S. virgínica (L.) Rollins (Virginian). — *Stem* hirsute, or glabrous above, 1–4 dm. high, usually depressed or *with divergent stiff branches; leaves lyrate-pinnatifid,* with oblong or linear divisions; flowering raceme rather close, in fruit much elongating; *petals white or pink-tinged,* 1.5–3 mm. long, oblanceolate to oblong or linear; *siliques stiff, broadly linear,* 2–2.5 cm. long, 1–2 mm. broad, flattish, on stout ascending pedicels, *valves faintly* 1-*nerved at base or nerveless;* seeds narrowly winged, about 1.5 mm. long. (*Cardamine* L.; *Arabis* Poir.; *Planodes* Greene) — Sandy or rocky soil, dry open woods, clearings and cult. fields, often a weed, Fla. to Tex. and Mex., n. to e. Va., s. O., s. Ind., s. Ill., Mo., se. Kans. and s. Calif. *Fl.* March–May; *fr.* April–June.

43. ÁRABIS L. Rock-Cress. Arabette (Que.)

Silique linear, flattened (terete in no. 3); placentae not thickened; the valves plane or convex, more or less 1-nerved in the middle, or longitudinally veiny. Seeds marginless or winged. Cotyledons accumbent or a little oblique. — Leaves seldom divided. Flowers white or purple (rarely yellowish). Large genus of Am. and Eurasia. (Name from the country, *Arabia,* according to Linnaeus.)*

a. Mature fruiting pedicels erect, ascending or divaricately spreading, not definitely descending nor deflexed; flowering pedicels at anthesis erect, ascending or merely divaricately spreading. . . *b.*

b. Mature siliques erect or ascending, often appressed to main stem, straight or inwardly falcate; fruiting pedicels erect or ascending. . . *c.*

c. Flowering stems matted, from a forking caudex, the branches divergent; valves of mature silique 1-nerved only at base; petals showy, white; boreal. . . . . 1. *A. alpina.*

c. Flowering stems erect or strongly ascending, simple or branching; valves of mature siliques 1-nerved to or above the middle. . . *d.*

d. Seeds definitely in only 1 row. . . *e.*

e. Cauline leaves spatulate to linear, not clasping; plant much branched from base. . . . . 2. *A. lyrata.*

e. Cauline leaves lanceolate to ovate, more or less clasping; stem simple below or with only 1 or 2 basal branches.

---

* Treatment based largely on those of Milton Hopkins in Rhodora, xxxix (1937) and of R. C. Rollins, ibid. xliii (1941).

Fruiting pedicels appressed or subappressed to rachis; petals lanceolate, oblanceolate or linear, yellowish- or greenish-white; lower and median cauline leaves entire or merely denticulate.

Mature siliques terete, 4–9.5 cm. long; style thick; seeds wingless or very narrowly winged. . . . . 3. *A. glabra.*

Mature siliques flat, often moniliform, 1.5–5 cm. long; style usually slender; seeds winged, broadly so above. . . . . 4. *A. hirsuta.*

Fruiting pedicels suberect to divergently ascending; petals broad, spatulate to obovate, white; lower and median cauline leaves serrate or dentate. . . . . 5. *A. patens.*

d. Seeds in 2 rows.

Fruiting pedicels merely ascending or divaricately spreading, finely stellate-pubescent or glabrous; stem finely appressed-pubescent at base with forking trichomes; petals pink or white. . . . 6. *A. divaricarpa.*

Fruiting pedicels strictly appressed or subappressed, glabrous; stem hirsute at base or glabrous.

Stem densely hirsute below with spreading hairs; petals yellowish-white, barely longer than sepals; siliques terete. . . . 3. *A. glabra.*

Stem glabrous throughout or sparsely hirsutulous at base; petals pink to white, twice as long as sepals; siliques flat. . . . 7. *A. Drummondi.*

b. Mature siliques divaricately spreading or curved outward or downward, often arcuate (mostly straight in no. 10), on ascending to divaricately spreading pedicels. . . *f.*

f. Stems and leaves glabrous.

Lower cauline leaves serrate-dentate to subentire; sepals nearly as long as petals; valves of siliques 1-nerved at base, rarely to middle; plant glaucous. . . . . 8. *A. laevigata.*

Lower cauline leaves sharply dentate to laciniate; sepals one-half length of petals; valves of siliques 1-nerved at least to middle or beyond; plant green. . . . . 9. *A. missouriensis.*

f. Stems and leaves pubescent. . . *g.*

g. Basal leaves finely stellate-pubescent on both surfaces; cauline leaves entire or subentire; petals pink to white. . . . . . 6. *A. divaricarpa.*

g. Basal leaves hirsute on both surfaces with simple hairs or, if stellate beneath, strigose above; larger cauline leaves serrate, dentate or sinuate; petals whitish to yellowish. . . *h.*

h. Seeds wingless; siliques finely and evenly stellate-pubescent (rarely glabrous). . . . . 10. *A. perstellata.*

h. Seeds definitely winged; siliques glabrous.

Lower cauline leaves serrate to dentate, ovate to oblong-lanceolate; siliques 2.5–4.5 cm. long. . . . . 5. *A. patens.*

Lower cauline leaves laciniate or lyrate-pinnatifid, lanceolate; siliques 6–9 cm. long. . . . . 9. *A. missouriensis.*

a. Mature fruiting pedicels pendulous or reflexed; flowering pedicels at anthesis reflexed or at least descending. . . . *i.*

i. Cauline leaves oblong to elliptic, tapering at base, the larger ones 1–2.5 cm. broad, denticulate and villous-hirsute; petals creamy-white; siliques 2.5–4 mm. broad; seeds broadly winged (wing about 0.75 mm. broad); plant of temperate area. . . . . 11. *A. canadensis.*

i. Cauline leaves linear, narrowly oblong or lanceolate, entire, auricled or sagittate at base, the larger ones 1.5–9 mm. broad, minutely stellate-pubescent or hirsutulous; petals purple, pink or white; siliques 1–2.5 mm. broad; seeds narrowly winged; boreal species. . . . . 12. *A. Holboellii.*

1. A. alpìna L. (alpine). — *Matted perennial with slenderly branching caudex; flowering stems* 1–3.5 dm. high, *branching below and there spreading-hirsute; leaves fleshy;* the radical oblanceolate to obovate-spatulate, petioled, *coarsely dentate or subentire, stellate-pubescent on both faces;* the cauline subamplexicaul, oblong to ovate-lanceolate; flowers in rather close racemes (rachis of mature leading racemes elongate); *petals white or whitish,* spatulate-oblanceolate, 7–9 *mm. long,* 2–4 *mm. broad,* delicately veined; or petals firm and greenish-white and coarsely veined in the local forma **phyllopétala** Fern. (leafy-petalled) of w. Nfld.; siliques glabrous, ascending, 4–7 cm. long, 1.5–2 mm. broad. — Damp basic or circumneutral rocks and gravels or springy slopes and meadows, Greenl. and e. Arct. Am., s. to w. Nfld., Gaspé Pen., Que., and Hudson Bay. *Fl.* July, Aug.; *fr.* July–Sept. (Eu.)

2. A. lyràta L. (lyre-shaped). — *Tufted* biennial or perennial, *slender, branching from base* and often above, 0.7–3.5 dm. high; flowering stems hirsute below with simple or bifurcate hairs; *radical leaves* spatulate to oblanceolate, 2–4 cm. long, *usually lyrate-pinnatifid or dentate,* hirsute below and often above; *cauline leaves* scattered, spatulate to linear, *tapering at base,* lyrate-pinnatifid to entire; racemes lax, with erect to spreading pedicels; sepals 1.5–3 mm. long, glabrous; petals white (rarely pink or purplish), 5–8 mm. long, 2–3 mm. broad, spatulate or oblanceolate; *siliques flattish,* spreading or loosely ascending, 2–4.5 *cm. long,* with style 0.65–1.25 mm. long, or siliques 1–2 cm. long with styles less than 0.5 mm. long and petals smaller in forma **parvisíliqua** M. Hopkins; *valves of silique* 1-*nerved to above the middle; seeds* about 1 mm. long, in 1 row, *wingless.* — Ledges, cliffs (basic or circumneutral), gravels and sands, Vt. to Ont. and Minn., s. to N.C., n. Ga., Tenn. and Mo.; also n. Alta. *Fl.* Apr., May; *fr.* May–July.

Var. **kamchàtica** Fisch. (of Kamtchatka). — *Stem and radical leaves* (except sometimes the petioles) *glabrous;* flowers smaller; style short or wanting. (Var. *glabra* (DC.) M. Hopkins) — Alaska to n. Wash., n. Mont. and Sask.; locally on n. shore of L. Sup. and in w. N.Y.

3. A. glàbra (L.) Bernh. (smooth), Tower-Mustard. — *Stoutish* biennial; *stem* erect, usually simple below, 0.6–1.2 m. high, *hirsute at base with spreading mostly simple hairs,* glabrous and glaucous above; radical leaves spatulate or oblanceolate, entire or irregularly dentate, 5–12 cm. long, those of first year finely stellate-pubescent on both surfaces; *cauline leaves imbricated at least below,* lanceolate to elliptic-oblong, *sagittate-amplexicaul,* mostly *glabrous* and entire, up to 12 cm. long; racemes virgate; *petals creamy or yellowish,* narrowly oblanceolate to linear, 2.5–6 mm. long; *siliques erect or suberect, terete,* slender, straight or nearly so, 4–9.5 *cm. long,* 0.7–1 *mm. thick, with short thick style;* valves of silique 1-nerved to middle or beyond; seeds irregular, in either 1 or 2 rows, about 1 mm. long, narrowly winged (rarely wingless). — Ledges, cliffs, thickets and fields, often a weed, Que. to Alaska, s. to N.C., W.Va., O., Ind., Ill., Ark. and Calif. *Fl.* May–July; *fr.* June, July. (Eurasia)

4. A. hirsùta (L.) Scop. (hirsute), var. **pycnocárpa** (M. Hopkins) Rollins (with crowded carpels). — In habit similar to no. 3; *stem* 1.5–8 dm. high, *slender, hirsute to summit* with spreading mostly simple hairs; *radical leaves* oblong or oblanceolate, 2–8 cm. long, *villous-hirsute to hirtellous on both faces; cauline leaves very numerous, imbricated or nearly so,* oblong to lanceolate, 1–4 *cm. long,* obtuse or subacute, the lower auriculate- or subsagittate-amplexicaul, *hirsute,* the upper merely sessile and less pubescent or smooth; racemes slender, becoming lax; *sepals herbaceous;* petals whitish or ochroleucous (rarely pinkish), lanceolate, 4–6 mm. long, 0.7–1 mm. broad; *siliques erect* or appressed, *flat, often moniliform,* 1.5–5 *cm. long,* 0.7–1 mm. broad; *style usually slender,* less than 1 mm. long; *seeds* in 1 row, (0.6–)1–1.25 *mm. long, broadly winged above.* — Cliffs, ledges and gravels (calcareous to circumneutral), Anticosti to Yuk., s. to N.B., N.E., Pa., rarely to nw. Ga., Mo., Kans., N.M., Ariz., and Calif. *Fl.* May, June; *fr.* June, July. — Variable representative of the Eurasian *A. hirsuta* (L.) Scop., which has more scattered leaves, plumper siliques nerved to tip, shorter and thicker style and less winged seeds.

Var. **glabràta** T. & G. (becoming smooth). — *Stem glabrous above the middle; cauline leaves only* 2–12, *remote or subremote, the middle and upper ones glabrous; sepals membranaceous;* siliques few. — Alta. and B.C. to N.M. and Calif.; sw. Wisc.

Var. **adpressípilis** (M. Hopkins) Rollins (with appressed hairs). — *Pubescence of stem strictly appressed or strigose, chiefly of bifurcate hairs;* cauline leaves glabrous or nearly so. — Ont. and Minn., s. to w. Va., s. Ind. and Mo.

5. A. pàtens Sulliv. (spreading). — Biennial to perennial, 3–6 dm. high; *stem* simple or branching, *spreading-hirsute throughout* with mostly simple hairs or glabrous above; radical leaves ovate or oblanceolate, 1.5–6 cm. long, toothed; *cauline leaves* ovate to oblong-lanceolate, 2–5 cm. long, auriculate-clasping, acute or acuminate, *the lower and median serrate or dentate, hirsute on both faces;* racemes loose; *petals white,* 5–7 mm. long, *broadly spatulate or obovate; siliques* 2.5–4.5 cm. long, 0.5–1 mm. broad, attenuate, *suberect to divergently ascending,* with a slender style, the valves strongly 1-nerved to or above the middle; seeds in 1 row, about 1.25 mm. long, narrowly winged. — Rocky shores and banks, se. Pa. to centr. O. and se. Ind., s. to w. N.C. and Tenn. *Fl.* Apr.–June; *fr.* May–Sept.

6. A. divaricárpa Nels. (spreading-fruited). — *Stem* erect, 2–9 dm. high, simple or branched, sparingly *appressed-pubescent at base with forking hairs; rosette-leaves* narrowly oblanceolate- to oblanceolate-spatulate, 2–6 cm. long, acute, usually dentate, minutely *stellate-pubescent; cauline leaves* narrowly oblong to linear-lanceolate, *strongly ascending, entire or subentire,* auriculate- or sagittate-based, *mostly glabrous;* flowers in loose racemes; *pedicels soon divergent, minutely stellate-pubescent or glabrous; petals pink or purplish* (rarely white), oblanceolate-spatulate, 5–8 mm. long; siliques glabrous, the lowest and mature ones loosely ascending to divergent, 2.5–9 cm. long, 1.25–3 mm. broad, with style only 0.2–0.7 mm. long; young seeds in 2 rows, when mature becoming somewhat uniseriate, 1–1.5 mm. in diameter, narrowly winged. (*A. brachycarpa* (T. & G.) Britt.) — Ledges, gravels and sands (basic to circumneutral), Gaspé Pen., Que., to Man., nw. to Yuk., s. to n. N.B., N.H., Vt., n. N.Y., n. O., s. Mich., s. Wisc., n. Ia., Neb., Colo. and Calif. *Fl.* May–July; *fr.* June–Aug.

Var. **stenocárpa** M. Hopkins (slender-fruited). — Siliques very slender, 0.7–rarely 1.25 mm. broad. — Ledges about Bic, Que.; also in Sask.

7. A. Drummóndi Gray (for its discoverer, Thomas Drummond, 1780–1835). — Resembling no. 6, often glaucous, *glabrous throughout* or rarely appressed-pubescent at base; radical leaves up to 9 cm. long, glabrous or merely with ciliate petioles; *flowers erect or strongly ascending; pedicels glabrous;* petals 5–10 mm. long; *siliques suberect to subappressed,* flattish, 4–10 cm. long, 1.5–3.3 mm. broad; seeds in 2 rows, about 1 mm. in diameter. — Basic or circumneutral ledges, gravels and thickets, se. Lab. to n. Alta. and s. B.C., s. to Nfld., N.S., N.E., Del., n. O., n. Ind., n. Ill., e. Ia., N.M., Ariz. and Calif. *Fl.* May–July; *fr.* May–Aug.

8. A. laevigàta (Muhl.) Poir. (smoothed). — *Glaucous; stem* 3–9 dm. high, simple or branching, *glabrous, averaging* 13 *internodes to first flower; rosette-leaves* (soon disappearing) spatulate-obovate to narrowly oblanceolate, *those of* 1st *year pilose,* of 2d year glabrous, *dentate or serrate,* 3–11 cm. long, 0.5–3 cm. broad; *cauline leaves* oblong-lanceolate to linear, *spreading or ascending,* 0.3–2 *dm. long,* sagittate- or auriculate-clasping, serrate-dentate or entire; racemes loose, the flowering pedicels ascending to divergent; *petals* whitish, 3–5 *mm. long, scarcely exceeding sepals,* spatulate to oblanceolate; siliques downwardly arcuate or straightish, recurved in maturity, 5–10 cm. long, 0.7–2.5 mm. broad, the *valves faintly 1-nerved at base or up to the middle; seeds* in 1 row, 1 *mm. long,* 0.5 mm. broad, winged. — Rich woods and slopes or shaded (chiefly calcareous) ledges, sw. Que. to Minn. and Colo., s. to Ga., Ala., Ark., and Okla. *Fl.* late March–July; *fr.* May–Sept.

Var. **Búrkii** Porter (for its discoverer, Isaac Burk, 1816–1893). — Cauline leaves linear to linear-lanceolate, entire to subdenticulate, sessile, not amplexicaul; valves of siliques 1-nerved to or beyond the middle. (*A. Burkii* (Porter) Small) — Bluffs and slopes, s. Pa. and W.Va. to e. Ky. and w. N.C.

9. A. missouriénsis Greene (of Missouri). — Smaller than no. 8, *green, glabrous throughout; stem* 2–5 dm. high, *averaging* 25 *internodes to first flower; rosette-leaves* lanceolate to spatulate, those of 1st year dentate to laciniate, persistent, *of 2d year strongly laciniate or lyrate-pinnatifid,* 2–9 cm. long, 5–15 mm. broad; *cauline leaves imbricate, appressed-ascending,* up to 8 *cm. long,* the lower resembling the basal but sagittate, the middle and upper lanceolate to linear; raceme less lax; *flowering pedicels strict; petals* ochroleucous, 6–8 *mm. long, twice as long as sepals; siliques at first erect, soon falcate and recurving,* 6–9 cm. long, 1.7–2 mm. broad; *valves 1-nerved one-half to two-thirds their length; seeds* 1.5–1.8 *mm. long, about* 1 *mm. broad.* (*A. viridis* Harger) — Circumneutral bluffs, ledges or rocky woods, sw. Me. to e. N.Y., s. to e. Pa.; n. Ga.; Mich.; Mo. to Okla. *Fl.* May–July; *fr.* June–Sept.

Var. **Deàmii** M. Hopkins (for its discoverer, Charles Clemon Deam, 1865–). — Short-hirsute nearly throughout. — Sandy or gravelly woods and bluffs, local, Wisc. to Ind. and Mo.

10. A. perstellàta E. L. Br. (very stellate). — *Stem* 1–7 dm. high, from biennial or perennial bases, *pubescent throughout with appressed to subappressed simple or stellate hairs; radical leaves* spatulate, oblanceolate or obovate, 1–15 cm. long, 0.5–6 cm. broad, *irregularly dentate to sinuate or lyrate-pinnatifid, finely stellate-pubescent at least beneath, strigose or stellate above; cauline leaves* similarly pubescent or upper surfaces glabrous, imbricate, lanceolate to narrowly obovate, 1–6 cm. long, auricled at base, *irregularly dentate;* flowering *pedicels hirsute,* ascending; *petals* white, pink or ochroleucous, 2–4 *mm. long,* linear to narrowly oblanceolate; *siliques subascending to divaricate,* straight or slightly curved, 1.5–4 cm. long, 0.7–1.25 mm. broad, *finely stellate-pubescent* (rarely glabrous), the valves nerveless or faintly 1-nerved at base; *seeds* in 1 row, *wingless,* about 1 mm. long. — Three vars.:

Perennial, with strong branching base and definitely perennial leafy basal offsets; stem and leaves gray with dense stellate pubescence throughout; flowering stems 1–5 dm. high; rosette-leaves lyrate-pinnatifid, 1–2.5 cm. long, 0.5–1.3 cm. broad; cauline leaves lyrate-pinnatifid to repand-dentate, 1–3 cm. long, 5–8 mm. broad, with acute basal auricles; petals roseate, 3–4 mm. long; fruiting pedicels 4–10 mm. long. . . . . **A. perstellata** (typical).

Biennial chiefly; stems 2–7 dm. high, with appressed or subappressed simple or forking hairs; radical leaves long-petioled, mostly dentate, thin and green, finely stellate beneath, strigose above, 3–15 cm. long and 1–6 cm. broad; cauline leaves similarly pubescent or upper surfaces glabrous, lanceolate to narrowly oblong, 2–6 cm. long, the basal auricles blunt to acute; petals white to creamy, 2–3 mm. long; fruiting pedicels 1–4 mm. long.

Siliques finely stellate-pubescent. . . . . Var. *Shortii.*

Siliques glabrous. . . . . Var. *phalacrocarpa.*

A. perstellàta (typical). — Wooded hillsides, Franklin Co., Ky. *Fl.* Apr.; *fr.* May.

Var. **Shórtii** Fern. (for its discoverer, Charles Wilkins Short, 1794–1863). *A. dentata* (Torr.) T. & G., not Clairv. — Rich woods, bluffs and calcareous ledges, N.Y. (local) to se. Minn. and e. S.D., s. to Va., Tenn., Ark. and e. Kans. *Fl.* Apr., May; *fr.* May–July.

Var. **phalacrocárpa** (M. Hopkins) Fern. (with bald carpels), *A. dentata*, var. M. Hopkins — Ia. to Ark.

11. A. canadénsis L. (Canadian), Sicklepod. — Biennial; stem erect, 3–9 dm. high, simple or sparingly branched above, sparsely hirsute at base, smooth above; rosette-leaves obovate to lanceolate, 2.5–13 cm. long, 1.5–4 cm. broad, serrate-dentate to runcinate, usually hirsute along midrib on both surfaces, soon wilting; *cauline leaves* approximate or subapproximate, *oblong-lanceolate to elliptic,* 2.5–12 cm. long, 0.5–2.5 *cm. broad, attenuate at base, often denticulate, the lowest villous-hirsute,* the upper hirsutulous to glabrous; raceme lax; *pedicels* becoming *pendulous and finally subgeniculate; petals creamy-white,* narrowly oblanceolate to oblong, *only slightly exceeding sepals,* 3–5 mm. long; *siliques falcate, pendulous* or recurved, 7–10 cm. long, 2.5–4 *mm. broad,* glabrous, reticulate-veiny, the valves definitely 1-nerved nearly or quite to the attenuate tip; *seeds* in 1 row, suborbicular, 1.25 mm. in diameter, *the broad wing cordate and about* 0.75 *mm. broad.* — Rich woods, thickets and rocky banks, centr. Me. (rare) to s. Ont., Minn. and Neb., s. to Ga., Ala., Ark., Okla. and Tex. *Fl.* Apr.–June; *fr.* June–Sept.

12. A. Holboéllii Hornem. (for its discoverer, Carl Peter Holbøll, 1795–1856). —*Stems* erect, 1–several, 1–7 dm. high, simple or often branched at base, *finely pubescent below,* glabrous and often glaucous above; *rosette-leaves* narrowly obovate to linear-oblanceolate, *entire or slightly dentate,* 1–8 cm. long, *minutely stellate-pubescent* with 2- and 3-forked canescent hairs; *cauline leaves oblong-lanceolate to narrowly oblong or almost linear,* remote or slightly overlapping, entire, 0.8–4 cm. long, 1.5–9 *mm. broad, subamplexicaul-sagittate, at least the lower densely pubescent;* racemes lax; *pedicels* stellate-pubescent to glabrous, *soon spreading or slightly descending, finally geniculate; sepals* herbaceous, oblong, 3–4.5 mm. long, 1–1.5 *mm. broad,* obtuse, purplish, with pale hyaline margin, the younger ones stellate-pubescent or glabrous; *petals purplish to white,* 3–10 *mm. long,* 1–2.25 *mm. broad,* the narrowly spatulate-obovate to -oblanceolate limb spreading; *siliques* arcuate or straightish, *secund,* slightly *to tightly reflexed to descending,* glabrous or puberulent, blunt or acutish, 2.5–8 cm. long, 1–2.5 *mm. broad,* the valves prominently 1-nerved at base, the nerve evanescent; *seeds* 0.75–1.5 mm. in diameter, *narrowly winged.* — Complex boreal species, with pronounced geographic vars. and discontinuous ranges. Ours are as follows:

Lower part of stem appressed-pubescent with fine dendritic or stellate hairs; stems 2–9 dm. high; rosette-leaves narrowly obovate to oblanceolate, 2–8 cm. long; cauline leaves 1.5–4 cm. long, 4–9 mm. broad; sepals 1–1.5 mm. broad; siliques 2.5–8 cm. long, 1.5–2.5 mm. broad.

Cauline leaves oblong-lanceolate to narrowly oblong, remote or slightly overlapping, flat; sepals promptly glabrate; petals 1.7–2.25 mm. broad; siliques slightly reflexed to descending, blunt, 2.5–6 cm. long, 1.5–2.5 mm. broad, the valves prominently 1-nerved at base; seeds 1–1.5 mm. broad. . . . *A. Holboellii* (typical).

Cauline leaves lanceolate to lance-linear, crowded, revolute-margined; sepals pubescent, rarely glabrate; petals 1–1.75 mm. broad; siliques strongly reflexed to appressed, acuminate, 3.5–8 cm. long, 1.75 mm. broad, the valves prominently nerved to middle or above; seeds 0.75–1.2 mm. broad. . . . Var. *retrofracta.*

Lower part of stem hirsute or hispid with long spreading simple and forking hairs; stems 1–3 dm. high; rosette-leaves 1–5 cm. long; cauline leaves 0.8–2.3 cm. long, 1.5–4.5 mm. broad; sepals 0.5–0.75 mm. broad; siliques straight, tightly appressed, acutish, 2.5–6 cm. long, 1–1.5 mm. broad, the 2 nerves prominent to or above middle. . . . . Var. *Collinsii.*

A. Holboéllii (typical). — Cold or shaded calcareous cliffs, local, Gaspé Co. to Rimouski and Charlevoix Cos., Que.; Yuk. and Alta. to s. B.C. and Wash. *Fl.* June–Aug.; *fr.* July–Sept. (Greenl.)

Var. **retrofrácta** (Graham) Rydb. (reflexed), *A. retrofracta* Graham — Dry bluffs, gravels and sands, local, Charlevoix Co., Que., to Algoma Distr., Ont., and n. Mich.; Alta. and B.C. to Sask., Colo., Utah, Nev., and Calif. *Fl.* late May, June; *fr.* June, July.

Var. **Collínsii** (Fern.) Rollins (for one of its discoverers, JAMES FRANKLIN COLLINS, 1863–1940), *A. Collinsii* Fern.; *A. pendulocarpa* sensu Hopkins, in part, not Nels. — Calcareous rock, gravel and sands, local, Rimouski Co., Que.; Man. to Alta., s. to S.D. and Wyo. *Fl.* June, July; *fr.* June–Aug.

FAM. 74. RESEDÀCEAE (MIGNONETTE FAMILY)

Herbs with unsymmetrical 4–7-merous small flowers, a fleshy 1-sided hypogynous disk between the petals and the (3–40) *stamens, which bears the latter. Calyx not closed in the bud. Capsule 3–6-lobed, 3–6-horned, 1-locular, with 3–6 parietal placentae, opening at the top before the seeds* (which are as in *Capparidaceae*) *are full grown.* — Leaves alternate, with only glands for stipules. Flowers in terminal spikes or racemes. A small and unimportant family nat. of the Old World, represented in gardens by the Mignonette (*Reseda odorata* L.).

1. RESÈDA L. MIGNONETTE

Petals 4–7, cleft, unequal. Stamens 12–40, on one side of the flower. (Name from *resedare, to calm*, in allusion to supposed sedative properties.)

1. R. LÙTEA L. (yellow). — Slender, 2–8 dm. high; *leaves irregularly pinnate-parted or bipinnatifid; flowers pale yellow;* sepals and petals 6; *stamens* 15–20. — Fields, roadsides and waste places, local and mostly casual, Me. to Mich., s. to Pa. and Mo. June–Sept. (Adv. from Eu.)

2. R. ÁLBA L. (white). — Similar; *leaves pinnately and rather regularly parted; flowers greenish-white;* petals 6; stamens 12–15. — Waste places and roadsides, mostly casual, N.E. to Ill. and Del. June–Oct. (Adv. from Eu.)

DYER'S ROCKET, R. LUTÈOLA L. (from the old generic name *Luteola*, yellowish), with *simple lanceolate leaves* and dense spikes, formerly cult. as a yellow dye, is occasionally found in waste. (Introd. from Eu.)

FAM. 75. SARRACENIÀCEAE (PITCHER-PLANT FAMILY)

Polyandrous and hypogynous bog-plants with hollow pitcher- or trumpet-shaped leaves, — comprising one genus in trop. S.Am., one in Calif., and the following Atl. N.Am. genus:

1. SARRACÈNIA L. PITCHER-PLANT

Sepals 5, with 3 bractlets at the base, colored, persistent. Petals 5, oblong or obovate, incurved, deciduous. Stamens numerous, hypogynous. Ovary compound, 5-locular, globose, crowned with a short style which is expanded at the summit into a very broad and petal-like 5-angled and 5-rayed umbrella-shaped body, the 5 delicate rays terminating under the angles in as many small hooked stigmas. Capsule with a granular surface, 5-locular, with many-seeded placentae in the axis, loculicidally 5-valved. Seeds anatropous, with a small embryo at the base of fleshy albumen. — Perennials, yellowish, green or purplish; the hollow leaves all radical, with a wing on one side, and a rounded arching hood at the apex. Scape naked, 1-flowered; flower nodding. (Named for Dr. *Michel Sarrasin de l'Étang*, 1659–1734, physician at the Court of Quebec, who sent our northern species to Europe.)

Leaves pitcher-shaped, curved, broadly winged. . . . . 1. *S. purpurea.*
Leaves slenderly trumpet-shaped, straight, almost wingless. . . . . 2. *S. flava.*

1. S. purpùrea L. (purple), SIDESADDLE-FLOWER, PITCHER-PLANT, HUNTSMAN'S-CUP, PETITS COCHONS (Que.), HERBE-CRAPAUD (Que.). — *Leaves pitcher-shaped,* spreading or loosely ascending, curved, *broadly winged;* the hood erect, open, reniform, covered with reflexed bristles; *flower subglobose,* on erect scape, commonly deep purple; the panduriform petals arched over the *greenish-yellow style.* — Very variable, with three pronounced variations. — The curious leaves usually partly filled with water and drowned insects.

Leaves glabrous (rarely a little hirsute) on the outside, elongate, the hollow portion mostly thrice as long as broad; wings of flattened-out hood extending only slightly beyond the margin of the narrowly winged pitcher.

Foliage more or less suffused with purple or red; sepals deep red-purple; petals and summit of stigma red. . . . . *S. purpurea* (typical).

Foliage pale green; sepals yellow-green; petals and summit of stigma yellowish. Forma *heterophylla.*

Leaves commonly densely hirsute (rarely glabrous) on the outside, the hollow portion usually less than thrice as long as broad; wings of flattened-out hood extending well beyond the margin of the broadly winged pitcher. . . . . Var. *venosa.*

S. purpùrea (typical) = subsp. *gibbosa* (Raf.) Wherry. — Sphagnous bogs and peaty barrens, Lab. to Mackenz., s. to Nfld., N.S., N.E., Del., Md., O., Ind., n. Ill., Minn., Man. and Sask. (the national flower of Nfld., there often called INDIAN-CUP or INDIAN-JUG). — Forma **heterophýlla** (Eat.) Fern. (variable-leaved), local. Late May–Aug.

Var. **venòsa** (Raf.) Fern. (veiny). — Fla. to La., n. to se. Va. and Tenn. May, June.

2. S. flàva L. (yellow), TRUMPETS, HUNTSMAN'S-HORN. — *Leaves long* (3–10 dm.) *and trumpet-shaped,* erect, *yellowish,* with an open mouth; the erect hood rounded, narrow at the base; wing almost none; *flower yellow;* the petals becoming long and drooping. — Wet pinelands and bogs, Fla. to Ala., n. to se. Va., Apr., May.

Hybrids of no. 2 and the var. of no. 1 occur. These are × **S. Càtesbaei** Ell. (named for MARK CATESBY, 1679 or '80–1749).

FAM. 76. DROSERÀCEAE (SUNDEW FAMILY)

Herbs of damp soil, with gland-tipped hairs on the leaves, with regular hypogynous pentamerous flowers, and withering-persistent calyx, corolla, and stamens, the anthers fixed by the middle and turned outward, and a 1-locular capsule with twice as many styles or stigmas as there are parietal placentae. — Calyx imbricated. Petals convolute. Seeds numerous, anatropous, with a short and minute embryo at the base of the albumen. Leaves circinate in bud, *i.e.,* rolled up from the apex to the base as in Ferns. Small family of insectivorous plants.

1. DRÓSERA L. SUNDEW. DAILY-DEW. ROSSOLIS (Que.)

Stamens 5. Styles 3, or sometimes 5, deeply 2-parted so that they are taken for 6 or 10, slender, stigmatose above on the inner face. Capsule 3 (rarely 5)-valved; the valves bearing the numerous seeds on a median band for their whole length. — Low perennials or biennials found on all continents; the leaves, in our species, all in a tuft at the base (often scattered in submersed plants), clothed with reddish gland-bearing bristles; the naked scape bearing the flowers (rarely solitary) in a 1-sided simple (or sometimes forking) raceme-like inflorescence, which nods at the undeveloped apex, so that the fresh-blown flower (which opens only in sunshine) is always highest. The plants yield a purple stain to paper. The glands of the leaves exude drops of a clear glutinous fluid, glittering like dew-drops (whence the name, from the Greek *droseros, dewy*).

a. Leaf-blades linear to orbicular, usually distinct from the petioles; plant not bulbous; petals white or pink, 3–8 mm. long, 2–4 mm. broad. . . *b.*

b. Blades linear to spatulate, much longer than broad, on slender mostly naked petioles. . . *c.*

c. Blades obovate- to linear-spatulate or -cuneate.

Stipules nearly free from bases of petioles; blades 0.5–2.5 cm. long; seeds ellipsoid-obovoid, with a close densely papillose testa. . . . 1. *D. intermedia.*

Stipules adnate, except at tip, to petioles; blades 1.5–5 cm. long; seeds fusiform, with the loosely alveolate testa prolonged at tips. . . . 2. *D. anglica.*

c. Blades linear, 1–6 cm. long; stipules adnate; seeds slenderly ellipsoid-obovoid, with close testa covered with shallow pits. . . . . . 3. *D. linearis.*

b. Blades suborbicular to obovate, cuneately tapering to the hairy petioles. . . *d.*

d. Blades suborbicular to broadly transverse-elliptic, on elongate slender petioles; petals white; seeds fusiform, with the loose testa prolonged at tips. . . . . 4. *D. rotundifolia.*

d. Blades cuneate-obovate, tapering into short broad petioles; petals pink or white; seeds ellipsoid to obovoid, with close testa.

Scapes and calyces glabrous; stipules elongate, nearly free; petals pink, 3–6 mm. long. . . . . 5. *D. capillaris.*

Scapes and calyces stipitate-glandular; stipules obsolete; petals white, 5–8 mm. long. . . . . 6. *D. brevifolia.*

a. Leaf-blades filiform, barely petioled; plant with woolly bulbous base; petals purple, 0.8–1.5 cm. long, 5–8 mm. broad. . . . . 7. *D. filiformis.*

1. **D. intermèdia** Hayne (intermediate). — *Leaf-blades spatulate,* 0.5–2.5 *cm. long, on* long ascending or divergent *slender naked petioles* in a rosette; *stipules nearly free;* scape glabrous or with minute sessile glands, 0.2–2.5 dm. high, 1–20-flowered, or forked and more floriferous in forma **corymbòsa** (DC.) Fern. (with a corymb); petals white, 3–6 mm. long; *seeds ellipsoid-obovoid, reddish-brown, with a close densely papillose coat.* (*D. longifolia* of ed. 7, and in part of L., name too ambiguous.) — Wet acid peat and sand, Nfld. to Ont., s. to N.S., N.E., S.C., se. Tenn., O., Ind. Ill. and Minn., and on the Coastal Plain to Fla. and e. Tex. June–Aug. Fig. 1114. — In very wet places and shallow water the caudex much prolonged and the leaves becoming scattered, forma **nàtans** Heuser (floating). (W.I.; Eurasia) — Sometimes hybridizes with no. 4.

1114. D. intermedia.

2. **D. ánglica** Huds. (English). — Stouter and stiffer than no. 1; *leaf-blades linear-spatulate to narrowly cuneate-obovate,* 1.5–5 *cm. long,* on naked erect or sparsely hairy petioles; *stipules adnate to petioles;* scapes 0.3–2 dm. high, stiffly erect, with 1–12 flowers; petals white, rarely pinkish, 5–6 mm. long; *seeds fusiform, blackish, with loosely alveolate testa prolonged at tips.* — Peaty and boggy spots, often calcareous, Nfld. and s. Lab. Pen. to Alaska, s. to Gaspé Co. and L. Mistassini, Que., Bruce Co., Ont., n. Mich., n. Wisc., Ida. and n. Calif. June–Aug. (Eurasia; Hawaii) Fig. 1115.

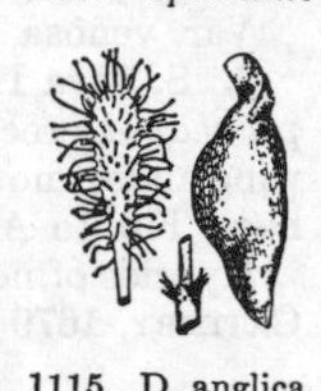

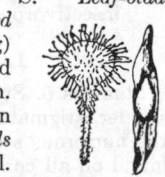

1115. D. anglica.

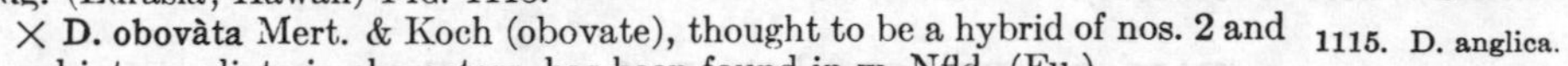

× **D. obovàta** Mert. & Koch (obovate), thought to be a hybrid of nos. 2 and 4 and intermediate in characters, has been found in w. Nfld. (Eu.)

3. **D. lineàris** Goldie (linear). — Similar to no. 2; *leaf-blades exactly linear,* 1–6 cm. long, 1.5–3 mm. wide, on erect naked petioles; *stipules adnate;* scapes 0.2–1.5 dm. high, 1–8-flowered; *seeds* black, *ellipsoid-obovoid, with a close testa densely covered with shallow pits.* — Marly bogs and wet limy shores, local, w. Nfld. to s. Alta., s. to Gaspé Co., Que., s. Aroostook Co., Me., Bruce Co., Ont., Mich., Wisc. and Minn. June–Aug. Fig. 1116.

1116. D. linearis.

4. **D. rotundifòlia** L. (round-leaved), Round-leaved S. — *Leaf-blade suborbicular or broadly transverse-elliptic, abruptly narrowed into the elongate slender pilose petiole;* scape (rarely forking) filiform, 0.5–3 dm. high, 2–25-flowered, or stiffly erect and shorter than to but slightly longer than leaves and 1–4 cm. high with only 1–3 flowers in forma **breviscàpa** (Regel) Domin (short-stemmed); petals (sometimes 6) white (rarely pink), 4–6 mm. long; *seeds fusiform, with loose testa prolonged at tips.* — Peaty or moist acid soils, Greenl. and Lab. to Alaska, s. to Nfld., N.S., N.E., n. Fla., Ala., Ill., Minn., Mont. and Calif. June–Aug. (Eurasia) Fig. 1117. — Forma *breviscapa* in bleak or alpine habitats, Lab. and mts. of Nfld. and Gaspé Pen., Que. (Eu.)

1117. D. rotundifolia.

Var. **comòsa** Fern. (with a tuft of hair). — Dwarf; *inflorescence* 1–few-flowered, *usually capitate; calyx crimson or roseate; petals greenish or crimson,* sometimes foliaceous; *carpels* and sometimes other parts of the flower *modified to green gland-bearing leaves.* — Very local, either in acid or calcareous peat, Gaspé Co., Que., to N.E. and n. N.Y.

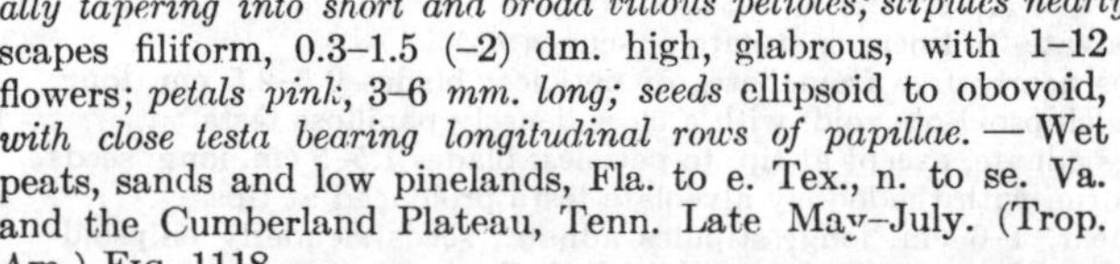

5. **D. capillàris** Poir. (hair-like), Pink S. — Rosettes compact, 2–6 cm. broad; *leaves* (including petioles) 1–3 *cm. long,* 2–7 mm. broad, *the cuneate-obovate blades gradually tapering into short and broad villous petioles; stipules nearly free, fimbriate;* scapes filiform, 0.3–1.5 (–2) dm. high, glabrous, with 1–12 flowers; *petals pink,* 3–6 *mm. long; seeds* ellipsoid to obovoid, *with close testa bearing longitudinal rows of papillae.* — Wet peats, sands and low pinelands, Fla. to e. Tex., n. to se. Va. and the Cumberland Plateau, Tenn. Late May–July. (Trop. Am.) Fig. 1118.

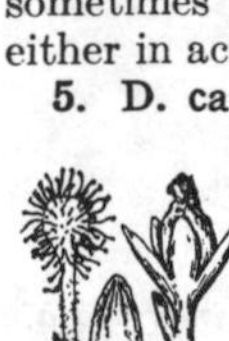

1118. D. capillaris.

6. **D. brevifòlia** Pursh (short-leaved). — Rosettes compact, 1–3.5 cm. broad; *leaves* cuneate-obovate, 0.5–1.7 *cm. long,* 2.5–6 mm. broad, *strongly cuneate to the short and broad villous petiolar base; stipules obsolete; scape* filiform, 1–8 (–10) cm. high, *its summit, with the rachis and calyces, stipitate-glandular; petals white,* 5–8 *mm. long; seeds* ellipsoid, *with longitudinal rows of shallow pits.* — Damp pinelands, ditches and low fields, Fla. to Tex., n. to se. Va., Tenn. and Ark. May, June. Fig. 1119. — The large flower closing at noon.

1119. D. brevifolia.

7. **D. filifórmis** Raf. (thread-like), Dew-thread. — *Base bulbous, woolly; leaves very long* (1–3 *dm.*) *and filiform,* erect, glandular throughout; *flowers numerous, purplish* (0.7–1.5 *cm. broad*); seeds ellipsoid-fusiform, covered

1120. D. filiformis.

with rows of shallow pits. — Damp sands of the Coastal Plain, s. N.J. to se. Mass. June–Sept. Fig. 1120. — A larger var. from n. Fla. to. La., n. to S.C.

FAM. 77. PODOSTEMÀCEAE (Riverweed Family)

Aquatics, growing on stones in running water, some with the aspects of Sea-weeds, *or others of* Mosses *or* Liverworts; *the minute naked flowers bursting from a spathe-like involucre, producing a 2–3-locular many-seeded ribbed capsule.* — An extraordinary, chiefly trop. group with thalloid or frondose vegetative growth, represented in e. N.Am. by a single species of the largely trop. genus

1. PODOSTÈMUM Michx. Riverweed

Flowers solitary, nearly sessile in a tubular sac-like involucre, destitute of floral envelopes. Stamens 2, borne on one side of the stalk of the ovary, with their long filaments united into one for more than half their length, and 2 short sterile filaments, one on each side; anthers 2-locular. Stigmas 2, subulate. Capsule pedicellate, oval, 8-ribbed, 2-locular, 2-valved. Seeds minute, very numerous, on a thick persistent central placenta, destitute of albumen. — Leaves 2-ranked. (Name from the Greek *pous, podos, foot,* and *stemon, stamen;* the two stamens being apparently raised on a stalk by the side of the ovary.)

1. P. ceratophýllum Michx. (horn-leaf), Threadfoot. — Leaves rigid or horny, dilated into a sheathing base, above mostly forked into filiform or linear lobes. — On rocks in streams, Ga. to Ark., n. to N.B., Me., s. Que. and se. Ont., locally abundant. July–Sept. — A small olive-green plant, of firm texture, resembling a Seaweed, tenaciously attached to loose stones by fleshy disks or processes in place of roots.

FAM. 78. CRASSULÀCEAE (Orpine Family)

Succulent herbs with perfectly symmetrical flowers: the petals and pistils equaling the sepals or calyx-lobes in number (3–30), *and the stamens the same or double their number,* — technically different from *Saxifragaceae* only in this complete symmetry, and in the carpels (in most of the genera) being quite distinct from each other. Also, instead of a perigynous disk, there are usually tiny scales on the receptacle, one behind each carpel. Fruit dry and dehiscent; the follicles opening down the ventral suture, many (rarely few)-seeded. Stipules none. Flowers usually cymose, small.

Sepals, petals and carpels 3 or 4; stamens 3 or 4; flowers solitary in leaf-axils; aquatic or subaquatic annuals. . . . . 1. *Tillaea.*
Sepals, petals and carpels 4–∞; stamens 8–∞; flowers cymose, cymose-racemose or cymose-paniculate; plants not aquatic.
Sepals, petals and carpels 4 or 5; stamens 8–10. . . . . 2. *Sedum.*
Sepals, petals and carpels 6–30; stamens 12–∞. . . . . 3. *Sempervivum.*

1. TILLAÈA L. Pigmyweed

Sepals, petals, stamens and carpels 3 or 4. Capsules 2–∞-seeded. Tiny tufted or matted aquatic or subaquatic annuals with slender stems, narrow connate-based entire leaves and tiny axillary flowers; the genus semicosmop. except in frigid areas. (Named for *Michelangelo Tilli,* 1655–1740, professor at Pisa.) Tillaeastrum Britt. Sometimes combined with the Old World genus Crassula L.

1. T. aquática L. (aquatic). — Filiform stems much branched from base, ascending and up to 1 dm. high or forming dense carpets up to 1.5 dm. across; leaves linear or linear-oblong, 2–7 mm. long, overtopping the solitary sessile to short-pedicelled axillary flowers; calyx about half as long as the greenish-white petals and the 8–10-seeded follicles, each of the latter with a scale at base. (Incl. *T. Vaillantii* of ed. 7, not Willd.) — Margins of pools and on fresh to tidal shores, Nfld. to lower St. Lawrence R., Que., s. along or near coast to Md.; La. to Tex. and Mex., inland n. to pools and depressions of Minn., Wyo., Utah and Wash. July–Oct. (Eu.; n. Afr.)

2. SÈDUM L. Stonecrop. Orpine

Sepals and petals 4 or 5, the mostly narrow petals distinct or barely united at base. Stamens 8–10, mostly perigynous. Follicles many-seeded, each follicle with a basal scale. — Mostly smooth and fleshy-leaved perennials (sometimes annuals) with chiefly alternate or imbricated leaves, and flowers in broad to one-sided cymes. Genus nearly cosmop., most species on rocky or dryish soils. (Name from *sedere, to sit,* alluding to the manner in which many species affix themselves to rocks or walls.)

*a.*Flowers perfect, purplish, pink, white or yellow; follicles divergent or ascending. . *b.*

*b.*Leaves of flowering stems or branches terete or subterete and linear-cylindric or thick-ovoid and tightly imbricated. . . *c.*

*c.*Leaves tightly imbricated, narrowly ovoid, thickened at base; flowers yellow; depressed and matted plant. . . . . 1. *S. acre.*

*c.*Leaves not tightly imbricated, linear-terete or subterete. . . *d.*

*d.*Plants annual, without perennating basal offsets.

Stems 5–15 cm. high, branching from near or below middle into elongate dichasial remotely flowered branches and branchlets; leaves flattish to subterete; petals yellow; mature follicles horizontally divergent. . . . . 2. *S. Nuttallianum.*

Stems 1–4.5 dm. high or long, simple up to the pink- or white-flowered crowded forking cyme; leaves linear-terete; follicles loosely ascending. . . . . 3. *S. pulchellum.*

*d.*Perennials with evergreen leafy basal offsets; cymes rather compact, corymbiform, terminating simple stems. . . *e.*

*e.*Petals yellow.

Leaves opposite or whorled. . . . . 4. *S. sarmentosum.*

Leaves alternate.

Sepals oblong-ovate, merely acutish, 2–3 mm. long; petals oblong, subacute, 6 mm. long, clear yellow. . . . . . . 5. *S. rupestre.*

Sepals lance-triangular, long-attenuate, 5–6 mm. long; petals oblong-acuminate, 7–10 mm. long, pale yellow or creamy. 6. *S. anopetalum.*

*e.*Petals white or pink-tinged. . . . . 7. *S. album.*

*b.*Leaves flat and dilated, spatulate to oblong or broader. . *f.*

*f.*Depressed or matted plants with prostrate basal offshoots and slender upcurving flowering stems 0.5–2 dm. high; leaves of flowering stems 0.5–3.5 cm. long. . *g.*

*g.*Leaves entire; petals white; native woodland plants.

Lower leaves mostly in whorls of 3; bracts of cyme dilated. . . . 8. *S. ternatum.*

Leaves all alternate; bracts of cyme slender or wanting. . . . . 9. *S. glaucophyllum.*

*g.*Leaves copiously dentate or serrate; petals pink or yellow; introd. and esc. from cult.

Leaves opposite, obovate, coarsely dentate; petals roseate. . . 10. *S. spurium.*

Leaves alternate, spatulate or oblanceolate, sharply serrate-dentate; petals yellow. . . . . 11. *S. Aizoön.*

*f.*Erect or ascending cespitose plants with coarse flowering stems 1.5–8 dm. high; larger elliptic, ovate or obovate leaves 3–10 cm. long. . . *h.*

*h.*Stem 3–8 dm. high; larger leaves 3–10 cm. long, mostly several-toothed; inflorescence corymbose or elongate-paniculate, 0.5–2 dm. broad; flowers deep rose-purple to pale pink; sepals barely one-third as long as petals; stamens about equaling petals; nectariferous scale longer than broad; introd. . . *i.*

*i.*Petals deep purple to roseate; stems 2–8 dm. high; leaves broad-oblong or elliptic, mostly several-toothed, alternate to whorled. . 12. *S. purpureum.*

*i.*Petals pale; leaves ovate to obovate or orbicular, mostly opposite; stems up to 5 dm. high.

Leaves ovate to ovate-lanceolate, regularly dentate; petals greenish-yellow. . . . . 13. *S. Telephium.*

Leaves oblanceolate or obovate to orbicular, entire or nearly so; petals pinkish-white. . . . . 14. *S. alboroseum.*

*h.*Stem 1.5–4(–6) dm. high; larger leaves 3–6 cm. long, entire or remotely dentate; corymb 2.5–8 (very rarely –14) cm. broad; flowers white or pale pink; sepals and stamens about half as long as petals; nectariferous scale broader than long; native. . . . . 15. *S. telephioides.*

*a.*Flowers dioecious, rarely polygamous, the staminate yellow, the pistillate yellow to purplish; petals very narrow; follicles closely approximate and erect; rhizome thick and deep-seated, the erect or nearly erect flowering stems annual; boreal or montane species. . . . . 16. *S. Rosea.*

1. S. ácre L. (pungent-tasting), Mossy S., Wallpepper, Love-entangle. — *Matted and creeping* evergreen, *forming moss-like freely rooting carpets; leaves closely imbricated, very thick, compressed-ovoid,* strongly spurred at base, blunt, 3–6 *mm. long; flowers in small cymes,* with small bracts; sepals oblong or ovate, 2–3 mm. long; *petals broadly lanceolate,* 5–10 mm. long, *yellow; follicles divergent,* long-beaked, 4–5 mm. long. — Rocks, walls and dry open places, Que. to Wash., s. to N.S., N.E., Va., O., Ind., Ill., etc. June, July. (Natzd. from Eu.)

2. S. Nuttalliànum Raf. (for Thomas Nuttall, 1786–1859). — *Erect or ascending annual* 5–15 *cm. high, mostly branching from near or below middle into dichasial elongate remotely flowered branches or branchlets;* leaves subterete to flattish, scattered, oblong, 4–6 mm. long; sepals ovate, unequal, 1.5–3 mm. long; *petals yellow,* slightly longer than sepals, oblong-lanceolate; *follicles widely divergent.* — Chert-barrens and other dry rock, sw. Mo., adj. Ark., Okla. and Tex. June, early July.

3. S. pulchéllum Michx. (beautiful and little), Rock-moss, Widow's-cross. — *Slender* winter-*annual; flowering stems ascending from decumbent bases,* 1–4.5 *dm. high or long, simple up to the cyme;* young plants with depressed branches bearing cuneate soon deciduous leaves; *leaves of flowering stems* very numerous, *linear-subterete to linear-spatulate,* 0.5–2.5 cm. long; *cyme with divergent to recurving closely flowered branches;* sepals lanceolate, 1.5–3 mm. long; *petals roseate or white,* 4–8 mm. long; *follicles loosely ascending,* 4–8 mm. long. — Dry to moist calcareous rock or thin soil, w. Va. to s. Ill., Mo. and Kans., s. to Ga., Ala., Ark. and Tex. Late May–early July.

4. S. sarmentòsum Bunge (producing lithe runners). — *Slender* perennial, *the lithe sterile and flowering stems creeping* and rooting at the nodes; up to 2.5 dm. long; *leaves in remote pairs or whorls of* 3, *oblong-lanceolate or narrowly oblanceolate,* entire, acutish, 1–2 cm. long; cyme few-flowered, leafy-bracted; sepals spurless, lance-oblong, bluntish, 3.5–5 mm. long; *petals yellow,* acute or mucronate, 5–8 mm. long; anthers acutish; follicles divergent, 5–6 mm. long. — Esc. from cult. to roadsides, waste ground and open woods, N.J. and e. Pa. to O. June. (Introd. from e. Asia)

5. S. rupéstre L. (of rock). — *Sterile stems extensively creeping and forking,* up to 5 dm. long, *with crowded alternate semiterete linear to narrowly oblanceolate apiculate leaves* about 1 cm. long; *flowering stems* up to 3 dm. high, *their leaves with broad basal spurs; corymb flattish or concave across top; sepals oblong-ovate, acutish,* 2–3 *mm. long; petals clear yellow, oblong, subacute,* 6 *mm. long;* follicles suberect, slender-beaked, papillose, 6–7 mm. long. (*S. reflexum* L.) — Roadsides, waste places or borders of woods, Mass. and N.Y., esc. from cult. June, July. (Introd. from Eu.)

6. S. anopétalum DC. (with upcurved petals). — Similar to no. 5; the *sterile stems* shorter, *more ascending, simpler, their leaves up to* 2 *cm. long* and *more terete; flowering stems* 1–2.5 *dm. high; sepals lance-triangular, long-attenuate,* 5–6 *mm. long; petals oblong-acuminate,* 7–10 *mm. long, pale yellow or creamy;* follicles more divergently ascending and longer. — Waste places, old fields, etc., esc. from cult., Lincoln Co., Me. and doubtless elsewhere. July, Aug. (Natzd. from se. Eu.)

7. S. álbum L. (white). — *Sterile stems* creeping and rooting, *forming nearly circular mats* 0.5–2 *dm. across, with alternate* thick or subterete *oblong or oval blunt leaves* 5–10 *mm. long;* flowering stems with similar leaves, ascending, 0.5–3 dm. high; *inflorescence a flat-topped corymb or low panicle* 1.5–7 *cm. broad,* the flowers slender-pedicelled; *sepals* spurless at base, *obovate, about* 1 *mm. long; petals white or pinkish, broadly lanceolate to obovate,* 3–4 *mm. long;* follicles ascending, 3–4 mm. long. — Spread from cult., e. Pa. to O. (Introd. from Eurasia)

8. S. ternàtum Michx. (in threes). — *Sterile branches* loosely spreading or prostrate, *with* 1 *or* 2 *remote pairs of cuneate-obovate entire flat leaves and a terminal rosette of about* 6 *similar crowded leaves* 1–2 cm. long and 0.7–1.5 cm. broad; *fertile branches arising from the sterile rosettes, upcurved,* 0.5–2 dm. high; *their lower leaves whorled and cuneate-obovate, the upper leaves much narrower and opposite or scattered; cyme leafy-bracted,* about 3-forked, with sessile flowers; sepals linear-oblong, equal, about 4 mm. long; *petals narrowly lanceolate, white,* 8–10 mm. long; anthers dark; follicles divergent. — Damp, often calcareous, rocks, mossy banks, brooksides, etc., N.Y. to Mich. and Ill., s. to Ga. and Tenn.; spread from cult. to damp roadsides and cool rocks farther north. April–June.

9. S. glaucophýllum R. T. Clausen (glaucous-leaved), Cliff-Stonewort. — Differing from no. 8 in its *alternate and spirally arranged glaucous leaves;* prostrate *sterile* decumbent *shoots* 0.5–5 *cm. long, with a dense terminal rosette;* the *rosette-leaves obovate- to oblong-spatulate and only* 4–16 *mm. long and* 2–7 *mm. broad; flowering stems* erect, *with* 30–50 flat *oblanceolate spreading leaves; cyme with* 3–several branches and *few or no narrow bracts; sepals* pale, *narrowly lanceolate,* 2.5–4.5 *mm. long; petals* white, *lance-acuminate,* 4–8 *mm. long.* (*S. Nevii* of ed. 7, not Gray) — Damp, chiefly calcareous, rock, Va. and W.Va. May–Aug.

10. S. spùrium Bieb. (false; previously confused with another species). — *Widely branching decumbent sterile stems* up to 1.5 dm. long; flowering stems erect, 1–2 dm. high; *leaves mostly opposite, obovate, coarsely dentate, 1.5–3 cm. long; corymb* rather dense, 2–8 cm. broad, *with foliaceous lanceolate or oblong papillose-margined bracts; sepals oblong-lanceolate blunt, papillose-tipped, 4–4.5 mm. long; petals roseate,* oblong to lanceolate, 7–10 mm. long. (*S. stoloniferum* of ed. 7, not Gmel.) — Rocky or sandy roadsides, banks or old fields, spread from cult., Nfld. to N.Y. and Pa. June–Aug. (Introd. and natzd. from Eurasia)

11. S. Aìzoön L. (for the genus *Aizoön*). — Resembling no. 10; *leaves alternate, spatulate to oblanceolate, sharply serrate; sepals unequal, linear to narrowly lanceolate,* not papillate; *petals yellow.* (Incl. *S. Ellacombianum* Praeger) — Locally spread from cult. (Introd. from Asia)

12. S. purpùreum (L.) Link (purple), Live-forever, Garden-O., Frogplant, Vit-toujours (Que.). — *Coarse erect perennial from fleshy carrot-like tubers,* the usually many *fleshy stems 2–8 dm. high; leaves* succulent, *broadly oblong or elliptic, coarsely dentate, green, alternate or in whorls of* 3, the larger ones 4–10 cm. long; *inflorescence a compact corymb or thyrsiform panicle* 0.7–2 *dm. broad,* the summit or the summits of secondary (rameal) corymbs rounded; *sepals hardly one-third as long as the purple-red to deep roseate,* wide-spreading *petals; nectariferous scales longer than broad; stamens about equaling petals;* follicles suberect. (*S. Fabaria* W. D. J. Koch) — Roadsides, open banks or open woods, spread from cult. and abundantly (often aggressively) natzd., Nfld. to Ont. and Wisc., s. to N.S., N.E., Md. and Ind. July–Sept. (Introd. and natzd. from Eu.)

13. S. Teléphium L. (for Telephus, son of Hercules), Garden-O., Live-forever. — Similar to no. 12 which may better be considered a var. of it; *stem rarely 5 dm. high; leaves mostly opposite, ovate or ovate-lanceolate, regularly toothed; petals greenish-yellow or creamy.* — Roadsides and borders of fields, locally spread from cult., Nfld. to Minn., s. to N.S., N.E., N.J., Pa. and Ind. Aug., Sept. (Introd. and natzd. from Eurasia)

14. S. alboròseum Baker (whitish pink), Garden-O. — Similar to nos. 12 and 13 and perhaps not specifically separable; stems often more spreading, up to 4 dm. high; *leaves glaucous, mostly opposite, oblanceolate to obovate-suborbicular, entire or merely undulate; inflorescence less compact, with longer pedicels; petals white with pink tinge.* — Spread from cult., Nfld. to Va. July–Sept. (Introd. from Asia)

15. S. telephioìdes Michx. (resembling *S. Telephium;* of which intricate complex it may be a geographic var.) — Resembling nos. 12–14; stems 1.5–4 (–6) dm. high, relatively slender; *leaves glaucous, obovate to obovate-oblong, petioled, entire or remotely dentate,* the larger ones 3–6 cm. long; *corymb* small, 2.5–8 (–14) *cm. broad; sepals and stamens about half as long as the white to pale pink petals; nectariferous scale quadrate, broader than long.* — Cliffs and knobs, w. N.Y. to s. Ill., s., especially on the mts., to Ga. Aug., Sept.

16. S. Ròsea (L.) Scop. (old name from the fragrant root, the *Rosea radix* of early apothecaries), Roseroot. — *Root thick and strong, suckering,* fragrant when bruised, with scaly crowns bearing numerous loosely ascending to erect leafy stems 0.3–4 dm. high; *leaves pale,* oblong, oval or obovate, dentate to entire, *the larger ones* 1–4 *cm. long,* rather *crowded,* spirally arranged to whorled; *flowers dioecious,* or rarely polygamous, *in a dense corymb* 1–6 cm. broad; *staminate flowers yellow or yellowish,* pistillate flowers commonly purple-tinged to full purple; *follicles* plump, *erect.* (*S. roseum* Scop.; *Rhodiola Rosea* L.) — Arctic reg., s. along rocky coast and sea-cliffs to Me.; and inland s. locally to Vt., N.Y. and ne. Pa.; Roan Mt., N.C. May–Aug. (Eurasia) — Probably including several so-called but scarcely separable species of w. N.Am.

Var. **Leèdyi** Rosend. & Moore (discovered in 1936 by John L. Leedy). — Leaves narrowly oblanceolate, many times longer than broad, entire or irregularly porrect-toothed. — Calcareous cliffs, se. Minn.

Other species are inclined to spread from cult. These should be sought in works on garden-plants.

3. SEMPERVÌVUM L. Houseleek

Calyx-lobes, petals and many-seeded carpels 6–many. Stamens usually twice as numerous. — Succulent perennials with imbricated leaves and cymose-paniculate yellow or purple flowers, nat. of Eurasia. (*Semper, ever,* and *vivus, alive,* from the tenacious vitality of the plants.)

1. S. tectòrum L. (of roofs or walls), Hens-and-chickens. — Leaves of the dense basal and lateral rosettes (on short thick offsets) ovate, acute, ciliate but otherwise glabrous; those of the stem more oblong, clammy-pubescent; flowers rose-purple. — Often planted, and persisting long after or esc. from cult. (Introd. from Eu.)

FAM. 79. SAXIFRAGÀCEAE (SAXIFRAGE FAMILY)

Herbs or shrubs of various aspect, distinguishable from Rosaceae *by having copious albumen in the seeds, opposite as well as alternate leaves, and usually no stipules, the stamens mostly definite, and the carpels commonly fewer than the sepals (the same no. in* Subfamily I), either separate or partly so, or all combined into one compound pistil. Calyx either free or adherent, usually persistent or withering away. Stamens and petals almost always inserted on the calyx. Ovules anatropous.

a. Herbs; fruit a capsule or follicle, with styles or tips of carpels distinct. . . *b.*
b. Flowers without gland-tipped staminodia alternating with the stamens, usually more than 1; leaves, if rosulate, toothed or lobed. . . *c.*
c. Ovary 5–7-locular, the 5 or 7 follicles dehiscent by circumscission below the spreading beaks. Subfam. I. PENTHOROIDEAE. . . . . 1. *Penthorum.*
c. Ovary 1–2 (rarely –3)-locular, fruits not circumscissile. Subfam. II. SAXIFRAGOIDEAE. . *d.*
d. Ovary 2 (rarely 3)-locular, with axile placentae or of as many distinct carpels. . . *e.*
e. Leaves ternately decompound; flowers polygamodioecious, in compound panicles of spikes or racemes; stamens 8 or 10; petals withering-persistent (marcescent). . . . . 2. *Astilbe.*
e. Leaves simple, unlobed or lobed; flowers perfect; petals mostly deciduous. . *f.*
f. Stamens 5; calyx-tube somewhat adherent to the ovary; leaves long-petioled, roundish, lobed.
Calyx-tube adherent only to base of ovary; petals marcescent; seeds wing-margined. . . . . 3. *Sullivantia.*
Calyx-tube adherent to nearly the whole ovary; petals deciduous; seeds not winged. . . . . 4. *Boykinia.*
f. Stamens 10; calyx-tube free from or adnate to ovary; leaves and habit various. . . . . 5. *Saxifraga.*
d. Ovary 1-locular, with 2 parietal or nearly basal placentae alternate with the stigmas. . . *g.*
g. Flowering stems simple, firm, bearing terminal spikes, racemes or thyrsoid panicles; basal leaves long-petioled; petals 5. . . *h.*
h. Petals entire; ovary and capsule elongate.
Flowers in a simple raceme; calyx nearly free from ovary; stamens 10; valves of capsule very unequal; placentae basal. . 6. *Tiarella.*
Flowers clustered on branches of the thyrse or panicle; calyx adherent to lower half of ovary; stamens 5; valves of capsule equal; placentae parietal. . . . . 7. *Heuchera.*
h. Petals pinnately fringed; ovary and capsule depressed; raceme slender, simple. . . . . 8. *Mitella.*
g. Flowering stems forking, soft and succulent, bearing axillary or cymose apetalous insignificant flowers. . . . . 9. *Chrysosplenium.*

b. Flowers with 5 palmately cleft gland-tipped staminodia alternating with the true stamens, single, terminating the naked or 1-leaved scape; leaves chiefly basal, entire. Subfam. III. PARNASSIOIDEAE. . . . . . 10. *Parnassia.*
a. Shrubs. . . *i.*
i. Fruit a capsule, partly or wholly free from the calyx-tube; leaves without stipules. . . *j.*
j. Leaves opposite; calyx adherent to base of capsule. Subfam. IV. HYDRANGEOIDEAE. . . *k.*
k. Stamens 20–40; capsules 3–10-locular, bursting at the sides or splitting into segments.
Erect shrubs; calyx-lobes conspicuous; petals 4, showy, convolute in the bud; styles 3–5. . . . . 11. *Philadelphus.*
High-twining vine; calyx-lobes small; petals 7–10 small, valvate in bud; style 1. . . . . 12. *Decumaria.*
k. Stamens 8 or 10; capsule 2-locular below, opening by a hole between the 2–4 divergent styles; marginal flowers often enlarged and sterile. . 13. *Hydrangea.*
j. Leaves alternate; calyx free from the 2-locular capsule. Subfam. V. ESCALLONIOIDEAE. . . . . 14. *Itea.*
i. Fruit a berry, surmounted by the marcescent calyx or its remnants, 1-locular, with 2 parietal placentae; leaves alternate, with or without stipules adnate to the petiole. Subfam. VI. RIBESIOIDEAE. . . . . 15. *Ribes.*

Subfam. I. PENTHOROÌDEAE

1. PÉNTHORUM L. Ditch-Stonecrop

Calyx-lobes 5 (–7). Petals rare, if any. Stamens 10. Pistils 5 (–7), united below, forming a 5-angled 5-horned and 5-locular capsule, which opens by the falling off of the beaks, many-seeded. — Upright weed-like perennials of e. N.Am. and Asia, with scattered leaves, and yellowish-green flowers loosely spiked along the upper side of the naked branches of the cyme. (Name from the Greek *pente, five,* and *horos, a mark,* from the quinary plan of the flower.)

1. P. sedoìdes L. (like Sedum). — Stoloniferous; the stem decumbent at base, simple to bushy-branched, 0.15–1 m. high; leaves elliptic or broadly lanceolate, finely serrate, acute at both ends. — Wet low grounds, Fla. to Tex., n. to s. N.B., N.E., sw. Que., s. Ont., Mich., Wisc., Minn. and Neb. July–Oct.

Subfam. II. SAXIFRAGOÌDEAE

2. ASTÌLBE Hamilton False Goatsbeard

Flowers dioeciously polygamous. Calyx 4–5-parted, small. Petals 4–5, spatulate, withering-persistent. Ovary almost free, many-ovuled; styles 2, short. Capsule 2-locular, separating into 2 follicles. Seed-coat loose and thin, tapering at each end. — Perennial herbs of e. N.Am. and s. and e. Asia, with twice or thrice ternately-compound ample leaves, cut-lobed and toothed leaflets, and small white or yellowish flowers in spikes or racemes, which are disposed in a compound panicle. (Name composed of the Greek *a, without,* and *stilbe, sheen,* because the foliage of the original Indian species was dull, in contrast with *Aruncus,* which it resembles.)

1. A. bitemàta (Vent.) Britt. (twice-ternately compound). — Somewhat pubescent (1–2 m. high); *leaflets mostly cordate;* petals minute or wanting in the pistillate flowers; stamens 10. — Woods in the mts., Va. and W.Va. to Ga. and Tenn. May–July. — Closely resembling *Aruncus* of the *Rosaceae.*

2. A. japónica (Morren & Dcne.) Gray (Japanese). — Much smaller than no. 1, 1.5–8 *dm. high; leaflets cuneate-oblanceolate;* flowers larger. — Spread from cult. to waste places and roadsides, N.E. and N.Y. (Introd. from Japan)

3. SULLIVÁNTIA T. & G.

Calyx campanulate, adhering below only to the base of the ovary, 5-cleft. Petals 5, oblanceolate, entire, acutish, withering-persistent. Stamens shorter than the petals. Capsule 2-beaked, many-seeded, opening between the beaks; seeds imbricated upward. — Low and reclined-spreading temp. N.Am. perennial herbs with rounded and cut-toothed or slightly lobed smooth leaves on slender petioles, and small white flowers in a branched loosely cymose panicle raised on a nearly leafless slender stem. Peduncles and calyx glandular; pedicels recurved in fruit. (Dedicated to the distinguished bryologist, *William Starling Sullivant,* 1803–1873, who discovered the original species.)

1. S. Sullivántii (T. & G.) Britt. (for the discoverer of the genus). — Stem slender, flexuous or decumbent, from a short oblong rhizome, 1–3.5 dm. high, naked or with 1 or 2 cauline leaves; *later radical leaves* reniform, 3–9 cm. broad, *cleft* one-sixth to one-third to base *into* broad *convex-sided* dentate *lobes; lower bracts* of panicle *round-obovate or broadly flabelliform,* 3–7-*toothed across the summit;* panicle 0.5–2.5 dm. long, lax; flowering *calyx* about 2 mm. long, *cleft one-third to half-way to base into triangular-lanceolate lobes; petals lanceolate or oblanceolate; base of fruiting calyx adnate to lower fifth to third of capsule; capsule* slenderly ovoid, *about* 4 *mm. long, broadest near bases of sinuses of calyx;* seeds 1.3–1.5 mm. long, winged. (*S. ohionis* T. & G.) — Wet limestone- and sandstone-cliffs, s. O., n. Ky., and se. Ind. June–Aug.

2. S. renifòlia Rosend. (with reniform leaves). — Differing from no. 1 in *straighter parallel sides of lobes of basal leaves;* bracts narrower, mostly 3-cleft; *calyx cleft only one-fourth to one-third its length into blunt oblong-ovate lobes;* petals ovate; *base of fruiting calyx adnate to lower third to half of capsule; capsule about 5 mm. long, broadest well below sinuses of calyx;* seeds 0.8–1.3 mm. long. — Similar habitats, sw. Wisc. and se. Minn. to nw. Ill. and ne. Mo. June–Aug.

4. BOYKÍNIA Nutt.

Calyx-tube turbinate, adherent to the 2-locular and 2-beaked capsule. Stamens 5, as many as the deciduous petals, these mostly convolute in the bud. Otherwise as in *Saxifraga.* — Perennial N. Am. and e. Asiatic herbs with alternate palmately 5–7-lobed or cut petioled leaves, and white flowers in cymes. (Dedicated to *Samuel Boykin*, early active botanist of Georgia.) THEROFON Raf.

1. **B. aconitifòlia** Nutt. (with leaves of *Aconitum*), BROOK-SAXIFRAGE. — Stem glandular (2–6 dm. high); leaves deeply 5–7-lobed. (*Therofon* Millsp.) — Wet rocks, banks of streams and rich woods, mts. of W.Va., Va. and Ky. to Ga. and Ala. June, July.

5. SAXÍFRAGA L. SAXIFRAGE

Calyx either free from or adhering to the base of the ovary, 5-cleft or -parted. Petals entire, imbricated in the bud, commonly deciduous. Styles 2. Capsule 2-beaked, 2-locular, opening down or between the beaks, or sometimes 2 almost separate follicles. — Chiefly perennial herbs of N. and S.Am. and Eurasia, with the radical leaves often clustered, those of the stem mostly alternate. (Name from *saxum*, a *stone*, and *frangere, to break;* the name early applied, through the doctrine of signatures, to European species bearing granular bulblets, which were supposed to dissolve urinary concretions.)

*a.*Ovary superior, entirely free from calyx-tube; calyx-lobes reflexed in fruit (flowers usually wanting in no. 1, these replaced by tufts of small leaves); plants with scapes and basal rosettes. . . *b.*

*b.*Rosette-leaves tapering to sessile or wing-petioled bases, lanceolate to ovate or obovate. . . *c.*

*c.*Flowers all or essentially all replaced by tufts of small leaves; Arctic-alpine. . . . . 1. *S. stellaris*, var. *comosa.*

*c.*Flowers normal, with calyx, petals and carpels; Alleghenian. . . *d.*

*d.*Petals uniform and equal, oval, ovate or oblong; follicles ribless. . . *e.*

*e.*Leaves with long winged petiole, abruptly dilated into an elliptic to deltoid-ovate or rounded blade 2–8 cm. long and 6–11-dentate on each margin.

Filaments clavate; follicle-bodies 4–5 mm. long. . . . . . 2. *S. caroliniana.*

Filaments slender-subulate; follicle-bodies 2.5–3 mm. long . . . 3. *S. Careyana.*

*e.*Leaves oblong or broadly oblanceolate, tapering gradually to the short winged petiole, the larger 1–3 dm. long and 12–40-toothed on each margin. . . . . 4. *S. micranthidifolia.*

*d.*Petals unequal, 3 large and cordate-based, 2 small and lanceolate; filaments subulate; follicles with longitudinal ribs. . . . . . 5. *S. Michauxii.*

*b.*Rosette-leaves slender-petioled, orbicular or reniform; petals unequal; follicles 5–6 mm. long; Newfoundland species. . . . . 6. *S. Geum.*

*a.*Ovary inferior or partly so, adnate to calyx-tube. . . *f.*

*f.*Leaves without lime-encrusted marginal or terminal pores. . . *g.*

*g.*Plants acaulescent, with basal rosettes of undulate, dentate or serrate leaves and soft flowering scapes (only exceptionally leafy). . . *h.*

*h.*Calyx-lobes ovate to broadly deltoid, ascending or only slightly spreading in fruit; leaves 1.5–8 cm. long; petals chiefly white.

Carpels 2; leaves crenate to sharply dentate.

Inflorescence at first broadly corymbiform, becoming an open panicle; its bracts linear, lanceolate or oblanceolate, many times shorter than the branches; calyx-tube obconic; Canadian and Alleghenian. . . . . 7. *S. virginiensis.*

Inflorescence spicate-racemose, 3–10-flowered; lower bracts oblong or ovate, nearly equaling or exceeding the flowers; calyx-tube hemispherical; Arctic-alpine. . . . . . . 8. *S. gaspensis.*

Carpels 3 or 4; leaves merely undulate or barely crenate; Ozarkian. 9. *S. texana.*

*h.*Calyx-lobes narrower, promptly reflexed; leaves 0.5–3 dm. long, remotely short-dentate; petals greenish, pearly, smoky or purple. . . 10. *S. pensylvanica.*

*g.*Plants with leafy flowering stems or with leafy basal offsets; leaves entire (but sometimes ciliate) or palmately lobed or with 3 terminal teeth, not regularly dentate. . . *i.*

*i.*Lower leaves reniform, cordate at base, filiform-petioled; basal offsets few and slender, arising from under ground, annual or biennial or represented by bulblets in the lower axils.

Stems loosely tufted, 0.1–1.5 dm. high, without bulblets in leaf-axils; flowers 1–5, pedicelled; calyx-lobes ovate; petals 3–5 mm. long. . . 11. *S. rivularis.*

Stems solitary or few, 1–2.5 dm. high, with pale bulblets in the lowest leaf-axils; upper cauline leaves subtending fascicles or branches of fascicles of purplish bulblets; flower solitary, terminal; calyx-lobes oblong; petals 6–12 mm. long. . . . . 12. *S. cernua.*

i. Leaves linear to cuneate-obovate, sessile or broad-petioled; basal offshoots numerous, trailing or matted, covered with imbricated marcescent leaves. . . *j.*

j. Leaves soft, flabelliform or cuneate-obovate, palmately 3–5-lobed; flowering stem with 1–5 leaves below inflorescence; petals white; capsule adnate nearly its whole length to calyx-tube. . . . . . 13. *S. cespitosa.*

j. Leaves firm, fleshy or rigid, linear to spatulate-oblanceolate, entire and mucronate or with 3 apical sharp teeth; flowering stem with 4–∞ leaves below inflorescence; petals yellow or yellowish-white, often dotted with red or orange; capsule adnate to calyx-tube only at base.

Leaves rigid, the basal crowded, cuneate-oblanceolate, with 3 cartilaginous sharp teeth; sepals 1–2.2 mm. long; petals yellowish-white. . . . . 14. *S. tricuspidata.*

Leaves fleshy, plano-convex, linear to oblong, obtuse, with a minute pore at tip below the short mucro; sepals 3–4 mm. long; petals bright yellow. . . . . 15. *S. aizoides.*

f. Leaves with lime-encrusted marginal or terminal pores on the upper surface.

Stolons ending in flat rosettes; rosette-leaves flat, elongate, serrulate with appressed cartilaginous teeth, with lime-encrusted pore at base of each tooth; flowering stem erect, paniculate; petals white, spreading. . . . 16. *S. Aizoön.*

Matted trailing branches with imbricated 4-ranked opposite short and keeled bristly-ciliate leaves, each with 1–3 terminal encrusted pores; peduncles 1-flowered; petals rose-lilac to purple, ascending. . . . 17. *S. oppositifolia.*

1. S. stellàris L. (star-like), var. **comòsa** Poir. (with a tuft of hair). — Rosettes solitary or 1–4 on slender stolons; *leaves thin, cuneate-oblanceolate to spatulate,* dentate-serrate above middle, 1–3.5 cm. long; scape slender, 0.2–2 dm. high; *inflorescence* a corymbiform raceme or panicle *of small leafy tufts replacing flowers,* a terminal normal flower sometimes present. (*S. foliolosa* R. Br.; *Spatularia fol.* Small) — Arct. reg., s. on mossy alpine rocks to nw. Nfld. and Mt. Katahdin, Me. July, Aug. (Eurasia) — Typical *S. stellaris,* with normal flowers on long pedicels, petals white with yellow basal spots, and short-beaked ovoid capsules 7–9 mm. long, reaches se. Lab. and is to be sought in Nfld. and e. Que.

2. S. caroliniàna Gray (of Carolina). — Viscid with glandular hairs; *leaf-blades* 3–6 *cm. long* (2–6 cm. broad), *coarsely 6–10-toothed on each margin,* rather abruptly or somewhat cuneately *contracted to long* hairy *petioles;* stem 3–4 dm. high; panicle ample; *petals ovate,* obtuse, white with two yellow spots; *filaments clavate; follicles* united only at the base, widely spreading, *their bodies* 4–5 *mm. long.* (*Micranthes* Small) — Limestone-cliffs and wet slopes, mts. of Va., N.C. and Tenn. May, June.

3. S. Careyàna Gray (for one of its discoverers, John Carey, 1791–1880). — Similar to no. 2, smoother; leaf-blades reniform to ovate; *petals oblong,* white with 2 yellow spots; *filaments slenderly subulate; follicles smaller* (*bodies* 2.5–3 *mm. long.*) (*Micranthes* Small) — Moist rocks and slopes, mts. of Va. and N.C. May, June.

4. S. micranthidifòlia (Haw.) Britt. (with leaves of *Micranthes*), Lettuce-S., Mountain-lettuce. — *Leaves* thin, *oblong or oblanceolate, tapering gradually to the barely petioled or wing-petioled base;* the larger *blades* 1–3 *dm. long, with* 12–40 *sharp teeth on each margin;* scape slender, 3–9 dm. high; panicle elongated, loosely flowered; pedicels slender; *calyx reflexed, entirely free, nearly as long as the oval obtuse* (*white*) *petals; filaments clavate; follicles* nearly separate, diverging, *narrow, pointed,* 6–8 mm. long. (*Micranthes* Small) — Brooksides, wet rocks and seeping banks, in the mts., Pa. and W.Va. to Ga. and Tenn. May, June.

5. S. Michaùxii Britt. (for its discoverer, André Michaux, 1746–1802). — Leaves spatulate-oblong, 5–15 cm. long, coarsely sharp-toothed; *stems* loosely pubescent, 2–5 dm. high, *with large leafy bracts* and a loose spreading corymbose or paniculate cyme; *petals* white, lanceolate, the 3 *larger ones with a cordate base* and a pair of yellow spots, the 2 smaller with a tapering base and no spots; *follicles* about 6 mm. long, *with slender longitudinal ribs.* (*S. leucanthemifolia* Michx., not La Pey.; *Spatularia petiolaris* and *Hydatica pet.* Small) — Cliffs, ledges and cool banks, mts. of Va. and W.Va. to Ga. and Tenn. June–Aug.

6. S. Gèum L. (for *Geum;* from similarity of basal leaves). — Loosely tufted, with elongate prostrate stems bearing terminal rosettes of *reniform or orbicular cartilaginous-margined crenate-dentate slender-petioled leaves* (blades 1–4 cm. in diameter); scape 1.5–3 dm. high, with a loose corymbiform panicle; sepals free nearly to base, oblong, 1.5–2.5 mm. long, strongly reflexed; *petals* white, with or without red or yellow dots, *slightly unequal*, spreading; filaments spatulate; capsule 5–6 mm. long, with short divergent horns. — Wet rocks and damp sheltered places, Nfld. (coll. many years ago by *Steinhauer*, exact locality unknown). June, July. (Local in w. Eu.)

7. S. virginiénsis Michx. (of Virginia), Early S. — Acaulescent, with thick cylindric rhizome and short caudex; *leaves* in a rosette, subcoriaceous to submembranaceous, ovate to oblong, 0.8–8 cm. long, *crenate or dentate*, often purple beneath, the broad petiole ciliate; scapes 1–7, soft, usually glandular-pubescent, 0.5–3.5 dm. high, the *bracts linear, lanceolate or oblanceolate, acute, much shorter than branches of the* at first densely corymbiform, finally loosely spreading *panicle* of cymes; *calyx-lobes* deltoid-ovate, 0.8–2.4 mm. long, *about equaling the obconic calyx-tube, ascending;* petals white, oblong or obovate, 4–5.5 mm. long (rarely green or wanting); capsule subinflated, 2.5–5.5 mm. long, of 2 nearly distinct follicles with divergent horns. Very variable; the following forms designated:

Petals white; stamens 10 or fewer.
  Flowers and fruits pedicelled; cymose branches of panicle elongating in fruit.
    Petals usually 5; stamens 10. . . . . *S. virginiensis* (typical).
    Petals numerous; stamens reduced. . . . . Forma *plena*.
  Flowers and fruits sessile in glomerules at tips of branches. . . . . . . Forma *glomerulata*.
Petals green or wanting.
  Petals green, pubescent. . . . . Forma *chlorantha*.
  Petals replaced by stamens, the latter 15. . . . . Forma *pentadecandra*.

S. virginiénsis (typical). — Dry or wet rocks and gravelly open or shaded slopes, Saguenay Co., Que., to Ont. and Minn., s. to N.B., N.E., Ga., Tenn. and Mo. Apr.–June. — Forma **plèna** Eames (double-flowered), local, Mass. to Pa.; forma **glomerulàta** Fern. (with glomerules), local, e. Mass.; forma **chlorántha** (Oakes) Fern. (green-flowered), local, e. Mass.; forma **pentadecándra** (Sterns) Fern. (fifteen-stamened), local, e. Mass. to s. N.Y.

8. S. gaspénsis Fern. (of Gaspé Pen., Que.). — Smaller than no. 6; *rosette-leaves narrowly cuneate-obovate,* narrowed to petiolar base, 1.5–3 cm. long, *acutely dentate above the long-cuneate base;* scape 1–7 cm. high, minutely glandular; *inflorescence spicate-racemose or corymbiform,* in maturity 1–2.5 cm. long, 3–10-*flowered; lowest bract oblong or ovate, nearly equaling to exceeding the subtended branch or flower; calyx-tube hemispherical,* lobes oblong or deltoid, 1.8–2 mm. long, ascending to spreading; *petals white,* lanceolate or narrowly elliptic, acute or subacute, 1.5–2 *mm. long;* capsules 4 mm. long, the follicles with short divergent beaks. — Basic rocks, alpine reg. of Shickshock Mts., Que.; n. Lab. July.

9. S. texàna Buckl. (Texan). — Resembling no. 6; *leaves subentire, undulate or barely crenate;* scape solitary, 0.5–1.5 dm. high; *corymb* of cymes rather *dense;* calyx-tube hemispherical, the oblong or ovate lobes ascending, 1.5–2 mm. long; petals oval or obovate, white; *carpels* 3 *or* 4. (*Micranthes* Small) — Acid rocks of prairies, glades and ravines, sw. Mo., Ark. and Tex. Late March, Apr.

10. S. pensylvánica L. (of Pennsylvania), Swamp-S., Wild-beet. — Rhizome thick, oblique; *rosette-leaves* leathery, *spatulate-oblong or narrowly spatulate-obovate to lanceolate or narrowly ovate,* with short broad more or less clasping petiole, 0.5–3 *dm. long,* remotely short-dentate, *the broad midrib conspicuous beneath;* scape soft, stoutish, 0.2–1.5 m. high, many-striate, glandular-pilose to villous, especially above, leafless to base of inflorescence; *panicle* at first dense, *becoming elongate,* much interrupted and lax, 0.3–6 *dm. long;* the lower spreading or spreading-ascending branches becoming 0.2–2 dm. long; bracts linear to lanceolate, much shorter than branches; or the lower 1 or 2 dilated, oval, one-half to two-thirds the length of branches and with purple petals in forma **fúltior** Fern. (supported); *calyx-lobes narrowly deltoid, soon reflexed,* 2–3 times as long as calyx-tube; *petals* linear-oblong, 2–3 mm. long, spreading, *yellowish-white, greenish-yellow, or* in forma **purpuripétala** (A. M. Johnson) House (purple-petaled) *purple;* filaments filiform-subulate; ovary adnate to short calyx-tube; green or purple; capsule 3–5 mm. long, with inflated follicles becoming free nearly to base. (*Micranthes* Haw.; incl. the possibly distinct *S. Forbesii* Vasey, and numerous recently proposed segregates of A. M. Johnson and others, these not yet convincingly differentiated) — Wet meadows, swamps, boggy thickets, prairies and seeping banks, s. Me. to Minn., s. to Va., W.Va., Ill. and Mo. Apr.–June. — Forma *fultior,* s. N.H.; forma *purpuripetala,* N.H. to N.Y. and N.J.

11. S. rivulàris L. (of rills), Alpine-brook-S. — Tufted, *delicate,* 0.1–1.5 dm. high; *rosette-*

leaves filiform-petioled, reniform-cordate, palmately 3–7-*lobed*, 0.3–2 cm. broad; *capillary stems* 1–3-leaved below the 1–5-*flowered* inflorescence, lowest flower long-pedicelled; calyx-lobes ovate, about equaling the hemispherical tube; petals oblong-obovate, white or purple-tinged, 3–5 mm. long; *filaments filiform; ovary adnate to calyx-tube;* capsule 4–7 mm. long, *follicles free only at summit.* — Arct. reg., s. to wet alpine areas of nw. Nfld., Mt. Washington, N.H., and Mont. July, Aug. (Eurasia)

12. S. cérnua L. (nodding). — *Stems erect,* simple or with ascending branches, 1–2.5 dm. high, *with pale bulblets at base, leafy; basal* and lower cauline *leaves cordate-reniform,* 5–7-lobed, *slender-petioled,* 1.2–2.5 cm. broad, *upper leaves* narrower and simple, or in forma **latibracteàta** (Fern. & Weath.) Polunin (broad-bracted) all or nearly all bracteal leaves broad and rounded, *subtending axillary clusters of small purple bulblets* or bulblet-bearing branches; flower terminal; petals white, 6–12 mm. long. — Arct. reg., s. to Nfld. (?), sea-cliffs and mts. of Gaspé Pen., Que., cliffs of Cook Co., Minn., and high mts. of Colo. Late June–early Sept. (Eurasia)

13. S. cespitòsa L. (in dense tufts). — The *closely matted persistent* superficial basal *stems* 0.1–2 dm. long, *covered with marcescent obovate-cuneate or flabelliform palmately* 3–5-*lobed ciliate soft leaves* 0.5–3 cm. long, the lobes linear or oblong, obtuse; flowering stems slender, erect, glandular, 0.1–2 dm. high, with 1–5 leaves; *flowers* 1–7, *loosely corymbed;* petals white, oblong, blunt, 4–6 mm. long, broader than calyx-lobes; *capsule* globose-ovoid, *adnate nearly its whole length to the campanulate calyx-tube.* (*Muscaria* Haw.) — Arct. reg., s. to calcareous gravels and ledges of w. Nfld., Anticosti I., Gaspé Pen. and Rimouski Co., Que., and Rocky Mts. to Colo. June–early Sept. (Eurasia)

14. S. tricuspidàta Rottb. (with three cusps). — *Leaves rigid, those of the basal shoots* crowded, *marcescent, cuneate-oblanceolate, with* 3 cartilaginous *apical teeth, middle tooth cuspidate;* flowering stem glabrous, with scattered entire leaves; *corymb loose;* calyx-lobes much longer than the scarcely accrescent tube, strongly ascending in fruit; petals creamy or yellowish-white, often red- or orange-dotted; capsule slender-conical, 5–7 mm. long, its free beaks 2–3 mm. long. (*Leptasea* Haw.) — Arct. reg., s. to rocks of Lab., Isle Royale, Mich., s. Man. and mts. of B.C. June–Aug. (Eurasia)

15. S. aizoìdes L. (similar to *Aizoön*), YELLOW MOUNTAIN-S. — Matted; basal offshoots superficial, 0.3–2 dm. long, trailing or decumbent, with somewhat imbricated *fleshy plano-convex linear to oblong obtuse leaves* 0.5–2.5 cm. long, *the tip with a minute pore below the short mucro;* some of the decumbent branches ascending at tip, bearing 1–20 *yellow often orange- or red-dotted flowers* in a simple or branching raceme; capsule globose-ovoid, adnate to calyx-tube, 6–9 mm. long, the beaks slightly divergent. (*Leptasea* Haw.) — Arct. reg., s. on calcareous gravels or cool and damp slopes to Nfld., e. Que., n. Vt., centr. and w. N.Y., s. Alta. and s. B.C. June–Sept. (Eurasia)

16. S. Aìzoön Jacq. (from similarity to *Aizoön*), var. **neogaèa** Butters (of the New World). — Stoloniferous, the *horizontal stolons terminated by rosettes; rosette-leaves flat,* leathery, marcescent, oblong, spatulate or narrowly obovate, *serrulate with appressed cartilaginous teeth, with a lime-encrusted pore on the upper surface at base of each tooth;* flowering stem erect, 0.5–5 dm. high, with alternate leaves; panicle racemiform to corymbiform; *petals white,* usually red-dotted, spreading; capsule adnate to calyx-tube, 4–6 mm. long. — Arct. reg., s. on exposed, chiefly calcareous, gravels and rocks, to Nfld., N.S., s. N.B., Mt. Katahdin, Me., Vt., n. N.Y., n. Mich., ne. Minn. and Sask. June–Sept. — With technical characters separating it from the typical European plant.

17. S. oppositifòlia L. (opposite-leaved), PURPLE MOUNTAIN-S. — Matted, with *subligneous trailing* crowded *branches with imbricated* 4-*ranked leathery opposite bristly-ciliate dorsally keeled* and ventrally concave *leaves* 2–5 *mm. long, each with* 1–3 *terminal lime-encrusted pores; peduncles* 1–5 cm. high, leafy, *terminated by a single flower;* calyx-lobes bristly-ciliate, as long as the tube; *petals rose-lilac, becoming violet,* or white in forma **albiflòra** (Lange) Fern. (white-flowered), *ascending;* capsule adnate to calyx-tube, 8–10 mm. long, the long beaks slightly divergent. (*Antiphylla* Fourr.) — Arct. reg., s. to calcareous gravels, ledges and cliffs of Nfld., Anticosti I. and Gaspé Pen., Que., n. Vt., James Bay, and high mts. of Wyo. and Wash. Late May–Aug. (Eurasia)

6. TIARÉLLA L. FALSE MITERWORT. FOAMFLOWER

Calyx campanulate, 5-parted. Petals 5, with claws. Stamens long and slender. Styles 2. Capsule membranaceous, 2-valved; the valves unequal. Seeds few, at the base of each parietal placenta, globular, smooth. — Perennials of temp. N.Am. and e. Asia; flowers white. (Name a diminutive from *tiara,* or *turban,* from the form of the pistil.)

1. T. cordifòlia L. (heart-leaved). — Leaves terminating the rhizome or on aestival runners, cordate-ovate, sharply to obtusely lobed and toothed, sparsely hairy above, downy beneath, the mature blades 5–10 cm. long; flowering scapes 1–3 dm. high (rarely leafy-bracted); racemes 3–15 cm. long, with glandular axis, in anthesis 2–3 cm. thick; *lower (longer) mature pedicels 7–13 mm. long;* flowers 2.5–5 mm. long, with oblong sepals and lanceolate entire white or pink-tinged petals; stamens 4–7 mm. long, the *filaments short-tipped;* in forma **parviflòra** Fern. (small-flowered) the flowers much smaller, the stamens only 2–3 mm. long; in forma **tridentàta** Lak. (three-toothed) the petals 3-toothed; upper valve of carpel one-half to three-fourths as long as lower valve, or in forma **subaequàlis** Lak. (subequal) of about the same length, the *mature capsule 8–12 mm. long, its valves gradually tapering* to the slender styles. — Rich woodlands, N.B. to s. Ont. and Mich., s. to N.S., N.E., and upland to N.C. and Tenn. Late April–July.

2. T. Whérryi Lak. (for its discoverer, Edgar Theodore Wherry, 1885–). — *Non-stoloniferous,* the rhizome often becoming stouter than in no. 1; leaf-blades mostly longer than broad; racemes more slender, *the longest pedicels becoming 6–10 mm. long;* petals narrower; *filaments slender-tipped; capsules 6–10 mm. long, their valves broadly rounded at summit.* (Incl. *T. cordifolia,* var. *collina* Wherry) — Rich wooded slopes and bluffs, Va. and Tenn. to Ga., Ala. and Miss. April, May.

7. HEÙCHERA L. Alumroot

Calyx 5-cleft. Petals 5, narrow. Styles 2, slender. Capsule 1-locular, with 2 parietal many-seeded placentae, 2-beaked, opening between the beaks. Seeds oval, with a rough (rarely smooth) and close seed-coat. — N.Am. perennials, with the round-cordate leaves principally long-petioled and basal, those on the stems, if any, alternate. Petioles with dilated margins or adherent stipules at their base. Flowers in small clusters borne in a narrow panicle, greenish or purplish. (Named for *Johann Heinrich Heucher,* 1677–1747, a German botanist.)*

a. Calyx in fresh anthesis 1.5–6.6 mm. long, regular or only slightly irregular, the tube (above the hypanthium) wanting or up to 1.6 mm. long; stamens long-exserted. . . *b.*
- *b.* Outside of regular flower villous with long white hairs; newly expanded calyx with tube at most 0.5 mm. long; capsule 3–5.5 mm. long. . . *c.*
  - *c.* Larger basal leaf-blades 0.7–3 dm. long, broadly cordate-ovate to -orbicular, acutely angulate-lobed and toothed; seeds echinate, about 0.8 mm. long. . . . . 1. *H. villosa.*
  - *c.* Larger basal leaf-blades 3–13 cm. long, reniform, definitely broader than long, with low rounded or obsolete lobes; seeds smooth or merely ridged, about 0.5 mm. long.
    - Petioles and flowering branches villous; freshly expanded calyx 1.5–2 mm. long, its tube obsolete; capsule 3–4.5 mm. long. . . . . 2. *H. parviflora.*
    - Petioles and flowering branches glandular-puberulent; freshly expanded calyx 2–4 mm. long, its tube 0.3 mm. long; capsule 4.2–5.5 mm. long. . . . . 3. *H. puberula.*
- *b.* Outside of slightly irregular flower minutely glandular-puberulent, without villosity; newly expanded calyx with tube 1–1.6 mm. long; capsule 5–7 mm. long.
  - Lower surface of leaf usually strigose; bractlets subulate, minute; freshly expanded calyx 2.5–5.5 mm. long, its barely oblique tube about 1 mm. long; capsule ovoid; seeds unsymmetrically ellipsoid, with 1 edge straight. . . . . 4. *H. americana.*
  - Lower surface of leaf glabrous; bractlets elliptic-lanceolate, evident; freshly expanded calyx 4.7–6.6 mm. long, its oblique tube 1.6 mm. long; capsule ellipsoid; seeds ovoid. . . . . 5. *H. hispida.*

a. Calyx in fresh anthesis 5–10 mm. long, irregular, the tube 2.5–5 mm. long; stamens included or but slightly exserted. . . *d.*
- *d.* Petioles and base of inflorescence puberulent to nearly glabrous; longer side of only slightly irregular calyx-tube of freshly expanded flower 2–4 mm. long; petals rhombic-spatulate, with undulate or toothed limb.
  - Bracts of inflorescence somewhat scarious; calyx-tube only slightly gibbous at base; styles and free beaks of carpels 6–8 mm. long during anthesis, long-exserted in fruit; seeds gibbous-ovoid. . . . . 6. *H. pubescens.*
  - Bracts herbaceous; calyx-tube strongly gibbous at base; styles and free beaks of carpels during anthesis 4–5 mm. long, wholly included in fruit; seeds ellipsoid. . . . . 7. *H. longiflora.*

---

* Treatment condensed in part from that of Rosendahl, Butters and Lakela, Monograph of the Genus Heuchera (1936).

d. Petioles and base of inflorescence hirsute; longer side of strongly irregular calyx-tube of freshly expanded flower 4–5 mm. long; petals spatulate, glandular-ciliate; seeds ovoid, with 1 or 2 flattened faces. . . . . 8. *H. Richardsonii.*

1. H. villòsa Michx. (long-hairy). — Acaulescent or nearly so from strong rhizome; *leaves* thin, *broadly cordate-ovate to -orbicular,* 0.7–3 *dm. long, acutely angulate-lobed and toothed,* coarsely strigose to glabrate above, more or less pubescent beneath, the petiole rusty-villous to glabrate; flowering stems rusty-villous, 2–8 dm. high; *bracts* of the loose to close panicle *linear to spatulate;* axis of inflorescence, peduncles and pedicels minutely glandular-hirsute or puberulent; flowers greenish to dull pink; *in anthesis the calyx* 1.5–3 *mm. long, regular, villous,* the *free upper half of calyx-tube then about* 0.5 *mm. long;* petals white to pinkish, twice or thrice length of calyx-lobes, often linear and twisted, or oblanceolate; stamens long-exserted; filamentous styles and free beaks of carpels together about 3 mm. long; capsule thick-ovoid, 3.5–5.5 mm. long, the beaks exserted; *seeds echinate,* fusiform, *about* 0.8 *mm. long.* Two well defined vars.:

Rhizome 5–9 mm. thick; petiole loosely villous; leaf-blade deeply and sharply lobed with the terminal lobe acutely triangular, lower surface glabrous to sparsely pubescent, the principal veins bordered by appressed stiff setae; flowering stem loosely villous; panicle lax; bracts linear, inconspicuous, mostly entire; bractlets narrowly lanceolate to subulate, glabrous or sparingly glandular-ciliate at tip. . . . . *H. villosa* (typical).

Rhizome 1–2 cm. thick; petiole densely shaggy-villous; leaf-blades with low and broadly triangular lobes much broader than long, lower surface hirsute, the principal veins bordered by long and soft villi; flowering stem shaggily villous; panicle closer, the axis usually hirsutulous; bracts conspicuous, oblong to spatulate, at least the lower toothed; bractlets short and broad, densely long-ciliate. . . . . Var. *macrorhiza.*

H. villòsa (typical). — Damp rocks and rich wooded slopes, uplands of Va. to S.C., Tenn. and se. Mo. June–Aug.

Var. **intermèdia** Rosend., Butt. & Lak. (intermediate). — Transitional between the preceding and the following, with relatively broad and low lobes of leaves and lax inflorescence, is possibly recognizable, from w. Va., W.Va. and s. O. to Ga. and Ala.

Var. **macrorhìza** (Small) Rosend., Butt. & Lak. (large-rooted), *H. macrorhiza* Small — W.Va. to se. Mo., s. to Ga. and Ala.

2. H. parviflòra Bartl. (small-flowered). — Smaller than no. 1, with short rhizomatous crown; *leaf-blades round-reniform, definitely broader than long,* 3–13 *cm. long,* up to 2.2 dm. broad, *with very low rounded lobes or lobing obscure,* thin but firm, the *lower surface velvety* with white pubescence and long-villous along the blades, the petioles villous; *flowering stems* 0.5–4.5 dm. high, *villous;* panicle lax; bracts rhombic-lanceolate, laciniate, usually very villous; axis of inflorescence, peduncles and pedicels glandular-hirsute to villous; flowers extremely villous, the *calyx at anthesis* 1.5–2 *mm. long, the tube obsolete;* styles and free beaks of carpels together 2 mm. long in anthesis; *capsule* 3–4.5 *mm. long,* ovoid; *seeds smooth* but slightly ridged, 0.5 mm. long. — Moist rocks and shaded cliffs, sw. Va., w. N.C. and e. Ky.

Var. **Rugélii** (Shuttlw.) Rosend., Butt. & Lak. (for its discoverer, FERDINAND RUGEL, 1806–1878). — *Leaves very thin* and membranaceous, *sparsely hirsute beneath* except on the more or less villous veins. — Similar habitats, W.Va. to s. Ill., s. to N.C., Tenn., n. Ala. and se. Mo. July–Sept.

3. H. pubérula Mackenz. & Bush (with short pubescence). — Similar to no. 2; *petioles* closely *glandular-puberulent,* leaf-blades mostly 2.5–15 cm. broad; *flowering stem glandular-puberulent;* bracts lanceolate, ciliate; axis of inflorescence, etc. puberulent; *calyx in anthesis* 2–4 *mm. long, with tube* 0.3 *mm. long;* fresh styles and free beaks of carpels together 3.5 mm. long; *capsule* 4.2–5.5 *mm. long,* lance-ovoid. — Shaded calcareous bluffs, Ky., Mo. and Ark. Sept., Oct.

4. H. americàna L. (American), ROCK-GERANIUM. — Caudex stout; *leaves* deeply cordate, suborbicular to round-ovate to rounded-pentagonal, 0.4–1.5 cm. long, with 5–9 rounded or obtuse lobes and many blunt teeth, the upper surface finely strigose to glabrous, *the lower surface usually somewhat strigose;* flowering stem glabrous to short-hirsute, 0.4–1 m. high; cylindric to slightly tapering inflorescence lax, with glandular-puberulent to -hirtellous axis; upper bracts and *bractlets subulate; flowers glandular-puberulent outside,* greenish or red-tinged; *calyx in young anthesis* 2.5–5.5 *mm. long, nearly regular, the tube* abruptly flaring from summit of hypanthium, *about* 1 *mm. long;* petals usually purplish, sometimes pale, obovate to spatulate or lanceolate; stamens and styles long-exserted in anthesis; *capsule ovoid,* 5–7 *mm. long,* with persistent styles nearly as long; *seeds* 0.5–0.7 mm. long, *unsymmetrically ellipsoid, with* 1 *edge straight,* echinate. — Very variable, with 4 vars. recognized:

Petioles glabrous, puberulent or only sparsely hirsute at summit.
Principal leaf-blades cordate at base, with narrow sinus and rounded bases, suborbicular to round-ovate, thickish; perianth in fresh anthesis 3–5.5 mm. long. *H. americana* (typical).
Principal leaf-blades subtruncate at base, with broad and open sinus and straight-sided bases, membranaceous; perianth in fresh anthesis 3–3.5 mm. long. Var. *subtruncata.*
Petioles densely hirsute or villous.
Petioles villous or villous-hirsute the villi relatively loose and up to 5 mm. long, with few or no glands; plants of the interior.
Perianth in anthesis 3–3.5 mm. long; calyx-tube nearly regular; petals 1–1.5 mm. long, narrowly elliptic to short-spatulate. . . . . . . Var. *interior.*
Perianth in anthesis 4–4.5 mm. long; calyx-tube oblique; petals 2–3 mm. long, oblanceolate. . . . . Var. *hirsuticaulis.*
Petioles densely glandular-hirsute, with the crowded glands 0.1–1 mm. long; flowering perianth regular, 3–4.5 mm. long; petals linear-oblanceolate; plant of e. Va. . . . . Var. *heteradenia.*

H. americàna (typical), incl. var. *brevipetala* Rosend., Butt. & Lak. — Loamy woods and shaded calcareous slopes and rocks, Ct. to s. Ont. and Mich., s. to Ga., Ala. and Okla. Late April–June.

Var. **subtruncàta** Fern. (somewhat truncate). — Calcareous wooded slopes, se. Va. Early June.

Var. **intèrior** Rosend., Butt. & Lak. (of the interior). — Ind. and Ill., s. to Tenn. and Ark. Late April–June.

Var. **hirsuticaùlis** (Wheelock) Rosend., Butt. & Lak. (with hirsute stems), *H. hirsuticaulis* (Wheelock) Rydb. — Ind., Ill. and Mo.

Var. **heteradènia** Fern. (with variable glands). — Calcareous bluffs, lower James R., Va. Late May, June.

5. H. híspida Pursh (stiffly short-hairy). — Intermediate between nos. 4 and 6, differing from the former in its usually *glabrous lower leaf-surfaces; bractlets elliptic-lanceolate;* freshly expanded perianth 4.7–6.6 mm. long, the *somewhat irregular calyx-tube* 1–1.6 *mm. long; petals* equaling to twice as long as calyx-lobes, rhombic-spatulate, *erose or undulate;* stamens unequal; capsule ellipsoid, 6.5–8 mm. long; *seeds ovoid.* — Calcareous ledges, mts. of Va. and W.Va.

6. H. pubéscens Pursh (hairy). — Acaulescent or sometimes with cauline leaves, from a stout caudex; *leaves with glandular-puberulent petioles;* the cordate-rotund blades 4–11 cm. long, their 5–7 well developed lobes with abruptly pointed teeth, *lower surface puberulent; flowering stems puberulent,* 2.5–7.5 dm. high, usually somewhat leafy; panicle elongate-conical, with few-flowered short-peduncled cymules; *bracts and bractlets* conspicuous, *lanceolate, somewhat scarious; calyx* glandular-puberulent with reddish dots, greenish or suffused with purple, *in fresh anthesis* 5–9 *mm. long,* broadly campanulate, *slightly irregular,* the oblique *calyx-tube only slightly gibbous at base, the longer side* 2.5–4 *mm. long; petals* somewhat exceeding calyx-lobes, *rhombic-spatulate, with undulate margin; stamens about equaling petals; styles and free beaks of carpels during anthesis* 6–8 *mm. long, long-exserted in fruit; seeds gibbous-ovoid,* about 0.7 mm. long, echinate. — Loamy woods and rock-crevices, Pa., Md., W.Va. and w. Va. Late May–Aug.

Var. **brachyándra** Rosend., Butt. & Lak. (short-stamened). — Stamens about length of calyx-lobes, included or barely exserted. — W. Md. and W.Va. to N.C. and Ky.

7. H. longiflòra Rydb. (long-flowered). — Differing from no. 6 in its often nearly glabrous petioles and *crenate blades with glabrous lower surface;* flowering stems glabrous or nearly so; *bracts herbaceous; calyx-tube strongly gibbous at base; styles and free beaks of carpels during anthesis only* 4–5 *mm. long, wholly included in fruit; seeds ellipsoid.* — Upland woods, W.Va. and Ky. to Ala.

8. H. Richardsònii R. Br. (for its discoverer, Sir John Richardson, 1787–1865). — Caudex stout, often multicipital; *petioles* more or less *hirsute* with gland-tipped pale hairs; leaf-blades cordate-rotund, up to 1.3 dm. long, with about 9 rounded to subtruncate lobes, the upper surface mostly glabrous, lower surface hispid; *flowering stems hispid; panicle slenderly cylindric,* the few-flowered cymules on short ascending peduncles; bracts and bractlets more or less scarious; *flowers* greenish, *irregular,* the perianth in fresh anthesis 5–10 mm. long, glandular-puberulent, the *calyx-tube strongly oblique and gibbous at base, its longer side in anthesis usually 4–5 mm. long; petals spatulate,* equaling or barely exceeding calyx-lobes, *glandular-ciliate;* stamens and styles slightly or hardly exserted; capsule ellipsoid, with included or barely exserted beaks; *seeds ovoid, with* 1 *or* 2 *flattened faces,* about 0.75 mm. long. Four vars. with us:

Capsules included; stamens barely exserted; petals glandular but not papillose.
  Flowering stems and petioles hispid with hairs at most 1.5 mm. long; leaf-blades 2–6 cm. broad. . . . . *H. Richardsonii* (typical).
  Flowering stems and petioles densely hispid with hairs 2–3.5 mm. long; leaf-blades 4–8 cm. broad. . . . . Var. *hispidior.*
Capsules more or less exserted; stamens definitely exserted; petals both papillose and glandular.
  Newly expanded perianths 6–10 mm. long, the calyx-tube strongly oblique. . Var. *Grayana.*
  Newly expanded perianths 5–7 mm. long, the calyx-tube only moderately oblique. . . . . Var. *affinis.*

H. Richardsònii (typical). — Dry rocky or sandy prairies and hills, Mackenz. and Alta., s. to n. Minn., S.D. and Colo. June, July.

Var. **hispídior** Rosend., Butt. & Lak. (more hispid). — Wisc. to N.D. and Wyo., s. to Neb. and Colo.

Var. **Grayàna** Rosend., Butt. & Lak. (in honor of Asa Gray, 1810–1888). — Mich. to Minn., s. to Ind., Ill., Mo. and Kans. May–July.

Var. **affìnis** Rosend., Butt. & Lak. (related). — Mich. and Wisc., s. to Ind., Ill. and Mo.

8. MITÉLLA L. Miterwort. Bishop's-cap

Calyx short, adherent to the base of the ovary, 5-cleft. Petals 5, slender, pinnately cleft. Stamens 5 or 10, included. Styles 2, very short. Capsule short, 2-beaked, 1-locular, with 2 parietal or rather basal several-seeded placentae, 2-valved at the summit. Seeds smooth and shining. — Low and slender N.Am. and e. Asiatic perennials, with round-cordate alternate slender-petioled leaves on the rhizome or stolons, and naked or 2–few-leaved flowering stems. Flowers small, in a simple slender raceme or spike. Fruit soon widely dehiscent. (Diminutive of *mitra, a cap,* alluding to the form of the young fruit.)

1. M. diphýlla L. (two-leaved), Coolwort. — *Rhizome stoutish,* without slender stolons; *basal leaves cordate-ovate, lobed, with prolonged terminal lobe,* pubescent beneath, sparsely so above; *flowering stem* erect, 1–4.5 dm. high, *with* 2 *opposite,* or sometimes alternate *sessile or subsessile leaves,* occasionally with a 3rd slightly above the paired ones in forma **triphýlla** Rosend. (three-leaved), or leaves definitely petioled in forma **oppositifòlia** (Rydb.) Rosend. (opposite-leaved); raceme stiffly erect, 0.5–2 dm. long, with 5–20 flowers on stoutish pedicels; *flowers white, the petals with obliquely ascending segments.* — Loamy and rocky woods, sw. Que. and N.H. to Minn., s. to upland of S.C., Tenn., Miss. and Mo. Late April–early June.

M. prostràta Michx. (prostrate), perhaps a hybrid of nos. 1 and 2, was based on a plant with rhizome more slender than in no. 1, leaves lobed and acute as in no. 1, the 4 alternate and remote cauline ones extending from base to summit of flowering stem, with lowest flower in axil of uppermost leaf. (*M. nuda,* forma *prostrata* (Michx.) Rosend.) — Local, reg. of Lake Champlain, Vt., and in sw. Ct.

2. M. nùda L. (naked, from the usually leafless scapes). — *Rhizome and elongate creeping stolons filiform; leaves cordate-suborbicular to reniform,* deeply and doubly crenate, strigose above, only sparsely pubescent beneath; *flowering scapes filiform, naked* or sparingly leafy, 0.3–2 dm. high; racemes 1.5–12.5 cm. long, the 2–13 remote flowers on filiform pedicels; *flowers greenish yellow, the petals with long divergent segments.* — Cool or mossy woods or swamps, s. Lab. to Mackenz., s. to Nfld., N.S., N.E., Pa. n. O., Mich., Wisc., Minn., N.D. and Mont. May–Aug. (Asia)

9. CHRYSOSPLÈNIUM L. Golden Saxifrage. Dorine (Que.)

Calyx-lobes 4–5, blunt, yellow or greenish within. Stamens 4–10, very short, inserted on a conspicuous disk. Styles 2. Capsule inversely cordate or 2-lobed, flattened, very short, 1-locular, with 2 parietal placentae, 2-valved at the top, many-seeded. — Low and small smooth herbs of cold or temp. reg., with tender succulent leaves, and small solitary or leafy-cymed flowers. (Name compounded of the Greek *chrysos, gold,* and *splen, the spleen;* probably from some reputed medicinal qualities.)

1. C. americànum Schwein. (American), Water-mat, Water-carpet. — Stems slender, *decumbent* and forking; *leaves principally opposite*, roundish or somewhat cordate, obscurely crenate-lobed; *flowers distant*, inconspicuous, *nearly sessile*, greenish, tinged with yellow or purple. — Springheads, rills and cold wet places, Gaspé Pen., Que. to Minn., s. to N.S., N.E., Md., upland to Ga., O., Ind. and Ia. Mar.–June.

2. C. ioénse Rydb. (of Iowa). — *Flowering stems erect*, arising from tips of stolons of preceding year; *leaves alternate*, the *lower round-reniform* and 7–11-lobed, *with nearly closed sinus*, the upper flabelliform and 6–9-lobed, *the 2 upper basally adnate to branches of cyme; bracts* 3–5-lobed, *yellow; flowers golden-yellow*, 3.5–4.5 mm. broad; *stamens* 5–8. (*C. tetrandrum* of ed. 7, not Fries) — Wet moss, Arct. Am.; ne. Ia.; Alta. May–July. (Asia)

Subfam. III. PARNASSIOÌDEAE

10. PARNÁSSIA L. Grass-of-Parnássus. Bog-stars

Sepals 5, imbricated in the bud, slightly united at the base, persistent. Petals 5, spreading, imbricated in the bud, a more or less cleft gland-bearing staminodium at the base of each. Stamens 5, alternate with the petals, persistent. Ovary 1-locular, with 4 projecting parietal placentae; stigmas 4, sessile. Capsule 4-valved, the valves bearing the placentae on their middle. Seeds very numerous, anatropous. Embryo straight; cotyledons very short. — Perennial smooth herbs of the N. Hemisph. with entire leaves, and solitary flowers on long scape-like stems, which often bear a single sessile leaf. Petals white, with greenish or yellowish veins. (Named from Mount Parnassus, the *Grass of Parnassus* of Dioscorides thought by some early botanists to have been *P. palustris.*)

a. Leaves membranaceous; calyx-lobes herbaceous, ascending in flower, appressed to fruit, slenderly nerved; petals shorter than to thrice length of calyx-lobes; staminodia with fan-shaped or obovate blades, cleft to or below the middle into 3–17 slender cilia. . . *b.*
- *b.* Leaves not cordate; petals 3–8 mm. long, at most once-and-a-half the length of calyx-lobes; anthers 0.8–1.6 mm. long; staminodia 0.5–1.8 mm. broad, with 3–7 cilia.
  - Scapes naked; petals shorter than or barely equaling calyx-lobes; staminodia 0.5–0.6 mm. broad, with 3–5 cilia; capsule depressed at summit. 1. *P. Kotzebuei.*
  - Scapes usually bearing a sessile leaf below or near the middle; petals once-and-a-half the length of calyx-lobes; staminodia 1–1.8 mm. broad, with 5–7 cilia; capsule rounded at summit. . . . . 2. *P. parviflora.*
- *b.* Leaves cordate at base; petals 8–15 mm. long, twice as long as calyx-lobes; anthers 1.5–2.3 mm. long; staminodia 2–2.5 mm. broad, bearing 9–17 cilia. . . . . 3. *P. palustris.*

a. Leaves coriaceous; calyx-lobes coriaceous, blunt, reflexed in fruit, with strong ribs; petals three or more times length of calyx-lobes; staminodia of 3–5 nearly distinct subulate segments or prongs.
- Leaves oval to suborbicular; petals sessile.
  - Staminodia shorter than to about equaling stamens; blades of largest leaves 2–7 cm. long; northern species. . . . . 4. *P. glauca.*
  - Staminodia much longer than stamens; blades of largest leaves 5–10 cm. long; southern species. . . . . 5. *P. grandifolia.*
- Leaves reniform; petals abruptly contracted to claws. . . . . . . 6. *P. asarifolia.*

1. P. Kotzebùei Cham. (for Count Otto von Kotzebue, 1787–1846). — *Leaves* thin, *deltoid- or rhombic-ovate, rounded at base, but tapering to the dilated petiole*, blades 0.6–3 cm. long; *scapes* 1–5, usually *naked*, furrowed, 0.2–2.8 dm. high; calyx-lobes herbaceous, narrowly oblong-lanceolate to oblanceolate, ascending in flower and fruit; *petals* oblong to elliptic, 3–7 mm. long, *shorter than to about equaling calyx-lobes; anthers* 0.8–1 *mm. long; staminodia* narrow-based; *the dilated blade* 0.5–0.6 *mm. broad, bearing* 3–5 *cilia*, distinctly shorter than stamens; *capsule ovoid, papery, depressed at summit.* — Wet calcareous rocks, n. Lab. to nw. Nfld. and Gaspé Pen., Que.; Hudson Bay; Alaska to Wyo. and ne. Asia; Greenl. July, Aug.

2. P. parviflòra DC. (small-flowered). — *Leaves thin, oval or oblong, slender-petioled*, blades 0.6–3.5 cm. long; *scapes* 1–25, 0.2–3.3 dm. high, usually *bearing a single leaf* commonly below or near the middle; calyx-lobes herbaceous, linear-lanceolate to narrowly oblong, ascending in flower and fruit; *petals* oblong or elliptic, 3.5–8 mm. long, *equaling to one-half longer than calyx-lobes; anthers* 1–1.6 *mm. long; staminodia cuneate*, 1–1.8 *mm. broad, bearing* 5–7 *unequal long cilia*, shorter than stamens; *capsule rounded at summit.* — Wet calcareous soils, Nfld. to Alaska, s. to C.B., P.E.I., Rimouski Co., Que., Ont., Mich., Wisc., S.D., Colo. and Utah. July, Aug.

3. P. palústris L. (of marshes), var. **neogaèa** Fern. (of the New World). — *Leaves* thin, ovate to suborbicular, *cordate-based,* 0.7–3 cm. long; *scapes* 1–many, 0.8–3.5 dm. high, usually *bearing a single sessile* ovate *leaf* below the middle; calyx-lobes subherbaceous, linear to lance-oblong, subacute, 4–11 mm. long, ascending in flower, barely spreading in fruit; *petals oval,* 8–15 *mm. long, twice length of calyx-lobes,* faintly 5–9-nerved with pale lines, *marcescent; anthers* 1.5–2.3 *mm. long; staminodia dilated upward to a greenish blade* 2–2.5 *mm. broad, bearing* 9–17 *slender cilia,* nearly to quite equaling stamens; capsule ovoid. — Wet basic or calcareous soils, Lab. to Alaska, s. to nw. Nfld., n. Mich., n. Minn., N.D., Wyo. and Oreg. July, Aug. — Typical **P.** PALÚSTRIS of Eurasia and the Behring Sea region of Alaska has rounded cauline leaf; shorter and firmer sepals; petals with more nerves, and staminodia slender-clawed.

4. P. glaùca Raf. (blue-green; from the foliage). — *Leaves coriaceous,* round-ovate to oblong, rounded to subcordate at base, *the larger blades* 2–7 *cm. long;* scapes naked or with 1 sessile leaf slightly above the base, 1–6 dm. high; *calyx-lobes oblong to oval,* rounded at the somewhat hooded tip, *coriaceous,* with broad hyaline margin, spreading in flower, *reflexed in fruit,* 2.5–5.5 mm. long, with 3–7 strong parallel ribs; petals oblong to oval, 1–1.8 cm. long, conspicuously nerved; anthers 1.2–2.8 mm. long; *staminodia cleft nearly to base into* 3 *stout lance-subulate prongs, shorter than to about equaling stamens;* capsule conic-ovoid, firm. (*P. caroliniana* of ed. 7 and most auth., not Michx.) — Wet calcareous soils, Nfld. to Man., s. to N.B., N.E., Pa., O., Ind., n. Ill., Ia. and S.D. Late July–Oct.

5. P. grandifòlia DC. (large-leaved). — Similar to no. 4, coarser; *larger leaves* 5–10 *cm. long;* anthers about 3 mm. long; *staminodia with* 3–5 *prongs,* these *nearly filiform and much exceeding stamens and nearly equaling petals.* — Wet calcareous rocks, shores and meadows, Fla. to Tex., n. to Va., W.Va., Tenn. and Mo. Aug.–Oct.

6. P. asarifòlia Vent. (with leaves like Wild Ginger, *Asarum*). — Similar, scapes angled, 2–5 dm. high; *leaves reniform,* slender-petioled, the larger 5–10 cm. broad; *petals* oblong-elliptic, 10–18 mm. long, *abruptly contracted to a claw; staminodia* 3-pronged, *mostly shorter than stamens.* — Bogs, wet woods and rocky banks, Va. and W.Va. to Ga. and Ala. Aug.–Oct.

Subfam. IV. HYDRANGEOÌDEAE

11. PHILADÉLPHUS L. Mock-orange or "Syringa"

Calyx-tube turbinate or campanulate; the limb 4–5-parted, spreading, persistent, valvate in the bud. Petals rounded or obovate, large. Styles united below or nearly to the top; stigmas oblong or linear. Capsule 3–4-locular, splitting at length into as many valves. Seeds very numerous, with a loose membranaceous coat prolonged at both ends. — N.Am., Asiatic and s. European shrubs, with opposite often toothed leaves, no stipules and solitary, racemose or cymose-clustered showy white flowers. (Said to be named for *Ptolemy Philadelphus,* King of Egypt, 283–247 B.C.)

Flowers 1–4, terminal.
  Stigmas distinct; flowering hypanthium and calyx-lobes glabrous outside, fruiting hypanthium 0.9–1.2 cm. long, the lobes erect; petals 2–3 cm. long. . . 1. *P. inodorus.*
  Stigmas united; flowering hypanthium strigose-hirsute, in fruit 4–6 mm. long, the lobes divergent; petals 1–2 cm. long. . . . . 2. *P. hirsutus.*
Flowers in a definite raceme of 2 or more nodes.
  Flowering calyx and lower leaf-surfaces glabrous. . . . . 3. *P. coronarius.*
  Flowering calyx and lower leaf-surfaces pubescent. . . . . 4. *P. pubescens.*

1. P. inodòrus L. (odorless). — Shrub 2 or 3 m. high; *leaves* narrowly ovate, acuminate, those of fertile branches 3–7 cm. long and 1.5–3.5 cm. broad, *glabrous on both surfaces or* remotely strigose above or *strigose-pilose on larger nerves beneath, obscurely denticulate; flowers* 1–4, *terminating branchlets; hypanthium and calyx-lobes glabrous; fruiting hypanthium tapering to base,* 0.9–1.2 *cm. long, the ovate acuminate calyx-lobes ascending in fruit; petals* round-obovate about 2 *cm. long; stigmas distinct.* — Rocky slopes, stream-banks and river-bluffs, Fla. and Ala. to Va. and Tenn.; spread from cult. northw. Late May, June.

Var. **grandiflòrus** (Willd.) Gray (large-flowered). — *Leaves* of fertile branchlets 5–9 *cm. long* and 2–5.3 *cm. broad, saliently serrate-dentate,* with principal veins more prominent beneath; *calyx-lobes broadly lanceolate, taper-pointed; petals* 2.5–3 *cm. long.* (*P. grandiflorus* Willd.) — Along streams, of similar range; spreading from cult. northw. Late May, June.

2. P. hirsùtus Nutt. (hairy). — Smaller than no. 1; *young branchlets appressed-hirsute; leaves* of fertile branchlets oblong or narrowly ovate, acuminate, sharply serrate, *densely pilose-hirsute beneath;* flowers smaller; *flowering hypanthium strigose-hirsute, in fruit* glabrescent, *campanulate,* 4–6 *mm. long, the calyx-lobes divergent; petals* obovate, 1–2 *cm. long; stigmas united* until dehiscence of capsule. — Rocky slopes, cliffs, river-banks, etc., Ga. and Ala., n. to N.C. and Ky. May, June.

3. P. coronàrius L. (suitable for a wreath). — Resembling no. 1; the oblong-ovate *leaves glabrous* except on veins beneath; *flowers 5–7 in definite racemes*, 2.5–3.5 cm. broad, *very fragrant;* calyx glabrous. — Roadside-thickets, occasional, spread from cult. June, July (Introd. from Eu.)

4. P. pubéscens Loisel. (hairy). — Differing from no. 3 in its closer bark; *leaves* more elliptic, *gray beneath with close pubescence; calyx pubescent; flowers odorless or only slightly fragrant.* (*P. verrucosus* and *P. latifolius* Schrad.) — Sandstone- and limestone-bluffs and river-banks, Tenn. and s. Ill., s. to Ala. and Ark. Late May, June.

Other species spread slightly from cult.; for these see works on cult. plants.

Two species of Deùtzia Thunb. (genus named in 1781 for *Jan van der Deutz*, patron of *Thunberg*), differing from *Philadelphus* in paniculate or paniculate-cymose inflorescences, calyx-teeth, petals and styles 5, stamens mostly only 10, are beginning to spread to thickets and roadsides from cult.: **D. scàbra** Thunb. (harsh) with stellate-pubescent and scabrous leaves, and **D. gràcilis** Sieb. & Zucc. (slender), more slender, with essentially glabrous foliage; both introd. from e. Asia.

12. DECUMÀRIA L. Climbing Hydrangea

Flowers all fertile. Calyx-tube turbinate, 7–10-toothed. Petals oblong. Stamens 20–30. Styles united into one, persistent. Stigma thick, 7–10-rayed. Capsule 10–15-ribbed, 7–10-locular, many-seeded, bursting at the sides, the thin partitions at length separating into numerous chaffy scales. — Smooth climbing shrubs, ours and one in China, with ovate or oblong entire or serrate leaves, no stipules, and numerous fragrant white mist-like flowers in compound terminal cymes. (Name said to be derived from *decimarius*, relating to tenths or tithes, referring to the often 10-merous flowers.)

1. D. bárbara L. (of Barbary; the species originally supposed to be African), Wood-vamp. — Leaves shining, sometimes pubescent; capsule with the persistent style and stigma urceolate, pendulous. — Rich woods and swamps, Fla. to La., n. to se. Va. and Tenn. May, June.

13. HYDRANGÈA L. Hydrangea

Calyx-tube hemispherical, 8–10-ribbed, adherent to the ovary; the limb 4–5-toothed. Petals ovate, valvate in the bud. Stamens 8–10, slender. Capsule 15-ribbed, 2-locular below, many-seeded, opening by a hole between the 2–4 diverging styles. — Am. and Asiatic shrubs, with opposite petioled exstipulate leaves. The marginal flowers of the compound cymes usually sterile and frequently radiant, then consisting merely of a showy membranaceous and colored flat and dilated 3- or 4-lobed calyx. (Name from the Greek *hydor*, *water*, and *aggeion*, *a vessel*, from the shape of the capsule.)

Flowers in flat- to round-topped corymbs. . . . . 1. *H. arborescens.*
Flowers in elongate panicles. . . . . 2. *H. paniculata.*

1. H. arboréscens L. (becoming tree-like), Wild H., Seven-bark. — Shrub up to 3 m. high; young branchlets glabrous or slightly pubescent; leaves ovate to ovate-oblong, pointed, serrate, glabrous, or pubescent beneath, *the long-petioled blades mostly* 0.6–1.8 *dm. long; corymb flat- or convex-topped.* — The following varieties and forms are recognized:

Lower surfaces of leaves glabrous, pale green.
  Principal leaf-blades broadly ovate to suborbicular, cordate or broadly rounded at base, the better developed ones two-thirds as broad to as broad as long.
    All or all but marginal flowers perfect. . . . . *H. arborescens* (typical).
    All or essentially all flowers sterile and radiant. . . . . Forma *grandiflora.*
  Principal leaf-blades gradually rounded to tapering at base, narrowly ovate to lance-elliptic or -oblong, the better developed ones one-third to two-thirds as broad as long.
    All or all but marginal flowers perfect. . . . . Var. *oblonga.*
    All or essentially all flowers sterile and radiant. . . . . Forma *sterilis.*
Lower surfaces of leaves whitened with minute pubescence; the principal blades broadly ovate to suborbicular, cordate or with broadly rounded base.
  All or all but marginal flowers perfect. . . . . Var. *Deamii.*
  All or essentially all flowers sterile and radiant. . . . . Forma *acarpa.*

H. arboréscens (typical). — Rich woods, calcareous rocky slopes, banks of streams, etc., Ga. to Okla., n. to s. N.Y., W.Va., O., Ind., Ill. and Mo. June, early July. — Forma **grandiflòra** (E. G. Hill) Rehd. (large-flowered), local.

Var. **oblónga** T. & G. (oblong). — Generally more common, n. Fla. to La. and Okla., n. to N.Y., W.Va., O., Ind., Ill. and Mo. — Forma **stérilis** (T. & G.) St. John (sterile), local.

Var. **Deàmii** St. John. (for its discoverer, CHARLES CLEMON DEAM, 1865–), *H. cinerea* Small. — Ga. to Okla., n. to W.Va., Ind., Ill. and Mo. — Forma **acárpa** (T. & G.) St. John (without carpels), local.

2. H. PANICULÀTA Sieb. (paniculate). — Shrub or small tree with pubescent branchlets; *leaves short-petioled, the oblóng to elliptic-ovate blades mostly* 4–12 *cm. long; panicle elongate*, up to 2.5 dm. long; sterile flowers showy. — Wooded swamp, abundant and freely seeding, Middlesex Co., Mass.; roadsides, etc., esc. from cult. elsewhere. Aug., Sept. (Introd. from Asia)

H. QUERCIFÒLIA Bartr. (oak-leaved), with large coarsely lobed leaves lanate beneath and elongate panicles, spreads from cult. northw. to Ct. (Introd. from the South)

SCHIZOPHRÁGMA (cleft wall) HYDRANGEOÌDES Sieb. & Zucc. (like *Hydrangea*), the CLIMBING HYDRANGEA of cult., climbing by aërial roots, with coarsely toothed rounded leaves, connate styles, the sterile flowers with a single large sepal, occasionally spreads from cult. (Introd. from Japan)

Subfam. V. ESCALLONIOÌDEAE

14. ÌTEA L. VIRGINIA-WILLOW. SWEET-SPIRES

Calyx 5-cleft, free from the ovary or nearly so. Petals 5, lanceolate, much longer than the calyx and longer than the 5 stamens. Capsule oblong, 2-grooved, 2-locular, tipped by the 2 united styles, 2-parted (septicidal) when mature, several-seeded. — Shrubs, ours and several in se. Asia, with simple alternate petioled exstipulate leaves and small white flowers in simple racemes. (Greek name of the Willow.)

1. I. virgínica L. (Virginian), TASSEL-WHITE. — Leaves deciduous, oblong, pointed, minutely serrate, at flowering time 5–8 cm. or more long; racemes (0.1–)0.5–2 dm. long, loose and open, the rachis and pedicels evident, or in the rare forma **abbreviàta** Fern. (abbreviated) the dense and compact racemes only 1–2.2 cm. long with hidden rachis and pedicels and the leaves only 1–2.5 cm. long. — Swamps, Fla. to Tex., n. to N.J., e. Pa., Ky., s. Ill., Mo. and Okla. May, June.

Subfam. VI. RIBESIOÌDEAE

15. RÌBES L. CURRANT. GADELLIER (Que.). GOOSEBERRY. GROSEILLIER (Que.)

Calyx 5-lobed, often colored; the tube adherent to the ovary. Petals 5, inserted in the throat of the calyx, small. Stamens 5, alternate with the petals. Ovary 1-locular, with two parietal placentae and 2 distinct or united styles. Berry crowned by the shriveled remains of the calyx. — Low N. and S.Am. and Eurasian sometimes prickly shrubs, with alternate palmately lobed leaves, which are plaited in the bud (except in subgen. *Symphocalyx*), often fascicled on the branches; the small flowers from the same fascicles, or from separate lateral buds. (Name said by Alphonse DeCandolle to come from the old Danish colloquial *ribs* for the Red Currant.)

a. Stems usually with firm spines at the nodes, at least the young stems usually bristly on the internodes; peduncles 1–5-flowered; pedicels not jointed at summit, the fruits not disarticulating from them; calyx with slender to campanulate tube, the lobes longer than broad. Subgen. GROSSULARIA (*Grossularia* Mill.), the GOOSEBERRIES. . . *b.*

b. Calyx-lobes much shorter than the campanulate tube; peduncle and pedicels elongate; ovary and fruit usually covered by glandless prickles. . . . 1. *R. cynosbati.*

b. Calyx-lobes equaling or longer than tube; ovary and fruit not prickly, either smooth, pubescent or bristly. . . . *c.*

c. Stamens much exceeding calyx-lobes during anthesis; peduncles or pedicels elongate.

Nodal spines 7–17 mm. long; flowers whitish, 9–12 mm. long; filaments 1–1.5 cm. long. . . . . 2. *R. missouriense.*

Nodal spines 2–5 mm. long; flowers greenish or purplish, 5–6 mm. long; filaments 4–7 mm. long. . . . . 3. *R. rotundifolium.*

c. Stamens shorter than to barely exceeding calyx-lobes; peduncles and pedicels short. . . *d.*

d. Calyx glabrous; fruit glabrous or only sparsely setose; indigenous. . . *e.*

e. Bark of middle and upper internodes of fruiting canes rarely bristly; leaf-blades and bracts of inflorescence not glandular; stamens distinctly exceeding petals. . . . . 4. *R. hirtellum.*

e. Bark of fruiting canes usually bristly; leaf-blades and bracts of inflorescence glandular; stamens and petals subequal.

Calyx-tube campanulate; ovary and fruit glabrous. . . . . . 5. *R. oxyacanthoides.*

Calyx-tube slenderly cylindric; ovary and fruit glandular-setose or glabrous. . . . . 6. *R. setosum.*

d. Calyx pilose; ovary closely pilose or glandular; fruit pubescent or glabrate; introduced. . . . . 7. *R. Grossularia.*

a. Stems with or without nodal spines and bristly cortex; peduncles bearing several–many-flowered racemes; pedicels jointed at summit, the fruits disarticulating from them. The CURRANTS. . . *f.*

f. New and old canes bristly and prickly; pendulous racemes and black fruits glandular-bristly; flowers saucer-shaped. Subgen. GROSSULARIOIDES. 8. *R. lacustre.*

f. New and old canes without bristles and prickles; racemes ascending to pendulous. Subgen. RIBESIA. . *g.*

g. Calyx saucer-shaped, rotate or campanulate, without elongate tube; leaves plicate in bud. . *h.*

h. Leaf-blades without resinous atoms; calyx flattened, saucer-shaped or rotate; fruit red (rarely yellow). . *i.*

i. Racemes ascending; ovary and fruit glandular-bristly; flower whitish to roseate. § HERITIERA. . . . . 9. *R. glandulosum.*

i. Racemes spreading to drooping; ovary and fruit glabrous; flowers purplish to green or yellowish. § RIBESIA.

Decumbent shrub; middle lobe of leaf deltoid; pedicels with capitate glands; flowers purplish. . . . . 10. *R. triste.*

Erect or ascending; middle lobe of leaf ovate; pedicels without glands; flowers yellowish or greenish. . . . . 11. *R. sativum.*

h. Leaf-blades, at least beneath, ovaries and fruits sprinkled with resinous atoms; calyx campanulate or cupuliform; fruit black. § EUCOREOSMA. . . *j.*

j. Racemes erect or ascending; calyx white, the cupuliform or open campanulate tube much shorter than the lobes. . . . . . . 12. *R. hudsonianum.*

j. Racemes spreading to pendulous; calyx campanulate, greenish to yellowish, the tube about equaling the lobes.

Bracts longer than pedicels; flowers yellowish and whitish, tubular-campanulate, glabrous, 8–10 mm. long. . . . . . . 13. *R. americanum.*

Bracts shorter than pedicels; flowers greenish, purplish or dull white, broadly campanulate, pubescent, 5–6 mm. long. . . . . 14. *R. nigrum.*

g. Calyx salverform, the long slender tube several times longer than the spreading lobes, golden-yellow; fruit yellow or black; leaves convolute in bud. Subgen. SYMPHOCALYX. . . . . 15. *R. odoratum.*

Subgen. GROSSULÀRIA (Mill.) Richard (GOOSEBERRIES)

1. R. cynósbati L. (dogberry), PRICKLY G., DOGBERRY. — Infra-axillary *spines slender*, 0.5–1 cm. long, or wanting in forma **inérme** Rehd. (unarmed); bark of internodes without bristles; *leaves* round-ovate, rounded or subcordate at base, *soft-pubescent;* racemes loose, long-peduncled, long-pedicelled, 2–6 cm. long; stamens and *undivided style* not longer than the *broadly campanulate* green *calyx; berries* large, *armed with long stiff prickles* or rarely smooth. (*Grossularia* (L.) Mill.) — Open loamy or rocky woods, w. N.B., w. Me. and s. Que. to Man., s. to w. N.C., n. Ala. and Mo. *Fl.* May, June; *fr.* July–Sept.

Var. **glabràtum** Fern. (becoming smooth). — Leaves promptly glabrate or only sparingly pilose on nerves beneath. — Western Va. and N.C. to O. and Mich.

Var. **átrox** Fern. (cruel). — Coarser, the fruiting (and young) canes heavily prickly with dark stiff bristles. — Manitoulin I., Ont.

2. R. missouriénse Nutt. (of Missouri), MISSOURI G. — *Spines* often long (7–17 mm.), *stout and red; peduncles long and slender; flowers white* or whitish; *filaments* capillary, 1–1.5

cm. long, generally connivent or closely parallel, soon *conspicuously longer than the oblong-linear calyx-lobes.* (*R. gracile* Pursh, not Michx.) — Open woods, thickets, fencerows etc., Ct. to Minn. and S.D., s. to Tenn., Ark. and Kans., rare eastw. *Fl.* April, May; *fr.* late June–Sept.

3. R. rotundifòlium Michx. (round-leaved). — Spines short (2–5 mm. long); *leaves* rather firm, sparingly pilose beneath, *mostly rounded at base; peduncles short; flowers greenish* or the lobes dull purplish; *filaments* slender, 4–7 *mm. long,* more or less *exceeding the narrowly oblong-spatulate calyx-lobes.* (*Grossularia* (Michx.) Cov. & Britt.) — Open rocky places and thickets, ascending to highest crests, w. Mass. to. W.Va., s. to mts. of N.C. *Fl.* April–early June; *fr.* June–Sept.

4. R. hirtéllum Michx. (bristly), FAUSSE-ÉPINE (Que.). — *Bark* freely exfoliating, that of new canes commonly prickly, *of fruiting canes mostly without prickles on the middle and upper internodes;* nodal spines 3–8 mm. long; *leaves* thin but firm, glabrescent and *without glands on the surfaces,* usually pilose on the nerves beneath, the blades cuneate to cuneate-truncate at base, *the petioles bearing long plumose trichomes;* peduncles very short; *bracts* ciliate, *without marginal glands;* flowers greenish-yellow to dull purple; *stamens exceeding petals;* ovary and fruit glabrous. (*Grossularia* (Michx.) Spach; *R. oxyacanthoides* of ed. 7 in large part, not L.) — Rocky or swampy woods and clearings, s. Lab. to e. Man., s. to Nfld., N.S., N.E., Pa., W.Va., n. O., n. Ind., n. Ill. and Minn. *Fl.* late April–July; *fr.* June–Sept.

Var. **saxòsum** (Hook.) Fern. (of stony places). — *Leaf-blades of fertile branches strongly rounded to cordate at base.* — W. Nfld. to s. Man., s. to N.S., s. N.B., se. Me., Mich., Minn. and N.D.

Var. **calcícola** Fern. (growing in lime). — *Leaves* subtruncate to cordate, *densely soft-pubescent beneath;* calyx often hirtellous. — Calcareous or other soils, s. Lab. Pen., Anticosti I. and Gaspé Pen., Que., to James Bay, s. to N.S., se. Mass. and s. R.I. (less characteristic), Ont., Mich. and Wisc.

5. R. oxyacanthoìdes L. (like *Oxyacantha,* hawthorn).— Resembling no. 4, but bark closer, the *internodes of the* fruiting *canes bristly; leaves* subtruncate to cordate at base, *with glands on the lower surface* mixed with pubescence; *bracts of inflorescence glandular-ciliate; calyx purplish, its tube campanulate and nearly or quite as broad as long;* stamens hardly longer than petals; fruit smooth. (*Grossularia* (L.) Mill.) — Hudson Bay to Yuk., s. to n. Mich., n. Minn., S.D. and Mont.

6. R. setòsum Lindl. (bristly). — Similar to no. 5; nodal spines up to 2 cm. long; *calyx whitish, its tube slenderly cylindric; ovary and fruit often glandular-hirtellous.* (*Grossularia* (Lindl.) Cov. & Britt.) — Rocky slopes, Alta. to Wyo., e. to Man., James Bay, Ont., n. Mich., Wisc. and Neb.

7. R. GROSSULÀRIA L. (old generic name derived from the French *groseille*), European G. — Resembling no. 4; bark of fruiting canes without bristles on middle and upper internodes; *nodal spines* stout, 1–1.5 *cm. long;* leaf-blades rounded to cordate at base; peduncles 1- or 2-flowered; *calyx pilose; ovary closely pilose or glandular; fruit pubescent* or rarely glabrate. (*R. Uva-crispa* and *reclinatum* L.; *Grossularia Uva-crispa* (L.) Mill.) — Spread from cult. to thickets and roadsides. (Introd. from Eu.)

Subgen. GROSSULARIOÌDES Jancz. (CURRANTS)

8. R. lacústre (Pers.) Poir. (of lakes), BRISTLY BLACK C. — *Young stems clothed with bristly prickles* and with weak thorns; *leaves* heart-shaped, 3–5-parted, *with* the *lobes deeply cut; racemes loosely spreading or drooping,* the *rachis, pedicels and ovary glandular-bristly;* calyx broad and flat; stamens and style not longer than the petals; *fruit bristly, purplish-black.* — Cold woods and swamps, Nfld. to Alaska, s. to N.S., n. N.E., w. Mass., N.Y., mts. to Tenn., n. O., Mich., Wisc., Minn., Colo., Utah and Calif. *Fl.* mid-May–Aug.; *fr.* July–Sept. — Bruised shrub and berries with skunk-like odor.

Subgen. RIBÈSIA Berl. (CURRANTS)

§ HERITIÈRA Jancz.

9. R. glandulòsum Grauer (glandular), SKUNK-C. — Stems reclining to ascending, smooth; leaves deeply cordate, smooth, 5–7-lobed, the ovate lobes acute and doubly serrate; *racemes ascending,* slender, *the bracts, pedicels, ovaries and red berries glandular-hispid; calyx flattish, rotate, whitish to roseate.* (*R. prostratum* L'Hér.) — Wet woods or clearings and rocky slopes, Lab. to Mackenz. and n. B.C., s. to Nfld., N.S., N.E., N.Y., mts. to N.C., n. O., Mich., Wisc., Minn. and Sask. *Fl.* May–Aug.; *fr.* June–Sept. — Bruised shrub and berries with odor of skunk.

§ Ribèsia Berl.

10. R. trìste Pall. (sad; probably from drooping racemes), Red C. — *Straggling or reclining*, the branches often rooting freely; *leaves* somewhat cordate, the mature blades 5–10 cm. broad, the *sides nearly parallel, the lobes mostly broad-deltoid*, glabrous to permanently white-tomentose beneath; *racemes borne on the old wood chiefly below the leafy tufts*, drooping, 3.5–9 cm. long; pedicels mostly glandular; *calyx smoke-color to purplish;* the segments broadly cuneate to subrhombic, as broad as or broader than long; petals broadly cuneate; disk a low broad pentagon; style deeply cleft; fruit mostly small and hard. (Incl. var. *albinervium* (Michx.) Fern.) — Cool woods, swamps and subalpine ravines, s. Lab. to Alaska, s. to Nfld., N.S., n. N.E., w. Mass., n. N.J., ne. Pa., centr. and w. N.Y., W.Va., s. Ont., Mich., Wisc., Minn., S.D. and Oreg. *Fl.* late April–July; *fr.* June–Aug. (Asia)

11. R. satìvum Syme (planted), Garden-C., Red C. — *Suberect; leaves* mostly cordate, slightly pubescent beneath or glabrate, the mature blades 3.5–6.5 cm. wide, *broadened upward,* 3–5-lobed, *the lobes mostly short-ovate; racemes borne chiefly among the leafy shoots,* spreading in anthesis, drooping in fruit, 3–5 (becoming 7) cm. long, the rachis glabrous, though often glandular; pedicels mostly glandless; *calyx yellow-green,* its segments oval and abruptly narrowed below the middle; petals narrowly cuneate; disks between the stamens and the slightly cleft style a high narrow ring with round-scalloped margin; fruit plump and juicy. (*R. vulgare* of ed. 7, not Lam.; *R. rubrum* of gardens, not L.) — Commonly cult., frequently spread to open woods and thickets. (Introd. and natzd. from Eu.)

§ Eucoreósma Jancz.

12. R. hudsoniànum Richards. (of the Hudson Bay region). — Erect or ascending; *leaves* reniform-ovate, cordate, 3–5-lobed, pubescent and *with resinous atoms beneath; racemes erect or ascending, with short caducous bracts; calyx* 4–5 *mm. long, white, the ascending lobes much exceeding the open-campanulate or cupuliform tube;* berries black. (*R. rigens* sensu Rouleau, not Michx., which is no. 9) — Swampy woods and rocky slopes, n. Que. to Alaska, s. to Mich., Wisc. ne. Ia., s. Man., Sask., Wyo., Utah and Oreg. *Fl.* May–early July; *fr.* July, Aug.

13. R. americànum Mill. (American), Wild Black C. — Ascending shrub up to 1.5 m. high; pubescent branchlets and 3–5-lobed doubly serrate suborbicular leaves resinous-dotted; *racemes drooping, downy,* the *elongate bracts persistent; flowers large, yellow and whitish; the tubular-campanulate glabrous calyx* 8–10 *mm. long,* the tube and lobes subequal; fruit black. (*R. floridum* L'Hér.) — Rich thickets and slopes, w. N.B. to Alta., s. to s. N.E., Del., w. Va., O., Ind., Ill., Mo., Neb. and N.M. *Fl.* late April–June; *fr.* June–Sept.

14. R. nìgrum L. (black), Black C. of gardens; Cassis (Que.). — Similar, but the *pubescent calyx* 5–6 *mm. long, the tube broadly campanulate, greenish-purple and dull whitish.* — Cult., and occasionally escaping to thickets, etc. (Introd. from Eu.)

Subgen. Symphócalyx Berl.

15. R. odoràtum Wendland f. (fragrant), Missouri or Buffalo-C. — Tall spineless shrub; *leaves* 3–5-lobed, rarely at all cordate, *convolute in bud;* racemes short; *flowers golden-yellow, spicy-fragrant; tube of salverform calyx three or four times longer than the oval lobes;* stamens short; berries black, or yellow in forma **xanthocàrpum** Rehd. (yellow-fruited). (*R. aureum* of ed. 7, not Pursh) — Rocky bluffs and slopes, S.D. to Tex., e. to Minn., Mo. and Ark.; and spread from cult. rather frequently e. to N.E., etc. *Fl.* late April–June; *fr.* June–Aug.

FAM. 80. HAMAMELIDÀCEAE (Witch-hazel Family)

Shrubs or trees with alternate simple leaves and deciduous stipules; flowers in heads or spikes, often polygamous or monoecious; the calyx adhering to the base of the ovary, which consists of 2 pistils united below and forming a 2-beaked 2-locular woody capsule opening at the summit, with a single bony seed in each locule, or several, only one or two of them ripening. — Petals inserted on the calyx, narrow, valvate or involute in the bud, or often none. Stamens twice as many as the petals and half of them sterile and changed into scales, or numerous. Seeds anatropous. Embryo large and straight, in scanty albumen; cotyledons broad and flat.

Leaves pinnately veined; flowers with definite calyx, perfect or polygamous, not in globular heads; capsules distinct, each locule with a single wingless bony seed.

Petals 4, linear; stamens 4, alternating with 4 staminodia; flowers few in axillary clusters. . . . . 1. *Hamamelis.*
Petals 0; stamens about 24; flowers in terminal spikes or heads. . . . . . . 2. *Fothergilla.*
Leaves palmately veined and lobed; flowers with rudimentary calyx and no petals, monoecious, in globular heads; capsules fused at base into a globular echinate head, each capsule with many fine winged seeds. . . . . 3. *Liquidambar.*

1. HAMAMÈLIS L. Witch-hazel

Flowers in small axillary clusters or heads, usually surrounded by a scale-like 3-parted involucre. Calyx 4-parted and with 2 or 3 bractlets at its base. Petals 4, liguliform, long and narrow, spirally involute in the bud. Stamens 8, very short; the 4 alternate with the petals anther-bearing, the others imperfect and scale-like. Styles 2, short. Capsule opening loculicidally from the top; the outer coat separating from the inner, which incloses the single large and bony seed in each locule, but soon bursts elastically into two pieces. — Tall shrubs or small trees of e. N. Am. and e. Asia, with straight-veined leaves and usually yellow perfect or polygamous flowers. (Ancient Greek name applied to the Medlar (*Mespilus*) or some similar tree.)

Flowering in autumn; petals 1.5–2 cm. long. . . . . 1. *H. virginiana.*
Flowering from midwinter to spring; petals 1–1.5 cm. long. . . . . 2. *H. vernalis.*

1. H. virginiàna L. (Virginian), Café du diable (Que.). — Coarse shrub or small tree up to 5 m. high, *not suckering; branchlets glabrous or sparsely pubescent and glabrate;* leaves obovate or oval to suborbicular, obliquely rounded or subcordate at base, wavy-toothed, 5–7-nerved, the mature ones 5–15 cm. long, glabrous beneath or sparsely pilose on the nerves; *flowers autumnal,* on short pedicels; *petals* yellow, 1.5–2 *cm. long; calyx-lobes* (fresh) *brownish-yellow inside;* capsules obovoid, 1–1.5 cm. long, densely pubescent, united nearly to middle with calyx-tube. — Dry or moist woods, s. Que. to Minn., s. to N.S., N.E., Ga., Tenn. and Mo. Sept.–Nov.

Var. parvifòlia Nutt. (small-leaved). — *Leaves thick and coriaceous,* mostly 3.5–10 cm. long, *densely stellate-tomentose and whitened to rufescent beneath.* — N.S. and N.E. to O. and mts. of Pa.; se. Va. to w. S.C. and La.

2. H. vernàlis Sarg. (vernal). — Shrub up to 2 or 3 m., *freely sprouting* from base; *branchlets densely stellate-tomentose;* leaves rounded-obovate to elliptic, cuneate to rounded at base, 5–12 cm. long, 4–6-nerved, pale or glaucous and glabrous or glabrate beneath, or permanently stellate-pubescent beneath in forma tomentélla Rehd. (minutely tomentose); *flowers in winter and early spring; petals* 1–1.5 *cm. long,* yellow, often reddish at base, or red throughout in forma cárnea Rehd. (flesh-colored); *calyx-lobes dark red within; flowers fragrant.* — Margins and shores of streams, s. Mo. and Okla. to La. and Ala. Jan.–Apr.

2. FOTHERGÍLLA Murr.

Flowers in a terminal ament-like spike, mostly perfect. Calyx campanulate, the summit truncate, slightly 5–7-toothed. Petals none. Stamens about 24, borne on the margin of the calyx in one row, all alike; filaments very long, thickened at the top (white). Styles 2, slender. Capsule adhering to the base of the calyx, 2-lobed, 2-locular, with a single bony seed in each locule. — Low shrubs of se. N. Am.; the oval or obovate leaves smooth, or hoary underneath, toothed at the summit; the flowers appearing rather before the leaves, each partly covered by a scale-like bract. (Dedicated to *John Fothergill,* 1712–1780, distinguished physician and botanist of London.)

1. F. Gárdeni Murr. (for its discoverer, Alexander Garden, 1730?–1791), Witch-alder. — Shrub up to 1 m. high; leaves obovate to elliptic, mostly 2–5 cm. long, tomentose beneath. — Sandy swamps, Ga. and Ala. to Va. Apr., May.

3. LIQUIDÁMBAR L. Sweet Gum

Flowers usually monoecious, in globular heads or aments; the staminate in a conical cluster, naked; stamens very numerous, intermixed with minute scales; filaments short. Pistillate inflorescence of many 2-locular 2-beaked ovaries subtended by minute scales in place of a calyx, all more or less cohering, and hardening in fruit, forming a spherical ament or head; capsules opening between the 2 subulate beaks. Styles 2, stigmatic down the inner side. Ovules many, but only one or two perfecting seeds. Seeds with a wing-angled seed-coat. — Aments in racemes, nodding, in the bud inclosed by a 4-leaved deciduous involucre. Two N. Am. and two Asiatic trees. (A mongrel name, from *liquidus, fluid,* and the Arabic *ambar, amber;* in allusion to the fragrant terebinthine juice or gum which exudes from the tree.)

1. L. Styracíflua L. (old generic name, meaning *flowering gum*), BILSTED. — Leaves rounded, deeply 5–7-lobed, smooth and shining, glandular-serrate, the lobes pointed. — Swampy woods, Fla. to Tex. and Mex., n. to s. Ct., se. N.Y., W.Va., s. O., s. Ind., s. Ill. and se. Mo. Apr., May. (Centr. Am.) — A large and beautiful tree, with fine-grained wood, the gray bark commonly with corky ridges on the branchlets, southward exuding a gum pleasant to chew. Leaves fragrant when bruised, turning deep crimson in autumn. The woody capsules filled mostly with abortive seeds, which resemble sawdust.

FAM. 81. PLATANÀCEAE (PLANE-TREE FAMILY)

Trees with watery sap, alternate palmately lobed leaves, sheathing stipules, and monoecious flowers in separate and naked spherical heads and with tiny and insignificant calyx and corolla; the fruit merely clavate 1-seeded nutlets, furnished with a ring of bristly hairs about the base. Only the following genus (of uncertain relationship).

1. PLÁTANUS L. SYCAMORE. BUTTONWOOD. PLANE-TREE

Staminate flowers of numerous stamens, with clavate little scales intermixed; filaments very short. Pistillate flowers in separate heads, consisting of inversely pyramidal ovaries mixed with tiny scales. Style rather lateral, subulate or filiform, simple. Nutlets coriaceous, small, tawny-hairy below, containing a single orthotropous pendulous seed. Embryo in the axis of thin albumen. — Large trees of N. Am. and Eurasia (India to se. Eu.), with the bark deciduous in broad thin brittle plates; dilated base of the petiole inclosing the bud of the next season. (The ancient name, from the Greek *platys, broad*, apparently referring to the large leaves.)

1. P. occidentàlis L. (western; *i.e.*, of Western Hemisphere). — Leaves broadly ovate, truncate or cordate at base, 3–5-lobed, with broad rounded shallow sinuses, or deeply lobed with long cuneate base in forma **attenuàta** Sarg. (tapering), the lobes broad, acuminate, serrate with long remote pointed teeth, under surface with pubescent nerves; pistillate heads 1 or 2, hanging on a long peduncle. — Rich soil, s. Me. to Ont. and Neb., s. to n. Fla., Ala., Miss. and Tex. Late Apr.–early June. — Our largest deciduous tree, up to 50 m. high, with trunks up to 4 m. in diameter.

Var. **glabràta** (Fern.) Sarg. (becoming smooth). — Leaves 3-lobed, mostly broader than long, broadly cuneate to subcordate, the sinuses rounded to acute, the entire lobes long-acuminate, glabrate. — Centr. Ia. and Mo. to w. Tex. and n. Mex.

FAM. 82. ROSÀCEAE (ROSE FAMILY)

Plants with regular flowers, numerous (rarely few) distinct stamens inserted on the calyx, and 1–many carpels, which are quite distinct, or (in the second tribe) united and combined with the calyx-tube. Ovules (anatropous) 1–few in each carpel; seeds almost always without albumen. Embryo straight, with large and thick cotyledons. Leaves alternate, with stipules; these sometimes caducous, rarely obsolete or wanting. — Calyx of 5 (3–8) sepals (the odd one superior) united at the base, often appearing double by a row of bractlets outside. Petals as many as the sepals (rarely wanting or by "doubling" becoming numerous), mostly imbricated in the bud and usually inserted with the stamens on the edge of a disk which lines the calyx-tube. Trees, shrubs or herbs.

a. Fruit dehiscent, a follicle or capsule, the superior ovary not inclosed in the calyx-tube nor adherent to it. Tribe I. SPIRAEEAE. . . *b.*

b. Leaves simple; stipules absent or caducous.

Carpels inflated, forming bladdery fruits, splitting into 2 valves; seeds plump, with crustaceous testa and copious albumen; leaves palmately nerved, the young with stipules; stamineal disk wanting. . . . . 1. *Physocarpus.*

Carpels not inflated, splitting down the ventral suture; seeds slender, with loose testa and no albumen; leaves mostly pinnately veined, without stipules; stamineal disk present. . . . . 2. *Spiraea.*

b. Leaves compound, stipulate.

Leaves 2–3-pinnate; carpels 3 or 4. . . . . 3. *Aruncus.*

Leaves simply odd-pinnate or 3-foliolate; carpels 5.

Suffruticose; leaves odd-pinnate, with 13–21 leaflets; flowers in dense panicles. . . . . 4. *Sorbaria.*

Herbaceous; leaves 3-foliolate; flowers loosely paniculate-corymbed. . 5. *Gillenia.*

a. Fruit or carpels indehiscent; leaves, at least when expanding, stipulate. . . *c.*

c. Ovary inferior, the 2–5 carpels usually connate, borne within and adnate to a cup-like or urn-like depression in the enlarged summit of the receptacle, the whole united to form a fleshy fruit (pome or berry); trees or shrubs, the stipules free from the petioles. Tribe II. POMEAE. . . *d.*

d. Mature carpels papery or soft-cartilaginous.

Carpels of the compound ovary as many as the styles, without false or partial partitions; leaves simple or pinnate; fruit a pome or berry-like. . . . . 6. *Pyrus.*

Carpels of the compound ovary subdivided by partial (false) partitions projecting inward from the back; leaves simple; fruit berry-like. . 7. *Amelanchier.*

d. Mature carpels hard and bony, 1–5, free or coherent in the pulpy fruit; spiny shrubs or trees with simple leaves.

Ovules in each locule of the carpel 1 or, if 2, unequal, 1 sessile and fertile, the other stalked and sterile; leaves deciduous, rarely evergreen, mostly serrate or dentate; mostly native. . . . . 8. *Crataegus.*

Ovules 2 in each locule, equal, fertile; leaves evergreen, crenulate; introd. . . . . 9. *Cotoneaster.*

c. Ovary or ovaries superior (in Tribes V and VI inclosed in the calyx-tube). . . *e.*

e. Ovaries several to many, borne on a broad to elongate receptacle, not inclosed by the calyx. . . *f.*

f. Carpels becoming dry achenes; plants herbaceous or, if woody, having bractlets at the sinuses of the calyx or the styles long and plumose. Tribe III. POTENTILLEAE. . . *g.*

g. Styles not elongate after anthesis, mostly deciduous. . . *h.*

h. Receptacle pulpy and greatly enlarged in fruit.

Bractlets of calyx similar to the narrow calyx-lobes; petals white or pinkish; receptacle becoming very juicy and edible. . . 10. *Fragaria.*

Bractlets much broader and longer than calyx-lobes; petals yellow; receptacle spongy, insipid and dryish. . . . . 11. *Duchesnea.*

h. Receptacle dry, not greatly enlarged in fruit. . . *i.*

i. Stamens 5.

Leaflets 3-toothed at apex; calyx flattish, without bractlets; stamens alternate with the yellowish petals. . . . . 12. *Sibbaldia.*

Leaflets dissected into linear segments; calyx turbinate, with bractlets; stamens opposite the white to purplish petals. 13. *Chamaerhodos.*

i. Stamens numerous.

Carpels 1-ovulate; calyx bracted or, if bractless, the flowers on a naked scape; stipules elongate.

Flowering stem scapose; bractlets small, at base of calyx and deciduous or wanting; petals yellow. . . 14. *Waldsteinia.*

Flowering stem bearing leaves; calyx with persistent bracts at sinuses; petals yellow, rarely white or purple. . . 15. *Potentilla.*

Carpels 2-ovulate; calyx without bractlets; stipules large, reniform; coarse plants with paniculate-cymose white or pink flowers and large pinnate leaves. . . . . 16. *Filipendula.*

g. Styles persistent and elongating after anthesis, plumose or jointed.

Shrubs with prostrate stems and simple leaves; flowers solitary at tips of scapes; receptacle flat. . . . . 17. *Dryas.*

Herbs with pinnate or lyrate basal leaves and upright leafy flowering stems; receptacle conical to cylindric. . . . . . 18. *Geum.*

f. Carpels ripening into drupelets; ovules 2, pendulous, but seed solitary. Tribe. IV. RUBEAE.

Carpels mostly numerous, ripening into juicy drupelets, crowded on a spongy receptacle (raspberries, blackberries, etc.); flowers mostly perfect and similar; herbaceous or mostly slightly woody. . . 19. *Rubus.*

Carpels few at the bottom of the calyx, nearly dry in fruit; creeping, with roundish leaves; showy flowers with white petals, sterile; fertile flowers apetalous, recurving and maturing fruit near the ground. . . . . 20. *Dalibarda.*

e. Ovaries 1-many, becoming achenes covered by the calyx, or drupes. . . *j.*

j. Ovary or ovaries (carpels) inclosed in the over-topping calyx; leaves compound or lobed. . . *k.*

k. Carpels 1–4, completely inclosed in the dry and firm calyx-tube, which is constricted or nearly closed at throat; herbs. Tribe V. POTERIEAE.

Leaves palmate or palmately lobed; calyx with bractlets alternating with the 4 lobes; petals none. . . . . 21. *Alchemilla.*

Leaves pinnate; calyx without bractlets; inflorescence spicate or spicate-racemose.
Calyx beset with hooked bristles, the 5-cleft limb closed after flowering and persistent; petals yellow. . . . . 22. *Agrimonia.*
Calyx not bristly, the 4 petaloid lobes white or purple, deciduous in fruit; petals none. . . . . 23. *Sanguisorba.*
k. Carpels numerous, lining the base or sides of the globose or urceolate fleshy calyx-tube, the latter ripening as a colored fruit suggesting a pome; petals large and showy; shrubs, often prickly, the pinnate leaves usually with decurrent stipules. Tribe VI. ROSEAE. . . . 24. *Rosa.*
j. Ovary free, solitary, becoming a drupe (plum, cherry, etc.); calyx deciduous, bractless. Tribe VII. PRUNEAE. . . . . 25. *Prunus.*

Tribe I. SPIRAÈEAE Camb.

1. PHYSOCÁRPUS Maxim. NINEBARK

Carpels 1–5, inflated, 2-valved; ovules 2–4. Seeds roundish, with a smooth and shining crustaceous testa and copious albumen. Stamens 30–40. Otherwise as *Spiraea.* — Shrubs, mostly N.Am., one in e. Asia, with simple palmately lobed leaves and umbel-like corymbs of white flowers. (Name from the Greek *physa, a pair of bellows,* and *karpos, fruit,* from the inflated carpels.)

1. **P. opulifòlius** (L.) Maxim. (with leaves of *Viburnum Opulus*). — Shrub, 1–3 m. high, with long branches, the old bark loose and separating in numerous thin layers; leaves roundish, somewhat 3-lobed and heart-shaped; pedicels and calyx glabrous to tomentulose; the purple-tinged membranaceous capsules usually 3, essentially glabrous, very conspicuous. (*Opulaster* Ktze.; *O. australis* Rydb.) — Shores, rocky banks and thickets, Que. to Hudson Bay, w. to Wisc. and s. to S.C., Tenn. and n. Ill.; often cult. and esc. May–July. Passing freely to

Var. **intermèdius** (Rydb.) Robins. (intermediate). — Capsules permanently pubescent. (*Opulaster intermedius* Rydb.) — Similar situations, w. N.Y. to Minn. and Colo., s. to Ind., Ill. and Ark.

2. SPIRAÈA L. SPIRAEA

Calyx 5-cleft, short, persistent. Petals 5, obovate, equal, imbricated in the bud. Stamens 10–50. Follicles 5–8, not inflated, few–several-seeded. Seeds linear, with a thin or loose coat and no albumen. — Shrubs of N. Hemisph. with simple leaves and white or rose-colored flowers in corymbs or panicles. (The Greek name, from *speira, a wreath.*)

Inflorescence an elongate panicle.
Leaves green both sides.
Branchlets of panicle puberulent or tomentulose. . . . . 1. *S. alba.*
Branchlets of panicle glabrous. . . . . 2. *S. latifolia.*
Leaves closely felted beneath with white or tawny tomentum. . . . . 3. *S. tomentosa.*
Inflorescence a compound corymb.
Calyx densely pubescent; petals pink; leaves long-acuminate. . . . . 4. *S. japonica.*
Calyx glabrous; petals white; leaves obtuse or merely acute or mucronate.
Leaves broadly oblong, coarsely toothed, not glaucous, the larger 2–5 cm. broad. . . . . 5. *S. corymbosa.*
Leaves narrowly oblong or oblanceolate, entire or nearly so, mucronate, glaucous beneath, 1–1.5 cm. broad. . . . . 6. *S. virginiana.*

1. **S. álba** Du Roi (white), MEADOW-SWEET. — Erect shrub 3–12 dm. high, with tough *yellowish-brown stems; leaves finely serrate,* lance-oblong, 5–7 cm. long, 1–1.8 cm. broad, rather firm in texture; *inflorescence thyrsoid, tomentulose;* flowers 6–8 mm. in diameter; petals suborbicular, white. (*S. salicifolia* of ed. 7, not L.) — Chiefly in low ground, sw. Que. and nw. Vt. to Sask., s. to Del., w. N.C., O., Ind., Ill., n. Mo. and N.D. June–Sept.

2. **S. latifòlia** (Ait.) Borkh. (broad-leaved), MEADOW-SWEET, THÉ DU CANADA (Que.). — Similar to no. 1; *branches red or purplish-brown; leaves* thin, obovate to oblanceolate, *coarsely toothed,* mostly 1.5–4 cm. broad; terminal or leading *panicles* mostly open-pyramidal, 0.5–3 dm. long, with elongate lower branches, *their branchlets glabrous,* secondary panicles more slender; *petals white or pale pink.* — Low grounds, Nfld. to Mich., s. to N.S., N.E., L.I. and interior N.C. June–Sept.

Var. **septentrionàlis** Fern. (northern). — Low, 1–7 dm. high; leading panicles dense, ovoid to cylindric, 1–10 cm. long, without elongate lower branches; flowers usually larger. — Nfld. and s. Lab. Pen. to Ung., s. to M.I., alpine reg. of Mt. Katahdin, Me., and White Mts., N.H.; mts. of n. Va.; Keweenaw Pen., Mich. July–Sept.

3. S. tomentòsa L. (tomentose), HARDHACK, STEEPLE-BUSH, THÉ DU CANADA (Que.). — *Branches and lower surface of* the ovate or oblong serrate *leaves very woolly;* panicle spire-like, the flowers very closely crowded in short spike-like racemes; *petals roseate*, or white in forma **albiflòra** Macbr. (white-flowered); follicles densely and permanently tomentulose. — Sterile low grounds and pastures, P.E.I. to the Laurentides, Que., and s. Ont., s. to N.S., N.E., L.I. and N.C. July–Sept.

Var. **ròsea** (Raf.) Fern. (rose-colored). — Flowers loosely clustered in distant groups on the branches of the panicle; follicles shorter-pubescent and glabrate. — Similar habitats, Ont. and Man., s. to e. Va., n. Ga., Tenn. and Ark. June–Aug.

4. S. JAPÓNICA L. f. (Japanese). — Stems 1 m. or more high; *leaves* 7–9 cm. long, glaucous beneath, *tapering* to base and *to acuminate tip, coarsely toothed;* corymbs broad, compound; *calyx densely pubescent; petals roseate.* — Spreading from cult. to thickets and roadsides, N.E. to Ind. and Tenn. July, Aug. (Introd. and natzd. from Asia)

5. S. corymbòsa Raf. (corymbose). — Stems erect, dark purple, simple or nearly so; *leaves oval or broadly oblong*, smoothish, of firm texture, *coarsely toothed* from near the middle to the rounded or obtuse apex, 2.5–5 cm. broad; *flowers white;* corymbs 4–10 cm. broad; *calyx glabrous; petals white.* — Rocky places and banks of streams in the mts., n. N.J. (and adj. N.Y.?), Pa. and W.Va. to Ga. and Ky., local. Late June, July.

6. S. virginiàna Britt. (of Virginia). — *Glabrous*, much-branched; *leaves lance-oblong, entire or subentire*, 1–1.6 cm. broad, often acute or acutish at the base; flowers white, about 6 mm. broad; *pedicels and glabrous calyx glaucous.* — Rocky places, W.Va. to w. N.C. and e. Tenn. June, July.

Several Old World species, introd. into cult., persist near old house-sites or spread locally to roadside-thickets; all described in works on cult. plants.

3. ARÚNCUS Adans. GOAT'S-BEARD

Dioecious. Carpels 3–4, splitting at the ventral suture. Flowers sessile or nearly so on the long spike-like branches of a large open panicle, the pistillate flowers reflexed in fruit. Petals small, white. — Tall, essentially herbaceous, N.Am. and Eurasian. Leaves 2–3-pinnate, the leaflets rather large, ovate-oblong. (*Aruncus, goat's-beard*, the name starting with Pliny and later applied to the European member of this genus.)

1. A. dioìcus (Walt.) Fern. (dioecious). — Stem erect, subsimple, bearing a few large compound petiolate leaves and a large pyramidal spicate panicle; leaflets 6–14 cm. long, green on both sides, sharply and somewhat doubly serrate, acuminate, the base mostly abrupt or subcordate, petiolulate, glabrous or pubescent beneath; calyx-lobes of staminate flowers thickish, deltoid; styles 0.5–0.8 mm. long; follicles olivaceous, semiovoid, 1.5–2 mm. long; seeds 1.5–2 mm. long. (*A. sylvester* of ed. 7, not Kostel.; *A. allegheniensis* Rydb.) — Rich woods and ravines, Catskill Mts., N.Y. (formerly), to Ky., s. to Ga. and Ala. Late May–July.

Var. **pubéscens** (Rydb.) Fern. (hairy). — Foliage thicker, grayer, either densely soft-pubescent beneath or glabrous; follicles subcylindric, 1.7–2.5 mm. long. (*A. pubescens* Rydb.) — Rich woods and limestone-bluffs, w. Ky., Ill. and Ia. to Ark. and Okla.

4. SORBÀRIA A. Br. FALSE SPIRAEA

Flowers perfect, paniculate, much as in *Spiraea.* Carpels mostly 5, opposite the calyx-lobes. Leaves regularly odd-pinnate, the leaflets lance-oblong, sessile, sharply serrate. Genus nat. of e. Asia. (Name from *Sorbus*, the Mountain-Ash, from the similar foliage.)

1. S. SORBIFÒLIA (L.) A. Br. (with leaves of *Sorbus*). — Suffruticose or nearly herbaceous, erect; leaves 1–4 dm. long, 13–21-foliolate; leaflets caudate-acuminate, with many straightish mostly simple veins springing from the midnerve; panicle ample, pyramidal, terminal; petals white. (*Spiraea* L.) — Common in cult., and esc. to waste land and copses, N.B. to Pa., Ind. and Minn. June, July. (Introd. and natzd. from Asia)

5. GILLÈNIA Moench INDIAN-PHYSIC

Calyx slender, somewhat constricted at the throat, 5-toothed; teeth erect. Petals 5, rather unequal, linear-lanceolate, inserted in the throat of the calyx, convolute in the bud. Stamens 10–20, included. Follicles 5, included, at first slightly cohering with each other, 2–4-seeded. — Perennial N. Am. herbs with almost sessile 3-foliolate leaves; the thin leaflets doubly serrate and incised. Flowers loosely paniculate-corymbed, pale rose-color or white. (Dedicated to an obscure German botanist and physician of the 17th century, *Arnold Gillen.*) PORTERANTHUS Britt.

1. G. trifoliàta (L.) Moench (with three leaflets), BOWMAN'S-ROOT. — Leaflets ovate-oblong, pointed, cut-serrate; *stipules small, subulate*, entire or slightly incised. — Rich woods, N.Y. to s. Ont. and Mich., s. to Ga. and Ala.; cult. and sometimes spreading northw. Late May–July.

2. G. stipulàta (Muhl.) Baill. (bearing stipules), AMERICAN-IPECAC. — Leaflets lanceolate, deeply incised; *stipules large and leaf-like*, doubly incised. — Woods, thickets and rocky slopes, Ga. and Ala. to e. Tex., n. to sw. N.Y., O., Ind., Ill., Mo. and Kans. May–July.

Tribe II. PÒMEAE B. & H.

6. PŸRUS L.

Calyx-like receptacle (hypanthium) urceolate, bearing 5 sepals (calyx-lobes). Petals roundish or obovate. Stamens numerous. Styles 2–5. Fruit a large fleshy pome, or smaller and berry-like, the 2–5 locules imbedded in the flesh, papery or cartilaginous, mostly 2-seeded. — Trees or shrubs, with showy flowers in corymbed or umbel-like cymes. (The classical name of the Pear-tree.) — A large genus, often subdivided, but with subgenera less strongly or constantly marked than our very few species would suggest: some Pears (true *Pyrus*) subglobose or depressed; some Apples (subgen. *Malus*) elongate and with grit-cells; some Mountain-Ashes (subgen. *Sorbus*) with simple leaves, one of them with pinnate leaves and large pyriform pomes with grit-cells; members of subgen. *Sorbus* hybridizing in the wild with members of true *Pyrus* and of subgen. *Aronia*. Cytologically there is no justification for separating them as genera and none in the wood-anatomy. Following DeCandolle, Eichler, Bentham & Hooker, Focke, Engler & Prantl, Gray and others of recognized good judgment, we maintain these confluent subgenera as a single true genus (comparable with *Potentilla, Rubus* and *Prunus*), including MALUS Mill., SORBUS L. and ARONIA Medic. Spelling often altered to the linguistically preferred PIRUS.

a. Leaves simple or merely lobulate except in some hybrids; flowers and fruits few in subumbellate or racemose inflorescences or, if paniculate-corymbose, in inflorescences rarely 6 cm. broad. . . *b.*

b. Trees or coarse shrubs often with spinescent branchlets; leaves not glandular along midrib; cyme simple and umbelliform or simply racemose; petals large (1–2 cm. long, 0.8–1.5 cm. broad); fruit 0.8–3 cm. or more in diameter. . . *c.*

c. Leaves involute in bud and before expanding; petals white; anthers reddish; orifice of concave receptacle partly or nearly closed by a disk-like cushion; styles free to base; flesh of fruit (in ours) with grit-cells. Subgen. PIROPHORUM (PEARS).

Leaves crenate; flowers 2.5–3 cm. broad; fruit obovoid or pyriform. . . 1. *P. communis.*

Leaves slenderly sharp-serrate; flowers 3–3.5 cm. broad; fruit subglobose. . . . . 2. *P. pyrifolia.*

c. Leaves convolute or folded lengthwise in bud and before expanding; petals roseate to white; anthers yellow or red; orifice of receptacle open; styles united at very base; fruit in ours subglobose and without grit-cells. Subgen. MALUS (APPLES). . . *d.*

d. Leaves convolute in bud and as they expand, not lobulate nor deeply cleft; anthers yellow; introd. and spread from cult. . . *e.*

e. Lower surface of leaves, petioles, young shoots and outside of persistent calyx with whitish or gray tomentum; leaves crenate-serrate. . . . . 3. *P. Malus.*

e. Leaves, petioles and calyx glabrous or promptly glabrate; leaves sharply serrate.

Young branchlets pubescent; calyx-lobes persistent; fruit about 2 cm. across. . . . . 4. *P. prunifolia.*

Young branchlets glabrescent; calyx-lobes deciduous; fruit barely 1 cm. across. . . . . 5. *P. baccata.*

d. Leaves longitudinally folded in bud and before expansion, dentate or serrate, those of at least the vigorous shoots often lobed or cleft; anthers red; native CRABS or CRAB-APPLES. . . *f.*

f. Calyx glabrous or sparsely pilose and glabrescent on outside.

Leaves of fertile branchlets oblong or narrowly elliptic, round-tipped, blunt or merely short-mucronate; colonial shrubs or small trees of Coastal Plain and southern range. . . . . 6. *P. angustifolia.*

Leaves of fertile branchlets broadly lanceolate, oval or ovate,

mostly acute; wide-ranging, chiefly inland and n. into Ontario, etc. . . . . 7. *P. coronaria.*
f. Calyx densely and permanently pannose-tomentose on outside. 8. *P. ioensis.*
b. Shrubs with slender ascending to spreading branches; leaves glandular along the midrib on upper side; inflorescence simple to more or less compound; petals mostly less than 1 cm. long and 7 mm. broad; fruit small and berry-like. Subgen. Aronia. . . *g.*
g. Lower surfaces of leaves, young shoots, rachis and pedicels soft-pubescent.
Sepals bearing stipitate glands; ripe fruit red, 5–7 mm. in diameter. . . 9. *P. arbutifolia.*
Sepals glandless or nearly glandless; ripe fruit purple or purple-black, 8–10 mm. in diameter. . . . . 10. *P. floribunda.*
g. Lower surfaces of leaves, new shoots, rachis and pedicels glabrous; fruit black. 11. *P. melanocarpa.*
a. Leaves odd-pinnate; cymes very compound; styles distinct; fruit berry-like. Subgen. Sorbus. . . *h.*
h. Winter-buds heavily glutinous, their outer scales glabrous; branchlets glabrous or nearly so; leaflets glabrous or glabrate beneath; branches of cyme and pedicels glabrous or glabrescent; native.
Inner bud-scales glabrous or merely ciliate; leaves yellow-green above; leaflets lanceolate to lance-oblong, long-attenuate, the larger ones three-and-a-half to five times as long as broad; flowers 5–6 mm. broad; fruits 4–6 mm. in diameter, not glaucous. . . . . 12. *P. americana.*
Inner bud-scales villous; leaves blue-green above; leaflets oblong to ob-long-ovate, short- to long-acuminate, the larger ones twice to thrice as long as broad; flowers about 1 cm. broad; fruits 8–12 mm. long, glaucous. . . . . 13. *P. decora.*
h. Winter-buds scarcely glutinous, white-villous; branchlets, axis of cyme and pedicels villous; lower leaf-surfaces pubescent; introd. . . . . 14. *P. Aucuparia.*

Subgen. Piróphorum Medic. (Pears)

1. P. commùnis L. (common), Common Pear. — Small tree, when wild often with spinescent branches; branchlets glabrous or glabrate; *leaves involute in the bud and on expanding,* round-ovate to elliptic, acuminate, *glabrous or promptly glabrate, with crenate margin; flowers* in umbelliform raceme, 2.5–3 *cm. broad; petals white,* broad-oblong; *anthers yellow; fruit pyriform to obovoid,* large, *the flesh with abundant grit-cells.* — Spread from cult. or from rejected cores to thickets, borders of woods and clearings. *Fl.* April, May. (Introd. from Eurasia)

2. P. pyrifòlia (Burm. f.) Nakai (with leaves of the pear; Burmann having mistaken it for a fig), Chinese or Japanese Pear. — Differing from no. 1 in its often tomentose young *leaves,* these glabrate, *finely sharp-serrate; flowers* larger (3–3.5 *cm. broad*), with broader petals; *fruit subglobose,* tardily softening. (Incl. *P. serotina* and *P. Lecontei* Rehd.) — Borders of woods and thickets, spread from cult., se. Va. *Fl.* April. (Introd. from e. Asia)

Subgen. Màlus (Mill.) Focke (Apples)

3. P. Màlus L. (the Greek name of the apple-tree), Apple, Pommier (Que.). — Small tree, when wild often with hard or spinescent short branchlets; *new growth* (*branchlets, buds, petioles, leaf-blades, etc.*) *white- or gray-tomentose, the tomentum somewhat persisting on lower leaf-surfaces, outside of calyx-lobes etc.; leaves* oblong-ovate, rounded to cordate at base, *crenate or crenate-serrate; inflorescence woolly,* the *stout pedicels* 1–2.8 *cm. long;* flowers pinkish-white; styles commonly pubescent below middle; fruit 3 cm. or more across, with impressed base and apex. (*Malus pumila* Mill.) — Roadsides, borders of woods, clearings, etc., spread from cult. *Fl.* late April–early June. (Introd. and natzd. from Eurasia)

4. P. prunifòlia Willd. (plum-leaved), Chinese Apple, Crab-Apple. — *Differing from no.* 3 *in its less pubescent branchlets; leaves* broader, *glabrate except for slight pubescence on veins beneath, sharply serrate; flowers white, on glabrous slender pedicels* 2–3.5 *cm. long; fruit* about 2 cm. broad, *crowned by the prominent beak-like calyx.* (*Malus* Borkh.) — Spread from cult. to road-sides and thickets, N.B., N.S. and N.E. to Pa. *Fl.* May, early June. (Introd. from e. Asia)

5. P. baccàta L. (berry-like), Siberian Crab. — Small tree *with glabrous or promptly glabrescent branchlets; glabrous* oblong-ovate *long-acuminate sharp-serrate leaves;* flowers white, with *oblong petals; calyx-lobes soon deciduous; fruit* subglobose, long-pedicelled, *about* 1 *cm. across,* red or yellow. (*Malus* Borkh.) — Spread from cult. to thickets and clearings. (Introd. from Asia)

6. P. angustifòlia Ait. (narrow-leaved), Wild Crab. — Trees up to 8 m. high or colonial shrubs; *flowering branchlets and leaves glabrous* or promptly glabrate; *leaves of fertile often spinescent branchlets oblong or narrowly elliptic, round-tipped, blunt or merely short-mucronate, tapering to base,* entire, crenate-dentate or serrate, *firm, often evergreen, in maturity* 2–7.5 *cm. long and* 0.8–3 *cm. broad;* sprout-leaves much larger, coarsely toothed or lobulate, up to 9 cm. long and 7 cm. broad; *pedicels glabrous;* calyx glabrous outside, tomentose within; petals roseate, fading to white; apple 2–3.5 cm. in diameter, yellowish-green, the sepals often deciduous. (*Malus* Michx.) — Woods, bottoms and thickets, Fla. to La., n. to Md., Va., W.Va., Ky. and Mo.

Var. **spinòsa** (Rehd.) Bailey (spiny). — *Branchlets and pedicels densely and rather persistently pannose;* young leaves pubescent; calyx glabrous or sparsely pilose outside. (*Malus ioensis,* var. Rehd.; *M. angustifolia,* var. *puberula* Rehd.) — N. to s. N.J., Ky. and se. Mo.

7. P. coronària L. (suitable for a wreath), WILD CRAB. — Differing from no. 6 in the *oval, ovate or ovate-lanceolate leaves of the fertile branchlets* more *acute and with rounded to cordate bases;* larger sprout-leaves often more lobulate. (*Malus* Mill.) — Three vars.:

Leaves of fertile branchlets roundish- or broad-ovate, five-eighths to four-fifths as broad as long, the mature ones 4–8.5 cm. long and 2.5–6.5 cm. broad, simply serrate or more or less lobulate.
- Hypanthium and sepals glabrous outside. . . . . . *P. coronaria* (typical).
- Hypanthium sparingly pilose outside; the sepals glabrate. . . . . Var. *dasycalyx.*

Leaves of fertile branchlets lance-ovate to broadly lanceolate, more acute to acuminate, one-third to half as broad as long, the mature ones 3.5–12 cm. long and 1.5–4 cm. broad, simply serrate. . . . . Var. *lancifolia.*

P. coronària L. (typical). — Bottoms, wooded slopes, thickets and clearings, centr. N.Y. and s. Ont. to s. Wisc., s. to Del., upland to N.C., Tenn. and Mo. — The trees with narrowest leaves somewhat lobulate are the scarcely worthwhile var. **elongàta** (Rehd.) Bailey (elongate).

Var. **dasýcalyx** (Rehd.) Fern. (with hairy calyx). — S. Ont. to Minn., s. to O., Ind. and Kans. — Perhaps partly a hybrid of nos. 7 and 8.

Var. **lancifòlia** (Rehd.) Fern. (lance-leaved), *Malus lancifolia* and *M. bracteata* Rehd. — N. Pa. to Ill., s. to upland of N.C. and to Tenn. and Mo.

8. P. ioénsis (Wood) Bailey (of Iowa), WILD CRAB. — Differing from nos. 6 and 7 in the *densely and permanently pannose-tomentose pale outer surface of hypanthium and sepals;* branchlets and expanding leaves pannose-tomentose; leaves of fertile branchlets narrowly to broadly oblong, elliptic or narrowly obovate, acute to obtuse, simply toothed to somewhat lobulate or subentire. (*Malus* Britt.) — Wisc., Minn. and Neb., s. to Ind., Ill., Ark. and Okla. — Very inconstant variations have been dignified by special names. The species evidently crossing with others, the commonest hybrid being

× **P. Soulárdi** Bailey (named in 1891 for JAMES G. SOULARD), SOULARD CRAB (*Malus* Britt.), a natural hybrid of nos. 3 and 8, either wild or spread from cult.

Subgen. ARÒNIA Reichenb. (CHOKEBERRIES)

9. P. arbutifòlia (L.) L. f. (with leaves of *Arbutus*), RED CHOKEBERRY. — Colonial shrub (spreading by subterranean offsets) 0.3–4 m. high or southw. becoming a slender tree up to 6 m. high; branches slender, loosely ascending; *new branchlets gray- or white-tomentose; leaves* broadly oblanceolate to narrowly obovate or subelliptic, tapering to base, acuminate or with abruptly pointed apex, *dark green and glabrous* (except for glandular midrib) *above, densely pannose-tomentose and pale beneath,* crenate-serrate, in maturity 2–9 cm. long and 0.5–4 cm. broad; *cyme* subsimple or with forking branches, flattish to convex above, 1.5–6 cm. broad, 2–25-flowered, *the rachis and pedicels tomentulose;* flowers about 1 cm. broad; *hypanthium tomentose, sepals bearing stipitate glands;* petals white or pink-tinged; *fruit* obovoid to subglobose, *bright or dull red,* 5–7 *mm. in diameter.* (*Aronia* Ell.; *Sorbus* Heynh.) — Low woods, thickets, swamps, damp pine-barrens, etc., Fla. to e. Tex., n. to N.S., s. N.E., N.Y., s. Ont., Mich. and Mo. *Fl.* April–mid-July (northw.); *fr.* Sept.–Nov. — Dwarf and subsimple individuals of pine-barrens and bogs and the largest-leaved individuals of more favorable habitats hardly merit the taxonomic distinction sometimes given them. — Hybridizes with no. 12, producing the rare × **P. hýbrida** Moench (hybrid) (× *Sorbus hybrida* Schneid.; × *Aronia hybrida* Zabel; × *Sorbaronia hybrida* Schneid.); also rarely with the introduced *P. Aria* (L.) Ehrh., producing the very local × **P. alpìna** Willd. (alpine) (× *Sorbus alpina* Heynh.; × *Sorbaronia alpina* Schneid.).

10. P. floribúnda Lindl. (full of flowers), PURPLE CHOKEBERRY. — Very similar to no. 9, 0.1–1 (–3) m. high; leaves, branchlets, rachis and pedicels often less densely pubescent; *sepals*

scarcely or not at all glandular; ripe fruit dark purple or purple-black, more juicy, 8–10 *mm. in diameter.* (*Aronia* Spach; *A. atropurpurea* Britt.; *A. prunifolia* (Marsh.) Rehd.; *P. prunifolia* Steud., not Willd.; *P. arbutifolia*, var. *atropurpurea* Robins.; *Sorbus arbutif.*, var. *atropurpurea* Schneid.) — Peats, low thickets, wet to dry clearings, etc., Nfld. and s. Lab. Pen. (mostly less than 1 m. high) to Algoma Distr., Ont., s. to N.S., N.E., Va. (mostly 1–3 m. high), s. Ont., O. and Ind. *Fl.* April–mid-July (northw.); *fr.* Sept.–Nov. — Hybridizes with nos. 11, 12 and 13.

11. P. melanocárpa (Michx.) Willd. (black-fruited), Black Chokeberry, Gueules noires (Que.). — Similar to no. 10 but *lower leaf-surfaces* (except sometimes the midrib), *new branchlets, rachis and pedicels glabrous; calyx and fruit essentially glabrous; fruit black,* juicy, 7–10 mm. in diameter; mostly lower than but quite as variable in foliage as nos. 9 and 10. (*Sorbus* Heynh.; *Aronia* Ell.; *Aronia nigra* Britt.) — Similar habitats or often in drier thickets and clearings or on bluffs and cliffs, Nfld. to nw. Ont. and Minn., s. to N.S., N.E., S.C. and Tenn. *Fl.* April–early July; *fr.* Aug.–Oct. Hybridizes with nos. 10, 12 and 14.

Subgen. Sórbus S. F. Gray (Mountain-Ash)

12. P. americàna (Marsh.) DC. (American), American Mountain-Ash, Roundwood, Dogberry, Missey-moosey, Cormier (Que.). — Small smooth-barked tree or coarse shrub *with glabrous or glabrate branchlets; winter-buds glutinous, glabrous or the inner ones ciliate; leaflets* (11–)13–15 (–17), *lanceolate or lance-oblong, taper-pointed,* sharply serrate, *membranaceous, yellow-green above,* glabrous or soon glabrate, *the larger ones three-and-a-half to five times as long as broad,* mostly 5–10 cm. long and 1.5–2.5 cm. broad; *cyme* corymbiform, rounded to flattish across top, *densely very many-flowered,* 0.8–2 dm. broad, the branches and pedicels glabrous or glabrescent; *flowers* white, 5–6 *mm. broad; fruit* subglobose, orange-red, 4–6 *mm. in diameter.* (*Sorbus* Marsh.; *Aucuparia* Nieuwl.) — Woods, Nfld. and Côte Nord, Que., to Minn., s. to N.S., N.E., n. N.J., Pa., Md., upland to Ga. and e. Tenn., W.Va., Mich. and n. Ill. *Fl.* late May–July; *fr.* late Aug.–Oct. (holding through winter). — Crosses with no. 10, producing × **P. Jáckii** (Rehd.) Fern. (for its discoverer, John George Jack, 1861–1949) = *Sorbaronia Jackii* Rehd.; and with no. 11, producing × **P. míxta** Fern. (mixed) = *Sorbus sorbifolia* (Poir.) Hedl. and *Sorbaronia sorbifolia* (Poir.) Schneid., not *P. sorbifolia* Cham.; and *Sorbus Sargentii* Dippel, not *P. Sargenti* Bean

13. P. decòra (Sarg.) Hyland (handsome), Mountain-Ash, Roundwood, Dogberry. — Differing from no. 12 in *villous inner bud-scales; leaves blue-green above,* whitish beneath; *leaflets oblong to narrowly oblong-oval, firm, rounded at tip to a short acumination; the larger ones only twice to thrice as long as broad,* usually 3.5–8 cm. long and 1.5–3 cm. broad; corymbose *cyme more open,* 6–16 cm. broad; *flowers about* 1 *cm. broad; fruits glaucous,* 8–12 *mm. long.* (*P. sitchensis* of ed. 7, not Piper; *Sorbus decora* (Sarg.) Schneid.; *Aucuparia subvestita* Nieuwl.) — Woods, rocky slopes and shores, s. Greenl. and Lab. to n. Ont. and se. Man., s. to N.S., n. N.E., nw. Mass., N.Y., n. O., n. Ind., Wisc. and Minn., ascending to subalpine areas. *Fl.* June, July (when growing with no. 12 flowering 10 days later); *fr.* Sept.–Nov. (holding through winter). — Hybridizes with no. 10, producing × **P. Arsénii** (Britt.) Arsène (named for its discoverer, Bro. Louis Arsène (Bigeul), 1875– , who inadvertently became author of the combination). (*Sorbus* Britt. and *Sorbaronia* G. N. Jones)

Var. **groenlándica** (Schneid.) Fern. (of Greenland). — Shrub or small tree; *leaflets membranaceous, tapering from near middle to prolonged acumination, merely paler green* (not whitish) *beneath,* very sharply toothed; cyme 6–9 cm. broad. (*Sorbus* G. N. Jones) — Greenl. and coastal Lab. to exposed to alpine habitats of Nfld., and alpine or subalpine reg. of Gaspé Pen., Que., and n. N.E. — Suggestive of no. 12 but with broad and short leaflets; the winter-buds, flowers and fruit of no. 13; foliage of fertile branches suggesting sprout-foliage of typical no. 13.

14. P. Aucupària (L.) Gaertn. (old generic name, meaning attractive to birds), European Mountain-Ash, Rowan. — Small tree differing from nos. 12 and 13 in its *white-villous nonglutinous winter-buds;* the *leaves with soft-pubescent lower surfaces;* the oblong leaflets blunt to short-acuminate, rather small; *branchlets and axis and branches of cyme white-villous,* at least *when young;* flowers and glaucous fruits much as in no. 13. (*Sorbus* L.) — Roadsides, borders of woods and other habitats near towns, spread from cult., Nfld. to Alaska, s. to N.S., N.E., Pa., O., Ind., Ill., Ia., etc. (Introd. from Eu.) — Hybridizes with no. 9 (see above); and with no. 11, producing × **P. fállax** (Schneid.) Fern. (deceitful) = *Sorbaronia* Schneid.

7. AMELÁNCHIER Medic. Juneberry. Sugarplum. Shadbush. Serviceberry. Sarviceberry. Poirier or Petites Poires (Que.)

Calyx 5-cleft. Petals obovate, oblong, oblanceolate or linear. Stamens many (mostly 20 in ours), short. Styles 5 (in ours), united below. Ovary 5-locular, each locule with 2 ovules but

with a projection growing from the back of each and forming a false cartilaginous partition, the sweet and edible berry-like small commonly juicy pome thus 10-locular with one seed in each locule (when all ripen). — Trees or shrubs of temp. N. Hemisph., with simple leaves and white or pink mostly racemose flowers. (Provençal name of a European species.)* — Species, especially in recently burned and newly populated clearings or other disturbed habitats, commingling and producing many perplexing hybrids, these mongrel offspring not here keyed.

*a.*Flowers or fruits racemose, with naked pedicels or only the lowest pedicels subtended by leaves; leaves conduplicate in bud, often longitudinally folded during anthesis, with rounded to cordate (only rarely subcuneate) bases; petioles in maturity (0.5–) 1–3 cm. long; summit of ovary low and rounded. . . *b.*

*b.*Summit of ovary densely pubescent in anthesis, somewhat so in fruit; leaves suborbicular, oblong or broadly ovate to obovate, with subtruncate, rounded or merely mucronate or short-pointed (not long-acuminate) apex. . . *c.*

*c.*Teeth of margin of mature leaves of fertile branches mostly (0–)2–5 per cm. . . *d.*

*d.*Leaf-blades usually broadly truncate or subtruncate; triangular sepals 2.5–3 mm. long, nearly as broad; species of Great Plains. . . . 1. ***A. alnifolia.***

*d.*Leaf-blades usually rounded or acute at apex; sepals 2–5.5 mm. long. . . *e.*

*e.*Veins or their branches extending into the teeth of the leaf. . . *f.*

*f.*Non-stoloniferous or essentially so; solitary-stemmed or slightly fastigiate shrubs or small trees. . . *g.*

*g.*Leaves rounded to subacute at apex, dentate, with usually open sinuses separating the teeth, the stronger veins mostly 12–15 pairs; sepals after falling of petals gradually arched-recurving from near middle. . . *h.*

*h.*Primary veins mostly 12–15 pairs, the upper mostly simple ones straight and running directly to the teeth; overwintering buds dull; petals linear or narrowly spatulate; shrubs 1–3 m. high.

Mature leaves green, with 4 or 5 teeth per cm.; lowest flowering pedicel 0.7–2.5 cm. long; calyx (hypanthium) 3.5–6 mm. in diameter below curve of sepals, the sepals 2–4 mm. long; petals 1–1.5 cm. long; anthers 0.6–0.8 mm. long; somewhat colonial and straggling wide-ranging chiefly non-calcicolous shrub. 2. ***A. sanguinea.***

Mature leaves glaucous, usually with 3 teeth per cm.; lowest flowering pedicel 2.7–4 cm. long; calyx (hypanthium) 7–9 mm. in diameter below curve of sepals, the sepals 3–5 mm. long; petals 1.6–2.2 cm. long; anthers 1–1.2 mm. long; stems solitary or few; calcicolous shrub of N. Y. and adj. Canada. . . . . 3. ***A. amabilis.***

*h.*Primary veins mostly 12 pairs, the upper ones curving upward and either simple or forking, their branches running into the teeth; leaf-margin with 3 or 4 teeth per cm.; overwintering buds lustrous; hypanthium 5–7.5 mm. in diameter; petals obovate or broadly oblanceolate; calcicolous shrub or small tree up to 7 m. high, of upper Great Lakes reg. . 4. ***A. huronensis.***

*g.*Leaves short-acuminate to apiculate and rounded above, serrate with 4 or 5 teeth per cm., the sinuses narrow and acute; stronger veins mostly 9–11 pairs, mostly forking at apex; sepals strongly recurving from base and tightly appressed to summit of fruit; hypanthium 4–5 mm. in diameter; petals 1–1.5 cm. long; lowest flowering pedicel 1.5–4 cm. long; arched-ascending shrub 3–8 m. high, wide-ranging northw. . 5. ***A. Wiegandii.***

*f.*Stoloniferous or spreading at or beneath surface of ground by elongate horizontal shoots and forming open or somewhat circular diffuse colonies.

Only slightly surculose, the solitary or slightly clustered strag-

---

* Treatment derived primarily from those of K. M. Wiegand in Rhodora, xiv. 118–162 (1912) and xxii. 146–151 (1920) and of Etlar L. Nielsen in Am. Midl. Nat. xxii. 160–208 (1939); as well as from shorter studies of Wiegand and the longer one, Am. Sp. Amel. of G. N. Jones in Ill. Biol. Mon. xx. no. 2 (1946). The figures show leaves, calyx after dropping of petals, fruits, and position of leaves in the expanding bud.

gling stems 1–3 m. high; mature leaves 3–6 cm. broad, dentate nearly or quite to base, the teeth with mostly broad and open sinuses; raceme lax, soon nodding, 4–7 cm. long, its axis and pedicels promptly glabrate; petals linear to narrowly spatulate, 1–1.5 cm. long. . . . . 2. *A. sanguinea.*

Strongly stoloniferous or surculose, loosely colonial, 0.3–8 m. high; mature leaves mostly 2–4 cm. broad, entire or toothed to near or slightly below middle, the teeth with narrow acute sinuses; racemes upright, 2–5 cm. long, densely silky-tomentose; petals oblong-obovate, 7–10 mm. long, about half as broad. . . . . 6. *A. humilis.*

e. Veins anastomosing below margin or forking, many of them not entering the teeth; low northern calcicolous colonial shrubs up to 1 m. high.

Expanding leaves densely pubescent, soon glabrate; mature blades 2.5–3 cm. broad, entire except near tapering short-acuminate or mucronate tip; raceme strict, dense, 3–5 cm. long; all pedicels naked, the lowest 1–1.4 cm. long; sepals 3.5–5.5 mm. long, soon recurving; petals elliptic to broadly oblong. . . . . . 7. *A. mucronata.*

Expanding leaves glabrous or but sparsely evanescent-pilose; mature blades 2–5 cm. broad, toothed essentially to base, rounded to subtruncate above; raceme open, 4–8 cm. long, the lower 1–3 pedicels usually subtended by a large long-petioled leaf, the lowest pedicel 1–3 cm. long; sepals 1.5–3.5 mm. long, ascending; petals narrowly oblanceolate. . . . . 8. *A. gaspensis.*

c. Teeth of margin of mature leaves of fertile branches mostly (5–)6–10 per cm. . *i.*

i. Stoloniferous or surculose colonial shrubs 0.3–1.5 m. high; overwintering buds 4–7 mm. long; leaves mostly rounded to tapering at base, oblong to suborbicular or obovate; racemes ascending to spreading; lowest pedicels 0.7–2 (–3.5) cm. long; sepals glabrous, ascending to recurving from near middle; eastern shrubs.

Expanding leaves densely white-tomentose beneath, soon glabrate, elliptic to oblong or suborbicular, obtuse or merely mucronate, sharply toothed along upper two-thirds of margin, veins obscure toward margin of blade; raceme rather close, erect, the pedicels and rachis pubescent during anthesis; lowest pedicels 7–15 mm. long; sepals 2.5–3 mm. long, soon reflexed from near middle; shrub of acid soils. . . . . 9. *A. stolonifera.*

Expanding leaves glabrous beneath, elliptic-oblong to oblong-obovate, acutish to rounded above, toothed to base, the lateral veins prominent beneath to tip; raceme spreading, lax, glabrous; lowest pedicels 1.5–3.5 cm. long; sepals 3–5 mm. long, erect or subascending; calcicolous shrub. . . . . 10. *A. Fernaldii.*

i. Non-stoloniferous straggling shrub or small tree up to 10 m. high; overwintering buds 6–13 mm. long; leaves cordate or subcordate, ovate or nearly so, short-acuminate, toothed nearly to base, the teeth with rounded sinuses; racemes drooping; lowest pedicels up to 4.5 cm. long; sepals tomentose above, at anthesis, soon reflexed from near base; northwestern. . . . . 11. *A. interior.*

b. Summit of ovary glabrous or only sparsely and temporarily pubescent; leaves blunt or with more or less prolonged acuminate apex. . . *j.*

j. Stoloniferous or surculose loosely colonial shrubs 0.2–2 m. high; flowering raceme compact, 1–4 cm. long; leaves only slightly developed at anthesis; mature leaves elliptic-oblong to oblong-obovate or suborbicular, blunt to acute; petals 3–7 mm. long. . . *k.*

k. Mature leaves elliptic-oblong, oblong-ovate or oblong-obovate, thin and membranaceous, grayish- or pale green, dull or but slightly lustrous above; hypanthium 2–5 mm. in diameter.

Expanding leaves heavily white-tomentose beneath, in maturity sparsely pilose; petioles permanently pilose; raceme close, 1–3 cm. long, the rachis and pedicels heavily white-pubescent; longest fruiting pedicels becoming 3–12 mm. long, pilose; hypanthium 2–3 mm. in diameter; sepals 1–2 mm. long, narrowly triangular, erect or but slightly spreading; se. species. . . . . . . 12. *A. obovalis.*

Expanding leaves sparingly pubescent, soon glabrate beneath, in maturity quite glabrous; petioles promptly glabrate; raceme lax, 2–4.5 cm. long, the rachis and pedicels promptly glabrate; longest fruiting pedicels 1–2 cm. long, glabrous; hypanthium 4–5 mm. in diameter; sepals 2–4 mm. long, lanceolate, attenuate, soon recurving; insular plant of se. Mass. . . . . 13. *A. nantucketensis.*

k. Mature leaves oblong, broadly elliptic or roundish, coriaceous, dark green and lustrous above, glabrous beneath, petioles glabrous; lowest fruiting pedicel 1–2.5 cm. long, glabrous; hypanthium 5–7 mm. in diameter; sepals erect; Nova Scotian. . . . . 14. *A. lucida.*

j. Fastigiate coarse and tall shrubs or arborescent; flowering raceme, pedicels, sepals and petals usually longer. . . *l.*

l. Racemes ascending; sepals ascending or in fruit somewhat divaricate or irregularly recurving; petals 7–12 mm. long; leaves elliptic-oblong, ovate or oblong-obovate, blunt or acute, rounded to but slightly cordate at base.

Young leaves heavily white-tomentose, becoming glabrate, only slightly developed at anthesis, without purple tinge; mature blade with 6–11 sharp teeth per cm.; flowering raceme dense, with tomentose rachis and pedicels; petals 7–10 mm. long; lower fruiting pedicels 1–2.2 cm. long; mature sepals ascending, tomentose above. 15. *A. canadensis.*

Young leaves sparsely pubescent beneath, often purple-tinged, well grown at anthesis; mature blade with 5–7 teeth per cm.; flowering raceme open, with rachis and pedicels less pubescent to glabrous; petals 9–12 mm. long; lower fruiting pedicels 2–3.5 cm. long; mature sepals ascending to recurving, glabrate. . . . . . . 16. *A. intermedia.*

l. Racemes often becoming pendulous; petals 1–1.8 cm. long; fruiting sepals tightly reflexed from base; leaves broadly to narrowly ovate, acuminate, cordate or rounded at base.

Overwintering buds 6–13 mm. long; leaves small and densely white-tomentose at anthesis, the pubescence somewhat persistent beneath to maturity, blades not purplish; petioles retaining slight pubescence; lowest pedicels 8–17 mm. long; fruits dryish, reddish-purple, insipid. . . . . 17. *A. arborea.*

Overwintering buds 9–17 mm. long; leaves half-grown at anthesis, glabrous or essentially so, purple or bronze; petioles glabrous; lowest pedicels becoming 2.5–5 cm. long; fruits juicy, sweet, blackish-purple. . . . . 18. *A. laevis.*

a. Flowers 1 or 2–4 in leafy-bracted fascicles; leaves imbricated in bud, or expanding promptly, becoming flat, glabrous or promptly glabrate, membranaceous, gradually tapering or cuneate to short (2–10 mm. long) petiole; ovary with subconical summit; fruit longer than thick; boreal shrub. . . . . . 19. *A. Bartramiana.*

1. A. alnifòlia Nutt. (alder-leaved). — *Stoloniferous* and colonial shrub 1–7 m. high; overwintering buds 5–8 mm. long, dull castaneous; *leaves* longitudinally folded at flowering time, then yellowish-tomentose beneath, soon expanded and glabrate, *the mature blade broadly elliptic to quadrate-rotund, broadly truncate or subtruncate across the summit,* 2.5–5 cm. broad, coarsely serrate-dentate *with* 2–5 *broad teeth per cm.; the prominent veins* curving upward, often forking, *they or their forks entering the teeth; racemes* few-flowered, close, erect, 1.5–3 cm. long in anthesis, the flowering rachis silky, *the lowest pedicel* 5–11 *mm. long; petals* 6–7.5 *mm. long;* hypanthium after falling of petals 3.5–4 mm. in diameter; the *triangular sepals* 2.5–3 *mm. long, nearly as broad,* their tips recurving; *summit of ovary tomentose;* fruit globose to obovoid, about 1 cm. in diameter, blue-purple, or in forma **álba** Nielsen (white) white, juicy and sweet. — Thickets, borders of woods and banks of streams, w. Ont. to Yuk., s. to nw. Ia., Neb., Colo. and Oreg. *Fl.* May; *fr.* July, Aug. Fig. 1121.

1121. A. alnifolia.

2. A. sanguínea (Pursh) DC. (blood-red; from the red branchlets). — *Arched-ascending to straggling* shrub 1–3 m. high, the solitary or few slender trunks *with red or reddish young branchlets; overwintering buds* slender, reddish-brown, 6–7 mm. long, *dull; leaves nearly or quite unfolded at flowering time,* then pale-tomentose beneath, soon glabrescent but often retaining slight pubescence on midrib beneath and slender petiole (1–2 cm. long); *mature blade elliptic-oblong to suborbicular, green, rounded to blunt or subacute apex,* 3–6 cm. long, with round to subcordate base, *the margin* coarsely *serrate-dentate to base,* the sharp *teeth* 4 *or* 5(–6) *per cm. and separated by usually open sinuses; primary veins mostly* 12–15 *pairs, the upper ones mostly simple, straight and running directly to the teeth; racemes* loose and open, soon arching, 4–7 cm. long in anthesis, the lowest pedicel 0.7–2.5 cm. long; *petals narrowly spatulate to linear,* 1–1.5 *cm. long; anthers* 0.6–0.8 *mm. long; hypanthium after falling of petals* 3.5–6 *mm.*

1122. A. sanguinea.

in diameter below curve of sepals (2–4 *mm. long*), *the latter soon recurving from the middle;* ovary tomentose above; fruit globose, dark purple, juicy. (*A. spicata* of ed. 7, not (Lam.) K. Koch) — Open woods, rocky slopes, river-banks, etc., in non-calcareous to slightly calcareous areas, s. Que. to Thunder Bay Distr., Ont., s. to n. and w. N.E., N.Y., mts. to n. N.C., n. O., Mich., Wisc. and n. Ia. *Fl.* May, June, *fr.* July, Aug. Fig. 1122. — Hybridizes with nos. 9, 18 and 19.

3. A. amábilis Wieg. (lovely). — Differing from no. 2 in somewhat *glaucous leaves;* teeth mostly 3 per cm.; *lowest flowering pedicels* 2.7–4 *cm. long; hypanthium below curve of sepals* (3–5 mm. long) 7–9 *mm. in diameter; petals* broader, 1.6–2.2 *cm. long; anthers* 1–1.2 *mm. long.* (*A. grandiflora* Wieg., not Rehd.; *A. sanguinea* sensu G. N. Jones, in part, not DC.) — Open woods, rocky (often calcareous) banks and shores, sw. Que. and s. Ont., s. into N.Y. *Fl.* May, early June; *fr.* Aug.

4. A. huronénsis Wieg. (of Lake Huron region). — *Fastigiate shrub or small tree* 3–7 m. high; *overwintering buds* 5–8 mm. long, dark purple to reddish-brown, *usually gummy; leaves nearly expanded at anthesis,* floccose-pubescent to glabrescent on the lower face, soon glabrate, *broadly oblong to suborbicular or slightly obovate,* 4–7 cm. long, *rounded to subacute above,* rounded to subcordate at base, coarsely serrate-dentate to below middle or nearly to base *with* 3 *or* 4 *teeth per cm.; primary veins mostly* 12 *pairs,* upcurving and simple or forking, *their tips entering the teeth; racemes loose* and open, 4–7 cm. long, the *lowest pedicel* 1.2–3 *cm. long; hypanthium* 5–7.5 *mm. in diameter,* the *narrow sepals* (3.5–5 *mm. long*) *strongly recurved after anthesis; petals broadly oblanceolate,* 1.2–1.8 *cm. long; anthers* 0.6–0.9 *mm. long;* ovary tomentose above; fruit subglobose, 5–8 mm. in diameter, sweet and juicy. (*A. sanguinea* sensu G. N. Jones, in part, not DC.) — Open woods, cliffs and shores, chiefly on trap or other basic rock, Manitoulin Dist. to Thunder Bay Distr., Ont., s. to Bruce Pen., Ont., n. Mich., Wisc. and Minn. *Fl.* late May–early July; *fr.* July, Aug. — Hybridizes with no. 18.

5. A. Wiegándii Nielsen (for Karl McKay Wiegand, 1893–1942, painstaking and accurate student of the genus). — Differing from no. 4 in its longer *overwintering buds* (*up to* 1 *cm. long*); *leaves bronze to purple on expanding* and in autumn, glabrous or promptly glabrescent, 3–6 cm. long, *short-acuminate to apiculate, with* 4 *or* 5 *very sharp teeth per cm.; primary veins mostly* 9–11 *pairs, chiefly forking at apex;* lowest pedicel 1.5–4 cm. long at anthesis; *hypanthium* 4–5 *mm. in diameter, the sepals tightly reflexed from base after anthesis; petals* linear to narrowly obovate. (*A. interior* sensu G. N. Jones, in part, not Nielsen) — Rocky slopes, banks of streams, sands, etc., Nfld. to Ont., s. to N.S., N.B., n. N.E., N.Y., n. Mich., n. Wisc. and n. Minn. *Fl.* late May–early July; *fr.* July, Aug. — Suggesting a small-leaved no. 18 but with shorter buds, fewer teeth, fewer veins, and summit of ovary heavily tomentose.

6. A. hùmilis Wieg. (low). — *Strongly surculose or with subterranean stolons, forming loosely scattered colonies* 0.3–8 *m. high;* overwintering buds ovoid, 4–9 mm. long, dull or slightly varnished; *leaves* partly or fully expanded *at flowering time,* then *densely white- or gray-tomentose beneath,* finally glabrescent or with pubescent petiole, elliptic, oblong, ovate or obovate; *those of fertile branchlets* 2.5–5 *cm. long,* 2–4 *cm. broad, broadly subacute to rounded above,* rounded to cordate at base, *coarsely serrate-dentate to below middle to entire, with* mostly 4 *or* 5(–6) *teeth per cm., separated by acute sinuses;* primary veins 7–13 pairs, forking near apex and entering teeth; *racemes erect,* 2–5 *cm. long, densely silky-tomentose; lowest pedicel* 9–13 *mm. long;* hypanthium 3–5 mm. in diameter; *sepals revolute from the middle* after anthesis; *petals oblong-obovate,* 7–10 *mm. long, about half as broad;* fruit globose to thick-ellipsoid, black, with a bloom, juicy and sweet. (*A. spicata* sensu G. N. Jones, in part, not K. Koch) — Very variable; hybridizes with nos. 9, 17 and 19. The following vars. recognized by Nielsen:

Leaves typically oblong but sometimes ovate.
- Racemes mostly 4–5 cm. long; hypanthium 4–5 mm. in diameter; leaves mostly coarsely dentate-serrate above middle, with 4 or 5 teeth per cm., rarely entire; overwintering buds glabrous. . . . . *A. humilis* (typical).
- Racemes 2.5–3 cm. long; hypanthium 3–3.5 mm. in diameter; leaves entire, or few-toothed at apex, then with about 6 teeth per cm.; buds pubescent. Var. *exserrata.*

Leaves typically elliptic, but occasionally ovate or obovate.
- Veins conspicuous and running into the teeth; racemes 2–5 cm. long, loose; hypanthium 3.5–4 mm. in diameter; sepals acute, 3 mm. long, longer than broad. . . . . Var. *campestris.*
- Veins forking and less prominent near margin; racemes 2–2.5 (–3) cm. long, very compact; hypanthium 3 mm. in diameter; sepals blunt, 2–2.5 mm. long, about as broad as long. . . . . Var. *compacta.*

A. hùmilis (typical). — Rocky or sandy shores or banks (often calcareous), sw. Que. to w. Ont., s. to Vt., Pa., O., Mich., Wisc., Minn. and S.D. *Fl.* May, early June; *fr.* July, Aug. FIG. 1123.

Var. **exserràta** Nielsen (without teeth). — Quartzite rock, se. Minn.

Var. **campéstris** Nielsen (of the plains). — Shores, Minn. and S.D.

Var. **compácta** Nielsen (compact). — Thickets and margins of woods, Minn. and S.D.

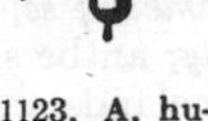

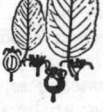

1123. A. humilis.

7. A. mucronàta Nielsen (mucronate). — Habitally similar to no. 6, about 1 m. high; *leaves* soon glabrate, *tapering to short-acuminate or mucronate apex; primary veins mostly* 9–14 *pairs, freely branching near margin but many of them not entering the teeth;* raceme strict, dense, 3–5 cm. long; all pedicels above the lowest (1–1.4 cm. long) without subtending leaf; sepals 3.5–5 *mm. long, soon recurving;* petals elliptic to broadly oblong, 8–9 mm. long. (*A. spicata* sensu G. N. Jones, in part, not K. Koch) — Basic rocks, n. Minn. and se. Man. *Fl.* June; *fr.* late Aug.

8. A. gaspénsis (Wieg.) Fern. & Weath. (of the Gaspé Peninsula). — *Colonial surculose* shrub rarely 1 m. high; *expanding leaves glabrous* or but sparsely pilose *beneath,* promptly glabrate; *mature blades broadly elliptic to broad-oblong or suborbicular, thinnish,* 3–6 cm. long, 1.5–5 *cm. broad, with rounded, subtruncate or short-pointed summit,* cordate to rounded at base, coarsely toothed nearly to base to subentire, the *teeth* 3–5(–6) *per cm.; primary veins* 6–13 *pairs, anastomosing* below margin, *many of them not entering teeth; racemes loose and open,* 4–8 *cm. long, with glabrous or nearly glabrous rachis and pedicels; lower* 1–3 *pedicels subtended by large leaves,* the lowest pedicel 1–3 cm. long; hypanthium 3–4 mm. in diameter, *sepals* 1.5–3.5 *mm. long, ascending; petals narrowly oblanceolate,* 6–9 *mm. long;* fruit about 1 cm. in diameter, blackish, glaucous, juicy and sweet. — Cliffs, ledges and shores (mostly calcareous), Gaspé Pen., Que., to James Bay, s. to n. N.B., n. Me., Algoma Distr., Ont., n. Mich. and ne. Minn. *Fl.* late June–mid-Aug.; *fr.* late July–Sept.

9. A. stolonífera Wieg. (stoloniferous). — Habitally similar to nos. 6 and 8, 0.3–1.5 m. high; overwintering buds 5–7 mm. long, dull reddish-brown; *expanding leaves half-grown at flowering time, then densely white-tomentose beneath, soon glabrate; mature blades dull green above, elliptic to oblong or suborbicular, obtuse or merely mucronate,* 2–5 *cm. long,* 2–3.5 *cm. broad, sharply toothed along upper two-thirds of margin with* (5–)6–8 *teeth per cm.; primary veins mostly* 7–9(–11) *pairs,* upcurved and *becoming indistinct toward margin of blade; racemes erect, dense,* 1.5–4 *cm. long,* the *pedicels and rachis pubescent during anthesis; lowest pedicel* 7–15 *mm. long;* hypanthium 3–4 mm. in diameter; *sepals* 2.5–3 *mm. long, soon reflexed from near middle,* their margins becoming revolute; *petals* oblong or narrowly obovate, 7–9 *mm. long,* or in forma **micropétala** (Robins.) Rehd. (with tiny petals) linear or narrowly spatulate and only 3–4 mm. long; summit of ovary densely tomentose; fruit blackish, juicy. (*A. spicata* sensu G. N. Jones, in part, not K. Koch) — Dry sterile (acid) rocky or sandy open habitats, Nfld. to Thunder Bay Distr., Ont., s. to N.S., N.E., L.I., Va., Mich. and Minn. *Fl.* May, early June; *fr.* July, Aug. FIG. 1124. — Hybridizes with nos. 2, 17, 18 and 19.

1124. A. stolonifera.

10. A. Fernáldii Wieg. (for one of its discoverers, 1873–1950). — In habit similar to nos. 6, 8 and 9, up to 1 m. high; *leaves glabrous from the first,* half-expanded during anthesis, in maturity *membranaceous, elliptic or oblong-obovate, acutish to rounded above,* 5–8 cm. long, 1.5–4.5 cm. broad, rounded at base, *sharply serrate to base,* the teeth (4–)6–10 per cm.; *primary veins* 7–13 *pairs, impressed above, prominent beneath to tips; racemes lax, spreading, glabrous,* 2–4 cm. long; *lowest pedicels* 1.5–3.5 cm. long; *hypanthium* 4–5 mm. in diameter, glabrous; *sepals* 3–5 *mm. long, erect or subascending; petals oval to broadly oblanceolate,* 8–11 *mm. long, half as broad;* fruit purple-black, 6–10 mm. in diameter. — Calcareous damp to dryish open barrens, thickets, ravines or shores, Nfld. and Anticosti to Charlevoix Co., Que., s. to C.B., P.E.I. and Gaspé Pen. and Rimouski Co., Que. *Fl.* late June, July; *fr.* Aug., Sept.

11. A. intèrior Nielsen (inland). — *Non-stoloniferous straggling* shrub or small tree up to 10 m. high; *overwintering buds* slenderly ovoid, *up to* 1.3 *cm. long; leaves broadly ovate to elliptic, acute or short-acuminate,* rounded to subcordate at base, *promptly glabrate,* 4–7 *cm. long,* 3–5 *cm. broad, finely serrate, with rounded sinuses,* 5–7 *teeth per cm.; primary veins mostly* 8–10 *pairs, anastomosing below margin; racemes loose, nodding, glabrous,* 3–7 cm. long; *lowest pedicel up to* 4.5 *cm. long;* hypanthium about 5 mm. in diameter; *sepals* 2.5–3.5 mm. long, tomentose above, *becoming* glabrate and *reflexed;* petals obovate, 1.5 cm. long; fruit globose, 6–8 mm. in diameter. — Hillsides and banks of streams, sw. Wisc., s. Minn. and adj. Ia. to s. S.D. *Fl.* June; *fr.* July.

12. A. obovàlis (Michx.) Ashe (obovate). — *Stoloniferous or surculose slender shrub forming* *more or less circular colonies,* 0.2–1.5 *m. high; overwintering buds reddish,* 5–8 *mm. long, varnished; leaves heavily white-tomentose beneath, unexpanded during anthesis; mature blades elliptic-oblong, oblong-ovate or oblong-obovate, blunt to acute,* 2–5.5 *cm. long,* 0.8–3 *cm. broad, thin, retaining minute pubescence beneath, pale and dull green above, sharply serrulate nearly to base, with* 6–9 *teeth per cm.; midrib prominent beneath; primary veins* 7–9 *pairs, evanescent toward margin; racemes expanding before leaves, compact,* 1–3 *cm. long,* the *rachis, pedicels and calyx white-tomentose; fruiting pedicels pilose, the longest becoming* 3–12 *mm. long; hypanthium* 2–3 *mm. in diameter; sepals* 1–2 *mm. long, erect or becoming divergent; petals narrowly elliptical,* 6–7 *mm. long;* anthers 0.6–0.7 mm. long; *summit of ovary glabrous;* fruit purple-black, 6–8 mm. long. — Pinelands and low woods, Ga. and Ala., n., chiefly on Coastal Plain, rarely among the mts., to s. N.J., se. Pa., e. Md. and interior Va. *Fl.* late March, April; *fr.* May, June.

13. A. nantucketénsis Bickn. (of Nantucket). — Resembling no. 12, differing in its *promptly glabrate or glabrous leaves becoming lustrous above,* 2–3 *cm. long* and 1.5–2 *cm. broad; raceme lax,* 2–4.5 *cm. long,* the *rachis and pedicels promptly glabrate; longest pedicels becoming* 1–2 *cm. long, glabrous in maturity; hypanthium* 4–5 *mm.* in diameter; sepals 2–4 *mm. long, lance-attenuate, soon recurving; petals linear-oblanceolate,* frequently involute, 3–7 mm. long, 1–2 mm. wide. (*A. canadensis* sensu G. N. Jones, in part, not Medic.) — Dry moors, pine-barrens and pond-margins, Nantucket and Martha's Vineyard Ids., Mass. *Fl.* May, early June; *fr.* late June, early July.

14. A. lùcida Fern. (lustrous). — Differing from no. 12 in *pale yellowish opaque overwintering buds; coriaceous* oblong, broadly elliptic or subrotund *leaves dark green and lustrous above,* 1.5–6.5 *cm. long* and 1–4 *cm. broad; sepals* 2–3.5 *mm. long,* erect; *fruiting raceme with glabrous rachis and pedicels, lowest fruiting pedicels* 1–2.5 *cm. long; hypanthium* 5–7 *mm. in diameter;* fruit larger. (*A. stolonifera,* var. *lucida* Fern.; *A. spicata* sensu G. N. Jones, in part, not K. Koch) — Sandy or peaty barrens, gravels and acid rock, N.S. *Fl.* unknown; *fr.* July, Aug.

15. A. canadénsis (L.) Medic. (Canadian). — *Fastigiate shrub* with usually several upright trunks up to 8 m. high, forming vase-like or alder-like clumps; *leaves scarcely half-grown at anthesis, then heavily white-felted beneath* and folded, *in maturity oblong, oblong-elliptic or narrowly oblong-obovate,* averaging 3–6 cm. long and 1.8–2.8 cm. broad, *rounded to barely mucronate or subacute at apex, rounded at base,* becoming glabrate, *finely serrate with* sharp teeth nearly or quite to base, *the teeth* 6–11 *per cm.* and with acute sinuses; *primary veins* 10–15 *pairs, becoming indistinct and anastomosing toward margin; racemes rather close, ascending,* 2.5–6 cm. long, *the rachis, pedicels and calyx tomentose; longer pedicels becoming* 1–2.2 *cm. long;* hypanthium 3–5 mm. in diameter, tomentose; *sepals tomentose within,* 1.5–3 mm. long, *erect or loosely spreading; summit of ovary glabrous* or nearly so; *petals* oblong-obovate to lance-linear, 7–10 *mm. long;* fruit blackish, juicy, sweetish. (*A. oblongifolia* T. & G.; *A. Botryapium* (L. f.) Borkh.) — Swamps, low grounds and thickets (non-calcareous), Ga. to se. and centr. Me., centr. N.H., sw. Que. and centr.-w. N.Y. *Fl.* late March (southw.)–early June (northw.); *fr.* June, July. FIG. 1125. — Hybridizes with nos. 9, 16, 17 and 18.

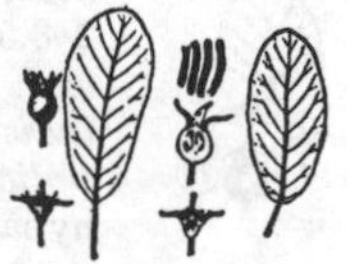
1125. A. canadensis.

Var. **subíntegra** Fern. (subentire). — *Leaves entire or with few teeth near the tip.* — Low pine-barrens of se. Va.

16. A. intermèdia Spach (intermediate). — Differing from no. 15 in the *often purplish coloring of the well-grown and only sparsely pubescent or glabrescent leaves at flowering time; mature blades* commonly *cordate and shorter and broader* (mostly 2.5–6 cm. long, 1.8–3 *cm. broad*), *usually short-acuminate, with* 5–7 *teeth per cm.; flowering raceme open, the rachis, pedicels and calyx nearly or quite glabrous; petals* 9–12 *mm. long; lower fruiting pedicels* 2–3.5 *cm. long; sepals ascending to recurving, glabrate.* (*A. canadensis* sensu G. N. Jones, in part, not Medic.) — Swamps, bogs (often calcareous), damp or dry thickets and shores, Nfld. and M.I. to n. Minn., s. to N.S., s. N.B., n. N.E., Va., upland of N.C., Mich. and s. Minn. *Fl.* May, June; *fr.* July, Aug.

17. A. arbòrea (Michx. f.) Fern. (tree-like). — *Fastigiate shrub* or tree up to 20 m. high (then with trunk up to 4 dm. in diameter); *overwintering buds* 6–13 *mm. long; expanding leaves green above, densely white-tomentose beneath, small and folded at flowering time; mature blades ovate to slightly obovate, acuminate, cordate* or rounded at base, *commonly* 4–10 *cm. long and* 2.2–5 *cm. broad, retaining some pilosity beneath and on the petioles,* sharply and often doubly serrate, *the long teeth mostly* 6–10 *per cm.;* primary veins mostly 11–17 pairs, these anastomosing and becoming indistinct; *racemes rather close, nodding,* 3–5 *cm. long; lower pedicels* 0.8–1.7 *cm. long;* hypanthium 2.5–3.5 mm. in diameter; *sepals* broadly oblong-triangular, obtuse or abruptly pointed, *soon strongly*

1126. A. arborea.

reflexed from base; petals linear or narrowly oblong, 10–14 *mm. long,* white; sometimes roseate; *fruit reddish-purple, dry, insipid.* (*A. canadensis* sensu Wieg., not Medic.; *A. canadensis,* var. *Botryapium* sensu ed. 7, not *A. Botryapium* (L.) Borkh.) — Rich woods, thickets and slopes, n. Fla. to La. and e. Okla., n. to sw. N.B., se. and centr. Me., sw. Que., s. Ont., n. Mich. and ne. Minn. *Fl.* late March (southw.)–early June (northw.); *fr.* June–Aug. Fig. 1126. — Hybridizes with nos. 6, 15, 18 and 19.

18. A. laèvis Wieg. (smooth). — Irregularly branching *fastigiate* shrub, or tree up to 13 m. high; *overwintering buds* 0.9–1.7 *cm. long; leaves half-grown at flowering time, usually reddish or purple-tinged, glabrous or essentially so; mature blades* dark green above, *glaucescent beneath,* elliptic, ovate, ovate-oblong or slightly obovate, *acuminate,* subcordate or rounded at base, those of fertile branches mostly 4–6 cm. long and 2.5–4 cm. broad; *the sharp callous-tipped almost subulate teeth with mostly rounded sinuses,* 6–8 *per cm.; primary veins* 12–17 pairs, *anastomosing near margin; racemes flexuous* or nodding, *nearly or quite glabrous,* 3–7 *cm. long; lowest pedicels becoming* 2.5–5 *cm. long; hypanthium* 2.5–5 mm. broad, glabrous; *sepals lanceolate to subulate,* 2.7–4 mm. long, *soon abruptly reflexed from base; summit of ovary glabrous; petals* linear-oblong, 1–2 *cm. long; fruit dark purple to blackish,* glaucous, *juicy, sweet.* — Dry to moist thickets, borders of woods, margins of swamps and clearings, Nfld. to Ont., s. to N.S., N.E., Del., Ga., O., s. Ind., n. Ill. and Ia., ascending to mountain-summits southw. *Fl.* late March (southw.)–mid-June (northw.); *fr.* June–Aug. Fig. 1127. — Hybridizes with nos. 2, 4, 5, 9, 15, 17, and 19.

1127. A. laevis.

Var. **nítida** (Wieg.) Fern. (shining). — Mature leaves dark green, lustrous above; teeth coarser and more prolonged. — Nfld. and N.S.

19. A. Bartramiàna (Tausch) Roemer (for William Bartram, 1739–1823, who sent seeds to European gardens), Mountain-J. — Cespitose-fastigiate slender shrub 0.5–2.5 m. high; leaves dull green, membranaceous (firmer in bleak habitats), *elliptical, elliptic-oblong or elliptic-oval, imbricated in bud, flat on expanding* (not conduplicate), glabrous except for slightly silky petiole, *blunt to acute, tapering or cuneate to very short* (2–10 *mm. long*) *petiole; blades of fertile branchlets* mostly 2–6 cm. long and 1.5–3.5 cm. broad, *closely toothed nearly to base* (sprout-leaves up to 7 cm. long and 6 cm. broad); *inflorescences of* 1–4 *flowers,* 1 *flower terminal, the others from leaf-axils; calyx campanulate* below, glabrous outside, the hypanthium 3–6 mm. in diameter; *sepals* triangular-subulate, 3–4 mm. long, *persistently tomentose on upper face,* ascending to spreading; petals oblong-oval, broadest at middle, 6–10 mm. long; *ovary densely tomentose at summit, conically tapering to pubescent style-base; fruit longer than thick,* 1–1.3 *cm. long,* becoming purple-black; seeds semiovate, curved, 4–5 mm. long, 2–3 mm. broad. (*A. oligocarpa* (Michx.) Roem.) — Peaty or boggy thickets, sphagnous bogs, bushy slopes and mountain-summits (ascending to subalpine areas), Lab. to Thunder Bay Distr., Ont., s. to Nfld., N.S., bogs and mts. of n. N.E. and Mass., ne. Pa., s. Ont., n. Mich., n. Wisc. and n. Minn. *Fl.* May (southw.)–Aug. (northw.); *fr.* July–Sept. Fig. 1128. — Hybridizes with nos. 2, 6, 9, 10, 15, 17 and 18.

1128. A. Bartramiana.

8. CRATAÈGUS L. Hawthorn. Red Haw. Thorn. Pommettes (Que.). Cenellier (Que.)

Calyx-tube campanulate or obconic, its limb 5-parted. Petals normally 5, deciduous. Stamens usually 5–20, in 1–3 series; filaments filiform; anthers oblong, white, yellow or red. Styles 1–5, distinct. Fruit a pome with 1–5 bony usually 1-seeded nutlets. Small trees or shrubs with usually crooked thorny branches and simple serrate or variously lobed leaves; those at the ends of vegetative shoots differently shaped, larger, and usually more deeply cut than on the flowering branchlets. — A large genus of the N. Hemisph., most abundant in ne. and centr. N.Am., where of great taxonomic difficulty. Some of the described species may be hybrids, and these and others here treated as varieties and forms require further study. (Name from the Greek *kratos, strength.*) *

a. Veins of the larger leaves running to the sinuses as well as to the points of the lobes. . . *b.*

* Treatment contributed by E. J. Palmer. The figures show characteristic flowers (in some), fruits, exposed nutlets and (in some) sculpturing of nutlets. The unnumbered binomials belong to trees or shrubs of very limited occurrence or of suspected hybrid origin.

b. **Leaves comparatively thin, becoming bright red or yellow in autumn, early deciduous; fruit 4–6 mm. thick; nutlets 3–5; thorns 2.5–6 cm. long; native species.**

Leaves of flowering branchlets often trilobate or with the lowest pair of lobes much enlarged, acuminate at apex and at points of lobes; fruit bright red, persistent on the branches until late in the season. Series 1. CORDATAE (see p. 769).

Leaves of flowering branchlets not trilobate, rounded or merely acute at apex and at points of lobes; fruit dull red, orange or greenish, falling early in the season. . . . . Series 2. MICROCARPAE (see p. 770).

b. **Leaves comparatively thick, persistent and remaining green until late in the season; fruit 6–9 mm. thick; nutlet single or rarely 2; thorns mostly 0.5–1.5 cm. long or rarely becoming elongated into spiny branchlets; cult. and esc. . . . . Series 3.** OXYACANTHAE (see p. 770).

a. **Veins of the leaves running only to the points of the lobes. . . *c.***

c. **Flowers single or rarely 2–5 in a cluster; calyx-lobes foliaceous, pectinate or deeply glandular-serrate; leaves small (mostly 0.5–1.5 cm. long), cuneate at base, serrate, crenate or dentate above the middle, unlobed or slightly lobed on vegetative shoots, rugose-veined. . . . . . Series 4.** PARVIFOLIAE (see p. 771).

c. **Flowers in cymes or corymbs; calyx-lobes not foliaceous, either entire or glandular-serrate; leaves usually larger. . . . *d.***

d. **Foliage and inflorescence conspicuously glandular; flowers in simple or but slightly branched mostly 4–8-flowered corymbs; fruit often bronze-red or dull greenish-yellow.**

Leaves of flowering branchlets variable, elliptic, oblong, or slightly ovate or obovate on the same plant, sometimes with 1–3 pairs of shallow lateral lobes above the middle, cuneate at base, mostly 1–2 cm. wide; primary veins strongly ascending or subparallel. Series 5. FLAVAE (see p. 771).

Leaves of flowering branchlets fairly uniform, usually rhombic or elliptic in outline, gradually or abruptly narrowed at the broad base, often with 4–5 pairs of sharp triangular lateral lobes or rarely unlobed, mostly 2.5–4 cm. wide; primary veins pinnate. . Series 6. INTRICATAE (see p. 771).

d. **Foliage and inflorescence eglandular or rarely with a few scattered glands; flowers in few- to many-flowered usually compound cymes or corymbs. . . *e.***

e. **Nutlets plane on the inner surface (except sometimes in Series 11). . *f.***

f. **Leaves of flowering branchlets mostly obovate, oblong or oblong-elliptic, broadest above or rarely about at the middle. . . *g.***

g. **Leaves relatively thin, those of the flowering branchlets variable and often asymmetric, mostly rhombic, oblong or oblong-obovate; primary veins strongly ascending, sometimes subparallel; petioles slender, slightly wing-margined above; fruit 0.4–1 cm. thick; nutlets usually 5. . . . . . . Series 7.** VIRIDES (see p. 773).

g. **Leaves thick or firm, those of the flowering branchlets fairly uniform in type; primary veins pinnate; petioles stoutish, often wing-margined to below the middle; fruit 0.8–1.5 cm. thick; nutlets 1–5.**

Leaves unlobed except rarely and slightly on vegetative shoots in some species, serrate or crenate; veins inconspicuous or only slightly impressed above; petioles short, usually less than one-fourth as long as the blades; fruit inedible, remaining hard and dry or rarely becoming mellow late in the season; nutlets 1–3, rarely 5. . . . . Series 8. CRUS-GALLI (see p. 774).

Leaves of flowering branchlets often more or less lobed and more deeply cut on vegetative shoots, sharply serrate or dentate; veins usually conspicuously impressed above; petioles often one-fourth to one-third as long as the blades; fruit fleshy, becoming edible; nutlets 2–5, usually 3–5. . . . Series 9. PUNCTATAE (see p. 778).

f. **Leaves of flowering branchlets not prevailingly obovate, oblong or oblong-elliptic, broadest at or below the middle. . . *h.***

*h.*Leaves of flowering branchlets mostly rhombic or elliptic or rarely oblong-ovate in outline, broadest about the middle, cuneate or abruptly narrowed at base.

Leaves of flowering branchlets usually acute or obtuse at apex and at points of lobes; leaves of vegetative shoots often broadly ovate or suborbicular, sometimes broader than long; nutlets plane on the inner faces. . . . . . Series 10. ROTUNDIFOLIAE (see p. 780).

Leaves of flowering branchlets usually acuminate at apex and points of the lobes; leaves of vegetative shoots mostly ovate, seldom as broad as long; nutlets sometimes pitted on the inner surface. . . . . Series 11. BRAINERDIANAE (see p. 783).

*h.*Leaves of flowering branchlets mostly ovate, oblong-ovate or deltoid in outline, broadest below or rarely at about the middle. *i.*

*i.*Young leaves roughened above with short appressed hairs; stamens usually 10 or fewer (rarely 20); fruit with a small sessile calyx.

Leaves thin, usually acuminate, the points of the lobes often reflexed; fruit usually bright red, succulent. Series 12. TENUIFOLIAE (see p. 785).

Leaves firm, pointed, the tips of the lobes not reflexed; fruit dull red or greenish, with thin dry or firm flesh. Series 13. SILVICOLAE (see p. 788).

*i.*Young leaves glabrous or pubescent; stamens 10 or 20; fruit with a large prominent calyx; nutlets usually 3–5. . *j.*

*j.*Foliage and inflorescence usually tomentose or pubescent with loose straight or matted hairs; petioles and midribs stout; fruit pubescent at least while young; nutlets usually 5. . . . . Series 16. MOLLES (see p. 795).

*j.*Foliage and inflorescence glabrous or pubescent; petioles and midribs slender; fruit glabrous; nutlets 3–5. . . *k.*

*k.*Leaves of flowering branchlets prevailingly oblong-ovate, finely and sharply serrate nearly to the rounded base; stamens usually 10 or less (rarely 20); styles and nutlets 2–4 (rarely 5); fruit mellow or soft when ripe. Series 15. COCCINEAE (see p. 793).

*k.*Leaves of flowering branchlets prevailingly ovate or deltoid, finely or coarsely serrate except sometimes near the base, usually lobed.

Leaves of flowering branchlets abruptly narrowed, truncate or rarely subcordate at base; flowers mostly 1.5–2 cm. wide; stamens 10 or 20; styles and nutlets 3–5; fruit with a large usually elevated calyx and thin dry flesh remaining hard and often green until late in the season. . . . . Series 14. PRUINOSAE (see p. 790).

Leaves of flowering branchlets usually truncate or subcordate at base; flowers 2–2.5 cm. wide; stamens 20; styles and nutlets 5; fruit with a wide nearly sessile calyx, brightly colored, with thick flesh, highly flavored and edible when ripe. . . . . . Series 17. DILATATAE (see p. 797).

*e.*Nutlets pitted on the inner surface; fruit usually 1 cm. or less thick.

Leaves of flowering branchlets mostly elliptic, rhombic or ovate; fruit bright red or orange. . . . . Series 18. MACRACANTHAE (see p. 798).

Leaves of flowering branchlets mostly obovate or oblong-obovate; fruit purplish-black. . . . . Series 19. DOUGLASIANAE (see p. 801).

Series 1. CORDÀTAE Beadle (see p. 768)

Leaves of flowering branchlets mostly rounded to subcordate at base; calyx deciduous from the mature fruit. . . . . 1. *C. Phaenopyrum.*

Leaves of flowering branchlets mostly narrowed or cuneate at base; calyx persistent or only partly and tardily deciduous from the ripe fruit. . . . . 2. *C. Youngii.*

1. C. Phaenópyrum (L. f.) Medic. (with the appearance of a pear), WASHINGTON THORN. — A small tree 7–8 m. high, with slender thorny branchlets and thin scaly bark; *leaves* mostly ovate, irregularly serrate, *often trilobate or with* 2–3 *pairs of spreading lateral lobes, the lowest pair enlarged,* thin but firm, glabrous or nearly so at maturity; petioles slender, one-third to three-fifths as long as the blades, eglandular; flowers 0.9–1.2 cm. wide, many in glabrous cymes; stamens about 20; anthers pale yellow; calyx-lobes deltoid, entire; *fruit* subglobose, 4–6 *mm. thick, bright red, lustrous, with* thin dry flesh and *a small deciduous calyx exposing the ends of the* 3–5 *nutlets.* — Thickets and open woods, Pa. to Fla., w. to Mo. and Ark. Often planted and esc. from cult. *Fl.* May, June; *fr.* Oct. FIG. 1129.

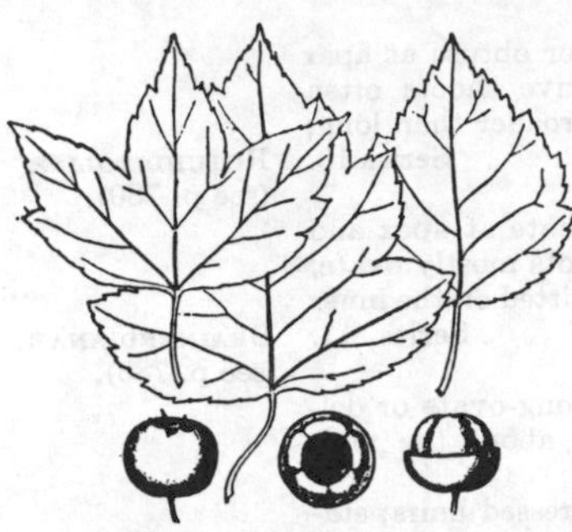

1129. C. Phaenopyrum.

2. C. Yoúngii Sarg. (named in 1923 for its discoverer, ROBERT C. YOUNG). — Very similar to no. 1; *leaves* averaging slightly smaller, those *of the flowering branchlets* mostly *narrowed* or *cuneate at base; fruiting calyx persistent or, if tardily and partially deciduous, not exposing the nutlets.* — Low wet woods and banks of streams, se. Va. to S.C. *Fl.* May; *fr.* Oct.

Series 2. MICROCÁRPAE Loud. (see p. 768)

Leaves of flowering branchlets spatulate or narrowly obovate, crenate or slightly lobed towards the apex, glabrous at maturity; fruit subglobose. . . . 3. *C. spathulata.*
Leaves broadly ovate, often as broad as or broader than long, deeply cut, with 2–3 pairs of lateral lobes often coarsely toothed or again cleft at the ends, pubescent; fruit oblong or obovoid. . . . . 4. *C. Marshallii.*

3. C. spathulàta Michx. (wedge-shaped; referring to the leaves). — A shrub or tree 5–7 m. high, with stoutish usually thorny branchlets and thin scaly bark; *leaves of flowering branchlets narrowly obovate,* mostly 1–2 cm. long, 0.5–1 cm. wide, with several coarse rounded teeth or small lobes above the middle or near the apex, gradually narrowed to the entire base, firm, glabrous at maturity, *with strongly ascending or nearly parallel* obscure *veins; petioles one-fourth to half as long as the blades;* flowers 6–8 mm. wide, many in compact glabrous corymbs; stamens about 20; anthers pale yellow; calyx-lobes deltoid, entire, persistent on fruit; *fruit subglobose,* 4–7 mm. thick, red, with thin mellow flesh and 3–5 nutlets. — Low woods along streams, Va. to Fla., w. to s. Mo. and e. Tex. *Fl.* May; *fr.* Oct. FIG. 1130.

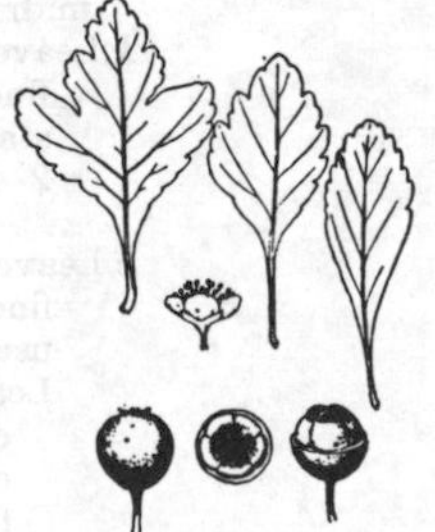

1130. C. spathulata.

4. C. Marshállii Egglest. (for HUMPHREY MARSHALL, 1722–1801, who first described it). — A shrub or small tree 6–8 m. high, with slender thorny or sometimes thornless branchlets pubescent while young and thin scaly bark; *leaves broadly ovate or deltoid-ovate,* sharply serrate and *deeply incised* with 2–3 pairs of lateral lobes often coarsely toothed or again cleft at the ends, pubescent while young; at maturity thin, glabrous above, often slightly pubescent along the veins beneath; *petioles* slender, *half to as long as or longer than the blades;* flowers 1–1.5 cm. wide, many in pubescent corymbs; stamens about 10; anthers red; calyx-lobes lanceolate, entire or nearly so; *fruit oblong or obovoid,* 5–9 mm. long, 4–8 mm. thick, bright red, with thin succulent flesh and 1–3 (usually 2) nutlets. — Open woods and low hills, usually along streams, se. Va. to Fla., w. along the Coastal Plain to e. Tex., and n. in the Miss. Val. to se. Mo. *Fl.* April; *fr.* Oct. FIG. 1131.

1131. C. Marshallii.

Series 3. OXYACÁNTHAE Loud. (see p. 768);
an Old World series; species introd. in our reg.:

5. C. MONÓGYNA Jacq. (having one ovary; from the solitary nutlet), ENGLISH HAWTHORN, AUBÉPINE (Que.). — A tree 7–8 m. high, with slender branches often armed with short (1–2 cm.) thorns and with dark slightly scaly bark; *leaves ovate to obovate, or on shoots* sometimes *deltoid, deeply incised,* trilobate or with 2–3 pairs of oblong lateral lobes, glabrous or nearly so, firm, deep green and *persistent until late in the season;* petioles slender, eglandular, from one-third to as long as the blades; flowers 1–1.5 cm. wide, many in glabrous corymbs; stamens about 20; anthers red; calyx-lobes deltoid, entire; *fruit* ellipsoidal or subglobose, 6–8 mm. thick, bright red, *with* thin flesh and *usually* 1 *nutlet.* — Cult. and often spreading to roadsides and borders of woods. *Fl.* May; *fr.* Oct. (Introd. from Eurasia and the Mediterr. reg.)

C. Oxyacántha L. (with sharp thorns), similar to no. 5 but with usually less deeply divided leaves and slightly larger *fruit with 2 nutlets,* is also commonly cult. and may be looked for as an escape.

Series 4. Parvifòliae Loud. (see p. 768);
a single species in our reg.:

6. C. uniflòra Muenchh. (one-flowered). — *A slender shrub* 0.6–1.5 m. high, with slender often flexuous thorny branchlets, villous while young; *leaves* obovate, oblong or elliptic, sharply or crenately serrate and sometimes obscurely lobed above the middle, especially on vegetative shoots, pubescent while young, *thick, the veins impressed above at maturity; petioles short, stout,* wing-margined above; *flowers* 1–1.5 cm. wide, *single or rarely* 2–5, on short tomentose pedicels; stamens 20 or more; anthers small, white or pale yellow; *calyx-lobes lanceolate, foliaceous,* pectinate or deeply glandular-serrate, *persistent on the fruit;* fruit 1–1.3 cm. thick, subglobose or slightly pyriform, greenish-yellow or dull red, with dry or mealy flesh and 3–5 (usually 5) nutlets. (*C. tomentosa* sensu Egglest., not L.) — Sandy or rocky banks and woods, N.Y. and Pa. to Fla., w. to the Ozark reg. and e. Tex. *Fl.* May; *fr.* Oct. Fig. 1132.

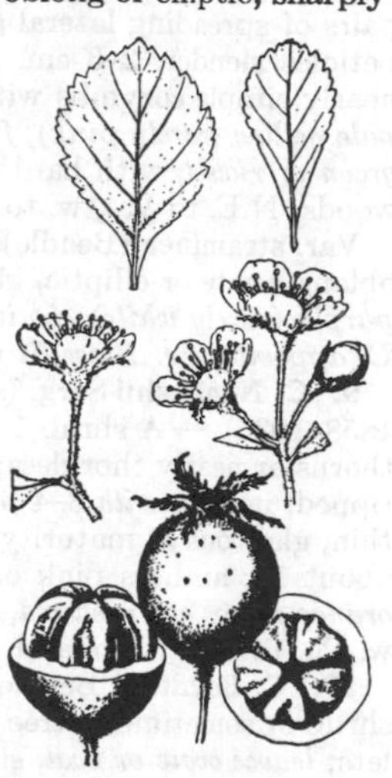

Series 5. Flàvae Loud. (see p. 768);
a single species native in our reg.:

7. C. flàva Ait. (yellow). — A tree 7–8 m. high, with flexuous thorny branchlets sometimes villous while young; *leaves* ovate, rhombic or elliptic, *glandular-serrate, often trilobate or asymmetrically lobed near the apex,* or with 2–3 pairs of lateral lobes on vegetative shoots, firm, yellowish-green, with the veins impressed above at maturity; *flowers* 1.6–1.7 cm. wide, mostly 5–7 *in simple or slightly branched* sparsely villous *corymbs;* stamens about 10 (rarely 20); anthers red; calyx-lobes finely glandular-serrate; *fruit oblong or pyriform, green or yellow,* 0.9–1.3 cm. thick, with a prominent calyx, thick dry flesh and 3–5 nutlets. (*C. aprica* sensu Egglest., not Beadle) — Open woods and low hills, usually in sandy or gravelly soil, se. Va. to Fla. *Fl.* May, June; *fr.* Oct.

1132. C. uniflora.

C. Evansiàna Sarg. (in honor of John Evans of Delaware Co., Pa., 1790–1862), with *leaves* oblong-elliptic or obovate, irregularly serrate and *glandular near the base, glabrous except for small tufts of tomentum in the axils of the strongly ascending veins beneath,* firm, lustrous above at maturity; flowers 4–10 in nearly simple slightly villous corymbs; fruit subglobose, with thin dry flesh and 5 nutlets, known from West Fairmont Park, Philadelphia, perhaps planted, of unknown origin, is probably a hybrid between a species of the *Flavae* and *C. viridis.*

Series 6. Intricàtae Sarg. (see p. 768)

a. Foliage and inflorescence glabrous or essentially so; fruit glabrous. . . *b.*
b. Leaves comparatively thin; fruit green, russet or red. . . *c.*
c. Leaves mostly ovate, broadest below the middle, often truncate or subcordate on vegetative shoots. . . *d.*
d. Leaves usually with 4–5 pairs of sharp spreading lateral lobes, the points acuminate and sometimes reflexed.
Anthers white or pale yellow (rarely pale pink); fruit obovoid or pyriform; nutlets 3–5, usually 3–4. . . . . 8. *C. intricata.*
Anthers pink or red (rarely white); fruit subglobose or short-oblong; nutlets 2–4, usually 2–3. . . . . 9. *C. Neobushii.*
d. Leaves usually with 3–4 pairs of broad shallow lateral lobes; fruit subglobose.
Flowers 1.3–1.5 cm. wide; petals not conspicuously cucullate; anthers small, pink or pale yellow. . . . . 10. *C. Boyntoni.*
Flowers 1.6–2 cm. wide; petals conspicuously cucullate; anthers large, yellow. . . . . 11. *C. foetida.*

c.Leaves mostly oblong-ovate or oval, broadest about the middle, acute, rounded or truncate at the base on vegetative shoots.
Leaves of flowering branchlets mostly narrowed or attenuate at base, often broadly ovate on vegetative shoots; nutlets 3–5, usually 3–4. 12. *C. rubella.*
Leaves of flowering branchlets mostly obtuse or rounded at base, broadly ovate to suborbicular on vegetative shoots; nutlets 2–3. . 13. *C. padifolia.*
b.Leaves thick or coriaceous at maturity; fruit bright yellow. . . . . . . 14. *C. fortunata.*
a.Foliage and inflorescence pubescent; fruit pubescent at least while young.
Leaves sharply serrate, with shallow lobes; anthers white or pale yellow; fruit subglobose or depressed-globose. . . . . 15. *C. biltmoreana.*
Leaves coarsely and doubly serrate with acute teeth, acutely lobed; anthers red; fruit ovoid or subglobose. . . . . 16. *C. Stonei.*

8. C. intricàta Lange (entangled). — An irregularly branched shrub 1–3.5 m. high, armed with slender thorns; *leaves mostly ovate or oblong-ovate*, serrate nearly to base, usually with 3–5 pairs of spreading lateral *lobes often reflexed at the points*, thin but firm, glabrous at maturity; petioles slender, 2–3 cm. long, usually glandular; flowers 1.3–1.7 cm. wide, few in glabrous nearly simple corymbs with conspicuous glandular bracts; stamens about 10; *anthers white or pale yellow* (*rarely pink*); *fruit* obovoid, oblong, or rarely subglobose, 0.9–1.3 cm. thick, *bronze-green or russet*, with hard dry flesh and 3–5 nutlets. (*C. apposita* Sarg.) — Thickets and open woods, N.E. to Va., w. to Mich. and Ind. *Fl.* May; *fr.* Oct.

Var. **stramínea** (Beadle) Palmer (straw-colored). — Leaves of the flowering branchlets mostly oblong-ovate or elliptic, glabrous except for a few caducous hairs when young; *anthers pink or purple* (*rarely white*); *fruit* about 1 cm. thick, *yellow or greenish-yellow*. (*C. straminea* Beadle; *C. apposita* var. *Bissellii* (Sarg.) Egglest.) — Ct. to Va., w. to Ky. and O.

9. C. Neobúshii Sarg. (a second species named for its discoverer, Benjamin Franklin Bush, 1858–1935). — A shrub 1–3 m. high, with irregular spreading branches armed with occasional thorns or nearly thornless; *leaves mostly ovate or rhombic*, sharply serrate, the lower teeth gland-tipped, usually *with* 3–4 *pairs of shallow lateral lobes*, or more deeply cut on vegetative shoots, thin, glabrous at maturity; flowers 1.6–1.8 cm. wide, 3–8 in usually glabrous corymbs; stamens about 10; anthers pink or rarely white; *fruit* subglobose or short-oblong, 0.9–1.2 cm. thick, *orange-red or greenish-red*, with hard dry flesh and 2–3 nutlets. — Thickets and rocky hillsides, w. Pa. to s. Ill., Mo. and Ark. *Fl.* May; *fr.* Oct.

10. C. Boŷntoni Beadle (for Frank Ellis Boynton, 1858– , its discoverer). — A stout shrub or sometimes a tree 6–8 m. high, with slender flexuous thorny or nearly thornless branchlets; *leaves ovate or oval*, simply or doubly serrate nearly to the base, usually with 3–4 pairs of small acuminate lateral lobes, or *on vegetative shoots sometimes as broad as long or broader* and more deeply cut; flowers 1.3–1.5 cm. wide, 3–10 in mostly glabrous corymbs; stamens about 10; anthers yellowish-white or pink; *fruit* subglobose or short-oblong, 0.8–1.3 cm. thick, *red, or green with red blotches*, with thin dry flesh and 2–3 nutlets. — Banks of streams and borders of woods, mostly in Piedmont reg., s. Pa. to N.C., Tenn. and Ala. *Fl.* May; *fr.* Oct.

11. C. foètida Ashe (ill-scented). — An irregularly branched shrub 1–4 m. high, with slender or stoutish usually thorny branchlets; *leaves* ovate or elliptic, coarsely serrate with broad shallow teeth and usually more or less lobed, or *broadly ovate to suborbicular and coarsely dentate on vegetative shoots*, firm, yellow-green, glabrous at maturity; petioles often glandular; *flowers showy but ill-scented*, 1.6–2 *cm. wide*, mostly 3–6 in compact glabrous corymbs, *with conspicuous glandular bracts;* stamens about 10; anthers pale yellow or rarely pink; *fruit* subglobose or depressed-globose, *sometimes thicker than long, green, dull orange or russet*, with thin dry flesh and 4–5 nutlets. — Thickets and rocky hillsides, s. Ont. and N.E. to Del. Pa. and Mich. *Fl.* May, June; *fr.* Oct.

12. C. rubélla Beadle (somewhat red). — An irregularly branched shrub 1–4 m. high or rarely a tree up to 5 m., with slender thorny or nearly thornless branchlets; leaves mostly elliptic or oval, finely serrate, usually with 1–3 pairs of very shallow lateral lobes above the middle, or sometimes ovate and more deeply divided on vegetative shoots, thin but firm, glabrous at maturity; *flowers* 1.2–1.6 *cm. wide*, mostly 3–6 *in small nearly simple* glabrous *corymbs;* stamens about 10; anthers pink; *fruit obovoid or oblong, red or orange-red*, 0.8–1.1 cm. thick, with thin dry flesh and 3–5 nutlets. — Rocky or open woods, Pa. to N.C. and Ala., w. to Ind. and Ky. *Fl.* April, May; *fr.* Oct.

13. C. padifòlia Sarg. (cherry-leaved). — A stout shrub or tree 5–6 m. high with slender thorny or nearly thornless branchlets; *leaves* ovate or oval, serrate nearly to the base with broad shallow teeth, *very slightly or obscurely lobed* except on vegetative shoots; flowers 1.5–1.8 cm. wide, mostly 3–6 in compact nearly simple glabrous corymbs; stamens about 10; anthers pink; *calyx-lobes deltoid*, entire or finely glandular-serrate; *fruit* subglobose, 0.9–1.2 cm. thick, *often slightly 5-angled, dull orange-red, with hard dry flesh* and 2–3 nutlets. — Rocky hillsides and open woods, s. Mo. and Ark. *Fl.* April; *fr.* Sept.

Var. **incarnàta** Sarg. (flesh-colored). — *Leaves* broadly ovate, *usually with distinct shallow lobes, often as broad as long on vegetative shoots; fruit* 1–1.4 cm. thick, dull red, *with thick juicy flesh* and 3–4 nutlets. — With the typical var.

14. C. fortunàta Sarg. (prosperous). — A shrub 2–3 m. high, with flexuous very thorny branchlets; *leaves* oval or ovate, sharply serrate nearly to the base, slightly lobed above the middle, or *on vegetative shoots* sometimes *nearly orbicular* and sharply lobed, glabrous, yellow-green, *thick*, and with *the veins impressed above at maturity;* flowers 1.5–1.7 cm. wide, mostly 3–8 in small nearly simple glabrous corymbs; stamens about 10; anthers pink; *fruit* short-obovoid or subglobose, 0.7–1.2 cm. thick, *bright yellow*, with a narrow calyx, thin mellow flesh and 2–3 nutlets. (*C. pallens* sensu Egglest. in part, not Beadle) — Pastures and open hillsides, Pa. and O. *Fl.* May; *fr.* Oct.

15. C. biltmoreàna Beadle (of Biltmore, N.C.). — A shrub 1–2 m. high or rarely a tree up to 5 m. high; *leaves* ovate or oval, *sharply serrate, with the lower teeth gland-tipped* and usually with 3–5 pairs of small triangular lateral lobes, or sometimes deeply and sharply lobed on vegetative shoots, *short-villous above and pubescent* along the veins *beneath while young*, firm, glabrous above at maturity; *petioles* slender, 1–3 *cm. long, pubescent and glandular; flowers* 1.5–1.8 cm. wide, mostly 3–7 *in pubescent corymbs;* stamens about 10; anthers white or pale yellow; *calyx-lobes* lanceolate, *usually deeply glandular-serrate, persistent on the fruit; fruit* subglobose or rarely short-obovoid, 0.9–1.2 cm. thick, dull orange or red, *pubescent at least while young*, with thin dry flesh and 3–5 nutlets. (*C. coccinea* sensu Egglest. in part, not L.; *C. intricata* sensu Egglest., not Lange) — Dry rocky woods and hillsides, Vt. to N.C., w. to Mo. and Ark. *Fl.* May; *fr.* Oct. Fig. 1133.

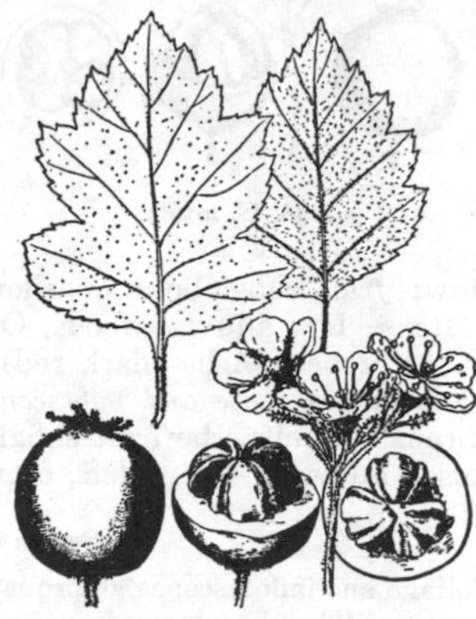

1133. C. biltmoreana.

16. C. Stònei Sarg. (for its discoverer, George Edward Stone, 1861–1941). — A shrub 1–2 m. high, with slender flexuous thorny branchlets usually pubescent when young; *leaves ovate, elliptic or rhombic, deeply serrate, glandular near the base*, usually with 3–4 pairs of small sharp lateral lobes, or *on vegetative shoots* more deeply cut and *with the lowest pair of lobes much enlarged, sparsely villous* above and along the veins beneath *while young*, becoming glabrate above; *flowers* 1.6–2 *cm. wide*, mostly 3–6 in nearly simple pubescent corymbs; stamens about 10; *anthers pink; calyx-lobes glandular-serrate or pectinate;* bracts conspicuously glandular; *fruit* obovoid or short-oblong, 1–1.5 cm. thick, *dull yellow or russet, pubescent toward the base while young*, with thin dry flesh and 3–5 nutlets. — Rocky woods and banks, w. Mass., N.Y. and Pa. *Fl.* May, June; *fr.* Oct.

C. pilòsa Sarg. (hairy), with *leaves ovate or deltoid*, more or less lobed, short-villous above while young, the *flowers* 1.8–2 cm. wide, 3–10 *in glabrous corymbs and with about* 20 *stamens* and red anthers, the subglobose and angled fruit 0.8–1 cm. thick and with 4–5 nutlets, is perhaps a hybrid between species of the *Intricatae* and *Pruinosae*, found at Lancaster, Mass.

Series 7. Vírides Beadle (see p. 768)

Leaves thin, scarcely lustrous above; fruit 5–8 mm. thick; nutlets normally 5. 17. *C. viridis.*
Leaves thick, lustrous above; fruit 0.8–1 cm. thick; nutlets 3–5. . . . . . 18. *C. nitida.*

17. C. víridis L. (green). — A tree sometimes 10–12 m. high, with slender unarmed or sometimes thorny branchlets and thin scaly pale gray bark over orange-brown inner bark; *leaves variable and often asymmetrical, thin, glabrous at maturity except for tufts of tomentum in the axils of the veins beneath*, on flowering branchlets mostly rhombic or oblong-elliptic, finely serrate above the middle or nearly to the base, on vegetative shoots often ovate and sharply serrate and sharply lobed or deeply cut toward the base; petioles slender, 1.2–5 cm. long; *flowers* 1.2–1.5 *cm. wide*, many in glabrous corymbs; stamens about 20; anthers small, pale yellow or rarely red; *fruit* subglobose, 5–8 *mm. thick, red or orange-red, with* thin juicy flesh and *usually* 5 *nutlets.* — Low wet or alluvial woods or rarely on calcareous slopes with seepage-water, Va. to Fla., w. to Ill., Mo. and e. Tex. *Fl.* April; *fr.* Oct. — Forma **padukénsis** Palmer

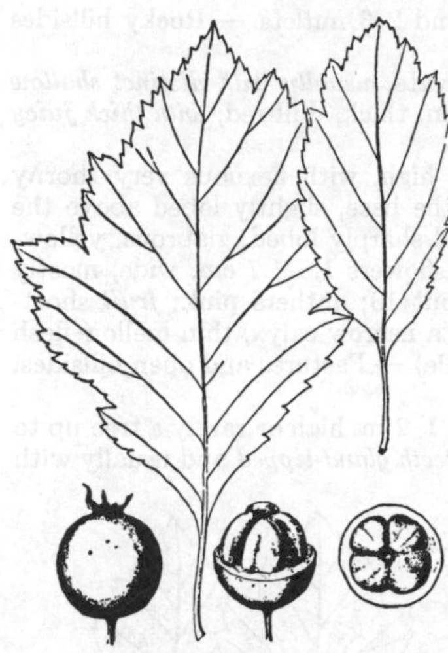

1134. C. nitida.

& Pickens (of Paducah, Ky.) has bright yellow fruit, found near Paducah, Ky. and perhaps elsewhere.

Var. **ovàta** (Sarg.) Palmer (ovate). — Leaves ovate or oblong-ovate, or sometimes nearly orbicular and only slightly lobed on vegetative shoots. (*C. ovata* Sarg.) — N.C. to Mo. and Ark.

Var. **lanceolàta** (Sarg.) Palmer (lanceolate). — Leaves of the flowering branchlets lance-elliptic or oblong-elliptic and on vegetative shoots often ovate or oblong-ovate and sharply lobed. — Ill. to Mo. and southw.

Var. **luténsis** (Sarg.) Palmer (growing in mud). — Leaves of flowering branchlets mostly oblong-obovate or elliptic, slightly longer than wide, deeply and irregularly serrate; terminal shoot-leaves often broadly ovate, slightly lobed. — W. Mo., Kans., and Okla.

18. C. nítida (Engelm.) Sarg. (shining). — A tree 10–12 m. high, with slender flexuous usually thorny branchlets and pale thin scaly bark; *leaves* oblong-lanceolate to elliptic, or sometimes oblong-ovate on vegetative shoots, sharply serrate and often slightly lobed above the middle, *firm or thick, glabrous, lustrous above at maturity; flowers* 1.5–1.7 *cm. wide*, many in glabrous corymbs; stamens about 20; anthers pale yellow; *fruit* subglobose or short-oblong, 0.8–1 *cm. thick*, with thick firm flesh and 3–5 nutlets. — Low alluvial woods, O. to Mo. and Ark. *Fl.* May; *fr.* Oct. Fig. 1134.

C. atrórubens Ashe (dark red), with *leaves* mostly ovate or oblong-elliptic, *seldom lobed, slightly villous above and pubescent beneath while young*, the flowers in slightly pubescent or glabrous corymbs, the fruit subglobose or short-obovoid and 0.8–1 cm. thick, orange-red or crimson and with 4–5 nutlets, occurs near St. Louis, Mo.

Series 8. Crùs-gálli Loud. (see p. 768)

a. Foliage and inflorescence glabrous or essentially so (except in forms and vars. of no. 19). . . *b.*

b. Leaves of flowering branchlets entire except for serrate margins, those of vegetative shoots unlobed or rarely obscurely lobed. . . *c.*

c. Leaves prevailingly obovate or oblong-obovate, on flowering branchlets mostly 1–2 cm. wide, on vegetative shoots larger but similar. . . *d.*

d. Leaves subcoriaceous, usually rounded or short-pointed at the apex; petioles stout, 4–8 mm. long, wing-margined nearly to the base; stamens 10 or fewer; nutlets usually 1–2. . . . . 19. *C. crus-galli.*

d. Leaves thin but firm, usually acutely pointed or acuminate; stamens 10–20; nutlets 1–4, usually 2–3.

Petioles slender, 0.6–2 cm. long, wing-margined above; fruit 7–9 mm. thick. . . . . 20. *C. pyracanthoides*, var. *arborea.*

Petioles stout, 4–8 mm. long, wing-margined nearly to base; fruit 0.9–1.2 cm. thick. . . . . 21. *C. Fontanesiana.*

c. Leaves prevailingly elliptic or oblong-obovate, relatively broad, on flowering branchlets mostly 1.5–2.5 cm. wide or, if narrower, with broadly obovate or suborbicular shoot-leaves. . . *e.*

e. Veins of leaves not noticeably impressed above; branchlets orange-brown or reddish at the end of the first season. . . *f.*

f. Terminal shoot-leaves often suborbicular, two to three times as broad as the average floral leaves.

Leaves of vegetative shoots merely serrate or dentate, usually obtuse or rounded at the apex; anthers red; fruit subglobose; nutlets 3–5. . . . . 22. *C. Reverchoni*, var. *discolor.*

Leaves of vegetative shoots often obscurely lobed above the middle, pointed or acuminate; anthers yellow; fruit ellipsoidal or rarely subglobose; nutlets 2–3. . . . . 23. *C. acutifolia.*

f. Terminal shoot-leaves mostly broadly ovate or oblong-elliptic, seldom more than twice as broad as the average floral leaves.

Leaves sharply pointed or acuminate; lobes of the terminal leaves, if present, usually above the middle. . . . . 24. *C. regalis.*

Leaves usually obtusely pointed or rounded at the apex; lobes of the terminal leaves, if present, usually near the base. . . 25. *C. Palmeri.*

*e.*Veins of leaves prominent, noticeably impressed above; branchlets pale olive-green or yellowish at the end of the first season. . . 26. *C. hannibalensis.*

*b.*Leaves of flowering branchlets sharply and irregularly serrate or dentate and sometimes slightly lobed, usually lobed on sterile shoots. . . *g.*

*g.*Leaves of flowering branchlets prevailingly obovate or oblong-obovate, sharply and doubly serrate, only rarely and slightly lobed.

Leaves thick or subcoriaceous; stamens about 10; fruit becoming succulent; nutlets 3–5. . . . . 27. *C. Canbyi.*

Leaves thin but firm; stamens about 20; fruit remaining hard and dry; nutlets 2–3. . . . . 28. *C. permixta.*

*g.*Leaves of flowering branchlets prevailingly oblong-obovate or rhombic, coarsely dentate and generally slightly lobed. . . . . . . 29. *C. schizophylla.*

*a.*Foliage and inflorescence somewhat pubescent (except in var. of no. 30). . *h.*

*h.*Leaves of flowering branchlets mostly cuneate-obovate, relatively small (0.8–1.5 cm. wide), reticulately veined; petioles 0.5–1 cm. long, wing-margined nearly to base. . . . . 30. *C. Engelmanni.*

*h.*Leaves of flowering branchlets mostly broadly obovate or oblong-obovate, relatively large (1.5–3 cm. wide), not reticulately veined; petioles 1–1.5 cm. long.

Leaves thin for the series, all deeply serrate with narrow acuminate teeth; anthers red; fruit subglobose, 1–1.3 cm. thick. . . . . . 31. *C. fecunda.*

Leaves thick, those of the flowering branchlets finely serrate; anthers yellow; fruit obovoid or ellipsoidal, 7–9 mm. thick. . . . . 24. *C. regalis*, var. *paradoxa.*

19. C. crùs-gálli L. (spur of a cock), COCKSPUR-THORN. — A tree 6–8 m. high, with wide-spreading branches, thorny flexuous branchlets and dark slightly scaly bark; *leaves* of flowering branchlets *mostly obovate*, sharply serrate except towards the cuneate base, at maturity *thick, shining above;* shoot-leaves often oblong-elliptic, sometimes twice as large and coarsely serrate or dentate; flowers 1–1.5 cm. wide, many in glabrous corymbs; stamens about 10, anthers pink or pale yellow; calyx-lobes linear-lanceolate, entire or nearly so; *fruit* short-oblong, slightly obovoid or rarely subglobose, 0.8–1 *cm. thick, often slightly 5-angled, greenish or dull red, with* thin dry flesh and 1–3, *usually* 1–2, *nutlets.* — Thickets and open ground, often in dry or rocky places, se. Can. to S.C., w. to Minn., e. Kans. and e. Tex. *Fl.* May, June; *fr.* Oct. FIG. 1135. — Forma **oblongàta** (Sarg.) Palmer (oblong), with oblong fruit 1 cm. thick and 1.2–1.5 cm. long, in Del. and Pa.; forma **truncàta** (Sarg.) Palmer (truncate), with the leaves mostly obtuse or slightly emarginate at the apex, Ill. and Mo.; forma **rùbens** (Sarg.) Palmer (reddening), with red flesh, Ontario Co., N.Y.

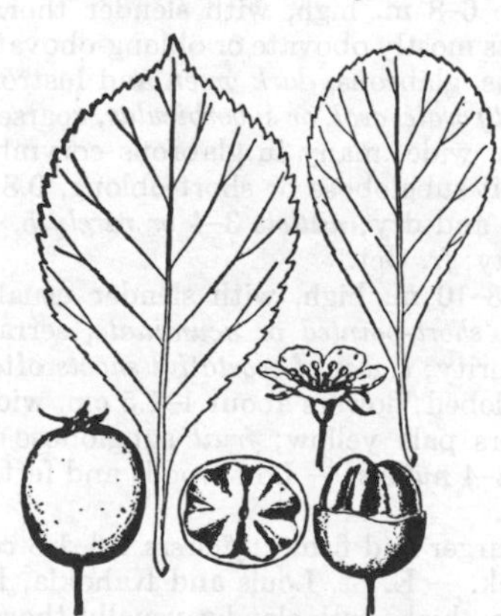

1135. C. crus-galli.

Var. **pyracanthifòlia** Ait. (with leaves of *Pyracantha*). — *Leaves* similar to those of the typical var. but *relatively narrower*, mostly 1–1.5 cm. wide on the flowering branchlets. — Occasional throughout the range of the typical var. and merging into it.

Var. **exígua** (Sarg.) Egglest. (small, insignificant). — *Leaves of vegetative shoots often obscurely lobed; fruit* short-oblong, 0.8–1 *cm. thick, bright red, often* 1-*seeded.* — Ct. to Ga., w. to Mo.

Var. **mácra** (Beadle) Palmer (meagre). — Similar to the last var., but the terminal leaves seldom lobed; fruit subglobose or short-oblong, 7–8 mm. thick. — S. Mo., southw. and eastw.

Var. **capillàta** Sarg. (slender). — Young leaves and inflorescence sparsely villous; flowers about 1 cm. wide. — Occasional, Del., Pa. and Mich.

Var. **leptophýlla** (Sarg.) Palmer (slender-leaved). — Young leaves sometimes sparsely hairy along the veins; *stamens about* 20; *nutlets* often 3–4.

Var. **barrettiàna** (Sarg.) Palmer (geographical name, for Barrett's Station, near St. Louis,

Mo.). — *Leaves rather thin, with prominent veins;* petioles rather slender, 5–7 mm. long; young branchlets orange-green or yellowish. — Ill., Mo. and Ark.

Var. **pachyphýlla** (Sarg.) Palmer (thick-leaved). — *Leaves thicker and averaging slightly larger* than in the typical var.; *fruit* 1.5 *cm. thick,* red, *with* usually 2–3 *nutlets.* — Ill. and Mo.

Var. **béllica** (Sarg.) Palmer (fierce). — *Leaves smaller than in the typical var.; branchlets* stoutish, flexuous and *very thorny;* flowering corymbs sometimes slightly short-pilose; *fruit rarely* 1 *cm. thick.* — Mo. and southw.

20. C. PYRACANTHOÌDES Beadle (resembling *Pyracantha*). — The typical var. is not found in our reg.; represented with us by

Var. **arbòrea** (Beadle) Palmer (tree-like). — A tree 8–10 m. high, with slender thorny or sometimes unarmed branchlets and dark gray scaly bark; *leaves obovate or oblanceolate, pointed* or rarely rounded *at the apex,* serrate except near the base, glabrous, firm, dark green and lustrous above; *flowers* 1–1.3 cm. wide, *mostly* 5–8 *in lax* glabrous *corymbs;* stamens 10 or 20; anthers pale yellow; *fruit* subglobose or short-oblong, 0.6–1 *cm. thick, red, with* thin mellow flesh and 2–3 *nutlets.* — Fertile or moist ground along streams, Ind. to Mo. and southw. *Fl.* April; *fr.* Oct.

C. ohioénsis Sarg. (of Ohio) has foliage similar to that of no. 20, the *flowering corymbs slightly hairy,* the anthers pink, the *fruit* obovoid *with persistent erect calyx-lobes,* found in Franklin Co., O.

21. C. Fontanesiàna (Spach) Steud. (named for RENÉ LOUICHE DESFONTAINES, French botanist, 1750–1833). — A tree 6–8 m. high; *leaves* obovate, lance-obovate or elliptic, sharply or doubly serrate, *often slightly lobed on vegetative shoots,* firm, glabrous, yellow-green at maturity; flowers 1.2–1.8 cm. wide, many in glabrous corymbs; stamens 10 or 20; anthers pink; *calyx-lobes* nearly entire, *persistent on fruit; fruit* subglobose or short-oblong, 1–1.5 cm. long and nearly as thick, green or red, *with* thin firm flesh and 2–3 *nutlets.* — Open ground and thickets, Ont. and Mich. to Pa. and O. *Fl.* May; *fr.* Oct.

C. ténax Ashe (tenacious), with similar leaves but with slightly larger flowers and fruit, may be distinct. — Ont. and Mich.

22. C. REVERCHÒNI Sarg. (named for JULIEN REVERCHON, 1837–1905). — The typical var. is not found in our range.

Var. **díscolor** (Sarg.) Palmer (of two colors). — A tree 6–8 m. high, with slender thorny branchlets and thin scaly bark; *leaves* of flowering branchlets mostly obovate or oblong-obovate, sharply serrate except near the base, firm to subcoriaceous, glabrous, *dark green* and lustrous *above, paler beneath; terminal leaves of vegetative shoots broadly ovate, oval, or suborbicular,* coarsely dentate and *sometimes obscurely lobed;* flowers 1.2–1.7 cm. wide, many in glabrous corymbs; stamens 10 or 20; anthers pink or rarely pale yellow; fruit subglobose or short-oblong, 0.8–1 cm. thick, dull red or green blotched with red; flesh thin and dry; *nutlets* 3–4 *or rarely* 5. — Fertile ground along streams, Mo., Ark. and Okla. *Fl.* May; *fr.* Oct.

23. C. acutifòlia Sarg. (with acute leaves). — A tree 8–10 m. high, with slender usually thorny branchlets; *leaves mostly oblong-obovate or elliptic, short-pointed or acuminate,* serrate except near the base, firm, yellow-green, glabrous at maturity; *leaves of vegetative shoots* often *broadly ovate to suborbicular,* coarsely toothed and slightly lobed; flowers about 1–1.5 cm. wide, several, in lax glabrous corymbs; stamens 10–15; anthers pale yellow; *fruit* subglobose or slightly obovoid, 0.7–1 cm. thick, *with* thin dry flesh and 2–4 *nutlets.* — Low woods and fertile ground along streams, Ind., Ill. and Mo. *Fl.* May; *fr.* Oct.

Var. **insígnis** (Sarg.) Palmer (well-marked). — Leaves larger and firmer; *flowers* 1.6–1.8 *cm. wide; pedicels sometimes slightly hairy;* fruit 1–1.5 cm. thick. — E. St. Louis and Kahokia, Ill.

24. C. regàlis Beadle (royal, splendid). — A tree 8–10 m. high, with slender usually thorny branchlets and scaly bark; *leaves* mostly elliptic or broadly obovate, sharply serrate or *on vegetative shoots coarsely dentate* and *with some of the teeth enlarged and lobe-like,* at maturity glabrous, firm but thin for the series; flowers 1.2–1.5 cm. wide, several, in glabrous corymbs; stamens about 10; anthers pale yellow; *fruit oblong or rarely subglobose,* 6–8 *mm. thick* and usually slightly longer, green or dull red, with thin dry flesh and 2–3 nutlets. — Fertile ground along streams, N.C. to Ind., Mo. and Ark. *Fl.* May; *fr.* Oct.

Var. **paradóxa** (Sarg.) Palmer (unexpected). — Young branchlets, foliage and inflorescence slightly pubescent; fruit ellipsoidal or oblong. — Mo., Kans. and Ark.

25. C. Pálmeri Sarg. (for its discoverer, ERNEST JESSE PALMER, 1875–). — A tree 8–10 m. high, with sparingly thorny or nearly thornless branchlets and pale gray bark; *leaves broadly obovate or elliptic,* serrate with shallow teeth, or *on vegetative shoots* sometimes *ovate or nearly orbicular* and *slightly lobed near the base,* firm but scarcely coriaceous; flowers 1.2–1.5 cm. wide, several, in lax glabrous corymbs; stamens about 10; anthers pale yellow; *fruit* subglobose,

6–8 *mm. thick*, pale dull red, with thin firm flesh and usually 3 nutlets. — Fertile uplands along small streams, Mo. and e. Kans. to Ark. and Okla. *Fl.* May; *fr.* Oct.

26. C. hannibalénsis Palmer (of Hannibal, Mo.). — A small tree or shrub *with* slender *pale* thorny *branchlets; leaves* mostly oblong-obovate elliptic, serrate, *yellowish-green*, firm to subcoriaceous and *with the veins impressed above;* flowers 1.4–1.6 cm. wide, 6–16 in glabrous corymbs; stamens about 10; anthers pale yellow; fruit short-oblong or slightly obovoid, 7–8 mm. thick, green or dull red, with thin dry flesh and 1–3, usually 2, nutlets. — Thickets in fertile calcareous soil, Ill., Ia. and Mo. *Fl.* May; *fr.* Oct.

C. vallícola Sarg. (inhabiting valleys). — Foliage and flowers similar to those of no. 26 but *with the flowering corymbs sometimes sparsely villous; fruit* subglobose or short-oblong, 1.2–1.6 *cm. thick.* — O.? to Mo.

C. persímilis Sarg. (very similar). — Somewhat similar to no. 26 in vegetative characters; stamens 10–20; anthers red; styles 3–4; *calyx-lobes glandular-serrate;* fruit short-oblong or subglobose. — Possibly a hybrid between species of the *Crus-galli* and *Macracanthae*, found in e. and centr. N.Y.

27. C. Cánbyi Sarg. (for its discoverer, William Marriott Canby, 1831–1904). — A shrub or small tree 5–7 m. high, with pale brown flexuous thorny branchlets; *leaves* mostly oblong-obovate or elliptic, pointed or rarely rounded at the apex, sharply serrate and *often slightly lobed on vegetative shoots*, glabrous, subcoriaceous at maturity; flowers 1–1.4 cm. wide, several, in glabrous corymbs; stamens 10–20; anthers pink; *fruit* oblong or subglobose, 7–9 mm. thick, *bright red, with thick juicy flesh and* 3–5 *nutlets.* — Thickets and banks of streams, e. Pa., Del. and Md. *Fl.* May; *fr.* Oct.

28. C. permíxta Palmer (much mixed). — A tree 5–6 m. high, with slender often flexuous thorny branchlets and dark gray scaly bark; *leaves oblong-obovate, elliptic or rhombic,* sharply serrate except near the base, or *on vegetative shoots with some of the teeth enlarged and lobe-like,* firm but *thin for the series*, glabrous; flowers 1–1.4 cm. wide, mostly 5–12 in compact glabrous corymbs, *stamens about* 20; anthers pink; *fruit* subglobose or obovoid, 0.8–1.2 cm. thick, *dark orange-red,* with thin firm flesh and 2–3 nutlets. (*C. intermixta* Sarg., not Beck) — Open woods and thickets, Ill. and Mo. *Fl.* May; *fr.* Oct.

29. C. schizophýlla Egglest. (with cut or split leaves). — A shrub or rarely a small tree 4 m. high, with very thorny flexuous branches and branchlets; *leaves* oblong-obovate, rhombic or elliptic, *coarsely and irregularly serrate or dentate*, and *more or less lobed on vegetative shoots*, firm but not coriaceous, deep green, *lustrous* and *with the veins impressed above at maturity; petioles slender,* 0.5–1.5 *cm. long*, wing-margined above, often slightly glandular; flowers 1.2–1.5 cm. wide, several in glabrous corymbs; stamens about 10; anthers pink; *calyx-lobes persistent on fruit; fruit* short-oblong or obovoid, 7–9 mm. thick, *bright red*, with thin dry flesh and usually 2 nutlets. — Thickets and open sandy ground, Martha's Vineyard, Mass. *Fl.* May, June; *fr.* Oct.

30. C. Engelmánni Sarg. (for George Engelmann, 1809–1884). — A stout shrub or small tree 6–7 m. high, with a broad flat-topped crown of stiff spreading branches, slender often flexuous thorny *branchlets villous while young*, and gray-brown scaly or fissured bark; *leaves* mostly obovate or oblong-obovate, or sometimes elliptic on vegetative shoots, usually *slightly pilose above or* on the veins *beneath while young*, subcoriaceous, *reticulately veined at maturity; petioles stout,* 3–8 *mm. long, villous*, wing-margined nearly to the base; *flowers* 1.2–1.5 cm. wide, several *in slightly villous corymbs;* stamens about 10; anthers pink or pale yellow; fruit subglobose or short-oblong, 6–8 mm. thick, dull crimson, with thin dry flesh and 1–3 nutlets. (*C. berberifolia*, var. (Sarg.) Egglest.) — Thickets and open rocky ground, Ill. to Okla. and southw. *Fl.* May; *fr.* Oct. — Forma **núda** Palmer (naked, without hairs) differs only in the young branchlets, foliage, and inflorescence being glabrous or essentially so, occasional with the typical form.

Var. **sinístra** (Beadle) Palmer (left, perverse). — *Inflorescence and petioles densely pilose-pubescent;* stamens 10 or rarely 20. — Mo., Tenn. and Ark.

31. C. fecúnda Sarg. (fruitful, prolific). — A tree 7–8 m. high, with slender thorny branchlets slightly villous while young; *leaves* obovate or elliptic, *sharply and deeply serrate nearly to the base,* or *on vegetative shoots coarsely and irregularly dentate and* sometimes *slightly lobed, short-villous above and slightly villous* on the veins *beneath while young,* glabrous, lustrous, and with the veins slightly impressed above at maturity; *petioles* slender, 0.6–1.5 cm. long, *at first pubescent* and *often glandular;* flowers 1.3–1.6 cm. wide, several in lax slightly villous corymbs; stamens 10 or rarely 20; anthers pink; *fruit* subglobose or short-oblong, 1–1.4 *cm. thick, slightly pubescent while young,* green or dull orange-red, with thick firm flesh and 2–3 nutlets. — Open woods, usually in moist or fertile ground along streams, Ill. and Mo. *Fl.* May; *fr.* Oct.

Series 9. PUNCTÀTAE Loud. (see p. 768)

a. Leaves of flowering branchlets prevailingly obovate or oblong-obovate, broadest above the middle, gradually narrowed below to the short wing-margined petiole. . b.

b. Terminal leaves of vegetative shoots prevailingly obovate, deeply cut or with narrow spreading lateral lobes near the apex; fruit 1.2–2 cm. thick. . . 32. *C. punctata.*

b. Terminal leaves of vegetative shoots prevailingly elliptic, not deeply cut but sometimes slightly lobed. . . c.

c. Leaves dull green above; flowering corymbs villous; nutlets 3–5. . . *d.*

d. Leaves of flowering branchlets finely serrate; flowers usually less than 2 cm. wide, in small mostly 5–10-flowered corymbs. . . . . . 33. *C. collina.*

d. Leaves sharply serrate or dentate; flowers 1.8–2.3 cm. wide, in many-flowered corymbs.

Leaves mostly acute, usually unlobed except sometimes on vegetative shoots; flowering corymbs lax. . . . . . 34. *C. verruculosa.*

Leaves pointed or rounded at apex, usually lobed; flowering corymbs rather compact. . . . . . 35. *C. Lettermani.*

c. Leaves slightly lustrous above; nutlets 2–3. . . *e.*

e. Leaves, young branchlets and inflorescence glabrous or nearly so; anthers pink.

Leaves green on both sides or only slightly paler beneath, pointed or acuminate, usually lobed on vegetative shoots; nutlets 1–3, usually 2. . . . . . 36. *C. disperma.*

Leaves decidedly paler beneath, mostly obtuse or rounded at the apex, only slightly if at all lobed even on vegetative shoots; nutlets 2–3. . . . . . 37. *C. peoriensis.*

e. Leaves, young branchlets and inflorescence villous or pubescent; anthers pale yellow. . . . . . 38. *C. incaedua.*

a. Leaves of flowering branchlets prevailingly oval or elliptic, broadest about the middle, gradually or abruptly narrowed to the slender petioles. . *f.*

f. Veins of the leaves slightly impressed above; flowering corymbs glabrous; calyx-lobes entire or nearly so; nutlets 3–5.

Terminal shoot-leaves oval or suborbicular; flowers 1.6–2.2 cm. wide; nutlets usually 5. . . . . . 39. *C. suborbiculata.*

Terminal shoot-leaves ovate or oblong-elliptic; flowers 1.5–1.7 cm. wide; nutlets 3–4. . . . . . 40. *C. Kellermanii.*

f. Veins of the leaves strongly impressed above; flowering corymbs more or less villous; nutlets usually 3. . . . . . 41. *C. celsa.*

32. C. punctàta Jacq. (dotted, punctate). — A tree 8–10 m. high, with stiff spreading branches, stout usually thorny branchlets and brownish-gray bark fissured on the trunk; *leaves* mostly obovate, serrate above the middle, usually slightly lobed toward the apex, or *on vegetative shoots sometimes oblong-elliptic* and *deeply lobed or laciniate,* roughened *with short appressed hairs above while young and more or less hairy* along the veins *beneath,* firm, *dull green,* and *with the veins distinctly impressed above at maturity; flowers* 1.3–2 cm. wide, *several, in pubescent corymbs;* stamens about 20; anthers red or yellow; calyx densely gray-pubescent; *fruit* subglobose or short-oblong, 1.2–2 *cm. thick, with thick mellow flesh* and 3–5 nutlets. — Open rocky ground, thickets and pastures, e. Can. and N.E. to e. Ky., Ind. and Ia. *Fl.* May; *fr.* Sept., Oct. FIG. 1136.

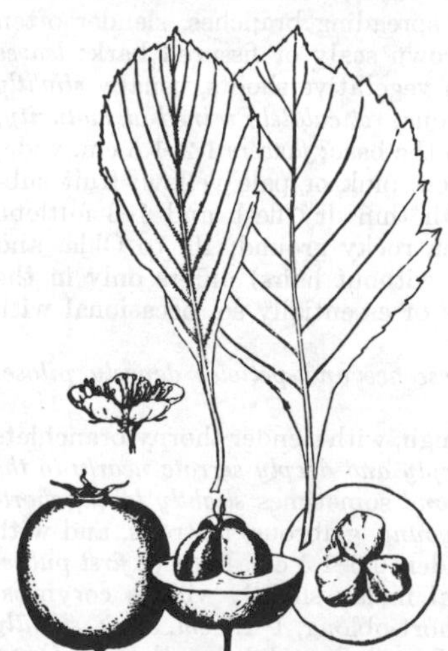
1136. C. punctata.

Var. **aùrea** Ait. (golden). — With *bright yellow fruit* and usually yellow anthers. — Occasional throughout the range of the species and often common northw.

Var. **canéscens** Britt. (gray, hoary). — Foliage and inflorescence densely gray-pubescent. — Ont. to Va., w. to O. and Ky.

Var. **microphýlla** Sarg. (small-leaved). — *Leaves smaller* (2–2.5 cm. long, 1.5–2 cm. wide); flowers 1–1.2 cm. wide, in compact few-flowered corymbs. — Pa. and O.

Var. **pausìaca** (Ashe) Palmer (olive-shaped). —

Foliage and inflorescence glabrous or nearly so, the *leaves slightly lustrous above;* flowers 1–1.5 cm. wide, with 10–20 stamens; nutlets 2–3. — Perhaps a hybrid with some species of the series *Crus-galli.* — Ont. to N.C., W.Va. and O., occasional.

33. C. collìna Chapm. (of the hills). — A tree 6–8 m. high or sometimes a stout shrub with stoutish thorny branchlets; *leaves* mostly obovate, serrate except near the base, *usually unlobed,* or on vegetative shoots sometimes broadly ovate or elliptic and slightly lobed above the middle, pubescent while young, dull green, firm and with the *veins slightly impressed above* at maturity; flowers 1.5–2 cm. wide, few, in nearly simple pubescent corymbs; stamens 10–15 or rarely 20; anthers pale yellow or pink; *fruit* subglobose, 0.8–1 *cm. thick,* dull red, pubescent while young, with thin firm flesh and 3–5 nutlets. — Open woods and thickets, usually in calcareous reg., Va. to S.C., w. to e. Kans. and Okla. *Fl.* April, May; *fr.* Oct.

Var. sórdida (Sarg.) Egglest. (dirty). — *Leaves comparatively thin, slightly lustrous above* at maturity; stamens about 20; anthers pink; calyx-lobes serrate or glandular-serrate; nutlets 2–4, usually 3. — S. Mo.

Var. collícola (Ashe) Palmer (inhabiting hills). — *Glabrous or nearly so throughout,* or sometimes with a few scattered hairs on the inflorescence; leaves subcoriaceous, slightly lustrous above; *fruit dull red or orange-red.* — Perhaps a hybrid with a species of the series *Crus-galli,* Va. and N.C. to Ky.

Var. sécta (Sarg.) Palmer (cut, divided). — Leaves mostly oblong-obovate or rhombic, *with small shallow lobes above the middle,* pubescent while young; *stamens* 5–10, *usually* 5. — S. Mo. and Ark.

Var. succíncta (Sarg.) Palmer (short). — Leaves mostly obovate, relatively narrow, slightly pubescent; *stamens about* 20; anthers yellow; *nutlets usually* 5. — S. Mo. and Ark.

34. C. verruculòsa Sarg. (rough, warty). — A tree 7–8 m. high with slender thorny branchlets villous while young; *leaves* oblong-obovate or elliptic, sharply serrate to below the middle, or *on vegetative shoots* sometimes ovate, *coarsely toothed and with* 2–3 *pairs of small lateral lobes,* short-villous above and slightly pubescent beneath while young, firm, yellow-green, *with the prominent veins impressed above* at maturity; flowers 1.8–2.3 cm. wide, several in loose villous corymbs; stamens about 20; anthers pink; fruit subglobose, 1–1.3 cm. thick, dark red, pubescent while young, with thick firm flesh and 3–5 nutlets. — Fertile open ground along streams, Ky. to Mo. and Ark. *Fl.* April, May; *fr.* Oct.

35. C. Lettermáni Sarg. (named for George Washington Letterman, 1841–1913). — A tree 6–7 m. high, with slender thorny branchlets tomentose while young; *leaves ovate or rhombic,* sometimes *broadly ovate on vegetative shoots,* sharply serrate, usually with 3–5 pairs of small shallow lateral lobes, short-villous above and slightly pubescent beneath while young; *flowers* 1.5–1.8 cm. wide, *several, in leafy-bracted pubescent corymbs;* stamens 10–15; anthers pale yellow; calyx-lobes pubescent, glandular-serrate; fruit subglobose, 1–1.4 cm. thick, with thick firm flesh and usually 5 nutlets. — Rich woods, uncommon, Mo. *Fl.* May; *fr.* Oct. — Perhaps a hybrid between a var. of *C. collina* and *C. mollis.* (*C. collina,* var. (Sarg.) Egglest.)

36. C. dispérma Ashe (two-seeded). — A tree 7–8 m. high, with slender glabrous usually thorny branchlets; *leaves* obovate or elliptic, or sometimes oblong-ovate on vegetative shoots, *finely serrate often only above the middle,* sometimes obscurely lobed near the apex, thick, *glabrous,* dark green and slightly *lustrous above at maturity, the veins slightly impressed;* flowers 1.3–1.8 cm. wide, mostly 5–12 in glabrous or sparsely villous corymbs; stamens about 10; anthers pink; fruit oblong or subglobose, 0.8–1 cm. thick, dull or bright red, with thin firm flesh and 2–3, *usually* 2, *nutlets.* — Open ground and borders of woods, Pa. to Va., w. to Ind. and Ia. *Fl.* May; *fr.* Oct. — Intermediate in characters between *C. crus-galli* and *C. punctata* and may have originated as a hybrid between them.

37. C. peoriénsis Sarg. (of Peoria, Ill.). — A tree 7–8 m. high, with slender glabrous thorny branchlets and dark gray-brown scaly bark; leaves mostly obovate, sharply serrate to below the middle, often obscurely lobed near the apex, or sometimes broadly elliptic or rhombic on vegetative shoots, glabrous, firm, dark green, lustrous, and with the veins impressed above at maturity; flowers 1–1.5 cm. wide, many in glabrous or sparsely villous corymbs; *stamens about* 10; *anthers red; calyx-lobes serrate or glandular-serrate;* fruit oblong or subglobose, 0.9–1.2 cm. thick, with thin hard flesh and 2–3, usually 3, nutlets. (*C. pratensis* Sarg. in part) — Open woods along streams, usually in moist argillaceous soil, n. and centr. Ill. *Fl.* May; *fr.* Oct.

38. C. incaèdua Sarg. (uncut). — A stout shrub or tree 6–7 m. high, with slender flexuous thorny branchlets villous while young; leaves mostly obovate or oblong-elliptic, sharply serrate to below the middle, or on vegetative shoots broadly oval or elliptic and coarsely serrate, short-villous above and pubescent beneath when young, subcoriaceous, glabrous, and with the veins impressed above at maturity; *flowers* 1.3–1.5 cm. wide, *several, in lax villous corymbs;* stamens about 10; anthers pale yellow; fruit short-oblong or subglobose, 0.8–1 cm. thick, green, dull yellow, or red, with thin hard flesh and 2–3, usually 2, nutlets. — Thickets and rocky open ground along streams, Ind., Mo. and Ark. *Fl.* May; *fr.* Oct. — Perhaps a hybrid between *C. collina* and *C. Calpodendron.*

C. pùberis Sarg. (downy), with *leaves* similar in outline and pubescence to those of no. 38, but *thinner and with* 4–5 *pairs of small spreading lateral lobes*, the *few flowers* 1.2–1.3 cm. wide and *in compact* villous *corymbs*, the *fruit* short-oblong, *bright red and pruinose, with* 3–4 *nutlets*, known from Tuscarora, N.Y.

39. C. suborbiculàta Sarg. (nearly orbicular; referring to the roundish terminal leaves). — A tree 6–7 m. high or sometimes a stout shrub with glabrous often flexuous thorny branchlets; *leaves mostly oblong-obovate to suborbicular*, sharply serrate and slightly or, on vegetative shoots, sometimes deeply lobed, glabrous, thin but firm, the veins impressed above at maturity; *flowers* 1.6–2 *cm. wide*, mostly 5–12 in glabrous corymbs; stamens about 20; anthers pink or rarely white; fruit subglobose or short-oblong, 1–1.5 cm. thick, dull red, with thin hard flesh and 3–5, *usually* 5, *nutlets*. — Calcareous hills and borders of woods, se. Can., N.Y. and Ct. *Fl.* May; *fr.* Oct.

C. recèdens Sarg. (retiring), is very similar, but with the young leaves and inflorescence somewhat villous, the nutlets 3–5, usually 3, found near Kutztown, Pa.

C. nitídula Sarg. (somewhat shining), is also similar to *C. suborbiculata* but with narrower oblong-obovate leaves and with few smaller (1.2–1.3 cm.) flowers in compact glabrous corymbs, the fruit subglobose or short-oblong, 1–1.3 cm. thick, in s. Ont. and near Port Huron, Mich.

C. Neobáxteri Sarg. (a second species named, in 1905, for M. S. Baxter, its discoverer), has leaves quite similar to those of *C. nitidula*, but the flowers 1.6–1.8 cm. wide in loose glabrous corymbs and the fruit short-oblong, bright red and pruinose with 3–4 nutlets, found near Tuscarora, N.Y.

40. C. Kellermánii Sarg. (for its discoverer, William Ashbrook Kellerman, 1850–1908). — A stout shrub or small tree with flexuous thorny glabrous branchlets; *leaves* oblong-obovate, finely serrate, usually *with* 4–5 *pairs of small acute lateral lobes above the middle*, glabrous, firm at maturity; flowers 1.3–1.5 cm. wide, mostly 6–10, in glabrous corymbs; *stamens about* 20; *anthers pink; fruit* short-oblong or subglobose, *slightly* 5-*angled*, 1.2–1.5 cm. thick, red or orange, punctate, with dry mealy flesh and 4–5 nutlets. — Thickets and pastures, n. and centr. O. *Fl.* May; *fr.* Oct. — Perhaps a hybrid between *C. punctata* and a species of the *Pruinosae.*

C. silvéstris Sarg. (of the forest), similar to *C. Kellermanii* but the larger dark yellowish green *leaves roughened above with short appressed hairs and slightly villous beneath while young, the* 5–8 *flowers in nearly simple villous corymbs*, the *fruit* about 1 cm. thick and *with a large prominent calyx* and usually 3 nutlets, found near London, Ont.

C. desuèta Sarg. (out of use), with *leaves* oblong-obovate or elliptic and *with* 4–6 *pairs of spreading or recurved lateral lobes*, glabrous or nearly so, the many flowers 1.2–1.5 cm. wide and in lax glabrous or slightly villous corymbs and with about 10 stamens and red anthers, the fruit short-oblong, 7–9 mm. thick, bright red, lustrous and with 3–4 nutlets, found in n. and centr. N.Y.

41. C. célsa Sarg. (high, elevated). — A stout shrub or tree 7–8 m. high, with slender thorny branchlets sometimes slightly villous when young; *leaves* obovate or rhombic, serrate, *with* 4–5 *pairs of small or obscure lateral lobes above the middle*, or sometimes broadly elliptic and more deeply divided on vegetative shoots, *slightly pubescent while young, firm, glabrous, lustrous and with the veins distinctly impressed above at maturity*, the many flowers 1.3–1.6 cm. wide, in villous corymbs; stamens about 20; anthers rose-color; fruit subglobose, 0.9–1.3 cm. thick, crimson, lustrous, with thin juicy flesh and usually 3 nutlets. — Thickets and hillsides, in fertile soil, s. Ont. and N.Y. *Fl.* May, June; *fr.* Oct. — Perhaps a hybrid between *C. punctata* and *C. succulenta.*

Series 10. Rotundifòliae Egglest. (see p. 769)

a. Leaves of flowering branchlets with rather uniform sharp spreading lobes. . . *b.*

b. Leaves of flowering branchlets mostly oblong-ovate or oval, short-pointed, often suborbicular on vegetative shoots. . . *c.*

c. Petioles and inflorescence glabrous or sparsely pubescent; fruit remaining hard until late in the season.

Inflorescence conspicuously bracteate; calyx-lobes glandular-serrate; fruit not pruinose. . . . . 42. *C. chrysocarpa.*

Inflorescence not conspicuously bracteate; calyx-lobes entire or nearly so; fruit sometimes pruinose. . . . . 43. *C. rotundata.*

*c.*Petioles and inflorescence villous or tomentose; fruit early ripening, succulent.
Leaves of vegetative shoots laciniately divided near the base; pubescence villous, often persistent along the veins beneath. . . . . 44. *C. irrasa.*
Leaves of vegetative shoots not laciniately divided; pubescence copious, of matted hairs; leaves soon glabrate. . . . . 45. *C. Faxoni.*
*b.*Leaves of flowering branchlets mostly elliptic, acuminate, attenuate at base, broader but not suborbicular on vegetative shoots.
Leaves of flowering branchlets seldom over 5 cm. long; flowers 1.5–2 cm. wide; fruit 0.9–1.2 cm. thick. . . . . 46. *C. Brunetiana.*
Leaves of flowering branchlets often 6–8 cm. long; flowers 2–2.3 cm. wide; fruit 1–1.3 cm. thick. . . . . 47. *C. Jonesae.*
*a.*Leaves of flowering branchlets with shallow, rounded or obscure lobes developed only above the middle of the blades or sometimes obsolete. . . *d.*
*d.*Leaves usually with 3–4 pairs of primary veins, the lowest pair at least strongly ascending; corymbs compound, mostly 8–12-flowered. . . *e.*
*e.*Leaves of flowering branchlets entire toward the usually cuneate or attenuate base; flowers 1.3–1.5 cm. wide.
Lobes of the leaves often rounded or sometimes obsolete; serration shallow or crenate; stamens about 20. . . . . 48. *C. Margaretta.*
Lobes of the leaves shallow but acute; serration sharp; stamens about 10. . . . . 49. *C. Dodgei.*
*e.*Leaves of flowering branchlets sharply serrate nearly to the abruptly narrowed base; flowers 2–2.3 cm. wide. . . . . 50. *C. Jackii.*
*d.*Leaves usually with 4–5 pairs of pinnate spreading primary veins; corymbs nearly simple, mostly 5–8-flowered. . . . . 51. *C. sicca.*

42. C. chrysocárpa Ashe (golden-fruited). — An intricately branched shrub or rarely a tree 5–6 m. high, with stoutish flexuous very thorny branchlets; *leaves* mostly *oval, elliptic or suborbicular,* serrate except near the base with *the lower teeth gland-tipped,* with 3–4 pairs of small triangular lateral lobes, roughened above with short appressed hairs while young, firm, yellow-green, glabrous, with the veins slightly impressed above at maturity; flowers 1.3–1.6 cm. wide, in loose villous corymbs; *stamens about* 10; anthers white or pale yellow; *calyx-lobes serrate or glandular-serrate;* fruit subglobose or short-oblong, 0.8–1 cm. thick, dark red or rarely yellow, remaining green until late in the season; nutlets 3–4. (*C. rotundifolia* var. *pubera* Sarg.; *C. rotundifolia* Moench, in part, not Lam.) — Thickets and rocky ground along streams, Nfld. and Que. to N.E. and N.Y., w. to Man., Colo. and N.M. *Fl.* May; *fr.* Oct. Fig. 1137. — Forma **rubéscens** (Sarg.) Palmer (reddening) has the foliage dark red while young and sometimes slightly red throughout the season; found in Que.

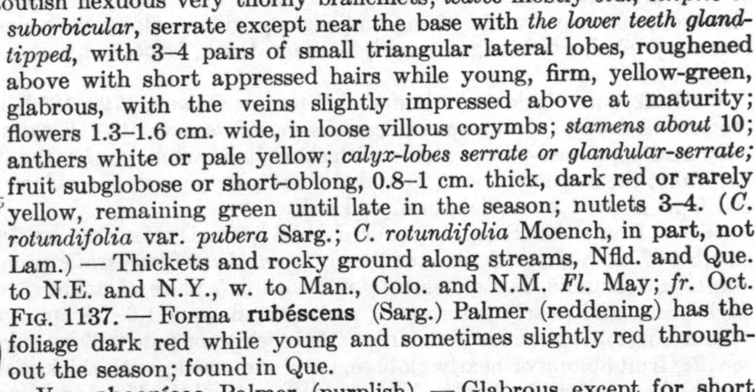

1137. C. chrysocarpa.

Var. **phoenícea** Palmer (purplish). — Glabrous except for short appressed hairs sometimes found on the upper side of the young leaves; *fruit dark red or purplish-red.* (*C. rotundifolia* Moench, in part, not Lam.) — Se. Can., N.E., and N.Y. to Wisc.

Var. **Bicknéllii** (Egglest.) Palmer (for its discoverer, Eugene Pintard Bicknell, 1859–1925). — *Leaves slightly larger and more deeply lobed* than in the other vars., with *points of the lobes often reflexed,* glabrous except on the upper surface while young; flowers in slightly villous corymbs; *anthers pink.* (*C. Bicknellii* Egglest.; *C. rotundifolia* var. *Bicknellii* Egglest.) — Nantucket Id., Mass.

Var. **caesariàta** (Sarg.) Palmer (covered with hairs). — Corymbs sparsely villous; *fruit oblong or obovoid, slightly pubescent at the ends while young;* nutlets 2–3; found near Albany and Wynantskill, N.Y.

C. Ideae Sarg. (named in 1905 for Mary Ellen Ide of St. Johnsbury, Vt.) with leaves similar in shape to those of *C. chrysocarpa,* the *flowers many in villous corymbs* and *with* about 10 stamens and *pink anthers,* the fruit subglobose and with thin flesh and 3–4 nutlets, is possibly a hybrid between *C. chrysocarpa* and *C. Brainerdi;* found near Concord, Vt.

C. Robinsòni Sarg. (for its discoverer, Charles Budd Robinson, 1871–1913) with *leaves mostly elliptic or rhombic,* narrowed at both ends, the flowers few in sparsely villous corymbs; the *fruit oblong or obovoid and 7–8 mm. thick, bright scarlet,* with juicy flesh and usually 3 *nutlets slightly pitted* on the inner surface, is perhaps a hybrid of similar relationship to the last, found near Pictou, N.S.

43. C. rotundàta Sarg. (rounded). — A stout shrub 3–4 m. high, with flexuous thorny branchlets; leaves ovate, oblong-elliptic or suborbicular, sharply serrate nearly to the base, usually with 3–4 pairs of shallow triangular lateral lobes, glabrous, firm at maturity; flowers, 1.3–1.5 cm. wide, few in compact glabrous corymbs; stamens 10 or less; anthers white; *fruit* subglobose or short-oblong, 1–1.5 cm. thick, *red or orange-red, sometimes with a slight bloom,* with thick flesh and 2–4 nutlets. — Thickets and pastures, Ont., N.Y. and Ct. *Fl.* May; *fr.* Oct.

44. C. irràsa Sarg. (unshorn). — A shrub spreading into thickets or sometimes a tree 4–5 m. high, with slender flexuous thorny branchlets; *leaves* ovate or elliptic, serrate nearly to the base and *deeply divided, with 4–5 pairs of acuminate spreading lateral lobes,* short appressed-pubescent above while young and *slightly pubescent beneath throughout the season;* flowers 1.3–1.5 cm. wide, mostly 6–12 in loose villous corymbs; stamens about 20; *anthers small, pale yellow;* calyx-lobes finely glandular-serrate; *bractlets early deciduous;* fruit oblong or subglobose, 0.8–1 cm. thick, bright red, with thin succulent flesh and 3–5 nutlets. — Rocky fields and borders of woods, Que. and N.Y. *Fl.* May; *fr.* Sept.–Oct.

Var. **Blanchárdi** (Sarg.) Egglest. (for its discoverer, William Henry Blanchard, 1850–1922). — *Flowers* slightly larger, *in compact villous corymbs;* anthers pink; *bractlets persistent through anthesis.* — Que., Vt. and N.Y.

45. C. Fáxoni Sarg. (for its discoverer, Charles Edward Faxon, 1846–1918). — Arborescent shrub 3–4 m. high, with numerous stems and slender thorny *branchlets villous while young; leaves* ovate, elliptic or rarely obovate, sharply serrate and with 4–6 pairs of small acute lateral lobes, *thickly coated above with short appressed hairs and densely pale-villous beneath while young,* firm, *nearly glabrous at maturity;* flowers 1–1.4 cm. wide, few in compact densely pubescent corymbs; stamens 5–10; anthers pale yellow; fruit oblong or nearly globose, 0.8–1 cm. thick, with thin dry flesh and 3–4 nutlets. (*C. irrasa* var. *Faxoni* (Sarg.) Egglest.) — Rocky pastures and borders of woods, Que., N.E. and N.Y. to Wisc. and Ill. *Fl.* May; *fr.* Sept., Oct.

Var. **praetermíssa** (Sarg.) Palmer (overlooked). — Leaves mostly oblong-elliptic or ovate, short-villous above while young and slightly villous along the veins beneath throughout the season; *fruit* short-oblong or ovoid, 8–9 mm. thick, *bright red, pubescent while young.* — Ferrisburg, Vt.

C. Oakesiàna Egglest. (in honor of William Oakes, 1799–1850) has the elliptic or broadly (*sometimes broader than long on vegetative shoots*) *leaves with 4–6 pairs of small acuminate lateral lobes* and appressed-villous above with short hairs while young, the flowers few in loose villous corymbs; the stamens about 20 and anthers pale yellow, the oblong or nearly globose orange-red fruit about 1 cm. thick, in n. and e. Vt.

46. C. Brunetiàna Sarg. (in honor of Abbé Louis Ovide Brunet, 1826–1876). — Arborescent shrub 6–7 m. high, with slender thorny *branchlets villous while young;* leaves oblong-obovate, elliptic or rhombic, acutely serrate and with 4–5 pairs of acute spreading lateral lobes, firm, dark green and nearly glabrous at maturity; flowers 1.6–1.8 cm. wide, mostly 5–10 in loose villous corymbs; stamens about 10; anthers white or pale yellow; *calyx-lobes coarsely glandular-serrate;* fruit oblong or nearly globose, 0.8–1 cm. thick, with thick mellow flesh and 3–4 nutlets. — Thickets and rocky banks, Nfld. to Ont., s. to N.S., Me. and Minn. *Fl.* May; *fr.* Sept.–Oct.

Var. **Fernáldi** (Sarg.) Palmer (named for its discoverer, Merritt Lyndon Fernald, 1873–). — Differing in the *less copious pubescence of the leaves* and inflorescence and in the slightly larger and more numerous *flowers with pink anthers.* — Que. and Me. to n. Mich.

47. C. Jónesae Sarg. (named in 1901 for its discoverer, Beatrix Jones, American landscape architect). — A tree 6–7 m. high or often an arborescent shrub with slender flexuous thorny branchlets villous while young; *leaves elliptic or oblong-obovate, 4–9 cm. long, 3–6 cm. wide,* sharply serrate, often glandular near the base and with 4–5 pairs of small acute spreading lateral lobes, sparsely short-villous above while young and pubescent beneath throughout the season, *thick* and *with the veins impressed above at maturity;* flowers 2–2.3 cm. wide, many in loose pubescent corymbs; stamens about 10; *anthers large,* pink; *calyx-lobes linear, entire;* fruit oblong or obovate, 0.9–1.2 cm. thick, bright red, with thick mealy flesh and 2–3, usually 3, nutlets. — Borders of streams and rocky banks, often near the coast, N.S., N.B., Que. and Me. *Fl.* June; *fr.* Oct.

48. C. Margarétta Ashe (named in 1900 for Margaret Henry Wilcox, later Mrs. Ashe). — A tree 6–7 m. high or sometimes a stout shrub with straight or flexuous thorny or nearly thornless branchlets; *leaves extremely variable in size and shape;* those of the flowering branchlets mostly oblong-elliptic, rhombic or short-obovate, *with shallow or crenate serration* and usually 2–4 pairs of shallow or rounded lateral lobes above the middle, or *on vegetative shoots often suborbicular or broader than long* and obscurely or deeply lobed and *sometimes deeply laciniate near the broad base,* glabrous or with a few short hairs above while young, with 3–4 pairs of ascending primary veins impressed above at maturity; flowers 1.2–1.5 cm. wide, mostly 4–10 in compact glabrous or rarely slightly villous corymbs; stamens about 20; anthers white or pale yellow; fruit subglobose, 0.9–1.2 cm. thick, dull red or orange-red, with thin mealy flesh and 2–4, usually 3, nutlets. — Thickets and rocky open ground, s. Ont. and Mich. to Ia., s. to Pa. and Mo. *Fl.* April, May; *fr.* Oct. — Forma **xanthocárpa** Sarg. (yellow-fruited), with the fruit bright yellow, at Steamboat Rock, Ia.

Var. **Brownii** (Britt.) Sarg. (for its discoverer, Addison Brown, 1830–1913). — Flowers 0.8–1 cm. wide; fruit 6–8 mm. thick. — Va. and Pa. to Ind. and Mo.

Var. **angustifòlia** Palmer (narrow-leaved). — Leaves of the flowering branchlets lance-elliptic or oblong-lanceolate, unlobed or slightly lobed near the apex, or on vegetative shoots often coarsely toothed or sharply lobed. — S. Mich. and Ind.

Var. **meiophýlla** (Sarg.) Palmer (smaller-leaved). — Leaves of the flowering branchlets smaller (1–3 cm. long, 0.8–2 cm. wide), mostly short-obovate or oval, usually lobed; flowers 6–12 in loose corymbs, with red anthers. — Near Mt. Victory, Harding Co., O.

49. C. Dódgei Ashe (for its discoverer, Charles Keene Dodge, 1844–1918). — An arborescent shrub or rarely a tree 5–6 m. high, with usually flexuous thorny branchlets; leaves oval, rhombic or broadly obovate, serrate except near the base, usually *with* 3–4 *pairs of small acute shallow lobes* above the middle, or on vegetative shoots sometimes ovate or suborbicular and with coarse teeth and more deeply lobed nearly to the base, glabrous or essentially so, firm, with the veins slightly impressed above at maturity; flowers 1.3–1.5 cm. wide, mostly 6–12 in compact glabrous corymbs; *stamens about* 10; anthers pale yellow; fruit subglobose or depressed-globose, 0.9–1.3 cm. thick, dull orange or red, with thin dry flesh and 2–3 nutlets. — Thickets and borders of woods, often in sandy soil, se. Can. and N.E. to Pa. and Wisc. *Fl.* May; *fr.* Sept.–Oct.

C. immànis Ashe (fierce) has the leaves mostly ovate or oblong-ovate and sharply lobed, the *flowers* 1.5–1.8 *cm. wide* and in slightly branched glabrous corymbs *with about* 20 *stamens and pink anthers*, the *fruit* subglobose or short-oblong and *angular*, 1–1.3 cm. thick, green or blotched with red, the *nutlets usually* 5. — Perhaps a hybrid between *C. Dodgei* and *C. pruinosa*, near Port Huron, Mich.

50. C. Jáckii Sarg. (for its discoverer, John George Jack, 1861–1949). — Arborescent shrub with slender flexuous very thorny branchlets; leaves oblong-obovate or oval, or on vegetative shoots sometimes ovate, serrate nearly to the base, slightly or obscurely lobed above the middle, glabrous except for a few short hairs above while young, thin but firm at maturity; *flowers* 1.5–1.7 *cm. wide*, many in slightly villous corymbs; *stamens* 5–10; anthers pale yellow; *calyx-lobes glandular-serrate;* fruit oblong or obovoid, about 1 cm. thick, dull dark red, with thin succulent flesh and 2–3 nutlets. — Calcareous ridges and banks of streams, s. Que. *Fl.* May; *fr.* Sept.

51. C. sícca Sarg. (dry). — Stout shrub or rarely a small tree 4–5 m. high, with slender thorny or nearly thornless branchlets; *leaves rhombic or elliptic*, sharply serrate, with 2–3 pairs of small or obscure lateral lobes above the middle, or sometimes oblong-obovate and more deeply divided on vegetative shoots, with short appressed hairs above and slightly villous along the veins beneath while young, thin but firm, glabrous or nearly so at maturity; *flowers* 1.5–2 cm. wide, *few in nearly simple* glabrous or slightly villous *corymbs;* stamens about 10; anthers white or pale yellow; *fruit* subglobose or short-oblong, *often angular*, 0.9–1.2 cm. thick, green or dull orange, with thin dry flesh and 3–5 nutlets. — Dry calcareous hills and rocky ground along streams, s. Mo. *Fl.* April, May; *fr.* Oct.

C. mercerénsis Sarg. (of Mercer Co., W.Va.) has the *leaves* oblong-obovate or elliptic, or *on vegetative shoots oblong-ovate to suborbicular*, sharply serrate, but otherwise *undivided or rarely with* 2–4 *pairs of shallow lateral lobes above the middle*, glabrous, thin but firm at maturity, the few flowers 1–1.3 cm. wide, in glabrous corymbs and with about 10 stamens and white anthers, the orange-red fruit subglobose and 1–1.2 cm. thick and with usually 3 nutlets. — Little-known and doubtfully placed in this series, Md., W.Va. and Tenn.

C. spíssa Sarg. (compact) has *leaves* similar to those of the last species but *larger and more deeply lobed* and slightly roughened above with short appressed hairs while young, the 5–10 flowers in compact nearly simple glabrous corymbs and with about 10 stamens and pink anthers, the *bright red fruit* oblong or nearly globose and 8–9 *mm. thick with* 3–4 *nutlets.* — Perhaps a hybrid between a species of the *Rotundifoliae* and *C. macrosperma*, Essex Co., N.Y.

Series 11. **Brainerdiànae** Egglest. (see p. 769)

Leaves with the teeth and points of the small sharp lobes acuminate, often reflexed; fruit with a small sessile calyx.

Young leaves roughened above with short appressed hairs; flowers 1.5–1.8 cm. wide; calyx-lobes linear-lanceolate.
Leaves of vegetative shoots mostly ovate or elliptic; fruit becoming soft. . 52. *C. Brainerdi.*
Leaves of vegetative shoots oval to suborbicular; fruit remaining firm or hard. . . . . 53. *C. Dunbari.*
Young leaves glabrous above; flowers 1.8–2.2 cm. wide; calyx-lobes narrowly deltoid; fruit becoming succulent. . . . . 54. *C. kingstonensis.*
Leaves with the teeth and points of the broad shallow lobes acute but not acuminate or reflexed; fruit with an elevated calyx, remaining hard and dry. . . 55. *C. Coleae.*

52. C. Brainérdi Sarg. (named for its discoverer, Ezra Brainerd, 1844–1924). — Shrub often flowering when only 2–3 m. high or sometimes a tree 6–7 m. high, with slender straight or flexuous thorny branchlets; *leaves* mostly ovate, oblong-ovate or elliptic, sharply serrate, *usually with* 4–6 *pairs of very small acute lateral lobes, the points acuminate and often reflexed,* glabrous except for short appressed hairs above while young, thin but firm, bluish-green at maturity; flowers 1.6–1.8 cm. wide, mostly 4–12 in glabrous corymbs; *stamens about* 20; anthers pink; *fruit oblong,* 8–9 *mm. thick, with rather thick flesh* and 2–5, usually 3, *nutlets sometimes slightly pitted on the inner surface.* — Thickets and pastures, Que., N.E. and N.Y. to Mich., and in the mts. of N.C. *Fl.* May, June; *fr.* Oct.

Var. **asperifòlia** (Sarg.) Egglest. (rough-leaved). — Leaves thicker, scabrate on the upper surface with more persistent hairs; pedicels of the flowers sometimes sparsely villous; stamens 10–20. — Se. Can., N.E. and N.Y.

Var. **scábrida** (Sarg.) Egglest. (rough to the touch). — Leaves of the flowering branchlets mostly elliptic or oblong-obovate, firm, dark green, scabrate above; anthers pink or pale yellow; fruit short-oblong or subglobose, mellow at maturity. — N.S., N.E., s. Ont. and N.Y. to Mich.

Var. **Egglestòni** (Sarg.) Robins. (for its discoverer, Willard Webster Eggleston, 1863–1935). — *Leaves of the flowering branchlets prevailingly oval,* slightly or obscurely lobed, obtuse or short-pointed *firm or thick at maturity;* fruit becoming mellow or slightly succulent. — N.S., N.E. and N.Y.

Var. **cyclophýlla** (Sarg.) Palmer (circular leaved). — Leaves mostly oval or elliptic or on vegetative shoots sometimes suborbicular, subcoriaceous at maturity; flowers 1.6–2 cm. wide, mostly 8–15 in loose glabrous corymbs; fruit 0.9–1 cm. thick, mellow when ripe. — N.H., Vt., and w. Mass.

C. rubrocárnea Sarg. (with red flesh) has the thin leaves ovate or oval, indented nearly to the base with 4–6 pairs of triangular lateral lobes and glabrous except for short appressed hairs on the upper surface while young and dark yellow-green at maturity, the mostly 6–12 flowers 1.3–1.6 cm. wide and in glabrous corymbs, with about 10 stamens and pink anthers, the fruit subglobose and 0.9–1 cm. thick, with thin *red flesh* and 2–3 *nutlets slightly pitted on the inner surface.* — Perhaps a hybrid between a species of this series and *C. macrosperma,* found at North Albany, N.Y.

C. shirleyénsis Sarg. (of Shirley, Mass.) has the thin but firm blue-green leaves elliptic, ovate or oblong-ovate and with 4–5 pairs of small acute lateral lobes above the middle, glabrous at maturity, the *flowers* 1.3–1.7 cm. wide and *in copiously pubescent corymbs,* with about 20 stamens and pink anthers; the orange-red fruit subglobose and 0.9–1 cm. thick with 3–4 nutlets. — Possibly a hybrid between a species of this series and one of the *Punctatae,* found at Shirley, Mass.

C. Hárryi Sarg. (named in 1908 for Henry T. Brown, its discoverer) has the leaves similar to those of the last species, the *flowers* 1.2–1.5 *cm. wide and in compound glabrous corymbs,* with 5–10 stamens and pink anthers, the short-oblong or obovoid fruit 7–9 mm. thick and with 2–4 nutlets. — Perhaps a hybrid of similar relationship to the last, known from Ontario and Livingston Cos., N.Y.

C. improvìsa Sarg. (unexpected) has the thin yellow-green leaves oblong-ovate and with short acute spreading lobes often reflexed at the tips, glabrous at maturity, the many flowers 1.5–1.7 cm. wide, in glabrous corymbs, with 10 stamens or less and red anthers, the orange-red oblong or obovoid fruit with thin succulent flesh and 2–3 nutlets. — Possibly a hybrid between species of the *Braincrdianae* and the *Coccineae,* found near Toronto, Ont.

C. Wébsteri Sarg. (named in 1905 for its discoverer, L. J. Webster) has the firm yellow-green *leaves* oval or elliptic and usually with 3–5 pairs of small acute lateral lobes above the middle and short-villous above and slightly villous along the veins beneath while young, *with the veins sharply impressed above at maturity,* the flowers in lax slightly villous corymbs and with 10 stamens or less and pink anthers, the bright red fruit short-oblong and 8–9 mm. thick and with 2–3 *nutlets slightly pitted on the inner surface.* — Probably a hybrid between species of the *Brainerdianae* and *Macracanthae* near Holderness, N.H.

C. pínguis Sarg. (fat) has the *large leaves* mostly oblong-ovate and with 4–5 pairs of shallow lateral lobes, short-villous above while young, but thin and glabrous at maturity, the many flowers about 1 cm. wide and in slightly villous corymbs, *with* 5–10 stamens and red anthers, the *calyx-lobes finely glandular-serrate,* the short-oblong or nearly globose fruit 0.8–1 cm. thick and with thin mellow flesh and 2–3 (usually 2) nutlets slightly pitted on the inner surface; near Grand Rapids, Mich.

53. C. Dunbári Sarg. (named in 1903 for its discoverer, John Dunbar). — A stout shrub 3–4 m. high, with slender usually thorny branches or becoming a small tree; *leaves* oval, ovate, or *on vegetative shoots sometimes nearly orbicular,* usually *with* 3–4 *pairs of very shallow lateral lobes,* glabrous except for short appressed hairs on the upper surface when young, thick, dark green at maturity; *flowers* 1.5–1.7 cm. wide, *in small compact glabrous corymbs;* stamens about 10; anthers red; *calyx-lobes glandular-serrate;* fruit subglobose or short-oblong, 0.9–1.5 cm. thick, with a slightly elevated calyx, thin dry flesh and 3–4 nutlets. — Rocky bluffs, near Rochester, N.Y. *Fl.* May; *fr.* Oct.

54. C. kingstonénsis Sarg. (of Kingston, Ont.). — A tree 7–8 m. high, with stoutish thorny branchlets; leaves elliptic or ovate, coarsely and irregularly serrate and slightly lobed above the middle, thin but firm, dark yellow-green, glabrous at maturity; *flowers* 1.8–2 *cm. wide,* in loose glabrous corymbs; stamens about 10; anthers red; *calyx-lobes narrowly deltoid, entire or slightly glandular-serrate; fruit* subglobose, 1–1.4 cm. thick, *dark red, with thick succulent flesh and* 4–5 *nutlets* concave or slightly pitted on the inner surface. — Fertile ground along streams, Kingston, Ont. *Fl.* May; *fr.* Oct. — Perhaps a hybrid between a species of this series and *C. dilatata.*

55. C. Cõleae Sarg. (named in 1902 for its discoverer, Emma Jane Cole, 1845–1910). — Arborescent shrub or a small tree 8–10 m. high, with flexuous thorny branchlets; leaves oblong-ovate or oval, irregularly serrate nearly to the base, usually with 3–5 pairs of small or obscure lateral lobes above the middle, thin but firm, glabrous, dark green at maturity; flowers 1.7–1.8 cm. wide, mostly 5–10 in glabrous corymbs; stamens 20 or rarely 10; anthers pink; *fruit* subglobose or short-oblong. 0.9–1.2 cm. thick, *with a broad prominent calyx,* rather thick firm flesh and 2–4, or rarely 5, nutlets. —Open woods and banks of streams, Pa. and Mich. *Fl.* May; *fr.* Sept., Oct.

C. Eatoniàna Sarg. (in honor of Amos Eaton, 1776–1842), with dark bluish green leaves ovate or elliptic and slightly or obscurely lobed above the middle, glabrous at maturity, the many *flowers* 1.3–1.5 cm. wide and *in loose glabrous corymbs,* the stamens about 20 with pink anthers, *the depressed-globose or short-oblong fruit* 0.9–1.2 cm. thick *with a broad prominent calyx,* thin dry flesh and 2–3 nutlets, found near Albany, N.Y.

C. Macaùleyae Sarg. (named in 1903 for Mary Elizabeth Macauley) has the *thin dull dark green* oval, rhombic or obovate (or sometimes ovate on vegetative shoots) *leaves* coarsely and irregularly serrate and usually slightly lobed above the middle and glabrous or nearly so at maturity, the flowers 1.2–1.5 cm. wide and in small compact glabrous corymbs, with about 20 stamens and pale yellow anthers, the subglobose *fruit* 0.9–1.2 cm. thick, and *with a narrow elevated calyx,* thin dry flesh and 4–5 nutlets, known from Ontario Co., N.Y.

Series 12. Tenuifòliae Sarg. (see p. 769)

a. Leaves glabrous except for short appressed hairs on the upper surface while young; flowering corymbs glabrous. . . *b.*

b. Flowers 1.3–1.8 cm. wide; calyx-lobes entire or nearly so.

Stamens 10 or fewer; fruit usually obovoid or oblong or, if subglobose, less than 1 cm. thick; nutlets plane or only slightly ridged on the back. . . 56. *C. macrosperma.*

Stamens about 20; fruit usually subglobose or short-oblong, 1–1.5 cm. thick; nutlets deeply grooved and ridged on the back. . . . . . 57. *C. basilica.*

b. Flowers 1.8–2 cm. wide; calyx-lobes conspicuously glandular-serrate.

Leaves of flowering branchlets mostly 4–6 cm. long, exclusive of the petioles, finely serrate; stamens about 20; fruit subglobose; nutlets usually 5. . . . . 58. *C. tortilis.*

Leaves of flowering branchlets mostly 5–8 cm. long, exclusive of the petioles; deeply serrate, stamens 5–10; fruit obovoid or oblong; nutlets usually 3. . . . . 59. *C. fretalis.*

a. Flowering corymbs more or less villous. . . *c.*

c. Leaves slightly divided, with shallow lobes; calyx-lobes glandular-serrate; fruit obovoid, with firm flesh; nutlets only slightly ridged on the back. . 60. *C. lucorum.*

c. Leaves deeply divided; calyx-lobes entire or nearly so; fruit subglobose, becoming succulent; nutlets strongly ridged on the back.

Leaves of flowering branchlets mostly elliptic, narrowed at the base, with 4–5 pairs of acute spreading lateral lobes. . . . . 61. *C. densiflora.*

Leaves of flowering branchlets mostly ovate, rounded at the base, with 5–6 pairs of narrow cuneate spreading lateral lobes. . . . . . . 62. *C. flabellata.*

56. C. macrospérma Ashe (large-seeded). — A tree 6–7 m. high or sometimes shrubby, the trunk and larger branches often angled or buttressed and covered with pale scaly bark; *leaves* mostly ovate or oval, sharply serrate, *with usually* 5 *pairs of broad triangular lateral lobes*, thin, glabrous except for short appressed hairs on the upper surface while young; petioles slender, 1.5–3 cm. long, sometimes slightly glandular; flowers 1.3–1.7 cm. wide, mostly 5–12 in glabrous corymbs; stamens 10 or fewer; anthers red; *fruit obovoid or oblong* 0.8–1.5 *cm. thick*, bright red, with thick mellow or succulent flesh and 3–5 nutlets plain or slightly ridged on the back. — Woods and thickets, usually in rocky ground, s. Can. and N.E. to the mts. of N.C. and Tenn., w. to Wisc. and n. Ill. *Fl.* May; *fr.* Sept. — A variable species with many varieties and forms:

Var. **acutíloba** (Sarg.) Egglest. (with sharp-pointed lobes). — *Leaves* averaging slightly larger than in the typical var., *deeply divided* with usually 5 pairs of acuminate spreading *lateral lobes often reflexed at the tips;* fruit usually less than 1 cm. thick. — Nfld., Que., N.E. and N.Y.

Var. **demíssa** (Sarg.) Egglest. (low). — A shrub 1–3 m. high; *leaves mostly ovate or deltoid, rounded, truncate or rarely subcordate at the base;* fruit subglobose or short-oblong, 6–9 mm. thick. — Se. Can. and N.E. to n. Ill.

Var. **matùra** (Sarg.) Egglest. (mature; early ripening). — *Leaves often truncate or subcordate at the base; fruit ripening in late Aug. or early Sept.* — Se. Can. and N.E. to N.Y.

Var. **pentándra** (Sarg.) Egglest. (with five stamens). — *Leaves mostly narrowed or rounded at the base; stamens* 5–8, *with large anthers;* fruit with firm mellow flesh and a slightly larger calyx. — Se. Can., N.E. and N.Y.

Var. **roanénsis** (Ashe) Palmer (of Roan Mountain, N.C.). — *Leaves deeply indented* with 4–5 pairs of acute spreading lateral *lobes, often acuminate but not reflexed* at the tips; fruit 0.7–1.2 cm. thick. — Sometimes resembling and grading into var. *acutiloba.* (*C. roanensis* Ashe) — Range of the typical var.

C. vittàta Ashe (provided with a chaplet), with the broadly ovate or oval leaves with 3–4 pairs of triangular lateral lobes, *the few flowers* 1.5–1.8 cm. wide, *in compact corymbs* and with 10 or fewer stamens and red anthers, the subglobose or short-oblong *fruit* 1–1.4 cm. thick and with thick mellow flesh and 3–4 nutlets *ripening in Sept.* is perhaps a hybrid between *C. macrosperma* and a species of the *Coccineae*, occurring in Berks Co., Pa.

C. mérita Sarg. (deserving; of good quality) is similar to the last species but with the mostly ovate *firmer leaves* acuminate and *more deeply indented* with 3–4 pairs of spreading lateral lobes, the *oblong or obovoid fruit* with thick firm flesh and *ripening in Oct.* is perhaps a hybrid of similar origin to the last, found in s. Mich.

C. Randiàna Sarg. (in honor of Edward Lothrop Rand, 1859–1924), has the oblong-obovate or oval (or *sometimes nearly orbicular on vegetative shoots*) *leaves* with 4–6 pairs of small lateral lobes, the few *flowers* 1.2–1.5 cm. wide and *in lax* glabrous *corymbs, with* 10 or fewer stamens and *dark red anthers*, the oblong fruit 8–9 mm. thick and with thin mealy flesh and 3–4 nutlets, is perhaps a hybrid between *C. macrosperma* and a species of the *Brainerdianae*, found on Mt. Desert I., Me.

57. C. basílica Beadle (royal, magnificent). — A tree 6–7 m. high or often shrubby, with slender flexuous thorny branchlets; *leaves* ovate or oval, sharply serrate nearly to the base, *usually with* 3–4 *pairs of broad shallow triangular lateral lobes, or* sometimes *more deeply divided on vegetative shoots*, glabrous except for short appressed hairs on the upper surface while young, firm, dark bluish-green at maturity; *flowers* 1.5–1.8 *cm. wide*, mostly 7–15 in glabrous corymbs; *stamens about* 20; anthers pink or red; *fruit* short-oblong or nearly globose, 1–1.3 *cm. thick*, bright red, *with thick* mellow or juicy *flesh* and 3–5 nutlets. (*C. alnorum* Sarg.) — Open woods and pastures, N.E. to s. Ont. and N.C. *Fl.* May; *fr.* Sept., Oct.

C. haemocárpa Ashe (with blood-red fruit), has the *leaves* similar to those of the last species but usually with more rounded or obscure lobes and *sometimes broadly ovate or deltoid and subcordate at the base on vegetative shoots*, the mostly 4–12 *flowers* 1.2–1.4 *cm. wide and in compact* glabrous *corymbs*, with about 20 stamens and pink anthers, the short-oblong or subglobose *often angular dark red* and *pruinose fruit* with firm hard flesh and 3–4 nutlets grooved and ridged on the back, is perhaps a hybrid between species of the *Tenuifoliae* and *Pruinosae* in W.Va. and N.C.

C. Férrissi Ashe (for its discoverer, JAMES HENRY FERRISS, 1849–1926) has the ovate or oblong-ovate firm dark green leaves deeply divided into 4–6 pairs of spreading *lateral lobes often reflexed at the tips with the lowest pair sometimes enlarged and separated by deep sinuses,* and *glossy above at maturity,* the flowers similar to those of the last species, the *obovoid or pyriform* bright red *fruit* 0.8–1.2 cm. thick and with thin mellow flesh and 4–5 nutlets, in n. Ill.

C. Schuéttei Ashe (for its discoverer, JOACHIM HEINRICH SCHUETTE, 1821–1908) has the thin but firm *leaves broadly ovate or deltoid* with 4–5 pairs of small acuminate lateral lobes and glabrous except for short appressed hairs on the upper surface while young and *with the veins distinctly impressed above at maturity, the few flowers* 1.3–1.5 cm. wide and *in nearly simple* glabrous *corymbs,* with about 20 stamens, the fruit short-oblong or nearly globose, about 1 cm. thick and with firm flesh and 4–5 nutlets, found near Green Bay, Wisc.

58. C. tórtilis Ashe (twisted). — A tree 7–8 m. high or sometimes shrubby, with stoutish flexuous branchlets armed with short stout thorns; *leaves* ovate, usually acuminate, finely serrate, with 3–5 pairs of acute lobes, glabrous, *thin but firm, bluish-green at maturity;* flowers 1.8–2.2 cm. wide, in compound glabrous corymbs; *stamens* 20 *or rarely* 10; anthers pink; *fruit* oblong or nearly globose, 0.9–1.2 cm. thick, *bright crimson, with thick succulent flesh and* 3–5, *usually* 5, *nutlets.* — Upland pastures and borders of woods, Wisc. and n. Ill. *Fl.* May, June; *fr.* Oct.

59. C. fretàlis Sarg. (of the straits; originally from the shore of L.I. Sound). — A tree 6–7 m. high or sometimes shrubby, with slender sparingly thorny branchlets; *leaves* ovate or oval, sharply serrate *with* 3–5 *pairs of small acute lateral lobes, or with the lowest pair* sometimes *enlarged and separated by deep sinuses on vegetative shoots,* glabrous except for short appressed hairs on the upper surface while young, thin but firm, yellow-green at maturity; *flowers about* 2 *cm. wide,* mostly 5–12 in loose glabrous corymbs, *with* 5 *or rarely* 6–8 *stamens* and red anthers; *calyx-lobes* lanceolate, *coarsely glandular-serrate;* fruit oblong or obovoid, 0.9–1 cm. thick, bright red, with slightly succulent flesh and usually 3 nutlets. — Open woods and rocky pastures, Mass. and Ct. *Fl.* May; *fr.* Sept. — Possibly a hybrid between species of the *Tenuifoliae* and *Coccineae.*

C. Knieskerniàna Sarg. (in honor of PETER D. KNIESKERN, 1800–1871), is quite similar to the last in foliage, flowers and fruit but with the 3–4 *nutlets depressed or pitted on the inner faces,* found near Herkimer, N.Y.

60. C. lucòrum Sarg. (of open woods). — An arborescent shrub or sometimes a tree 7–8 m. high, with slender thorny branchlets sometimes slightly villous when they first appear, soon glabrous; leaves ovate, oblong-ovate, or rarely elliptic, deeply serrate, with 3–4 pairs of triangular lateral lobes; *flowers* 1.5–1.8 cm. wide, *mostly* 4–10 *in compact* slightly villous or rarely glabrous *corymbs,* with 5–10 or rarely 20 stamens and pink anthers; *calyx-lobes lanceolate, glandular-serrate or pectinate, sometimes only near the apex;* fruit obovoid or oblong, 0.8–1.5 cm. thick, with mellow or juicy flesh and 4–5 nutlets. — Borders of woods and banks of streams, occasional. Ct. to Ill. *Fl.* May; *fr.* Sept. — Perhaps a hybrid between species of the *Tenuifoliae* and *Coccineae.*

61. C. densiflòra Sarg. (closely or densely flowered). — A stout shrub 4–5 m. high, with slender branchlets armed with stout thorns 5–7 cm. long; *leaves* oval or oblong-ovate, sharply serrate nearly to the base, *with* 4–5 pairs of triangular *lateral lobes acuminate and often reflexed at the points,* glabrous except for short appressed hairs on the upper surface while young; *flowers* 1.2–1.4 cm. wide, mostly 5–10 *in compact pubescent corymbs;* stamens 10 or fewer; anthers pink; *calyx-lobes glandular-serrate;* fruit oblong or subglobose, 1–1.3 cm. thick, dark or purplish-red, with soft pulpy flesh and 3–4 nutlets. — Limestone-ridges and -hillsides, s. Que. *Fl.* May; *fr.* Sept.

C. apiomórpha Sarg. (pear-shaped) is similar to no. 61, but with the leaves less deeply cut, many smaller flowers in sparsely villous corymbs, and obovoid fruit with firm flesh and 3–5 nutlets, found in n. Ill.

C. lemingtonénsis Sarg. (of Lemington, Vt.), is an arborescent shrub 4–5 m. tall and forming thickets, the *leaves* broadly ovate or oval, or *on vegetative shoots sometimes nearly orbicular,* with 4–5 pairs of small acuminate lateral lobes, the few flowers 1.3–1.5 cm. wide and in compact slightly villous corymbs, the bright red fruit short-oblong or subglobose and 0.7–1 cm. thick, with succulent flesh and usually 3 nutlets, in Vt. and N.H.

62. C. flabellàta (Spach) Kirchn. (fan-like). — An arborescent shrub or rarely a tree 5–6 m. high, with slender glabrous thorny branchlets; *leaves* ovate or rhombic, sharply serrate nearly to the base, with 4–6 pairs of small acuminate lateral lobes, or *on vegetative shoots* often *broadly ovate to nearly orbicular* and *deeply laciniate with narrow triangular lobes often reflexed at the acuminate points,* short-pilose above and sometimes villous along the veins beneath while young, firm, glabrous or nearly so at maturity; flowers 1.5–1.8 cm. wide, many in more or less pubescent corymbs; *stamen* 10 *or fewer;* anthers pink; *fruit oblong or subglobose,* 0.9–1 *cm. thick,* crimson, with thick mellow flesh and 3–5 nutlets. (*C. crudelis* Sarg.) — Thickets and open woods, usually in rocky ground, along the St. Lawrence River, Que. *Fl.* May; *fr.* Sept.

Var. **Grayàna** (Egglest.) Palmer (in honor of ASA GRAY, 1810–1888). — Very similar in foliage and flowers; *stamens about* 20; *fruit subglobose or short-oblong,* 1–1.2 *cm. thick, often slightly angular.* (*C. Grayana* Egglest.) — Se. Can., N.E. and N.Y.

C. ínsolens Sarg. (unusual), similar to the last var. but with smaller leaves less deeply cut, slightly villous inflorescence and smaller oblong fruit, is perhaps a hybrid between the above var. and *C. macrosperma,* known from W. Concord, Vt.

Series 13. SILVÍCOLAE Beadle (see p. 769)

a. Leaves of flowering branchlets mostly 5–6 cm. long, 2.5–4 cm. wide, or on vegetative shoots about one-third longer; fruit obovoid, oblong or subglobose, mostly 1.2 cm. or less thick (except in no. 66). . . *b.*

b. Leaves with shallow or acute triangular lobes, not deeply incised except rarely and near the base on vegetative shoots. . . *c.*

c. Fruit subglobose or oblong, full and rounded at the base, slightly or not at all pruinose. . . *d.*

d. Leaves thin at maturity; flowers 1.2–1.5 cm. wide.

Leaves finely serrate with narrow acuminate teeth; flowers mostly 3–7 in compact corymbs. . . . . 63. *C. iracunda.*

Leaves more coarsely serrate with acute broad-based teeth; flowers mostly 6–12 in loose corymbs. . . . . 64. *C. brumalis.*

d. Leaves thick or firm at maturity; flowers 1.5–2 cm. wide. . . *e.*

e. Leaves of flowering branchlets mostly short-ovate or deltoid, often truncate or subcordate at base.

Leaves of flowering branchlets with broad shallow lobes; flowers mostly 5–10 in the corymb; fruit 0.9–1.2 cm. thick. . . . . 65. *C. stolonifera.*

Leaves of flowering branchlets with sharp spreading lobes; flowers mostly 6–12 in the corymb; fruit 1.2–1.6 cm. thick. . . . . 66. C. *beata.*

e. Leaves of flowering branchlets mostly ovate or oval, acute, rounded, or rarely truncate at base. . . . . 67. *C. populnea.*

c. Fruit usually obovoid, narrowed at the base, conspicuously pruinose. . 68. *C. levis.*

b. Leaves deeply incised, sometimes half way to the midrib on vegetative shoots. . . . . 69. *C. filipes.*

a. Leaves relatively large, often 6–7 cm. long and 5–6 cm. wide on flowering branchlets; fruit obovoid or oblong, mostly 1–1.5 cm. thick.

Flowers in nearly simple mostly 6–10-flowered corymbs; fruit about 1 cm. thick; nutlets 4–5. . . . . 70. *C. gravis.*

Flowers many in compound corymbs; fruit 1–1.5 cm. thick; nutlets 3–4, usually 4. . . . . 71. *C. compta.*

63. C. iracúnda Beadle (irascible). — An arborescent shrub or tree 5–6 m. high, with thorny often flexuous branchlets and ashy gray scaly bark; *leaves* of flowering branchlets mostly ovate, very sharply and finely serrate, with 2–4 pairs of small or obscure lateral lobes, or *on vegetative shoots broadly ovate or deltoid* and more *deeply divided,* thin, glabrous except for short appressed hairs on the upper surface while young; *flowers* 1.3–1.5 cm. wide, mostly 3–7 *in nearly simple* glabrous *corymbs;* stamens about 10; anthers pink or purplish; fruit subglobose, 0.8–1 cm. thick, red or green blotched with red, with thin hard flesh and 3–5, usually 3, nutlets. (*C. populifolia* sensu Egglest., not Walt.) — Low flat or open woods, Va. to Ga. and Ky. *Fl.* April, May; *fr.* Oct.

Var. **silvícola** (Beadle) Palmer (inhabiting the forest). — *Leaves mostly broad-ovate or deltoid,* more finely serrate and less distinctly lobed, *the blades often as broad as long or broader on vegetative shoots.* (*C. silvicola* Beadle) — Pa. to S.C. and La.

64. C. brumàlis Ashe (of the winter; from the long-persistent fruit). — An arborescent shrub or a small tree 6–8 m. high, with slender thorny branchlets and gray-brown slightly scaly bark; *leaves* ovate, oval, or *on vegetative shoots* sometimes *deltoid or nearly orbicular,* serrate nearly to the base with broad-based acuminate teeth, with 4–5 pairs of broad shallow or sharp triangular lateral lobes, glabrous except for short appressed hairs above while young, thin but firm at maturity; flowers 1.3–1.5 cm. wide, mostly 6–12 in glabrous corymbs; stamens 10 or fewer; anthers pink or purplish; fruit subglobose or short-oblong, 0.9–1.2 cm. thick, with thin dry or mellow flesh and 3–5 nutlets strongly grooved and ridged on the back. — Thickets and open woods, s. N.E. and N.Y. to Mich. and Ky. *Fl.* May; *fr.* Oct.

65. C. stolonífera Sarg. (bearing stolons or suckers). — A shrub 2–3 m. high, spreading into thickets, with slender often flexuous thorny branchlets; *leaves mostly ovate or deltoid*, serrate nearly to the base, with 5–6 pairs of triangular acuminate lateral lobes, or *on vegetative shoots* more deeply divided and *with the lowest pair of lobes enlarged and separated by deep sinuses*, short-villous on the upper surface while young, firm or thick, glabrous at maturity; flowers 1.5–1.8 cm. wide, mostly 5–10 in glabrous corymbs; stamens 10 or less; anthers pink; *fruit* subglobose or short-oblong, 0.9–1.2 cm. thick, bright red, *with a broad* shallow calyx, firm or mellow flesh and 3–4 nutlets. (*C. populnea* Egglest. in part, not Ashe) — Borders of woods and on hillsides, s. Ont. and N.Y. to Pa. and Del. *Fl.* May; *fr.* Sept., Oct.

C. Béckwithae Sarg. (for its discoverer, FLORENCE E. BECKWITH, 1843–1929), has leaves and flowers similar to those of *C. stolonifera*, but *fruit* subglobose or short-obovoid and *with a small narrow calyx*, thin dry flesh and 4–5 nutlets. (*C. silvicola* var. *Beckwithae* (Sarg.) Egglest.; *C. filipes* Egglest. in part, not Ashe) — N.H., Vt. and N.Y.

66. C. beàta Sarg. (rich). — A shrub or tree 6–7 m. high, with slender often flexuous thorny branchlets; *leaves* of the flowering branchlets mostly ovate or oblong-ovate, sharply and deeply serrate nearly to the base, with 3–5 pairs of sharp spreading acuminate lateral lobes, or *on vegetative shoots often deltoid or suborbicular and sometimes broader than long*, glabrous except for short appressed hairs on the upper surface while young, firm, bluish-green at maturity; *flowers* 1.5–2 cm. wide, *many in* glabrous *corymbs; stamens* 15–20; *anthers large*, red; fruit subglobose or short-oblong, 1.2–1.6 cm. thick, crimson, with thick mellow flesh and usually 5 nutlets. — Thickets and open woods, s. Ont. and N.Y. *Fl.* May; *fr.* Sept., Oct.

Var. **ópulens** (Sarg.) Palmer (rich). — Leaves similar but thicker; *flowers smaller, few; fruit slightly angled*, with 3–4, usually 4, nutlets. — Centr. and w. N.Y.

67. C. popúlnea Ashe (poplar-like). — An arborescent shrub or a tree 6–7 m. high, with slender usually thorny branchlets; leaves mostly ovate or oval, sharply serrate nearly to the base, usually with 4–5 pairs of broad shallow lateral lobes, glabrous, firm or thick, dark green at maturity; flowers 1.5–1.7 cm. wide, mostly 5–12 in glabrous corymbs; stamens 10 or fewer; anthers pink; fruit oblong or subglobose, 0.9–1.2 cm. thick, dull or bright red, with firm mellow flesh and 3–4 nutlets. — Open woods and moist ground along streams, N.Y. and Pa. to Mich. and O. *Fl.* May; *fr.* Sept., Oct.

68. C. lévis Sarg. (lightly armed). — An arborescent shrub 3–4 m. high, with slender thorny branchlets and dark gray slightly scaly bark; leaves mostly ovate or oval, finely serrate, with 3–4 pairs of small acuminate lateral lobes, glabrous, bluish-green, thin but firm at maturity; *flowers* 1.4–1.6 cm. wide, *mostly* 4–10 *in nearly simple* glabrous *corymbs;* stamens 10 or fewer; anthers pink; *fruit obovoid or pyriform, gradually narrowed below the middle*, 0.8–1 cm. thick, *dull crimson, pruinose*, with thin mellow flesh and 3–4 nutlets. — Open woods and pastures, s. Ont., N.E. and N.Y. *Fl.* May; *fr.* Sept., Oct.

69. C. fílipes Ashe (with thread-like pedicels). — An arborescent shrub or small tree 6–7 m. high; *leaves ovate or deltoid*, finely serrate *deeply indented* with 4–5 pairs of triangular acuminate lateral lobes, glabrous, thin but firm at maturity; flowers 1.6–1.8 cm. wide, mostly 5–8 in nearly simple glabrous corymbs; stamens 5–10; anthers rose-color; *fruit obovoid or pyriform*, 0.8–1 cm. thick, *bright red*, sometimes *slightly pruinose*, with thin mellow flesh and 4–5 nutlets. — Thickets and borders of woods, Pa., Mich. and Wisc. *Fl.* May; *fr.* Oct.

70. C. grávis Ashe (heavily laden). — A shrub or small tree 6–7 m. high, with slender often flexuous thorny branchlets; *leaves* ovate or oval, serrate with broad-based acuminate teeth, *with 4–5 pairs of shallow lateral lobes*, or *on vegetative shoots broadly ovate to nearly orbicular and more distinctly lobed*, glabrous except for short appressed hairs on the upper surface while young, firm or thick, yellowish-green at maturity; flowers 1.5–1.8 cm. wide, usually 5–10 in glabrous corymbs; stamens 10 or fewer; anthers pink or purplish; fruit obovoid or short-oblong, 0.8–1 cm. thick, with thin dry flesh and 3–5 nutlets. — Hillsides and borders of woods, usually in fertile soil, s. Ont., N.Y. and Pa. to Mich. and Ind. *Fl.* May; *fr.* Oct.

71. C. cómpta Sarg. (adorned). — A shrub or tree 6–7 m. high; *leaves* mostly ovate or oblong-ovate, sharply serrate, with 3–5 pairs of acute spreading lateral lobes, or *on vegetative shoots broadly ovate or deltoid*, sometimes *as broad as long or broader*, with short appressed hairs above while young, glabrate, thick or coriaceous, dull blue-green at maturity; flowers 1.5–2 cm. wide, many in glabrous corymbs; stamens 10 or fewer; anthers red; *calyx-lobes glandular-serrate; fruit* obovoid or oblong, 1–1.3 cm. thick, *bright red, slightly pruinose*, with thick firm or mellow flesh and 3–4 nutlets. — Thickets and open woods along streams, s. Ont. and N.Y. to Pa. and Mich. *Fl.* May; *fr.* Oct.

C. iteràta Sarg. (repeated; substitute for an earlier incorrectly repeated name) has the *irregularly serrate* and slightly or obscurely lobed glabrous thin but firm *leaves ovate, oval, or suborbicular* and *with the veins impressed above at maturity*, the *many flowers* about 1.5 cm. wide and *in loose* glabrous *corymbs*, the fruit subglobose with thin dry flesh and 3–4 nutlets, found near Richmond, N.Y.

C. mèdia Sarg. (intermediate) has the leaves similar to those of the last in shape but thicker, the *few flowers* 1.3–1.5 cm. wide and *in* glabrous *corymbs*, the fruit short-oblong or subglobose and about 1.2 cm. thick, with thick mellow flesh and 3–5 nutlets, found near Oxford, Ct.

C. xanthophýlla Sarg. (yellow-leaved) has the *relatively large* mostly *broadly ovate or deltoid leaves* slightly indented with 3–4 pairs of broad shallow lateral lobes, *thick*, *yellow-green* and glabrous *at maturity*, the mostly 6–12 flowers 1.5–1.8 cm. wide, the *calyx-lobes glandular-serrate*, the fruit short-oblong or subglobose and 1–1.3 cm. thick, with thick mellow flesh and 3–4 nutlets, found near Buffalo, N.Y. and Scranton, Pa.

Series 14. PRUINÒSAE Sarg. (see p. 769)

a. Leaves and inflorescence glabrous or essentially so from the first (except in forms and vars. of nos. 72 and 73). . *b.*
b. Fruit subglobose or depressed-globose (sometimes slightly narrowed at the base while young). . . *c.*
c. Leaves of flowering branchlets seldom over 4 cm. wide, bluish-green or dark green at maturity. . . *d.*
d. Leaves with broad triangular or obscure lobes; petioles one- to two-thirds as long as the blades. . . *e.*
e. Leaves thick at maturity; flowers 1.8–2.2 cm. wide.
Leaves of flowering branchlets mostly rounded or abruptly narrowed at the base, more than half-grown when the flowers open; bracts and bractlets inconspicuous and early deciduous. . . 72. *C. pruinosa.*
Leaves of flowering branchlets rounded, truncate, or subcordate at the base, half-grown or less when the flowers open; bracts and bractlets conspicuous and more persistent. . . 73. *C. Mackenzii.*
e. Leaves thin but firm at maturity; flowers 1.5–1.7 cm. wide. . . 74. *C. Gattingeri.*
d. Leaves with sharp acuminate spreading lobes; petioles half to nearly as long as the blades. . . 75. *C. leiophylla.*
c. Leaves of the flowering branchlets often 5–6 cm. wide, mature leaves yellowish-green or dull green.
Leaves distinctly or sharply lobed, mostly truncate or subcordate at base; fruit with a broad elevated calyx. . . 76. *C. rugosa.*
Leaves with shallow or obscure lobes, mostly rounded or abruptly narrowed at base; fruit with a narrow slightly elevated calyx. . . 77. *C. disjuncta.*
b. Fruit pyriform, obovoid or oblong or, if nearly globose, with the leaves deeply divided. *f.*
f. Leaves more or less lobed but not deeply cut.
Leaves thick at maturity, sharply lobed, the veins distinctly impressed above; flowers 1.6–1.8 cm. wide. . . 78. *C. Porteri.*
Leaves thin but firm at maturity, with shallow or obscure lobes; the veins not noticeably impressed above; flowers 1.4–1.6 cm. wide. . . 79. *C. compacta.*
f. Leaves deeply divided, laciniate at least on vegetative shoots. . . 80. *C. Jesupi.*
a. Young leaves and sometimes the inflorescence more or less villous (forms or vars. of nos. 72 and 73 may be sought here). . . 81. *C. locuples.*

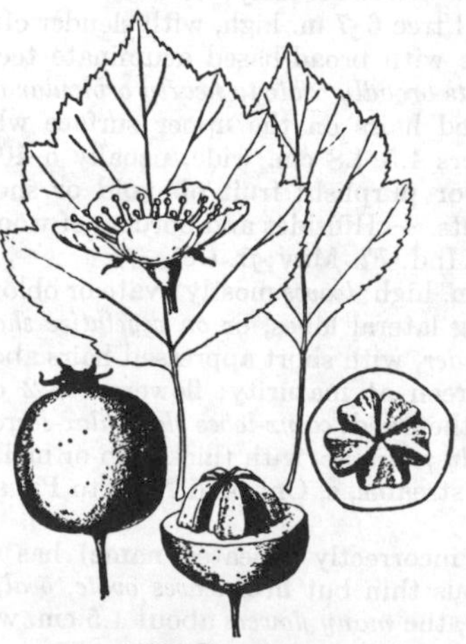
1138. C. pruinosa.

72. C. pruinòsa (Wendl.) K. Koch (frosty, with a bloom). — A tree 7–8 m. high or sometimes an arborescent shrub, with thorny intricate branchlets and dark gray scaly bark; *leaves of flowering branchlets* prevailingly ovate, *rounded or abruptly narrowed at the base*, sharply and irregularly serrate, usually with 3–5 pairs of shallow or obscure lobes, or on vegetative shoots sometimes broadly ovate, more deeply divided, and with truncate or rarely subcordate bases, glabrous, bluish-green, firm or thick at maturity; flowers 1.7–2 cm. wide, mostly 5–10 in glabrous corymbs; stamens about 20; anthers pink or rarely creamy-white; calyx-lobes lanceolate, entire or slightly serrate; *fruit* subglobose or short-oblong, *often slightly angled*, 0.9–1.3 cm. thick, *dull crimson or green and dark-dotted*, *pruinose*, *with a prominent elevated calyx*, thin dry flesh and 4–5 relatively large nutlets. — Thickets and rocky ground, Nfld., se. Can. and N.E. to N.C., w. to Wisc., Ky. and n. Ark. *Fl.* May; *fr.* Oct. FIG. 1138. — Forma **angulàta** (Sarg.) Palmer (angled). —

Fruit conspicuously 5-angled and flattened at the base, occasional with the typical form.

Var. **díssona** (Sarg.) Egglest. (differing). — *Stamens about* 10; *calyx of the fruit* broad but *nearly sessile.* — With the typical var. throughout the n. part of its range.

Var. **latisépala** (Sarg.) Egglest. (with broad calyx-lobes). — Leaves often undivided or obscurely lobed; *calyx-lobes deltoid or gradually narrowed from a broad base;* anthers white or yellowish. — Se. Can. and N.E. to Pa. and O.

Var. **brachýpoda** (Sarg.) Palmer (short-pedicelled). — *Flowers* 1.5–1.7 cm. wide, mostly 5–7 *in nearly simple compact corymbs; pedicels* 1–1.5 *cm. long;* stamens 10 or rarely 20, with white or pale yellow anthers; fruit green or slightly tinged with red, with a nearly sessile calyx. — Ky. to Mo. and Ark.

Var. **delawarénsis** (Sarg.) Palmer (of Delaware). — Similar to the last var. but with the *pedicels longer* and the *calyx-lobes serrate or glandular-serrate; fruit dark red, not pruinose.* — Del. and e. Pa.

C. formòsa Sarg. (beautiful) has the thick dull dark green leaves ovate or oblong-ovate and usually with 3–4 pairs of acute spreading lateral lobes, the 4–10 flowers 1.8–2.2 cm. wide and in glabrous corymbs and with 15–20 stamens and white anthers, the *bright red* and *pruinose* oblong or nearly globose *fruit* 0.9–1 cm. thick and *with a broad nearly sessile calyx,* thin flesh and 5 nutlets; perhaps a hybrid between species of the *Silvicolae* and *Pruinosae,* found in n. and centr. N.Y.

C. gaùdens Sarg. (joyous) has the elliptic or oblong-ovate *leaves finely serrate* and usually with 4–5 pairs of shallow lateral lobes, firm and glabrous at maturity and *with conspicuously long petioles,* the few flowers 1.4–1.6 cm. wide and with about 20 stamens and red anthers, the obovoid or nearly globose fruit with thin succulent flesh and 3–5 nutlets; Allegheny Co., Pa.

C. franklinénsis Sarg. (of Franklin Co., O.), is very similar to the last but with about 10 stamens and the fruit dark red with thin hard flesh; O. and Ky.

C. glareòsa Ashe (growing in gravel) has the *relatively large* and *sharply and deeply serrate* subcoriaceous and glabrous *leaves* ovate or oval and usually obscurely lobed above the middle or *on vegetative shoots* sometimes *nearly orbicular, dentate* with *the lower teeth gland-tipped* and with *the veins distinctly impressed above* at maturity, the petioles glandular, the several flowers 2–2.4 cm. wide and in glabrous or slightly villous corymbs, with about 20 stamens, pale yellow anthers and glandular-serrate calyx-lobes, the *bright red* subglobose *fruit with a large elevated calyx,* thin dry flesh *and* usually 4 *nutlets;* perhaps a hybrid between *C. pruinosa* and *C. succulenta* var. *michiganensis,* found near Port Huron, Mich.

C. durobrivénsis Sarg. (of Rochester, N.Y.; from the Latin name of Rochester, England) has the *leaves* similar to those of the last species but thinner and *with the veins only slightly impressed above,* the flowers slightly smaller in glabrous corymbs and with about 20 stamens and red anthers, the *fruit with* a broad slightly elevated calyx and usually 5 *nutlets;* perhaps a hybrid between *C. pruinosa* and *C. suborbiculata,* in n. and centr. N.Y.

73. C. Mackénzii Sarg. (named for its discoverer, Kenneth Kent Mackenzie, 1877–1934). — An arborescent shrub or a small tree 5–6 m. high, with slender usually very thorny branchlets; *leaves* mostly *ovate or deltoid, often nearly as broad as long on vegetative shoots,* sharply and deeply serrate, with 3–4 pairs of acute triangular lateral lobes, firm, glabrous, bluish-green at maturity; *flowers* 1.7–2 cm. wide, *few in nearly simple glabrous corymbs;* stamens about 20; anthers pink; fruit subglobose, 1–1.6 cm. thick, sometimes thicker than long, often angular, dull red, pruinose, with dry or mealy flesh and 3–5 relatively large nutlets. — Thickets, glades and open woods, Ky. to se. Ia. and Okla. *Fl.* April, May; *fr.* Oct.

Var. **bracteàta** (Sarg.) Palmer (provided with bracts). — *Leaves* mostly ovate, *slightly lobed or rarely unlobed;* fruit 0.9–1.2 cm. thick. — S. Mo., Kans., Okla. and Ark.

Var. **áspera** (Sarg.) Palmer (roughened). — Similar to the last var. but the *leaves short-villous above while young* and somewhat *pubescent along the veins beneath* throughout the season; *inflorescence* more or less *pubescent.* (*C. aspera* Sarg.) — S. Mo.

C. virélla Ashe (somewhat green) has the leaves similar in outline to those of the last var. but sharply lobed and short-villous above and often slightly pubescent along the veins beneath while young, the few *flowers* 1.6–1.8 cm. wide *in slightly villous or glabrous corymbs,* the fruit 0.9–1.3 cm. thick and with a prominent elevated calyx and 3–5 nutlets; Pa. to O. and Ill.

C. platycárpa Sarg. (broad-fruited) has the more or less lobed leaves ovate or deltoid and glabrous except for a few hairs above and along the veins beneath while young, the few flowers 2–2.2 cm. wide and in nearly simple glabrous corymbs and with about 20 stamens and large pink or red anthers, the bright red or orange-red fruit subglobose or depressed-globose and often broader than long with a slightly elevated calyx and thick mellow flesh and 3–5 nutlets; s. Ill., Mo. and Ark.

74. C. Gattíngeri Ashe (named for its discoverer, AUGUSTIN GATTINGER, 1825–1903). — An arborescent shrub or a small tree 6–7 m. high, with slender flexuous usually very thorny branchlets; *leaves* mostly *ovate or deltoid,* finely serrate, usually *with* 4–5 *pairs of broad or acute lateral lobes and an often elongated terminal lobe,* glabrous, thin but firm at maturity; flowers few, 1.5–1.7 cm. wide, in glabrous corymbs; stamens about 20; anthers pink; fruit subglobose or sometimes slightly attenuate at base, 0.9–1 cm. thick, dull red, with thin dry flesh and 3–5 nutlets. — Thickets and open woods, Pa. to Tenn. and Ark. *Fl.* May; *fr.* Oct.

Var. **rígida** Palmer (rigid). — Branchlets stout and rigid, conspicuously flexuous; thorns stout, 1–2 cm. long; flowers 1.2–1.5 cm. wide. — Ind. and Ky.

75. C. leiophýlla Sarg. (smooth-leaved). — An arborescent shrub or rarely a tree 5–6 m. high; *leaves ovate or deltoid, sharply serrate nearly to the base, deeply indented* with 3–5 pairs of acute spreading lateral lobes, *or on vegetative shoots* sometimes broader than long and *deeply laciniate,* glabrous, thin but firm at maturity; flowers 1.7–2 cm. wide, mostly 4–7 in glabrous corymbs; stamens about 20; anthers pink or creamy white; fruit subglobose, 1–1.3 cm. thick, with a prominent elevated calyx, thin dry or mealy flesh and 2–3 nutlets. — Thickets and woods along streams, s. Ont. and N.Y. *Fl.* May; *fr.* Oct.

76. C. rugòsa Ashe (rugose). — An arborescent shrub or small tree 6–7 m. high, with slender usually thorny branchlets; *leaves broadly ovate or deltoid, often broader than long on vegetative shoots,* serrate, and with 3–4 pairs of shallow or acute lateral lobes, thick, glabrous, *yellowish-green* at maturity; *flowers* 1.6–2 cm. wide, usually 5–8 *in nearly simple* glabrous *corymbs;* stamens about 20; anthers white or rarely pink; fruit subglobose or short-oblong, 1–1.5 cm. thick, dull red or green blotched with red, with a prominent slightly elevated calyx, dry mealy flesh and 2–3 nutlets. — Thickets along streams, N.Y. and Pa. to Ky., Ia. and Mo. *Fl.* April, May; *fr.* Oct.

77. C. disjúncta Sarg. (distinct). — An arborescent shrub or a small tree with slender thorny branchlets; *leaves* mostly ovate, *rounded or abruptly narrowed at base,* acuminate, or on vegetative shoots broader and truncate to subcordate at base, sharply serrate with acute or acuminate teeth, *usually with* 2–4 pairs of *shallow or obscure lateral lobes,* dull green, glabrous, thin but firm at maturity; flowers 1.7–2 cm. wide, in nearly simple corymbs; stamens 10 or rarely 20; anthers white or pink; *fruit* subglobose, often slightly angled, 1–1.4 cm. thick, green or dull red, *with a slightly elevated calyx,* thin dry flesh and 2–5 nutlets. — Upland thickets and along small streams, w. Ky. to s. Mo. *Fl.* April, May; *fr.* Oct.

78. C. Pórteri Britt. (for THOMAS CONRAD PORTER, 1822–1901). — An aborescent shrub or a tree 3–4 m. high; *leaves* ovate or elliptic, finely serrate, *with* 3–4 *pairs of sharp acuminate spreading lateral lobes,* glabrous, *bluish-green, lustrous, coriaceous* at maturity; flowers 1.5–1.8 cm. wide, few in lax glabrous corymbs; stamens about 20; anthers white or pale yellow; *fruit pyriform or obovoid,* 0.8–1 cm. thick, *bright red,* with a narrow elevated calyx, thin dry flesh and 3–5 nutlets. — Borders of woods and thickets, Pa. and adj. O. *Fl.* May; *fr.* Oct.

Var. **caeruléscens** (Sarg.) Palmer (becoming bluish). — *Leaves* ovate or broadly ovate, *thin but firm,* dull bluish-green; *fruit dull red,* pruinose. — E. Mass.

C. Mílleri Sarg. (named in 1923 for its discoverer, JOHN MILLER), has the ovate or oblong-ovate leaves with 3–5 pairs of acuminate spreading lateral lobes and sparsely short-villous on the upper surface while young, but becoming glabrate and bluish-green and thin but firm at maturity, the mostly 8–15 *flowers* 1.4–1.6 cm. wide and *in lax corymbs,* with about 10 stamens and pink anthers, the dull red fruit obovoid and 0.8–1 cm. thick with a slightly elevated calyx, thin dry flesh and 3–5 nutlets; n. Pa. and ne. O.

C. crawfordiàna Sarg. (named for Crawford Co., Pa.) has the glabrous and *finally yellowish-green* ovate or elliptic *leaves deeply serrate* and usually slightly indented with 3–5 pairs of shallow lateral lobes, the *mostly* 5–8 *flowers* 1.5–1.8 cm. wide and *in lax corymbs,* with about 20 stamens and white or pink anthers, the obovoid or pyriform fruit 0.9–1 cm. thick and with a narrow elevated calyx, thin dry flesh and 3–5 nutlets; N.Y., Pa. and O.

C. littoràlis Sarg. (of the shore) has the ovate or rhombic finely serrate glabrous and *subcoriaceous leaves* usually with 3–4 pairs of broad shallow lateral lobes and *with the veins deeply impressed above* at maturity, the few flowers 1.7–1.8 cm. wide and in nearly simple glabrous corymbs, *with about* 20 *stamens and large yellow anthers,* the obovoid or nearly globose *fruit* 0.9–1.2 cm. thick and *with a prominent elevated calyx,* thin hard flesh and 3–4 nutlets; e. Ct.

79. C. compácta Sarg. (compact). — An arborescent shrub 3–4 m. high, forming thickets; *leaves* elliptic, rhombic, or oblong-ovate, *relatively small,* finely serrate, *unlobed or but slightly lobed* above the middle, glabrous except for short appressed hairs on the upper surface while young, thin but firm at maturity, *the lowest pair of primary veins strongly ascending;* flowers mostly 5–9, 1.2–1.5 cm. wide, in compact glabrous corymbs; stamens about 20; anthers pink or white; *fruit* short-oblong or nearly globose, *usually slightly narrowed at the base,* 0.8–1.3 cm. thick, with thin dry flesh and 3–5 nutlets. — Thickets and pastures, often in sandy or sterile ground, s. Ont., and Mich. to Pa. and O. *Fl.* May; *fr.* Oct.

80. C. Jésupi Sarg. (for Henry Griswold Jesup, 1826–1903). — An arborescent shrub or a small tree 3–6 m. high; *leaves* ovate or oblong-ovate, *coarsely and irregularly serrate, deeply indented with* 3–4 pairs of *acute or acuminate* lateral *lobes,* or on *vegetative shoots* often *deeply laciniate,* thin but firm, glabrous, yellow-green at maturity; flowers few, 1.5–1.8 cm. wide, in lax nearly simple corymbs; stamens 10 or fewer; anthers pink; fruit obovoid or short-oblong, 1–1.3 cm. thick, dull or dark red, with a slightly elevated calyx, thin dry flesh and 3–4 nutlets. — Rocky pastures and hillsides, Vt. to Ct., Pa. and O. *Fl.* May; *fr.* Oct.

81. C. lócuples Sarg. (opulent). — A tree 7–8 m. high, with stoutish often flexuous thorny branchlets; *leaves* ovate or oval, *deeply serrate and slightly indented* with 3–4 pairs of shallow or acute lateral lobes, *or on vegetative shoots* sometimes *deeply divided and laciniate toward the base, copiously short-pilose above while young* and more or less *pubescent beneath throughout the season,* firm, yellow-green at maturity; *flowers* 2–2.4 cm. wide, 7–10 *in pubescent corymbs;* stamens about 20; anthers pink or pale yellow; *calyx-lobes glandular-serrate;* fruit subglobose, 1–1.2 cm. thick, with a broad slightly elevated calyx, thin dry flesh and 4–5 nutlets. — Open woods and along streams, O. to Ky. and Mo., occasional. *Fl.* April, May; *fr.* Sept., Oct. — Probably a hybrid between a species of the *Pruinosae* and *C. mollis.*

C. lécta Sarg. (choice), has the mostly elliptic or oblong-ovate *leaves short-pilose above while young* and *slightly pubescent along the veins beneath,* finely serrate, usually slightly lobed above the middle, firm, dark and blue-green at maturity, the few flowers 1.5–1.7 cm. wide and in slightly villous corymbs, with about 20 stamens and yellowish-white anthers, the short-obovoid and angular orange-red fruit 1.2–1.4 cm. thick. — Allegheny Co., Pa.

Series 15. Coccíneae Loud. (see p. 769)

a. Leaves of flowering branchlets mostly elliptic, oval, or oblong-ovate, broadest about the middle, seldom cordate at base even on vegetative shoots; fruit obovoid or oblong (sometimes nearly globose in var. of no. 83). . . *b.*
- *b.* Young branchlets, petioles and inflorescences glabrous or villous; fruit glabrous (except sometimes in no. 84 and var. of no. 83). . . *c.*
  - *c.* Leaves flat or plane; calyx-tube glabrous or sparsely villous; fruit bright red, with thin dry flesh.
    - Leaves of flowering branchlets elliptic or oblong-obovate, narrowed toward the acute or rounded base; fruit obovoid or oblong, noticeably longer than thick. . . . . . . 82. *C. Holmesiana.*
    - Leaves of flowering branchlets mostly oblong-ovate or oval, rounded or rarely acute at base; fruit oblong, only slightly longer than thick. . . 83. *C. pedicellata.*
  - *c.* Leaves often slightly cupped (concavo-convex); calyx-tube villous; fruit dull dark red, with thick flesh, sometimes slightly pubescent at the ends. 84. *C. Pringlei.*
- *b.* Young branchlets, petioles, and inflorescences tomentose; fruit pubescent at the ends. . . . . 85. *C. anomala.*

a. Leaves of flowering branchlets mostly ovate, broadest below the middle, on vegetative shoots broadly ovate to deltoid, truncate or subcordate at base; fruit subglobose or short-oblong, 1–1.6 cm. long and as thick or thicker.
- Petioles stoutish; calyx-tube usually pubescent; stamens about 10. . . . 86. *C. pennsylvanica.*
- Petioles slender; calyx-tube glabrous; stamens about 20. . . . . . 87. *C. Putnamiana.*

82. C. Holmesiàna Ashe (for Joseph Austin Holmes, 1859–1915). — A tree or often an arborescent shrub with slender usually thorny branchlets; *leaves* elliptic, ovate or oblong-ovate, *usually narrowed or acute at the base except on vegetative shoots,* sharply and deeply serrate, usually with 4–6 pairs of shallow or small acute lateral lobes, glabrous except for short appressed hairs on the upper surface while young, thin but firm, yellow-green at maturity; flowers 1.5–1.8 cm. wide, many in glabrous corymbs; stamens about 10; anthers pink or red; calyx-lobes glandular-serrate; *fruit obovoid or oblong,* 1–1.2 cm. thick, bright red, with a broad shallow calyx, thin firm flesh and usually 3 nutlets. — Open woods and thickets, usually in fertile ground, se. Can. and N.E. to Pa. and Minn. *Fl.* May; *fr.* Sept., Oct.

Var. **víllipes** Ashe (with hairy pedicels). — Petioles and inflorescence more or less villous. — Range of the typical var. and often commoner northw.

Var. **magniflòra** (Sarg.) Palmer (large-flowered). — Leaves very slightly lobed or sometimes

unlobed; flowers 1.8–2 cm. wide, mostly 7–10 in glabrous corymbs, with about 20 stamens; nutlets 3–5. — Near Chicago, Ill.

Var. amícta (Ashe) Palmer (clothed). — Leaves of flowering branchlets oblong-ovate or oval, with 3–5 pairs of shallow lateral lobes; flowers 1.6–1.8 cm. wide, in sparsely villous corymbs; stamens 5–8. — O. and n. Ill.

Var. chippewaénsis (Sarg.) Palmer (of Chippewa, Ont.). — Petioles beset with stalked glands; flowers mostly 12–15 in villous corymbs, with about 20 stamens; nutlets 3–5. — S. Ont.

83. C. pedicellàta Sarg. (pedicelled). — A tree 6–8 m. high or sometimes a stout shrub with slender glabrous thorny branchlets; leaves of flowering branchlets mostly oblong-ovate or oval, sharply serrate nearly to the base, more or less indented with 4–5 pairs of acute spreading lateral lobes, or on vegetative shoots often ovate and more deeply divided, roughened above with short villous hairs while young, glabrous, thin but firm at maturity, flowers 1.8–2 cm. wide, many in slightly villous corymbs; stamens about 10; anthers pink or red; calyx-lobes glandular-serrate; *fruit oblong or obovoid*, 0.7–1 cm. thick, *bright red*, with a broad nearly sessile calyx, thin flesh and 3–5 nutlets. (*C. coccinea* of auth. in part) — Thickets and banks of streams, N.E., Ont. and N.Y. to Ind. and n. Ill. *Fl.* May; *fr.* Sept.

Var. álbicans (Ashe) Palmer (becoming white). — Leaves relatively broader, sharply lobed; *flowers* 1.7–1.8 cm. wide, 5–10 *in glabrous or nearly glabrous corymbs; fruit* short-oblong to nearly globose, *dark red, pruinose*, with firm mellow flesh and 4–5 nutlets. (*C. albicans* Ashe) — Me. to N.Y. and Pa., w. to n. Ill.

Var. Ellwangeriàna (Sarg.) Egglest. (in honor of GEORGE ELLWANGER, 1816–1906). — Leaves roughened above with short appressed hairs while young and slightly pubescent along the veins beneath; *inflorescence* many-flowered, *pubescent; fruit* oblong, 1–1.5 cm. thick, sometimes *slightly pubescent at the ends*, crimson, lustrous, with thick mellow flesh and 3–5 nutlets. — Ont., N.Y. and Pa. to Mich. — Possibly a hybrid with a species of the *Molles*.

Var. Robesoniàna (Sarg.) Palmer (in honor of WILLIAM R. ROBESON, on whose estate it was discovered). — Leaves relatively large, mostly ovate or oblong-ovate, deeply serrate and sharply or deeply lobed; flowers 1.6–1.8 cm. wide; few in slightly villous corymbs; fruit obovoid, with a small sessile calyx, thick juicy flesh and 4–5 nutlets. — N.E. to Pa., s. Ont. and n. Ill.

C. Híllii Sarg. (for its discoverer, ELLSWORTH JEROME HILL, 1833–1917) has the *branchlets villous when they first appear*, the ovate or oval leaves usually slightly indented with 3–5 pairs of small lateral lobes and short-villous above and pubescent along the veins beneath while young, becoming firm and glabrous or nearly so and yellow-green at maturity, the many *flowers* 1.8–2 *cm. wide and in lax villous corymbs*, with about 20 stamens and pink anthers, the *obovoid crimson fruit with thick firm flesh* and 4–5 nutlets; O. and n. Ill.

C. illecebròsa Sarg. (enticing) has the elliptic or ovate and sharply serrate more or less lobed leaves glabrous except for short appressed hairs on the upper surface while young and thin but firm at maturity, the *petioles glandular*, the few *flowers* 1.8–2.2 *cm. wide* and glabrous corymbs, *with* 20–25 *stamens and dark red anthers*, the obovoid crimson *fruit* 1–1.2 cm. thick and *with thick succulent flesh* and 3–5 nutlets, is perhaps a hybrid between *C. pedicellata* and *C. dilatata*, growing near Kingston, Ont.

C. Letchworthiàna Sarg. (in honor of WILLIAM P. LETCHWORTH on whose farm it was discovered in 1904) has the oval or elliptic (or *on vegetative shoots broadly ovate to suborbicular*) *leaves* with 3–4 pairs of acute lateral lobes and glabrous except for short villosity on the upper surface and along the veins beneath while young, *firm or thick* and *with the veins impressed above* at maturity, the mostly 8–12 flowers 1.8–2 cm. wide and in glabrous or nearly glabrous corymbs, with about 10 stamens and pink anthers, the *short-obovoid or oblong fruit* 1–1.2 cm. thick and *with thin firm flesh* and 2–4 nutlets, is perhaps a hybrid between *C. pedicellata* and *C. suborbiculata*, found near Portage, N.Y.

84. C. Príngleì Sarg. (for its discoverer, CYRUS GUERNSEY PRINGLE, 1838–1911). — A tree 7–8 m. high, with slender often flexuous branchlets villous when they first appear; *leaves* ovate or oval, *drooping on the branches, often slightly cupped or concave below*, sharply serrate, usually slightly indented with 3–5 pairs of small acuminate lateral lobes, with short appressed hairs above and slightly pubescent along the veins beneath while young, thin, yellow-green, glabrous or nearly so at maturity, *flowers* many, 1.8–2 *cm. wide*, in villous corymbs; stamens 5–10; anthers pink or red; *fruit* oblong or obovoid, 0.8–1.2 cm. thick, *dull dark red*, with a small sessile calyx, firm thick flesh and 3–5 nutlets. — Thickets and woods along streams, N.E. and Ont. to n. Ill. *Fl.* May; *fr.* Sept., Oct.

Var. exclùsa (Sarg.) Egglest. (separated; from the species with which it had been confused). — *Leaves* broadly ovate or *on vegetative shoots nearly orbicular*, more coarsely serrate and with small but more distinct lobes; flowers 2–2.2 cm. wide, in tomentose corymbs; *fruit bright red*. — Vt. and N.Y.

Var. **lobulàta** (Sarg.) Egglest. (having small lobes). — *Leaves* ovate or oblong-ovate, *sharply and deeply indented with 4–6 pairs of acute lateral lobes, these often reflexed at the tips;* fruit oblong, 0.8–1 cm. thick, with thick mellow flesh. — N.E., N.Y. and s. Ont.

85. C. anómala Sarg. (anomalous, unusual). — An arborescent shrub or a small tree 6–7 m. high, with stoutish often flexuous thorny branchlets tomentose when they first appear, *leaves* elliptic or oblong-ovate, *sharply and deeply serrate,* slightly indented with 4–6 pairs of small acuminate lateral lobes, roughened with short appressed hairs on the upper surface and *pubescent* along the veins *beneath while young,* thin, light yellow-green at maturity; *petioles* slender, *pubescent while young, often glandular;* flowers 1.5–1.7 cm. wide, many in lax pubescent corymbs; stamens 10 or fewer; anthers red; *fruit* oblong or obovoid, 0.8–1.2 cm. thick, ***crimson,*** *pale-dotted, usually pubescent at the ends,* with a large nearly sessile calyx, thick juicy flesh and 4–5 nutlets. — Rocky banks and open woods, s. Que., n. N.E. and N.Y. *Fl.* May; *fr.* Sept., Oct. — Possibly a hybrid between species of the *Coccineae* and the *Molles.*

86. C. pennsylvánica Ashe (of Pennsylvania). — A tree 9–10 m. high, with slender thorny branchlets villous when they first appear; leaves oval or ovate, on vegetative shoots broadly ovate or deltoid, sharply serrate, with 4–6 pairs of small acute spreading lateral lobes, sparsely short-villous above while young and slightly pubescent along the veins beneath throughout the season, thin but firm, yellow-green, glabrate above at maturity; *flowers* 1.6–1.8 cm. wide, *many in villous corymbs;* stamens about 10; anthers small, white; *calyx-lobes laciniately glandular-serrate; fruit subglobose or broader than long,* 1.3–1.6 cm. thick, with a broad sessile calyx, thin firm flesh and 4–5 nutlets. — Open woods and thickets, in moist fertile soil, Del., Pa. and W.Va. *Fl.* May; *fr.* Sept., Oct.

87. C. Putnamiàna Sarg. (in honor of Gen. RUFUS PUTNAM, 1738–1824). — A tree 8–10 m. high or sometimes an arborescent shrub with slender thorny glabrous branchlets; *leaves broadly ovate or deltoid,* sharply serrate, usually with 3–4 pairs of shallow or acute lateral lobes, *glabrous except for a few short villous hairs above and along the veins beneath while young,* thin but firm, bright yellow-green at maturity; flowers few 1.6–1.8 cm. wide, in glabrous corymbs; stamens about 20; anthers red; *calyx-lobes entire or nearly so; fruit* subglobose, 1.2–1.5 cm. thick, *often slightly angular, bright red, with* a broad nearly sessile calyx, *thick firm flesh* and 4–5 nutlets. — Thickets and open woods along streams, O., Ky., Ind. and s. Ill. — *Fl.* May; *fr.* Oct.

C. confragòsa Sarg. (very rough; application not explained) has the mostly ovate and sharply serrate leaves with 4–6 pairs of small acute or acuminate lateral lobes and glabrous at maturity, the petioles often glandular, the *flowers* (1.6–2 cm. wide) in loose glabrous corymbs and *with* about 20 stamens, pink anthers and *lanceolate broad-based calyx-lobes deeply glandular serrate above the middle,* the short-oblong to subglobose fruit with thick mealy flesh and usually 5 nutlets; near Sarnia, Ont.

C. aùlica Sarg. (princely) is similar to the last but with the leaves more coarsely serrate and slightly villous along the veins beneath, the flowers with 10 or fewer stamens, the nutlets 3–5, in s. Ont.

C. Haberèri Sarg. (for its discoverer, JOSEPH V. HABERER, 1855–1925) has the mostly *broadly ovate to deltoid or suborbicular leaves* sharply serrate and with 4–5 pairs of shallow or acute spreading lateral lobes and glabrous except for short appressed hairs on the upper surface while young, thin, yellow-green at maturity, the few flowers 1.4–1.5 cm. wide and in nearly simple glabrous corymbs, with about 10 stamens and rose-colored anthers and glandular-serrate calyx-lobes, the *oval or short-obovoid fruit* 1–1.3 cm. thick and *with soft succulent flesh* and 3–5 nutlets; N.Y. and O.

C. corúsca Sarg. (gleaming; from lustrous branches) has the branchlets slightly villous when they first appear but soon glabrous, the broadly ovate or oval *leaves often as broad as long on vegetative shoots,* sharply serrate and rather deeply indented with 4–6 pairs of acute spreading lateral lobes and glabrous except for short appressed hairs on the upper surface while young, thin, *with the veins impressed above at maturity,* the many flowers 1.6–1.8 cm. wide and in slightly pubescent corymbs, with about 20 stamens, red anthers and glandular-serrate calyx-lobes, the oblong or obovoid, bright red fruit with a narrow slightly elevated calyx and 4–5 nutlets, is perhaps a hybrid between a species of the *Coccineae* and *C. mollis,* found in Lake Co., Ill.

Series 16. MÓLLES Sarg. (see p. 769)

a. Leaves yellow-green or dark green, firm or thick, not velvety beneath at maturity. . . *b.*

b. Corymbs many (10–20)-flowered; fruit 1.3–1.8 cm. thick, with edible flesh. . . *c.*

c. Foliage and inflorescence copiously tomentose, the midribs and petioles appearing thickened; leaves of vegetative shoots mostly truncate or cordate at the base or, if rounded in forms, with coarse shallow teeth. 88. ***C. mollis.***

c. Foliage and inflorescence thinly tomentose or villous, the midribs and petioles not appearing thickened; leaves of vegetative shoots narrowed, rounded or rarely truncate at base. . . *d.*

d. Leaves mostly ovate or oblong-ovate, noticeably longer than wide except sometimes on vegetative shoots; stamens about 10. . . . . . 89. *C. submollis.*

d. Leaves mostly short-ovate to suborbicular, only slightly longer than wide, or sometimes wider than long on vegetative shoots.

Leaves dark green; flowers 1.6–1.8 cm. wide; stamens about 20; fruit ripening in Oct.; nutlets usually 5. . . . . . . 90. *C. canadensis.*

Leaves yellow-green; flowers 1.8–2 cm. wide; stamens about 10; fruit ripening in Aug. or Sept. . . . . 91. *C. arnoldiana.*

b. Corymbs few (6–12)-flowered; fruit 1–1.2 cm. thick, with thin scarcely edible flesh. . . . . 92. *C. noelensis.*

a. Leaves blue-green, coriaceous or subcoriaceous, velvety beneath at maturity. 93. *C. lanuginosa.*

88. C. móllis (T. & G.) Scheele (soft). — A tree 10–12 m. high, with stout sparingly thorny or nearly thornless branchlets villous while young, and thick deeply fissured brownish-gray bark on the trunk; *leaves relatively large*, variable in shape, *mostly ovate or deltoid*, sharply or coarsely serrate, usually with 4–5 pairs of lateral lobes, or *on vegetative shoots sometimes deeply laciniate, thickly coated with short appressed hairs on the upper surface and densely white-tomentose* especially along the veins *beneath while young*, firm, yellow-green, glabrous above and often slightly pubescent beneath at maturity; flowers 2–2.3 cm. wide, many in tomentose corymbs; stamens about 20; *anthers small*, yellowish- or rarely pink; *calyx-tube densely tomentose, the lobes glandular-serrate; fruit* subglobose or rarely oblong or obovoid, 1.3–1.8 cm. thick, scarlet or bright crimson, *pubescent at least toward the ends*, with a broad shallow calyx, thick mellow flesh and normally 5 nutlets. — Open woods, usually in alluvial or fertile ground, s. Ont. and Mich. to Ala., w. to Minn., e. Okla., Ark. and Miss. *Fl.* April, May; *fr.* Aug.–Oct. — Forma **dumetòsa** (Sarg.) Palmer (of thickets) has the leaves narrowly ovate or elliptic, pointed at the apex and narrowed or rounded at the base, slightly lobed or nearly undivided, from Ind. to Mo. and ne. Okla.

Var. **sèra** (Sarg.) Egglest. (ripening late). — *Leaves* slightly thinner, ovate or oblong-ovate, *rounded or abruptly narrowed at the base; fruit usually obovoid*, rarely subglobose, 1–1.3 cm. thick, *ripening late*. — Sometimes well marked, but passing into the typical var., s. Ont. and Mich. to n. Ill.

89. C. submóllis Sarg. (somewhat soft). — A tree 8–10 m. high or sometimes an arborescent shrub with flexuous thorny branchlets villous when they first appear and slightly scaly brownish-gray bark; leaves ovate or oval, coarsely serrate and slightly divided with 4–5 pairs of shallow or obscure lateral lobes, thickly covered with short appressed hairs above and tomentose beneath while young, thin but firm, dark green, glabrous above at maturity; *flowers* many, 2–2.2 cm. wide, *in loose tomentose corymbs; stamens* 10 *or fewer;* anthers white or pale yellow; calyx-tube tomentose, the lobes glandular-serrate; *fruit pyriform or obovoid*, 1.2–1.5 cm. thick, bright red, slightly pubescent at the ends, with mellow flesh and usually 5 nutlets. (Incl. *C. champlainensis* Sarg.) — Wooded hillsides and open fertile ground, E. Can., N.E. and N.Y. *Fl.* May; *fr.* Sept.

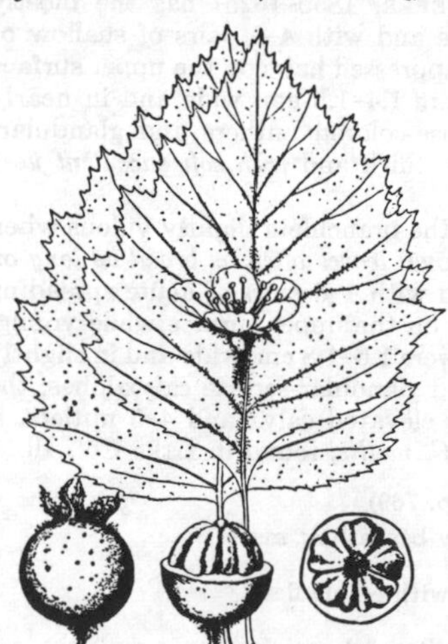
1139. C. arnoldiana.

90. C. canadénsis Sarg. (of Canada). — A tree 8–9 m. high, with slender flexuous very thorny branchlets villous while young; *leaves ovate to suborbicular*, sharply serrate, with 4–5 pairs of shallow lateral lobes, short-pilose above and somewhat tomentose beneath while young, *dark green*, glabrous above *at maturity;* flowers many, 1.6–1.8 cm. wide, in loose tomentose corymbs; stamens about 20; *anthers small, yellowish white; fruit subglobose or short-oblong*, 1.2–1.4 cm. thick, crimson, slightly villous at the ends, with mellow flesh and usually 5 nutlets. — Wooded calcareous ridges along the St. Lawrence R., Que. *Fl.* May; *fr.* Sept.

91. C. arnoldiàna Sarg. (for the Arnold Arboretum, where first recognized). — A tree 7–9 m. high, with stoutish flexuous thorny branchlets villous when they first appear; leaves mostly ovate or broadly ovate, sharply serrate and indented sometimes only above the middle with 3–5 pairs of small acute lateral lobes, short-villous on the upper surface and pubescent beneath while young, thin but firm, dark yellow-green, glabrous above and slightly pubescent along the veins beneath at maturity; flowers many, 2–2.3 cm. wide, in lax tomentose corymbs; *stamens about* 10; *anthers large,* pale yellow; *fruit* subglobose, 1.4–1.6 cm. thick, *bright crimson with pale dots,* pubescent at the ends, *with* thick mellow flesh and *usually* 3–4 *nutlets.* — Wooded banks, e. Mass., and Ct. *Fl.* May; *fr.* Aug.–Sept. FIG. 1139.

92. C. noelénsis Sarg. (of Noel, Mo.). — A tree 7–8 m. high, with a round top of wide-spreading branches and stout thorny branchlets villous when they first appear; *leaves* of flowering branchlets mostly oval or oblong-ovate, sharply and coarsely serrate and obscurely lobed, short-pilose above and slightly pubescent beneath while young, *thin but firm, dark yellow-green,* glabrous above *at maturity; flowers* 2–2.3 cm. wide, *mostly* 5–12 *in thinly tomentose corymbs;* stamens 10 or fewer; anthers red or sometimes pale yellow; *fruit* subglobose, 1–1.2 cm. thick, *orange-red or scarlet, with* a broad slightly elevated calyx, *thin firm flesh and* 3–5 *nutlets.* — Ozark reg., s. Mo. and n. Ark. *Fl.* April; *fr.* Sept.

93. C. lanuginôsa Sarg. (covered with wool). — A tree 8–9 m. high or sometimes an arborescent shrub *with stout flexuous very thorny branchlets* villous when they first appear *and with many compound thorns on the larger branches; leaves ovate to suborbicular,* sharply and deeply serrate, sometimes slightly or obscurely lobed with 2–4 pairs of shallow lateral lobes, or *occasionally deeply incised near the base on vegetative shoots,* copiously short-pilose above and densely tomentose beneath while young, *thick or coriaceous, dark blue-green, scabrous above and velutinous beneath at maturity;* flowers 2–2.3 cm. wide, many in compact tomentose corymbs; stamens about 20; anthers large, pink or red; *fruit* subglobose, 1–1.4 cm. thick, *dark crimson, pubescent,* with thick firm flesh and usually 5 nutlets. — Thickets and calcareous hills, sw. Mo. and adj. Kans. to centr. Ark. *Fl.* April; *fr.* Oct.

C. dispéssa Ashe (spread out) has the branchlets at first pubescent, the *elliptic or oval* finely serrate *leaves undivided or rarely slightly lobed on vegetative shoots* and sparsely short-pilose above while young and pubescent along the veins beneath throughout the season, thin but firm at maturity, the mostly 5–10 flowers in lax pubescent corymbs and with about 20 stamens and pale yellow anthers, the short-oblong or nearly globose fruit 1–1.2 cm. thick and with thin firm flesh and 4–5 nutlets. (*C. pyriformis* Britt.) — S. Mo.

C. Kellóggii Sarg. (named in 1903 for its discoverer, JOHN H. KELLOGG) with the *relatively* small mostly *broadly ovate to suborbicular* sharply serrate *leaves* indented with 4–5 pairs of small rounded lateral lobes and short-pilose above while young and more or less pubescent along the veins beneath throughout the season, firm and dark yellow-green and *with* the *veins distinctly impressed above at maturity,* the few flowers 1.4–1.6 cm. wide and in villous corymbs, with about 20 stamens and pink or white anthers, the subglobose yellow or dark red fruit 1.5–1.7 cm. thick and with thin mealy flesh and usually 5 nutlets, is probably a hybrid between *C. Margaretta* and *C. mollis,* found in Ind. and Mo.

C. latebrôsa Sarg. (of a hiding place; from the woodland habitat) with the oblong-elliptic or oval finely serrate and *unlobed leaves* (*or* these sometimes *with a few shallow asymmetric lobes above the middle*) short-villous above while young and slightly pubescent along the veins beneath throughout the season, the *few flowers in compact densely tomentose corymbs,* with about 10 stamens and pale yellow anthers, the short-oblong or subglobose fruit about 1 cm. thick and pubescent at the ends, with thin firm flesh and 3–5 nutlets, is perhaps a hybrid between *C. noelensis* and *C. collina,* found near Noel, Mo.

Series 17. DILATÀTAE Sarg. (see p. 769)

Young leaves short-pilose above, sometimes slightly villous on the veins beneath, not crisped at the edges, rarely subcordate except on vegetative shoots; inflorescences glabrous or slightly villous; fruit subglobose or slightly obovoid. 94. *C. dilatata.*

Leaves glabrous or essentially so from the first, crisped at the edges, often subcordate on flowering branchlets; inflorescences glabrous; fruit subglobose or broader, often angular. . . . . 95. *C. coccinioides.*

94. C. dilatàta Sarg. (widened). — A tree 7–8 m. high or sometimes shrubby with slender often flexuous thorny branchlets; *leaves* broadly ovate or deltoid-ovate, sharply and deeply serrate, with 3–5 pairs of triangular lateral lobes, *short-pilose above and often slightly pubescent beneath while young,* thin but firm, glabrous, dark yellow-green at maturity; *flowers* 2.2–2.4 *cm. wide, many in* glabrous or sparsely villous *corymbs;* sta-

1140. C. coccinioides.

mens about 20; anthers large, rose-color; calyx-lobes glandular-serrate; *fruit* subglobose, 1–1.5 cm. thick, bright red, *with thick mellow or succulent flesh* and usually 5 nutlets. (*C. coccinioides* var. *dilatata* (Sarg.) Egglest.) — Borders of woods and hillsides, usually in fertile soil, se. Can. and N.E. *Fl.* May; *fr.* Sept.

95. C. coccinioides Ashe (resembling *C. coccinea*).— A tree 6–7 m. high or sometimes a stout shrub with slender usually very thorny branchlets; *leaves* broadly ovate or deltoid, sharply and deeply serrate, with 4–5 pairs of triangular lateral lobes, *glabrous except for a few caducous hairs sometimes found on the upper surface while young*, thin but firm, yellow-green, with *the edges often crisped* at maturity; *flowers 2.4–2.6 cm. wide, mostly 4–7 in compact* glabrous *corymbs;* stamens about 20; anthers large, pink or rose-color; calyx-lobes laciniately glandular-serrate; *fruit* subglobose or depressed-globose, *often angular*, 1.3–1.7 cm. thick, *with* a very broad nearly sessile calyx, *thick firm juicy flesh* and 5 nutlets. — Thickets and calcareous hills, s. Ill. to se. Kans., Okla. and Ark. *Fl.* May; *fr.* Oct. FIG. 1140.

Series 18. MACRACÁNTHAE Loud. (see p. 769)

a. Leaves prevailingly ovate or oval, broadest at or below the middle, usually lobed at least on vegetative shoots. . . *b.*

b. Leaves dark or bright green or if yellow-green, lustrous above; young branchlets glabrous or rarely slightly villous. . . *c.*

c. Leaves thick or coriaceous and lustrous above at maturity; stamens 10 or 20. . . *d.*

d. Inflorescence usually many-flowered; nutlets 2–3.

Flowers 1.2–1.8 cm. wide, in more or less branched but not conspicuously lax corymbs. . . . . 96. *C. succulenta.*

Flowers 1–1.2 cm. wide, in conspicuously lax corymbs. . . . . . 97. *C. laxiflora.*

d. Inflorescence usually 6–8-flowered; nutlets 3–4. . . . . 98. *C. chadsfordiana.*

c. Leaves thin but firm, scabrate above at maturity; stamens 5–10. . . . 99. *C. divida.*

b. Leaves dull yellow-green above, often irregularly lobed above the middle or sometimes lobed only on vegetative shoots; young branchlets tomentose. 100. *C. Calpodendron.*

a. Leaves prevailingly obovate or oblong-obovate, broadest at or above the middle, unlobed except rarely and slightly on vegetative shoots.

Flowers usually many in compound corymbs; calyx-lobes entire or finely glandular-serrate. . . . . 101. *C. laetifica.*

Flowers 2–8, in nearly simple clusters; calyx-lobes pectinately glandular-serrate. . . . . 102. *C. Vailiae.*

96. C. succulénta Link (succulent; referring to the fruit). — A tree 7–8 m. high or sometimes a stout shrub with slender often flexuous and usually thorny branchlets glabrous or rarely slightly villous when they first appear; *leaves* elliptic, rhombic or ovate, sharply serrate except near the base, usually indented above the middle with 4–5 pairs of shallow or short acute lateral lobes, with short appressed hairs on the upper surface and sometimes slightly pubescent beneath while young, *thick or coriaceous, glabrous, lustrous,* and *with the veins appressed above at maturity;* flowers many, 1.3–1.8 cm. wide, in glabrous or slightly villous corymbs; stamens about 20; anthers pink or rarely white; calyx-lobes glandular-serrate; *fruit* subglobose, 0.7–1.2 cm. thick, glabrous, *bright red and succulent when ripe;* nutlets 2–3. (*C. macracantha* var. *succulenta* (Link) Egglest.) — Thickets, pastures and borders of woods, usually in dry or rocky ground, se. Can., N.E., N.Y. and Pa., w. to Ia. *Fl.* May, June; *fr.* Sept., Oct. FIG. 1141.

Var. **macracántha** (Lodd.) Egglest. (with large thorns). — *Stamens about* 10, with white, pale yellow or rarely pink anthers; *fruit* 1 cm. thick or less, *remaining hard and dry until late in the season; thorns* usually more numerous, *up to 7–8 cm. long.* (*C. macracantha* Lodd.) — Often difficult to distinguish from the typical var., se. Can. and N.E. to Minn.

Var. **michiganénsis** (Ashe) Palmer (of Michigan). — *Leaves* oval or rarely oblong-obovate, *rounded or abruptly pointed at the apex, very slightly or obscurely lobed,* glabrous above and slightly pubescent beneath at maturity; flowers 1.2–1.3 cm. wide, with about 20 stamens and pink anthers. — Near Port Huron, Mich.

Var. **pertomentòsa** (Ashe) Palmer (very tomentose). — *Leaves* oval, rhombic, or rarely ovate or slightly obovate, slightly or obscurely lobed, finely appressed-villous above while young and pubescent beneath throughout the season, *at maturity coriaceous, dark green,* glabrous, *and with the veins distinctly impressed above, much paler and velvety-pubescent beneath;* flowering corymbs tomentose; stamens about 10; *fruit* subglobose, 8–9 mm. thick, bright red, *usually pubescent at the ends,* with thin flesh and 2–3 or rarely 4 nutlets. (*C. pertomentosa* Ashe) — Ill., Ia., Mo. and e. Kans.

Var. **occidentàlis** (Britt.) Palmer (western). — *Leaves* elliptic, rhombic, or rarely oblong-obovate, usually acute at both ends, green on both sides or *only slightly paler beneath,* firm to subcoriaceous, *glabrous or nearly so at maturity;* flowers many in villous corymbs, with about 10 stamens; *fruit* subglobose or short-oblong, 7–9 *mm. thick,* bright red, *pubescent at the ends.* — Mich. to Man. and Mont., s. to Neb. and Colo.

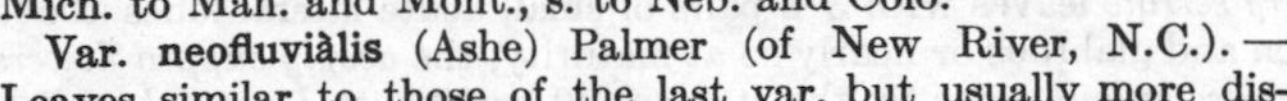

Var. **neofluviàlis** (Ashe) Palmer (of New River, N.C.). — Leaves similar to those of the last var. but usually more distinctly lobed, glabrous or nearly so; flowers 1.3–1.5 cm. wide, many in glabrous or slightly villous corymbs, with 10 or rarely more stamens; fruit 6–8 mm. thick, remaining dry or becoming mellow late in the season, glabrous. (*C. neofluvialis* Ashe; *C. macracantha* var. *neofluvialis* (Ashe) Egglest.) — N.E., s. Ont. and N.Y. to N.C., w. to Mo.

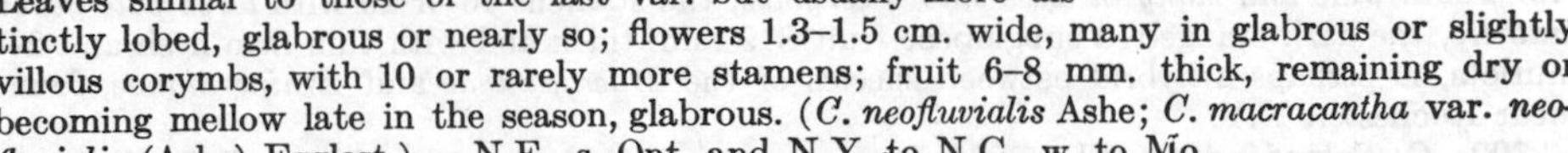

1141. C. succulenta.

C. **integrìloba** Sarg. (entire-lobed) with the oval or rarely oblong-obovate *leaves deeply and irregularly serrate* and slightly lobed above the middle and *glabrous except for a few short villous hairs on the upper surface while young,* dark green and thin but firm, with the veins impressed above at maturity, the *many flowers* 1.5–1.7 cm. wide and *in pubescent corymbs, with* about 10 stamens and *large red anthers,* the *calyx-lobes linear-lanceolate, entire* and *usually persistent on fruit,* the subglobose fruit 0.8–1 cm. thick and bright scarlet with thin mellow flesh and 2–3 nutlets is perhaps a hybrid between species of the *Macracanthae* and *Punctatae,* found in s. Que.

C. **árdua** Sarg. (difficult) with leaves similar to those of the last but more finely and evenly serrate, the *flowers slightly larger* and with 10–20 stamens and white or red anthers, the *calyx-lobes sometimes glandular-serrate,* is perhaps a hybrid of similar origin to the last, near Toronto, Ont.

C. **menandiàna** Sarg. (named for Menands, near Albany, N.Y.) *with thinner* and less distinctly lobed *leaves,* the *flowers in lax glabrous corymbs* and the subglobose succulent fruit 1–1.2 cm. thick, may also belong here.

97. C. laxiflòra Sarg. (loosely flowered). — A tree 7–8 m. high, with slender glabrous flexuous thorny branchlets; *leaves* oval, oblong-elliptic, or oblong-obovate, finely serrate except near the base, usually indented with 3–5 pairs of shallow lateral lobes above the middle, glabrous or nearly so, *subcoriaceous, dark green and lustrous above at maturity; flowers many,* 1–1.2 *cm. wide, in lax slender-branched villous corymbs;* stamens about 20; anthers pale yellow; *calyx-lobes glandular-serrate;* fruit subglobose or short-oblong, 7–8 mm. thick, bright red, with a narrow elevated calyx, thin dry flesh and 2–3 nutlets. — Open woods along streams, n. Ill. *Fl.* May; *fr.* Oct.

C. **laurentiàna** Sarg. (for the St. Lawrence R. reg.) with leaves similar in outline, but *more coarsely and irregularly serrate and more distinctly lobed,* the many *flowers* 1.4–1.6 *cm. wide and in* lax *hoary-tomentose corymbs,* with about 10 stamens and pink anthers, the oblong or obovoid fruit 0.8–1.1 cm. thick and with thin flesh becoming succulent and 4–5 nutlets, is probably a hybrid between *C. succulenta* and *C. Brunetiana,* occurring along the St. Lawrence R., Que.

98. C. chadsfordiàna Sarg. (for Chadsford, Delaware Co., Pa.). — An arborescent shrub 2–3 m. high, with slender flexuous thorny branchlets; leaves mostly ovate or rhombic, sharply serrate and indented with 4–5 pairs of small triangular lateral lobes mostly above the middle, *glabrous or nearly so,* yellow-green and *lustrous above at maturity; flowers mostly* 6–8, 1.6–2 *cm. wide, in nearly simple glabrous corymbs;* stamens about 10; anthers white or pale yellow; calyx-lobes nearly entire; *fruit* subglobose or short-oblong, 1–1.2 *cm. thick, with thin firm flesh* and usually 3–4 nutlets. — Thickets and wooded slopes, w. Mass. and Pa. *Fl.* May; *fr.* Sept., Oct. — Perhaps a hybrid between species of the *Macracanthae* and *Pruinosae.*

C. **spatiòsa** Sarg. (ample), is similar but with the *leaves* mostly broadly ovate or rhombic (or *on vegetative shoots often as broad as long or broader* and *deeply divided*), found near New London, Ct.

C. putàta Sarg. (pruned; application not explained) with oval, elliptic or broadly ovate finely serrate and usually slightly lobed leaves, the *flowers* 1.5–1.7 cm. wide and *with about* 20 *stamens* and pink anthers, the *orange-red fruit* 7–9 *mm. thick* and with thin dry flesh and 2–3, usually 3, nutlets, is found near Scranton, Pa.

C. membranàcea Sarg. (thin, referring to the leaves) has the mostly elliptic or oblong-ovate *coarsely and irregularly serrate leaves* indented with 3–5 pairs of small acute lateral lobes mostly above the middle, these glabrous and *thin at maturity*, the many flowers 1.8–2 cm. wide and in glabrous or slightly villous corymbs, the fruit 0.9–1 cm. thick and with thick flesh becoming succulent and usually 3 nutlets, about Middlebury, Ct.

99. C. dívida Sarg. (separated). — A tree 8–9 m. high or often a stout shrub with slender straight or flexuous thorny branchlets; *leaves* mostly oblong-obovate, sharply serrate and slightly indented with 4–5 pairs of small acute lateral lobes, glabrous except for short appressed hairs on the upper surface while young, *thin but firm, bluish-green at maturity;* flowers 1.5–2 cm. wide, many in glabrous corymbs; stamens 5–10; anthers pink or pale yellow; fruit subglobose or short-oblong, 0.9–1.2 cm. thick, with thin mellow flesh and 2–3 nutlets. — Borders of woods and argillaceous hills, usually in moist soil, s. Ont., N.Y. and n. Ill. *Fl.* May; *fr.* Oct.

C. Làneyi Sarg. (named in 1902 for its discoverer, C. C. Laney) with the elliptic or oblong obovate *sharply and irregularly serrate* leaves with 3–5 pairs of small acute lateral lobes above the middle and yellowish green and glabrous or nearly so at maturity, the *deeply cupped flowers* 1.5–2 cm. wide and *many in loose villous corymbs*, the stamens 10 or 20 with *large pale yellow anthers*, the dark orange-red subglobose fruit 8–9 mm. thick and with thin firm flesh and 2–4 nutlets, is perhaps a hybrid between species of the *Macracanthae* and *Brainerdianae*, found near Rochester, N.Y.

100. C. Calpodéndron (Ehrh.) Medic. (urn-tree; referring to the shape of the fruit). — A tree 5–6 m. high or often an arborescent shrub, with slender sparingly thorny or nearly thornless branchlets tomentose while young and dark brownish-gray bark becoming thick and furrowed on old trunks; *leaves ample*, mostly ovate, oblong-elliptic, or rhombic, coarsely serrate except near the base, usually *with* 3–5 *pairs of shallow often asymmetric lateral lobes*, short-villous above while young and usually pubescent beneath throughout the season, firm, dull yellow-green and with the veins impressed above at maturity; flowers 1.2–1.5 cm. wide, many in tomentose corymbs; stamens about 20; anthers pink or rarely white; *calyx-tube pubescent, the lobes glandular-serrate or pectinate; fruit oblong or obovoid,* rarely nearly globose, 7–9 *mm. thick, pubescent while young,* bright red or orange-red, *with thin sweet flesh becoming succulent,* and 2–3 nutlets deeply pitted on the inner surface. (*C. Chapmani* var. *Plukenetii* Egglest.; *C. tomentosa* of many auth.) — Open woods and thickets, usually along small rocky streams, s. Ont. and N.Y. to Ga. and Ala., w. to Minn. and Mo. *Fl.* May, June; *fr.* Oct.

Var. **microcárpa** (Chapm.) Palmer (small-fruited). — *Leaves* similar to those of the typical var. but *with the primary veins closer together and more deeply impressed above;* fruit subglobose, 5–8 mm. thick. (*C. Chapmani* (Beadle) Ashe) — Va. and N.C. to Mo. — Sometimes well-marked, but intermediate forms are difficult to distinguish.

Var. **globòsa** (Sarg.) Palmer (globose). — *Leaves often obscurely lobed or unlobed; flowers fewer in more compact corymbs; fruit subglobose,* 7–9 mm. thick, with thin flesh remaining hard and dry until late in the season. (*C. globosa* Sarg.) — Ky. to Mo., Kans. and e. Tex.

Var. **mollícula** (Sarg.) Palmer (soft). — A shrub 2–3 m. high; *leaves* obscurely lobed or undivided, *velvety-pubescent beneath;* flowers 1.5–1.7 cm. wide, with 10 or fewer stamens and usually yellow anthers. — S. Mo. and Ark.

Var. **hispídula** (Sarg.) Palmer (minutely hispid). — *Leaves more sharply serrate than in the typical var.*, with narrow acuminate teeth and small acuminate spreading lateral lobes, *hispidulous beneath at maturity;* fruit obovoid or short-oblong, pubescent while young. — S. Mo. and se. Kans.

C. Whittakèri Sarg. (named in 1925 for Page Whittaker, on whose farm it was discovered) has leaves similar to those of *C. Calpodendron*, the *flowers* 1.5–2 *cm. wide and in small compact densely tomentose corymbs*, the short-oblong or nearly globose *fruit* 1–1.2 *cm. thick*, found near Olney, Ill. — Described as a possible hybrid between *C. Calpodendron* and *C. mollis.*

101. C. laetífica Sarg. (making glad). — An arborescent shrub or a small tree with stoutish flexuous thorny glabrous branchlets; *leaves obovate or oblong-obovate, finely serrate* except near the cuneate base, *otherwise entire, glabrous*, dark yellowish-green, *lustrous above* at maturity; *flowers* 1.2–1.5 cm. wide, *many in compact glabrous corymbs;* stamens 10 or 20; anthers pink; calyx-lobes glandular-serrate; *fruit* subglobose, 0.9–1.3 *cm. thick, dark red,* with thin slightly succulent flesh and 2–3 nutlets with small cavities on the inner faces. — Open woods and borders of swamps, Pa. and O. *Fl.* May; *fr.* Sept., Oct. — Probably a hybrid between species of the *Macracanthae* and *Crus-galli.*

C. simulàta Sarg. (resembling), with the ovate, oval or elliptic *leaves sharply serrate nearly to the base but otherwise undivided* and with short appressed hairs above and *pubescent along the veins beneath while young,* thin but firm and yellowish-green at maturity, *the mostly* 5–15 *flowers* 1.6–1.8 cm. wide and *in slightly villous corymbs,* with about 20 stamens and dark red anthers and coarsely glandular-serrate calyx-lobes, the short-oblong or subglobose *fruit* 7–8 mm. thick and *with* thin dry flesh and *usually* 3 *nutlets with shallow cavities on the inner faces* is probably a hybrid between *C. Calpodendron* and *C. Palmeri,* found near Joplin, Mo.

C. nùda Sarg. (naked; without hairs) has similar but *glabrous and thinner leaves,* the *many flowers* 1.3–1.5 cm. wide and *in glabrous corymbs,* about 10 stamens and pale pink or white anthers, the *nutlets usually* 2 *with irregular depressions on the inner faces,* is perhaps a hybrid between a var. of *C. succulenta* and a species of the *Crus-galli,* known in Taney Co., Mo.

102. C. Vailiae Britt. (named for its discoverer, Anna Murray Vail, 1863–). — An arborescent shrub or a small tree 4–5 m. high, with slender usually thorny *branchlets tomentose while young; leaves* elliptic, oblong-obovate, or *on vegetative shoots sometimes broadly ovate to suborbicular, coarsely serrate or dentate, otherwise undivided,* roughened on the upper surface with short appressed hairs while young and pubescent along the veins beneath throughout the season, thick, dark green, slightly lustrous above at maturity; *petioles stout,* 0.5–1 *cm. long, wing-margined often nearly to the base; flowers* 1–1.5 cm. wide, *mostly* 2–6 *in nearly simple tomentose clusters;* stamens about 20; anthers white or red; *calyx-lobes glandular-pectinate, usually persistent on* the *fruit; fruit* subglobose or obovoid, 0.8–1 cm. thick, red or orange-red, *pubescent* at least *while young,* with thin firm flesh and 3–5 nutlets usually pitted on the inner faces. — Thickets and rocky ground along streams, s. Va. and N.C. to s. Mo., occasional. *Fl.* May; *fr.* Oct. — Probably a hybrid between *C. Calpodendron* and *C. uniflora.*

Series 19. Douglasiànae Egglest. A single species in our reg. (see p. 769)

103. C. Douglásii Lindl. (for David Douglas, botanical explorer, 1798–1834). — A tree 10–12 m. high or sometimes an arborescent shrub, with slender glabrous thorny or thornless branchlets and brownish-gray scaly bark; *leaves oblong-obovate, oval or elliptic,* serrate except near the base, *usually indented above the middle or only near the apex with* 2–4 *pairs of broad shallow or acute lateral lobes,* glabrous or with a few short appressed hairs on the upper surface while young, firm, dark green, lustrous above at maturity; flowers 1–1.3 cm. wide, mostly 5–12 in glabrous corymbs; stamens 10 or fewer; anthers white or pink; *fruit* short-oblong or rarely subglobose, 0.8–1 cm. thick, *turning from dark wine-color to black* when fully ripe, with thick succulent flesh and 3–5 nutlets slightly ridged on the outer and pitted on the inner faces. — Open woods and rocky banks, sw. Ont. and n. Mich. to the Pacific coast and widely distributed in the Rocky Mts. *Fl.* May; *fr.* Aug., Sept. Fig. 1142.

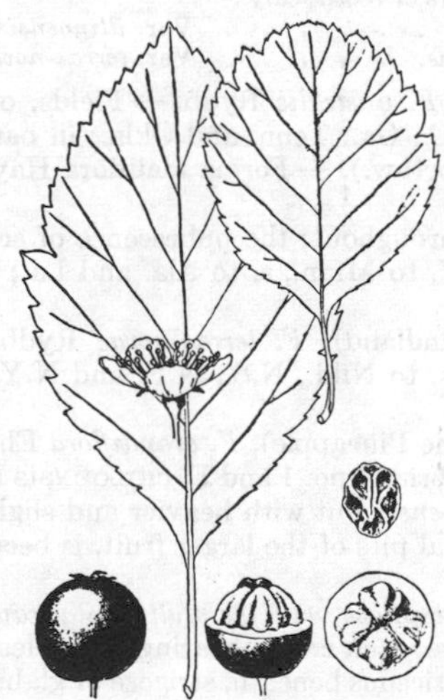
1142. C. Douglasii.

9. COTONEÁSTER Ehrh.

Calyx small, adherent to the 2–5 carpels, the 5 lobes short, persistent as teeth. Styles free, stigmatic at the slightly enlarged summit. Carpels at maturity bony, 1-seeded. Fruit small, berry-like, mealy. — Much branched shrubs nat. of Eurasia and n. Afr., with small alternate usually coriaceous and often evergreen leaves, and small white cymose flowers. (Name New Latin, implying resemblance to the quince, from *cotonea, quince,* and the suffix *-aster, a kind of.*) Including Pyracantha Roem.

1. C. Pyracántha (L.) Spach (fire-thorn), Fire-Thorn. — Shrub, armed with slender spreading purple spines; leaves elliptic-oblanceolate, crenate-serrate, coriaceous, 3–6 cm. long; fruit globose, scarlet. — Attractive shrub, used for formal hedges, etc., esc. from cult. and estab., Fla. to La., n. to Pa. (Introd. from Eu.)

Tribe III. Potentílleae B. & H.

10. FRAGÀRIA L. Strawberry. Fraisier (Que.)

Flowers nearly as in *Potentilla*, but in varying degrees polygamodioecious. Styles deeply lateral. Receptacle in fruit much enlarged and conical, becoming pulpy and usually scarlet, bearing the minute dry achenes scattered over or slightly imbedded in its surface. — Rosulate N. and S.Am. and Eurasian perennials, usually with stolons, and with white to pink cymose flowers on scapes. Leaves radical; leaflets 3, obovate-cuneate, coarsely serrate; stipules cohering with the base of the petioles, which with the scapes are usually hairy. (Name from the Latin *fraga*, the *strawberry*, so called from the fragrance of the fruit.)

Inflorescence umbelliform or a rounded to flattish cyme, with subequal primary branches; calyx-lobes appressed or connivent about the young fruit; achenes in pits of the pulpy receptacle.
- Crown mostly single, short-lived, at the tip of a thick rhizome; superficial runners regularly developed; leaflets petiolulate, the larger 1.5–10 cm. long. . . 1. *F. virginiana.*
- Crowns numerous, erect, long-lived, forming a multicipital caudex; superficial runners wanting or rarely developed; leaflets sessile, the larger 1–3 cm. long. 2. *F. multicipita.*

Inflorescence soon racemose or irregularly racemiform, the primary branches of the cyme quite unequal, the leading one prolonged as the axis of a raceme; calyx-lobes loosely spreading or reflexed; achenes superficial or nearly so. . . 3. *F. vesca.*

1. F. virginiàna Duchesne (Virginian). — Leaves, scapes and runners from a *subsimple caudex at end of a simple thickish rhizome;* old leaves firm or coriaceous, new ones thinner, plane or hardly rugose; *leaflets short-petiolulate,* the terminal one cuneate-obovate, the lateral with rounded outer margin, *sharply toothed;* flowers dioecious or polygamous, the strictly pistillate much smaller than the staminate, 0.6–2.5 cm. broad, 2–many in umbelliform cymes; calyx-lobes 4–11 mm. long; fruits 0.5–2 cm. in diameter, subglobose to ovoid; *achenes* 1.3–1.6 *mm. long, in deep pits*; pulp juicy. Very variable; the best marked variations are the following:

Scapes and petioles villous or hirsute with loosely spreading to divaricate hairs.
- Terminal leaflets of larger leaves 1.5–10 cm. long, with 4–8 pairs of teeth; calyx-lobes 4–8 mm. long. . . . . *F. virginiana* (typical).
- Terminal leaflets of larger leaves 5–10 cm. long, with 8–15 pairs of teeth; calyx-lobes 5–10 mm. long. . . . . Var. *illinoensis.*

Scapes and petioles appressed- or strigose-pubescent to glabrous. . . . . . Var. *terrae-novae.*

F. virginiàna (typical), incl. *F. canadensis* Michx. and *F. australis* Rydb. — Fields, open slopes and borders of woods, Nfld. to Alta., s. to N.S., N.E., Ga., Tenn. and Okla.; in barely separable forms westw. and southwestw. *Fl.* Apr.–July (northw.). — Forma **maliflòra** Haynie (with apple-flowers) has petals pink or pink-striped.

Var. **illinoénsis** (Prince) Gray (of Illinois). — Coarser throughout; the pubescence of scape and petiole denser. (*F. Grayana* Vilmorin) — From w. N.Y. to Minn., s. to Ala. and La.; esc. from cult. eastw. to N.E.

Var. **térrae-nòvae** (Rydb.) Fern. & Wieg. (of Newfoundland), *F. terrae-novae* Rydb. — Commoner northw. and at high alts., se. Lab. to n. Ont., s. to Nfld., N.S., N.E. and N.Y. — Occasionally crosses with no. 3 in Que. and n. Me.

× **F. Ananássa** Duchesne (from *Ananas* or *Ananassa*, the Pineapple), *F. grandiflora* Ehrh., not Crantz, the Cultivated Strawberry, considered a hybrid of no. 1 and F. chiloensis (L.) Duchesne and strongly resembling *F. virginiana*, var. *illinoensis* but with heavier and slightly rugose foliage, coarser and blunter teeth and more superficial pits of the larger fruit, is becoming freq. along roads and railroads. (Esc. from cult.)

2. F. multicípita Fern. (with many crowns). — *Densely cespitose, with the multicipital caudex forking into* 4–40 *erect long-lived crowns;* runners none or rare; each crown bearing 3 or 4 leaves with cuneate *sessile or subsessile leaflets* 1–3 *cm. long,* these sericeous beneath, strigose or glabrate above, very sharply toothed; scapes silky, shorter than leaves, with 1–2(–4) umbellate flowers and fruits; calyx as in no. 1, its lobes 3–4 mm. long; fruit subglobose or ovoid, 5–10 mm. long; *achenes* 1 *mm. long,* in pits. — Gravels along R. Ste. Anne des Monts, Gaspé Co., Que. *Fr.* July.

3. F. vésca L. (weak), Woodland S., Sow-teat S., Fraisier à vaches (Que.). — Crowns 1–several; runners numerous; *leaflets* firm, dark green and *strongly veined above,* deeply toothed; *petiole and scape with mostly loose or widely spreading pubescence; inflorescence* 1–9-flowered, *becoming racemiform or paniculate by elongation of the axis above the earlier flowers;* flowers 1–1.5 cm. broad, with white petals, or petals pink in forma **ròsea** Rostrup (rose-colored); *calyx spreading or reflexed;* fruit ovoid-conic to subglobose, red, or yellowish or whitish in forma **álba** (Ehrh.) Rydb. (white), rather dry and insipid; *achenes superficial,* 1–1.3 mm. long. — Rocky woods and openings, Nfld. to Mich. and W.Va.; indig. in Nfld. and e. Que., chiefly introd. from Eu. elsewhere. *Fl.* May–Aug.

Var. **americàna** Porter (American), Sow-teat S. — *More slender; leaves thin, pale green; pubescence* of scapes as well as pedicels and sometimes the petioles *more or less closely appressed, often sparse;* fruit slenderly conical to subcylindric-ovoid, sometimes subglobose. (*F. americana* Britt.) — Wooded slopes or rocky banks, Gaspé Pen., Que., to n. Alta., s. to N.S., N.E., upland Va., n. Ill., Mo., Neb. and N.M.

11. DUCHÈSNEA Sm. Indian Strawberry

Calyx 5-parted, the lobes alternating with much larger foliaceous spreading 3-toothed appendages. Petals 5, yellow. Receptacle in fruit spongy but not juicy. Flowers otherwise as in *Fragaria.* — Perennial herb nat. of Asia, with leafy stolons and 3-foliolate leaves similar to those of the true strawberries. (Dedicated to *Antoine Nicolas Duchesne*, 1747–1827, an early monographer of *Fragaria.*)

1. D. índica (Andr.) Focke (of India). — Fruit red, insipid. — Waste ground, grassy places, etc., n. Fla. to Okla., n. to s. Ct., N.Y., O., Ind. and Ia. April–June. (Natzd. from Asia)

12. SIBBÁLDIA L.

Calyx flattish, 5-cleft, with 5 bractlets. Petals 5, linear-oblong, minute. Stamens 5, alternate with the petals, inserted into the margin of the woolly disk which lines the base of the calyx. Achenes 5–10; styles lateral. — Low and depressed arct. and boreal perennials. (Dedicated to Sir *Robert Sibbald*, 1641–1722, first professor of medicine at Edinburgh.)

1. S. procúmbens L. (lying on the ground). — Leaflets 3, cuneate, 3-toothed at the apex; petals yellow. — Greenl. to Alaska, s. to alpine gravels and meadows of nw. Nfld., Shickshock Mts., Gaspé Pen., Que., White Mts., N.H., Colo., Utah and Calif. Late June–Aug. (Eurasia)

13. CHAMAÈRHODOS Bunge

Calyx turbinate, 5-cleft, without bractlets. Petals 5, obovate, white or purplish, about as long as the calyx-lobes. Stamens 5, opposite the petals. Carpels 5–20; styles decidedly lateral or basilar, articulated near the base. Ovule solitary, ascending. — Erect pubescent essentially herbaceous plants of N.Am. and Siberia, with 3-foliolate leaves; the leaflets cleft into linear segments. (Name from the Greek *chamai, on the ground, low, dwarf*, and *rhodon, a rose.*)

1. C. Nuttállii Pickering (for its discoverer, Thomas Nuttall, 1786–1859). — Glandular-pubescent; root woody; stem erect, 1–3 dm. high, often with ascending branches, leafy; flowers small, crowded in small rounded cymes. (*C. erecta* of ed. 7, not (L.) Bunge) — Sandy and gravelly soil, arid plains, etc., Yuk. to Colo., e. to w. Man., N.D. and w. Minn. Late May–July.

Var. **keweenawénsis** Fern. (for Keweenaw Peninsula, Mich.). — Villous as well as glandular. — Dry gravelly bluffs, Keweenaw Co., Mich.

14. WALDSTEÌNIA Willd.

Calyx-tube inversely conical; the limb 5-cleft, with 5 often minute and deciduous bractlets. Petals 5. Stamens many, inserted into the throat of the calyx. Achenes 2–6, minutely hairy; the terminal slender styles deciduous from the base by a joint. Seed erect; radicle inferior. — Low perennial herbs of n. temp. reg., with chiefly radical 3–5-lobed or -divided leaves, and small yellow flowers on bracted scapes. (Named in honor of *Francis Adam*, 1759–1823, Count of *Waldstein-Wartenburg*, a German botanist.)

1. W. fragarioìdes (Michx.) Tratt. (like *Fragaria*), Barren Strawberry. — Low; leaflets 3, broadly cuneate, cut-toothed; scapes several-flowered; calyx-lobes 2–7.5 mm. long; petals mostly 5–10 mm. long and 3–6 mm. broad. — Woods, thickets and clearings, Carleton Co., N.B.; s.-centr. Me. and sw. Que. to Ont. and Minn., s. to uplands of Ga. and Tenn. and locally to Ind. and Mo. Late April–June. Passing southw. into

Var. **parviflòra** (Small) Fern. (small-flowered). — Calyx-lobes 3–4.5 mm. long; petals 2.5–4 mm. long, 1–1.5 mm. wide. (*W. parviflora* Small) — Mts. of Pa. and Ky. to Tenn. and Ga.

15. POTENTÌLLA L. Cinquefoil. Five-finger

Calyx flat, deeply 5-cleft, with as many bractlets at the sinuses, thus appearing 10-cleft. Petals 5, usually roundish. Stamens few to many. Achenes many, collected in a head on the dry mostly pubescent or hairy receptacle; styles lateral or terminal, deciduous. Radicle superior. — Herbs, or rarely shrubs, mostly of N. Hemisph., with compound leaves, and solitary or cymose

flowers; their parts rarely in fours. (Name a diminutive from *potens*, powerful, originally applied to *P. anserina*, from its once reputed medicinal powers.)*

*a.*Carpels and achenes pubescent. . . *b.*
*b.*Shrub; leaves pinnate, with 5–7 approximate entire leaflets; petals yellow; style attached below middle of achene, enlarged upward, only slightly overtopping achene. . . . . 1. *P. fruticosa.*
*b.*Herbaceous or merely suffruticose at base; leaflets 3; petals white; style filiform, prolonged.
Leaves coriaceous, evergreen; leaflets narrowly cuneate, entire except at 3–5-toothed apex; flowers in stiff terminal cymes; achenes densely hairy; style basal. . . . . 2. *P. tridentata.*
Leaves herbaceous; leaflets obovate to rounded-ovate, with regularly toothed margins; flowers 1–few on filiform branches; achenes sparsely hairy; style terminal. . . . . 3. *P. sterilis.*
*a.*Carpels and achenes glabrous. . . *c.*
*c.*Suffruticose, with prolonged woody stems rooting in water or marsh; leaves not rosulate, pinnate; petals purple; disk thick and hairy; style lateral. . 4. *P. palustris.*
*c.*Herbaceous, annual or biennial, or, when perennial, often with flowering stems arising from crowns or the crowns bearing rosettes of leaves. . . *d.*
*d.*Glandular-villous erect perennial; leaves pinnate, with many (in ours 7–11) leaflets; petals white or creamy; stamens 25–30, in 5 groups; style subbasal, thickened below middle. . . . . 5. *P. arguta.*
*d.*Rarely if ever glandular-villous; petals yellow; stamens fewer; style borne at or well above middle of carpel. . . *e.*
*e.*Style thickened at base or below the middle, slenderly conical; flowers mostly numerous (1–few in nos. 6 and 9), in cymes. . . *f.*
*f.*Lower leaf-surfaces or branches of inflorescence tomentose or lanate with implexed soft hairs (rarely glabrous or glabrate in age); perennials with crowns long retaining marcescent old stipules or their remnants. . . *g.*
*g.*Style short and much thickened, shorter than to but slightly longer than mature carpel. . . *h.*
*h.*Basal leaves ternate or digitate.
Densely cespitose; basal leaves mostly ternate, with regularly toothed ovate, obovate or broadly oblong leaflets; flowers 1–few, long-peduncled, 1–2 cm. broad; petals much exceeding calyx-lobes. . . . . 6. *P. nivea.*
Loosely cespitose or with erect or ascending stems terminating in many-flowered cymes; leaflets of basal leaves 5–7, cuneate; flowers 1–1.5 cm. broad; petals shorter than to barely exceeding calyx-lobes.
Cauline and radical leaves very dissimilar, with revolute incised or irregularly toothed margins; tomentum of lower leaf-surface very dense, not mixed with long trichomes; style papillose at base. . . . . 7. *P. argentea.*
Cauline and radical leaves similar, with flat regularly toothed margins; tomentum of lower surface sparse, mixed with long trichomes; style epapillose. . . . . 8. *P. canescens.*
*h.*Basal leaves pinnate.
Flowering stems 1–9 cm. long, 1 (rarely–3)-flowered. . . . 9. *P. ustica-pensis.*
Flowering stems 1.5–5 dm. high, many-flowered.
Basal leaves oblong to oblong-obovate, very much longer than broad, with 7–15 leaflets. . . . . 10. *P. pensyl-vanica.*
Basal leaves suborbicular to round-obovate, with 5–7 leaflets. . . . . 11. *P. pectinata.*
*g.*Style short-conic above the base, then slender and prolonged to twice or thrice length of mature carpel. . . *i.*
*i.*Basal leaves pinnate, with 5–11 leaflets, the lower pairs of leaflets remote.
Bractlets and calyx-lobes subequal; tomentum of leaves with silvery-glistening hairs intermixed. . . . . 12. *P. Hippiana.*
Bractlets much narrower and barely half as long as calyx-lobes; tomentum opaque. . . . . 13. *P. effusa.*

---

* The figures show a characteristic leaf for many species, a portion of stem of a few, rhizomes of nos. 30 and 31, calyx of most species, petal of some, and (for most species) an enlarged achene.

i. Basal leaves digitate or subpalmate, the 5–7(–9) leaflets approximate. . . . . 14. *P. gracilis.*

f. Plants (except no. 22) without true tomentum, often hirsute or pilose, frequently glandular, biennial (sometimes annual) or short-lived perennial, with few or no loosely disposed marcescent stipules (glabrescent var. of no. 14 might be sought here). . . *j.*

j. Perennial with nearly leafless cyme; basal leaves digitate, with 5–7 leaflets; flowers 1.5–2.5 cm. broad; petals showy, pale yellow; stamens 30; mature calyx 1–1.5 cm. high. . . . . 15. *P. recta.*

j. Biennial, annual or short-lived perennial, with very leafy inflorescence; basal leaves pinnate or, if digitate, with only 3–5 (rarely 7) leaflets; flowers 3–10 mm. broad; petals small, deep yellow; stamens 5–20; calyx smaller (if as large, the plant with 3-foliolate basal leaves). . . *k.*

k. Basal leaves pinnate, with 5–11 leaflets. . . *l.*

l. Lower leaves with 2 or 3 approximate pairs of leaflets; inflorescence paniculate-cymose; achenes smooth, without corky enlargement. . . . . 16. *P. rivalis.*

l. Lower leaves with 2–5 distant pairs of leaflets; achenes longitudinally ribbed, the ventral suture with a corky enlargement.

Leaves all pinnate, with 7–11 leaflets; inflorescence paniculate-cymose. . . . . 17. *P. paradoxa.*

Lower leaves pinnate, with 5–7 leaflets; middle and upper ones ternate; branches of inflorescence racemose-elongate. 18. *P. Nicolletii.*

k. Basal leaves digitate, with 3–5 leaflets. . . *m.*

m. Petals cuneate or narrowly obovate; achenes smooth, 0.5–0.8 mm. long.

Leaves all ternate; flowers or flowering branches often borne near base of stem; stamens 10–15. . . . . . . 19. *P. millegrana.*

Lower leaves quinate, or ternate with lower leaflets deeply cleft; flowers mostly from upper half of stem; stamens 5. 20. *P. pentandra.*

m. Petals broadly obovate to rounded; achenes longitudinally ribbed, 0.8–1.3 mm. long.

Basal leaves ternate; foliage more or less hirsute or pilose, not tomentulose beneath; fruiting calyx finally 0.8–1.3 (–1.7) cm. high, its bracteoles acutish. . . . . . 21. *P. norvegica.*

Basal leaves chiefly quinate (rarely with 4 or 7 leaflets); foliage sparsely tomentulose beneath and with longer pilosity; fruiting calyx finally 5–8 mm. high, its bracteoles blunt. . . . . 22. *P. intermedia.*

e. Style slender at base, subterminal or lateral; flowers solitary, or few in scarcely leafy corymbs. . . *n.*

n. Leaves digitate or ternate; style slenderly clavate or slightly enlarged upward to the stigmatic tip (except in no. 23), subterminal. . *o.*

o. Densely cespitose, with few–many tough crowns usually covered with marcescent dark stipules and terminated by rosettes of leaves; flowering stems rising from the rosette, bearing 1–15 long-peduncled flowers. . . *p.*

p. Radical leaves ternate, their stipules with broad blunt or abruptly tipped auricles; flowering stem 1–3-flowered; alpine species.

Flowering stems 1–3.5 cm. high; calyx 5–8 mm. broad, 3–4 mm. high; petals narrowly obovate. . . . . 23. *P. Robbinsiana.*

Flowering stems 0.3–2 dm. high; calyx 1–1.8 cm. broad, 6–10 mm. high; petals round-obovate. . . . . . . 24. *P. hyparctica.*

p. Radical leaves mostly quinate, their stipules with prolonged ovate to linear auricles; flowering stems 1–several-flowered; chiefly lowland plants.

Caudex multicipital, the strongly ascending crowns densely invested below with marcescent stipules; stipules of radical leaves membranaceous, glabrous or sparsely short-pubescent, with ovate or lanceolate auricles; receptacle depressed, becoming subconic; plant of Nfld. . . . . . . 25. *P. Crantzii.*

Caudex with prostrate offshoots, bearing only few loose old stipules; stipules of radical leaves coriaceous, densely strigose-villous, with prolonged linear or linear-lanceolate auricles; receptacle hemispherical; more southern. . . . 26. *P. verna.*

*o.*Flowering stems elongate, alternately leafy, arising from rhizomes or crowns without densely rosulate leaves, bearing slender-peduncled flowers, mostly singly from the axils or from opposite the leaves. . . *q.*

*q.*Stems erect or loosely ascending, paniculate-branched, not rooting at nodes or tips; cauline leaves sessile or subsessile, their stipules resembling leaflets, deeply incised. . . . . . . 27. *P. erecta.*

*q.*Stems trailing or at first erect and finally trailing or rooting at tips; cauline leaves petioled, their stipules quite unlike the leaflets. . *r.*

*r.*Stems from a deep non-tuberous root; leaves essentially glabrous; flowers 1–2 cm. broad, on peduncles 3–17 cm. long; bracteoles lanceolate, oblong or elliptic; calyx-lobes ovate.

Flowers mostly 4-merous; stems soon dichotomously branching from the lower and middle (flowering) nodes. 28. *P. anglica.*

Flowers mostly 5-merous; stems rarely with more than incipient branches from the flowering nodes. . . . . . 29. *P. reptans.*

*r.*Stems from tuberous-thickened rhizomes; leaves pubescent beneath; flowers 1–1.5 cm. broad, on peduncles 1–9 cm. long; bracteoles linear or linear-lanceolate; calyx-lobes lanceolate to lance-ovate.

Rhizome praemorse, cylindric, 0.5–2 cm. long, 4–8 mm. thick; stolons without tuberous enlargements at tip; stem at flowering time 0.1–1.5 dm. high, soon prostrate, filiform; cauline leaves at flowering time not well-expanded; stipules of basal leaves with oblong-lanceolate flat auricles; first flower usually borne from node above first well developed internode. . . . . 30. *P. canadensis.*

Rhizome irregularly enlarged, up to 8 cm. long, 0.5–2 cm. thick; tips of stems late in autumn bearing tuberous enlargements; stem at flowering time erect or ascending, up to 2–5 dm. high then greatly prolonging, arching, forking and rooting at tips; cauline leaves well-expanded at flowering time; stipules of basal leaves with linear-lanceolate long-attenuate usually inrolling auricles; first flower usually from node above second well-developed internode. . . . . 31. *P. simplex.*

*n.*Leaves pinnate, elongate; style filiform, lateral.

Leaflets silvery-silky beneath, at least the younger lustrous; stolons, peduncles, petioles and rachises more or less villous; achene thick-ovoid to subglobose, corky, dorsally sulcate. . 32. *P. anserina.*

Leaflets white-tomentose beneath with opaque hairs or glabrate or glabrous, not lustrous; stolons, etc. glabrous or soon glabrate; achene laterally compressed, firm, rounded on back, not sulcate. 33. *P. Egedii.*

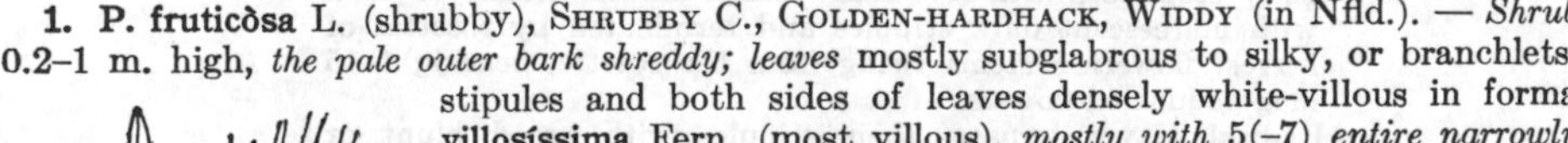

1. P. fruticòsa L. (shrubby), SHRUBBY C., GOLDEN-HARDHACK, WIDDY (in Nfld.). — *Shrub* 0.2–1 m. high, *the pale outer bark shreddy; leaves* mostly subglabrous to silky, or branchlets, stipules and both sides of leaves densely white-villous in forma **villosíssima** Fern. (most villous), *mostly with* 5(–7) *entire narrowly oblong to lanceolate or oblanceolate leaflets* 2–9 mm. broad and 1–3 cm. long; expanded flower 1.5–3 cm. broad; calyx 0.8–1.5 cm. high, villous; bracteoles narrowly oblong to elliptic or lanceolate, equaling to much exceeding the thinner deltoid acuminate calyx-lobes; petals yellow, rounded to broadly elliptic; anthers 2–3 mm. long; *carpels densely villous; the short upwardly enlarged styles mostly hidden in the villosity,* attached below middle of carpel. (*Dasiphora* Rydb.) — Wet or dry open ground, s. Lab. to Alaska, s. to Nfld., N.S., N.E., n. N.J., Pa., O., Ind., n. Ill., n. Ia., S.D., N.M., Ariz. and Calif. June–Oct. (Eurasia) FIG. 1143.

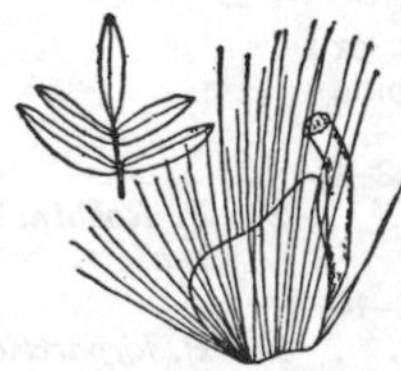

1143. P. fruticosa.

Var. **tenuifòlia** Lehm. (slender-leaved). — Depressed or low (up to 4 dm. high); leaflets linear or linear-lanceolate, strongly revolute, 1–3 mm. broad, only 0.5–1.5 cm. long; flowers 1–1.5 cm. broad; calyx 0.5–1 cm. high, bracteoles narrowly linear; anthers 1.5–2 mm. long. — Exposed barrens, gravels and ledges, Lab. to Wash., s. to Nfld., Gaspé Pen., Que., Minn., Colo., Nev. and Calif. (Eurasia)

2. P. tridentàta Ait. (three-toothed), THREE-TOOTHED C. — Perennial *with* extensively creeping subterranean stems and *depressed superficial ligneous branches;* flowering stems ascending,

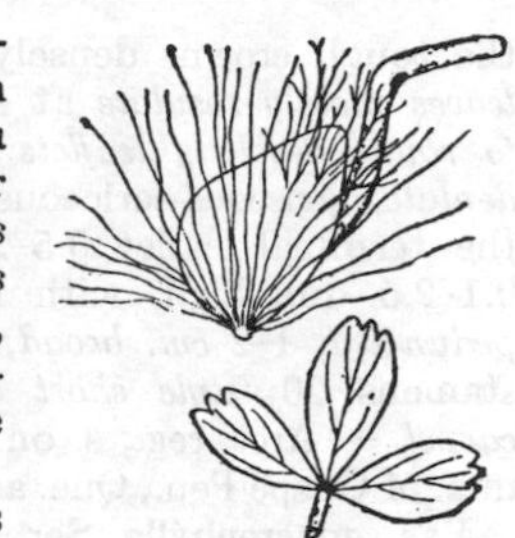

1144. P. tridentata.

0.2–3 dm. high; *leaves* palmate, *with 3 coriaceous evergreen cuneate-oblong leaflets entire except for the* 3 (*or* 5) *-toothed tips*, bright green and nearly glabrous, or hirsute on both surfaces in forma **hirsutifòlia** Pease (hirsute-leaved); *cymes stiff, few–many-flowered; petals white*, or roseate in forma **auròra** Graustein (like the dawn); carpels, *achenes* and receptacle *densely hairy; style filiform, basal.* (*Sibbaldiopsis* Rydb.) — Dry open sterile rocky, gravelly, sandy or peaty soils, Greenl. and Lab. to Mackenz., s. to Nfld., N.S., N.E., N.Y., along exposed mts. to Ga., s. Ont., Mich., Wisc., ne. Ia. and N.D. Late May–Oct. Fig. 1144.

3. P. stérilis (L.) Garcke (sterile; on the supposition that it was

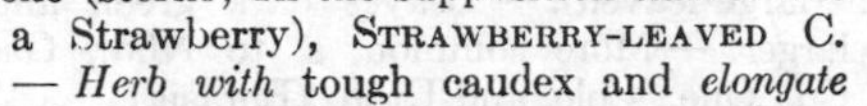

a Strawberry), Strawberry-leaved C. — *Herb with* tough caudex and *elongate stoloniform leafy branches;* stems and petioles long-villous; *radical leaves with 3 coarsely crenate-serrate obovate to rounded blunt leaflets; flowers* 1–3 *on filiform short branches,* 1–1.5 cm. broad; *petals white,* obcordate, equaling or slightly exceeding the villous calyx-lobes; *carpels* and achenes *sparsely pilose; style filiform, terminal.* (*P. Fragariastrum* Ehrh.) — Thickets, clearings and rocky slopes, rare, se. Nfld. Fig. 1145. — A remarkable species, separated from *Fragaria* only by its dry receptacle and the pubescent carpels. (Eu.)

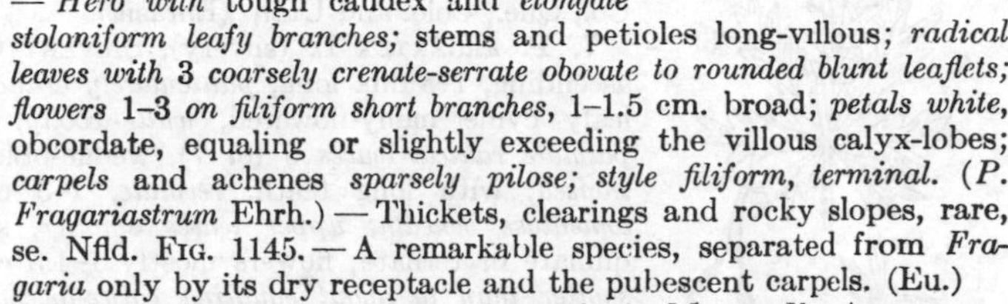

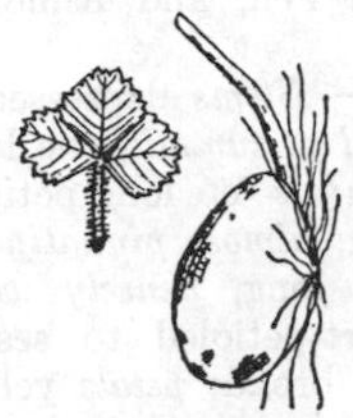

1145. P. sterilis.

4. P. palústris (L.) Scop. (of marshes), Marsh-F., Argentine rouge or Comaret (Que.). — Stems stout, *ascending from a decumbent rooting ligneous base,* 1–6 dm. high, glabrous below, more or less minutely pilose or glandular above, 1–many-flowered; leaves pinnate, the 5–7 *oblong-lanceolate to oblanceolate* acutish to obtuse *leaflets green and glabrous or nearly so,* or silvery-sericeous in forma **subserícea** (Becker) Wolf (somewhat silky) *above,* glaucous and puberulent to sericeous beneath; *terminal leaflet* of larger primary leaves *one-sixth to one-half as broad as long,* 2–10 cm. long, 0.7–3.8 cm. wide; *petals purple,* shorter than the broad calyx-lobes, somewhat persistent; *disk thick and hairy; achenes glabrous; style lateral; receptacle hairy,* becoming large and spongy. (*Comarum* L.) — Inundated meadows, margins of streams, swales, etc., Greenl. and Lab. to Alaska, s. to Nfld., N.S., N.E., n. N.J., Pa., centr. O., n. Ind., n. Ill., n. Ia., Wyo. and Calif. June–Aug. (Eurasia) Fig. 1146. — Forma *subsericea* chiefly in exsiccated areas and developed late in summer.

1146. P. palustris.

Var. **parvifòlia** (Raf.) Fern. & Long (small-leaved). — Similar, smaller; branches 1–4-flowered; *leaflets elliptic to cuneate-obovate,* subtruncate or rounded at tip; *the terminal one one-half to two-thirds as broad as long,* 1.3–4.5 *cm. long,* 0.9–2.5 cm. broad. — Greenl. and Lab. Pen. to Alaska, s. to Nfld., and Sable I., N.S. July, Aug.

Var. **villòsa** (Pers.) Lehm. (soft-pubescent). — Often coarser; branches few–many-flowered, with the petioles, peduncles, bractlets, etc., *densely* (*usually glandular-*) *villous; leaflets villous or densely sericeous, oblong-elliptic to narrowly obovate,* rounded above, *one-third to three-fifths as broad as long;* the terminal 3.3–10 cm. long, 1.4–3 cm. broad. — Nfld. to Man., s. to N.S., Me., Mass., N.Y., Mich. and Minn. (Eu.)

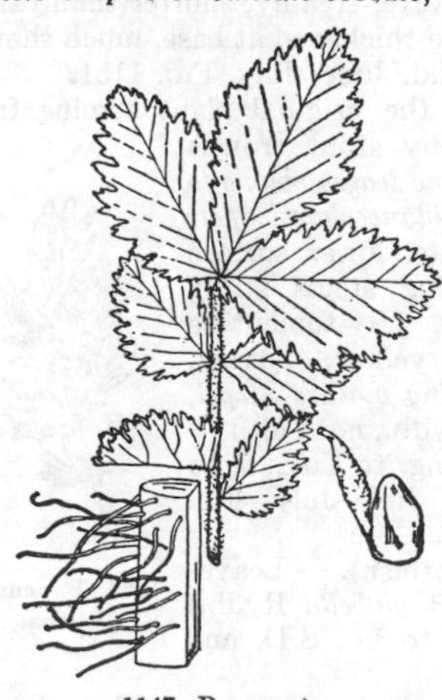

1147. P. arguta.

5. P. argùta Pursh (acute; from teeth of leaflets), Tall C. — *Erect glandular-villous* perennial, 0.3–1 *m. high,* rather coarse, the clammy pubescence brownish; *basal leaves pinnate, with* 7–11 oval to ovate cut-serrate *leaflets downy beneath;* cyme strict, rather close, corymbiform; *petals* broadly obovate to suborbicular, *whitish or creamy; stamens* 25–30 (*usually* 30), *borne in 5 groups* on a thick glandular disk; *style subbasal, thickened near middle.* (*Drymocallis* Rydb.; *P. agrimonioides* Rydb.) — Rocky, bushy or alluvial soils or prairies, N.B. to n. B.C., s. to D.C., W.Va., O., Ind., Ill., Mo., Okla. and Colo. June–Aug. Fig. 1147.

6. P. nívea L. (for the *snow-white* lower leaf-surfaces). — *Densely cespitose* perennial,

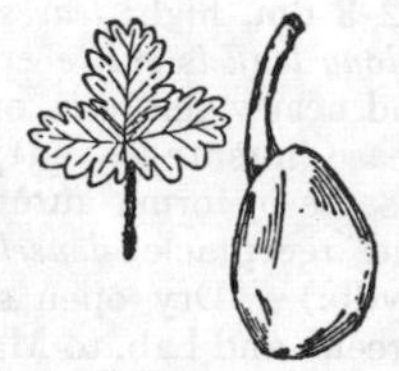
1148. P. nivea.

the tough crowns densely invested with marcescent dark stipules; *leaves mostly rosulate* at summits of crowns, *3-foliolate; the obovate to rounded-oblong leaflets regularly and deeply crenate- to elongate-dentate*, griseous-sericeous above, *densely white-tomentose beneath*, the terminal leaflet 0.5–2.5 cm. long; flowering stems ascending, 0.1–2.5 dm. high, with few reduced leaves; *flowers 1–few, long-peduncled, 1–2 cm. broad; petals* yellow, *much exceeding calyx-lobes;* stamens 20; *style short and much thickened, shorter than mature carpel.* — Arct. reg., s. on calcareous rocks and gravels to Nfld. and mts. of Gaspé Pen., Que. and of Colo. June, July. (Eurasia) FIG. 1148.

Var. **macrophýlla** Ser. (large-leaved). — Leaves dark green and glabrous above, often larger. — More common, s. to Nfld., Gaspé Pen. and Rimouski Co., Que., Colo. and Utah. (Eurasia)

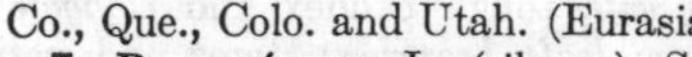

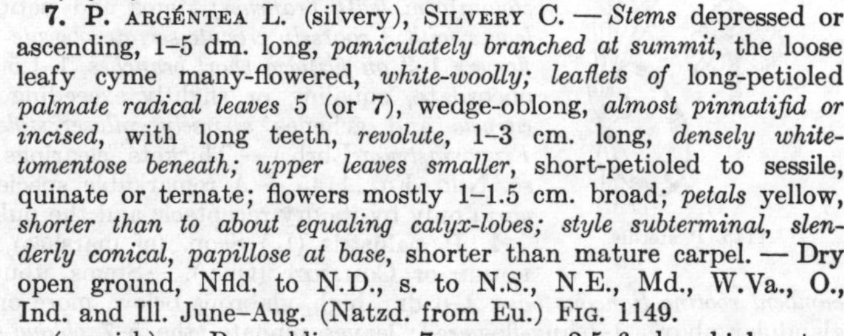

1149. P. argentea.

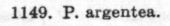

7. P. ARGÉNTEA L. (silvery), SILVERY C. — *Stems* depressed or ascending, 1–5 dm. long, *paniculately branched at summit*, the loose leafy cyme many-flowered, *white-woolly; leaflets of* long-petioled *palmate radical leaves* 5 (or 7), wedge-oblong, *almost pinnatifid or incised*, with long teeth, *revolute*, 1–3 cm. long, *densely white-tomentose beneath; upper leaves smaller*, short-petioled to sessile, quinate or ternate; flowers mostly 1–1.5 cm. broad; *petals* yellow, *shorter than to about equaling calyx-lobes; style subterminal, slenderly conical, papillose at base*, shorter than mature carpel. — Dry open ground, Nfld. to N.D., s. to N.S., N.E., Md., W.Va., O., Ind. and Ill. June–Aug. (Natzd. from Eu.) FIG. 1149.

8. P. CANÉSCENS Bess. (grayish). — Resembling no. 7, coarser, 2.5–6 dm. high; stem grayish-lanate; *cauline and radical leaves very similar*, pale, the radical with the 5–7 oblong-oblanceolate *leaflets grayish beneath and with the sparse tomentum mixed with long trichomes, the margin not revolute, regularly toothed; style epapillose.* (*P. inclinata* of many auth., not Vill.) — Roadsides and old fields, sw. Que. and Ont., s. to Ct., Pa., Ind. and Mich. June–Aug. (Natzd. from Eu.) FIG. 1150.

1150. P. canescens.

9. P. usticapénsis Fern. (of Burnt Cape, Nfld.). — *Dwarf;* crown densely covered with old blackish stipules; *leaves* radical, short-petioled, 1–2.5 *cm. long, pinnate; leaflets* 3–5, *densely villous-lanate above*, densely white-tomentose and lanate beneath, deeply incised; *flowering stems filiform, decumbent*, lanate, *subscapose*, 1–9 *cm. long*, 1 (*rarely* –3)*-flowered;* calyx villous-lanate, 5–6 mm. high; the bracteoles oblong, obtuse, shorter than narrowly ovate calyx-lobes; petals rounded-obovate, creamy, shorter than calyx-lobes; stamens 20; style subterminal, papillose-thickened at base, much shorter than ripe carpel. — Dry limestone-gravel, Burnt Cape, n. Nfld. June, July. FIG. 1151.

1151. P. usticapensis

10. P. pensylvánica L. (of Pennsylvania; a misnomer, the original plant coming from shores of Hudson Bay). — Densely cespitose, with 1–many stout crowns covered with marcescent brown stipules; *basal and lower cauline leaves oblong to oblong-obovate, longer than broad, pinnate, with* 7–15 *oblong to oblanceolate leaflets* divided three-fourths to middle into linear-oblong segments, lower surface gray-tomentose; upper gray-green, often strigose; flowering stems stiffly ascending, 1.5–5 dm. high, gray-pilose; cyme rather close, gray-tomentose and strigose; bracteoles lanceolate, calyx-lobes ovate; petals yellow, obovate, about equaling calyx-lobes; *style shorter than to about equaling mature carpel*, slenderly conical, glandular at base. (Incl. *P. strigosa* sensu Rydb., not Pall.) — Dry plains, bluffs and calcareous rocks, Ung. to Yuk., s. to n. Mich., Minn., Neb., N.M. and Calif. June–July. FIG. 1152.

1152. P. pensylvanica.

Var. **glabràta** (Hook.) S. Wats. (smoothish). — Leaves nearly to quite glabrous on both faces. (*P. glabella* Rydb.) — Thunder Bay Distr., Ont., to Alta., s. to Ia., S.D. and Wyo.

Var. **bipinnatífida** (Dougl.) T. & G. (twice deeply cleft). — Leaves whitish-tomentose below, silvery-silky above, the leaflets cut nearly to middle into linear segments. (*P. bipinnatifida* Dougl.) — W. Ung. to Alta., s. to Minn., N.D. and Colo.

1153. P. pectinata.

11. P. pectinàta Raf. (comb-like). — Similar to and perhaps an extreme of no. 10; *basal leaves suborbicular to round-obovate*, pinnate (but often by crowding of leaflets *appearing subdigitate*), *the* 5–7 *oblanceolate to oblong-obovate leaflets deeply cleft* into oblong to linear-lanceolate segments; flowering stems slender, arched-ascending to erect, glabrous to canescent-pilose. (*P. pensylvanica* in large part of ed. 7; *P. litoralis* Rydb.) — Calcareous or basic rocky, sandy or gravelly open soils, s. Lab. to Alaska, s. to Nfld. and lower St. Lawrence and Baie des Chaleurs, Que.; n. C.B.; coast of Me. and N.H. June–Sept. Fig. 1153.

12. P. Hippiàna Lehm. (named in 1830 for the friend of Lehmann, Charles Friedrich Hipp). — Cespitose, with stout brown-scaly crowns; leaves and stem densely white-tomentose and *silvery-silky throughout; basal leaves pinnate, with the* 5–11 *cuneate-oblong leaflets incisely toothed* at least at apex, *diminishing uniformly down the rachis;* cyme open; *bractlets and calyx-lobes subequal;* carpels 10–30; *style prolonged,* two or three times length of mature carpel. — Plains, dry slopes and gravels, n. Mich. to Alta., s. to Neb., N.M. and Ariz.; casual in grassland eastw. June, July. Fig. 1154.

1154. P. Hippiana.

1155. P. effusa.

13. P. effùsa Dougl. (loosely spreading). — In habit like no. 11, *tomentose throughout* and *with* scattered *opaque villosity; basal leaves interruptedly pinnate, the* 5–11 *coarsely incised or dentate leaflets* cuneate-oblong, *the alternate ones smaller; bracteoles much narrower than and barely half as long as calyx-lobes.* — Dry plains and slopes, Man. to Alta., s. to w. Minn., Neb. and N.M. June, July. Fig. 1155.

1156. P. gracilis.

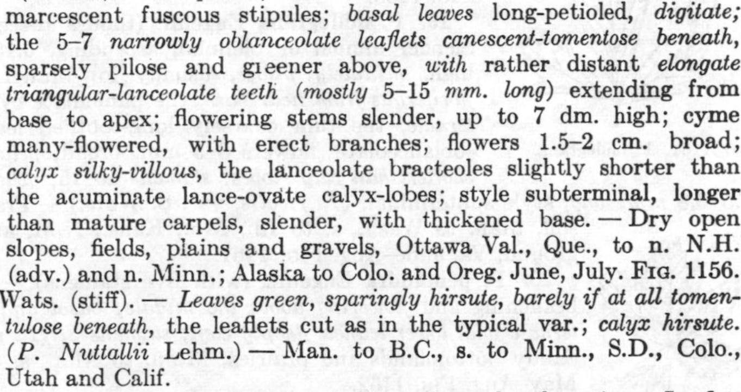

14. P. grácilis Dougl. (slender). — Cespitose, with several thick crowns covered with marcescent fuscous stipules; *basal leaves* long-petioled, *digitate;* the 5–7 *narrowly oblanceolate leaflets canescent-tomentose beneath,* sparsely pilose and greener above, *with* rather distant *elongate triangular-lanceolate teeth* (*mostly* 5–15 *mm. long*) extending from base to apex; flowering stems slender, up to 7 dm. high; cyme many-flowered, with erect branches; flowers 1.5–2 cm. broad; *calyx silky-villous,* the lanceolate bracteoles slightly shorter than the acuminate lance-ovate calyx-lobes; style subterminal, longer than mature carpels, slender, with thickened base. — Dry open slopes, fields, plains and gravels, Ottawa Val., Que., to n. N.H. (adv.) and n. Minn.; Alaska to Colo. and Oreg. June, July. Fig. 1156.

Var. **rígida** (Nutt.) S. Wats. (stiff). — *Leaves green, sparingly hirsute, barely if at all tomentulose beneath,* the leaflets cut as in the typical var.; *calyx hirsute.* (*P. Nuttallii* Lehm.) — Man. to B.C., s. to Minn., S.D., Colo., Utah and Calif.

Var. pulchérrima (Lehm.) Fern. (very handsome). — *Leaflets* of radical leaves narrowly oblong-obovate to -oblanceolate, *very densely white-tomentose beneath,* appressed-pilose above, *approximately crenate or crenate-serrate, the teeth mostly* 2–5 *mm. long;* stem and *calyx white-villous.* (*P. pulcherrima* Lehm.) — Occasionally in fields, e. to N.H. (Adv. from w. of our range.)

1157. P. recta.

15. P. récta L. (upright). — *Stem* erect, very leafy, 1.5–7 dm. high, loosely *hirsute,* from a perennial base; *basal leaves* on long hirsute petioles, *digitately* 5–7-*foliolate; leaflets oblanceolate,* 3–14 cm. long, *with* 7–17 *prolonged* narrowly deltoid *teeth, more or less hirsute on both surfaces,* paler beneath; *cyme* stiffly erect, *standing above the principal foliage; flowers* on erect stalks, 1.5–2.5 *cm. across; calyx* hirsute, *becoming* 1–1.5 *cm. high;* petals obcordate, deeply emarginate, pale (rarely deep) yellow, equaling or exceeding calyx-lobes; *stamens mostly* 30 (or 25); style shorter than mature carpel. (Incl. *P. sulphurea* Lam.) — Dry fields and roadsides, Nfld. to Ont. and Minn., s. to N.S., N.E., Va., Tenn., Ark. and se. Kans., rapidly spreading. Late May–Aug. (Natzd. from Eu.) Fig. 1157.

16. P. rivàlis Nutt. (of brooksides). — *Annual or biennial,* rather slender, 1–5 dm. high, *softly villous, with paniculate-cymose very leafy flowering summit and branches; lower leaves pinnate, with* 2 *or* 3 *closely approximate pairs of leaflets or a single pair with the terminal leaflet* 3-*parted;* leaflets cuneate-obovate or -oblong; cauline leaves with 3 or 5 leaflets; flowers

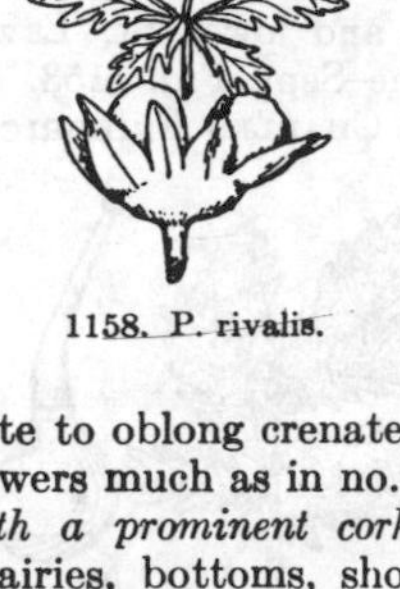

1158. P. rivalis.

4–8 mm. broad; mature calyx 5–8 mm. high, pilose; petals tiny, cuneate; stamens 5–20; *achenes smooth.* — Prairies, bottoms and waste places, Man. to B.C., s. to Mo., Kans., Mex. and Calif. June–Aug. Fig. 1158.

17. P. paradóxa Nutt. (paradoxical or strange; in having prominent corky enlarged bases of carpels). — Annual, biennial or short-lived perennial; *stems decumbent* or ascending, often stout, leafy, *subvillous; leaves all pinnate,* oblong-obovate, *with* 2–5 *distant pairs of small* obovate to oblong crenate-dentate *leaflets; cyme open-paniculate,* leafy; flowers much as in no. 15; stamens 20; *achenes longitudinally ribbed, with a prominent corky enlargement along the ventral suture.* — Prairies, bottoms, shores and damp places, Ont. to B.C., s. to w. N.Y., nw. Pa., O., Ill., Mo., Kans. and N.M. Late May–Sept. (E. Asia) Fig. 1159.

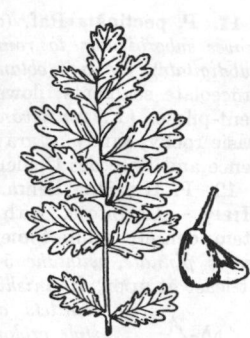

1159. P. paradoxa.

18. P. Nicollétii (S. Wats.) Sheld. (for its discoverer, Joseph Nicholas Nicollet, 1786–1843, pioneer explorer of Upper Mississippi region). — Similar to no. 17, slender; *lower leaves pinnate, with* 5–7 sharply toothed *leaflets; middle and upper leaves ternate and digitate; branches of inflorescence racemose-elongate;* stamens 10–15. — River-banks and prairies, local, Minn. and N.D. to Mo. and Kans. June–Aug. Fig. 1160.

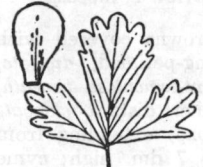

1160. P. Nicolletii.

19. P. millegrâna Engelm. (thousand-seeded). — Slender annual or biennial, ascending, 0.15–1.2 m. high, *minutely pilose,* usually low-branching and

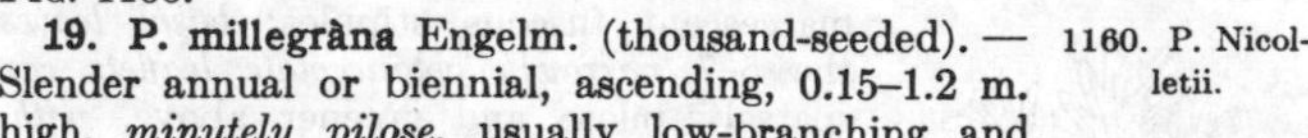

floriferous from near base, the paniculate cyme very leafy; *leaves all ternate,* the thin coarsely long-toothed leaflets cuneate-oblong to -oblanceolate; flowers 3–5 mm. broad; *petals* yellow, *cuneate, much shorter than calyx-lobes; stamens* 10–15; *achenes smooth,* pale brown, 0.5–0.8 *mm. long;* style subterminal. (*P. rivalis,* var. S. Wats.) — Prairies, bottoms and open soil, Man. to Wash., s. to Ill., Mo., Kans., N.M. and Calif.; adv. in waste ground, Va. June–Aug. Fig. 1161.

1161. P. millegrana.

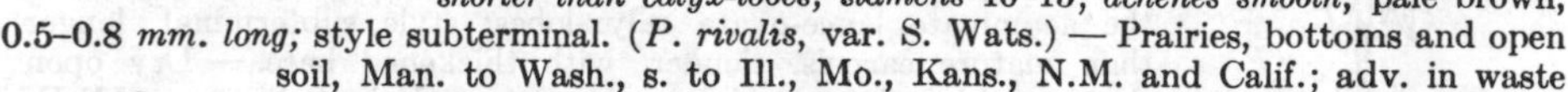

20. P. pentándra Engelm. (with five stamens). — Similar, *hirsute,* chiefly branching and *flowering above the middle; basal and lower leaves quinate, or ternate with lower leaflets deeply cleft; stamens* 5. (*P. rivalis,* var. S. Wats.) — Sandy bottomlands and prairies, Minn. to Alta., s. to Ark. and Kans. Late May–Aug. Fig. 1162.

1162. P. pentandra.

21. P. norvégica L. (Norwegian). — Erect or ascending annual or biennial (rarely short-lived perennial) 1–9 dm. high; *stem hirsute with stiff mostly spreading hairs,* often with shorter pubescence intermixed; *lower leaves* long-petioled, 3-*foliolate; leaflets* obovate to oblanceolate, coarsely serrate, usually more or less hirsute, otherwise *green;* upper leaves sessile, often with narrow leaflets; inflorescence a leafy cyme; *calyx in fruit enlarging to* 0.8–1.3 (–1.7) *cm. high, its bracteoles acutish;* petals obovate, mostly shorter than calyx-lobes; stamens 15–20; style slenderly conical at base, subterminal, about equaling mature carpel; *achenes longitudinally ribbed,* 0.8–1.3 *mm. long.* (Incl. var. *hirsuta* (Michx.) Lehm.; *P. monspeliensis* L.) — Thickets, clearings, roadsides and waste places, Greenl. and Lab. to Alaska, s. beyond our limits. June–Oct. (Eu.) Fig. 1163. — Passing northw. into

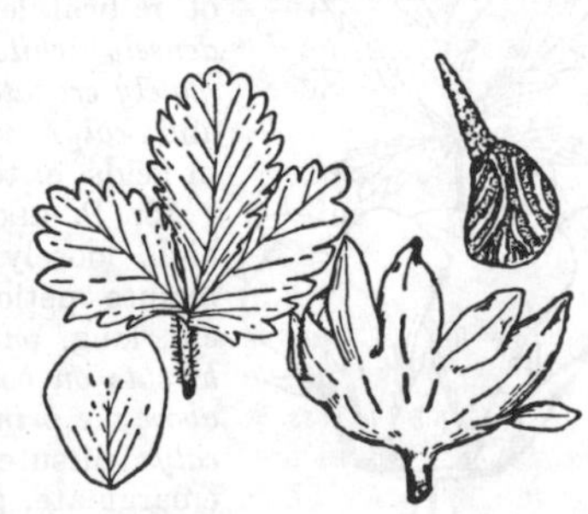

1163. P. norvegica.

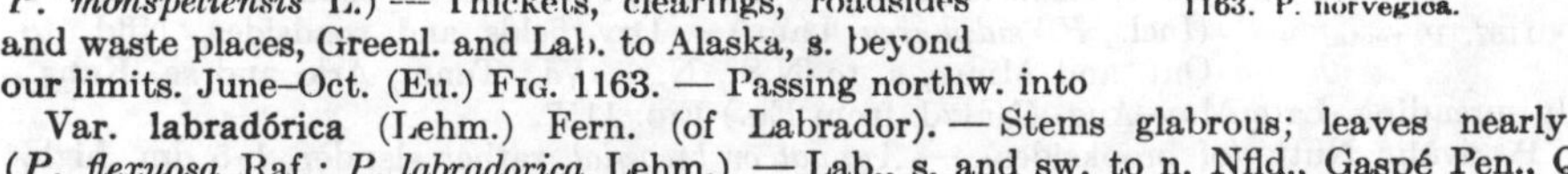

Var. **labradórica** (Lehm.) Fern. (of Labrador). — Stems glabrous; leaves nearly so. (*P. flexuosa* Raf.; *P. labradorica* Lehm.) — Lab., s. and sw. to n. Nfld., Gaspé Pen., Que., mts. of n. N.E., and Bruce Pen., Ont.

22. P. intermèdia L. (intermediate). — Resembling nos. 7 and 20; *stems* up to 7 dm. high, *grayish-tomentulose; basal leaves* mostly *with* 5 (*rarely* 4 *or* 7) *oblanceolate to nar-*

rowly obovate deeply dentate leaflets green above but *grayish-villous and very minutely tomentulose beneath;* cyme diffuse, leafy; *calyx* villous, *in maturity* 5–8 *mm. high, its bracteoles blunt;* achenes longitudinally ribbed, about 1 mm. long, about equaled by the basally conic and papillose-thickened style. — Roadsides and waste places, local but rapidly spreading, Nfld. to Mich. and Va. Late May–Aug. (Adv. and natzd. from Eu.) Fig. 1164. — Nondescript plants sometimes develop abundant midsummer and autumnal basal foliage with 3, 4 or 5 broader and lower-toothed leaflets, the middle one sometimes petiolulate. These need special study.

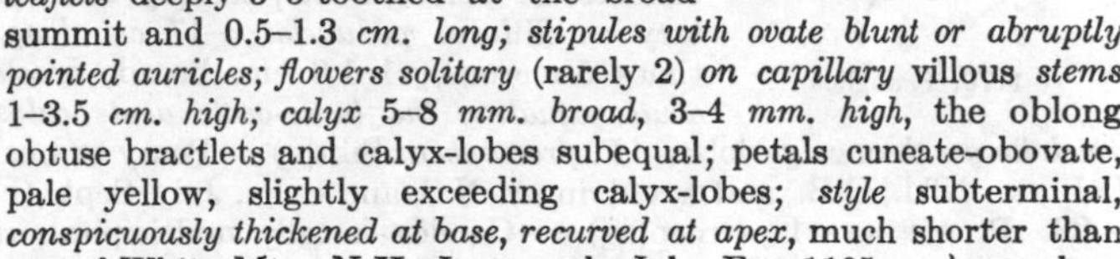

1164. P. intermedia.

23. P. Robbinsiàna Oakes (for its discoverer, James Watson Robbins, 1801–1879). — *Densely tufted dwarf, alpine,* with crowns covered with marcescent stipules; *leaves 3-foliolate,* with the cuneate-obovate nearly glabrous *leaflets* deeply 3–5-toothed at the broad summit and 0.5–1.3 *cm. long; stipules with ovate blunt or abruptly pointed auricles; flowers solitary* (rarely 2) *on capillary* villous *stems* 1–3.5 *cm. high; calyx* 5–8 *mm. broad,* 3–4 *mm. high,* the oblong obtuse bractlets and calyx-lobes subequal; petals cuneate-obovate, pale yellow, slightly exceeding calyx-lobes; *style* subterminal, *conspicuously thickened at base, recurved at apex,* much shorter than mature carpel. — Alpine reg. of White Mts., N.H. June, early July. Fig. 1165. — Anomalous species, with some characters of one series but with style of another.

1165. P. Robbinsiana.

24. P. hypárctica Malte (subarctic), var. **elàtior** (Abrom.) Fern. (taller). — Coarser than no. 23, softly pubescent below, villous above; *flowering stems* 0.3–2 *dm. high,* 1–3 (–5)-*flowered; leaflets subcoriaceous,* pilose when young, glabrate, broadly obovate, the terminal often nearly as broad as long, 1–1.5 cm. long, obtusely toothed; *flowers* 1.5–2 *cm. broad;* petioles pilose; auricles of stipules lance-ovate, blunt; *calyx* 1–1.8 *cm. broad, the elliptic bracteoles about as long as the oblong-ovate blunt calyx-lobes;* petals obcordate, emarginate; *style somewhat clavate,* enlarged below the dilated stigma, much shorter than mature carpel. (*P. emarginata* Pursh, not Desf.) — S. Greenl. and Baffin I. to Keewatin, s. to Lab. and Ung.; calcareous schists, alpine reg. of Shickshock Mts., Gaspé Pen., Que. Late June–Aug. Fig. 1166. — Southern representative of *P. hyparctica* of the Arctic, which has lance-attenuate stipules and more spreading pubescence.

1166. P. hyparctica, v. elatior.

25. P. Crántzii (Crantz) G. Beck (named for — and by — its discoverer, Heinrich Johann Nepomuk Crantz, 1722–1799). — Densely to loosely *cespitose, with many* depressed or ascending *crowns heavily covered with marcescent dark stipules; leaves* chiefly rosulate at summits of crowns, *mostly quinate* (sometimes ternate), digitate, *with* slender *sparsely pilose petioles and membranaceous veiny brown stipules with glabrous or short-pubescent ovate to lanceolate auricles; leaflets* obovate, deeply incised-serrate above the middle, cuneate and entire below, *sparsely long-villous, green,* 0.5–3.5 cm. long; flowering stems arched-ascending, 1–15-flowered, 0.3–2 dm. high; flowers long-peduncled, 1.5–2.5 cm. broad; calyx spreading-pilose; bracteoles oblong to elliptic, shorter than to equaling the lance-ovate calyx-lobes; petals obovate, emarginate, pale yellow, with orange base, exceeding calyx-lobes; *receptacle at first depressed, becoming subconic.* (*P. alpestris* Hall. f.; *P. maculata* E. Mey.) — Arct. reg., s. to calcareous gravels, peats, meadows and sea-cliffs of w. Nfld. and Ung. June–Aug. (Eu.) Fig. 1167.

1167. P. Crantzii.

26. P. vérna L. (vernal). — Similar to no. 25; crowns loosely spreading and *prostrate, with few if any residual old stipules; stipules of radical leaves coriaceous, densely strigose-villous, with prolonged linear or linear-lanceolate auricles;* leaflets mostly 1–2 cm. long, these and petioles gray-strigose; flowering stems strongly long-pilose; petals deep yellow; *receptacle hemispherical.* — Grassy roadsides, local, s. Ct. May, June. (Adv. from Eu.) Fig. 1168.

1168. P. verna.

27. P. erécta (L.) Räuschel (erect). — *Stems* slender, *erect or ascending,* from a subtuberous caudex 1-3 cm. thick, 0.5–3 dm. high, *usually panicu-*

lately branched, very leafy; radical leaves slender-petioled, with 3 (4 or 5) cuneate-obovate subtruncate few-toothed sessile leaflets, evanescent; *cauline leaves sessile* with oblong-lanceolate incised smooth leaflets *nearly equaled by the deeply incised stipules; flowers solitary*, slender-peduncled, opposite the leaves or in the forks, 4-*merous*, about 1 cm. broad; bracteoles linear-oblong, much narrower than oblong-ovate calyx-lobes; petals obcordate, about equaling calyx-lobes, yellow with orange base; style subterminal, slender, about equaling the mature *rugose carpel.* (*P. Tormentilla* Neck.) — Mossy spots, se. Nfld., local; adv. in e. Mass. June–Aug. (Eurasia; Azores) Fig. 1169.

1169. P. erecta.

28. P. ánglica Laicharding (English). — *Caudex* stout, *terminating a long root;* stems slender, at first ascending, *soon trailing*, up to 7 dm. long, *mostly forked*, with lash-like tips often rooting late in the season and producing vegetative buds; lower *leaves* long-petioled, either 5-, 4- or 3-nate; upper leaves shorter-petioled, similar, *glabrous or glabrescent;* stipules of lower leaves with lance-ovate acute auricles; leaflets oblong or cuneate- or oblong-ovate, sharply toothed at summit; *flowers* solitary *on* filiform *peduncles* 3–17 *cm. long, mostly* 4-*merous* (sometimes 5-merous), 1–1.8 cm. broad; *bracteoles oblong or lanceolate, acute, equaling the lance-ovate acute calyx-lobes; petals* obcordate, *much longer* than calyx-lobes. (*P. procumbens* Sibth.) — Peaty or bushy slopes and openings, s. Lab.; se. Nfld.; C.B.; and natzd. in sw. N.S. and e. Pa. July–Sept. (Eu. and Azores) Fig. 1170.

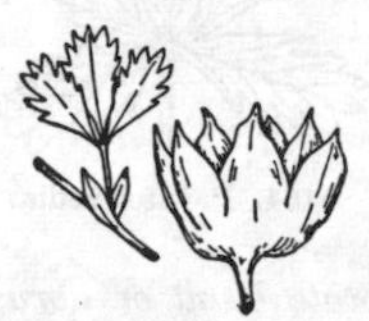
1170. P. anglica.

29. P. réptans L. (creeping). — Caudex usually multicipital; *stems repent*, elongate, up to 1 m. long, *producing adventitious plants*, simple or with incipient branches from the flowering nodes; *radical leaves with* 5 *or* 7 (sometimes 4 or 3) *leaflets*, long-petioled; the *cauline similar*, short-petioled; stipules of radical leaves with lanceolate auricles, of cauline shorter and ovate or oblong; leaflets oblong-obovate to -oblanceolate, crenate-serrate; *flowers* 5-*merous* on filiform peduncles 3–10 cm. long, 1.5–2 cm. broad; bracteoles elliptic to oblong, blunt, about equaling the acute calyx-lobes. — Lawns, roadsides and waste places, local, N.S. and e. Mass. to sw. Ont., s. to Va. May–July. (Adv. from Eu.) Fig. 1171.

1171. P. reptans.

30. P. canadénsis L. (of Canada). — *Rhizome* short, *praemorse*, in well-developed plants 0.5–2 *cm. long*, 4–8 *mm. thick; stems, petioles, lower leaf surfaces, calyx, etc., silky-pilose with appressed or loosely ascending soft pubescence; stems* at flowering time 0.1–1.5 dm. high, *soon becoming prostrate, flagelliform or filiform,* 0.3–1 *mm. thick*, not bearing tuberous enlargements at tips; *cauline leaves during anthesis not well expanded* (those subtending flowers then 0.5–3 cm. long, including petiole); leaflets narrowly cuneate-obovate, 5 (sometimes 3), coarsely and deeply 5–15-toothed around the rounded summit, entire below middle; middle leaflet of largest leaves 1.5–4 cm. long; *stipules of basal leaves with oblong-lanceolate flat auricles*, of mature primary cauline leaves mostly 3-cleft and 4–12 (–15) mm. long; *first flowers usually borne from node above the first well developed internode*, 1–1.5 cm. broad, on filiform peduncles 1–9 cm. long; bracteoles linear or linear-lanceolate; petals rounded at summit or retuse, deep yellow, or cream-color in forma **ochroleùca** (Weath.) Fern. (cream-colored). (*P. pumila* Poir.) — Dry open soil, w. N.S. and s. Me. to sw. Ont., s. to L.I., S.C. and O. March–June. Fig. 1172.

1172. P. canadensis.

Var. **villosíssima** Fern. (most villous). — Pubescence of young growth, stems, petioles etc., *long-villous, loosely spreading to reflexed;* mature leaflets up to 6 cm. long (*P. caroliniana* of ed. 7, not Poir.). — Md. to O., s. and sw. to Ga., Tenn. and Mo.; casually adv. in e. Mass.

31. P. símplex Michx. (unbranched), Old-field-C. — Coarser than no. 30; *rhizome irregularly enlarged, often nodose or moniliform-thickened, up to* 8 *cm. long* and 0.5–2 cm. thick; *stem* hirsute or villous-hirsute with spreading hairs, *at first erect or ascending, up to* 2–5 *dm. high, then greatly prolonging* (to 1.2 m.), *arching, forking and rooting at tip* (*producing tubers*, the young rhizomes of following years), 1–3 *mm. thick at base; cauline leaves during anthesis mostly well-expanded* (those subtending flowers then 1.5–10 cm. long, including petiole); *leaflets* green and more or less strigose-pubescent or barely whitened beneath, *narrowly obovate, narrowly elliptic or oblanceolate; the middle ones* of larger leaves 1.5–7.5 cm. long, 9–27-*toothed for three-fourths their length; stipules of basal leaves with linear-lanceolate usually inrolling auricles*, of mature primary cauline leaves 0.7–3 cm. long;

first flower usually from node above second well-developed internode. (*P. canadensis* of ed. 7, not L.) — Dry or moist fields, thickets, open woods, etc., N.S. and sw. N.B. to s. Ont. and Minn., s. to N.C., Tenn., s. Mo. and Okla. April–June. Fig. 1173.

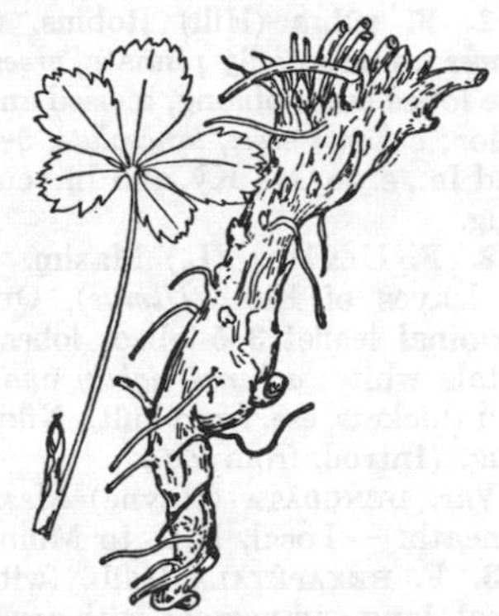
1173. P. simplex.

Var. **calvéscens** Fern. (becoming bald). — Similar, but stem, petioles, etc., glabrous or appressed-strigose. — Nfld. to Minn., s. to N.S., N.E., S.C., Ill. and Okla.

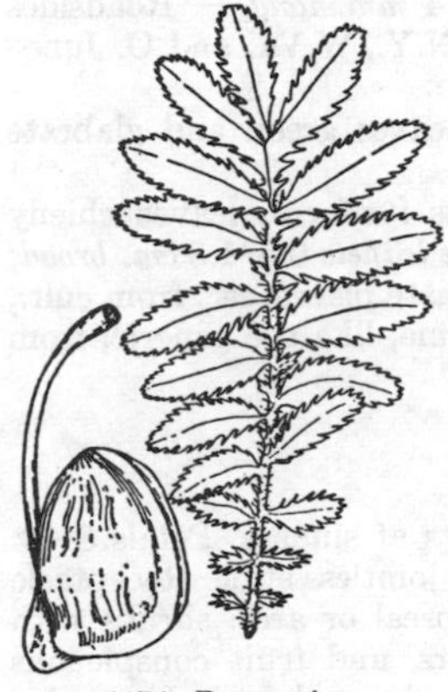
1174. P. anserina.

Var. **argyrísma** Fern. (silvery). — Leaves densely silvery-silky beneath; stems densely spreading-villous. — Pa. to Tenn. and Ill.

32. P. anserina L. (of geese), Silverweed, Argentine, Richette (Que.). — Spreading by slender many-jointed runners; the *stolons, peduncles, petioles and rachises* more or less *pubescent with ascending or loosely spreading hairs;* leaves all radical, interruptedly pinnate; *leaflets* oblong, oblanceolate or obovate, sharply serrate, *silky-tomentose* beneath, or *at least the younger lustrous,* green and glabrous or nearly so above, or silvery-silky on both sides in forma **serícea** (Hayne) Hayek (silky), this minor form identical with var. *concolor* Ser. and *Anserina concolor* and *argentea* Rydb.; peduncles elongated; flower 1–2.5 cm. broad; bracteoles often cleft; *achenes thick-ovoid to subglobose, more or less corky, dorsally sulcate.* — Gravelly or sandy shores or banks, Nfld. to Alaska, s. to N.S., N.B., n. N.E., N.Y., Ind., n. Ill., Ia., N.M. and Calif. June–Aug. (Eurasia) Fig. 1174.

33. P. Egédei Wormsk. (for Hans Egede, 1686–1758, father of modern Greenland). — Similar to no. 32; *stolons, petioles, rachises, etc. glabrous; leaves glabrous or opaquely grayish-tomentulose and glabrate beneath,* 1.5–8 cm. long, simply pinnate, with 2–4 remote pairs of narrowly obovate leaflets (larger ones 0.5–2 cm. long; *bracteoles simple; achenes laterally compressed, firm, rounded on back, not sulcate.* — Greenl., Baffin I. and n. Lab., n. Que. and Keewatin. Fig. 1175. — Southw. passing into smaller plants of

1175. P. Egedei.

Var. **groenlándica** (Tratt.) Polunin (of Greenland), Argentine. — Stolons, etc. glabrous or remotely pubescent; *leaves interruptedly pinnate,* 0.3–5 dm. long, *densely white-tomentose beneath with opaque* or but slightly sericeous *hairs,* the larger of the 7–31 oblong, oblanceolate or obovate leaflets 1–6 cm. long; flowers 1–3 cm. broad; bracteoles mostly simple. (*P. pacifica* Howell; *P. anserina,* var. *groenlandica* Tratt. and var. *grandis* T. & G.; *Argentina pacifica* and *litoralis* Rydb.) — Seacoast, s. Greenl. and Lab. to L.I.; Hudson Bay; Pacific coast, Alaska to Calif. May–Aug. (E. Asia)

16. FILIPÉNDULA Mill.

Flowers perfect or polygamous. Calyx (4–)5-parted. Petals (4–)5, short-clawed. Stamens 20 or more, almost hypogynous, the disk obscure. Carpels 5-15, free, 2-ovuled, mostly 1-seeded, indehiscent, compressed, sometimes twisted. — Perennial herbs of n. temp. reg., with pinnate leaves and panicled cymose flowers. Stipules reniform. (Name from *filum, a thread,* and *pendulus, hanging,* in allusion to the roots of no. 3.)

Leaves with large leaflets, the larger (terminal) one 0.5–2 dm. broad; carpels glabrous.
- Terminal leaflet 1–2 dm. broad, 7–9-parted; petals pink; carpels straightish, erect. 1. *F. rubra.*
- Terminal leaflet 0.5–1.5 dm. broad, 3–5-lobed; petals white; carpels spirally imbricated. 2. *F. Ulmaria.*

Leaves with very numerous pectinate leaflets, the larger only 1–1.5 cm. broad; carpels pubescent. 3. *F. hexapetala.*

1. F. rùbra (Hill) Robins. (red), QUEEN-OF-THE-PRAIRIE. — Glabrous, 6–25 dm. high; *leaves* interruptedly pinnate, *green* and *scarcely paler beneath;* terminal leaflet large, 7–9-parted; the lobes lance-oblong, incised and toothed; lateral leaflets also cut; petals deep peach-blossom-color; *achenes erect, lanceolate,* 5–8 *mm. long, glabrous.* — Meadows and prairies, Pa. to Mich. and Ia., s. to Ga., Ky. and Ill.; cult. and esc. n. and e. to N.S., N.E., N.Y., etc. Late June–early Aug.

2. F. ULMÀRIA (L.) Maxim. (old generic name, from fancied resemblance of the leaflets to leaves of Elm, *Ulmus*), QUEEN-OF-THE-MEADOW. — *Leaves canescent-tomentose beneath;* terminal leaflet 3–5-lobed; lobes ovate, doubly serrate; the lateral leaflets mostly unlobed; petals white; *achenes tightly and spirally imbricated, the group* 2.5–4 *mm. long.* — Roadsides and thickets, esc. from cult., Nfld. and e. Que. to N.S., N.E., N.J., N.Y., W.Va. and O. June–Aug. (Introd. from Eu.)

Var. DENUDÀTA (Hayne) Maxim. (denuded; without hair). — Leaves green and glabrate beneath. — Local, N.E. to Minn. (Introd. from Eu.)

3. F. HEXAPÉTALA Gilib. (with six petals), DROPWORT. — Roots fusiform; leaves chiefly basal, long and narrow, with *very many lacerate or pectinate lanceolate leaflets* 0.5–1.5 *cm. broad;* flower-buds pink; petals white; *carpels pubescent.* — Roadsides and waste places, esc. from cult., Nfld. to N.Y. June, early July. (Introd. from Eu.) — Colloquial name, like the generic, from the fusiform ("drop"-like) enlargements of the roots.

17. DRỲAS L. DRYAS

Calyx saucer-shaped to campanulate, 8–10-cleft, without bractlets at sinuses. Petals 8–10. Stamens many. Achenes numerous, tipped by the persistent plumose jointless style. Receptacle flat, dry. Seed ascending, basal. — Depressed matted or trailing boreal or arct. shrubs with simple coriaceous leaves whitened beneath, solitary scapose flowers, and fruit conspicuous through elongation of the feathery styles. (Named for the mythological dryads, wood-nymphs, from the resemblance of the foliage to tiny oak-leaves.)

1. D. integrifòlia Vahl (entire-leaved). — Branches horizontally trailing, or forming close mats; *leaves* elliptic, deltoid-ovate or lanceolate, short petioled; *the blades* revolute, *entire or few toothed at base,* 0.5–2 cm. long, 2–7 *mm. broad,* rugose, dark green and lustrous above, or strongly whitened and canescent-pilose in forma **canéscens** (Simmons) Fern. (grayish-white); scapes 1–11 cm. high; *calyx* canescent, often with coarse dark trichomes at base, the *linear to lanceolate lobes* 4–8 mm. long; *petals creamy-white,* elliptic, rounded at summit, 1–1.3 cm. long; plumes of mature (fruiting) styles smoky-white or tinged with bronze or purple, inclined to twist together, 2–4 cm. long. — Arct. Am., s. on calcareous rocks and gravels to w. Nfld., Mingan Ids., Anticosti I., eastern Gaspé Pen., Shickshock Mts. and Lake Mistassini, Que., Hudson Bay reg. and Alta. June, July.

2. D. Drummóndii Richards. (for its discoverer, THOMAS DRUMMOND, 1780–1835). — Coarser; *leaves* elliptic to narrowly obovate, *coarsely dentate to tip,* 1–4.5 cm. long, 0.5–2.3 *cm. broad,* canescent-arachnoid and becoming glabrate or quite glabrous above; scapes 0.3–2.5 dm. high; *calyx* heavily stipitate-glandular, *with oblong to lance-ovate lobes; petals orange,* ascending; plumes of fruit whitish to buff, 3.5–5.5 cm. long. — Calcareous cliffs, talus and river-gravels, w. Nfld.; Anticosti I. and Gaspé Pen., Que.; Slate I., Thunder Bay Distr., Ont.; Mackenz. to Alaska, s. to Mont. and Oreg. June–early Aug.

18. GÈUM L. AVENS. BENOÎTE (Que.)

Calyx campanulate or deeply 5-cleft, usually with 5 small bractlets at the sinuses. Petals 5. Stamens many. Achenes numerous, crowded on a conical or cylindrical dry receptacle, the long persistent styles forming hairy or naked and straight or jointed tails. Seed erect; radicle inferior. — Perennial herbs of cold and temp. reg., with pinnate or lyrate leaves. (A name used by Pliny.)

a. Styles jointed and geniculate near or above the middle; the lower segment, after falling of the stigmatose upper one, uncinate and stiff; stems with spreading to ascending branches and dilated leaves. . . *b.*

b. Calyx-tube saucer-shaped, the lobes soon reflexed; petals spreading, gradually narrowed at base; lower half of style glabrous or merely glandular, upper half not plumose; flowers ascending. . . *c.*

c. Calyx usually with bractlets at the sinuses; head of carpels sessile in the calyx; upper segment of style usually hispid above the joint. §EUGEUM. . . *d.*

*d.*Key to flowering material. . . *e.*

*e.*Petals white to pale or greenish-yellow, oblong to spatulate, 1–4.5 mm. broad, if yellow shorter than calyx-lobes. . . *f.*

*f.*Peduncles filiform, minutely pilose or puberulent; long hairs, if present, scattered; petals yellow and short or white and 5–9 mm. long by 1–4.5 mm. broad (except in late depauperate flowers).

Cauline leaves not rapidly reduced in size from base to summit of stem, their tips and teeth mostly acute; petals white, 5–9 mm. long, 2–4.5 mm. broad, about equaling to longer than calyx-lobes. . . . . 1. *G. canadense.*

Cauline leaves rapidly reduced in size toward summit of stem, their tips and teeth blunt; petals pale or greenish-yellow, 2–4 mm. long, 1–2 mm. broad, much shorter than calyx-lobes. . 2. *G. virginianum.*

*f.*Peduncles stout and short, copiously hirsute with divergent or reflexed crowded hairs 1–2 mm. long; petals white, 2–5 mm. long, 1–2 mm. broad, much smaller than calyx-lobes. . . . 3. *G. laciniatum.*

*e.*Petals orange or deep yellow, suborbicular or broadly obovate, 3–9 mm. broad, equaling or longer than calyx-lobes.

Terminal segment of some or all basal leaves cuneate-obovate to -oblanceolate, incised; calyx-lobes lanceolate or lance-ovate, 5–9 mm. long; petals 5–10 mm. long, 5–9 mm. broad; base of style glandless. . . . . 4. *G. aleppicum.*

Terminal segment of all (except sometimes in var.) basal leaves cordate-reniform or -suborbicular, dentate; calyx-lobes broadly deltoid, 2.5–5 mm. long; petals 3.5–5 (–7) mm. long, 3–5 (–6) mm. broad; base of style minutely glandular. . . . . 5. *G. macrophyllum.*

*d.*Key to fruiting material. . . *g.*

*g.*Denuded receptacle glabrous or only minutely pubescent, the pits and scars evident.

Peduncles thick, mostly 1–5 cm. long, copiously hirsute with divergent or reflexed crowded hairs 1–2 mm. long; calyx-lobes lanceolate or lance-ovate, 5–11 mm. long; fruiting head 1.7–2.5 cm. in diameter, the drab or brownish styles not glandular. . 3. *G. laciniatum.*

Peduncles filiform, mostly longer, minutely puberulent, long hairs if present scattered; calyx-lobes broadly deltoid, 2.5–5 mm. long; fruiting heads 1.2–1.8 cm. in diameter, the usually purple styles minutely glandular at base. . . . . 5. *G. macrophyllum.*

*g.*Denuded receptacle copiously hirsute, the pits and scars hidden. . . *h.*

*h.*Basal leaves chiefly simple or ternately divided, the margins lobed, serrate or dentate; middle and upper leaves simple or 3-cleft, serrate or dentate; peduncles filiform; fruiting heads spherical; the upper styles loosely ascending or spreading, only tardily reflexed.

Cauline leaves not rapidly reduced in size from base to summit of stem, their tips and teeth mostly acute. . . . . . . 1. *G. canadense.*

Cauline leaves rapidly reduced in size from base to summit of stem, their tips and teeth obtuse. . . . . 2. *G. virginianum.*

*h.*Basal leaves chiefly interruptedly pinnate, with 5–9 incised leaflets and interspersed smaller ones; middle and upper leaves similar, incised; peduncles coarse, clavate at tip; fruiting heads finally globose-obovoid; the styles soon tightly reflexed-appressed and imbricated. . . . . 4. *G. aleppicum.*

*c.*Calyx without bractlets; head of carpels raised on a stipe above the calyx; upper segment of style glabrous; petals yellow to cream-color, 1.5–2 mm. long. §Stylipus. . . . . 6. *G. vernum.*

*b.*Calyx campanulate, the usually purple lobes erect or ascending; petals erect, abruptly contracted to a claw; base and summit of style plumose; flowers nodding, fruits erect. §Caryophyllata. . . . . 7. *G. rivale.*

*a.*Styles not jointed, wholly persistent, not geniculate nor uncinate; calyx turbinate-campanulate; stems subscapose, with reduced cauline leaves, simple or forking toward summit. §Sieversia.

Basal leaves with many cuneate subuniform segments; flowers nodding; bractlets longer than calyx-lobes; petals erect, oblong, stramineous to purplish; style plumose nearly to tip, slender and flexuous. . . . 8. *G. triflorum.*

Basal leaves with a large suborbicular or reniform terminal segment and very small lateral segments; flowers ascending; bractlets shorter than calyx-lobes; petals spreading, round-obovate, deep yellow or orange; style plumose only at base, stiff and straight. . . . . 9. *G. Peckii.*

§ Eugèum T. & G.

1. **G. canadénse** Jacq. (Canadian). — Stem 0.2–1.2 m. high, glabrous to sparingly hirsute, rather slender, often minutely pubescent or glandular-puberulent at summit; *leaves of basal tufts long-petioled,* the petioles smooth or sparsely hairy, *simple and undivided or with* 3–5 (rarely –7) *rhombic serrate leaflets;* lower cauline leaves similar, short-petioled to sessile, mostly with 3 leaflets; *upper ternately cleft or simple, sharply serrate and acute;* stipules ovate-oblong, 1–2 cm. long, subentire or cleft; *peduncles filiform, minutely pilose to glandular-puberulent;* calyx-lobes lanceolate to lance-ovate, acuminate, 4–10 mm. long; *petals white, oblong,* 5–9 *mm. long,* 2–4.5 *mm. broad, about equaling to longer than calyx-lobes; fruiting head spherical,* with 30–160 carpels, 1.2–2 cm. in diameter; *the upper segment of the style ascending or spreading, only tardily reflexed; denuded receptacle densely white-villous.* — Our most variable species, with several recognized vars. and forms:

a. Upper stigmatose segment of style conspicuously (though sparsely) bearded at base with stiff white hairs; body of carpel usually sparsely appressed-pubescent as well as long-setiferous; outer surface of calyx-lobes and petioles and stems glabrate or with a few long hairs. . . *b.*

b. Carpels 30–60, when mature with mostly broad-ovate to obovate bodies 2.5–3 mm. long; peduncles (except for glands, when present) typically fine-puberulent or with few scattered longer hairs; leaves membranaceous or thin, the cauline mostly glabrous above.

Outer surface of calyx-lobes and peduncles glandless. . . . . . . *G. canadense* (typical).

Outer surfaces of calyx-lobes and peduncles beset with articulate gland-tipped trichomes. . . . . Forma *glandulosum.*

b. Carpels 60–160, their mature bodies narrowly obovate to cuneate and 3–4 mm. long; peduncles often with coarser pubescence; leaves firmer, mostly strigose-pilose above.

Outer surface of calyx-lobes and peduncles glandless. . . . . . Var. *camporum.*

Outer surface of calyx-lobes and peduncles beset with gland-tipped articulate trichomes. . . . . Forma *adenophorum.*

a. Upper stigmatose segment of style minutely and sparsely short-hispidulous; body of carpel hispid above, otherwise glabrous; peduncles puberulent or minutely pilose, with or without short glands; outer surface of calyx-lobes, petioles and stem merely puberulent to glabrate.

Upper cauline leaves simple, ovate-rotund to rhombic, short-petioled to sessile, much shorter than the fruiting peduncles; these mostly 3–8 cm. long; calyx-lobes 5–10 mm. long; carpels and achenes 75–120. . . Var. *Grimesii.*

Upper cauline leaves mostly compound, ovate to obovate, long-petioled, longer than the axillary divergent fruiting peduncles; these 1–3 cm. long; calyx-lobes 4–6 mm. long; carpels and achenes 50–60. . . . . Var. *brevipes.*

G. canadénse (typical), incl. *G. Meyerianum* Rydb. — Rich thickets and borders of woods, N.B. to Minn., s. to N.S., s. N.E., L.I., S.C., Tenn., Mo. and Kans. June–early Aug. Fig. 1176. — Forma **glandulòsum** Fern. & Weath. (glandular), occasional.

Var. **campòrum** (Rydb.) Fern. & Weath. (of the plains), *G. camporum* Rydb. — Open woods, thickets, fields and roadsides, Que. to N.D., s. to N.S., N.E., N.C., Ala., Okla. and Tex. — Forma **adenóphorum** Fern. & Weath. (bearing glands), local.

Var. **Grìmesii** Fern. & Weath. (for Earl Jerome Grimes, 1893–1921, its discoverer). — N.Y. to Ind. and Ga.

1176. G. canadense.

Var. **brévipes** Fern. (short-stalked). — Alluvial woods, Nottoway R. and tributaries of Roanoke R., Va.

2. **G. virginiànum** L. (Virginian). — Stem (at least below) and petioles hirsute; *radical leaves firm,* ovate, simple or 3–7-foliolate; the blades or their segments *obtuse, dentate; cauline leaves rapidly reduced in size toward summit of stem,* lanceolate to ovate, simple or cleft, *blunt and with obtuse teeth,* pilose; stipules incised or lobed, the larger 1.5–5.5 cm. long; peduncles elongate, minutely pilose; *petals pale or greenish-yellow,* oblong or narrowly obovate, 2–4 *mm. long,* 1–2 *mm. broad, much smaller than calyx-lobes* (4–6 mm. long); fruiting head spherical,

1177. G. virginianum.

1.5–2 cm. in diameter; denuded receptacle densely hirsute. (*G. flavum* (Porter) Bickn.) — Dry woods and thickets or rocky banks, centr. Mass. to Ind., s. to S.C. and Tenn. June–Aug. Fig. 1177.

3. G. laciniàtum Murr. (slashed). — Rather coarse, up to 1 m. high; stem villous or hirsute; *basal leaves diverse,* with simple and rounded to pinnate blades with several incised leaflets; *cauline leaves* simple and lobed or 3-parted or -divided, *strongly toothed or incised; peduncles* stout, mostly 1–5 cm. long, *copiously hirsute with crowded divergent or reflexed hairs 1–2 mm. long; petals white,* 2–5 mm. long, 1–2 mm. broad, *much smaller than the lanceolate to lance-ovate calyx-lobes* (5–11 *mm. long*); *carpels very numerous,* their closely *imbricated* mature styles nearly all reflexing, the dense fruiting head 1.7–2.5 cm. in diameter, the pale-brown to drab or olivaceous achenes glabrous; *denuded receptacle glabrous or nearly so.* (*G. virginianum* of ed. 7, not L.) — Damp thickets, meadows and roadsides, N.S. to s. Ont. s. to Mass., N.J., Md., O. and Ind. June, July. Fig. 1178.

1178. G. laciniatum.

Var. trichocàrpum Fern. (hairy-fruited). — Achenes bristly at summit. — Commoner southw., N.S. to s. Ont. and Minn., s. to N.C., Ill., Mo. and e. Kans.

4. G. aléppicum Jacq. (of Aleppo). — Somewhat hairy, 0.3–1.5 m. high; basal leaves with large rounded terminal leaflet or the 5–9 leaflets cuneate-obovate to -oblanceolate and incised, with interspersed small leaflets; *cauline leaves with* 3–5 acute, *mostly incised or deeply cut rhombic-ovate or oblong acute leaflets; peduncles stiff, clavate at summit; calyx-lobes lanceolate to lance-ovate, 5–9 mm. long; petals orange or deep yellow, suborbicular to broadly obovate, 5–10 mm. long, 5–9 mm. broad; fruiting heads finally globose-ovoid,* 1.4–2.3 cm. broad, *the glandless styles soon tightly reflexed-appressed and imbricated;* body of achene long-villous; *denuded receptacle long-hirsute.* — Eurasia. Represented with us by

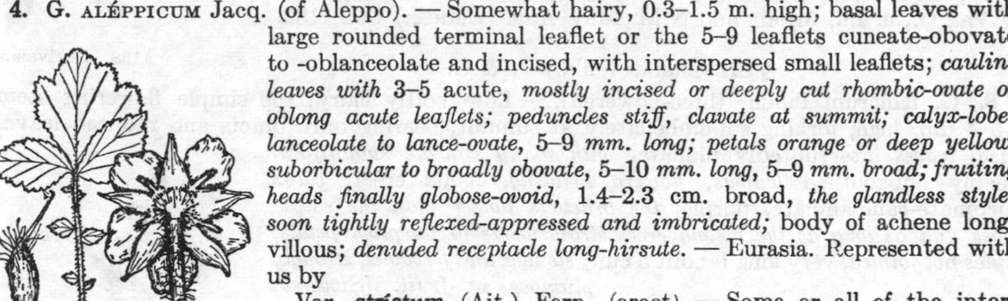

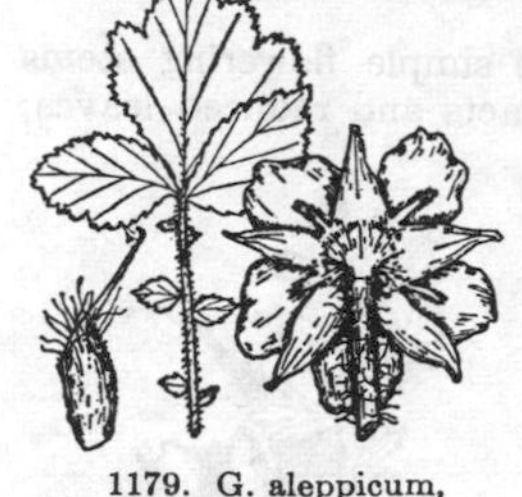
1179. G. aleppicum, v. strictum.

Var. stríctum (Ait.) Fern. (erect). — Some or all of the interruptedly pinnate basal leaves with incised cuneate-obovate to -oblanceolate terminal segments; *sides of achene only short-pilose to glabrous.* (*G. strictum* Ait.) — Thickets, meadows and clearings, Gaspé Pen., Que., to B.C., s. to N.S., N.E., N.J., Pa., W.Va., O., Ind., Ill., Ia., Neb., N.M. and Mex. June–Aug. Fig. 1179.

5. G. macrophýllum Willd. (large-leaved). — Bristly-hairy below, 0.3–1 m. high; *radical leaves with a cordate-reniform or -suborbicular dentate terminal segment; lateral leaflets of cauline leaves 2–4,* small; *the terminal round and large,* 3-cleft, *with broadly cuneate blunt lobes; peduncles filiform,* minutely puberulent or with scattered long hairs; *calyx-lobes broadly deltoid, 2.5–5 mm. long; petals* bright yellow, 3.5–5 *mm. long,* 3–5 *mm. broad; base of purple style minutely glandular; fruiting heads globose,* 1.2–1.8 *cm. in diameter; denuded receptacle glabrous or merely short-hispid.* — Rich woods, damp thickets and openings, se. Lab. to Alaska, s. to Nfld., N.S., n. Me., mts. of N.E. and N.Y., n. Mich., Wisc., Minn., Mont., Id. and Calif. June–Aug. (E. Asia; n. Eu.) Fig. 1180.

1180. G. macrophyllum.

Var. perincìsum (Rydb.) Raup (much incised). — *Terminal segment of basal leaves* usually more *incised or sharply toothed; cauline leaves smaller, with narrower and more incised leaflets;* petals up to 7 mm. long and 6 mm. wide. (*G. perincisum* Rydb.) — Similar habitats, Yuk. to Calif., e. to Ung. and Lake Superior reg. of Ont. and Mich.

× **G. púlchrum** Fern. (handsome), a natural hybrid between nos. 5 and 7, with habit and foliage of no. 7, purple calyx with spreading lobes 4–5 mm. long, golden-yellow obovate but definitely clawed spreading petals and carmine styles, is local from Que. to Alta., s. to Vt. and Id.

× **G. aurantìacum** Fries (orange), a similar hybrid of nos. 4 and 7, is very local in n. N.Y.

G. urbànum L. (of towns), resembling no. 5, with acuminate leaflets of cauline leaves rhombic, narrow petals barely longer than the lanceolate calyx-lobes and long-hirsute receptacle, is spreading in dooryards and on shaded roadsides, locally, about towns of e. Mass. and e. Pa. (Adv. from Eu.)

§ Stỳlipus (Raf.) T. & G.

6. G. vérnum (Raf.) T. & G. (vernal). — Stems arched-ascending, smooth or slightly pubescent, 2.5–6 dm. high; basal leaves roundish-cordate, 3–5-lobed, or some of them pinnate, with lobes cut; peduncles capillary, subumbellately disposed; *calyx* small, *without bractlets*, the lobes 1.5–3 mm. long; *head of carpels and fruit raised on a stipe above the calyx; petals yellow to cream-color*, 1.5–2 *mm. long; upper segment of style glabrous;* fruiting head 0.8–1.4 cm. in diameter; the achenes few, loosely divergent. (*Stylipus* Raf.) — Rich woods and openings, N.Y. to s. Ont. and Mich., s. to Md., W. Va., Tenn., Mo. and se. Kans. Late Apr.–June. Fig. 1181.

1181. G. vernum.

§ Caryophyllàta Ser.

7. G. rivàle L. (of brook-sides), Water- or Purple A.; Chocolate-root. — Stems nearly simple, several-flowered, 0.3–1.2 m. high; basal leaves lyrate and interruptedly pinnate, those of the stem few, 3-foliolate or 3-lobed; *flowers nodding; calyx purple*, or yellowish to green in forma **virèscens** Lilja (greenish), *campanulate, with erect purple lobes; petals erect*, buff or smoky-yellowish to purplish, *abruptly contracted to claw*, the blade retuse; head of fruit stalked, erect; *base and summit of style plumose.* — Wet meadows, bogs and peaty slopes, se. Lab. to B.C., s. to Nfld., N.S., N.E., n. N.J., Pa., W.Va., O., n. Ill., Minn., and N.M. May–Aug. (Eurasia) Fig. 1182.

1182. G. rivale.

§ Sievérsia (Willd.) T. & G.

8. G. triflòrum Pursh (three-flowered). — Low, softly hairy; the simple flowering stems 1.5–4 dm. high, forking subumbellately at summit, bearing only bracts and reduced leaves; *basal leaves* interruptedly pinnate, *with many cuneate subuniform* deeply cut-toothed *leaflets; flowers nodding*, fruits erect; *calyx* turbinate-campanulate, purple, *the bractlets longer than the calyx-lobes and as long as the oblong erect stramineous to purplish petals; styles not jointed*, very long (about 5 cm.) *slender and flexuous, strongly plumose in fruit.* (*Sieversia* R. Br.) — Calcareous rocks, gravels and prairies, Watertown, N.Y. (formerly); shores of L. Huron, Ont., to Alta., s. to n. Ill., Ia., Neb. and Mont. Apr.–June. Fig. 1183.

1183. G. triflorum.

9. G. Péckii Pursh (for its discoverer, William Dandridge Peck, 1763–1822). — Smoothish; *root-leaves rounded-reniform*, radiate-veined, 5–12 cm. broad, doubly or irregularly cut-toothed and obscurely 5–7-lobed, with a set of minute leaflets down the long petiole; stems 1.5–4 dm. high, 1–5-flowered; *bractlets minute; petals yellow, round-obovate* and more or less obcordate, exceeding the calyx (1 cm. long), *spreading; styles naked* except at the base. (*Sieversia* R. Br.) — Damp peats, gravels and cliffs, White Mts., N.H.; Brier I., w. N.S. June–Sept. Fig. 1184.

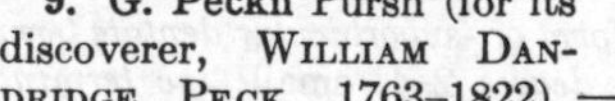

1184. G. Peckii.

Tribe IV. Rùbeae Dumort.

19. RÙBUS L. Bramble

Calyx 5(3–7)-parted, without bractlets. Petals mostly 5, deciduous. Stamens numerous. Carpels usually many, collected on a spongy or succulent receptacle, becoming small drupelets; styles nearly terminal. — Perennial herbs, or more often somewhat shrubby often prickly plants of N. Hemisph. and mts. of S.Am., with white (rarely reddish) flowers, and usually edible fruit. (The Roman name, kindred with *ruber, red.*)

Unarmed; leaves simple or pedately 3- or 5-foliolate.
- Flowering stems herbaceous, 0.1–4 dm. high, very slender, arising from creeping subterranean or superficial prolonged bases; leaves 3- or 5-foliolate or if simple only 1–3 and the solitary flowers dioecious.
  - Leaves suborbicular, simple; flowers solitary, terminal, dioecious; calyx cleft nearly to base into ovate short-tipped lobes; petals white, spreading; mature fruit amber-color, becoming yellowish, when fully ripe dropping quickly from the dry receptacle. . . . . Subgen. I. Chamaemorus (see p. 819).
  - Leaves 3–5-foliolate; flowers 1–7, some or all hermaphrodite; calyx cleft

partly to base, with linear lobes; petals erect, white to roseate; fruit red, only tardily separating from the spongy receptacle. . . .Subgen. II. CYLACTIS (see p. 819).

Shrubby, the leafy stems mostly 1–2 m. high, from close crowns; leaves simple, lobed; calyx-lobes caudate-tipped; petals broad, spreading; fruit depressed, readily falling from the dry receptacle. . . . . .Subgen. III. ANAPLOBATUS (see p. 820).

Armed with prickles or bristles (unarmed in a few species); canes slightly woody, annual, biennial or of few years' duration; leaves mostly compound (simple in rare cases); fruit subglobose to elongate.

Leaves of both primocanes (1st year's canes) and fruiting canes (floricanes) pinnate or simple; fruit readily falling intact from the dry receptacle, red or black. . . . . .Subgen. IV. IDAEOBATUS (see p. 820).

Leaves of primocanes digitately 3–7-foliolate; flowers in racemes, panicles or corymbs or solitary; petals showy, wide-spreading; fruit tardily separating from the fleshy receptacle (often drying without falling), usually falling intact, black or blackish. . . . . .Subgen. V. EUBATUS (see p. 822).

Subgen. I. CHAMAEMÒRUS (Ehrh.) Focke CLOUDBERRY A single species.

1. R. Chamaemòrus L. (old generic name for this species, meaning ground-mulberry), BAKED-APPLE-BERRY, BAKED-APPLE, CHICOUTÉ (Que.). — Extensively creeping herb; the *upright branches* 1–3 dm. high, simple, 1–3-*leaved; leaves round-reniform*, more or less 5-lobed, coriaceous, serrate; *flower solitary*, terminal, *dioecious; calyx cleft nearly to base, its ovate or ovate-oblong leathery lobes blunt or mucronate-tipped; petals* white, obovate, *wide-spreading*, 1–1.8 cm. long; *fruit* at first closely embraced in calyx, *finally* with calyx-lobes loosely spreading, red-tinged, then amber-color, finally (when ripe) *yellowish* and soft, of very large drupelets with stones 4–5 mm. long, *quickly dropping from the dry receptacle.* — Acid peats, Greenl. and Lab. to Alaska, s. to Nfld., N.S., coastal reg. of e. Me.; mts. of w. Me. and n. N.H., Montauk Pt., L.I.; Man., Alta. and B.C. *Fl.* June–early Aug.; *fr.* July, Aug. (Eurasia)

Subgen. II. CYLÁCTIS (Raf.) Focke DWARF RASPBERRY. PLUMBOY

Central leaflet broadly ovate or ovate-rhombic, tapering about equally to acute apex and base, mostly becoming 2.5–10 cm. long; flowers 1–7; vegetative branches sometimes prolonged to flagelliform rooting tips and forming filiform subligneous old stems.

Leaves membranaceous, opaque; calyx-lobes 3–7 (rarely –10) mm. long; petals white or pale pink, 6–10 mm. long, 1.5–3 mm. broad; vegetative shoots usually much prolonged into flagelliform tips. . . . . 2. *R. pubescens.*

Leaves subcoriaceous, sublustrous above; calyx-lobes 7–12 mm. long; petals roseate, 1–1.7 cm. long, 3–7 mm. broad; vegetative shoots short, rarely prolonged. . . . . 3. *R. arcticus.*

Central leaflet cuneate-obovate, broadly rounded at summit, subcoriaceous, lustrous, the mature ones 1–3.5 cm. long; flowers solitary, as in no. 3 or larger; vegetative shoots very short. . . . . 4. *R. acaulis.*

2. R. pubéscens Raf. (hairy), CATHERINETTES (Que.). — *Stems slender, trailing* or loosely ascending, *bearing* erect *herbaceous flowering branches* 1–4 dm. high, slightly pubescent above, *the vegetative shoots much prolonged and ending in flagelliform rooting tips; leaves* pedately 3- or 5-foliolate, *membranaceous, opaque; the central leaflet* rhombic-ovate to -lanceolate, *tapering about equally to the acute apex and base*, coarsely and doubly serrate, glabrous or nearly so, 4–10 cm. long; flowers 1–7; calyx-lobes 3–7 (–10) mm. long, linear; *petals white or pale pink*, or deep roseate in forma **roseiflòrus** (Peck) House (rose-flowered), erect, 6–10 *mm. long*, 1.5–3 *mm. broad;* fruit dark red, only tardily separating from the spongy receptacle, the richly juicy drupelets large. (*R. triflorus* Richards.; *R. americanus* Britt.) — Damp slopes, rocky shores, low thickets, etc., Lab. to n. B.C., s. to Nfld., N.S., N.E., n. N.J., Pa., O., Ind., Wisc., ne. Ia., S.D. and Colo. *Fl.* May–July (–Aug., alpine); *fr.* June–Aug. — Hybridizes with no. 3.

Var. **pilosifòlius** A. F. Hill (hairy-leaved). — Leaves closely pilose and velvety to touch beneath. — Nfld. to N.E., w. to Mont., local.

3. R. árcticus L. (arctic). — Similar to no. 2, *rarely with prolonged vegetative shoots*, the flowering stems (0.5–3.5 dm. high) chiefly rising from subterranean rhizomes or slender crowns; *leaflets firm, sublustrous*, more broadly ovate, less attenuate at tip, simply or doubly dentate,

2.5–6 cm. long; flowers 1–3; *calyx-lobes 7–12 mm. long; petals roseate, 1–1.7 cm. long, 3–7 mm. broad;* fruit with rather few coarse drupelets. (*R. paracaulis* Bailey) — Se. Lab. to Alaska, s. in damp peats or in gravel (often calcareous) to Nfld. and St. P. et Miq., Côte Nord, Mingan Ids. and Gaspé Pen., Que., Man. and s. Alta. *Fl.* July, Aug.; *fr.* late July–early Sept. (N. Eurasia) — Hybridizes with nos. 2 and 4.

4. R. acaùlis Michx. (stemless). — Plant usually smaller than no. 3; *flowering stems all from subterranean filiform bases, 1–10 cm. high;* leaves coriaceous, lustrous; the *leaflets cuneate-obovate and broadly rounded at summit,* doubly dentate, the larger ones 1–3.5 cm. long; flower solitary, as large as or larger than in no. 3; fruit of more numerous and smaller drupelets, with smaller stones, richly flavored. (*R. arcticus,* var. *grandiflorus* Ledeb.) — Lab. to Alaska, s. on damp soils to Nfld., St.P. et Miq., Anticosti I. and Gaspé Pen., Que., n. Minn., Man., Sask., Colo. and s. B.C. *Fl.* June–early Aug.; *fr.* July–late Aug. (E. Asia) — Hybridizes with no. 3.

Subgen. III. Anaplóbatus Focke (see p. 819)

Calyx-lobes heavily covered with dark gland-tipped hairs; petals roseate (except in albino); fruit dry and insipid. . . . . 5. *R. odoratus.*

Calyx-lobes canescent-pilose or villous, rarely with some short pale glands; petals white; fruit juicy and of rich flavor. . . . . 6. *R. parviflorus.*

5. R. odoràtus L. (fragrant), Purple-flowering Raspberry, Thimbleberry, Chapeaux rouges (Que.). — *Shrubby,* unarmed, 1–2 m. high, *the branches, stalks and calyx more or less bristly with long or short dark glandular-clammy hairs;* leaves 3–5-lobed, mostly 1–3 dm. broad, glabrous or nearly so to pilose above, pilose along nerves or glabrate to velvety beneath, the lobes ovate- or deltoid-acuminate, dentate-serrate; *peduncles 3–75-flowered; flowers* showy, 3–6 *cm. broad;* calyx-lobes slender-tipped; *petals* elliptic-oblong to rounded, *rose-purple,* or white and with foliage pale in forma **albiflòrus** House; *fruit* broad, low-hemispherical, red, *dryish and rather insipid.* (*Rubacer* Rydb.) — Thickets and borders of woods, s. Que. and s. Ont., s. to N.S., N.E., L.I., Ga. and Tenn. *Fl.* June–Sept.; *fr.* July–Sept. — Excessively variable in pubescence.

Var. **columbiànus** Millsp. (in acknowledgment of comparisons made in the herbarium of Columbia College). — Lobes of leaf lanceolate, incised and sharply double-serrate; flowers 2–3 cm. broad. (*Rubacer columbianum* Rydb.) — Local, n. W.Va.

6. R. parviflòrus Nutt. (small-flowered; a misnomer), Thimbleberry. — Similar to no. 5, more or less glandular, but stems scarcely bristly; lobes of leaves ovate, coarsely toothed; peduncles 1–15-flowered; *calyx-lobes canescent-pilose or villous, rarely with short pale glands; petals white; fruit juicy and of rich flavor.* (*Rubacer* Rydb.) — Thickets and borders of woods, Bruce Pen., Ont., to Minn.; Black Hills, S.D.; Alta. and s. Alaska, s. to mts. of Mex., Ariz. and Calif. *Fl.* June, July; *fr.* Aug., Sept. — Polymorphic species, highly variable in degree of pubescence and glandularity.

Subgen. IV. Idaeóbatus Focke (see p. 819) Raspberry

Fruiting stems annual; leaves with 2 or 4 remote pairs of lance-attenuate leaflets; flower 3–4 cm. broad, with conspicuously clawed spreading petals; fruit suggesting a strawberry, 2–3 cm. in diameter. § 1. Rosaefolii. . . . . . 7. *R. illecebrosus.*

Fruit chiefly on 2-year-old canes; leaves of fruiting canes with 3 leaflets; flowers much smaller, the petals spreading-ascending; fruit smaller. § 2. Idaeanthi.

Canes, pedicels and calyx densely villous with long red gland-tipped hairs; calyx-lobes becoming erect and enveloping the developing fruit. . . . . 8. *R. phoenicolasius.*

Canes prickly, bristly, smooth or minutely pubescent, if glandular-hairy with the hairs pale and not crowded; calyx-lobes spreading or reflexed under developing fruit.

Canes erect, not rooting at tip, arising from subterranean rhizomes and stolons, bristly to smoothish, with or without hooked prickles; leaves of primocanes, if quinate, pinnate; calyx often bristly, its lobes (excluding caudate tips) about equaling petals; fruit red (rarely yellowish), the drupelets not separated by bands of tomentum. . . . . . . 9. *R. idaeus.*

Canes and branches long-arching, finally rooting at tips, arising from crowns, with hooked prickles and no bristles, heavily glaucous; leaves of new canes, if quinate, digitate; calyx scarcely if ever bristly, its lobes much exceeding petals; fruit purple-black (rarely pale), the immature drupelets separated by bands of white tomentum. . . . . 10. *R. occidentalis.*

§ 1. Rosaeifòlii Focke

7. R. illecebròsus Focke (enticing), Strawberry-Raspberry. — Slender *annual green angular* glabrous *canes* with flat-based curved prickles, arising from the subterranean rhizomes and stolons; *leaves pinnate, with 5–9 lance-attenuate leaflets, the pairs remote,* the rachis prickly; *flowers* from upper axils, 3–4 *cm. broad,* fragrant; calyx-lobes slender-tipped; *petals white, spreading, conspicuously clawed; fruit* scarlet, 2–3 *cm. in diameter, suggesting a strawberry.* — Roadsides, thickets and old house-sites, N.S., N.E. and N.Y.; originally cult. *Fl.* July–Sept.; *fr.* Aug.–Oct. (Introd. from Japan)

§ 2. Idaeánthi Focke

8. R. phoenicolàsius Maxim. (with purple-red hairs), Wineberry. — *Canes* biennial, elongate, arching, *rooting at tip, densely covered, like the pedicels, petioles and calyx, with long gland-tipped reddish hairs;* leaflets 3, strongly white-tomentose beneath and with reddish veins, the terminal round-cordate; panicles terminal, several-flowered; *calyx-lobes lance-attenuate, much exceeding the small petals, soon covering the developing fruit,* finally loosely spreading; fruit red. — Roadsides, thickets and open woods, Mass. to Ind., s. to Va. and Ky., originally cult. *Fl.* late May–July; *fr.* late June–Sept. (Introd. from e. Asia)

9. R. idaèus L. (of Mt. Ida), Raspberry, Framboisier (Que.). — *Canes upright,* biennial, *not rooting at tip,* arising from subterranean suckers and stolons, *prickly, bristly or smoothish; leaves of primocanes pinnate* (rarely simple), *with* 3–7 mostly oblong-ovate to -lanceolate cut-serrate pointed *leaflets,* the lateral ones sessile, white or gray beneath; leaves of fruiting canes mostly ternate; flowers drooping, from upper axils or in small clusters, about 1 cm. broad; *calyx often bristly, its lobes* (excluding slender tip) *about equaling the narrow white subascending petals; fruit* depressed-globose to conic-thimbleshaped, *red* (rarely yellow), *falling intact from the dry receptacle.* — Highly variable circumboreal type. We have the following:

a. Inflorescence without glands or minute bristles.
  Prickles of primocanes strong and obviously broadened at base. . . . *R. idaeus* (typical).
  Prickles wanting; canes smooth. . . . . Forma *inermis.*
a. Inflorescence bearing glands and minute bristles; primocanes usually bearing slender bristles and often stipitate glands. . . *b.*
  b. Bark of primocanes glabrous or at most glaucous beneath the bristles, in age becoming lustrous. . . *c.*
    c. Prickles mostly strong and obviously broadened at base. . . . . . Var. *aculeatissimus.*
    c. Prickles (when present) bristleform and not much thickened at base. . . *d.*
      d. Leaves of the primocanes with oblong to ovate acuminate leaflets; of fruiting canes with 3 (rarely 5) similar but shorter leaflets.
        Primocanes bristly. . . . . Var. *strigosus.*
        Primocanes without bristles. . . . . Forma *tonsus.*
      d. Leaves of primocanes with 3 short-ovate to suborbicular round-tipped or blunt leaflets; of fruiting canes simple and rounded or at most 3-lobed. . . . . Var. *Egglestonii.*
  b. Bark of primocanes cinereous-tomentulose beneath the prickles. . . *e.*
    e. Many of the prickles stout and broad-based. . . . . Var. *heterolasius.*
    e. Prickles all bristleform. . . *f.*
      f. Leaves with distinct lanceolate to ovate acuminate leaflets.
        Leaflets lanceolate to ovate, coarsely toothed or merely lobed. . . Var. *canadensis.*
        Leaflets narrowly lanceolate, caudate-tipped, those of primocanes cleft nearly or quite to base. . . . . Var. *caudatus.*
      f. Leaves with simple rounded or round-lobed blades or with 3 barely separate rounded leaflets. . . . . Var. *eucyclus.*

R. idaèus (typical). — Roadsides, thickets and neighborhoods of settlements, Nfld. to Ont., s. to N.S., s. N.E., N.Y., O., Mich., Minn. and S.D., spread from cult. *Fl.* late May–July; *fr.* late June–Oct. (Introd. and natzd. from Eu.) — Forma **inérmis** Kaufmann (unarmed), with prickleless canes, is apparently indig. on the Magdalen Ids. and in Minn.

Var. **aculeatíssimus** Regel & Tiling (most prickly), *R. melanolasius* Focke — Western N.Am., e., locally, to Mich. (Asia)

Var. **strigòsus** (Michx.) Maxim. (beset with straight stiff hairs), *R. strigosus* Michx. —

Thickets, clearings and borders of woods, Lab. to B.C., s. to Nfld., N.S., N.E., w. Va., O., n. Ind., Wisc., Minn., Neb. and Wyo. — Forma **tónsus** Fern. (sheared), with primocanes without bristles, occasional, Nfld. to n. N.E.; forma **álbus** (Bailey) Fern. (white), with fruit amber-white, rare.

Var. **Egglestònii** (Blanch.) Fern. (for its discoverer, Willard Webster Eggleston, 1863–1935), *R. Egglestonii* Blanch. — Local, Vt.

Var. **heterolásius** Fern. (variously hairy). — Coastal reg., sw. N.B. and Me.

Var. **canadénsis** Richards. (Canadian), *R. carolinianus* and *subarcticus* Rydb. — Lab. to Alaska, s. to Nfld., N.S., N.E., locally to mts. of N.C., Mich., Wisc., S.D. and Colo. (E. Asia)

Var. **caudàtus** (Robins. & Schrenk) Fern. (tailed). — Se. Nfld., local.

Var. **eucỳclus** Fern. & Weath. (well-rounded). — Very local, n. Gaspé Co., Que. — This and var. *Egglestonii* N.Am. counterparts of the rare European var. *anomalus*, reversionary variants with rounded leaves and leaflets and mostly sterile.

× **R. negléctus** Peck (neglected), a series of plants, obviously crosses between *R. idaeus*, var. *strigosus* and no. 10, is occasional where the ranges of the two coincide.

10. R. occidentàlis L. (western; *i.e.*, of the Western Hemisphere), Black Raspberry, Thimbleberry, Framboise noire (Que.). — *Heavily glaucous; canes and their branches long-arching, rooting at tips, armed* like the pedicels *with hooked prickles, not bristly,* arising from crowns; *leaves of primocanes* ternate, or quinate and *digitate,* strongly whitened beneath, coarsely double-serrate, the lateral somewhat stalked; *calyx* pannose-tomentose, *scarcely if ever bristly, its lobes much exceeding the small petals; fruit purple-black,* or yellowish or amber in forma **pállidus** (Bailey) Robins. (pale), compact, hemispherical, *the drupelets separated by bands of white tomentum.* — Rich thickets, ravines and borders of woods, Que. to Minn., s. beyond our limits, with close allies extending into the Tropics. *Fl.* late April–July; *fr.* June–Aug.

R. triphýllus Thunb. (three-leaved), with procumbent canes pilose when young, pinnate ternate leaves, long-stalked rhombic-rotund terminal leaflet and purple incumbent petals, has appeared in vacant lots in Boston, Mass. (Adv. from Japan)

Subgen. V. Eùbatus Focke * (see p. 819) Blackberries, Mûrier (Que.)

a. Inflorescence a terminal panicle, its branches and pedicels strongly armed with stout and firm flat prickles; plants coarse, trailing, reclining, arching or climbing, the crowns often perennial; the strongly branching canes with flattened prickles.
  Leaves white-felted beneath; leaflets not deeply dissected; calyx-lobes unarmed, without long appendages; petals uncleft. . . . . § 3. Discolores (see p. 824).
  Leaves greenish or but slightly grayish beneath; leaflets cleft into lacerate segments; calyx-lobes prickly, with prolonged appendages; petals 3-lobed. § 4. Sylvatici (see p. 824).

a. Inflorescence of 1–many flowers, if several-flowered corymbose, cymose or racemose, if slightly paniculate without strong prickles on the pedicels; crowns and canes biennial; prickles terete. . . *b.*
  b. Floricanes and mature primocanes habitually trailing, creeping, low-arching or forming domes (care is necessary to exclude normally ascending species with floricanes depressed through weight of snow); tips of primocanes or their branches prolonging along surface of ground, commonly tip-rooting; flowers in corymbs or short racemes or solitary. . . *c.*
    c. Floricanes and mature primocanes trailing or very low-arching and trailing, the flowering shoots erect or suberect from the prostrate floricane, the branching thus appearing unilateral. . . *d.*

* Taxonomically a most difficult group. Our few original wide-ranging, essentially unvarying and ancient species have greatly commingled since extensive clearing of the land and have crossed, producing somewhat localized but rapidly spreading offspring which largely breed true, either as fertile or apomictic or allopolyploid (often even cleistogamous) abruptly established incipient "species." As in Europe, the latter will greatly increase with time and new crossings. In the earlier sections (§§ 3–8) the present treatment is probably too liberal; in the later sections (§§ 9–14) fewer minor and perhaps only temporary trends or local phases or states are accepted as fixed species, for it was soon found that the author of most of them was following his stated personal philosophy: "when I write 'new species' (or *species nova*) I do not use the term in its old formal final sense." For many very local and other proposed "species" not here treated see L. H. Bailey's numerous publications in Gentes Herbarum. In collecting, careful note should always be made of habit: whether the new canes (*primocanes*) and the flowering and fruiting canes (*floricanes* or the preceding year's primocanes) are prostrate, doming or arching and tip-rooting, or with no rooting tips. Without such careful records and adequate material of both primocanes and floricanes identification is nearly impossible.

d. Primocanes and usually floricanes with strong mostly scattered prickles. . . *e.*

e. Leaves somewhat evergreen, persistent over winter, coriaceous, glabrous; leaflets lanceolate to oblong, those of primocanes only 1–3 cm. broad; primocanes slender, terete, bearing glandular bristles among the hard curved prickles; pedicels glabrous except for the glandular bristles; plants of s. Coastal Plain. . . . . . § 5. VEROTRIVIALES (see p. 824).

e. Leaves deciduous (evergreen in some members of § *Hispidi* with broad leaflets and pilose pedicels), glabrous or pubescent beneath; leaflets of primocanes ovate, elliptic, rhombic or obovate, mostly broader; primocanes angled or corrugated or, if terete, not glandular; pedicels mostly pubescent.

Canes with whitish bloom, terete, with prickles not restricted to the angles; fruit with a bloom. . . . . § 6. TRIVIALES (see p. 825).

Canes becoming ligneous, without bloom, the stronger ones angled, with prickles mostly along the angles; longer pedicels (0.5–)3–12 cm. long; calyx-lobes (excluding terminal slender beak-like tip sometimes present) 4–12 mm. long; petals 1–2.5 cm. long, 3–18 mm. broad, usually one-half to three-fourths as broad as long; flowers mostly from mid-April to early July; fruit without bloom, juicy and mostly of rich flavor, early-ripening (mostly mid-May to Aug.), 1–2 cm. in diameter. . . . . § 7. FLAGELLARES (see p. 825).

d. Primocanes and closely trailing floricanes hispid or bristly, often over the surfaces, without any or many strong prickles; longest pedicels 0.5– rarely 4 cm. long; calyx-lobes 2.5–6 mm. long; petals 5–12 mm. long, 2–6 mm. broad, mostly much less than half as wide as long; flowers mostly from late May to early Sept.; fruit late-ripening (mostly from mid-Aug. to Oct.), 6–15 mm. in diameter, mostly sour or inferior. . . . . §8. HISPIDI (see p. 834).

c. Primocanes and usually the floricanes doming or high-arching 0.5–2 m. or more above surface of ground, the tips of branches of the former soon reaching the ground and then trailing and often tip-rooting; flowers mostly from April–Aug.; longer pedicels 0.5–8 cm. long; calyx-lobes 3–12 mm. long; petals 5–15 mm. long, 2–10 mm. broad; fruit inferior to rich and juicy, promptly ripening (ripe from June to early Sept.), 0.6–2 cm. in diameter. . . . . § 9. THOLIFORMES (see p. 838).

b. Floricanes and primocanes erect to arched-ascending or becoming depressed over winter, not usually tip-rooting nor with trailing tips; inflorescences racemiform or corymbiform. . . *f.*

f. Canes and other parts bearing bristles or slender prickles and usually no broad-based claw-like prickles, the canes 0.2–1.5 m. high (rarely taller); leaflets mostly plicate, not very flat; inflorescences corymbiform to short-racemiform; calyx-lobes with blades 3–7.5 mm. long; petals 0.7–1.5 cm. long, 2–8 mm. broad; relatively low soft-stemmed or barely ligneous shrubs with inferior subglobose fruit 0.7–1.5 cm. in diameter; mostly northward; flowers mid-June–mid-Aug.; fruit late July–Sept. . § 10. SETOSI (see p. 846).

f. Canes not bristly, either bearing broad-based and often hooked prickles or prickleless; the floricanes subligneous and dark-barked, of medium height to tall; inflorescences racemose or corymbiform; fruit usually of good to superior quality. . . *g.*

g. Primocanes closely canescent-pilose, relatively low, stiff and rigidly prickly; leaflets white- or gray-tomentulose beneath, firm to subcoriaceous, those of primocanes mostly only 3–10 cm. long and obtuse to short (not long) -acuminate; flowers 1–few in leafy-bracted cymes, in latest spring and early summer; fruit from June-Sept.; section of southern range. . . . . § 11. CUNEIFOLII (see p. 851).

g. Primocanes not closely canescent-pilose, of low or medium to tall stature and diverse habit; leaflets green to grayish beneath, mostly membranaceous, the terminal ones of primocanes mostly longer and often acuminate; flowers in racemes or corymbs, mostly earlier than in § 11. . . *h*.

h. Canes glabrous or essentially so, without or with some straight prickles; leaves glabrous above, green and glabrous or but slightly pubescent beneath (not velvety to touch); rachis of inflorescence and pedicels smooth or nearly so, without glands; flowers in June and July; fruit ripe in Aug. and early Sept.; northern and montane. § 12. CANADENSES (see p. 852).

h. Canes more or less pubescent when young, or glabrous, mostly with strong and hooked or broad-based prickles; leaves pubescent on one or both surfaces; rachis and pedicels mostly pubescent or glandular; wide-ranging sections.

Inflorescence and young parts conspicuously stipitate-glandular; primary inflorescence a definite elongate raceme or a corymbiform (or sometimes subpaniculate) raceme; flowers May–July; fruit Aug., Sept. . . . . § 13. ALLEGHENIENSES (see p. 853).

Inflorescence and young growth glandless (rarely with few inconspicuous glands); inflorescences short-racemose or corymbiform; flowers late April–mid-July; fruit June–Sept. (–Oct.). . . . . § 14. ARGUTI (see p. 856).

§ 3. DISCOLÒRES P. J. Muell. (see p. 822) Two European species natzd.:

Canes canescent toward tips; panicle thyrsiform-subcylindric, its prickles falcate. 11. *R. procerus*.
Canes nearly or quite glabrous; panicle, when well developed, pyramidal, its prickles straight. . . . . 12. *R. bifrons*.

11. R. PROCÈRUS P. J. Muell. (elongate), HIMALAYA-BERRY. — Coarse, with climbing, arching or trailing freely branched and quadrate *flat-sided canes* more or less *canescent toward tips, bearing* broad-based *curving* strong *flat prickles; leaves strongly whitened beneath with dense felty tomentum;* primocane-foliage 5-foliolate, the terminal long- and prickly-stalked leaflet ovate, double-serrate; *inflorescence a thyrsiform subcylindric panicle, with very strong falcate flat prickles up to* 1 *cm. long; calyx* canescent-tomentose, *its unarmed lobes without long appendages;* petals rounded-ovate, short-clawed, roseate to white; fruit subglobose or slightly elongate, with large black drupelets. — Roadsides and waste places, spread from cult., Del. to Va. (Introd. from Eu.)

12. R. BÌFRONS Vest (having two kinds of leaf; the lower quinate, the upper ternate). — Similar to no. 11; coarse *canes glabrous; prickles straight; well developed inflorescence pyramidal, its strong prickles straight.* — Roadsides, waste places and old house-sites, Fla. to La., n., locally, to R.I., Tenn. and Mo. *Fl.* May–July; *fr.* July, Aug. (Introd. from Eu.)

§ 4. SYLVÀTICI P. J. Muell. (see p. 822) One European species naturalized:

13. R. LACINIÀTUS Willd. (slashed), CUT-LEAVED BLACKBERRY. — More slender, with glabrescent canes, armed with broad-based curved prickles; *leaves greenish or but slightly grayish beneath, the leaflets pinnately cleft into small leaflets and segments;* inflorescence corymbiform or short-paniculate, armed with falcate flat prickles; *calyx armed with sharp slender prickles, its lobes with long narrow foliaceous appendages; petals 3-lobed.* — Roadsides, borders of groves and waste places, esc. from cult., Mass. to Mich., and southw. *Fl.* June–Aug.; *fr.* July–Sept. (Cult. and natzd., of Old World origin.)

§ 5. VEROTRIVIÀLES Bailey (§ *Triviales* Rydb., not P. J. Muell.; § *Persistentes* Fern.) (see p. 823)

Chiefly of the southern Coastal Plain, represented with us by

14. R. triviàlis Michx. (ordinary), SOUTHERN DEWBERRY. — Trailing or low-arching and soon depressed, with *slender* tough *terete canes, tip-rooting; primocanes* and usually floricanes *bearing glandular bristles among the hard curved prickles; leaves of primocanes coriaceous, somewhat evergreen;* the usually 5 *glabrous leaflets oblong to lanceolate,* coarsely toothed, 2–10 cm. long, 1–3 cm. broad; leaflets of floricanes smaller; *flowers 1–few on erect* filiform *bristly or glandular otherwise glabrous long pedicels;* calyx glabrous or glabrescent on back or glandular; petals obovate, 7–10 mm. broad; fruit subglobose to elongate, black. (Incl. *R. rubrisetus* Rydb. and

R. continentalis (Focke) Bailey) — Low to dry grounds of Coastal Plain, Fla. to Tex., n. to Md., Mo. and Okla. *Fl.* April, May; *fr.* late May, June.

§ 6. TRIVIÀLES P. J. Muell. (see p. 823) European section; one species introd. with us:

15. R. CAÈSIUS L. (bluish-gray), European DEWBERRY. — Low-arching or trailing, *with glaucous or pruinose terete canes* bearing weak prickles; primocane-leaves 3-foliolate, the thin slightly pubescent leaflets broad- or rhombic-ovate; inflorescence a corymb; pedicels glandular and sparingly bristly; calyx-lobes abruptly long-acuminate, reflexed in anthesis but appressed to the fruit; petals suborbicular or broad-ovate, white; *fruit* black, *with a bloom.* — Spreading from cult., local. (Introd. from Eu.)

§ 7. FLAGELLÀRES Bailey (§ *Procumbentes* Rydb.) (see p. 823) DEWBERRIES. A large group of eastern North America.

a. Pedicels of the 1–15(–20) flowers and fruits nearly erect, all but the short central one elongating to 2–12 cm.; mostly southern, a few reaching s. Can. . . *b.*

b. Canes becoming strongly woody, 3–6 mm. thick at base; leaves mostly firm (except in shade); well-developed corymbs (excluding those of weaker shoots and attenuated tip) mostly 2–9(–20)-flowered; longer pedicels 2–12 cm. long. . . *c.*

c. Pedicels, rachis, petioles and usually the young stem glandless or essentially so. . . *d.*

d. Mature leaflets of both primocane and floricane glabrous or merely appressed-pilose or puberulent on veins beneath, not soft nor velvety to touch. . . *e.*

e. Pedicels and petioles of bracteal leaves glabrous to closely appressed-pilose or puberulent, rarely with any spreading trichomes.

Terminal leaflet of primocane-foliage ovate to subelliptic, abruptly terminated by a narrow acumination 0.5–2 cm. long; simple upper bracts of corymb ovate, obovate or lanceolate, mostly much longer than broad.

Canes simply forking, with recurving prickles 1–4 mm. long; terminal leaflet of primocane-foliage usually abruptly contracted or shouldered below the acumination; petals white; wide-ranging in dry soil. . . . . 16. *R. flagellaris.*

Canes intricately branching and entangled, with straight divergent prickles 4–6 mm. long; terminal leaflet gradually tapering; fresh petals roseate; of wooded swamps, se. Va. 17. *R. plexus.*

Terminal leaflet of primocane-foliage roundish-elliptic to suborbicular, short-acuminate or merely acutish; simple bracteal leaves of corymb round-ovate, about as broad as long. . 18. *R. felix.*

e. Pedicels and petioles (at least at anthesis) of bracteal leaves densely spreading-villous.

Leaves submembranaceous; leaflets serrate or sharply dentate; terminal leaflet or simple blade of bracts of corymb acute to acuminate, 2–7 cm. long; larger corymbs 2–15(–20)-flowered, their pedicels becoming 2–7 cm. long.

Canes almost prickleless, the remote prickles (when present) subulate and 1–2.5 mm. long; petioles and pedicels unarmed or with remote bristles 1–2 mm. long; leaflets coarsely dentate or dentate-serrate with broad simple or but slightly notched teeth; larger corymbs 2–6-flowered. . . . . 19. *R. nefrens.*

Canes with copious strong prickles often 2–4 mm. long; petioles of primocane-leaves, upper axis of flowering branchlets and pedicels often similarly prickly; leaflets sharply and often deeply double-serrate; larger corymbs 2–15(–20)-flowered. . 20. *R. tetricus.*

Leaves coriaceous, their leaflets coarsely blunt-dentate; terminal leaflet or simple bract of flowering branchlet obtuse or bluntish, 2–4 cm. long; larger corymbs 2–6-flowered, their lower pedicels 2.5–4 cm. long. . . . . 21. *R. celer.*

d. Mature leaflets of both primocane and floricane pilose on surfaces beneath, soft or velvety to touch. . . *f.*

f. Principal leaflets or simple bracteal leaves mostly one-half to essentially as broad as long, with regular or nearly regular teeth 3–5 mm. long. . . *g.*

g. Upper simple broad bracts overtopping or only slightly overlapped by their subtended pedicels.

Leaves sharply and doubly serrate; terminal primocane-leaflet elliptic-oval; principal leaflets of floricanes cuneate at base; pedicels scarcely villous. . . . . 22. *R. meracus.*

Leaves dentate or serrate-dentate; terminal primocane-leaflet broadly ovate; principal leaflets of floricanes rounded at base; flowering pedicels copiously villous. . . . . 23. *R. occidualis.*

g. Upper single broad bracts greatly overtopped by their subtended pedicels.

Principal terminal leaflets of primocane- and floricane-leaves suborbicular to round-ovate, nearly or quite as broad as long. 24. *R. Jaysmithii.*

Principal terminal leaflets of primocane- and floricane-leaves definitely much longer than broad.

Pedicels copiously spreading-villous during anthesis. . . . 25. *R. roribaccus.*

Pedicels glabrous or essentially so.

Leaves of primocane 3-foliolate, coarsely dentate, leaflets merely acutish; the terminal one subtruncate or subcordate at base, broadly ovate. . . . . . . 26. *R. temerarius.*

Leaves of primocane 5-foliolate, finely sharp-serrate; leaflets caudate-acuminate, the terminal one elliptic-oval. . . . 27. *R. injunctus.*

f. Principal leaflets or simple bracteal leaves mostly less than half as broad as long, lacerate-serrate with many teeth 6–8 mm. long; upper pedicels greatly overtopping their subtending bracts. . . 28. *R. Maltei.*

c. Pedicels, rachis or petioles and often young canes glandular (when old or weathered often merely glutinous). . . *h.*

h. Lower faces of leaves glabrous; appressed hairs only on nerves.

Primocanes with glandular hairs among the many slender prickles; their leaflets mostly 5.

Pedicels only slightly to barely overtopping their subtending bracts; calyx bristly; petioles of primocane-leaves copiously glandular, the terminal leaflet with strong shoulders and abrupt tip; plant of se. Va. and southw. . . . . 29. *R. iniens.*

Pedicels mostly twice to thrice length of subtending bracts; calyx scarcely bristly; petioles of primocane-leaves glabrous or nearly so, the terminal leaflet regularly tapering to acuminate tip; plant of n. Ind. . . . . 30. *R. profusiflorus.*

Primocanes glandless (except sometimes at expanding tip), glabrous, remotely prickly; primocane-leaflets 3.

Terminal primocane-leaflet abruptly long-acuminate; pedicels mostly overtopping bracts; plant of Ky. . . . . . . 31. *R. kentuckiensis.*

Terminal primocane-leaflet only gradually tapering to subacute tip; pedicels mostly shorter than bracts; plant of N.J. and e. Md. . . . . 32. *R. depavitus.*

h. Lower faces of leaves soft-pilose or velvety to touch.

Terminal leaflet of primocane-leaves with long narrowly acuminate tip.

Pedicels shorter than to barely longer than their subtending bracts; upland species, N.Y. to Va. . . . . 33. *R. invisus.*

Pedicels mostly much overtopping subtending bracts.

Leaflets of primocanes 3, the petioles glandular; axis of flowering branchlets and calyx glandular; plant of O., Ind., Ky. and Tenn. . . . . 34. *R. Rosagnetis.*

Leaflets of primocanes mostly 5, the petioles nearly glandless; axis of flowering branchlets and calyx essentially glandless; plant of e. Va. . . . . 35. *R. clarus.*

Terminal leaflet of primocane-leaves gradually tapering to merely acute apex.

Primocane-leaves 3-foliolate, grayish-pubescent above, the round-ovate leaflets nearly as broad as long; species of Ind. . . . . 36. *R. centralis.*

Primocane-leaves 5-foliolate, glabrous or promptly glabrate above, the median ovate to subelliptic leaflets much longer than broad; plant of Md. . . . . 37. *R. redundans.*

b. Canes relatively slender, mostly flexible and only slightly woody, 1–5 mm. thick near base (sometimes remaining ascendant but subsimple and arching to rooting tips, then separated from § *Tholiformes* by the slender subsimple canes); leaves membranaceous; flower usually 1 (exceptionally 2 or 3) on erect filiform pedicel (weakened or terminal 1-flowered branchlets of normally 3–∞-flowered species of the parallel "*b*" not here included). . . *i*.

i. Petioles and pedicels appressed-pilose, puberulent or glabrate, mostly glandless.

Floricanes very slender, 1–3 mm. thick, arching to trailing; primocanes suberect, prickleless or with remote bristleform weak prickles only 0.5–2 mm. long. . . . . 38. *R. Enslenii.*

Floricanes and mature primocanes relatively stout, 2.5–5 mm. thick, commonly with stout hooked prickles 1.5–3 mm. long, the floricanes closely trailing. . . . . 39. *R. Baileyanus.*

i. Petioles (young) and pedicels spreading-villous, often with divergent stipitate glands.

Primocane-leaves mostly 3-foliolate, the lower leaflets subsessile.

Terminal leaflet of most of the 3-foliolate leaves of flowering shoots one-half to three-fourths as broad as long; canes subrigid, 2–3.5 mm. thick toward base, reddish-brown or darker, with broad-based hooked prickles mostly 1.5–4 mm. long; Mass. southw. . . . . 40. *R. scambens.*

Terminal leaflet of most 3-foliolate leaves of flowering shoots one-third to barely half as broad as long; canes flexible, cord-like, subherbaceous, greenish, tardily brownish, rarely 3 mm. thick, their few acicular prickles 1–2 mm. long; se. Va. southw. . . . . . 41. *R. leviculus.*

Primocane-leaves mostly 5-foliolate, the median leaflets on petiolules 0.6–1.8 cm. long.

Petioles, axis of flowering shoot and pedicels heavily pubescent; leaflets soft-pubescent; upland species. . . . . 42. *R. particularis.*

Petioles, axis of flowering shoot and pedicels promptly glabrate; leaflets glabrate beneath; Coastal Plain species. . . . . . . 43. *R. longipes.*

a. Pedicels of the mostly 2–20(–40) flowers chiefly loosely ascending to somewhat spreading (only exceptionally erect), mostly 0.5–4 (rarely –6) cm. long; mostly of Can. and N. states. . . *j*.

j. Primocanes with diverse prickles, some coarse and stout, others more slender and with many setae scattered over the surface. Such plants, sometimes placed in this section, should be sought in § *Hispidi*.

j. Primocanes without fine bristles scattered over the surfaces. . . *k*.

k. Leaves soft-pilose or velvety on surface beneath. . . *l*.

l. Terminal leaflet of more typical primocane-leaves abruptly short-acuminate, the blade often broadest near or above the middle.

Terminal leaflet of primocane-leaves round-ovate to suborbicular below the abrupt tip, essentially as broad as long.

Margins of primocane-leaflets sharply double-serrate; northeastern. . . . . 44. *R. arenicola.*

Margins of primocane-leaflets coarsely dentate; Kentuckian. . . . 45. *R. Whartoniae.*

Terminal leaflet of primocane-leaves more narrowly ovate, one-half to five-sevenths as broad as long, sharply serrate; southeastern. . 46. *R. hypolasius.*

l. Terminal leaflet of more typical primocane-leaves gradually tapering from ovate or ovate-elliptic base to long narrow acumination. . *m*.

m. Terminal leaflet of primocane-leaf round-cordate, excluding the apical acumination as broad as long; northwestern. . . . . . 47. *R. foliofiorus.*

m. Terminal leaflet of primocane-leaf narrowly to broadly ovate, one-half to three-fourths as broad as long, gradually rounded to only shallowly cordate at base. . . *n*.

n. Pedicels and petioles of floricane-leaves and axis of flowering shoot copiously glandular; prickles of cane acicular, 1–2 mm. long; plant of s. Mich. . . . . 48. *R. vagus.*

n. Pedicels, petioles and axis of flowering shoot glandless or with only occasional glands among the trichomes; prickles of canes mostly larger. . . *o*.

o. Canes bearing 60–100 or more broad-based prickles per dm.; axis of flowering shoot and pedicels copiously prickly; plant of s. Mich. . . . . 49. *R. conabilis.*

o. Canes bearing 0–50 prickles per dm.; axis of flowering shoot and pedicels prickleless or prickly. . . *p*.

p. Leaflets jagged-serrate or incised, many of the lance-attenuate teeth 4–8 mm. long; plant of Great Lakes reg. . . . 50. *R. michiganensis.*

p. Leaflets dentate with large teeth, or regularly serrate with the triangular (but often double) teeth mostly 2–3.5 mm. long; plants of Atl. states. . . *q.*

q. Leaflets serrate, with the teeth mostly 2–3.5 mm. long, the sharp teeth of floricane-leaflets only 1.5–4 mm. broad at base.

Terminal leaflet of primocane-leaves narrowly ovate, gradually arching to tip from below the middle and merely rounded to subcordate at base; villous pedicels mostly with scattered stipitate glands intermixed.

Pedicels and axis of flowering shoot with numerous fine prickles; southeastern. . . . . 51. *R. Grimesii.*

Pedicels and axis of flowering shoot prickleless or with only 1 or 2 prickles; of s. N.E. . . . . 52. *R. curtipes.*

Terminal leaflet of primocane-leaves broadly ovate, tapering from near or above middle to the caudate tip, cordate-based; strongly villous pedicels glandless or nearly so and mostly prickleless. . . . . . . 53. *R. cordifrons.*

q. Leaflets coarsely lobulate-dentate, the teeth mostly 3–7 mm. long, those of floricane-leaflets blunt and 4–6 mm. broad at base; southeastern. . . . . . . 54. *R. imperiorum.*

k. Leaves glabrous or merely pilose along veins beneath, not velvety to touch. . . *r.*

r. Pedicels glandless or essentially so, glabrescent, puberulent or incurved-pilose.

Leaves opaque; the leaflets flattish, with veins not impressed above. 55. *R. recurvicaulis.*

Leaves sublustrous; the leaflets with plicate or puckered margins, their principal veins impressed above and very prominent and riblike beneath. . . . . 56. *R. plicatifolius.*

r. Pedicels obviously glandular and spreading villous.

Leaflets of both primocane and floricane gradually long-acuminate, sharply and mostly doubly serrate; corymb 5–17-flowered, all but lowest pedicels subtended by reduced bracts. . . . . . 57. *R. licens.*

Leaflets of primocane abruptly short-tipped, of floricane blunt or short-tipped, the teeth of both kinds of leaves dentate or dentate-serrate; corymbs (1–) 5-flowered.

Canes slender, 2–3 mm. in diameter, prickly; floricane-leaflets acute; flowering shoot spreading-villous and glandular; flowers not overtopping the leafy bracts. . . . . 58. *R. indianensis.*

Canes stout, 4–8 mm. thick, smooth or with caducous small prickles; floricane-leaflets obtuse; flowering shoot glabrous; flowers much overtopping their greatly reduced bracts. . . . 59. *R. jactus.*

16. R. flagellàris Willd., at least sensu Bailey (like a whip-lash). — *Very long-creeping and prostrate,* the canes eventually flat-trailing and *usually rooting at tips. Primocanes up to* 6 *mm. thick at base,* at first ascending, later depressed, becoming 2–5 m. long, slender toward tips, *glabrous* and with firm subulate but thick-based scattered and slightly recurved prickles 1–4 mm. long along the angles; *leaves* on slender *subglabrous or merely puberulent or appressed-pilose* prickleless or sparsely prickly *petioles; leaflets* 3 or 5, serrate or dentate, *glabrous or with the veins beneath appressed-pilose;* the *terminal leaflet ovate to subelliptic,* rounded to base, *rather abruptly contracted* (often shouldered) *to an acuminate tip* 0.5–2 *cm. long,* the petiolule elongate, the lateral leaflets subsessile and often asymmetrical, all *glabrous or merely appressed-pilose on nerves beneath. Floricanes* trailing, woody and tough, often reddish or purplish; *flowering branchlets* erect, *those near base producing corymbs of* (2–)3–9 (*sometimes more*) *flowers on nearly erect filiform glabrous to merely puberulent or appressed-pilose* (rarely remotely villous) prickleless or sparsely setose-prickly *pedicels, the lower pedicels* 3–9(–12) *cm. long;* corymbs toward tip of floricane often reduced to 2 (or 1) flowers; *bracteal leaves* 3-foliolate or simple, *glabrous or nearly so beneath and on glabrescent petioles, the terminal leaflet or the simple blade ovate, obovate or lanceolate, longer than broad;* calyx pubescent; *petals* white, oblong to elliptic-oval or obovate, 1–1.5 *cm. long,* 5–10 *mm. broad; fruit* relatively *large,* globose or slightly elongate, *mostly* 1–1.5 *cm. in diameter, usually of rich flavor.* (*R. villosus* of ed. 7, not Ait.; *R. procumbens* sensu many auth., not Muhl. as validated by Bart.; *R. geophilus* Blanch., thinner- and paler-leaved plants with most jagged toothing; *R. alacer* Bailey; *R. fecundus* Bailey, form with floricane-leaflets merely subacute) — Highly variable but scarcely divisible species of dry fields, openings and borders of thickets, w.-centr. Me. and sw. Que. to s. Ont. and Minn., s. beyond our limits. *Fl.* May, June; *fr.* June–Aug.

17. R. pléxus Fern. (entangled). — *Intricately branching and forking*, prostrate; *primocanes* at first erect and strongly angled, soon becoming prostrate and subterete; these *and floricanes bearing straight horizontally divergent broad-based subulate hard prickles 4–6 mm. long; terminal glabrate leaflets or simple bracts* narrowly ovate and *gradually tapering to acuminate tips;* corymbs 3–5-flowered, the minutely pilose sparsely prickly *erect pedicels 3–4 cm. long at anthesis; fresh petals roseate,* 1.3 cm. long, 9 mm. wide. — Wooded swamps, Princess Anne Co., Va. *Fl.* late April, early May; *fr.* unknown.

18. R. fèlix Bailey (fruitful). — Differing from no. 16 in *less acuminate leaflets; those of primocane relatively broader, the terminal one round-elliptic to suborbicular, the simple bracteal blades of the corymb round-ovate and about as broad as long.* (Suggesting illustration of type of *R. flagellaris* Willd., the designation of a new species perhaps not felicitous; *R. maniseesensis* Bailey) — Dry open woods, clearings and banks, s. N.E. to se. Va.

19. R. néfrens Bailey (unable to bite; presumably meaning unable to scratch). — Resembling no. 16 but *with flowering pedicels and petioles of bracteal leaves densely spreading-villous; canes with remote subulate prickles only 1–2.5 mm. long; petioles unarmed or with remote bristles 1–2 mm. long;* primocane-leaves firm-membranaceous, the terminal ovate leaflet acuminate-tipped; *floricane-leaflets merely acute, all leaflets coarsely dentate or dentate-serrate with broad simple or but slightly notched teeth;* larger corymbs with 2–6 flowers on *nearly prickleless pedicels* up to 7 cm. long. (? *R. Steelei* Bailey; *R. multifer* Bailey) — Dry open soil or thin woods, N.Y. to Minn., s. to Va., Tenn. and Okla. *Fl.* May, early June; *fr.* mid-June, July.

20. R. tétricus Bailey (harsh). — Strongly *resembling no. 16 but with densely spreading-villous flowering pedicels and petioles of bracteal leaves; canes with broad-based strong prickles 2–4 mm. long* interspersed with shorter subulate ones; *leaflets with narrow sharp and mostly deeply cleft teeth; petioles, rachis and pedicels commonly with prickles up to 4 mm. long; larger corymbs 2–20-flowered.* — Dry open soil, sw. Me. to Mich., s. to upland Va., O. and Ind. *Fl.* late May, June; *fr.* July, early Aug.

21. R. cèler Bailey (swift; from its rapid growth). — Trailer, differing from nos. 19 and 20 in its more *slender* (*3–4 mm. thick*) *greenish canes* with few short subulate prickles; *leaves coriaceous, gray-green, blunt-dentate; those of flowering branchlets narrowly* obovate, *bluntish, the terminal one or the simple bract only 2–4 cm. long; corymbs* (1–)*2–6-flowered, the longer erect spreading-villous* nearly prickleless *pedicels 2.5–4 cm. long;* petals about 1 cm. long, 5 or 6 mm. wide. — Dry open soil, Md., D.C. and Va. *Fl.* late April, early May; *fr.* late May, June. — Perhaps better placed in the prevailingly 1-flowered series under 2nd "*b*".

22. R. meràcus Bailey (unmixed). — Strongly resembling no. 16 and perhaps a pubescent phase of it; canes slender, about 3 mm. thick, recurved-prickly; *leaves soft-pubescent beneath,* otherwise much as in no. 16, sharply (often doubly serrate); *primocane-leaves* mostly 5-foliolate, the *terminal leaflet elliptic-oval and long-acuminate; pedicels* 3–6, *suberect, glabrous or nearly so,* the lowest becoming 4.5–6.25 cm. long, *the upper equaled or nearly equaled by the subtending bracts.* — Dry slopes, sw. Mich.

23. R. occiduàlis Bailey (western). — Habit of preceding; canes with few to many recurving hard prickles; *leaves soft-pilose beneath, coarsely dentate or serrate-dentate,* with numerous recurving prickles on petiole and petiolule; *primocane-leaves* 3- or 5-foliolate; the *terminal leaflet broadly ovate, with subtruncate or subcordate base; terminal leaflet of floricane-leaves nearly elliptic, merely acutish; erect copiously villous pedicels mostly equaled or overtopped by subtending bracts,* lowest pedicel prolonged. — Dry prairies, open woods and fields, Ill. and Ia., s. to Kan., Mo. and Okla. *Fl.* May, early June.

24. R. Jaysmíthii Bailey (for Stanley Jay Smith, 1915–). — Trailing canes about 5 mm. thick; *leaves soft-pilose beneath; primocane-leaflets mostly* 5, broad, *obtuse or merely acute,* not caudate-tipped; *terminal leaflet suborbicular or round-ovate, about as broad as or broader than long, the petiole mostly prickleless;* floricane-leaflets broadly ovate, bluntish; *pedicels erect, sparingly pilose, mostly prickleless, much overtopping the subtending mostly simple bracts;* petals about 5 mm. broad; fruit up to 2 cm. across. — Dry open soil, w. Ct. to s. Ont., s. to n. Va.

25. R. roribáccus (Bailey) Rydb. (dewberry). — Canes rather slender, 3 or 4 mm. in diameter; *leaves soft-pilose or velvety beneath; primocane-leaves* 3- or 5-foliolate, double-serrate; *their terminal leaflets ovate, rounded or subcordate at base, long-acuminate; petioles of floricane-leaves villous, the terminal leaflet elliptic to slightly obovate;* erect *pedicels* mostly 3–11, *villous during anthesis, the uppermost ones much longer than their small subtending bracts,* the lowest broad bracteal one up to 7.5 cm. long; fruit of fine quality, about 1.5 cm. across. (? *R. frustratus*

Bailey) — Dry woods and openings, e. Mass. to sw. Que., s. to Pa. and Ky. *Fl.* June, early July; *fr.* July, Aug.

26. R. temeràrius Bailey (ill-advised; because originally misidentified). — Similar to no. 25 but *pedicels essentially glabrous; leaflets coarsely dentate, merely acute; primocane-leaflets broadly ovate, the terminal leaflet broadly subtruncate or subcordate at base,* the petiole with numerous acicular prickles; axis of *flowering branchlet pilose; leaflets mostly elliptic; erect glabrous* but bristly *pedicels* about 5 cm. long. — Upland of w. Va.

27. R. injúnctus Bailey (not united, because it "strongly suggests" another species "but distinguishes itself by" stated characters). — Slender trailer, the canes 2.5–4 mm. thick, with straightish retrorse slender prickles about 3.5 mm. long; *primocane-leaves* 5-foliolate, *soft-pilose beneath; the leaflets caudate-acuminate, finely double-serrate,* the *terminal long-petiolulate leaflet elliptic-oval;* terminal acute floricane-leaflet narrowly rhombic-obovate; *corymbs 2–5-flowered, the pilose or villous erect pedicels mostly much longer than their subtending bracts; petals oblanceolate, distant, 5 or 6 mm. broad.* — Rocky slopes, n. W.Va.

28. R. Máltei Bailey (for MALTE OSCAR MALTE, 1880–1934, Swedish-Canadian botanist). — Differing from our other dewberries in its *very narrow and jagged or lacerate-toothed caudate-tipped leaflets;* canes slender, with few setae; *leaves tomentulose beneath; primocane-leaves* 3- or 5-foliolate, *the terminal long-stalked leaflet narrowly ovate, the others lance-oblong; floricane-leaflets lance-oblong to narrowly rhombic,* the *petioles and petiolules spreading-villous; flowers 2–5 on greatly prolonged erect pilose pedicels;* fruit 1–1.5 cm. long. — Sw. Que. and s. Ont., s. to s.-centr. N.Y.

29. R. íniens Bailey (starting on a journey; because of rapidly creeping canes). — Strong trailer with *canes covered by long acicular prickles 3 mm. long, these intermixed with abundant short prickles and stipitate glands; leaves glabrous or glabrescent beneath; primocane-leaves* 3- or 5-foliolate, *with heavily setose and glandular petiole and terminal petiolule;* the *terminal leaflet* obovate, *strongly shouldered and abruptly acuminate; lower (or median in 5-foliolate leaves) leaflets distinctly petiolulate; flowering branchlets copiously glandular on axis, petioles and pedicels;* terminal leaflet and simple bracts subrhombic-elliptic; flowers 1–4, their *erect very glandular pedicels about equaling the subtending bracts; calyx bristly and glandular;* fruit elongate, 1.5 cm. long. — Sandy fields, se. Va.

30. R. profusiflòrus Bailey (profusely flowering). — *Less glandular* than no. 29; *longer and less crowded prickles of primocane 4 mm. long; petioles and petiolules of primocane-leaves glandless; the terminal leaflet elliptic-ovate, only gradually tapering to acuminate tip;* suberect glandular and bristly *pedicels much longer than subtending bracts,* up to 6.5 cm. long; *calyx* glandular, *scarcely bristly.* — Sandy open soil, n. Ind.

31. R. kentuckiénsis Bailey (of Kentucky). — *Canes slender, about 3 mm. thick, glabrous and glandless,* with scattered short bristles; *leaves soft-pubescent beneath, on glabrous and mostly glandless petioles; primocane-leaflets* 3, the lower subsessile, the petiolulate *terminal one rotund-ovate and subtruncate below the abruptly acuminate tip; flowering branchlets glabrous,* remotely short-bristly; erect *pedicels* 3–5, *glandular, much overtopping subtending bracts;* flower rather small, the petals 1–1.2 cm. long and about 5 mm. broad. — Upland of Ky.

32. R. depavìtus Bailey (trampled down). — Coarser than no. 31, habitally resembling nos. 16 and 36; *canes* 5–6 mm. thick, *glabrous, with remote* hard *prickles* about 4 mm. long; *leaves glabrescent beneath; primocane-leaves 3-foliolate,* the *petiole glandless,* the *leaflets gradually tapering to acuminate tips, terminal leaflet elliptic-ovate;* flowering branchlet glabrous; *floricane-leaves with mostly elliptic leaflets;* nearly erect *pedicels glandular, mostly shorter than subtending bracts; calyx often glandular;* petals 1.2–1.5 cm. long, about 8 mm. broad. — Borders of dry woods, N.J. to ne. Md. — Very close to no. 33.

33. R. invìsus (Bailey) Britt. (hateful or detested). — Habitally much like no. 16; canes glabrous, with few divergent or recurved acicular prickles; *leaves soft-pubescent beneath; primocane-leaves 3- or 5-foliolate, the terminal ovate leaflet tapering gradually to long acuminate tip; flowering branchlet glabrescent or slightly glandular, remotely setose; floricane-leaves with narrowly ovate acute leaflets;* suberect *pedicels glandular, mostly shorter than subtending bracts.* (Incl. *R. Masseyi* Bailey and *R. Macdanielsii* Bailey, too difficult to distinguish) — Dry fields and borders of woods, interior N.Y. to upland Va.

34. R. Rosagnètis Bailey (named in 1943 for its discoverer, Sister ROSE AGNES GREENWELL). — *Canes* 3–4 mm. thick, *copiously armed with broad-based and hooked prickles; leaves soft-pilose beneath; primocane-*leaves 3-foliolate, with strongly armed glandless or glandular petioles; *leaflets* sharply double-serrate, *the terminal one ovate and tapering to long acumination; flowering axis and pedicels copiously glandular;* the *strongly ascending pedicels mostly overtopping their subtending bracts;* petals 1.5–2 cm. long, 1 cm. broad. (Incl. *R. Deamii* Bailey which seems to differ chiefly in glandless petioles of primocane-leaves; *R. Gordonii* Bailey) — Dry fields, rocky slopes and thickets, O. and Ind., s. to Tenn.

35. R. clàrus Bailey ("clear; now clarified" after confusion with no. 36). — Very long-running, with *canes* 4–6 *mm. thick, with scattered slender prickles* 3–5 *mm. long; leaves soft-tomentulose beneath; primocane-leaves* 3–5-foliolate, pilose but glabrate above, densely velvety beneath, *the petiole sparsely glandular to glandless; leaflets tapering to long acuminate tip,* the terminal leaflet ovate; *flowering branchlets villous and sparsely glandular;* flowers 1–6 on nearly *erect villous and stipitate-glandular pedicels* up to 6 cm. long, *the upper pedicels well overtopping the subtending bracts; fruit* of richest quality, 1.5–2 *cm. in diameter.* — Sandy thickets, clearings and borders of woods, e. Va. *Fr.* June.

36. R. centràlis Bailey (central). — Differing from nos. 34 and 35 in the weakly armed canes; *leaflets of primocane-leaves rotund-ovate, nearly as broad as long, gradually tapering to short tips, coarsely dentate* as in nos. 19 and 23, *broadly rounded to subcordate at base, grayish-pilose above,* densely soft-pilose beneath, the *petiole pilose; floricane-leaflets mostly oblong;* suberect *glandular and villous pedicels shorter than or but slightly longer than subtending bracts;* petals 7 or 8 mm. broad; fruit 1.5–2 cm. long. — Wooded slopes and openings, Ind.

37. R. redúndans Bailey (superfluous or redundant; because originally illustrated as no. 32, the author changing his mind). — Resembling no. 36; *with more strongly armed canes, the strong prickles* 3 *or* 4 *mm. long; primocane-leaves 5-foliolate, promptly glabrate above; terminal leaflet ovate, two-thirds as broad as long, median leaflets more elliptic.* — Ne. Md. and se. Pa.

38. R. Enslènii Tratt. (for Aloysius Enslen, Austrian botanical explorer of our southeastern states early in 19th century). — The erect but *finally low-arching to trailing canes slenderly cord-like,* 1–3 *mm. thick, prickleless or with remote bristleform weak prickles* 0.5–2 *mm. long; primocanes* glabrous, *flexible, their leaves glabrescent to slightly soft-pilose beneath,* 3-foliolate, *on glabrous or minutely pilose petioles; leaflets thin,* acuminate, serrate; *floricanes varying from ascending and down to* 2 *dm. long to trailing and* 1 *m. or more long;* leaves 3-foliolate or simple, the terminal leaflet elliptic-ovate to slightly obovate; *flower* 1 (exceptionally 2 or 3), *terminal, its glabrescent or appressed-pilose and glandless* (or very sparsely glandular) *pedicel* 1–7 cm. long; *flower* large, *with broad* white *petals* 1.2–1.5 *cm. long;* fruit plump and juicy. (Incl. *R. tenuicaulis* Bailey) — Dry or moist open woods, thickets, etc., e. Mass. to s. Ind., s. beyond our limits. *Fl.* mid-April (southw.)-late June (northw.); *fr.* June–Aug.

39. R. Baileyànus Britt. (for Liberty Hyde Bailey, 1858– , vigorous author of voluminous work on the genus). — Much *coarser than no.* 38; the *canes trailing* and longer, 2.5–5 *mm. thick, bearing stout hooked prickles* 1.5–3 *mm. long;* leaves firmer and darker, glabrescent or merely pilose on veins beneath; primocane-leaves 3- or 5-foliolate, the leaflets coarsely double-serrate; pedicels glandless. (Incl. *R. connixus, Housei* and *Sailori* Bailey) — Dry to moist sandy open woods, thickets and clearings, Mass. to s. Ont. and Ind., s. beyond our limits.

40. R. scámbens Bailey (becoming bow-legged). — Low-arching and trailing; *canes firm, soon becoming reddish-brown or purplish,* 2–3.5 *mm. thick at base, with scattered broad-based hooked prickles mostly* 1.5–4 *mm. long; floricane-leaves 3-foliolate, with the lower leaflets subsessile,* often *plush-like beneath; leaflets caudate-acuminate,* sharply double-serrate, *the terminal one ovate and one-half to three-fourths as broad as long,* the laterals nearly as large but oblique; *flowering shoots more or less stipitate-glandular, their petioles commonly so;* 3-foliolate leaves of flowering shoots with elliptic-oval terminal leaflet at least half as broad as long, the simple bracteal leaves broader; *flower* usually *single, on villous and often glandular pedicel.* — Open woods, thickets and clearings, se. Mass to se. Va. *Fl.* late April–June; *fr.* May–Aug.

41. R. levículus Bailey (very small). — More *delicate* than no. 40; the *flexible and cord-like very prolonged greenish to pale brown canes almost prickleless or with remote bristle-like straight or slightly curved prickles only* 1–2 *mm. long; leaves very thin,* at first pilose but becoming glabrescent beneath; *primocane-leaves* 3-foliolate, *with* the coarsely serrate-dentate *leaflets mostly less than half as broad as long and attenuate,* the lower leaflets cuneate to subsessile base, the terminal one narrowly rhombic- or elliptic-ovate; principal leaves of flowering shoots similar and coarsely serrate-dentate; flowers solitary on sparsely spreading-villous and stipitate-glandular pedicels; fruit very rich and juicy, about 1.5 cm. in diameter. — Thin woods, bottoms and fertile thickets, e. S.C. to se. Va. *Fl.* late April; *fr.* June.

42. R. particulàris Bailey (special or apart from others, "not closely related"). — Relatively coarse, the *strongly armed canes up to* 5 *mm.* thick and bearing stout straight prickles up to 3 or 4 mm. long; *leaves soft-pilose beneath; primocane-leaves* 5-foliolate, *with densely villous petiole and petiolules;* the *median leaflets on elongate petiolules,* elliptic-ovate, these and the broadly elliptic terminal one abruptly caudate-acuminate; *stipules narrowly linear,* 1–1.5 *cm. long; floricane with densely villous flowering shoots,* their leaflets oblong to narrowly elliptic-obovate;

flowers 1 or 2 on erect strongly pubescent pedicels only 1–1.5 cm. long in anthesis. — Upland of W.Va.

43. R. lóngipes Fern. (long-stalked; from the petiolules). — Resembling no. 42 in its *strong canes and prickles but the latter often claw-like; leaves promptly glabrate, as are the at first villous and glandular pedicels; primocane-leaves 5-foliolate, the median* subrhombic-obovate *leaflets long-petiolulate,* the terminal one broadly elliptic and abruptly acuminate; stipules of larger leaves lance-linear, 1.5–2 cm. long; *floricane-leaflets* mostly narrowly obovate and *blunt;* flower 1, on *promptly glabrate pedicel* 1.5–3.5 *cm. long.* — Sandy slopes near Meherrin R., Southampton Co., Va. *Fr.* mid-June.

44. R. arenícola Blanch. (sand-dweller). — Diffusely trailing; primocanes with relatively numerous (35–150 per dm.) *subulate to needle-like slightly recurving prickles* 3–5 *mm. long; primocane-leaflets* 3 or 5, *soft-pilose or velvety beneath,* the more typical ones *round-ovate to nearly suborbicular, with abrupt apical acumination;* the terminal one 3–7 cm. long, *closely double-serrate;* floricane-leaflets smaller, obovate to ovate, coarsely toothed, soft-pubescent beneath, somewhat so above; *corymb closely leafy-bracted, bracts subtending most of the loosely ascending pilose and often prickly rather short* (1–4 *cm.*) *pedicels; bracts often simple,* with fine sharp teeth rarely 3 mm. broad at base; petals large, 7–10 mm. broad; fruit subglobose. (Incl. the hardly separable *R. obsessus* Bailey) — Dry open soil sw. N.S. and sw. Me. to N.Y. and s. N.E. *Fl.* late May–July (–mid-Aug.); *fr.* late June–early Sept. — Spelling corrected from the original *R. arenicolus* of Blanchard.

45. R. Whartòniae Bailey (named in 1943 for its discoverer, Mary E. Wharton). — Habitally similar to no. 44; prickles stouter; *primocane-leaflets coarsely dentate, the terminal one with about* 25 *broad-based low simple or slightly notched teeth; flowers* 3–5, *much overtopping bracteal leaves;* petals 5–7 mm. broad. — Dry shaly soil, e. Ky.

46. R. hypolásius Fern. (hairy beneath). — Superficially suggesting no. 16 but *leaves downy beneath* and shorter pedicels spreading ascending; *canes* stout, *intricately branched,* the main axis up to 5 or 6 mm. thick, *strongly armed with horizontally divergent subulate hard prickles up to* 6 *mm. long; primocane-leaflets* 3 or 5, *scabridulous with sparse hairs above, velvety beneath;* the *terminal leaflet of the main axis ovate to ovate-elliptic, abruptly acuminate,* 4.5–7.5 *cm. long and* 2.5–5 *cm. broad, sharply double-serrate;* floricane-leaves subcoriaceous, blue-green; *corymb with* several flowers on *mostly arching glandless villous pedicels up to* 2 *or* 3 *cm. long, these rarely overtopping the bracts.* — Wet pinelands and peaty swales, se. Va. to Cape May, N.J. *Fr.* June.

47. R. folioflòrus Bailey (with leaves and flowers). — Strongly suggesting no. 44 but with *primocane-leaflets more gradually attenuate to the acumination,* their margins less lobulate and teeth more numerous; *terminal leaflet deeply cordate, round-ovate, lateral leaflets as broad as long;* bracteal leaves greatly overtopping the short (*up to* 2 *cm. long*) *pedicels.* — Dry sand or gravel, se. Minn.

48. R. vàgus Bailey (rambling). — Suggesting no. 44; *canes very slender,* 2–3 *mm. thick,* with *subulate-acicular straight* scattered *prickles* 1–2 *mm. long; primocane*-leaves 3–5-foliolate, rather soft-pilose beneath; the *terminal leaflet* ovate, *tapering from near middle* to acuminate tip, shallowly cordate, with very short double serration; *flowering shoots, petioles and pedicels stipitate-glandular;* terminal leaflet of bract rhombic; *corymb* up to 12-flowered, *most flowers much overtopping bracts, the glandular* and setose *pedicels* mostly divergent and 1.5–2 cm. long; flower small, petals 8 mm. long and only 3 or 4 mm. broad; fruit longer than thick, about 1 cm. broad. — Low woods, s. Mich. — Perhaps better placed in § *Hispidi.*

49. R. conábilis Bailey (difficult). — Strongly *woody canes* 4–5 mm. thick, freely branching, *with* 60–100 *or more strong straight prickles* (*up to* 8 *mm. long*) *per dm.;* petioles very prickly; *primocane-leaflets acuminate from above the middle,* strongly ribbed, soft-pilose beneath, the terminal one ovate and broadly rounded at base; *flowering shoot and arched-ascending to spreading pedicels* (up to 2.8 cm. long) *very prickly, the upper pedicels much exceeding their bracts;* fruit 1–1.5 cm. long. — Meadows and low fields, s. Mich.

50. R. michiganénsis (Card) Bailey (of Michigan). — *Canes with very few setiform prickles* (0–20 *per dm.*)*; leaflets* soft-pilose beneath, *jagged- or incised-serrate with lance-attenuate teeth often* 4–8 *mm. long* and narrow acute sinuses, the tips attenuate-acuminate; flowering shoot villous; *corymb leafy, the simple oval upper bracts exceeding the arched-ascending* villous and often setulose rather *long* (2–4.5 *cm.*) *pedicels;* globular fruit up to 1.5 cm. in diameter, of good quality. — Sands and other light soils, s. Mich., n. Ind. and s. Wisc.

51. R. Grìmesii Bailey (for its discoverer, Earl Jerome Grimes, 1893–1921). — Freely forking tough trailing canes 2–4 mm. thick, with *usually* 10–20 *broad-based* soon uncinate *prickles per dm.;* leaves soft-pilose beneath; *terminal leaflet of primocane narrowly ovate, tapering from near or below middle to long acumination, with short* double *serration,* the *base rounded;*

flowering shoots villous and setulose; corymbs 2–7-flowered; *pedicels* arched-ascending to divergent, 1–3.5 cm. long, *with stipitate glands and setae among the spreading villi; petals* 1.5 *cm. long,* 0.8–1.2 *cm. broad;* fruit 1.5 cm. in diameter, juicy and rich. — Old fields and borders of dry woods, Md. and D.C. to se. Va. *Fl.* April, early May; *fr.* June.

52. R. cúrtipes Bailey (short-stalked; from the pedicels). — Very similar to no. 51; *axis of flowering shoot and short pedicels* (1–2 *cm. long*) *prickleless or with only* 1 *or* 2 *setae and usually glandless* (or "sometimes . . . one or two indifferent glands . . . A well-marked species"). — Open habitats, e. Mass. — Perhaps eventually to be merged with no. 51.

53. R. cỏrdifrons Bailey (with heart-shaped leaf; *i.e.* leaflet). — Very strong and intricately forking low-arching and long-trailing and root-tipping, with *canes up to 8 mm. thick at base* but down to 2 mm. at rooting tips; *canes with scattered broad-based prickles; terminal leaves membranaceous, downy beneath; terminal primocane-leaflet broadly ovate and cordate or subcordate,* in maturity 3–7 cm. broad, *tapering from near or above middle to caudate tip, sharply double-serrate; flowering axis and pedicels villous but mostly without bristles or glands;* corymb 2–7 (at tips of canes often 1)-flowered, the arched-ascending or spreading villous *pedicels* 1–4 cm. long and *mostly overtopped by the bracts.* — Sandy thickets, borders of woods and clearings, N.J. and Pa., s. to Va. and W.Va. *Fl.* early May–mid-June; *fr.* June, July. — A handsome double-flowered plant with stamens largely altered to petals is forma **pleniflỏrus** Fern. (double-flowered).

54. R. imperiỏrum Fern. (of the Dominions). — In habit, stout canes, prickles and pubescence like no. 52; *broad primocane-leaflets coarsely dentate, the dentations* 3–5 *mm. long and broad, the terminal petiolule* 3–3.5 *cm. long; floricane-leaflets lobulate-dentate, the low rounded teeth* 3–7 *mm. long and* 4–6 *mm. broad, the terminal leaflet and simple bracts broadly rhombic;* leafy corymb 2–8-flowered; petals 1.2 cm. long, 8 mm. broad; fruit 1.2 cm. in diameter. — Grassy borders of pine-woods, se. Va. *Fl.* April, early May; *fr.* June.

55. R. recurvicaủlis Blanch. (with recurving stems). — Trailing or soon recurving and prostrate and tip-rooting, up to 2.5 m. long (or in the smallest extremes the primocanes short or tardily developed), *with relatively few* (*or none* in var. *inarmatus* Blanch.) *slenderly subulate straightish prickles; primocane-leaves* firm, *opaque, glabrous* on both faces *or pilose only on nerves beneath,* 5 (more rarely 3)-*foliolate;* the ovate to rhombic-elliptic or oval *leaflets acuminate,* unequally serrate or serrate-dentate, *flat, their principal nerves not impressed above nor unusually prominent beneath;* terminal leaflet subcordate or rounded at base, 3–12 cm. long; floricane-leaves and lower bracts 3-foliolate or 3-lobed, the leaflets from broad-ovate to narrowly obovate; *inflorescences short racemes or racemiform corymbs, leafy-bracted only at base,* mostly 3–15(–40)-flowered; *all but the lowest spreading-ascending to divaricate* glabrescent, puberulent or incurved pilose *pedicels* (1–5 cm. long) *subtended by reduced narrow bracts,* the pedicels unarmed or (in *R. armatus* (Fern.) Bailey) armed with needle-like prickles; calyx sparsely pubescent or glabrescent, its lobes loosely spreading or ascending in fruit; petals 4–9 mm. broad; fruit of rich quality, 1–1.5 cm. in diameter. (Incl. the smallest extremes, *R. oriens* and *rhodinsulanus* of Bailey, at least the former chiefly riparian plants of which the primocanes are often destroyed by action of frost, shifting stones or ice and freshet; also *R. fandus, aptatus, uvidus, bretonis, polybotrys, botruosus* of Bailey, and some others) — Dry or gravelly open habitats, s. Nfld. to Ont., s. to N.S., s. N.E., N.Y., Mich. and Wisc. *Fl.* June–Aug.; *fr.* mid-July–Sept.

56. R. plicatifỏlius Blanch. (with leaves plaited). — Differing from no. 55 in its often fulvous-green *leaves with sublustrous* and (when fresh) *strongly plicate leaflets puckered at the edge, the main ribs impressed* (*forming furrows*) *above and very prominent beneath.* (Incl. *R. semierectus* Blanch.) — Dry open soil, N.S. and se. Me. to s. Que., N.Y. and Ct. *Fl.* June–Aug.; *fr.* Aug., Sept.

57. R. lìcens Bailey (unrestrained). — Habitally like nos. 55 and 56; canes rather slender, bearing about 20 prickles per dm.; *leaves membranaceous, with sharp simple or double serration,* the *leaflets tapering to long-acuminate tips; corymb racemiform* as in no. 55, with 5–17 *glandular-villous pedicels, most of the pedicels much exceeding the small bracts.* — Openings, thickets and borders of woods, sw. Que. to s. Mich.

58. R. indianénsis Bailey (of Indiana). — Slender, the short-prickly *canes* 2–3 *mm. in diameter;* primocane-leaves 3-foliolate, the glabrous or glabrescent *oval short-acuminate leaflets coarsely dentate-serrate; flowering shoots and petioles spreading-villous; flowers* 1–4 *on arched ascending or divergent glandular and villous pedicels* 1.5–2.5 cm. long; *foliaceous bracts overtopping pedicels.* — Dry soil, s. Ind.

59. R. jáctus Bailey (thrown out or spreading). — *Stout canes smooth or with caducous prickles,* 4–8 *mm. in diameter;* smooth *primocane-leaflets* ovate, *abruptly tipped, dentate; flowering shoots glabrous; floricane-leaves with elliptic-oblong or -obovate blunt leaflets; flowers* 3–5 *on* arched-ascending *glandular pedicels* up to 3 cm. long, *these subtended by very small bracts.* — Nw. Vt., little known; perhaps of some other §.

§ 8. HÍSPIDI Rydb. (see p. 823)

a. Leaves of primocane coriaceous and lustrous (when fresh), dark green or purple- or bronze-tinged, evergreen or nearly so (at least strongly marcescent).

Primocanes soon closely prostrate, 1–5 mm. in diameter, covered by few (or no) to very numerous slender bristles and glands; primocane-leaflets 3 (rarely 5), not prickly beneath; terminal leaflet obovate to rhombic-ovate, blunt-toothed; axis of inflorescence without long bristles; wide-ranging continental species. . . . . 60. *R. hispidus.*

Primocanes arching, soon depressed and vigorously forking, stout, covered with very many hard subulate prickles 3–4.5 mm. long, as well as some glands and bristles; primocane-leaflets with hooked prickles on midrib beneath, 3 or 5; terminal leaflet obovate, elliptic or ovate, sharper-toothed; axis of inflorescence and pedicels prickly; northeastern coastwise species. . . . . 95. *R. arcuans.*

a. Leaves of primocanes membranaceous to subcoriaceous, opaque or nearly so, dark to pale green, sharply toothed or dentate. . . *b.*

b. Leaflets of upper floricane-leaves broadly rounded or obtuse at apex. . . *c.*

c. Primocane-leaflets blunt or merely abruptly short-pointed. . . *d.*

d. Inflorescence a corymb or irregular cyme borne well above the leaves of the flowering shoots; leaflets sharp-serrate. . . *e.*

e. Pedicels glandless.

Primocane-leaflets mostly 5; southeastern.

Leaves submembranaceous, glabrous or promptly glabrate on both surfaces; primocanes with some stipitate glands among the longer bristles and prickles. . . . . . 61. *R. vigil.*

Leaves subcoriaceous, permanently pilose on both surfaces; primocanes glandless. . . . . 62. *R. ambigens.*

Primocane-leaflets mostly 3; canes and pedicels glandless; mid-western. . . . . 63. *R. kalamazoensis.*

e. Pedicels copiously glandular; primocane-leaflets 3 or 5; northern. . 64. *R. Rowleei.*

d. Inflorescence a leafy-bracted interrupted raceme, its strongly glandular and bristly pedicels overtopped by the 3-foliolate and simple bracteal leaves; leaflets dentate; West Virginian. . . . . . . 65. *R. vagulus.*

c. Primocane-leaflets tapering to acute or acuminate tip.

Primocane-leaves 3-foliolate, pubescent beneath, merely acute, 2–3.5 cm. long, broadly ovate; most pedicels subtended by attenuate bracts barely 1 mm. broad; West Virginian. . . . . . . 66. *R. vegrandis.*

Primocane-leaves 3- or 5-foliolate; leaflets glabrous, long-acuminate, mostly 5.5–8 cm. long, ovate-lanceolate or narrowly obovate; most pedicels subtended by elliptic or oblong bracts 2–5 mm. broad; species of Vt. . . . . 67. *R. cubitans.*

b. Leaflets of upper floricane-leaves gradually tapering to acute or acutish tips. *f.*

f. Primocane-leaflets abruptly short-tipped. . . . *g.*

g. Terminal leaflet of 3-foliolate bracteal leaves about half as broad as long.

Leaves glabrous beneath except along nerves, subcoriaceous; northeastern. . . . . 68. *R. trifrons.*

Leaves pubescent beneath, membranaceous; mid-western.

Primocane-leaflets gradually rounded or tapering at base, sharply double-serrate; canes and petioles essentially glandless. . . . 69. *R. plus.*

Primocane-leaflets cordate or subcordate, blunt-serrate; canes and petioles copiously stipitate-glandular. . . . . . . 70. *R. distinctus.*

g. Terminal leaflet of 3-foliolate bracteal leaves nearly as broad as long, firm.

Canes glandless; primocane-leaflets 5 (or 3); plant of n. Wisc. . . 71. *R. Fassettii.*

Canes copiously stipitate-glandular; primocane leaflets 3; plant of montane W. Va. . . . . 72. *R. Huttonii.*

f. Primocane-leaflets gradually tapering to acute or acuminate tips. . . *h.*

h. Leaves glabrous or merely pilose on nerves beneath. . . *i.*

i. Leaflets jagged or lacerate-toothed, with many sinuses 3–5 mm. deep; ribs deeply impressed above; calyx glandular; local in se. Vt. . . 73. *R. Blanchardianus.*

i. Leaflets regularly serrate, double-serrate or serrate-dentate, the sinuses mostly 1–3 mm. deep. . . *j.*

j. Canes remotely prickly or bristly, the surface of the cane clearly evident between the bristles; petioles smooth to sparsely prickly. . . *k.*

k. Terminal primocane-leaflet merely acute, not long-acuminate; leaflets usually only 3; plant of n. Mich. . . . . 74. *R. compos.*

k. Terminal primocane-leaflet acuminate or caudate-tipped.

Primocane glandless or only exceptionally gland-bearing, from nearly prickleless to prickly, the coarsest prickles merely subulate-setiform. . . . . 75. *R. tardatus.*

Primocane copiously glandular.

Canes (especially primocanes) merely bristly and glandular, 2.5–3.5 mm. in diameter; leaves pale green, thin, dull, with deeply impressed ribs. . . . . 76. *R. jacens.*

Canes with many broadish-based hard subulate-conic prickles, stout, 3–7 mm. thick; leaves dark green, glossy, the ribs not impressed. . . . . 77. *R. biformispinus.*

j. Canes very densely hirsute or bristly, the surface obscured; petioles of primocane-leaves densely clothed. . . *l.*

l. Primocane-leaves 5 (sometimes 3)-foliolate; pedicels copiously setiferous; northeastern.

Leaflets of both primocane and floricane only one-half to one-third as broad as long, caudate-tipped; inflorescence a slender raceme, its lower foliaceous bracts all simple. . . . . 78. *R. Parlinii.*

Leaflets broader, rarely less than half as broad as long, acute or acuminate, scarcely caudate; inflorescence a simple to much-branched corymb with lower bracts 3-foliolate. . . . 91. *R. adjacens.*

l. Primocane-leaves 3-foliolate; West Virginian.

Canes slender, about 2 mm. thick; petioles glandless; axis of inflorescence and pedicels with remote short prickles. . . 79. *R. Davisiorum.*

Canes stout, 7–8 mm. thick; petioles glandular; axis of inflorescence and pedicels copiously long-bristly. . . . . 80. *R. zaplutus.*

h. Leaves soft-pubescent or velvety to touch beneath. . . *m.*

m. Canes copiously glandular amongst the bristles.

Petiolule of terminal primocane-leaflet only 4 or 5 mm. long; rachis of inflorescence and pedicels with long spicules; plant of ne. N.Y. . . . . 81. *R. paganus.*

Petiolule of terminal primocane-leaflet 1.5–3.3 cm. long; rachis of inflorescence minutely stipitate-glandular or short-setulose.

Terminal primocane-leaflet on a petiolule 2–3.3 cm. long, the median petiolules (of 5-foliolate leaves) 0.6–2 cm. long; longer leaflets of flowering shoots 5–7 cm. long; plant of se. Mass. . 85. *R. laevior.*

Terminal primocane-leaflet on a petiolule 1.5–2 cm. long, the median petiolule (of 5-foliolate leaves) 6–8 mm. long; longer leaflets of flowering shoots 3–4 cm. long; plant of centr. N. Y. 82. *R. furtivus.*

m. Canes glandless; petiolule of terminal primocane-leaflet 1 cm. long; pedicels minutely glandular; plant of interior N. B. . . . . . 83. *R. emeritus.*

60. R. híspidus L. (bristly). — Trailing, up to 2.5 m. long; canes woody, the stronger portions 2–5 mm. thick; *primocanes with* 300–2000 *bristles and intermixed glands per dm.* on the median and terminal growth; *leaves coriaceous, mostly evergreen, lustrous, dark green, often purple- or bronze-tinged* (especially beneath), *with* 3 (rarely 5) *obtuse to abruptly short-pointed obovate to rhombic-ovate or suborbicular blunt-toothed leaflets;* petioles of principal new primocane-foliage with 100–500 bristles; *terminal leaflet* 2.5–7 *cm. long,* 2–5.5 *cm. broad;* floricanes somewhat less bristly, the leaves smaller, the terminal leaflet 1.5–6 cm. long; inflorescence a lax corymb, cyme or irregular raceme; pedicels spreading or ascending, minutely pilose, with or mostly without stiffish bristles; calyx pubescent, mostly without bristles or glands; *petals* 5–9 mm. *long,* 2–5 mm. broad; fruit seedy, tardily blackening. (Var. *major* Blanch., in part) — Moist or dry open soil, ditches, swales and open woods, P.E.I. to Algoma Distr., Ont., s. to N.S., N.E., Del., Md. and upland to N.C. *Fl.* June–early Sept.; *fr.* mid-Aug.–Oct. Passing into

Var. **obovàlis** (Michx.) Fern. (inverted oval). — More slender; *canes weaker, mostly* 1–2 *mm. in diameter, quite smooth or with* 1–200 *bristles per dm.; petioles of primocane-foliage smooth or with* 1– *rarely* 100 *bristles; terminal leaflet* 1.5–4.5 *cm. long,* 1–3.5 *cm. broad.* — More often in damp habitats, Que. to Wisc., s. to N.S., N.E., L.I., N.C., O., Ind. and Ill.

61. R. vígil Bailey (alert). — Trailing like no. 60, the *canes with* abundant *stiff and resistant bristles mixed with stipitate glands; primocane-foliage submembranaceous, with 5 or 3 abruptly pointed sharply toothed broadly obovate to obliquely ovate dark green leaflets,* the larger leaflets 4–6.5 cm. long; *floricane-leaflets mostly obtuse or rounded at summit,* those at base of loose corymb with terminal leaflet 3–6 cm. long; *pedicels* pilose, with or without a few bristles, 1–4 cm. long, mostly *subtended* by bladeless *linear-lanceolate bracts* 1–2 *mm. broad* and 3–13 mm. long; petals 9–11 mm. long, 4–6 mm. broad; fruit of poor quality. — Wet woods, thickets and swales, D.C. to N.C. *Fl.* June; *fr.* Aug.

62. R. âmbigens Fern. (doubtful). — Resembling no. 61; *expanding foliage white-pilose beneath; primocanes glandless; their subcoriaceous and fulvous leaves* 5-*foliolate, pilose on both surfaces*, the larger leaflets 6–7.5 cm. long. — Wet peaty savannas, pinelands and clearings, se. Va. *Fl.* late May, early June.

63. R. kalamazoénsis Bailey (of Kalamazoo, Mich.). — Resembling no. 60, but with stronger prickles; *leaves thin, dull green, with* 3 (sometimes 5) *blunt glabrous sharply serrate leaflets; canes without glands; pedicels glandless or nearly so, bearing numerous spicules.* — Marshes, s. Mich.

64. R. Rowleei Bailey (for its discoverer, Willard Winfield Rowlee, 1861–1923). — Resembling the three preceding but with *pedicels copiously glandular;* primocanes with mixture of glands, bristles and prickles; *primocane-leaflets* 3 or 5, *the terminal one elliptic.* — Oswego Co., N.Y.

65. R. vágulus Bailey (wandering). — Differing from nos. 61 and 62 in *more abundant glands on the primocane and the flowering shoots; primocane-leaflets* 3, glabrous; the *terminal leaflet broadly elliptic-obovate, with short bluntish tip, coarsely dentate;* 3-foliolate *floricane-leaves with subacute elliptic leaflets, the upper bracteal leaves simple and equaling the very bristly and glandular suberect pedicels of the almost virgate leafy raceme;* fruit very small. — Swamps, W.Va.

66. R. vegrándis Bailey (not very large). — Similar to nos. 61 and 62, but canes with recurving subulate prickles, without bristles; *leaves* dull green, *pubescent beneath; leaflets ovate, acute,* sharply serrate, the larger *one* (at flowering stage) *of the* 3 *primocane-leaflets* 3 *or* 4 *cm. long;* pedicels sparsely spiculose and glandular; petals about 8 mm. long and 4–5 mm. broad. — Sandy soil, ne. W.Va.

67. R. cûbitans Blanch. (lying down). — Similar to no. 61; *primocanes with few stiff bristles,* their *subcoriaceous leaves* 3- or 5-foliolate; *leaflets ovate-lanceolate or narrowly obovate, very long-acuminate, mostly* 5.5–8 *cm. long; floricane-leaves similar but leaflets rounded at summit,* those at base of raceme with terminal leaflet 2.5–3 cm. long; *raceme elongate, with axis* 0.5–1.5 *dm. long; ascending pedicels mostly* 3–8 *cm. long,* unarmed and *subtended by elliptic or oblong bracts* 2–5 *mm. broad.* — Dry woods and clearings, local, se. Vt. *Fl.* late May, early June.

68. R. trîfrons Blanch. (with three leaves; *i.e.* leaflets). — Resembling no. 60 but *with thinner and more acute leaflets and sharp serration; primocane* trailing or low-arching in maturity, relatively stout, with 100–500 *recurving setae or subulate prickles per dm.; primocane-leaflets subcoriaceous, dark green, opaque, with* 3 *or* 5 *broadly ovate to obovate sharp-tipped glabrous leaflets one-half to four-fifths as broad as long, their veins not prominent,* petioles strongly bristly, terminal leaflet 3–8 cm. long; *floricane-leaflets* smaller, mostly *acute, sharply toothed;* inflorescence much as in no. 60, pedicels and calyx sparsely if at all glandular-hispid; *calyx-lobes* 3.5–5 *mm. long; petals* 5–9 *mm. long,* 2.5–4 *mm. broad;* fruit of poor quality. (Incl. *R. alter* Bailey and *R. harmonicus* Bailey) — Rocky, gravelly or peaty open woods and clearings, N.S. to e. Ont. and e. N.Y., s. to s. N.E. *Fl.* late June–early Aug.; *fr.* late Aug., Sept.

Var. **pûdens** (Bailey) Fern. (modest). — Slender, the primocanes and petioles nearly bristleless or with few scattered setae. (*R. pudens* Bailey) — C.B. to e. Vt. and se. Mass.

69. R. plús Bailey (more — but not the last). — Resembling no. 68, but the *membranaceous leaves pubescent beneath;* perhaps a western extreme of no. 68. — Dry thickets, s. Mich.

70. R. distínctus Bailey (distinct). — Coarser than no. 68, the *stout primocanes with many stipitate glands mixed with the divergent bristles and prickles; petioles glandular; primocane-leaflets* 3, *soft-pubescent beneath, broadly round-ovate, the terminal one cordate or subcordate at base;* floricane-leaflets narrow and acute; pedicels glandular. — Low ground, s. Mich.

71. R. Fasséttii Bailey (for its discoverer, Norman Carter Fassett, 1900–). — In habit like no. 68; *primocanes without glands* but with scattered setae or slender prickles; *primocane-leaves subcoriaceous,* of 3 or 5 leaflets; *terminal leaflet round-ovate,* acute, *rounded-subcordate at base; terminal leaflet of* 3-*foliolate round-ovate bracteal leaves three-fourths as broad as long.* — Low sandy soil, n. Wisc.

72. R. Huttònii Bailey (named in 1947 for its discoverer, Eugene E. Hutton, Jr.). — Differing from no. 68 in the *copiously stipitate-glandular slender canes; primocane-leaves coriaceous but opaque, subpersistent,* 3-*foliolate;* the leaflets abruptly short-acuminate, glabrous except on nerves beneath, sharply serrate; *terminal leaflet of* 3-*foliolate bracts elliptic-ovate, acute, the simple upper bracts nearly overtopping the corymb.* — Mts. of W.Va.

73. R. Blanchardiànus Bailey (for its discoverer, William Henry Blanchard, 1850–1922, pioneering and self-sacrificing student of eastern North American *Rubi*). — Characterized by *jagged or lacerate-toothed leaflets with acute sinuses up to* 5 *mm. deep;* primocanes retrorse-bristly and stipitate-glandular; primocane-leaves 3- or 5-foliolate, glabrous, with narrowly ovate acuminate leaflets; corymb short, only the lowest pedicels subtended by foliaceous bracts; calyx glandular. — Windham Co., Vt.

74. R. cómpos Bailey ("master of its domain"). — Forming extensive carpets; *primocanes glandless, with scattered* divergent *bristles; primocane-leaves sublustrous,* dark green, 3 (sometimes 5)*-foliolate, subcoriaceous,* their petioles remotely bristly; *their terminal leaflets rhombic-ovate to -obovate, acute,* 4–7 cm. long, serrate or serrate-dentate; *floricane-leaflets acute; small raceme corymbiform,* the pilose *pedicels* mostly 8–12 *mm. long.* — Open ground, n. Mich. *Fr.* late Aug.

75. R. tardàtus Blanch. (tardy). — Low-arching and trailing, forming extensive implicated carpets; *primocanes* 2–6 *mm. thick, from nearly prickleless to prickly with* 1 *or* 2–180 *acicular prickles* (1–4 *mm. long) per dm., mostly glandless; primocane-leaves* mostly 5*-foliolate, deep green, submembranaceous, glabrous,* with smooth to sparsely prickly petioles, *acuminate to caudate-tipped, elliptic-oblong to rhombic-obovate,* double-serrate; the terminal one 4–11 cm. long, its petiolule 0.5–2.5 cm. long; floricane-leaves relatively large; corymb simple to very compound, the lower glandular *pedicels* 1.5–5 *cm. long;* calyx glandular; petals relatively large, 10–12 mm. long, 3–6 mm. broad. (Incl. *R. rixosus* Bailey, the plants with caudate-tipped primocane-leaflets; and *R. Weatherbyi* Bailey, the almost prickleless plants). — Dry to wet thickets, swales, meadows and clearings, P.E.I. to Algoma Distr., Ont., s. to N.S., N.E., Del., Pa. and Mich. *Fl.* late June–mid-Aug.; *fr.* late Aug., Sept.

76. R. jàcens Blanch. (prostrate). — Differing from no. 75 in its *copiously glandular,* as well as bristly, and more *slender canes* (2.5–3.5 *mm. thick*), these rarely so much branching and prolonged; *primocane-leaves pale green and chartaceous, with lateral ribs impressed above and very prominent beneath,* their *leaflets oblong-ovate* and 6–7 cm. long by 2–3.5 cm. broad; raceme short or corymbiform; *pedicels* with few or no glands or bristles, 1–2 *cm. long; calyx not glandular; petals* 7–10 *mm. long,* 2.5–4 mm. broad. — Dry pastures and clearings, s. N.H. to e. N.Y. and Pa. *Fl.* late June, July; *fr.* Aug., Sept.

77. R. biformispìnus Blanch. (with two forms of prickles). — Coarse trailer; *primocanes* up to 3 m. long, 3–7 *mm. thick,* nearly terete, *with* 40–150 *stout subulate-conic prickles per dm. mixed with more slender prickles and glandless and glandular fine bristles and hairs; primocane-leaves dark green, subcoriaceous, glabrous or with veins beneath sparsely pilose,* 5*-foliolate,* the petioles glandular and prickly; *primocane-leaflets ovate to oblong-obovate,* taper-pointed, closely sharp-serrate; *the terminal one* 7–11 *cm. long, on a petiolule* 1.5–3 *cm. long;* leaflets of floricanes much smaller and narrower, merely acute to bluntish; *inflorescence racemose or racemose-corymbiform,* with axes 4–15 cm. long, leafy-bracted only at base; pedicels loosely ascending or spreading, pilose and often glandular-hispid, 0.5–4 cm. long, mostly subtended by oblong or lance-elliptic bracts; calyx sparsely pubescent or glabrescent on back, glandless or sparingly glandular; petals 4–6 mm. broad. (Incl. *R. particeps* Bailey) — Rocky or sandy clearings, pastures or borders of woods, Que. to N.Y., s. to N.S., s. N.E. and ne. Pa. *Fl.* June, July; *fr.* July–Sept.

78. R. Parlínii Bailey (for its discoverer, John Crawford Parlin, 1863–1948). — *Primocanes densely retrorse-bristly and stipitate-glandular,* ascending, soon reclining, *becoming* 6–8 *mm. thick;* primocane-leaves 5 (or 3)-foliolate, with hispid and glandular petiole; *primocane-leaflets opaque,* glabrous but *with midrib bristly beneath, narrowly lance-oblong or ovate, caudate-acuminate, the terminal one* 8–12 *cm. long and* 4–5 cm. *broad; raceme elongate, most of the long* (4–5.5 *cm.*) *glandular pedicels subtended by lance-acuminate simple foliaceous bracts.* — Sw. Me. to e. Ct.

79. R. Davisiòrum Bailey (for its discoverers, Hannibal Albert, 1899– and Tyreeca Stemple Davis, 1902–). — Very slender trailer with *canes about* 2 *mm. thick;* primocanes closely covered with fine setae and glands up to 3 mm. long; *primocane-leaves* 3*-foliolate on very long hispid petioles, subcoriaceous and lustrous;* the *leaflets narrowly elliptic-ovate, acute,* 4–5 *cm. long,* sharply serrate; *racemes slender and lax; the erect nearly prickleless pedicels glandular and up to* 5 *cm. long,* only the lowest leafy-bracted; *calyx* 5 or 6 mm. long, *glandless and prickleless; petals* 8–10 *mm. long,* 2–4 *mm. broad.* — Woods and openings, Preston Co., W.Va.

80. R. zaplùtus Bailey (very rich; for the "strikingly vigorous roaming" habit). — *Very stout,* trailing for 2–3 m. and strongly branching; *primocanes* 7–8 *mm. thick, densely long-setose and glandular,* the *setae up to* 6 *or* 7 *mm. long; primocane-leaves* 3*-foliolate, dull, on glandular and hirsute petioles; terminal leaflet elliptic-ovate, acuminate,* 9 or 10 cm. long, on a petiolule 1.5 cm. long; lateral leaflets often lobed; *flowering shoots and their petioles and pedicels copiously long-setose;* flowers racemose-corymbose; calyx glandular and hispid. — Preston Co., W.Va.

81. R. pagànus Bailey (of the country). — *Primocanes slender, closely covered with glands and* ~~~~*ded fine bristles 3 or 4 mm. long; primocane-leaves* mostly *5-foliolate, soft-pubescent bene*~~ *5–7 cm. long; terminal leaflet rhombic-obovate, acuminate, on a very short* (4–5 *mm.*) *petiolule;* flowering shoots with larger leaves, these extending well into the corymb; *axis and pedicels spiculose; calyx very glandular.* — Open ground, St. Lawrence Co., N.Y.

82. R. furtìvus Bailey (concealed; the "plant . . . likely to remain furtive or concealed in grass and low herbage"). — *Primocanes subulate-bristly and glandular,* about 4 mm. thick; *primocane-leaves soft-pubescent beneath, 5-foliolate; leaflets narrowly elliptic-ovate, caudate-acuminate,* sharply serrate; *terminal leaflet with petiolule 1.5–2 cm. long; flowering shoots glandular, their largest leaflets 3 or 4 cm. long; corymbiform raceme with glandular rachis and pedicels,* only the lowest pedicels leafy-bracted; calyx glandular; petals 10 mm. long, 4 mm. broad. — Flat "on ground in open sun", centr. N.Y.

83. R. emèritus Bailey (unfit for service). — Resembling no. 75; *canes, petioles and axis of flowering shoot glandless; leaves dull, soft-pubescent beneath; primocane-leaflets* 5, oblong-oval or slightly obovate, *acute, the terminal one about 6 cm. long and on a petiolule 1 cm. long; pedicels* of corymbiform raceme *minutely glandular;* calyx glandular; *petals* 10–12 mm. long, *about* 4 *mm. broad.* — Dry open ground, w.-centr. N.B. *Fl.* late June, July.

§ 9. Tholifórmes Fern. (see p. 823)

a. Primocanes bearing evident gland-tipped hairs, mostly setiferous, without or with hard prickles. . . *b.*

b. Lower surfaces of leaves soft-pilose, more or less velvety to touch. . . *c.*

c. Primocanes with 1000–5000 soft purple hairs and gland-tipped bristles with few prickles intermixed on a dm. of young growth; leaflets of primocane-foliage rhombic-ovate to obovate, abruptly short-pointed; the terminal ones 2.5–6.5 cm. long, two-thirds to nine-tenths as broad, on heavily glandular petiolules only 0.7–2 cm. long. . . . . . . 84. *R. permixtus.*

c. Primocanes with numerous strong prickles mixed with stiffish bristles and glands; leaflets of primocane-foliage more ovate-acuminate; terminal leaflet of primary leaves (not of side-branches) 6–12 cm. long, mostly proportionately narrower, on only sparsely glandular to glandless petiolules 1.5–4.3 cm. long. . . *d.*

d. Blades of calyx-lobes (excluding terminal appendage) 3–4 mm. long, 1.5–3 mm. broad; petals 4–5 mm. broad; terminal leaflets of primocane-foliage 3–5.5 cm. broad.

Glands and prickles of primocane numerous and about equally dispersed; petioles, pedicels and calyces only minutely pilose to glabrescent (except for stalked glands); short corymbiform raceme or open cyme with setose pedicels; terminal leaflet of primocane-foliage one-half to five-sevenths as broad as long; plant of Cape Cod. . . . . 85. *R. laevior.*

Glands of primocanes very few, mostly confined to youngest growth; petioles, etc. villous-tomentose; pedicels of elongate raceme mostly without prickles; terminal leaflet of primocane-foliage barely half as long; plant of upper Ct. Val. . . . . 86. *R. fraternalis.*

d. Blades of calyx-lobes 4–7 mm. long, 3–4.5 mm. broad; petals 6–12 mm. broad; terminal leaflet of primary primocane-foliage 4–8.5 cm. broad. . . *e.*

e. Primocanes with many hundreds of glands, bristles and prickles per dm.; northern species.

Rachis of inflorescence, pedicels, calyx and petiolules sparsely pilose to glabrescent (except for stipitate glands); lower pedicels bearing 3–6 conspicuous overlapping bracteoles; middle leaflets of primocane-foliage 1–1.2 dm. long; plant of Ontario basin. . 87. *R. bracteoliferus.*

Rachis, etc. densely villous-tomentose; pedicels bractless or rarely with a single bracteole; middle leaflets of primocane-foliage mostly 5–9 cm. long.

Primocanes with the often fascicled stalked glands more numerous than the stout prickles; inflorescence elongate-racemose, with strongly prickly rachis; bladeless upper bracts entire or merely toothed; pedicels mostly 1–2(–3) cm. long; Nova Scotian. . . . . 88. *R. adenocaulis.*

Primocanes with strong prickles usually more abundant than scattered glands; inflorescence corymbose, cymose or corymbose-racemose, with prickleless or merely setose rachis; bladeless upper bracts mostly deeply cleft; many pedicels 3–8 cm. long; plant of n. N. H. . . . . 89. *R. aculiferus.*

*e.*Primocanes with relatively few prickles and only remote glands; s. Appalachian. . . . . 90. *R. Boyntoni.*

*b.*Lower surfaces of leaves glabrous, glabrescent or only minutely pubescent, not soft-pilose nor velvety to touch. . . *f.*

*f.*Primocanes usually bearing no broad-based stout prickles, strongly bristly with slender setae or needle-like prickles. . . *g.*

*g.*Foliage of primocanes dark green, lustrous; leaflets abruptly pointed, 3–8 cm. long.

Primocanes slender, 1–4 mm. thick, rarely 1.5 m. long, the surfaces readily visible between the scattered (100-500 per dm.) bristles; leaves membranaceous to subcoriaceous. . . . . . 68. *R. trifrons.*

Primocanes coarse, 3–8 mm. thick, up to 2.5 m. long, the surfaces mostly hidden by the purple or colored densely overlapping bristles (3000–5000 per dm.); leaves coriaceous. . . . 91. *R. adjacens.*

*g.*Foliage of primocane paler green, thinner and duller; leaflets long-acuminate, the terminal one 7–14 cm. long.

Primocanes gray to fuscous with dense coat of fine bristles and stipitate glands; the acicular bristles mostly less than 2 mm. long, horizontally spreading; leaves chartaceous, on copiously glandular petioles, the nerves pilose beneath; pedicels mostly 1–2 cm. long; petals 0.5–1 cm. long, 3–4 mm. broad. . . . . . . 92. *R. tholiformis.*

Primocanes greener, with more scattered reflexed prickles up to 3 mm. long and few glands; leaves firm, on essentially glandless petioles, the nerves beneath glabrous or nearly so; pedicels mostly 2.5–5 cm. long; petals 1.2–1.5 cm. long, 5–9 mm. broad. . . . 93. *R. spiculosus.*

*f.*Primocanes bearing broad-based prickles.

Foliage of primocanes membranaceous, pale or medium green, scarcely lustrous; terminal leaflet cordate-ovate to oval-oblong, taper-pointed, 6–11 cm. long and 3–8.5 cm. broad, finely serrate, on petiolule 1.5–3 cm. long; inflorescence mostly without prickles. . . . 77. *R. biformispinus.*

Foliage of primocanes coriaceous, dark green and lustrous.

Primocane-leaflets with prolonged caudate tips 2–3 cm. long; terminal leaflet cordate, ovate, 1–1.2 dm. long, 6–7 cm. broad; inflorescence not long-bristly; northwestern. . . . . 94. *R. grandidens.*

Primocane-leaflets with abrupt tips about 1 cm. long; terminal leaflet cuneate-obovate to ovate-rotund, 3–8 cm. long, 2.5–5.5 cm. broad, on petiolule 0.5–2 cm. long; inflorescence copiously prickly; northeastern. . . . . 95. *R. arcuans.*

*a.*Primocanes glandless or essentially so, mostly (except in nos. 96 and 119) without fine setae but with hard prickles. *h.*

*h.*Lower surfaces of leaves soft-pilose, more or less velvety to touch. *i.*

*i.*Fuller inflorescences with 1–4 flowers and fruits; southeastern. *j.*

*j.*Terminal primocane-leaflet ovate or lanceolate, regularly fine-serrate, with rounded or cordate base; pedicels essentially glandless.

Prickles acicular, scarcely stout-based; leaflets of median primocane-leaves ovate-lanceolate, one-third as broad as long, the terminal petiolule 5 mm. long; canes 1.5–3.3 mm. in diameter. . . . . 96. *R. obvius.*

Prickles broad-based and strong; leaflets of median primocane-leaves ovate, one-half to two-thirds as broad as long, the terminal petiolule 1.5–3 cm. long; main axis of canes 4–10 mm. in diameter.

Leaves firm or subcoriaceous, sparsely strigose above with hairs scarcely 0.5 mm. long, or glabrate; principal primocane-leaves 3-foliolate; leaflets of bracteal leaves blunt or subacute. . . 97. *R. Akermani.*

Leaves membranaceous, with strigae of upper surface 1 mm. or more long; principal primocane leaves 5-foliolate; leaflets of bracteal leaves acute. . . . . 98. *R. subinnoxius.*

*j.*Terminal primocane-leaflet broadly rhombic-obovate, deeply jagged-lacerate, cuneate at base; pedicels heavily glandular. . . . . . 99. *R. pernagaeus.*

*i.*Fuller inflorescences with (4–)5–15 or more flowers and fruits. . . *k.*

*k.*Pedicels obviously spreading-pilose to densely villous. . . *l.*

*l.*Pedicels definitely setiferous or prickly. . . *m.*

*m.*Prickles of primocane rather distant, mostly 1–2 cm. or more apart, 2–5 mm. long; leaves membranaceous; prickles of primocane-petioles 5–15. . . *n.*

*n.*Terminal primocane-leaflet deeply cordate at base; terminal leaflet or simple blade of bracteal leaves broadly cordate-ovate; inflorescence overtopped by foliaceous bracts; northwestern. 100. *R. minnesolanus.*

*n.*Terminal primocane-leaflet rounder or only shallowly cordate at base; terminal leaflet or simple blade of bracteal leaves not cordate-ovate; inflorescence overtopping the foliaceous bracts; southeastern.

Primocanes with straight horizontal subulate prickles 2–3 mm. broad at base; leaflets of primocane mostly 3, with low dentation; terminal leaflet cordate; petiole of floricane-leaf, pedicels (1–2.5 cm. long) and calyx glandless. . . . . 101. *R. Sewardianus.*

Primocanes with recurving deltoid-subulate prickles 3–6 mm. broad at base; leaflets of primocane 5, with long sharp serration; terminal leaflet not cordate; petiole of floricane-leaf, pedicels (2–4 cm. long) and calyx glandular. . . . . . 102. *R. cathartium.*

*m.*Prickles of stronger parts of primocane approximate, frequently clustered, very hard, mostly less than 1 cm. apart, the larger only 5–8 mm. long; leaves firm; prickles of petioles of larger primocane-leaves 10–25; plant of s. N. E. . . . . . . 103. *R. multispinus.*

*l.*Pedicels prickleless or at most with inconspicuous and tiny spicules among the pubescence. . . *o.*

*o.*Terminal primocane-leaflet barely half as broad as long, usually gradually narrowed to base; plant of w. Ky. . . . . . 104. *R. decor.*

*o.*Terminal primocane-leaflet more than half as broad as long, broadly rounded to cordate at base.

Terminal primocane-leaflet two-thirds as broad as long, deeply cordate, serrate-dentate; median paired leaflets on petiolules 1.5 cm. long; plant of n. N. Y. . . . . 105. *R. satis.*

Terminal primocane leaflet only slightly more than half as broad as long, not deeply cordate, sharp-serrate; the median paired leaflets on petiolules 2–15 mm. long.

Leaflets regularly doubly serrate, not lacerate; terminal primocane-leaflet tapering from near base to acumination; bracteal leaves not lobed; pedicels often glandular. . 106. *R. arundelanus.*

Leaflets lacerate or jagged-toothed; terminal primocane-leaflet tapering from near middle to acumination; some bracteal leaves lobed; pedicels glandless. . . . . . 201. *R. recurvans.*

*k.*Pedicels glabrous or only inconspicuously pilose. . . *p.*

*p.*Primocane-leaflets mostly 3, abruptly short-acuminate; prickles acicular, about 2 mm. long; northwestern. . . . . 107. *R. Rosendahlii.*

*p.*Primocane-leaflets mostly 5, gradually tapering to long acumination, broadened at base, 4–5 mm. long. . . *q.*

*q.*Inflorescence cymose-corymbiform, broader than long; floricane-leaflets blunt or subacute, shallowly serrate-dentate; Missourian. 108. *R. missouricus.*

*q.*Inflorescence racemose or racemose-corymbose, as long as broad; leaflets of floricane acuminate, sharply serrate; eastern.

Terminal primocane-leaflet 5–7 cm. broad, three-fifths to two-thirds as wide as long; longer pedicels 2–2.5 cm. long; calyx-lobes promptly reflexed; plant of s.-centr. N. Y. . . . . 109. *R. pityophilus.*

Terminal primocane-leaflet 2.5–3.7 cm. broad, one-third to scarcely half as broad as long; longer pedicels 2–5 cm. long; calyx-lobes scarcely reflexed in fruit; plant of se. Mass. . . 110. *R. paludivagus.*

*h.*Lower surfaces of leaves glabrous or merely pilose along nerves beneath. . . *r.*

*r.*Rachis and pedicels copiously prickly or bristly.

Primocane-leaflets oblong-lanceolate, less than half as broad as long, gradually tapering at tip; bracts of raceme coarsely dentate. . . 111. *R. Porteri.*

Primocane-leaflets broadly ovate to suborbicular, the terminal one three-fifths to almost as broad as long, abruptly tipped; bracts sharply serrate.

Prickles of primocane very stout and often crowded, broad-based, up to 5–8 mm. long; terminal primocane-leaflet ovate, three-fifths to two-thirds as broad as long; pedicels glandless. . . . . 103. *R. multispinus.*

Prickles of primocane lance-subulate, 2–3 mm. long; terminal leaflet suborbicular, nearly as broad as long; pedicels glandular. . . . 112. *R. eflagellaris.*

*r.*Rachis and pedicels prickleless or rarely with 1–few very short bristles. . *s.*

*s.*Nerves of lower side of leaf glabrous or essentially so.

Primocanes bearing 0–40 prickles per dm.; primocane-leaves membranaceous, pale green, the terminal leaflet narrowly ovate and less than to about half as broad as long; inflorescence typically a raceme. 113. *R. multiformis.*

Primocanes with 150–300 prickles per dm.; primocane-leaves subcoriaceous, dark green, the terminal leaflet broadly ovate to obovate and definitely more than half as broad as long; inflorescence typically corymbiform or loosely cymose. . . . . . . 114. *R. severus.*

*s.*Nerves of lower side of leaf definitely pilose. . *t.*

*t.*Prickles of the canes broad-based and hard, 2–3 mm. wide at base.

Leaflets of primocane abruptly narrowed to short acuminate tips 5–10 mm. long; eastern.

Terminal primocane-leaflet about half as broad as long, its petiolules with many stout prickles; leaflets of bracts subacute. . . . . 115. *R. novanglicus.*

Terminal primocane-leaflet nearly as broad as long, its petiole unarmed or with few prickles; leaflets or blade of bracts sharply acute. . . . . 116. *R. noveboracus.*

Leaflets of primocane gradually tapering to caudate-acuminate tips 1.5–2 cm. long; bracteal leaflets acuminate; plant of s. Mich. . . . . 117. *R. Schoolcraftianus.*

*t.*Prickles of primocane setiform or merely slender-subulate; plants of Mich. and Wisc.

Prickles subulate, 40–75 per dm.; primocane-leaflets gradually tapering, about half as broad as long; bracts mostly simple, broadly ovate; pedicels heavily pubescent. . . . . . . 118. *R. complex.*

Prickles and bristles 300–600 per dm.; primocane-leaflets abruptly tipped, the terminal one four-fifths as broad as long; bracts with 3 narrow slender-tipped leaflets; pedicels smoothish. . . . 119. *R. setospinosus.*

84. R. permíxtus Blanch. (confused). — Arching, up to 7 dm. high, the canes then recurving and tip-rooting or trailing; *primocanes densely covered with soft purple plush-like hairs, mixed with stipitate glands and* relatively *few slender prickles* (1000–5000 *hairs, bristles and prickles per dm. on young growth*); *primocane-leaflets* pale and soft-pubescent (*almost velvety*) *beneath, rhombic-ovate to obovate, abruptly short-pointed; the terminal leaflet* 2.5–6.5 *cm. long, two-thirds to nine-tenths as broad, on heavily glandular petiolules only* 0.7–2 *cm. long;* floricanes more depressed; their leaflets narrowly ovate to obovate, sharply double-serrate, 1.5–5.5 cm. long; *inflorescence racemose, the rachis heavily glandular; bracts mostly elliptic to lanceolate or ovate, often lobed or cleft; pedicels* spreading-ascending, mostly 0.5–2.5 cm. long, *villous-tomentose*, more or less setose and glandular; *calyx villous* and glandular; petals oblong-spatulate, 7–10 mm. long, about 4 mm. broad; fruit about 1 cm. in diameter. — Thickets and openings, sw. Me. to centr. N.Y. *Fl.* June, early July; *fr.* Aug., Sept.

85. R. laèvior (Bailey) Fern. (smoother). — Low-arching or extensively creeping, tip-rooting; *primocanes with scattered* needle-like *prickles, bristles and stipitate glands; primocane-foliage* green on both sides; *the 5 leaflets* oblong to narrowly oval or barely obovate, *acuminate, the terminal leaflets* 6–9 *cm. long* and *on glandless or very sparsely glandular petiolules* 2.3–3.5 *cm. long;* primocane-foliage submembranaceous, with acuminate leaflets mostly 4–8 cm. long; *inflorescence corymbose, corymbose-racemose or loosely cymose, the rachis, pedicels and calyces only minutely pilose to glabrescent;* pedicels mostly 1.5–5 cm. long, setose; *bladeless bracts linear-*

lanceolate, entire; fruit slightly elongate, inferior. — Borders of dry shady woods, Barnstable Co., Mass. *Fr.* Aug.

86. R. fraternàlis Bailèy (closely related). — In habit similar to no. 85; *primocanes with needle-like prickles and only few glands, these chiefly on young growth; primocane-foliage heavily gray-tomentose beneath, the petioles, rachis, pedicels and calyx villous-tomentose; leaflets* narrowly oval, *long-acuminate; the terminal about* 1 *dm. long, barely half as broad,* closely double-serrate, *on glandless petiolule* 3–4 *cm. long; raceme* elongate, often subpaniculate, *the tomentose rachis, pedicels and calyx copiously glandular, mostly without prickles;* petals oblong, nearly 1.5 cm. long, 4–5 mm. wide; fruit globose. — (*R. fraternus* Brain., not Gremli) — Sandy thickets and openings, Ct. Val., N.H. and Vt.

87. R. bracteolíferus Fern. (bearing small bracts). — Coarse, strongly arching or doming and tip-rooting; primocanes 6–10 mm. in diameter, *with many hundreds of strong conic-subulate prickles, bristles and stalked glands per dm.;* leaves velvety beneath, the 5 *ovate* double-serrate *leaflets long-acuminate; terminal leaflet* 1–1.2 *dm. long,* 7–8 *cm. broad,* cordate, *on glandular and prickly petiolule* 3.5–4.3 *cm. long;* floricane-foliage large, the ovate leaflets long-acuminate; *raceme lax, the rachis* 4–11 cm. long and *only sparsely pilose but glandular and prickly;* lower bracts resembling leaflets of foliage-leaves, *upper bracts abruptly reduced; pedicels arcuate-ascending,* 2–4 cm. long, glandular, *the lower bearing* 3–6 *imbricated bracteoles; calyx* short-pilose to glabrescent, *its* glandular-hispid *lobes* 7–10 *mm. long;* fruit subglobose, 1.5 cm. in diameter. — Ontario basin, N.Y. *Fr.* Aug.

88. R. adenocaùlis Fern. (glandular-stemmed). — Arching and forming intricate domes, with the canes and their recurving branches reaching the ground, then trailing and tip-rooting, up to 2 m. long; *primocanes* 5–8 mm. in diameter, *very densely glandular, with crowded or fascicled stipitate glands and fewer straight* broad-based slender *prickles; primocane-foliage firm, soft-pilose beneath,* on glandular and prickly petioles 4–12 cm. long; *leaflets* oval or ovate to oval-obovate, *acuminate; the terminal one* subcordate, 5–9 *cm. long,* 3.5–8.5 cm. broad, on glandular and prickly petiolule 2–4 cm. long; leaflets of floricane-foliage narrowly oval, acuminate; *elongate raceme with rachis, pedicels and calyces densely villous-tomentose, the rachis strongly armed; upper bracts entire or subentire; pedicels* subascending, 1–2(–3) *cm. long;* calyx-lobes 4–6 mm. long; petals elliptic-obovate, 1–1.4 cm. long, 5–10 mm. broad; fruit globose-ovoid, 1.5–2 cm. long. — Rocky or peaty openings and borders of spruce-woods, sw. N.S. *Fl.* July; *fr.* Aug., Sept.

89. R. aculíferus Fern. (bearing prickles). — Resembling no. 88; young primocane at first simple and ascending to 1–2 m., then arching, freely branching and forming intricate domes with branches up to 2 or 3 m. long, their ends trailing and rooting; *main stems with many conic-subulate straight prickles and few glands; primocane-foliage* submembranaceous; the *terminal leaflet* 5–8 *cm. broad, on an essentially glandless* prickly *petiolule* 1.5–3.5 cm. long; leaflets of floricanes often obovate, coarsely toothed; *inflorescence corymbose, cymose or coyrmbose-racemose, with prickleless or merely setose rachis; bladeless upper bracts mostly deeply cleft; many pedicels* 3–8 *cm. long, divergent;* fruit about 1 cm. in diameter, of superior quality. — Borders of dry woods, roadsides and clearings, n. N.H. *Fl.* July, early Aug.; *fr.* late Aug., Sept.

90. R. Boỳntoni Ashe (named in 1903 for its discoverer, FRANK ELLIS BOYNTON). — Similar to no. 77, more arching, with some ascending and other trailing and rooting branches; *primocanes sparsely glandular,* remotely prickly; *leaves velvety-pubescent beneath; primocane-leaves* 5-*foliolate, the terminal broadly cordate-ovate* finely double-serrate acuminate *leaflet* about 1 dm. long and *on a petiolule* 3.5–4 *cm. long; inflorescence racemose or subcorymbose,* leafy-bracted at base, *with copiously glandular* and often prickly *ascending pedicels* 2–5 cm. long; calyx glandular-hispid, its lobes reflexed in fruit; petals very large; fruit elongate. — Uplands of w. Va. and N.C. *Fl.* May; *fr.* late June–Aug.

91. R. ádjacens Fern. (neighboring). — Arching or doming, finally depressed; young *primocanes* at first ascending, then *arching, the tips and branch-tips finally trailing,* sometimes rooting, *becoming* 1–2.5 *m. long,* 3–8 *mm. thick, densely covered with overlapping* purple or colored *long reflexed setae* (3000–5000 *per dm.*) and glands; *primocane-foliage coriaceous, dark green, shining, glabrous,* quinate or ternate, the setose and glandular petiole 6–12 cm. long; *leaflets obovate or rhombic, abruptly short-acuminate,* serrate-dentate; the *terminal leaflet* rhombic-obovate, rounded-cuneate at base, 4–8 cm. long, 2.5–5 cm. broad, *its pilose, glandular and setose petiolule* 0.5–1.8 *cm. long;* leaflets of floricane-foliage narrowly cuneate-obovate, subcoriaceous, acute or acutish, àcutely serrate; inflorescence corymbose, corymbose-racemose or loosely cymose, with the pilose rachis and pedicels bristly and glandular; middle bracts lanceolate, simple or incised; pedicels arcuate-ascending, mostly 1.5–2.5 cm. long; calyx pilose, more or less glandular-setose, its lobes 2.7–5(–6) mm. long; petals narrow, 7–12 mm. long, 2–5 mm. wide; fruit of poor quality, about 1 cm. in diameter. (*R. signatus* and *Fulleri* Bailey) — Peaty or sandy thickets and clearings, s. Que. to Wisc., s. to N.S., Me., Mass., N.Y. and n. Ind. *Fl.* July; *fr.* mid-Aug., Sept.

92. R. tholifórmis Fern. (dome-shaped). — In habit like nos. 88 and 89; canes 3–7 mm. thick, elongating to 1 m., often rooting at tip; *primocanes densely gray- or fuscous-glandular and setose; the bristles divergent and only* 1–2 *mm. long, subrigid; primocane-foliage* quinate or ternate, *chartaceous, pale green, dull, glabrous,* with elliptic-oval leaflets mostly tapering to base and apex, doubly fine-serrate, *the ribs beneath prominent and pilose;* the *terminal leaflet* 7–11 cm. long, 3.5–6.5 cm. broad, somewhat rounded to base, *on a glandular and setose petiolule* 1.2–2.4 *cm. long;* leaflets of floricane-foliage acute, narrowly serrate-dentate; *inflorescence* corymbose or corymbose-racemose, *with densely villous rachis* and often trifid upper bracts; *pedicels* subascending, 1–2 *cm. long, gray-villous and glandular,* sometimes setose; calyx villous and glandular-setose, with short soon reflexed lobes; *petals* narrow, 0.5–1 *cm. long,* 3–4 *mm. broad;* fruit subglobose, 1–1.3 cm. in diameter, of fine quality. — Sandy alluvium, sand-plains and damp thickets, Coös Co., N.H. *Fl.* July; *fr.* Aug., Sept.

93. R. spiculðsus Fern. (covered with spicules). — Habit of no. 92; canes elongating to 2 m., frequently tip-rooting; *primocanes green and glabrous, openly retrorse-setose, with* 100–500 straight *reflexed setae* (and only scattered glands) per dm.; *primocane-foliage* firm, pale green, glabrous, *with ribs* less prominent beneath and *glabrous or nearly so,* the *petioles and petiolules glandless;* terminal leaflet subcordate, long-acuminate, 7–14 cm. long, 3.5–8 cm. broad, on bristly petiolule 1–3 cm. long; floricane-leaflets narrowly ovate or rhombic-obovate, acuminate, coarsely serrate; racemes lax, often subcorymbiform, the pilose rachis 3–8 cm. long, the lanceolate bracts often incised; *pedicels loosely divergent,* pilose and glandular, scarcely setose, mostly 2.5–5 *cm. long;* calyx-lobes 6–7 mm. long; *petals* showy, narrowly obovate, 1.2–1.5 *cm. long,* 5–9 *mm. broad;* fruit subglobose, 1.3 cm. in diameter. — Thickets and borders of woods, n. and centr. N.H. *Fl.* July; *fr.* late Aug., Sept.

94. R. grándidens Bailey (large-toothed). — Somewhat intermediate between nos. 77 and 95; primocanes high-doming, then the tips arching and finally trailing, about 6 mm. thick, with stout-based prickles 4 mm. long mixed with fine and short bristles and glands; primocane-leaves coriaceous, dark green, lustrous, 5-foliolate; *leaflets with caudate tips* 2–3 *cm. long; terminal and median leaflets broadly ovate, cordate or subcordate; the terminal* 1–1.2 *dm. long and* 6–7 *cm. broad,* on petiolules 2.5 cm. long; petiolules of median leaflets 1.5 cm. long; floricane-leaves small, narrowly oval, acute; racemose-corymbiform *inflorescence with glandular but prickleless pedicels* 1.5–2.5 cm. long. — Damp sand, n. Wisc.

95. R. árcuans Fern. & St. John (arching). — Strongly suggesting no. 91, doming, arching or sprawling; *primocanes with numerous hard broad-based hooked prickles* 3–4.5 mm. long, mixed with relatively few or no bristles and glands; *primocane-foliage coriaceous, dark green, lustrous,* 3- or 5-foliolate; *the obovate to suborbicular leaflets abruptly tipped, coarsely toothed; terminal leaflet* 3–8 *cm. long,* 2.5–5.5 cm. broad, the midrib beneath armed with hooked prickles, *the prickly petiolule* 0.5–2 *cm. long; raceme* interrupted, simple or paniculate, 0.6–3.2 dm. long; *the axis and pedicels* densely pilose, *usually much armed,* sometimes glandular; pedicels 1.5–4 cm. long; calyx prickly or unarmed, its lobes ascending in fruit; petals spatulate, 7–11 mm. long, 4–8 mm. broad; fruit globose, small. (Incl. *R. provincialis, vigoratus, Bicknellii* and *mananensis* Bailey) — Sandy soil, P.E.I. and N.S. sw. near coast to se. Mass. *Fl.* July, early Aug.; *fr.* late Aug., Sept.

96. R. óbvius Bailey (not wholly obvious, through contradiction as to leaflets, between original description and illustration). — Primocanes up to 1 m. high, doming, with recurving and prostrate-tipped branches; *canes rather slender,* 1.5–3.3 *mm. thick, with scattered acicular prickles* 3–4 mm. long; *primocane-leaves mostly 5-foliolate,* on long remotely acicular petioles; *leaflets narrowly ovate-lanceolate, about thrice as long as broad,* long-acuminate, the *terminal one rounded to base and on a petiolule* 5 *mm. long; flowering shoots* finely pubescent, 1–3-*flowered;* simple or trifoliolate *bracts acuminate* and exceeding the flowers; calyx-lobes long-pointed; petals oblong-spatulate, 1.6 cm. long, 7–8.5 mm. broad. — Borders of woods, ne. Md.

97. R. Akermáni Fern. (in honor of Alfred Akerman, 1876–). — Coarser, *forming intricate domes* up to 1.3 m. high, with long-arching and finally tip-rooting *branches up to* 3 *m. or more long; primary canes* 4–10 *mm. thick,* remotely *armed with hard recurving prickles* 3–6 *mm. long and with flattened bases* 2–6 *mm. broad; leaves firm to coriaceous, soft-pubescent beneath,* the *upper surface with strigae barely* 0.5 *mm. long* or glabrate; *primocane-leaves 3-foliolate, with ovate* regularly toothed acuminate *leaflets; terminal leaflet cordate or rounded at base,* 5–8 cm. long, 2.5–5.5 cm. broad, *on strongly armed villous petiolules* 1.5–3 *cm. long; flowering shoots* with small leaves *with obtuse to merely acutish* elliptic to *cuneate-obovate leaflets,* the terminal leaflet 1.5–6 cm. long; flowers 1 or 2–4 in a compact corymb, equaled by the bracts; pedicels ascending, villous, glandless and mostly unarmed, 0.7–1.8 cm. long; calyx pilose, unarmed; fruit nearly 2 cm. in diameter. — Argillaceous thickets or peaty or sphagnous swamps, Brunswick and Greensville Cos., Va. *Fr.* mid-June, early July.

98. R. subinnóxius Fern. (almost harmless). — Similar to no. 97, the great *domes up to 2 m. high,* the *branches and long* (up to 2 m.) *branchlets more pendulous; primocanes and branches unarmed or with* very *remote prickles* 2–4 *mm. broad at base; leaves membranaceous, the strigae of the upper surface mostly* 1 *mm. long; leaflets mostly* 5, *abruptly acuminate;* petiolule of terminal leaflet 2.5–3 cm. long; petals oblong-obovate, 1 cm. long, 7 mm. broad; *branches and branchlets of floricane* very intricate, *pendulous; leaflets acute or acuminate;* fruit 1.5 cm. in diameter. — Wet or dry thickets, Isle of Wight and Southampton Cos., Va. *Fl.* late April, early May; *fr.* June.

99. R. pernagaèus Fern. (of the land of hams, from the type-region, Smithfield, Va.). — Much smaller than nos. 97 and 98; *primocane* 2–3.5 *mm. in diameter,* with scattered subulate prickles; *primocane-leaves* 3- or 5-foliolate, somewhat velvety beneath; *the terminal leaflet cuneate-obovate, coarsely lobulate and incised* (at anthesis 2.5–3.5 cm. long, doubtless becoming longer); flowers 1–3; *flowering pedicels loosely ascending, heavily glandular, mostly bracteolate; calyx glandular; petals rose-tinged,* 6–8 mm. long, 4 mm. wide. — Dry thickets, local, Isle of Wight Co., Va. *Fl.* early April. — Little known; fuller material and study needed.

100. R. minnesotànus Bailey (of Minnesota). — Forming dense domes up to nearly 1 m. high, the arching branches reaching the ground and trailing or tip-rooting; *stronger parts of primocane with scattered subulate straight prickles* 2–3 *mm. long; primocane-leaves* 3- or 5-foliolate, *membranaceous,* glabrous above, *soft-pubescent beneath; leaflets broadly ovate,* long-acuminate, *sharply double-serrate;* the *terminal one deeply cordate,* 8–13 cm. long, 6–9 cm. broad; petiole remotely unguiculate-prickly; *flowering shoots with large 3-foliolate or simple cordate-ovate leafy bracts,* these *mostly equaling or exceeding the* racemose-corymbiform *inflorescence; pedicels* ascending, 1.5–4 cm. long, *remotely prickly;* calyx pilose, the subulate-tipped lobes 7 mm. long; petals 1.5 cm. long, about 1 cm. broad; fruit 1–1.5 cm. in diameter. — Se. and centr. Minn.

101. R. Sewardiànus Fern. (in honor of Walter Seward, 1860–1932, founder of the Seward Forest of the Univ. of Va.). — Forming dense mounds, the *primocanes* and their finally intricate divergent branches and trailing-tipped branchlets up to 2 m. long, *remotely armed with horizontally divergent straight subulate prickles* 3–5 *mm. long and* 2–3 *mm. broad at base; primocane-leaves mostly 3-foliolate,* membranaceous, *tomentulose beneath;* the ovate-acuminate *leaflets with low serrate-dentate teeth;* the *terminal leaflet cordate at base* and 8–12 cm. long by 5–7.5 cm. broad, the *unguiculate-armed petiolule* 2.5–3 cm. long; *floricanes* inextricably and divergently much branched, *their prickles becoming unguiculate; floricane-leaflets narrowly elliptic-oval, acuminate, doubly serrate;* short racemose *corymb* 2–8-flowered, *overtopping the bracts; pedicels* divergently ascending, 1–2.5 *cm. long, retrorsely villous* and prickly; calyx pilose; fruit 1.5–1.8 cm. in diameter. — Dry clearings and thickets, Brunswick Co., Va. *Fr.* June.

102. R. cathártium Fern. (of buzzards; from their nesting near the type-colony). — Differing from no. 101 in its longer and more flexible *primocanes with obliquely deltoid*-subulate *recurved or unguiculate* hard *prickles* 3–6 *mm. broad* at base; *primocane-leaves* mostly *5-foliolate, the leaflets slenderly sharp-serrate,* the *terminal leaflet rounded at base,* the petiolule 1.5–3 cm. long; *floricanes subsimple and more rigid; petiole of floricane-leaves, pedicels* (2–4 *cm. long*) *and calyces glandular.* — Dry thickets and clearings, Brunswick Co., Va. *Fr.* late June.

103. R. multispìnus Blanch. (many-prickled). — Very coarse and intricately recurved-branching and repeatedly forking doming plant with long-arching and tip-trailing or -rooting branches and branchlets; *stronger parts of stout primocane with* 4 *or more rows of very strong and hard broad-based prickles up to* 5–8 *mm. long, these often approximate in groups and rarely* 1 *cm. apart; leaves firm and thick, strongly uncinate-prickly on midrib beneath,* usually strongly pubescent beneath; *primocane-leaves* 3- or 5-foliolate, *often fulvous-green;* the *leaflets* rather *abruptly acuminate,* sharply double-serrate; *terminal leaflet broadly ovate, the larger ones* 8–15 *cm. long by* 4–9.5 *cm. broad,* broadly rounded to subcordate at base, the very prickly petiolule 1.5–5 cm. long; *petiole with* 10–25 *recurved prickles;* flowering shoots with *small corymbiform-racemose clusters of* (3–)5–8 *very prickly-* and short-*pedicelled flowers standing above the small bracteal leaves;* fruit up to 2 cm. long, juicy. — Dry to damp thickets and clearings, chiefly coastwise, Knox Co., Me. to centr. and s. Ct. *Fl.* June; *fr.* Aug., Sept.

104. R. dècor Bailey (ornamental). — Grayish, doming to 6 dm. high; primocane with scattered subulate divergent prickles; *primocane-leaves 3-foliolate or imperfectly 5-foliolate; the leaflets barely half as broad as long, gradually tapering or subcuneate at base;* the terminal one *on a petiolule nearly* 2 *cm. long,* the lateral subsessile, soft-pubescent beneath; primocane-leaflets elliptic-oblong to narrowly obovate, acutish; *corymb 4–6-flowered*, the *ascending pubescent pedicels* up to 3 cm. long; calyx pilose; petals narrowly obovate, about 1.8 cm. long and 8–10 mm. broad. — Borders of dry woods, w. Ky.

105. R. sátis Bailey (enough or sufficient, if not satisfactory). — *High-doming, up to 1 m. high*, with trailing and root-tipping branchlets; primocanes with remote broad-based prickles; *primocane-leaves 5-foliolate*, on remotely prickly glabrescent petioles; *leaflets broadly ovate*, rounded to base, *abruptly acuminate, serrate-dentate*, the *median on petiolules* 1.5 *cm. long; the strongly cordate terminal leaflet two-thirds as broad as long and on a petiolule* 3.5 *cm. long; inflorescence* racemose-corymbiform, 3–7-flowered, *the axis and nearly prickleless suberect pedicels spreading-pilose; foliaceous bracts* sharply toothed, *the simple elliptic-lanceolate upper ones acuminate;* fruit small and seedy. — Fields and borders of ditches, Jefferson Co., N.Y.

106. R. arundelànus Blanch. (for Arundel, York Co., Me.). — Dense domes up to 1 m. and more high; *primocanes with scattered subulate hard prickles; primocane-leaves* 5-foliolate, *thickish, with often plicate margins*, light green, sparingly pilose to glabrous above, *tomentose or velvety beneath; leaflets ovate, long-acuminate, slightly more than half as broad as long, sharply double-serrate, rounded at base; median leaflets on petiolules 2–15 mm. long; petiolule of terminal leaflet 2–3.5 cm. long; petiole and petiolules villous; corymb leafy-bracted, with* 4–10 or more flowers, with almost prickleless *villous pedicels;* calyx pilose; petals broadly oval, 1.5 cm. long, half as broad; fruit about 1 cm. in diameter. (Incl. *R. Janssonii* and *R. prosper* Bailey) — Dry clearings, thickets and borders of woods, Knox Co., Me., to e. N.Y., s. to Mass., s. R.I. and s. Ct. *Fl.* June, early July; *fr.* Aug., early Sept. Passing northw. into

Var. Jeckylànus (Blanch.) Bailey (for Stevenson's Dr. Jekyll, because of its variable character). — *Branches and branchlets less pendulous; terminal leaflet of primocane-foliage cordate,* its petiolule 1–2.5 cm. long, *scarcely plicate; pedicels more glandular.* — With the species and passing into it, sw. Me.

107. R. Rosendáhlii Bailey (for its discoverer, Carl Otto Rosendahl, 1875–). — Doming, up to 5 dm. high, *slender; primocanes 3–4 mm. thick, with very remote slenderly deltoid-subulate prickles about 2 mm. long; primocane-leaves 3-foliolate*, the broad *leaflets abruptly short-acuminate*, glabrous or slightly pubescent beneath; *terminal leaflet broadly oval*, usually *more than half as wide as long*, finely double-serrate; floricane-leaflets small, those of 3-foliolate bracts obovate, the *simple upper bracts ovate and acute;* racemose corymb with smooth axis and *naked pedicels; calyx-lobes 5 mm. long, smoothish, soon reflexed;* petals narrowly oblong, 8–9 mm. long; *fruit* subglobose, 8 *mm. in diameter.* — Sandy soil, se. Minn. *Fl.* June; *fr.* July.

108. R. missoùricus Bailey (of Missouri). — Doming to 1 m. high, in dense clumps; *primocanes* about 5 *mm. thick, with scattered subulate* barely curved *prickles* 4 or 5 mm. long; *primocane-leaflets* 5, *narrowly elliptic-ovate, acuminate, serrate-dentate*, the terminal one about half as broad as long; *lower leaves of fertile shoots with obovate bluntish leaflets; bracteal leaf* 3-foliolate, *with narrowly oval acute leaflets; cyme as broad as long, loosely forking, with reduced bracts;* petioles of floricane and pedicels (1–2.5 cm. long) mostly unarmed; *calyx-lobes* 7 mm. long, *essentially glandless, ascending in fruit;* petals 9 or 10 mm. long, 3–4 mm. broad; fruit nearly equaled by calyx. — Open habitats, w. Mo.

109. R. pityóphilus S. J. Smith (lover of pines). — Arched-ascending to 5 dm. or more high; *primocanes* stout, *the stronger parts up to 1 cm. thick, with scattered straightish lance-subulate prickles; primocane-leaflets* 5; *the ovate, subcordate and acuminate terminal one three-fifths to two-thirds as broad as long* (5–7 cm. wide); *leaflets and simple bracts of flowering shoot sharply serrate, acuminate; inflorescence racemose-corymbose, the lower flowers greatly exceeded by the bracts; longer pedicels 2–2.5 cm. long; calyx-lobes soon reflexed; petals obovate, about* 1.5 *cm. long and* 1 *cm. broad.* — In pine-woods, s.-centr. N.Y.

110. R. paludívagus Fern. (wandering over the bog). — *Primocanes* at first simple, *soon intricately branching, 3–6 mm. thick*, at first puberulent; becoming glabrate, *stiffly armed with subulate prickles* 5 mm. long and 2–3 mm. broad at base; *primocane-leaflets* 5, *strigose-pilose above*, velvety beneath, *oblong or narrowly lance-ovate*, serrate, acuminate, *the terminal one one-third to less than half as broad as long (2.5–3.7 cm. wide); membranaceous floricane-leaflets acuminate; pedicels loosely ascending*, minutely pilose, *the longer ones 2–5 cm. long*, mostly unarmed; *calyx-lobes* 3–5 mm. long, unarmed, *scarcely reflexed in fruit; petals 1–1.5 cm. long, 6–7 mm. broad; fruit 1.5 cm. in diameter, of rich quality.* — Wet thickets and bogs, Cape Cod, Mass. *Fl.* mid-June, early July; *fr.* late Aug.

111. R. Pórteri Bailey (for its discoverer, Thomas Conrad Porter, 1822–1901). — *High-mounding*, with many overarching and finally tip-trailing glabrous branches *with numerous lance-subulate recurving prickles (60–120 per dm.) 2 mm. long; primocane-leaflets 5, firm, pale green, prominently ribbed, glabrous on both sides, oblong-lanceolate, acuminate, one-third to two-*

fifths as broad as long, the terminal one 8–9 cm. long and about 3 cm. broad, the *petiole with many prickles;* lower *bracteal leaf* 3-foliolate; the upper simple and oval, *dentate; axis and pedicels of* the *long raceme* bristly *with fine divergent prickles* 2–3 *mm. long; calyx bristly;* fruit seedy and insipid, about 1 cm. in diameter. — Thickets and open woods, Pocono Plateau, ne. Pa.

112. R. eflagellàris Bailey (without a whip-lash). — Strongly resembling no. 103 but the *terete primocane with fewer lance-subulate prickles only* 2–3 *mm. long; primocane-leaflets broader,* the *terminal one suborbicular-ovate, three-fifths to two-thirds as broad as long;* corymbs mostly overtopping leaves, the *bracts mostly inconspicuous; pedicels glandular* as well as prickly. — Sandy soil, centr. Ct.

113. R. multifórmis Blanch. (of many forms). — Mounding, the canes arching; their branches and branchlets soon trailing and greatly prolonged, tip-rooting; *canes glabrous, with* 0–40 *lance-subulate prickles per dm.; primocane-leaves glabrous, membranaceous, pale green,* 5-foliolate, *on smooth or only sparsely prickly petioles; leaflets coarsely and subsimply serrate, corrugated, with long caudate-acuminate tips; terminal leaflet narrowly ovate, less than to about half as broad as long* (6–15 cm. long, 2.5–5 cm. broad), the smooth or sparsely prickly petiolule 1.5–4 cm. long; *inflorescence long-racemose or racemose-corymbiform, many-flowered; most* of the *pedicels subtended by reduced bracts; pedicels very slender, smooth* or only remotely short-bristly, *the longer ones* 2.5–7 *cm. long;* calyx glandless and prickleless, becoming reflexed; petals 8–12 mm. long, 4–7 mm. broad; fruit finally juicy, inferior. — Wet to dry woods, thickets, clearings, etc., N.S. and s. N.B. to Vt. *Fl.* late June–early Aug.; *fr.* late Aug., early Sept.

114. R. sevèrus Brain. (harsh). — Habitally like no. 113; primocanes with 150–300 *stout horizontal to arched-recurving prickles per dm.; primocane-leaves subcoriaceous, dark green above,* coarsely double-serrate; *leaflets ovate to obovate,* acuminate; *the terminal one* 6–13 *cm. long by* 3.5–8 *cm. broad,* its strongly armed petiolule 1.5–4 cm. long; leaflets of floricane elliptic-obovate; *inflorescence loosely paniculate, corymbiform or open-cymose;* rachis remotely bristly; *filiform pedicels* 2–5 *cm. long,* without or with few bristles and glands; *calyx-lobes* 4–8 mm. long, *reflexed in fruit;* petals 9–11 mm. long, 4–5 mm. broad; fruit subglobose. — Thickets and clearings, N.H. and e. Vt. *Fl.* late June–early Aug.; *fr.* late Aug., early Sept.

115. R. novánglicus Bailey (of New England). — With habit and *coriaceous dark lustrous leaves* somewhat as in no. 114, but *broad-based prickles* more *scattered; primocane-leaflets abruptly short-acuminate and pilose along nerves beneath; inflorescence an open broad raceme with unarmed pedicels* up to 4 cm. long; *calyx not reflexed in fruit;* petals 1.2 cm. long, 5–6 mm. broad. — Dry "hard" land, se. Ct. *Fl.* June.

116. R. noveboràcus Bailey (of New York). — Resembling no. 113 but forming *dense mounds* 1 *m. high; primocane-leaves* more *approximate, very firm,* 3–7-*foliolate, pubescent on nerves beneath; leaflets finely double-serrate, abruptly* slender-*tipped; terminal one round-ovate, nearly as broad as long,* 7–9 cm. long; *inflorescence short and compact, racemose-corymbiform,* its foliaceous bracts acute; *pedicels* finely pilose, with very short bristles, *up to* 3 *cm. long; calyx-lobes reflexed* under the large (2 cm. long) fruit. — Dry soil, s.-centr. N.Y.

117. R. Schoolcraftiànus Bailey (in honor of Henry Rowe Schoolcraft, 1793–1864). — Strongly resembling no. 113, but *with stouter and broad-based* (5 *mm. wide at base*) *prickles; leaves pilose on veins beneath; terminal leaflet more than half as broad as long; pedicels spreading-pilose,* the longer ones 2.5 cm. long. — Woods and borders of swamps, s. Mich.

118. R. cómplex Bailey (closely connected with). — Habit of the several preceding species; *prickles of canes merely subulate,* 2–4 mm. long, divergent, 40–75 *per dm.; primocane-leaves* 3- or 5-foliolate, *finely pubescent on nerves above and beneath; leaflets* broadly ovate to oblong-ovate, gradually tapering to tip, *coarsely serrate-dentate, about half as broad as long* (4–6 cm.); *flowering shoots spreading-pubescent,* ending in a leafy-bracted raceme; *bracts mostly simple, ovate, acute, longer than the pubescent pedicels;* petals 1 cm. long, 5 mm. broad; fruit globose, up to 1 cm. in diameter. — Sandy open places, Mich.

119. R. setospinòsus Bailey (with bristly prickles). — *Stout, doming to* 1 *m. high; primocanes up to* 1 *cm. in diameter with* 300–600 *mixed bristles and slender prickles per dm.; leaves* glabrous above, *pilose on nerves beneath;* 5-foliolate *primocane-leaves* on very prickly petioles and with strongly setose petiolules; *terminal leaflet round-ovate, abruptly acuminate, four-fifths as broad* (about 1 cm.) *as long; small leafy corymb with slender smooth pedicels overtopped by the 3-foliolate bracts with narrowly lanceolate to oblong acuminate leaflets.* — Dry open soil, w. Wisc.

§ 10. Setòsi Bailey (see p. 823)

a. Lower surfaces of leaflets glabrous (except for pilosity on veins). . . *b.*

b. Primocanes bearing hundreds or thousands of spreading to reflexed essentially uniform soft bristles per dm., these not stiff and resistant to the touch. . . *c.*

*c.*Leaflets of floricane-leaves acute or subacute, not broadly rounded at summit.

Terminal primocane- and floricane-leaflets one-third to three-fifths as broad as long. . . . . 120. *R. setosus.*

Terminal primocane- and floricane-leaflets three-fourths as broad as long. . . . . 121. *R. Lawrencei.*

*c.*Leaflets of floricane-leaves broadly rounded at summit. . . . . 122. *R. rotundior.*

*b.*Primocanes bearing some hard and piercing slenderly subulate prickles, either by themselves or mixed with glands and fine bristles. . . *d.*

*d.*Axis of primocane with more or less crowded prickles and bristles (and often glands), 200–1000 or more per dm.; these or their scars mostly evident on floricanes. . . *e.*

*e.*Pedicels and axis of inflorescence copiously long-setose with firm glandless prickles and few or no intermixed glands; wide-ranging northeastward. . . . . 123. *R. Groutianus.*

*e.*Pedicels and axis of inflorescence copiously glandular; West Virginian. . . *f.*

*f.*Inflorescence strongly corymbiform; pedicels armed with long flexuous gland-tipped hairs. . . . . 124. *R. discretus.*

*f.*Inflorescence a simple raceme; pedicels heavily glandular but scarcely bristly.

Primocane-leaflets lance-attenuate, less than half as broad as long; pedicels mostly subtended by leafy bracts. . . . . 125. *R. angustifoliatus.*

Primocane-leaflets ovate to elliptic-obovate, more than half as broad as long, abruptly tipped; pedicels mostly subtended by stipule-like bracts. . . . . 126. *R. nocivus.*

*d.*Axis of primocane with fewer (3–150 per dm.) scattered or remote uniform prickles, these (except in no. 128) not mixed with shorter bristles and stipitate glands. . . *g.*

*g.*Terminal primocane-leaflet ovate, ovate-lanceolate or ovate-oblong, two-fifths to three-fifths as broad as long, tapering to acuminate tip; inflorescence racemose to corymbiform.

Axis of primocane, petioles and axis of inflorescence glandless or essentially so; pedicels glandless or only sparsely glandular; wide-ranging northward. . . . . 127. *R. vermontanus.*

Axis of primocane, petioles, petiolules, axis of inflorescence and pedicels copiously glandular; northwestern. . . . . 128. *R. regionalis.*

*g.*Terminal primocane-leaflet broadly obovate, abruptly short-tipped, three-fourths to four-fifths as broad as long; inflorescence broadly corymbose.

Axis of inflorescence and pedicels long-setose; lower bracteal leaves simple, narrowly oblong to lanceolate and jagged-toothed; midwestern. . . . . 129. *R. Wheeleri.*

Axis of inflorescence and pedicels mostly unarmed; lower bracts broadly ovate, regularly serrate; northeastern. . . . 130. *R. univocus.*

*a.*Lower surfaces of leaflets soft-pilose to subvelutinous to touch; prickles or bristles of primocanes scattered, 25–200 per dm. (1000 or more and closely crowded only in no. 132 with stipitate glands intermixed). . *h.*

*h.*Primocanes and main axes of flowering branchlets glandless or essentially so (except in no. 138). . . *i.*

*i.*Principal leaflets of primocane tapering to acute to long-acuminate tips; leaflets of upper leaves of flowering shoots or lower bracteal leaves mostly acute. . . *j.*

*j.*Prickles of primocane acicular to slenderly subulate, mostly 3–5 mm. long and less than 1.5 mm. broad at base. . *k.*

*k.*Pedicels glandless or essentially so. . . *l.*

*l.*Inflorescence definitely overtopping the basal foliaceous bracts or the uppermost leaves. . . *m.*

*m.*Axis of inflorescence prickleless; pedicels prickleless or merely with fine setae mostly less than 1 mm. long.

Leaflets of primocane with broad-based acumination 5–12 mm. long; leaflets of bracteal leaves merely acute or acutish; inflorescence strongly overtopping the bracteal leaves; petals 3–5 mm. broad; plant of s. N.E., s. N.Y. and e. Pa. 131. *R. semisetosus.*

Leaflets of primocane narrow-based, subcaudate tip 1–1.5 cm. long; leaflets of bracteal leaves long-acuminate; inflorescence but slightly overtopping bracteal leaves; petals 5–8 mm. broad; plant of N.S. and se. Me. . . . . 132. *R. ortivus.*

*m.*Axis of inflorescence and pedicels prickly with stiff bristles 2–3 mm. long; midwestern.

Inflorescence leafy-bracted at base; pedicels strongly ascending, the lower ones 4–6 or more cm. long. . . . . . . 133. *R. dissensus.*

Inflorescence without leafy bracts; pedicels divergent, up to 2 cm. long. . . . . 134. *R. Schneideri.*

*l.*Inflorescence nearly or quite equaled or overtopped by the leafy bracts or upper leaves. . . *n.*

*n.*Axis of inflorescence prickleless or with rare and tiny setae; pedicels 1–2 cm. long; principal leaves of flowering shoot with sharply acute leaflets.

Prickles of primocane very remote, 25–40 per dm., acicular.

Inflorescence with several leafy bracts, largely overtopped by them; plant of N.S. and e. Me. . . . . . . 132. *R. ortivus.*

Inflorescence without leafy bracts rising beyond petiole of the upper leaf; midwestern. . . . . . . 135. *R. mediocris.*

Prickles of primocane 100–300 per dm., unguiculate; inflorescence leafy-bracted; western. . . . . 136. *R. uniformis.*

*n.*Axis of inflorescence or pedicels more or less prickly with setae up to 2 or 3 mm. long (or prickleless); inflorescence leafy-bracted at base; pedicels 1.5–6 cm. long; principal leaves of flowering shoot with blunt or merely subacute leaflets; eastern. . . . . 137. *R. hispidoides.*

*k.*Pedicels copiously stipitate-glandular.

Prickles of primocane weak and bristleform; axis of inflorescence and petioles of bracteal leaves stipitate-glandular; plant of n. New England. . . . . 138. *R. perinvisus.*

Prickles of primocane hard, subulate-unguiculate; axis of inflorescence and petioles of bracteal leaves glandless; plant of se. Pennsylvania. . . . . 139. *R. Benneri.*

*j.*Prickles of primocane hard and subulate, many of them 6–9 mm. long and with bases 2–3 mm. broad; inflorescence prickless or only sparsely setose; plant of s. New England. . . . . 140. *R. ascendens.*

*i.*Principal leaflets of primocanes barely subacute or abruptly short-tipped; leaflets of flowering shoots blunt or merely acutish.

Axis of inflorescence and long (up to 3 cm.) pedicels prickleless; eastern. 141. *R. Clausenii.*

Axis of inflorescence and short (up to 1.5 cm.) pedicels prickly; midwestern. . . . . 142. *R. jejunus.*

*h.*Primocanes and main axes of flowering branches stipitate-glandular.

Canes densely covered with thousands of closely crowded slender bristles and glands per dm.; inflorescence with several foliaceous bracts, the leaflets or simple blades half as broad as long; plant of centr. New York. . . . . 143. *R. beatus.*

Canes with 100–200 firm subulate subremote prickles per dm.; inflorescence with leafy bract only at base, its leaflets a third to two-fifths as broad as long; West Virginian. . . . . 144. *R. racemiger.*

120. R. setòsus Bigel. (bristly). — *Primocanes* slender, 0.3–1.5 m. high or long, *arched-ascending to erect, densely covered with hundreds or thousands of spreading to reflexed pale nearly uniform bristles; these scarcely resistant to touch,* neither rigid nor pungent, 1.5–4 mm. long; *primocane-leaves* 3- or mostly 5-foliolate, *on more or less bristly petioles;* the *leaflets lance- or oblong-ovate to slightly obovate or subrhombic, pale green, dull, plicate,* double-serrate (or sometimes dentate), mostly *acutish to acuminate,* terminal leaflet distinctly petiolulate; *floricanes reclining to upright, covered with fuscous bristles;* the branching mostly sublateral or unilateral and prolonged on reclining or partially buried canes, more symmetrical and shorter on upright canes; *floricane-leaflets narrowly obovate to oblong-rhombic, acutish to acuminate, one-third to three-fifths as broad as long; inflorescence a simple and racemose short corymb or a branching corymbiform panicle, its axis and pedicels setose;* the lower bracts leaf-like, the upper greatly reduced; calyx-lobes usually bristly, 3.5–7.5 mm. long; petals 7–10 mm. long, 2–4 mm. broad;

fruit inferior, dryish, subglobose, about 1 cm. in diameter. (Incl. *R. nigricans* Rydb., an erect and symmetrically branching state; and *R. junior, significans, dissimilis, Ribes* (erect and symmetrically branching state), *notatus, Boottianus, condignus* and *tectus* Bailey, states which, as individual collections, show some tendencies but which in a series of hundreds of nos. show little constancy, making up what the author of most of them would call a "small various bramble".) — Swamps, swales, damp thickets, etc., s. Que. to Wisc., s. to N.B., s. N.E., Md. and W.Va. *Fl.* late June–early Aug.; *fr.* late Aug., Sept.

121. R. Lawréncei Bailey (for its discoverer, George Hill Mathewson Lawrence, 1910–). — Differing from no. 120 in its *broader leaflets; terminal leaflet of primocane broadly rhombic-oval, about three-fourths as broad as long, on very short petiolule; terminal floricane-leaflet of similar proportions, elliptic-obovate.* (Incl. *R. exter* and *R. udus* Bailey) — Low thickets and swamps, w. Me. to Wisc. and Pa. — Suggesting derivation, in part, from *R. hispidus; R. udus* is transitional to no. 120.

122. R. rotúndior Bailey (rounder). — Similar to no. 121 but nearly glandless; *terminal primocane-leaflet round-ovate, abruptly tipped; floricane-leaflets broadly obovate, strongly rounded at summit; axis of inflorescences and pedicels with few or no setae.* — Sw. Que. to Wisc.

123. R. Groutiànus Blanch. (for the Grout family, pioneer settlers in the type-region). — Habitally close to no. 120; *primocanes with* 200–1000 *or more stiff bristles and acerose acicular to slenderly subulate firm prickles per dm., the longer prickles 3–7 mm. long; leaflets flat, scarcely plicate,* the petiole usually very prickly; pedicels and axis of racemose to corymbose inflorescence bearing long glandless setae. (Incl. *R. electus* Bailey, state with glands relatively abundant on pedicels, and *R. gulosus* Bailey, form with narrowest leaflets.) — Moist to dry thickets, clearings and swamps, N.B. to s. Ont., s. to N.Y. and ne. Pa. *Fl.* mid-June–early Aug.; *fr.* late Aug., Sept.

124. R. discrètus Bailey (separate). — Resembling coarser states of no. 123; *primocanes* 4 or 5 dm. high, *copiously stipitate-glandular and with long subulate prickles;* primocane-leaves with 5 leaflets; terminal leaflet ovate, doubly serrate, abruptly long-acuminate; the *prickly petiole stipitate-glandular; corymb with strongly ascending long pedicels strongly armed with long gland-tipped divergent hairs.* — Swamps of e. W.Va.

125. R. angustifoliàtus Bailey (with narrow leaflets). — *Primocanes densely beset with short recurved prickles, bristles and stipitate glands; the prickly- and glandular-petioled leaf with* 5 petiolulate *lance-attenuate leaflets; upper leaflets barely half as broad as long; inflorescence loosely racemose, the ascending glandular pedicels mostly subtended by* 3-foliolate or simple *bracts resembling the foliage-leaves.* — Low ground, W.Va.

126. R. nocìvus Bailey (hurtful). — Differing from no. 123 in its *densely stiff-bristly or prickly primocane strongly glandular; petiole and petiolule of leaf densely glandular and retrorse-prickly; elliptic-obovate or ovate leaflets abruptly tipped, more than half as broad as long; raceme strongly glandular, the divergent pedicels mostly subtended by small or nearly obsolete stipule-like bracts.* — Known only from e. W.Va. (presumably in neighboring states).

127. R. vermontànus Blanch. (of Vermont). — Habitally like nos. 120 and 123 but *with sparser armature; primocanes* erect to depressed, 0.2–1.5 m. high, *with very few to* 150 *firm slender acerose prickles per dm., these without interspersed glands and usually without finer bristles; floricane-leaves* 5-foliolate, their petioles more or less prickly; *leaflets ovate-lanceolate, ovate or ovate-oblong, tapering to acuminate apex, two-fifths to three-fifths as broad as long,* sharply double-serrate, *thinnish, dull or barely sublustrous, glabrous above,* pale to deeper green and *glabrous beneath; inflorescence varying from a simple raceme to a forking corymb;* the *pedicels unarmed* or rarely with few short setae, *mostly glandless* (occasionally with few glands); lower bracts foliaceous, upper reduced and stipule-like; petals 1–1.5 cm. long, 4–8 mm. broad; fruit about 1 cm. in diameter, juicy but sour. (Incl. *R. junceus* Blanch., plants "with a flimsy look"; *R. viridifrons* Bailey, i.e., *R. vermontanus,* var. *viridifolius* Blanch., with dark and sublustrous leaves and with some glands on pedicels, a "guise" which "has no separate range or habitat. . . In a single collecting . . . one may distinguish both kinds." — *Bailey; R. navus, R. unanimus* and *R. superioris* Bailey; *R. spectatus* Bailey, with setose pedicels; *R. apparatus* Bailey, with as few prickles as *R. junceus; R. malus* Bailey, with foliage and glandular pedicels as in *R. viridifrons* but collected in N.B. instead of Vt.; *R. offectus* Bailey, suggesting *R. junceus* with some setae on a few pedicels; *R. suppar* Bailey, like *R. viridifrons* but with dull leaves; *R. quebecensis* Bailey, nearly unarmed; *R. Deaneanus* Bailey; *R. singulus* Bailey, with leaflets slightly narrower than in *R. suppar*) — Dry to moist thickets, clearings, etc., plastic but forming a consistent series of interlocked minor trends, Nfld. to Ont., s. to N.S., s. N.E., N.Y., n. Pa., Mich., Wisc. and Minn. *Fl.* mid-June–mid.-Aug.; *fr.* late July–Sept.

128. R. regionàlis Bailey (belonging to a region). — Strongly suggesting no. 127 and presumably an extreme of it; *differing in having copious glands on the primocane, petioles, rachis and pedicels; petals illustrated as rather smaller.* — N. Wisc.

129. R. Wheèleri Bailey (in honor of CHARLES FAY WHEELER, 1842–1910). — With *canes remotely armed* as in no. 127; *corymb as heavily spiculose* as in no. 123; *primocane-leaves sublustrous above; the terminal leaflet rounded-obovate, abruptly short-tipped, three-fourths to four-fifths as broad as long; corymb as broad as long,* the *lower foliaceous bracts narrowly oblong and entire to lance-acuminate and lacerate;* petals 7 or 8 mm. broad. — Swamps and damp thickets, s. Mich.

130. R. unívocus Bailey (of one voice; apparently meaning that everyone agrees). — Coarse, with erect primocanes but *mostly depressed floricanes,* the few setiform hard prickles scattered; *primocane-leaves suggesting those of no.* 129; *lower long-petioled foliaceous bracts with simple or 3-foliolate broad and regularly serrate blades; corymb neither spiculose nor glandular.* — Dry clearings, old fields and roadsides, N.B., s. Que. and Ont., s. to N.S., and se. Me.

131. R. semisetòsus Blanch. (half bristly). — *Canes more woody than in no.* 120, *more like those of no.* 127, the *prickles* 80–150 *per dm.* on the glandless primocanes; *primocane-leaves* much as in no. 127, but *soft-pubescent beneath, the leaflets with broad-based acumination; racemes or racemose corymbs well overtopping the floricane-foliage,* the ascending *pedicels with or without scattered minute setae* mostly *less than* 1 *mm. long; bracteal leaves* simple or 3-foliolate, *acute or acutish;* calyx-lobes often glandular; petals 3–5 mm. broad; fruit inferior, 6 or 8 mm. in diameter. (Incl. *R. Bigelovianus* Bailey) — Dryish meadows, swampy thickets, etc., s. N.E., se. N.Y. and e. Pa. *Fl.* mid-June–early July; *fr.* Aug., Sept.

132. R. ortìvus Bailey (eastern). — *Erect* or nearly so, slender, 2–6 *dm. high,* the *canes* unarmed below, *remotely short-setose above; primocane-leaflets* sublustrous and *deep green above, grayish-tomentulose beneath,* lanceolate, *with subcaudate tip* 1–1.5 *cm. long; bracteal leaves crowded, nearly equaling or overtopping the corymbiform raceme, the leaflets or narrow simple bracts acuminate; pedicels stipitate-glandular;* oblong calyx-lobes 5–6 mm. long, abruptly caudate-tipped; *petals* 5–8 *mm. broad.* — Dry open soil and dryish meadows, w. N.S. and se. Me. *Fl.* July; *fr.* Sept.

133. R. dissénsus Bailey (disagreeing). — Erect or arching, the hard *canes with stiff subulate prickles up to* 6 *mm. long;* petioles of primocane-leaves very prickly; *leaves dull green,* glabrous above, *soft-pubescent beneath;* the *terminal leaflet elliptic-ovate, short-acuminate; axis and suberect pedicels of corymb strongly armed with setae* 2–3 *mm. long,* the *lower pedicels up to* 6 *cm. or more long; lower bracteal leaves dilated,* trifoliolate or simple; *calyx-lobes* 3–5 *mm. long; petals* 7–9 mm. long, 2–4 *mm. broad.* — Borders of marshes, s. Mich.

134. R. Schneìderi Bailey (named in 1941 for its discoverer, RICHARD K. SCHNEIDER). — *Differing from no.* 133 in *its less prickly petioles; primocane-leaflets long-acuminate; inflorescence without leafy bracts, the short* (*up to* 2 *cm. long*) *divergent pedicels glandular and all subtended by minute bracts.* — Damp depressions among sand-hills, n. Ill.

135. R. mediòcris Bailey (mediocre). — *Similar to no.* 133 *but with few prickles; median primocane-leaflets with longer petiolules; corymb without foliaceous bracts, overtopped by the foliage; pedicels only about* 1 *cm. long, divergent, illustrated as without setae* (described as having them). — Borders of swamps, s. Mich.

136. R. unifórmis Bailey (uniform; a remarkable character in *Rubus*). — *In the copious* (100–300 *per dm.*) *reflexed hard subulate-unguiculate prickles suggesting no.* 133, *but the leaflets soft-pubescent beneath;* primocane-leaves with entire *stipules* 2–3 *cm. long; leaflets plicate,* ovate or ovate-elliptic, *caudate-acuminate; corymb compact, leafy-bracted at base, the short divergent pedicels pubescent but unarmed.* — W. Wisc.

137. R. hispidoìdes Bailey (resembling *R. hispidus*). — Ascending to reclining, 0.6–1.8 m. high; *primocane with remote firm acicular prickles up to* 4 *mm. long; primocane-leaves* dull, soft-pubescent beneath; *terminal leaflet elliptic-ovate to -obovate, abruptly acuminate, doubly serrate-dentate; floricane-leaves* with remotely acicular petioles, *the broadly elliptic leaflets only subacute or obtuse; racemes and corymbs with setose axis and ascending pedicels up to* 6 *cm. long;* calyx-lobes with blades 4–6 mm. long; *petals* 9–15 *mm. long,* 3–8 *mm. broad;* fruit 1–1.3 cm. in diameter. — Swampy thickets and woods or borders of bogs, e. Mass. *Fl.* late June, July; *fr.* mid-Aug., Sept. — Originally described from etiolated woodland specimens, the species at borders of bogs becomes stout, up to 1.8 m. high, with numerous and intricate racemiform corymbs borne the full length (up to 4 dm.) of the vigorous fertile shoots, the pedicels elongating to 6 cm.

138. R. perinvìsus Bailey (unseen or unknown). — *Strongly suggesting no.* 132, having similarly sparse prickles, slender stipules, leaflets with prolonged slender tips, etc., *but the upper half of the flowering shoot* (*petioles, axis and pedicels*) *copiously stipitate-glandular.* — Moist or dry thickets, P.E.I. and N.B. to sw. Me. — Probably only a glandular state of no. 132.

139. R. Bénneri Bailey (for its discoverer, WALTER MACKINNETT BENNER, 1888–).— *Resembling no.* 127, *but with leaves soft-pubescent beneath;* up to 2 m. high; primocane-leaflets abruptly long-acuminate; *petioles of floricane-leaves glandless;* racemose-corymbose *inflorescence with glandless axis but densely glandular pedicels.* — Damp thickets, e. Pa.

140. R. ascéndens Blanch. (ascending). — Canes 0.6–1.2 m. high, *with* 60–100 *strong subulate prickles per dm., these often* 6–9 *mm. long and with bases* 2–3 *mm. broad; leaves tomentose beneath;* terminal leaflet of primocane-foliage ovate to elliptic, acuminate, yellow-green, its petiolule 1.2–2.5 cm. long; *raceme or corymb* leafy-bracted below, its *axis and pedicels unarmed or only sparingly armed.* — Open thickets, fields and borders of woods, s. N.E. *Fl.* June, early July; *fr.* late Aug., Sept.

141. R. Clausénii Bailey (for its discoverer, Robert Theodore Clausen, 1911–). *Differing from no.* 140 *in its blunt to abruptly short-tipped leaflets;* prickles only 3–4 mm. long; *leaflets of primocane-foliage rounded-elliptic to elliptic-obovate, the terminal one three-fifths as broad as long; bracteal leaflets similar* but smaller; *inflorescence overtopped by bracts, its prickleless pedicels up to* 3 *cm. long; petals* narrowly obovate, 1.3–1.4 cm. long, 5–8 *mm. broad.* — Dry to wet peaty soil, n. N.J.

142. R. jejùnus Bailey (starved). — *More slender than no.* 141, *the canes with more slender prickles* 2–3 *mm. long;* rhombic-oval or elliptic-obovate *leaflets subacute; axis of inflorescence and short* (*up to* 1.5 *cm. long*) *pedicels prickly; petals* 7 mm. long, 3.5–5 *mm. broad;* fruit 7–8 mm. broad, of few drupelets. — Wooded swamps, s. Mich.

143. R. beàtus Bailey (prosperous). — *Canes* much as in no. 123, *densely covered with close-packed sharp acicular prickles, bristles and stipitate glands, many hundreds or some thousands per dm.,* 0.4–1 m. high; *leaves soft-pubescent to touch beneath; terminal primocane-leaflet cordate-ovate,* narrowly acuminate, sharply serrate, the *petioles densely setose and glandular; flowering branchlet and its petioles, pedicels and calyces similarly setose and glandular; raceme elongate, leafy-bracted to above middle, the lower pedicels up to* 8 *cm. long* and often forking. — Sandy and mossy low ground, Oswego Co., N.Y.

144. R. racèmiger Bailey (bearing racemes). — Very *similar to no.* 143, *but canes clearly evident between the scattered subulate prickles* (100–200 *per dm.*)*; primocane-leaves with glandless remotely prickly petioles;* primocane-leaflets mostly narrower; *raceme leafy-bracted only at base, its more divergent pedicels only about* 2 *cm. long.* — Upland W.Va.

§ 11. Cuneifòlii Bailey (see p. 823) Sand-B.

Leaflets subtruncate or abruptly short-tipped, coriaceous, obovate, white-pannose beneath. . . . . 145. *R. cuneifolius.*
Leaflets acute or acuminate, membranaceous to barely subcoriaceous, obovate, elliptic, oblong or ovate, gray-tomentose beneath.
Leaflets of primocanes chiefly 3; terminal leaflet broadly rounded at base, mostly five-eighths to practically as broad as long, coarsely dentate. . . 146. *R. Longii.*
Leaflets of primocane chiefly 5; terminal leaflet cuneate or subcuneate at base, rarely more than half as broad as long, sharply serrate. . . . . 147. *R. sejunctus.*

145. R. cuneifòlius Pursh (cuneate-leaved). — *Stiff,* erect to arching *fiercely prickly,* 3–9 *dm. high; primocanes ashy-tomentose when young, with many divergent or recurving broad-based strong pale prickles up to* 6 *mm. long; primocane-leaves* 3- or 5-foliolate, with arcuate-prickly petioles; *leaflets subcoriaceous, glabrate or glabrous and green and furrowed above, white-pannose beneath, cuneate-obovate, with subtruncate to broadly rounded summit, often abruptly short-pointed, coarsely dentate-serrate; terminal leaflet* 2–5 *cm. long; floricane* stiffly bushy-branched, *its leaflets cuneate-obovate;* flowers 1–∞, on variously prickly or unarmed pedicels; calyx-lobes 6–10 mm. long; *petals* 1–1.5 cm. long, 7–10 *mm. broad;* fruit rather dry, sweet, of good flavor, about 1 cm. in diameter. — Dry sandy or rocky soil of Coastal Plain and outer Piedmont, Fla., n. to se. Pa., N.J. and centr. Ct. *Fl.* May–mid-July; *fr.* July–Sept.

Var. **subellípticus** Fern. (nearly elliptical). — Leaflets rotund- to elliptic-obovate, more rounded at base. — Del. to se. Va.

146. R. Lóngii Fern. (for one of its discoverers, Bayard Long, 1885–). — Mostly *taller or more arching to diffuse than no.* 146; primocanes often branching, up to 1.8 m. long, occasionally tip-rooting; *longer prickles* 6–9 *mm. long; primocane-leaves mostly 3-foliolate, copiously strigose-pilose above when young, loosely cinereous-tomentulose beneath; the leaflets elliptic, ovate or elliptic-obovate, acuminate, membranaceous, scarcely furrowed above, coarsely dentate, the divergent broad teeth with a triangular callous tip; terminal leaflet* 3.5–8.5 *cm. long,* 2.5–5.5 *cm. broad;* upper bracts of inflorescence often simple, broad, dentate, 1.5–7 cm. long; fruit rich and juicy. (Incl. *R. pascuus* and *R. acer* Bailey; *R. uliginosus* Fern.) — Dry to wet argillaceous, siliceous or peaty soil, se. Va. to L.I. *Fl.* May, early June; *fr.* late June, July.

147. R. sejúnctus Bailey (separated). — Like no. 146, with canes erect or arching, sometimes tip-rooting, up to 1.3 m. long; *primocane-leaves mostly 5-foliolate; leaflets narrowly oblong- to elliptic-obovate, acuminate, firm, promptly glabrate and impressed-nerved above, cinereous-tomentose beneath, sharply serrate, with slender callous tips; terminal leaflets* 4.5–5.5 *cm. long,* 2–2.7 *cm. broad; bracteal leaves mostly 3-foliolate, with sharply serrate leaflets* 1–3 *cm. long.* — Dry to wet peaty or argillaceous or siliceous soil, se. Va. to se. N.C. *Fl.* May; *fr.* June, early July.

§ 12. Canadénses Bailey (see p. 824)

*a.*Canes prickleless (very rarely with remote small prickles); petiole of primocane-leaf unarmed or nearly so; flowers racemose (up to 25), on naked pedicels; wide-ranging. . . . . 148. *R. canadensis.*

*a.*Canes definitely prickly; petioles of primocane well armed; northern. . . *b.*

*b.*Principal inflorescence (excluding lower secondary ones) mostly 6–15-flowered, racemose or racemose-corymbose. . . *c.*

*c.*Bracteal leaflets or leaves toothed essentially to base.

Prickles of primocanes 2–10 per dm.; midrib of terminal primocane-leaflet unarmed; median leaflets cuneate or subcuneate at base, their petiolules 3–13 mm. long; pedicels of cylindric raceme 1–3.5 cm. long, unarmed. . . . . 149. *R. amicalis.*

Prickles of primocanes 30–100 per dm.; midrib of terminal primocane-leaflet usually somewhat prickly beneath; median leaflets broadly rounded at base, their petiolules 0.6–3 cm. long; pedicels of broad-topped racemose corymb mostly 2–7 cm. long, frequently armed. 150. *R. elegantulus.*

*c.*Bracteal leaves or leaflets nearly or quite entire toward the cuneate base; prickles of primocane 50–200 per dm. . . . . 151. *R. miscix.*

*b.*Principal inflorescences strongly corymbiform, 2–few-flowered.

Coarser prickles of primocane acicular or slenderly subulate; corymb barely or not at all exceeding the leaves, the stiff pedicels mostly 0.5–3 cm. long; wide-ranging northw. . . . . 152. *R. Kennedyanus.*

Coarser prickles of primocane broad-based; corymb greatly overtopping leaves, its filiform pedicels mostly 3.5–6 cm. long; local in Me. . . . 153. *R. multilicius.*

148. R. canadénsis L. (Canadian), Smooth B. — *Canes smooth and without* (or with exceptional and very remote) *prickles,* 0.5–3 m. high, erect or high-arching; *primocane-leaves with smooth petioles* or exceptionally with 1 or 2 small prickles; *leaflets* (3–)5(–7), ovate or ovate-lanceolate, long-acuminate, *glabrous on both surfaces, the median and terminal ones long-petiolulate, the terminal with broadly rounded to cordate base; floricanes strongly ascending* (unless lodged by heavy snow); *primary inflorescences elongate racemes of numerous (up to 25) flowers, standing well above the basal foliaceous bracts; most bracts reduced and stipuliform,* the rachis minutely pilose; petals 1.2–2 cm. long, 6–12 mm. broad; fruit globose to thimble-shaped, –12 mm. long, dryish or in some colonies juicy and of rich quality. (*R. Millspaughii* Britt.; *R. Randii* (Bailey) Rydb., the smallest extreme) — Thickets and clearings, Nfld. to w. Ont. and Minn., s. to N.S., N.E., Pa., upland to n. Ga. and Tenn. *Fl.* June, July; *fr.* Aug., early Sept.

149. R. amicàlis Blanch. (friendly). — Very similar to no. 148 and perhaps a phase of it; *differing in having scattered slender prickles on the canes (2–10 per dm.);* petioles of *primocane-leaves* smooth or only sparsely prickly; *the unarmed leaflets* relatively narrow, *the median ones tapering to a cuneate or subcuneate base and on short petiolules (3–13 mm. long); raceme* rather short, *its lower bracteal leaves or leaflets toothed to base, the pedicels 1–3.5 cm. long.* — N.B. and N.S. to sw. Me.

150. R. elegántulus Blanch. (neatish). — *Erect to low-arching or,* after heavy snows, *trailing; primocanes armed with firm subulate prickles, these 30–100 or more per dm.; petiole of primocane-leaf usually with a few recurved prickles; leaflets* much as in no. 148, *the ovate or ovate-lanceolate median ones broadly rounded at base, their petiolules 0.6–3 cm. long; terminal ovate leaflet* long-petiolulate, *commonly with a few prickles along midrib beneath; raceme corymbiform, broad across the top, its* simple or forking *slender ascending often slightly armed pedicels mostly 2–7 cm. long;* otherwise much like no. 148. (*R. canadensis,* var. Farw.) — Centr. Nfld. to s. Que., s. to N.B., N.E., n. N.J. and ne. Pa. *Fl.* June, early July; *fr.* Aug., early Sept.

151. R. míscix Bailey (changeable). — Much more prickly than no. 150, the *acicular prickles* 35–200 *per dm.* on new primocane; *petiole of primocane-leaf very prickly; midrib of terminal leaflet usually armed beneath; raceme usually overtopped by the bracteal leaves; leaflets or simple blades of bracteal leaves entire near base (one-fourth to one-third their length).* (*R. peculiaris* Blanch., not Sampaio) — Se. and centr. to sw. Me.

152. R. Kennedyànus Fern. (for its discoverer, RAE BALDWIN KENNEDY, 1879–). — *Canes* 0.3–1.5 dm. high, *with scattered straight acicular prickles;* plant suggesting nos. 148 and 150 but *with few-flowered* (2–7) *corymbs; petiole of primocane-leaf somewhat armed; the broadly ovate terminal leaflet with rounded to cordate base; corymbs mostly overtopped by the bracteal leaves, the stiff pedicels mostly* 0.5–3 *cm. long.* (Incl. *R. acridens, R. quaesitus* and *R. ulterior* Bailey) — Nfld. to Minn.

153. R. multilícius Bailey (with many threads; from the very slender pedicels). — Suggesting slender no. 148 but the *primocane strongly armed with broad-based hard triangular-subulate divergent prickles; petiole, petiolules and midribs beneath of primocane-leaflets very prickly;* leaflets caudate-acuminate, lanceolate or narrowly ovate, glabrous; *raceme corymbiform, broad across the summit, elevated high above the bracts,* the *very slender filiform arched-ascending to spreading pedicels mostly* 3.5–6 *cm. long; calyx-lobes narrowly lanceolate, slender-tipped.* — Dry open soil, Piscataquis to Lincoln Co., Me.

§ 13. ALLEGHENIÉNSES Bailey (see p. 824)

a. Axis of primocane (except sometimes at expanding tip) without or essentially without stipitate glands on the internodes or among the prickles. . . *b.*

b. Primary inflorescences (not always those of lower shoots) cylindric and elongate racemes two to four times as long as thick (mostly 1–3 dm. long and 2.5–10 cm. thick). . . *c.*

c. Primocanes with broad-based prickles mostly 3–10 mm. long (prickles wanting in one var. of no. 154); leaflets regularly and rather finely double-serrate; the terminal primocane-leaflet broadly to narrowly ovate, with cordate or broadly rounded base; calyx-lobes ovate or ovate-lanceolate.

Prickles of lower half of primocane distant (or wanting), not crowded, 2–4 mm. broad at base, narrowed to lance-subulate straight summits. . . . . 154. *R. allegheniensis.*

Prickles of lower half of primocane crowded, confluent or clustered, the larger ones 5–9 mm. broad at base and subdeltoid. . . . . . . 155. *R. pugnax.*

c. Primocanes with slender acicular prickles 100–200 per dm.; primocane-leaflets coarsely and subsimply incised, lanceolate or lance-ovate and subcuneate or only slightly rounded at base; calyx-lobes lance-attenuate. . . . . 156. *R. flavinanus.*

b. Primary inflorescences racemose-subcorymbose, broadest or with flowers crowded at summit, mostly one-half as broad to essentially as broad as long. . . *d.*

d. Terminal primocane-leaflet narrowly oblong- or ovate-lanceolate, merely rounded to base, less than half as broad as long; expanded flower about 2 cm. or less across; fruit 8 or 9 mm. thick. . . . . . . 157. *R. saltuensis.*

d. Terminal primocane-leaflet broadly ovate, broadly rounded or cordate at base, three-fifths to four-fifths as broad as long; flower 2–4 cm. broad; fruit 1.5–2 cm. long. . . *e.*

e. Leaflets of primocane with long-attenuate to caudate tips, the tip of the terminal leaflet 1.5–2.5 cm. long; flowers 3–4 cm. broad; wide-ranging. . . . . 158. *R. alumnus.*

e. Leaflets of primocane merely acute or abruptly short-tipped, the narrow tip of the terminal leaflet only 4–10 mm. long.

Blades and leaflets of bracteal leaves acute; flowers standing well above the foliaceous bracts; primocane-leaflets 5; median primocane-leaflets half to three-fifths as broad as long, definitely petiolulate; flowers about 2 cm. broad; plant of s. N.E. . . . . . 159. *R. paulus.*

Blades and leaflets of bracteal leaves obtuse, equaling or overtopping the upper flowers; primocane-leaflets 3, the broad paired ones subsessile; flowers 2–3 cm. broad; midwestern. . . . . . . 160. *R. attractus.*

a. Axis of primocane bearing several to very many long stipitate glands among the prickles. . . *f.*

f. Inflorescence regularly flowering above the basal bracteal leaves. . . *g.*

g. Inflorescence a regular elongate cylindric raceme, usually without basal forking. . . *h.*

h. Prickles of primocane firm and strong, not mixed with glandless bristles; canes angled or corrugated.

Prickles subulate-acicular; terminal primocane-leaflet cordate; upper bracteal leaves with obtuse leaflets. . . . . 161. *R. ravus.*

Prickles with broad deltoid bases; terminal primocane-leaflet merely rounded at base; upper bracteal leaves sharply acute. . . . 162. *R. nuperus.*

h. Prickles acicular and bristleform, densely intermixed; leaflets of bracts acuminate; canes terete. . . . . 163. *R. reravus.*

g. Inflorescence corymbiform, broadest toward summit, often forking below. . *i.*

i. Inflorescence essentially unarmed, at most with a rare weak and short spicule or two.

Leaflets sparsely pilose beneath, their principal veins (beneath) merely pilose to glabrescent, their petioles glabrous or only sparsely pilose and gland-bearing; leaflets of both primocane and floricane acuminate; wide-ranging northeastw. . . . . . . 164. *R. glandicaulis.*

Leaflets densely velutinous beneath, their principal veins and petioles densely spreading-villous and glandular; floricane-leaflets obtuse or bluntish; local in se. Vt. . . . . 165. *R. frondisentis.*

i. Inflorescence armed with several to many hooked or straight prickles or elongate setae; leaflets acuminate.

Primocanes bearing acicular to slenderly subulate scattered prickles; petioles glabrous to but slightly pubescent, without or with few glands; leaflets sparingly pilose and with appressed-pilose to glabrous veins beneath; wide-ranging northeastw. . . . . . 164. *R. glandicaulis.*

Primocane with crowded deltoid-subulate prickles 2–4 mm. broad at base; petioles heavily armed and copiously stipitate-glandular; leaflets densely velvety and with spreading-villous and coarsely prickly glandular midrib beneath; local in ne. N.H. . . . . 166. *R. sceleratus.*

f. Inflorescence of few axillary flowers much exceeded by foliaceous bracts and terminating in a tuft of simple leaves. . . . . 167. *R. inclinis.*

154. R. alleghenìénsis Porter (of the Alleghenies), SOW-TEAT B. — *Erect or high arching,* mostly 1–3 m. high, armed *with scattered broad-based lanceolate to lance-subulate prickles or prickles wanting,* the young primocanes often ridged or angled and finely pubescent, the expanding tip often glandular; *primocane-leaves* mostly 5-foliolate (sometimes 3- or 7-foliolate), *on pubescent and mostly stipitate-glandular* unarmed or armed *petioles; leaflets soft-pilose or velvety beneath,* pilose above when young; *terminal leaflet ovate, cordate or broadly rounded at base, doubly serrate, with long attenuate or caudate tip,* long-petiolulate; the *median paired leaflets petiolulate* and mostly narrower; floricanes simply or intricately branching; *principal racemes cylindric or subcylindric, mostly* 1–3 *dm. long and* 2.5–10 *cm. thick,* the *rachis and* spreading-ascending to divergent *pedicels copiously stipitate-glandular* and either unarmed or armed; *calyx-lobes ovate to ovate-lanceolate,* often with elongate tips; *petals* 1–2 *cm. long,* 4–12 *mm. broad; fruit globose to thimble-shaped, up to* 2 *cm. long,* either dryish or juicy and of rich flavor. — Common and wide-ranging species with several recognizable vars. and forms, of which the following are best marked:

a. Primocane and usually lower internodes of floricane bearing broad-based stout prickles mostly 5–10 mm. long; the canes often angulate, rather coarse, up to 3 m. high. . . *b.*

b. Bracteal leaves and leaflets elongate and acute. . . *c.*

c. Well-developed flowering or fruiting racemes mostly 5–10 cm. thick, with the longer pedicels mostly 3–9 cm. long; calyx-lobes (excluding slender tips) 5–8 mm. long (in forma *calycosus* foliaceous and up to 7 cm. long); petals 1.2–2 cm. long, 5–12 mm. broad. . . *d.*

d. Rachis of raceme and pedicels unarmed or with few remote fine bristles among the many glands; petioles and petiolules unarmed or remotely armed. . . . *e.*

e. Foliaceous bracts of raceme 1–5, subtending only the lowest flowers.

Calyx-lobes not foliaceous, the blades 5–8 mm. long; fruit plump and juicy.

Fruit purple-black. . . . . *R. alleghenìensis* (typical).

Fruit amber-white. . . . . Forma *albinus.*

Calyx-lobes changed to thin leaf-like often toothed structures up to 7 cm. long; fruit small and dry. . . . . Forma *calycosus.*

e. Foliaceous bracts of raceme 6–15, subtending and usually overtopping most of the single or fascicled pedicels. . . . . Forma *suffultus.*

d. Rachis of raceme strongly armed with stout prickles; some or most pedicels usually bearing 3–8 hard prickles; petioles and petiolules strongly armed. . . . . Var. *neoscoticus.*

c. Well-developed flowering or fruiting raceme 2.5–5 cm. thick, with longer pedicels 1–2 cm. long; calyx-lobes (excluding slender tips) 3.5–5 mm. long; petals 1 cm. long and 4–7 mm. broad. . . . . Var. *plausus*.

b. Bracteal leaves or leaflets round or roundish-elliptic, rounded at summit. . Var. *populifolius*.

a. Primocanes and floricanes prickleless or with only rare prickles 1–3 mm. long on exceptional internodes; the canes slender and terete, 2–5 mm. thick and 0.3–1.5 m. high; racemes 3–7 cm. in diameter, mostly without prickles. . . Var. *Gravesii*.

R. allegheniénsis (typical). — Canes up to 1 cm. or more thick at base; terminal primocane-leaflet 0.7–2 dm. long, 3.5–11 cm. broad; lower elongate pedicels often forking. (Incl. *R. nigrobaccus, auroralis, longissimus, virginianus, separ, uber, marilandicus, Fernaldianus* and *Rappii* Bailey) — Dry clearings and thickets, usually abundant, N.B. and s. Que. to Minn., s. to N.S., N.E., L.I., Md., upland N.C. and Tenn., and Mo. *Fl.* May–July; *fr.* Aug., Sept. — Forma **albìnus** (Bailey) Fern. (albino), scarce; forma **calycòsus** Fern. (with enlarged calyx), locally abundant, propagated by basal offsets; forma **suffúltus** Fern. (supported or subtended, incl. *R. fissidens* Bailey, a less defined state), locally, N.H. to N.Y. and Ct.

Var. **neoscóticus** (Fern.) Bailey (Nova Scotian). — As robust as the typical var., more prickly. (*R. glandicaulis* var. Fern.; *R. pennus* Bailey) — Common in N.S., infreq. w. to N.Y.

Var. **plaûsus** Bailey (praiseworthy). — Chiefly in very dry soil, se. Me. to sw. Que., w. to Mich. and Ind., s. to s. N.E. and n. N.J.

Var. **populifòlius** Fern. (poplar- or aspen-leaved). — Borders of woods, Berkshire Co., Mass.

Var. **Gràvesii** Fern. (for its discoverer, Charles Burr Graves, 1860–1936). — In its usually prickleless and slender canes a well-marked extreme (*R. Gravesii* (Fern.) Bailey) of woods and thickets, w. Me. to ne. N.Y., s. to Md.

155. R. púgnax Bailey (pugnacious). — Differing from coarser states of no. 154 in having the lower half of the *glabrous primocanes with crowded or confluent subdeltoid hard prickles, the larger ones 5–9 mm. broad at base; petioles, petiolules and midribs (beneath) of primocane-foliage copiously prickly, the margins of the leaflets with mostly simple teeth;* racemes 0.5–1 dm. long. — Upland pastures, Hartford Co., Ct.; reports from e. Mass. to centr.-w. N.Y. need confirmation.

156. R. flavinànus Blanch. (yellowish and small). — *Slender primocanes 4–9 dm. high, yellowish,* glabrous, *their median and upper internodes bearing 100–200 acicular often hooked prickles per dm.; primocane-leaves* yellow-green, 5-foliolate, *glabrous above, pale and velvety beneath; petiole, petiolules and midribs beneath strongly armed; leaflets lanceolate to lance-ovate, coarsely and subsimply incised-serrate, subcuneate toward base;* terminal leaflet 7–10 cm. long, 3–4.5 cm. broad, long-acuminate; *racemes cylindric,* often branching near base, the densely villous rachis and pedicels copiously stipitate-glandular, the longer simple pedicels 2–3 cm. long; *calyx-lobes narrowly lanceolate and prolonged;* petals about 1.2 cm. long, 4–6 mm. broad; fruit small and inferior. — Dry upland (at about 550 m.), Windham Co., Vt. *Fl.* June; *fr.* Aug.

157. R. saltuénsis Bailey (of bushy pastures). — Similar to no. 156, but *primocane with only 20–30 straight prickles per dm.; petiole, petiolules and midrib beneath of leaflets scantily armed or the latter essentially unarmed; leaflets with fine double serration; terminal leaflet with broadly rounded subcordate base; inflorescence racemose-corymbiform, broadest at summit, the rachis and long (up to 5.5 cm.) pedicels only sparsely stipitate-glandular; calyx-lobes ovate, short-tipped.* — Shaded thickets, Tolland Co., Ct.

158. R. alúmnus Bailey (nursling). — *As coarse as larger plants of* no. 154, *the canes similarly corrugated and prickly;* the *terminal and median broadly ovate long-acuminate leaflets strongly cordate* or rounded at base and *with caudate acumination* 1.5–2.5 *cm. long; raceme usually more compact, with the upper flowers more crowded, or tending to be corymbiform and one-half to nearly as thick as long; flowers* 3–4 *cm. broad;* the succulent *fruit* 1.5–2 *cm. long.* (Incl. *R. Rosa, impos* and *apianus* Bailey) — Dry pastures, old fields and clearings, in part spread from cult., N.S. and Me. to sw. Que., w. to Minn., s. to s. N.E., Va., Ky. and Mo. — Source of several horticultural vars.

159. R. paûlus Bailey (little). — Rather slender, 0.6–2 m. high, *with scattered straight subulate-acerose prickles* and angled canes; *primocane-leaves* 5-foliolate, *the terminal and petiolulate median leaflets ovate, with acute tips up to 1 cm. long and rounded to subcordate bases,* finely double-serrate; *bracteal leaves and leaflets acute, mostly ovate; short corymbiform raceme well overtopping the larger bracts;* the glandular pedicels 1–3 cm. long; calyx-lobes lanceolate; flowers about 2 cm. broad. (Incl. *R. licitus* Bailey, state with pubescent primocane-petiole) — R.I. to se. N.Y.

160. R. attráctus Bailey (attractive). — Canes subterete, with scattered straightish *subulate prickles 2–3 mm. long; primocane-leaves 3-foliolate,* with glabrous petiole and petiolule; *terminal leaflet ovate, merely acute,* broadly rounded at base, *coarsely serrate-dentate,* 5–6 cm. long; *paired* *leaflets obliquely rhombic, subsessile; bracteal leaves or leaflets mostly obtuse and broadly oblong to narrowly obovate, the upper about equaling or overtopping the short corymbiform raceme; flowers 2–3 cm. broad,* with broad petals; *calyx-lobes ovate, not prolonged* at tip. — Dry sands, sw. Que. to s. Mich.

161. R. ràvus Bailey (grayish-yellow or tawny). — *Suggesting small extremes of* no. 154, *but with the angular primocane bearing many stipitate glands interspersed among the* subulate-acicular *divergent prickles; primocane-leaves with strongly stipitate-glandular and prickly petiole and petiolules; the long-acuminate cordate-ovate doubly serrate terminal leaflet grayish with velvety pubescence beneath; cylindric raceme* strongly glandular, the *lower flowers subtended by 3-foliolate bracts with narrowly obovate blunt leaflets;* calyx-lobes lance-attenuate; *corolla scarcely 2 cm. broad;* petals about 8 mm. long and 4–5 mm. broad. — Dry soil, Knox Co., Me. to se. Vt.

162. R. nùperus Bailey (fresh or new). — *Resembling strong and tall much-branched states of no.* 154, up to 3 m. high and diffusely branched and tangled; the *primocane with numerous stipitate glands and remote deltoid-subulate hard prickles;* petiole and petiolules of *primocane-leaves* stipitate-glandular, *the terminal* ovate acuminate *leaflet rounded at base; bracts of cylindric raceme with acuminate leaflets;* calyx-lobes lance-attenuate; petals 1–1.2 cm. long. — Low grounds, s.-centr. N.Y.

163. R. reràvus Bailey (another *R. ravus*). — *Coarser than no.* 161, *with terete canes up to 1 cm. in diameter; primocanes densely covered by acicular prickles, setae and stipitate glands closely intermixed; primocane-leaves with similarly clothed petiole and petiolules;* raceme simple or with lower pedicels forking; *calyx-lobes lance-attenuate with prolonged tips; petals narrowly oblanceolate,* 1.4–1.8 *cm. long.* — Mts. of Garrett Co., Md.

164. R. glandicaùlis Blanch. (with glandular stem). — High arching (or through weight of snow low-arching to depressed), up to 2 m. high; *primocanes coarse, angled, with acicular or slenderly subulate prickles mixed with stipitate glands; primocane-leaves with glabrous to sparsely pilose and glandular petioles and petiolules; leaflets acuminate, thinly pilose beneath,* glabrous or glabrescent above, *the principal veins beneath appressed-pilose or glabrescent, the terminal leaflet ovate or oval and cordate or broadly rounded at base; inflorescences racemose-corymbose, the lower pedicels often becoming forked elongate branches,* the *bracteal leaves or leaflets acute or acuminate; rachis and pedicels unarmed or with few setae;* calyx-lobes lanceolate, often prolonged; *petals* oblanceolate to oblong-obovate, 0.8–1.5 *cm. long,* 5–8 *mm. broad; fruit subglobose to subcylindric,* 0.8–1.5 cm. long, of good quality. (Incl. *R. frondisentis* sensu Brain. & Peitersen, not Blanch.; *R. montpelierensis* Blanch.; *R. acadiensis* Bailey) — Dry to moist thickets and clearings, N.B. to sw. Que., s. to N.S., n. N.E. and ne. N.Y. *Fl.* June, July; *fr.* Sept.

R. abbrèvians Blanch. (short or abbreviated), a low plant of upland Vt. with fewer glands and nearly terete slender canes, the leaves described as either velutinous or glabrous beneath, is presumably a mixture of no. 164 and some other species.

165. R. frondisèntis Blanch. (leafy bramble). — Erect, 0.9–1.5 m. high; *primocanes slender, terete in maturity, densely covered with acicular prickles, setae and intermixed stipitate glands; primocane-leaves firm,* yellow-green above, *densely gray-velvety beneath, with the principal veins beneath spreading-villous and glandular;* leaflets 3 or 5, ovate, acuminate, broadly rounded at base; inflorescence similar to that of no. 164; *bracteal leaves heavy and thick, the simple blades or leaflets ovate to obovate or rhombic, obtuse, with coarse dentation.* — Local, Windham Co., Vt.

166. R. sceleràtus Brainerd (vicious or accursed). — Strong, *high-arching, freely forking, making intricately entangled domes 1.8–3 m. high; primocanes closely armed with broad-based deltoid-subulate fulvous prickles* (2–4 *mm. broad at base*) *intermixed with more slenderly subulate prickles and stipitate glands; primocane-leaves with similarly armed petioles and petiolules;* the ovate acuminate coarsely double-toothed *leaflets* glabrous above, *fulvous-tomentose beneath; the midrib beneath spreading-villous, coarsely prickly and glandular; inflorescence* a very leafy corymbiform raceme or open panicle, *with mostly forking ascending branches, the axis and branches or pedicels armed with firm setae;* calyx-lobes ovate-lanceolate; fruit subglobose about 8 mm. in diameter. — Clearings, terraces of Androscoggin R., ne. N.H.

167. R. inclìnis Bailey (inclined or leaning). — Slender, with ascending angled primocanes and arching or leaning floricanes; primocanes with remote slender prickles, rufous pubescence and many stipitate glands; leaves dull; those of primocane soft-pubescent beneath, with glandular and prickly petiole and petiolules; terminal leaflet ovate, subcordate; *inflorescence a few long-pedicelled flowers in the axils of the median long-petioled simple or 3-foliolate acute simply toothed bracts, with simple leaves produced well above the flowers.* — Marshland, s.-centr. N.Y.; an anomalous plant.

§ 14. ARGÙTI Rydb. (see p. 824)

a. Upper 3 leaflets of fully developed primocane-leaves twice to thrice as long as broad, acuminate. . . *b.*

*b.*Teeth (or many of them) of primocane- and floricane-leaflets simply sharp-serrate, the longer ones 3–6 mm. long. . . . . 168. *R. argutus.*

*b.*Teeth (or most of them) of primocane- and floricane-leaflets doubly serrate and rarely more than 2–4 mm. long. . . *c.*

*c.*Primocane-leaflets mostly 5 (or 7), acuminate; floricane-leaflets mostly acuminate. . . *d.*

*d.*Lower surfaces of primocane- and floricane-leaflets densely velvety or soft-pubescent to touch.

Larger primocane-leaflets, bracteal leaves or terminal leaflets mostly only one-third as broad as long; fruit amber-white. . . . . . 169. *R. louisianus.*

Larger primocane-leaflets two-fifths to half as broad as long; simple bracts or terminal bracteal leaflets two-fifths to three-fifths as broad as long; fruit purple-black. . . . . 170. *R. jugosus.*

*d.*Lower surfaces of primocane- and floricane-leaflets sparsely pilose to glabrescent, hard (not velvety) to touch. . . . . 171. *R. Blakei.*

*c.*Primocane-leaflets mostly 3, these and floricane-leaflets merely acute or acutish. . . . . 172. *R. fatuus.*

*a.*Upper leaflets of fully developed primocane-leaves mostly less than twice as long as broad. . . *e.*

*e.*Primary inflorescences cylindric racemes, not broadened across the top. . . *f.*

*f.*Terminal (and usually the lower) primocane-leaflets with continuous margin, not lobulate. . . *g.*

*g.*Leaves densely tomentose beneath, velvety to touch; pedicels, rachis, etc. villous or tomentose; expanding leaves of primocane-tip white-tomentose. . . *h.*

*h.*Terminal leaflet of 5-foliolate leaves oblong- to elliptic-ovate, only slightly more than half as broad as long, gradually long-attenuate, glabrous or promptly glabrate above; fruit 1–1.5 cm. thick.

Ribs of leaves slender, often hidden and not prominent beneath; longer pedicels of leading racemes filiform, 2–5 cm. long; fruit globose, about 1 cm. long. . . . . 173. *R. floricomus.*

Ribs strong and conspicuous beneath; longer pedicels stoutish, 0.5–2 cm. long, unarmed; fruit short-cylindric, about 1.5 cm. long. . . . . 174. *R. amnicola.*

*h.*Terminal leaflet of 5-foliolate leaves broadly cordate- or subcordate-ovate, much more than half as broad as long, abruptly acuminate. . . *i.*

*i.*Leaves strigose-pilose above; pedicels stoutish; corolla 3.5–4 cm. broad; fruit becoming 2–3 cm. long, of rich quality.

Calyx-lobes spreading or only slightly arched-recurving at and after anthesis; fruit subglobose; erect or high-arching shrub of N.S., N.E. and L.I. . . . . 175. *R. facetus.*

Calyx-lobes abruptly reflexed at and after anthesis; fruit ellipsoid to thimble-shaped; intricate shrub, with arching branches nearly reaching the ground, of N.Y. to Mich., s. to mts. of nw. N.J. . . . . 176. *R. bellobatus.*

*i.*Leaves glabrous above; pedicels filiform; corolla 2 cm. broad; fruit globose, 1 cm. long; western.

Leaves membranaceous, their veins not prominent beneath; terminal primocane-leaflet on petiolule 4–6 cm. long; raceme much elongate, its lower pedicels 3–4.5 cm. long; mid- and north-western. . . . . 177. *R. bractealis.*

Leaves becoming hard and coriaceous, their veins very prominent beneath; terminal primocane-leaflet on petiolule 1–3 cm. long; raceme short, its lowest pedicels up to 2 cm. long; south-western. . . . . 178. *R. mollior.*

*g.*Leaves in maturity merely slightly pilose beneath and not velvety to touch (mature no. 178 might be sought here); expanding leaves merely gray-green; terminal primocane-leaflet cordate- or subcordate-ovate.

Pedicels mostly unarmed, the longer ones 3–5 cm. long; longer racemes 12–20-flowered; petiolule of terminal primocane-leaflet 3–4.5 cm. long. . . . . 179. *R. avipes.*

Pedicels prickly, the longer ones 3 cm. long; racemes 8–10-flowered; petiolule of terminal primocane-leaflet 2 cm. long. . . . . . 180. *R. associus.*

*f.*Terminal (and often other) primocane-leaflets angulate-lobulate.

Prickles subulate; primocane-leaves 5-foliolate; plants of W.Va. or Ky.

Racemes of lateral branches solitary, compact; the pedicels mostly exceeded by the bracts. . . . . 181. *R. condensiflorus.*

Racemes of lateral branches numerous, subpaniculately disposed, loose and open; pedicels much exceeding bracts. . . . . 182. *R. congruus.*

Prickles deltoid, broad-based; primocane-leaflets 3 or 5; plant of s.-centr. N.Y. . . . . 183. *R. independens.*

*e.*Primary inflorescence subcorymbose-racemose, with short axis, corymbosely broadened across the top; or flowers only 1–3 at tips of branches. . . *j.*

*j.*Flowers (1–)2–several in well-developed inflorescences. . . *k.*

*k.*Margin of terminal primocane-leaflet continuous, not lobulate, with relatively short teeth and shallow sinuses (mostly 1–5 mm. deep). . . *l.*

*l.*Canes becoming woody, up to 3 m. high, stout or slender, erect or ascending, mostly prickly; terminal primocane-leaflets (3.5–)6–20 cm. long; corollas 2–3.5 cm. broad; fruits mostly 1–3 cm. long. . . *m.*

*m.*Most of the pedicels subtended by small stipule-like bracts (all or nearly all foliaceous in an aberrant form of no. 184); large foliaceous bracts usually only 1–3 or 4. . . *n.*

*n.*Pedicels all or nearly all unarmed (occasionally 1 or 2 of them slightly armed); veins of primocane-leaves not strongly impressed above; fruit usually of good quality. . . *o.*

*o.*Bracteal leaves or leaflets prevailingly acute; terminal primocane-leaflet definitely longer than broad, often long-acuminate. . . *p.*

*p.*Terminal primocane-leaflet ovate to elliptic, with rounded to cordate base.

Primocane-leaves (except lowest) mostly 5-foliolate and quite unlike leaves of flowering shoots; terminal leaflet acuminate; wide-ranging, through most of our area. . 184. *R. pensilvanicus.*

Primocane-leaves mostly 3-foliolate, resembling foliage of flowering shoots; terminal leaflet elliptic, merely acutish; low (up to 6 dm.) shrub of e. Md. . . . . . 185. *R. libratus.*

*p.*Terminal primocane-leaflet rhombic-obovate, cuneate at base; small-leaved plant of Md. . . . . 186. *R. subsolanus.*

*o.*Bracteal leaves or leaflets mostly obtuse, elliptic-oblong to oblong-obovate; terminal primocane-leaflet round-ovate or obovate, abruptly tipped, about four-fifths as broad as long; tangled shrub with pendulous branchlets, of coast of s. N.E. and N.Y. . . . . 187. *R. insulanus.*

*n.*Pedicels all or nearly all strongly armed; subcoriaceous small leaves with straightish nerves impressed above; fruit poor, often bitter; stiff shrub of N.E. . . . . 188. *R. barbarus.*

*m.*Most (at least more than half) of the pedicels subtended by foliaceous bracts. . . *q.*

*q.*Stouter parts of primocane with 4 or more rows of broad-based (4–6 mm.) often crowded or approximate hooked prickles in groups rarely 1 cm. apart; intricately branching canes with long-arching and tip-trailing branchlets; shrub of s. N.E. . 103. *R. multispinus.*

*q.*Primocanes with scattered and usually straighter or more slender prickles; branchlets not usually trailing. . . *r.*

*r.*Foliaceous bracts of leading inflorescences 7–12, subtending all or nearly all pedicels.

Simple bracteal leaves narrowly ovate to lanceolate, long-attenuate; pedicels unarmed; eastern. . . . . . 184. *R. pensilvanicus,* forma *phyllophorus.*

Simple bracteal leaves broadly ovate to reniform, lobulate, short-pointed; pedicels armed; northwestern. . . . . 189. *R. latifolilus.*

*r.*Foliaceous bracts of leading inflorescences fewer, subtending only the lower and median pedicels. . . *s.*

*s.*Primocanes soon forming long-arching or doming shrubs with drooping to tip-trailing branches and branchlets. . . *t.*

t. Larger bracts or terminal leaflets of bracts mostly 4–7 cm. long.
Upper flowers equaled or overtopped by the foliaceous bracts; prickles of primocane subulate, nearly straight; wide-ranging. . . . . 190. *R. frondosus*.
Upper flowers greatly overtopping the foliaceous bracts; prickles of primocane broad-based and hooked; southeastern. . . . . 191. *R. floridus*.
t. Larger bracts or terminal leaflets of bracts 1.5–2.5 cm. long; pedicels mostly overtopping the bracts; shrub of se. cypress-swamps. . . . . 192. *R. cupressorum*.

s. Primocanes stiffly erect or but slightly arching at summit. . . *u*.
u. Terminal primocane-leaflets 7–10 cm. long; larger bracts or their terminal leaflets 4–6 cm. long; corymbs 3–12-flowered; pedicels glandless; shrubs of the interior.
Terminal primocane-leaflet ovate, broadest toward base; prickles of primocane subulate or acicular. 193. *R. pratensis*.
Terminal primocane-leaflet oblong-elliptic, broadest near middle; prickles broad-based. . . . 194. *R. Bushii*.
u. Terminal primocane-leaflet 4–7 cm, long; larger bracts or their terminal leaflets 1.5–3.5 cm. long; corymbs 2–4-flowered; pedicels often bearing some short glands; small plants of southeastern states.
Floricanes 2–7 dm. high, with strongly ascending branches, erect; leaflets of floricane ovate, acute. . 195. *R. pauxillus*.
Floricanes 0.6–1.5 m. high, with widely divergent to horizontal branches, arching above; leaflets of floricane narrowly cuneate-obovate or -subrhombic, bluntish. . . . . 196. *R. defectionis*.

l. Canes subherbaceous or weak and very slender, not strongly woody in age, 1.5–5 mm. in diameter, 0.2–1 (–1.5) m. high; terminal primocane-leaflets mostly 4–9 (–11) cm. long; corollas 1.5–2 cm. broad; fruit 0.7–1.5 cm. long. (no. 195 and slender plants of other preceding species might be looked for here). . . *v*.
v. Terminal primocane-leaflet broadly cordate-ovate, nearly as broad as long, the larger ones 7–11 cm. long.
Leaves serrate-dentate, the teeth of primocane-leaves often as broad as long; canes unarmed or with very remote subulate prickles. . . . . 197. *R. Brainerdii*.
Leaves sharply serrate, the teeth longer than broad; canes with several prickles per dm. . . . . 198. *R. perfoliosus*.
v. Terminal primocane-leaflet scarcely cordate, more narrowly ovate, definitely longer than broad, 4–9 cm. long.
Leaflets of primocane-foliage and blades or leaflets of bracts sharp-pointed, with prominent and very sharp teeth. 199. *R. pauper*.
Leaflets of primocane-foliage and blades or leaflets of bracts only acute or acutish, with small and merely acute or acutish teeth. . . . . 200. *R. perpauper*.
k. Margin of terminal primocane-leaflet rather definitely but irregularly 1-∞-lobulate, with often deep sinuses. . . *w*.
w. Primocane-leaflets prevailingly 5. . . *x*.
x. Preponderant inflorescences exceeded by or nearly equaled by the numerous bracteal leaves; canes or their branches and branchlets soon arching or recurving, the tips nearly or quite reaching the ground and sometimes tip-rooting. . . . . . 201. *R. recurvans*.
x. Preponderant inflorescences with most of the pedicels raised or borne well above the 1–3 basal bracteal leaves; canes not long-arching nor with branches reaching the ground.
Canes terete; terminal primocane-leaflet subrhombic; most flowers borne high above bracteal leaves. . . . . . 202. *R. praepes*.
Canes angled; terminal primocane-leaflet tapering from cordate-ovate base; bracts of lower pedicels exceeding the inflorescence. . . . . 203. *R. Cardianus*.

w.Primocane-leaflets prevailingly 3; terminal primocane-leaflet round-ovate. . . . . 204. *R. wisconsinensis.*

j.Flower solitary on an erect strongly armed axillary pedicel; primocane-leaves 3-foliolate; small erect plant with simple floricane; southeastern. 205. *R. dissitiflorus.*

168. R. argûtus Link (sharply serrate). — Erect or high-arching and stiff, up to 2.5 m. high; *primocanes* angled and furrowed when young, *with scattered subulate straight or slightly arching hard prickles mostly* 4–10 *mm. long, the stouter prickles broadening at base to a width* of 2–5 *mm.; primocane-leaflets mostly* 5-*foliolate,* the glabrous petiole with a few unguiculate prickles; *leaflets narrowly oblong-lanceolate, acuminate, velvety beneath; the terminal one about one-third as broad as long and gradually narrowed to base, with mostly sharp simple coarse teeth* 3–6 *mm. long; full-grown primocane-leaflets usually with* 3–4 (–5) *teeth per cm. of median margin;* floricanes with leaflets of flowering shoots coarsely toothed; *racemes few-flowered,* often corymbiform, *the rachis unguiculate-prickly,* the *narrow bracteal leaves and leaflets coarsely and mostly simply toothed,* the densely pilose pedicels mostly unarmed; calyx-lobes lanceolate or lance-ovate; petals 1.2 cm. long, 6 mm. broad; fruit globose to thick-cylindric, 1 cm. thick. (Incl. *R. incisifrons* Bailey) — Dry or moist thickets and borders of woods, Ga. and Ala. to Ark., n. to se. Mass., N.J., Pa., Ky. and s. Ill. *Fl.* May, early June; *fr.* late June–early Aug.

169. R. louisiànus Berger (of Louisiana). — Differing from no. 168 in the very fine teeth of the primocane-leaflets, these only 2–3 mm. long; *simple blades and terminal leaflets of bracteal leaves only one-third as broad as long; fruit amber-white.* — Local, S.C. to La., n. to e. Md., Va. and Ky.

170. R. jugòsus Bailey (of the mountains; from type-region). — Differing from no. 169 in the broader and shorter *primocane-leaflets; the terminal one two-fifths to half as broad as long, rounded to subcordate at base, the fine teeth mostly* 5–8 *per cm. of median margin; simple blade or terminal leaflet of bracteal leaf two- to three-fifths as broad as long; fruit purple-black.* (Incl. *R. ablatus* and *R. suus* Bailey and probably *R. subtractus* Bailey, an extreme suggesting possible influence of no. 168) — Dry thickets and borders of woods, S.C. to Ala. and La., n. to N.E., N.J., Pa., Ky. and Mo. *Fl.* May, June; *fr.* late June–Sept.

171. R. Blàkei Bailey (for its discoverer, Sidney Fay Blake, 1892–). — *Similar to no.* 170 *but the narrow gradually acuminate primocane-leaflets only thinly pilose to glabrescent beneath;* the corymbiform-racemose inflorescence with ascending pedicels; fruit subglobose to short-cylindric, sour and insipid. — Dry or moist thickets, meadows and roadsides, Ga. to Ark., n. to Mass., Ill. and Mo. *Fr.* July, Aug.

172. R. fátuus Bailey (foolish; perhaps from the diffuse habit). — *Primocanes slender, with* scattered *subulate prickles* 2–4 *mm. long; primocane-leaves* 3-*foliolate,* velvety beneath; *the leaflets oblong or oblong-lanceolate, merely acute, the terminal one* 8–10 *cm. long* and 3–4 *cm. broad;* floricanes diffusely branched, their leaves similar to those of primocanes; corymbiform racemes few-flowered, nearly overtopped by the bracts; flowers 3 cm. broad. — Thickets, D.C.

173. R. florícomus Blanch. (covered with flowers). — *Erect* or but slightly arching, 1–2.3 m. high; primocanes glabrous, furrowed, often branching, with straight strong prickles along the angles; *primocane-leaves glabrous or promptly glabrate above, densely velvety-tomentose beneath, the expanding ones white-velvety,* the glabrous petioles and petiolules stoutly armed; *terminal leaflet of* 5-*foliolate leaves oblong- to elliptic-ovate, about half as broad as long,* 10–15 cm. long, 5–8 cm. broad, *its ribs not prominent beneath, gradually long-acuminate,* with mostly simple sharp teeth; *leaflets of* 3-*foliolate bracts acuminate; raceme leafy-bracted only toward base, most of the filiform* unarmed or armed villous *pedicels subtended by reduced stipule-like bracts;* the *longer pedicels* 2–5 *cm. long;* flowers 2.5–3.5 cm. broad; calyx-lobes promptly reflexed; *fruit globose, about* 1 *cm. in diameter.* (*R. par* Bailey) — Thickets, clearings and swamps, N.E., n. N.J. and e. Pa. to N.C. *Fl.* late May–early July; *fr.* Aug., Sept.

174. R. amnícola Blanch. (growing by a river). — Differing from no. 173 in the more strongly arching canes; *leaflets with the ribs impressed above and very prominent* and straight *beneath,* scarcely hidden in the tomentum; *racemes more crowded, their longest rather thick unarmed pedicels only* 0.5–2 *cm. long; fruit short-cylindric,* about 1.5 cm. long. — Savannas, river-meadows and damp thickets, sw. N.S. to sw. Me. *Fl.* late June, July; *fr.* Aug.

175. R. facètus Bailey (elegant). — *Differing from nos.* 173 and 174 *in its very great stature; primary canes* 1–2 *cm. in diameter and up to* 4.5 *m. high, erect or stiffly arching; primocane-leaves strigose-pilose above; terminal leaflet cordate- or subcordate-ovate, abruptly long-acuminate,* 10–15 cm. long, 7–10.5 *cm. broad,* this and the similar median leaflets long-petiolulate; corolla 3.5–4 cm. broad; *calyx-lobes tardily spreading, only slightly recurved in fruit; fruit subglobose, becoming*

2–2.5 *cm. in diameter*, very rich, with grape-like flavor. — Damp thickets, pond-margins and disturbed soil, w. N.S. and centr. Me. to s. N.E. and L.I. *Fl.* June, early July; *fr.* Aug., Sept.

176. R. bellóbatus Bailey (beautiful bramble), KITTATINNY BLACKBERRY. — *Differing from no.* 175 *in its lesser stature* (*up to* 2 *m. or more*) and *more slender canes; branches of canes long-arching nearly to ground; racemes with many foliaceous bracts; calyx-lobes promptly and strongly reflexed; fruit ellipsoid, distinctly longer than thick,* 2–3 *cm. long.* — Thickets and borders of woods, sw. Que. to Mich., s. to Vt., N.Y. and nw. N.J.; cult. and locally spread elsewhere.

177. R. bracteàlis Bailey (with bracts). — Only about 1 m. high; with *leaves of primocane glabrous or glabrate above*, soft-pubescent beneath; *terminal and median leaflets broadly cordate-ovate*, abruptly acuminate, 8–10 *cm. broad, long-petiolulate; raceme long and slender, all but the lowest filiform pedicels subtended by small bracts; calyx promptly reflexed; corolla about* 2 *cm. broad; fruit nearly globular, about* 1 *cm. in diameter.* — S. Wisc.

178. R. móllior Bailey (softer). — *Stiff, erect,* up to 1.7 m. high, with broad-based arching hard prickles up to 6 mm. long; *primocane-leaves becoming very thick and coriaceous, with the veins impressed above and very prominent and rib-like beneath,* the *upper surface glabrous but scabridulous, the lower at first soft-pubescent but becoming hard to the touch; leaflets ovate-oblong or oval, the terminal one on a short petiolule* 1–3 *cm. long; raceme rather short, its lowest pedicels* 1–2 *cm. long;* corolla 2–3 cm. broad; fruit short-oblong. — S. Mo. and Kans., s. to Ark. and Okla.

179. R. ávipes Bailey (bird-footed; from the pedate leaf). — Erect, rather slender, up to 2 m. high, strongly resembling no. 154 but glandless; *prickles of primocane remote,* 2–4 *mm. long; primocane-leaves* merely grayish on unfolding, at first soft-pubescent but *soon glabrescent or merely slightly pilose beneath; terminal primocane-leaflet cordate- or subcordate-ovate, on a petiolule* 3–4.5 *cm. long;* median leaflets long-petiolulate; *racemes elongate, the longer ones* 12–20-*flowered; pedicels unarmed, the lower ones* 3–5 *cm. long; flowers* 2.5–3.5 *cm. broad;* calyx-lobes soon reflexed; fruit globose or subovoid, scarcely 1 cm. in diameter. — Thickets and clearings, w. N.E. to Mich. and Ill.

180. R. assòcius Hanes (associating with; other local plants). — Differing from no. 179 in its more abundant prickles; *terminal primocane-leaflet on shorter petiolule* (*about* 2 *cm. long*); *median leaflets short-petiolulate; raceme* 8–10-*flowered, the rachis and pedicels prickly, the longer pedicels about* 3 *cm. long.* — Borders of swamps, Kalamazoo Co., Mich.

181. R. condensiflòrus Bailey (closely flowered). — Erect or arching, up to 2 m. high, the furrowed canes with *scattered subulate prickles* 3–4 mm. long; *primocane-leaves* glabrous or glabrescent above, soft-pubescent beneath, 5-*foliolate; terminal leaflet more or less dentate-lobulate,* cordate-ovate, the other leaflets narrow and tapering to base; *lateral racemes single, rather short, the simple lower pedicels overtopped by the foliaceous bracts;* flower 2–3 cm. broad; calyx-lobes soon reflexed. — Shale-region of Ky.

182. R. cóngruus Bailey (agreeing). — *More stiffly erect* than no. 181; *lateral branchlets of floricane bearing several more or less paniculately branched racemes, these forming a subpaniculate thyrsiform group of racemes; pedicels mostly much exceeding bracts.* — Dry woods and thickets, W.Va. and Ky.

183. R. indepéndens Bailey (independent). — Erect, up to 1 m. high, *with deltoid prickles* 6–10 *mm. long;* primocane-leaflets 3 or 5; the *terminal leaflet* cordate-ovate, acuminate, *somewhat lobulate;* racemes standing well above bracteal leaves, the lower pedicel sometimes forking, the *pedicels only* 1–2 *cm. long; corolla* about 2 *cm. broad.* — Chemung Co., N.Y.

184. R. pensilvánicus Poir. (Pennsylvanian). — Erect or ascending usually firm- and finally purple-stemmed plant with canes 0.5–3 m. high, freely branching; *primocanes with stout straight or slightly arching remote prickles; primocane-leaves* mostly 5-foliolate, *soft- or velvety-pubescent beneath,* on more or less prickly petioles; *terminal leaflet narrowly to broadly ovate, acuminate, with broadly rounded to cordate base,* doubly serrate, *the mostly submembranaceous blade becoming* 0.6–2 *dm. long and* 4–10 *cm. broad,* its petiolule elongate, the *principal veins slender and partially obscured by the pubescence beneath; racemes corymbiform or subcorymbiform, broadest above the middle, the axis densely pilose or villous; pedicels* elongate and *pilose, mostly unarmed, some rarely armed, mostly subtended by small stipule-like bracts, only the lower* 1–4 *subtended by foliaceous bracts,* or all or nearly all (including the upper ones) with foliaceous bracts in the rare forma **phyllóphorus** Fern. (bearing leaves); flowers showy, 2–3 or more cm. broad; calyx-lobes soon reflexed; *fruit squarish-subglobose to thick-ellipsoid,* 1–3 *cm. long, usually very juicy and of fine quality,* rarely bitter. (Incl. many local phases which can scarcely be distinguished: *R. Andrewsianus, orarius, pergratus* and *philadelphicus* of Blanchard; *R. laudatus* Berger; and *R. abactus, difformis, Hanesii, honorus, insons, litoreus, ozarkensis, parcifrondifer, pubifolius, rosarius* and *uniquus* of Bailey) — Thickets, borders of woods, clearings and (luxuriantly) recent burns, sw. Nfld. to s. Ont. and Minn., s. to N.S., N.E., Va., upland to Ala., Tenn., Ark. and Okla. *Fl.* May (southw.)–mid-July (northw.); *fr.* July (southw.)–Sept. (northw.).

185. R. libràtus Bailey (horizontal; from the sometimes lodged floricanes). — Described as differing from small plants of no. 184 in its *mostly 3-foliolate primocane-leaves*, these with *elliptic-oblong merely acutish leaflets resembling those of the bracteal leaves.* — Dry woods, e. Md.

186. R. subsolànus Bailey (eastern). — Suggesting low and small-leaved phases of no. 184, with slender prickles; *leaflets of primocane cuneate at base; the terminal one rhombic-obovate, about 1 dm. long; terminal leaflets of flowering shoots narrowly obovate,* 2–3 cm. long; slender *naked pedicels shown as mostly strongly ascending* (also described as divaricate); *calyx-lobes not reflexed;* corolla 1.5–1.8 cm. broad. — Damp thickets and fields, e. Md.

187. R. insulànus Bailey (of islands). — *Diffusely branched,* up to 1 m. high, *with intricately tangled pendulous branches and branchlets;* canes with numerous straight prickles; *primocane-leaves* rather short-petioled, *glabrous and sublustrous above, thin-pilose and glabrescent beneath, with the ribs prominent; terminal leaflet round-ovate or -obovate, abruptly short-acuminate, 7–8 cm. long* and 5 or 6 cm. broad, subcordate; corymbs few-flowered; *bracteal leaves or leaflets mostly obtuse, elliptic-oblong to oblong-obovate;* ascending pedicels short; calyx-lobes soon reflexed; fruit subglobose, about 1 cm. in diameter. — Sands of coast, islands and mainland, Block I., R.I. to s. Ct. and e. L.I.

188. R. bárbarus Bailey (barbarous). — Rather *rigidly erect, with hard stiff canes* 5–9 dm. high, *bearing hard subulate straight prickles mostly* 5–10 *mm. long; leaves subcoriaceous, scabridulous and with impressed nerves above, soft-pubescent and with prominent straight nerves beneath; primocane-leaves* on petioles with hooked prickles, the petiolules similarly prickly, *the terminal* oblong- or elliptic-ovate *coarsely toothed leaflet* 3.5–7 *cm. long; leaflets of flowering shoots* often larger but *similar;* corymbiform raceme 5–12-flowered, the *divergent densely pilose pedicels all or partly armed;* calyx-lobes promptly reflexed; corolla up to 2 cm. broad; *fruit subglobose,* 5–10 mm. in diameter, *inferior* (*often bitter*). (*R. conanicutensis* and *originalis* Bailey) — Clearings, thickets and borders of dry woods, s. N.H. to se. N.E. and s. N.J. *Fr.* Aug.–Oct.

189. R. latifolìolus Bailey (with broad leaflets). — *Primocanes arching, their drooping and finally tip-trailing branches and branchlets forming mounds* about 1 m. high, *terete,* with subacicular prickles 2–3 mm. long; *primocane-leaves* with very prickly petioles and petiolules; *terminal leaflet broadly cordate-ovate,* abruptly acuminate, closely double-serrate, glabrous above, *sparsely pilose beneath, the ribs prominent beneath; racemiform corymb very leafy-bracted, all the bristly pedicels overtopped by the large bracts; simple upper bracts broadly ovate, lobulate; calyx-lobes tardily reflexed;* fruit subglobose, small, seedy. — Borders of woods, se. Minn. — Perhaps better placed in § *Tholiformes.*

190. R. frondòsus Bigel. (leafy). — Tall (up to 2.5 m.), the *arching canes with branches and branchlets often reaching the ground,* soon terete, *with straight subulate prickles; primocane-leaves* with slightly prickly to smooth petioles and petiolules; *terminal leaflet* broadly ovate, 0.9–1.5 *dm. long,* subcordate, abruptly acuminate, *sparsely pilose or glabrate above, velvety beneath; short and compact corymbs leafy-bracted to or above the middle, the bracts overtopping the flowers, the larger simple coarsely toothed bracts or their terminal leaflets mostly* 4–7 *cm. long;* flowers 3–10 on *unarmed densely pubescent short pedicels;* corollas 2.5–3.5 cm. broad; calyx-lobes soon reflexed; fruit subglobose, 1–1.5 cm. thick. (Incl. *R. heterogeneus* Bailey, an unusual form) = Thickets and borders of woods, Mass. to Ind. and Va. *Fl.* May, June; *fr.* late June–Sept.

191. R. flóridus Tratt. (flowering). — *Similar to no.* 190; *canes more corrugated* (up to 3 m. high), *with broader-based and more hooked prickles; terminal primocane-leaflet narrowly ovate, long-acuminate; corymbs* (1–)2–7-*flowered, the upper flowers standing high above the bracts, the latter with mostly narrower blades;* pedicels elongate, often armed. — Borders of woods, Va. and southw. *Fl.* late April, May; *fr.* June, early July.

192. R. cupressòrum Fern. (of cypresses). — *Erect,* 1–2 m. high, *with widely arching branches and intricate branchlets; primocane with stout broad-based prickles* 4–8 or more mm. long; *petioles of primocane-leaves with stout hooked prickles; terminal leaflet elliptic-ovate to slightly obovate,* long-acuminate, 7–10 cm. long, *strigose-pilose above and along nerves beneath; corymb* (1–)2–7-*flowered, short and compact, all but the lowest unarmed short pedicels overtopping the bracts; longer bracteal leaves or their terminal leaflets only* 1.5–2.5 *cm. long;* calyx-lobes 3–4 mm. long, soon reflexed; fruit ellipsoid, small. — Borders of cypress-swamps, se. Va.

193. R. praténsis Bailey (of meadows; evidently meaning prairies). — *Stiffly erect,* 1 m. or so high, the primocanes *with subulate prickles* 2–5 mm. long; *primocane-leaves* smooth or sparingly short-pilose above; *terminal leaflet cordate-ovate, long-acuminate,* about 1 dm. long, *sharply toothed, velvety beneath; corymb* 5–8-flowered, *the middle and lower flowers subtended by bracts with simple blades or with terminal leaflet* 4–5 *cm. long; the short upper pedicels mostly overtopped by the lower bracts; corolla about* 2 *cm. broad;* fruit globose or subglobose, about 1 cm. in diameter. — Prairies, Mo. — *R. limulus* Bailey of s. Mich. seems hardly separable; said to differ in coarser toothing of leaves, and more prickly canes becoming lodged over winter.

194. R. Búshii Bailey (for Benjamin Franklin Bush, 1858–1937). — *Differing from no.* 193 *in stouter prickles; terminal primocane-leaflet oblong-elliptic, the petiole with broad hooked prickles; corymb with* 3–12 *flowers.* (Incl. *R. sertatus* and *R. senilis* Bailey) — Dry open slopes, prairies and borders of woods, w. W.Va. to w. Mo.

195. R. pauxíllus Bailey (little). — *Slender, erect,* 2–7 *dm. high, the terete canes* with slender prickles as long as the diameter, *the branches of the floricane strongly ascending; leaves soft-pubescent beneath; primocane-leaves 3-foliolate,* short-pilose on upper surface; *terminal leaflet subcordate-ovate,* 4–6 *cm. long; corymb* 2–4-*flowered; larger simple blades or terminal leaflets of bracts* 1.5–2.75 *cm. long, oblong-ovate, acute; filiform pedicels erect, glandless or minutely glandular;* flowers 2 cm. broad; calyx-lobes about 5 mm. long, soon reflexed. — D.C. and e. Va. to Ga. and Tenn.

196. R. defectiònis Fern. (of eclipse; from the type-region, Eclipse, Va.). — *Larger than no.* 195; *canes up to* 1.5 *dm. high,* the *floricanes with divergent to horizontal or recurving branches; later primocane-leaves 5-foliolate,* glabrate above, with deeply impressed nerves; larger *terminal leaflets* 5–6.5 cm. long, *rounded at base, elliptic-oval to slightly obovate; larger simple blades or terminal leaflets of bracteal leaves* 2–3.5 cm. long, *narrowly cuneate-obovate or subrhombic;* fruit subglobose, 1–1.5 cm. in diameter. — Bases of bluffs along James R., Nansemond Co., Va. *Fr.* June.

197. R. Brainérdii Rydb. (for its discoverer, Ezra Brainerd, 1844–1924). — *Weak, diffusely ascending to reclining; the subherbaceous to barely woody canes* about 1 m. high, *about* 4 *mm. in diameter, unarmed or with very remote subulate-acicular prickles* 2–4 *mm. long; primocane-leaves* glabrous above, velvety beneath; *terminal roundish-ovate cordate-based abruptly acuminate leaflet nearly as broad as long* (*mostly* 7–11 *cm. long and* 6–9 *cm. broad*), closely and doubly *serrate-dentate,* with *the larger teeth as broad as long;* corymbs mostly 3–7-flowered, the short ascending pedicels densely villous and mostly equaled by the bracts; *corolla about* 2 *cm. broad;* calyx-lobes soon reflexed; fruit subglobose, about 1 cm. in diameter. (*R. sativus* Brain. in part, not as to basonym) — Dry to moist soil, Vt. to Ct.

198. R. perfoliòsus Bailey (very leafy). — Differing from no. 197 in *more erect* habit; *canes* often lower and *stouter, with frequent prickles, teeth of leaves sharper, narrower than long, the leaf-tips sharper and more prolonged; fruit* 1.5 *cm. thick.* — Dry, often rocky soil, Essex Co. to Tompkins Co., N.Y.

199. R. paùper Bailey (poor or meagre). — Differing from no. 197 in its *often mounding and prostrate or ascending canes only* 2–3 *mm. in diameter; leaflets very sharply serrate and sharp-tipped, the terminal primocane-leaflet narrowly ovate and much longer than broad.* (Incl. plants of various habits but probably phases of one species: *R. humilior* and *polulus* Bailey) — Thin, dry soil, Ct., N.Y. and Pa. — Perhaps a series of starved states of normally coarser species.

200. R. perpaùper Bailey (very poor or meagre). — As small and weak as no. 199; *leaflets of primocane-foliage and blades or leaflets of bracts only acute or acutish, with small and merely acutish teeth.* (Incl. *R. pauperrimus* and perhaps *R. tantulus* Bailey) — Ct. to Pa. and Mich. — Perhaps phases of members of § *Setosi.*

201. R. recúrvans Blanch. (recurving). — *Arching, with the tips of canes and branches strongly recurving and nearly touching the ground or tip-trailing or sometimes tip-rooting, very intricately branched;* canes becoming terete, with scattered slender straight hard prickles; *primocane-leaves* mostly 5-foliolate, *thin,* soft-pubescent beneath; *the leaflets jagged- or lacerate-serrate,* acuminate, from lanceolate to ovate; terminal leaflet narrowly to broadly ovate, with rounded to cordate base, up to 12 cm. long; *inflorescences* very variable; *the more prevalent and typical ones conspicuously leafy-bracted and often covered or exceeded by the narrow jagged-toothed bracteal leaves; pedicels filiform, elongate;* corolla 1.5–3 cm. broad; fruit subglobose to ellipsoid, mostly 1–1.5 cm. long. (Incl. *R. Rossbergianus* Blanchard, *R. Wiegandii* Bailey and *R. heterophyllus* sensu Am. auth., not Willd.) — Dry to moist thickets, clearings or borders of woods, se. Que. to Minn., s. to N.S., N.E., Va., Ill. and Mo. *Fl.* May, June; *fr.* July–Sept.

202. R. praèpes Bailey (nimble). — *Erect,* with spreading summit; *terete canes up to* 2 *m. high,* with scattered acicular-subulate prickles; *primocane-leaves* 5; *terminal leaflet rhombic-ovate, irregularly lobulate-dentate,* long-petiolulate; corymbs with 2 or 3 rather small basal bracts, *most of the pedicels subtended by tiny bractlets;* corolla about 2 cm. broad. — Thin woods and clearings, Ky.

203. R. Cardiànus Bailey (in honor of Fred Wallace Card, 1863–1941). — *Erect,* 1 m. or more tall, *with stout spreading or horizontal branches, the canes angled* and bearing stoutish prickles; *primocane-leaves 5-foliolate, with coarse jagged double toothing; terminal leaflet cordate-ovate, tapering gradually from base to apex; inflorescences with* 2 *or* 3 *long basal bracts, the several upper naked pedicels subtended by tiny bracts but equaled or overtopped by the lower foliaceous ones; flowers* 3–4 *cm. broad;* fruit 1.2–1.5 cm. in diameter. — Thickets, Tompkins Co., N.Y.

204. R. wisconsinénsis Bailey (of Wisconsin). — *Erect to reclining, with spreading to nearly procumbent branches,* the *canes soon terete, bristly; primocane-leaves mostly* 3 (sometimes 5) *-foliolate,* the *terminal round-ovate leaflet jagged-toothed and lobulate; inflorescence compact, the lower armed pedicels subtended by leafy bracts, these nearly or quite overtopping the upper flowers;* corolla about 2 cm. broad; globose fruit hardly 1 cm. in diameter. — Clearings and fields, Wisc.

205. R. dissitiflórus Fern. (remotely flowered). — *Erect,* 4–5 dm. high; *slender tough terete canes simple, strongly armed with broad-based hooked prickles; primocane-leaves 3-foliolate, the petiole and petiolule unguiculate-armed;* the *leaflets coarsely rounded-dentate; terminal leaflet* elliptic-ovate, acuminate, 5–6 *cm. long; floricane unbranched* except for flowering shoots; leaves similar to but smaller than primocane-foliage; *flowers solitary on erect* slender villous and prickly *axillary pedicels 2–4 cm. long; broad pilose calyx-lobes 6–7 mm. long, appressed to the small globose fruit (scarcely* 1 *cm. thick*). — Calcareous bluffs by James R., Isle of Wight Co., Va., little known. *Fr.* mid-June.

Kérria DC. (for *John Bellenden Ker,* 1764–1842) japónica (L.) DC. (Japanese), alternate-leaved shrub with green branches, 5 yellow petals and 5–8 dryish drupe-like achenes, and Rhodótypos Sieb. & Zucc. (with character of a rose) scándens (Thunb.) Makino (climbing; an unfortunate name), an upright shrub with opposite leaves and 4-merous flowers with large sepals, white petals and 4 black drupe-like dry fruits, are commonly cult. and occasionally spread to thickets, etc. (Introd. from e. Asia)

20. DALIBÁRDA Kalm False Violet

Calyx deeply 5–6-parted, 3 of the divisions larger and toothed. Petals 5, sessile, deciduous. Stamens many. Ovaries 5–10, becoming nearly dry seed-like drupes; styles terminal, deciduous. — Low unarmed perennial of e. N.Am., with creeping and densely tufted stems or root-stocks, and roundish-cordate crenate leaves on slender petioles. Flowers of 2 kinds, a few upright long-peduncled usually sterile ones with white petals, and numerous fertile apetalous ones on short curved peduncles. (Named for *Thomas François Dalibard,* 1703–1779, a French botanist of the time of Linnaeus.) Sometimes placed in Rubus.

1. D. rèpens L. (creeping), Robin-run-away. — Downy; sepals of the petaliferous flowers spreading, of the cleistogamous ones converging and inclosing the fruit. — Woods, Que. and s. Ont., s. to N.S., Me., Mass., nw. Ct., mts. of N.C., n. O. and Mich. June–Aug.

Tribe V. Poterìeae Reichenb.

21. ALCHEMÍLLA L. Lady's-mantle

Calyx-tube inversely conical, contracted at the throat; limb 4-parted with as many alternate accessory lobes or bractlets. Petals none. Stamens 1–4. Pistils 1–4; the slender style arising from near the base; achenes included in the tube of the persistent calyx (hypanthium). — Low herbs of cool reg., with palmately lobed or compound leaves, and small corymbed greenish flowers. (From *Alkemelyeh,* the Arabic name, having reference to the silky pubescence of some species.)

*a.*Small annual with equably leafy stem up to 1 dm. high; leaves flabelliform, 2–6 mm. broad; flowers axillary; stamen 1. Subgen. Aphanes. . . . . 1. *A. microcarpa.*
*a.*Coarser perennials with stout base and large long-petioled basal leaves and paniculate inflorescences; stamens 4. Subgen. Eualchemilla. . . *b.*
*b.*Leaves silvery-silky beneath, divided to base into distinct leaflets. . . . 2. *A. alpina.*
*b.*Leaves green, lobed, rarely cleft to the middle. . . *c.*
*c.*Pubescence of stem and petioles spreading. . . *d.*
*d.*Pedicels and branchlets of inflorescence glabrous; hypanthium glabrous to very sparsely hirsute.
Stem hairy nearly up to the pedicels; mature hypanthium 1–1.5 mm. high, campanulate, faintly ribbed; leaves glabrous above. . . 3. *A. pratensis.*
Stem hairy only at base, glabrous above; mature hypanthium 1.5–2 mm. high, turbinate, prominently ribbed; leaves strigose-pilose above. . . . . 4. *A. filicaulis.*
*d.*Pedicels, branchlets and hypanthium hirsute; upper surface of leaves pilose.
Mature hypanthium campanulate, 1–1.5 mm. high, only obscurely ribbed; pedicels glabrous. . . . . 5. *A. monticola.*
Mature hypanthium turbinate, 1.5–2 mm. high, prominently ribbed; pedicels hirsute. . . . . 6. *A. minor.*
*c.*Pubescence of stem and petioles appressed, stem glabrous above; mature hypanthium campanulate, faintly ribbed, glabrous, 1–1.5 mm. high. . 7. *A. alpestris.*

Subgen. Áphanes (L.) Rothmaler

1. A. microcárpa Boiss. & Reut. (minute-fruited), Parsley-piert. — *Tiny annual* up to 1 dm. high, simple or freely branching from base; *leaves flabelliform*, deeply cleft, pilose, 2–6 *mm. broad; flowers fascicled in the axils of dilated stipules;* the mature hypanthium minutely pubescent, 0.5–1 mm. long, barely 0.5 mm. thick, with connivent hairy calyx-lobes. (*A. arvensis* of ed. 7, not Scop.; *Aphanes australis* Rydb.) — Sandy open pastures, old fields and roadsides, Del. and D.C. to Ga., Ala. and Tenn. May, June. (Natzd. from Eu.)

The coarser A. arvénsis (L.) Scop. (of cult. fields), with leaves 0.5–1.5 cm. broad, hypanthium 1.5–2 mm. long and calyx-lobes erect, has been recorded from N.S. (Adv. from Eu.)

Subgen. Eualchemílla (Focke) Buser

2. A. alpìna L. (alpine), Alpine L. — Perennial with elongate root and 1–several crowns with pale brown sheathing stipules and long-petioled *palmate* radical *leaves silvery-silky beneath; leaflets* 5–7, *distinct* or nearly so, oblong, sharply toothed at summit; flowering stems 1–2.5 dm. high, with few small leaves and a silvery panicle of glomerulate flowers; hypanthium and outer faces of calyx-lobes heavily appressed-pubescent; disk broad; stamens 4. — Dry shingle and gravel, Belle-Rivière, Miquelon (*Arsène*). July. (Greenl.; Eu.)

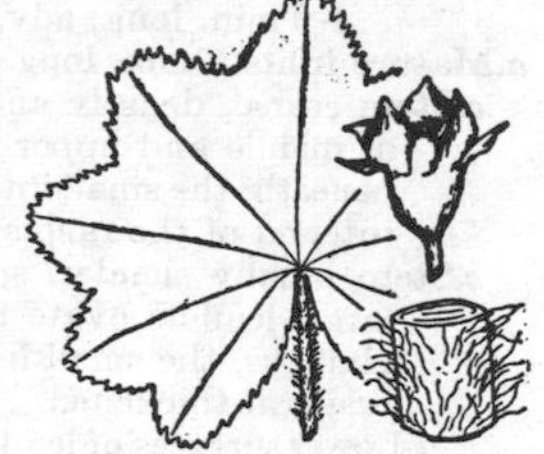
1185. A. pratensis.

3. A. praténsis F. W. Schmidt (of meadows). — Stout-based perennial; stems several from the crown, 1–5 dm. high, *spreading-hirsute nearly to summit; basal leaves* slender-petioled, reniform or suborbicular, 0.5–1.2 dm. broad, with 5–9 rounded-oblong or short-ovate serrate-margined lobes, *the upper surface glabrous;* cauline leaves smaller, with sessile clasping stipular bases; cymules very many in a finally loose panicle; *mature hypanthia campanulate, rounded at base, faintly ribbed,* 1–1.5 *mm. high, glabrous;* calyx-lobes narrowly deltoid-ovate, larger than the alternating bractlets, about 1 mm. long; stamens 4; disk broad. — Roadsides, fields and thickets, a troublesome weed, w. N.S.; local, Me. to N.Y. June–Oct. (Natzd. from Eu.) Fig. 1185.

4. A. filicaùlis Buser (thread-stemmed). — Similar to no. 3; *stem glabrous above the spreading-hirsute base; leaves strigose-pilose above;* cymules fewer; *mature hypanthia turbinate-campanulate* or elongate-obconic, *prominently ribbed,* 1.5–2 *mm. long, glabrous* or with 1–few trichomes; *calyx-lobes* 1.2–2 *mm. long.* (*A. vulgaris,* var. Fern. & Wieg.) — Cool or wet rocks, brooksides, etc., sea-level to subalpine areas, w. Nfld. and e. Saguenay Co., Que. June–Sept. (Eu.) Fig. 1186.

5. A. montícola Opiz (of mountains). — As in no. 3, but *leaves pilose above; hypanthia campanulate, only obscurely ribbed,* 1–1.5 *mm. long, copiously pilose.* — Fields and lawns, a bad weed, se. Nfld.; casual in e. Mass. June–Oct. (Adv. from Eu.) Fig. 1187.

1186. A. filicaulis.

1187. A. monticola.

1188. A. minor.

1189. A. alpestris.

6. A. mìnor Huds. (smaller). — As in no. 4, but *stem spreading-hirsute to summit,* the large *hypanthia* (and commonly the pedicels) *copiously spreading-hirsute.* (*A. vulgaris,* var. *vestita* (Buser) Fern. & Wieg.) — Cold or wet calcareous slopes and stream-margins. se. Lab. Pen. and w. Nfld. June–Sept. (Eu.) Fig. 1188.

7. A. alpéstris F. W. Schmidt (of high mountains). — Similar to nos. 3 and 5; *stem and petioles appressed-pubescent, the stem glabrous above;* leaves essentially glabrous; hypanthia as in no. 3. (*A. vulgaris,* var. *grandis* Blytt) — Local, e. Que. June–Sept. (Adv. from Eu.) Fig. 1189.

22. AGRIMÒNIA L. Agrimony. Cocklebur. Harvest-lice

Calyx-tube (hypanthium) turbinate or hemispherical, the throat beset with hooked bristles, indurated in fruit and inclosing 2 achenes; the limb 5-cleft, closed after flowering. Petals 5, yellow. Stamens 5–15. Styles terminal. — Perennial widely distributed genus of herbs, with interruptedly pinnate leaves, crenate-serrate leaflets, and small spicate-racemose flowers. Bracts 3-cleft. (Name a corruption of *Argemone*, a plant mentioned by Pliny.)

*a.*Mature fruits (including beak) 6–8 (–10) mm. long; coarse plants without fusiform-thickened roots. . . *b.*
*b.*Rachis of inflorescence minutely glandular-puberulent, occasionally with remote divergent long hairs interspersed; hooked bristles of fruit mostly wide-spreading from a thin horizontal flange, the short outer bristles often reflexed. . . . . 1. **A. gryposepala.**
*b.*Rachis non-glandular, densely pilose or hirsute; hooked bristles of fruit ascending and connivent from an insignificant flange.
Leaves scattered, reaching well up to the inflorescence; leaflets long-acuminate, glandular-granulose beneath and minutely pilose; rachis of inflorescence minutely pilose; calyx-lobes much longer than broad; petals about 3 mm. long; hypanthium glabrous to minutely pilose; native species. . . . . 2. **A. striata.**
Leaves subapproximate on lower half of stem; leaflets not long-acuminate, densely villous and non-glandular beneath; rachis spreading-hirsute; sepals about as broad as long; hypanthium densely long-hirsute; petals 4–6 mm. long; adv. weed. . . . . 3. **A. Eupatoria.**
*a.*Mature fruits 5 mm. long or less. . . *c.*
*c.*Stem coarse, densely and divergently long-hirsute or villous; larger leaflets of middle and upper leaves lance-acuminate, copiously glandular-dotted beneath; the small interspersed leaflets very unequal, often 3–5 pairs on an interval of the rachis; roots not fusiform-thickened. . . . . 4. **A. parviflora.**
*c.*Stem usually slender, sparsely if at all spreading-villous (except at base), larger leaflets ovate to obovate or oblong, if copiously glandular-dotted glabrous, the small interspersed leaflets 1 or 2 pairs on an interval; roots fusiform-thickened. . . *d.*
*d.*Lower surfaces of leaflets and rachis of inflorescence glabrous or essentially so; mature hypanthium round-based, not deeply furrowed. . . . . 5. **A. rostellata.**
*d.*Lower surfaces of leaflets and rachis of inflorescence pubescent; mature hypanthium furrowed, campanulate to turbinate.
Stems (at least below) and rachises of lower leaves with long horizontally divergent hairs; larger leaflets of principal leaves 5 (rarely 7), mostly broadly rounded at summit; hypanthium of mature fruit about as broad as long, 1.5–3 mm. long. . . . . 6. **A. microcarpa.**
Stems and rachises of leaves with loose ascending or incurving pubescence; larger leaflets of principal leaves 5–9, acute or at least not broadly rounded at summit; mature hypanthium longer than to about as broad as long, 2.5–4 mm. long. . . . . 7. **A. pubescens.**

1. A. gryposépala Wallr. (having hooked sepals). — Tall, 0.3–1.8 m. high, with long fibrous roots; stem sparsely spreading-hirsute; leaves remote, extending to bases of inflorescences; larger *leaflets* 5–9, oblong-oblanceolate to narrowly obovate or rhombic, 1.5–5 cm. broad, coarsely dentate, mostly thin, *glandular-dotted beneath, glabrous* except for the sparsely hirtellous veins, the small leaflets 1–3 unequal pairs on the intervals of the rachis; *rachis of inflorescence minutely glandular, with remote long divergent hairs intermixed; fruits 6–8 mm. long, the hooked bristles spreading from a thin horizontal flange, the short outer bristles often reflexed* and then covering the shallowly grooved hypanthium. — Thickets and borders of woods, P.E.I. and s. Que. to N.D., s. to N.S., N.E., N.C., Tenn., Mo. and e. Kans.; also s. B.C. to Calif. July, Aug. Fig. 1190.

1190. A. gryposepala.

1191. A. striata.

2. A. striàta Michx. (grooved). — Similar to no. 1, 0.3–2 m. high, stiffer; *leaflets firmer*, more veiny, lanceolate to narrowly obovate, *long-acuminate, the larger 1–4 cm. broad*, coarsely serrate-dentate, *veiny, somewhat fine-pilose beneath* as well as *glandular-dotted; rachis of inflorescence appressed-pilose; hypanthium glabrous to short-pilose; calyx-lobes much longer than broad; bristles of fruits ascending or connivent from a narrow base, the mature hypanthium turbinate and deeply furrowed.* — Thickets and borders of woods, Nfld. to B.C., s. to N.S., N.E., n. N.J., Pa., W.Va., O., Wisc., Ia., Neb., N.M. and Ariz. July–Sept. Fig. 1191.

3. A. Eupatòria L. (from an old application of the name *Eupatorium*). — Stem 0.3–1 m. high, usually densely hirsute at base; *leaves subapproximate on lower half of plant; leaflets* oblong to narrowly obovate, *not long-acuminate*, dark-green and strigose above, *densely cinereous-villous and nonglandular beneath, the larger 1–2.5 cm. broad; inflorescences on scapiform peduncles, the rachis spreading-hirsute; hypanthium densely long-hirsute; calyx-lobes nearly as broad as long; petals 4–6 mm. long;* fruit 7–10 mm. long. — Waste places and old fields, local, Mass., Wisc. and Minn. (Adv. from Eu.) Fig. 1192.

1192. A. Eupatoria.

4. A. parviflòra Ait. (small-flowered). — *Stout* and tall, up to 2 m. high, from long fibrous roots; *stem densely and divergently long-hirsute or villous; larger leaflets* of middle and upper leaves 11–15, *lance-acuminate*, sharply serrate, firm, veiny, *copiously glandular-dotted beneath; the smaller interspersed leaflets very unequal, often 3–5 pairs on the intervals of the hirsute rachis; fruits 4–5 mm. long;* the hooked bristles borne on a horizontal flange, spreading to ascending, the outer ones strongly spreading; hypanthium turbinate, with deep rounded grooves. — Damp thickets, rocky slopes, etc., Fla. to e. Tex., n. to w. Ct., N.Y., s. Ont., O., Ind., Ill., and Neb. Aug., Sept. Fig. 1193.

1193. A. parviflora.

5. A. rostellàta Wallr. (with a small beak). — *Slender, with fusiform roots; stem nearly glabrous,* 0.2–1 m. high, with ascending filiform branches; *leaflets* of principal leaves 5–7, *elliptic to narrowly obovate,* with coarse blunt teeth, thin, *glabrous* or essentially so, intermediate small leaflets 1–3; *fruits* 3.5–4.5 *mm. long; the bristles ascending,* or the outer spreading, *the longer ones overtopped by the dome-like beak of calyx-lobes; hypanthium round-based, hemispherical, grooveless or only shallowly grooved,* smooth or glandular-granulose. — Open woods, Ga. to Okla., n. to centr. Mass., Ct., N.Y., O., Ind., Ill., Mo. and e. Kans. Late July–Sept. Fig. 1194.

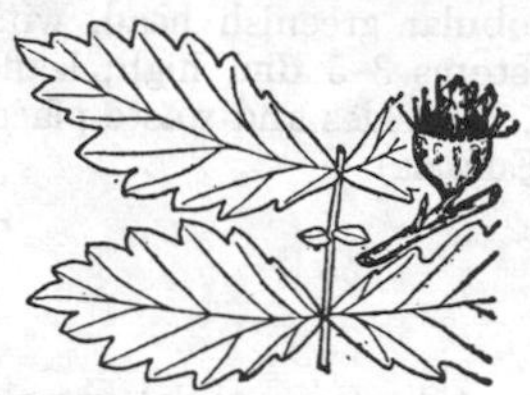
1194. A. rostellata.

6. A. microcárpa Wallr. (tiny-fruited). — Resembling no. 4; the *slender stem divergently long-villous below,* 0.4–1 m. high; *rachises of principal leaves horizontally long-villous; larger leaflets* 5 (rarely 7), narrowly obovate to elliptic, *rounded at summit,* coarsely dentate, thin, glabrous or sparsely strigose above, pilose beneath; rachises of inflorescences filiform, becoming 1.5–4 dm. long, minutely pilose, remotely flowered; *hypanthium of mature fruit* obconic to short-campanulate, *about as broad as long, 1.5–3 mm. long,* the bristles ascending or the *outer* slightly spreading. (Incl. *A. platycarpa* Wallr.) — Woods, Fla. to e. Tex., n. to N.J., e. Pa., W.Va. and Ky. July, Aug. Fig. 1195.

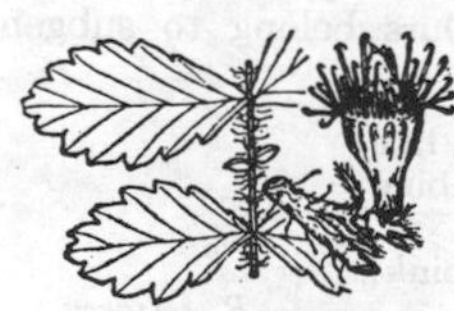
1195. A. microcarpa.

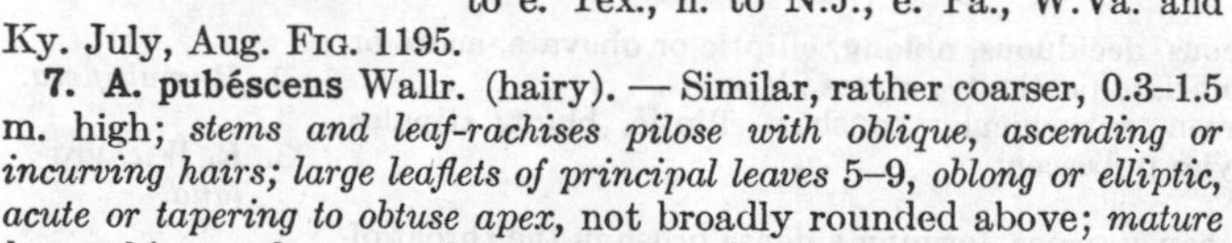

7. A. pubéscens Wallr. (hairy). — Similar, rather coarser, 0.3–1.5 m. high; *stems and leaf-rachises pilose with oblique, ascending or incurving hairs; large leaflets of principal leaves 5–9, oblong or elliptic, acute or tapering to obtuse apex,* not broadly rounded above; *mature hypanthium as long as or longer than broad, 2.5–4 mm. long.* (*A. mollis* (T. & G.) Britt.; *A. Bicknellii* (Kearney) Rydb.) — Rich woods or about shaded calcareous ledges, Ga. to Okla., n. to Mass., N.Y., s. Ont., Mich., Ill., Minn. and e. Kans. Mid-July–Sept. Fig. 1196.

1196. A. pubescens.

23. SANGUISÓRBA L. Burnet

Calyx with a turbinate tube, constricted at the throat, persistent; the 4 broad petal-like spreading lobes imbricated in the bud, deciduous. Petals none. Stamens 4–12 or more, with flaccid filaments and short anthers. Pistils 1–3; the slender terminal style tipped by a tufted or brush-like stigma. Achene (commonly solitary) inclosed in the 4-angled dry and thickish calyx-tube. Seed suspended. — Chiefly perennial herbs of N. Hemisph., with unequally pinnate leaves, stipules adherent to the petiole, and small often polygamous or dioecious flowers crowded in a dense head or spike at the summit of a long and naked peduncle, each bracteate and 2-bracteolate. (Name from *sanguis, blood,* and *sorbere, to drink up, to absorb,* from reputed styptic properties in folk-medicine.) POTERIUM L., in part.

Principal leaflets 2.5–10 cm. long; stamens 4.
  Spikes white or whitish; filaments long-exserted, dilated upward. . . . . . 1. *S. canadensis.*
  Spikes purple; filaments included or barely exserted, filiform. . . . . . . 2. *S. officinalis.*
Principal leaflets 1–2.5 cm. long; stamens 12 or more . . . . 3. *S. minor.*

1. S. canadénsis L. (Canadian), CANADIAN B., HERBE À PISSER (Que.). — Stem 0.3–2 m. high, simple or branching; leaflets of lower leaves 7–17, narrowly to broadly oblong-lanceolate or -ovate, obtuse, coarsely serrate-dentate, mostly petiolulate, cordate to rounded at base, 2.5–10 cm. long, without or sometimes with stipels; *spikes whitish,* elongate-cylindric, in maturity mostly 0.5–2 dm. long; *filaments* 4, *spatulate or clavate, exserted.* — Peaty or boggy soils, Lab. to Mich., s. to Nfld. (common) and interruptedly to C.B., N.E., L.I., s. N.J., n. Del., mts. to Ga., O., Ind. and Ill. Late June–Oct. — Extreme plants with short and broad leaflets have been erroneously referred to the western var. *latifolia* Hook. or *S. stipulata* Raf. (*S. sitchensis* C. A. Mey.)

2. S. OFFICINÀLIS L. (of the shops; from early repute in medicine), BURNET-BLOODWORT. — Habit of no. 1; *spikes deep red or purplish,* ovoid, ellipsoid or thick-cylindric, 1–3.5 cm. long; *filaments filiform, included or barely exserted.* — Locally in low fields and roadside-thickets, Me. and Minn.; and on Pacific slope. June–Oct. (Adv. from Eurasia)

3. S. MÌNOR Scop. (smaller), GARDEN-B. — Stamens 12 or more in the lower flowers of the globular greenish head, with drooping capillary filaments, the upper flowers pistillate only; stems 3–5 dm. high; leaflets small, ovate, deeply cut. (*Poterium Sanguisorba* L.) — Fields, roadsides and waste places, local, N.S. to Ont., s. to Va. and Tenn. May–July. (Adv. from Eurasia)

Tribe VI. RÒSEAE Griseb.

24. RÒSA L. ROSE

Calyx-tube (receptacle) urceolate to globose, contracted at the mouth, becoming fleshy in fruit. Petals (in typical flowers) 5, obovate or obcordate, inserted with the many stamens into the edge of the hollow thin disk which lines the receptacle, the latter bearing the numerous carpels below or at base. Ovaries hairy, becoming bony achenes in fruit. — Shrubs, usually prickly, mostly with odd-pinnate leaves, and stipules chiefly adnate to the petiole; flowering stalks, foliage, etc., often bearing aromatic glands. Many of the species highly variable and freely crossing, often indeterminable from incomplete specimens. Ours belong to subgen. *Eurosa* Focke. (The ancient Latin name.)*

a. Styles united into a protruding column about equaling the stamens and, like them, long-exserted from the throat of the receptacle; stems mostly climbing or trailing. § SYNSTYLAE.
  Leaflets mostly 3; stipules entire or nearly so; flowers 4–8 cm. broad, pink; native. . . . . 1. *R. setigera.*
  Leaflets mostly 7 or 9; stipules deeply toothed; flowers 2–5 cm. broad, white; introd.
    Leaflets membranaceous, deciduous, oblong, elliptic or obovate, acute or obtuse; stipules fimbriate-pectinate; styles glabrous. . . . . . 2. *R. multiflora.*
    Leaflets firm, evergreen, suborbicular to short-elliptic, blunt; stipules jagged-dentate; styles pubescent. . . . . 3. *R. Wichuraiana.*
a. Styles free, much shorter than stamens, forming a dense brush in the throat of the receptacle. . . *b.*
  b. Stipules cleft nearly to base into lance- or linear-attenuate pectinate caducous halves; leaves evergreen, of 5–9 very hard and thick narrowly obovate leaflets; flowers terminating villous-tomentose branchlets and closely subtended by dissected bracts; trailing or climbing; introd. § BRACTEATAE. 4. *R. bracteata.*
  b. Stipules adnate one-third to three-fourths their length to the petiole, persistent; plants mostly not creeping or climbing. . . *c.*
    c. Calyx-lobes divergent to ascending after flowering, promptly deciduous in fruit or, if persistent, the shrub with glandular-aromatic foliage. . . *d.*

---

* Our species, both native and introduced, freely hybridize, with the result that the characters which evidently mark the isolated populations become hopelessly confused where two or more species commingle. Only the clearer-cut species and varieties are here included. Many scores of recently proposed "species" are omitted until their relative stability is better demonstrated.

d. Carpels and achenes lining the inner wall as well as base of the receptacle; introd. shrubs. . . *e.*

e. Stem with intermixed bristles and slender (subulate) prickles; infrastipular prickles wanting or scarce; stipules narrow; flower solitary (rarely 2 or 3). § GALLICANAE. Represented by . . . 5. *R. gallica.*

e. Stem with many stout prickles (only rarely a little bristly); infrastipular prickles mostly strong; stipules dilated; flowers often corymbed. § CANINAE. . . *f.*

f. Leaves glandular beneath, the teeth doubled and glandular or gland-tipped. . . *g.*

g. Leaflets tomentose; prickles straight, nearly subulate above the broad base. . . . . 6. *R. tomentosa.*

g. Leaflets finely pilose to glabrous beneath; prickles (or at least the larger ones) stout and hooked.

Leaflets heavily glandular on both faces, pilose beneath, strongly gummy-aromatic; sepals persistent until full ripening of fruit; styles pubescent. . . . . . . 7. *R. Eglanteria.*

Leaflets not strongly glandular above, only slightly fragrant; sepals promptly deciduous; styles glabrous or nearly so. . 8. *R. micrantha.*

f. Leaves glandless on both faces, glabrous, not aromatic, their sharp often simple teeth glandless. . . . . 9. *R. canina.*

d. Carpels and achenes confined to the bottom of the receptacle; infrastipular prickles usually present; receptacle and fruit often bristly; indigenous species of acid or non-calcareous soils. § CAROLINAE. . *h.*

h. Stipules mostly strongly herbaceous, thin, upwardly dilated, the wide-spreading adnate portion of the pair 3–10 mm. broad, the free tips dilated and porrect; leaflets mostly lustrous above, sharply serrate; shrubs of dry to moist or inundated habitats.

New canes with many hundreds or thousands of slender crowded bristles per dm., with or without few longer acicular harder prickles extending nearly to summit; adnate lower half of stipules 5–18 mm. long; leaflets submembranaceous, toothed nearly to base, those of flowering shoots 1–5 cm. long and 4–12 (–18) mm. broad; northeastern. . . . . 10. *R. nitida.*

New canes with stout broad-based straight to curving hard prickles below, and similar infrastipular prickles, with few (if any) acicular prickles; adnate lower half of stipules 1–4 cm. long; leaflets firm to subcoriaceous, mostly larger; wide-ranging. . . . 11. *R. virginiana.*

h. Stipules firmer, linear or with parallel sides, either spreading or with connivent margins, the adnate halves of the pair 0.5–2 (–3) mm. broad, the short and narrow free tips ascending to divergent; leaflets dull to but slightly lustrous; wide-ranging and southern.

Infrastipular prickles stout-based, subulate to hooked; adnate portion of stipules 1–2.5 cm. long; leaflets closely appressed-serrate, each margin with 12–25 fine teeth above the middle; mostly tall shrub of damp or swampy soil. . . . . . . 12. *R. palustris.*

Infrastipular prickles slenderly acicular and straight; adnate portion of stipules 0.5–1.5 cm. long; leaflets coarsely serrate, each margin with 5–15 teeth above the middle; low shrub of dry habitats. . . . . 13. *R. carolina.*

c. Sepals erect to divergent after flowering, persistent in fruit; pedicels and receptacle usually smooth; carpels lining inner walls as well as base of the receptacle; foliage not glandular aromatic. . . *i.*

i. Flower solitary, without a bract at base of pedicel; petals yellow, white or pink; leaves very small; introd. § PIMPINELLIFOLIAE. Slender stem with abundant needle-like prickles; leaflets ovate to orbicular, 0.5–2 cm. long. . . . . 14. *R. spinosissima.*

i. Flowers solitary or in corymbs, the base of the pedicel bearing a dilated bract; petals pink to rose-purple (white only in albinos); leaves and leaflets larger; native and introd. § CINNAMOMEAE. . . *j.*

j. Branches heavily tomentose among the crowded prickles and bristles; prickles pubescent; leaflets strongly rugose, the strongly reticulate veins elevated on lower surface; expanded flower 7–12 cm. broad, its petals rose-purple (or white); introd. . . . . 15. *R. rugosa.*

j. Branches and prickles (when present) glabrous or essentially so; expanded flowers smaller, pink. . . *k.*

k. Infrastipular prickles commonly present, clearly different from prickles of internodes of cane.

Leaflets oblong, finely serrate; stipules glandless; flowers mostly double; sepals glandless, 2–2.5 cm. long; introd. shrub up to 2 m. high. . . . . 16. *R. cinnamomea.*

Leaflets oblong-obovate, coarsely serrate; stipules glandiferous; sepals glandular, 1–1.5 cm. long; nw. native 2–9 dm. high. . 17. *R. Woodsii*, var. *Fendleri.*

k. Infrastipular prickles wanting or not differentiated from prickles of internodes of cane. . *l.*

l. Sepals widely divergent or reflexed in maturity. . . *m.*

m. Leaves and petioles glabrous or promptly glabrate; the leaflets dark green and sublustrous above; stipules scarcely or only sparsely glandular-ciliate. . . . . . . 18. *R. johannensis.*

m. Leaves and petioles copiously pubescent, the leaflets pale green and dull above. . . *n.*

n. Stipules strongly ciliate with stipitate glands, glandular-pubescent on the back; upper half of flowering cane without setae; northeastern.

Stipules 1–2 cm. long; sepals 0.9–1.5 cm. long. . . 19. *R. Williamsii.*

Stipules 2–3.5 cm. long; sepals 1.8–2.5 cm. long . . . 20. *R. Rousseauiorum.*

n. Stipules glandless, pubescent; petiole, leaf-rachis and lower leaf-surfaces pilose; flowering canes with scattered setae extending to summit; southwestern. . . . 21. *R. conjuncta.*

l. Sepals porrect in fruit, forming a loose beak at summit of the receptacle. . . *o.*

o. Prickles extending nearly or quite to summit of flowering stem, acicular.

Petioles, leaf-rachises or margins of stipules or all of them glandular. . . . . 22. *R. acicularis.*

Petioles, leaf-rachises and margins of stipules glandless. 23. *R. arkansana.*

o. Prickles, if any, confined to base of stem or only scattered above, not extending far into flowering portion. . . . . . 24. *R. blanda.*

§ Synstỳlae DC.

1. R. setígera Michx. (bearing bristles; a misnomer), Climbing R., Prairie-R. — *Climbing, leaning or trailing,* the *canes* prolonging to many meters and *bearing remote broad-based prickles,* or prickles wanting in forma **inérmis** Palmer & Steyerm. (unarmed); *leaflets* of flowering branchlets 3 *or* 5, *ovate or ovate-oblong, sharply serrate, acuminate,* up to 1 dm. long, *lustrous above, glabrous or merely pilose on nerves beneath; stipules* with *entire or merely ciliate* margins; *flowers several in a corymb, roseate,* fading to whitish, 4–8 *cm. broad;* pedicels and receptacle glandular-hispid; sepals reflexed in anthesis, lance-attenuate, 1.2–1.6 cm. long, deciduous; *styles united into a column about equaling stamens;* fruit red, subglobose, 8–12 mm. long. — Open woods, thickets, clearings and banks, Fla. to Tex., n. to interior N.Y., W.Va., O., Ind., Ill., Mo. and e. Kans.; natzd. to s. N.E., e. N.Y., Mich., etc. Late May–July.

Var. **tomentòsa** T. & G. (tomentose). — *Leaflets dull above, tomentose beneath.* (*R. rubrifolia* Ait.) — W. Ga. to e. Tex., n. to s. Ont., O., n. Ind., Ill. and Neb.; natzd. e. to s. N.E. — Forma **serèna** (Palmer & Steyerm.) Fern. (serene) has the canes prickleless (var. *serena* Palmer & Steyerm.).

2. R. multiflòra Thunb. (many-flowered). — Differing from no. 1 in more trailing or arching habit; *leaflets mostly* 7 *or* 9, 2–4 *cm. long,* obtuse to acute; *stipules fimbriate-pectinate; flowers abundant in pyramidal inflorescences,* 2–4 *cm. broad, mostly white; sepals* 5–8 *mm. long; style glabrous;* fruit smaller. (*R. polyantha* Sieb. & Zucc.) — Clearings, roadsides and borders of woods, s. N.E. southw. May, June. (Introd. and natzd. from e. Asia)

3. R. Wichuraiàna Crépin (for its discoverer, Max Wichura, 1817–1866), Memorial R. — Semievergreen trailer and climber, differing from no. 2 in smaller *rounder and firmer leaflets; stipules jagged-dentate; flowers* fewer, 4 *or* 5 *cm. broad; styles pubescent.* — Locally spread from cult. to open banks, clearings and borders of woods, Va., O., etc. (Introd. from Asia)

§ Bracteàtae Thory

4. R. bracteàta Wendl. (bracted), Macartney R. — *Depressed evergreen with very broad-based paired infrastipular prickles; branchlets densely tomentose* and slenderly prickly and stipitate-glandular; *stipules about* 5 *mm. long, cleft essentially to base into slender segments; leaflets*

5–9, *hard and coriaceous*, lustrous above, *narrowly obovate, blunt, finely crenate-dentate*, 1.3–2.5 cm. long; *flowers* 1–few, *closely subtended by large dissected bracts*, 5–7 *cm. broad, white; receptacle tomentose.* — Dry open soil, Fla. to Tex., n. to Va. (Introd. and natzd. from China)

§ Gallicànae DC.

5. R. gállica L. (French), French R. — *Stems arising singly from subterranean rootstocks, slender*, 0.5–1.5 m. high, *covered with mixed bristles, stipitate glands and subulate prickles; stipules narrow*, the free lanceolate tips spreading; *leaflets* (3–)5(–7), *very thick or coriaceous, roundish-ovate or elliptic, rugose above, pubescent beneath, obtuse or subacute*, serrate-dentate; *flowers* 1 (rarely 2 or 3) *on stout erect glandular pedicels*, 4–7 cm. broad, *crimson or pink* (often double); receptacle and reflexed finally deciduous sepals glandular-hispid. — Roadside-thickets, N.E. and N.Y. June, July. (Introd. and natzd. from Eu.)

× **R. centifòlia** L. (hundred-leaved; application not clear), the Cabbage-R., differing from no. 5 in stouter and taller growth, with coarser prickles, thinner and longer leaflets and nodding flowers, sometimes spreads from cult. (Introd. from the Caucasus)

§ Canìnae DC.

6. R. tomentòsa Sm. (tomentose). — Upright shrub 0.7–2 m. high, the branches arching; *stems armed with broad-based stout conic-subulate but compressed straight prickles* 5–10 *mm. long;* infrastipular prickles similar; stipules broad, abruptly slender-tipped, glandular; *leaflets* 5–7, *oval or narrowly ovate, tomentose and glandular beneath, with doubly serrate glandular-margined teeth;* petiole and rachis tomentulose and glandular; flowers 1–few, on glandular-hispid pedicels; sepals and *subglobose fruit with long subrigid stipitate glands.* — Roadside-thickets, P.E.I. (Introd. and natzd. from Eu.)

7. R. Eglantèria L. (Latinization of the old English and French name), Sweet-Brier, Eglantine. — *Coarse shrub* (sometimes tree-like) up to 3.5 m. high, *the stout new stems bearing many broad-based flattened recurving prickles* 7–13 *mm. long, these often supplemented by straight slender prickles of varying lengths; leaflets* 7–9, *glandular-scurfy on both faces, resinous-aromatic, pilose beneath, suborbicular to elliptic*, 1–3 cm. long, *blunt to acutish*, glandular-toothed; *flowers* pink, 3–5 *cm. broad*, the short pedicels glandular-hispid; *sepals subpectinate with stipitate glands*, reflexed, *subpersistent, dropping tardily from the scarlet or orange fruit; styles pubescent.* (*R. rubiginosa* L.) — Thickets, clearings, roadsides, etc., throughout our range and beyond. Mid-May–July. (Introd. and natzd. from Eu.)

8. R. micrántha Sm. (small-flowered). — Differing from no. 7 in its *uniform arching lanceolate prickles* at most 1 cm. long, *without slender ones intermixed; leaflets* 5–7, *more ovate, acuminate, nearly glandless above, not strongly fragrant; flowers smaller* (2–3 *cm. broad*), *paler, often white; styles glabrous; sepals promptly deciduous.* (*R. rubiginosa*, var. *micrantha* Lindl.) — Similar habitats, less common, Anticosti to s. Ont. and Wisc., s. to N.S., N.E., N.C., Ky. and Tex.; often passing into no. 7. (Introd. and natzd. from Eu.)

9. R. canìna L. (of a dog; mean), Dog-R. — Resembling no. 8, with *broader hooked prickles* 3–8 *mm. long; leaflets glandless, glabrous or nearly so, with glandless sharp teeth; flowers* white or pink, 4–5 *cm. broad, on smooth pedicels; receptacle smooth; sepals pinnatifid, glabrous on back*, tightly reflexed in anthesis, *promptly falling* from the ellipsoid scarlet fruit. — Dry banks, thickets and open fields, local, N.S. to w. N.Y., s. to Va. and Tenn. Mid-May–July. (Introd. and natzd. from Eu.)

§ Carolìnae Crépin

10. R. nítida Willd. (shining). — Slender shrub 0.2–1 m. high, from slender stoloniferous rhizomes; *canes* 2–5 *mm. thick* (excluding bristles); *new canes densely covered with many hundreds or thousands of dark purple fine spreading or slightly reflexed bristles per dm., these with or without a few longer acicular hard straight prickles, the bristles ascending to tip of cane; bristles fuscous or blackish on old canes, often extending to* (but less numerous) *flowering tips; stipules dilated, herbaceous, the flaring adnate portion* 5–18 *mm. long* and 3–10 mm. broad, the shorter *free tips porrect; leaflets* 7–9, *submembranaceous, lustrous and dark green above*, sharply *fine-serrate nearly to base*, narrowly elliptic or oblong-oval, *those of flowering shoots* 1–4(–5) *cm. long* and 4–12 (–18) *mm. broad; flowers* pink, solitary, or few in corymbs, 4–6 cm. broad, *in evening with fragrance of Convallaria;* sepals long-attenuate, with or without slightly dilated tips, reflexed in anthesis, ascending in young fruit, finally deciduous from the dark red subglobose fruit; *larger mature achenes cinnamon-brown*, ellipsoid, 3–5 mm. long, 1.5–2 mm. broad. — Bogs, wet thickets, pond-margins, etc., in acid soil, Nfld. to s. Que., s. to N.S. and s. N.E. June (southw.)–early Sept. (northw.)

11. R. virginiàna Mill. (Virginian). — Shrub up to 2 m. high, *scarcely stoloniferous; canes 4–10 mm. thick at base*, there variously prickly; *upper half of cane with strong thick-based straightish to recurving or hooked infrastipular prickles, or prickleless; stipules upwardly dilated, herbaceous, the wide-spreading adnate portion of the pair 1–4 cm. long and 3–10 mm. broad the free broad and herbaceous tips porrect*, glandular-ciliate or eciliate; *leaflets* 5–11, *firm to subcoriaceous, lustrous and dark green above*, essentially glabrous, elliptic to narrowly elliptic-obovate, mostly acute, gradually rounded to cuneate at base, *the coarse acute serration extending along the upper three-fourths of the margin*, 5–17 *teeth on each margin above the middle;* mature blades 2–6 cm. long and 1–3 cm. broad; *flowers* pink, *mostly in corymbs* (sometimes single), 5–7 *cm. broad*, on more or less stipitate-glandular pedicels; receptacle stipitate-glandular or smooth; *sepals reflexed in anthesis*, spreading or ascending in fruit, soon deciduous from the red fruit; *achenes obovoid, tan-brown or pale castaneous*, becoming 3.5–4 mm. broad. (*R. lucida* Ehrh.) — Damp to dry thickets, clearings, swamps and shores, Nfld. to s. Ont., s. to N.S., N.E., Va., upland N.C. to Ala., Tenn. and Mo. June (southw.)–Aug. (northw.). — Forma **nanélla** (Rydb.) Fern. (very dwarf) is the *smallest extreme*, compact, 1.5–4 dm. high, *with leaflets* 0.8–2 *cm. long and* 4–13 *mm. broad*, the prickles and stipules proportionally short (*R. nanella* Rydb.), on barrens, rocks and sands, Nfld. to s. N.J.

12. R. palústris Marsh. (of marshes). — Up to 2.5 m. high, from prolonged stoloniform rhizomes; canes up to 1.2 cm. thick at base, there bearing stout and hard pale conical prickles up to 1 cm. broad at base and 1 cm. long; *median and upper internodes slender*, terete, *prickleless; infrastipular prickles stout-based, subulate or unguiculate*, 2–8 mm. long; *stipules firm, trough-like or with connivent margins; the adnate portion linear*, 1–2.5 *cm. long and* 0.5–3 *mm. broad*, the short free tips ascending or divergent; *leaflets* 5–9, *dull green*, oblong or elliptic, *minutely pubescent beneath*, acute at both ends or obtusish, mostly 2–6 cm. long, *minutely appressed-serrate nearly to base*, 12–25 *teeth on each margin above the middle;* flowers in corymbs or single (*R. floridana* Rydb.), 4–5.5 cm. broad; fruit depressed-globose to ellipsoid (*R. dasistema* Raf.), glandular-hispid or smooth; *achenes light brown, narrowly subcuneate-obovoid*, 2.5–3.5 *mm. long*, 1.5–2 *mm. broad.* (*R. carolina* L. (1762), not L. (1753); *R. pensylvanica* Michx.) — Swamps, wet thickets and shores, Fla. to Ark., n. to N.S., N.B., s. Que., Algoma Distr., Ont., Mich. Wisc. and Minn. — June (southw.)–late Aug. (northw.).

13. R. carolìna L. (Carolinian). — Low and *slender*, 2–9 dm. high; *canes usually borne singly from stolons*, 2–6 *mm. thick at base*, there often bearing short acicular scattered prickles; *middle and upper internodes with few or no prickles; infrastipular prickles acicular, firm, straight* or nearly so, *horizontally divergent or slightly reflexed*, 4–9 mm. long (rarely wanting); *stipules as in no.* 12, *the halves of the adnate pair more spreading*, 0.5–1.5 *cm. long and* 0.5–2 *mm. broad; leaflets* 5–9, *elliptic, narrowly ovate-lanceolate or narrowly obovate*, acute or obtuse, *submembranaceous to firm, dull or barely sublustrous above;* the blades mostly 1.5–4 cm. long, *with* 5–15 *coarse serrations on each margin above the middle;* rachis glabrous, or in forma **glandulòsa** (Crépin) Fern. (glandular) stipitate-glandular and the leaflets more or less glandular-toothed (var. *glandulosa* Farw.; *R. serrulata* sensu Rydb., not Raf.); *flowers mostly solitary*, 3.5–5.5 *cm. broad;* calyx stipitate-glandular or smooth; fruit red, subglobose; *achenes castaneous, obovoid or semi-obovoid* (*straight on one side*), 4–5 *mm. long.* — Dry sandy, rocky or open habitats or thin woods, Fla. to Tex., n. to N.S., sw. Me., N.H., Vt., N.Y., s. Ont., Mich., Wisc., Minn., and Neb. Mid-May (southw.)–early July (northw.).

Var. **villòsa** (Best) Rehd. (villous). — Similar; *leaflets soft-pubescent beneath.* (*R. Lyonii* Pursh) — N.H. to Minn., s. to Ga., La. and Okla.

Var. **grandiflòra** (Baker) Rehd. (large-flowered). — *Leaflets obovate or broadly oval, obtuse or subacute, firm*, sublustrous; *flowers* 4–7 *cm. broad.* (*R. obovata* Raf.) — Sw. Me. to s. Ont., Wisc. and Ia., s. to Ga., Tenn. and Ark.

§ PIMPINELLIFÒLIAE DC.

14. R. SPINOSÍSSIMA L. (mostly spiny), SCOTCH or BURNET-R. — Slender, *very prickly with straight or recurved acicular prickles and innumerable shorter bristleform ones;* stipules very narrow; *leaves* 2.5–8 *cm. long, with* 7–13 *dryish ovate to orbicular serrate leaflets* 0.5–2 *cm. long; solitary flowers on naked pedicels*, white, yellow or pink, 2–4 cm. broad; sepals erect at summit of blackish fruit. — Roadside-thickets, spread from cult. Late May–early July. (Introd. and natzd. from Eu.)

§ CINNAMÒMEAE DC.

15. R. RUGÒSA Thunb. (rugose). — *Coarse*, in dense clumps up to 2 m. high; *canes very prickly and bristly as well as tomentose*, the mixed prickles straight and acerose or setiform; *petioles and lower surfaces of upwardly dilated stipules tomentose; leaflets* 5–9, *dark green and* *deeply furrowed above, pale and with prominent reticulation beneath*, elliptic to oblong-obovate, the toothed margin recurving; *flowers* 1–few, rose-purple (or white), *when fully expanded* 7–12 *cm. broad*, on short bristly pedicels; the *smooth* brick-red *fruit up to* 2.5 *cm. in diameter, capped by the long* (2.5–4.5 *cm.*) slender-tipped *sepals.* — Roadsides, seashore-thickets, dune-sands, etc., Que. to Minn. s. to N.S., N.E., N.J. and lower Great Lakes. June–Sept. (Introd. and natzd. from e. Asia)

16. R. CINNAMÒMEA L. (from fragrance supposed to suggest cinnamon), CINNAMON-R. — *Forming dense thickets by subterranean shoots*, up to 2 m. high; the *slender flexuous red branches armed with broad-based straight or slightly recurving infrastipular pale prickles; leaflets* narrowly elliptic, 2–4 cm. long, *dull green, pubescent*, paler beneath, sharply and finely serrate; *flowers mostly double*, 4 or 5 cm. broad, pinkish-purple; *glandless sepals* 2–2.5 cm. long; fruit rarely formed in the wild, scarlet, 1.2–1.5 cm. in diameter. (*R. spinosissima* L. in small part only) — Roadsides, fencerows, etc., originally derived from cult. shrubs. June, July. (Introd. and natzd. from Eurasia)

17. R. Woódsii Lindl. (for JOSEPH WOODS, 1776–1864, English student of the genus), **var. Féndleri** (Crépin) Rydb. (for its discoverer, AUGUST FENDLER, 1813–1883). — *Small shrub* 2–9 *dm. high*, with reddish-brown, or later grayish, bark; *infrastipular prickles acicular or subulate, straight*, or arching from base; *stipules glandular and pubescent on back*, often glandular-serrulate, the pair 4–7 mm. broad; *leaflets* 5–9, *elliptic-obovate, coarsely serrate, thin*, pubescent beneath, 1–3 cm. long; *flowers* pink, solitary or corymbed, *about* 3 *cm. broad; sepals* 1–1.5 *cm. long, often glandular*, erect or spreading in fruit; receptacle smooth; *fruit* subglobose, red, *about* 8 *mm. in diameter.* — Prairies and dry slopes, w. Ont. and Minn. to B.C., s. to Mo., Neb. and n. Mex. June. — Typical and taller, more glabrous and glandless *R. Woodsii* more western.

18. R. johannénsis Fern. (of the St. John River, Me. and N.B.). — *Canes* 0.3–1 *m. high*, the adult unarmed or with basal setae or rare prickles, the latter straight and broad-based; *branches lustrous, glabrous*, unarmed or slightly bristly, mostly purplish; *stipules dilated*, 1.5–4 *cm. long, glabrous, entire or sometimes glandular-dentate*, the free limbs lance-ovate; *petiole and leaf-rachis glabrous or promptly glabrate; leaflets* 5–9, oval or narrowly obovate, coarsely serrate, *dark green and sublustrous above, glabrous* or merely pilose on nerves beneath, 1.5–5.5 cm. long; flowers solitary or in corymbs, 4–6 cm. broad, rosy-pink, or white in forma **albìna** Fern. (albino); *sepals* more or less glandular, caudate-appendiculate, 2–5.5 *cm. long, after anthesis divergent or strongly reflexed, persistent;* fruit smooth, orange-red, subglobose, 1–1.5 cm. in diameter; *carpels lining inner walls and base of receptacle;* achenes obovoid, medium brown, 4.5–5 mm. long. — Gravelly shores or rocky banks, chiefly riparian, Que., N.B. and n. Me. June, July.

19. R. Williámsii Fern. (for one of its discoverers, EMILE FRANCIS WILLIAMS, 1859–1929). — Smaller than no. 18, up to 5 dm. high; *stipules* 1–2 *cm. long, glandular-pulverulent beneath, copiously ciliate with stipitate glands; petiole and leaf-rachis minutely pilose and usually glandular; leaflets* 5–7, *pale green and dull*, cuneate-obovate, mostly rounded to truncate at summit, *short-pilose on both surfaces, glandular beneath along the nerves*, 1–3.5 cm. long; flowers 3.5–5 cm. broad; *sepals tightly reflexed in fruit*, 0.9–1.5 *cm. long; fruit ovoid to pyriform*, 1–1.3 cm. long, 7–8 mm. thick; *achenes ellipsoid*, 3–3.5 *mm. long.* — Calcareous rocks, e. Gaspé Co. to Rimouski Co., Que. July.

20. R. Rousseauiòrum Boivin (for its discoverers, JACQUES ROUSSEAU, 1905– , and ZÉPHIRIN ROUSSEAU, 1900–). — *Much coarser* than no. 19, 1–2.25 *m. high, with stout canes densely retrorse-setose below* (or slightly among) *the branches; stipules* 2–3.5 *cm. long, the marginal red stipitate glands often crowded, the teeth of the free blades often appearing glandular-pectinate;* leaflets 5–9, sharply serrate above the middle, 2–4.5 cm. long; *sepals* 1.8–2.5 *cm. long;* fruit 8–12 mm. thick. — Marshes and borders of the sea, ascending to 300 m. alt., St. Lawrence R. and Gulf, Que. Late June, July.

21. R. conjúncta Rydb. (connected). — Differing from no. 20 in the remotely *setose axis of the lower* (about 5 dm. high) *flowering cane; stipules glandless;* leaflets elliptic to narrowly obovate; flowers more numerous, in terminal corymbs. — Bluffs and open banks, nw. Mo.

22. R. aciculàris Lindl. (with needle-like prickles). — *Canes* 3–12 dm. high, *densely bristly and acicular-prickly below, the prickles and bristles extending densely or interruptedly to the flowering summit* and branchlets; *stipules* dilated, *glandular-ciliate and resiniferous; leaf-rachis usually glandular; leaflets* 3–7, ovate or elliptic, obtuse or acute, gradually rounded to subcordate at base, 1.5–5 cm. long, *opaque*, simply serrate, glabrous above, smoothish or downy and *often resinous-puberulent beneath;* flowers solitary (sometimes few in a small corymb), 4–6 cm. broad, roseate; sepals erect in fruit, often glandular on the back; *fruit smooth, ellipsoid or* *slenderly pyriform, contracted to a neck below the beak of sepals,* 1.5–2 cm. long; achenes pale brown, 4–5.5 mm. long. — Thickets, rocky (often acid) slopes, w. N.E. (local), e. N.Y., Mich., Wisc., Minn., S.D. and Colo. Mid-June, July. (E. Asia) — Passing insensibly into

Var. **Bourgeauiàna** Crépin (for its discoverer, Eugène Bourgeau, 1813–1877). — *Fruit subglobose to oblate, strongly rounded to both ends, the neck less developed.* (*R. Bourgeauiana* Crépin) — L. Mistassini, Que. to Yuk. and B.C., s., locally to coast of sw. Me., s. Vt., s. Ont., Mich., Wisc., Minn., etc.

23. R. arkansàna Porter (for the Arkansas R., Colo.). — Canes low, 2–5 dm. high, the shrub *resembling no. 22; differing in non-glandular foliage; rachis and* obtuse to acute sharply serrate *leaflets glabrous on both surfaces,* the *leaflets sublustrous above; inflorescences chiefly few–many-flowered corymbs; fruit subglobose,* 1–1.5 cm. in diameter, beaked by sepals 1–1.5 cm. long; achenes plump-ellipsoid, 5–5.5 mm. long. — Rocky slopes, thickets and dry prairies, Wisc. and Minn. to Colo. and Kans.— More generally eastw. as

Var. **suffúlta** (Greene) Cockerell (supported or propped-up; because the original material bore supernumerary reduced leaflets from between the stipular lobes). — *Petiole, rachis and lower leaf-surface soft-pilose.* (*R. suffulta, pratincola* and *heliophila* Greene; *R. Bushii* Rydb.) — Dry thickets, rocky slopes, sands, etc., ne. N.Y. to Alta., s. to D.C., Ind., Wisc., Mo., Kans., Tex. and N.M. May–Aug.

24. R. blánda Ait. (mild; from its lack of prickles).— *Canes* 0.07 (on wind-swept subalpine rock)–2 m. high, the vigorous new ones often acerose-prickly below, *the prickles wanting or sometimes few and scattered above, not paired and infrastipular; stipules pilose to glabrescent, the margins and blades entire, undulate or dentate* (rarely glandular-ciliate); *petiole; leaf-rachis and lower leaf-surface tomentulose, pilose or subglabrous; the upper surface pale green, glabrous and dull; leaflets* 5–7 (–9), *elliptic to oblong-obovate,* acute or obtuse, serrate, (0.7–)2–6 cm. long, in the rare forma **angústior** Vict. & Rolland (narrower) linear or linear-lanceolate and long-acuminate and with margins entire or merely undulate; flowers single or in corymbs, rosy-pink, or white in forma **álba** (Schuette) Fern. (white), 4–6 cm. broad; *pedicels and receptacle glabrous, often glaucous; sepals* glabrous or stipitate-glandular, *mostly* 1–2 *cm. long; fruit* subglobose to ovoid or obovoid, 1–1.5 cm. in diameter, *capped by the persistent erect beak of sepals;* achenes ellipsoid-ovoid, 4.5–5 mm. long. (*R. Solanderi* Tratt.; *R. subblanda* Rydb.) — Dry to moist rocky (calcareous to circumneutral) slopes, shores, etc., Anticosti and Mingan Ids., Que., to Man., s. to N.B., n. and w. N.E., Pa., n. O., Ind., Ill., Mo. and Neb. June–early Aug. — Many so-called vars. have been described; these seem to be trivial forms or hybrids.

Tribe VII. Prùneae B. & H.

25. PRÙNUS L. Plum, Cherry, etc.

Calyx 5-cleft; the tube campanulate, urceolate or tubular-obconical, deciduous after flowering. Petals 5, spreading. Stamens 15–20 or more. Pistil solitary, with 2 pendulous ovules. Drupe fleshy, with a bony stone. — Small trees or shrubs of N. Hemisph. and Andean S. Am., with mostly edible fruit. (The ancient Latin name of the plum.)*

a. Flowers chiefly solitary, umbellate or corymbed, from scaly overwintering buds, expanding before or with the leaves. . . *b.*

b. Ovary and fruit covered by bloom or pubescence; fruit with 2 opposite longitudinal shallow furrows or lines; flesh more or less fibrous; stone somewhat compressed, longer than broad. . . *c.*

c. Flowers 1–several in an umbel from solitary old axillary buds; terminal bud lacking; the slender elongate pedicels only tardily separating from the mature fruit; ovary and fruit glabrous, with a bloom; fruit not readily broken into halves; stone not deeply corrugated and sculptured; leaves convolute or conduplicate in bud. Subgen. I. Prunophora. . . *d.*

d. Flowers 1 or 2; leaves convolute in bud; stone slightly sculptured; introd. species. § Euprunus. . . . . 1. *P. insititia.*

d. Flowers 3–several (exceptionally fewer); leaves mostly conduplicate in bud (convolute in nos. 8 and 9); stone smoothish; indigenous. § Prunocerasus. . . *e.*

e. Teeth of leaves not glandular, sharply acute or sharp-pointed; calyx-lobes mostly glandless. . . *f.*

f. Leaves ovate, oval, obovate or rounded, merely acute or acutish to obtuse or truncate, crenate or dentate, the teeth sharp-pointed; dense shrubs with blue (sometimes purple or yellow) fruit.

---

* Much aid derived from Hedrick's Plums of New York (1911).

Leaves ovate, oval or obovate, longer than broad, acute to obtuse. . . . . 2. *P. maritima.*

Leaves suborbicular, subtruncate. . . . . 3. *P. Gravesii.*

f. Leaves lanceolate to ovate, elliptic or ovate-oblong, long-acuminate, sharply serrate; trees or open shrubs with red, purple or yellow fruit.

Leaves gradually acuminate; flowers 1–1.5 cm. broad; calyx-lobes oblong-ovate, blunt, only tardily reflexed; style 4–7 mm. long; fruit about 1 cm. in diameter, dark purple; stone turgid. 4. *P. alleghaniensis.*

Leaves abruptly long-acuminate; flowers 1.8–3 cm. broad; calyx-lobes lance-attenuate, promptly reflexed; style 7–12 mm. long; fruit red or yellow, 2–3 cm. in diameter; stone flattened. 5. *P. americana.*

e. Teeth of young leaves glandular or gland-tipped, rounded, crenate or depressed, the glands, if deciduous, leaving callous remnants; fruit red or yellow. . . *g.*

g. Calyx-lobes glandular on margin; flowers 1.2–3 cm. broad; leaves conduplicate in bud and soon flat (except in no. 8), mature blades mostly 6–12 cm. long and 3–6 cm. broad; trees with trunks up to 3 dm. in diameter, or large shrubs. . . *h.*

h. Flowers 2–3 cm. broad, the white petals soon becoming roseate; calyx-lobes 3.5–5 mm. long, promptly reflexed; leaves abruptly acuminate, their teeth relatively coarse and spreading. 6. *P. nigra.*

h. Flowers 1.2–1.5 cm. broad, the petals mostly unchanging in color; calyx-lobes 1.5–3 mm. long, tardily reflexed; leaves gradually tapering to tip, their teeth relatively fine and appressed.

Leaves conduplicate in bud, becoming flat; flowers chiefly borne on one-year-old slender branches, with or after expansion of leaves; stone pointed at both ends. . . . . . 7. *P. hortulana.*

Leaves convolute in bud, commonly longitudinally folded in maturity; flowers borne on short spurs before or at beginning of leaf-expansion; stone obliquely truncate at base. . 8. *P. Munsoniana.*

g. Calyx-lobes glandless; flowers 6–9 mm. broad; leaves convolute in bud, often longitudinally folded, the mature blades 2.5–8 cm. long and 0.5–2 cm. broad; branchlets spinescent; shrub or small tree with trunks rarely 1 dm. in diameter. . . . . . 9. *P. angustifolia.*

c. Flowers 1 or 2, sessile or nearly so, from clustered buds along the branches; ovary and fruit tomentulose; fruit readily separated into halves through the sutures; stone deeply sculptured and furrowed; leaves conduplicate in bud. Subgen. II. Amygdalus. Represented with us by . . . . 10. *P. Persica.*

b. Ovary and fruit glabrous and without bloom, the fruit globose or nearly so, without longitudinal furrows; flowers 1 or few and corymbose or sometimes in short and conspicuously bracted racemes. Subgen. III. Cerasus. . . *i.*

i. Inflorescence umbellate or corymbose, without obvious foliaceous bracts; leaves elongate, tapering or only gradually rounded to base. . . *j.*

j. Low and slender shrubs; leaves entire or subentire toward base, the teeth not gland-tipped. . . *k.*

k. Branches and branchlets erect or strongly ascending; stone ellipsoid to orbicular or subglobose, rounded at base. . . *l.*

l. Leaves narrowly oblanceolate, mostly tapering from near or above middle to prolonged base and acuminate apex, when full-grown one-sixth to one-third as broad as long, subcoriaceous, barely glaucescent beneath, appressed-serrate; stone 6–8 mm. broad, apiculate; northern. . . . . 11. *P. pumila.*

l. Leaves elliptic to oblong or narrowly obovate, blunt or merely short-acute, two- to three-fifths as broad as long, whitish beneath.

Leaves of fertile branchlets (not the leading shoots) mostly 3–7 cm. long and up to 3 cm. broad, firm-membranaceous, with low crenation; fruit about 1 cm. in diameter, astringent; stones 5–6 mm. broad, acute; northern and eastern. . . . . . 12. *P. susquehanae.*

Leaves of fertile branchlets mostly 2–4 cm. long and 0.8–2 cm. broad, becoming coriaceous, with acute teeth; fruit 1.5 cm. in diameter, sweet; stone 6–8 mm. broad, round above or barely apiculate; western. . . . . 13. *P. Besseyi.*

k. Branches trailing, forming mats up to 2 m. across; leaves pale beneath, spatulate-oblanceolate to -obovate, submembranaceous, rounded to acutish at apex, mostly one-third to one-fourth as broad as long; fruit about 1 cm. in diameter; stone ellipsoid, acute at both ends, 4.5–6 mm. broad; northeastern. . . . . 14. *P. depressa.*

j. Trees or coarse shrubs; leaves toothed to base, the young teeth gland-tipped. . . *m.*

m. Flowers 1.2–1.6 cm. broad; leaves membranaceous, gradually acute or acuminate, glabrous or promptly glabrate, finely toothed; fruit 5–7 mm. in diameter; stone 4–5 mm. long; native. . . . . . 15. *P. pensylvanica.*

m. Flowers 2.5–3.5 cm. broad; leaves firmer and thicker, abruptly pointed, coarsely toothed; fruit 1.5–2.5 cm. in diameter; stone about 1 cm. long; introd.

Leaves loose or somewhat drooping, soft, pubescent on nerves beneath; inner scales of buds widely divergent or reflexed below the pedicels; calyx-tube constricted below the entire lobes; fruit sweetish. . . . . 16. *P. avium.*

Leaves porrect, firm, glabrous beneath; inner scales of bud erect; calyx-tube not constricted below the toothed lobes; fruit sour. 17. *P. Cerasus.*

i. Inflorescence a short 4–10-flowered raceme, the lower pedicels subtended by foliaceous bracts; leaves roundish, broadly rounded to cordate at base. . . . . 18. *P. Mahaleb.*

a. Flowers in elongate racemes terminating new leafy branchlets of the season. Subgen. IV. Padus.

Leaves with blunt callous teeth; calyx-limb persistent in fruit, its narrow lobes acute. . . . . 19. *P. serotina.*

Leaves with sharp teeth; calyx-limb deciduous, its broad lobes blunt. . . . 20. *P. virginiana.*

Subgen. I. Prunóphora Focke (Plums)

§ Euprùnus Koehne

1. P. insitítia L. (used in grafting), Damson or Bullace. — Somewhat spinescent large shrub or small tree *with tomentose young branchlets; flowers mostly* 2 *together*, expanding with the leaves, 2–2.5 *cm. broad, on pubescent pedicels; leaves elliptic to elliptic-obovate, convolute in the bud, becoming* 3–7 *cm. long and* 1.5–4.5 *cm. broad, obtuse or subacute, closely crenate-serrate*, at first soft-pubescent, becoming glabrate and dull above and *slightly pubescent beneath; calyx-*lobes oblong, blunt, ascending, *longer than the tube;* petals white, elliptic or narrowly obovate; *fruit blue-black*, with a bloom, subglobose, 1.5–2 cm. long; *stone only slightly compressed, somewhat sculptured.* — More or less natzd. in thickets and borders of woods through much of our area, spread originally from cult. *Fl.* late April, May. (Introd. and natzd. from Eu.)

P. spinòsa L. (spiny), the Sloe of Europe, differing in more spiny branches, mostly solitary flowers only 1–1.5 cm. broad and on glabrous pedicels, the leaves oblong and much smaller, **P. doméstica** L. (domesticated), the common old-fashioned Plum of horticulture, a larger tree with larger leaves more pubescent beneath, the fruit larger and the stones smoother, and **P. cerasìfera** Ehrh. (bearing cherries), the Cherry-Plum of hort., a small tree with glabrous branchlets, acuminate sharply serrate ovate leaves, often solitary flowers and red (or yellow) cherry-like fruit with round stones, all spread from cult. (Introd. from Eu.)

§ Prunocérasus Koehne

2. P. marítima Marsh. (maritime), Beach-P. — Low and *straggling or ascending densely branched shrub* 0.3–2.5 m. high; branchlets at first pubescent, becoming glabrate; overwintering *buds acute;* flowers expanding before the leaves, in many fascicles of 2 or 3; *leaves conduplicate in bud, ovate, elliptic or obovate, acute or obtuse, with sharp-pointed crenate teeth;* the short-petioled *blades* with 2 or more basal glands, *soft-pubescent beneath*, glabrate above, 3–6 *cm. long,* 1.5–3.5 *cm. broad; flowers* 1.2–2 *cm. broad,* white; *pedicels pubescent; calyx* pubescent, *with oblong obtuse lobes; fruit* 1.3–2.5 cm. in diameter, *blue-purple or full purple*, or in forma **flàva** G. S. Torr. (yellow) yellowish; stone turgid, acute to truncate at base. — Sandy soil along or near the coast, Knox Co., Me. to e. Pa. and Del. *Fl.* April–June; *fr.* Sept., Oct.

3. P. Gràvesii Small (for its discoverer, Charles Burr Graves, 1860–1936). — *Smaller,* up to 1.2 m. high; *leaves suborbicular, truncate or merely broadly retuse and apiculate,* 2–4 cm. long, *sharply serrate; flowers larger* (*up to* 3 *cm. broad*), with rounder petals; *fruit blacker, only* 1–1.5 *cm. in diameter; stone plumper.* — Sandy and gravelly ridges by L.I. Sound, se. Ct. *Fl.* late May; *fr.* early Sept.

4. P. alleghaniénsis Porter (of the Alleghanies), SLOE or ALLEGHANY-P. — Low tree up to 5 m. high, with trunks up to 2 dm. in diameter, or a straggling shrub; *branchlets* pubescent but *glabrate, becoming reddish-brown; leaves lanceolate, oblong-ovate or narrowly obovate, gradually acuminate,* 4–9 cm. long, 2–4.5 cm. broad, *sharply serrate,* slightly pubescent or *glabrescent beneath; flowers* 1–1.5 *cm. broad,* slender-pedicelled; *calyx-lobes oblong-ovate, blunt, only tardily reflexed; style* 4–7 *mm. long,* usually hidden among the stamens; *fruit dark purple, about* 1 *cm. in diameter; stone turgid.* — Thickets and borders of woods. Ct. to centr.-w. Pa., and along mts. to w. Va. *Fl.* late April, May; *fr.* Aug., Sept.

5. P. americàna Marsh. (American), WILD P. — Coarse shrub, or tree up to 8 m. high and with dark shaggy-barked trunks up to 3 dm. in diameter; *branchlets* often spinescent, *glabrous or glabrate; leaves* lance- or oblong-ovate to broadly ovate or obovate, *abruptly long-acuminate,* glabrous above, *slightly pubescent to glabrate beneath, sharply* (*often doubly*) *serrate, the fully grown ones* 5–12 *cm. long; petioles mostly without terminal glands; flowers* in fascicles of 2–5, 1.8–3 *cm. broad; calyx glabrous or nearly so, its lance-attenuate lobes promptly reflexed;* petals white; *style* 7–12 *mm. long,* often exceeding stamens; *fruit red to yellow,* 2–3 *cm. in diameter; stone compressed.* — Thickets, borders of woods, stream-banks and fence-rows, Fla. to N.M., n. to w. N.E., N.Y., s. Ont., Mich., Wisc., Minn., s. Man., Wyo. and Utah, often spread from cult. *Fl.* late April–early June; *fr.* late Aug., Sept. — Inland and westw. passing insensibly into

Var. **lanàta** Sudw. (woolly). — *Branchlets and petioles pubescent, the latter without or with apical glands; leaf-blades densely soft-pubescent beneath.* (Incl. *P. lanata* Mackenz. & Bush, *P. arkansana* Sarg. and *P. americana,* var. *mollis* of ed. 7, not T. & G.). — Ind. to Ia., s. to Tenn., Ark., Tex. and Mex.

6. P. nìgra Ait. (black, from the dark branches), CANADA P., "POMEGRANATE," GUIGNIER and PRUNIER SAUVAGE (Que.). — Very similar to no. 5, the arborescent form mostly smaller and with trunks rarely 2 dm. in diameter; *leaves with coarse divergent rounded or blunt gland-tipped teeth, the petiole usually with* 2 *apical glands; calyx-lobes glandular-serrate,* 3.5–5 mm. long, acuminate, abruptly reflexed; *petals often becoming roseate; fruit light red, orange-red or yellowish,* slightly *elongate, almost without bloom.* — Thickets, stream-banks, borders of woods, etc., Que. to Man., s. to N.S. (introd.), N.E., Va., upland to Ga., W.Va., n. O., n. Ind., n. Ill. and Ia. *Fl.* mid-April (southw.)–early June (northw.); *fr.* Aug.–Oct.

7. P. hortulàna Bailey (of the garden), WILD-GOOSE-P. — Small tree or coarse shrub with gray-brown *bark containing abundant red cork-cells; branchlets glabrous, reddish-brown;* overwintering buds obtuse; *leaves conduplicate in bud, becoming flat,* narrowly ovate to ovate- or lance-oblong, *gradually tapering to acuminate tip,* glabrous above, *glabrate or merely pubescent along the nerves beneath, finely crenate-serrate with gland-tipped teeth;* the slender *petiole with* 2–6 *glands* toward tip; *flowers borne chiefly on the prolonged slender branches of preceding year,* 1.2–1.5 *cm. broad; calyx-lobes* 1.5–3 *mm. long, glandular-serrate, only tardily reflexed;* fruit red to yellow, globose or nearly so, 2–3 cm. in diameter; *stone pointed at both ends, reticulate.* — Bottomlands, thickets and borders of woods, s. Ind. to Ia., s. to n. Ala., w. Tenn., Ark. and Okla. *Fl.* late March–early May; *fr.* Aug.–Oct.

8. P. Munsoniàna Wight & Hedrick (for THOMAS VOLNEY MUNSON, 1843–1913), WILD-GOOSE-P. — Differing from no. 7 in having the *leaves convolute in bud and usually longitudinally folded to maturity,* the fine serrations tipped with red glands; *flowers borne chiefly on short lateral spurs;* fruit ripening earlier; stone obliquely truncate at base. — Rich thickets, O. to Kans., s. to Okla. and Tex. — In its convolute leaves approaching the next.

9. P. angustifòlia Marsh., CHICKASAW P. — *Shrub, rarely a small tree* up to 4 m. high, *with* frequently spinose *reddish-brown glabrous branchlets; leaves convolute in bud, mostly longitudinally folded or trough-like,* the mature blades 2.5–8 cm. long and 0.5–2 cm. broad, lanceolate to lance-oblong, *the appressed teeth gland-tipped,* dull, glabrous, the short red petioles with 1 or 2 red glands at summit; flowers expanding with the leaves, 6–9 mm. broad, creamy-white, in nearly sessile fascicles; *calyx* glabrous without, its short *lobes glandless;* fruit red or yellowish, subglobose, 1–2 cm. in diameter; stone plump, both ends obtuse or summit acutish. — Dry thickets and borders of woods, Fla. to Tex., n. to N.J., Md., Ky. and Mo., and as a relic of cult. elsewhere. *Fl.* late March, April; *fr.* June, July. — Westward passing into

Var. **Wátsoni** (Sarg.) Waugh (named in 1894 for its discoverer, LOUIS WATSON), SAND-P. — Lower, usually less spinescent; leaves elliptic to elliptic-oblong often blunter, coriaceous, usually with more prominent teeth; flowering and fruiting later. — Sand-ridges and plains, se. Neb. to Tex.

Subgen. II. Amýgdalus (L.) Focke (Peach)

10. P. Pérsica (L.) Batsch (old generic name), Peach. — Small tree; *leaves lance-oblong, attenuate, serrate, conduplicate in bud; flowers subsessile, appearing before the leaves,* from scaly buds, pink, 2.5–3.5 cm. broad; calyx-lobes pubescent; ovary tomentulose; *fruit* subglobose, *tomentulose, 3–6 cm. or more in diameter; stone deeply sculptured.* — Roadside-thickets, etc., spread from cult., N.E. to Mich. and southw. *Fl.* late March–May; *fr.* July–Oct. (Introd. from Asia)

Subgen. III. Cérasus Pers. (Cherry)

11. P. pùmila L. (dwarfish), Sand-C. — *Small shrub with erect or strongly ascending branches and branchlets,* 0.3–2.5 m. high; *leaves narrowly oblanceolate, one-sixth to one-third as broad as long, tapering gradually from near or above the middle to the narrowly cuneate base and slender petiole* (1–1.5 cm. long) *and to the acuminate apex;* the blade mostly 4–7.5 cm. long and 0.6–1.5 cm. broad, *coriaceous, dark green and lustrous above, slightly paler but scarcely glaucous beneath, appressed-serrate* above the middle; deciduous *stipules* 1.2–1.5 *cm. long, thread-like, fimbriate or setose;* flowers umbellate, from 2 or 3 buds, about 1.5 cm. broad; pedicels filiform, *erect or ascending; fruit purple-black, about* 1 *cm. in diameter, astringent; stone subglobose, apiculate at tip, rounded at base, 6–8 mm. broad.* — Dunes and sands or calcareous rocky shores, Ont., s. to St. Lawrence basin from N.Y. to Minn. *Fl.* May, early June; *fr.* July–Sept.

12. P. susquehánae Willd. (of the Susquehanna region, Pa.), Sand-C. — Differing from no. 11 in its ascending or erect branches (*lower 0.2–1 m. high*); *leaves elliptic to oblong or narrowly obovate, firm-membranaceous, pale green above, glaucous or grayish beneath, with gradually tapering to rounded bases, blunt to merely subacute,* on petioles mostly 0.5–1 cm. long; *blades of fertile branchlets two- to three-fifths as broad as long, mostly* 3–7 *cm. long and* 1.5–3 *cm. broad, low-crenate above middle; stipules 5–8 mm. long, scarcely fimbriate; stone* ellipsoid, 5–6 mm. broad, *acute.* (*P. cuneata* Raf.; *P. pumila,* var. *susquehanae* Jaeger) — Sandy or other acid dry to wet open habitats, sw. Me. and sw. Que. to se. Man., s. to L.I., Va., O., Ind., Ill. and Minn. *Fl.* May, June; *fr.* July–Sept.

13. P. Bésseyi Bailey (for Charles Edwin Bessey, 1845–1915), Sand-C. — Differing from no. 12 in its *coriaceous* smaller *acutely toothed leaves, the blades of fertile branchlets mostly* 2–4 *cm. long* and 0.8–2 *cm. broad; stipules firm and subpersistent; fruit* 1.5 *cm. in diameter, sweet* and very fleshy; *stone* 6–8 *mm. broad, rounded at both ends or barely apiculate at summit.* — Sandhills, open plains, rocky slopes or shores, Man. to Wyo., s. to Minn., Kans. and Colo. *Fl.* late April, May; *fr.* July–Oct.

14. P. depréssa Pursh (depressed), Sand-C., Gogoune (Que.). — *Depressed, forming prostrate or low-doming mats* up to 2 m. broad; the *new reddish shoots highly lustrous, often freely rooting; leaves spatulate-oblanceolate to narrowly obovate, submembranaceous, pale green above, whitened beneath, obtuse or subacute, with appressed teeth;* the *blades of fertile branches one-third to one-fourth as broad as long,* tapering to *petioles* 3–10 *mm. long; stipules flexuous,* 4–8 *mm. long, barely or not at all setose;* flowers 1.2–1.5 cm. broad; *fruit* red-purple to purple-black, *about* 1 *cm. in diameter, acid but of rich quality; stone ellipsoid, acute at both ends,* 4.5–6 *mm. broad.* (*P. pumila,* var. Bean) — Gravelly or sandy beaches in calcareous or basic soils or on calcareous pavement or slopes, Gaspé Pen. to L. Mistassini, Que., w. to Manitoulin Distr., Ont., s. to rivers of N.B., n. N.E., w. Mass., Delaware R., Pa., N.Y. and Wisc. *Fl.* May–July; *fr.* late July–Sept.

15. P. pensylvánica L. f. (Pennsylvanian), Bird-, Pin- or Fire-C., Cerises d'été or Petit Merisier (Que.). — *Tree* up to 12 m. high *or coarse shrub, with thin* (*easily removed*) *reddish-brown bark; leaves* oblong-lanceolate to narrowly ovate or subelliptic, *gradually acute to acuminate, membranaceous, finely and sharply serrate,* lustrous above, *green and glabrous beneath; flowers* in umbels or corymbs, expanding with the leaves, 1.2–1.6 *cm. broad,* white; *fruit globose, light red,* 5–7 *mm. in diameter,* with thin acid flesh; *stone subglobose,* 4–5 *mm. long.* — Dry woods, recent burns and openings, s. Lab. Pen. to B.C., s. to Nfld., N.S., N.E., L.I., Va., upland to N.C., Tenn., Ill., Ia., S.D. and Colo. *Fl.* late March (southw.)–early July (northw.); *fr.* July–Sept.

16. P. Àvium L. (of birds), Sweet C., Mazzard, Cerisier de France (Que.). — Tree with ashy-gray branches; *leaves soft, inclined to droop,* becoming 0.7–1.5 dm. long and up to 7 cm. broad, *abruptly acuminate, pubescent on nerves beneath, coarsely and doubly dentate; flowers* 2.5–3.5 *cm. broad, the pedicels subtended by reflexed or wide-spreading enlarged inner scales of the over-wintering buds; calyx-tube constricted at summit,* the *lobes entire; fruit* 2–2.5 *cm. in diameter,* subglobose, deeply cordate at base, red to purplish-black (sometimes yellowish), *with sweet pulp; stone* ellipsoid, blunt, smooth, *about* 1 *cm. long.* — Roadside-thickets, borders of woods, etc., spread from cult., N.S. to s. Ont. and southw. *Fl.* April, May; *fr.* June, July. (Introd. and natzd. from Eurasia)

17. P. Cérasus L. (Classical name of the cherry, which was brought into Europe from Cerasus or Keresoon), Sour or Pie-C. — Differing from no. 16 in *more ascending or porrect and firmer smaller leaves* (up to 1 dm. long and 5 cm. broad) *glabrous beneath* and with smaller teeth; *inner scales of overwintering buds erect at anthesis; calyx-tube not constricted above,* the *lobes serrate; fruit* red, *sour;* stone rounder. — Roadside-thickets and borders of woods, P.E.I. to Mich. and southw. (Introd. and natzd. from Asia)

18. P. Màhaleb L. (Arabic name), Mahaleb, Perfumed C., St. Lucie-C. — Small tree *with open branching and aromatic bark; young branchlets tomentulose; leaves round-ovate, broadly rounded to cordate at base, abruptly short-tipped or obtuse,* crenate-dentate, *glandular between the teeth,* the glabrescent *blade* 2.5–6 *cm. long;* inflorescences *short* 4–10-*flowered corymbiform racemes* of very fragrant white flowers about 1.5 cm. broad, *the lower pedicels subtended by small leaflike bracts; fruit* subglobose or ovoid, *nearly black,* 6–10 *mm. long.* — Roadsides, borders of woods, rocky banks, etc., N.E. to s. Ont. and southw., spread from cult. *Fl.* April, May; *fr.* July. (Introd. and natzd. from Eurasia)

Subgen. IV. Pàdus (Moench) Koehne (Cherry)

19. P. seròtina Ehrh. (late-ripening), Black or Rum-C., Cerises d'automne (Que.). — Tree up to 30 m. high, with dark bark, the branches reddish-brown, the *inner bark aromatic; leaves* lance-oblong to oblong-ovate, acuminate, *taper-pointed or acute,* firm to coriaceous, *crenate-serrate with blunt callous teeth,* the blades 3.5–15 cm. long, *dark green and lustrous above,* pale green beneath, *the broad midrib prominent beneath* and often villous; *racemes slender* and elongate (6–14 *cm. long*), the divergent *pedicels* 3–10 *mm. long;* flowers 7–10 mm. broad; *calyx with narrow acute lobes, persistent in fruit; fruit* globose, 7–10 mm. in diameter, *dark red, becoming purple-black, sweetish or bitter.* (*P. virginiana* sensu many auth., not L.) — Dry woods and fence-rows, Fla. to Tex. and Mex., n. to N.S., N.B., s. Que., s. Ont., Minn. and N.D. *Fl.* late April (southw.)–June (northw.); *fr.* July–Sept.

20. P. virginiàna L. (Virginian), Choke-C., Cerisier (Que.). — *Large shrub or small tree,* differing from no. 19 in its *non-aromatic bark; leaves* ovate or obovate, *thin, sharply serrulate, glabrous beneath* or merely tufted in the axils, or in forma **Deàmii** G. N. Jones (for Charles Clemon Deam, 1865–) copiously pubescent beneath and the young shoots and the rachis of the compact raceme tomentose, the *midrib slender;* raceme often thicker; *calyx-lobes blunt, the limb deciduous in fruit; fruit* deep red, becoming red-purple, or in forma **leucocárpa** (S. Wats.) Haynie (white-fruited) amber-colored to whitish or yellow, *acid and astringent.* (*P. nana* Dur.) — Thickets, shores and borders of woods, Nfld. to Sask., s. to N.S., N.E., upland of N.C., Tenn., Mo. and Kans. (vars. farther west). *Fl.* late April (southw.)–early July (northw.); *fr.* Aug.–Oct.

P. Pàdus L. (name used by Theophrastus, from *Padus,* the River Po), European Bird-C., differing from no. 20 in larger flowers (up to 1.5 cm. broad), the calyx-tube pubescent within, the stone strongly sculptured, spreads from cult. to roadside-thickets, as do several others locally. These should be sought in horticultural works.

FAM. 83. LEGUMINÒSAE (Pulse Family)

Plants with papilionaceous or sometimes regular flowers, 10 (*rarely* 5 *and sometimes many*) *monadelphous, diadelphous or rarely distinct stamens, and a single simple free pistil becoming a legume in fruit. Seeds mostly without albumen. Leaves alternate, with stipules,* simple or *usually compound.* One of the sepals inferior (*i.e.* next the bract); one of the petals superior (*i.e.,* next the axis of the inflorescence). — A very large family.

Subfam. I. MIMOSOÌDEAE

Flowers regular, small, in dense heads, with stamens overtopping the perianth. Corolla valvate in aestivation, often united into a 4–5-lobed cup, hypogynous, as are the (often very numerous) exserted stamens. Embryo straight. Leaves twice pinnate. — Often separated as Fam. Mimosaceae.

Trees or shrubs; stamens very numerous.
Shrub with hirsute branches; heads (in ours) 1–2 cm. in diameter, yellow or salmon-colored; sepals nearly distinct; legumes 4–8 cm. long, about 1 cm. broad. . . . . 1. *Acacia.*

Tree with smooth branchlets; heads about 5 cm. in diameter, pink or lilac; sepals united below into a cup; legumes 1–1.8 dm. long, 1.8–2.5 cm. broad. 2. *Albizzia.*
Herbs; stamens 5–10.
Petals distinct; legume smooth, flat. . . . . 3. *Desmanthus.*
Petals united below into a cup; legume prickly, thickish. . . . . 4. *Schrankia.*

SUBFAM. II. CAESALPINIOÏDEAE

Corolla imperfectly or not at all papilionaceous, sometimes nearly regular, imbricated in the bud, the upper or odd petal inside and inclosed by the others. Stamens 10 or fewer, commonly distinct, inserted on the calyx. Seeds anatropous, often with albumen. Embryo straight. — Often separated as Fam. CASSIACEAE.

*a.*Leaves simply or doubly pinnate; flowers not papilionaceous. . . *b.*
*b.*Trees with simply or doubly pinnate leaves without petiolar gland; flowers dioecious or polygamous; the calyx with tubular base, petals narrow, equal.
Thornless tree; leaves bipinnate, with ovate leaflets; flowers long-stalked, in open racemes; legume with thick woody valves . . . . 5. *Gymnocladus.*
Thorny (rarely thornless); leaves pinnate or bipinnate, with oblong leaflets; flowers short-stalked, in small spiciform racemes; legume with thin flat valves. . . . . 6. *Gleditsia.*
*b.*Herbs or low shrubs, with simply pinnate leaves, the petiole bearing 1 or more obvious glands; flowers perfect; calyx split to base; petals broad and showy, often somewhat unequal. . . . . 7. *Cassia.*
*a.*Leaves simple, rounded; flowers pink, papilionaceous, appearing before the leaves; tree. . . . . 8. *Cercis.*

SUBFAM. III. PAPILIONOÏDEAE

Calyx of 5 sepals more or less united, often unequally so. Corolla inserted into the base of the calyx, of 5 irregular petals (or very rarely fewer) more or less distinctly *papilionaceous, i.e.*, with the upper or odd petal (*vexillum* or *standard*) larger than the others and inclosing them in the bud, usually turned backward or spreading; the two lateral ones (*wings*) oblique and exterior to the two lower, which last are connivent and commonly more or less coherent by their anterior edges, forming the *carina* or *keel*, which usually incloses the stamens and pistil. Stamens 10, very rarely 5, inserted with the corolla, monadelphous, diadelphous (mostly with 9 united into a tube which is cleft on the upper side, and the tenth or upper one separate) or occasionally distinct. Ovary 1-locular, sometimes 2-locular by an intrusion of one of the sutures, or transversely 2–many-locular by cross-division into *articles;* style simple; ovules amphitropous, rarely anatropous. Cotyledons large, thick or thickish; radicle incurved. — Leaves simple or simply compound, the earliest ones in germination usually opposite, the rest alternate; leaflets quite entire or sometimes toothed. Flowers perfect. — Often separated as Fam. FABACEAE.

*a.*Stamens (10) distinct (distinct except at monadelphous base in no. 24, with corolla reduced to 1 petal). . . *b.*
*b.*Leaves palmately 3-foliolate or simple; herbs with yellow, bluish or white flowers.
Legume plump or inflated, roundish to subcylindric. . . . . 9. *Baptisia.*
Legume flat, linear. . . . . 10. *Thermopsis.*
*b.*Leaves pinnate, with numerous leaflets; flowers white.
Smooth-barked tree; flowers in pendulous panicles; legumes flat. . . 11. *Cladrastis.*
Low herb; flowers in erect racemes; legume terete, moniliform. . . . 12. *Sophora.*
*a.*Stamens monadelphous or diadelphous (9 and 1, rarely 5 and 5) but distinct nearly to base in no. 24. . . *c.*
*c.*Anthers of 2 forms; stamens monadelphous; legume promptly dehiscent; leaves digitate, simple, or rarely phyllodial. (Nos. 23 and 41, with anthers of 2 forms, have indehiscent legumes.) . . *d.*
*d.*Calyx subequally 5-lobed; legume inflated; herbs with simple leaves. 13. *Crotalaria.*
*d.*Calyx 2-lipped; legume flattened; shrubs, or herbs with palmate leaves of several leaflets. . . *e.*
*e.*Shrubs; leaves simple or 1–3-foliolate; corolla yellow.
Leaves with definite blades; branches smooth.
Branches terete, leafy to summit, the leaves simple; seed estrophiolate. . . . . 14. *Genista.*

Branches prominently 4-angled, vivid green and often naked above; leaves 3-foliolate or often with 1 terminal leaflet; seeds strophiolate. . . . . 15. *Cytisus.*

Leaves reduced to pungent petioles; stem hairy, spiny; seed strophiolate. . . . . 16. *Ulex.*

*e.*Herbs (with us); leaves palmate, with numerous elongate leaflets; flowers blue, purple, pink or white. . . . . . 17. *Lupinus.*

*c.*Anthers uniform (of 2 forms but the legume indehiscent in nos. 23 and 41). . . *f.*

*f.*Leaves digitately (rarely pinnately) 3-foliolate; leaflets denticulate or serrulate; stamens diadelphous; legumes small, 1–few-seeded, often inclosed in the calyx or curved or coiled. . . *g.*

*g.*Inflorescence capitate; legume membranaceous, straight or but slightly curved; petals adherent to stamen-tube, more or less persistent in fruit. . . . . 18. *Trifolium.*

*g.*Inflorescence racemose or racemose-spicate; legume coriaceous; petals free from stamens, deciduous in fruit.

Legume ovoid, straight or nearly so, wrinkled; racemes slender, more or less 1-sided. . . . . 19. *Melilotus.*

Legume scythe-shaped, spiral or tightly coiled; racemes mostly with spirally arranged flowers. . . . . 20. *Medicago.*

*f.*Leaves 1–several-foliolate (exceptionally simple); leaflets entire. . . *h.*

*h.*Leaves pinnately (rarely palmately) 1–several-foliolate; plants not climbing by tendrils nor twining (except in no. 31, a woody vine with many leaflets). . . *i.*

*i.*Legume dehiscent or, if indehiscent, plump and not articulated. . . *j.*

*j.*Flowers in umbels, loosely capitate or solitary in axils; herbs.

Flowers in clover-like heads subtended by many-cleft involucre; calyx loose, becoming vesicular, inclosing the legume. . . 21. *Anthyllis.*

Flowers 1–few; calyx close, not bladdery; legumes long-exserted. 22. *Lotus.*

*j.*Flowers in racemes, spikes or spiciform heads. . . *k.*

*k.*Leaves, calyces or legumes glandular- or waxy-dotted; stamens mostly monadelphous at least at base; legumes short, plump, indehiscent (or only tardily dehiscent), mostly 1-seeded. . . *l.*

*l.*Leaves pinnately 3-foliolate or palmately 3–5-foliolate; flowers in racemes or spikes; corolla strongly papilionaceous; alternate anthers large and small. . . . . 23. *Psoralea.*

*l.*Leaves pinnate, with several leaflets; flowers spicate-racemose to spicate-capitate; corolla but slightly to scarcely papilionaceous; anthers uniform. . . *m.*

*m.*Shrubs; flowers spicate-racemose; corolla of 1 petal (the standard) wrapped around the stamens and style; legume exserted from calyx. . . . . 24. *Amorpha.*

*m.*Herbs or half-shrubs; flowers spicate or capitate; corolla of 5 petals, the standard borne from bottom of calyx, the other 4 borne on the sheath of filaments; legume included in mature calyx.

Corolla imperfectly papilionaceous; stamens 10 (or 9); the keel- and wing-petals borne on the middle of the sheath of filaments. . . . . 25. *Dalea.*

Corolla barely or scarcely papilionaceous; stamens 5; the keel- and wing-petals borne at summit of the sheath of filaments. . . . . 26. *Petalostemum.*

*k.*Leaves (pinnate), etc., not glandular-dotted; flowers racemose; stamens mostly diadelphous; legumes dehiscent (indehiscent, fleshy and plum-like in a few species of no. 33), several-seeded. . . *n.*

*n.*Standard orbicular to round-ovate or -obovate. . . *o.*

*o.*Herbs. . . *p.*

*p.*Leaves odd-pinnate; wing-petals more or less coalescing to the keel.

Keel-petals with a lateral spur or gibbosity; legumes 4-angled. . . . . 27. *Indigofera.*

Keel-petals spurless; legumes flat. . . . . . . 28. *Tephrosia.*

*p.*Leaves even-pinnate; wing-petals and keel free. . . . 29. *Sesbania.*

*o.*Trees, shrubs or woody climbers. . . *q.*

*q.*Flowers violet, purple, pink or white; legume firm, not inflated.

Erect trees or shrubs; legume flat, thin, with one edge margined. . . . . 30. *Robinia.*
Twining woody climber; legume terete or tumid, not margined. . . . . 31. *Wisteria.*
*q.*Flowers yellow; legume membranous, bladdery-inflated. 32. *Colutea.*
*n.*Standard relatively narrow, not round-ovate nor orbicular; legume subcylindric, turgid or inflated. . *r.*
*r.*Ovary and legume not muricate nor prickly; legume (except in a few species of no. 33 with fleshy plum-like fruits) promptly dehiscent; calyx 5-toothed.
Keel not tipped by a sharp appendage or point. . . . 33. *Astragalus.*
Keel tipped by a sharp appendage or point. . . . . 34. *Oxytropis.*
*r.*Ovary and nearly indehiscent legume muricate or prickly; 2 upper lobes of calyx shorter and partly united. 35. *Glycyrrhiza.*
*i.*Legume indehiscent (or very tardily and irregularly dehiscent), if of more than 1 fertile segment (article) with transverse joints (except in no. 42, the peanut). . . *s.*
*s.*Leaves odd-pinnate, the many leaflets not stipellate.
Calyx 2-lipped; corolla yellowish and reddish, with navicular keel; stamens equally diadelphous (5 and 5); legume readily separable into quadrate flat articles. . . . . . . 36. *Aeschynomene.*
Calyx subequally 5-toothed; corolla purple, pink or white; stamens unequally diadelphous; articles of legume not flat and quadrate.
Inflorescence umbellate; articles of legume subcylindric, 4-angled. . . . . 37. *Coronilla.*
Inflorescence racemose; articles flat, rounded to oval, reticulate. . . . . 38. *Hedysarum.*
*s.*Leaves with (1–)2–4 leaflets. . . *t.*
*t.*Leaflets 3 (rarely only 1). . . *u.*
*u.*Leaflets stipellate; calyx 2-lipped; flowers uniform, purple or pink to white; legume with (1–)2–several flattish articles covered with minute hooked hairs (adhering to animals and clothing). . . . . 39. *Desmodium.*
*u.*Leaflets not stipellate; calyx subequally 5-toothed; legume with 1 fertile non-adhesive article.
Flowers mostly of two sorts, the petaliferous ones purple to white, in racemes, the apetalous ones very fertile; stamens diadelphous (9 and 1), with uniform anthers; calyx persistent in fruit. . . . . 40. *Lespedeza.*
Flowers uniform, orange or yellow (rarely whitish), 1–few; stamens monadelphous, with anthers of 2 sorts; calyx deciduous from fruit. . . . . 41. *Stylosanthes.*
*t.*Leaflets 2 or 4; corolla yellow; stamens monadelphous.
Soft-stemmed annual; leaflets obovate or elliptic, 2–5 cm. long; calyx-tube filiform, the 4 upper membranaceous lobes united; legumes (peanuts) developed at or beneath surface of ground, with reticulate and ropy cortex. . . . . . 42. *Arachis.*
Wiry freely forking perennial; leaflets narrow and small; calyx 5-toothed, the tube not elongate; flowers and small fruits covered by leafy bracts, from the axils and tips of the branches. . . . . 43. *Zornia.*
*h.*Leaves either 3-foliolate (pinnate and with many leaflets in no. 46), the stems herbaceous, twining (at least at tip when reaching supports) or trailing (rarely erect); or leaves pinnate with few–many leaflets and terminating in tendrils, the plants then usually climbing by means of tendrils; flowers on axillary peduncles. . . *v.*
*v.*Leaves abruptly pinnate, terminated by a tendril or bristle; stamens diadelphous; keel not strongly curved; seeds globular to barrel-form.
Wings coherent with keel; style filiform, bearded with a tuft or ring of hairs at apex. . . . . 44. *Vicia.*
Wings free or nearly so; style somewhat dilated and flattened upward, bearded down the inner face. . . . . 45. *Lathyrus.*
*v.*Leaves odd-pinnate, without tendrils; plants mostly twining or trailing; keel strongly curved or coiled. . . *w.*
*w.*Leaflets 5–several; keel slender, scythe-shaped, at length coiled. . 46. *Apios.*
*w.*Leaflets 3. . . *x.*

x.Ovules and seeds several in each legume (from petaliferous flowers, the apetalous basal or subterranean cleistogamous flowers of no. 54 producing 1-seeded indehiscent fruits); corolla usually tinged with or wholly blue, purple, pink or white, rarely wholly yellow. . . *y*.

y.Style bearded. . . *z*.

z.Style bearded lengthwise on the upper surface.

Flowers numerous in racemes or panicles; keel spirally coiled; seeds round-reniform to plumply sublunate. . 47. *Phaseolus*.

Flowers 1–few in pedunculate heads or short racemes; keel not spirally coiled.

Flowers 0.6–2.5 cm. long; the standard arched, not much longer than other petals; keel strongly incurved; legume sessile or subsessile in the calyx.

Wing-petals auricled at base; legumes 1.5–3 cm. long; seeds reniform to subglobose, with rounded ends and back; cult. annual. . . . . 48. *Vigna*.

Wing-petals not auricled at base; legumes 2–10 cm. long; seeds with truncate ends and angled back; native. . . . . 49. *Strophostyles*.

Flowers 4–6 cm. long; the standard erect, flat, very much longer and broader than the other petals; keel scythe-shaped; legume raised out of the fruiting calyx on a slender stipe; stems erect or twining. . . . 50. *Clitoria*.

z.Style bearded at summit about the stigma.

Peduncles mostly shorter than leaves, 1–4-flowered; calyx-teeth linear-setaceous; standard much larger than other petals, spurred at base of back; legume narrowly linear; native perennial. . . . . 51. *Centrosema*.

Peduncles elongate, flowers numerous in interrupted spiciform racemes; calyx-teeth deltoid; standard about equaling other petals, spurless; legume lunate-oblong; cult. annual. . . . . 52. *Dolichos*.

y.Style beardless.

Erect cult. annual; flowers few in axillary fascicles; corolla less than twice length of long-villous calyx; ovary and legume densely long-villous. . . . . 53. *Glycine*.

Twining or prostrate perennials; flowers in peduncled racemes or interrupted spikes; corolla twice to thrice length of smooth to short-hairy calyx.

Flowers of 2 kinds, petaliferous and apetalous; calyx of petaliferous flowers ebracteolate, tubular, with subequal teeth. . . . . 54. *Amphicarpa*.

Flowers all petaliferous; calyx with basal bracteoles, bilabiate, deeply cleft.

Slender native herbs with twining or non-twining minutely pubescent stems; petioles rarely 4 cm. long; leaflets firm, entire, rarely 6 cm. long; racemes interruptedly linear-cylindric, with ovate to lance-linear inconspicuous bracts; legumes finely pubescent. . 55. *Galactia*.

Coarse more or less woody introduced climber, rusty-villous; petioles elongate; leaflets membranaceous, entire or lobed, mostly 1–2 dm. long; racemes dense, columnar or slenderly pyramidal, with slender-tipped bracts conspicuously overtopping unexpanded flowers; legumes long-villous. . . . . . . 56. *Pueraria*.

x.Ovules and seeds 1 or 2 in each short legume; corolla orange or yellow; plants twining, trailing or erect; keel scythe-shaped. 57. *Rhynchosia*.

Subfam. I. MIMOSOÌDEAE

1. ACÀCIA Mill.

Flowers perfect or polygamous, regular, small, capitate or spicate. Sepals 4 or 5, nearly distinct or united into a 4- or 5-toothed campanulate cup. Petals as many, narrow. Stamens ∞, exserted. Legume oblong to linear, compressed or turgid. — Shrubs or trees (mostly armed), chiefly trop. and subtrop., with bipinnate or (in certain Austral. species) vertically expanded phyllodial leaves. (Ancient Greek name, *Acacia*, of an Egyptian species.)

1. A. angustíssima (Mill.) Ktze. (most narrow), var. **hírta** (Nutt.) Robins. (harsh). — Unarmed hirsute undershrub; pinnae 8–14 pairs and leaflets mostly 18–40 pairs (both less numerous on young shoots); flowers in yellow or salmon-colored paniculate globose heads 1–2 cm. in diameter; legume flat, 4–8 cm. long, about 1 cm. broad. — Dry bluffs, rocky woods and prairies, Tex. and Ark., n. to s. Kans. and sw. Mo. June–Oct.

2. ALBÍZZIA Durazzini

Flowers perfect or polygamous. Calyx tubular, 5-dentate. Petals united for more than half their length into a tubular somewhat salverform corolla. Stamens numerous; the filaments much elongated. Legume narrowly oblong, the valves neither twisted nor elastically spreading. — Unarmed trees nat. of Mex., Asia, Afr. and Austral., with bipinnate leaves. (Dedicated to *Filippo degli Albizzi*, who, two centuries ago, introduced this genus into European cult.) Originally spelled ALBIZIA, corrected to ALBIZZIA by Bentham.

1. A. JULIBRÍSSIN Durazzini (modification of Persian name), SILK-TREE, "MIMOSA". — Flowers pink or lilac, in several tassel-like globular clusters about 5 cm. in diameter at the ends of slender naked peduncles; legumes flat, 1–1.8 dm. long, 1.8–2.5 cm. broad. — Frequently cult., and natzd. on roadsides and in thickets and borders of woods, n. to se. Va., Ky. and Ind. June–Aug. (Introd. and natzd. from Asia and Afr.)

3. DESMÁNTHUS Willd.

Flowers perfect or polygamous, regular. Calyx campanulate, 5-toothed. Petals 5, distinct. Stamens 5 or 10. Legume flat, membranaceous or somewhat coriaceous, several-seeded, 2-valved, smooth. — Herbs of warm. reg. of N.Am. (1 in Madagascar), with twice-pinnate leaves of numerous small leaflets and with one or more glands on the petiole, setaceous stipules, and axillary peduncles bearing a head of small greenish-white flowers. (Name composed of the Greek *desme, a bundle*, and *anthos, flower*.)

1. D. illinoénsis (Michx.) MacM. (of Illinois), PRAIRIE-MIMOSA, PRICKLE-WEED. — Nearly glabrous perennial, erect, 3–24 dm. high; pinnae 6–15 pairs; leaflets 20–30 pairs; peduncles 2.5–7.5 cm. long; stamens 5; *legumes numerous in dense globose heads, oblong or lanceolate, strongly curved,* 1.5–2.5 *cm. long,* 4.5–5.5 *mm. wide,* 2–6-seeded. (*Acuan* Ktze.) — Prairies and alluvium, Ala. to Tex., n. to O., Ind., Ill., Minn., N.D. and Colo. June–Aug.

2. D. leptólobus T. & G. (slender-lobed). — Similar, scabrous; peduncles shorter; *legumes linear, straight, pedately ascending,* 3–7 *cm. long,* 1.5–4 *mm. wide,* several-seeded. — Prairies and bluffs, Tex. to s. Mo. and Kans. June–Aug.

4. SCHRÁNKIA Willd. SENSITIVE BRIER

Flowers polygamous, regular. Calyx minute, 5-toothed. Petals united into a funnelform 5-cleft corolla. Stamens 10–12, distinct, or the filaments united at base. Legumes long and narrow, rough-prickly, several-seeded, 4-valved, *i.e.*, the two narrow valves separating on each side from a thickened margin. — Perennial Am. herbs, closely related to the true Sensitive Plants (*Mimosa*); the procumbent stems and petioles recurved-prickly, with twice pinnate sensitive leaves of many small leaflets, and axillary peduncles bearing round heads of small rose-colored flowers. (Named for *Franz von Paula von Schrank*, a German botanist, 1747–1835.)

1. S. Nuttállii (DC.) Standl. (for its discoverer, THOMAS NUTTALL, 1786–1857), CAT-CLAW. — Prickles hooked; pinnae 4–6 pairs; *leaflets elliptical, reticulated* with strong veins beneath; *legumes* oblong-linear, nearly terete, short-pointed, densely prickly, 3–7 *cm. long.* (*S. uncinata* of ed. 7. not Willd.; *Leptoglottis* DC.) — Sandy pinelands, fields and woods, Ala. to Tex., n. to N.C., Ill. and Neb. June–Sept.

2. S. microphýlla (Dryander) Macbr. (small-leaved). — *Leaflets oblong-linear, scarcely veined; legumes* more slender, 8–15 *cm. long,* taper-pointed, sparingly prickly. (*S. angustata* T. & G.; *Leptoglottis* Britt.) — Dry sandy soil, Fla. to Tex., n. to Va. and Ky. June–Sept.

Subfam. II. CAESALPINIOÏDEAE

5. GYMNÓCLADUS Lam. KENTUCKY COFFEE-TREE. CHICOT (Que.)

Flowers dioecious or polygamous, regular. Calyx elongated-tubular below, 5-cleft. Petals 5, oblong, equal, inserted on the summit of the calyx-tube. Stamens 10, distinct, short, inserted with the petals. Legume oblong, flattened, hard, pulpy inside, several-seeded. Seeds flattish. —

Tall unarmed trees, ours and one in China, with rough bark, stout branchlets, and large unequally twice-pinnate leaves. Flowers whitish, in terminal racemes. (Name from the Greek *gymnos, naked,* and *clados, a branch,* alluding to the stout branches for many months destitute of foliage.)

1. G. dioíca (L.) K. Koch (dioecious). — Leaves 6–9 dm. long, with several large partial leafstalks bearing 7–13 ovate stalked leaflets, the lowest pair with single leaflets; stipules wanting; legume 1.5–2.5 dm. long, 3–4 cm. broad; seeds over 1.3 cm. across. — Rich woods, centr. N.Y. to S.D., s. to Tenn., Mo. and Okla.; natzd. from Del. to Va. and cult. northw. to Canada. May, June.

6. GLEDÍTSIA L. Honey-Locust

Flowers polygamous. Calyx short, 3–5-cleft, the lobes spreading. Petals as many as the sepals and equaling them, the two lower sometimes united. Stamens 3–10, distinct, inserted with the petals on the base of the calyx. Legume flat, 1–many-seeded. Seeds flat. — Thorny (rarely thornless) trees of Am., Asia and Afr., with abruptly once or twice pinnate leaves, and inconspicuous greenish flowers in small spikes. Thorns above the axils. (Simplified and Latinized name from that of *Johann Gottlieb Gleditsch,* 1714–1786, a botanist contemporary with Linnaeus.)

1. G. triacánthos L. (three-thorned), Honey-shuck. — *Thorns stout,* or wanting in forma **inérmis** (Pursh) Schneid. (unarmed), *commonly triple or compound; leaflets lanceolate-oblong,* somewhat serrate; *legumes linear,* elongated (2–4.5 dm. long), often twisted, *filled with sweet pulp* between the *oval seeds.* — Rich woods, w. N.Y. to S.D., s. to Fla. and Tex.; common in cult. and becoming estab. northeastw. to N.S. and N.E. May, June.

× **G. texàna** Sarg. (of Texas), thought to be a hybrid with the next, has legumes rarely 1.5 dm. long, without pulp.

2. G. aquática Marsh. (growing in water), Water-Locust. — *Thorns slender, mostly simple* on the branches; *leaflets ovate or oblong; legumes 4–5 cm. long,* mostly 1(–3)-seeded, *without pulp; seeds orbicular.* — River-swamps, Fla. to Tex., n. to N.C., s. Ind., s. Ill. and se. Mo. — Smaller tree than no. 1.

7. CÁSSIA L. Senna

Sepals 5, scarcely united at base. Petals 5, only slightly unequal, spreading. Stamens 5–10, unequal and some of them often imperfect, spreading; anthers opening by 2 pores or chinks at the apex. Legumes many-seeded, often with cross-partitions. — Herbs (in the United States) or ligneous plants of wide range in warm reg., with simply and abruptly pinnate leaves, and mostly yellow flowers. (An ancient name of some aromatic plant.)

a. Larger leaflets 0.7–3.5 cm. broad, not sensitive; stipules deciduous; 3 upper anthers imperfect; flowers or racemes axillary and terminal. . . *b.*
b. Leaflets 4–10 pairs, oblong, lanceolate or ovate; legumes flat or flattish. . . *c.*
c. Leaflets oblong or elliptic, mucronate, 5–10 pairs; young legumes pubescent.
Stipules linear-setaceous; gland near base of petiole slenderly clavate, stalked; ovary long-villous; segments of legumes as long as broad. . 1. *C. hebecarpa.*
Stipules linear-lanceolate; gland at base of petiole short-cylindric to conic; ovary strigose-hirsute; segments of legumes much shorter than broad. . . . . 2. *C. marilandica.*
c. Leaflets ovate or ovate-lanceolate, acuminate, 4–6 pairs; legumes glabrous. 3. *C. occidentalis.*
b. Leaflets 2 or 3 pairs, obovate or cuneate; legumes 4-sided. . . . . . . 4. *C. Tora.*
a. Leaflets smaller, sensitive to touch; stipules persistent, striate; anthers all perfect; flowers or small fascicles mostly supra-axillary.
Basal petiolar gland saucer-shaped or discoid; pedicels 1–2.5 cm. long; calyx about 1 cm. long; petals 1–2 cm. long; stamens 10. . . . . . . 5. *C. fasciculata.*
Basal gland stalked; pedicels 2–4 mm. long; calyx 3–4 mm. long; petals 4–8 mm. long; stamens 5. . . . . 6. *C. nictitans.*

1. C. hebecárpa Fern. (hairy-fruited), Wild S. — Root perennial; stem 9–12 dm. high, *sparsely villous above; stipules linear-setaceous,* caducous; *leaflets 5–9 pairs, oblong, obtuse;* petiole with a *slender clavate gland* near the base; *ovary spreading-villous with implexed pale hairs; legumes* linear, slightly curved, *flat,* at first hairy, 6.5–12 cm. long, *their segments as long as broad; seed flat, quadrate-orbicular.* (*C. marilandica* of most auth., not L.; *Ditremexa marilandica* sensu Britt. & Rose) — Alluvial soil, N.E. to Wisc., s. to w. N.C. and Tenn. July, Aug.

Var. **longípila** E. L. Braun (long-hairy). — Petioles densely long-villous near the glands; calyx conspicuously long-villous. — Dry slopes, e. Ky.

2. C. marilándica L. (of Maryland), Wild S. — Similar, *glabrous* or nearly so; *stipules linear-lanceolate;* petiolar *gland short-cylindric to conic-ovoid;* leaflets oblong or oblong-lanceolate, acutish; *ovary strigose-hirsute; legumes thickish,* 5–9 cm. long, *their segments much shorter than broad; seed plump, oblong-obovoid,* twice as long as thick. (*C. Medsgeri* Shafer; *Ditremexa Medsgeri* Britt. & Rose) — Dry roadsides and thickets, Pa. to Ia. and Kans., s. to Fla. and Tex. July, Aug.

3. C. occidentàlis L. (western), Coffee-Senna, Styptic-weed. — Annual; *leaflets* 4–6 *pairs, ovate-lanceolate, acute;* an ovoid gland at the base of the petiole; *legumes* long-linear (12 cm. long), with a tumid border, *glabrous.* — Waste places, cult. land, shores, etc., Fla. to Tex., n. to Va., Ind., Ill., Ia. and e. Kans. Aug., Sept. (Natzd. from the Tropics)

4. C. Tòra L. (an East Indian name), Sicklepod. — Annual; *leaflets* 3 *or rarely* 2 *pairs, obovate, obtuse,* with an elongated gland between those of the lower pairs or lowest pair; *legumes* slender, 4-*sided,* about 1.5 dm. long, curved. (*Emelista* Britt. & Rose) — Rich soil, waste land, etc., Fla. to Tex. and Mex., n. to Pa., Ind., Mich., Ill., Mo. and e. Kans. July–Sept. (Trop. reg.)

5. C. fasciculàta Michx. (in bunches), Partridge-Pea, Prairie-Senna, Golden Cassia. — *Annual,* suberect, 1.5–9 dm. high; *branches usually simple,* ascending; pubescence minute, subappressed, usually scanty; *stipules striate, persistent;* leaflets 10–15 pairs, linear-oblong, oblique at the base, with a cup-shaped gland beneath the lowest pair; *flowers* (*large*) *on slender pedicels,* 2 or 3 of the showy (1–2 cm. long) yellow, or white in forma **Jénseni** Palmer & Steyerm. (named in 1935 for its discoverer, L. C. Jensen), petals often with a purple spot at base; *anthers* 10, *elongated, unequal* (4 of them yellow, the others purple); style slender; legumes 2.5–5 cm. long and 4–4.5 mm. broad, with segments 3–4.5 mm. broad; the 4–13 seeds 3.5–5 mm. long, 2.5–4 mm. broad. (*Chamaecrista* Greene; *Cassia Chamaecrista* of ed. 7, not L.) — Sandy open soil, Fla. to Tex., n. to Mass., N.Y., s. Ont., O., Ind., Wisc., Minn. and S.D. July–Sept. — Two wholly aberrant sports from se. Va. are forma **transmutàta** Fern. (changing over) with inflorescences changed to forking glomerules of strongly ribbed bracts and bractlets; and forma **mutàta** Fern. (changing) with leaves odd-pinnate, of 3 or 5 leaflets, the large terminal leaflet composed of 2 fused smaller ones.

Var. **robústa** (Pollard) Macbr. (stout). — *Hirsute with spreading hairs;* leaflets 9–18 pairs; petals 1.5–2 cm. long. (*Chamaecrista* Pollard) — Fla. to La., n. to Va., O., Ill. and Mo.

Var. **depréssa** (Pollard) Macbr. (depressed). — *Diffuse, hirsute; leaflets* 6–10 *pairs; petals* 1.2–1.7 *cm. long.* (*Chamaecrista depressa* (Pollard) Greene) — Fla. to se. Mo.

Var. **macrospérma** Fern. (large-seeded). — Plants 1–1.7 m. high, the stems pilose-hirsute or glabrate; *pedicels hirsute; legumes mostly* 4–8.5 *cm. long,* 5–10 *mm. broad,* with hirsute sutures, the segments 5.5–7 mm. broad; *seeds* 4–10, *mostly* 5.5–7.5 *mm. long* and 4–6 mm. wide. — Tidal marshes and estuaries, Va.

6. C. níctitans L. (winking; from the opening and closing of the leaves), Wild Sensitive Plant. — Leaflets 10–20 pairs, oblong-linear; *flowers very small, on very short pedicels; anthers* 5, *nearly equal;* style short; *legumes finely pilose with short incurved-appressed hairs.* (*Chamaecrista procumbens* (L.) Greene) — Sandy soil, Ga. to Tex., n. to Mass., s. Vt., N.Y., O., Ind., Ill., Mo. and Kans. July–Sept.

Var. **hebecárpa** Fern. (hairy-fruited). — *Legumes villous-hirsute* with spreading hairs up to 1 mm. long. — Sandy coastal bluffs and dunes, e. Va. and e. N.C.

Var. **leiocárpa** Fern. (smooth-fruited). — *Legumes glabrous.* — Bell Co., Ky.

8. CÉRCIS L. Redbud. Judas-tree

Calyx 5-toothed. Corolla imperfectly papilionaceous; standard smaller than the wings and inclosed by them in the bud; the keel-petals larger and not united. Stamens 10, distinct, declined. Legume oblong, flat, many-seeded, the upper suture with a winged margin. Embryo straight. — Trees of N. Hemisph., with rounded cordate simple leaves, caducous stipules, and red-purple flowers in umbel-like clusters along the branches of the last or preceding years, appearing before the leaves, acid to the taste. (The ancient name of the oriental *Judas-tree.*)

1. C. canadénsis L. (Canadian), Redbud. — Leaves short-pointed, pubescent beneath when young, more or less so in maturity, or glabrous in forma **glabrifòlia** Fern. (glabrous-leaved); corolla reddish, or white in forma **álba** Rehd. (white); legumes nearly sessile above the calyx. — Rich woods and ravines, nw. Fla. to Tex. and ne. Mex., n. to s. Ct. (local), se. N.Y., Pa., s. Ont., s. Mich. and s. Wisc. Late March–May. — Large shrub or small tree, often cult.

Subfam. III. PAPILIONOÌDEAE

9. BAPTÌSIA Vent. FALSE INDIGO

Calyx 4- or 5-toothed. Standard not longer than the wings, its sides reflexed; keel-petals nearly separate and, like the wings, straight. Stamens 10, distinct. Legumes stalked in the persistent calyx, roundish or subcylindric, inflated, pointed, many-seeded. — Perennial N.Am. herbs with palmately 3-foliolate (rarely simple) leaves, which generally blacken in drying, and racemose flowers. (Name from the Greek *baptizein, to dye,* from the economic use of some species, which yield a poor indigo.)

*a.*Racemes very numerous, terminating the fine branchlets; leaflets 0.6–4 cm. long; flowers 1–1.6 cm. long; petals yellow; body of legume 0.5–1.5 cm. long. 1. *B. tinctoria.*

*a.*Racemes 1–few, terminal, later seemingly axillary by prolongation of branches; leaflets, flowers and legumes mostly larger. . . *b.*

*b.*Leaves, ovaries and legumes glabrous. . . *c.*

*c.*Stipules slender, shorter than petioles, early deciduous; petals white; legume slender-stiped, abruptly slender-beaked, beak 3–8 mm. long.

Slender, divergently branched; flowers 1.5–1.8 cm. long; legumes erect, linear-cylindric, their stipes barely exserted from the calyces. . . 2. *B. alba.*

Coarse, with ascending branches; flowers 2.2–3 cm. long; legumes drooping, ellipsoid-ovoid, their stipes long-exserted. . . . . . 3. *B. leucantha.*

*c.*Stipules broad, longer than petioles, persistent; petals indigo-blue; legume with upwardly enlarged stipe, tapering to a falcate beak 0.8–2 cm. long.

Petioles 0.5–2 cm. long; bracts 1–1.2 cm. long; wings and keel 2.5 cm. long; legume 1–1.5 cm. thick, its stipe not exceeding calyx. 4. *B. australis.*

Petioles 1–4 mm. long; bracts 7–9 mm. long; wings and keel 2.7–3 cm. long; mature legume 1.5–2.5 cm. broad, its stipe twice length of calyx. 5. *B. minor.*

*b.*Leaves (young) or ovaries and legumes (at least when young) pubescent.

Pedicels 5–8 mm. long, their small subtending bracts caducous; petals yellow. . . . . 6. *B. cinerea.*

Pedicels 1–3.5 cm. long, their large subtending bracts persistent; petals cream-color. . . . . 7. *B. leucophaea.*

1. B. tinctòria (L.) R. Br. (used for dyeing), WILD INDIGO, RATTLEWEED, HORSEFLY-WEED. — Smooth and slender, blue-glaucous when young, bushy-branched, 3–9 dm. high; leaves almost sessile, the *leaflets* narrowly *cuneate or with slightly concave sides* below the broad summit, *the larger ones* 0.6–1.8 *cm. long and* 0.5–1 *cm. broad (drying black)*; stipules and bracts minute, deciduous; *racemes numerous,* 0.3–1 (–1.5) dm. long, loosely few-flowered, *terminating the branchlets; flowers* 1–1.3 *cm. long, the petals yellow; legumes* long-stipitate, *their bodies* 0.5–1 *cm. long,* strongly rounded at base and summit. (*B. Gibbesii* Small) — Dry open woods and clearings, Fla. to Va. and less characteristically to N.Y. and R.I.; W.Va. to Ind. and Minn. Late May–Sept.

Var. **crèbra** Fern. (common). — Larger; *leaflets of primary leaves* 1.5–4 *cm. long,* 0.8–1.8 *cm. broad, cuneate or with convex sides;* flowers 1.3–1.6 cm. long; *mature legumes* 0.8–1.5 *cm. long,* attenuate to tip. — S.C. to La., n. to sw. Me., s. N.H., s. Vt., N.Y., s. Ont., s. Mich., Ill. and se. Minn. — Hybridizes with nos. 2 and 3.

Var. **projécta** Fern. (thrust forward). — As in var. *crebra,* but the primary raceme 2–4.5 dm. long; flowers 1.3–1.6 cm. long. — Along the mts., Pa. to w. Va.

2. B. álba (L.) R. Br. (white). — *Smooth,* 3–9 dm. high, *the branches slender and widely spreading; petioles slender; stipules and bracts minute* and deciduous; *leaflets oblong or oblanceolate, the larger* 2.5–5 *cm. long (drying green)*; *racemes* slender, *on long naked peduncles,* becoming 2–5 dm. long; *flowers* 1.5–1.8 *cm. long; petals white; legumes erect, linear-cylindric,* 1.5–4 cm. long, 6–9 *mm. thick, their stipes barely exserted.* — Dry pineland, Fla. to Eastern Shore, Va., and Tenn. May, June.

× **B. pinetòrum** Larisey (of the pines), a hybrid of no. 2 with *B. tinctoria,* var. *crebra,* occurs in Accomac Co., Va.

3. B. leucántha T. & G. (white-flowered), WHITE OR PRAIRIE-F. — Stout, *smooth,* glaucous, up to 1.5 dm. high, *with ascending branches; leaflets oblong-cuneate,* obtuse, *the larger* 3–6.5 *cm. long; stipules slender, shorter than petioles, earlier deciduous;* racemes stout; *flowers* short-pedicelled, 2.2–3 *cm. long; petals white; legumes drooping, ellipsoid-ovoid,* 1.3–1.8 cm. thick, *their slender stipes long-exserted, the slender beak* 4–8 *mm. long.* — Woods, prairies and river-banks, Miss. to Tex., n. to sw. Ont. (*Goldie*), O., Mich., Wisc., Minn. and Neb. May–July.

× **B. Deàmii** Larisey (for CHARLES CLEMON DEAM, 1865–), a hybrid of no. 3 with *B. tinctoria,* var. *crebra,* occurs in Ind.

4. B. austrālis (L.) R. Br. (southern), Blue F. — *Smooth,* stout, up to 1.6 m. high, *with ascending branches* or lower branches spreading; leaves slender-petioled, the median on *petioles* 0.5–2 *cm. long; leaflets* oblong-cuneate, obtuse, *the larger* 4–8 *cm. long,* firm; *stipules* lanceolate to narrowly ovate, *longer than petioles, persistent;* raceme elongated, coarse, erect; *bracts* 1–1.2 *cm. long;* flowers 2–3 cm. long; *petals dull indigo-blue; wing-petals and keel* 2.5 *cm. long,* paler; *legumes* ovoid-ellipsoid, ascending or divergent, 3–4 cm. long, 1–1.5 *cm. broad, narrowed below to a thick stipe and above to a falcate beak* 0.8–2 *cm. long, stipe not exceeding calyx.* — Rich woods and alluvial thickets, Ga. to Pa., W.Va., s. Ind. and Ky.; cult. and spread to alluvium of Vt. and N.Y. May, June.

5. B. mìnor Lehm. (smaller). — Lower than no. 4, divaricately branched; *petioles* 1–4 *mm. long;* stipules and leaves smaller, grayish-green, coriaceous, the *larger leaflets only* 1.5–4 *cm. long; bracts* 7–9 *mm. long; wings and keel* 2.7–3 *cm. long;* mature *legume* 1.5–2.5 *cm. broad, stipe twice length of calyx.* (*B. vespertina* Small, at least in part) — Rocky prairies, ravines and open woods, Mo. and Kans. to Tex. May, June. Hybridizes with no. 7.

6. B. cinèrea (Raf.) Fern. & Schub. (ashy). — Sometimes soft-hairy, usually minutely *pubescent* when young, erect, 6–9 dm. high, *with divergent branches;* leaves almost sessile; leaflets cuneate-lanceolate or obovate, 4–9 cm. long; lower stipules lanceolate and persistent; on the branches often small, subulate and caducous; racemes many-flowered; *pedicels* 5–8 *mm. long; bracts subulate, mostly deciduous; corollas yellow; legumes ovoid-ellipsoid,* taper-pointed, *minutely pubescent.* (*B. villosa* sensu Nutt., not *Sophora villosa* Walt., basonym) — Sands and sandy woods, se. Va. (very rare) to e. S.C. April, May.

7. B. leucophaèa Nutt. (cream-colored). — Coarse, *hairy,* 2–8 dm. high, *with divergent branches;* leaves nearly sessile, with firm spatulate or narrowly oblong-obovate leaflets; *stipules conspicuous, one-third as large as leaflets, persistent; bracts of raceme large, foliaceous, persistent; pedicels* slender, 1–3.5 *cm. long,* loosely ascending to divergent; flowers 2–3 cm. long; *petals cream-color; legume* ovoid or ellipsoid, 3–5 cm. long, *pubescent,* tapering to a short thick stipe and long falcate beak. (*B. bracteata* of ed. 7, not Muhl.) — Prairies, sands and open woods, Ark. to e. Tex., n. to s. Mich., Wisc., Minn. and Neb.; sometimes spread from cult. eastw. May.

Var. **glabréscens** Larisey (becoming smooth). — Stem and leaves glabrous. — La. and Tex. n. to Wisc. and w. Ill.

× **B. bícolor** Greenm. & Larisey (two-colored), a hybrid of no. 7 with no. 5, occurs in Mo., Kans. and Okla.

10. THERMÓPSIS R. Br.

Legume sessile or short-stipitate in the calyx, flat, linear, straight or curved. Otherwise nearly as *Baptisia.* — Perennial N. Am. and Asiatic herbs with palmately 3-foliolate leaves and foliaceous stipules not blackening in drying, and yellow flowers in terminal racemes. (Name from the Greek *thermos, the lupine,* and *opsis, appearance.*)

1. T. móllis (Michx.) M. A. Curtis (soft), Bush-pea. — Finely appressed-pubescent, 0.4–1.5 m. high; leaflets rhombic-lanceolate, 2.5–7.5 cm. long; stipules narrow, mostly shorter than the petiole; raceme elongated; legumes narrow, short-stipitate, somewhat curved, 2–10 cm. long. — Dry woods and ridges, mts. of Va. to Ga., Tenn. and Ala.; Essex Co., Mass., introd. and natzd. (*Birrell*) May, June.

11. CLADRÁSTIS Raf. Yellowwood. Virgilia

Calyx 5-toothed. Standard large, roundish, reflexed; the distinct keel-petals and wings straight, oblong. Stamens 10, distinct; filaments slender, incurved above. Legume short-stalked above the calyx, linear, flat, thin, marginless, 4–6-seeded, at length 2-valved. — Handsome trees, ours and three in Asia, with yellow wood (yielding a dye), smooth bark, nearly smooth pinnate leaves of 7–11 oval or ovate leaflets, and ample panicled racemes (2.5–5 dm. long) of showy white flowers drooping from the ends of the branches. Stipules obsolete. Base of petioles hollow, inclosing the leaf-buds of the next year. Bracts minute and fugacious. (Name said to be from the Greek *clados, a branch,* and *thraustos, brittle,* the latter Rafinesquian etymology.)

1. C. lùtea (Michx. f.) K. Koch (yellow). — Sometimes 17 m. high; legumes 7.5–10 cm. long. — Rich woods and calcareous bluffs, s. Ind., s. Ill. and Ky. to w. N.C., n. Ga., n. Ala., and Mo.; also in cult. and often spreading from planted trees n. to Mass. May, early June.

12. SOPHÒRA L.

Calyx campanulate, shortly 5-toothed. Standard rounded; keel nearly straight. Stamens distinct or nearly so. Legume coriaceous, stipitate, terete, more or less constricted between

the seeds, indehiscent. Seeds subglobose. — Shrubby, or ours an herbaceous perennial, of N. Am. and Asia, the leaves pinnate with numerous leaflets, and flowers white or yellow in terminal racemes. (Said by Linnaeus to be the ancient name of an allied plant.)

1. S. serícea Nutt. (silky). — Silky-canescent, erect, 3 dm. high or less; leaflets oblong-obovate, 6–12 mm. long; flowers white; legumes few-seeded. — Prairies and dry hills, S.D. to Wyo., s. to Kans., Tex., Ariz. and Mex. May, June. — Likely to be confused with *Astragalus*.

13. CROTALÀRIA L. Rattlebox

Calyx 5-cleft, scarcely 2-lipped. Standard large, cordate; keel scythe-shaped. Sheath of the monadelphous stamens cleft on the upper side; 5 of the anthers smaller and roundish. Legume inflated, subcylindric, many-seeded. — Herbs of warm reg., with simple leaves. Flowers yellow. (Name from the Greek *crotalon, a rattle;* the loose seeds rattling in the coriaceous inflated pods.)

a.Stems erect or ascending; leaves (or all but the earliest ones) elongate, tapering to base, the longer ones 0.3–2 dm. long; inversely sagittate stipules usually conspicuous; seeds 2–5 mm. broad. . . *b.*

b.Slender, 1–5.5 dm. high, with linear to oblong or oblanceolate leaves (3–8 cm. long) tapering to both ends; raceme 2–6-flowered; seeds 2–3 mm. broad.

Stems and calyx loosely villous or hirsute; peduncles 1–4 cm. long. . . 1. *C. sagittalis.*

Stems and calyx minutely strigose; peduncles 3–12 cm. long. . . . . 2. *C. Purshii.*

b.Coarse, 0.5–2 m. high, with obovate round-topped leaves mostly 0.6–2 dm. long; raceme 1.5–4.5 dm. long, many-flowered; seeds 4–5 mm. broad. . 3. *C. spectabilis.*

a.Stems prostrate or trailing; leaves oval to suborbicular, strongly rounded at both ends, the larger ones 1–3 cm. long; stipules mostly small or wanting. . 4. *C. angulata.*

1. C. sagittàlis L. (like an arrow-head). — Annual (southw. often perennial), *loosely villous or hirsute, erect or ascending,* simple to bushy-branched, up to 4 dm. high; *decurrent inversely sagittate stipules usually conspicuous;* lower *leaves* oval, *the upper linear to oblong-lanceolate, tapering to base and apex, the larger* 3–7 *cm. long; peduncles* terminal and axillary, 1–4 *cm. long* below the 2–4 flowers; *calyx loosely villous or hirsute, exceeding the corolla; legume* 1.5–2.7 *cm. long;* seeds obliquely reniform, 2.3–3 mm. broad. — Dry sandy or gravelly soil, Fla. to Tex. and Mex., n. to Mass., s. Vt., s. N.Y., s. O., s. Mich., Wisc., Minn. and S.D. June–Sept. (W.I.)

Var. **oblónga** Michx. (oblong). — All leaves, including the bracteal ones, broadly elliptic-oblong. — Fla. to Mex., locally n. to se. Va.

2. C. Púrshii DC. (for its discoverer, Frederick Traugott Pursh, 1774–1820). — Similar, perennial, *simple or sparingly simple-branched, erect,* up to 5.5 dm. high; *stem and calyx minutely strigose;* principal (upper) leaves lanceolate or oblanceolate, 3–8 cm. long; *lower and median leaves lance-oblong and* 6–15 *mm. wide; peduncles* 3–12 *cm. long below the* 2–6 *finally remote flowers, the bractlets borne mostly toward the summit; corolla equaling or exceeding the calyx; legume* 2.5–4 *cm. long;* seeds 2–2.5 mm. broad. — Dry sands and pinelands, Fla. to Tex., n. to se. Va. June–Sept.

Var. **bracteolífera** Fern. (bearing bractlets). — *Bushy-branched, the branches often forking; leaves linear or narrowly linear-lanceolate, the broader ones* 2–7 *mm. wide; longer peduncles with* 7–14 *scattered flowers or bractlets, the latter extending down to the base.* — Bogs and savannas of Coastal Plain, Fla. to se. Va.

3. C. spectábilis Roth (showy). — *Coarse* annual, with *canescent* angulate *stem* 0.5–2 *m. high; leaves cuneate-obovate,* canescent-tomentulose beneath, *the primary ones* 0.6–2 *dm. long,* the upper narrower and smaller; *racemes becoming* 1.5–4.5 *dm. long,* with many long-pedicelled showy flowers (1.5–2.5 cm. long) in the axils of acuminate persistent bracts; legumes 3–5 cm. long; seeds 4–5 mm. broad. — Fields and roadsides, southern states, n. to e. Va. and se. Mo. Sept.–Nov. (Introd. from the Tropics)

C. retùsa L. (shallowly notched), similar to no. 3 but with linear-subulate bracts, introd. in the South, has once been collected in N.J.

4. C. angulàta Mill. (angled; from the not very prominent angles of the stems), Rabbit-bells. — Perennial, *the trailing stems and branches forming carpets up to* 8 *dm. broad;* stems villous to strigose; *leaves oval to suborbicular, strongly rounded at both ends, mostly* 1–3 *cm. long; stipules usually small or wanting;* peduncles 3–15 cm. long below the finally remote flowers; corolla equaling or exceeding calyx; legume 1.8–3 cm. long; *seeds* 1.8–2 *mm. broad.* (*C. rotundifolia* of recent auth., not (Walt.) Poir.) — Sandy pinelands, Fla. to La., n. to se. Va. Late May–Sept. (Trop. Am.)

14. GENÍSTA L. Woad-waxen. Whin

Calyx 2-lipped. Standard oblong-oval, spreading; keel oblong, straight, deflexed. Stamens monadelphous, the sheath entire; 5 alternate anthers shorter. Legumes mostly flat and several-seeded. — Shrubby plants nat. of Eu., n. Afr. and w. Asia, with simple leaves and yellow flowers. (Name from the Celtic *gen, a bush.*)

1. G. tinctòria L. (used in dyeing), Dyer's Greenweed. — Low, not thorny, with striate-angled erect branches; leaves lanceolate; flowers in spiciform racemes. — Dry sterile soil, s. Me. to Mass., locally to D.C. and Mich. June, July. (Natzd. from Eu.)

15. CÝTISUS L. Broom

Calyx campanulate, with 2 short broad lips. Petals broad, the keel obtuse and slightly incurved. Stamens monadelphous. Legume flat, much longer than the calyx. Seeds several, with a strophiole at the hilum. — Shrubs nat. of Eu., n. Afr. and w. Asia, with stiff green branches, leaves mostly digitately 3-foliolate and large bright yellow flowers. (The ancient Greek name of a plant, probably a *Medicago.*)

1. C. scopàrius (L.) Link (broom-like), Scotch B. — Glabrous or nearly so, 1–2 m. high; leaflets small, obovate, often reduced to a single one; flowers solitary or in pairs, on slender pedicels, in the axils of the old leaves, forming leafy racemes along the upper branches; style very long and spirally incurved. — Sandy roadsides, open woods and barrens, N.S.; sw. Me. to w. N.Y., s. to Va., W.Va. and Ga. May, June. (Natzd. from Eu.)

16. ÙLEX L. Furze. Gorse

Calyx deeply 2-lipped. Standard ovate; wings and keel oblong, of about equal length. Stamens monadelphous. Legume short-oblong. — Low densely branched shrubs nat. of Eu. and Afr., with spine-like phyllodial leaves. (An ancient name, used by Pliny for some not certainly identified plant.)

1. U. europaèus L. (European). — Calyx large, yellow, tomentulose. — Sometimes cult. as a sand-binder, now locally natzd. on sands from se. Mass. to Va. and W.Va. May–Sept. (Introd. and natzd. from Eu.)

17. LUPÌNUS L. Lupine

Calyx very deeply 2-lipped. Sides of the standard reflexed; keel falcate, pointed. Sheath of the monadelphous stamens entire; anthers alternately oblong and roundish. Legume oblong, flattened, often knotty by constrictions between the seeds. Cotyledons thick and fleshy. — Herbs of Am., Eu. and Mediterr. reg., with palmately 1–18-foliolate leaves, stipules adnate to base of the petiole and showy flowers in terminal racemes or spikes. (The ancient name of the lupine, from *lupus, wolf,* because of a belief that it destroys the soil.)

Stems smooth to minutely pubescent, if villous with the villi sparse; leaves glabrous or short-strigose beneath; lower lip of calyx 4–6 mm. long.
- Flowering stems 2–7 dm. high; lower leaves with 7–11 leaflets 1.5–5 cm. long; raceme 0.7–3 dm. long. . . . . 1. *L. perennis.*
- Flowering stems 0.6–1.2 m. high; lower leaves with 12–18 leaflets 6–13 cm. long; raceme 3–7 dm. long. . . . . 2. *L. polyphyllus.*

Stems densely long-villous; leaves densely silky-villous beneath; lower lip of calyx 8–10 mm. long. . . . . 3. *L. nootkatensis.*

1. L. perénnis L. (perennial), Wild L. — *Stems several from creeping subterranean caudices,* **2–7** *dm. high,* minutely pubescent to glabrous; *leaves* long-petioled, *the lower with* 7–11 oblanceolate *leaflets* 1.5–5 *cm. long,* minutely strigose beneath, the lower petioles often with long villi; *racemes* 0.7–3 *dm. long,* the young tips with bracts 3–11 mm. long; rachis and pedicels minutely silky to pilose-hirtellous; flowers blue-purple, or flesh-pink or roseate in forma **ròseus** Britt. (rose-color), or white in forma **leucánthus** Fern. (white-flowered); legumes densely sericeous-villous to pilose-hirsute, 5–6-seeded. — Dry open woods, clearings and openings, sw. Me. to N.Y., s. to Fla. April–July.

Var. **occidentàlis** S. Wats. (western). — Upper half of stem and upper (as well as lower) petioles with sparse spreading long villi: flowers blue-purple, or white in forma **albiracèmus** A. H. Moore (with white raceme). — Similar habitats, N.Y. to s. Ont. and Minn., s. to Md., W.Va., n. O., Ind. and n. Ill.

2. L. polyphýllus Lindl. (many-leaved). — *Stout* perennial 0.6–1.2 *m. high, smooth or* *nearly so; lower leaves with* 12–18 oblanceolate acute *leaflets* 6–13 *cm. long; raceme* 3–7 *dm. long;* the blue-purple flowers about 1.5 cm. long: *lower lip of calyx* 4 *mm. long.* — Dry roadsides and banks, P.E.I., N.S. and n. N.E. Late June–early Aug. (Introd. and natzd. from nw. Am.)

3. L. nootkaténsis Donn (of Nootka Sound, Vancouver Island). — Similar to no. 2, 4–7 *dm. high,* the stout *stem densely long-villous; leaves densely silky-villous, the lower with* 5–9 oblanceolate *leaflets* 3–7 *cm. long; racemes* 1–3 *dm. long;* flowers 1.5–2 cm. long; *lower lip of calyx* 8–10 *mm. long.* — Roadsides and open banks, Avalon Pen., Nfld., to n. N.E. Late May–July. (Introd. and natzd. from nw. Am.)

18. TRIFÒLIUM L. Clover. Trefoil. Trèfle (Que.)

Calyx persistent, 5-cleft, the teeth usually bristleform. Corolla mostly withering or persistent; the claws of all the petals, or of all except the oblong or ovate standard, more or less united below with the stamen-tube; keel short and obtuse. Tenth stamen more or less separate. Legume small and membranous, often included in the calyx, 1–6-seeded, indehiscent, operculate or opening by one of the sutures. — Herbs of temp. reg. Leaves mostly palmately (sometimes pinnately) 3-foliolate; leaflets usually toothed. Stipules united with the petiole. Flowers in heads or head-like racemes. (Name from *tres,* three, and *folium,* a leaf.)

a. Flowers sessile or subsessile in the dense head, the middle and upper ones not strongly reflexed in age. . . *b.*
b. Heads on naked peduncles, not subtended by paired leaves or stipules; leaves all alternate. . . *c.*
c. Heads cylindric, cylindric-ovoid or long-conical; individual flowers without bracteoles; calyx long-villous, its subequal setiform teeth much longer than the campanulate uninflated and merely 10-nerved tube.
Petioles of lower and median leaves shorter than the narrowly oblanceolate or linear-oblong leaflets; corolla pale, marcescent, overtopped by the calyx-teeth; heads drab or grayish. . . . . 1. *T. arvense.*
Petioles of lower and median leaves longer than the obovate leaflets; corolla usually scarlet or deep red, soon shrivelling, exceeding the brown calyx. . . . . 2. *T. incarnatum,* var. *elatius.*
c. Heads subglobose to globose-ovoid; flowers subtended by bracteoles; calyx short-pilose to glabrous, bilabiate, the shorter teeth shorter than the finally inflated or vesicular and strongly reticulate tube.
Perennial, with repent stems; leaflets elliptic to obovate; bracteoles lanceolate, equaling the calyx; corolla not resupinate. . . . . 3. *T. fragiferum.*
Annual or biennial, with ascending branches; leaflets cuneate; bracteoles truncate, much shorter than calyx; corolla resupinate. . . . . 4. *T. resupinatum.*
b. Heads subtended by a pair of opposite leaves or their dilated stipules, either sessile or slightly elevated above them. . . *d.*
d. Heads cylindric or ellipsoid, 7–10 mm. thick; calyx-teeth linear-subulate, the upper ones much shorter than the urceolate to subglobose fruiting tube; corolla barely exceeding calyx, persistent. . . . . . . 5. *T. striatum.*
d. Heads subglobose, ovoid or obovoid, 1.5–3.5 cm. thick; calyx-teeth setaceous, all or at least the lower much longer than the campanulate to obconic tube; corolla much exceeding calyx. . . *e.*
e. Perennial or biennial; stems and petioles glabrous or only sparsely pilose; stipules gradually tapering to tip; calyx with 10-nerved campanulate or ovoid tube and unequal teeth, the upper and lateral teeth about equaling the tube.
Stipules oval, the free portion broadly triangular, tapering to a subulate tip; leaflets oval to obovate; calyx-tube pubescent, rim of throat glabrous. . . . . 6. *T. pratense.*
Stipules lance-acuminate; leaflets narrowly elliptic or oblong; calyx-tube glabrous, rim of throat hairy. . . . . 7. *T. medium.*
e. Annual; stems and petioles copiously spreading-villous; stipules abruptly contracted to a bristle-tip much longer than free blade; calyx with 20-nerved obconic lustrous-villous tube and subequal teeth all much exceeding it. . . . . 8. *T. hirtum.*
a. Flowers distinctly pedicellate, in loosening heads; pedicels reflexed in age. . . *f.*
f. Corolla white, roseate or purple; heads 1.2–4.5 cm. in diameter; lower petioles longer than leaflets. . . *g.*
g. Heads 1.5–4.5 cm. in diameter; flowers 8–13 mm. long; calyx scarcely 2-lipped, its teeth linear-subulate to subaristate, glabrous or appressed-pilose; standard not prolonged into a subulate tip. . . *h.*

h. Stems widely creeping or with long basal runners.
Peduncles rising from the creeping stems, scapiform; heads 1.5–3 cm. in diameter; calyx-teeth shorter than tube. . . . . . . 9. *T. repens.*
Peduncles axillary on ascending stems; principal heads about 3 cm. in diameter; calyx-teeth twice as long as tube. . . . . . . 10. *T. stoloniferum.*
h. Stems erect or ascending, without basal runners. . . *i.*
i. Stems elongate, from relatively soft bases; leaflets obovate or obovate-elliptic; peduncles mostly longer than subtending leaves; pedicels and calyx glabrous or only sparingly pilose.
Stems and petioles villous; heads 2.5–4.5 cm. in diameter; flowers 10–13 mm. long; calyx 7–8 mm. long, its teeth many times longer than the saucer-like tube. . . . . 11. *T. reflexum.*
Stems and petioles glabrous; heads 1.5–3.5 cm. in diameter; flowers 6–11 mm. long; calyx 3.5–5 mm. long, its teeth only slightly longer than the campanulate tube. . . . . 12. *T. hybridum.*
i. Stems short, cespitose, from a subligneous crown and thick root; leaflets narrowly oblong to oblanceolate; peduncles strongly pilose, mostly shorter than subtending leaves; pedicels and calyx densely pilose. . . . . 13. *T. virginicum.*
g. Heads 1–1.7 cm. in diameter; flowers 4.5–6 mm. long; calyx 2-lipped, with softly ciliolate veiny lanceolate teeth much longer than tube; standard prolonged into a subulate tip. . . . . 14. *T. carolinianum.*
f. Corolla yellow, becoming brown and marcescent; heads 0.5–1.5 cm. thick; petioles mostly shorter than leaflets.
Heads densely many-flowered, 0.8–1.5 cm. thick; corollas conspicuously striate-sulcate in age.
Leaflets all sessile or subsessile; stipules elongate, not dilated at base; style and legume subequal; seed globose. . . . . . . 15. *T. agrarium.*
Terminal leaflet distinctly petiolulate; stipules short, rounded at base; style many times shorter than legume; seed ovoid. . . . . . . 16. *T. procumbens.*
Heads loosely 3–15-flowered, 5–8 mm. thick; corollas not striate-sulcate. 17. *T. dubium.*

1. T. arvénse L. (of cult. fields), Rabbit-foot-, Old-field- or Stone-C. — Pubescent branching annual 1–4.5 dm. high; *lower petioles mostly shorter than the narrowly oblanceolate to linear-oblong leaflets;* free blades of stipules linear-setiform; *heads* short-peduncled, *ovoid-cylindric,* 1–4 cm. long, 8–14 mm. thick, *gray or drab; the long-villous 10-nerved sessile campanulate calyces* crowded, spreading, *their setiform teeth much longer than the tube and the pale rose,* or white in forma albiflòrum Sylvén (white-flowered), *marcescent corolla.* — Dry roadsides and fields, Que. and Ont., s. throughout our range. May–Oct. (Natzd. from Eu.)

2. T. incarnàtum L. (blood-red), var. elàtius Gibelli & Belli (tall), Crimson or Italian C. — Erect soft-pubescent annual up to 6 dm. high; *lower and median leaves with very long petioles and with cuneate-obovate* usually emarginate *leaflets;* free blades of stipules ovate, dentate; peduncles elongate; *heads slenderly conical, becoming cylindric,* 2–7 cm. long; *calyx fulvous-villous,* the strongly 10-ribbed campanulate tube slightly shorter than the linear-setiform subequal teeth; *corolla scarlet or deep red* (rarely white), *exceeding calyx-teeth.* — A cult. forage plant, occasionally spreading to waste ground, roadsides, etc. May–July. (Introd. from Eu.)

3. T. fragíferum L. (bearing strawberries), Strawberry-headed C. — *Perennial with repent stems* 1–3 dm. long; long-petioled leaves with *broadly elliptic to narrowly obovate leaflets,* their marginal nerve-tips parallel and prominent; *heads* on scapiform peduncles, *subglobose,* in fruit becoming 1.2–2 cm. thick; *flowers crowded, subsessile, the outer subtended by lanceolate bracteoles as long as the bilabiate calyx; calyx* in flower *with the spreading upper teeth shorter than the tube, in fruit vesicular and reticulate;* corolla roseate, about 6 mm. long. — Weed in lawns, Tompkins Co., N.Y. June–Sept. (Adv. from Eu.)

4. T. resupinàtum L. (upside-down), Reversed C. — Smooth *annual or biennial, with ascending stems* and branches up to 3 dm. high; *leaflets cuneate;* peduncles axillary; heads globose, in fruit 1–1.5 cm. in diameter; *bracteoles truncate, much shorter than the calyx;* fruiting calyx vesicular, reticulate; *corolla* purplish, 4–6 mm. long, twice or thrice length of calyx, *resupinate, the standard turned outward.* — Lawns, fields and roadsides, local, Mass. to Ill., s. to Ala. and e. Kans. May–Sept. (Adv. from Eu.)

5. T. striàtum L. (striped), Knotted C. — Pubescent annual with depressed or ascending stems and branches up to 7 dm. long; lower leaves long-petioled; stipules abruptly pointed; leaflets cuneate-obovate; *heads short-cylindric or ellipsoid,* 7–10 *mm. thick, subtended by 2 opposite* *leaves or their dilated stipules*, terminating short branches or sessile in upper axils, sometimes paired; *calyx-tube* becoming inflated, *urceolate to subglobose*, strongly ribbed, pale, *longer than the upper linear-subulate erect teeth; corolla* pale reddish, *barely exceeding calyx, persistent.* — Fields and roadsides, local, se. Mass. to N.J. May–Sept. (Adv. from Eu.)

6. T. praténse L. (of meadows), Red C. — *Biennial or* short-lived *perennial* with ascending stems 0.5–4 dm. high, glabrous or sparingly pilose, or densely villous or hirsute in forma pilòsum (Griseb.) Hayek (soft-hairy); leaves long-petioled; *stipules oval, the free blades broadly triangular, tapering to a subulate tip; leaflets oval to cuneate-obovate*, 1–3 cm. long, 0.5–1.5 cm. broad; *heads* subglobose, ovoid or obovoid, dense, 1.2–3 cm. long, *subtended by a pair of leaves*, either sessile or short-peduncled; flowers sessile; *calyx-tube pubescent*, 10-nerved, *campanulate or ovoid, about equaled by the setaceous upper and lateral teeth, the throat glabrous within;* corolla roseate, or creamy-white in forma leucochràceum Aschers. & Prantl (whitish), much exceeding calyx; *legume opening by a lid.* — Roadsides, clearings and turf, Lab. to B.C., s. beyond our limits. May–Sept. (Natzd. from Eu.) — Less common than

Var. satìvum (Mill.) Schreb. (sown), Cultivated Red C. — Coarser, longer-lived; stems 3–8 dm. high; larger leaflets 3–7 cm. long, 1.5–3.5 cm. broad; heads 3–4 cm. long; corollas roseate, or whitish in forma flàvicans (Vis.) Hayek (yellowing). — Extensively cult. and freely natzd. throughout. (Introd. and natzd. from Eu.)

7. T. mèdium L. (intermediate), Zigzag C. — More slender than no. 6, with few glabrous stems; *stipules lance-acuminate; leaflets narrowly elliptic or oblong*, the larger 2.5–6 cm. long and 0.8–2 cm. broad; heads sessile or peduncled above the paired upper leaves, 2–4.5 cm. long; *calyx-tube glabrous outside but with a ring of hairs at the inner margin of the throat;* corolla purplish, longer than in no. 6; *legume 2-valved.* — Fields, open slopes, borders of woods and roadsides, local, e. Que. and N.B. to Mass. June–Aug. (Natzd. from Eu.)

8. T. hírtum All. (rough-hairy). — *Annual*, 1–4 dm. high, *densely long-villous; stipules narrow, abruptly contracted to a bristle-tip longer than the free blades;* leaves petioled, the narrowly cuneate-obovate leaflets 0.8–2 cm. long; head globose, sessile, subtended by paired leaves or their stipules, 1.5–2.5 cm. in diameter; flowers crowded, sessile; *calyx-tube obconic, lustrous-villous, 20-nerved, much shorter than the subequal teeth;* corolla purplish, slender, exceeding calyx. — Roadsides and fields, local, Prince Edward Co., Va. May–July. (Adv. from Eu.)

9. T. rèpens L. (creeping), White C. — Smooth *perennial, with long repent stems;* stipules pale, with abruptly tipped lanceolate free lobes; leaves long-petioled; leaflets inversely cordate or shallowly notched; *peduncles scapiform*, elongate; *heads subglobose*, 1.5–3 *cm. in diameter*, the pedicels elongating and recurving in age; *calyx-teeth shorter than the tube*, shorter than the white or rose-tinged corolla; legume 3–4-seeded; seeds globose-reniform. — Grasslands, roadsides and open pastured woods, throughout our range and beyond. May–Oct. (Introd. and natzd. from Eu.) — The unusual forma phyllánthum (Ser.) Fiori & Béguinot (leafy-flowered) has the pedicels much elongate and the calyx-lobes more or less altered to leaves.

10. T. stoloníferum Eat. (with runners), Running Buffalo-C. — *Perennial with ascending flowering stems* 1–4 dm. high, *sending out long basal runners; flowering stem scapose below, with 2 large leaves toward the summit*, their stipules with ovate-oblong blades, their obovate leaflets 2.5–4.5 cm. long; leaves of the runners with lanceolate stipules and mostly smaller leaflets; *peduncles short, from the upper axils; heads* subglobose, 2.5–3.5 *cm. in diameter; calyx-teeth twice as long as the glabrous tube;* corolla white, tinged with purple, exceeding calyx. — Open woodlands and prairies, W.Va. to S.D., s. to Ky., Mo. and e. Kans., local. May–Aug.

11. T. refléxum L. (bent back), Buffalo-C. — *Upright annual or biennial*, branching from base, the stems villous toward summit, 1–5 dm. high; *leaves* alternate, *with* acuminate ovate stipules, *long pubescent petioles* and oblong-obovate finely toothed leaflets; *heads globose*, 2.5–4.5 *cm. in diameter, the pubescent pedicels recurving in age; flowers* 10–13 *mm. long; calyx* pilose, 7–8 *mm. long, its subulate teeth many times longer than the saucer-like tube;* corolla with roseate standard, the other petals white, or rarely all white. — Borders of sandy woods, fields, roadsides, etc., Fla. to Tex., n. to Va., Ky., Ill., Mo. and se. Kans. May–Aug.

Var. **glàbrum** Lojacono (smooth). — Stem and branches nearly or quite glabrous; pedicels and calyx glabrous. — Ala. to Okla., n. to w. N.Y., s. Ont., Ind., Ill., Ia. and S.D.

12. T. hýbridum L. (hybrid; from an early misconception), Alsike C. — Glabrous *perennial with* arcuate-*ascending to erect* hollow and soft *stems* 3–8 dm. high; leaves long-petioled; free blades of stipules lance-attenuate; leaflets oval to cuneate-obovate, the larger ones 2.5–6 cm. long and 2–3.5 cm. broad; peduncles longer than leaves; *heads dense*, 2–3.5 *cm. in diameter;* flowers 8–11 mm. long; *calyx* about equaling pedicels, 3.5–5 *mm. long, its teeth only slightly longer than the campanulate tube;* corolla pink and white or roseate throughout, becoming dull-brown; legume stipitate, 2–4-seeded; seeds lenticular, subreniform. (*T. fistulosum* Gilib.) —

Much cult., spreading to roadsides and clearings, Nfld. to B.C., s. into the n. states. June–Oct. (Introd. from Eu.) — More generally represented by

Var. ÉLEGANS (Savi) Boiss. (elegant). — Smaller throughout; *stems firmer and slender,* 1.5–6 *dm. high; larger leaflets* 1.5–3.5 *cm. long and* 1–2.5 *cm. broad; heads* 1.5–2.5 *cm. in diameter; flowers* 6–8 *mm. long.* — Roadsides, clearings and fields, throughout our range and beyond. May–Oct. (Natzd. from Eu.)

13. T. virgínicum Small (Virginian). — *Densely cespitose, from a heavy root and firm caudex; stems very short, up to* 1 *dm. long, pilose;* leaves pilose, long-petioled, *the narrowly oblong to oblanceolate firm leaflets* 2–7 *cm. long; peduncles very pilose, mostly shorter than subtending leaves;* heads globose, 2–3 cm. in diameter; *pedicels and calyx pilose;* corolla white. — Shaly slopes, local, mts. of Md., s. Pa., W.Va. and Va. May, June.

14. T. caroliniànum Michx. (of Carolina). — Slender, procumbent, with single or tufted more or less pubescent stems 0.5–2.5 dm. high; leaves long-petioled, with cuneate-obovate slightly notched leaflets 0.3–1.5 cm. long; *stipules ovate, foliaceous, veiny;* peduncles elongate; *heads* depressed-globose, lax, 1–1.7 *cm. in diameter;* flowers 4.5–6 mm. long; *calyx 2-lipped, with softly ciliolate veiny lanceolate teeth much longer than the tube;* corolla purplish or whitish, slightly exceeding calyx, the *standard prolonged into a subulate tip;* legume 4-seeded. — Sandy fields, prairies and roadsides or on granitic rock, Fla. to Tex., n. to S.C., sw. Mo. and se. Kans.; formerly on ballast and waste northw. Apr.–June.

15. T. AGRÀRIUM L. (of fields),* YELLOW or HOP-C. — Smoothish annual with ascending stems up to 4.5 dm. high; leaves short-petioled; *stipules elongate, cohering with the petiole for half its length, not dilated at base; leaflets* oblong-obovate, truncate or emarginate, *all essentially sessile;* peduncles mostly longer than leaves; *head* dense, cylindric-ovoid, 1–2 cm. long, 1–1.5 *cm. thick; corolla yellow, in age strongly striate-sulcate* and becoming warm-brown; *style and legume subequal;* seeds globose. (*T. aureum* Poll.) — Roadsides, waste places and dry fields, throughout our range, commonest eastw. June–Sept. (Natzd. from Eu.)

16. T. PROCÚMBENS L. (trailing), LOW HOP-C. — Similar to no. 15; stems often depressed or with wide-spreading branches, up to 3 dm. high; *stipules ovate, much shorter than petiole, broadly rounded at base; leaflets* cuneate-obovate, *the terminal distinctly petiolulate,* the lateral sessile or nearly so; heads 0.5–1.5 cm. long, 0.8–1.2 cm. thick; *style many times shorter than legume;* seeds ovoid. — Roadsides, waste places and old fields, Que. to N.D., s. to N.S., N.E., Ga., Miss., Ark. and Kans.; also Pacific slope. May–Sept. (Natzd. from Eu.)

17. T. DÙBIUM Sibth. (doubtful). — Similar to no. 16, smaller throughout; stipules obliquely ovate, one side broadly rounded; leaflets narrowly cuneate-obovate, emarginate, 3–12 mm. long, the terminal with longer petiolule than the lateral; peduncles filiform; *head 3–15-flowered lax,* 5–8 *mm. thick; standard* about 11-nerved, *scarcely or not at all striate in age;* style very short; seeds ovoid. — Dry roadsides, waste places and old fields, Fla. to Tex., n. to N.S., N.E., N.Y., s. Ont., Mich., Wisc. and B.C., chiefly eastw. Apr.–Oct. (Natzd. from Eu.)

Species of the Old World genus ONÒNIS L. (classical name), with 3-foliolate or simple stipulate leaves, the blades denticulate, the flowers mostly solitary in the axils, the keel-petal with a pointed beak, the stamens all united, occasionally occur on ballast and in waste, perhaps not persistent; especially O. SPINÒSA L. (spiny), a small suffruticose plant with spinescent branchlets, oblong leaflets and purplish flowers. (Adv. from Eu.)

19. MELILÒTUS Mill. MELILOT. SWEET CLOVER. VIEUX GARÇONS (Que.)

Flowers much as in *Trifolium* but in slender spike-like racemes, small. Corolla deciduous, free from the stamen-tube. Legume ovoid, coriaceous, wrinkled, longer than the calyx, scarcely dehiscent, 1–2-seeded. — Annual or biennial herbs nat. of Old World, fragrant in drying, with pinnately 3-foliolate leaves. (Name from the Greek *meli, honey,* and *lotos,* some leguminous plant.)

Petals yellow.
Corolla 4.5–6 mm. long; fruit 2.5–6 mm. long.
Fruit 2.5–3.5 mm. long, glabrous, prominently reticulate; standard much longer than wing-petals. . . . . 1. *M. officinalis.*
Fruit 4.5–6 mm. long, obscurely reticulate; petals subequal. . . . . 2. *M. altissima.*
Corolla 2–2.5 mm. long; fruit about 2 mm. long, honeycomb-reticulate. . . . 3. *M. indica.*
Petals white; fruit 3–4 mm. long, reticulate. . . . . 4. *M. alba.*

*** The technical characters and specific names of nos. 15–17 are very differently interpreted but not uniformly agreed upon by some recent European students. Until they reach agreement the long-established names are retained.**

1. M. officinàlis (L.) Lam. (of the shops; various preparations from it formerly used in medicine and for flavoring), Yellow M., Trèfle d'odeur jaune (Que.). — Upright, usually tall; *leaflets* obovate-oblong, obtuse, *closely serrate; corolla yellow,* 5–6 *mm. long; fruit* 2.5–3.5 *mm. long,* glabrous or glabrate, *prominently reticulate.* — Waste or cult. ground, Que. to B.C., s. beyond our range. Late May–Oct. (Natzd. from Eu.)

2. M. altíssima Thuill. (very tall). — Similar; *leaflets* linear- to lance-oblong, *subentire or remotely toothed; fruit* gibbous, 4.5–6 *mm. long, pubescent, obscurely reticulate.* — Roadsides and waste places, local, C.B. to N.Y. and Pa. June–Sept. (Adv. from Eu.)

3. M. índica (L.) All. (of India). — Low; *leaflets* cuneate-oblanceolate or -obovate, *truncate or emarginate,* toothed above the middle; *corolla yellow,* 2–2.5 *mm. long; fruit* gibbous, about 2 *mm. long, alveolate.* — Roadsides, grassland and waste ground southw., n. locally to N.S. and Minn. May–Oct. (Adv. from Eurasia)

4. M. álba Desr. (white), White M. — Tall; leaflets narrowly obovate to oblong, serrate, truncate or emarginate; *corolla white,* 4–5 mm. long; fruit 3–4 mm. long, somewhat reticulate. — Rich soil, roadsides, etc., throughout our range. May–Oct. (Natzd. from Eu.)

20. MEDICÀGO L. Medick

Flowers nearly as in *Melilotus.* Legume 1–several-seeded, falcate, incurved or variously coiled. — Leaves pinnately 3-foliolate; leaflets toothed; stipules often cut. Genus nat. of Old World. (*Medice,* the name of *Alfalfa,* because it came to the Greeks from Media.)

a. Flowers 6–12 mm. long; legume loosely spiral-coiled, the spiral open at center, or merely falcate to straightish, unarmed; ascending perennials.
  Corolla blue-violet or purple; legume spiraled through 2 or 3 revolutions. . . 1. *M. sativa.*
  Corolla yellow; legume falcate to nearly straight. . . . . 2. *M. falcata.*
a. Flowers 2–5 mm. long, yellow; fruit reniform or coiled into a spiral with closed center; low or diffuse annuals. . . b.
  b. Flowers numerous in dense elongate spiciform racemes; legume subreniform, slightly spiral only at summit, 1.5–3 mm. long, unarmed. . . . . . 3. *M. lupulina.*
  b. Flowers 1–8; legume strongly spiraled, larger, armed with slender spines. . . c.
    c. Leaflets broadly obcordate, as broad as or broader than long, upper surface often with a dark blotch; spines strongly arched, somewhat conforming to curve of legume. . . . . 4. *M. arabica.*
    c. Leaflets cuneate-obovate or -oblong, narrower than long, unmottled; spines straightish (except sometimes at uncinate tip), widely divergent.
      Stipules pectinate, the blade as slender as the linear-filiform lateral segments; seeds separated by partition-walls. . . . . 5. *M. hispida.*
      Stipules dentate to entire (except at base), with broad blade; seeds not separated by partition-walls.
        Pilose; base of stipule short-dentate; leaflets rounded, subtruncate or barely emarginate at summit. . . . . 6. *M. minima.*
        Glabrous; base of stipule lacerate; leaflets deeply emarginate. . . . 7. *M. laciniata.*

1. M. satìva L. (sown), Alfalfa, Lucerne, Luzerne or Lentine (Que.). — Ascending *deep-rooted perennial,* branching from base, up to 1 m. high; leaflets oblanceolate to oblong, sharply serrate at truncate apiculate or retuse apex; peduncles about equaling the subcapitate raceme; *flowers* 7–12 *mm. long; corolla blue-violet to purple,* or white in forma Álba Benke (white); *legume loosely spiraling through* 2 *or* 3 *revolutions.* — Important fodder-crop; more or less natzd. or persistent along roadsides, in old fields and waste, throughout. May–Oct. (Introd. and natzd. from Old World)

2. M. falcàta L. (sickle-shaped). — Similar to no. 1; leaflets linear to cuneate-oblong; *flowers* 6–8 *mm. long; corolla yellow; legume straight or falcate.* — Waste places and roadsides, local, Mass. to Man., s. to Del. and Mich. May–Sept. (Adv. from Eu.)

3. M. lupulìna L. (hop-like; from the spiciform racemes), Black M., Nonesuch. — Decumbent or prostrate annual; the branching stems up to 4 dm. long, pubescent especially below; stipules ovate-lanceolate, few-toothed; leaflets suborbicular- to broadly obovate-cuneate; peduncles slender, pilose or glabrate; *flowers* 3–4 *mm. long, crowded in spiciform racemes;* corolla yellow, longer than villous calyx; *fruit subreniform,* with closed sinus, pilose or glabrate, reticulate, 1.5–3 *mm. long,* becoming black, *unarmed.* — Roadsides and waste places, throughout. March–Dec. (–Feb.). (Natzd. from Eu.) — Passing to

Var. glandulòsa Neilr. (glandular). — Peduncles and fruits with divergent stalked glands. — Often as abundant, throughout most of our range. (Natzd. from Eu.)

4. M. arábica (L.) Huds. (Arabian), Spotted M. — Depressed or spreading annual, with glabrous or but sparsely pubescent branches up to 6 dm. long; stipules broadly semisagittate; *leaflets broadly obcordate, as broad as or broader than long,* 1–3.5 cm. broad, *usually with a dark blotch near center of upper surface;* peduncles filiform, 2–5-flowered; flowers 4–5 mm. long, yellow; fruit subglobose, 6–9 mm. in diameter, the *legume* with 4–7 spirals, *its faces smooth,* the dorsal edge thick, the suture marked by 2 nerves, these depressed between prominent lateral nerves; *spines* in a double row, *arching and somewhat conforming to the curve of the fruit; seeds separated by partitions.* — Waste places, infrequent, Fla. to Tex. and Mex., n. to N.B., N.E., Okla. and Wash. Apr.–Oct. (Adv. from Old World)

5. M. HÍSPIDA Gaertn. (hispid), BUR-CLOVER. — Similar to no. 4, smaller; *stipules pectinate, the blade as slender as the linear-filiform lateral segments; leaflets cuneate-obovate, unmottled; fruit* slightly smaller, *deeply reticulated, with a thin keeled edge, the spines nearly straight and widely divergent.* — Waste places, Fla. to Tex. and Mex., n. to Que., Mich., Neb., Mont. and B.C. Apr.–Oct. (Adv. from Eu.)

6. M. MÍNIMA (L.) Desr. (tiny), BUR-CLOVER. — Similar to no. 5, *pilose* throughout; *stipules* lanceolate, *short-dentate at base; leaflets* of upper leaves cuneate-oblong or -linear, *rounded, truncate or barely emarginate at tip; peduncles 1–3 cm. long, much exceeding subtending leaves, with several* flowers and *fruits; spirals,* including bristles, 5–7 *mm. in diameter.* (Var. *elongata* Rochel) — Waste places and old fields, local, Ct. to Va. May–Oct. (Adv. from Eu.)

Var. COMPÁCTA Neyraut (compact). — Peduncles much shorter than leaves, mostly 1–2-flowered, with chiefly solitary fruits 5–7 mm. in diameter. — Similar habitats, local, Mass. to Fla. (Adv. from Eu.)

Var. LONGISÈTA DC. (with long bristles). — *Fruits,* including long bristles, 8–10 *mm. in diameter.* — Similar habitats, local, Mass. to Va. (Adv. from Eu.)

7. M. LACINIÀTA Mill. (slashed). — Like no. 6, but *glabrous; base of stipule lacerate; leaflets deeply emarginate.* — Waste places, wool-waste, etc., local, Me. and Mass. May–Oct. (Adv. from Eu.)

Numerous other species are found sporadically in waste but rarely persist.

21. ANTHÝLLIS L. LADY'S-FINGERS

Calyx 5-toothed, loose, persistent and somewhat vesicular in age. Corolla yellow to crimson. Keel blunt or short-pointed. Legume mostly stalked, included in the calyx, nearly or quite indehiscent, 2–several-seeded. — Herbs or low shrubs nat. of the Mediterr. reg., with pinnate leaves and large loose clover-like heads. (Ancient Greek plant-name.)

1. A. VULNERÀRIA L. (an old generic name, from the ancient use of plant as a vulnerary or astringent). — Pubescent, 2–3 dm. high; leaflets mostly 5–13 (on the basal leaves often fewer and sometimes reduced to a solitary enlarged terminal leaflet); heads ovoid or subglobose, involucrate, yellow to red. — In clover-fields and waste ground, widely distributed but local, Que. and Ont., s. to N.E., Pa., O., Mich., Mo. and N.D. June, July. (Adv. from Eu.)

22. LÒTUS L. BIRDSFOOT-TREFOIL. BASTARD INDIGO

Calyx-teeth nearly equal. Petals free from the diadelphous stamens; standard ovate or roundish, its claw often remote from the others; wings obovate or oblong; keel incurved. Legume linear, compressed or somewhat terete, sessile, several-seeded. — Herbs of N. Am. and Old World, with pinnate leaves (in ours 1–5-foliolate and with gland-like stipules) and small yellow or reddish flowers in umbels or solitary upon axillary leafy-bracteate peduncles. (Ancient Greek plant-name, used in many senses; restricted by Linnaeus to these clover-like plants.) Including HOSACKIA Dougl.

1. L. CORNICULÀTUS L. (horned). — Diffuse many-stemmed perennial, with *pinnately 5-foliolate leaves, the basal pair of leaflets simulating stipules; flowers* yellow, *in capitate umbels on very long erect peduncles.* — Fields, roadsides and waste places, local, Nfld. to Minn., s. to Va. and O. June–Sept. (Adv. from Eu.)

2. L. americànus (Nutt.) Bisch. (American), PRAIRIE-TREFOIL. — Annual, more or less *silky-villous or subglabrous,* 1.5–5 dm. high; leaves nearly sessile, the 1–3 (rarely 4 or 5) *leaflets* ovate to lanceolate; stipules gland-like; *peduncles only slightly overtopping the leaves,* 1 (rarely 2) *-flowered,* bracteate with a single leaflet. (*Hosackia* Piper; *Acmispon* Rydb.) — Sandy or dry prairies and glades, Man. and w. Minn. to the Pacific, s. to Ark., Okla., Tex., N.M., Ariz. and Calif.; adv. e. to w. N.Y. and Va. June–Sept.

23. PSORÀLEA L. SCURF-PEA

Calyx 5-cleft, persistent, the lower lobe longest. Stamens diadelphous or sometimes monadelphous. Legume seldom longer than the calyx, thick, often wrinkled, indehiscent, 1-seeded. —

Perennial herbs of N. and S. Am., Afr. and Austral., usually sprinkled all over or roughened (especially the calyx, legumes, etc.) with glandular dots or points. Leaves mostly 3–5-foliolate. Flowers spicate or racemose, white or mostly blue-purplish. Root sometimes tuberous and farinaceous. (Name from the Greek *psoraleos, scabby*, from the glands or dots.)

*a.*Calyx 2–4 mm. long. . . *b.*

*b.*Leaves (at least the upper) pinnately compound, the 3 lanceolate to lance-ovate leaflets 0.5–6 cm. broad; fruits spirally or transversely corrugated.

Leaflets lance-ovate, 2–6 cm. broad; peduncles shorter than subtending leaves; bracts of spike, calyx and fruits without glands. . . . . 1. *P. Onobrychis.*

Leaflets lanceolate to narrowly oblong, 0.5–1.5 cm. broad; peduncles much exceeding leaves; bracts, etc., often glandular-dotted. . . 2. *P. psoralioides.*

*b.*Leaves palmate, the 3 or 5 linear to oblanceolate leaflets 1–13 mm. broad, coarsely dotted; fruits coarsely dotted, plane.

Calyx-lobes obtuse; petals white, only the keel with a purple spot; fruits globose. 3. *P. lanceolata.*

Calyx-lobes acute; petals blue or purple; fruits compressed-ovoid. . . 4. *P. tenuiflora.*

*a.*Calyx 8–16 mm. long. . . *c.*

*c.*Silvery-silky; spike interrupted, of 1–3 distant whorls; calyx narrowed to base, its lowest lobe twice to thrice length of others; corolla scarcely exceeding longest calyx-lobe. . . . . 5. *P. argophylla.*

*c.*Not silvery; pubescence various, not silky; calyx gibbous at base, its lobes less dissimilar; corolla clearly exceeding calyx.

Calyx conspicuously punctate, essentially glabrous. . . . . . . 6. *P. cuspidata.*

Calyx not obviously punctate, pubescent.

Stem villous-hirsute with long widely spreading hairs. . . . . . 7. *P. esculenta.*

Stem canescent with minute incurved pilosity. . . . . . 8. *P. canescens.*

1. P. Onóbrychis Nutt. (from resemblance to *Onobrychis*). — Nearly smooth and *free from glands*, erect, 1–1.5 *m. high, from a stoloniferous base;* leaves pinnately 3-foliolate; the *lance-ovate* and *pointed leaflets* 5–11 cm. long, 2–6 *cm. broad*, sometimes minutely punctate; *stipules and bracts subulate; peduncles shorter than leaves*, the slender racemes 5–10 cm. long; calyx about 3 mm. long, with short deltoid lobes; corolla 6–7 mm. long, bluish; *fruit* obliquely compressed-ovoid, about 1 cm. long, *obliquely corrugated*, often tuberculate. (*Orbexilum* Rydb.) — Rich woods, thickets and clearings, O. to Ill., s. to w. Va., Tenn. and e. Mo.

P. stipulàta T. & G. (with stipules), with *diffuse stems, ovate-elliptic leaflets, ovate stipules, short-peduncled head-like racemes with ovate bracts*, originally on limestone-rocks of the Ohio R. in Ky. and Ind., is apparently extinct.

2. P. psoralioides (Walt.) Cory (resembling *Psoralea*, the plant originally placed in *Trifolium*), Sampson's Snakeroot. — Slender, *from a* long *fusiform tap-root;* stems smoothish, erect, 3–9 dm. high; *leaflets lanceolate to narrowly oblong, bluntish*, 3–7 cm. long, 0.5–1.5 *cm. broad;* stipules subulate; *peduncles much overtopping the leaves;* the dense spiciform raceme 2–5, elongating to 10 cm. long; *bracts* suborbicular, *abruptly caudate, conspicuously glandular-dotted*, 6–10 mm. long, the cauda 3–5 mm. long; calyx 3–5 mm. long, with triangular-ovate lobes, the ventral lobe 1.6–2.6 mm. long, dorsal lobes 1–2 mm. long; corolla lilac-purple; *fruits suborbicular*, 4–5 *mm. long*, obliquely transverse-wrinkled, *heavily dotted.* (*P. pedunculata* (Mill.) Vail; *Orbexilum pedunculatum* Rydb.) — Dry open woods, clearings and fields, se. Va. to Ga. May–July.

Var. **eglandulòsa** (Ell.) F. L. Freeman (without glands). — *Less* conspicuously *dotted to* nearly or quite *glandless*, more pubescent; *bracts* ovate-acuminate, 5–8 *mm. long;* calyx 3–6 mm. long, ventral lobe 2.4–3.8 mm. long, dorsal lobes 2–3 mm. long. — Interior Ga. to e. Tex., n. to Va., O., Ind., s. Ill., Mo. and se. Kans.

3. P. lanceolàta Pursh (lanceolate). — Stems 1.5–6 dm. high, mostly bushy-branched, glabrous or nearly so; *leaves palmately* 3–5-*foliolate*, yellowish-green, densely and coarsely punctate; *leaflets linear to oblanceolate*, 1–13 *mm. broad; flowers in spiciform racemes* 1–3 *cm. long; calyx* 2–3 *mm. long*, glandular-dotted, *with deltoid obtuse lobes;* petals white or violet-tinged, 5–6 mm. long; *fruits globose*, about 5 mm. long, *smooth.* (*Psoralidium* Rydb.) — Dry prairies and hills, N.D. and Sask. to Alta. and Wash., s. to se. Kans., N.M. and Ariz. May–July.

4. P. tenuiflòra Pursh (slender-flowered; meaning sparsely flowered). — Bushy-branched, up to 1 m. high, minutely hoary-pubescent when young; leaves palmately 3–5-foliolate; leaflets linear to oblong-oblanceolate, obtuse, 5–8 mm. broad, glandular-dotted; *racemes* 1.5–4 cm. long, *slender, becoming lax, with flowers mostly* 1 *or* 2 *at a node* and *about* 5 *mm. long; calyx* 2–2.5 mm. long, *with acute lobes;* corolla bluish, or white in forma **álba** Steyerm. (white); *fruit compressed-ovate*, glandular-dotted, 6–8 mm. long. (*Psoralidium* Rydb.) — Dry prairies, open woods and rocky banks, Ind. to Neb., s. to Mo., Tex., N.M. and Ariz. June–Sept.

Var. **floribúnda** (Nutt.) Rydb. (full of flowers). — Coarser; *racemes up to* 1 *dm. long, with* 2–4 *flowers at a node; calyx* 3 *mm. long; corolla* 6–7 *mm. long.* (*P. floribunda* Nutt.; *Psoralidium floribundum* Rydb.) — Dry prairies, bluffs and open woods, Ill. to Minn. and Neb., s. to Ark. and Tex.

5. P. argophýlla Pursh (silvery-leaved). — *Silvery silky-white throughout,* from creeping rhizomes, 2.5–6 dm. high, divergently branched; leaves palmately lobed, *leaflets elliptic-lanceolate; spike interrupted, of* 1–3 *distant whorls; calyx narrowed to base, its lowest attenuate lobe* 8–10 *mm. long,* twice to thrice *as long as upper lobes;* corolla bluish or purple, scarcely exceeding longest calyx-lobe; fruit ovoid, silky. (*Psoralidium* Rydb.) — Dry prairies, Wisc. to Sask., s. to Mo., Okla. and N.M. Late June–early Aug.

6. P. cuspidàta Pursh (sharp-tipped). — *Strigose to glabrate,* from a strong parsnip- or turnip-shaped or moniliform tap-root, 1.5–6 dm. high, divergently branched; *leaves palmately* 5(–7)-*foliolate,* the leaflets elliptic to narrowly obovate; spike dense, thick-cylindric, 2–6 cm. long; *calyx gibbous at base,* 1–1.4 *cm. long, strigose and glandular-punctate, its* lance-acuminate *lobes only slightly unequal;* corolla blue; *fruit* oval, long-beaked, *rupturing about the middle.* (*Pediomelum* Rydb.) — Dry plains and calcareous hills, Minn. to Mont., s. to Ark. and Tex. June, July.

7. P. esculénta Pursh (edible), Breadroot. — Resembling no. 6, but *villous-hirsute* throughout; stem 1–4 dm. high; leaflets lance- to obovate-oblong; *calyx hirsute, without glandular dots,* up to 1.6 cm. long. (*Pediomelum* Rydb.) — Dry prairies, rocky woods and calcareous hills, Man. and Wisc. to Alta., s. to Mo., Tex. and N.M. May–July. — This, and presumably no. 6, was the Pomme blanche or Pomme de Prairie of voyageurs across the plains.

8. P. canéscens Michx. (grayish-pubescent), Buckroot. — Bushy-branched, 4–9 dm. high; the stem and divergent branches *minutely canescent-pilose;* leaves with petiolulate elliptic-obovate leaflets; *calyx strigose-villous;* the lower lobe narrow, longer than the lance-acuminate upper ones; corolla blue or violet, changing to green. (*Pediomelum* Rydb.) — Sandy woods, Fla. and Ala., n., locally, to se. Va. June, July.

24. AMÓRPHA L.

Calyx inversely conical, 5-toothed, persistent. Standard (the other petals entirely wanting!) wrapped around the stamens and style. Stamens 10, monadelphous at the very base, otherwise distinct. Legume oblong, longer than the calyx, 1–2-seeded, roughened, tardily dehiscent. — Shrubs or herbs of warm temp. and trop. N.Am. with odd-pinnate leaves; the leaflets marked with minute dots, usually stipellate, the midvein excurrent. Flowers violet or purple, crowded in terminal spiciform racemes. (Greek *amorphos, deformed,* from the absence of four of the petals.)*

Low shrubs, usually less than 1 m. high; leaves nearly sessile, petioles usually shorter than width of lowest leaflets.
- Plant densely to sparsely villous; calyx-tube and legume white-villous; style and beak nearly as long as legume. . . . . 1. *A. canescens.*
- Plant glabrous or only minutely and sparsely pubescent; legume glabrous or essentially so, longer than the style and beak, resinous-dotted.
  - Inflorescence paniculate, 1–2.5 dm. long, of many branches; upper calyx-lobes two-thirds as long as tube, lower equaling it or longer. . . . . 2. *A. brachycarpa.*
  - Inflorescence of single (rarely 2 or 3) racemes 3–8 cm. long, terminating the branches; calyx-lobes about half as long as tube. . . . . 3. *A. nana.*

Taller shrubs, mostly 1–4 m. high; leaves distinctly petioled, petioles longer than width of lowest leaflets.
- Legumes dotless or essentially so; branchlets glabrous; leaflets thin, lustrous above. . . . . 4. *A. nitens.*
- Legumes coarsely resinous-dotted; branchlets minutely to strongly pubescent at least when young; leaflets becoming firm, dull above. . . . . . . 5. *A. fruticosa.*

1. A. canéscens Pursh (grayish-pubescent), Leadplant. — Undershrub rarely 1 m. high, *densely white-villous,* or greener and sparsely pubescent in forma **glabràta** (Gray) Fassett (becoming smooth); *leaves subsessile, with* 15–51 *crowded* elliptic-oblong or -lanceolate *nearly sessile leaflets* 4–10 *mm. wide;* racemes paniculate-clustered, 0.8–2.5 dm. long; *calyx-tube and legume white-villous; style and beak nearly as long as legume.* — Dry sandy prairies and hills, Mich. to Sask., s. to Ind., Ill., Ark., Tex. and N.M. Late May–Aug.

2. A. brachycárpa Palmer (short-fruited). — Slender shrub 6–9 dm. high, *nearly or quite*

* Treatment derived partly from that of E. J. Palmer in Journ. Arn. Arb. xii. 157–197 (1931).

glabrous; leaves nearly sessile, with 21–45 oblong *crowded leaflets 4–8 mm. wide; panicle* of spiciform racemes 1–2.5 *dm. long; calyx* 4–5 mm. long, *its upper lobes two-thirds as long as tube, the lower equaling it or longer; fruit* obliquely obovate, glabrous, conspicuously resinous-dotted, *longer than style and beak*. — Barrens and glades, Mo. May–Aug.

3. A. nàna Nutt. (dwarf), FRAGRANT FALSE INDIGO. — Undershrub 3–9 *dm. high, glabrous* or nearly so; leaves nearly sessile, the firm green oblong to oval or narrowly obovate leaflets 4–8 mm. broad; *racemes solitary* (rarely 2 or 3) *at the tips of branches*, 3–8 *cm. long; calyx-lobes about half as long as tube;* fruit obliquely ovate, dotted, short-beaked. (*A. microphylla* Pursh) — Dry prairies, Man. and Sask., s. to Ia. and Kans. June, July.

4. A. nìtens F. E. Boynt. (shining). — *Shrub* 1–3 *m. high, glabrous* or essentially so; *leaves long-petioled*, the 9–19 oblong or oblong-ovate *leaflets* rounded at each end, *thin, lustrous above*, sparsely pubescent beneath, 1–2 cm. broad; *racemes usually solitary* (rarely 2 or 3), *up to* 2.5 *dm. long;* calyx glabrous except for ciliate rounded or obtuse short lobes; *fruit* curved, narrowly oblong, glabrous, *dotless*. — Thickets and banks of streams, Ga. to Ark., n. to s. Ill. May, June.

5. A. fruticòsa L. (shrubby), FALSE OR BASTARD INDIGO, INDIGO-BUSH. — *Shrub up to 4 m. high*, more or less *pubescent to glabrate;* leaves long-petioled, *the* 13–35 *or more leaflets firm, opaque*, 1–6 cm. long; *inflorescence a panicle of several erect racemes; fruits* 6–8 mm. long, often curved, *with large resinous dots*. — Heteromorphous; the following, of a dozen named variations, seem most recognizable:

Branches, foliage and calyx grayish-pilose or minutely appressed-pubescent to glabrate.
  Leaflets mostly ovate or oblong, rounded or obtuse at base; pubescence of young growth spreading.
    Leaflets 13–25, averaging once-and-a-half to twice as long as broad, not crowded. . . . . *A. fruticosa* (typical).
    Leaflets 21–35 or more, averaging two or three times as long as broad, often crowded.
      Leaflets mostly 1–2 cm. long, pubescent; fruits straight or straightish. Var. *tennesseensis*.
      Leaflets 2–5 cm. long, sparsely pubescent to glabrate; fruits curved. . Var. *oblongifolia*.
  Leaflets mostly elliptic or obovate, narrowed to base; pubescence appressed. . Var. *angustifolia*.
Branches, foliage and calyx copiously villous with tawny or rufescent hairs; leaflets 2.5–6 cm. long. . . . . Var. *croceolanata*.

A. fruticòsa (typical). — River-banks, rich thickets, etc., n. Fla. to La., n. to s. Pa., W. Va., O., s. Mich., Wisc. and Kans.; cult. and esc. ne. to N.E. and N.Y. May, June.

Var. tennesseénsis (Shuttlew.) Palmer (of Tennessee), *A. tennesseensis* Shuttlew. — Fla. to Tex., n. to Ky., Ill., Mo. and Kans.

Var. oblongifòlia Palmer (oblong-leaved). — Mo. and Ark.

Var. angustifòlia Pursh (narrow-leaved), *A. fragrans* Sweet — Tex. to Ariz. and n. Mex., n. to Wisc., Minn., Man. and Sask.

Var. croceolanàta (P. W. Wats.) Schneid. (with saffron-colored wool), *A. croceolanata* P. W. Wats. — Fla. to La., n. to Ga., Ky., s. Ill. and Mo.

25. DÀLEA Juss. DALEA

Calyx 5-cleft or -toothed. Corolla imperfectly papilionaceous; petals all with claws; the standard cordate, inserted in the bottom of the calyx; the keel and wings borne on the middle of the monadelphous sheath of filaments, which is cleft down one side. Stamens 10, rarely 9. Fruit membranaceous, 1-seeded, indehiscent, inclosed in the persistent calyx. — Mostly herbs of N. and S.Am., more or less glandular-dotted, with minute stipules; the small flowers in terminal spikes or heads. (Named for *Samuel Dale*, 1659–1739, an English botanist.) PAROSELA Cav.

1. D. alòpecuroides Willd. (resembling *Alopecurus;* in the young spikes), FOXTAIL-D. — Erect *annual*, 3–6 dm. high; leaflets 19–35, glabrous, linear-oblong; flowers light rose-color or whitish, in cylindrical spikes; bracts ovate-lanceolate, acuminate, deciduous; calyx very villous, with long slender teeth. (*Parosela Dalea* Britt.; *P. alopecuroides* Rydb.) — Bottomlands, roadsides and fields, Ill. to S.D., s. to Ala. and N.M.; adv. e. to N.Y. July–Sept.

2. D. enneándra Nutt. (with nine stamens). — Erect *perennial*, 3–12 dm. high, branching; leaflets 5–13, linear, 4–6 mm. long; spikes loosely flowered; bracts conspicuous, persistent, almost orbicular and very obtuse; petals white; calyx densely villous, the long teeth beautifully plumose. (*Parosela* Britt.) — Dry soil, Ia. to N.D., s. to Miss., Tex. and N.M. May–Aug.

26. PETALOSTÈMUM Michx. PRAIRIE-CLOVER

Calyx 5-toothed. Corolla indistinctly papilionaceous; petals all with filiform claws, 4 of them nearly alike and spreading, borne on the top of the monadelphous and cleft sheath of filaments, alternate with the 5 anthers; the fifth (standard) inserted in the bottom of the calyx, cordate or oblong. Fruit membranaceous, inclosed in the calyx, indehiscent, 1–2-seeded. — Chiefly perennial N.Am. herbs, upright, glandular-dotted, with crowded odd-pinnate leaves, minute stipules, and small flowers in very dense terminal and peduncled heads or spikes. (Name, often but not originally spelled *Petalostemon*, combined of the two Greek words for petal and stamen, alluding to the peculiar union of these organs in this genus.) KUHNISTERA Lam.

a. Calyx and bracts glabrous or nearly so on outer surfaces (but sometimes ciliate). . . *b.*
b. Lateral leaflets rarely more than 4 pairs; petals white. . . *c.*
c. Spikes globose or short-ovoid, 1–1.7 cm. in diameter. . . . . . . 1. *P. multiflorum*
c. Spikes elongate, cylindric, becoming two to five times as long as thick.
Spikes dense and uninterrupted in maturity; bracts much exceeding calyces. . . . . 2. *P. candidum.*
Spikes loosening and subflexuous in maturity; bracts barely exceeding calyces. . . . . 3. *P. occidentale.*
b. Lateral leaflets mostly 10–13 pairs; petals roseate, fading to white. . . . 4. *P. foliosum.*
a. Calyx and bracts pubescent on the outer surfaces; petals usually roseate to purple. . . *d.*
d. Silky-villous throughout; leaflets 13–19, narrowly elliptic or oblong; spikes elongating to 3–12 cm., loosening. . . . . 5. *P. villosum.*
d. Glabrous or only sparsely (rarely densely) villous; leaflets 3–7, linear; spikes very dense, 1–5 cm. long.
Calyx densely villous-tomentose. . . . . 6. *P. purpureum.*
Calyx silvery-silky, at least on nerves. . . . . 7. *P. pulcherrimum.*

1. P. multiflòrum Nutt. (many-flowered). — *Glabrous throughout,* stems 2–6.5 dm. high, *corymbosely branched;* leaf-rachis 1.2–2 cm. long, the 3–9 linear to narrowly oblong leaflets 4–9 mm. long; *spikes globose or short-ovoid,* 1–1.7 *cm. in diameter; bracts* subulate-setaceous (except lowest), *much shorter than calyces;* petals white, oval or elliptic, abruptly contracted at base, standard reniform. (*Kuhnistera* Heller) — Dry prairies and rocky slopes, w. Ia. and Neb., s. to Ark. and Tex. June–Sept.

2. P. cándidum (Willd.) Michx. (shining white), WHITE P. — *Glabrous,* stems 3–7 dm. high; leaf-rachis 0.8–3 cm. long, the 5–7 linear-lanceolate to oblong *leaflets* 0.9–3 *cm. long; spikes* terminating long ascending branches, *very dense,* cylindric, *two to five times as long as thick,* becoming 2–10 cm. long; *bracts* slender-tipped, much *exceeding the calyces,* conspicuously spreading in the young spikes; petals white (*Kuhnistera* Ktze.) — Prairies, Ind. to Sask., s. to Miss., La. and Tex. June, July.

3. P. occidentàle (Gray) Fern. (western). — Differing from no. 2 in its linear or linear-oblanceolate *leaflets rarely more than* 1.5 *cm. long; spikes* more slender, *loosening in maturity; bracts barely exceeding calyces.* (*P. oligophyllum* (Torr.) Rydb.) — Dry prairies and bluffs, Minn. to Alta., s. to La., Tex. and Mex. June–Aug.

4. P. foliòsum Gray (leafy). — *Glabrous,* 2–6 dm. high; leaf-rachis 3–5 cm. long; *leaflets mostly* 20–27, obtuse, apiculate, oblong, 8–13 mm. long; *spikes* short-peduncled, *dense,* cylindric, 1.6–4.5 *cm. long;* the slender-tipped lanceolate *bracts much overtopping the buds; petals roseate,* fading to whitish. (*Kuhnistera* Ktze.) — Rocky hills, glades and river-banks, n. Ill. to Tenn. July–Sept.

5. P. villòsum Nutt. (long-hairy), SILKY P. — *Soft-downy or silky-villous all over; leaflets* 13–19, *narrowly elliptic or oblong,* 6–12 mm. long; *spikes* short-peduncled, slenderly cylindric, *soft-villous, loosening and elongating to* 3–12 *cm. long;* petals roseate. (*Kuhnistera* Ktze.) — Sandy hills and prairies, Mich. to Sask. and Mont., s. to Tex. and N.M. July–Sept.

6. P. purpùreum (Vent.) Rydb. (purple). — Stems upright, from erect crowns, 0.3–1 m. high, *sparingly villous to glabrous,* or densely villous in forma **pubéscens** (Gray) Fassett (hairy); *leaflets* 3–5, narrowly linear, 0.8–2 cm. long; *spikes* short-peduncled, *globose-ovoid, becoming short-cylindric* and 1–5 *cm. long, very dense;* bracts equaling or slightly surpassing the densely villous-tomentose calyx; petals roseate to crimson. (*Kuhnistera* MacM.) — Prairies and dry hills, Ind. to Sask. and Mont., s. to w. Tenn., Ark., Tex. and N.M.; adv. e. to N.Y. June–Sept. — Forma **arenàrium** F. C. Gates (of sand) is a depressed extreme of open sands with radiating crowns, divaricate stems and appressed leaves.

7. P. pulchérrimum Heller (most handsome). — Similar to no. 6, mostly lower; *leaflets* 3–7, 2–4 *cm. long;* spikes longer-peduncled; *calyx-tube glabrous or glabrate, the nerves silvery-sericeous.* — Prairies and barrens, Mo. and Ark. to Tex. and N.M. June–Aug.

27. INDIGÓFERA L. Indigo

Calyx small, obliquely toothed. Standard broad and rounded, sessile or with short claw; wing-petals oblong, slightly coalescing to the keel; keel-petals with a lateral gibbosity or pouch. Stamens 10, monadelphous, but 1 free above the middle of the tube. Legume plump, globose to linear- or quadrate- cylindric, septate between the few estrophiolate seeds. — Chiefly trop. shrubs or herbs with bifurcate appressed (malpighiaceous) hairs, pinnate (rarely digitate or simple) leaves, flowers purplish or red in axillary racemes. (Name from *indigo* and *fero, to bear.*)

1. I. leptosépala Nutt. (with slender sepals). — Perennial herb, up to 1 m. high, branching from base, decumbent or loosely spreading; leaves with 7–9 strigose-canescent narrowly obovate or cuneate broad-tipped leaflets; racemes overtopping leaves; calyx with prolonged unequal lance-attenuate lobes; corolla barely 1 cm. long; legume linear-cylindric, 4-angled, straight. — Sandy and gravelly soils, e. Kans. to w. Tex. and n. Mex. June, July.

28. TEPHRÒSIA Pers. Hoary Pea

Calyx about equally 5-cleft. Standard roundish, usually silky outside, turned back, scarcely longer than the coherent wings and keel. Stamens monadelphous or diadelphous. Legume linear, flat, several-seeded, 2-valved. — Am., Afr. and Austral. perennial herbs, with odd-pinnate leaves and white or purplish racemose flowers. Leaflets mucronate, veiny. (Name from the Greek *tephros, ash-colored* or *hoary.*) Cracca L., not Benth.

Stems stiffly ascending, simple, leafy to summit, white-pubescent; raceme or panicle dense, barely peduncled. . . . . 1. *T. virginiana.*
Stems loosely ascending or reclining, often branched below, with few scattered leaves; raceme at first subcapitate, becoming interruptedly spiciform, with remote flowers and legumes, very long-peduncled.
  Stem, rachis and petiolules strigose-hispid; leaflets strigose-sericeous beneath; pedicels filiform; upper calyx-lobes deltoid-lanceolate. . . . 2. *T. hispidula.*
  Stem, rachis and petiolules spreading-pilose or villous; leaflets appressed-pilose beneath; upper calyx-lobes lance-subulate. . . . . . . 3. *T. spicata.*

1. T. virginiàna (L.) Pers. (Virginian), Goat's-rue, Catgut, Rabbit's-pea. — *Silky-villous* with whitish hairs at least when young; *stem erect and simple,* 3–6 dm. high, *leafy* to the top, the upper *internodes with* few to many *long spreading villi;* rachis of leaf spreading-villous; *leaflets* 17–29, linear-oblong to narrowly elliptic, pilose to silky beneath, *green and glabrous or sparsely strigose above;* flowers large and numerous, clustered in a terminal *ellipsoid dense raceme or panicle,* yellowish-white marked with purple; *legumes heavily shaggy-villous.* (*Cracca* L.) — Dry sandy woods and openings, Fla. to Tex., n. to s. N.H., Mass., N.Y., s. Ont., s. Mich., s. Wisc., Mo. and Okla. Mid-May–early Aug. — Westward passing freely into

Var. **holoserícea** (Nutt.) T. & G. (wholly silky). — Silvery silky-villous throughout, often very densely so; *leaflets silvery-silky above.* — Sandy woods, plains and bluffs, Mich. to Minn. and S.D., s. to Ark., Okla. and Tex.

Var. **glàbra** Nutt. (hairless). — Internodes of *stem and rachises of leaves without long villi, the pubescence very short* and hirtellous *or wanting; leaflets minutely pubescent to glabrous beneath; legumes densely appressed-pubescent, not long-villous.* — Sandy pine-barrens and dry woods, Ga. to w. Va. and N.J.

2. T. hispídula (Michx.) Pers. (with short, straight hairs). — Slender, the *strigose-hispid to glabrate stems* arising from a crown with thick roots, reclining or ascending; leaves with 6–14 pairs of *lanceolate, lance-oblong or oblanceolate acutish to round-tipped* mucronate *leaflets strigose-sericeous beneath, the rachis and petiolules strigose;* peduncles slender, axillary, much elongate; flowers at first closely clustered, becoming scattered; pedicels filiform; *upper calyx-lobes deltoid-lanceolate,* 1.5–3 mm. long; flower 1.5–2 cm. long; corolla at first whitish or flesh-color, changing to red; legume 3–4 cm. long, 4–5 mm. broad, strigose. (*Cracca* Ktze.) — Pinelands and peaty or sandy openings, Fla. to La., n. to N.C. and, locally or formerly, e. Va. June–Sept.

3. T. spicàta (Walt.) T. & G. (spicate). — Similar; the straggling to ascending rather zigzag *stems* very densely *rusty-pilose or -villous; leaflets oblong-obovate, broadly rounded-subtruncate to the* mucronate *tip,* pilose beneath and often above, *the rachis and petiolules spreading-pilose;* pedicels stoutish; *upper calyx-lobes lance-subulate,* 2.5–3.5 *mm. long;* legumes closely appressed-pilose. (*Cracca* Ktze.) — Sandy woods and openings, Fla. to La., n. to Ky. and Del. Late May–Sept.

Var. **semitónsa** Fern. (partly shaved). — Greener, the stems openly or sparsely pilose; rachis subglabrous; leaflets glabrous or nearly so above; legumes remotely pilose. — Similar habitats, n. to se. Va. and Ky.

29. SESBÀNIA Scop.

Calyx campanulate, equally toothed. Standard large, round. Stamens diadelphous. Ovary many-ovuled; legume long. — Herbs or shrubs of warm reg., with long even-pinnate leaves. Flowers on axillary peduncles or lateral racemes. (Name latinized from the earlier SESBAN Adans., said to be of Arabic origin.)

1. S. exaltàta (Raf.) Cory (tall). — Erect annual, 0.7–3 m. high; leaflets 12–25 pairs, narrowly oblong; corolla pale yellow, often spotted; legumes 2 dm. in length, narrow, with thickened margins. (*S. macrocarpa* Muhl.) — Fields and low grounds, Ala. to Tex., n. to Mo. and Okla.; adv. from se. Pa. to se. N.Y. July–Oct.

30. ROBÌNIA L. LOCUST

Calyx short, 5-toothed, slightly 2-lipped. Standard large and rounded, turned back, scarcely longer than the wings and keel. Stamens diadelphous. Legume linear, flat, several-seeded, at length 2-valved. — N.Am. trees or shrubs, often with spines for stipules. Leaves odd-pinnate, the ovate or oblong leaflets stipellate. Flowers showy, in drooping axillary racemes. (Named for *Jean Robin*, 1550–1629, herbalist to Henry IV of France, and his son, *Vespasian Robin*, 1579–1662, who first cultivated the Locust-tree in Europe.)

Branchlets glabrous or pubescent, not bristly, strongly glandular nor viscid.
  Branchlets glabrous; leaflets glabrous or promptly glabrate; flowers white; ovary and legume glabrous. . . . . 1. *R. Pseudo-Acacia.*
  Branchlets pubescent; leaflets permanently pubescent beneath; flowers purple; ovary and legume hispid. . . . . 2. *R. Elliottii.*
Branchlets bristly, strongly glandular or viscid; legume bristly.
  Branchlets and peduncles clammy-viscid; flowers pale pink, 2 cm. long, crowded (8–20) in compact racemes; leaflets 6–12 pairs; tree. . . . . . 3. *R. viscosa.*
  Branchlets and peduncles long-bristly; flowers roseate or purple, 2–3 cm. long, few (3–8) in loose racemes; leaflets 3–7 pairs; shrub. . . . . 4. *R. hispida.*

1. R. Pseùdo-Acàcia L. (old generic name, *false Acacia*), BLACK L., FALSE ACACIA. — Large shrub, or usually a coarse-barked *tree* up to 25 m. high; branches with spines; *branchlets glabrous or soon glabrate;* leaflets 3–10 pairs, elliptic to ovate, *racemes* 1–2 *dm. long, rather closely flowered; flowers white, fragrant,* 1.5–2.3 cm. long; lateral calyx-lobes deltoid, shorter than tube; *ovary and legume glabrous.* — Woods and thickets, Ga. to La., n. to Pa., W.Va., s. Ind., Ia. and Okla.; natzd. n. to N.S., Que. and Ont. May, June. — Much planted for valuable timber and for ornament, now natzd. in many countries.

2. R. ELLIÓTTII (Chapm.) Ashe (for STEPHEN ELLIOTT, 1771–1830). — *Shrub* up to 1.5 m. high; slender branches with tiny spines, the short *branchlets pilose* and crowded toward summit; leaves with 5–7 pairs of elliptic *leaflets pubescent beneath; raceme lax, with* 3–10 *rose-purple flowers* 2–2.5 cm. long; *legume hispid.* — Borders of woods and thickets ne. Md., esc. from cult. May. (Introd. from the South)

3. R. viscòsa Vent. (sticky), CLAMMY L. — *Branchlets and leaf-stalks clammy; flowers crowded in short racemes,* roseate-tinged, nearly inodorous; legume glandular-hispid. — Dry woods, Pa., and W.Va. to Ga. and Ala.; esc. from cult. northw. May, June.

4. R. híspida L. (beset with bristles), BRISTLY L., MOSSY L., ROSE-ACACIA. — Shrub 0.5–3 m. high, stoloniferous; *branches and peduncles bristly* with brown hairs; leaflets 3–6 pairs, rounded-oblong, glabrous or nearly so, 2–5.5 cm. long, 1.5–4.5 cm. broad, strongly mucronate; *racemes lax and open,* 3–8-*flowered; flowers rose-purple,* 2.5–3 cm. long, *inodorous, with bristly pedicels and calyx; legume bristly.* — Dry woods, thickets and slopes, Va. and Tenn., southw.; cult. and esc. northw. May, June.

Var. FÉRTILIS (Ashe) Clausen (fruitful). — Leaflets elliptic, acute or acutish, 0.7–2.5 cm. broad; flowers 2–2.5 cm. long. (*R. fertilis* Ashe) — Spread from cult. n. to Ct. (Introd. from the South)

31. WISTÈRIA Nutt. WISTERIA

Calyx campanulate, somewhat 2-lipped; upper lip of 2 short teeth, the lower of 3 longer ones. Standard roundish, large, turned back, with 2 callosities at its base; keel falcate; wings doubly auricled at the base. Stamens diadelphous. Legumes elongated, thickish, knobby, stipitate, many-seeded, at length 2-valved. Seeds large. — Ovate-lanceolate leaflets numerous. Racemes of large and showy lilac to violet (or white) flowers. Twining shrubs of N.Am. and e. Asia. (Dedicated to Professor *Caspar Wistar*, 1760–1818, distinguished anatomist of Philadelphia.) KRAUNHIA Raf. WISTARIA Spreng. (a later spelling).

Ovary and legume glabrous; pedicels 6–10 mm. long.
  Raceme 4–12 cm. long; pedicels and calyx with few, if any, clavate glands. . 1. *W. frutescens.*
  Raceme 1.5–3 dm. long; pedicels and calyx with abundant clavate glands. . 2. *W. macrostachya.*

Ovary and legume pilose or velvety; pedicels 1–2.5 cm. long.
  Leaflets 7–13; raceme thick-cylindric, 1.5–2 dm. long; flowers opening nearly simultaneously, 2.3–2.6 cm. long, on pedicels becoming 1.5–2.5 cm. long. . 3. *W. sinensis.*
  Leaflets 13–19; raceme slender, 2–5 dm. long; flowers gradually expanding from base to apex of raceme, 1.5–2 cm. long, on pedicels 1–2 cm. long. . . 4. *W. floribunda.*

1. W. frutéscens (L.) Poir. (shrubby). — *Leaflets* 9–15, oblong to narrowly ovate, minutely pubescent, becoming glabrate, *when mature* 2–6 cm. long, 1–2 *cm. wide; racemes* dense, 4–12 *cm. long; bracts* lanceolate to narrowly ovate, 6–10 *mm. long*, caducous; the ascending to spreading *pedicels pilose with uniform hairs*, rarely with clavate glands; *calyx with few if any clavate glands;* flowers 1.5–2 cm. long; the petals lilac; legume glabrous, 0.5–1 dm. long. (*Kraunhia* Raf.) — Borders of wooded swamps and banks of streams, Ala. and Fla., n. to se. Va. and, esc. from cult., Md. Late Apr., May.

2. W. macrostáchya Nutt. (large-spiked). — Coarser; *leaflets* more rounded at base, *abruptly narrowed to acuminate tips*, when mature 4–8 cm. long, 2–4.5 *cm. broad; racemes* 1.5–3 *dm. long;* the spreading *pedicels and* the *calyx with abundant clavate glands* mixed with the hairs; bracts 0.8–1.5 cm. long. (*Kraunhia* Small) — Swamps and rich woods, La. and Tex., n. to Ky., s. Ill. and Mo. Late Apr., May. — Perhaps better treated as a var. of no. 1.

3. W. SINÉNSIS Sweet (Chinese), CHINESE W. — High-twining; *leaflets* 7–13, *silky when expanding*, becoming glabrate, oblong- or lance-ovate; *racemes thick-cylindric*, 8–10 *cm. in diameter*, becoming lax, 1.5–2 *dm. long; flowers* 2.3–2.6 *cm. long*, violet-blue, slightly fragrant, *all expanding nearly simultaneously, on filiform pedicels* 1.5–2.5 *cm. long; ovary and legume velvety.* — Open woods, roadside-tangles and abandoned gardens, s. N.E. to Va. and Ill. April, May. (Introd. and natzd. from Asia)

4. W. FLORIBÚNDA (Willd.) DC. (full of flowers), JAPANESE W. — Similar to no. 3; *leaflets* 13–19; *raceme more slender*, 5–7 *cm. in diameter*, 2–5 *dm. long; flowers* 1.5–2 *cm. long*, very fragrant, *gradually expanding from base to apex of raceme;* pedicels 1–2 cm. long. — Open woods, old homesteads, etc., local, Mass. to La. April–June. (Introd. and natzd. from Asia)

32. COLÙTEA L. BLADDER-SENNA

Calyx 5-toothed. Corolla showy; standard suborbicular, flat, with 2 folds or callosities above the claw; the wings short-clawed, falcate; the broad keel incurved, its claws united. Stamens in a tube, 1 of them free. Legume stipitate, many-seeded, membranaceous, bladdery-inflated, indehiscent or 2-valved at apex. — Shrubs nat. of Eurasia, with odd-pinnate leaves, small stipules and few large yellow to red flowers in axillary racemes. (Classical name of some plant with inflated fruit.)

1. C. ARBORÉSCENS L. (becoming tree-like). — Shrubs 3 or 4 m. high; leaflets 9–13, elliptic to narrowly obovate, 1.5–3 cm. long; flowers yellow, 2 cm. broad; legumes semiovoid, 6–8 cm. long. — Roadside-thickets and about towns, occasional, esc. from cult., Mass. westw. and southw. June–Sept. (Introd. from Eu.)

33. ASTRÁGALUS L. MILK-VETCH

Calyx 5-toothed. Corolla usually long and narrow; standard narrow, equaling or exceeding the wings and blunt keel, its sides reflexed or spreading. Stamens diadelphous. Legume several–many-seeded, various, mostly turgid, one or both sutures usually projecting into the locule, either slightly or so as to divide the cavity lengthwise into two. — Chiefly herbs (ours perennials) of N. Hemisph. with odd-pinnate leaves and spicate or racemose flowers. Mature legumes

are usually necessary for certain identification of the species. (The ancient Greek name of a leguminous plant, as also of the ankle-bone.)*

Key to Flowering Material

a. Calyx-tube 1.5–3.5 mm. long. . . *b.*

b. Flowering stems elongating, with evident internodes and remote leaves (very abbreviated only in dwarfed plants of no. 16 with their leaflets only 2–10 mm. long, violet corolla and black-hairy calyx with teeth shorter than tube). . . *c.*

c. Flowering stems erect to decumbent, from the crown or crowns of a deep root, not creeping and forming a mat. . . *d.*

d. Racemes in anthesis (before shrivelling of lower corollas and evident enlargement of ovary) linear-cylindric, 0.5–2 cm. in diameter, 3-∞ times as long as thick, remotely flowered. . . *e.*

e. Calyx (including teeth) 2–4 mm. long, usually pubescent; corolla white, ochroleucous or merely tipped with purple (if purple throughout, the calyx only about 2 mm. long); stems minutely strigose at least above; species of the West. . . *f.*

f. Calyx-tube 1.5–2 mm. long, strigose with minute straight pubescence; corolla 5–8 mm. long.

Middle and upper stipules dark brown or green, deltoid, merely acute or acuminate; peduncles 1–2 cm. long; calyx sparsely strigose (or glabrate); corolla ochroleucous or merely pink-tipped. . . . . 1. *A. tenellus.*

Middle and upper stipules pale brown, abruptly caudate-tipped; peduncles 4–12 cm. long; calyx densely white-strigose; corolla purple. . . . . 2. *A. gracilis.*

f. Calyx-tube 3 mm. long, densely pilose with soft incurving hairs; corolla 8–10 mm. long, whitish. . . . . 6. *A. flexuosus.*

e. Calyx (including teeth) 4–5 mm. long, glabrous; corolla purple; stems glabrous; plant of Nfld. . . . . 15. *A. eucosmus,* var. *facinorum.*

d. Racemes in anthesis ovoid to thick-cylindric, 1.5–3 cm. in diameter, shorter than to twice, rarely thrice, as long as thick, rather closely flowered. . . *g.*

g. Leaflets of principal leaves 1.5–3 cm. broad; principal stipules 1–2 cm. long; racemes hardly elongating after anthesis; calyx glabrous; corolla yellowish, becoming brown. . . . . 18. *A. glycyphyllos.*

g. Leaflets 2–10 mm. broad; principal stipules 2–9 mm. long; racemes conspicuously (except in no. 4) elongating and loosening after anthesis; calyx pubescent (if rarely glabrous, the corolla purple). . . *h.*

h. Stems diffuse, the numerous basal leaves shorter than and with much smaller leaflets than the upper; raceme compact, only slightly loosening after anthesis; ovaries sessile, the old flowers and fruits mostly upcurved. . . . . 4. *A. distortus.*

h. Stems stiffly erect or ascending, without numerous small basal leaves; racemes evidently or conspicuously elongating after anthesis; ovaries and legumes stalked in the calyx (sessile only in no. 15, with old flowers and fruits strongly reflexed). . . *i.*

i. Ovaries (and legumes) sessile in the calyx, usually densely villous-hirsute with black (or white) hairs; corolla (except in albinos) deep-purple. . . . . 15. *A. eucosmus.*

i. Ovaries (and legumes) stipitate within the calyx, sparsely strigose or glabrous (white-pilose only in the white-flowered no. 10). . . *j.*

j. Stems, leaves, stipules, calyces and ovaries copiously white-pilose; corolla white. . . . . 10. *A. scrupulicola.*

*** One of the largest genera of flowering plants, the North American species alone treated by M. E. Jones under 30 sections and by Rydberg as an equal number of "genera". The fundamental grouping depends on the degree of intrusion of sutures in the fruit. Our comparatively limited number of species may be recognized largely on superficial characters; the accompanying keys are, therefore, quite artificial. The illustrations show fruiting inflorescences and legumes or only one of these. — By some recent botanists the genus made to include the traditionally separated genus *Oxytropis*.**

*j.*Stems, leaves and stipules strigose to glabrate or glabrous; calyces and ovaries strigose, often with black hairs, to glabrate. . . *k.*

*k.*Corolla white, 6–8 mm long; ovary (and legume) strongly rounded below the style. . . . . 11. *A.* ***Robbinsii.***

*k.*Corolla violet or purple, 9–12 mm. long; ovary (and legume) gradually tapering to the style. . . *l.*

*l.*Leaflets glabrous on both sides or merely strigose on midribs beneath. . . . . 14. *A. Jesupi.*

*l.*Leaflets cinereous-strigose on both surfaces when young and permanently so beneath.

Stipe of legume included in calyx-tube; plant of Nfld. and Lab. . . . . 13. *A. Fernaldi.*

Stipe distinctly exserted; plant of n. N.E. . . . . 12. *A. Blakei.*

*c.*Flowering stems creeping, at least at base, and forming an entangling stoloniferous mat.

Pedicels soon divergent; ovary stalked, partially 2-locular by slight intrusion of ventral suture; principal stipules 4–6 mm. long. . . . 16. *A. alpinus.*

Pedicels erect or suberect; ovary sessile, 1-locular, the ventral suture not intruded; principal stipules 2–4 mm. long. . . . . . 9. *A. stragulus.*

*b.*Flowering stems very short or wanting, usually overtopped by the basal leaves; the latter canescent, with firm leaflets 0.5–2 cm. long; corolla yellow; the white-hairy calyx with teeth much longer than the tube. . . . 5. *A. lotiflorus.*

*a.*Calyx-tube 3.5–10 mm. long. . . *m.*

*m.*Calyx-tube 3.5–6 mm. long. . . *n.*

*n.*Principal stipules 1–2 cm. long; leaflets of principal leaves 2–6 cm. long; calyx-tube and ovary glabrous. . . . . 7. *A. frigidus,* var. *gaspensis.*

*n.*Principal stipules 3–8 mm. long; leaflets mostly smaller; calyx-tube pubescent. . . *o.*

*o.*Corollas 8–12 mm. long. . . *p.*

*p.*Raceme not conspicuously loosening and elongating after anthesis; corollas white, cream-color or greenish; ovaries glabrous, sessile in the calyx; fruits ascending.

Raceme dense; wing-petals with a long basal auricle; ovary 2-locular. . . . . 19. *A. canadensis.*

Raceme lax; wing-petals with short basal auricle; ovary 1-locular. 8. *A. neglectus.*

*p.*Raceme loosening and conspicuously elongating after anthesis; corollas violet or purple; ovaries pubescent; fruits reflexed or loosely spreading. . . *q.*

*q.*Ovaries (and legumes) sessile in the calyx, densely villous-hirsute with black (or white) hairs; corollas deep purple, 6–8 mm. long. 15. *A. eucosmus.*

*q.*Ovaries (and legumes) stipitate within the calyx, sparingly strigose-hispid; corollas violet or blue-purple, 9–12 mm. long.

Leaflets glabrous on both sides or merely strigose on midribs beneath; calyx whitish, papery; ovary sparsely strigose; plant of Ct. Val. . . . . 14. *A. Jesupi.*

Leaflets cinereous-strigose on both surfaces when young and permanently so beneath; calyx blackish, herbaceous; ovary copiously strigose; plant of Nfld. and Lab. . . . . 13. *A. Fernaldi.*

*o.*Corollas 13–18 mm. long. . . *r.*

*r.*Ovary glabrous or merely strigose-hispid. . . *s.*

*s.*Stems 0.3–1.5 m. high; peduncles shorter than to but slightly overtopping their subtending leaves; flowers divergent or reflexed during anthesis, whitish or cream-color (sometimes pinkish in no. 17); ovary glabrous or sparsely strigose. . . *t.*

*t.*Raceme lax; corollas white or pinkish; ovary 1-locular or only partially 2-locular.

Ovary stipitate in the calyx, linear; fruits reflexed. . . . 17. *A. racemosus.*

Ovary sessile, ovoid; fruits strongly ascending. . . . . . 8. *A. neglectus.*

*t.*Raceme dense; corollas creamy to greenish-white; ovary sessile, completely 2-locular, ellipsoid-ovoid; fruits erect. . . . 19. *A. canadensis.*

*s.*Stems 1.5–4 dm. high; peduncles mostly much longer than their subtending leaves; flowers strongly ascending in a dense raceme, purple, rarely white; ovary subsessile, oblong, densely white-strigose. . . . . 20. *A. striatus.*

*r.*Ovary densely villous or woolly, sessile, ovoid, 2-locular; stems slender, 0.5–3 dm. long; racemes short, dense, thick.

Flowers strongly ascending; calyx-tube 4–6 mm. long; ovary ripening into a villous dehiscent legume 6–9 mm. long. . . . 21. *A. goniatus.*
Flowers divergent; calyx-tube 6–7 mm. long; ovary ripening into a pilose to glabrate fleshy tardily dehiscent fruit 1.5–2 cm. long. . 24. *A. plattensis.*

*m.*Calyx-tube 7–10 mm. long. . . *u.*

*u.*Subacaulescent, the very short branches crowded, the long peduncles therefore subscapiform; calyx strigose, with stiff appressed hairs; corolla purple; ovary 1-locular. . . . . 3. *A. missouriensis.*

*u.*Stems elongate, 1–5 dm. long; calyx pilose or villous; ovary 2-locular. . . *v.*

*v.*New stipules 5–8 mm. long; stems pilose or villous, not spreading-hirsute. . . *w.*

*w.*Ovary glabrous; corolla 1.8–2.5 cm. long.

Corolla purple; leaflets grayish. . . . . 22. *A. caryocarpus.*
Corolla cream-color or merely with bluish tips; leaflets green. . 23. *A. mexicanus*, var. *trichocalyx.*

*w.*Ovary pilose; corolla 1.4–1.6 cm. long, cream-color or tipped with blue or purple. . . . . 24. *A. plattensis.*

*v.*New stipules 1.3–1.6 cm. long; stems densely and horizontally long-hirsute; corolla cream-color or with bluish tips, 1.5–2 cm. long; ovary villous-hirsute. . . . . 25. *A. tennesseensis.*

Key to Fruiting Material

*a.*Fruit sessile or essentially so within the calyx. . . *b.*

*b.*Fruit dry, a readily dehiscent legume. . . *c.*

*c.*Legume less than 1 cm. long. . . *d.*

*d.*Stems ascending or erect, not forming a prostrate mat; leaflets chiefly 1–2.5 cm. long; racemes several–many-flowered. . . *e.*

*e.*Fruiting raceme lax, elongate, the fruits divergent or reflexed.

Leaflets linear, 1–3 mm. wide; middle and upper stipules abruptly caudate-tipped; calyx-tube 1.5–2 mm. long, white-strigose; body of legume ellipsoid-ovoid, white-strigose, strongly cross-ribbed. 2. *A. gracilis.*
Leaflets oblong, 2–9 mm. wide; stipules blunt; calyx-tube 2.5–5 mm. long, black-hairy (rarely glabrous); body of legume ovoid, black-villous or glabrous, not cross-ribbed. 15. *A. eucosmus.*

*e.*Fruiting raceme dense, thick-cylindric to subglobose, the fruits ascending.

Legumes thick-cylindric or slenderly ovoid, finely white-strigose. 20. *A. striatus.*
Legumes ovoid, long-villous. . . . . 21. *A. goniatus.*

*d.*Stems trailing and freely branching and interlocking, forming a dense carpet; leaflets elliptic, 0.2–1 cm. long; racemes 1–10-flowered; legumes inflated, ovoid, black-strigose. . . . 9. *A. stragulus.*

*c.*Legumes 1–3 cm. long. . . *f.*

*f.*Subacaulescent, the very short branches crowded, the peduncles therefore subscapiform; legumes firm, lance-cylindric to slenderly ovoid or ellipsoid, acuminate.

Peduncles all elongate; calyx-tube 7–10 mm. long, strigose with stiff appressed hairs, the teeth much shorter; legumes cross-ribbed beneath the pubescence. . . . . 3. *A. missouriensis.*
Peduncles elongate or sometimes nearly wanting; calyx-tube 2–3 mm. long, white-pilose, the teeth as long or longer; legumes not cross-ribbed, frequently in the leaf-axils. . . . . 5. *A. lotiflorus.*

*f.*Stems elongate, 0.1–1.5 m. high, the peduncles obviously axillary. . . *g.*

*g.*Legumes strongly ascending, in a rather dense raceme, thick-cylindric to ovoid.

Legumes 1-locular, inflated-ovoid, 1–1.4 cm. thick. . . . 8. *A. neglectus.*
Legumes 2-locular, with a longitudinal partition, subcylindric to slenderly-ovoid, 3–7 mm. thick, scarcely inflated.

Peduncles mostly longer than subtending leaves; fruiting racemes 2–8 cm. long; legumes 0.7–1 cm. long, 3–4 mm. in diameter, white-strigose. . . . . 20. *A. striatus.*
Peduncles mostly shorter than subtending leaves; fruiting racemes 0.3–2 dm. long; legumes 1–2 cm. long, 4–7 mm. in diameter, glabrous. . . . . 19. *A. canadensis.*

g.Legumes divergent, arched-ascending or reflexed. . . *h*.
h.Legumes lanceolate or lance-cylindric, glabrous or merely white-strigose, 1.5–3 cm. long; prairie and Alleghenian species.
Stems diffuse; the numerous basal leaves shorter than and with much smaller leaflets than the upper; racemes compact, rachis in fruit 0.5–4 cm. long; legumes glabrous, partially 2-locular. 4. *A. distortus.*
Stems strongly ascending; basal leaves few or none, similar to the upper; racemes lax, rachis in fruit 4–12 cm. long; legumes white-strigose, strictly 1-locular. . . . . 6. *A. flexuosus.*
h.Legumes ovoid, black-villous (rarely glabrous), 7–12 mm. long; boreal and montane species. . . . . 15. *A. eucosmus.*
b.Fruits fleshy, indehiscent or tardily splitting at tip, 2-locular. . . *i*.
i.New stipules 5–8 mm. long; stems pilose or villous, not spreading-hirsute.
Fruit glabrous.
Leaflets grayish; calyx white-strigose; corolla purple; fruit 1.5–2 cm. in diameter. . . . . 22. *A. caryocarpus.*
Leaflets green; calyx villous; corolla cream-color; fruit 2–3 cm. in diameter. . . . . 23. *A. mexicanus*, var. *trichocalyx.*
Fruit pilose, 1–1.6 cm. in diameter. . . . . 24. *A. plattensis.*
i.New stipules 1.3–1.6 cm. long; stems densely and horizontally long-hirsute; fruit villous-hirsute. . . . . 25. *A. tennesseensis.*

a.Fruit stipitate within the calyx. . . *j*.
j.Leaflets of principal leaves 2–6 cm. long, 0.6–3 cm. broad; principal stipules 1–2 cm. long; calyx-tube glabrous, 3–6 mm. long; corollas cream-color, yellowish or brownish; legumes glabrous, subterete or strongly inflated.
Fruiting racemes compact, with linear-cylindric falcate ascending or merely spreading 2-locular legumes 2–4.5 cm. long (excluding stipes). 18. *A. glycyphyllos.*
Fruiting racemes lax, with inflated ovoid reflexed partially 1-locular legumes 1.5–2.2 cm. long (excluding stipes). . . . . 7. *A. frigidus*, var. *gaspensis.*
j.Leaflets and stipules mostly smaller; calyx-tube pubescent (if glabrate only 1.5–2 mm. long); legumes laterally compressed, flattish to triangular or rounded in cross-section. . . *k*.
k.Flowering stems erect to decumbent, from a crown or crowns, not depressed and forming an intricate mat; fruiting racemes 3–18 cm. long, lax, the distant or subdistant fruits divergent or loosely reflexed. . *l*.
l.Peduncles 1–2 cm. long; calyx-tubes 1.5–2 mm. long; legumes flat, 1-locular, their bodies 8–15 mm. long, glabrous, transversely reticulate; western species. . . . . 1. *A. tenellus.*
l.Peduncles (in fruit) 4–18 cm. long; calyx-tubes 2.5–5 mm. long; legumes turgid (elliptic to triangular or rounded in cross-section), their bodies partially 2-locular by a narrow false septum intruded slightly between the halves, 1–3 cm. long. . . *m*.
m.Legumes pubescent (at least strigose), not deeply grooved, with papery or membranous valves; boreal and northeastern species. . . *n*.
n.Stems, leaves, stipules and calyces copiously white-pilose; legumes white-strigose, their bodies 1.5–2.5 cm. long, stipes 5–7 mm. long. 10. *A. scrupulicola.*
n.Stems, leaves and stipules strigose to glabrate or glabrous; calyces and fruits strigose, often with black hairs, rarely glabrate, their bodies 1–2 cm. long. . . *o*.
o.Legumes strongly rounded below the style, the body 1–1.5 cm. long; stipe 2–4 mm. long, scarcely exserted; plant of Vt. . 11. *A. Robbinsii.*
o.Legumes gradually tapering to the style (beak), the body 1–2 cm. long. . . *p*.
p.Leaflets cinereous-strigose on both surfaces when young and permanently so beneath; calyx blackish, herbaceous.
Stipe of legume 4–7 mm. long, exserted from calyx; plant of n. N.E. . . . . 12. *A. Blakei.*
Stipe 1–3 mm. long, not exserted; plant of Nfld. and Lab. 13. *A. Fernaldi.*
p.Leaflets glabrous on both sides or merely strigose on midribs beneath; calyx papery, whitish except for sparse black strigae; stipe 5 mm. long, barely exserted; plant of Ct. Val. 14. *A. Jesupi.*

*m.*Legumes glabrous, deeply grooved, coriaceous, the cross-section compressed-triangular with concave sides; western species. . . . 17. *A. racemosus.*

*k.*Flowering stems depressed, freely branching, forming an intricate mat; fruiting racemes 1–7 cm. long, the approximate fruits strongly reflexed and overlapping. . . . . 16. *A. alpinus.*

1. A. tenéllus Pursh (delicate), Loose-flowered M. — Stems numerous, slender, ascending, 2–4.5 dm. high, *minutely strigose at least above,* branching; *middle and upper stipules* connate, *dark brown or green,* deltoid, *acute or acuminate;* leaflets 9–21, linear or narrowly oblong, pale green, 1–2 cm. long, glabrous or sparingly strigose beneath; *peduncles 1–2 cm. long; racemes linear-cylindric,* lax and upon expansion *remotely flowered; calyx-tube 1.5–2 mm. long, sparsely strigose* or glabrate, the teeth lance-subulate, shorter; *corolla ochroleucous or merely pink-tipped,* 6–8 mm. long; *legumes stalked, flat, 1-locular, the glabrous transversely reticulate body 8–15 mm. long* and 3–4.5 mm. wide. (*Homalobus* Britt.; *H. stipitatus* Rydb.) — Calcareous gravelly shores bluffs and plains, Yuk. to N.M., e. to Man. and Minn. June, July. Fig. 1197.

1197. A. tenellus.

2. A. grácilis Nutt. (slender). — Similar, up to 6 dm. high; middle and upper *stipules pale brown, abruptly caudate-tipped;* leaflets 7–15, in remote pairs, firm, linear to narrowly cuneate-oblong, 1–2.5 mm. long, 1–3 mm. wide; *peduncles 4–12 cm. long; calyx densely white-strigose,* the tube 1.5–2 mm. long; *corolla purple,* 5–8 mm. long; *legumes sessile, ellipsoid-ovoid, plump, 5–9 mm. long, strongly cross-ribbed beneath the close white pubescence, 1-locular.* (*A. parviflorus* MacM., not Lam.; *Microphacos parviflorus* and *gracilis* Rydb.) — Dry prairies, hills and barrens, Mont. to N.M., e. to Neb., Kans., Okla. and Tex. May–July. Fig. 1198.

1198. A. gracilis.

3. A. missouriénsis Nutt. (of old Missouri). — Cespitose, *subacaulescent;* leaves crowded, ascending, the 9–15 elliptic *leaflets* 5–10 mm. long, *whitish-strigose on both surfaces; peduncles subscapiform, 3–13 cm. long,* canescent; racemes short, hemispherical to ovoid, few-flowered, 3.5–4 mm. broad; flowers subsessile to short-pedicelled; *calyx-tube 7–10 mm. long, strigose with stiff whitish hairs,* the teeth much shorter; *corolla purple, 1.5–2 cm. long; legume sessile* in the calyx, *firm, lance- to slenderly oblong-cylindric,* acuminate, *1.5–2.5 cm. long, cross-ribbed* beneath the strigose pubescence, *both sutures prominent.* (*Xylophacos* Rydb.) — Rocky or sandy plains and bluffs, Alta. to N.M., e. to Man., Minn., Neb., Kans., Okla. and Tex. Late April–June. Fig. 1199.

1199. A. missouriensis.

4. A. distórtus T. & G. (twisted; from the legumes). — *Stems* numerous, *diffuse or loosely ascending,* 0.5–4.5 dm. long, *subglabrous; lower leaves numerous, shorter than and with smaller leaflets than the upper;* upper leaves *with* 13–25 *oblong to cuneate-obovate green and glabrous or glabrate blunt or retuse leaflets* up to 1 cm. long; *raceme compact, only slightly loosening after anthesis;* flowers spreading; *calyx-tube 2–3 mm. long,* slightly longer than the lanceolate teeth; corollas lilac, purple or white, about 1 cm. long; *legumes sessile,* lance-cylindric to slenderly ovoid, glabrous, *lunate,* 2–3 cm. long, *loosely spreading or ascending,* appearing *falsely 2-locular by the incurving of the sutures* but without a septum. (*Holophacos* Rydb.) — Sandy or rocky (calcareous) prairies, slopes and glades, Ill. and se. Ia. to e. Kans., s. to Miss., La. and Tex.; also dry rocky slopes, w. Md., ne. W.Va. and nw. Va. May, June. Fig. 1200.

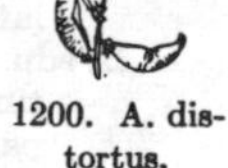

1200. A. distortus.

5. A. lotiflòrus Hook. (with flowers of *Lotus*), Low M. — Cespitose, *hoary with strigose pubescence, the stems almost wanting* or up to 1 dm. long; leaves ascending, with 2–10 remote pairs of firm elliptic or oblong leaflets 0.5–2 cm. long; *earlier flowers sessile or short-stalked in the axils, few,* others (often later) in larger racemes on peduncles up to 1 dm. long; *calyx-tube 2–3 mm. long, white-pilose, the subulate teeth as long or longer; corolla yellow or cream-color,* 8–11 mm. long; *legumes sessile, coriaceous, obliquely lance-ovoid,* acuminate, 1.5–2.5 cm. long, *often lunate,* the back more or less impressed, the acute ventral suture nearly straight. (*Batidophaca lotiflora* and *cretacea* Rydb.) — Dry calcareous gravel, bluffs and plains, B.C. to N.M., e. to Man., w. Minn., w. Ia., w. Mo., Okla. and Tex. April, May. Fig. 1201.

1201. A. lotiflorus.

6. A. flexuòsus Dougl. (flexuous). — *Stems slender,* ascending, 3–6 dm. high, *ashy-puberulent; leaflets* 7–12 remote pairs, firm, *linear to oblong-oblanceolate,* 0.5–1.5 cm. long, grayish-strigose beneath; peduncles 3–10 cm. long; *racemes* linear-cylindric, *lax, in fruit 4–12 cm. long; calyx-tube* 3 mm. long,

1202. A. flexuosus.

densely pilose with incurving hairs, the lance-deltoid teeth much shorter; *corolla whitish*, 8–10 mm. long; *legumes nearly sessile* in the calyx, *slenderly lance-cylindric*, 1.2–2.2 cm. long, sparsely strigose, straight or falcate, 1-locular. (*Pisophaca flexuosa* and *elongata* Rydb.) — Dry gravels, bluffs and prairies, Alta. to N.M., e. to Man., w. Minn., Neb. and Kans. June, July. Fig. 1202.

7. A. frígidus (L.) Gray (of cold regions). — Stems stout, smooth, solitary or few, simple, up to 4 dm. high; *stipules herbaceous*, ovate, 1–1.7 *cm. long;* leaves with 7–17 *oblong to oval thin* obtuse *leaflets* 1.5–4 *cm. long;* peduncles 5–13 cm. long; *racemes thick, rather lax; calyx* campanulate, pale, glabrous or sparsely strigose, 5–7 *mm. long*, the margins and low teeth black-ciliolate; corolla creamy or yellowish, 1.3–1.7 mm. long; *legumes inflated, membranaceous, ellipsoid or ovoid, contracted* at base *to a long stipe, the body* 1.5–2.3 *cm. long*, minutely pubescent. — Calcareous soils, Eu. — A variable boreal type, represented with us by

1203. A. frigidus, v. gaspensis.

Var. **gaspénsis** (Rousseau) Fern. (of the Gaspé Pen., Que.). — Mostly taller, 0.2–1.4 m. high; *leaflets* 1.5–6 *cm. long*, 0.7–3 *cm. broad; calyx* 3.5–6 mm. long, glabrous, *its margin and teeth white-ciliolate;* corolla 1.1–1.4 cm. long; *body of legume glabrous*. — Calcareous alluvium of wooded river-banks, Bonaventure Co., Que. Late June–early Aug. Fig. 1203.

8. A. negléctus (T. & G.) Sheld. (overlooked), Cooper's M. — Stems few–∞ from crowns, nearly smooth, erect or ascending, branching, 0.3–1.3 m. high; stipules narrowly deltoid, 2–5 mm. long; leaflets 5–10 subdistant pairs, oblong or elliptic, obtuse or emarginate, somewhat strigose beneath, glabrous above, 1–3 cm. long; *peduncles shorter than* to barely exceeding *subtending leaves; flowering racemes* ellipsoid or thick-cylindric, *lax;* calyx-tube about 4 mm. long, pilose, the lanceolate teeth half as long; *corolla white or tinged with violet*, or lemon-yellow in forma **limònius** (Farw.) Fern. (lemon), 1.2–1.4 *cm. long; wing-petals with a short basal auricle; legumes sessile* in the calyx, *erect, inflated-ovoid*, 1.5–2 cm. long, 1–1.4 *cm. thick, glabrous*, 1-*locular*. (*Phaca* T. & G.) — Calcareous gravels, talus and cliffs, e. Ont. to Minn., s. to w. N.Y., O., s. Mich., s. Wisc. and n. Ia. June, July. Fig. 1204.

1204. A. neglectus.

9. A. strágulus Fern. (forming a carpet). — Depressed, the *filiform freely branching and repent stems forming carpets*, the ascending filiform branchlets glabrous or sparsely strigose; *stipules* 2–4 *mm. long;* leaves divergent, the 4–9 pairs of *elliptic retuse leaflets* 0.2–1 *cm. long*, glabrous, or strigose beneath; peduncles filiform, 0.2–9 cm. long; *flowers solitary or* 2–10 *in close racemes* 1–4 cm. long, divergent; calyx-tube campanulate, 2.5–3 mm. long, black-strigose, the slender teeth half as long; *corolla lilac*, 8–12 *mm. long; legumes sessile, erect, inflated*, 1-*locular*, oblong-ovoid, apiculate, black-strigose, 6–8 *mm. long*, 2.5–4 mm. thick. (*Phaca* Rydb.) — Sandy and turfy calcareous shores, Pistolet Bay, Nfld. July, Aug. Fig. 1205.

1205. A. stragulus.

10. A. scrupulícola Fern. & Weath. (growing on broken rock). — *Stems, leaves, stipules, calyces and ovaries copiously white-pilose;* stems numerous, erect or ascending, 1.5–3 dm. high; lower stipules connate, orbicular to broadly ovate, 4–6 mm. long, the upper lance-attenuate; leaflets 4 or 5 remote pairs, oblong-lanceolate, obtuse, ascending, 1–2.5 cm. long; peduncles 3–10 cm. long; *racemes lax*, 7–17-flowered, in anthesis 2–5 cm., *in fruit elongating to* 6–10 *cm. long;* flowers ascending or spreading; calyx-tube 2.5–3 mm. long, the lance-subulate teeth shorter; *corolla white*, 7–9 *mm. long; legumes arcuate-recurving*, elliptic-lanceolate, falcate, compressed, 2–3.2 *cm. long*, acute, *attenuate at base to a stipe* 5–7 *mm. long, short-pilose*, nearly 1-locular but with a very narrow partial septum. —Dry calcareous cliffs and talus, nw. Gaspé Co., Que. June, early July. Fig. 1206.

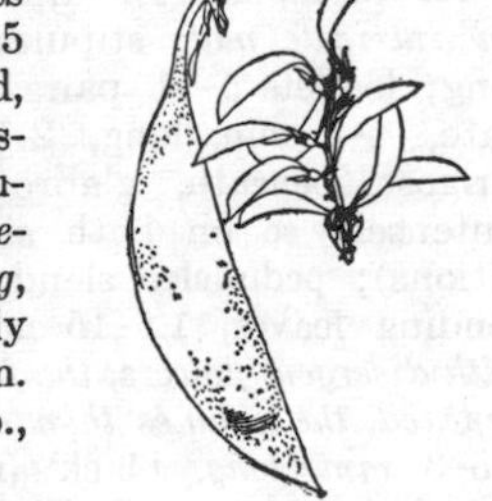
1206. A. scrupulicola.

11. A. Robbínsii (Oakes) Gray (for its discoverer, James Watson Robbins, 1801–1879). — Green and nearly glabrous or sparsely strigose; stems erect, 1.5–4 dm. high; stipules lance-ovate, 4–5 mm. long; *leaflets* 2–5 remote pairs, oblong, rounded or retuse at tip, *green and glabrous above, sparingly strigose beneath*, 0.8–2.2 cm. long; peduncles 0.6–1.5 dm. long, erect; *racemes in anthesis* subglobose to thick-cylindric, *rather dense*, 1.5–3 cm. long, *in fruit lax*

1207. A. Robbinsii.

and elongate to 0.6–1.5 *dm.;* flowers ascending or spreading; calyx-tube 3–3.5 mm. long, black-strigose, the lance-deltoid teeth shorter; *corolla white,* 6–8 *mm. long; legumes* spreading or recurving, compressed, straight or slightly falcate, *the oblong body* 1–1.5 *cm. long,* sparingly black-strigose, *rounded to the short beak and to the mostly included stipe* (2–4 *mm. long*), slightly 2-locular through a narrow partial septum. (*Atelophragma* Rydb.) — Dry calcareous ledges, Winooski R., Vt. (probably extinct). May, June. Fig. 1207.

12. A. Blàkei Egglest. (for its discoverer, Joseph Blake, 1814–1888). — Similar to no. 11; *leaflets cinereous-strigose on both surfaces when young,* later glabrate above but strigose beneath; *calyx herbaceous,* black-strigose; *corolla violet or blue-purple,* 9–12 *mm. long; legumes attenuate to style and stipe, the body* 1.5–2 *cm. long, triangular in cross-section, the exserted stipe* 4–7 *mm. long.* (*Atelophragma* Rydb.) — Calcareous ledges, cliffs and talus, local, n. Me. and n. Vt. May–early July. Fig. 1208.

1208. A. Blakei.

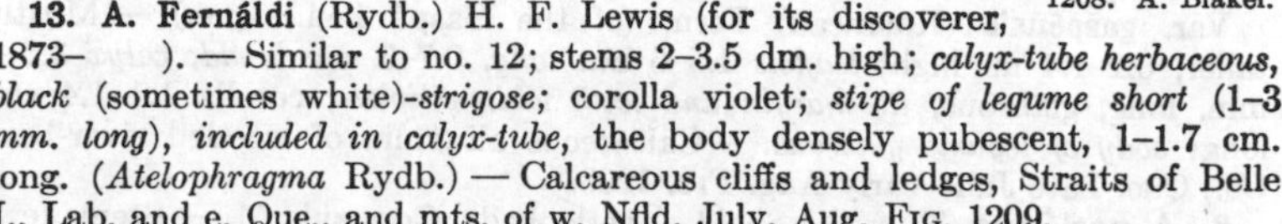

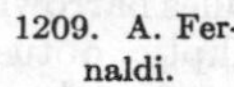

1209. A. Fernaldi.

13. A. Fernáldi (Rydb.) H. F. Lewis (for its discoverer, 1873–). — Similar to no. 12; stems 2–3.5 dm. high; *calyx-tube herbaceous, black* (sometimes white)*-strigose;* corolla violet; *stipe of legume short* (1–3 *mm. long*), *included in calyx-tube,* the body densely pubescent, 1–1.7 cm. long. (*Atelophragma* Rydb.) — Calcareous cliffs and ledges, Straits of Belle I., Lab. and e. Que., and mts. of w. Nfld. July, Aug. Fig. 1209.

14. A. Jésupi (Egglest. & Sheld.) Britt. (for its discoverer, Henry Griswold Jesup, 1826–1903). — Similar to nos. 12 and 13, 2.5–6 dm. high; *leaflets glabrous on both sides or merely strigose on midribs beneath; calyx papery, whitish,* sparsely black-strigose, the tube 3–3.8 mm. long; *corolla violet or purple,* 9–12 *mm. long; legumes* attenuate at both ends, *the stipe barely exserted* (about 5 mm. long), the sparsely strigose body 1.5–2 cm. long. (*Atelophragma* Rydb.) — Rocky banks of Conn. R., N.H. and Vt. May, early June. Fig. 1210.

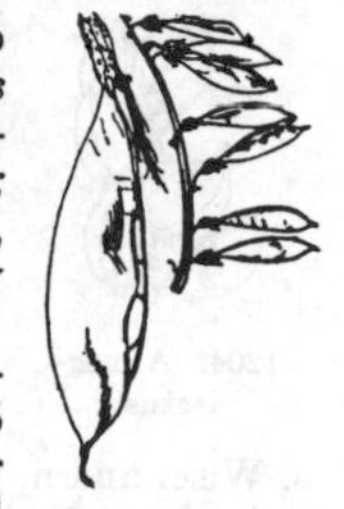

1210. A. Jesupi.

15. A. eucósmus Robins. (elegant). — Stems few to many from ascending caudices, 1–6 dm. high, glabrous or strigose; *stipules* ovate, *blunt,* 3–5 *mm. long; leaflets* 4–8 pairs, narrowly to broadly *oblong or elliptic,* obtuse, 0.6–2.5 cm. long, 2–9 mm. wide, green and smoothish to canescent, usually strigose beneath and in arid habitats also above; peduncles longer than subtending leaves, 0.3–1.6 dm. long; *flowering raceme* at first dense, *soon elongating* and in fruit lax and secund, becoming 0.3–1.5 dm. long; *calyx-tube* 2.5–5 *mm. long, black-hairy,* the teeth much shorter; *corolla deep-purple,* or white in forma **albìnus** Fern. (like plaster), 6–8 *mm. long; legumes sessile or nearly so, obliquely ellipsoid to ovoid, reflexed,* 7–12 *mm. long, densely* black (rarely white) *-villous-hirsute to -strigose,* nearly 1-locular, but with narrow partial partition. (*Atelophragma elegans* Rydb.) — Calcareous gravel and ledges, Baffin I. to Alaska, s. to Nfld., Gaspé Pen., Que., n. N.B., n. Me. and Rocky Mts. to Colo. June, July. Fig. 1211.

1211. A. eucosmus.

Var. **facìnorum** Fern. (of Exploits River, Nfld.) — Stems, leaves and calyx-tube glabrous; legumes sparsely white-strigose. — Ledges, Exploits R., Nfld.

16. A. alpìnus L. (alpine). — *Stems depressed or creeping at base, freely branching, forming an intricate mat;* stipules lanceolate to deltoid, 3–6 mm. long; leaflets 5–11 pairs, oblong, elliptic, or narrowly obovate, 4–5 mm. long, 2–8 mm. broad, white-strigose beneath, glabrous to strigose above (intensely so on both sides under arid conditions); peduncles slender, exceeding the subtending leaves, 1.5–15 cm. long; *raceme short, with divergent flowers, the fruiting calyces strongly deflexed, the legumes then overlapping; calyx-tube* 2.5–3 *mm. long,* black (rarely white)-hirsute, the hairy teeth shorter; corolla blue-violet (either pale or deep), 1.1–1.5 cm. long; *legumes tightly reflexed, stipitate, plump, densely* black (rarely white) *-hairy,* the lance-ovoid to lanceolate straightish to slightly falcate body 9–12 mm. long. (*Atelophragma* Rydb.) — Arct. reg., s. to n. Nfld., n. Wisc., S.D. and Colo. June–Aug. (Eurasia) Fig. 1212.

1212. A. alpinus (typical).

1213. A. alpinus, v. labradoricus.

1214. A. alpinus, v. Brunetianus.

Var. **labradóricus** (DC.) Fern. (of Labrador). — Looser and often coarser; *calyx-tube*

2–2.5 *mm. long, sparsely black-strigose; legumes subcylindric-oblanceolate, black-strigose, straight.* (*A. labradoricus* DC.; *Atelophragma lab.* Rydb.) — Calcareous shores and gravels, Nfld. to L. St. John and Drummond Co., Que. July, Aug. Fig. 1213.

Var. **Brunetiànus** Fern. (in honor of Ovide Brunet, 1826–1877). — Similar; the *lance-subcylindric legumes falcate.* (*A. Brunetianus* Rousseau) — Calcareous ledges and gravels, Restigouche R. system, Que. and N.B.; St. John R. system, N.B. and Me.; Kennebec R., Me.; Conn. R., N.H. and Vt. June–Sept. Fig. 1214.

17. A. racemòsus Pursh (racemose). — Stems several, erect, stoutish, 0.3–1 m. high, strigose or glabrate; stipules 5–8 mm. long, connate, acuminate; *leaflets* 6–13 pairs, linear to elliptic, 1.5–3.5 *cm. long, canescent beneath;* peduncles 5–15 cm. long; racemes thick-cylindric, rather lax, in fruit elongating to 6–15 cm.; flowers divergent to nodding; *calyx-tube* gibbous-campanulate, 4–5 *mm. long, white-strigose; corolla white or pinkish, about* 1.5 *cm. long; legumes* stipitate, *pendent or loosely spreading,* slender, plump, *glabrous, coriaceous, triangular-compressed, with concave sides,* the body 2–2.5 cm. long. (*Tium* Rydb.) — Dry prairies, shores and slopes, Minn. to Sask., s. to Okla. and N.M. Late May, June. Fig. 1215.

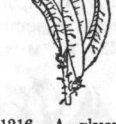

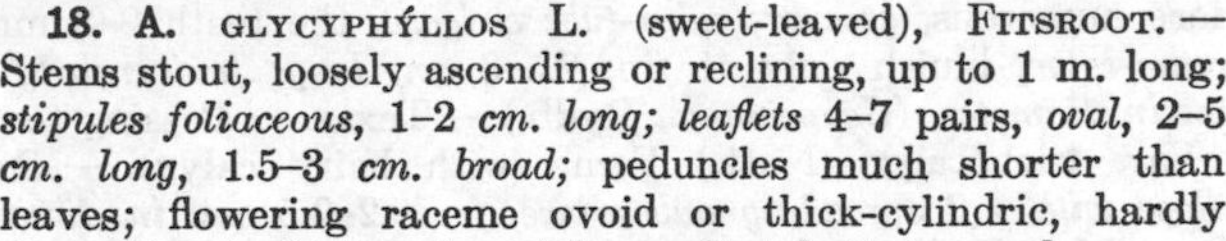

1215. A. racemosus.

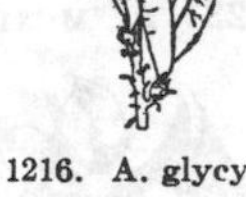

1216. A. glycyphyllos.

18. A. glycyphýllos L. (sweet-leaved), Fitsroot. — Stems stout, loosely ascending or reclining, up to 1 m. long; *stipules foliaceous,* 1–2 *cm. long; leaflets* 4–7 pairs, *oval,* 2–5 *cm. long,* 1.5–3 *cm. broad;* peduncles much shorter than leaves; flowering raceme ovoid or thick-cylindric, hardly elongating in fruit; *calyx glabrous, its tube* 3–4 *mm. long; corolla yellowish, becoming brown,* 1–1.3 cm. long; *legumes* short-stipitate, glabrous, *linear-cylindric, falcate,* 2–4.5 *cm. long,* 2-*locular, spreading or loosely ascending.* (*Hedyphylla* Rydb.) — Roadsides and waste grounds, local, Mass. to Ind. May–July. (Adv. from Eu.) Fig. 1216.

19. A. canadénsis L. (Canadian). — Rather coarse, 0.3–1.6 m. high, somewhat pubescent or glabrate; stipules deltoid to lance-attenuate, 0.5–1 cm. long; leaflets 7–15 pairs, oblong or elliptic, glabrous above, strigose beneath, 1.5–4 cm. long; *peduncles much shorter than subtending leaves; racemes dense, thick-cylindric, in anthesis* 2.5–3.5 *cm. thick, in fruit* 0.3–2 dm. long, *dense,* 1.5–3 *cm. thick;* flowers loosely spreading or reflexed, *calyx-tube* 4.5–6 *mm. long, minutely strigose to glabrous;* the slender *teeth* 1–3 mm. long, *less than half as long as tube; corolla creamy to greenish-white,* 1.2–1.5 cm. long; *legumes erect, sessile, crowded,* thick-cylindric, thick-walled, *glabrous* and cross-reticulate or rugulose, 1–2 *cm. long,* 4–7 *mm. thick, completely* 2-*locular.* — Shores and rich thickets, sw. Que. and L. Champlain, Vt., to Hudson Bay, w. to B.C., s. to Va., W.Va., Ark., Tex. and Colo. Late June–Aug. Fig. 1217. — Passing insensibly into

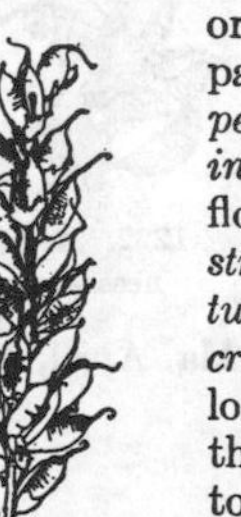

1217. A. canadensis.

Var. **longilobus** Fassett (long-lobed). — Racemes similar; *calyx-teeth* 2.5–5.5 *mm. long, more than half the length of the tube* (4–5 mm. long). — Wisc., Minn. and Ia.

Var. **caroliniànus** (L.) Jones (of Carolina), Rattle-vetch. — *Flowering raceme looser,* 2–2.5 *cm. thick; fruiting raceme less crowded,* 1.7–2 *cm. in diameter; calyx-teeth more than half as long as tube,* 2–3.7 mm. long. (*A. carolinianus* L.) — Md. and W.Va. to Ga. and Tenn.

20. A. striàtus Nutt. (with parallel lines). — Ascending or decumbent, 1–4 dm. high, cinereous with minute appressed pubescence or glabrate; stipules scarious, deltoid, about 5 mm. long; *leaflets* 6–12 pairs, narrowly oblong, 0.5–2 cm. long, *strigose* especially *beneath; peduncles much shorter than subtending leaves; racemes dense, obovoid in anthesis,* with flowers ascending, *thick-cylindric or ellipsoid in fruit;* calyx-tube 4–6 mm. long, strigose, the slender teeth half or two-thirds as long; *corollas* purple, creamy or white, 1.5–1.8 *cm. long; legumes erect,* crowded, *slenderly ellipsoid,* 7–10 *mm. long, densely white-strigose,* 2-*locular,* triangular-compressed. (*A. adsurgens* Hook., not Pall.; *A. Chandonnettii* Lunell, the creamy-flowered form.) — Dry prairies and ridges, Man. to Wash., s. to w. Minn., Neb., w. Kans. and N.M. Late May–July. Fig. 1218.

1218. A. striatus.

21. A. goniàtus Nutt. (angled; from the three-angled legumes). — Slender, tufted, *diffuse,* the sparsely strigose to glabrous stems 0.5–3 dm. long; leaflets 6–10 pairs, oblong to linear, 0.4–2 cm. long; peduncles elongate, mostly overtopping the subtending leaves; *racemes dense,* in anthesis obovoid to subglobose, *in fruit nearly spherical,* the flowers strongly ascend-

1219. A. goniatus.

ing; *calyx-tube 4–6 mm. long,* hirsute, the slender teeth half as long; corollas purple, 1.5–2 cm. long; *legumes erect, ovoid to subglobose, long-villous, 6–9 mm. long, 2-locular.* (*A. hypoglottis* of Am. auth., not L.) — Damp prairies, bottoms and meadows, Man. to Alta. and Wash., s. to Ia., Kans., N.M. and n. Calif. May–July. Fig. 1219. — Related to and perhaps referable to the Eurasian *A. danicus* Retz.

22. A. caryocárpus Ker (nut-fruited; from the indehiscent fruit, originally described as "not unlike a stinted walnut"), Ground-plum, Buffalo-bean. — *Stems* decumbent, minutely *appressed-pubescent,* 1–3.5 dm. long; *stipules* deltoid, 5–7 *mm. long; leaflets* 7–10 pairs, elliptic, oblanceolate or narrowly oblong, 0.5–2 cm. long, *grayish;* peduncles shorter than subtending leaves; *racemes* rather dense, *in anthesis subglobose to ellipsoid; calyx-tube* 8–10 *mm. long,* pilose; *corolla purple,* 1.8–2.2 cm. long; *fruit glabrous, ovoid-globular, at first succulent,* becoming thick-walled and corky when dry, *subsessile, indehiscent, 2-locular,* 1.5–2 *cm. in diameter.* (*A. crassicarpus* Nutt.?; *Geoprumnon crassicarpum* Rydb.) — Prairies, Man. to Sask. and Mont., s. to Ark., Okla., Tex. and N.M. April, May. Fig. 1220. — The unripe fruits of this and the succeeding species resemble green plums (whence the popular name) and are eaten raw or cooked.

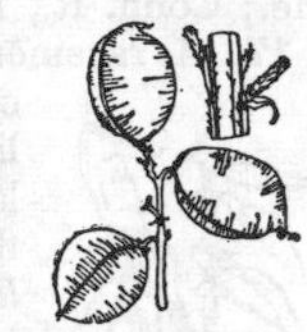

1220. A. caryocarpus.

23. A. mexicànus A. DC. (Mexican), Ground-plum. — Similar to no. 24; *leaflets* often more numerous, *greener; calyx-tube strigose,* the teeth 2–3 mm. long; *corolla cream-color,* bluish only at tip, 2.5–3 *cm. long; fruit nearly globular,* 2.5–3 *cm. in diameter.* (*Geoprumnon* Rydb.) — Texas. — Passing to

Var. **trichócalyx** (Nutt.) Fern. (with hairy calyx). — *Calyx-tube white-pilose with soft spreading pubescence; fruit* 2–2.5 *cm. in diameter.* (*Geoprumnon trichocalyx* Rydb.) — Rocky prairies, knobs and barrens, Tex. and La., n. to Mo. and s. Ill. April, May.

1221. A. plattensis.

24. A. platténsis Nutt. (of the Platte region), Ground-plum. — Similar; *loosely villous;* stipules lance-deltoid, 3–5 mm. long; leaflets obovate to oblong, 0.5–1.5 cm. long; *calyx-tube pilose; corolla cream-color* or tipped with purple, 1.4–1.6 *cm. long; fruits pilose,* 1–1.6 *cm. in diameter.* (*Geoprumnon* Rydb.) — Prairies, Tex. to Ala., n. to S.D., Minn. and Ill. May, June. Fig. 1221.

25. A. tennesseénsis Gray (of Tennessee), Ground-plum. — *Stems* depressed or decumbent, *densely and horizontally long-hirsute; stipules* deltoid or ovate, 1.3–1.6 *cm. long;* calyx villous; corolla cream-color or with bluish tips, 1.5–2 cm. long; *fruit villous-hirsute, slenderly ovoid,* 2.4–3 cm. long. (*Geoprumnon* Rydb.) — Calcareous barrens and cedar-glades, s. Ill., Tenn. and Ala. April, May. Fig. 1222.

1222. A. tennesseensis.

34. OXÝTROPIS DC.

Keel tipped with a sharp projecting point or appendage; otherwise as in *Astragalus.* Legume often more or less 2-locular by the intrusion of the ventral suture. — Our species mostly low nearly acaulescent perennials, with tufts of often numerous very short stems from a hard and thick crown or rootstock covered with scaly adnate stipules; pinnate leaves of many leaflets; peduncles scape-like, bearing a head or short spike of flowers. Genus of N. Hemisph. (Name from the Greek *oxys, sharp,* and *tropis, keel.*) Spiesia Neck. Aragallus Neck. — By some united with the technically distinguished and overwhelmingly variable genus *Astragalus;* here maintained on traditional lines.

a. Stipules strongly adnate to petioles; flowers and fruits ascending or spreading, not secund; corolla 1–2 cm. long. . . *b.*

b. Leaflets opposite, subopposite or alternate, not fascicled; legumes exserted from calyces. . . *c.*

c. Walls of legumes coriaceous or firm; leaflets 9–23, subcoriaceous, linear or narrowly lanceolate, 1–4 cm. long; prairie species. . . . . 1. *O. Lambertii.*

c. Walls of legumes membranaceous, easily crushed; leaflets 13–35, thin, broadly lanceolate to elliptic, mostly shorter; northern and eastern species. . . *d.*

d. Bracts of inflorescence and legume viscid.

Leaflets 30–35, oblong, obtuse, loosely strigose-villous on both surfaces; corolla yellowish-white, purple-maculate on keel; legume 1–1.3 cm. long, white-villous; plant of Gaspé Pen. . . . . . 2. *O. gaspensis.*

Leaflets 31–51, narrowly lanceolate or lance-linear, acutish, sparsely strigose or glabrescent; corolla reddish-violet; legume 1.3–1.5 cm. long, black-strigose; plant of L. Sup. reg. . . . . 3. *O. ixodes.*

*d.*Bracts and legumes not viscid; corolla rose-purple to violet (often drying bluish). . . *e.*

*e.*Larger leaves 1–3 dm. long, with leaflets 1–3 cm. long; scapes mostly 1.5–4 dm.high; spikes elongating, 3–12 cm. long; bracts 5–11 mm. long, strongly pubescent on back; calyx-teeth 1.5–3 mm. long.

Stipules villous-hirsute, becoming glabrate, their free blades 0.6–1.8 cm. long; legumes 1.5–2 cm. long; seeds 1.8–2 mm. broad. . . 4. *O. johannensis.*

Stipules permanently silky-villous below, their free glabrous blades 3–9 mm. long; legumes 1 cm. long; seeds 1–1.2 mm. broad. . . 5. *O. chartacea.*

*e.*Larger leaves 2–9 cm. long, with leaflets 2–8 mm. long; scapes 0.15–1.2 dm. high; spikes subcapitate, 1.5–3 cm. long; bracts 3–5 mm. long, glabrous or sparingly pubescent; calyx-teeth 0.5–1.5 mm. long. 6. *O. terrae-novae.*

*b.*Leaflets mostly in fascicles of 3 or 4 along the rachis; legumes only slightly surpassing calyces. . . . . 7. *O. splendens.*

*a.*Stipules nearly free from petioles; flowers and fruits reflexed, the latter becoming secund; corolla violet, at most 1 cm. long. . . . . 8. *O. foliolosa.*

1. O. Lambértii Pursh (for Aylmer Bourke Lambert, 1761–1842, from whose cultivated plants Pursh described it), Locoweed, Crazyweed. — Tufted, more or less *silky with fine appressed hairs;* leaves rather stiff, the 9–23 *subcoriaceous linear or linear-lanecolate* often curved *leaflets* 1–4 *cm. long;* scapes stoutish, up to 2.5 dm. high; spikes 3–15 cm. long, elongating sometimes to 2 dm.; bracts firm, lance-attenuate; calyx-tube silky, 6–7 mm. long, the deltoid-subulate teeth about 2 mm. long; corolla purple, drying bluish, about 2 cm. long; *legumes coriaceous* or firm, lance-cylindraceous, *erect,* long-beaked, 2–3 *cm. long,* silky or strigose, partly 2-locular by intrusion of the ventral suture. (*Spiesia* Ktze.; *Aragallus* Greene; *O. involuta* (Nels.) K. Schum.) — Dry prairies, calcareous gravels and bluffs, Minn. and Man. to Mont., s. to Mo., Okla. and Tex. May–July.

2. O. gaspénsis Fern. & Kelsey (of the Gaspé Pen., Que.). — Multicipital; leaves strongly ascending; *leaflets* 30–35, oblong, mostly obtuse, 0.8–2 cm. long, thin, loosely strigose-villous, *viscid;* scapes 0.5–2 dm. high, shorter than or barely exceeding leaves, white-pilose; *spikes dense,* in anthesis 2–5 cm. long, rarely lengthened in fruit; *bracts* lanceolate, herbaceous, 5–10 mm. long, *viscid and somewhat verrucose;* calyx campanulate, white-villous, the tube 4–5 mm. long, the deltoid viscid teeth 1–2 mm. long; *corollas yellowish-white,* 1–1.6 *cm. long, purple-maculate on the keel; legumes* short-ovoid, 1–1.3 cm. long, *viscid* and white-villous, thin-walled, nearly 2-locular; seeds 1.6–2 mm. long. — Calcareous cliffs and talus, nw. Gaspé Co., Que. June, July.

3. O. ixòdes Butt. & Abbe (sticky). — Branches of the multicipital caudex 2–10 mm. thick; stipules adnate, subglutinous, white with green veins, white-hirsute, the deltoid-ovate free blades with caudate appendages, or merely acuminate in forma **ecaudàta** Butt. & Abbe (without tails); *leaflets* 31–51, *narrowly lanceolate to lance-linear, acutish, sparsely strigose to glabrescent;* scapes about equaling leaves, sparsely strigose-villous, with darker viscid hairs above; spikes dense, 0.6–2 cm. long, lengthening to 5–6 cm.; bracts linear-lanceolate, exceeding calyx, with viscid warts among the hairs; *calyx hirsute with mixed white and black hairs,* the tube 5–6 mm. long, the lanceolate viscid teeth 3–4 mm. long; *corolla deep reddish-violet,* yellowish at base, 1.5 cm. long; *legume* 1.3–1.5 cm. long, *black-strigose* and glandular-viscid. — Slaty cliffs, Cook Co., Minn. and adj. Ont. Late June, early July.

4. O. johannénsis Fern. (of the St. John River, Me. and N.B.). — Multicipital; *stipules* whitish, membranaceous, villous-hirsute when young, later *glabrate* or merely bristly-ciliate at tip, the lanceolate or ovate *free blades* 0.6–1.8 *cm. long; leaves* 0.5–3 *dm. long,* the petiole and rachis appressed-pubescent, the 15–31 linear-lanceolate to oblong sericeous to glabrate *leaflets* 0.6–3 *cm. long; scapes up to* 3.3 *dm. high; spikes often elongating and loosening with age,* 2–11 *cm. long; bracts* lance-attenuate, 5–11 *mm. long, herbaceous, silky-villous on back, becoming glabrate; calyx*-tube 5–7 mm. long, *the lanceolate* pubescent *teeth* 1.5–3 *mm. long; corolla* 1.5–2 *cm. long, purple or purple-violet* (rarely albino), the *vexillum* 7–10 *mm. broad; legumes* strongly ascending, thin-walled, thick-cylindric or obliquely lance-ovoid, 1.5–2 *cm. long; seeds as high as broad,* 1.8–2 *mm. across.* (*O. campestris,* var. *johannensis* Fern.) — Calcareous rocks and gravels, w. Nfld. to James Bay, s. to St. Paul's I., N.S., St. John R., N.B. and Me., and Lévis Co., Que. June, July.

5. O. chartàcea Fassett (papery). — Similar to no. 4; *stipules silky-villous* with long hairs, *only their short* (3–9 *mm. long*) *free blades glabrous,* ciliate; leaflets silky-pilose; scapes silky-pilose or -villous; *bracts villous below; legumes* subascending, 1 *cm. long; seeds* 1–1.2 *mm. wide.* — Sandy lake-shores, n. Wisc. June, July.

6. O. térrae-nòvae Fern. (of Nfld.). — Dwarf, loosely matted; *stipules with glabrous* or glabrate *deltoid-acuminate free blades only* 2–7 *mm. long; leaves* 2–9 *cm. long*, the 13–25 sparsely sericeous to glabrate lanceolate to elliptic *leaflets* 2–8 *mm. long; scapes* divergent or ascending, 0.15–1.2 *dm. long; spike subcapitate*, few-flowered, 1.5–3 *cm. long; bracts* subchartaceous, *glabrous or nearly so*, 3–5 *mm. long; calyx* black-pilose or with a few white hairs, *its deltoid teeth* 0.5–1.5 *mm. long;* corolla purple or violet; the *vexillum* 1.2–1.5 cm. long, with the obcordate blade 5–8 *mm. broad;* legumes slender-ovoid, plump, black-pilose, excluding the short beak 1.2–1.7 mm. long; seeds slightly shorter than broad, 1.8–2.2 mm. broad. — Hudson Straits, s. on calcareous rocks and gravel to w. Nfld.; Mackenzie Distr. July, Aug.

7. O. spléndens Dougl. (splendid). — *Villous or loosely sericeous;* leaves strongly ascending; the very numerous lanceolate *leaflets mostly in fascicles of* 3 *or* 4 *along the rachis*, 1–2.5 cm. long; scapes 1–3 dm. high; spike compact or loose below or in age, 3–18 cm. long; *calyx villous, its slender teeth nearly equaling the tube;* corolla purple to bluish, 12–15 mm. long; legume villous, about 1 cm. long, but slightly exserted, 2-locular. (*Spiesia* Ktze.; *Aragallus* Greene; *O. Richardsonii* (Hook.) K. Schum.) — Prairies, meadows and draws, w. Minn. and Man. to Alta., s. to N.D. and N.M. June, July.

8. O. foliolòsa Hook. (with many leaflets). — Loosely tufted; *leaves* 2–10 cm. long, *with nearly free* lanceolate pilose *stipules;* leaflets 15–29, in approximate pairs, narrowly ovate to elliptic, 2–12 mm. long, pilose; scapes 2–15 cm. high, pilose; spikes compact, 1–3 cm. long, with the 2–10 *flowers soon divergent and later reflexed and secund;* calyx black-pilose; its tube 2.6–3.5 mm. long, truncate, about equaled by the lance-subulate lobes; *corolla deep violet*, with pale base, 8–10 *mm. long*, the vexillum 4.5–5 mm. broad; *legumes reflexed*, stipitate, subcylindric, 1–1.5 cm. long, 3–5 mm. broad, black-hirsute. — Hudson Straits to Alaska, s. on basic rock and gravel to n. and w. Nfld., Gaspé Pen., Que., and Colo. July, Aug.

35. GLYCYRRHÌZA L. Licorice

Calyx with the two upper lobes shorter or partly united. Anther-locules confluent at the apex, the alternate stamens smaller. Fruit ovate or oblong-linear, compressed, scarcely dehiscent, few-seeded. The flower, etc. otherwise as in *Astragalus.* — Long perennial root sweet (whence the name, from the Greek *glycys, sweet*, and *rhiza, root*); herbage glandular-viscid; leaves odd-pinnate, with minute stipules; flowers in axillary spikes, white or bluish. Genus of temp. N. and S.Am., s. Eurasia, Afr. and Austral.

1. G. lepidòta (Nutt.) Pursh (scaly), Wild L. — Tall (6–9 dm. high); leaflets 15–19, oblong-lanceolate, mucronate-pointed, sprinkled with little scales when young, and with corresponding dots when old; spikes peduncled, short; flowers whitish; fruits oblong, beset with hooked prickles, suggesting burs of *Xanthium.* — Prairies, meadows and rich shores, w. Ont. to Wash., s. to nw. Mo., Tex. and Mex.; adv. eastw. June–Aug.

Var. glutinòsa (Nutt.) S. Wats. (glutinous). — Stem beset with heavy glands and small setae; foliage glutinous. — Old fields, e. Va. (Introd. from the West)

Onóbrychis (Greek name for the plant) viciifòlia Scop. (vetch-leaved), *O. sativa* Lam., Sainfoin, a perennial with pinnate leaves of many oblong leaflets, long-peduncled spikes of pink flowers and strongly reticulate round fruits, is a sporadic adventive but perhaps not long persistent. (Introd. from Eu.)

36. AESCHYNÓMENE L. Sensitive-joint-Vetch

Calyx 2-lipped; the upper lip 2-, the lower 3-cleft. Standard roundish; keel navicular. Stamens diadelphous in two sets of 5 each. Legume flattened, composed of several easily separable articles. — Plants of trop. and warm reg.; leaves odd-pinnate, with several pairs of leaflets, sometimes sensitive to stimulation, as if shrinking from touch (whence the name, Greek for *ashamed*).

1. A. virgínica (L.) BSP. (Virginian). — Erect bristly annual up to 2 m. high; leaflets very numerous, those of the principal leaves 1–2.5 cm. long, linear-oblong; racemes few-flowered; flowers yellowish and reddish, 1.3–1.5 cm. long; legumes strongly torulose, 3.5–7 cm. long, with stipe 1–2.3 cm. long, the pustulate-hirsute segments with thick unwrinkled walls, the lowest segment 0.8–1.5 cm. long and 5–7 mm. broad; seeds 4.5–6 mm. long. — Fresh to brackish tidal shores, s. N.J. and e. Md. to s. Va. Aug.–Oct.

37. CORONÍLLA L. Crown-Vetch

Calyx 5-toothed. Standard orbicular; keel incurved. Stamens diadelphous, 9 and 1. Legume terete or 4-angled, jointed; the articles subcylindric. — Glabrous herbs or shrubs, nat. of w. Eurasia and n. Afr., with pinnate leaves, and the flowers in umbels terminating axillary peduncles. (Diminutive of *corona, a crown;* alluding to the inflorescence.)

1. C. vària L. (variable). — A perennial herb with ascending stems; leaves sessile; leaflets 15–25, oblong; flowers rose-color; legumes coriaceous, 3–7-jointed, the 4-angled articles 6–8 mm. long. — Roadsides, waste places and old house-sites, originally cult., now abundantly but locally estab., N.E. to S.D., s. to Va., W.Va., Ky. and Mo. June–Aug. (Introd. and natzd. from Eu.)

38. HEDÝSARUM L.

Calyx 5-cleft, the lobes narrow and nearly equal. Keel nearly straight, obliquely truncate, not appendaged, longer than the wings. Stamens diadelphous, 9 and 1. Legume flattened, composed of several equal-sided separable roundish articles connected in the middle. — Perennial herbs (rarely shrubs) of N. Hemisph.; leaves odd-pinnate. (Greek name for some plant, *hedysaron*, taken up by Linnaeus for this and allied genera.)

Calyx-teeth deltoid or lanceolate, not subulate-tipped; articles of fruit glabrous, with broad irregularly quadrate areolae. . . . . 1. *H. alpinum.*
Calyx-teeth lance- or linear-subulate; articles of fruit minutely pubescent, with elongate transverse areolae. . . . . 2. *H. Mackenzii.*

1. H. alpìnum L. (alpine), var. **grandiflòrum** Rollins (large-flowered). — Stems few–many, from brown caudices and a large root, erect or arching, 1–6 dm. high; leaves numerous, with 13–23 oblong glabrous or minutely pubescent leaflets; stipules scaly, opposite the petioles, united at base; raceme 1-sided, 3–15 cm. long, of many deep purple to pink deflexed *flowers* 1.6–2 *cm. long; calyx*-tube finely pubescent to glabrous, the *teeth deltoid to lanceolate; keel* much longer than the standard, *its halves* 4–5.5 *mm. broad; articles* 2–4, *in fruit* 4.5–10 mm. long, *glabrous*, thin-margined, *the surfaces with broad irregularly quadrate areolae.* — Calcareous rocks and gravels, s. Lab. and n. and w. Nfld.; Alta., B.C., Yuk. and Alaska. July, Aug. — Typical *H. alpinum* European.

Var. **americànum** Michx. (American). — Often taller, up to 1 m. high; leaflets oblong to lanceolate; *flowers* 1–1.5 *cm. long*, pink or magenta, or rarely white in forma **albiflòrum** Fern. (white-flowered); *half-keel* 3–4 *mm. broad.* (*H. boreale* of ed. 7, not Nutt.; *H. americanum* (Michx.) Britt.) — Similar habitats, Nfld. to Alaska, s. to Gaspé Pen., Que., and n. N.B., St. John R. syst., N.B. and Me., mts. of n. Vt., n. shore of L. Sup., Man. and B.C. June–Aug.

2. H. Mackénzii Richards. (for its discoverer, Sir Alexander Mackenzie, 1822–1892). — Stems few, 1–4 dm. high; *leaflets* 7–15; racemes compact, 2–8 cm. long; *calyx* somewhat canescent, *with linear- to lance-subulate teeth* equaling or longer than tube; corolla 1.5–2.3 cm. long, deep purple, the half-keel 5–7 mm. broad; *articles canescent-strigulose, with prominent elongate transverse areolae.* — Calcareous gravels and slopes, local, w. Nfld.; Anticosti I.; Arct. nw. Am., s. to Man., Alta. and B.C. July. (Siberia)

39. DESMÒDIUM Desv. Tick-trefoil. Tick-clover. Beggar's-ticks

Calyx bilabiate, upper lip bifid, the lower (usually deeply) tridentate. Standard orbicular to ovate; short-unguiculate wings slightly adherent (by a small flap, not usually recognizable except in relatively fresh young material) to the falcate, mostly apically truncate, long-unguiculate keel-petals. Stamens monadelphous (in only 3 of our species) or diadelphous, 9 and 1, at least above. Loment stipitate, (1–)2–multiarticulate, its lower suture usually more deeply indented than the upper, generally separating into 1-seeded, mostly uncinulate-pubescent indehiscent articles. — Perennial herbs in our region, with stipulate, pinnately 3 (rarely 1 or 5)-foliolate leaves and stipellate leaflets. Flowers in axillary or terminal, often paniculately compound, racemes. — Herbaceous to almost arborescent plants of cool-temp. to trop. reg. throughout, excepting w. U.S. (Pacific slope), Eu. and N.Z. (Name from the Greek *desmos, a bond* or *chain*, from the connected articles of the loment.)* Meibomia Adans.

a. Stamens monadelphous; pedicel mostly equaled or exceeded by the loment-stipe or, if not exceeded by it (as in no. 3), the stipe at least more than three times as long as the calyx; isthmi between the articles narrow (about 1 mm.) and superior, *i.e.*, the lower suture deeply invaginated, the upper straight or nearly so; primary bracts lance-linear or linear (rarely narrowly ovate). Ser. Americana. . . *b.*

b. Inflorescence not axillary; flowers rose to purple (rarely white).

---

* Treatment contributed by Bernice G. Schubert. In measuring the loment-articles the length of each article has been taken parallel to the long axis of the loment, the width of each article perpendicular to the long axis of the loment. The figures of the loments are, in most cases, approximately natural size.

Flowering stem leafless, arising from the base; stipules deciduous; pedicels slender; loment-stipe (5–)10–22 mm. long. . . . . 1. *D. nudiflorum.*

Flowering stem arising from a whorl of leaves; stipules persistent; pedicel stout; loment-stipe shorter (4–10 mm.). . . . . 2. *D. glutinosum.*

*b.*Inflorescences axillary and terminal; flowers white; pedicels slender; loment-stipe 5–9 mm. long. . . . . 3. *D. pauciflorum.*

*a.*Stamens diadelphous (9 and 1) at least above; pedicel usually exceeding the loment-stipe in length; the stipe rarely more than twice as long as the calyx; isthmi between the articles broader and more nearly central; the sutures subequally to equally angled or curved; primary bracts ovate or ovate-acuminate. . . *c.*

*c.*Stipules conspicuous, ovate-attenuate and cordate to semicordate at base. Ser. STIPULATA. . . *d.*

*d.*Plants trailing; stems rather slender and flexuous; some long tapering (not hooked) trichomes present at least on stem and petioles; leaflets orbicular to ovate.

Terminal leaflets chiefly orbicular; flowers purple (drying bluish); pedicels 6–13 mm. long; loment-articles more or less evenly pubescent throughout, 5–7.5 mm. long, 4–5 mm. broad, the margins neither folded back nor revolute. . . . . 4. *D. rotundifolium.*

Terminal leaflets chiefly ovate, obtuse to acute, cordate at base; flowers creamy to white; pedicels 10–22 mm. long; loment-articles densely pubescent on the sutures, sparsely, if at all, on the surfaces, 7–10 mm. long, 5–8 mm. broad, with margins often alternately involute and revolute. . . . . 5. *D. ochroleucum.*

*d.*Plants upright; stems stout; leaflets ovate, ovate-oblong or -lanceolate.

Inflorescences axillary and terminal, usually much branched; pubescence of rachis of inflorescence chiefly of long soft and spreading trichomes; stipules regularly reflexed at maturity; pedicels 8–14 mm. long; flowers pinkish, becoming green; loment-articles usually semirhombic. . 6. *D. canescens.*

Inflorescence terminal, rarely branched; the pubescence of its rachis with hooked and multicellular hairs predominating; stipules not regularly reflexed at maturity; pedicels 12–23 mm. long; flowers white; loment-articles oval to orbicular in outline. . . . . . . 7. *D. illinoense.*

*c.*Stipules linear to lance- or ovate-attenuate. . . *e.*

*e.*Loments 1–3(–4)-articulate, at least the lower suture curved; plants with small bracts (not more than 3 mm. long) and small (to 6 mm. long) flowers. Ser. PAUCIARTICULATA. . . *f.*

*f.*Plants erect, with linear, linear-oblong or -lanceolate leaflets.

Leaflets linear-oblong, 0.7–1.4 cm. broad; leaves sessile to very short-petiolate; petioles when present not more than 3 mm. long; pedicels stout, 1.5–4 mm. long. . . . . 8. *D. sessilifolium.*

Leaflets linear to linear-lanceolate, 0.3–0.7 cm. broad; leaves petiolate; petioles 0.4–1.8 cm. long; pedicels to 11 mm. long.

Upper suture of loment-articles convex (*i.e.,* curved); pedicels 4–8 mm. long; plants of savannas and swaley grounds. . . . . . 9. *D. tenuifolium.*

Upper suture of loment-articles slightly concave or invaginated; pedicels 6–11 mm. long; plants of dry sandy pinelands. . . . 10. *D. strictum.*

*f.*Plants prostrate to ascending or erect, with ovate, ovate-oblong or almost orbicular leaflets. . . *g.*

*g.*Leaflets elliptic- to lance-ovate, large (the terminal leaflet 5–6.7 cm. long), thick; stipules and bracts very early deciduous; inflorescence lax; pedicels 6–17 mm. long. . . . . 11. *D. rigidum.*

*g.*Leaflets chiefly ovate, rounded-ovate, oblong or orbicular, much smaller (terminal leaflet 0.9–3.1 cm. long); stipules mostly persistent.

Plants erect; leaflets rather thin; inflorescence stout.

Leaves nearly sessile, petioles up to 1.5 cm. long; stems and petioles mostly pilose; pedicels 4–9 mm. long. . . . . 12. *D. ciliare.*

Leaves mostly petiolate, petioles 1.2–2.7 cm. long; plants essentially glabrous; pedicels (6–)8–19 mm. long. . . . . . 13. *D. marilandicum.*

Plants prostrate, trailing, glabrous to puberulent throughout; leaflets thickish; inflorescence lax. . . . . 14. *D. lineatum.*

*e.*Loments multiarticulate (*i.e.*, usually with 4 or more large articles). . *h.*

*h.*Plants with long stipules, large showy primary bracts (to 12 mm. long), large flowers (6.5–13.5 mm. long) and short-stipitate to sessile loments. Ser. LONGIBRACTEATA.

Leaves short-petiolate (petioles 0.8–2.5 cm. long); petiolules up to 3 mm. long; leaflets thickish, oblong to ovate-lanceolate, obtuse or acutish and with conspicuous venation; stipules 4.5–9.5 mm. long; stipels 2–4 mm. long; both sutures of loment-articles curved. . . 15. *D. canadense.*

Leaves longer-petiolate (petioles 4–9.7 cm. long); petiolules 3–5 mm. long; leaflets thin, ovate-acuminate, glaucous beneath; stipules 11–20 mm. long; stipels 5–9 mm. long; both sutures of loment-articles regularly angled. . . . . 16. *D. cuspidatum.*

*h.*Plants with smaller primary bracts and flowers and longer-stipitate loments. Ser. STIPITATA. . . *i.*

*i.*Pedicels mostly short (3–12 mm. long) and stout; leaflets densely tomentose to glabrescent (except on veins) beneath, but not glaucous. . . *j.*

*j.*Plants with coriaceous leaflets, glabrous or densely pubescent or, if not coriaceous, at least with conspicuously reticulate venation. . . *k.*

*k.*Leaflets moderately to densely tomentose beneath, soft-velvety to touch; upper surface moderately soft-pilose, not prominently reticulate.

Terminal leaflet usually rhombic to deltoid, acute to cuneate or truncate at base, width generally at least two-thirds the length; leaflets thick; articles chiefly rhombic or with the upper suture somewhat angled. . . . . . . 17. *D. viridiflorum.*

Terminal leaflet (except of the uppermost leaves) elliptic-ovate, mostly rounded at base, width about one-half the length; leaflets thinner; articles rounded above (*i.e.*, the upper suture curved rather than angled. . . . . . 18. *D. Nuttallii.*

*k.*Leaflets glabrescent to moderately pilose beneath and strongly reticulate.

Leaflets uncinulate-pubescent along midrib and veins, otherwise essentially glabrous beneath; terminal leaflet obtuse to acute at apex; stipules early deciduous. . . . . . 19. *D. Fernaldii.*

Leaflets mostly appressed-pilose on both surfaces (with uncinulate pubescence only above); stipules usually persistent through flowering; terminal leaflet obtuse and emarginate. 20. *D. glabellum.*

*j.*Plants with thinner leaflets, less conspicuously reticulate and rather inconspicuously to densely pilose. . . *l.*

*l.*Leaflets thin, with appressed pubescence sparse to moderate; terminal leaflet mostly lanceolate and twice to many times longer than broad; stems sparsely if at all pilose, usually only uncinulate-puberulent. . . . . 21. *D. paniculatum.*

*l.*Leaflets mostly thicker and abundantly to densely appressed-pilose on both surfaces; terminal leaflet usually elliptic or rhombic to ovate and about twice as long (sometimes longer) as broad; stems, petioles and usually rachis of inflorescence moderately to very densely spreading-pilose as well as uncinulate-puberulent.

Plants erect to spreading; stipules linear- to lance-attenuate. 22. *D. perplexum.*

Plants prostrate and trailing, with stiffish (not flexuous) stems; stipules ovate- to lance-attenuate. . . . . 23. *D. humifusum.*

*i.*Pedicels longer (10–19 mm.), slender and lax; leaflets essentially glabrous, glaucous beneath. . . . . 24. *D. laevigatum.*

Ser. AMERICÀNA Schub.

1. D. nudiflòrum (L.) DC. (with naked flowering scape). — Ascending to erect, *the slender simple sterile stem* (or several, from the base) *up to approximately 2 dm. high, with the leaves regularly whorled at its apex, slightly to three times exceeded by the flowering scape; this* also arising from the base and *ordinarily simple and leafless*, or in forma **foliolàtum** (Farwell) Fassett (leafy) with leaves scattered, or in forma **personàtum** Fassett (masked) with leaves subverticillate or the flowering stem branched and leafy (resembling no. 3); *stipules deciduous;* leaflets finely puberulent and remotely pilose above, glaucous and moderately pilose beneath; the *terminal leaflets* mostly *rhombic to obovate,* acute to short-acuminate at apex, 4.5–12 cm. long, 3.5–8 cm. wide; the lateral leaflets oblique, with the broader side half-ovate, 3.8–10 cm. long, 3–5.5 cm. wide; the *rose to purple flowers,* or white in forma **Dúdleyi** House (for William Russell Dudley, 1849–1911), *borne on slender puberulent pedicels* 10–23 *mm. long, these mostly equaled or exceeded by the puberulent loment-stipe* (10–22 *mm. long);* loments 1–4-articulate; *sinuses between the articles broadly inverted-U-shaped;* articles with surfaces uncinulate-puberulent and sutures glabrous, 7–12 mm. long, 4–5 mm. wide. (*Meibomia* Ktze.) — In rich or dry woods, sw. Me. to sw. Que., w. to e. Minn., s. to n. Fla., s. Miss., centr. La. and e. Tex. *Fl.* July, Aug. Fig. 1223.

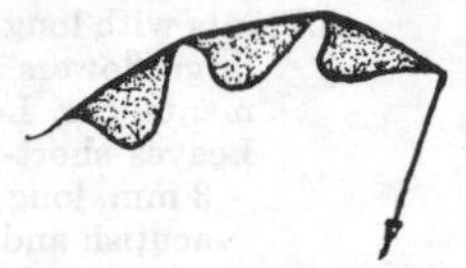

1223. D. nudiflorum.

2. D. glutinòsum (Muhl.) Wood (glutinous; for the uncinulate-puberulent rachis of the inflorescence). — Erect, *simple-stemmed, to* 11 *or more dm. high, from a slender branching root to* 3.5 *dm. long;* the *leaves trifoliolate and whorled at the base of the terminal inflorescence,* or scattered along the stem in forma **Chandonnétii** (Lunell) Schub. (for its discoverer, Zephyrin Leonard Chandonnet, 1848–1916), or unifoliolate and scattered in forma **unifoliolàtum** Schub. (unifoliolate); *terminal leaflet ovate or nearly orbicular, with an abruptly acuminate apex,* 7–13 cm. long, 6–12.3 cm. wide; the lateral leaflets oblique and more elongate, 8–11.8 cm. long, 4–7 cm. wide; *the lance- or linear-attenuate persistent stipules* 1–10 *mm. long;* the purple (drying bluish) or very rarely white flowers on *stout short pedicels* 3–8 *mm. long, these equaled or slightly exceeded by the loment-stipe;* loment 1–4-articulate, its *sinuses* between the articles *very slenderly inverted-U-shaped;* articles with sutures and stipe essentially glabrous at maturity and the surfaces of the articles uncinulate-puberulent, 8–11 mm. long, 4–7 mm. wide. (*Meibomia* Ktze.; *D. grandiflorum* of ed. 7, not (Walt.) DC.; *D. acuminatum* (Michx.) DC.) — Dry or rocky woods, centr. Me. to sw. Que., w. to n. Minn. and se. Neb., s. to Fla., e. Tenn., centr. La. and ne. Tex.; Mex. *Fl.* July, Aug. Fig. 1224.

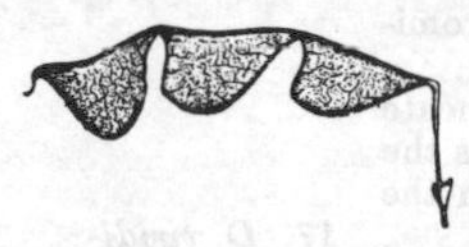

1224. D. glutinosum.

3. D. paucifiòrum (Nutt.) DC. (few-flowered). — *Simple or much branched,* 1.5–5.5 *dm. high;* stem very slender, arising from a filiform to thick usually branched and nodulose root-system, or roots with more elongate stouter thickenings at intervals and ends, up to 2.5 dm. long; stem and branches moderately uncinulate-puberulent and sparsely pilose; the *early deciduous stipules* slenderly lanceolate to oblong and filiform to acute at apex, puberulent and ciliate, 1.5–5 mm. long; petioles with pubescence like that of stem, 5–7.5 cm. long; the leaf-rachises similar but shorter, 0.6–1.3 cm. long; leaflets appressed-pilose above and beneath with long white trichomes, paler on the lower surface, ciliate; the *terminal leaflets* obovate to *rhombic, acute or short-acuminate at apex and acutish at base,* 5–9 cm. long, 4–6.5 cm. wide; the oblique lateral leaflets with the narrower half semielliptic, the wider semirhombic; *inflorescence of usually unbranched short mostly axillary racemes; the lanceolate primary bracts rather long-persistent,* 2–4 *mm. long; corolla white, to* 6.5 *mm. long;* the persistent stamen-tube 3–4 mm. long at fruiting period; the stipitate 1–3-articulate loment borne on a glabrous to *uncinulate-puberulent stipe* 5–9 *mm. long; articles more or less inverted-triangular in outline,* uncinulate-puberulent on surfaces and occasionally on sutures, with straight trichomes frequent at isthmi, 9–14 mm. long, 6–8 mm. wide; *sinuses between the articles inverted-V-shaped.* (*Meibomia* Ktze.) — Rich woods and wooded banks, w. N.Y. to ne. Ia., s. to n. Fla., nw. Ala., s. La. and e. Tex. *Fl.* June–Aug.

Ser. Stipulàta Schub.

4. D. rotundifòlium DC. (round-leaved). — *Stem prostrate,* trailing, slender and *densely soft- and white-pilose,* or only sparsely uncinulate-puberulent in forma **glabràtum** (Gray) Schub. (glabrate), to more than 1 m. long, from a branching root up to 5 mm. thick and 3 or more dm. long; *leaflets chiefly orbicular, the terminal with cuneate base,* up to about 5 cm. long and wide, *appressed- to spreading-pilose on both surfaces; stipules ovate-acuminate,* obliquely cordate, ciliate and somewhat pilose on outer surface, reflexed at maturity and 5–12 *mm. long; flowers purple, on stout pedicels* 6–13 *mm. long; loments* (3–)4–6-*articulate with subrhombic to elliptic uncinulate-puberulent articles* 5–7.5 *mm. long and* 4–5 *mm. wide.* (*Meibomia* Ktze.; *M. Michauxii* Vail) — Dry woods, Fla. to Tex., n. to Mass., s. Vt., N.Y., s. Ont. and Mich. *Fl.* July–Sept.

5. D. ochroleùcum M. A. Curtis (yellowish-white; referring to the flowers). — The somewhat spreading-pilose and uncinulate-puberulent *terete procumbent stem* from a frequently branching root; leaves occasionally unifoliolate; *leaflets thick, with prominent reticulate venation and fine sparse uncinulate puberulence above and below* (straight white trichomes only along the midrib beneath); *the terminal ones cordate- to rhombic-ovate, obtuse or less often acute, 3.7–7.5 cm. long and 2.2–5.3 cm. wide;* the lateral ones obliquely ovate, mostly obtuse and about one-fourth shorter; *stipules* similar to those of no. 4 but *merely puberulent on outer surface; flowers creamy or white, fading yellow, on pedicels 1.2–2 cm. long; loments 2–4-articulate with essentially glabrous strongly reticulate articles; these densely uncinulate-puberulent only on the sutures, suborbicular to subrhombic in outline and seemingly contorted by irregular folding of margins near or at the isthmi, 7–10 mm. long and 5–8 mm. wide.* (*Meibomia* Ktze.) — Sandy to loamy woods, Ga. and Tenn. to Del. and Md. Fig. 1225.

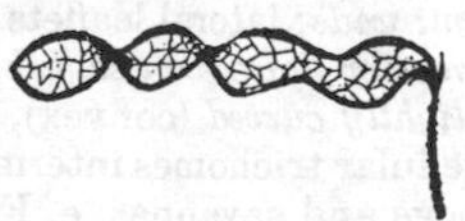

1225. D. ochroleucum.

6. D. canéscens (L.) DC. (canescent, *i.e.*, gray or hoary). — *Arising from a long* (1–4 *dm.*) *brown tap-root; the stem much branched and variously pubescent but predominantly spreading-pilose with characteristic tapering trichomes longer than its diameter; terminal leaflet ovate, with truncate to rhombic or rounded base and acute to gradually acuminate apex, 5–13 cm. long and 3–10 cm. wide;* lateral leaflets similar but somewhat narrower in proportion to their length; *stipules* deltoid- to ovate-acuminate or -attenuate, oblique and *with one side almost lobed at base, usually reflexed at maturity,* 5–13 mm. long, the long pilosity of the outer surface not usually persisting; *inflorescence axillary and terminal, lax and diffuse,* with pubescence of the rachis similar to that of stem; the *flowers pinkish, soon green,* or white in forma **albinum** Fern. (albino), *on pedicels 8–14 mm. long;* loments 1–6-articulate; articles 6–13 mm. long and 4–7 mm. wide, with uncinulate pubescence intermixed with multicellular hairs, these particularly dense on isthmi and sutures; the lower suture acutish and the isthmi narrow, or obtuse and the isthmi broad. (*Meibomia* Ktze.; incl. *D. canescens,* var. *hirsutum* (Hook.) Robins. of ed. 7 (*Meib. canescens,* var. *hirsuta* Vail); var. *villosissimum* T. & G.) — Dry sandy woods and fields, sw. Mass. to sw. Ont., w. to sw. Wisc., Iowa and Neb., s. to Fla., sw. Ala., ne. Miss., La. and e. Tex. Fig. 1226.

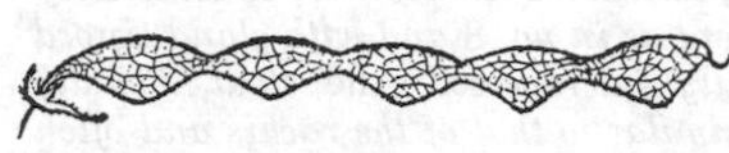

1226. D. canescens.

7. D. illinoénse Gray (of Illinois). — *Roots reddish; stem slender, little branched,* upright, with various pubescence as in no. 6 but the *pilosity usually sparse or lacking;* leaflets not reaching as great size as in no. 6 but similar; *stipules obliquely ovate-acuminate, with mostly truncate base, persistent but not characteristically reflexed,* 10–15 mm. long; *inflorescence somewhat virgate, not usually branched,* with *pubescence* of rachis similar to that of stem but *with multicellular (often glandular) trichomes predominating;* the *white flowers on pedicels 12–23 mm. long; loments 2–7-articulate, with mostly suborbicular articles 4–7 mm. long* and *3.5–4.5 mm. wide,* densely uncinulate-pubescent with straight multicellular trichomes intermixed, and *with broad isthmi.* (*Meibomia* Ktze.; *D. Tweedyi* Britt.) — Dry soils, n. Ohio and s. Mich. to Ia. and se. Neb., s. to Mo., Okla. and Tex. Fig. 1227.

1227. D. illinoense.

Ser. Pauciarticulàta Schub.

8. D. sessilifòlium (Torr.) T. & G. (with sessile leaves). — Erect, to 1.2 m. high, from a moderately branched root-system; stem terete to subangulate, finely uncinulate-puberulent and moderately uncinate-pubescent with stout hooked hairs, lacking straight tapering trichomes except when immature or just below nodes; *leaves sessile or with a petiole 2–3 mm. long; leaf-rachis 3–8 mm. long; terminal leaflet elliptic-lanceolate to very narrowly ovate, narrowed to the obtusish apiculate apex, rounded at base,* ciliate, rather prominently reticulate, somewhat pilose above and moderately uncinulate-pubescent, densely pilose and much paler beneath, with uncinulate pubescence on midrib, 3.7–7.3 *cm. long and* 0.7–1.4 *cm. wide; lateral leaflets* similar, smaller, 2.9–6.4 *cm. long,* 0.6–1.2 *cm. wide;* stipules lance-attenuate, 4–9.5 mm. long; inflorescence racemose-paniculate, its rachises uncinulate-puberulent and -pubescent and with some sparsely scattered multicellular trichomes; *stout pedicels* 1.5–4.5 *mm. long,* their pubescence similar to that of the rachis; the pinkish to lavender corolla up to 6 mm. long; loment 1–4-articulate, stipitate; *stipe to 3 mm. long; articles broadly curved above and rounded beneath,* densely uncinulate- to uncinate-pubescent throughout, 4–6 *mm. long,* 3–4.5 *mm. wide.* (*Meibomia* Ktze.) — Dry sandy soils, rather local, se. Mass. to Mich., Ill., s. Mo. and e. Kans., s. to se. Pa., w. Va., e. S.C., sw. La. and e. Tex.

9. D. tenuifòlium T. & G. (slender-leaved). — Simple-stemmed or branched from the base, up to 1 m. high; root rather slender and with some irregular thickenings; stem slender, terete to subangulate, with pubescence similar to that of no. 8; leaves short-petiolate; *petioles* mostly glabrous, 0.4–1.7 *cm. long;* the leaf-rachis somewhat stiffly ascending-pilose, 0.4–0.8 cm. long; *terminal leaflet linear, obtuse* and mucronulate, dark and glabrous to puberulent above, paler, uncinulate-puberulent and appressed ascending-pilose beneath, 4.1–6.2 *cm. long,* 0.3–0.6 *cm. wide;* lateral leaflets similar and only slightly smaller; inflorescence as in no. 8, with *slightly smaller flowers on stiffish pedicels 4–8 mm. long; loment* 1–3-*articulate; articles with upper sutures slightly curved* (convex), *the lower more so and deeply indented,* uncinulate-pubescent with multicellular trichomes intermixed, 3.5–4.5 *mm. long* and 3 *mm. wide.* (*Meibomia* Ktze.) — Sphagnous bogs and savannas, e. Fla. and sw. Ala., n., locally, to se. Va.

10. D. stríctum (Pursh) DC. (very straight). — Mostly erect, often simple-stemmed, little branched below the inflorescence; root long, slender (less than 5 mm. thick), somewhat branched; stem terete to angulate, pubescent as the two preceding species, becoming glabrescent below; in general aspect *similar to no.* 9 *but with slightly shorter leaflets and slender stiff pedicels* 6–11 *mm. long; the upper suture of the loment-articles slightly concave; the articles* pubescent as in no. 9, 4–6 *mm. long and* 3–4 *mm. wide.* (*Meibomia* Ktze.) — Sandy soil of pinelands of Coastal Plain, Fla. to La., n. to N.J.

11. D. rígidum (Ell.) DC. (stiff). — Erect, slender, often with several stems arising from the stoutish branched root; pubescence of stem as in the preceding species, with long tapering trichomes only near the nodes; leaves short-petiolate, *petioles* 1–2 *cm. long;* leaf-rachis 0.6–1.1 cm. long; *stipules* narrowly ovate to lanceolate, attenuate, *very early deciduous,* 3–6 mm. long; *terminal leaflet slenderly to broadly ovate, rounded to truncate at base, mostly obtuse at apex, prominently reticulate, thick,* abundantly pilose beneath, less so above and with moderate uncinulate pubescence, 5–6.7 *cm. long,* 2–3.3 *cm. wide; lateral leaflets similar,* considerably smaller; *inflorescence often very lax, pubescent as in no.* 8 *and with gland-tipped trichomes of various types often abundantly interspersed;* the *stout, usually outwardly curved pedicels with pubescence similar to that of the rachis and often with gland-based trichomes,* 6–17 *mm. long; the pale rose to almost white corolla drying yellow; articles with the lower suture more deeply curved than the often darker upper one, uncinate-puberulent throughout and usually with slender multicellular trichomes intermixed,* 4–4.5 *mm. long,* 2.5–4 *mm. wide.* (*Meibomia* Ktze.) — Dry sandy woods, se. Mass. to s. Mich. and Ind., s. to se. S.C., Fla., La. and e. Tex. Fig. 1228.

1228. D. rigidum.

12. D. ciliàre (Muhl.) DC. (ciliate). — Ascending to erect slender-stemmed plant often branched from base; root thick, usually branched, up to about 2 dm. long; the subangulate strongly lineate *stem usually spreading-pilose with long slender tapering trichomes* intermixed with shorter hooked ones; *leaves short-petiolate; petioles* sulcate, *long-pilose chiefly on margins of the groove,* 0.1–1.5 *cm. long;* leaf-rachis similar to petiole, 0.2–0.9 cm. long; *the lance-attenuate stipules persistent,* 2.5–5 mm. long; *terminal leaflet elliptic to narrowly ovate to rhombic, obtuse, abundantly pilose above, less abundantly so beneath, about* 1–3 *cm. long and* 0.5–1.7 *cm. wide;* lateral leaflets similar; uncinulate-puberulent rachis of inflorescence usually also spreading-pilose, at least when young; pedicels 4–9 mm. long; loments not deeply indented on upper suture but more or less curved to form an arc; articles 4–5.5 mm. long and 2.7–4 mm. wide. (*Meibomia* Blake; *D. obtusum* of ed. 7) — Dry sandy woods and clearings, Fla. to Tex. and Mex., n. to s. N.E., N.Y., O., Mich. and s. Mo. (W.I.)

Var. **lancifòlium** Fern. & Schub. (lance-leaved). — With lanceolate to lance-ovate leaflets. — Pine-barrens, se. Va.

13. D. marilándicum (L.) DC. (of Maryland). — A meter or more high, from a subligneous simple or branched slender root; *stem glabrous or sparsely to moderately uncinulate-puberulent; petioles with pubescence similar to that of stem,* 1.2–2.7 *cm. long; leaf-rachis similar or with sparse pilosity,* 0.5–1 *cm. long;* the ovate, ovate-rhombic or suborbicular terminal *leaflets* obtuse, *dark and essentially glabrous above,* paler and somewhat pilose beneath, 1.9–2.4 *cm. long,* 1–1.7 *cm. wide; lateral leaflets elliptic to deltoid-ovate,* 1.5–2.5 *cm. long and* 1–1.6 *cm. wide;* rachis of inflorescence moderately uncinulate-puberulent; the pedicels (6–)8–19 mm. long; loments similar to those of no. 12; the articles 4–5.5 mm. long and 3–4 mm. wide. (*Meibomia* Ktze.) — Dry open woods, s. N.E. to sw. Ont., Mich., Ill. and Mo., s. to S.C., Tenn., Ark., Okla. and e. Tex.

14. D. lineàtum DC. (marked with lines; *i.e.* on the stem). — Uncinulate-puberulent terete and lineate *stem trailing,* sometimes branched from base (often forming carpets), to 7 dm. or more in length; root reddish-brown, smooth, essentially straight, to 1 cm. thick and 3 dm. long; *stipules* lance-attenuate or obliquely ovate-attenuate, *moderately persistent,* 2–5 mm. long; leaflets ciliate, somewhat uncinulate-puberulent and pilose above and beneath; *terminal leaflet ovate to rhombic, obovate or orbicular,* 0.9–2.6 *cm. long,* 1.3–2.9 *cm. wide;* lateral leaflets similar,

1229. D. lineatum.

about 0.8–2.2 cm. long and wide; *pedicels 8–16 mm. long;* loments 1–4-articulate, straight to slightly curved; articles deeply indented on the lower suture, less so on the upper, 3.5–5 mm. long, 2.5–3.5 mm. wide. (*Meibomia* Ktze.; *M. arenicola* Vail; *D. aren.* (Vail) F. J. Herm.) — Sandy pinelands, Fla. to Tex., n. on Coastal Plain to se. Md. Fig. 1229.

Ser. Longibracteàta Schub.

15. D. canadénse (L.) DC. (of Canada). — Erect, one to several stems to 13 dm. high, arising from the simple or branched slender (to 0.5 cm. thick) brown root (about 1.5–2 dm. long); *the terete stem smooth below, becoming lineate, uncinulate-puberulent and spreading-pilose above; the linear-lanceolate to lance-attenuate and ciliate stipules often pilose on the outer surface* and minutely puberulent, 4.5–9.5 *mm. long; stipels lanceolate to linear-lanceolate, puberulent and ciliate,* 2–4 *mm. long; petioles* lineate, shallowly sulcate, *more or less pilose throughout,* 0.8–2.5 *cm. long; leaf-rachis* similar, 0.95–1.5 *cm. long; petiolules* 1.5–3 *mm. long; leaflets* thick, ciliate, *somewhat pilose above, abundantly so* and paler *beneath;* the *terminal* one elliptic to rhombic or almost orbicular at base of plant, becoming ovate to lance-ovate above, short-acute to obtuse at apex, 4.8–10.5 *cm. long,* 2–3.8 *cm. wide* (about half-way up the stem); *lateral leaflets similar, smaller,* 4.5–8.2 *cm. long,* 1.9–2.8 *cm. wide; rachis of inflorescence with* dense uncinulate puberulence and some scattered multicellular trichomes as well as *abundant pilosity of long tapering trichomes; primary bracts very conspicuous before flowering, lance- to ovate-attenuate, ciliate, pilose and uncinulate-puberulent on outer surface,* 4–10 *mm. long,* soon deciduous; the densely uncinulate-puberulent *pedicels* often bearing glutinous-based or -tipped multicellular trichomes, 5–6 (–9.5) *mm. long;* flowers rose-purple, changing to blue, the *corolla* 8.5–13.5 *mm. long;* the 1–5-articulate loments short-stipitate; *sutures of loment-articles curved to straight above, curved to obtusely angled below; articles densely uncinulate-pubescent throughout,* 5–7 *mm. long,* 4–5 *mm. wide.* (*Meibomia* Ktze.) — Open woods, N.S. to s. Sask., s. to N.E., Md., w. Va., W.Va., O., Ind., Ill., Mo. and Okla. *Fl.* July, Aug. Fig. 1230.

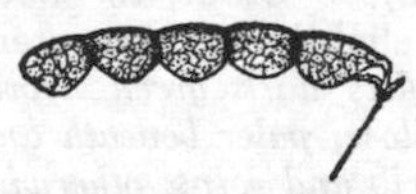
1230. D. canadense.

16. D. cuspidàtum (Muhl.) Loud. (made pointed, *i.e.*, the leaflets). — Erect to 1 m. or more high; root to 1.5 dm. or longer; *stem glabrous or occasionally with minute puberulence; the slenderly and obliquely ovate-attenuate stipules* 1.1–2 *cm. long; stipels linear-attenuate or subulate, glabrous to puberulent,* 5–9 *mm. long; petioles essentially glabrous,* with a broad shallow groove 4–9.7 *cm. long; leaf-rachis* similar, 1.8–3.5 *cm. long; petiolules dark,* rugose, with a few scattered hooked or tapering trichomes, 3–5 *mm. long; terminal leaflet ovate-acuminate,* more or less rounded to cuneate at base, *glabrous except for a few trichomes along midrib and veins above and beneath, lower surface* paler and *glaucous,* 8–14 *cm. long,* 4.2–6.7 *cm. wide;* rachis of inflorescence strongly ribbed and moderately uncinulate-puberulent; *primary bracts* ovate-attenuate, ribbed, *essentially glabrous on outer surface, inconspicuously ciliate,* not long persistent, 6–12 *mm. long; pedicels* puberulent to finely pilosulous with multicellular often glutinous-tipped hairs and occasionally some hooked trichomes interspersed, *often almost glabrate at maturity,* 3.5–7 *mm. long;* the roseate flowers 7–8 mm. long; *loments* 1–7*-articulate, the sutures often more or less evenly indented above and below,* the *isthmi* broad, *more densely uncinulate-puberulent than the reticulate surfaces of the articles,* which are 7–10 *mm. long and* 4–5 *mm. wide.* (*Meibomia* Ktze.; *D. bracteosum* (Michx.) DC.; *D. grandiflorum* (Walt.) DC., not of ed. 7 and auth. generally) — Rich woods and wooded banks, s. N.H. and Vt. to Mich., s. to Fla., Ark. and e. Tex. Fig. 1231.

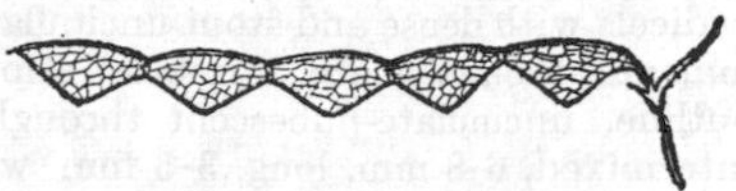
1231. D. cuspidatum.

Var. **longifòlium** (T. & G.) Schub. (long-leaved). — Differing in having *stem, petioles and leaf-rachises moderately spreading-pilose as well as densely uncinulate-puberulent; the leaflets somewhat ciliate and moderately to abundantly pilose on both surfaces; rachis of the inflorescence and pedicels also with some pilosity and the latter remaining puberulent even at maturity; the primary bracts conspicuously ciliate and puberulent on the outer surface and the loment-articles more densely uncinulate-puberulent than in the typical variety.* (*D. bracteosum,* var. *longifolium* (T. & G.) Robins.; *D. longifolium* (T. & G.) Smyth; *Meibomia* Vail) — No. O., s. Mich., s. Wisc., Ill. and Neb., s. to Ala., La. and Kans.

Ser. Stipitàta Schub.

17. D. viridiflòrum (L.) DC. (green-flowered), Velvety T. — The erect, usually simple, ridged and grooved stem somewhat pilose and uncinulate-pubescent when young; *terminal* *leaflets mostly rhombic* or deltoid, *with* truncate base, acute to acuminate or obtuse, moderately to abundantly spreading-pilose with short straight or hooked trichomes or both and *impressed veins above, densely tomentose with soft spreading white trichomes beneath, width usually at least two-thirds the length,* 5.2–11.8 *cm. long,* 3.6–8.8 *cm. wide; lateral leaflets similar,* 4–10.2 *cm. long,* 2–6.5 *cm. wide; pedicels with* spreading uncinulate puberulence and *some pilosity,* 3–8 mm. long; *flowers* pinkish or rose, *turning green after anthesis,* to 8.5 mm. long; loment stipitate, 2–6 (usually 4–5)-articulate; stipe 3–6 mm. long; *articles more or less rhombic with the upper suture regularly angled,* moderately to densely uncinulate-pubescent throughout, 5–9 *mm. long,* 3–3.5 *mm. wide.* (*D. viridiflorum* of ed. 7 in part; *Meibomia* Ktze.) — Chiefly in dry woods and clearings, n. Fla. to e. Tex., n. to Del. and inland only to Ark. and Tenn.

18. D. Nuttállii (Schindl.) Schub. (for THOMAS NUTTALL, 1786–1859). — Erect or ascending, simple or branched from base, up to about 1.5 m. tall, from a thickish branched root to 3 dm. long; very similar in appearance to no. 17 but differing in having the *leaflets* tomentulose beneath, *the terminal ones* (except of the uppermost leaves) *elliptic-ovate and rounded at base, one-half as wide as long;* the loment-stipe 2.5–4 mm. long; *the loment-articles with the upper suture curved rather than angled* and almost rounded and deeply indented below, 4–7 *mm. long,* 3–4.5 *mm. wide.* (*D. viridiflorum* of ed. 7 in part; *Meibomia* Schindl.) — Chiefly in dry sandy open woods, N.Y. to Ind., s. to n. Fla., Ala. and Ark. FIG. 1232.

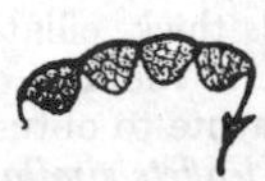

1232. D. Nuttallii.

19. D. Fernáldii Schub. (for MERRITT LYNDON FERNALD, 1873–1950). — Stout, to about 1.3 m. tall, from a slender branching root to 4 dm. long; the terete to slightly angulate stem finely puberulent and sparsely to densely uncinulate-pubescent; *leaflets dark green above, moderately uncinulate-pubescent and also somewhat short- and straight-pilose, paler beneath and uncinulate-pubescent* (rarely also somewhat pilose) *along prominent midrib and veins, otherwise essentially glabrous; terminal leaflet rhombic to ovate, obtuse to acute at apex,* cuneate to rounded at base, (4.5–)6–8 (–9.3) *cm. long,* 2.5–5 *cm. wide;* lateral leaflets elliptic to ovate-elliptic, often acute at apex and truncate at base, 4–6 cm. long and 1.8–3.5 cm. wide; *stipules* lance-attenuate, striate, pilose, *very early deciduous,* 2–4 mm. long; loment-stipe 2.5–4 mm. long; loment 1–5-articulate; articles uncinulate-pubescent throughout, somewhat deltoid, with the upper suture slightly angled, the lower obtuse, 5.5–8 mm. long, 3.5–5 mm. wide. (*D. rhombifolium* of ed. 7 and auth. generally, in part, not (Ell.) DC). — Sandy woods, se. Va. to S.C., La., and Newton Co., Tex. (*Cory*).

20. D. glabéllum (Michx.) DC. (smooth). — Spreading to erect, to 1.5 m. or more high, from a thickish branched root almost 4 dm. long; the lineate *stems* minutely and rather densely uncinulate-puberulent and usually at least sparsely *spreading-pilose; stipules not long persistent,* 2–4 *mm. long; leaflets with* rather *prominent reticulate venation,* ciliate, somewhat uncinulate-puberulent and moderately to abundantly *appressed-pilose above, paler and pilose beneath; terminal leaflets* narrowly to broadly ovate to nearly elliptic or rhombic, broadly acute or obtuse and *slightly emarginate,* 3–8 cm. long, 1.8–5.5 cm. wide; lateral leaflets mostly elliptic and obtuse, 2.1–6 cm. long, 1–3.5 cm. wide; bracts small (to 3 mm. long), not persistent; pedicels with dense and stout uncinulate puberulence and occasional pilosity, 3.5–12 mm. long; loments 1–5-articulate, stipitate; stipe 3–5(–7) mm. long; articles triangular to rhombic in outline, uncinulate-pubescent throughout with fine puberulence or multicellular trichomes intermixed, 6–8 mm. long, 3–5 mm. wide. (*Meibomia* Ktze.; *D. Dillenii* Darl. in part) — Dry sandy woods, se. Mass. to Mich. and Ill., s. to S.C., Ala., La. and Tex.

21. D. paniculàtum (L.) DC. (paniculate). — *Stem slender,* to 6 or more dm. high, *glabrous to puberulent,* often several arising from the base; *the linear to lance-attenuate stipules* 3–6 *mm. long; petioles* 1.4–5.3 *cm. long* and minutely puberulent to moderately pilose; the linear-lanceolate to lance-ovate, acute to acuminate leaflets rounded at base; the terminal ones 4.3–10 cm. long, 1–2.3 cm. wide, the lateral leaflets somewhat smaller; *inflorescence* paniculately branched and often very diffuse, its *rachis minutely puberulent to moderately uncinulate-pubescent* and only occasionally with scattered pilosity; *bracts small and early deciduous; pedicels* ascending, minutely puberulent to uncinulate-puberulent, 4–11 *mm. long;* loments stipitate, 1–5-articulate, uncinulate-puberulent to -pubescent throughout; their *articles triangular to subrhombic,* 5.5–7 mm. long, 3.5–4.5 mm. wide. (Incl. vars. *angustifolium* and *pubens* T. & G.; *Meibomia* Ktze.) — Clearings and borders of dry woods, N.H. to Ont., Ia. and Neb., s. to Fla., Ala., Miss., La. and Tex.

Var. **epetiolàtum** Schub. (lacking petioles). — *With broader, sessile to short-petiolate leaflets and rounded loment-articles.* — Sphagnous bogs, damp clearings and sandy pine- and oak-woods, Coastal Plain of se. Va. and e. N.C.; Colorado Co., Tex.

22. D. perpléxum Schub. (perplexing; in reference to the taxonomic confusion over the identity of the group). — Slender to stout erect to spreading *stem* 6 dm. or more high, uncinulate-puberulent and usually *with abundant spreading or upwardly directed pilosity*, from a rather thick root about 1 dm. long; *the slender long-attenuate stipules persistent;* petioles and leaf-rachis pilose; *leaflets* elliptic-ovate to ovate and *chiefly acute* with appressed pilosity on the upper as well as the paler lower surface; inflorescence axillary and (chiefly) terminal, diffuse, its rachis uncinulate-puberulent and usually with at least some spreading pilosity interspersed; pedicels not usually more than 1 cm. in length; loments stipitate, 2–5-articulate, the articles chiefly rhombic in outline and uncinulate-puberulent throughout. (*D. Dillenii* Darl. in part) — Sandy woods, centr. Me. to Wisc. and s. rather generally throughout our range and beyond. Fig. 1233. — A very variable species showing close relationship to no. 20 and to no. 21 in some of its phases, as well as to no. 23 in its form with small ovate leaflets.

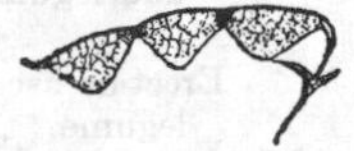
1233. D. perplexum.

23. D. humifùsum (Muhl.) Beck (spread on the surface of the ground). — *Prostrate,* with angulate to subterete stiffish, strongly lineate *stems* 0.9—1.8 m. in length, *sparsely to densely spreading-pilose with long tapering trichomes* and somewhat uncinulate-puberulent; *stipules obliquely ovate- to lance-acuminate,* sometimes long-attenuate, *ciliate, 4.5–8 mm. long; petioles densely spreading-pilose,* 2.8–4.8 cm. long; *leaf-rachis similar,* 1–1.5 cm. long; *terminal leaflets ovate to rhombic-ovate,* moderately appressed-pilose above and beneath, with occasional uncinulate puberulence on midrib and chief lateral veins above, 4.7–6.6 cm. long, 3–5.1 cm. wide; *lateral leaflets similar or becoming nearly orbicular, with truncate base,* 3.6–6 cm. long, 2.1–4 cm. wide; *bracts soon deciduous; pedicels finely uncinulate-puberulent, to 9 mm. long; purple corolla to about* 9.5 *mm. long;* loment 3–4-articulate; articles deltoid, with the upper suture essentially straight, or rhombic with the upper suture angled, 6–8 mm. long, 4–5 mm. wide. (*Meibomia* Ktze.) — Dry sandy woods, s. N.E. to Penn., Md. and Del.; Mo. Fig. 1234.

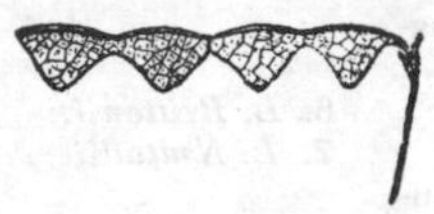
1234. D. humifusum.

24. D. laevigàtum (Nutt.) DC. (smooth). — Erect, simple to much branched, 4.6–12 dm. high, from a thick straight slightly branched root 2 or more dm. long; stem terete, lineate, glabrous to finely and minutely puberulent; *stipules very early deciduous;* the rather thick *leaflets glabrous to sparsely puberulent above and beneath or also short-pilose on the midrib and veins of the glaucous lower surface,* with somewhat revolute margins and slightly ciliate base; *terminal leaflets ovate, acute to acuminate,* mucronulate, *rounded to acute at base,* 4.3–8.4 cm. long, 2.8–5.2 cm. wide; *lateral leaflets elliptic-oblong or -ovate with obtuse apex and truncate or rounded base,* 3.2–7.4 cm. long, 2.5–4 cm. wide; corolla deep rose to purple; *the slender lax uncinulate-puberulent pedicels 10–19 mm. long;* the 2–5-articulate *loment with a mostly glabrous stipe 6–6.5 mm. long;* the reticulate articles half-rhombic to almost rhombic in outline and 5–7 mm. long, 3.5–4 mm. wide. (*Meibomia* Ktze.) — Dry sandy woods and clearings, N.Y. to Ind. and Mo., s. to n. Fla., Tenn., La. and e. Tex. Fig. 1235.

1235. D. laevigatum.

40. LESPEDÈZA Michx. Bush-Clover

Calyx 5-cleft; the lobes nearly equal, slender. Stamens diadelphous (9 and 1); anthers all alike. Legume of a single 1-seeded article (sometimes jointed, with the lower article empty and stalk-like), oval or roundish, flat, indehiscent, reticulated. — Herbs or half-shrubs of N. Am., e. Asia and Austral., with pinnately 3-foliolate leaves not stipellate, the fine and straightish veins horizontally divergent from the midrib. Flowers often cleistogamous, in summer and autumn. (Dedicated to *Vincente Manuel de Céspedes,* Spanish Governor of East Florida during the explorations there of Michaux late in the 18th century; the name later misspelled, probably by Michaux's editor, as *de Lespedez.*)*

a. Stipules subulate-setaceous or subulate-tipped; bracts slender and minute; calyx-lobes setaceous or with setaceous tips; perennials. . . *b.*

b. Flowers (except in no. 5) obviously of 2 kinds; the larger violet-purple or with preponderance of purple tones, perfect but often not fruitful, racemose or panicled; the smaller pistillate and freely fertile ones mostly apetalous, in small clusters by themselves or intermixed with the petaliferous ones; calyx mostly shorter than fruit. . . *c.*

* Our species needing critical study, the most-used characters often seeming relatively unimportant. Hybridization evidently frequent.

c. Petaliferous flowers on elongate filiform peduncles two to four times overtopping the subtending leaves. . *d.*

d. Stems soft-downy with short spreading hairs.

Trailing or reclining; calyx 1.5–3 mm. long, much shorter than corolla and legume. . . . . 1. *L. procumbens.*

Erect or ascending; calyx 6–9 mm. long, nearly equaling corolla and legume. . . . . 2. *L. Manniana.*

d. Stems glabrate to appressed-pubescent. . . *e.*

e. Flowers of 2 kinds; the petaliferous ones rarely 1 cm. long, borne at tips of very elongate peduncles; legumes 3–7 mm. long.

Stems prostrate or trailing; stipules mostly 2–4.5 mm. long; keel, wings and standard subequal. . . . . 3. *L. repens.*

Stems ascending; stipules mostly 5–8 mm. long; keel much longer than wing-petals and standard. . . . . 4. *L. violacea.*

e. Flowers all showy, 1.2–1.7 cm. long, in elongate slender racemes; legumes 7–10 mm. long. . . . . 5. *L. Thunbergii.*

c. Petaliferous flowers on stouter peduncles, some or all of the peduncles as short as or shorter than the subtending leaves. . . *f.*

f. Calyx 4–6 mm. long, villous; fruit 5–9 mm. long, villous-hirsute, long-attenuate; inflorescences frequently with some elongate peduncles.

Leaflets densely villous beneath. . . . . 6. *L. Brittonii.*

Leaflets sparingly villous to appressed-pubescent beneath. . . . 7. *L. Nuttallii.*

f. Calyx 3–5 mm. long; fruit 3.5–7 mm. long, roundish to elliptic-attenuate, minutely strigose (pilose in no. 8); inflorescence with few or no prolonged peduncles. . . *g.*

g. Calyx and legumes pilose or villous-hirsute; lower leaf-surfaces densely pilose, velvety to touch. . . . . 8. *L. Stuevei.*

g. Calyx and legumes strigose or strigillose; lower leaf-surfaces strigose to glabrescent.

Leaflets linear or linear-oblong; petaliferous inflorescences mostly sessile or subsessile. . . . . 9. *L. virginica.*

Leaflets oval, rounded or short-oblong; petaliferous inflorescences often short-peduncled. . . . . 10. *L. intermedia.*

b. Flowers uniform (rarely some cleistogamous), in heads or spikes or in axillary clusters; corollas creamy or whitish with purple blotch; calyx usually equaling or exceeding legume. . . *h.*

h. Heads or racemes many-flowered, borne at summits of stems and branches; calyx copiously pilose; indigenous; apetalous flowers none or rare. . . *i.*

i. Heads thick-cylindric to subglobose, dense, with crowded or imbricated flowers, 1–2 cm. thick. . . *j.*

j. Heads very dense; the strongly appressed calyces 7–13 mm. long, closely overlapping, with the inner ones mostly hidden, greatly exceeding the legume; peduncles very short, usually much shorter than subtending leaves. . . . . 11. *L. capitata.*

j. Heads racemose-spiciform, cylindric, the ascending to divergent lower flowers not hiding those above; mature calyx 4–10 mm. long, nearly equaling or only slightly exceeding legume; peduncles often much longer than subtending leaves.

Principal leaves with petiolule of terminal leaflet 4–10 mm. long and conspicuously with denser and more pilose summit; leaflets rounded-obovate to linear-lanceolate, 0.6–3.5 cm. broad; calyx 6–10 mm. long; bracteoles 2–4 mm. long. . . . . . . 12. *L. hirta.*

Principal leaves with petiolule of terminal leaflet 0.5–4 mm. long and not conspicuously modified at summit; leaflets linear, 2–6 mm. broad; mature calyx 4–6.5 mm. long; bracteoles 1–2 mm. long. . . . . 13. *L. angustifolia.*

i. Spikes slender-cylindrical, interrupted, 5–8 mm. thick; leaflets linear or linear-oblong. . . . . 14. *L. leptostachya.*

h. Racemes of 1–several petaliferous flowers borne in axils of most leaves up and down the slender stems; calyx glabrescent or sericeous; apetalous flowers frequent; introd. as crops.

Stems stiffly erect and virgate; leaflets narrowly cuneate; racemes of petaliferous flowers subsessile, 1–4-flowered. . . . . 15. *L. cuneata.*

Stems loosely ascending; leaflets oblong or oblong-lanceolate; racemes of petaliferous flowers long-peduncled, the flowers numerous. . . . 16. *L. daurica.*

a. Stipules and bracts broad and scarious; calyx-lobes broad; low annuals suggesting small clovers.

Flowers and fruits in tiny axillary clusters, their subtending leaves not bristly-ciliate. . . . . 17. *L. striata.*

Flowers and fruits in elongate leafy-bracted spiciform racemes, the bracteal leaves bristly-ciliate. . . . . 18. *L. stipulacea.*

1. L. procúmbens Michx. (trailing). — *Stems trailing or reclining*, up to 1.3 dm. long, *covered with dense soft spreading pubescence; leaflets downy*, oval, rounded to base and apex, not more than twice as long as wide, the larger 1.2–2.5 cm. long and 0.7–1.5 cm. broad; *petaliferous flowers 2–8 at summit of elongate filiform peduncles much longer than the subtending leaves; calyx* 1.5–3 *mm. long;* keel-petals purplish, equaling the wings; legumes suborbicular to elliptic, 3–7 mm. long, minutely strigose. — Dry sandy or rocky woods and clearings, Fla. to Tex., n. to s. N.H., Mass., N.Y., O., Ind., s. Wisc., Ia. and se. Kans. Mid-Aug.–early Oct. — Crosses with nos. 3, 7, 8 and 12. — Perhaps better treated as a var. of no. 3.

Var. ellíptica Blake (elliptic). — Leaflets narrowly elliptic-oblong, about 4 times as long as wide, the larger ones 1.8–3.5 cm. long and 4–9 mm. wide. — Ala. to Mo., n. to e. Mass. and Ind.

2. L. Manniàna Mackenz. & Bush (named in 1902 in honor of CAMERON MANN). — *Erect or ascending*, 3–7 dm. high, the rather slender stems appressed-pubescent or slightly pilose; leaves mostly short (0.5–1.5 cm.) -petioled, the *linear-oblong to narrowly elliptic thick leaflets strigose-pubescent beneath;* peduncles various, many of them elongate; *calyx* 6–9 *mm. long, about equaling the corolla and* the strigose *legume.* — Barrens and dry open woods, Mich. to Ark. and Tex.

3. L. rèpens (L.) Bart. (creeping). — Habitally like **no. 1**, but more slender, the *stems glabrate to minutely appressed-pubescent; stipules* subrigid, *mostly* 2–4.5 *mm. long;* leaflets round-oval to obovate or oblong, minutely appressed-silky to glabrate beneath. — Sandy or rocky open woods, thickets and openings, Fla. to Tex., n. to Ct., N.Y., O., Ind., s. Wisc., Ia. and e. Kans. Late May–Sept. — Crosses with nos. 1 and 12.

4. L. violàcea (L.) Pers. (violet). — *Stems upright* or spreading, slender, branched, 2–7 dm. high, rather *sparsely leafy* and sparingly pubescent; stipules setaceous, mostly 5–8 mm. long; *leaflets thin, broadly oval or oblong*, finely appressed-pubescent beneath; those of the cauline leaves mostly 2–5 *cm. long*, 1.2–2.2 *cm. broad; peduncles very slender, loosely few-flowered*, mostly longer than the leaves; petals 6–8 mm. long, the keel often the longest; *legume ovate*, 4–6 mm. long, minutely strigose. (*L. prairea* Britt.) — Dry woods, thickets and openings, Fla. to Tex., n. to s. N.H., s. and w. Vt., N.Y., O., s. Mich., s. Wisc. and e. Kans. July–Sept. — Crosses with no. 7 and others.

5. L. THUNBÉRGII (DC.) Nakai (in honor of CARL PETER THUNBERG, 1743–1828). — Stems numerous, diffusely arched-ascending, up to 3 m. high, much branched above the middle; leaves on petioles up to 7 cm. long; leaflets oblong, deep green above, whitish with appressed pilosity beneath; *racemes long and slender, curving, up to 2 dm. long; flowers* 1.2–1.7 *cm. long*, purple; *legume* 7–10 *mm. long.* — Spreading from cult., Mass. and southw. Aug.–Oct. (Introd. from e. Asia)

6. L. Brittònii Bickn. (in honor of NATHANIEL LORD BRITTON, 1859–1934). — *Densely cinereous-velvety or -tomentose; stems loosely ascending or arching*, 6–13 *dm. long;* leaves mostly short-petioled, *the thick oblong or lance-elliptic leaflets velvety beneath*, cinereous-pilose or glabrate above, the principal ones 1.5–4 cm. long; inflorescences numerous along the upper half of the stem or on short lateral branches; *peduncles various*, some shorter than the leaves, others elongate; calyx 4–5 mm. long; corolla 6–8 mm. long, pink and purple, the standard deeper purple at base; *legume tomentose*, sharply acute or acuminate, up to 8 mm. long. — Dry soil, Mass. and N.Y. to Md. Aug., Sept. — Perhaps a hybrid of nos. 1 and 7.

7. L. Nuttállii Darl. (for its discoverer, THOMAS NUTTALL, 1786–1859). — *Stems erect*, stoutish, 6–12 dm. high, *villous;* leaves mostly long (1–3 cm.)-petioled; the *oval leaflets* glabrous or glabrate above, *appressed-pubescent or sparingly villous beneath*, the principal ones 2.5–4 cm. long; *peduncles stoutish, of various lengths; calyx* 4–6 *mm. long, villous;* corolla pink or purple; *legumes* 5–9 *mm. long, villous-hirsute, long-attenuate.* — Dry woods and openings, s. N.H. to Mich., s. to S.C., Tenn., Mo. and s. Kans. Aug.–Oct. — Crosses with nos. 1, 4, 9 and others.

8. L. Stuèvei Nutt. (named in 1818 for its discoverer, Dr. W. STUEVE of Bremen). — *Stem* upright-spreading, 3–12 dm. high, very leafy, *downy with spreading pubescence*, simple or with few densely flowered *wand-like branches; leaves crowded*, short-petioled; *the elliptical* firm *leaflets woolly or velvety beneath* and sometimes above, mostly 1–2.5 *cm. long;* peduncles all short, *the crowded racemes mostly appearing sessile or subsessile; calyx* 3–5 mm. long, *pilose*, much shorter than the *villous-canescent legume.* — Dry soil, Ala. to Tex., n. to Mass., s. Vt., N.Y., O., Ind., Ill., Mo. and se. Kans. Aug., Sept. — Crosses with nos. 4 and 9.

Var. **angustifòlia** Britt. (narrow-leaved). — Leaflets linear or linear-oblong. (*L. neglecta* Mackenz. & Bush) — Less common, Ga. to Tex., n. to N.J., Md., Ind. and Mo. — Perhaps a hybrid of nos. 8 and 9.

× **L. acuticárpa** Mackenz. & Bush (acute-fruited) seems to be a cross of no. 4 with no. 9 or 10.

× **L. simulàta** Mackenz. & Bush (similar) seems to be a cross of no. 11 with no. 9 or 10.

9. L. virgínica (L.) Britt. (Virginian). — *Stems* upright, 3–11 dm. high, *wand-like* or with few erect branches, *minutely appressed-pubescent* or glabrate, or spreading-pubescent in forma **Deàmii** M. Hopkins (for Charles Clemon Deam, 1865–); leaves very crowded; *the principal* cauline ones with slender rather long petioles; their thickish *linear or linear-oblong leaflets* 1.5–4 *cm. long,* 3–7 *mm. broad, finely appressed-pubescent;* flowers on *very crowded short peduncles;* keel shorter than the standard; *calyx strigose,* 3–5 *mm. long,* shorter than the strigillose legume. — Dry open woods, thickets and barrens, Fla. to Tex., n. to s. N.H., Mass., N.Y., s. Ont., Mich., Wisc. and e. Kans. Late July–Sept. — Crosses with nos. 1 and 8.

10. L. intermèdia (S. Wats.) Britt. (intermediate). — Stems erect or ascending, 1.5–7 dm. high, slightly appressed-pubescent or glabrate, or densely spreading-pilose in forma **Háhnii** (Blake) M. Hopkins, (named in 1924 for its discoverer, W. L. Hahn); leaves mostly with slender long (1.5–3 cm.) petioles; *the oval to oblong firm leaflets finely appressed-pubescent or glabrate,* those of the cauline leaves 1.5–4 cm. long; peduncles of various lengths, mostly very short, a few sometimes nearly equaling the leaves; *calyx* 3–5 *mm. long,* much shorter than the strigillose legume. (*L. frutescens* of ed. 7, not *Hedysarum frutescens* L., the basonym.) — Dry open woods and thickets, Fla. to Tex., n. to sw. Me., s.-centr. N.H., Vt., N.Y., s. Ont., Mich., Wisc. and e. Kans. July–Sept.

11. L. capitàta Michx. (in heads). — Stems stiffish and relatively stout, simple below or with few upright branches, 0.6–1.2 m. high; petioles short; leaflets thickish; *heads* of flowers *subglobose, subsessile or on peduncles shorter than leaves, very dense; calyces* 7–13 *mm. long,* villous, *closely overlapping, the inner ones hidden, greatly exceeding the legume;* minute cleistogamous and apetalous flowers hidden among the others; corolla creamy-white, the standard with a purple spot at base. — One of our most variable species, crossing with nos. 9, 10, 12 and 13; the following vars. recognized:

Lower surfaces of leaves with closely appressed or sericeous pubescence.
  Leaflets oblong, elliptic, oval or obovate.
    Leaflets oblong or narrowly elliptic; heads crowded and very short-peduncled among the upper leaves.
      Leaves brilliantly silvery beneath, grayish and lustrous above; upper heads densely aggregated and mostly hiding the subtending leaves. . *L. capitata* (typical).
      Leaves opaque or only slightly lustrous beneath, green above; subtending leaves often exceeding the heads. . . . . Var. *vulgaris.*
    Leaflets broadly elliptic-oval to rounded-obovate; some or all peduncles scattered and equal or exceeding the subtending leaves. . . . . Var. *calycina.*
  Leaflets lanceolate to lance-linear, usually sericeous beneath. . . . . . Var. *stenophylla.*
Lower surfaces of oblong to narrowly obovate leaflets velvety-pilose with dense dull cinereous pubescence; inflorescence leafy; northeastern. . . . . . . Var. *velutina.*

L. capitàta (typical), var. *sericea* H. & A. — Dry open soil, Fla. to Tex., n. to s. N.E., L.I., N.J., Pa., Tenn., Wisc., Minn. and Neb., common southw. and westw. Late July–Sept. — Passing into

Var. **vulgàris** T. & G. (common), *L. capitata* of ed. 7. — W.-centr. Me. and sw. Que. to Minn. and Neb., s. to upland of N.C., and Mo.

Var. **calycìna** (Schindl.) Fern. (with large calyx). — Pinelands and borders of woods, se. Va. to Fla., w. to e. Tex.

Var. **stenophýlla** Bissell & Fern. (narrow-leaved), var. *longifolia* of ed. 7, not *L. longifolia* DC., basonym. — Mass. to s. Wisc., s. to N.J. and n. Mo. — Upper surfaces of leaves green, or silvery-silky in forma **argèntea** Fern. (silvery).

Var. **velùtina** (Bickn.) Fern. (velvety), *L. Bicknellii* House — Centr. Me. to e. N.Y. and n. N.J.

12. L. hírta (L.) Hornem. (stiffly-pubescent). — Stems erect or ascending, up to 1.5 m. high, rather stout, with mostly spreading pubescence; *leaves definitely petioled;* leaflets rounded-obovate or round-elliptic to narrowly lanceolate or even linear, the *terminal leaflet* of the principal leaves *on a petiolule* 4–10 *mm. long, this conspicuously thickened and with denser pubescence at summit; petaliferous flowers in cylindric spiciform racemes, not densely imbricated, loosely ascending or divergent; bracteoles* 2–4 *mm. long; calyx* very hairy, 6–10 *mm. long;* corolla whitish, with purple base; apetalous flowers often in separate inflorescences. — Very variable, often crossing with vars. of no. 11 and with other species. Our vars. are as follows:

Peduncles mostly overtopping their subtending leaves; racemes relatively open, in fruit 1.5–4.5 cm. long.
Leaflets rounded-obovate to oblong-ovate, the terminal one of the primary leaves 1–3.5 cm. broad.
Stem villous or copiously pilose; leaves pubescent beneath, at least on the veins, with spreading or spreading-ascending hairs; terminal leaflet of primary leaves 2–6 cm. long, 1.5–4 cm. broad. . . . . *L. hirta* (typical).
Stem densely short-pubescent; leaves grayish, silvery beneath and sometimes above with minute sericeous puberulence; terminal leaflet of primary leaves 1–2.7 cm. long, 1–2 cm. broad; southeastern. . . . . . Var. *appressipilis*.
Leaflets narrowly oblong to linear, 6–10 mm. broad; southern.
Leaflets oblong, the larger 2–3 cm. long and 7–10 mm. broad; calyx 6–8 mm. long. . . . . Var. *longifolia*.
Leaflets narrowly linear, the larger 3–7 cm. long and 6–8 mm. broad; calyx 8–10 mm. long. . . . . Var. *intercursa*.
Peduncles mostly much shorter than their subtending leaves, producing a virgate leafy inflorescence; racemes relatively compact in fruit, only 1–2.5 cm. long; leaflets oblong, velutinous or sericeous, the larger ones 3–6 cm. long and 1–2 cm. broad; northern. . . . . Var. *dissimulans*.

L. hírta (typical). — Dry soils, sw. Me. to s. Ont., s. to Ga., Ala., Ark. and e. Tex. July–Oct. — Crosses with nos. 1, 3, 9, 11 and 13.

Var. **appressípilis** Blake (with appressed hairs). — Pinelands and barrens, Fla. to e. N.C., and less characteristically to se. Va.

Var. **longifòlia** (DC.) Fern. (long-leaved), *L. longifolia* DC.; *L. capitata*, var. *longifolia* T. & G.; *L. hirta*, var. *oblongifolia* Britt. — Local, s. N.J. to La. — Little known; thought by some to be a hybrid with no. 13.

Var. **intercúrsa** Fern. (running between). — Bogs, swales and wet thickets, se. Va. — Very puzzling, with inflorescences of no. 12, leaves of no. 13, forming transition between them.

Var. **dissímulans** Fern. (dissembling). — Dry, often argillaceous soil, centr. Me. to s. and w. N.E.; Wisc. — Most perplexing, with habit of no. 11 and nearly or quite exserted legume of no. 12.

13. L. angustifòlia (Pursh) Ell. (narrow-leaved). — Very slender, stiffly ascending, 0.3–1.2 m. high, with minute appressed silky, or velutinous in forma **subvelùtina** Fern. (somewhat velvety), pubescence; leaves subsessile or short-petioled, with *firm linear leaflets* 2–6 *mm. broad; petiolule of terminal leaflet of principal leaves* 0.5–4 *mm. long, not conspicuously modified at summit; flowers* much as in no. 12, smaller, *in* long-stalked *compact racemes* 0.8–2 *cm. long; bracteoles* 1–2 *mm. long; calyx* 4–6.5 *mm. long, barely longer or even shorter than legume.* — Sandy barrens and openings, Fla. to La., n. to se. Mass. Aug., Sept.

14. L. leptostáchya Engelm. (slender-spiked). — In habit like no. 13; linear leaflets strongly silky beneath; the *spikes slender-cylindric, interrupted,* 2–4 cm. long, 5–8 *mm. thick,* on peduncles as long as the leaves; *legume* ovate, about 3 mm. long, *about equaling the calyx.* — Prairies, n. Ill., s. Wisc., e. Minn. and n. Ia. July, Aug.

15. L. cuneàta (Dumont) G. Don (wedge-shaped). — Short-lived perennial, the *erect often upright-branched* stems up to 1.5 m. high; *leaves crowded,* gray-green, or silvery-silky beneath, *with narrowly cuneate retuse or truncate small leaflets;* bracts and bracteoles ovate-lanceolate; *flowers white with purple veins, in* 1–4*-flowered small clusters* in the leaf-axils; legume minutely ciliate. — Much cult. in the South, spread to roadsides, etc., n. to Pa., Tenn. and Mo. Sept., Oct. (Introd. and natzd. from e. Asia)

16. L. daùrica Schindl. (of Davuria or Mongolia). — More loosely ascending and lower than no. 15; *leaflets glabrate, oblong or oblong-lanceolate, obtuse to subacute; flowers numerous in long-peduncled* axillary *racemes;* corolla white, with standard purple-marked. — Cult. southw., spreading locally to roadsides and borders of woods, n. to Del. (Introd. from e. Asia)

17. L. striàta (Thunb.) H. & A. (with parallel lines), Japanese Clover. — *Diffusely branched* decumbent subpubescent *annual;* petioles very short; leaflets oblong-obovate, 1.2 cm. long or less; *peduncles very short,* with 1–5 pinkish flowers; legume tiny, little exceeding calyx. — Roadsides and dry open soil, common southw., extending n. to N.J., Pa., O., Ind., Ill., Mo. and Kans. July–Oct. (Introd. and natzd. from e. Asia)

18. L. stipulàcea Maxim. (bearing stipules), Korean Clover. — Similar to no. 17; the *flowers and fruits in elongate leafy-bracted spiciform racemes* 1–3.5 cm. long, the *bracteal leaves bristly-ciliate.* — Cult. southw., now rapidly spreading in dry open soil, n. to Pa., W.Va. and Ia. July–Oct. (Introd. and natzd. from e. Asia)

41. STYLOSÁNTHES Sw. Pencil-flower

Calyx early deciduous; tube slender and stalk-like; limb unequally 4–5-cleft, the lower lobe more distinct. Corolla and monadelphous stamens inserted at the summit of the calyx-tube; standard orbicular; keel incurved. Anthers 10, in two series. Style filiform, its upper part deciduous, the lower incurved or hooked, persistent on the 1–2-articled short reticulated fruit; the lower article when present empty and stalk-like. — Low perennials of warmer reg. of Am., Asia and Afr., branched from the base, with wiry stems, pinnately 3-foliolate leaves, and small mostly yellow flowers in terminal heads or short spikes. (Name composed of the Greek *stylos, a column*, and *anthos, a flower*, from the stalk-like calyx-tube.)

Erect or stiffly ascending; leaflets of middle and upper leaves lanceolate, 1.5–4 cm. long; stipular bases of uppermost leaves (subtending fascicles of flowers) or primary bracts usually bristly-hispid, their leaflets bristly-ciliate. . . . 1. *S. biflora.*
Depressed or loosely ascending; leaflets of middle and upper leaves oval, elliptical or broadly oblanceolate, 0.5–2.3 cm. long; stipular bases of uppermost or bracteal leaves smooth, their blades eciliate. . . . . 2. *S. riparia.*

1. S. biflòra (L.) BSP. (two-flowered). — *Erect or ascending, stiff*, from a stoutish caudex; stems 1.5–5 dm. high, *simple or with stiffly ascending branches*, finely pubescent or bristly at summit; *leaflets of the middle and upper leaves lanceolate*, subulate-tipped, 1.5–4 *cm. long;* upper *leaves crowded at base of flower-fascicles; stipular base of the uppermost* (or primary) *involucre commonly bristly on the surface, the leaflets bristly-ciliate;* corolla yellow or orange. — Dry woods, thickets and openings, Fla. to Tex., n. to se. N.Y., N.J., Pa., W.Va., O., Ind., s. Ill. and se. Kans. June–Sept.

Var. **hispidíssima** (Michx.) Pollard & Ball (most hispid). — Stem copiously spreading-hispid or -hirsute throughout. — La. to Ariz., n. to Ind., Ill., Mo. and Okla.; s. N.J. to N.C.

2. S. ripària Kearney (of river-banks). — *Depressed or loosely branched*, slender, the stems minutely pubescent in lines, 1–3.5 dm. high; *leaflets of middle and upper leaves oval, elliptic or broadly oblanceolate*, 0.5–2.3 *cm. long*, glabrous or nearly so, with smooth (rarely a little ciliate) margins; *stipular base of uppermost* or involucral *leaf smooth, the leaflets eciliate;* corolla yellow or orange, or milk-white in forma **ochroleùca** Fern. (creamy). — Dry woods and openings, Fla. to Tex., n. to N.J., e. Pa., W.Va., s. Ind., s. Ill., Mo. and Okla. June–Sept.

Var. **setífera** Fern. (bristly). — Stems more or less hirsute-setose with divergent hairs 1 or 2 mm. long; corolla whitish. — Local, Southampton Co., Va.

42. ÁRACHIS L. Peanut

Calyx-tube filiform; lobes membranaceous, the 4 upper connate, the lowest distinct and slender. Corolla and monadelphous stamens inserted at summit of calyx-tube; standard suborbicular; wing-petals oblong, free; keel incurved, beaked. Style filiform; ovary 2–3-ovulate, after anthesis reflexed on the elongating rigid stipe and becoming buried underground; the mature legume indehiscent, with reticulate and ropy cortex, torulose. Seeds 1–3, ovoid-cylindric. — Low herbs, nat. of S.Am., with abruptly pinnate leaves with mostly 4 leaflets (rarely odd-pinnate), and yellow flowers in axillary sessile spikes. (Name a contraction of the earlier *Arachidna*, originally used for a clover with recurving and subterranean fruiting heads.)

1. A. hypogaèa L. (beneath the ground), Common P. — Decumbent; leaflets obovate or elliptic, 2–5 cm. long; corolla 2–2.5 cm. long. — Roadsides, waste places and old fields, Del. and Kans. and southw., esc. from cult. July–Sept. (Introd., originally from S.A.)

43. ZÓRNIA Gmel.

Calyx bilabiate, 5-toothed, the tube not elongated. Corolla yellow. Stamens monadelphous. Ovary sessile. — Prostrate wiry-stemmed perennials of trop. and warm reg., with long tough root. (Named for *Johannes Zorn*, 1739–1799, a German apothecary.)

1. Z. bracteàta (Walt.) Gmel. (bearing bracts). — Prostrate intricately branched stems forming broad carpets; leaves 4-foliolate. — Dry sandy woods and openings, Fla. to Tex., n. to se. Va. June–Sept.

44. VÍCIA L. Vetch. Tare. Vesce (Que.)

Calyx 5-cleft or 5-toothed, the 2 upper teeth often shorter, or the lowest longer. Wings of the corolla adhering to the middle of the keel. Stamens more or less diadelphous (9 and 1); the orifice of the tube oblique. Style filiform, hairy all round or only on the back at the apex

or beneath the stigma. Legume usually laterally compressed, 2-valved, 2-several-seeded. Seeds globular to barrelform. Cotyledons very thick, remaining under ground in germination. — Herbs of N. Hemisph. and temp. S. Am., mostly more or less climbing by the tendril at the ends of the pinnate leaves. Stipules half-sagittate. Flowers or peduncles axillary. (The classical Latin name.)

a. Peduncle many times shorter than leaflets or wanting; flowers 1–6, 0.6–3.5 cm. long; style bearded on the lower side beneath the stigma. . *b.*

b. Tendrils all simple; flowers 6–8 mm. long, solitary; legume 1.5–2.5 cm. long, sessile in the calyx; seeds strongly quadrate, verrucose; annual. 1. ***V. lathyroides.***

b. Tendrils all (or chiefly) forking; flowers 1–3.5 cm. long, mostly 2–5; legume 2.5–7 cm. long (sometimes shorter in the perennial no. 5, with 2–5 large flowers, stipitate legume and smooth globose seeds); seeds globose or orbicular, smooth. . . *c.*

c. Calyx-teeth subequal, all linear-subulate to narrowly lanceolate; seeds globose or subglobose to reniform; annuals or biennials. . *d.*

d. Flower 2.6–3.5 cm. long, with yellowish corolla; calyx-teeth one-third to half as long as glabrous tube. . . . . 2. ***V. grandiflora.***

d. Flower 1–3 cm. long, with purple corolla; calyx-teeth nearly as long as to longer than pubescent tube.

Flower 1.8–3 cm. long; corolla purple with violet wings; legume more or less torulose; seeds 5 mm. broad. . . . . 3. ***V. sativa.***

Flower 1–1.8 cm. long; corolla uniformly purple; legume plane; seeds 3 mm. broad. . . . . 4. ***V. angustifolia.***

c. Calyx-teeth very unequal, the upper shortest and triangular.

Leaflets mostly 4–8 pairs; flowers 0.8–1.5 cm. long, deep blue, striped with purple; legume stipitate in the calyx, 1.8–3 cm. long, with smooth sutures; seeds globose; perennial. . . . . 5. ***V. sepium.***

Leaflets 1–3 pairs; flowers 1.5–2.5 cm. long, purple; legume sessile, 5–7 cm. long, with pectinate-fringed sutures; seeds compressed; annual. 6. ***V. narbonensis.***

a. Peduncle definite, nearly as long as to much longer than leaflets; flowers 1–very many, 0.3–2 cm. long; style pubescent at summit or all around, not bearded beneath the stigma. . . *e.*

e. Flowers 1–8, 3–8 mm. long. . . *f.*

f. Legume nearly equally rounded from both sutures to the blunt tip, beakless, 1–1.3 cm. long. . . . . 7. ***V. tetrasperma.***

f. Legume strongly oblique at tip, tapering to a beak. . . *g.*

g. Flowers 3–4 mm. long; legume 6–10 mm. long, pubescent, 2-seeded. . 8. ***V. hirsuta.***

g. Flowers 5–8 mm. long; legume 2–3 cm. long, glabrous, 4–8-seeded.

Leaflets 6–12, elliptic; flowers 2–8; calyx-teeth lance-subulate, the longer equaling the tube. . . . . 9. ***V. ludoviciana.***

Leaflets 4–6, linear; flowers solitary; calyx-teeth deltoid, shorter than tube. . . . . 10. ***V. micrantha.***

e. Flowers 3–40 or more, 0.8–2 cm. long, if less than 8 (in well-developed racemes) 1.5–2 cm. long. . . *h.*

h. Fully developed inflorescences equaling or overtopping subtending leaves, many-flowered; flowers 1–1.5 cm. long; leaflets mostly 12–24, not prominently veiny. . . *i.*

i. Racemes lax; flowers mostly scattered, 8–12 mm. long; corolla white, with keel tipped with blue; calyx-teeth deltoid, about as broad as long. . . . . 11. ***V. caroliniana.***

i. Racemes dense; flowers strongly overlapping, 1–1.5 cm. long; corolla blue, violet or violet and white (all white only in albinos); at least the lower calyx-teeth lance-attenuate to linear-acicular. . . *j.*

j. Upper side of calyx gradually rounded, not gibbous, at base, the lower teeth lance-attenuate, the upper very short and broad; limb of standard as long as the claw; hilum one-third to one-fourth as long as circumference of seed; perennial. . . . . 12. ***V. Cracca.***

j. Upper side of calyx gibbous or saccate at base, the lower teeth linear-acicular; the upper lanceolate; limb of standard less than half as long as claw; annual or biennial; hilum at most one-sixth as long as circumference of seed.

Plant spreading-villous; lower calyx-teeth long-ciliate, 2–4 mm. long. . . . . 13. ***V. villosa.***

Plant glabrous or appressed-pilose; lower calyx-teeth glabrescent, 1–2 mm. long. . . . . 14. *V. dasycarpa.*

h. Fully developed inflorescences shorter than subtending leaves, 3–9-flowered; flowers 1.5–2 cm. long; leaflets 8–18, with lateral veins elevated beneath. . . . . 15. *V. americana.*

1. V. LATHYROÌDES L. (similar to *Lathyrus*). — Low usually tufted slender annual up to 2 dm. high; leaves with 4–8 narrowly obovate leaflets and a *simple tendril;* stipules entire, semisagittate; *flowers solitary,* sessile, 6–8 *mm. long;* lance-subulate equal calyx-teeth about equaling conic tube; corolla violet; *legume* flattened, 1.5–2.5 *cm. long; seeds quadrate, verrucose.* — Sandy grassland, local, Nantucket I., Mass., to Va. May, June. (Adv. from Eu.) FIG. 1236.

1236. V. lathyroides.

2. V. GRANDIFLÒRA Scop. (large-flowered). — Annual, with stems up to 6 dm. long; leaves with 6–14 linear or oblong leaflets and slender forking tendrils; stipules semisagittate; *flowers* 1 or 2, subsessile, 2.6–3.5 *cm. long; calyx* pilose, the subequal lanceolate *teeth much shorter than glabrous tube; corolla yellow or yellowish,* often suffused with lilac or black dots or with the standard purple; legume compressed, smooth, becoming black, 3.5–5 cm. long; seeds barrelform, 3–3.5 mm. long. — Roadsides, cult. fields and open woods, Del. to e. Va. Apr.–June. (Natzd. from Eu.) FIG. 1237.

1237. V. grandiflora.

3. V. SATÌVA L. (sown), SPRING-V. — Annual (or winter-annual), *pubescent,* becoming glabrate; the stem simple or branched at base; leaves essentially uniform; leaflets 4–8 pairs, oblong to oblong-obovate, truncate to emarginate and mucronate at apex, 1.5–3 cm. long, 5–13 mm. broad; *flowers* chiefly in twos in the upper axils, 1.8–3 *cm. long, showy,* purple and rose-color; calyx 1–1.5 cm. long; *legume* pubescent when young, *torulose,* 4–8 cm. long, 7–8 mm. wide; *seeds* compressed-globose, 5 *mm. broad.* — Cult. for forage, occasionally spread to roadsides and waste places, Nfld. to Minn. and southw. July–Sept. (Introd. from Eu.) FIG. 1238.

Var. LINEÀRIS Lange (linear). — Leaflets of all but basal leaves linear and emarginate or apiculate. — Waste places, old fields, etc., local, Que. to Va. and westw. (Introd. from Eu.)

1238. V. sativa.

4. V. ANGUSTIFÒLIA Reichard (narrow-leaved), COMMON V. — Similar, *glabrous or glabrate; leaflets* 2–5 (rarely 6) pairs, those *of the lower leaves oblong and truncate, of the upper linear- to lance-attenuate,* mucronate, 1.5–3 cm. long, 1–4 mm. broad; *flowers smaller* (1–1.8 *cm. long*); calyx 7–11 mm. long; *legume plane,* 4–5.5 cm. long, 5–7 mm. broad; *seeds* 3 *mm. broad.* — Waste ground, roadsides, etc., e. Can. to Mich. June–Oct. (Natzd. from Eu.) FIG. 1239.

Var. SEGETÀLIS (Thuill.) W. D. J. Koch (growing as a weed in wheat-fields). —*Leaflets of upper leaves oblong to oblong-obovate, truncate or emarginate* and mucronate at apex, 2–9 *mm. broad.* — Roadsides and waste places, nearly throughout our range. March–Oct. (Natzd. from Eu.)

Var. UNCINÀTA (Desv.) Rouy (bearing tendrils).—*Leaflets of upper leaves narrowly linear, truncate or abruptly narrowed,* 1–2 *mm. broad.* — Nfld. to Va., local. (Natzd. from Eu.)

1239. V. angustifolia.

5. V. SÈPIUM L. (of hedges). — *Perennial,* pilose or glabrate, with filiform reddish stolons and moniliform tubers; stems angular, up to 1 m. long; leaves with 4–8 oval round-tipped or emarginate leaflets; *stipules coarsely toothed,* semilunate; *peduncles very short, with* 2–6 *short-pedicelled flowers* 8–15 *mm. long; calyx-teeth very unequal, the upper shortest and triangular; corolla deep blue, striped with purple; legume stipitate,* linear, obliquely long-beaked, 1.8–3 cm. long, 6–7 mm. broad, black; seeds globose-barrelform, 3–4 mm. broad, the hilum extending two-thirds around the circumference. — Roadsides and old fields, local, Nfld. to Ont., s. to N.S., N.B. and n. N.E. June–Sept. (Natzd. from Eu.) FIG. 1240.

Var. MONTÀNA W. D. J. Koch (of mountains). — Leaflets elliptic-lanceolate or narrowly ovate, tapering to acutish tips. — Que. and Me., local. (Natzd. from Eu.)

1240. V. sepium.

6. V. NARBONÉNSIS L. (of Narbonne), NARBONNE V. — *Coarse annual;* the lower leaves without tendrils, the upper with them short and simple or forking; *leaflets* 1–3 *pairs,* oval or elliptic, mostly 3–7 cm. long and 2–4 *cm. broad,* entire, *fleshy;* peduncle short or almost wanting; *flowers* 1–5, 1.5–2.5 *cm. long;* calyx-teeth unequal; *corolla purplish* (drying black); *legume sessile,* plump, 5–7 cm. long, *pectinate along the sutures;* seeds compressed, with flat sides, orbicular, 5–6 mm. in diameter, with short oblong hilum. — Cult. southw., occasionally estab. as a weed in D.C. and Md. April–July. (Introd. from Eu.) FIG. 1241.

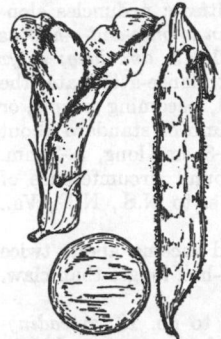

1241. V. narbonensis.

V. FÀBA L. (old generic name of Etruscan origin), BROAD BEAN, *a coarse erect annual* up to 2 m. high, *without tendrils, the large obtuse oval leaflets in* 1–3 *pairs, the large white corolla blackish-blotched, the legume plump* and *the large seed compressed,* is much cult. in Nfld. and Canada; occasionally spontaneous. (Introd. from Eu.)

7. V. TETRASPÉRMA (L.) Moench (four-seeded). — Annual with branching often matted capillary stems up to 5 dm. long; leaves with 2–5 *pairs of oblong round-tipped* mucronulate *leaflets* and simple or forking tendrils; stipules entire; *peduncles filiform,* longer than the leaflets, 1–2-*flowered; flowers* 7–8 *mm. long;* calyx-teeth unequal, lower lanceolate, upper triangular; corolla lilac, with deeper veins; *legume* glabrous, pale brown, 1–1.3 *cm. long, subequally rounded from both sutures to the blunt beakless tip;* seeds 3–5, subglobose, 1.5–2.5 mm. in diameter, hilum extending one-fifth around the circumference. — Waste places, old fields, roadsides, etc., Nfld. to Ont., s. to N.S., N.E., Fla. and Miss. May–Sept. (Natzd. from Eu.) FIG. 1242.

1242. V. tetrasperma.

Var. TENUÍSSIMA Druce (slenderest). — *Leaflets* of upper leaves *linear, attenuate to slender tips.* — Local, N.E. and N.Y. (Natzd. from Eu.)

8. V. HIRSÙTA (L.) S. F. Gray (hirsute). — Resembling no. 7; elongate lower lobe of stipule with setiform segments; *peduncle* 3–6-*flowered; flowers* 3–4 *mm. long; calyx with equal linear-subulate teeth* all longer than tube; *legume obliquely attenuate to beak, hirsute,* becoming black, 6–10 mm. long, 2-*seeded;* hilum extending one-third around circumference of seed. — Roadsides and waste places, Nfld. to B.C., s. beyond our limits. Apr.–Sept. (Natzd. from Eu.) FIG. 1243.

1243. V. hirsuta.

9. V. ludoviciàna Nutt. (of Louisiana). — Annual, with slender stems up to 1 m. long; leaves with 3–6 *pairs of oblong to elliptic round-tipped* or emarginate *leaflets* up to 2.5 cm. long; *peduncles* shorter than to exceeding leaves, 2–8-*flowered; flowers* 6–8 *mm. long; calyx-teeth unequal, the lower lance-subulate and equaling the pilose tube,* the upper shorter and broader; corolla blue-violet; *legume* narrowly oblong, *glabrous,* obliquely beaked, 2–3 *cm. long,* 4–8-*seeded;* seed suborbicular, flattened, the thread-like hilum extending one-third around the circumference. — Rich woodlands and thickets, Fla. to Tex., n. to sw. Mo. and Okla. Apr., May. FIG. 1244.

1244. V. ludoviciana.

10. V. micrántha Nutt. (tiny-flowered). — As slender as no. 9, stems up to 5 dm. long; *leaves with* 2 *or* 3 *pairs of linear elongate leaflets* and long simple or forking tendrils; *peduncle* shorter than leaves, 1-*flowered;* flower 5–8 mm. long; *calyx-teeth triangular, shorter than tube;* legume broadly linear-falcate. — Woods, thickets and glades, Fla. to Tex., n. to Tenn. and Mo. Apr., May. FIG. 1245.

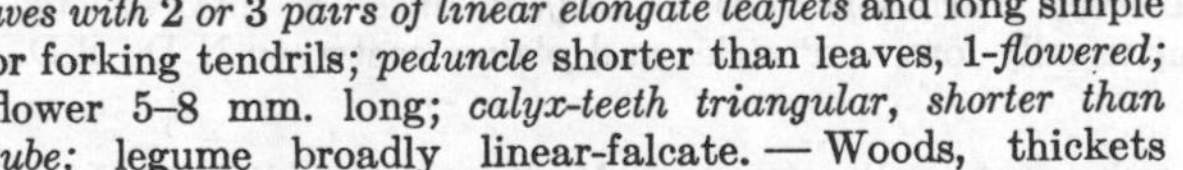

11. V. caroliniàna Walt. (of Carolina), WOOD-V. — Slender perennial, 0.3–1.5 m. high; leaflets mostly 6–12 pairs, oblong, obtuse, mucronulate; peduncles elongate, with *many loosely disposed pedicelled flowers* 8–12 *mm. long; calyx-teeth triangular, about as broad as long; corolla white, the keel tipped with blue* (rarely blue throughout); legume stipitate, narrowly oblong, obliquely long-beaked, 2–3 cm. long; seed compressed-subglobose, 3–4 mm. in diameter, the hilum extending three-fourths around the circumference. — Rich woods, thickets and shores, N.Y. and s. Ont. to Minn., s. to Ga., Ala., Miss., La. and Okla. Apr.–June. FIG. 1246.

1245. V. micrantha.

1246. V. caroliniana.

12. V. CRÁCCA L. (Latin name applied by Rivinius to this plant, the Italian name being *Cracca,* a French name *Vesce craque*), TUFTED V., CANADA-PEA, JARGEAU (Que.). — Perennial with striate angular *stems* up to 2 m. long,

1247. V. Cracca.

appressed-pubescent; leaves with 8–12 pairs of appressed-pilose to glabrate, or in forma SERÍCEA (Peterm.) G. Beck (silky) silky and silvery-lustrous, oblong to linear mucronate leaflets; stipules semisagittate; peduncles elongate, with many-flowered 1-sided *racemes shorter than to once-and-a-half as long as subtending leaf; flowers* crowded, reflexed, 1–1.3 *cm. long; calyx gradually rounded at base on upper side,* its lower teeth lance-attenuate, the upper ones very short and broad; corolla blue-violet, becoming purple, or white in forma ÁLBIDA (Peterm.) Gams (whitish), limb of standard about equaling claw; legume linear-lanceolate, glabrous, 2–3 cm. long, 5–7 mm. broad; hilum extending one-third to one-fourth around circumference of seed. — Fields, thickets and shores, Nfld. to B.C., s. to N.S., N.E., Va., Mich. and Ill. May–Aug. (Natzd. from Eu.) FIG. 1247.

Var. TENUIFÒLIA (Roth) G. Beck (slender-leaved). — Peduncle and raceme about twice length of subtending leaf; flowers 12–15 mm. long; limb of standard one-half longer than claw. — Locally in cult. land, Wisc. (Natzd. from Eu.)

13. V. VILLÒSA Roth (long-hairy), HAIRY or WINTER-V. — Similar to no. 12, *spreading-villous, annual or biennial; flowers* 1.4–1.5 *cm. long; calyx-tube gibbous at base on upper side, its lower teeth* linear-acicular, *long-villous and* 2–4 *mm. long,* upper teeth lanceolate; corolla violet and white, or wholly white in forma ALBIFLÒRA (Schur) Gams (white-flowered), *limb of standard less than half as long as claw;* legume oblong, obliquely beaked, 2–3 cm. long, 7–10 mm. broad; hilum extending around one-seventh of the circumference of the globular seed. — Cult. as a forage-crop, spread to roadsides and fields through most of our range. May–Oct. (Introd. and natzd. from Eu.) FIG. 1248.

1249. V. dasycarpa.

1248. V. villosa.

14. V. DASYCÁRPA Ten. (hairy-carpelled). — Differing from no. 13 in its merely *appressed-pubescent or glabrate* stem and foliage; flowers fewer and less reflexed, 1–1.5 cm. long; *lower calyx-teeth* 1–2 *mm. long, glabrescent.* — Roadsides, fields and waste places, Me. to Mont., s. to Ga., Mo. and Calif. May–Oct. (Natzd. from Eu.) FIG. 1249.

15. V. americàna Muhl. (American). — *Perennial,* with glabrous or glabrate stems up to 1 m. long; *leaflets* 4–9 pairs; those of middle and upper leaves oblong-ovate or elliptic, *obtuse,* rarely acute, *mostly* 1.5–3.5 *cm. long* and 0.6–1.4 *cm. broad, the lateral veins prominent and rib-like beneath in drying; fully developed inflorescence shorter than subtending leaves; flowers* 3–9, 1.5–2 *cm. long;* calyx-teeth unequal, the lower lance-attenuate, the others short and broad; corolla bluish-purple; legume 2.5–3.5 cm. long; seeds subglobose, 4 mm. long, hilum extending one-fifth around circumference. — Damp or gravelly shores, thickets or meadows, w. Que. to se. Alaska, s. to Va., O., Ind., Ill., Ark., Kans., N.M. and Ariz. May–July. FIG. 1250. — Passing westw. into

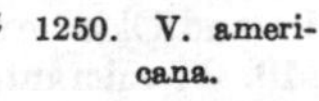
1250. V. americana.

Var. **truncàta** (Nutt.) Brewer (truncate; cut off square). — *Leaflets* of upper leaves broadly oblong, 5–13 *mm. broad, conspicuously truncate.* (*V. oregana* Nutt.) — Commoner in w. N. Am., eastw. to e. Man., Minn. and Mo.

Var. **angustifòlia** Nees (narrow-leaved). — *Leaflets linear or linear-oblong,* 1–4 *mm. wide;* plant usually low. — Prairies and plains, eastw. to N.D., S.D., Ia., Kans. and Okla.

LÉNS Mill. (from shape of seed) CULINÀRIS Medic. (of the kitchen), *L. esculenta* Moench, LENTIL, superficially resembling *Vicia* no. 8, but with 5-parted calyx cleft nearly to base, the style hairy along the upper surface, the seeds lenticular, is occasional in waste but scarcely persistent. (Introd. from Old World)

The CHICK-PEA, CÍCER L. (classical name) ARIETÌNUM L. (similar to a ram's head; from fancied resemblance of seed), an erect annual with odd-pinnate leaves of several incised-dentate small leaflets, solitary geniculate-stalked axillary white or pink-tinged flowers with free wing-petals, turgid legume 1.5–2 cm. long, and 1 or 2 large wrinkled seeds pointed at one end, is casual in waste places. (Introd. from sw. Asia)

45. LÁTHYRUS L. VETCHLING. WILD PEA. GESSE (Que.)

Style dilated and flattish (not grooved) above, hairy along the inner side (next the free stamen). Sheath of the filaments scarcely oblique at the apex. Otherwise nearly as in *Vicia.* — Herbs of N. Hemisph. and temp. S. Am. and Afr., our species perennial and mostly smooth plants. (*Lathyros*, a leguminous plant of Theophrastus, the name often said to be composed of the prefix, *la*, *very*, and *thuros*, *passionate*, the original plant reputed to be an aphrodisiac.)

a.Leaflets of principal leaves 4 or more. . . . *b*.
  b.Stipules broadly ovate, somewhat hastate, with 2 basal lobes, nearly as large as the fleshy leaflets. . . . . 1. *L. japonicus*.
  b.Stipules semisagittate to semicordate, each with 1 basal lobe, mostly smaller than the scarcely fleshy leaflets. . . . *c*.
    c.Principal stipules semisagittate, 1–10 mm. broad, prolonged into sharp tips and basal lobes; leaflets firm, their petiolules more or less pilose above; lateral calyx-teeth triangular or triangular-lanceolate; corolla violet or purple (rarely white).
      Leaflets of larger leaves 4–10, elliptic, lanceolate or oblanceolate to linear; peduncles and inflorescences (2–9-flowered) nearly equaling to overtopping the subtending leaves. . . . . 2. *L. palustris*.
      Leaflets of larger leaves 10–12, elliptic to ovate; peduncles and inflorescences (5–19-flowered) much exceeded by the subtending leaves. . . 3. *L. venosus*.
    c.Principal stipules semicordate, entire or more or less toothed at base, 7–15 mm. broad; leaflets membranaceous, ovate or elliptic, glabrous throughout (including petiolules); lateral calyx-teeth ovate; corolla yellowish-white. . . . . 4. *L. ochroleucus*.

a.Leaflets of mature leaves 2. . . . *d*.
  d.Stipules sagittate-ovate or -lanceolate, each with 2 prolonged basal lobes; corolla bright yellow. . . . . 5. *L. pratensis*.
  d.Stipules semisagittate, each with 1 prolonged basal lobe; corolla violet, purple, pink or white. . . . *e*.
    e.Stem wingless, glabrous; roots tuberous; leaflets oblong or elliptic, 1.5–3.5 cm. long; flowers about 1.5 cm. long, roseate, fragrant. . . . . . 6. *L. tuberosus*.
    e.Stem winged; roots not tuberous; flowers odorless. . . . *f*.
      f.Petioles wingless; peduncles 1–3-flowered; flowers 6–13 mm. long; legumes 1.5–4 cm. long; annual or biennial.
        Flowers 6–10 mm. long; calyx-teeth lance-linear, long-attenuate, much exceeding the tube; legumes smooth. . . . . . . 7. *L. pusillus*.
        Flowers 9–13 mm. long; calyx-tube nearly equaling the broadly lanceolate acute teeth; legumes covered with pustular-based long hairs. . . . . 8. *L. hirsutus*.
      f.Petioles winged; peduncles 4–10-flowered; flowers 1.5–2.5 cm. long; legumes 5–10 cm. long; perennials.
        Leaflets lanceolate, oblong or oval, mostly 4–9 cm. long; flowers about 2.5 cm. long; legumes 6–9 cm. long, with smooth dorsal suture. . . 9. *L. latifolius*.
        Leaflets narrowly lanceolate, mostly 1–1.5 cm. long; flowers about 1.5 cm. long; legumes 5–7 cm. long, the dorsal suture with toothed ridges. . . . . 10. *L. sylvestris*.

1. L. japónicus Willd. (Japanese), Beach-Pea, Pois de mer (Que.). — *Fleshy* perennial *with widely creeping and forking slender rhizomes; stems* 0.1–1.5 m. or more long, *stiffly branching; stipules foliaceous, broadly ovate, nearly or quite as large as the leaflets*, somewhat hastate, *with 2 basal lobes;* leaflets 4–10, 1–7 cm. long, 0.5–4 cm. broad; peduncles arching or straight, with 3–10 purple or purple and violet or bluish flowers 1.2–3 cm. long; legume firm-chartaceous, 3–7 cm. long. — Highly variable circumpolar species, with several geographic vars.:

a.Stems 1–3 (rarely –10) dm. long, 0.5–2.5 mm. thick (in dried specimens); leaflets thinnish, submembranaceous, green and not strongly glaucous, the better developed ones (on each plant) 1–4(–5) cm. long, 0.7–2.5 cm. broad; tendrils mostly simple; peduncles filiform, 0.5–1.5 mm. thick, often equaling or exceeding the subtending leaves; corolla 1.8–3 cm. long; legumes 3–5 cm. long.
  Plant glabrous or essentially so throughout. . . . . *L. japonicus* (typical).
  Plant more or less pubescent; the stem, lower leaf-surfaces, peduncles, pedicels and calyx all or nearly all densely pilose. . . . . Var. *aleuticus*.
a.Stems 0.2–1.5 m. or more long, 2–5 mm. thick (in dried specimens); leaflets subcoriaceous or heavily fleshy, glaucous, the better developed ones (on each plant) 2–7 cm. long, 1.5–4 (in the rare forma *acutifolius* down to 0.5) cm. wide; tendrils mostly forking; peduncles stoutish, 1–2 mm. thick, definitely shorter than the subtending leaves; corolla 1.2–2.5 cm. long; legumes 3–6.5 cm. long. . . . *b*.
  b.Stems and leaves glabrous or sparsely pilose and glabrate; rachis of raceme, pedicels and calyx-tube glabrous. . . . *c*.

*c.*Corolla 1.5–2.5 cm. long, the standard 1–2 cm. broad; legumes mostly 4–6.5 cm. long; seeds (dried) mostly 4–5 mm. in diameter.
Leaflets elliptic or obovate, blunt, 1.5–4 cm. broad. . . . . . . Var. *glaber.*
Leaflets elliptic-lanceolate, acute, 0.5–1 cm. broad. . . . . . . Var. *glaber*, forma *acutifolius.*
*c.*Corolla 1.2–1.45 cm. long, the standard 0.5–0.8 cm. broad; legumes 3–4.8 cm. long; seeds 3.5–4 mm. in diameter. . . . . Var. *parviflorus.*
*b.*Stem (at least above), lower leaf-surfaces, peduncles, pedicels and calyces densely pilose. . . . . Var. *pellitus.*

L. japónicus (typical). — Gravelly sea-shores, Greenl., Lab. and Nfld. July, Aug. (Hudson Bay; Alaska to Oreg. and Japan; Chile)

Var. **aleùticus** (Greene) Fern. (of the Aleutian Islands), *Pisum maritimum* L., as to Lapland plant. — Coasts, Greenl. to Alaska, s. to Nfld., M.I. and James Bay. Late June–Aug. (N. Eurasia) — The albino, with white corolla, is forma **albinus** Fern.

Var. **glàber** (Ser.) Fern. (smooth), *L. maritimus* Bigel. and var. *glaber* Eames — Gravelly or sandy coast and shores, Lab., Nfld. and lower St. Lawrence, Que., to N.J., inland on L. St. John, Que., Oneida L., N.Y., L. Simcoe, Ont., and Great Lakes, the corolla blue-violet, or in forma **spectábilis** Fassett (showy) crimson. June–Sept. (B.C. to Calif.; Eurasia) — Forma **acutifòlius** (Bab.) Fern. (acute-leaved) local in Nfld. (Nw. Eu.)

Var. **parviflòrus** Fassett (small-flowered). — Shores of L. Nipissing, Ont., L. Erie, O., and L. Winnipeg, Man.

Var. **pellìtus** Fern. (clad in fur). — Coast, Nfld. and Gulf of St. Lawrence, Que., to N.J., inland on L. St. John, Que., on L. Champlain, and locally on Great Lakes. June–Sept. — Forma **cándidus** Fern. (white) has white corollas.

2. L. palústris L. (of marshes), Vetchling. — Perennial with slender creeping rhizomes; stems winged or wingless, 0.1–1.2 m. long; *stipules obliquely lanceolate to ovate, sharp-pointed at both ends;* principal leaves with 2–5 pairs of ovate to linear firm leaflets; *peduncles* slender, 2–9-*flowered*, the *inflorescence nearly equaling to overtopping the subtending leaf;* flowers 1–2.5 cm. long; *lateral calyx-teeth triangular-lanceolate;* corolla purple to violet. — Polymorphous circumboreal species; several vars. with us:

Stem relatively stoutish, winged (rarely wingless), excluding the wings 1.5–3 mm. in diameter below lowest peduncle; larger leaves with 2–5 pairs of elliptic to lanceolate or oblanceolate leaflets 3–8.5 cm. long and 0.7–2.3 cm. broad; flowers 1.5–2.5 cm. long.
Stems, leaves, calyx and legume essentially glabrous. . . . . *L. palustris* (typical).
Stems, leaves, etc., finely pubescent. . . . . Var. *macranthus.*
Stem comparatively slender, 0.5–1.5 mm. in diameter below lowest peduncle; larger leaves with 2 or 3 (rarely -5) pairs of leaflets 1.5–5.5 cm. long and 0.15–1.7 cm. broad; flowers 1–1.8 cm. long.
Stem winged (rarely wingless), 1–6(–8) dm. high; leaflets 2–5 pairs, linear, lanceolate or narrowly oblong, 1.5–5.5 cm. long and 1.5–9 mm. broad; peduncles 2–5-flowered; flowers 1.3–1.8 cm. long.
Stem, leaves, etc., glabrous or nearly so. . . . . Var. *linearifolius.*
Stem, leaves, etc., pubescent. . . . . Var. *pilosus.*
Stem wingless or but slightly winged, up to 1 m. high; leaflets 2–4 pairs, ovate, obovate, elliptic or broadly lanceolate, 2–4.5 cm. long and 6–17 mm. broad; peduncles 3–9-flowered; flowers 1–1.6 cm. long.
Glabrous; leaflets 2 or 3 pairs, broadest at or below the middle, ovate, elliptic or broadly lanceolate; flowers 3–9. . . . . . . Var. *myrtifolius.*
Minutely pilose; leaflets 2–5 pairs, broadest near tip, cuneate-elliptic, broadly subtruncate and retuse at summit; flowers 4. . . . Var. *retusus.*

L. palústris (typical). — Shores, damp thickets and meadows, Anticosti I. to Alaska, s. to N.S., N.E., L.I., (adv. in N.C.), centr. N.Y., O., Ind., Ill. and Mo. June–Sept. (Eurasia)

Var. **macránthus** (T. G. White) Fern. (large-flowered), *L. macranthus* (T. G. White) Rydb. — Similar habitats, w. Nfld. to N.D., s. to N.S., N.E., w. N.Y. and Mich.; Alaska to Oreg. (Asia)

Var. **linearifòlius** Ser. (linear-leaved). — Rather local, Que. to B.C., s. to n. N.Y., Mich., Wisc. and Minn. (Eurasia)

Var. **pilòsus** (Cham.) Ledeb. (hairy). — Damp sands, gravels, shores and marshes, Lab. to Alaska, s. to Nfld., N.S., N.E., N.Y., Mich., Wisc., Minn. and Oreg. (Asia)

Var. **myrtifòlius** (Muhl.) Gray (with leaves of Myrtle), *L. myrtifolius* Muhl. — Meadows, shores and damp thickets, Que. to Minn., s. to Mass., N.C., Tenn. and Mo. — Forma **pállidus** Farw. (pale) has whitish corollas.

Var. **retùsus** Fern. & St. John (notched at rounded apex). — Local, St. P. et Miq.; Sable I., N.S.

3. L. venòsus Muhl. (veiny). — Mostly stouter than no. 2; stem 4-angled, ridged and striate, up to 2 m. long; stipules semisagittate; *leaflets* of principal leaves 10 *to* 12, *ovate or elliptic,* veiny beneath, 1.5–6.5 cm. long, 1–3 cm. broad; *peduncles one-half to two-thirds length of subtending leaf, 5–19-flowered;* flowers 1.5–2 cm. long, purplish; upper calyx-lobes short and convergent, the 3 lower linear-lanceolate. — Three vars.:

Stipules of the larger leaves ovate-lanceolate, 2–3.5 cm. long, 4–10 mm. broad; flowers 5–14; plant glabrous or nearly so. . . . . *L. venosus* (typical).

Stipules linear-lanceolate, the larger 0.8–2 cm. long and 1.5–5 mm. broad; flowers 5–19.

Plant glabrous or sparsely hirtellous; leaflets ovate to lanceolate. . . . . Var. *meridionalis.*

Plant copiously hirtellous; leaflets elliptic. . . . . Var. *intonsus.*

L. venòsus (typical). — Rich woods, thickets and banks of streams, w. N.J. to Wisc., s. to Va. and W.Va. May, June.

Var. **meridionàlis** Butt. & St. John (southern). — Wooded slopes, w. Va. to Ark., s. to Ga., La. and e. Tex.

Var. **intónsus** Butt. & St. John (unshaved), *L. Rollandii* Vict. & Rousseau — Dry or sandy soils, Gaspé Pen., Que.; s. Ont. to Sask., s. to W.Va., Tenn. and Mo.

4. L. ochroleùcus Hook. (yellowish-white). — Perennial with slender rhizomes, *glabrous throughout;* stem slender, wingless, up to 1 m. high; *stipules semicordate, membranaceous, half as large as the* 4 or 6 (rarely 8) *thin* ovate or elliptic blunt *leaflets, the larger stipules 7–15 mm. broad; peduncles* slender, *much overtopped by the subtending leaves,* 5–10-flowered; *lateral calyx-teeth ovate; corolla yellowish-white,* 1.5–1.8 cm. long. — Dry or moist woods, slopes and rocky banks, w. Que. to n. B.C., s. to w. Vt., N.Y., centr. Pa., n. O., Ind., n. Ill., Ia., S.D., Wyo., Ida. and Oreg. May–July.

5. L. PRATÉNSIS L. (of meadows), YELLOW VETCHLING. — Slender perennial with creeping filiform rhizomes; stems loosely ascending or straggling, much branched, up to 7 dm. long; *stipules sagittate-ovate or -lanceolate, each with* 2 *prolonged basal lobes; leaflets* 2, lanceolate to linear-oblong, acute, bright green, 2–8 *mm. broad,* with a prolonged simple or forking tendril; *peduncles prolonged,* 4–10-flowered; calyx-teeth all slender; *corolla bright yellow,* 1.3–2 cm. long. — Springy slopes, meadows, shores and roadsides, Nfld. to Ont., s., locally, to N.S., N.E., N.Y., O., Mich. and Ill. June–Aug. (Natzd. from Eu.; possibly native in w. Nfld.)

6. L. TUBERÒSUS L. (with tubers), TUBEROUS VETCHLING. — Slender perennial; the rootstocks bearing numerous tubers; stems glabrous; *leaves and stipules thin; petioles and tendrils filiform;* the 2 oblong *leaflets* 1.5–3.5 *cm. long;* peduncles filiform, 3–6-flowered; the *fragrant pink to violet flowers* about 1.5 cm. long. — Fields, meadows and roadsides, local, Vt. to s. Ont. and Wisc. June–Aug. (Introd. and natzd. from Eu.)

7. L. pusíllus Ell. (very small). — Low *annual or biennial,* freely branching at base; *stems winged,* up to 7 dm. long; *stipules* semisagittate, *with prolonged erect lance-falcate sharp-pointed blade* and shorter pointed basal lobe; *leaflets* 2, linear or narrowly lanceolate, with forking tendril; *peduncle* rather short, 1–3-*flowered; calyx-tube much shorter than the lance-linear long-attenuate teeth; corolla* 6–10 *mm. long,* purplish, *with obovate or elliptic standard; legume* linear, *glabrous.* — Woods, meadows, ditches and damp gravels, Fla. to Tex., n. to N.C., Mo. and se. Kans. April, May.

8. L. HIRSÙTUS L. (hirsute). — Coarser than no. 7; stipules ascending to spreading; leaflets oblong or lanceolate; *peduncle much prolonged,* loosely 2–3-flowered; *calyx-tube nearly equaling the broadly lanceolate acute teeth; corolla* 9–13 *mm. long, with suborbicular standard; legume* broadly linear, *long-pubescent with pustular-based hairs.* — Roadsides and borders of fields and thickets, Va. to Ala. and Miss. May, June. (Natzd. from Eu.)

9. L. LATIFÒLIUS L. (broad-leaved), EVERLASTING OR PERENNIAL PEA, POIS VIVACE (Que.). — Tall high-climbing *perennial with broadly winged stems; leaves and stipules coriaceous and veiny;* petioles winged; *leaflets* 2, *lanceolate, oblong or oval, mostly* 4–9 *cm. long;* peduncles stiff, 4–10-flowered; *flowers* odorless, purple, pink or white, *about* 2.5 *cm. long; legumes* 6–9 *cm. long, with smooth dorsal suture; seeds tuberculate.* — Esc. from cult. to roadsides, thickets and waste places, N.E. to Ind., s. to Va., Mo. and Kans. June–Sept. (Introd. from Eu.)

10. L. SYLVÉSTRIS L. (of woods), EVERLASTING OR PERENNIAL PEA. — Similar to no. 9; *leaflets narrowly lanceolate, mostly* 1–1.5 *cm. long; flowers about* 1.5 *cm. long; legumes* 5–7 *cm. long, the dorsal suture with toothed ridges; seeds smoothish or obscurely pebbled.* — Roadsides and waste places, local, Que. to Mich. and Conn. June, July. (Introd. from Eu.)

The familiar SWEET PEA, L. ODORÀTUS L. (fragrant), an annual with winged stem, 2 leaflets, peduncles 1–3-flowered, the fragrant flowers 2 cm. or more long, of varying colors, spreads to waste ground but scarcely persists. (Introd. from s. Eu.)

PÌSUM L. (ancient name) SATÌVUM L. (planted), the GARDEN-PEA, an annual with foliaceous calyx-lobes, strongly dilated style, mostly white corollas and subinflated legumes, and var. ARVÉNSE (L.) Poir. (of cult. fields), the FIELD-PEA, with smaller pink or purple and green flowers and smaller legumes and seeds, spread from cult. but rarely persist. (Introd. from Eurasia)

46. ÀPIOS Medic. GROUNDNUT. WILD BEAN. POTATO-BEAN

Calyx somewhat 2-lipped, the 2 lateral teeth being nearly obsolete, the upper very short, the lower one longest. Standard very broad, reflexed; the long scythe-shaped keel strongly incurved, at length coiled. Stamens diadelphous. Legume straight or slightly curved, linear, elongated, thickish, many-seeded. — Perennials of e. N. Am. and China, twining and climbing over bushes; the rootstocks with tuberous enlargements. Leaflets 3–9, ovate-lanceolate, obscurely stipellate. Flowers in dense and short often branching racemes. (Name *apios*, Greek for *pear;* from the somewhat pyriform tuberous enlargements of the rootstock.)

1. A. americàna Medic. (American), PÉNACS or PATATES EN CHAPELET (Que.). — *Rootstocks moniliform, the tuberous enlargements numerous;* stems and leaves glabrous, or pilose in forma **pilòsa** Steyerm. (hairy); racemes compact, strongly rounded at summit, the mature denuded rachis 3–17 cm. long; *calyx* 4–5 *mm. long; corolla purple-brown and mauve,* violet-scented, or clavate and closed and scarcely exserted in forma **cleistógama** Fern. (pollinated without expanding), *standard unappendaged* at summit. (*A. tuberosa* Moench; *Glycine Apios* L.) — Rich thickets, N.B. to Minn. and Colo., s. to N.S., N.E., L.I., Fla., La. and Tex. July–Sept.

Var. **turrígera** Fern. (bearing towers). — Racemes loosely lanceolate- or ovoid-attenuate, with prolonged tips; mature denuded rachis 1–2 dm. long. — Local, S.C. to La. and Okla., n. to se. Va., Ill. and e. Kans.

2. A. Priceàna Robins. (for its discoverer, SARAH FRANCES PRICE, 1849–1903). — *Tuber solitary, up to* 1–2 *dm. thick; calyx* 8–10 *mm. long; corolla* much larger, *greenish-white,* with purple tips, the *standard bearing a fleshy knob at apex.* (*Glycine* Britt.) — Woods and thickets, Ky. and Tenn. July–Sept.

47. PHASÈOLUS L. KIDNEY-BEAN

Calyx 5-toothed or 5-cleft, the two upper teeth often shallower. Keel and style coiled. Stamens diadelphous. Stigma oblique or lateral. Legume scythe-shaped, several–many-seeded, tipped with the hardened base of the style. Cotyledons thick and fleshy, rising out of the ground in the cult. species, nearly unchanged in germination. — Twining herbs of warm reg. with pinnately 3-foliolate stipellate leaves. Flowers racemose, produced in summer and autumn. (The ancient name of the Kidney-Bean.)

1. P. polystáchios (L.) BSP. (many-spiked), WILD BEAN. — Perennial, twining or trailing; *leaflets* ovate to roundish, short-pointed, 4–10 cm. long, *firm or subcoriaceous, minutely scabridulous above,* softly subvelutinous beneath, not quickly wilting, adherent when fresh; small purple flowers on slender pedicels in simple or forking elongate long-stalked racemes; *rachis usually short-hispid; calyx* (dry) relatively thin, *its veins and veinlets evident;* legumes 4–7 cm. long, drooping, 4–5-seeded; *seeds strongly flattened on both sides, black or black and gray,* 5–10 mm. long, 5–6.5 mm. broad. — Dry pine- or oak-woods or sandy thickets, Fla. to Tex., n. to s. N.J., W.Va., O., Ind., s. Ill., Ia. and Neb. July–Sept.

Var. **aquilònius** Fern. (northern). — *Leaves submembranaceous,* promptly wilting, *smooth and glabrous or glabrescent above,* less pilose beneath, the leaflets up to 1.3 dm. long; *rachis* usually *with inflexed pilosity; calyx* (dry) of thicker texture, *its veins obscure or invisible; seeds strongly biconvex, reddish-black,* slightly smaller. — S. Ct. and se. N.Y. to Del. and upland of N.C.

Several cult. species tend to spread but hardly to persist in our climate: P. VULGÀRIS L. (KIDNEY-B.) and its var. HÙMILIS Alef. (BUSH-B.); P. COCCÍNEUS L. (SCARLET RUNNER) and P. LIMÉNSIS Macfad. (LIMA B.).

48. VÍGNA Savi

Habit and floral characters nearly as in *Phaseolus*, but the keel merely arcuate, not coiled at the tip; wing-petals with prominent auricle at base. — Twining herbs (ours annual) of trop.

and warm-temp. reg. with pinnately 3-foliolate leaves. (Dedicated to *Dominico Vigna*, Italian scientist of the 17th century.)

1. V. sinénsis (L.) Endl. (Chinese), Cow-Pea, Black-eyed Pea. — Annual; leaflets broadly ovate, often very oblique or sometimes slightly contracted above an obtusely hastate base; flowers few, loosely subcapitate at the end of the long stiffish peduncle, purplish; legumes 1.5–3 dm. long; seeds reniform to subglobose, with a dark ring around the oblong hilum. — Cult. and natzd. or adv. by roadsides and in fields and thickets, Fla. to Tex., n. to Del., Ind., Ill. and Mo. (Introd. from Asia)

49. STROPHOSTỲLES Ell. Wild Bean

Keel of the corolla with the included stamens and style elongated, strongly incurved, not spirally coiled. Legume linear, terete or flattish, straight or nearly so. Seeds quadrate or oblong with truncate ends, mealy-pubescent or glabrate; hilum linear. Otherwise as *Phaseolus*. — Stems prostrate or climbing, more or less retrorsely hairy. Stipules and bracts striate. Perennials of warm reg., sometimes merged with *Phaseolus*. (Name from the Greek *strophe, a turning*, and *stylos, a style*)

Leaves green, glabrous or only remotely strigose; calyx-tube glabrous or nearly so; standard 1–2 cm. broad; legume 3.5–10 cm. long; seeds 3–12 mm. long, scurfy.
Principal leaflets ovate, either lobed or unlobed, the terminal one 2–8 cm. long and 0.8–6.5 cm. broad; legumes 3.5–10 cm. long; seeds 6–12 mm. long; root annual. . . . . 1. *S. helvola.*
Principal leaflets oblong to narrowly ovate, the larger 2–5 cm. long and 0.4–2 cm. broad; legumes 3.7–7 cm. long; seeds 3–6 (rarely –10) mm. long; root perennial. . . . . 2. *S. umbellata.*
Leaves gray with abundant silky-strigose pubescence; calyx-tube hairy; standard 5–8 mm. broad; legumes 2–3.5 cm. long; seeds 2–4 mm. long, lustrous-black. 3. *S. leiosperma.*

1. S. hélvola (L.) Ell. (yellowish). — *Annual; stem branching*, trailing or twining, 0.2–2 m. long, *green and glabrous or spreading-pilose*, mostly horizontally branched at base; principal leaves *with 3-lobed and more or less panduriform or quite unlobed ovate and acuminate leaflets green and glabrous or sparsely strigose, the larger* (of each plant) 2–6.5 cm. long and 0.8–4 *cm. broad;* mature peduncles 0.5–2 dm. long; *calyx-tube glabrous or nearly so;* corolla pink or purple, turning greenish, 8–13 mm. long, the *standard* 1–2 *cm. broad; legumes* terete, 3.5–8 *cm. long; seeds* 4–8, *oblong*, covered with a deciduous felty drab coat, 6–9.5 *mm. long*, the hilum 4–5 mm. long. — Damp thickets and shores, Fla. to Tex., n. to e. Mass., sw. Que., s. Ont., Mich., Wisc., Minn. and S.D. June–Oct.

Var. missouriénsis (S. Wats.) Britt. (of Missouri). — Climbing 3–10 m.; *principal leaflets* rounded-ovate to rhombic-ovate, unlobed, *with sides gradually rounded to blunt or barely acutish tips*, the terminal one 4–8 cm. long and 3–6.5 cm. broad; flowers 1–1.5 cm. long; legume 5–10 cm. long; *seed* 8–12 *mm. long*, hilum 5–7 mm. long. (*S. missouriensis* (S. Wats.) Small) — River-banks and alluvial thickets, Fla. to Ark., n. to Pa., s. Ill., Mo. and Kans.

2. S. umbellàta (Muhl.) Britt. (umbelled). — *Perennial;* stems more slender, their young tips retrorse-pilose; *leaflets oblong to narrowly ovate*, subcoriaceous, the larger (of each plant) 2–5 cm. long and 0.4–2 *cm. broad*, strigose-pilose to glabrous beneath; mature peduncles 1–2.5 dm. long; *legumes* 3.7–5.5 *cm. long; seeds* quadrate-short-oblong to subcubical, 3–4.5 (rarely –6) *mm. long* and 2–3 *mm. thick*, scurfy. — Sandy woods, clearings and fields, Fla. to Tex., n. to L.I., s. Ind., Ill., Mo. and Okla. July–Oct.

Var. paludígena Fern. (dweller in marshes). — *Glabrous* or glabrescent; *legumes* 5–7 *cm. long; seeds* quadrate-oblong, 5–10 *mm. long*, 3.5–5 *mm. thick*. — Fresh to brackish tidal marshes of Chesapeake Bay drainage, D.C. and Va.

3. S. leiospérma (T. & G.) Piper (smooth-seeded). — *Annual; stem and leaves gray with abundant silky-strigose pubescence; leaflets linear to narrowly ovate; the larger* 2–4.5 cm. long, 0.3–2 *cm. broad;* mature peduncles 3–12 cm. long; *calyx-tube pubescent; standard* 5–8 *mm. broad; legumes* 2–3.5 *cm. long; seeds square*, 2–4 *mm. long, shining*. (*S. pauciflora* (Benth.) S. Wats.) — Dry, mostly sandy, soils, Miss. to Tex., n. to Ind., Wisc., Minn., Neb. and Colo. July–Oct.

50. CLITÒRIA L. Butterfly-Pea

Standard of expanded flower much larger than the rest of the corolla, erect, rounded, notched at the top, not spurred on the back; keel small, shorter than the wings, incurved, acute. Stamens monadelphous below. Legume linear-oblong, flattish, knotty, several-seeded, pointed with the

base of the style. — Erect or twining perennials of warm reg., with mostly pinnate 3-foliolate stipellate leaves and very large as well as small and later cleistogamous flowers. Peduncles of early inflorescences 1–3-flowered; bractlets opposite, striate. (Derivation from the small keel, suggesting the mammalian *clitoris*.)

1. C. marìana L. (of Maryland). — Low, ascending or twining, smooth; leaflets oblong-ovate or ovate-lanceolate: stipules and bracts subulate; peduncles short; the showy pale blue flowers 5–6 cm. long. (*Martiusia* Small) — Dry soil, Fla. to Ariz., n. to s. N.Y., W.Va., s. O., s. Ind., s. Ill. and Ia. June–Aug. — Fresh seed viscid and adhesive.

51. CENTROSÈMA (DC.) Benth. BUTTERFLY-PEA

Corolla, etc., much as in *Clitoria*, but the spreading standard with a spur-shaped projection on the back near the base; keel broad. Legume long and linear, flat, pointed with the subulate style, many-seeded, thickened at the edges, the valves marked with a raised line on each side next the margin. — Twining perennials of warm reg., with 3-foliolate stipellate leaves and large showy flowers. (Name from the Greek *centron, a spur*, and *sema, a standard*.) BRADBURYA Raf.

1. C. virginiànum (L.) Benth. (Virginian). — Rather rough with minute hairs; leaflets of most well-developed leaves ovate, tapering gradually to a subacuminate apex, very veiny, shining; peduncles 1–4-flowered; calyx-teeth linear-subulate; corolla violet, 2.5–3.5 cm. long; legume straight, 7–14 cm. long. (*Bradburya* Ktze.) — Sandy woods and fields, Fla. to Tex., n. to s. N.J., Tenn. and Ark. July, Aug.

Var. ellípticum (DC.) Fern. (elliptic). — Leaflets of most well developed leaves ovate-oblong to elliptic, blunt or rounded at tip. — Fla. to La., n., locally, to se. Va. and s. Ky.

52. DÓLICHOS L.

Calyx campanulate, with deltoid teeth, the upper pair united nearly or quite to the apex. Standard orbicular, with incurved auricles at base; keel strongly incurved, beaked. Style thickened above, bearded. Flowers fasciculate-racemose. Legumes linear and falcate, or oblong-lunate, compressed; seeds several. — Herbs of warm reg., with 3-foliolate pinnate leaves, small stipules, nodose rachis and caducous bracts. (The Greek *dolichos, long*, a word also employed by Theophrastus as the name of some kind of pulse.)

1. D. LÁBLAB L. (native East Indian name), HYACINTH-BEAN. — Stoutish twining annual 3–6 m. in length; leaflets large, deltoid-ovate; flowers purple to white; legumes 2 cm. broad. — Often cult. for ornament and in trop. countries for its seeds; tending to esc., D.C. to O. and Ga. Aug.–Oct. (Introd. from India)

53. GLYCÌNE L.

Calyx with the 2 upper lobes united to or above the middle. Standard suborbicular, sub-auriculate at base, not inflexed; wings narrow, slightly adhering to short obtuse keel. Stamens all anther-bearing. Style slender-tipped, beardless. Legume linear to falcate, compressed, cellulose-septate between the estrophiolate globose seeds. — Herbs nat. of Old World, with stalked leaflets (mostly 3), and small purplish or pale flowers solitary or fascicled on the continuous rachis. (Name from the Greek *glycys, sweet*.)

1. G. MÁX (L.) Merr. (old name), SOY-BEAN. — Erect, bushy-branched brown-villous annual; leaflets ovate, 0.7–1.5 dm. long; flowers inconspicuous, in fascicles; legumes pendulous, long-stalked, 4–8 cm. long, villous; seeds 2–4, green to black. (*G. Soja* (L.) Sieb. & Zucc.) — Roadsides, old fields, etc., southw., n. to Del., Mich. and Ill., esc. from cult. Aug., Sept. (Introd. from e. Asia)

54. AMPHICÁRPA Ell. HOG-PEANUT

Flowers of 2 (or 3) kinds; those of the racemes from the upper branches perfect; those near the base and on filiform creeping branches with the corolla none or rudimentary, and few free stamens, but fruitful; reduced flowers of slightly different form sometimes also on aerial racemes. Calyx about equally 4 (rarely 5)-toothed. Stamens diadelphous. Legumes of the upper flowers, when formed, somewhat scimitar-shaped, stipitate, 3–4-seeded; of the lower ones commonly subterranean and fleshy, obovate or pear-shaped, indehiscent, ripening usually but one large seed. — Low and slender e. Am. and Asiatic perennials; the twining stems clothed with brownish hairs. Leaves pinnately 3-foliolate; leaflets rhombic-ovate, stipellate. Petals mostly purplish. Bracts persistent, round, partly clasping, striate, as well as the stipules.

(Name from the Greek *amphi, of both kinds*, and *carpos, fruit;* the allusion to the two kinds of fruits.) FALCATA J. F. Gmel.

1. A. bracteàta (L.) Fern. (bracted). — Stems capillary, retrorsely *appressed-pubescent or sparingly hirsute; leaflets* minutely strigose on both surfaces, *commonly* 2–6 (rarely –8) *cm. long;* racemes nodding, simple (rarely forked), with 2–15 *pale lilac* (sometimes purple) *to white flowers* 0.9–1.3 *cm. long;* floral bracts 2–2.5 mm. long; calyx-tube 4–5 mm. long, glabrous or strigose; blades of keel-petals longer than claws; *aerial legumes strigose or smooth on the sides,* 1.5–3 *cm. long,* with beak (style) 1–3 mm. long; subterranean fruits strigose. (*A. monoica* (L.) Ell.) — Damp woodlands, Que. to Man. and Mont., s. to N.S., N.E., L.I., Fla., La. and Tex. Aug., Sept. — Passing insensibly into

Var. **comòsa** (L.) Fern. (bearded), PITCHER'S H. — *Coarser* throughout; *stems and petioles densely villous-hirsute* with mostly reflexed sordid hairs; *leaflets more coarsely pubescent,* often softly so beneath, *mostly* 5–10 *cm. long;* racemes simple or forking; *flowers often deeper purple,* 1.1–1.6 *cm. long;* bracts longer; calyx-tube slightly longer; blades of keel-petals about equaling claws; *aerial legumes villous-hirsute,* 2–4 *cm. long,* with beak 2–5 mm. long. (*A. Pitcheri* T. & G.; *Glycine comosa* L., not *Falcata comosa* of most recent auth.) — Richer, often calcareous or alluvial soil, Me. to N.D., s. to Va., Tenn. and Tex.

55. GALÁCTIA P. Br. MILK-PEA

Keel scarcely incurved. Stamens diadelphous or nearly so. Legumes linear, flat, several-seeded (a few of them rarely subterranean and fleshy or deformed). — Low mostly prostrate or twining perennial herbs of warm reg. Leaflets usually 3, stipellate. Flowers in somewhat interrupted or knotty racemes, purplish; in summer. (Name from *gala, milk,* Patrick Browne originally stating that it has "milky branches".)

Peduncle and raceme stiff, 2–15 cm. long; flower-buds straight or but slightly curved at tip, their basal bracts ovate.
  Stems depressed, rarely twining except at tips, minutely puberulent; leaflets short-strigose to glabrous beneath; calyx strigillose to glabrous, 6–9 mm. long; keel-petals 10–14 mm. long. . . . . 1. *G. regularis.*
  Stems and branches intricately twining, divergently or retrorsely pilose; leaflets pilose beneath; calyx spreading-pilose, 4–5.5 mm. long; keel-petals 6–10 mm. long. . . . . 2. *G. volubilis.*
Peduncle and raceme filiform, flexuous above, the well-developed ones 0.7–3 dm. long; stems twining, minutely strigillose; flower-buds with prolonged falcate beak, their basal bracts linear- or lance-subulate; keel-petals 9–10 mm. long. 3. *G. Macreei.*

1. G. regulàris (L.) BSP. (according to rule, Linnaeus evidently thinking it typical of *Dolichos,* with which he placed it). — *Stem prostrate or depressed,* twining (if at all) only at tip, *minutely puberulent; leaflets* elliptical or ovate-oblong, firm, *short-strigose* to glabrous beneath; *peduncles and rachises* of racemes *stiff, minutely puberulent,* 2–11 *cm. long; fully developed* but unexpanded *calyces strigillose to glabrous, straight or nearly so,* 6–9 *mm. long, their basal bracts ovate;* corolla violet-purple; *keel-petals* 10–14 *mm. long;* legume canescent-strigose, 2–5 cm. long. — Dry sandy soil, Fla. to La., n. to se. N.Y. and e. Pa.; recorded (but no specimens seen) from Tenn., Mo. and Kans. June–Aug.

2. G. volùbilis (L.) Britt. (twining). — *Stems and branches intricately twining, pilose with divergent to reflexed hairs; leaflets* oval to oval-oblong, the larger 1–3 cm. broad, *minutely pilose beneath,* glabrous or glabrate above; *peduncles and rachises stiff, pilose, mostly* 3–15 *cm. long,* floriferous nearly to base, the true peduncle only 1 mm.–3.5 cm. long, *the groups of flowers* 0.5–2 *cm. apart; full-grown calyces just before expansion slightly curved, spreading-pilose,* 4–5.5 *mm. long, their basal bracts ovate;* corolla pink or roseate, essentially unicolorous; *keel-petals* 6–7 *mm. long;* legumes densely pilose, 2–5.5 cm. long. — Dry thickets and borders of woods, Fla. to Tex., n. to L.I., N.J., e. Pa., W.Va., Ind. and Kans. July, Aug.

Var. **mississippiénsis** Vail (of the Mississippi Valley). — *Leaflets strigose-pilose above; keel-petals* 6–10 *mm. long.* — Ala. and Miss., n. to Ky., Ill., Mo. and se. Kans.

3. G. Macreèi M. A. Curtis (for its discoverer, JAMES FERGUS MCREE, 1794–1869). — *Filiform stem* and branches intricately twining, *minutely retrorse-strigillose;* leaflets oblong, the larger 0.5–2 cm. broad, smooth, or strigose beneath; *peduncles and flexuous rachises filiform, retrorsely strigillose to glabrous, mostly* 0.7–3 *dm. long,* flowering well above the base, the true peduncles 3–7 cm. long, *the groups of flowers* 1.5–4 *cm. apart; full-grown* unexpanded *calyx* subappressed-pilose, *falcate-beaked,* 6–10 *mm. long, its basal bracts linear- or lance-subulate;* corolla pink, with deep purple eye; *keel-petals* 9–10 *mm. long;* legumes minutely strigose, 3–7 cm. long. — Damp or wet thickets, low woods and pond-margins, Fla. to Tex., n. on Coastal Plain to se. Va. July, Aug.

56. PUERÀRIA DC.

Calyx with the two upper lobes united. Keel ascending or arcuate at tip, about equaling wings; the standard suborbicular or obovate. Stamens monadelphous, with the axillary one free at base. Legume flattish, continuous or with internal partitions, the ovules numerous, the seeds suborbicular or transversely ovoid, compressed. — High-climbing ligneous to herbaceous plants with pinnately 3-foliolate leaves, the leaflets entire to palmately lobed, the stipules herbaceous, stipels subulate. Racemes simple or compound, axillary, the purple flowers fascicled at nodes of the rachis, the small bracts deciduous, the bracteoles tiny. Small Asiatic genus. (Named for *M. W. Puerari*, Swiss botanist, 1765–1845.)

1. P. lobàta (Willd.) Ohwi (lobed), Kudzu-vine. — Very high-climbing, the twining stems becoming ligneous and up to 2.5 cm. thick; new growth pilose; leaflets entire or coarsely lobed, up to 1.8 dm. long; racemes peduncled, elongate, with caducous long-tipped bracts; corollas reddish-purple, with rich fragrance of grapes. (*P. Thunbergiana* (Sieb. & Zucc.) Benth.) — Borders of woods and fields, se. states (rapidly spreading), n. to Pa. and Tenn., rarely flowering n. of Va. Late Aug., Sept. (Introd. and natzd. from e. Asia) — Originally introd. for its farinaceous tuberous roots and for fiber; much used as a quick-growing ornamental climber.

57. RHYNCHÒSIA Lour.

Stamens diadelphous. Ovules only 2. Legume 1–2-seeded, flat, 2-valved. — Perennial herbs of warm reg., with leaves pinnately 3-foliolate, or with a single leaflet, not stipellate. Flowers yellow, racemose or clustered. (Name from the Greek *rhynchos, a beak*, from the shape of the keel.)

Stems trailing or twining, spreading- or reflexed-pilose.
  Racemes close, 1–4 cm. long, short-peduncled. . . . . 1. *R. difformis.*
  Racemes open, 0.5–3 dm. long, long-peduncled. . . . . 2. *R. latifolia.*
Stems stiffy erect, tomentulose-pilose with appressed-ascending hairs; axillary
  racemes close, 1–4 cm. long. . . . . 3. *R. tomentosa.*

1. R. diffórmis (Ell.) DC. (with two forms — of leaves). — *Trailing or twining*, the slender *stems and branches with spreading or reflexed pilosity;* earliest leaves with a single reniform leaflet, the later with 3 *rounded to ovate or elliptic leaflets* 2–5 *cm. long; racemes close,* 1–4 *cm. long, few-flowered, short-peduncled, shorter than the petioles;* calyx 8–11 mm. long, about equaling the yellow corolla, 4-parted, the upper lobe 2-cleft; legume oblong, with falcate tip, 1.5–2 cm. long. (*R. tomentosa* sensu recent auth., not (L.) H. & A.) — Dry sandy woods and clearings, Fla. to Tex., n. to Va. and se. Mo. Late June–Aug.

2. R. latifòlia Nutt. (broad-leaved). — Coarser, with shorter dense pilosity; *leaflets* of principal leaves 3–7 *cm. long; racemes long-peduncled, elongate and open, becoming* 0.5–3 *dm. long; calyx* 1.1–1.5 *cm. long.* — Woods, thickets and sandy prairies, La. and Tex., n. to s. Mo. and Okla. Late June–Aug.

3. R. tomentòsa (L.) H. & A. (tomentose; with close soft pubescence). — *Stems erect,* 1.5–9 dm. high, *tomentulose-pilose with appressed-ascending pubescence;* leaflets 3, oblong to oval, canescent-tomentose beneath, 2–7 cm. long; *racemes dense, subsessile, shorter than the petioles;* calyx 5–9 mm. long; legumes blunt or short-beaked. (*R. erecta* (Walt.) DC.) — Dry sandy woods and clearings, Fla. to Tex., n. to Del., Md., D.C. and Tenn. Late June–Sept.

FAM. 84. LINÀCEAE (Flax Family)

Herbs (rarely shrubs) with the regular and symmetrical hypogynous flowers 4–6-merous throughout, strongly imbricated calyx and convolute petals, 5 stamens monadelphous at base, and an 8–10-seeded capsule having twice as many locules as there are styles.

Flowers 5-merous. . . . . 1. *Linum.*
Flowers 4-merous. . . . . 2. *Millegrana.*

1. LÌNUM L. Flax. Lin (Que.)

Sepals (usually persistent), petals, stamens and styles 5, regularly alternate with each other. Capsule of 5 united carpels (into which it splits in dehiscence), 5-locular; with 2 seeds hanging from the summit of each locule, which is partly or completely divided into two by a false partition projecting from the back of the carpel, the capsule thus becoming 10-locular. Seeds

anatropous, mucilaginous, flattened, containing a large embryo with plano-convex cotyledons. — Herbs, nearly world-wide, with tough fibrous cortex, simple and sessile entire leaves without stipules but often with glands in their place, and with corymbose or panicled flowers. Corolla usually ephemeral. (The classical name of Flax.)

*a.*Fruiting pedicels 1–4 cm. long; petals blue or white; sepals without marginal glands. . . *b.*

*b.*Leaves alternate, very numerous; petals blue, 1 cm. or more long; capsules 5–10 mm. high.

Pedicels erect; sepals long-acuminate, the inner ciliolate-serrulate; stigmas linear; annual, mostly simple at base. . . . . 1. *L. usitatissimum.*

Pedicels arching and spreading; sepals blunt or short-mucronate, not ciliolate; stigmas capitate; perennial, usually with several stems from the crown. . . . . 2. *L. Lewisii.*

*b.*Leaves chiefly opposite, 2–8 pairs below the inflorescence; petals white, with yellow base, 2–5 mm. long; capsules 2–3 mm. high. . . . . . . 3. *L. catharticum.*

*a.*Fruiting pedicels 0.5–10 mm. long; petals yellow; inner sepals often glandular-margined. . . . *c.*

*c.*Outer as well as inner sepals coarsely glandular-serrate, 4–10 mm. long; styles united below; petals 8–15 mm. long; leaves with a pair of stipular (basal) glands.

Sepals soon deciduous; capsule with dark cartilaginous thickenings at base; individual carpels blunt or merely subacute. . . . . 4. *L. rigidum.*

Sepals persistent; capsule without basal thickenings; individual carpels or valves sharply pointed. . . . . 5. *L. sulcatum.*

*c.*Outer sepals entire, 1.5–4 mm. long; styles free to base; petals rarely 1 cm. long; leaves without stipular glands. . . . *d.*

*d.*Capsule ovoid, rounded to conical at summit, 2–3 mm. high.

Middle and upper leaves linear, 1–2 mm. broad, tapering to a subulus 0.5–1 mm. long; capsule rounded at summit, beakless, the individual carpels bluntish or with short incurved tip. . . . . 6. *L. floridanum.*

Middle and upper leaves narrowly oblong to elliptic, 1–6 mm. broad, the subulus shorter; capsule pointed, the individual carpels or their valves tapering to erect points. . . . . 7. *L. intercursum.*

*d.*Capsule oblate, strongly depressed, much lower than broad, 1–2 mm. high. . . . *e.*

*e.*All but lowest leaves alternate, firm, nonviscid; stem terete; inflorescence corymbiform; outer sepals lance-ovate to lance-attenuate, firm, 2–3.5 mm. long.

Inner sepals glandular-erose or ciliate; leaves firm; branches stiffly fastigiate. . . . . 8. *L. medium.*

Inner sepals with smooth margins or with few promptly deciduous glands; leaves thinnish; branches filiform, loosely ascending or spreading. . . . . 9. *L. virginianum.*

*e.*All or all but the upper leaves of main stem opposite, submembranaceous, viscidulous; stem striate-angled; inflorescence paniculate; outer sepals elliptic, herbaceous, 1.5–2.5 mm. long. . . . . . . 10. *L. striatum.*

1. L. usitatíssimum L. (most useful), Common F. — Erect *annual* up to 7.5 dm. high, corymbosely branched above; leaves linear-lanceolate, numerous; flowers and fruits on *long erect pedicels; sepals long-acuminate, the inner with serrulate margins; petals blue, 1 cm. or more long; stigmas linear; capsule* globose-ovoid, 7–10 *mm. high.* — Waste places, railroad-yards, etc., a casual weed. June–Sept. (Introd. from Eu.) Fig. 1251.

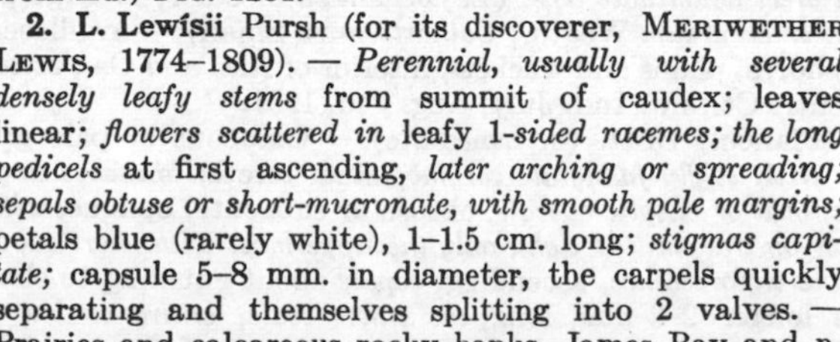

2. L. Lewísii Pursh (for its discoverer, Meriwether Lewis, 1774–1809). — *Perennial, usually with several densely leafy stems* from summit of caudex; leaves linear; *flowers scattered in* leafy *1-sided racemes; the long pedicels* at first ascending, *later arching or spreading; sepals obtuse or short-mucronate, with smooth pale margins;* petals blue (rarely white), 1–1.5 cm. long; *stigmas capitate;* capsule 5–8 mm. in diameter, the carpels quickly separating and themselves splitting into 2 valves. — Prairies and calcareous rocky banks, James Bay and n.

1252. L. Lewisii.

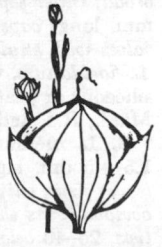

1251. L. usitatissimum.

Ont. to Alaska, s. to Wisc., Tex., n. Mex. and s. Calif. June, July. Fig. 1252. — A polymorphic group, in need of careful study. An isolated representative (needing further study) in Pendleton Co., W.Va. (*Allard*).

L. grandiflòrum L. (large-flowered), Flowering F., an annual with red petals 1–2 cm. long, spreads locally from gardens. (Introd. from n. Afr.)

3. L. cathárticum L. (cathartic), Fairy-F. — Delicate annual or winter-annual with crowded basal leaves and often decumbent base; *stems* 1–many, *filiform*, 0.5–3 dm. high *with* 2–8 *pairs of opposite elliptic leaves below the loose leafy panicle; pedicels* erect, capillary, *in fruit becoming* 1–2.5 *cm. long;* sepals lance-oblong, acuminate, 2–3.5 mm. long; petals longer, white, with yellow base; capsule depressed-ovoid, splitting promptly into 10 valves. (*Cathartolinum* Small) — Calcareous slopes and openings, w. Nfld.; locally natzd. from Eu. in old fields and along calcareous ditches, etc., C.B. to Ont., s. to centr. N.S., s. N.B., n. N.E. and centr. N.Y. June–Aug. Fig. 1253.

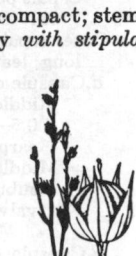

1253. L. cathartieum.

4. L. rígidum Pursh (stiff). — Glaucous, stiff, bushily fastigiate-branched, the *rigid branches angled; leaves* linear, erect, usually *with stipular glands;* flowers scattered along the branches, large; *sepals* lance-attenuate, rigid, 6–10 *mm. long, all glandular-serrate,* soon *deciduous; petals yellow,* 1–1.5 *cm. long; styles united below; capsule ellipsoid,* its 5 firm carpels *with brown cartilaginous thickenings at base, blunt or merely subacute.* (*Cathartolinum* Small) — Dry open soil, Minn. and Man. to Alta., s. to Mo. (adv.), Tex. and n. Mex. June, July. Fig. 1254.

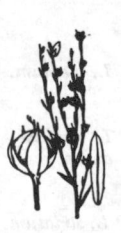

1254. L. rigidum.

5. L. sulcàtum Riddell (furrowed). — Similar to no. 4, less compact; stems mostly simple below, 1.5–7.5 dm. high; *leaves* linear, mostly *with stipular glands; branches deeply sulcate; sepals* broadly *lanceolate,* rigid, *persistent, all glandular-serrate, the outer* 4–7 *mm. long;* petals yellow, 8–12 mm. long; *capsule globose-ovoid or slightly depressed,* 2.5–3 dm. high, *without basal thickenings, promptly splitting into* 10 *valves, the individual carpels* (*or their valves*) *tapering to erect beaks.* (*Cathartolinum* Small) — Dry prairies and calcareous rocks and sands, e. Mass. (rare and local) to Man., s. to Ga., Ala., Ark. and Tex. July–Sept. Fig. 1255.

1255. L. sulcatum.

6. L. floridànum (Planch.) Trel. (of Florida). — Slender, 3–8 dm. high, fastigiate-corymbose at summit; stems solitary, terete; *leaves very numerous* (50–150), firm, appressed-ascending; *the middle and upper ones linear, with a subulus* 0.5–1 *mm. long,* 0.6–1.8 cm. long, 1–2 *mm. broad, not showing internal venation by transmitted light;* flowers very short-pedicelled, on spiciform branches; *outer sepals* 3–4 *mm. long,* broadly lanceolate, sharp-pointed, little exceeding *the inner ovate glandular-ciliolate* ones; petals yellow, about 1 cm. long; *capsule globose-ovoid, rounded at summit,* 2.5–3 *mm. long,* tardily dehiscent, *the firm individual carpels or their valves blunt or with short incurved tip.* (*Cathartolinum* Small) — Damp pinelands and low woods, Fla. to La., n. to se. Va. and s. Ill. July, Aug. Fig. 1256.

1256. L. floridanum.

7. L. intercúrsum Bickn. (running between). — Similar to no. 6, coarser, 1–7 dm. high; *leaves fewer* (15–60) *below the corymb,* thinner, *with definite internal venation shown by transmitted light; the middle and upper ones narrowly oblong to elliptic, with very short subulus,* 0.8–2 cm. long, 1–6 *mm. broad; inner sepals lance-ovate, outer* slightly longer (2–3 *mm. long*); petals 5–7 mm. long; *capsule ovoid, pointed,* 1.5–2.5 *mm. long, promptly splitting into* 10 *valves with straight erect acuminate tips.* (*L. floridanum* of ed. 7, in large part; *L. floridanum,* var. *intercursum* Weath.; *Cathartolinum* Small) — Argillaceous, siliceous or peaty shores, plains and thickets, interior of Ala. and Ga., n. to se. Mass., R.I. and centr. Ct.; nw. Ind. July, Aug. Fig. 1257.

1257. L. intercursum.

8. L. mèdium (Planch.) Britt. (intermediate). — Glaucous or pale green, 1.5–4.5 dm. high, with *stiffly fastigiate corymb;* stem terete, *branches angled; leaves firm, elliptic, oval or elliptic-obovate,* obtuse or subacute, opaque, *without obvious veins shown by transmitted light, only the uppermost with short subulate tips,* 25–40 *below the inflorescence,* ascending; *sepals* about equaling to shorter than capsule, the longer 2–3 mm. long; *the inner ovate, entire or but sparingly glandular-ciliate;* petals 5–8 mm. long; *capsule depressed or oblate,* 1.5–2 mm. high, 2–2.5 mm. broad, promptly splitting into 10 blunt valves. — Dry or moist open soil, s. Ont., nw. Pa. and n. O. July, Aug. Fig. 1258.

1258. L. medium.

Var. **texànum** (Planch.) Fern. (Texan). — Mostly taller, up to 1 m. high; *leaves more crowded, 30–150 below the inflorescence, linear or linear-lanceolate to lance-elliptic,* subtranslucent, *with evident veins shown by transmitted light, all but the lowermost with prolonged subulate tips; sepals* mostly exceeding capsule; *the inner* ovate to lanceolate, *copiously glandular-ciliolate; the longer 2.5–5 mm. long;* capsule often slightly larger, splitting into the 5 carpels or tardily into 10 valves. — Dry or damp sterile open soil, Fla. to Tex., n. to sw. Me., ne. Mass., s. Vt., e. Pa., W.Va., s. Mich., n. Ind., Ill., Mo. and Okla. June–Aug.

9. L. virginiànum L. (Virginian). — Green or only slightly glaucous, 1.5–8 dm. high, with *loose corymb of elongate filiform* often flexuous *spreading-ascending branches; leaves thinnish,* green, elliptic-lanceolate to narrowly oblong, alternate, the lowest spatulate and opposite; flowers scattered, distinctly pedicelled; *sepals* ovate to broadly lanceolate, nearly or quite *without marginal glands; capsule oblate or depressed-globose.* (*Cathartolinum* Small) — Open woods, thickets and clearings, Mass. to s. Ont., s. to Ga. and Ala. July, Aug. Fig. 1259.

1260. L. striatum.

1259. L. virginianum.

10. L. striàtum Walt. (furrowed; from the striate-angled stem). — Stems usually decumbent at base, 2–12 dm. high, *striate-angled* above and *on the branches of the usually elongate and slender loose panicle; leaves* thin, *viscidulous, elliptic, the lower 5–10 (–14) pairs opposite, the upper 5–25 and those of the axis of the panicle alternate; flowers* numerous, *short-stalked,* on the spreading-ascending branches; *outer sepals elliptic, herbaceous,* 1.5–2.5 mm. long, inner smaller and often glandular-toothed; petals 5–7 mm. long; *capsule depressed,* about 2 mm. broad. (*Cathartolinum* Small) — Damp sands, peats and low woods, Fla. to Tex., n. to s. N.E., e. N.Y., Pa., W.Va., s. Mich., s. Ill., Mo. and Okla. Late June–Aug. Fig. 1260. — Fruiting plant habitally resembling *Lechea.*

Var. **multíjugum** Fern. (with many pairs). — *Cauline leaves and lower leaves of panicle-axis all opposite.* — Cape Cod and islands of se. Mass.; Block I., R.I.

2. MILLEGRÀNA Adans. All-seed

Sepals (toothed), petals, stamens, and styles 4. Capsule of 4 almost 2-locular carpels, each carpel 4-seeded. Seeds without albumen. — A tiny annual, nat. of Eu. and Afr., with filiform simple stems or forking branches, opposite leaves, and small corymbiform cymes. Corolla fugacious. (Name from *mille, thousand,* and *granum, seed.*) Radiola Roth.

1. M. Radìola (L.) Druce (the old generic name). — The only species. — Roadsides, old fields and ditches near the coast, N.S. July–Sept. (Natzd. from Eu.)

FAM. 85. OXALIDÀCEAE (Wood-Sorrel Family)

Plants with regular 5-merous 10–15-androus flowers. Ovary superior, 5-locular, the carpels 2–∞-ovuled, usually distinct above, loculicidal. — Ours low herbs with sour watery juice and delicate impunctate palmate alternate or radical leaves with 3 obcordate leaflets.

1. ÓXALIS L. Wood-Sorrel. Lady's-Sorrel

Sepals 5, persistent. Petals 5, sometimes united at base, withering after expansion. Stamens 10, usually monadelphous at base, alternately shorter. Styles 5, distinct. Capsule oblate, prismatic, cylindric or subulate, membranaceous; valves persistent, being fixed to the axis by the partitions. Seeds pendulous from the axis, anatropous, their outer coat loose and separating. Embryo large and straight, in fleshy albumen; cotyledons flat. — Several species produce small cleistogamous flowers, precociously fertilized in the bud and particularly fruitful; the ordinary flowers being often dimorphous or even trimorphous in the relative length of the stamens and styles. Semicosmop. genus. (Greek name for sorrel, from *oxys, sour.*)*

a. Stemless; leaves and scapes all basal; petals white to purple.

Creeping by slender rhizomes; flowers solitary; tips of sepals plane. . . . 1. *O. montana.*

Bulbous; flowers in umbels; tips of sepals with orange callosities. . . . 2. *O. violacea.*

* Treatment of leafy-stemmed species chiefly derived from that of K. M. Wiegand in Rhodora, xxvii. 113–124 and 133–139 (1925).

*a.*Leafy-stemmed; petals yellow. . . *b.*

*b.*Flowers 5–11 mm. long, mostly homogamous, *i.e.* relative length of style and stamens constant (rarely heterogamous in nos. 5 and 6). . . *c.*

*c.*New leafy flowering stems repent, rooting at the nodes, from slender tap-roots, without subterranean stolons; stipules broad, brownish or purplish. 3. *O. corniculata.*

*c.*New leafy flowering stems ascending, not rooting above the sometimes persistent and decumbent bases; stipules narrow or obsolete. . . *d.*

*d.*Flowers 1–5, in umbels; fruiting pedicels curved, horizontally arched or deflexed, the capsules erect; subterranean stolons wanting (except in no. 5). . . *e.*

*e.*Lower internodes of flowering stems appressed-strigose with white hairs or glabrous; styles 1–2 mm. long (rarely a little longer).

Capsules hoary from base to apex, mostly 1.5–2.5 cm. long; sepals 3.5–7 mm. long; old decumbent bases without subterranean stolons; leaflets 1–2 cm. broad; stipules oblong. . . . . 4. *O. stricta.*

Capsules glabrous or only remotely pilose, 8–12 mm. long; sepals 2–4.5 mm. long; old filiform decumbent bases bearing autumnal subterranean stolons; leaflets 3–12 mm. broad; stipules obsolete. 5. *O. filipes.*

*e.*Lower internodes with loosely spreading sordid pilosity; styles mostly 3–4 mm. long. . . . . . 6. *O. florida.*

*d.*Flowers 1–9 in cymose umbels; fruiting pedicels straight, erect or ascending, not deflexed; autumnal fleshy subterranean stolons frequent; stipules obsolete or nearly so. . . . . 7. *O. europaea.*

*b.*Flowers 1.2–1.8 cm. long, trimorphic as to relative length of stamens and styles; fruiting pedicels rarely deflexed; stipules nearly obsolete.

Leaflets 2–5 cm. broad, mostly with a narrow purple margin; capsules 7–10 mm. long . . . . 8. *O. grandis.*

Leaflets 5–12 mm. broad, without definite purple margin; capsules 1–1.7 cm. long. . . . . 9. *O. recurva.*

1. O. montàna Raf. (of the mountains), COMMON W., WOOD-SHAMROCK, PAIN DE LIÈVRE (Que.). — *Creeping*, with pale scaly rhizomes; leaves and scapes basal; leaflets broadly obcordate, 1.2–3 cm. broad; earlier *scapes* equaling or exceeding leaves, 1-*flowered; the later ones short, recurving and bearing small cleistogamous flowers;* sepals with plane tips; *petals oblong or oblong-obovate*, deeply notched, *white or whitish, with lilac or red lines*, or purple or roseate throughout in the local forma **rhodántha** Fern. (rose-flowered); capsule depressed. (*O. Acetosella* of ed. 7, not L.) — Damp woods, s. Nfld. and e. Que. to Man., s. to N.S., n., centr. and w. N.E., Pa., e. O., Mich., Wisc. and Minn., and along the mts. to N.C. and Tenn. Petaliferous *fls.* late May–Aug.

2. O. violàcea L. (violet), VIOLET W. — Nearly glabrous; *base bulbous*, scaly; leaves radical, on glabrous petioles; *scapes* umbellately *several-flowered*, 1.2–2.5 dm. high, exceeding the leaves; *sepals tipped by orange callosities;* petals purple or violet, or white in forma **álbida** Fassett (white). (*Ionoxalis* Small) — Woods, shaded slopes, gravelly banks and prairies, Fla. to N.M., n. to Mass., N.Y., O., Ind., Wisc., Minn., N.D. and Colo. April–July (–Oct.). — Sometimes producing beneath the bulb a long vertical icicle-like water-storage organ or fleshy root.

Var. **trichóphora** Fassett (bearing hairs). — Petioles bearing long pale often gland-tipped hairs. — Locally replacing the glabrous plant, Ark. and Mo. to w. Va., s. Pa. and w. Vt.

O. INTERMÈDIA A. Richard (intermediate), a coarse species with scaly central bulbs setting off abundant stalked lateral bulbs, the leaflets inversely deltoid and much broader than high (the larger ones 3–8 cm. broad), the violet flowers numerous in umbels, spreads freely inside and outside a greenhouse in Essex Co., Mass.; presumably elsewhere. (Introd. from W.I.)

3. O. CORNICULÀTA L. (horned), CREEPING L. — *Stems* from a slender tap-root, without stolons, *prostrate and often rooting at many nodes; stipules broad, brown or purple;* leaflets comparatively small, green, purple or bronze; peduncles shorter than to exceeding leaves, 1–5-flowered; flowers 4–8 mm. long, with yellow petals; *pedicels reflexing in fruit;* capsules prismatic-cylindric, 0.8–2.5 cm. long, closely puberulent or slightly viscidulous-pilose. (*O. repens* Thunb.; *Xanthoxalis corniculata* and *Langloisii* Small) — Waste places, roadsides and fields, northw. chiefly about hot-houses, Fla. to Tex. and Mex., n. to Nfld., N.E., N.Y., Ill., N.D., etc. April–Nov. (Semicosmop. weed) FIG. 1261.

1261. O. corniculata.

4. O. strícta L. (erect). — Gray-green, ascending or becoming decumbent and matted, from a tap-root, *without subterranean stolons; stems and branches appressed-strigose with whitish hairs; stipules oblong*, firm, pale; leaflets 1–2

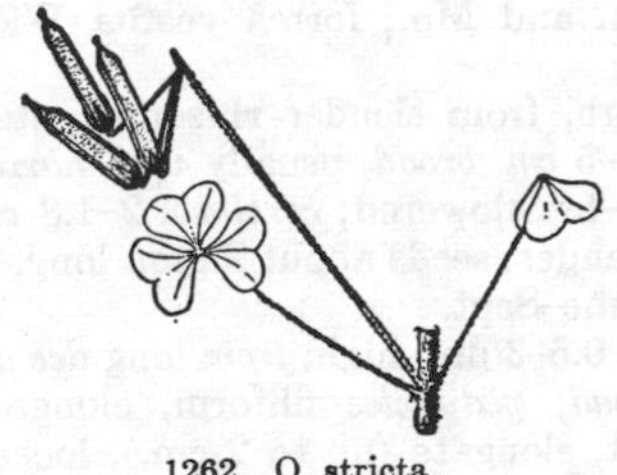

1262. O. stricta.

cm. broad; peduncles strigose-pubescent, mostly equaling or overtopping leaves, umbellately 1-4-flowered; *pedicels* 1–2.5 cm. long, *deflexed in fruit; flowers* 7–11 *mm. long,* homogamous, the yellow petals sometimes red at base; *sepals* 3.5–7 *mm. long; styles* 1–2 *mm. long; capsules* prismatic-cylindric, mostly 1.5–2.5 *cm. long, hoary-puberulent;* seeds 1–1.3 mm. long. (*Xanthoxalis* Small) — Dry open soil, P.E.I. and s. Me. to s. B.C., s. to Fla., Gulf States and Mex. May–Oct. Fig. 1262. — Forma **viridiflòra** (Hus) Fern. is an occasional variation with green petals.

5. O. fílipes Small (slender-stalked). — Flowering *stems* erect or depressed, 1–4 dm. long, *appressed-pubescent or glabrous,* from persistent filiform but wiry decumbent bases, often *with slender subterranean stolons; stipules obsolete* or nearly so; *leaflets* 3–12 *mm. broad;* peduncles exceeding leaves, with 2–5 umbellate yellow sometimes heterogamous flowers 7–10 mm. long; *pedicels deflexed in fruit; sepals* 2–4.5 *mm. long; styles about* 2 *mm. long; capsule* cylindrical, *glabrous or nearly so,* 8–12 *mm. long;* seeds 1–1.2 mm. long. (*Xanthoxalis* Small) — Sandy thickets, borders of woods and fields, Fla. to La., n. to Ct., Tenn. and Mo. April–Sept.

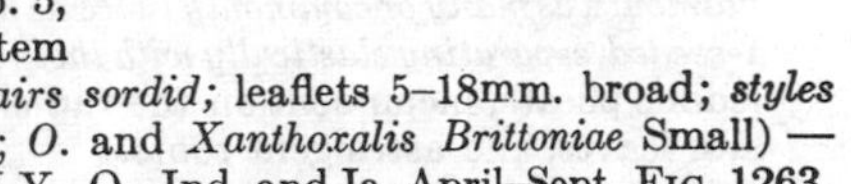

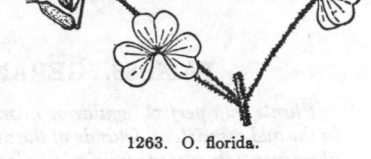

1263. O. florida.

6. O. flórida Salisb. (flowering). — Resembling no. 5, *non-stoloniferous; lower* (and often all) *internodes* of stem *and petioles loosely spreading-pilose or villous, the hairs sordid;* leaflets 5–18 mm. broad; *styles mostly* 3–4 *mm. long.* (*O. filipes* of ed. 7, not Small; *O.* and *Xanthoxalis Brittoniae* Small) — Dry sandy or rocky soil, Fla. to Ark., n. to s. N.E., N.Y., O., Ind. and Ia. April–Sept. Fig. 1263.

7. O. europaèa Jord. (European), Pain d'oiseau (Que.), Sûrette (Que.). — Erect or depressed, up to 6 dm. high, simple to freely branched, *annual, or perennial with fleshy subterranean stolons;* stems glabrous to densely spreading-villous; *stipules obsolete or nearly so;* leaflets green to purple, 0.6–4 cm. broad; peduncles mostly exceeding leaves, glabrous, pilose or villous; *flowers* 1–9 *in cymose umbels,* orange-yellow, or in forma **pallidiflòra** Fern. (pale-flowered) pale lemon-color; *pedicels straight, ascending or spreading, not deflexed, in fruit;* sepals 3–5 mm. long; styles 2–4 mm. long; capsules 0.5–1.5 cm. long. (*O. corniculata* of ed. 7, largely, not L.) — Our most polymorphous species; the following variations have been designated:

Upper leaf-surfaces glabrous.
- Hairs of pedicels appressed, scarcely viscid.
  - Stems glabrous or sparsely strigose-pubescent. . . . . . *O. europaea* (typical).
  - Stems appressed-villous. . . . . . Forma *pilosella.*
- Hairs of pedicels spreading, usually viscid.
  - Stems nearly or quite glabrous. . . . . . Forma *cymosa.*
  - Stems villous. . . . . . Forma *villicaulis.*

Upper leaf-surfaces pubescent.
- Hairs of pedicels appressed, scarcely viscid.
  - Stems villous. . . . . . Var. *Bushii.*
  - Stems glabrous or appressed-pubescent. . . . . . Var. *Bushii,* f. *subglabrata.*
- Hairs of pedicels spreading, usually viscid; stems villous. . . . . . Var. *Bushii,* f. *vestita.*

O. europaèa (typical). — Weed of fields, cult. ground, roadsides, etc., e. Que. to N.D. and Colo., s. to N.S., N.E., Ga., Tenn., Okla. and Ariz. May–Oct. (Introd. in Eu.) — Forma **pilosélla** Wieg. (somewhat hairy), occasional; forma **cymòsa** (Small) Wieg. (cymose), *O. cymosa* and *O. rufa* and *Xanthoxalis cymosa* and *rufa* Small, very common, s. to Fla. and Tex.; forma **villicaùlis** Wieg. (hairy-stemmed), N.S. to Minn., s. to Va., Tenn. and Ill. Fig. 1264.

1264. O. europaea, f. villicaulis.

Var. **Búshii** (Small) Wieg. (for its discoverer, Benjamin Franklin Bush, 1858–1937), *O.* and *Xanthoxalis Bushii* Small — W. Ont. and Minn., s. to Ind., Ill., Mo. and Kans.; e. Mass. to sw. N.B. — Forma

subglabràta Wieg. (nearly glabrous), Ind. and Wisc. to Ia. and Mo.; forma **vestìta** Wieg. (clothed), Ind., Ill. and Mo.; e. Mass. and Va.

8. O. grándis Small (large). — Coarse, up to 1.2 m. high, from slender rhizomes; stem sparingly loose-pubescent; stipules nearly obsolete; *leaflets* 2–5 *cm. broad*, usually *with narrow purple margin;* peduncles but slightly overtopping leaves, 1–few-flowered; *corolla* 1.2–1.8 *cm. long*, yellow; pedicels not deflexed; *capsules* 7–10 *mm. long*, slender; seeds about 2 mm. long. — Woods and shaded slopes, Pa. to s. Ill., s. to Ga. and Ala. June–Sept.

9. O. recúrva Ell. (turned back). — *Flowering stems* erect, 0.5–3 dm. high, *from* long *decumbent filiform bases, spreading-villous; leaflets* 5–12 *mm. broad; peduncles* filiform, elongate, *appressed-pubescent*, 2–7-flowered; pedicels strigose-pubescent, elongate (up to 2 cm.), loosely arched-ascending, rarely deflexed; *petals* yellow, 13–18 *mm. long*, glabrous; *capsules* 1–1.3 *cm. long*. (*Xanthoxalis* Small) — Dry sandy or rocky woods and openings, Fla. to Miss., n. to N.C. and Mo. April–Aug.

Var. **macrántha** (Trel.) Wieg. (large-flowered). — Stems and petioles shaggily spreading-villous; peduncles loosely pilose; petals often hairy outside; capsules slightly larger. (*O. Priceae* Small; *Xanthoxalis macrantha, hirsuticaulis* and *Priceae* Small) — Fla. to Miss., n. to Ky.

FAM. 86. GERANIÀCEAE (Geranium Family)

Plants with perfect regular or nearly regular 5-merous hypogynous flowers. Sepals imbricated in the bud, persistent. Glands of the disk 5, *alternate with the petals. Stamens, counting the sterile filaments, as many or commonly twice as many as the sepals. Ovary deeply lobed; carpels 2-ovuled,* 1-*seeded, separating elastically with their long styles, when mature, from the elongated axis.* Cotyledons plicate, incumbent on the radicle. — Our species herbs with lobed or divided stipulate leaves, and astringent roots.

Leaves palmately lobed or divided; anthers 10 (rarely 5); ripe carpels plump, dehiscent on the inner suture, the separated stylar portions merely recurving, nearly glabrous on the inner side; seeds often reticulate. . . . . 1. *Geranium.*

Leaves pinnate or pinnatifid; anthers 5; ripe carpels sharp-pointed at base, tardily dehiscent, the freed stylar portions spiraling, bearded on inner side; seeds smooth. . . . . 2. *Erodium.*

1. GERÀNIUM L. Cranesbill

Stamens 10 (rarely 5), all with perfect anthers, the 5 longer with glands at their base (alternate with the petals). Ripe carpels dehiscent on the inner suture, the separated stylar portions arched-recurving and smooth along the inner side. — Stems forking. Leaves palmately lobed. Peduncles 1–∞-flowered. Genus widespread in temp. and trop. reg. (An old Greek name, from *geranos, a crane;* the long fruit-bearing beak thought to resemble the bill of that bird.)*

*a.*Leaves variously palmate-lobed, not divided into distinct leaflets; carpel-bodies only tardily separating from or permanently attached to styles. . . *b.*

*b.*Petals 1.4–2.3 cm. long, 0.7–1.3 cm. broad, much exceeding calyx; terminal beak of mature style-column 5–10 mm. long; anthers 2–3 mm. long; perennials with thick crowns and stout rhizomes.

Pedicels and style-columns not obviously glandular; pedicels erect in fruit. 1. *G. maculatum.*

Pedicels and style-columns glandular-pilose; pedicels nodding in fruit. . 2. *G. pratense.*

*b.*Petals smaller, shorter than to exceeding calyx; beak of mature style-column 0–6 mm. long; anthers 0.5–0.8 mm. long; annuals or biennials (nos. 3, 6 and 10 perennial) with tap-roots. . . *c.*

*c.*Sepals prominently awned or subulate-tipped, the tips 0.7–3 mm. long; seeds with reticulate surfaces. . *d.*

*d.*Peduncles all 1-flowered or terminated by 1 pedicel. . . . . . 3. *G. sibiricum.*

*d.*Peduncles all or nearly all with 2 or more pedicels. . . *e.*

*e.*Fruiting pedicels much longer than calyx.

Beak of mature style-column 2.5–6 mm. long; plants annual or biennial, without strong rhizome; seed coarsely reticulate.

Pedicels with minute appressed glandless pubescence; carpel-bodies glabrous. . . . . 4. *G. columbinum.*

Pedicels densely glandular-pilose; carpel-bodies hairy. . . . 5. *G. Bicknellii.*

* Treatment based largely on that in Rhodora, xxxvii. 295–301 (1935).

Beak less than 1.5 mm. long; plant perennial with stout rhizome; seed very minutely reticulate. . . . . 6. *G. nepalense*, var. *Thunbergii.*

*e.*Fruiting pedicels shorter than to slightly longer than calyx; beak of mature style-column 1–2 mm. long. . . *f.*

*f.*Carpel-bodies short-hirsute with horizontally spreading hairs; lobes of middle and upper leaves acute; seed strongly reticulate or pitted with subuniform thick-walled square to rounded areolae. 7. *G. dissectum.*

*f.*Carpel-bodies long-villous with ascending hairs; leaf-lobes obtuse; seeds loosely reticulate with irregular elongate thin-walled areolae or obscurely and subuniformly reticulate.

Larger mature sepals broadly ovate, 5–8 mm. wide, 5-nerved; seed subspherical, 2–2.7 mm. in diameter, fainly reticulate with about 50 rows of small square to rounded areolae. . . 8. *G. sphaerospermum.*

Larger mature sepals 3–4.5 mm. wide, 3-nerved; seed oblong, 1–1.5 mm. thick, with 20–35 irregular rows of loosely elongate areolae. . . . . 9. *G. carolinianum.*

*c.*Sepals awnless, at most with minute callous tips; seeds smooth or only minutely granular. . *g.*

*g.*Carpel-bodies not cross-wrinkled, finely pubescent; style-column beakless.

Sepals 5–8 mm. long; petals twice as long, deeply notched; carpel-bodies 3–3.5 mm. long; fruiting style-column 1.1–1.5 cm. long. . 10. *G. pyrenaicum.*

Sepals 2.5–4 mm. long; petals about as long, shallowly notched; carpel-bodies 2 mm. long; fruiting style-column 6–9 mm. long. . 11. *G. pusillum.*

*g.*Carpel-bodies conspicuously cross-wrinkled or smooth, glabrous; style-column terminated by a filiform beak 1–2 mm. long. . . . . 12. *G. molle.*

*a.*Leaves divided into 3–5 distinct pinnatifid leaflets, terminal leaflet stalked; carpel-bodies promptly separating from the style. . . . . 13. *G. Robertianum.*

1. G. maculàtum L. (mottled), Wild or Spotted C. — Erect, 2–5.5 dm. high, *from a stout rhizome;* the stem scapose below the pair of *large* (0.5–1.5 *dm. across*) rounded-pentagonal *deeply* 5-*parted leaves* with cuneate lobed and incised segments; basal leaves similar, long-petioled; inflorescence a terminal corymb of few-many large flowers; *pedicels minutely pubescent, scarcely glandular, erect in* flower and *fruit;* sepals about 1 cm. long, slender-tipped, ciliate and sparsely pubescent; *petals* entire, rosy-purple, or white in forma **albiflòrum** (Raf.) House (white-flowered), 1.4–2.3 *cm. long,* about 1 cm. broad; *style-column minutely pubescent, not glandular,* 2–3 cm. long. — Woods, thickets and meadows, centr. Me. to Man., s. to s. N.E., L.I., Ga., Tenn., Mo. and Kans. Late April–June. Fig. 1265.

1265. G. maculatum.

2. G. praténse L. (of meadows). — Similar, tall, 3–7 dm. high; leaves 5–7-parted, the segments more sharply incised; *pedicels and style-column heavily glandular-hirsute;* sepals hirsute; petals bluish-purple; *fruiting pedicels recurved.* — Fields and roadsides, Lab., Nfld. and Que. to N.S. and Me., rarely to Mass. and N.Y. June–Aug. (Natzd. from Eu.) Fig. 1266.

1266. G. pratense.

G. ibèricum Cav. (of Iberia), Iberian C., similar to nos. 1 and 2, with *long-villous erect pedicels* and sepals, and violet petals, and G. sanguíneum L. (bloody), Blood-red C., with low wiry slender stems, scattered leaves and scattered long peduncles with *solitary* deep crimson (or white) *flowers,* are tending to spread from cult. (Both introd. from Eu.)

3. G. sibíricum L. (Siberian), Siberian C. — *Weak diffusely branched perennial,* with villous leafy stems 3–10 dm. long; *leaves* 3–8 cm. broad, 3(–5)-*parted,* the segments broadly lanceolate or rhombic, sharply cut-toothed, acute; *peduncles* scattered, mostly solitary, 1-*flowered,* hispidulous; *petals about equaling sepals* (6–7 *mm.*

long), lilac to white with violet markings; carpel-bodies finely pubescent, style-column 1–1.5 cm. long; seeds reticulate. — Roadsides, fields and waste places, locally abund., N.Y. and Pa. to Ill. Aug., Sept. (Natzd. from Eurasia) FIG. 1267.

1267. G. sibiricum.

4. G. COLUMBÌNUM L. (of a dove, the old English name being *Dove's-foot*), LONG-STALKED C. — Diffuse annual, 1.5–8 dm. high; *leaves* 5–7-parted, 2–4 cm. broad, the segments *cut into linear lobes; peduncles and pedicels* filiform, *elongate, minutely appressed-pubescent, from most of the axils, pedicels deflexed in fruit;* sepals awned, the blades 0.5–1.2 cm. long; petals purple, retuse, about equaling sepals; *carpel-bodies glabrous;* style-column 1.5 cm. long, with a *filiform beak* 2.5–5 *mm. long.* — Roadsides, fields, etc., N.Y. to Ind., s. to N.C. and Tenn.; S.Dak. June–Sept. (Natzd. from Eu.) FIG. 1268.

1268. G. columbinum.

5. G. Bicknéllii Britt. (for EUGENE PINTARD BICKNELL, 1859–1925). — Simple or commonly much branched annual or biennial, 1.5–6 dm. high; stem spreading-hirsute; *hairs of upper internodes unequal, some long, others short and sometimes gland-tipped;* leaves angulate-rotund, 2–7 cm. broad, deeply 5-parted, the cuneate segments deeply oblong-lobed; *peduncles in forks of stem, loosely hirsute and with shorter often glandular hairs intermixed; pedicels distinctly longer than calyx, with mixed long glandless and short glandular hairs; sepals broadly lanceolate or oblong-ovate,* awn-tipped, their bodies 4.5–8 mm. long and *unequally ciliate-hirsute on margin and nerves;* petals roseate, slightly exceeding calyx; carpel-bodies long-villous with ascending hairs; *style-column* in maturity 1.5–2.3 cm. long, *with a filiform beak* 3–5 *mm. long;* seeds thick-cylindric, clearly reticulated with elongate areolae. — Open woods, clearings and disturbed soil, Nfld. to Alta., s. to N.S., Me., Mass., w. Ct., N.Y., n. Ind., Ill. and Ia. July–Sept. FIG. 1269.

1269. G. Bicknellii.

Var. **lóngipes** (S. Wats.) Fern. (long-stalked). — *Pubescence of upper internodes, peduncles and pedicels more uniformly short and gland-tipped; sepals with shorter gland-tipped ciliation.* — W. Que. to w. Ont., n. Mich. and Minn.; Yuk. and Alaska to Sask., Colo., Utah and n. Cal.

6. G. NEPALÉNSE Sweet (of Nepal), var. THUNBÉRGII (Sieb. & Zucc.) Kudo (for its discoverer, CARL PETER THUNBERG, 1743–1828). — Perennial *with stoutish oblique rhizome; stems* slender, *decumbent at base,* lengthening to 4 dm., *retrorsely hirsute; leaves* on retrorse-hirsute petioles, *the larger blades with* 3 or 5 narrowly cuneate-obovate *coarsely toothed segments;* axillary peduncles hairy, up to 8 cm. long, the usually 2 *slender pedicels spreading-villous; sepals lanceolate, spreading-pubescent,* with terminal mucro about 1 mm. long; petals violet, broadly obovate, about equaling or slightly exceeding sepals; fruit about 1.7 cm. long, hirtellous, its short beak puberulous; *seed nearly smooth or very finely reticulate.* — Garden-weed, locally abund., Middlesex Co., Mass. (Adv. from e. Asia)

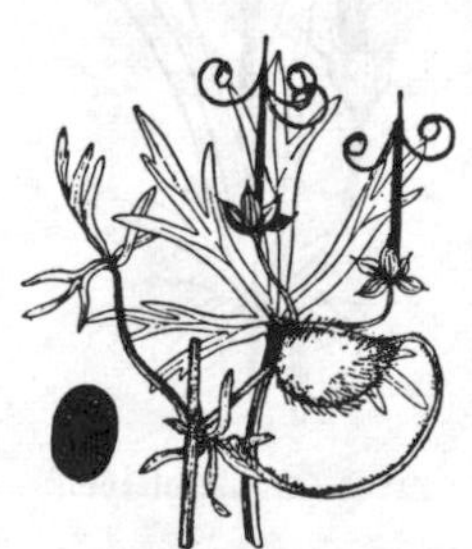
1269a. G. dissectum.

7. G. DISSÉCTUM L. (dissected). — Similar to no. 5, loosely branching; *lobes* of segments *of upper leaves linear, acute;* fruiting pedicels shorter than to slightly exceeding calyces; *carpel-bodies short-hispid with divergent hairs; seed subglobose, strongly reticulate or pitted with subuniform small thick-walled square or rounded areolae.* — Roadsides and waste places, local, Mass. to Mich. and N.C.; Pacific slope. Apr.–Aug. (Adv. from Eu.) FIG. 1269a.

8. G. sphaerospérmum Fern. (spherical-seeded). — Annual or biennial, simple or with ascending branches, 1–4.5 dm. high; *stem minutely retrorse-pilose;* leaves reniform-suborbicular, 2–7 cm. broad, with 5 deeply cleft and blunt-lobed cuneate segments; *flowers in crowded terminal umbelliform corymbs; peduncles and paired pedicels minutely retrorse-pilose, shorter than fruiting calyx; sepals broadly ovate,* awned, 5-*nerved,* the larger *in maturity* 5–8 *mm.*

wide, with pilose-hirsute nerves; petals roseate, about equaling sepals; carpel-bodies long-villous with ascending hairs; mature style-column 1–1.5 cm. long, pilose-hirsute, with *subulate beak* 1–1.5 *mm. long; seeds subspherical,* 2–2.7 *mm. in diameter, faintly reticulate with about* 50 *rows of small square to rounded thin-walled areolae.* — Calcareous gravel and rich clearings, Ont. to Sask. and Wash., s. to n. N.Y., S.D., Mont. and n. Calif. Late May–Sept. Fig. 1270.

1270. G. sphaerospermum.

9. G. caroliniànum L. (of Carolina). — Similar to no. 8, often more bushy-branched, 1–8 dm. high; internodes, petioles, peduncles and pedicels *densely retrorse-hirsute with subuniform hairs mostly less than* 0.5 *mm. long, mixed (on pedicels) with short glands;* leaves with obtuse linear-oblong ultimate lobes; *inflorescences peduncled, solitary in the upper forks or loosely aggregated as* 4–12-*flowered terminal corymbs;* the lower peduncles elongate, the upper abbreviated; *pedicels shorter than calyx; sepals narrowly ovate,* 3-*nerved,* short-pilose, *with villous-ciliate nerves, in maturity* 3–4.5 *mm. wide;* petals pale; mature style-column 1.2–1.8 cm. long, hirtellous and glandular; beak 1–2 mm. long; *seeds oblong,* 1–1.5 *mm. thick, with* 20–35 *irregular loose rows of elongate areolae.* (*G. Langloisii* Greene) — Dry rocky woods, fields and waste places, Fla. to s. Calif., n. to Mass., Ct., N.Y., W.Va., s. Mich., Ill., Mo., Kans., Wyo., Ida. and s. B.C. May–July. Fig. 1271.

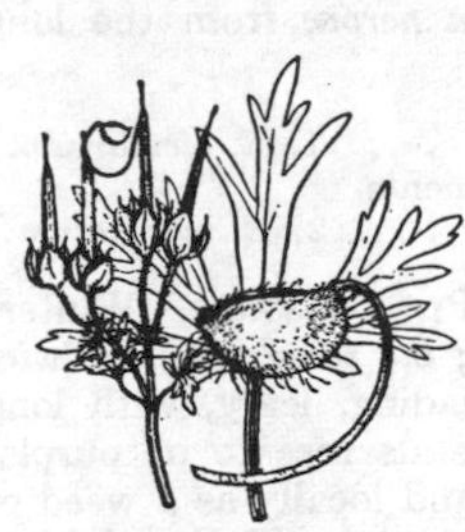

1271. G. carolinianum.

Var. **confertiflòrum** Fern. (with crowded flowers). — *Pubescence more loosely villous-hirsute, the hairs mostly* 1 *mm. long; inflorescences mostly crowded* 5–25-*flowered terminal corymbs.* — Dry rocky or sandy soil, s. Me. to Minn., s. to s. N.E., L.I., Del., uplands of N.C. and Tenn., and Mo. May, June.

10. G. PYRENÀICUM Burm. f. (of the Pyrenees). — *Perennial* with short and thick scaly caudex; stems simple below, bushy-branched above, 2–8 dm. high, minutely glandular, sparsely retrorse-pilose; *leaves* orbicular or reniform, 5–7-*cleft two-thirds in to base, the lobes oblong; peduncles scattered* in upper axils; *sepals awnless,* 5–8 *mm. long,* puberulent; *petals* rose-colored, *about twice as long as sepals, deeply notched; carpel-bodies* 3–3.5 *mm. long, minutely puberulent; fruiting style-column beakless,* 1.1–1.5 *cm. long;* seeds minutely granular. — Roadsides and waste places, near Quebec, Que., rarely southw. June–Aug. (Adv. from Eu.) Fig. 1271a.

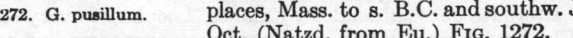

1271a. G. pyrenaicum.

11. G. PUSÍLLUM L. (very small). — Annual or biennial, slender, weak, the branching stems 1–6 dm. long; leaves reniform, 5–7-parted two-thirds to base, with oblong obtuse or acutish lobes; 2-flowered peduncles from most axils, the elongate pedicels minutely glandular; *sepals callous-tipped,* 2.5–4 *mm. long; petals about as long, shallowly notched;* stamens 5; *carpel-bodies* 2 *mm. long, strigose; fruiting style-column beakless,* 6–9 *mm. long;* seeds smooth. — Roadsides, fields and waste places, Mass. to s. B.C. and southw. June–Oct. (Natzd. from Eu.) Fig. 1272.

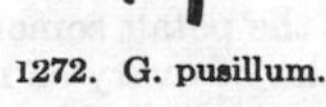

1272. G. pusillum.

12. G. MÓLLE L. (soft), DOVESFOOT-C. — Weak spreading annual, resembling no. 10, soft-pubescent; leaves reniform, cleft to middle, the segments lobed; sepals 3–4 mm. long, downy; stamens 10; *carpel-bodies conspicuously cross-wrinkled,* or smooth in forma PREUSCHÓFFII Abromeit (named about 1890 for its discoverer J. PREUSCHOFF), *glabrous;* mature *style-column with a filiform beak* 1–2 *mm. long.* — Lawns and waste places, N.S. to s. B.C. and southw. June–Aug. (Adv. from Eu.) Fig. 1273.

1273. G. molle.

13. G. Robertiànum L. (for St. Robert), HERB-ROBERT, HERBE À L'ESQUINANCIE (Que.). — Sparsely hairy, diffuse, strong-scented annual; *leaves divided into* 3–5 *distinct pinnatifid leaflets, terminal*

1274. G. Robertianum.

leaflet stalked; sepals awned; petals purple, or white in forma **albiflòrum** (G. Don) House (white-flowered), long-clawed, exceeding sepals; *carpel-bodies promptly separating from styles, wrinkled;* seeds smoothish. (*Robertiella* Hanks) — Rocky woods, ravines and gravelly shores, Nfld. to Man. s. to N.S., N.E., L.I., Md., W.Va., O., Ind. and Ill. Late May–Oct. (Eurasia; n. Afr.) FIG. 1274.

2. ERÓDIUM L'Hér. STORKSBILL

Leaves in ours pinnate or pinnatifid. Peduncles mostly bearing several flowers in an umbel. Upper petals slightly smaller than the others. Antheriferous stamens 5. Ripened carpels sharp-pointed below, at most tardily dehiscent, the stylar portion when freed spirally twisting below, bearded on the inner side. Seed smooth. — Old World and w. Am. herbs. (Name from the Greek *erodios, a heron;* from the long beak of the fruit.)

Leaflets sessile; awns of sepals bristle-tipped; filaments not toothed. . . . . . 1. *E. cicutarium.*
Leaflets short-stalked; awns of sepals not bristle-tipped; antheriferous filaments 2-toothed at base. . . . . 2. *E. moschatum.*

1. E. CICUTÀRIUM (L.) L'Hér. (resembling *Cicuta*), ALFILERIA, PIN-CLOVER. — Winter-annual or biennial, making rosettes of pinnate overwintering leaves; *the sessile leaflets twice pinnatifid into linear fine segments;* stems loosely ascending or spreading, leafy, with long axillary peduncles of umbellate flowers; *sepal-awns bristle-tipped;* petals roseate or purple; *filaments without teeth at base.* — Fields, roadsides and waste places, and locally as a weed of sandy farm-lands, Que. to Mich. and Ill., s. to N.S., N.E., Va., Tenn., Ark., Tex. and Mex. Mar.–Nov. and sporadically all winter. (Natzd. from Eu.)

2. E. MOSCHÀTUM (L.) L'Hér. (smelling of musk). — Coarser; *leaflets short-petiolulate, merely pinnatifid,* the bluntish teeth rarely extending half-way to midrib; *sepals* glabrous or sparsely hirtellous, *their awns without bristles; filaments with a tooth on each side of base.* — Waste places, infreq., e. Mass. to N.Y. and Del.; abund. on Pacific slope. (Adv. from Eu.)

Var. PRAÈCOX Lange (early). — Leaflets deeply and often bipinnately cleft; sepals strongly viscid-pubescent. — Waste, often from woolen-mills, Gaspé Pen., Que., to Ct. (Adv. from Eu.)

Other Old World species are casuals and may become estab.

FAM. 87. ZYGOPHYLLÀCEAE (CALTROP FAMILY)

Herbs (or southward woody plants) with opposite (or alternate), in our species abruptly pinnate, undotted leaves, and perfect regular mostly 5-merous flowers. Stamens free, essentially hypogynous, in ours twice as many as the petals. Pistil of several united 1–few-ovuled carpels. Ovules anatropous, with superior micropyle and large straightish embryo. — Chiefly tropical.

Carpels 4 or 5, each with 3–5 ovules, prickly. . . . . 1. *Tribulus.*
Carpels 8–12, each with 1 ovule, tuberculate. . . . . 2. *Kallstroemia.*

1. TRÍBULUS L.

Sepals and petals (4–)5. Filaments slender, unappendaged; those before the petals sometimes slightly united with them, the alternate ones subtended by glands. Locules of ovary as many as the petals, 3–5-ovulate. — Chiefly confined to trop. and warm reg.; ours spreading annuals. (Latin name of the *caltrop,* the form of which is suggested by the prickly fruit.)

1. T. TERRÉSTRIS L. (terrestrial), CALTROP. — Branched from the base; leaflets 5–7 pairs; flowers small, short-peduncled; petals pale yellow; mature carpels crested and armed with 2–4 spreading prickles. — Dry waste places and open sandy ground, Fla. to Tex., n. to s. N.Y., O., Mich., Ill., Ia., S.D., etc. June–Sept. (Natzd. from Old World)

2. KALLSTROÈMIA Scop.

Sepals, petals and stamens as in *Tribulus.* Locules of the ovary twice as many as the petals, each 1-ovuled, becoming 1-seeded nutlets, dorsally rounded, smooth or tuberculate but not

prickly, at maturity falling away from the persistent stylar axis. — Diffuse trop. and subtrop. annuals, ours unusually northern. (In honor of *Kallstroem,* obscure contemporary of Scopoli.)

1. K. intermèdia Rydb. (intermediate). — Prostrate, grayish-hirsute; leaflets 4 or 5 pairs, oblong, obtuse, about 1 cm. long; flowers 9–15 mm. in diameter; *sepals subulate; petals* yellow, 7–10 *mm. long; fruit* depressed-ovoid, *beaked with a stoutish columnar style* 5–6 *mm. long and exceeding body of fruit.* (*K. maxima* of ed. 7, not (L.) T. & G.) — Dry open soil, Ill. to Colo., s. to Mo., Tex. and Mex. June–Sept.

2. K. hirsutíssima Vail (most hirsute). — Similar to no. 1; leaflets 3 or 4 pairs; *sepals narrowly lanceolate; petals* 5 *or* 6 *mm. long; beak more conic, shorter than body of fruit,* 3–4 *mm. long.* — Waste places and sands, w. Mo. to Colo., s. to Tex. and Mex.

FAM. 88. RUTÀCEAE (RUE FAMILY)

***Plants with simple or compound leaves dotted with pellucid glands and abounding in a pungent or bitter-aromatic acrid volatile oil, producing hypogynous mostly regular* 3–5-*merous flowers, the stamens as many or twice as many as the sepals (rarely more numerous); the* 2–5 *pistils separate or combined into a compound ovary of as many locules, raised on a prolongation of the receptacle (gynophore) or glandular disk.* Embryo large, usually in fleshy albumen. Styles commonly united or cohering. Fruit usually a capsule, berry or samara. Stipules none. — A large family, chiefly of the Old World and the S. Hemisph.**

Trees or lignescent shrubs, with distinct leaflets.
Flowers dioecious; leaves odd-pinnate, with several leaflets.
Leaves alternate; branches and petioles prickly; fruits fleshy follicles with solitary heavy seeds. . . . . 1. *Xanthoxylum.*
Leaves opposite; branches and petioles not prickly; fruit a drupe. . . . . 2. *Phellodendron.*
Flowers polygamous or perfect; leaves 3-foliolate; fruit indehiscent.
Branches without prickles; petioles slender; flowers small, greenish-white, in compound peduncled cymes; fruit an orbicular samara. . . . . 3. *Ptelea.*
Branches bright green, with strong green spines; petioles winged; flowers 3–5 cm. broad, white, subsessile on old wood; fruit a many-seeded orange-like berry. . . . . 4. *Poncirus.*
Herbaceous or suffruticose, the leaves pinnatifid or simple; flowers perfect; fruit a many-seeded 4–5-lobed capsule. . . . . 5. *Ruta.*

1. XANTHÓXYLUM Gmel. PRICKLY ASH. FRÊNE ÉPINEUX (Que.). CLAVALIER (Que.)

Flowers dioecious. Sepals 4 or 5, obsolete in one species. Petals 4 or 5, imbricated in the bud. Stamens 4 or 5 in the staminate flowers, alternate with the petals. Pistils 2–5, separate, but their styles connivent or slightly united. Follicles thick and fleshy, 2-valved, 1–2-seeded. Seed-coat crustaceous, black, smooth and shining. Embryo straight, with broad cotyledons. — Shrubs or trees of N.Am. and Asia, with mostly pinnate alternate leaves, usually the stems and often the petioles prickly. Flowers small, greenish or whitish. (From the Greek *xanthos, yellow,* and *xylon, wood.*) ZANTHOXYLUM L.

1. X. americànum Mill. (American), NORTHERN P., TOOTHACHE-TREE, CLAVALIER (Que.). — Aromatic shrub; leaves and *flowers in sessile axillary umbellate clusters,* the flowers yellow-green, expanding before the leaves; *leaflets* 2–4 pairs and an odd one, ovate-oblong, *downy when young;* calyx none; petals 4–5; pistils 3–5, with slender styles; follicles short-stalked. — Rich woods and river-banks, w. Que. to N.D., s. to Ga., Ala., Mo. and Okla. Apr., May. — Forma impùniens Fassett (not punishing) has prickleless branches.

2. X. Clàva-Hérculis L. (Hercules'-club), SOUTHERN P., HERCULES'-CLUB. — Small tree with very sharp prickles, those of the trunk with broad pyramidal bases; *leaflets* 3–8 *pairs and an odd one, ovate or ovate-lanceolate, oblique,* shining above, glabrous; *flowers in an ample terminal cyme;* sepals and petals 5; pistils 2–3, with short styles; follicles sessile. — Sandhills and dry woods and thickets, Fla. to Tex., n. to e. Va. and Ark. Apr., May.

2. PHELLODÉNDRON Rupr. CORK-TREE

Flowers dioecious. Sepals and petals 5–8, imbricate in the bud. Staminate flowers with rudimentary ovary and 5 or 6 stamens longer than the petals; anthers 2-lobed at base. Pistillate flowers with 5 or 6 staminodia and a 5-locular subglobose ovary; stigma 5-lobed, thick. Fruit a subglobose drupe with 5 1-seeded small stones. Embryo straight, with oblong cotyledons. —

Aromatic trees, nat. of e. Asia, with prickleless trunks, odd-pinnate opposite leaves, and yellowish-green paniculate or corymbose flowers. (From the Greek, *phellos, cork*, and *dendron, tree.*)

1. P. JAPÓNICUM Maxim. (Japanese). — Tree up to 10 m. high, with thin brown bark and reddish branchlets; leaves whitish-pubescent beneath; the 9–13 ovate-oblong or obliquely ovate leaflets abruptly acuminate, mostly 4–6 cm. broad; panicle open, 5–10 cm. broad; drupes glabrous. — Roadsides and borders of woods, e. Pa. May–July. (Introd. from e. Asia)

Other species are likely to spread from cult.

3. PTÈLEA L. SHRUBBY TREFOIL. HOP-TREE

Flowers polygamous. Sepals 3–5. Petals 3–5, imbricated in the bud. Stamens as many. Ovary 2-locular; style short; stigmas 2. Fruit a 2-locular and 2-seeded samara, winged all around, nearly orbicular. — N. Am. shrubs, with 3-foliolate leaves and greenish-white small flowers in compound terminal cymes. (The Greek name of the Elm, by Linnaeus applied to this genus with similar fruit.)

1. P. trifoliàta L. (three-leaved), STINKING ASH, WAFER-ASH. — Tall shrub or small tree; branchlets glabrous or essentially so; leaflets ovate, pointed, glabrous or promptly glabrate, or permanently pilose beneath in forma **pubéscens** (Pursh) Voss (hairy); fruits orbicular to reniform. — Alluvial thickets, rocky slopes and gravels, sw. Que. and N.Y. to s. Ont., and Neb., s. to Fla., Ala., La. and Tex.; cult. and spreading in N.E. Late May–early July.

Var. **móllis** T. & G. (soft). — Branchlets densely pubescent; leaflets firmer, often smaller and less pointed, permanently soft-pubescent beneath. (*P. tomentosa* Raf.) — N.C. to Ga., w. to Okla., Ariz. and Mex.; dunes of L. Mich.

4. PONCÌRUS Raf. TRIFOLIATE ORANGE

Flowers perfect. Sepals 4–7, nearly distinct, triangular. Petals 5, oblong-obovate, imbricate in bud, spreading. Stamens 8–10, free. Ovary 6–8-locular, pubescent, the ovules in 2 rows in each locule. Fruit an orange-like berry. — Green-branched shrubs nat. of Asia, or small trees with coarse green spines, the wing-petioled alternate leaves with 3 crenate small leaflets. Flowers subsessile on the old wood, before the leaves, white, fragrant. (*Poncire*, the French name for some *Citrus.*)

1. P. TRIFOLIÀTA (L.) Raf. (three-leaved). — Flowers 3–5 cm. broad; fruit downy, lemon-yellow, globose, 3–5 cm. in diameter, fragrant. — Much used in the South for hedges, esc. to borders of woods, etc., Fla. to Tex., n. to e. Va. April, May. (Introd. from e. Asia)

5. RÙTA L. RUE

Flowers perfect, 4–5-merous. Calyx persistent. Petals yellow, the sides and apex strongly inrolled, the margin denticulate or ciliate-dentate. Stamens 8–10, inserted about the base of the torus, the alternate ones smaller. Capsule 4–5-lobed, dehiscent at the summit, many-seeded. — Heavy-scented herbs or undershrubs nat. of Eurasia, with alternate simple or variously compound leaves. (The ancient name.)

1. R. GRAVÈOLENS L. (strong-smelling), COMMON R., HERB-OF-GRACE. — Suffruticose, glaucous, 0.3–1 m. high; leaves thickish, 2–3-pinnatifid, ultimate lobes or divisions obovate-cuneate; petals denticulate. — Formerly cult. for medicinal and aromatic qualities, now locally estab. in old pastures, on roadsides, etc., Vt. to Va., W.Va. and Mo. June, July. (Introd. from Eu.)

FAM. 89. SIMARUBÀCEAE (QUASSIA FAMILY)

Trees and shrubs with floral structure much as in the Rutaceae *but the foliage destitute of pellucid dots.* — Chiefly trop.

1. AILÁNTHUS Desf. TREE-OF-HEAVEN

Flowers polygamous. Calyx regular, 5-parted, the lobes imbricated. Petals 5, infolded-valvate. Stamens in staminate flowers 10, in perfect flowers 2–3, in pistillate flowers none. Disk lobed. Ovary 2–5-parted, becoming in fruit 1–5 narrowly oblong membranaceous samaras (1-seeded in the middle). — Handsome trees of rapid growth, nat. of Asia and Austral. Leaves odd-pinnate. Flowers small, green or yellowish, in ample terminal panicles, especially the staminate of unpleasant odor. (Name said to be from a vernacular Moluccan designation, meaning *tree-of-heaven*, in allusion to the height in the native habitat.)

1. A. altíssima (Mill.) Swingle (tallest), Copal-tree. — Leaves 3–6 dm. long, 11–25-foliolate; leaflets oblong or lanceolate, acuminate, 8–18 cm. long; samaras 3–5 cm. long. (*A. glandulosa* Desf.) — Spread from cult. by basal suckers as well as by seed, Mass. to s. Ont. and Ia. and southw., often too aggresive. June, July. (Introd. and natzd. from Asia) — Rapid-growing tree.

FAM. 90. MELIÀCEAE (Mahogany Family)

Trees or shrubs with hard wood and with alternate dotless pinnate leaves without stipules, paniculate inflorescences and perfect mostly 5 (*or* 6)*-merous small regular flowers. Sepals imbricated in bud; petals imbricated or convolute in the bud. Stamens monadelphous; anthers 2-locular, introrse. Locules of ovary usually of same number as petals, the base surrounded by a ring or cup-like disk; styles and stigma united into one.* — Trop. and subtrop. group of great economic importance.

1. MÉLIA L.

Petals 5 or 6, narrowly spatulate, spreading. Stamen-tube cylindrical, with 10–12-toothed orifice and many included sessile anthers. Ovary with a pair of superposed ovules in each locule. Drupe 5–6-locular or by abortion 1-locular, with thin flesh and a single seed in each locule of the bony putamen; embryo in thin fleshy albumen. — Trees and shrubs nat. of Asia and Austral. (Greek name of Ash.)

1. M. Azédarach L. (aboriginal name), Pride-of-India, China-tree, "Mahogany". — Tree up to 12 m. high; flowers pale lilac, fragrant; drupes yellow, subglobose, 1.5–2 cm. in diameter, bitter and astringent. — Thickets and borders of woods, Fla. to Tex., n. to e. Va. and Okla., spread from cult. May, June. (Introd. and natzd. from Asia)

FAM. 91. POLYGALÀCEAE (Milkwort Family)

Plants with irregular hypogynous flowers, 4–8 *diadelphous or monadelphous stamens, their* 1*-locular anthers opening at the top by a pore or chink; the fruit a* 2*-locular and* 2*-seeded capsule.*

1. POLÝGALA L. Polygala. Milkwort

Flower commonly very irregular. Calyx persistent, of 5 sepals, of which 3 (the uppermost and the 2 lowest) are small and often greenish, while the two lateral or inner (called *wings*) are much larger and colored like the petals. Petals 3, hypogynous, connected with each other and with the stamen-tube, the middle (lower) one keel-shaped and often crested on the back. Stamens 6 or 8; their filaments united below into a split sheath, or into 2 sets, cohering more or less with the petals, free above; anthers 1-locular. Ovary 2-locular, with an anatropous ovule pendulous in each locule; style prolonged and curved; stigma various. Fruit a small loculicidal 2-seeded capsule, usually rounded and notched at the apex, much flattened contrary to the very narrow partition. Seeds usually arillate. Embryo large, straight, with flat and broad cotyledons, in scanty albumen. — Bitter plants of most temp. and trop. reg. (low herbs in temp. reg.), with simple usually entire often dotted leaves, and no stipules. (An old name composed of the Greek *polys, much,* and *gala, milk,* applied by Dioscorides to some low shrub reputed to increase lactation.)*

a. Creeping perennial; the principal leaves approximate at tips of short branches, elliptic or oval, 1–4.7 cm. long; showy flowers 1–4, 1.5–2.3 cm. long; sepals deciduous, their wings deciduous; small cleistogamous flowers and suborbicular fruits scattered on subterranean branches. . . . . 1. *P. paucifolia.*

a. Upright annuals with single stems, or more or less tufted perennials or biennials without long creeping stems; leaves mostly narrower and shorter; flowers numerous, smaller; sepals and their wings persistent; cleistogamous subterranean flowers only in no. 2. . . . *b.*

b. Flowers purple, pink, white or greenish, not changing color, in simple racemes; sepals not decurrent on the pedicels; pedicels wingless. . . . *c.*

c. Flowers loosely racemose, spreading or recurving on slender pedicels; stems usually several from a biennial to perennial root; minute cleistogamous flowers and their plump capsules in 1-sided prostrate racemes about the base after the fading of the petaliferous flowers. . . . . . . 2. *P. polygama.*

* Treatment prepared with aid from that of S. F. Blake in N. Am. Fl. xxv. pt. 4 (1924).

c. Flowers spicate-racemose, ascending to spreading; no cleistogamous inflorescences. . . d.

d. Stems several from a thick and tough caudex and perennial root; leaves alternate, lanceolate to ovate (linear in no. 4), 0.8–7 cm. long, 0.1–3 cm. broad; racemes dense, white.

Leaves lanceolate to ovate, 1.3–7 cm. long, 0.25–3 cm. broad, irregularly serrulate; wings suborbicular, 3–3.7 mm. long, 2.2–3 mm. broad; capsule suborbicular, 3–4.3 mm. broad. . . . . . 3. *P. Senega.*

Leaves linear, 0.8–2.5 cm. long, 1–1.5 mm. wide, entire; wings elliptic, 2.5–3 mm. long, about 1.5 mm. broad; capsule elliptic, 1.3–1.6 mm. broad. . . . . 4. *P. alba.*

d. Stems solitary, annual; leaves mostly smaller and narrower (if sometimes broadened upward whorled), entire; racemes purple, pink or green (exceptionally white). . . . e.

e. Stem glaucous; leaves linear-subulate or -involute, soon deciduous; cylindric racemes usually with persistent fruits at base and a tuft of flowers at tip; keel conspicuously crested; claws of the true petals united into a long and slender cleft tube much surpassing the wings; aril not obviously lobed, equitant. . . . 5. *P. incarnata.*

e. Stem not glaucous; leaves flat, persistent; fruits soon dropping from base of raceme; keel only inconspicuously crested; true petals not longer than wings; aril 2-lobed. . . f.

f. Raceme dense, ovoid, subcapitate or thick-cylindric, rounded at summit (apiculate in no. 9), 4–16 mm. thick; wings 2.1–6.3 mm. long. . g.

g. Bracts soon deciduous from the base of the elongating rachis. . h.

h. Wings 4.8–6.3 mm. long, 2.5–3.5 mm. broad, about twice length of keel; aril half the length of the seed or longer, 1–1.3 mm. long. . . . . 6. *P. sanguinea.*

h. Wings 2.7–4.7 mm. long, 1.3–2.2 mm. broad; aril 0.4 mm. long, much shorter than seed (0.9–1.1 mm. long).

Wings 2.7–3.6 mm. long, 6-nerved; capsule 1.8–2 mm. broad. . . . . 7. *P. mariana.*

Wings 3.5–4.7 mm. long, 3–5-nerved; capsule 1.3–1.5 mm. wide. . . . . 8. *P. Harperi.*

g. Bracts persistent as subulate hooks after the falling of fruit from the elongating rachis. . . i.

i. Leaves alternate, the upper linear and subulate-tipped.

Racemes 4–6 mm. thick, soon loosening; wings 2–2.5 mm. long, 1–1.3 mm. broad, barely longer than keel; aril 3–4 mm. long. . . . . 9. *P. Nuttallii.*

Racemes 6–14 mm. thick; wings 3–6.3 mm. long, 1.7–3.5 mm. broad, much longer than keel.

Racemes dense, the rose-purple to green or whitish flowers tightly overlapping; pedicels 1–1.5 mm. long; wings ovate or oval, 9-nerved, 2.5–3.5 mm. broad; aril 1–1.3 mm. long, with linear lobes, three-fourths length of seed. . . . . 6. *P. sanguinea.*

Racemes quickly loosening, bright pink to rose-purple; the keel and upper petals yellow-tipped; pedicels 1–2 mm. long; wings elliptic, 5–7-nerved, 1.7–2.9 mm. broad; aril 0.3–0.4 mm. long, with obovate lobes, one-fourth length of seed. . . . . 10. *P. Curtissii.*

i. Leaves (at least the lower) whorled, blunt; aril 0.9–1.2 mm. long, three-fourths length of seed.

Peduncle 0.1–4 mm. long; racemes 0.7–2 cm. thick; bracts 1.5–3 mm. long; wings deltoid, acuminate, subulate-tipped, 2.5–5.5 mm. long, 2.5–3.6 mm. wide. . . . . 11. *P. cruciata.*

Peduncle 0.7–8 cm. long; racemes 7–12 mm. thick; bracts 1 mm. long; wings ovate to oblong, merely acutish or short-tipped, 2.8–3.2 mm. long, 1.5–2.3 mm. wide. . . 12. *P. brevifolia.*

f. Raceme open to dense, slenderly cylindric, lanceolate or conic, tapering to tip, 2–4.5 mm. thick, greenish, whitish or dull purple; bracts deciduous; wings 0.9–2.6 mm. long; at least the lower leaves commonly whorled. . . . . 13. *P. verticillata.*

b. Flowers bright yellow to orange, often changing (when dried) to paler yellow, green or blackish; sepals decurrent on the winged pedicels.

Racemes single at tips of stems and branches, capitate-spicate, 0.9–2 cm. thick, orange-yellow, becoming paler in drying. . . . . 14. *P. lutea.*
Racemes numerous in a terminal cyme, darkening on drying, the individual racemes 7–13 mm. thick.
Basal leaves elliptic to obovate, obtuse, 0.7–2 cm. long; cauline leaves subequal to summit of stem; wings 2.9–3.5 mm. long; seed ellipsoid, hairy, with evident aril. . . . . 15. *P. ramosa.*
Basal leaves linear to narrowly lanceolate, attenuate, 0.3–1.5 dm. long; cauline leaves greatly reduced upward; wings 2.3–3 mm. long; seed hemispheric, glabrous, with minute aril. . . . . 16. *P. cymosa.*

1. P. paucifòlia Willd. (few-leaved), FRINGED POLYGALA, FLOWERING WINTERGREEN, BIRD-ON-THE-WING. — Perennial with slender subterranean rhizome and stolons; flowering stems short (7–10 cm. high); *lower leaves* small and *scale-like*, scattered; the *upper ovate, petioled, crowded* at the summit, glabrous or merely ciliate or pilose on nerves beneath, or densely canescent-pilose in forma **vestìta** Fern. (clothed); *flowers* 1–4, 1.5–2.3 *cm. long*, rose-purple, or white and with foliage paler in forma **álba** Wheelock (white); wings obovate, rather shorter than the fringe-crested keel; stamens 6; caruncle of 2 or 3 subulate lobes longer than the seed; through summer and autumn bearing minute cleistogamous flowers and suborbicular fruits on subterranean branches. (*Triclisperma* Nieuwl.) — Woods, in light soil, Gaspé Co., Que., to Man., s. to s. N.B., N.E., inland Va., mts. of Ga. and Tenn., n. Ill. and Minn. May–early July. FIG. 1275.

1275. P. paucifolia.

2. P. polỳgama Walt. (polygamous). — *Stems* mostly *numerous from the biennial or perennial root*, simple or forking, ascending to depressed, very leafy, 1–4.5 dm. high, *bearing 1-sided racemes of minute cleistogamous flowers and plump pale fruits from the base in summer and autumn;* rosette-leaves spatulate, cauline very numerous and narrowly oblanceolate to linear, mostly 2–4 cm. long, or in the local forma **obovàta** Blake (obovate) obovate or elliptic and only 0.8–2 cm. long; racemes *very loose*, finally elongating to 8–15 cm.; *the divergent flowers* 4–6 *mm. apart*, 5–6 *mm. long*, rose-pink to rose-purple, *on pedicels* 1.5–3.5 *mm. long; the broadly obovate wings* longer than the keel and *much exceeding the* plump ovoid *capsule* (3–4 mm. long); seeds ellipsoid, pilose, with conspicuous aril. (Var. *macrospora* Chodat) — Dry sandy woods and openings, Fla. to N.J. and Ky. June, July.

Var. **obtusàta** Chodat (blunt). — *Racemes* becoming 2–12 cm. long, *closer-flowered; the flowers* 3–5 (–6) mm. long, *mostly* 1–4 *mm. apart, on pedicels* 0.5–2 *mm. long;* wings strongly rounded above, shorter than to exceeding capsule. — Dry open soil, Fla. to Tex., n. to N.S., centr. Me., centr. N.H., s. Vt., sw. Que., n. N.Y., s. Ont., Mich., Wisc. and Minn. June–Aug. FIG. 1276. — Late in the season often branching or bearing short racemes of cleistogamous flowers and fruits resembling those of the basal racemes but deeper-colored. Plants with pale lilac petals are forma **pállida** Britt. (pale); those with white petals forma **albiflòra** House (white-flowered).

1276. P. polygama, v. obtusata.

3. P. Sénega L. (old generic name, from use of the plant among the Seneca Indians), SENECA-SNAKEROOT. — *Stems several from a thick crown* and stout root, simple, 1–5 dm. high; *leaves* numerous, all but the bracteal lower ones linear-lanceolate to lance-elliptic or -ovate, *acuminate*, 1.3–7 cm. long, 2.5–15 *mm. wide, the subcrustaceous margin serrulate;* racemes terminal, solitary, dense, 6–7 mm. thick, with white or whitish flowers; *wings* suborbicular, 3–3.3 mm. long, 2.2–2.5 *mm. wide; capsules* plump, rounded, 2.5–3.5 *mm. long*, 3–3.8 *mm. wide*, persistent; seeds about 2.5 mm. long, pilose. — Dry rocky or gravelly, chiefly calcareous, areas, Que. to Alta., s. to N.B., ne. Me., w. N.E., Ga., Tenn., Ark. and S.D. May–July. FIG. 1277.

1277. P. Senega.

Var. **latifòlia** T. & G. (broad-leaved). — Simple or branching, 3–5 dm. high; *leaves* thinner and broader, *the upper oval or lance-ovate, more strongly serrulate* 1.5–3.5 *cm. broad;* racemes 5–9 mm. thick; *wings* 3–3.7 mm. long, 2.8–3 *mm. wide; capsules*, 3.5–4.2 *mm. long*, 4–4.3 *mm. wide;* seeds 3–3.5 mm. long. — Rich woods, calcareous bluffs and slopes, Pa. and Del. to Minn. and S.D., s. to N.C., Tenn. and Mo.

4. P. álba Nutt. (white). — Resembling no. 3, more slender, 2–3.5 dm. high; *leaves* mostly *linear, firm*, cuspidate, 0.8–2.5 *cm. long*, 1–1.5 *mm. broad, smooth, entire*, soon deciduous;

1278. P. alba.

peduncles elongate; raceme slenderly lance-cylindric, acuminate, 4–8 mm. thick, 2–8 cm. long; corolla white, with green center and often purple crest; *wings elliptic,* 2.5–3 mm. *long, about* 1.5 *mm. broad,* about equaling keel; *capsule elliptic,* 2.5–3 mm. long, 1.3–1.6 *mm. wide.* — Dry plains and gravelly hills, Wash. to Mex., e. to Minn., Neb., Kans., Okla. and Tex. May–Aug. Fig. 1278.

5. P. incarnàta L. (flesh-colored). — *Glaucous,* the annual stem sulcate, stiff, very slender, simple or with few erect branches, 1.5–6 dm. high; *leaves linear-subulate or involute,* scattered, subappressed, *soon deciduous; racemes* dense, *subcylindric, some or all of the fruits persistent below the tuft of terminal* flesh-colored *flowers; claws of the true petals united into a long slender cleft tube much surpassing the wings; keel conspicuously crested; aril* equitant, *not obviously lobed.* — Dry open soil, Fla. to e. Tex. and Mex., n. to L.I., N.J., Pa., s. Ont., s. Mich., s. Wisc., Ia. and Neb. June–Nov. Fig. 1279.

1279. P. incarnata.

6. P. sanguínea L. (blood-red). — Annual, simple to bushy-branched, 0.7–4 dm. high, very leafy to top with linear mucronulate ascending leaves; *racemes thick-cylindric to capitate, rounded at summit, very dense, with the flowers tightly overlapping,* 6–14 *mm. thick,* the fruits dropping promptly; bracts persistent or deciduous along the old axis; *pedicels* 1–1.5 *mm. long;* flowers pink to rose-purple at summit, or greenish in forma **viridéscens** (L.) Farw. (greenish) = *P. viridescens* L., or whitish in forma **albiflòra** (Wheelock) Millsp. (white-flowered); *wings ovate or oval,* 4.8–6.3 *mm. long,* 2.5–3.5 *mm. broad,* broadly rounded above, 9-*nerved, twice as long as keel; aril* 1–1.3 *mm. long, with* 2 *linear* appressed *lobes, three-fourths as long as* the pyriform-subglobose pilose *seed.* — Sterile meadows and other moist open acid soils, N.S. to s. Ont. and Minn., s. to s. N.E., L.I., S.C., Tenn., La. and Okla. Late June–Oct. Fig. 1280.

1280. P. sanguinea.

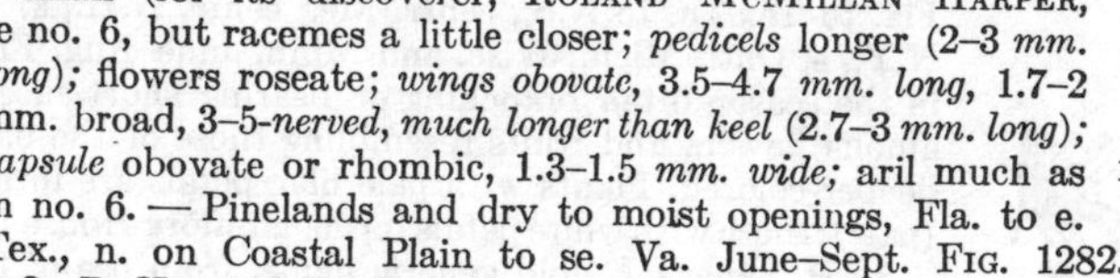

7. P. mariàna Mill. (of Maryland). — Resembling no. 5; capitate *racemes* obtuse or apiculate, 6–11 mm. thick, *with flowers distinct, not closely overlapping,* pink, purplish or greenish; *bracts caducous; wings elliptic,* 2.7–3.6 *mm. long,* 1.3–2.2 *mm. wide,* 6-*nerved, only slightly longer than keel; capsule* suborbicular, 1.8–2 *mm. broad; aril* 0.4 *mm. long,* with oblong cellular lobes, *one-third length of seed.* — Dry to moist sands, clays and peats, chiefly on Coastal Plain, Fla. to Tex., n. to N.J., Del., Md. and Ky. June–Oct. Fig. 1281.

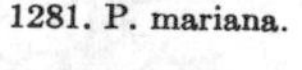

1281. P. mariana.

8. P. Hárperi Small (for its discoverer, Roland McMillan Harper, 1878–). — Like no. 6, but racemes a little closer; *pedicels* longer (2–3 *mm. long*); flowers roseate; *wings obovate,* 3.5–4.7 *mm. long,* 1.7–2 mm. broad, 3–5-*nerved, much longer than keel* (2.7–3 *mm. long*); *capsule* obovate or rhombic, 1.3–1.5 *mm. wide;* aril much as in no. 6. — Pinelands and dry to moist openings, Fla. to e. Tex., n. on Coastal Plain to se. Va. June–Sept. Fig. 1282.

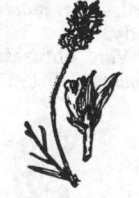

1282. P. Harperi.

9. P. Nuttállii T. & G. (for its discoverer, Thomas Nuttall, 1786–1859). — Similar to no. 6, smaller throughout, 0.5–2.5 dm. high; *racemes cylindric,* greenish-white to dull purple, apiculate, finally rounded at summit, 4–6 *mm. thick,* rather dense, soon loosening; *bracts persistent* after fall of fruit as upcurved hooks; *wings* obovate, 2–2.5 *mm. long,* 1–1.3 *mm. wide, barely longer than keel;* aril much as in nos. 6 and 7. — Dry open soil and pinelands, Ga. to Miss., n. to s. N.E., s. N.Y., N.J., e. Pa., Ky. and Ark. June–Oct. Fig. 1283.

1283. P. Nuttallii.

10. P. Curtíssii Gray (for its discoverer, Allen Hiram Curtiss, 1845–1907). — Similar to no. 6; *racemes quickly loosening,* thick-cylindric, 9–13 mm. thick, bright pink to rose-purple, *the keel and upper petals yellow-tipped; bracts persistent; pedicels* 1–2 *mm. long; wings elliptic,* 5–7-*nerved,* 3–6 mm. long, 1.7–2.9 *mm. broad, nearly twice as long as keel; capsule* obovate, 2.5–3 mm. long, *half as long as wings;* aril 0.3–0.4 mm. long. — Dry to moist pinelands and openings, Ga. to La., n. on Coastal Plain to Del. and Md. and in the interior n. to w. Va., W.Va., O. and Ky. June–Oct. Fig. 1284.

1284. P. Curtissii.

11. P. cruciàta L. (cross-shaped; from the whorls of four leaves). — Simple to bushy-branched annual, 1–5 dm. high, the *primary axis with* 5–12 *leaf-*

bearing nodes; leaves chiefly in whorls of 3 or 4, *linear-spatulate or linear-oblanceolate, firm, the larger ones* 1.5–3(–4) *mm. wide; racemes dense, thick-cylindric to ellipsoid, subsessile or on peduncles up to 4 cm. long, the principal ones* before falling of lower flowers (1–) 1.5–4.5 *cm. long and* 1.2–2 *cm. thick*, purple to greenish or whitish; *bract persistent* after fall of fruit, 2–3 *mm. long; wings deltoid-cordate, longer than broad,* 3.5–5.5 *mm. long, with subulate tip* 1.5–3 *mm. long; seed ellipsoid, faintly rugulose;* its aril about 1 mm. long, its linear lobes about as long as seed. (Var. *cuspidata* (H. & A.) Wood; var. *ramosior* Nash; *P. ramosior* (Nash) Small) — Wet pinelands, savannas, peats and sands, Fla. to e. Tex., n. on or near Coastal Plain to Va. and inland n. to Ky. July–Oct. Northward passing to

Var. **aquilònia** Fern. & Schub. (northern). — Simple or *with few more divergent branches,* 0.5–2.5 dm. high, the *primary axis with* 3–5 (–6) *leaf-bearing nodes; leaves spatulate to narrowly oblanceolate, herbaceous, the larger ones* (2–) 3–7 *mm. wide; racemes sessile or on very short* (*up to* 5 *mm.*) *peduncles,* the *leading raceme* before falling of flowers 0.7–3.5 cm. long and 0.7–1.5 *cm. thick; bracts* 1.5–2 *mm. long; wings about as wide as long, their blades* 2.5–4 *mm. long,* purplish to greenish, or in forma **álba** (Oakes) Fern. & Schub. (white) whitish, *with awn* 0.5–1 *mm. long; seed ellipsoid-obovoid, coarsely rugulose.* (*P. cruciata* of ed. 7, not L.) — Damp peat, sands, sterile meadows, rarely far from coast, and at borders of saline marshes, Va. to s. Me.; n. O. to Minn., s. to mts. of Ala. and Tenn. Fig. 1285.

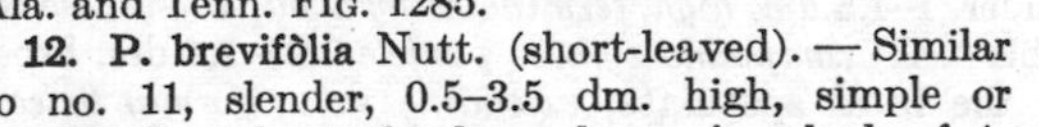

1285. P. cruciata, v. aquilonia.

1286. P. brevifolia.

12. P. brevifòlia Nutt. (short-leaved). — Similar to no. 11, slender, 0.5–3.5 dm. high, simple or branched; at least the *lower leaves in whorls* of 4 or 5, linear-oblanceolate; *peduncles* 0.7–8 *cm. long; racemes* 7–12 *mm. thick,* rose-purple; *bracts about* 1 *mm. long; wings ovate to oblong, merely acutish or short-mucronate,* 2.8–3.2 *mm. long,* 1.5–2.3 *mm. wide.* — Sandy swamps, Fla. to Miss.; N.J. July–Sept. Fig. 1286.

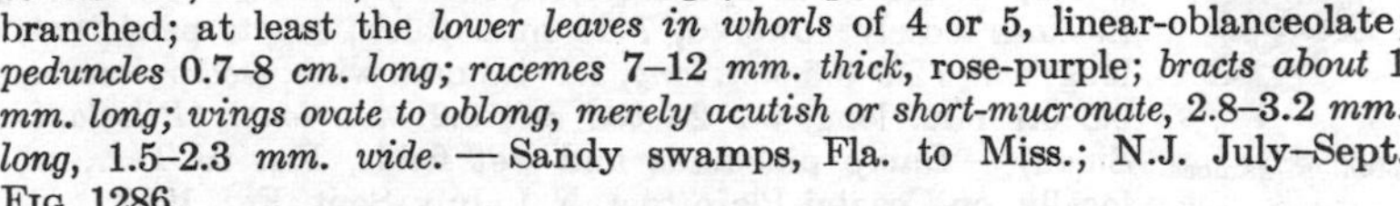

13. P. verticillàta L. (whorled). — Slender annual 0.5–4 dm. high; *leaves of at least* 1–3 *lower primary* and sometimes of other *nodes in whorls* of 3's–7's, linear to narrowly lanceolate, acute; *racemes* peduncled, *white, greenish or purplish, linear-cylindric to lance-conic, attenuate at summit,* the floriferous portion (after falling of basal fruits) 0.5–5 cm. long, 2–4.5 *mm. thick; bracts caducous;* wings obovate to oval, 0.9–2.6 mm. long, shorter than to about equaling capsule; keels yellow or yellowish; capsule oval, 0.9–2.2 mm. long; aril with 2 oblong to narrowly obovate lobes, one-third to three-fourths the length of the pubescent seed. — Polymorphic, with 5 defined vars.:

a. Leaves of all or most of the primary nodes whorled; racemes lance-conic, the floriferous part 0.5–2 cm. long; wings shorter than capsule. . . *b.*

b. Stems 1.5–4 dm. high, with ascending branches; lower leaves in 3's–5's, the upper or sometimes all but those of lower nodes scattered, linear, rather firm, the larger 1.5–3 cm. long and 0.8–3 mm. broad; peduncles 2–7 cm. long; racemes lax, white or purplish, the floriferous portion 1–2 cm. long; mature pedicels 0.5–1 (–2) mm. long; capsule about 1.5 mm. long. . . . *P. verticillata* (typical).

b. Stems 0.5–3 dm. high, with spreading or spreading-ascending branches; leaves in 3's–7's, or the uppermost and rameal opposite or alternate, broadly linear to narrowly lanceolate, thinnish, the larger 1–2.5(–3) cm. long and 1.5–3.5 mm. broad; peduncles 0.5–4 cm. long; racemes dense, white to greenish, the floriferous portion 0.5–1.2 cm. long; pedicels 0.1–0.3 mm. long.

Plant 0.5–2 dm. high; larger leaves 1–2.5 cm. long; capsules 1–1.6 mm. long. . . . . Var. *isocycla.*

Plant 1–3 dm. high; larger leaves 2–3 cm. long; capsules 1.6–2.3 mm. long. . . . . Var. *sphenostachya.*

a. Leaves (except of the 1–3 lowest primary nodes) scattered and alternate; stem 1.5–4 dm. high, with ascending branches; racemes white to purplish, linear-cylindric to slenderly lanceolate, loosely flowered, becoming interrupted at base, the floriferous portion 1–5 cm. long. . . . *c.*

c. Wings about equaling capsule, 1–1.5 mm. long, whitish, greenish or purplish. . . . . Var. *ambigua.*

c. Wings much exceeding capsule, 2–2.6 mm. long, milk-white. . . . . . Var. *dolichoptera.*

P. verticillàta L. (typical), *P. Pretzii* Pennell. — Moist to dryish sterile open habitats, s. Me. to s. Ont. and Mich., s. to s. N.E., Va. and La. June–Oct. Fig. 1287.

1287. P. verticillata.

Var. **isocýcla** Fern. (with equal cycles), *P. verticillata* sensu Pennell, not primarily of L. — Dry or moist sterile habitats, Mass. to s. Man., s. to Fla., Ala., La., Tex., Colo. and Utah.

Var. **sphenostáchya** Pennell (wedge-spiked). — O. to N.D., s. to Mo., Okla. and Tex.

Var. **ambígua** (Nutt.) Wood (doubtful), *P. ambigua* Nutt. — Sw. Me. to n. N.Y., w. to Ill. and Mo., s. to Ga., Ala., La., Okla. and Tex.

Var. **dolichóptera** Fern. (long-winged). — Se. Va.; Mo., Ark. and Okla.

14. P. lùtea L. (yellow), YELLOW M., YELLOW BACHELOR'S-BUTTON. — Biennial, forming a *rosette of obovate to oblong round-tipped leaves;* stems 1–several, ascending, 0.6–4.8 dm. high, simple or forking; cauline leaves alternate, spatulate-lanceolate to narrowly obovate, 1.5–4.5 cm. long; *racemes peduncled, bright orange-yellow* (rarely paler yellow), *becoming paler in drying,* at first depressed, becoming *rounded-ovoid to thick-cylindric,* 0.9–2 cm. thick; *pedicels winged* by decurrence of the sepals; wing-petals obliquely elliptic, 5–7 mm. long, 2.5–3.3 mm. broad; linear lobes of aril nearly equaling seed. (*Pilostaxis* Small) — Damp or wet sandy or peaty soil, Coastal Plain, Fla. to La., n. to L.I., N.J. and e. Pa. Mid-May–Oct. FIG. 1288.

1288. P. lutea.

15. P. ramòsa Ell. (branching). — Biennial; *rosette-leaves spatulate to elliptic, obtuse,* 0.7–2 *cm. long; stems* 1–few, slender, 1–4.5 *dm. high, terminated by a sulphur-yellow* (after drying dark green to blackish) *compound cymose panicle* 0.25–1.4 dm. broad; *cauline leaves* linear, or the basal spatulate, *extending subuniformly to corymb;* individual racemes loosely flowered, 7–11 mm. thick; bracts persistent, about 1.5 mm. long; *wings* 2.9–3.5 *mm. long,* 1.1–1.3 mm. wide; *seed pubescent,* 0.6–0.7 mm. long, *the 2-lobed aril* 0.2 *mm. long.* (*Pilostaxis* Small) — Damp pinelands and wet fields, Fla. to Tex., n., locally, on Coastal Plain to s. N.J. July–Sept. FIG. 1289.

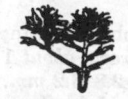

1289. P. ramosa.

16. P. cymòsa Walt. (cymose). — Similar to no. 15; *basal leaves linear to narrowly lanceolate, attenuate,* 0.3–1.5 *dm. long; stem* solitary, distended at base, 0.45–1.2 *m. high; the bracteiform leaves* attenuate, *greatly reduced upward; wings oval,* 2.3–3 *mm. long; seed* 0.8–0.9 mm. long, *glabrous,* minutely areolate, *with minute unlobed aril.* (*Pilostaxis* Small) — Wet pinelands, swales and ditches, Fla. to La., n., locally, on Coastal Plain to Del. July–Sept. FIG. 1290.

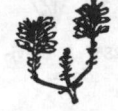

1290. P. cymosa.

FAM. 92. EUPHORBIÀCEAE (SPURGE FAMILY)

Plants usually with a milky acrid juice; the monoecious or dioecious flowers mostly apetalous, sometimes achlamydeous (occasionally polypetalous or gamopetalous); the ovary free and usually 3-locular, with one or sometimes two ovules hanging from the summit of each locule; stigmas or branches of the style as many or twice as many as the locules; fruit commonly a 3-lobed capsule, the lobes or carpels separating elastically from a persistent axis and elastically 2-valved; seed anatropous; embryo straight, almost as long as and the flat cotyledons mostly as wide as the fleshy or oily albumen. Stipules often present. — A vast family in the warmer parts of the world; most numerously represented in northern countries by the genus *Euphorbia,* which has very reduced flowers within a calyx-like involucre.

a. Flowers with a calyx, not in the base of an involucre. . . *b.*
b. Seeds and ovules 1 in each locule. . . *c.*
c. Flowers in di- or trichotomous cymose panicles; calyx corolla-like, that of staminate flowers salverform; plant very prickly, with large palmate leaves. . . . . 1. *Cnidoscolus.*
c. Flowers in spikes, racemes or glomerules, or 1–few in leaf-axils. . . *d.*
d. Stamens 5–innumerable; styles mostly divided. . . *e.*
e. Anthers inflexed in bud; plants stellate-downy or scurfy.
Flowers chiefly in terminal spikes or spiciform racemes; ovary and capsule 2–4-locular. . . . . 2. *Croton.*
Flowers few, scattered on the branchlets; ovary and capsule 1-locular. . . . . 3. *Crotonopsis.*
e. Anthers erect in bud; plants not scurfy. . . *f.*
f. Slender plants with small simple leaves; stamens 8–20, with simple filaments; capsule small.

Staminate calyx 3-parted; stigmas not cleft; casual weed. . . 4. *Mercurialis.*
Staminate calyx 4-parted; stigmas cut-fringed; native. . . . 5. *Acalypha.*
f. Coarse plant with large roundish and palmately divided leaves; stamens very abundant, with forking filaments; terminal racemes in panicles; capsule covered with soft spines; garden-escape. . . . . 6. *Ricinus.*
d. Stamens 2 or 3; styles simple.
Flowers racemose; calyx-lobes valvate in bud; plant pubescent. . 7. *Tragia.*
Flowers spicate; calyx-lobes imbricate in bud; plant glabrous. . . 8. *Stillingia.*
b. Seeds and ovules 2 in each locule; flowers axillary.
Small annual herb with apetalous flowers; stamens 3. . . . . . . 9. *Phyllanthus.*
Shrub with staminate and usually pistillate flowers petaliferous; stamens 5 or 6. . . . . 10. *Andrachne.*
a. Flowers all without calyx, included in a cup-shaped and calyx-like involucre which surrounds a group of many staminate flowers (each of a single naked stamen) and a central pistillate flower with 3-lobed pistil. . . . . 11. *Euphorbia.*

1. CNIDÓSCOLUS Pohl Spurge-Nettle

Flowers monoecious, rarely dioecious, in a terminal open forking cyme; pistillate ones usually in the lower forks. Calyx corolla-like, in the staminate flowers often salverform, 5-lobed; in the pistillate 5-parted, imbricated or convolute in the bud. Petals 0. Glands of the disk opposite the calyx-lobes. Stamens 10–30, monadelphous at base. Ovary mostly 3-locular; styles 3, united below their summits, once or twice forked. Capsule separating into 3 two-valved carpels. Seed carunculate. — Perennial herbaceous or shrubby plants, chiefly of trop. Am., with alternate mostly long-petioled palmately veined leaves, and stipules, armed with stinging bristles. (Name from the Greek *cnide*, *nettle*, and *scolops*, *prickle* or *sting.*)

1. C. stimulòsus (Michx.) Gray (tormenting or stinging), Tread-softly, Bull-Nettle, "Stinging Nettle". — Herbaceous, from a long perennial root, branching, 1.5–6 dm. high; leaves roundish-cordate, 3–5-lobed nearly to the base, on long petioles; the divisions entire or acutely toothed, cut, or even pinnatifid, often discolored; flowers white, fragrant, 1.8 cm. long or more; filaments 10, monadelphous only at the woolly base, the outer set almost distinct. (*Jatropha* Michx.) — Dry sandy woods, fields and sandhills, Fla. to Tex., n. to Va. June–Sept.

2. CRÒTON L. Croton

Flowers monoecious, rarely dioecious, mostly in terminal spike-like racemes or spikes. *Stam. Fl.* Calyx 5 (rarely 4–6)-parted; the divisions lightly imbricated or nearly valvate in the bud. Petals usually present, as many, but mostly small or rudimentary, hypogynous. Glands or lobes of the disk as many as and alternate with the petals. Receptacle usually hairy. Stamens 5 or more; filaments with the anthers inflexed in the bud. *Pist. Fl.* Calyx 5–10-cleft or -parted, nearly as in the staminate flowers, but petals none or minute rudiments. Ovary 3 (rarely 2–4) -locular with a single ovule in each locule; styles as many, from once to thrice 2-cleft. Capsule separating into 2–4 2-valved 1-seeded carpels. Seeds carunculate. — Stellate-downy, scurfy or hairy and glandular plants of warm reg., mostly strong-scented; the pistillate flowers usually at the base of the staminate spike or cluster. Leaves alternate, or sometimes imperfectly opposite, with or without obvious stipules. (*Croton, a tick*, the Greek name of the Castor-oil-plant, of this family, from similarity of the seed to a tick, but not to the so-called "Croton-bug", a roach of European origin, which first invaded southeastern New York coincidentally with the construction of the Croton aqueduct.)

Monoecious; staminate flowers with petals.
Leaves toothed; calyx of staminate flowers 4-parted; the petals 4, without alternating glands; pistillate flowers with minute petals, the 3 styles 2-cleft. 1. *C. glandulosus*, var. *septentrionalis.*
Leaves entire; staminate flowers with glands alternating with petals; pistillate flowers without petals.
Pistillate flowers capitate-crowded at base of terminal staminate spike; calyx 7–12-parted; styles 3, thrice 2-parted; stamens 7–14.
Seeds ventrally flattened, lenticular and orbicular; trichomes of inflorescence brownish. . . . . 2. *C. capitatus.*
Seeds little compressed, longer than broad; trichomes tawny-yellow to white. . . . . 3. *C. Lindheimeri.*
Pistillate flowers on recurved peduncles, calyx 5-parted; stigmas 2, sessile, 2-parted; stamens 3–8. . . . . 4. *C. monanthogynus.*

Dioecious; staminate flowers apetalous; styles twice or thrice dichotomously 2-parted. . . . . 5. *C. texensis.*

1. C. glandulòsus L. (glandular), var. **septentrionàlis** Muell. Arg. (northern). — Rough-hairy and glandular somewhat umbellately branched annual 1–6 dm. high; *leaves* oblong, *obtusely toothed,* the base with a saucer-shaped gland on each side; staminate flowers in a spike, with the pistillate ones capitate-clustered at base, terminal or sessile in the forks; *staminate flowers with 4-parted calyx, 4 petals, a 4-rayed disk and 8 stamens; pistillate flowers with 5-parted calyx, minute rudimentary petals, and 3 bifid styles;* capsule about 5 mm. in diameter. — Sandy woods, fields and waste places, Fla. to Tex., n. to Del. (adv. to N.J. and Pa.), Ind., Ill., Ia. and Kans. July–Oct. — Typical *C. glandulosus* broader-leaved and more southern or tropical.

2. C. capitàtus Michx. (in heads), Hogwort, Woolly C. — Densely soft-woolly and somewhat glandular branching annual up to 2 m. high; leaves long-petioled, lance-oblong to elliptic or oval, entire, rounded to subcordate at base; *trichomes of inflorescence brownish; pistillate flowers capitate-crowded at base of staminate spike, with 7–12-parted calyx, no petals, and the 3 styles twice or thrice 2-parted; staminate flowers with 5-parted calyx, 5 glands alternating with 5 densely fimbriate lance-obovate petals, and 7–14 stamens;* capsule 7–9 mm. in diameter; *seeds ventrally flattened, lenticular, orbicular.* — Dry open soil and waste places, Ga. to Tex., n. to s. N.Y., sw. O., Ind., Ill., Ia. and Kans. June–Oct.

3. C. Lindheimeri (Engelm. & Gray) Wood (for its discoverer, Ferdinand Jacob Lindheimer, 1801–1879). — Differing from no. 2 in its generally more tapering and often larger leaves; in the *whiter or paler vestiture of the inflorescence;* in the *barely compressed somewhat barrel-shaped seeds.* (*C. Engelmanni* Ferguson) — Dry pinelands, fields and waste places, Fla. to Tex., n. to Ga., Mo. and Kans. Aug.–Oct.

4. C. monanthògynus Michx. (with 1 pistillate flower), Prairie-tea. — Whitish-stellate or rusty-glandular annual 1.5–6 dm. high, the slender stem often umbellately 3–4-forked below, then repeatedly 2–3-forked or alternately branched; leaves oblong to ovate, entire, longer than petioles; *flowers in the forks; the staminate few at summit of a short erect peduncle, with unequally 3–5-parted calyx, 3–5 alternating spatulate petals and glands and 3–8 stamens; pistillate flowers* solitary or few *on short recurved peduncles, with equally 5-parted calyx, 5 glands but no petals and 2 sessile 2-parted stigmas;* ovary 2-locular but often by abortion 1-locular and 1-seeded; capsule 3–4 mm. long. — Calcareous openings or barrens and in waste places, Ga. to Tex., n. to se. Va., O., Ind., Ill., Ia. and Kans. June–Sept. (Mex.)

5. C. texénsis (Klotzsch) Muell. Arg. (Texan), Skunkweed. — Dioecious closely canescent-stellate dichotomously branched or spreading annual 0.3–1.5 dm. high; staminate plant smaller and narrower-leaved than pistillate; leaves narrowly lance-oblong to linear, entire; *staminate flowers* in racemes 1–3 cm. long, *with calyx equally 5-parted, petals none,* stamens 8–12; pistillate flowers 1–4, with triangular calyx-segments, the *styles twice or thrice dichotomously 2-parted;* capsule 4–6 mm. in diameter, stellate-tomentose and somewhat muricate. — Dry soil, Ala. to Ariz., n. to Ill., S.D. and Wyo.; sometimes adv. e. to N.E. and Del. May–Oct. (Mex.)

3. CROTONÓPSIS Michx. Rushfoil

Flowers monoecious, in very small terminal or lateral spikes or clusters, the lower fertile. *Stam. Fl.* Calyx equally 5-parted. Petals 5, spatulate. Stamens 5, opposite the petals; filaments distinct, inflexed in the bud, enlarged at the apex. *Pist. Fl.* Calyx unequally 3–5-parted. Petals none. Glands (petal-like scales) 5, opposite the sepals. Ovary 1-locular, simple, 1-ovuled, bearing a twice or thrice forked style. Fruit dry and indehiscent, 1-seeded. Seed without caruncle. — Slender low N.Am. annuals, with short-petioled linear or elliptic-lanceolate leaves, which are green and smoothish above but silvery-hoary with stellate hairs and scurfy with brownish scales underneath. (*Croton* and *opsis, appearance,* from likeness to *Croton.*)

1. C. ellíptica Willd. (elliptical). — Stem 1–4 dm. high; *petals and filaments of staminate flowers hardly exceeding sepals;* ovary and *fruit smooth* or merely scurfy. (*C. linearis,* in part, of ed. 7, not Michx.) — Dry sandy soil, n. Fla. to e. Tex., n. to s. Ct., N.J., e. Pa., Md., s. Ind., s. Ill., s. Mo. and se. Kans. July–Sept.

2. C. lineàris Michx. (linear). — Usually taller, up to 8 dm. high; *petals and filaments of staminate flowers longer than sepals; fruits spiny* at summit. — Sands and rocky barrens, Fla. to Tex., n. to Va., s. Ill. and se. Mo. July–Sept.

4. MERCURIÀLIS L. Mercury

Dioecious or monoecious. Flowers apetalous, in interrupted axillary spikes. Stamens 8–20, distinct. Calyx small, green, globose in bud, 3-parted. Carpels 2(–3). — Herbs nat. of Eurasia and n. Afr., with opposite pinnately veined leaves. (A plant-name used by Pliny and meaning *belonging to the god Mercury.*)

1. M. ÁNNUA L. (annual), BOYS-AND-GIRLS. — Weak erect leafy-stemmed annual; leaves lanceolate or ovate-lanceolate, crenate-serrate; carpels hispid. — Waste places and ballast-ground, local, Que. to O., and s. beyond our range. July–Nov. (Adv. from Eu.)

5. ACALỲPHA L. THREE-SEEDED MERCURY. COPPERLEAF

Flowers monoecious; the staminate very small, clustered in spikes; the few or solitary pistillate flowers at the base of the same spikes, or sometimes in separate ones. Calyx of the staminate flowers 4-parted and valvate in bud; of the pistillate 3–5-parted. Corolla none. Stamens 8–16; filaments short, monadelphous at base; anther-locules separate, long, often worm-shaped, hanging from the apex of the filament. Styles 3, the upper face or stigmas cut-fringed (usually red). Capsule separating into 3 globular 2-valved carpels, rarely of only one carpel. — Herbs (ours annuals), or in the tropics often shrubs, resembling Nettles or Amaranths, of temp. and trop. reg.; leaves alternate, petioled, with stipules. Clusters of staminate flowers with a minute bract; the pistillate surrounded by a large and leaf-like cut-lobed persistent bract. (Name from the Greek *Acalephe, nettle,* as explained by Linnaeus.)*

Leaves not cordate; fruit smooth or merely pubescent; seeds nearly smooth.
  Primary leaves mostly ovate to ovate-rhombic, their petioles one-third to nearly the length of the glabrous or remotely long-hairy blades; pistillate bracts deeply 5–7(–9)-lobed. . . . . 1. *A. rhomboidea.*
  Primary leaves mostly ovate-lanceolate to linear, their petioles at most half as long as the pubescent (sometimes glabrous) blades; pistillate bracts with 9–15 lobes or teeth.
    Stems with at least some long and spreading hairs, often villous; primary leaves lanceolate, tapering to tip, their petioles one-third to half the length of blades; pistillate bracts deeply cut into mostly lanceolate very acute lobes, hispid, usually not glandular. . . . . 2. *A. virginica.*
    Stems with incurved or ascending fine pubescence; primary leaves lance-oblong to linear, mostly obtuse or abruptly contracted at apex, the petioles rarely more than one-fourth length of blades; pistillate bracts shallowly cut into ovate or broadly deltoid teeth, sparsely beset with whitish or reddish glands. . . . . 3. *A. gracilens.*
Leaves cordate-ovate; fruit echinate with soft bristly green projections; seeds rough-wrinkled. 4. *A. ostryaefolia.*

1. A. rhomboidea Raf. (rhombic). — Stem simple or widely branching, incurved-puberulent or glabrate, rarely sparsely villous, 0.1–1 m. high; *primary leaves lance- to rhombic-ovate,* tapering to blunt apex, 2–9 cm. long, coarsely crenate, essentially *glabrous or with remote* coarse *appressed* white *hairs,* the nerves often puberulous, *petioles one-third to essentially as long as blade; pistillate bracts* strigillose on nerves and margins, sometimes long-ciliate, sparsely beset with long-stipitate deciduous whitish glands or glabrous, *with* 5–7(–9) *deep oblong to lanceolate lobes;* staminate spike 4–10 mm. long, usually not exceeding bract; seeds 1.6–1.8 mm. long. (*A. virginica,* chiefly, of ed. 7) — Roadsides, fields, borders of woods, etc., s. N.S., centr. Me. and sw. Que. to w. Ont., Minn. and Neb., s. to Fla., Ala., Ark. and Okla. July–Oct.

Var. **Deàmii** Weath. (for its discoverer, CHARLES CLEMON DEAM, 1865–). — Coarse throughout; broadly ovate *primary leaves* 7.5–10.5 cm. long, *drooping; seeds 2.5–3 mm. long.* — Damp woods, banks and roadsides, s. O. and s. Ind.

2. A. virgínica L. (Virginian). — Similar; *stem* and branches densely puberulent or pubescent with short incurved hairs and *usually with some spreading hairs, often villous; primary leaves lanceolate* (narrowly to broadly), *the petioles one-third to half the length of the blades; the latter* 2–8 cm. long, remotely crenate, *tapering to blunt apex,* finely pubescent or glabrate beneath; *pistillate bracts cut one-third to three-fifths their breadth into* 9–15 *sharply acute lanceolate lobes, usually glandless, mostly pubescent;* staminate spike usually 1 (rarely –2) cm. long, equaling or exceeding bract; seeds 1.4–1.8 mm. long, borne in all valves of capsule. (*A. digynea* Raf.) — Dry soils, Ga. to Tex., n. to Mass., Ind., Ill., Mo. and Kans. Aug.–Oct.

3. A. grácilens Gray (slender). — Stem 1–4.5 dm. high, the lower branches (when present) arcuate-ascending, *with short incurved or ascending pubescence,* rarely with spreading hairs; *primary leaves oblong to oblong-lanceolate, mostly obtuse or abruptly contracted at apex, the petioles one-eighth to one-fourth the length of blade; pistillate bracts approximate, cut a fourth to half their breadth into* 9–11 *broadly oblong, triangular-ovate or broadly deltoid teeth, sparsely beset with* long-

---

* Treatment derived largely from that of C. A. WEATHERBY in Rhodora, xxix. 193–204 (1927).

stipitate whitish *glands*, red sessile glands, or both; staminate spike 5–15 mm. long; seeds 1.3–2 mm. long, usually borne in all valves of capsule. — Dry sterile soils, Fla. to Tex., n. to s. N.H., Mass., N.Y., O., Ind. and Wisc. Aug.–Oct.

Var. **Fràseri** (Muell. Arg.) Weath. (for its discoverer, JOHN FRASER, 1750–1811). — More densely and coarsely pubescent; branches few and slender; *leaves linear or linear-lanceolate, their petioles* very short (*about one-tenth the length of blade*); *staminate spike* 2–4 *cm. long; pistillate bracts* with red sessile glands only, *often becoming distant below;* seeds about 1.8 mm. long. — Ga. to Tex., n. to Ill. and Okla.

Var. **monocócca** Engelm. (with one locule). — *Branches stiffly ascending*, often crowded, densely pubescent; *leaves linear to linear-lanceolate; staminate spike* 5–12 *mm. long; only* 1 *valve of capsule seed-bearing;* seeds 1.8–2 mm. long. — Ark. and Tex., n. to Mo. and Kans.

4. A. ostryaefòlia Riddell (with leaves of hop-hornbeam, *Ostrya*). — Stem pubescent, 2–8 dm. high, with loosely ascending branches; *leaves cordate-ovate*, thin, long-petioled, recurving, finely and closely serrate-dentate, abruptly acuminate; staminate spikes short, axillary; *pistillate spikes mostly terminal and elongate, their bracts* deeply *cut into linear lobes; fruit echinate with soft bristly green projections; seeds* about 2 mm. long, *rough-wrinkled.* — Thickets, roadsides, waste places and cult. fields, Fla. to Tex. and Mex., n. to se. Va., s. O., s. Ind., s. Ill., sw. Ia. and Kans. (old records from N.J.). July–Oct. (W.I.)

6. RÍCINUS L. CASTOR-OIL-PLANT. CASTOR-BEAN. PALMA CHRISTI

Flowers in racemose or panicled clusters, the pistillate above, the staminate below. Calyx 5-parted. Stamens very numerous, with repeatedly branching filaments. Styles 3, united at base, each bifid and plumose, red. Capsule very large, covered with soft spine-like processes, 3-lobed, with compressed ellipsoid lustrous-silvery and mottled seeds about 1 cm. long. — A tall stately annual (southw. perennial) with very large alternate peltate and palmately 5–11-cleft leaves often 3–9 dm. broad. (Named for the Mediterranean sheep-tick, *Ricinus*, because of resemblance of seed to a tick.)

1. R. COMMÙNIS L. — Cult. for ornament, and sometimes spreading to waste ground. (Introd. from the Tropics) — The large and handsome seeds dangerously poisonous to eat.

7. TRÀGIA L.

Flowers monoecious, in racemes, apetalous. *Stam. Fl.* Calyx 3–5 (chiefly 3)-parted, valvate in the bud. Stamens (in ours) 2 or 3; filaments short; anther-locules united. *Pist. Fl.* Calyx 3–8-parted, persistent. Style 3-cleft or 3-parted; the branches 3, simple. Capsule 3-locular, 3-lobed, bristly, separating into three 2-valved 1-seeded carpels. Seeds not carunculate. — Erect or climbing plants (ours perennial herbs) of warm reg., pubescent or hispid, sometimes stinging, with mostly alternate stipulate leaves; the small-flowered racemes terminal or opposite the leaves; the staminate flowers above, the few pistillate at the base, all with small bracts. (Named for the early herbalist, *Hieronymus Bock*, 1498–1554, Latinized *Tragus.*)

Leaves short-petioled to sessile, oblanceolate or oblong to lance-ovate, 1–7 cm. long; capsule 5–10 mm. broad; seeds 3–4 mm. long.
  Leaves oblanceolate to obovate-oblong, acute at base; stamens 2. . . . . 1. *T. urens.*
  Leaves lanceolate to triangular-ovate, with truncate to cordate base; stamens 3. 2. *T. urticifolia.*
Leaves long-petioled, ovate, deeply cordate, 5–12 cm. long; capsule 1.2–1.6 cm. broad; seeds about 5 mm. long. . . . . 3. *T. cordata.*

1. T. ùrens L. (burning or stinging). — Deep-rooted; stems erect, simple or loosely branched, 1.5–4.5 dm. high, softly hairy; *leaves oblanceolate to obovate-oblong, acute at base*, coarsely toothed above or subentire, 1.5–5 cm. long, *sessile or short-petioled;* staminate flowers only 1–5, the calyx usually 4-parted; *stamens* 2; capsule 7–9 mm. broad, hispid; seeds globose, marbled, 3.5–4 mm. in diameter. (Var. *lanceolata* Michx.; var. *innocua* (Walt.) Pax) — Dry sandy woods and clearings, e. Va. to Fla. and Ala. June–Sept. — Plant usually far from innocuous.

2. T. urticifòlia Michx. (nettle-leaved). — *Erect or reclining* or slightly twining, hirsute with stinging hairs; *leaves ovate-lanceolate or triangular-lanceolate*, or the lower ovate, *all somewhat cordate or truncate at base*, coarsely cut-toothed, *short-petioled;* staminate calyx usually 3-parted and stamens 3; capsule 7–10 mm. broad. (*T. nepetaefolia* of ed. 7, not Cav.; incl. *T. ramosa* of ed. 7, not Torr.) — Dry sandy soil, Fla. to Ariz., n. to Va. (acc. to *Pursh*), Mo., Kans. and Colo. June–Sept.

3. T. cordàta Michx. (heart-shaped). — *Twining*, somewhat hirsute; *leaves deeply cordate,* ovate, mostly narrowly acuminate, sharply serrate, 5–12 cm. long, all but the uppermost *long-petioled; capsule* 1.2–1.6 *cm. broad; seeds about* 5 *mm. long.* (*T. macrocarpa* Willd.) — Open woods and bluffs, Fla. to Tex., n. to Ga., s. Ind., s. Ill. and Mo. July–Sept.

8. STILLÍNGIA Garden

Flowers monoecious, aggregated in a terminal spike. Petals and glands of the disk none. Calyx 2–3-cleft or -parted; the divisions imbricated in the bud. Stamens 2 or 3; anthers adnate, turned outward. Style thick; stigmas 3, diverging, simple. Capsule 3-locular, 3-lobed, 3-seeded. Seed carunculate. — Smooth upright plants of warm parts of Am. and islands of Pacific and Indian Oceans, with the alternate leaves mostly 2-glandular at base; the pistillate flowers few at the base of the dense staminate spike (rarely separated); the bract for each cluster with a large gland on each side. (Named for Dr. *Benjamin Stillingfleet*, 1702–1771, English naturalist.)

1. S. sylvática L. (of woods), QUEEN'S-DELIGHT, QUEEN'S-ROOT. — Deep-rooted; stems several from the crown, herbaceous, 3–9 dm. high; leaves almost sessile, elliptic to narrowly obovate, obtuse, mostly 2–4.5 cm. broad, closely crenate, the teeth with deciduous spinules; staminate spike becoming open, the glands at bases of flowers evident, saucer-shaped; seeds 6–8 mm. long. — Sandy woods and openings, Fla. to Tex., n. to se. Va. May–July.

Var. **salicifòlia** Torr. (willow-leaved). — Leaves rhombic to lanceolate, acuminate to both ends, mostly 1–3 cm. broad; staminate spike denser, the basal glands less evident. (*S. salicifolia* (Torr.) Raf.) — Ark. to Kans. and Tex.

9. PHYLLÁNTHUS L.

Flowers monoecious, axillary. Calyx usually 5–6-parted, imbricated in the bud. Petals none. Stamens mostly 3, erect in the bud, often united. Ovules 2 in each locule of the ovary. Capsule depressed; each carpel 2-valved, 2-seeded. Seeds not carunculate. — Leaves alternate, 2-ranked, with small stipules. Plants of warm reg. (Name composed of the Greek *phyllon, leaf,* and *anthos, flower,* because the flowers in a few species are borne upon leaf-like dilated branches.)

1. P. caroliniénsis Walt. (of Carolina). — Annual, low and slender, branched; leaves obovate or oval, short-petioled; flowers commonly 2 in each axil, almost sessile, one staminate, the other pistillate; calyx 6-parted; stamens 3; styles 3, each 2-cleft; glands of the disk in the pistillate flowers united into a cup. — Gravelly banks and damp sandy soil, Fla. to Tex. and Mex., n. to Pa., W.Va., O., Ind., Ill., Mo. and se. Kans. June–Sept. (W.I.)

10. ANDRÁCHNE L.

Flowers monoecious, pedicellate, the staminate petaliferous, fasciculate; the pistillate often petaliferous, usually solitary in the axils. Stamens and calyx-segments 5–6. Fruit dry, splitting into three 2-valved carpels. — Shrubs and undershrubs of trop. and warm reg., with many ascending leafy branches. Leaves oval or obovate, entire. (Classical Greek for the purslane.)

1. A. phyllanthoides (Nutt.) Muell. Arg. (resembling *Phyllanthus*). — Nearly glabrous shrub up to 1 m. high; stems and ascending simple branches lithe; leaves broadly obovate, membranaceous, 1.5 cm. long, shortly petiolate; pedicels capillary, 7–14 cm. long; petals in the staminate flowers about as long as the obovate calyx-segments, in the pistillate flowers obsolescent. (*Savia* Pax & Hoffm.) — Dry hills and rocky barrens, s. Mo. to Tex. May–Oct.

11. EUPHÓRBIA L. SPURGE

Flowers monoecious, borne in a somewhat cupuliform outer structure (the *cyathium*) resembling a calyx or corolla with united lobes. Staminate flowers few to many at base of the cyathium, each with a single free or adnate basal *bracteole*, the filament-like pedicellate stalk (*androphore*) jointed at base and terminated by a single roundish anther; the single pistillate flower pedicellate and naked or with a rudimentary 3-lobed calyx at base of the 3-locular sessile ovary, the pedicel (*gynophore*) elongating and exserted; styles 3, each 2-cleft. The group of flowers surrounded by an *involucre* (within the cyathium) which usually projects as 5 somewhat foliaceous *lobes* (modified leaves). Margin or summit of the cyathium bearing 4 or 5 nectariferous *glands*, with or without colored and often petaloid marginal *appendages;* one of the glands (*fifth gland*) often wanting or greatly modified, its space then occupied by the arching gynophore. Seed often carunculate. Inflorescence cymose, with umbelliform branching and bearing a solitary flower at the summit of the main axis, the di- or trichotomous rays greatly

overtopping it and bearing cyathia, or in subgenus *Chamaesyce* the central terminal cyathium suppressed and the branching more lateral. — Plants with milky acrid juice; the genus of most strikingly dissimilar subgenera but all with similarly modified inflorescences (our species all herbaceous), many hundreds of species dispersed over most temp. and trop. reg. (the herbs of temp. areas often becoming weeds). (Named for *Euphorbus*, physician of King Juba of Numidia.)*

a. Glands of cyathium without petaloid appendages; stems erect or ascending; leaves essentially symmetrical, alternate, opposite or whorled; stipules wanting or mere glands. . . *b.*

- **b.** Cyathia irregularly clustered at summit of stem, with opposite leaves below the clusters; glands 1–few, cup-shaped. Subgen. POINSETTIA.
  - Scabrous-pubescent; bracts or bracteal leaves green or sometimes pale at base; cyathia nearly sessile, the gland or glands short-stalked. . . 1. ***E. dentata.***
  - Glabrous or essentially so; bracts of inflorescence often red or red-based; cyathia about as long as their pedicels, the gland or glands sessile. . 2. ***E. heterophylla.***
- **b.** Cyathia in umbelliform cymes (pleiochasia) subtended by a whorl of leaves, with opposite leafy bracts (floral leaves) at the forks; cauline leaves mostly alternate; glands of involucre flat or convex, entire or crescentic. Subgen. ESULA. . . *c.*
  - **c.** Glands transversely oval or barely subreniform, obtuse. . . *d.*
    - **d.** Leaves entire, minutely downy beneath; strong perennial with rhizome 1–2 cm. thick; capsule 6–8 mm. high; seeds obovoid or subglobose, 3–4 mm. long. . . . . 3. ***E. purpurea.***
    - **d.** Leaves serrulate; annual, biennial or barely perennial (by offsets), slender; capsule 2–3 mm. high; seeds 1.3–2 mm. long. . . *e.*
      - **e.** Capsule verrucose; seeds compressed or lenticular; floral leaves broad-based. . . *f.*
        - **f.** Leaves pubescent beneath, acute; rays of terminal umbel mostly 5; seeds about 2 mm. long, with a circular depression at the chalaza; introd. northeastw. . . . . 4. ***E. platyphylla.***
        - **f.** Leaves glabrous; rays of terminal umbel mostly 3; seeds without depression at the chalaza; indigenous.
          - Floral leaves cordate-clasping; glands red or scarlet, all present; seeds 1.7–2 mm. long, nearly smooth. . . . . . . 5. ***E. obtusata.***
          - Floral leaves truncate to subcordate at base; glands yellow, one of them replaced by a small lobe or tuft of hairs; seeds 1.3–1.5 mm. long, prominently reticulate. . . . . . . 6. ***E. dictyosperma.***
      - **e.** Capsule smooth; seeds ovoid-subglobose, 2 mm. long, areolate; rays of terminal umbel 5; floral leaves narrowed to base; introd. . . 7. ***E. Helioscopia.***
  - **c.** Glands of cyathium crescent-shaped or 2-horned; leaves entire. . . *g.*
    - **g.** Cauline leaves alternate, often crowded; floral leaves 0.5–1.5 cm. long; cyathia 1–4 mm. high; capsule 1.5–3.5 mm. high; seeds 1.2–2.5 mm. long. . . *h.*
      - **h.** Perennial by strong horizontal widely creeping rootstocks and adventitious shoots; cyathia 2.5–4 mm. high; seeds smooth, 1.7–2.5 mm. long. . . *i.*
        - **i.** Cauline leaves narrowly linear to linear-filiform or -spatulate, 1–2 cm. long, 0.5–3 mm. wide; floral leaves 4–6 mm. long. . . 8. ***E. Cyparissias.***
        - **i.** Cauline leaves broadly linear to oblong-lanceolate, longer and wider; larger floral leaves 1–2 cm. long.
          - Median cauline leaves 2–10 mm. wide; larger floral leaves 1–1.3 cm. long; cyathia 2.5–3 mm. high. . . . . 9. ***E. Esula.***
          - Median cauline leaves 1–2.5 cm. wide; larger floral leaves 1.2–2 cm. long; cyathia 3–4 mm. high. . . . . . . 10. ***E. lucida.***
      - **h.** Annuals; cyathia 1–2 mm. high; seeds pitted or sculptured, 1.2–1.8 mm. long. . . *j.*

*** Treatment of subgen. *Esula* based largely on that of Section *Tithymalus* by J. B. S. NORTON in Mo. Bot. Gard. Ann. Rep. xi. 85–144, plates 11–52 (1900); of subgen. *Chamaesyce* derived chiefly from the monograph by L. C. WHEELER in Rhodora, xliii (Contrib. Gray Herb. cxxxvi (1941)). In the illustrations from the latter treatment the parts are numbered as follows: 1, branch, × ½; 2, cyathium with mature capsule, × 3; 3, summit of capsule, showing styles, × 3; 4, glands from above, × 6; 5, styles, × 6; 6, basal view of seed, with the raphe above, × 3; 7, lateral view of seed, with raphe at left, × 3; 8, seed, with raphe median, × 3; 9, leaf, × 1; 10, node, showing stipules, × 2.**

j.Floral leaves linear-lanceolate, similar to cauline leaves; cyathia 1 mm. high. . . . . 11. *E. exigua.*

j.Floral leaves ovate-lanceolate to subrhombic, broadly ovate or reniform, contrasting with the cauline leaves; cyathia 1.5–2 mm. high. . . *k.*

k.Floral leaves broader than long; seeds carunculate, finely pitted or reticulate.

Cauline leaves obovate, obtuse or retuse; floral leaves often connate at base; glands with slender horns twice as long as breadth of body; seeds finely pitted; indigenous. . . . 12. *E. commutata.*

Cauline leaves linear, sharp-pointed; floral leaves distinct; glands with blunt horns as short as breadth of body; seeds clearly reticulated; introd. . . . . 13. *E. segetalis.*

k.Floral leaves longer than broad; seeds ecarunculate, with few coarse pits.

Median cauline leaves obovate to roundish, obtuse or retuse, distinctly petioled; floral leaves bluntish; cocci of capsule with broad and thin dorsal keels. . . . . 14. *E. Peplus.*

Median cauline leaves oblanceolate, tapering to subsessile base, acute or acutish; floral leaves acute; cocci of capsule without evident keels. . . . . 15. *E. falcata.*

g.Cauline leaves opposite, strongly decussate, in subremote pairs; floral leaves 2–6 cm. long, acuminate; cyathia 3–4 mm. high; capsules 8–12 mm. long; seeds 4–5 mm. long. . . . . 16. *E. Lathyris.*

a.Glands of cyathia usually with petaloid broad to narrow appendages (if these wanting, the depressed annual plants with definite stipules and asymmetrical leaves). . . *l.*

l.Perennials (nos. 17 and 18 annuals) with symmetrical entire leaves; cyathia mostly 5-lobed, with 5 glands; flowers in terminal umbels or cymes or solitary and long-stalked in leaf-axils. Subgen. Agaloma. . . *m.*

m.Annuals with stipules or stipular glands.

Cauline leaves linear or linear-lanceolate, with stipular glands; flowers in forks of upper branchlets, the slender subtending leaves green. . 17. *E. hexagona.*

Cauline leaves oval, ovate or broadly oblong, with deciduous lanceolate stipules; uppermost leaves and bracts of the terminal umbels dilated and with broad white petaloid margins. . . . . 18. *E. marginata.*

m.Perennials; stipules wanting. . . *n.*

n.Main axis or primary stem subterranean, crowning a long vertical root, bearing a dense crown of finally di- or trichotomous slender prostrate or loosely ascending branches; lowest pair of cauline leaves subtending lowest fork of branch; cyathia long-stalked in the forks.

Appendages much narrower than body of gland, purple, greenish or whitish. . . . . 19. *E. Ipecacuanhae.*

Appendages many times longer than breadth of body of gland, dilated and petal-like, white or pink. . . . . 20. *E. arundelana.*

n.Main axis or primary stem above ground, erect or ascending, solitary or few. . . *o.*

o.Flowers long-stalked and solitary in the forks of the rays and raylets of the open leafy umbel, the subtending leaves little reduced in size. . . *p.*

p.Leaves very thin and membranaceous, slender-petioled, those of main axis wanting or often deciduous; cyathia (including glands) 3–5 mm. broad, with narrow appendages; plant of Appalachian Upland. . . . . 21. *E. mercurialina.*

p.Leaves firm, sessile or subsessile; cauline leaves present and persistent.

Cyathia (including glands) 5–10 mm. broad, with broad and long clear white appendages; roots deeply descending, vertical; plant of montane, Piedmont and Coastal Plain regs. southward. . . . . 22. *E. zinniiflora.*

Cyathia 3–6 mm. broad, with purplish, fuscous or dull white appendages; roots or rootstocks horizontal, nearly superficial; plant of Coastal Plain and Piedmont. . . . . 23. *E. marilandica.*

o.Flowers clustered in terminal umbelliform cymes, their foliaceous bracts greatly reduced.

Leaves thin, membranaceous or submembranaceous, on slender petioles mostly 3–7 mm. long, 9–15 leaves or scars below the cyme; se. plant. . . . . 24. *E. apocynifolia.*

Leaves firm to subcoriaceous, sessile or on much shorter thick petioles, 25–75 or more leaves or scars below the cyme; wide-ranging species. . . . . 25. *E. corollata.*

l.Annuals (with us) with leaves oblique or inequilateral at base, all opposite; stipules usually well developed and persistent; stems slender, prostrate to ascending, mostly much branched; cyathia small, in upper axils or in axillary small glomerules. Subgen. CHAMAESYCE. . . q.

q.Ovaries and capsules pubescent.

Capsules minutely appressed-pubescent; styles 0.4–0.7 mm. long, upper third or half of styles bifid.

Cyathia not split down the back; facets of seed with 3 or 4 transverse ridges; wide-ranging. 26. *E. supina.*

Cyathia split down the back; facets of seed smooth or minutely pebbled; chiefly western. . . . . . . 27. *E. humistrata.*

Capsules spreading-villous, especially along the angles; styles 0.2–0.3 mm. long, bifid nearly to base; facets of seed with several transverse wrinkles; chiefly southern. . . . . . . 28. *E. Chamaesyce.*

q.Ovaries and capsules glabrous. . . r.

r.Facets of seeds smooth and plane; leaves entire; plant glabrous or essentially so. . . s.

s.Leaves oblong to linear, mostly 0.5–3 cm. long; capsules 2–3.5 mm. long; seeds 1.3–2.6 mm. long and 1–1.9 mm. thick. . . t.

t.Appendages of glands wanting or at most twice as high as the body of the gland; cyathia bearing (0–)5–17 staminate flowers. . u.

u.Glands nearly or quite unappendaged; capsules 3–3.5 mm. long; seeds compressed-ovoid, 2–2.6 mm. long, 1.6–1.9 mm. broad; littoral plant of Atl. coast and Great Lakes. . . . 29. *E. polygonifolia.*

u.Glands with definite appendages; capsules 2 mm. long; seeds not compressed, 1.3–2 mm. long; southeastern or inland species.

Seeds roundish-ovoid, 1.5–1.9 mm. long and 1.3–1.5 mm. thick; se. coastal species. . . . . . . 30. *E. ammannioides.*

Seeds narrowly ovoid, 1.3–1.6 mm. long and 1 mm. thick; western species. . . . . 31. *E. Geyeri.*

t.Appendages conspicuous, oblong-ovate, much higher than body of gland; cyathia bearing 29–48 staminate flowers; capsule 2–2.5 mm. long; seeds broadly ovoid, 1.5–2 mm. long and 1.1–1.4 mm. thick; southwestern. . . . . 32. *E. missurica.*

s.Leaves subcordate, 2–7 mm. long; capsules 1.2 mm. long; seeds 1 mm. long and 0.5 mm. thick. . . . . 33. *E. serpens.*

r.Facets of seeds roughened or transversely wrinkled; leaves more or less serrate. . . v.

v.Leaves serrate along both margins, at least on the outer margin continuously from base to subacute or narrowed apex; glands stalked; styles slender, 0.6–1 mm. long; facets of seeds with vermiform or slightly wrinkled or rippled surfaces.

Stem erect or obliquely ascending, often simple or subsimple below, glabrous or merely with pilose lines; leaves mostly 1–3.5 cm. long; capsules 1.9–2.3 mm. long; seeds obtusely angled. . 34. *E. maculata.*

Stem depressed or but slightly ascending, freely branching from base, pilose; leaves 0.5–2 cm. long; capsules 1.6–1.9 mm. long; seeds sharply angled. . . . . 35. *E. vermiculata.*

v.Leaves serrate only at rounded tip and sometimes near base; glands sessile; styles 0.3–0.5 mm. long, thick or clavate; facets of seeds with transverse ridges or wrinkles.

Leaves ovate or obovate to broadly oblong; involucres 0.8–1.2 mm. in diameter; cyathia bearing 5–18 staminate flowers; glands flat; seeds minutely pitted and with short cross-ridges. . . . . . . 36. *E. serpyllifolia.*

Leaves linear-oblong; involucres 0.6–0.9 mm. in diameter; cyathia bearing 4(1–5) staminate flowers; glands depressed in middle; seeds with 5 or 6 sharp transverse ridges. . . . . . . 37. *E. glyptosperma.*

Subgen. Poinsettia (Graham) House

1. E. dentàta Michx. (toothed). — Erect or ascending *strigose-hispid* annual 0.2–1.2 m. high, with ascending branches; *leaves* of primary nodes and summit *lance- to rhombic-ovate, or in forma* **cuphospérma** (Engelm.) Fern. (hump-seeded) *linear to lanceolate, dentate,* slender-petioled, opposite or alternate, *the upper and bracteal leaves often pale (especially at base); cyathia nearly sessile,* with 5 oblong sharply dentate lobes and 1 or more short-stalked broad glands; seeds globose-ovoid, tuberculate-roughened. (Incl. *E. cuphosperma* (Engelm.) Boiss.; *Poinsettia dentata* and *cuphosperma* Small) — Dry open soil, thin woods, railroad-banks and waste places, N.Y. to Minn., S.D. and Wyo., s. to Va., La., Tex. and Mex., now largely as an adv. weed. July–Sept.

2. E. heterophýlla L. (various-leaved), Painted-leaf. — *Glabrous* or essentially so; *primary cauline and upper leaves alternate, oval and often panduriform,* the rameal either broad or narrow and unlobed, *very thin, the uppermost and bracteal leaves often red or red-based; cyathia about as long as their pedicels,* the 5 lobes incised; the 1 or more *glands sessile; seeds subglobose,* tuberculate-roughened. (*Poinsettia* Klotzsch & Garcke) — Damp sandy soil, Fla. to Tex., n., partly as a weed of waste places and railroad-banks, to Va., Ind., Wisc., Minn. and S.D. Aug., Sept. Passing insensibly into

Var. **graminifòlia** (Michx.) Engelm. (grass-leaved). — Primary cauline and upper leaves linear or narrowly lanceolate, mostly unlobed. (*Poinsettia pinetorum* Small) — Fla. to Tex., n. to Ind., Wisc. and Minn.

Subgen. Ésula Pers. Wolf's-milk

3. E. purpùrea (Raf.) Fern. (purple; from color of cyathium). — Stout *perennial, with thick rootstock (1–2 cm. thick)* retaining rounded scars as in *Polygonatum; stem* 0.6–1.3 m. high, *up to 1 cm. thick,* with slender flowering branches from upper axils; *cauline leaves oblanceolate to lanceolate, or the upper becoming oblong-oval,* 0.5–1 cm. long, 1.3–3 *cm. broad, white-pilose or glabrate beneath, entire;* terminal umbel with 5–8 slender ascending rays; *floral leaves* broadly ovate, cordate or truncate-based, obtuse or emarginate, 2–3 *cm. broad; cyathia* 3–4 *mm. high,* the obovate or obcordate incurved lobes often purple above; *glands transversely reniform-oblong, about* 1 *mm. broad; capsule* subglobose, 7–9 *mm. broad; seeds* ovoid to subglobose, 3–4 *mm. long,* brown, with depressed-conic caruncle. (*E. Darlingtonii* Gray; *Tithymalus Darlingtonii* Small) — Rich or swampy woods and thickets, local, N.J. to O., s. to Del., Md. and w. N.C. July–Sept.

4. E. platyphýlla L. (broad-leaved). — Erect *annual,* 1.5–8 dm. high; upper *cauline leaves oblanceolate-spatulate to lanceolate-oblong, acute,* cordate at base, 0.5–1.5 cm. broad, minutely serrulate, mostly *with scattered hairs beneath;* floral ones triangular-ovate, subcordate; umbel 5-rayed; *cyathia* 2 mm. high, *with ciliate lobes* and large sessile glands; *styles* longer than the ovary, *united at base, slightly 2-cleft;* capsule depressed-globose, 2.5 mm. long, scarcely sulcate, covered with depressed warts; *seeds* compressed, about 2 *mm. long, with a circular depression at the chalaza.* — Shores, thickets and waste places, sw. Que. and Ont., s. to Vt., N.Y., O., Mich., Ill. and Mo. June–Aug. (Natzd. from Eu.)

5. E. obtusàta Pursh (bluntish). — Erect *annual,* 1.5–9 dm. high; *leaves oblong-spatulate,* minutely serrulate, *smooth, all obtuse;* upper ones cordate at base; *floral leaves* ovate, dilated, *cordate-clasping,* barely mucronate; umbel once or twice divided into 3 (rarely -5) rays, then into 2; *cyathia about* 1 *mm. high, with naked lobes* and small *stipitate* transversely oblong *red or scarlet glands; styles distinct,* longer than the ovary, erect, 2-cleft to the middle; capsule 3 mm. high, beset with long warts; *seeds lenticular, blackish,* 1.7–2 mm. long, *with thin flat caruncle, nearly smooth.* (*Tithymalus* Klotzsch & Garcke) — Damp woods, bottoms, fields, roadsides, etc., S.C. to e. Tex., n. to Pa., O., Ind., Ill., Ia. and Neb. May–July.

6. E. dictyospérma Fisch. & Mey. (with netted or reticulate seed). — Stem erect, 2–4.5 dm. high, resembling no. 5; differing in the *truncate-based or subcordate floral leaves; glands sessile, yellow, one of them replaced by a small lobe or tuft of hairs;* styles shorter than the ovary, spreading or recurved; *seeds delicately reticulated,* 1.3–1.5 *mm. long.* (*E. arkansana* and var. *missouriensis* Norton; *Tithymalus arkans.* Klotzsch & Garcke) — Prairies, barrens and roadsides, Minn. to Wash., s. to Ala., La., Tex. and Mex. May–July. (S. Am.)

7. E. Helioscópia L. (ancient name, turning toward the sun; from the whorl of leaves at summit of stem), Wartweed, Réveille-matin (Que.). — Annual, stems ascending, 0.5–6 dm. high; *leaves all obovate* and very rounded or retuse at the end, *finely serrate,* smooth or a little hairy, those of the stem cuneate; *umbel divided into* usually 5 *rays,* then into 3, or at length simply forked; *floral leaves narrowed to base; glands orbicular or elliptical, yellow, stalked;*

capsule smooth and even; seeds with coarse honeycomb-like reticulations. (*Tithymalus* Hill) — Waste places and dry open soil, e. Que. to Sask., abund.; more locally s. to N.S., N.E., Md., O. and Ill. July–Sept. (Natzd. from Eu.)

8. E. Cyparíssias L. (name used by Pliny for some related plant), Cypress S. — *Densely tufted from extensively creeping* and forking rope-like *rootstocks;* the sterile and fertile upright stems 1–7 dm. high, densely leafy; *cauline leaves narrowly linear to linear-filiform or -spatulate,* pale green, 1–2 *cm. long,* 0.5–3 *mm. wide;* umbel several-rayed, compact; *floral leaves* broadly ovate, 4–6 *mm. long,* yellowish when young, often becoming purplish or red in age; cyathia 3 mm. high, their lunate glands waxy-yellow; capsule depressed-globose (often undeveloped with us), 3 mm. long; *seeds brownish-gray,* nearly 2 mm. long, *smooth, with flattish caruncle.* (*Tithymalus* Hill) — Roadsides, old fields, neglected cemeteries, etc., originally cult., through most of our area and beyond. April–Aug. (Introd. and natzd. from Eu.)

9. E. Ésula L. (old name from the Celtic *esu,* sharp; from the acrid milky juice), Wolf's-milk, Leafy S. — Taller (3–9 dm. high) than no. 8, with more slender rootstock and less crowded stems; *principal cauline leaves broadly linear to narrowly oblong-lanceolate or -oblanceolate,* 2–10 *mm. broad,* less crowded; many upper branches floriferous; terminal umbel open; *larger floral leaves* reniform, 1–1.3 *cm. long,* yellow-green; *cyathia* 2.5–3 *mm. high;* glands strongly 2-horned, yellow or green; capsule warty; *seeds ellipsoid-ovoid,* about 2 mm. long, *yellow-brown,* with yellow emarginate caruncle. (*Tithymalus* Hill; and incl. *E. virgata* Waldst. & Kit.) — Sandy banks, old fields, roadsides, etc., Que. to Alta., s. to N.S., N.E., Pa., Ind., Ill., Ia., Neb., etc. May–Sept. (Natzd. from Eu.)

10. E. lùcida Waldst. & Kit. (shining; from the foliage). — Resembling no. 9, much coarser, with rootstock up to 2 cm. thick; stems up to 1.3 m. high; *lustrous cauline leaves* 1–2.5 *cm. broad; larger floral leaves* 1.2–2 *cm. long; cyathia* 3–4 *mm. high; seeds* 2.5 *mm. long.* (*Tithymalus* Klotzsch & Garcke) — Old fields, roadsides, etc., local, centr. N.Y. and s. Ont. to Ia. (Natzd. from Eu.)

11. E. exígua L. (small). — Slender erect to depressed *annual* 0.5–4 dm. high, with ascending branches from near middle; *leaves linear, with sessile often cordate bases,* very narrow; *floral leaves* similar to cauline leaves, *linear-lanceolate, acute; cyathia* 1 *mm. high,* with 2-horned yellow glands; *capsule* 2–2.5 *mm. long, smooth and shining; seeds* 1.5 *mm. long,* 4-angled, tuberculate, with small caruncle. (*Tithymalus* Hill) — Waste places, roadsides, etc., local, C.B. to w. N.Y. and W.Va. July–Oct. (Adv. from Eu.)

12. E. commutàta Engelm. (changeable), Wood-S. — Annual (or perennial by new basal sprouts); stems branched from a commonly decumbent base, 1.5–4 dm. high; *leaves* obovate, *obtuse or retuse,* the upper all *sessile;* the *upper floral leaves* roundish-dilated, *broader than long, often connate at base;* umbel 3-forked; cyathia 1–2 mm. high, whitish; *glands with slender horns twice as long as breadth of body; capsule obtusely angled,* 3 mm. long; *seeds ovoid, pitted all over,* 1.8 mm. long. (*Tithymalus* Klotzsch & Garcke) — Along streams, on shady slopes or about calcareous rocks, Pa. to Minn., s. to Fla., Ala., Tenn., Mo., Okla. and Tex. April–July.

13. E. segetàlis L. (growing in grain-fields). — Similar to no. 12, up to 1 m. high; differing in having *linear sharp-pointed rather crowded and divergent cauline leaves;* inflorescence becoming loose, open and broad; *floral leaves distinct;* cyathia 1.5 mm. high, the yellow *glands with blunt horns as short as breadth of body;* capsule 3.5 mm. long; *seeds* 2.5–3 *mm. long, clearly reticulated.* — Local in waste places, Pa. and southw. (Adv. from Eu.)

14. E. Péplus L. (name used by Pliny and Dioscorides for this or a similar plant), Petty S. — Habitally resembling no. 12; *median cauline leaves obovate to roundish, blunt or retuse, with distinct petioles;* the many slender branches fertile at tip; larger inflorescences forking; *floral leaves ovate, blunt; cyathia* 1.5 *mm. high, the large yellow glands with prolonged narrow horns; capsule* 2 mm. long, *its* 3 *cocci with broad and thin dorsal keels; seeds* 1.5 mm. long, *greenish,* 6-*sided, the* 2 *inner faces each with a longitudinal furrow, the* 4 *outer faces each with* 3–5 *roundish pits, caruncle wanting.* (*Tithymalus* Hill) — Weed of cult. and waste ground, Nfld. to Wisc. and Ia., s. to N.S., N.E., Md., W.Va., O. and Ind. June–Nov. (Natzd. from Eu.)

15. E. falcàta L. (sickle-shaped; from the glands). — Habitally like nos. 12 and 14 but the *median cauline leaves oblanceolate, acute or acutish and tapering to subsessile base; floral leaves acute,* concealing the cyathia; cyathia 1.5 mm. high, the *glands with short and blunt horns; capsule* 2 mm. long, *without keels; seeds* 4-*sided, each side with reddish-brown transverse pits.* — Roadsides, railroad-banks, cult. or disturbed soil, becoming freq., Pa. and O., s. to Va. and W.Va. July–Sept. (Natzd. from Eu.)

16. E. Láthyris L. (name used by Pliny "for a kind of wolf's-milk"), Caper-S., Mole-plant. — Annual (or by sprouting from near base of old stem becoming slightly perennial), erect, up to 1.5 m. high; *cauline leaves opposite, strongly decussate,* narrowly oblong to lanceolate with cordate clasping base, firm and thick; *floral leaves cordate,* lance- or oblong-ovate, *long-acuminate, 2–6 cm. long; cyathia 3–4 mm. long;* glands strongly 2-horned; *capsule 8–12 mm. long; seeds 4–5 mm. long,* wrinkled, with a helmet-like caruncle. (*Tithymalus* Hill) — Roadsides, waste places, etc., originally spread from cult., s. N.E. to O., s. to N.C. May–Sept. (Introd. and natzd. from Eu.)

Subgen. Agalòma (Raf.) House

17. E. hexágona Nutt. (six-angled; inappropriate name published as of Nuttall by Sprengel to whom Nuttall trustingly sent material, Nuttall himself slightly later formally describing it as *E. heterantha* (diverse-flowered) because of separation of the sexes in the cyathia!). — Slightly pubescent *slender* erect *bushy-branched annual* 0.15–1.5 m. high; *leaves very thin, pale green, the primary opposite ones linear-lanceolate, entire, the later and rameal ones very narrow; main axis and branches with scattered cyathia, most of the latter with staminate flowers only,* a few with pistillate flowers, the 5 *glands with triangular-ovate green appendages;* capsule smooth; seeds ovoid, tuberculate. (*Zygophyllidium* Small) — Sandy shores, bottoms and open ground, Minn. to Mont. and southw.; casual weed eastw. July–Sept.

18. E. marginàta Pursh (margined), Snow-on-the-mountain. — Erect *annual with broadly oblong, oval or ovate pale leaves with lanceolate deciduous stipules; uppermost leaves and leafy bracts with broad white petaloid margins;* inflorescence a *terminal umbel, usually 3-rayed; cyathia 5-lobed, the glands with broad white appendages.* (*Lepadena* Nieuwl.; *Dichrophyllum* Klotzsch & Garcke) — Minn. to Mont. and southw.; extensively cult. and spread to waste places e. to Atl. coast. June–Oct.

19. E. Ipecacuánhae L. (early Virginian name, the plants having the emetic properties of Brazilian *Ipecacuanha*), Wild Ipecac, Ipecac-s., Carolina Ipecac. — *Main stem or stems subterranean, arising from summit of a thick vertical root* (up to 1 m. long) *and at summit forking into many slender at first simple but soon dichotomously and diffusely forking branches* up to 3 dm. long; the *fertile branches naked or with few scales up to the paired leaves at the lowest fork;* succeeding forks with similar sessile or subsessile oval, obovate, oblong or linear fleshy green to purple entire blunt leaves up to 5.5 cm. long and to 3 cm. broad; *cyathia solitary at the forks, on elongate ascending slender peduncles* 1.3–4.5 *cm. long,* the cup about 1.5 mm. high, its 5 transverse oblong glands each with a very narrow whitish, yellow, greenish or purple appendage; capsule 4–5 mm. broad. (*Tithymalopsis* Small) — Dry sands, pinelands and barrens, Coastal Plain and outer Piedmont, Fla. to N.J., locally to L.I., formerly (125 years ago) to centr. Ct. April, May (sporadically to July). — Hopelessly variable, plants of all leaf-forms and varying in color from green to crimson or purple and with the narrow appendages of the glands of different colors occurring in indiscriminate mixture, without evident difference of habitat. The plants with linear leaves seem perhaps to be less succulent, drying more quickly than those with broader leaves. These may be separated as forma **lineàris** (Moldenke) Fern. (linear).

20. E. arundelàna Bartlett (for Anne Arundel Co., Md.). — Like no. 19 and as variable in leaf-shape and coloring, often pubescent; differing in *dilated white or pink oval or obovate appendages several times longer than the breadth of the body of the gland.* — Sands of N.J. and e. Md. — Perhaps a hybrid of nos. 19 and 23.

21. E. mercurialìna Michx. (resembling *Mercurialis*). — *Perennial, with* 1–few slender erect stems 1.5–7 dm. high with 2 *or* 3 *simple or once-forking long rays* subtended by 2–4 leaves; *leaves membranaceous, oval,* the larger 3–9 cm. long, pale beneath, *on slender petioles* 3–10 *mm. long, blades of the main axis often wanting or caducous;* bracteal leaves large; flowers 1–3 at the forks, on filiform ascending peduncles; *cyathia (including the* 5 *narrowly white-appendaged glands)* 3–5 *mm. broad,* about 2 mm. high; capsule 5–6 mm. broad. (*Tithymalopsis* Small) — Rich woods, bottoms and calcareous slopes, Fla. and Ala., n. to D.C. and Ky. Late April, May.

22. E. zinniiflòra Small (with flowers of [very tiny] *Zinnia*). — *Differing from no.* 21 *in* smaller *linear, oblong or oval firmer* round-tipped *sessile or subsessile leaves mostly 2–5 cm. long and green on both sides; cyathia broader than high, including the large clear white roundish appendages 5–10 mm. across.* (*Tithymalopsis* Small) — Sandy or rocky open woods and clearings, Fla. to e. Tex., n. to D.C., n. Va., W.Va., Ky. and Ark. June–Sept.

23. E. marilándica Greene (of Maryland).—Habit much as in nos. 21 and 22 but *with stout horizontal shallowly seated or superficial rootstock; leaves* green to purplish, *glaucescent,* firm, oblong-obovate to linear, those at base of the 2–6-rayed di- or trichotomous leafy cyme opposite or whorled; *cyathia, including the purplish, fuscous or dull white appendages, 3–6 mm. broad.* (*Tithymalopsis* Small) — Sandy soil, Coastal Plain and Piedmont, Del. and Md. to S.C. May–Aug.

24. E. apocynifòlia Small (with leaves of *Apocynum*). — Perennial with *slender horizontal rootstock;* slender leafy erect *stems* 1.5–7 dm. high, *terminated by an open umbelliform much-forked cyme with greatly reduced or tiny floral leaves* much as in no. 25; *leaves of main axis* 9–15, *thin to membranaceous,* oblong, oval or oblong-obovate, rounded at apex, *mostly* 2.5–6 *cm. long, on slender definite petioles mostly* 3–7 *mm. long,* those at base of umbel whorled; *cyathia* 1.5 *mm. high, including the white roundish appendages* 4–5 *mm. broad; longer pedicels* 2–8 *mm. long;* capsule 5 mm. broad. (*Tithymalopsis* Small) — Open woods and clearings, Fla. and Ala., n. to se. Va. July–Sept.

25. E. corollàta L. (with corollas), FLOWERING S., TRAMP'S S., WILD HIPPO. — *Stems* 1–several from strong and deep perennial root, *up to* 1 *m. or more high, glabrous, with* 25–75 *or more firm sessile or subsessile* oval, oblong or linear obtuse *glabrous leaves up to the umbellate,* usually *very floriferous* 3–7-forked and again 2–5-forked *inflorescence;* the *bracteal leaves greatly reduced;* the *longer pedicels* (at the forks) 7–25 *mm. long; cyathia* 1–1.5 mm. high, *including the conspicuous white suborbicular to obovate appendages* 7–10 *mm. broad;* capsule 3.5–4.5 mm. broad. (*Tithymalopsis* Small) — Dry open woods, clearings, fields, roadsides, etc., Fla. to Tex., n. to N.Y., s. Ont., Mich., Wisc., Minn. and Neb., and casually adv. in N.E. June–Oct.

Var. móllis Millsp. (soft). — Similar but upper half of stem and lower surfaces of leaves soft-pubescent. — Ala. to Tex., n. to Va., N.C. and Ind.

Var. paniculàta (Ell.) Boiss. (with panicles). — Cyathia more crowded on the branchlets of the cyme, including the relatively small appendages 5–7 mm. broad; the longer pedicels only 0.5–5 mm. long. (*E. paniculata* Ell.; *Tithymalopsis panic.* Small) — Ga. and Ala. to e. Tex., n. to se. Va., Ind., Ill., Ark. and Okla.

Subgen. CHAMAESỲCE Raf.

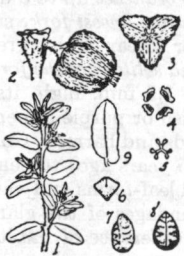

26. E. supìna Raf. (lying on its back), MILK-PURSLANE. — Slender, prostrate or ascending, branching from near base, forming mats 1–9 dm. across; young stems villous or tomentulose; leaves 4–17 mm. long, elliptic-ovate to oblong, serrulate to subentire, sparsely villous, glabrate, often purple-mottled; cyathia mostly on short crowded lateral branches; *involucres about* 0.8 *mm. in diameter,* villous; glands transversely elongate, 0.15–0.25 mm. long, with narrow white appendages; staminate flowers 2–5; crisp-pubescent gynophore barely exserted; *ovary and capsule strigose-pubescent; styles mostly* 0.4 *mm. long, cleft one-third to one-fourth to base, thickish;* capsule subacutely 3-angled, 1.4 mm. long; quadrangular *seed* about 1 mm. long, *the facets with low transverse ridges.* (*Chamaesyce* Moldenke; *E. maculata* of ed. 7, not L., as to type; *C. macul.* sensu Small and others, not as to type) — Dry open soil, roadsides, etc., generally common, s. Que. and s. Ont. to N.D., s. beyond our limits. June–Sept. FIG. 1291.

1291. E. supina.

27. E. humistràta Engelm. (carpeting the ground). — Stiffer and often coarser than no. 26; primary leaves oval to oblong-ovate, those of branches and branchlets narrower, entire or remotely serrulate, strongly inequilateral at base; *involucres* 0.6–0.8 *mm. in diameter;* glands with white or pink appendages; *styles* 0.7 *mm. long, cleft to middle, slender; seeds nearly smooth.* (*Chamaesyce* Small) — Sandy bottoms, river-banks, etc., sw. O. to Ill., Mo. and e. Kans., s. to Ala., La. and ne. Tex.; adv. along railroads, Va. Aug.–Oct. FIG. 1292.

1292. E. humistrata.

28. E. CHAMAESỲCE L. (name used by Dioscorides for some prostrate plant, literally ground-fig). — Differing from no. 26 in its mostly uniform-colored leaf-surfaces, the blades mostly 4–8 (rarely –11) mm. long, the petioles shorter; involucres 0.6–0.9 mm. in diameter; glands usually depressed in middle, the appendages as wide as to twice as wide as gland; *styles* 0.2–0.3 *mm. long, cleft nearly to base; capsule sharply angled, with long divergent hairs especially near the angles; seeds sharply angled, the facets with low transverse wrinkles.* (*E. prostrata* Ait.; *Chamaesyce malaca* Small) — Open gravelly or sandy waste places, roadsides, etc., Fla. to Tex. and Mex., n. to se. Va. and Mo. June–Oct. (Natzd. from trop. Am.) FIG. 1293.

1293. E. Chamaesyce.

29. E. polygonifòlia L. (with leaves of knotgrass, *Polygonum*),

SEASIDE-S. — Glabrous, slender, the subsimple to branching matted or ascending stems 0.1–2.5 dm. long; *leaves pale green, oblong-linear to narrowly oblong-lanceolate, 0.6–1.6 cm. long, entire,* usually *mucronate, the midrib prominent beneath;* peduncles in forks, as long as or longer than petioles; *cyathia solitary* at the upper nodes; involucre 1–1.4 mm. in diameter; *glands transverse, broadly oval to subrotund* or sometimes double and suggesting the figure 8, *shallowly cupped, the appendage a mere rudiment or wanting; staminate flowers* 5–14 or wanting; ovary and capsule glabrous; *styles* 0.7–1 *mm. long, cleft to middle; capsules 3–3.5 mm. long,* with rounded lobes; *seeds compressed-ovoid,* subtruncate at base, subacute, 2–2.6 *mm. long,* 1.6–1.9 *mm. broad, the facets smooth* under low magnification. (*Chamaesyce* Small) — Sandy or gravelly beaches above high tide, or dune-hollows of the coast, M.I., P.E.I. and e. N.B., s. to Ga.; beaches and dune-hollows of Lakes Ont., Erie, Mich. and s. Huron. July–Oct. FIG. 1294.

1294. E. polygonifolia.

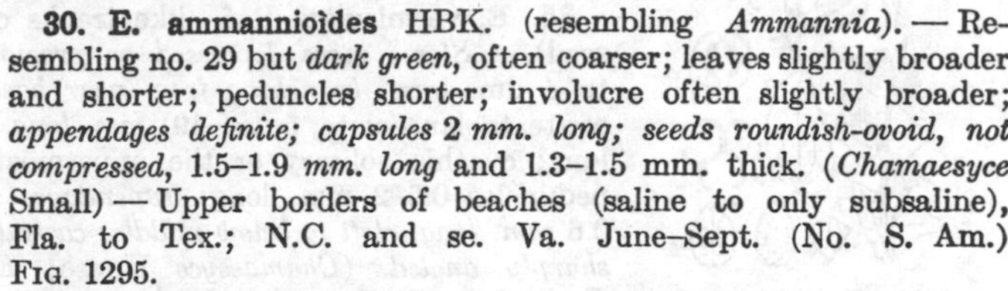
1295. E. ammannioides.

30. E. ammannioìdes HBK. (resembling *Ammannia*). — Resembling no. 29 but *dark green,* often coarser; leaves slightly broader and shorter; peduncles shorter; involucre often slightly broader; *appendages definite; capsules 2 mm. long; seeds roundish-ovoid, not compressed,* 1.5–1.9 *mm. long* and 1.3–1.5 mm. thick. (*Chamaesyce* Small) — Upper borders of beaches (saline to only subsaline), Fla. to Tex.; N.C. and se. Va. June–Sept. (No. S. Am.) FIG. 1295.

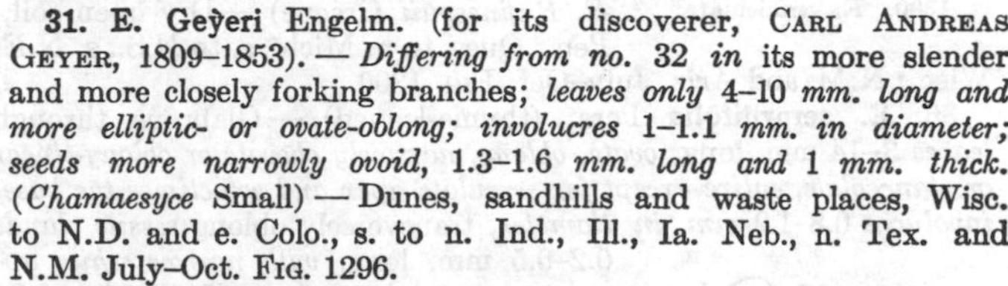
1296. E. Geyeri.

31. E. Geỳeri Engelm. (for its discoverer, CARL ANDREAS GEYER, 1809–1853). — *Differing from no.* 32 *in* its more slender and more closely forking branches; *leaves only* 4–10 *mm. long and more elliptic- or ovate-oblong; involucres 1–1.1 mm. in diameter; seeds more narrowly ovoid, 1.3–1.6 mm. long and 1 mm. thick.* (*Chamaesyce* Small) — Dunes, sandhills and waste places, Wisc. to N.D. and e. Colo., s. to n. Ind., Ill., Ia. Neb., n. Tex. and N.M. July–Oct. FIG. 1296.

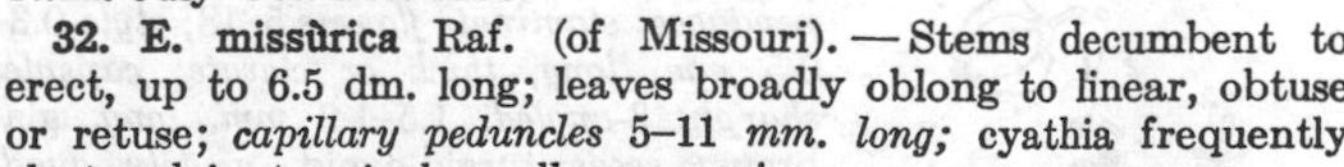

32. E. missùrica Raf. (of Missouri). — Stems decumbent to erect, up to 6.5 dm. long; leaves broadly oblong to linear, obtuse or retuse; *capillary peduncles 5–11 mm. long;* cyathia frequently clustered in terminal small cymes; *involucres 1.7–1.9 mm. in diameter; appendages prolonged, oblong-ovate,* white or pink; *staminate flowers* 29–48; styles 0.7–0.9 mm. long, cleft about half-way to base; *capsule 2–2.5 mm. long, obtusely angled; seeds broadly ovoid, 1.5–2 mm. long and 1.1–1.4 mm. thick, scarcely angled.* (*E. zygophylloides* Boiss.; *Chamaesyce zyg.* and *Nuttallii* Small) — Dry sands, fields, roadsides, etc., w. Mo. and Kans. to Ark. and Tex. FIG. 1297. — More commonly as

1297. E. missurica.

Var. **intermèdia** (Engelm.) L. C. Wheeler (intermediate). — *Leaves linear, truncate, mostly emarginate;* branchlets thicker; *peduncles thickish, 2–4 mm. long; seeds angled.* (*E. petaloidea* Engelm.; *Chamaesyce pet.* Small) — Dunes, sandhills, dry fields, roadsides, etc., w. Minn. to e. Mont., s. to Tex. and N.M. June–Sept.

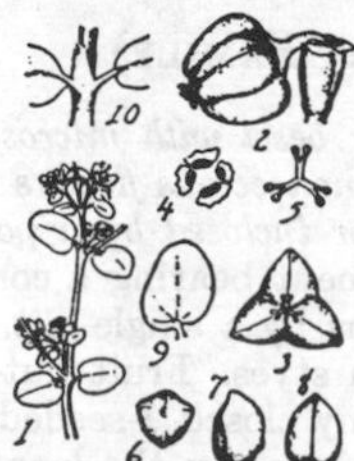
1298. E. serpens.

33. E. sèrpens HBK. (crawling). — Very leafy, slender and prostrate or sometimes repent glabrous annual; *leaves roundish-ovate or -oblong, subcordate, 2–7 mm. long,* entire, the *petiole less than* 1 *mm. long;* peduncles up to 2 mm. long; *involucres* 1 *mm. long and* 1 *mm. in diameter;* transversely oblong *glands about* 2 *mm. long,* yellowish-white, with narrow white appendages; staminate flowers 5–10; *styles clavate, 0.2 mm. long; capsule 1.2 mm. long; seeds* ovoid, *about* 1 *mm. long and* 0.5 *mm. thick, smooth.* (*Chamaesyce* Small) — Alluvial or rich soil, s. Ont. to Mont., s. to Ala., La., Tex., N.M. and Mex.; as a weed of fields and waste places e. to N.E., N.J., Ga. and Fla. July–Oct. (S. Am.) FIG. 1298.

34. E. maculàta L. (spotted; from the purplish spot on the leaf), EYEBANE. — *Stem* simple or mostly branched, *erect or ascending*, 0.08–1 m. high, crisp-pubescent at young tips, *soon glabrate* and firm; *leaves* oblong, oblong-lanceolate or lance-falcate, 0.8–3.5 *cm. long, serrulate;* cyathia solitary or clustered; peduncle 0.5–5 mm. long; involucres 0.7–1 mm. in diameter; *glands long-stipitate*, circular to broadly and transversely elliptical, 0.1–0.3 *mm. in diameter; appendages rudimentary or very narrow, white to reddish;* staminate flowers 5–11; *styles* 0.6–1 *mm. long*, cleft above the middle; *capsules* 1.9–2.3 *mm. long; seeds obtusely angled*, 1.1–1.6 *mm. long and* 0.9–1.1 *mm. thick, the facets with finely rippled surface.* (***E. nutans*** Lag.; *E. Preslii* Guss.; *E. hypericifolia* sensu most auth., not L.; *Chamaesyce Preslii* Arthur) — Dry open soil, waste places, cult. fields, roadsides, etc., Fla. to Tex. and Mex. n. to N.E., N.Y., s. Ont., Mich., Wisc., Minn. and N.D. June–Oct. FIG. 1299.

1299. E. maculata.

1300. E. vermiculata.

35. E. vermiculàta Raf. (like tracks of worms; from surface of seed). — *Stems more depressed or procumbent than in no.* 34, up to 4 dm. long, *branching from near base, pilose or hirsute; leaves* ovate to lanceolate, 5–15(–19) *mm. long*, the petioles about 1 mm. long; cyathia solitary or the uppermost in small leafy cymules; peduncles 0.5–2 mm. long; *appendages white, definite; style about* 0.6 *mm. long, cleft to below middle; capsules* 1.6–1.9 *mm. long; seeds sharply angled.* (*Chamaesyce* House; *E. hirsuta* (Torr.) Wieg.; *E. Rafinesquii* Greene) — Dry open soil, waste places, etc., Gaspé Pen., Que., to n. Mich., s. to N.S., s. N.E., N.J., Pa., O., Ind. and s. Wisc.; N.M. and Ariz. July–Oct. FIG. 1300.

36. E. serpyllifòlia Pers. (thyme-leaved). — Glabrous throughout; prostrate to erect; *leaves* 3–14 mm. long, *ovate, oblong, narrowly obovate or oblong-linear or -lanceolate, entire except for serrulate apex and sometimes the base; involucre* 0.8–1.2 *mm. in diameter;* transversely oblong *sessile glands* 0.2–0.5 mm. long, *with narrow white appendages; staminate flowers* 5–18; *style* 0.3–0.5 *mm. long, thick or clavate; capsules sharply* 3*-angled*, 1.5–1.9 *mm. long*, glabrous; *seeds* turgid-ovoid, *acutely quadrangular*, 1–1.4 *mm. long, their facets slightly pitted and with short cross-ribs.* (*Chamaesyce* Small) — Sandy or alluvial soil, B.C. to Mex., e. to nw. Mich., Minn., w. Ia., w. Mo., N.M. and w. Tex.; adv. along railroads e. to N.B. July–Oct. FIG. 1301.

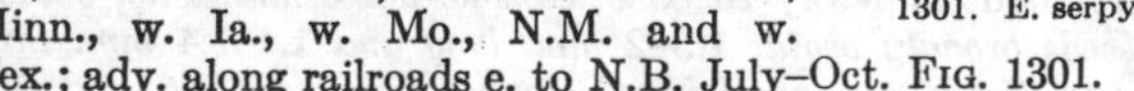

1301. E. serpyllifolia.

1302. E. glyptosperma.

37. E. glyptospérma Engelm. (with engraved seed). — Differing from no. 36 in mostly narrowly oblong leaves; *involucres* 0.6–0.9 *mm. in diameter; staminate flowers mostly* 4 (1–5); *glands* averaging shorter and *depressed in middle; facets of seed with* 5 *or* 6 *sharp elongate transverse ridges.* (*Chamaesyce* Small) — Dry open soil, N.B. and n. Me. to B.C., s. to N.H., Vt., N.Y., s. Ont., O., Ind., Ill., Mo., Tex., Mex. and Calif. June–Oct. FIG. 1302.

FAM. 93. CALLITRICHÀCEAE (WATER-STARWORT FAMILY)

Low slender and usually tufted chiefly aquatic herbs (glabrous or beset with microscopic stellate scales) with entire to undulate opposite obovate to linear leaves, monoecious flowers (solitary or 2 or 3 together in the axil of the same leaf) wholly naked or inclosed by a pair of membranaceous bracts. Staminate flower a single stamen, the filament bearing a cordate 4-locular anther, which by confluence becomes 1-locular and opens by a single slit. Pistillate flower a single 4-locular ovary, bearing 2 distinct filiform styes. Fruit nut-like, compressed, 4-lobed, 4-locular, separating at maturity into as many closed 1-seeded carpels. Seeds pendulous; embryo slender, straight or slightly curved, nearly the length of the oily albumen.

1. CALLÍTRICHE L. Water-Starwort. Water-Chickweed

The only genus, nearly cosmop. (Name from the Greek *callos, beautiful*, and *thrix, hair*, from the slender stems.) — Species mostly polymorphic, the terrestrial states often quite different in appearance from the aquatic; our species in need of careful study.

*a.*Flowers without bracts at base; leaves uniform.
Submersed aquatic perennial with prolonged stems; leaves linear, from a broad base, retuse or notched at apex, 4–12 mm. long, 1-nerved; fruit nearly sessile, flattened, 1.5–2.5 mm. long. . . . . 1. *C. hermaphroditica.*
Terrestrial tiny annuals; leaves obovate to oblanceolate or spatulate, narrowed at base, obtuse, 2–4 mm. long, 3-nerved; fruit stalked to subsessile, 0.5–0.7 mm. long.
Pedicels curving, rarely 1 mm. long; fruit above ground, plump, with rounded sides. . . . . 2. *C. deflexa*, var. *Austini.*
Pedicels straight, elongating to 4 mm. and thus burying the flat-sided thick-edged fruits. . . . . 3. *C. Nuttallii.*
*a.*Flowers 2-bracted at base; leaves in submersed plants often di- or trimorphic, the floating ones obovate and 3–7-nerved, the submersed often linear (those of terrestrial plants linear to oblong). . . *b.*
*b.*Fruit 1.5–2 mm. broad, with prominent thin wings; floating leaves 3–8 mm. broad, 5–7-nerved by branching of the lateral nerves. . . . . . 4. *C. stagnalis.*
*b.*Fruit much smaller, wingless or barely wing-margined above; floating leaves mostly smaller, 3 (rarely 5)-nerved. . . *c.*
*c.*Fruit definitely longer than broad, obovoid to ellipsoid, narrowed to base, its carpels sharply keeled or very narrowly winged above; styles (promptly deciduous) shorter than young fruit. . . . . 5. *C. palustris.*
*c.*Fruit round or round-obovate, as broad as long, rounded at base, its carpels with rounded edges; styles (deciduous) as long as to much longer than the young fruit.
Stems compressed-filiform, more slender than submersed leaves, in aquatic state much prolonged; fruit 1–1.4 mm. long. . . . . . 6. *C. heterophylla.*
Stems strongly flattened, much broader than submersed leaves, in aquatic state rarely 1.5 dm. long; fruit 0.5–1 mm. long. . . . . 7. *C. anceps.*

1. C. hermaphrodítica L. (hermaphrodite). — Submersed perennial with prolonged slender very leafy stems; *leaves* fulvous, *uniform, all oblong-linear, broad-based*, with rounded to retuse tip, *with the single nerve prominent beneath*, 4–12 mm. long; *flowers without subtending bracts; styles* long and *reflexed*, deciduous; *fruit circular*, 1.5–2.5 *mm. across*, its *broadly winged carpels* separated by a deep notch. (*C. autumnalis* L.) — Quiet waters, often calcareous or brackish, Greenl. and s. Lab. to Alaska, s. to Nfld., N.B., w. Vt., n. and w. N.Y., Mich., Wisc., Minn., Sask., Colo., Utah and Calif. Aug.–Dec. (Eurasia)

2. C. defléxa A. Br. (turned back), var. **Aústini** (Engelm.) Hegelm. (for its discoverer, Coe Finch Austin, 1831–1880). — Tiny *terrestrial annual* with simple to much-branching slender stems 0.5–3.5 cm. long; *leaves uniform, obovate to oblanceolate*, narrowed to base, 2–4 *mm. long, 3-nerved, obtuse; flowers without subtending bracts; fruits* distinctly *short-pedicelled* (*pedicel arching, rarely* 1 *mm. long*) to subsessile, *plump, broader than long*, 0.5–0.7 *mm. long*, deeply notched at summit and base; spreading *styles much shorter than fruit*, subpersistent. (*C. Austini* Engelm.; *C. terrestris* sensu DC., probably not Raf.) — Damp earth, fallow fields, footpaths, etc., local, Fla. to Tex. and Mex., n. to Mass., Ct., se. N.Y., N.J., e. Pa., W.Va., O., s. Ind., Ill. and Mo. April–July.— Typical S. Am. *C. deflexa* differs in slightly longer pedicels.

3. C. Nuttállii Torr. (for its discoverer, Thomas Nuttall, 1786–1859). — Like no. 2; leaves often smaller and more spatulate; flowers sessile, but their *mature pedicels prolonging to* 4 *mm. and burying underground the maturing fruit;* styles prolonged and appressed-recurving; *fruits with flat to concave sides and wire-like margins.* — Fallow fields and damp soil, Ala. to Tex., n. to Ky. and Ark. April, May.

4. C. stagnàlis Scop. (of pools). — Aquatic perennial with elongate stems and diverse foliage; *submersed leaves linear-lanceolate or -oblanceolate, spatulate or somewhat like the floating ones*, when linear 1-nerved and deeply emarginate, when broader 3-nerved; *floating leaves* dark green, obovate, crenulate, 3–8 *mm. broad and up to* 2 *cm. long* (3)–5–7-*nerved by forking of lateral nerves; fruit* suborbicular, 1.5–2 *mm. broad, with prominent thin wings;* styles spreading, deciduous.— Running or quiet waters, rapidly spreading, St. Lawrence R., Que.; Mass. to Md. Aug.–Dec. (Natzd. from Eu.)

5. C. palústris L. (of marshes). — *Pale green* aquatic (or stranded) perennial with submersed stems elongate and delicate (emersed plants in small mats with reduced and rarely dilated oblong leaves); lower submersed leaves linear, 1-nerved, shallowly emarginate; transitional leaves oblanceolate; *floating leaves* (often crowded in a rosulate tuft) petioled, *narrowly obovate or broadly spatulate, tapering above to rounded summit,* 3-nerved; *fruits definitely longer than broad, narrowly obovate to ellipsoid, tapering to base,* 1–1.4 mm. long, about equaled by the erect deciduous styles; *the carpels sharply keeled or narrowly winged above,* with a wide groove between them. (*C. verna* L., later name) — Springheads, rills, quiet waters and wet shores, Greenl. and Lab. to Alaska, s. to Nfld., N.S., N.E., w. Va., W.Va., Mich., Ill., Minn., Neb., N.M., Ariz. and Calif. April–Nov. (Eurasia)

6. C. heterophýlla Pursh (diverse-leaved). — Similar to no. 5, often darker green and coarser; stems compressed-filiform, in deep water much elongating; *floating leaves flabelliform-obovate, subtruncate to broadly sloping to obtuse summit,* 3–5-nerved; *fruits round-obovate to suborbicular, rounded at base,* 1–1.4 *mm. long, with rounded margins,* the finally deciduous styles longer. — Quiet waters and mud, Fla. to Tex. and Mex., n. to Nfld., s. Lab. and s. Can. across the continent. April–Dec. — Consisting of three or more strains much in need of critical study.

7. C. ánceps Fern. (two-edged). — Similar to no. 6, but pale green, the *stems flat and broadly winged, much broader than submersed leaves,* simple or branching, in submersed state 1–10 (rarely –15) *cm. long;* leaves mostly linear and submersed, very slender; *floating leaves gradually rounded to summit; fruit* 0.5–1 *mm. long, the short styles promptly deciduous.* — Shallow pools and wet shores, Greenl. and Lab. Pen. to Nfld. and e. Que., often in alpine areas; Mt. Mansfield, Vt. July–Sept.

FAM. 94. BUXÀCEAE (Box Family)

Perennial herbs or more often trees or shrubs with simple opposite or alternate usually evergreen leaves, watery juice and small greenish monoecious or dioecious apetalous flowers; sepals imbricated or none; stamens opposite the sepals or indefinite; carpels 3; *ovary 3-locular; styles* 3, *simple; ovules (in ours) geminate in the locules, suspended, the raphe dorsal.* — A small family, chiefly of trop. and warm reg.

1. PACHYSÁNDRA Michx.

Flowers monoecious, in naked spikes. Calyx 4–5-parted. Petals none. *Stam. Fl.* Stamens 4, separate; filaments long-exserted, thick and flat; anthers oblong-linear. *Pist. Fl.* Styles thick, subulate, recurved, stigmatic down their whole inner side. Capsule deeply 3-horned, 3-locular, splitting into 3 at length 2-valved 2-seeded carpels. — Nearly glabrous low and procumbent perennial herbs or subshrubs of N. Am. and e. Asia, with matted creeping rhizomes and alternate ovate or obovate coarsely toothed leaves narrowed at base into a petiole. Flowers each 1–3-bracted, the upper staminate, a few pistillate ones at base, unpleasantly scented; sepals greenish or purplish; filaments white (their size and thickness giving the name, from the Greek *pachys, thick,* and *andros,* used for *stamen*).

1. P. procúmbens Michx. (upturned from base), Allegheny-spurge. — Stems 1.5–2.3 dm. long, bearing several approximate leaves at the summit on slender petioles, and a few many-flowered spikes along the base; the intervening portion naked, or with a few small scales. — Rich calcareous woods, Ky. to w. Fla. and La.; cult. and sometimes esc. northw. March–May.

The Asiatic P. terminàlis Sieb. & Zucc. (terminal), with smaller cuneate-obovate leaves and smaller whitish flowers in terminal spikes, is much cult. and tends to esc.

FAM. 95. EMPETRÀCEAE (Crowberry Family)

Low shrubby evergreens with the narrow and rigid foliage, aspect and compound pollen suggesting Heaths, *and the drupaceous fruit somewhat resembling that of* Arctostaphylos, *but the divided or laciniate stigmas, etc., of some* Euphorbiaceae. — By some considered an apetalous and degenerate ally of the *Ericaceae* but better kept in the Order *Sapindales;* comprising three genera, two within the limits of this work, the third very doubtfully so.

Flowers scattered and solitary or few in the axils. Sepals 3, petaloid. . . . . 1. *Empetrum.*
Flowers collected in terminal heads. Calyx none. . . . . 2. *Corema.*

1. ÉMPETRUM L. Crowberry. Camarine (Que.)

Flowers polygamous, monoecious or dioecious, scattered and solitary or few in the axils of the leaves, inconspicuous, scaly-bracted. Calyx of 3 spreading and somewhat petal-like sepals. Stamens 3. Style very short; stigma 6–9-rayed. Fruit a juicy berry-like drupe, with 6–9 seed-like nutlets, each containing an erect anatropous seed. — Besides our species one other (*E. rubrum*) subantarctic. (An ancient name, from the Greek *en, upon,* and *petros, a rock.*)

Branchlets or margins of expanding leaves glandular, the latter not tomentose; mature leaves divergent, soon reflexed; fruit black (rarely purple). . . 1. *E. nigrum.*
Branchlets and margins of expanding leaves white-tomentose; plant not glandular; leaves ascending to divergent, rarely (and then very tardily) reflexed; fruit pink, red or purple-black.
 Fruit 5–9 mm. in diameter, red to purplish-black; seeds 2–2.4 mm. long; leaves soon loosely divergent, those of leading shoots 4–6.5 mm. long. . . . 2. *E. atropurpureum.*
 Fruit 3–5 mm. in diameter, pink or light red; seeds 1.2–1.5 mm. long; leaves crowded, ascending, tardily divergent, those of leading shoots 2.5–4 mm. long. . . . . 3. *E. Eamesii.*

1. E. nìgrum L. (black), BLACK C., CURLEWBERRY, CORBIGEAU or GRAINES À CORBIGEAUX (Que.). — Procumbent and spreading, with branches creeping in humus; *branchlets* slender, *glabrous or minutely glandular or pilose with viscid and sordid hairs; leaves* linear to narrowly elliptic, *glabrous or merely glandular-pulverulent,* 2.5–7 mm. long, divergent, *soon reflexed; drupes black,* or purple in forma **purpùreum** (Raf.) Fern. (purple), or whitish in forma **leucocárpum** L. M. Neuman (white-fruited), with watery juice and stone-like seeds 1.5–3 mm. long. — Arct. reg., s. in peaty soil to Nfld., N.S., coast of Me., alpine areas of n. N.E. and n. N.Y., n. Mich., n. Minn., s. Alta. and n. Calif.; isolated on e. L.I. *Fr.* July–Nov. (Eurasia)

2. E. atropurpùreum Fern. & Wieg. (dark purple), PURPLE C. — Trailing *branchlets white-tomentose* at least when young, *not glandular; leaves* linear-oblong, *tomentose or arachnoid on expanding,* at first ascending, *finally loosely spreading, those of the leading shoots 4–6.5 mm. long; drupes 5–9 mm. in diameter, red or purple-black, opaque,* with *seeds 2–2.4 mm. long.* (*E. nigrum,* var. *andinum* in part, of ed. 7, not (Phil.) DC.) — Granitic or acidic gravel and sands, s. coast of Lab. Pen. to Lake Mistassini, Que., s. to M.I., P.E.I. and s. coast of N.S.; mts. of n. N.E. *Fr.* July–Nov.

3. E. Eàmesii Fern. & Wieg. (for EDWIN HUBERT EAMES, 1865–1948, who first pointed out its characters), ROCKBERRY. — Closely prostrate; the *young branchlets white-tomentose; leaves crowded, ascending and imbricated,* in age slightly spreading, narrowly elliptic- or spatulate-oblong to oblong-linear, *very coriaceous and lustrous,* round-tipped, *those of leading shoots 2.5–4 mm. long; drupes 3–5 mm. in diameter, pink or light red,* with thin *translucent* skin and nearly colorless pulp; *seeds 1.2–1.5 mm. long.* — Exposed sands and siliceous gravels and rocks, se. Lab. Pen., Nfld., St. P. et Miq. and N.S. *Fr.* July–Nov.

2. CORÈMA D. Don BROOM-CROWBERRY

Flowers dioecious or polygamous, in terminal heads, each in the axil of a scaly bract and with 5 or 6 scarious imbricated bractlets but no proper calyx. Stamens 3, rarely 4. Style slender, 3 (or rarely 4–5)-cleft; stigmas narrow, often toothed. Drupe small, with 3 (rarely 4–5) nutlets. — Diffusely branched little shrubs, with subverticillate narrowly linear heath-like leaves; our species and another in Portugal and on the Azores. (Name the Greek *corema, a broom,* from the bushy aspect.)

1. C. Conràdii Torr. (for its discoverer, SOLOMON WHITE CONRAD, 1779–1831), POVERTY-GRASS. — Shrub, 1.5–6 dm. high, diffusely branched, nearly smooth; drupe very small, dry and juiceless when ripe. — Sandy pine-barrens, sandhills and siliceous rocks, ?Nfld.; M.I., P.E.I. and N.S. along or near coast to se. Mass.; Shawangunk Mts., N.Y. and N.J.; Pine Barrens of N.J. March–early May. — The staminate plant handsome in flower, on account of the tufted purple filaments and brown-purple anthers.

CERATÌOLA (Greek *ceras, horn*) ERICOÌDES Michx. (heath-like), an erect shrub with linear revolute falsely tubular divergent leaves 1 cm. long and crowded axillary flowers, recorded in 1842 by *Edward Tuckerman* from N.S., is not now known from n. of S.C.

FAM. 96. LIMNANTHÀCEAE (FALSE MERMAID FAMILY)

Herbaceous plants with perfect regular 3–6-merous slightly perigynous symmetrical flowers, the persistent sepals valvate. Glands alternate with the petals. Stamens distinct. Carpels nearly distinct, with a common style, 1-ovuled, at length fleshy and indehiscent, not beaked, separating from a very short axis. Embryo straight; cotyledons very thick; radicle very short. — Low tender annuals, with alternate pinnate exstipulate leaves.

1. FLOÈRKEA Willd. FALSE MERMAID

Sepals 3. Petals 3, shorter than the calyx, oblong. Stamens 6. Ovaries 3, opposite the sepals, united only at the base; the style rising in the center; stigmas 3. Fruit of 3 (or 1 or 2) roughish fleshy achenes. Seed anatropous, erect. — Small and inconspicuous herb, with minute solitary flowers on axillary peduncles. (Named for *Heinrich Gustav Floerke*, 1764–1835, a German botanist.)

1. F. proserpinacoìdes Willd. (like *Proserpinaca*). — Leaflets 3–5, lanceolate, sometimes 2–3-cleft. — Rich low or alluvial woods or wet calcareous rocks, w. N.S.; sw. Que. and Vt. to s. Ont. and N.D., s. to Del., Va. and Tenn. Apr.–June. — Taste slightly pungent.

FAM. 97. ANACARDIÀCEAE (CASHEW FAMILY)

Trees or shrubs with resinous or milky acrid juice, dotless alternate leaves, and small often polygamous regular 5-merous flowers, but the ovary 1-locular and 1-ovuled, with 3 styles or stigmas. — Petals imbricated in the bud. Fruit mostly drupaceous. Seed without albumen, borne on a curved stalk which rises from the base of the locule. Stipules none. Some species pervaded by an exceedingly active poisonous (irritating) principle.

Leaves simple; pedicels elongating after flowering and the abortive ones becoming plumose; fruit glabrous, becoming gibbous, with remains of styles lateral. 1. *Cotinus.*
Leaves compound; pedicels neither much elongating nor plumose; drupes globose or ovoid, with terminal styles. . . . . 2. *Rhus.*

1. CÓTINUS Duham. SMOKE-TREE

Calyx 5-parted. Petals 5. Stamens 5, inserted under edge of disk. Fruit dry, glabrous, becoming gibbous, the remains of the 3 styles becoming lateral. Pedicels much elongating after flowering, those of aborted flowers becoming plumose-villous. — Shrubs or small trees, one of e. N.Am., one of Asia and s. Eu., with strong-smelling yellow wood, simple deciduous entire alternate leaves and ample loose finally plumose terminal panicles of polygamous or dioecious yellowish flowers. (Name used by Pliny for the wild olive.)

1. C. obovàtus Raf. (obovate), AMERICAN S., CHITTAM-WOOD. — Erect shrub or small tree up to 10 m. high, glabrous or nearly so; *leaves* membranaceous, obovate to oval or elliptic, *sparsely silky beneath when young, becoming* 1–2 *dm. long; flowers dioecious,* mostly abortive; flowering *panicles* 0.7–2 dm. long, *in fruit up to 3 dm. long,* with many plumose sterile pedicels accompanying the filiform fruiting ones; *fruit* obliquely reniform, reticulate, 4.5–5 *mm. long.* (*Rhus cotinoides* and *C. americanus* Nutt.) — Calcareous rocky woods and bluffs, nw. Ala. and adj. Tenn.; sw. Mo., nw. Ark. and e. Okla.; sw. Tex. *Fl.* April, May; *fr.* June–Sept.

C. COGGÝGRIA Scop. (modification of the name *Coccygea* used by Theophrastus), *Rhus Cotinus* L., the Old World SMOKE-TREE, with glabrous leaves 3–8 cm. long, flowers perfect and fruits 3–4 mm. long, is commonly planted, sometimes escaping. (Introd. from Eurasia)

2. RHÙS L. SUMAC. POISON IVY

Calyx small, 5-parted. Petals 5. Stamens 5, inserted under the edge or between the lobes of a flattened disk in the bottom of the calyx. Fruit small and indehiscent, a sort of dry drupe. — Leaves compound. Flowers greenish-white or yellowish. Shrubs or trees of warm or temp. reg. (The old Greek and Latin name.)

a. Inflorescences terminal, pyramidal or thyrsiform erect panicles 0.5–4.5 dm. long; drupe covered with red acid hairs; leaves odd-pinnate; trees or erect shrubs. § SUMAC. . . *b.*
b. Petioles wingless; leaflets thin, much paler beneath, sharply serrate.
Branches, petioles, inflorescences and drupes densely velvety-villous with long hairs. . . . . 1. *R. typhina.*
Branches and petioles glabrous or merely puberulent; drupes with short appressed hairs. . . . . 2. *R. glabra.*
b. Petioles wing-margined between the firm entire or but slightly toothed leaflets; branches and petioles puberulent. . . . . 3. *R. copallina.*
a. Inflorescences small simple or clustered spikes expanding before or with the leaves, or later axillary panicles. . . *c.*
c. Flowers expanding from scaly-bracted aments formed in autumn and opening before or at expansion of leaves, aments either simple or in clusters; drupes red, hairy; low shrubs with ternate leaves. § LOBADIUM. . . . 4. *R. aromatica.*

c. Flowers in axillary panicles; drupes dun-colored to whitish, the thin epicarp finally breaking and exposing the fibrous mesocarp. § TOXICODENDRON. . . *d.*

d. Leaves pinnate, with 7–13 leaflets; coarse shrub or small tree. . . . . . 5. *R. Vernix.*

d. Leaves 3-foliolate; slender shrubs or climbing vines.

Stems strongly woody and prolonged, bushy and much branched, trailing, leaning or high-climbing and then with aërial clinging roots; leaves alternately scattered along the branches; leaflets acuminate to acute tips; drupes glabrous (only rarely pubescent). . . . 6. *R. radicans.*

Stems erect, simple or sparsely upright-branched, woody for only 0.5–6 dm. above the creeping subterranean stoloniferous base, never climbing; leaves approximate on long erect petioles at summit of stems and branches, thus appearing falsely verticillate.

Petioles and lower leaf-surfaces glabrous; terminal leaflet broadly ovate to suborbicular, with abrupt acute tip; fruit usually glabrous; n. and upland shrub. . . . . 6. *R. radicans,* var. *Rydbergii.*

Petioles and lower leaf-surfaces velvety-tomentulose; terminal leaflet elliptic, rhombic or obovate, blunt or with rounded tip; fruit usually pubescent; shrub of s. Coastal Plain. . . . . 7. *R. Toxicodendron.*

§ SÙMAC DC.

1. R. typhìna L. (like *Typha*, cat-tail; from the velvety branches), STAGHORN-S., VELVET S., VINAIGRIER (Que.). — Shrub or tree, 1–10 m. high, with orange-colored wood; *branches and petioles densely velvety-hairy;* leaflets 11–31, pale beneath, oblong-lanceolate, pointed, serrate; flowers polygamous, rarely dioecious, in dense terminal pyramidal to ovoid panicles 0.5–2 dm. long; *drupes velvety-villous with long spreading red hairs.* (*R. hirta* (L.) Sudw.) — Dry, rocky or gravelly soil, Gaspé Pen., Que., to s. Ont. and Minn., s. to N.S., N.E., N.C., Ky., Ill. and Ia. June, July. Apparently hybridizes with the next species. — Forma **laciniàta** (Wood) Rehd. (deeply slashed) has leaflets and bracts more or less deeply and laciniately toothed; a frequent form, at least in some cases pathological and with inflorescence transformed in part into contorted bracts, is *Datisca hirta* L.; forma **dissécta** Rehd. (dissected) has leaves bipinnatifid to bipinnate.

2. R. glàbra L. (glabrous), SMOOTH S. — Similar to no. 1, but a *glabrous glaucous* shrub 0.5–3(–5) m. high; panicles more often dioecious, the staminate open and up to 4.5 dm. long, the perfect or pistillate dense and smaller; *drupes covered with minute appressed hairs.* — Dry soil, centr. Me. and sw. Que. to s. B.C., s. beyond our range. June, July. — Forma **laciniàta** (Carr.) Robins. (deeply slashed) has leaves laciniately bipinnatifid or bipinnate. — This species segregated by Greene into 29 so-called species.

Var. **boreàlis** Britt. (northern). — Branches short-pilose or puberulent. (*R. borealis* (Britt.) Greene; *R. pulvinata* Greene) — Local, N.E. to s. Man., s. to N.C., Ind. and Minn. — Probably a hybrid of nos. 1 and 2.

3. R. copallìna L. (from the exudation of copal-like atoms), DWARF OR SHINING S., WING-RIB-S. — Shrub or small tree up to 10 m. high, *with ashy-puberulent branches, petioles, panicle-rachis, etc.; leaves* 1.5–3.5 dm. long, *the rachis wing-margined between the* 11–23 *lance- or linear-oblong entire leaflets; these* lustrous above, oblique at base, *attenuate to both base and apex,* mostly 4–9 cm. long and 1–2 cm. broad; panicle dense, or in the local forma **frondòsa** Fern. (leafy) interspersed with foliage-leaves; drupes short-pubescent. — Dry woods and openings, Fla. to se. N.Y. July–Sept.

Var. **latifòlia** Engler (broad-leaved). — Low; *the* 5–13 *broadly oblong to narrowly ovate leaflets strongly rounded to base* on the upper side, mostly 1.5–4 cm. broad. — Dry thickets and openings, s. Me. to n. Ill., s. to Ga., Ala., La. and e. Tex.

§ LOBÁDIUM DC.

4. R. aromática Ait. (aromatic), FRAGRANT S., LEMON-S., POLECAT-BUSH. — Straggling to upright shrub with ascending branches, *pungently fragrant when bruised,* ascending to 2 m. high; *leaves 3-foliolate,* the terminal (and often the lower) leaflet dentate above the middle; *aments formed in late summer, expanding before or with leaves, usually crowded in short clusters,* their loosening bracts with a bare area between the villous base and summit; flowers pale yellow; *fruit* compressed-globose, *closely covered with long acid soft red hairs;* stone compressed but with convex sides, 3.8–4.5 mm. long. — A highly variable species, represented with us by 4 vars.:

Terminal leaflet elliptic- to rhombic-ovate, tapering about equally to subcuneate base and acute or acutish apex, 3–9 cm. long; flowers preceding expansion of leaves.
Leaves more or less soft-pubescent when young, becoming glabrate; branchlets puberulent to glabrate. . . . . *R. aromatica* (typical).
Leaves permanently appressed-pubescent above, velvety beneath; branchlets tomentulose. . . . . Var. *illinoensis*.
Terminal leaflet more flabelliform-obovate, with cuneate base and rounded summit; flowers expanding with the leaves.
Upright, up to 2 m. high; leaves sparsely pilose to glabrate beneath, the terminal leaflet 2.5–6 cm. long. . . . . Var. *serotina*.
Depressed and low, with very slender branches; leaves velvety beneath, the terminal leaflet 1.5–3.5 cm. long. . . . . Var. *arenaria*.

R. aromática (typical), *Schmaltzia* Desv.; *R. canadensis* Marsh., not Mill.; *R. crenata* (Mill.) Rydb., not Thunb. — Dry rocks, sands and open woods, sw. Que. and w. Vt. to ne. Kans., s. to upland of nw. Fla., n. Ala., Miss., n. La. and e. Tex. *Fl.* April, May; *fr.* late May–July.

Var. **illinoénsis** (Greene) Rehd. (of Illinois). — Ill. and Mo.

Var. **serôtina** (Greene) Rehd. (late), *R. trilobata*, var. Barkley — Sands, barrens, rocks and open woods, Ill. to Neb., s. to Ark., Okla. and ne. Tex.

Var. **arenària** (Greene) Fern. (growing in sand), *R. trilobata*, var. Barkley — Dunes of n. O., n. Ind. and ne. Ill. *Fl.* May; *fr.* June, July.

§ Toxicodéndron (Mill.) Gray

5. R. Vérnix L. (varnish; Linnaeus supposing this species to be the Asiatic source of lacquer), Poison S., Poison Elder or Poison Dogwood. — Coarse *shrub or small tree* 2–7 m. high, *with* gray smoothish bark and glaucous and *glabrous branchlets; leaves* odd-pinnate, *with* 7–13 *obliquely ascending entire oblong-obovate to elliptic acuminate leaflets; panicles axillary*, spreading or pendulous, up to 2 dm. long; fruit whitish or drab, globular. (*Toxicodendron* Ktze.) — Wooded swamps, Fla. to e. Tex., n. to sw. Me., s. N.H., w. Vt., sw. Que., N.Y., s. Ont., s. Mich., s. Wisc. and se. Minn. *Fl.* May–July; *fr.* Aug.–Nov. (conspicuous all winter). — The acid waxy atoms on foliage, flowers and fruits are as poisonous to most skins as those of the two following species; the handsome orange to scarlet autumnal foliage dangerously tempting.

6. R. radìcans L. (rooting; from the aerial roots), Poison Ivy, Mercury or Markry, Cow-itch, Herbe à la puce (Que.), Bois de chien (Que.). — Upright, trailing or high-climbing *shrub, the branches when climbing with aërial clinging roots; leaflets* 3, *acuminate to acute tips*, glabrous to pilose beneath, the terminal leaflet longer-stalked than the lateral; panicles axillary or from axils of past years, ascending or divergent; flowers yellowish-green; fruits subglobose, dun-colored to whitish, the thin and dry epicarp readily breaking and exposing the fibrous mesocarp. (*Toxicodendron* Ktze.; *T. vulgare* Mill.) — A perplexingly variable species for which Greene, Nieuwland and others have published more than 30 specific names. Our vars. and forms are the following:

Stems strongly woody and prolonged, bushy and much branched, erect, leaning, trailing, or high-climbing and then forming aërial clinging roots; leaves alternately scattered along the branches; terminal leaflet narrowly to broadly ovate, scarcely rotund, gradually acuminate.
Leaves firm to subcoriaceous, on petioles 2–10 (rarely –18) cm. long; leaflets mostly entire, the terminal one 3.5–10(–14) cm. long; fruiting panicles dense, 1.5–5(–7) cm. long; erect, leaning or climbing.
Fruit glabrous.
Leaves glabrous beneath or pilose to hispid only along the midrib and bases of veins. . . . . *R. radicans* (typical).
Leaves pilose on the lower surface. . . . . Forma *hypomalaca*.
Fruit pubescent. . . . . Forma *malacotrichocarpa*.
Leaves membranaceous, on petioles mostly 6.5–2 dm. long; leaflets coarsely dentate, undulate or entire, the terminal one mostly 1.1–2 dm. long; fruiting panicle more open, 3.5–8 cm. long; high-climbing.
Lower leaf-surface and petiole glabrous. . . . . Var. *vulgaris*.
Lower leaf-surface soft-pubescent.
Petioles glabrous. . . . . Forma *intercursa*.
Petioles villous-tomentose. . . . . Forma *Negundo*.

Stems woody for only 0.5–6 dm. above the creeping subterranean stoloniferous bases, simple or very sparsely upright-branched, with no aerial roots; leaves approximate at summit of stem and branches, appearing falsely verticillate, on erect petioles 0.5–2.3 dm. long; leaflets membranaceous to subcoriaceous (in exposed habitats), dentate, undulate or rarely entire, the terminal one broadly ovate to suborbicular (rarely narrower) and abruptly acuminate, 0.4–1.5 dm. long, the veins and veinlets of the lower surface often hirtellous. . . . . Var. *Rydbergii.*

R. radìcans (typical), *R. Toxicodendron* of ed. 7, in large part, not L.; *R. Toxicodendron*, var. *microcarpa* Michx.; *Toxicodendron radicans* Ktze. — Thickets, open woods, sandy or rocky places and fence-rows, often too abundant, Que. to Great Lakes reg., s. to N.S., N.E., L.I., Fla. and Ky. *Fl.* May–July; *fr.* Aug.–Nov. (and all winter). — Forma **hypomálaca** Fern. (soft-pubescent beneath), N.Y. to Ky.; forma **malacotrichocárpa** (A. H. Moore) Fern. (with fruits soft-hairy) (= *R. littoralis* Mearns and *Toxicodendron rad.*, var. *littorale* Barkley), Me. to Va. and probably elsewhere.

Var. **vulgàris** (Michx.) DC. (common). — Wooded swamps and bottomlands, Fla. to e. Tex., n. to s. Me., Mass., N.Y. and Okla. — Forma **intercúrsa** Fern. (running between), Pa. to Va.; forma **Negúndo** (Greene) Fern. (for *Acer Negundo*, box-elder), Fla. to e. Tex., n. to Va., O., Ind., Ill. and Ia.

Var. **Rydbérgii** (Small) Rehd. (for Per Axel Rydberg, 1860–1931), *R. Rydbergii* Small — Woods and rocky slopes, Gaspé Co., Que. to s. B.C., s. to N.S., n. and w. N.E., mts. of w. Va., Mich., n. Ill., Minn., w. Kans., Tex., N.M. and Ariz.

7. R. Toxicodéndron L. (old generic name, poison tree), Poison Oak. — *Stem slender*, erect, simple or with very few *erect* branches, *woody for* 0.5–6 *dm., not climbing nor with aërial roots*, spreading by subterranean stolons; *leaves on* long erect *velvety-tomentose petioles*, mostly near summit of stem and often appearing falsely verticillate; *leaflets elliptic, rhombic or obovate*, pilose above, *velvety-tomentose beneath, obtuse or rounded above*, variously lobed, suggesting oak-leaves, or unlobed in forma **elobàta** Fern. (unlobed); *fruit greenish to buff, pubescent*, or glabrous in forma **leiocárpa** Fern. (smooth-fruited), rather large. (*R. quercifolia* (Michx.) Steud.; *Toxicodendron quercifolium* Greene; *T. Toxicodendron* Britt.) — Dry barrens, pinelands and sands, n. Fla. to e. Tex., n. to N.J., Md., Tenn., s. Mo. and e. Okla. *Fl.* May, June; *fr.* Aug.–Nov. (and all winter).

FAM. 98. CYRILLÀCEAE (Cyrilla Family)

American shrubs or small trees with alternate entire thickish leaves, no stipules, and (4–)5-*parted small regular and perfect flowers. Stamens hypogynous,* 5 *or* 10, *when* 5 *alternate with the petals. Ovary* 2–5-*locular; locules* 1–4-*ovuled.* Petals (white or roseate) imbricated or convolute in bud, sessile or unguiculate. Fruit a small corky drupe or tardily dehiscent capsule. Flowers racemose-spicate.

1. CYRÍLLA Garden Leatherwood. Black Ti-ti

Petals sessile. Stamens 5, attached with the petals under a disk; anthers somewhat sagittate. Ovary 2–3-locular; ovules anatropous or half-anatropous; cotyledons terete, small; radicle superior. — Leaves oblanceolate to narrowly obovate, coriaceous, evergreen or nearly so. Handsome shrubs or small trees of trop. and warm-temp. Am. (Named in honor of *Domenico Cirillo*, 1734–1799, professor of medicine at Naples.)

1. C. racemiflòra L. (with flowers in racemes), He-Huckleberry. — Freely branched shrub or small tree up to 6 (farther s. to 10) m. high; branchlets glabrous; leaves coriaceous, lustrous above, oblanceolate to narrowly obovate or elliptic, round-tipped, blunt or subacute, in age clearly reticulate beneath; racemes slenderly cylindric, often crowded, 0.3–1.5 dm. long; pedicels soon horizontally divergent; sepals lanceolate or lance-ovate, sharp-pointed, 1–1.8 mm. long; petals milk-white, oblong-lanceolate, about 5 mm. long; drupes yellow, ovoid or conic, 2–3 mm. high, tipped by style and persistent stigmas. — Sandy swamps, Fla. to Tex., n. on Coastal Plain to se. Va. June, early July. — Autumnal foliage scarlet or orange, gorgeous when mingled with the bright yellow fruit.

Var. **subglobòsa** Fern. (nearly globular). — Leaves relatively thin, the veins beneath hardly reticulate; sepals narrowly ovate, 1 mm. long; drupes depressed, subglobose, broader than long, with a deep longitudinal furrow, the style and stigmas very short. — Local, se. Va.

FAM. 99. AQUIFOLIÀCEAE (Holly Family)

Trees or shrubs with small axillary 4–9-merous flowers, a minute calyx free from the 4–9-locular ovary and the 4–9-seeded berry-like drupe; the stamens as many as the divisions of the almost or quite 4–9-petaled corolla and alternate with them, usually attached to their very base. Corolla imbricated in the bud. Anthers opening lengthwise. Stigmas 4–9 or united into one, nearly sessile. Seeds suspended and solitary in each locule, anatropous, with a minute embryo in fleshy albumen. Leaves simple, mostly alternate. Flowers white, greenish or yellowish, mostly polygamodioecious.

Flowers 4–9-merous; calyx persistent in fruit; petals oblong or broader, white in ours, slightly united at base; stamens adnate to base of corolla. . . . . 1. *Ilex.*
Flowers 4–5-merous; calyx often obsolete; petals linear, yellowish, free; stamens free. . . . . 2. *Nemopanthus.*

1. ÌLEX L. Holly. Houx (Que.)

Calyx 4–9-toothed. Petals 4–9, separate or united only at the base, oval or obovate, obtuse, spreading. Stamens 4–9. The berry-like drupe containing 4–9 small nutlets. — Leaves alternate. Pistillate flowers inclined to be solitary, and the staminate or partly staminate flowers to be clustered in the axils. Trees and shrubs of trop. and temp. reg. (The ancient Latin name of the Holly-Oak, rather than of the Holly.)

*a.*Calyx-segments and petals of pistillate or polygamous flowers in 4's or 5's (rarely 6's); drupes red or purple (rarely yellow); nutlets grooved or ribbed on the back. . . *b.*
*b.*Leaves coriaceous and evergreen; flowers 4-merous. § Aquifolium.
Leaves spiny-margined or -tipped; calyx-segments acute, ciliate; drupe 7–10 mm. in diameter. . . . . 1. *I. opaca.*
Leaves crenate-margined; calyx-segments rounded, scarcely ciliate; drupe 5–8 mm. in diameter. . . . . 2. *I. vomitoria.*
*b.*Leaves deciduous; flowers 4–6-merous. § Prinoides.
Fruiting pedicels 1.2–2.2 cm. long. . . . . 3. *I. longipes.*
Fruiting pedicels mostly much shorter.
Calyx-segments ciliate; leaves ovate to lanceolate, acuminate, sharply serrate. . . . . 4. *I. montana.*
Calyx-segments eciliate; leaves oblong to oblanceolate or obovate, blunt to acutish, crenate or blunt-serrate.
Leaves oblong, strongly rugose beneath; fruits solitary, dull, on pedicels 0.7–2 cm. long. . . . . 5. *I. Amelanchier.*
Leaves oblanceolate to obovate, not rugose; drupes mostly clustered, lustrous, on shorter pedicels. . . . . 6. *I. decidua.*
*a.*Calyx-segments and petals of pistillate flowers in 6's–8's (rarely 9's); drupes red (rarely yellow) or black; nutlets smooth and even. § Prinos. . . *c.*
*c.*Leaves deciduous; drupes red or yellow.
Leaves appressed-pubescent beneath, at least on the veins, or glabrous, sharply serrate; calyx-segments obtuse, pubescent and strongly ciliate. 7. *I. verticillata.*
Leaves glabrous, or spreading-pilose on veins beneath, appressed-serrulate; calyx-segments acute or acutish, glabrous, eciliate or only sparsely ciliate. . . . . 8. *I. laevigata.*
*c.*Leaves coriaceous, evergreen; drupes black or dark purple.
Young twigs velutinous-puberulent; leaves mostly blunt, crenate or crenate-serrate above the middle. . . . . 9. *I. glabra.*
Young twigs glabrous or viscid-puberulent; leaves sharply acute, mostly spinescent-serrate above the middle. . . . . 10. *I. coriacea.*

§ Aquifòlium Gray

1. I. opàca Ait. (dull or opaque; as contrasted with Eurasian species), American H. — *Leaves* oval to elliptic-lanceolate, *coriaceous, evergreen, with remote spiny teeth,* or entire or with only 1 or 2 small teeth in forma **subíntegra** Weath. (nearly entire), 0.4–1 *cm. long;* flowers clustered along the bases of the puberulent young branchlets or in the axils, 4-merous; *calyx-segments acute, ciliate; drupes* 7–10 *mm. in diameter,* red, or yellow in forma **xanthocárpa** Rehd. (yellow-fruited); nutlets grooved on the back. — Moist woodlands, Fla. to Tex., n. to e. Mass., se. N.Y., e. Pa., W.Va., s. O., n. Ky., s. Ill., se. Mo. and Okla. May, June. — Tree up to 30 m. high.

2. I. vomitòria Ait. (causing vomiting; the stimulating brew emetic when drunk to excess), CASSINA, YAUPON. — Stiffly and divergently branched shrub or small tree with whitish-gray close bark; *leaves* lance-oval or elliptic, coriaceous, evergreen, *crenate*, 1–4.5 *cm. long;* flower-clusters nearly sessile; *calyx-segments rounded, scarcely ciliate;* drupes 5–8 mm. in diameter; nutlets grooved on back. (*I. Cassine* of many authors, not L.) — Sandy woods and clearings, Fla. to Tex., n. to se. Va. and n. Ark. May, June. — The dried leaves much used and recommended (in spite of the name) as tea, like Asiatic Tea containing an appreciable amount of caffeine.

§ PRINOÌDES Gray

3. I. lóngipes Chapm. (long-stalked). — Large shrub; *leaves thin, elliptic to broadly lanceolate*, obtuse to abruptly pointed, low-serrate with sharp teeth to dentate, *glabrous* or with midrib pilose beneath, 3–9 cm. long; *staminate pedicels* 1.2–1.8 *cm. long, pistillate up to* 2.3 *cm. long;* drupe globose, 7–10 mm. in diameter, red, or yellow in forma **Vantrómpi** M. Brooks (named in 1940 for its discoverer, H. O. VAN TROMP); nutlets prominently ribbed. — Woods, bogs and rocky slopes, Fla. to La., n. to N.C., W.Va. and Tenn. May, June.

4. I. montàna T. & G. (of mountains), MOUNTAIN-WINTERBERRY, MOUNTAIN-H., LARGE-LEAVED H., HULVER. — Tall shrub or small tree with slender glabrous branchlets; *leaves oblong-lanceolate to ovate*, membranaceous, acuminate, *sharply serrate*, 0.6–1.4 dm. long, glabrous, often clustered at tips of fruiting spurs; peduncles of pistillate flowers and fruits short; flowers 4 or 5 (rarely 6) -merous; *calyx ciliate; drupes* about 1 *cm. in diameter*, scarlet; *nutlets striately many-ribbed on back.* (*I. monticola* Gray) — Rich wooded slopes and mountain-sides, e. to w. N.Y., s. on the upland to Ala. and Tenn. June. (In Japan, as the barely separable var. *macropoda* (Miq.) Fern.)

Var. **móllis** (Gray) Britt. (soft). — Leaves softly pilose beneath, at least along the nerves. (*I. monticola*, var. *mollis* (Gray) Britt.; *I. dubia* sensu Loesener, not *Prinos dubius* G. Don) — Sw. Mass. and nw. Ct. to W.Va., s. to Ga. and Tenn.

5. I. Amelánchier M. A. Curtis (for *Amelanchier;* from resemblance of leaf). — Shrub 1–2 m. high, with slightly pubescent branchlets; *leaves oblong*, blunt, leathery, 4–8 cm. long, *dull above, prominently rugose* and slightly pubescent *beneath*, serrulate; staminate flowers 6–9; *pistillate flowers solitary, on pedicels* 0.7–2 *cm. long; drupe dull*, scarlet, 0.8–10 mm. in diameter; nutlets grooved on back. (*I. dubia* BSP., not Weber) — Sandy swamps, very rare, Ga. to La., n. to se. Va. (acc. to *Loesener*).

6. I. decídua Walt. (deciduous), POSSUM-HAW. — Small tree or shrub; *leaves cuneate-oblong, oblanceolate or lance-obovate*, submembranaceous, becoming thickish, *crenate or obtusely serrate, pilose along the midrib beneath;* peduncles of staminate inflorescence longer than petioles, of pistillate shorter; flowers 4- or 5-merous; *calyx-teeth acute, smooth; drupes* 6–8 mm. in diameter, scarlet, *lustrous, short-stalked, mostly clustered;* nutlets many-ribbed. — Low woods, thickets and bottoms, Fla. to Tex., n. to Md., s. Ind., s. Ill., Mo. and se. Kans. April, May.

§ PRÌNOS (L.) Gray

7. I. verticillàta (L.) Gray (whorled; from the clusters of axillary flowers), BLACK ALDER, WINTERBERRY, APALANCHE OR AULNE BLANCHE (Que.). — Shrub 1–4 m. high, rarely tree-like and with trunks 1 dm. in diameter; *leaves* lanceolate to round-obovate, serrate, *appressed-pilose or downy or glabrous beneath, dull above*, deciduous; *staminate flowers* clustered, 2–10, all *short-stalked*, 4–6-merous; pistillate flowers short-stalked, 5–8-merous; *calyx-segments obtuse*, mostly pubescent and *ciliate;* drupes bright red (rarely yellow), 5–7 mm. in diameter. — Highly variable:

Leaves thickish, opaque to transmitted light, not pellucid-puncticulate, pilose at least on the nerves beneath, the blades lanceolate to rounded-obovate, 2–10 cm. long; bark of two-year-old branchlets dark or fuscous-brown, rarely pale.

Lower surfaces of leaves minutely appressed-pilose or downy along principal nerves. . . . . *I. verticillata* (typical).

Lower surfaces minutely downy on tissue as well as nerves. . . . . . . Var. *padifolia*.

Leaves thickish and glabrous throughout, or membranaceous and appearing pellucid-puncticulate under a lens; bark of two-year-old branchlets whitish-gray.

Branchlets divergent, rather distant; leaves membranaceous, pellucid-puncticulate under a lens, glabrous or sparsely downy on primary veins beneath; blades oblanceolate to elliptic-obovate, 3–8 cm. long, 1–4 cm. broad. . . Var. *tenuifolia*.

Branchlets strongly ascending, more approximate or fastigiate; leaves firm, less puncticulate, glabrous throughout; blades lanceolate to lance-elliptic, 2–5 cm. long, 0.7–2 cm. broad. . . . . Var. *fastigiata.*

I. verticillàta (typical). — Swamps, pond-margins and damp thickets, Nfld. to Minn., s. to N.S., N.E., L.I., Ga., Tenn. and Ill. June–Aug. — The rare forma **chrysocárpa** Robins. (yellow-fruited) has yellow fruit.

Var. **padifòlia** (Willd.) T. & G. (with leaves of *Padus*). — N.S. to Minn., s. to s. N.E., Ga. and Mo.

Var. **tenuifòlia** (Torr.) S. Wats. (thin-leaved), *I. bronxensis* Britt. — Nfld. to Ont., s. to N.S., N.E., N.Y., n. N.J. and Mich.

Var. **fastigiàta** (Bickn.) Fern. (with branches erect and near together), *I. fastigiata* Bickn. — Nfld.; N.S.; se. Mass. and s. Conn.

8. I. laevigàta (Pursh) Gray (smooth), SMOOTH WINTERBERRY. — Similar to no. 7, up to 4 m. high; *leaves* lance- to oval-elliptic, 3–9 cm. long, *lustrous above, glabrous beneath or with wide-spreading pilosity on the veins,* appressed-serrulate; *staminate flowers 1 or 2, on slender peduncles* 0.6–2 *cm. long;* pistillate flowers and scarlet drupes (7–10 mm. in diameter) on shorter but definite peduncles; *calyx-segments acute or acutish, not at all or only sparsely ciliate.* — Wooded swamps, s. Me. to N.Y., s. to n. Ga., mostly at low alts. Mid-May–early July. — Forma **Hérveyi** Robins. (for ELIPHALET WILLIAMS HERVEY, 1834–1925) has bright yellow fruit.

9. I. glàbra (L.) Gray (smooth), INKBERRY, BITTER GALLBERRY. — Shrub 0.15–3 m. high, with *ashy-puberulent twigs; leaves* coriaceous, evergreen, lustrous, cuneate-lanceolate to oblong, *mostly blunt, crenate or crenate-serrate above the middle,* 1.5–5 *cm. long,* 0.7–2 *cm. wide;* staminate flowers several, on long peduncles; pistillate flowers solitary; calyx-teeth blunt; *drupes finally black, mostly solitary in the axils, firm, dry, persistent.* — Low sandy or peaty soil, Fla. to La., n. to Mass.; Isle au Haut, Me.; N.S. May (southw.)–Aug. (N.S.).

10. I. coriàcea (Pursh) Chapm. (leathery), LARGE or SWEET GALLBERRY. — Similar to no. 9, up to 4 m. high; *branchlets glabrous or viscid-puberulent; leaves* obovate, oblanceolate or oval, coriaceous, *sharply acute, mostly spinescent-serrate above the middle,* 3.5–8 cm. long, 1.5–3.5 *cm. broad;* staminate flowers crowded, on very short peduncles; *drupes* 1–5 *in an axil, soft and pulpy when ripe, dropping in autumn,* said to be edible. (*I. lucida* of ed. 7, not (Ait.) T. & G.) — Sandy woods and swamps, Fla. to La., n. to Great Dismal Swamp, se. Va. May.

2. NEMOPÁNTHUS Raf. MOUNTAIN-HOLLY. CATBERRY

Flowers polygamodioecious. Calyx in the staminate flowers of 4–5 minute deciduous teeth, in the pistillate ones obsolete. Petals 4–5, oblong-linear, spreading, distinct. Stamens 4–5; filaments slender. Drupe with 4–5 bony nutlets. — A much-branched shrub, with ashy-gray bark, alternate deciduous entire or slightly toothed smooth leaves on slender petioles. Flowers yellowish, on long slender axillary peduncles, solitary or sparingly clustered. (Name from the Greek *nema, a thread, pous, foot,* and *anthos, flower.*)

1. N. mucronàta (L.) Trel. (abruptly short-tipped), FAUX HOUX (Que.). — Erect, 0.3–3 m. high; bark gray; leaves elliptic-oblong to narrowly obovate, thin, slightly paler beneath; drupes red, or pale yellow in forma **chrysocárpa** (Farw.) Fern. (yellow-fruited). — Damp woods, thickets and swamps, Nfld. to Minn., s. to N.S., N.E., upland Va., W.Va., O., Ind. and n. Ill. May, June.

FAM. 100. CELASTRÀCEAE (STAFF-TREE FAMILY)

Shrubs with simple leaves and small regular flowers, the sepals and the petals both imbricated in the bud, the 4 or 5 perigynous stamens as many as the petals and alternate with them, inserted on a disk which fills the bottom of the calyx and sometimes covers the ovary. Seeds arillate. Ovule anatropous; styles united into one. Fruit 2–5-locular, free from the calyx. Embryo large, in fleshy albumen; cotyledons broad and thin. Stipules minute and fugacious. Pedicels jointed.

Leaves opposite; flowers solitary or in axillary cymes; branches 4-angled; shrubs or small trees.
  Ovary 3–5-locular; capsule lobed; seed inclosed in a scarlet or orange aril; ours deciduous-leaved. . . . . 1. *Euonymus.*
  Ovary 2-locular; capsule oblong, not lobed; seed with a white lacerate basal aril; dwarf evergreen. . . . . 2. *Pachystima.*
Leaves alternate, deciduous; flowers in panicles or racemes; fruit a globular 3-valved orange or yellow capsule; branches terete; high-twining woody vine. 3. *Celastrus.*

1. EUÓNYMUS L. Spindle-tree

Flowers perfect. Sepals 4 or 5, united at the base, forming a short and flat calyx. Petals 4–5, rounded, spreading. Stamens short, borne on the edge or face of a broad and flat 4–5-angled disk, which coheres with the calyx and is stretched over the ovary and more or less adhering to it. Style short or none. Capsule 3–5-lobed, 3–5-valved, loculicidal. Seeds 1–4 in each locule, inclosed in a red aril. — Shrubs of N. Am., Eurasia and Austral. with 4-sided branchlets, opposite serrate leaves, and loose axillary pedunculate cymes of small flowers. (Ancient Greek name, from *eu*, *good*, and *onoma*, *name*, but used ironically, the plants having had the bad reputation of poisoning cattle.) Also spelled Evonymus.

Flowers 4-merous; ovules ascending, 2 in each locule; fruit deeply lobed, not tuberculate.
 Leaves pubescent beneath; flowers 7–15, purple; fruit purple; aril scarlet. . . 1. *E. atropurpureus.*
 Leaves glabrous; flowers 2–5, yellow-green; fruit pink; aril orange. . . . 2. *E. europaeus.*
Flowers 5-merous; fruit tuberculate, shallowly lobed; seeds 4–10 in each locule.
 Leaves acuminate or pointed, sessile, thickish, lustrous; petals clawed, distant; upright loosely branching shrub. . . . . 3. *E. americanus.*
 Leaves obtuse, petioled, thin, opaque; petals clawless, approximate; low procumbent shrub. . . . . 4. *E. obovatus.*

1. E. atropurpùreus Jacq. (dark purple), Burning-bush, Wahoo. — Large shrub or small tree 2–8 m. high; *leaves* petioled, oblong-oval, acuminate, *pubescent beneath, the terminal* (larger) 7–14 *cm. long,* finely serrate; *peduncles 7–15-flowered; flowers purple,* chiefly 4-merous; ovules ascending, 2 in each locule; *fruits* purple, about 1.5 cm. broad, deeply 4-lobed, *smooth; seed* brown, *with a scarlet aril.* — Rich woods and thickets, Ont. to Mont., s. to e. Va., Ala., Tenn., Ark. and Okla.; cult. and somewhat natzd. northeastw. June, July.

E. Fortùnei (Turcz.) Hand.-Maz. (for its discoverer, Robert Fortune, 1817–1880), an evergreen climbing shrub with aërial rootlets, coriaceous elliptic leaves evidently veiny beneath, dense long-peduncled cymes and smooth globose fruit, tends to spread from cult. (Introd. from Asia)

2. E. europaèus L. (European), European S. — Similar to no. 1; *leaves* smaller (*the larger 4–8 cm. long*), *glabrous,* crenulate; *peduncles 2–5-flowered; flowers yellow-green; fruit pink; seeds with orange aril.* — Roadsides and waste places, occasional, Mass. to Wisc. and southw., spread from cult. Late May, June. (Introd. and natzd. from Eu.)

E. alàtus (Thunb.) Sieb. (winged), Winged S., with opposite branches corky-winged, the fruit of 2–4 distinct locules, spreads from cult. (Introd. from e. Asia)

3. E. americànus L. (American), Strawberry-bush, Bursting-heart. — Shrub, low, upright or straggling, 1–2 m. high; *leaves almost sessile, thickish,* bright green, ovate to oblong-lanceolate, *acute or pointed;* parts of the greenish-purple flowers mostly in 5's; *petals distinctly clawed; capsules rough-warty, depressed,* crimson when ripe; the aril and dissepiments scarlet. — Rich woods and ravines, Fla. to Tex., n. to se. N.Y., Pa., W.Va., s. Ind., s. Ill., Mo. and Okla. May, June.

4. E. obovàtus Nutt. (obovate), Running Strawberry-bush. — *Trailing, with repent branches;* flowering branches 3–6 dm. high; *leaves thin, opaque,* obovate or oblong, *obtuse, petioled;* flowers and fruits similar to those of no. 3; the *petals without claws,* nearer together. — Rich dry to damp woods, thickets and slopes, w. N.Y. and s. Ont. to Mich. and Ill., s. to Tenn. and Mo. May, June.

2. PACHÍSTIMA Raf.

Flowers perfect. Sepals and petals 4. Stamens 4, on the edge of the broad disk lining the calyx-tube. Ovary free; style very short. Capsule small, oblong, 2-locular, loculicidally 2-valved. Seeds 1 or 2, inclosed in a white membranaceous many-cleft aril. — Low N. Am. evergreen shrubs, with smooth serrulate coriaceous opposite leaves and very small green flowers solitary or fascicled in the axils. (Name from the Greek *pachys*, *thick*, and *stigma*.) Generic name also spelled by Raf. as *Paxistima* and *Pachystima*.

1. P. Cánbyi Gray (for its discoverer, William Marriott Canby, 1831–1904), Cliff-green, Mountain-lover. — Leaves linear to linear-oblong or oblong-obovate, obtuse, 6–25 mm. long; pedicels very slender, often solitary, shorter than the leaves; fruit 4 mm. long. — Calcareous rocks and slopes, mts. of w. Va. and W.Va.; Carter Co., Ky.; Highland and Adams Cos., O. May–Sept.

3. CELÁSTRUS L. STAFF-TREE. SHRUBBY BITTERSWEET

Flowers polygamodioecious. Petals (crenulate) and stamens 5, inserted on the margin of a cupuliform disk which lines the base of the calyx. Capsule globose, orange-color and berry-like, 3-locular, 3-valved, loculicidal. Seeds 1 or 2 in each locule, erect, inclosed in a pulpy scarlet aril. — Leaves alternate. Flowers small, greenish, in raceme-like clusters terminating the branchlets. Twining shrubs of e. N.Am., e. and s. Asia and Austral. (*Celastros*, an ancient Greek name for some evergreen tree.)

1. C. scándens L. (climbing), WAXWORK, CLIMBING BITTERSWEET, BOURREAU DES ARBRES (Que.). — Twining shrub up to 18 m. high, trunks up to 2.5 cm. thick; *leaves ovate-oblong, finely serrate, pointed;* capsule orange or orange-yellow, upon splitting exposing the scarlet to crimson seeds. — Thickets, river-banks and woods, s. Que. to s. Man., s. to s. N.E., L.I., Ga., Ala., La. and Okla. Late May, June.

2. C. ORBICULÀTUS Thunb. (round). — *Leaves suborbicular to broadly obovate, with crenate teeth; flowers and fruits in small axillary cymes.* — Roadsides, fence-rows and thickets, N.Y. to Va. and southw. (Introd. and natzd. from e. Asia)

FAM. 101. STAPHYLEÀCEAE (BLADDERNUT FAMILY)

Shrubs or small trees with opposite chiefly pinnate stipulate leaves and perfect flowers. Stamens as many as and alternate with the petals, borne outside a large disk. Fruit (in ours) a bladdery inflated 2–3-horned capsule. Seeds (in ours) with scanty albumen and straight embryo. —Chiefly Asiatic.

1. STAPHYLÈA L. BLADDERNUT

Calyx deeply 5-parted, the lobes erect, whitish. Petals 5, erect, spatulate. Pistil of 3 several-ovulate carpels united in the axis, their long styles lightly cohering. Capsule large, inflated, 3-locular, at length bursting at the summit; the locules containing 1–4 bony anatropous seeds. Cotyledons broad and thin. — Upright shrubs or small trees of temp. N. Am. and Eurasia, with opposite pinnate leaves of 3 or 5 serrate leaflets and white flowers in drooping raceme-like clusters terminating the branchlets. Stipules and stipels deciduous. (Name from Greek *staphyle, a bunch of grapes.*)

1. S. trifòlia L. (three-leaved). — Shrub or small tree, with greenish striped branches; leaflets elliptic to ovate, acuminate, serrulate, the terminal long-stalked, the lateral subsessile or short-petiolulate; fruit ellipsoid, pyriform or subglobose, often 3-lobed at summit. — Rich thickets and borders of woods, sw. Que. to Minn., s. to Mass., Ct., Ga., Ala., Ark. and Okla. Mid-Apr.–early June.

FAM. 102. ACERÀCEAE (MAPLE FAMILY)

Trees and shrubs with watery saccharine sap, opposite simple and palmately lobed or more rarely palmately or pinnately divided leaves, small regular mostly polygamous or dioecious sometimes apetalous flowers. Ovary 2-locular, 2-lobed; ovules 2 in each locule. Embryo coiled or folded; *cotyledons long and thin.* — Chiefly trees of temp. reg.

1. ÀCER L. MAPLE

Flowers polygamodioecious. Calyx colored, 5 (rarely 4–12)-lobed or -parted. Petals either none or as many as the lobes of the calyx, equal, with short claws (if any), inserted on the margin of a perigynous or hypogynous disk. Stamens 3–12. Ovary 2-locular, with a pair of ovules in each locule; styles 2, long and slender, united only below, stigmatic down the inner side. The back of each carpel bearing a wing, converting the fruit into two 1-seeded at length separable samaras or keys. — Trees or sometimes shrubs of N. Hemisph., with palmately lobed or rarely pinnate leaves and small flowers. Pedicels not jointed. (Latin name of the maple.)

a. Leaves simple, palmately lobed or cleft; flowers variously unisexual to essentially perfect, often bearing a basal disk within. . . *b.*

b. Inflorescence a panicle or elongate raceme. . . . *c.*

c. Inflorescence a thyrse-like panicle, pubescent during anthesis; petals linear to narrowly spatulate, inconspicuous; stamens borne inside the disk; samaras strongly ribbed over the seed; bark of trunk and branches brown or darker, not striped. § SPICATA.

Panicle heavy, pendulous; calyx 4–5 mm. long; styles deeply cleft; leaves heavy, glabrous and prominently ribbed beneath; samaras 3–5 cm. long, the wing 1–2 cm. broad; trunk single. . . . . 1. *A. Pseudo-Platanus.*

Panicle slender, ascending; calyx 1–2 mm. long; styles barely notched; leaves membranaceous, pubescent and not prominently ribbed beneath; samaras 1.5–2.5 cm. long, the wing 5–10 mm. broad; trunks usually in clumps (shrubby). . . . . 2. *A. spicatum.*

*c.*Inflorescence a slender pendulous glabrous raceme; petals obovate, conspicuous; stamens inserted outside the disk; samaras ribless over the seed; bark of young trunk and branches green, with longitudinal pale or dark stripes. § Macrantha. . . . . 3. *A. pensylvanicum.*

*b.*Inflorescence corymbose, subumbellate, fascicled or glomerulate. . . *d.*

*d.*Flowers appearing with or little before the leaves, on filiform pedicels, yellowish; stamineal disk well-developed; mature pairs of samaras with broadly U-shaped or open sinus, the long axis of the seed divergent. . . *e.*

*e.*Inflorescence peduncled, the flowers on glabrous ascending pedicels; sepals distinct; petals conspicuous; stamens inserted on middle of disk; samaras (including wings) nearly horizontally spreading, 3.5–5.5 cm. long, the flattened seed-bearing base 1–1.5 cm. long, wing 1–1.6 cm. broad and scarcely narrowed at base; petioles with milky juice. § Platanoidea. . . . . 4. *A. platanoides.*

*e.*Inflorescence sessile, the flowers on filiform hairy pendulous pedicels; sepals united below into a tube; petals wanting; stamens inserted on inner margin of disk; samaras strongly upcurving, 1.5–4 cm. long, the plump seed-bearing base 6–10(–12) mm. long; wing 4–11 mm. broad, the inner side narrowed at base; petioles without milky juice. § Saccharina. . . *f.*

*f.*Pedicels of staminate flowers elongating to 5–10 cm.; calyx 4–6 mm. long; ovary and fruit glabrous; bark deeply furrowed, gray to blackish; northern or upland trees.

Leaf-blades flat, with pale or glaucous lower surfaces, those of fertile branches with open sinuses; full-grown stipules not covering axillary buds; petioles of vigorous terminal leaves only gradually enlarged to base, without foliar outgrowths; fresh branchlets brown; bark of trunk and branches gray. . . . . . 5. *A. saccharum.*

Leaf-blades with drooping margins, green or fulvous beneath, with nearly or quite closed sinuses; full-grown stipules inclosing axillary buds; petioles of vigorous terminal leaves abruptly enlarged at base, often bearing foliar outgrowths; fresh branchlets orange-brown; bark of trunk and branches darker. . . . . . . 6. *A. nigrum.*

*f.*Pedicels of staminate flowers elongating to 3–5 cm.; calyx 1.5–4 mm. long; ovary and young fruit pubescent; leaves broadly rounded or shallowly cordate to subtruncate at base, pale and pubescent beneath; bark (except on old bases of trunks) smooth, whitish, suggesting *Fagus;* trees of southern Coastal Plain and outer Piedmont. 7. *A. barbatum.*

*d.*Flowers red to yellow, appearing long before the leaves, the staminate subsessile or but short-pedicelled in capitate or subcapitate clusters; the hermaphrodite and pistillate in sessile umbels, with fruiting pedicels elongating; disk rudimentary or wanting; mature pairs of samaras with V-shaped sinus, the long axis of the seed ascending. § Rubra.

Petals about equaling sepals; ovaries and young fruit glabrous; samaras 1.5–5 cm. long, glabrous, deep red to yellow or green, seed-bearing body 5–9 mm. long; leaves rarely cleft more than half-way to base; branchlets spreading or ascending. . . . . 8. *A. rubrum.*

Petals wanting; ovaries and young fruits white-villous; samaras 4–8 cm. long, usually hairy at base, pale, seed-bearing body 1–1.5 cm. long; leaves cleft one-half to two-thirds of the way to base; branchlets pendulous. . . . . 9. *A. saccharinum.*

*a.*Leaves pinnate, with 3–7(–9) leaflets; flowers dioecious, the staminate fascicled, the pistillate racemose; disk wanting. § Negundo. . . . . . 10. *A. Negundo.*

§ Spicàta Pax

1. A. Pseùdo-Plátanus L. (false plane-tree), Sycamore-M. — Tree with *heavy dark green leaves glaucous, glabrous,* and *strongly ribbed beneath;* the glabrous blades 5-lobed, coarsely serrate, up to 1.6 dm. broad; *panicle* elongate, *thyrsiform*, pubescent, *pendulous*, with stoutish pedicels; flowers greenish; *calyx* 4–5 *mm. long;* petals linear; *styles deeply cleft; stamens attached inside the disk; samaras* 3–5 *cm. long*, strongly ribbed over the seed, *the wing* 1–2 *cm. broad.* — Much planted, freely establishing seedlings and sometimes estab. in fence-rows, on roadsides, etc. *Fl.* May, June; *fr.* June–Sept. (Introd. from Eu.)

A. GÍNNALA Maxim. (a native name), a shrub or low tree with lustrous oblong or ovate small leaves deeply cleft near base, panicles of fragrant whitish flowers, and glabrous nearly parallel erect samaras 2–2.5 cm. long, is locally estab. from Me. to Ct. and w. N.Y. (Introd. from Asia)

2. A. spicàtum Lam. (spiked), MOUNTAIN-M., PLAINE BÂTARDE, ÉRABLE or FOUÉREUX (Que.). — Tall shrub or bushy small tree with drab flaky or furrowed bark; *panicle slender*, subcylindric, *ascending*, pubescent; flowers greenish, insignificant; *calyx* 1–2 *mm. long; petals linear-spatulate; styles barely notched; leaves thin, downy beneath,* 3 (or slightly 5)-lobed, coarsely serrate, the lobes taper-pointed; *samaras red* (sometimes yellow), showy, 1.5–2.5 *cm. long*, the wing 5–10 *mm. broad, strongly ribbed over the seed.* — Cool woods, Nfld. to Sask., s. to N.S., N.E., n. N.J., Pa., uplands to n. Ga. and Tenn., O., Mich., Wisc. and ne. Ia. *Fl.* late May–early Aug.; *fr.* July–Oct. — Leaves brilliant red in late summer and autumn.

§ MACRÁNTHA Pax

3. A. pensylvánicum L. (Pennsylvanian), STRIPED M., MOOSEWOOD, WHISTLEWOOD, BOIS BARRÉ (Que.). — Slender tree with green, pale- to dark-striped smooth bark; *leaves* (up to 2 dm. broad) 3-lobed at apex, *finely and sharply double-serrate, the short lobes taper-pointed* and also serrate; *racemes drooping*, loose, slender, *glabrous; petals obovate*, pale green to yellowish, much *larger than sepals; stamens inserted outside disk; samaras ribless over the seed,* 2–3.3 cm. long, the wings loosely ascending to divergent. — Rich cool woods, Gaspé Pen., Que., to Man., s. to N.S., N.E., Pa., uplands to n. Ga. and Tenn., O. and Mich. *Fl.* May, June; *fr.* June–Sept. — Leaves turning yellow in autumn. By woodsmen in Me. called MALEBERRY; supposed by them to be the male of VIBURNUM ALNIFOLIUM!

§ PLATANOÌDEA Pax

4. A. PLATANOÌDES L. (like plane-tree, *Platanus*), NORWAY M. — Close-barked tree with spreading-ascending branches; *leaves* green to bronze, glabrous, 5-lobed, up to 1.8 dm. broad, sinuately sharp-toothed, *their petioles with milky juice; corymbs* appearing with the leaves, *glabrous, peduncled,* the showy yellow-green *flowers on ascending to spreading pedicels; sepals distinct;* petals larger; *stamens inserted on middle of disk; samaras almost horizontally divergent,* 3.5–5.5 *cm. long, the flattened seed-bearing base* 1–1.5 *cm. long, the wing* 1–1.6 *mm. broad and scarcely narrowed at base.* — Much planted; seedlings abundantly thriving in hedge-rows, road-side-thickets, etc. *Fl.* April, May; *fr.* May–Aug. (Introd. from Eu.)

A. CAMPÉSTRE L. (of fields), HEDGE-M., a shrub or small tree, with dull green leaves pubescent beneath and only 0.5–1 dm. broad, with 3–5 entire obtuse lobes and milky juice, pubescent corymbs, stamens inserted at inner margin of disk, flat-based strongly ribbed samaras up to 3 cm. long, sometimes spreads from cult., hardly natzd (Introd. from Eu.)

§ SACCHARÌNA Pax (SUGAR-MAPLES) (*Saccharodendron* Nieuwl.)

5. A. sáccharum Marsh. (sugar), ROCK- or SUGAR-M., ÉRABLE À SUCRE (Que.). — Large tree, the *older gray bark* becoming *deeply furrowed;* branches spreading, new branchlets brown; *mature leaves of fertile branches flat, firm,* roundish in general outline, cordate or subcordate, *with open sinus,* 5–3-lobed, 0.8–2 dm. broad, dark green above, *glabrous and pale beneath,* or glaucous in forma **glaûcum** (Schmidt) Pax (gray-blue), few-lobulate, at least the terminal lobe broad-oblong and shouldered below the acuminate tip; *their full-grown stipules not covering the axillary buds; leaves of leading shoots* larger, more cordate and even with closed sinuses, *their glabrous petioles only gradually enlarged at base, without foliar outgrowths,* or in forma **Rugélii** (Pax) Palmer & Steyerm. (named for its discoverer, FERDINAND RUGEL, 1806–1878) with the coriaceous 3-lobed leaves of fertile branches more reniform and 0.6–1.5 dm. broad and with the lobes entire or barely shouldered and the lateral lobes prolonged and divergent; flowers expanding with the leaves, pendulous on filiform hairy *pedicels elongating to* 5–10 *cm.; calyx tubular,* 4–6 *mm. long,* yellowish; *petals* 0; *style-branches* 5–6 *mm. long; samaras* 2.5–4 *cm. long;* the plump *seed-bearing bases divergent,* 8–12 *mm. long;* the ascending *wing* 6–11 *mm. broad,* the inner side narrowed at base. (*A. barbatum* of Sarg., not Michx.; *A. saccharinum* Wang., not L.; *A. saccharophorum* K. Koch; *Saccharodendron barbatum* sensu Nieuwl., not *A. barbatum* Michx., basonym) — Rich, mostly hilly woods, Gaspé Pen., Que., to e. Man., s. to N.S., N.E., Del., uplands to Ga. and Ala., Miss., Ark. and ne. Tex. *Fl.* late April–early June;

fr. June–Sept. — Leaves yellow to scarlet in autumn. This and the next highly valued for their hard wood and as the chief sources of maple-syrup and -sugar. — Forma **cónicum** Fern. (conical), with strongly ascending crowded branches and dense conical outline, local, N.H.

Var. **Schnéckii** Rehd. (for its discoverer, JACOB SCHNECK, 1843–1906). — Petioles and veins of lower leaf-surfaces densely villous. — S. Ind., Ill. and Mo.

6. A. nìgrum Michx. f. (black), BLACK M. — *Bark darker,* usually with narrower and shallower furrows than in no. 5; *young branchlets orange-tinged; leaves* deeper green, *weaker; the margins, especially of the lower lobes, drooping; petioles and yellow-green to fulvous lower surfaces permanently pubescent* (rarely glabrate); *blades* on fruiting branches rotund to subreniform, deeply cordate *with narrow or closed sinus, mostly* 3 (rarely 5)*-lobed, the divergent lobes entire or only slightly lobulate; full-grown stipules inclosing axillary buds; petioles of vigorous leading shoots abruptly enlarged at base, often bearing foliar outgrowths;* seed-bearing base of samara (8–)10–12 mm. long. (*A. saccharum,* var. *nigrum* (Michx. f.) Britt.; *Saccharodendron* Small) — Rich calcareous or alluvial woods, sw. Que. and w. N.H. to Minn. and S.D., s. to Ga., Ala. and La. *Fl.* May, June; *fr.* June–Sept. — As valuable for timber and sugar as no. 5.

7. A. barbàtum Michx. (bearded; from the long beard in the summit of the flower), SOUTHERN SUGAR-MAPLE, SUGAR-TREE. — Tree with *thin whitish-gray bark becoming furrowed only in age,* the trunks up to 7 dm. in diameter; branchlets grayish, with purple tinge; young branchlets of the season 1–2 mm. thick, densely short-pilose; or in forma **floridànum** (Chapm.) Fern. (of Florida) young branchlets and petioles glabrous; *mature leaves minutely pilose to glabrescent beneath; those of fertile shoots* (not the vigorous leaders) 3–9.5 (av. 6.7) *cm. long,* 3.5–11 (*av.* 8) *cm. broad,* the middle lobe 2–5.5 (*av.* 3.3) cm. long, *the base truncate, rounded or with broad open sinus,* the petioles 0.5–1 mm. thick near middle, densely short-pilose, or glabrous in forma *floridanum; flowering pedicels elongating to* 3–5 *cm.*; *flowering calyx* (including hypanthium) greenish to greenish-yellow, 1.5–2.5 *mm. long, with conspicuous long white beard projecting from throat; ovary long-setose;* style 1–2.5 mm. long, stigmas 1.5–5 mm. long; anthers 1–1.5 mm. long; samaras 1.5–3 cm. long, the mature locules 5–10 (av. 6.5) mm. long and 4–6.5 (av. 5.4) mm. thick, the mature wings 1–2.2 (av. 1.6) cm. long and 4.5–9 (av. 7) mm. broad. (*A. floridanum* (Chapm.) Pax; *Saccharophorum floridanum* (Chapm.) Small) — Rich calcareous or alluvial woods, Fla. to e. Tex., n. to e. Va. and se. Mo. *Fl.* April, early May; *fr.* June, rarely–Sept. — Passing northward into

Var. **Lóngii** Fern. (for one of its discoverers, BAYARD LONG, 1885–   ). — Trunks up to 1.2 m. in diameter, the old bark exfoliating in broad sheets; *young branchlets* of the season 2–3 *mm. thick,* often *densely villous; mature leaves densely velvety beneath, those of the fertile shoots* 7.5–13 (av. 10) *cm. long and* 7–17 (*av.* 13) *cm. broad,* slightly cordate to subtruncate at base; *the* 3 *lobes narrowly oblong-ovate, long-attenuate,* entire or only remotely and obtusely lobulate, *the middle lobe* 3.5–8 *cm. long,* or in forma **platýlobum** Fern. (broad-lobed) the *blades* definitely *cordate with the broad-oblong to oblong-obovate lobes coarsely and acutely lobulate; petioles* 1.5–2 mm. thick at middle, *heavily velutinous; flowering calyx* 3–4 *mm. long;* style 2.5–3 mm. long, stigmas 5–6 mm. long; anthers 1.5–2 mm. long; *samaras* 2.5–3.5 *cm. long,* the mature locules 8–10.5 (av. 9.25) mm. long and 5–7 (av. 6.5) mm. thick, the mature wings 1.7–2.5 (av. 2.1) cm. long and 9–11 (av. 10) mm. broad. — Calcareous slopes, James R., James City Co., and bottomlands of creeks tributary to Roanoke R., Brunswick Co., Va.

§ RÙBRA Pax (RED MAPLES)

8. A. rùbrum L. (red), RED, SCARLET, SOFT or SWAMP-M., PLAINE or PLAINE ROUGE (Que.). — Medium-sized tree with *spreading to ascending* smoothish *branches;* branchlets red; *flowers dark red to scarlet,* or yellowish in forma **pallidiflòrum** (K. Koch) Fern. (pale-flowered), appearing long before the leaves in at first dense clusters, the staminate in capitate clusters, the hermaphrodite and pistillate in sessile umbels with fruiting pedicels elongate; *petals linear-oblong; ovaries and young fruits glabrous; samaras* 1.5–2.5 *cm. long; the seed-bearing bases* ascending and 5–9 *mm. long,* with V-shaped sinus; young leaves more or less pilose to tomentose, mature ones glabrate, or leaves permanently pubescent beneath in forma **tomentòsum** (Desf.) Dansereau (downy), pale beneath; those of the fertile branches subcordate, suborbicular to deltoid-ovate, mostly 0.6–1.5 dm. broad, with 3 or 5 acuminate coarsely toothed ascending lobes, the middle lobe 4–8 cm. long, the upper lateral ones 2–5 cm. long. (*Rufacer* Small) — Swamps or uplands, Nfld. and Gaspé Pen., Que., to Man., s. beyond our range. *Fl.* March–May; *fr.* May–July. — The very rare forma **breviramúsculum** Vict. (with short branchlets) has long wand-like ascending branches fringed with abundant branchlets only 2–5 cm. long, these with leaves only up to 7 cm. across. — Typical *A. rubrum* has foliage red to yellow in autumn. — Passing freely into

Var. **trìlobum** K. Koch (three-lobed). — Mature leaf-blades of fertile branches reniform to ovate, rounded to somewhat cuneate at base, 3–10 cm. broad, with 3 short (or even obsolete) terminal small-toothed lobes, middle lobe 1–5 cm. long, upper laterals 0.5–2 (–3.5) cm. long; fruit much as in typical *A. rubrum.* (Var. *tridens* Wood; *Rufacer carolinianum* (Walt.) Small) — Swamps and low woods, Fla. to Tex., n. to N.S., N.B., N.E., s. Que., N.Y., W.Va., Tenn., s. Ill. and Mo.

Var. **Drummóndii** (H. & A.) Sarg. (for its discoverer, THOMAS DRUMMOND, 1780–1835). — Foliage as in typical *A. rubrum,* often deeper-cleft, more coriaceous, more or less permanently pubescent beneath; *samaras* 2.7–4(–5) *cm. long.* (*Rufacer Drummondii* Small) — Bottomlands, sandy woods and swamps, Fla. to Tex., n. to N.J., s. Ind., s. Ill., and Mo.

9. A. sacchàrinum L. (sugary), SILVER, WHITE, SOFT or RIVER-M., PLAINE BLANCHE (Que.). — More slender and open than no. 8, *with pendulous branchlets;* flowers greenish-yellow or -reddish, earlier than in no. 8; *petals* 0; *ovaries and young fruits white-villous; samaras* 4–8 *cm. long,* usually hairy at base, pale, *the seed-bearing body* 1–1.5 *cm. long; leaves very deeply* 5-*lobed* with rather acute sinuses, silvery-white (downy when young) beneath; the divisions narrow, cut-lobed and toothed. (*Argentacer* Small) — River-banks and bottomlands, N.B. to Ont. and Minn., s. beyond our limits. *Fl.* Feb.–early May; *fr.* April–June. — Leaves yellow in autumn. — Much planted.

§ NEGÚNDO K. Koch

10. A. Negúndo L. (aboriginal name), BOX-ELDER, ASH-LEAVED M., ÉRABLE À GIGUÈRE (Que.). — Small tree with *green,* glabrous *twigs; leaves* of fertile branches *with* 3–5(–7) petiolulate very veiny and (when young) pubescent *leaflets,* the mature pubescent to glabrate beneath; terminal leaflet elliptic to obovate, lateral narrower and coarsely few-toothed or entire; leaves of vigorous tips and sprouts with more numerous often lobed leaflets; *flowers greenish, dioecious,* slightly before the leaves, *the staminate* fascicled and *pendulous on filiform pedicels, the pistillate racemose; petals and disk* 0; *samaras* 2.5–3.5 cm. long, yellowish, *strongly ascending,* the seed prolonged. (*Negundo fraxinifolium* Nutt.) — River-banks, Fla. to Tex., n. to w. N.E., N.Y., s. Ont. and se. Minn.; much cult. and natzd. e. to Maritime Provinces and e. Que. *Fl.* April, May; *fr.* June–Oct.

Var. **violàceum** (Kirsch.) Jaeg. (violet). — Differing only in its glabrous *glaucous branchlets.* (*Negundo Nuttallii* (Nieuwl.) Rydb.) — Mich. to Mont., s. to Mo., Kans. and Colo.; cult. and estab. eastw. to N.E. and N.J.

Var. **texànum** Pax (Texan). — *Branchlets puberulent; lateral leaflets* of fertile branches lanceolate to elliptic-oblong, *entire or but slightly crenate-dentate.* — Tex. to w. N.C., n. to Okla., Mo., s. Ind. and s. O.

Var. **intèrius** (Britt.) Sarg. (of the interior). — *Branchlets puberulent; leaflets* ovate to oblong, *coarsely lobulate,* long-petiolulate. (*Negundo interius* (Britt.) Rydb.) — Man. to Mont., s. to nw. Mo., Okla., N.M. and Ariz.

FAM. 103. HIPPOCASTANÀCEAE (BUCKEYE FAMILY)

Trees or shrubs with exstipulate opposite palmate leaves, the leaflets straight-veined. Flowers polygamous, irregular, showy, in a terminal thyrse or panicle; petals 5 (*or* 4), *brightly colored or white, unguiculate. Fruit a leathery capsule,* 3-*locular and* 2- or 3-*seeded, or by abortion* 1-*seeded.* Cotyledons very thick and fleshy, their contiguous faces coherent, remaining under ground in germination; plumule 2-leaved; radicle curved. — A single characteristic genus.

1. AÈSCULUS L. HORSE-CHESTNUT. BUCKEYE

Calyx tubular, 5-lobed, often oblique or gibbous at base. Petals 4–5, more or less unequal, with claws, nearly hypogynous. Stamens 7 (rarely 6 or 8); filaments long, slender, often unequal. Style 1; ovary 3-locular, with 2 ovules in each locule. Fruit a leathery capsule, 3-locular and 3-seeded or usually by abortion 2- or 1-locular and 2- or 1-seeded, loculicidally 3-valved. Seed very large, with thick shining coat and a large round pale scar. — Trees or shrubs of N. Hemisph. Flowers often polygamous, most of them with imperfect pistils and sterile; pedicels jointed. Seeds farinaceous, but imbued with a bitter and narcotic principle. (The ancient name of some Oak or other mast-bearing tree.)

Calyx campanulate, essentially regular; petals villous-ciliate, not glandular-margined, yellow, creamy-white or reddish.

Petals subequal, all exceeded by the stamens; young fruit prickly. . . . . 1. *A. glabra.*

Petals very dissimilar, at least the upper pair about equaling or exceeding the stamens; fruit lepidote.

Calyx and pedicels glandular; petals dark yellow; longer petiolules 1–7 mm. long; large tree of the interior. . . . . 2. *A. octandra.*

Calyx and pedicels glandless; petals pale yellow or creamy, at least the upper ones purple-striped at base; longer petiolules 0.7–1.5 cm. long; small tree or shrub of Coastal Plain and outer Piedmont. . . . . 3. *A. sylvatica.*

Calyx tubular, oblique or gibbous at base; petals glandular-margined, very unequal, at least the upper exceeding the upper stamens, red or red and yellow; fruit smooth.

Leaflets lance-elliptic to lance-obovate, glabrous or promptly glabrate beneath, the mature 2–6 cm. broad. . . . . 4. *A. Pavia.*

Leaflets broadly elliptic to oblong-obovate, velvety to tomentulose beneath, the mature 4–8 cm. broad. . . . . 5. *A. discolor.*

1. A. glàbra Willd. (smooth), OHIO B. — Small tree, up to 10 m. high; leaflets 5 (rarely 6 or 7), elliptic to obovate, acuminate, cuneate at base, glabrous or quickly glabrate beneath, or permanently pubescent in forma **pállida** (K. Koch) Fern. (pale); *calyx campanulate, nearly regular; petals subequal,* upright, *shorter than the stamens, villous-ciliolate, yellowish; fruit prickly,* at least when young. — Woods and bottoms, w. Pa. to Ia. and Neb., s. to Ala., Miss., Ark. and Okla. Apr., May. Often cult.; hybridizes with nos. 2, 4 and 5.

Var. **leucodérmis** Sarg. (white-barked). — Often larger, up to 20 m. high; bark whitish; *leaflets strongly whitened beneath.* — Ind. to Ia., s. to Ark.

Var. **Sargéntii** Rehd. (in honor of CHARLES SPRAGUE SARGENT, 1841–1927). — Small tree or large shrub; *leaflets* 6 *or* 7, *lanceolate,* long-acuminate, pubescent beneath. (Var. *arguta* of ed. 7, not *A. arguta* Buckl., basonym). — O. to Ia., s. to Miss. and Okla.

2. A. octándra Marsh. (with eight stamens), YELLOW B., SWEET B. — Large shrub or tree, up to 30 m. high; *leaflets* 5, *on short petiolules only* 1–7 *mm. long,* glabrate, or permanently tomentulose beneath in the rare forma **vestìta** (Sarg.) Fern. (clothed); *calyx and pedicels glandular-villous; petals very dissimilar, at least the upper pair equaling or exceeding the stamens, dark yellow throughout; fruit prickleless, closely coated with pale brown scales.* — Woods, w. Pa. to Mich. and Ia., s. to Ga. May, early June. — Hybridizes with nos. 1 and 4.

3. A. sylvática Bartr. (of the woods). — Small tree or coarse shrub, differing from no. 2 in the leaflets more generally minutely velutinous beneath when young and on *longer petiolules* (0.7–1.5 *cm. long*); *calyx and pedicels glandless; petals pale yellow or creamy, the upper or all purple-veined at base; fruit with dark lines separating the larger appressed scales.* (*A. neglecta* Lindl.) — Rich woods and banks of streams, Coastal Plain and outer Piedmont, Fla. and Ala., n. to Meherrin R., se. Va. Late March–early May.

4. A. Pàvia L. (old generic name, in honor of PETER PAAW who died in 1617 in Leyden), RED B. — Shrub or small tree; *leaflets* 5, *lance-elliptic to lance-obovate, glabrous or promptly glabrate beneath, the mature* 2–*rarely* 6 *cm. broad; calyx tubular, strongly gibbous on one side at base,* dark red; *petals glandular-ciliate,* bright *red,* or with red tips, *strongly unequal,* the upper exceeding the adjacent stamens; fruit smooth. — Woods and thickets, Fla. to La., n. to Va. and W.Va. April. — Hybridizes with nos. 1 and 2, and with other spp. in cult., including the Horse-Chestnut.

5. A. díscolor Pursh (particolored). — Similar to no. 4; *leaflets broadly elliptic to oblong-obovate, velvety to tomentulose beneath, the mature* 4–8 *cm. broad;* calyx red, red and yellow or yellow-green. — Low or rich woods and thickets, Ga. to Tex., n. to N.C., Ky. and Mo. April, May. — Hybridizes with no. 1.

A. HIPPOCÁSTANUM L. (old generic name, literally HORSE-CHESTNUT), MARRONNIER of Que., with large resinous buds, spreading white petals red-marked at base, long-exserted stamens and large prickly fruit and large seeds, is often self-seeding from cult. trees. (Introd. from se. Eu.)

FAM. 104. SAPINDÀCEAE (SOAPBERRY FAMILY)

Trees, shrubs or rarely herbaceous climbers with exstipulate chiefly alternate and compound leaves. Flowers often polygamous, mostly unsymmetrical. Stamens commonly more numerous than the petals, rarely twice as many. Embryo curved or convolute, rarely straight; cotyledons thick and fleshy. — Large family, chiefly woody climbers in the tropics.

Trees or upright shrubs; leaves simply pinnate; flowers nearly regular; fruit a globose or 2–3-lobed berry. . . . . 1. *Sapindus.*

Herbaceous, climbing by tendrils; leaves biternate; flowers zygomorphic; fruit a bladdery capsule. . . . . 2. *Cardiospermum.*

1. SAPÍNDUS L. Soapberry

Flowers regular, polygamous. Sepals 4–5, imbricated in 2 rows. Petals 4–5, with a scale at the base. Stamens 8–10, upon the hypogynous disk. Ovary 3-locular, with an ascending ovule in each locule. Fruit a globose or 2–3-lobed berry, 1–3-seeded. Seed crustaceous, globose. — Trees or shrubs of trop. and warm reg. with alternate abruptly pinnate leaves and small flowers in terminal or axillary racemes or panicles. (Name a contraction of *sapo indicus*, *Indian soap*, having reference to the saponaceous character of the berries.)

1. **S. Drummóndi** H. & A. (for its discoverer, Thomas Drummond, 1780–1835). — Tree, 6–18 m. high; leaflets 4–9 pairs, obliquely lanceolate, sharply acuminate, entire, 3.7–7.5 cm. long; the rachis of the leaf not winged; flowers white, in a large panicle; fruit mostly globose, 1.2 cm. in diameter. — Limestone-bluffs, slopes and creek-sides, La., Tex. and Mex., n. to sw. Mo., Kans., N.M. and Ariz. June.

2. CARDIOSPÉRMUM L. Balloon-vine. Heart-seed

Flowers zygomorphic. Sepals 5 or, more commonly, by union of 2 of them, reduced to 4 in unequal pairs. Petals 4, alternating with sepals, each bearing from near the base an irregular claw-like or wing-like appendage, the appendages of the upper petals larger. Disk extrastamineal, with a gland opposite each of the upper petals. Stamens 8, deflexed. Seeds without an aril, black, with a light-colored cordate scar. — Herbaceous or slightly woody climbers of warm reg. of Am., with biternate alternate leaves and tendrils and bladdery capsules. (Greek *cardia, heart*, and *sperma, seed*.)

1. C. Halicácabum L. (old generic name, as *Halicacabus*). — Annual; leaflets cut-toothed and petiolulate; petals whitish, obovate; fruit subglobose or obovoid, 2–3.5 cm. long. — Moist thickets and waste places, Fla. to Tex., n. to N.J., Pa., O., Ill., Mo., etc. July–Sept. (Introd. from Trop. Am.)

Koelreutèria (for *Joseph Gottlieb Koelreuter*, 1733–1806) paniculàta Laxm. (panicled), Pride-of-India or China-tree, a round-topped tree with bipinnate leaves and broad panicles of yellow flowers, followed by bladdery fruits, is beginning to spread from cult. (Introd. from e. Asia)

FAM. 105. BALSAMINÀCEAE (Touch-me-not Family)

Herbs or undershrubs with bland watery juice, alternate simple exstipulate leaves, irregular flowers, and petaloid imbricated spurred calyx. Stamens 5, with short flat filaments and introrse more or less connivent anthers. Ovary 5-locular. Seeds without albumen; embryo straight. — Ours succulent annuals.

1. IMPÀTIENS L. Balsam. Jewelweed. Snapweed. Touch-me-not

Sepals apparently only 4; the anterior one notched at the apex (probably two combined); the posterior one (appearing anterior as the flower hangs on its pedicel) largest and forming a usually spurred sac. Petals 2, 2-lobed (each a pair united). Filaments appendaged with a scale on the inner side, the 5 scales connivent over the stigma; anthers introrse. Capsule with evanescent partitions and a thick axis bearing several anatropous seeds; valves 5, coiling elastically and projecting the seeds in dehiscence. — Leaves ovate or oval to lanceolate, coarsely toothed, petioled. Flowers axillary or panicled, often of two sorts, *viz.*, the larger ones which seldom ripen seeds; and very small cleistogamous ones which are fertilized early in the bud, their floral envelopes never expanding but forced off by the growing capsule and carried upward on its apex. Large genus, mostly of trop. and subtrop. Asia and Afr., a few in temp. Eurasia and Am. (Name the Latin for *impatient*, from the sudden bursting of the capsules when touched, whence also the popular appellation.)

Leaves alternate, elliptic, crenate or crenate-serrate; flowers orange, yellow or creamy, with or without deeper mottling; capsules slenderly clavate.
  Spur of saccate sepal strongly curved; flower, excluding spur, 1.5–3 cm. long; leaves crenate, bluntish, the peduncles in axils of the scattered upper ones.
    Flower pale yellow; spur bent at right angles, one-fifth to one-fourth as long as the sac, the latter broader than long. . . . 1. *I. pallida.*
    Flower orange to reddish or whitish; the spur gradually bent parallel with the sac and one-third to one-half its length, sac longer than broad. . 2. *I. capensis.*
  Spur directed straight back; flower, excluding spur, less than 1.5 cm. long;

leaves crenate-serrate, acuminate, crowded with the long and overtopping peduncles at summits of stem and branches. . . . . 3. *I. parviflora.*

Leaves opposite, elongate, lanceolate or oblanceolate, finely serrate; flowers usually purple, numerous, on terminal and subterminal long peduncles; capsules clavate-obovoid. . . . . 4. *I. glandulifera.*

1. I. pállida Nutt. (pale), PALE TOUCH-ME-NOT or SNAPWEED. — Stems freely branching, 0.5–2.5 m. high, glaucous or pale green; leaves glaucous or blue-green, elliptical, crenate, mostly exceeding their petioles; inflorescences few-flowered from the scattered upper axils, their bracts lanceolate; *flower canary-yellow* and sparingly spotted with brownish-red or unspotted, or in forma **speciòsa** Jennings (showy) creamy-white, or in forma **dichròma** Steyerm. (two-colored) with upper and lateral petals white but the posterior ones and the saccate sepal yellow; the *saccate sepal dilated and very obtuse, broader than long, with the short spur (one-fifth to one-fourth its length) bent at right angles.* — Wet or springy places, often in shade and chiefly in calcareous areas, sw. Nfld. to Sask., s. to N.S., N.E., Ga., Tenn., Mo. and Kans. July–Sept.

2. I. capénsis Meerb. (of the Cape; wrongly thought by its author to have been introduced into European gardens from the Cape of Good Hope), SPOTTED TOUCH-ME-NOT or SNAPWEED, LADY'S-EARRINGS, "CELANDINE" or "SOLENTINE". — Similar, less glaucous; leaves usually smaller; bracts of inflorescence linear-subulate; *flower orange,* with crimson spots or variously colored (see forms below), *the saccate sepal longer than broad, its spur one-third to half its length and bent back parallel with it;* small, immature or poorly developed plants producing minute cleistogamous flowers. (*I. biflora* Walt.; *I. fulva* Nutt.) — Wet or springy places, even in acid or subacid swamps, Nfld. to Alaska, s. to Fla., Ala., Ark. and Okla. June–Sept. — Including *I. Nortonii* Rydb., with unusually prolonged saccate sepal, an extreme of Mo., Kans. and Okla. needing further observation. — The following, among the very numerous flower-variations, have been named: forma **immaculàta** (Weath.) Fern. & Schub. (unspotted), flower orange and unspotted or only slightly spotted at throat; forma **citrìna** (Weath.) Fern. & Schub. (lemon-colored), flower lemon-yellow with crimson spots; forma **albiflòra** (Rand & Redf.) Fern. & Schub. (white-flowered), flower whitish or cream-color with scattered pink or brownish-red spots; forma **Peàsei** (A. H. Moore) Fern. & Schub. (for ARTHUR STANLEY PEASE, 1881–), flower cream-color but appearing pink from the heavy and coalescent pink spots; forma **platýmeris** (Weath.) Fern. & Schub. (broad-lobed), flower orange and spotted, with the basal lobe of the petals as large as and overlapping the terminal lobe (much smaller in other forms).

3. I. PARVIFLÒRA DC. (small-flowered). — Branching annual up to 1 m. high; *leaves* elliptic, *acuminate, crenate-serrate, becoming crowded at the summits of stems and branches and subtending long ascending peduncles; flower small,* 6–12 *mm. long, without the long straight spur, lemon-yellow.* — Shaded waste ground and barnyards, P.E.I. and locally in Que. July–Sept. (Adv. from Eurasia)

4. I. GLANDULÍFERA Royle (bearing glands). — Tall and coarse, up to 2 m. high; *leaves opposite or whorled, elongate-lanceolate to oblanceolate, long-acuminate, sharply and copiously serrate; peduncles elongate,* many-flowered, *subumbellate* from the crowded terminal axils; *flowers magenta, red or purple,* or in forma PALLIDIFLÒRA (Hook. f.) Weath. (pale-flowered) pale pink with brownish or reddish spots, 3–4 *cm. long,* with broad petals and short recurving spur; capsule obovoid. — Waste grounds and roadside-thickets, N.S., s. N.B., n. N.E., Que. and Ont., becoming occasional, esc. from cult. Aug., Sept. (Introd. from Asia)

I. BALSÁMINA L. (old generic name), the garden BALSAM, with simple pubescent stem, elongate slender-toothed alternate leaves much overtopping the variously colored large axillary flowers with broad petals, and with pubescent capsules splitting from the apex, esc. from cult. (Introd. from Asia)

FAM. 106. RHAMNÀCEAE (BUCKTHORN FAMILY)

Shrubs or small trees with simple leaves, small and regular flowers (sometimes apetalous), with the 4 or 5 perigynous stamens as many as the valvate sepals and alternate with them, accordingly opposite the petals! Drupe or capsule with only one erect seed in each locule, not arillate. Petals folded inwards in the bud, hooded or concave, inserted with the stamens into the edge of the fleshy disk which lines the short tube of the calyx and sometimes unites it to the lower part of the 2–5-locular ovary. Ovules solitary, anatropous. Stigmas 2–5. Embryo large, with broad cotyledons, in sparse fleshy albumen. Flowers often polygamous, sometimes dioecious. Leaves mostly alternate; stipules small or obsolete. Branches often thorny. — Slightly bitter and astringent; the fruit often mucilaginous, commonly rather nauseous or cathartic.

Calyx and disk free from the ovary; flowers greenish or greenish-white; fruit a drupe.
- Petals sessile, entire, as long as calyx-lobes; drupe with thin flesh and a 2-locular bony putamen; high-twining vine. . . . . 1. *Berchemia.*
- Petals clawed, notched at apex, or none; drupe juicy and berry-like, with 2–4 separate nutlets; non-twining trees and shrubs. . . . . 2. *Rhamnus.*

Calyx with disk adherent to base of ovary; flowers white or lilac; fruit dry, 3-lobed, splitting into 3 carpels; shrubs. . . . . 3. *Ceanothus.*

1. BERCHÈMIA Neck. Supple-Jack. Rattan-vine

Calyx with a very short and roundish tube; its lobes equaling the 5 oblong sessile acute petals, longer than the stamens. Disk very thick and flat, filling the calyx-tube and covering the ovary. Drupe ellipsoid, with thin flesh and a bony 2-locular putamen. — Woody high-climbing twiners of se. N.Am., s. and e. Asia and e. Afr. with the pinnate veins of the leaves straight and parallel, the small greenish-white flowers in small panicles. (Presumably named for *Berthout van Berchem*, 18th century Dutch botanist.)

1. B. scándens (Hill) K. Koch (climbing). — Glabrous; leaves oblong-ovate, acute, scarcely serrulate; drupe blue, ellipsoid, about 8 mm. long, the short style deciduous. — Climbing high over trees in rich or low woods, Fla. to Tex., n. to Va., Tenn. and s. Mo. May.

2. RHÁMNUS L. Buckthorn

Calyx 4–5-cleft; the tube campanulate, lined by the disk. Petals small, short-clawed, notched at the end, wrapped around the short stamens or sometimes none. Ovary free, 2–4-locular. Drupe berry-like (black), containing 2–4 separate seed-like nutlets of cartilaginous texture. — Shrubs or small trees, mostly of temp. reg., with loosely pinnate-veined leaves, and greenish perfect, polygamous or dioecious flowers in axillary or supra-axillary cymes. (The ancient Greek name.)

Flowers dioecious or polygamodioecious, expanding essentially with the leaves, without a common peduncle; nutlets and seeds often grooved on the back; raphe dorsal; winter-buds scaly. § Eurhamnus.
- Calyx-lobes and stamens 5; petals wanting; seed flat, scarcely grooved on back; low shrub. . . . . 1. *R. alnifolia.*
- Calyx-lobes, petals and stamens 4; seed plump, deeply grooved down the back; tall shrubs or trees.
  - Long shoots ending in spines; branches often spine-like, glabrous; leaves elliptic or subovate, glabrous; drupe 3–4-seeded, the groove of the seed deep and narrow. . . . . 2. *R. cathartica.*
  - Long shoots and branches without spines, often puberulent; leaves ovate-lanceolate, more or less pubescent beneath; drupe 2-seeded, the groove broad and open. . . . . 3. *R. lanceolata.*

Flowers perfect, appearing after the leaves; seeds notched at base but not furrowed on back; raphe lateral; winter-buds naked, hairy. § Frangula.
- Umbels of flowers and fruits peduncled; pedicels pubescent. . . . . . 4. *R. caroliniana.*
- Umbels sessile; pedicels glabrous or glabrate. . . . . 5. *R. Frangula.*

§ Eurhámnus Griseb.

1. R. alnifòlia L'Hér. (alder-leaved). — *Low shrub*, 1.5–8 dm. high, with simple or slightly forking upright gray branches; leaves oval, serrate, nearly straight-veined; *flowers apetalous*, polygamodioecious, the *calyx-lobes and stamens* 5; drupes 3-seeded; the *seeds flat, scarcely grooved on the back.* — Swamps, low woods and meadows (often calcareous), Nfld. to B.C., s. to N.S., n. and w. N.E., n. N.J., Pa., W.Va., O., n. Ind., n. Ill., Minn., Neb., Wyo. and Calif. May–July.

2. R. cathártica L. (cathartic), Common B., Nerprun (Que.). — Small tree or coarse shrub, *with spine-tipped long shoots and branches*, the latter glabrous; *leaves elliptic or subovate, glabrous;* flowers polygamodioecious, the *calyx-lobes, petals and stamens* 4; *drupe* 3–4-*seeded,* the plump *seeds with a deep and narrow dorsal groove.* — Open woods, pastures and fence-rows, early introd. for hedges, quickly scattered by birds and now often appearing as if native eastw., Que. to Minn., s. to N.S., N.E., Va., O., Ill. and Mo. May, June. (Introd. and natzd. from Eu.)

3. R. lanceolàta Pursh (lanceolate). — Tall unarmed shrub; *leaves oblong-lanceolate* and acute, or on flowering shoots oblong and obtuse, finely serrulate, minutely downy beneath; the yellowish-green flowers of two forms on distinct plants: both perfect, one with short pedicels clustered and with a short included style, the other with pedicels oftener solitary, style ex-

serted; petals deeply notched; ***fruit 2-seeded, seeds with broad open groove.*** **— Thickets and borders of woods, Ala. to Tex., n. to s.-centr. Pa., W.Va., s. O., Ind., s. Wisc., s. Ia. and Neb. May.**

Var. glabràta Gleason (glabrate). — Leaves glabrous or promptly glabrate beneath. — Moist woods, Ky. to Neb., s. to Tenn. and Ark.

§ Frángula S. F. Gray

4. R. caroliniàna Walt. (of Carolina), Carolina B., Indian Cherry. — Shrub or small tree up to 11 m. high, without thorns; ***young branchlets puberulent; leaves*** **elliptic or oblong, obscurely serrulate,** ***the mature*** **4–15** ***cm. long, usually slightly pubescent but glabrate beneath; flowers perfect,*** **5-merous,** ***some*** **solitary, others** ***in short-peduncled umbels; drupe 3-seeded,*** **the** ***seeds not grooved on the back.*** **— Rich woods and sheltered slopes, Fla. to Tex., n. to w. Va., W.Va., s. O., s. Ind., s. Ill., Mo. and Neb. May, June.**

Var. móllis Fern. (soft). — Leaves permanently velvety beneath. — S. Ind. to Tenn., Mo. and Tex.

5. R. Frángula L. (old generic name), Alder-B. — Shrub or small tree up to 6 m. high; young branchlets pubescent; ***leaves*** **short-oblong to obovate,** ***the mature 3–7 cm. long; flowers in*** **small** ***sessile umbels, with glabrous pedicels,*** **perfect, 5-merous;** ***drupe 2-seeded;*** **seeds not furrowed. — Fence-rows and thickets, local, s. Que., Ont. and Minn., s. to N.S., N.E., N.J., O., Ind. and Ill. May–July. (Natzd. from Eu.) — Recently and very rapidly spreading; likely to become obnoxious.**

3. CEANÒTHUS L. Redroot

Calyx 5-lobed, incurved; the lower part cohering with the thick disk to the ovary, the upper separating across in fruit. Petals hooded, spreading, on slender claws longer than the calyx. Filaments elongated. Fruit 3-lobed, dry and splitting into its 3 carpels when ripe. — Shrubby N. Am. plants; flowers in small umbel-like clusters, forming dense panicles or corymbs at the summit of naked flowering branches or peduncles; calyx and pedicels colored like the petals. (An obscure name used by Theophrastus, surely not for this genus.)

Peduncles developed from old buds chiefly below the leafy shoots of the season; large shrub up to 4 m. high. . . . . 1. *C. sanguineus.*
Peduncles borne on leafy shoots of the season.
Leaves ovate or ovate-oblong, acute or acuminate (rarely blunt); peduncles elongate, mostly axillary, naked or few-bracted at summit. . . . . . 2. *C. americanus.*
Leaves narrowly elliptic or elliptic-lanceolate, obtuse to barely acute; peduncles short, mostly terminating regular leafy branches. . . . . . 3. *C. ovatus.*

1. C. sanguíneus Pursh (blood-red), Wild-lilac. — ***Large shrub,*** **up to 4 m. high;** ***branchlets glabrous; leaves elliptic, ovate, obovate or suborbicular, rounded at tip,*** **rounded or subcordate at base, crenate-serrate, more or less pubescent when young,** ***on petioles*** **0.7–2** ***cm. long; peduncles borne from buds of the preceding year, mostly below the leafy shoots of the season;*** **the** ***panicle elongate, with scattered corymbs;*** **capsule obovoid, nearly smooth and crestless. — Dry rocky crests, bluffs and borders of woods, Keweenaw Co., Mich.; Black Hills, S.D.; s. B.C. to nw. Mont. and n. Calif. June.**

2. C. americànus L. (American), New Jersey Tea.— Low shrub, up to 1 m. high; the ***flowering branches herbaceous, new each year,*** **erect;** ***leaves ovate or ovate-oblong, acuminate or acute,*** **3-ribbed, serrate, cordate or rounded at base, green and glabrous above, gray and more or less** ***pilose beneath, mostly*** **5–10** ***cm. long and*** **2.5–6** ***cm. broad; peduncles axillary, naked or with few subterminal bracts,*** **floriferous only toward the summit, usually 2–10 on a branch;** ***flowering thyrse*** **1.5–4** ***cm. in diameter, in fruit*** **2.5–5** ***cm. in diameter.*** **— Dry open woods and gravelly or rocky banks, Fla. to Ala., n. to centr. Me., s. Que., s. Ont. and s. Man. Late May–Sept. — Leaves considered one of best substitutes for tea during the Am. Revolution.**

Var. intermèdius (Pursh) K. Koch (intermediate). — More intricately and slenderly branched, the upper branches more ligneous and smoother; ***leaves ovate-oblong or lance-ovate,*** **chiefly 2–6** ***cm. long and*** **1–3.5** ***cm. broad;*** **peduncles filiform, tending to cluster from the upper axils;** ***flowering thyrse*** **1–2.5** ***cm. in diameter, in fruit*** **1.5–3** ***cm. thick.*** **(*C. intermedius* Pursh) — Sandy pinelands and openings, Fla. to La., n., locally, to Cape Cod, Mass., Cecil Co., Md., W.Va., s. Ill. and Mo.**

Var. Pítcheri T. & G. (for its discoverer, Zina Pitcher, 1797–1872). — ***Leaves*** **broadly oblong-ovate or oval,** ***blunt or round-tipped, pilose above,*** **velvety beneath; the branches densely villous. — N. Ga. to Tex., n. to Ind., Ill., Ia. and Kans.**

3. C. ovàtus Desf. (ovate). — Low shrub up to 1 m. high; the ligneous upright branches slender, puberulent or pilose; *leaves narrowly elliptic or elliptic-lanceolate, blunt to subacute,* crenate-serrate (*the young teeth gland-tipped*), *those of the fertile branchlets* 2.5–6 *cm. long and* 1–2.5 *cm. broad,* sparingly pilose beneath, or branchlets and lower leaf-surfaces sordid-villous in forma **pubéscens** (S. Wats.) Soper (pubescent) or *C. pubescens* (S. Wats.) Rydb.; *peduncles borne chiefly at tips of regularly leafy branchlets,* the thyrse short and corymbiform; capsules without crests. — Sandy or rocky plains, prairies and slopes, w. Me. and w. Que. to Man., s. to Mass., w. Ga., Ala., Ark. and Tex., local eastw. Late Apr.–early July.

FAM. 107. VITÀCEAE (VINE FAMILY)

Shrubs with watery acid juice, usually climbing by tendrils, with small regular greenish commonly polygamous flowers, a minute or truncated calyx, its limb mostly obsolete, and the stamens as many as the valvate petals and opposite them. Berry 2-locular, usually 4-seeded. **Petals 4–5, very deciduous, hypogynous or perigynous. Filaments slender; anthers introrse. Style short or none; stigma slightly 2-lobed; ovary 2-locular, with 2 erect anatropous ovules from the base of each locule. Seeds bony, with a minute embryo at the base of the hard albumen. Stipules deciduous. Leaves alternate, palmately veined or compound; tendrils and inflorescences opposite the leaves.**

Bark close, covered with lenticels; pith white; inflorescence a dichotomous or umbelliform cyme; petals expanding, free from one another, dropping singly; seeds trigonous.
- Leaves simple or pinnately compound; tendrils all with slender tips; ovary surrounded by a free nectariferous or glanduliferous disk.
  - Cyme dichotomous; flowers 5-merous; disk with entire or crenulate margin. . . 1. *Ampelopsis.*
  - Cyme umbelliform; flowers 4-merous; disk deeply 4-lobed. . . . . 2. *Cissus.*
- Leaves (or some of them) digitate; some or all tendrils with adhesive disks; ovary adnate to obscure disk. . . . . 3. *Parthenocissus.*

Bark mostly loosening and freely exfoliating in ropy shreds, without lenticels; pith brown; inflorescence paniculate; petals separating only at base, falling without expanding; seeds mostly pyriform. . . . . 4. *Vitis.*

1. AMPELÓPSIS Michx.

Flowers mostly 5-merous and perfect. Petals expanding, free. Disk cup-shaped, free from ovary except at base, the margin entire or barely crenate. Berry dry or with scant pulp. Seeds trigonous-obovoid. Tendrils few, mostly in the inflorescence. — Climbing or erect N.Am. and Asiatic shrubs with close bark bearing lenticels, the pith white; leaves thin, deciduous; inflorescence a dichotomous cyme. (Name from the Greek *ampelos, the vine,* and *opsis, appearance.*)

1. A. cordàta Michx. (heart-shaped). — Nearly glabrous; *leaves heart-shaped* or truncate at the base; coarsely and sharply toothed, acuminate, *not lobed;* panicle small and loose; style slender; berries of the size of a pea, 1–3-seeded, bluish or greenish. (*Cissus Ampelopsis* Pers.) — Rich woods and bottoms, Fla. to Tex. and Mex., n. to Va., s. O., s. Ind., s. Ill. and se. Neb.; cult. and sometimes natzd. n. to Mass. May–July.

2. A. BREVIPEDUNCULÀTA (Maxim.) Trautv. (short-peduncled). — Similar to no. 1 but *pubescent;* the longer-acuminate *leaves 3-lobed;* berries lilac, finally becoming bright blue. (*A. heterophylla* Sieb. & Zucc.) — Thickets and shores, N.E. to O., s. beyond our limits. (Introd. from Asia)

3. A. arbòrea (L.) Koehne (of trees; from its high-climbing habit), PEPPER-VINE. — Nearly glabrous, high-climbing or bushy and rather upright; *leaves twice pinnate or ternate,* the leaflets cut-toothed; flowers cymose; calyx 5-toothed; disk very thick, adherent to the ovary; berries black, obovoid. (*Cissus* Des Moulins) — Swampy woods, Fla. to Tex., n. to e. Md., s. Ill., Mo. and Okla. June–Aug.

2. CÍSSUS L. POSSUM-GRAPE

Flowers 4-merous and perfect or polygamomonoecious. Petals expanding, free. Disk a deeply 4-lobed cup free from ovary except at base. Berry dry, 1–2-seeded. Seeds trigonous-obovoid. — Fleshy mostly tuberous-rooted climbing shrubs (or herbs) of trop. to warm-temp. reg., with close bark with lenticels and white pith; leaves fleshy, deciduous or evergreen; inflorescence an umbelliform cyme. (*Cissos,* Greek name of the Ivy.)

1. C. incìsa (Nutt.) Des Moulins (sharply cut), MARINE-VINE. — A stout vine, with somewhat *succulent* deeply *3-parted or pinnately 3-foliolate leaves*, the leaflets ovate or obovate, cuneate, coarsely and irregularly toothed; inflorescence suggesting a compound umbel. — Rocky or sandy open woods and bluffs, Fla. to Tex., n. to Mo. and se. Kans. May–July.

3. PARTHENOCÍSSUS Planch. VIRGINIA CREEPER. WOODBINE. VIGNE-VIERGE (Que.)

Calyx slightly 5-toothed. Petals concave, thick, expanding before they fall. Disk none. — N.Am. and Asiatic woody climbers, with digitate or palmately lobed leaves; leaflets 3–7, rather coarsely serrate, or blade cordate and merely lobed. Inflorescence cymosely compound. Tendrils branched, their tips twining or affixing themselves by enlarged terminal adhesive disks. (Name from the Greek *parthenos, virgin,* and *cissos, ivy;* based on the French name *Vigne-Vierge* or the English *Virginia Creeper.*) PSEDERA Neck.

Leaves all with 5(3–7) distinct leaflets.
  Tendrils terminating in adhesive disks; fresh foliage opaque above; inflorescence usually of 25–200 or more flowers in panicled groups of cymes, the central axis of the group prolonged, the lower divergent to ascending branch solitary; fruit 5–7 mm. in diameter. . . . . 1. *P. quinquefolia.*
  Tendrils without adhesive disks; fresh foliage lustrous above; inflorescence usually 10–60-flowered, without elongate central axis, usually with a pair of subequal divergent branches; fruit 8–10 mm. in diameter. . . . . . 2. *P. inserta.*
Leaves of basal sprouts mostly 3-foliolate, of fertile branches merely 3-lobed. . 3. *P. tricuspidata.*

1. P. quinquefòlia (L.) Planch. (five-leaved). — *Climbing by means of adhesive disks* of the 3–8-branched tendrils; *leaflets* 0.4–1.5 dm. long, distinctly petiolulate, oblong-obovate to elliptic, acuminate, coarsely toothed, *when fresh pale green and dull above,* glaucescent beneath, glabrous, or pubescent in forma **hirsùta** (Donn) Fern. (stiffly hairy); *cymes usually approximate and with prolonged central axis, forming a paniculate inflorescence with solitary ascending to divergent lower branch and* 25–200 *or more flowers; fruits* 5–7 *mm. in diameter,* 1–3-seeded. — Woods and rocky banks, Fla. to Tex. and Mex., n. to se. Me., s. N.H., Vt., sw. Que., N.Y., Ind., Ill. and Minn.; spread from cult. elsewhere. June–Aug. — Many minor horticultural forms have been named.

2. P. insérta (Kerner) K. Fritsch (inserted). — Loosely climbing, leaning or trailing; *tendrils with* 3–5 *slender-tipped twining branches without disks;* fresh foliage lustrous above, green beneath, glabrous, or in forma **dùbia** Rehd. (doubtful) pubescent; the leaflets mostly 5–12 cm. long, or in forma **macrophýlla** (Lauche) Rehd. (large-leaved) up to 2 dm. long and 1 dm. broad; *inflorescences* mostly scattered (not crowded) in elongate panicles or solitary, *without a prolonged central axis, consisting of a pair of subequal divergent few-flowered branches; fruit* 8–10 *mm. long;* seeds 3–4, larger. (*P. vitacea* (Knerr) Hitchc.) — Woods, thickets and banks, Que. to Man. and Mont., s. to N.S., N.E., Pa., O., Ind., Ill., Mo., Kans., N.M., Ariz. and Calif. June, July.

3. P. TRICUSPIDÀTA (Sieb. & Zucc.) Planch. (with three cusps), BOSTON IVY. — Extensively branching and high-climbing, clinging to supports by adhesive disks of the abundant tendrils; *leaves* of basal trailing offshoots 3-foliolate, *of the fruiting stems 3-lobed, cordate,* lustrous; fruit with a bloom. — Locally esc. from cult., Mass. to O. (Introd. from Asia)

4. VÌTIS L. GRAPE. RAISIN OR VIGNE (Que.)

Flowers polygamodioecious (some plants with perfect flowers, others staminate and with at most a rudimentary ovary), 5-merous. Calyx very short, usually with a nearly entire border or none at all. Petals separating only at base and falling off without expanding. Hypogynous disk of 5 nectariferous glands alternate with the stamens. Berry pulpy. Seeds mostly pyriform, with beak-like base. — Plants mostly climbing by the coiling of naked-tipped tendrils; mostly in temp. reg. of N. Hemisph. Flowers in a compound thyrse, very fragrant; pedicels mostly umbellate-clustered. Leaves simple, mostly rounded and cordate. (The classical Latin name.)

a. Bark of main stem and of branches (after 1st year) shreddy and exfoliating, without distinct lenticels; pith interrupted by diaphragms within the nodes; tendrils, when present, forked; seeds pyriform. § EUVITIS. *b.*
b. Leaves covered beneath with a permanent and continuous rusty felt, the lower leaf-surface not exposed in age; a tendril or inflorescence borne at each of 3–7 or more successive nodes. . . . . 1. *V. Labrusca.*

*b.*Leaves with thin or flocculent loosening and finally deciduous pubescence, or glabrate or glabrous; tendrils or inflorescences intermittent, none opposite each third leaf (except sometimes in nos. 6 and 9). . *c.*

*c.*Leaves toward and at maturity strongly whitened, cinereous or rufescent and in young condition often tomentose beneath.

Grapes 1–2.5 cm. in diameter; seeds 7–8 mm. long, 5–6 mm. broad; pubescence of lower leaf-surface a thin scarcely flocculent web; southwestern species. . . . . 2. *V. Lincecumii.*

Grapes 4–12 mm. in diameter; seeds 4–7 mm. long, 3–4 mm. broad; pubescence of lower leaf-surface flocculent or deciduous; southern, central and eastern species.

Young sprouts and tips terete, glabrous or soon glabrate; leaf-blades of fruiting branches 0.7–2 dm. broad, suborbicular in general outline, shallowly to very deeply lobed. . . . . 3. *V. aestivalis.*

Young sprouts and tips angled, permanently close-pubescent; leaf-blades of fruiting branches 0.6–1.5 dm. broad, ovate, with prolonged triangular summit, unlobed or with 2 short shoulders. . . 4. *V. cinerea.*

*c.*Leaves toward and at maturity green or at most gray-green and glabrous or merely short-pilose or hirtellous beneath, floccose pubescence, if any, soon disappearing or persisting only in axils of veins beneath. . . *d.*

*d.*Bark promptly exfoliating; leaf-blades of fertile branches rounded- to prolonged-ovate, with tapering tips; the mature blades as long as or longer than broad, 0.4–1.8 dm. broad; tendrils well developed on fruiting branches, the plants climbing; mature thyrse 3–15 cm. long; grapes acid, or sweet when fully ripe. *e.*

*e.*Blades of most leaves on fertile branches with porrect and prolonged tapering lobes 1–4 cm. long.

Diaphragms (dividing pith) 4–5 mm. thick; new branchlets bright red when fresh; leaf-blades scarcely ciliolate, those of fruiting branches mostly 4–9 cm. broad; rachis of thyrse copiously hirtellous; grapes 5–8 mm. in diameter, black, without bloom. . . 5. *V. palmata.*

Diaphragms 0.8–2 mm. thick; new branchlets green or dull red; leaf-blades ciliolate, those of fruiting branches mostly 7–15 cm. broad; rachis of thyrse glabrous or nearly so; grapes 8–12 mm. in diameter, with heavy bloom. . . . . 6. *V. riparia.*

*e.*Blades on fertile branches unlobed or merely with short shoulder-like lobes. . . *f.*

*f.*New branchlets permanently close-tomentulose. . . . . 4. *V. cinerea.*

*f.*New branchlets glabrous or promptly glabrate. . . *g.*

*g.*Expanding leaves glabrous or only lightly pilose; inflorescences or tendrils intermittent, none opposite each third leaf; axis of thyrse not lanate; grapes 3–9 mm. in diameter, black, without or barely with bloom, sweet when ripe; seeds 4–5 mm. long.

Petioles and lower faces of leaves glabrous or promptly glabrate; branchlets terete. . . . . 7. *V. vulpina.*

Petioles and veins of lower surfaces of leaves permanently pilose; branchlets angled. . . . . 8. *V. Baileyana.*

*g.*Expanding leaves densely felted, soon glabrate; an inflorescence or tendril usually borne in 3–5 or more successive axils; axis of thyrse floccose-lanate; grapes 1.2–1.7 cm. in diameter, purple-black, with bloom, sharply acid; seeds 6.5–8 mm. long. . . 9. *V. novae-angliae.*

*d.*Bark tardily exfoliating; leaf-blades of fertile branches reniform to deltoid-ovate, abruptly slender-tipped to subtruncate; the mature blades broader than long, 3–10 cm. broad; tendrils wanting or only at tips of fertile branches, the plants shrubby or sprawling; mature thyrse 2–7 cm. long; grapes sweet.

Young branchlets and leaves glabrous or only sparsely pilose and promptly glabrate; leaves of fruiting branches reniform; seeds 4–5 mm. long. . . . . 10. *V. rupestris.*

Young branchlets and leaves arachnoid-pilose or tomentose, the deltoid-ovate to suborbicular blades permanently somewhat pilose beneath; seeds 5–6 mm. long. . . . . 11. *V. acerifolia.*

*a.*Bark of branches and main stem close, not exfoliating, abundantly dotted with lenticels; pith continuous, without separating diaphragms within the nodes; tendrils simple; leaves orbicular to deltoid-ovate, firm, lustrous; fruit thick-skinned, musky and sweet when ripe; seeds oblong. § MUSCADINA. . . . 12. *V. rotundifolia.*

§ Euvìtis Planch.

1. V. Labrúsca L. (early Latin name of the wild vine), Fox-G. — High-climbing, coarse; *branchlets and young leaves densely felted with whitish to rusty or reddish tomentum, the latter persisting on the lower leaf-surfaces;* leaves thick, strongly veined, those of the fertile portion of the flowering branches quadrate-orbicular or -ovate, with more or less forward-pointing lobes or shoulders, in maturity 0.5–1.8 dm. broad, the margin coarsely dentate, covered beneath with a blanket of tomentum, of which the separate hairs are visible under a lens; an *inflorescence or tendril at each of* 3–7 (*or more*) *successive nodes;* thyrse compact, with flocculent-tomentose axis and mostly 20 or fewer quickly dropping purple or purple-black, or amber-white to russet or pinkish in forma **álba** (Prince) Fern. (white), finally sweetish *grapes* 1.5–2.5 *cm. in diameter;* seeds 5–8 mm. long, 4–6 mm. broad. — Wet or dry thickets and borders of woods, s. Me. to s. Mich., s. to s. N.E., Pa., upland to Ky. and Tenn., more rarely on Coastal Plain to Ga. *Fl.* mid-May–early July; *fr.* Sept., Oct. — Source of the Concord, Champion, Chautauqua, Moore's Early and other popular cult. strains and, by crossing, many scores more; these, constituting V. labruscàna Bailey, spread from cult.

Var. **subedentàta** Fern. (almost without teeth). — Leaves of fertile portions very densely felted with close (scarcely separable) shining tomentum, the margins with poorly developed shoulders and very low teeth. — Coastal Plain and Piedmont, S.C. to se. N.Y.

2. V. Lincecúmii Buckl. (for its discoverer, Gideon Lincecum, 1793–1874), Post-oak-G. — Resembling no. 3; *diaphragms* dividing pith 2–3 *mm. thick;* leaves of fertile branches quadrate-rotund to ovate, as broad as long, 1–1.5 dm. broad, obscurely shouldered or shallowly to deeply 3–5-lobed; *tomentum of lower surface forming a loose scarcely flocculent thin* cinereous to rufescent *mesh;* petiole tomentose; *thyrse in fruit becoming* 1–2.5 *dm. long,* the rachis tomentose; *grapes* 1–2.5 *cm. in diameter,* purple-black, with slight bloom; *seeds* 7–8 *mm. long,* 5–6 *mm. broad.* — Woods, thickets and glades, Miss. to Tex., n. to s. Ind. and Mo. *Fl.* May, June; *fr.* July–early Sept.

Var. **glaùca** Munson (blue-green). — Petiole glabrous or glabrate; lower leaf-surface pale and glabrate; fruit more glaucous. — Mo., se. Kans. and Ark. to Tex.

3. V. aestivàlis Michx. (of summer), Summer- or Pigeon-G. — High-climbing; *young branchlets and petioles with rusty or reddish persistent to flocculent-deciduous tomentum or velutinous pilosity; diaphragms* dividing pith 3–4 *mm. thick;* leaves of fertile branches rounded-ovate, cordate, slightly longer than broad, unlobed or merely shouldered to deeply 3–5-lobed, 0.7–2 dm. broad, *lower surface with subpersistent but loose and flocculent tomentum, the prominent ribs tomentose to velutinous-hispid;* thyrse in maturity 0.5–1.8 dm. long; *grapes* 5–12 *mm. in diameter,* black, with a thin bloom, persistent; *seeds* 5–7 *mm. long,* 4–5 *mm. broad.* — Dry woods and thickets, Ga. to Tex., n. to Mass. and N.Y. (rare) and to O., Mich. and Wisc. *Fl.* late May–early July; *fr.* Sept., Oct. — Passing freely into

Var. **argentifòlia** (Munson) Fern. (silvery-leaved), Summer- or Silverleaf-G. — *Branchlets and petioles glabrous or glabrate,* often glaucous; *leaves glaucous or strongly whitened beneath, soon glabrate* or with merely the ribs beneath short-hirtellous. (*V. bicolor* of ed. 7, perhaps not LeConte; *V. argentifolia* Munson; *V. Lecontiana* House) — Similar habitats, often ascending to higher alts., N.H. to s. Minn., s. to se. Va., Ala., Tenn., Mo. and Kans.

4. V. cinèrea Engelm. (ashy), Graybark- or Pigeon-G. — Lax high climber; *growing tips and branchlets angled, permanently close-pubescent with ashy-white or gray hairs; leaves of fertile branches* 0.6–1.5 *dm. broad, ovate, with prolonged and tapering tips, either unlobed or with short shoulders* or, more rarely, with 2 or 4 prolonged lobes; upper surface floccose, becoming glabrate; *lower surface canescent-pilose or grayish-floccose,* the looser hairs somewhat deciduous; *petioles canescent; thyrse canescent or gray-floccose,* 6–15 cm. long, rather open; grapes 4–9 mm. in diameter, blackish, with slight bloom, finally sweet; seeds 4–5 mm. long. (Incl. var. *canescens* Bailey) — Rich low thickets, bottoms and banks of streams, Fla. to e. Tex., n. to se. Va., s. O., Ind., Ill., Ia. and Neb. *Fl.* June, early July; *fr.* Sept., Oct. — Passing freely into

Var. **floridàna** Munson (of Florida), Pigeon-G. — *Pubescence rufescent or rusty;* grapes black, acid. (*V. Simpsoni* Munson; *V. austrina* Small) — Similar habitats, more common eastw., Fla. to Tex., n. to e. Va. and Ark. *Fl.* June; *fr.* Sept., Oct.

5. V. palmàta Vahl (palmate-leaved), Red, Cat- or Catbird-G. — Slender, high-climbing; shoots and flowering *branchlets* herbaceous, *bright red* when fresh; *diaphragms* 4–5 *mm. thick; leaves of fertile branches* ovate, *long-acuminate, with two prolonged acuminate erect lobes,* with broad open sinus, glabrous, *mostly* 4–9 *cm. broad, the margins without ciliation;* leaves of vegetative shoots larger and with 3–5 long acuminate lobes with broad rounded sinuses; *thyrse* 0.5–1.5 dm. long, rather open, *its rachis and branches copiously hirtellous; grapes* 5–8 mm. in diameter, *black, without bloom,* finally sweet; seeds 4.5–5 mm. long. (*V. rubra* Michx.) —

Margins of ponds or sloughs or in low woods, La. and Tex., n. to s. Ind., Ill. and se. Ia. *Fl.* mid-June, July; *fr.* Oct.

6. V. ripària Michx. (of river-banks), RIVER-BANK- or FROST-G., RAISIN SAUVAGE (Que.). — High-climbing; *new branchlets green or dull red,* glabrous or pubescent and glabrate; *diaphragms* 0.8–2 *mm. thick; leaves of fertile branches mostly* 7–15 *cm. broad, ciliolate-margined,* glabrous or glabrate, with glabrous or glabrate petiole, *cordate-ovate, with prolonged acuminate apex, coarse acuminate teeth and* 2 or more *erect and prolonged lobes* 1–4 *cm. long;* leaves of vegetative sprouts similar or more deeply palmate-lobed; *thyrse* 4–12 cm. long, *its axis and branches glabrous or nearly so; grapes* crowded, 8–12 *mm. in diameter, with heavy bloom,* acid; seeds about 5 mm. long. (*V. vulpina* of ed. 7, not L.) — River-banks and rich thickets, Que. to Man. and Mont., s. to N.B., N.E., n. Va., W.Va., Tenn., Mo., Tex. and N.M. *Fl.* mid-May–early July; *fr.* Aug., Sept.

Var. **praècox** Engelm. (early ripening), JUNE-G. — Thyrse only 4–6 cm. long; grapes 6–7 mm. in diameter, without bloom, sweet; seeds about 4 mm. long. (*V. vulpina*, var. Bailey) — Sw. Ill. to centr. Mo. *Fl.* Apr.; *fr.* June.

Var. **syrtícola** (Fern. & Wieg.) Fern. (dweller on sand-dunes), DUNE-G. — Petioles and lower faces of mature leaves copiously pilose. — Sand-dunes of Great Lakes and adjacent dry soil, N.Y. to Mich. and Ind. — Tendrils or inflorescences sometimes in several successive axils.

7. V. vulpìna L. (foxy), WINTER-, FROST- or CHICKEN-G. — High-climbing, with stout trunks; *branchlets terete,* glabrous or soon glabrate; diaphragms 2–6 mm. thick; *leaves of fertile branches cordate-acuminate,* with a deep basal sinus, coarsely and sharply toothed, *unlobed or merely with angled shoulders,* firm, lustrous, *glabrous or promptly glabrate* (except for axillary tufts), mostly 0.5–1.5 dm. broad; thyrse loose and open, up to 1.3 dm. long; fruiting pedicels about 5 mm. long; *grapes black and shining,* at first very acid, becoming sweet after frost, 3–9 mm. in diameter; seeds about 5 mm. long. (*V. cordifolia* Michx.) — River-banks, bottomlands and rich thickets, Fla. to Tex., n. to se. N.Y., Pa., W.Va., O., Ind., Ill., Mo. and e. Kans. *Fl.* mid-May–mid-June; *fr.* Sept., Oct.

8. V. Baileyàna Munson (in honor of LIBERTY HYDE BAILEY, 1858–), POSSUM-G. — Similar to no. 7; differing in *angled* or striate *branchlets* mostly pubescent when young; *petioles and lower faces of leaf-blades permanently pilose;* the leaf-blades more rounded and less prolonged, with smaller teeth; fruiting pedicels mostly less than 3 mm. long; grapes crowded in the thyrse, ripening earlier. — River-banks and rich thickets, e. Va. to Ky. and Mo., s. to Ala. and Ark. *Fl.* late May–mid-June; *fr.* late Aug.–Oct.

9. V. nòvae-ángliae Fern. (of New England), NEW ENGLAND G. — Vigorous climber; young tips reddish-tomentose; *diaphragms* 1.5–2 *mm. thick; expanding leaves densely felted, soon glabrate; blades* of fertile branches *quadrate- to round-ovate,* acuminate, *unlobed or merely shouldered,* deltoid-acuminate, mostly 0.7–1.8 dm. broad; *an inflorescence or tendril usually in* 3–5 *or more successive axils; thyrse* 4–10 cm. long, compact, *its axis floccose-lanate; grapes* 1.2–1.7 *cm. in diameter,* purple-black, *with bloom,* sharply acid; *seeds* 6.5–8 *mm. long.* — Alluvial or rich thickets, N.E., e. N.Y., N.J. and Pa. *Fl.* June, July; *fr.* Sept.

10. V. rupéstris Scheele (of rocks), SAND- or SUGAR-G. — Usually *low and bushy,* the *bark tardily exfoliating;* diaphragms 2–3 mm. thick; *tendrils wanting or only at tips of fertile branches; leaves* firm, shining, trough-like, those *of fruiting branches glabrous or promptly glabrate, reniform,* subtruncate to abruptly slender-tipped, coarsely toothed, or deeply and irregularly lobed and toothed in forma **dissécta** (Eggert) Fern. (dissected), *the mature* 3–10 *cm. broad;* thyrse 2–5 cm. long; grapes globose to depressed, about 1 cm. in diameter, black, with or without bloom, soon dropping, sweet; *seeds* 4–5 *mm. long.* — Sandy banks, shores, hills, etc., s. Pa. and D.C. to Mo., s. to w. N.C., Tenn., Ark. and Tex. *Fl.* May; *fr.* July, Aug.

11. V. acerifòlia Raf. (maple-leaved), BUSH-G. — Similar to no. 10; *young parts arachnoid-pilose or tomentose; leaves more deltoid-ovate to suborbicular, permanently somewhat pilose beneath;* grapes with heavier bloom, more persistent; *seeds* 5–6 *mm. long.* (*V. Longii* Prince) — Ravines and sandy shores, Tex., n. to sw. Mo., Kans. and se. Colo. *Fl.* May; *fr.* July, Aug.

§ MUSCADÌNA Planch.

12. V. rotundifòlia Michx. (round-leaved), MUSCADINE, SCUPPERNONG, BULLACE-G. — *Bark close, not exfoliating, abundantly dotted with small lenticels; pith continuous; tendrils simple;* high-climbing; *leaves orbicular to deltoid-ovate, firm,* lustrous, with broad blunt teeth, those of fertile branches 5–10 cm. broad; thyrse 2–4 cm. long; *grapes* promptly falling, 1.2–2.5 *cm. in diameter,* purple-black to bronze, without bloom, the skin tough, the heavy flesh musky; *seeds oblong,* 7–8 *mm. long.* —Woods, thickets, sandhills and shores, Fla. to Tex., n. to s. Del., Va., W.Va., s. Ind., se. Mo. and Okla. *Fl.* June; *fr.* Sept., Oct. — On inundated bottomlands the branches often producing long drooping roots (these formed during freshets).

FAM. 108. TILIÀCEAE (Linden Family)

Trees (rarely herbs) with the mucilaginous properties, fibrous bark, valvate calyx, etc., of the Mallow Family; *but the sepals deciduous, petals imbricated in the bud, the stamens usually polyadelphous, and the anthers 2-locular.* — Represented in northern reg. by the single genus:

1. TÌLIA L. Linden. Basswood

Sepals 5. Petals 5, spatulate-oblong. Stamens numerous; filaments cohering in 5 groups with each other (in European species) or with the base of a spatulate petal-like body situated opposite each of the real petals. Pistil with a 5-locular ovary and 2 half-anatropous ovules in each locule, a single style, and a 5-toothed stigma. Fruit dry and woody, indehiscent, globular, becoming 1-locular and 1–2-seeded. Embryo in hard albumen; cotyledons broad and thin, 5-lobed, crumpled. — Fine trees of N. Temp. reg. with soft and white wood, very fibrous and tough inner bark, more or less cordate and serrate alternate leaves (oblique and often truncate at the base), deciduous stipules, and cymes of flowers each hanging on an axillary peduncle which is united to a ligulate membranaceous bract. Flowers cream-color, honey-bearing, fragrant. (The classical Latin name.)*

Leaves glabrous or the young pubescent beneath with simple promptly deciduous hairs.
- Expanding leaves glabrous except for tufts in axils of lateral veins; peduncles and pedicels glabrous; northern species. . . . . . . 1. *T. americana.*
- Expanding leaves villous, soon glabrate; peduncles and pedicels pubescent; southern species. . . . . 2. *T. floridana.*

Leaves permanently more or less stellate-pubescent or densely felted beneath.
- Lower surfaces of leaves green, with scattered stellate or loosely enmeshed hairs. . . . . 3. *T. neglecta.*
- Lower surfaces whitened with felt-like or crowded and stellate pubescence. . 4. *T. heterophylla.*

1. T. americàna L. (American), Basswood or Whitewood, Bois blanc (Que.). — Large tree with glabrous twigs; *leaves glabrous from the first,* with tufts in the axils of many of the lateral veins; the blades cordate to obliquely subtruncate at base, abruptly acuminate, *sharply serrate with slender-tipped teeth, blades* of fertile shoots mostly 0.7–2 *dm. long; petiole, pedicels* and bract of inflorescence *glabrous;* the bract stalked or sessile, often tapering to base; flowers with staminodia; fruit ellipsoid to globose. (*T. glabra* Vent.) — Rich woods, s. Que. to Man., s. to N.B., N.E., Ala., Tenn., Ark. and Tex. Late June–early Aug.

2. T. floridàna (V. Engler) Small (of Florida). — Smaller tree than no. 1; *expanding leaves white-villous or -tomentose,* promptly glabrate, *in maturity* 0.5–1.2 *dm. long,* with short curved teeth; young peduncle and *pedicels pubescent;* bract round-based. — Rich woods, w. Fla. to Tex., n. to se. Va., s. Ind., s. Ill. and Mo. June.

Two European species, **T. europaèa** L. (European), a tree with glabrous leaves only 0.5–1 dm. long, the flowers without staminodia; and **T. platyphýllos** Scop. (broad-leaved), Large-leaved L., with similar flowers and with leaves sparsely pubescent beneath, spread from cult. to rubbish and roadsides. (Introd. from Eu.)

3. T. neglécta Spach (overlooked). — Resembling no. 1; *leaves greenish to grayish beneath,* more or less *loosely pubescent with stellate and simple hairs,* often only sparsely so, the blades 0.7–2dm. long. (*T. Michauxii* sensu Sarg., not Nutt.) — An inconstant and rather nondescript series, perhaps better treated as variations of no. 1. — Sw. Que. to Minn., s. to s. N.E., N.C., s. Ill. and Mo.

4. T. heterophýlla Vent. (diverse-leaved), White B. — *Leaves strongly whitened beneath with close felt-like tomentum or crowded stellate pubescence;* the blades ovate to ovate-oblong, acuminate, 0.8–2 dm. long, sharply toothed; *pedicels pubescent.* (Including var. *Michauxii* (Nutt.) Sarg. and *T. monticola* Sarg.) — Rich woods, n. Fla. to Ala., n. to N.Y., W.Va., O., Ind., s. Ill. and e. Mo. June, early July.

T. petiolàris DC. (petioled), Pendent Silver L., a tree with pendulous branchlets, the young ones tomentulose, the suborbicular leaves white-felted beneath, sometimes spreads from cult. to waste places. (Introd. from Eurasia)

* The characters assigned in a monograph of the American species in 1918 are so inconstant that it is evident that the genus demands careful restudy before it can be satisfactorily treated. The following treatment is wholly tentative.

FAM. 109. MALVÀCEAE (MALLOW FAMILY)

Herbs or shrubs, with alternate stipulate leaves and regular flowers, the calyx valvate and the corolla convolute in the bud, numerous stamens monadelphous in a column and united at base with the short claws of the petals, 1-locular anthers and reniform seeds. Sepals 5, united at base, persistent, often involucellate with a whorl of bractlets forming a sort of exterior calyx. Petals 5. Anthers reniform, opening along the top. Pistils several, the ovaries united in a ring or forming a several-locular capsule. Seeds with little albumen; embryo curved, the leafy cotyledons variously doubled up. — Mucilaginous non-poisonous plants, with tough bark and palmately-veined leaves. Flower-stalks with a joint, axillary.

a. Carpels 5–20 or more, in a single series around a central axis, 1–few-ovulate; stamineal column antheriferous at summit. Tribe MALVEAE. . . *b.*
b. Stigmas along the inner faces of the style-branches; carpels reniform and indehiscent, 1-ovulate. . . *c.*
c. Flowers perfect; calyx subtended by an involucel or naked (in some species of no. 3 with denticulate petals). . . *d.*
d. Petals obcordate or emarginate, with plane margins; carpels beakless.
Involucel of 2 or 3 distinct bractlets. . . . . . . 1. *Malva.*
Involucel of 6–9 or more bractlets often united at base. . . . . 2. *Althaea.*
d. Petals with truncate erose-denticulate summits; carpels beaked. . . . 3. *Callirhoë.*
c. Flowers dioecious; calyx naked; petals entire; stamens few (15–20); carpels beakless. . . . . 4. *Napaea.*
b. Stigmas terminal, capitate; carpels 1–9-ovulate. . . *e.*
e. Calyx subtended by 2 or 3 (sometimes deciduous) bractlets; ovules and seeds 1–4, ascending. . . *f.*
f. Carpels tardily dehiscent from above, 1–4-seeded, without transverse partition.
Tall perennials with maple-like leaves mostly 0.8–1.5 dm. broad; flowers 3–5 cm. broad, roseate; basal half of capsule not reticulate. 5. *Iliamna.*
Low annual or perennial; leaves simple or pedately parted, up to 4 cm. broad; petals smaller, yellow, scarlet or brick-red; basal half of capsule reticulate. . . . . 6. *Sphaeralcea.*
f. Carpel with a transverse partition separating the 2 seeds. . . . 7. *Modiola.*
e. Calyx naked; ovules and seeds 1 and pendulous or 2–9 with at least the lower pendulous.
Carpels 1-seeded.
Petals bluish to lavender; carpels radiate-divergent in a depressed and radiate capsule, which promptly breaks up, the partitions becoming obliterated, the firm dorsal portion of each carpel embracing the horizontal seed. . . . . . 8. *Anoda.*
Petals yellow or orange to whitish; carpels not divergent, their walls firm and enduring, covering the pendulous seed. . . . . . 9. *Sida.*
Carpels 2–9-seeded, with long divergent beaks. . . . . . . 10. *Abutilon.*
a. Carpels forming a 5 (in ours)-locular loculicidal capsule with no central column; column of stamens antheriferous much of its length but naked at the 5-toothed apex. Tribe HIBISCEAE.
Locules 1-seeded. . . . . 11. *Kosteletzkya.*
Locules 2–several-seeded. . . . . 12. *Hibiscus.*

Tribe MÁLVEAE Endl.

1. MÁLVA L. MALLOW

Calyx with a 3-leaved involucel at the base, like an outer calyx. Petals obcordate. Styles numerous, stigmatic down the inner side. Fruit depressed, separating at maturity into as many 1-seeded and indehiscent round-reniform blunt carpels as there are styles. Radicle pointing downward. Herbs, nat. of Eurasia and n. Afr. (An old Latin name, from the Greek *malache*, or *moloche*, indicating the emollient leaves.)

a. Flowers fascicled in the axils of many leaves, surpassed by the long petioles; leaf-blades round-cordate or reniform and merely obtusely lobed; annual or biennial. . . *b.*
b. Flower 3–5 cm. in diameter; bractlets (at base of calyx) oblong to ovate-lanceolate; stem erect. . . . . 1. *M. sylvestris.*
b. Flower much smaller; bractlets linear to linear-lanceolate. . . *c.*
c. Petals twice as long as calyx, about 1 cm. long; carpels round-margined, smooth or but slightly reticulated on the back.

Stems trailing or procumbent; leaves only obscurely lobed; flowers mostly pedicelled; carpels not reticulated. . . . . 2. *M. neglecta.*
Stems erect; leaves shallowly but definitely 5–7-lobed; flowers mostly sessile; carpels obscurely reticulate. . . . . 3. *M. verticillata.*

*c.*Petals scarcely to barely exceeding calyx; carpels acute-margined, rugose-reticulate; stems ascending.
Claws of petals glabrous; calyx enlarged and widely spreading under the fruit, veiny-reticulate; carpels glabrous, with thin-margined and denticulate angles. . . . . 4. *M. parviflora.*
Claws of petals bearded; calyx barely enlarged, mostly closed over fruit, scarcely reticulate; carpels at first tomentulose, becoming glabrate, with acute wingless angles. . . . . 5. *M. rotundifolia.*

*a.*Flowers only from upper axils, surpassing subtending leaves, forming a terminal racemose-paniculate inflorescence; petals many times longer than calyx; cauline leaves cleft or dissected; erect perennials.
Pubescence of spreading simple hairs. . . . . 6. *M. moschata.*
Pubescence of short stellate hairs. . . . . 7. *M. Alcea.*

1. M. sylvéstris L. (of woodlands), High M. — *Erect* biennial 2–8 dm. high, *hirsute;* leaves long-petioled, the *blades sharply 5–7-lobed, the lobes triangular;* flowers slender-stalked in the axils; *corolla 3–5 cm. in diameter,* rose-purple with deeper veins; carpels wrinkled-veiny. — Roadsides and waste places, Que. to N.D. and southw., esc. from cult. May–July. (Introd. from Eu.)

Var. mauritiàna (L.) Boiss. (of Mauritius). — *Glabrous* or nearly so; *lobes of leaf very broad and rounded;* petals deeper-colored. (*M. mauritiana* L.) — N.E. to N.D. and southw. (Introd. from Eu.)

2. M. neglécta Wallr. (overlooked), Common M., Cheeses, Amours (Que.). — *Stems procumbent* from a deep biennial root; *leaves* round-cordate, on very long petioles, crenate, *only obscurely lobed;* flowers fascicled in the axils, more or less pedicelled; *petals* pale lilac or white, *twice as long as calyx; carpels* about 15, puberulent and *rounded but not reticulate on back.* (*M. rotundifolia* of ed. 7, not L.) — Weed of barnyards, waste places, etc., throughout. Apr.–Oct. (Natzd. from Eu.)

3. M. verticillàta L. (whorled; from the flowers clustered in leaf-axils). — *Erect* smoothish annual up to 2 m. high; *leaves crenately* 5–7-*lobed,* long-petioled; *flowers* small, *sessile or subsessile,* purple to white, crowded in the axils; calyx tending to close in fruit, accrescent, finally thin and veiny; *carpels* at maturity *smoothish or only obscurely veined* with weak transverse simple ridges starting at the edge but not reaching the obscure midnerve. — Local weed of gardens and waste places, Que. to Ia., s. to Pa. and Del. July–Sept. (Adv. from Eu.)

Var. críspa L. (curled or crisped), Curled M. — Leaves crisped. (*M. crispa* L.) — Waste places and roadsides, Que. to Ill. and Ct. (Adv. from Eu.)

4. M. parviflòra L. (small-flowered). — Glabrous or sparsely hairy erect or ascending branching annual; leaves angulate-lobed, long-petioled; flowers axillary, short-pedicelled; *calyx enlarged and widely spreading under the fruit, veiny-reticulate; petals* whitish, *about equaling calyx, with glabrous claw; carpels glabrous, with thin-margined and denticulate angles.* — Waste places, local, Que. to N.J.; N.D. to B.C., s. to Mo., Tex., N.M. and Mex. June–Oct. (Natzd. from Eu.)

5. M. rotundifòlia L. (round-leaved). — Similar to no. 4; *petals with bearded claws; calyx barely enlarged, mostly closed over fruit, scarcely reticulate; pedicels* mostly longer, inclined to be *reflexed in fruit; carpels at first tomentulose,* becoming glabrate, *with acute wingless mostly entire margins.* (*M. pusilla* Sm.; *M. borealis* Wallm.) — Waste places and roadsides, local, Mich. and Ind. to the Pacific and southw., June–Sept. (Natzd. from Eu.)

6. M. moschàta L. (musky), Musk-M. — *Perennial* with ascending firm stems 2–7 dm. high; *pubescence chiefly of* scattered *divergent simple hairs;* basal leaves simple and rounded, unlobed or shallowly cleft, lower cauline leaves similar to basal, upper leaves shallowly 5-cleft into broad rhombic simply cleft lobes, or in the commoner forma heterophýlla (Vis.) Hayek (various-leaved) the lower leaves merely crenate or with 5 broad cuneate lobes and the upper blades 5-cleft with segments dissected into linear laciniae, or in the common forma laciniàta (Desr.) Hayek (slashed) all the cauline leaves with the 5 lobes dissected into pinnatifid linear segments, faintly musk-scented; *inflorescence terminal, racemose-paniculate; petals* pink or white, 2–3 *cm. long; fruit downy.* — Fields and roadsides, often in old gardens; one or more forms abundant from Nfld. to n. N.E., less common westw. to Ont., s. to Del., Md., Tenn. and Neb. June, July (–Oct.). (Natzd. from Eu.)

7. M. Álcea L. (from the genus *Alcea,* with which this species was early associated). —

Similar, with *short stellate pubescence; cauline leaves only once 5-parted or -cleft,* the lobes incised; large flowers as in the last; *fruit smooth;* bractlets of involucre ovate. — Roadsides, local, N.E. to Pa., O. and Mich. (Natzd. from Eu.)

2. ALTHAÈA L.

Calyx surrounded by a 6–9-cleft involucel. Otherwise as in *Malva,* and like it nat. of Eurasia and n. Afr. (Old Greek and Latin name, from the Greek *althaino, heal,* in allusion to its emollient properties.)

1. A. OFFICINÀLIS L. (of the shops), MARSHMALLOW. — Stem erect, 6–12 dm. high; leaves ovate or slightly cordate, toothed, sometimes 3-lobed, velvety-downy; peduncles axillary, many-flowered; flowers pale rose-color. — Borders of saline or fresh marshes, very local, Co. Deux-Montagnes, Que.; Ct. to Va.; also locally westw. to Mich. and Ark.; formerly cult. Aug.–Oct. (Introd. from Eu.) — Perennial root yielding the original non-synthetic mucilaginous marshmallow-paste.

A. RÒSEA Cav. (rosy), the HOLLYHOCK of gardens, sometimes persists after cult. (Introd. from Eurasia)

3. CALLÍRHOË Nutt. POPPY-MALLOW

Calyx either naked or with a 3-leaved involucel at its base. Petals cuneate and truncate (usually red-purple). Styles, etc. as in *Malva.* Carpels 10–20, straightish, with a short empty beak, separated within from the 1-seeded locule by a narrow projection, indehiscent or partly 2-valved. Radicle pointing downward. The N. Am. counterpart of Old World *Malva.* (Name of several characters and of several fountains in Greek mythology.)

Calyx subtended by a 3-leaved involucel.
  Leaves triangular or hastate, coarsely crenate; peduncle short, umbellately few–several-flowered; carpels not rugose. . . . . 1. *C. triangulata.*
  Leaves rounded, 3–7-parted or -cleft; peduncle elongate, 1-flowered; carpels rugose-reticulated at maturity.
    Stem hirsute, at least some of the pubescence spreading; lobes or divisions of lower leaves toothed or dissected. . . . . 2. *C. involucrata.*
    Stem mostly appressed-pubescent or glabrate; divisions of leaves entire. 3. *C. Papaver.*
Calyx without subtending involucel.
  Radical leaves oblong- to deltoid-cordate; flowers corymbose; calyx and summit of carpels strigose-pubescent. . . . . 4. *C. alcaeoides.*
  Radical leaves round-cordate; flowers subracemose; calyx and carpels nearly or quite glabrous. . . . . 5. *C. digitata.*

1. C. triangulàta (Leavenw.) Gray (triangular). — Stellate-pubescent; stems nearly *erect,* 6 dm. high, from a fusiform root; *leaves triangular or halberd-shaped,* or the lowest rather cordate, coarsely *crenate;* the upper incised or 3–5-cleft; flowers panicled, short-pedicelled, purple; involucel as long as the 5-cleft 5-nerved calyx; *carpels not rugose.* — Dry woods and prairies, Ala. to Tex., n. to N.C., Ind., Ill. and Wisc. May–Aug.

2. C. involucràta (T. & G.) Gray (with an involucre). — *Hirsute or hispid,* the *procumbent* stems from a deep napiform root; *leaves rounded, 5–7-parted or -cleft into linear or lanceolate incisely toothed segments;* peduncles elongated; *bractlets of involucel 3, linear or oblong,* about half as long as and *close to* the 5-parted *calyx;* calyx-lobes lanceolate, 3–5-nerved; petals red or purplish, 2–3 cm. long; *carpels indehiscent, rugose-reticulate.* — Plains, N.D. to Wyo., s. to Mo., Okla., Tex. and Utah; adv. in waste places e. to O. June–Aug.

Var. **Búshii** (Fern.) Martin (for its discoverer, BENJAMIN FRANKLIN BUSH, 1858–1937). — *Erect, retrorsely hirsute* and minutely stellate-pubescent; *leaf-segments oblong to broadly obovate; bractlets ovate.* (*C. Bushii* Fern.) — Ozark reg. of sw. Mo. and n. Ark.

3. C. Papàver (Cav.) Gray (from *Papaver,* the Poppy; Cavanilles seeing a resemblance of our plant to *Papaver Rhoeas*). — Similar to no. 2; pubescence of procumbent or loosely ascending stem mostly *strigose; segments of leaves entire,* linear or oblanceolate and cuneate; *involucel slightly remote from calyx,* its bractlets linear; calyx-lobes 3-nerved. — Sandy or rocky woods, glades and prairies, Fla. to Tex., n. to Ga. and Mo. June–Aug.

4. C. alcaeoìdes (Michx.) Gray (like *Alcaea*). — *Strigose-pubescent;* stems slender, 3 dm. high, erect from a perennial root; *lower leaves triangular-cordate,* incised, the upper 5–7-parted, laciniate, the uppermost divided into linear segments; flowers rose-color or white, corymbose, on slender peduncles; involucel none; calyx 5-parted; *carpels* strongly rugose, *strigose-pubescent* at least at summit. — Barrens and plains, Ill. and Ky. to Neb., s. to Ala., Mo., Okla. and Tex. May–Aug.

5. C. digitàta Nutt. (digitate). — *Sparsely hirsute or glabrous*, erect; *leaves* few, *round-cordate*, 5–7-parted, the cauline commonly with linear divisions; peduncles subracemose, long, filiform; flowers red-purple to white; *involucre none;* calyx 5-parted; carpels rugose. — Dry plains and barrens, Mo. and Kans. to Tex.; adv. ne. to Ind. and Ill. Apr.–July.

4. NAPAÈA L. Glade-Mallow

Calyx naked at the base, 5-toothed. Petals entire. Flowers dioecious; the staminate destitute of pistils, with 15–20 anthers; the pistillate with a short column of usually antherless filaments. Styles 8–10, stigmatic along the inside. Fruit depressed-globular, separating into as many reniform 1-seeded beakless scarcely dehiscent carpels as styles. Radicle pointing downward. — Tall roughish perennial herb, with very large 9–11-parted lower leaves, the pointed lobes pinnatifid-cut, and small white panicled flowers. (From the Greek *nape, a glade*, or, poetically, *a nymph of the glades.*)

1. N. dioìca L. (dioecious). — Stems nearly simple, 1.5–3 m. high; lower surfaces of leaves strigose-pilose, with or without admixture of short stellate hairs, in age often glabrate, or in forma **stellàta** Fassett (star-like) the pubescence chiefly of short stellate hairs. — Moist ground and roadsides, Pa. to Minn., s. to Va., O., Ind., Ill. and Ia., rare; spread from cult. elsewhere. June, July.

5. ILIÁMNA Greene

Calyx with involucel of 3 narrow bracts. Petals densely villous along margins of claws, white, pink or purple. Column stout, at most half as long as petals, coarsely stellate-hirsute toward base. Carpels oblong, thin-walled, smooth on sides, bordered near the back by a band of coarse stellate hairs; the back coarsely hirsute, dehiscent from apex toward base two-thirds down the ventral margin, attached to central axis by a stout subpersistent fiber. Seeds 2–4 in each carpel, reniform or reniform-ovate. — Stout perennials with ligneous rhizomes and maple-like leaves, mostly of w. N. Am., two species eastern. (Name unexplained.)

1. I. remòta Greene (remote; from its western congeners). — Erect, bushy-branched, densely stellate-pubescent, 1–1.7 *m. high; leaves* aceriform, 5–7-lobed, *up to* 1.5 *dm. or more broad, with terminal lobe deltoid, crenate, with obtuse sinuses;* flowers clustered in upper axils and subspicate, up to 5 cm. broad, delicately fragrant; calyx densely pubescent, its caudate-acuminate lobes 1–1.5 cm. long; petals rose-color. (*Phymosia* Britt.; *Sphaeralcea* Fern.) — Open woods, gravels and shores, Kankakee R., Ill.

2. I. Còrei Sherff (for Earl Lemley Core, 1902–). — *Smaller, up to* 1 *m. high; leaves less than* 1 *dm. broad, with the lobes more prolonged, the terminal lobe oblong-lanceolate; the teeth dentate-serrate, with acute sinuses.* — Rocky wooded mountain-slopes, local, w. Va.

6. SPHAERÁLCEA St. Hil. False Mallow. Globe-Mallow

Calyx with 2 or 3 bractlets or none. Petals usually more or less notched or emarginate, small, yellow to magenta. Column one-half to nearly as long as petals, glabrous or sparsely pubescent. Styles 5 or more; stigmas capitate; carpels 1–4-seeded, closely stellate-pubescent, beakless or with smooth empty summit, at length dehiscent above, the basal non-dehiscent part reticulate on the sides; the reniform seeds ascending and the radicle pointing downward. — Small genus of Am. and Afr. (Name from the Greek *sphaira, a sphere*, and *alcea, a mallow;* from the commonly spherical fruit.)

1. S. angústa (Gray) Fern. (narrow), Yellow F. — *Annual, slightly hairy*, erect, 1–5.5 dm. high; *leaves lance-oblong or linear, with scattered fine callous teeth;* flowers glomerulate or solitary in the upper axils, on short peduncles; bractlets and stipules setaceous; *petals yellow, scarcely exceeding the calyx; carpels* 5 *or* 6, *reniform*, smooth, at length 2-valved. (*Malvastrum* Gray; *Sidopsis hispida* Rydb.) — Gravelly and rocky barrens, hills and prairies, Ill. and Ia. to Ala., Mo. and Kans. July, Aug.

2. S. coccínea (Pursh) Rydb. (scarlet), Red F. — *Perennial, low and hoary*, 0.5–3 dm. high; *leaves* 2–4 *cm. broad, 5-parted or -pedate;* flowers in short spikes or racemes; *calyx-lobes lance-triangular,* 3–5 *mm. long; petals coppery-scarlet or brick-red*, 1–1.5 cm. long, *much exceeding calyx; carpels* 10 or more, *reticulated on the sides*, about 3 mm. high, tardily and incompletely dehiscent, with 1 seed. (*Malvastrum* Gray) — Sandy plains, Man. and w. Ia. to Tex. and westw. May–Aug.

7. MODÌOLA Moench

Calyx with a 3-leaved involucel. Petals obovate. Stamens 10–20. Stigmas capitate. Carpels 14–20, reniform, pointed, and at length 2-valved at the top; the cavity divided into two by a cross-partition, with a single seed in each locule. — Humble procumbent or creeping Am. and Afr. annuals or biennials, with cut leaves and small purplish to orange flowers solitary in the axils. (Name from *modiolus*, the *nave of a wheel*, from the shape of the fruit.)

1. M. caroliniàna (L.) G. Don. (of Carolina). — Hairy; leaves 3–5-cleft and incised; fruit hispid at the top. — Low grounds, roadsides, etc., Fla. to Tex., Mex. and Calif., n. to Va., rarely adv. n. to Mass. April–June. (Trop. Am.)

8. ANÒDA Cav.

Calyx explanate under the fruit. Petals bluish to lavender. Stigmas capitate; carpels 5–20, mostly radiate-divergent in a depressed capsule which breaks up at maturity, the partitions or sides of the carpels evanescent or obliterated, the firmer dorsal portion embracing the horizontal seed. — Annuals, chiefly nat. of Mex. and S. Am. (Name a Ceylonese colloquial one for *Abutilon*, taken over by Cavanilles and applied to this genus.)

1. A. CRISTÀTA (L.) Schlecht. (crested). — Villous-hirsute, erect, up to 1 m. high; leaves ovate, angulate-lobed and coarsely dentate; flowers axillary, on filiform pedicels; calyx becoming flat, with 5 narrowly ovate long-acuminate lobes, in fruit 2.5–3 cm. across; petals blue-violet to lavender, 2–2.5 cm. long, 1.5–2 cm. broad; carpels dark green, long-awned, hirsute, their backs separated by pale bands; seed pubescent. — Waste places and roadsides, se. Pa.; Ia. and Mo. to Tex. Aug.–Oct. (Adv. from sw. U.S. and Mex.)

Var. BRACHYANTHÈRA (Reichenb.) Hochr. (short-flowered). — Petals blue-violet, shorter than to but slightly exceeding calyx. — Weed of cult. fields and waste places, se. Va. (Adv. from S.Am.)

9. SÌDA L.

Calyx naked at the base, 5-cleft. Petals entire, usually oblique. Styles 5 or more, tipped by capitate stigmas; the ripe fruit separating into as many 1-seeded carpels, which are closed, or commonly 2-valved at the top, and which tardily separate from the axis. Seed pendulous. Embryo abruptly bent; the radicle pointing upward. Wide-ranging genus of warm reg. (A name used by Theophrastus for some related plant.)

Leaves palmately 3–7-cleft; calyx terete at base; petals white. . . . . . . 1. *S. hermaphrodita.*
Leaves uncleft; calyx angled; petals yellow.
  Petiole slender and elongate; leaf-blade oblong-lanceolate to -ovate; carpels 5, each 2-beaked. . . . . 2. *S. spinosa.*
  Petiole short or wanting; carpels 8–12, each with a single point.
    Leaves rhombic-oblong or cuneate-ovate to oblanceolate; carpels smooth. . 3. *S. rhombifolia.*
    Leaves narrowly linear to oblong; carpels reticulate-rugose on sides.
      Leaf-blades lance- or linear-oblong; carpels rounded at awnless summit to a blunt incurved tip. . . . . 4. *S. inflexa.*
      Leaf-blades mostly narrowly linear; carpels with 2 erect horns at summit. 5. *S. Elliottii.*

1. S. hermaphrodita (L.) Rusby (hermaphrodite; Linnaeus originally contrasting it with *Napaea dioica*), VIRGINIA MALLOW. — A smooth tall (1.2–3 m. high) perennial; *leaves 3–7-cleft;* the lobes oblong, pointed and toothed; *flowers white, umbellate-corymbed*, 2.5 cm. wide; carpels 10, pointed. — Glades and river-banks, rare, Pa. to s. Mich., s. to Md., w. Va. and e. Tenn.; cult. and adv. n. to Mass. July–Oct.

2. S. SPINÒSA L. (spiny), PRICKLY MALLOW. — *Annual* weed, minutely and softly pubescent, low (0.2–1 m. high), much branched; *leaves ovate-lanceolate or oblong*, serrate, rather *long-petioled;* peduncles axillary, 1-flowered, shorter than the petiole; *flowers yellow*, small; *carpels* 5, combined into an ovoid fruit, *each splitting at the top into 2 beaks.* — Waste places and open ground, Mass. to Mich., Neb. and southward, where common. June–Oct. — A little tubercle at the base of the leaves on the stronger plants gives the specific name, but it cannot be called a spine. (Natzd. from the Tropics)

3. S. RHOMBIFÒLIA L. (rhombic-leaved). — Ascending *annual* up to 1 m. high; *leaves short-petioled to subsessile, rhombic-oblong or cuneate-ovate to oblanceolate*, obtuse, cinerous-puberulent beneath, serrulate, entire at base; peduncles more or less elongate; calyx cinereous, in maturity with 5–10 callous-thickened nerves at base; petals pale yellow, often red at base; *carpels* 8–12, *smoothish, subulately 1-awned.* — Waste places and roadsides, Fla. to Tex., n. to se. Va. (formerly on ballast to N.J.). July–Oct. (Adv. from the Tropics)

4. S. infléxa Fern. (incurved). — A minutely puberulent erect *perennial* 6–12 dm. high; *leaves lance- or linear-oblong*, 0.4–2 cm. broad, serrate, short-petioled; *flowers yellow*, rather large (petals 1.5 cm. long), *mostly in terminal small corymbs; calyx villous-hirsute on nerves at* *base; carpels* 9–10, *slightly and abruptly incurved-pointed,* puberulent, the sides horizontally ribbed and reticulate, forming a depressed fruit. — Sandy pinelands and openings, se. Va. Aug.–Oct.

5. S. Elliòttii T. & G. (for its discoverer, STEPHEN ELLIOTT, 1771–1830). — Differing from no. 4 in smoother often more branching stem; *leaves narrowly linear,* 1.5–7 *mm. broad; flowers mostly solitary in axils or in small corymbs on peduncles up to* 2.5 *cm. long; calyx at most strigose on ribs at base; carpels with two short erect horns at summit* of dorsal ribs, glabrous or nearly so at summit, weakly cross-ribbed on sides. — Sandy soil, nw. Fla. and Ala., n. to S.C., Tenn. and se. Mo. Aug.–Oct.

10. ABÙTILON Mill. INDIAN MALLOW

Carpels 2–9-seeded, at length 2-valved. Radicle ascending or pointing inward. Otherwise as in *Sida.* Genus of trop. and subtrop. reg. (Name, *Aubutilun,* given by Avicenna.)

1. A. THEOPHRÁSTI Medic. (for THEOPHRASTUS, 370–285 B.C., Greek scientist), VELVET-LEAF, BUTTER-PRINT, PIE-MARKER. — Tall annual 6–12 dm. high; leaves roundish-cordate, taper-pointed, velvety; peduncles shorter than the petioles; corolla yellow; carpels 12–15, hairy, beaked. (*A. Avicennae* Gaertn.; *A. Abutilon* Rusby) — Waste places, vacant lots and in cult. fields, N.E., westw. and southw. Aug., Sept. (Natzd. from India)

Tribe HIBÍSCEAE Endl.

11. KOSTELÉTZKYA Presl SEASHORE-MALLOW

Capsule depressed, with a single seed in each locule. Otherwise as *Hibiscus.*— Shrubs or, as in the case of our single species, perennial herbs, of warm parts of Am., Mediterr. reg. and n. Afr. (Named for *Vincenz Franz Kosteletzky,* 1801–1887, Bohemian botanist.)

1. K. virgínica (L.) Presl (Virginian). — *Minutely stellate-puberulent, greenish* or barely cinereous, somewhat scabrous, 0.5–1.5 m. high; *leaves* gray-green, the lower *cordate-rotund to -ovate,* angulate or coarsely toothed, *upper and bracteal ones* more *lanceolate and with divergent (hastate) basal lobes; pedicels capillary, often equaling or exceeding bracteal leaves;* flowering calyx minutely puberulent, 8–13 mm. high, its *linear-subulate bracteoles* 6–10 *mm. long; petals* roseate, 3.2–4.5 *cm. long,* 2–3 cm. broad; column (including styles) 1.5–2.5 cm. long; *carpels copiously villous-hirsute with hairs* 1.5–2 *mm. long.* — Brackish to nearly fresh marshes and shores, La. to Fla., n. to se. Va. Late July–Sept.

Var. **aquilònia** Fern. (northern). — Smaller in all parts; *upper leaves often broader; flowering calyx* 6–10 *mm. high, its bractlets* 2.5–6 *mm. long; petals* 1.8–3 *cm. long,* 1–1.6 *cm. broad; column* 0.65–1.5 *cm. long; carpels sparsely hispid-setose with setae* 0.5–1.5 *mm. long.* — Va. to L.I.

Var. **altheaefòlia** Chapm. (with leaves of *Althaea*). — Coarser; *gray pubescence longer, harsh, stellate-hirsute or -tomentose; leaves heavily pubescent with plush-like pubescence, often without hastate bases; pedicels coarse and short, rarely equaling their subtending leaves; calyx heavily pubescent; carpels densely villous-hirsute.* — Fla. to Tex., n. to Del. (W.I.)

12. HIBÍSCUS L. ROSE-MALLOW

Calyx involucellate at base by a circle of numerous bractlets, 5-cleft. Column of stamens long, bearing anthers for much of its length. Styles united; stigmas 5, capitate. Fruit a 5-locular loculicidal capsule. Seeds several or many in each locule. — Herbs or shrubs of warm reg., usually with large and showy flowers. (An old Greek and Latin name for some large Mallow.)

*a.*Calyx herbaceous, not inflated, closely applied to or filled by the capsule. . . *b.*

*b.*Shrub with rhombic-ovate 3-lobed glabrous leaves. . . . . 1. *H. syriacus.*

*b.*Perennial herbs. *c.*

*c.*Stems and lower surfaces of leaves canescent to tomentose; leaf-blades rounded to cordate at base. . *d.*

*d.*Leaves glabrous or glabrate above, pannose beneath; ovary and capsule glabrous.

Median cauline leaves (below inflorescence) broadly ovate to rounded, commonly 3-lobed; peduncles all or nearly all leafless or united close to base with subtending petiole; petals pink or purple (white in albino); branches of exserted style usually heavily pubescent; capsule subglobose with depressed or broadly rounded summit, blunt or abruptly short-tipped. . . . . 2. *H. palustris.*

Median cauline leaves narrowly ovate to lanceolate, unlobed or the median and lower ones tricuspidate; 1–several peduncles usually fused for a third to three-fourths their length to subtending petiole; petals white or creamy; branches of exserted style glabrous or remotely hispid; capsule conic-ovoid, tapering to erect beak. . 3. *H. Moscheutos.*

d. Leaves closely pubescent above, loosely and coarsely stellate-tomentose beneath; capsule short-cylindric, subtruncate or rounded at apex, heavily long-villous and tomentose. . . . . 4. *H. lasiocarpos.*

c. Stem and often hastate median and upper leaves glabrous. . . . . . 5. *H. militaris.*

a. Calyx bladdery-inflated, soon becoming scarious, closed over the globular capsule; annual with principal leaves 5-parted. . . . . 6. *H. Trionum.*

1. H. syrìacus L. (Syrian), Shrubby Althea of gardens, Rose-of-Sharon. — Tall smooth *shrub; leaves rhombic- or cuneate-ovate,* pointed, cut-toothed or lobed; corolla usually roseate. — Roadsides and thickets, Fla. to Tex., n. to Ct., N.Y., O. and Mo. July–Sept. (Introd. from Asia)

2. H. palústris L. (of swamps), Swamp-R., "Marsh-Mallow", Mallow-Rose, Sea-Hollyhock. — *Mousy-smelling* perennial, 1–2.5 m. high; *stem and lower surface of leaves canescent with close pannose puberulence,* the *upper leaf-surface dark green and glabrous or promptly glabrate; median cauline leaves usually broadly ovate to roundish and commonly 3-lobed,* 7–18 cm. long and 4.5–11.5 cm. broad, sometimes as broad as or broader than long; *peduncles all or nearly all leafless or united close to base with the subtending petiole* (*the joint or node* 0.5–2 *cm. below the calyx*); *petals pink to purple,* or in forma **Péckii** House (for Charles Horton Peck, 1833–1917) creamy-white, usually with red or crimson base; stamineal column averaging 1.4 cm. thick; *style* (measured from summit of ovary) averaging 4.4 cm. long, *its exserted half with usually densely-pubescent branches; capsule subglobose, with depressed or rounded summit, blunt or abruptly short-tipped,* 2–2.5 *cm. high,* glabrous. (*H. Moscheutos,* in part, of ed. 7; *H. Moscheutos,* var. *purpurascens* Sweet) — Saline, brackish or fresh marshes, e. Mass. to e. N.C.; inland from w. New York to s. Ont., s. Mich., n. Ind. and ne. Ill. Late July–early Oct. — Perhaps better as a geographic var. of no. 3.

3. H. Moscheùtos L. (from its odor suggesting Musk-Rose), Swamp-R., Mallow-Rose, Wild Cotton. — *Differing from no. 2 in narrower leaves; the median ones narrowly ovate to lanceolate,* 0.8–2.2 dm. long and 3–9 cm. broad, *unlobed or the middle and lower ones tricuspidate; one to several peduncles usually fused for one-half to three-fourths their length to the subtending petiole* (*the node* 1–5 *cm. below the calyx*); *petals white or creamy* with purple or red base; stamineal column averaging 2 cm. thick; *style* averaging 6 cm. long, *its exserted branches glabrous* (rarely remotely hispid); *capsule conic-ovoid, tapering to erect beak,* 2.5–3 *cm. long,* glabrous. (*H. oculiroseus* Britt.) — Marshes, Fla. and Ala., n. to e. Md., Va., W.Va., s. O. and s. Ind. July–Sept. — Apparently crosses with nos. 2 and 4.

4. H. lasiocárpos Cav. (hairy-fruited). — *Leaves* broadly to narrowly ovate, *soft-pubescent upon both surfaces, the upper surface bearing many simple* or subsimple *hairs; bractlets ciliate;* petals white or rose-color, crimson-blotched at base; *capsule short-cylindric, subtruncate, densely villous-hirsute.* — Marshes, Ga. to Tex., northw. in Miss. basin to Ky., Ind., Ill., and Mo. July–Sept.

5. H. militàris Cav. (soldierly), Halberd-leaved R. — *Smooth throughout;* lower leaves ovate-cordate, toothed, 3-lobed; *upper leaves commonly hastate;* peduncles slender; corolla 5–7.5 cm. long, flesh-color, with purple base; *fruiting calyx* becoming oblong-campanulate and at length ovoid, *loosely inclosing* the puberulent to glabrous *capsule,* the lobes incumbent; *seeds hairy.* — Wet rivèr-banks and wooded swamps, Fla. to Tex., n. to s. Pa., W.Va., s. O., Ind., Ill., Minn., and Neb. July–Sept.

6. H. Triònum L. (old generic name), Flower-of-an-hour. — A low rather *hairy annual;* upper leaves 3-parted, with lanceolate divisions, the middle one much the longest; fruiting *calyx inflated, membranaceous, 5-winged, with numerous dark ciliate nerves; corolla sulphur-yellow,* with a blackish eye, ephemeral. — Cult. and waste ground, mostly casual, throughout. July–Sept. (Adv. from Eu.)

H. esculéntus L. (edible), Okra or Gumbo, an annual with lobed rounded leaves, spathiform calyx splitting down one side and deciduous from near base, large yellowish corolla, and finger-like capsules with many mucilaginous seeds, is spread from cult. to waste ground but rarely persistent. (Introd. from Africa)

Gossýpium L. (late Latin name for the Cotton plant) **herbàceum** L. (herbaceous), the common Cotton of cult., annual with 3-lobed leaves, involucel of 3 large cordate and laciniate foliaceous bractlets, undivided style with 5 sessile stigmas, creamy petals, subglobose cap-

sule, and white-woolly seeds 6–7 mm. broad, spreads to waste ground northw. to se. Va. and is casual in waste n. to N.E.; probably not long persistent.

FAM. 110. THEÀCEAE (TEA OR CAMELLIA FAMILY)

Trees or shrubs with alternate simple feather-veined leaves and no stipules, the regular flowers hypogynous and polyandrous, the sepals and petals both imbricated in aestivation, the stamens more or less united at the base with each other (monadelphous or 3–5-adelphous) and with the base of the petals. **Anthers 2-locular, introrse. Fruit a woody 3–5-locular loculicidal capsule. Seeds few, with little or no albumen. Embryo large, with broad cotyledons. — A family with showy flowers, the most familiar of which are the well-known CAMELLIA and the more important TEA plant.**

1. STEWÀRTIA L.

Sepals 5, rarely 4 or 6, ovate or lanceolate. Petals 5, rarely 6, obovate, crenulate. Stamens monadelphous below. Capsule 5-locular. Seeds 1 or 2 in each locule, crustaceous, anatropous, ascending. Radicle longer than the cotyledons. — Shrubs or small trees of e. N. Am. and e. Asia, with membranaceous deciduous oblong-ovate serrulate leaves soft-downy beneath, and large short-peduncled flowers solitary in their axils. (Named for *John Stuart*, 1713–1792, — or as formerly often written *Stewart* — third Earl of Bute.) STUARTIA L'Hér.

1. S. Malachodéndron L. (generic name used by Mitchell), SILKY CAMELLIA. — Petals 5, white, 2.5 cm. long; sepals ovate; stamens with dark purple filaments and blue anthers; style 1; stigma 5-toothed; capsule globular, blunt; seeds not margined. — Rich woods, w. Fla. to w. La., n. to Va., Tenn. and Ark. June.

2. S. ovàta (Cav.) Weath. (ovate), MOUNTAIN-CAMELLIA. — Leaves larger, 1.3–1.5 dm. long; sepals acute; petals often 6; styles 4, distinct; capsule angled, pointed; seeds wing-margined. (*S. pentagyna* L'Hér.) — Rich woods and stream-banks, Ga. and Ala., n. to Va. and Ky. May–Aug.

FAM. 111. GUTTÍFERAE (ST. JOHN'S-WORT FAMILY)

Herbs or shrubs, with opposite entire dotted mostly sessile leaves and no stipules, usually regular hypogynous flowers, the petals mostly oblique and convolute in the bud, and many or few stamens sometimes collected in 3 or more clusters or bundles. Capsule 1-locular with 2–5 parietal placentae and as many styles, or 3–7-locular by the union of the placentae in the center; dehiscence mostly septicidal. **Sepals 4 or 5, imbricated in the bud, herbaceous, persistent. Petals 4 or 5, mostly deciduous. Styles persistent, at first sometimes united. Seeds numerous, small, anatropous, with no albumen. — Plants usually smooth. Flowers solitary or cymose. HYPERICACEAE.**

Sepals 4, in 2 very unequal pairs; petals 4; stamens many, distinct. . . . . 1. *Ascyrum.*
Sepals 5; petals 5; stamens many to few, often in 3 or 5 clusters. . . . . . 2. *Hypericum.*

1. ÁSCYRUM L. ST. PETER'S-WORT

Sepals 4; the two outer very broad and leaf-like; the inner much smaller. Petals 4, oblique, very deciduous, convolute in the bud. Stamens numerous; the filaments distinct and scarcely in clusters. Capsule strictly 1-locular, 2–4-valved. — Low rather shrubby smooth pale green plants of N. Am. and the Himalayas, with nearly solitary light yellow flowers. (Ancient Greek name for some plant probably of this family.)

Leaves broadly oblong, the upper cordate or clasping, 1–2 cm. broad; petals obovate, 7–10 mm. broad; styles 3 or 4. . . . . 1. *A. stans.*
Leaves linear- to oblong-oblanceolate, narrowed at base; petals oblong, 1.5–4 mm. broad; styles 2. . . . . 2. *A. Hypericoides.*

1. A. stáns Michx. (standing upright), ST. PETER'S-WORT. — Stems erect or suberect, 2-edged, 3–10 dm. high; *leaves broadly oblong, the upper cordate or clasping,* mostly 2–4 cm. long and 1–2 *cm. broad,* round-tipped, firm; *outer sepals* round-cordate, 1–2 *cm. long, nearly as broad,* strongly cordate; *inner sepals lanceolate,* 7–14 *mm. long; petals* showy, *obovate,* longer than sepals, 7–10 *mm. broad; styles* 3 *or* 4. — Moist or dry sandy woods, meadows and barrens, Fla. to Tex., n. to L.I., N.J., e. Pa. and Ky. Late July–Sept.

2. A. Hypericoìdes L. (old generic name, like *Hypericum*), ST. ANDREW'S CROSS. — *Leaves*

linear- to oblong-oblanceolate, *narrowed at base*, the principal ones 2–9 *mm. broad; outer sepals oblong, elliptic or ovate*, at most subcordate, 3–10 *mm. broad; inner sepals minute or obsolete; petals oblong*, about equaling the outer sepals, 1.5–4 *mm. broad; styles* 2. — A complex species; 3 well defined varieties with us:

Leaves linear-oblanceolate or linear-oblong; the principal ones 0.5–2 cm. long, 2–4(–5) mm. broad, with crowded axillary fascicles; outer sepals 5–11 mm. long, 3–5 mm. broad; ascending shrub. . . . . *A. Hypericoides* (typical).

Leaves oblong-oblanceolate, the principal ones 4–9 mm. broad.
Decumbent or diffuse, with numerous slender stems 1–2(–3) dm. long, flowering from tips and the uppermost axils; larger leaves 1–2.3 cm. long, without long axillary branches; outer sepals elliptic, oval or oblong-oblanceolate, rounded at base, 5–11 mm. long, 3–7 mm. broad. . . . . Var. *multicaule*.
Erect or ascending, with 1–few stems 3–9 dm. high, simple or sparsely branched below, with flowering (often elongate) branches from most of middle and upper axils; primary leaves in distant pairs, 2–3 cm. long; outer sepals broadly ovate, often subcordate, 10–15 mm. long, 7–10 mm. broad. . Var. *oblongifolium*.

A. Hypericoìdes (typical). — Dry sandy soil, Fla. to Tex. and Mex., n. to e. Va. and Tenn. July–Sept. (W.I.)

Var. **multicaùle** (Michx.) Fern. (many-stemmed). — Dry sand and rock, Ga. to e. Tex., n. to Nantucket I., Mass., N.J., Pa., W.Va., Ky., s. Ind., s. Ill., Mo. and se. Kans.

Var. **oblongifòlium** (Spach) Fern. (oblong-leaved). — Sandy woods and thickets, Fla. to Miss., n. to e. Md., w. Tenn. and se. Mo.

2. HYPERÌCUM L. St. John's-wort. Millepertuis (Que.)

Sepals 5, usually subequal. Petals 5, oblique, convolute in the bud (except in § *Elodea*). Stamens frequently united or clustered in 3–5 fascicles; capsule 1-locular or 3–5-locular. Seeds usually cylindrical. — Herbs or shrubs, mostly of the N. Hemisph. with cymose yellow, flesh-colored or purplish flowers. (*Hypericon*, the ancient Greek name.)

a.Petals yellow or orange, convolute in bud; stamens numerous or few, distinct, or united merely at base into 3–5-clusters, without intervening glands; flowers terminal or in terminal cymes. . . *b*.
b.Stamens numerous (mostly 20–100 or more); leaves linear to oval; roots perennial. . . *c*.
c.Capsules 3–5-locular or nearly so by inward projection of the placentae. . *d*.
d.Styles 5, wide-spreading above the united base, ending in thick or capitate stigmas; flowers 5–7 cm. broad; capsule 2–3 cm. high, 5-locular; large herb. § Roscyna. . . . . 1. *H. pyramidatum*.
d.Styles 3 (sometimes 2 or 4), 5 only in nos. 5 and 8 (shrubs); flowers mostly smaller (if as large, plant shrubby); capsules smaller. . . *e*.
e.Stamens in 3–5 definite clusters; styles 3, distinct and usually divergent, with capitate stigmas; surfaces of herbaceous capsules marked with elongate oil-vesicles; herbs, the leaves, sepals or petals with dark dots or lines. § Euhypericum. . . *f*.
f.Much branched, usually with the leaves of the branches subtending suppressed branchlets; sepals linear-lanceolate, attenuate, without prominent elongate oil-vesicles; petals orange-yellow, at most black-dotted on margin only; seeds roughened by coarse reticulation. . . . . 2. *H. perforatum*.
f.Simple or with simple branches; sepals blunt or merely acute, with prominent rib-like oil-vesicles; seeds smoothish or only minutely reticulate.
Petals pale yellow, streaked with dark lines; flowers crowded, very numerous; capsules 3–5 mm. long. . . . . . . 3. *H. punctatum*.
Petals copper-yellow, dotted (if at all) only on margin; flowers few; capsules 8–10 mm. long. . . . . 4. *H. mitchellianum*.
e.Stamens rarely in definite clusters; styles united into a long sharp beak, finally becoming distinct, with elongate or minute stigmas;

capsules firm, usually without superficial oil-vesicles; shrubs or herbs, mostly without dark dots. § MYRIANDRA. . *g.*

g. Shrubs, mostly with elongate flower-bearing branches and without evident stolons; stamens promptly deciduous, the capsules naked at base. . . *h.*

h. Sepals (or the larger ones) foliaceous, 5–30 mm. long; capsules 3–7 mm. thick, 7–15 mm. long (excluding beaks).

Corymbs terminal, comparatively open; capsule ovoid, usually of 5(3–6) carpels, 7–10 mm. long (excluding beaks), 4–7 mm. thick. . . . . 5. *H. Kalmianum.*

Corymbs both terminal and in upper axils, contracted; capsules subcylindric to lance-ovoid, of 3 carpels, 8–15 mm. long, 3–5 mm. thick.

Sepals shorter than petals and capsules, 5–7 mm. long; corolla 1.5–2.5 cm. broad. . . . . 6. *H. spathulatum.*

Sepals often much longer than petals and capsules, the larger 2–3 cm. long; corolla 2.5–4 cm. broad. . . . . 7. *H. frondosum.*

h. Sepals 1.5–4(–5) mm. long, not foliaceous; capsules 1.5–3 (rarely –5) mm. thick, 3.5–6.5(–7) mm. long. . . *i.*

i. Principal leaves subtending axillary fascicles; capsules lanceolate or slender-conic, 1.5–3 mm. thick, of 3–5 separate carpels. . . . . 8. *H. densiflorum.*

i. Principal leaves without axillary fascicles; capsules ovoid, 4.5–6 mm. thick, 1-locular, with 3 inwardly projecting placentae. 9. *H. nudiflorum.*

g. Herbs with slender stoloniferous bases and simple stems (except for axillary sterile branches); stamens persistent, forming a marcescent mass at base of capsule. . . . . 10. *H. adpressum.*

c. Capsules 1-locular, with 3 (or 4) parietal placentae. . . *j.*

j. Styles united below into a beak; stigmas minute, elongate; capsules conical, ellipsoid or globose, 4–7 mm. long (excluding beak); plant not virgate. § BRATHYDIUM. . . *k.*

k. Herb with filiform rhizome and numerous filiform stolons; leaves and sepals membranaceous to herbaceous; capsule ellipsoid or slenderly ovoid, thin-walled, not angled. . . . . 11. *H. ellipticum.*

k. Shrubby or suffruticose, without slender rhizome and stolons; leaves and sepals firm to coriaceous; body of capsule broadly ovoid to subglobose, hard-walled, 3-angled.

Sepals 6–13 mm. long, very unequal, the larger ones 2–8 mm. broad; corolla 1.5–2 cm. broad. . . . . 12. *H. dolabriforme.*

Sepals 3–5 mm. long, the larger ones 1.5–3 mm. broad; corolla 1–1.5 cm. broad. . . . . 13. *H. sphaerocarpum.*

j. Styles free to base, with capitate stigmas; capsules ovoid to lanceolate, 4–5 mm. long (excluding beak); plants virgate (resembling *Linum*). § BRATHYS.

Stems, leaves and sepals glabrous. . . . . 14. *H. denticulatum.*

Stems and leaves copiously pilose; sepals ciliate. . . . . . . 15. *H. setosum.*

b. Stamens 5–12, if –20 the flower very small (nos. 19 and 20) or the leaves usually subulate or linear-subulate, the branching fastigiate and the root annual. § BRATHYS. . . *l.*

l. Leaves flat, linear or lanceolate to elliptic, oval or rounded, spreading; stem simple or loosely branched, from a slightly perennial base. . *m.*

m. Ultimate bracts of inflorescence dilated, resembling the foliage-leaves. 16. *H. boreale.*

m. Ultimate bracts of inflorescence setaceous to linear-subulate. *n.*

n. Leaves orbicular, ovate-deltoid or rounded-oblong, the upper cordate at base or clasping.

Diffusely branched; leaves (except sometimes the upper) ovate-oblong to short-elliptic, rounded at tip; capsule short-ellipsoid. 17. *H. mutilum.*

Simple or nearly so, virgate, the cyme with ascending branches; middle and upper leaves ovate-deltoid, tapering to tip; capsule slender-conical. . . . . 18. *H. gymnanthum.*

n.Leaves lanceolate or linear to narrowly obovate, merely sessile or slightly petioled (if clasping, with lance-attenuate outline). . . *o*.

o.Leaves (at least the upper) lanceolate, tapering from base to apex; the larger upper ones 5–7-nerved at base, 5–15 mm. broad; mature sepals 4–7 mm. long. . . . . 19. *H. majus.*

o.Leaves linear, linear-oblong or linear-oblanceolate (rarely narrowly obovate), rounded at tip, 3 (rarely 5)-nerved at base, the larger ones 1–10 mm. broad; sepals 2.5–4.5 mm. long.

Leaves linear to linear-oblanceolate or rarely narrowly obovate; mature capsules (excluding styles) 3–6.5 mm. long, regularly developing. . . . . 20. *H. canadense.*

Leaves linear-oblong; mature capsules 3–4.5 mm. long, many of them not developing. . . . . 21. *H. dissimulatum.*

l.Leaves subulate or subulate-linear, appressed or strongly ascending; stem fastigiate-branching; root annual.

Leaves 6–20 mm. long; capsules ovoid, slightly exceeding calyx. . . . 22. *H. Drummondii.*

Leaves smaller, scale-like; capsules lance-subulate, strongly exceeding calyx. . . . . 23. *H. gentianoides.*

a.Petals flesh-color to mauve-purple, imbricated in bud; stamens mostly 9, strongly triadelphous, with 3 large orange glands alternating with the 3 bundles of stamens; flowers clustered in the axils and at summit of stem. § ELODEA.

Leaves clasping or cordate at base; filaments united at base. . . . . . 24. *H. virginicum.*

Leaves (except sometimes the uppermost) tapering to the petioled or sessile base; filaments united to above the middle. . . . . 25. *H. tubulosum.*

§ RÓSCYNA (Spach) Endl.

1. H. pyramidàtum Ait. (pyramidal), GREAT S. — Stems stout, *herbaceous*, 0.6–2 m. high, with 2–4-angled branches; leaves ovate-oblong, partly clasping, the larger 7–10 cm. long; *petals* narrowly obovate, 2.5–3 *cm. long; stamens very numerous, 5-adelphous; styles* 5, *united into a beak below, divergent above, with capitate stigmas; capsule conical,* 2–3 *cm. high, its* 5 *placentae turned far back into the locules; seeds with winged raphe.* (*H. Ascyron* of Am. auth., not L.) — Rich thickets, river-meadows, etc., ne. Me. to Man., s. to w. N.E., N.J., n. Md., O., Ind., Ill., Mo. and ne. Kans. July–Sept. FIG. 1303.

1303. H. pyramidatum.

§ EUHYPERÌCUM Boiss.

2. H. PERFORÀTUM L. (perforated), COMMON S., PERTUISANE (Que.). — Stems tough, 3–9 dm. high, much branched, usually clustered from a depressed base, 2-edged, producing leafy basal offshoots; *leaves* elliptic- to linear-oblong, with pellucid dots, those *of the branches subtending abbreviated branchlets; cymes leafy, very floriferous; sepals linear-lanceolate, attenuate*, plane; *petals orange-yellow*, occasionally black-dotted on the margin, twice as long as sepals; *stamens* numerous *in* 3–5 *tufts; styles* 3, *separate; seeds blackish*, lustrous, *rough with coarse reticulations.* — Dry pastures, roadsides and neglected fields, a weed difficult to eradicate, Nfld. to B.C. and s. beyond our limits. June–Sept. (Natzd. from Eu.) FIG. 1304.

1304. H. perforatum.

3. H. punctàtum Lam. (dotted). — Conspicuously marked with both black and pellucid dots; stem terete, 0.3–1.2 m. high, sparingly branched; *leaves oblong, rounded at tip,* the base either subclasping or sessile, or leaf oblanceolate and tapering to narrow base in forma **subpetiolàtum** (Bickn.) Fern. = *H. subpetiolatum* Bickn. (somewhat petioled); *flowers crowded; petals pale yellow, marked with dark lines and dots, about twice as long as the oblong bluntish or acute sepals; capsule* 3–5 *mm. high, covered with elongate oil-vesicles.* — Thickets, borders of woods and damp openings, Fla. to Tex., n. to s. Que., s. Ont. and Minn. June–Sept. FIG. 1305.

1305. H. punctatum.

Var. **pseudomaculàtum** (Bush) Fern. (false *H. maculatum*). — Uppermost and bracteal

1306. H. mitchellianum.

leaves subdeltoid and acutish. (*H. pseudomaculatum* Bush) — Fla. to Tex., n. to s. Va., w. Tenn., Ill., Mo. and Okla.

4. H. mitchelliànum Rydb. (of Mt. Mitchell, North Carolina), MOUNTAIN-S. — Resembling no. 3; stem only 3–6 dm. high; leaves elliptic, 3–4 cm. long, obtuse, glandular-punctate along margin beneath; *cymes few-flowered; sepals* 5 *mm. long,* lanceolate, *blunt,* with prominent linear oil-vesicles; *petals* oval, *copper-yellow, with gland-dots chiefly confined to margin;* styles subulate, about equaling ovary; *capsule* ovoid, 8–10 *mm. long.* (*H. graveolens* of ed. 7, not Buckl.) — Damp slopes, mts. of Va., N.C. and Tenn. June–Aug. FIG. 1306.

§ MYRIÁNDRA (Spach) Endl.

5. H. Kalmiànum L. (for its discoverer, PEHR KALM, 1715–1779), KALM'S S. — Slender shrub 2–6 dm. high, with papery whitish bark and ascending 4-edged branches, with 2-edged branchlets; leaves rather crowded, linear to oblanceolate, mostly 3–4.5 cm. long; flowers 1–10 in *terminal open corymbs,* 2–3.5 cm. across; *sepals or some of them foliaceous,* oblong to elliptic, 5–15 *mm. long; capsules ovoid,* usually *with* 5 (sometimes 3, 4 or 6) *carpels and styles,* 7–10 *mm. long* (excluding beaks), 4–7 *mm. thick.* — Rocky and sandy soil, chiefly near the Great Lakes, Ottawa R., Que., to w. Ont., s. to w. N.Y., O., Ind. and Ill. July, Aug. FIG. 1307.

1307. H. Kalmianum.

6. H. spathulàtum (Spach) Steud. (spatula-like), SHRUBBY S. — Coarser, up to 2.5 m. high; leaves narrowly oblong, mostly 3–7 cm. long, obtuse, narrowed at base; *corymbs both terminal and in the* 1–3 *upper pairs of axils, contracted, the inflorescences thus interruptedly cylindric;* sepals 5–7 mm. long; *styles* 3 (rarely 4); *capsules subcylindric to lance-ovoid, of* 3 *carpels,* 8–15 *mm. long,* 3–5 *mm. thick.* (*H. prolificum* sensu Am. auth., not L.) — Dry or damp sandy or rocky thickets, pastures and slopes, se. N.Y. to Ont. and Minn., s. to Ga., Ala. and Ark.; cult. and locally esc. in Mass. July–Sept. FIG. 1308.

1308. H. spathulatum.

7. H. frondòsum Michx. (full of leaves; from the large sepals). — Widely branching shrub up to 1 m. high, resembling no. 6; *sepals very unequal, often inclosing capsule, some of them ovate or oblong and* 2–3 *cm. long; corolla* 2.5–4 *cm. broad,* the firm petals becoming reflexed; stamens very numerous; capsule slenderly conic-ovoid, not lobed, 9–12 mm. high. (*H. aureum* Bartr.) — Limestone-barrens and

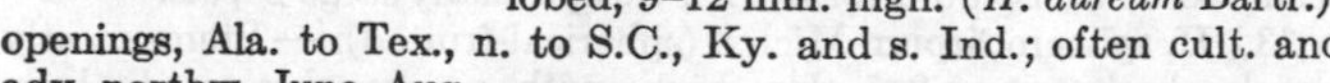
openings, Ala. to Tex., n. to S.C., Ky. and s. Ind.; often cult. and adv. northw. June–Aug.

8. H. densiflòrum Pursh (densely flowered). — Exceedingly branched above, 0.5–2 m. high, the branches slender, with linear to linear-elliptic *acute leaves and axillary fascicles; flowers* 1.2–1.7 *cm. in diameter* and numerous in crowded compound cymes; *sepals* firm, *not foliaceous,* 2–5 *mm. long; capsules lanceolate to slender-conic, the body* 3.5–6.5 *mm. long,* 1.5–3 *mm. thick, of* 3 *distinct carpels, with* 3 *styles.* — Swamps and wet acid soil, Fla. to Tex., n. to L.I., N.J., W.Va., s. Ind. and s. Mo.; rarely adv. northw. July–Sept. FIG. 1309.

1309. H. densiflorum.

Var. **lobocárpum** (Gattinger) Svenson (with lobed fruit). — Bark brown; *leaves* linear-lanceolate to narrowly oblong, *obtuse; flowers* very crowded; *styles* mostly 4 or 5; *capsule of* 4 or 5 *distinct carpels.* — Dry or moist acid soils, oak-barrens, etc., w. N.C. to se. Mo. and La. Late June–Aug.

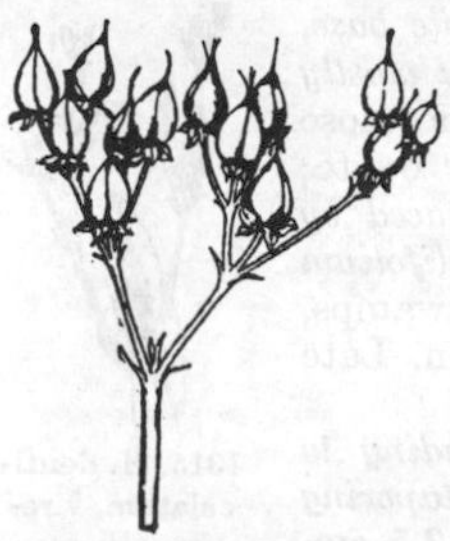
1310. H. nudiflorum.

9. H. nudiflòrum Michx. (naked-flowered; from the inflorescence having only small leaves). — Ligneous below, 0.9–2 m. high, sending up long subherbaceous brown-barked 4-angled flowering branches; *leaves oblong to oval-lanceolate,* obtuse, thin, pale green, 4–7 cm. long, *minutely punctate beneath, all or nearly all without axillary fascicles;* corymbs terminal, open, dichotomous, with minute lance-subulate bracts; flowers 1.5–2 cm. broad; sepals firm, linear to oblong, 2–5 mm. long; styles 3; *capsule ovoid,* 4.5–6.5 mm. long (excluding beak), 4–5 *mm. thick,* 1-*locular, with* 3 *inwardly projecting placentae.* — Damp sandy woods and thickets or swamps, Fla. to Tex., n. to se. Va. and Tenn. Late June, July. FIG. 1310.

1311. H. adpressum.

10. H. adpréssum Bart. (appressed; from the crowded axillary leaves), CREEPING S. — *Herbaceous; the simple stem* 3–8 dm. high, *from a creeping freely stoloniferous base,* rather slender, or when long immersed prominently spongy-thickened below as forma **spongiòsum** (Robins.) Fern. (spongy), obscurely 4-angled below and 2-edged above; *leaves ascending,* linear-lanceolate and acute to oblong and obtuse, *subtending axillary fascicles;* cyme terminal, open, leafy at base; flowers 1.5–2 cm. broad; sepals lanceolate to ovate; *stamens marcescent, persisting as a mass at base of the ovoid* 1-locular *capsule* (3–6 mm. high), the latter with 3 inwardly projecting placentae. — Damp sands, peats and pond-margins, Ga. to La., n. to se. Mass., W.Va., Jasper Co., Ind., and s. Ill. Late July–early Sept. FIG. 1311.

§ BRATHÝDIUM (Spach) Endl.

11. H. ellípticum Hook. (elliptical). — Stem simple, herbaceous, 2–5 dm. high, obscurely 4-angled, from a *slender creeping stoloniferous base; leaves spreading, elliptical-oblong,* obtuse, usually narrower toward the subclasping base, thin; cyme nearly naked, rather few-flowered, borne above the foliage, or in forma **foliòsum** Vict. (leafy) much overtopped by ascending leafy branches; petals bright yellow, 6–10 mm. long; *sepals oblong; capsule ellipsoid or slenderly ovoid, obtuse, 1-locular;* seeds minutely striate. — Damp sandy or gravelly shores or wet places, Nfld. to Man., s. to N.S., N.E., Md., W.Va., O., Ind., Ill., Ia. and N.D. Late June–Aug. FIG. 1312. — Submersed plants with simple sterile stems and membranous round to ovate leaves are forma **submérsum** Fassett (submersed).

1312. H. ellipticum.

12. H. dolabrifórme Vent. (shaped like an axe). — *Stems* branched from the decumbent base, *woody below,* 1.5–5 dm. high, terete; *leaves linear-lanceolate,* widely spreading, veinless; cyme leafy, few-flowered; *sepals oblong or ovate-lanceolate,* about the *length of the very oblique petals* (1–1.3 *cm. long*), *the larger ones* 2–8 *mm. broad; capsule* 1-*locular, ovoid-conic,* 4–7 *mm. high, strong-walled* and -beaked. — Rocky woods, glades and barrens, Ga. to Ky. and se. Mo. June–Aug. FIG. 1313. — *H. Bissellii* Robins., once found in w.-centr. Ct. but not rediscovered, seems to be an isolated form with exceptionally large sepals.

1313. H. dolabriforme.

13. H. sphaerocárpum Michx. (spherical-fruited). — Stems mostly simple, herbaceous, 3–5 dm. high, with a *somewhat woody base,* angled with 4 very narrow salient lines; leaves narrowly oblong to nearly linear, 3–7 cm long, sessile, with a somewhat clasping base; cyme naked, compound, usually many-flowered; *sepals ovate,* 3–5 *mm. long, the larger ones* 1.5–3 *mm. broad; corolla* 1–1.5 *cm. broad; capsules depressed-globular or ovoid, firm;* seeds large, oblong, very rough-pitted. (*H. cistifolium* sensu Coult. and ed. 7, not Lam.) — Rocky woods, barrens and shores, Ala. and Miss., n. to O., Ind., s. Wisc., Ia. and e. Kans. July–Sept. FIG. 1314.

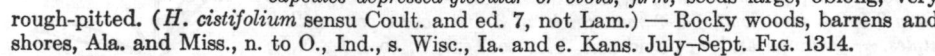

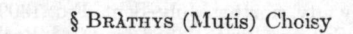

1314. H. sphaerocarpum.

§ BRÀTHYS (Mutis) Choisy

14. H. denticulàtum Walt. (small-toothed; from marginal teeth on petals), COPPERY S. — Glabrous throughout; *stems* slender, herbaceous, strict or strongly ascending, often in clumps with surculose bases, simple below, *sharply 4-angled,* 2.5–6.5 dm. high; *leaves nearly erect, narrowly to broadly ovate, oval or obovate, rounded to sessile base, blunt to subacute, thick and firm, the larger ones* 0.8–2(–2.5) *cm. long and mostly* 0.6–1.7 *cm. broad; inflorescence compound,* naked, the scattered flowers racemose along the ascending branches; sepals herbaceous, oblong to narrowly ovate; *petals coppery-yellow,* 8–10 mm. long; *capsule* 1-locular, ovoid, *embraced by the sepals;* styles distinct, short, with capitate stigmas. (Var. *ovalifolium* (Britt.) Blake; *H. angulosum* Michx.) — Sandy or argillaceous shores, swamps, ditches or pine-barrens, Coastal Plain, N.J. to S.C.; Coffee Co., Tenn. Late June–Aug.

Var. **recògnitum** Fern. & Schub. (studied anew). — *Leaves ascending to divergent, narrowly oblong or oblong-elliptic to narrowly obovate,* thinner, *tapering to acute or acutish base, with acute to obtuse tip,* the larger blades 1.5–3.5 cm.

1315. H. denticulatum, v. recognitum.

long and 4–10 mm. broad. (*H. denticulatum* sensu recent Am. auth., not Walt.) — Wet or dry oak-barrens or bogs, Ga. to Miss., n. in the Piedmont and upland areas and at inner margin of Coastal Plain to se. Va., s. W.Va., Ky. and s. Ind. FIG. 1315.

Var. **acutifòlium** (Ell.) Blake (acute-leaved). — Similar to last var.; *leaves or at least the middle and upper ones narrowly linear or linear-lanceolate, broad-based,* all or *at least the upper tapering to firm nearly subulate tips and* 1–3.5 *cm. long by* 1.5–7 *mm. broad.* (*H. virgatum* Lam.) — Pine- or oak-barrens and margins of pools, n. Fla., n., locally, to s. Va. and Tenn. — Often taller, up to 10.5 dm. high.

15. H. setòsum L. (covered with bristles). — Similar, *scabrous-tomentose or -pilose;* leaves oval to narrowly oblong, closely appressed, their margins revolute; *sepals ciliate;* petals yellow. (*H. pilosum* Walt.) — Wet pinelands and ditches, Fla. to La., n. to se. Va. June–Sept. FIG. 1316.

1316. H. setosum.

16. H. boreàle (Britt.) Bickn. (northern). — Perennial; *the stems decumbent and leafy-bracted at base,* slender, 5–40 cm. high (rarely submersed and very elongate); *leaves elliptic,* rounded at tip, *sessile,* 3–20 mm. long, 3–5-nerved; cymes leafy-bracted, *all the bracts foliaceous and broad;* pedicels short; sepals linear, blunt, shorter than the *rounded short-ellipsoid capsule* (3–5 *mm. long*). — Damp peat, sand and shallow water, Nfld. to Ont. and Minn., s. to N.S., N.E., L.I., se. and w. Va., W.Va., O., Ind., Ill., and e. Ia. July–Sept. FIG. 1317. — Hybridizes with no. 20. — Submersed plants with elongate flexuous simple sterile stems and roundish slightly or hardly punctate thin leaves are forma **callitrichoìdes** Fassett (like *Callitriche*).

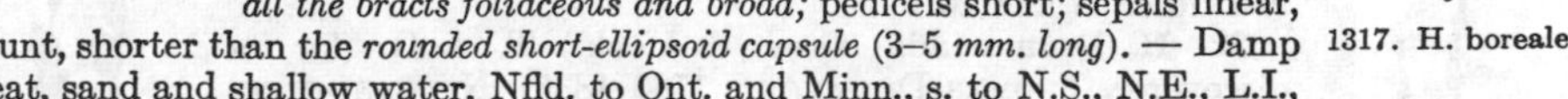

1317. H. boreale.

17. H. mùtilum L. (cut-off; the Linnaean type being merely a cut-off fragment of a plant!). — Stem slender, weak, 3–9 dm. high, widely branching at least above, perennial with leafy-bracted decumbent bases; *leaves ovate to narrowly oblong,* partly clasping, 5-nerved, *the upper and rameal ovate- to lance-deltoid* or tapering from base to obtuse tip; cyme in well-developed plants diffuse, somewhat leafy-bracted, *the ultimate bracts linear-setaceous;* flowers about 4 mm. broad; sepals linear-lanceolate, acute; *capsules* 2.5–3.5 *mm. long, short-ellipsoid, rounded at summit.* — Low grounds, Fla. to Tex., n. to s. N.E., centr. N.Y., O., Ill., Mo. and Kans. July–Sept. FIG. 1318. — Hybridizes with nos. 18, 19 and 20.

Var. **parviflòrum** (Willd.) Fern. (small-flowered). — Leaves all elliptic and gradually rounded to summit. — Extending n. to N.S., N.B., Que., s. Ont., Mich., Wisc. and Minn.

1318. H. mutilum.

Var. **latisépalum** Fern. (with broad sepals). — Sepals broadly lanceolate or oblong, foliaceous. — Low grounds, Fla. to Tex., n. to tidal mud of s. N.J.

18. H. gymnánthum Engelm. & Gray (with naked flowers). — Almost simple, with strict stem and branches, 3–9 dm. high; *leaves clasping, deltoid-cordate,* acute or obtuse; cyme naked, the floral leaves reduced to small subulate bracts; capsules slender-conical, pointed, 4–5 mm. long, slightly exceeding the lance-acuminate sepals. — Sandy, muddy or peaty low grounds, Fla. to Tex., n., locally, to L.I., N.J., Pa., W.Va., O., Ill., Mo. and e. Kans. Late June–Sept. FIG. 1319. — Hybridizes with no. 17.

1319. H. gymnanthum.

19. H. màjus (Gray) Britt. (larger). — Perennial by short leafy offshoots; stems solitary or tufted, erect, rather stout, 1–7 dm. high; *upper leaves chiefly* 5–7-*nerved at the rounded or subcordate* sessile or clasping *base, lanceolate,* acute or bluntish, 1.5–4.5 cm. long, 5–15 mm. broad; cymes essentially naked, the bracts slender; *sepals lance-attenuate,* 4–7 mm. long; capsule conic-ellipsoid, bluntish. — Wet or dry open soil, N.B. and Que. to s. B.C., s. to N.E., L.I., N.J., n. Del. (formerly), Pa., n. O., Ind., Ill., Ia., Neb., Colo. and Wash. July–Sept. FIG. 1320. — Hybridizes with nos. 17 and 20.

1320. H. majus.

20. H. canadénse L. (Canadian). — Perennial by short leafy offshoots; stems slender, 0.5–7.5 dm. high; *leaves* 1–3-*nerved, linear to linear-oblanceolate, rounded at tip, narrowed to the* sessile or subpetiolar *base,* 1–4 cm. long, 1–6 mm. broad; cymes naked except for the linear-setaceous bracts; *sepals* linear-lanceolate, blunt or acutish, 2.5–4.5 *mm. long;* petals yellow, oblong to oblanceolate, broadly round-tipped; *capsule* conical, *red or purplish,* 4–6.5

1321. H. canadense.

1322. H. dissimulatum.

mm. long. — Wet or dry, often exsiccated, places, Nfld. to Man., s. to Ga., Ala., Ill. and Ia. July–Sept. Fig. 1321. — Hybridizes with nos. 16, 17 and 19.

Var. **magninsulàre** Weath. (of Grand Manan Island). — *Petals more lanceolate, tapering to tip, pale lemon-yellow, tinged with red.* — Swampy ground, Grand Manan I., N.B.

Var. **galiifórme** Fern. (with form of *Galium*). — Dwarf, 0.8–3 dm. long or high, *weak* and reclining to erect; *leaves narrowly obovate or oblanceolate, petioled, the larger ones 5–15 mm. long and 2–4 mm. wide; sepals 1.5–2 mm. long; capsules ovoid, rounded at summit, 3–3.5 mm. long.* — In *Sphagnum* or on wet shores, se. Va. Sept.

21. H. dissimulàtum Bickn. (disguised). — Resembling nos. 19 and 20, either virgate or somewhat diffusely branched above; *leaves linear-oblong*, round-tipped, 0.5–3 cm. long, 1.5–10 mm. broad, sessile but not clasping; cymes as in nos. 19 and 20 or more diffuse; *sepals 2.5–4.5 mm. long; capsules 3–4.5 mm. long, often not filled-out.* — Peaty or wet sandy open soil, N.S. to e. N.Y., s. to se. Va. July–Sept. Fig. 1322. — Perhaps an unusually constant and recurrent hybrid of nos. 16 or 17 and 20.

1323. H. Drummondii.

1324. H. gentianoides.

22. H. Drummóndii (Grev. & Hook.) T. & G. (for its discoverer, Thomas Drummond, 1780–1835), Nits-and-lice. — Annual; stem and the mostly alternate bushy branches rigid, erect, 1.5–8 dm. high; *leaves linear-subulate*, nearly erect, *1-nerved*, 6–20 mm. long; *flowers scattered* along the upper part of the leafy branches, *short-pedicelled; capsule ovoid, barely as long as or but slightly exceeding the calyx.* (*Sarothra* Grev. & Hook.) — Dry or moist sandy or rocky soil, Fla. to Tex., n. to Md., W.Va., O., Ind., s. Ill., Ia. and se. Kans. July–Sept. Fig. 1323.

23. H. gentianoìdes (L.) BSP. (gentian-like), Orange-grass, Pineweed. — Annual; stem and bushy branches thread-like, wiry, 1–6 dm. high; *leaves minute subulate scales, appressed; flowers minute, mostly sessile* and scattered along the erect branches; *capsules lance-subulate*, acute, *much longer than the calyx.* (*Sarothra* L.) — Sandy, sun-baked or rocky soil, Fla. to Tex., n. to s. Me., s. N.H., s. Vt., N.Y., s. Ont., O., Ind. and Wisc. July–Oct. Fig. 1324.

§ Elodèa (Juss.) Choisy (Triadenum Raf.)

24. H. virgínicum L. (Virginian), Marsh-S. — Stoloniferous; stem simple or bushy-branched, 2–8 dm. high; *leaves* oblong or oblong-ovate, round-tipped, *sessile and cordate or clasping*, the principal ones 2–7 cm. long, often purplish; *flowers in small clusters in the upper axils and terminal;* mature sepals lanceolate, acute, 5–7 mm. long; *filaments free except near base, the 3 fascicles alternating with 3 large orange glands; petals flesh-color or mauve;* capsule tapering to summit, the *mature styles 2–3 mm. long.* (*Triadenum* Raf.) — Wet sands, bogs and swamps, Fla. to N.S., s. N.E., N.Y., O., Ind. and Ill. July, Aug.

Var. **Fràseri** (Spach) Fern. (for its discoverer, the younger John Fraser, 1799–1860?). — *Mature sepals oblong or elliptic, blunt,* 2.5–5 *mm. long;* capsule often rounded at tip, the *mature styles* 0.5–1(–2) *mm. long.* (*Triadenum Fraseri* (Spach) Gleason) — Nfld. and s. Lab. Pen. to Man., s. to N.S., N.E., centr. Pa., Ind., Mich., Minn. and Neb. Fig. 1325.

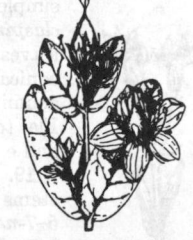

1325. H. virginicum, v. Fraseri.

1326. H. tubulosum, v. Walteri.

25. H. tubulòsum Walt. (tubular; from the united filaments), Marsh-S. — Similar, greener; *leaves* oblong or oblong-oblanceolate, the principal ones 4–15 cm. long, *sessile or nearly so,* the upper and sometimes all rounded to cordate at base; sepals oblong to linear-lanceolate, obtuse or acute; *filaments united to above the middle.* (*Triadenum* (Walt.) Gleason; *T. longifolium* Small) — Cypress- and gum-swamps, Fla. to La., n. to se. Va., s. O., s. Ind. and Mo. Late Aug., Sept.

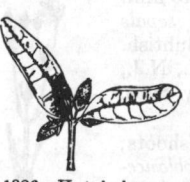

Var. **Wálteri** (Gmel.) Lott (in honor of Thomas Walter, ?1740–1789). — *Leaves tapering to petioles.* (*H. petiolatum* Walt., not L.; *Triadenum Walteri* (Gmel.)

Gleason; *T. petiolatum* Britt.) — Similar habitats, Fla. to Tex., n. to Md. (?N.J.), W.Va., s. Ind. and se. Mo. Fig. 1326.

FAM. 112. ELATINÀCEAE (Waterwort Family)

Low and bland herbs with opposite or sometimes verticillate simple dotless leaves with membranaceous stipules between them, small regular axillary flowers, the persistent or marcescent sepals and petals imbricated in the bud. **Stamens same number as petals and alternate with them or twice as many. Ovary with as many locules as the sepals, placentae axile. Styles introrsely stigmatose or stigmas sessile. Valves of capsule alternate with dissepiments. Seeds, oblong-cylindric, straight or curved, the crustaceous testa nearly filled by the cylindric embryo; cotyledons thick and short.**

Flowers 2–4-merous (2- or 3- in ours); sepals membranaceo-herbaceous, obtuse, without midrib; capsules globose or depressed, membranaceous, often bursting irregularly; dwarf aquatic or amphibious creeping glabrous annuals or subperennials. . . . . 1. *Elatine.*

Flowers 5-merous; sepals pointed, with thickened midrib and scarious margins; capsule firm, ovoid; terrestrial annuals or perennials, mostly pubescent. . . 2. *Bergia.*

1. ELATÌNE L. Waterwort

Sepals 2–4, obtuse. Petals 2–4, hypogynous. Stamens as many, rarely twice as many as petals. Styles or sessile capitate stigmas 2–4. Capsule membranaceous, 2–4-locular, several–many-seeded, 2–4-valved; the partitions left attached to the axis, or evanescent. — Dwarf plants, often rooting at the nodes, in temp. reg. of most lands. (Classical name of a low and creeping plant, transferred to this genus.)

Seeds mostly straight, with obtuse cross-ribs separating the elliptical round-ended pits; flowers 2-merous; plant closely matted, with ascending branchlets only 0.2–5 cm. long; leaves cuneate-obovate to oblong, rounded at summit, 0.7–5 mm. long. . . . . 1. *E. minima.*

Seeds with acute cross-ribs separating the 6-sided angular-ended pits; flowers 3- or 2-merous; closely matted or with elongate branches; leaves 2–15 mm. long.

- Leaves obovate to broadly spatulate, rounded at summit, 3–8 mm. long, the larger 1.5–5 mm. broad; seeds borne from lower half of central axis, ascending, curved, slenderly cylindric, the pits 20–30 in each row; stems densely matted; the crowded ascending branches 1–5 cm. long. . . . . . . 2. *E. americana.*
- Leaves linear, lanceolate, oblong or narrowly spatulate, 2–15 mm. long, 0.5–3 mm. broad; seeds borne along entire length of thickened central axis, divergent.
  - Stems loosely matted or with submersed branches up to 2 dm. long; leaves 2.8–15 mm. long, 0.5–3 mm. broad, commonly truncate to emarginate; seeds elongate, curved, with 16–25 pits in each row, the larger pits 0.08–0.1 mm. long. . . . . 3. *E. triandra.*
  - Stems densely matted, the ascending branchlets 0.5–2.5 cm. long; leaves 1.5–4 mm. long, 0.7–1.8 mm. broad, with rounded, tapering or barely notched tips; seeds straightish, thick-cylindric, with 9–15 pits in each row, the larger pits 0.12–0.14 mm. long. . . . . 4. *E. brachysperma.*

1. E. mínima (Nutt.) Fisch. & Mey. (smallest). — Creeping, *forming small mats rarely* 1 *dm. broad,* the erect or strongly ascending fleshy *branchlets* 0.2–5 *cm. high; leaves cuneate-obovate to oblong,* sessile or obscurely petioled, *rounded at summit,* 0.7–5 *mm. long,* 0.3–3 *mm. broad;* flowers sessile, solitary in the axils; sepals, petals and stamens 2; *seeds borne at base of central axis, vertical,* thick-cylindric or barrel-shaped, *mostly straight,* 220–280 μ *thick, with straight and definite longitudinal ribs and* 15–18 *cross-ribs, the elliptical round-ended pits* 15–18 *in each row.* — Sandy, peaty or (rarely) muddy margins and shores of ponds and deadwaters, Nfld. and Saguenay Co., Que., to Minn., s. to Md., ne. Va., s. Mich. and Wisc. July–Oct.

2. E. americàna (Pursh) Arn. (American). — *Forming prostrate mats* up to 2 dm. across; the subascending branchlets 1–5 cm. long; *leaves obovate to broadly spatulate, rounded or shallowly notched at summit,* 3–8 *mm. long, the larger* 1.5–5 *mm. broad;* flowers 3-merous, with pinkish petals; *seeds borne at base of central axis, erect,* slender-cylindric, *usually curved,* 140–190 μ *thick, with meandering or obscure longitudinal ribs, the* 20–30 *6-sided angular-ended pits in each row* *separated by acute cross-ribs.* (*E. triandra*, var. (Pursh) Fassett) — Muddy (often tidal) shores and margins of ponds and streams, N.B. and Que. to e. Va.; also Mo. and Okla. July–Oct.

3. E. triándra Schkuhr (with three stamens). — Matted and *with linear, narrowly lanceolate or narrowly spatulate, mostly truncate or emarginate leaves* 2.8–6.5 mm. long and 0.5–3 *mm. wide*, or branches elongating to 1–2 dm. and flaccid with leaves up to 1.5 cm. long in forma **submérsa** Seubert (submerged); flowers 3-merous; *seeds borne along entire length of thickened central axis, horizontally divergent, slender and curved*, resembling those of no. 2, *the larger of the* 16–25 *pits in each row* 0.08–0.1 *mm. long.* — Shallow water and shores, Wisc. to Alta. and Wash., s. to Tex., n. Mex. and s. Calif.; introd. in Somerset Co., Me. Aug.–Oct. (Eurasia)

4. E. brachyspérma Gray (short-seeded). — In *tiny mats* 1–5 cm. broad, *the ascending branchlets* 0.5–2.5 *cm. long; leaves* narrowly oblong to oval, *narrowed or rounded to apex*, 1.5–4 *mm. long*, 0.7–1.8 *mm. broad;* flowers 2- or 3-merous; capsule depressed; the *short-oblong thick seeds with* 9–15 *pits in each irregular row, the larger pits* 0.12–0.14 *mm. long* and separated by acute cross-ribs. (*E. triandra*, var. (Gray) Fassett) — Shallow water and shores, rare, O. and Ill. to Oreg., s. to Tex., Ariz. and Calif. July–Oct.

2. BÉRGIA L.

Sepals 5, acuminate, with thickened midnerve and scarious margins. Petals 5. Stamens 5 or 10. Capsule of firm texture. — Diffuse or ascending plants, chiefly trop. (Named for *Peter Jonas Bergius*, 1723–1817, a Swedish botanist.)

1. B. texàna (Hook.) Seubert (Texan). — Branched from the base; branches 1–4 dm. long; flowers scarcely peduncled in the axils of the lance-oblong serrulate leaves. — Swamps and wet banks, Ark. to Tex. and s. Calif., n. to s. Ill., Mo., S.D. and Wash. June–Sept.

FAM. 113. TAMARICÀCEAE (Tamarisk Family)

Shrubs or trees (rarely perennial herbs) with alternate entire thickish or scale-like exstipulate leaves and regular perfect or rarely dioecious flowers. Sepals 4–6, free nearly or quite to base and imbricated. Petals of same number, either distinct or united, inserted outside a hypogynous disk. Stamens 4–∞, with free or united filaments; anthers introrse, 2-locular. Ovary single, 1-locular, with 3–5 styles and 3–5 parietal placentae. Capsule with valves of same number as styles. Seeds few–∞, mostly without endosperm; embryo straight. Small family nat. of warm reg. of the Old World.

1. TÁMARIX L. Tamarisk

Sepals 4–6. Petals distinct nearly or quite to base. Stamens 4–12, nearly or quite distinct. Ovary ovoid-attenuate, with 3 or 5 short thick styles. Placentae basal, with many ovules. Leaves minute and scale-like. — Mediterr. and Asiatic trees and shrubs with scaly branches and spicate pink or white small flowers. (The classical name.)

1. T. gállica L. (French), French T. — Glabrous shrub or small tree with flexuous brown branches and savin-like pale green or glaucous foliage; flowers numerous in terminal spikes. — River-banks, bars, and saline marshes, Mo. and Kans., and in roadside-thickets and waste places, s. states and casually esc. from cult. n. to Mass. and Ind. May–Sept. (Introd. and natzd. from s. Eu.)

FAM. 114. CISTÀCEAE (Rockrose Family)

Low shrubs or herbs with regular or nearly regular flowers, distinct and hypogynous mostly indefinite stamens, a persistent calyx, a 1-locular 3–5-valved capsule with as many parietal placentae borne on the middle of the valves, and orthotropous albuminous seeds. Sepals 5; the two external much smaller and bract-like or sometimes wanting; the 3 others slightly twisted in the bud. Petals mostly 3 or 5, convolute in the opposite direction from the calyx in the bud. Anthers short, innate, on slender filaments. Style single or none. Ovules few or many, on slender stalks. Embryo long and slender, straightish or curved, in mealy albumen; cotyledons narrow. Leaves simple and mostly entire, the lower opposite and the upper alternate or all opposite or all alternate.

Bases of plants without overwintering rosulate branches; petals 5 (or sometimes none in no. 1), crumpled in the bud, yellow, fugacious; style usually definite; stamens very numerous (at least in petaliferous fls.); capsule 1-locular.

Leaves dilated, not imbricated; flowers of two sorts, the terminal and earlier ones with broad petals; the later with petals reduced or wanting, these flowers cleistogamous and highly fertile; style short or wanting. . . . . . 1. *Helianthemum*.

Leaves scale-like or needle-like, imbricated; flowers uniform, axillary, with small narrow petals; style slender, elongate. . . . . 2. *Hudsonia*.

Bases of plants usually with overwintering rosulate leafy offsets; petals 3, flat in the bud, reddish, withering-persistent, small, not showy; stamens 3–25; style short or none; capsule partly 3-locular, each of the imperfect partitions bearing a pair of ovules. . . . . 3. *Lechea*.

1. HELIÁNTHEMUM Mill. Rockrose

Capsule 1-locular. Embryo curved in the form of a hook or ring, or plicate. Flowers in many N. Am. (and in some Eu.) species of two sorts, viz., *primary* or earlier ones, with large yellow petals, indefinitely numerous stamens and many-seeded capsules; and *secondary* or later ones, which are much smaller and cleistogamous and with small petals or none, 3–10 stamens, and much smaller 3–few-seeded capsules. — The large flowers opening only once, in sunshine, and casting their petals by the next day. Genus extensive in Am., Eurasia and Afr. (Name from the Greek *helios, the sun,* and *anthemon, flower.*) Including Halimium (Dunal) Spach and Crocanthemum Spach

Early petaliferous flowers solitary (rarely 2), soon overtopped by lateral branches; later cleistogamous flowers 1–few on each branchlet, their larger capsules 3–5 mm. in diameter; seeds truncate above, papillate or pebbled.

Stems mostly 2–4 dm. high, with strongly ascending branches and branchlets; cleistogamous flowers several, very unequal in size; seeds longer than thick, with long papillae. . . . . 1. ***H. canadense.***

Stems 0.5–2.5 dm. high, with widely divergent branches and branchlets; cleistogamous flowers solitary and large, at tips and forks of leafy branchlets; seeds nearly as thick as long, with low pebbling. . . . . . 2. ***H. dumosum.***

Early petaliferous flowers 2–15 in terminal corymbs, these rarely overtopped by the later branches; mature cleistogamous capsules uniform, crowded, 2–3 mm. in diameter; seeds rounded, finely reticulate.

Stems clustered at the tips of multicipital caudices, without creeping rootstock, mostly 2–6 dm. high. . . . . 3. ***H. Bicknellii.***

Stems scattered, many of them arising singly from subterranean elongate rootstocks, mostly 1–3 dm. high. . . . . 4. ***H. propinquum.***

1. H. canadénse (L.) Michx. (Canadian), Frostweed. — Subcespitose, the numerous hoary-pubescent erect or arching *stems* from a multicipal caudex, at first simple and glutinous, *mostly 2–4 dm. high; leaves* lance-oblong to oblanceolate, tapering at base, pale beneath, *green above; branches and branchlets ascending; terminal petaliferous flower* solitary (rarely 2), 2.5 cm. broad, *soon overtopped by the prolonging upper branches,* their capsules 6 mm. long; the small mostly *apetalous cleistogamous flowers glomerulate at tips of 1–few-flowered ascending branchlets and often scattered and sessile below in the leaf-axils; their capsules unequal, those from the terminal flowers* larger and *3–4 mm. in diameter; seeds* somewhat *thimble-shaped, truncate at one end,* rounded at other, *much longer than thick, strongly papillate.* (*Crocanthemum* Britt.) — Dry open woods, clearings and barrens, N.S.; s. Me. to sw. Que., s. Ont. and Wisc., s. to N.C., Ky., Miss. and Mo. *Petaliferous fls.* mid-May–early July. — Late in autumn crystals of ice shoot from the cracked bark at the base of this and no. 3, whence the popular name.

Var. **sabulònum** Fern. (of sands). — Stems few, *decumbent or loosely ascending; leaves oblong-elliptic,* often canescent above; *cleistogamous flowers* mostly *long-pedicelled, in loose corymbs terminating upper branches, uniform, in maturity 4–5 mm. in diameter.* — Dunes and open sands, local, se. Mass. and Oneida L., N.Y., to se. Va.

2. H. dumòsum (Bickn.) Fern. (bushy). — *Forming loosely ascending to depressed mounds* 0.5–2.5 *dm. high;* the *stems soon stiffly and almost horizontally branching,* with coarser and more elongate pubescence than in no. 1 often intermixed with reddish glandular hairs; leaves mostly smaller than in no. 1, crowded on the divergent branches; *petaliferous flowers solitary at tips of stem and main branches,* paler and earlier than in no. 1; sepals commonly with red papillae among the pale hairs; *cleistogamous flowers solitary and large, at tips and forks of leafy branchlets; seeds nearly as thick as long, pebbled.* (*Crocanthemum* Bickn.) — Dry sands, barrens and open woods, se. Mass.; Block I., R.I.; L.I. *Petaliferous fls.* late May, June.

3. H. Bicknéllii Fern. (for the student of the group in North America, Eugene Pintard Bicknell, 1859–1925), Frostweed. — Habit of no. 1, the many clustered stems up to 6 dm.

high; *leaves* oftenest *canescent above;* terminal *petaliferous flowers* 2–15 *in a loose* (often racemiform) *corymb,* this usually *not surpassed by the later appressed short branches which are crowded with* canescent cleistogamous flowers producing *uniform capsules* 2–3 *mm. in diameter; seeds rounded, finely reticulate.* (*H. majus* sensu BSP. and *Crocanthemum majus* sensu Britt., not *Lechea major* L. upon which their binomials rest) — Dry rocky, sandy or argillaceous open woods, clearings and plains, centr. Me. to Minn., S.D. and Colo., s. to Md., mts. of N.C., O., Ind., Ill., Mo. and Kans. *Petaliferous fls.* June, July.

4. H. propínquum Bickn. (near). — *Very slender, with widely creeping subterranean rootstocks sending up solitary and scattered or slightly clustered wiry erect stems* 1–2(–3) *dm. high; leaves linear-*, to narrowly oblong-*oblanceolate;* petaliferous flowers in lax racemiform corymbs, long-pedicelled; cleistogamous flowers and fruits much as in no. 3, densely crowded on erect branches which often overtop the vernal inflorescence; seeds nearly as in no. 3. (*Crocanthemum* (Bickn.) Bickn.) — Dry open sand and barrens, se. Mass. to Md. and D.C. *Petaliferous fls.* late May–July.

2. HUDSÒNIA L. Hudsonia

Petals much larger than calyx. Style long and slender; stigma minute. Capsule terete, inclosed in the calyx, strictly 1-locular, with 1 or 2 seeds attached near the base of each nerve-like placenta. Embryo coiled into the form of a closed hook. — Bushy heath-like low N. Am. shrubs, covered with the small subulate or scale-like alternate persistent downy or villous leaves, producing numerous small but showy usually bright yellow flowers crowded along the upper part of the branches. (Named in honor of *William Hudson*, 1730–1793, English botanist.)

1. H. ericoìdes L. (heath-like), Golden-heather. — Somewhat downy but *greenish; leaves* villous but *not densely involved in pubescence, the* linear-subulate *blades essentially free and prolonged; flowers on slender naked pedicels* 4–10 *mm. long,* bright yellow, or whitish in forma **leucántha** Fern. (white-flowered); *ovary pubescent.* — Dry sands, pinelands and acidic rocks, Nfld.; P.E.I.; N.S.; se. and s. Me. and centr. N.H., interruptedly to N.J. and Del.; reports from Va. and N.C. needing verification; ascending from sea-level to bleak mountain-tops. Late May–early July. — Represented on mts. of N.C. by the closely related *H. montana* Nutt.

2. H. tomentòsa Nutt. (tomentose), Beach-heath, Poverty-grass. — *Hoary with villous tomentum; leaves densely implicated in pubescence, closely imbricated, their tips not projecting; flowers sessile* or on very short leafy-bracted pedicels; petals sulphur-yellow; *ovary glabrous or nearly so.* — Dunes and sand-blows, Lake St. John, Que., to n. Alta., s. to s. Ont., nw. Ind., Ill., Minn. and Sask.; and chiefly near the coast, e. and s. Gaspé Pen., Que., M.I. and e. N.B. and from s. and sw. Me. and centr. N.H. to N.C. May–July.

Var. **intermèdia** Peck (intermediate). — Leaf-tips more evident, often projecting; flowers on naked pedicels 1.5–7 mm. long. — S. Lab. Pen., Que., to n. Alta., s., locally, to P.E.I., w. Me., centr. N.H., e. Cape Cod, Mass., w. Vt., ne. N.Y., Mich., nw. Ind. and n. Ill. — Needs critical study; looking like a hybrid of nos. 1 and 2, but extending 2000 mi. nw. of range of no. 1.

3. LÉCHEA L. Pinweed

Sepals 5, in 2 distinct series; the 2 outer sepals narrow, foliaceous, often inconspicuous; the 3 inner ones imbricated in bud, broad, scarious. Petals 3, reddish, mostly shorter than sepals, imbricated, flat in bud, marcescent, rarely seen expanded. Stamens 3–25 (mostly 5–15). Ovary short-stipitate; style very short or none; stigmas 3, dark red, fimbriate-plumose; placentae 3, broad and valve-like, each bearing a pair of erect subsessile ovules, one on each side of the posterior face. Capsule 3-valved. Embryo slender, nearly straight to curved, in hard endosperm. — Low N. Am. suffruticose or herbaceous perennials, mostly with overwintering basal leafy shoots, alternate or falsely verticillate small cauline leaves, and abundant small (rarely expanded except in early morning) drab to brownish or reddish pyriform to subglobose paniculate flowers only 2–4 mm. broad. (Named for *Johan Leche*, 1704–1764, Swedish botanist.)*

a. Outer sepals nearly equaling to exceeding the inner ones. . . *b.*

b. Leaves of basal shoots ovate or elliptic; cauline leaves lanceolate or oblanceolate to elliptic or ovate.

Leaves of basal shoots 0.8–1.5 cm. long; pubescence of stem spreading-pilose or villous; fruiting calyx depressed-globose, the outer sepals about equaling the inner ones; seeds lustrous, inequilateral, the embryo clearly visible. . . . . 1. *L. villosa.*

---

* Treatment condensed from that of A. R. Hodgdon in Rhodora, xl. (1938).

Leaves of basal shoots 4–7 mm. long; pubescence of stem appressed or incurving; fruiting calyx pyriform, the outer sepals exceeding the inner ones; seeds opaque, equilateral, the embryo not clearly visible. . . 2. *L. minor.*

*b.*Leaves of basal shoots and of fruiting stem narrowly linear; fruiting calyx depressed-globose, the outer foliaceous sepals equaling to much exceeding the inner. . . . . 3. *L. tenuifolia.*

*a.*Outer sepals shorter than the inner ones. . . *c.*

*c.*Leaves of basal shoots thin, lance-elliptic to oblong-ovate, less than three times as long as broad; fruits scattered-racemose on the secondary branches; capsules pyriform, long-pedicelled, 1–1.2 mm. broad; seeds 1–3. 4. *L. racemulosa.*

*c.*Leaves of basal shoots firm to thin, linear to elliptic-lanceolate, three to six times as long as broad; fruits scattered or crowded; capsules 1.3–2.1 mm. broad; seeds 2–6. . . *d.*

*d.*Leaves of basal shoots thick, dull green, pilose over the lower surface, lanceolate to lance-elliptic, 1.5–3.5 mm. broad; axis of basal shoots densely pilose to tomentulose; panicle usually branched from well below the middle, often broadly subpyramidal. . . . . . 5. *L. maritima.*

*d.*Leaves of basal shoots thin, bright green above, pilose beneath only on midrib and margin, linear-lanceolate or -oblanceolate to narrowly lance-elliptic, 0.7–2(–2.5) mm. broad; axis of basal shoots subappressed- to spreading-pilose; panicle usually branching from above middle, cylindric, spire-like or slenderly pyramidal. . . *e.*

*e.*Seeds 2–4, dorsiventrally compressed (2-sided) or obscurely 3-sided, nearly equilateral, elliptic-oblong or enlarged at base, smooth at maturity.

Plant gray-green; stem and crowded strongly ascending or suberect branches strongly pilose; calyx strongly spreading-pilose. . . . 6. *L. stricta.*

Plant bright green; stem and ascending to broadly spreading branches sparingly pilose; calyx sparingly appressed-pilose. . . 7. *L. Leggettii.*

*e.*Seeds 4–6, inequilateral and prominently 3-sided, dorsally convex, with 2 subequal lateral faces meeting in a straight ventral keel, at maturity with a white reticulate membranaceous covering. . . . . . . 8. *L. intermedia.*

1. L. villòsa Ell. (villous). — Caudex low, simple; *basal shoots* simple or branching, up to 1 dm. long, *their axes densely villous;* their crowded mostly whorled *leaves bright green, broadly elliptic-ovate,* glabrous above, villous beneath, 0.8–1.5 *cm. long; fruiting stems* few, erect, 1.5–9 dm. high, brownish-purple, *villous;* cauline leaves oblanceolate to lance-elliptic, bright green, villous below, sparsely spreading-pilose above, 1–3 cm. long; panicle closely subcylindric to broadly pyramidal, often very densely glomerate-flowered; *calyx depressed-globose,* dark brown, with pale brown obconic base; *inner sepals subacute,* 1.3–1.6 mm. long, deeply concave, *prominently keeled,* the keel conspicuously pilose; *exterior sepals* linear or narrowly lanceolate, *about equaling inner; capsule depressed-globose; seeds* mostly 3, bright brown to clear yellow, *lustrous, inequilateral,* strongly convex dorsally, concave or flattened ventrally, *the embryo clearly visible* through the transparent endosperm. — Dry sandy or gravelly open woods and clearings, Fla. to e. Tex., n. to s. N.H., s. Vt., e. and w. N.Y., s. Ont., Mich., n. Ill., Mo. and Kans. *Fr.* July–Nov.

2. L. mìnor L. (smaller). — Caudex simple or with erect branches; *basal shoots* rarely 5 cm. long, *subappressed-pilose; their* ovate or broadly elliptic more or less pilose *leaves* mostly in 3's, 4–7 *mm. long; fruiting stems* 1–several, 2–5 dm. high, *their pubescence subappressed or incurving;* cauline leaves bright green, elliptic-ovate to lanceolate, spreading-pilose on both faces, often in 3's or 4's; panicle compact, leafy, slenderly subcylindric to subglobose; *calyx pyriform,* with short obconic base, 1.6–2 mm. long, 1.2–1.5 mm. thick, on pedicels 1–2 mm. long; *inner sepals* brownish-drab, subappressed-pilose, *obtuse,* abruptly narrowed at base; *exterior sepals* bright green, *equaling or exceeding inner; capsule ellipsoid-ovoid to elongate-ellipsoid, the valves unequal; seeds* 2 or 3, light to dark brown, *opaque, nearly equilateral,* dorsiventrally compressed but slightly convex dorsally and flattened beneath or obscurely 3-sided and conspicuously angled ventrally, *the embryo not clearly visible.* — Sandy woods, clearings and dry pond-borders, n. Fla. to La., n. to se. Mass., sw. N.H., se. N.Y., N.J., Pa., O., s. Ont., s. Mich., nw. Ind. and ne. Ill. *Fr.* Aug.–Nov.

3. L. tenuifòlia Michx. (slender-leaved). — Caudex reduced, hardly branched; *basal shoots* up to 7 cm. long, *slender, often numerous and much branched, forming dense mats,* the axis subappressed-pilose; *the crowded leaves narrowly lanceolate or oblancèolate to linear,* 4–6 *mm. long, less than* 1 *mm. broad, bright green* and glabrous above, *subappressed- or slightly spreading-pilose beneath; fruiting stems* 1–several, 1.2–4 dm. high, *capillary,* minutely pilose; *cauline leaves scattered, linear or linear-subulate,* 0.8–2 cm. long, 1–1.5 *mm. broad;* rameal leaves smaller and more persistent; *panicle openly spreading-branched from below the middle,* subglobose to broadly spire-like, the fruits mostly subappressed and racemose on the branchlets; *calyx subglobose, completely covering capsule,* 1.6–1.9 mm. long; inner sepals spreading-pilose, dull green, subacute, with a hard keel; *outer sepals 2–3 mm. long,* equaling to much exceeding inner; *capsule ovoid,* 1.3–1.5 mm. thick; seeds 2–5, opaque, yellowish- to reddish-brown, plump, broadened and greatly thickened toward base, 0.7–1 mm. long. — Dry sandy or rocky open woods and slopes, upland of S.C. to sw. Me.; La. and Tex., n. to Ind., Wisc. and e. Minn. *Fr.* late June–Nov.

Var. **occidentàlis** Hodgdon (western). — More robust, with *branches erect; leaves dull, densely pilose beneath; fruiting calyx* 1.9–2.2 *mm. long,* the outer sepals prolonged; capsule equaling inner sepals; seeds mostly 2, 1–1.2 mm. long. — Dry woods, rocks and sands, nw. Ill.; s. Neb. to n. Tex.

4. L. racemulòsa Michx. (with small racemes). — Caudex short, mostly simple; *basal shoots* up to 8 cm. long, ascending, finally often branched, their axes spreading-pilose, *their thin bright green oblong-ovate to elliptic-lanceolate leaves 4–6 mm. long and* 1.5–2.5 *mm. broad;* fruiting stems 1–4.5 dm. high, minutely and sparsely pilose; cauline leaves elliptic-lanceolate to narrowly oblanceolate, those of the branches linear; panicles loosely to compactly branched, or 1 central short panicle subglobose, the *long-pedicelled fruits racemosely scattered on the secondary branches; calyx slenderly pyriform,* subappressed-pilose, 1.8–2 mm. long, about half as broad; inner sepals narrowly elliptic to spatulate, longer than the linear or lanceolate outer ones; *capsule* equaling or exceeding calyx, *slenderly ellipsoid or ovoid,* 1–1.2 *mm. broad,* often inequilateral; *seeds* 1–3 (mostly 2), dark brown, 1–1.3 mm. long, ellipsoid-ovoid, nearly equilateral, the embryo not visible through the opaque endosperm. — Sandy open woods, barrens and fields, upland of Ga. and Ala., n. on upland and Coastal Plain to se. N.Y., N.J., e. and s. Pa., s. O. and s. Ind., locally to nw. Ind. and e. Mo. *Fr.* Aug.–Nov.

5. L. marítima Leggett (maritime). — *Caudex* simple to multicipital, *with stout nearly erect branches; basal shoots* simple, up to 1 dm. long, *the stout axis densely pilose to subtomentose; their thick dull green* imbricated *lanceolate to elliptic leaves three to five times as long as broad and pilose over the lower surface;* fruiting stems stout, erect to strongly inclined; cauline leaves whorled and caducous below the panicle, alternate and more persistent in the panicle, narrowly oblanceolate, 0.7–2.5 cm. long, 1.2–4 mm. broad; *panicle subpyramidal to subcylindric or 1-sided,* 1.5–3.5 dm. high, *its branches starting from near the base to about the middle of the stem;* fruits mostly subappressed, on secondary or tertiary branches; calyx pyriform to ovoid, 1.6–2.1 mm. long, longer than pedicels, subappressed-pilose; inner sepals acute or subacute, outer much shorter than to equaling them; capsule 1.3–2.1 mm. long, subglobose to ovoid; seeds 2–5(–6), light brown, with smooth dull surface, 0.8–1.2 mm. long, the embryo faintly seen through the subtransparent endosperm. Three vars.:

Panicles thick-subcylindric to broadly subpyramidal, their branches mostly from below middle of stem; calyx mostly pyriform to obconic or subglobose, the outer sepals much shorter than the inner; seeds 2–4(rarely –5).

Seeds 3–4(–5), obscurely 3- or 2-sided, ventrally convex, 0.8–1(–1.1) mm. long, much narrower; panicles mostly uniformly branching at about the middle of stems. . . . . *L. maritima* (typical).

Seeds 2, dorsally convex, ventrally flattened to slightly concave, 1–1.2 mm. long, 0.7–0.8 mm. broad; panicles branching chiefly from near bases of stems, often 1-sided and persistent to 2nd year. . . . . Var. *virginica.*

Panicles slenderly subcylindric, their branches mostly from above middle of stem; calyx slightly depressed-globose, the outer sepals often nearly equaling the inner; seeds 4–6. . . . . Var. *subcylindrica.*

L. marítima (typical). — Dunes and sandy flats near coast and slightly inland, s. Me. to n.-centr. N.H. (Crawford Notch) and w. Mass., s. to e. Md. *Fr.* Aug.–Nov. — Including var. *interior* Robins.

Var. **virgínica** Hodgdon (Virginian). — Dunes and open sands, e. Md. and e. Va. *Fr.* Sept.–Nov.

Var. **subcylíndrica** Hodgdon (nearly cylindric). — Sandy coast, e. N.B. *Fr.* late Aug.–Oct.

6. L. strícta Leggett (upright). — *Gray-green;* basal shoots simple, 2–3 cm. long, their axis strongly spreading- or subappressed-pilose to villous, their crowded lanceolate or lance-elliptic leaves 0.7–1.2 cm. long; *fruiting stems* 1–several, 2.5–4.5 dm. high, erect or strongly ascending, *at first subtomentose, becoming strongly pilose; cauline leaves subappressed, narrowly oblanceolate, acute,* 1.3–2 cm. long, strongly pilose beneath on midrib and margin; *panicle* branching from above middle of stem, the *branches erect or strongly ascending, strongly pilose; calyx* subglobose, *strongly pilose with spreading pubescence,* 1.6–1.8 mm. long, 1.2–1.3 mm. broad; inner sepals ovate or obovate, obtuse, the outer linear and rarely two-thirds as long; capsule subglobose; seeds 3 or 4, smooth, narrowly ovoid and equilateral, or dorsiventrally compressed and inequilateral. — Sandy woods, prairies and fields, Wisc. and Minn., s. to nw. Ind., n. Ill., Ia. and ne. Neb.; near Belleville, Ont. and in w. N.Y. *Fr.* July–Oct.

7. L. Leggéttii Britt. & Hollick (for close student of the genus, William Henry Leggett, 1816–1882). — Bright green; caudex mostly simple; basal shoots up to 8 cm. long, their axes subappressed-pilose, their scarcely or but slightly overlapping acute lanceolate or oblanceolate leaves 4–10 mm. long and sparsely pilose on midrib and margin beneath; *fruiting stem* mostly solitary, 2.5–8 dm. high, *sparsely subappressed-pilose to glabrate;* cauline leaves linear or narrowly oblanceolate, 1–2 cm. long, attenuate to slender petioles about 1 mm. long; panicle subcylindric to subglobose; *calyx* subglobose to slenderly pyriform, 1.6–2 mm. long, *sparingly subappressed-pilose;* inner sepals prominently 3-veined, not keeled, longer than the lanceolate outer ones; capsule subglobose or ellipsoid, 1.3–1.5 mm. broad; *seeds* 2 *or* 3(–4), 1–1.25 mm. long, brown, the embryo visible. — Three vars.:

Panicle slenderly ovoid to subcylindric, principal branches subequal, the ultimate ones greatly reduced, the fruits crowded or clustered; fruiting calyx brown, cuneate-obovoid; seeds 3 or 4, 1–1.1 mm. long, frequently thickened dorsiventrally and keeled. . . . . *L. Leggettii* (typical).

Panicle subcylindric to subglobose, its branches diminishing upward, the ultimate ones several cm. long, bearing racemes of scattered fruits; fruiting calyx brownish- to reddish-purple, pyriform; seeds 2 or 3, 1.1–1.25 mm. long, dorsiventrally compressed or 3-sided.

Fruiting stems at most 6 dm. high; panicles branching from near the middle of stem, subcylindric to slenderly ellipsoid; calyx slenderly pyriform; seeds mostly 3, frequently inequilateral and 3-sided. . . . . Var. *moniliformis.*

Fruiting stems up to 8 dm. high; panicles branching well above the middle, broadly and irregularly ovoid to globose; seeds mostly 2, dorsiventrally compressed and equilateral. . . . . Var. *ramosissima.*

L. Leggéttii (typical). — Dry woods, pond-shores and fields, mts. of Va. to Mass., R.I., Ct., se. N.Y., N.J., centr. Pa. and ne. O. *Fr.* late July–Oct.

Var. **monilifórmis** (Bickn.) Hodgdon (resembling a chain of beads). — Dry or damp open woods, sands and peats, Md., N.J., L.I. and ids. of se. Mass.; w. L. Erie reg., s. Ont. and O., to s. Mich. and n. Ill.

Var. **ramosíssima** Hodgdon (most branched). — Sandy open woods, peats and fields, chiefly of Coastal Plain, Va. to Fla. and La.

8. L. intermèdia Leggett (intermediate). — Caudex mostly simple; *basal shoots* up to 7 cm. long, *their axes subappressed-pilose,* their elliptic- to oblong-lanceolate leaves 3–8 mm. long and sparingly subappressed-pilose on midrib and margin beneath; *fruiting stems* 1–many, up to 6 dm. high, *sparingly subappressed-pilose;* cauline leaves oblanceolate, pointed, 1–2.5 cm. long; panicles subcylindric to spire-like, their spreading-ascending to suberect branches short; *calyx appressed-pilose,* ovoid to globose or sometimes pyriform, 1.9–2.3 mm. long, 1.9–2.1 mm. broad; inner sepals 3–5-nerved, not keeled, the linear outer ones one-half to three-fourths as long; *capsules ovoid to depressed-globose,* 1.8–2.1 *mm. broad; seeds* 4–6, pale to medium brown, *when mature with a thin adherent puckering coat,* 1.1–1.3 mm. long, 3-*sided and inequilateral, dorsally convex, with* 2 *subequal lateral faces meeting in a straight ventral keel.* — Three vars.:

Calyx and mature capsules depressed-globose; the capsule conspicuous, equaling to exceeding the calyx; interior sepals broadly ovate, obtuse or subobtuse, 5–3-veined; panicles mostly openly branched; capsules opening in mid- to late autumn. . . . . *L. intermedia* (typical).

Calyx and mature capsules globose to broadly ellipsoid; calyx slightly pyriform, exceeding the capsule; interior sepals elliptic-ovate and subacute, 3-veined; plants densely and closely branched; capsules opening in early autumn.

Basal leaves sparsely subappressed- to spreading-pilose on midrib and margin beneath from the first, becoming nearly or quite glabrous on the margin, with the midrib remaining slightly pilose, frequently 4-ranked, in conspicuous approximate to remote whorls, elliptic-lanceolate, uniformly narrowed to the acute apex and base, three (frequently) to five times as long as broad. Var. *juniperina.*

Basal leaves at first strongly subappressed- to spreading-pilose beneath, retaining a prominent pilosity on midrib and margin, mostly 2-ranked in crowded obscure whorls or scattered, lanceolate to oblanceolate, more than four times as long as broad. . . . . Var. *laurentiana*.

L. intermèdia (typical). — Dry sterile soil, mts. of Va. to N.S., P.E.I., e. N.B., centr. Me., N.H., Vt., N.Y., s. Ont., Mich., Wisc. and ne. Minn.; sw. S.D. and nw. Neb. *Fr.* July–Nov.

Var. **juniperìna** (Bickn.) Robins. (savin-like). — E. and n. N.H. to C.B.

Var. **laurentiàna** Hodgdon (of the St. Lawrence Valley). — St. Lawrence and Ottawa drainage, s. Que., s. Ont. and n. N.Y.

FAM. 115. VIOLÀCEAE (Violet Family)

Herbs (with us), with a somewhat irregular 1-spurred or gibbous corolla of 5 *petals,* 5 *hypogynous stamens with adnate introrse anthers connivent over the pistil, and a 1-locular 3-valved capsule with* 3 *parietal placentae.* Sepals 5, persistent. Petals imbricated in the bud. Stamens with their short and broad filaments continued beyond the anther-locules, and often coherent with each other. Style usually clavate, with the simple stigma turned to one side. Valves of the capsule bearing the several-seeded placentae on their middle; after opening each valve as it dries folding firmly together lengthwise, projecting the seeds. Seeds anatropous, with a hard seed-coat, and a large straight embryo nearly as long as the albumen; cotyledons flat. — Leaves alternate, with stipules. Flowers axillary, nodding.

Sepals not auricled; petals (in ours) equal in length; stamens united into a sheath. 1. *Hybanthus.*
Sepals auricled; lower petal spurred; stamens distinct, the two lower spurred. . 2. *Viola.*

1. HYBÁNTHUS Jacq. Green Violet

Petals nearly equal (or in extralimital species very unequal) in length, but the lower one larger and gibbous or saccate at the base, and more notched than the others at the apex. Stamens (in ours) completely united into a sheath inclosing the ovary and bearing a broad gland on the lower side. Style hooked at the summit. — Perennials, chiefly of the Tropics or warm-temp. reg., with stems leafy to the top and few small greenish-white flowers on short recurved axillary pedicels. (Name from the Greek *hybos, hump-backed,* and *anthos, flower.*) Solea Spreng. Ionidium Vent. Cubelium Raf.

1. H. cŏncolor (T. F. Forst.) Spreng. (of one color). — Strong-based; the solitary or clustered erect stems 3–9 dm. high, pubescent, or glabrous or nearly so in forma **subglàber** (Eames) Zenkert (almost glabrous); leaves oblong, acuminate to base and apex, entire, or the later ones toothed; flowers from median axils; capsule plump, about 2 cm. long. (*Solea* Gingins; *Ionidium* B. & H.; *Cubelium* Raf.) — Rich woods, calcareous slopes and ravines, etc., Ga. to Miss. and Kans., n. to Ct., N.Y., s. Ont., Mich. and Wisc. April–early June.

2. VÌOLA L. Violet

Petals somewhat unequal, the lower one spurred at base. Stamens closely surrounding the ovary, often slightly cohering with each other; the two lower ones bearing spurs which project into the spur of the corolla. Besides the conspicuous petaliferous mostly vernal flowers others are produced in all our sections but §§ 1 and 7, either on peduncles or stolons (sometimes concealed underground), these (the cleistogamous flowers or *cleistogenes*) not developing petals nor expanding but self-fertilized in the bud and very fruitful. — The closely allied species of the same section, when growing together, often hybridizing, producing offspring very confusing to the student of the group, the hybrids commonly displaying more or less intermediate characters and marked vegetative vigor but impaired fertility, successive generations raised exclusively from seed from cleistogamous capsules gradually (in eight to twelve generations) eliminating, in ideal Mendelian fashion, all traces of the blend and reverting to the characters of the two original parents of the cross. Large nearly cosmop. and variable genus, ours herbaceous. (*Viola* the classical name.)*

*a.*Key to petaliferous flowering material. . . *b.*
*b.*Stemless plants, the leaves and peduncles (at least of petaliferous flowers) arising directly from the rhizome or from stolons. . . . *c.*

---

* Treatment derived largely from the studies of Ezra Brainerd. With the eventual study of various types abroad some changes of names may be necessary.

*c.*Rhizome fleshy, erect, praemorse; leaves pedately cleft into firm segments or lobes; flowers all petaliferous, flat and nearly rotate; petals all beardless; style thick-clavate, the orifice on one side toward summit; cleistogamous aestival flowers wanting, the capsules all from petaliferous flowers and on long erect peduncles. . . . . . . 1. *V. pedata.*

*c.*Rhizomes erect, oblique or nearly horizontal; leaves various; petaliferous vernal flowers with some or all petals more or less porrect; lateral petals bearded in the blue- or violet-flowered species (sometimes not in white-flowered species); style enlarged upward to a truncate or laterally beaked summit; cleistogamous apetalous and very fruitful flowers (*cleistogenes*) usually numerous. *d.*

*d.*Flowers commonly blue, violet, purple or lilac (except in albinos); rhizome not producing stolons (except in nos. 29 and 30). . . *e.*

*e.*Rhizome thick and fleshy, 3–10 mm. thick, not producing stolons; flowers not specially fragrant. . . *f.*

*f.*Principal leaves merely crenate, dentate or serrate, not deeply lobulate nor divided. . . *g.*

*g.*Expanding leaves, petioles and peduncles glabrous (or petioles or peduncles only sparsely pilose or puberulent but plant otherwise glabrescent). . . *h.*

*h.*Leaves reniform, cordate-ovate or broadly cordate-deltoid, regularly toothed, much more than half as broad as long. . . *i.*

*i.*Marginal teeth of leaf low (rarely more than 2 mm. high), mostly as broad as to much broader than high or long. . . *j.*

*j.*Beard of lateral petals or much of it clavate, with rounded knob- or hood-like tips; spurred petal shorter than the lateral petals, beardless at base; corolla blue-violet, usually with darker center, often much overtopping leaves; plant of wet places. . . . . . 2. *V. cucullata.*

*j.*Beard of lateral petals not strongly knobbed, mostly with slender tips; corolla richer violet to purplish, not darkened toward center (except in no. 4); spurred and lateral petals subequal. . . *k.*

*k.*Spurred petal beardless at base. . . *l.*

*l.*New leaves at flowering time with margins gradually rounded to bluntish or short-deltoid apex; corolla rich violet with pale center (or albino), the flowers rarely overtopping the leaves; young cleistogenes on prostrate or low-arching peduncles; wide-ranging. . . . . . . 3. *V. papilionacea.*

*l.*New leaves at flowering time gradually tapering or attenuate to acutish triangular apex.

Flowers pale violet, darkened toward center, their peduncles shorter than to but slightly overtopping leaves; western. . . . . . 4. *V. missouriensis.*

Flowers deep violet, with pale center, greatly overtopping leaves; southeastern. . . . . 5. *V. Langloisii.*

*k.*Spurred petal bearded or villous at base. . . *m.*

*m.*Young leaves at flowering time broadly ovate, nearly as broad as to broader than long, blunt or with short deltoid tip; sepals blunt or round-tipped.

Broader petals 5–8 mm. wide; young cleistogenes on mostly ascending peduncles, their young capsules green; petioles and peduncles smooth; northern, transcontinental species of shores and open ground. . . . . 6. *V. nephrophylla.*

Broader petals 7–13 mm. wide; young cleistogenes on mostly depressed or prostrate peduncles, soon purplish; petioles and peduncles granulose above; e. woodland species. . . . . . 7. *V. latiuscula.*

*m.*Young leaves at flowering time narrowly ovate, tapering to acutish tip; sepals acute or acutish; broader petals 4–8 mm. wide; young cleistogenes mostly on loosely ascending slender peduncles; petioles and peduncles smooth; wide-ranging. . 8. *V. affinis.*

i. Marginal teeth prolonged, those near base of blade 2–6 mm. long, the blade pectinate with teeth much longer than broad; flowers mostly overtopping leaves; young cleistogenes on erect peduncles; coastwise species. . . 9. *V. pectinata.*

h. Leaves lanceolate, oblong or narrowly deltoid, less than half as broad to nearly as broad as long, at most subcordate, their bases often incised or coarsely toothed; young cleistogenes on erect peduncles.

Leaf-blades lance-oblong or narrowly lanceolate, twice to thrice as long as broad; flowers nearly or quite overtopping the leaves. . . . . 17. *V. sagittata.*

Leaf-blade nearly triangular, one-half to nearly as broad as long; flowers soon overtopped by the leaves. . . . 16. *V. emarginata.*

g. Expanding leaves or petioles and usually peduncles definitely pubescent. . . *n.*

n. Leaf-blades at flowering time pubescent beneath or petioles pubescent below. . . *o.*

o. Sepals or their auricles ciliate. . . *p.*

p. Spurred petal strongly villous at base. . *q.*

q. New leaves at flowering time cordate-ovate, the petioles mostly much longer than the blades; young cleistogenes mostly on short prostrate or recurving peduncles; plants of damp or mesophytic habitats.

Petioles and lower surfaces of young leaves densely villous; some leaf-blades uncleft, others lobed; margins of sepals sparingly ciliate; southern species extending northw. in ne. states. . . . . 18. *V. triloba.*

Petioles and lower surfaces of young leaves sparsely hirsute or hispidulous; leaves all unlobed; margins of sepals closely ciliate; northern species. . . . . 11. *V. septentrionalis.*

q. New leaves at flowering time not cordate-ovate, their petioles (except in no. 17) shorter than to but slightly exceeding the blades; young cleistogenes on erect peduncles; plants chiefly of dry open habitats.

Blades orbicular, reniform or roundish-elliptic, with deep sinus, broadly rounded above, not lobulate below; southern, n. to s. Va. . . . . . . 13. *V. villosa.*

Blades of new leaves at flowering time oblong-ovate, obtuse to subacute, truncate to subcordate and often lobulate toward base; wide-ranging, into s. Can. . . . . 15. *V. fimbriatula.*

p. Spurred petal glabrous or only very sparsely hairy at base; later leaves cordate-ovate, strongly villous beneath and at summit of petiole; cleistogenes on prostrate short peduncles; wide-ranging. . . . . 10. *V. sororia.*

o. Sepals and auricles glabrous, not ciliate.

Leaves cordate-ovate, with regularly low-toothed margin, only sparingly if at all pubescent; petioles pubescent; cleistogenes on prostrate or low-arching short peduncles.

Leaf-blade broadly ovate to subreniform, nearly as broad as or broader than long, the lower surface and petiole granulose; eastern woodland plant. . . . 7. *V. latiuscula.*

Leaf-blade narrowly ovate or ovate-oblong, much longer than broad, smooth; petioles spreading-villous below; chiefly on shores, damp rock or meadow northward. . . . . 12. *V. novae-angliae.*

Leaves with lanceolate to lance-oblong blades merely undulate above, twice to thrice as long as broad, hastate-lobulate near base, pilose beneath; cleistogenes on erect peduncles; plant of dry open habitats. . . . . . 17. *V. sagittata.*

n. Leaf-blades at flowering time glabrous beneath, strigose above, cordate-ovate.

Leaves strongly ascending, the blades much shorter than

the villous-hirsute or hispidulous petioles; sepals definitely ciliate; wide-ranging northw. . . . . 11. *V. septentrionalis.*

Leaves spreading or depressed, the blades at flowering time longer than to about half as long as glabrous petiole; sepals glabrous or only rarely with a few cilia; southern, n. to Ct., etc. . . . . 14. *V. hirsutula.*

f. Principal leaves lobed or divided. . . *r.*

r. Leaves pubescent. . . *s.*

s. Blades lanceolate to oblong or oblong-ovate, hastately lobulate near base; cleistogenes on erect peduncles; plants of dry open habitats.

Leaf-blades (above the lobulate base) oblong-ovate, longer to barely shorter than the petiole, copiously pubescent; sepals ciliate. . . . . 15. *V. fimbriatula.*

Leaf-blades (above the lobulate base) lance-oblong or narrowly lanceolate, mostly on very long slender petioles, sparsely pubescent; sepals not ciliate. . . . . . 17. *V. sagittata.*

s. Blades cordate-ovate, some or all palmately or pedately lobed, the young villous, on densely villous petioles; cleistogenes on depressed or low-arching short peduncles; plants of woodland.

Leaves both unlobed and lobed; the lobed ones chiefly 3-lobed, with the central lobe broadly oblong or oblong-ovate, the 2 basal lobes broad and with strongly recurving inner margin, all lobulate. . . . . 18. *V. triloba.*

Leaves all deeply divided into 5–11 more uniformly narrow lobes, these often further dissected; basal sinus usually broader and with the inner margins less recurving. . . . 19. *V. palmata.*

r. Leaves glabrous or essentially so. . . *t.*

t. Blades lanceolate, oblong or narrowly deltoid, more or less hastately incised near base; cleistogenes on erect peduncles; plants of dry open habitats.

Leaf-blades nearly triangular, one-half to nearly as broad as long; flowers soon overtopped by the leaves. . . . 16. *V. emarginata.*

Leaf-blades lance-oblong or narrowly lanceolate, twice to thrice as long as broad; flowers nearly or quite overtopping the leaves. . . . . 17. *V. sagittata.*

t. Blade of leaf broader, palmately or pedately lobed or dissected. . . *u.*

u. Auricles of the sepals prolonged, much longer than broad, acute, acutish or toothed.

Leaves all or essentially all divided into 5–15 narrow lobes or segments, the base scarcely to barely cordate; coastwise from Me. to N.C. . . . . 20. *V. Brittoniana.*

Leaves partly unlobed, partly lobed, the lobed ones with 3–9 simple or divided lobes, cordate, the middle lobe broad; southeastern, reaching n. to se. Va.

Leaves coriaceous, the lobed ones with 3–9 or more lobes and divisions; simple blades few. . . . . 21. *V. esculenta.*

Leaves membranaceous, the lobed ones with mostly 3 short lobes; simple blades abundant. . . . . . 22. *V. chalcosperma.*

u. Auricles broad and low, mostly broader than long, with rounded to subtruncate summits. . . *v.*

v. Spurred petal glabrous at base.

Leaves with broadly rounded to cordate base, the lobes blunt to subacute; spurred petal broad and flattish; lateral petals densely bearded; cleistogenes on prostrate to loosely arching peduncles; e. woodland species. . . . . 23. *V. Stoneana.*

Leaves truncate at base, the lobes acute or acutish; spurred petal narrow and inrolled; lateral petals sparsely bearded; cleistogenes on erect peduncles; prairie species. . . . . 24. *V. viarum.*

v. Spurred petal villous at base.

Leaves cordate at base, many of them unlobed, the others shallowly to deeply lobed, membranaceous; cleistogènes rarely showing at flowering time; southeastern. . . . . 25. *V. septemloba.*

Leaves cuneate to subtruncate at base, flabelliform, all divided nearly to base into many ascending narrow coriaceous segments; young cleistogamous flowers evident at flowering time; inland and western.

Cleistogenes on prostrate subterranean fleshy peduncles; plant of calcareous barrens, Ky. and Tenn. . . . . 26. *V. Egglestoni.*

Cleistogenes on slender erect peduncles; prairie species. . . . . 27. *V. pedatifida.*

e. Rhizome thread-like or cord-like, not strongly thickened, 1–4 mm. thick, more or less stoloniferous (except in no. 28, with deep and nearly closed leaf-sinus, the spur about equaling the blade of the spurred petal).

Nonstoloniferous; leaves strigose-hirsute above, the deep sinus nearly closed; flower not strongly fragrant, its spur half to two-thirds as long as the blade of the spurred petal; n. woodland plant. . . . . 28. *V. Selkirkii.*

Stoloniferous; leaves glabrous or minutely pubescent, with open sinus; flowers often richly fragrant, the spur much shorter than blade of spurred petal.

Glabrous; ovary glabrous; style terminated by a broad disk; subalpine and alpine native. . . . . . . 29. *V. palustris.*

Minutely pubescent; ovary pubescent; style slender, with hooked tip; spread from cult. . . . . 37. *V. odorata.*

d. Flowers white or yellow. *w.*

w. Flowers white with purple lines, expanding with or after the leaves; rhizome filiform to slenderly cord-like, stoloniferous (except in no. 33); cleistogenes evident at flowering time, on simple peduncles above ground. . *x.*

x. Leaves reniform to cordate-ovate, with deep basal sinus; flowers usually fragrant. . . *y.*

y. Stoloniferous; new leaves at flowering time mostly ovate or if reniform, the flowers with lateral petals bearded.

Leaves glabrous, at anthesis 1–3 cm. long, ovate or reniform; flower 7–10 mm. long, very fragrant; upper petals broadly obovate, the lateral ones beardless; young cleistogenes on erect peduncles; plant of wet places. . . . . 30. *V. pallens.*

Leaves more or less pubescent on one or both faces, at anthesis mostly 3–6 cm. long, chiefly ovate; flower 1–1.5 cm. long, only slightly fragrant; young cleistogenes on prostrate peduncles.

Young leaves pubescent beneath; petioles and peduncles green, the petioles ascending; upper petals not strongly reflexed, broad; lateral petals bearded; plant of damp or wet habitats northw. . . . . 31. *V. incognita.*

Young leaves glabrous beneath; petioles spreading, these and peduncles reddish; upper petals strongly reflexed, narrow; lateral petals beardless; plant of rich (mostly dry) woods, southw. and n. into s. Can. . . . 32. *V. blanda.*

y. Nonstoloniferous; leaves reniform to orbicular; petals all beardless; northern woodland plant. . . . . 33. *V. renifolia.*

x. Leaves linear-lanceolate to oblong or ovate, tapering at base to barely subcordate; cleistogenes on erect peduncles.

Leaves lanceolate, gradually tapering into petiole, the blade one-third to one-twentieth as broad as long. . . . . . 34. *V. lanceolata.*

Leaves oblong to ovate, gradually rounded to subcordate at base, the blade two-fifths to quite as broad as long. . . . 35. *V. primulifolia.*

w. Flowers yellow, expanding before or at earliest unrolling of leaves, the latter roundish to elliptical; rhizome thick and fleshy, without stolons; style beakless; cleistogenes not evident at flowering time; plant of rich woodland. . . . . 36. *V. rotundifolia.*

b. Stems elongate, leafy, with flowers borne from leaf-axils. . . *z.*

z. Stipules bract-like, narrow and elongate, entire or fringed; style clavate or slender; apetalous cleistogamous flowers in summer and autumn; native perennials. . . . a.

a. Flowers yellow, creamy-white or white with violet tinge on back. . b.

b. Stipules entire or merely crenulate (sharp-toothed in no. 39), at least the lower ones often scarious; style upwardly clavate, bearded at summit. . . c.

c. Petals yellow; upper stipules herbaceous. . . d.

d. Leaves (or some of them) deeply 3-parted; plant glabrous; yellow petals with violet tinge outside; rhizome subligneous; southern. . . . . 38. *V. tripartita.*

d. Leaves undivided. . . e.

e. Stem and leaves glabrous; leaves narrowly hastate-ovate or narrowly ovate, at summit of stem; stipules often sharply toothed, 2–5 mm. long; yellow petals violet on back; rhizome pale and fleshy; southern. . . . . . . 39. *V. hastata.*

e. Stem and leaves pubescent or glabrescent; leaves broad-ovate; stipules entire or crenate, much longer; petals yellow, not violet-backed; rhizome subligneous; wide-ranging.

Expanding leaves and upper half of stem densely pubescent; basal leaves usually wanting; cauline leaves heavy, strongly veiny; stipules semiovate. . . . . . 40. *V. pubescens.*

Expanding leaves and upper half of stem thinly pubescent or glabrescent; basal leaves commonly 1–3; cauline leaves less veiny; stipules lanceolate or narrowly semiovate. . . . . 41. *V. pensylvanica.*

c. Petals white, often more or less violet-tinged on back (especially in age); stipules narrowly lanceolate, slender-tipped, scarious or subscarious and pale; plant sparsely pubescent to glabrous.

Rhizome thickish, not stoloniferous, ascending, the crown 0.5–1 cm. thick, densely covered with roots; wide-ranging, chiefly eastw. . . . . 42. *V. canadensis.*

Rhizome slender and cord-like, with long subterranean nearly horizontal stolons, the cord-like crowns with few roots; western. . . . . 43. *V. rugulosa.*

b. Stipules sharply toothed or fringed (often entire in the northern and subalpine violet-flowered no. 46), herbaceous; flowers violet (white only in no. 44 and albinos); spur two or more times as long as thick; style slender.

Petals creamy- or milk-white; leaves membranaceous, with rounded crenations; stipules toothed their whole length, the larger ones 1–2.5 cm. long; auricles of sepals 2–4 mm. long. . . . 44. *V. striata.*

a. Petals violet, purple or lilac; leaves firm, with low and broad flat-topped crenations; stipules toothed mostly toward base or entire, rarely more than 1.3 mm. long; auricles rarely 1.5 mm. long. . . g.

g. Lateral petals bearded; spur less than 8 mm. long; tip of style bent downward, somewhat pubescent near summit. . . h.

h. Stems tufted, ascending; the branches not forming stolons; rameal leaves mostly ovate; northern and wide-ranging.

Leaves membranaceous, the upper cauline blades reniform-ovate, with rounded or abrupt blunt tips; flowers pale violet; wide-ranging southward, n. into s. Can. . . . 45. *V. conspersa.*

Leaves firm to subcoriaceous, the upper blades ovate, gradually narrowed above; flowers deep violet to violet-blue; northern and transcontinental. . . . . 46. *V. adunca.*

h. Stems soon prostrate and developing rooting stolons; rameal leaves orbicular or round-reniform; southern. . . . 47. *V. Walteri.*

g. Lateral petals beardless; spur slender, 0.9–1.8 cm. long; style straight, with glabrous tip; upper leaves ovate, tapering to tip; petals lilac with a central violet spot; e. woodland species. 48. *V. rostrata.*

z. Stipules leaf-like, with dilated summit, pectinate or pinnatisect toward base; style enlarged upward into a globose hollow summit with a wide orifice on the lower side; Pansies; introd. annuals or short-lived perennials without rhizomes.

Stipules pinnatisect at base; upper leaves crenate-serrate.

Petals twice or thrice length of sepals, variously whitish, yellow and violet. . . . . 49. *V. tricolor.*

Petals about equaling to but slightly longer than sepals, pale yellow. 50. *V. arvensis.*

Stipules palmately pectinate at base; upper leaves entire or nearly so; petals bluish-white to creamy, nearly twice as long as sepals. . . . 51. *V. Kitaibeliana*, var. *Rafinesquii.*

*a.*Key to cleistogamous or fruiting material. . . **i.**

i.Stemless (or merely with stolons in a few species), the leaves and peduncles arising from the rhizome or stolons. . . **j.**

j.Cleistogamous apetalous flowers and their fruits wanting; fruiting peduncles erect and subequal; mature but not overripe capsule with 1 valve tipped by a clavate style; leaves variously palmate-cleft; plant of dry habitats. . . . . 1. *V. pedata.*

j.Cleistogamous apetalous flowers (cleistogenes) and their fruits developed, usually more fruitful and often on shorter peduncles than petaliferous flowers; style, if persistent, not clavate. . . **k.**

k.Rhizome thick and fleshy, 3–10 mm. in diameter; no stolons (except subterranean fruitful ones in the depressed-leaved no. 36). . . **l.**

l.Leaves crenate, dentate or serrate, not deeply lobulate nor divided. . . **m.**

m.Cleistogenes borne singly (if rarely clustered, on erect peduncles); leaves ascending or if flat on ground rarely 6.5 cm. long. . . **n.**

n.Leaves cordate at base, the sinus deep. **o.**

o.Plants essentially glabrous (pubescence, if any, not on foliage). . . **p.**

p.Young cleistogenes sagittate-lanceolate, the green capsule ovoid-cylindric and only slightly exceeding the sepals; auricles of sepals pointing straight back, mostly 2–6 mm. long; cleistogenes on erect peduncles; peduncles of petaliferous flowers mostly overtopping foliage; plant of cool wet habitats. . . . . . 2. *V. cucullata*

p.Young cleistogenes plumper, the more ovoid capsule once-and-a-half to four times as long as sepals; auricles shorter, usually more appressed. . . **q.**

q.Cleistogenes on mostly ascending to erect or arched-ascending filiform green peduncles.

Leaves firm to subcoriaceous, gradually rounded above, their teeth strongly flattened; sepals obtuse; capsule green; seeds not dotted with purple, buff, becoming deep brown, the pale caruncle prolonged; northern transcontinental plant of shores and open ground. . 6. *V. nephrophylla.*

Leaves membranaceous, gradually tapering to narrowly triangular tip, their teeth obliquely dome-shaped; sepals acute or acutish; capsule green or purple-mottled; seeds usually dotted with purple, when pale-colored soon becoming dark; pale caruncle not prolonged; eastern and southern woodland plants.

Auricles of sepals spreading or reflexed, acutish or angulate, 2 mm. long; capsule 6–12 mm. long; southeastern. . . . . 5. *V. Langloisii.*

Auricles appressed, broad and rounded, shorter; capsule 5–9 mm. long; wide-ranging eastw. . 8. *V. affinis.*

q.Cleistogenes mostly on prostrate or nearly prostrate fleshy peduncles, these often buried until maturity, then becoming elevated; capsule often mottled with purple; seeds pale, often purple-dotted, often becoming dark in maturity. . . **r.**

r.Leaves broadly cordate-ovate, the larger ones nearly as broad as to broader than long; auricles broad and rounded, appressed, about 1 mm. long; caruncle short and broad.

Leaves with rounded sides, with gradual convex curve to the blunt or short tip; wide-ranging. . . 3. *V. papilionacea.*

Leaves with less rounded sides or with a concave curve below the acute or acutish tip. . . . . 7. *V. latiuscula.*

r.Leaves cordate-deltoid, tapering (not arching) from base to apex, narrower than to barely as broad as long; auricles elongate, not appressed, 1–2 mm. long; caruncle elongate; western. . . . . 4. *V. missouriensis.*

o. Plants definitely pubescent. . . **s.**

s. Leaves ascending to erect, the later ones tapering to acute or subacute apices, long-petioled; wide-ranging eastw. or northw. . . **t.**

t. Sepals (at least toward base) and their auricles ciliate; mature leaves as broad as to much broader than long.

Mature leaves dark green, thickish, the larger ones (on each mature plant) 6–13 (av. 9) cm. broad, with 21–52 (av. 32) prominent teeth on each half; summit of petiole and adjacent lower surface of leaf villous; cleistogenes on fleshy prostrate peduncles at first partially buried but finally arching above ground; auricles broad, appressed, 0.5–1 mm. long; seeds plump-obovoid, 1.75–2.5 mm. long excluding the short caruncle, 1.2–1.5 mm. thick; wide-ranging southw. and eastw. . . . . 10. *V. sororia.*

Mature leaves pale green, membranaceous, the larger ones 2–11 (av. 5.5) cm. broad, with 10–26 (av. 17) low teeth on each half; summit of petiole and adjacent lower leaf-surface usually sparsely hispid; auricles narrow or lobed not appressed, 1.5–3 mm. long; seeds slenderly obovoid, 1.5–2 mm. long, excluding the prolonged caruncle (–0.5 mm. long), 0.8–1 mm. thick; northern. . . . . 11. *V. septentrionalis.*

t. Sepals and their auricles glabrous; most leaves narrowly cordate-ovate or triangular, the longest mature ones mostly 3.5–7 (av. 5) cm. long and 2–5 (av. 3.4) cm. wide, with 6–12 (av. 10) teeth on each half; seeds plump-ovoid, 1.7–2 mm. long excluding prolonged caruncle (0.3–0.4 mm. long), 1–1.6 mm. thick; northern. . . . . 12. *V. novae-angliae.*

s. Leaves forming flat or spreading rosettes 0.5–1.5 dm. across, rounded or obtuse above, short-petioled; southern.

Leaves closely short pilose on both faces; cleistogenes on ascending slender (but short) peduncles; capsule green, ovoid-cylindric, 9–10 mm. long. . . . . . 13. *V. villosa.*

Leaves glabrous beneath, strigose-hispid above; cleistogenes on depressed often fleshy peduncles; capsules purple-marked, plump-ovoid, 6–8 mm. long. . . . 14. *V. hirsutula.*

n. Leaves truncate or subtruncate to barely subcordate at base, without deep sinus, the blades lanceolate or oblong to deltoid; young cleistogenes on erect peduncles; their auricles narrow, prolonged and erect or divergent. . . **u.**

u. Leaves regularly 10–16-toothed along each margin nearly to apex of the deltoid blade, not hastately lobed, the lower uniform teeth 2–6 mm. long; coastwise, Mass. to Va. . . 9. *V. pectinata.*

u. Leaves with very short or suppressed teeth above the more or less hastate base, there often lacerate into 2–6 sharp prolonged teeth or lobes; wide-ranging. . . **v.**

v. Leaf-blades (above lobulate base) triangular to oblong-ovate, the larger (on each plant) 1.5–8 cm. broad, mostly overtopping the peduncles.

Leaves and petioles pilose, the blades oblong-obovate; sepals and auricles ciliate. . . . . 15. *V. fimbriatula.*

Leaves and petioles glabrous, the blades triangular; sepals and auricles glabrous. . . . . 16. *V. emarginata.*

v. Leaf-blades (above lobulate base) lance-oblong to narrowly lanceolate, glabrous or minutely pubescent, the larger ones 1–3.5 cm. broad, nearly or quite equaled by the peduncles; sepals and auricles glabrous. . . . . . . 17. *V. sagittata.*

m. Cleistogenes mostly 2-several on a peduncle.

Leaves ascending, pilose on both surfaces, with oblong blades hastate-lobulate at base; peduncles erect or nearly so, terminated by 3-flowered umbels; plant of dry open soil. 15. *V. fimbriatula*, form.

Leaves spreading or flat on the ground, roundish-elliptic or oval to subrotund, with deep sinus, glabrous above; peduncles prostrate or subterranean, stoloniform, with long-stalked flowers in lax racemes; plant of deciduous woods. . . . 36. *V. rotundifolia.*

l.Leaves or some of them lobed or divided. . . **w.**

w.Leaves cuneate to truncate at base, not definitely cordate. . . **x.**

x.Some or all leaves cleft, with the lobes at base of elongate broader terminal blade, others unlobed.

Leaves much longer than broad or, if in autumn nearly as broad, the basal lobes rarely more than 1 cm. long and the auricles or the sepals narrow, not appressed and 2–4 mm. long; seeds (incl. caruncle) 1.4–1.8 mm. long; eastern. . . Species 15–17 under **v** above.

Leaves as broad as to broader than long, with basal lobes often 1–4 cm. long; auricles broad and rounded or truncate, hardly 1 mm. long; seeds (incl. caruncle) 1.9–2 mm. long; western. . . . . 24. *V. viarum.*

x.All or essentially all leaves flabelliform and palmately divided nearly to base into many narrow segments; inland or western.

Cleistogenes on prostrate fleshy subterranean peduncles, the mature fruit finally arching above ground; seeds, including caruncle (0.3–0.5 mm. long), 2.7–3.3 mm. long; plant of calcareous barrens, Ky. and Tenn. . . . . . 26. *V. Egglestoni.*

Cleistogenes on slender ascending peduncles; seed, including caruncle (0.2 mm. long), 2 mm. long; prairie species. . . 27. *V. pedatifida.*

w.Leaves definitely cordate at base. . **y.**

y.Leaves pubescent.

Blades of leaf both unlobed and lobed; lobed ones chiefly 3-lobed, with the central lobe broadly oblong or oblong-ovate, the 2 basal lobes broad and with strongly recurving inner margin; seeds slenderly obovoid, 2.3–2.5 mm. long. . 18. *V. triloba.*

Blades all or nearly all deeply divided into 5–11 more uniformly narrow lobes, these often more dissected; basal sinus usually broader and with inner margins less recurving; seeds plump-obovoid, 2 mm. long. . . . . 19. *V. palmata.*

y.Leaves glabrous. . . **z.**

z.Auricles of sepals of cleistogenes prolonged, much longer than broad, acute, acutish or toothed, 2–5 mm. long.

Later leaves all or essentially all divided into 5–15 narrow lobes or segments; mature erect petioles 0.5–3 dm. high; seeds 0.9–1 mm. thick; chiefly coastwise, Me. to N.C. . 20. *V. Brittoniana.*

Later leaves partly unlobed, partly lobed, the lobed ones with the middle lobe very broad; mature spreading or ascending petioles 0.3–2 (rarely –2.5) dm. high; southern.

Leaves coriaceous, the lobed ones with 3–9 or more lobes and divisions; simple blades few; seeds 1.8–1.9 mm. long, 1–1.3 mm. thick, with prominent caruncle. . . 21. *V. esculenta.*

Leaves membranaceous, the few lobed ones with mostly 3 short lobes; simple blades abundant; seeds 1.5–1.6 mm. long, 0.9–1 mm. thick, with very short caruncle. 22. *V. chalcosperma.*

z.Auricles broad and low, mostly broader than long (about 1 mm.), rounded to subtruncate.

Later leaves all cleft nearly or quite to base, 0.7–1.8 dm. broad; cleistogenes on prostrate fleshy colored peduncles, the capsules plump-ovoid and purplish; seeds 1.8–2 mm. long, 1.1–1.4 mm. thick; plant of rich deciduous woods, Pa. and O. to Va. . . . . 23. *V. Stoneana.*

Later leaves uncleft or cleft, 2.5–7 cm. broad; cleistogenes on slender green ascending peduncles, the capsules ovoid-cylindric, green; seeds 1.5–2 mm. long, 1.4–1.5 mm. thick; plant of pineland and swamps, southeastw. . . . 25. *V. septemloba.*

k.Rhizome thread-like or cord-like, not strongly thickened, 1–4 mm. thick, usually stoloniferous except in nos. 28 and 33. . **a.**

a.Leaves reniform to cordate-ovate, with deep basal sinus. . **b.**

b.Nonstoloniferous or essentially so; leaves pubescent on one or both faces; northern.

Principal leaves ovate, with nearly closed sinus, membranaceous, in maturity 1.5–4.5 cm. broad; auricles of calyx glabrous; seeds, incl. short caruncle, 1.5–1.9 mm. long, 1–1.1 mm. thick. 28. *V. Selkirkii.*

Principal leaves reniform or orbicular, with open sinus, firm, in maturity (1.5–)3–8 cm. broad; auricles ciliate; seeds, incl. elongating caruncle (0.3–0.5 mm.), 1.9–2.4 mm. long, 1–1.4 mm. thick. . . . . . 33. *V. renifolia.*

b.Stoloniferous, the stolons filiform or slenderly cord-like. . c.

c.Extensively creeping, with cord-like leafy stolons; leaves closely pubescent on both surfaces; capsule densely pubescent; seeds creamy-white, including the long caruncle (1 mm.), 3.4–4 mm. long, 1.75–2 mm. thick; introd. . . . . . 37. *V. odorata.*

c.Slightly creeping, the thread-like stolons leafless or only sparsely leafy at tip; capsule glabrous; seeds grayish or buff to blackish, including the short caruncle (0–0.2 mm.), 1–2.1 mm. long, 0.7–1.3 mm. thick; native. . d.

d.Seeds 1–1.4 mm. long, 0.7–0.8 mm. thick, soon blackish; leaves cordate-ovate, glabrous; capsule ellipsoid, green; northern or upland species. . . . . . 30. *V. pallens.*

d.Seeds 1.5–2.1 mm. long, 1–1.3 mm. thick, grayish, buff or brown; capsule plump-ovoid, often purplish or reddish-brown. e.

e.Leaves cordate-rotund or -reniform to ovate, glabrous, 2–4 cm. broad; stolons cord-like, 1–1.5 mm. thick, stiffish; capsule green or greenish; seeds, including very short caruncle, 1.5–1.7 mm. long, 1 mm. thick, gray; arctic-alpine and high northern. . . . . . 29. *V. palustris.*

e.Leaves cordate-ovate, pubescent or glabrate, the larger (of each plant) 2.5–10 cm. broad; stolons filiform, flexuous; capsule usually purple-brown; seeds, including evident caruncle (0.1–0.2 mm.), 1.6–2.1 mm. long, 1–1.3 mm. thick, yellowish to brown; non-alpine.

Leaves membranaceous, rugulose, more or less pubescent on veins beneath, ascending; petioles and bases of peduncles green; plant of wet places and cool mountain-slopes, northw. . . . . 31. *V. incognita.*

Leaves firmer, flat, glabrous beneath, soon spreading; petioles and bases of peduncles usually tinged with red; plant of rich (mostly dry) woods southw. and n. into s. Can. . . . . 32. *V. blanda.*

a.Leaves linear-lanceolate to oblong or ovate, tapering at base to barely subcordate; cleistogenes on erect peduncles.

Leaves lanceolate, gradually tapering into petiole, the blade one-third to one-twentieth as broad as long. . . . . . 34. *V. lanceolata.*

Leaves oblong to ovate, gradually rounded to subcordate at base, the blade two-fifths to quite as broad as long. . . . . 35. *V. primulifolia.*

i.Leafy-stemmed, the stems elongating, with flowers borne from the leaf-axils. . . f.

f.Stipules bract-like, narrow and elongate, entire or fringed; native perennials. . g.

g.Leaves (at least some of them) 3-parted; rhizome subligneous, bearing long tough roots; sepals 4–8 mm. long, thin; capsule trigonous-ovoid, green, 1–1.2 mm. long; seeds drab or pale brown, plump-obovoid, 2.5–2.8 mm. long, 1.5–2 mm. thick; southern. . . . . 38. *V. tripartita.*

g.Leaves all simple. h.

h.Fruiting stems naked except at summit, slender, glabrous or essentially so; leaves at summit of stem, thin, glabrous, hastate-triangular, tapering gradually to tip, 2–5 cm. broad at base; stipules thin, attenuate, 2–10 mm. long, entire or toothed; seeds 2–2.5 mm. long; rhizome whitish, fleshy; southern. . . . . . . 39. *V. hastata.*

h.Fruiting stem leafy or, if naked below, pubescent and with large rounded-ovate pubescent leaves and large herbaceous stipules; rhizome subligneous or slender and elongate. . . i.

i.Stems erect or ascending, 1–few, with 0–few erect basal leaves; lower stipules scarious, the upper more herbaceous or scarious and entire or merely dentate or undulate; larger mature leaves mostly 4–10 cm. long and 2.5–12 cm. broad, pilose or hirtellous to glabrate; seeds 1.5–3 mm. long, 1.2–1.8 mm. thick. . . j.

j.Thick-leaved; upper leaves broadly ovate to roundish, blunt-tipped, coarsely dentate, with broad open sinus; upper stipules foliaceous; capsules densely woolly or glabrous, 0.7–1.4 cm. long; seeds 2–3 mm. long, 1.3–1.8 mm. thick.

Summit of stem and lower (sometimes upper) leaf-surface closely pubescent, the leaves heavy and rugulose; basal leaves usually 0; stipules semiovate. . . . . 40. *V. pubescens.*

Summit of stem and leaves glabrescent or only sparsely pubescent, the leaves thinner and not rugulose; basal leaves usually 1–3; stipules lanceolate or narrowly semiovate. . . 41. *V. pensylvanica.*

j.Thin-leaved, the leaves with deep sinuses; upper leaves narrowly ovate, tapering to prolonged tips, closely serrate or dentate; stipules more scarious; capsules not woolly, 5–9 mm. long; seeds 1.5–2.2 mm. long, 1.2–1.5 mm. thick.

Rhizome thickish, not stoloniferous, ascending, the crowns 0.5–1 cm. thick, densely covered with strong roots; seeds 1.5–2 mm. long, 1.2–1.5 mm. thick; wide-ranging. . . . 42. *V. canadensis.*

Rhizome slender, much elongate and forking into subterranean stolons, the cord-like crowns with few roots; seeds 2–2.2 mm. long, 1.2–1.25 mm. thick; northwestern. . . . 43. *V. rugulosa.*

i.Stems tufted or rosulate with rosettes of basal leaves, loosely ascending or spreading; stipules herbaceous, fimbriate to entire; larger mature leaves (except in no. 44) small, glabrous or puberulent; seeds 1.5–2.4 mm. long, 0.8–1.4 mm. thick. . . k.

k.Leaves long-acuminate, those of primary axes 3–7 cm. long, their teeth rounded-crenate; stipules 1–2.5 cm. long, their longer teeth 3–5 mm. long; sepals closely ciliolate, their auricles about 2 mm. long. . . . . 44. *V. striata.*

k.Leaves rounded, blunt or (in no. 48) sometimes short-acuminate, those of primary axes 1–4.5 cm. long, their teeth flattened; stipules rarely 1.5 cm. long, their longer teeth 0.1–3 mm. long or wanting; sepals eciliate or essentially so, their auricles 1 mm. or less long. . . l.

l.Stems prostrate and trailing, often perennial and with proliferating rooting tips; leaves orbicular or round-reniform, the veins bordered by dark green; capsules purple, globose-ovoid; southern. . . . . 47. *V. Walteri.*

l.Stems loosely spreading or ascending, not rooting at tips; leaves uniformly colored; capsules ovoid, green or brown. . . m.

m.Leaves roundish, reniform or ovate, all blunt; stipules entire or with longer teeth up to 2 mm. long; seeds 0.9–1.1 mm. thick.

Leaves membranaceous, glabrous or nearly so, the upper cauline blades reniform-ovate, with rounded or abrupt blunt tips; plant of mesophytic or moist habitats southw. and n. into s. Can. . . . . . . 45. *V. conspersa.*

Leaves firm to subcoriaceous, puberulent or glabrous, the upper cauline blades ovate and gradually narrowed above; plant of dry to alpine habitats northw. . . . 46. *V. adunca.*

m.Leaves ovate, membranaceous, the upper gradually acuminate; longer teeth of stipules 3 mm. long; seeds 1.1–1.4 mm. thick; plant of rich (mostly calcareous) woods. . . 48. *V. rostrata.*

f.Stipules leaf-like, with dilated summit, pectinate or pinnatisect toward base; cleistogamous flowers and fruits wanting; introd. annuals or short-lived perennials without rhizomes. . . n.

n.Stipules pinnatisect at base; upper leaves and middle lobe of stipule crenate-serrate.

Lower and median leaves rounded or cordate at base. . . . . 49. *V. tricolor.*

Lower and median leaves gradually narrowed to base. . . . . 50. *V. arvensis.*

n.Stipules palmately pectinate at base; upper leaves and middle lobe of stipule entire or nearly so. . . . . 51. *V. Kitaibeliana*, var. *Rafinesquii.*

GROUP I. PLANTS STEMLESS, THE LEAVES AND PEDUNCLES OR INFLORESCENCES DIRECTLY FROM THE RHIZOME OR STOLONS (for GROUP II see p. 1040).

§ 1. *Style clavate, beakless, obliquely concave at summit; stigma within a small protuberance near the center of the cavity.* BIRDFOOT-V. FIG. 1327.

1. V. pedàta L. (foot-like), PANSY-V. — Tufted from a *fleshy praemorse erect rhizome; leaves* spreading or erect, *3-divided, the lateral divisions 3–5-parted or -cleft into linear or narrowly spatulate firm* often apically toothed *segments;* leaves of early spring and late autumn often less cut; *flowers* all alike (*none cleistogamous*) *on erect peduncles, flat and nearly rotate,* 1.5–3 cm. broad; *petals all beardless, the upper petals dark violet,* the others pale to deep lilac-purple; the orange tips of the stamens conspicuous at the center of the flower; *capsules green, all from petaliferous flowers early in the season, on erect subequal peduncles; mature capsule with 1 valve retaining the clavate style* until overripe. (Var. *bicolor* Pursh) — Dry sunny openings, in argillaceous or siliceous soil, upland S.C. to La., n. to s. Ct. and se. N.Y. (very rare), N.J., Pa., w. Va., Ky., Ind., Ill. and Mo. Late March (southw.)–early June (northw.).

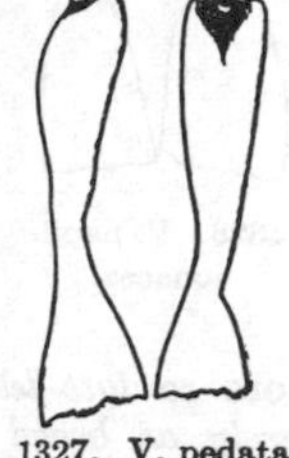

1327. V. pedata.

Var. **linearíloba** DC. (linear-lobed). — Leaves dissected nearly to base into linear or narrowly oblanceolate segments as in the typical var., or the segments cuneate-obovate and only slightly divided in forma **cuneatíloba** Brainerd (cuneate-lobed), or leaves all or nearly all narrowly flabellate-obovate with the margin shallowly cleft or merely dentate or entire in forma **ranunculifòlia** (Juss.) Fern. (buttercup-leaved); *flower pale blue, or bluish-violet,* or in forma **ròsea** A. L. Sanders (roseate) rose-pink, or in forma **álba** (Thurb.) Britt. (white) white. (Vars. *inornata* Greene and *concolor* Holm) — Of much wider range, Fla. to e. Tex., n. to s. N.H., Mass., N.Y., s. Ont., Mich., Wisc., Minn. and Kans. April–early June.

§ 2. *Style dilated upward in a vertical plane, capitate, with a conical beak on the lower side; stigma within the tip of the beak.* STEMLESS VIOLETS. . . *a.* FIG. 1328.

a. Rhizome fleshy and thickened, without stolons: petals violet-blue to purple, the lateral ones bearded. BLUE VIOLETS.

2. V. cucullàta Ait. (hooded; from the inrolled young leaves). — Tufted from a stout rhizome, with 1–several crowns; leaves ascending, glabrous or essentially so, long-petioled, cordate-ovate, crenate; *petaliferous flowers often much overtopping leaves,* blue-violet, usually *dark toward center,* or with petals irregularly blue and white in forma **Thurstònii** (Twining) House (for its discoverer CHARLES ORION THURSTON, 1857–), or white in forma **albiflòra** Britt. (white-flowered); *spurred petal shorter than lateral petals, beardless; beard of lateral petals containing many clavate or round-tipped trichomes; sepals with lobed auricles 2–6 mm. long and pointed straight back; cleistogenes on long erect peduncles, sagittate-lanceolate, their ovoid-cylindric green capsules only slightly exceeding the sepals, 1–1.5 cm. long;* seeds 1.4 mm. long, becoming black. — Wet meadows, springy swamps, bogs, etc., Nfld. to Thunder Bay Distr., Ont. and Minn., s. to N.S., N.E., L.I., Va., upland to Ga., Tenn., Ark. and Neb. April–July. — Hybridizes with no. 3, producing × **V. Bisséllii** House (for CHARLES HUMPHREY BISSELL, 1857–1925); with no. 6, producing × **V. inséssa** House (fixed); with no. 8, producing × **V. consòcia** House (interrelated); with no. 10, producing × **V. conturbàta** House (confused); with no. 11, producing × **V. melissaefòlia** Greene (*Melissa*-leaved); with no. 15, producing × **V. Porteriàna** Pollard (for THOMAS CONRAD PORTER, 1822–1901); with no. 17, producing × **V. festàta** House (in festal attire); with no. 18, producing × **V. Greenmáni** House (for JESSE MORE GREENMAN, 1867–); with no. 19, producing × **V. Ryòniae** House (for ANGIE M. RYON, 1867–1948?); with no. 20, producing × **V. notábilis** Bicknell (notable); and with no. 24.

Var. **microtìtis** Brainerd (tiny-eared). — Auricles of sepals only 1–2 mm. long. — Nfld., M.I., N.B. and N.S.

3. V. papilionàcea Pursh (butterfly-like). — Differing from no. 2 in its often larger broadly cordate-ovate *leaves* in maturity 4–12 cm. broad, *these mostly overtopping* (rarely much overtopped by) *the rich violet white-centered petaliferous flowers,* or flower whitish or pale grayish with blue veins in the frequent and much cult. forma **albiflòra** Grover (white-flowered) = *V. Priceana* Pollard (CONFEDERATE V.); *spurred petal* beardless and *about equaling lateral petals, the latter with slender beard;* outer *sepals* ovate-lanceolate, *the rounded auricles short and appressed; cleistogenes on prostrate or low-arching at first partly subterranean but finally arched-ascending to erect short peduncles; capsule ellipsoid,* green to purple, 1–1.5 cm. long, *much longer than sepals;* seeds brown or brown-dotted, about 2 mm. long, *the caruncle very short.* (Incl. *V. domestica* Pollard, the coarsest and weedy or domesticated extreme, and *V. pratincola*

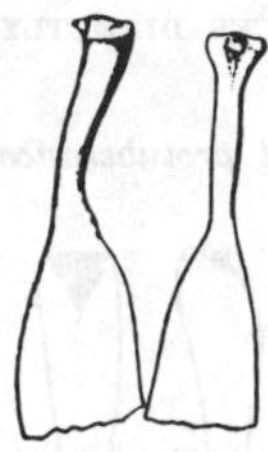
1328. V. papilionacea.

Greene) — Damp woods, meadows, roadsides, door-yards, etc., centr. Me. and s. Que. to N.D. and Wyo., s. beyond our limits. Late March–early June. FIG. 1328. — One of the commonest species, freely hybridizing with many others: with nos. 2, 4 and 6; with no. 8, producing × **V. filicetòrum** Greene (of ferny places); with no. 10, producing × **V. nàpae** House (probably intended to mean of a wooded dell); with no. 14, producing × **V. cordifòlia** (Nutt.) Schwein. (heart-leaved); with no. 15, producing × **V. abérrans** Greene (aberrant); with no. 16, producing × **V. Greènei** House (for EDWARD LEE GREENE, 1843–1915); with nos. 17, 18 and 19; with no. 20, producing × **V. insólita** House (unusual); with no. 23; with no. 27, producing × **V. Bernárdi** Greene (named in 1898 for ARTHUR BERNARD SAUNDERS).

4. V. missouriénsis Greene (of Missouri). — Differing from no. 3 in the more *cordate-deltoid leaves tapering nearly straight from base to apex and narrower than to barely as broad as long; petaliferous flowers pale violet, darkened toward center; auricles of sepals elongate,* 1–2 mm. long, *not appressed; seeds with prolonged caruncle.* — Low woods, thickets and bottoms, Ind. to Ia. and Neb., s. to Ky., Ark. and Tex. April, May. — Hybridizes with nos. 3 and 10.

5. V. Langloisii Greene (for the Louisiana botanist, AUGUSTE BARTHÉLEMY LANGLOIS, 1832–1900). — In small tufts from horizontal thick rhizomes; *leaves much as in no. 4, membranaceous,* up to 6 cm. long, *with rounded dome-like teeth, greatly overtopped by the flowers; these deep violet with pale center; cleistogenes at first sagittate, in maturity with ellipsoid capsules 6–12 mm. long; sepals narrowly lance-attenuate, the spreading or reflexed acute or angled auricles about 2 mm. long.* — Low woods, damp slopes and borders of streams, Fla. to e. Tex., n. to se. Va. and Okla. Late March, April.

6. V. nephrophýlla Greene (with kidney-shaped leaves). — Tufted, *with mostly vertical crowns,* in habit resembling no. 3; *leaves firm to subcoriaceous, gradually rounded above,* cordate-ovate to reniform, *nearly as broad as to broader than long,* becoming 2–6 cm. long, their *teeth broadly flattened; petaliferous flowers nearly equaling to overtopping leaves, warm violet,* or white in forma **albínea** Farw. (white); *spurred petal villous; broader petals 5–8 mm. wide; sepals ovate, oblong or lanceolate, obtuse or round-tipped; cleistogenes on mostly ascending to arched-recurving filiform green peduncles;* the *capsules green,* short-ellipsoid; *seeds buff, becoming olive-brown, with prolonged caruncle.* — Gravelly (often calcareous) shores, slopes, bogs and open low grounds, Nfld. to B.C., s. to P.E.I., N.B., n. and w. N.E., n. N.Y., s. Ont., Mich., Wisc., n. Ia., N.D., N.Mex., Ariz. and s. Calif. May–July. — Hybridizes with nos. 2 and 3; with no. 8, producing × **V. subaffinis** House (nearly *V. affinis*) and with nos. 10 and 27.

7. V. latiúscula Greene (broadish). — Crowns 1–few, ascending, with few leaves on erect *more or less granulose-angled petioles; leaf-blades broadly cordate-ovate or -deltoid, mostly broader than long, in maturity 4–12 cm. broad, with slightly rounded to concave sides below the acutish apex,* serrate-dentate; *petaliferous flowers very large and full, rich violet; spurred petal somewhat villous at base; larger petals 7–13 mm. wide; sepals bluntish,* with short rounded auricles; *cleistogenes on short fleshy purplish prostrate peduncles,* the purple-flecked ovoid or ellipsoid capsule 8–12 mm. long; seeds often purple-dotted, pale, usually dark in maturity. — Open deciduous woods, Vt. and N.Y. to Va. April–June. — Hybridizes with no. 10; and with no. 18, producing × **V. Slavínii** House (named in 1924 for B. H. SLAVIN).

8. V. affinis Le Conte (near or related). — *Resembling no. 3, but with somewhat narrower leaves gradually tapering to acutish tips, their teeth obliquely dome-shaped,* the full-grown blades 2–8 cm. long; *sepals acute or acutish; auricles rounded, appressed; spurred petal bearded,* the broader lateral petals 4–8 mm. wide; *cleistogenes mostly on loosely ascending or spreading filiform green peduncles; capsule* purple-dotted or green, glabrous or minutely pubescent, 5–9 *mm. long;* seeds pale and purple-dotted or dark, the caruncle short. — Meadows, damp thickets and slopes, low woods and shores, sw. Que. and Vt. to Wisc., s. to s. N.E., Ga., Ala., Tenn. and Ark. April–early June. — Hybridizes with nos. 2 and 3; with no. 10, producing × **V. cónsona** House (harmonious); with no. 11, producing × **V. champlainénsis** House (of L. Champlain); with no. 14, producing × **V. consobrìna** House (related); with no. 15, producing × **V. Hollíckii** House (for ARTHUR HOLLICK, 1857–1933); with no. 16; with no. 17, producing ×**V. disséensa** House (discordant); with no. 18, producing × **V. Mílleri** Moldenke (presumably for WALDRON DEWITT MILLER, 1879–1929); with no. 19, producing × **V. díscors** House (discordant); with no. 20, producing × **V. Davísii** House (named in 1924, presumably for WILLIAM T. DAVIS).

9. V. pectinàta Bickn. (comb-like). — *In foliage suggesting no. 8 but the leaves pectinate with teeth much longer than broad, those on lower half 2–6 mm. long; petaliferous flowers mostly equaling* *or overtopping leaves*, rich violet, with conspicuous white throat; sepals narrow and acuminate; their *prolonged auricles much longer than broad, acutish or toothed; cleistogenes on erect peduncles, their auricles divergently ascending and 2–5 mm. long; capsules green, subcylindric*, 1–1.3 cm. long; *seeds* drab to pale brown, *excluding the caruncle* (0.3–0.4 *mm.*) 1.5 *mm. long and* 1 *mm. thick.* — Peaty meadows and thickets along or not far from coast, Mass. to Va. May, June. — Hybridizes with no. 20.

10. V. soròria Willd. (sisterly; resembling other species). — *In aspect like no.* 3 *but petioles and lower faces of expanding* cordate-ovate *leaves densely villous*, the *mature leaves dark green and thickish, the larger ones* 6–13 *cm. broad and with* 21–52 *prominent teeth on each margin;* petaliferous flowers about equaling to overtopping leaves, violet or lavender, or white in forma **Béckwithae** House (for FLORENCE E. BECKWITH, 1843–1929); *spurred petal glabrous or only sparsely hairy;* outer *sepals* ovate-oblong, usually obtuse, *more or less ciliate toward base and on the short and broad appressed auricles; cleistogenes on fleshy prostrate at first buried but later arching peduncles;* capsules usually purplish, ovoid; *seeds* buff to brown, *plump-ovoid, excluding the short caruncle* 1.75–2.5 *mm. long and* 1.2–1.5 *mm. thick.* — Meadows, low woods, moist slopes, etc.; sw. Que. to Minn. and S.D., s. to N.C., Ky., Mo. and Okla. Late March–June. — Hybridizes with nos. 2, 3, 4, 6, 7, and 8; with no. 11, producing × **V. montívaga** House (wandering over the mountains); with no. 14, producing × **V. pérpera** House (defective); with no. 15, producing × **V. Fernáldii** House (for MERRITT LYNDON FERNALD, 1873–1950); with no. 17; with no. 18, producing × **V. populifòlia** Greene (poplar-leaved); with no. 19, producing × **V. Peckiàna** House (for CHARLES HORTON PECK, 1833–1917); with nos. 20 and 27.

11. V. septentrionàlis Greene (northern). — In aspect similar to small plants of nos. 3 and 10; *petioles* (often purple-based) *and lower surfaces of young leaves sparsely hirsute or hispidulous; mature broadly cordate-ovate bluntish leaves membranaceous, pale green, the larger ones becoming* 2–11 *cm. broad and with* 10–26 *low teeth on each margin;* petaliferous flowers deep violet to pale lilac, or white in forma **álba** Vict. & Rousseau (white), *all the petals more or less pubescent; sepals* mostly obtuse, *closely ciliolate nearly to tip; the narrow or angled ciliate auricles conspicuous,* 1.5–3 *mm. long; cleistogenes* on horizontal to arching short peduncles, *their subglobose capsules often purplish* and 6–10 mm. long; *seeds slenderly obovoid, excluding the long caruncle* (–0.5 *mm.*) 1.5–2 *mm. long and* 0.8–1 *mm. thick.* — Open woods (often coniferous), clearings, alluvial thickets, etc., Nfld. to s. B.C., s. to N.S., N.E., N.Y., upland to Va. and Tenn., Mich., Wisc., Ia., Neb. and Wash. May, June. — Hybridizes with nos. 2, 8 and 10; and with no. 15, producing × **V. párca** House (sparse), and with no. 19.

Var. **grísea** Fern. (gray). — Strongly suggesting no. 12 in its *narrowly deltoid-ovate leaves* but *these heavily pilose-hirsute beneath and somewhat so above;* the calyx as in no. 11. — Dry sandy soil, n. Mich.

12. V. nòvae-ángliae House (of New England). — Differing from no. 11 in the *mostly narrowly cordate-ovate, triangular or ovate-oblong leaves only minutely pubescent to glabrescent;* the *larger mature blades mostly* 3.5–7 *cm. long and* 2–5 *cm. broad, with only* 6–12 *teeth on each half; petioles villous only at base; sepals and auricles without cilia; seeds plumper,* 1–1.6 *mm. thick.* — Gravels, wet rocks, shores and meadows, rather local, N.B. to Minn. Late May, June.

13. V. villòsa Walt. (villous). — *Depressed, forming small rosettes* with a simple strongly rooting oblique subligneous rhizome; *leaves spreading or flat on ground, their orbicular, reniform or roundish-elliptic blades with a narrow, deep sinus, pilose on both surfaces, obtuse or rounded at apex,* with very low flat teeth; petioles short, pilose; *petaliferous flower* violet, *its globose spur* and the lower petals *bearded; sepals and auricles ciliate; cleistogenes* on slender short ascending peduncles, *their ovoid-cylindric green capsules* 9–10 *mm. long;* seeds becoming blackish. — Dry woods and clearings, Fla. to Tex., n. to s. Va., Ark. and Okla. Late March, April.

14. V. hirsùtula Brainerd (slightly hirsute). — *Resembling no.* 13 *but leaves more round-ovate, with broader sinus, quite glabrous* and often purple *beneath, strigose-hispid above; petioles glabrous; peduncles of petaliferous flowers glabrous,* mostly overtopping leaves; *sepals mostly eciliate; cleistogenes on fleshy depressed* (finally upcurved) *peduncles, their ovoid purplish capsules* 6–8 *mm. long; seeds buff or pale brown.* — Moist to dry woods, thickets and clearings, Ga. and Ala., n. to Ct., N.Y., W.Va. and s. Ind. April, May. — Hybridizes with nos. 3, 8 and 10; with no. 15, producing × **V. redácta** House (restored); with no. 16, producing × **V. dimíssa** House (sent apart); with no. 17; with no. 18, producing × **V. díssita** House (segregated); with no. 19, producing × **V. ràvida** House (grayish); and with no. 23.

15. V. fimbriátula Sm. (slightly fringed). — *Leaves* strongly rosulate, spreading to erect, *pilose on both surfaces,* or glabrescent in forma **glabràta** Pennell (glabrate); *blades triangular to oblong-ovate, low-crenate, obtuse to subacute, unlobed or subhastately lobed at base, the sinus subtruncate to subcordate,* the *larger thick mature blades* 1.5–8 *cm. broad; pilose petioles shorter than* *to twice length of blade;* petaliferous flowers violet-purple, or white in forma **albéscens** Farw. (whitened), on erect pubescent peduncles, or in the rare forma **umbelliflòra** Fern. (with umbelled flowers) with the usually single flowers replaced by umbels of 3 and the leaves deeply lacerate at base; *cleistogenes on erect peduncles,* their oblong-ovoid *green capsules* 6–10 *mm. long; sepals and divergent acute auricles ciliate;* seeds becoming brownish, 1.75–1.8 mm. long, 1–1.2 mm. thick. (*V. sagittata*, var. *ovata* (Nutt.) T. & G.) — Dry sterile woods, clearings and fields, N.S. to Minn., s. to N.E., L.I., n. Fla., Ala., La. and Okla. April, May. — Hybridizes with nos. 2, 3, 8, 10, 11 and 14; with no. 16, producing × **V. errática** House (erratic); with no. 17, producing × **V. abúndans** House (abundant); with no. 18, producing × **V. Robinsoniàna** House (for Benjamin Lincoln Robinson, 1864–1935); with no. 19, producing × **V. convícta** House (proven) and with no. 20, producing × **V. Mulfórdae** Pollard (named in 1902 for A. Isabel Mulford).

16. V. emarginàta (Nutt.) LeConte (emarginate). — *Very similar to no.* 15 *but glabrous; leaves triangular, one-half to nearly as broad as long, only slightly if at all blunt-lobulate at base, dark green, the slender erect glabrous petioles two to four times as long as the blade; sepals and auricles glabrous;* petals blue-violet, frequently emarginate (whence the name); cleistogamous capsules 8–14 mm. long. — Dry to moist open woods, slopes and clearings, Ga. to Tex., n. to Mass., N.Y., W.Va., O., Mo. and Kans. Late March–early May. — Hybridizes with nos. 3, 8, 10, 14 and 15; with no. 17, producing × **V. céstrica** House (of Chester Co., Pa.); with nos. 18 and 19; with no. 20, producing × **V. Holmiàna** House (for Theodor Holm, 1854–1932); and with nos. 23 and 25.

Var. **acutíloba** Brainerd (acutely lobed). — Later leaves sharply lacerate at base, the lobes 1–2.5 cm. long. — Dry mixed or deciduous woods and clearings, D.C. to se. Va.

17. V. sagittàta Ait. (arrow-shaped; from the leaf), Arrow-leaved V. — *More slender than no.* 16, *equally glabrous or minutely pubescent; leaf-blades lance-oblong to narrowly lanceolate, twice or thrice as long as broad, the later ones commonly with sharply lobed hastate base, the larger mature ones* 1–3.5 *cm. broad, nearly or quite equalled by the erect peduncles of the petaliferous flowers; sepals and auricles glabrous; petals rich violet-purple with broad white center.* — Damp to dry siliceous or argillaceous open woods, sterile meadows and clearings, Mass. to Minn., s. to Ga., La. and e. Tex. April–June. — Hybridizes with nos. 2, 3, 8, 10, 14, 15 and 16; with no. 18, producing × **V. caesariénsis** House (of New Jersey); with no. 19, producing × **V. mistùra** House (a mixture); with no. 20, producing × **V. marylándica** House (of Maryland); and with no. 27.

18. V. tríloba Schwein. (three-lobed). — *Very similar to no.* 10 and by some considered a variant of it, *but the spurred petal strongly villous at base; leaves cordate-ovate or some or all of them* 3–5*-lobed, with the middle segment broad, the basal segments lunate and with strongly recurved inner margin, the lower surfaces and petioles villous; seeds more slenderly obovoid than in no.* 10, 2.3–2.5 mm. long including the long caruncle (0.4–0.5 mm.), 1.3–1.5 mm. thick. — Rich woods, bottoms, shaded (mostly calcareous) ledges, etc., s. and w. N.E. to Ind., s. to Ala. and Mo. April, May. — Hybridizes with nos. 2, 3, 7, 8, 10, 14, 15, 16 and 17; with no. 19, producing × **V. Angéllae** Pollard (discovered in 1899 by Lillie Angell); and with nos. 20 and 23.

Var. **dilatàta** (Ell.) Brainerd (dilated). — Divided leaves cleft nearly to base into 5–7 or more narrow segments, the sinuses very sharp and deep. — Fla. to e. Tex., n. to Pa., s. Ind., Ill., Mo. and Okla.

19. V. palmàta L. (palmate). — *Strongly suggesting the var. of no.* 18, *but the leaves essentially all deeply divided into* 5–11 *uniformly narrow lobes or segments, the base of the blade subtruncate; seeds plumper,* 2 *mm. long, the caruncle not prolonged.* — Rich deciduous woods, shaded calcareous ledges, etc., s. N.H. to s. Ont. and Minn., s. to Fla. and Miss. Possibly should include as vars. nos. 10 and 18. — Hybridizes with nos. 2, 3, 8, 10, 11, 14, 15, 17 and 18; and with no. 20, producing × **V. Eàmesii** House (for Edwin Hubert Eames, 1865–1948).

20. V. Brittoniàna Pollard (for Nathaniel Lord Britton, 1859–1934). — *Superficially resembling no.* 19 *but leaves, peduncles, calyx, etc. glabrous;* crowns 1–few; first *leaves* merely dentate; the others *erect, on long petioles, the blade deeply divided into* 5–15 *narrow elongate lobes or segments, the base subtruncate to shallowly subcordate; flowers rich violet with conspicuous white center; sepals* acuminate, *glabrous, with prolonged narrow auricles* 2–5 *mm. long; cleistogenes on tall slender erect peduncles, the cylindric-ovoid green capsule* 1–1.8 *cm. long; seeds* pale brown, *excluding the short* (0.2 *mm.*) *caruncle* 1.3–1.5 *mm. long,* 0.9–1 *mm. thick.* — Sandy or peaty soil, coast of s. Me. to coast and mts. of N.C. April–June. — Hybridizes with nos. 2, 3, 8, 9, 10, 15, 16, 17, 18 and 34.

21. V. esculénta Ell. (esculent). — Rhizome and *thick crowns often becoming red in age; leaves partly simple and ovate, partly palmately lobed, thick and succulent or subcoriaceous,* rugulose when fresh, *glabrous; the lobed ones with* 3–9 *lobes, the lateral lobes falcate, the median broadly rhombic to deltoid;* petiole glabrous, elongate, divergent or ascending; *petaliferous flowers* large, *pale violet,* the spurred petal slightly villous, other petals 6–12 mm. broad; *sepals eciliate, with long emarginate auricles; cleistogenes on erect or ascending stoutish peduncles; capsule* cylindric-ovoid, green or purple-flecked, 1–1.5 *cm. long; seeds* buff to brown, *including the prominent caruncle* 1.8–1.9 *mm. long,* 1–1.3 *mm. thick.* — Dry to swampy woods and thickets, Coastal Plain, Fla. to se. Va. April.

V. Lovelliàna Brainerd (named in 1910 for its discoverer, Phoebe Lovell), *differing from no.* 21 *in minutely hoary-pubescent lower surface of leaf and summit of petiole; petals deep violet; ciliate auricles short and rounded and cleistogenes on fleshy prostrate peduncles;* a species of the Gulf Coastal Plain, has been reported from Mo.

22. V. chalcospérma Brainerd (coppery-seeded). — *Differing from no.* 21 *in more slender gray or brown crowns; leaves membranaceous, erect; the abundant early ones unlobed, cordate-ovate and broadly flattened-crenate; the few later lobed blades shallowly and bluntly 2- or 4-lobed toward base; petaliferous flowers lilac-purple,* mostly overtopping leaves; *cleistogamous capsules ellipsoid,* 7–11 *mm. long; seeds* coppery to darker, *including the very short caruncle* 1.5–1.6 *mm. long and* 0.9–1 *mm. thick.* — Cypress-swamps, wooded bottoms and rich slopes, Coastal Plain, Fla. to se. Va. April.

23. V. Stoneàna House (for Witmer Stone, 1866–1939). — Habitally *suggesting the var. of no.* 18 *but glabrous,* the 1–few crowns with long-petioled erect leaves; *larger or later blades cleft nearly or quite to base into long segments,* cordate or rounded at base, *becoming* 0.7–1.8 *dm. broad; petaliferous flowers* large, violet or lilac-purple, *darkened toward center; spurred petal glabrous, broad and flattish; auricles low and rounded; cleistogenes on fleshy prostrate or low-arching colored peduncles; capsule plump-ovoid, purplish; seeds* buff to olive-brown, including short caruncle (0.2 mm. long), about 2 mm. long, 1.1–1.4 *mm. thick.* — Deciduous woods, N.J. and Pa. to Va. and Ky. April–early June. — Hybridizes with nos. 3, 14, 16 and 18.

24. V. viàrum Pollard (of roads; originally found by a railroad). — *With aspect of no.* 18 *but glabrous; principal leaves truncate at base, broadly deltoid, with hastately and deeply incised base,* the lobes acute or acuminate; *petals* deep violet; *the spurred one narrow, emarginate and rolled into a tube, the lateral ones converging about the spurred petal; cleistogenes on slender ascending green peduncles, their cylindric-ovoid capsules green;* seeds nearly black. — Dry or moist open woods, prairies and roadsides, Ia., Mo., Kans. and Okla. Late April, May.

25. V. septémloba LeConte (seven-lobed). — Distinguished from small plants of nos. 23 and 24 by *villous spurred petal; leaves deeply cordate, many of them unlobed, the others shallowly to deeply lobed, membranaceous, the larger ones becoming* 2.5–7 *cm. broad; petaliferous flowers mostly overtopping leaves, blue-violet with broad white center,* the lateral petals villous at base; *cleistogenes not showing at time of vernal flowering,* later developed, *on erect green peduncles; their* cylindric-ovoid *green capsules* 1.2–1.5 cm. long; *seeds* buff to brown, plump-obovoid, 1.5–2 mm. long, 1.4–1.5 *mm. thick, the caruncle inconspicuous.* — Pinelands, sandy woods, clearings and swamps, Coastal Plain, Fla. to La., n. to se. Va. Late April, May.

26. V. Egglestòni Brainerd (for Willard Webster Eggleston, 1863–1935). — *In aspect suggesting var. of no.* 18 *but the plant glabrous,* the *deeply cleft leaves truncate to flabellate-cuneate at base;* principal leaves with many narrow coriaceous segments; corolla violet-purple, the spurred and lateral petals bearded at base; auricles of sepals broad and low; *cleistogenes on prostrate subterranean* (finally emerging) *fleshy peduncles; capsule ellipsoid, green or yellowish; seeds, including prolonged caruncle* (0.3–0.5 *mm.*) 2.7–3.3 *mm. long.* — Calcareous barrens of Ky. and Tenn.

27. V. pedatífida G. Don (pedately cleft). — Crowns 1–few from erect rhizome; *all or nearly all ascending or erect minutely pubescent or glabrate firm leaves truncate to cuneate at base, suborbicular to flabelliform and divided to base into* few to many *erect segments;* petaliferous flowers mostly overtopping leaves, rich violet; *cleistogenes on slender ascending peduncles; seeds, including caruncle* (0.25 *mm.*) 2 *mm. long,* 1–1.1 *mm. thick.* — Prairies and dry openings, n. O. to Alta., s. to Mo., Okla., N.M. and Ariz. April–early June. — Hybridizes with nos. 3, 6, 10 and 17.

a.Rhizome slender and thread- or cord-like, not strongly thickened, 1–4 mm. in diameter, either with or without stolons; petals white (pale violet only in nos. 28 and 29). . . *b.*

b.Petals violet, all beardless; spur and blade of spurred petal subequal; plant usually not stoloniferous; leaves with nearly closed slender sinus. Great-spurred V.

28. V. Selkírkii Pursh (for Thomas Douglas, Earl of Selkirk, 1771–1820). — *Delicate rosulate plant with slender elongate rhizome,* only exceptionally with stoloniform roots; *leaves loosely spreading on slender glabrous petioles, the membranaceous cordate-ovate blades glabrous* *below, strigose above, the deep slender sinus often overlapped by the rounded inner margins of the basal lobes;* mature blade 1.5–4.5 cm. broad; *petaliferous flower violet, about 2 cm. broad, the spur as long as the blade of the spurred petal; all petals beardless; auricles of calyx glabrous; cleistogenes on ascending slender peduncles,* the *subglobose capsule* often dotted with purple; seeds, including short caruncle 1.5–1.9 mm. long, 1–1.1 mm. thick. — Rich woods, shaded or cool rocky (often calcareous) slopes, etc., s. Greenl. and Lab. to Alaska, s. to Nfld., N.S., n. and w. N.E., Pa., s. Ont., Mich., Wisc., Minn., Colo. and s. B.C. May–July. (N. Eurasia)

b.Petals white (or in no. 29 often lilac-purple); plants stoloniferous (except in no. 33 with reniform pubescent leaves); spur much shorter than blade of spurred petal; leaves from reniform or orbicular to linear-lanceolate. Alpine Marsh-V. (no. 29) and White Violets.

29. V. palústris L. (of marshes), Alpine Marsh-V. — *Glabrous creeping plant with strongly scaly elongate forking subterranean rhizome and leafy stolons; leaves cordate-rotund or -reniform to ovate, glabrous,* becoming 2–4 cm. broad; *stolons cord-like,* 1–1.5 *mm. thick, stiffish; flower lilac-purple,* or white with lilac spur in forma **albiflòra** Neum. (white-flowered), 1–1.5 *cm. broad, very fragrant, the short spur rounded; capsules green; seeds, including the very short caruncle* 1.5–1.7 *mm. long,* 1 *mm. thick,* grayish. — Subarct. reg., s. to subalpine and alpine brooksides and damp slopes to Nfld., Gaspé Pen., Que., n. N.E., Colo., Utah and Oreg. Late June–Aug. (Eurasia)

30. V. pállens (Banks) Brainerd (pale). — *With very slender, flexuous filiform stolons; leaves cordate-ovate to reniform, glabrous, membranaceous, blunt, at anthesis* 1–3 cm., in maturity becoming 1–8 cm. long, crenate with low flattened teeth, *on sparsely hirtellous or glabrous petioles; petaliferous flowers mostly overtopping subtending leaves, white,* with purple veins, 7–10 *mm. long, very fragrant, the upper petals obovate, the lateral ones beardless; sepals linear-lanceolate; cleistogenes on shorter filiform ascending peduncles; capsules green, ellipsoid; seeds becoming black,* 1–1.4 *mm. long,* 0.7–0.8 *mm. thick.* — Wet or springy woods, thickets, slopes and openings, Lab. to Alaska, s. to Nfld., N.S., N.E., L.I., Del., Pa., W.Va., upland to S.C. and Ala., O., Ind., n. Ill., n. Ia., N.D. and Mont. April–July. — Hybridizes with no. 34, producing × **V. sublanceolàta** House (nearly lanceolate); and with no. 35, producing × **V. mollícula** House (softish).

31. V. incógnita Brainerd (unknown; until recognized). — *Coarser than no.* 30, *forming intricate carpets; the pubescent leaves at anthesis* 3–6 *cm. long, chiefly ovate, the larger ones at maturity* 2.5–10 *cm. broad; ascending petioles and lower surfaces of membranaceous rugulose leaves soft-pubescent when young, the upper surface glabrous or nearly so, the petioles mostly green; flowers* 1–1.5 *cm. long, not especially fragrant, on pubescent peduncles; lateral petals bearded, the obovate upper ones not strongly reflexed;* sepals ovate-lanceolate; *cleistogenes on prostrate or low-arching peduncles, their* plump *capsules usually purplish; seeds,* including caruncle (0.1–0.2 mm.), 1.9–2.1 *mm. long,* 1–1.25 *mm. thick, buff, becoming olivaceous.* — Wet to dryish woods, thickets and openings, se. Lab. to Thunder Bay Distr., Ont., s. to Nfld., N.S., N.E., N.Y., Mich., Wisc., Minn. and N.D. May–early July. — Apparently hybridizes with no. 33.

Var. **Fórbesii** Brainerd (for Fayette Frederick Forbes, 1851–1935). — Nearly or quite glabrous except for scattered hairs on upper leaf-surface. — Often more common, Nfld. to Minn., s. to N.S., N.E., L.I., Del. and Pa. — Strongly simulating no. 32.

32. V. blánda Willd. (mild). — *Differing from no.* 31 *in less matted growth, in the more flattened and depressed leaves firmer and glabrous except for scattered strigae on the upper surface (especially toward base) when young,* the *petioles and bases of peduncles reddish and glabrous; narrow upper petals strongly reflexed, the lateral petals beardless; seeds shorter and plumper,* 1.6–1.9 *mm. long,* 1.25–1.3 *mm. thick.* — Rich, chiefly deciduous, woods, sw. Que. and N.H. to Minn., s. to s. N.E., Del., Md., upland to Ga. and Tenn., O., Ind., Ill. and Wisc. April, May.

33. V. renifòlia Gray (kidney-leaved). — *Non-stoloniferous, with crowns* 2–4 *mm. thick; leaves widely spreading to ascending, reniform to orbicular or round-ovate, cordate at base, when young whitened with soft persistent villosity on both sides as well as along petiole,* the nearly suppressed teeth very broad and flat; *mature blades* 1.5–8 *cm. broad;* flowers white with purple stripes, 1–1.5 cm. long, the *petals all beardless; auricles of sepals ciliate; peduncles of cleistogenes prostrate or arching, villous above, the sepals and auricles copiously ciliate, the subglobose capsule usually purple; seeds* warm brown, *including the prominent caruncle* (0.3–0.5 *mm.*) 1.9–2.4 *mm. long.* — Cool woods, wooded swamps, rocky (often calcareous) slopes, etc., n. Nfld. (local) to Minn., s. to n. N.E., w. Mass., e. N.Y. and s. Ont. May–July. — Much more restricted geographically than

Var. **Brainérdii** (Greene) Fern. (for Ezra Brainerd, 1844–1924, painstaking student of the genus). — Leaves glabrous or promptly glabrate and bright green above, glabrescent beneath, the petioles similarly glabrescent; peduncles and sepals glabrescent. (*V. Brainerdii* Greene) — Similar habitats and often ascending to subalpine barrens northeastw., s. Lab. to Alaska, s. to Nfld., N.S., n. N.E., nw. Ct., n. and n.-centr. N.Y., L. Huron and Thunder Bay Distr., Ont., n. Mich., n. Wisc., n. Minn., S.D., Colo. (at 2500 m.) and mts. of s. B.C.

34. V. lanceolàta L. (lanceolate), LANCE-LEAVED V. — *Stoloniferous, glabrous, the often alternately leafy superficial stolons frequently (late in the season) forming close mats and bearing many alternate cleistogenes; leaves lanceolate to oblanceolate, gradually tapering into the margined often red-based petioles; mature obscurely crenate blades* 2.5–12 *cm. long and* 0.7–2.5 *cm. broad; white petaliferous* purple-veined flowers often equaling or overtopping expanding leaves, their peduncles 0.2–1.7 dm. high, the *lateral petals* usually *beardless; sepals lanceolate, acute; cleistogenes nodding on elongate erect peduncles from the crowns or on shorter peduncles from the axils of the stolons,* their cylindric-ovoid *capsules green;* seeds becoming olivaceous, plump-obovoid, acutish-based, 1.4–1.5 mm. long, 0.9–1 mm. thick. — Damp to inundated argillaceous, siliceous or peaty open ground or in slight shade, Fla. to e. Tex., n. to C.B., N.B., N.E., s. Que., s. Ont., Mich., Wisc., Minn. and Neb. Late March (southw.)–early July (northw.). — Hybridizes with nos. 20 and 30; and with no. 35, producing × **V. modésta** House (modest). — Southw. passing into

Var. **vittàta** (Greene) Weath. & Grisc. (from resemblance of leaves to fronds of *Vittaria*). — *Stolons* mostly subterranean, usually *without scattered leaves and cleistogenes but eventually producing terminal leafy crowns; later leaves with linear-lanceolate blades* 0.6–3 *dm. long and* 0.5–1.5 *cm. broad,* often more denticulate and sometimes sparsely pubescent beneath; peduncles of vernal flowers mostly 0.7–2.5 dm. high. (*V. vittata* Greene) — Wet pinelands, sands and peats, Fla. to e. Tex., n. to borders of ponds and pond-holes of s. N.J. and inland from the Gulf to s. Tenn. March, April.

35. V. primulifòlia L. (primrose-leaved). — Similar to no. 34, *glabrous;* the filiform *rhizomes chiefly terminating in crowns of leaves, the superficial stolons late in the season (mostly later than in no.* 34) *developing scattered alternate leaves but few or no axillary cleistogenes; leaf-blades elliptic-oblong to ovate, cordate or subcordate above the junction with the broadly decurrent summit of the petiole, the later ones* 2.5–12.5 cm. long, 1.7–8 cm. broad, *averaging four-fifths as broad as long;* flowers and fruits as in no. 34; seeds slightly larger (1.5–1.7 mm. long). — Wet to dryish open shores, meadows, thin woods, etc., Fla. to e. Tex., n. to N.J., e. Pa. and Okla. April–June. — Northw. passes into

Var. **acùta** (Bigel.) T. & G. (acute; name not very appropriate). — *Glabrous leaves gradually rounded to subtruncate above the broad summit of the petiole, the later ones* 1.7–8.5 cm. broad, *averaging one-half as wide as long.* — Fla. to e. Tex., n. to N.S., s. N.B., centr. Me., sw. Que., s. Ont., Mich. and Minn. April–June.

Var. **villòsa** Eat. (villous). — *Leaves* as in var. *acuta* but *villous on one or both surfaces or along veins beneath.* — Fla. to La., n. to N.J. and e. Pa.

§ 3. *Style enlarged upward, abruptly capitate, beakless; stigma within a small orifice on the lower side of the summit; petals yellow.* FIG. 1329.

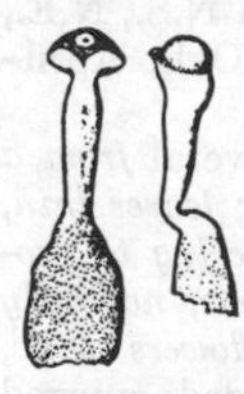
1329. V. rotundifolia.

36. V. rotundifòlia Michx. (round-leaved), ROUND-LEAVED or EARLY YELLOW V. — *Rhizome thick and fleshy; leaves barely unrolling and minutely pubescent at flowering time, soon widely spreading from summit of rhizome and lying flat on the ground, in maturity roundish-elliptic to oval or subrotund,* 4–12 cm. long, with shallow slender sinus, crenulate, *subcoriaceous and glabrous at least above; petaliferous flowers bright yellow,* the three lower petals brown-veined, the lateral ones bearded; *cleistogenes not evident at vernal flowering time, later borne on prostrate or partially subterranean stoloniform branches, the* (1–)2–*several long-stalked cleistogenes forming open racemes;* capsules ovoid, 6–10 mm. long, purple-dotted; seeds pale. — Rich, chiefly deciduous, woods, n.-centr. Me. to s. Ont., s. to s. N.E., Del., Pa., W.Va. and O. and along mts. to n. Ga. and Tenn. April, May.

§ 4. *Style neither clavate nor capitate, ending in a small beak pointing downward; petals violet or white.* FIG. 1330.

37. V. ODORÀTA L. (fragrant), ENGLISH or SWEET V. — *Spreading by long cord-like superficial leafy stolons; leaves* broadly cordate-ovate, *finely pubescent; petaliferous flowers richly fragrant,* warm violet, or in forma ALBIFLÒRA Oborny (white-flowered) white; *capsules closely pubescent, purple, plump-ovoid; seeds, including the very long caruncle* (1 *mm.*) 3.4–4 *mm. long,* 1.75–2 *mm. thick.* — Spread from cult. to roadsides, open woods, etc. (Introd. and natzd. from Eu.)

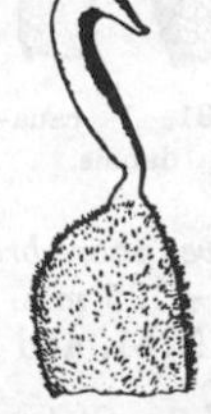
1330. V. odorata.

GROUP II. PLANTS WITH LEAFY STEMS; PETALIFEROUS FLOWERS AXILLARY.

§ 5. *Style capitate, beakless, bearded at summit; spur short; stipules entire or finely toothed, the lower ones more or less scarious; petals yellow or white.* LEAFY YELLOW VIOLETS and CANADA-V. (nos. 42 and 43). FIG. 1331.

38. V. tripartita Ell. (three-parted). — *Stem erect, slender, glabrous,* 1.5–4.5 dm. high, usually single *from a subligneous rhizome, leafy only toward summit; leaves pubescent beneath, some of them deeply 3-parted into narrow elongate segments or leaflets,* the upper simple leaves, when present, rhombic- to ovate-lanceolate and *plicate when fresh,* or leaves all simple and rhombic to narrowly ovate in forma **glabérrima** (DC.) Fern. (most glabrate); *stipules thin, green,* toothed or entire; *petaliferous flowers on long erect filiform axillary peduncles;* sepals linear-lanceolate, ciliate; *petals yellow, the upper ones often purplish on back,* the three lower bearded; *capsule trigonous-ovoid, green,* 1–1.2 *cm. long,* usually pubescent; cleistogenes on crowded short peduncles in upper axils; *seeds drab or pale brown, plump-obovoid,* 2.5–2.8 *mm. long,* 1.5–2 *mm. thick.* — Rich woodlands, n. Fla. and Ala., n. to N.C. and s. O. Late April, May.

39. V. hastàta Michx. (hastate), HALBERD-LEAVED YELLOW V. — Habitally *similar to no.* 38, *the very slender glabrous stem* 1–3 *dm. high, from a pale and fleshy rhizome; leaves narrowly hastate-ovate or ovate,* 2–4, at summit of stem, thin, *tapering to acute tip, flat when fresh,* slightly serrate, 2–5 cm. broad at base; *stipules thin, attenuate,* 2–10 *mm. long, often sharply toothed; yellow petals violet on back;* capsule glabrous; *seeds* 2–2.5 *mm. long.* — Rich deciduous woods, Fla. and Ala., n. to Pa., W.Va. and n. O. Late April, May.

40. V. pubéscens Ait. (pubescent), DOWNY YELLOW V. — *Softly villous stems* 1–few, 1–4.5 dm. high, *usually leafless at base,* occasionally with 1 basal leaf; *cauline leaves pale green,* few toward summit of stem, *very broadly ovate or roundish, blunt-tipped, heavy, veiny, the expanding ones densely soft-pubescent, the fully grown ones* 5–12 *cm. broad and with broad open sinuses, the coarse dentations mostly* 15–23 *on each half; stipules ovate-oblong or semiovate, entire or merely undulate, soft-pubescent, the upper ones herbaceous, the lower scarious; petaliferous flowers yellow,* purple-veined, on filiform peduncles; sepals narrowly lanceolate, acute; cleistogenes clustered on short axillary peduncles; *capsules* ovoid, *white-woolly;* seeds pale, 2.5–2.9 mm. long, 1.5–1.8 mm. thick. — Rich deciduous woods, centr. Me. and sw. Que. to Minn. and S.D., s. to Del., Pa., upland Va., Tenn., Mo. and Neb. May, early June.

Var. **Péckii** House (for CHARLES HORTON PECK, 1833–1917). — Quite similar but foliage darker and capsules glabrous. — Local, through range of typical form.

41. V. pensylvánica Michx. (Pennsylvanian), SMOOTH YELLOW V. — *Differing from no.* 40 *in its more slender and soon glabrescent stems often tufted and bearing* 1–3 *basal leaves; expanding leaves thinly pilose, becoming glabrate; cauline leaves thinner, deeper green, less veiny, in maturity* 2–10 *cm. broad, usually with* 8–15 *teeth on each half; stipules lanceolate to narrowly semiovate; capsule densely white-tomentose.* (*V. eriocarpa* and *scabriuscula* Schwein.) — Damp woods, wooded bottoms, cool shaded rocky slopes, etc., centr. Ct. to s. Minn., s. to Ga., Ala., Ark. and Okla. April, May.

Var. **leiocárpa** (Fern. & Wieg.) Fern. (smooth-fruited). — *Capsules glabrous.* — Generally more northern or of cooler habitats southw., e. Gaspé Co., Que. to Man., s. to N.S., N.E., Del., Md., upland of N.C. (to 5000 ft. alt.) and Tenn., O., Ind., Ill., Mo. and Okla. April–July.

42. V. canadénsis L. (Canadian), CANADA-V., TALL WHITE V. — *Stems* 1–several *from a thickish subligneous rhizome,* minutely pubescent to glabrate, up to 4 dm. high; *leaves thin,* cordate-ovate *with deep sinus, the upper narrowly ovate and tapering to prolonged tip, closely serrate or dentate; stipules more or less scarious, narrowly lanceolate, slender-tipped;* slender-peduncled axillary *petaliferous flowers white, often violet-tinged on the back, especially in age;* lateral petals bearded; spurred petal yellow at base and dark-lined; sepals narrowly lance-acuminate; capsules subglobose, 6–10 mm. long; *seeds* brown, 1.5–2 *mm. long,* 1.2–1.5 *mm. thick.* — Deciduous woods, sw. Que. and w. N.H. to Mont., s. to Pa. and Md., upland to S.C., Ala., Tenn., Ia., S.D., Colo. and Utah. Late April–July, sporadically –Oct. FIG. 1331.

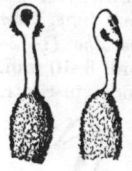

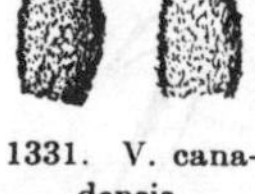

1331. V. canadensis.

43. V. rugulòsa Greene (slightly wrinkled). — *Differing from no.* 42 *in the elongate cord-like subterranean stoloniferous rhizome and slender crowns; upper leaves more broadly ovate and more abruptly tipped; stipules more scarious; seeds* 2–2.2 *mm. long,* 1.2–1.25 *mm. thick.* — Mixed or coniferous woods, calcareous bluffs, etc., n. Alta. and B.C. to N.M. and Ariz., e. to Man., Minn., Wisc. and Ia. May, June.

§ 6. *Style not capitate, slender; spur at least twice as long as thick; stipules herbaceous, mostly fringe-toothed; petals* (*except in no.* 44) *violet or bluish.* Leafy Blue Violets and (no. 44) Cream-V. Fig. 1332.

1332. V. conspersa.

44. V. striàta Ait. (with fine lines), Cream-V. — Basal leaves rosulate in spring, cordate-subrotund, soon shrivelling; *stems tufted, loosely spreading to ascending,* in vernal anthesis simple and 1–3 dm. high, later elongating to 6 or 7 dm. and often much branched; *principal cauline leaves membranaceous, cordate-ovate, those of primary axis 3–7 cm. long,* usually *long-acuminate, with rounded crenations; stipules herbaceous,* 1–2.5 *cm. long, their longer teeth* 3–5 *mm. long; petaliferous vernal flowers creamy to milk-white; auricles of sepals* 2–4 *mm. long;* capsules ovoid, glabrous, 4–6 mm. long; seeds pale brown, 1.4–2.6 mm. long. — Low woods and meadows (chiefly calcareous), Ga. to Ark., n. to N.Y., s. Ont., O., Mich. and Wisc.; sometimes spread from cult. elsewhere. April–June.

45. V. conspérsa Reichenb. (sprinkled; from the dotted lower leaf-surface), American Dog-V. — Habitally similar to no. 44; stems at flowering time 0.5–2 dm. long; *leaves* membranaceous, *glabrous or nearly so, the upper cauline blades reniform-ovate, at anthesis* 1.5–3 *cm. long, with rounded or abrupt blunt tips, the teeth strongly flattened; stipules ovate-lanceolate, with longer teeth up to* 2 *mm. long;* flowers pale violet, or white in forma **Masònii** (Farw.) House (named in 1917 for its discoverer, E. W. Mason), mostly raised high above the leaves; *lateral petals bearded, spur less than* 8 *mm. long; tip of style bent downward, somewhat pubescent; cleistogenes abundant in upper axils* of the finally firm and enlarged leaves; *seeds* buff to brown, *slenderly obovoid,* including caruncle (0.2 mm.), 1.7–2 *mm. long,* 1–1.1 *mm. thick.* — Meadows, damp woods, bottomlands, etc., Gaspé Pen., Que. to Thunder Bay Distr., Ont., and Minn., s. to N.S., N.E., Md., upland (up to 1800 m. alt.) to Ga., Ala., and Tenn. May–early July. Fig. 1332. — Hybridizes with no. 46; and with no. 48, producing × **V. Malteàna** House (for Malte Oscar Malte, 1880–1933).

46. V. adúnca Sm. (hooked; from the not unusual spur). — Similar to no. 45, *smaller and lower, densely puberulent, more densely tufted; leaves subcoriaceous, the upper ones ovate and gradually narrowed to apex; stipules linear-lanceolate, entire, subentire or short-toothed at base; flowers deep violet or violet-blue,* or in forma **albiflòra** Vict. & Rousseau (white-flowered) white; *seeds,* including caruncle (0.2–0.4 mm.) 1.5–2 mm. long, 0.8–1 *mm. thick.* (*V. arenaria* of ed. 7, not DC.) — Dry sterile open woods, pastures, rocky slopes, etc., Côte Nord, Que., to Alaska, s. to n. N.E., ne. Mass., N.Y., s. Ont., Mich., Wisc., Minn., S.D., Colo. and Calif. Mid-May–July. — Hybridizes with no. 45.

Var. mìnor (Hook.) Fern. (smaller). — *Glabrous or at most with scattered hairs on upper leaf-surface; leaves smaller, less tapering to rounded; stipules mostly subentire.* (Var. *glabra* Brainerd; *V. labradorica* Schrank) — Wet to dry rocks, sands, woods, swamps or openings, Greenl. and Lab. to Alaska, s. to Nfld., N.S., n. N.E., n. N.Y., Ont., n. Mich., n. Wisc., N.D., Colo. and Calif., with us ascending to alpine areas.

47. V. Wálteri House (for Thomas Walter, ?1740–1789). — *Young crowns and round-reniform to orbicular leaves* suggesting those of no. 45; the *crowns soon sending out long trailing stolons, these rooting at tip and establishing new crowns;* leaves scabrous, their veins often with dark borders (when fresh); *stipules deeply lacerate;* vernal petaliferous flowers blue-violet; *cleistogenes on long axillary peduncles, produced from spring to autumn; capsules subglobose,* 5–6 mm. long, often purplish; brown seeds, including caruncle (0.2 mm. long), 1.7–2.4 mm. long, 1–1.4 mm. thick. — Dry woods and rocky slopes, Fla. to Tex., n. to W.Va. and Ky. Late April, May.

48. V. rostràta Pursh (beaked; from the long spur), Long-spurred V. — *Differing from no.* 45 *in gradually acuminate upper leaves;* longer teeth of stipules 3 mm. long; *petals lilac-purple with darker spot near center,* the *lateral petals beardless; spur very slender,* 0.9–1.8 *cm. long; style straight, glabrous at tip.* — Rich, often calcareous, woods, sw. Que. and Vt. to Wisc., s. to Ct., n. N.J., Pa., W.Va. and upland to Ga. and Ala. April–early June.

§ 7. *Style much enlarged upward into a globose hollow summit with a wide orifice on the lower side; stipules large and leaf-like; ours annual or short-lived perennials, introd.* Pansies. Fig. 1333.

49. V. trìcolor L. (three-colored), Pansy, Heart's-ease, Johnny-jump-up, Ladies'-delight. — Stems angled, soft and brittle, 1–3 dm. high; lowest leaves roundish; *upper leaves oblong, cordate, crenate-serrate; stipules pinnatisect at base; petals twice or thrice length of sepals,* variously yellow, white or purple; capsule ovoid; seeds brown. — Persistent after cult. and as a weed of fields, etc. (Introd. from Eu.)

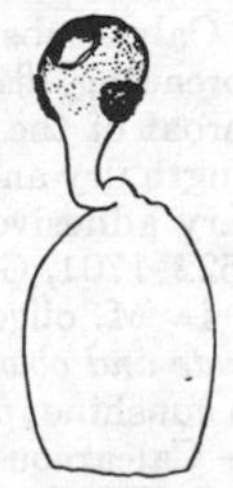

1333. V. tricolor.

50. V. arvénsis Murr. (of fields), Wild Pansy. — Similar to no. 49 but smaller; *leaves cuneate at base; petals about equaling to but slightly longer than sepals, pale yellow; capsule globose;* seeds, including caruncle 1.5–1.7 mm. long, 0.8–1 mm. thick. — Fields and roadsides, through our area and beyond. April–Sept. (Natzd. from Eu.)

51. V. Kitaibeliàna R. & S. (for Paul Kitaibel, 1757–1817), var. Rafinésquii (Greene) Fern. (for Constantine Samuel Rafinesque-Schmaltz, 1783–1840), Field-Pansy. — *Slender, simple or branched from base,* the *bruised roots with fragrance of wintergreen; leaves small,* the basal rounded, *the cauline obovate to linear-oblanceolate and entire, attenuate to base; stipules palmately pectinate at base; flowers* small, 7–10 *mm. long, the obovate bluish-white to creamy petals nearly twice as long as sepals; seeds,* including caruncle (0.2–0.25 mm. long), 1.3–1.5 *mm. long,* 0.8–0.9 *mm. thick.* (*V. Rafinesquii* Greene) — Dry fields, roadsides, etc., Ga. to Tex., n. to N.Y., O., Mich., Ill., Ia. and Kans. April, May. (Natzd. from Eurasia)

FAM. 116. PASSIFLORÀCEAE (Passion-flower Family)

Herbs or woody plants, climbing by tendrils, with perfect flowers, 5 *monadelphous stamens, and a stalked* 1-*locular ovary free from the calyx, with* 3 *or* 4 *parietal placentae, and as many clavate styles.*

1. PASSIFLÒRA L. Passion-flower

Calyx of 5 sepals united at the base; the throat crowned with a double or triple fringe. Petals 5, on the throat of the calyx. Filaments united into a tube which sheaths the long stalk of the ovary, separate above; anthers large, fixed by the middle. Berry (often edible) many-seeded. Leaves alternate, generally palmately lobed, with stipules. Peduncles axillary, jointed. Large genus of trop. and warm-temp. Am., Asia, Austral. and Madagascar, ours perennial herbs. (An adaptation of *flos passionis,* a translation of *fior della passione,* the popular Italian name early applied to the flower from a fancied resemblance of its parts to the implements of the crucifixion.)

1. P. lùtea L. (yellow). — Stem, when young, more or less pilose, slender; *leaves obtusely* 3-*lobed at the summit, the lobes entire;* petioles glandless; *flowers greenish-yellow,* 1.5–2.5 *cm. broad,* the *calyx pilose-hirsute at base; fruit* about 1.2 *cm. in diameter.* — Thickets and borders of woods, Fla. to se. Pa. June–Sept.

Var. **glabriflòra** Fern. (smooth-flowered). — Glabrous throughout, or only the stems rarely pilose. — W.Va. and O. to Ill., Mo., se. Kans. and southw.

2. P. incarnàta L. (flesh-colored), Apricot-vine. — Pubescent; *leaves* 3–5-*cleft, the lobes serrate,* the base bearing 2 glands; *flower large* (5–8 *cm. broad*), *nearly white, with a triple purple and flesh-colored crown;* involucre 3-leaved; *fruit* ellipsoid, 3.5–6 *cm. long.* — Sandy thickets and open soil, Fla. to Tex., n. to Md., sw. Pa., W.Va., s. O., s. Ind., s. Ill., s. Mo. and Okla. June–Sept. — Fruit called Maypops.

FAM. 117. LOASÀCEAE (Loasa Family)

Herbs with a rough or stinging pubescence, no stipules, the calyx-tube adherent to a 1-*locular ovary with* 2 *or* 3 *parietal placentae;* — represented with us only by the genus.

1. MENTZELIA L.

Calyx-tube cylindrical or clavate; the limb 5-parted, persistent. Petals 5 or 10, regular, spreading, flat, convolute in the bud, deciduous. Stamens inserted with the petals on the throat of the calyx. Styles 3, more or less united into 1; stigmas terminal, minute. Capsule at length dry and opening at the summit. Seeds flat, anatropous. — Stems erect. Leaves alternate, very adhesive by the barbed pubescence. Small Am. genus. (Dedicated to *Christian Mentzel,* 1622–1701, German botanist.)

1. M. oligospérma Nutt. (few-seeded), Stickleaf. — Much branched, 3–9 dm. high; *leaves ovate and oblong,* cut-toothed or angled, often petioled; *flowers yellow,* 1.5–2 *cm. broad,* opening in sunshine; *petals* 5, wedge-oblong, pointed; stamens 20 or more; capsule small, about 9-seeded. — Calcareous hills and banks, Ill. to S.D. and Colo., s. to La., Tex. and Mex. June–Aug.

2. M. decapétala (Pursh) Urban & Gilg (with ten petals). — Larger in all its parts; *leaves elongate-lanceolate,* sharply and coarsely dentate; *flowers white or pale yellow,* 7–12 *cm. broad,* opening in the evening; *petals* 10, lanceolate; stamens abundant; seeds numerous. (*Nuttallia* Greene) — Rocky hillsides and dry prairies, nw. Ia. to Sask. and Alta., s. to Okla., Tex. and Nev. July–Sept.

FAM. 118. CACTÀCEAE (Cactus Family)

Fleshy and thickened plants, mostly without green leaves, globular or columnar and many-angled or flattened and jointed, usually with prickles, spines or stinging bristles. Flowers solitary, sessile or with the base of the ovary prolonged; sepals and petals numerous, imbricated in several rows, the bases adherent to the 1-locular ovary. Stamens numerous, inserted on the inside of the tube or cup formed by the union of the sepals and petals. Style 1; stigmas several.

Branching or jointed plants; the segments or internodes separated by joints or articulations, flattened to terete, bearing the flowers along the margins of the newer segments. . . . . 1. *Opuntia.*
Globose or ovoid plants covered with spine-bearing tubercles; flowers borne from between the tubercles. . . . . 2. *Mamillaria.*

1. OPÚNTIA Mill. Prickly Pear. Indian Fig

Sepals and petals spreading, regular, the inner roundish and not united into a prolonged tube. Stem composed of segments or internodes separated by articulations, bearing small scarious or scale-like mostly deciduous leaves spirally arranged, with clusters (*areoles* or *pulvini*) of short hairs and barbed bristles (*glochids*) and often longer spines in their axils. Flowers in our species yellow, opening in sunshine for two or more days. Large genus of arid soils of trop. and temp. Am. (some natzd. in Old World). (A name used by Theophrastus for some plant, certainly not this.)

Segments or internodes of stem strongly flattened and tightly articulated, separating only under pressure, the mature ones 0.4–2.5 dm. long; fruit juicy, not covered with long spines.
Roots fibrous, not thickened.
Segments without spines or with mostly single spines from some areolae; wide-ranging. . . . . 1. *O. humifusa.*
Segments with 3–8 very unequal spines from the areolae; western. 2. *O. tortispina.*
Roots tuberous-thickened; spines 1–4; southwestern. 3. *O. macrorhiza.*
Segments turgid, the terminal one loosely articulated and easily separated, 1.5–5 cm. long; fruit dry, spinescent. . . . . 4. *O. fragilis.*

1. O. humifùsa Raf. (spreading on the ground). — Prostrate or spreading from branching *fibrous roots,* forming carpets up to 1 m. or more across, pale to deep green, often glaucous; *the larger segments suborbicular, broadly obovate or narrowly oblong-obovate, strongly flattened,* 0.4–2.5 *dm. long,* with quickly deciduous brown leaves 4–9 mm. long; *glochids reddish-brown to paler,* 2–5 *mm. long; spines wanting or* 1 (sometimes 2) and 1–5 *cm. long, light brown to whitish; flower* yellow, 5–9 *cm. broad,* sometimes with a red star-shaped eye, the *petals* 8–12; *fruit pulpy, edible,* green to dull purple, obovoid or somewhat clavate-based, 2–5 cm. long, often with some glochids on the tube; seeds about 5 mm. in diameter, compressed. (*O. vulgaris* of ed. 7, not Mill.; *O. Opuntia* Karst.; *O. Rafinesquii* Engelm.; *O. macrarthra* Gibbes) — Dry sands and rocks, Mass. to Minn., s. to S.C., upland Ga., Ala., Miss., Mo. and Okla. June, July. — Polymorphous but not clearly separated into geographic vars.

2. O. tortispìna Engelm. (with twisted spines). — Differing from no. 1 in having the areoles provided *with 3–8 very unequal divergent spines.* — Colo. and N.M., e. to Wisc., O., and n. Ky.

3. O. macrorhìza Engelm. (large-rooted). — Differing from no. 1 in its *tuberous-thickened roots, the larger ones* 5–7.5 *cm. in diameter;* areoles spineless or with 1–4 *very unequal yellow to brown spines, the longer one up to* 2.5 *cm. long; fruit slenderly obovoid, inedible.* — Tex. to Kans. and sw. Mo.

4. O. fràgilis (Nutt.) Haw. (fragile). — Decumbent, *forming mounds* 1–4 *dm. across and up to* 2 *dm. high, with plump to subterete segments* 1.5–5 *cm. long, the terminal one readily disarticulating; pulvini white-woolly; spines* 1–7, *strongly divergent, up to* 2 *cm. long; flower about* 5 *cm. broad; fruit dry, ovoid, spiny,* 1.5–2 *cm. long.* — E. Man. to B.C., s. to Ill., Ia., Kans. nw. Tex. and Ariz.

2. MAMILLÀRIA Haw.

Flowers about as long as broad, the tube campanulate or funnel-shaped. Ovary often hidden between the bases of the tubercles, naked, the succulent berry exserted. Seeds yellowish-brown to black, crustaceous. — Globose or ovoid plants, covered with spine-bearing cylindrical, ovoid or conical tubercles, the flowers from distinct woolly or bristly areoles at their base. (Name from *mamilla*, a *nipple*, referring to the tubercles.) Neomammillaria Britt. & Rose. Coryphantha (Engelm.) Lemaire

1. M. vivípara (Nutt.) Haw. (sprouting on the parent plant). — Single or tufted, 2.5–12 cm. high, the almost terete tubercles bearing bundles of 5–8 reddish-brown spines (2 cm. long or less) surrounded by 15–20 grayish ones in a single series, all straight and rigid; flowers red or purple, with fringed sepals and lance-subulate petals; berries ovoid, green; seeds pitted, light brown. (*Coryphantha* Britt. & Rose; *Neomammillaria* Britt. & Rose) — Granite ledges and dry plains, Man. to Alta., s. to w. Minn., Kans. and Colo.

FAM. 119. THYMELAEÀCEAE (Mezereum Family)

Shrubs with acrid and very tough (not aromatic) bark, entire leaves and perfect flowers with a regular and simple colored calyx, bearing usually twice as many stamens as its lobes and free from the 1-locular and 1-ovuled ovary, which forms a berry-like drupe in fruit with a single suspended anatropous seed. Embryo large; albumen little or none.

Calyx tubular, without spreading lobes; stamens and style exserted. . . . . . 1. *Dirca.*
Calyx-lobes (4) spreading; stamens included; style short or none. . . . . . 2. *Daphne.*

1. DÍRCA L. Leatherwood. Moosewood. Bois de plomb (Que.)

Calyx corolla-like, tubular-funnelform, truncate, the border wavy or obscurely about 4-toothed. Stamens inserted on the calyx above the middle, the alternate ones longer. Style filiform. Drupe ovoid to ellipsoid or globose, green, yellowish or red. — A much branched single-trunked Am. shrub, with jointed branchlets, oval-obovate alternate leaves on very short petioles, the bases of which conceal the buds of the next season. Flowers light yellow, preceding the leaves, 3 or 4 in a cluster from a bud of as many dark-hairy scales, the latter forming an involucre from which soon after proceeds a leafy branchlet. (Named for *Dirce*, wife of *Lycus*, who, after her brutal murder, changed into the fabulous fountain Dirce.)

1. D. palústris L. (of swamps), Wicopy, Rope-bark. — Shrub, 1–3.3 m. high; the wood white, soft and very brittle; but the fibrous bark remarkably tough (used by the Indians to make thongs, whence one of the popular names). — Rich deciduous or mixed woods, N.B. to Ont. and Minn., s. to n. Fla. and La. Apr., May.

2. DÁPHNE L. Mezereum

Calyx salverform or somewhat funnelform. Anthers nearly sessile on the calyx-tube. Stigmas capitate. Drupe red. — Hardy low shrubs nat. of Eurasia and n. Afr. (Mythological name of the nymph transformed to Apollo's dismay into a Laurel.)

1. D. Mezèreum L. (modification of the Arabic name), Daphne, Bois gentil (Que.). — Shrub, 3–9 dm. high, with purple-roseate (rarely white) flowers in lateral clusters on shoots of the preceding year, before the lanceolate smooth leaves. — Roadsides, thickets and old lime-quarries, Nfld. to s. Ont., s. to N.S., N.E., N.Y., O. and doubtless elsewhere. Apr., May. (Introd. and natzd. from Eu., often cult.)

FAM. 120. ELAEAGNÀCEAE (Oleaster Family)

Shrubs or small trees with silvery- to golden- or reddish-scurfy leaves and perfect or dioecious flowers; further distinguished from the Mezereum Family by the erect or ascending albuminous seed and the calyx-tube which becomes pulpy and berry-like in fruit, strictly inclosing the achene.

Leaves alternate; flowers perfect; stamens 4. . . . . 1. *Elaeagnus.*
Leaves opposite; flowers dioecious; stamens 8. . . . . 2. *Shepherdia.*

1. ELAEÁGNUS L. Oleaster

Calyx cylindric-campanulate above the persistent cylindrical or globose base, the limb valvately 4-cleft, deciduous. Stamens 4, in the throat. Style linear, stigmatic on one side.

Fruit drupe-like, with an ellipsoid 8-striate stone. — Leaves alternate, entire and petioled, and flowers axillary and pedicellate; shrubs of N. Am. and Eurasia. (From *elaia, the olive*, and *agnos*, the Greek name of the Chaste-tree, *Vitex Agnus-castus.*)

1. E. commutàta Bernh. (changeable), SILVERBERRY, BOIS D'ARGENT, CHALEF (Que.). — A stoloniferous unarmed shrub, 0.3–4 m. high, the younger branches covered with ferruginous scales; *leaves* elliptic to lanceolate, undulate, *silvery-scurfy on both sides* and more or less ferruginous; *flowers* numerous, *deflexed,* silvery without, pale yellow within, fragrant; *fruit* round-ovoid, *silvery, dry and mealy,* edible, 8–10 mm. long. (*E. argentea* Pursh, not Moench) — Dry calcareous slopes, Gaspé Pen., Que., to Alaska, s. to sw. Que., Minn., S.D. and Utah. June, July.

2. E. UMBELLÀTA Thunb. (with umbels). — *Leaves soon green above; flowers ascending* and slender-tubed, the *tube much longer than the lobes; drupes juicy, reddish or pink,* 6–8 mm. long, *on pedicels about* 1 *cm. long.* — Spreading from cult. to thickets and roadside-banks, Me. to N.J. and Pa. (Introd. from e. Asia)

E. MULTIFLÒRA Thunb. (many-flowered), similar to no. 2, but with *calyx-tube and limb about equal,* the *fruits* 1–1.5 *cm. long* and *on pedicels* 1.5–2.5 *cm. long,* has begun to spread from cult. (Introd. from e. Asia)

E. ANGUSTIFÒLIA L. (narrow-leaved), OLEASTER OR RUSSIAN OLIVE, similar to no. 1 but with silvery branches, leaves green above, perfect flowers with style covered by disk, and sweet drupes, is spreading from cult. (Introd. from Eurasia)

2. SHEPHÉRDIA Nutt.

Flowers dioecious; the staminate with a 4-parted calyx (valvate in the bud) and 8 stamens, these alternating with as many processes of the thick disk; the pistillate with an urceolate 4-cleft calyx, inclosing the ovary (the orifice closed by the teeth of the disk) and becoming berry-like in fruit. Style slender; stigma 1-sided. — Leaves opposite, entire, deciduous; the small flowers nearly sessile in their axils on the branches, clustered, or the pistillate ones solitary. N. Am. shrubs. (Named for *John Shepherd*, 1764?–1836, for many years curator of the Liverpool Botanic Garden.) LEPARGYREA Raf.

1. S. canadénsis (L.) Nutt. (Canadian), SOAPBERRY. — Shrub 0.3–2 m. high; *leaves elliptical or ovate, nearly naked and green above,* silvery-downy and scurfy with rusty scales beneath; fruit yellowish-red, or pale yellow in forma **xanthocárpa** Rehd. (yellow-fruited), nauseous. (*Lepargyrea* Greene) — Calcareous rocks and banks and on sandy shores, Nfld. to Alaska, s. to N.S., centr. Me., Vt., n. and w. N.Y., n. O., n. Ind., n. Ill., n. Minn., S.D. and N.M. Apr.–June.

2. S. argéntea Nutt. (silvery), BUFFALO-BERRY. — Somewhat thorny, 1–6 m. high; *leaves cuneate-oblong, silvery on both sides;* fruit ovoid, scarlet, acid and edible. (*Lepargyrea* Greene) — Banks of streams, Man. to Alta., s. to Ia., Kans. and N.M. Apr., May.

FAM. 121. LYTHRÀCEAE (LOOSESTRIFE FAMILY)

Herbs or shrubs with mostly opposite entire leaves, no stipules, the calyx inclosing but free from the 1–4*-locular many-seeded ovary and membranous capsule, and bearing the* 4–7 *deciduous petals and* 4–14 *stamens on its throat, the latter inserted below the petals. Style* 1; *stigma capitate or rarely 2-lobed.* Flowers axillary or whorled, rarely irregular, perfect, sometimes dimorphous or even trimorphous, those on different plants with filaments and style reciprocally longer and shorter. Petals sometimes wanting. Capsule often 1-locular by the early breaking away of the thin partitions; placentae in the axis. Seeds anatropous, without albumen. — Branches usually 4-sided.

*a.*Flowers regular or nearly so. . . *b.*

*b.*Calyx short, campanulate to globular. . . *c.*

*c.*Herbs with opposite leaves; flowers uniform; petals 0 or 4, very small and promptly deciduous; capsule indehiscent, irregularly bursting or septicidal. . . *d.*

*d.*Calyx without appendages; petals 0; capsule globular, indehiscent; small aquatic with flowers solitary in axils. . . . 1. *Peplis.*

*d.*Calyx usually with an appendage in each sinus; petals 4 or 0; capsule dehiscent; terrestrial or paludal plants.

Flowers 1 (rarely –3) in an axil; capsule with 3 or 4 valves, septicidal. . . . . 2. *Rotala.*

Flowers (1–)3–several in an axil; capsule globular, 2–4-locular, irregularly bursting. . . . . 3. *Ammannia.*

*c.*Shrub with arching woody or corky stems and opposite or whorled leaves;

flowers showy, trimorphous, with stamens of different lengths; petals and calyx-teeth 5–7; capsule loculicidal. . . . . 4. *Decodon.*
*b.*Calyx tubular, cylindric; petals usually 6; stamens mostly 6 or 12. . . . 5. *Lythrum.*
*a.*Flowers irregular and unsymmetrical, the calyx spurred or enlarged at base on one side; petals 6, unequal; stamens 6–14 (mostly 11) in 2 sets. . . . . 6. *Cuphea.*

1. PÉPLIS L. Water-Purslane

Petals none in ours. Calyx without appendages in ours. Stamens 4. Capsule globular, indehiscent, 2-locular. — N.Am. and Eurasian aquatics (sometimes terrestrial), rooting in the mud, with opposite leaves and very small greenish flowers solitary in their axils. (Name used by Pliny for Purslane or *"porcilaca"*.) Didiplis Raf.

1. P. diándra Nutt. (with 2 stamens; a misnomer). — Leaves when submersed elongated, thin, closely sessile by a broad base, when emersed shorter and contracted at base; calyx with broad triangular lobes; style very short; capsules very small. (*Didiplis* Wood) — Shallow waters and their margins, Fla. to Tex., n. to Va., O., Ind., and Wisc. June–Aug.

2. ROTÀLA L.

Petals 4 (in ours). Capsule-valves (under a strong lens) transversely and closely striate. Genus of warm reg. (Name an incorrect diminutive of *rota*, a wheel; from the whorled leaves of the original species.)

1. R. ramòsior (L.) Koehne (very branching), Tooth-cup. — Plant low, simple to diffusely branched or depressed, *rarely 2 dm. high; larger leaves* 1.5–4(–5) *mm. broad, definitely petioled; fruit* 2–4 mm. long, 2–2.3 *mm. broad;* bractlets subulate, 0.5–1.4 mm. long. — Sandy shores and damp depressions, Fla. to Tex., n. to s. N.H., Mass., N.Y. and Mo.; sands of s. Mich. and n. Ind. to Minn.; Wash. and Oreg. July–Oct.

Var. **intèrior** Fern. & Grisc. (of the interior). — More robust, *up to* 4.5 *dm. high; larger leaves* 5–10 *mm. wide, subsessile or short-petioled; fruit* 3.5–5 mm. long, (3.2–)3.8–4.5 *mm. broad;* bractlets linear-lanceolate, 1.6–2.5(–4) mm. long. — Fla. to La., n. to s. N.Y., W.Va., O., Ill., Ia. and Kans.

3. AMMÁNNIA L.

Flowers small, in 3–many-flowered axillary cymes. Calyx globular or campanulate, 4-angled, 4-toothed, usually with a little horn-shaped appendage at each sinus. Petals 4 (purplish), small and deciduous, sometimes wanting. — Low and inconspicuous smooth herbs of warm reg., with opposite narrow leaves. (Named for *Paul Ammann*, 1634–1691, German botanist.)

Leaves all or all but the very lowest auriculate-cordate at base; calyx-teeth prominent; style filiform, elongate; seeds reddish-brown.
Inflorescences or flowers mostly stalked; fruiting calyx 2–3 mm. in diameter. 1. *A. auriculata.*
Inflorescences sessile or nearly so; fruiting calyx 3–5 mm. in diameter. . . . 2. *A. coccinea.*
Leaves all tapering to base or only the middle and upper auriculate-clasping; inflorescences sessile or nearly so; fruiting calyx 5–6 mm. in diameter, its teeth very short; style thick and short; seeds whitish-brown. . . . . 3. *A. teres.*

1. A. auriculàta Willd. (with ear-lobes). — Erect or with ascending branches, 1–9 dm. high; *leaves* linear or linear-lanceolate, *long-attenuate, all but the very lowest auriculate-cordate; cymes loosely* 3–15*-flowered, peduncled or with the flowers pedicelled; calyx* 1.5–2 *mm. long, with prominent teeth,* 8-nerved, in fruit becoming subglobose; petals violet to white, soon dropping; stamens 4–8; *style filiform, equaling or longer than ovary; fruiting calyx* 2–3 *mm. in diameter; seeds reddish-brown.* — Pond-borders, swamps and ditches, Miss. to N.M. and Mex., n. to Ind. and S.D. July–Oct. (Trop. reg.)

2. A. coccínea Rottb. (scarlet). — Stouter than no. 1, ascending or depressed; leaves all auricled, less attenuate; *cymes closely* 3–5*-flowered, nearly or quite sessile; calyx* 2.5–5 *mm. long, in fruit* 3–5 *mm. in diameter.* — Wet shores and muddy (often alkaline) places, Fla. to Tex. and Mex., n. to O., Ill., Minn., Neb., Mont. and Wash. Aug.–Oct. (Tropics)

3. A. tères Raf. (terete or quill-like). — Erect, simple or with few erect branches, 1–9 dm. high, fleshy; *leaves oblong or oblanceolate, obtuse or subacute, the lower tapering at base, the middle and upper auricled, the longer ones* 2.5–8 *cm. long; cymes* closely 3–7-flowered, *sessile or essentially so; calyx-teeth very short and broad;* petals pink, minute, fugacious; *style thick and much shorter than ovary; fruiting calyx* 5–6 *mm. in diameter; seeds whitish-brown.* (*A. Koehnei* Britt.) — Swamps and tidal marshes, Fla. to Miss., locally n. to n. N.J. July–Sept.

Var. exauriculàta Fern. (without ear-lobes). — Decumbent at base; leaves broadly oblanceolate to spatulate, 2–3 cm. long, all narrowed to base; petals wanting or unknown. — Tidal marshes of North Landing and James Rivers, se. Va.

4. DÉCODON J. F. Gmel. SWAMP-LOOSESTRIFE

Calyx with 5–7 erect teeth and as many longer and spreading horn-like processes at the sinuses. Stamens exserted, of two lengths. Capsule globose, 3–5-locular, loculicidal. — Perennial N.Am. herb or slightly shrubby plant, with opposite or whorled leaves, and axillary clusters of trimorphous flowers. (Name from the Greek *deca, ten,* and *odous, tooth,* from the summit of the calyx.)

1. D. verticillàtus (L.) Ell. (*whorled*), WATER-WILLOW, WATER-OLEANDER. — Smooth or downy; stems recurved, 6–25 dm. long, 4–6-sided; leaves lanceolate, nearly sessile, opposite or whorled, the upper with clustered short-pedicelled flowers in their axils; petals 5, cuneate-lanceolate, magenta, 1.2 cm. long; stamens 10, half of them shorter. — Swamps and shallow pools, Fla. to La., n. to centr. Me., s. N.H., Mass., N.Y., s. Ont. and s. Ill. July, Aug. — Bark of submersed part of stem spongy-thickened; arching branches rooting at tips and thence starting new arching stems, the plant thus rapidly spreading.

Var. laevigàtus T. & G. (smooth). — Stems, lower leaf-surfaces and pedicels glabrous. — Va. to Tenn., n. to N.S., Me., centr. N.H., n. Vt., sw. Que., n. N.Y., O., Wisc. and Minn.

LAGERSTROÈMIA L. (for *Magnus Lagerstroem,* ?–1759) INDICA L. (East-Indian), CRAPE-MYRTLE, a small tree or shrub with oval to obovate leaves, showy paniculate flowers with large crisped roseate to white petals, much planted in the South, spreads about old house-sites n. to Md. and Ind. (Introd. from Asia)

5. LỲTHRUM L. LOOSESTRIFE

Calyx cylindrical, striate, 5–7-toothed, with as many little processes in the sinuses. Petals 5–7. Stamens as many as the petals or twice the number, inserted low down on the calyx-tube. Capsule subcylindrical, 2-locular. — Slender herbs of warm or temp. reg., with pink or magenta (rarely white) flowers in summer. (From *lytron,* a name used by Dioscorides for our no. 5.)

Flowers small, solitary and sessile or only short-stalked in the axils of at least the upper leaves.
  Leaves linear to lanceolate, tapering or narrowed to base.
    Annual, flowering from near the base upward; base of ovary without a thickened ring. 1. *L. Hyssopifolia.*
    Perennial by stoloniform basal offshoots, the lower axils without flowers.
      Principal cauline leaves opposite, linear or linear-oblanceolate, 1.4–4 mm. broad; calyx-teeth and narrowly triangular appendages subequal. . . 2. *L. lineare.*
      Principal or primary cauline leaves mostly alternate, lanceolate to linear-elliptic, 5–12 mm. broad; appendages of calyx subulate, much longer than the teeth. . . . . 3. *L. lanceolatum.*
  Leaves oblong-ovate to lanceolate, cordate or rounded at base; perennial. . . 4. *L. alatum.*
Flowers showy, trimorphous, cymose in axils of whorled or opposite upper leaves, forming interrupted spikes; tall perennials.
  Leaves rounded to cordate at base; plant usually pubescent; calyx-lobes much shorter than the subulate appendages. . . . . 5. *L. Salicaria.*
  Leaves attenuate at base; plant glabrous; calyx-lobes and appendages subequal. . . . . 6. *L. virgatum.*

1. L. Hyssopifòlia L. (old generic name, Hyssop-leaved). — *Annual,* loosely branched or simple, 0.5–6 dm. high, pale, *flowering from near base to summit;* leaves oblong-linear, obtuse, 1–3 cm. long, 1–7 mm. broad; flowers with 4–6 calyx-teeth, stamens and small pale purple petals; appendages longer than calyx-lobes; stamens included. — Marshes and sterile soil, mostly near the coast, s. Me. to N.J. and Pa.; locally in O.; also Pacific coast. June–Sept. (Eu.) — Perhaps originally adv. with us.

2. L. lineàre L. (linear). — *Perennial* by creeping basal offshoots; *stem* slender, 0.3–1.3 m. high, pale green, bushy at summit, *with 2 margined angles; leaves chiefly opposite, linear or linear-oblanceolate, those of the primary stem 1.4–4 mm. broad,* those of the flowering branches smaller; petals whitish or pale-lilac; flowers with 6 included stamens and a long style, or stamens exserted and style short; *calyx-teeth and narrow triangular appendages subequal; ovary* on a thick stipe, *without a basal ring.* — Brackish to saline marshes, Fla. to Tex., n. to L.I. July–Sept.

3. L. lanceolàtum Ell. (lanceolate). — Greener than no. 2, mostly coarser; *leaves of primary*

***stem mostly alternate, lanceolate to linear-elliptic,* 5–12 *mm. broad,* those of the flowering branches smaller and often opposite; *calyx-teeth much shorter than the slenderly subulate appendages;* petals purple; *a thick ring at base of ovary.* — Swamps, damp sandy thickets, etc., Fla. to Tex., n., locally, to se. Va. and Ark. June–Sept.**

4. L. alàtum Pursh (winged). — Tall and wand-like *perennial; branches with margined angles; leaves oblong-ovate to linear-lanceolate, acute, with a cordate or rounded base,* the upper mostly *alternate; calyx* 4–7 *mm. long;* petals rather large, deep purple; *stamens of the short-styled flowers exserted; fleshy hypogynous ring prominent.* — Swamps, meadows, prairies and ditches, Ont. and n. N.Y. to B.C., s. to Ga., La. and Tex.; adv. in N.E., N.J., etc. June–Sept.

5. L. SALICÀRIA L. (old generic name, *like a willow*), SPIKED L., SALICAIRE or BOUQUET VIOLET (Que.). — *More or less downy* and tall; *leaves* lanceolate, *cordate at base,* sometimes whorled in threes; flowers magenta, trimorphous in the relative lengths of the stamens and style; *calyx and bracts* greenish, somewhat *pubescent,* the calyx-lobes much shorter than the subulate appendages. — Wet meadows, river-floodplains, etc., locally abundant, often too aggressive in choking out native vegetation, Nfld. and Que. to Minn., s. to N.S., N.E., Va., W.Va., O., Ind. and Mo. June–Sept. (Natzd. from Eu.)

Var. TOMENTÒSUM (Mill.) DC. (with implexed hairs). — *Calyx and bracts white-tomentose.* — Similar habitats, Que. and Ont., s. into N.E. and N.Y. (Natzd. from Eu.)

Var. GRACÍLIOR Turcz. (more slender). — *Glabrous* or essentially so; leaves rounded to cordate at base; *spike slender and loosely* or often remotely *flowered.* — Local, N.E. (Introd. from Asia)

6. L. VIRGÀTUM L. (wand-like). — Similar, *glabrous throughout; leaves narrowed to the* sessile or short-petioled *base;* the calyx-lobes shorter than or equaling the appendages. — Local, N.H. and Mass., spread from cult. (Introd. from Eu.)

6. CÙPHEA P. Br. CUPHEA

Calyx tubular, 12-ribbed, gibbous or spurred at the base on the upper side, 6-toothed at the apex and usually with as many little processes in the sinuses. Ovary with a curved gland at the base next the spur of the calyx, 1–2-locular; style slender; stigma 2-lobed. Capsule ovoid or ellipsoid, few-seeded, early ruptured through one side. Flowers solitary or racemose, stalked. Large chiefly trop. Am. genus. (Name from the Greek *cyphos, gibbous,* from the shape of the calyx.)

1. C. petiolàta (L.) Koehne (with petioles), CLAMMY C., BLUE WAXWEED. — Annual, very viscid-hairy, branching; leaves ovate-lanceolate; petals ovate, short-clawed, purple; seeds flat. (*Parsonsia* Rusby) — Dry open soil, Ga. to La., n. to s. N.E., s. N.Y., O., Ind., Ill. and Ia.; adv. elsewhere. July–Oct.

FAM. 122. NYSSÀCEAE (SOUR GUM FAMILY)

Trees or shrubs with simple broad alternate and exstipulate leaves and dioecious greenish flowers borne at the summit of axillary peduncles.* STAM. FL. *numerous, the calyx small and 5-parted, the petals as in pistillate flower or none; stamens* 5–12, *often* 10, *inserted on the outside of a convex disk; no pistil.* PIST. FL. *solitary or* 2–8, *sessile in a bracted cluster, much larger than the staminate flowers; petals very small and fleshy, deciduous, or often wanting; stamens* 5–10, *with perfect or imperfect anthers; style elongated; drupe ovoid or ellipsoid.

1. NÝSSA L. TUPELO. PEPPERIDGE. SOUR GUM

Characters of the family; ovary 1-locular; style simple. — Few species in e. N. Am. and se. Asia. (The name of a Nymph; the genus "so called because it [the original species] grows in the water.")

Leaves of fertile branches 1–3 dm. long; fertile flowers solitary; fruits 2–3 cm. long. . . . . 1. *N. aquatica.*
Leaves of fertile branches 2.5–15 cm. long; fertile flowers 2 or more on each peduncle; fruits less than 1.5 cm. long. . . . . 2. *N. sylvatica*

1. N. aquática L. (aquatic), COTTON-GUM. — A large tree; *leaves* oblong or ovate, sometimes slightly cordate at base, long-petioled, entire or angulate-toothed, pale and downy-pubescent beneath, at least when young, 1–3 *dm. long; pistillate flower solitary* on a slender peduncle; *fruit ellipsoid, blue,* 2.5 *cm. or more in length,* resembling olives. — Inundated swamps, Fla. to e. Tex., n. to Va., s. Ind., s. Ill. and Mo. Apr., May. — Wood soft; that of the roots very light and spongy.

2. N. sylvática Marsh. (of the woods), BLACK GUM. — Small or middle-sized tree, with stiffly horizontal branches; *leaves adjacent to flowers and fruits obovate to elliptic*, frequently with short abrupt tip, *firm to subcoriaceous, lustrous above*, smooth and glabrate or glabrous beneath, 3–10 *cm. long*, 2–6 *cm. broad;* fruiting peduncles (1.5–)3–6.5 cm. long; fruits 1–1.2 cm. long, acid; *staminate pedicels* 0.5–4 *mm. long.* — Low acid woods, swamps and shores, centr. Me. to Mo., s. to s. N.E., L.I., Fla., Tex. and Mex. Apr.–June.

Var. **biflòra** (Walt.) Sarg. (two-flowered). — Base of trunk, when submersed, swollen; *leaves* coriaceous, *those adjacent to flowers and fruits oblanceolate, narrowly obovate or narrowly oblong-lanceolate, round-tipped, obtuse or subacute*, lustrous above, 2.5–9 cm. long, 1.5–3.5 *cm. broad;* fruiting peduncles 1–3.5 cm. long; fruits 1.2–1.5 cm. long, bitter; staminate pedicels 3–5 mm. long. (*N. biflora* Walt.) — Inundated swamps and damp sands, Fla. to e. Tex., n. to Del., Md. and interior S.C., Ga., Ala. and Miss.

Var. **dilatàta** Fern. (expanded). — *Mature leaves* of fertile shoots *broadly oval to round-oblong*, 5–8 *cm. broad*, silky-pilose at base beneath; fruiting peduncles 1.5–5 cm. long; staminate pedicels 3–8 mm. long. — Fla. to Tex., n. to e. Va., Tenn., Ark. and Okla.

Var. **caroliniàna** (Poir.) Fern. (of Carolina). — *Leaves submembranaceous to thin-coriaceous; those of fertile shoots rhombic-oval to rhombic-obovate, tapering subequally to acuminate base and apex*, papillose beneath, 8–15 cm. long, 3.5–8 cm. broad; fruiting peduncles 1.5–5 cm. long; staminate pedicels 2–4.5 mm. long. — Chiefly in uplands of the interior, se. Pa. to s. Ont. and Mo., s. to se. Va., N.C., Tenn., n. Miss. and Tex.

FAM. 123. MELASTOMATÀCEAE (MELASTOMA FAMILY)

Plants with opposite 3–7-ribbed leaves and definite stamens, the anthers opening by pores at the apex; otherwise much as in the Onagraceae. — All tropical, except the genus

1. RHÉXIA L. DEERGRASS. MEADOW-BEAUTY

Hypanthium urceolate, adherent to the ovary below and continued above it, persistent, 4-cleft at the apex. Petals 4, convolute in the bud, oblique, inserted with the 8 stamens on the summit of the calyx-tube. Anthers long, 1-locular, inverted in the bud. Style 1; stigma 1. Capsule 4-locular, with 4 many-seeded placentae projecting from the central axis. Seeds (in ours) coiled like a snail-shell, without albumen. — Low perennial often bristly e. N.Am. herbs with showy cymose flowers in summer; the petals falling early in the day. (A name used by Pliny for some unknown plant.)*

a. Anthers oblong, straight, spurless, 1.5–2 mm. long; flowers sessile, among terminal leaves; stem glabrous; leaves oval, at most 2.5 cm. long, bristly-ciliate. . . . . 1. *R. ciliosa.*

a. Anthers elongate, linear-subulate or -cylindric, tapering, curved, with a minute spur on the back at attachment of filament, 5–10 mm. long; flowers peduncled; stem more or less bristly (except in no. 2 and sometimes no. 3); leaves longer. . . *b.*

b. Leaves entire or only remotely serrate; calyx-lobes longer than neck of hypanthium; bristles of hypanthium not gland-tipped; stem glabrous. . . 2. *R. aristosa.*

b. Leaves regularly serrulate; calyx-lobes shorter than to about equaling neck of hypanthium; bristles of hypanthium (when present) gland-tipped; stems pubescent or glabrous. . . *c.*

c. Tuberous-rooted; stolons when present soft; seeds 0.65–0.8 mm. long. 3. *R. virginica.*

c. Non-tuberous, the bases developing horizontally spreading or creeping subligneous stoloniform stems; seeds 0.5–0.6 mm. long.

Neck of fruiting hypanthium longer than body; stems subterete, not obviously 4-angled. . . . . 4. *R. mariana.*

Neck of fruiting hypanthium about equaling to shorter than body; stems 4-angled, especially above.

Mature hypanthium (excluding calyx-lobes) 7.5–10 mm. long, its body 4–5 mm. in diameter; seeds with low rounded pebbling, the surface appearing relatively uniform. . . . . 5. *R. interior.*

Mature hypanthium 9–14 mm. long, its body 5.5–8 mm. in diameter; seeds with prominent thin ridges and slender papillae. . . . . 6. *R. ventricosa.*

1. R. ciliòsa Michx. (ciliate). — *Stem* from a tough base, *square*, slender, simple or slightly forking above, *glabrous*, 1.5–7.5 dm. high, with many pairs of *oval, bristly-ciliate leaves* 1–2.5

---

* Treatment derived in part from that of FERNALD & GRISCOM in Rhodora, xxxvii. 169–173 (1935).

cm. long; flowers sessile among terminal leaves; hypanthium glabrous; calyx-lobes bristly, non-glandular; *petals ascending,* roseate, 1–2 cm. long. — Damp pine-barrens, peats and sands, Fla. to La., n. to se. Va. July–Sept.

2. R. aristòsa Britt. (awned; referring to the petals). — *Stem from a fusiform tuber, glabrous,* slightly wing-angled, 3–6 dm. high, simple or with ascending branches; *leaves sessile,* lanceolate to linear-oblong, *entire or only remotely toothed, firm,* glabrous or nearly so; *hypanthium with the non-glandular bristles mostly below the throat; calyx-lobes longer than neck; petals* roseate, *aristate at summit;* seeds 0.4–0.5 mm. long, with low but coarsely pebbled ridges. — Wet pine-barrens, local, N.J. and Del.; S.C. and Ga. July–early Sept.

3. R. virgínica L. (Virginian). — *Stem from a tuber,* simple or branched, 0.1–1 m. high, 1–4.5 mm. in diameter above the often spongy base, *with 4 wing-angles* 0.1–rarely 1 *mm. wide,* sparsely hispid or glabrous, usually bristly at the nodes; *leaves oval to lance-oblong, rounded at base, sessile, sharply serrulate,* commonly bristly, especially above, the larger ones 0.5–3 cm. broad; *those at base of inflorescence* 0.5–2 *cm. wide, their longer teeth* 0.5–1.2 *mm. long; hypanthium* commonly glandular-hispid, *its neck in maturity much shorter than the body;* petals crimson; *seeds* 0.65–0.8 *mm. long, strongly papillose.* — Peats and wet sands and gravels, N.S. to s. Ont. s. to Ga., Ala., Tenn. and Mo.; ascending to 700 m. alt. July–Sept.

Var. septemnérvia (Walt.) Pursh (seven-nerved). — Mostly much *coarser; stem* commonly much branched, 0.6–1.6 m. high, 5–8 *mm. thick, its conspicuous thin wings* 1–2 *mm. wide;* larger *leaves* 2–4 cm. broad; *those at base of inflorescence* 1.5–3 *cm. broad, their longer teeth* 1–1.5 *mm. long.* — Low woods, swamps and borders of rills, Fla. to La., n. on Coastal Plain to se. Va. Late July, Aug.

4. R. mariàna L. (of Maryland). — Non-tuberous, the base forming *slender horizontally spreading or creeping subligneous stoloniform stems,* sending up solitary or scattered *subcylindric hirsute flowering stems* 2–6 dm. high; *leaves* lanceolate to elliptic, *narrowed to petioles,* subglabrous to hirsute; *hypanthium* glandular-hirsute to glabrous, *its neck when mature longer than the body; petals pale rose to whitish,* 1.2–2 *cm. long; seeds about* 0.5 *mm. long, rather sharply muriculate.* — Damp sands and peats of Coastal Plain, Fla. to e. Pa., L.I. and se. Mass.; inland from Ga. to Va. and Ky. Late June–early Sept.

Var. leiospérma Fern. & Grisc. (smooth-seeded). — Similar; *seeds with obsolete or depressed papillae.* — La. and Tex., n. to se. Mo., s. Ill. and s. Ind.

Var. purpùrea Michx. (purple). — Stouter; *leaves lanceolate, more villous-hirsute; petals crimson,* 1.5–2.5 *cm. long; seeds with conspicuous pebbling.* (*R. Nashii* Small) — Wet pine-barrens, savannas and bogs, La. to Fla., n. to se. Va. July–Sept.

5. R. intèrior Pennell (inland). — Base as in no. 3; *stem 4-angled,* especially above, 2–6 dm. high, hirsute; *leaves* lanceolate to narrowly elliptic, *sessile,* resembling those of no. 2; *mature hypanthium* (excluding calyx-lobes) 7.5–10 *mm. long, its body* 4–5 *mm. in diameter* and about as long as neck; *seeds* 0.5–0.6 mm. long, *with low rounded pebbling.* — Pond-shores, wet ground and prairies, sw. Mo. and se. Kans. June, July (rarely –Sept.)

6. R. ventricòsa Fern. & Grisc. (bellied-out). — Similar to no. 4, coarser, 2.5–8 dm. high; leaves elliptic-lanceolate to oblong-ovate, sessile or subsessile, hispid; *mature hypanthium* 9–14 *mm. long, the base* 5.5–8 *mm. in diameter, the tube more constricted below the flaring summit; seeds with prominent thin ridges and slender evenly spaced papillae* (giving a falsely alveolate appearance under magnification). — Argillaceous soils (dry or moist), se. Va. and e. N.C. Late June–Sept.

FAM. 124. HYDROCARYÀCEAE (Water-Chestnut Family)

Aquatic annuals with a whorl or pair of rhombic floating leaves with inflated petioles; the submersed leaves remote, with capillary segments; flowers borne among the floating leaves; calyx-tube short, inclosing base of ovary; fruit indehiscent, large, with 2 or more strong spines, 1-locular, 1-seeded. Calyx-limb 4-parted, the segments persistent and becoming spinescent. Cotyledons unequal, one large and full, the other scale-like, the smaller one raised above ground in germination.

1. TRÀPA L. Water-nut. Water-Chestnut

Characters of the family. Small genus nat. of Eurasia. (Name abridged from *calcitrapa, a caltrop,* in allusion to the spreading points of the fruit.)

1. T. nàtans L. (floating), **Water-Caltrop.** — Fruit 4-horned; seed edible. — Quiet streams and ponds, locally too abund. and aggressive, Mass. and N.Y. to Md. July–Sept. (Introd. and natzd. from Eurasia)

FAM. 125. ONAGRACEAE (Evening-Primrose Family)

Herbs with 4-merous (sometimes 2–3- or 5–6-merous) perfect and symmetrical flowers; the tube of the calyx adhering to the 2–4-locular ovary, its lobes valvate in the bud or obsolete; the petals convolute in the bud, sometimes wanting; and the stamens as many or twice as many as the petals or calyx-lobes, inserted on the summit of the calyx-tube. Style single, slender; stigma 2–4-lobed or capitate. Pollen-grains often connected by cobwebby threads. Seeds anatropous, small, without albumen. — Mostly herbs, with opposite or alternate leaves. Stipules none or glandular.

*a.*Parts of flowers in fours or fives or more numerous. . . *b.*
*b.*Fruit a many-seeded dehiscent capsule, usually loculicidal. . . *c.*
*c.*Calyx-limb divided to summit of ovary, persistent.
Petals 4–6; stamens twice as many; capsule elongated. . . . . . . 1. *Jussiaea.*
Petals 4 or 0; stamens 4; capsule relatively short. . . . . . . 2. *Ludwigia.*
*c.*Calyx-tube or deeply cleft limb deciduous from summit of capsule; petals 4; stamens 8.
Capsule slender, promptly splitting, when ripe, into recurving membranous valves; seeds usually with a silky coma at summit; petals (in ours) pink, purple or white; lower leaves often opposite. 3. *Epilobium.*
Capsule stout or broad to slender, firm, the hard valves when opening, rarely recurved; seeds naked; petals mostly yellow; leaves alternate. 4. *Oenothera.*
*b.*Fruit indehiscent, dry, 1–4-seeded.
Calyx-tube (hypanthium) obconical; filaments appendaged at base; fruit 3–4-ribbed or -angled. . . . . 5. *Gaura.*
Calyx-tube (hypanthium) filiform; filaments unappendaged at base; fruit 8-ribbed. . . . . 6. *Stenosiphon.*
*a.*Parts of flowers in twos; fruit indehiscent, bristly; leaves opposite. . . 7. *Circaea.*

1. JUSSIAÈA L. Primrose-willow. Water-Primrose

Calyx-tube elongated, not at all prolonged beyond the ovary; the lobes 4–6, herbaceous and persistent. — Herbs, chiefly trop., with mostly entire and alternate leaves and axillary yellow flowers in summer; submersed roots often distended into "swimming bladders". (Dedicated to *Bernard de Jussieu,* 1699–1776, the founder of the "Natural System of Botany".)

Plant annual, without prolonged base; decurrent wings running down the stem from leaf-bases. . . . . 1. *J. decurrens.*
Plant perennial, with prolonged creeping base; no wings on stem.
Stem, leaves, hypanthium and calyx-segments glabrous or nearly so; calyx-segments 7–12 mm. long; petals 0.7–1.3 cm. broad; fruit 2.5–4 cm. long. . 2. *J. repens,* var. *glabrescens.*
Stem (above) of fertile shoots, leaves (beneath), hypanthium and calyx-segments villous-hirsute; calyx-segments 0.6–2 cm. long; petals 0.9–2.5 cm. broad.
Petioles of principal leaves 1.5–4 cm. long; peduncles becoming 2.5–6 cm. long; calyx-segments 1.5–2 cm. long; petals 2.2–3 cm. long, 2–2.5 cm. broad. 3. *J. Michauxiana.*
Petioles of principal leaves 0.5–2.5 cm. long; peduncles becoming 1–2 cm. long; calyx-segments 6–13 mm. long; petals 1.2–2 cm. long, 9–15 mm. broad. 4. *J. uruguayensis.*

1. J. decúrrens (Walt.) DC. (decurrent). — *Annual with clustered roots; stem* up to 2 m. high, subsimple or freely ascending-branched, the internodes *with decurrent wings running down from leaf-bases,* the filiform branches with greatly reduced leaves; *larger leaves* lanceolate, *sessile or subsessile,* 5–18 cm. long, membranaceous; flowers from many axils, on *peduncles only* 1–10 *mm. long; hypanthium and fruit obconic, winged or angled, becoming* 1–2 *cm. long;* calyx-segments and *petals* 6–10 *mm. long; seeds in several rows in each locule.* — Wet places, ditches, etc., Fla. to Tex., n. to Va., W.Va., s. Ind., s. Ill., Mo. and Okla. May–Oct. (Trop. Am.)

2. J. rèpens L. (repent), var. **glabréscens** Ktze. (smoothish). — *Perennial with creeping or floating and rooting stems,* the simple to forking flowering branches *glabrous or sparingly pilose; principal leaves* glabrous or glabrescent, oblong-lanceolate to spatulate, mostly 3–9 cm. long and *on slender petioles up to 4 cm. long; mature peduncles* 3–8 *cm. long, smooth or smoothish; fruit tough, slenderly cylindric,* 2.5–4 *cm. long; calyx-segments* glabrous or sparingly pubescent, *about* 1 *cm. long; seeds in* 1 *row in each locule.* (*J. diffusa* of ed. 7, not Forsk.) — In water or muddy places, Ala. to Tex. and Mex., n. to s. Ind. (introd.), s. Ill., Mo. and Kans.; ditches, etc., presumably adv., lower Del. R., N.J. and Pa., and Wicomico Co., Md. June–Oct. — Representative of a pan-trop. type.

3. J. Michauxiàna Fern. (for its discoverer, André Michaux, 1746–1802). — Coarser than no. 2, the long trailing bases producing abundant aquatic or subaquatic roots; the *flowering branches erect,* simple or slightly forking, *up to* 9 *dm. long, the upper internodes densely villous-hirsute with long spreading hairs; principal leaves* oblong-lanceolate to -oblanceolate, *villous-hirsute on the veins beneath, the blade* 8–11 cm. long and 2–3 *cm. broad,* the petiole 1.5–4 cm. long; *peduncle, hypanthium, fruit and backs of calyx-segments villous;* mature peduncle 2.5–6 cm. long; *fruit* 1.4–2 *cm. long; calyx-segments* 1.5–2 *cm. long; petals* 2.2–3 *cm. long,* 2–2.5 *cm. broad.* (*J. grandiflora* Michx., not R. & P.) — Shallow water and mud, Fla. to La., n. to N.C.; natzd. or spread from cult. n. to e. Pa. July–Oct.

4. J. uruguayénsis Camb. (of Uruguay). — Smaller in most parts than no. 3, similarly pubescent; leaves of flowering branches oblanceolate to lance-elliptic, 0.8–2.5 cm. broad, their *petioles only* 0.5–2.5 *cm. long; peduncles lengthening to* 1–2 *cm.; calyx-segments* 6–13 *mm. long; petals* 1.2–2 *cm. long and* 9–15 *mm. broad.* — Ponds and quiet waters, locally n. to se. N.Y. Aug.–Oct. (Introd. from. S.Am.)

J. leptocárpa Nutt. (slender-fruited), a *bristly* plant inclined to branch above, *with lanceolate or oblanceolate* acute *leaves* (*upper ones* 3–5 *mm. wide*), pedicels up to 1.5 cm. long, *calyx-segments and petals mostly* 6, *subequal; fruit linear-cylindric* and 1.8–4.5 cm. long and 2.5–3 *mm. thick,* the calyx-segments 5–8 mm. long, is reported in our range while this is in press. — Fla. to Tex. and Mex., n. to Ga. and se. Mo. (Trop. and warm-temp. Am.)

2. LUDWÍGIA L. False Loosestrife

Calyx-tube not at all prolonged beyond the ovary; the lobes 4, usually persistent. Capsule cubical or subglobose to turbinate or cylindrical, many-seeded. — Perennial herbs, chiefly of trop. and warm reg., with axillary to capitate flowers through summer and autumn. (Named for *Christian Gottlieb Ludwig,* 1709–1773, Professor of Botany at Leipsic.)* Ludvigia L. (excluded spelling)

*a.*Leaves alternate; flowering stems erect. . . *b.*

*b.*Flowers pedicelled; petals yellow, conspicuous, 0.8–1.6 cm. long, nearly equaling to exceeding the foliaceous calyx-lobes; capsules nearly cubical but rounded at base, 5–8 mm. in diameter, opening by terminal pores; autumnal basal shoots wanting; roots fascicled, often fusiform. . . *c.*

*c.*Leaves petioled, lanceolate, acute or acutish at each end; stem and leaves glabrous or minutely pubescent; petals about equaling calyx-lobes; capsule longer than pedicel. . . . . 1. *L. alternifolia.*

*c.*Leaves sessile, mostly linear, oblong or oblong-ovate, obtuse; pedicel as long as or longer than capsule.

Plant glabrous or minutely puberulent; principal leaves linear to narrowly oblong-lanceolate, 3–7 mm. broad; calyx-lobes reflexed after anthesis. . . . . 2. *L. virgata.*

Plant villous-hirsute; leaves oblong to oblong-ovate, mostly 5–12 mm. broad; calyx-lobes erect to merely spreading. . . . . 3. *L. hirtella.*

*b.*Flowers sessile or subsessile; petals inconspicuous, greenish or purplish or wanting (showy, yellow and up to 8 mm. long in no. 8); capsules cylindric or turbinate to subglobose, smaller, the valves separating from the terminal disk; autumnal prostrate basal offshoots with short and broad leaves frequent; roots scattered, fibrous. . . *d.*

*d.*Fruit cylindric, 1.5–2.3 mm. in diameter; calyx-lobes 1–2 mm. long. . . 4. *L. glandulosa.*

*d.*Fruit obconic or obpyramidal to subspherical, mostly thicker; calyx-lobes 1.5–4 mm. long. . . *e.*

*e.*Hypanthium 1–2 mm. long; leaves obovate, 0.5–2.5 cm. long. . . . 5. *L. microcarpa.*

*e.*Hypanthium 2.4–10 mm. long; leaves linear, lanceolate or oblong, longer. . . *f.*

*f.*Branches wing-angled; bractlets of hypanthium reaching nearly to or beyond sinuses between calyx-lobes; fruit turbinate-campanulate, glabrous.

---

* Treatment largely derived from that of Fernald & Griscom in Rhodora, xxxvii. 173–177 (1935).

Fruiting hypanthium 4–7 mm. long, longer than broad, with rounded angles. . . . . 6. *L. polycarpa.*
Fruiting hypanthium 2.4–4 mm. long, as broad as long, with wing-angles. . . . . 7. *L. alata.*
*f.*Branches scarcely if at all winged; bractlets wanting or much shorter than hypanthium, or fruit hemispherical.
Fruit slenderly campanulate to elongate-obconic, much longer than broad, glabrous or minutely papillate; petals yellow, often equaling calyx-lobes. . . . . 8. *L. linearis.*
Fruit hemispheric to subglobose, pubescent; petals wanting or minute.
Stems and leaves glabrous or minutely appressed-pubescent; hypanthium minutely pubescent, its bractlets very short. . . 9. *L. sphaerocarpa.*
Stems, leaves and hypanthium velvety-pilose; bractlets nearly reaching sinuses of calyx. . . . . . . 10. *L. pilosa.*
*a.*Leaves opposite; stems depressed or subascending, often prostrate and repent or floating. . . *g.*
*g.*Flowers distinctly pedicelled, clavate or slenderly turbinate, with calyx-lobes 2.5–6 mm long; petals yellow; capsule clavate or clavate-campanulate.
Pedicels 5–12 mm. long; flowering hypanthium 3.5–6 mm. long, fruiting 8–10 mm. long; expanded calyx 1.2–1.3 cm. broad; calyx-lobes 3.5–6 mm. long; petals broad, equaling calyx-lobes. . . . . 11. *L. brevipes.*
Pedicels 1–2.5 mm. long; flowering hypanthium 2–3 mm. long, fruiting, only slightly longer; expanded calyx 5.5–6.5 mm. broad; calyx-lobes 2.5–3 mm. long; petals elliptic-oblong, shorter than calyx-lobes. . . 12. *L. lacustris.*
*g.*Flowers sessile or subsessile, broadly turbinate, campanulate or urceolate, with calyx-lobes 1–2.5 mm. long; petals minute, purplish to green, or usually wanting; capsule campanulate.
Hypanthium 4–6 mm. long, without green longitudinal bands but often with 1 or more narrow and elongate bractlets borne well above the base. 13. *L. natans.*
Hypanthium 2–4.5 mm. long, with 4 green longitudinal bands; the bractlets, when present, basal and minute. . . . . . . 14. *L. palustris,* vars.

1. L. alternifòlia L. (alternate-leaved), SEEDBOX. — *Glabrous* or nearly so, erect, branched, 0.2–1 m. high; *roots fascicled, often fusiform; leaves petioled*, alternate, *lanceolate, pointed at both ends; flowers* distinctly *short-pedicelled; calyx-lobes foliaceous*, ovate to broadly lanceolate; *petals yellow, about equaling calyx-lobes; capsules cubical* but rounded at base, 5–8 *mm. thick, opening by terminal pores*, wing-angled, *longer than the short pedicels.* — Swamps, Fla. to e. Tex., n. to Mass., N.Y., s. Ont., s. Mich., Ill., Ia. and Kans. June–Aug. FIG. 1334.

1334. L. alternifolia.

Var. **pubéscens** Palmer & Steyerm. (hairy). — Summit of stem, leaves and calyx densely pubescent with minute spreading hairs. — La. and e. Tex., n. to s. Ind., s. Ill., s. Mo. and Kans.

Var. **linearifòlia** Britt. (linear-leaved). — Leaves linear; calyx-lobes linear-lanceolate. — W.Va., local.

2. L. virgàta Michx. (wand-like). — Stem or stems arising from a decumbent tuberous-rooted base, 3–8 dm. high, *simple or with slender ascending branches, glabrous or minutely puberulent; leaves linear to linear-oblong*, suberect, *sessile, blunt, mostly* 3–7 *mm. wide; flowers long-pedicelled, in slender bracted wand-like racemes; petals longer than* oblong-lanceolate *soon reflexed calyx-lobes;* style 7–10 mm. long, often conspicuous; capsule shorter than pedicel, quadrate-globose. — Bogs, savannas and wet pineland of Coastal Plain, n. Fla. to se. Va. Late June–Aug.

3. L. hirtélla Raf. (stiffly short-hairy). — Similar to no. 1, *simple* or sparingly branched, 0.2–1 m. high, *villous-hirsute; leaves sessile, oblong to oblong-ovate*, or the upper lanceolate, *blunt at both ends;* petals exceeding calyx-lobes; *capsule shorter than pedicel, crowned by erect or spreading calyx-lobes.* — Pine-barrens and sandy or peaty swamps, Fla. to e. Tex., locally n. to N.J. and Ky. July–Sept. FIG. 1335.

1335. L. hirtella.

4. L. glandulòsa Walt. (bearing glands; from the 4 glands in place of petals). — Bushy-branched, 0.15–1.5 m. high; *leaves* of upright stems thin, *oblong- or spatulate-lanceolate, tapering to short petioles*, those of basal offshoots short and elliptic; petals minute or wanting; *fruit cylindric*, 4–9 mm. long, 1.5–2.3 *mm. thick;* basal

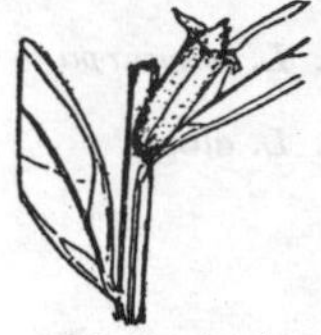
1336. L. glandulosa.

bractlet minute; *calyx-lobes* 1–2 *mm. long.* — Swamps, wet woods and shallow water, Fla. to e. Tex., n. to se. Va., s. Ind., s. Ill., se. Mo. and se. Kans. Late June–Sept. FIG. 1336.

5. L. microcárpa Michx. (tiny fruited). — Very slender, depressed or loosely ascending; the *stems angled,* 1–5 dm. high; *leaves obovate,* 0.5–2.5 *mm. long,* tapering to petioles; flowers sessile in the axils, apetalous; *hypanthium broadly obpyramidal,* 1–2 *mm. high,* shorter than the broad deltoid-ovate calyx-lobes. — Ditches, wet thickets, etc., Fla. to La., n. to N.C., Tenn. and se. Mo. July–Oct. (W.I.)

1337. L. polycarpa.

6. L. polycárpa Short & Peter (many-fruited). — Stoutish, 1.5–8 dm. high, mostly with ascending *wing-angled branches;* leaves narrowly lanceolate, acute at both ends, those of basal shoots short and oblong-spatulate; petals minute or wanting; *fruit* somewhat quadrate, turbinate-*campanulate,* 4–7 mm. long, 3–6 mm. thick, the *lance- to linear-attenuate bractlets nearly or quite as long;* calyx-lobes narrowly deltoid to broadly lanceolate. — Pond-shores and open wet places, sw. Me. to Ct.; sw. Ont. and O. to Minn., s. to Tenn., Mo. and Kans. July–Sept. FIG. 1337.

7. L. alàta Ell. (winged). — Slender, 0.7–1.5 m. high, loosely branching; *stem wing-angled;* leaves as in no. 4; *fruit depressed, turbinate-campanulate,* 2.4–4 *mm. long, about as broad, with wing-angles, the oblanceolate bractlets nearly as long;* calyx-lobes deltoid, as long as hypanthium. — Brackish or tidal swales and marshes, Fla. to La., n. to se. Va. Aug., Sept. FIG. 1338.

1339. L. linearis.

8. L. lineàris Walt. (linear). — Slender, 1.5–8 dm. high, simple or with ascending angled branches; *leaves narrowly linear or linear-lanceolate,* those of basal offshoots elliptic; *petals yellow,* narrow, *often equaling the lance-deltoid calyx-lobes; fruit elongate-turbinate or slenderly campanulate,* 4-sided, 5–10 mm. long, *naked or with bristle-like short bractlets;* seeds sausage-shaped. — Wet pine-barrens, ditches and swamps, Fla. to e. Tex., n. to N.J. and Tenn. July–Sept. FIG. 1339.

1338. L. alata.

9. L. sphaerocárpa Ell. (spherical-fruited). — Erect, 0.3–1 m. or more high, with ascending branches (stems thickly spongy when in water); leaves linear-lanceolate to narrowly oblong, those of the long basal shoots elliptic to narrowly obovate; *calyx subglobose, minutely pilose;* petals minute or wanting; *fruit spherical or subspherical, pilose,* 2.5–4.6 *mm. long,* naked or with bristle-like basal bractlets; *calyx-lobes* deltoid, *about equaling fruit; seeds globose or broadly obovoid.* — A polymorphous species:

Rameal leaves strongly reduced, glabrous or pubescent, linear or lanceolate.
  Mature fruits broader than high, 2.5–3.2 mm. long, 2.8–4 mm. broad.
    Branches and narrowly linear-lanceolate and attenuate leaves glabrous. . . . *L. sphaerocarpa* (typical).
    Branches and lanceolate acute leaves pubescent. . . . . Var. *jungens.*
  Mature fruits mostly longer than broad, 3.5–4.6 mm. long, 3.2–4 mm. broad. Var. *macrocarpa.*
Rameal leaves scarcely smaller than the narrowly oblong subacute and pubescent primary ones; stem pubescent; fruit about 3 mm. long and broad. . . . . Var. *Deamii.*

1340. L. sphaerocarpa.

L. sphaerocárpa (typical). — Sandy and peaty pond-margins and swamps, Fla. to La., n., locally, to L.I. and R.I. Late July–Sept. FIG. 1340.

Var. **júngens** Fern. & Grisc. (connecting). — Se. Va. to N.J. and e. Pa.

Var. **macrocárpa** Fern. & Grisc. (large-fruited). — N.J. and se. N.Y. to e. Mass.

Var. **Deàmii** Fern. & Grisc. (for its discoverer, CHARLES CLEMON DEAM, 1865–). — Porter Co., Ind., and s. Mich.

10. L. pilòsa Walt. (soft hairy). — Resembling no. 9, but *velvety-pilose* on young stem-tips and leaves of fertile stems; leaves lanceolate to oblong; *mature fruits* 5–6 *mm. broad,* usually *with a long bractlet;* seeds elongate. — Ditches, pools,

1341. L. pilosa.

bottomlands, and wet pine-barrens, Fla. to e. Tex., n. to se. Va. Late July–Sept. Fig. 1341. — Stems and branches often arching and tip-rooting in autumn; prostrate autumnal basal shoots glabrescent.

11. L. brévipes (Long) Eames (short-stalked). — Stems repent or floating and forming mats, glabrous or nearly so; emersed leaves firm, opposite, oblong-oblanceolate, narrowed to the blunt apex and tapering at base, 1–2.5 cm. long; submersed leaves flaccid, narrower and longer (up to 4 cm. long); *flowers on pedicels* 5–12 *mm. long, in anthesis* slenderly turbinate, 4-angled, 3.5–6 *mm. long,* with 2 basal bractlets; *calyx-lobes* lanceolate to lance-ovate, 3.5–6 *mm. long; petals* yellow, *broadly elliptic, about equaling calyx-lobes;* style 1–2 mm. long, surrounded by the 4 lobes of the stylopodium; *fruit* clavate-subcylindric, 8–10 *mm. long.* (*L. arcuata,* in part, of ed. 7, not Walt.; *Ludwigiantha* Long) — Damp sands and shallow pools of Coastal Plain, local, N.J. to n. Fla. June–Sept. Fig. 1342.

1342. L. brevipes.

12. L. lacústris Eames (of lakes). — Similar; more compact, forming prostrate mats, or when submersed elongating to 1 m. and with flaccid linear-lanceolate leaves up to 4.5 cm. long; *flowers on pedicels* 1–2.5 *mm. long, in anthesis* 2–3 *mm. long,* with *calyx-lobes* ovate to ovate-lanceolate and 2.5–3 *mm. long; petals elliptic-oblong, much shorter than calyx-lobes;* fruits smaller. — Sandy pond-shores and -margins, s. R.I. and s. Ct. Aug., Sept. (Cuba?) Fig. 1343.

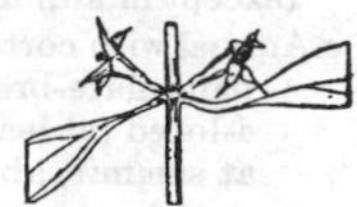
1343. L. lacustris.

13. L. nàtans Ell. (floating). — Smooth, forming repent mats (or floating); leaves opposite, petioled, elliptic to obovate, blunt to subacute, firm, 1–4 cm. long (in submersed state flaccid and up to 6 cm. long and 3 cm. broad); *flowers sessile or subsessile,* turbinate, *usually with narrow and elongate bractlets borne on the body of the hypanthium; petals minute or wanting; fruit quadrate-campanulate or -turbinate,* 4–6 *mm. long;* the broad calyx-lobes slender-tipped, 1.5–2.5 mm. long. (*Isnardia* Ktze.) Wet places or shallow water, Fla. to Tex., n., locally, to N.C., Tenn. and s. Mo. June–Sept. (Bermuda, and a var. in the W.I. and Mex.) Fig. 1344.

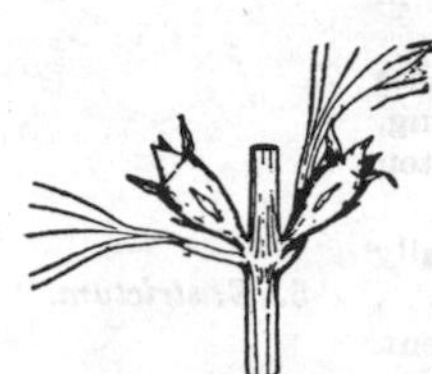
1344. L. natans.

14. L. palústris (L.) Ell. (of marshes), Water-purslane. — Similar; *hypanthium with* 4 *green longitudinal bands* terminating well below the sinuses; body of fruit whitish and corky, 4–5.3 mm. long; *bractlets, when present, basal and minute.* (*Isnardia* L.) — Europe; represented with us by

Var. **americàna** (DC.) Fern. & Grisc. (American). — *Fruit darker and less corky,* 2–4.5 *mm. long,* 1.8–3.5 *mm. thick, the longitudinal green bands reaching or nearly reaching sinuses; calyx-lobes broadly deltoid.* — Wet open places, shores or shallow water, N.S. and N.B. to Minn., s. to Ga., La. and Tex.; also e. Wash. to Mex. Late June–Sept. (Bermuda) Fig. 1345. — Completely submersed plants with sterile branches and flaccid translucent leaves are forma **elongàta** Fassett (elongate).

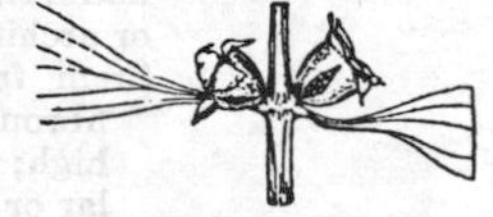
1345. L. palustris, v. americana

Var. **nàna** Fern. & Grisc. (dwarf). — Dwarf; *leaves* long-petioled, 1–2.5 *cm. long; fruits* 2–3 *mm. long,* 1.4–2 *mm. broad, with* triangular to broad-lanceolate *acuminate calyx-lobes.* — Low grounds, Fla. to Tex. and Mex., n. to Del. (W.I.; S.Am.)

3. EPILÓBIUM L. Willow-herb

Calyx-tube scarcely or not at all prolonged beyond the ovary; limb 4-cleft or -divided. Petals 4, violet, magenta, pink or white (in ours). Capsule slender, many-seeded. Seeds usually with a tuft (*coma*) of long hairs at the summit. — Mostly perennial herbs of cool or temp. reg., chiefly with nearly sessile leaves. (Name from the Greek *epi, upon,* and *lobon, a capsule;* from the perianth surmounting the capsule.)

a. Calyx cleft essentially to summit of ovary; corolla slightly irregular, with entire petals; the latter 1–3 cm. long, widely spreading; stamens and style successively declined, the former in a single series; flowers racemose; perennials with strong crowns. § Chamaenerion.
  Stems solitary or few, erect, 0.1–2 m. high; leaves membranaceous, green above, reticulate-veiny beneath, lanceolate, the longer ones 0.3–2 dm.

long; petals 1–2 cm. long; style pilose at base, in maturity longer than stamens; seeds oblong, 1–1.3 mm. long. . . . . 1. *E. angustifolium.*

Stems numerous, tufted, depressed or arched-ascending, rarely 4.5 dm. long; leaves thick and fleshy, whitish or glaucous, not veiny, elliptic-ovate to lanceolate, 2–8 cm. long; petals 1.8–3 cm. long; style glabrous, much shorter than stamens; seeds fusiform, 2 mm. long. . . . . . . 2. *E. latifolium.*

a. Calyx-tube slightly prolonged above ovary; corolla regular; petals notched, rarely (except in no. 3) more than 1 cm. long, with ascending bases; stamens and style erect or ascending, the former of 2 or more series; flowers corymbed, panicled or few in the upper axils; perennials, biennials or annuals. § LYSIMACHION. . . *b.*

b. Plant long-villous with divergent hairs, spreading by stout rope-like subterranean stolons; leaves sessile, partly clasping; petals 1–2 cm. long; stigma 4-parted; introd. species. . . . . 3. *E. hirsutum.*

b. Plant glabrous to short-pubescent, not long-villous, with filiform stolons or stolons none; leaves not clasping; petals 3–10 mm. long; stigma entire (except in annual no. 4); indigenous. . . *c.*

c. Annual with cortex at base cracking and exfoliating, usually very slenderly paniculate-branched; locules of capsule 6–10-seeded; stigma often 4-lobed (at least in larger flowers); seeds obovoid, rounded and beakless at summit, abruptly contracted just above base, its coma caducous. . 4. *E. paniculatum.*

c. Perennial (or biennial), often with basal offsets, stolons, autumnal rosettes or scaly bulbs; stigma entire or but shallowly notched; seeds more numerous, often tapering above to a collar or short beak, gradually narrowed to base; coma more persistent (wanting or rudimentary in no. 17). . . *d.*

d. Stem terete, without decurrent lines running down from the leaf-bases (except in no. 9); cauline leaves linear, lanceolate or narrowly oblong, entire or merely undulate (few-toothed in no. 9), the margins often revolute. . . *e.*

e. Stem, leaves and capsules grayish-velvety with short horizontally divergent pubescence. . . . . 5. *E. strictum.*

e. Stem, leaves and capsule glabrous to minutely incurved-pubescent. . . *f.*

f. Median and upper cauline leaves linear to lanceolate, the larger ones of each plant 1–10 cm. long; calyx-segments tapering to acute or subacute tips; seed with short and thick to barely evident neck. . . *g.*

g. Leaves closely and evenly pubescent above with minute incurved hairs; tips of stems or branches and buds before flowering erect or arching, not strongly nodding.

Stem freely upright-branching to simple, arising from a fibrous-rooting non-stoloniferous base; calyx 3–4.5 mm. high; petals 4–6.5 mm. long; seeds tapering to evident collar or neck. . . . . 6. *E. leptophyllum.*

Stem simple (except for axillary fascicles) or with few erect branches, arising from decumbent prolonged and rooting simple base or from loosened scaly bulbs, propagating late in the season by elongate filiform stolons with bulbous tips; calyx 4.5–7 mm. long; petals 7.5–10 mm. long; seeds essentially without a collar. . . . . 7. *E. nesophilum.*

g. Leaves glabrous above or with few remote hairs; tips of stems and pedicels before flowering nodding.

Flowering stems arising from tips of slender stolons of the preceding year or from loosened scaly bulbs, late in the season sending out filiform stolons; uppermost internodes and capsules, if pubescent, uniformly so; leaves with often revolute entire to shallowly undulate margins, acute to blunt; calyx sparsely pubescent. . . . . 8. *E. palustre.*

Flowering stems arising from basal rosettes of round-tipped leaves, without stolons; internodes and capsule glabrous except for restricted lines of hairs; leaves with flat ciliate denticulate margins, round-tipped; calyx glabrous. . . 9. *E. davuricum.*

f. Median and upper cauline leaves elliptic-oblong, obtuse or round-tipped, 0.8–2 cm. long; calyx-segments with broadly rounded tips; seed with long slender neck. . . . . 10. *E. Pylaieanum.*

d. Stem terete or 4-angled, with decurrent lines running down from the leaf-bases; cauline leaves lanceolate to elliptic or ovate, membranaceous to subcoriaceous, flat (not revolute-margined), toothed or, if entire, the stems ascending from slender elongate tufted basal shoots. . *h.*

h. Stems solitary (rarely 2 or more), erect or ascending, without elongate slender scaly basal offshoots, the bases in summer or autumn bearing sessile or short-stalked leafy rosettes or fleshy and scaly bulbiform offsets; principal cauline leaves more or less toothed. . . *i.*

i. Stem terete, very slender, 0.4–3 dm. high.

Leafy basal rosette present at flowering time, its round-tipped leaves crowded; cauline leaves remote, linear, 1–2 mm. broad; seed 1.5 mm. long. . . . . 9. *E. davuricum.*

Leafy basal rosette developed in autumn, represented at flowering time by dry scale-like vestiges; cauline leaves lance-elliptic to oval, 3–8 mm. broad; seed about 1 mm. long. . . 11. *E. leptocarpum.*

i. Stem 4-angled, 0.03–1 m. high. . . *j.*

j. Coma of long slender hairs; seed smoothish or with low rounded to deltoid pebbling. . . *k.*

k. Cauline leaves elongate-lanceolate, gray-green, distinctly petioled, rugose-veiny, the larger ones 30–75- or more-toothed; seeds blackish, with cinnamon-colored coma. . . 12. *E. coloratum.*

k. Cauline leaves lanceolate, oblong, elliptic or ovate, scarcely rugose-veiny, mostly fewer-toothed, not grayish; seeds fulvous (rarely blackish), with white or whitish coma. . . *l.*

l. Middle and upper internodes of stem minutely incurved-pilose on sides as well as angles; calyx-lobes pilose; plant with simple to paniculate-branched stem 0.03–1 m. high.

Leaves firm-membranaceous (except in deep shade), full green, mostly opposite, with lateral veins prominent beneath, sessile or short-petioled, lanceolate, oblong or ovate, rounded to cordate at base; flowers 4–9 mm. long; petals pink to lilac; fruiting pedicels 0.2–1.5 cm. long; coma persistent. . . . . . 13. *E. glandulosum.*

Leaves thin and flaccid, pale green, inclined to be alternate, elliptic to narrowly ovate, tapering to definite petioles up to 1 cm. long, rarely rounded at base and sessile; flowers 3–6 mm. long; petals white or whitish; fruiting pedicels 0.5–3 cm. long; coma caducous. . . 14. *E. ciliatum.*

l. Middle and upper internodes of stem glabrous or promptly glabrate on sides; calyx-lobes glabrous or promptly glabrate; stem simple or with short erect branching, 1–4.5 dm. high; lateral nerves of leaves delicate.

Flowering from near base to summit, stem fastigiately short-branched; leaves thin to flaccid, mostly alternate, the larger 3–7 cm. long, tapering to petioles; plant of w. Nfld. . . . . 15. *E. scalare.*

Flowering only at summit of simple stem; leaves firm, thickish, mostly opposite, the larger 1.5–5 cm. long, rounded to sessile bases; plant of Lab., Nfld. and Gaspé. . . . . 16. *E. Steckerianum.*

j. Coma wanting or merely rudimentary; seed heavily covered with rows of whitish elongate trichome-like papillae; estuarine species. . . . . 17. *E. ecomosum.*

h. Stems tufted or matted, arising from slender stoloniform creeping basal offshoots, the bases at flowering time often sending out elongate slender scaly and leafy stolons; leaves entire or shallowly few-toothed, their lateral nerves not prominent. . . *m.*

m. Seeds distinctly papillate or pebbled (under magnification); flowers 6–10 mm. long, with lilac petals; median cauline leaves elliptic- to oblong-ovate, 1–2.5 cm. broad, tapering to blunt tips. . 18. *E. Hornemanni.*

*m.*Seeds smooth; flowers smaller, white or, if lilac or purplish, the median leaves rounded to tip and only 0.5–1 cm. broad.
Flowering stems erect, 1–5.2 dm. high; median cauline leaves ovate to oblong-elliptic, repand-dentate, tapering to blunt tip, 1.5–6 cm. long, 0.6–2 cm. broad; flowers 1–several from the upper axils; petals milk-white. . . . . 19. *E. alpinum.*
Flowering stem curving, 0.3–1.5 dm. high: median cauline leaves oblong, essentially entire, broadly rounded at tip, 0.8–2 cm. long, 5–10 mm. broad; flowers 1–3, terminal; petals pink or purple. . . . . 20. *E. anagallidifolium.*

§ Chamaenèrion (Ludw.) Tausch

1. E. angustifòlium L. (narrow-leaved), Great W., Fireweed, Wickup. — *Stem* or stems arising from a strong perennial crown, *erect*, smooth, 0.1–2 m. high, very leafy; lowest *leaves* scale-like; *the principal and very numerous ones* alternate, *membranaceous, green above*, pale and *reticulate-veiny beneath*, 0.3–2 *dm. long*, nearly entire; *raceme or racemes elongate, bracted;* calyx cleft to summit of ovary; *flower-buds becoming reflexed, then ascending; petals* magenta or pink (rarely white), 1–2 *cm. long, entire, spreading, slightly unequal; style longer than stamens, declined, hairy at base; stigmas slender, elongate, soon revolute;* stamens in 1 series; capsules 5–8 cm. long, canescent; *seeds oblong*, 1–1.3 *mm. long.* (*Chamaenerion* Scop.) — Very variable boreal species; the following varieties and forms with us:

*a.*Leaves long-attenuate at apex. . . *b.*
*b.*Plant tall, 0.5–2 m. high; median cauline leaves 0.7–2 dm. long; mature raceme 1.7–7.5 dm. long; petals 1.3–2 cm. long. . . *c.*
*c.*Median leaves 0.5–3.5 cm. broad, the secondary nerves not prominent; racemes open, with small bracts or, if leafy-bracted, with the lower bracteal leaves at most 2.5 cm. broad. . . *d.*
*d.*Petals purplish or roseate. . . . . *E. angustifolium* (typical).
*d.*Petals white or whitish.
Sepals whitish. . . . . Forma *albiflorum.*
Sepals red. . . . . Forma *spectabile.*
*c.*Median leaves 2–5.5 cm. broad, the secondary nerves very prominent beneath; inflorescence leafy-bracted, the lower bracteal leaves 2.3–3.5 cm. broad. . . . . Var. *macrophyllum.*
*b.*Plant low, 1–5 dm. high; median cauline leaves 3–7 cm. long; mature raceme 0.4–1.3 dm. long; petals 1–1.5 cm. long. . . . . Var. *intermedium.*
*a.*Leaves merely acutish or bluntish, not long-attenuate at tip; racemes and flowers much as in typical var. . . . . Var. *platyphyllum.*

E. angustifòlium (typical). — Recent clearings, burned woodlands, damp ravines, etc., subarct. Am., s. to Nfld., N.S., N.E., Md., mts. of N.C., W.Va., n. O., centr. Ind., n. Ill., n. Ia., Black Hills, S.D., Ariz. and Calif. July–Sept. (Eurasia) — Forma **albiflòrum** (Dumort.) Haussk. (white-flowered) occasional; forma **spectàbile** (Simmons) Fern. (showy), local.

Var. **macrophýllum** (Haussk.) Fern. (large-leaved). — Local, w. Nfld. and M.I.; Alaska.

Var. **intermèdium** (Wormsk.) Fern. (intermediate). — Greenl. and Lab. to Nfld., e. Que., L. Mistassini and James Bay.

Var. **platyphýllum** (Daniels) Fern. (broad-leaved). — Shickshock Mts., Que.; n. Minn. to B.C., s. to N.M.

2. E. latifòlium L. (broad-leaved), River-beauty. — *Depressed densely matted* perennial *with* the *upwardly arching crowded stems* from thick crowns, 1–4.5 *dm. long; leaves thick and fleshy, whitish or strongly glaucous, not veiny*, elliptic-ovate to lanceolate, 2–8 *cm. long; flowers 1–few*, racemosely borne *in the upper leaf-axils;* calyx often purple-tinged; *petals* showy, 1.8–3 *cm. long*, purple or roseate, or in forma **leucánthum** (Ulke) Fern. (white-flowered) white or whitish; *style glabrous*, declined, *much shorter than stamens; the stigmas short and thick, not becoming revolute; seeds fusiform*, 2 mm. long. (*Chamaenerion* Spach) — River-gravels, margins of streams and damp slopes, Greenl. and Arct. Am., s. to Nfld., Gaspé Pen., Que., James Bay, Black Hills, S.D., Colo. and Wash. Late June–early Sept. (Eurasia)

§ Lysimáchion Tausch

3. E. hirsùtum L. (hirsute). — *Creeping by thick rope-like rhizome and subterranean stolons; stem, leaves,* etc., *villous-hirsute with long spreading hairs;* stem freely branched, subterete, 0.7–2 m. high; *leaves sessile, somewhat clasping, soft and hairy*, lanceolate to lance-oblong, with falcate teeth; *flowers from the upper axils, forming racemes;* calyx-tube about 2 mm. long, the segments apiculate; *petals* purple, obcordate, 1–2 *cm. long,* subascending; *stigma 4-cleft;* seeds obovoid, rounded above, about 1.5 mm. long. — Waste places, roadside-thickets, meadows, etc., s. Que. and s. Ont., s. to s. N.E., N.Y., O., Mich. and n. Ill. July–Sept. (Natzd. from Eu.)

4. E. PANICULÀTUM Nutt. (panicled). — *Annual,* 1.5–8 dm. high, glabrous or glabrate, *slenderly dichotomous-paniculate, the pale dry cortex soon exfoliating at base; leaves opposite or alternate* and subtending tiny axillary fascicles, narrowly lanceolate, *sparingly denticulate,* often folded, mostly petioled; flowers scattered, on often bracted short pedicels; calyx 5–6 mm. long; petals purple, notched, 6–8 mm. long; *stigma 4-lobed; capsule fusiform, prominently beaked, with strong ribs,* 1.2–2.5 cm. long, *on pedicels* 1–5 *mm. long; seeds* 6–10 *in each locule, obovoid, rounded and beakless at summit, strongly constricted above the base,* the *coma caducous.* — Open woods, clearings, prairies and rocky slopes, B.C. to s. Calif., e. to Man., N.D., S.D., Colo. and Ariz.; adv. in w. Que. June–Sept.

Var. **subulàtum** (Haussk.) Fern. (awl-shaped). — More slender; *calyx and petals* 3–5.5 *mm. long, subequal; stigma less lobed; fruiting pedicels* filiform, 0.5–2 *cm. long;* capsules often shorter (down to 1 cm.). — B.C. to Calif., e. to Sask., Black Hills, S.D., Colo. and Ariz.; dry calcareous talus, rocks, open woods, etc., near L. Huron, Ont.; casual on railroad-ballast elsewhere. June–Sept.

5. E. stríctum Muhl. (pressed together). — Erect, simple or paniculately branched at summit, with often crowded axillary fascicles and short branches, propagating late in the season by filiform stolons; *stem* 0.15–1.35 m. high, terete, *with the leaves, pedicels and capsule grayish-velvety with crowded horizontally divergent pubescence; leaves oblong-lanceolate to linear,* entire or undulate, with revolute margins; petals pink, 7–9 mm. long, broadly obovate, notched; fruiting pedicels rather short; seed about 2 mm. long, rounded above to very short neck, the coma often becoming dingy. (*E. densum* Raf.; *E. molle* Torr., not Lam.) — Bogs, mossy thickets, meadows, etc., Gaspé Pen., Que., to Minn., s. to N.S., N.E., n. Va., n. O., centr. Ind. and n. Ill. July–Sept.

6. E. leptophýllum Raf. (narrow-leaved). — *Minutely hoary-pubescent with incurved hairs,* 0.2–1.1 m. high, simple to densely upright-branched, usually with short leafy axillary fascicles, *from branching root, rarely, if ever, stoloniferous; leaves linear or linear-lanceolate,* 1–3 mm. wide, *attenuate, canescent above,* the lateral veins inconspicuous, the entire or barely undulate *margins revolute,* or in forma **umbròsum** (Haussk.) Fern. (of shade) the thin and flat primary leaves 4–10 mm. broad and with evident lateral nerves; *calyx* 3–4.5 *mm. high, canescent,* the segments acutish; *petals* pink to whitish, 4–6.5 *mm. long;* pedicels one-fourth to half as long as canescent capsules; *seed* about 1.5 mm. long, *tapering to an evident short neck.* (*E. densum* of ed. 7, not Raf.; *E. lineare* of auth., not Muhl.) — Low grounds, Gaspé Pen., Que., to Alta., s. to N.S., N.E., n. Va., W.Va., O., Ind., Ill., Mo., Kans. and Colo. July–Sept.

7. E. nesóphilum Fern. (island-lover). — Coarser than no. 6; *stem* simple (except for axillary fascicles) or with few erect branches, canescent above with minute incurved hairs, 1.5–8.5 dm. high, *arising from prolonged decumbent simple bases and from loosened scaly bulbs, propagating late in the season by filiform stolons tipped by small bulbs;* leaves lanceolate to lance-linear, attenuate, minutely incurved-pubescent; *calyx* 4.5–7 *mm. high, with* minutely incurved-pilose *bluntish* lanceolate *segments; petals* pink, 7.5–10 *mm. long;* longer fruiting pedicels 1–4 cm. long; *seeds with nearly obsolete collar, the coma as if arising directly from summit of seed.* — Bogs, swales, damp thickets, dune-hollows, etc., Nfld., Anticosti and M.I. Mid-July–Sept.

Var. **sabulonénse** Fern. (of Sable Island). — Leaves oblong-lanceolate, blunt or merely acutish, not attenuate to tip; longest fruiting pedicels 1–1.5 cm. long. — Sable I., N.S.

8. E. palústre L. (of marshes). — *Stems* erect, simple or branching, *arising from decumbent tips of last year's filiform stolons or from loosened scaly bulbs,* commonly *after flowering sending out filiform stolons;* internodes of stem terete, the lower glabrescent, the upper more or less incurved-pilose; *tips of stems before flowering nodding; leaves linear to oblong-lanceolate,* narrowed to sessile bases, *the apex usually callus-tipped, entire, glabrescent or with the often revolute margin crisp-puberulent; flower-buds at first nodding,* the pedicels and flowers becoming erect; flowers 4–8 mm. long; petals deeply notched, lilac, violet, pink or white; *calyx-segments lanceolate, acutish, sparsely pilose;* capsules 3–9 cm. long, on pedicels 1–5 cm. long; *seeds fusiform,* 1.5–2 mm. long, *attenuate to short neck and to acute base.* — Polymorphic boreal species. We recognize the following varieties:

a. **Principal leaves lanceolate, 5–15 mm. broad, tapering from near base to apex, membranaceous (but not flaccid) to firm, ascending or spreading, often with axillary fascicles or definitely branched; fruiting pedicels mostly exceeded by subtending leaves.**

Stem simple or branching above; branches, when developed in the lower axils, short and sterile. . . . . *E. palustre* (typical).

Stem branching from lowest nodes; the prolonged branches strongly ascending, nearly equaling the main axis and floriferous. . . . . Var. *longirameum.*

*a.*Principal leaves linear, linear-lanceolate or narrowly oblong, 1–8 mm. wide; fruiting pedicels often or mostly equaling or overtopping the subtending upper leaves. . . *b.*

*b.*Stem 0.5–8 dm. high, simple or with axillary fascicles or with many commonly erect branches; leaves flaccid, linear or linear-lanceolate, tapering gradually to tip, the longer ones in remote pairs, 2–8 mm. wide, 3–7 cm. long. . . . . Var. *grammadophyllum.*

*b.*Stem 0.5–5.5 dm. high, mostly simple, rarely with axillary tufts; principal leaves linear to linear-oblong, 1–5 mm. wide, 1–3.5 cm. long, mostly obtuse, strongly ascending. . . . *c.*

*c.*Pairs of leaves distant or subdistant, firm to coriaceous, extending to summit of stem, the latter commonly 1–5.5 dm. high.

Elongate median and upper leaves 5–17 pairs, mostly longer than internodes, their blunt tips often overtopping the bases of those above; flowers, when fully developed, on gently arching to ascending pedicels, 4–6(–8) mm. high; petals white to pink; capsules up to 6 cm. long; stem up to 5.5 dm. high. . . . . Var. *oliganthum.*

Elongate median and upper leaves 3–9 pairs, mostly shorter than internodes, the upper ones tapering to acutish tips; flowers, when first expanded on strongly divergent to drooping pedicels, 5–8 mm. high; petals deep lilac or purple; capsules rarely 4 cm. long; stem rarely more than 3 dm. high. . . . . Var. *lapponicum.*

*c.*Pairs of lower and median leaves strongly overlapping, thin-membranaceous or flaccid; expanded flowers on slightly arched to ascending pedicels, 4–5 mm. high; petals white or pink; capsules rarely 4.5 cm. long; stem 0.5–2 dm. high. . . . . Var. *labradoricum.*

E. palústre (typical). — Bogs, mossy meadows and damp slopes, Greenl. and Lab. to Alaska, s. to Nfld., N.S., n. N.E., e. Mass., w. Ct., Ont. and Wyo. July, Aug. (Eurasia)

Var. **longiràmeum** Fern. & Wieg. (long-branched). — Swamps, wet places and shores, Lab. and Côte Nord, Que.

Var. **grammadophýllum** Haussk. (engraved-leaved; from pellucid lines often present), var. *mandjuricum* sensu Am. auth., not Haussk. and *E. wyomingense* Nels. — Wet rocks, mossy swales, etc., sw. Greenl. and s. Lab. to Alaska, s. to Nfld., N.S., n. N.E., ne. Mass., n. N.Y., Minn., S.D., Colo. and Oreg.

Var. **oligánthum** (Michx.) Fern. (few-flowered), var. *monticola* of Am. auth., not Haussk.; *E. oliganthum* Michx. — Peaty or mossy soils, Lab. to Alta., s. to Nfld., N.S., N.E., N.Y., Ont. and Minn.

Var. **lappónicum** Wahlenb. (of Lapland). — Damp limestone-barrens and peaty swales, w. Nfld. and Anticosti; Alaska. (N. Eu.)

Var. **labradóricum** Haussk. (of Labrador). — Low grounds, wet rocks, etc., Greenl. and Lab. Pen., s. to Nfld. and alpine areas of Gaspé Pen., Que., and of Mt. Washington, N.H.

9. E. davùricum Fisch. (of Dahuria). — *Stems* few or solitary *from a compact sessile basal rosette of crowded* deep green *spatulate-oblong entire round-tipped leaves only* 0.5–1.5 *cm. long, without basal stolons*, simple or slightly branched above, very slender, 0.5–3 dm. high, *terete*, or subangulate at the glabrescent base, *the upper internodes with fine incurved hairs borne in lines; cauline leaves remote, the upper often alternate;* the principal ones linear, 2–5 cm. long, 1–2 *mm. broad, remotely denticulate, with flat ciliate margins, broadly rounded at tip; flowers* few, 4–6 mm. long, *before anthesis strongly nodding;* the fruiting pedicels erect; *calyx glabrous; petals whitish*; seed slenderly fusiform, obviously attenuate at summit and base, 1.5 mm. long. — Subarct. Am., coming s. to damp limestone-barrens and bogs of nw. Nfld., Anticosti and L. Mistassini, Que., shores of James Bay, etc. Mid-July, Aug. (Eurasia)

10. E. Pylaieànum Fern. (in honor of its discoverer, Auguste Jean Marie Bachelot de la Pylaie, 1786–1856). — Small plant with filiform subterranean rhizomes; stem simple or sparingly branched, very slender, 0.3–1.7 dm. high, terete, minutely pilose; *principal cauline leaves* 5–10 pairs, *oblong or elliptic, with rounded or blunt tips, the larger ones* 0.8–2 *cm. long and* 2.5–5 *mm. broad*, entire, *sublustrous above; buds erect*, apiculate; flowers 3–5 mm. high; *calyx-segments oblong, broadly rounded at tip;* petals white or pink; *seed* 2 mm. long, *slenderly fusiform, with a very slender neck* 2 *mm. long.* — Bogs and wet acid rock, s. Nfld. Aug., Sept.

11. E. leptocárpum Haussk. (slender-fruited). — Very small; the slender *terete stem incurved-pilose,* scarcely 1 dm. high, with many ascending branches; *leaves 1–1.5 cm. long,* 2–3 *mm. wide, broadly lanceolate,* with few low teeth, obtuse to subacute, *the winged petiole with decurrent lines running down from base, lateral nerves obscure;* flowers 3 mm. long, pale; calyx-segments acute; capsules glabrous, 1–2 cm. long, on glabrous pedicels 0.5–1 cm. long; seeds fusiform, attenuate to both ends, 1 mm. long, thinly papillose. — Pacific N. Am.; represented with us by

Var. **Macoùnii** Trel. (for John Macoun, 1831–1920). — Coarser; the simple or *subsimple stem* 0.5–3 *dm. high,* with vestiges of old rosettes at base during flowering time; *cauline leaves* lance-elliptic to oval, 1–3.5 *cm. long,* 3–8 *mm. wide; flowers* 4–5 *mm. long,* the petals often pinkish; *capsule up to* 5 *cm. long.* — Damp, often calcareous, spots, local, n. Nfld. to B.C., s. to N.S., Bruce Pen., Ont., Man., Ida. and Wash. Late June–Sept.

12. E. coloràtum Biehler (colored; from the red nerves of the leaves). — *Stem* erect or strongly ascending, bushy-branched at summit (simple only in small plants), up to 1 m. high, glabrous below, *minutely pubescent above, at least in lines, with incurved pale hairs,* in autumn developing sessile leafy basal rosettes; *leaves elongate-lanceolate,* long-acuminate, membranaceous, *gray-green, rugulose-veiny,* divergent to ascending, distinctly short-petioled, *closely and irregularly serrulate; the larger ones* of each plant 0.4–1.5 dm. long and 0.5–2.5 cm. broad, *with 30–75 or more teeth;* panicle usually intricately branched, with greatly reduced leafy bracts and numerous approximate flowers of varying degrees of development; calyx short, with the pedicels and capsule canescent; petals pink, 3–5 mm. long; *seed* about 1.5 mm. long, *abruptly rounded to almost collarless summit;* mature *coma cinnamon-colored.* — Low grounds, springy slopes, etc., s. Que. to Minn. and S.D., s. to N.S., N.E., L.I., Ga., Ala., Tenn., Ark. and Kans. July–Oct.

13. E. glandulŏsum Lehm. (bearing glands). — *Stem* generally solitary, 0.05–1 m. high, *obtusely 4-angled,* glabrous below, *minutely incurved-pilose and often glandular above, the sides as well as the angles pubescent,* simple to much branched above the middle, in autumn producing above or below ground sessile or subsessile rosettes or subglobose pale and fleshy bulbs; *leaves* narrowly to broadly lanceolate, elliptic, oblong or ovate, *full green, shallowly denticulate, flat* and membranaceous to firm, *with plane surfaces and the lateral nerves prominent beneath, sessile or short-petioled; flowers* crowded among the upper leaves or in open panicles, 4–9 *mm. long; pedicels* erect, *in fruit* 0.2–1.5 *cm. long; calyx minutely pilose; petals lilac or pink;* seeds 1.2–1.8 mm. long, oblong, rounded or gradually tapering above to short hyaline neck, variously pebbled with rounded to acutish papillae; *coma white or whitish.* — A perplexingly variable complex greatly in need of prolonged and cautious checking. The following varieties are tentatively recognized:

Leaves rather crowded and overlapping, not greatly decreasing in size into the crowded inflorescence, the pedicels and lower halves of the stiffly ascending capsules covered by the leaves; stem simple or with few erect branches; leaves rounded to sessile or subsessile bases; flowers 6–9 mm. long; seeds 1.4–1.8 mm. long. *E. glandulosum* (typical).

Leaves more remote, conspicuously decreasing in size into the looser and more open inflorescence, the pedicels and capsules not strongly covered by subtending leaves; flowers 4–8 mm. long; seeds 1.2–1.5 mm. long.

- Median cauline leaves cordate-subclasping, ovate- or broadly lance-acuminate; principal internodes 4–10 cm. long; stem simple or sparingly branched above; fleshy bulbiform pale basal offsets frequently developed before end of anthesis. Var. *cardiophyllum.*
- Median cauline leaves rounded, barely subcordate or tapering at base, ovate-lanceolate to narrowly lanceolate or oblong; median internodes mostly shorter; inflorescence often more branched.
  - Leaves narrowly ovate to lanceolate, acuminate to acutish; inflorescence of large plants open-paniculate.
    - Median leaves ovate to ovate-lanceolate. . . . . Var. *adenocaulon.*
    - Median leaves elongate-lanceolate. . . . . Var. *occidentale.*
  - Leaves elliptic-oblong to oblong-lanceolate, rounded at summit; stem simple or subsimple. . . . . Var. *brionense.*

E. glandulŏsum (typical). — Coast of s. Lab., s. to coasts of Nfld., Que. and N.B.; Alaska to Oreg. July–Sept. (E. Asia)

Var. **cardiophýllum** Fern. (heart-leaved). — Damp thickets, etc., s. Lab. to Alaska, s. to Nfld., e. Que., n. Vt., n. N.Y., n. Minn., Colo. and Oreg. — Erroneously identified with *E. boreale* Haussk.

Var. **adenocaùlon** (Haussk.) Fern. (glandular-stemmed), *E. adenocaulon* Haussk. — Damp thickets, etc., generally common, Nfld. to Alaska, s. to N.S., N.E., Del., Md., W.Va., O., Ind., Ill., n. Ia., Neb., Colo., etc. June–Sept.

Var. **occidentàle** (Trel.) Fern. (western), *E. occidentale* (Trel.) Rydb. — Nfld. to B.C., s. to N.S., n. N.E., n. N.Y., Ont., Mich., Wisc., S.D., Colo. and Calif.

Var. **brionénse** Fern. (of Brion I.). — Brion I., M.I.

14. E. ciliàtum Raf. (having marginal hairs). — Resembling weak and etiolated states of no. 13; stem very slender, simple or slightly branched, 0.3–4 dm. high; *leaves pale green, flaccid or thin-membranaceous, elliptic, oval or narrowly ovate, with faint lateral veins, the upper* and sometimes all *scattered and alternate,* 1–5 cm. long, tapering into slender *petioles* 2–10 *mm. long* (rarely sessile); flowers 3–6 mm. long; calyx sparsely pilose; *petals white or pale pink; capsules on* ascending to arching *pedicels* 0.5–3 *cm. long; seeds* fulvous to fuscous, 0.9–1.5 *mm. long, with prominent rows of deltoid subacute papillae; coma caducous.* (*E. americanum* Haussk.; *E. adenocaulon,* var. *perplexans* Trel.) — Damp spots, springy places, wet rocks, etc., Nfld. to B.C., s. to N.S., N.E., Pa., s. Ont., Wisc., N.M., Ariz. and Calif. July–Sept.

15. E. scalàre Fern. (of a ladder; from the type-locality, "John Kanes's Ladder"). — Rhizome elongate, bearing pale fleshy bulbiform subterranean buds; *stem* erect, 0.5–3.5 dm. high, simple or *usually with many short erect fastigiate flowering branches, the sides of the internodes glabrous* but with 2 elevated pilose lines; *leaves pale green, fleshy to flaccid, glabrous,* crowded, *mostly alternate,* spreading to reflexed, elliptic to narrowly oblong-ovate, *narrowed to petioled and to subacute apices, the larger blades* 3–7 *cm. long,* shallowly repand-dentate; *flowers borne from base to summit,* erect, 5–6 mm. long, on pedicels 2–3 mm. long; *calyx-lobes* acute, *glabrate,* slightly exceeded by the purple petals; *capsules erect, partly covered by leaves, glabrate;* seeds with pellucid collar, gradually tapering to acute base, papillose, 1.4 mm. long. — Wet slaty cliffs, nw. Nfld. July, Aug.

16. E. Steckeriànum Fern. (named in 1918 for its discoverer, ADOLPH STECKER, Moravian missionary in Labrador late in the 19th century). — *Rhizome* short and thick, late in the season *sending out stalked elongate bulbiform fleshy buds; stem strict, simple,* 4-angled, *glabrous* except for decurrent pilose lines, 1–4.5 dm. high; *leaves* 5–15 pairs, *all opposite or the uppermost scattered, subcoriaceous,* strongly ascending, often overlapping, *oblong-ovate to oblong-lanceolate, subequal, the largest* 1.5–5 *cm. long and* 0.6–2 *cm. broad, blunt,* repand-dentate, *rounded to sessile base; flowers* erect, 4.5–6 mm. long, *in upper axils,* the lower equaled or exceeded by subtending leaves, very short-pedicelled; *calyx glabrate;* petals purplish or white; *capsule glabrate;* seeds papillate. — Springy or wet mossy banks and glades, Lab., n. Nfld. and Shickshock Mts., Gaspé Pen., Que. July, Aug. — Sometimes mistaken for the cordilleran *E. Drummondii* Haussk. or *E. saximontanum* Haussk.

17. E. ecomòsum (Fassett) Fern. (without coma). — Plant superficially resembling *E. glandulosum,* var. *adenocaulon;* stem simple to paniculately branched, 0.5–7 dm. high; *seeds grayish, abruptly rounded at base, covered with approximate high and crest-like ridges of pale trichome-like papillae; coma wanting or rudimentary.* — Fresh tidal flats of the St. Lawrence River, Quebec. July–Sept.

18. E. Hornemánni Reichenb. (for JENS WILKEN HORNEMANN, 1770–1841). — *Stems tufted or in mats from short and slender creeping branches of a many-forked crown,* arched-ascending to suberect, slender, simple or slightly forking, *smooth, except for minutely pilose lines on upper internodes,* 0.5–4.5 dm. high; *principal cauline leaves elliptic- to oblong-ovate,* 1.5–5 *cm. long and* 1–2.5 *cm. broad, tapering to blunt tip* and to definite petiole, *membranaceous, faintly nerved, shallowly and remotely denticulate;* flowering tip arched when young, soon ascending; *flowers few–several from upper axils,* 6–10 *mm. high; petals lilac or pink* (rarely white), *much exceeding calyx-lobes;* capsules longer than to about equaling the ascending pedicels; *seed* elongate-fusiform, about 1 mm. long, *its surface distinctly pebbled* (under magnification). (*E. alpinum* sensu some auth., not L.) — Damp rocks, margins of rills, etc., Arct. Am., s. to Nfld., Gaspé Pen., Que., C.B., Mt. Katahdin, Me., White Mts., N.H., Adirondack Mts., N.Y., Colo., Utah, Nev. and Calif. July, Aug. (Eurasia)

19. E. alpìnum L. (alpine). — *Stiffer* than no. 18; up to 5.2 dm. high; *leaves firmer,* the median ovate to oblong-elliptic, 1.5–6 cm. long and 0.6–2 cm. broad; *flowers* 3–6 *mm. high, with milk-white* (sometimes pink-tipped) *petals; seed* 1.2–1.5 *mm. long, smooth.* (*E. lactiflorum* Haussk.) — Margins of rills, etc., Arct. Am., s. to n. Nfld. and alpine areas of Gaspé Pen., Que., Mt. Katahdin, Me., White Mts., N.H., Colo., Utah, Nev. and Calif. July, Aug. (Eurasia)

20. E. anagallidifòlium Lam. (pimpernel-leaved). — *Dwarf, the flowering stems arching,* eventually ascending, 0.3–*finally* 1.5 *dm. high,* with spreading or decumbent leafy basal shoots; *leaves narrowly elliptic to oblong, round-tipped, obscurely petioled, essentially entire, the median* *ones 0.8–2 cm. long and 5–10 mm. broad; flowers 1–3, terminal,* often nodding, 4–7 mm. high, *with pink or purple petals; pedicel nearly equaling to longer than capsule, at first arching, later erect; seed smooth,* about 1.5 mm. long. (*E. alpinum* in part of L.) — Arct. Am., s. in damp moss or on wet rock to n. Nfld., alpine areas of Gaspé Pen., Que., and Mt. Katahdin, Me., Colo., Utah, Nev. and Calif. July, Aug. (Eurasia)

4. OENOTHÈRA L. Evening-Primrose

Calyx-tube prolonged beyond the ovary, deciduous; the lobes 4, at first ascending and united, later reflexed and more or less separated. Petals 4. Stamens 8; anthers mostly linear and versatile. Capsule 4-valved, many-seeded. Seeds naked or with an obscure membranaceous crest. — Leaves alternate or rarely all basal, those of first year often in basal rosettes. Petals yellow, white or roseate. Styles often of different lengths in the same species (sometimes obsolete). A complex and very plastic Am. genus. (Name used by Theophrastus for a species of *Epilobium.*)*

a.Capsules slenderly cylindric, subcylindric or corniculate, many times longer than thick, if 4-sided not prominently winged. . *b.*

b.Stigma with 4 linear-cylindric lobes; calyx-tube slender; flowers somewhat vespertine or nocturnal. . *c.*

c.Seeds in 2 (rarely more) rows in each locule; flower-buds erect; fresh petals yellow, often reddish in age. . *d.*

d.Ovules and seeds horizontal in the locules, in 2 or more rows, the seeds angled; erect or ascending mostly stiff biennials. Subgen. 1. Oenothera proper (*Onagra*). . *e.*

e.Tips of calyx-lobes in the unexpanded bud closely connivent or parallel as a tube, the tips of the reflexed lobes pointing directly back and mostly 2–6(–10) mm. long. . *f.*

f.Petals 1–2.5 cm. long; calyx-lobes 1–2.5 cm. long; styles mostly 0.3–1.7 cm. long; wide-ranging native. . . . . . . 1. *O. biennis.*

f.Petals 3–6 cm. long; calyx-lobes 2.2–5 cm. long; styles mostly 1.8–5 cm. long.

Ovaries and capsules villous, ascending; petals 4–6 cm. long; locally spread from cult. . . . . 2. *O. grandiflora.*

Ovaries and capsules glabrous or essentially so, their bases divergent from the axis of the spike; petals 3–4 cm. long; native along mts., southw. . . . . 3. *O. argillicola.*

e.Tips of calyx-lobes in unexpanded bud not closely connivent, standing apart; tips of reflexed lobes deflected above the basal auricle or appendage and 1–3 (rarely –5) mm. long.

Petals linear, only 1–3 mm. wide; leaves membranaceous, minutely pilose beneath, spreading, spreading-ascending or reflexed; bracts soon deciduous, leaving naked fruiting spike; mature capsule 4.5–7 mm. thick at base. . . . . . . 4. *O. cruciata.*

Petals obovate, broader; leaves thickish to fleshy, stiff, mostly ascending, strigose to glabrous beneath; bracts persistent, the fruiting spike leafy-bracted; mature capsule 6–10 mm. thick at base. . . . . 5. *O. parviflora.*

d.Ovules and seeds ascending; seeds subcylindric or slenderly obovoid, pitted, not sharply angled; capsules slenderly cylindric; annual, biennial or perennial. Subgen. 2. Raimannia. . *g.*

g.Flowers in axils of entire or only sinuate upper leaves, in fruit forming spikes.

Cauline leaves and bracts merely acute or acutish; tips of calyx-lobes subacute, densely spreading-villous; filaments flattened, 7–10 mm. long; fruiting spike open; capsules 1.5–3.5 cm. long; plant of coastal sands, southeastw. . . . . 6. *O. humifusa.*

Cauline leaves sharply acuminate; tips of calyx-lobes slenderly attenuate, appressed-strigose; filaments filiform, 1–1.5 cm. long; fruiting spike dense; capsules 1–2 cm. long; prairie species, natzd. eastw. . . . . 7. *O. rhombipetala.*

g.Flowers in axils of sinuate-pinnatifid leaves, these persistent on the leafy-bracted interrupted fruiting summits. . . . . 8. *O. laciniata.*

---

* The treatment of subgenera 2–8 based largely on the studies by Philip A. Munz in Am. Journ. Bot. xvi–xxii (1929–1935) and in Bull. Torr. Bot. Cl. lxiv (1937).

*c.*Seeds in 1 row in each locule, smooth, slenderly obovoid to lance-linear; flower-buds nodding; petals white, becoming reddish; perennial with creeping subterranean rhizome and exfoliating white epidermis; northwestern. Subgen. 3. ANOGRA. . . . . 9. *O. Nuttallii.*

*b.*Stigma disk-like, barely 4-lobed; calyx-tube funnelform, 4-angled, the lobes prominently keeled; stamens unequal, the longer ones alternate with the petals. Subgen 4. CALYLOPHIS. . . . . 10. *O. serrulata.*

*a.*Capsules clavate, obovoid, ovoid, ellipsoid or roundish, 4-angled or -winged. . . *h.*

*h.*Capsule clavate to obovoid, 4-angled, usually tapering below to a sterile pedicel-like base or stipe; seeds clustered in the locules; leafy-stemmed; flowers diurnal. . . *i.*

*i.*Flower-buds erect (or young axis recurving in no. 15); calyx-tube filiform; fresh petals yellow; capsule 4-angled, not prominently ribbed; seeds clustered in each locule, not in distinct rows; wide-ranging e. of Great Plains. Subgen. 5. KNEIFFIA. . . *j.*

*j.*Cauline leaves linear-filiform, 0.5–1 mm. broad; lower bracts shorter than capsule; calyx-tube 1.5–2 mm. long; petals 3–5 mm. long; capsule ellipsoid, with ridge-like wings; stigma-lobes short and thick; annual. § PENIOPHYLLUM. . . . . 11. *O. linifolia.*

*j.*Cauline leaves broadly linear to oblanceolate, oblong or ovate, broader; lower bracts often exceeding capsule; calyx-tube 0.4–2.5 cm. long; petals 0.5–3 cm. long; capsules narrowly winged; stigma-lobes elongate, linear; perennials. § EUKNEIFFIA. . *k.*

*k.*Inflorescence compact at summits of stems and branches, erect in bud; calyx-tube 0.5–2.5 cm. long, the lobes 0.5–2 cm. long; petals 1–3 cm. long; anthers 4–8 mm. long. . . *l.*

*l.*Capsule without gland-tipped hairs, variously pubescent, clavate.

Mature capsule clavate, sessile or short-stipitate; calyx-tube 1.5–2.5 cm. long, the divaricate subulate tips mostly 2–3 mm. long; anthers 5–8(–10) mm. long. . . . . 12. *O. pilosella.*

Mature capsule with clavate to ovoid, obovoid or subglobose body narrowed to a definite stipe; calyx-tube 0.5–1.5 cm. long, the scarcely to slightly divaricate tips 0.5–2.5 mm. long; anthers 4–6 mm. long. . . . . 13. *O. fruticosa.*

*l.*Capsule bearing gland-tipped hairs, its short-stipitate body ellipsoid or quadrate-oblong. . . . . 14. *O. tetragona.*

*k.*Inflorescence of scattered flowers, the axis and tip drooping in bud, finally straightening; calyx-tube 4–8 mm. long, lobes 5–8 mm. long; petals 5–9 mm. long; anthers 1.2–2 mm. long; stipe of clavate to ellipsoid capsule 2–4 mm. long. . . . . 15. *O. perennis.*

*i.*Flower-buds drooping or nodding; calyx-tube enlarged upward; petals white or roseate; capsule strongly 8-ribbed; seeds clustered, in more than 2 rows in the locules; western and southern; introd. eastw. Subgen. 6. HARTMANNIA. . . . . 16. *O. speciosa.*

*h.*Capsule ovoid, ellipsoid or suborbicular, strongly winged; seeds in 1 or 2 definite rows in each locule; flowers vespertine; western.

Caulescent, with alternate entire or repand leaves: petals yellow; fruit ellipsoid to suborbicular, 5–8 cm. long; seeds in 1 row, crested. Subgen. 7. MEGAPTERIUM. . . . . 17. *O. missouriensis.*

Acaulescent, the rosulate leaves sinuate or runcinate-pinnatifid; petals pale yellow to white or roseate; fruit ovoid to ellipsoid, 2–3.5 cm. long; seeds in 2 rows, crestless. Subgen. 8. LAVAUXIA. . . . . 18. *O. triloba.*

Subgen. 1. OENOTHÈRA proper (ONAGRA Ludwig) EVENING-PRIMROSES *

1. O. biénnis L. (biennial). — Rather stout; rosettes of first year developing many elongate leaves and strong fleshy roots; stem of second year erect, simple or branching, especially above, 0.15–1.5 m. high, the green to purple-tinged stem more or less villous or hirsute to glabrous;

* A hopelessly confused and freely hybridizing group, early introduced into Europe and there cultivated and, like other plants of the garden, intermixed; then spreading to waste or open ground. The types of several species, described from European material, not wholly clarified; and further confusion added by the publication of many scores of "mutants" or "elementary species" as true "species". (Compare other strongly apomictic groups of similar behavior, such as *Rubus*, subgen. *Eubatus*, and European *Taraxacum officinale* and *Hieracium*.) These, for the most part, not specially considered here.

leaves ascending or spreading, lanceolate to oblong- or (rarely) ovate-lanceolate, repand-denticulate, acute or acuminate; *bracts of the prolonged spikes* lanceolate, shorter than to slightly longer than capsules, *finally usually deciduous and leaving the fruiting spike bractless;* calyx-tube 1.8–4.4 cm. long; *calyx-lobes in unexpanded buds with slender closely connivent or parallel tips* 2–6 *mm. long, these forming a tube at base; body of calyx more or less villous,* the tips strongly pubescent; *reflexed expanded calyx-lobes* 1–2.5 *cm. long, their tips arching or extended straight back; petals* obovate, yellow (in age often purplish), 1–2.5 *cm. long;* anthers 3–11 mm. long; *style* (0–)0.3–1.7 *cm. long,* the lobes of the stigma 2–8 mm. long; *capsule lance-cylindric, tapering at tip, strongly appressed-ascending,* 1–3.5 *cm. long; mature seed* 1.2–1.8 *mm. long,* about 0.8 *mm. broad.* — Polymorphic, transcontinental; the best marked vars. are as follows:

Surface of calyx-lobes, ovaries and capsules evident, the trichomes scattered or wanting; leaves membranaceous, soft-pilose beneath, spreading or loosely ascending.
  Bracts of inflorescence finally deciduous, leaving a naked fruiting spike.
    Body of calyx more or less villous; capsule villous or hirsute, 1.5–3.5 cm. long. . . . . *O. biennis* (typical).
    Body of calyx glabrous or essentially so, often viscid; glabrescent capsule 1–2.5 cm. long. . . . . Var. *nutans.*
  Bracts persistent in fruit; otherwise much as in typical var. . . . . . . Var. *pycnocarpa.*
Surface of calyx-lobes, ovaries and capsules hidden by dense canescent or whitish appressed or strigose-villous pubescence; leaves firm, strongly ascending.
  Pubescence of capsule and calyx of closely appressed short strigae. . . . . Var. *canescens.*
  Pubescence of capsule and calyx with many loosely ascending to spreading long villi. . . . . Var. *hirsutissima.*

O. biénnis (typical), incl. *O. comosa, grandifolia, Hazelae, novae-scotiae, parva* and *Royfraseri* R. R. Gates; *O. Victorini* Gates & Catcheside; and several others. — Dry open soil, Nfld. and Côte Nord, Que., to se. Alta., s. to N.S., N.E., L.I., n. Fla., Tenn., Ark., N.D. and Id. June–Oct.

Var. **nùtans** (Atkinson & Bartlett) Wieg. (nodding; from the drooping of the wilting flowers). — Very definite in its glabrous parts. (*O. nutans* Atkinson & Bartlett) — N.Y. to Ga.

Var. **pycnocárpa** (Atkinson & Bartlett) Wieg. (with crowded fruits). — In its persistent bracts suggesting no. 5, but otherwise as in no. 1. (*O. pycnocarpa* Atkinson & Bartlett) — N.E. to Minn. and southw.

Var. **canéscens** T. & G. (canescent). — Strongly marked by its appressed canescence. (*O. eriensis, niagarensis* and *repandodentata* R. R. Gates) — S. Que. and e. N.B. to se. Alta., s. to w. N.Y., s. Ont., O., Ill., Mo. and Okla. — Passing insensibly into

Var. **hirsutíssima** Gray (most hirsute). — More wide-spread than the preceding. (*O. strigosa* (Rydb.) Mackenz. & Bush) — S. Que. to w. B.C., s. to P.E.I., N.B., s. N.E., N.J., Pa., Mich., Ill., Kans., Tex., N.M., Ariz. and Mex.

2. O. grandiflòra Ait. (large-flowered). — *Differing from no.* 1 *in its longer flowers; the broadly obovate petals* 4–6 *cm. long; calyx-tube* 3–5 *cm. long, body of calyx villous, the lobes* 2.5–5 *cm. long, the slender tips* (3–)5–10 *mm. long; anthers* 0.7–1.5 *cm. long; style* (0.7–)1.8–5 *cm. long, the lobes of the stigma up to* 1 *cm. long.* (Incl. *O. Lamarckiana* Ser.) — Nat. s. of our range, spread from cult. to fields, etc., Que. and southw. July–Nov. (Introd. and natzd. via Eu.)

3. O. argillícola Mackenz. (growing in clay). — *Glabrous* or essentially so *throughout,* up to 1.5 m. high; *leaves very numerous, linear-lanceolate, long-attenuate,* 5–15 *mm. wide,* ascending; *petals* 3–4 *cm. long,* broadly obovate; *calyx-tube* 3.5–4.5 *cm. long; body of calyx* much as in nos. 1 and 2, *glabrous or nearly so, the lobes* 2.2–3.5 *cm. long,* with tips 3–6 mm. long; *anthers* 8–12 *mm. long; style* 2.2–3.2 *cm. long,* with stigma-lobes 2.5–6.5 mm. long; *capsules abruptly divergent at base,* gradually upcurved. — Dry shale, slate and clay, upland of Pa., Va. and W.Va. Late June–Oct.

4. O. cruciàta Nutt. (cross-shaped). — Habitally like no. 1, simple or with ascending branches, 0.15–1 m. high, the *stem commonly reddish and villous-strigose; cauline leaves lanceolate, acute, repand-denticulate,* mostly 1–2.5 cm. wide, *membranaceous or submembranaceous,* minutely pilose beneath, *remote* (*internodes prolonged*); *bracts* shorter than to twice as long as ovaries, *promptly deciduous, leaving a naked interrupted fruiting spike; petals linear,* 1–3 *mm. broad,* 5–12 *mm. long;* calyx-tube 2.5–3.5 cm. long; *body of calyx with scattered villi,* the *free tips distinct to base; expanded and reflexed calyx-lobes* 1–4.5 cm. long, *with a prominent corniculate appendage below the deflected tip, the latter* 1.5–3.5 *mm. long;* anthers 5–7 mm. long; *style* 0.35–1.2 *cm. long,* the lobes of the stigma 1.5–6 mm. long; capsule much as in no. 1, 2–3.5 cm. long, 4.5–7 mm. thick at base, loosely spreading-villous; mature seed 1.6–2 mm. long, 1–1.5 mm. broad. — Dry open soil, N.E. to Mich. July–Sept.

Var. **sabulonénsis** Fern. (of Sable I., N.S.). — *Stem* 3–3.5 dm. high, simple, *canescent-strigose with appressed short hairs; leaves oblong, bluntish,* mostly 1.5–2.5 cm. broad, *crowded, scarcely dentate; body of calyx sparsely and minutely pilose; capsules* 8–10 *mm. thick,* loosely villous.— Sand-dunes, Sable I., N.S. July, Aug.

Var. **stenopétala** (Bickn.) Fern. (narrow-petalled). — *Very slender,* 1–3.5 dm. high; *stem cinereous with crowded minute appressed pubescence; leaves remote, narrowly lanceolate or oblanceolate, acute, repand-denticulate,* 0.4–1.5 *cm. broad, closely cinereous-strigose beneath with short strigae; calyx and capsule similarly cinereous-strigose.*— Nantucket and Martha's Vineyard Ids., Mass.

5. O. parviflòra L. (small-flowered). — Differing from no. 1 in the *simple or rarely branching stem* 1–8 dm. high, *strigose-puberulent, usually beset at least above with longer spreading hairs on enlarged reddish pustular bases; cauline leaves narrowly lanceolate, ascending,* shallowly repand-denticulate, *rather fleshy* (when fresh) *and thick, strigose to glabrate beneath, passing without evident interruption into the persistent foliaceous bracts;* lower flowers much overtopped by their bracts; calyx-tube 1.5–3 cm. long, the body of *unexpanded calyx* more or less *strigose-villous, the tips free to base* and 1–5 mm. long; expanded and *reflexed calyx-lobes* 0.65–2.3 *cm. long, with an evident auricle below the slightly deflected tip,* somewhat as in no. 4; *petals obovate,* 1.2–2 cm. long; anthers 4–7.5 mm. long; *style* (0–)0.45–2.5 *cm. long,* the stigma-lobes 2.5–6 mm. long; capsules subfusiform-cylindric, 2–4 cm. long, more or less scattered-villous; *seeds mostly* 1.7–2.2 *mm. long* and 1–1.5 mm. thick. (*O. muricata* L.; *O. angustissima,* var. *quebecensis* R. R. Gates) — Gravelly shores, talus (often calcareous), sands and dry openings, Nfld. and Côte Nord, Que. to James Bay and Thunder Bay Distr., Ont., s. to N.S., N.E. and N.Y. July–Sept.

Var. **angustíssima** (R. R. Gates) Wieg. (narrowest). — Stem, leaves, calyx and capsules glabrous or essentially so. (*O. angustissima* R. R. Gates) — Que. to w. N.Y., s. to D.C. and W.Va.

Var. **Oakesiàna** (Robbins) Fern. (for its discoverer, William Oakes, 1800–1848). — *Stem, lower leaf-surfaces, backs of calyx-lobes and capsules closely and minutely puberulent with short appressed hairs,* rarely with longer strigae; leaves narrowly to broadly lanceolate; *auricle near summit of calyx-lobe often longer and more conical.* (*O. Oakesiana* Robbins; *O. Tidestromii* Bartlett) — Sands along coast and rarely inland, Plymouth and Worcester Cos., Mass., to Northampton Co., Va. July–Oct.

Subgen. 2. Raimánnia (Rose) Munz

6. O. humifùsa Nutt. (spreading on the ground), Seabeach-E. — *Hoary with short dense appressed pubescence;* stem simple or divergently branched, ascending to depressed, 1–7.5 dm. high, *suffrutescent,* the older epidermis peeling off; *cauline leaves* firm, oblong-lanceolate or -oblanceolate, 1–4 cm. long, *sparingly repand-dentate or entire, merely acute or acutish;* basal leaves pinnatifid; *bracteal leaves scarcely reduced; flowers nocturnal, in axils of upper leaves, forming an open spike in fruit; calyx-*tube 2–3 cm. long; *lobes* distinct, 0.8–1.2 cm. long, *the densely spreading-villous tips subacute;* petals obovate, yellow, changing to reddish, 0.8–1.5 cm. long; *filaments flattened,* 7–10 mm. long; *capsules* cylindrical, sessile, curved, densely strigose, 1.5–3.5 *cm. long; seeds* oblong-ovoid, about 1.2 mm. long, *obscurely pitted.* (*Raimannia* Rose) — Sandy seabeaches and dune-hollows, Fla. to La., n. to N.J. June–Sept. (Bermuda)

7. O. rhombipétala Nutt. (with rhombic petals). — Differing from no. 6 in more generally erect stem up to 1 m. high; *cauline leaves sharply acuminate,* 2–7 cm. long; bracts 5–15mm. long; *flowers forming dense spikes; calyx-lobes* mostly united in anthesis, often reddish, 1–1.8 cm. long, *their slenderly attenuate tips appressed-strigose; petals* rhombic-obovate, 1.3–2.5 *cm. long; filaments filiform,* 1–1.5 *cm. long; capsules* strigose, 1–2 *cm. long; seeds finely pitted.* (*Raimannia* Rose) — Sandy prairies, Fla. to e. Tex., n. to Ga., Ind., Wisc., Minn. and N.D.; natzd. eastw. to N.Y. and N.J. June–Oct.

8. O. laciniàta Hill (slashed). — *Greener* than nos. 6 and 7, depressed or ascending, simple or branching, *more or less strigose-villous or -hirsute; leaves sinuate-pinnatifid* (rarely subentire), 2–10 *cm. long; bracts similar, much longer than ovaries and capsules, persistent;* calyx-tube 1.5–3.5 cm. long; the lobes 6–12 mm. long, with free tips 1–2 mm. long; petals yellow or whitish, becoming reddish, 0.5–1.8 cm. long; filaments 5–10 mm. long; anthers 2–5 mm. long; capsules 2–3 cm. long; *seeds strongly pitted.* (*Raimannia* Rose) — Sandy open ground, N.J. to N.D., s. to Fla., La. and Tex.; adv. northeastw. to N.Y. and N.E. May–Oct.

Var. **grandiflòra** (S. Wats.) Robins. (large-flowered). — Calyx-tube 2.5–5 cm. long, the lobes 2–3 cm. long and with free tips 2–5 mm. long; petals 2–3.5 cm. long; filaments 1.5–2 cm. long; anthers 8 mm. long. — Tex., N.M. and ne. Mex., ne. to Mo. and Kans.

Subgen. 3. Anogra (Spach) Jepson

9. O. Nuttállii Sweet (for its discoverer, Thomas Nuttall, 1786–1859), White E. — Perennial *with subterranean rhizomes;* stems erect, 1.5–12 dm. high, often branched above, *with white exfoliating epidermis; leaves* pale, linear to oblong-oblanceolate, 2–10 cm. long, entire or remotely denticulate, *glabrous above,* strigose beneath, sessile or petioled; *flower-buds nodding;* flowers with disagreeable odor; *calyx-tube* 2–4 cm. long, enlarged at base, *glandular;* lobes glandular, adhering, 2–3 cm. long; *petals white, becoming roseate,* obcordate, suborbicular, 1.5–2.5 cm. long; *capsules glandular,* 2–3 cm. long; *seeds slenderly obovoid, about 2 mm. long, nearly smooth.* (*Anogra* Nels.; *O. pallida* of ed. 7, not Lindl.) — Dry plains and prairies, Man. to B.C., s. to Wisc., Minn., Neb. and Colo. June–Sept.

Subgen. 4. Calýlophis (Spach) Munz

10. O. serrulàta Nutt. (finely serrate). — *Suffrutescent* to herbaceous, *canescent,* slender, simple to much branched and cespitose, 1–5 dm. high; leaves linear to linear-lanceolate, 2–6 cm. long, subentire or shallowly denticulate, not much reduced upward; flowers in upper axils; *calyx-tube funnelform, 4-angled,* 0.5–1.5 cm. long, *with broadly dilated strongly nerved throat,* the ovate *prominently keeled lobes* 3–10 *mm. long;* petals yellow, 0.3–2.5 cm. long; *filaments of 2 lengths, the longer ones alternating with petals; stigma disk-like, shallowly 4-lobed,* 1–3 *mm. across; capsule* cylindric, 1–2.5 cm. long, 1–1.5 *mm. thick; seeds sharply angled,* acute at base, broadened upward. (*Meriolix* Walp.; *M. intermedia* Rydb.) — Sandy or rocky soil, Man. to Mont., s. to Wisc., Mo., Tex. and N.M.; locally adv. e. to N.E. May–Sept.

Subgen. 5. Kneìffia (Spach) Munz Sundrops

§ Peniophýllum (Pennell) Munz

11. O. linifòlia Nutt. (flax-leaved). — Erect simple or diffusely branched *slender annual or biennial* 1–4.5 dm. high, *glabrous except for the strigillose-puberulent inflorescence; cauline leaves linear-filiform,* 0.5–1 *mm. wide; open-spiciform inflorescence peduncled, the greatly abbreviated bracts mostly shorter than the capsules; calyx-tube* 1.5–2 *mm. long,* filiform; *the lobes* reflexed in pairs, 1.5–2 *mm. long, puberulent; petals* yellow, obcordate, 3–5 *mm. long; stigma with thick lobes, about* 1 *mm. across; capsule ellipsoid,* sessile, 4–6 mm. long, *with ridge-like wings.* (*Kneiffia* Spach; *Peniophyllum* Pennell) — Prairies and dry slopes, Ga. to Tex., n. to s. Ill., Mo. and Kans. May–July.

§ Eukneìffia Munz

12. O. pilosélla Raf. (finely pilose). — Perennial, erect, simple or with few ascending upper branches, 0.3–1 m. high, the *stem copiously spreading-villous with divergent hairs up to* 2 *or* 3 *mm. long;* cauline leaves oblong-lanceolate to -ovate, 3–13 cm. long, 1–4 cm. broad, bluntish, minutely and remotely denticulate, more or less covered with long villi or glabrescent above; *flowers* few at tips of stem and branches, *in axils of only slightly reduced leaves; calyx-tube* filiform, 0.8–2.5 *cm. long;* body of unexpanded *calyx* green or purplish, more or less villous; *lobes* united in 2's or 4's in anthesis, *with divergent villous-hirsute free tips mostly* 2–3 *mm. long;* petals obcordate, 1.3–3 cm. long; *anthers* 5–8(–10) *mm. long; capsule* thick-clavate, *sessile or short-stipitate* (stipe sometimes up to 5 mm. long), spreading-villous. (*Kneiffia* Heller; *O. pratensis* (Small) Robins.; *O. fruticosa,* var. *hirsuta* Nutt.) — Open woods, low prairies and meadows, s. Ont. to Mich. and Ill., s. to w. Pa., W.Va., O., Ind., Ill. and Ark.; natzd. e. to N.E., e. Pa. and Va. June, July.

13. O. fruticòsa L. (shrubby; a misnomer). — Perennial herb; stems erect, ascending or (in two vars.) diffusely clustered or cespitose; first year's rosettes with ovate to spatulate petioled leaves; cauline leaves lanceolate, oblong or linear; inflorescence much as in no. 12; *calyx-tube* 0.5–1.5 *cm. long; the body appressed- to spreading-pubescent, the hairs not gland-tipped; the calyx-lobes reflexed in 4's, their scarcely to barely divaricate tips* 0.5–2.5 *mm. long; anthers* 4–6 *mm. long; capsule clavate, tapering to a slender stipe; the body oblanceolate to ellipsoid or obovoid-subglobose, its pubescence not gland-tipped.* — Polymorphous; the following vars. recognized with us:

Stems ascending to erect, simple to branched above, mostly 3–10 dm. high.
Slender tips of calyx-lobes connivent or not unguiculate, at most about 1 mm. long.
Body of capsule 6–11 mm. long, much longer than thick, with appressed

strigose canescent pubescence or with divergent pubescence; longer stipes mostly 5–15 mm. long.

Pubescence of calyx, capsule and stipes appressed-strigose. . . . . ***O. fruticosa*** **(typical).**

Pubescence spreading. . . . . **Var.** ***linearis.***

Body of capsule 3–6 mm. long, one-half to three-fourths as thick as long, covered by crisp pilosity; stipes 0.5–5 mm. long. . . . . **Var.** ***microcarpa.***

Slender tips of calyx-lobes free, arching, 1–2.5 mm. long; body of calyx appressed-strigose; body of capsules 4–10 mm. long, slenderly obconic, spreading-villous, on villous stipes up to 1.5 cm. long. . . . . **Var.** ***unguiculata.***

Stems depressed, freely branching from base, forming mats 1–2 dm. deep.

Branches, leaves, calyx and capsule canescent with appressed strigae; body of capsule 5–8 mm. long. . . . . **Var.** ***humifusa.***

Branches, leaves, calyx and capsule copiously spreading-hirsute; body of capsule 7–12 mm. long. . . . . **Var.** ***Eamesii.***

O. fruticòsa (typical), var. *vera* Hook.; *Kneiffia* Raimann — Fresh to brackish marshes, swales, meadows, fields and borders of woods, Fla. to Okla., n. to s. N.E., N.Y., Tenn. and Mo. May–Aug.

Var. **lineàris** (Michx.) S. Wats. (linear), incl. *O. linearis* Michx., *O. longipedicellata* (Small) Robins. and *O. frut.*, forma *angustifolia* Léveillé — Similar habitats, n. to s. N.E., L.I., N.J., Md., Mich. and Mo.

Var. **microcárpa** Fern. (tiny-fruited). — Moist or dry sterile soil, e. S.C. to serpentine-barrens of Cecil Co., Md.

Var. **unguiculàta** Fern. (with claws). — Dry to moist woods and openings, se. to sw. Va., s. to S.C.

Var. **humifùsa** Allen (spreading on the ground). — Sands and dryish coastal marshes, Stratford Co., Ct.; e. L.I.; fallow field in pineland, Sussex Co., Va., 1936, the station destroyed by cult.

Var. **Eàmesii** (Robins.) Blake (for its discoverer EDWIN HUBERT EAMES, 1865–1948), *O. linearis*, var. Robins. — Dry saline meadows, Stratford Co., Ct.; specimens transitional to the last on L.I.

14. O. tetragòna Roth (4-angled).— Differing from nos. 12 and 13 in having gland-tipped hairs on calyx and capsule or these merely viscid but glabrous; body of capsule more generally oblong-quadrate or quadrate-ellipsoid, abruptly narrowed to short stipe or sessile base. — Very variable; the following vars. recognized:

a.**Ovary and capsule with glandular hairs only. . . .** ***b.***

b.**Leaves not glaucous beneath, the cauline lanceolate to lance-linear, mostly 0.5–1.5(–2) cm. broad.**

Stems and branches spreading-villous to glabrescent; capsule oblong, subsessile or short-stipitate. . . . . ***O. tetragona*** **(typical).**

Stems and branches appressed-pubescent; capsule oblanceolate to obovoid and tapering to stipe. . . . . **Var.** ***longistipitata.***

b.**Leaves paler or glaucous beneath, the cauline broadly ovate to oval or lanceolate. . .** ***c.***

c.**Plant glabrous throughout; leaves oval to broadly ovate, the principal ones 2–4 cm. broad, with 1–3(–4) low teeth per 2 cm. of margin, mostly without axillary fascicles; petals 2–3 cm. long. . . . .** **Var.** ***Fraseri.***

c.**Plant more or less pubescent; leaves lanceolate to lance-ovate, the larger ones 1–2 cm. broad, with 4–6 teeth per 2 cm. of margin, mostly subtending axillary fascicles; petals 1–3 cm. long.**

Stem, leaves, calyx and capsule with spreading hairs. . . . . . **Var.** ***hybrida.***

Stem, leaves, calyx and capsule more or less appressed-strigose. . . **Var.** ***latifolia.***

a.**Ovary and capsule with mixed glandular and non-glandular hairs; leaves densely subvelutinous with crowded strigae, narrowly ovate to lanceolate, mostly subtending axillary fascicles. . . . .** **Var.** ***velutina.***

O. tetragòna (typical), *O. fruticosa*, var. *ambigua* Nutt.; *O. hybrida*, var. *ambigua* Blake — Swales, meadows and damp thickets, N.Y. to Ill., s. to Ga. and Tenn., natzd. e. to e. Me. and N.S. June–Aug.

Var. **longistipitàta** (Pennell) Munz (with long stipes). — Ct. to Mich., s. to Ga. and Tenn.

Var. **Fràseri** (Pursh) Munz (for its discoverer, JOHN FRASER, 1750-1811). Incl. *O. glauca* Michx. and *O. fruticosa*, var. *Fraseri* Hook. — Upland of Va. and W.Va. to (?)O., s. chiefly along the mts. to N.C. and e. Tenn. May–Aug.

Var. hýbrida (Michx.) Fern. (hybrid; a reasonable guess in this genus). Incl. var. *Fraseri*, forma *hybrida* Munz and *O. hybrida* Michx. — Upland of N.C. and Tenn.; outer Piedmont and inner Coastal Plain, se. Va. Late May–July.

Var. latifòlia (Rydb.) Fern. (broad-leaved). Incl. *Kneiffia latifolia* Rydb. and *O. tetragona*, var. *Fraseri*, forma *latifolia* Munz — Pa. and W.Va., s. to w. N.C. and e. Tenn.

Var. velùtina (Pennell) Munz (velvety). — Sands of L.I.; dry to damp sandy or argillaceous soil, e. Va. June, July.

15. O. perénnis L. (perennial). — Perennial, *usually retaining basal rosettes;* stems 1–several, erect or ascending, 0.5–8 dm. high, simple or *with* ascending *strigose-puberulent to glabrescent branches;* rosette-leaves oblanceolate or spatulate, obtuse; *cauline leaves linear-lanceolate to oblanceolate, entire or nearly so, strigulose or glabrescent,* the upper gradually reduced to narrow bracts; *inflorescence slender, interrupted, in larger mature plants becoming* 1–3 *dm. long, in bud with recurved tip, glandular-puberulent; calyx* often reddish, the slender *tube* 4–8 *mm. long,* the *lobes* 5–8 *mm. long; petals* 5–9 *mm. long,* obcordate; *anthers* 1.2–2 *mm. long; stigma-lobes thick,* 1.5–2 *mm. long; capsule* ellipsoid- or oblong-*clavate,* 4–10 *mm. long and* 3–3.5 *mm. thick, the stipe* 2–4 *mm. long.* (*O. pumila* L.) — Dry to moist open ground, Nfld. to e. Man., s. to N.S., N.E., L.I., Del., e. Va., upland to Ga., W.Va., O., Ind. and Mo. Late May–Aug.

Var. rectípilis Blake (with straight hairs). — Hairs of stem and branches spreading. — Local, ne. N.B.; Niagara reg. of Ont.

Subgen. 6. HARTMÁNNIA (Spach) Munz

16. O. speciòsa Nutt. (showy), WHITE E. — Erect to spreading *perennial with creeping subterranean rootstock,* often suffrutescent; slender stems and branches pubescent; *leaves oblong-lanceolate to linear, repand-dentate or subpinnatifid; flowers nodding in bud,* few to several in upper axils, *vespertine; calyx-tube enlarged upward,* 1–2 cm. long, the long-acuminate narrow lobes 1.5–3 cm. long, coherent, with free tips; *petals white to roseate,* obcordate, 2–4 cm. long; *capsule clavate-obovoid, strongly 8-ribbed,* the fertile body 3–5 mm. thick and attenuate to a slender base; *seeds clustered in more than 2 rows in each locule.* (*Hartmannia* Small) — Prairies and plains, Mo. and Kans. to Tex. and Mex.; spread, partly through cult. and natzd. from Ill. to La., eastw. to Va., N. and S.C., Ga. and Fla. May–July.

Subgen. 7. MEGAPTÈRIUM (Spach) Munz

17. O. missouriénsis Sims (of Missouri). — *Perennial from deep root,* the short *leafy stems* more or less *hoary; leaves lanceolate to lance-linear, green, acuminate to apex and petiole; flowers erect in bud; calyx-tube enlarged at summit,* 5–12 *cm. long,* the lobes 2–4 cm. long; *petals yellow,* often drying paler or reddish, 3–6 *cm. long; anthers linear,* 1.2–2 *mm. long; stigma-lobes linear,* 5–7 *mm. long; fruit ellipsoid to suborbicular,* 5–8 *cm. long, the wings broad; seeds in* 1 *row, crested.* (*Megapterium* Spach) — Rocky or sandy soil, Mo. and Kans. to Tex. May–Sept.

Subgen. 8. LAVAÙXIA (Spach) Jepson

18. O. tríloba Nutt. (three-lobed). — *Acaulescent winter-annual or biennial, with a rosette of oblanceolate runcinate-pinnatifid leaves up to* 3 *dm. long; flowers basal,* axillary; *calyx-tube gradually expanding upward,* 2–15 cm. long; the lobes distinct or united, 1–2 cm. long; *petals round-obovate, pale yellow to whitish or roseate,* 1–4 cm. long; anthers 5–9 mm. long; *fruit ovoid to ellipsoid,* 2–3.5 *cm. long, strongly winged; seeds in* 2 *rows, crestless.* (*Lavauxia* Spach) — Dry, often calcareous, ground, w. Va. to Kans., s. to Ala., Tenn., Okla. and Tex. Late April, May.

5. GAÙRA L. GAURA

Calyx-tube much prolonged beyond the ovary, deciduous; the lobes 4 (rarely 3), reflexed. Petals clawed, unequal or turned to the upper side. Stamens mostly 8, often turned down, as is also the long style. A small scale-like appendage before the base of each filament. Stigma 4-lobed, surrounded by a ring or cup-like border. Fruit hard and nut-like, 3–4-ribbed or -angled, indehiscent or nearly so, usually becoming 1-locular and 1–4-seeded. Seeds naked. — Small Am. herbs; leaves alternate, sessile. Flowers rose-color or white, changing to reddish in fading, in spikes or racemes, in our species quite small (so that the name, from the Greek *gauros, superb,* does not seem always appropriate).

Ovary and capsule sessile or nearly so.

Plant villous, downy or strigose; cauline leaves oblong- to ovate-lanceolate, the larger ones 1–5 cm. wide; fruit 4-angled to base.

Calyx-segments 6–8 mm. long; petals 5 mm. long; anthers linear, 2.5–3 mm. long; capsule 5–7 mm. long, ovoid-fusiform. . . . . 1. *G. biennis.*
Calyx-segments 1.5–3 mm. long; petals 1.5–2 mm. long; anthers oval, 1 mm. long; capsule 6–10 mm. long, slenderly fusiform. . . . . . . 2. *G. parviflora.*
Plant minutely canescent to glabrate; cauline leaves linear to narrowly lanceolate, the larger ones 4–8 mm. wide; fruit terete at base. . . . . 3. *G. coccinea.*
Ovary and capsule narrowed below to stipe-like terete base 1.5–3 mm. long. . 4. *G. filipes.*

1. G. biénnis L. (biennial). — Biennial or winter-annual, *villous and downy,* up to 3.5 m. high, with divergently branched panicle of spikes; rosette-leaves oblanceolate, 1–3 dm. long; *cauline leaves oblong-lanceolate, tapering at both ends,* remotely denticulate, the larger ones 3–10 cm. long and 1–2.5 cm. broad, *glabrate,* frequently with axillary fascicles; *spikes* wand-like, *bearing both gland-tipped and non-glandular hairs; calyx-tube* slender, 5–7 *mm. long; calyx-segments* 6–8 *mm. long,* reddish, *reflexed in pairs; petals* white, becoming pink, *about* 5 *mm. long; anthers linear,* 2.5–3 *mm. long; fruits sessile, ovoid-fusiform, obtusely angled to base,* 5–7 *mm. long.* — Damp shores, meadows, etc., w. Que. to Minn., s. to Va., w. N.C., Tenn. and Mo.; adv. in N.E. June–Oct.

Var. **Pítcheri** Pickering (for Zina Pitcher, 1797–1872). — Stems merely short-pubescent; leaves canescent; inflorescence without glands. — Ill. to Neb., s. to Ark., Okla. and Tex.

2. G. parviflòra Dougl. (small-flowered). — Less branched than no. 1, the 1–few and wand-like spikes more strict; *cauline leaves lanceolate to lance-ovate, up to 5 cm. broad; calyx-tube* 1.5–3 *mm. long,* minutely puberulent, or in forma **glàbra** Munz (smooth) glabrous; *calyx-segments* greenish to reddish, 1.5–3 *mm. long, reflexed separately during anthesis; petals* pink, 1.5–2 *mm. long; anthers oval,* 1 *mm. long; fruits slenderly fusiform,* 6–10 *mm. long,* glabrous, or in forma **lachnocárpa** Weath. (hairy-fruited) spreading-puberulent. — Dry prairies, open places and waste ground, Tex. and n. Mex., n. to Ind., Ill., Ia., S.D., Wyo., Ida. and Wash.; casual weed e. to N.E. June–Oct.

3. G. coccínea Pursh (scarlet). — *Perennial* with several stems 1–3 dm. high, *canescent, either puberulent or glabrate,* densely leafy; *leaves linear to narrowly lanceolate, the larger primary ones* 4–8 *mm. wide;* flowers in simple *sessile spikes* up to 1 dm. long, *terminating stems and branches;* calyx-tube 5–8 mm. long, slenderly funnelform; calyx-segments separately reflexed, 5–8 mm. long; petals white to red, 3–6 mm. long; *anthers* 2–3 *mm. long; stigmatic lobes* 0.25 *mm. long,* rounded; *fruit* sessile, 5–7 mm. long, the 4-angled body *tapering into the* obpyramidal *terete base.* — Dry prairies, roadsides and waste places, Alta. to Calif., e. to Man., Minn., Mo., Okla. and Tex.; spreading as a casual weed e. to N.Y. May–Aug.

4. G. FÍLIPES Spach (slender-stalked). — *Perennial* with long erect stems up to 2 m. high, *nearly smooth* or somewhat puberulent, *openly and paniculately branched at summit,* the *branches very slender;* leaves crowded below, linear or nearly so; calyx-tube canescent-strigose, 3–5 mm. long; calyx-segments separately reflexed, 5–7 mm. long; petals 4–5 mm. long, white, becoming pink; *anthers* 1.5–2 *mm. long; stigmatic lobes* 0.4–0.8 *mm. long; fruit clavate-obovoid, acutely angled,* the body 4–5 mm. long, *narrowed to a slender stipe-like base* 1–3 *mm. long.* — Sandy pine-barrens, S.C. to Fla. and Ala. Represented with us by

Var. **màjor** T. & G. (larger). — Calyx-segments 8–10 mm. long; petals 6 mm. long. — Dry rocky or sandy woods or barrens, w. Fla. to La., n. to Ky. and s. Ind. July–Sept.

6. STENOSÌPHON Spach

Calyx prolonged beyond the ovary into a filiform tube. Fruit 1-locular, 1-seeded. Habit of *Gaura;* only one species. (Name from the Greek *stenos, slender,* and *siphon, a tube.*)

1. S. linifòlius (Nutt.) Britt. (flax-leaved). — Slender, 6–12 dm. high, glabrous, leafy; leaves narrowly lanceolate to linear, pointed, entire, much reduced above; flowers numerous in an elongated spike, white, 1.2 cm. long; fruit pubescent, ovoid, 8-ribbed, 2.5–4 mm. long. — Dry calcareous hills and prairies, Neb. and Colo., s. to Ark., Tex. and Mex. July–Sept.

7. CIRCAÈA L. Enchanter's Nightshade

Calyx-tube slightly prolonged, the end filled by a cup-shaped disk, deciduous; lobes 2, reflexed. Fruit indehiscent, small and bur-like, bristly with hooked hairs, 1–2-locular; locules 1-seeded. — Low perennials of temp. and cool areas of the N. Hemisph., with opposite leaves on slender petioles, and small whitish to roseate flowers in racemes, produced in summer. (Named for *Circe,* the enchantress, the plant-name originating with Dioscorides and transferred by later writers to the present modest — except for clinging fruits — genus.)

Stem firm; leaves dark green above, firm, oblong-ovate, at most subcordate, on subterete petioles; calyx-lobes 1.8–2.6 mm. broad; disk cup-like, prolonged about 0.5 mm. above perianth; mature fruit with 3–5 corrugations on each face, including bristles 3.5–6 mm. thick. . . . . 1. *C. quadrisulcata*, var. *canadensis.*

Stem rather weak; leaves pale green, mostly flaccid, ovate, cordate or subcordate (rarely only rounded to base), on channeled or margined petioles; calyx-lobes 0.8–1.7 mm. broad; disk inconspicuous; fruit not corrugated, 1–3 mm. thick.

Rhizome cord-like, 1–3 mm. thick, not tuberous-thickened; calyx-lobes 1.2–1.7 mm. broad; petals 2.3–3.5 mm. long; anthers 0.5–0.8 mm. long; fruit 2-locular, 1.5–3 mm. thick. . . . . 2. *C. canadensis.*

Rhizome tuberous-thickened; calyx-lobes 0.8–1.2 mm. broad; petals 1.2–3 mm. long; anthers 0.2–0.3 mm. long; fruit 1-locular, 1–1.5 mm. thick. . . 3. *C. alpina.*

1. C. quadrisulcàta (Maxim.) Franch. & Sav. (four-furrowed), var. **canadénsis** (L.) Hara (Canadian). — *Stem firm,* glabrous below, with slender horizontal rhizome and filiform stolons, 0.2–1 m. high; *leaves dark green* above, rather firm, *oblong-ovate,* 4.5–15 cm. long, *shallowly undulate-dentate,* rounded or merely subcordate at base, *on subterete petioles;* leading racemes becoming 0.7–2.5 dm. long in fruit; *calyx-lobes* 1.8–2.6 *mm. broad; disk cup-like, prolonged about* 0.5 *mm. above the perianth; petals* broadly obovate, as long as or longer than broad, cuneate at base, 2.5–3.6 *mm. long;* anthers 0.7–1 mm. long; *stigma subcapitate, shallowly 2-lobed; mature pedicels strongly reflexed,* 4–10 mm. long; ripe *fruit* compressed-pyriform, *with* 3–5 *corrugations on each face,* including the strong hooked bristles 3.5–6 *mm. thick.* (*C. lutetiana* of Am. auth., not L.; *C. latifolia* Hill) — Rich woods, thickets and ravines, N.S. and w. N.B. to s. Ont. and N.D., s. to Ga., Tenn., Mo. and Okla. Late June–early Aug. (Typical *C. quadrisulcata* is Asiatic.)

2. C. canadénsis Hill (Canadian). — *Rhizome slender,* scarcely tuberous-thickened; *stem succulent* and weak, 1.5–4.5 dm. high, usually branching; *leaves pale green,* flaccid, ovate, *coarsely sharp-dentate,* cordate or subcordate, 2.3–11 cm. long, *on channeled or margined petioles;* leading racemes becoming 0.3–1 dm. long in fruit; *calyx-lobes* 1.2–1.7 *mm. broad; disk inconspicuous, scarcely prolonged;* petals 2.3–3.5 mm. long; anthers 0.5–0.8 mm. long; *stigma deeply cleft; mature pedicels spreading or slightly reflexed,* 3–5 mm. long; ripe *fruit* clavate to slender-pyriform, *not corrugated,* unequally *2-locular,* including the long and soft trichomes 1.5–3 *mm. thick.* (*C. intermedia* Ehrh.) — Rich or alluvial woods, Gaspé Pen. to L. St. John, Que., and w. to Minn., s. to N.S., s. Me., s. N.H., Ct., Pa., mts. of Va. and W.Va. July–early Sept. (Eu.)

Var. **virginiàna** Fern. (Virginian). — Rhizome stout and firm, 2.5–3 mm. thick; stem firm, simple; *leaves firm, shallowly dentate, definitely cordate;* bases of pedicels conspicuously dark purple; *backs of calyx-lobes villous.* — Rich woods, local, se. Va. June, early July.

3. C. alpìna L. (alpine). — Differing from no. 2 in its *tuberous-thickened rhizome;* stems 0.4–3 dm. high; leaves mostly smaller, 1–6.5 cm. long; leading racemes becoming 1.5–7 cm. long; *calyx-lobes* 0.8–1.2 *mm. broad;* petals 1.2–3 mm. long; *anthers* 0.2–0.3 *mm. long; fruit* on pedicels 2–6 mm. long, *1-locular,* including the very short hairs 1–1.5 *mm. thick.* — Cool moist woods and openings, Lab. to Alaska, s. to Nfld., N.S., N.E., s. N.Y., mts. to Ga. and Tenn., Mich., n. Ill., Ia., S.D., Colo., Utah and Wash. Mid-June–Sept. (Eurasia)

FAM. 126. HALORAGÀCEAE (Water-Milfoil Family)

Aquatic or paludal plants (at least in northern countries) with the inconspicuous symmetrical (perfect or unisexual) *flowers sessile in the axils of leaves or bracts, calyx-tube adherent to the ovary, which consists of 2–4 more or less united carpels, the styles or sessile stigmas distinct.* Limb of the calyx obsolete or very short in perfect or pistillate flowers. Petals small or none. Stamens 1–8. Fruit indehiscent, 3–4-locular, with a single anatropous seed suspended from the summit of each locule. Embryo in the axis of fleshy albumen; cotyledons minute.

Carpels 4, separating into distinct nutlets; stamens 4 or 8; leaves whorled or alternate. . . . . 1. *Myriophyllum.*

Carpels 3, united into an indehiscent 3-angled fruit; stamens 3; leaves alternate. 2. *Proserpinaca.*

1. MYRIOPHÝLLUM L. Water-Milfoil

Flowers monoecious or polygamous. Calyx of the staminate flowers 4-parted, of the pistillate 4-toothed. Petals 4 or none. Stamens 4–8. Fruit nut-like, 4-locular, deeply 4-lobed; stigmas 4, recurved. — Perennial aquatics, the genus nearly cosmop. Leaves often whorled; those under water pinnately parted into capillary divisions. Flowers sessile, chiefly in the axils of the upper leaves, usually above water in summer; the uppermost staminate. (Name from the Greek *myrios*, *numberless*, and *phyllon*, *leaf;* alluding, like Milfoil, to the innumerable divisions of the leaves.)

*a.*Stamens 8; petals quickly deciduous; fruits 1.5–3 mm. long, their mericarps rounded on back and without dorsal ridges; submersed leaves definitely whorled. . . *b.*

*b.*Upper floral leaves or bracts entire; bracteoles at base of flower entire or merely serrulate; stems whitened in drying. . . *c.*

*c.*Submersed leaves 3–12 mm. long, with stiffish divisions; inflorescences simple or forking, with scattered flowers; upper bracts promptly deciduous; bracteoles serrulate; fruits 1.5–2 mm. long, nearly cubical. . . 1. *M. alterniflorum.*

*c.*Submersed leaves 1–3 cm. long, flaccid; inflorescences simple; flowers whorled; upper bracts persistent; bracteoles entire; fruits 2.3–3 mm. long, subglobose.

Submersed leaves with 6–11 pairs of divisions, upper foliage-leaves pinnate; bracts of spike spatulate-obovate or oblong, rarely 3 mm. long; fruit slenderly 4-sulcate, mericarps smooth. . . . . . 2. *M. exalbescens.*

Submersed leaves with 3–7 pairs of divisions; upper leaves subentire or short-pectinate; bracts linear-oblanceolate, conduplicate, falcate, 0.3–1 cm. long; fruit very broadly sulcate, mericarps rugose. . . . 3. *M. magdalenense.*

*b.*Upper bracts pinnate, like submersed leaves or smaller; bracteoles palmately 7-cleft; stem scarcely whitened in drying. . . . . 4. *M. verticillatum.*

*a.*Stamens 4; petals often persistent; mericarps with 1 or 2 dorsal ridges or, if not ridged, only 0.7–1.2 mm. long; submersed leaves verticillate, subverticillate or scattered. . . *d.*

*d.*Submersed leaves capillary-pinnate, with numerous segments; stems flexuous, when submersed elongate. . . *e.*

*e.*Carpels or mericarps with 1 or 2 dorsal ridges, in maturity 1–2.5 mm. long. . . *f.*

*f.*Submersed leaves mostly verticillate or subverticillate; flowers in emersed leafy-bracted spikes; bracts entire or pectinate; petals 1.5–3 mm. long; anthers 0.8–2.5 mm. long; fruits 1–2 mm. long. . . *g.*

*g.*Submersed leaves with 5–10 pairs of capillary divisions; bracteoles ovate to lanceolate, acuminate; mericarps with 2 smooth dorsal ridges.

Emersed leaves and bracts lanceolate to spatulate or elliptic, 1.5–5 mm. wide; bracteoles 1–1.3 mm. long; mature fruit subglobose, 1–1.5 mm. long, minutely papillose, mericarps rounded on the sides. . . . . 5. *M. heterophyllum.*

Emersed leaves and bracts linear to lanceolate, 1–2 mm. wide; bracteoles 0.6–0.7 mm. long; mature fruit 1.6–2 mm. long, deeply sulcate, with rectangular sinuses and laterally compressed smoothish mericarps. . . . . 6. *M. hippuroides.*

*g.*Submersed leaves with about 5 pairs of capillary divisions; bracteoles obtusely triangular; mericarps with 2 prominently tuberculate dorsal ridges. . . . . 7. *M. pinnatum.*

*f.*Submersed leaves mostly scattered, rarely subverticillate, with flowers in submersed axils, neither spicate nor bracteate; petals 1 mm. long; anthers 0.4 mm. long; fruits 2–2.5 mm. long, dorsal ridges tuberculate or armed. . . . . 8. *M. Farwellii.*

*e.*Carpels rounded on back, smooth or only minutely roughened, not dorsally ridged, in fruit 0.7–1.2 mm. long. . . . . 9. *M. humile.*

*d.*Submersed or lower leaves merely minute simple scales; the seemingly naked stems scapiform; spike remotely alternate-flowered. . . . . 10. *M. tenellum.*

Supplementary Key to Sterile or Immature Specimens

*a.*Leaves elongate, at least the submersed ones pinnate; stems flexuous or prostrate and creeping. . . *b.*

*b.*Leaves all or nearly all in definite whorls. . . *c.*

*c.*Stems commonly whitened in drying. . . *d.*

d. Primary submersed leaves 3–12 mm. long, 3–14 mm. broad, stiffish; spikes with upper flowers scattered or in 2's, their bracts quickly deciduous. . . . . 1. *M. alterniflorum.*

d. Primary submersed leaves 1–5 cm. long, 1–4 cm. broad, flaccid; flowers whorled, in terminal spikes; bracts persistent.

Leaves with 6–11 pairs of divisions; floral bracts and upper submersed leaves spatulate-obovate to oblong-cochleiform, 0.8–1.8 mm. long. . . . . 2. *M. exalbescens.*

Leaves with 3–7 pairs of divisions; upper emergent ones elongate-oblanceolate to linear; bracts linear-oblanceolate, conduplicate, upcurved at tip, 0.3–1 cm. long. . . . . 3. *M. magdalenense.*

c. Stem not whitened in drying.

Floral bracts and emersed leaves pinnate or pectinate. . . . . . 4. *M. verticillatum.*

Floral bracts and emersed leaves entire or merely serrate.

Bracts and emersed leaves lanceolate or oblanceolate to elliptic, 1.5–5 mm. wide; bracteoles at base of flower 1–1.3 mm. long; petals 1.5–3 mm. long. . . . . 5. *M. heterophyllum.*

Bracts linear or lanceolate, 1–2 mm. wide; bracteoles 0.6–0.7 mm. long; petals 1.5 mm. long. . . . . 6. *M. hippuroides.*

b. Leaves whorled, subverticillate or alternate on the same plant. . . *e.*

e. Plant with all leaves submersed and capillary-pinnate.

Stems horizontally arching far below surface of water; leaves usually with minute black spicules in axils of capillary divisions; dried plants usually green. . . . . 8. *M. Farwellii.*

Stems reaching nearly or quite to surface; leaves usually without black spicules; dried plants usually reddish. . . . . 9. *M. humile.*

e. Plant with some or all leaves emersed.

Emersed leaves with middle portion flat and lanceolate, sharply or pectinately serrate; bracteoles at base of flower about 1 mm. long, sinuate-toothed; petals 1.5–2 mm. long. . . . . 7. *M. pinnatum.*

Emersed leaves uncleft or pectinate, blade of middle portion slenderly linear or linear-spatulate; bracteoles 0.5 mm. long, with a pair of median teeth; petals 1 mm. long. . . . . 9. *M. humile.*

a. Leaves wanting or merely minute scales on lower half of the stiffly erect stems. 10. *M. tenellum.*

1. M. alterniflòrum DC. (alternate-flowered). — Very slender, forking, up to 1 m. or more long, stem usually whitened in drying; *submersed leaves* in whorls of 3–5, 3–12 *mm. long,* 3–14 *mm. wide, with* 3–7 *pairs of* capillary *stiffish divisions; spikes* emersed, *loosely flowered; lower flowers* pistillate, *verticillate in* 3's; *upper flowers* staminate, some or all of them *alternate; upper bracts entire,* 1–2 *mm. long, soon deciduous;* lower bracts pectinate or pinnatifid; *bracteoles* 2, minute, *broadly ovate or suborbicular, serrulate,* 0.15–2 *mm. long,* firm, dark-margined; petals of staminate flowers boat-shaped, 2–2.5 mm. long; stamens 8, anthers 2 mm. long; *fruit nearly cubical,* 1.5–2 *mm. long, with subcylindric red-toothed mericarps.* (Incl. var. *americanum* Pugsley, the clones with smallest leaves) — Lakes, ponds and streams, Nfld. to Alaska, s. to N.S., Me., Mass., Ct., n. N.Y., n. Mich. and n. Minn. June–Sept. (Greenl.; Eu.) Fig. 1346.

1346. M. alterniflorum.

2. M. exalbéscens Fern. (becoming pale). — *Stems* simple or forking, purple, *when dry becoming white; leaves* in 3's or more commonly in 4's, 1.2–3 cm. long, *with* 6–11 *pairs of* capillary flaccid or slightly stiffish *divisions; spikes almost naked, with verticillate flowers,* lower flowers pistillate, upper staminate; *bracts rarely equaling fruit, spatulate-obovate or oblong-cochleiform,* the lower serrate, *the upper entire;* bracteoles ovate, entire, 0.7–1 mm. long; *petals* oblong-obovate, concave, 2.5 *mm. long;* anthers 8, oblong, 1.2–1.8 mm. long; *fruits subglobose, very slenderly 4-sulcate,* 2.3–3 *mm. long; the mericarps rounded on the back, smooth or rugulose.* (*M. spicatum* of Am. auth., not L.) — Ponds, pools and quiet streams, often brackish or calcareous, se. Lab. to Alaska, s. to Nfld., N.S., N.E., Md., W.Va., O., Ind., ne. Ill., Minn., nw. Kans., Ariz. and s. Calif. July–Sept. Fig. 1347.

1347. M. exalbescens.

3. M. magdalenénse Fern. (of the Magdalen Islands). — Similar to no. 2;

1348. M. magdalenense.

leaves with 3–7 *pairs of* capillary *flaccid divisions;* the upper emergent or *amphibious leaves elongate-oblanceolate or linear, short-pectinate or subentire; bracts similar to upper leaves, conduplicate, upcurved at apex,* entire or pectinate, 0.3–1 *cm. long; petals* ovate-oblong, 1.5 *mm. long; fruit* 3 mm. long, *very broadly 4-sulcate,* the *mericarps* dorsally *rugose.* — Shallow sandy ponds, M.I. July–Sept. Fig. 1348.

4. M. verticillàtum L. (whorled). — Stems simple or with few elongate branches up to 2.5 m. long, in autumn producing turbinate crowded winter-buds 1–2 cm. long; *leaves in 4's and 5's;* the *submersed ones* 0.8–4.5 cm. long, *with* 9–13 opposite or alternate *pairs of* capillary flaccid *divisions* 0.5–2.8 cm. long; *emersed leaves and bracts smaller* and *with coarser divisions or* merely *pectinate-pinnate;* flowers in whorls of 4–6, hermaphrodite or the lower pistillate, the upper staminate; *bracteoles palmately 7-lobed,* 0.5 mm. long; petals (merely rudiments in pistillate flowers) spoon-shaped, obtuse, 2.5 mm. long; anthers 8, about 2 mm. long; *fruit subglobose,* 2–2.5 mm. long, *deeply 4-furrowed,* the brown *smooth mericarps rounded* on back. — Highly variable circumpolar species, the typical Old World form (var. *pinnatifidum* Wallr.) with bracteal leaves elongate, about 10 times as long as flowers. We have

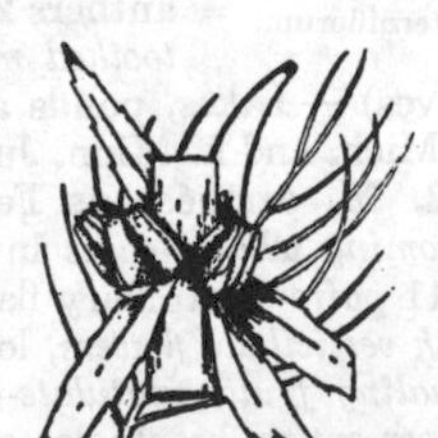

Var. **pectinàtum** Wallr. (comb-like). — Lower usually submersed leaves 1.5–4.5 cm. long; bracts very short, once to thrice length of flowers. — Shallow waters, chiefly in argillaceous or calcareous reg., Nfld. to B.C., s. to N.S., N.E., Del., Md., Great Lakes states and Utah. Late June–Sept. (Eurasia) Fig. 1349.

1349. M. verticillatum, v. pectinatum.

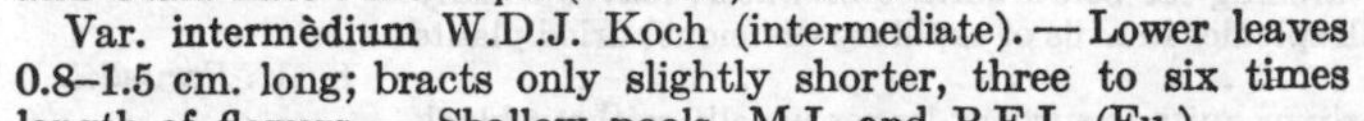

Var. **intermèdium** W.D.J. Koch (intermediate). — Lower leaves 0.8–1.5 cm. long; bracts only slightly shorter, three to six times length of flowers. — Shallow pools, M.I. and P.E.I. (Eu.)

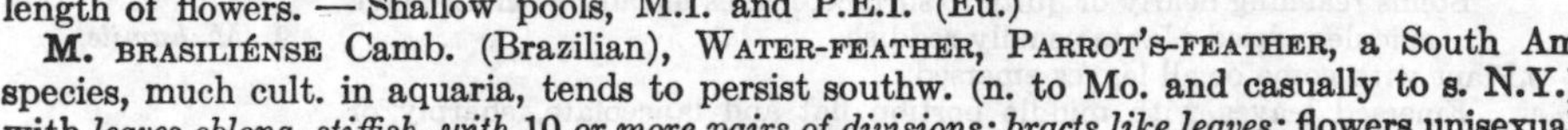

M. brasiliénse Camb. (Brazilian), Water-feather, Parrot's-feather, a South Am. species, much cult. in aquaria, tends to persist southw. (n. to Mo. and casually to s. N.Y.), with *leaves oblong, stiffish, with* 10 *or more pairs of divisions; bracts like leaves;* flowers unisexual; *bracteoles filiform,* 2–3-*cleft.* (*M. proserpinacoides* Gillies)

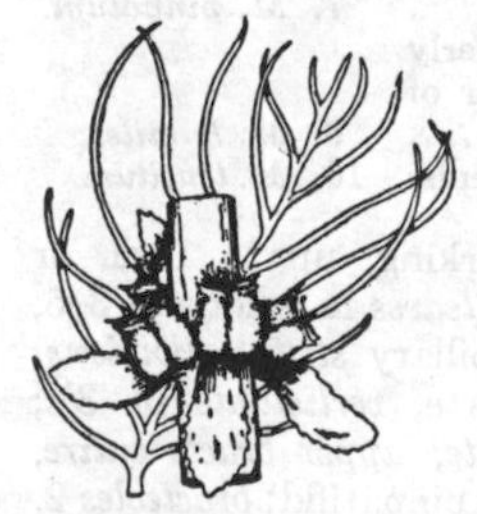
1350. M. heterophyllum.

5. M. heterophýllum Michx. (diverse-leaved). — *Stems stoutish; leaves* in 4's and 6's; *the submersed* pinnate ones 2–5 cm. long, *with* 7–10 *pairs of divisions; the amphibious pinnatisect; the emersed and the bracts firm, lanceolate or lance-spatulate to elliptic, entire or serrate,* 0.4–3 cm. long, 1.5–5 *mm. broad;* spikes 0.3–3.7 dm. long; flowers in whorls of 4–6, hermaphrodite or lower pistillate and upper staminate; *bracteoles ovate, acuminate,* serrate, 1–1.3 *mm. long,* 0.5–0.7 mm. broad; *petals* (of staminate flowers) *acutish,* 1.5–3 *mm. long; the* 4 *anthers* 1–2.5 *mm. long; fruit subglobose,* 1–1.5 *mm. long and wide,* minutely papillose; *each mericarp 2-ridged on back but rounded on sides,* prominently beaked. — Ponds and streams, Fla. to Tex. and N.M., n. to N.D., Ont. and sw. Que. June–Sept. Fig. 1350.

6. M. hippuroìdes Nutt. (resembling *Hippuris,* mare's-tail). — Similar to no. 5; *leaves* in whorls of 4–6 or slightly scattered; *the median amphibious ones pinnate but smaller* than the submersed ones; *the upper emersed ones forming plume-like spikes, linear or lanceolate,* pectinate, serrate or subentire, 0.5–2 cm. long, 1–2 *mm. broad;* flowers in upper axils, perfect, or lower pistillate and upper staminate; *bracteoles long-acuminate,* coarsely serrate, 0.6–0.7 *mm. long; petals* 1.5 *mm. long; anthers* 4, 0.9–1.3 *mm. long; fruit* 1.6–2 *mm. long, deeply sulcate, with rectangular sinuses and laterally compressed smoothish mericarps with* 2 *nerve-like angles along the dorsal ridge.* (*M. verticillatum,* var. *Cheneyi* Fassett) — Shallow waters, N.Y. to Wisc.; Wash. to Calif. June–Oct. Fig. 1351.

1351. M. hippuroides.

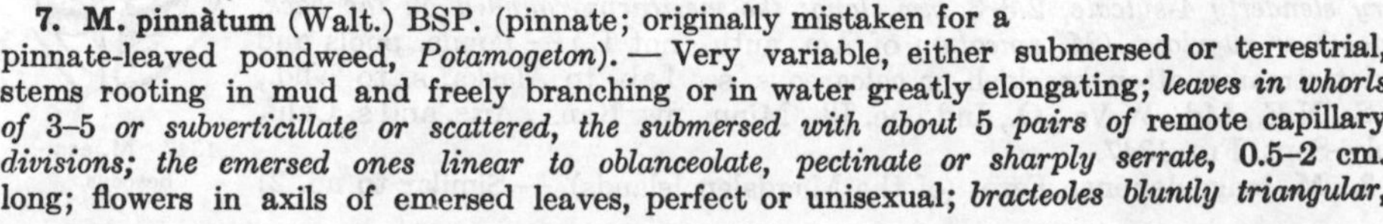

7. M. pinnàtum (Walt.) BSP. (pinnate; originally mistaken for a pinnate-leaved pondweed, *Potamogeton*). — Very variable, either submersed or terrestrial, stems rooting in mud and freely branching or in water greatly elongating; *leaves in whorls of* 3–5 *or subverticillate or scattered, the submersed with about* 5 *pairs of* remote capillary *divisions; the emersed ones linear to oblanceolate, pectinate or sharply serrate,* 0.5–2 cm. long; flowers in axils of emersed leaves, perfect or unisexual; *bracteoles bluntly triangular,*

1352. M. pinnatum.

1 mm. long; petals purplish, 1.5–2 mm. long, rounded above, with slender claw; anthers 4, 0.8–1.1 mm. long; *fruit pale, cubic-ovoid,* 1.3–1.8 *mm. long, mericarps with flat sides and 2 tuberculate dorsal ridges.* (*M. scabratum* Michx.) — Peaty or muddy shores or in shallow waters, Fla. to Tex., n. to s. N.E., W.Va., Ky., Ill., and Ia. June–Oct. (W.I.) FIG. 1352.

8. M. Farwéllii Morong (for its discoverer, OLIVER ATKINS FARWELL, 1868–1944). — *Stems* with *very slender* flexuous branches 3–6 dm. long, *horizontally arching and fruiting far beneath surface of water; leaves uniform, very delicate, crowded, verticillate, subverticillate and alternate,* 1–2.5 cm. long, *with slenderly capillary rachis;* the 5–8 pairs of slender *divisions usually black-spiculate in their axils; flowers in middle axils,* perfect; *bracteoles lanceolate, hyaline,* about 0.4 mm. long; *petals rhombic,* 1 *mm. long,* with black ciliate-serrulate margin; the 4 anthers 0.4 mm. long; *fruit* cubic-ovoid, 2–2.5 *mm. long,* the flat-sided *mericarps with dorsal rows of tubercles and thick hooks.* — Ponds and slow streams, Gaspé Co., Que. (to subalpine areas), to Ont., s. to N.S., n. N.E., centr. N.Y., n. Mich. and n. Minn. June–Sept. FIG. 1353.

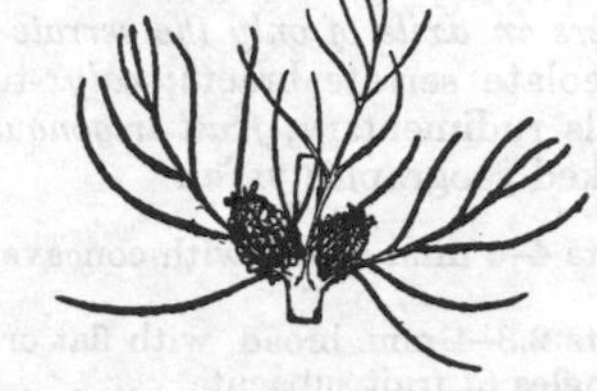
1353. M. Farwellii.

9. M. hùmile (Raf.) Morong (lowly). — Exceedingly variable, occurring in three strikingly dissimilar forms due to change of water-depths; the TERRESTRIAL or subterrestrial typical FORM *with* stems rooting in shores, the depressed branches 1–8 cm. long, the *linear to linear-spatulate and uncleft to pinnatifid alternate leaves* 2–10 *mm. long* and 0.3–4 *mm. broad,* the divided ones with 2–4 pairs of subopposite short segments; the WHOLLY SUBMERGED forma **capillàceum** (Torr.) Fern. (hair-like) with stem elongating to 1 m. or more, *with* the *subverticillate or scattered* finely pinnatifid *leaves* 1–3.5 *cm. long* and 1.5–3 cm. broad, and *with* 4–8 *pairs of capillary divisions;* forma **nàtans** (DC.) Fern. (floating) similar but with emersed inflorescences with bracts like the leaves of typical shore-form; flowers mostly perfect; *bracteoles narrowly deltoid-ovate,* about 0.5 mm. long; *petals* 1 *mm. long; anthers* 4, *about* 0.6 *mm. long; fruit nearly cubical,* 0.7–1.2 *mm. long, with round-backed* smooth or minutely roughened *slender-cylindric mericarps.* — Sandy, peaty or muddy shores and ponds, N.S. to e. N.Y., s. to Pa. and e. Md. June–Oct. FIG. 1354.

1354. M. humile.

10. M. tenéllum Bigel. (delicate). — *Stems* numerous from creeping base, *appearing leafless and scape-like, thickish,* some sterile, others floriferous at the simple or rarely corymbose-forked summit, 0.25–5.5 dm. high, *bearing remote alternate minute simple leaves; spikes remotely alternate-flowered; bracts entire, blunt,* 1–4 mm. long; *bracteoles linear-triangular,* about 1 mm. long; petals reddish, oblong, about 2 mm. long; stamens 4, anthers 1.2–1.7 mm. long; *fruit truncate-ovoid,* deeply 4-furrowed, 1–1.2 *mm. long, lustrous,* smooth or minutely rugulose. — Shallow margins of ponds and pools in sand, granitic gravel, mud, and peat, Nfld. to Ont., s. to N.S., N.E., L.I., N.J., e. Pa., centr. N.Y., Mich. and Minn. July–Oct. FIG. 1355.

1355. M. tenellum.

2. PROSERPINÀCA L. MERMAID-WEED

Flowers perfect. Calyx-tube 3-sided, the limb 3-parted. Petals none. Stamens 3. Stigmas 3, cylindrical. Fruit bony, 3-angled, 3-locular, 3-seeded, nut-like. — Low e. Am. perennial herbs, with the stems creeping at base, alternate leaves, and small flowers sessile in the axils, solitary or 2–5 together, in summer. (Name applied by Pliny to a *Polygonum,* meaning *pertaining to Proserpina;* transferred to the present genus presumably because of the ready adaptation of the plants to different habitat-conditions.)

Leaves subtending the flowers merely serrate; submersed pinnatifid leaves 1.5–6 cm. long, with 8–14 pairs of divisions 0.5–3 cm. long. . . . . 1. *P. palustris.*
Leaves uniform or nearly so, all pinnatifid or pinnatisect.
Median portion of blade 1–4 mm. broad, the 7–12 pairs of divisions 2–3.5 mm. long; fruit 2.3–3.6 mm. broad. . . . . 2. *P. intermedia.*

Median portion of blade 0.2–1 mm. broad, the 4–9 pairs of divisions 2–7.5 mm. long; fruit 2–2.8 mm. broad. . . . . 3. *P. pectinata.*

1. P. palústris L. (of marshes). — Repent, with ascending summit or branches, or suberect, 0.1–1 m. long, the base usually submersed (at least in spring), the summit becoming emersed; *submersed leaves* rufescent, sessile, finely pinnatifid, 1.5–6 *cm. long, with* 8–14 *pairs of* linear-filiform *divisions* 0.5–3 *cm. long,* commonly bearing minute black axillary spicules; *median portion linear,* 0.7–1 *mm. wide;* amphibious leaves petioled, pinnatisect, 1.7–7 cm. long, with lanceolate middle portion 1.5–9 mm. broad; *emersed leaves* (or those produced on terrestrial plants) *lanceolate to oblanceolate,* 1.5–8.5 cm. long, 0.2–1.4 cm. broad, *serrate; flowers in axils of only the serrate leaves,* solitary or in clusters of 2–5, subtended by minute lanceolate serrate bracts; *calyx*-tube 3-angled; its *lobes* ovate to deltoid, *obtuse or acutish;* petals rudimentary; *fruit trigonous-urceolate, or pyramidal.* — Very variable, with three well-marked geographic vars.:

Fruits 4–6 mm. broad, with concave sides and thin or winged angles. . . . *P. palustris* (typical).
Fruits 2.3–4 mm. broad, with flat or rounded sides, wingless.
Angles of fruit subacute. . . . . Var. *crebra.*
Angles rounded or nearly obsolete. . . . . Var. *amblyogona.*

P. palústris (typical), var. *latifolia* Schindl.; *P. platycarpa* Small — Shallow water and shores, Fla. to La., n. to se. Va. and se. Mo., and more locally to s. N.E. June–Oct. (W.I.)

Var. **crèbra** Fern. & Grisc. (common). — N.S. to Minn., s. to Ga. and Okla. July–Oct. (Mex.; Centr. Am.)

Var. **amblyogòna** Fern. (with rounded angles). — Ga. to Tex., n. to L. Huron, Ont., Mich. and se. Mo. July–Oct.

2. P. intermèdia Mackenz. (intermediate). — Intermediate between *P. palustris,* var. *crebra* and no. 3; *leaves nearly uniform,* 1.7–4 cm. long, *with* linear to lanceolate *median portion* 1–4 *mm. broad* and 7–12 pairs of firm *divisions* 2–3.5 *mm. long;* fruit intermediate, 3.7–4.3 mm. long, 2.3–3.6 mm. broad. — Similar habitats, either with one or both of the others or isolated, local, sw. N.S.; e. Mass. to se. Va. July–Sept.

3. P. pectinàta Lam. (comb-like). — Stem *very slender,* repent, with ascending rufescent summit 0.5–3 dm. high; *leaves all pinnatifid,* 1.3–2.3 *cm. long, with linear median portion* 0.2–1 *mm. wide* and 4–9 *pairs of slender rather firm divisions* 2–7.5 *mm. long,* sometimes bearing minute black spicules; flowers solitary, rarely in 2's or 3's, in the middle and upper axils; *calyx-lobes acuminate; fruit urceolate, with rounded or obtuse angles,* 3–4 mm. long, 2–2.8 *mm. broad.* — Peaty or muddy quagmires, savannas and shallow water, chiefly on Coastal Plain southward, Fla. to Tex., n. to N.S., sw. Me. and Tenn. June–Oct.

FAM. 127. HIPPURIDÀCEAE (Mare's-tail Family)

Flowers perfect or polygamous. Calyx entire. Style filiform, stigmatic down one side, received in the groove between the lobes of the large anther. Fruit nut-like, 1-locular. — Perennial aquatics with entire leaves in whorls, the minute flowers sessile in the axils. A single genus:

1. HIPPÙRIS L. Mare's-tail

Flaccid or fleshy plants of cool reg. of N. and S. Hemisph. (Greek name, meaning horse-tailed, from the Greek *hippos, a horse,* and *oura, a tail,* applied to a water-plant.)

1. H. vulgàris L. (common). — Stems erect, from creeping rhizomes, hollow, simple (rarely forking), cylindric, 1.5–8.5 dm. high; *leaves* 6–12 *in a whorl, linear-attenuate,* firm and thick, 0.4–3.5 cm. long, or flaccid and up to 6 cm. long in the submersed forma **fluviátilis** (Coss. & Germ.) Glück (of rills or brooks); flowers in middle and upper axils; calyx tubular or barrel-shaped, subtruncate at summit, adherent to ovary, in maturity 1.5–2.5 mm. long. — Shallow pools and margins of streams and lakes (forma *fluviatilis* in deep or running water), Greenl. to Alaska, s. to Nfld., N.S., n. N.E., centr. N.Y., Ind., n. Ill., Minn., Neb. and N.M. June–Sept. (Eurasia)

2. H. tetraphýlla L. f. (4-leaved). — Stems 1–4 dm. high; *leaves* 3–6 *in a whorl, oblanceolate, elliptic or oblong-obovate, obtuse,* 0.4–1.5 cm. long. (*H. vulgaris* L., var. *maritima* sensu Am. auth., not Wahlenb.) — Saline or brackish marshes, Lab. to Gaspé Pen., Que., and James Bay; Alaska. (Eurasia)

FAM. 128. ARALIÀCEAE (Ginseng Family)

Herbs, shrubs or trees with much the same characters as Umbelliferae, *but with usually more than 2 styles, and the fruit a few–several-locular drupe.* Albumen mostly fleshy. Petals 5, epigynous, not inflexed. Stamens 5, epigynous, alternate with the petals.

Leaves compound.
  Leaves alternate, decompound, the ultimate divisions pinnate; carpels 5; fruit blackish. . . . . 1. *Aralia.*
  Leaves whoried, palmately 3–7-foliolate; carpels 2 or 3; fruit red or yellow. 2. *Panax.*
Leaves simple, roundish, palmately lobed or entire, alternate.
  Densely prickly deciduous upright shrub; carpels 2; fruit red, 2-seeded. . . 3. *Oplopanax.*
  Smooth trailing or climbing shrub attaching by aërial roots; carpels 5; fruit black, 3–5-seeded. . . . . 4. *Hedera.*

1. ARÀLIA L.

Flowers polygamous. Petals imbricated in the bud. Ovary 5-locular; ovules solitary, anatropous, suspended in the locules. — Leaves compound or decompound. Flowers white or green, in umbels or panicles. Qualities aromatic. Small genus of N. Am. and of Asia to Austral. (Name from the French-Canadian *Aralie,* the original specimens sent by the Quebec physician, Sarrasin, to Tournefort under that name.)

Umbels numerous in a large compound panicle or raceme.
  Shrub or tree; stem and leaf-stalks prickly; umbels in a large broad panicle. . 1. *A. spinosa.*
  Herb, without prickles; umbels in a racemose panicle. . . . . . . 2. *A. racemosa.*
Umbels 2–∞ in a corymb.
  Stem bristly at base, leafy. . . . . 3. *A. hispida.*
  Stem smooth, subscapose, with a single long-stalked leaf at base. . . . . 4. *A. nudicaulis.*

1. A. spinòsa L. (spiny), Angelica-tree, Hercules'-club, Devil's-walking-stick, "Prickly Ash". — *Shrub or tree* up to 10 m. high, the *stout trunk and stalks of* the large decompound *leaves coarsely prickly;* leaflets ovate, pointed, serrate, pale beneath; *umbels in a large compound panicle.* — Bluffs, rich woods and riverbanks, Fla. to Tex., n. to N.J., Pa., w. N.Y., O., Ind., s. Ill. and Ia.; esc. from cult. n. to s. N.E., centr. N.Y. and Mich. July–Sept.

2. A. racemòsa L. (racemose), Spikenard, Petty Morrel, Life-of-Man. — *Herbaceous,* prickleless, 0.5–3 m. high, *widely branched; leaflets cordate-ovate,* pointed, doubly serrate, slightly downy; *umbels racemose-paniculate; styles united.* — Rich woods and thickets, Gaspé Pen., Que., to Man., s. to N.S., N.E., Ga., Ala., Miss., Mo. and ne. Kans. June–Aug. — Well known for its spicy-aromatic large roots.

3. A. híspida Vent. (with straight hairs), Bristly Sarsaparilla, Dwarf-elder. — *Stem* 2–9 dm. high, *bristly, leafy,* terminating in a peduncle bearing several umbels; leaves twice pinnate; leaflets oblong-ovate, acute, cut-serrate. — Rocky and sandy open woods and clearings, Nfld. and s. Lab. Pen. to Man., s. to N.S., N.E., w. N.C., W.Va., O., Ind., Ill. and Minn. June–Aug.

4. A. nudicaùlis L. (naked- or leafless-stemmed), Wild Sarsaparilla, Salsepareille (Que.). — *Stem scarcely rising out of the ground, smooth, bearing a single long-stalked leaf* (2–4 dm. high) *and a shorter naked scape,* with 2–7 umbels; leaflets oblong-ovate or oval, pointed, serrate, 5 (rarely more) on each of the 3 divisions. — Woodlands, Nfld. and s. Lab. Pen. to B.C., s. to N.S., N.E., Ga., Tenn., Ill., Mo., Neb., Colo. and Ida. May–July. Forma **viréscens** Vict. & Rousseau (greenish) has the carpels changed to leaves. — The long horizontal aromatic roots a substitute for officinal Sarsaparilla.

2. PÀNAX L. Ginseng

Flowers dioeciously polygamous. Umbel solitary, simple, terminal. Carpels 2–3. — Herbaceous perennials of e. N. Am. and e. Asia, springing from thickish roots or tubers, the erect simple stems bearing a solitary whorl of 3 palmate leaves. (Name from the Greek *pas, all,* and *akos, cure,* that is, *all healing,* a panacea, from the repute of the plants in China.)

1. P. quinquefòlium L. (five-leaved), Ginseng, Sang. — *Root large and spindle-shaped, often forked,* 1–2 dm. long, aromatic; stem about 3 dm. high; *leaflets long-stalked,* mostly 5, large and thin, obovate-oblong, pointed; styles mostly 2; *fruit bright red.* — Rich and cool woods, Que. to Man., s. to n. Fla., Ala., La. and Okla. June, July. Formerly much sought for the root, which is purchased by the Chinese and extensively employed by them in their medicine, as is also a similar and even more highly prized Asiatic species.

2. P. trifòlium L. (three-leaved), Dwarf G., Ground-nut. — *Root or tuber globular*, deep in the ground; stem 1–2 dm. high; *leaflets* 3–5, *sessile* at the summit of the leaf-stalk, narrowly oblong, obtuse; styles usually 3; *fruit yellowish.* — Rich woods and damp clearings, P.E.I. to Minn., s. to N.S., N.E., n. Ga., O., Ind., Ia. and Neb. Apr.–June.

3. OPLÓPANAX (T. & G.) Miq.

Flowers perfect or polygamous. Umbels numerous in simple or compound racemes or paniculately disposed. Calyx-margin narrow or obsolete, obscurely crenate-lobed. Carpels 2. — Stout sometimes arborescent shrubs of N. Am. and e. Asia, very prickly. Leaves simple, long-petioled, the limb suborbicular and palmately lobed. (Name from the Greek *hoplon, weapon*, and *Panax*.)

1. O. hórridus (Sm.) Miq. (prickly), Devil's-club. — Coarse shrub, thickly beset with stramineous prickles; leaves 1–3 dm. in diameter, with 5–13 deltoid acute lobes, the margin sharply and unevenly serrate, the ribs prickly beneath. (*Fatsia* B. & H.; *Echinopanax* Dcne. & Planch.) — Low rocky woods and sheltered cliffs, Thunder Bay Distr., Ont.; I. Royale and adj. ids., n. Mich.; Alaska to Mont. and Calif. July, Aug. (Jap.)

4. HÉDERA L. Ivy

Flowers perfect. Umbels solitary or racemose at tips of branches. Calyx 5-toothed. Petals 5, valvate. Carpels 5. — Evergreen creeping or climbing shrubs nat. of Eurasia and n. Afr. with aërial roots and alternate petioled palmately lobed to entire coriaceous leaves. (The ancient Latin name.)

1. H. Hélix L. (old generic name, meaning twining), English I. — Leaves 3–5-lobed, dark green, often with paler markings; berries black. — Open woods, Va. and southw. (Introd. and natzd. from Eu., commonly cult.)

FAM. 129. UMBELLÍFERAE (Parsley Family)

Herbs with small flowers in umbels (or rarely heads), the calyx entire or 5-toothed, its tube wholly adhering to the 2-locular and 2-ovuled ovary, the 5 petals and 5 stamens inserted on the disk which crowns the ovary and surrounds the base of the 2 styles. Fruit of 2 seed-like dry carpels (called *mericarps*) cohering by their inner face (the *commissure*), when ripe separating from each other and usually suspended from the summit of a slender prolongation of the axis (*carpophore*); each carpel marked lengthwise with 5 *primary ribs*, and often with 4 intermediate (*secondary*) ones; in the *interstices* or *intervals* are commonly oil-tubes (*vittae*), longitudinal canals containing aromatic oil (these best seen in sections made across the fruit). Seed suspended from the summit of the locule, anatropous. Stems usually hollow. Leaves alternate, often compound, the petioles usually expanded or sheathing at base. Umbels usually compound, the secondary ones being termed *umbellets;* the bracts which often subtend the general umbel form the *involucre*, and those of the umbellets the *involucels*. The frequently thickened base of the styles is called the *stylopodium*. — A large and difficult family, some of the species harmless and aromatic or supplying food, others with very poisonous properties.

Synoptic Key to Genera

a. Seed-cavity with a thick layer of ligneous strengthening cells around the seed; fruits without distinct oil-tubes, usually with an oil-bearing continuous layer beneath the epidermis; ours weak herbs with simple (or proliferous) umbels. Subfam. I. Hydrocotyloideae.

Leaves orbicular or reniform; umbels several-flowered; mericarps with 5 simple ribs and plane sides. . . . . 1. *Hydrocotyle.*

Leaves ovate; umbels 2–4-flowered; mericarps with 5 primary and 4 secondary ribs, the marginal and sometimes the intermediate ribs with lateral branches. . . . . 2. *Centella.*

a. Seed-cavity with soft tissue surrounding the seeds; oil-tubes variously developed or absent. . . *b*.

b. Styles arising from a ring-like disk; fruits sessile, covered with hooked prickles, scales or tubercles, not definitely ribbed. Subfam. II. Saniculoideae.

Leaves palmate; flowers greenish or yellowish, in irregular umbels; fruits covered with hooked prickles. . . . . 3. *Sanicula.*

Leaves simple; flowers bluish or white, in dense bracteate heads; fruits scaly or tuberculate. . . . . 4. *Eryngium.*

*b.*Styles terminating the 2 stylopodia; umbels chiefly compound; fruits smooth or variously roughened, usually ribbed or winged. Subfam. III. Apioideae. . . *c.*

*c.*Fruit with only the primary ribs prominent, *i.e.*, 3 dorsal ribs on each carpel prominent, secondary ribs weaker or not evident. . . *d.*

*d.*Inner face of seed concave or furrowed. . . *e.*

*e.*Each rib of the fruit supported by well developed strengthening tissue; fruit (in ours) linear-cylindric to spindleform or slenderly ovoid. . . *f.*

*f.*Fruit linear-cylindric to slenderly oblanceolate or broadly lanceolate, glabrous, strigose-pubescent or appressed-hispid.

Fruit linear-cylindric to lanceolate, rounded at base.

Petals obovate; fruit with prominent obtuse ribs; oil-tubes conspicuous, 1 in each interval. . . . . . 5. *Chaerophyllum.*

Petals acuminate; fruit obscurely ribbed or ribless; oil-tubes wanting or obscure. . . . . 6. *Anthriscus.*

Fruit oblanceolate or spindleform, tapering to long-attenuate base, the ribs appressed-hispid, the slender bases upwardly hispid. . . . . 7. *Osmorhiza.*

*f.*Fruit lance-ovoid, densely covered with spreading long and short prickles. . . . . 8. *Torilis.*

*e.*Each rib with weakly developed or no strengthening tissue beneath; fruit ovoid to reniform or globose.

Fruit globose or subglobose, not readily separating into halves, smooth or obscurely ribbed, with conspicuous calyx-teeth. . . 9. *Coriandrum.*

Fruit laterally compressed, ovoid to reniform, the mericarps separating in maturity, slenderly ribbed, calyx-teeth obsolete.

Low weak vernal perennial with deep-seated round tuber; umbel 2–4-rayed, subtended by a ternately compound leaf; fruit reniform or suborbicular, the curving mericarps separated by a broad plane area. . . . . 10. *Erigenia.*

Erect annual or biennial; umbel few–many-rayed, with deciduous involucre of linear-lanceolate short bracts or without involucre; fruit ovoid, the marginal ribs of the mericarps contiguous and parallel.

Coarse, with mottled stem; leaves with lanceolate pinnatifid leaflets; involucre of linear-lanceolate bracts; fruit smooth, with prominent undulate ribs; oil-tubes none. . . . 11. *Conium.*

Slender, not mottled; leaf-segments linear or filiform; involucre none; fruit tuberculate or bristly, the ribs hidden; oil-tubes present. . . . . 12. *Spermolepis.*

*d.*Inner face of seed plane or nearly so or convex, not concave. . *g.*

*g.*Fruit laterally compressed or flattened (as shown in cross-section), the whole fruit distinctly broader than thick. . . *h.*

*h.*Leaves simple, entire, perfoliate. . . . . 13. *Bupleurum.*

*h.*Leaves compound or, if simple, not perfoliate. . . *i.*

*i.*Petals yellow or orange; leaves toothed or incised.

Biennial, with finely incised leaf-segments; flowers all pedicelled; calyx-teeth obsolete; petals greenish-yellow; stylopodium cushion-like. . . . . 14. *Petroselinum.*

Perennials, with coarser leaves; central flower of each umbellet sessile; calyx-teeth prominent; petals orange-yellow; stylopodium wanting. . . . . 15. *Zizia.*

*i.*Petals white (or yellow in no. 20, with entire leaflets). . *j.*

*j.*Oil-tubes solitary in all the intervals. . . *k.*

*k.*Primary involucre none.

Leaves decompound; calyx-teeth prominent; fruit ellipsoid-oblong to reniform; stylopodium depressed. . . 16. *Cicuta.*

Leaves 3-foliolate; calyx-teeth obsolete; fruit slenderly oblong-cylindric; stylopodium slenderly conical. . . 17. *Cryptotaenia.*

*k.*Primary involucre well developed.

Lower leaves palmate, subcoriaceous, with strongly decurrent lanceolate finely cartilaginous-toothed leaflets. 18. *Falcaria.*

Lower leaves pinnate, with filiform divisions of the membranous leaflets. . . . . 19. *Carum.*

*j.*Oil-tubes 3 or more in some or all intervals, or 0. . *l.*

*l.*Petals yellow; leaflets entire. . . . . 20. *Taenidia.*

*l.*Petals white; leaflets toothed or incised. . . *m.*

*m.*Primary involucre wanting (rarely a single bract). . . *n.*
*n.*Leaves pinnate, the larger forming a basal rosette. . . 21. *Pimpinella.*
*n.*Leaves (or some of them) ternately divided, not rosulate.
Oil-tubes 0; leaflets lanceolate or ovate. . . . . 22. *Aegopodium.*
Oil-tubes 3 in the intervals; leaflets and segments linear. . . . . 23. *Perideridia.*
*m.*Primary involucre well developed; leaves all pinnate (rarely simple).
Fruit ovoid to ellipsoid; ribs prominent; oil-tubes 1–3 in the intervals. . . . . 24. *Sium.*
Fruit nearly globose; ribs inconspicuous; oil-tubes numerous in a continuous series. . . . . 25. *Berula.*
*g.*Fruit nearly or quite round in cross-section or dorsally compressed. . . *o.*
*o.*Fruit not at all or barely compressed; ribs filiform to winged, the marginal usually not conspicuously winged or, if so, not broader than the dorsal ones. . . *p.*
*p.*Ribs low, much lower than height of body of carpel; stylopodium usually prominent. . . *q.*
*q.*Oil-tubes solitary in the intervals. . . . *r.*
*r.*Petals white.
Primary involucre present (or, if wanting, leaves reduced to septate hollow petioles).
Ascending annuals with compound umbels.
Leaves with numerous capillary segments or reduced to hollow petioles; seed subterete. . . . 26. *Ptilimnium.*
Leaves with few linear or lanceolate segments or, if simple, flat; seed dorsally flattened, with plane face. . . . . 27. *Cynosciadium.*
Creeping perennial with filiform subterranean stems; leaves reduced to hollow septate petioles; umbel simple. . . . . 28. *Lilaeopsis.*
Primary involucre none; annual, with 2–3-ternately compound leaves with many-cleft segments. . . . . . 29. *Aethusa.*
*r.*Petals yellow; leaf-segments capillary.
Perennial; sheathing base of petiole of larger leaves 3–10 cm. long; marginal ribs of fruit like the dorsal. . . 30. *Foeniculum.*
Annual or biennial; sheathing base of petiole of larger leaves 1–3 cm. long; marginal ribs of fruit broadened into thin wings. . . . . 31. *Anethum.*
*q.*Oil-tubes 2–6 in the intervals; coarse perennials, with ternately compound leaves and broad leaflets. . . . . 32. *Ligusticum.*
*p.*Ribs nearly or quite as high as body of carpel; stylopodium low or wanting.
Coarse mostly maritime plants with inflated obcordate petiolar sheaths; petals whitish; styles arched-recurving in fruit, persistent; stylopodium low but definite. . . . . 33. *Coelopleurum.*
Slender inland plants with tight tapering petiolar sheaths; petals yellow or purple; styles ascending in fruit, soon deciduous; stylopodium obsolete. . . . . 34. *Thaspium.*
*o.*Fruit strongly compressed dorsally, with wing-margins. . . . *s.*
*s.*Oil-tubes not conspicuous on back of fruit or, if conspicuous, slender and extending nearly or quite the length of the intervals. . . *t.*
*t.*Stylopodium evident, conical or depressed. . . *u.*
*u.*Petals white (sometimes purplish).
Leaves twice to thrice compound.
Stylopodium conical; slender plants with thin finely divided leaves. . . . . 35. *Conioselinum.*
Stylopodium depressed, the crenate disk prominent; stout plants with coarse leaves and leaflets. . . 36. *Angelica.*
Leaves simply pinnate or palmate or reduced to quill-like petioles. . . . . 37. *Oxypolis.*
*u.*Petals yellow.
Slender annual or biennial; leaf-segments capillary; involucre none. . . . . 31. *Anethum.*

Coarse perennial; leaflets coarse, rhombic-obovate; involucre conspicuous. . . . . 38. *Levisticum.*

*t.*Stylopodium obsolete or essentially so. . . *v.*

*v.*Umbel dense, subglobose, the rays scarcely elongate; ribs of fruit all forming wings; acaulescent perennial of the Plains. . . . . 39. *Cymopterus.*

*v.*Umbel open, with prolonged rays; ribs of fruit capillary. . *w.*

*w.*Principal leaves ternately compound; slender or low perennials with terete uncorrugated stems.

Leafy-stemmed; leaflets and leaf-segments broad.

Leaflets entire; fruit with distinct dorsal ribs; oil-tubes 1 or 2 in the intervals. . . . . 40. *Pseudotaenidia.*

Leaflets serrate; dorsal ribs obscure on the depressed backs of the fruit, the corky border very prominent; oil-tubes numerous around the seed and through the pericarp. . . . . 41. *Polytaenia.*

Acaulescent; leaflets capillary or very fine. . . . . 42. *Lomatium.*

*w.*Leaves simply or doubly pinnate; coarse and tall biennial with corrugated stem. . . . . 43. *Pastinaca.*

*s.*Oil-tubes conspicuous on the back of the large flat fruit, obclavate, extending only one-half to two-thirds the length of the fruit; stout hairy perennials with coarse leaves. . . . . 44. *Heracleum.*

*c.*Fruit with secondary ribs the most prominent, winged and armed with barbed or hooked prickles, the low primary filiform ribs bristly.

Involucre of short simple bracts or none; bristles of fruit mostly barbless; inner face of seed furrowed. . . . . 8. *Torilis.*

Involucre of more or less conspicuous pinnate bracts; bristles of fruit retrorsely barbed at tip; inner face of seed broadly concave or nearly plane. . . . . 45. *Daucus.*

Artificial Key based on Superficial Characters

*a.*Leaves all simple. . . *b.*

*b.*Leaves sessile, perfoliate, entire. . . . . 13. *Bupleurum.*

*b.*Lower leaves petioled or elongate, not perfoliate. . . . *c.*

*c.*Leaves with definite blades. *d.*

*d.*Flowers in dense bracted heads; leaves elongate and spinulose-margined or lanceolate or ovate. . . . . 4. *Eryngium.*

*d.*Flowers in simple or compound umbels; leaves not spinulose.

Leaves orbicular or reniform; umbels simple. . . . . . 1. *Hydrocotyle.*

Leaves ovate to linear.

Stem creeping under ground; leaves ovate, long-petioled; umbel simple, 2–4-flowered. . . . . 2. *Centella.*

Stem ascending or above ground, leafy; umbels compound, many-flowered.

Leaves ovate to lanceolate, toothed.

Leaves fascicled in old axils, membranaceous; umbel with primary involucre; flowers white; plant of estuaries. . 24. *Sium.*

Leaves alternate, coriaceous; umbel without primary involucre; flowers yellow or purple; plants of woods and thickets, with rhizomes. . . . . 34. *Thaspium.*

Leaves linear or narrowly oblanceolate, entire; involucre wanting; plant of southern bogs and wet pineland, with bulbiform base and fusiform roots. . . . . 37. *Oxypolis.*

*c.*Leaves reduced to hollow septate petioles or phyllodia.

Stem filiform, creeping in brackish or saline mud, bearing clavate round-tipped 3–18-jointed leaves 1–20 cm. long; umbels simple. . 28. *Lilaeopsis.*

Stem ascending or reclining, above ground; leaves tapering, linear-filiform; umbels compound.

Stem thickish, quill-like, upright, 0.8–1.2 m. high; principal leaves 1–2 dm. long; umbel with involucre of filiform bracts; fruit 4–8 mm. long, retuse at both ends; bog-plant of Del. . . . . 37. *Oxypolis.*

Stem slender, weak, 1–6 dm. long; principal leaves 1–4 cm. long; involucre 0 or 1 or 2 short bracts; fruit 1.2–2 mm. long, rounded to base and apex; river-bank plant of Appalachian region. . . . 26. *Ptilimnium.*

*a.*Some or all of the leaves compound. . . *e.*

*e.*Perfect flowers sessile or essentially sessile in glomerules, interspersed with stalked staminate ones; the fruits covered with hooked prickles; basal leaves long-stalked, with broad digitate blades; inflorescence irregularly paniculate. . . . . 3. *Sanicula.*

*e.*Flowers all or nearly all perfect, mostly pedicelled; fruits smooth (except for ribs) or, if prickly, the leaves finely dissected. . . *f.*

*f.*Fruits with spreading prickles, barbs or tubercles.

Primary involucre 0 or of short simple bracts.

Leaf-segments broad, incised.

Carpels and fruit echinate, tipped by a long smooth beak; rays of umbel glabrous. . . . . 6. *Anthriscus.*

Carpels and fruit covered to the obscure beak with hooked bristles; rays of umbel pubescent. . . . . 8. *Torilis.*

Leaf-segments filiform or nearly so. . . . . . . 12. *Spermolepis.*

Primary involucre of long pinnate bracts. . . . . 45. *Daucus.*

*f.*Fruits smooth or only with appressed or strigose pubescence. . . . *g.*

*g.*Axils of upper leaves bearing clustered bulblets. . . . . . 16. *Cicuta.*

*g.*Axils without bulblets. . . *h.*

*h.*Leaflets or segments of larger leaves filiform or slenderly linear, at most 1 mm. broad. . . *i.*

*i.*Stem leafy.

Primary involucre of numerous filiform bracts; lowland annuals. 26. *Ptilimnium.*

Primary involucre wanting or of only 1 or 2 short bracts.

Flowers white (rarely roseate); styles slender and persistent at summit of fruit.

Leaves all pinnate; biennial with long thick tap-root; stem 2–7.5 dm. high, branching below middle; fruits strongly ribbed, the styles wide-spreading; weed of northern fields and roadsides. . . . . 19. *Carum.*

Upper leaves ternate; perennial with fascicled tuberous-thickened roots; stem 0.7–1.2 m. high, branching only toward summit; fruits only faintly ribbed, their styles recurving; native of rich woods, prairies and low grounds, Central states. . . . . 23. *Perideridia.*

Flowers yellow; styles short and stout, or wanting in fruit.

Perennial; petiolar sheaths of larger leaves 3–10 cm. long; fruit with marginal and dorsal ribs all filiform, the short styles evident. . . . . 30. *Foeniculum.*

Annual or biennial; petiolar sheaths of larger leaves 1–3 cm. long; fruit thin-winged, the minute styles deciduous. 31. *Anethum.*

*i.*Acaulescent deep-rooted perennial of prairies and plains; naked scape 0.3–3 dm. high; leaves ternately decompound. . . . 42. *Lomatium.*

*h.*Leaflets or chief segments of larger leaves linear to ovate, obovate or suborbicular, more than 1 mm. broad. . . *j.*

*j.*Leaflets of compound leaves 3 (rarely only 2). . . *k.*

*k.*Leaflets toothed, sessile or only short-stalked; roots fibrous.

Leaves membranaceous, finely double-serrate; umbel very irregular; petals white; fruits linear- to oblong-subcylindric, beaked by the subulate stylopodia. . . . . . . 17. *Cryptotaenia.*

Leaves firm, simply toothed; umbels regular; fruit strongly flattened; stylopodium low or obsolete.

Leaflets linear-lanceolate, with cartilaginous mucronate-serrulate margins, strongly decurrent, the larger leaflets mostly divided nearly to base; involucre of filiform bracts; petals white; fruit narrowly oblong, with equal filiform ribs. . . . . 18. *Falcaria.*

Leaflets lanceolate to ovate, not strongly decurrent; involucre 0 or a rudiment only; petals yellow or purple; fruit elliptic or ovate to suborbicular.

Central perfect flower and fruit of each umbellet sessile; fruits with filiform ribs. . . . . 15. *Zizia.*

Flowers all pedicelled; fruits with thin dorsal and lateral wings. . . . . 34. *Thaspium.*

*k.*Leaflets entire, linear to linear-spatulate or oblanceolate; petals white to purple.

Leaflets sessile; involucre of several filiform bracts; fruit ovoid-flask-shaped, with slender neck; annual with fibrous roots. . . . . 27. *Cynosciadium.*

Leaflets with slender petiolules about equaling blade; involucre 0 or a rudiment; fruit ellipsoid, rounded at summit; perennial with fascicle of spindle-shaped fibres. . . . . 37. *Oxypolis.*

*j.*Leaflets more than 3. . . *l.*

*l.*Leaflets entire.

Leaves pinnate; leaflets elongate; petals white; plants of wet or peaty habitats, with fibrous or fusiform roots.

Leaflets linear, at most 1.5 mm. wide; leading umbels 3–9 cm. broad, with involucre of several narrow bracts; annual, with fibrous roots. . . . 27. *Cynosciadium.*

Leaflets linear to oblong, 0.2–4 cm. broad; leading umbels 0.6–1.6 dm. broad, without involucre or with only 1 or 2 small bracts; perennial with thick fusiform roots. . . 37. *Oxypolis.*

Leaves ternately decompound; leaflets lanceolate, ovate or elliptical; petals yellow or creamy; perennials of dry woods, thickets and slopes, with heavy deep root.

Petals yellow; fruit about 4 mm. long, flattened laterally, wingless. . . . . 20. *Taenidia.*

Petals creamy; fruit 4–6 mm. long, flattened dorsally, with thin broad wing. . . . . 40. *Pseudotaenidia.*

*l.*Leaflets toothed or incised. . . *m.*

*m.*Scapose, the leafless flowering stem arising from among basal leaves.

Umbel subtended by a ternately compound leaf; fruits with equal filiform ribs; delicate woodland herb, with subglobose tuber. . . . . 10. *Erigenia.*

Umbel exinvolucrate; fruits broadly winged; firm-leaved plants of the Plains, with deep tap-roots.

Involucel cup-like, deeply cleft, the segments foliaceous; umbel dense, subglobose; 1 or 3 dorsal ribs of fruit flattened into wings; stylopodium obsolete. . . . . . 39. *Cymopterus.*

Involucel of distinct slender bractlets; umbel open in fruit; dorsal ribs of fruit filiform; stylopodium definite. 42. *Lomatium.*

*m.*Leafy-stemmed. . . *n.*

*n.*Larger leaves simply or doubly pinnate, the petiole not ternately forking at summit. . . *o.*

*o.*Involucre of many persistent linear or lanceolate bracts. . . *p.*

*p.*Principal leaves simply pinnate, the linear to narrowly ovate or oblong sessile or subsessile lateral leaflets copiously toothed from base to apex; involucral bracts herbaceous; petals white; fruit laterally compressed; the plump mericarps wingless, 1.8–3 mm. long; plants of marshland and water.

Stem coarsely corrugated, with stiffly ascending branches (except in submersed states), 0.3–2 m. high; leaflets of principal cauline leaves 9–17, firm, 0.3–1.5 dm. long, closely serrate; fruit ovoid to ellipsoid, prominently ribbed, with tightly recurving styles. . . . . 24. *Sium.*

Stem slender, finely corrugated, ascending or reclining, 1–9 dm. long; principal cauline leaflets membranaceous, 0.5–4 cm. long.

Branches few, strongly ascending; principal cauline leaves with 3–9 simply toothed leaflets; fruit (rarely developed) ellipsoid, prominently ribbed with tightly recurving styles. . . . . 24. *Sium.*

Branches numerous, loosely divergent; leaflets of principal cauline leaves 9–19, lacerate or deeply cleft; fruit subglobose, obscurely ribbed, with spreading styles. . . . . 25. *Berula.*

*p.*Principal leaves bipinnate; the rhombic-obovate coarse leaflets mostly petiolulate, coarsely few-toothed only at summit, entire below; involucral bracts broadly scarious-margined; petals greenish-yellow; fruit dorsally compressed, winged, 5–7 mm. long; escape from cult. . . . . 38. *Levisticum.*

*o.*Involucre none or of few filiform mostly deciduous bracts. . . *q.*

*q.*Relatively slender smooth or smoothish plants with terete firm stems; cauline leaves with leaflets or ultimate segments 0.1–4 cm. broad; oil-tubes mostly extending the length of the fruit, visible only on the back or invisible on both faces. . . *r.*

*r.*Leaves mostly in a basal rosette, simply pinnate, with 4–8 pairs of coarsely toothed oblong or obovate leaflets; strong-rooted perennials of fields, etc. . 21. *Pimpinella.*

*r.*Leaves mostly cauline; plants with fibrous or fusiform-thickened roots or tap-roots.

Leaflets linear to oblong or narrowly obovate, 0.2–4 cm. broad, entire except for few coarse teeth above middle; stem 0.5–1.8 m. high, with fascicled slender fusiform roots; plant of peats, bogs and wet places. . . . . 37. *Oxypolis.*

Leaflets deeply cleft or dissected, the segments narrow; plants of drier habitats.

Stem and petioles scabrous-puberulent; cauline leaves with segments mostly 3–10 mm. broad; petals yellow; fruits with a broad thick corky wing continuous around the depressed and only obscurely ribbed back; perennial, with deep tap-root, of dry habitats, Miss. basin. . . . . 41. *Polytaenia.*

Smooth and glabrous; cauline leaves with segments mostly 0.5–2 mm. broad; petals white (or roseate); fruit without broad wings.

Flowers unsymmetrical, the outer with 2 or more petals much larger than the others, 4–6 mm. long, deeply cleft; fruit subglobose, its nearly nerveless halves separated with difficulty; annual, spread from cult. . . . 9. *Coriandrum.*

Flowers symmetrical, with essentially uniform obcordate petals 1–2 mm. long; fruit compressed, ovate or oblong, with filiform ribs, the mericarps readily separated when ripe.

Stem low-branching, 2–7.5 dm. high; cauline leaves pinnate, with a pair of finely dissected leaf-like stipules below the petiolar sheath; fruit prominently ribbed, with wide-spreading styles; biennial with long tap-root; weed of fields northw. . . 19. *Carum.*

Stem branching only near summit, 0.7–1.2 m. high; upper leaves ternate, without leafy stipules; fruit faintly ribbed, with tightly recurving styles; perennial with fascicled finger-like roots, native of rich woods, prairies, etc., Central states. . . 23. *Perideridia.*

*q.*Coarse, with angulate-corrugated soft or hollow stems; larger leaflets 0.3–1 dm. or more broad; oil-tubes of one or both faces of the very thin scale-like fruits conspicuous, linear-oblanceolate tò obclavate, only one-fourth to three-fourths length of fruit.

Glabrous biennial with thick tap-root (parsnip); petals yellow, essentially uniform; fruits 4–7 mm. long, the dorsal oil-tubes filiform and nearly as long as fruit, the commisural linear-oblanceolate and about three-fourths as long. . . . . . 43. *Pastinaca.*

Hairy perennial with stout root; petals white, very unequal in the outer flowers; fruits 0.8–1.5 cm. long, the obclavate dorsal oil-tubes one-third to two-thirds the length of the fruit. . . . . . 44. *Heracleum.*

*n.*Larger leaves with petiole forking into 3 primary divisions, the leaves therefore ternately compound. . . *s.*

*s.*Fruit clavate, strigose-setose on the angles, 1–2.5 cm. long, attenuate into long slender upwardly barbed bases. . 7. *Osmorhiza.*

*s.*Fruit not long-attenuate at base, glabrous or merely finely pubescent. . . *t.*

t. Leaves membranaceous, the principal leaflets deeply dissected into narrow segments. . . *u.*

u. Leaves chiefly at base of naked scape-like flowering stem, the latter with a ternately divided leaf subtending the 2–4-rayed umbel; vernal delicate plant with deep-seated subglobose tuber. . . . 10. *Erigenia.*

u. Stem with alternate leaves; involucre, when present, of several slender bracts; mostly later-flowering plants. . . *v.*

v. First or central umbel quickly overtopped by the upper branches. . . *w.*

w. Umbels simple or with only 2–4 rays, some often sessile at the nodes (without primary or common peduncle); ovary and fruit linear to linear-lanceolate; vernal-flowering annuals.

Rays of umbel strongly ascending; bractlets of involucels oblong to narrowly obovate; stylopodium depressed-conic. . . . . . 5. *Chaerophyllum.*

Rays of umbel soon wide-spreading; bractlets linear-acicular or wanting; stylopodium subulate. . . . . 6. *Anthriscus.*

w. Umbels with 5 or more rays.

Primary involucre of several narrow (finally deciduous) bracts; stem with viscid usually purple blotches and spots; ribs of plumpish ovoid fruit undulate; tall biennial with taproot; a weed. . . . . 11. *Conium.*

Primary involucre wanting or rarely of 1 or 2 bracts; stems without viscid spots; ribs of fruit not undulate.

Rays of umbel very unequal, several of them twice or thrice shorter than the others; involucels drooping, mostly longer than pedicels; petals of outer flowers very unequal; annual or biennial weed. 29. *Aethusa.*

Rays of umbel subequal, mostly elongate; involucels inconspicuous and spreading or short or wanting.

Involucels of numerous narrowly ovate short reflexed bractlets; petals white; fruits linear-cylindric, lustrous, with subulate beaks (stylopodia). . . . 6. *Anthriscus.*

Involucels of filiform short bractlets or wanting; petals yellow; fruits ovate, ovoid or oblong; stylopodium cushion-like or wanting.

Glabrous throughout; petals greenish-yellow, about 0.6 mm. long; fruit 2–3 mm. long, with filiform ribs, the stylopodium cushion-like; garden annual or biennial, escaped. . . . 14. *Petroselinum.*

Bearded at the nodes; petals bright-yellow, larger; fruit about 6 mm. long, with broad thin wings; stylopodium wanting. . . . . . 34. *Thaspium.*

v. First or central umbel rarely and then but slightly overtopped by branches.

Simple nearly to summit; some leaves ternately divided, others pinnate, with prolonged linear segments; stipular bases of petioles close; rays of central umbel 3–10 cm. long; fruit equally slender-ribbed; roots fascicled, finger-like. . . . . 23. *Perideridia.*

Branching near or below middle; leaves all ternate, with shorter leaflets; stipular bases inflated; rays of central umbel 0.5–5 cm. long; fruit winged; roots stout, not fascicled. . . 35. *Conioselinum.*

t. Leaves firm to subcoriaceous or fleshy; the principal

leaflets merely toothed, not divided into narrow segments. . . *x*.

x.Stem glabrous or only minutely pilose, mostly firm and cylindric; larger leaflets of lower leaves 0.1–1.5 dm. long, lanceolate to ovate or oblong; flowers relatively small; fruits plump or somewhat flattened, not scale-like; oil-tubes, if conspicuous on back of fruit, slender and extending quite the length of the intervals. . . *y*.

y.Upper sheaths (in and about inflorescence) shorter than the blade-bearing portion of leaf. . . *z*.

z.Sheaths close, relatively slender, tapering or merely rounded or notched at summit; stem slender; first or central umbel with 2–21 glabrous or barely pubescent rays; involucels wanting or of slender bractlets much shorter than pedicels.

Petals yellow (rarely purple); stylopodium wanting.

Central perfect flower and fruit of each umbellet sessile; fruits with filiform ribs. 15. *Zizia*.

All flowers pedicelled; fruits with thin dorsal and lateral wings. . . . . 34. *Thaspium*.

Petals white or whitish; stylopodium prominent.

Styles shorter than to barely longer than the low-conical stylopodia; involucels of linear bractlets; fruit dorsally compressed, 4–10 mm. long; rhizome or deep root stout. . . . . 32. *Ligusticum*.

Styles filiform, much longer than the stylopodia; fruit laterally compressed, 2–4 mm. long; stoloniferous or with fascicled finger-like roots.

Loosely stoloniferous; leaflets of radical leaves 9; petals with prolonged linear incurved tips; stylopodium conical; escape from cult. . . . . 22. *Aegopodium*.

Non-stoloniferous, the roots a fascicle of finger-like or tuberiform fleshy fibers; leaflets of radical leaves mostly very numerous; petals without prolonged tips; stylopodium depressed; native of damp or wet places. . . . . . 16. *Cicuta*.

z.Sheaths inflated, strongly veiny, with strongly rounded and obcordate prolonged summits; stem coarse; first or central umbel with 20–50 closely scabrous-puberulent rays; involucels of numerous lanceolate to oblong or broader bractlets nearly equaling to exceeding the pedicels; petals whitish; fruit barely compressed, ellipsoid, with strong subequal corky ribs; maritime and subalpine, with coarse root. . . . . 33. *Coelopleurum*.

y.Upper sheaths (in and about inflorescence) bladeless or much longer than their reduced blades; stem stout; petals whitish; fruit dorsally compressed, winged, 4–10.5 mm. long, the depressed stylopodium with a strongly crenate disk. . . 36. *Angelica*.

x.Stem hairy, coarse, angulate-corrugated, soft; larger leaflets of lower leaves 1.5–6 or more dm. long and broad; marginal flowers of the umbellets 0.8–1.5 cm. broad, with very unequal white petals; fruits thin and scale-like, obcordate, 0.8–1.4 cm. long; oil-tubes obclavate, superficially conspicuous, much shorter than intervals. . . . . 44. *Heracleum*.

Subfam. HYDROCOTYLOÌDEAE

1. HYDROCÓTYLE L. Water-Pennywort. Navelwort

Calyx-teeth obsolete. Carpels with 2 of the ribs enlarged and often forming a thickened margin; oil-tubes none, but usually a conspicuous oil-bearing layer beneath the epidermis. — Low mostly smooth paludal or aquatic perennials of temp. and warm reg. with slender creeping stems, round, peltate or reniform leaves, and scale-like stipules. Flowers small, whitish, in simple umbels or clusters, which are either single or proliferous (one above another), appearing all summer. (Name from the Greek *hydor, water,* and *cotyle, a flat cup,* the peltate leaves of several species being somewhat cup-shaped.)

Leaves peltate.
  Inflorescence a simple (rarely proliferating) loose subglobose umbel, in maturity 1.3–3 cm. in diameter, the (12–)20–100 or more filiform pedicels radiating in all directions; fruits deeply notched at base. . . . . . . 1. *H. umbellata.*
  Inflorescence simple or forking, moniliform, of hemispherical verticillate umbels 3–15 mm. in diameter, the 2–20 flowers and fruits merely divergent or ascending; fruits rounded to only shallowly notched at base. . . . . . 2. *H. verticillata.*
Leaves with deep sinuses, not peltate.
  Flowering stems and branches prostrate and repent; umbels definitely peduncled.
    Aquatic or subaquatic, with fleshy stems, petioles and peduncles; petioles mostly 0.4–4 dm. long, leaf-blades 1–6 cm. broad; peduncles much shorter than petioles; umbels hemispherical; fruits stalked. . . . . 3. *H. ranunculoides.*
    Terrestrial, with filiform stems, petioles and peduncles; petioles 0.5–2 cm. long, leaf-blades 5–10 mm. broad; peduncles equaling or exceeding petioles; umbels spherical; fruits sessile. . . . . 4. *H. sibthorpioides.*
  Flowering stems ascending from creeping bases, bearing filiform superficial stolons; umbels sessile or very short-stalked. . . . . 5. *H. americana.*

1. H. umbellàta L. (with umbels). — *Stems* freely branching and elongate, fleshy, *creeping and rooting at nodes; leaves erect, on long petioles; the blades* suborbicular to reniform, *centrally peltate,* coarsely low-crenate, mostly 1–7.5 cm. broad; peduncles erect, nearly equaling to overtopping leaves; *umbel terminal* (rarely proliferating), *loosely subglobose, becoming* 1.3–3 *cm. in diameter,* the filiform *widely radiating pedicels* (12–)20–100; *fruits* reniform, *with deeply notched base,* 1–2 mm. high, 2–3 mm. broad, the remote dorsal ribs obtuse. — Pond-shores, ditches and low grounds, Fla. to Tex. and Mex., n. to Mass. and w. N.S., N.Y., ne. O. and s. Mich. July–Sept.; *fr.* Aug.–Nov. (Trop. Am.)

2. H. verticillàta Thunb. (whorled). — In habit similar to no. 1, glabrous; *inflorescences* simple or with 2–several erect branches, *moniliform,* usually overtopped by the leaves; *the* 1–12 *or more whorls or small hemispherical umbels of* 2–20 *spreading or ascending sessile or subsessile flowers; fruit* depressed, much broader than long, 1–2.5 mm. high, 2–4 mm. broad, *broadly rounded to subtruncate or barely notched at base,* the distant acute ribs prominent. — Swamps, shores and low grounds, Fla. to Tex. and Mex., n. to se. Mass., Mo., Okla., Utah and Oreg. June–Aug.; *fr.* Aug.–Nov.

Var. **Fetherstoniàna** (Jennings) Mathias (named in 1932 for its discoverer, Edith Fetherstone). — Petioles sparsely and minutely hirsute, only 1.5–9 cm. long; fruits more or less emarginate at base. — By woodland-stream, Wyoming Co., N.Y., local.

Var. **triradiàta** (A. Richard) Fern. (three-rayed; described from plants with 3-forked scapes). — Inflorescence often forking (more often simple northward); *flowers on distinct pedicels* 1–10 *mm. long;* fruit often shallowly emarginate at base. (Var. *racemosa* (Moc. & Sessé) Mathias; *H. Canbyi* and *H. australis* C. & R.) — Damp soils, Fla. to Tex. and Mex., n. to se. Mass. and Calif. (Trop. Am.)

3. H. ranunculoìdes L. f. (resembling *Ranunculus*). — *Fleshy aquatic or subaquatic* plant *with* widely *creeping or floating stems* sending out long roots at the nodes; *petioles erect, mostly* 0.4–4 *dm. high, the leaf-blades* 3–7*-cleft,* the lobes crenate, the sinus open; *peduncles much shorter than petioles,* 1–6 *cm. long, soon recurving; umbels* dense, *hemispheric; fruits stalked,* brownish, rounded or short-oblong, 2–3 *mm. long, the stylopodia and styles* 0.6–1 *mm. long,* the ribs obscure. — Shallow water and shores, Fla. to Tex. and Mex., n. to se. Pa., W.Va., Ark., Okla. and Oreg. May–Aug.; *fr.* July–Oct.

4. H. sibthorpioìdes Lam. (resembling *Sibthorpia*), Lawn-W. — Delicate, the *filiform stems creeping and rooting at the nodes; petioles filiform,* 0.5–2 *cm. long; the blades* reniform,

5–10 *mm. broad*, with crenate margin and deep sinus; *peduncles filiform, equaling or exceeding petioles; umbel capitate,* 2–4 *mm. in diameter;* the *pale sessile fruits* 1–1.5 *mm. long*, the stylopodia and styles less than 0.5 mm. long. (*H. rotundifolia Lam.*) — Lawns, roadsides, etc., Pa. and Del. to Ind. and Ky. Apr.–Sept. (Natzd. from Asia and Afr.)

1356. H. americana.

5. H. americàna L. (American). — Propagating by slender tuberiferous stolons; stems filiform, *branching and creeping; leaves thin*, round-reniform, *not peltate, crenate-lobed* and the lobes crenulate, shining; few-flowered *umbels axillary and almost sessile;* fruit less than 2 mm. broad; intermediate ribs prominent; no oil-bearing layer; seed-section broadly oval. — Meadows, damp woods, etc., Nfld. to Minn., s. to N.S., N.E., L.I., Md., upland of N.C., and Tenn. Late June–Sept. FIG. 1356.

2. CENTÉLLA L.

Calyx-teeth obsolete. Petals white, imbricated in bud. Carpels 7–9-ribbed and somewhat reticulate. — Creeping perennials of N. and S. Am., Afr., s. Asia and Austral. reg., with simple ovate to reniform leaves. Umbels subtended by 2 conspicuous bracts. (Name a diminutive of *centum;* from the rounded coin-like leaves.)

1. C. erécta (L. f.) Fern. (erect). — Leaves with ascending petioles 0.05–3 dm. long; the cordate-ovate fleshy repand-dentate blade 1.5–8 cm. long; scapes filiform, ascending to arching, 0.1–1 dm. long; fruits 3.5–5 mm. broad. (*C. asiatica* of ed. 7, not (L.) Urban; *C. repanda* (Pers.) Small and *C. floridana* (C. & R.) Nannf.) — Wet or exsiccated sand or clay or margins of pools, Fla. to Tex. and Mex., n. on Coastal Plain to Del. May–Oct. (W.I.; Centr. Am.)

Subfam. SANICULOÏDEAE

3. SANÍCULA L. SANICLE. BLACK SNAKEROOT

Calyx-teeth manifest, persistent. Fruit globular to ellipsoid; the carpels not separating spontaneously, ribless, thickly clothed with hooked prickles. — Perennial or biennial rather tall glabrous herbs of N. and S.Am., Eurasia and Afr., with few palmately lobed or parted leaves, those from the base long-petioled. Umbels irregular or compound, more or less paniculate, the flowers (greenish or yellowish) capitate in the umbellets, perfect, and with staminate ones intermixed or in separate umbellets. Involucre and involucels few-leaved. (Name said to be from *sanare, to heal.*)

Styles much exceeding bristles of fruit; principal branches of inflorescence strongly ascending; staminate flowers in separate umbellets or mixed with the perfect ones; plants perennial, with short rhizomes.
- Sepals of staminate flowers rigid, lance-subulate, 1–2 mm. long; fruits sessile, 5–8 mm. long. . . . . 1. *S. marilandica.*
- Sepals of staminate flowers soft, deltoid, 0.5–1 mm. long; fruits stipitate, their bodies 2.5–4 mm. long. . . . . 2. *S. gregaria.*

Styles not exceeding bristles; principal branches divergent; staminate flowers mixed with the pistillate; biennial, the base shriveling after fruiting (no. 5 with tuberous roots perhaps perennial).
- Pedicels of staminate flowers three or four times length of calyx; fruit elongate, 5.5–7 mm. long, with beak of sepals prolonged beyond bristles. . . . . 3. *S. trifoliata.*
- Pedicels of staminate flowers once or twice length of calyx; fruit rounded, 3–6 mm. long, its beak hidden by bristles.
  - Roots fibrous; longer branches of inflorescence twice or thrice forked, apparently dichotomous through shortening of middle ray; fruits stipitate, 3–5 mm. long; styles included within calyx. . . . . 4. *S. canadensis.*
  - Roots tuberous; longer branches once (rarely twice) 3-forked, middle ray elongate; fruits sessile, 5–6 mm. long; styles exceeding calyx. . . . . 5. *S. Smallii.*

1357. S. marilandica, × ca. 1.

1. S. marilándica L. (of Maryland). — Perennial with thick rhizome, 0.3–1.2 m. high; *leaves thickish*, the basal long-petioled, mostly borne on rhizome, 1–3 dm. across, mostly with 5 cuneate obovate to elliptic unequally serrate leaflets, lateral leaflets often deeply cleft; cauline leaves similar but smaller, the upper nearly sessile, the lower with sessile or short-petiolulate leaflets; *inflorescence with* 2–several stiff *ascending subequal* usually 3-forked *rays* and a shorter simple central ray; *staminate flowers* mixed with the perfect ones or in separate umbellets; *their sepals lance-subulate, rigid,* 1–2 *mm. long; fruits* 3–8 in each umbellet, ovoid, *sessile,* 5–8 *mm. long; styles long-exserted,* curving. — Thickets, shores, meadows and open woods, n. Nfld. to Hudson Bay and B.C., s. to N.S., N.E., Va., upland of nw. Fla., Great Lakes states, n. Kans. and Colo. May–July. FIG. 1357.

Var. **petiolulàta** Fern. (with petiolules). — Leaflets of 1 or 2 lower cauline leaves on petiolules 1.5–5 cm. long. — Dry sandy pineland, se. Va. to S.C.

2. S. gregària Bickn. (herding together). — More slender, 3–8 dm. high; *leaves membranaceous,* mostly smaller, the leaflets sharply and finely serrate; *inflorescence* more leafy, its *rays capillary; staminate flowers* tiny, *their herbaceous deltoid sepals* 0.5–1 *mm. long; fruits stipitate, their bodies* 2.5–4 *mm. long.* — Rich woods and thickets, w. N.B. to Minn., s. to N.S., N.E., n. Fla., Ala., Mo. and e. Kans. Late Apr.–July. FIG. 1358.

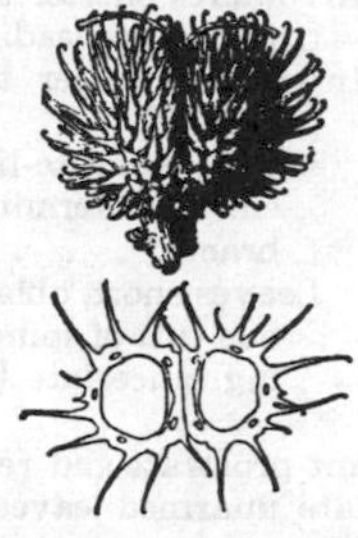

1358. S. gregaria, × 4.

3. S. trifoliàta Bickn. (with 3 leaflets). — *Biennial,* 3–8 dm. high; *leaves thin,* 3-*foliolate; lateral leaflets* of the lower leaves often cleft, *of the upper mostly uncleft, rhombic-obovate,* coarsely double-serrate; inflorescence with large subtending leaves; primary rays mostly alternate, spreading-ascending, elongate; *staminate flowers* few, mixed with perfect ones, *their pedicels three or four times length of calyx;* sepals firm, with incurved subulate tips; *fruits ellipsoid-fusiform,* sessile, 5.5–7 *mm. long, with the beak of sepals overtopping the bristles.* — Rich woods, w. N.B. to Minn., s. to N.E., Va. and Tenn. June, July. FIG. 1359.

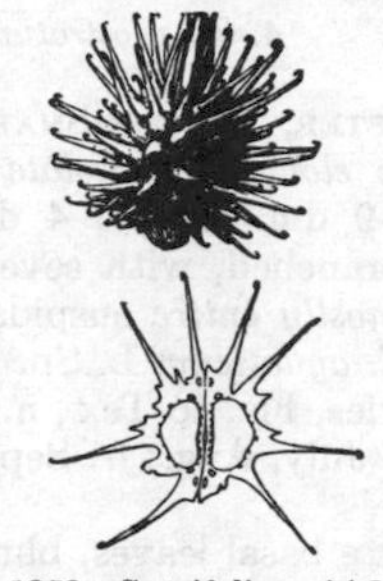

1359. S. trifoliata, × 3.

4. S. canadénsis L. (Canadian; a misnomer). — Biennial, 0.3–7.5 dm. high, *divergently and seemingly dichotomously branched, the longer branches twice or thrice forked, with central simple ray very short;* lower and median leaves 3-foliolate, with lateral leaflets deeply cleft and sharply and doubly serrate with acerose teeth or often incised; larger leaflets of petioled leaves 3.5–8 cm. long and 1.5–4(–5) cm. broad, of lower subsessile leaves (at lower fork of stem) 3–7 cm. long; *upper leaves much reduced, the bracteal very small and commonly paired; staminate flowers* mixed with perfect ones, few, *on pedicels only slightly longer than calyx,* their sepals firm and subulate-tipped; triads of fruits 7–9 mm. broad, the individual *fruits stipitate with* rounded *bodies* 3–4 *mm. long, their sepals overtopped by bristles, the styles included.* — Dry open woods, Fla. to Tex., n. to s. N.H., e. and se. Mass., centr. Conn., N.J., Pa., W.Va., O., Ky., Mo. and Okla. May–July.

Var. **grándis** Fern. (large). — Coarser, up to 2 m. high; larger leaflets of petioled leaves 5.5–13 cm. long and 2.5–8 cm. broad, of lower subsessile leaves 4.5–12 cm. long; triads of fruits 1–1.5 cm. broad. — Rich woods, w. Vt. to Minn. and Neb., s. to N.C., Tenn., Mo., Okla. and Tex. FIG. 1360.

1360. S. canadensis, v. grandis, × 3.

Var. **floridàna** (Bickn.) H. Wolff (of Florida). — Smaller throughout, 1.5–9 dm. high; *larger leaves* 2–8 *cm. broad, their abruptly cuneate firm leaflets spinulose-toothed.* — Dry sandy woods, Fla. to Miss., n. to se. Va. June, July.

5. S. Smállii Bickn. (for its discoverer, JOHN KUNKEL SMALL, 1869–1938). — Biennial or perennial, with *tuberous-thickened roots,* 3–7.5 dm. high; *leaves coriaceous, pale beneath,* 3-foliolate; *leaflets* rhombic-obovate, *obtuse to subacute,* broadly acerose-toothed, the lateral leaflets often cleft; inflorescence with few elongate rays, the *longer branches once* (rarely twice) 3-*forked, the simple middle ray elongate;* staminate flowers about half as long as their thick pedicels, few, mixed with perfect ones; *fruits sessile,* rounded, 5–6 *mm. long, their short sepals overtopped by the styles.* — Rich woods and thickets, Fla. to Tex., n. to s. Va., Tenn. and se. Mo. May, June.

4. ERÝNGIUM L. ERYNGO

Calyx-teeth prominent, rigid and persistent. Styles slender. Fruit ovoid or obovoid, covered with little hyaline scales or tubercles, with no ribs, and usually 5 slender oil-tubes on each carpel. — Chiefly perennials of temp. and trop. reg., with usually coriaceous toothed, cut, or prickly leaves, and blue or white bracted flowers closely sessile in dense heads. (*Eryngion,* classical name for some prickly plant.)

Plant erect, coarse, with hard more or less spinescent to bristly-margined leaves; leading heads becoming 1–4.5 cm. long.

Involucres shorter than the half-breadth of the head, their bracts mostly entire; mature leading heads 1.5–2.5 cm. long. . . . . 1. *E. yuccifolium.*

Involucres longer than the half-breadth of the head, their bracts mostly pinnatifid.

Leaves elongate-linear to -lanceolate; leading heads 1–1.5 cm. long, without enlarged terminal bractlets; involucre with linear spreading or reflexed bracts. 2. *E. aquaticum.*

Leaves short, oblanceolate; leading heads 2.5–4.5 cm. long, with a conspicuous tuft of spinescent enlarged terminal bractlets; involucre with ascending lanceolate bracts. . . . . 3. *E. Leavenworthii.*

Plant prostrate and repent, with filiform stems and herbaceous ovate to lanceolate unarmed leaves; heads 4–9 mm. long. . . . . 4. *E. prostratum.*

1. E. yuccifòlium Michx. (with leaves of *Yucca*), RATTLESNAKE-MASTER, BUTTON-SNAKE-ROOT. — Stiffly erect, the coarse stem 0.5–1.8 dm. high; *lower leaves elongate-lanceolate or broadly linear, hard,* more or less slenderly spinescent-margined, 1.5–9 dm. long, 1–4 dm. broad, *prominently parallel-veined;* inflorescence elongate, irregularly branched, with several ovoid-globose *heads becoming* 1.5–2.5 *cm. long, with* ovate-lanceolate *mostly entire* cuspidate *bracts shorter than the half-diameter of the head,* and similar bractlets. (*E. aquaticum* L. Sp. Pl. ed. 2, in part, not ed. 1) — Dry or moist open woods, thickets and prairies, Fla. to Tex., n. to N.J. and s. Ct. (perhaps introd.), O., Mich., Wisc., Minn. and e. Kans. July, Aug.; *fr.* Sept.–Nov.

E. PLÀNUM L. (flat), with turnip-like root, firm cordate oblong to ovate basal leaves, bluish inflorescence, with the heads 1–1.5 cm. long and exceeded by the involucre, is beginning to spread from cult. (Introd. from Eurasia)

2. E. aquáticum L. (aquatic). — Slender, erect, 0.4–1.5 m. high; the *firm lower leaves elongate-linear to -lanceolate,* on long fistulous petioles; the *blades* 0.5–3.5 cm. broad, entire or with hooked teeth, *pinnately veined;* upper leaves sessile, spiny-toothed or laciniate; *heads* many in a terminal irregular umbel, *the leading ones becoming* 1–1.5 *cm. long; involucre of linear spreading or reflexed bracts* nearly as long as the head. (*E. virginianum* Lam.) — Fresh to brackish marshes, streams, ponds and bogs, Fla. to Tex., n. to N.J. Late July–Sept.; *fr.* Sept.–Nov.

3. E. Leavenwórthii T. & G. (for its discoverer, MELINES CONKLIN LEAVENWORTH, 1796–1862). — Stout, 4–9 dm. high; *lowest cauline leaves broadly oblanceolate, spinosely toothed, the rest* sessile and *deeply and palmately parted into narrow incisely pinnatifid spreading pungent segments; heads* ovoid-ellipsoid, 2.5–4.5 *cm. long, with pinnatifid spinose ascending* long *bracts* and 3–7 cuspidate *bractlets, the terminal ones very prominent and resembling the bracts.* — Dry soil, e. Kans. and Ark. to Tex. Aug.–Oct.

4. E. prostràtum Nutt. (prostrate). — *Prostrate, with filiform* forking and *repent stems* 1–6 dm. long; *leaves membranaceous; the basal ones* slender-petioled, *unarmed,* oblong to ovate, entire or bluntly lobulate, 1.5–7 cm. long; cauline leaves mostly smaller, falsely verticillate; *heads mostly solitary at nodes,* slender-peduncled, thick-cylindric, obtuse, 4–9 mm. long, blue, soon fading; the involucre of reflexed narrow bracts; *fruit subglobose or hemispherical, about* 1 *mm. broad, broader than high,* covered with subsessile papillae and *capped by broad loosely ascending marcescent calyx-segments.* — Damp sands, shores, etc., Fla. to e. Tex., n. to se. S.C., sw. Ga., w. Ky., se. Mo. and e. Okla. Late June–Nov.

Var. **disjúnctum** Fern. (separated). — *Fruits mostly higher than broad, mostly obconic,* 0.6–0.9 *mm. long,* 0.5–0.6 *mm. broad.* — Sandy pond-shores, local, se. Va.

Subfam. APIOÌDEAE

5. CHAEROPHÝLLUM L.

Calyx-teeth obsolete. Fruit slenderly ovoid to linear-lanceolate, notched at base, with short beak or none and equal ribs; oil-tubes solitary in the intervals; seed-face more or less deeply grooved; stylopodium low-conic. — Annuals of N. Hemisph., with ternately decompound leaves, pinnatifid leaflets with small lobes, mostly no involucre, involucels of many bractlets, and white flowers. (Name from the Greek *chairo, rejoice,* and *phyllon, a leaf,* alluding to the agreeable odor of the foliage.)

Lobes of leaflets oblong, lanceolate or oblanceolate; larger simple umbels or umbellets with 1–10 fruits; stylopodia erect or slightly divergent, distinct to base.
Fruiting pedicels 1–6, filiform, nearly uniform in thickness; intervals of fruit broader than ribs; leaf-surfaces glabrous. . . . . 1. *C. procumbens.*
Fruiting pedicels 3–10, thickish, clavate; ribs (except in var. *floridanum*) broader than intervals; leaves (especially lower) pilose. . . . . . . 2. *C. Tainturieri.*
Lobes of leaflets linear; larger simple umbels or umbellets with 6–15 fruits; stylopodia connivent or convergent. . . . . 3. *C. texanum.*

1. C. procùmbens (L.) Crantz (lying flat). — Slender, simple, or loosely branched below, 1.5–5.5 dm. high, the *weak stem glabrous* except sometimes at base; *lobes of glabrous leaflets oblong,* bluntish or round-tipped; umbels simple or with 2 or 3 slender ascending rays; bractlets of involucels elliptic to narrowly obovate; *fruits* 1–6, *on filiform pedicels;* their glabrous bodies narrowly oblong, 6–10 mm. long, 1.5–2 mm. thick, contracted above into a thick neck below *the erect stylopodia.* — Rich low woods and shady places, N.Y. and s. Ont. to Ia., s. to Ala., Miss., Ark. and e. Kans. Apr.–June. Fig. 1361.

1361. C. procumbens.

Var. **Shòrtii** T. & G. (for its discoverer, Charles Wilkens Short, 1794–1863). — Fruit more broadly oblong to narrowly ovate, glabrous or minutely pubescent, 4.5–6.5 mm. long, 2–2.5 mm. broad, not contracted at summit. (*C. Shortii* (T. & G.) Bush) — Similar habitats, w. Pa. to Ind., s. to w. Va., Tenn. and La. — Often earlier flowering than typical *C. procumbens.*

2. C. Tainturièri Hook. (for L. F. Tainturier des Essarts, who sent plants of Louisiana to Sir William Hooker from 1824–1836). — Stiffer, erect or spreading, 2–7 dm. high; *stem densely pilose, especially at base; leaves mostly pilose, the lobes of the leaflets lanceolate to oblanceolate,* acutish; bractlets often reflexed in fruit; *fruits* 3–10, *on clavate pedicels, their* glabrous lance-oblong tapering *bodies with ribs broader than the intervals.* — Open woods, roadsides and waste places, Fla. to Tex., n. to Va., s. O., s. Ind., Mo. and Kans. Apr.–June.

Var. **floridànum** C. & R. (of Florida). — Intervals broader than ribs of fruit. (*C. floridanum* (C. & R.) Bush) — Fla. to se. Va. and Mo.

3. C. texànum C. & R. (of Texas). — Erect, 2–6 dm. high, simple or with stiffly ascending branches, rather densely short-pubescent below; *lobes of leaflets linear;* bractlets often reflexing; *fruits mostly* 6–15, *on clavate pedicels,* lance-oblong, 5–8 mm. long, tapering above but scarcely beaked, the ribs broader than the intervals; *stylopodia connivent.* — Prairies, open woods, barrens and waste places, Mo. and Kans. to Tex. Apr.–June.

6. ANTHRÍSCUS Bernh. Chervil

Fruit linear-cylindric to lance-ovoid, notched at base, beaked, glabrous, smooth or muricate, without ribs (but the beak ribbed); oil-tubes none; stylopodium subulate, seed-face sulcate. — Resembling *Chaerophyllum* in vegetative characters, nat. of Old World. (The ancient Roman name.)

Fruits lance-ovoid, sharply muricate. . . . . 1. *A. scandicina.*
Fruits linear-cylindric, smooth.
Lateral umbels sessile or short-peduncled at the nodes of the stem, rays pubescent; body of fruit twice as long as slender beak. . . . 2. *A. Cerefolium.*
Lateral and terminal umbels long-peduncled, rays glabrous; body of fruit many times longer than beak. . . . . 3. *A. sylvestris.*

1. A. scandicìna (Web.) Mansf. (resembling *Scandix*). — Slender, 1–6 dm. high, essentially glabrous; leaves 3-pinnatisect, the obtuse ultimate segments pinnatifid, lower leaves long-petioled, upper sessile; *umbels short-peduncled with* few *glabrous rays,* mostly without involucres, but with involucels of 4 or 5 linear bracts; *fruit ovoid-lanceolate, the sharply muricate bodies about thrice length of the smooth beak.* (*A. vulgaris* Pers., not Bernh.) — Waste places, as yet local, se. Va. May; *fr.* June. (Adv. from Eu.)

2. A. Cerefòlium (L.) Hoffm. (old generic name; wax-leaf). — Very slender, 2–8 dm. high; stem pubescent near the nodes; *umbels sessile or short-peduncled* at the nodes and terminal; *rays minutely pubescent; body of linear-cylindric fruit twice as long as slender beak.* — Roadsides and waste places, local, Que. to Pa., originally cult. May–Aug. (Introd. from Eu.)

3. A. sylvéstris (L.) Hoffm. (of woodland). — Coarse, up to 1 m. high, *pubescent at base; umbels all peduncled; rays glabrous; body of lustrous lance-linear fruit many times longer than beak.* — Fields and waste places, se. Nfld. and vicinity of Montreal, Que., to N.J. Late May–Aug. (Natzd. from Eu.)

7. OSMORHÌZA Raf. Sweet Cicely

Calyx-teeth obsolete. Fruit with prominent caudate attenuation at base and equal ribs. — Glabrous to hirsute or villous perennials of N. and S.Am. and e. Asia, with thick aromatic roots, ternately compound leaves, ovate variously toothed leaflets, few-leaved involucres, and white flowers in few-rayed and few-fruited umbels. (Name from the Greek *osme, a scent*, and *rhiza, a root.*) Washingtonia Raf.

Rays of umbellules subtended by an involucel of simple narrow spreading to reflexed bractlets (these more or less deciduous at ripening of fruit); wide-ranging to our southern limits.
Stipules ciliate-hispid, otherwise glabrous; bractlets of involucel linear-attenuate, 3–8 mm. long; stylopodium and connivent styles 0.7–1.5 mm. long; vertical roots only slightly fleshy, rather fibrous and covered by rootlets, rank-tasting, only slightly sweet. . . . . 1. *O. Claytoni.*
Stipules tomentose-felted at least toward margin; bractlets of involucel lance-attenuate, 0.5–1 cm. long; stylopodium and distinct to spreading styles 2–4 mm. long; vertical roots fleshy, fusiform, sweet and aromatic, the larger ones 3–10 mm. thick. . . . . 2. *O. longistylis.*
Rays of umbel mostly without involucels; plants northern.
Body of fruit 8–12 mm. long, rounded or short-beaked at tip; stylopodium depressed, broader than high, with the styles 0.3–0.5 mm. long. . . . 3. *O. obtusa.*
Body of fruit 11–17 mm. long, with a conical beak about 2 mm. long; stylopodium higher than broad, conical, with the styles about 1 mm. long. . . . 4. *O. chilensis.*

1. O. Clàytoni (Michx.) C. B. Clarke (for John Clayton, ? — 1773, pioneer Virginia botanist), Sweet Jarvil. — *Stem* rather slender, 3–9 dm. high, *villous-pubescent*, from *rather slenderly fusiform and fibrous rank-tasting roots; leaves* 2–3-ternate, *crisp-hairy; leaflets* mostly 4–7 cm. long, *acuminate*, crenate-dentate and *often much cleft, rachis and racheolas horizontally hispid; stipules glabrous on back, ciliate-margined;* primary involucre of 0–2 bracts; rays of umbel 3–5; *involucels of linear-attenuate bractlets* 3–6(–8) *mm. long; petals with a very short incurved point; stylopodium and closely appressed-connivent styles* 0.7–1.5 *mm. long;* body of fruit 1–1.3 cm. long, the caudate basal appendages 5–7 mm. long. — Woods and wooded slopes, Gaspé Pen., Que., to s. Sask., s. to N.S., N.E., w. L.I., inland Va., upland N.C. and Ala., e. Kans. and nw. Ark. May, June; *fr.* June–Aug.

2. O. longistỳlis (Torr.) DC. (long-styled), Anise-root. — *Coarser than no.* 1, up to 1.2 m. high, *the stem glabrous or essentially so except at the nodes*, from somewhat *fleshy and often carrot-like strongly anisate and sweet roots up to* 1 *cm. thick; leaves* coarser, *their leaflets broader and less cleft, sparingly hirsutulous to glabrescent on veins beneath; stipules densely pilose near margin;* primary involucre of 1–several bracts up to 1.5 cm. long; rays of umbel 3–6; *involucels of several lanceolate bractlets* 0.5–1 *cm. long; petals larger, with a long incurved point; stylopodium and spreading or remote ascending styles* 2–4 *mm. long;* body of fruit 1.2–1.5 cm. long, *the seed-face more deeply and broadly concave than in no.* 1. — Rich, often alluvial, woods and thickets, rarely (if ever) with no. 1, Gaspé Pen., Que., to s. Alta., s. to N.S., N.E., Va., Ky., Mo., w. Okla. and ne. N.M. May, June; *fr.* June–Aug. Fig. 1362.

1362. O. longistylis.

Var. **brachỳcoma** Blake (with short beard). — Stems, petioles, and at least bases of branches *densely puberulent with spreading hairs averaging* 0.5 *mm. long.* — More local, Ct. to s. Ont. and s. Mich., s. to upland Va., W.Va., O. and Ind.

Var. **villicaùlis** Fern. (with villous stem). — Stem, petioles and at least bases of branches *densely villous with spreading hairs* 0.5–2 *mm. long; backs of stipular leaf-bases densely villous.* (*O. villicaulis* (Fern.) Rydb.) — Coastal Plain of Va. to n. Ala., sw. to e. Tex., n. to s. Ct., s. N.Y., s. O., s. Mich., Ill. and s. Mo., the common var. southeastward.

3. O. obtùsa (C. & R.) Fern. (blunt). — Stems glabrous or sparingly pubescent, 1.5–7 dm. high; leaves 2–3-ternate, more or less crisp-pubescent; leaflets 1.5–6 cm. long, acuminate, the teeth mucronate; *umbels* naked or obsoletely involucrate, *with* 3–5 *naked finally very divergent rays; fruit on divergent long pedicels, the enlarged portion* 8–12 *mm. long, rounded or short-beaked at tip; stylopodium depressed, broader than high, with the styles* 0.3–0.5 *mm. long.* — Woodlands, Nfld. and s. Lab. Pen. to n. N.B.; Ont. and n. Mich. to ne. Minn.; s. Alaska and n. B.C., s. to Black Hills, S.D., N.M. and Calif. Late June, July; *fr.* July–Sept.

4. O. chilénsis H. & A. (of Chile). — Similar, usually taller (4–10 dm. high); *umbels with* 3–7 *ascending-spreading rays; fruit on ascending pedicels,* 11–17 *mm. long, with a conical beak* 2 *mm. long;* stylopodium conical, with the styles about 1 mm. long. (*O. nuda* Torr.; *O. divaricata*

(Britt.) Suksd.) — Woodlands and clearings, Nfld. to Alaska, s. to N.S., n. Me., n. N.H., Ont., n. Mich., n. Wisc., S.D., Ariz. and Calif. Late May–July; *fr.* late June–Aug. (Temp. w. S.Am.)

8. TORÌLIS Adans. HEDGE-PARSLEY

Calyx-teeth short, triangular, persistent. Fruit bristly with hooked prickles or warty, the primary ribs not so prominent as the secondary. — Erect slender caulescent annuals nat. of Old World, with bipinnate leaves, compound umbels, and dense heads of white flowers, the involucres and involucels of linear bracts. (Etymology uncertain.)

1. T. JAPÓNICA (Houtt.) DC. (Japanese). — *Umbels* open, loose, *long-peduncled,* raised above the leaves; *prickles evenly distributed on the fruit.* (*T. Anthriscus* of ed. 7, not Bernh.) — Open woods and waste places, N.Y. to Ia., s. to Fla. and Tex. June–Aug. (Natzd. from Eurasia)

9. CORIÁNDRUM L. CORIANDER

Fruit nearly globose, not at all narrowed at the commissure; ribs filiform or acutish. Seed dorsally compressed, somewhat concave on the inner face. — Slender glabrous herbs nat. of Old World, with pinnately dissected leaves, compound umbels, no involucre, few-parted involucels, and white or roseate unequal petals. (The classical name.)

1. C. SATÍVUM L. (sown). — Lower leaves pinnate, the leaflets flabelliform, many-cleft, cuneate at the base, upper leaves deeply cut into linear segments. — Waste places, chiefly spread from cult. June, July. (Introd. from Eurasia)

10. ERIGENÌA Nutt. HARBINGER-OF-SPRING. PEPPER-AND-SALT

Calyx-teeth obsolete. Petals obovate or spatulate, flat, entire, white. Fruit didymous, laterally flattened, the carpels incurved at top and bottom, nearly reniform, with 5 very slender ribs and 1–3 small oil-tubes in the intervals. — A small glabrous vernal plant with a simple stem, bearing 1 or 2 twice- or thrice-ternately divided leaves and a few-flowered leafy-bracted umbel. (Name from the Greek, meaning *born in the spring.*)

1. E. bulbòsa (Michx.) Nutt. (bulbous). — Stem 1–2.3 dm. high, from a subglobose tuber; leaf-segments linear-oblong; fruit 2 mm. long, 3 mm. broad. — Deciduous woods, etc., s. Ont. and w. N.Y. to Wisc., s. to Ala., Miss. and Mo. Feb.–May.

11. CONÌUM L. POISON HEMLOCK

Fruit somewhat flattened at the sides, glabrous, with prominent wavy ribs; oil-tubes none, but a layer of secreting cells next the seed, the face of which is deeply and narrowly concave. — Poisonous biennials nat. of Eurasia and Afr. with spotted stems, large decompound leaves with lanceolate pinnatifid leaflets, involucre and involucels of narrow bracts, and white flowers. (*Coneion*, the Greek name of the *Hemlock*, by which Socrates and various criminals were put to death at Athens.)

1363. C. maculatum.

1. C. MACULÀTUM L. (mottled or spotted). — A large branching herb, in waste places, Que. to Ia., s. beyond our limits. June–Aug. (Natzd. from Eu.) — Notoriously poisonous, fatal to eat. FIG. 1363.

12. SPERMÓLEPIS Raf.

Involucre none but involucels present. Flowers small, in pedunculate compound irregular umbels. Stylopodium small, conical. Fruit thin-walled; oil-tubes 1 in each interval, 2 on the commissural side. — Slender smooth branching Am. annuals with finely dissected leaves with filiform or narrowly linear segments, umbels with unequal rays, ovoid fruits tuberculate or bristly. (Name from the Greek *sperma, seed,* and *lepis, scale,* alluding to the scurfy or bristly fruit.) LEPTOCAULIS Nutt. — Sometimes placed in the genus *Apium.*

1. S. inérmis (Nutt.) Math. & Const. (unarmed). — Stem geniculate, 3–5 dm. high; leaf-segments linear-filiform; *fruit merely warty;* oil-tubes many. (*S. patens* (Nutt.) Robins.) — Sandy soil and barrens, Ind. to Neb., s. to Ark. and Tex. May, June.

2. S. echinàta (Nutt.) Heller (prickly). — Similar in habit; *fruit bristly;* oil-tubes 6. — Dry prairies and rocky barrens, Miss. to Ariz. and Calif., n. to Mo. April–June.

ÀPIUM (ancient Greek name) GRAVÈOLENS L. (strong-smelling), CELERY, is a casual escape, but hardly persistent. (Introd. from Eu.)

13. BUPLEÙRUM L. Thoroughwax

Calyx-teeth obsolete. Fruit oblong, with very slender ribs, no oil-tubes, depressed stylopodium, and seed-face somewhat concave. — Smooth annual of a large chiefly Old World group, with ovate perfoliate entire leaves, no involucre, involucels of 5 very conspicuous ovate mucronate bractlets, and yellow flowers. (Name from the Greek *bous, an ox,* and *pleuron, a rib.*)

1. B. rotundifòlium L. (round-leaved). — Glaucous, up to 6 dm. high; lower leaves oblong-lanceolate to obovate, upper ones ovate; fruit oblong-ovate, about 3 mm. long, purplish or brown, with filiform ribs. — Roadsides and fields, Ala. to Mo., etc., rarely northw. to N.E., N.Y., W.Va., Ind. and S.D. May, June. (Adv. from Eu.)

14. PETROSELÌNUM Hoffm. Parsley

Calyx-teeth obsolete. Petals greenish-yellow, with attenuate incurved points. Fruit ovate, glabrous, laterally compressed; carpels pentagonal, the primary ribs filiform, subequal; oil-tubes solitary in the intervals; stylopodium cushion-like. — Chiefly biennials nat. of Eu. and Mediterr. reg., with ternately pinnate decompound leaves, toothed leaf-segments, compound umbels, few-parted or no involucres, and several-many-parted or no involucels. (Name from the Greek *petra, a rock,* and *selinon, parsley.*)

1. P. críspum (Mill.) Mansf. (crinkled), Common P. — Leaflets small, ovate 3-cleft or -toothed. (*P. hortense* Hoffm.) — Commonly cult., occasional as an esc. (Introd. from Mediterr. reg.)

15. ZÍZIA W. D. J. Koch

Calyx-teeth prominent. Fruit ovate to oblong, glabrous, with filiform ribs. Oil-tubes large and solitary in the broad intervals, and a small one in each rib; stylopodium wanting; seed terete. — Smooth N. Am. perennials, with mostly *Thaspium*-like leaves, no involucre, involucels of small bractlets, yellow flowers, and the central fruit of each umbellet sessile. Flowering in spring. (Named for *Johann Baptist Ziz,* a Rhenish botanist, 1779–1829.)

Basal leaves all or nearly all simple, ovate to suborbicular. . . . . 1. *Z. aptera.*
Basal leaves all or nearly all ternately compound.
Leaves coriaceous, coarsely dentate or low-serrate; rays of umbel filiform, 2–11, the longest becoming 3–11 cm. long; fruit ovoid to orbicular, 2–3.5 mm. long. . . . . 2. *Z. trifoliata.*
Leaves membranaceous, the leaflets serrate; rays of umbel coarser, 6–20, the longest becoming 2–5 cm. long; fruit oblong, 3–4 mm. long. . . . . 3. *Z. aurea.*

1. Z. àptera (Gray) Fern. (without wings). — *Radical leaves* mostly long-petioled, cordate, *ovate or rounder,* crenately toothed, very rarely lobed or divided; *cauline leaves simply ternate or quinate,* with the coriaceous ovate to lanceolate leaflets serrate, incised or sometimes parted; *longer rays of umbel lengthening to* 1.5–6 *cm.*; *fruit* ovate, 3 *mm. long.* (*Z. cordata* sensu W. D. J. Koch and later auth., not *Smyrnium cordatum* Walt. upon which the name rests) — Woods, thickets, meadows and prairies, R.I. to B.C., s. to Ga., Ala., Mo., Colo., Utah and Oreg.; perhaps adv. in Laurentide Mts., Que. *Fl.* April–June; *fr.* July–Sept. — Often confused with *Thaspium trifoliatum.*

2. Z. trifoliàta (Michx.) Fern. (three-leaved). — Slender, up to 8 dm. high; lower leaves all or nearly all ternate, with 3–9 *dentate or low-serrate coriaceous leaflets,* rarely simple; cauline leaves coriaceous, once or twice ternate, greatly reduced upward; *umbels* long-stalked, *the* 2–11 *filiform rays very unequal, the longer ones lengthening to* 3–11 *cm.*; *fruit ovoid or orbicular,* 2–3.5 *mm. long.* (*Z. Bebbii* (C. & R.) Britt.) — Rich woods, especially along the mts., Va. and W.Va. to Ga. and Tenn. *Fl.* May, June; *fr.* July–Sept.

3. Z. aùrea (L.) W. D. J. Koch (golden), Golden Alexanders. — Stems from elongate base, 0.3–1 m. high; *leaves membranaceous,* all (except the uppermost) 2–3-*ternate,* the radical long-petioled; *leaflets* ovate to lanceolate, acuminate, *sharply serrate,* or rounded-elliptical to obovate or suborbicular and obtuse in forma **obtusifòlia** (Bissell) Fern. (blunt-leaved); *rays of umbel* 6–20, *mostly* (all but 2 or 3) *subequal, stoutish, the longer lengthening to* 2–5 *cm.*; *fruit oblong,* 3–4 *mm. long.* — Meadows, shores, damp thickets and wet woods, Que. to Sask., s. to N.B., N.E., Ga., Tenn., Mo., Okla. and Tex. *Fl.* Apr.–June; *fr.* July–Sept. Fig. 1364.

1364. Z. aurea.

16. CICÙTA L. Water-Hemlock

Calyx-teeth prominent. Fruit ovoid to nearly globose, glabrous, with strong flattish corky ribs (the lateral largest); oil-tubes conspicuous, solitary; stylopodium depressed; seed nearly terete. — Very poisonous plants of n. temp. reg. with pinnately or ternately compound leaves and serrate leaflets, involucre usually none, involucels of several slender bractlets, and white flowers. (The ancient Latin name of the poisonous Hemlock, a deadly Old World herb.)

Axils of upper leaves bearing clustered bulblets; leaflets linear. . . . . . 1. *C. bulbifera.*
Axils without bulblets; leaflets linear-lanceolate to ovate-oblong.
  Fruits ellipsoid, ovoid or subglobose, with subequal prominent round-backed pale ridges alternating with dark grooves; leaflets lanceolate to ovate-oblong. 2. *C. maculata.*
  Fruits reniform to cordate-ovoid, the lateral ribs prominent, the others obscure; leaflets linear-lanceolate. . . . . 3. *C. Victorinii.*

1. C. bulbífera L. (bearing bulbs). — Slender, from fascicled tuberous-thickened roots, 0.15–1 m. or more high; leaves 2–3-pinnate (the lower often ternate); leaflets linear, sparsely slender-toothed, 1–7 cm. long; *upper axils bearing fascicled bulblets;* umbels 1–few, 2–8 cm. broad; fruit (rarely maturing) reniform, 1.5–2 mm. long, with low and broad round-backed subequal ribs and slender intervals. — Swamps and wet thickets, Nfld. to B.C., s. to N.S., N.E., Va., O., Ind., Ill., Minn., Neb., Mont., Ida. and Oreg.

2. C. maculàta L. (mottled or spotted), Spotted Cowbane, Musquash-root, Beaver-poison, Carotte à Moreau (Que). — Biennial, with a fascicle of fleshy tuberiform finger-like roots; stem often mottled with purple below, relatively stout, 0.6–2.2 m. high; leaves 2–3-pinnate, the lower long-petioled; *leaflets narrowly lanceolate to lance-oblong,* acuminate, *thin,* the midribs and primary veins evident beneath, the margins *sharply serrate with lance-attenuate teeth; larger leaflets of lower leaves* 0.6–3(–4) *cm. broad;* leading umbels 5–14 cm. broad; *fruits ellipsoid or ovoid,* rarely subglobose, 3–4 mm. long, with alternating rounded ribs and dark furrows, usually with *the marginal ribs confluent until maturity.* — Meadows, swales, low thickets and prairies, Gaspé Pen., Que., to e. Man., s. to N.S., N.E., Md., upland to N.C., Tenn., Mo. and Tex. June–Sept. Fig. 1365. — The tuber-like roots, resembling small sweet potatoes and with the fragrance of parsnips, are deadly poisonous.

1365. C. maculata.

Var. **Curtíssii** (C. & R.) Fern. (for its discoverer, Allen Hiram Curtiss, 1845–1907). — Darker-green; *leaflets thicker, with veinlets evident beneath,* broadly oblong-lanceolate to narrowly oblong-ovate, *coarsely crenate or dentate-serrate with broad-based teeth, the larger leaflets* 3–5.5 *cm. broad;* leading umbels 0.8–1.8 dm. broad; *fruits subglobose or reniform-globose,* 2–3 mm. high, usually *with a dark furrow separating the marginal ribs.* (*C. Curtissii* and *C. mexicana* C. & R.) — Ditches, stream-margins, wet thickets and swampy woods, Fla. to Tex. and ne. Mex., n. to N.J., W.Va. and Tenn., evidently passing northw. into typical *C. maculata.*

3. C. Victorínii Fern. (for its discoverer, Conrad Kirouac, Frère Marie-Victorin, 1885–1944). — Similar to no. 2, only 3–6 dm. high; *leaves* biternate, *with* dentate-serrate *linear-lanceolate segments;* umbels 3–8 cm. broad; *fruits reniform or cordate-ovoid,* 3.5–4 mm. long, *the lateral ribs prominent, the others obscure.* — Fresh tidal flats of the St. Lawrence R., Que. *Fr.* Aug., Sept.

17. CRYPTOTAÈNIA DC. Honewort. Wild Chervil

Calyx-teeth obsolete. Fruit linear- to oblong-subcylindric, subulate-beaked, glabrous, with obtuse equal ribs; oil-tubes solitary in the intervals and beneath each rib; stylopodium slender-subulate; seed-face plane.—Glabrous perennials of N.Am., Eurasia and Afr., with thin 3-foliolate leaves, no involucre, involucels of minute bractlets or none, and white flowers. (Name from the Greek *cryptos, hidden,* and *tainia, a fillet,* referring to the concealed oil-tubes.)

1. C. canadénsis (L.) DC. (Canadian). — Plants 0.3–1 m. high; leaflets large, ovate, 5–10 cm. long, pointed, doubly serrate, often lobed; umbels irregular and unequally few-rayed; pedicels very unequal; fruit 4–6 mm. long, often curved. (*Deringa* Ktze.) — Rich woods and thickets, w. N.B. to Man. s. to N.E., Ga., Ala., Ark. and Tex. June–Sept. Fig. 1366. (Closely related plant in e. Asia)

1366. C. canadensis.

18. FALCÀRIA Bernh. SICKLEWEED

Calyx 5-dentate, the tube of the sterile flowers wanting, of the fertile cylindric. Petals arching, deeply emarginate, the lobes reflexed. Fruit laterally compressed, oblong, each mericarp with 5 equal filiform ribs; oil-tubes solitary in the intervals. Seed subterete. — Smooth oriental and s. European perennials with ternate pinnati- or palmatisect leaves with narrow decurrent cartilaginous-toothed lobes, involucres and involucels of slender leaves and white sterile and fertile flowers. (Name from *falcarius, pertaining to a sickle,* from the arching lobes of the petals.)

1. F. VULGÀRIS Bernh. (common). — Glaucous, 0.3–1 m. high, from a deep fusiform root; leaves subcoriaceous, with linear-lanceolate mucronate-serrulate segments, the lower palmate, the upper pinnate; fruit 4–5 mm. long. (*F. sioides* (Wibel) Aschers.; *F. Rivini* Host) — Local weed of fields, N.Y. to Neb., s. to Pa., Mo. and Kans. *Fl.* Aug., Sept., *fr.* Oct. (Adv. from Eu.)

19. CÀRUM L. CARAWAY

1367. C. Carvi.

Calyx-teeth small. Fruit ovate or oblong, with filiform or inconspicuous ribs; oil-tubes solitary; stylopodium conical; seed-face plane or nearly so. — Smooth erect slender biennials of N. Hemisph. with fusiform or tuberous roots, pinnate leaves, involucre and involucels wanting or of few to many bracts, and white (rarely pink) flowers. (Modification of the old Latin name *Careum.*)

1. C. CÁRVI L. (old officinal name), CARAWAY. — Leaves with filiform or narrowly linear divisions; stem 2–7.5 dm. high, widely branching. — Neglected fields, abund. northw., Nfld. to Alta., s. to N.S., N.E., Pa., Ill., etc. May–July. (Natzd. from Eu.) FIG. 1367. — Forma RHODOCHRÁNTHUM A. H. Moore (with rose-colored flowers) with pinkish petals, and forma ATRÓRUBENS Lange (dark red) with them deep purple, are local forms. — Fruits ("CARAWAY-SEEDS") a familiar seasoning.

20. TAENÍDIA Drude YELLOW PIMPERNEL

Fruit short-oblong, flattened laterally, wingless, glabrous; oil-tubes mostly 3 in the intervals; seed subterete but the face slightly concave. Involucre and involucels mostly wanting. Flowers yellow. — Glabrous glaucous e. Am. perennial, with ternate leaves. (Name from the Greek *tainidion, a little band,* in reference to the small scarcely prominent ribs.)

1. T. integérrima (L.) Drude (quite entire). — Slender, 5–10 dm. high; leaves 2–3-ternate; leaflets lanceolate to ovate, entire; fruit oblong, 4 mm. long. — Dry rocky or gravelly woods and thickets, w. Que. to Minn., s. to Ga., Ala., Miss., La. and Tex. May–July. FIG. 1368.

1368. T. integerrima.

21. PIMPINÉLLA L.

Fruit oblong to ovate, glabrous, with slender equal ribs, numerous oil-tubes, and depressed or cushion-like stylopodium. — Smooth perennials, nat. of Eurasia, with involucre and involucels scanty or none; ours with pinnate rosulate leaves and white flowers. (Ancient name (originally *pipinella*) for these and similar plants.)

1. P. SAXÍFRAGA L. (stone-breaker), BURNET-SAXIFRAGE. — Perennial 3–9 dm. high, corymbosely branched; leaves mostly near base of plant, the rosulate lowest ones with 4–8 pairs of coarsely and variously toothed nearly sessile oblong to obovate leaflets; umbels flat-topped, without involucres; fruit oblong, 2 mm. long; stylopodium cushion-like. — Roadsides, fields and shores, local, Nfld. to Del., Pa. and Ind. June–Aug.; *fr.* Aug., Sept. (Natzd. from Eu.)

P. ANÌSUM L. (old generic name), ANISE, an annual with long-petioled simply pinnate to ternately divided leaves, yellowish flowers and ovate fruit, esc. from cult. (Introd. from Eu.)

22. AEGOPÓDIUM L. GOUTWEED

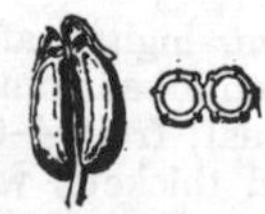

1369. A. Podagraria.

Fruit ovate, glabrous, with equal filiform ribs and no oil-tubes; stylopodium conical and prominent; seed nearly terete. — Coarse glabrous perennials nat. of Eurasia, with creeping rhizome, ternate leaves, sharply toothed ovate leaflets, and rather large naked umbels of white flowers. (Name from the Greek *aix, goat,* and *podion, a little foot,* probably from the shape of the leaflets.)

1. A. Podagrària L. (old generic name, derived from *podagra, gout*). — Waste places and roadsides, Nfld. to Mich., s. to N.S., N.E. and N.C. June–Aug. (Natzd. from Eu.) Fig. 1369. — A form with white-variegated leaves cult. for ornament.

23. PERIDERÍDIA Reichenb.

Calyx-teeth prominent. Fruit ovate or oblong, glabrous, with equal filiform ribs; oil-tubes 1–5 in the intervals; stylopodium conical, with long recurved styles; seed-face broadly concave, with a central longitudinal ridge. — Small N.Am. genus; ours perennial with tuberous-thickened roots, pinnately compound leaves, involucels of numerous narrowly lanceolate acuminate bractlets and long-peduncled umbels of white flowers. (Name said to be from the Greek *peri, around,* and *derris,* a *leathern coat.*) Eulophus Nutt.

1370. P. americana.

1. P. americàna (Nutt.) Reichenb. (American). — Erect, 0.7–1.2 m. high; radical and lower cauline leaves large, 1–2-pinnately compound, with leaflets cut into short narrow segments; upper cauline leaves ternate, with narrowly linear elongated leaflets; fruit 4–6 mm. long. (*Eulophus* Nutt.) — Low grounds, prairies and rich woods, O. to sw. Mich. and Ill., s. to Tenn., Ark. and se. Kans. July. Fig. 1370.

24. SÌUM L. Water-parsnip

Calyx-teeth minute. Fruit ovoid to ellipsoid, glabrous, with prominent corky nearly equal ribs; oil-tubes 1–3 in the intervals; stylopodium depressed; seed-face plane. — Smooth perennials of N. Hemisph. and Afr., with mostly pinnate leaves and serrate or pinnatifid leaflets, involucre and involucels of numerous narrow bracts, and white flowers. (From *sion*, the Greek name of some paludal plant.)

1371. S. suave.

1. S. suàve Walt. (fragrant). — Earliest submersed leaves of 1st year's rosette very thin and twice or thrice pinnately dissected into linear or linear-filiform segments; emersed basal leaves long-petioled, pinnate, with 5–17 linear, lanceolate or lance-oblong copiously serrate leaflets 3–15 cm. long; *cauline leaves* similar, gradually reduced upward, *the lower and median strongly ascending, with* 5–17 *firm to membranous leaflets* 4–15 *cm. long; stems* erect, 0.3–2 m. high, *strongly corrugated or angled*, branching chiefly above the middle, *the corrugated branches strongly ascending;* larger *umbels* 4–11 cm. broad, *with* 10–25 *angled rays; pedicels stiffly ascending; petals roundish, equaling the stamens;* styles rather short; fruit 2.5–3 mm. long, with prominent ribs. (*S. cicutaefolium* Schrank) — Meadows, wet thickets, muddy banks, etc., Nfld. to B.C., s. to N.S., N.E., Fla., O., Ind., Ill., Mo., e. Kans., Colo., Utah, Nev. and Calif. July–Sept. (E. Asia) Fig. 1371. — Readily distinguished from the deadly poisonous *Cicuta maculata* by simply pinnate leaves and corrugated stems. — Forma **Carsònii** (Durand) Fassett (for its discoverer, Joseph Carson, 1808–1876), weak, creeping, floating, trailing or loosely ascending stems 1–6 dm. long, the leaves with 3–9 oblong to ovate crenate or serrate leaflets 1–4 cm. long or when submersed with filmy and narrower lacerate leaflets, the fruit rarely if ever maturing (*S. Carsonii* Durand), a miscellaneous series of submersed or partially submersed weak forms; forma **fasciculàtum** Fassett (bunched), repent or erect with basal leaves reduced to a single terminal lanceolate to ovate coarsely serrate leaflet, the cauline leaves mostly with 1 similar leaflet and fascicled in old axils, the fascicle-bases hardening into corms, the fruit not maturing, growing in estuary of Cathance R., Me.

2. S. floridànum Small (of Florida). — Similar but *stem terete*, only shallowly furrowed in drying, flexuous, *branching from near or below the middle, the terete branches divergent or loosely ascending; leaves membranaceous, divergent,* the larger with 3–11 leaflets 2–9 cm. long; *larger umbels* 1–7 *cm. broad, with* 7–15 *filiform rays; pedicels filiform, curving or arching,* prolonged; *petals elliptic, shorter than stamens; styles prolonged.* — Rich bottomlands, se. Va. to Fla. June–Sept.

25. BÉRULA Hoffm. Water-parsnip

Calyx-teeth minute. Fruit emarginate at base, glabrous; carpels nearly globose, with very slender inconspicuous ribs and thick corky pericarp; oil-tubes numerous and contiguous about the seed-cavity; seed terete. — Smooth aquatic N. Am. and Eurasian perennials, with simply pinnate leaves and variously cut leaflets, usually conspicuous involucre and involucels of narrow bracts, and white flowers. (Latin name of some aquatic plant.)

1372. B. pusilla.

1. B. pusílla (Nutt.) Fern. (tiny). — Stem slender, 2–9 dm. high, erect to reclining or loosely divergent-branched; leaflets 5–9 pairs, linear to oblong or ovate, serrate to cut-toothed, often laciniately lobed, sometimes crenate, 2–8 cm. long; fruit scarcely 2 mm. long. (*B. erecta* sensu Coville, not *Sium erectum* Huds., basonym) — Swamps and streams, Ont. to B.C., s. to Mich., Ill., Minn., Okla., N.M. and Calif. July–Sept. Fig. 1372.

26. PTILÍMNIUM Raf. Mock Bishop's-weed

Fruit ovoid, glabrous; carpels with dorsal ribs filiform to broad and obtuse, the lateral ribs very thick and corky, those of the two carpels closely contiguous and forming a dilated obtuse or acute corky band; oil-tubes solitary; stylopodium conical; seed nearly terete. — Smooth N. Am. annuals, usually with involucre of slender foliaceous bracts or wanting, involucels of prominent or minute bractlets, and white flowers. (Name unexplained by Rafinesque, presumably from the Greek *ptilon, a feather* or *down*, and *limne, mud*, in allusion to the finely divided leaves and the habitat.)

Leaves divided into capillary segments; involucre of several bracts.
  Styles shorter than to equaling the stylopodia.
    Leading umbel with 2–16 rays; petals 0.5–1 mm. broad; calyx-teeth deltoid, much shorter than stylopodium, 0.07–0.15 mm. long; fruits 2–3.2 mm. long, their dorsal ribs narrower than the intervals. . . . . 1. *P. capillaceum.*
    Leading umbel with 12–30 rays; petals 1–2 mm. broad; calyx-teeth subulate, half as long as to nearly equaling stylopodium, 0.3–0.6 mm. long; fruits 1–2 mm. long, their dorsal ribs broader than the intervals. . . 2. *P. Nuttallii.*
  Styles two to four times as long as stylopodia; fruits 2.5–5 mm. long. . . 3. *P. costatum.*
Leaves reduced to slender septate hollow petioles; involucre wanting or of a single bract. . . . . 4. *P. viviparum.*

1. P. capillàceum (Michx.) Raf. (hair-like). — Stem slender, 0.1–1.9 m. high, usually loosely branched; leaves with numerous filiform divisions; *leading umbels* 2–16 *rayed;* involucre of filiform often cleft or parted bracts, the involucels commonly developed; *petals* 0.5–1 *mm. broad; calyx-teeth rudimentary or deltoid* and 0.07–0.15 *mm. long, much shorter than the stylopodia;* styles shorter than stylopodia; *fruits* 2–3.2 *mm. long, their dorsal ribs narrower than the intervals,* the corky marginal ribs confluent as a broad band. — Brackish to fresh marshes, Fla. to Tex., n. to s. N.E., se. N.Y., s. Ill. and Mo. July–Oct. Fig. 1373.

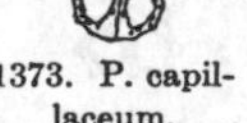

1373. P. capillaceum.

2. P. Nuttállii (DC.) Britt. (for its discoverer, Thomas Nuttall, 1786–1859). — Similar to no. 1; leaves with fewer and longer segments; *leading umbels* 12–30-*rayed;* involucral bracts mostly simple; *petals* broadly obcordate, 1–2 *mm. broad; calyx-teeth subulate, half as long as to nearly equaling stylopodia,* 0.3–0.6 *mm. long; fruits* 1–2 *mm. long, their dorsal* and lateral *ribs* subequal, *broader than the intervals.* — Damp prairies, glades and shores, Ala. to Tex., n. to Ill., Mo. and Kans. June–Aug.

3. P. costàtum (Ell.) Raf. (ribbed). — Coarser, 0.6–1.8 m. high; leaves with crowded segments (often appearing subverticillate); leading umbels with 10–30 rays; involucral bracts short, simple; flowers nearly as in no. 2; *styles prolonged, two to four times as long as stylopodia; fruits* 2.5–5 *mm. long,* their dorsal ribs slender. (Incl. *P. missouriense* C. & R.) — Swamps and wet barrens, Ga. to La., n. to N.C., Ky., s. Ill. and Mo. July–Oct.

4. P. vivíparum (Rose) Mathias (with the young well developed on the parent plant). — Stem slender and weak, simple or but slightly branched, ascending or reclining, 1–6 dm. long; *leaves terete, nodose, hollow, in autumn often bearing axillary bulblets; umbels without involucres* (or with 1 tiny bract), 1–2 *cm. broad,* 3–6-rayed; involucels minute; petals tiny; calyx-teeth sharp; *fruits* 1.2–2 *mm. long, with slender ribs.* — Along Potomac R., mts. of w. Md., ne. W.Va. and n. Va. Aug.–Oct.

27. CYNOSCIÁDIUM DC.

Calyx-teeth distinct. Fruit short, glabrous, scarcely flattened; lateral ribs forming a corky margin; stylopodium conical. — Slender N.Am. annuals with narrow simple or palmately or pinnately divided leaves. Involucre and involucels present. Petals white. (Name from the Greek *cyon, dog,* and *sciadion, a sunshade,* a fanciful designation referring to the umbels.)

1. C. pinnàtum DC. (pinnate). — Slender, 0.2–8 dm. high, subsimple or with ascending branches; *cauline leaves* linear or linear-lanceolate, *simple or pinnate with small scattered segments;* umbels mostly 3–9 cm. broad; *fruit* ellipsoid, 2.5–3.5 mm. long, *not strongly constricted* *below the* prolonged *stylopodia.* — Wet places, borders of pools, etc., La. and Tex., n. to sw. Mo. and se. Kans. June–Aug.

2. C. digitàtum DC. (with fingers). — Similar to no. 1; most of the *cauline leaves digitate; fruit constricted to a slender beak below the stylopodia.* — Wet places, Miss. to Tex., n. to se. Mo. and Okla.

28. LILAEÓPSIS Greene

Calyx-teeth small. Fruit globose or slightly flattened laterally; dorsal ribs filiform, the lateral thick and corky; oil-tubes solitary in the intervals, 2 on the commissure. — Dwarf creepers of temp. N. and S.Am., Austral. and N.Z., with hollow cylindrical or subulate nodose petioles (phyllodia) in place of leaves, simple few-flowered umbels and white flowers. (Named from its resemblance to *Lilaea.*) CRANTZIA Nutt., not Scop.

1. L. chinénsis (L.) Ktze. (of China; Linnaeus mistakenly thinking it was found there). — Upright or arching clavate broadly round-tipped *phyllodia* 1–6 *cm. long*, 3–6-*jointed*, scattered and solitary along the creeping filiform stem; *fruits* pyriform or constricted at base, about 2 mm. long, *the thick lateral wings forming a corky margin.* (*L. lineata* Greene) — In mud of brackish marshes and tidal shores along the coast, w. N.S. to Fla. and Miss. June–Sept.

2. L. carolinénsis C. & R. (of Carolina). — Much coarser, extensively creeping or floating, forming broad mats; *phyllodia* mostly 1–2 *dm. long*, linear- to oblanceolate-spatulate, 4–11 *mm. broad*, 10–18-*jointed;* peduncles much shorter; *fruits with nearly uniform ribs.* — Shallow pools and ponds, very local, coast of se. Va. to S.C.; La. April, May. (Temp. e. S.Am.)

29. AETHÙSA L. FOOL'S-PARSLEY

Calyx-teeth obsolete. Fruit ovoid-globose, slightly flattened dorsally; carpels with 5 thick sharp ribs; oil-tubes solitary in the intervals, 2 on the commissure. — Poisonous annual nat. of Eu., with twice- or thrice-ternately compound leaves, their divisions pinnate; ultimate segments small and many-cleft, no involucre, long narrow involucels, and white flowers. (Name from the Greek *aithon*, *glistening*, in allusion to the bright or shining foliage, probably in translation of the Swedish vernacular name *glis.*)

1374. A. Cynapium.

1. A. CYNÀPIUM L. (old generic name). — A fetid poisonous herb, in waste or cult. grounds, local, N.S. to s. Ont. and Minn., s. to Del., Pa., and O. June–Aug. (Natzd. from Eu.) FIG. 1374.

30. FOENÍCULUM Mill. FENNEL

Fruit ellipsoid, glabrous, with prominent ribs and solitary oil-tubes. — Stout glabrous aromatic perennials, nat. of Old World, with leaves dissected into numerous filiform segments, no involucre nor involucels, and large umbels of yellow flowers. (The Latin name, from *foenum*, *hay*, presumably from the abundant thread- or straw-like segments of the leaves.)

1. F. VULGÀRE Mill. (common). — Stem up to 2 m. high, glaucous; *petiolar sheaths of larger leaves* 3–10 *cm. long;* fruit about 5 mm. long. — Dry fields and roadsides, Ct. to Mich., Neb. and southw. June–Sept. (Introd. and natzd. from Eu.)

31. ANÈTHUM L. DILL

Petals yellow. Fruit ellipsoid, somewhat flattened dorsally, the lateral ribs winged. Involucre and involucels none. — Slender caulescent annuals nat. of Old World, with finely divided leaves and compound umbels of yellow flowers. (*Anethon*, ancient Greek name of the *dill*, thought to come from *aithein*, *blaze*, in allusion to the pungent seeds.)

1. A. GRAVÈOLENS L. (strong-smelling). — Erect, glabrous, usually branched, 3–10 dm. high; leaves finely dissected, fennel-like, the *petiolar sheaths of larger leaves* 1–3 *cm. long.* — Roadsides and waste places, local, Ct. to Minn., Pa. and N.J. July, Aug. (Introd. and natzd. from Asia)

32. LIGÚSTICUM L. LOVAGE

Fruit ellipsoid or ovoid, flattened laterally if at all, glabrous; carpels with prominent equal acute ribs and broad intervals; oil-tubes 2–6 in the intervals, 6–10 on the commissure. Stylopodium conical. — Smooth perennials of temp. and cold areas, from large aromatic roots, with large ternately compound leaves, mostly no involucre, involucels of narrow bractlets, and white flowers in large many-rayed umbels. (Name from *ligusticus*, of the country *Liguria*, where the officinal *Lovage* of the gardens abounds.)

1375. L. scothicum.

1. L. canadénse (L.) Britt. (Canadian; a misnomer), NONDO, ANGELICO. — Stem stout, branched, 1–2 m. high; *leaves very large*, 3–4-*ternate; leaflets broadly oblong*, 5–12 *cm. long, coarsely serrate; fruit ovate*, 4–6 mm. long; seed with angled back. — Rich woods, s. Pa. and W.Va. to Ga., Ala. and Mo. May–Aug.

2. L. scóthicum L. (Scotch), SCOTCH L. — Stem simple, 3–6 dm. high, red or purple below; leaves biternate; *leaflets* fleshy, ovate, 2.5–5 *cm. long, coarsely toothed; fruit narrowly oblong* 8–10 *mm. long;* seed with round back. — Saline marshes and rocks along the coast from Greenl. and Lab. to s. N.Y. June–Sept. (Eu.) FIG. 1375.

33. COELOPLEÙRUM Ledeb.

Fruit globose to ellipsoid, with prominent nearly equal thick corky ribs (none of them winged); oil-tubes solitary in the intervals and under the ribs, 2–4 on the commissure; styles arched-recurving in fruit. Seed loose in the pericarp. — Stout glabrous (or inflorescence puberulent) maritime perennials of cool n. reg., with twice- or thrice-ternate leaves on very large inflated petioles, few-leaved deciduous involucre, involucels of numerous oblong to linear-lanceolate bractlets (often conspicuous or even leaf-like), and greenish-white flowers in many-rayed umbels. (Name from the Greek *coilos, hollow*, and *pleuron, a rib.*)

1376. C. lucidum.

1. C. lùcidum (L.) Fern. (shining). — Stem stout, 0.3–1.5 m. high, often with gummy spots; petiolar sheath with broadly obcordate summit; leaflets ovate, irregularly cut-serrate, 3–8 cm. long; central umbel with 20–50 rays, these scabrous-puberulent or hirtellous; fruit 4–7 mm. long. (*C. actaeifolium* C. & R.; *Angelica lucida* L.) — Rocky and gravelly coast, s. Greenl. and Lab. to s. N.Y.; also subalpine meadows, Gaspé Pen., Que. June–Sept. — Forma **frondòsum** Fern. (leafy) has the involucels changed to broad leafy bracts. (Perhaps nw. Am. and ne. Asia) FIG. 1376.

34. THÁSPIUM Nutt. MEADOW-PARSNIP

Calyx-teeth conspicuous. Fruit ovoid to ellipsoid, slightly flattened dorsally; carpels with 3 or 4 or all the ribs strongly winged; oil-tubes solitary in the intervals, 2 on the commissure. Stylopodium wanting; styles long, ascending in fruit or deciduous. — N. Am. perennials with ternately divided leaves (or the lower simple) and broad serrate or toothed leaflets, greenish, yellow or purple flowers, and all the fruits pedicelled. (Name a play upon *Thapsia*, the name of a related genus, so called from the peninsula of Thapsus.)

1. T. trifoliàtum (L.) Gray (three-leaved). — Caudex slender, 3–5 mm. thick; *stems* 1–3, slender, *glabrous*, 2–6 dm. high; *basal leaves firm*, long-petioled, *simple and ovate to rotund*, deeply cordate, crenate or dentate, chiefly 1.5–5 cm. long, *or* less often *ternate* and with the stalked ovate to lanceolate leaflets 1.5–4 cm. long; *cauline leaves ternate or simple*, dentate or serrate, those at first fork of stem with the terminal leaflet (or the simple blade) 2–4.5 cm. long; umbels 1.5–3(–4) cm. broad, with 4–11 rays; petals greenish to purple; fruit rounded-ellipsoid, 3–4 mm. long, all the ribs equally winged (or dorsal ribs sometimes suppressed). (*T. aureum*, var. *atropurpureum* (Desv.) C. & R.) — Woods, Fla. to La., n. to N.J. and e. Pa., upland of Ky., and s. Mo. *Fl.* Apr.–June; *fr.* Aug.–Oct.

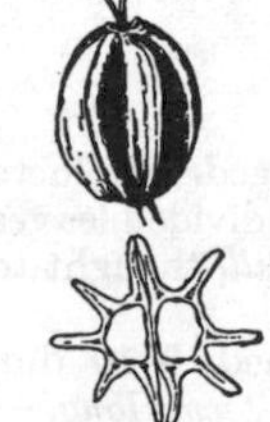

1377. T. trifoliatum, v. flavum.

Var. **flàvum** Blake (yellow). — Coarser; rhizome knotty, 0.5–1.5 cm. thick; stems 0.3–1.5 m. high; simple basal leaves (when present) 3–10 cm. long; middle leaflet of basal and larger cauline leaves (sometimes quinate) 3–8 cm. long; umbel 3–9 cm. broad; petals yellow (rarely purple); fruit 4–5 mm. long. (*T. aureum* sensu Nutt., not as to nomenclatural source of name.) — Rich woods and thickets, calcareous bluffs, etc., N.Y. to Minn., s. to Ala., Ark. and s. Kans. *Fl.* May–July; *fr.* late July–Sept. FIG. 1377. — Often confused with *Zizia aptera*.

2. T. barbinòde (Michx.) Nutt. (with bearded nodes). — Loosely branched, 0.4–1.2 m. high, the green or glaucous *stems hairy at the nodes; basal leaves* long-petioled, *twice or thrice ternate, membranaceous*, cauline leaves similar but shorter-petioled to sessile; *leaflets* ovate to lanceolate, coarsely toothed or cleft, *those of the leaf at first fork of stem* 9–15, mostly 0.5–5 *cm. broad;* flowers pale yellow; fruit broadly oblong, about 6 mm. long and 4 mm. broad, with mostly 7 prominent wings. — Rich woods, thickets and talus, N.Y. to Minn., s. to Ga., Ala., Miss., Ark. and Okla. *Fl.* Apr.–June; *fr.* late July–Oct.

Var. **angustifòlium** C. & R. (narrow-leaved). — Leaflets of basal leaves cut into linear or oblanceolate segments; *leaflets* at first fork of stem 15–75, oblanceolate, 2–10 *mm. wide.* — S. Ont. to Ill. and Ky.; se. Va.

3. T. pinnatífidum (Buckl.) Gray (pinnately cleft). — Similar to no. 2 and perhaps a var. of it; leaves three or four times divided, *the* 150–300 *narrow leaflets of the lower leaves dissected into linear segments; fruit* 3–4 *mm. long,* 2–3 *mm. broad.* — Rich woods, w. N.C. and Ky. to Ala.

35. CONIOSELÌNUM Hoffm. Hemlock-parsley

Fruit thick-oval, flattened dorsally, glabrous, the lateral ribs extended into broad wings; seed slightly concave on the inner face. — Slender glabrous N.Am. and Eurasian perennials, with finely twice- or thrice-pinnately compound leaves, few-leaved involucre or none, involucels of elongated (in ours) linear-setaceous bractlets, and white flowers. (Name a compound of *Conium* and *Selinum,* from its resemblance to these genera.)

1. C. chinénse (L.) BSP. (of China; Linnaeus mistakenly thinking it from there). — Plant 0.1–1.5 m. high; *petiolar sheaths* of leaves *dilated, with scarious margins;* leaflets pinnatifid; leading umbels 0.3–1.5 dm. broad; involucels of few linear-filiform short bracts; *fruit* ellipsoid, *distinctly longer than broad,* 4.5–6 *mm. long,* lateral wings as broad as seed, *dorsal wings thin and as broad as thickness of seed;* oil-tubes 2–3 in the intervals, sometimes 1 or 4. — Thickets, open slopes and (southward) wet woods, Lab. to w. Ont., s. to Nfld., N.S., N.E., Pa., upland to N.C., O., Ind., Ill. and Mo. July–Sept. Fig. 1378.

1378. C. chinense.

2. C. pùmilum Rose (dwarf). — Smaller, 0.1–7.5 dm. high; *sheaths lance-tapering,* firm, purple; leading umbels 2–5 cm. broad; *fruits suborbicular, only slightly longer than broad,* 3–4 *mm. long; dorsal wings low and rounded.* — Serpentine-barrens, Lab. and Nfld. July–Sept.

Other forms in Nfld. and on the Shickshock Mts. must await fuller material and detailed study.

36. ANGÉLICA L. Angelica

Fruit strongly flattened dorsally; primary ribs very prominent; the lateral ones extended into broad distinct wings, forming a double-winged margin to the fruit; oil-tubes 1–several in the intervals or indefinite, 2–10 on the commissure. — Stout perennials of N.Am., Eurasia and N.Z., with ternately or pinnately compound leaves, the uppermost reduced to tubular sheaths or with sheaths much longer than blades, large terminal umbels, scanty or no involucres, small many-leaved involucels, and white or greenish flowers. (Named *angelic* from cordial and medicinal properties of some species.)

Sheaths of upper blade-bearing petioles slenderly tubular, linear-cylindric to lanceolate, 0.5–1 cm. in diameter, not prominently veined; largest (central) umbels 0.6–1.5 dm. broad; seeds adherent to pericarp; oil-tubes distinct, 1–6 in each interval.
  Fruit pubescent; oil-tubes mostly 3–6 in each interval. . . . . 1. *A. venenosa.*
  Fruit glabrous; oil-tubes 1 (exceptionally 2–3) in each interval.
    Summit of stem and rays of umbel densely puberulent. . . . . 2. *A sylvestris.*
    Summit of stem and rays of umbels glabrous. . . . . 3. *A. triquinata.*
Sheaths of upper blade-bearing petioles inflated, ellipsoid or ovoid, 2–5 cm. in diameter, coarsely veiny; largest (central) umbels 1–3 dm. broad; seeds loose in pericarp; oil-tubes 25–30, continuous around the seed.
  Dorsal ridges of fruit wire-like, much lower than breadth of the thin marginal wings, the latter little if at all overtopping the stylopodium. . . . 4. *A. atropurpurea.*
  Dorsal ridges thin and wing-like, equaling the lateral, their summits overtopping the stylopodium. . . . . 5. *A. laurentiana.*

1. A. venenòsa (Greenway) Fern. (very poisonous). — Relatively slender; the terete firm *stem* 0.6–1.8 m. high, *closely pilose at summit;* basal and lower cauline leaves twice pinnately or ternately divided; their *thick leaflets lanceolate to oblong,* 2–7 cm. long, serrate; *upper leaves* rapidly reduced to *linear-cylindric or lanceolate tubular sheaths,* with or without small blades; central umbel 0.6–1.5 dm. broad, hemispherical, its horizontal or ascending rays canescent-pilose; *fruits pubescent*, 4–8 mm. long, 3–5 mm. broad, the dorsal and intermediate ribs prominent, the thin lateral ones as broad as the body; *oil-tubes* 3–6 (rarely only 1 or 2) *in the intervals*, 6–10 on the commissural side. (*A. villosa* (Walt.) BSP., not Lag.) — Dry woods, thickets and openings, Ct. to Mich. and Ill., s. to Fla., Ala., Miss. and Mo. *Fl.* July–Sept.; *fr.* Aug.–Oct.

2. A. sylvéstris L. (of woodland). — Resembling no. 1; *glabrous except for densely puberulent peduncles and rays of umbels; leaflets thin, ovate or lanceolate, acuminate,* sharply serrate; *fruits glabrous; oil-tubes solitary in the intervals.* — Old fields and roadsides, C.B. *Fl.* July, Aug.; *fr.* Sept., Oct. (Natzd. from Eu.)

3. A. triquinàta Michx. (with three groups of five leaflets), Filmy A. — Similar to no. 1; *stem* slender, purple, *glabrous to summit,* 0.6–1.8 m. high; *leaflets* thinnish, ovate to lanceolate, irregularly toothed, *mostly jagged-serrate, long-acuminate, up to* 1.5 *dm. long; rays of umbel glabrous; fruit glabrous.* (*A. Curtisii* Buckl.) — Woods and thickets, chiefly along the mts., Pa. and W.Va. to N.C. *Fl.* Aug., Sept.; *fr.* Sept., Oct.

4. A. atropurpùrea L. (dark purple), Alexanders. — Very stout, the usually *purple* or purple-stained glabrous *stem* 1–3 m. high, 2–4 cm. in diameter at base; lower leaves 2–3-ternately divided, the pinnate segments of 5–7 serrate oblong-lanceolate to ovate *mostly distinct* glabrous *leaflets* up to 1.5 dm. long; *upper leaves* reduced, *their inflated coarsely veiny ellipsoid to ovoid basal sheaths* 2–5 *cm. broad; umbels spherical or subspherical,* the 20–46 rays minutely puberulent; the central umbel 1–3 dm. in diameter; *fruit* 5.5–7.5 mm. long, *the dorsal ridges wire-like and much lower than the breadth of the thin marginal wings, the latter little if at all overtopping the stylopodium;* seeds loose in the pericarp; oil-tubes 25–30, continuous around the seed. — Rich thickets, bottomlands, swamps, etc., s. Lab. to Wisc., s. to Nfld., N.S., N.E., Del., Md., W.Va., O., Ind. and Ill., ascending to subalpine reg. of Que. and n. N.E. *Fl.* late May (southw.)–Sept. (northw.); *fr.* July–Oct. Fig. 1379.

1379. A. atropurpurea.

Var. **occidentàlis** Fassett (western). — Leaves minutely pilose beneath, scabridulous above. — W. Wisc., e. Minn. and ne. Ia.

5. A. laurentiàna Fern. (of the Gulf of St. Lawrence). — Mostly greener than no. 4, 1–2 m. high; *terminal segments of the lower leaves strongly confluent; umbels hemispherical; fruits* 5.5–10.5 *mm. long; the dorsal ridges wing-like, equaling the lateral, their summits overtopping the stylopodium.* — Rich thickets, calcareous shores and mountain-ravines, Straits of Belle Isle, Que., and n. Nfld. *Fl.* July, Aug.; *fr.* Aug., Sept.

37. OXÝPOLIS Raf. Hog-fennel

Calyx-teeth evident. Fruit ovate to obovate, flattened dorsally; dorsal ribs filiform, the lateral broadly winged and closely contiguous and strongly nerved next to the body (giving the appearance of 5 dorsal ribs); oil-tubes solitary in the intervals, 2–6 on the commissure; stylopodium short, thick-conical. — Glabrous erect Am. paludal herbs with fusiform roots, involucels present, and flowers white. (Name from the Greek *oxys, sharp,* and *polios,* white; from the subulate involucels and white petals.)

Leaves with definite leaflets; primary involucre wanting or but slightly developed.
  Leaves pinnate, mostly with 5–11 sessile leaflets. . . . . 1. *O. rigidior.*
  Leaves simple, or palmate with 2 or 3 long-stalked narrow leaflets. . . . . 2. *O. ternata.*
Leaves reduced to slender rush-like nodose petioles; involucre well developed. . 3. *O. Canbyi.*

1. O. rigídior (L.) C. & R. (stiffer or rather stiff), Cowbane, Water-Dropwort. — Stem 0.5–1.8 dm. high; roots 1–2 dm. long, their fusiform enlargements 2–8 cm. long; *leaves simply pinnate,* the larger ones *with* 3–15 narrowly obovate, oblong or lanceolate subcoriaceous coarsely 3–several-toothed or entire *leaflets* 2.5–15 cm. long and 0.4–4 *cm. broad;* leading umbel 0.6–1.6 dm. broad, *without primary involucre or with only* 1*-few slender bracts;* fruits ellipsoid to ovoid, more than half as thick as broad, 3–7 mm. long, the wings about as broad as the coarsely ribbed body; oil-tubes mostly slender. — Bogs, swamps, wet woods and damp rocks, Fla. to La., n. to N.Y., O., Mich., Wisc. and Minn. Aug., Sept.; *fr.* Sept.–Nov. Fig. 1380. — Poisonous.

1380. O. rigidior.

Var. **ambígua** (Nutt.) Robins. (doubtful). — Leaflets narrowly linear, attenuate, the larger 9–15 cm. long and 2–4 mm. broad. (Var. *longifolia* (Pursh) Britt.) — N.J. and Del., and rarely to Mo. and La.

2. O. ternàta (Nutt.) Heller (in threes). — Very slender, 3–9 dm. high; base bulbiform, the fusiform roots 1–1.5 cm. long; *leaves slender, petioled, simple, or palmate with* 2 *or* 3 *petiolulate linear to oblanceolate leaflets* with blades 4–12 cm. long and 1–10 mm. broad; umbels without primary involucre or with 1 or 2 slender bracts, their filiform rays elongate; fruit ellipsoid, 3.5–5 mm. long. — Damp pinelands and bogs, rare, Fla. to se. Va. Aug.–Oct.; *fr.* Nov., Dec.

3. O. Cánbyi (C. & R.) Fern. (for its discoverer, WILLIAM MARRIOTT CANBY, 1831–1904).— Stem hollow, 0.8–1.2 m. high; *leaves reduced to filiform terete septate petioles* (*phyllodia*); the lower and median leaves 1–2 dm. long, with dilated base only 0.5–1 cm. long; *involucre of many filiform bracts;* fruit suborbicular to broad-oblong, 4–8 mm. long, retuse at both ends, the slenderly conical stylopodium nearly overtopped by the broad corky lateral wings. (*O. filiformis,* var. C. & R.) — Bogs, Sussex Co., Del., local, perhaps extinct.; Lee Co., Ga. *Fr.* Aug., Sept.

38. LEVÍSTICUM Hill LOVAGE

Calyx-teeth obscure. Petals greenish-yellow. Fruit oblong, rounded at each end, strongly ribbed, the lateral ribs moderately winged; oil-tubes solitary in the intervals, 2 on the commissure; seed flattish on the inner face. — Stout perennial herb nat. of Eu., with branched stems, large bipinnate leaves with rhombic-obovate and compound conspicuously involucrate umbels. (Name said to be a corruption of *Ligusticum.*)

1. L. OFFICINÀLE W. D. J. Koch (of the shops). — Essentially glabrous; leaflets coarsely toothed toward the apex, entire at the cuneate base; fruit 5–7 mm. long. (*L. Levisticum* Karst.) — Cult. for the aromatic qualities, especially of its seeds, and now occasionally found as a local esc. — May–July (Introd. from Eu.)

39. CYMÓPTERUS Raf.

Calyx-teeth more or less prominent. Fruit usually globose, with all the ribs conspicuously winged; oil-tubes 1–several in the intervals, 2–8 on the commissure. Stylopodium depressed. Seed-face slightly concave. — Mostly low (often cespitose) glabrous N.Am. perennials, from a thick elongated root, with more or less pinnately compound leaves, with or without an involucre, with prominent involucels and white flowers (in ours). (From the Greek *cyma, a wave,* and *pteron, a wing,* referring to the often undulate wings.)

1. C. acaùlis (Pursh) Raf. (stemless). — Low (1–2 dm. high), with a short erect caudex bearing leaves and peduncles at the summit, glabrous; rays and pedicels very short, making a compact cluster; involucre none; involucel of a single palmately 5–7-parted bractlet; fruit globose, 6–8 mm. in diameter; wings rather corky; oil-tubes 4–5 in the intervals. — Dry open soil, nw. Minn. to Alta. and e. Oreg., s. to Okla. and Colo. April–June.

40. PSEUDOTAENÍDIA Mackenz.

Calyx-teeth short, thickish. Petals creamy. Fruit thickish, strongly compressed dorsally, oblong to obovate; carpels obcompressed, with slender dorsal ribs and broad somewhat corky lateral wings. Oil-tubes mostly solitary in the intervals. — Glabrous erect perennial, with twice- to thrice-ternate leaves, entire leaflets and exinvolucrate compound umbels. (Name from the Greek *pseudos, false,* and *Taenidia,* to which this genus possesses a marked habital resemblance.)

1. P. montàna Mackenz. (of the mountains). — Slender, erect, 5–8 dm. high; root slightly thickened; leaves chiefly on lower half of plant, their petioles broad and clasping; leaflets elliptical to lance-ovate or -oblong, entire, thin; umbels 3–15-rayed; involucels none or inconspicuous; fruit 4–6 mm. long. — Argillaceous or shaly wooded slopes, mts. of w. Md., e. W.Va. and w. Va. Apr.–July; *fr.* Aug.–Oct.

41. POLYTAÈNIA DC.

Calyx-teeth conspicuous. Fruit obovate to oval, much flattened dorsally; dorsal ribs small or obscure in the depressed back, the lateral with broad thick corky closely contiguous wings forming the margin of the fruit; oil-tubes 12–18 about the seed and many scattered through the thick corky pericarp. — A perennial mostly glabrous herb, with twice-pinnate leaves (upper opposite and 3-cleft), the segments cuneate and incised, no involucre, narrow involucels, and bright yellow flowers in spring. (Name from the Greek *poly-, many,* and *tainia, a fillet,* alluding to the numerous oil-tubes.) PLEIOTAENIA C. & R.

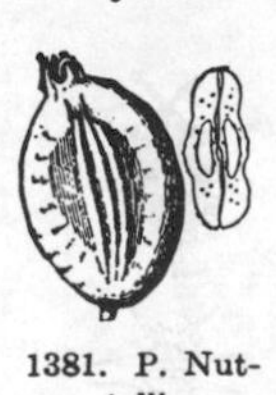
1381. P. Nuttalli.

1. P. Nuttállii DC. (for its discoverer, THOMAS NUTTALL, 1786–1859), PRAIRIE-PARSLEY. — Plant 0.5–1 m. high; pedicels and involucels pubescent; fruit 0.5–1 cm. long. — Dry prairies, open woods and glades, Ind. to N.D., s. to Ala., La. and Tex. *Fl.* Apr.–June; *fr.* June–Aug. FIG. 1381.

42. LOMÀTIUM Raf.

Fruit flattened dorsally, oblong to nearly orbicular, laterally winged; oil-tubes usually many. Roots fusiform. Leaves dissected. Involucre none. — Am. perennials of dry ground, nearly or quite acaulescent. Petals yellow or white. (Name from the Greek *lomation, a little border,* referring to the winged fruit.) COGSWELLIA Spreng.

1. **L. orientàle** C. & R. (eastern; occurring farther east than most of the genus). — *Puberulent to glabrous;* scape 1–3 dm. high (rarely with a median leaf); *leaves bipinnate,* with small oblong segments; flowers white or pink; *involucels glabrous* or nearly so; *fruit* glabrous, about 5 mm. long, nearly round, *its dorsal ribs indistinct.* (*Cogswellia* (C. & R.) Jones) — Dry or gravelly plains, Minn. to Mont., s. to Ia., Kans., N.M. and Ariz. Apr.–June.

2. **L. foeniculàceum** (Nutt.) C. & R. (resembling fennel, *Foeniculum*). — *Scape and involucels villous; leaves ternately divided,* the pinnate divisions with finely dissected leaflets, *ultimate segments filiform; bractlets broadly scarious-margined, villous; flowers yellow; fruit* 5–8 mm. long, elliptic to rounded, the *dorsal ribs prominent.* (Incl. *L. daucifolium* (Nutt.) C. & R.) — Prairies and limestone-barrens and glades, Man. to Mo. and Tex. Apr.–June.

43. PASTINÀCA L. PARSNIP. PANAIS (Que.)

Calyx-teeth obsolete. Fruit oval, very much flattened dorsally; dorsal ribs filiform, the lateral extended into broad wings, which are strongly nerved toward the outer margin; oil-tubes small, solitary in the intervals, 2–4 on the commissure; stylopodium depressed. — Tall stout glabrous biennials of a small Eurasian genus, with pinnately compound leaves, mostly no involucre nor involucels, and yellow flowers. (The Latin name, from *pastino,* to prepare the ground for planting of the vine.)

1. **P. SATÌVA** L. (sown). — Stem grooved; leaflets ovate to oblong, cut-toothed. — Waste places, roadsides, etc., throughout. May–Oct. (Introd. and natzd. from Eu.) FIG. 1382.

1382. P. sativa.

44. HERACLÈUM L. COW-PARSNIP

Fruit obovate, as in *Pastinaca,* but with a thick conical stylopodium, and the conspicuous obclavate oil-tubes extending scarcely below the middle. — Tall stout boreal perennials with large compound leaves, broad umbels, deciduous involucre, many-leaved involucels, white or purplish flowers, and obcordate petals, of which the outer ones are commonly larger and 2-cleft. (Dedicated to *Hercules,* Pliny having thought species no. 2 of the highest medicinal importance.)

1383. H. maximum.

1. **H. máximum** Bartr. (largest), MASTERWORT. — Woolly; stem grooved, 1–2.8 m. high; *leaves ternate,* the lowest with larger leaflets 1.5–6 or more dm. long and broad, irregularly *cut-toothed.* (*H. lanatum* Michx.) — Rich or low ground, Lab. to Alaska, s. to Nfld., N.S., N.E., mts. of Ga., O., Ind., Ill., Mo., Kans., N.M. and Calif.; ascending to subalpine areas. June–Aug. FIG. 1383.

2. **H. SPHONDÝLIUM** L. (an old generic name), ELTROT or HELTROT, HOG-WEED. — Spreading-pubescent and somewhat scabrous; *leaves pinnate;* leaflets 3–7, ovate or rounded, mostly 5–10 cm. long, coarsely and rather *bluntly toothed.* — Fields and roadsides, a rank-smelling weed, se. Nfld.; locally in waste places, C.B. to N.Y. June–Aug. (Natzd. from Eurasia)

Var. ANGUSTIFÒLIUM Jacq. (narrow-leaved). — Leaves with elongate acute leaflets up to 2 dm. long, these deeply cleft into acute segments. — Waste places about Boston, Mass. (Adv. from Eurasia)

45. DAÙCUS L. CARROT

Fruit oblong, flattened dorsally; stylopodium depressed; carpel with 5 slender bristly primary ribs and 4 winged secondary ones, each of the latter bearing a single row of barbed prickles; oil-tubes solitary under the secondary ribs, 2 on the commissural side. — Bristly Am., Eurasian and Afr. annuals or biennials with pinnately decompound leaves, foliaceous and cleft involucral bracts and compound umbels which become strongly concave. (The ancient Greek name.)

1384. D. Carota.

1. **D. CARÒTA** L. (old generic name for carrot), WILD C., QUEEN ANNE'S-

LACE, DEVIL'S-PLAGUE. — Biennial; stem bristly, 0.3–1.6 m. high; ultimate leaf-segments lanceolate and cuspidate; *primary umbel* 6–10 *cm. broad*, the rays numerous; *leaves of involucre simply pinnate*, with long linear attenuate segments; flowers white or whitish, or all roseate to purplish in forma RÒSEUS Millsp. (rosy), the central one usually dark purple, or this wanting in forma EPURPURÀTUS Farw. (without purple). — Dry fields and waste places, a pernicious weed, Que., and w. and s. beyond our limits. May–Oct. (Natzd. from Eu.) FIG. 1384.

2. D. pusillus Michx. (very small). — Similar, merely hispidulous, 1.5–6 dm. high; leaves more finely dissected; *primary umbels* 2–6 *cm. broad*, with short rays; *leaves of involucre bipinnatifid.* — Barrens, prairies and dry hills, Fla. to Calif., n. to S.C., Mo., se. Kans. and B.C. April–June.

FAM. 130. CORNÀCEAE (DOGWOOD FAMILY)

Shrubs, trees or herbs with opposite or rarely alternate simple leaves; the calyx-tube adherent to the 1–2-locular ovary, its limb minute; the petals (valvate in the bud) and as many stamens borne on the margin of an epigynous disk in the perfect flowers; style one; a single anatropous ovule hanging from the top of the locule; the fruit a 2-seeded drupe; embryo nearly as long as the albumen, with large foliaceous cotyledons. — Bark bitter and tonic.

1. CÓRNUS L. CORNEL. DOGWOOD. CORNOUILLER (Que.)

Flowers perfect (or in some foreign species dioecious). Calyx minutely 4-toothed. Petals 4, oblong, spreading. Stamens 4; filaments slender. Style slender; stigma terminal, flat or capitate. Drupe small, with a 2-locular and 2-seeded stone. — Leaves opposite (except in one species), entire. Flowers small, in open naked cymes, or in close heads surrounded by a corolla-like involucre. Am., Eurasian and Afr. genus. (Latin name from *cornu, a horn;* alluding to the hardness of the wood, the European *C. sanguinea* having long been used for skewers by butchers, whence *Skewerwood* in English provinces and DAGWOOD from the Old English *dagge*, a dagger or sharp pointed object.)

a. Flowers greenish or purple, in a close cyme or head, surrounded by a usually 4-bracted white or pink corolla-like involucre; fruit red. . . *b.*
b. Herbs, the low flowering stems arising from slender cord-like rhizomes; flowers short-pedicelled; drupes soft, very numerous, in globular clusters. Subgen. ARCTOCRANIA.
Leaves dissimilar, short-petioled, with pinnate venation; the upper much larger than the lower, ovate or rhombic, acute or acutish, the larger 2.5–9 cm. long; flowers greenish to purplish; flesh of fruit insipid. . 1. *C. canadensis.*
Leaves nearly uniform, sessile, palmately or subpalmately veined, elliptic, mostly obtuse, 1.5–5 cm. long, in several distant pairs; flowers dark purple; flesh of fruit slightly acid. . . . . 2. *C. suecica.*
b. Trees; flowers sessile; drupes hard, sessile. Subgen. BENTHAMIDIA. . . . 3. *C. florida.*
a. Flowers white or creamy, in open cymes; involucre none; fruit spherical, blue or white. Subgen. THELYCRANIA. . . *c.*
c. Leaves opposite, not crowded into terminal pseudoverticillate clusters, their petioles 0.5–1.5(–2.5)cm. long. . . *d.*
d. Cymes broad and flattish-topped or only slightly rounded, compact; leaves 1.5–13 cm. broad. . . *e.*
e. Pith of branches 1–2 years old white.
Branches red or purple-brown; leaves elliptic, ovate or lanceolate, appressed-pubescent or glabrous (if short-woolly beneath, in var. of no. 4, there whitened and the fruit white), the lateral veins 3–6 pairs.
Leaves glaucous or whitened and finely pubescent beneath; cyme flattish-topped, its branches minutely pubescent; anthers cream-color; fruit white; northern. . . . . 4. *C. stolonifera.*
Leaves green and glabrous beneath; cyme round-topped, its branches glabrous; anthers blue; fruit blue; southern. . . . 5. *C. foemina.*
Branches greenish, usually with purple blotches; leaves broadly ovate to rotund, woolly beneath, with (6–)7–9 pairs of veins; fruit bluish. 6. *C. rugosa.*
e. Pith of branches 1–2 years old brown or drab (rarely white in no. 7, with scabrous leaves).
Leaves harsh (rarely smooth) above, pilose-woolly beneath; fruit white.

Petals 4.5–6 mm. long; drupes 5–6 mm. in diameter. . . . . . 7. *C. Drummondi.*
Petals 2.5–4 mm. long; drupes 3–4 mm. in diameter. . . . . 8. *C. Priceae.*
Leaves smooth above, minutely appressed-pubescent beneath; fruit blue.
Leaves rounded at base, green beneath, the lower surfaces minutely pubescent with brown or reddish or grayish hairs. . . 9. *C. Amomum.*
Leaves cuneate at base, glaucous beneath, the lower surfaces with white or colorless minute pubescence. . . . . 10. *C. obliqua.*
d. Cymes elongate, loosely paniculate, with bright red pedicels; branches gray; fruit white; leaves 1–4 cm. broad. . . . . 11. *C. racemosa.*
c. Leaves alternate but by crowding at the tips of the branches, appearing falsely verticillate, the slender petioles mostly 2–5 cm. long; fruit blue-black, on red pedicels. . . . . 12. *C. alternifolia.*

Subgen. Arctocrània Endl.

1. C. canadénsis L. (Canadian), Dwarf Cornel, Bunchberry, Crackerberry, Pudding-berry, Quatre-temps (Que.). — Extensively *creeping* by slender forking subterranean rhizomes; flowering stems slender, erect, *herbaceous*, 0.5–3 dm. high; the remote lower 1 or 2 pairs of leaves small, the upper pairs of *pinnately veined ovate-oblong to rhombic unequal leaves* somewhat crowded and appearing falsely whorled; *the blades 2–9 cm. long, tapering to acuminate apex and subcuneate base, with definite but short petiole*, or pairs of leaves all scattered in forma **elongàta** Peck (lengthened); peduncle terminal, erect; *flowers greenish-white to creamy*, sometimes pink-tipped, the dense umbelliform *cyme surrounded by* usually 4 elliptic- to round-ovate acute white or purple-tipped, or in forma **purpuráscens** (Miyabe & Tatewaki) Hara (purplish) roseate throughout, petaloid deciduous *bracts* 0.7–2.5 *cm. long* and 0.5–1.8 cm. broad; capitate cymes of fruit 2–3.5 cm. in diameter; drupes globular, bright red, with insipid soft flesh. (*Cynoxylon* Schaffner; *Chamaepericlymenum* Aschers. & Graebn.) — Woods, thickets and damp openings, s. Greenl. and Lab. to Alaska, s. to Nfld., N.S., N.E., n. N.J., Pa., Md., W.Va., O., n. Ind., n. Ill., Minn., S.D., N.M. and Calif. *Fl.* May–July; *fr.* late July–Oct. (Ne. Asia) — Innumerable minor forms have been noted.

× **C. unalaschkénsis** Ledeb. (of Unalaska), *C. canadensis × suecica*, a hybrid of nos. 1 and 2, occurs in w. Nfld.; also Greenl. and Alaska.

2. C. suècica L. (Swedish). — Differing from no. 1 in more *uniform sessile round-based palmately or subpalmately veined elliptic* mostly *obtuse leaves in* 3–6 *distant pairs*, the *blades* 1.5–5 *cm. long;* peduncle filiform, often nodding in fruit and then frequently overtopped by elongate axillary branches; *flowers dark purple;* involucral bracts obtuse or subacute, 5–12 mm. long, 3–9 mm. broad; *drupe ellipsoid*, with juicier and more acid or sweeter flesh. — Peaty, turfy or rocky habitats, Greenl. and Lab. to Alaska, s. to Nfld., St. P. et. Miq., M.I., ne. N.S. and lower St. Lawrence, Que. *Fl.* late June, July; *fr.* Aug., Sept. (N. Eurasia)

Subgen. Benthamídia Spach

3. C. flórida L. (flowering), Flowering D. — *Large shrub or tree* up to 12 m. high; branchlets greenish; leaves ovate, acute or acuminate, tapering or rounded to the short petiole, dark green above, pale and glabrous or silky beneath; *flowers* greenish-white or creamy, *sessile, subtended by* 4, or 6–8 or more in forma **pluribracteàta** Rehd. (many-bracted), *obovate*, emarginate, subtruncate or pointed, *white*, or rose-purple in forma **rùbra** (Weston) Palmer & Steyerm. (red), often unequal deciduous *bracts* 1.5–7 *cm. long* (these, unexpanded, covering the flower-buds over winter); *drupes ellipsoid*, beaked by the persistent calyx-lobes, about 1 cm. long, dark red, or yellow in forma **xanthocárpa** Rehd. (yellow-fruited), hard, *in small heads.* (*Cynoxylon* Raf.) — Acidic woods, Fla. to Tex. and Mex., n. to sw. Me., s. N.H., s. Vt., N.Y., s. Ont., O., Ind., Ill., Mo. and Kans. *Fl.* March–June; *fr.* Aug.–Nov.

Subgen. Thelycrània Endl.

4. C. stolonífera Michx. (bearing stolons), Red Osier, Harts rouges or Poison (Que.).— Loosely spreading to ascending stoloniferous shrub with the osier-like *branches of recent years deep red*, smooth, some of them often prostrate; *pith white, usually a third the diameter of the branch; leaves* broadly ovate to ovate- or oblong-lanceolate, with 5–7 pairs of veins, dark green and glabrous or appressed-pubescent above, *glaucous and appressed-pilose to glabrous beneath;* cyme flattish-topped; corollas ovoid in bud; *fruit white or lead-colored,* rarely with bluish flush; stone rounded below, as broad as or slightly broader than high. — Shores and thickets, Nfld. and s. Lab. Pen. to Yuk., s. to N.S., N.E., D.C., W.Va., O., Ind., Ill., Ia., Neb., N.M., Ariz. and Calif. *Fl.* May–Aug.; *fr.* July–Oct. — Southward often spread (by birds) from cult. — **Forma rèpens** Vict. (creeping) is depressed and rooting at the nodes, forming carpets 4–8 dm. across, with leaves only 3–5 cm. long.

Var. Baileyi (Coult. & Evans) Drescher (for LIBERTY HYDE BAILEY, 1858–). — *Branches browner,* the young branchlets pubescent; *leaves densely soft-pilose beneath* with spreading and curving hairs; stone broader than high. (*C. Baileyi* Coult. & Evans) — St. Lawrence valley, Ont., to Alaska, s. to n. and w. N.Y., nw. Pa., O., Ind., n. Ill., Minn. and S.D.

5. C. foèmina Mill. (female; so considered by Miller, as opposed to the "male" no. 3), STIFF D. — Small tree or erect shrub up to 5 m. high; *branches of recent years reddish or brown,* smooth; *pith white,* slender, less than one-third diameter of branch; *leaves* broadly lanceolate to narrowly elliptic-ovate, tapering to narrow and prolonged tip and at base to short petiole, *glabrous* or nearly so, dark green above, *slightly paler beneath,* with 4 or 5 pairs of veins; *cymes round-topped,* rather open, *on peduncles* 2.5–7 *cm. long;* corollas subcylindric in bud; *anthers bluish; fruit bluish;* stone longer than broad, slightly furrowed. (*C. stricta* Lam.) — Swamps, wet woods and bottomlands, Fla. to e. Tex., n. to e. Va., s. Ind. and se. Mo. *Fl.* May, June; *fr.* Aug.–Oct. — Frequently confused in the herbarium with *C. racemosa.*

C. GLABRÀTA Benth. (becoming glabrous) of the Pacific slope spreads from cult. Differing from no. 5 by branches sooner becoming gray; the cymes flat-topped, 2–3 cm. broad; fruit white or very pale blue; stone subglobose, ribless. (Introd. from w. N. Am.)

6. C. rugòsa Lam. (wrinkled), ROUND-LEAVED D., BOIS DE CALUMET (Que.). — Coarse ascending shrub up to 3 m. high; *branches of recent years greenish, usually blotched* with purple; pith white; *leaves broadly ovate to suborbicular,* abruptly pointed, 3–13 cm. broad, broadly rounded at base, or in forma **eucỳcla** Fern. (well rounded) of C. B. membranaceous and orbicular to orbicular-obovate and scarcely apiculate, *woolly beneath, with* (6–)7–9 *pairs of veins;* cymes dense, flat-topped, on stout peduncles 1–4 cm. long; *fruit light blue,* rarely white; stone subglobose. (*C. circinata* L'Hér.) — Dry woods and rocky slopes, Gaspé Co., Que., to Man., s. to N.S., N.E., w. Va., W.Va., O., Ind., n. Ill. and ne. Ia. — *Fl.* late May–July; *fr.* Aug.–Oct.

× **C. Slavínii** Rehd. (named in 1910 for its discoverer, B. H. SLAVIN), a hybrid of nos. 4 and 6, has the branches purple; leaves narrowly ovate and tapering at summit, subglabrous to somewhat woolly beneath; fruit pale blue to white. — Infrequent, N.Y. to Wisc.

7. C. Drummóndi Meyer (for its discoverer, THOMAS DRUMMOND, 1780–1835). — Upright shrub or small tree up to 15 m. high; *branches gray, with brown* (rarely white) *slender pith; leaves* ovate to elliptic, with prolonged tips, *scabrous* (rarely smooth) *above, pilose-woolly beneath;* corymbs round-topped, pubescent, 4–7 cm. broad; corolla subcylindric in bud; petals 4.5–6 mm. long; *fruit white,* 5–6 *mm. in diameter;* stone subglobose. (*C. asperifolia* of auth., not Michx.) — Shores and damp woods and thickets, Miss. to e. Tex., n. to s. Ont., O., Ind., Ill., Ia. and Neb. *Fl.* late May, June; *fr.* Aug.–Oct.

8. C. Prìceae Small (for its discoverer, SARAH FRANCES PRICE, 1849–1903), MISS PRICE'S D. — Similar to no. 7; branches redder; leaves very scabrous, the oval to round-ovate blades with prolonged tips; *petals* 2.5–4 *mm. long; drupes* white, *only* 3–4 *mm. in diameter.* — River-banks and bluffs, Ky. and Tenn. *Fl.* late May, early June; *fr.* July, Aug. — Perhaps better considered as *C. Drummondi,* forma *Priceae* (Small) Rickett.

9. C. Amòmum Mill. (Latin name of some shrub), "RED WILLOW". — Upright or spreading shrub up to 3 m. high, with reddish-brown to grayish bark and tawny pith; young growth sericeous with brown hairs; *leaves ovate or broadly elliptic, with broadly rounded bases and abrupt short tips, green or rufescent beneath, usually with reddish* (sometimes colorless or pale) *hairs on the nerves* and commonly on the surface, *those of the fertile branches* 3–8 cm. broad, *averaging more than half as broad as long;* cyme 4–7 cm. broad, its peduncle and pedicels silky-pilose; corolla in bud cylindric; *fruit blue or bluish-white;* stone subglobose. (*Svida* Small, in part) — Swamps and damp thickets, s. Me. to Ind., s. to s. N.E., Ga. and Ala. *Fl.* June, July; *fr.* Aug.–Oct.

10. C. oblìqua Raf. (oblique), SILKY D. — Similar to no. 9 and often confused with it; *leaves lanceolate to narrowly ovate-oblong, with tapering or gradually narrowed bases and gradually acuminate tips, whitish or glaucous beneath,* the lower surface with appressed white hairs, *blades of the fertile branches* 1.5–5 cm. broad, *averaging less than half as broad as long.* (*C. Purpusi* Koehne) — Swamps and damp thickets, N.B. to N.D., s. to s. N.E., N.J., Pa., W.Va., Ky., Ark. and Okla. *Fl.* mid-May–mid-July; *fr.* late May–Oct.

× **C. arnoldiàna** Rehd. (from growing in the Arnold Arboretum), a hybrid of nos. 10 and 11, taller than the latter, with its slender gray branches; elongate acuminate leaves mostly 6–10 cm. long, and somewhat paniculate cyme, is occasional from N.E. to Pa., O. and Mo.

11. C. racemòsa Lam. (racemose). — Ascending shrub up to 2(–3) m. high, with *slender gray branches* and slender pale brown pith, the pith of the young branchlets often white; *leaves oblong-lanceolate, elliptic or narrowly ovate,* long-acuminate, tapering to short (mostly 3–10 mm. long) petioles, *glaucous beneath, glabrous or appressed-pubescent,* the blade 2.5–7(–10) cm. long and 1–4 *cm. broad; cymes paniculate, nearly or quite as high as broad* (3–6 cm.), *open, with bright red branches;* petals recurving, 3–4 mm. long; *fruit white,* on red pedicels, depressed, 5–7 mm. high; stone obliquely subglobose. (*C. paniculata* L'Hér.; *C. candidissima* Marsh., not Mill.; *Svida foemina* sensu Small, not *C. foemina* Mill.) — Dry to moist open habitats, centr. Me. to s. Ont. and Minn., s. to s. N.E., Del., Md., w. Va., (?"Ga."), Ky., Mo. and Okla. *Fl.* late May–early July; *fr.* July–Oct.

12. C. alternifòlia L. f. (alternate-leaved), PAGODA-D., GREEN OSIER. — Shrub or small tree up to 8 m. high, the *branches* often grouped *in irregular horizontal platforms,* glabrous, those of recent years *greenish,* with slender white pith; *leaves alternate but on fertile branches subapproximate and appearing as if in whorls,* elliptic- to rhombic-ovate, acuminate, *tapering to slender petioles mostly* 2–6 *cm. long,* pale beneath, with 5 or 6 pairs of veins; cyme flat-topped, 3–6 cm. broad; fruit bluish-black, with a bloom, or yellow in forma **ochrocárpa** Rehd. (yellowish-fruited), hard; stone subglobose. (*Svida* Small) — Dry woods and rocky slopes, Nfld. to s. Ont. and Minn., s. to N.S., N.E., Ga., Ala. and Mo. *Fl.* May–early July; *fr.* late July–Sept.

× **C. acadiénsis** Fern. (of Acadia), a hybrid of nos. 4 and 12, from C.B., has purple-brown branches with drab pith, elliptic or oval short-pointed opposite leaves crowded in false whorls, small cymes and fleshy blue drupes.

Various cult. species spread from local plantings; these should be sought in works on cult. plants.

Subclass II. METACHLAMÝDEAE, GAMOPÉTALAE

FAM. 131. CLETHRÀCEAE (WHITE ALDER FAMILY)

Shrubs or trees with stellate pubescence, regular flowers in panicles or racemes, calyx free from ovary, corolla of distinct petals, anthers extrorse in bud and opening by pores at base and inverted in flower, superior ovary 3-locular, style 3-cleft, 3-valved capsule with 2-cleft valves, many-seeded porrect placentae remaining attached to the columella, seeds with transparent cellular coat. Sepals five. Stamens ten. Pollen-grains simple. A single small genus:

1. CLÈTHRA L. WHITE ALDER. SUMMER-SWEET

Sepals imbricated in the bud. Petals obovate-oblong. Anthers sagittate, erect in the bud, becoming inverted. Style slender. Capsule 3-valved, many-seeded, inclosed in the calyx.— Shrubs or trees of trop. reg. and temp. e. Am., e. Asia and Madeira, with alternate serrate deciduous leaves, and white flowers in terminal hoary racemes. Bracts deciduous. (Ancient Greek name of Alder, applied to this very different genus because of some similarity in foliage.)

1. C. alnifòlia L. (alder-leaved), SWEET PEPPERBUSH. — Shrub 1–3 m. high; *leaves* 3.5–7 cm. long, *cuneate-obovate, sharply serrate,* entire toward the base, prominently straight-veined, smooth, green both sides; *racemes upright, usually panicled;* petals white, or pink in the local forma **ròsea** Rehd. (rose-colored); *bracts shorter than the flowers;* filaments smooth. — Swamps, damp thickets and sandy woods, Fla. to e. Tex., n. to s. Me., s. N.H., Mass., se. N.Y. and e. Pa. July–Sept. — Flowers deliciously fragrant.

2. C. acuminàta Michx. (taper-pointed). — A tall shrub or small tree; *leaves oval or oblong, pointed,* thin, finely serrate, 7–15 cm. long, pale beneath; *racemes solitary, flexuous or drooping; bracts longer than the flowers;* filaments and capsules hairy. — Woods and acid rocks, mts. of Va. and W.Va. to Ga. and Tenn. July, Aug.

FAM. 132. PYROLÀCEAE (WINTERGREEN FAMILY)

Herbaceous or nearly so from perennial slender rootstocks and mostly with evergreen simple leaves, or saprophytes or root-parasites without chlorophyll, with distinct or united petals, mature anthers mostly inverted and with basal pores, ovary and capsule 5 (or 4)*-locular, placentae and seed much as in* Clethraceae. Sepals and petals mostly 5 (5–2 in *Monotropa*). Stamens 10. Pollen-grains simple or compound. Small family of N. Hemisph.

Slender-stemmed herbaceous or subherbaceous mostly green-leaved plants with slender subterranean rootstocks; pollen in tetrads.
Stem leafy; flowers in a corymb; filaments dilated, hairy; style nearly obsolete. 1. *Chimaphila.*

Stem scapose, leafy only near base; flowers in a raceme or solitary; filaments slender, glabrous; style definite.
Flower solitary; petals widely spreading, valves of capsule smooth-margined. . . . . . 2. *Moneses.*
Flowers in a raceme; petals arching; valves of capsule with cobwebby margins. . . . . . 3. *Pyrola.*
Fleshy-stemmed saprophytes or root-parasites without chlorophyll, the bases thickened, the leaves scaly; pollen simple.
Corolla of 4–6 distinct petals; calyx represented by 2–5 bracts. . . . . . 4. *Monotropa.*
Corolla gamopetalous; calyx of 5 regular sepals.
Coarse clammy-pubescent plant 3–9 dm. high; corolla globose-urceolate, shallowly 5-toothed; anthers 2-awned on back. . . . . . 5. *Pterospora.*
Small glabrous plant up to 1 dm. high; corolla open-campanulate, deeply 5-lobed; anthers awnless. . . . . . 6. *Monotropsis.*

1. CHIMÁPHILA Pursh PIPSISSEWA. WINTERGREEN. WAXFLOWER. HERBE À CLÉ (Que.)

Petals 5, concave, orbicular. Stamens 10; filaments enlarged and hairy in the middle; anthers as in *Pyrola,* but more or less conspicuously 2-horned. Style nearly immersed in the depressed summit of the globular ovary; stigma broad and orbicular, disciform, the border 5-crenate. Capsule, etc., as in *Pyrola,* but splitting from the apex downward. — Low nearly herbaceous plants of the N. Hemisph., with long creeping subterranean shoots, and thick shining leaves somewhat whorled or scattered along the short ascending stems; the flowers pink, roseate or white, on a terminal peduncle. (Name from the Greek *cheima, winter,* and *philein, to love,* in allusion to one of the popular names, WINTERGREEN.)

1. C. UMBELLÀTA (L.) Bart. (umbellate). — Leafy, 1–2.5 dm. high; leaves dark green, not mottled, cuneate, blunt, obtusely toothed, about 3.5 cm. long; peduncle umbellately 2–8-flowered; calyx-lobes ovate, longer than broad; petals flesh-color; anthers violet; stigma 1.9–2.3 mm. broad. Eurasia. — Represented with us by

Var. **cisatlántica** Blake (on this side of the Atlantic), PRINCE'S PINE, PIPSISSEWA. — Mostly larger, up to 3 dm. high; *leaves* 3–7 cm. long, 1–2.2 cm. broad, *definitely veined beneath, acutely toothed,* mucronulate; flowers more or less racemose; *calyx-lobes usually broader than long; capsule* 5–6 *mm. in diameter;* stigma 2.3–2.5 mm. broad. (*C. corymbosa* Pursh) — Dry woods, Gaspé Pen., Que., to w. Ont., s. to N.S., N.E., Ga., O., Mich., ne. Ill. and Minn. July, Aug.

Var. **occidentàlis** (Rydb.) Blake (western). — Coarser than the preceding; *leaves very obscurely veined or essentially veinless beneath,* cuneate or cuneate-obovate, up to 9 cm. long, 1–2.5 cm. broad; *calyx-lobes longer than broad; capsule* 6–7.5 *mm. in diameter.* (*C. occidentalis* Rydb.) — Dry woods, B.C. and se. Alaska to Calif., Utah and Colo.; n. Mich.

2. C. maculàta (L.) Pursh (mottled or spotted), SPOTTED W. — Stem 1–2.5 dm. high; *leaves lanceolate or ovate-lanceolate, obtuse at the base,* remotely toothed, *the upper surface variegated with white;* peduncles 1–5-flowered, stiffly erect in fruit; flower 2 cm. broad, very fragrant, with white petals. — Dry woods, s. N.H. to s. Ont., Mich. and ne. Ill., s. to Ga., Ala. and Tenn. June–early Aug.

2. MONÈSES Salisb. ONE-FLOWERED PYROLA

Petals 5, orbicular. Filaments subulate, naked; anthers as in *Pyrola,* but conspicuously 2-horned. Stigma large, peltate, with 5 narrow and conspicuous radiating lobes. (Flowers occasionally tetramerous.) — Intermediate between *Pyrola* and *Chimaphila,* a single boreal species. (Name formed of the Greek *monos, single,* and *hesis, delight,* from the attractive solitary flower.)

1. M. uniflòra (L.) Gray (one-flowered). — A small perennial; the rounded and veiny serrate thin leaves 1–3 cm. long, clustered at the ascending apex of creeping subterranean shoots; the 1–2-bracted scape 3–13 cm. high, bearing a fragrant waxy-white or rose-tinged terminal flower 1–2 cm. wide. — Cool mossy woods, Lab. to Alaska, s. to Nfld., N.S., N.E., Pa., upland W.Va., Mich., Wisc., Minn., Colo., Utah and Oreg. June–Aug. (Eurasia)

3. PÝROLA L. PYROLA. WINTERGREEN. SHINLEAF

Calyx 5-parted, persistent. Petals 5, concave and more or less converging, deciduous. Stamens 10; filaments naked; anthers extrorse in the bud, but in the flower commonly inverted by the inflexion of the apex of the filament, more or less 4-locular, opening by a pair of pores at the blunt or somewhat 2-horned base (by inversion the apparent apex). Stigma 5-lobed or 5-rayed.

Capsule depressed-globose, 5-lobed, 5-valved from the base upward (loculicidal). Seeds minute, innumerable, with a very loose cellular-reticulated coat. — Low and smooth perennial herbs of N. Hemisph., with creeping subterranean shoots, bearing a cluster of petioled evergreen basal leaves, and a simple raceme of ascending, spreading or nodding flowers on an upright more or less scaly-bracted scape. (Name a diminutive of *Pyrus*, the Pear-tree, from resemblance in the foliage.)

*a.*Raceme 1-sided; corolla campanulate, longer than broad; capsule 3–5 mm. broad, shorter than the long straight style. . . . . 1. *P. secunda.*

*a.*Raceme spiral; corolla subglobose or short-campanulate, broader than long; capsule 4–9 mm. broad, with very short straight or long upwardly curved style. . . *b.*

*b.*Corolla subglobose; petals nearly meeting at tip, 3–5 mm. long; stamens equally connivent around the short (0.5–1.3 mm. long) straight style. 2. *P. minor.*

*b.*Corolla with spreading or loosely converging petals 4–11 mm. long; stamens declined-ascending; style upwardly curved, in maturity 3–11 mm. long. . . *c.*

*c.*Pedicels ascending during anthesis; petals acuminate; anthers remaining erect and extrorse, not inverted, in anthesis. . . . 3. *P. oxypetala.*

*c.*Pedicels divergent or arched-recurving during anthesis; petals rounded at summit; anthers becoming inverted in anthesis. . . *d.*

*d.*Leaf-blade usually shorter than petiole or wanting, opaque and coriaceous; calyx-lobes rounded or obtuse, 0.4–2 mm. long; corolla greenish-white, with converging petals. . . . . 4. *P. virens.*

*d.*Leaf-blade (except in deep shade) nearly equaling to longer than petiole, lustrous or, if opaque, thin and membranaceous; calyx-lobes lanceolate to ovate, acutish or mucronate, 1.5–7 mm. long; corolla milk- or cream-white to crimson, with loosely spreading petals. . . *e.*

*e.*Leaf-blade opaque and thin, usually longer than petiole; bracts of raceme linear-subulate; calyx-lobes 1.5–2 mm. long, less than one-fourth the length of the petals. . . . . 5. *P. elliptica.*

*e.*Leaf-blade lustrous and coriaceous (if dull, the plant pink-flowered), usually as short as petiole; bracts of raceme lanceolate to ovate or obovate; calyx-lobes 1.6–7 mm. long, one-third to three-fifths as long as petals. . . *f.*

*f.*Filaments flat; anthers 2–3.6 mm. long, mostly mucronate at base, their locules definitely constricted above into necks; calyx-lobes coriaceous, chartaceous or scarious; bracts of scape oblong-lanceolate to ovate.

Calyx-lobes firm, 3–5-nerved nearly to tip; petals milk-white or creamy (rarely pink-tinged), thick and leathery, scarcely veiny, not obviously contracted to claw; leaf-blades leathery and lustrous, not cordate. . . . . 6. *P. rotundifolia.*

Calyx-lobes thin, membranaceous or scarious, nerveless or obscurely nerved; petals crimson to white, rather thin, conspicuously veiny (at least when dry), contracted to a claw; leaf-blade, when leathery and lustrous, with cordate base. . . 7. *P. asarifolia.*

*f.*Filaments almost filiform at summit; anthers 1.7–2.3 mm. long, muticous at base, their locules barely constricted above; calyx-lobes herbaceous to petaloid; bracts of scape short-oblong to obovate. . . . . 8. *P. grandiflora.*

1. P. secúnda L. (one-sided), ONE-SIDED P. or W. — Caudex forking; *basal bracts ciliate,* lanceolate, *strongly involute, acute, firm;* leaves rosulate or somewhat scattered; blade longer than petiole, subcoriaceous, lustrous, elliptic to ovate, narrowed to mucronate tip, or rounded at summit in forma **eucýcla** Fern. (well rounded), crenate-serrate, 1.5–6 cm. long; scapes 1–2 dm. high, 2–5-bracted; *raceme strongly 1-sided,* 6–20-flowered, in anthesis 2–5, in fruit up to 8.5 cm. long; *bracts of raceme ciliate,* equaling or shorter than the spreading or reflexed pedicels; *calyx-lobes ciliate-serrulate,* obtuse, 0.6–1.2 mm. long; *corolla campanulate, as long as or longer than broad; petals oblong, erose,* greenish-yellow, 3.5–5.5 mm. long, *with* 2 small *tubercles at base; stamens equally connivent about pistil; style* slender, *straight,* exserted, tipped by the 5-lobed broad peltate stigma, in maturity 5–9 mm. long; *ovary with* 10 *small tubercle-like bodies at base; capsule depressed,* 3–5 *mm. broad.* (*Ramischia secunda* (L.) Garcke) — Dry or moist woods, Nfld. to Alaska, s. to N.S., N.E., Va., n. O., Ind., n. Ill., ne. Ia., S.D., N.M. and Calif. June–Aug. (Eurasia) FIG. 1385.

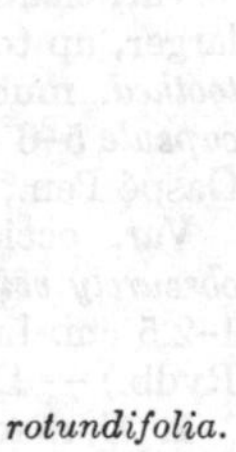

1385. P. secunda.

Var. **obtusàta** Turcz. (blunt). — Basal bracts oblong to ovate, only slightly involute, membranaceous; leaf-blades membranaceous, scarcely lustrous, ovate to orbicular, often rounded above, crenate, 0.8–3 cm. long; scape 0.5–1.5 dm. high; raceme 2–10-flowered, in anthesis 0.5–3, in fruit up to 4 cm. long; petals creamy-white; mature style 4–6 mm. long. — Greenl. and n. Lab. to Alaska, s. in mossy or boggy (often calcareous) woods and on wet rocks, to Nfld., N.S., n. N.E., w. Mass., n. and w. N.Y., Mich., Minn., Colo. and Calif. June–Aug. (Asia)

1386. P. minor.

2. P. mìnor L. (smaller). — Basal bracts oblong to obovate, cuspidate; leaves rosulate or but slightly scattered, the *blade* short-oblong to orbicular, thin, crenate, *dull*, 1–4.5 cm. long, shorter than to equaling petiole; scape 0.5–2.6 dm. high, naked or with 1 or 2 bracts; *raceme cylindrical*, 6–17-flowered, in anthesis 1.3–7 cm. long, longer in fruit; pedicels spreading or arched-recurving; calyx-lobes entire; *corolla subglobose; petals* bright white to pink, 3–5 mm. long, *their tips connivent; stamens equally connivent around the short* (0.5–1.3 *mm. long*) *straight style; anthers not constricted below the pores; stigma peltate;* capsule 4–5.5 mm. broad. (*Erxlebenia minor* (L.) Rydb.) — Greenl. and Lab. to Alaska, s. in cool thickets and woods to Nfld., N.S., n. N.E., n. Mich., Minn., N.M. and Calif. Late June–early Aug. (Eurasia) Fig. 1386.

3. P. oxypétala Austin (with sharp petals). — Basal bracts lanceolate, dentate; *leaves* rosulate; *blade narrowly ovate*, 1.5–2.5 cm. long, coriaceous, shorter than petiole; scape 2-bracted, about 2 dm. high; raceme cylindric, 3–4 cm. long, 7–9-flowered; *pedicels ascending; calyx-lobes elongate-triangular, acute,* 1–1.5 *mm. long; petals* converging, greenish, *lance-oblong, acuminate,* 6–7 mm. long; stamens declined-ascending; *anthers remaining erect and extrorse,* not inverted in flower, mucronate, 2-horned, 3–3.5 mm. long; *style straightish,* about 4 mm. long. — Wooded hill near Deposit, Delaware Co., N.Y., June 1, 1860, *Austin.* Fig. 1387. — A remarkable and wholly anomalous plant (other members of the genus having anthers inverted), to be most carefully sought.

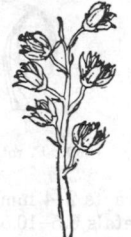

1387. P. oxy-petala.

1388. P. virens.

4. P. virens Schweigger (greenish). — Basal bracts remote, lance-subulate to oblanceolate; *leaf-blades* reniform, suborbicular or ovate, rounded to base and apex, rather numerous (4–11), *thick and opaque,* obscurely crenate, *slender-petioled,* the larger 1.5–3.5 cm. broad; *scape* 0.7–3 dm. high, *naked,* rarely with 1 or 2 bracts; raceme cylindric, (1–)2–13-flowered; *calyx-lobes* ovate-oblong to deltoid-ovate, entire, usually as long as or longer than broad, 0.8–1.7 *mm. long, rounded at tip or obtuse;* corolla open-campanulate; *petals* spreading-ascending, *convergent, oblong to oval,* obtuse, *greenish-white,* 4–6.5 mm. long, 2.5–4 mm. broad; stamens declined-ascending; anthers constricted below pores, 2.3–3.3 mm. long; style declined, upwardly curved at apex, with a scalloped ring below the stigma; capsule 5.5–9 mm. long; mature style 4–7 mm. long. (*P. chlorantha* Sw.) — Dry or dryish coniferous woods and thickets, Lab. to Alaska, s. to Nfld., N.S., Me., Mass., Wisc., S.D., Ariz. and Oreg. June–early Aug. (Eurasia) Fig. 1388. — Forma **paucifòlia** Fern. (few-leaved) has leaf-blades mostly cuneate at base and truncate or subtruncate at summit or flabelliform-obovate, few (1–7 or even wanting), the larger 0.7–2.5 cm. broad and long; calyx-lobes broadly deltoid, mostly broader than long, 0.5–1 mm. long; petals 4–6.5 mm. long; anthers 1.5–2.6 mm. long, found in dry or moist woods from N.S. and P.E.I. to Minn., s. to n. and w. N.E. and mts. of Pa.

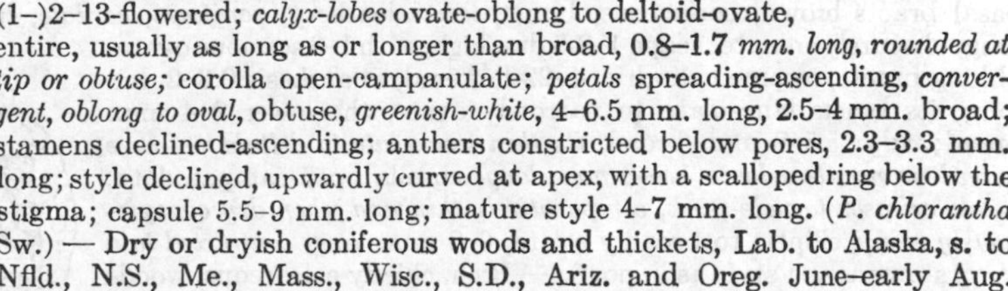

Var. **convolùta** (Bart.) Fern. (rolled up; referring to the twisted stem). — Coarser than typical *P. virens;* leaf-blades rounded at base, the larger 2–4.5 cm. broad; calyx-lobes 1.2–2 mm. long; petals 6.5–9 mm. long, 3.5–6 mm. broad; anthers 3–4 mm. long; mature style 8–10 mm. long. — Dry woods, se. and centr. Me. to s. Ont., s. to s. N.E., Del., Md., W.Va., Ind., Wisc., Minn. and Neb. Late May–early July.

1389. P. elliptica.

5. P. ellíptica Nutt. (elliptic), Shinleaf, Wild Lily-of-the-valley. — Basal bracts chartaceous, oblong, obtuse; leaves rosulate; *blades thin and membranaceous, dull, usually longer than the petioles,* elliptic, oblong or obovate, commonly rounded at summit, plicate-serrulate or crenate, 2–8 cm. long; *scape* 1.2–3 dm. high, *naked or with 1 or 2 lance-subulate bracts; raceme* cylindric, 3–21-flowered, *with bracts linear-subulate; calyx* 4–5.5 mm. in diameter, the triangular to ovate acute *lobes* 1.5–2 *mm. long;* corolla open-campanulate, milk-white or creamy, rarely pink-tinged, fragrant; petals spreading, obovate, 7–10 mm. long; stamens declined-ascending; style declined and upcurved, in maturity 7–10 mm. long. — Dry or moist woods, Nfld. to B.C., s. to N.S., N.E., Pa., W.Va., O., Ind., n. Ill., Ia., S.D. and N.M. June–Aug. (Japan) Fig. 1389.

6. P. rotundifòlia L. (round-leaved), Wild Lily-of-the-valley., Muguet des bois (Que.). — Bracts of elongate caudex brown, coriaceous, oblong to elliptic, blunt; *leaves* rosulate, *thick and leathery, evergreen, lustrous;* the blades suborbicular to rounded-obovate, 1.8–5 cm. long, the revolute margin obscurely crenate-serrulate; *petiole margined;* scape 1–3 dm. high, with 1–7 conspicuous brownish scarious oblong or oblong-lanceolate bracts; raceme cylindric, 3–13-flowered, in anthesis 2–9 cm. long; its lowest bracts lanceolate or oblong, scarious, brown, 1–2 mm. broad; pedicels spreading or arching; *calyx* 5–7 *mm. broad; its firm oblong-lanceolate to oblong-obovate lobes* 1.6–3 *mm. long, usually* 3–5-*nerved nearly to tip;* corolla open-campanulate, very fragrant; its creamy or milk-white (rarely pinkish) *petals thick and leathery, scarcely veiny, not obviously contracted to a claw,* broadly rounded, 5–7 mm. long, 4–6 mm. broad; *filaments flat; anthers* 2–2.7 *mm. long, usually mucronate* at base, the very short necks straight or slightly oblique; style deflexed at base, upwardly arched, in maturity 5–10 mm. long, with a distinct ring below the 5 lobes of the stigma. — Damp sandy or gravelly thickets, barrens and sphagnous bogs. Greenl. to Nfld., e. Que. and N.S. July, Aug. (Eurasia)

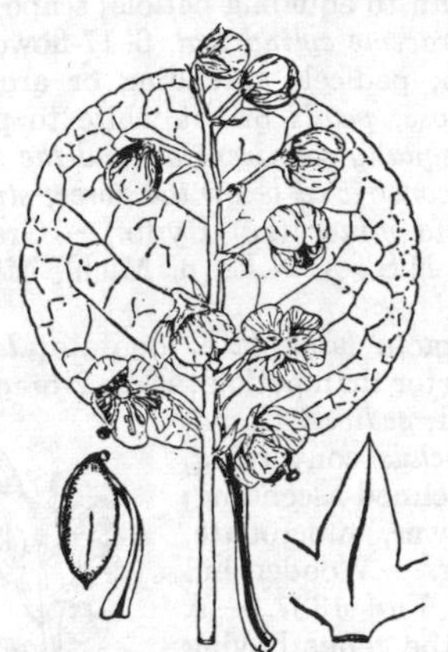

1390. P. rotundifolia, v. americana.

Var. **americàna** (Sweet) Fern. (American). — Leaf-blades 2.5–8 cm. long; scape up to 4 dm. high; raceme 5–21-flowered, in anthesis 0.25–2 dm. long, its lowest bracts 2–4 mm. broad; calyx 6.3–10 mm. broad, its oblong to rhombic lobes 2.5–4.5 mm. long; petals 6.5–10.5 mm. long, 3.5–8 mm. broad; anthers 2.7–3.6 mm. long, with prominent arching necks; style 7–11 mm. long. (*P. americana* Sweet) — Woods and clearings, s. Gaspé Pen., Que., to Ont., s. to N.S., N.E., N.C., Ky., n. Ind., Mich., Wisc. and Minn. June–mid-Aug. Fig. 1390.

7. P. asarifòlia Michx. (with leaves of *Asarum*), Pink P. or W. — Extensively creeping; basal bracts brown, coriaceous, blunt; *leaves* leathery, *reniform, cordate, lustrous,* 2–6.5 cm. long, obscurely crenate; scape 1–3.5 dm. high, with 1–3 scarious ovate-oblong bracts; raceme cylindric, 4–22-flowered, in anthesis 0.3–2 dm. long; its bracts lanceolate to narrowly ovate, the lower 2–4 mm. broad; *calyx* 5.5–8 mm. broad, *its thin lanceolate to deltoid or ovate nerved or nerveless lobes* 1.8–3 *mm. long;* corolla open-campanulate; *petals crimson to pale pink, comparatively thin and when dry conspicuously veiny,* elliptic to broadly ovate, 6–8 mm. long, *contracted to a claw;* stamens and style as in no. 6. — Rich, chiefly calcareous, woods and thickets, Nfld. to Yuk., s. to P.E.I., n. N.E., centr. and w. N.Y., n. Ind., Mich., Wisc., S.D. and N.M. Late June–early Aug. Fig. 1391.

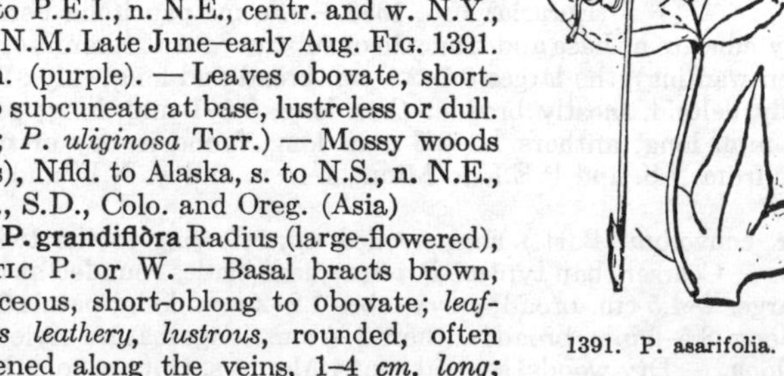

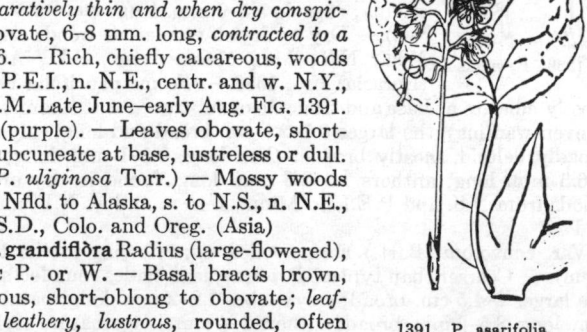

1391. P. asarifolia.

Var. **purpùrea** (Bunge) Fern. (purple). — Leaves obovate, short-elliptic or orbicular, rounded to subcuneate at base, lustreless or dull. (Var. *incarnata* (Fisch.) Fern.; *P. uliginosa* Torr.) — Mossy woods and swamps (chiefly calcareous), Nfld. to Alaska, s. to N.S., n. N.E., w. Mass., N.Y., n. Ind., Minn., S.D., Colo. and Oreg. (Asia)

8. P. grandiflòra Radius (large-flowered), Arctic P. or W. — Basal bracts brown, coriaceous, short-oblong to obovate; *leaf-blades leathery, lustrous,* rounded, often whitened along the veins, 1–4 *cm. long; scape* 0.5–1 (rarely –2) *dm. high, with* 1–3 brown membranaceous *short-oblong to oblong-obovate bracts; raceme* cylindric, 2–9-flowered, in anthesis 1.5–7 cm. long; *its bracts* scarious, brown, *broad-lanceolate; calyx-lobes herbaceous or petaloid, purplish,* oblong, broadly lanceolate or oblong-ovate, *with erose scarious margins, distinctly* 5–7-*nerved,* 3–4.5 mm. long; corolla as in no. 6, *the thick* crimson to cream-white *rounded petals* 6–11 *mm. long; stamens* declined-ascending, 4–5.5 *mm. long; filaments* slender, *almost filiform at summit; anthers* 1.7–2.3 mm.

1392. P. grandiflora.

long, *muticous at base, barely constricted to necks above; style* deflexed at base, in maturity 3–6 *mm. long.* — Arct. reg., s. on damp slopes to mts. of w. Nfld., alpine reg. of Shickshock Mts., Que., n. Ont., n. Man. and Mackenz. Mid-June–Sept. (N. Eurasia) FIG. 1392.

4. MONÓTROPA L. INDIAN-PIPE. PINESAP

Calyx of 2–5 lanceolate bract-like scales, deciduous. Corolla of erect spatulate or cuneate scale-like petals which are gibbous or saccate at the base and tardily deciduous. Stamens 8 or 10; filaments subulate; anthers becoming 1-locular. Style columnar; stigma discoid, 4–5-rayed. Capsule ovoid, 8–10-grooved, the 4–5 locules loculicidal; the very thick placentae covered with innumerable minute seeds which have a very loose coat. — Low and fleshy herbs of N. Hemisph., tawny, reddish or white, parasitic on roots or growing on decomposing vegetable-matter; the clustered stems springing from a ball of matted fibrous rootlets, furnished with scales or bracts in place of leaves, 1–several-flowered; the summit at first nodding, in fruit erect. (Name composed of the Greek *monos, one,* and *tropos, turn,* the summit of the flowering stem being turned to one side.)

Plant essentially inodorous, 1-flowered; calyx of 2–4 irregular scales or bracts; anthers transverse, opening equally by 2 chinks; style short and thick. § EUMONOTROPA. . . . . 1. *M. uniflora.*

Plant commonly aromatic-fragrant; flowers several in a scaly raceme, the terminal one usually 5-merous, the others 3- or 4-merous; bract-like sepals mostly as many as the petals; anthers opening by a continuous line into 2 very unequal valves; style longer than ovary, hollow. § HYPOPITYS. . . . . . . 2. *M. Hypopithys.*

§ EUMONÓTROPA Gray

1. M. uniflòra L. (one-flowered), INDIAN-PIPE, CORPSE-PLANT, CONVULSION-ROOT, FITS-ROOT. — Smooth, at first waxy-white, flesh-pink or rarely (in autumn) reddish, quickly blackening in drying, 0.5–3 dm. high; bud and flower nodding, fruit erect; stigma naked. — Woodland humus, Nfld. to se. Alaska, s. beyond our range. June–Sept. (Mex., Asia)

§ HYPÓPITYS Gray

2. M. Hypópithys L. (old generic name from the Greek *hypo, under,* and *pitys, pine*), PINESAP, FALSE BEECH-DROPS. — Somewhat pubescent or downy, tawny, yellowish or (in autumn) sometimes red, darkened in drying, 1–4 dm. high; young raceme drooping, soon becoming erect; capsule globular or ovoid; stigma ciliate. (*Hypopithys lanuginosa* (Michx.) Nutt.) — Woodland humus, Nfld. to B.C., south beyond our limits. June–Oct. (Mex., Eurasia) — A complex species, needing careful study; the variations of pubescence of parts and of color not clearly supporting the differentiation of species as sometimes maintained. The specific name, as used consistently by Linnaeus, often altered to *Hypopitys.*

5. PTERÓSPORA Nutt. PINE-DROPS

Calyx 5-parted. Corolla ovoid, urceolate, persistent. Stamens 10. Style short; stigma 5-lobed. Capsule globose, depressed, 5-lobed, 5-locular, loculicidal, but the valves cohering with the columella. Seeds very numerous, ovoid, tapering to each end, the apex expanded into a broad reticulated wing many times larger than the body of the seed. — A stout and simple purplish-brown clammy-pubescent root-parasitic herb 3–9 dm. high; the wand-like stem furnished towards the base with scattered lanceolate scales in place of leaves, above bearing many nodding white to red flowers in a long bracted raceme. (Name from the Greek *pteros, a wing,* and *spora, seed;* alluding to the singular wing borne by the seeds.)

1. P. andromedèa Nutt. (like *Andromeda;* from the nodding reddish to white flowers), GIANT BIRD'S-NEST. — Hard argillaceous or basic soils in dry, chiefly pine, woods, very local and probably largely extinct with us, P.E.I. to B.C., s. to Vt., N.Y., Mich., Wisc. and Rocky Mts. to Mex. June–Aug.

6. MONOTRÓPSIS Schwein. SWEET PINESAP

Calyx of 5 oblong-lanceolate acute scale-like sepals, erect, persistent. Corolla persistent, rather fleshy, slightly 5-gibbous at the base. Stamens 10; anthers much shorter than the filaments, fixed near the summit, awnless, with two sac-shaped locules. Capsule ovoid, 5-locular, with a short and thick style and a large 5-angular stigma. Seeds innumerable. — Low and smooth purplish-brown e. N. Am. plants, 0.5–1 dm. high, with the aspect of *Monotropa*, scaly-bracted, the flowers several in a terminal spike, at first nodding, light rose-color, fragrant. (Name from *Monotropa* and *opsis, appearance*, from resemblance to that genus.)

1. M. odoràta Ell. (fragrant), PYGMY-PIPES. — Sandy, chiefly pine, woods, Md. and W.Va. to Ga. and Ala. Late March–early May. — Fresh plant deliciously aromatic, usually hidden by fallen leaves.

FAM. 133. ERICÀCEAE (HEATH FAMILY)

Trees or shrubs with regular or nearly regular flowers; stamens as many or twice as many as the 4–5 lobes or petals of the corolla, free from it or borne at its base and with a well developed hypogynous disk; anthers upright, 2-locular, opening by terminal pores. Ovary 2–10-locular. Fruit capsular, baccate or drupaceous. Embryo minute, in fleshy albumen. Pollen in tetrads. Large family, represented with us by 2 subfamilies.

a. Calyx free from the superior ovary (or in *Gaultheria* the ripened ovary appearing falsely inferior through extension around it or its base of the fleshy calyx, the fruit becoming falsely baccate). Subfam. I. RHODODENDROIDEAE. . . *b.*

b. Leaves dilated or needle-like; corolla deciduous. . . *c.*

c. Fruit a septicidal capsule; the valves (except in no. 4) in dehiscence separating from the persistent placentiferous columella. Tribe RHODODENDREAE. . . *d.*

d. Flowers developed from scaly buds; capsule longer than broad. . . *e.*

e. Bud-scales or bracts caducous; anther-locules opening by a hole or chink at summit. . . *f.*

f. Corolla of 5 distinct white petals; leaves evergreen, densely woolly beneath; flowers in terminal umbelliform clusters; capsule opening from base upward. . . . . 1. *Ledum.*

f. Corolla campanulate, subglobose or funnelform, of united petals; capsule opening from apex to base.

Flowers mostly 5-merous; corolla campanulate or funnelform; leaves deciduous or evergreen. . . . . 2. *Rhododendron.*

Flowers 4-merous; corolla globose-campanulate; leaves deciduous. . . . . 3. *Menziesia.*

e. Bud-scales firm and persistent; anther-locules opening lengthwise; leaves evergreen; capsule opening from apex to base.

Corolla of 5 distinct white petals; stamens 10, exserted. . . . 4. *Leiophyllum.*

Corolla gamopetalous, deeply 5-cleft, roseate to whitish; stamens included. . . . . 5. *Loiseleuria.*

d. Flowers from the axils of coriaceo-foliaceous persistent bracts or from leaf-axils; capsule subglobose, dehiscent from above.

Corolla saucer-shaped, 10-saccate toward base, the pouches receiving the anthers in bud; leaves oblong to linear, alternate, opposite or whorled. . . . . 6. *Kalmia.*

Corolla ovoid or urceolate, without basal pouches; leaves narrowly linear, crowded, heath-like. . . . . 7. *Phyllodoce.*

c. Fruit a loculicidal short capsule, the valves usually bearing the partitions, which separate from the persistent placentiferous columella, or fruit a drupaceous berry. . . *g.*

g. Fruit capsular, either a dry capsule subtended by the dry calyx or (in no. 17) surrounded by the enlarged fleshy calyx or its base, the fruit thus falsely baccate. Tribe ANDROMEDEAE. . . *h.*

h. Calyx remaining dry and the fruit a dehiscent capsule; plants without "wintergreen" flavor. . . *i.*

i. Anthers opening only at summit; stigma mostly entire; corolla cylindric, urceolate, ovoid or subglobose. . . *j.*

j. Calyx-lobes or sepals valvate or open in bud, not overlapping. . . *k.*

k. Locules of anther awned. . . *l.*

l. Awns projected forward; seeds ovoid or ellipsoid, firm.

Leaves entire, linear to narrowly oblong, revolute, evergreen; flowers in small terminal umbels; corolla globose-urceolate; sutures of capsule not thickened; seed smooth and lustrous; boreal. . . . . . . 8. *Andromeda.*

Leaves crenate, lance-oblong to elliptic or oval, flat, deciduous; umbel-like fascicles or racemes of open campanulate flowers from axils; sutures of capsule keeled; seeds angled, opaque; southeastern. . . . . 9. *Zenobia.*

*l.*Awns reflexed; broad leaves evergreen; panicles and racemes terminal and axillary; corolla urceolate; sutures of capsule not thickened; seeds scobiform, with loose coat. . . . 10. *Pieris.*

*k.*Locules of anther awnless; leaves deciduous or evergreen; capsule with thickened sutures; seeds scobiform, with loose testa. . . . . 11. *Lyonia.*

*j.*Calyx-lobes or sepals imbricated or overlapping in young bud. . . *m.*

*m.*Leaves dilated, not imbricated; ascending shrubs or trees; corolla cylindric to conic-urceolate; anther-locules awnless or with ascending awns; inflorescence racemose or paniculate. . . *n.*

*n.*Racemes simple or panicled, the flowers not in the axils of dilated leaves; anthers linear to oblong, blunt; valves of capsule entire; large shrubs or tree, the petioled and toothed leaves not scurfy; of temperate and warm reg.

Shrub with axillary and terminal racemes or groups of racemes; corolla glabrous; capsule depressed-globose. 12. *Leucothoë.*

Tree with large panicle terminating new branchlets; corolla minutely pubescent; capsule ovoid-pyramidal. . 13. *Oxydendrum.*

*n.*Racemes simple, the flowers in the axils of reduced subsessile entire small foliage-leaves; anther-locules tapering to a tubular beak; the 5 valves of the depressed capsule eventually separating into 2 layers, the dehisced capsule thus 10-valved; small boreal and paludal shrub. . 14. *Chamaedaphne.*

*m.*Leaves tiny, bristle-like, imbricated; depressed moss- or heath-like arctic-alpine shrubs, with axillary flowers; corolla campanulate; anther-locules with a recurved awn; valves of capsule 2-cleft. . . . . 15. *Cassiope.*

*i.*Anthers opening their entire length, unappendaged; stigma 5-lobed; corolla salverform, with long slender tube; trailing broad-leaved subherbaceous plant. . . . . 16. *Epigaea.*

*h.*Calyx becoming enlarged and fleshy, wholly or partly covering the fleshy capsule, the fruit thus being a false berry; plants with "wintergreen" flavor, ours weakly suffruticose, with extensively creeping rhizomes and stolons, or trailing. . . . . 17. *Gaultheria.*

*g.*Fruit a fleshy drupe or drupaceous berry subtended by the calyx; corolla urceolate to globular or barrel-shaped; stamens twice as many as corolla-lobes, included; ours trailing shrubs with evergreen or marcescent leaves. Tribe Arbuteae. . . . . 18. *Arctostaphylos.*

*b.*Leaves scale-like or needle-like, very small, imbricated, opposite or in whorls, persistent; corolla of 4 (rarely 5) marcescent lobes persistent over the fruit, the lobes convolute in bud; stamens twice as many as corolla-lobes; fruit a capsule; heaths. Tribe Ericeae.

Calyx brightly colored, greatly overtopping corolla. . . . 19. *Calluna.*

Calyx not brightly colored, much smaller than corolla. . . . 20. *Erica.*

*a.*Calyx-tube adnate to whole length of ovary which in fruit becomes a berry or drupe crowned with the calyx-teeth; corolla of united (though sometimes prolonged) petals, the disk epigynous; anthers erect, introrse, their locules partly separated or prolonged into a tubular appendage with apical pore; pollen in tetrads; shrubs or small trees with scaly buds and alternate leaves. Subfam. II. Vaccinioideae.

Ovary 10-locular, 10-ovuled, the ovules become hard nutlets, the fruit baccate-drupaceous; leaves mostly glandular-dotted or punctate; flowers in lateral racemes, the corolla tubular to campanulate; anthers awnless. . 21. *Gaylussacia.*

Ovary 5-locular or by development of false partitions becoming 10-locular, with many fine or fewer coarse seeds, the fruit a berry; leaves mostly glandless and non-punctate; flowers variously disposed, the corolla strongly gamopetalous to deeply cleft; anthers awnless or awned. . 22. *Vaccinium.*

Subfam. RHODODENDROÌDEAE

Tribe Rhododéndreae Spreng.

1. LÈDUM L. Labrador-tea

Calyx 5-toothed, very small. Corolla of 5 obovate and spreading petals. Capsule 5-locular, splitting from the base upward, many-seeded; placentae borne on the summit of the columella. — Low boreal shrubs with the alternate entire leaves usually clothed with white to rusty wool underneath, persistent, the margins revolute; herbage fragrant when bruised. Flowers white, small, in terminal umbel-like clusters. (*Ledon*, the ancient Greek name of species of *Cistus* yielding aromatic resin, which the aroma of *Ledum* was thought to resemble.)

1. L. groenlándicum Oeder (of Greenland). — Erect, 1 m. or less high; *leaves oblong or linear-oblong*, 2–5 cm. long, very obtuse; *stamens 5–7*; *capsule slender, subcylindric.* — Peaty soils, Greenl. and Lab. to Alaska, s., especially in bogs, to Nfld., N.S., N.E., n. N.J., Pa., O., Mich., Wisc., Minn., Alta. and Wash. May, June, rarely to Aug. — Forma **denudàtum** Vict. & Rousseau (denuded) has the thinnish and flat leaves without dense tomentum beneath.

2. RHODODÉNDRON L. Rhododendron. Rosebay. Azalea

Calyx mostly small or minute. Stamens sometimes as few as the corolla-lobes, more commonly twice as many, usually declined; anther-locules opening by a round terminal pore. Capsule 5-locular, 5-valved, many-seeded. Seeds scale-like. — Shrubs or small trees of N. Hemisph. (one Austral.) of diverse habit and character, with alternate entire or toothed leaves and large and showy flowers in umbelled clusters from terminal buds. (Greek name, meaning *rose-tree.*)*

a. Leaves coriaceous, evergreen; stamens (5–)10–20. Subgen. Eurhododendron (Rhododendrons).
  Large shrubs or small trees; leaves 0.5–2.5 dm. long, not scurfy; corollas 3–6 cm. broad; filaments pubescent at base; capsules not scurfy. § Leiorhodium.
    Leaves narrowed to acutish base and apex, usually pubescent beneath; corolla 3.5–4 cm. broad; ovary and capsule glandular. . . . . . 1. *R. maximum.*
    Leaves rounded to base and apex, glabrous beneath; corolla about 6 cm. broad; ovary and capsule rusty-pubescent. . . . . 2. *R. catawbiense.*
  Dwarf matted shrub; leaves 0.5–2 cm. long, scurfy; corollas 1–2 cm. broad; filaments glabrous; ovary and capsule scurfy. § Lepipherum. . . . 3. *R. lapponicum.*

a. Leaves chartaceous to membranaceous, deciduous; stamens 5–10. Subgen. Anthodendron (Azaleas). . . *b.*
  b. Key to flowering material. . . *c.*
    c. Buds (expanding into inflorescences) of few and early caducous scales; corolla strongly irregular, with short or hardly any tube, anteriorly divided to base, the limb equaling the 10 stamens and style. § Rhodora. 4. *R. canadense.*
    c. Buds of numerous much imbricated scales; corolla with conspicuous elongated funnelform tube; style and 5 stamens long-exserted. § Pentanthera. . . *d.*
      d. Flowers appearing before or with the young leaves. . . *e.*
        e. Corolla bright yellow, orange or scarlet, about 5 cm. broad, the tube much shorter than the lobes. . . . . 5. *R. calendulaceum.*
        e. Corolla pink, roseate or white, rarely creamy, 2–4.5 cm. broad, the tube about equaling to longer than the lobes. . . *f.*
          f. Shrubs mostly 1–5 m. high, not strikingly colonial; corolla-buds without restricted rows of glands, tube pubescent within.
            Leaves glabrous beneath or merely with strigose midrib; flowers barely if at all fragrant. . . . . 6. *R. nudiflorum.*
            Leaves softly canescent-tomentulose or -pilose beneath; flowers very fragrant.
              Corolla-tube gradually dilated upward, about as long as lobes; filaments about twice as long as tube. . . . . . . 7. *R. roseum.*
              Corolla-tube slender, abruptly broadened at summit, about twice as long as lobes; filaments thrice as long as tube. . 8. *R. canescens.*

---

*Treatment of Subgen. *Anthodendron* derived largely from that of *Rehder, The Azaleas of North America,* in Wilson & Rehder, Monograph of Azalea (1921).

f.Shrubs 0.2–1 m. high, colonial, with abundant subterranean stolons; corolla-buds with slender rows of stalked glands, glabrous within. 9. *R. atlanticum.*

d.Flowers expanding after the leaves.

Corollas white or suffused with pink.

Young branchlets bristly or strigose; leaves bristly or hirsute along midrib beneath; calyx-lobes 1–2 mm. long; style 4.5–6 cm. long; ovary appressed- or ascending-pilose, rarely glandular.

Leaves of fertile branches 0.7–2 cm. broad, green or whitish beneath; pedicels 0.5–1 (very rarely –1.5) cm. long; corolla-tube gradually dilated at summit, 1.5–2.5 cm. long, about once-and-a-half length of lobes, pubescent within above middle; style pubescent below middle. . . . . 10. *R. viscosum.*

Leaves of fertile branches 1.5–3.8 cm. broad, green both sides; pedicels 1–2.3 cm. long; corolla-tube slenderly cylindric nearly to summit, 2.5–3.5 cm. long, about twice length of lobes, glabrous within; style glabrous, or minutely pubescent only at base. . . . . 11. *R. serrulatum.*

Young branchlets glabrous; leaves glabrous; calyx-lobes 3–6 mm. long; corolla-tube 2.5–3.5 cm. long; style 6–9 cm. long; ovary glandular-villous. . . . . 12. *R. arborescens.*

Corollas red, orange or yellow; calyx-lobes 1–3 mm. long; young branchlets glabrous or but slightly strigose. . . . . 13. *R. cumberlandense.*

b.Key to fruiting material. . . *g*.

g.Capsules 0.7–1.5 cm. long, glaucous-puberulent, sparsely if at all setose; style 1.5–2 cm. long; leaves oblong or elliptic, blunt, closely pubescent (rarely glabrous) beneath; branches strongly ascending. . . . . . 4. *R. canadense.*

g.Capsules 0.8–2.5 cm. long, if less than 1.5 cm. long strongly setose; style 4–9 cm. long; leaves, if closely pubescent beneath, acute or acuminate and the branches widely spreading. . . *h*.

h.Leaves glabrous or essentially so beneath, except for strigose bristles along the midrib or nerves. . . *i*.

i.Branchlets more or less strigose-setose or bristly; midrib of leaf beneath strigose or hirsute; style 4.5–7 cm. long.

Shrubs 0.5–5 m. high, not strongly colonial; branchlets rarely glandular; capsules covered with long subappressed bristles.

Pedicels glandless or nearly so; calyx-lobes long-ciliate with non-glandular hairs. . . . . 6. *R. nudiflorum.*

Pedicels copiously glandular; calyx-lobes glandular-ciliate.

Floral winter-buds with 8–12 round-tipped or merely mucronate scales; leaves of fertile branches 0.7–2 cm. broad, green or whitened beneath; pedicels 0.5–1 (rarely –1.5) cm. long; style pubescent below middle; capsule lance-cylindric. . 10. *R. viscosum.*

Floral winter-buds with 15–20 mucronate to aristate scales; leaves of fertile branches 1.5–3.8 cm. broad, green beneath; pedicels 1–2.3 cm. long; style glabrous or minutely pubescent only at base; capsule slenderly ovoid. . . . . 11. *R. serrulatum.*

Shrubs 0.2–1 m. high, freely stoloniferous and colonial; branchlets and capsules with abundant spreading stalked glands. . . . 9. *R. atlanticum.*

i.Branchlets glabrous; leaves glabrous, including midrib; capsule densely glandular-villous; style 6–9 cm. long; shrub up to 6 m. high. 12. *R. arborescens.*

h.Leaves canescent-tomentulose or pilose beneath.

Capsule with wholly or mostly non-glandular hairs; southern species.

Calyx-lobes 1–4 mm. long, glandular-ciliate; style 6–7 cm. long. . 5. *R. calendulaceum.*

Calyx-lobes barely 1 mm. long, usually glandless; style 4–6 cm. long. . . . . 8. *R. canescens.*

Capsule copiously glandular-setose; calyx-lobes barely 1 mm. long, glandular-ciliate; style 4–5 cm. long; northern and upland species. 7. *R. roseum.*

Subgen. EURHODODÉNDRON Endl.

§ LEIORHÒDIUM Rehd.

1. R. máximum L. (greatest), GREAT LAUREL, ROSEBAY. — Shrub or straggling tree up to 10 m. high, with pubescent young branchlets; *leaves* evergreen, very thick, oblong or oblong-obovate, 0.8–2.5 *dm. long, acute at each end,* dark green above, *paler and* often *closely tomentose beneath;* flowers numerous in clusters; pedicels viscid; *corollas* campanulate, 3.5–4 *cm. broad,* rose-pink and white, or wholly white in forma **álbum** (Pursh) Fern. (white), or deep pink to purple in forma **purpùreum** (Pursh) Fern. (purple), greenish in the throat on the upper side, and spotted with yellow or red; ovary and capsule glandular. — Damp woods, swamps and pond-margins, common in the uplands southward, local northw., Ga. and Ala. to N.S. (formerly), N.E., N.Y., s. Ont. and O., extending out to inner Coastal Plain or outer Piedmont from e. Va. northw. June, July.

2. R. catawbiénse Michx. (from Catawba R., N.C.), MOUNTAIN-ROSEBAY, PURPLE LAUREL. — Shrub up to 3 m. high or small tree to 6 m.; *leaves* 0.7–1.5 *dm. long, rounded to both ends, glabrous beneath; corolla about* 6 *cm. broad,* rose- or lilac-purple, with broad uncleft lobes, or the lobes divided into narrow segments in forma **tomopétalum** Rehd. (with dissected petals); *ovary and capsule rusty-pubescent.* — Rocky summits, rocky slopes, woods and stream-banks, upland and Piedmont, Va. and W.Va. to Ga. and Ala. May, June. — Much cult.

§ LEPÍPHERUM Don

3. R. lappónicum (L.) Wahlenb. (of Lapland), LAPLAND ROSEBAY. — Dwarf, *forming depressed* aromatic *mats* 1–6 *dm. across; branchlets scurfy; leaves* thick, *scurfy-rugulose,* elliptic to broadly oblanceolate or narrowly obovate, 0.5–2 *cm. long; corollas royal-purple,* 1–2 *cm. broad; filaments glabrous,* exserted; *capsules* 4–8 *mm. long,* with the erect fruiting pedicels *scurfy.* — Arct. and subarct. reg., s. to (often magnesian and calcareous) barrens and cliffs of Nfld.; alpine barrens of Shickshock Mts., Gaspé Pen., Que., Mt. Katahdin, Me., Presidential Range, N.H., and Adirondack Mts., N.Y.; sandstone-cliffs, Dells of Wisconsin R., Wisc. Late May, June (southw.)–Sept. (Nfld.). (Eurasia)

Subgen. ANTHODÉNDRON Endl.

§ RHODÒRA G. Don

4. R. canadénse (L.) Torr. (Canadian), RHODORA. — Shrub *with strongly ascending branches,* 0.15–1 m. high; *inflorescences mostly expanding before leaves,* from buds with few and early-caducous scales; leaves oblong, gray-green or glaucous, more or less pubescent, or dark green and shining above and barely pilose beneath in the western Nova Scotian forma **viridifòlium** Fern. (green-leaved), 2–6 cm. long, with revolute margins; *flowers* 2–several; *corolla* pale to deep rose-purple, or white in forma **albiflòrum** (Rand & Redf.) Rehd. (white-flowered), 1.5–3 cm. long, *with very short tube, bilabiate, anteriorly divided to base, the posterior lip 3-lobed, the anterior of* 2 oblong-linear *nearly distinct recurving petals; stamens* 10, *about equaling style and corolla; style* 1.5–2 *cm. long; capsule* 0.7–1.5 *cm. long, glaucous-puberulent.* (*Rhodora* L.) — Bogs, damp thickets, acid barrens and rocky summits and slopes, Nfld. and e. Saguenay Co., Que., to n. N.Y., s. to N.S., N.E., n. N.J. and ne. Pa. March–July (–Sept.).

R. VÀSEYI Gray (for GEORGE RICHARD VASEY, who discovered it in 1878), PINKSHELL-A., taller shrub, with pale roseate flowers before the leaves, the latter glabrous or nearly so and acuminate, the corolla not divided to base, the exserted style 2.5–3 cm. long, stamens 5–7, has spread from cult. in Mass. (Introd. from south of our range.)

§ PENTANTHÈRA G. Don

5. R. calendulàceum (Michx.) Torr. (colored like *Calendula*), FLAME- or YELLOW A. — Shrub up to 5 m. high, loosely branching, the *young branchlets pubescent* or strigillose; *flowers* from tips of last year's branchlets, *before or with expanding leaves; leaves* elliptic, oblong or narrowly obovate, membranaceous, 4–8 cm. long, acute, *closely short-pilose beneath; calyx-lobes* 1–4 *mm. long, glandular-ciliate; corolla* glandular outside, *about* 5 *cm. broad, deep orange, or scarlet to bright red* and with an orange blotch in forma **aurántium** (Lodd.) Rehd. (orange), or *yellow or yellowish* with darker blotch in forma **cròceum** (Michx.) Rehd. (saffron-colored), essentially odorless, the *tube much shorter than the lobes;* style 6–7 cm. long; capsules lance-cylindric, 1.5–3 cm. long, finely pubescent and setulose. (*Azalea* Michx.; *R. luteum* Schneid., not Sweet) — Open woods, Ga. and Ala., n. to sw. Pa., W. Va. and se. O. May, June.

R. JAPÓNICUM (Gray) Suringar (Japanese), with obtuse leaves glabrous beneath or merely setulose on midrib, corolla more campanulate and 5–6 cm. broad, salmon- to brick-red, or yellow in forma AÙREUM Wilson (golden), and glandless without, the stamens not exserted, has spread from cult. in Ct. and n. N.J. (Introd. from Japan)

6. R. nudiflòrum (L.) Torr. (naked-flowered; from flowering before the leaves expand), PINXTER-FLOWER, PURPLE HONEYSUCKLE, ELECTION-PINK. — Shrub up to 3 m. high; *branch-*

lets strigose-setose; flowers essentially odorless, expanding before the leaves; leaves elliptic to oblong or oblong-obovate, abruptly acuminate or acute, rarely obtuse, 3–8 cm. long, *strigose on midrib beneath,* pilose along midrib above, green both sides; *pedicels strigose-pilose* (rarely slightly glandular); *calyx-lobes* 0.5–2 *mm. long, ciliate with* usually *glandless hairs; corolla pale pink, pilose outside,* or *white in* forma **álbum** (Ait.) Rehd. (white), or *deep roseate to purplish* and somewhat glandular outside *in* forma **glandíferum** (Porter) Fern. (bearing glands), *with* funnelform slender *tube* 1.5–2.5 cm. long and *slightly flaring above and longer than the* widely spreading *lobes;* style 4.5–7 cm. long, much exceeding the exserted stamens; *capsule* 1–2.5 cm. long, *setose.* (*Azalea* L.; *A. lutea* L., in part; *A. periclymenoides* Michx.) — Woods, thickets and swamps, S.C. and Tenn. to Mass., N.Y. and s. O. March–May.

7. R. ròseum (Loisel.) Rehd. (rosy), Honeysuckle, Early A., Election-pink. — In habit like no. 6; *branchlets canescent-pilose* and more or less strigose; *leaves canescent-tomentulose or short-pilose beneath,* more or less pilose above; *flowers very fragrant; pedicels with stipitate glands* among the villosity; *calyx-lobes* rarely 1 mm. long, *glandular-ciliate; corolla stipitate-glandular* and villous, bright pink to whitish; *the tube* 1.5–2 cm. long and *broadening upward, pubescent inside, about equaling the lobes; stamens about twice length of corolla-tube;* style 4–5 cm. long; *capsule copiously glandular-setose.* (*R. canescens,* in part, of ed. 7, not Sweet; *R. prionophyllum* Millais; *Azalea prionophylla* Small) — Dry woods, thickets and rocky banks, sw. Me. to sw. Que., s. and sw. to uplands of Va., Tenn. and Mo. May, June.

8. R. canéscens (Michx.) Sweet (hoary with gray pubescence). — Similar to no. 7; leaves mostly oblanceolate to narrowly obovate or oblong, up to 11 cm. long; *pedicels villous, rarely glandular; calyx-lobes usually glandless; corolla-tube slender, abruptly broadened at summit, about twice as long as lobes; filaments about thrice length of corolla-tube;* style 4–6 cm. long; *capsule non-glandular.* (*Azalea* Michx.) — Wooded swamps, dry woods and stream-banks, Fla. to e. Tex., n. to n. Del., ne. Md., s. O. and s. Ill. April, early May.

9. R. atlánticum (Ashe) Rehd. (from near the Atlantic), Dwarf A. — *Widely stoloniferous, forming colonies,* 0.2–0.6(–1) *m. high, the erect stems simple below; young branches* more or less strigillose and *glandular-hirsute; leaves* cuneate-obovate to elliptic or oblong, acutish to rounded at tip, green to bluish-glaucous, glabrous or glabrescent, or minutely setulose when young, bristly-ciliate, *the midrib beneath often hirsute or strigillose; flowers* in dense clusters *before or at unfolding of leaves, strongly fragrant; pedicels glandular-hirsute; calyx-lobes* 1–4 mm. long, *glandular; corolla* funnelform, purple to white, 3–4.5 cm. long, equally broad, *with lines of stipitate glands outside, glabrous within,* the tube longer than the spreading lobes; *capsule* lance-cylindric, 1.5–2 cm. long, *glandular-hirsute.* — Highly variable. The following forms are named:

Corolla purple to pink throughout.
- Corolla glabrous outside except for scattered rows of straight stipitate glands; pedicels spreading-glandular. . . . . *R. atlanticum* (typical).
- Corolla minutely pilose outside, the glands mixed with long villi; pedicels villous and more sparsely glandular. . . . . Forma *neglectum.*

Corolla white or whitish, at most pink, purple, salmon or yellow at base or outside.
- Corolla white, suffused outside or on tube with pink or purple, the tube and pedicels merely glandular-hispid; buds white or pink.
  - Lobes of corolla uncleft. . . . . Forma *confusum.*
  - Lobes dissected into linear-spatulate segments or altered to stamens. . . Forma *tomolobum.*
- Corolla white, suffused with yellow and salmon-orange tints, commonly pilose on surface as well as glandular; buds yellowish. . . . . Forma *luteo-album.*

R. atlánticum (typical). — Pine-barrens and sandy open woods, e. S.C. to Del. and ne. Md. April, May. — Forma **negléctum** (Ashe) Rehd. (overlooked), similar range, n. to s. N.J. and ne. Md.; forma **confùsum** Fern. (confused), e. S.C. to e. Md., our commonest form, April–early June; forma **tomólobum** Fern. (with dissected lobes), local, Nansemond Co., Va.; forma **lùteo-álbum** (Coker) Fern. (yellow-white), e. S.C. to Del.

10. R. viscòsum (L.) Torr. (sticky), Swamp-Honeysuckle, Clammy A. — Branching shrub 0.2–3(–5) m. high; *floral winter-buds with* 8–12 *round-tipped or merely short-mucronate scales;* branchlets strigose or hirsute; *leaves* narrowly ovate or elliptic-obovate to oblong-oblanceolate, acute to obtuse, *those of fertile branches* 1.5–6 cm. long, 0.7–2(–2.5) *cm. broad,* bristly-ciliate, glabrous beneath except for setulose midrib, green both sides or barely glaucescent beneath, or strongly whitened beneath and sometimes above in forma **glaùcum** (Lam.) Voss (blue-green); *pedicels* 0.5–1(–1.5) *cm. long,* glandular-hirsute and pubescent; *calyx-lobes* semicircular to ovate,

0.5–2 mm. long, *glandular-ciliate; corolla* white or with pink stripes along the lobes outside, or intense rose-purple in forma **rhodánthum** Rehd. (rose-flowered) or intense pink to carmine and with leaves glaucous in forma **ròseum** Hollick (roseate) = forma *rubescens* (Sweet) Rehd.; *the cylindric glandular and pilose tube gradually dilated upward, 1.5–2.5 cm. long, about equaling to once-and-a-half as long as the* ovate-oblong acuminate *lobes, pubescent inside above middle of tube; style* 4–5 cm. long, *pubescent below middle; capsule lance-cylindric,* strigose-setose, 1–2 cm. long. — Swamps, thickets and damp clearings, sw. Me. to ne. O., s. to S.C. and e. Tenn. June–Aug. (–Sept.)

11. R. serrulàtum (Small) Millais (finely saw-toothed). — Loosely branched shrub 1.5–7 m. high, differing from no. 10 in more strongly strigose reddish to brown young branchlets; *floral winter-buds with* 15–20 *mucronate to aristate dark-bordered scales; leaves* elliptic to obovate or oblanceolate, green both sides, those *of fertile branches* 2.5–8 cm. long and 1.5–3.8 *cm. broad,* serrulate-ciliate; *pedicels* 1–2.3 *cm. long,* villous and glandular; *corolla-tube slenderly cylindric nearly to summit,* 2.5–3.5 *cm. long, about twice length of lobes, glabrous within; style glabrous or minutely pubescent only at base; capsule slenderly ovoid.* — Swamps and low sandy woods, Fla. to Miss., n. to se. Va. June, early July.

12. R. arboréscens (Pursh) Torr. (tree-like), Smooth A. — Similar to no. 11, but *branchlets and lower leaf-surfaces glabrous* or essentially so throughout; leaves obovate to elliptic or oblong-oblanceolate, 1.3–3 cm. broad, bright green above, green to glaucous beneath; *petioles* 3–7 *mm. long, essentially glabrous;* pedicels rarely glandular; *calyx-lobes* linear-oblong to ovate, 3–6 *mm. long; corolla-tube* slenderly cylindric, *barely dilated at summit,* 2.5–3 *cm. long,* usually glabrous except for stipitate glands, *pubescent within, much longer than lobes; style* 6–9 *cm. long,* glabrous or minutely pubescent at base; *ovary and ovoid capsule densely glandular-villous* with reddish glands. — Swamps, stream-banks and rocky woods, Pa. to Ky., s. to Ga. and Ala. June, July.

13. R. cumberlandénse E. L. Br. (of the Cumberland Plateau). — Suggesting no. 5 but not flowering until after maturing of leaves, more compact; *new branchlets glabrous beneath the sparse strigose bristles; floral winter-buds with glabrous scales,* the *outer scales aristate;* dark green *leaves* more crowded, *glabrous* beneath except for strigose midrib; *pedicels glandless; corolla* red, orange or yellow, with orange-yellow blotch, the outside *with subsessile glands* and non-glandular pubescence; *style and filaments carmine;* filaments less pubescent or glabrous; capsule thicker and shorter. — Oak-woods, Cumberland Plateau and Mts., Ky. and w. Va. June, July.

3. MENZIÈSIA Sm.

Calyx small and flattish, 4-toothed or 4-lobed. Corolla cylindraceous-urceolate, soon campanulate. Stamens included; anther-locules opening by an oblique pore. Capsule ovoid, woody, 4-locular, 4-valved, many-seeded. Seeds narrow, with a loose coat. — Low N. Am. and e. Asiatic shrubs, the straggling branches and the alternate leaves usually hairy and ciliate with rusty rather chaff-like bristles. Flowers small, developed with the leaves, in terminal clusters, greenish-white and purplish, nodding. (Named for *Archibald Menzies,* 1754–1842, who from Vancouver's voyage brought the original species from the Northwest Coast.)

1. M. pilòsa (Michx.) Juss. (hairy), Minnie-bush. — More or less chaffy; *leaves* obovate-oblong, prominently glandular-mucronate, *strigose-hirsute* especially above; *filaments glabrous; capsule beset with short gland-tipped bristles; seeds merely apiculate.* — Upland woods, Pa. and W.Va. to Ga. and Tenn. May, June.

4. LEIOPHÝLLUM (Pers.) Hedw.fil. Sand-myrtle

Calyx 5-parted. Corolla of obovate-oblong spreading petals. Style filiform. Capsule 2–3-locular, splitting from the apex downward, many-seeded. — Low branched e. N.Am. evergreens with the aspect, foliage, etc., of *Ledum,* but the crowded leaves sometimes opposite and scarcely petioled. Flowers small, white, in terminal umbelliform clusters. (Name formed of the Greek *leios, smooth,* and *phyllon, leaf.*) Dendrium Desv.

1. L. buxifòlium (Berg.) Ell. (with leaves of box, *Buxus*). — Shrub, 1–9 dm. high; leaves oval or oblong, apiculate, smooth and shining, 3–13 mm. long; pedicels glandless. — Sandy pine-barrens, N.J. May, June.

Var. **Hùgeri** (Small) Schneid. (for its discoverer, Arthur Middleton Huger, 1842–1925). — Pedicels glandular. (*Dendrium Hugeri* Small) — Sandy pine-barrens, N.J.; Coastal Plain to mountain-crests, N.C. and S.C.; e. Ky.

5. LOISELEÙRIA Desv. Alpine Azalea

Calyx 5-parted, nearly as long as the campanulate regular corolla. Stamens not declined. Style short. Capsule ovoid, 2–3-locular, many-seeded, 2–3-valved; valves 2-cleft from the apex; placentae borne on the middle of the columella. — A small depressed shrubby circumpolar evergreen, much-branched and tufted, smooth, with coriaceous opposite revolute-margined elliptical leaves on short petioles. Flowers small, white or rose-color, 2–5 in a cluster. (Named for *Jean Louis Auguste Loiseleur-Delongchamps*, 1774–1849, French botanist.) Chamaecistus Oeder

1. L. procúmbens (L.) Desv. (lying flat). — Greenl. to Alaska, s. to peaty or rocky exposed habitats of Nfld. and (locally) N.S., and alpine reg. of Que., Me., N.H. and Alta. June–Aug. (Eurasia)

6. KÁLMIA L. Laurel (of America)

Calyx 5-parted. Corolla shallowly 5-lobed, crateriform. Filaments long and filiform. Capsule globose, 5-locular, many-seeded. — Evergreen mostly smooth N. Am. shrubs, with alternate, opposite or ternate entire coriaceous leaves, naked buds, and showy flowers. (Dedicated to *Pehr Kalm*, 1716–1779, a pupil of Linnaeus, who traveled and collected extensively in America.)

Branchlets terete; leaves flat; pedicels glandular or puberulent; calyx without conspicuous ring at base, herbaceous, its lobes not overlapping.
Coarse shrub or small tree; corymbs terminal; pedicels viscid-pilose to glabrous, ascending in fruit; corolla 1.5–3 cm. broad. . . . . 1. *K. latifolia.*
Slender shrub; corymbs lateral; pedicels not pilose, recurving in fruit; corolla 0.8–1.3 cm. broad. . . . . 2. *K. angustifolia.*
Branchlets 2-edged; leaves revolute-margined; pedicels glabrous; calyx with conspicuous ring at base, papery, its lobes overlapping. . . . . 3. *K. polifolia.*

1. K. latifòlia L. (broad-leaved), Mountain-L., Calico-bush, Spoonwood, "Ivy", Ivy-bush. — Coarse shrub or small tree; *leaves mostly alternate*, sometimes opposite or ternate, *flat*, coriaceous, evergreen, narrowly to broadly elliptical, narrowed to both ends, or in the rare forma **obtusàta** Rehd. (blunt) rounded to both ends and mucronate, 0.3–1 dm. long, on *petioles* 0.5–2 *cm. long; corymbs terminal;* bracts ovate to lanceolate, firm, viscid-puberulent; *pedicels viscid-pilose*, 2–3.5 cm. long, *ascending in fruit; calyx viscid; corolla* pink, or deep crimson in forma **rùbra** K. Koch (red), or almost chocolate-purple within in forma **fuscàta** Rehd. (dusky), or white in forma **álba** (Mouillefert) Rehd. (white), with saucer-shaped acutely 5-lobed limb 1.5–3 *cm. across*, or in forma **polypétala** (Nicholson) Rehd. (many-petaled) divided into linear to filiform petals; *capsule* depressed, 5–7 *mm. broad*, thick-walled, *glandular*, tardily dehiscent; *style* 1.3–2 *cm. long;* seeds oblong. — Rocky or gravelly woods and clearings, sometimes swamps, in acid or sterile areas, w. Fla. to La., n. to N.E., N.Y., O. and Ind. Late May–mid-July.

Var. **laèvipes** Fern. (smooth-stalked). — Pedicels smooth, glabrous or merely with sessile viscid spots. — Coastal Plain and outer Piedmont, Ala. to e. Md.

2. K. angustifòlia L. (narrow-leaved), Lambkill, Sheep-, Pig- or Dwarf L., Wicky. — Slender shrub up to 1.7 m. high; branches terete, strongly ascending; *leaves* commonly *opposite* or ternate, flat, thinnish, evergreen, oblong to elliptic-lanceolate, glabrous, or puberulent and glabrate, when young ferruginous, on short petioles; *corymbs lateral* (often appearing verticillate), scattered or crowded; *pedicels glandular-puberulent, recurving in fruit; calyx glandular-puberulent; corolla* deep rose-pink or crimson, or white in forma **cándida** Fern. (white), 0.8–1.3 *cm. broad; capsule depressed-globose*, 2.5–4.5 *mm. broad, puberulent; style* 4–7 *mm. long.* — Dry or wet sterile soil, old pastures and barrens, Lab. to Man., s. to Nfld., N.S., N.E., Va., mts. of Ga., and Mich. Late May–Aug.

Var. **carolina** (Small) Fern. (Carolinian). — *Leaves permanently pale-puberulent beneath; calyx puberulent, not glandular.* (*K. carolina* Small) — Swamps and sandy woods, Va. to S.C. and Tenn. May, June (–Oct.)

3. K. polifòlia Wang. (with leaves of *Polium*), Pale or Bog-L. — Slender *straggling shrub with 2-edged branches; leaves* opposite (rarely in 3's), *sessile or nearly so, firm,* lanceolate or linear, *with blunt callous tip,* 0.7–3.5 cm. long, lustrous-green above, *conspicuously whitened beneath,* often revolute, *the midrib prominent beneath; corymbs terminal,* umbelliform, becoming racemose; bracts like leaves but smaller; *pedicels glabrous,* filiform, loosely ascending, erect in fruit; *calyx glabrous, with a conspicuous ring at base, with papery overlapping lobes; corolla* deep pink to crimson, 1.2–2 *cm. broad; capsule globose-ovoid, glabrous, promptly dehiscent, its valves*

2-cleft; seeds linear. — Peaty soils, Lab. to Alaska, s., chiefly in bogs, to Nfld., N.S., N.E., n. N.J., Pa., Mich., Minn., Id. and Oreg. Mid-May–July.

7. PHYLLÓDOCE Salisb.

Corolla 5-toothed, urceolate or campanulate. Stamens 10; anthers pointless, shorter than the filaments. Capsule 5-locular, 5-valved, many-seeded. — Low circumpolar or alpine heath-like evergreen undershrubs, clothed with crowded linear and obtuse rough-margined leaves. Flowers nodding on solitary or umbelled peduncles at the summit of the branches. — Sometimes united with *Bryanthus*, a Siberian genus with 4-parted umbelled flowers. (*Phyllodoce*, a sea-nymph mentioned by Virgil.)

1. **P. caerùlea** (L.) Bab. (sky-blue; a misnomer). — Calyx pubescent; corolla cylindric-urceolate, 5-toothed, purplish, smooth, 7–8 mm. long; style included. (*Bryanthus taxifolius* Gray) — Arct. reg., s. to alpine rocks and peat of Nfld., Que., Me. and N.H. June–Aug. (Eurasia) — Fresh corolla never bluish with us but said to be sometimes so in n. Eu., with us becoming bluish only on fading.

Tribe ANDROMÉDEAE Drude

8. ANDRÓMEDA L. ANDROMEDA

Calyx of 5 nearly distinct sepals, valvate in bud, but soon wide-spreading. Corolla globose-urceolate. Stamens 10, the filaments unappendaged, each locule of the anther with a slender ascending awn. Capsule obovoid or depressed-globose, 5-locular and 5-valved, the sutures not thickened; the many-seeded placentae attached next to the summit of the columella. Seeds ovoid or ellipsoid, with a smooth and lustrous firm coat. — Two slender boreal pale evergreen shrubs, with firm revolute narrow leaves and small umbels of pink or white flowers. (The northern and Eurasian *A. Polifolia* fancifully named by Linnaeus for *Andromeda* of Greek mythology.)

1393. A. glaucophylla.

1. **A. glaucophýlla** Link (with blue-green leaves), BOG-ROSEMARY. — Low shrub with elongate creeping base, the ascending glaucous stems 1–7 dm. high; *leaves* linear to narrowly oblong, soon becoming revolute, *whitened beneath with close minute puberulence;* bud-scales glaucous; *flowers on thickish curved pedicels rarely twice their length;* calyx-lobes whitish, usually spreading; *capsule depressed, turban-shaped, glaucous.* — Bogs, peats, margins of pools, etc., sw. Greenl. and Lab. to e. Man., s. to Nfld., N.S., N.E., n. N.J., Pa., W.Va., O., Ind., Wisc. and Minn. May–July (–Sept.) FIG. 1393.

9. ZENÒBIA D. Don ZENOBIA

Calyx barely 5-parted, thickish, with the thin margins valvate in the bud. Corolla broadly campanulate, with 5 blunt lobes. Stamens 10; filaments naked, abruptly dilated at base; anthers lanceolate, each locule surmounted by a pair of slender ascending awns. Capsule depressed-globose, obtusely 5-lobed, somewhat keeled at the dorsal sutures; the placentae on the middle of the short columella. Seeds angled. — Coriaceous-leaved (deciduous with us) shrub of se. U.S., the white flowers in umbelliform fascicles of racemes from axillary buds of the preceding year. (*Zenobia*, queen of Palmyra in the 3rd century of our era.)

1. **Z. pulverulénta** (Bartr.) Pollard (covered with powder). — Shrub up to 3 m. high, the branchlets and foliage whitened with bloom; leaves of fertile branches elliptic-oblong or oval, obtuse, shallowly crenate, 2–7 cm. long; flowers nearly clear white; corolla as broad as or broader than long, 6–8 mm. long. In forma **nítida** (Michx.) Fern. (shining) leaves as in typical form but green both sides; branchlets barely if at all glaucous. In forma **nùda** (Vent.) Fern. (naked) leaves lance-oblong to narrowly ovate, acute at both ends, prominently toothed, green both sides (*Z. cassinefolia* (Vent.) Pollard) — Damp sandy or peaty pine-barrens, se. Va. to ne. S.C. June. — Fresh and handsome flowers deliciously fragrant.

10. PÌERIS D. Don PIERIS

Calyx deeply 5-parted, its firm thick-edged ovate lobes valvate in the bud. Corolla ovoid-urceolate, 5-toothed. Stamens 10; anther-locules each with a deflexed awn on the back at junction with the unappendaged filament. Capsule globose, not thickened at the sutures; placentae borne on summit of columella; seeds scobiform, with loose cellular coat. — Evergreen shrubs or small trees of e. N. Am. and Asia, with coriaceous broad mostly toothed leaves, the white flowers expanding in spring in terminal and axillary racemes or panicles formed the preceding summer, the pedicels 2–3-bracteolate. (*Pieris*, a muse.)

1. P. floribúnda (Pursh) B. & H. (full of flowers), FETTER-BUSH. — Very leafy, 5–15 dm. high; *young branchlets, etc. strigose-hairy; leaves* lanceolate-oblong, acute or acuminate, ciliate-serrulate, *glandular-dotted beneath*, 4–6 cm. long; *racemes crowded in short terminal panicles*, densely flowered. (*Andromeda* Pursh) — Moist hillsides, in the Alleghenies from Va. to Ga. May.

11. LYÒNIA Nutt.

Similar to *Andromeda*. Filaments hairy and often toothed or appendaged; anthers oblong, unappendaged. Capsule 5-angled, the dorsal sutures with a thickened ridge which usually divides in dehiscence of the capsule; the placentae borne both upon the columella and the walls of the locules. Seeds scobiform, with a loose thin testa. — Am. and E. Asiatic shrubs with fascicled, racemose or panicled white or roseate flowers. (Named for *John Lyon*, 17?–1818, early American botanist and explorer of the southern Alleghenies.)

Leaves rigidly coriaceous and evergreen; branches strongly angled; flowers in axillary clusters of racemes. . . . . 1. *L. lucida.*
Leaves thinner, deciduous; branches terete.
  Inflorescences umbellate-racemose; corollas nodding, cylindric-ovoid, white to pink, 0.8–1.3 cm. long. . . . . 2. *L. mariana.*
  Inflorescence paniculate; globose- or ovoid-urceolate white corollas 2.5–5 mm. long. . . . . 3. *L. ligustrina.*

1. L. lùcida (Lam.) K. Koch (shining), FETTER-BUSH. — Glabrous *evergreen* shrub up to 2 m. high, the green *branchlets sharply 3-angled; leaves firm-coriaceous*, dark green, *lustrous*, oblong-ovate to lanceolate, *entire, with a conspicuous nerve next the revolute margin;* flowers in *clusters of nodding racemes from axils of persistent leaves of preceding year;* calyx-lobes rigid, purplish; *corolla ovoid, 6–8 mm. long*, white or pale pink, or deep pink or reddish in forma **rùbra** (Lodd.) Rehd. (red); filaments with a pair of bristle-like small appendages at summit; fruit subglobose. (*L. nitida* (Bartr.) Fern.; *Desmothamnus lucidus* Small) — Low woods and peaty thickets, Fla. to La., n. to Va. April, May.

2. L. mariàna (L.) D. Don (of Maryland), STAGGER-BUSH. — Slender upright shrub, usually with stems naked below and with strongly ascending branches above, 0.2–2 m. high; *leaves* borne on new shoots, *deciduous*, oblong, oval or narrowly obovate, 2–8 cm. long, glabrous or merely pilose on the nerves beneath, or lower surface pilose in forma **vestìta** (Rehd.) Parks (clothed); *inflorescences umbellate-racemose, fascicled along leafless old branches;* flowers long-pedicelled, nodding; *corolla cylindric-ovoid*, 0.8–1.3 *cm. long*, white or pinkish; filaments usually 2-toothed near apex; *capsule ovoid-pyramidal*, truncate at the contracted apex, surrounded by the finally appressed calyx-lobes. (*Neopieris* Britt.) — Peaty or sandy pinelands and open woods or low grounds, Fla. to e. Tex., n. to s. R.I., s. Ct., se. N.Y., N.J., e. Pa., w. Tenn. and Ark. May, June. — Foliage said to poison lambs and calves.

3. L. ligustrìna (L.) DC. (resembling privet, *Ligustrum*), MALEBERRY or MALE BLUEBERRY. — Shrub 0.4–4 m. high, with deciduous lanceolate, oblong, elliptic or obovate serrulate to entire leaves; racemes borne in panicles terminating the branches or from upper axils; corollas whitish, globose- or ovoid-urceolate, 2.5–5 mm. long; filaments flat, not appendaged; capsule globose or subglobose, 2–5 mm. long. — A polymorphous wide-ranging species:

Leaves of fertile branches 2.5–9 cm. long, 1–5 cm. broad, lanceolate or oblanceolate to ovate or broadly elliptic, acuminate.
  Flowers 2.5–3.5 (–4) mm. long; fruits 2.5–3 mm. long.
    Leaves thickish, rugulose, opaque, oblanceolate, obovate, elliptic or oval, abruptly short-acuminate, more or less setulose or pilose beneath or glabrate; branchlets pilose to glabrous; panicles often dense, without leafy bracts along their branches. . . . . *L. ligustrina* (typical).
    Leaves membranaceous, lustrous above, plane, lance- to narrowly ovate-elliptic, long-acuminate, mostly glabrous; branchlets glabrous or nearly so; panicles loose, their branches with several leafy bracts. . . . . . Var. *salicifolia.*
  Flowers 4–5 mm. long; fruits 3.5–5 mm. long; leaves subcoriaceous, broadly elliptic to ovate or obovate, short-acuminate. . . . . Var. *capreaefolia.*
Leaves of fertile branches 1.3–4.5 cm. long, 0.8–2.5 cm. broad, oblong to narrowly obovate, round-tipped, blunt or merely subacute or mucronate, glabrous or nearly so; branches of panicle leafy-bracted; flowers 2.5–3.5 mm. long; fruits 2.5–3 mm. long. . . . . Var. *foliosiflora.*

L. ligustrìna (typical). — Shrub 0.5–3 dm. high. (*Xolisma* Britt.; *Arsenococcus* Small) — Wet or dry thickets, S.C. and n. Ga. to N.E., N.Y., W.Va. and Ky. May–July. — Foliage poisonous to young stock. More pubescent forms are mistaken for the more southern var. *pubescens* Gray, which is low, with blunt or merely acutish tomentulose leaves and cinereous-puberulent branchlets.

Var. **salicifòlia** (Wats.) DC. (willow-leaved). — Tall, up to 4 m. high. (Var. *foliosiflora* of ed. 7, not (Michx.) Fern.) — Damp thickets, swampy woods and low pinelands, Fla. to La., n. to Va., Ky., Ark. and Okla. — Forma **nanélla** Fern. (very dwarf) of e. Virginian bogs is only 1.5–2 dm. high, its mature leaves only 1–3 cm. long and 0.5–1.4 cm. broad, the capsules only 2–3 mm. in diameter.

Var. **capreaefòlia** (Wats.) DC. (with leaves of *Salix Caprea*). — Tall, up to 4 m. high. — Swampy woods, Fla. to e. Tex., n. to se. Va., w. N.C., Tenn. and Ark.

Var. **foliosiflòra** (Michx.) Fern. (with leaves among the flowers). — Low glabrous or glabrescent compact shrub 0.4–2 m. high. (*Xolisma foliosiflora* Small) — Low pinelands, shrub-bogs and savannas, Fla. to se. Va.

12. LEUCÓTHOË D. Don Leucothoë. Fetter-bush

Calyx of 5 nearly distinct sepals, imbricated in the bud. Stamens 10; anthers naked, or the locules with 1 or 2 erect awns at the apex, opening by a pore. Capsule depressed, more or less 5-lobed, 5-locular, 5-valved, the sutures not thickened; the many-seeded placentae borne on the summit of the short columella. Seeds mostly pendulous. — Am. shrubs with petioled and serrulate or crenate leaves and white scaly-bracted flowers in spiciform axillary or terminal racemes. (*Leucothoë*, daughter of Orchamus, King of Babylon, referred to by Ovid.)

Racemes sessile, dense, not 1-sided, with persistent bracts, in the axils of thick and shining evergreen leaves; calyx not bracteolate; anthers awnless; stigma 5-rayed. Subgen. Euleucothoë.
- Leaves abruptly to gradually pointed, with petioles at most 1 cm. long; bracts broadly ovate or rounded; sepals broadly ovate, imbricated, blunt. . . 1. *L. axillaris.*
- Leaves taper-pointed, with petioles mostly 1–1.5 cm. long; bracts narrow, acuminate; sepals ovate-oblong, acuminate, scarcely imbricated. . . . . . 2. *L. editorum.*

Racemes 1-sided, open, with deciduous bracts, mostly terminating branchlets; leaves membranaceous, deciduous; calyx 2-bracteolate below the imbricated sepals; anther-locules 1–2-awned from the apex; stigma simple. Subgen. Eubotrys.
- Racemes recurving; sepals ovate; anthers 2-awned; capsule 5-lobed; seeds winged. . . . . 3. *L. recurva.*
- Racemes straight, divergent to erect; anthers 4-awned; capsule unlobed; seeds wingless. . . . . 4. *L. racemosa.*

Subgen. Euleucóthoë Drude

1. L. axillàris (Lam.) D. Don (axillary). — Depressed shrub 0.5–2 m. high, with loosely spreading branches; coriaceous lustrous dark green evergreen *leaves* of flowering branches oblong-lanceolate to elliptic or oval, *abruptly pointed*, more or less cartilaginous- or spinulose-toothed often only toward apex, *the petioles* 0.4–1 *cm. long;* the blades 0.3–1.3 dm. long, five-twelfths to four-fifths as broad; racemes 1–several in an axil, up to 7 cm. long; *bracts broadly ovate to orbicular, blunt; sepals broadly ovate, imbricated, blunt.* — Low woods, Fla. to Miss., n. to se. Va. (Oct.–Dec.–) Feb.–May.

Var. **âmbigens** Fern. (uncertain). — Leaves lanceolate or narrowly lance-ovate, with prolonged acuminate tips, the blades of fertile branches mostly 0.6–1.4 dm. long and only one-fourth to three-eighths as broad. — Fla. to se. Va.

2. L. editòrum Fern. & Schub. (of highlands), Dog-hobble, Switch-ivy. — Similar to no. 1; *leaves* up to 1.5 dm. long, ovate-lanceolate, *taper-pointed*, closely serrulate with ciliate-spinulose teeth, the *petioles* 1–1.5 *cm. long; bracts lance-acuminate; sepals ovate-oblong, acuminate, not imbricated.* (*L. Catesbaei* sensu Gray and most auth., not *Andromeda Catesbaei* Walt., basonym) — Moist woods along the mts., Va. to Ga. and Tenn. May.

Subgen. Eubòtrys (Nutt.) Gray

3. L. recúrva (Buckley) Gray (turned back). — *Branches and racemes recurved-spreading; leaves* lanceolate or ovate, taper-pointed, *deciduous; sepals ovate; anther-locules* 1-*awned; capsule* 5-*lobed; seeds flat and cellular-winged.* — Dry hills, mts. of Va. to Ala. Apr. — Lower and more straggling than the next.

4. L. racemòsa (L.) Gray (with racemes). — Shrub up to 4 m. high, with ascending branches; *leaves oblong to oval-lanceolate, acute,* serrulate, 2–7.5 cm. long, somewhat pubescent when young; *racemes* mostly solitary, *ascending or divergent,* 2–5 (–7) cm. long; *sepals ovate-lanceolate; anther-locules each 2-awned; capsules not lobed; seeds angled and wingless.* — Moist thickets, pond-shores, etc., Fla. to La., n. to Mass., se. N.Y. and e. Pa. May, June.

Var. **projécta** Fern. (thrust forward). — Racemes 1-several, the longer 0.6–2 dm. long. — Del. to Ga. and Tenn.

13. OXYDÉNDRUM DC. Sorrel-tree. Sourwood

Calyx of 5 almost distinct sepals, valvate in the bud. Corolla ovoid, puberulent. Stamens 10; anthers fixed near the base, linear, awnless, the locules tapering upward. Capsule 5-locular, 5-valved; the many-seeded placentae at the base of the locules. Seeds slender, the thin and loose reticulated coat extended at both ends into subulate appendages. — A tree with deciduous oblong-lanceolate pointed soon smooth serrulate leaves on slender petioles, and white flowers in long one-sided racemes clustered in an open panicle terminating the branches of the season. Bracts and bractlets minute, deciduous. Foliage acid (whence the name, from the Greek *oxys, sour,* and *dendron, tree*).

1. O. arbòreum (L.) DC. (becoming a tree), Titi. — Woods, Fla. to La., n. to Pa., W.Va., O. and Ind. June, July. — Leaves turning scarlet in autumn, then very handsome.

14. CHAMAEDÁPHNE Moench Leather-leaf. Cassandra

Calyx of 5 distinct sepals or segments. Stamens 10; anther-locules tapering into a tubular beak, awnless. Capsule depressed, 5-locular, many-seeded. Seeds flattened, wingless. — Low and much-branched circumpolar shrub, with nearly evergreen and coriaceous leaves which are scurfy, especially underneath. Flowers white, in the axils of the upper small leaves, forming small 1-sided leafy racemes. (From the Greek *chamai, on the ground,* and *daphne, laurel.*) Cassandra D. Don

1. C. calyculàta (L.) Moench (with an outer calyx). — Leaves oblong to oblong-lanceolate, mostly 2.5–5 cm. long, about thrice as long as wide; sepals broadly deltoid-lanceolate to narrowly deltoid-ovate, acuminate, one-third to half as long as the urceolate corolla. — Eurasia; represented with us by

Var. **angustifòlia** (Ait.) Rehd. (narrow-leaved). — Similar, up to 1 m. or more high; *leaves two-fifths to one-fourth as broad as long; acute or acutish sepals* about one-third as long as subcylindric corolla. — Peaty swales, bogs, pond-margins, etc., Nfld. to Alaska, s. to N.S., N.E., L.I., Ga., O., n. Ind., n. Ill., Wisc., n. Ia., Alta. and B.C. March–July. (E. Asia)

Var. **latifòlia** (Ait.) Fern. (broad-leaved). — Depressed or ascending, 1–6 dm. high; *leaves oblong to oblong-elliptic, rounded at summit, 1–3 cm. long, about half as broad; sepals broad- or rounded-ovate, blunt,* one-third to half as long as the urceolate-cylindric corolla. — Peaty barrens and bogs, Lab. Pen. to Mackenz., s. to Nfld. (where common), N.S. and n. N.E.

15. CASSÌOPE D. Don

Calyx without bractlets, of 4 or 5 nearly distinct ovate sepals imbricated in the bud. Corolla open-campanulate, 4–5-lobed or -cleft. Stamens 8 or 10; anthers fixed by the apex; the ovoid locules each opening by a large terminal pore and bearing a long recurved awn behind. Capsule 4–5-locular; placentae many-seeded, pendulous from the summit of the columella. Seeds smooth and wingless. — Small arctic or alpine evergreen plants with scale-like or needle-like leaves and solitary white or rose-colored flowers nodding on slender erect peduncles. (Named for *Cassiope,* mother of *Andromeda.*) Harrimanella Coville

1. C. hypnoìdes (L.) D. Don (resembling *Hypnum,* a moss). — Tufted and procumbent, moss-like, 1–12 cm. high; leaves acerose, loosely imbricated; corolla 5-cleft; style short and conical. (*Harrimanella* Coville) — Arct. reg., s. to mossy alpine areas of Nfld., Que., Me., N.H. and (?) N.Y. June–Aug. (Eurasia)

16. EPIGAÈA L. Ground-laurel. Trailing Arbutus

Corolla salverform, its tube hairy inside, as long as the ovate-lanceolate scale-like nearly distinct sepals. Stamens 10, with slender filaments; anthers oblong. Style slender, its apex (as in *Pyrola*) forming a sort of ring or collar around and partly adnate to the 5 small lobes of the stigma. Capsule depressed-globular, 5-lobed, 5-locular, many-seeded. — A prostrate or trailing scarcely shrubby plant with evergreen and reticulated oval to suborbicular and cordate-based alternate leaves on slender petioles, and with rose-colored to white spicily fragrant subdioecious and dimorphous flowers in small axillary and terminal clusters, from scaly bracts; a second species in Japan. (Name composed of the Greek *epi, upon,* and *gaea, the earth,* from the trailing growth.)

1. E. rèpens L. (running on the ground), MAYFLOWER, FLEUR DE MAI (Que.). — Mature leaves scabrous and persistently setose or pilose on both surfaces. — Sandy to peaty woods or clearings, Fla. to Miss., n. to s. N.E., se. N.Y., Pa., W.Va. and O. March–May.

Var. **glabrifòlia** Fern. (smooth-leaved). — Leaves (except sometimes the nerves beneath) glabrous or soon glabrate on both surfaces. — Lab. to Sask., s. to Nfld., N.S., N.E., Pa., upland to N.C. and Tenn., Mich., Ill. and Ia. April–July. — The rare forma **plèna** Rehd. (double) has "double" flowers.

17. GAULTHÈRIA L. AROMATIC WINTERGREEN. CREEPING SNOWBERRY

Corolla cylindrical-ovoid or urceolate to campanulate, 4- or 5-toothed or -cleft. Stamens 8 or 10, included. Capsule depressed, many-seeded, inclosed when ripe by the calyx or its base, which thickens and becomes fleshy, so as to appear as a globular to ellipsoid berry! — Shrubs or almost herbaceous plants of N. and S. Am. and of Asia to Austral. with alternate evergreen leaves and axillary nearly white flowers; pedicels with 2 bractlets. (Dedicated to *Jean-François Gaultier* — also spelled *Gaulthier, Gauthier* and *Gautier,* ?1708–1756 — naturalist and court-physician at Quebec.) Including CHIOGENES Salisb.

1. G. procúmbens L. (lying flat), TEABERRY, CHECKERBERRY, MOUNTAIN-TEA, IVRY-LEAVES, PETIT THÉ DES BOIS (Que.). — Stems slender and extensively creeping on or below the surface; *the flowering branches ascending, leafy at summit,* 5–15 cm. high; *leaves* at first tender, soon hard and coriaceous, 1.5–5 *cm. long, elliptic to narrowly obovate,* narrowed to base, or suborbicular and rounded at base and summit in forma **suborbiculàta** Fern. (nearly round), *obscurely serrate;* flowers few, solitary or in small racemes in the axils, nodding; *corolla cylindric-ovoid to urceolate, 5-toothed; stamens* 10 with arching white hairy filaments; *anthers with 2 double horns,* bursting outwardly their whole length; ovary roundish, depressed, 5-angled, borne above the 10-toothed reddish glandular disk; fruits red, formed by the *fleshy glabrous calyx surrounding the capsule,* often holding the shriveled corolla, or the enlarged corolla becoming fleshy and pink and persistent in forma **accrèscens** Fern. & Hodgdon (enlarging), subglobose to obovoid, or elongate and subcylindric in forma **elongàta** Svenson (elongate). — Sterile woods and clearings, Nfld. to Man., s. to N.S., N.E., Ga., Ala., Wisc. and Minn. July, Aug. — Plant aromatic.

2. G. hispídula (L.) Bigel. (finely rough-hairy), CREEPING SNOWBERRY, MOXIEPLUM, CAPILLAIRE, MAIDENHAIR-BERRY, PETIT THÉ OR OEUFS DE PERDRIX (Que.). — *Trailing and creeping matted delicate evergreen with* barely woody stems and *abundant* ovate firm *thyme-like revolute leaves* 0.5–1 *cm. long and bristly beneath; flowers* mostly solitary in the axils, on short nodding peduncles, *4-merous; corolla campanulate, deeply 4-cleft; stamens* 8, inserted on 8-toothed disk, *with* broad and short filaments, and *pointed but awnless anther-locules* opening by a large median chink; *berry bright white,* juicy, delicately acid and aromatic, *the fleshy accrescent calyx surrounding its base.* (*Chiogenes* T. & G.) — Mossy (oftenest coniferous) woods and knolls, and southw. in bogs, Lab. to B.C., s. to Nfld., N.S., N.E., Pa., upland to N.C., Mich., Wisc., Minn. and Ida. April, May (–Aug. in mts.).

Tribe ARBÙTEAE Drude

18. ARCTOSTÁPHYLOS Adans. BEARBERRY

Corolla with a short revolute 5-toothed limb. Stamens 10, included; anthers with 2 reflexed awns on the back near the apex, opening by terminal pores. — Circumpolar and Am. shrubs with alternate leaves and scaly-bracted flowers in terminal racemes or clusters. (Name composed of the Greek *arctos, a bear,* and *staphyle, a bunch of grapes,* the Greek of the popular name.)

Leaves entire, evergreen, leathery, not clearly veiny; branchlets finely pubescent; drupe dry and mealy. . . . . 1. *A. Uva-ursi.*
Leaves crenate, marcescent or deciduous, clearly veiny; branchlets glabrous; drupe juicy.
  Leaves strongly rugose, subcoriaceous, ciliate toward base and on petiole with long bristles; drupes purple to purple-black; nutlets 2–3.6 mm. broad. . . 2. *A. alpina.*
  Leaves plane or only slightly rugose, submembranaceous, not ciliate; drupes scarlet; nutlets 1.6–2.2 mm. wide. . . . . 3. *A. rubra.*

1. A. Ùva-úrsi (L.) Spreng. (old generic name, meaning bear's grape), COMMON B., KINNIKINICK, MEALBERRY, HOG-CRANBERRY, BOUSSEROLE OR RAISIN D'OURS (Que.). — Trailing shrub with long flexible branches covered with papery reddish to ashy exfoliating bark; *new branchlets minutely tomentulous and viscid, becoming glabrate; leaves* short-petioled, obovate to spatulate, *coriaceous,* lustrous above, *entire,* rounded at summit, 0.9–3 cm. long, *evergreen; flowers* 5–12 *in dense terminal racemes; bracts* lanceolate to lance-deltoid, *thick and firm;* corolla urceolate, pale pink, white or white with pink tips, about 5 mm. long, with very short rounded lobes; *anthers with* 2 *appendages* as long as the short broad-based filaments; *drupe* dull red, *dry and mealy; stone* solitary, 3–7 mm. in diameter, *consisting of* 5–10 *wholly or partially fused nutlets,* or compressed nutlets distinct. (*Uva-ursi Uva-ursi* (L.) Britt.) — Arct. reg., s. on exposed rock and sands to w. Nfld., Que., S.D., N.M. and Wash. May–early July. (Eurasia)

Var. **coáctilis** Fern. & Macbr. (felted). — *Young branchlets not viscid, minutely and permanently white-tomentulose.* — Exposed rocks, crests, downs and sands, Nfld. to Yuk., s. to N.S., N.E., L.I., Va., Ind., n. Ill., Colo. and n. Calif. Apr.–June (rarely Oct.). Forma **heterochròma** Fern. (variously colored), chiefly autumn-flowering, with the autumnal corollas often longer and subcylindric, carmine to deep purple, vernal flowers similar or like those of the typical form, in open sandy woods, e. Cape Cod and Nantucket, Mass. Oct., Nov. and May.

Var. **adenótricha** Fern. & Macbr. (with glandular hairs). — Branches and petioles viscid-villous, the *pubescence intermixed with black stipitate glands.* — Côte Nord, Que., to B.C., s. to n. Minn. and Mont.

2. A. alpìna (L.) Spreng. (alpine), Alpine B. — Matted and trailing, with brittle branches covered with papery bark; *branchlets glabrous,* slender; persistent *inner scales of winter-buds obovate, rounded at tip; leaves* slender-petioled, *subcoriaceous, deeply veiny-rugose, crenate-serrate, highly marcescent,* becoming whitened when old; *blade* obovate or oblanceolate, 1–3 *cm. long, ciliate* at base and along petiole *with bristles* 1–2 *mm. long; flowers* 2 *or* 3 *from terminal scaly buds,* expanding with leaves; *corolla* urceolate, *white, with short greenish lobes; bracts oval, thin; anthers unappendaged* or with rudimentary appendages; *drupe purple to purplish-black, juicy,* edible; *nutlets* 5, *distinct,* 2.7–4.6 *mm. long,* 2–3.6 *mm. broad.* (*Mairania alpina* (L.) Desv.; *Arctous alpina* (L.) Niedenzu) — Arct. reg., s. on bare rock and gravel to Nfld. and alpine areas of Que., Me. and N.H. May, June. (Eurasia)

3. A. rùbra (Rehd. & Wils.) Fern. (red). — Similar to no. 2, but with more ascending branchlets; *inner persistent scales of buds* lanceolate to ovate, *acuminate; leaves submembranaceous,* deciduous or barely marcescent, *flat or but slightly rugose,* the oblanceolate to cuneate-obovate *non-ciliate blade* 1.3–4.5 *cm. long; berries scarlet,* very juicy; *nutlets* 2.5–3 *mm. long,* 1.6–2.2 *mm. broad.* (*Arctous rubra* (Rehd. & Wils.) Nakai; *Arctous erythrocarpa* Small) — Calcareous gravel and rocks, nw. Nfld.; Anticosti I., Mingan Ids. and L. Mistassini, Que., to Alaska and Alta. (Asia)

Tribe ErícEAE Drude

19. CALLÙNA Salisb. Heather. Scotch Heather. Ling

Calyx of 4 colored sepals. Corolla much shorter and less conspicuous than the calyx, both becoming scarious and persistent. Stamens 8, distinct; anthers with a pair of deflexed appendages on the back, the locules each opening by a long chink. Capsule 4-locular, 4-valved. — Evergreen undershrub, nat. of Eu. and s. Asia, with no scaly buds, opposite and minute leaves (mostly extended at base into 2 sharp auricles) crowded and imbricated on the branches. Flowers axillary or terminating very short shoots and crowded on the branches, forming close mostly one-sided spikes or spike-like racemes, rose-colored or sometimes white, small, bracted by 2 or 3 pairs of leaves, the innermost of which are more or less scarious. (Named from the Greek *callunein, to brush or sweep,* brooms being made of its twigs.)

1. C. vulgàris (L.) Hull (common). — Peaty or damp sandy spots, always of small extent, very locally from Nfld. to Mich., s. to N.S., N.E., N.J. and mts. of W.Va. July–Nov. (Originally introd. from Eu., now natzd.)

20. ERÌCA L. Heath

Calyx of 4 short sepals. Corolla ovoid, globular, campanulate or tubular, 4-toothed or -lobed, marcescent about the capsule. Stamens 8. Capsule free, 4-locular, opening down the back of the locule, each locule several-seeded. — Large genus nat. of Europe and sw. Asia and Mediterr. reg., very largely S. African; low and much branched shrubs with small entire revolute leaves in whorls or scattered, the flowers in terminal or axillary clusters, usually drooping. (The ancient Greek and Latin name.)

Anthers included, 2-awned near junction with filament.
 Principal leaves in whorls of 4, without axillary fascicles. . . . . 1. *E. Tetralix.*
 Principal leaves in 3's, subtending axillary fascicles. . . . . 2. *E. cinerea.*
Anthers exserted, awnless. . . . . 3. *E. vagans.*

1. E. Tétralix L. (name used by Theophrastus, four spirals), Cross-leaved H. — Straggling or loosely ascending, with erect branches; *leaves chiefly in* 4's, stiffly ciliate, often whitish-downy; flowers barrel-shaped, 6–8 mm. long, deep pink, in terminal clusters; *awns nearly as long as anther, entire.* — Peaty clearings, open woods and margins of bogs, very local, Franklin Co., Me.; Cape Cod and Nantucket I., Mass.; s. N.J.; Grant Co., W.Va. Late July–Oct. (Natzd. from Eu.)

2. E. cinèrea L. (ashy), Scotch H. — Primary leaves in whorls of 3, very slender, *with dense axillary fascicles;* flowers purple, in terminal or axillary clusters; *awns very short, toothed.* — Moors and open woods, local, Nantucket I., Mass. July–Oct. (Introd. from Eu.)

3. E. vàgans L. (wandering), Cornish H. — Leaves in 3's or 4's, slenderly linear; *flowers axillary,* abundant, *forming subterminal elongate inflorescences; corolla* whitish or pink, campanulate, *changing to globular; anthers exserted, unappendaged.* — Open pine-woods, local, Nantucket I., Mass. July–Oct. (Introd. from Eu.)

Subfam. II. VACCINIOÌDEAE

21. GAYLUSSÀCIA HBK. Huckleberry

Corolla tubular, ovoid or campanulate; the border 5-cleft. Stamens 10; anthers awnless; locules tapering upward into more or less of a tube, opening by a chink at the end. Fruit a berry-like drupe, containing 10 seed-like nutlets. — Branching N. and S. Am. shrubs with the aspect of *Vaccinium,* commonly sprinkled with resinous dots; the flowers (pale, tinged with purple or red) in lateral and bracted racemes. (Named for *Louis Joseph Gay-Lussac,* 1778–1850, French chemist.)

Leaves evergreen, coriaceous, somewhat serrate, not resinous-dotted; racemes in leaf-axils. . . . . 1. *G. brachycera.*
Leaves deciduous, entire, sprinkled more or less with resinous or waxy atoms; racemes from axils of preceding year.
 Leaves thick or subcoriaceous, upper face shining and dark green; new growth glandular-hairy; bracts of raceme foliaceous, persistent; ovary and fruit glandular-pubescent. . . . . 2. *G. dumosa.*
 Leaves thinner, dull and paler; fruit glabrous; bracts of raceme small and deciduous.
  Resin-dots confined to lower surface of leaf; racemes lax and open, with pedicels mostly 1–2.5 cm. long; corolla campanulate to subglobose. . . . 3. *G. frondosa.*
  Resin-dots abundant on both leaf-surfaces; flowers short-pedicelled, in short sessile racemes; corolla conic-ovoid, becoming cylindraceous. . . . . 4. *G. baccata.*

1. G. brachýcera (Michx.) Gray (short-horned), Box-H. — *Evergreen,* the branches ascending from subterranean creeping stems, 2–4 dm. high, *glabrous or barely puberulent; leaves oval, finely crenate-toothed, thick and coriaceous, not resinous-dotted;* racemes short, nearly sessile, axillary, of few crowded flowers; bracts and bractlets scaly, caducous; corolla white or flesh-color, cylindraceous-campanulate; anthers shorter than ciliate filaments. — Sandy woods and slopes, local, Del. to W.Va., s. to Tenn. *Fl.* April, May; *fr.* Aug.

2. G. dumòsa (Andr.) T. & G. (bushy), Dwarf H. — Flowering stems slender, 1–5 dm. high, from creeping subterranean base, *leaves* cuneate-oblanceolate to oblong-obovate, conspicuously *mucronate, glandular beneath, glandless* or soon becoming so *on the lustrous upper surface; racemes leafy-bracted,* elongate; *the bracts persistent, glandular beneath,* rarely so above, oval, as long as pedicels; corolla campanulate, white or pink; *ovary and black fruit glandular-pubescent.* — Dry barrens and pinelands, Fla. to Miss., n. to N.J., e. Pa. and Tenn. *Fl.* May, June; *fr.* Aug., Sept.

Var. **Bigeloviàna** Fern. (for its discoverer, Jacob Bigelow, 1787–1879). — Leaves and bracts more generally and persistently glandular above as well as beneath. — Sphagnous bogs and wet peats, Nfld. and e. N.B. to L.I. and less typically to e. Va. *Fl.* June–Aug.; *fr.* Aug.–Oct.

3. G. frondòsa (L.) T. & G. (leafy; from the bracteate racemes), Dangleberry, Blue-tangle. — Slender, up to 2 m. high; young branchlets glabrous or merely puberulent; leaves oblong to oval-obovate, mostly obtuse and retuse, *pale, finely pubescent and glaucous beneath,* in maturity 2.5–7 cm. long, *resinous-dotted only beneath; racemes lax and open, the slender pedicels up to* 2.5 *cm. long* and longer than the oblong or linear deciduous bracts; corolla campanulate to subglobose; anthers with long tubular tips; *fruit dark blue with pale bloom, sweet and juicy.* — Dry woods and clearings, Fla. to La., n. to s. N.H., Mass. and se. N.Y. *Fl.* April–June; *fr.* July–Sept.

4. G. baccàta (Wang.) K. Koch (berry-bearing), BLACK H., GUEULES NOIRES (Que.). — Much branched, stiff, *slightly pubescent* when young, 0.3–1 m. high; *leaves* oval, oblong-ovate or oblong, thickly clothed and at first *clammy, as well as the flowers, with shining resinous globules;* racemes short, clustered, one-sided; pedicels about the length of the flowers; *bracts and bractlets reddish;* corolla ovoid-conical, or at length cylindrical with an open mouth; fruit black, without bloom, sweet. (*G. resinosa* T. & G.) — Dry or moist woods, thickets and clearings, Nfld. to Sask., s. to N.S., N.E., L.I., Ga. and La. *Fl.* May–July; *fr.* Aug.–Oct. — In forma **glaucocárpa** (Robins.) Mackenz. (fruit with a bloom) the fruit blue with a bloom, often larger and juicier; in forma **leucocárpa** (Porter) Fern. (white-fruited) fruit white to pinkish, somewhat translucent.

22. VACCÌNIUM L. BLUEBERRY. BILBERRY. CRANBERRY

Corolla various in shape, the limb 4- or 5-cleft. Stamens 8 or 10; anthers awned on back or awnless, the locules opening by a terminal pore. Berry 4- or 5-locular, many-seeded, or 8- or 10-locular by false partitions extending from the back of each locule to the placenta. Large and polymorphic genus of more or less woody plants, the aggregate group cosmop., the subgenera in restricted and very typical geographic areas. (Ancient name, presumably from the Latin *vaccinus, of cows,* as reflected in the germanic folk-name, *Kuhteke.*)

a. Corolla open-campanulate to cylindric, urceolate or globose, its lobes erect or spreading-ascending; leaves membranaceous to firmer, lanceolate, oblanceolate, ovate, obovate or broader, deciduous, if rarely evergreen broad and flat; anthers included (exserted only in subgen. 2, with flowers and fruits in leafy-bracted racemes or panicles). . . *b.*

b. Flowers and fruits in axils of leafy-bracted racemes or in axils of regular foliage-leaves; anthers 2-awned on back. . . *c.*

c. Flowers and fruits on filiform pedicels in bracted racemes or panicles; corolla open-campanulate, with 5 long spreading lobes; stamens or style exserted; ovary and fruit falsely 10-locular; chiefly southern. . . *d.*

d. Flowers and fruits articulated with pedicels; anthers included, their awns more than half the length of the tubular tips; berry black, dry, with hard stone-like seeds; coarse shrub or small tree with coriaceous often evergreen leaves. Subgen. 1. BATODENDRON. . . . . . 1. *V. arboreum.*

d. Pedicels not articulated, elongate; anthers exserted, their awns much shorter than the tubular tips; berries green to amber, purple or blackish, juicy, with soft seeds; slender shrubs with deciduous leaves. Subgen. 2. POLYCODIUM.

Bracts of racemes much smaller than foliage-leaves, 2–10 mm. broad; mature leaves (3–)5–9 cm. long. . . . . 2. *V. stamineum.*

Bracts one-half to nearly as large as foliage-leaves, 1–2 cm. broad; mature leaves 1.5–5 cm. long. . . . . 3. *V. caesium.*

c. Flowers and fruits solitary in leaf-axils or 1–3 drooping from scaly axillary buds; corolla slenderly campanulate to subglobose, 4- or 5-toothed; stamens and style included; ovary and fruit 4- or 5-locular; northern. Subgen. 3. EUVACCINIUM. . . *e.*

e. Leaves coriaceous, obovate to elliptic-oblong or roundish; flowers 1–few from a scaly bud, nearly sessile, 4-merous; corolla white or pink-tipped. . . . . 4. *V. uliginosum*, var. *alpinum.*

e. Leaves membranaceous or nearly so; flowers solitary on definite pedicels from the axils of new leaves, 5-merous; corolla pink to red. . . *f.*

f. Branches terete; leaves serrate or serrate-dentate, 0.5–1.5(–1.8) cm. broad; shrubs 0.5–7 dm. high; the corollas lasting until development of the leaves.

Depressed or subcespitose, with slender close-barked tufted branches, 0.5–3 dm. high; leaves narrowly obovate, cuneate-oblanceolate or -spatulate; corolla slenderly ellipsoid, 2–4 mm. thick; berries light blue. . . . . 5. *V. cespitosum.*

Ascending, with bark of coarse branches exfoliating, 0.2–7 dm. high; leaves elliptic; corolla ellipsoid-urceolate, 4–5 mm. thick; berries dark blue or black. . . . . . . 6. *V. nubigenum.*

*f.*Branches 4-angled; leaves mostly 1.5–3.5 cm. broad; upright or straggling shrubs 0.3–2 m. or more high.

Leaves sharply serrulate, bright green; flowers expanding with or after the leaves, their pedicels 6–12 mm. long; berries black. . 7. *V. membranaceum.*

Leaves entire or shallowly undulate, pale green; flowers expanding before leaves, on pedicels 1–5 mm. long; berries blue, with a bloom. . . . . 8. *V. ovalifolium.*

*b.*Flowers and fruits in dense terminal or lateral clusters from scaly buds separate from the leaf-buds; pedicels short, spreading or ascending; corolla 4- or 5-toothed, tubular or campanulate to subglobose-urceolate; filaments hairy; anthers awnless, included; ovary and berry 8- or 10-locular or incompletely so by a projection from the back of each carpel. . . *g.*

*g.*Ascending or upright shrubs with deciduous lanceolate to broader leaves; flowers 5-merous; berries blue or black. Subgen. 4. CYANOCOCCUS. . . *h.*

*h.*Mature leaves 0.5–4 cm. long and 3–15(–20) mm. broad; shrubs 1–9 dm. (in the se. no. 9 up to 3.5 m.) high. . . *i.*

*i.*Leaf-margins with few to no bristle-tipped teeth. . . *j.*

*j.*Leaves lustrous above, green both sides, 1–4 cm. long, 0.4–2 cm. broad; corollas slenderly campanulate to cylindric, nearly or quite twice as long as thick, milk-white to pink; berries black, 5–8 mm. in diameter; species of se. Coastal Plain.

Leaves membranaceous, lustrous and glabrous on both faces or merely pilose on midrib beneath, rounded to base; loosely branching or straggling tall shrub up to 3.5 m. high. . . 9. *V. Elliottii.*

Leaves firm, glandular-pubescent beneath, cuneate-oblanceolate or -obovate; compact low shrub 1–6 dm. high. . . . 10. *V. tenellum.*

*j.*Leaves opaque or barely sublustrous above, pale-pubescent or strongly whitened beneath; corolla globose-urceolate or ellipsoid, nearly or quite as thick as long, greenish to pink- or purple-tinged; berries blue with a bloom, 7–10 mm. in diameter (rarely black); n. or wide-ranging species.

Branchlets and lower leaf-surface copiously velvety-pilose with spreading hairs; leaves oblong-lanceolate to elliptical, 0.5–1.5 cm. broad, entire; corolla 4–6 mm. long; berries sour. 11. *V. myrtilloides.*

Branchlets and lower leaf-surface glabrous or only sparsely appressed-pilose; leaves obovate, oval or elliptic, 1.5–3.5 cm. broad, entire or serrulate-dentate, glaucous beneath; corolla 5–8mm. long; berries sweet. . . . . 12. *V. vacillans.*

*i.*Leaf-margins (at least when young) with slender bristle-tipped teeth. . . *k.*

*k.*Leaves bright green on both sides, narrowly lanceolate to oblong, 0.7–3.5 cm. long, 3–15 mm. wide, very closely spinulose-serrulate; corolla milk-white to pink-tinged; berries light to dark blue (usually with bloom). . . . . 13. *V. angustifolium.*

*k.*Leaves gray-green or blue-green, more or less glaucous beneath.

Leaves narrowly lanceolate to oblong, becoming 5–15 mm. broad, very closely spinulose-serrulate; corolla milk-white to pinkish, tubular-campanulate or barrel-shaped; berries black, without bloom. . . . . 13. *V. angustifolium*, var. *nigrum.*

Leaves obovate, oval or elliptic, 1.5–3.5 cm. broad, sparsely spinulose-serrulate; corolla greenish-white, with or without purple tinge, globose-urceolate; berries blue, with bloom. . . 12. *V. vacillans*, var. *crinitum.*

*h.*Mature leaves mostly 3–8 cm. long and 1.5–4 cm. broad; shrubs 1–4 m. high, of relatively southern range (not in our more northern areas). . . *l.*

*l.*Leaves glabrous or but slightly pubescent beneath, half-grown at flowering time; calyx usually glaucous, with acute or merely bluntish lobes; corolla ovoid to cylindric-urceolate, white to pinkish, 5–12 mm. long; berries blue to blue-black, with bloom.

Leaves of fertile branches 4–8 cm. long and 2–4 cm. broad, entire or serrulate, glabrous or slightly pubescent; corolla 6–12 mm.

long, white or pinkish; berries mostly 6–12 mm. in diameter; shrub 1–4 m. high, wide-ranging. . . . . 14. ***V. corymbosum.***

Leaves of fertile branches mostly 3–5 cm. long and 1.5–2 cm. broad, entire, glabrous and glaucous; corolla 4–6 mm. long, dull white; berries 5–8 mm. in diameter; shrub 1–3 m. high, chiefly coastwise. . . . . 15. ***V. caesariense.***

l. Leaves downy to woolly beneath, entire, unexpanded at flowering time; calyx not glaucous, its lobes obtuse or rounded; corolla turgid-ovoid to ellipsoid, yellowish- or greenish-white tinged with red, 5–8 mm. long; berries polished black, without bloom, 5–8 mm. in diameter. . . . . 16. ***V. atrococcum.***

g. Creeping or trailing narrow- and coriaceous-leaved evergreens, forming prostrate carpets; flowers 4- or 5-merous; berries red or black.

Branchlets ligneous; leaves green both sides; flowers in small axillary clusters, 5-merous, with deciduous bracts; corolla white or pink; stamens 10; berries black, sweet; southern. Subgen. 5. Herpothamnus. . . . . 17. ***V. crassifolium.***

Branchlets subligneous; leaves paler beneath and bearing scattered dark bristles; flowers in terminal clusters, 4-merous, with nearly persistent reddish bracts; corollas pink to reddish; stamens 8; berries red, acid; northern. Subgen. 6. Vitis-Idaea. . . . . 18. ***V. Vitis-Idaea,*** var. ***minus.***

a. Corolla deeply 4-parted or -cleft, with reflexed linear lobes, nodding on long filiform pedicels; anthers exserted, awnless, with very long terminal tubes; berries red to dark purple, brown, blackish or sometimes pale.

Upright shrub with broad deciduous membranaceous serrate leaves; flowers solitary, axillary, jointed with the pedicel; berries insipid or sweetish. Subgen. 7. Oxycoccoides. . . . . 19. ***V. erythrocarpum.***

Trailing lithe-stemmed evergreens with narrow coriaceous often revolute entire leaves; flowers solitary or in small racemes from terminal buds, not jointed with pedicel; berries acid. Subgen. 8. Oxycoccus.

Leaves ovate-oblong to ovate or triangular, strongly whitened beneath; pedicels springing from terminal short rachis, each with 2 colored bractlets 1–2.5 mm. long at or below middle; filaments half as long as anthers. 20. ***V. Oxycoccos.***

Leaves oblong-elliptic, blunt, only slightly paler beneath; pedicels springing from a rachis terminated by a leafy shoot and bearing toward the tip 2 flat foliaceous bractlets 4–10 mm. long; filaments scarcely a third the length of the anthers. . . . . 21. ***V. macrocarpon.***

Subgen. 1. Batodéndron (Nutt.) Gray

1. V. arbòreum Marsh. (tree-like), Farkleberry or Sparkleberry. — Coarse shrub or small tree up to 9 m. high and with trunk up to 3 dm. in diameter; *leaves* obovate to oblong, entire to denticulate, *lustrous above, thick and leathery, southward quite evergreen;* flowers in loose leafy-bracted racemes, the bracts smaller than and often of different shape from the foliage-leaves; *pedicels jointed;* corolla white, open-campanulate; *anthers included, their awns more than half the length of the tubular tips;* style exserted; *berry black, hard, dry, with hard stone-like seeds.* (*Batodendron* Nutt.) — Sandy woods, thickets and clearings, Fla. to Tex., n. to Va., s. Ind., s. Ill., s. Mo. and Okla. *Fr.* Aug.–Oct. — Passing insensibly into

Var. **glaucéscens** (Greene) Sarg. (somewhat glaucous). — Bracts and foliage similar and subequal. (*Batodendron glauc.* Greene; *B. andrachneforme* Small) — S. Ill. and s. Mo. to Ala., Okla. and Tex.

Subgen. 2. Polycòdium (Raf.) Sleumer

2. V. stamíneum L. (with prominent stamens), Deerberry, Squaw-Huckleberry. — Diffusely branched shrub 0.3–3 m. high, the young branchlets and leaves more or less pubescent; the latter persistently so on the lower surface, ovate to oval or oblong, (3–)5–9 cm. long, pale, glaucous or whitened beneath; *flowers in loose bracted racemes, pendulous on filiform pedicels; the acute to obtuse bracts much smaller than the foliage-leaves, 2–10 mm. broad;* calyx glabrous or nearly so; *corolla* greenish-white to purplish (blackish in drying), *open-campanulate, with 5 spreading lobes; anthers much exserted, their awns much shorter than the tubular tips; berry juicy,* tough-skinned, *green to amber, purple or blue,* with or without bloom, with soft seeds, dropping promptly; *bracts of raceme caducous.* (*Polycodium* Greene) — Dry woods, thickets and clearings, Fla. to La., n. to se., centr. and w. Mass., N.Y., s. Ont., O., Ind. and Mo. *Fr.* July–Sept. — A polymorphous species with no seeming constancy in shape and size of leaves or color and flavor of fruit. Passing into

Var. **melanocárpum** Mohr (black-fruited). — Leaves pubescent beneath; *calyx densely white-tomentose;* berries black-purple. (*V. melanocarpum* Mohr) — N.Y. to N.C. and s. Mo.

Var. **intèrius** (Ashe) Palmer & Steyerm. (inland). — Young branchlets pubescent; *mature leaves glabrous or merely with midrib pilose beneath;* calyx glabrous. — Md. to Kans., s. to Va., Ark. and Okla.

Var. **negléctum** (Small) Deam (overlooked). — Branchlets and *leaves* glabrous, the latter *green to slightly glaucous beneath;* fruit green to dark purple. (*V. neglectum* (Small) Fern.) — N.Y. to N.J., Pa., W.Va., O., Ind., Mo. and e. Kans.

3. V. caèsium Greene (bluish-green), DEERBERRY, SQUAW-HUCKLEBERRY. — Smaller than no. 2, 0.2–5 (rarely –10) dm. high, pubescent to glabrate; *leaves* whitened or green beneath, mostly blunt, *becoming* 1.5–5 *cm. long; bracts* of raceme *one-half to nearly as large as the foliage-leaves and* similar to them in outline, 1–2 *cm. broad,* much exceeding the elongate pedicels. — Barrens and oak- or pine-scrub, Fla. to La., n. to mts. of w. Pa. and W.Va.

Subgen. 3. EUVACCÌNIUM Gray, the BILBERRIES

4. V. uliginòsum L. (of swamps), var. **alpinum** Bigel. (alpine), ALPINE B. — *Stiff and stoutish depressed or ascending,* 0.2–6 dm. high, *with strong woody branches;* branchlets and leaves glabrous to puberulent; *leaves coriaceous, obovate, oblong or roundish, dull green above, pale and reticulate beneath,* in maturity 0.5–2 cm. long; *flowers 1–few from scaly buds* at expansion of leaves, subsessile or on pedicels 0.1–3.5 mm. long, 4-merous; corolla urceolate, white or pink-tinged; *berry* blue to blackish, *with a bloom, sweet.* — Arct. reg., s. to rocky or dry peaty barrens and upper slopes of mts. of Nfld., C.B., n. N.E. (rarely down to sea-level), n. N.Y., n. Mich. and n. Minn. *Fr.* late July–Sept. — Smaller throughout, with shorter petioles and pedicels and with the horns of the anthers relatively longer than in typical Eurasian and nw. Am. *V. uliginosum.*

5. V. cespitòsum Michx. (forming tufts), DWARF B. — *Dwarf tufted* depressed or ascending *shrub with slender close-barked branches terete and* 0.5–3 *dm. high; leaves membranaceous, smooth and shining, narrowly obovate, cuneate-oblanceolate or -spatulate, serrate,* in maturity 1–4 cm. long and 0.4–2 cm. broad; *flowers solitary in leaf-axils, 5-merous,* nodding on short pedicels; *corolla deep pink to coral-reddish, slenderly ellipsoid, 2–4 mm. thick; berries light blue, with bloom.* — Gravelly or rocky shores, openings, woods, etc., Lab. to Alaska, s. to Nfld., N.B., n. N.E., n. N.Y., n. Mich., n. Wisc., n. Minn., Colo. and Calif., ascending to alpine areas and descending to sea-level. *Fr.* July, Aug.

6. V. nubígenum Fern. (born of the clouds). — *More woody* than no. 5, *the ascending terete woody stems and branches* 0.2–7 *dm. high, with exfoliating thin bark; leaves elliptic,* becoming 1–3.5 cm. long, *firmer, slightly lustrous, prominently veiny beneath;* flowering with expanding leaves; *corolla ellipsoid-urceolate, 4–5 mm. thick; berries blue-black,* with bloom, of rich grape-like flavor. — Rocky, peaty or heathy slopes and barrens, sea-level to subalpine areas, n. Nfld.; subalpine reg., Shickshock Mts., Gaspé Pen., Que. *Fr.* Aug., early Sept.

7. V. membrànaceum Dougl. (membrane-like; from the thin leaves). — Erect or loosely ascending *shrub* 0.3–1.5 *m. high, with* 4-*angled branchlets and exfoliating bark; leaves ovate, acute or pointed,* thin, *closely and sharply serrulate, bright green,* becoming 2–7 cm. long; *flowers expanding with or after the leaves, on pedicels* 6–12 *mm. long;* corolla pinkish, subglobose-urceolate; *berry* depressed, *purplish to black, without or with only slight bloom,* of good quality. — Thickets and borders of woods, Bruce Pen., Ont. and ne. Mich.; Black Hills, S.D.; s. Alta. and s. B.C., s. to Calif. *Fr.* Aug., early Sept.

8. V. ovalifòlium Sm. (oval-leaved), MATHERS (Nfld.). — *Straggling shrub with 4-angled branchlets* and exfoliating bark, 0.3–1.7 m. high; *leaves elliptical, obtuse, firm, dull above,* pale beneath, *entire or remotely undulate,* becoming 1.5–5 cm. long; *flowers expanding before leaves, on pedicels* 1–5 *mm. long;* corolla subglobose, pinkish; *berries blue, with a bloom,* often of poor or disagreeable flavor with us. — Thickets, open woods and peaty slopes, sometimes ascending in subalpine ravines, se. Lab.; Nfld.; Gaspé Pen. and Laurentide Mts., Que.; Algoma Distr., Ont. and n. Mich.; Alaska to Ida. and Oreg. *Fr.* late July–early Sept.

Subgen. 4. CYANÓCOCCUS Gray *, the BLUEBERRIES, BLUET or BLEUET (Que.)

9. V. Elliόttii Chapm. (in honor of STEPHEN ELLIOTT, 1771–1830), ELLIOTT'S B. — *Straggling or loosely branched* slender shrub 1–3.5 *m. high, with* minutely pilose to glabrous *slender divergent branches; leaves* elliptic, *rounded to base, membranous, glossy-green on both surfaces,* glabrous, or pilose along nerves beneath, 1–3.5 cm. long, 0.5–1.7 cm. broad, crenate or serrulate; *flowers* 3–8 *in short mostly terminal racemes, on long slender pedicels,* expanding as leaves unfold; calyx-lobes broad, triangular; *corolla* pink to milk-white, *slenderly subcylindrical,* 5–6 mm. long, *with tiny recurving teeth; long-pedicelled berries black or black-purple,* 5–8 mm. in diameter, dryish. (*Cyanococcus* Small) — Woods, thickets and swamps, Fla. to La., n. to se. Va. and Ark. *Fr.* June, July.

10. V. tenéllum Ait. (slender). — *Dwarf* colonial shrub 1–6 *dm. high, with few strongly ascending glabrous or minutely pilose branches* or almost unbranched; *leaves cuneate-oblanceolate to -obovate,* 1.5–3 *cm. long,* 0.4–1.5 *cm. broad, with reddish glandular hairs on minutely pilose to glabrate lower surface; flowers in dense fasciculate racemes from old leaf-axils,* before or at expanding of leaves; calyx-lobes ovate to triangular; *corolla slenderly subcylindric,* pink to milk-white or merely pink-tipped, 5–8 mm. long; berries small, black, rather dry. (*Cyanococcus* Small) — Dry sandy woods and clearings, Fla. to Miss., n. to se. Va. and Tenn. *Fr.* June, July.

11. V. myrtilloìdes Michx. (like *V. Myrtillus*), SOUR-TOP- or VELVET-LEAF-B. — Low, 2–6 (–9) dm. high, with *new branchlets densely velvety or divergently pilose-hirsute; leaves oblong-lanceolate to elliptical,* in maturity 2–4 (rarely –5) cm. long and 0.5–1.5 (rarely –2.5) *cm. broad, downy beneath and often above, entire;* close racemes terminating the branchlets, the flowers expanding when the leaves are half-grown; *corolla* globose-urceolate or short-campanulate, 4–6 mm. long, *greenish* or tinged with purple; *berries* 7–10 mm. in diameter, blue, with a heavy bloom, or whitish in forma **chiocóccum** (Deane) Fern. (with snowy berry), *sour.* (*V. canadense* Kalm) — Moist woods, swamps and clearings, Nfld. (local) and e. Saguenay Co., Que., to B.C., s. to N.S., n. and w. N.E., Pa., mts. to Va., n. O., w. Ind., n. Ill., ne. Ia., Sask. and Mont. Fr. July–Sept.

12. V. vacíllans Torr. (vacillating), LOW B., EARLY SWEET B., SUGAR-HUCKLEBERRY. — Low, 3–9 dm. high, *glabrous,* with stiff greenish or brown-tinged branches; *leaves* obovate, oval or elliptic, *pale or dull,* glaucous beneath, *in maturity subcoriaceous* and 1.5–5 cm. long, 1.5–3.5 *cm. broad, entire to serrulate-dentate* (especially toward summit); racemes terminating branchlets or from old axils, flowering when the leaves are partly grown; calyx often reddish; *corollas greenish, greenish and purple or wholly purplish, globose-urceolate,* 5–8 mm. long; *berries dark blue,* with faint bloom, 6–9 mm. in diameter, *very sweet.* (Not *V. vacillans* Kalm., *nomen nudum; V. Torreyanum* Camp) — Dry open woods, thickets and clearings, Ga. to Mo., n. to w. N.S., s. Me., s. N.H., Vt., N.Y., s. Ont., O., Mich., Ill. and ne. Ia. *Fr.* late June–Sept.

Var. **crinìtum** Fern. (hairy). — Young branchlets and lower surfaces of leaves pubescent; fruit blue, with bloom. — Throughout much of the range, perhaps partly of hybrid origin.

Var. **missouriénse** Ashe (of Missouri). — Similar to the last; fruit black, without bloom. — Ind. to Ark. and e. Kans.

13. V. angustifòlium Ait. (narrow-leaved), LOW SWEET or LATE SWEET B.; SWEET HURTS. — Dwarf *intricately branched* and depressed *freely stoloniferous* shrub 0.5–3.5 *dm. high, with smooth or minutely pilose new branchlets; leaves narrowly lanceolate,* 0.7–2 *cm. long,* 3–8 *mm. wide, bright green and glabrous on both sides, closely and minutely spinulose-serrulate;* flowers in glomerulate racemes from tips of branchlets or from old leaf-axils; *corollas milk-white* or tinged with pink, *globose-urceolate or ellipsoid-campanulate,* 5–6 *mm. long; berries blue, with heavy bloom, sweet,* 5–8 mm. in diameter. (*V. pensylvanicum,* var. *angustifolium* Gray) — Dry open barrens, peats and rocks, Lab. and e. Que. to Minn., s. to Nfld., C.B., mts. of n. N.E. and N.Y., and locally in bogs to s. N.B. and s. N.E. *Fr.* Aug., Sept. — Passing southw. into

Var. **laevifòlium** House (smooth-leaved). — *Coarser,* up to 6 dm. high; *leaves lanceolate to oblong,* 1.5–3.5 *cm. long and* 8–15 *mm. wide, glabrous* on both sides or minutely pilose at base of midrib beneath; *corolla cylindric-campanulate,* 6–10 *mm. long; berries blue, with bloom* (rarely black), or whitish in forma **leucocárpum** (Deane) Rehder (white-fruited), sweet, 6–15 *mm. in diameter.* (*V. pensilvanicum* Lam., not Mill.; *V. Lamarckii* Camp) — Nfld. and Côte Nord, Que., to Sask., s. to N.S., s. N.E., L.I., Del., Md., upland to Va., W.Va., O., Ind., n. Ill. and ne. Ia. *Fr.* June–Sept.

Var. **hypolásium** Fern. (hairy beneath). — Similar to last var. but with *leaves hairy beneath*

* Species freely hybridizing, their classification difficult, contradictory and confused. The hybrids or apparent hybrids not here included.

especially along the nerves, *with spreading pubescence.* — Nfld. to Hudson Bay, s. to N.S., Mass., upland to Va., Mich., Wisc. and Minn.

Var. **nìgrum** (Wood) Dole (black). — Of stature and size and shape of foliage of var. *laevifolium* but the firm *leaves blue-green, glaucous beneath;* flowers and fruits as in the latter, but *berries usually black without bloom.* (*V. pensilvanicum*, var. Wood; *V. nigrum* Britt.; *V. Brittonii* Porter) — Nfld. (rare) to n. Ont., s. to N.S., N.E., N.J., Pa., mts. to W.Va., n. Ind., Wisc. and Minn.

V. pállidum Ait. (pale), a slender *low shrub* (*mostly* 2.5–5 *dm. high*), *with glabrous leaves thinnish* and *closely serrate*, the blades becoming 3–5 *cm. long and* 1.5–2.5 *cm. broad; the corollas subcylindric, greenish-white or pink-tinged and* 4–6 *mm. long, the dryish berries* 5–7 *mm. across*, extends n. in dry woods from Ga., Ala. and Ark. to se. Va., W.Va. and Mo. (reported n. to N.Y. and Pa.).

14. V. corymbòsum L. (bearing corymbs), Highbush-B. — Shrub up to 4 m. high (southw.), forming compact or open clumps; *leaves green on both sides, entire, half-grown at flowering time*, elliptic-lanceolate to ovate, becoming 4–8 cm. long and 2–4 *cm. broad, the lower side pubescent along the nerves,* the margin eciliate; flowers rather densely clustered; *calyx usually glaucous and glabrous, its lobes acute or acutish; corollas bright white or milk-white,* sometimes pink-tinged, *ovoid to cylindric-urceolate,* 6–12 *mm. long and* 4–6 *mm. thick; berries blue to blue-black, with a bloom,* 6–12 *mm. in diameter,* sweet and juicy. — Swamps, low woods or even dry uplands, N.S. to s. Que., w. to Wisc., s. beyond our limits. *Fr.* late June–early Sept.

Var. **albiflòrum** (Hook.) Fern. (white-flowered). — Similar; *green leaves finely serrulate or serrulate-ciliate.* (Var. *amoenum* sensu Gray, not *V. amoenum* Ait., basonym) — Similar range, usually less abundant.

Var. **glàbrum** Gray (glabrous). — *Leaves glaucous and nearly or quite glabrous beneath, ciliate-serrulate.* (Var. *pallidum* sensu Gray, not *V. pallidum* Ait., basonym) — Occasional throughout the range.

15. V. caesariénse Mackenz. (of (New) Jersey), Highbush-B., New Jersey B. — Generally lower than no. 14 but up to 3 m. high; *leaves quite glabrous,* 3–5(–7) *cm. long* and 1.5–2 *cm. broad,* glaucous beneath, entire; *corolla dull white,* 4–6 *mm. long,* 3–4 *mm. in diameter; berries dark blue,* with bloom, 5–8 *mm. in diameter.* — Swamps, peaty thickets and bogs, chiefly on or near Coastal Plain, n. Fla. to s. Me., inland to centr. N.Y.

16. V. atrocóccum (Gray) Heller (black-berried), Black Highbush-B. — Differing from the two preceding in having the *heavily downy or woolly entire leaves unexpanded at flowering time; calyx not glaucous, its lobes obtuse or rounded; corolla turgid-ovoid to ellipsoid, yellowish- or greenish-white tinged with purple,* 5–8 *mm. long; berries polished black, without bloom,* or whitish in forma **leucocóccum** Deane (white-fruited), 5–8 mm. in diameter. — Swamps, low woods and barrens, n. Fla. to Ark., n. to N.E., N.Y., s. Ont. and Ind. *Fl.* and *fr.* a week to ten days earlier than in nos. 14 and 15 in same regions.

Subgen. 5. Herpothámnus (*Herpothamnus* Small, as a genus)

17. V. crassifòlium Andr. (thick-leaved), Creeping Blueberry. — *Prostrate* and more or less repent, *forming carpets* up to 2 m. across, from nodose-thickened bases; *stems slender, ligneous; leaves elliptic, coriaceous, evergreen,* 0.5–1.5 cm. long, 3–9 mm. broad, *the revolute margin glandular-crenate; flowers few in small racemes shorter than subtending leaves, the small bracts deciduous; calyx 5-lobed; corolla* subglobose, white and pink, *the 5 short lobes spreading; stamens* 10; *berry black or black-purple, lustrous,* juicy, *sweet* and bland, with large seeds. (*Herpothamnus* Small) — Sandy or peaty pinelands and barrens, e. S.C. to se. Va. *Fr.* late Aug.–Oct.

Subgen. 6. Vìtis-Idaèa (Moench) W. D. J. Koch

18. V. Vìtis-Idaèa L. (grape of Mt. Ida), var. **mìnus** Lodd. (smaller), Mountain-Cranberry, Rock-C., Cowberry, Lingen, Lingberry, "Pomme-de-Terre" (Que.). — Extensively *creeping evergreen with lithe slender stems and branches; leaves* evergreen, *coriaceous,* narrowly obovate, 0.5–1.8 cm. long, 4–9 mm. broad, smooth and *lustrous above, paler and dotted beneath* with blackish bristly points; *flowers 4-merous, in terminal clusters; corolla* campanulate, *pink or reddish;* stamens 8; *berries red, acid and slightly bitter.* — Rocky or dry peaty acid soil, subarct. Am., s. to Nfld., N.S., N.E., L. Sup. reg. of Ont., n. Minn., Man., Sask., Alta. and B.C. *Fr.* autumn, overwintering and superior in taste at melting of snow. (Greenl.; e. Asia) — Smaller throughout than in typical *V. Vitis-Idaea* of Eu.

Subgen. 7. Oxycoccoìdes (B. & H.) Sleumer

19. V. erythrocárpum Michx. (red-fruited), Mountain-Cranberry, Bearberry. — Woody *divergently branched shrub* 0.3–2.5 *m. high*, with exfoliating bark; *leaves oblong- to ovate-lanceolate, acute, membranaceous*, up to 7.5 cm. long and 1–3 cm. broad, *closely serrate; flowers solitary in the axils; corolla deeply 4-cleft nearly to base into recurving pink or white narrow lobes; anthers long-exserted, awnless, with very long terminal tubes; berry* red to brownish or darker, or black in forma **nigrum** Allard (black), *insipid to sweetish.* (*Hugeria* Small) — Thickets, rocky woods, slopes or summits, mts. of Ga. and Tenn. to W.Va. and w. Va. *Fr.* Aug., Sept. — Closely related species in e. Asia; worthy the thought of those who separate subgen. *Oxycoccus* as a genus!

Subgen. 8. Oxycóccus (Hill) Gray, the Cranberries or Atocas (Que.)

20. V. Oxycóccos L. (old generic name, as *Oxycoccus*, meaning sour berry), Small C. — Stem very slender and creeping, the branches almost capillary and ascending; *leaves ovate-oblong to ovate or triangular*, 3–8 *mm. long*, 1–3 *mm. broad, strongly revolute, conspicuously whitened or glaucous beneath; pedicels* 1–4, *springing from a terminal* (rarely proliferating) *short rachis* at most 3 or 4 mm. long *and bearing near or below the middle* 2 lanceolate or lance-ovate often involute *colored bractlets* (1–2.5 *mm. long*); roseate *corolla-segments* 5–6 *mm. long; filaments one-half as long as anthers; berry* 5–8 *mm. in diameter*, pale and speckled, becoming red when ripe, or whitish when ripe in forma **leucocárpum** Aschers. & Magnus (white-fruited). (*Oxycoccus* MacM.; *O. microcarpus* Turcz.; *O. quadripetalus* Gilib.) — Boggy or peaty soil, Greenl. and Lab. to Alaska, s. to Nfld., N.S., N.E., L.I., N.J., Pa., upland to Va. and W.Va., n. O., Mich., Wisc., Minn., Man., Alta. and Wash. *Fr.* Aug.–Oct. (holding over winter). (Eurasia)

Var. **ovalifòlium** Michx. (oval-leaved). — Coarser; *leaves* 6–15 *mm. long* and 2–8 *mm. broad, only slightly revolute; pedicels* 2–10 *from a rachis* often 5–10 *mm. long; corolla-segments* 6–8 *mm. long; berry* 8–10 *mm. in diameter.* (Var. *intermedium* Gray; *Oxycoccus ovalifolius* (Michx.) A. E. Porsild) — Nfld. to B.C., s. to N.S., N.E., N.Y., mts. of N.C., Mich., Minn., Sask. and Oreg.

21. V. macrocárpon Ait. (large-fruited), Large or American C. — Stem usually stouter than in no. 20, very elongate and intricately forking, the flowering branches ascending; *leaves oblong-elliptic*, blunt or rounded at tip, 6–17 mm. long, 2–8 mm. broad, pale or but slightly whitened beneath, flat or slightly revolute; *pedicels* 1–10, *springing from an elongate rachis* 1–3 *cm. long, this terminated by a long leafy shoot;* the *pedicel bearing toward the tip 2 flat leaf-like bractlets* 4–10 *mm. long; corolla-segments* 6–10 *mm. long; filaments scarcely one-third the length of the anthers; berry* 1–2 *cm. thick*, globose, ellipsoid, obovoid or pyriform. (*Oxycoccus* Pers.) — Open bogs, swamps and wet shores, Nfld. to Minn., s. to N.S., N.E., L.I., N.C., W.Va., O., Ind., Ill. and (rarely) Ark. *Fr.* Sept.–Nov., holding through winter. — Many forms of fruit perpetuated in cult.

FAM. 134. DIAPENSIÀCEAE (Diapensia Family)

Low perennial herbs or suffruticulose tufted plants, glabrous or nearly so, with simple leaves, no stipules, regular 5-merous flowers (except the 3-locular ovary), stamens adnate to the corolla and sometimes monadelphous (those opposite its lobes, when present, reduced to staminodia); pollen simple; loculicidal capsule and seeds of Ericaceae. Flowers solitary or racemose. Style 1, with 3-lobed stigma. — Distinguished from the *Ericaceae* chiefly by the insertion of the stamens upon the corolla.

Dwarf evergreens with small entire crowded coriaceous leaves; staminodia 0; filaments adnate to campanulate corolla up to sinuses; anthers 2-locular; flowers solitary. Tribe Diapensieae.

Flowers (or at least fruits) on scape-like peduncles; sepals coriaceous; anther-locules blunt, obliquely dehiscent. . . . . 1. *Diapensia.*

Flowers sessile on short leafy branches; sepals thin; anther-locules awned at base, transversely dehiscent. . . . . 2. *Pyxidanthera.*

Acaulescent herb, with creeping rhizome and long-petioled roundish crenate leaves; staminodia 5; stamens monadelphous; anthers 1-locular; flowers in a terminal spike. Tribe Galacineae. . . . . 3. *Galax.*

Tribe DIAPENSÌEAE Gray

1. DIAPÉNSIA L.

Corolla campanulate, 5-lobed; lobes rounded. Filaments broad and flat, adherent to the corolla up to the sinuses, short; anthers adnate, of 2 ovoid locules, diverging below. Capsule inclosed in the calyx, cartilaginous; locules few-seeded. — Boreal, growing in very dense convex tussocks, the stems covered below by persistent imbricated cartilaginous narrowly spatulate mostly opposite leaves, and terminated by a 1-flowered peduncle 3-bracted under the calyx; a 2nd species Himalayan. Corolla white, 1.5–2 cm. wide. (An ancient name of the Sanicle, *Sanicula*, strangely applied by Linnaeus to this boreal plant.)

1. D. lappónica L. (of Lapland). — Leaves 5–15 mm. long; peduncle at length 1.5–3 cm. long. — Arct. reg., s. to bare ledges and gravel of Nfld. and alpine reg. of Que., n. N.E. and n. N.Y. June, July. (Eurasia)

2. PYXIDANTHÈRA Michx.

Prostrate and creeping, with narrowly oblanceolate and awl-pointed leaves mostly alternate on the sterile branches and somewhat hairy near the base. Flowers solitary and sessile, very numerous, white or rose-color. — A single or possibly 2 species. (Name from the Greek *pyxis, a small box*, and *anthera*, new Latin for *anther;* the anther opening as if by a lid.)

1. P. barbulàta Michx. (with a little beard), FLOWERING-MOSS, PYXIE. — Leaves 3–8 mm. long. — Sandy pine-barrens, N.J.; se. Va. to S.C. March–May.

Tribe GALACÍNEAE Gray

3. GÀLAX L. WANDFLOWER

Calyx imbricate, persistent. Petals hypogynous, obovate-spatulate, rather erect, deciduous. Filaments united into a 10-toothed tube, slightly adhering to the base of the petals, the 5 teeth opposite the petals naked, the alternate ones shorter and bearing roundish anthers which open across the top. Style short. Capsule ovoid, 3-locular; columella none. Seeds numerous, the cellular loose coat tapering to each end. — Evergreen herb, with a thick matted tuft of scaly creeping rhizomes beset with fibrous red roots and sending up round-cordate crenate-toothed and veiny shining leaves (3–16 cm. wide) and a slender naked scape 3–8 dm. high, this bearing a virgate spike or raceme of small and minutely bracted white flowers. (Name from the Greek *gala, milk*, presumably from the milk-white flowers.)

1. G. aphýlla L. (leafless; from the naked scape), BEETLEWEED. — Open woods, Va. and W.Va. to Ga. and Ala.; cult. and locally esc. n. to Mass. Late May–July.

FAM. 135. PRIMULÀCEAE (PRIMROSE FAMILY)

Herbs with simple (rarely dissected) leaves and regular perfect flowers, the stamens as many as the lobes of the gamopetalous (rarely polypetalous) corolla (corolla none in Glaux) *and inserted opposite them* (on the tube or base), *and a 1-locular ovary with a central free placenta rising from the base and bearing several or many seeds.* Calyx free from the ovary, or in *Samolus* partly adherent. Stamens 4 or 5, rarely 6 or 8. Style and stigma one. Seeds with a small embryo in fleshy albumen. Ovules amphitropous, except in *Hottonia.*

a. Leaves entire or merely toothed. . . *b.*
  b. Ovary wholly free from the calyx. Tribe PRIMULEAE. . . *c.*
    c. Scapose, with terminal umbels; leaves forming a basal rosette. . . *d.*
      d. Calyx tubular, with erect teeth; corolla-lobes merely spreading; stamens distinct, included.
        Corolla-tube equaling to exceeding calyx, open at throat, the lobes rotate and obcordate; style filiform, elongate; perennials. . . . 1. *Primula.*
        Corolla short, with short tube exceeded by calyx, the throat constricted; style very short; annuals or biennials. . . . . . 2. *Androsace.*
      d. Calyx deeply cleft, its lobes and the corolla-lobes usually reflexed; filaments monadelphous below, anthers forming an exserted cone. . 3. *Dodecatheon.*
    c. Leafy-stemmed, with flowers variously disposed, rarely in definite umbels. . . *e.*
      e. Corolla rotate or nearly so, deeply cleft, not persistent (except as a shriveled cap over the capsule in no. 7). . . *f.*

f. Corolla longer than calyx; leaves (or most of them) opposite or whorled. . *g*.
g. Corolla white or yellow; capsule splitting vertically; stems mostly ascending or erect (creeping in *Lysimachia Nummularia*); ours perennial.
Stem leafy throughout, the leaves opposite or whorled; corolla yellow, 5- or 6-parted; calyx 5- or 6-parted. . . . . . . 4. *Lysimachia*.
Stem with scale-like alternate leaves below and whorled thin leaves at summit, with slender-peduncled white mostly 7-parted flowers from their axils. . . . . 5. *Trientalis*.
g. Corolla scarlet or blue (rarely white); capsule circumscissile, the top falling off as a lid; prostrate branching annual. . . . . 6. *Anagallis*.
f. Corolla shorter than calyx, 4–5-cleft, often persisting as a withered cap over the circumscissile capsule; leaves alternate; small annual. . 7. *Centunculus*.
e. Corolla wanting; calyx campanulate, petaloid, pink or pink and white, persistent around the pointed and dehiscent capsule; succulent opposite-leaved plants with nearly sessile axillary flowers. . . . . . 8. *Glaux*.
b. Ovary adnate at base to base of calyx; slender alternate-leaved, with lax racemes of white campanulate flowers. Tribe Samoleae. . . . . . . 9. *Samolus*.
a. Leaves pinnately dissected, scattered below, crowded above at base of peduncles; flowers whorled, in a raceme. Tribe Hottonieae. . . . . 10. *Hottonia*.

Tribe Primùleae Endl.

1. PRÍMULA L. Primrose. Cowslip. Primevère (Que.)

Calyx tubular, angled, 5-cleft. Corolla enlarging above the insertion of the stamens; the 5 lobes often notched or obcordate. Capsule many-seeded, splitting at the top into 5 valves or 10 teeth. — Low mostly boreal or alpine perennial herbs (only one or two in s. S. Am.), with single or tufted and simple scapes, the flowers in an umbel. (Name a diminutive of *primus*, from the flowering of the true primrose in *early spring*.)*

Leaves mostly dentate or crenate, often farinose beneath; limb of corolla 0.8–2 cm. broad, lilac, pink or white (rarely bluish); lobes 3–7 mm. broad, emarginate or shallowly obcordate; mature capsules thick-cylindric to ellipsoid, 2–5 mm. in diameter; seeds dark brown or fulvous.
Comparatively stout; scape 0.6–3 mm. thick at summit; involucral bracts usually saccate or gibbous at base, 5–14 mm. long; calyx urceolate-campanulate, in maturity 5.5–11 mm. long, 3–6 mm. thick; stigma and anthers not exserted from throat of corolla; capsule ellipsoid, 2.5–5 mm. in diameter, its valves splitting into linear halves 1.5–2 mm. wide. . . . . . . 1. *P. laurentiana*.
Comparatively slender; scape 0.3–1.8 mm. thick at summit; involucral bracts usually not saccate at base, 2–6 mm. long; calyx more slender, in maturity 3–6 mm. long, 2–3.5 mm. thick; stigma or tops of anthers exserted from throat of shrivelled corolla; capsule cylindric, 2–3 mm. in diameter, its valves splitting into linear halves 0.5–1 mm. wide.
Leaves usually yellow-farinose beneath; calyx commonly farinose; seeds angulate and truncated, strongly rugose. . . . . 2. *P. intercedens*.
Leaves green, rarely farinose beneath; calyx usually efarinose; seeds rounded, smooth or nearly so. . . . . 3. *P. mistassinica*.
Leaves entire or undulate, green both sides, slender-petioled; limb of corolla 5–9 mm. broad, white or violet; lobes 1.6–4 mm. broad, deeply cleft; mature capsules slenderly cylindric-attenuate, 1.8–2.1 mm. thick; seeds pale brown or yellowish, smooth. . . . . 4. *P. egaliksensis*.

1. **P. laurentiàna** Fern. (of the Gulf and River St. Lawrence), Bird's-eye-P. — *Leaves white- or yellowish-farinose beneath*, or efarinose in forma **chlorophýlla** Fern. (green-leaved), oblanceolate, spatulate or narrowly rhombic, *dentate*, up to 13 cm. long and 3 cm. broad; *scape* 1–4.5 dm. high, stoutish, 0.6–3 *mm. thick at summit*, often farinose above; *involucral bracts* lance-subulate or involute, *usually saccate or gibbous at base*, 5–14 *mm. long;* pedicels up to 5 cm. long, rather stout; *calyx urceolate-campanulate, usually farinose, in maturity* 5.5–11 *mm. long* and 3–6 mm. thick, the lobes about equaling the tube; *corolla* lilac, its tube but slightly exserted; *the limb* 9–13 *mm. broad*, with broadly obcordate lobes 3.5–5.5 mm. wide; style and anthers included within the yellow throat; *capsule ellipsoid*, barely exceeding to twice length of calyx, 2.5–5 *mm. in diameter*, its linear half-valves 1.5–2 *mm. broad;* seeds angulate, conspicuously reticulate;

* Treatment derived from that in Rhodora, xxx. 59–77 and 85–104 (1928).

fresh flowers with fragrance of jonquils; roots musky. (*P. farinosa* of ed. 7, not L.; *P. farinosa*, var. *macropoda* Fern.) — Ledges, cliffs and meadows, chiefly calcareous, s. Lab., Nfld. and e. Que. to N.S. and e. and n.-centr. Me. June–Aug.

2. P. intercèdens Fern. (standing between). — More slender; *leaves yellow-farinose beneath*, firm, up to 7 cm. long, cuneate, scarcely petioled; *scape* up to 2.5 dm. high, 0.5–1.8 *mm. thick at* the often farinose *summit; involucral bracts scarcely saccate at base*, 3–6 *mm. long;* pedicels stiff, lengthening to 0.6–3 cm.; *calyx* turbinate-campanulate, *commonly farinose, in fruit* 4–6 *mm. long; corolla* lilac, its tube slightly exserted, *its limb* 1–1.5 *cm. broad*, the obcordate *lobes* 4–7 *mm. broad; stigma or tips of anthers slightly exserted* at least from shriveled corollas; *capsule cylindric*, one-and-a-half times length of calyx, 2–3 *mm. thick, its* linear *half-valves* 0.5–1 *mm. broad; seeds angulate and truncated, strongly rugose.* (*P. farinosa*, var. *americana* Torr.) — Calcareous shores, Ont., n. Mich. and n. Minn. June, July. — Apparently crosses with the next.

3. P. mistassínica Michx. (of Lake Mistassini), Bird's-eye-P. — *Very slender, efarinose* or slightly farinose; *leaves dentate*, oblanceolate to cuneate-obovate, *sessile or short-petioled*, up to 7 cm. long and 1.6 cm. broad; *scape filiform*, 0.3–1.4 *mm. thick* at summit, 0.3–2.1 dm. high; *involucral bracts* usually *non-saccate at base*, 2–6 *mm. long;* pedicels filiform, becoming 0.5–3.5 cm. long; *calyx slenderly campanulate*, 3–6 mm. long, 2–3.5 *mm. thick; corolla-tube* yellow, *exserted;* the pale pink, lilac or bluish-purple, or white in forma **leucántha** Fern. (white-flowered), *limb* 0.8–2 *cm. broad, with conspicuous yellow eye*, with broad cuneate-obcordate lobes; *top of stigma or tips of anthers exserted* in old flowers; *capsule subcylindric*, once-and-a-half to twice length of calyx, 2–3 *mm. thick, the half-valves* 0.5–0.9 *mm. wide; seeds rounded, nearly smooth.* — Calcareous or argillaceous rocks, shores and meadows, s. Lab. to e. Alaska, s. to Nfld., N.S., centr. Me., n. Vt., n. Mich., Wisc., centr. Ia., s. Alta. and s. B.C. May–Aug.

Var. **noveboracénsis** Fern. (of New York). — Leaves obovate or broadly oblanceolate; *limb of corolla only* 0.8–1 *cm. broad, without conspicuous yellow eye.* — Similar habitats, nw. and centr. N.Y., s. Ont., s. Mich. and n. Ill.

4. P. egaliksénsis Wormsk. (of Egalik Fjord, Greenl.), Greenland P. — *Leaves thin, efarinose, ovate, oblong, obovate or spatulate, entire* or obscurely undulate, *slender-petioled*, up to 5 cm. long and 1.5 cm. broad; *scapes strict*, efarinose, pale-green, 0.1–2.4 dm. high; *involucral bracts dilated and gibbous-saccate at base*, 2.5–7 mm. long; *pedicels strict*, in fruit *becoming very unequal*, the longest 0.5–5 cm. long; *calyx* pale green, cylindric to slenderly campanulate, 3.5–6 mm. long, *cleft only one-third or one-fourth to base into glandular-ciliate lobes;* corolla with white limb 5–9 mm. broad, or limb violet or deep lilac and whole plant darker or suffused with purple in forma **violàcea** Fern. (violet), *the short cuneate lobes only* 1.6–4 *mm. broad and deeply cleft; capsule slender-cylindric, tapering above, becoming twice or thrice the length of calyx*, 7–13 *mm. long*, 1.8–2.1 *mm. thick; seeds pale, smooth.* — Meadows and wet calcareous shores, s. Greenl. and Lab. to Alaska, s. to n. Nfld., e. Saguenay Co., Que., James Bay, Alta. and B.C. Late June–Aug.

P. vèris L. (of spring) or *P. officinalis* Hill, the European Cowslip, with large yellow or purple corollas, is occasionally and locally estab. (Esc. from cult.)

2. ANDRÓSACE L.

Calyx 5-cleft; tube short. Corolla salverform or funnelform; the tube shorter than the calyx; limb 5-parted. Capsule 5-valved. — Small herbs of N. Hemisph., with clustered radical leaves and very small solitary or umbelled flowers. (A name used by Pliny for some unidentified plant.)

Involucral bracts elliptic to narrowly ovate or lanceolate. . . . . 1. *A. occidentalis.*
Involucral bracts linear. . . . . 2. *A. septentrionalis.*

1. A. occidentàlis Pursh (western). — Dwarf annual; *rosette-leaves* linear to oblong or elliptic-lanceolate, 0.5–2 cm. long, 1–5 mm. broad, stiffly *white-pubescent above with mostly simple hairs; scapes* 1–∞, 1–7 *cm. high*, scabrous with stellate pubescence; *involucral bracts elliptic to narrowly ovate or lanceolate; pedicels* erect or ascending, *stellate-pubescent*, 0.5–3 cm. long; calyx-tube pale, campanulate, the bright green but pubescent lobes deltoid to lance-subulate. — Dry sands and gravels or rocky woods, w. Ont. to B.C., s. to Ind., Ill., Ark., Tex., N.M. and Ariz. Late March–May.

2. A. septentrionàlis L. (northern). — Similar; *rosette-leaves* linear-lanceolate to oblanceolate, *often with stellate pubescence*, entire or jagged-toothed, 1–4 cm. long; *scapes* 1–several, 0.3–3 *dm. high*, often unequal, erect or arched-ascending; *involucral bracts linear-subulate; pedicels* ascending or the lateral spreading and arched, *soon glabrate*, 1–5.5 cm. long; calyx turbinate-campanulate, stramineous to green, with short deltoid lobes. — Arct. reg., s., locally, to gravels and cliffs (chiefly calcareous) of w. Nfld., n. Gaspé Co., Que., James Bay, and Rocky Mts. to N.M. June–Aug. (Eurasia)

3. DODECÁTHEON L. American Cowslip

Calyx deeply 5-cleft, the divisions lanceolate. Corolla with a very short tube and thickened throat; the divisions long and narrow. Filaments short, monadelphous at base; anthers long and linear, approximate in a slender cone. — Perennial smooth herbs with fibrous roots, a cluster of basal leaves and a simple naked scape involucrate with small bracts at the summit and bearing an umbel of showy flowers, these nodding on slender pedicels. Corolla lilac to pale pink or violet, or sometimes white. N. Am. genus, mostly western. (Name from the Greek *dodeca, twelve,* and *theos, god,* given by Pliny to the primrose, which was believed to be under the care of the twelve superior gods.)*

1. D. Meàdia L. (a generic name given by Catesby in honor of Dr. Richard Mead, 1673–1754, English physician and patron of Catesby), Shooting-star. — Scape 1.5–5 dm. high; *bases of leaves marked with red,* especially when fresh, or without red in the rare forma **sèdens** Fassett (sitting), narrowly elliptic-oblong to oblanceolate, tapering to petiole, apex blunt or rounded; *umbel* 4–125-*flowered;* the young pedicels erect, recurving in anthesis, ascending in fruit; *calyx-lobes at expansion of flower* 3–7 (*mostly* 4–5) *and in fruit* 4–9 (*mostly* 5.5–7) *mm. long; fresh corollas lilac to pale pink,* or white in forma **álbum** Macbr. (white); tube of filaments 1–2 mm. long; *anthers* 6.5–10 (*mostly* 7–8) *mm. long; capsule dark reddish-brown, ovoid to barrel-shaped,* twice to thrice as long as thick, 10.5–18 mm. long, *with firm ligneous wall* 130–325μ *thick.* — Open woods, meadows, moist slopes and prairies, D.C. to w. Pa., thence nw. to w. Wisc., s. on upland to n. Ga., n. Ala., Ark., nw. La. and e. Tex.; forma *sedens* from Tenn. and s. Ill. to Tex. April–June.

Var. **brachycárpum** (Small) Fassett (short-fruited). — Leaf-bases marked with red, or red wanting in forma **pállidum** Fassett (pale); flowers few(1–14); calyx-lobes at anthesis 2.5–5 (mostly 3.5–4), in fruit 4–4.5 mm. long; anthers 4–7 (mostly 5.5–6.5) mm. long; capsules 7.5–10 mm. long. — Sw. Va. to w. Mo., s. to n. Ala., w. Ark. and Tex.

Var. **Frénchii** Vasey (for its discoverer in 1871, George Hazen French). — Leaves ovate or elliptic, cordate or abruptly contracted to petiole; otherwise like the typical var., into which it clearly passes in favorable conditions. — Under overhanging cliffs, very local, sw. Wisc. and s. Ill.

2. D. amethýstinum Fassett (amethyst-colored). — More slender than most no. 1, 1–3 dm. high; *bases of leaves rarely marked with red; umbel* 1–10 (*rarely* –18)-*flowered; calyx-lobes of expanding flowers* 2–5 (*mostly* 3), *in fruit* 3–6 (*mostly* 4–5) *mm. long; fresh corolla-lobes deep red-purple,* or white in forma **margaritàceum** Fassett (pearly), reflexed, or in forma **Strícklerae** (Fern.) Fassett (named in 1937 for its discoverer Alice Strickler) directed forward; *anthers* 5–7.5 *mm. long; capsule cylindric,* mostly more than thrice as long as thick, *pale brown to yellowish, with thin papery walls* 35–120 μ *thick.* — Damp or shaded calcareous cliffs and banks along streams, lower Susquehanna and Schuylkill Rivers, e. Pa.; ?ne. W.Va.; Ohio R., Ky.; Miss. R., sw. Wisc. and adj. Ill., Ia. and Minn. ("driftless area") and ne. Mo. Late April–early June.

4. LYSIMÁCHIA L. Loosestrife

Calyx 5–6-parted, imbricate or valvate in bud. Corolla 5–6-parted, rotate, convolute in bud, or in § *Steironema* each division separately convolute or involute about its stamen. Filaments distinct or nearly so on a ring at base of corolla, or monadelphous at base; anthers ovoid to slender. Capsule few–many-seeded. — Leafy-stemmed perennials of temp. reg., with entire opposite or whorled leaves and (at least with us) yellow or orange sometimes purple-dotted corollas. (In honor of King *Lysimachus* of Thrace, or from *lysis, a release from,* and *mache, strife;* tradition telling of *Lysimachus,* chased by a maddened bull, in desperation seizing a plant of *Loosestrife* and pacifying the bull by waving the plant before him.)

a. Divisions of corolla entire (or glandular-ciliate only in no. 2), neither cuspidate nor rolled about their stamens; filaments commonly monadelphous at base, without staminodia; anthers oblong or oval; flowers on ascending or divergent peduncles or pedicels; leaves often punctate. . *b.*

b. Flowers borne singly from the axils of leaves or bracts, if racemose the racemes terminal or terminating leafy branches; corolla with lanceolate to

* Treatment largely copied from that of N. C. Fassett in Am. Midl. Nat. xxxi. 455–486 (1944).

ovate divisions, the sinuses without teeth; stamens monadelphous at base, often unequal. . . *c.*

c. Stems erect or ascending, with linear to ovate elongate leaves; flowers in terminal panicles or racemes or long-pedicelled from the axils; corollas 1–2 cm. broad. § LYSIMASTRUM. . . *d.*

d. Flowers in leafy panicles or whorled in the upper axils; corolla plain yellow, without dark markings.

Calyx 4–5 mm. long, its segments dark-margined; corolla-lobes entire. . . . . 1. *L. vulgaris.*

Calyx 7–10 mm. long, green throughout; corolla-lobes glandular-ciliolate. . . . . 2. *L. punctata.*

d. Flowers in racemes or borne in the leaf-axils; corolla dark-dotted or streaked.

Flowers all or at least the lower from the axils of regular foliage-leaves.

Stem loosely hairy or smooth; leaves whorled (rarely opposite or alternate); flowers all in axils of foliage-leaves. . . . . . 3. *L. quadrifolia.*

Stem glabrous; leaves opposite or in whorls of 3–5; flowers in terminal racemes, the lowest from axils of upper foliage-leaves. . 4. ×*L. producta.*

Flowers in terminal bracted racemes; stem smooth; leaves opposite (rarely alternate). . . . . 5. *L. terrestris.*

c. Stems creeping; leaves roundish; flowers 2–3 cm. broad, solitary in the axils of ordinary leaves; corolla with dots, its divisions broadly ovate. § NUMMULARIA. . . . . 6. *L. Nummularia.*

b. Flowers in short spiciform racemes terminating peduncles from the lower axils; corolla deeply cleft into linear often purple-dotted divisions, with or without a small tooth in each sinus; filaments distinct, equal. § NAUMBERGIA. . . . . 7. *L. thyrsiflora.*

a. Divisions of corolla erose-denticulate above, cuspidate-tipped, each separately involute or convolute around its stamen; filaments distinct or nearly so, with 5 slender staminodia alternating with the fertile stamens; anthers linear, becoming arcuate; flowers nodding; leaves not punctate. § STEIRONEMA. . . *e.*

e. Leaves membranaceous or submembranaceous, linear to ovate, at least the lower petioled or, if sessile (no. 11), ciliate at base, the pinnate veins evident. . . *f.*

f. Principal leaves ovate or broadly lanceolate, with rounded to subcordate bases. . . *g.*

g. Petioles smooth or smoothish, or strongly ciliate only at base; so. species.

Stems or flowering branches reclining or creeping, often rooting at the nodes; flowers 8–12 mm. broad; calyx-segments 3–5 mm. long, exceeded by capsule. . . . . 8. *L. radicans.*

Stems and branches ascending; flowers 1.3–2.5 cm. broad; calyx-segments 5–8 mm. long, much exceeding capsule. . . . . . . 9. *L. tonsa.*

g. Petioles long-ciliate to summit; corolla 1.5–2.8 cm. broad; capsule and calyx subequal; wide-ranging continental species. . . . . . 10. *L. ciliata.*

f. Principal leaves linear to narrowly lanceolate or oblong, gradually tapering at base.

Flowering stems slender, arising from slender cord-like stolons and rhizomes; middle and upper leaves sessile or subsessile, pale beneath, bristly-ciliate at base; calyx-segments nearly nerveless. . . . . 11. *L. lanceolata.*

Flowering stems coarser, without basal stolons; middle and upper leaves petioled, green both sides, the blades rarely ciliate; calyx-segments nerved. . . . . 12. *L. hybrida.*

e. Leaves firm, linear, sessile or nearly so, the veins obscure, the blades without ciliation. . . . . 13. *L. quadriflora.*

§ LYSIMÁSTRUM Duby

1. L. VULGÀRIS L. (common), GARDEN-L. — Coarse, up to 1.2 m. high, softly viscid-pilose, paniculate-branching above; leaves opposite or whorled, lanceolate to lance-ovate or elliptic, 8–13 cm. long, mostly 1.5–4 cm. broad; the middle and upper short-petioled, acuminate; *flowers* 1.5–2.5 cm. broad, *in leafy panicles; calyx-segments* 3.5–5 *mm. long, dark-margined; corolla plain yellow,* without dark markings, its lobes entire; glandular filaments united to near middle. — Roadsides, swampy thickets and flats, Que. and Ont., s. to N.S., N.E., Md., O. and Ill. June–Sept. (Introd. and natzd. from Eu.)

2. L. PUNCTÀTA L. (dotted), GARDEN-L. — Similar; leaves ovate, bluntish; flowers merely whorled in upper axils or on short branches; *calyx-segments* 5.5–9 *mm. long, green throughout;*

corolla-lobes glandular-ciliolate. — Roadsides and waste ground, Nfld. to Ill., s. to N.S., N.E., N.Y., N.J. and Pa. June–Sept. (Introd. and natzd. from Eu.)

3. L. quadrifòlia L. (four-leaved), Whorled L. — Loosely hairy or smooth; stem simple, 2–8 dm. high; *leaves whorled* in 4's or 5's (sometimes in 2's, 3's, 6's or 7's or even alternate), lanceolate to lance-ovate, sessile or subsessile, 3–9 cm. long, 1–2.5 cm. broad; *flowers on long capillary peduncles in axils of unreduced foliage-leaves;* calyx-segments 2.5–5 mm. long; *corolla* 1–1.5 cm. broad, *dark-dotted or streaked,* its lobes ovate-oblong; filaments unequal, conspicuously monadelphous. — Dry or moist open woods, thickets and shores, Me. to s. Ont. and Wisc., s. to s. N.E., L.I., Ga., Ala., Tenn. and Ill. May–Aug. — Hybridizes with no. 5, producing no 4.

4. × L. prodúcta (Gray) Fern. (elongated). — Stem smooth, 0.5–1 m. high, simple or slightly branched; leaves opposite or in whorls of 3–5, lanceolate to ovate-lanceolate; *flowers in terminal bracted racemes* (0.5–5 dm. long), *the lower from the axils of the upper foliage-leaves;* calyx-segments 3–5 mm. long; corolla as in no. 3, 1–2 cm. broad, the lobes ovate-oblong to oblong-lanceolate. (*L. foliosa* Small) — Damp thickets and shores, Me. and sw. Que. to Mich., s. to s. N.E., S.C. and O. — A widely distributed and often abundant hybrid of nos. 3 and 5.

5. L. terréstris (L.) BSP. (terrestrial; Linnaeus thinking the form with elongate bulblets a terrestrial mistletoe), Yellow or Swamp-L., Swamp-candles. — *Smooth;* stems 0.2–1 m. high, simple or loosely branched, very leafy, from stoloniferous bases; lowest *leaves* small, middle and upper *opposite,* rarely alternate, narrowly to broadly *lanceolate, acute or acutish,* 3.5–10 cm. long, 0.4–1.6 cm. broad; *racemes conspicuously bracted, terminating the main axis* (often wanting in a sterile state with moniliform axillary bulblets), 0.5–2.5 dm. long, usually overtopping the branches (when developed); flowers slender-stalked, divergent or ascending; calyx-segments 2–3.5 (–5) mm. long; corolla dotted or streaked, its lobes lance-oblong. — Low grounds and wet shores, Nfld. to James Bay and Minn., s. to N.S., N.E., Ga., Ky. and Ia. June–Aug. — Hybridizes with no. 3, producing no. 4; also with no. 7.

Var. **ovàta** (Rand & Redf.) Fern. (egg-shaped). — Leaves elliptic-ovate, blunt, mostly 1.5–4 cm. long; branches wholly or partly overtopping the raceme. — Swamps, local, Mt. Desert I., Me.; Martha's Vineyard I., Mass.

A frequent hybrid of nos. 5 and 7, combining in various degrees their distinctive characters, occurs from Que. and M.I. to N.E. and N.Y.

§ Nummulària (Gilib.) Endl.

6. L. Nummulària L. (old generic name, coin-like, from the round leaves), Moneywort, Monnayère (Que.). — Smooth; stems *creeping* and trailing; *leaves* opposite, *roundish,* 1.5–3 cm. long; *flowers 2–3 cm. broad, solitary in the axils;* calyx-segments ovate, 5–9 mm. long; *corolla* dotted (at least when dry), its *divisions broadly ovate;* filaments slightly monadelphous. — Damp roadsides, grasslands and shores, Nfld. to Ont., s. to N.S., N.E., Ga., Mo. and Kans. June–Aug. (Introd. and natzd. from Eu.)

§ Naumbérgia (Moench) Duby (*Naumbergia* Moench, as a genus)

7. L. thyrsiflòra L. (with flowers in a thyrse), Tufted L. — *Stems* from creeping rhizomes, smooth (or loosely scurfy above when young), 2–8 dm. high, *simple;* all but lower (reduced) *leaves lanceolate to elliptic, acuminate, attenuate to base,* 0.5–1.5 dm. long, 1–5.5 cm. wide, pale beneath, *the axils of* 1–4 *lower or middle pairs bearing short-peduncled spike-like racemes* of light yellow dotted flowers; calyx-segments 1.5–3.5 mm. long; *corolla very deeply* 5(–7)*-parted into linear divisions* 3–5 mm. long, with or without a small tooth in each sinus; *filaments distinct, equal.* (*Naumbergia* Reichenb.) — Cold swamps, springy marshes and bogs, Saguenay Co., Que., to James Bay, nw. to Alaska, s. to N.S., N.E., n. N.J., Pa., W.Va., O., Ind., n. Ill., Mo., Colo. and Calif. May–July. (Eurasia)

§ Steironèma (Raf.) Gray (*Steironema* Raf., as a genus)

8. L. radìcans Hook. (rooting). — *Stem* slender, *soon reclined,* the elongated branches often rooting at the nodes; *leaves* membranaceous, *lanceolate to lance-ovate,* the principal ones with blades 3–9 cm. long, rounded at base and *with slender naked or nearly naked petioles;* rameal leaves reduced; *flowers* nodding, 8–12 *mm. broad; calyx-segments* 3–5 *mm. long; corolla-segments erose-dentate, cuspidate; capsule exceeding calyx.* (*Steironema* Gray) — Wet woods, bottomlands and swamps, Miss. to Tex., n. to w. Tenn. and Mo.; local, e. Va. June–Aug.

9. L. tónsa Wood (shaved). — *Erect,* smooth, 2–7 dm. high, subsimple or with ascending branches; *leaves with glabrous petioles; the ovate to broadly lanceolate* thin acuminate *blades rounded at base,* 2.5–10 *cm. long;* peduncles whorled, in the upper axils or on the branches; *flowers* 1.3–2.5 *cm. broad; calyx-segments* 5–8 *mm. long;* corolla-segments erose at summit;

capsule much shorter than calyx. (*Steironema* Bickn.; *S. intermedium* Kearney, not Janka) — Dry open woods and bluffs, Va. to Ga. and Ark. June, July.

10. L. ciliàta L. (with marginal hairs). — Erect, simple or branched, 0.1–1.2 m. high, from a slender rhizome; *leaves ovate-lanceolate to ovate,* acuminate, *broadly rounded to subcordate at base,* the blades 3–15 cm. long, *all on long ciliate-fringed petioles;* flowers on filiform axillary peduncles, mostly whorled, 1.5–2.8 cm. broad; *calyx-segments* firm, subulate-tipped, 5–10 *mm. long;* corolla yellow, unspotted, its broad segments cuspidate and erose-dentate; capsule exceeding to shorter than mature calyx. (*Steironema* Raf.; *S. pumilum* Greene) — Low grounds, thickets, rich woods and shores, L. St. John, Que., to B.C., s. to N.S., N.E., Fla., Tex. and Colo. June–Aug.

11. L. lanceolàta Walt. (lance-shaped). — *Stems slender and firm, from elongate cord-like or filiform rhizomes and stolons,* 0.5–7 dm. high, simple or with ascending branches (the latter often abbreviated); *basal leaves often rosulate, oblong, elliptic or rounded; middle and upper leaves* linear to lanceolate or narrowly oblong, attenuate and *bristly-ciliate at base, sessile, paler beneath;* peduncles filiform, from upper axils; flowers 1.2–2 cm. broad; *calyx-segments firm,* lance-acuminate, 3.5–7 mm. long, *their nerves not evident;* corolla-segments suborbicular, erose, cuspidate; filaments equaling or longer than anthers; capsule shorter than calyx. (*L. heterophylla* Michx.; *Steironema heterophyllum* Raf.) — Dry or moist open woods, thickets, gravels and shores, Fla. to La., n. to Pa., O., s. Mich. and Wisc. June–Aug.

12. L. hýbrida Michx. (hybrid; name given through misconception). — *Stems stoutish* and soft *at base, without slender rhizomes and stolons,* 0.2–1.5 m. long, ascending or, when elongate, often decumbent or reclining, *the autumnal basal rosettes sessile or subsessile; cauline leaves* linear-lanceolate to oblong, *mostly petioled, green both sides, the petiole but rarely, the blade somewhat, ciliate;* flowers 1.5–2.5 cm. broad; *calyx-segments herbaceous,* lance-acuminate, 4.5–10 mm. long, 3-*nerved;* corolla-segments obovate, erose, cuspidate; filaments shorter than to about equaling anthers; capsules nearly equaling calyx. (*Steironema lanceolatum* of ed. 7, including var. *hybridum* (Michx.) Gray) — Swamps and wet shores, sw. Que. to w. Ont. and N.D., s. to N.E., Fla., Miss. and Tex. July, Aug.

13. L. quadriflòra Sims (four-flowered). — *Stems erect,* 4-angled, slender, 2–9 dm. high, from a thickish caudex, simple or branched; *cauline leaves sessile, narrowly linear, elongated,* 3–9 cm. long, smooth and shining, *rather rigid,* obtuse, the margins often a little revolute, *the veins obscure;* the lowest leaves oblong or spatulate; corolla longer than the calyx, the lobes conspicuously pointed; fruiting calyx 5–7 mm. long, the firm segments obscurely nerved. (*Steironema* Hitchc.) — Calcareous bogs, swales and shores, Ont. and Man., s. to w. N.Y., w. Va., Ky., Ill. and Mo. July, Aug.

5. TRIENTÀLIS L. Chickweed-wintergreen

Corolla spreading, flat, without tube. Filaments slender, united in a ring at the base; anthers oblong, revolute after flowering. Capsule few-seeded. — Low and smooth Eurasian and N. Am. perennials with simple erect stems, bearing a few alternate usually minute and scale-like leaves below, and a whorl of thin veiny leaves at the summit. Peduncles 1 or more, very slender, bearing a delicate white and star-shaped flower. (A Latin name, meaning the third part of a foot, alluding to the height of the plant.)

1. T. boreàlis Raf. (northern), Star-flower. — Stem 0.3–2.6 dm. high, spreading by slender elongate stolons (these rarely from the upper axils); leaves lanceolate, acuminate, somewhat unequal, in a single whorl, or in forma **ramòsa** Vict. (branching) with axillary leafy branches; petals finely pointed. (*T. americana* Pursh) — Woodlands and peaty slopes, ascending to subalpine reg., Lab. to Sask., s. to Nfld., N.S., N.E., Va., W.Va., O., Ind., Ill. and Minn. May–Aug. (subalpine).

6. ANAGÁLLIS L. Pimpernel

Corolla rotate, with almost no tube; the divisions broad. Stamens 5; filaments bearded. Capsule membranaceous, many-seeded. — Low spreading or procumbent herbs, nat. of Old World, mostly annuals with opposite or whorled entire leaves and solitary flowers on axillary peduncles. (The ancient Greek name, probably from *ana, again,* and *agallein, to delight in,* from its habit of closing the flowers in cloudy weather but opening them again in sunshine.)

1. A. arvénsis L. (of cultivated fields), Common or Scarlet P. — Leaves ovate, sessile, shorter than the peduncles; petals obovate, obtuse, fringed with minute teeth and stalked glands. — Waste sandy fields, June–Aug. — Flowers variable in size, scarlet or rarely white, or in forma caerùlea (Schreb.) Baumg. (sky-blue) petals blue and often nearly or quite destitute of ciliation, quickly closing at the approach of bad weather; whence the English popular name of "POOR MAN'S WEATHER-GLASS". (Natzd. from Eu.)

7. CENTÚNCULUS L. CHAFFWEED

Corolla rotate, with an urceolate short tube, usually withering on the summit of the capsule (which is like that of *Anagallis*). Stamens 4 or 5; filaments beardless. — Small annuals of temp. and trop. reg., with entire leaves and solitary inconspicuous flowers in their axils. (Latin name of some plant, diminutive of *cento*, patchwork.)

1. **C. mínimus** L. (smallest). — Stems ascending, 3–8 cm. long; leaves ovate, obovate, or spatulate-oblong; flowers nearly sessile, the parts mostly in fours. — Damp sand and mud, Fla. to Tex., Mex. and Calif., n. to Ga., and very locally to Del. and N. S. (Sable I.) and P. E. I.; in the interior n. to O., Ill., Minn., Sask., etc. June–Aug. (Semicosmop.)

8. GLAÙX L. SEA-MILKWORT

Calyx campanulate, 5-cleft, brightly colored; lobes ovate or oblong, petal-like. Corolla wanting. Stamens 5, on the base of the calyx, alternate with its lobes. Capsule 5-valved, few-seeded. — A low and leafy fleshy perennial of N. Hemisph., with opposite entire sessile leaves, and solitary nearly sessile (white, pink or lavender and crimson) flowers in their axils. (An ancient Greek name, from *glaucos, sea-green.*)

1. **G. marítima** L. (of the sea). — Diffusely branched (rarely simple); the branches prostrate, loosely ascending or sometimes erect, 0.3–3.3 dm. high; *leaves linear to narrowly oblong*, the larger 3–12 mm. long and 1.5–6 *mm. broad, bluntly pointed;* flowers 3–5 mm. long; *mature capsule* 2–3 *mm. long,* 2–2.5 *mm. broad.* — Saline or brackish shores, marshes and sands, Gaspé Pen., Que., to Va.; subsaline soil, Sask. to B.C., s. to N.M. and Oreg. June, July. (Eurasia)

Var. **obtusifòlia** Fern. (blunt-leaved). — *Erect*, 0.5–4 dm. high, simple or with few erect branches; *leaves oval or broadly oblong*, the principal ones 8–15 mm. long and 4–8 mm. broad, *with rounded tips; mature capsule* 2.5–4 *mm. broad.* — Nfld. to lower St. Lawrence, Que., s. to N.J.; James Bay; Pacific coast. (E. Asia)

Tribe SAMÒLEAE Endl.

9. SÁMOLUS L. WATER-PIMPERNEL. BROOKWEED

Calyx 5-cleft. Corolla somewhat campanulate, 5-cleft. True stamens 5, on the corolla-tube, included. Capsule globose, 5-valved at the summit, many-seeded.— Smooth semicosmop. herbs, with alternate entire leaves and small white flowers. (Ancient name of probable Celtic origin, said to refer to curative properties of this genus in diseases of cattle and swine.)

1. **S. parvìflòrus** Raf. (small-flowered). — Slender, 0.5–6 dm. high, often becoming diffuse; leaves obovate or spatulate, obtuse, the basal rosulate; racemes elongate, often panicled; pedicels 1–2 cm. long, spreading; capsule 2–3 mm. long. (*S. floribundus* HBK.) — Shallow water and wet soils, Fla. to Calif. and Mex., n. to P.E.I., N.B., s. Que., s. Ont., Mich., Ill. and B.C. May–Sept. (Trop. Am.)

Tribe HOTTONÌEAE Endl.

10. HOTTÒNIA L. FEATHERFOIL. WATER-VIOLET

Calyx 5-parted, the divisions linear. Corolla with a short tube, limb 5-parted. Stamens 5, included. Capsule many-seeded, 5-valved; the valves cohering at the base and summit. Seeds anatropous. — Short-lived Eurasian and N. Am. aquatic herbs with the erect hollow flower-stems almost leafless. Flowers white or whitish, whorled at the joints, forming an interrupted raceme. (Named for *Petrus Hotton*, 1648–1709, professor at Leiden.)

1. **H. inflàta** Ell. (inflated). — Leaves dissected into thread-like divisions, scattered on the floating and rooting stems and crowded at the base of the cluster of peduncles, which are strongly inflated between the joints; pedicels short. — Pools and ditches, local, Fla. to Tex., n. to N.E., N.Y., O., Ind., s. Ill. and Mo. April–June. — Winter-annual, germinating and growing in autumn and winter, reaching the surface and flowering in spring.

FAM. 136. PLUMBAGINÀCEAE (LEADWORT FAMILY)

Herbs with regular 5-merous flowers, a plaited calyx, the 5 stamens opposite the separate petals or the lobes of the corolla, and the free ovary 1-locular, with a solitary ovule hanging from a long cord which rises from the base of the locule. — Represented in our flora by two genera:

Flowers in tiny spikelets borne in 1-sided spikes, these in panicles or corymbs; leaves with dilated blades. 1. *Limonium.*
Flowers in capitate glomerules; leaves slender, without differentiation of blade and petiole. . . . . 2. *Armeria.*

1. LIMÒNIUM Mill. SEA-LAVENDER. MARSH-ROSEMARY

Flowers scattered or loosely spiked on 1-sided branches, 2–3-bracted. Calyx funnelform, dry and membranaceous, persistent. Corolla of 5 nearly or quite distinct petals, with long claws, the 5 stamens attached to their bases. Styles 5, rarely 3, separate. Fruit membranous and indehiscent, within the persistent calyx. Embryo straight, in mealy albumen. — Seaside perennials with thick and petioled radical leaves; the naked flowering stems or scapes branched into panicles. (*Leimonion*, the ancient Greek name, presumably from *leimon, a marsh.*) STATICE Willd., not L.

1. L. caroliniànum (Walt.) Britt. (of Carolina). — Caudex thick and woody, very astringent; leaves oblong, spatulate or obovate-lanceolate, 1-ribbed, tipped with a deciduous bristly point, petioled; scape much branched, panicled, 1.5–6 dm. high; spikelets 1- or 2-flowered; flowers lavender; *calyx-tube quite glabrous; the lobes* 0.5–1 *mm. long*, triangular to oval, with teeth in the sinuses. (Incl. var. *angustatum*, largely, *sensu* Blake, not *L. angustatum* (Gray) Small as to type, the latter a local species of the Florida Keys with linear-lanceolate leaves ending in cusps 2 mm. long.) — Salt-marshes, Fla. to Miss., n. to s. N.Y. and rarely to se. N.H. where (except for strictly glabrous calyx) seeming to pass into the next. July–Oct.

2. L. Náshii Small (for GEORGE VALENTINE NASH, 1864–1921). — Stouter, often with more compact inflorescence; *calyx-tube pubescent at base and often part way up* 1 *or more angles, the lobes* 1–1.7 *mm. long.* (Incl. var. *trichogonum* Blake; *L. trichogonum* Blake) — Salt marshes, Fla. to Tex., n. to Nfld. and Gulf and R. St. Lawrence, Que. July–Oct. — N.H. specimens with strictly glabrous calyx-tube are embarrassingly transitional to the preceding. — Plants with white flowers have been called forma **albiflòrum** (Raf.) House (white-flowered). Species needs restudy.

2. ARMÈRIA Willd. THRIFT

Flowers closely glomerulate-capitate, mixed with scarious bracts, the outer of which form the involucre, the 2 outermost of them prolonged below their insertion into a sleeve-like sheath about the summit of the scape. Calyx dry and scarious, in ours pale and lustrous. Petals barely united at base, usually pink. Styles pubescent at base. — Densely tufted, with marcescent leaves without distinction of blade and petiole, these in ours crowded and narrowly linear or linear-subterete. Genus most abundant in the cooler half of the N. Hemisph. and the Andes. (Name said to be of Celtic origin.)

1. A. labradórica Wallr. (of Labrador). — Dense tussocks with grayish-green thickish linear leaves 1–3-nerved, 3.5–8 cm. long and 0.7–2 mm. broad; *scapes* 1–several, erect, 0.1–3 dm. high, *pubescent;* glomerule hemispherical, 1.5–2.5 cm. in diameter; outer bracts lance-ovate, acuminate, brownish, the inner broadly elliptic; the fruiting (innermost) bracts subtending the 2- or 3-flowered spikelets, cuneate-suborbicular or obovate; *calyx* 6–7 mm. long, with funnelform limb, the *lobes tipped by a cusp* 0.4–0.5 *mm. long*, the obconic tube densely pilose on the 10 ribs and on the connecting cross-ribs; petals pink to purple. — Greenl. and Baffin I., s. to calcareous barrens along Straits of Belle Isle, Nfld., and shore of James Bay. July, Aug.

Var. **submùtica** (Blake) H. F. Lewis (nearly awnless), FOXFLOWER. — *Calyx-lobes merely acuminate or with minute cusp up to* 0.2 *mm. long; scapes glabrous.* — Greenl. and Baffin I., s. to limestone- and serpentine-barrens of Nfld. and of Mt. Albert, Gaspé Pen., Que. July, early Aug.

FAM. 137. SAPOTÀCEAE (SAPODILLA FAMILY)

Trees or shrubs, mostly with a milky juice, simple and entire alternate leaves (often rusty-downy beneath), small and perfect regular flowers usually in axillary clusters; the calyx free and persistent; the fertile stamens commonly as many as the lobes of the hypogynous short corolla and opposite them, inserted on its tube along with one or more rows of appendages and scales (or sterile stamens); anthers turned outward; ovary 4–12-locular, with a single anatropous ovule in each locule; seeds large. Albumen mostly none; but the large embryo with thickened cotyledons. Style single, pointed. — A small mostly trop. family.

1. BUMÈLIA Sw. IRONWOOD

Calyx 5-parted. Corolla 5-cleft, with a pair of internal appendages at each sinus. Fertile stamens 5; anthers sagittate. Sterile stamens 5, petal-like, alternate with the lobes of the corolla. Ovary 5-locular. Fruit small, resembling a cherry, black, containing a large ovoid and erect seed with a roundish scar at its base. — Flowers small, white, in fascicles from the axils of the leaves. Branches sometimes spiny. Leaves often fascicled on short spurs. Wood very hard. Trees and shrubs of warm temp. and trop. Am. (The ancient Greek name of Ash.)

Branchlets and lower surfaces of leaves woolly. . . . . 1. *B. lanuginosa*, var. *oblongifolia*.

Branchlets glabrous or essentially so; leaves sericeous-puberulent to glabrous beneath.

Mature leaves 6–15 cm. long; fruit obovoid or ellipsoid. . . . . 2. *B. lycioides*.

Mature leaves 2–6 cm. long; fruit subglobose. . . . . . . 3. *B. Smallii*.

1. B. lanuginòsa (Michx.) Pers. (woolly), var. **oblongifòlia** (Nutt.) R. B. Clark (oblong-leaved), FALSE BUCKTHORN. — Spiny, 3–18 m. high; *leaves oblong-obovate or cuneate-obovate, rusty-woolly beneath*, obtuse, 2.5–9 cm. long; *clusters* 6–12-*flowered, pubescent; fruit globular.* — Sandy woods, La. and e. Tex., n. to s. Mo. and s. Kans. July.

2. B. lycioìdes (L.) Gaertn. f. (resembling *Lycium*), SOUTHERN BUCKTHORN. — Spiny, 3–9 m. high; *leaves of fertile branches elliptic-oblong to narrowly obovate, tapering to a blunt but subacuminate apex, silky*, becoming *glabrate, in maturity* 6–15 *cm. long* and 2–3.5 *cm. broad; flower-clusters densely many-flowered, glabrous; fruit ovoid.* — Rich woods and river-banks, Fla. to La., n. to s. Va., Ky. and s. Ill. June, July.

Var. **virginiàna** Fern. (Virginian). — *Leaves of fertile branches oblanceolate*, 1.3–2.8 *cm. broad, strongly rounded at summit.* — Rich woods and river-shores, se. Va. to Ga.; s. Ind. Late June, July.

3. B. Smállii R. B. Clark (for JOHN KUNKEL SMALL, 1869–1938). — Smaller than no. 2; the coriaceous and *strongly reticulate leaves becoming* 2–6 *cm. long;* the *fruit subglobose.* — Low woods, La. and e. Tex., n. to se. Mo.

FAM. 138. EBENÀCEAE (EBONY FAMILY)

Trees or shrubs with alternate entire leaves and polygamous regular flowers which have a calyx free from the 3–12-*locular ovary; the stamens* two to four *times as many as the lobes of the corolla, often in pairs before them, their anthers turned inward, and the fruit a several-locular berry; ovules* 1 *or* 2, *suspended from the summit of each locule.* Seeds anatropous, mostly single in each locule, large and flat, with a smooth coriaceous integument; the embryo shorter than the hard albumen, with a long radicle and flat cotyledons. Styles wholly or partly separate. Wood hard and dark-colored. No milky juice. — A small family, chiefly trop.

1. DIOSPỲROS L. PERSIMMON

Calyx 4–6-lobed. Corolla 4–6-lobed, convolute in the bud. Stamens commonly 16 in the staminate flowers and 8 in the pistillate, in the latter imperfect. Berry large, globular, surrounded at base by the thickish calyx, 4–8-locular, 4–8-seeded. — Flowers dioeciously polygamous, the pistillate axillary and solitary, the staminate smaller and often clustered. Trees of trop. and warm reg. (Name from the Greek *Dios, of Jove*, and *pyros, grain.*)

1. D. virginiàna L. (Virginian), COMMON P. — Leaves thickish, ovate-oblong, smooth or nearly so; peduncles very short; calyx 4-parted; corolla pale yellow, thickish, between campanulate and urceolate, 1–1.5 cm. long in the pistillate flowers, much smaller in the staminate; styles 4, two-lobed at the apex; ovary 8-locular. — Dry woods, old fields and clearings, Fla. to Tex., n. to s. N.E. (local), se. N.Y., Pa., W.Va., O., Ind., Ill., s. Ia. and e. Kans. May, June. — Tree, 6–30 m. high, with very hard blackish wood; the plum-like fruit 2–4 cm. in diameter, exceedingly astringent when green but yellow to reddish, sweet and edible when ripe or on exposure to frost.

Var. **pubéscens** (Pursh) Dippel (hairy). — Branchlets villous or densely tomentose; leaves pubescent beneath. — Fla. to Ark., n. to Va., s. Ill. and s. Ia. — An early-ripening extreme with depressed fruits up to 7.5 cm. across, occurring in Mo., Ark., Kans. and Okla., has been designated var. **platycárpa** Sarg. (broad-fruited).

FAM. 139. SYMPLOCÀCEAE (Sweetleaf Family)

Shrubs or trees with alternate simple exstipulate leaves and perfect regular flowers; the calyx adherent to lower part of the 3-locular ovary, its 5 lobes valvate or imbricate; the corolla of 3–11 (5 in ours) petals united merely at base; the stamens in several series, free or united at base, the anthers short and innate; style 1; fruit drupe-like or baccate, 1–5-locular, the locules 1-seeded. Seeds anatropous, with short cotyledons. Corolla with a cluster of stamens at the base of each petal. Ovules 2–4, pendulous. Pubescence simple. — A small family of warm parts of Am., Asia and Australia.

1. SÝMPLOCOS Jacq. Sweetleaf

Characters of the family. — Flowers in ours in short axillary sessile clusters or racemes, yellow. Large genus of most (except Afr.) trop. and warm reg. (Name from the Greek *symplocos, connected;* from the union of the stamens.)

1. S. tinctòria (L.) L'Hér. (used as a dye), Horse-sugar or Yellowwood. — Large shrub or small tree up to 6 m. high; leaves elongated-oblong, acute, obscurely toothed, thickish, almost evergreen, minutely pubescent and pale beneath, 7–15 cm. long and 3–6 cm. broad; flowers 6–14, in close and bracted clusters, fragrant; drupes cylindric-ellipsoid, about 1 cm. long. — Woods, swamps and bottomlands, Fla. to Ark. and Tex., n. to Del. April, May. — Leaves sweetish, greedily eaten by cattle.

Var. **pygmaèa** Fern. (dwarf). — Dwarf shrub 0.3–1.3 m. high; mature leaves elliptic-ovate, acute at both ends, 2–5.5 cm. long and 1–2.5 cm. wide. — Dry pine-barrens, se. Va.

FAM. 140. STYRACÀCEAE (Storax Family)

Shrubs or trees with stellate pubescence; the leaves alternate, simple, exstipulate; the regular flowers perfect; calyx adherent to the ovary or at least to its base, entire or with 4–8 valvate teeth or lobes; the corolla with 4–8 petals free or united at base; the stamens in a single series, of same number or two to four times as many as corolla-lobes, adnate to corolla-tube or free; anthers elongate; style 1; fruit a drupe or capsule. — Seeds anatropous; embryo surrounded by endosperm, the slender radicle as long as or longer than the flat cotyledons. Ovules 2–∞ in each locule. A small family of warm reg.

Calyx adherent to whole surface of ovary; corolla 4-lobed; fruit 2–4-winged. . . 1. *Halesia.*
Calyx adherent only to base of ovary; corolla 5 (rarely 4–8)-parted; fruit globular, wingless. . . . . 2. *Styrax.*

1. HALÈSIA Ellis Snowdrop- or Silverbell-tree

Calyx inversely conical, 4-toothed; the tube 4-ribbed. Petals 4, united at base, or oftener to the middle, into an open campanulate corolla, convolute or imbricated in the bud. Stamens 8–16; filaments united into a ring at base and usually a little adherent to the base of the corolla; anthers linear-oblong. Ovules 4 in each locule. Fruit large and dry, bony within. Seeds single, cylindrical. — Shrubs or small trees of e. N.Am. and China, with large and veiny pointed deciduous leaves; the snowy-white flowers drooping on slender pedicels, in clusters or short racemes, from axillary buds of the preceding year. Pubescence partly stellate. (Named for *Stephen Hales*, 1677–1761, author of "Vegetable Statics", etc.) Mohrodendron Britt.

1. H. carolìna L. (of Carolina), Opossum-wood, Shittim-wood. — Leaves oblong-ovate; fruit 4-winged, 3–4 cm. long. — Rich woods and banks of streams, Fla. to e. Tex., n. to Va., W.Va., s. O., s. Ind., s. Ill., se. Mo. and Okla. May. — Often cult. and occasionally spreading.

2. STÝRAX L. Storax

Calyx truncate, somewhat 5-toothed. Corolla 5 (rarely 4–8)-parted, large, the lobes mostly soft-downy. Stamens twice as many as the lobes of the corolla; filaments flat, united at the base into a short tube; anthers linear. Fruit globular, its base surrounded by the persistent calyx, dry, often 3-valved. Seed globular, erect, with a hard coat. — Shrubs or small trees of warm parts of N.Am. and Eurasia, with commonly deciduous leaves, and axillary or leafy-racemed white and showy flowers on drooping peduncles, produced in spring. (The ancient Greek name of the tree which produces *storax.*)

Racemes 0.5–1.5 dm. long, tomentose, many-flowered; leaves white-tomentose beneath, the leaders 0.7–2 dm. long. . . . . 1. *S. grandifolia.*
Racemes 1–5 cm. long, scurfy-canescent or glabrous, few-flowered; leaves scurfy or glabrous beneath, the leaders 2–10 cm. long. . . . . 2. *S. americana.*

1. S. grandifòlia Aít. (large-leaved). — Shrub 1–3.5 m. high; *leaves* obovate, acute or short-acuminate, *white-tomentose beneath*, 0.5–2 dm. long; *flowers mostly in elongated racemes;* corolla 1.5 cm. long, convolute-imbricated in bud. — Woods, Fla. to La., n. to s. Va., Tenn. and Ark.; reported from s. O. May.

2. S. americàna Lam. (American), MOCK-ORANGE. — Shrub 1–4 m. high; *leaves* oblong to elliptic, acute at both ends, 2–10 cm. long, *smooth or barely pulverulent beneath; flowers axillary or in 3–4-flowered racemes;* corolla valvate in the bud. — Along streams, in cypress-swamps, etc., Fla. to La., n. to se. Va., s. O., s. Ind., s. Ill. and se. Mo. May.

Var. pulverulénta (Michx.) Perkins (covered with powder). — Lower; *leaves* elliptic to obovate, *sparingly puberulent above, scurfy beneath;* corolla-lobes often more convolute in bud. (*S. pulverulenta* Michx.) — Swamps and banks of streams, Fla. to Tex., n. to s. Va. and Ark.

FAM. 141. OLEÀCEAE (OLIVE FAMILY)

Trees or shrubs with opposite and pinnate or simple leaves, a 4-cleft (or sometimes obsolete) calyx, a regular 4-cleft or nearly or quite 4-petalous corolla, sometimes apetalous; the stamens only 2 (rarely 3 or 4); the ovary 2-locular, with 2 (rarely more) ovules in each locule. Seeds anatropous, with a large straight embryo in hard fleshy albumen, or without albumen.

Fruit a dry indehiscent winged samara; leaves usually pinnate in ours; flowers apetalous in ours; trees or shrubs. Tribe FRAXINEAE. . . . . 1. *Fraxinus.*
Fruit a capsule, drupe or few-seeded berry; leaves simple; shrubs or small trees.
 Fruit a loculicidal capsule. Tribe SYRINGEAE. . . . . 2. *Syringa.*
 Fruit a drupe or berry. Tribe OLEINEAE.
  Corolla wanting; calyx minute or obsolete; flowers in scaly ament-like clusters from axils of preceding year; leaves deciduous. . . . . . 3. *Forestiera.*
  Corolla present; calyx well developed.
   Corolla with linear elongate nearly separate petals; panicles drooping, from lateral buds. . . . . 4. *Chionanthus.*
   Corolla funnelform, with broad lobes; panicles erect or merely spreading.
    Panicles scaly-bracted, from the old axils; flowers dioecious or polygamous; stigma capitate; leaves coriaceous, evergreen; drupe 1-locular. . . . . 5. *Osmanthus.*
    Panicles open, terminal, erect; flowers perfect; stigma 2-lobed; leaves tardily deciduous; drupe-like berry 2-locular. . . . . . 6. *Ligustrum.*

Tribe FRAXÍNEAE Gray

1. FRÁXINUS L. Ash. FRÊNE (Que.)

Flowers dioecious, polygamous or monoecious. Calyx small and 4-cleft, toothed, entire or obsolete. Petals 4, or altogether wanting in our species. Stamens 2, sometimes 3 or 4; anthers linear or oblong, large. Style single; stigma 2-cleft. Fruit 1–2-locular, flattened (sometimes 3-angled in no. 4), 1–2-seeded. Cotyledons elliptical; radicle slender. — Timber-trees (some shrubby) of N. Hemisph., with petioled mostly pinnate leaves; the small flowers in crowded panicles or racemes from the axils of the preceding year's leaves. (The classical Latin name.) — Ours belong to § FRAXINÁSTER DC.

a. Calyx present, persistent as a lobed cup at base of fruit; flowers dioecious; wing of fruit narrowed or becoming obsolete at base of fruit; anthers linear or linear-oblong, apiculate; leaflets 5–11, definitely petiolulate. Sub-§ MELIOIDES. . . *b.*
 b. Fruits with a definite subterete body terminated by an obtuse to subtruncate (rarely acute) wing 3.5–10.5 mm. broad. . . *c.*
  c. Petiolules of middle and lower mature leaflets wingless nearly their entire length, 0.3–2 cm. long.
   Leaflets strongly whitened beneath, the longest blades 0.5–1.5 dm. long; petiolules 0.3–1 (rarely –1.5) cm. long; mature fruits (1–) 2.5–5 cm. long, body 0.3–1.6 cm. long, wing 2–7(–10) mm. broad; anthers with apiculate tips 0.2–0.4 mm. long. . . . . 1. *F. americana.*
   Leaflets green or fulvous beneath, the longest blades 0.9–2 dm. long, petiolules 0.5–2 cm. long; mature fruits 4–7.5 cm. long, body 1–3 cm. long, wing 6–12 mm. broad; anthers with slender terminal cusps 0.5–1 mm. long; branchlets, petioles and leaf-rachises velvety. . . 2. *F. tomentosa.*
  c. Petiolules of middle and lower mature leaflets winged nearly to base with decurrent blade-tissue, 1–5(–9) mm. long; wing decurrent down the upper half of the subterete body of the fruit. . . . 3. *F. pennsylvanica.*

*b.*Fruits winged to the acute base, with the body concealed, rhombic, elliptic or oblanceolate, 0.5–2 cm. broad at the middle or above; leaflets with slender petiolules. . . . . 4. *F. caroliniana.*

*a.*Calyx deciduous, a mere ring or wanting; flowers perfect, polygamous or dioecious; wing of the oblong fruit extending without obvious narrowing to the rounded base; anthers oblong, cordate, blunt; leaflets 7–11, sessile or short petiolulate. Sub-§ Bumelioides.

Branchlets 4-angled, often winged; leaflets petiolulate; flowers perfect, with small deciduous calyx. . . . . 5. *F. quadrangulata.*

Branchlets terete; leaflets sessile; flowers polygamous or dioecious, without calyx. . . . . 6. *F. nigra.*

Sub-§ Melioìdes Endl.

1. F. americàna L. (American), White A., Franc-frêne (Que.). — *Branchlets, petioles and leaf-rachises glabrous; leaflets* 5 or 7 (or 9), *on slender petiolules* 0.3–1.5 *cm. long,* ovate- or lance-oblong, acuminate to blunt, *whitish or pale* and glabrous or sparsely pilose *beneath,* entire, undulate or serrate, 0.5–1.5 dm. long; *anthers* linear or linear-oblong, *with an apiculation* 0.2–0.4 *mm. long; fruits* 2.5–5 *cm. long, the body* (inclosing seed) 0.7–1.6 *cm. long,* thick-cylindric or slenderly ellipsoid, terminated by an oblanceolate to linear-cuneate stramineous, or purple in forma **iodocàrpa** Fern. (purple-fruited), *wing twice to thrice as long as the body and* 4–7 *mm. broad.* — Rich upland- to lowland-woods, Que. to Minn., s. to N.S., N.E., L.I., Fla. and Tex. Apr., May. Fig. 1394. — Large and very valuable tree with gray furrowed bark, smooth gray branchlets (the young ones lustrous), rust-colored buds and durable whitish wood.

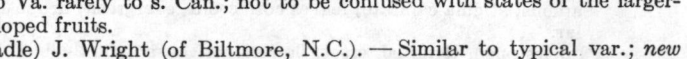

1394. F. americana (typical).

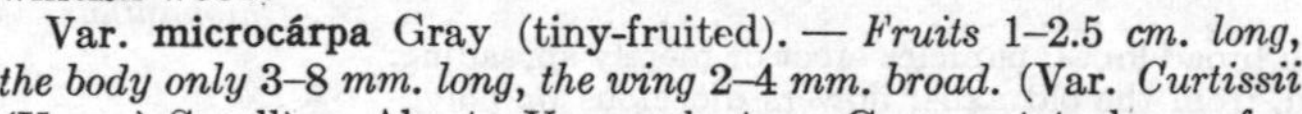

Var. **microcàrpa** Gray (tiny-fruited). — *Fruits* 1–2.5 *cm. long, the body only* 3–8 *mm. long, the wing* 2–4 *mm. broad.* (Var. *Curtissii* (Vasey) Small) — Ala. to Va. rarely to s. Can.; not to be confused with states of the larger-fruited tree with undeveloped fruits.

Var. **biltmoreàna** (Beadle) J. Wright (of Biltmore, N.C.). — Similar to typical var.; *new branchlets, petioles and leaf-rachises velvety-tomentulose;* leaflets strongly whitened beneath, on slender petiolules; *fruits with less distinction of body and wing,* 6–7.25 *mm. broad.* (*F. biltmoreana* Beadle) — Rich woods, N.J. to s. Ill. and Mo., s. to Ga. and Ala. Fig. 1395.

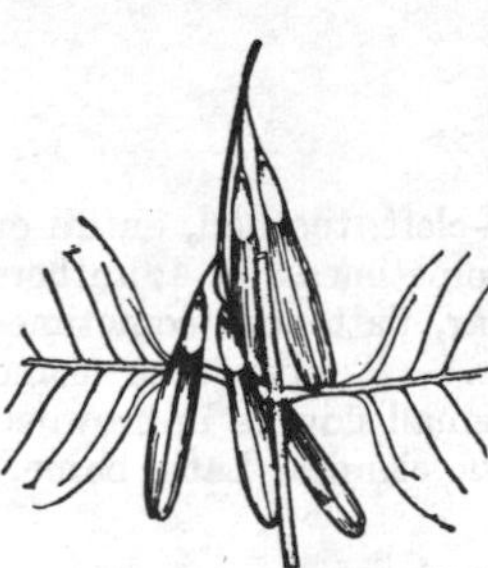

1395. F. americana, v. biltmoreana.

2. F. tomentòsa Michx. f. (with dense and soft pubescence), Red or Pumpkin-A. — *New branchlets, petioles and leaf-rachises velvety-tomentulose; leaflets* 7–9, *on slender petiolules* 0.5–2 *cm. long,* elliptic to ovate-lanceolate, long-acuminate, entire or undulate, subcoriaceous, lustrous above, more or less rufescent-tomentulose to glabrate beneath, the *longer blades* 0.9–2 *dm. long; anthers* linear, *with slender terminal cusps* 0.5–1 *mm. long; fruits* linear-oblong, 4–7.5 *cm. long, the vaguely defined body* 1–3 *cm. long,* the narrowly decurrent *wing* 6–12 *mm. broad; fruiting calyx* 2–5 *mm. long.* (*F. profunda* Bush; *F. Michauxii* Britt.) — Inundated swamps and bottoms, nw. Fla. to La., n. to se. and nw. N.Y., O., Ind., s. Ill., and Mo. April, May. Fig. 1396. — Small to very tall (up to 40 m.) tree.

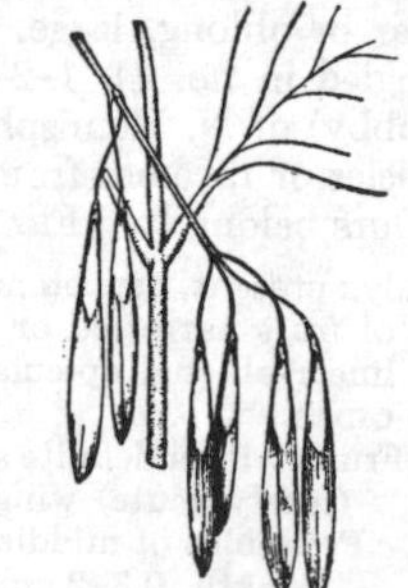

1396. F. tomentosa.

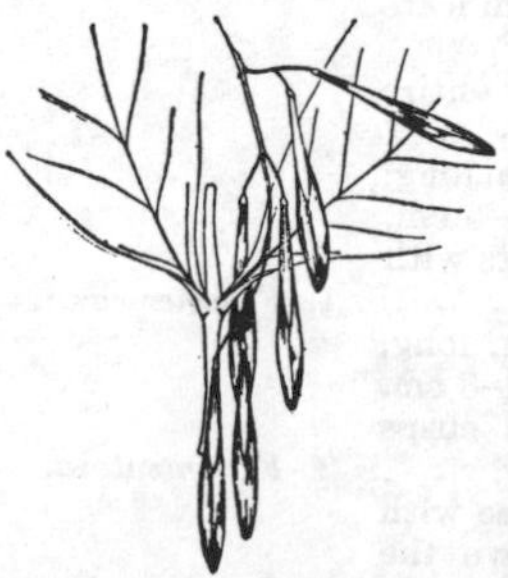

1397. F. pennsylvanica.

3. F. pennsylvánica Marsh. (of Pennsylvania), Red A., Frêne rouge (Que.). — *Branchlets velvety-tomentose; petioles, rachises and lower leaf-surfaces fulvous-pubescent; leaflets* 5–9, ovate or oblong-lanceolate, taper-pointed, entire or undulate, rarely toothed, *gradually narrowed at base and decurrent along the upper half of the short petiolule; mature samaras* stramineous or brownish, or red in forma **erythrocàrpa** Vict. & Rolland (red-fruited), 4–7.5 *cm. long,*

3–6(–7.5) mm. broad, the slender terete body 1.7–3 cm. long with only its upper half bordered by the decurrent base of the long *linear-oblanceolate wing.* (*F. Darlingtoniana* Britt.) — Low grounds and river-banks, Ala. to La., n. to warm valleys of N.E., sw. Que., N.Y., s. Ont., Mich., Wisc. and Minn. May. Fig. 1397. — Middle-sized tree; inner face of outer bark of branches red or cinnamon-color when fresh. Passing into

Var. **Aùstini** Fern. (for its discoverer, Coe Finch Austin, 1831–1880). — Similar but *leaflets usually toothed;* mature *samaras* 2.8–3.7(–4) *cm. long, their bodies only* 1–1.7 *cm. long* and *spatulate wings* 4–11 mm. broad. — Banks of streams, Que. to Man., s. to N.S., N.E., Va., O., Ill. and Ia. Passing into

Var. **subintegérrima** (Vahl) Fern. (almost entire), Green A. — *Branchlets, petioles and leaf-rachises glabrous;* leaflets toothed, green on both sides, glabrous or sparsely pale-villous on nerves beneath; samaras 2.7–4.5 cm. long, their bodies 1–2 cm. long, the spatulate or oblanceolate wings 4–6 mm. broad. (Var. *lanceolata* (Borkh.) Sarg.; *F. lanceolata* Borkh.; *F. viridis* Michx. f.) — Banks of streams, Que. to Sask. and Mont., s. to N.E., Ga., Ala., La. and Tex.

4. F. caroliniàna Mill. (Carolinian), Water-A., Pop-A. — Shrubby tree rarely 10 m. high, with trunks rarely 1.5 dm. in diameter, *branchlets terete; leaves with terete petioles; leaflets 5–7,* ovate-lanceolate to oblong or roundish, membranaceous, acute at base, acute or obtuse at apex, serrate-dentate or entire, *green beneath, those of fertile branches* 4–12 cm. long and *on slender petiolules* 0.3–2 *cm. long; fruits broadly 2-winged and flat or spoon-shaped or 3-winged* to base, *rhombic, elliptic or oblanceolate, acute at base,* 2.5–5.5 cm. long, 0.5–2 *cm. broad,* with small calyx. Our most variable species; the following vars. and forms recognized:

Samaras broadly oblong-oblanceolate to rhombic or subelliptic, 1–2 cm. broad, 2.5–4.5 cm. long.
  Petiole, rachis and lower leaf-surface glabrous or only sparsely pilose along nerves. . . . . *F. caroliniana* (typical).
  Petiole, rachis and lower leaf-surface tomentose. . . . . . . Forma *pubescens.*
Samaras oblanceolate, 0.5–1.3 cm. broad, 3–5.5 cm. long.
  Samaras broadly oblanceolate to spatulate, 1–1.3 cm. broad above middle, 1–3 mm. broad at 5 mm. above base.
    Leaves glabrous or essentially so. . . . . Var. *oblanceolata.*
    Leaves with tomentose petiole, rachis and lower surface. . . . . . Forma *hypomalaca.*
  Samaras narrowly oblanceolate, 5–9 mm. broad above middle, only 1–2 mm. broad at 5 mm. above base.
    Branchlets, petioles, rachis and lower leaf-surface glabrous or essentially so. Var. *cubensis.*
    Branchlets, petioles, rachis and lower leaf-surface densely short-velvety. . Forma *lasiophylla.*

F. caroliniàna (typical). — Inundated swamps, bottomlands and wet shores, Fla. to e. Tex., n. on or near Coastal Plain to Va. and Ark. April, May. Fig. 1398. — Forma **pubéscens** (M. A. Curtis) Fern. & Schub. (pubescent), var. *pubescens* (M. A. Curtis) Fern. and var. *Rehderiana* (Lingelsh.) Sarg., local throughout.

Var. **oblanceolàta** (M. A. Curtis) Fern. & Schub. (oblanceolate). — Less common, Fla. to se. Va. — Forma **hypomálaca** Fern. & Schub. (soft beneath), local.

Var. **cubénsis** (Griseb.) Lingelsh. (of Cuba). — Cuba and Fla. and locally northw. — Forma **lasiophýlla** Fern. & Schub. (hairy-leaved), local in Southampton Co., Va. and probably farther south.

1398. F. caroliniana.

Sub-§ Bumelioìdes Endl.

5. F. quadrangulàta Michx. (4-angled), Blue A. — *Branchlets square* (at least on vigorous shoots), glabrous; leaflets 7–11, short-stalked, oblong-ovate or lanceolate, pointed, sharply serrate, green both sides; *flowers perfect, the minute calyx promptly deciduous; anthers cordate, oblong, blunt; fruit oblong, blunt and of the same width at both ends* or slightly narrowed at the base, often notched at the apex, 2.5–5 cm. long, 6–15 mm. wide. — Dry or moist rich woods, s. Ont., Mich. and Wisc., s. to Ala., Ark. and Okla. March, April. Fig. 1399. — Large timber-tree, the inner bark yielding a blue color to water.

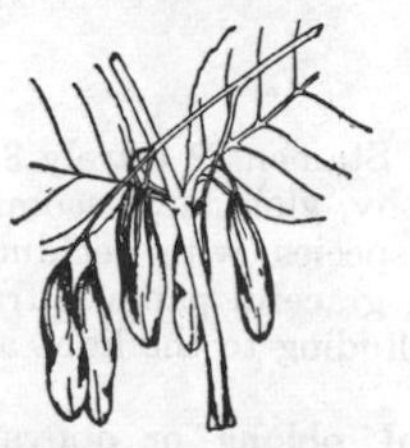

1399. F. quadrangulata

6. F. nigra Marsh. (black), Black A., Frêne noir (Que.). — Similar to no. 5; *branchlets terete,* glabrous; *leaflets* 7–11; *the lateral ones sessile,* oblong-lanceolate, tapering to point, serrate, obtuse

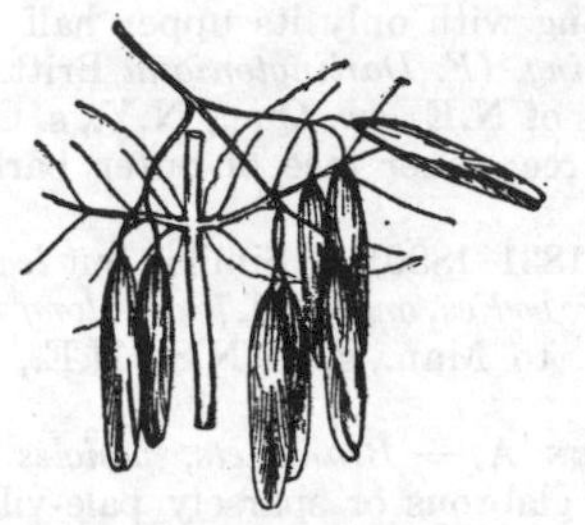
1400. F. nigra.

or rounded at base, green and smooth both sides, when young often rusty beneath; *flowers* naked, *polygamous or dioecious; fruits linear-oblong or narrowly elliptical,* blunt at both ends. — Swamps and shores, Nfld. to Man., s. to N.S., N.E., Del., Va., W.Va., O., Ind., Ill., Ia. and N.D. May, June. Fig. 1400. — Small to middle-sized tree with pale gray rather flaky bark and very tough fissile wood. Bruised foliage with odor of *Sambucus.*

Tribe Syríngeae G. Don

2. SYRÍNGA L. Lilac. Lilas (Que.)

Corolla salverform, much exceeding the 4-toothed calyx, pale violet to roseate or white. Ovary 2-locular; ovules 2 in each locule, pendulous. — Upright shrubs nat. of Eurasia, with simple opposite ovate or lanceolate leaves and numerous flowers in thyrsoid or pyramidal panicles. (Name from the Greek *syrinx, a pipe,* from the use of the branchlets of *Philadelphus,* to which the name originally belonged, for pipes, the name later transferred to the present wholly different genus.)

1. S. vulgàris L. (common), Common L. — *Leaves ovate,* acuminate, entire, *truncate or subcordate at base,* slender-petioled; corolla lilac-purple, rarely white. — Long popular in cult. and not rarely found in a wild state. May, June. (Introd. from Eu.)

2. S. pérsica L. (Persian), Persian L. — More slender, *with lance-acuminate leaves tapering at base* and with smaller panicles of pale lilac flowers. — Cult. and less frequently esc. (Introd. from Asia)

Forsýthia (for *William Forsyth,* 1737–1804) suspénsa (Thunb.) Vahl (suspended), with showy yellow vernal flowers, pendulous hollow branchlets and many of the leaves 3-parted, and F. viridíssima Lindl. (very green), with ascending branchlets with laminated pith and narrower uncleft leaves, spread slightly or persist about old places but are scarcely established. (Both introd. from Asia)

Tribe Oleíneae Gray

3. FORESTIÈRA Poir.

Calyx of 4 minute sepals. Anthers oblong. Ovary ovoid, 2-locular, with 2 pendulous ovules in each locule; style slender; stigma somewhat 2-lobed. Drupe small, ovoid, 1-locular, 1-seeded. — Shrubs of N. and S. Am., with opposite and often fascicled deciduous leaves and small flowers from the axils of the preceding year. Peduncles of pistillate flowers short, 1–3-flowered. (Named for *Forestier,* physician of Saint-Quentin, deceased about 1820, Poiret's first teacher in botany.) Adelia P. Br., not L.

1. F. acuminàta (Michx.) Poir. (with tapering tip), Swamp-Privet. — Glabrous, somewhat spinescent, 1.5–3 m. high; *leaves* thin, *oblong-ovate or ovate-lanceolate, acuminate* at both ends, often serrulate; *drupe elongated-ellipsoid,* usually pointed. (*Adelia* Michx.) — Wet river-banks, ponds and swamps, Fla. to Tex., n. to S.C., s. Ind., Ill., e. and s. Mo. and se. Kans. Late March–early May.

2. F. ligustrìna (Michx.) Poir. (Privet-like). — *Leaves oblong to narrowly obovate, blunt,* more or less short-pubescent, 1–4 cm. long; *drupe short-ovoid,* sessile. — Rocky slopes, barrens and glades, Fla. and Ala., n. to Ga. and Ky. Aug.

4. CHIONÁNTHUS L. Fringe-tree

Calyx 4-parted, very small, persistent. Petals barely united at base. Stamens 2 (rarely 3 or 4), on the base of the corolla, very short. Stigma notched. Drupe fleshy, globular, becoming 1-locular, 1–3-seeded. — Low trees or shrubs, ours and an e. Asiatic species, with deciduous and entire petioled leaves, and delicate flowers in loose and drooping graceful panicles from lateral buds. (Name from the Greek *chion, snow,* and *anthos, flower;* alluding to the light and snow-white clusters of flowers.)

1. C. virgìnica L. (Virginian), Old-Man's-beard. — Leaves oval, oblong or obovate-lanceolate; flowers on slender pedicels; petals 2–2.5 cm. long, narrowly linear, acute, varying to 5 or 6 in number; drupe purple, with a bloom, ovoid, 1–1.8 cm. long. — Damp woods, thickets or bluffs, Fla. to Tex., n. to N.J., e. Pa., W.Va., s. O., s. Mo. and Okla.; spreading from cult. northw. May, June.

5. OSMÁNTHUS Lour. Devilwood

Calyx 4-cleft, short, persistent. Corolla small, 4-cleft, its broad obtuse lobes imbricated in bud. Stamens 2 (or rarely 4), inserted on the short corolla-tube, included; anthers ovate. Style short; stigma small and entire. Ovules a pair in each locule. Drupe 1-seeded, ovoid or globose. — Evergreen shrubs or small trees with small dioecious or polygamous flowers in scaly-bracted panicles from the old axils; e. N.Am., Polynesian and e. and s. Asiatic. (Name from the Greek *osme, odor*, and *anthos, flower;* from the fragrant flowers.)

1. O. americànus (L.) Gray (American), Wild Olive. — Tree up to 15 m. high, or large shrub; bark whitish; branchlets, leaves, etc., glabrous; heavily coriaceous entire lance-oblong lustrous leaves 0.7–1.5 dm. long, exceeding the panicles of dull white flowers; drupe dark purple, 1–1.5 cm. long. (*Amarolea* Small) — Rich woods, swamps and bluffs, Fla. to La., n., locally, to Cape Henry, Va. April, early May.

6. LIGÚSTRUM L. Privet

Calyx campanulate or short-tubular, 4-toothed. Stamens 2, inserted on the tube of the corolla, included or barely exserted. Berry 2-locular, 1–2-seeded, blackish, hard. — Shrubs nat. of Old World, with entire leaves and small white flowers in panicles terminating branches and branchlets. (The classical name of no. 1.)

Branchlets glabrous or barely puberulent; leaves glabrous; calyx glabrous.
- Leaves deciduous; branchlets puberulent or glabrous; corolla-tube scarcely to barely as long as the limb; anthers much shorter than corolla-lobes. . . . 1. *L. vulgare.*
- Leaves semievergreen; branchlets glabrous; corolla-tube about thrice length of lobes; anthers as long as corolla-lobes. . . . . 2. *L. ovalifolium.*

Branchlets strongly pubescent; leaves pubescent (at least on midrib) beneath; corolla-tube much longer than lobes.
- Calyx glabrous; anthers much shorter than corolla-lobes. . . . . 3. *L. amurense.*
- Calyx pubescent; anthers and corolla-lobes subequal. . . . . 4. *L. obtusifolium.*

1. L. vulgàre L. (common), Privet or Prim. — Tall shrub (up to 5 m. high) with *minutely puberulent or glabrescent branchlets; leaves firm, glabrous, oblong- or ovate-lanceolate,* 3–6 cm. long, deciduous; panicles dense, 3–6 cm. long; corolla-tube shorter than limb; *anthers shorter than limb;* fruit black, lustrous, ovoid to subglobose, 6–8 mm. long. — Thickets and open woods, s. Me. to s. Ont., s. to N.C., O. and Mich., long estab. in open woods from N.E. to Pa. June, July (Introd. and natzd. from Eu.)

2. L. ovalifòlium Hassk. (oval-leaved), California P. — Similar to no. 1; *branchlets glabrous; leaves more lustrous* and firmer, *elliptic-ovate,* acute, *subevergreen;* panicles mostly longer; *corolla-tube thrice length of limb; anthers and corolla-lobes subequal.* — Cult. and spread to roadsides, thickets, etc., southw. (Introd. and natzd. from Japan)

3. L. amurénse Carr. (of Amur River), Amur P. — Resembling no. 1 but *branchlets densely pilose; leaves* oblong or elliptic, *pubescent on the midrib beneath;* panicles pubescent, rather open; *calyx glabrous* (except near pedicel); *corolla-tube much longer and anthers much shorter than lobes.* — Thickets and open woods, se. Va. and southw. June. (Introd. and natzd. from ne. Asia)

4. L. obtusifòlium Sieb. & Zucc. (obtuse-leaved). — Low shrub differing from no. 3 in mostly blunter leaves, *pubescent calyx,* and *subequal anthers and corolla-lobes.* — Thickets, borders of woods and roadsides, e. Pa. and southw. (Introd. and natzd. from Japan)

Several other Asiatic species may spread from cult.; these should be sought in works on cult. plants.

FAM. 142. LOGANIÀCEAE (Logania Family)

Herbs, shrubs or trees with opposite and entire leaves and stipules or with a stipular membrane or line between the leaves, and with regular 4–5-merous 4–5-androus perfect flowers, the ovary free from the calyx; a connecting group between *Gentianaceae, Apocynaceae* and *Scrophulariaceae* (from all of which it is known by the stipules) and *Rubiaceae,* from which it differs in the free ovary; our representatives of the family are related most nearly to the *Rubiaceae*, to which, indeed, they have sometimes been appended.

Woody twiner (rarely prostrate) with coriaceous evergreen leaves; corolla large, yellow, its lobes imbricated in bud; stigmas 2, each 4-parted. Tribe GELSEMIEAE. . . . . 1. *Gelsemium.*
Herbs (with us) with annual leaves; stigma entire or 2-lobed. Tribe LOGANIEAE.
Leaves broad; flowers unilaterally spicate in terminal scorpioid inflorescences, 5-merous; corolla-lobes valvate in bud.
Perennial with tubular-funnelform red and yellow corolla 3–5 cm. long; style single, jointed in the middle; capsule circumscissile above the cup-like persistent base. . . . . 2. *Spigelia.*
Annual with small urceolate white corolla only slightly longer than calyx; styles 2, short, converging and united at summit, with a common stigma; capsule mitre-shaped or 2-horned, opening down the inner side of each horn. . . . . 3. *Cynoctonum.*
Leaves awl-shaped or linear; inconspicuous white flowers in axils and on tips of branches, 4-merous; corolla-lobes imbricated in bud. . . . . 4. *Polypremum.*

Tribe GELSEMÌEAE G. Don

1. GELSÈMIUM Juss. YELLOW JESSAMINE

Corolla open-funnelform. Stamens 5, with oblong sagittate anthers. Divisions of stigma linear. Capsule elliptical, flattened contrary to the narrow partition, 2-locular, septicidally 2-valved. Seeds many or several, winged. Embryo straight, in fleshy albumen; the ovate flat cotyledons much shorter than the slender radicle. — Smooth twining or trailing shrubby plants, one of e. N.Am., the other of e. Asia, with ovate or lanceolate leaves, minute deciduous stipules, and showy yellow dimorphous flowers. (*Gelsemino*, the Italian name of the Jessamine.)

1. G. sempérvirens (L.) Ait. f. (evergreen), EVENING-TRUMPET-FLOWER. — Stems slender, wiry, freely branched and tangling, high-twining or forming broad repent carpets; leaves short-petioled, linear-lanceolate to elliptic-ovate (sometimes on a single plant), lustrous, subpersistent; flowers in short axillary clusters; pedicels scaly-bracted; corolla 2.5–4 cm. long, deliciously fragrant; capsule pointed. — Dry to wet woods, thickets and sands, Fla. to Tex., n. to se. Va. and Ark. Late March–early May.

Tribe LOGANÌEAE Mart.

2. SPIGÈLIA L. PINKROOT. WORMGRASS

Corolla tubular-funnelform, 5-lobed at the summit. Stamens 5; anthers linear. Style slender, hairy above. Capsule short, 2-locular, twin, laterally flattened, separating at maturity from a persistent base into 2 carpels which open loculicidally, few-seeded. — Chiefly herbs of N. and S. Am., with opposite leaves united by stipules, and the flowers spiked in one-sided cymes. (Named for *Adrian Spiegel*, latinized *Spigelius*, 1578–1625, who was perhaps the first to give directions for preparing an herbarium.)

1. S. marilándica L. (of Maryland), INDIAN-PINK. — Stems simple and erect, 3–6 dm. high, from a perennial root; leaves sessile, ovate-lanceolate to broadly ovate, acute; spike simple or forked, short; corolla 3–5 cm. long, red outside, yellow within; tube four times the length of the calyx, the lobes lanceolate; anthers and style exserted. — Rich woods, Fla. to Tex., n. to Md., O., Ind., Mo. and Okla. May, June.

3. CYNÓCTONUM J. F. Gmel. MITERWORT

Corolla little longer than the calyx, somewhat funnelform. Stamens 5, included. Ovary slightly adnate at base to the bottom of the calyx, 2-locular. Capsule exserted, strongly 2-horned or mitriform, opening down the inner side of each horn, many-seeded. — N. and S. Am. and se. Asiatic annual smooth herbs, with small stipules between the leaves, and small white flowers spiked along one side of the branches of a terminal peduncled cyme. (Greek *cyon, dog*, and *cteinein, to kill.*) MITREOLA R. Br. — Plants easily mistaken for *Houstonia* of the *Rubiaceae*.

1. C. Mitrèola (L.) Britt. (old generic name, referring to the miter-like fruit). — Stem simple or *loosely slender-branched*, 1–7.5 dm. high; *leaves* thin, elliptic to lance-oblong, *acuminate, petioled*, 2–8 *cm. long;* cymes on long slender peduncles, their branches loosening in fruit; *corolla* *twice as long as calyx; fruiting calyces slightly distant;* fruits 3–4 mm. high. — Wet sandy or argillaceous shores, ditches and meadows, Fla. to e. Tex., n. to se. Va., Tenn. and Ark. June–Nov.

2. C. sessilifòlium (Walt.) J. F. Gmel. (sessile-leaved). — Stem *stiffly erect,* simple or with few erect branches; *leaves* firm, oval or elliptic, *sessile, blunt to abruptly tipped,* 1–2 cm. long; cymes more compact; *corolla only slightly longer than calyx; fruiting calyces approximate.* — Savannas, damp shores and low woods, Fla. to e. Tex., n. to se. Va. July–Oct.

4. POLYPRÈMUM L.

Calyx 4-parted; the divisions subulate from a broad scarious-margined base. Corolla almost rotate, bearded in the throat. Stamens 4, very short; anthers globular. Style very short; stigma ovoid, entire. Capsule ovoid, a little flattened, notched at the apex, 2-locular, loculicidally 2-valved, many-seeded. — A smooth diffuse much branched small annual or perennial, with narrowly linear or subulate leaves connected at base by a slight stipular line; the small flowers solitary and sessile in the forks and at the ends of the branches; corolla inconspicuous, white. (Name altered from the Greek *polypremnos, many-stemmed.*) — Plants easily mistaken for *Hedyotis* of the *Rubiaceae.*

1. P. procúmbens L. (prostrate). — Dry sandy fields, roadsides and openings, inclined to be a weed, Fla. to Tex., n. to L.I., N.J., e. Pa. and se. Mo. June–Oct. (Trop. Am.)

Buddlèja (for *Adam Buddle,* 1660–1715) Dàvidi Franch. (for Armand David, 1826–1900), a cult. shrub with lower leaf-surfaces white-tomentose, and spiciform inflorescences of cymules of tubular lilac orange-throated glabrous flowers, is becoming natzd. n. to Md. (Introd. from China)

FAM. 143. GENTIANÀCEAE (Gentian Family)

Smooth herbs with a colorless bitter juice, opposite (or whorled) and sessile entire and simple leaves (alternate and petioled, compound or crenate in nos. 9 and 10) *without stipules, regular flowers with the stamens as many as the lobes of the corolla, which are convolute (rarely imbricated and sometimes valvate) in the bud, a 1-locular ovary with 2 parietal placentae, or nearly the whole inner face of the ovary ovuliferous; the fruit usually a 2-valved and septicidal many-seeded capsule.* Calyx persistent. Corolla mostly withering-persistent, the stamens inserted on its tube. Seeds anatropous, with a minute embryo in fleshy albumen. — Bitter-tonic plants.

*a.*Leaves opposite or whorled, simple, entire, sessile (except sometimes the radical); aestivation of corolla never valvate. Subfam. I. Gentianoideae. . . *b.*
*b.*Lobes of the corolla convolute in bud. . . *c.*
*c.*Style filiform, usually deciduous from capsule; anthers mostly twisting or curving in age.
Parts of flower in 4's–13's; corolla rotate; anthers soon recurved, arcuate or revolute; stigmas clavate or spatulate; capsule globose or ovoid. 1. *Sabatia.*
Parts of flower in 4's or 5's; corolla salverform; anthers spirally twisting; stigmas oblong to flabelliform; capsule ellipsoid to fusiform. 2. *Centaurium.*
*c.*Style short and thick, persistent, or none; anthers remaining straight. . . *d.*
*d.*Corolla without nectariferous pits or glands at base.
Corolla funnelform, campanulate or salverform, the sinuses with or without plaits or appendages; calyx with elongate tube; stamens borne on corolla-tube. 3. *Gentiana.*
Corolla rotate, the acute divisions with a pair of scale-like appendages at base; calyx deeply parted; stamens on base of corolla. 4. *Lomatogonium.*
*d.*Corolla with a large pit or nectariferous gland at the base of each lobe.
Corolla 4-parted, rotate, with a fringed glandular spot on each lobe; corolla-lobes sinistrorsely convolute in bud. . . . . 5. *Swertia.*
Corolla 4–5-cleft, campanulate, usually 4–5-spurred at base; corolla-lobes dextrorsely convolute in bud. . . . . 6. *Halenia.*
*b.*Lobes of corolla imbricate in bud, without appendages or glands.
Calyx 4-parted; corolla deeply 4-cleft, slenderly campanulate; stigma nearly sessile. . . . . 7. *Bartonia.*
Calyx of 2 foliaceous sepals; corolla 4-lobed, ellipsoid-campanulate; style definite. . . . . 8. *Obolaria.*
*a.*Leaves alternate, mostly petioled, compound or crenate; aestivation of corolla induplicate-valvate; seed-coat crustaceous; paludal or aquatic perennials. Subfam. II. Menyanthoideae.
Leaves 3-foliolate; corolla bearded inside. . . . . 9. *Menyanthes.*
Leaves simple, rounded; corolla naked, or bearded on margin only. . . 10. *Nymphoides.*

Subfam. I. GENTIANOÏDEAE

1. SABÀTIA Adans. SABATIA

Calyx 5–12-parted, the lobes slender. Corolla 5–12-parted, rotate. Stamens 5–12; anthers soon recurved. Style 2-cleft or -parted, slender. — N. Am. herbs with slender stems, and mostly cymose-panicled handsome (white or yellow to rose-purple) flowers in summer or autumn. (Dedicated to *Liberato Sabbati*, Italian botanist who published in 1745.) SABBATIA Salisb.

*a.*Calyx-lobes and corolla-lobes usually 5 (exceptionally 4, 6 or 7), corolla-lobes 0.5–2 cm. long; annuals, biennials or non-stoloniferous perennials. § PENTAPETALA. . *b.*

*b.*Branches of inflorescence all opposite; stems somewhat 4-angled or ridged; flowers in a panicle or corymb of cymes. Sub-§ ANGULARES. . . *c.*

*c.*Principal leaves linear to ovate-lanceolate, firm, 1–3-nerved, not clasping; stems not wing-angled; calyx-lobes linear-acicular; corolla-lobes spatulate to narrowly obovate-oblong; style-branches 2–3.5 mm. long. . . *d.*

*d.*Perennial with no rosette-leaves at flowering time; stem simple up to the corymb; middle and upper leaves acuminate; flowers chiefly solitary at tips of elongate branches of cyme; corolla white, drying creamy-yellow, its lobes 1–2 cm. long. . . . . 1. *S. difformis.*

*d.*Biennial, usually with basal rosette of obovate leaves persisting to flowering time; stem simple or branching; middle and upper leaves blunt or merely acutish; corolla white and becoming irregularly suffused with saffron when dry, or roseate.

Corollas white, suffused when dry with diffuse saffron, the lobes 5–13 mm. long; flowers mostly in small cymules at tips of the branches. . . . . 2. *S. paniculata.*

Corollas pink (or permanently white), their lobes 1–1.8 cm. long; flowers both clustered and solitary at branch-tips. . . . . . 3. *S. brachiata.*

*c.*Principal leaves cordate-ovate (or lower often suborbicular), membranaceous, clasping, 3–7-nerved; stem wing-angled; corolla roseate (rarely white), with broadly oblong-obovate lobes; style-branches 3.5–6 mm. long; biennial, when well developed freely branched. . . . . . . 4. *S. angularis.*

*b.*Branches of inflorescence all or nearly all alternate; stems less definitely 4-angled (except in no. 5); flowers borne singly at tips of peduncles and branches. . . *e.*

*e.*Calyx-tube wing-angled up to the sinuses; corolla-lobes oval; leaves chiefly subcordate-clasping, narrowly deltoid-ovate; prairie annual or biennial. Sub-§ CAMPESTRES. . . . . 5. *S. campestris.*

*e.*Calyx-tube without wing-angles; corolla-lobes elliptic to obovate; leaves narrower, not clasping; eastern species. Sub-§ CAMPANULATAE. . . *f.*

*f.*Calyx-lobes foliaceous, mostly as broad as and often longer than corolla-lobes; branching both alternate and opposite; leaves membranaceous, oblong to narrowly obovate, obtuse. . . . . 6. *S. calycina.*

*f.*Calyx-lobes firmer, slender, much narrower than corolla-lobes; branching alternate; leaves firm, narrower.

Annual with solitary stem and tap-root; primary leaves narrowly elliptic to spatulate, tapering to base and apex, usually darkened in drying; lateral peduncles mostly bractless; herbaceous calyx-lobes linear-lanceolate, usually shorter than corolla-lobes. . . . 7. *S. stellaris.*

Cespitose perennial with short woody caudex; primary leaves linear to oblong or narrowly oval, obtuse, with broad sessile base, not blackening; lateral peduncles usually 1–2-bracted; calyx-lobes linear-setaceous, firm, often nearly equaling to exceeding corolla-lobes. . . . . 8. *S. campanulata.*

*a.*Calyx-lobes and corolla-lobes (7–)8–13; corolla-lobes 1.5–3 cm. long; perennials, often stoloniferous. § PLEIENTA.

Primary leaves with mucronate tip; calyx-lobes firm, linear-setaceous, subulate-tipped, 0.5–1.5 mm. broad; calyx-tube nerveless; corolla-lobes cuneate-obovate, rounded or emarginate at tip, 0.6–1.5 cm. broad, the basal yellow spot of each lobe 2.5–5 mm. broad. . . . . 9. *S. Kennedyana.*

Primary leaves blunt or acutish; calyx-lobes herbaceous, 1–3 mm. broad; calyx-tube distinctly nerved or corrugated; corolla-lobes narrowly oblong-spatulate to oblanceolate, acutish or obtuse, 4–9 mm. broad, the basal yellow spots 0.7–2.3 mm. broad. . . . . 10. *S. dodecandra.*

§ PENTAPÉTALA Blake

Sub-§ ANGULÀRES Blake

1. S. difförmis (L.) Druce (of two forms; from the original specimen having some flowers with 6 corolla-lobes). — *Perennial, with basal rosettes rarely present at flowering time; stem* firm, 0.4–1 m. high, *unbranched up to the corymbiform terminal* opposite-branched *cyme; leaves* in remote pairs (rarely in 3's), firm, the lower oblong or narrowly ovate and rounded at base, *the middle and upper lance-acuminate; flowers* 5–45, *chiefly solitary at the tips of the slender pedicels;* calyx-lobes linear-filiform; *corolla white, with* 5 oblong-oblanceolate *lobes* 1–2 *cm. long,* drying yellowish. (*S. lanceolata* (Walt.) T. & G.) — Bogs, sandy swamps, peaty shores and wet pine-barrens, Fla. to Tenn. and N.C. and more locally to s. N.J. July–Sept. FIG. 1401.

1401. S. difformis.

2. S. paniculàta (Michx.) Pursh (with panicles). — *Biennial, with basal rosette* of obovate leaves usually *persisting to flowering time;* stem wire-edged, simple or loosely forking, 2–5 dm. high; *cauline leaves oblong to narrowly lanceolate, blunt or subacute,* with only the midrib prominent beneath; cyme corymbose-paniculate, with 1–8 pairs of ascending opposite branches; *flowers mostly clustered in small terminal cymules; corolla white when fresh, becoming irregularly suffused with saffron in age, the* 5 *spatulate-oblanceolate lobes* 5–13 *mm. long;* style-branches 2–3.5 mm. long. — Dry to moist pinelands, clearings and fields, Ala. and Fla., n. to se. Va. July, Aug. FIG. 1402.

1402. S. paniculata.

3. S. brachiàta Ell. (with divergent branches). — Like no. 2, but stem less rigid, 1.5–6 dm. high; *flowers both clustered and solitary at tips of branchlets; corolla roseate, with yellow or greenish eye,* or white and unchanging in forma **cándida** Fern. (white), *the* mostly 5 narrowly *oblong-obovate lobes* 1–1.8 *cm. long.* — Dry to moist pinelands, oak-woods and clearings, Fla. to La., n. to se. Va., s. Ind. and Mo. Mid-June–early Aug. FIG. 1403.

1403. S. brachiata.

4. S. angulàris (L.) Pursh (with angles), ROSE-PINK, BITTER-BLOOM. — Biennial, with subpersistent rosette of obovate leaves; *stem strongly 4-angled,* simple or usually bushy- or pyramidal-branched, 2–9 dm. high; cauline *leaves membranaceous,* the lower suborbicular or round-oblong; *the middle and upper cordate-ovate, clasping, 3–7-nerved; fruiting calyx-tube* 1–3 *mm. long,* the lobes 4–13 mm. long; *corolla* roseate, or white in forma **albiflòra** (Raf.) House (white-flowered), with yellow eye, delicately fragrant, *with* 5 *round-tipped obovate lobes* 1–2 *cm. long, their yellow basal spots oblong or ovate and entire or slightly lobed,* or in forma **cleistántha** Fern. (with hidden flowers) the minute pale tubular corolla unexpanding; *style-branches* 3.5–6 *mm. long.* — Open woods, clearings, prairies and fields, Fla. to La. and Okla., n. to se. N.Y., s. Ont., Mich., Wisc. and Mo.; casually adv. in Mass. July–Sept. FIG. 1404.

1404. S. angularis.

Sub-§ CAMPÉSTRES Blake

5. S. campéstris Nutt. (of the Plains). — Similar to no. 4, lower, 1–3.5 dm. high, with few *simple or forking mostly alternate branches; calyx with* short *wing-angles extending up to the sinuses; its lobes* lanceolate, 3-nerved, 1–2.5 *cm. long; corolla-lobes oval,* 1–2 cm. long, *their yellow basal spots* 3–6-*lobed with* 1 *or* 2 *middle lobes longest; fruiting calyx-tube* 5–8 *mm. high.* — Prairies and fields, Ill. and Ia. to Ark., Okla. and Tex.; rarely adv. in N.E. July–Sept. FIG. 1405.

1405. S. campestris.

Sub-§ CAMPANULÀTAE Blake

6. S. calycina (Lam.) Heller (with conspicuous calyx). — Perennial, with slender hard caudex; stems weak, simple or with *either opposite or alternate divergent branches,* 1–4.5 dm. high; *leaves membranaceous, oblong to narrowly obovate, round-tipped or obtuse;* flowers borne singly; *calyx-lobes foliaceous,* oblanceolate, *nearly as broad as to broader and often longer than the* pale pink to whitish spatulate or narrowly obovate thin *corolla-lobes.* — Shaded river-swamps and bottoms, Fla. to Tex., n. to se. Va. and Ark. Late June, July. FIG. 1406.

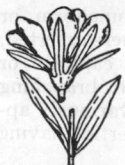

1406. S. calycina.

1407. S. stellaris.

7. S. stellàris Pursh (star-like), SEA-PINK, MARSH-PINK. — *Annual or biennial with shallow tap-root; stem solitary,* 0.3–5 dm. high, unbranched or *with* loosely ascending *mostly alternate* simple or forked *branches; primary leaves narrowly elliptic to spatulate, narrowed to acute base and apex,* usually darkened in drying; *lateral peduncles mostly bractless; calyx-lobes herbaceous, linear-lanceolate,* usually shorter than corolla; corolla-lobes 4–7, crimson-pink, or white in forma **albiflòra** Britt. (white-flowered), elliptical to oblong-obovate, 0.8–1.7 cm. long, *the yellow or green and yellow basal spots more or less toothed or lobed* and with brown or reddish borders. (*S. amoena* (Raf.) G. Don) — Saline or brackish marshes and meadows, Fla. to La., n. to se. Mass. July–Oct. FIG. 1407.

1408. S. campanulata.

8. S. campanulàta (L.) Torr. (bell-shaped). — Similar to no. 7; *perennial with woody caudex; stems usually 2–several,* 1.5–6 dm. high; *primary leaves oblong or narrowly oval to linear, obtuse, broadly sessile,* not darkened in drying; lateral peduncles naked or with 1–2 bracts; *calyx-lobes linear-setaceous,* firm, 1–2 cm. long; corolla-lobes pink or pale crimson, or white in forma **albìna** Fern. (whitish), elliptic to obovate, 1–1.7 cm. long, the ovate to obovate basal *yellow spots entire or shallowly lobed.* — Damp sands and peats, upland and Piedmont reg., Ga. and Ala., n. to Va. and Ind., and on or near Coastal Plain to se. Mass. July–Sept. FIG. 1408. — Southward on the Coastal Plain merging into

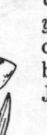

Var. **grácilis** (Michx.) Fern. (slender). — More slender; *leaves linear, the upper narrowly so; peduncles mostly leafy-bracted; calyx-lobes setaceous,* 6–14 *mm. long; corolla-lobes* 6–14 *mm. long.* (*S. gracilis* Michx.) — Wet sands and peats of Coastal Plain, n. to se. Va.

§ PLEIÉNTA (Raf.) Blake

9. S. Kennedyàna Fern. (in honor of GEORGE GOLDING KENNEDY, 1841–1918), PLYMOUTH GENTIAN, LARGE S. — *Stoloniferous perennial* 2.5–8 dm. high; *stolons slender,* superficial, *with terminal rosettes of* oblanceolate *acuminate leaves* 3–8 cm. long; *cauline leaves lance-acuminate, firm,* sessile or subclasping, *mucronate; flowers* 1–24, long-peduncled, 3–7 *cm. broad,* delicately fragrant; *calyx-tube nerveless, the linear-subulate lobes* 0.5–1.5 *mm. broad and hyaline-margined; corolla-lobes* roseate, or white in forma **cándida** Fern. (white), 7–13, *narrowly cuneate-obovate,* or in forma **eucỳcla** Fern. (well-rounded) of Yarmouth Co., N.S., *broadly obovate* and with overlapping margins, *rounded or emarginate at summit,* 1.5–3 cm. long, 0.6–1.5 *cm. broad;* the brown-margined *basal yellow spots subtruncate, 3-lobed and* 2.5–5 *mm. broad.* — Sandy and peaty margins of fresh ponds, Tusket Valley, Yarmouth Co., N.S.; e. Norfolk Co. to Bristol and Barnstable Cos., Mass.; e. Kent and Washington Cos., R.I. Late June–Sept. (rarely –Nov.). FIG. 1409.

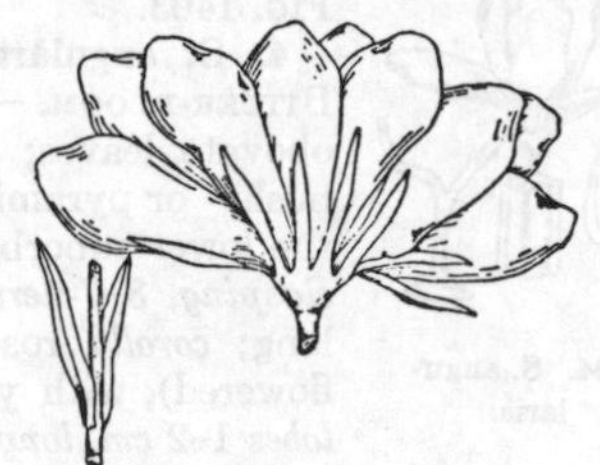
1409. S. Kennedyana.

10. S. dodecándra (L.) BSP. (having twelve stamens), LARGE MARSH- OR SEA-PINK. — Less stoloniferous, 0.8–6 dm. high; *cauline leaves* oblong to lanceolate, *bluntish or acute, not mucronate; calyx-tube distinctly nerved or corrugated, the herbaceous lobes* 1–3 *mm. broad; corolla-lobes narrowly oblong-spatulate to oblanceolate, acutish or obtuse,* 4–9 *mm. broad,* the *basal yellow spots* 0.7–2.3 *mm. broad.* — Saline, brackish or rarely fresh marshes and meadows, Fla. to La., n. on or near Coastal Plain to s. Ct. July–Sept. FIG. 1410.

1410. S. dodecandra.

2. CENTAÙRIUM Hill CENTAURY

Calyx 4–5-parted, the divisions slender. Corolla funnelform or salverform, with slender tube and 4–5-parted limb. Anthers exserted, erect, spirally twisting. Style slender, single; stigma capitate or 2-lipped. — Low and small N.Am., Eurasian and Afr. branching annuals, chiefly with rose-purple or reddish flowers in summer. (An old name, variously applied by the herbalists, from the Latin *Centaurus, Centaur,* its medicinal properties having been discovered by the Centaur.) ERITHREA Neck. ERYTHRAEA Borkh.

Flowers in spikes. . . . . 1. *C. spicatum.*
Flowers in cymes or panicles.
  Flowers in definite terminal cymes, all nearly sessile. . . . . 2. *C. umbellatum.*

Flowers loosely paniculate or paniculate-cymose, all pedicelled.
Corolla-lobes 3–5 mm. long; anthers oblong.
Cauline leaves ovate-oblong to oval; pedicels shorter than calyx. . . . 3. *C. pulchellum.*
Cauline leaves linear or linear-lanceolate; pedicels equaling or exceeding calyx. . . . . 4. *C. texense.*
Corolla-lobes 7–10 mm. long; anthers linear. . . 5. *C. calycosum.*

1. C. spicàtum (L.) Fern. (spiked). — Stem strictly upright, 1–4 dm. high; the *flowers sessile and scattered along one side of the simple or rarely forked spiciform branches;* leaves oval and oblong, rounded at base, acutish; tube of the rose-colored or whitish corolla scarcely longer than the calyx, the lobes oblong. — Borders of sandy salt-marshes, Nantucket I., Mass.; Dorchester Co., Md.; Norfolk Co., Va. June–Sept. (Natzd. from Eu.)

2. C. umbellàtum Gilib. (umbelled), Centaury. — Stem upright, 1–5 dm. high, *corymbosely branched* above; leaves oblong or elliptical, acutish, the basal rosulate, the uppermost linear; *cymes clustered, flat-topped, the flowers all nearly sessile;* tube of the purple-roseate corolla not twice the length of the oval lobes. — Damp sands and pond-margins, Sable I., N.S.; meadows, borders of woods, waste places, etc., local, Que. to Mich., s. to Ga. and Ind. Aug., Sept. (Natzd. from Eu.)

3. C. pulchéllum (Sw.) Druce (beautiful). — Similar to no. 2, 0.5–3 dm. high, *simple or branched from near the base, most of the plant forming a loosely dichotomous cyme; basal leaves not rosulate,* cauline oblong to oblong-lanceolate; *flowers short-pedicelled; corolla-lobes* lanceolate, *3–4 mm. long; capsule about one-half longer than calyx.* — Fields and wet places, N.Y. to Ill., s. to e. Va. Late June–Aug. (Natzd. from Eu.)

4. C. texénse (Griseb.) Fern. (Texan). — Similar to the preceding, but more diffusely forked; *cauline leaves linear or linear-lanceolate, the upper reduced to subulate bracts; pedicels equaling or exceeding the calyx;* corolla-tube twice the length of the lance-oblong lobes. — Dry soil and calcareous barrens, Mo. to Tex. June–Aug.

5. C. calycòsum (Buckley) Fern. (with prominent calyx). — *Simple or corymbose-branched,* 1–6 dm. high; leaves oblong to lance-linear; pedicels equaling or exceeding the calyx; *corolla-tube nearly equaled by the oblong or oval lobes (7–10 mm. long); anthers linear.* — Damp soil along streams, Tex. to Mo. (there possibly adv.). June–Aug.

3. GENTIÀNA L. Gentian

Corolla 4–5-lobed, often with intermediate plaited folds which bear appendages or teeth at the sinuses. Stamens inserted on the tube of the corolla. Style short or none; stigmas 2, persistent. Capsule ellipsoid, 2-valved, the innumerable seeds either borne on placentae at or near the sutures, or in most of our species covering nearly the whole inner face of the capsule. — Flowers solitary or cymose, showy, in late summer and autumn. Large genus of cool and temp. reg. (Name from *Gentius,* king of Illyria, who, according to Pliny, discovered the plant, *i.e.,* its medicinal virtue.)*

a. Flowers usually 4-merous, mostly on distinct peduncles or pedicels and without basal bracts; corolla with more or less spreading lobes, without plaits, prolonged teeth or secondary lobes in the sinuses; annuals or biennials. . . *b.*
b. Flower large; calyx 0.7–4.5 cm. long, cleft about one-half way to the base; corolla deep sky-blue or violet-blue to white, 1.2–6 cm. long; its lobes rounded at summit, fringed or erose-toothed, 0.3–1.8 cm. broad; seeds squamulose- or papillose-roughened. Fringed Gentians, § Crossopetalum. . . *c.*
c. Upper leaves ovate to ovate-lanceolate, broadly rounded or subcordate at base; upper half of corolla-lobes fimbriate with a uniformly long fringe 2–6 mm. long; capsule distinctly stipitate. . . . . . . 1. *G. crinita.*
c. Upper leaves linear, linear-lanceolate, narrowly oblong or spatulate; fringe of corolla-lobes of varying lengths, often long below but of short triangular teeth less than 2 mm. long above; capsule sessile to stipitate. . . *d.*
d. Upper leaves linear or linear-lanceolate, acute or acutish, the pairs extending well up the stem; calyx with all 4 lobes acute to attenuate. . . *e.*
e. Elongate leaves (above the basal loose rosette) 8–13 pairs below the

---

* The figures show a flower, often a leaf or leaves, in fig. 1412 the papillose keel of the calyx-lobe, in fig. 1413 a stamen, in some the ovary or capsule, in fig. 1418 the bearding of the corolla, and in several the enlarged seed.

primary peduncle; calyx-lobes (except in a rare form) with strongly papillose-scabrous keels, the longer lobes 1–3 cm. long; corolla 2–5.5 cm. long; its lobes 1–1.8 cm. broad, after anthesis becoming spreading-ascending, the longer marginal teeth 2–4 mm. long; ovary nearly or quite sessile. . . . . 2. *G. procera.*

e. Elongate leaves (above basal rosette) 2–7(–9) pairs below the primary peduncle; keels of calyx-lobes smooth or merely granulose, longer lobes 0.5–1.7 cm. long; corolla 2–4.5 cm. long; its lobes 3–12 mm. broad, after anthesis strongly ascending, the longer marginal teeth wanting or up to 2 mm. long. *f.*

f. Fringe of lower half of corolla-lobes of slender teeth 1–2 mm. long; keels of calyx-lobes granulose-roughened.

Filaments more or less ciliate near the middle; teeth around summits of corolla-lobes rarely 0.5 mm. long. . . . . 3. *G. Macounii.*

Filaments glabrous; teeth around summits of corolla-lobes mostly 0.5–1.5 mm. long.

Corollas of primary flowers 2–3 cm. long; their lobes 3–5 mm. broad, with summit-teeth 0.5–rarely 1 mm. long; ovary barely stipitate, tapering to a thick neck below the stigmas. 4. *G. gaspensis.*

Corollas of primary flowers 2.5–4.5 cm. long; their lobes 1–1.2 cm. broad, with summit-teeth mostly 1–1.5 mm. long; ovary stipitate, with definite slender style. . . . . 5. *G. Victorinii.*

f. Fringe of lower half of corolla-lobes wanting or of few deltoid teeth rarely 1 mm. long; keels of calyx-lobes smooth. . . . . 6. *G. tonsa.*

d. Upper leaves oblong, spatulate or oblanceolate, obtuse or round-tipped, mostly crowded on lower third of plant; primary peduncle two-fifths to nine-tenths the entire height of plant; broad calyx-lobes obtuse or merely subacute; corolla-lobes merely dentate. . . . . 7. *G. nesophila.*

b. Flower small; calyx 0.3–1.3 cm. long, cleft one-half to three-fourths to base; corolla light violet, lilac or creamy-white, 0.5–2.3 cm. long, its acute or tapering entire or merely denticulate lobes 1.5–4.5 mm. broad; seeds with smooth close coats. FELWORTS, § ENDOTRICHA. *g.*

g. Corolla-lobes acute or obtuse, without aristate tips, strongly fimbriate-crowned at base (at the throat). . . . . 8. *G. Amarella.*

g. Corolla-lobes bristle-tipped, without a crown at base.

Leaves oblong to oblanceolate, the 2–5 pairs very remote; pedicels often many times longer than flowers; calyx with very unequal foliaceous lobes, 2 of them oblong or ovate; capsule sessile. . . . 9. *G. propinqua.*

Leaves (except the basal) ovate or ovate-lanceolate, with rounded or clasping bases, the 5–16 pairs not very remote; pedicels rarely equaling the flowers; calyx with subequal to unequal lance-subulate to oblong-lanceolate lobes; capsule slender-stalked. . . . 10. *G. quinquefolia.*

a. Flowers 5-merous; corollas with thin membranous teeth, appendages or plaits between the firm incurved to spreading lobes; capsules stipitate; perennials, often with clustered stems. § PNEUMONANTHE. . *h.*

h. Flowers long-peduncled and solitary, without basal bracts; corolla open-funnelform, with wide-spreading azure-blue to pink or whitish lobes; anthers not connected; seeds wingless. . . . . 11. *G. autumnalis.*

h. Flowers mostly clustered at summit of stem or in upper axils, sessile or very short-peduncled, with 1 or 2 basal bracts; corollas with erect or ascending blue-violet to yellowish or white lobes; anthers cohering in a ring or short tube (or free in nos. 12 and 13); seeds winged (except in no. 19). CLOSED GENTIANS. . . *i.*

i. Anthers free or promptly separating; seeds restricted to region of the sutures; leaves firm or rigid, the principal ones 1.5–5 cm. long.

Leaves 13–19 pairs below the inflorescence; calyx-tube about 1 cm. long; corolla 3.5–5 cm. long, the lobes twice as long as the 2-cleft appendages. . . . . 12. *G. puberula.*

Leaves 7–13 pairs below the inflorescence; calyx-tube 5–7 mm. long; corolla 2–3 cm. long, the lobes only slightly exceeding the laciniate appendages. . . . . 13. *G. affinis.*

i. Anthers often cohering in a ring, tube or cone around the style; seeds covering inner walls of the capsule; leaves membranaceous to subcoriaceous, mostly longer. . . *j.*

j. Calyx-lobes distinctly ciliate. . . *k.*

k. Corolla with nearly truncate summit; the firm true lobes nearly obsolete, narrowed at summit, the broader intervening thin prolongations of the membranous bands forming a fimbriate-dentate border. 14. *G. Andrewsii.*

k. Corolla-lobes broad, rounded, 2–10 mm. long, as broad as or broader than the intervening simple to cleft appendages. . . *l.*

l. Calyx-tube pubescent, 0.7–1.5 cm. long, the lobes 2–8 mm. long; corolla 3–4 cm. long. . . . . 15. *G. decora.*

l. Calyx-tube glabrous, mostly longer, the lobes mostly longer; corolla 2.5–5 cm. long. *m.*

m. Leaves lanceolate to lance-ovate, acuminate; involucre of terminal inflorescence of 4–6 leaves, the 2 outer leaves (except in dwarf individuals) 5–10 cm. long and 2–4 cm. broad; calyx-lobes herbaceous, oblong to obovate, finally spreading; corolla 2.5–4 cm. long, scarcely opening, its lobes incurved. . . 16. *G. clausa.*

m. Leaves elliptic to oblong or linear, acute or obtuse, not acuminate; upper involucre of 2–4 leaves, the 2 outer leaves 3–7 cm. long and 0.5–2.5 cm. broad; calyx-lobes firm, linear to oblanceolate, strongly ascending; corolla 3–5 cm. long, open at summit, its lobes erect or ascending. . . *n.*

n. Stem glabrous; leaves oblong to narrowly elliptic or linear, mostly 4–8 cm. long; corolla cylindric-oblanceolate, 1–1.5 cm. broad at the slightly open summit, the lobes only slightly if at all longer than the thin appendages.

Leaves oblong to narrowly elliptic; outer involucral bracts similar, 0.7–2 cm. broad; calyx-lobes 2–5 mm. broad. . 17. *G. Saponaria.*

Leaves linear or linear-lanceolate; outer involucral bracts 4–6 mm. broad; calyx-lobes 1–2 mm. broad. . . . . 18. *G. cherokeensis.*

n. Stem puberulent; leaves lanceolate to broadly oblong-ovate, mostly 1.5–7 cm. long; corollas upwardly enlarged, campanulate, wide-open, 1.5–2.5 cm. broad at summit, the lobes much longer than the appendages. . . . . . . 19. *G. Catesbaei.*

j. Calyx-lobes smooth-margined or at most scabrous; appendages much shorter than corolla-lobes. . . *o.*

o. Principal leaves narrowly obovate or obovate-oblong, blunt; bracts similar, overtopping the pale striped corollas; corolla-lobes triangular, acute; seeds wingless. . . . . 20. *G. villosa.*

o. Principal leaves linear to lance-ovate, acute; flowers standing well above the involucral bracts; corolla-lobes round-tipped; seeds winged. . . *p.*

p. Calyx with narrow acute sinuses, the ovate to cordate lobes spreading or reflexed; corolla yellowish-white or greenish. . . . 21. *G. flavida.*

p. Calyx with broad rectangular sinuses and linear, oblanceolate or oblong lobes; corolla blue-violet (white only in albinos).

Middle and upper leaves with rounded or cordate bases; involucral bracts cordate, ovate, 1.5–3 cm. broad; corolla-lobes three to five times as long as the appendages. . . . . . . 22. *G. rubricaulis.*

Middle and upper leaves like the lower, all narrowed to base; involucral bracts linear, lanceolate or narrowly oblong, narrowed to base, 0.3–1.5 cm. broad; corolla-lobes once-and-a-half to thrice length of appendages. . . . . 23. *G. linearis.*

§ Crossopétalum Froel.

1. G. crinìta Froel. (having long hairs or fringe), Fringed G. — Biennial or annual; stem simple or with upright branches, 0.1–1 m. high, with 1–176 (!) peduncled flowers; lowest *leaves* spatulate; *the middle and upper ovate to ovate-lanceolate,* 0.7–2.5 cm. broad, *broadly rounded to subcordate at base;* peduncles 2–12 cm. long; *calyx* acuminate in bud, 1.6–4 cm. long, *with sharply keeled lobes and tube; the lobes conspicuously unequal,* the 2 shorter (inner) oblong to ovate and hyaline-margined, *the 2 outer lance-attenuate;* corolla campanulate-funnelform, deeply 4-lobed, 3.5–6 cm. high; *the violet-blue,* or white in forma **albìna** Fern. (whitish), *cuneate-obovate lobes wide-spreading in sunshine, conspicuously fringed around the broad summit with slender teeth 2–6 mm. long;* nectariferous glands 4, at base of corolla; *capsule* spindle-shaped or lanceolate, *with a distinct stipe* and slender beak; seeds conspicuously papillose. (*Anthopogon* Raf.) — Meadows, brooksides, wet thickets and low woods, centr. Me. to s. Man., s. to Ga., O., Ind. and n. Ia. Late Aug.–Nov. Fig. 1411. — Seeds often all blowing to new areas and the plants disappearing from former stations.

1411. G. crinita.

1412. G. procera.

2. G. procèra Holm (tall), FRINGED G. — Similar to no. 1, 1.5–5 dm. high, with 1–24 flowers; *leaves* 8–13 *pairs below the primary peduncle; the median and upper linear-lanceolate, acute,* 0.2–1.2 cm. broad; peduncles 0.2–1.5 dm. long; *calyx* 1.5–4.5 cm. long, *the* longer *lobes* linear-lanceolate and 1–3 cm. long, the shorter broader and with membranous margins, *all acuminate and with strongly papillose-scabrous keels,* or keels smooth in the rare Mich. forma **laevícalyx** Fern. (with smooth calyx); *corolla* 2–5.5 *cm. long;* the blue-violet *lobes* spreading in anthesis, *finally spreading-ascending,* 1–1.8 *cm. broad, fringed on the sides with slender teeth 2–4 mm. long, dentate across the summit;* ovary and capsule nearly or quite sessile. (*Anthopogon* Rydb.) — Boggy prairies, sandy swamps, borders of sloughs, wet calcareous rocks, etc., w. N.Y. and s. Ont. to Alaska, s. to O., Ind., n. Ill., Ia. and N.D. Mid-Aug.–Oct. FIG. 1412.

3. G. Macoùnii Holm (for its discoverer, JOHN MACOUN, 1831–1920). — Smaller than no. 2; stem 1.5–4 dm. high, simple or with few erect branches, 1–8-flowered; *leaves* 2–8 *pairs below the long primary peduncle* (a third to half the height of the plant), 1.5–4 mm. wide; *calyx* 1.5–2.5 cm. long, *with granulose-roughened keels; corolla* 2.3–4 cm. long; the oblong-obovate finally suberect *lobes with few slender marginal teeth* 1–2 *mm. long,* the summit-teeth short and deltoid; *filaments ciliate near the middle;* ovary subsessile or with short thick stipe. (*Anthopogon* Rydb.) — Boggy prairies, subsaline marshes and shores, Hudson Bay, Ung., to B.C., s. to w. Ia., N.D. and Mont. Late July–Sept. FIG. 1413.

1413. G. Macounii.

1414. G. gaspensis.

4. G. gaspénsis Vict. (of Gaspé Peninsula). — Similar to no. 3; calyx purplish; *corollas* 2–3 *cm. long, their lobes* 3–5 *mm. wide and with summit-teeth* 0.5–*rarely* 1 *mm. long; filaments glabrous; ovary tapering to a thick neck, subsessile.* — Brackish to fresh marshes, mouth of Bonaventure R., Gaspé Pen., Que. Aug., Sept. FIG. 1414.

5. G. Victorínii Fern. (for its discoverer, JOSEPH-LOUIS-CONRAD KIROUAC, FRÈRE MARIE-VICTORIN, 1885–1944). — Larger than nos. 3 and 4; stem 1–5 dm. high, 1–30-flowered; *elongate leaves* 4–10 pairs below primary peduncle (one-sixth to one-third height of plant), 2–7 *mm. wide;* longer calyx-lobes subulate-acuminate; *corollas* 2.5–4.5 *cm. long; their lobes* 1–1.2 *cm. broad, with summit-teeth mostly* 1–1.5 *mm. long; filaments glabrous; ovary stipitate, with definite slender style* 1–2 *mm. long.* — Fresh to brackish tidal estuary of the St. Lawrence R., Que. Late July–early Sept. FIG. 1415.

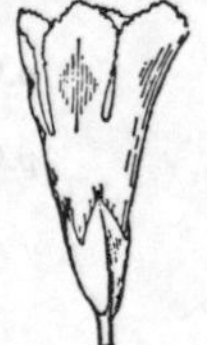
1415. G. Victorinii.

6. G. tōnsa (Lunell) Vict. (shaved). — Like no. 3, but *calyx with smooth keels; fringe of lower half of corolla-lobes wanting or of few deltoid teeth rarely* 1 *mm. long;* filaments glabrous. (*Anthopogon* Rydb.) — Brackish or fresh marshes and shores, James Bay to Alta., s. to Minn. and S.D. Late June–Sept. FIG. 1416. — Often mixed with no. 3; perhaps a variation of it.

1416. G. tonsa.

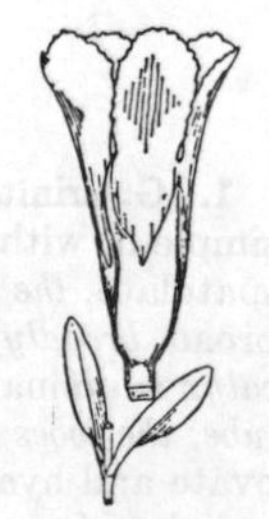

7. G. nesóphila Holm (island-loving). — Biennial or annual, 0.2–2.3 *dm. high,* commonly branching from the base, 1–20-flowered; *leaves rather crowded* near the base in branching plants, more scattered in simple ones, *somewhat fleshy,* glaucous, *oblong, spatulate or oblanceolate, obtuse;* the median ones (0.5–) 1–5 cm. long, 0.2–1.5 cm. broad; peduncles (0.3–)1–11 cm. long, strongly angled; *calyx* 0.7–3 cm. long; *its lobes* subequal, *with smooth keels, the 2 broader lobes blunt or merely acute; corolla* 1.2–5 cm. long, the blue-violet to creamy *lobes merely dentate;* capsule sessile or subsessile. — Calcareous turf, gravels or rocks, n. and w. Nfld.: Mingan Ids. and Anticosti I., Que. Mid-July–early Sept. FIG. 1417.

1417. G. nesophila.

§ ENDÓTRICHA Froel.

8. G. Amarélla L. (old generic name, referring to the bitter taste), FELWORT. — Annual or biennial, 0.5–5 dm. high, simple or *with strongly ascending nerve-angled branches;* radical leaves oblanceolate to spatulate-obovate, obtuse, often persistent; cauline leaves lanceolate to narrowly ovate, acuminate, 0.6–5 cm. long, 0.3–2 cm. broad at base, 3–7-nerved; *upper branches usually short, lower longer and often much elongated;* pedicels 0.1–2 cm. long; *calyx* 0.3–1.3 cm. long, *cleft three-fourths to base* into slightly unequal lanceolate lobes; *corolla* funnelform or salverform, 1–1.7 *cm. long,* its slender *tube exserted, fringed at the throat;* the light violet or lilac, or creamy-white in forma **Michauxiàna** Fern. (for André Michaux, 1746–1802), oblong-lanceolate entire *acute to obtusish lobes* one-fourth to one-third the length of the tube, expanded in the morning; *capsule sessile or subsessile;* seeds smooth. (*G. Amarella*, var. *acuta* (Michx.) Herder; *G. acuta* Michx.; *G. plebeja* Cham.; *Amarella acuta* Raf.) — Damp turf, gravel, beaches, dunes, wet (mostly calcareous) rocks, etc., Lab. to Alaska, s. to w. Nfld., M.I., N.B., n. Me., Smuggler's Notch, Vt., n. Minn., S.D., N.M. and Calif. Mid-July–Sept. Fig. 1418. — Characteristic circumboreal type, either one variable species or innumerable minor ones.

1418. G. Amarella.

9. G. propínqua Richards. (related to; in this case to *G. Amarella*). — Lower than no. 8, 0.3–2.5 dm. high; *usually with the basal branches more numerous, elongate and slender;* the 2–5 pairs of oblong to lanceolate leaves very remote; pedicels often much longer than flowers; *calyx-lobes very unequal, the broader ones ovate to oblong and foliaceous; corolla* 1–2 cm. long, *without crown at throat, its* ovate-lanceolate acuminate lavender *lobes bristle-tipped; capsule sessile.* (*Amarella* Greene) — Turfy or gravelly calcareous slopes and shores, se. Lab. and adjacent e. Que. and nw. Nfld.; w. Can., Alaska and ne. Asia. Mid-July–Sept. Fig. 1419.

1419. G. propinqua.

10. G. quinquefòlia L. (five-leaved), Stiff G., Ague-weed, Gall-of-the-earth. — Annual or biennial, 0.3–8 dm. high, with habit of no. 8; stem and branches wing-angled; lower cauline *leaves* oblong and obtuse; *middle and upper* 4–15 *pairs ovate-lanceolate, clasping by rounded bases,* 0.5–8 cm. long, 0.2–3.5 cm. broad, rather thin, with 3–7 nerves prominent beneath; *pedicels rarely equaling flowers; calyx* 3–9 mm. long, *cleft* to below middle *into subequal lance-subulate lobes; corolla* pale violet-blue or lilac, or greenish-white in forma **lutéscens** Fern. (whitish), 1–2 *cm. long,* funnelform, its triangular-ovate *lobes bristle-tipped; capsule slender-stalked.* (*Gentianella* Small) — Rich woods, wet gravelly banks, damp fields, etc., sw. Me. to w. N.Y., s. to n. Fla. and Tenn. Late Aug.–Nov. Fig. 1420.

1420. G. quinquefolia.

Var. **occidentàlis** (Gray) Hitchc. (western). — *Calyx-lobes subfoliaceous,* linear to oblong-lanceolate, longer; corolla 1.5–2.5 cm. long, with longer lobes. (*Amarella occidentalis* Greene; *Gentianella occidentalis* Small) — S. Ont., s. to Tenn., La. and Mo.

§ Pneumonánthe Link

11. G. autumnàlis L. (autumnal), Pine-barren G. — Very slender, the few or solitary smooth arcuate to suberect simple or 2–3-forked stems 2–6 dm. high; *leaves very narrowly linear or linear-oblanceolate,* thick, 7–15 pairs; *flower slender-peduncled,* naked at base or with linear bracts; calyx-lobes similar to upper leaves; corolla indigo-blue, rarely white or white and green in forma **albéscens** Fern. (whitish), or white and blue in the rare forma **albocaerùlea** Fern. (white and blue), or purple or lilac in the very unusual forma **Porphýrio** (J. F. Gmel.) Fern. (for *Porphyrio*, the Purple Gallinules), the obconic *tube brown-spotted within, the limb strongly spreading to rotate,* the ovate or broad-oblong *lobes* 1.3–2 cm. long and *much longer than the dissected pleats;* anthers distinct; seeds wingless. (*G. Stoneana* Fern.; *G. Porphyrio* J. F. Gmel., technically applicable only to forma *Porphyrio*) — Pine-barrens and borders of pineland, N.J.; se. Va. to S.C. Sept., Oct. Fig. 1421.

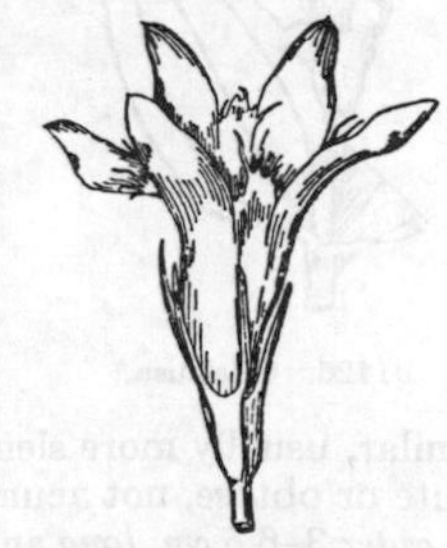
1421. G. autumnalis.

12. G. pubérula Michx. (closely short-pubescent). — Perennial; *stems* 1–few, 2–5 dm. high, *puberulent,* slender, rather rigid, *with* 13–19 *pairs of* linear-lanceolate to oblong-ovate *stiff revolute- and scabrous-margined leaves* mostly 3–5 cm. long; flowers crowded at summit and in upper axils; *calyx-tube about* 1 *cm. long,* the *lobes rigid and linear-lanceolate to subulate; corolla open-funnelform,* 3.5–5 cm. long, *the* blue-purple *spreading lobes twice as long as the 2-cleft appendages; anthers separate or promptly separating; seeds restricted to region of sutures,* winged. (*Dasystephana* Small) — Dry calcareous prairies, barrens and sandy ridges, Ga. to Kans., n. to w. N.Y., s. Ont., Mich., Wisc., Minn. and N.D. Sept., Oct. Fig. 1422.

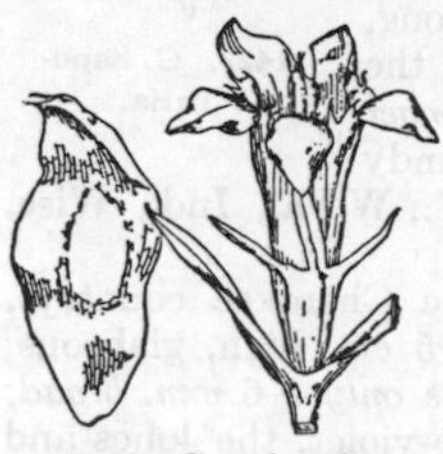
1422. G. puberula.

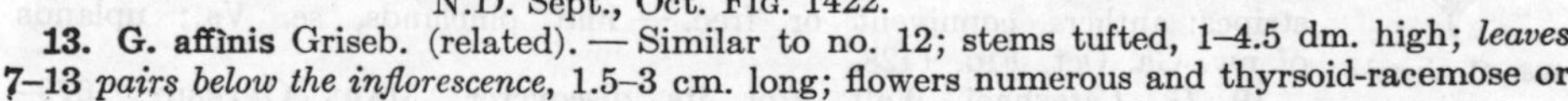

13. G. affìnis Griseb. (related). — Similar to no. 12; stems tufted, 1–4.5 dm. high; *leaves* 7–13 *pairs below the inflorescence,* 1.5–3 cm. long; flowers numerous and thyrsoid-racemose or few or almost solitary; *calyx-tube* 5–7 *mm. long; the lobes unequal,* the longest ones rarely equaling the tube, the shortest often minute; *corolla* blue or bluish, 2–3 *cm. long, slenderly funnelform,* the spreading *lobes only slightly exceeding the laciniate appendages.* (*Dasystephana* Rydb.) — Damp soil, Pacific slope and Rocky Mts., e., locally, to w. Minn. and S.D. July, Aug. Fig. 1423.

1423. G. affinis.

14. G. Andrewsii Griseb. (of Henry C. Andrews, English botanical artist and engraver, who published from 1794–1830), Closed G. — Perennial with stoutish caudex and coarse long roots; stems usually tufted, smooth, ascending, 3–8 dm. high; lowest leaves reduced to small connate bracts; *median and upper leaves* gradually larger, *lanceolate to oblong-ovate, acuminate, narrowed to the sessile base; the uppermost* 0.3–1.1 dm. long, *forming a 4–6-leaved involucre;* flowers 3.5–4.5 cm. long, sessile, 2–5 in a terminal cluster and some often in the upper axils; calyx-lobes herbaceous, ciliate, lanceolate to ovate, mostly spreading, shorter than the chartaceous tube; *corolla* subcylindric, blue-violet, or pinkish in forma **rhodántha** Rouleau & Kucyniak (rose-flowered), or white in forma **albiflòra** Britt. (white-flowered), *with nearly truncate summit, the broader intervening thin prolongations of the membranous bands forming a fimbriate-dentate border;* capsule distinctly exserted, long-stipitate; seeds oblong, whitish-brown, lustrous, 2–3 mm. long, broadly winged. (*Dasystephana* Small) — Meadows, damp prairies, low thickets, etc., Ga. to Ark., n. to e. Mass., Vt., sw. Que., s. Ont., Man. and Sask. Aug.–Oct. Fig. 1424.

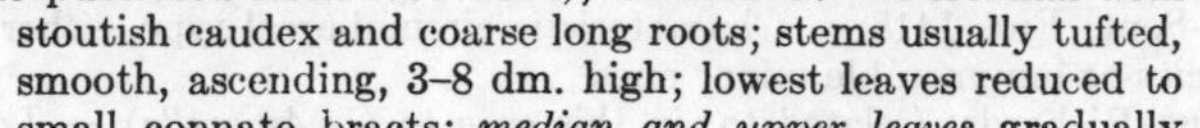
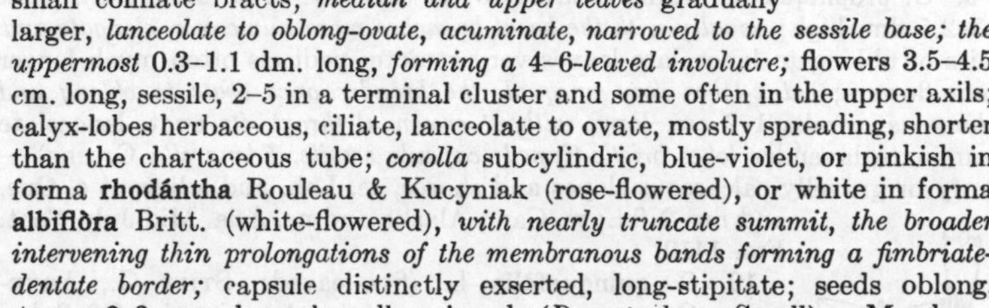

1424. G. Andrewsii.

15. G. decòra Pollard (beautiful). — Resembling no. 16; *stems* slender, 3–5 dm. high, *minutely puberulent above; leaves* 6–8 pairs below the inflorescence, *elliptic-lanceolate to narrowly ovate, acuminate to tip and base,* mostly 4–9 cm. long; involucre of terminal inflorescence of 2–4 leaves; flowers all terminal or in 1 or 2 pairs of upper axils; *calyx-tube pilose,* 0.7–1.5 *cm. long,* the subulate to narrowly lanceolate ciliate *lobes* 2–8 *mm. long; corolla* 3–4 *cm. long,* its erect lobes longer than the broad appendages. (*Dasystephana* Small) — Moist woods, uplands of W.Va., Va., and Ky. to n. Fla. Sept., Oct. Fig. 1425.

1425. G. decora.

16. G. claùsa Raf. (closed), Closed, Blind or Bottle-G. — Similar to no. 14, sometimes smaller; *leaves lanceolate to ovate-lanceolate, acuminate; involucre of terminal inflorescence of* 4–6 *leaves; the* 2–4 *outer leaves* (except in dwarf specimens) subequal, 5–10 *cm. long,* 2–4 *cm. broad; calyx-lobes herbaceous,* ciliate, *oblong to obovate, finally spreading; corolla* 2.5–4 cm. long, porcelain-blue changing to blue-violet, *scarcely opening; the roundish lobes slightly incurved,* 2–8 mm. long, *as broad as or broader than the intervening* 2–3*-cleft appendages.* — Borders of rich woods and thickets, stream-banks, meadows, etc., N.E. to Minn., s. to Md., uplands of N.C. and Tenn., and Mo. Late Aug.–early Oct. Fig. 1426.

1426. G. clausa.

17. G. Saponària L. (from resemblance of foliage to that of *Saponaria*), Soapwort-G. — Similar, usually more slender; *stem glabrous; leaves oblong to narrowly elliptic,* acute or obtuse, not acuminate, mostly 4–8 cm. long; *involucre of 2–4 leaves, the outer* 3–6.5 *cm. long* and 0.7–2 *cm. broad; calyx-lobes firm, oblong to oblanceolate, ascending; corolla cylindric-oblanceolate,* 3–5 cm. long, 1–1.5 *cm. in diameter at the slightly open summit;* the roundish to subacute purplish *lobes erect, only slightly longer than the appendages.* (*Dasystephana* Small) — Glades, sandy swamps, shores and bogs, Ga. to e. Tex., n. to N.Y., W.Va., Ind., Wisc. and Minn. Sept., Oct. Fig. 1427.

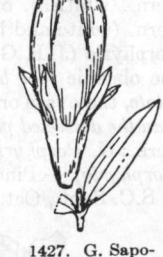

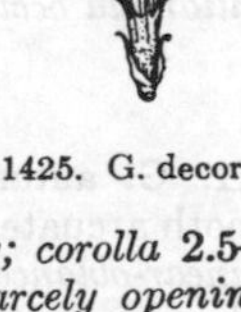
1427. G. Saponaria.

18. G. cherokeénsis (W. P. Lemon) Fern. (of the Cherokee country). Cherokee G. — More slender than no. 17; stem 2.5–5 cm. high, glabrous; *leaves linear to linear-lanceolate; outer involucral bracts only* 4–6 *mm. broad; calyx-lobes linear-lanceolate,* 1–2 *mm. broad;* corolla blue-violet, the lobes and pleats subequal, the tube and inside of throat with conspicuous dark blue stripes; anthers connivent or free. — Flat pinelands, se. Va.; uplands of nw. Ga. Oct. Fig. 1428.

1428. G. cherokeensis.

19. G. Càtesbaei Walt. (for its discoverer, Mark Catesby, 1679–

1749), Catesby's G., Sampson's Snakeroot. — In habit like no. 17; *stems puberulent*, slender, 3–8 dm. high, simple or loosely branched; *leaves* firm, narrowly lanceolate to oblong-ovate, acute or acutish, *mostly 3–7 cm. long and 1–3 cm. broad;* flowers 1–7 at summit of stem, 1–3 on the branches, involucrate; calyx-lobes linear to narrowly lanceolate, erect, equaling or longer than tube; *corolla* deep blue to blue-violet or violet-purple, *campanulate, upwardly enlarged*, 1.5–2.5 *cm. in diameter at the wide-open summit*, the *lobes only slightly longer than the appendages.* (*Dasystephana parvifolia* and *latifolia* Small; *G. Elliottii* Chapm., not *G. Elliottea* Raf.; *G. parvifolia* (Chapm.) Britt.) — Damp sands, peats, wet woods and pinelands, Fla. and Ala., n. on Coastal Plain to Del. Sept., Oct. Fig. 1429.

1429. G. Catesbaei.

1430. G. villosa.

Var. **nummulariaefòlia** Fern. (with leaves of *Lysimachia Nummularia*). *Leaves membranaceous, blunt to round-tipped, elliptic-oval, only* 1–1.8 *cm. long; calyx-lobes broadly oblanceolate.* — Sphagnous bogs, very local, se. Va.

20. G. villòsa L. (long-hairy; a misnomer, the misdescribed type being glabrous), Sampson's Snakeroot. — Perennial with 1–several loosely ascending smooth stems 1–6 dm. high; *principal leaves narrowly obovate to obovate-oblong, blunt*, all narrowed to base, 4–10 cm. long, *the uppermost or involucral leaves overtopping the clustered flowers; calyx-lobes linear, smooth*, unequal, much longer than the tube; *corolla greenish-white to purplish-green, striped, with loosely ascending acute triangular lobes* much exceeding the oblique appendages; *seeds wingless.* (*Dasystephana* Small) — Open woods and pinelands, Fla. to La., n. to N.J., Pa., s. O. and s. Ind. Late Aug.–Oct. Fig. 1430.

21. G. flàvida Gray (yellowish), Yellowish G. — Stout perennial with smooth stem; *leaves ovate-lanceolate, acuminate, cordate-clasping*, the upper 7–11 cm. long; flowers 3–4.5 cm. long, crowded in terminal and sometimes axillary clusters; *calyx with acute sinuses, the ovate to cordate smooth herbaceous lobes spreading or reflexed; corolla yellow-white or greenish*, cylindric-campanulate, *open at summit*, with erect ovate lobes much longer than the broad erose appendages. (*Dasystephana* Britt.) — Damp woods, prairies and meadows, N.C. to Ark., n. to Pa., s. Ont., Mich., Ill., Minn. and Man. Mid-Aug.–early Oct. Fig. 1431.

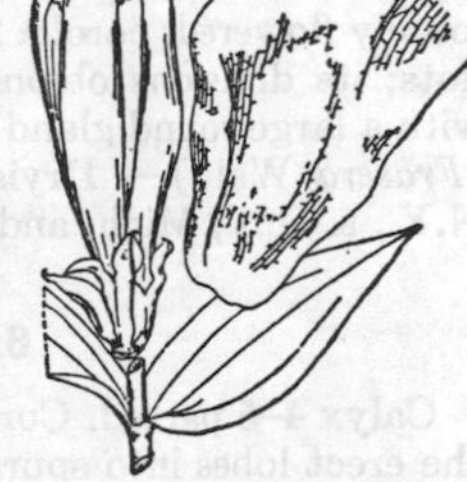

1431. G. flavida.

22. G. rubricaùlis Schwein. (red-stemmed), Closed G. — Perennial; stems few–several, commonly tinged with red, smooth, 3–8 dm. high; *leaves* 5–10 *well-developed pairs* below the inflorescence; *the middle and upper ones* lanceolate to ovate, *rounded to cordate at base; involucral leaves ovate, cordate,* 1.5–3 *cm. broad;* inflorescence terminal, or sometimes 1 or 2 flowers axillary; *calyx with broad rectangular sinuses and oblong to narrowly obovate smooth* or barely scabrous-margined *herbaceous lobes; corolla* blue-violet, 2.5–3.5 cm. long, subcylindric, *with erect or slightly incurved ovate lobes three to five times as long as the appendages;* seeds distinctly winged. (*G. linearis*, var. *latifolia* Gray; *G. Grayi* Kusnezow; *Dasystephana Grayi* Britt.) — Low meadows, woods and shores, local in Charlotte Co., N.B., and s. Somerset Co., Me.; s. Ont. and Mich. to Minn. and Neb. Late July, Aug. Fig. 1432.

1432. G. rubricaulis.

23. G. lineàris Froel. (linear; referring to the leaves), Closed G. — Similar to no. 22; stems 1.5–7 dm. high, slender; *leaves* 7–12 (–15) *well-developed pairs* (rarely 3's) below the inflorescence; *the middle and upper linear, lanceolate or narrowly oblong, narrowed to base; involucral leaves similar,* 0.3–1.5 *cm. wide;* inflorescences terminal and often in upper axils; *calyx-lobes linear or lanceolate*, smooth or barely scabrous; *corolla* porcelain-blue to blue-violet, or white in forma **Blanchárdii** Fern. (for its discoverer, William Henry Blanchard, 1850–1922), subcylindric or thick-clavate, 3–5.5 cm. long, *with erect or slightly incurved rounded lobes once-and-a-half to thrice length of appendages.* (*Dasystephana* Britt.) — Bogs, wet meadows, swampy woods, wet ledges, etc., chiefly in siliceous and granitic districts, Grand Falls of Hamilton R., Lab. Pen.,

1433. G. linearis.

to L. Sup., s. to s. N.B., Me., Mass., N.Y., n. Pa., Md., W.Va. and Minn., ascending to **1575 m.** (ca. 5250 ft.) alt. Late July–mid-Sept. Fig. 1433.

4. LOMATOGÔNIUM A. Br.

Acute divisions of the showy corolla with a pair of scale-like appendages at base. Stamens inserted at base of corolla. Style none; stigmas decurrent. — Small annuals or biennials of boreal reg. (Name from the Greek *loma*, *hem*, and *gone*, here used in the sense of *pistil*, alluding to the shape of the stigmas.) Pleurogyne Esch.

1. L. rotàtum (L.) Fries (wheel-shaped). — Stem 0.3–2.5 dm. high, simple to bushy-branched, with strongly ascending capillary branches; *leaves fleshy*, the lower spatulate or oblong-obovate; *the middle and upper* 0.4–3 cm. long, *linear-lanceolate and acuminate*, or oblong-lanceolate and obtuse in forma **americànum** (Griseb.) Fern. (American), or oval and obtuse in forma **ovalifòlium** Fern. (oval-leaved); sepals 2–5, similar to median leaves, nearly equaling to much exceeding corolla; corolla-segments porcelain-blue, or white in forma **albiflòrum** Polunin (white-flowered), lanceolate to elliptic, 0.5–1.5 cm. long; ovary oblong, subacuminate, 0.5–1.7 cm. long, with stigmas one-half to three-fourths as long. (*Pleurogyne rotata* (L.) Griseb.) — Turfy or sandy seashores, Lab. and Nfld. to the lower St. Lawrence, Que., s. to se. Me.; Hudson Bay to Alaska, s. to Colo. July–Sept. (Greenl. and n. Eu. to ne. Asia)

5. SWÉRTIA L. Columbo

Calyx deeply 4–5-parted. Corolla with lobes sinistrorsely convolute in bud, rotate, with a fringed glandular spot on each lobe. Filaments subulate, usually monadelphous at base; anthers oblong, versatile. Style persistent; stigma 2-lobed. Capsule oval, flattened, 4–14-seeded. Seeds large and flat, wing-margined. — Tall and showy Am., Eurasian and Afr. herbs with thick root, upright mostly simple stems, whorled leaves and numerous peduncled flowers in open cymes disposed in an ample elongated panicle. (Named for the Dutch herbalist, *Emanuel Sweert*, born in 1552.) Frasera Walt.

1. S. caroliniénsis (Walt.) Ktze. (of Carolina). — Smooth biennial or triennial, 1–2.5 m. high; leaves mostly in fours, lance-oblong, the lowest spatulate, veiny; panicle pyramidal, loosely flowered; corolla 2–3 cm. broad, light greenish-yellow, marked with small brown-purple dots; its divisions oblong, mucronate, longer than the narrowly lanceolate calyx-lobes, each with a large round gland below the middle; capsule much flattened parallel with the flat valves. (*Frasera* Walt.) — Dryish meadows, rich woods and calcareous slopes, Ga. to La., n. to w. N.Y., s. Ont., Mich. and Wisc. May, June.

6. HALÈNIA Borkh. Spurred Gentian

Calyx 4–5 parted. Corolla without folds or fringe, usually prolonged at the base underneath the erect lobes into spurs, which are glandular in the bottom. Stigmas 2, sessile, persistent on the oblong flattish capsule. Seeds rather numerous, oblong. — Small and upright Am. and Eurasian herbs, with yellowish or purplish panicled-cymose flowers. (Named for *Jonas Halenius*, pupil of Linnaeus, who in 1750 described the first recognized species.) Tetragonanthus Gmel.

1. H. defléxa (Sm.) Griseb. (bent downward; referring to the spurs). — Leafy annual or biennial, 1–9 dm. high, simple or branched above; *leaves* 3–5-nerved, the lowest oblong-spatulate and petioled; the others *oblong-lanceolate to ovate, acuminate; internodes mostly elongate*, those above lowest flowering nodes 1–12 cm. long; flowers in terminal and axillary umbelliform cymes (sometimes solitary); corolla greenish or bronze, 0.8–1.5 cm. long, acutely 4-lobed, with curved-descending spurs, or spurs of many or all flowers lacking in forma **heterántha** (Griseb.) Fern. (variable-flowered). — Damp and cool woods, se. Lab. to s. B.C., s. to Nfld., N.S., Me., n. N.H., w. Mass., N.Y., s. Ont., Mich., n. Ill., Minn., S.D. and n. Mont.; mts. of Mex. July–Sept.

Var. **Brentoniàna** (Griseb.) Gray (named in 1836 for its discoverer, Mary G. Brenton). — More *compact*, 0.3–1.5 dm. high, *with shortened internodes; upper and bracteal leaves ovate, with* strongly rounded to *cordate sessile bases; flowers purple or purple-tinged*, broader and shorter (*rarely more than* 1 *cm. long*). — Exposed mossy, turfy or gravelly slopes and coasts, se. Lab. Pen., Nfld., M.I. and C.B

7. BARTÒNIA Muhl.

Stamens short. Capsule thick-cylindric, flattened, pointed by a large persistent at length 2-lobed stigma. Seeds minute, innumerable, covering the whole inner surface of the capsule. —

Small Am. annuals or biennials with filiform stems and tiny subulate scales in place of leaves. Flowers small, peduncled. (Dedicated to *Benjamin Smith Barton*, 1766–1815, Philadelphia botanist.)

Leaves chiefly opposite on stem below the inflorescence; stigma (including the divergent basal lobes) 1.5–2.3 mm. long. . . . . 1. *B. virginica.*
Leaves chiefly alternate or scattered below the inflorescence; stigma 0.8–1.5 mm. long. 2. *B. paniculata.*

1. B. virgínica (L.) BSP. (Virginian). — *Stem wiry*, 0.4–3 dm. high, *yellowish-green*, stiffly erect or slightly flexuous, rarely twining, simple or commonly *with mostly opposite* simple to dichotomous strongly ascending short *branches* (0.5–4.5 cm. long); *scale-like leaves subulate, chiefly opposite, yellowish-green*, persistent, those below the lower branches 1.5–4.5 mm. long; *calyx yellowish, with firm keeled subulate-lanceolate* almost distinct *lobes* 2–4.5 mm. long; *corolla* yellowish or stramineous, 2.6–4.5 mm. long, *with* oblong to lanceolate blunt or acuminate *often erose lobes;* anthers yellowish; *stigma* (including decurrent bases) 1.5–2.3 *mm. long.* — Dry, exsiccated or wet acid soil, Fla. to La., n. to St. P. et Miq., N.S., se. Me., sw. Que., n. N.Y., W.Va., O., Mich., n. Ill. and Minn. July–Sept.

2. B. paniculàta (Michx.) Muhl. (panicled), Screw-stem. — Similar; *stem* less rigid, 0.4–4.5 dm. high, *often flexuous*, sometimes twining, the alternate or opposite simple or forking branches mostly 2–6 cm. long; *leaves scattered or alternate*, greenish-yellow to purple, 1–2.5 *mm. long; corolla* 2–7 *mm. long, with entire lobes;* anthers yellow to purple; *stigma* 0.8–1.5 *mm. long.* — Our most variable species:

Calyx cleft nearly or quite to base; corolla-lobes lanceolate to narrowly oblong, sharply acuminate or at least acute.
Plant yellowish-green, rarely purplish; inflorescence thyrsoid or a simple raceme; leaves and calyx-lobes firm, subulate to linear-lanceolate, yellowish-green or at most purple-tipped; flowers 2.5–5 mm. long; corolla-lobes mostly creamy, lance-acuminate, 0.7–1.5 mm. broad; anthers mostly yellowish. . *B. paniculata* (typical).
Plant purplish or fulvous; inflorescence a simple raceme, rarely subpaniculate; leaves and calyx-lobes fleshy or herbaceous; the latter deeper green to purple, lanceolate to oblong; flowers 3.8–6 mm. long; corolla-lobes often purple-tipped or watery-white, lance-oblong, 1.2–2 mm. broad; anthers mostly purple. . . . . Var. *intermedia.*
Calyx cleft (at least on one side) only two-thirds to three-fourths way to base; its lobes herbaceous, oblong to ovate; corolla-lobes petaloid, oblong to narrowly ovate, blunt or merely acutish, 1–2 mm. broad.
Racemes dichotomous to simple; pedicels clavate; 2 or 3 calyx-lobes distinct to base; corolla 3–5 mm. long, creamy-white; anthers mostly yellowish. . . Var. *sabulonensis.*
Racemes mostly simple; pedicels filiform; calyx-tube 1–2 mm. long, its lobes not distinct to base; corolla 4–7 mm. long, often purplish; anthers mostly purple. . . . . Var. *iodandra.*

B. paniculàta (typical). — Wet peat and sand, Fla. to La., n. to N.S., s. N.E., s. N.Y., Ky., Ark. and Okla. Aug.–Oct.

Var. **intermèdia** Fern. (intermediate). — N.S. and, less characteristically, s. to N.J. Aug., Sept.

Var. **sabulonénsis** Fern. (of Sable Island). — St. P. et Miq. and N.S., local. Aug., Sept.

Var. **iodándra** (Robins.) Fern. (with purple anthers), *B. iodandra* Robins. — Wet peat, Nfld. and C.B. Aug., Sept.

8. OBOLÀRIA L. Pennywort

Calyx of 2 spatulate spreading sepals resembling the leaves. Corolla withering-persistent; the lobes oval-oblong, or with age spatulate, imbricated in the bud. Stamens inserted at the sinuses of the corolla, short. Style short, persistent; stigma 2-lipped. Capsule ovoid, 1-locular, the locule cruciform; the seeds covering the whole face of the walls. — A low and very smooth purplish-green perennial 6–15 cm. high, with a simple or sparingly branched stem, opposite cuneate-obovate leaves; the dull white or purplish flowers solitary or in groups of three, terminal and axillary, nearly sessile, in spring. (Name from *obolos*, a small Greek coin, from the thick rounded leaves.)

1. O. virgínica L. (Virginian). — Herbaceous and fleshy; the lower leaves scale-like; flowers 1–1.5 cm. long. — Moist hardwoods and thickets, Fla. to Tex., n. to N.J., Pa., W.Va., O., Ind. and s. Ill. March–May.

Subfam. II. MENYANTHOÌDEAE

9. MENYÁNTHES L. Buckbean. Bogbean. Herbe à canards (Que.)

Calyx 5-parted. Corolla short-funnelform, 5-cleft, deciduous, the whole upper surface white-bearded. Style slender, persistent; stigma 2-lobed. Capsule bursting somewhat irregularly, many-seeded. Seed-coat hard, smooth and shining. — A boreal perennial herb with a thickish creeping rootstock sheathed by the membranous bases of the long petioles, which bear 3 oval or oblong leaflets; the flowers racemed on the naked scape (1–3 dm. high), white or slightly reddish. (Name used by Theophrastus, from *menyein, disclosing,* and *anthos, a flower,* later applied to this genus with flowers progressively expanding in the raceme.)

1. M. trifoliàta L. (three-leaved). — Corolla 1.5–3 cm. across, the upper surfaces white or often pink or roseate, the long beard covering nearly all the upper faces of the lobes. — Eurasia and Pacific N.Am. Represented with us by

Var. **minor** Raf. (smaller). — Corolla 1.5–2 cm. broad, the upper face white, or colored only at tips; beard shorter, most abundant on lower half of lobes. — Quagmires, shallow water and pond-margins, Lab. to Alaska, s. to Nfld., N.S., N.E., Del., Md., w. Va., W.Va., centr. O., Ind., Ill., Mo., Neb. and Wyo. Late Apr.–mid-July (rarely –Sept.).

10. NYMPHOÌDES Hill Floating-heart

Calyx 5-parted. Corolla almost rotate, 5-parted, the divisions bearing a glandular appendage near the base. Style short or none; stigma 2-lobed, persistent. Capsule few–many-seeded, at length bursting irregularly. Seed-coat hard. — Perennial Am., Eurasian and Afr. aquatics, with floating leaves on very long petioles which, in most species, bear near the summit the umbel of polygamous flowers, often along with a cluster of short and spur-like roots; flowering all summer. (Name from *Nymphaea* and *eidos, appearance.*) Limnanthemum Gmel.

Rhizomatous; flowering stems simple; calyx 3–8 mm. long; corolla white, at most 2 cm. broad, the segments not fringed; capsule 3–9 mm. long, scarcely beaked; seeds without fringe.
  Floating leaves cordate-ovate, 1.5–5 cm. broad; spur-like roots frequent near umbel; corolla 0.5–1 cm. broad; capsule ovoid-subglobose, 3–5 mm. long; seeds smooth. 1. *N. cordata.*
  Floating leaves suborbicular to reniform, 0.4–1.5 dm. broad; spur-like roots infrequent; corolla about 2 cm. broad; capsule elongate, 6–9 mm. long; seeds roughened. . . . . 2. *N. aquatica.*
Extensively creeping, branching; calyx 1–1.3 cm. long; corolla yellow, 2–3 cm. broad, its segments fringed; capsule 1.2–2.5 cm. long, strong-beaked; seeds fringed. . . . . 3. *N. peltata.*

1. N. cordàta (Ell.) Fern. (heart-shaped). — Rhizome stout; *floating leaves cordate-ovate* subcoriaceous, 1.5–5 *cm. broad, the filiform smooth petioles* frequently *bearing spur-roots* among or near the flowers; pedicels filiform, 0.5–6 cm. long; *calyx* 3–5 *mm. long; corolla* white, with a crest-like yellow gland at base of each oval lobe, 0.5–1 *cm. broad;* style none; *capsule ovoid-subglobose, little exceeding calyx,* 3–5 *mm. long,* beakless; *seeds smooth.* (*N. lacunosa* of recent authors, not (Vent.) Ktze., which by typification is a synonym of no. 2.) — Ponds and slow streams, Nfld. to Ont., s. to N.S., N.E., Fla. and La. July–Sept.

2. N. aquática (Walt.) Ktze. (aquatic). — Coarser; *floating leaves suborbicular to reniform,* much heavier in texture, 0.4–1.5 *dm. broad,* the stouter *purple-glandular petioles usually without spur-roots* at summit; pedicels 2–8 cm. long; *calyx* 4–8 *mm. long; corolla* 1–2 *cm. broad; capsule elongate, much exceeding calyx,* 6–9 *mm. long; seeds glandular-roughened.* — Ponds and streams, Fla. to Tex., n., locally, to s. N.J. June–Aug.

3. N. peltàta (Gmel.) Ktze. (shield-shaped), Yellow F. — Coarse, *extensively creeping, branching;* petioles without spur-roots; blades suborbicular, coarsely undulate-dentate; *calyx* 1–1.3 *cm. long; corolla yellow,* 2–3 *cm. broad, its segments somewhat fringed; capsule strongly beaked,* 1.2–2.5 *cm. long; seeds with fringe-like margin.* — Quiet waters, cult. and locally spreading in s. states, n. to Hudson valley, N.Y., and Mo. July–Sept. (Introd. from Eu.)

FAM. 144. APOCYNÀCEAE (Dogbane Family)

Plants almost all with milky acrid juice, entire chiefly opposite leaves without stipules, regular 5-merous and 5-androus flowers; the 5 lobes of the corolla convolute and twisted in the bud; the filaments distinct, inserted on the corolla, and the pollen glandular; calyx free from the two ovaries, which (in our genera) are distinct (forming follicles), though their styles or stigmas are united into one. Seeds amphitropous or anatropous, with a large straight embryo in sparse albumen, often bearing a tuft of down (comose). — Chiefly trop. acrid-poisonous plants.

Leaves alternate; erect herbs with bluish flowers, the corolla-tube bearded inside. 1. *Amsonia.*
Leaves opposite.
Plant creeping or trailing; flowers solitary in the axils; corolla with salverform limb 1.5–5 cm. broad; seeds without coma. . . . . 2. *Vinca.*
Plant erect, ascending, depressed or twining; flowers in cymes; corolla campanulate or funnelform, much smaller; seeds comose.
Woody twiner; calyx glandular within; corolla funnelform, not appendaged, greenish-yellow; filaments slender. . . . . 3. *Trachelospermum.*
Erect or depressed herbs; calyx not glandular; corolla campanulate, white to deep pink, appendaged within; filaments broad and flat. . . . . . 4. *Apocynum.*

1. AMSÒNIA Walt.

Calyx small. Corolla with a slender funnelform tube; the limb divided into long linear lobes. Stamens inserted on the tube, included; anthers obtuse at both ends. Ovaries 2; style 1; stigma rounded, surrounded by a cup-like membrane. Capsules (follicles) 2, long and slender, many-seeded. Seeds cylindrical, abrupt at both ends, packed in one row. — Perennial N. Am. and Japanese herbs, with *alternate leaves*, and pale blue flowers in terminal panicled cymes. (Named for Dr. *Amson*, physician of Gloucester, Virginia, in 1760, and friend of John Clayton.)

Principal leaves lanceolate to ovate or elliptic, relatively few, 1–6 cm. broad; corolla-lobes pubescent outside.
Leaves membranaceous, dull, 1–6 cm. broad; follicles firm, scarcely torulose, tardily dehiscent, erect. . . . . 1. *A. Tabernaemontana.*
Leaves subcoriaceous, lustrous above, 1–2.5 cm. broad; follicles thin-walled, torulose, promptly dehiscent, loosely ascending or drooping. . . . . . 2. *A. illustris.*
Principal leaves linear, very numerous, 0.5–5 mm. broad; corolla-lobes glabrous outside. . . . . 3. *A. ciliata*, var. *filifolia.*

1. A. Tabernaemontàna Walt. (an old generic name, in honor of Jacobus Theodorus Tabernaemontanus, German herbalist of the 16th century). — Stems clustered, pubescent, becoming glabrate, or glabrous, 0.3–1 m. high, the upper branches after the flowering elongating and much overtopping the central inflorescence; *leaves subremote, elliptic-oblong to ovate,* 3–6 *cm. broad, membranaceous, opaque,* glabrous or glabrate; *inflorescence close, short-peduncled; calyx glabrous; corolla* salverform, the tube 6–8 mm. long; the *lobes* wide-spreading to reflexed, narrowly oblong to lanceolate, 5–8 mm. long, *pubescent outside; follicles erect, continuous, firm-walled,* abruptly acuminate, 8–10 cm. long, *tardily dehiscent.* — Rich woods, river-banks, etc., Ga. to La. and e. Okla., n. to se. Va., Tenn., s. Ill., Mo. and e. Kans.; cult. and natzd. ne. to Mass. Apr., May.

Var. **salicifòlia** (Pursh) Woodson (willow-leaved). — Lower (3–5 dm. high); *leaves lanceolate,* glabrous or glabrate and *glaucous beneath,* 1–3 *cm. broad; inflorescence loose,* relatively *few-flowered.* — Ga. to Tex., n. to se. Va., s. Ind., Ill. and Mo.

Var. **Gattíngeri** Woodson (for its discoverer, Augustin Gattinger, 1825–1903). — Stems 3–10 dm. high; *leaves lanceolate, green both sides,* 1–3 cm. broad; *inflorescence dense, many-flowered.* — Ga. to Tex., n. to Tenn., Ill., Mo. and Kans.

2. A. illústris Woodson (lustrous). — Differing from tall plants of no. 1 in its *subcoriaceous lanceolate leaves lustrous above; calyx strigillose-hirsute; follicles loosely spreading or drooping, thin-walled, torulose, promptly dehiscent and readily breaking into 1-seeded articles.* — Sand-bars and gravelly shores, Mo. and Kans. to Tex. May, June.

3. A. ciliàta Walt. (ciliate), var. **filifòlia** Wood (with thread-like leaves). — Stems 0.2–1 m. high, pilose or glabrate; *leaves very numerous, crowded, narrowly linear,* 0.5–5 *mm. broad; inflorescence long-peduncled,* dense; calyx glabrous or sparsely hairy; *corolla glabrous outside;* follicles continuous. (Var. *tenuifolia* (Raf.) Woodson) — Sandhills, barrens and rocky shores, Ga. to Tex. and Mex., n. to N.C. and s. Mo. May, June.

2. VÍNCA L. Periwinkle. Pervenche (Que.)

Calyx-lobes acuminate. Corolla-tube funnelform; the limb salverform. Stamens inserted below the throat; filaments short. Style slender. Capsules short-cylindric. Seeds rough. —

Smooth trailing hardy plants (or in the Tropics tender annuals), nat. of Old World, with firm leaves and axillary flowers. (Name abbreviated from Pliny's *Vincapervinca,* the ancient name reflected in the colloquial Italian *Pervinca,* the French *Pervenche* and the English *Periwinkle.*)

1. V. mìnor L. (smaller), Common P., "Myrtle". — Spreading by creeping stems; leaves glossy, ovate to oblong, 1.5–3 cm. long; flowers peduncled; *calyx-teeth linear-lanceolate,* 2–3 *mm. long, glabrous; corolla* blue-violet (rarely white), 1.5–3 *cm. broad,* with truncate lobes; follicles 2–2.5 cm. long, abruptly beaked. — Roadsides, etc., spread from cult. March–June. (Introd. from Eu.)

2. V. herbàcea Waldst. & Kit. (herbaceous). — Branches herbaceous, from a stoutish base; *leaves lanceolate or narrowly elliptic, subherbaceous, scabrous;* flowers long-stalked; *calyx-teeth linear-lanceolate,* 2–3 *mm. long; corolla* blue-violet, 2.5–3 *cm. across.* — Shaded terraces of Ct. R., Franklin Co., Mass., originally spread from cult. May, June. (Introd. from Eu.)

3. V. màjor L. (larger), Large P. — Trailing, with ascending branches up to 2 m. high; *leaves ovate, ciliolate,* 3.5–8 *cm. long;* pedicels much shorter than leaves; *calyx-lobes linear,* 1–1.5 *cm. long, ciliate; corolla* blue or violet, 3.5–5 *cm. broad;* follicles 3.5–4 cm. long, gradually tapering. — Roadsides and borders of woods, originally spread from cult., Va. to Ga. Apr., May. (Introd. and natzd. from Eu.)

3. TRACHELOSPÉRMUM Lemaire Climbing Dogbane

Calyx with 3–5 glands at its base inside. Stamens included; anthers sagittate, with an inflexed tip. Follicles 2, slender, many-seeded. Seeds oblong. — Twining plants, all but ours of trop. and subtrop. Asia, more or less woody, with small flowers in cymes. (Name from the Greek *trachelos, a neck,* and *sperma, seed,* upon the supposition that the seed was beaked.)

1. T. difförme (Walt.) Gray (of two forms; from the dissimilar leaves). — Nearly herbaceous and glabrous; leaves oval-lanceolate to obovate, pointed, thin; calyx-lobes taper-pointed; corolla pale yellow; follicles slender-cylindric, 1.5–2.5 dm. long. — Swamps and low grounds, twining high over shrubs and trees, Fla. to Tex., n. to Del., D.C., s. Ind., s. Ill., Mo. and Okla. June, July.

4. APÓCYNUM L. Dogbane. Indian Hemp

Calyx-lobes acute. Corolla campanulate, tubular or ovoid, bearing 5 triangular appendages below the throat opposite the lobes. Stamens on the very base of the corolla; filaments shorter than the sagittate convergent anthers, which slightly adhere to the stigma. Style none; stigma large, ovoid, slightly 2-lobed. Fruit of 2 long and slender follicles. Seeds with a tuft of long silky hairs (coma) at apex. — Perennial herbs of N. Hemisph., with branching stems, opposite often mucronate-pointed leaves, a tough fibrous bark, and small and pale cymose flowers on short pedicels. (Ancient name of the Old World Dogbane, composed of the Greek *apo, far from,* and *cyon, a dog.*)

Leaves reflexed or wide-spreading; corolla pink or pink-striped, campanulate, 4–9 mm. long, the lobes more or less spreading or reflexed; seeds 2.5–4 mm. long.
- Inflorescences both terminal and axillary, the cymes flowering simultaneously; flowers soon nodding; corollas 5–10 mm. broad, their lobes soon recurving; seeds 2.5–3 mm. long. . . . . 1. *A. androsaemifolium.*
- Inflorescences terminal and at ends of leafy branches, the central cyme flowering much earlier than the others; flowers ascending or spreading; corollas 3.5–6 mm. broad, their lobes not recurving; seeds 3–4 mm. long. . . 2. *A. medium.*

Leaves ascending or wide-spreading; corolla greenish- to milk-white, tubular to ovoid, pentagonal, 2–4 mm. long, with ascending lobes; cymes terminal, mostly compact, the flowers ascending to spreading; seeds 4–6 mm. long.
- Leaves of primary axis (except sometimes the lowest) definitely petioled; small bracts of inflorescence scarious; fully grown follicles 1.2–2 dm. long, falcate; coma of seed 2.5–3 cm. long. . . . . 3. *A. cannabinum.*
- Leaves of primary axis sessile or nearly so; small bracts of inflorescence green and herbaceous; fully grown follicles 4–10 cm. long, straightish; coma of seed 0.8–2 cm. long. . . . . 4. *A. sibiricum.*

1. A. androsaemifòlium L. (with leaves of *Androsaemum*), Spreading D., Herbe à la puce (Que.). — Stems 1–5 dm. high below the lowest of the *loosely ascending often dichotomous branches;* leaves ovate-oblong to ovate, mucronate-tipped, slender-petioled, *loosely spreading or drooping, dark green* and glabrous *above,* paler and glabrous to pubescent beneath; *cymes both terminal and axillary, flowering simultaneously,* the central usually largest; *the flowers* *spreading-ascending to nodding,* fragrant; *bracts oblong or lanceolate, subpersistent;* calyx rarely half as long as the corolla-tube; *corolla campanulate, pink, with deeper stripes in the tube,* 6–9 *mm. long,* 5–10 *mm. broad, the lobes* prominently flaring and *finally recurved;* follicles slenderly linear-cylindric, attenuate to tip, 0.7–2 dm. long; *seeds* slenderly fusiform, 2.5–3 *mm. long.* — Dry thickets and borders of woods, Nfld. to Alaska, s. to N.S., N.E., upland of N.C., W.Va., O., centr. Ind., Ill., Ark., Neb., N.M. and n. Mex. June–Aug.

2. A. mèdium Greene (intermediate). — Similar to no. 1; branches ascending or spreading; leaves firm, ovate-oblong to elliptic; *cymes terminal or at the tips of elongate leafy branches, the central one flowering first; bracts linear-attenuate, caducous; flowers spreading or slightly nodding; corolla pink-tinged to white,* 4–7 mm. long, 3.5–6 *mm. broad, urceolate-campanulate to short-tubular, the spreading lobes not recurving; seeds* 3–4 *mm. long.* — Dry or moist ground, rocky shores, etc., Nfld. to B.C., s. to N.S., N.E., Va., Tenn., Mo., Tex. and N.M. June–Aug. — A polymorphous series, combining traits of no. 1 and nos. 3 and 4. Several names have been assigned the different phases, many of the latter evidently hybrids of nos. 1 and 3.

3. A. cannàbinum L. (hemp-like), Indian Hemp. — Glabrous or nearly so, 3–6 dm. high, with ascending branches; *leaves* mostly ascending, ovate to lanceolate, glabrous to sparingly pubescent beneath, *those of the primary axis* (except sometimes the lowest) *narrowed at base to distinct petioles* 2–7 *mm. long,* those of the branches often subsessile; *cymes* terminal, *of mostly ascending flowers, the central cyme flowering first;* bracts scarious, linear-attenuate, caducous; *calyx* glabrous, *its lobes about equaling corolla-tube; corolla greenish-white, tubular to ovoid, pentagonal,* 2–4 *mm. long, with erect lobes;* fully grown *follicles* 1.2–2 *dm. long, falcate; seeds* 4–6 *mm. long, with coma* 2.5–3 *cm. long.* — Open ground, thickets and borders of woods, w. Que. to Alta. and Wash., s. to w. and s. N.E., Fla., Ala., Miss., La., Tex., N.M., Ariz. and Calif. June–Aug.

Var. **pubéscens** (Mitchell) A. DC. (pubescent). — *Upper part of stem, leaves, branches of cyme and calyx white-pubescent.* — Se. Mass. to s. Ont. and N.D., s. to Ga., Ala., Miss. and e. Tex.

Var. **nemoràle** (G. S. Mill.) Fern. (of the woods). — *Glabrous* or nearly so; *leaves mostly spreading or drooping on elongate petioles* 1–1.5 *cm. long.* — Open woods, local, n. Va. and W.Va.

4. A. sibíricum Jacq. (Siberian; Jacquin supposing the plant in cult. to have come from Asia). — Similar to no. 3, erect or ascending, up to 5 dm. high, or depressed and forming prostrate mats in forma **arenàrium** (F. C. Gates) Fern. (growing in sand), glabrous throughout, with ascending branches; *leaves sessile or nearly so, those of primary axis rounded to cordate at base* and oblong to oblong-lanceolate or narrowly ovate, 1–4 cm. broad; *bracts herbaceous, green; corolla milk-white; follicles* 4–10 *cm. long; coma of seed* 0.8–2 *cm. long.* (*A. hypericifolium* Ait.) — Rocky or gravelly soil, often along streams, Nfld. to B.C., s. to N.S., N.E., Va., W.Va., O., Ind., Ill., Mo. and Tex. June–Aug.

Var. **cordígerum** (Greene) Fern. (bearing hearts). — *Leaves of primary axis ovate to oval-oblong, deeply cordate and clasping,* 2.5–5.5 *cm. broad.* — Chiefly on shores, w. N.E. to Sask., s. to O., Ind., Mo. and Kans.

FAM. 145. ASCLEPIADÀCEAE (Milkweed Family)

Plants with milky juice and opposite or whorled (rarely scattered) entire leaves; the follicles, seeds, anthers (connected with the stigma), latex, etc., as in the preceding family, from which they differ in the commonly valvate corolla and in the singular connection of the anthers with the stigma, the cohesion of the pollen into wax-like or granular masses (pollinia), etc., as explained under the typical genus, *Asclepias.*

a. Pollen cohering into 2 waxy masses or pollinia; crown without widely divergent long appendages; follicles fusiform, lanceolate or ovoid; stems herbaceous. . . *b.*

b. Anthers tipped with an inflexed or sometimes erect scarious membrane, the locules lower than the top of the stigma; pollinia suspended. Tribe Cynancheae. . . . *c.*

c. Stem erect or merely decumbent; crown of 5 hooded fleshy bodies.

Each of the 5 hoods of the crown with a salient crest; leaves alternate. 1. *Asclepiodora.*

Each hood of crown without a crest; leaves opposite, alternate or whorled. . . . . 2. *Asclepias.*

c. Stem twining; leaves opposite.

Corolla campanulate, with erect lobes; crown of 5 distinct thin oblong, truncate or abruptly tipped bodies terminated by a 2-cleft tail or awn. 3. *Ampelamus.*

Corolla rotate, spreading; crown a fleshy 5–10-lobed ring or disk. . . 4. *Cynanchum.*

b. Anthers with short (if any) scarious tip, borne on the margin of or close under the disk of the stigma; pollinia horizontal; twining broad-leaved herbs. Tribe Gonolobeae. . . . . 5. *Gonolobus.*

a. Pollen granulose, loosely aggregated in 2 masses in each anther-locule; crown with 5 filiform spreading appendages nearly equaling lobes of rotate corolla; follicles linear-cylindric; woody climber. Tribe PERIPLOCEAE. . . . . 6. *Periploca.*

Tribe CYNÁNCHEAE Endl.

1. ASCLEPIODÒRA Gray SPIDER-MILKWEED

Resembling *Asclepias;* but the corolla-lobes ascending or spreading; the hoods destitute of a horn, widely spreading and somewhat incurved, slipper-shaped and laterally compressed, the cavity divided at the apex by a crest-like partition. — Umbels solitary and terminal or corymbed, loosely flowered. Follicles ovoid, often somewhat muricate with soft spinous projections. Small Am. genus. (Name from *Asclepias,* and *doron* or *dorea, the gift of Aesculapius.*)

1. **A. víridis** (Walt.) Gray (green), ANTELOPE-HORN. — Almost glabrous; stems 3–7 dm. high; leaves alternate, short-petioled, ovate-oblong to lanceolate, 3–13 cm. wide; umbels several in a cluster, short-peduncled; flowers large (2–3 cm. broad), green, with a purplish crown. — Dry woods, fields and prairies, Fla. to N.M., n. to S.C., s. O., s. Ill., Mo. and Neb. May, June.

2. ASCLÈPIAS L. MILKWEED. SILKWEED

Calyx persistent; divisions small, reflexed. Corolla deeply 5-parted; divisions valvate in bud, deciduous. Crown of 5 hooded bodies seated on the tube of stamens, each with (or sometimes without) an incurved horn. Stamens 5, inserted on the base of the corolla; filaments united into a tube which incloses the pistil; anthers adherent to the stigma, each with 2 vertical locules tipped with a membranaceous appendage, each locule containing a flattened pyriform and waxy pollen-mass; the two contiguous pollen-masses of adjacent anthers forming pairs which hang by a slender prolongation of their summits from 5 cloven glands which grow on the angles of the stigma (extricated from the locule by insects, and directing copious pollen-tubes into the point where the stigma joins the apex of the style). Ovaries 2, tapering into very short styles; the large depressed 5-angled fleshy stigmatic disk common to the two. Follicles 2, one of them often abortive, soft, ovoid or lanceolate. Seeds anatropous, flat, margined, usually (not in our no. 6) bearing a tuft of long silky hairs *(coma)* at the hilum, downwardly imbricated all over the large placenta, which separates from the suture at maturity. Embryo large, with broad foliaceous cotyledons in thin albumen. — Perennial Am. and Afr. herbs; peduncles terminal or lateral and between the usually opposite petioles, bearing simple many-flowered umbels in summer. (The Greek name of *Aesculapius,* to whom the genus is dedicated.) — Species frequently hybridizing. Including ACERATES Ell.

a. Hoods with an incurved horn arising from the cavity. . . *b.*

b. Hoods sessile, broader or at least not attenuate at base, the horn or crest conspicuous; anther-wings broadest and usually angulate-truncate and salient at base. . . *c.*

c. Hirsute; juice of stem not obviously milky; leaves irregularly alternate, only in small part opposite, linear-lanceolate to oblong; hoods orange or yellow. . . . . 1. *A. tuberosa.*

c. Glabrous or softly pubescent; juice of stem often milky; leaves mostly opposite or whorled; hoods purple to white or greenish, orange or yellow only in no. 2. . . *d.*

d. Corolla bright red to purple; hoods orange, purple or purplish. . . *e.*

e. Umbels 1–few, distant or irregularly disposed; leaves transversely veined; hoods 4.5–7 mm. long, exceeding the anthers; follicles erect, at tips of deflexed pedicels.

Leaves glabrous, linear to ovate, tapering to long acuminate tips; follicles glabrous.

Leaves linear to elongate-lanceolate, 0.2–2.5 cm. broad; hoods orange-yellow to scarlet. . . . . 2. *A. lanceolata.*

Leaves broadly lanceolate to ovate, the larger 2.5–6.5 cm. broad; hoods red or purplish. . . . . 3. *A. rubra.*

Leaves minutely downy beneath, elliptical to ovate-oblong, abruptly short-pointed; corolla and hoods red or purple; follicles canescent-tomentose. . . . . 4. *A. purpurascens.*

e. Umbels several, the upper usually forming a more or less definite broad corymb; leaves with ascending veins; hoods 2–3 mm. long, equaling anthers; follicles lance-fusiform, minutely pubescent, on ascending pedicels. . . . . 5. *A. incarnata.*

*d.*Corolla and hoods greenish, yellowish, white or merely purplish-tinged. . . *f.*

*f.*Leaves lanceolate to ovate, obovate, elliptic or oblong, 0.1–1.8 dm. broad. . . *g.*

*g.*Leaves membranaceous, the principal ones gradually attenuate to long slender tips. . *h.*

*h.*Larger leaf-blades 1–5.5 cm. broad, 5–15 cm. long; corolla-lobes 2.5–6 mm. long; hoods 2–4.5 mm. long.

Leaves 8–15 pairs, becoming somewhat approximate toward summit of stem; corolla-lobes 2.5–4 mm. long; hoods 2–2.5 mm. high; follicles drooping, on recurved pedicels; seeds without coma; southern paludal plant. . . . . 6. *A. perennis.*

Leaves 2–5 pairs or whorls, those of at least 1 node often in whorls; corolla-lobes 4–6 mm. long; hoods 3–4.5 mm. high; follicles erect, on ascending pedicels; seeds with coma; plant of upland woods. . . . . 7. *A. quadrifolia.*

*h.*Larger leaf-blades 4–10 cm. broad, 1.2–3 dm. long; corolla-lobes 6–10 mm. long; hoods 4–6 mm. high; follicles erect at tips of strongly deflexed pedicels. . . . . 8. *A. exaltata.*

*g.*Leaves firm to coriaceous, blunt or merely short-pointed, not attenuate at tip (except in the coriaceous-leaved no. 12). . *i.*

*i.*Leaves petioled (petiole often very short). . *j.*

*j.*Leaves (except when very young) glabrous or promptly glabrate.

Leaves tapering to slender petioles; flowers 8–12 mm. long, bright white, with purple center; hoods globular-ventricose; follicles lance-fusiform, long-attenuate, 1–2 cm. thick. . . . . 9. *A. variegata.*

Leaves rounded or cordate at base, with very short and thick petioles; flowers 1.5–2 cm. long, not clear white; hoods ovate to obovate; follicle lance-ovoid, 2.5–3 cm. thick.

Leaves oblong to oblong-ovate, 2.5–7 cm. broad, the larger 1.2–2 dm. long; corolla purplish to whitish. 10. *A. Sullivantii.*

Leaves orbicular to broadly oval, rigidly coriaceous, 5–12 cm. broad, the larger 7–12 cm. long; corolla greenish. . . . . 11. *A. latifolia.*

*j.*Leaves minutely tomentulose beneath. . *k.*

*k.*Stem slender to coarse, 2–6 dm. high; leaves with slender ascending veins, or veins nearly hidden; flowers about 1 cm. long; hoods yellowish or whitish; follicles plane, about 2 cm. thick.

Stem slender; leaves oval, ovate or broadly lanceolate, tapering to blunt or subacute apex, 3–9 cm. long, 1–5 cm. broad, with slender ascending veins; hoods yellowish. . . . . 12. *A. ovalifolia.*

Stem coarse; leaves squarish- or quadrangular-oblong, broadly rounded to truncate or retuse at summit, 4–12 cm. long, 3–9 cm. broad, the veins hidden; hoods whitish. . . . . 13. *A. arenaria.*

*k.*Stem coarse, up to 2 m. high; leaves much larger, with strong transverse veins; flowers 1.2–2.5 cm. long; hoods greenish or whitish to purplish; follicles usually softly spinulose, thicker.

Pedicels, etc. minutely pubescent; hoods 3–4 mm. high, obtuse. . . . . 14. *A. syriaca.*

Pedicels, etc. densely white-tomentose; hoods 1–1.5 cm. high, their summits prolonged into a lance-lingulate appendage. . . . . 15. *A. speciosa.*

*i.*Leaves sessile and clasping; umbels usually terminating a long terminal peduncle.

Leaves oblong or elliptical, rounded at summit, with lingulate margins, mostly 0.7–1.5 dm. long; flowers 13–18 mm. long. 16. *A. amplexicaulis.*

Leaves cordate-ovate to -lanceolate, tapering from base to narrow tip, flat, 3–8 cm. long; flowers 12–14 mm. long. . 17. *A. Meadii.*

*f.*Leaves narrowly linear, with revolute margins, or filiform.
Stems from crowns with fibrous roots, from deep-seated rhizomes, 3–9 dm. high; leaves narrowly linear, with revolute margins, whorled or opposite. . . . . 18. *A. verticillata.*
Stems from woody bases, 1–1.5 dm. high; leaves filiform or filiform-linear, crowded, spirally arranged. . . . . 19. *A. pumila.*
*b.*Hoods sessile, with a narrow wholly adnate internal crest terminating in a minute horn; anther-wings widening to the broadly rounded base and conspicuously auriculate just above it; leaves narrowly linear; umbels subsessile or short-peduncled. . . . . 20. *A. stenophylla.*
*a.*Hoods with neither horn nor crest; flowers greenish. . . *l.*
*l.*Crown borne on a short column, definitely shorter than the anthers; leaves very numerous, mostly alternate or spirally arranged, linear to narrowly lanceolate; umbels mostly lateral, slender-peduncled.
Leaves slenderly linear to filiform; pedicels villous-tomentose; hoods emarginate, with broad basal auricles. . . . . 21. *A. auriculata.*
Leaves linear to narrowly lanceolate; pedicels puberulent or hirtellous; hoods rounded at summit, without basal auricles.
Pedicels and follicles minutely puberulent with incurved hairs; flowers about 7 mm. long; Coastal Plain species. . . . . . . 22. *A. longifolia.*
Pedicels and follicles hirtellous with spreading hairs; flowers 8–8.5 mm. long; inland species. . . . . 23. *A. hirtella.*
*l.*Crown sessile, about equaling anthers; leaves 8–24, many or all of them opposite, oval to lanceolate or linear; lateral umbels (when present) sessile or on very short thick peduncles.
Stem and leaves minutely pilose or glabrate; umbels mostly lateral, subsessile; expanded flowers 9–13 mm. long. . . . . 24. *A. viridiflora.*
Stem and leaves spreading-hirsute; umbel terminal, peduncled; expanded flowers 6–8 mm. long. . . . . 25. *A. lanuginosa.*

1. A. tuberòsa L. (tuberous), Butterfly-weed, Pleurisy-root, Chigger-flower. — *Roughish-hairy,* 3–9 dm. high; stems ascending or decumbent, very leafy, branching at the summit, and bearing umbels in a terminal corymb, or scattered in racemes along the branches; *leaves from linear to oblong-ovate, sessile or slightly petioled;* divisions of corolla oblong, greenish-orange to red, or yellow and with orange hood in forma **bìcolor** Standl. (two-colored); *hood orange* to deep red, or pale yellow and with pale petals in forma **lùtea** Clute (yellow), scarcely longer than the nearly erect and slender subulate horns; follicles pubescent, erect at tops of deflexed pedicels, lance-fusiform, 7–12 cm. long. (Incl. the vaguely separable *A. decumbens* L.) — Dry open soil, Fla. to Tex. and Ariz., n. to s. N.H., s. Vt., N.Y., s. Ont., Mich., Wisc., Minn., Neb. and Colo. June–Sept. — Our most variable species.

2. A. lanceolàta Walt. (lanceolate). — *Glabrous,* up to 1.5 m. high; *leaves elongate-lanceolate, long-acuminate, thick and firm,* the larger ones 1–2.5 cm. broad, with midrib prominent beneath and *with transverse veins;* umbels 1–3 at summit of stem, 5–12-flowered; divisions of red corolla oblong; *hoods bright orange-yellow to scarlet,* broadly oblong, 5–7 mm. high, obtuse, much exceeding the incurved horn; follicles smooth, lance-attenuate, 8–9 cm. long, erect at tips of deflexed pedicels. — Brackish to fresh marshes along the coast, Fla. to N.J. June–Aug.

Var. **paupércula** (Michx.) Fern. (stunted). — Leaves elongate-linear, 0.2–1 cm. wide; hoods deep orange to scarlet, or in forma **flaviflòra** Fern. (yellow-flowered) clear yellow. — Wet pinelands, swamps and fresh marshes, Fla. to Tex., n. to Del. Late June–Aug.

3. A. rùbra L. (red). — *Glabrous;* the firm *broadly lanceolate to ovate leaves tapering from a rounded to cordate base to the long slender tip,* 0.5–1.8 dm. long, *the broader* 2.5–6.5 *cm. broad, with conspicuous transverse veins;* umbels 1–few, irregularly scattered, many-flowered; *divisions of corolla and hoods* (*about* 6 *mm. long*) *purplish-red; horn long, slender, straightish;* follicles much as in no. 2. — Wet pinelands, bogs and peats, Fla. to Tex., n. to L.I., N.J. and e. Pa. June, July.

4. A. purpuràscens L. (purplish), Purple M. — Stem rather slender, 1 m. or less high; *leaves elliptical or ovate-oblong,* the upper taper-pointed, *minutely velvety-downy underneath,* smooth above, *contracted at base into a short petiole; pedicels* shorter than the peduncle, *three to four times the length of the dark purple lance-ovate divisions of the corolla;* hoods oblong, abruptly narrowed above, 4.5–7 mm. high; *the horn broadly scythe-shaped, with a narrow and abruptly horizontal point;* follicles erect, at tips of deflexed pedicels, canescent-tomentose, lance-ovoid, 1–1.5 dm. high. — Dry to damp woods, thickets and openings, s. N.H. to s. Ont., Minn. and S.D., s. to N.C., Tenn., Miss., Ark. and Okla. Late May–July.

5. A. incarnàta L. (flesh-colored), Swamp-M. — *Stems* solitary or clustered, 0.3–1.5 m. high, *very leafy,* with two downy lines above and on the branches of the inflorescence, with little milky juice; *leaves* rather numerous, *oblong-lanceolate to ovate, with fine ascending veins; umbels* *usually several, forming a broad corymb; flowers pink to rose-purple* (rarely whitish); *hoods* 2–3 *mm. high, equaling the anthers; follicles* lance-fusiform, minutely hairy, 5–9 cm. long, *on ascending pedicels.* Very variable:

Stems and lower surfaces of leaves glabrous or only very sparsely pilose.
Stems 0.6–1.5 m. high; leaves 10–18 pairs on the main stem, narrowly to broadly oblong-lanceolate, tapering to tip, the principal ones 1–2 dm. long. *A. incarnata* (typical).
Stems 3–5 dm. high; leaves 7–11 pairs, ovate-oblong, obtuse or subacute, 4.5–6.5 cm. long. . . . . Var. *neoscotica.*
Stems and lower leaf-surfaces copiously short-pilose; leaves 11–21 pairs, broadly lanceolate to elliptic, often more rounded at base, the larger 0.9–1.8 dm. long; flowers often paler. . . . . Var. *pulchra.*

A. incarnàta (typical). — Swamps, wet thickets and shores, Que. to Man. and Wyo., s. to N.S., N.E., L.I., S.C., Tenn., La., Tex. and N.M. Late June–Aug. — Forma **albiflòra** Heller (white-flowered) has whitish flowers.

Var. **neoscótica** Fern. (of Nova Scotia). — Siliceous gravelly shores, N.S. July, early Aug.

Var. **púlchra** (Ehrh.) Pers. (beautiful), *A. pulchra* Ehrh. — Swamps, wet thickets and shores, N.S. to w. N.Y. and W.Va., s. to Ga. July–Sept. — Forma **cándida** Fern. (white) has whitish flowers.

6. A. perénnis Walt. (perennial). — Stems often *decumbent and rooting at base*, then erect, 3–9 dm. high; *leaves 8–15 pairs, often subapproximate toward summit of stem, membranaceous,* lance- to ovate-attenuate, narrowed to slender petioles, nearly or quite glabrous, 1–4 *cm. broad,* 1–1.5 *dm. long;* umbels 1–few in the upper axils, on filiform peduncles mostly longer than the pedicels; flowers white; *corolla-lobes* 2.5–4 *mm. long; hoods* 2–2.5 *mm. high, shorter than the needle-shaped horn; follicles pendulous, on recurved pedicels,* glabrous; *seeds without coma.* — Wet woods and argillaceous swamps, Fla. to Tex., n. to S.C., s. Ind., s. Ill. and se. Mo. Late June–Sept.

7. A. quadrifòlia Jacq. (four-leaved). — Stem slender, 3–8 dm. high, mostly leafless below, bearing usually *one or two whorls* of four in the middle and one or two pairs *of ovate or ovate-lanceolate* taper-pointed petioled *leaves* (0.5–1 dm. long); pedicels slender; *corolla-lobes pale pink,* oblong, 4–6 *mm. long; hoods* 3–4.5 *mm. high,* white, elliptic-ovate, the incurved horn short and thick; *follicles slenderly lance-attenuate,* smooth, *on erect pedicels.* — Dry woods, N.H. to s. Ont. and Minn., s. to N.C., Ala., se. Kans. and Ark. May–early July.

8. A. exaltàta L. (very tall). — Smooth, 0.5–1.5 m. high; *leaves* 7–11 pairs, ovate, or *the middle and upper oval-lanceolate, membranaceous,* pale beneath, pointed at both ends, *tapering to slender acuminate tips,* short-petioled, *the larger* 4–10 *cm. broad and* 1.2–3 *dm. long;* umbels terminal and from 1–several upper axils, shorter than subtending leaves, lax, with numerous loosely spreading to nodding pedicels 2–5 cm. long; *corolla-lobes* ovate-oblong, greenish, 6–10 *mm. long; hoods white,* truncate, the margins 2-toothed at summit, 4–6 *mm. high, the horns with a long projecting subulate point; follicles* lance-fusiform, attenuate to base and apex, *erect, at the tips of strongly deflexed pedicels,* 1–1.5 dm. long. (*A. phytolaccoides* Pursh) — Rich woods and clearings, s. Me. to Minn., s. to Ga., Ky., Ill. and Ia. June, July.

9. A. variegàta L. (variegated). — Glabrous or nearly so, 3–9 dm. high; *leaves* (4–6 pairs), *ovate, oval or obovate,* pale beneath, blunt or short-pointed, *tapering into definite petioles;* umbels rounded; pedicels finely hairy; *flowers white, with purplish center,* 8–12 *mm. long;* divisions of corolla ovate; *hoods orbicular-ventricose,* entire, the horn semilunate with a horizontal point; follicles slenderly lance-fusiform, long-attenuate, 1–2 cm. thick. (*Biventraria* Small) — Woods, Fla. to Tex., n. to s. Ct., se. N.Y., N.J., Pa., W.Va., Ky., s. Ill., se. Mo. and Okla. May–July.

10. A. Sullivántii Engelm. (for its discoverer, William Starling Sullivant, 1803–1873). — *Very smooth throughout,* 0.6–1.5 m. high, from creeping rhizome; *leaves oblong or oblong-ovate with cordate base, nearly sessile but short-petioled, the larger* 2.5–7 *cm. broad and* 1.2–2 *dm. long;* corolla purplish to whitish, 1.5–2 cm. long; *hoods obovate,* entire, *obtusely 2-eared at base outside;* anther-wings 2-toothed at base; *follicle* nearly glabrous, *slenderly and obliquely ovoid,* 2.5–3 cm. *thick,* obscurely soft-spinous towards tip. — Rich low grounds and prairies, s. Ont. to Minn., s. to O., Ind., Ill., Mo. and e. Kans. June, July.

11. A. latifòlia (Torr.) Raf. (broad-leaved). — Farinose-puberulent when young, *soon* quite *green and glabrous,* 4–8 dm. high; *leaves rigidly coriaceous,* in subapproximate pairs, *orbicular to broadly oval,* often emarginate, 5–12 *cm. broad, the larger* 7–12 *cm. long, subcordate,* very short-petioled, with straightish transverse veins; *umbels mostly lateral,* short-peduncled, densely many-flowered; *flowers greenish,* 1.5–2 *cm. long;* corolla-lobes ovate; hoods barely equaling anthers, broad, with transverse entire summit; crest falciform-triangular, its apex extended into a subulate horn; follicles turgid-ovoid, smooth, 6–8 cm. long, 3 cm. thick. (*A. Jamesii* Torr.) — Dry prairies and plains, w. Mo. and Kans. June–Aug. (Adv. from farther west and southwest)

12. A. ovalifòlia Dcne. (oval-leaved). — *Slender, 0.2–6 dm. high; leaves* 3–7 pairs, short-petioled, *oval, ovate or broadly lanceolate, tapering to blunt or subacute apex,* soft-downy beneath, 3–9 *cm. long,* 1–5 *cm. broad, with slender ascending veins;* umbels sessile or short-peduncled, loosely 10–18-flowered; *flowers about* 1 *cm. long; hoods yellowish,* oblong, obtuse, entire, twice or thrice length of anthers, with a small horn, about the length of the oval greenish-white (purplish outside) corolla-lobes; *follicle* thick-lanceolate, downy, *plane, about* 2 *cm. thick.* — Oak-openings and prairies, Man. to Alta., s. to n. Ill., Ia. and Neb. June–July.

13. A. arenària Torr. (of sand). — *Stem stoutish,* decumbent or ascending, *tomentose,* 2–6 *dm. high;* principal *leaves squarish- or quadrate-oblong with broadly rounded, truncate or retuse summit,* 4–12 cm. long, 3–9 cm. broad, very thick, tomentulose beneath; umbels very short-peduncled, much shorter than subtending leaves, dense; *flowers greenish, about* 1 *cm. long;* the *whitish hood* about as broad as long, surpassing the anthers, truncate at base and summit, the latter oblique and notched on each side; follicle tomentulose, with plane surface. — Sandy soil, Neb. and e. Kans. to Colo., s. to Tex. and N.M. June, early July.

14. A. syrìaca L. (Syrian; early carried from e. Am. to s. Eu. and supposed by Linnaeus to have come from the Orient), COMMON M. or SILKWEED, HERBE À COTON or COCHONS DE LAIT (Que.). — *Stem coarse,* up to 2 m. high, finely soft-pubescent above, from creeping rhizomes; *leaves lance-oblong to broadly oval,* 1–2.6 *dm. long,* 0.4–1.8 *dm. broad, rounded or tapering to the short thick petiole,* grayish-tomentulose beneath, with strong transverse nerves; umbels many-flowered, rounded, with *minutely pubescent pedicels; corolla-lobes* 6–9 *mm. long,* dull purple, or in forma **leucántha** Dore (white-flowered) whitish; *hoods* 3–4 *mm. high, ovate, obtuse, with a tooth on each side of the stout claw-like horn; follicles slenderly ovoid,* 2.5–3.5 *cm. thick,* tomentose, remotely echinate with soft conical processes 1–3 mm. long, or surface plain in forma **inérmis** Churchill (unarmed). — Thickets, roadsides, dry fields, etc., w. N.B. to Sask., s. to Ga., Tenn., Ia. and Kans. June–Aug. — Flowers heavily fragrant; young shoots and firm well-grown pods cooked as vegetables.

Var. **kansàna** (Vail) Palmer & Steyerm. (of Kansas). — Follicles densely covered with subulate-filiform processes 3–10 mm. long. (*A. kansana* Vail) — Ia., Neb., Mo. and Kans.

15. A. speciòsa Torr. (showy). — Coarse, up to 2 m. high; *leaves subcordate*-oval to -oblong, very short-petioled, as large as in no. 14; new growth, *pedicels, etc., densely white-tomentose;* corolla-lobes purplish, ovate-oblong, 9–13 mm. long; *hoods* 1–1.5 *cm. long,* with a short inflexed horn, *the truncate summit abruptly produced into a very long lanceolate-lingulate appendage; follicles* obliquely ovoid, *white-woolly,* with numerous prolonged processes. — Prairies and openings, Minn. to B.C., s. to Mo., Okla., Tex., N.M., Ariz. and Calif. June–Aug.

16. A. amplexicaùlis Sm. (clasping the stem). — *Smooth,* 0.3–1 m. high, *glaucous; leaves* sessile, *clasping,* firm, *oblong or elliptical, rounded at summit,* mucronate, *undulate or wavy-margined,* the larger ones 0.7–1.5 dm. long; umbel *solitary* (rarely 2 or 3) *on an elongate erect usually naked peduncle,* many-flowered; *flowers* 13–18 *mm. long;* corolla greenish-purple or cream-color, the lobes oblong; *hoods* flesh-color, *erosely truncate and somewhat toothed at the broad summit,* hardly exceeding anthers and shorter than the falcate-subulate horn; *anther-wings* 2-*horned at base;* follicles slenderly fusiform, long-attenuate, minutely hirtellous or glabrous, 0.8–1.5 cm. long, on strongly recurved or contorted pedicels. — Dry open soil, Fla. to Tex., n. to s. N.H., Mass., N.Y., W.Va., O., Mich., Wisc., s. Minn. and Neb. May–July.

17. A. Meàdii Torr. (for its discoverer, SAMUEL BARNUM MEAD, 1799–1880). — Slender, 4–9 dm. high, glabrous, glaucous; *leaves* 3–4 *distant pairs, cordate-ovate to -lanceolate, clasping, tapering from base to narrow apex,* 3–8 *cm. long;* umbel solitary at tip of long peduncle; *flowers* 12–14 *mm. long;* corolla greenish-yellow; *hoods rounded-truncate at summit,* with a sharp tooth at each margin, somewhat exceeding the stout horn; follicles fusiform, 7–8 cm. long, glabrous. — Dry prairies, rare and local, nw. Ind., sw. Wisc., Ill. and Ia.; reported from ne. Kans. June.

18. A. verticillàta L. (whorled). — *Stems* slender, simple or sparingly branched, 3–9 *dm. high, from a fibrous root,* very leafy to the summit; *leaves* linear, with revolute margins, 3–6 *in a whorl; umbels small,* lateral and terminal; *divisions of the corolla ovate, greenish-white; hoods roundish-oval, about half the length of the incurved claw-shaped horns;* follicles slenderly-fusiform, on erect pedicels. — Dry woods and open sterile soil, Fla. to Tex. and Mex., n. to Mass., N.Y., s. Ont., Mich., Wisc., s. Man. and s. Sask. June–Sept.

19. A. pùmila (Gray) Vail (dwarf). — Similar; *low* (1–1.5 dm. high) *and many-stemmed from a woody caudex; leaves* much *crowded, spirally arranged,* filiform or filiform-linear. — Dry plains, Ia. to Mont., s. to Ark., Tex. and N.M. July, Aug.

20. A. stenophýlla Gray (narrow-leaved). — Puberulent, but foliage glabrous; stems slender, 0.3–1 m. high; leaves narrowly linear, the upper alternate, lower opposite; umbels several, short-peduncled, 10–15-flowered; corolla-lobes oblong, greenish; *hoods* whitish, equaling the anthers, conduplicate-concave, *with a minute horn exserted from the 2-lobed apex; anther-wings broadly rounded at base and conspicuously auriculate-notched just above it; follicles erect, on slender pedicels.* (*Acerates angustifolia* Dcne.) — Dry prairies and rocky ground, S.D. to Colo., s. to Mo., Okla., N.M. and Ariz. June, July.

21. A. auriculàta (Engelm.) Holz. (with ear-like appendages). — Resembling no. 20; stems 0.5–1 m. high, 1–few, glabrous, or puberulent above; *leaves narrowly linear to filiform,* 1–2 dm. long, very numerous, *alternately and spirally arranged;* umbels numerous, lateral, on peduncles about as long as the *villous-tomentose* slender *pedicels; calyx villous-tomentose; crown borne on a very short column, the emarginate hoods auricled at base and much shorter than the anthers;* follicles lance-ovoid, attenuate, 6–8 cm. long. (*Acerates* Engelm.) — Dry plains, e. Neb. to Colo., s. to Tex., N.M. and Ariz. July.

22. A. longifòlia Michx. (long-leaved). — Stems decumbent or ascending, glabrous or minutely pubescent with incurved hairs; *leaves* numerous, linear or linear-lanceolate, alternate or irregularly subapproximate, *smooth or barely scabrous; umbels* terminal and lateral, 1–10, *on slender puberulent or minutely incurved-pilose peduncles; pedicels and follicles similarly pubescent; flowers about* 7 *mm. long;* petals externally purple toward apex, with narrow white border; *hoods* shorter than anthers, *rounded at summit, without basal auricles;* follicles slenderly lanceolate, slender-tipped, about 1 dm. long. (*Acerates* Ell.; *Acerates floridana* sensu Hitchc., not as to type of Lam.) — Damp pinelands and siliceous or argillaceous swales, Fla. to Tex., n., locally, to se. Va.; Sussex Co., Del. (formerly). June.

23. A. hirtélla (Pennell) Woodson (covered with short and stiff hairs). — Similar to no. 22; *pubescence* of peduncles, pedicels and follicles *spreading-hispid; leaves* crowded, *more scabrous; flowers* 8–8.5 *mm. long;* petals greenish, with broader white border. (*Acerates* Pennell) — Prairies, fields and waste places, W.Va. to Wisc., s. to Ala., Ark. and e. Kans. June–Aug.

24. A. viridiflòra Raf. (green-flowered), Green M. — *Minutely pilose or glabrate;* stems 3–8 dm. high, leaning or ascending; *leaves* 8–24, *many or all opposite,* oval, obovate or oblong, obtuse or merely short-tipped, very thick; *umbels subsessile, axillary,* dense and globose; expanded *flower* 9–13 *mm. long; crown sessile, about equaling anthers;* follicles slenderly lance-attenuate, 8–11 cm. long. (*Acerates* Eaton) — Dry woods and openings, Fla. to Tex. and N.M., n., somewhat locally, to Mass., se. N.Y., Pa., Ky. and sw. Kans. June–Aug.

Var. **lanceolàta** (Ives) Torr. (lanceolate). — Leaves lanceolate to narrowly ovate, acute or acuminate, mostly 1.5–4.5 cm. broad. (*Acerates viridiflora,* var. *Ivesii* Britt.) — Dry sands and prairies, s. Ont. to Man. and Wyo., s. to Ky., La., Tex. and N.M.; se. Pa. to se. Va. Passing into

Var. **lineàris** (Gray) Fern. (linear). — Leaves elongate-linear or linear-lanceolate, 0.2–1 cm. broad. (*Acerates viridifl.,* var. Gray) — W. Ont. and Man., s. to La., Okla. and N.M.

25. A. lanuginòsa Nutt. (woolly). — Spreading-hirsute; stems 1–2.5 dm. high; leaves 5–8 pairs, lanceolate to lance-ovate; *umbel solitary, terminal, slender-peduncled; flowers* 6–8 *mm. long;* hoods purplish, nearly equaling anthers. (*Acerates* Dcne.) — Prairies, Wisc. to Mont., s. to n. Ill., Ia., Kans. and Wyo. Late May, June.

3. AMPÉLAMUS Raf. Sandvine

Crown of 5 free membranaceous blades which are truncate or obscurely lobed at the apex, where they bear a pair of flexuous awns united at base. Anthers nearly as in *Asclepias;* pollen-masses oblong, obtuse at both ends, fixed below the summit of the stigma to the descending glands. Follicles elongate-ovoid to lanceolate, smooth. Seeds with a coma, as in *Asclepias.* — A perennial twining herb, smooth, with opposite cordate-ovate and pointed long-petioled leaves and small whitish flowers in raceme-like clusters on slender axillary peduncles. (Name from the Greek *ampelos, a vine,* and a Rafinesquean ending, together said to mean "Sand Vine".) Ampelanus (alteration of spelling) of the original auth.; Enslenia Nutt., not Raf.; Gonolobus sensu Vail, not Michx.

1. A. álbidus (Nutt.) Britt. (whitish), Honeyvine. — Climbing, 3–4 m. high; leaves 3.5–12 cm. wide. (*Gonolobus laevis* of ed. 7, not Michx.) — Rich sandy or alluvial thickets and disturbed soils, w. Ala. to Tex., n. to e. Pa., W.Va., O., Ind., Ill., Ia. and Neb. July–Sept.

4. CYNÁNCHUM L.

Crown flat, simple. Anthers, smooth follicles and seeds much as in *Asclepias.* — Herbs, nat. of Old World, often twining. (An ancient name for some plant supposed to be poisonous to dogs, from the Greek *cyon, dog,* and *anchein, to strangle.*) VINCETOXICUM Medic., Moench and others, not Walt.

1. C. nìgrum (L.) Pers. (black). — *Twining,* nearly smooth; leaves ovate or lance-ovate; *flowers* small, *dark purple,* in an axillary cluster, on a peduncle shorter than the leaves; *corolla pubescent within;* follicles lance-fusiform, 4.5–8 cm. long. — Waste places, roadsides, thickets and fields, Me. to O., Pa. and Kans. June–Sept. (Natzd. from Eu.) — Flowers with heavy fragrance.

2. C. Vincetóxicum (L.) Pers. (ancient name, meaning rope-like poison). — *Suberect,* 3–6 dm. high; leaves ovate-lanceolate; *flowers greenish-white; corolla glabrous.* — Esc. from cult., w. N.Y. and s. Ont. (Introd. from Eu.)

Tribe GONOLÓBEAE Endl.

5. GONÓLOBUS Michx. ANGLE-POD

Corolla rotate, sometimes reflexed-spreading; the lobes convolute in the bud. Crown small, annular or cupuliform, in the throat of the corolla. Anthers partly hidden under the flattened stigma, opening transversely. Pollen-masses 5 pairs, horizontal. Follicles turgid, mostly muricate with soft warty projections, sometimes ribbed. Seed with a coma. — Am. herbs or shrubs with opposite cordate leaves and corymbose-umbelled greenish or dark purple flowers on peduncles arising from between the petioles. Our species belong to the typical section with the crown simple and unappendaged and the corolla nearly veinless. (Name from the Greek *gonia, an angle,* and *lobos, a capsule,* from the angled fruit.)* VINCETOXICUM Walt., not Moench.

a. Follicles costate-angled, not muricate; calyx glabrous or only slightly pubescent toward tips of lobes; crown low, 10-lobed, at base of anther-column; pollinia slenderly obovoid, attached to the retinaculum by caudicles at least 0.2 mm. long; anther-sacs inconspicuous, with narrow slits.
- Flower-buds short-conical, abruptly acuminate; calyx practically glabrous; corolla-lobes broadly lanceolate, usually pubescent but glabrate within, 5–7 mm. long, about twice length of calyx-lobes. . . . . 1. *G. suberosus.*
- Flower-buds conical, gradually acute or acuminate; calyx glabrous or its lobes ciliolate toward apex; corolla-lobes linear-lanceolate, glabrous within, three or four times length of calyx-lobes. . . . . 2. *G. gonocarpos.*

a. Follicles muricate, not costate-angled; calyx pubescent; crown cup-shaped, as high as or higher than anther-column; pollinia semilunate or oblong, attached to the retinaculum by caudicles less than 0.2 mm. long; anther-sacs obvious, with open slits. . . *b.*

b. Flower-buds bluntly ovoid; corolla rotate. . . . . 3. *G. carolinensis.*

b. Flower-buds oblong-conical; limb of corolla ascending. . . . *c.*

c. Crown of thin texture, the long bifid lobes overtopping the anther-column.
- Corolla-lobes whitish, imbricate or only slightly contorted in bud, 8–12 mm. long, 1.5–2.5 mm. wide; longer teeth of crown-lobes usually subulate. . . . . 4. *G. Baldwynianus.*
- Corolla-lobes brownish-purple, strongly contorted in bud, 10–15 mm. long, 3–6 mm. broad; longer teeth of crown-lobes flat. . . . . . 5. *G. decipiens.*

c. Crown fleshy, as long as or slightly longer than anther-column.
- Corolla-lobes broadly linear, 13–15 mm. long, 2–2.5 mm. broad, dark purple. . . . . 6. *G. Shortii.*
- Corolla-lobes slenderly linear, 6–12 mm. long, 1.5–2 mm. broad, greenish-fuscous outside, purple on inner face. . . . . 7. *G. obliquus.*

1. G. suberòsus (L.) R. Br. (corky). — Stem hirsute; leaves narrowly ovate, with a rounded sinus, abruptly short-acuminate, glabrate, or minutely pilose beneath, 6–15 cm. long, 3.5–9.5 cm. broad; umbels much shorter than petioles, 3–9-flowered; *corolla short-conic in bud, then abruptly acuminate, twisted, its lobes* fuscous-green without, purple-brown and *pubescent or glabrate within, ovate to triangular-lanceolate, acute, 5–7 mm. long, about twice length of the glabrous calyx-lobes; crown low,* 10-*lobed; follicles wing-angled, smooth,* lance-ovoid, 1–1.3 dm. long; seeds tuberculate. (*Vincetoxicum* Britt.) — Rich thickets and river-banks, Fla. to e. Va. Aug., Sept.

2. G. gonocárpos (Walt.) Perry (angular-fruited). — Similar to no. 1; leaves with deeper sinus, ovate to rotund, the larger ones 1–2 dm. long and 0.5–2 dm. broad; *corolla conical in bud, gradually acute to acuminate, hardly twisted;* the greenish-brown or fuscous linear-lanceo-

* Treatment based on that of L. M. PERRY in Rhodora, xl. 281–287 (1938).

late *corolla-lobes 7–10 mm. long, three or four times length of ciliate-tipped calyx-lobes; follicles with prominent but scarcely winged angles.* (*Vincetoxicum* Walt.; *Gonolobus macrophyllus* Michx.) — Rich thickets and bottomlands, Ga. to Tex., n. to se. Va., s. Ind., s. Ill., s. Mo. and Okla. June, July.

3. G. carolinênsis (Jacq.) Schultes (of Carolina). — *Stems puberulent* and more or less hirsute; leaves minutely pubescent, broadly ovate to rounded, with deep narrow sinus, the basal lobes sometimes overlapping; *flower-bud ovoid; corolla rotate,* chocolate-purple, 1.5–2.5 cm. broad, *its elliptic-oblong round-tipped lobes* 3.5–6 *mm. broad;* crown fleshy, cup-shaped, 10-crenate; *follicle lance-attenuate, slenderly muricate,* 1–1.6 dm. long; seeds short-tuberculate at base. (*Odontostephana* E. J. Alex.; *G. hirsutus* Michx.; *Vincetoxicum hirsutum* Britt.) — Rich thickets, Fla. to Miss., n. to Del., Md., w. Va. and Tenn. June–Aug.

4. G. Baldwyniànus Sweet (named by the English author, in his own Sweet way, for WILLIAM BALDWIN, 1779–1819). — Hirsute; leaves broadly ovate, mostly 8–14 cm. long; peduncles about equaling petioles; umbels 10–20-flowered; *corolla whitish; its ascending narrowly oblong to spatulate lobes imbricated in bud,* 8–12 *mm. long,* 1.5–2.5 *mm. broad; crown of thin texture, deeply cleft into* 5 *usually emarginate lobes half as long as the pair of usually subulate processes in each sinus.* (*Vincetoxicum* Britt.; *Odontostephana* E. J. Alex.) — Rich woods and thickets, Ga. and Ala. to s. Mo. and Okla. May, June.

5. G. decípiens (E. J. Alex.) Perry (deceitful or puzzling). — Hirsute, resmebling no. 3; flowers very numerous; *corolla with ascending brownish-purple lobes* 1–1.5 *cm. long,* 3–6 *mm. broad, these strongly contorted in bud; crown* much as in no. 4, *its longer teeth flat.* (*Odontostephana* E. J. Alex.; *Vincetoxicum carolinense* in part of ed. 7, not Britt.) — Rich thickets, rocky (calcareous) woods and banks of streams, S.C. to La., n. to (?) Md., Mo. and Okla. May, June.

6. G. Shórtii Gray (for its discoverer, CHARLES WILKINS SHORT, 1794–1863). — Resembling nos. 5 and 7; *corolla-lobes broadly linear,* 1.3–1.5 *cm. long,* 2–2.5 *mm. broad, chocolate-purple; crown fleshy, as long as or longer than anther-column;* the alternate teeth thinner, narrower and longer. (*Vincetoxicum* Britt.; *Odontostephana* E. J. Alex.) — Rich woods, local, Pa. to Ky. and Ga. May, June.

7. G. oblìquus (Jacq.) Schultes (oblique; from the ascending corolla-lobes). — Hirsute; leaves abruptly acuminate, rounded- to cordate-ovate; umbel many-flowered; *corolla conical in bud; its slenderly linear ascending lobes* 6–12 *mm. long,* 1.5–2 *mm. broad, greenish-fuscous outside, purple on inner face.* (*Vincetoxicum* Britt.; *Odontostephana* E. J. Alex.) — Rich thickets, borders of woods and rocky slopes, Pa. to s. Ill., s. to Ga., Tenn. and Mo. May–July.

Tribe PERIPLÒCEAE B. & H.

6. PERÍPLOCA L. SILKVINE

Corolla rotate, with long lobes. Crown adnate to base of corolla, bearing 5 long slender spreading appendages. Anthers bearded on the back. Pollen granulose, loosely aggregated in 2 masses in each anther-locule. Follicles linear-cylindric, divaricate. — Woody shrubs nat. of the Old World, often climbing, with opposite leaves and loose terminal and axillary cymes of livid or blackish flowers. (Name from the Greek *peri, around,* and *plocein, to entwine.*)

1. P. GRAÈCA L. (Greek). — Corolla about 2.5 cm. across, its lobes pubescent above. — Frequently cult. and locally esc. to thickets, Fla. to Okla., n. to s. N.E., N.Y. and Kans. June–Sept. (Introd. from s. Eu.)

FAM. 146. CONVOLVULÀCEAE (CONVOLVULUS FAMILY)

Chiefly twining or trailing herbs (or in the Tropics sometimes ligneous), often with some milky juice, with alternate leaves (or scales) and regular 4- or 5-androus flowers; a calyx of 4 or 5 imbricated sepals; a 4- or 5-plaited or -lobed corolla convolute or twisted in the bud (imbricate in the parasitic no. 7); a 2 (rarely 3)-locular ovary (or in tribe I of 2 or 4 separate carpels), with a pair of erect ovules in each locule, the locules sometimes doubled by a false partition between the seeds, so becoming 4-locular; the embryo large, curved or coiled in mucilaginous albumen. Fruit a globular or plump 2–6-seeded capsule. Flowers mostly showy, on axillary peduncles (small and paniculate or glomerulate in no. 7); pedicels articulated, often bracted. — Many are cult. for ornament and one, the SWEET POTATO, for its edible farinaceous roots; those of several species are cathartic, *e.g.* JALAP.

a. Ovary and capsule of 2 or 4 distinct or nearly distinct carpels; styles 2, basilar; embryo large, with cotyledons; creeping herbs. Tribe I. DICHONDREAE.
Corolla deeply 5-cleft; pistils 2, each 1-seeded. . . . . 1. *Dichondra.*

*a.*Ovary and capsule of united carpels. . . *b.*
*b.*Green-leaved plants, not parasitic, rooting in soil. Tribe II. CONVOLVULEAE. . . *c.*
*c.*Styles deeply 2-cleft or 2-divided; plant prostrate or diffuse, not twining.
Branches of style simple; stigmas capitate. . . . . 2. *Breweria.*
Branches of style each 2-cleft; stigmas linear-filiform. . . . . 3. *Evolvulus.*
*c.*Style undivided or 2-cleft only at apex; plants often twining. . . . *d.*
*d.*Style undivided; stigmas capitate, globose or thickened-ovoid.
Stigma 1, capitate or of 2 or 3 globular lobes. . . . . . 4. *Ipomoea.*
Stigmas 2, thickened-ovoid but flattened. . . . . . . 5. *Jacquemontia.*
*d.*Style undivided or merely cleft at apex; stigmas 2, linear-filiform to subulate or slenderly ovoid. . . . . 6. *Convolvulus.*
*b.*Never green; variously colored twining annual parasites soon free from the ground, with leaves reduced to scales; embryo filiform, coiled, without cotyledons. Tribe III. CUSCUTEAE. . . . . 7. *Cuscuta.*

Tribe I. DICHÓNDREAE B. & H.

1. DICHÓNDRA Forst.

Calyx 5-parted. Corolla broadly campanulate. Stamens included. Styles, ovaries and utricular 1–2-seeded capsules 2, distinct. Stigmas thick. — Small and creeping perennial herbs of trop. and warm reg., soft-pubescent, with reniform entire leaves and axillary 1-flowered bractless peduncles. Corolla small, yellowish or white. (Name from the Greek *di, two,* and *chondros, a grain,* from the fruit.)

1. D. rèpens Forst. (creeping), var. **carolinénsis** (Michx.) Choisy (of Carolina). — Leaves round-reniform, pubescent, green both sides; corolla not exceeding calyx. (*D. carolinensis* Michx.) — Wet pinelands, low grounds and, as a weed, in lawns, Fla. to Tex., n. to se. Va. and Ark. June–Sept. (W.I.) — The species, with geographic vars., pantrop. or of warm reg.

Tribe II. CONVOLVÙLEAE B. & H.

2. BREWÈRIA R. Br.

Styles 2 (rarely 3), simple and distinct or united into one toward base or nearly or quite to summit; stigmas depressed-capitate. Otherwise much as in *Convolvulus* and *Evolvulus.* — Perennial prostrate or diffusely spreading herbs of warm reg.; flowers in ours (of subgen. *Stylisma*) small, in summer and autumn; corolla more or less hairy or silky outside. (Named for an English amateur and gardener, *Samuel Brewer,* 1670–1743.) Incl. STYLISMA Raf.

Leaves oblong-lanceolate to elliptic; peduncles terminated by slender short subulate-tipped bracts; style divided nearly to base into 2 long branches.
Backs of sepals nearly or quite glabrous; capsule glabrous except at summit. 1. *B. humistrata.*
Backs of sepals and capsules densely appressed-pubescent. . . . . . 2. *B. Michauxii.*
Leaves narrowly linear or linear-oblanceolate; peduncles terminated by foliaceous elongate paired bracts resembling short foliage-leaves; style cleft only toward summit into branches up to 3 mm. long or merely notched or subentire. . 3. *B. Pickeringii.*

1. B. humistràta (Walt.) Gray (stretched out on the ground). — Slender much branched minutely pilose trailing or slightly twining stem up to 2 m. or more long; *leaves slightly pubescent to glabrescent, oblong to elliptic-obovate, broadly rounded to cordate at base,* with obtuse and mucronate apex; *peduncles* equaling to much exceeding subtending leaves, 1–7-flowered, *with short subulate glabrous terminal bracts;* pedicels much shorter than to equaling calyx; *sepals lance-ovate, long-acuminate, nearly or quite glabrous,* 6–8 mm. long; *corolla white,* 1.5–2.5 cm. high; *filaments hairy; styles divided to near base; capsule* exserted, plump-ovoid, *glabrous except at summit.* (*Stylisma* Chapm.) — Dry sandy woods and clearings of Coastal Plain, Fla. to e. Tex., n. to se. Va. and Ark. July, Aug.

2. B. Michaùxii Fern. & Schub. (for its discoverer, ANDRÉ MICHAUX, 1746–1802). — Resembling no. 1; *leaves copiously pubescent; peduncles* 1–3(–5)-*flowered, the terminal bracts more elongate and downy; pedicels shorter or flowers almost sessile; sepals and capsule densely pubescent; corolla pink or purple.* (*B. aquatica* of ed. 7 and most auth., not *Convolvulus aquaticus* Walt., basonym) — Pinelands, sands and pond-margins, Fla. to Tex., n. to N.C. and (fide *House* and *Small*) s. Va.

3. B. PICKERÍNGII (Torr.) Gray (in honor of CHARLES PICKERING, 1805–1878). — Freely branching, with soft-pubescent stems; *leaves narrowly linear to linear-oblanceolate,* 1–6 *mm.*

broad, the primary ones 3–6 *cm. long, all gradually tapering to base; peduncles terminated by a pair of foliaceous and elongate bracts resembling short foliage-leaves and subtending the* 1–3-*flowered inflorescence; central or solitary flower sessile, the lateral ones on pedicels only* 1–4 *mm. long; sepals ovate, blunt, heavily villous,* 5–6 *mm. long;* corolla white; *filaments glabrous or glabrescent; style unequally 2-cleft at summit,* the longer branch 2–3 mm. long; *capsule pubescent.* — The typical var. known only from sandhills near Wilmington, N.C.; the species represented with us by two vars.:

Var. **caesariénsis** Fern. & Schub. (of (New) Jersey). — *Flowers* 1–5, *all pedicelled, the lateral pedicels* 0.5–1.5 *cm. long; corolla* 1–1.4 *cm. high; style deeply cleft.* — Sands of Pine Barrens, N.J., local. Mid-July–late Aug.

Var. **Pattersòni** Fern. & Schub. (in honor of Harry Norton Patterson, 1853–1919). — Pubescence rather *more cinereous; sepals acute; style subentire or shallowly notched or cleft, the longest branch up to* 1–1.5 *mm. long.* — Dry prairies and sands, Henderson Co., Ill. and Muscatine Co., Ia.; (?) Kans.; Okla. and Tex. July, Aug.

3. EVÓLVULUS L.

Calyx of 5 sepals, naked at base. Corolla open-funnelform or almost rotate. Styles 2, each 2-cleft; stigmas obtuse. Capsule 2-locular, the locules 2-seeded. — Low and small herbs or suffrutescent plants of N. and S. Am. and Afr., mostly diffuse, never twining (hence the name, from *evolvere*, to unroll, in contrast with *Convolvulus*).

1. E. Nuttalliànus R. & S. (for Thomas Nuttall, 1786–1859). — Many-stemmed from a somewhat woody base, dwarf, *silky-villous all over;* leaves crowded, broadly lanceolate, sessile, or the lower oblong-spatulate and short-petioled; *flowers almost sessile* in the axils; *corolla* purple, *about* 1 *cm. broad.* (*E. argenteus* Pursh, not R. Br.) — Sterile plains and prairies, N.D. to Colo., s. to Mo., Kans., Tex. and Ariz. May–July.

2. E. alsinoìdes L. (like *Alsine*). — *Very slender, villous or hirsute; leaves oval or oblong-lanceolate, subtending elongate filiform peduncles with slender refracted pedicels; corolla about* 5 *mm. across.* — Waste ground, Mo. (Adv. from farther south)

4. IPOMOÈA L. Morning-glory

Calyx not bracteate at base, but the outer sepals commonly larger than the inner. Corolla salverform to nearly campanulate; the limb entire or slightly lobed. Style undivided; stigma capitate or of 2 or 3 globular lobes. Capsule globose or subglobose, 4–6 (or by abortion fewer)-seeded, 2–4-valved. — A complex and very large genus of most trop. and warm-temp. reg. (Name from the Greek *ips, a worm,* and *homoios, resembling;* from the twining habit.)

Corolla salverform or with slender funnelform tube; stamens and style exserted; flower scarlet, orange or red (rarely white); annuals. § 1. Quamoclit.
- Leaves pinnately divided into parallel linear segments. . . . . . . 1. *I. Quamoclit.*
- Leaves entire or palmately parted. . . . . 2. *I. coccinea.*

Corolla funnelform to nearly campanulate, contorted in bud; stamens and style not exserted. § 2. Euipomoea.
- Lobes of stigma and locules of capsule 3; sepals herbaceous, narrow and attenuate or acute to acutish; annuals with roseate, purple, blue or white corollas.
  - Leaves 3-lobed or unlobed; sepals drawn out into long linear tips; corolla 3–4.5 cm. long, the limb sky-blue when fresh (quickly changing to pink or purple). . . . . 3. *I. hederacea.*
  - Leaves unlobed; sepals with short acute or acutish tips; corolla 4.5–7 cm. long, variously colored. . . . . 4. *I. purpurea.*
- Lobes of stigma 0 or 2; capsule 2-locular; sepals firm; corollas white or purplish.
  - Deep-rooted trailing perennial; sepals glabrous or nearly so, obtuse, 1.2–2 cm. long; corolla 4.5–8 cm. long, 7–10 cm. broad. . . . . 5. *I. pandurata.*
  - Slender twining or creeping annual; sepals hirsute, bristle-tipped, 0.9–1.4 cm. long; corolla 1.5–2 cm. long, 1–2 cm. broad. . . . . 6. *I. lacunosa.*

§ 1. Quámoclit Meissn., Cypress-vines

1. I. Quámoclit L. (native Mexican name). — Capillary-stemmed twining annual; *leaves pinnately dissected into linear-filiform divergent segments* 1–2 cm. long; *peduncles* 1- *or* 2-*flowered,* small-bracted at summit (at junction with pedicel); *sepals* appressed to corolla-tube, *short-mucronate; corolla* with very slender tube, 3.5–4 *cm. long,* with rotate limb, scarlet-red (rarely white); capsule globose-ovoid. (*Quamoclit* Britt.; *Q. pennata* (Desr.) Voigt) — Locally spread from cult., n. to Va. and Mo. Aug.–Oct. (Introd. and adv. from Trop. Am.)

2. I. coccínea L. (scarlet), Red Morning-glory. — Differing from no. 1 in its *simple cordate-ovate* entire or angulate-margined thin *leaves;* flowers 2–several on thicker peduncle and pedicels; *sepals abruptly long-awned;* corolla 2–3 cm. long. (*Quamoclit* (L.) Moench) — Thickets, roadsides, waste places, etc., Fla. to Ariz., n. to Mass., N.Y., W.Va., O., Mich., Ill., Mo. and Kans. Aug.-Oct. (Natzd. from Trop. Am.)

§ 2. Euipomoèa Gray, Morning-glories

3. I. hederàcea (L.) Jacq. (resembling *Hedera*, English Ivy). — Slender *retrorsely hairy twining annual; leaves deeply cordate,* broadly ovate to suborbicular in outline, *deeply 3-lobed, the lobes acuminate; peduncles 1–3-flowered mostly shorter than petioles; flowers sessile or short-pedicelled; sepals with broad densely hairy blades, abruptly tapering to very prolonged linear tips; fresh corolla sky-blue,* quickly changing to rose-purple, with white tube, 3–4.5 *cm. long.* (*Pharbitis* Choisy) — Weed of cult. fields, roadsides and waste places, Fla. to Ariz., n. to N.E., N.Y., O., Ind., Ill., Minn. and N.D. July–Oct. (Natzd. from Trop. Am.)

Var. integriúscula Gray (nearly entire). — Leaves simply cordate-ovate, unlobed. — N. to Md. and Kans.

4. I. purpúrea (L.) Roth (purple), Common M. — *Coarser than no.* 3, pubescent to glabrescent, climbing high; *leaves broadly cordate-ovate, unlobed; peduncles elongate, umbellately* (1–) 2*-several-flowered, the pedicels mostly elongate; calyx hirsute or setose below, the narrow sepals acute or acutish; corolla* 4.5–7 *cm. long,* purple, red. bluish, white or variegated. (*Pharbitis* Voigt) — Roadsides, borders of fields and cult. ground, originally spread from cult. July–Oct. (Introd. and natzd. from Trop. Am.)

5. I. panduràta (L.) G. F. W. Mey. (fiddle-shaped), Wild Potato-vine, Man-of-the-earth. — *Stem nearly smooth to slightly pilose, strong and freely forking,* often purplish, trailing or slightly twining, *from a deep and farinaceous vertical root* (said sometimes to weigh 4–11 kg.); *leaves cordate-ovate or panduriform, short-pilose on one or both surfaces, or* in forma **leviúscula** Fern. (smoothish) *glabrous on both faces; peduncles* mostly *longer than petioles,* 1–5-flowered; *pedicels stout, ascending; sepals glabrous or essentially so, coriaceous, oblong-ovate, obtuse,* 1.2–2 *cm. long; corolla* open-funnelform, *white, with purple in the tube,* 4.5–8 *cm. long,* 7–10 *cm. broad.* — Dry open or partly shaded soil, Fla. to Tex., n. to Ct., N.Y., s. Ont., O., s. Mich., Ill., Mo. and Kans. June–Sept.

6. I. lacunòsa L. (with air-spaces; from the loosely reticulate venation of the leaves). — Smoothish *slender small creeping or twining thin-leaved annual;* leaves entire or angulate-lobed, cordate, acuminate; peduncles short, 1–3-flowered; *sepals sparsely long-hirsute or -ciliate,* lance-oblong, *bristle-tipped,* 0.9–1.4 *cm. long; corolla* sharply 5-lobed, 1.5–2 *cm. long,* 1–2 *cm. broad,* white, or in forma **purpuràta** Fern. (purplish) mauve-purple. — Damp to dryish thickets, meadows, fields and roadsides, Ga. to Tex., n. to Pa., O., Ind., Ill., Mo. and Kans. Aug.–Oct.

5. JACQUEMÓNTIA Choisy

Resembling *Ipomoea* but with the 2 stigmas thickened-ovoid and dorsiventrally compressed. Flowers in cymes, corymbs or glomerules. — A small chiefly trop. and subtrop. group, standing between *Ipomoea* and *Convolvulus.* (Named for *Victor Jacquemont,* 1801–1832, botanical explorer.) Incl. Thyella Raf.

1. J. tamnifòlia (L.) Griseb. (with leaves of *Tamus* or *Tamnus*). — Erect or slightly twining *fulvous-hirsute annual;* leaves long-petioled, cordate-ovate; *dense glomerules* terminating stem and long axillary peduncles, these with an involucre of lanceolate to narrowly ovate bracts; sepals linear- or lance-subulate, ferruginous-hirsute; corolla blue or violet, funnelform, about equaling calyx. — Fields, local, se. Va. and southw. Aug.–Oct. (Adv. from farther south or from the Tropics)

6. CONVÓLVULUS L. Bindweed. Liseron (Que.)

Corolla funnelform to campanulate. Stamens included. Style undivided or merely cleft at apex; stigmas 2, linear-filiform to subulate or slenderly ovoid. Capsule globose or subglobose, 2-locular or imperfectly 4-locular by false partitions between the 2 seeds, or by abortion 1-locular, mostly 2- or 4-valved. Herbs or somewhat shrubby twining, erect or prostrate plants of temp. and trop. reg. (Name from *convolvere, to entwine.*) Incl. Volvulus Medic. and Calystegia R. Br.

*a.*Calyx subtended by 2 broad leafy bracts; stigmas ovoid to ellipsoid. § 1. Calystegia. . . *b.*

*b.*Plants erect or merely twining at tip; flowers 1–4 from lower and median axils, white; basal leaves much smaller than well-developed upper ones; petiole short to barely half as long as leaf-blade. . . . . 1. *C. spithamaeus.*

*b.*Plants twining or trailing, the elongate stems forking and high-climbing or creeping; flowers numerous, from many axils, white or pink; basal leaves only slightly reduced; petioles one-fourth to quite as long as leaf-blade. . . *c.*

*c.*Bracts 1.5–3.5 cm. long; corolla (4–)5–8 cm. high; stamens (2–)2.5–3.3 cm. long; wide-ranging, chiefly native. . . . . 2. *C. sepium.*

*c.*Bracts 0.8–1.5(–2.5) cm. long; corolla 2.5–4.5 cm. high; stamens 1.2–1.6 cm. long (or changed to petals); introd. or spread from cult.

Plant glabrous; petioles one-half to quite as long as leaf-blade; hastate base three-fourths as broad as to broader than length of blade; flower with stamens. . . . . 3. *C. Wallichianus.*

Plant soft-pubescent; petioles one-fourth to one-third as long as blade; sagittate-hastate base about one-third as broad as length of blade; stamens often altered to petals (flower double). . . . . . . 4. *C. pellitus.*

*a.*Calyx without bracts; stigmas filiform. § 2. STROPHOCAULOS.

Perennial, trailing or twining; corolla 1.5–2 cm. high; weedy plant. . . . . 5. *C. arvensis.*

§ 1. CALYSTÈGIA (R. Br.) Gray

1. C. spithamaèus L. (a span high), Low B. — Plant glabrescent to downy; *stem simple or with few ascending branches from near base, usually not twining,* 0.7–5 *dm. high; the lowest leaves very small; the others nearly uniform and broadly to narrowly oblong or oblong-oval,* with cordate or subtruncate to tapering bases, the *mature blades* 3–8 cm. long *on petioles usually much shorter than the blade; flowers* 1–4, *erect, from lower axils, the leafy axis extending well above them; corolla white,* 3.5–7 cm. high. (*C. stans* Michx.) — Sandy or rocky open soil, thin woods, etc., sw. Me. and s. Que. to Thunder Bay Distr., Ont., and Minn., s. to Del., Md., D.C., upland Va., Tenn. and Ia. May–July (–Aug.). — Passing through the transitional *C. Purshianus* Wherry into

Var. **pubéscens** (Gray) Fern. (pubescent). — *Stems prolonged* (*up to* 1 *m.*) *and spiraling or twining above* the floriferous area; *leaves more scattered and longer-petioled,* cordate-based, *mostly velutinous,* the upper ones smaller. (Var. *Catesbeianus* (Pursh) Tryon; *C. Catesbeianus* (Pursh) Ell.) — Mts. of w. Va. to w. Ga. and n. Ala.

2. C. sèpium L. (of hedges), HEDGE-B., WILD MORNING-GLORY, GLOIRE DU MATIN (Que.). — *Twining or creeping, extensively branching,* glabrous to pubescent; *leaves cordate- or deltoid-ovate to narrowly sagittate-lanceolate, long-petioled; flowers from many axils, on 4-angled* short to prolonged divergent or ascending *peduncles;* paired *bracts* cordate, ovate (rarely subrotund), 1.5–3.5 *cm. long; corolla white to roseate,* 4–8 *cm. high,* nearly or quite as broad; *stamens* 2–3.3 *cm. long.* — Variable species of temp. reg., represented with us by the following:

Upper two-thirds (or blade) of principal leaves above the sinus (excluding lowest leaves) broadly ovate to ovate-triangular to broadly oval, mostly two-thirds as broad as to broader than long, usually 3–10 cm. broad.

Sinus of leaf U- or V-shaped, with sloping or divergent sides; corolla roseate or white.

Leaves broadly ovate, pointed, longer than broad, with angulate, truncate or rounded basal lobes; bracts acute or acutish, 2–3.5 cm. long; corolla 4.5–8 cm. high, white to roseate. . . . . *C. sepium* (typical).

Leaves round-oval to suborbicular, with rounded summit and lobes; bracts blunt, 1.5–2 cm. long; corolla white, 4–4.5 cm. high. . . . . . . Var. *dumetorum.*

Sinus of leaf quadrangular, the sides nearly parallel; corolla white. . . . . Var. *fraterniflorus.*

Upper two-thirds of blade of principal leaves lance-ovate to lanceolate, long-attenuate, one-fifth to rarely half as broad as long, mostly 1–3 cm. broad. . Var. *repens.*

C. sèpium (typical). — Leaves and stem glabrous or glabrescent; corolla white, more commonly with us as forma **coloràtus** Lange (colored) with roseate corolla. (Incl. vars. *americanus* Sims and *communis* Tryon, the leaves of both readily matched in Eurasian specimens.) Forma **malachophýllus** Fern. (soft-leaved) has the leaf-surfaces, petioles and stems soft-pilose, the plant chiefly coastal. — Thickets, shores, etc., Nfld. to B.C., s. to Fla., Ala., Mo., N.M. and Oreg. Mid-May–Sept. (Eurasia; N.Z.)

Var. DUMETÒRUM Pospichal (of bramble-thickets). — Waste places and roadsides, w. N.S. (Adv. or natzd. from Eu.)

Var. **fraterniflòrus** Mackenz. & Bush (with brotherly flowers; these sometimes in pairs). —

Very similar to the white-flowered typical *C. sepium* but with quadrangular leaf-sinus. — Thickets, roadsides, fields and waste places, Pa. to N.D., s. to Va., Ky. and Ark.

Var. **rèpens** (L.) Gray (repent). — Our narrowest-leaved extreme. (Incl. *C. Nashii* House) — Shores, beaches, dune-hollows, etc., Fla. to Tex., n. along coast to e. N.B. and inland n. to O., Ind. and Wisc.

3. C. Wallichiànus Spreng. (for its discoverer, Nathaniel Wallich, 1786–1854). — Slender *glabrous* twining perennial; *leaves on long slender petioles nearly or quite as long as leaf-blade, membranaceous, broadly hastate-lobed at base, there nearly as broad as to broader than the* oblong or narrowly ovate *blade;* peduncle equaling or overtopping leaf; *bracts cordate-ovate,* 0.8–1.6 *cm. long; corolla* 2.5–3.5 *cm. high,* roseate; *stamens* 1.5–1.6 *cm. long.* — Weed of cult. land, se. Pa. (Adv. from Asia)

4. C. pellìtus Ledeb. (clothed with a hairy skin). — Differing from no. 3 in its *close soft pubescence on stem, leaves and bracts; leaves firm, only slightly hastate, narrowly sagittate, with broad shallow or subtruncate sinus,* about *one-third as broad as long, on short petioles (one-fourth to one-third as long as blade); bracts rounded at base; corolla* 4–5 cm. long, pink, *simple in the typical form, very double and with many narrowly lacerate and overlapping segments and no stamens nor style in* forma anéstius Fern. (homeless) = *Calystegia pubescens* Lindl. — Typical single-flowered plant casual in waste ground, e. Mass. (adv. from Asia); forma *anestius* freq. in old or fallow fields, on roadsides or in waste places, e. Me. to Mich., s. to Va., Tenn. and Mo. (Introd. as a cult. plant, sent as a double-flowered curiosity from China to England a century ago, now natzd. with us)

§ 2. Strophocaùlos (Don) Choisy

5. C. arvénsis L. (of fields), Field-B. — *Slender trailing or slightly twining* pubescent to glabrate *perennial,* with freely branching stems up to 1 m. long; *leaves* long-petioled, very variable, *from ovate to oblong or almost linear, with hastate, cordate or sagittate bases;* peduncles mostly 1-flowered; bracts minute, remote; *corolla* 1.5–2 *cm. long,* white or reddish-tinged. — Old fields and waste places from s. Que. w. and s. beyond our range. June–Sept. (Natzd. from Eurasia) — The typical form has the sagittate- or hastate-ovate leaf with acute basal lobes; forma cordifòlius Lasch (heart-leaved) has the broad blade cordate and with rounded basal lobes; forma auriculàtus Desr. (eared) has linear-oblong to lanceolate blades with acute auricles.

Tribe Cuscùteae B. & H.

7. CUSCÙTA L. Dodder. Love-vine. Strangle-weed

Calyx 5 (rarely 4)-cleft or of 5 (or 4) sepals. Corolla globular-urceolate, campanulate or short-tubular, the border 5 (rarely 4)-cleft, imbricate. Stamens with a scale-like often fringed appendage (the *scale*) at base. Ovary 2-locular, mostly 4-ovuled; styles distinct, or rarely united. Capsule mostly 4-seeded. Embryo spirally coiled in the rather fleshy albumen, sometimes with a few alternate scales (belonging to the plumule); germination occurring in the soil. — Leafless annual herbs of most trop. and temp. reg., with filiform yellowish or reddish stems, bearing a few minute scales in place of leaves; on rising from the ground becoming entirely parasitic on the bark of herbs and shrubs on which they twine and to which they adhere by means of suckers developed on the surface in contact. Flowers small, cymose-clustered, mostly white, usually produced in summer and autumn. (Name supposed by some to be of Arabic derivation.) *

a. Stigmas capitate or dilated; styles distinct. Subgen. 1. Grammica; capsule remaining closed at maturity or finally irregularly rupturing (not circumscissile). § Cleistogrammica. . . *b.*

b. Calyx gamosepalous; ovary and capsule subglobose to globose-ovoid. . . *c.*

c. Ovary and capsule depressed-globose, of thin texture, without thickened summit; withered corolla basal (or in no. 2 capping the capsule); inflorescence dense or globose. . . *d.*

d. Flowers 4-merous (rarely 5-merous); lobes of corolla erect; stamens included or barely exserted; margins of scales only shallowly fringed.

Corolla-lobes triangular, acute, as long as or longer than the thick tube; scales 2-cleft or reduced to a few teeth; styles shorter than capsule; withered corolla persisting at base of capsule. . . . . . 1. *C. Polygonorum.*

---

* Many details derived from T. G. Yuncker's The Genus Cuscuta, Mem. Torr. Bot. Cl. xviii[2]. (1932), and from other studies by him.

Corolla-lobes ovate, obtuse, much shorter than slender tube; scales regularly short-toothed along margin and summit; styles equaling or longer than capsule; withered corolla capping summit of capsule. 2. *C. Cephalanthi.*

*d.*Flowers 5-merous; lobes of corolla acute, spreading, with often inflexed tips; stamens exserted; scales with long fringe; capsule protruding from marcescent corolla.

Flowers 1.5–2 mm. long; corolla-lobes lance-acuminate; scales reaching slightly above middle of corolla-tube, their longer fringe about one-fifth as long as blade; seeds about 1 mm. long. . . . 3. *C. pentagona.*

Flowers 2–3 mm. long; corolla-lobes triangular-ovate; scales reaching summit of corolla-tube, their longer fringe half as long as blade; seeds 1.5 mm. long. . . . . 4. *C. campestris.*

*c.*Ovary and capsule subglobose, with thickened stylopodium; the capsule capped by the marcescent corolla; flowers in rather open panicled cymes. . . *e.*

*e.*Corolla-lobes with inflexed acute or acutish tips; calyx-lobes triangular, acute or acutish; flowers granulate or papillose.

Flowers 5-merous, 2–5 mm. long; calyx about half as long as corolla-tube; scales large, deeply and regularly fringed, not bifid. . . 5. *C. indecora.*

Flowers 4-merous, about 2 mm. long; calyx equaling corolla-tube; scales bifid, consisting of few-toothed narrow halves. . . . 6. *C. Coryli.*

*e.*Corolla-lobes ovate, obtuse, reflexed in fruit, the tips not inflexed; flowers smooth or glandular. . . . *f.*

*f.*Corolla-lobes spreading; ovary globose or globose-ovoid, not prolonged at summit; capsule beakless; flowers 2–4 mm. long.

Lobes of corolla wide-spreading in anthesis; scales oblong, fringed from near base to summit, nearly equaling corolla-tube; styles two-thirds as long as capsule; seeds about 1.5 mm. long. . . . 7. *C. Gronovii.*

Lobes or corolla only slightly spreading in anthesis; scales bifurcate, long-fringed at summit, about half as long as corolla-tube; styles one fourth as long as capsule; seeds 2–2.8 mm. long. . . . 8. *C. megalocarpa.*

*f.*Corolla-lobes erect; ovary flask-shaped, with beaked summit; capsules long-pointed; scales oblong, half as long as corolla-tube; seeds about 2.4 mm. long; flower 4–6 mm. long. . . . . 9. *C. rostrata.*

*b.*Calyx of distinct sepals surrounded by 2 or more similar bracts; globose to ovoid beaked or beakless capsule capped by the marcescent corolla. . . *g.*

*g.*Flowers pedicelled in loose panicles; ovate bracts and sepals appressed, cuspidate; seeds 1.4 mm. long. . . . . 10. *C. cuspidata.*

*g.*Flowers sessile in dense masses; seeds 1.7–2.6 mm. long.

Bracts and sepals broadly oval to suborbicular, rounded at summit, appressed; corolla-lobes obtuse; ovary and capsule globose-ovoid; seeds about 2.6 mm. long. . . . . 11. *C. compacta.*

Bracts and sepals narrowly oblong or lanceolate, with recurved-spreading slender tips; ovary and capsule flask-shaped; seeds about 1.7 mm. long. . . . . 12. *C. glomerata.*

*a.*Stigmas filiform, elongate; styles distinct. Subgen. 2. Succuta; capsule regularly circumscissile toward base. § Eucuscuta. . . *h.*

*h.*Flowers 4 (rarely 5)-merous; lobes of calyx and corolla obtuse; scales bifid, fringeless. . . . . 13. *C. europaea.*

*h.*Flowers 5-merous; lobes of calyx and corolla acute; scales fringed.

Flowers yellowish; calyx-lobes broadly ovate; corolla-lobes ascending or soon conforming to capsule; stamens included. . . . . . 14. *C. Epilinum.*

Flowers whitish or pinkish; calyx-lobes triangular; corolla-lobes spreading; stamens exserted. . . . . 15. *E. Epithymum.*

Subgen. 1. Grámmica (Lour.) Engelm.

§ Cleistográmmica Engelm.

1. C. Polygonórum Engelm. (of Polygonums). — *Stems slender, orange-colored*; inflorescences *compact glomerules* 0.5–1 *cm. in diameter; flowers* 2–2.5 mm. long, subsessile, mostly *4-merous;* calyx-lobes narrowly ovate, obtuse, equaling or exceeding corolla-tube; *corolla-lobes triangular-lanceolate, acute, as long as or longer than thick tube, erect; scales* oblong, *2-cleft or reduced to a few teeth; stamens* included, *borne directly from the sinuses; styles shorter than depressed-globose ovary; capsule* exserted, *with persistent withered corolla at base; seeds* yellowish-brown, about

1.3 *mm. long.* (*C. obtusiflora* of ed. 7, not HBK.; *C. chlorocarpa* Engelm.) — Low ground, on *Polygonum*, *Lycopus*, *Penthorum* and other herbs, sw. Que. to Minn. and Neb., s. to s. N.E., L.I., Del., Md., Tenn., Mo. and Kans. July–Sept.

2. C. Cephalánthi Engelm. (of *Cephalanthus*). — Coarse yellow stems high-climbing; glomerules larger and commonly more crowded or confluent than in no. 1; flowers 4 (sometimes 3 or 5)-merous; *calyx much shorter than* rather *slender corolla-tube*, its narrowly ovate *lobes obtuse; corolla-lobes ovate, obtuse, much shorter than tube; scales narrowly oblong, shallowly short-toothed from near base to subtruncate apex; stamens borne below and barely reaching sinuses of corolla; styles equaling or longer than capsule; marcescent corolla capping capsule; seeds* light brown, *about* 1.6 *mm. long.* — Low grounds, on various shrubs and coarse herbs, N.S. to s. B.C., s. to s. N.E., Va., Tenn., Mo., Tex. and N.M. Aug., Sept.

3. C. pentagòna Engelm. (5-angled). — *Stems pale, very slender* and low, the globular glomerules scattered or slightly confluent; *flowers 5-merous, whitish,* 1.5–2 *mm. long,* somewhat glandular, pedicelled; *calyx nearly inclosing corolla-tube, its short roundish lobes overlapping at the sinuses; corolla-lobes lance-acuminate, wide-spreading, their slender elongate tips inflexed; stamens arising below sinuses, exserted in flower;* oblong *scales reaching slightly above middle of corolla-tube, their longer fringe about one-fifth as long as blades; capsule protruding from marcescent corolla; seeds* 1 *mm. long.* (*C. arvensis* Beyrich) — Dry open soil, on many herbaceous or slightly ligneous hosts, Fla. to s. Calif., n. to s. N.E., se. N.Y., e. Pa., Md., s. Ont., Ind., s. Wisc., s. Minn., S.D. and Mont. June–Oct.

4. C. campéstris Yuncker (of prairies). — Coarser than no. 3, the glomerules larger; *flowers* 2–3 *mm. long,* short-pedicelled; calyx-lobes broader; *corolla-lobes* narrowly *triangular-ovate; stamens borne at sinuses; scales reaching summit of corolla-tube, their longer fringe half as long as blade; seeds about* 1.5 *mm. long.* — On various herbs, including cult. species, partly adv., S.C. to Tex., Mex. and Calif., n., locally in the East, to Mass. and sw. Que., and more generally to Ind., Ill., Minn., Man., Sask., Colo. and Wash. June–Oct. (Trop. Am.)

5. C. indecòra Choisy (not ornamental). — Relatively coarse; *flowers 5-merous,* 2–2.5 mm. long, *fleshy, papillose, shorter than the slender pedicels,* forming rather open panicles; *calyx much shorter than corolla-tube, with narrowly triangular-ovate acute or acutish lobes; campanulate corolla with spreading-ascending to erect triangular lobes with inflexed acute tips;* stamens half as long as corolla-lobes; *scales* oblong, *equaling corolla-tube, deeply and regularly fringed;* styles equaling or longer than the depressed-globose ovary; *capsule thickened at summit,* wrapped by the marcescent corolla; seeds about 1.7 mm. long. — Damp pinelands, bottoms and sandy openings, on various herbs and shrubs, Fla. to Tex. and Mex., n. to se. Va., N.D. and Colo. July, Aug. (W. Ind.)

Var. **neuropétala** (Engelm.) Hitchc. (with nerved petals). — Flowers 2–5 mm. long; corolla much thicker-campanulate, the tube nearly or quite equaled by the calyx-lobes. — Fla. to Tex. and Mex., n. to Ill., Minn., S.D., Wyo., Ida. and Calif. (W.I.)

6. C. Córyli Engelm. (of *Corylus*). — Similar to no. 5; *flowers mostly* 4 (rarely 5)*-merous, about* 2 *mm. long; calyx-lobes equaling or exceeding corolla-tube; corolla* slenderly campanulate, *with erect lanceolate lobes* with incurved tips; *scales bifid, of few-toothed narrow halves, shorter than corolla-tube;* withered corolla soon cast off from the *depressed capsule;* seeds 1.5 mm. long. — On various shrubs and herbs, on sandy pond-shores, bottoms and low or dry grounds, s. N.E. to Mont., s. to N.C., Tenn., Mo., Tex., N.M. and Ariz. July–Sept.

7. C. Gronòvii Willd. (for Jan Fredrik Gronovius, 1690–1762, teacher of Linnaeus and author of *Flora Virginica* (1739)). — Stems slender to coarse; *flowers in* loosely to densely *panicled cymes,* 2–4 *mm. long, mostly 5-merous,* short- or long-pedicelled, often glandular; *calyx about half as long as the thick-campanulate corolla-tube;* the *ovate-oblong to suborbicular lobes obtuse,* overlapping below; *corolla-lobes spreading, ovate, obtuse,* reflexed in age; *scales* oblong, *fringed from base to summit, nearly equaling corolla-tube, the upper fringe prolonged; styles two-thirds as long as globose-ovoid capsule, the latter enveloped by the withered corolla* (or rarely crowned by it); *seeds about* 1.5 *mm. long.* (Including var. *latiflora* Engelm. = var. *Saururi* (Engelm.) MacM.) — Low grounds, parasitic on very many hosts, Que. to Man., s. to N.S., N.E., L.I., Fla., Ala., La., Tex., N.M. and Ariz. July–Oct.

8. C. megalocárpa Rydb. (large-fruited). — *Differing from no. 7 in globose-glomerulate clusters more scattered; lobes of corolla only slightly spreading in anthesis; calyx much shorter; scales much shorter than corolla-tube, bifurcate; styles about one-fourth as long as large* (3–6 *mm. long*) *globose-ovoid umbonate-topped capsule; seeds* 2–2.8 *mm. long.* (*C. curta* (Engelm.) Rydb., a later binomial) — On various shrubs and herbs, sometimes cult. crops, Minn. to Mont., s. to Colo. and Utah. July, Aug.

9. C. rostràta Shuttlew. (beaked). — *Differing from no. 7 in its more delicate and whiter* *flowers 4–6 mm. long and on shorter pedicels; corolla-lobes at first erect, the tube with slender thickened lines running down from the bases of the stamens, the tube thus appearing angled,* the lobes finally reflexed; *scales* oblong, *half as long as corolla-tube; ovary flask-shaped, the summit beaked; capsule beaked,* enveloped by the marcescent corolla; *seeds about* 2.4 *mm. long.* — On various herbs and shrubs, mts. of W.Va. to S.C. and Tenn. Late July–Sept.

10. C. cuspidàta Engelm. (cuspidate). — Stems rather slender; *flowers* 5-merous, 3–5 *mm. long, in* loose to dense *bracted panicled clusters; calyx of distinct* or nearly distinct *sepals subtended by* 1 *or more broad bracts; sepals broadly ovate to suborbicular, cuspidate, appressed,* half as long as corolla-tube; *corolla-lobes oblong to lanceolate, acute,* spreading, shorter than the cylindric tube; oblong scales long-fringed, shorter than corolla-tube; stamens short, in sinuses of corolla; styles much longer than subglobose ovary; withered corolla capping capsule; *seeds about* 1.4 *mm. long.* — Dry to wet prairies, on various herbs, s. Ind. to N.D., Colo. and Utah, s. to La. and Tex. July–Oct.

11. C. compácta Juss. (compact). — Rather coarse; *flowers* 4–5 mm. long, *sessile in dense scattered to confluent subglobose glomerules;* the 3–5 *bracts and sepals orbicular or nearly so, concave,* slightly crenate, *closely appressed; corolla-*tube cylindric, becoming urceolate, the oblong *blunt lobes* spreading to reflexed; stamens very short; scales long-fringed; *ovary globose-ovoid, subconical at the thickened summit;* capsule similar, the summit surrounded by the withered corolla; *seeds* about 2.6 *mm. long.* — On a great variety of shrubs and coarse herbs in low woods and thickets, Fla. to Tex., n. to s. N.H., Mass., se. N.Y., N.J., e. Pa., D.C., Va., s. Ind., Ill., Mo. and Okla. July–Oct.

12. C. glomeràta Choisy (with glomerules). — *Differing from no.* 11 *in the white or stramineous depressed-globose or oblate dense glomerules,* these soon obliterating the winding stems; *bracts numerous, oblong or lanceolate, scarious, with slender recurved tips; sepals similar to bracts, their tips arched-recurving; corolla-lobes narrowly oblong to lanceolate, acute or acutish,* spreading; *ovary and capsule flask-shaped,* the withered corolla carried at summit of capsule; *seeds about* 1.7 *mm. long.* — On many coarse herbs, mostly *Compositae,* sw. Mich. to S.D., s. to Miss., Mo., Kans. and Tex. July–Sept.

Subgen. 2. Succuta (Des Moulins) Yuncker

§ Eucuscùta Engelm.

13. C. europaèa L. (European). — Stems thread-like, greenish to reddish-yellow; *flowers* yellowish, 2–3 *mm. long,* in dense subglobose glomerules, 4 (rarely 3 or 5)*-merous;* calyx campanulate, its obtuse lobes appressed-ascending; *corolla* campanulate, becoming urceolate, *its ovate blunt lobes erect or incurved; scales very short, bifid, not fringed,* barely reaching insertion of filaments; *styles and filiform stigmas shorter than ovary; capsule* subglobose, *circumscissile near base;* seeds about 1.5 mm. long. — On many herbaceous or shrubby hosts; with us casual in a hedge-row, Oxford Co., Me. July, Aug. (Adv. from Eu.)

14. C. Epílinum Weihe (on flax), Flax-D. — *Stem* slender, *nearly simple, pale green; flowers yellowish,* about 3 mm. long, 5*-merous; calyx-lobes fleshy, ovate, acutish, as long as corolla-tube; corolla-lobes ovate, acutish, ascending; scales bifid, with short marginal teeth; styles soon bowed outward, the slender stigmas converging;* capsule depressed-globose, circumscissile; *seeds about* 1.2 *mm. long.* — Very injurious to cult. flax, Que. to Mich., s. to Del. and Pa. June–Aug. (Adv. from Eu.)

15. C. Epíthymum Murr. (on thyme), Clover-D. — *Stems* very slender, *red, freely branching, making dense entangling masses; calyx red or pink, with* 5 *thin triangular acute lobes; corolla with triangular acute spreading lobes; scales roundish, fringed;* stamens about equaling corolla-lobes; *seeds about* 1 *mm. long.* — Very injurious in clover-fields, N.E. to Wash., s. through much of our range. June–Oct. (Natzd. from Eu.)

FAM. 147. POLEMONIÀCEAE (Polemonium Family)

Herbaceous or suffruticose, with alternate or opposite leaves, mostly regular 5-merous and 5-androus flowers, the lobes of the corolla convolute in the bud, a 3-locular ovary and 3-lobed style; capsule 3-locular, 3-valved, loculicidal, few–many-seeded; the valves usually breaking away from the triangular central column. Seeds amphitropous, the coat frequently mucilaginous when moistened and emitting spiral threads. Embryo straight, in the axis of copious albumen. Calyx persistent, imbricated. Corolla with a 5-parted border. Anthers introrse. — Plants without poisonous or medicinal properties, many ornamental and in cult.*

* Many details drawn from the studies of Edgar T. Wherry in Bartonia, xi–xxiii (1929–1945) and from his other papers, which see for further subdivisions of species.

Leaves pinnate, with several broad leaflets, alternate; calyx herbaceous, with broad lobes, accrescent in fruit; corolla open-campanulate to nearly rotate; filaments declined; perennials. . . . . 1. *Polemonium.*
Leaves simple or dissected; calyx firm, with narrow lobes, not accrescent; corolla salverform to funnelform or campanulate; filaments not declined.
 Stamens equally inserted in or below throat or in sinuses of corolla; leaves divided. . . . . 2. *Gilia.*
 Stamens unequally inserted on tube of corolla; leaves simple, entire.
  Leaves alternate; flowers slender, in leafy-bracted glomerules; calyx-tube not ruptured by expanding capsule; seeds, when wet, exuding mucilage and spiraling threads (spiricles); annual. . . . . 3. *Collomia.*
  Leaves opposite; flowers in cymes, the rotate limb of the corolla showy; calyx-tube ruptured by expanding capsule; seeds not emitting spiricles; perennials (some annuals in cult.). . . . . 4. *Phlox.*

1. POLEMÒNIUM L. GREEK VALERIAN. JACOB'S-LADDER

Stamens equally inserted at the summit of the very short tube of the open-campanulate or short-funnelform corolla; filaments declined, hairy-appendaged at base. Calyx herbaceous, with broad lobes, accrescent in fruit. Capsule few–several-seeded. — Perennials of cool and temp. reg. of N. Hemisph., with pinnate leaves and blue to purple mostly bractless corymbed flowers. (Ancient name, by some thought to be for *Polemon*, early Athenian philosopher, by others derived from the Greek *polemos, war.*)

*a.*Larger leaves with 15–29 leaflets; inflorescence a thyrsiform or contracted panicle; pedicels much shorter than calyx; stem 0.2–1 m. high; lowest bracteal leaves (subtending lowest flowering branch) with 3–7 greatly reduced and crowded leaflets. . . *b.*
*b.*Leaflets oblong-lanceolate to narrowly lance-linear, closely sessile, those of the larger leaves 19–29; stamens hardly to barely exserted.
 Rhizome deep, ascending; summit of stem bearing stipitate glands; leaflets oblong-lanceolate, up to 1 cm. broad; crowded leaflets of lowest bract of inflorescence 3–7; calyx-lobes oval, obtuse; corolla blue, 1.5–2 cm. high; introd. and spread from cult. . . . . 1. *P. caeruleum.*
 Rhizome horizontal, very slender and creeping; summit of stem without elongate glands; leaflets lance-linear, 2–7 mm. broad; crowded leaflets of lowest bract 9–13; calyx-lobes deltoid-oblong to -lanceolate, acutish; corolla 1–1.5 cm. high, violet; native northwestw. . . . . . . 2. *P. occidentale*, var. *lacustre.*
*b.*Leaflets narrowly ovate or ovate-lanceolate, short-petiolulate, those of larger leaves 15–21; stamens long-exserted; native eastw. . . . . . 3. *P. Van-Bruntiae.*
*a.*Larger leaves with 11–17 leaflets; inflorescence openly branching, bearing remote long-peduncled corymbs with pedicels often longer than calyx; stems loosely ascending, slender, rarely 5.5 dm. high; lowest bracteal leaves (subtending lowest flowering branch) with 9–11 leaflets nearly as large as in median foliage-leaves; natives.
 Leaflets acute, entire; corolla blue, 1–1.6 cm. long; style about equaling corolla; stamens included; wide-ranging. . . . . 4. *P. reptans.*
 Leaflets obtuse, often bluntly lobulate at summit; corolla red-purple, 7–8 mm. long; stamens long-exserted; style obsolete; local in se. Pa. . . . 5. *P. Longii.*

1. P. caerùleum L. (blue). — *Deep-rooting, strong, erect,* with simple stem *up to 1 m. high; lower and longer leaves with* 19–29 *narrowly oblong-lanceolate closely sessile and obliquely round-based* nearly approximate *leaflets (the pairs rarely* 1 *cm. apart at base) up to* 1 *cm. broad; inflorescence virgate or slenderly thyrsoid, with very short branches, the lowest branch subtended by a greatly reduced short blade with* 3–7 *crowded leaflets; glandular and viscid pedicels shorter than calyx; calyx short-pubescent at base, its lobes oval, obtuse; corolla blue* (rarely white), 1.5–2 *cm. long; stamens and style scarcely or barely exserted.* — Spread from cult. to roadsides and waste places, e. Can. and N.E. to Minn. and doubtless elsewhere. (Introd. from Eurasia)

2. P. occidentàle Greene (western), var. **lacústre** (Wherry) Lakela (of lakes). — *Differing from no.* 1 *in its very slender and horizontally creeping rhizome; summit of stem* viscid and pilose but *without elongate glands; leaflets lance-linear,* 2–7 *mm. wide;* the *lowest bracteal leaf with* 9–13 *leaflets; panicle more branching and open,* with long ascending branches; *calyx-lobes deltoid-oblong to -lanceolate, acutish; corolla bright violet, strongly whitened at center,* 1–1.5 *cm. high.* — Arbor-vitae swamp, St. Louis Co., Minn. Late June, early July. — Eastern representative of a western species.

3. P.Ván-Brúntiae Britt. (named in 1892 for Mrs. CORNELIUS VAN BRUNT). — *Differing from no.* 1 *in its horizontal rhizome; lower and longer leaves with* 15–21 *more distant (mostly* 1–2 *or rarely* –3.5 *cm. apart) narrowly ovate to ovate-lanceolate short-petiolulate leaflets; pedicels short-hirtellous, scarcely viscid; calyx with long divergent trichomes at base, its lobes lanceolate, acute; corolla blue-purple,* 1.2–1.5 *cm. long;* stamens and style much exserted. — Wooded swamps, bottoms, sphagnous bogs and mossy glades, Vt. and N.Y. to Md. and W.Va., often at high alts. Mid-June–late July.

4. P. réptans L. (creeping; a misnomer). — *Tufted, the diffuse and low-branching stems* 1.5–5.5 *dm. high,* glabrous or sparsely villous; *basal and longer leaves with* 11–17 *lanceolate to oval thin* acute *leaflets in remote pairs; lowest bracteal leaf with* 9–11 *leaflets nearly as large as in cauline leaves; inflorescence with long ascending slender branches terminated by few-flowered corymbs; pedicels often as long as calyx; calyx-lobes lanceolate to lance-triangular, acute or acutish; corolla deep blue,* 1–1.6 *cm. high, completely overtopping the stamens; style included or barely exserted.* — Rich woods and bottoms, N.Y. to Minn., s. to interior Ga., Ala., Miss., Mo. and Okla.; cult. and esc. elsewhere. Mid-April–June.

5. P. Lóngii Fern. (for its discoverer BAYARD LONG, 1883–). — *Differing from no.* 4 *in higher branching; longer leaves with* 11–13 *oblong or oblong-obovate round-tipped often apically lobulate leaflets; calyx* 5–6 mm. long, *with broadly triangular lobes; corolla reddish-purple,* 7–8 *mm. high; stamens long-exserted; style wanting.* — Alluvial woods, Montgomery Co., Pa., not collected since 1925; to be sought elsewhere.

2. GÍLIA R. & P.

Calyx firm, with narrow segments or lobes, not accrescent, the tube scarious and ruptured by expanding capsule. Corolla campanulate to funnelform or salverform, regular. Stamens equally inserted at or below the throat of the corolla, the naked filaments not declined. Ovules and seeds solitary to numerous. Leaves variously divided. Large genus, chiefly annuals and biennials, mostly of w. N. Am., a few in S. Am. (Named in 1794 for *Felipe Gil*, a Spanish botanist thus honored by Ruiz & Pavon.)

1. G. rùbra (L.) Heller (red), STANDING-CYPRESS. — Tall simple-stemmed biennial up to 1 m. or more high; leaves crowded, dissected into filiform segments; slender thyrse 2–5 dm. long, leafy at base, naked above; calyx-lobes triangular-lanceolate, acute; corolla slenderly funnelform, red, pink or white, 2.5–3.5 cm. long; stamens included or barely exserted; ovules and seeds numerous. (*Ipomopsis* Wherry) — Spread from cult. and natzd. in pastures, on roadsides, etc., n. to N.E., N.Y., O., Mich., Ill. and Mo. June–Aug. (Introd. and natzd. from s. U.S.)

A number of annual small species of *Gilia* and of *Navarretia* R. & P. (with spinulose unequal sepals, the tube not ruptured by expanding capsule) are sporadic in hay- or clover-fields or in waste places, all adv. from w. N. Am. but apparently not long-persistent.

3. COLLÒMIA Nutt.

Calyx-tube subherbaceous, scarcely accrescent, not ruptured by expanding capsule. Corolla tubular-funnelform or salverform, with unequally inserted stamens protruding from the open orifice. Ovules and seeds 1–many, the seeds, when moistened, exuding mucilage and spiraling threads (spiricles). Annuals or biennials with simple or cut sessile alternate leaves, the flowers in leafy-bracted glomerules; a small N.Am. genus. (Name from the Greek *colla, gluten,* from the mucilaginous seeds.)

1. C. lineàris Nutt. (linear). — Simple or branching and in age often spreading, 1–5 dm. high; leaves linear- to oblong-lanceolate; calyx-lobes triangular-lanceolate, acute; corolla about 1 cm. long, lilac-purple to whitish, very slender, with small limb. (*Gilia* Gray) — Dry open soil, gravelly shores, etc., sw. Gaspé Pen., Que., and adj. n. N.B.; Thunder Bay Distr., Ont., to B.C., s. to Wisc., Minn., Neb., N.M., Ariz. and Calif.; and as an adv. weed from N.S. to Ia., s. to N.J., Pa. and Mo. June–Aug.

4. PHLÓX L. PHLOX

Calyx somewhat prismatic or plaited and angled. Corolla with a long tube. Stamens very unequally inserted in the tube of the corolla, included or slightly exserted. Capsule ovoid, with sometimes 2 or 3 ovules but usually ripening only a single seed in each locule. — Perennials (except a few southern species) with chiefly opposite and sessile or sometimes petioled entire leaves, the floral ones often alternate. Flowers cymose, mostly bracted; the open clusters terminal or crowded in the upper axils. (Greek for *flame*, an ancient name of *Lychnis*, transferred to this North American (and slightly Asiatic) genus.) Most of our species are cult.

*a.*Stems ligneous or subligneous, trailing or depressed, often with fascicles or short leafy branchlets in leaf-axils at anthesis; leaves persistent or mostly so, hard and stiffish, linear to subulate; cymose flowers long-pedicelled. § 1. SUBULATAE. . . *b.*

*b.*Lobes of corolla entire, erose or very shallowly emarginate; stamens included; style 0.5–3 mm. long; ovules 1–3 in each locule; se. species of Coastal Plain and Piedmont. . . . . 1. *P. nivalis.*

*b.*Lobes of corolla usually deeply notched; stamens partially exserted; style 4–12 mm. long; ovule 1 in each locule.

Stems with very numerous and crowded nodes; principal leaves 0.8–2 cm. long, subtending crowded axillary fascicles; notch of corolla-lobes about 1 mm. deep; densely matted; inland and upland species, on Coastal Plain only northw. . . . . 2. *P. subulata.*

Stem with few nodes 1 cm. or more apart; leaves 2–6 cm. long, with few or no axillary fascicles; notch of corolla-lobe 3–7 mm. deep; diffuse plant west of Alleghenies. . . . . 3. *P. bifida.*

*a.*Herbaceous; the flowering stems erect to decumbent, with flat or flattish mostly deciduous leaves without conspicuous axillary fascicles at anthesis; flowers rather short-pedicelled. . . *c.*

*c.*Stamens and style at most reaching middle of corolla-tube; style much shorter than villous fruiting calyx. § 2. DIVARICATAE. . *d.*

*d.*Stems spreading or ascending from decumbent often elongate and rooting bases or with elongate decumbent sterile shoots; leaves oblong, lanceolate or ovate, obtuse or barely acute; inflorescence becoming loose and open in maturity; calyx densely stipitate-glandular. . . . . . . 4. *P. divaricata.*

*d.*Stems erect or ascending, without decumbent sterile shoots; inflorescence compact to open; calyx glandless (sometimes glandular in no. 5).

Leaves narrowly lanceolate to linear, attenuate to sharp tip; inflorescence finally becoming open, its branches standing well out from the basal leaves; calyx glandless or glandular; wide-ranging northw. into s. Can. . . . . 5. *P. pilosa.*

Leaves broadly linear to oblong-lanceolate or ovate, blunt or merely acutish; cyme compact, sessile, embraced by leafy involucre; calyx glandless; southern. . . . . 6. *P. amoena.*

*c.*Stamens (at least the upper) and style reaching the orifice of the corolla-tube or more or less exserted; style equaling or exceeding the fruiting calyx. . *e.*

*e.*Leaves with obscure lateral veins and smooth to slightly roughened eciliate margins; anthers deep yellow. § 3. OVATAE. *f.*

*f.*Loosely stoloniferous, with long creeping to flagelliform repent leafy runners; rosette-leaves spatulate-obovate; viscid-villous pedicels nearly or quite equaling calyx. . . . . 7. *P. stolonifera.*

*f.*Ascending to erect, without repent stolons, the basal offshoots at most decumbent and upcurving; basal leaves mostly narrower and often not evergreen. . . *g.*

*g.*Flowering shoots arising from leafy tips of decumbent stems, with 3–7 pairs of leaves below the corymb.

Lowest tufted leaves ensiform, evergreen; cauline leaves linear, ensiform or narrowly lanceolate and sessile; inflorescence densely glandular. . . . . 8. *P. Buckleyi.*

Lowest tufted leaves elliptic to ovate, marcescent; lower and median cauline leaves oblanceolate to elliptic or obovate and petioled; inflorescence glandless. . . . . 9. *P. ovata.*

*g.*Flowering shoots chiefly arising from leafless crowns or branches of rhizomes, with 6–15 or more pairs of leaves below the corymb. . . *h.*

*h.*Stems green or at most vaguely suffused with purple, with erect to short-decumbent bases; inflorescence a corymb or long-branched panicle; calyx-lobes at anthesis nearly or quite as long as tube.

Upper and bracteal leaves linear or narrowly lanceolate; nodes of flowering stem 12–23; calyx 5–8.5 mm. long. . . . . . 10. *P. glaberrima.*

Upper and bracteal leaves ovate or lanceolate; nodes 5–12; calyx 8–12 mm. long. . . . . 11. *P. carolina,* var. *triflora.*

*h.*Stem usually with strong red or purple dots, arising from tip of slender elongate rootstock; inflorescence cylindric, short-branched, usually at least twice as long as thick; calyx-lobes at anthesis much shorter than tube. . . . . 12. *P. maculata.*

e. Leaves with lateral veins evident beneath, conspicuously reticulate or areolate (by transmitted light), the young margins bristly-ciliolate; anthers pale yellow or white; stems erect, clustered. § 4. PANICULATAE.
Nodes below inflorescence 14–25 or more; inflorescence usually glandless; corolla-tube commonly pubescent; 1 or more anthers exserted. 13. *P. paniculata.*
Nodes 7–15; inflorescence copiously glandular; corolla-tube glabrous; anthers included. . . . . 14. *P. amplifolia.*

§ 1. SUBULÀTAE Peter

1. P. nivàlis Lodd. (snowy), TRAILING P. — *Stems subligneous, forming tufts or freely branching prostrate mats up to* 6 *dm. or more across; nodes crowded, bearing linear-subulate hard and somewhat evergreen squarrose-divergent leaves* 1–2.5 cm. long, *these with dense axillary fascicles;* flowering branchlets ascending; cyme few-flowered; the *densely pilose slender pedicels one-half to twice as long as calyx; calyx-tube* villous, 4.5–6 *mm. long,* about equaling linear-subulate lobes; *corolla* white or pale pink, or in forma **roseiflòra** Fern. (rosy-flowered) deep rose or magenta, with dark eye, the *tube long-exserted,* the limb 2–3 cm. broad; *corolla-lobes entire, erose or very shallowly emarginate; stamens included; style* 0.5–3 *mm. long; ovules and seeds* 1–3 *in each locule.* — Dry sandy or gravelly pine- or oak-lands, barrens and scrub, Fla. to Tex., n., chiefly on Coastal Plain and Piedmont, to s. Va., w. N.C., n. Ga. and centr. Ala. Late March–early May.

2. P. subulàta L. (awl-shaped), MOSS- or MOUNTAIN-P., MOSS-PINK. — Habitally similar to no. 1, often forming broader mats; *principal leaves* 0.8–2 *cm. long; pedicels and calyx glandless* or rarely glandular; *calyx* 6–9.5 *mm. long; corolla smaller; the tube* 8.5–16 *mm. long;* the *lobes* 8–12 *mm. long and* 4.5–12.5 *mm. broad, with emargination* 1–2 *mm. deep,* roseate or purplish, or white in forma **albiflòra** Britt. (white-flowered); *stamens partially exserted; style elongate, up to* 1.2 *cm. long; ovule and seed single in each locule.* (Var. *ciliata* (Brand) Wherry) — Dry sandy, gravelly or rocky soil, L.I. to nw. N.Y., w. to s. Mich., s. to N.J. and (inland from Coastal Plain) to Md., upland to N.C. and Tenn.; spread from cult. and natzd. in N.E. April–early July.

Var. **Brittònii** (Small) Wherry (for NATHANIEL LORD BRITTON, 1859–1934). — More delicate; *leaves mostly* 5–10 *mm. long; hairs of inflorescence usually gland-tipped; calyx* 4–7(–7.5) *mm. long; corolla-lobes* 4–9 *mm. long and* 2–6 *mm. broad,* lavender-white, or in forma **austràlis** (Wherry) Fern. (southern) deep purple. — Mts. of Md. to L. Erie, Norfolk Co., Ont., s. to N.C. and Tenn.

3. P. bífida Beck (2-cleft). — Subligneous at base, *diffusely ascending and much branched,* forming tussocks 1–2 *dm. high; stem and branches densely pilose or puberulent,* with or without glands, the *nodes subdistant* (*mostly* 1–3.5 *cm. apart*); *leaves narrowly linear, long-attenuate,* 2–6 *cm. long, with few or no axillary fascicles; flowers few, on elongate filiform pedicels, often scattered;* calyx-lobes subulate, about equaling tube; *lobes of* pale purple *corolla deeply* 2 (or 3)-*cleft to below middle, the divergent segments oblong or linear* and 3–7 *mm. long.* (Var. *glandifera* Wherry) — Dry cliffs, bluffs, sandhills, dunes, etc., s. Mich. to centr. Ia., s. to Ky., Mo. and e. Okla. Late April–early June. — Passing into

Var. **cedària** (Brand) Fern. (Brand's very modern Latin from "Cedar Barrens" of Tenn.). — Glabrous or very thinly pubescent, glandless. (Var. *Stellaria* (Gray) Wherry; *P. Stellaria* Gray) — Local on cliffs and rock-barrens, s. Ind. to Tenn. and Ark.

§ 2. DIVARICÀTAE Peter

4. P. divaricàta L. (strongly divergent; from the branching of the inflorescence), BLUE P. — *Stems spreading or ascending from a decumbent base, loosely villous; the sterile basal shoots often prolonged, prostrate and rooting;* upright flowering stems 1.5–5 dm. high, slender; *leaves* in few remote pairs, *sparsely villous, oblong, lanceolate or narrowly ovate, obtuse or barely acute; cyme* corymbose-paniculate, *loosely flowered, with soon divergent branches; calyx densely glandular, its slender subulate-tipped teeth* longer than the tube; *lobes of corolla* blue to purple, or white in forma **albiflòra** Farw. (white-flowered), *commonly obcordate or emarginate,* equaling or longer than the tube. (Var. *canadensis* (Sweet) Wherry) — Damp to dry open woods, rocky slopes, etc., nw. Vt. and sw. Que. to Mich. and ne. Ill., s., mostly w. of Coastal Plain, to S.C. and n. Ala.; cult. and locally natzd. elsewhere. April–early June. — The remarkable forma **Coùlteri** Fern. (for its discoverer, JOHN MERLE COULTER, 1851–1928), covering a large area of glacial debris near Crawfordsville, Ind., has the corolla-lobes only 3–5 mm. long, 0.5–1 mm. broad, and abruptly long-acuminate.

Var. **Laphàmii** Wood (for INCREASE ALLEN LAPHAM, 1811–1875). — Differing in its almost regularly unnotched corolla-lobes. — Coastal Plain of n. Fla. to se. Tex., n., locally, to se. Va., n. Ala., Wisc., Minn. and Neb.

5. P. pilòsa L. (soft-pubescent). — *Stems nearly erect from a subligneous crown,* 2–7.5 dm. high, usually *pubescent; leaves narrowly lanceolate to linear, attenuate to sharp tips,* the longer ones 3–10 cm. long, the uppermost often subamplexicaul; *cymes at first close, becoming open; pedicels and calyx copiously stipitate-glandular; calyx-teeth narrowly linear-attenuate or subulate, with almost awn-like tips; corolla* roseate to pink or violet, *its* obovate *lobes entire.* (Incl. var. *virens* (Michx.) Wherry) — Dry open woods, sandhills, openings, prairies, etc., Fla. to Tex., n. to sw. Ct., n. N.J., w. N.Y., s. Ont., s. Mich., e. Ia., Mo. and e. Kans. May–early July.

Var. **fúlgida** Wherry (lustrous; from the pubescence). — Differing in having the inflorescence whitish-villous with glandless pubescence; the rare forma **albiflòra** (MacM.) Standl. (white flowered) with white corolla. — Wisc. to Sask., s. to Ind., Ill., n. and w. Mo. and ne. Kans.

6. P. amoèna Sims (charming). — Habitally similar to no. 5, usually lower; *leaves broadly linear to oblong-lanceolate or -ovate, obtuse or barely acute, 2–5 cm. long,* those of sterile shoots often ovate; *cyme mostly compact and sessile, embraced by broad leafy bracts; calyx glandless, its linear to subulate teeth rarely awned;* lobes of purple, pink or rarely white corolla obovate and entire or rarely notched. (Var. *Walteri* (Gray) Wherry) — Dry open woods, rocky slopes, barrens, etc., n. Fla. to e. Miss., n. to e. S.C., w. N.C., Tenn. and s. Ky. May, June.

§ 3. Ovàtae Wherry

7. P. stolonífera Sims (bearing stolons), Creeping P. — *Basal leafy offshoots forming prostrate and repent villous stolons with* several pairs of *spatulate-obovate or -elliptic leaves; rosette-leaves similar, evergreen, petioled,* up to 7 cm. long and 2.3 cm. broad; *flowering stems ascending,* 1–3 dm. high, villous above, *with few distant nodes; corymb few-flowered, viscid-pilose or -villous;* slender *pedicels nearly or quite as long as calyx; calyx-lobes linear-subulate, about equaling tube; corolla* violet or purple, rarely white, *its tube 2–2.5 cm. long,* the round-obovate mostly entire *lobes 1–1.7 cm. long; uppermost anthers slightly exserted.* — Moist woods and bottoms, Pa. to s. O., s. in upland to w. N. and S.C. and n. Ga.; occasionally spread from cult. elsewhere. Mid-April–early June.

8. P. Búckleyi Wherry (for its discoverer, Samuel Botsford Buckley, 1809–1884), Buckley's P., Swordleaf-P. — *Decumbent slender stem terminated by a rosette of crowded evergreen (but the 2nd year marcescent) firm linear-ensiform leaves* up to 1.5 dm. long; *flowering stem erect, from center of old rosette,* 1.5–3 dm. high, with 3–7 nodes, *glandular above; cauline leaves sessile, similar to rosulate ones* or narrowly lanceolate; *corymb* dense to slightly open, *densely glandular; pedicels short; calyx* 7–13 mm. long, the *short-awned lobes about one-third as long as tube; corolla* bright purple; *the glandular tube 1.7–2.3 cm. long,* the obovate or suborbicular lobes 8–12 mm. long; anthers reaching summit of corolla-tube; style-tips barely exserted. — Thin woods and shaly slopes, sw. W.Va. and adj. Va. May–July.

9. P. ovàta L. (ovate), Mountain-P. — *Flowering stems* 1.5–6 dm. high, *arising from leafy tips or from axils of decumbent basal branches, with 3–7 pairs of leaves below the inflorescence; lowest tufted leaves elliptic to ovate, marcescent; lower and median cauline leaves oblanceolate to elliptic or obovate, the lower* 4–12 cm. long and *petioled, the upper sessile; corymb* at first crowded, soon loosening, *mostly glandless;* pedicels chiefly shorter than calyx, glabrous to pubescent; *caylx-lobes mostly* 8.5–12 *mm. long;* corolla reddish-purple to deep pink, its tube 1.7–2.3 cm. long, the lobes more than half as long; stigmas reaching or exceeding anthers. — Open woods, thickets and meadows, se. Pa. to ne. Ind., s. to Piedmont and upland N.C. and n. Ala. Mid-May–early July.

10. P. glabérrima L. (very smooth). — *Stems 1–few from an upright crown,* 0.25–1.7 *m. high,* slender, erect, glabrous or sparsely pubescent, *with* (8–)12–23 *nodes below the inflorescence;* lowest few leaves reduced and scale-like; most *cauline leaves narrowly to broadly linear to narrowly lanceolate, long-attenuate, the larger ones,* 0.35–2 *dm. long,* the uppermost and bracteal much reduced; corymb close, or in larger plants with elongate rays; *pedicels short; calyx* 6.5–8.5 *mm. long, the lance-attenuate subulate-tipped lobes* 2.5–4 *mm. long; corolla* reddish-purple to violet, rarely to white, its tube 1.6–2.5 cm. long, the *lobes* 0.7–1 *cm. long.* (Var. *melampyrifolia* (Salisb.) Wherry) — Swampy woods, thickets, swales, etc., Ga. and nw. Fla. to e. Tex., n. to se. Va., central N.C., w. Ky., se. Mo. and Ark. May, June. Passing westw. into

Var. **intèrior** Wherry (inland). — Calyx 5–7(–7.5) mm. long, its broader and less attenuate lobes 1.5–3 mm. long. — Nw. O. to se. Wisc. and Ill., s. to s. Ky. and ne. Mo.

11. P. carolìna L. (of Carolina), var. **triflòra** (Michx.) Wherry (three-flowered; from the smaller cymes). — Similar to no. 10; *stems either several in a crown or scattered from axils of old decumbent branches,* 2–7.5 *dm. high, puberulent or pilose* (sometimes glabrous), *with* 5–12 *nodes* below the inflorescence; *principal leaves linear to oblong-lanceolate, the longest rarely more than* 1 *dm. long,* the bracteal ones cordate at base; *fruiting pedicels often as long as calyx; calyx* slenderly campanulate, 8–12 *mm. long, with lance-linear attenuate lobes* 4–6 *mm. long; corolla* bright red-purple to pink (rarely white), *its lobes* 1–1.4 *cm. long.* — Wet to dry open woods, thickets and clearings, N.C. to Ala., n. to Md., Ky. and s. Ind. May, June. — Northern representative of a polymorphic s. species, not too sharply separated from no. 10.

12. P. maculàta L. (spotted), WILD SWEET WILLIAM. — Smooth or barely hispidulous; the *stem arising from the tip of a slender horizontal rootstock, purple-spotted,* or unspotted and white- or pale-flowered in forma **immaculàta** Fern. (unspotted), 3–9 *dm. high, with* 7–15 often remote *nodes* below the inflorescence; *lower and median leaves lanceolate to linear, long-attenuate,* the upper often broader, *the bracteal lanceolate to lance-ovate and rounded or cordate at base; panicle thyrsiform, subcylindric, short-branched,* 0.5–4 dm. long; pedicels short; *calyx about* 7 *mm. long, with short triangular-lanceolate lobes;* corolla purple or red-violet (white or pale in forma *immaculata*), the tube 1.7–2.6 cm. long, the lobes 0.6–1.1 cm. long. (Var. *odorata* (Sweet) Wherry) — Meadows, bottomlands, stream-banks, etc., sw. Que., N.Y. and sw. Ct. to se. Minn., s. to s. N.J., n. Del., n. Md., upland to Va., Tenn. and e. Mo.; spread from cult. in N.E. and e. Can. Late May–Aug.

Var. **purpùrea** Michx. (purple). — Usually taller, 0.4–1.5 m. high, with spotted stem and purple flowers, or stem unspotted and corolla white in forma **cándida** (Michx.) Fern. (white); nodes 15–30; leaves more nearly uniform and only gradually changing up the stem, the bracteal ones often narrow. (Var. *pyramidalis* (Sm.) Wherry) — The common extreme southw., N.C. and Tenn., n. to Md., s. Pa., W.Va., s. O., Ind. and Ia. June–Sept.

§ 4. PANICULÀTAE Peter

13. P. paniculàta L. (panicled), FALL-P., PERENNIAL P. — Strong perennial, soon forming clumps 0.5–2 m. high; *stem* glabrous, or pubescent above, *with* 14–25 (–40) *nodes; leaves* opposite or the upper ones somewhat alternate, *elliptic- to oblong-lanceolate or -ovate, very short-petioled or sessile, the young margins bristly-ciliolate;* panicle large, pyramidal-corymbose, with very short glabrous or pubescent pedicels; *calyx* 6–10 *mm. long, with awn-pointed lobes; corolla* pinkish-purple (varying in cult. to many shades or to white), *its mostly pubescent tube* 1.5–2.5 *cm. long,* the lobes 7–12 mm. long; *anthers creamy,* included or some of them slightly exserted. — Rich open woods, thickets, bottoms, etc., centr. N.Y. to n. Ia., s. to interior N.C., nw. Ga., n. Miss. and Ark.; spread from cult. elsewhere. July–Oct. (or killing frosts).

14. P. amplifòlia Britt. (ample-leaved), BROADLEAF-P. — *Differing from no.* 13 *in its relatively few* (7–15) *distant pairs of very thin leaves; these scabrous-hirtellous above, broadly rhombic-ovate to elliptic-lanceolate,* usually *abruptly contracted to narrow base; inflorescence copiously glandular; corolla-tube glabrous,* the limb usually paler than in no. 13; anthers creamy, varying to yellow. — Rocky wooded slopes and banks of streams, ascending high on mts., sw. Va. to s. Ind., w. to e. Mo., s. to w. N.C., Tenn. and Ala. Mid-June–Aug.

P. DRUMMÓNDII Hook. (for its discoverer, THOMAS DRUMMOND, 1780–1836), DRUMMOND-P., an annual with glandular-viscid pubescence, oblong or lanceolate often clasping leaves (the upper ones mostly alternate), is a familiar plant of gardens and sometimes spreads to waste lots. (Introd. from e. Tex.)

FAM. 148. HYDROPHYLLÀCEAE (WATERLEAF FAMILY)

Herbs, commonly hairy, with mostly alternate leaves, regular 5-*merous and* 5-*androus flowers, in aspect between the foregoing and the next family; but the ovary entire and* 1-*locular, with* 2 *parietal* 4–*many-ovuled placentae, or rarely* 2-*locular by the union of the placentae in the axis; style* 2-*cleft, or* 2 *separate styles; fruit a* 2-*valved* 4–*many-seeded capsule.* Seeds mostly reticulated or pitted. Embryo small, in copious albumen. Flowers chiefly blue, purple or white, in 1-sided cymes or false racemes, which are mostly bractless and coiled from the apex when young, as in the Borage Family. — A small family of plants of no marked properties; some cult. for ornament.

a. Leaves cut-toothed, lobed or pinnate; style 2-cleft; ovary and capsule 1-locular; seeds pitted or reticulated, their albumen cartilaginous. Tribe 1. HYDROPHYLLEAE. . . *b.*

b. Ovary lined with the dilated and fleshy placentae which inclose the ovules and 4 seeds like an inner pericarp. . . *c.*

c. Perennial or biennial, with alternate leaves; flowers in scorpioid cymes; stamens exserted. 1. *Hydrophyllum.*

c. Annuals with the lower or all the leaves opposite; flowers solitary in the axils; stamens included.

Calyx with reflexed appendages at the sinuses. . . . . 2. *Nemophila.*

Calyx unappendaged. . . . . 3. *Ellisia.*

*b.*Ovary with narrow parietal placentae which in fruit project more or less inward; flowers in scorpioid cymes. . . . . 4. ***Phacelia.***

*a.*Leaves entire; styles 2, distinct; ovary and capsule 2-locular, their placentae often projecting from the axis far into the locules; seeds striate-ribbed, their albumen fleshy. Tribe 2. HYDROLEAE. . . . . 5. ***Hydrolea.***

Tribe 1. HYDROPHÝLLEAE R. Br.

1. HYDROPHÝLLUM L. WATERLEAF

Calyx 5-parted, sometimes with a small appendage in each sinus, early open in the bud. Corolla campanulate, 5-cleft; the lobes convolute in the bud; the tube furnished with 5 longitudinal linear appendages, opposite the lobes, forming a nectariferous groove. Stamens and style mostly exserted; filaments more or less bearded. Ovary bristly-hairy (as is usual in the family); the placentae soon free from the walls except at the top and bottom. Capsule ripening 1–4 seeds, spherical. — N.Am. perennials, with petioled ample leaves, and white or bluish-purple cymose-clustered flowers. (Name formed of the Greek *hydor, water,* and *phyllon, leaf;* the original species with very watery stems and petioles.)

*a.*Perennial, with scaly-toothed rhizomes; calyx naked at the sinuses or with minute teeth in no. 3, scarcely altered in fruit; stamens long-exserted. § EUHYDROPHYLLUM. . . *b.*

*b.*Leaves pinnate or pinnatifid, ovate to oblong; calyx-lobes copiously hirsute or hispid, without teeth in sinuses.

Rough-hairy; divisions of leaves 9–13, obtuse; peduncle shorter than petiole; cyme globose. . . . . 1. ***H. macrophyllum.***

Smoothish; divisions of leaves 3–7, acute or acuminate; peduncle usually longer than petiole; cyme opening. . . . . 2. ***H. virginianum.***

*b.*Leaves palmately lobed, rounded; calyx-lobes glabrous or nearly so, often with minute teeth in sinuses. . . . . 3. ***H. canadense.***

*a.*Biennial; calyx appendaged with a reflexed lobe at each sinus, enlarging under the fruit; stamens only slightly exserted; cauline leaves palmately lobed. § DECEMIUM. . . . . 4. ***H. appendiculatum.***

§ EUHYDROPHÝLLUM Gray

1. H. macrophýllum Nutt. (large-leaved). — *Rough-hairy; leaves oblong, pinnate and pinnatifid; the divisions* 9–13, *ovate, obtuse,* coarsely cut-toothed; basal leaves 2–3.5 dm. long; *peduncle shorter than the petiole;* calyx-lobes lanceolate-pointed from a broad base, very hairy; flowers 1 cm. or so long, crowded in a globular cluster; anthers short-oblong. — Rich woods, W.Va. and w. Va. to Ill., s. to Ga. and Ala. May, June.

2. H. virginiànum L. (Virginian), JOHN'S-CABBAGE. — *Smoothish,* 2–7 dm. high; *leaves pinnately divided; the divisions* 5–7, *ovate-lanceolate or oblong, pointed,* sharply cut-toothed, the lowest mostly 2-parted, the uppermost confluent; *peduncles longer than the petioles* of the upper leaves, forked; calyx-lobes narrowly linear, bristly-ciliate; corollas lavender to white, up to 1 cm. long. (Incl. *H. patens* Britt.) — Rich woods and damp clearings, Que. to Man., s. through w. N.E. to Va. and to Tenn., n. Ark. and e. Kans. May–Aug. — Forma **simplicifòlium** Fern. (simple-leaved), a remarkable "root-sport" discovered by *Pringle* at Charlotte, Vt., in 1879, needs closer observation: leaves simple with short obtuse lobes, only 4–9 cm. long; calyx only 3–4 mm. long, with appressed pubescence; corolla only 5–6 mm. long, with very narrow lobes.

Var. **atránthum** (E. J. Alex.) Constance (dark-flowered). — Divisions of leaves more often 7–9; corolla dark violet, the lobes longer than the tube. — Mts. of W.Va., w. Va. and N.C.

3. H. canadénse L. (Canadian). — *Nearly smooth,* 2–7 dm. high; *leaves* 0.5–2.5 dm. broad, *palmately* 5–7-*lobed, rounded,* cordate at base, unequally toothed, those from the base sometimes with 2–3 small and scattered lateral leaflets; *peduncles mostly shorter than the petioles,* forked, the nearly white flowers on very short pedicels; calyx-lobes linear-subulate, nearly smooth, often with minute teeth in the sinuses. — Damp rich woods, sw. Vt. and w. Mass. to s. Ont., s. to n. Ala. and e. Mo. June, July.

§ DECÉMIUM (Raf.) Gray

4. H. appendiculàtum Michx. (having appendages). — Hairy; cauline leaves palmately 5-lobed, rounded; the lobes toothed and pointed, the lowest pinnately divided; cymes rather loosely flowered; *filiform pedicels and calyx bristly-hairy, the latter with a small reflexed lobe in each sinus; stamens little exserted.* (*Decemium* Brand) — Rich woods, Ont. and Minn., s. to sw. Pa., Tenn., Mo. and e. Kans. May, June.

2. NEMÓPHILA Nutt.

Corolla campanulate or almost rotate; lobes convolute in the bud; tube mostly with 10 small folds or scales inside. Placentae (each bearing 2–12 ovules), capsule and seeds as in *Hydrophyllum.* — Diffuse fragile N.Am. annuals, with opposite or partly alternate pinnatifid or lobed leaves, and 1-flowered peduncles. (Name from the Greek *nemos, glade,* and *philein, to love.*)

1. N. micrócalyx (Nutt.) Fisch. & Mey. (with tiny calyx). — Small, roughish-pubescent; stems diffusely spreading, 0.5–4.5 dm. long; leaves parted or deeply cleft into 3–5 roundish or wedge-obovate sparingly cut-lobed divisions, the upper all alternate; peduncles opposite the leaves, shorter than the long petioles; flowers minute; corolla white, longer than the calyx; placentae each 2-ovuled; capsule 1–2-seeded. — Moist woods, w. Fla. to Tex., n. to e. Va., Tenn., se. Mo., Ark. and e. Okla. April, May.

3. ELLÍSIA L.

Corolla campanulate or cylindraceous, not longer than the calyx, 5-lobed above; the lobes imbricated or convolute in the bud, the tube with 5 minute appendages within. Placentae (each 2-ovuled), fruit and seeds much as in *Hydrophyllum.* — Delicate and branching N. Am. annuals, with lobed or divided leaves, the lower opposite, and small whitish flowers. (Named for *John Ellis,* 1710?–1776, distinguished naturalist and correspondent of Linnaeus.) MACROCALYX Trew

1. E. nyctélea L. (nocturnal). — Minutely or sparingly roughish-hairy, divergently branched, 1–4 dm. high; leaves pinnately parted into 7–13 lanceolate or linear-oblong sparingly cut-toothed divisions; peduncles solitary, supra-axillary or opposite the leaves, 1-flowered; calyx-lobes lanceolate, pointed, about the length of the cylindraceous corolla, in fruit ovate-lanceolate, 1 cm. long; capsules pendulous. (*Nyctelea* Britt.) — Rich low woods, stream-banks and disturbed soil, w. N.J. and e. Pa. to N.C.; Man. to Alta., se. and s. to Ind., Ill., Ark., Okla. and Colo.; casual adv. ne. to N.E. April–July.

4. PHACÈLIA Juss. SCORPION-WEED

Corolla open-campanulate, 5-lobed. Filaments slender, often (with the 2-cleft style) exserted; anthers ovoid or oblong. Ovary with 2 linear placentae adherent to the walls, the two often forming an imperfect partition in the ovoid 4–many-seeded capsule. Ovules 2–30 on each placenta. — N. and S.Am. herbs, with simple, lobed, or divided leaves, and often handsome (blue, purple, or white) flowers in scorpioid raceme-like cymes (to which the name, from the Greek *phacelos, a fascicle,* doubtless alludes).

a. Corolla without scales or lamellae at the bases of the filaments (scales, when present, remote from filaments). § COSMANTHUS. . . *b.*
b. Lobes of corolla fimbriate; ovules and seeds only 2 on each placenta.
Calyx-segments linear or linear-lanceolate, with divergent long marginal hairs, the bases in fruit arching away from the capsule. . . . 1. *P. Purshii.*
Calyx-segments oblong to spatulate, with ascending marginal hairs, the bases closely appressed to the fruit. . . . . 2. *P. fimbriata.*
b. Lobes of corolla not fimbriate; ovules and seeds 2–8 on each placenta. . . . *c.*
c. Corolla rotate to broadly campanulate, 5–15 cm. in diameter; filaments pilose; fruiting pedicels ascending or merely spreading. . . *d.*
d. Calyx-segments definitely ciliate with spreading hairs, subequal; corolla-tube with linear-triangular scales remote from the filaments.
Summit of stem and pedicels spreading-hirsute; calyx-segments divergent in flower, about a third as long as corolla. . . . . . 3. *P. hirsuta.*
Summit of stem and pedicels minutely canescent-pilose or sometimes strigose; calyx-segments appressed to and one-half to three-fourths as long as corolla. . . . . 4. *P. gilioides.*
d. Calyx segments eciliate or with few ascending basal and marginal trichomes, unequal; corolla 8–14 mm. broad, with no scales in the tube and the lobes entire. . . . . 5. *P. dubia.*
c. Corolla tubular-campanulate, with erect limb 2–4 mm. broad; filaments glabrous; fruiting pedicels strongly recurving. . . . . 6. *P. ranunculacea.*

a.**Corolla with conspicuous lamellae at the bases of the filaments, rotate-campanulate.**

Flowers and fruits long-pedicelled, the fruiting pedicels divergent or recurving; filaments mostly long-exserted; ovules and seeds 2 on each placenta.

§ EUPHACELIA. . . . . 7. *P. bipinnatifida.*

Flowers and fruits subsessile, the very short pedicels appressed-ascending; filaments only slightly exserted; ovules and seeds several on each placenta.

§ EUTOCA. . . . . 8. *P. Franklinii.*

§ COSMÁNTHUS (Nolte) Gray

1. P. Púrshii Buckl. (for its discoverer, FREDERICK TRAUGOTT PURSH, 1774–1820), MIAMI-MIST. — Sparsely hairy annual or biennial with erect or ascending simple to bushy-branched stem 1.5–5 dm. high; cauline leaves with 5–9 acute oblong to lanceolate lobes; racemes soon loosening, the filiform pedicels ascending or finally spreading; *calyx-segments linear to linear-lanceolate, with long divergent marginal hairs, their bases in fruit strongly arching away from the capsule; corolla* bluish, rarely white, about 1 cm. broad, the *lobes fimbriate; ovules and seeds* 2 *on each placenta.* — Rich woods, clearings and fields, sometimes becoming a weed, Ala. to Okla., n. to Pa., W.Va., O., Ind., Ill. and Wisc.; casually adv. to N.E. April–June.

2. P. fimbriàta Michx. (fringed). — Weaker, nearly glabrous, 1–3.5 dm. high; cauline leaves with 5–9 usually blunter lobes; *calyx-segments oblong to spatulate, with ascending marginal hairs, the bases closely appressed to the fruit;* corolla white to lilac. — Upland woods, Va. to Ala. May, June.

3. P. hirsùta Nutt. (stiffly hairy). — Annual or biennial, commonly branching from base, the *stems* 1–4 dm. high, *spreading-hirsute* as well as more or less finely pubescent; principal cauline leaves oblong, divided into linear or oblong blunt segments, strigose; *pedicels spreading-hirsute; calyx segments* linear-oblong, *divergent in flower*, appressed in fruit, *about one-third as long as corolla, hirsute with long slender spreading-ascending pale trichomes;* corolla blue-violet, 1–1.5 cm. broad, its lobes entire-margined; fruiting calyx 5.5–8 mm. high; ovules and seeds 6 on each placenta; seeds rugulose. — Calcareous rocks, low woods, prairies and river-banks, sw. Mo. and Ark. to Tex.; casual adventive eastw. May, June.

4. P. gilioìdes A. Brand (like *Gilia*). — Rather stiffly branched, *canescent or grayish*, the *stem and pedicels minutely pilose; principal segments of cauline leaves linear or linear-lanceolate,* copiously strigose; *calyx-segments appressed* in anthesis, linear or linear-oblong, *strigose on the back and with spreading-ascending slender or barely pustular-based whitish cilia*, in maturity becoming 5–7 mm. long; *corolla* white to bluish, 5–8 *mm. broad*, its lobes undulate to dentate, the tube with linear-triangular scales; ovules and seeds 6 on each placenta; seeds with deep transverse furrows. (*P. hirsuta* largely of sw. botanists, not Nutt.) — Prairies, calcareous barrens and open woods, Mo., e. Kans., Okla. and Ark. May, June.

5. P. dùbia (L.) Trel. (doubtful; Linnaeus placing it with doubt in *Polemonium*). — Annual or winter-annual, resembling nos. 3–5, simple or with basal or higher arching or diffuse branches, up to 4 dm. high; basal rosette with simple, pinnatifid or pinnate leaves; *cauline leaves pinnately cleft; racemes promptly elongating, the rachis and pedicels minutely appressed-pilose; calyx-segments unequal, the outer enlarged, eciliate or merely with few basal ascending trichomes, lanceolate to lance-oblong or -linear, sometimes spatulate, in fruit becoming* 4.5–7.5 *mm. long* and about twice as long as capsule; *corolla* white to lilac, becoming bluish in drying, open-campanulate, 8–14 mm. broad, *with entire lobes and no scales in throat;* filaments hairy; lower fruiting pedicels 3–25 mm. long. — Rich woods, thickets and clearings, Ga. and Tenn., n. to Del., Pa. and e. O., rarely to centr. N.Y. Late March–June.

6. P. ranunculàcea (Nutt.) Constance (resembling *Ranunculus*). — Weak annual or biennial, simple or branching from base, 1–2 dm. high; lowest leaves simple and ovate, or with 1 or 2 pairs of oval leaflets; cauline leaves with 5–7 thin oblong or obovate often cleft leaflets or segments; *racemes 2–6-flowered, the long slender pedicels soon recurving;* calyx-segments linear; *corolla tubular-campanulate, with erect bluish limb* 2–4 *mm. broad*, the tube naked within; *filaments glabrous;* capsule depressed-globose; ovules and large seeds 2 on each placenta. (*P. Covillei* S. Wats.) — Rich woods and alluvium, upper Potomac R., Md., D.C. and Va.; Wabash R., Ind., to w. Ill. and Ark. April, May. — Probably of more general range; easily overlooked or mistaken for *Ellisia nyctelea.*

§ EUPHACÈLIA Gray

7. P. bipinnatífida Michx. (twice pinnately cleft). — Biennial; stem upright, hairy, 2.5–6 dm. high; leaves long-petioled, pinnately 3–5-divided; the divisions ovate or oblong-ovate, acute, coarsely and often sparingly cut-lobed or pinnatifid; racemes elongated, loosely many-flowered, glandular-pubescent; *pedicels about the length of the calyx*, spreading or recurved; *corolla* bright blue, 1–1.5 cm. broad, *with 5 pairs of conspicuous lamellae subtending the long-exserted hairy filaments;* ovules and seeds 2 on each placenta. — Rich woods, shaded rocks and alluvium, Ga. and Ala. to Ark., n. to Va., W.Va., O., Ind., s. Ill. and Mo.; cult. and sometimes adv. northw. May, June.

Var. **Plúmmeri** Wood (named in 1847 for its discoverer, John Plummer). — Corolla smaller, 6–12 mm. broad; stamens and style little exserted. (Var. *brevistyla* (Buckl.) Gray; *P. brevistyla* Buckl.) — Ala. to N.C., Ky. and se. Ia.

§ Eùtoca (R. Br.) Gray

8. P. Franklínii (R. Br.) Gray (in honor of Sir John Franklin, 1786–1847). — Soft-hairy; stem erect, 2–6 dm. high, rather stout; leaves pinnately parted into many lanceolate or oblong-linear lobes, which are crowded and often cut-toothed or pinnatifid; *racemes* short, *dense*, crowded, *spiciform, the ascending pedicels very short;* calyx-lobes linear; corolla bluish to white, rotate-campanulate; *stamens only slightly exserted, the filaments subtended by pairs of lamellae; ovules and seeds several on each placenta.* — Dry sands, gravels, burned lands and disturbed soils, n. Ont. to Mackenz. and Yuk., s. to n. Mich., n. Minn., Wyo. and Idaho. June–Aug.

Tribe 2. Hydròleae R. Br.

5. HYDRÒLEA L.

Corolla 5-cleft. Filaments dilated at base. Capsule globular, with very large and fleshy many-seeded placentae, thin-walled, 2–4-valved or bursting irregularly. Seeds minute, striate-ribbed. — Herbaceous or scarcely shrubby, growing in water or wet places of trop. and warm reg., often having spines in the leaf-axils, and clustered blue flowers. (Name unexplained, doubtless in part from the Greek *hydor, water*, in allusion to the aquatic habitat.) Nama L., in part.

1. H. uniflòra Raf. (one-flowered). — *Glabrous throughout;* stem ascending from a creeping base; *leaves lanceolate*, tapering to a very short petiole; *flowers in small axillary leafy-bracted clusters; divisions of calyx lance-ovate, equaling the corolla* and the irregularly bursting globose capsule. (*H. affinis* Gray) — Swampy woods, sloughs and shallow water, Miss. to Tex., n. to s. Ind., s. Ill., se. Mo. and Okla. June–Sept.

2. H. quadriválvis Walt. (four-valved; a misleading name). — Similar, but *villous-hispid above; divisions of calyx linear or linear-lanceolate.* — Swampy woods, alluvium and shallow pools, Fla. to La., n. to se. Va. July–Sept.

3. H. ovàta Nutt. (egg-shaped). — Hirsute or puberulent; *leaves ovate;* flowers in *terminal leafy panicles; calyx hirsute, with lanceolate divisions shorter than the corolla.* — Swamps and bayous, Ga. to Tex., n. to se. Mo. June–Sept.

FAM. 149. BORAGINÀCEAE * (Borage Family)

Chiefly rough-hairy herbs, with alternate entire or shallowly toothed leaves, and symmetrical flowers with a 5-parted calyx, a regular 5-lobed corolla (except in Echium), 5 *stamens inserted on its tube, a single style and a usually deeply 4-lobed ovary* (as in Labiatae), *forming in fruit* 4 *seed-like 1-seeded nutlets*, or separating into two 2-seeded or four 1-seeded nutlets. Albumen none. Cotyledons plano-convex; radicle pointing to the apex of the fruit. Stigmas 1 or 2. Calyx valvate, the corolla imbricated (in *Myosotis* convolute) in the bud. Flowers mostly on one side of the branches of a reduced cyme, simulating a spike or raceme, which is rolled up from the end, and straightens as the flowers expand (circinate or scorpioid), often bractless. — A rather large family of non-poisonous mucilaginous and slightly bitter plants; the roots of some species yielding a red dye.

a. Ovary unlobed, surmounted by the style; stigma peltate-annular, forming a complete ring, surmounted usually by a tip or appendage. Subfam. Heliotropioideae. . . . . 1. *Heliotropium.*

a. Ovary 4-parted from above into 1-locular 1-ovuled divisions forming nutlets around base of style; stigma not annular, terminal. Subfam. Boraginoideae. . . *b.*

* The key to genera and species based partly on that of I. M. Johnston in Contrib. Gray Herb., n. s., lxx. pt. 1 (1924).

b. Attachment of nutlet surrounded by a tumid annular rim, strongly convex or plug-like, leaving a pit upon the flat or low-convex gynobase. Tribe ANCHUSEAE. . . *c.*

c. Stamens dorsally appendaged, closely crowded around the style; corolla rotate. . . . . 2. *Borago.*

c. Stamens unappendaged, included within the more or less tubular corolla. . . *d.*

d. Throat of corolla without appendages but with villous spots. . . . . . 3. *Nonea.*

d. Throat of corolla with appendages (faucal appendages) formed by intrusions of tissue.

Corolla thick-tubular, with campanulate throat, the short lobes erect or recurving at tip; faucal appendages lanceolate, acute, with denticulate margins. . . . . 4. *Symphytum.*

Corolla funnel- or salverform, with ill-defined throat, the elongate lobes spreading or divergent; faucal appendages deltoid or oblong, blunt, usually hairy.

Tube of corolla bent near middle, limb oblique or slightly irregular. . . . . 5. *Lycopsis.*

Tube of corolla straight, limb regular. . . . . 6. *Anchusa.*

b. Attachment of nutlet without an annular rim, flat or concave, leaving no pit on the gynobase. . . *e.*

e. Stigmas 2 or style bifid. Tribe LITHOSPERMEAE. . . *f.*

f. Corolla irregular, oblique; stamens unequal. . . . . 7. *Echium.*

f. Corolla regular or essentially so; stamens equal.

Lobes of corolla acute or acuminate, erect; style long-exserted, protruded as buds expand; anthers sagittate. . . . . 8. *Onosmodium.*

Lobes of corolla rounded or obtuse, ascending, spreading or recurving; style short-exserted or included, never protruding until flower is fully expanded; anthers oblong. . . . . 9. *Lithospermum.*

e. Stigma solitary and simple. . . *g.*

g. Nutlets attached near apex, strongly divergent, the anterior faces forming broad angles with the floral axis. Tribe CYNOGLOSSEAE. . . . 10. *Cynoglossum.*

g. Nutlets attached near base or middle, erect or nearly so and parallel, the anterior faces essentially parallel; calyces regular. Tribe ERITRICHIEAE. . *h.*

h. Flowers of the slender raceme mostly without leafy bracts; corolla-lobes convolute in bud; nutlets smooth, sessile, basally attached. . 11. *Myosotis.*

h. At least the lower flowers leafy-bracted; corolla-lobes imbricate in bud; nutlets rugose, wrinkled or echinate (sometimes smooth in no. 15), obliquely attached. . . *i.*

i. Fruiting calyx greatly enlarged, strongly veiny, with 5 broad flat lobes, irregularly toothed. . . . . 12. *Asperugo.*

i. Fruiting calyx not greatly enlarged nor veiny, of 5 elongate entire lobes. . . *j.*

j. Nutlets smooth, granular or wrinkled; throat of corolla open, with or without transverse folds. . . *k.*

k. Corolla yellow or orange, unappendaged; gynobase elongate; cotyledons 2-lobed. . . . . 13. *Amsinckia.*

k. Corolla blue, pink or white, with or without transverse folds in the throat; gynobase flat or merely convex; cotyledons unlobed.

Corolla white, its throat very short and shallow, its tube shorter than to barely exceeding calyx; nutlets with median dorsal keel; style shorter than nutlets. . . . 14. *Plagiobothrys.*

Corolla blue or pink (exceptionally white), with cylindric or funnelform throat, the tube surpassing calyx; nutlets lacking a dorsal median keel, overtopped by the style. . 15. *Mertensia.*

j. Nutlets armed with prickles at least on the upturned rim of the back; throat of corolla closed by 5 scales.

Pedicels and fruiting calyces erect or ascending; nutlets equaling the subulate gynobase, attached for nearly their whole length along the ventral keel, with no areola. . . . . 16. *Lappula.*

Pedicels and fruiting calyces reflexed; nutlets twice surpassing the stout gynobase, attached above the middle, with a deltoid or ovoid areola. . . . . 17. *Hackelia.*

Subfam. HELIOTROPIOÌDEAE

1. HELIOTRÒPIUM L. Turnsole. Heliotrope

Corolla salverform or funnelform, unappendaged, more or less plaited in the bud. Anthers nearly sessile. Style short; stigma conical or capitate. Fruit separating into 2 indurated 2-locular and 2-seeded closed carpels, or more commonly into 4 one-seeded nutlets. — Herbs or low shrubby plants of warm-temp. and trop. reg.; leaves entire; fl. in summer. (The ancient name, from the Greek *helios, the sun*, and *trope, a turn;* ancient writers believing that it turns toward the sun in flowering.)

Fruits 4-lobed, separating into 4 1-locular 1-seeded nutlets.
  Flowers in 1-sided scorpioid spikes, bractless.
    Hoary-pubescent; leaves oval. . . . . 1. *H. europaeum.*
    Glabrous; leaves spatulate to linear. . . . . 2. *H. curassavicum.*
  Flowers scattered in axils of linear leaves or bracts. . . . . 3. *H. tenellum.*
Fruits 2-lobed, separating into 2 2-locular 2-seeded carpels with a pair of empty false locules.
  Annual; leaves ovate or oval, long-petioled; flowers in single elongate spikes. 4. *H. indicum.*
  Perennial; leaves oblong-lanceolate, sessile or barely petioled; cyme at first dense, the 3 or 4 branches later loosening. . . . . 5. *H. amplexicaule.*

1. H. europaèum L. (European). — Erect annual, 1.5–8 dm. high, *hoary-pubescent; leaves oval*, long-petioled; lateral *spikes* single, the terminal in pairs, *scorpioid, bractless;* calyx spreading in fruit, hairy; corolla white or bluish, rarely 4 mm. broad; *nutlets* 4, *1-seeded, tuberculate.* — Roadsides and waste places, Fla. to Ala., n. to N.J. (locally to Mass.). June–Sept. (Natzd. from Eu.)

2. H. curassávicum L. (of Curaçao), Seaside-H. — Apparently annual, *glabrous;* stems ascending or forming carpets; *leaves lance-linear or spatulate,* thickish, pale, almost veinless; spikes in pairs; flowers white or bluish. — Sandy seashores and borders of fresh or saline marshes, Fla. to N.M. and Mex., n. to Del. and casually on ballast northw. June–Oct. (Trop. Am.) Fig. 1434.

1434. H. curassavicum.

3. H. tenéllum (Nutt.) Torr. (delicate). — Stem 1.5–4 dm. high, paniculately branched, slender, strigose-canescent; *leaves narrowly linear,* with revolute margins; *flowers in the upper axils or terminal and bractless,* white. (*Lithococca* Small) — Dry open limy soil, Ala. to Tex., n. to Ky., Mo. and se. Kans. June–Aug. Fig. 1435.

1435. H. tenellum.

4. H. índicum L. (of India), Turnsole. — Coarse hairy annual up to 7 dm. high; leaves petioled, ovate or oval; *spike single,* elongate, *scorpioid, bractless;* corolla bluish; *fruit 2-lobed, separating into* 2 *2-locular 2-seeded carpels,* mitriform, *with an empty false locule before each seed-bearing one.* (*Tiaridium* Lehm.) — Waste places, rich bottoms, etc., Fla. to Tex., n. to Va., W.Va., Ky., s. Ill., Mo. and Okla. July–Nov. (Natzd. from Asia) Fig. 1436.

1436. H. indicum.

5. H. amplexicaùle Vahl (clasping the stems). — *Stems decumbent from perennial base,* arching, up to 3 or 4 dm. high, spreading-hirsute; *leaves oblong-lanceolate,* dentate, hirsute beneath, *narrowed to sessile or short-petioled base; cyme at first dense,* bristly, *the 3 or 4 close scorpioid branches* soon loosening and *elongating;* corolla lilac; fruit similar to that of no. 4. — Waste places, roadsides, etc., Fla. to Tex. and Calif., n. to N.J. and O. May–Aug. (Natzd. from S.Am.)

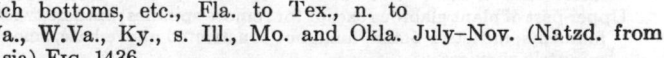

Coldénia (for *Cadwallader Colden*, 1688–1775) Nuttállii Hook. (for Thomas Nuttall, 1786–1859), a depressed slender annual with long internodes and dichotomous opposite divergent naked branches, falsely verticillate small rhombic-ovate petioled canescent leaves, the small pink or white flowers crowded at the forks and tips, the style 2-cleft, the 4 oblong-ovate nutlets united only at center, with a linear ventral scar, has occurred as a railroad-weed in Mo. (Adv. from farther west)

Subfam. BORAGINOÌDEAE

Tribe ANCHÙSEAE Gürke

2. BORÀGO L. BORAGE

Corolla rotate, with very short tube, the throat closed by short scales; the 5 acute lobes imbricate in the bud. Calyx deeply 5-cleft. Filaments very short, produced above into an erect dilated appendage; anthers connivent, forming an erect cone around the simple style as in *Solanum*. Nutlets erect, ovoid or ellipsoid, excavated at the basal point of attachment to the low gynobase. — Hairy annuals or biennials nat. of Eurasia; flowers on recurving pedicels in a loose cyme. (The old name, presumably of folk-origin.)

1. B. OFFICINÀLIS L. (of the shops). — Coarse, harshly villous-hirsute, 3–5 dm. high; principal leaves obovate to oblong, the lower petioled; corolla clear blue; anthers dark. — Cult. and long-persistent or spreading to waste places, occasional, Nfld. to Ont. and N.D., s. to N.S., N.E., Va., Tenn. and Ill. July–Sept. (Introd. from Eu.)

3. NÒNEA Medic.

Corolla with long cylindric tube and spreading limb, the throat unappendaged but with villous spots; the lobes obtuse. Calyx obliquely campanulate, cleft one-fourth to one-half its length into broad lobes, ampliate below. Filaments very short, unappendaged; anthers oblong, included or barely exserted. Style 2-cleft or with 2 thick stigmas. Nutlets plump, reticulate, with a broad concave areola. — Herbs, nat. of Old World, the cymes with elongating leafy-bracted more or less secund branches, with deflexing ampliate fruiting calyces. (Named for *Johann Philipp Nonne*, 1729–1772, physician of Erfurt.)

1. N. LÙTEA (Desr.) DC. (yellow). — Stem branching from base; *corollas mostly yellow; nutlets nearly erect, brown*, obscurely rugose, *longer than thick*, attached near base. — Infrequent weed of cult. and waste ground, N.Y. to Ky. July–Sept. (Adv. from s. Eu.)

N. RÒSEA (Bieb.) Link (roseate), similar but with rose-colored corollas, and with strongly oblique conspicuously rugose lead-colored nutlets as broad as long, has been found on waste land in Me.; **N.** VESICÀRIA (L.) Reichenb. (bladdery), with blackish-red corollas and laterally attached nutlets with coarse papillose-dentate annular basal rim, is reported as adv. about New York. (Adv. from Eu.)

4. SÝMPHYTUM L. COMFREY. CONSOUDE (Que.)

Corolla 5-toothed, the short teeth spreading. Stamens included; anthers elongated. Style filiform. Nutlets erect, fixed by the large hollowed base, which is finely toothed on its margin. — Coarse perennial herbs, nat. of Old World, with thickened bitterish mucilaginous roots; the nodding raceme-like clusters either single or in pairs. (Ancient Greek name from *symphyton, grown together*, for its reputed healing virtues, as indicated by the decurrent leaves.)

Many or all the trichomes of stem and inflorescence broad-based firm spinules.
- Upper part of plant copiously bristly or hispid; anthers and filaments subequal; nutlets 4 mm. long, constricted above base, strongly areolate and tuberculate. . . . . 1. *S. asperum.*
- Upper part of plant glabrous except for remote spinules; anthers twice length of filaments; nutlets 4–5 mm. long, not constricted above base, smooth or only minutely rugulose. . . . . 2. *S. uliginosum.*

Pubescence pilose-hispid, without broad-based spinules.
- Upper part of stem and inflorescence densely pilose-hispid; leaves broadly decurrent; nutlets 4.5 mm. long, not constricted above the base. . . . 3. *S. officinale.*
- Upper part of stem, etc. sparsely hirsute; leaves scarcely or but slightly decurrent; nutlets 3.5 mm. long, constricted above the base. . . . . 4. *S. tuberosum.*

1. S. ÁSPERUM Lepechin (harsh), PRICKLY C. — Root thick and branching; *stems branching*, up to 1.5 m. high, *copiously uncinate-spinulose*, especially above; *leaves* harsh, *barely or not at all decurrent;* calyx 3–5 mm. long in flower, elongating to 1–1.7 cm. in fruit; *corolla roseate, changing to bluish*, 1–1.5 cm. long, *its lobes erect; anthers and filaments subequal; nutlets 4 mm. long, constricted above the base, strongly areolate and tuberculate.* (*S. asperrimum* Donn) — Roadsides and waste places (also cult.), Nfld. to Ont., s. to N.S., N.E., Md. and Mich. Late May–Aug. (Introd. from the Caucasus)

2. S. ULIGINÒSUM Kern. (growing in marshes). — Similar to no. 1 but *glabrous except for*

1437. S. officinale.

scattered uncinate pale spinules; corolla violet; *anthers twice as long as filaments; nutlets 4–5 mm. long, not constricted above base, smooth or minutely rugulose, lustrous.* — Locally in roadside-ditches, Madison Co., N.Y. June–Aug. (Adv. from Eu.)

3. S. OFFICINÀLE L. (of the shops), COMMON C. — Coarse, up to 1 m. high, branching and *densely pilose-hispid above; lower leaves larger than the next succeeding ones; upper leaves broadly decurrent by wings along the stems; corolla whitish, yellowish or dull purple, 1.5–1.7 cm. long, the lobes recurved at top; nutlets 4.5 mm. long, smooth or faintly areolate, not constricted above the entire-rimmed base.* — Damp roadsides and waste places (and often cult.), Nfld. to w. Ont., s. to N.S., N.E., Ga., Tenn. and La. Late May–Sept. (Introd. from Eu.) FIG. 1437.

4. S. TUBERÒSUM L. (tuberous). — *Root tuberous-thickened,* nodose; *stems 2–5 dm. high,* with short upper branches, *sparsely hirsute; lower leaves smaller than next succeeding ones, upper scarcely or but slightly decurrent; corolla* yellowish, 1.2–1.6 *cm. long; nutlets* 3.5 *mm. long,* minutely tuberculate or rugulose, *constricted above the toothed annulate base.* — Locally in ditches and fields, Me. to Ct. Late May–July. (Adv. from Eu.)

5. LYCÓPSIS L. BUGLOSS. CHAUDRONNETTE (Que.)

Corolla funnelform, with curved tube and slightly unequal limb; the throat closed with 5 convex obtuse bristly scales opposite the lobes. Stamens and style included. Nutlets rough-wrinkled, erect, fixed by a hollowed-out base. — Annuals, nat. of Old World. (Name from the Greek *lycos, a wolf,* and *opsis, appearance.*)

1. L. ARVÉNSIS L. (of cult. fields), SMALL B. — Very rough-bristly, 1–6 dm. high; leaves lanceolate; flowers in leafy raceme-like clusters; calyx as long as the tube of the small blue corolla. — Dry or sandy fields and waste places, Nfld. to Ont., s. to N.S., N.E., Va., O. and Neb., frequent in e. Can., local southw. June–Aug. (Natzd. from Eu.) FIG. 1438.

1438. L. arvensis.

6. ANCHÙSA L. ALKANET

Corolla funnelform, with straight tube often enlarged at summit, usually with hairy scales closing the throat; the regular limb spreading, with 5 narrow lobes. Stamens included. Nutlets angular, wrinkled, attached by their broad concave bases. — Coarse hairy biennials or perennials, nat. of Old World; the blue flowers all subtended by leafy bracts, forming a cyme of 1-sided spikes. (The Greek name.)

Bracts ovate or oblong; calyx-lobes narrowly triangular or lanceolate; limb of corolla 5–10 mm. broad; nutlets horizontal. . . . . . 1. *A. officinalis.*
Bracts and calyx-lobes linear-lanceolate; limb of corolla 1.2–2 cm. broad; nutlets erect. . . . . . . 2. *A. azurea.*

1. A. OFFICINÀLIS L. (of the shops), COMMON A. — Biennial with stiffly hirsute stems up to 8 dm. high; leaves hispid, narrowly oblanceolate to lanceolate, the basal petioled, the upper sessile; panicle terminal, of elongating and often forking spikes; *bracts ovate or oblong; calyx* tubular, hispid, the tube and *lanceolate or narrowly triangular lobes subequal; corolla* 5–10 *mm. broad, with merely pubescent faucal appendages; nutlets horizontal,* ovoid, 2 mm. high, 3 mm. long. — Roadsides and waste places, local, Me. to O., Mich. and N.J. Late May–Oct. (Adv. from Eu.)

2. A. AZÙREA Mill. (sky-blue). — Coarser; basal leaves broadly oblanceolate or narrowly obovate; *bracts linear-lanceolate, resembling the strongly white-bristly very long calyx-lobes* (*finally* 1–1.5 *cm. long, many times exceeding tube*); *limb of corolla* 1.2–2 *cm. broad, with long-penicillate faucal appendages; nutlets erect,* ellipsoid-trigonous, strongly reticulate. — Roadsides and waste places, local, N.E. to Ia. and westw. Late May–Aug. (Adv. from Eu.)

A. BARRELLIÈRI (All.) Vitm. (for JACQUES BARRELLIER, 1606–1673), similar to no. 2, with mature calyces 5–7 mm. long, bracts lanceolate and corollas only 3–6 mm. broad, has appeared sporadically in Ct. (Adv. from Eu.)

Tribe Lithospérmeae Gürke

7. ÉCHIUM L. Viper's Bugloss. Vipérine (Que.)

Corolla with a cylindraceous or funnelform tube; lobes rounded, spreading. Stamens mostly exserted, unequal. Style threadform. Nutlets roughened or wrinkled, fixed by a flat base. Biennials, nat. of Old World. (A plant-name used by Dioscorides, from *echis, a viper;* from a fancied resemblance of the nutlets to a viper's head.)

1439. E. vulgare.

1. E. vulgàre L. (common), Blueweed, Blue Devil. — Rough bristly biennial; stem erect, 3–9 dm. high; cauline leaves linear-lanceolate, sessile, their slender spreading pubescence (when present) non-pustular at base or only slightly so; flowers showy, in short lateral clusters disposed in a close narrow elongate thyrse 3–7 cm. thick; corolla brilliant blue, or pink in forma ròseum F. Zimm. (rose-colored), or white in forma albiflòrum R. Hoffm. (white-flowered). — Roadsides, dry fields and waste places, often an obnoxious weed, Que. to w. Ont., s. through our range and beyond. June–Sept. (Natzd. from Eu.) Fig. 1439.

Var. pustulàtum (Sibth. & Sm.) Coincy (blistered). — Pubescence of leaves coarse and spreading, with conspicuously pustular bases; inflorescence a loose and broad thyrse 0.7–3 dm. thick. — Local, N.J. to W.Va. (Adv. from Eu.)

8. ONOSMÒDIUM Michx. False Gromwell. Marble-seed

Divisions of calyx linear and erect. Corolla tubular or tubular-funnelform (the sinuses minutely hooded-inflexed), the 5 acute lobes converging or barely spreading. Anthers oblong-linear or sagittate, mucronate, inserted in the throat. Style filiform, much exserted. Nutlets bony, ovoid, erect; the scar not hollowed out. — Chiefly perennial Am. herbs, coarse and often hairy, with oblong and sessile rib-veined leaves, and white, greenish or yellow flowers in at length elongated and erect leafy raceme-like clusters; fl. in summer. (Named from a likeness to the genus *Onosma,* meaning *smell of an ass.*)

Corolla clear yellow to orange; its lobes narrowly lance-acuminate, twice or thrice as long as broad. . . . . 1. *O. virginianum.*
Corolla sordid or greenish-white or -yellowish; its lobes triangular, merely acute, little longer than wide.
 Stem silky, strigose or hirsute, dull.
  Nutlets with base rounded, if weakly constricted the plant silky-hirsute.
   Closely and finely grayish-pubescent, 4–6 dm. high; calyces 5–7 mm. long; nutlets conspicuously pitted, slightly constricted at base. . . . 2. *O. molle.*
   Spreading-pubescent, 0.4–1 m. high; calyces 8–12 mm. long; nutlets unpitted, rounded and not at all constricted at base. . . . 3. *O. occidentale.*
  Nutlets strongly constricted at base; plant coarsely spreading-hirsute. . . . 4. *O. hispidissimum.*
 Stem essentially glabrous and lustrous below. . . . . 5. *O. subsetosum.*

1. O. virginiànum (L.) A. DC. (Virginian). — Clothed with harsh and rigid appressed short bristles; stems rather slender, 3–8 dm. high; leaves narrowly oblong or oblong-lanceolate, 3–9 cm. long, the lower narrowed at base; *corolla clear yellow to orange, its lance-acuminate lobes twice or thrice as long as broad;* nutlets 2–2.8 mm. long, irregularly pitted, not conspicuously constricted at base. — Pinelands, dry sandy woods and open sands, Fla. to La., n., locally, to Mass. and N.Y. May–July. Fig. 1440.

1440. O. virginianum.

2. O. mòlle Michx. (soft). — Finely grayish-pubescent, 4–6 dm. high; the lowest leaves oblanceolate, the others ovate to ovate-lanceolate, conspicuously veined, acutish; calyx silky as well as sparingly hirsute, 5–7 mm. long; *corolla greenish-white or -yellowish, its triangular lobes nearly as broad as long; nutlets* 3 mm. long, *conspicuously pitted, slightly constricted at base.* — Calcareous barrens and dry openings, Tenn. and Ky. Late May–July.

3. O. occidentàle Mackenz. (western). — Coarser, 0.4–1 m. high; the *pubescence of the stem more spreading; calyces 8–12 mm. long; nutlets rounded and not at all constricted at base, unpitted.* — Sandy, gravelly or rocky prairies, glades and open woods, Minn. to Sask., s. to w. Ill., Mo., Okla., Tex. and N.M. Late May–July.

4. O. hispidíssimum Mackenz. (most hispid). — Stout, erect, 10–12 dm. high, green, coarsely hirsute throughout; cauline leaves ovate, 8–10 cm. long, 1.8–4 cm. wide, the pubescence mostly spreading; *nutlets* about 3 mm. long, *with a very short neck or constriction at the base*, brownish-tinged.— Dry calcareous rocky or gravelly prairies, banks and glades, N.Y. and s. Ont. to Minn. and Neb., s. to N.C., Tenn., La. and Tex. June, July. Fig. 1441.

Var. **macrospérmum** Mackenz. & Bush (large-seeded). — Nutlets shining-white, 4 mm. long. — Local, Ill. and Mo.

5. O. subsetòsum Mackenz. & Bush (somewhat bristly). — *Stem* erect, *nearly or quite glabrous to the middle*, 6 dm. or more in height; leaves narrowly lanceolate, finely appressed-pubescent upon both surfaces, 6–8 cm. long, 1–1.5 cm. wide; *nutlets* brownish-white, *not constricted at the base*, usually pitted, 2–3 mm. long. — Dry calcareous barrens, glades and bluffs of Ozark reg., Mo. and Ark. May, June.

1441. O. hispidissimum.

9. LITHOSPÉRMUM L. Gromwell. Puccoon. Grémil (Que.)

Throat of corolla naked, or with a more or less evident transverse fold or scale-like appendage opposite each lobe; the limb 5-cleft. Anthers oblong, almost sessile, included. Nutlets smooth or roughened, mostly bony or stony; scar nearly flat. — Herbs (genus semicosmop.) with thickish and commonly red roots and sessile leaves; flowers solitary and as if axillary, or spicate and leafy-bracted, sometimes dimorphous as to insertion of stamens and length of style. (Name formed of the Greek *lithos, stone*, and *sperma, seed*, from the hard nutlets.)

a.Annual or biennial, with small tap-root; throat of white corolla without evident folds or appendages; nutlets rough-wrinkled, gray, dull. . . . . . . 1. *L. arvense.*

a.Perennial, with stout or subligneous base; throat of corolla (except sometimes in no. 5) with 5 folds or appendages; nutlets smooth or slightly pitted, white to pale brown, lustrous. . . *b*.

b.Leaves lanceolate or narrowly ovate, membranaceous, with lateral veins evident beneath; corolla small, 2–6 mm. long, greenish-white to pale yellow. . . *c*.

c.Basal (rosulate) leaves wanting at flowering time; cauline leaves lanceolate; nutlets 2–3.5 mm. long; roots not tuberous.

Pubescence closely appressed; leaves ovate-lanceolate, 2–4.5 cm. broad, remote; nutlets 2.8–3.5 mm. long. . . . . 2. *L. latifolium.*

Pubescence loosely appressed to spreading; leaves lanceolate, 0.6–2 cm. broad, crowded; nutlets 2–2.8 mm. long. . . . . 3. *L. officinale.*

c.Basal rosette-leaves persistent, narrowly obovate; cauline leaves elliptic, oval or narrowly obovate, blunt; nutlets about 2 mm. long; roots tuberous-thickened. . . . . 4. *L. tuberosum.*

b.Leaves linear to narrowly lanceolate or oblong, firm, without evident lateral veins; corolla 0.9–2.5 cm. broad, orange or yellow, with long tube. . *d*.

d.Flowers dimorphic; the earlier and terminal salverform, with corolla-limb 0.9–2 cm. broad and lobes fimbriate, slender tube 1.5–3.5 cm. long; the later axillary ones small or cleistogamous; fruiting pedicels mostly arching; nutlets with a collar below the basal constriction. . . . . . 5. *L. incisum.*

d.Flowers uniform, slenderly funnelform, orange-yellow, with tube one-half to twice longer than calyx; fruiting pedicels erect; nutlets without basal collar. . . . *e*.

e.Hispid or hirsute with scattered stiff hairs; stems in maturity 0.15–1 m. high; mature calyx-lobes 1–1.5 cm. long, hispid or hirsute; corolla-limb 1.5–2.5 cm. broad, the tube pubescent within at base; nutlets 3.5–4.5 mm. long.

Well developed leaves mostly 35–45 below the inflorescence, hispid with pustular-based hairs; calyx-lobes strongly keeled; limb of corolla thickish, opaque, the veins anastomosing above the throat. 6. *L. croceum.*

Well developed leaves mostly 15–25 below the inflorescence, hirsute with slender-based hairs; calyx-lobes flat; limb of corolla translucent (to transmitted light), the veins continuous from throat into the lobes. . . . . 7. *L. caroliniense.*

e.Hoary with dense soft pilosity; stems 1–4.5 dm. high; mature calyx-lobes 6–8 mm. long, finely appressed-pubescent; corolla-limb 1–1.5 cm. broad, the tube glabrous within at base; nutlets 2–3 mm. long. 8. *L. canescens.*

1. L. arvénse L. (of planted fields), Corn-G., Bastard Alkanet. — *Annual or biennial*, slender, minutely roughened and hoary, 2–7 dm. high; *leaves* linear or lanceolate *without lateral veins; corolla white, scarcely longer than calyx, without evident folds in throat; nutlets*

1442. L. arvense.

gray, prominently rugose.— Weed of sandy fields and roadsides, N.S. to B.C., s. to Fla., La. and Calif. Apr.–June. (Natzd. from Eu.) Fig. 1442.

2. L. latifòlium Michx. (broad-leaved).— Perennial with tough base; stems 0.4–1 m. high, *harsh, with tightly appressed short hairs; leaves scattered,* narrowly *ovate,* taper-pointed, *2–4.5 cm. broad, thin, with lateral veins prominent beneath; corolla* greenish-white to pale yellow (rarely pinkish), small, with tube *shorter than calyx,* and with 5 distinct pubescent crests in throat; *nutlets white, lustrous,* 2.8–3.5 mm. long. — Rich woods and thickets, N.Y. to Minn., s. to Tenn., Ark. and e. Kans. May, June. Fig. 1443.

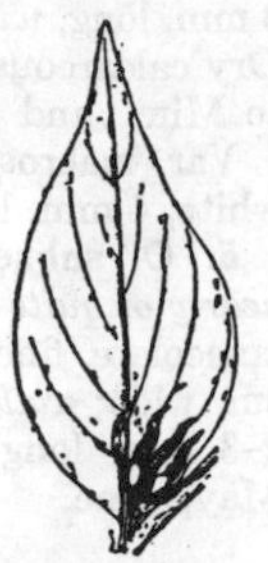

1443. L. latifolium.

3. L. officinàle L. (of the shops), European G.— Similar to no. 2; *pubescence loosely appressed to spreading; leaves lanceolate,* very numerous, *somewhat overlapping, 0.6–2 cm. broad; corolla slightly exceeding calyx; nutlets 2–2.8 mm. long.*— Roadsides, pastures and open places, e. Que. to Ont., s. to N.E., N.J., N.Y., O., Ind. and Ill. June, July. (Natzd. from Eu.) Fig. 1444.

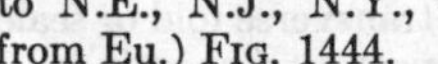

1444. L. officinale.

4. L. tuberòsum Rugel (tuberous).— Slender perennial with *tuberous-thickened roots and a basal rosette of narrowly obovate* ribbed *leaves* harsh on both faces; stem 2–7 dm. high, forking above, hirsute; *cauline leaves* remote, *elliptic, oval or narrowly obovate,* blunt; corolla yellowish-white, 3–5 mm. long; *nutlets* buff, globose, *about 2 mm. in diameter.*— Calcareous wooded slopes and bluffs, Fla. and Ala., n. to Ga. and Ky. Apr.–June.

5. L. incìsum Lehm. (sharply cleft).— Erect or diffusely branched from base, 1–5 dm. high, minutely rough-strigose and hoary; *leaves linear, firm, veinless; flowers* of two sorts, pedicelled; *the earlier showy,* yellow, *salverform, with* limb 0.9–2 cm. broad, *fimbriate lobes and* slender *tube 1.5–3.5 cm. long,* the throat appendaged; *the later small,* with unappendaged corolla, *or apetalous and cleistogamous; fruiting pedicels mostly recurving; nutlets strongly constricted above a basal collar,* lustrous-white to buff. (*L. angustifolium* Michx., not Forsk.)— Dry prairies and open soils, s. Ont. to B.C., s. to Ind., Ill., Mo., Tex. and n. Mex. Apr.–July.

6. L. cròceum Fern. (saffron or golden-yellow), Puccoon.— Erect from deep perennial purple-staining root; stems 1.5–6 dm. high, pilose; *leaves nearly uniform, linear-oblong or lanceolate, usually* 33–45 *below the inflorescence, papillose-hispid with hairs about* 1 *mm. long;* flowering *racemes* crowded; *fruiting ones* elongate, *strict,* 1–2 dm. long, *with* harsh lanceolate to ovate *imbricated bracts; flowers short-pedicelled; calyx-lobes strongly keeled, papillose-hispid in maturity, 1–1.5 cm. long; corolla funnelform, orange,* pilose, with exserted *tube bearded at base within,* limb 1.5–2 cm. broad, throat with low appendages, *the opaque lobes with veins* (when cleared) *distinctly anastomosing at base; nutlets* white, lustrous, *3.5–4 mm. long.*— Sands, gravels and dry woods, Ont. to Mont., s. to w. N.Y., O., Ind., Ill., Mo. and Kans. Late May–July.

7. L. caroliniénse (Walt.) MacM. (of Carolina), Puccoon.— Similar to no. 6; stems 0.2–1 m. high, *with 15–25 well developed leaves below the inflorescence; leaves* linear-lanceolate to oblong, *their pilose-hispid pubescence scarcely or barely papillose-based; fruiting racemes* more loosely ascending, 1–3 dm. long, *their spreading bracts becoming remote; calyx-lobes flai,* lanceolate to oblong, *hirsute with non-papillose long hairs;* the slender midrib villous-hispid; *corolla* of thinner texture, the limb 1.5–2.5 cm. broad, *the throat with elongate appendages,* the outside silky, the tube minutely pilose within, *the veins continuing without evident anastomosing into the lobes; fruiting calyx on slender pedicels 3–8 mm. long.*— Sandhills, pine-barrens and dry sandy woods, Fla. to e. Tex. and Mex., n. to se. Va., Ark. and Okla. Late May, June.

8. L. canéscens (Michx.) Lehm. (hoary), Puccoon, Indian-paint.— Similar to nos. 6 and 7 but dense *pubescence hoary, soft and very fine,* without papillose bases; *stems 1–4.5 dm. high; lower leaves short, the upper well-developed ones 8–30 below inflorescence;* racemes fewer, strongly scorpioid-secund; *mature calyx-lobes flat, finely appressed-pubescent, 6–8 mm. long; corolla* paler, 1–1.5 cm. broad, *with tube glabrous within at base; nutlets 2–3 mm. long.*— Dry or sandy open woods, prairies, etc., s. Ont. to Sask., s. to Ga., Ala., Miss. and Tex. Apr.–early June. Fig. 1445.

1445. L. canescens.

Tribe Cynoglósseae Gürke

10. CYNOGLÓSSUM L. Hound's-tongue. Beggar's-lice. Langue de chien (Que.)

Corolla funnelform, the tube about equaling the 5-parted calyx; lobes rounded. Stamens included. Nutlets depressed or convex, oblique, fixed near the apex to the base of the style, roughened all over with short barbed or hooked prickles. — Coarse herbs of temp. and subtrop. reg., with petioled lower leaves; the mostly panicled (so-called) racemes naked above, usually bracted at base. (Name from the Greek *cynos, of a dog*, and *glossa, tongue;* from the shape and texture of the leaves.)

Leafy to summit, canescent with fine soft hairs; calyx-lobes oval, subfoliaceous; corolla mauve-purple; nutlets with flat depressed back bordered by a rim, ascending, overtopped by the firm beak-like style. . . . . 1. *C. officinale.*

Leafy only below the middle, long-pedunculate above, spreading-hirsute or villous; calyx-lobes smaller; corolla pale blue, lilac or white; nutlets horizontal, rounded on back, rimless, hiding the minute delicate style.

Larger petioled leaves 5–11 cm. broad; calyx in anthesis 3.5–4.5 mm. long; corolla 1–1.2 cm. broad; nutlets 5.5–7 mm. long. . . . . 2. *C. virginianum.*

Larger petioled leaves 3–8 cm. broad; calyx in anthesis 2–2.5 mm. long; corolla 6–8 mm. broad; nutlets 3.5–4.5 mm. long. . . . . 3. *C. boreale.*

1. C. officinàle L. (of the shops), Common H. — Biennial, *clothed with short soft hairs, leafy,* panicled above; upper leaves lanceolate, closely sessile by a rounded or slightly cordate base; racemes nearly bractless; *corolla reddish-purple* (rarely white); *nutlets* flat on the broad upper face, somewhat margined, 5–7 *mm. long, ascending, overtopped by the firm beak-like style.* — Calcareous pastures, roadsides and waste places, locally abundant, Que. to the Pacific and s. beyond our range. May–July. — An obnoxious weed with mousey odor, the fruits firmly clinging to fleeces of sheep and to clothing, etc. (Natzd. from Eu.) Fig. 1446.

1446. C. officinale.

2. C. virginiànum L. (Virginian), Wild Comfrey. — Perennial, *roughish with spreading bristly hairs;* stem simple, *few-leaved,* 3–8 dm. high; cauline leaves lance-oblong, clasping by a deep cordate base; *racemes few and corymbed, raised on long naked peduncles,* bractless; *calyx* in anthesis 3.5–4.5 *mm. long; corolla pale lilac to white,* 1–1.2 *cm. broad, with suborbicular lobes and closed sinuses; nutlets* strongly echinate, compressed, orbicular-obovoid, cuneate at base, 5.5–7 *mm. long,* becoming whitish-gray. — Open deciduous woods, w. Fla. to e. Tex., n. to sw. Ct., N.J., Pa., W.Va., O., Ind., s. Ill. and Mo. Late Apr.–early June.

3. C. boreàle Fern. (northern), Northern Wild Comfrey. — Similar to no. 2 but more slender; stems villous-hispid at base, appressed-pubescent above; only the upper cauline leaves clasping; *calyx* in anthesis 2–2.5 *mm. long; corolla blue,* 6–8 *mm. broad, the lobes oblong-ovate, the sinuses open; nutlets* compressed, pyriform-obovoid, 3.5–4.5 *mm. long,* less inclined to become whitish in ripening. — Rich woods and thickets, Anticosti I., Que., to Ont., s. to N.S., Me., N.H., n. Ct., N.Y., Ind., Wisc. and Ia; also s. B.C. May, June.

Tribe Eritrichìeae Gürke

11. MYOSÒTIS L. Scorpion-grass. Forget-me-not

Corolla-tube about the length of the 5-toothed or 5-cleft calyx, the throat with 5 small and blunt arching appendages opposite the rounded lobes; the latter convolute in the bud. Stamens included, with very short filaments. Nutlets compressed. — Low and mostly soft-hairy herbs of temp. reg., with entire leaves, those of the stem sessile, and with small flowers in naked racemes, which are entirely bractless or occasionally with small leaves next the base, prolonged and straightened in fruit. (Name composed of the Greek *myos, of a mouse,* and *ous, ear,* from the short and soft leaves in some species.)

a. Hairs on calyx few, short, straight, closely appressed, eglandular; plants of wet soil.

Stem angled, stoloniferous; inflorescence usually bractless; calyx-lobes commonly shorter than tube; corolla 6–9 mm. broad; style overtopping nutlets, about equaling calyx-tube. . . . . 1. *M. scorpioides.*

Stem terete, non-stoloniferous; main branches of inflorescence usually bracted at base; calyx-lobes and -tube subequal; corolla 3–6 mm. broad; style much exceeded by nutlets and calyx-tube. . . . . 2. *M. laxa.*

*a.*Hairs (at least some of them) of calyx hooked or gland-tipped; plants of drier habitats. . . *b.*

*b.*Calyx unequally cleft, usually 2-lipped; corolla inconspicuous, white, 1–2 mm. broad.

Stiffly ascending; principal leaves 2–10 mm. broad; central fruiting racemes 0.3–1.8 dm. long; pedicels erect, curved at tip, 1–5 mm. long, the lowest on main axis 0.5–2 cm. apart; fruiting calyx persistent, its base with few if any hooked bristles; nutlets 1–1.3 mm. broad. . . . . . 3. *M. verna.*

Loosely ascending; principal leaves 0.5–1.7 cm. broad; central fruiting racemes 1.2–4.7 dm. long; pedicels divergently ascending from base, 3–10 mm. long, the lowest on main axis 2–5 cm. apart; fruiting calyx promptly deciduous, its base and tube with hundreds of upcurved strongly hooked bristles; nutlets about 2 mm. broad. . . . . . 4. *M. macrosperma.*

*b.*Calyx subequally cleft, nearly or quite regular; corolla 1.5–8 mm. broad, blue (sometimes yellow in no. 8).

Limb of corolla flat and rotate, 5–8 mm. broad. . . . . 5. *M. sylvatica.*

Limb spreading-ascending, concave, 1.5–4 mm. broad.

Pedicels as long as or longer than calyx; calyx-lobes spreading. . . . 6. *M. arvensis.*

Pedicels shorter than calyx; calyx-lobes erect.

Floriferous to near base, lower flowers in leaf-axils; corolla blue, its tube about equaling the calyx. . . . . 7. *M. stricta.*

Floriferous only above the leaves, the racemes naked; corolla at first yellow, changing to blue, its tube exceeding the calyx. . . . . . 8. *M. versicolor.*

1. M. scorpioìdes L. (like a scorpion), True F. — Perennial; stems angled, succulent, ascending from an oblique creeping base, 3–7 dm. high, loosely branched, smoothish; leaves rough-pubescent, oblong-lanceolate or linear-oblong; inflorescence wholly above the leaves; *calyx-lobes shorter than the tube; limb of corolla* 6–9 *mm. broad*, sky-blue, with a yellow eye; *style overtopping the nutlets.* — In quiet water or wet ground, Nfld. to Ont., s. to N.S., N.E., Ga., Tenn. and La.; also Pacific Am. May–Oct. (Natzd. from Eu.) Fig. 1447.

1447. M. scorpioides.

2. M. láxa Lehm. (loose). — Perennial, non-stoloniferous; *stems terete*, very slender, decumbent; pubescence all appressed; leaves lanceolate-oblong or somewhat spatulate; lowest flowers often leafy-bracted; *calyx-lobes as long as the -tube; limb of corolla* 3–6 *mm. broad*, pale blue; *style shorter than nutlets.* — In water and wet ground, Nfld. to Ont., s. to N.S., N.E., Ga. and Tenn.; Pacific N. Am. May–Aug. (Eurasia; Chile) Fig. 1448.

1448. M. laxa.

3. M. vérna Nutt. (vernal). — Simple or with stiff upright branches, 1–4 dm. high; principal leaves 2–10 mm. broad, strigose-hirsute; *corolla white*, 1–2 *mm. broad; racemes in maturity* elongating to 0.3–1.8 *dm. long; fruiting pedicels* 1–5(–6) *mm. long, erect and nearly parallel with rachis*, the tip curving, *the lowest* 0.5–2 *cm. apart; fruiting calyx bilabiate*, 4–6 mm. long, *persistent* on the pedicel, *the tube with few straight or slightly hooked short bristles, the base with mostly reflexed and appressed strigae; nutlets* obovoid, 1–1.3 *mm. broad*, the strophiole 0.4–0.5 mm. broad. (*M. virginica* sensu BSP. and others, not *Lycopsis virginica* L., basonym) — Rocky woods, dry or moist banks, prairies and openings, N.E. to Minn., s. to n. Fla., Tenn., Okla. and Tex.; Ida. to s. B.C., s. to Wyo. and Calif. April–June. (So. S. Am.) Fig. 1449.

1449. M. verna.

4. M. macrospérma Engelm. (large-seeded). — Differing from no. 3 in its lax or loosely branching stem (1.5–)2.5–8 dm. high; principal leaves 0.5–1.7 cm. broad; *central raceme elongating to* 1.2–4.7 *dm. long; fruiting pedicels* 3–10 *mm. long, loosely spreading-ascending from base, the lowest* 2–5 *cm. apart; fruiting calyx* 5.5–9 mm. long, *promptly disarticulating from tip of pedicel, the tube covered to base with hundreds of strongly hooked upcurving long bristles* (fruiting calyces adhering to passing animals); *nutlets* rounded-obovoid, *about* 2 *mm. broad*, the strophiole 0.5–0.8 mm. broad. (*M. virginica*, var. *macrosperma* (Engelm.) Fern.) — Loamy calcareous woods and bottomlands, Fla. to e. Tex., n. to Md. and D.C., Ky., s. Ind., s. Ill. and Mo. Late April, May.

5. M. sylvática Hoffm. (of the woods), Garden-F. — Tufted annual or biennial, becoming 2–5 dm. high, with oblong leaves; inflorescences naked, becoming elongate and lax; pedicels slender, equaling or exceeding the subequally lobed *oblique calyx; corolla-limb flat and rotate, 5–8 mm. broad,* pale and clear blue (or pink or white) with yellow eye. — Commonly cult., spreading and locally natzd., Gaspé Pen., Que., to Ont., s. to N.E., N.Y. and Mich. June, July. (Introd. from Eu.)

1450. M. arvensis.

6. M. ARVÉNSIS (L.) Hill (of cult. ground). — Annual or biennial, erect or ascending, spreading-hirsute, 1–4.5 dm. high; leaves oblong-lanceolate, acutish; *racemes naked at base; calyx* closing or *with erect subequal lobes* in fruit, spreading-pubescent *with minutely hooked or gland-tipped basal hairs, in maturity mostly shorter than the pedicels; corolla with spreading-ascending blue limb rarely* 4 *mm. broad.* — Waste places, roadsides and fields, Nfld. to Minn., s. to N.S., N.E., N.J., Pa. and W.Va. June–Aug. (Natzd. from Eu.) Fig. 1450.

7. M. STRÍCTA Link (upright). — Stems 0.5–2 dm. high, *branching from the base, with the oblong or lanceolate blunt leaves pubescent with mixed straight and divergent hooked hairs; some scattered flowers produced from the lower axils; pedicels* very short, mostly *with* some *divergent hooked hairs;* corolla blue, the tube barely equaling the calyx; *style rarely equaling the nutlets.* — (*M. micrantha* sensu ed. 7, not Pall.) — Roadsides, old fields, etc., local, s. Que. and s. Ont., s. to s. N.E., Va., O., Ind., and Ia. May–Aug. (Natzd. from Eu.) Fig. 1451.

8. M. VERSÍCOLOR (Pers.) Sm. (variously colored). — Slender, 1–3 dm. high, *mostly simple at base, often branched above; stems and leaves uniformly pubescent with ascending straightish hairs; upper leaves somewhat pointed; racemes* loose, mostly *naked at base; pedicels appressed-pubescent,* shorter than calyx; corolla pale yellow, changing to blue, then to violet, the tube exceeding the calyx; *style distinctly longer than the nutlets.* — Fields, local, N.S., Mass. and N.Y. to Va. April–July. (Adv. from Eu.) Fig. 1452.

1451. M. stricta.

1452. M. versicolor.

12. ASPERÙGO L. Madwort

Corolla with short tube slightly enlarged above, and with spreading limb, smaller than the conspicuous calyx which in fruit becomes greatly enlarged and with deeply cleft roundish lobes. Stamens included. Nutlets granulated. — Low annual nat. of Old World, with harsh slender stems, oblong or spatulate leaves, and few axillary flowers on short recurved pedicels. (Name from *asper, rough,* referring to the harsh leaves.)

1. A. PROCÚMBENS L. (ascending from a prostrate base). — Waste places and new seeding, mostly casual and not long-persistent, Mass. to Minn. s. to Del., D.C. and O. May–July. (Adv. from Eu.) Fig. 1453.

1453. A. procumbens.

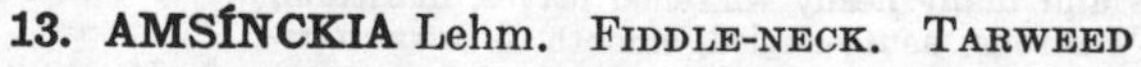

13. AMSÍNCKIA Lehm. Fiddle-neck. Tarweed

Corolla salverform or tubular-funnelform. Style filiform. Nutlets rough, dull, ovoid-trigonous, attached below the middle. Rough-hairy N. and S.Am. annuals, with oblong or narrower leaves and scorpioid-spicate yellow flowers, at least the lowest leafy-bracted. (In honor of *Wilhelm Amsinck,* a burgomaster of Hamburg, who early in the 19th century gave important support to the botanical garden of that city.)

1. A. BARBÀTA Greene (bearded). — Decumbent, loosely branched, 3–6 dm. high; leaves lanceolate or ovate-lanceolate, repand-denticulate; the lower flowers mostly bractless; *calyx-lobes marginally bearded with soft white hairs;* corolla pale yellow; *nutlets* trigonous-ovoid, *carinate, sharply muriculate,* rugose toward tip. (*A. lycopsoides* of ed. 7, not Lehm.) — Fields, henyards and waste places, somewhat sporadic, N.E. to O. June, July. (Adv. from B.C.)

2. A. SPECTÁBILIS Fisch. & Mey. (showy). — Similar; *calyx-lobes without long marginal hairs; nutlets granulate and sharply muricate, blackish,* 1–2 mm. long. — Casual in fields and waste places, Mass. to Ill. (Adv. from Calif.)

Other species appear occasionally in fields and waste places in our range, but are not usually long-persistent. These include A. Menzièsii (Lehm.) Nels. & Macbr. (for its discoverer, Archibald Menzies, 1754–1842), similar to no. 2, but with larger (2 mm. or more long) *transversely rugose pale nutlets*, the *corolla-limb broadly flaring;* A. Douglasiàna A. DC. (for its discoverer, David Douglas, 1799–1834), similar to the last but leaves generally entire, fruiting calyx at most 8 mm. long, *corolla more tubular and with narrow limb*, and *nutlets irregularly rugulose, scarcely 3 mm. long;* A. intáctilis J. F. Macbr. (not to be touched), coarser than the last but *fruiting calyx* 1–1.2 *cm. long, nutlets* 4 *mm. long, strongly rugose toward the apex.* These and others, adv. from western fields, are likely to increase.

14. PLAGIOBÒTHRYS Fisch. & Mey.

Calyx 5-parted, lax in fruit. Corolla white, short-salverform to rotate, slightly constricted at the more or less appendaged throat; tube not exceeding calyx, shorter than the roundish lobes. Filaments very short, inserted on middle of tube; oblong anthers included. Style short; stigma capitate. Nutlets trigonous, oblique, acutely cristulate-muricate or rugose, dorsally and ventrally keeled, incurved-ascending, on a short stipe; stipes united in pairs or distinct. — Am. herbs with alternate leaves, or lower leaves opposite. (Name from the Greek *plagios, oblique*, and *bothros, a pit.*)

1. P. scopulòrum (Greene) I. M. Johnston (of rocks). — *Prostrate or diffuse basally branching* annual *up to* 2.5 *dm. high*, strigose; *leaves* linear or linear-oblanceolate, 2–4 *cm. long*, 1.5–3 mm. broad; loosely flowered *simple racemes bracted to summit;* calyx slightly accrescent, in maturity with linear ascending lobes 2–4 mm. long; *corolla white, very small; nutlets lanceolate*, asymmetrical, acute, rounded at base, 1.5–2.3 *mm. long*, rugulose with anastomosing low ridges — Saline flats and damp depressions or shores, Sask. and Mont. to N.M., e. to w. Minn. June–Aug.

2. P. híRTUS (Greene) I. M. Johnston (harsh), var. figuRàTUS (Piper) I. M. Johnston (adorned with figures). — *Erect* strigose annual *up to* 7 *dm. high;* linear *acute leaves up to* 1.5 *dm. long;* forking summit bearing elongate *bractless* solitary or paired finally erect *racemes* up to 2 dm. long; *corolla* 5–10 *mm. broad, with conspicuous yellow appendages in throat; nutlets ovoid*, 1–1.5 *mm. long*, obliquely rugose *with forking ridges*, the lateral scar ovate and one-fourth the length of the nutlet. — Low fields, local, s. Mich. June–Aug. (Adv. from Pacific slope)

15. MERTÉNSIA Roth Lungwort

Corolla longer than the deeply 5-cleft or 5-parted calyx, naked, or with 5 small glandular folds or appendages in the open throat. Anthers oblong or sagittate. Style long and filiform. Nutlets ovoid, fleshy when fresh, smooth or wrinkled, obliquely attached by a prominent internal angle; the scar small. — Smooth or soft-hairy perennial herbs of N. Hemisph., with pale and entire leaves, and handsome purplish-blue (rarely white) flowers in loose and short panicled or corymbed raceme-like clusters, only the lower one leafy-bracted; pedicels slender. (For *Franz Karl Mertens*, 1764–1831, a distinguished German botanist.)

Erect large-leaved plants of inland habitats; corolla slender-tubed, 1–2.5 cm. long; nutlets dull, obtusely angled, becoming wrinkled when dry. § Eumertensia.
- Glabrous, very glaucous; corolla naked at throat; filaments filiform, longer than the exserted anthers. . . . . 1. *M. virginica.*
- Pubescent, greener; corolla with crests at the throat; filaments broad and short, the anthers included. . . . . 2. *M. paniculata.*

Prostrate, with trailing branches and many fleshy whitened leaves, maritime; corolla 3–9 mm. long; nutlets lustrous, acute-angled, smooth. § Steenhammera. . . . . 3. *M. maritima.*

§ Eumerténsia Gray

1. M. virgínica (L.) Pers. (of Virginia), Virginian Cowslip, Bluebells, Roanoke-bells. — Very smooth, glaucous, 2–7 dm. high; leaves of basal shoots long-petioled, with elliptic or ovate blades up to 2 dm. long; middle and upper cauline leaves short-petioled or sessile, ovate to oblong; flowers in 1-sided scorpioid cymes; *corolla* blue, pink when young (rarely white), trumpet-shaped, the slender *tube* 1–2 *cm. long, densely hairy within at base*, the spreading shallowly undulate limb *almost or wholly naked at throat; anthers exserted;* nutlets 3 mm. long, finally roughened. — Rich woods, clearings and bottomlands, N.Y. and s. Ont. to e. Minn., s. to S.C., Ala., Ark. and e. Kans. Late March–early June. Fig. 1454.

1454. M. virginica.

2. M. paniculàta (Ait.) G. Don (panicled). — Erect, *more or less rough-hairy*, 2–7.5 dm. high, loosely branched above; basal leaves long-petioled, the lance-elliptic to cordate-ovate harsh blades up to 2 dm. long; cauline leaves sessile, lance- to ovate-acuminate; cymes at first dense, becoming lax; *corolla* blue (rarely white), pink in bud, the *tube* 4.5–7 *mm. long, the throat with conspicuous crests; filaments broad and short, anthers included.* — Rich, usually damp, thickets, woods and shores, James Bay, Que., to Alaska, s. to n. Mich., Wisc., ne. Ia., Minn., Mont., Ida. and Wash. June–Sept.

§ STEENHÁMMERA (Reichenb.) Gray

3. M. marítima (L.) S. F. Gray (of seashores), SEA-L., OYSTERLEAF. — Spreading or decumbent, smooth, glaucous; leaves ovate, obovate or spatulate, fleshy, the upper surface becoming papillose; *corolla* rose-pink, becoming pale blue, or white in forma **albiflòra** Fern. (white-flowered), *campanulate-funnelform,* 3–9 *mm. long,* twice length of calyx, *with crested throat; filaments longer and narrower than anthers; nutlets smooth and lustrous, acute, utricular.* (*Pneumaria* Hill) — Gravel-beaches, rocks and sands of the coast, Greenl. to Alaska, s. to Mass., shores of James Bay and B.C. June–Sept. (Arct. and n. Eurasia) FIG. 1455.

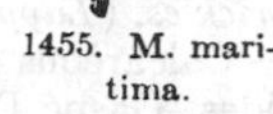
1455. M. maritima.

16. LÁPPULA Moench STICKSEED. BARDANETTE (Que.)

Corolla salverform, short, imbricated in bud, stamens included. Nutlets slenderly attached their whole length along the well developed median ventral keel, triangular or compressed, the back or dorsal margin usually armed with prickles which are barbed at apex. Gynobase subulate, much longer than broad. Style usually overtopping the nutlets. — Rough-hairy or grayish annuals found in many regions, with bracteate scorpioid racemes and short erect axillary pedicels and blue or white corollas. (Name a diminutive of *lappa,* a *bur.*)

1456. L. echinata.

1. L. ECHINÀTA Gilib. (prickly). — Erect, 1.5–6 dm. high, hispid; principal leaves linear to narrowly lanceolate; *nutlets* rough-granulate or tuberculate on the back, *the margins with a double row of slender distinct prickles,* or these irregularly distributed over most of the back. (*L. Myosotis* Moench) — Roadsides and waste places, Nfld. to Alaska, s. to N.S., N.E., N.J., Pa., W.Va., Ky., Ill., Mo., Kans. and Tex. June–Sept. (Natzd. from Eu.) FIG. 1456.

2. L. REDÓWSKII (Hornem.) Greene (in honor of D. REDOWSKY, Russian botanist who published early in the 19th century), var. OCCIDENTÀLIS (Wats.) Rydb. (western). — Erect, 1.5–6 dm. high, at length diffuse; *nutlets* irregularly and minutely sharp-tuberculate, *the margins armed with a single row of stout flattened prickles* sometimes confluent at base. — Dry open soil, w. U.S. and Can., spreading eastw. as a weed of roadsides and waste places to Mich. and Mo. and sporadically to N.E. May–Sept. (Patagonia) FIG. 1457.

1457. L. Redowskii, v. occidentalis.

17. HACKÈLIA Opiz STICKSEED. BEGGAR'S-LICE

Corolla as in *Lappula.* Nutlets attached by a large oblique submedian ovate or deltoid areola; ventral keel on only upper half of nutlet. Gynobase pyramidal, broader than high. Style overtopped by nutlets. — N.Am. and Eurasian biennials or perennials with naked or sometimes bracted racemes, the slender extra-axillary pedicels recurved or deflexed in fruit. (Named in 1839 for *Paul Hackel,* Bohemian scientist.)

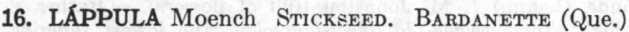

Cauline leaves membranaceous, lanceolate to oblong-ovate, spreading; branches of inflorescence loosely divergent, bracted; corollas 1.5–3 mm. broad; marginal prickles of nutlets nearly subulate.
Unseparated group of fruits globose; nutlets subequally prickly over the whole back. . . . . 1. *H. virginiana.*
Unseparated group of fruits broadly pyramidal; nutlets only marginally prickly or with few median prickles. . . . . 2. *H. americana.*
Cauline leaves firm, linear to narrowly oblanceolate, strongly ascending; branches of inflorescence erect or nearly so, bractless; corollas 4–6 mm. broad; marginal prickles flat, lanceolate. . . . . 3. *H. floribunda.*

1. H. virginiàna (L.) I. M. Johnston (Virginian). — Stem 3–12 dm. high; *radical leaves round-ovate or cordate,* slender-petioled; cauline 0.5–2.5 dm. long, ovate-oblong to oblong-lanceolate, acuminate at both ends; loosely paniculate racemes divaricate, with small bracts; *pedicel and flower* each about 2 mm. long, *frequently supra-axillary; nutlets*

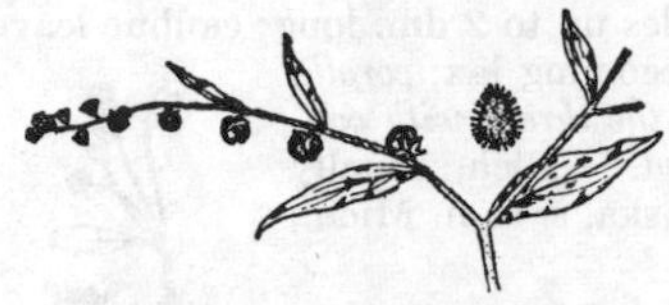
1458. H. virginiana.

of the globose fruit equally short-glochidiate over the whole back. (*Lappula* Greene) — Rich woods and thickets, s. Que. and Me. to S.D., s. to Ga., Ala., La. and Okla. Late June–Sept. Fig. 1458.

2. H. americàna (Gray) Fern. (American). — Similar to no. 1; *basal leaves narrowly ovate-lanceolate or narrowly obovate,* usually shriveled at flowering time; cauline narrowly to broadly lanceolate to oblanceolate, 0.1–1.5 dm. long; racemes loosely ascending to divergent, their lower bracts lanceolate, their upper linear-acicular and minute; *pedicels generally borne opposite the bracts;* corollas 1.5–2 mm. broad; *fruit pyramidal, the deltoid nutlets* 1–1.5 mm. broad, *with few unequal or no dorsal prickles.* (*Lappula deflexa*, var. *americana* (Gray) Greene; *L. americana* Rydb.) — Calcareous rocky or gravelly woods, bluffs and talus, spreading to roadsides, Gaspé Pen., Que., and adj. ne. N.B.; nw. Que. to B.C., s. to n. Vt. Bruce Pen., Ont., n. Mich., Wisc., ne. Ia., ne. Kans., Mont., Ida. and Wash. Late June, July. Fig. 1459.

1459. H. americana.

3. H. floribúnda (Lehm.) I. M. Johnston (full of flowers). — Rather *strict,* 0.3–1.5 m. high; *leaves oblong- to linear-lanceolate,* the lowest tapering into margined petioles; *racemes* numerous, commonly geminate and in fruit rather *strict;* corolla blue, sometimes white, 0.5–1 cm. in diameter; *nutlets* 4–6 *mm. long,* scabrous and *margined with a close row of flat-subulate prickles.* (*Lappula* Greene) — Damp thickets and shores, Pacific slope and Rocky Mts., e. to w. Ont. and Minn. June, July. Fig. 1460.

1460. H. floribunda.

FAM. 150. VERBENÀCEAE (Vervain Family)

Herbs or shrubs, *with opposite leaves, more or less 2-lipped or irregular corolla, and didynamous stamens, the 2–4-locular fruit dry or drupaceous, usually splitting when ripe into as many 1-seeded indehiscent nutlets;* resembling the following family, but *the ovary not deeply 4-lobed, the style therefore terminal,* and the plants seldom aromatic or furnishing a volatile oil. Seeds with straight embryo and little or no albumen. — A large family in the Tropics, sparingly represented in cool regions.

Flowers in simple and terminal spikes or heads; stamens included, didynamous; fruits (in ours) dry, separating into nutlets. Tribe Verbeneae.
- Calyx tubular, plicately 5-angled, 5-toothed; corolla salverform, equally or subequally 5-lobed; stigma mostly 2-lobed; fruit (schizocarp) separating into 4 nutlets. . . . . 1. *Verbena.*
- Calyx campanulate or compressed and 2-keeled, 2–4-cleft; corolla 2-lipped; stigma capitate; fruit splitting into 2 nutlets. . . . . 2. *Lippia.*

Flowers in axillary cymes; stamens exserted, equal; fruit a drupe. Tribe Viticeae. . . . . 3. *Callicarpa.*

Tribe Verbèneae Spreng.

1. VERBÈNA L. Vervain

Calyx 5-toothed, one of the teeth often shorter than the others. Corolla tubular, often curved, salverform; the border somewhat unequally 5-cleft. Stamens included; the upper pair occasionally without anthers. Style slender; stigma mostly 2-lobed. — Flowers sessile, in single or often panicled spikes, bracted, produced all summer. Plants of temp. and trop. reg. (The Latin name for any sacred herb; derivation obscure.)* — The species present numerous spontaneous hybrids.

a. Sterile lobe of style not protruding beyond the stigmatic surface; ovary entire at apex or barely emarginate, the style apical; schizocarp not constricted along the lines of cleavage; anthers unappendaged; calyx 1.7–5 mm. long, rarely more than twice length of schizocarp and not contorted beyond it; mostly erect annuals or perennials with small corollas (in ours 2–9 mm. broad). § Verbenaca. . . *b.*

b. Spike close or crowded, becoming short-cylindric (in ours 2–5 cm. long), in

* Treatment derived from that of L. M. Perry in Ann. Mo. Bot. Gard. xx. 239–392 (1933).

compound open cymes. Ser. PACHYSTACHYAE. The 4 angles of the tall stem prominent; leaves in few remote pairs, cuneate at base, acuminate, incised-serrate; calyx pubescent; corolla purple. . . . . 1. *V. brasiliensis.*

*b.*Spikes slender, filiform to pencil- or tail-like, with distant to approximate fruiting calyces, single or in simple cymes or panicles. Ser. LEPTOSTACHYAE. . . *c.*

*c.*Spikes panicled at the apices of stems and branches; fruiting calyces 1.7–3 mm. long; corolla-limb 2–4.5 mm. broad. . . *d.*

*d.*Spikes filiform or very slender, with remote fruiting calyces. . . *e.*

*e.*Leaves obtuse to subacute, pinnatifid or deeply incised; calyx truncate, with minute subulate teeth, glandular; corollas purple or blue, with limb 3.5–4 mm. broad; annuals.

Mature calyx 2–2.5 mm. long, twice as long as bract; corolla purple; mature schizocarp nearly as broad as long. . . . . . 2. *V. officinalis.*

Mature calyx 3 mm. long, about equaled by bract; corolla bluish; mature schizocarp twice as long as broad. . . . . . . 3. *V. riparia.*

*e.*Leaves acuminate, ovate to lanceolate, regularly toothed; calyx not glandular, with deltoid to lanceolate lobes; corolla white, pink or bluish, with limb about 2 mm. broad; perennials.

Calyx-lobes lanceolate, in maturity connivent and forming an acute beak beyond the schizocarp; stigmatic surface subtended by 2 sterile lobes of the style; nutlets 1–1.3 mm. long, reticulate above. 4. *V. scabra.*

Calyx-lobes deltoid, with subulate tips, about equaling and not connivent above the schizocarp; stigmatic surface subtended by 1 sterile lobe of style; nutlets 1.5–2 mm. long, not reticulate. . 5. *V. urticifolia.*

*d.*Spike pencil-like, with mostly imbricated fruiting calyces; corollas violet-blue (rarely pink), with limb 3–4.5 mm. broad; mature calyx overtopping the schizocarp, its teeth incurved. . . . . . 6. *V. hastata.*

*c.*Spikes 1 or in 3's at tips of stems or branches, often pencil-like, with appressed-ascending short floral bracts, if several to a branch (in no. 10) with divergent bracts much exceeding calyx; fruiting calyces 2.5–5 mm. long; corolla-limb 2.5–9 mm. broad, purple or pink (rarely white). . . *f.*

*f.*Leaves regularly toothed, not pinnatifid; spikes stiffly pencil-like, with appressed-ascending subulate or subulate-lanceolate bracts rarely much exceeding calyx; corolla-limb 5–9 mm. broad; stiffly ascending plants. . . *g.*

*g.*Stems glabrous or sparsely strigose; leaves narrowly lanceolate to oblanceolate, the broader 3–15 mm. broad; corolla-limb 5–6 mm. broad. . . . . 7. *V. simplex.*

*g.*Stems copiously pubescent; leaves broader, the larger 1.5–6 cm. broad.

Stem subterete, stoutish, hirsute; leaves oblong to broadly obovate or rounded-elliptic, pilose beneath; spike stiff, the white-pilose calyces imbricated; corolla-limb 8–9 mm. broad; nutlets readily separable, reticulate above. . . . . 8. *V. stricta.*

Stem 4-angled, slender, puberulent; leaves oblong to oblanceolate, scabrous-hirtellous beneath; spike slender, greatly interrupted; calyx glandular-hirtellous; corolla-limb 5 mm. broad; nutlets separated with difficulty, pitted above. . . . . 9. *V. carnea.*

*f.*Leaves pinnately incised or lobed; spike lax, with divergent lanceolate or linear bracts much exceeding calyx; corolla-limb 2.5–3 mm. broad; low and diffusely branching. . . . . 10. *V. bracteata.*

*a.*Sterile lobe of style protruding well beyond the stigmatic surface; ovary becoming definitely lobed at summit, the style apparently inserted in the depression between the lobes; schizocarp constricted along the lines of cleavage; connective of upper anthers mostly appendaged; low and often prostrate or decumbent perennials; flowering inflorescence corymbose, later elongating; fruiting calyx (in ours) 8.5–13 mm. long, its prolonged tips constricted or contorted; corolla 0.8–1.5 cm. broad. § GLANDULARIA.

Bracts mostly shorter than the glandular-hirsute calyx; corolla-tube about twice length of calyx. . . . . 11. *V. canadensis.*

Bracts mostly longer than the villous-hirsute calyx; corolla-tube about one-half longer than calyx. . . . . 12. *V. bipinnatifida.*

§ VERBENÀCA Schauer

Ser. 1. PACHYSTÁCHYAE Schauer

1. V. BRASILIÉNSIS Vellozo (Brazilian). — Strong erect perennial 1–2 m. high; the 4 *angles of the stem prominent; leaves in few remote pairs, cuneate at base,* the lower narrowly obovate, *the median and upper* lance-*acuminate* and *sharply incised-serrate; cyme open and compound,* with long ascending branches with linear-lanceolate leaves; *spikes fascicled at tips of slender branches and branchlets, dense, elongating in fruit to 2–5 cm.; calyx pubescent;* small *corolla purple;* nutlets trigonous, about 2 mm. long, strongly striate, reticulate at apex. — Waste places, fields and roadsides, Fla. to La., n. to se. Va. and Ark. *Fl.* July–Sept. (Natzd. from S. Am.)

Ser. 2. Leptostáchyae Schauer

2. V. officinàlis L. (of the shops), European V. — *Annual,* glabrous or nearly so, *loosely branched,* 3–9 dm. high; *leaves* oblong to narrowly ovate or obovate, 2–7 cm. long, strigillose; *the lower pinnatifid or 3–5-cleft,* with incised divisions; *spikes* loosely paniculate, *becoming filiform and remotely flowered* in maturity; *calyx glandular, becoming 2–2.5 mm. long, truncate,* with minute subulate teeth, *twice as long as bract;* corolla purple, its limb about 4 mm. broad; *mature schizocarp nearly as broad as long;* nutlets scarcely 2 mm. long, striate below, reticulate above. — Waste places and roadsides, Fla. to La., n. to Mass. (rare), N.Y., W.Va. and Tenn. June–Oct. (Natzd. from Eu.) Fig. 1461.

1461. V. officinalis.

3. V. ripària Raf. (of river-banks). — Coarser, up to 1.5 m. high; lower and median *leaves 0.4–1.2 dm. long;* mature *calyx 3 mm. long, about equaled by the bract;* corolla light blue; *mature schizocarp twice as long as broad.* (*V. urticifolia,* var. *riparia* Britt.) — Rich thickets and banks of streams, local, N.J. to w. Va. and N.C. July–Sept.

4. V. scàbra Vahl (harsh). — Erect perennial, 0.5–1.5 m. high, with habit of no. 5, simple or paniculately branched, the stem hispidulous; *leaves* narrowly ovate, 3–13 cm. long, petioled, serrate-dentate, acuminate, *harshly scabrous above;* spikes filiform, interrupted; *fruiting calyx* 2.5–3 mm. long, hispidulous, *ovoid,* longer than the bracts, *with the lanceolate lobes connivent as an acute beak prolonged above the schizocarp;* corolla white, pink or bluish, its limb about 2 mm. across; *stigmatic surface subtended by 2 sterile obtuse lobes of the stigma; nutlets 1–1.3 mm. long, reticulate above.* — Swamps, rich low woods and shores, Fla. to n. Mex. and Calif., n. to se. Va. June–Sept. (W.I.) Fig. 1462.

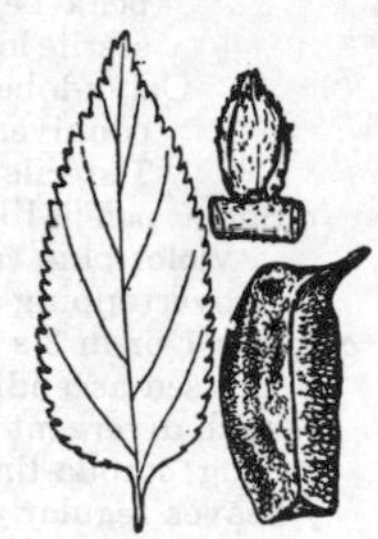
1462. V. scabra.

5. V. urticifòlia L. (nettle-leaved), White V. — Erect, paniculate-branched, 0.5–1.5 m. high, the stem hirtellous to glabrate; *leaves* oblong-ovate to lanceolate, acuminate, 0.8–2 dm. long, coarsely (often doubly) crenate-serrate, *the lower surface strigose-hirsute on the veins with stiff hairs up to 1.3 mm. long,* or glabrate; slender *spikes* interrupted, *usually stiffly ascending, the axis strigose; mature calyx strigose, 2–2.3 mm. long, with deltoid* subulate-tipped *lobes, about equaling or shorter than mature schizocarp;* the bracts 1–1.5 mm. long; corolla white, limb 2 mm. across; *stigmatic surface subtended by 1 sterile lobe of style; nutlets 2 mm. long, corrugated on back.* — Rich thickets and borders of woods, sw. Que. and w. Me. to s. Ont. and S.D., s. to s. N.E., L.I., n. Fla., Ala., La. and Tex. June–Sept. Fig. 1463.

1463. V. urticifolia.

Var. **leiocárpa** Perry & Fern. (with smooth carpels). — *Leaves minutely velutinous beneath, the hairs hardly 0.3 mm. long; flowering branches filiform, loosely ascending or divergent, puberulent; mature calyx* 1.7–2 mm. long, *puberulent,* its bract 0.5–1 mm. long; *nutlets 1.5 mm. long,* shining and *plane* on *back.* — Rich thickets, Ct. to S.C. and Tenn.

6. V. hastàta L. (halberd-shaped; the original material having such basal leaves), Blue V., Simpler's-joy. — Erect, 0.4–1.5 m. high, branched above; leaves lanceolate or lance-oblong to narrowly ovate, 0.5–1.8 dm. long, acuminate, petioled, coarsely incised-serrate, sometimes hastately lobed at base, scabrous-pubescent; *spikes strict,* pencil-like, usually numerous in a panicle, *compact, with mostly imbricated fruiting calyces;* bracts lance-subulate, mostly shorter than the pubescent *calyx* (2.5–3 *mm. long*); *calyx with finally incurving* subulate-tipped *lobes overtopping the schizocarp; corolla-limb violet-blue,* or pink in forma **ròsea** Cheney (rose-colored), 3–4.5 *mm. broad;* nutlets 2 mm. long, nearly smooth or faintly striate. — Swales, damp thickets and shores, N.S. and N.B. to s. B.C., s. to s. N.E., L.I., nw. Fla., e. Tenn., Mo., Tex., N.M., Ariz. and Calif. July–Sept. Fig. 1464.

1464. V. hastata.

7. V. símplex Lehm. (simple). — Erect perennial 2–6 dm. high, with *glabrous or sparsely*

1465. V. simplex.

strigose stem; leaves narrowly lanceolate to oblanceolate, tapering at base, 3–10 cm. long, *the larger* 3–15 *mm. broad,* serrate, reticulate and often scabrous above, glabrate or strigillose on both surfaces; spikes solitary or few, erect, pencil-like; *fruiting calyx* 4–5 mm. long, with acuminate lobes, *usually longer than the lance-subulate bract; corolla* purple, its tube barely exserted, *its limb* 5–6 *mm. broad;* nutlets 2.5 mm. long, reticulate above. (*V. angustifolia* Michx., not Mill.) — Dry or sandy soil, sw. Que. and w. N.H. to Ont., Minn., and Neb., s. to Fla., Ala., Miss., La. and Kans. May–Sept. Fig. 1465.

8. V. strícta Vent. (strict) Hoary V. — *Stems* stoutish, erect, *subterete,* simple or branched above, 0.3–1.2 m. high, *densely hirsute; leaves oblong to broadly obovate or rounded-elliptic, pilose beneath,* 6–10 cm. long, 3–6 *cm. broad; spikes* as in no. 7 or *denser,* the *white-pilose calyces imbricated; corolla* purple, or rosy-pink in forma **roseiflòra** Benke (rosy-flowered) or white in forma **albiflòra** Wadmond (white-flowered), its *limb* 8–9 *mm. broad; nutlets* 2.5 mm. long, *reticulate above.* — Prairies and barrens, s. Ont. to Mont., s. to Tenn., Ark., Okla., Tex. and N.M.; adv. e. to N.E., s. N.Y. N.J. and Del. June–Sept. Fig. 1466.

1466. V. stricta.

9. V. cárnea Medic. (flesh-colored). — *Stem slender, quadrate, scabrous-puberulent,* 2–8 dm. high, simple or with arcuate branches; *leaves oblong to oblanceolate,* round-tipped, serrate-dentate, *scabrous on both faces, the larger* 1.5–3 *cm. broad; spike* slender, *greatly interrupted in fruit; calyx glandular-hirtellous,* about 5 mm. long, with acute lobes; corolla flesh-color or white, its limb 5 mm. broad; *schizocarp* 3 mm. long, *ellipsoid, of closely united or tardily separable nutlets,* the summit dome-shaped. (*Stylodon carolinienis* (Walt.) Small) — Sandy woods and openings, Fla. to Tex., n. to e. Va. May–Sept. Fig. 1467.

1467. V. carnea.

10. V. bracteàta Lag. & Rodr. (with bracts). — Diffusely branched, with decumbent or ascending hirsute stems and branches; *leaves* 1–6 cm. long, *pinnately incised or lobed,* hirsute; *spikes* terminal, sessile, thick, *elongate,* rather closely flowered, *with divergent* hirsute lanceolate to linear *bracts much longer than calyces;* calyx 3–4 mm. long, hirsute, the short lobes connivent over the schizocarp; *corolla* purple, its *limb* 2.5–3 *mm. broad;* nutlets 2 mm. long, reticulate above. (*V. bracteosa* Michx.) — Sandy prairies, fields and waste places, Fla. to Tex. and Mex., n. to Va. (adv. to N.E.), s. Ont., Mich., Wisc., Minn., N.D., Alta. and s. B.C. June–Sept. Fig. 1468.

1468. V. bracteata.

§ Glandulària Schauer

11. V. canadénsis (L.) Britt. (Canadian), Rose-V. — Stem *decumbent,* rooting at nodes, spreading-hirsute to glabrous; *leaves* ovate, 3–9 dm. long, *incised or incised-pinnatifid or 3-cleft,* glabrous or appressed-pubescent; *spikes peduncled,* at first broadly corymbose, later elongate; *bracts shorter than or about equaling calyx; calyx glandular-hirsute,* in maturity 1–1.3 cm. long, with subulate-setaceous lobes; *corolla-tube about twice length of calyx,* its flesh-color (rarely white) to reddish-purple *limb* 1–1.5 *cm. broad;* gland at tip of anther-connective minute; schizocarp constricted at maturity along line of cleavage. (*Glandularia* Small) — Sandy or rocky prairies and barrens, roadsides, etc., Fla. to Tex., n. to se. Va., Ill., Ia., Kans. and Colo. Apr.–Oct. Fig. 1469. — Often cult.

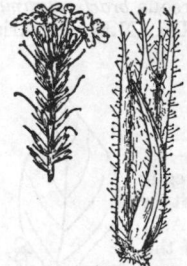

1469. V. canadensis.

12. V. bipinnatífida Nutt. (twice pinnately dissected). — Similar, hispid-hirsute; *leaves* bipinnately parted or 3-parted into more

1470. V. bipinnatifida.

or less bipinnatifid divisions, *the ultimate segments linear; bracts mostly longer than the villous-hirsute calyx; mature calyx* 8.5–10 *mm. long; corolla-tube about one-half longer than calyx.* — Dry prairies and plains, Ala. to Ariz. and Mex., n. to Mo. and S.D. May–Aug. Fig. 1470.

2. LÍPPIA L.

Calyx often flattened, 2–4-toothed or 2-lipped. Corolla 2-lipped, upper lip notched; lower much larger, 3-lobed. Stamens included. Style slender; stigma obliquely capitate. Plants of trop. and warm-temp. Am. and Afr. (Dedicated to *Agostino Lippi*, Italian naturalist, 1678–1704.) Phyla Lour.

Leaves tapering to acuminate base and apex, 5–11-toothed on each side to below middle, on elongate ascending flowering branches. . . . . 1. *L. lanceolata*.
Leaves blunt or round-tipped, cuneate or spatulate, with 1–7 teeth on each side toward summit, on mostly trailing or depressed stems.
Stems repent, very elongate; leaves 3–7-toothed on each margin; mature fruiting spikes 4.5–8 mm. thick, their broad bracts gradually tapering above. . 2. *L. nodiflora*.
Stems trailing (rarely repent) from stout ligneous caudex; leaves 1–2-toothed on each margin; mature fruiting spikes 7–11 mm. thick, their bracts abruptly pointed. . . . . 3. *L. cuneifolia*.

1. L. lanceolàta Michx. (lance-shaped), Fog-fruit. — Decumbent or procumbent, green, *with elongate ascending flowering branches; leaves lanceolate, acuminate to tip and base, the larger* 0.8–1.5 *cm. broad,* 5–11-*toothed on each margin to below the middle;* peduncles filiform, from the middle or upper (but rarely the uppermost) axils, twice to thrice length of subtending leaf; the solitary short head with pink, bluish or white flowers subtended by *broad-ovate pointed bracts.* (*Phyla* Greene) — Wet brackish to fresh sand, Fla. to La., n. to s. N.J. May–Sept.

1471. L. lanceolata, v. recognita.

Var. **recógnita** Fern. & Grisc. (studied again). — Coarser; *leaves oblong-lanceolate to narrowly elliptic or oval, the larger* 1–4 *cm. broad.* — Bottomlands, ditches, etc., Fla. to s. Calif., n. to Pa., s. Ont., Mich., Wisc., Minn. and Neb. June–Oct. Fig. 1471.

2. L. nodiflòra (L.) Michx. (flowering from the nodes). — *Stems repent,* elongating to 1 m. or more, with short ascending branches, greenish to cinereous; *leaves spatulate to cuneate-obovate,* rounded at summit, *with* 3–7 *teeth on each margin chiefly above the middle; the larger* 1.5–4.5 cm. long, 0.7–2 *cm. broad; peduncles two to four times length of subtending leaves; mature fruiting spikes* cylindric, 0.8–2.5 cm. long, 4.5–8 *mm. thick, their bracts gradually tapering above.* (*Phyla* Small) — Low grounds, Fla. to Tex. and Mex., n. to se. Va., se. Mo., Okla., etc. June–Oct. (Trop. Am.) Fig. 1472.

1472. L. nodiflora.

3. L. cuneifòlia (Torr.) Steud. (with wedge-shaped leaves). — Stems *procumbent* (not repent) *from a stout ligneous caudex,* canescent; *leaves rigid, linear-cuneate to cuneate-oblanceolate, with* 1 *or* 2 *pairs of sharp teeth near summit on each side; the larger* 1.5–4 cm. long, 3–8 *mm. wide;* peduncles shorter than to twice length of subtending leaves; *mature fruiting heads* ellipsoid to thick-cylindric, 1–2 cm. long, 7–11 *mm. thick, their* cinereous *bracts abruptly pointed.* — Low places in the plains, S.D. and Wyo., s. to e. Kans., Okla., Tex., N.M. and Ariz. May–Sept.

Tribe Vitíceae Spreng.

3. CALLICÁRPA L.

Calyx 4–5-toothed. Corolla tubular-campanulate, 4–5-lobed, nearly regular. Stamens 4, nearly equal, exserted; anthers opening at the apex. Style slender, thickened upward. — Shrubs or trees of warm parts of N. Am., e. Asia and Australia, with scurfy pubescence and small flowers. (Name formed of the Greek *callos, beauty,* and *carpos, fruit.*)

1. C. americàna L. (American), French Mulberry, Beauty-berry. — Shrub 0.5–2 m. high; *leaves* ovate-oblong with a tapering base, acuminate, toothed, *whitish-tomentose beneath;* cymes many-flowered; calyx obscurely 4-toothed; *corolla bluish;* fruit pinkish or violet-color. — Rich woods and thickets, Fla. to Tex., n. to Md., Tenn., Ark. and Okla. June–Aug. Fig. 1473.

1473. C. americana.

FAM. 151. LABIÀTAE (Mint Family)

Chiefly herbs, ordinarily with square stems, opposite aromatic leaves, more or less 2-lipped corolla, didynamous stamens or these only two, and a deeply 4-lobed ovary, which forms in fruit 4 little seed-like nutlets or achenes, surrounding the base of the single style in the bottom of the persistent calyx, each filled with a single erect seed. Nutlets smooth to roughish and fixed by their base, except in the first four genera. Albumen mostly none. Embryo straight (except in *Scutellaria*); radicle at the base of the fruit. Upper lip of the corolla 2-lobed or sometimes entire; the lower 3-lobed. Stamens inserted on the tube of the corolla. Style 2-lobed at the apex. Flowers axillary, chiefly in cymose clusters, these often aggregated in terminal spikes or racemes. — Foliage mostly dotted with small glands containing a volatile oil, upon which depends the warmth and aroma of the plants of this large and well-marked family.

a. Ovary merely 4-lobed; nutlets rugose-reticulated, the areola (scar of attachment) large, very oblique or even on the inner face of the nutlet. . . *b.*

b. Limb of corolla nearly regular or merely oblique, its 5 similar lobes subequal.

Calyx campanulate, deeply and equally 5-cleft; the throat of corolla campanulate, the obovate lobes not declined; stamens included; ovary 5-lobed. . . . . 1. *Isanthus.*

Calyx oblique, the upper teeth elongate, the lower very short; corolla with declined oblong lobes; stamens long-exserted; ovary 4-lobed. . . . 2. *Trichostema.*

b. Limb of corolla irregular, seemingly unilabiate, the upper lip either split down or very short; stamens exserted from the cleft.

Corolla with a very short and almost truncate upper lip, the spreading lower lip with 2-cleft or emarginate middle lobe. . . . . 3. *Ajuga.*

Corolla deeply cleft between the 2 small lobes of the upper lip, these lobes united (one on each side) with the lateral lobes of the declined lower lip, the middle lobe of the latter much the larger. . . . . 4. *Teucrium.*

a. Ovary deeply 4-parted; nutlets smooth, rugose, papillate or reticulated, the areola basal and small. . . *c.*

c. Stamens declined toward or resting on the lower lip of the corolla, 2 long and 2 short; corolla declined, the upper lip with 4 entire lobes, the lower lip saccate and abruptly deflexed at the contracted and callous base. . . 5. *Hyptis.*

c. Stamens not declined and resting on the lower lip. . *d.*

d. Stamens 4, ascending and parallel; calyx bilabiate, closed in fruit, with a helmet-like projection on the upper side, the rounded lips entire; corolla bilabiate, the upper lip usually arched. . . . . 6. *Scutellaria.*

d. Stamens ascending or spreading, extended straight forward; calyx usually open in fruit, without a projection on the upper side. . . *e.*

e. Stamens and style included in the tube of the small white corolla; white-woolly plant; flowers in axillary clusters; calyx tubular, cylindraceous, the 5 or 10 rigid teeth claw-like. . . . . 7. *Marrubium.*

e. Stamens and style exserted beyond the corolla-tube; calyx-teeth not strongly hooked. . *f.*

f. Fertile stamens 4, all with 2-locular anthers; upper lip of corolla galeate or concave. . . *g.*

g. Upper (inner) pair of stamens longer than the lower, ascending or diverging; calyx 13- or 15-nerved. . . *h.*

h. Anthers separated or distant (not approximate in pairs), their locules nearly parallel.

Tall plants with dense spikes of small flowers; stamens divergent or distant, exserted, the upper pair declined and crossing the ascending lower pair. . . . . 8. *Agastache.*

Low stoloniferous plant with few large (2–3.5 cm. long) corollas; stamens all ascending, not exceeding corolla. . . 9. *Meehania.*

h. Anthers approximate in pairs; their locules strongly divergent; filaments ascending, not exserted. . . *i.*

i. Calyx oblique at throat, with subequal narrow teeth.

Stem erect; flowers crowded in dense interrupted spiciform terminal clusters; locules of anthers merely divergent. . 10. *Nepeta.*

Stem weak, trailing or creeping; flowers in axils of petioled roundish to kidney-shaped leaves; anthers of perfect flowers with locules at right angles, each pair forming a cross. . . . . 11. *Glechoma.*

*i.*Calyx equal at summit, the upper tooth much broader than the others. . . . . 12. *Dracocephalum.*

*g.*Upper pair of stamens shorter than lower (except in no. 17), both pairs parallel and ascending under the galeate or concave upper lip of the corolla; calyx 5–10-nerved, not strongly 2-lipped (except in no. 13). . . *j.*

*j.*Calyx deeply bilabiate, 10-nerved, reticulate-veiny, closed in fruit, the upper lip flat and 3-toothed. . . . . . . 13. *Prunella.*

*j.*Calyx not strongly bilabiate, open in fruit, the lobes or teeth subequal. . . *k.*

*k.*Calyx membranaceous, inflated or enlarged after flowering, obscurely nerved but somewhat veiny, 4- or 5-toothed; throat of showy corolla inflated; the upper lip concave, not galeate; filaments villous.

Calyx nearly equally 5-toothed; corolla gradually inflated upward; anthers uniform, with nearly parallel locules; perennial with narrow leaves. . . . . . . 14. *Physostegia.*

Calyx deeply 4-cleft; corolla slender-tubed, inflated into a ventricose throat; anthers with divergent locules, the contiguous ones of the upper stamens sterile and connate; annual or biennial with deeply cordate ovate leaves. . . 15. *Synandra.*

*k.*Calyx firm, distinctly 5- or 10-nerved or striate, 5- or 10-toothed. . . *l.*

*l.*Calyx funnelform-dilated at summit, 10-nerved, the 5-toothed border spreading; nutlets rounded at top. 16. *Ballota.*

*l.*Calyx tubular to turbinate, 5- or 10-nerved. . . *m.*

*m.*Hard-stemmed tall perennials; cauline leaves deeply cleft or incised, all much longer than the capitate-verticillastrate flowers; calyx-teeth at length divergent, subulate-spinescent; whitish upper lip of small corolla densely woolly. . . . . 17. *Leonurus.*

*m.*Soft-stemmed; cauline leaves merely crenate, dentate or serrate; calyx-teeth ascending; upper lip of corolla glabrous or merely pilose. . . *n.*

*n.*Anthers opening by transverse valves, the inner valve of each locule hirsute-ciliate; calyx-teeth spinescent; nutlets rounded at summit; annual weeds. . . . 18. *Galeopsis.*

*n.*Anthers opening by longitudinal valves, the valves not ciliate.

Stamens not deflexed after anthesis; nutlets truncate at summit; flowers crowded in axils of well-developed leaves; introduced weeds or wanderers from cult. . . . . 19. *Lamium.*

Stamens all (or the lower pair sometimes) deflexed to the sides of the throat or contracted after anthesis; nutlets rounded at summit; flowers interruptedly spicate, the upper or all bracteal leaves reduced; chiefly indigenous. . . . . 20. *Stachys.*

*f.*Fertile stamens 2 or, if 4, the upper pair shorter and the upper lip of corolla neither galeate nor concave. . . *o.*

*o.*Antheriferous stamens 2, ascending and parallel, the upper pair wanting or rudimentary; anthers really or apparently 1-locular; corolla strongly 2-lipped. . . *p.*

*p.*Anther with an elongate filament-like connective articulated with the filament, a fertile locule at the ascending end and sometimes 1 at the lower end; calyx bilabiate, its lower lip 2-cleft; flowers mostly interrupted-spicate, with upper bracteal leaves reduced. . . . . 21. *Salvia.*

*p.*Anther with both locules fertile and similar, contiguous and divergent, connate or confluent at their junction, thus becoming 1-locular; flowers verticillastrate-capitate, with leafy outer bracts.

Calyx elongate-tubular, nearly regular, with 5 subequal teeth; middle lobe of lower lip of corolla larger or longer than the lateral. . . . . 22. *Monarda.*

Calyx ovoid-tubular, 2-lipped, with subulate or bristleform teeth; middle lobe of lower lip of corolla narrower than the rounded lateral lobes. . . . . 23. *Blephilia.*

*o.*Antheriferous stamens 2 or 4, upper pair shorter or wanting; anthers 2-locular; corolla only moderately to scarcely 2-lipped, the upper lip neither galeate nor curved. . . *q.*

*q.*Flowers in more or less crowded clusters or whorls or spicate. . . *r.*

*r.*Corolla more or less 2-lipped; the upper lip erect, entire or emarginate. . . *s.*

*s.*Stamens (at least the lower pair) ascending or arching, commonly converging or ascending parallel under the upper lip. . . *t.*

*t.*Calyx gibbous on the lower side; fertile stamens only 2, the others wanting or sterile; small-leaved annuals with flowers loosely clustered. . . . . 24. *Hedeoma.*

*t.*Calyx not strongly gibbous below; fertile stamens 4.

Calyx flattish on upper side; tube of corolla curved upward; flowers loosely clustered in axils of thin dentate broad leafy bracts. . . . . . . 25. *Melissa.*

Calyx tubular or campanulate; tube of corolla straight.

Herbaceous, with flat opposite leaves. . . . . 26. *Satureja.*

Suffruticose or woody, with linear revolute-margined fascicled leaves. . . . . 27. *Conradina.*

*s.*Stamens distant and straight, often divergent, never connivent nor curved. . . *u.*

*u.*Antheriferous stamens 4. . . *v.*

*v.*Calyx naked in the throat; leaves entire (sometimes toothed in no. 29).

Inflorescence a spike of crowded clusters of blue-purple flowers; garden-escape. . . . . . . 28. *Hyssopus.*

Inflorescence an open or dense corymb of dense heads or heads in upper axils; corollas white to purple; indigenous. . . . . 29. *Pycnanthemum.*

*v.*Calyx bearded in the throat.

Upright herb, with terminal corymbs of cylindric or ellipsoid spikes with large imbricated purple bracts; calyx nearly equally toothed. . . . 30. *Origanum.*

Prostrate half-shrub; flowers crowded at ends of branches, with minute bracts; calyx distinctly 2-lipped. . . . . 31. *Thymus.*

*u.*Antheriferous stamens 2; calyx strongly bearded in throat, with subequal teeth; upright herb with serrate ovate leaves and corymbs of clustered flowers. . . . 32. *Cunila.*

*r.*Corolla regular or essentially so, almost equally 4-lobed.

Fertile stamens 2, the others wanting or rudimentary; flowers white, clustered in the axils of reduced upper leaves; plants slenderly stoloniferous or tuberiferous. . . . 33. *Lycopus.*

Fertile stamens 4; flowers purplish, bluish or white, spicate or clustered in upper axils.

Corolla-tube included; anthers not versatile, the locules parallel and distinct; perennials, spreading by stolons and basal offsets; flowers in axillary clusters or in interrupted spikes or heads. . . . . 34. *Mentha.*

Corolla-tube exserted; anthers versatile, the divergent locules becoming confluent; annual, non-stoloniferous, with close slender spikes of tiny flowers. . . . 35. *Elsholtzia.*

*q.*Flowers in loose terminal racemes or panicles, slender-pedicelled; calyx enlarged and declined in fruit, becoming bilabiate.

Perennial with hard rhizome; corolla yellowish to purplish, many times longer than flowering calyx, the stamens long-exserted.

Fertile stamens 2; corolla not viscid. . . . 36. *Collinsonia.*

Fertile stamens 4; corolla viscid. . . . . 37. *Micheliella.*

Annual with fibrous roots; corolla white, minute, barely exceeding flowering calyx; stamens barely exserted.

Fertile stamens 4; anthers with parallel locules. . . . 38. *Perilla.*

Fertile stamens 2; anthers with divaricate locules. . . . 39. *Mosla.*

1. ISÁNTHUS Michx. FALSE PENNYROYAL

Calyx equally 5-lobed, enlarged in fruit. Corolla little longer than the calyx; the border campanulate, with obovate lobes. Stamens slightly didynamous, incurved-ascending, scarcely exceeding the corolla. — A low annual, clammy-pubescent, with nearly entire lance-oblong to -linear 1–3-nerved leaves, and small pale blue flowers on axillary 1–3-flowered peduncles. (Name from the Greek *isos, equal,* and *anthos, flower,* referring to the almost regular corolla.)

1. **I. brachiàtus** (L.) BSP. (with arm-like branches). — Corolla about 5 mm. long, little exceeding calyx. — Dry, often calcareous, soil, sw. Que. and Vt. to Minn., s. to Ga., Tenn., Ark. and Tex. July–Sept. FIG. 1474.

1474. I. brachiatus.

2. TRICHOSTÈMA L. BLUECURLS

Calyx campanulate, deeply 5-cleft; the 3 upper teeth elongated and partly united, the 2 lower very short. Stamens with very long curved capillary filaments; anther-locules divergent and at length confluent. — Low N. Am. annuals, somewhat clammy-glandular and balsamic, branched, with entire leaves, and mostly solitary 1-flowered pedicels terminating the branches and becoming lateral by the production of axillary branchlets, and the flower appearing to be reversed, *i.e.,* the short teeth of the calyx upward, etc. Corolla blue, varying to pink, rarely white, small; fl. in summer and autumn. (Name composed of the Greek *thrix, hair,* and *stema, stamen,* from the capillary filaments.)

Leaf-blades oblong to rhombic-lanceolate. . . . . 1. *T. dichotomum.*
Leaf-blades linear. . . . . 2. *T. setaceum.*

1. **T. dichótomum** L. (regularly forking in pairs), BASTARD PENNYROYAL. — Stem 1–8 dm. high, bushy-branched, viscid-pilose with slender hairs and abundant long-stalked glands; *leaves oblong to rhombic-lanceolate; corolla* 4–6 *mm. long,* its lower lobe oblong and broader than the others. — Dry open soil, s. Me. to Mich., s. to s. N.E., L.I., N.C., Ky. and Mo. Aug.–Oct. FIG. 1475.

1475. T. dichotomum.

Var. **pubérulum** Fern. & Grisc. (minutely pubescent). — Stem merely puberulent, the branches similar or with minute glands. — Fla. to Tex., n., locally, to se. Va., Tenn. and Mo.

2. **T. setàceum** Houtt. (bristle-like). — Stem 1–3 dm. high, more slender and less branched than in no. 1, puberulent; *leaves linear, glabrous or nearly so; corolla* 6–10 *mm. long.* (*T. lineare* Walt.) — Dry sands, local, Fla. to La., n. to sw. Ct., centr. Pa. and s. O. Aug., Sept.

3. ÁJUGA L. BUGLEWEED

Calyx 5-toothed. The large and spreading lower lip of the corolla with the middle lobe emarginate or 2-cleft. Stamens as in *Teucrium,* but anther-locules less confluent. Herbs, nat. of Old World. (From the Greek *a-, without,* and *zugos* (Latin *jugum*), *yoke,* from the seeming absence of a yoke-fellow to the lower lip of the corolla.)

Leaves broad, merely undulate or dentate; corollas bluish; perennials.
  Freely stoloniferous, glabrous or but slightly pubescent. . . . . 1. *A. reptans.*
  Non-stoloniferous, tufted, copiously soft-pubescent. . . . . 2. *A. genevensis.*
Leaves deeply cleft into linear hirsute segments; corollas yellow; annual or short-lived perennial. . . . . 3. *A. Chamaepitys.*

1. A. RÉPTANS L. (creeping). — *Perennial,* 1–2.5 dm. high, smooth or but slightly pubescent, *with* copious *creeping stolons;* leaves obovate or spatulate, sometimes sinuate, the cauline sessile, the floral approximate, subtending several sessile bluish to purple flowers. — Locally by roadsides and in fields, Nfld. to Wisc., s. to Pa. and O. May–July. (Natzd. from Eu.)

2. A. GENEVÉNSIS L. (of Geneva). — Similar; the stems copiously soft-pubescent, *tufted, not stoloniferous;* leaves somewhat pubescent; flowers larger. — Locally in fields, waste places, etc., N.E. to Ont., s. to Pa., O. and Ill. May–July. (Natzd. from Eu.) FIG. 1476.

3. A. CHAMAÈPITYS (L.) Schreb. (old generic name), YELLOW B. — *Annual or short-lived perennial,* bushy-branched, 0.5–2 dm. high, *strongly villous-hirsute; leaves crowded, cleft nearly to base into linear stiff segments; flowers in axillary pairs much shorter than leaves; corolla yellow,* the lower lip brown-spotted at base. — Dry banks and old fields, e. Va. May–Sept. (Natzd. from Eu.)

1476. A. genevensis.

4. TEÙCRIUM L. Germander. Germandrée (Que.)

Calyx 5-toothed. Corolla with the 4 upper lobes nearly equal, oblong, turned forward, so that there seems to be no upper lip; the lower lobe much larger. Stamens 4, exserted from the deep cleft between the 2 upper lobes of the corolla; anther-locules confluent. Large genus of temp. reg. (*Teucrion*, name used by Dioscorides for some related plant.)

Leaves merely dentate or serrate; flowers in terminal racemes; perennials with stoloniferous rootstocks.
Racemes cylindric, spiciform; calyx densely pubescent.
Calyx canescent with felt-like (pannose) coat, the 3 upper lobes obtuse (or middle one acutish). . . . . 1. *T. canadense.*
Calyx viscid-villous, the 3 upper lobes acute (or middle one acuminate). . . 2. *T. occidentale.*
Racemes 1-sided, slender, with paired flowers; calyx glabrous or glabrate. . 3. *T. Scorodonia.*
Leaves pinnatifid into narrow segments; flowers in axillary verticils; annual. . 4. *T. Botrys.*

1. T. canadénse L. (Canadian), American G., Wood-Sage. — Stem stiff, erect simple or branching, 0.2–1 m. high, appressed-pubescent; *leaves lanceolate to lance-ovate*, serrate, 1.5–3 cm. broad, tapering to rounded at base, short-petioled, thick, hoary beneath, *the upper surface papillose beneath the dense appressed pubescence;* raceme spiciform, elongate or wand-like, with crowded to slightly separated whorls of about 6 flowers; *calyx canescent-pannose, the 3 upper lobes obtuse, or the middle one acutish;* corolla 1–2 cm. long, purplish, pink or creamy. (Var. *littorale* (Bickn.) Fern.; *T. littorale* Bickn.) — Shores, thickets and woods, Fla. to Tex., n. near the coast to e. N.B. and N.S., and up the Miss. embayment to Okla. July–Sept.

Var. **virgínicum** (L.) Eat. (Virginian). — Mostly taller, 0.5–1.3 m. high; *leaves lanceolate to ovate,* 2–6 *cm. broad,* rounded or narrowed at base, *green and glabrous or only sparingly appressed-pubescent, but scarcely papillose above.* (*T. canadense* of ed. 7) — Rich low woods and thickets, s. Que. and w. N.E. to Minn. and Neb., s. to Ga., Ala., Miss. and Tex. — Corolla white in forma **albiflòrum** House (white-flowered).

2. T. occidentàle Gray (western). — *Stem villous,* 3–9 dm. high; *leaves lance- to ovate-oblong,* white-villous beneath; *calyx and bracts villous with viscid hairs and with shorter capitate or stipitate glands;* the upper calyx-lobes acute or the middle one acuminate; corolla 8–12 mm. long. — Alluvial soil, Me. to B.C., s. to Pa. (adv.?), O., Ind., Ill., Mo., Kans., N.M. and Calif. July, Aug.

Var. **boreàle** (Bickn.) Fern. (northern). — Stem more closely pubescent; calyx and bracts with few or no capitate glands amongst the often viscid hairs; corolla slightly longer. (*T. boreale* Bickn.) — Similar situations, s. Que. and Me. to Wash., s. to s. N.E., N.Y., O., Ind., Ill. and Neb. July, Aug.

3. T. Scorodònia L. (old generic name for this plant), Wood-Sage, Germander-Sage. — Stems ascending from a freely creeping rootstock, villous, 2–5 dm. high, simple or with few erect branches; leaves deltoid-lanceolate to -ovate, crenate, rugose, cordate or truncate at base, short-petioled; flowers pale yellow, paired in slender 1-sided racemes; *upper tooth of the glabrous or glabrate calyx large and recurved.* — Locally estab. in Ont. and O. (Adv. from Eu.)

4. T. Bótrys L. (old generic name, meaning *a bunch of grapes*), Cut-leaf G. — Erect or decumbent annual 1–3 dm. high; *leaves* long-petioled, rhombic-ovate, 1–2 cm. long, *divided into* few linear or oblong *segments; flowers* slender-pedicelled, *in axillary verticils;* corolla reddish-purple; calyx gibbous-campanulate, the lower side saccate, the subequal deltoid teeth short. — Dry limy pastures, locally abundant, w. Mass. and Ct. to Ont. and O. June–Aug. (Natzd. from Eu.)

5. HÝPTIS Jacq.

Calyx ovoid or tubular, with 5 subequal teeth. Corolla declined, the upper lip with 4 entire lobes, the lower lip saccate, abruptly deflexed at the contracted and callous base. The 4 stamens declined or resting on the lower lip, didynamous. Disk entire or with a gland on the anterior side. Nutlets ovoid or oblong. — A very large genus chiefly of trop. Am., either herbaceous or shrubby, the flowers in ours in spikes of scattered capitate glomerules. (Name from the Greek *hyptios, resupinate,* in reference to the deflexed lower lip of the corolla.)

1. H. mutábilis (A. Richard) Briq. (changeable), var. **spicàta** (Poit.) Briq. (spicate). — Annual, branching, up to 2 m. high, the stem harsh with muricate hairs and finer villosity; leaves reniform-ovate, acuminate, unequally serrate; spikes of numerous small axillary glomerules subtended by foliaceous bracts; calyx tubular, in anthesis 1.5, in fruit to 6 mm. long;

corolla about 4 mm. long, lavender-blue; nutlets about 1.5 mm. long. — Waste places and roadsides, Fla. to se. Va. Sept., Oct. (Adv. from Trop. Am.)

6. SCUTELLÀRIA L. Skullcap

Calyx campanulate in flower, splitting to the base at maturity, the lips entire, the upper usually falling away. Corolla with an elongated curved ascending tube, dilated at the throat; the upper lip entire or barely notched, the lateral lobes mostly connected with the upper rather than the lower lip; the lower lobe or lip spreading and convex, notched at the apex. Stamens ascending under the upper lip; anthers approximate in pairs, ciliate or bearded, those of the lower stamens 1-locular (halved), of the upper 2-locular and cordate. — Bitter perennial herbs, not aromatic, the short peduncles or pedicels chiefly opposite, 1-flowered, often in 1-sided axillary racemes or terminal spikes or racemes; fl. in summer; the genus occurring in all but the colder reg. (Name from *scutella, a dish*, in allusion to the appendage of the fruiting calyx.)*

a. Stem arising from firm base, usually without stolons (in no. 1 with few superficial stolons tardily developed); racemes terminal or in terminal or leafy-bracted panicles, not 1-sided; corollas 1.4–3 cm. long. . . *b.*

b. Cauline leaves dentate or dentate-serrate, oblong- to round-ovate; bracts of racemes or panicles greatly reduced and unlike foliage-leaves. . . *c.*

c. Principal leaves cordate-ovate, with deep basal sinus, villous to glabrescent, on long slender petioles; lateral lobes of corolla almost equaling the short upper lip. . . . . 1. *S. ovata.*

c. Principal leaves not cordate or, if shallowly so, not villous; lateral lobes of corolla much shorter than the galeate upper lip. . . *d.*

d. Principal leaves rhombic-ovate, 3–6 cm. long, cuneate to short winged petioles, only 2–5 pairs below the inflorescence; corolla 1.4–2 cm. long. 2. *S. elliptica.*

d. Principal leaves ovate or oval, scarcely rhombic, 4–14 cm. long.

Stem glabrous (rarely pubescent), with only 4–6 pairs of thin (rarely thick and firm) foliage-leaves (not bracts), the upper blades largest; raceme simple, its lower internodes 1.5–8 cm. long; corolla 2–3.5 cm. long. . . . . 3. *S. serrata.*

Stem pubescent, with 8–15 or more pairs of firm leaves below the panicle of racemes, the median blades largest; floriferous nodes of racemes mostly less than 1 cm. apart; corolla 1.5–2.5 cm. long. . 4. *S. incana.*

b. Cauline leaves (or all but lower ones in no. 5) entire or nearly so; the median and upper linear, narrowly oblong, lanceolate or oblanceolate ones passing gradually into the lower elongate bracts.

Leaves of lowest nodes slender-petioled, with usually dentate blades much broader than the middle and upper ones; flowers in axils of upper rapidly reduced leaves or in terminal simple or panicled racemes with middle and upper bracts much shorter than flowers; wide-ranging species. 5. *S. integrifolia.*

Leaves all narrowly oblanceolate, entire, subsessile or very short-petioled; flowers in axils of only gradually reduced nearly uniform leaves, the middle and upper exceeded or about equaled by the bracts; endemic in Mo. . . . . 6. *S. Bushii.*

a. Stem arising from creeping rhizomes derived from subterranean stolons; flowers or often 1-sided racemes from axils of well-developed leaves or terminal; corollas 0.5–2.5 cm. long. . . *e.*

e. Stolons filiform or but slightly thickened, not tuberiferous; larger leaves regularly toothed, 1–12 cm. long. . . *f.*

f. Stem pubescent at least on the angles; calyx uniformly puberulent or pilose; nutlets wingless, uniformly pebbled or tuberculate, on obscure thick gynobase. . . *g.*

g. Pubescence of stem (angles), petioles and lower leaf-surfaces (when present) incurved-ascending; principal leaves ovate, on slender petioles 0.2–3.5 cm. long. . . *h.*

h. Lower internodes of principal racemes 1–3.5 cm. long; corolla 1–2 cm. long.

Principal leaves cordate-ovate, obtuse or short-tipped, on petioles 1–3.5 cm. long; lateral lobes of corolla almost equaling short upper lip. . . . . 7. *S. saxatilis.*

Principal leaves oblong-ovate, acuminate, rounded or subcordate

---

* Some characters derived from studies of C. W. Penland in Rhodora, xxvi. 61–79, 2 pl. (1924), E. C. Leonard in Contrib. U. S. Nat. Herb. xxii, pt. 10 (1927), and Carl Epling in Univ. Calif. Pub. Bot. xx., no. 1, 3 pl. (1942).

at base, on petioles 2–9 mm. long; lateral lobes of corolla much shorter than the arched upper lip. . . . . 8. *S. Churchilliana.*

*h.*Lower internodes of the chiefly lateral racemes 2–10 cm. long; principal leaves ovate, acuminate, rounded or subcordate at base, on petioles 0.5–3 cm. long; corolla 5–9 mm. long. . . . . . 9. *S. lateriflora.*

*g.*Pubescence of stem, petioles and lower leaf-surfaces recurving; leaves subsessile or with thick petioles at most 4 mm. long, mostly oblong; corolla 1.5–2.5 cm. long. . . . . . 10. *S. epilobiifolia.*

*f.*Stem glabrous; calyx glabrous except for hirtellous nerves; leaves glabrous, ovate, subsessile or short-petioled; corolla 0.8–1.1 cm. long; nutlets with broad disk-like wing, on a slender gynobase. . . . 11. *S. nervosa.*

*e.*Stolons producing thick moniliform tubers; leaves entire or remotely toothed, 0.5–2 cm. long. . . . . 12. *S. parvula.*

1. S. ovàta Hill (ovate). — Stem simple or forking, rarely (late in season) with superficial stolons, from erect and 1.5–8.5 dm. high to depressed and low (down to 0.5 dm.), the middle and upper *internodes villous to pilose-hirtellous*, with or without some gland-tipped hairs; *leaves cordate-ovate, with deep basal sinus*, more or less pubescent to glabrescent, *on long slender petioles*, the blades 1–13 cm. long, crenate-serrate; racemes terminal, simple, forking or in panicles, the rachis pilose to villous or hirsute; bracts of raceme ovate, the lower merging into foliage-leaves; *calyx glandular*, in anthesis 3–4 mm., in fruit 6–8 mm. long; *corolla* blue with whitish lower lip, 1–2.3 cm. long, the slender tube flaring to 5 mm. thick at the throat, *the lateral lobes almost equaling the short upper lip; nutlets* buff to orange, covered *with conical papillae.* — Highly variable but only 3 vars. sufficiently definite for recognition:

Stem erect, 1.5–8.5 dm. high; leaves green, the larger blades 4–13 cm. long and 2.5–11 cm. broad; corolla 1–2.3 cm. long.

Leaves narrowly ovate, thick and firm, heavily pubescent, the larger ones 4–7.5 cm. long and 2.5–6 cm. broad, their petioles 1–5 cm. long; bracts 0.8–2 cm. long, drying dark, the upper overtopping calyx. . . . . . . *S. ovata* (typical).

Leaves broad-ovate, thinnish to thickish, pilose to glabrescent, the larger ones 6–13 cm. long and 5–11 cm. broad, their petioles 2.5–8 cm. long; bracts much shorter, all but the lowest shorter than calyx, greenish. . . . . . Var. *versicolor.*

Stem depressed or sprawling, weak, 0.5–3 dm. long; leaves often purplish, the round-tipped blades 1–4 cm. long and 0.8–3 cm. broad; bracts like a continuation of the foliage-leaves; corolla about 1 cm. long. . . . . . Var. *rugosa.*

S. ovàta (typical), var. *bracteata* Blake; *S. versicolor*, var. *bracteata* Benth. — Rocky woods, bluffs, creek-banks and limestone-barrens, La. to Tex. and e. Mex., n. to Tenn., s. Ill., Mo. and Kans. June, July. (Selected small individuals have been called subsp. *mississippiensis* Epl.) — Passing freely into

Var. **versícolor** (Nutt.) Fern. (differently colored), *S. versicolor* Nutt. — Calcareous soils, wooded slopes, river-thickets (where large), limestone-barrens (where smaller- and thicker-leaved), etc., S.C. to La., n. to Md., W.Va., O., Ind., Wisc. and Minn. May–July (–Aug.) Fig. 1477. — Minor variations include subspp. *virginiana*, *calcarea* and *pseudovenosa* Epl.

Var. **rugòsa** (Wood) Fern. (wrinkled). — Rocky bluffs and shores, local, W.Va. and w. Va. to Mo. and Ark. (Incl. subspp. *rugosa* (Wood) Epl. and *rupestris* Epl.)

1477. S. ovata, v. versicolor.

2. S. ellíptica Muhl. (elliptic), Hairy S. — *Stems* 1–few from a firm base, 1.5–7 dm. high, *pilose with short soft curving hairs, the longest internodes* 3.5–15 *cm. long; only* 2–5 *pairs of leaves or nodes below the inflorescence;* the lower slender-petioled blades ovate to deltoid, smaller than those above; *the median and upper blades* more *rhombic-oval*, dentate, *blunt, cuneate to short petioles*, 3–6 *cm. long* and 2–3.5 cm. broad, sparsely strigose-hirsute above, very minutely pilose along veins beneath; racemes terminal, 6–20-flowered, or 3–5(–7) in a panicle, the rachis curling-pilose; bracts of all but the lowest flowers spatulate to oblong-obovate, entire, exceeding calyx; lowest flowers 1–2 cm. apart; calyx 2.5–3, in fruit 6–8 mm. long, glandular-hirsute; *corolla* 1.4–2 *cm. long*, pubescent, the tube pale, the unequal lips blue-violet; nutlets subglobose, dark brown, about 2 mm. in diameter, with many concentric rings of rosulate slightly flattened papillae. (*S. ovalifolia* Pers.; *S. pilosa* Michx.) — Dry woods and thickets, nw. Fla. to e. Tex., n. to se. N.Y., n. N.J., Pa., W.Va., s. O. and Mo. Late May–early July.

Var. **hirsùta** (Short) Fern. (hirsute). — *Stem with many straight divergent hairs* 1–2 *mm.*

long; median and upper leaves more deltoid-ovate. — Sw. Pa. to s. Mich., s. to se. Va., n. Ga., s. Ala., s. Miss. and se. Tex.

3. S. serràta Andr. (saw-toothed). — Habitally suggesting no. 2; *stems* 2–7 dm. high, *slender, glabrous or sparsely pubescent* toward summit; *leaves 4–6 pairs, thin, ovate or ovate-oblong,* acuminate, serrate or serrate-dentate, *narrowed to short petioles, the upper blades largest, glabrous* or with few scattered hairs, *the larger blades* 6–12 *cm. long* and 2.5–7 cm. broad; *raceme* 1, rarely 3, terminal, *open, its lower internodes* 1.5–8 *cm. long;* bracts lanceolate to lance-oblong, the lower longer, the upper shorter than flowers; calyx puberulent; *corolla* 2–3 *cm. long, abruptly upcurved from base,* blue, the throat about 1 cm. in diameter, the lips subequal; nutlets about 2 mm. in diameter, covered with blunt-pointed papillae. — Rich woods and bluffs, se. N.Y. and w. N.J. to W.Va., s. to S.C. and Tenn. May, June.

Var. **montàna** (Chapm.) Penland (of the mountains). — *Stem and leaves glandular-pubescent; corolla* 2.5–3.5 *cm. long.* (*S. montana* Chapm.) — Local, Ga., Tenn., Ala. and Mo.

4. S. incàna Biehler (hoary). — *Stem* 0.3–1.2 m. high, simple to *paniculate-branched above, canescent-pilose* (except toward base); *leaves* 8–15 *or more pairs,* ovate or ovate-oblong, *slender-petioled, the median blades largest,* 3–7 *cm. broad,* lowest rounded to shallowly cordate at base, the upper rarely cordate, crenate, *whitish with minute pubescence beneath,* green and glabrous or but sparsely pubescent above; *racemes panicled, their internodes mostly less than* 1 *cm. long;* all but lowest bracts linear to lanceolate and short; calyx densely pubescent; *corolla* canescent, blue, 1.5–2.5 *cm. long, the tube enlarging to* 7 *mm. in diameter at the throat;* nutlets 1.5 mm. in diameter, covered with truncate papillae. (*S. canescens* Nutt.) — Dry woods, thickets and clearings, n. N.J. to sw. N.Y., w. to Ia., s. to N.C., s. Ala., Ark. and e. Kans. June–Sept.

Var. **punctàta** (Chapm.) Mohr (dotted). — Less pubescent; *leaves oblong-ovate,* the larger ones 2.5–6.5 cm. broad, *glabrous beneath* except along nerves. (*S. punctata* (Chapm.) Leonard) — Va. to Mo., s. to nw. Fla. and Ala. — Probably scarcely more than a glabrous or glabrescent form.

5. S. integrifòlia L. (entire-leaved). — *Stems* rather slender, 1–several from subligneous base, simple or with arched-ascending branches above, *often with abbreviated axillary branches,* 3–7 dm. high, *densely incurved-pilose,* with or without some glands; *leaves* 3–8 pairs below inflorescence; *the lowest ones ovate, slender-petioled and crenate; the successively narrower middle and upper ones mostly oblong to linear-lanceolate and entire,* narrowed to short petioles, punctate, *minutely hoary,* 2–4 *cm. long,* 0.4–1.2 *cm. broad;* inflorescence a terminal raceme or leafy elongate panicle of 6–40-flowered racemes; all but lowest flowers in axils of much reduced foliaceous linear-lanceolate bracts, the lowest internodes 1–2.5 cm. long; pedicels short; calyx 2.5–3.5, becoming 5–7 mm. long, puberulent; corolla 2–2.5 cm. long, purple-blue and whitish, or in forma **rhodántha** Fern. (rose-flowered) rose-pink, the large lips subequal; *nutlets* subglobose, blackish, 1.2–1.6 *mm. in diameter,* covered with rosulate flattish papillae. — Borders of woods, thickets and clearings, n. Fla. and Ala., n. to se. Mass., centr. Ct., se. N.Y., n. N.J., s. Pa., W.Va., s. O., Ky. and Mo. Late May–July.

Var. **híspida** Benth. (hispid). — Stem with numerous straighter and divergent hairs; transition from lower to median leaves more gradual. — Fla. to e. Tex., n. to se. Va., Tenn. and Ark.

Var. **multiglandulòsa** Kearney (with many glands). — Stems 0.7–3.5 dm. high, simple or with few erect branches; lower internodes closely covered with capitate glands and short spreading or retrorse hispidity, upper internodes with divergent capitate glands mixed with pilosity; leaves nearly uniform, subelliptic, 1–4.5 cm. long, about a third as broad; calyx glandular. (*S. multiglandulosa* (Kearney) Small) — Interior Ga., n. to peaty shores and bogs of se. Va.

6. S. Búshii Britt. (for its discoverer, Benjamin Franklin Bush, 1858–1937). — Suggesting no. 5; the numerous closely puberulent-pilose stems 1.5–4 dm. high, from a strong ligneous caudex; *leaves uniform, oblanceolate, entire, subsessile or very short-petioled; flowers in axils of only gradually reduced leaves, the middle and upper ones exceeded or about* equaled by the bracts. — Glades and calcareous knobs, s. Mo. May, June.

7. S. saxátilis Riddell (of rocks). — Spreading by filiform subterranean stolons and rhizomes; *stem weak,* ascending, simple to bushy-branched, slender, 1–5 dm. long, *nearly glabrous except for scattered upcurved pilosity on angles below; leaves ovate* to ovate-lanceolate or deltoid, *cordate-based, coarsely crenate, obtuse,* thin, sparsely pubescent to glabrous; the larger blades 4–5 cm. long, *on slender petioles* 1–3.5 *cm. long;* racemes single or loosely panicled, lax and open, often 1-sided, with narrowly ovate to lanceolate bracts; *lower internodes of racemes* 1–3.5 *cm. long; calyx glandular-pilose; corolla* 1.2–1.9 cm. long, light blue-violet and white, *the nearly glabrous lateral lobes almost equaling the smooth upper lip;* nutlets 1 mm. in diameter, tuberculate. — Rocky woods, talus-slopes and bluffs, n. Del. to sw. Pa. and sw. to s. Ind., s. to w. S.C. and Tenn. June–Sept.

Var. **pilòsior** Benth. (more hairy). — *Stems and petioles more pilose; leaves acutish, with sharper teeth.* (Var. *arguta* Gray; *S. arguta* Buckl.) — Among the mts., w. Va., w. N.C., e. Tenn. and nw. Ga.

8. S. Churchilliàna Fern. (for its discoverer, JOSEPH RICHMOND CHURCHILL, 1845–1933). — Slenderly stoloniferous; *stem* ascending, simple or forking above, 2–6 dm. high, *the angles minutely pilose with incurved-ascending hairs; leaves lance- or oblong-ovate, acuminate, rounded or subcordate at base,* coarsely crenate or dentate, thin, minutely and sparsely pilose on nerves beneath, 2–8 cm. long, 0.7–3.5 *cm. broad, on slender upwardly pilose petioles* 2–9 *mm. long;* flowers axillary or 2–20 in axillary remotely flowered 1-sided leafy-bracted racemes, the *internodes of lateral racemes* 1–2.5 *cm. long; calyx in anthesis* 2.5–4 *mm. long,* sparingly pubescent; *corolla* blue-violet, pilose, 1–2 *cm. long.* — Sandy or gravelly shores or alluvial thickets, Rimouski and Kamouraska Cos., Que., s. to w. N.B. and s. Me. July–Sept. PLATE III, FIG. 1.

9. S. laterìflòra L. (lateral-flowered), MAD-DOG S. — Perennial by filiform rhizomes and stolons; *stem* 0.1–1 m. high, simple or freely branched, smooth, or *minutely pilose on angles with incurved-ascending hairs; leaves ovate,* acuminate, coarsely serrate or serrate-dentate, rounded to subcordate at base, thin, glabrous or at most minutely pilose on veins beneath and margin, 1.8–12 cm. long, 0.6–6 *cm. broad, on upwardly pilose slender petioles* 0.5–3 *cm. long; racemes 1-sided, axillary* and terminal (occasionally reduced to 1 flower), usually with 2–44 flowers in the axils of progressively smaller leafy bracts, *internodes of lateral racemes mostly* 2–7 (rarely –10) *mm. long; calyx* 2–2.7, *becoming in fruit* 3–4 *mm. long,* puberulent; *corolla* slender, blue-violet, or in forma **rhodântha** Fern. (rose-flowered) pink, or in forma **albiflòra** (Farw.) Fern. (white-flowered) white, 5–9 *mm. long,* pilose; nutlets pale, 1.4–1.7 mm. high, obtusely pebbled. — Alluvial thickets, meadows, swampy woods, etc., Ga. to s. Calif., n. to Nfld., Anticosti I., Que., n. Ont., Man., Sask., s. Alta. and s. B.C. June–Sept. PLATE III, FIG. 2.

10. S. epilobiifòlia A. Hamilton (with leaves of Willow-herb), COMMON S. — Perennial with slender creeping rhizome and mostly subterranean slightly thickened stolons; *stem* ascending, simple or forking, 0.1–1 m. high, *minutely pilose on angles above with recurved hairs; leaves oblong-lanceolate to oblong-ovate,* crenate, rounded to cordate at base, 1–8 cm. long, 0.3–3 cm. broad, often somewhat rugose, veiny, *minutely pubescent* (often canescent) *beneath with recurving hairs, sessile or on a thick petiole at most* 4 *mm. long;* flowers axillary or in the axils of reduced leaf-like bracts and forming 1-sided interrupted racemes, very short-pedicelled; *calyx* 3.5–4.5, *in fruit* 5–6.5 *mm. long,* minutely pilose; *corolla* blue-violet with whitish throat and tube, pilose, 1.5–2.5 *cm. long,* or pink in forma **ròsea** (Rand & Redf.) Fern. (roseate), or white in forma **albiflòra** (Millsp.) Fern. (white-flowered); *nutlets* pale olive, 1.5–2 mm. long, *with broad low pebbling.* (*S. galericulata* of Am. auth., not L.) — Gravelly, sandy or rocky shores, meadows, swampy thickets, etc., Nfld. and s. Lab. Pen. to James Bay, thence nw. to Alaska, s. to N.S., N.E., L.I., Del., s. Pa., W.Va., O., Ind., Ill., Mo., Kans., N.M., Ariz. and Calif. June–Sept. PLATE III, FIG. 3. — Differing consistently from the Eurasian *S. galericulata* in the low pebbling of the nutlets; consisting of several geographic vars. yet to be worked out.

11. S. nervòsa Pursh (prominently nerved). — Rhizomes and stolons filiform; *stem* slender, erect or ascending, *weak,* simple or slightly forking below, acutely angled, *glabrous,* 2–5 dm. long; lowest *leaves* rounded, cordate; the others *ovate, obtuse, remotely crenate-dentate, subsessile* or very short-petioled, *very thin,* when young copiously strigose on upper surface, *with nerve-like veins prominent beneath,* 2–5 *cm. long,* 1–3.25 cm. broad; *flowers solitary in the axils,* on slender hirtellous pedicels 2–7 mm. long; *calyx* 2.5–3.5, in fruit 5–6 *mm. long, hirtellous on the veins; corolla* pale bluish, 0.8–1.1 *cm. long; nutlets* yellowish or buff, *on a slender gynobase, the muricate-papillate summit surrounded by a broad and thin disk.* — Moist woods and thickets, e. Pa. to centr. O., s. Ind. and se. Ill., s. in upland to w. N.C. and e. Tenn.; also nw. La. May–early July. FIG. 1478.

1478. S. nervosa.

Var. **calvifòlia** Fern. (with hairless leaves). — Similar, the leaves glabrous above. — No. N.J. to s. Ont., s. at low alts. to e. Va., s. O., s. Ind. and w. Tenn.

Var. **ambígua** (Nutt.) Fern. (doubtful). — Much smaller and stiffer; leaves thick and hard, the longer strigose cauline ones entire or barely dentate, narrowly ovate, 1.5–2.5 cm. long, 8–10 mm. broad. (*S. ambigua* Nutt.) — Dry rocky woods, local, D.C. to s. O., Ky. and Tenn.

12. S. pârvula Michx. (very small). — Plant strongly stoloniferous and *producing moniliform tubers* with fusiform to ellipsoid white segments 0.5–2 cm. long and 2–14 in a chain; stem simple or more commonly forking from base, slender, 0.8–3 dm. high, pubescent to glabrate; *leaves* ovate to ovate-oblong, *entire or sparingly toothed;* the median and upper ones sessile or subsessile, 0.5–2 *cm. long,* 0.3–1.4 cm. broad, pilose to glabrate; flowers axillary, the pedicels about equaling the calyx; calyx pubescent, 2–3, in fruit 4–5 mm. long; corolla slender, bluish,

Plate III

Fig. 1, Scutellaria Churchilliana; fig. 2, S. lateriflora; fig. 3, S. epilobiifolia.

0.5–1 cm. long; *nutlet brown, about 1 mm. broad, with a smooth marginal band surmounted by a rosulate crown of finger-like processes or pebbling, usually with a ring of drooping processes between the marginal band and the gynobase.* — Three confluent vars.:

Stem pilose, at least along the angles; lowest leaves rounded, on petioles 0.5–1.5 cm. long; median and upper leaves ovate to ovate-oblong, 0.3–1.4 cm. broad, pubescent beneath, flat; nutlets surmounted by a crown of finger-like processes.
Stem evenly pilose with divergent to recurved hairs intermixed with longer glands; lower leaf-surfaces with sessile glands among the long trichomes; pedicels retrorse-hirtellous. . . . . *S. parvula* (typical).
Stem pilose on angles with appressed-ascending hairs usually intermixed with glands; lower leaf-surfaces with long glandular hairs; pedicels ascending-pubescent. . . . . Var. *australis*.
Stem minutely puberulent or glabrate; lowest leaves on petioles 3–11 mm. long; median and upper leaves 0.2–1 cm. broad, glabrous or minutely puberulent, their margins revolute; nutlets more often with papillae reduced to short rounded pebbling. . . . . Var. *Leonardi*.

S. pårvula (typical). — Dry or moist sandy or gravelly, chiefly calcareous soils, sw. Que. to Wisc. and Ia., s. to Ga., Ala., Miss., Ark. and e. Tex. May–July.

Var. **austràlis** Fassett (southern), *S. australis* (Fassett) Epling. — N. Fla. to e. Tex., n. to s. Ct., L.I., D.C., W.Va., Ind., Ill., Mo. and Kans.

Var. **Leónardi** (Epling) Fern. (for EMERY CLARENCE LEONARD, 1892–), *S. parvula*, var. *ambigua* sensu Fern. and *S. ambigua* sensu Leonard, not Nutt. — Centr. Me. to N.D., s. to se. Va., w. N.C., Tenn., La. and e. Tex.

7. MARRÙBIUM L. HOREHOUND

Calyx-teeth more or less spiny-pointed and spreading at maturity. Upper lip of the corolla erect, notched; the lower spreading, 3-cleft, its middle lobe broadest. Stamens 4. — Whitish-woolly bitter-aromatic perennials, nat. of Old World, branched at the base, with rugose and crenate or cut leaves, and many-flowered axillary whorls. (A name used by Pliny, from the Hebrew *marrob*, a bitter juice.)

1479. M. vulgare.

1. M. VULGÀRE L. (common), COMMON H. — Stems ascending; leaves round-ovate, petioled, crenate-toothed; whorls capitate; calyx with 10 recurved teeth, the alternate ones shorter; corolla small, white. — Waste places, freq. southw., locally n. to N.E., N.Y., s. Ont., etc. May–Sept. (Natzd. from Eu.) FIG. 1479.

8. AGÁSTACHE Clayt. GIANT HYSSOP

Calyx tubular-campanulate, 15-nerved, oblique, 5-toothed, the upper teeth rather longer than the others. Upper lip of corolla nearly erect, 2-lobed; the lower 3-cleft, with the middle lobe crenate. Stamens 4, exserted; the upper pair declined, the lower and shorter pair ascending, so that the pairs cross; anther-locules nearly parallel. — N.Am. and e. Asiatic perennial tall herbs, with petioled serrate leaves, and small flowers crowded in interrupted terminal spikes in summer. (From the Greek *agan, much* and *stachys, an ear of corn* (*wheat*), in reference to the numerous spikes.)

Leaves glabrous or essentially so; calyx-teeth ovate, blunt; corolla yellowish. . 1. *A. nepetoides*.
Leaves finely pubescent beneath; calyx-teeth lanceolate, acute; corolla purple or bluish.
Stem grayish-pubescent; leaves not white beneath; corolla purple. . . . . 2. *A scrophulariaefolia*.
Stem glabrous; leaves white beneath; corolla blue. . . . . 3. *A. Foeniculum*.

1. A. nepetoìdes (L.) Ktze. (resembling Catnip, *Nepeta*), YELLOW G. — Stem stout, 0.7–1.5 m. high, sharply 4-angled, *smooth* or nearly so; leaves ovate, somewhat pointed, coarsely crenate-toothed, 5–12 cm. long; spikes 3–12 cm. long, crowded with the ovate pointed bracts; *calyx-teeth ovate, rather obtuse, little shorter than the pale greenish-yellow corolla.* — Rich thickets and borders of woods, sw. Que. to w. Ont. and S.D., s. to Ga., Ky., Mo. and Kans. July–Sept.

2. A. scrophulariaefòlia (Willd.) Ktze. (with leaves of Figwort, *Scrophularia*), PURPLE G. — Stem (obtusely 4-angled) and lower surface of the ovate or somewhat cordate acute leaves slightly *pubescent;* spikes 0.5–5 dm. long; *calyx-teeth lanceolate, acute, shorter than the purplish corolla;* otherwise like the preceding. — Rich woods and thickets, Vt. and e. Mass. to w. Ont., s. to N.C., Ky., Mo. and Kans. July–Sept. FIG. 1480.

1480. A. scrophulariaefolia.

Var. **móllis** (Fern.) Heller (soft). — Stems and lower leaf-surfaces densely villous. — Local, Vt. to Minn., s. to Ct., N.Y., O. and Mo.

3. A. Foenículum (Pursh) Ktze. (from its odor resembling that of Fennel), Blue G. — Smooth, but the ovate acute *leaves glaucous-white underneath* with minute down; *calyx-teeth lanceolate, acute; corolla blue.* (*A. anethiodora* Britt., based on the unidentifiable *Hyssopus anethiodorus* Nutt.) — Dry thickets, plains and barrens, Ont. to Mackenz., s. to Ill., Ia., S.D., Colo. and Wash.; adv. eastw. in fields to Que., N.E. and Del. June–Sept. — Foliage with the scent of anise.

9. MEEHÁNIA Britt.

Calyx rather obliquely 5-toothed, 15-nerved. Corolla ample, expanded at the throat; the upper lip flattish or concave, 2-lobed, the lower 3-cleft, the middle lobe largest. Stamens 4, ascending, the lower pair shorter; anther-locules parallel. — Low stoloniferous herb, with pale purplish flowers. (Named for *Thomas Meehan*, 1826–1901, Philadelphian botanist.)

1. M. cordàta (Nutt.) Britt. (heart-shaped). — Low, with slender runners, hairy; leaves broadly cordate, crenate, petioled, the floral shorter than the calyx; whorls few-flowered, at the summit of short ascending stems; corolla hairy inside, 2–3.5 cm. long; stamens shorter than the upper lip. (*Cedronella* Benth.) — Rich woods and shaded banks, w. Pa. to Ill., s. to N.C. and Tenn. June.

10. NÉPETA L. Catmint. Herbe à chats (Que.). Chataire (Que.)

Calyx tubular, often incurved. Corolla dilated in the throat; the upper lip erect, rather concave, notched or 2-cleft; the lower 3-cleft; the middle lobe largest, either 2-lobed or entire; locules of the anthers merely divergent. — Perennial herbs with flowers crowded toward summits of stems and branches, nat. of Old World. (The Latin name, thought to be derived from *Nepete*, an Etruscan city.)

1. N. Catària L. (old generic name), Catnip. — Erect perennial, 0.3–1 m. high; *stem whitish-downy; principal leaves ovate* to ovate-oblong, long-petioled, *minutely white-pubescent beneath, coarsely dentate; calyx* urceolate, *very pubescent;* corolla whitish, dotted with purple. — Weed of dooryards, roadsides and waste places, throughout our range and beyond. June–Sept. (Natzd. from Eu.)

2. N. grandiflòra Bieb. (large-flowered). — Similar to no. 1, *glabrate or glabrous; principal leaves oblong, finely crenate;* calyx more slender. — Roadsides, local, s. Que. and e. N.Y. July–Sept. (Natzd. from the Caucasus)

11. GLECHÒMA L. Ground-Ivy. Lierre terrestre (Que.)

Differing from *Nepeta* in having the flowers in the axils of regular foliage-leaves; the 4 locules of the 2 anthers of the perfect flowers at right angles, thus forming a cross. — A small genus, nat. of Old World, with creeping and trailing herbaceous stems and petioled suborbicular to ovate leaves. (Old Greek name *glechon*, for pennyroyal.)

1481. G. hederacea, v. micrantha.

1. G. hederàcea L. (ivy-like), Gill-over-the-ground, Run-away-Robin. — Extensively creeping; stems and leaves smoothish to minutely pubescent; leaves petioled, round-reniform, crenate, or in forma acutíloba Neuman (acute-lobed) sharp-toothed, green or purplish on one or both sides; corolla 1.6–2.2 cm. long, purplish-blue. (*Nepeta* Trevisan; *N. Glechoma* Benth.) — Roadsides, yards and damp shaded places, Nfld. to Ont., s., somewhat locally, to N.S., N.E., Va., Tenn. and Mo. April–July (Natzd. from Eu.)

Var. micrántha Moricand (small-flowered). — Corolla 1–1.5 cm. long. (*Nepeta hed.*, var. *parviflora* Druce) — Usually more common, Nfld. to Ont., s. to N.S., N.E., Ga., Ala., Mo. and Kans. (Natzd. from Eu.) Fig. 1481.

12. DRACOCÉPHALUM L. Dragonhead

Calyx tubular, 13–15-nerved, 5-toothed. Upper lip of the corolla slightly arched and notched; the lower 3-cleft, with its middle lobe largest and 2-cleft or notched at the end. — Whorls few-many-flowered in leafy or bracted terminal heads or interrupted spikes; plants of N. Hemisph.

(Name from the Greek *dracon, a dragon,* and *cephale, head,* alluding to the form of the corolla in the original species.) MOLDAVICA Adans.

1. D. parviflòrum Nutt. (small-flowered). — Annual or biennial, with erect stems 1–8 dm. high, solitary or clustered at base, simple or branching; *leaves* lanceolate or lance-ovate, *sharply cut-toothed,* petioled; *whorls of flowers crowded into a terminal head with spinescent-tipped and sharp-toothed imbricated bracts,* the light blue to violet *corolla only slightly exserted* from the calyx; upper tooth of the calyx ovate. (*Moldavica* Britt.) — Rocky or gravelly calcareous soils, often in recent clearings, e. Que. to Yuk., s. to n. and w. N.Y., Mich., Wisc., Ia., Neb., N.M. and Ariz.; adv. elsewhere eastw. to N.E. June–Aug. FIG. 1482.

1482. D. parviflorum.

2. D. MOLDÁVICA L. (old generic name of Tournefort; the plant first known from Moldavia). — *Leaves* oblong or narrowly triangular, obtuse, *with blunt dentation; whorls of flowers scattered in axils of upper prolonged leaves; corolla* 2.5 *cm. long, much exceeding calyx.* (*Moldavica* Britt.; *M. punctata* Moench) — Waste or cult. ground, local, N.E. to Wisc. and Neb. June–Oct. (Adv. from Eu.)

3. D. THYMIFLÒRUM L. (thyme-flowered). — Stem 0.1–4 dm. high, simple or nearly so, thinly pubescent; lower *leaves* triangular-ovate, long-petioled; *upper ones ovate-lanceolate to lance-oblong,* short-petioled; *whorls in prolonged interrupted spike; bracts narrow, subentire; calyx* 6–7 *mm. long, its teeth subequal in length,* the upper much broader than the remaining narrowly lanceolate ones; *corolla* purplish, *scarcely exceeding calyx.* — Fields, open woods and waste places, as yet local, Mass., Ont., etc. June, July. (Adv. from Eu.)

13. PRUNÉLLA L. SELFHEAL. HERBE AU CHARPENTIER (Que.)

Calyx tubular-campanulate, somewhat 10-nerved, naked in the throat, closed in fruit; upper lip broad, truncate. Corolla ascending, slightly contracted at the throat and dilated at the lower side just beneath it, 2-lipped; upper lip erect, arched, entire; the lower reflexed-spreading, 3-cleft, its lateral lobes oblong, the middle one rounded, concave, denticulate. Filaments 2-toothed at the apex, the lower tooth bearing the anther; anthers approximate in pairs, their locules diverging. — Low nearly cosmop. perennials, with nearly simple stems, and 3-flowered clusters of flowers sessile in the axils of round and bract-like membranaceous floral leaves, imbricated in a close spike or head. (Name of uncertain and much disputed origin, often written *Brunella,* which was a pre-Linnaean form.)

1. P. vulgàris L. (common), HEAL-ALL, CARPENTER-WEED. — Stems mostly tufted or loosely ascending from leafy-tufted bases, 0.5–6 dm. high, simple or branched; bracts of inflorescence mostly bristly-ciliate in our vars. — A widely dispersed species of the N. Hemisph., represented with us by the following vars. and forms:

a. Principal or median cauline leaves ovate to ovate-oblong, rounded at base, two-fifths to two-thirds (averaging one-half) as broad as long.
  Stems and leaves only sparingly and not conspicuously pilose.
    Corolla bluish, violet or lavender. . . . . *P. vulgaris* (typical).
    Corolla white. Forma *albiflora.*
  Stems, petioles and often the lower surfaces of leaves densely villous-hispid with white pubescence. . . . . Var. *hispida.*
a. Principal or median cauline leaves lanceolate to oblong, gradually narrowed or cuneate at base, one-fifth to one-half (av. one-third) as broad as long. . . *b.*
b. Ribs and margins of calyx-lobes ciliate with long spreading villi.
  Calyx green or at most with purple-tinged margins.
    Corolla bluish, violet or lavender. . . . . Var. *lanceolata.*
    Corolla white. . . . . Forma *candida.*
  Calyx purple.
    Corolla bluish or violet. . . . . Forma *iodocalyx.*
    Corolla pink. . . . . Forma *rhodantha.*
b. Ribs and calyx-lobes glabrous or only very short-ciliolate. . . . . Var. *Rouleauiana.*

P. VULGÀRIS (typical). — Grasslands, roadsides, waste ground, etc., Nfld. to B.C., s. beyond our limits. May–Sept. Forma ALBIFLÒRA (Bogenhard) Britt. (white-flowered), local. (Natzd. from Eurasia) — When repeatedly mowed, cropped or trampled the plants become densely matted, depressed and small-leaved.

Var. HÍSPIDA Benth. (hispid). — Roadsides, clearings, groves, etc., N.C. and Ky. to Fla. and La. (Natzd. from Eurasia)

Var. **lanceolàta** (Bart.) Fern. (lanceolate), var. *elongata* Benth. — Open woods, thickets, roadsides, etc., Nfld. to Alaska, s. to N.S., N.E., L.I., N.C., Tenn., Mo., Kans., N.M., Ariz. and Calif. May–Oct. (E. Asia) — Forma **cándida** Fern. (white), occasional; forma **iodócalyx** Fern. (with purple calyx), frequent; forma **rhodántha** Fern. (rose-flowered), local.

Var. **Rouleauiàna** Vict. (in honor of ERNEST ROULEAU, 1916–). — Plant glabrous; ribs and lobes of calyx glabrous, glabrescent or merely with very short cilia; otherwise like preceding var. — Fresh tidal shores (estuary) of lower St. Lawrence R., Que.

14. PHYSOSTÉGIA Benth. FALSE DRAGONHEAD

Calyx obscurely 10-nerved, short-tubular or campanulate, more or less enlarged and slightly inflated in fruit. Corolla funnelform, with a much inflated throat, 2-lipped; upper lip erect, nearly entire; the lower 3-parted, spreading, small, its middle lobe larger, broad and rounded, notched. — Smooth N. Am. perennials, with upright virgate stems, and sessile linear, lanceolate or oblong mostly toothed leaves. Flowers large and showy, rose or flesh-color variegated with purple, opposite, scattered or crowded in simple or panicled terminal leafless spikes. (Name from the Greek *physa*, *a bladder*, and *stege*, *a covering;* in allusion to the calyx, which is at length somewhat inflated.)

*a.*Leaves (especially the median and upper) with sharp-tipped or acute serration. . *b.*

*b.*Principal leaves lanceolate to oblanceolate or oblong, except in dwarf colonies 1–4 cm. broad, scarcely rigid; spikes mostly 0.3–2 dm. long, with the pairs of calyces subapproximate or overlapping.

Corolla 2–3 cm. long; leaves 2–4 cm. broad and not greatly reduced below the inflorescence or, if only 1–2 cm. broad, the upper usually much reduced; terminal spikes of larger plants becoming 1–2 dm. long; eastern and midwestern. . . . . 1. *P. virginiana.*

Corolla 0.8–1.5 cm. long; larger leaves 1–2 cm. broad, not much reduced below inflorescence; terminal spike 3–15 cm. long; northwestern. . . 2. *P. parviflora.*

*b.*Principal leaves linear or narrowly lanceolate, rigid, 3–10 mm. wide, greatly reduced upward; spike very slender, 1–3.5 dm. long, remotely flowered, the lower calyces or empty bracts becoming 1–5 cm. apart; corolla 2.5–3 cm. long. . . . . 3. *P. angustifolia.*

*a.*Leaves with undulate or dentate margins. . *c.*

*c.*Leaves membranaceous, the lower 4–6 pairs on slender elongate petioles; upper sessile leaves greatly reduced; the slender interrupted spike appearing long-peduncled, solitary or with strictly erect secondary spikes, in well-developed plants 1–4 dm. long. . . . . 4. *P. denticulata.*

*c.*Leaves firm, sessile or only the very lowest petioled. . *d.*

*d.*Larger leaves 1–3 cm. broad; lateral branches of panicle, if developed, arched-ascending; corolla 2–3.3 cm. long; southern or southeastern.

Principal leaves narrowly obovate to broadly spatulate-oblanceolate, with rounded summits, only the remote and greatly reduced upper ones attenuate; spikes remotely flowered, the pairs of calyces distant; corolla 2.5–3.3 cm. long. . . . . 5. *P. obovata.*

Principal leaves oblong, acute or subacute, not greatly reduced above; spikes closely flowered, the fruiting calyces overlapping; corolla 2–2.5 cm. long. . . . . 6. *P. aboriginorum.*

*d.*Larger leaves 3–12 mm. broad, all but the lowest attenuate, rapidly reduced upward; lateral branches of inflorescence, if developed, erect; flowers remote; corolla 1–1.5 cm. long; southwestern. . . . 7. *P. intermedia.*

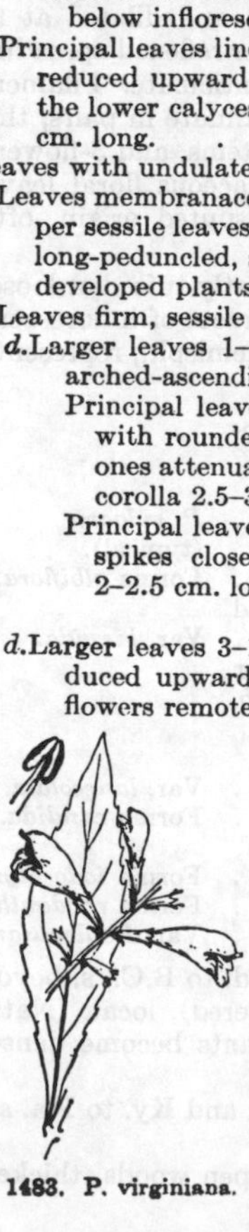

1483. P. virginiana.

1. **P. virginiàna** (L.) Benth. (Virginian). — Stoloniferous (stolons rarely collected), the smooth stem (0.2–)0.4–1.5 m. high, simple or paniculately branched at summit; *leaves submembranaceous to firm, lanceolate,* often copiously *sharp-serrate,* 4–12 cm. long, 1–2.5 *cm. wide, the uppermost much reduced;* the virgate spike (or panicled spikes) *rather closely flowered, in fruit with overlapping calyces,* the longer spikes becoming 0.6–2 dm. long; *calyx* tubular-campanulate, with sharp-tipped short triangular teeth, *viscid-glandular and usually minutely hirtellous-puberulent; corolla* purple or roseate, or rarely whitish in forma **cándida** Benke (white), 2–3 *cm. long.* (*Dracocephalum* L.) — River-banks, damp thickets and waste places, s. Que. to Minn., s. to N.B., N.E., Del., Md., upland to N.C., Tenn. and Mo., often cult. and mostly an escape eastw. June–Sept. FIG. 1483. — Passing to

Var. **granulòsa** (Fassett) Fern. (composed of grains). — Low, 2–6 dm. high; *leaves* 2–8 *cm. long,* 5–15 mm. wide, *with* 4–10(–17) pairs of *low and depressed teeth; calyx glandular-puberulent.* — Tidal shores of St. Lawrence R., Que., and Susquehanna R. and head of Chesapeake Bay, Pa. and Md.; fresh shores, Kennebec R., Me., and Lake Champlain, Vt. and N.Y.

Var. **speciòsa** (Sweet) Gray (showy). — The coarsest extreme, with sharply toothed *leaves* broadly lanceolate or oblanceolate to oblong and 2–4 *cm. broad, more gradually or sometimes scarcely reduced above; calyx with fewer or no glands.* (*P. speciosa* Sweet) — O. to N.D., s. to w. N.C., Tenn. and La.; spread from cult. e. to N.E. and Pa.

2. P. parviflòra Nutt. (small-flowered). — Similar to no. 1, 0.3–1 m. high; *leaves elliptic-lanceolate, tapering at base,* acuminate, firm (or membranaceous in deep shade), *the larger ones* 1–2 *cm. broad, not much reduced upward;* inflorescence simple or with erect branches, the *terminal spikes* 3–15 *cm. long,* dense; *corolla* 0.8–1.5 *cm. long.* (*Dracocephalum Nuttallii* Britt.) — Shores, damp thickets and swamps, Wisc. to n. B.C., s. to Ind., Ill., Mo., Kans., Colo. and Oreg. Late July–Sept.

3. P. angustifòlia Fern. (narrow-leaved). — Stem *stiffly erect,* 0.3–1 m. high; *leaves rigid, grayish or pale, linear or narrowly lanceolate, sessile,* very numerous (16–22 pairs), the lower obtuse or subacute; *the median ones attenuate* and *appressed-acerose-serrate,* 4–10 *cm. long,* 3–10 *mm. wide, the upper* leaves *strongly reduced; spikes slender,* virgate, *solitary or with few erect laterals,* 1–3.5 *dm. long, remotely flowered, the lower calyces or the often numerous lower bracts becoming* 1–5 *cm. apart; corolla* pale to deep purple or whitish, 2.5–3 *cm. long.* — Prairies, low grounds and thickets, Ill. to Tenn., Miss., La. and Tex. June–Sept.

4. P. denticulàta (Ait.) Britt. (with small teeth). — Strongly stoloniferous; the slender smooth stem 0.4–1 m. high; *leaves membranaceous, full green, the* 4–6 *lower pairs slender-petioled,* distant, *oblong or oblong-lanceolate,* blunt to acute, *undulate-margined or low-dentate; the median sessile leaves* narrowly oblong, lanceolate or oblanceolate, *acutish,* all but the reduced uppermost ones *narrowed at base and* 0.7–2 *cm. broad, dentate; spike* very slender and *interrupted, appearing long-peduncled,* in well developed plants *becoming* 1–4 *dm. long and flexuous, lateral spikes (if present) erect; corolla* bright magenta, 1.8–2.5 *cm. long, the summit of the throat* 6–9 *mm. across.* — River-swamps and bottomlands, se. Va. to Fla. June, early July.

5. P. obovàta (Ell.) R. K. Godfrey (obovate). — Generally coarser than no. 4, 0.5–1 m. high; all but the remote and greatly reduced uppermost *firm leaves narrowly obovate to broadly spatulate-oblanceolate, with broadly rounded summit,* undulate or dentate, *mostly* 1–3 *cm. broad; spikes remotely flowered, the laterals when present on arched-ascending* branches, the pairs of fruiting calyces very distant; *corolla* 2.5–3.3 *cm. long, the summit of the throat* 1–1.5 *cm. across.* (*P. Digitalis* Small) — Low pinelands, thickets and swamps, Fla. to Tex., n. to N.C. and s. Mo. June–Sept.

6. P. aboriginòrum Fern. (wrongly constructed name, intended to mean *of the aborigines,* from the type-station, Indian Creek, Va.). — Coarser than no. 4, *the stout stem paniculate-branched above; leaves firm, sessile,* or the lowermost on short and broad petioles, *oblong, acute or acutish, not greatly reduced above,* undulate-sinuate, 1.7–2.5 *cm. broad; lateral spikes on arched-ascending peduncles;* the *spikes densely flowered, with imbricated fruiting calyces,* 0.4–2 dm. long; *corolla* 2–2.5 *cm. long, the summit of the throat* 8–10 *mm. across.* — Swamps and stream-margins, local, se. Va. June, July.

7. P. intermèdia (Nutt.) Engelm. & Gray (intermediate). — Slender, 0.3–1.5 m. high; *leaves linear-lanceolate to linear,* thickish, *pale, the larger* 3–12 *mm. broad, repand-denticulate, all but the lowest attenuate,* the uppermost greatly reduced; spike very slender, much interrupted, the lateral ones, if developed, strictly erect, the rachis 0.5–3.5 dm. long; *corolla* 1–1.5 *cm. long.* (*Dracocephalum* Nutt.) — Damp barrens, prairies and swamps, Ky. to Kans., s. to Ala., La. and Tex. June, July.

15. SYNÁNDRA Nutt.

Calyx campanulate, inflated, membranaceous, irregularly veiny. Corolla with a long tube, much expanded above and at the throat; the upper lip slightly arched, entire; the lower spreading and 3-cleft, with ovate lobes, the middle one broadest and notched at the end. Filaments hairy; anthers approximate in pairs under the upper lip; the two upper each with one fertile and one smaller sterile locule, the latter locules cohering (whence the name; from the Greek *syn, together,* and *aner* (*andr*), *man,* here used for *anther*). A single species.

1. S. hispídula (Michx.) Baill. (stiffly hairy). — Hairy biennial, 3–6 dm. high; lower leaves long-petioled, broadly ovate, cordate, crenate, thin; the floral sessile, gradually reduced to bracts, each with a single sessile flower; corolla 3–4 cm. long, yellowish-white. — Wet woods, damp thickets and stream-banks, Va. and W.Va. to Ill., s. to Tenn. June.

16. BALLÒTA L. Fetid Horehound

Calyx nearly funnelform; the 10-ribbed tube expanded above into a spreading regular border, with 5–10 teeth. Anthers exserted beyond the tube of the corolla, approximate in pairs. Otherwise much as in *Marrubium*. Small genus nat. of Old World. (The Greek name, of uncertain origin.)

1. B. nìgra L. (black), Black Horehound. — More or less hairy, but green, erect; the root perennial; leaves ovate, toothed; whorls many-flowered, dense; calyx-teeth 5, longer than the tube of the purplish corolla. — Waste places, local, s. N.E. and N.Y., s. to N.J., Pa. and Md. June–Oct. (Adv. from Eu.)

17. LEONÙRUS L. Motherwort. Agripaume (Que.)

Calyx 5-nerved, with 5 nearly equal teeth. Upper lip of the corolla oblong and entire, somewhat arched; the lower spreading, 3-lobed; its middle lobe larger, narrowly oblong-obovate, entire; the lateral ones oblong. — Upright herbs, nat. of Old World, with cut-lobed leaves, and close whorls of flowers in their axils; in summer. (Name from the Greek *leon, a lion*, and *oura, tail*, i.e., *Lion's-tail.*)

1. L. Cardìaca L. (old generic name, *for the heart*), Common M. — Tall perennial; *leaves long-petioled; the lower rounded; palmately lobed, the floral cuneate at base, subentire or 3-cleft*, the lobes lanceolate; *upper lip of* the pale purple *corolla bearded.* — Waste places, Que., westw. and southw. beyond our range. June–Aug. (Natzd. from Eu.)

2. L. Marrubiástrum L. (old generic name, *with flowers of horehound*). — Tall biennial, with elongated branches; *cauline leaves oblong-ovate, coarsely toothed; corolla* whitish, shorter than the calyx-teeth, the *tube naked within;* lower lip rather erect. — Waste places, etc., Pa. and Del. to Fla. and Ill., local. June–Aug. (Adv. from Eu.)

3. L. sibíricus L. (Siberian). — Tall biennial; *leaves 3-parted, the divisions 2–5-cleft, or deeply 3–7-cleft and incised;* corolla purplish, twice as long as the calyx, the upper lip fornicate, the lower little spreading. — Waste grounds, Fla. to La., locally n. to Que., Ky., and Ia. June–Aug. (Adv. from Eurasia)

18. GALEÓPSIS L. Hemp-Nettle. Gratte (Que.)

Calyx about 5-nerved, with 5 somewhat equal teeth. Corolla dilated at the throat; upper lip ovate, arched, entire; the lower 3-cleft, spreading, the lateral lobes ovate, the middle one rounded to inversely cordate; palate with 2 nipples at the sinuses. — Annuals, nat. of Eurasia and n. Afr., with spreading branches, and several–many-flowered whorls in the axils of floral leaves which are nearly like the lower ones. (Name composed of the Greek *gale, a weasel*, and *opsis, appearance*, from some fancied likeness of the corolla to the head of a weasel.)

Stems swollen under the nodes, bristly-hairy. . . . . 1. *G. Tetrahit.*
Stems not swollen under the nodes, covered with soft appressed hairs. . . . 2. *G. Ladanum.*

1. G. Tetràhit L. (old generic name meaning 4-parted). — Branching *bristly-hirsute* annual, the *stem swollen below the joints* and, when old, disarticulating there; *leaves ovate, mostly rounded at base, acuminate,* coarsely serrate; *calyx-teeth in fruit* 7.5–11 *mm. long; corolla about* 2 *cm. long,* commonly white or suffused with purple, rarely purple throughout; the middle lobe rarely longer than broad, usually rounded, rarely emarginate, the margins flat; nipples at base of lip large. — Waste places, roadsides, cult. fields, etc., Nfld. to Alta., s. to N.S., N.E., N.Y., Mich., Wisc., Ia., etc. June–Sept. (Natzd. from Eurasia)

Var. bífida (Boenn.) Lej. & Court. (two-cleft). — *Leaves cuneate at base; fruiting calyx with teeth* 5–8 *mm. long; corolla* purplish, or white in forma albiflòra House (white-flowered), 1.3–1.6 *cm. long,* more slender than in the preceding, with smaller lower lip; *its middle lobe oblong, more or less emarginate,* the margins often revolute; nipples smaller and narrower. — Common weed, Nfld. to Alaska, s. to N.S., N.E., N.C., W.Va., O., Mich., S.D., etc. (Natzd. from Eu.)

Var. arvénsis Schlecht. (of fields). — *Leaves rounded-ovate, blunt; flowers as in* var. *bifida.* — Seashore-litter, Gaspé Co., Que. (Adv. from Eu.)

2. G. Ládanum L. (old generic name). — Small annual with arched-ascending branches, the *stems canescent with appressed pubescence, not enlarged at joints;* leaves linear or lanceolate, more or less downy, entire or obscurely serrate; corolla red or roseate, often spotted with yellow, much exceeding calyx; calyx-teeth shorter than in no. 1. — Waste ground, local, Que. to Mich., s. to N.E., N.J. and Ind. July–Sept. (Adv. from Eu.)

Var. LATIFÒLIA (Hoffm.) Wallr. (broad-leaved). — *Leaves ovate-oblong, sharply toothed; summit of plant glandular.* — Local and perhaps casual, St. P. et Miq. to N.E. (Adv. from Eu.)

19. LÀMIUM L. DEAD-NETTLE

Calyx tubular-campanulate, about 5-nerved, with 5 nearly equal awl-pointed teeth. Corolla dilated at the throat; upper lip ovate or oblong, arched, narrowed at the base; the middle lobe of the spreading lower lip broad, notched at the apex, contracted as if stalked at the base; the lateral ones small, at the margin of the throat. — Decumbent herbs, nat. of Eurasia and n. Afr., the lowest leaves small and long-petioled, the middle cordate and doubly toothed, the upper bracteal ones subtending the whorled flower-clusters. (Old Latin name of a nettle-like plant mentioned by Pliny.)

Annuals or short-lived biennials, without creeping stems; corolla 0.3–1.5 cm. long, its very slender tube without a ring of hairs inside (except in the purple-bracted no. 3).

Upper or bracteal leaves sessile, reniform; flowers 6–10 or more in a whorl; calyx 5–6.5 mm. long, with connivent teeth and without a bractlet at base. . . . 1. *L. amplexicaule.*

Upper or bracteal leaves petioled, ovate to subrhombic; flowers 3–7 in a whorl; calyx 6–9.5 mm. long, with spreading teeth and usually with a short linear bract at base.

Leaves pale green; the upper and bracteal ones ovate-deltoid or subrhombic, deeply and doubly toothed, 2–4.5 cm. long; inside of corolla-tube naked; nutlets 2.3–3 mm. long. . . . . 2. *L. hybridum.*

Leaves deep green or purplish, all cordate-ovate; the upper and bracteal ones crowded, purple, shallowly dentate, 1–2.5 cm. long; corolla-tube with a ring of hairs inside; nutlets about 2 mm. long. . . . . . 3. *L. purpureum.*

Perennials with creeping stems bearing offshoots; corolla 1.8–3 cm. long, its stoutish tube with a ring of hairs inside.

Upper and bracteal leaves deltoid-ovate, crenate-dentate, frequently with a pale blotch bordering the midrib; corolla roseate or whitish, the ring of internal hairs transverse. . . . . 4. *L. maculatum.*

Upper and bracteal leaves ovate-lanceolate, coarsely serrate, long-acuminate; corolla creamy-white, the ring of internal hairs oblique. . . . . . . 5. *L. album.*

1. L. AMPLEXICAÙLE L. (clasping), HENBIT. — Annual or winter-annual, branching freely from base; the arched-ascending stems glabrous, slender, remotely leafy, 0.5–3.5 dm. high; *leaves round-ovate to reniform,* crenately incised, pilose above, 0.8–4 cm. broad, the lower long-petioled, the median short-petioled; *the upper larger, sessile and clasping; flowers* 6–10 *or more in a whorl,* in axils of upper leaves; *calyx* pilose-hirsute, 5–6.5 *mm. long, with* long lance-acuminate ciliate *connivent teeth; corolla* 1.2–1.5 cm. long, *with very slender and straight tube* and pink or purple lips, the hairy upper lip oval and entire, *the inside of tube naked,* or in forma CLANDESTÌNUM (Reichenb.) G. Beck (hidden) the very small unopening corolla tubular and round-tipped; nutlets smooth and shining, brown or olivaceous, pebbled or mottled with white, clavate-obovate, trigonous, 2–2.4 mm. long. — Waste places, new seeding, roadsides, fallow fields, etc., common southw., more locally n. to Nfld., se. Lab., s. N.B., N.E., N.Y., s. Ont. Mich., etc. March–Nov. (Natzd. from Eu.)

2. L. HÝBRIDUM Vill. (hybrid; on the unsupported assumption that it is a cross of nos. 1 and 3), HENBIT. — Stem less tufted than in no. 1, 1–5 dm. high; *leaves pale green,* glabrous or setulose, all petioled, the small remote cordate-ovate obtuse crenate-dentate lower ones long-petioled; the *upper ones* larger, 2–4.5 *cm. long, subremote to approximate, ovate-deltoid to subrhombic,* acute or obtuse, *subtruncate to attenuate to the short petiole, incised, with the large teeth commonly cleft or irregularly notched;* flowers 3–7 in a whorl, subtended by short linear bracteoles; *calyx* green or slightly dark-angled, 6.5–9.5 *mm. long, with* lance-acuminate *finally divergent teeth; corolla* purplish, 0.7–1.5 cm. long, *with* slender straight *tube* gradually *enlarged to the throat and naked inside,* the hairy upper lip entire; nutlets grayish or olivaceous, lustrous, 2.3–3 mm. long. — Cult. soil, waste places, etc., Nfld. to N.E., locally s. to N.C. May–Oct. (Natzd. from Eu.)

3. L. PURPÙREUM L. (purple), PURPLE D. — Annual; hollow stem 1–3 dm. high; *leaves deep green or purplish, all cordate-ovate, obtuse, shallowly and rather regularly crenate,* setulose, the lower long-petioled; *the upper ones crowded* and short-petioled, 1–2.5 *cm. long,* red or purple;

flowers 3–6 in a whorl, subtended by a short linear bractlet; calyx purple-angled, hairy, 6–8 mm. long, with lance-attenuate bristly lobes; *corolla* purplish, 0.8–1.5 cm. long, the slender straight *tube with a ring of hairs inside*, upper lip hairy and entire; *nutlets about* 2 *mm. long.* — Roadsides and waste places, local, Nfld., N.S. and N.E. to Mich., s. to S.C., W.Va., O., Ind., Ill. and Mo. April–Oct. (Natzd. from Eu.)

4. L. MACULÀTUM L. (spotted). — *Perennial* with creeping stems and basal offshoots; flowering stems often forking, 1.5–5 dm. high; *leaves* all petioled, ovate, crenate-dentate, often doubly so, *frequently with a whitish blotch bordering the midrib,* the lower ones cordate or cordate-ovate, *the upper triangular ones* 2–6 *cm. long* and becoming remote; flowers 6–15 in a whorl, naked or each subtended by a small bractlet; calyx 7–8, becoming 13–15 mm. long, the unequal lobes finally divergent; *corolla rose-purple,* or white in forma LÁCTEUM (Wallr.) G. Beck (milky), 1.8–2.5 cm. long, the cylindric *tube exceeding the calyx and with an internal transverse ring of hairs,* the upper lip strongly arched and with 2 dorsal ridges. — Roadsides, waste ground, etc., esc. from cult., N.E. to s. Ont., s. to N.C. and Tenn. April–Oct. (Introd. from Eu.)

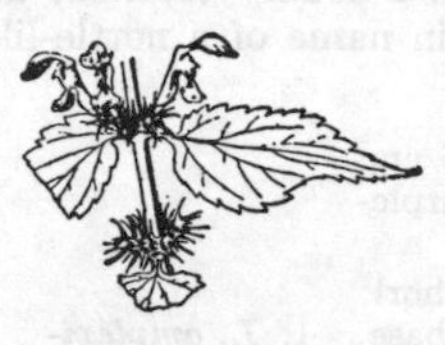
1484. L. album.

5. L. álbum L. (white), SNOWFLAKE. — Creeping-based perennial; the *hirsute stems* 1.5–6 dm. high; *leaves* all petioled, *ovate to ovate-lanceolate, coarsely serrate; the upper long-acuminate,* 0.3–1 dm. long; flowers many in a whorl, commonly subtended by a minute subulate bract; calyx about 1 cm. long, often purple, with spreading lobes; *corolla creamy-white,* 2–3 *cm. long, the tube about equaling calyx and with an internal ring of oblique hairs,* the strongly arched upper lip with 2 dorsal ridges. — Waste places, roadsides and old lawns, often spread from cult., Que. to Minn., s. to N.E. and Va. May–Oct. (Introd. from Eu.) FIG. 1484.

20. STÁCHYS L. HEDGE-NETTLE. ÉPIAIRE (Que.)

Corolla not dilated at the throat; upper lip erect or rather spreading, often arched, entire or nearly so; the lower usually longer and spreading, 3-lobed, with the middle lobe largest and nearly entire. Stamens 4, ascending under the upper lip (often reflexed on the throat after flowering); anthers approximate in pairs. Nutlets obtuse, not truncate. — Whorls 2–many-flowered, approximate in a terminal raceme or spike (whence the name, from the Greek *stachys,* an *ear of wheat* or *spike*). Genus wide-spread.

a. Lower leaf-surfaces and calyces tomentose or lanate with whitish felt.
 Plant biennial; stem pilose-tomentose; principal leaves rounded to cordate at base, obviously dentate, silky-pilose above, tomentose beneath. . . . 1. *S. germanica,* var. *italica.*
 Plant perennial; stem and leaves heavily lanate, the latter narrowed to base, the dentation hidden by the felt. . . . . 2. *S. olympica.*
a. Lower leaf-surfaces and calyces glabrous, setose, pilose or puberulent, not tomentose. . . *b.*
b. Annuals or biennials with slender fibrous roots and no stolons; leaves 1–4.5 cm. long.
 Stem long-hirsute; leaves cordate or round at base, hirsute; calyx-lobes curved; corolla pink. . . . . 3. *S. arvensis.*
 Stem minutely pilose to glabrate; leaves attenuate at base, glabrous or nearly so; calyx-teeth straight; corolla yellow. . . . . . . 4. *S. annua.*
b. Perennials with mostly creeping bases, rhizomes or stolons, the latter often becoming thickened and more or less tuberiform; leaves 0.15–2 dm. or more long. . . *c.*
c. Sides of middle and upper internodes of stem pilose, puberulent or villous over the surfaces (in typical var. of no. 5 sometimes almost glabrous); lower leaf-surfaces and surfaces of calyx pilose or puberulent. . . *d.*
d. Lower and median leaves sessile, subsessile or on petioles at most one-eighth as long as blade; blades narrowly to broadly lanceolate to narrowly ovate, rounded, subtruncate or only shallowly cordate at base; calyx-teeth lance-subulate, about equaling tube; transcontinental northern species. . . . . 5. *S. palustris.*
d. Lower and median leaves long-petioled, the petioles one-fifth to quite as long as the deeply cordate to broad-ovate to ovate-oblong blade.
 Stem and petioles heavily viscid-villous; calyx copiously long-glandular, its lance-subulate teeth about equaling tube; casual adventive eastward. . . . . 6. *S. sylvatica.*
 Stem and petioles minutely puberulent and often hirsute, without

obvious glands; calyx minutely pilose, sometimes hirsute, not prominently glandular, the deltoid-based teeth much shorter than tube; primarily of Appalachian Upland. . . . . 7. *S. Riddellii.*

*c.*Sides of middle and upper internodes glabrous or rarely with remote setae; lower leaf-surfaces glabrous, the pubescence, if present, spreading-hirtellous and along the veins; surfaces of calyx glabrous or minutely pilose or merely setose along the angles and margins. . . *e.*

*e.*Leaves membranaceous (firmer in exposed habitats), veiny, dark green above, broadly linear to ovate; the principal ones 0.5–2 dm. long, 1.2–9 cm. wide, acuminate, prominently toothed, the lowest prolonged lateral nerves arching forward 2–13 mm. from the margin; nutlets 2.1–2.8 mm. long. . . *f.*

*f.*Calyx minutely pilose over the surface; leaves ovate to broadly elliptic-oblong or oval, the principal ones 3.5–9 cm. broad; calyx-teeth deltoid.

Leaves glabrous beneath, glabrous or remotely short-strigose above, sessile or nearly so; axis of spike minutely appressed-pilose to glabrate; calyx-teeth broadly deltoid, erect, with short subulate tips. . . . . 8. *S. latidens.*

Leaves copiously hispid on veins and veinlets beneath, strigose above with long and short hairs, distinctly petioled; axis of spike retrorse-hispid; calyx-teeth narrowly triangular, soon spreading, with prolonged subulate tips. . . . . 9. *S. Clingmanii.*

*f.*Calyx glabrous except for setae sometimes along the nerves; leaves linear-lanceolate to narrowly ovate, the principal ones 1.2–6 cm. broad; calyx-teeth lance-attenuate, soon strongly curved. . . . 10. *S. tenuifolia.*

*e.*Leaves firm, only obscurely veined, pale green, narrowly linear to narrowly oblong; the larger ones 1.5–7 cm. long, 2–15 mm. broad, blunt or merely acutish, entire or with low flattened teeth, the lowest prolonged lateral nerves within 0.3–1 mm. from the often revolute margin; nutlets 1.5–1.8 mm. long. . . . . 11. *S. hyssopifolia.*

1. S. germánica L. (German), var. itálica (Mill.) Briq. (Italian). — *Biennial,* erect, 0.4–1 m. high, the *stem villous-tomentose* with grayish hair; *leaves oblong or oblong-lanceolate,* thick, *crenate-dentate, rugulose, rounded or cordate at base,* grayish above, whiter and *tomentose beneath,* the lower petioled, the upper sessile; whorls of flowers dense, somewhat scattered, leafy-bracted; *calyx* densely lanate, *its acerose teeth evident;* corolla lanate on the back, roseate or purple, its tube included. — Waste places, fields and roadsides, local, R.I. to s. Ont., where a troublesome weed. June–Sept. (Natzd. from Eu.)

2. S. olýmpica Poir. (Olympian), Woolly H., Lamb's-ears. — Matted *perennial* with *coarse stem and leaves densely felted with whitish implexed silky wool; leaves* oblong-elliptic, *narrowed to base, the small teeth hidden; spike dense,* white-woolly, the *bracts mostly short and hidden; calyx-teeth hidden in wool;* corolla purple, heavily woolly without, the tube included. (*S. lanata* Jacq.) — Pastures and roadsides, region of Bruce Pen., Ont. July–Sept. (Natzd. from the Caucasus)

3. S. arvénsis L. (of fields). — Small *annual;* the decumbent or ascending *stem* often branching from base, 0.5–4 dm. high, *loosely long-hirsute; leaves cordate,* suborbicular to ovate, all but the bracteal *slender-petioled,* strongly crenate, *hirsute,* 1–3.5 cm. long, 0.5–2.3 cm. broad; *spikes* leafy at base, *crowded at tip,* lax below, becoming 0.2–2 dm. long, *with no bracteoles at base of calyx; calyx long-hirsute, with straight* lance-deltoid sharp-tipped *teeth; corolla pink,* spotted with purple, *only slightly exceeding calyx,* with transverse hairs in throat; the *rounded upper lip entire;* nutlets 1.8–2 mm. long. — Waste places and grasslands, rather local, N.E., N.Y. and Pa. June–Oct. (Adv. from Eu.)

4. S. ánnua L. (annual). — Small *annual; stem* erect or with ascending branches, *minutely pilose,* 1–4 dm. high; *leaves glabrous or nearly so,* with long-ciliate slender petioles (or upper sessile), oblanceolate to lanceolate or elliptic-oblong, attenuate at base, obtuse, crenate, or the upper acute and entire, 1.5–4.5 cm. long, 0.7–1.8 cm. wide; leafy spike becoming 0.5–2 dm. long, with 3–12 remote whorls; *flowers subtended by minute bractlets; calyx minutely hairy and gland-dotted above,* hirsute at base, heavily bearded at throat, the campanulate *tube* becoming 2–4.5 mm. long and *about equaled by the curved lanceolate* spinose-tipped *teeth; corolla yellow,* with few dark spots, 1.2–1.8 cm. long, the *upper lip crenate and dilated,* the lower 3-lobed. — Waste places and weedy fields, local, N.E., N.Y. and N.J. July–Sept. (Adv. from Eu.)

5. S. **palústris** L. (of marshes), Woundwort. — Rank-smelling perennial with creeping rootstock, bearing subterranean stolons terminated by crisp fusiform whitish tubers; *stems* *pubescent on the sides as well as on the angles*, simple or loosely branching, 0.2–1 m. high; *leaves sessile, subsessile, or rarely with* prolonged *petioles, lanceolate to narrowly ovate, rounded, subtruncate or shallowly cordate at base*, crenate or crenate-serrate, *pubescent on the surface beneath*, appressed- or strigose-hispid above; spike with divergent or reflexed leafy bracts at base, nearly or quite naked above, becoming 0.5–2.5 dm. long, the lowest whorls in fruit remote, the upper approximate; *calyx closely pubescent*, the slenderly campanulate tube in maturity 3–5 mm. long, *the lance-subulate teeth about equaling the tube;* corolla rose-purple mottled with paler and darker tones, 1.2–1.5 cm. long, pubescent without, the large 3-lobed lower lip much longer than the upper; *nutlets dark brown*, sublustrous, ellipsoid-obovoid, 1.8–2.2 *mm. long*, 1.2–1.8 mm. broad. — Wide-ranging circumpolar species, the typical subspecies with its vars. introd. from Eu., the native subspecies with vars. parallel with those of Europe:

a. Calyx closely viscid-pilose, many gland-tipped hairs mixed with short glandless ones, the latter rarely more than 1 mm. long. . . *b.*
b. Leaves lanceolate, acuminate, the principal ones four to seven times as long as broad, 5–12 cm. long. . . *c.*
c. Leaves sessile or nearly so.
Sides of stem closely pubescent with appressed hairs very much shorter than the long retrorse-hirsute pubescence of the angles. . *S. palustris* (typical).
Sides of stem densely retrorse-hirsute with prolonged hairs as long as on the angles. . . . . Var. *segetum.*
c. Leaves (or principal ones) definitely slender-petioled; pubescence of stem as in typical var. . . . . Var. *petiolata.*
b. Leaves oblong or oblong-ovate, obtuse or merely subacute, only two or three times as long as broad, the principal ones 2.5–8 cm. long. . . . . . Var. *elliptica.*
a. Calyx long-hirsute as well as short-pilose, the shorter pubescence largely hidden by long whitish and glandless subsetiform trichomes mostly 1.5–3 mm. long. . . *d.*
d. Leaves sessile or on very short and inconspicuous thick petioles. . *e.*
e. Principal leaves oblong to oblong-ovate or oval, blunt or merely subacute, three-tenths to one-half as broad as long, 3.5–12 cm. long, 1.5–5 cm. broad; chiefly western. . . . . Var. *pilosa.*
e. Principal leaves narrowly to broadly lanceolate or narrowly oblong, acuminate, one-seventh to one-third as broad as long, 2–13 cm. long, 0.4–3 cm. broad; chiefly eastern and northern. . . *f.*
f. Pubescence of sides of stem many times shorter than that of angles, or minute and with long hairs of angles wanting.
Angles of stem retrorsely long-hirsute, sides short-pilose; leaves lanceolate to oblong, the principal ones 4–13 cm. long and 0.8–3 cm. broad, strongly hispid on the veins and pilose on the surfaces beneath; boreal and transcontinental. . . . . Var. *nipigonensis.*
Angles of stem with few or no long hairs, sides minutely and retrorsely appressed-pilose-puberulent; leaves narrowly lanceolate, the principal ones 2–5.5 cm. long and 0.4–1.5 cm. broad, sparsely and minutely pubescent to glabrous beneath; plant of Upper Great Lakes region. . . . . Var. *macrocalyx.*
f. Pubescence of sides of stems very dense and long, as long as and not readily distinguished from that of angles; northeastern. . . . Var. *homotricha.*
d. Leaves slender-petioled, the petioles of the principal ones 1–1.4 cm. long, the oblong to lance-ovate acuminate blades mostly 6–12 cm. long and 2–4 cm. broad; angles of stem with relatively few or no long hairs. . . . . Var. *phaneropoda.*

S. PALÚSTRIS (typical). — Ditches, wet roadsides, low meadows and waste places, Nfld. to s. Ont., s. to N.S., N.E. and N.Y. July–Sept. (Natzd. from Eu.)

Var. SÉGETUM (Mutel) Grogn. (of corn-fields). — Waste places, old fields, shores, etc., ne. N.B. to e. Me. (Natzd. from Eu.)

Var. PETIOLÀTA Clos (petioled). — Meadows and shores, local, Nfld. to s. Ont. (Natzd. from Eu.)

Var. ELLÍPTICA Clos (elliptic). — Waste places, ballast-grounds, shores and fields, Gaspé Pen., Que., to N.Y. and N.J. (Natzd. from Eu.)

Var. **pilòsa** (Nutt.) Fern. (soft-hairy). — Shores, stream-margins, meadows, etc., B.C. to Ariz., e. to James Bay, Que., Wisc., Ill. and Neb., and adv. along railroads and in waste places e. to N.E. Late June–Sept.

Var. **nipigonénsis** Jennings (of Lake Nipigon). — Low grounds, Que. to Alaska, s. to N.E., N.Y., O., Mich., Ill., Ia., N.M. and Ariz.

Var. **macrócalyx** Jennings (with large calyx). — Shores, Thunder Bay Distr., Ont., to n. Wisc. and n. Minn.

Var. **homótricha** Fern. (with uniform hairs). — Shores, meadows and fields, Me. and s. Que. to Minn., s. to Ct., N.Y., O. and Ill.

Var. **phaneròpoda** Weath. (with evident footstalks). — Local, O. to Wisc. and Ia.

S. Nuttállii Shuttlew. (for Thomas Nuttall, 1786–1859), a coarse plant with sides of stem minutely glandular-puberulent, the sessile or subsessile cordate-based narrowly ovate to oblong-elliptic acuminate leaves nearly or quite glabrous beneath except for hispid nerves, the glandular and short-pilose calyx with deltoid-acuminate teeth, has been reported from Virginia. It occurs from N.C. to Ark. and Okla.; its status with us needing verification.

6. S. sylvática L. (of the woods). — Rank-smelling, *heavily glandular* and stoloniferous; *stem* 0.3–1 m. high, *the upper internodes viscid-villous on the sides* as well as the angles; *leaves mostly on long slender glandular and villous petioles often equaling the cordate-ovate crenate* and densely strigose-pilose *leaves;* spike interrupted, with short bracts, the flowers 6–10 in a whorl; *calyx copiously glandular-pilose, its lanceolate subulate-tipped teeth about equaling the tube;* corolla deep purple, variegated with white, the tube well-exserted. — Waste places, N.Y. and Pa. June–Sept. (Adv. from Eu.)

S. Siebòldii Miq. (in honor of Philipp Franz von Siebold, 1796–1866, famous Dutch botanist), Japanese or Chinese Artichoke, Knotroot or Crosnes, cult. for its edible tubers, with slender-petioled cordate leaves, the pubescence of stem, petioles and foliage minute and glandless, the small whitish to red corollas with tube included in calyx, spreads from cult. but is hardly estab. (Introd. from e. Asia)

S. officinàlis (L.) Trev. (of the shops), with long-petioled mostly obtuse narrowly oblong-ovate coarsely dentate cordate-based lower leaves, the short-petioled to sessile smaller cauline ones only 1 or 2 remote pairs, the dense long-peduncled spike with showy purple flowers, spreads from cult. but is hardly estab. (Introd. from Eu.)

7. S. Riddéllii House (in honor of its discoverer, John Leonard Riddell, 1807–1865). — Tall perennial with large tubers; stem 0.4–1.2 m. high, *long-hirsute on the angles, puberulent or minutely pilose on the sides; leaves* (except the upper) *long-petioled, cordate-ovate, membranaceous,* crenate-dentate, acuminate, glabrous except for long strigae above, *hispid on the nerves beneath; spike slender* and interrupted, all but the lowest whorls with only small bracts; *calyx minutely pilose, sometimes hirsute,* barely if at all glandular, *the deltoid teeth much shorter than the tube;* corolla about 1 cm. long, purplish, with deeper mottling. (*S. cordata* Riddell, not Gilib.) — Wooded bottomlands, rich thickets and shaded low ground, w. Md. to Ill., s. to N.C. and Tenn. June, July.

8. S. làtidens Small (with broad teeth). — *Stem* stout, 3–9 dm. high, *glabrous on the sides,* the angles smooth, muricate or hirsute; *leaves ovate-oblong to elliptic,* acuminate, coarsely crenate-dentate, *cordate at base, sessile or short-petioled;* the blades mostly 1–1.3 dm. long and 3–5 cm. broad, *glabrous or only remotely short-strigose above, glabrous beneath; spike stiff,* interrupted, becoming 0.5–1.5 dm. long, only the lowest whorl with long bracts; *the axis minutely appressed-pilose or glabrate; calyx* short-campanulate, *finely pubescent, the broadly deltoid teeth erect and with short subulate tips.* — Brooksides and meadows, D.C. and W.Va., s. along the mts. to Ga. and Tenn. June–Aug.

9. S. Clingmánii Small (for Clingman's Dome, N.C., and its discoverer). — Strongly stoloniferous, with elongate tubers; *stem* 0.6–1.8 m. high, rather *weak, copiously bristly-hirsute on the angles, glabrous on the sides; leaves* (except the upper) *on bristly petioles; the thin blades elliptic-ovate to -oblong or narrowly -obovate,* acuminate, coarsely dentate, *copiously pilose-hispid along veins and veinlets beneath, strigose-hispid above* with long and short strigae, the principal blades 0.8–1.5 dm. long and 3–9 cm. broad, *with broadly rounded bases;* spike open and *flexuous-moniliform,* becoming 1–3 dm. long, all but the lowest whorls short-bracted, *the flexuous axis retrorse-hispid; calyx puberulent or minutely pilose, its narrowly triangular teeth with long-subulate soon spreading tips.* (*S. Nuttallii* of ed. 7, not Shuttlew.) — Alluvial woods, bottoms and swales, e. Va. to Ill., s. to N.C. and Tenn. June, early July.

10. S. tenuifòlia Willd. (thin-leaved). — Extensively creeping; the erect simple or branching *stems glabrous* or at most sparsely hirtellous *on the sides, smooth, roughened or hirsute on the angles,* 0.2–1.3 m. high; *leaves linear or linear-lanceolate to narrowly ovate,* the principal ones 0.8–6 cm. broad, *glabrous, or hirsute on one or both faces, with rounded, subtruncate or tapering base,* taper-pointed, sharply dentate to serrate; spike interrupted; *calyx glabrous, or bristly along the angles, the lance-attenuate teeth soon outwardly curving.* — Very variable; the following are the best marked vars.:

Bracts at base of glomerules of spike not ciliate (except rarely at base); calyx glabrous throughout; leaves glabrous on both faces, the principal ones with slender petioles 0.5–3 cm. long.

Leaf-blades lanceolate to ovate-oblong or narrowly ovate, two to five times as long as broad, the principal ones 4–10 cm. long and 1.2–6 cm. broad; spike stiff; wide-ranging continental plant. . . . . *S. tenuifolia* (typical).

Leaf-blades narrowly oblong, about six times as long as broad, the median ones 1–1.5 dm. long and 1.6–2 cm. wide; spike flexuous; plant of se. Va. . Var. *perlonga.*

Bracts bristly-ciliate; calyx often bristly on the angles; leaves often hispid on veins beneath, often strigose above, sessile, or the lower ones short-petioled.

Principal leaves broadly oblong to narrowly ovate, one-third to three-fifths as broad as long, 5–15 cm. long, 2–6 cm. broad. . . . . Var. *platyphylla.*

Principal leaves narrowly lanceolate to narrowly lance-oblong, one-sixth to one-fourth as broad as long, 3–10 cm. long, 0.8–2.4 cm. broad. . . . . Var. *hispida.*

S. tenuifòlia (typical). — Rich bottomlands, shores, low woods and meadows, N.Y. to Minn., s. to S.C., Tenn., La. and e. Tex. June–Sept.

Var. **perlónga** Fern. (very long). — Wooded bottomlands, Meherrin R., se. Va. July, Aug.

Var. **platyphýlla** Fern. (broad-leaved), var. *aspera* sensu ed. 7, not *S. aspera* Michx., basonym. — Low woods, rich shores and meadows, s. Que. to Minn., s. to e. Va., w. N.C., O., Ind. and Ill. June–Aug.

Var. **híspida** (Pursh) Fern. (hispid), incl. *S. ambigua* of ed. 7, in part, not *S. ambigua* (Gray) Britt. as to basonym. — Meadows, swamps, low woods and shores, s. N.H., e. Mass. and R.I. to e. S.C.; Mich. to Minn., s. to Ind. and n. Ill. June–Sept.

11. S. hyssopifòlia Michx. (hyssop-leaved). — Stoloniferous, with whitish stolons thickened and crisp; *stems slender*, simple to freely branched, 1–7.7 dm. high, *glabrous; leaves linear to linear-oblong, obtuse to subacute, entire* or obscurely repand along the margin, sessile or short-petioled, 1.5–7 *cm. long*, 2–9 *mm. wide*, often with a basal tuft of hairs, *otherwise glabrous, pale green and firm, with only the midrib evident beneath, the lower lateral nerves following forward* within 0.3–1 *mm. from the* often revolute *margin; spikes* leafy below, short-bracted above, straight, of 1 whorl or oftenest 0.2–1.3 *dm. long* and of 2–8 remote whorls, *the upper bracts linear; calyx glabrous* or rarely with remote bristles on angles, the campanulate tube 3–4 mm. long in maturity, *the lance-attenuate teeth about equaling the tube, becoming erect or connivent in fruit;* corolla pink mottled with purple and white, 1–1.5 cm. long, the straight dilated pubescent upper lip shorter than the 3–5-lobed drooping lower one; *nutlets* blackish, ellipsoid to subglobose, 1.5–1.8 *mm. long.* — Sandy, gravelly or peaty shores and bogs, se. Mass. to centr. and s. Ct. and se. N.Y., s. on or near the Coastal Plain to S.C. and in the mts. of Va. and N.C.; s. Mich. and nw. Ind. June–Oct. Passing freely to

Var. **ambígua** Gray (doubtful). — *Stems* more regularly *retrorse-hispid on angles; leaves* 5–15 *mm. broad,* often more tapering at tip, *regularly depressed-serrate; upper bracts* of spike *broadly lanceolate to ovate; calyx minutely pilose to glabrate,* more often setose on the angles, *the lance-triangular teeth often spreading-ascending in fruit.* (*S. aspera* Michx.; *S. ambigua* (Gray) Britt., not Sm.; *S. Grayana* House) — Damp sands, swamps and prairies, n. Fla. to e. Mo., n. to Pa., Ky., Ill. and Ia.; adv. in centr. Me. June–Aug.

21. SÁLVIA L. Sage

Upper lip of calyx 3-toothed or entire, the lower 2-cleft. Corolla deeply 2-lipped, ringent; upper lip straight or falcate, entire or barely notched; the lower spreading or pendent, 3-lobed, its middle lobe larger. Stamens on short filaments, jointed with the elongated transverse connective, one end of which, ascending under the upper lip, bears a linear 1-locular (half-) anther, the other, usually descending, generally bearing an imperfect or deformed (half-) anther or none at all (both halves perfect in no. 1). — Flowers mostly large and showy, in spiked, racemed, or panicled whorls. Large genus of warm and temp. reg. (The old Latin name of Sage.)

a. Leaves chiefly in a basal rosette, membranaceous, obovate, sinuate, repand or lyrate-pinnatifid, their surfaces glabrous or only sparsely strigose-hirsute; flowering stem scapose or with only 1 or 2 remote pairs of well-developed coarsely dentate thin leaves; both anther-locules polliniferous. . . . . 1. *S. lyrata.*

a. Leaves cauline or, if chiefly rosulate, thick, not lyrate, densely pubescent; lower anther-locule wanting or rudimentary. . . *b.*

b. Leaves linear, lanceolate or narrowly oblong, mostly less than 2 cm. broad; upper lip of calyx entire.

Perennial with leaves short-petioled; calyx obscurely bilabiate, cleft about one-fourth its length; corolla 1.5–2.5 cm. long, tube long-exserted. . 2. *S. azurea*, var. *grandiflora.*

Annual with primary leaves long-petioled; calyx deeply bilabiate, cleft about halfway to base; corolla about 1 cm. long, barely exserted. . 3. *S. reflexa.*

b. Leaves broadly oblong, ovate or rounded, 2.5–10 or more cm. broad; upper lip of calyx 3-toothed. . . c.

c. Pedicels nearly as long as fruiting calyx.

Principal leaves deltoid- to rhombic-ovate, cuneately tapering into winged petioles, glabrescent; calyx-teeth triangular-ovate; native. 4. *S. urticifolia.*

Principal leaves cordate-lyrate, slender-petioled, pubescent; calyx-teeth lance-subulate; introd. . . . . 5. *S. verticillata.*

c. Pedicels many times shorter than fruiting calyx.

Leaves chiefly in a basal rosette, the cauline only 1–3 pairs; bracts cordate-ovate, much shorter than flower; upper lip of calyx not recurved. . . . . 6. *S. pratensis.*

Leaves mostly cauline, 6–8 pairs; bracts suborbicular, about equaling calyces; upper lip of calyx recurved. . . . . . 7. *S. sylvestris.*

1. S. lyràta L. (lyre-shaped), LYRE-LEAVED S., CANCER-WEED. — *Rosulate perennial; the basal obovate leaves petioled, membranaceous, the vernal ones sinuate, repand or lyrate-pinnatifid;* those of summer and autumn (holding over winter) mostly subentire, *their surfaces glabrous or sparsely strigose-hirsute,* often purple-shaded; scape naked or with 1 or 2 pairs of leaves, simple or with virgate branches; whorls of flowers 3–10, becoming distant; bracts oblong-linear, not longer than calyx; *broad upper lip of calyx truncate,* with 3 widely separated sharp teeth; violet *corolla* 2–3 *cm. long,* exserted, its short and straight upper lip much shorter than the broad lower one; *both forks of connective bearing fertile anthers;* nutlets fuscous, obovoid, 2–2.3 mm. long. — Sandy open woods and clearings, Fla. to e. Tex., n. to sw. Ct., se. N.Y., N.J., Pa., W.Va., O., Ind., Ill., Mo. and Okla. April–June.

2. S. azùrea Lam. (sky-blue), var. **grandiflòra** Benth. (large-flowered), BLUE S. — Erect perennial 0.5–1.5 m. high, *cinereous with short recurving pubescence, leafy;* lower leaves lanceolate or oblong, obtuse, denticulate or serrate, tapering to petioles; *upper leaves* narrower, *often linear, entire;* inflorescence spike-like, the whorls crowded above, tomentulose-sericeous; *calyx cleft about a quarter its length, obscurely bilabiate, the broad upper lip entire; corolla deep blue* (rarely white), 1.5–2.5 *cm. long, with prominently exserted tube.* (*S. Pitcheri* Torr.) — Dry prairies and openings, Minn. to Neb., s. to Ky., Ark. and Tex.; esc. from cult. e. to N.E. July–Sept.

3. S. refléxa Hornem. (turned back; from the reflexed leaves). — *Annual,* often bushy-branched, the puberulent to glabrous stem 1.5–6 dm. high; *leaves lanceolate to lance-linear,* irregularly serrate to entire, spreading or reflexed, *glabrescent above, the larger ones up to* 5 *cm. long and tapering to slender petioles* 0.5–1.5 *cm. long;* flowers 1–3 in a whorl, forming a slender interrupted spike; *calyx deeply bilabiate, cleft about halfway to base, the upper lip entire; corolla blue, only slightly exserted.* (*S. lanceaefolia* of ed. 7, not Poir.) — Dry open soil, Wisc. to Mont., s. to Ark., Tex. and Mex.; adv. e to Mich., O., W.Va. and N.J. May–Oct.

4. S. urticifòlia L. (nettle-leaved). — Perennial; the stems arising from elongate bases, 3–6 dm. high, viscid and villous to glabrate; *leaves rhombic- to triangular-ovate,* coarsely serrate, *cuneately tapering into decurrent wings of the petioles, glabrescent;* spike interrupted, the distant whorls numerous and several-flowered; *pedicels nearly equaling the calyx; calyx-lips divergent, the upper with* 3 *triangular-ovate teeth,* the lower 2-cleft; corolla blue and white, 1–1.5 cm. long, twice length of calyx; style bearded. — Woods, Fla. to La., n. to Pa. and O. May–July.

5. S. VERTICILLÀTA L. (whorled). — *Stem pilose-hispid,* 2–9 dm. high; *leaves rounded or round-ovate, long-petioled, often lyrately lobed at base,* irregularly crenate, *pubescent on both sides;* inflorescence simple or paniculate, interrupted; *pedicels about equaling the tubular villous calyx; calyx-teeth lance-subulate;* corolla about twice length of calyx, its tube barely exserted. — Roadsides, fields and waste places, local, N.E. to Pa. and Md. June–Oct. (Adv. from Eu.)

6. S. PRATÉNSIS L. (of meadows). — Stout-based perennial *with a rosette of long-petioled oblong or oblong-ovate blunt* crenate *rugulose-bullate leaves glabrous above* but pubescent along the nerves beneath; *stem* pubescent, *with only* 1–3 *pairs of well developed leaves;* inflorescence spiciform or paniculate, with remote whorls; *bracts cordate-ovate, much shorter than flowers,* more or less reflexed; *calyces subsessile, viscid, with connivent teeth;* corolla thrice length of calyx, bluish. — Waste places, pastures, fields, etc., Mass. to Ill. and Pa. May, June. (Natzd. from Eu.)

7. S. SYLVÉSTRIS L. (of woodland). — *Stem regularly leafy to summit; leaves oblong-lanceolate,* crenate, *rugose,* glabrous above, pubescent beneath, rounded to cordate at base, the lower petioled, *the upper sessile; whorls subapproximate or slightly separated; bracts orbicular, colored, about as long as calyces; calyx* ovoid-tubular, the rounded *upper lip recurved* and shortly 3-toothed; corolla twice length of calyx. — Waste places, old fields, etc., local, Mass. to s. Ont., Wisc. and Neb., s. to N.J., Pa., Ky. and Colo. June–Sept. (Natzd. from Eu.)

Numerous Old World species occur casually with us but rarely persist. The common SAGE of the kitchen-garden, S. OFFICINÀLIS L. (of the shops), and the SCARLET S., S. SPLÉNDENS Ker. (shining), a South American species, are frequent on dumps, etc.

22. MONÁRDA L. MONARDA. HORSEMINT

Calyx 15-nerved, usually hairy in the throat. Corolla clongated, with a slightly expanded throat; lips linear or oblong, somewhat equal; the upper erect or arched, entire or slightly notched; the lower spreading, 3-lobed at the apex, its lateral lobes ovate and obtuse, the middle one narrower and slightly notched. Stamens elongated, ascending, inserted in the throat of the corolla. — Odorous N. Am. erect herbs, with entire or toothed leaves, and large attractive flowers in a few verticils closely surrounded by bracts. (Dedicated to *Nicolás Monardes*, author of many tracts upon medicinal and other useful plants, especially those of the New World, in the latter half of the 16th century.)*

*a.*Glomerules solitary and terminal or sometimes 2 or 3; stamens and style conspicuously exserted beyond the linear straight acutish upper lip of the corolla. § EUMONARDA. *b.*

*b.*Corolla glabrous or its tube with short spreading or curving hairs, the upper lip not bearded. . . *c.*

*c.*Principal leaves definitely petioled. . . *d.*

*d.*Bracts and bractlets eciliate or merely ciliolate with short hairs; horizontally divergent bristles at summit of calyx-tube wanting or few and less than 0.5 mm. long.

Bracts vividly red-tinged; calyx 9–14 mm. long, nearly or quite naked in throat, its teeth 1–2 mm. long; corolla vermilion to nearly scarlet, 3–4.5 cm. long. . . . . 1. *M. didyma.*

Bracts white or white-tinged; calyx 6–10 mm. long, bearded in the throat, its teeth about 1 mm. long; corolla dull white to flesh-color, dark-spotted, 1.5–3 cm. long. . . . . 2. *M. clinopodia.*

*d.*Bracts and bractlets stiffly pectinate-ciliate, with straight divergent hairs 1–1.3 mm. long, dark purple or purplish; summit of mature calyx with a dense crown of stiffly divergent bristles 1–1.5 mm. long; corolla deep red-purple. . . . . 3. *M. media.*

*c.*Principal or all leaves sessile or essentially so.

Internodes mostly shorter than to once-and-a-half length of subtending leaves; larger leaves ovate, 2.3–5.5 cm. broad; glomerule (without corollas) 2–3 cm. broad; corolla-tube included in calyx; tip of upper lip bearded. . . . . 4. *M. Russeliana.*

Internodes mostly twice to thrice length of subtending leaves; larger leaves narrowly ovate to lanceolate, 1–2.5 cm. broad; glomerule 1–2 cm. broad; corolla-tube long-exserted, upper lip not bearded. . 5. *M. virgata.*

*b.*Corolla closely pubescent with incurved pilosity, tip of upper lip bearded.

Leaves deep green, membranaceous, ovate, the sides gradually curving to the acuminate tips, larger blades 2.5–6 cm. broad; bracts dark purple or purplish; corolla deep red-purple. . . . . 3. *M. media.*

Leaves grayish, firm (except in deep shade), narrowly triangular-ovate to lanceolate, the straightish sides tapering from base to apex, larger blades 1–4 cm. broad; bracts pale green to grayish or lilac-tinged; corolla pink or lilac. . . . . 6. *M. fistulosa.*

*a.*Glomerules axillary or interrupted-spicate; stamens not exceeding the falcate upper lip of the short-tubed corolla. § CHEILYCTIS. *e.*

*e.*Perennial with hard forking crowns, annual in a southwestern var.; leaves subtending the glomerules and the bracts without long awn-tips; calyx-teeth triangular to lanceolate, barely subulate-tipped; corolla yellowish, strongly dotted with purple-brown. . . . . 7. *M. punctata.*

*e.*Annual without strong crowns; bracteal leaves and bracts of glomerule ending in long awns; calyx-teeth awn-like; corolla whitish to pinkish, not at all or but lightly spotted.

Foliar bracts densely pubescent, white to purple on upper surface, their bases erect and forming a cup around the calyces, their summits outwardly curved. . . . . 8. *M. citriodora.*

Foliar bracts nearly or quite glabrous, green or greenish, straight, divergent from base. . . . . 9. *M. pectinata.*

---

* Some characters derived from the study of MCCLINTOCK & EPLING in Univ. Calif. Pub. Bot. xx. no. 2 (1942).

§ Eumonárda Benth.

1. M. dídyma L. (twin; used by Linnaeus for plants with stamens in pairs), Oswego-tea, Bee-balm. — Stem 0.5–1.5 m. high, acutely angled, simple or ascending-branched above, glabrous or sparsely villous and glabrate, usually villous at nodes; leaves petioled, with dark green membranous cordate-ovate to ovate-lanceolate serrate blades with long spreading villous-hirsute bearding along the principal veins beneath, the larger blades 6–14 cm. long and 3–6.5 cm. broad; glomerules solitary at tips of stem and branches, rarely a 2nd one below, 2–4 cm. broad (without corollas); *bracteal leaves red or red-based, inner red bracts linear-bristleform; calyces* 9–14 mm. long, *glabrous or minutely pubescent, glabrous or nearly so in the throat,* the subulate teeth 1–2 mm. long; *corolla glabrous or nearly so,* 3–4.5 *cm. long, vermilion to nearly scarlet;* stamens inserted 2–4 mm. within throat of corolla. — Rich woods, thickets and bottomlands, N.Y. to Mich., s., chiefly in the uplands, to Ga. and Tenn.; esc. from cult. e. to Que., N.E., L.I. and N.J. Late June–Aug.

2. M. clinopódia L. (from resemblance of the bracteal leaves to those of *Clinopodium,* now *Pycnanthemum incanum*). — Similar to no. 1; *leafy bracts white or white-tinged; calyx* 6–10 *mm. long, bearded in the throat, the teeth about* 1 *mm. long; corolla dull white to flesh-pink, dark-spotted,* 1.5–3 *cm. long; stamens inserted* 1–2 *mm. within the throat.* — Moist woods, thickets, ravines and stream-banks, N.Y. to Ill., s. to Md., and along mts. to Ala. and Ky., and s. Ind.; spread from cult. in N.E. Late June–early Sept.

3. M. mèdia Willd. (intermediate). — In habit like nos. 1 and 2; *bracts and bractlets stiffly pectinate-ciliate with long horizontally divergent hairs* 1–1.3 *mm. long, dark purple or purple-red; calyces heavily long-bearded in the throat and usually with a horizontally divergent crown of many straight hairs* 1–1.5 *mm. long; corolla deep red-purple, soft-pubescent, with the tip of the upper lip often bearded.* (Incl. *M. fistulosa,* var. *rubra* Gray) — N.Y. to Ont. and Ind., s. to upland of N.C. and Tenn.; much cult. and commonly esc. to roadside-thickets, etc., e. to N.E. Mid-July–early Sept.

4. M. Russeliàna Nutt. (in honor of "Dr. Russel" who, in 1819, aided Nuttall in his Arkansas work). — *Stem* simple or nearly so, *slender,* strict, *glabrous* (rarely with scattered divergent hairs), 2–5 *dm. high,* from slender rhizome; *leaves sessile* or nearly so, *cordate-ovate,* firm, appressed-hirsute on both surfaces, *longer than to two-thirds length of internode above, the larger ones* 5–10 cm. long and 2.3–5.5 *cm. broad; glomerule* (without corollas) 2–3 *cm. broad;* outer foliaceous bracts often purple-tinged; calyces 9–14 mm. long, with aristate teeth 2–4 mm. long, densely hairy in the throat; *corolla white to pale purple,* purple-dotted, 2.5–4 cm. long, *its tube included in calyx, tip of upper lip bearded.* (*M. Bradburiana* Beck) — Dry bluffs, ravines, open woods and thickets, Ky. to w.-centr. Ill. and Ia., s. to n. Ala., Ark. and ne. Tex. May.

5. M. virgàta Raf. (wand-like). — More slender than no. 4, 3–8.5 dm. high; *leaves a third to half as long as internode above, the larger blades* 4–6 cm. long and 1–2.5 *cm. broad; glomerules* 1–2 *cm. broad; corolla with slender tube long-exserted out of calyx, upper lip not bearded.* (*M. Russeliana* Nutt., 1837, not Nutt. ex Sims, 1824) — Open woods and slopes, Ky. and Tenn. to Okla. and ne. Tex. May, June.

6. M. fistulòsa L. (tubular), Wild Bergamot. — Stem 0.5–1.5 m. high, simple or often branching above, the *upper internodes more or less villous or hirsute; leaves gray-green, firm* (except in deep shade), *narrowly triangular-ovate to -lanceolate,* or sometimes cuneate at base, *the straightish* serrate-dentate *sides tapering to tip,* definitely petioled, *spreading-hirsute or -villous with long hairs along the nerves beneath, the larger blades* 1–4 *cm. broad and on petioles* 0.5–1.5 *cm. long;* uppermost leaves and foliaceous *bracts often pink-tinged;* glomerules mostly terminal; *calyx-tube densely bearded in throat* and often with short and stiff hairs divergent from bases of lobes; *corolla lilac or pink,* or in forma **albéscens** Farw. (whitish) white, 2–3 cm. long, *closely pubescent with soft incurved hairs, tip of upper lip bearded.* — Dry thickets, clearings and borders of woods, sw. Que. and w. N.E. to Minn., s. to Md., upland to Ga. and Ala., La. and e. Tex.; spread from cult. elsewhere. July, Aug. Passing into

Var. **móllis** (L.) Benth. (soft). — *Lower surfaces of leaves puberulent, minutely short-pilose or glabrescent.* (*M. mollis* L.) — Sw. Me. to Sask., s. to Va., upland to Ga. and Ala., Miss., La., Tex. and ne. Mex. Passing into

Var. **menthaefòlia** (Graham) Fern. (mint-leaved). — *Stiffer,* the mostly *simple* to subsimple *stem* 3–7.5 *dm. high, softly curved-pubescent to* glabrate; leaves much as in last, but subsessile or with petioles up to 5 mm. long; corolla 2.5–3.5 cm. long. (*M. menthaefolia* Graham) — Man. to Mackenz. and B.C., s. to Minn., S.D., Coahuila (Mex.), N.M. and Ariz.

§ Cheilýctis Benth.

7. M. punctàta L. (dotted), Dotted M., Horsemint. — *Perennial with tough crowns;* the simple to branching *stems* 0.2–8.5 dm. high, *minutely pilose-puberulent; leaves* petioled, lanceolate, narrowed at base, *minutely pilose or puberulent beneath,* the primary ones 3–8 cm. long and 6–17 mm. wide; *glomerules axillary,* (1–)2–7 *in interrupted spikes;* leaves subtending glomerules pointed but not awn-tipped; bracts recurved, purplish-lilac or whitish, entire but ciliate at base; *calyces* 6–9 mm. long, *with narrowly triangular teeth; corolla cream-color to yellowish with purple spots, with falcate upper lip longer than the stamens.*— Dry sandy soils on or near Coastal Plain, La. to Fla., n. to L.I. Late July–Oct.

Var. **villicaùlis** Pennell (hairy-stemmed). — *Stem densely short-villous or pilose* with recurving hairs; *leaves closely grayish pilose to tomentulose beneath;* bracts whitish to pale lilac; calyces 5–7.5 mm. long. — Dry sandy soil, w. Vt. to Minn., s. to N.J., Tenn., Mo. and Okla.

Var. **lasiodónta** Gray (hairy-toothed). — Perennial or *annual,* 2–4 dm. high, minutely canescent-pilose; *leaves* minutely pubescent on both surfaces, *the larger blades* 2.5–5 *cm. long; calyx* 4.5–7 mm. long, finely pubescent, *the broadly triangular teeth silky-villous with long hairs.* (Var. *occidentalis* (Epl.) Palmer & Steyerm.) — Open ground, sw. Ill. to Kans., s. and sw. to Tex., Chihuahua (Mex.) and Ariz.

8. M. citriodòra Cerv. (with odor of *Citrus*), Lemon-Mint. — Somewhat resembling no. 7, *annual,* 2–8 dm. high, minutely pubescent; *leaves* oblong to lanceolate, puberulent but glabrate, *awn-tipped; foliar bracts* oblong, *densely pubescent, their bases ascending and forming a cup around the calyces, their upper halves outward-curved, ending in long slender awns; calyx-teeth long-awned,* 2–7 *mm. long,* finally spreading; *corolla* white to pink, *lightly dotted with purple.* — Limestone-barrens and slopes, Mo. and Kans. to ne. Mex. and w. Tex.; adv. e. to Mich., Tenn. and Ga. June–Aug.

9. M. pectinàta Nutt. (comb-like). — Lower (1.5–3 dm. high) and more slender, more freely branched; *foliar bracts nearly or quite glabrous, straight and flat, divergent from base, greenish;* corolla unspotted. — Dry soil, e. Neb. to Colo. and Utah, s. to w. Tex., N.M. and Ariz. June, July.

23. BLEPHÍLIA Raf.

Calyx 13-nerved, naked in the throat; upper lip with 3 awned teeth, the lower with 2 nearly awnless teeth. Corolla inflated in the throat, nearly equally 2-lipped; upper lip erect, entire; the lower spreading, 3-cleft, its lateral lobes ovate and rounded, larger than the oblong and notched middle one. Stamens exserted (the upper pair minute or none). — Perennial N.Am. herbs, with nearly the foliage, etc., of *Monarda;* the small pale bluish-purple flowers crowded in axillary and terminal globose whorls; in summer. (Name from the Greek *blepharis, the eyelash,* in reference to the hairy fringe of the bracts and calyx-teeth.)

1. B. ciliàta (L.) Benth. (with marginal hairs). — Somewhat downy, 3–9 dm. high; *leaves* of flowering stem *almost sessile, oblong-ovate or -lanceolate, narrowed at base,* whitish-downy underneath; those of vegetative offshoots long-petioled, blunt, nearly elliptic or oval; outer bracts ovàte, acute, colored, ciliate, as long as the calyx; corolla hairy. — Dry woods, thickets and openings, Vt. to Wisc. and Ia., s. to Ga., Miss., Mo. and e. Tex. June–Aug. — Forma **albiflòra** House (white-flowered) has white corollas.

2. B. hirsùta (Pursh) Benth. (with straight hairs), Wood-Mint. — Taller, hairy throughout; *leaves long-petioled, ovate, pointed, rounded or cordate at base;* the lower floral ones similar, the uppermost and the *bracts linear-subulate,* shorter than the long-haired calyx; corolla pale, with darker purple spots. — Moist shady places, w. Que. and Vt. to Minn., s. to Ga., Tenn., Mo. and e. Tex. June–Sept.

Var. **glabràta** Fern. (becoming hairless). — Stem and leaves becoming glabrous. — Sw. Vt.

24. HEDEÒMA Pers. Mock Pennyroyal

Calyx ovoid or tubular, gibbous on the lower side near the base, 13-nerved, bearded in the throat, 2-lipped; upper lip 3-toothed, the lower 2-cleft. Upper lip of corolla flat, notched at the apex, the lower spreading and 3-cleft. Posterior stamens 0 or sterile. — Low odorous Am. annuals, with small leaves and loose axillary clusters of flowers (in summer) often forming terminal leafy racemes. (Name altered from *hedyosmon,* an ancient name of Mint, from the Greek *hedys, sweet,* and *osme, scent.*)

Leaves oblong-ovate, petioled, commonly toothed; sterile filaments manifest. . 1. *H. pulegioides.*
Leaves linear, entire, hispid, ciliate, sessile or nearly so: sterile filaments minute or obsolete. . . . . 2. *H. hispida.*

1. H. pulegioìdes (L.) Pers. (like *Mentha Pulegium*), PENNYROYAL of America, PUDDING-GRASS. — Erect, branching, *hairy;* whorls few-flowered; *upper calyx-teeth triangular*, the lower setaceous-subulate; corolla bluish, pubescent, scarcely exserted, 3–5 mm. long; taste and odor nearly of the true *Pennyroyal* (*Mentha Pulegium*) of Eu. — Dry soil, s. Que. to Minn. and S.D., s. to N.S., N.E., Fla., Tenn., Ark. and e. Kans. July–Sept.

2. H. híspida Pursh (with short stiff hairs). — Mostly low; *leaves linear, crowded, almost glabrous*, somewhat hispid-ciliate; *bracts spreading or reflexed;* upper flowers rather crowded; *calyx-teeth all subulate*, equaling the bluish corolla. — Plains, sands and rocky banks, N.Y. and Ont. to Sask., s. to Miss., La. and Tex.; adv. e. to w. N.E. May–July.

H. ACINOÌDES Scheele (resembling *Acinos*), similar to no. 2, minutely pubescent, slender, with slender-petioled thin oblong to ovate leaves, slender calyx and long-exserted almost filiform corolla-tube, is appearing as a weed in sw. Mo. (Adv. from farther south)

25. MELÍSSA L. BALM

Calyx with the upper lip flattened and 3-toothed, the lower 2-cleft. Corolla with a recurved-ascending tube. Stamens 4, curved and conniving under the upper lip. Otherwise nearly as *Satureja.* — Clusters few-flowered, loose, one-sided, with few and mostly ovate bracts resembling the leaves. Small Eurasian genus. (Name from the Greek *melissa, a bee;* the flowers yielding abundance of honey.)

1. M. OFFICINÀLIS L. (of the shops), COMMON B. — Upright, branching, perennial, pubescent; leaves broadly ovate, crenate-toothed, lemon-scented; corolla nearly white. — Roadsides, open woods and waste, esc. from cult., local, N.E. to O., s. and sw. to Fla., Ark. and Kans. June–Sept. (Introd. from Eu.)

26. SATURÈJA L. SAVORY. CALAMINT. SARRIETTE (Que.)

Calyx tubular to campanulate, 10–13-nerved, naked or hairy in the throat. Corolla with a straight tube and an inflated throat, distinctly 2-lipped; the upper lip erect, flattish, entire or notched; the lower spreading, 3-parted, the middle lobe usually largest. Stamens 4, somewhat ascending. — Herbs (ours) of warm or temp. reg., with mostly purplish or whitish flowers produced all summer; inflorescence various. (The ancient Latin name, *savory.*) Including CLINOPODIUM L. and CALAMINTHA Lam.

Flowers in loose or dense cymes, without long-ciliate subtending bracts.
 Calyx campanulate, naked in the throat; annual, with linear leaves. . . 1. *S. hortensis.*
 Calyx cylindric to tubular, hairy in the throat.
  Stem, branches and leaves pubescent.
   Stoloniferous, with ovate leaves; axillary cymes of flowers open, surpassing the subtending bracts; calyx minutely pubescent to glabrescent. . 2. *S. Calamintha.*
   Non-stoloniferous, with elliptic or oblong leaves; axillary cymes dense, overtopped by the bracts; nerves of calyx bristly-ciliate. . . 3. *S. Acinos.*
  Stem, branches and leaves glabrous; subsessile cymes of long-pedicelled flowers from axils of leafy bracts.
   Non-stoloniferous, the decumbent base sending off erect sprouts; nodes of stem bearded; primary leaves oval to elliptic or oblong-lanceolate, the larger ones 2.5–5 cm. long and 5–17 mm. broad; corolla 1 cm. long; filaments stout, twice to thrice length of anther; one style-branch much shorter than the other. . . . . 4. *S. glabella.*
   Stoloniferous, the repent stolons leafy; nodes of stems beardless or nearly so; leaves of flowering stem linear to linear-oblanceolate, the larger ones 1–2.5 cm. long and 1–5 mm. broad; corolla smaller; slender filaments four or five times length of anther; both style-branches elongate. . 5. *S. arkansana.*
Flowers in subglobose dense clusters in upper axils, accompanied by many long-ciliate linear-subulate bracts; calyx nearly naked in throat. . . 6. *S. vulgaris.*

1. S. HORTÉNSIS L. (of the gardens), SUMMER-S. — *Annual*, simple or erect-branching, up to 4.5 dm. high, the stem minutely pilose; *leaves* gray-green, *lance-linear*, blunt, glandular-punctate; *flowers in dense 1-sided axillary cymes; calyx tubular-campanulate, naked at throat, 10-nerved, with subequal* lanceolate ciliate *teeth;* corolla but slightly exserted, pale purplish; *branches of style nearly equal, slender*. — Roadsides and waste places, esc. from cult. Aug.–Oct. (Introd. from Eu.)

2. S. CALAMÍNTHA (L.) Scheele (old generic name meaning *beautiful mint*), BASIL-THYME. — *Stoloniferous, stiffly branching perennial* with pubescent stem up to 9 dm. high, commonly with tufts of reduced leaves in the axils; *leaves rhombic- or elliptic-ovate,* obtuse or rounded at tip, rounded to subcordate at base, petioled, subentire to crenate; *axillary cymes with slightly elongate and somewhat open axis, mostly surpassing the subtending bracts*, the flowers pedicelled; *calyx with* straight 11–13-nerved *tube, minutely pubescent to glabrescent*, 4.5–10 mm. long, strongly bilabiate, *bearded in the constricted throat*, the teeth of the lower lip elongate; corolla about twice length of calyx, lilac to violet. — Excessively variable Eurasian species, the following vars. (all natzd. from Eu.) with us:

a.Foliage-leaves membranaceous, the principal ones 4–7 cm. long and on petioles 1.5–2.5 cm. long; cymes loose, the lowest ones overtopped by the foliaceous bracts; pedicels elongate; calyx 7–10 mm. long, barely puberulent to glabrescent; corolla 1.5–2 cm. long. . . . . 1. *S. Calamintha* (typical).
a.Foliage-leaves thick and firm, the principal ones 1–3 cm. long and on petioles 3–8 mm. long; cymes all surpassing the subtending bracts; calyx 4.5–6 mm. long; corolla 1–1.5 cm. long. . . *b*.
b.Cymes lax and open, on peduncles 0.7–3 cm. long; stem long-villous; primary leaves 1.5–3 cm. long, on petioles 5–8 mm. long; calyx glabrescent. . . Var. *nepetoides*.
b.Cymes compact, on peduncles 2–7 mm. long; primary leaves 1–1.5 cm. long, on petioles 3–5 mm. long.
Stem and branches long-villous; calyx puberulent-hirtellous. . . . Var. *Nepeta*.
Stem and branches cinerous-puberulent; calyx barely puberulent or glabrescent, the glands therefore evident. . . . . Var. *glandulosa*.

S. Calamíntha (typical). — Local in woods and clearings, se. Va. (Natzd. from Eu.)

Var. nepetoìdes (Jord.) Briq. (like *Nepeta*). — Waste places, roadsides, etc., D.C., Va. and Tenn. July–Oct. (Natzd. from Eu.)

Var. Népeta (L.) Briq. (for *Nepeta*, Catnip), *S. Nepeta* (L.) Scheele. — Roadsides, pastures, old fields and waste places, N.C. to Ark., n. to Md. and Ky. July–Oct. (Natzd. from Eu.)

Var. glandulòsa (Riquien) Briq. (bearing glands). — Similar habitats, Ga. and Ala., n. to Md. and Tenn. (Natzd. from Eu.)

3. S. Ácinos (L.) Scheele (the ancient name of a fragrant herb), Mother-of-Thyme. — *Nonstoloniferous* annual or short-lived perennial, *commonly branching from the decumbent base*, 1.5–4 dm. high, the *stem with reflexed soft pubescence; leaves elliptic or oblong*, the bracteal similar to the others; flowers in few-flowered axillary *cymes*, these *much shorter than the subtending bracts; calyx contracted above the gibbous base*, tubular, bilabiate, *the* 13 *nerves coarsely bristly-ciliate;* small purplish corolla exserted 3–5 mm. from calyx; style-branches unequal, the upper subulate, the longer lower branch flattened. — Old fields, pastures, roadsides, etc., sw. Que. and s. Ont., s. to N.E., N.J., Pa. and Mich. June–Sept. (Natzd. from Eu.)

4. S. glabélla (Michx.) Briq. (smooth). — Essentially *glabrous perennial with* often *decumbent base* and erect basal sprouts; *stems* ascending or diffuse, simple or with strongly ascending branches, 2.2–6.5 *dm. long, bearded at the nodes; lower primary leaves oval to* elliptic or oblong-lanceolate, 2.5–5 *cm. long and* 5–17 *mm. broad*, entire or sparingly toothed, short-petioled, *with only the midrib prominent;* upper and bracteal leaves similar but smaller; *cymes subsessile in axils of leaf-like bracts, the flowers on slender pedicels*, the *bractlets* 5–15 *mm. long;* calyx bearded in the throat; the slender tube nearly smooth, 5 mm. long, the teeth lance-attenuate; corolla bluish-purple, about 1 cm. long; *filaments thickish, twice to thrice length of anther; style-branches very unequal;* nutlets obscurely reticulate. — Damp calcareous cliffs, bluffs, gravels and river-silts, Ky., Tenn., s. Mo. and Ark. June–Aug.

5. S. arkansàna (Nutt.) Briq. (of Arkansas). — Smaller than no. 4, *freely stoloniferous, the repent stolons with short ovate to elliptic leaves; stems* slender, mostly erect, 0.5–4 dm. high, simple or erect-branching, *scarcely bearded at the nodes; cauline leaves linear-oblanceolate*, nearly or quite entire, *the larger ones* 1–2.5 *cm. long and* 1–5 *mm. broad; cymes* few–many-flowered, *their bractlets* 3–5 *mm. long;* calyx more strongly ribbed; corolla smaller; *filaments very slender, four or five times length of anther; style-branches both elongate*, strongly curling. (*S. glabra* sensu ed. 7; *S. glabella*, var. *angustifolia* (Torr.) Svenson) — Calcareous rock, banks, shores and barrens, e. Ont. to Minn., s. to w. N.Y., O., Ind., Ill., Ark. and Tex. June–Oct. — Plant with odor of Pennyroyal.

6. S. vulgàris (L.) Fritsch (common), Basil. — *Perennial with creeping bases;* the *flowering stems* ascending, simple or branching, *covered with spreading to reflexed villosity; leaves elliptic-ovate*, obtuse, with *the midrib and lateral nerves prominent beneath, densely villous or almost velvety beneath, copiously strigose-villous above*, the margins nearly entire or slightly dentate; *cymes densely many-flowered*, the lower and usually all exceeded by the foliaceous bracts, *subglobose, with elongate long-ciliate linear-subulate bracts and bractlets subtending flowers*, the common peduncle almost wanting; pedicels densely villous, up to 5 mm. long; *calyx tubular*, *long-villous above, the tube gradually incurved,* 5–6 mm. long, strongly 13-nerved, *scarcely bearded* at throat; the lobes 3–4 mm. long, ovate to lanceolate, bristle-tipped; *corolla purple-red,* exserted about 8 mm. out of calyx, the tube gradually passing into the throat; *branches of style very unequal, the long and flattened lower branch often concealing the short subulate upper one.* — Eurasian; represented with us by

Var. DIMINÙTA (Simon) Fern. & Wieg. (diminished). — Pubescence of the typical form; ovate leaves 1.5–3 cm. long, densely pubescent on both faces; double cymes only 1–2 cm. broad, barely exceeded by or surpassing the bracteal leaves; corolla red-purple. — Roadsides, pastures, etc., Jefferson Co., N.Y. (Natzd. from Eu.)

Var. **neogaèa** Fern. (of the New World), DOGMINT. — Stems 1.5–6 dm. high; *leaves* elliptic to elliptic-ovate, *the larger ones 2–5 cm. long, glabrous or only sparsely strigose above, hirtellous to villous on nerves,* more rarely on surfaces, *beneath; corolla with slender tube, lilac-pink to whitish.* — Woods, thickets, rocky or alluvial shores, etc., Nfld. to Man., s. to N.S., N.E., Del., upland to N.C., Tenn., Ind., Wisc. and Minn.; Colo. and N.M. June–Sept.

27. CONRADÌNA Gray

Calyx much as in *Satureja.* Corolla with a slender and straight tube somewhat shorter than the calyx, abruptly bent backward at the throat, deeply bilabiate, ringent; the upper lip concave and incurved, emarginate or retuse; the lower lip dependent, contracted at base, deeply 3-lobed, the lateral lobes roundish, the emarginate middle one dilated. Stamens arched-ascending under upper lip, parallel; locules of anther at length separated on a transversely dilated connective, their base with a tuft of fine villi. Leaves linear, revolute, 1-ribbed. — Small N. Am. genus of suffruticose to woody low plants with stiff fascicled leaves. (In honor of *Solomon White Conrad,* Philadelphia botanist, 1779–1831.)

1. **C. verticillàta** Jennison (whorled). — Diffuse low shrub, the lower branches rooting, the bark finally exfoliating; leaves 1.5–2 cm. long, glabrous and glandular-pitted above, minutely hoary beneath; flowers 2–6 in upper axils or terminal; calyx 6–7 mm. long, strongly 13-nerved, hirsute; corolla lavender, strongly spotted within. — Sandy and gravelly shores and open woods, local, Ky. and Tenn. May.

28. HYSSÒPUS L. HYSSOP

Calyx tubular, 15-nerved, equally 5-toothed, naked in the throat. Corolla short, 2-lipped; upper lip erect, flat, obscurely notched; the lower 3-cleft, with the middle lobe larger and 2-cleft. — Perennial herb, nat. of Eurasia, with virgate simple branches, lanceolate or linear entire leaves, and blue-purple flowers in small clusters crowded in a spike. (The ancient Greek name.)

1. **H. OFFICINÀLIS** L. (of the shops). — Dry pastures and roadsides, locally abund., s. and w. N.E. to Ont., Minn. and southw. July–Oct. (Introd. from Eu.)

29. PYCNÁNTHEMUM Michx. MOUNTAIN-MINT. BASIL

Calyx about 13-nerved, naked in the throat. Corolla short, more or less 2-lipped; the upper lip straight, nearly flat, entire or slightly notched; the lower 3-cleft, its lobes all ovate to oblong and obtuse. Lower pair of stamens rather longer than the upper; anther-locules parallel. — Perennial upright N. Am. herbs with a pungent mint-like flavor, corymbosely branched above, the floral leaves often whitened; the many-flowered whorls dense, crowded with bracts, and usually forming terminal heads or close cymes. Corolla whitish or purplish, the lips mostly dotted with purple. *Fl.* summer and early autumn. Varies, like *Mentha,* with the stamens exserted or included in different flowers. (Name composed of the Greek *pycnos, dense,* and *anthemon,* a *flower;* from the compact inflorescence.)* KOELLIA Moench

a. Calyx nearly regular, its teeth of essentially uniform length. . *b.*

b. Calyx-teeth 1–5 mm. long, lanceolate to subulate, with sharp or pointed tips. . . *c.*

c. Calyx-teeth and inner bracts rigidly awned, at most hirtellous or scabrous (or with 1 divergent long trichome in no. 3); leaves short-petioled, either ovate, the larger ones 1–3 cm. broad or, if oblong, narrowly lanceolate or linear, blunt. . . *d.*

d. Calyx-teeth 1.3–2.2 mm. long; leaves elliptic- or oblong-ovate, the larger

---

* Some characters taken from the *Study of Pycnanthemum* by GRANT & EPLING in Univ. Calif. Pub. Bot. xx. no. 3 (1943).

ones 1–3 cm. broad and with 5–7 pairs of lateral veins; corolla-tube about 4.5 mm. long. . . . . 1. *P. setosum.*

*d.*Calyx-teeth 2.3–5 mm. long; corolla-tube 4.5–7 mm. long.

Leaves narrowly oblong to oblong-lanceolate or almost linear, blunt, gray-green, entire or nearly so, the larger ones 0.5–1.5 cm. broad and with 3–5 pairs of lateral veins arising at or below the middle of the blade; calyx-teeth unappendaged. . . . . . 2. *P. flexuosum.*

Leaves ovate to ovate-lanceolate, acuminate, full green, serrate-dentate, the larger ones 2–3 cm. broad and with 5–8 pairs of lateral veins, the uppermost veins from well above middle of blade; some or all of calyx-teeth with a flexuous multicellular divergent trichome below the tip. . . . . 3. *P. monotrichum.*

*c.*Calyx-teeth merely acuminate (or, if sharp-pointed, not rigid), hirtellous to pilose; leaves linear to lanceolate, tapering to tip, only 0.15–1.5 cm. wide.

Stem glabrous or rarely a little pilose on angles; leaves linear, sessile, 1.5–5.5 mm. wide; calyx-teeth sharp-tipped, glabrescent. . . . . 4. *P. tenuifolium.*

Stem finely pubescent; leaves lanceolate to lance-linear, the larger ones 0.5–1.5 cm. wide, definitely petioled; calyx-teeth merely acuminate, pubescent. . . . . 5. *P. Torrei.*

*b.*Calyx-teeth 0.7–1 mm. long, deltoid, not sharp-tipped. . . *e.*

*e.*Sides of stem glabrous, angles pubescent; leaves sessile, lanceolate to lance-linear; bracts glabrous. . . . . 6. *P. virginianum.*

*e.*Sides of upper half of stem pubescent; primary leaves short-petioled; bracts puberulent to pilose. . . . *f.*

*f.*Primary leaves lanceolate, the larger ones 0.8–2 cm. broad, tapering at base, entire or shallowly dentate.

Sides of upper internodes minutely pilose to glabrescent; lower surfaces of leaves glabrous or glabrate, only the stronger veins short-pilose; heads on branches of corymb often single and stalked or 2–3 of them verticillate or subspicate. . . . . 7. *P. verticillatum.*

Sides of upper internodes as well as angles densely hoary-villous; lower surfaces of leaves hoary; glomerules rarely verticillastrate, often crowded and nearly sessile. . . . . 8. *P. pilosum.*

*f.*Primary leaves ovate or broadly ovate-lanceolate, rounded to subcordate at base, glabrous or nearly so, the larger ones 1.5–4 cm. broad, serrate; glomerules mostly solitary at tips of elongate peduncles or verticillate. . . . . 9. *P. muticum.*

*a.*Calyx bilabiate, the upper and the lower lips definitely unequal, the shorter one two-thirds to one-third as long as the other. . . *g.*

*g.*Glomerules somewhat flattened from above or corymbiform, loosening in age to expose the axes, either terminating stem and branches or verticillate and slightly peduncled. . . *h.*

*h.*Bracts green, not whitened.

Leaves lance-acuminate, the larger ones 1–2 cm. broad; calyx-teeth with long multicellular loosely ascending to spreading trichomes much exceeding the close short pubescence. . . . . 10. *P. clinopodioides.*

Leaves oval, bluntish, the larger ones 2.5–3.5 cm. broad; calyx-teeth merely minutely hirtellous. . . . . 11. *P. umbratile.*

*h.*Bracts strongly whitened, at least above. . . . *i.*

*i.*Upper internodes and petioles copiously short-villous or spreading-pilose; lower surfaces or midribs of leaves hirsute; inner bracts and calyx-teeth bearing many long flexuous multicellular divergent trichomes; longer calyx-teeth 1.5–3 mm. long; corolla deep pink to purple, lateral lobes of lower lip spreading-ascending. . . . . 12. *P. pycnanthemoides.*

*i.*Upper internodes and petioles canescent with minute incurving hairs or puberulent to glabrescent, spreading hairs sparse or none; lower surfaces of leaves not hirsute; longer calyx-teeth 1.5–2(–2.5) mm. long; corolla white or slightly pink-tinged, lateral lobes of lower lip and middle lobe nearly parallel.

Inner bracts and calyx-teeth with or without long divergent multicellular trichomes toward their tips; calyx-teeth subulate to lance-acuminate, slender-tipped; eastern. . . . . . . 13. *P. incanum.*

Inner bracts and calyx-teeth without subterminal slender trichomes; calyx-teeth ovate, deltoid or oblong, obtuse; southwestern. . . . 14. *P. albescens.*

*g.*Glomerules dense, subspherical, closely sessile in axils of upper green glabrous leaves. . . . . 15. *P. montanum.*

1. P. setòsum Nutt. (bearing bristles). — Stem 4–8 dm. high, puberulent; *leaves elliptic- or oblong-ovate,* acute to bluntish, *green, with low scattered teeth, the primary blades* 2.5–6 cm. long and 1–3 *cm. broad and with* 5–7 *pairs of lateral veins,* on petioles 1–2 mm. long; glomerules loosely corymbose, solitary at tips of branches or verticillate in upper axils; *inner bracts and subequal calyx-teeth rigidly awned,* scabrous; *calyx-teeth* 1.3–2.5 *mm. long; corolla* white or pinkish, *its tube about* 4.5 *mm. long.* (*P. aristatum* Michx., illegitimate name) — Dry to moist sandy woods or marshes, N.J. and Md. to n. Fla. July–Sept.

2. P. flexuòsum (Walt.) BSP. (flexuous). — *Grayish-green; stem* 0.35–1 m. or more high, minutely puberulent, *commonly with short or suppressed axillary branchlets; leaves narrowly oblong to oblong-lanceolate or oblong-linear, blunt, entire or nearly so, the larger ones* 2.5–5.5 cm. long and 0.5–1.5 *cm. broad, with* 3–5 *pairs of lateral veins arising at or below middle of blade,* the petioles 1–2 mm. long; inflorescence as in no. 1; *calyx-teeth* 2.3–5 *mm. long, the rigid awns stramineous; corolla-tube* 4.5–7 *mm. long.* (*P. hyssopifolium* Benth.) — Dry sandy or argillaceous woods and clearings, Fla. and Ala., n. to Va., D.C. and Tenn. July–Sept.

3. P. monòtrichum Fern. (one-haired). — In habit similar to no. 1, the puberulent stem 0.6–1.2 m. high, usually with elongate finally flowering branches; *leaves ovate or ovate-lanceolate, acuminate, full green, serrate-dentate, the larger ones* 4–5.5 cm. long and 2–3 *cm. broad, with* 5–8 *pairs of lateral veins, the uppermost veins from well above middle of blade; calyx-teeth lance-aristate,* 2.5–3 *mm. long, most or all of them bearing a single* (rarely more) *prolonged flexuous multicellular divergent trichome below the tip.* — Dry sandy woods and clearings, se. Va. July–Sept.

4. P. tenuifòlium Schrad. (slender-leaved). — *Stem glabrous,* 0.3–1 m. high, *usually bearing abundant sterile simple or forking axillary branches; leaves linear, sessile,* 1.5–5.5 *mm. wide, tapering to tip,* firm and sometimes revolute, *entire;* corymb crowded or open, simple or made up of several to many secondary corymbs; *appressed lance-attenuate bracts and glabrescent lance-subulate calyx-teeth* 1–2 mm. long, *with firm sharp tips.* (*P. flexuosum* sensu BSP., not *Origanum flexuosum* Walt., basonym) — Dry to moist siliceous or argillaceous open ground, thickets and bogs, Ga. to Tex., n. to N.E., n. N.Y., s. Ont., O., s. Mich., Wisc. and Minn. June–Sept.

5. P. Tòrrei Benth. (for John Torrey, 1796–1873). — *Stem* more loosely branched than in no. 4, the branches elongating, *finely pubescent; leaves thin* and membranaceous, *distinctly short-petioled, lanceolate to lance-linear, the larger ones* 0.5–1.5 *cm. wide, nearly or quite glabrous; glomerules gathered into capitate clusters at tips of stem and upper branches; outer divergent or reflexed bracts prolonged; calyx-teeth lance-acuminate, pilose.* — Dry, often fertile, woods and thickets, s. Ct. and e. L.I. to W.Va., s. to e. Va. and upland to Ga.; s. Ill. to Kans. and Ark. Late June–Sept.

Var. **lèptodon** (Gray) Boomhour (slender-toothed). — *Stem villous or spreading-pilose; leaves hirsute on veins beneath; calyx spreading-pilose.* (*P. leptodon* Gray) — W.Va. and w. Va. to O., Mo. and mts. of N.C.

6. P. virginiànum (L.) Durand & Jackson (Virginian). — In habit resembling no. 4, less densely branched; *stems with glabrous sides and pubescent angles; leaves lanceolate to lance-linear,* entire, *rounded to sessile base, glabrous or minutely pubescent beneath,* the larger ones 6–11 mm. broad; *corymb* much as in no. 4, *made up of few–many convex-topped secondary corymbs; foliaceous bracts glabrous, green; calyx-teeth triangular, less than* 1 *mm. long, acuminate, pubescent at tip.* — Gravelly shores, meadows, dry to wet thickets, etc., centr. Me. to N.D., s. to s. N.E., L.I., Va., upland to w. N.C. and e. Tenn., O., Ind., Ill., Mo. and e. Kans. July–Sept.

7. P. verticillàtum (Michx.) Pers. (whorled). — *Stem minutely pilose* to glabrescent *on sides* as well as angles, its branches prolonged and mostly flowering late in season; *leaves lanceolate, short-petioled, glabrous to glabrate beneath, with only the principal veins short-pilose,* the broader blades 1–1.5 cm. broad, *tapering at base; corymb open,* with foliaceous bracts canescent with minute hairs, *the glomerules terminating slender stalks or* 2 *or* 3 *of them in upper axils; calyx-teeth narrowly triangular, acuminate, barely* 1 *mm. long.* — Dry to moist thickets, clearings swales or (southward) wet peat, sw. Que. to Mich., s. to s. N.E., L.I., Va. and w. N.C. July–Sept.

8. P. pilòsum Nutt. (hairy). — *Sides and angles of stems and lower surfaces of leaves hoary-pilose or -villous; leaves lanceolate,* entire or shallowly dentate, acuminate, *tapering to short petioles,* the broader blades 0.8–2 cm. broad; *corymb of* few–several *convex-topped dense small corymbs of nearly sessile crowded glomerules,* the latter rarely verticillastrate; *calyx-teeth hardly*

1 *mm. long*, acuminate, *sharp-tipped, canescent-pilose.* — Dry to moist woods, thickets and clearings, sw. Ont. and s. Mich. to Ia., s. to Tenn., Ark. and Okla.; adv. from e. Mass. to e. Pa. July–Sept.

9. P. mùticum (Michx.) Pers. (awnless). — Habit much as in no. 7; *stem minutely hoary-pilose* to puberulent; *leaves ovate to broadly ovate-lanceolate, acuminate, rounded to subcordate at* subsessile to short-petioled *base*, firm, shallowly serrate-dentate, glabrous, or minutely pilose beneath, *the larger ones* 1.5–4 *cm. broad; foliaceous bracts whitened with minute velvety pubescence;* corymb as in no. 7, the *glomerules solitary at tips of branches or verticillastrate; calyx* minutely pilose, *the narrowly triangular acuminate but scarcely sharp-tipped teeth less than* 1 *mm. long.* — Dry (or southw. wet) woods, thickets and clearings, sw. Me. to Mich. and Ill., s. to L.I., n. Fla., Tenn., La. and e. Tex. July–Sept.

10. P. clinopodioìdes T. & G. (like *Clinopodium*). — *Stem pubescent with* short curving and longer *spreading hairs; leaves lanceolate, acuminate, tapering to petioles* mostly 3–6 mm. long, sharply serrate to subentire, *pale green, membranous,* more or less hirtellous especially on veins beneath, *the primary blades* 1–2 *cm. broad; foliaceous bracts* similar to foliage-leaves, *green; glomerules hemispherical, either terminating branches or verticillastrate; calyx* short-pubescent, *bilabiate, the free teeth of the upper lip about two-thirds as long as the acuminate lower teeth* (1–1.5 mm. long), *these and the inner bracts usually appendaged near tip by long flexuous divergent multicellular trichomes.* — Dry to moist wooded slopes, thickets and shores, rather local, e. Mass. to W.Va., rarely s. to n. Del. and along mts. to N.C. July–Sept.

11. P. umbràtile Fern. (of deep shade). — Differing from no. 10 in its *curved-pilose pubescence of stem without longer and straight hairs; leaves oval, bluntish,* dark green, essentially entire, *broadly rounded to petiole,* barely puberulent on midrib beneath, *the larger ones* 2.5–3.5 *cm. broad; corymb terminal,* hemispherical, subtended by elliptic large green bracts; *inner bracts and* subulate-tipped *calyx-teeth merely short-hirtellous, without long trichomes.* — Bottomland-woods, Blackwater R., se. Va.

12. P. pycnanthemoìdes (Leavenw.) Fern. (resembling *Pycnanthemum*, the species originally considered a separate genus). — Tall, up to 1.5 m. high; the *stem* with loosely ascending, elongate and eventually flowering branches, *pilose with incurved hairs and with some divergent straighter ones; leaves* ovate-oblong, acuminate, *coarsely toothed, the lower and median ones* dark green above, *grayish and minutely pilose beneath, the spreading hairs of the midrib barely* 0.5 *mm. long;* larger blades 6–9 cm. long and 2.5–4 cm. broad, on petioles 4–10 mm. long; *upper leaves and foliaceous bracts whitened-puberulent above;* glomerules hemispherical, becoming lax and broad, terminating long branches or at 2 or 3 upper nodes; *inner* elongate *bracts and teeth of canescent calyx bearing numerous elongate flexuous multicellular divergent trichomes; tips of calyx unequal,* the lower pair 1.5–2.5 mm. long, the upper ones much shorter; *corolla* relatively large and showy, *roseate to purple, its dark-spotted lower lip with the lateral lobes spreading-ascending.* (*P. Tullia* Benth.) — Woods and thickets, w. Va. and W.Va. to s. Ill., s., principally in mts., to n. S.C., n. Ga. and Tenn. July, Aug.

Var. **viridifòlium** Fern. (green-leaved). — *Spreading pubescence of stem more abundant; primary foliage-leaves greener beneath* and more strongly hirsute, *the abundant hairs along the midrib up to* 1 *mm. long,* the larger blades 6.5–10 cm. long and 3–4.5 cm. broad, on petioles up to 1.5 cm. long; long divergent trichomes of inner bracts and calyx-teeth less numerous. (*P. viridifolium* (Fern.) Grant & Epl.) — Rich woods and thickets, e. Va. to w. Va., s. to e. S.C., Ga. and Ala. July, Aug.

13. P. incànum (L.) Michx. (gray). — *Upper internodes of stem cinerous-pilose with curving crowded hairs intermixed with some longer and straighter divergent ones,* the loosely ascending branches soon prolonged and eventually flowering; *leaves ovate to ovate-oblong,* remotely toothed, *hoary-pilose beneath with copious elongate hairs,* lower and median *blades* green above, *the upper and bracteal* ones *canescent above;* larger primary blades somewhat rounded to the petiole, 4.5–11 cm. long and 1.5–5.5 cm. broad; glomerules loosely hemispherical, terminating branches or in the upper axils; *calyx closely canescent, the inner bracts and the lance-acuminate slender-tipped teeth with or without long flexuous subterminal trichomes,* the lower calyx-teeth 1–1.5 mm. long; *corolla white to pale lilac,* purple-spotted, *the lobes of the lower lip nearly parallel.* — Dry woods and thickets, s. N.H. to w. N.Y., sw. to s. Ill., s. to N.C. and Tenn. July–Sept. Passing southw. into

Var. **Loomísii** (Nutt.) Fern. (named in 1834 in honor of its discoverer, Dr. H. Loomis). — *Upper internodes densely canescent with minute incurved hairs,* without or with few spreading longer ones; *lower surfaces of leaves minutely pilose or pruinose; leaves ovate-lanceolate to broadly lanceolate, tapering at base, the larger* 1.5–3 *cm. broad; calyx densely canescent, the lower teeth* 1.5–2 *mm. long; summits of bracts and calyx-teeth with long divergent multicellular trichomes.* (*P. Loomisii* Nutt.) — Woods and thickets, N.J. to s. Ill., s. to n. Fla. and Ala. Passing into

Var. **pubérulum** (Grant & Epl.) Fern. (puberulent). — Similar to preceding but *puberulence of internodes sparse and very minute to almost wanting; calyx sparsely puberulent,* not densely canescent. (*P. puberulum* Grant & Epl.) — Fla. to La., n. to Va., W.Va. and s. O.

14. P. albéscens T. & G. (whitish). — Habitally similar to last var. of no. 13; upper internodes puberulent or minutely incurved-pilose, with or without spreading hairs; *leaves lanceolate to ovate,* all but the lower strongly whitened, *the larger ones* 3–7 *cm. long and* 1–2.5 *cm. broad,* pale and minutely pubescent beneath; bracteal ones strongly whitened; glomerules relatively small; *calyx* densely canescent-pannose, *without long multicellular trichomes, the ovate, deltoid or oblong teeth blunt,* the lower ones 1–1.3 mm. long. — Dry woods and thickets, nw. Fla. to e. Tex., n. to Mo. and e. Okla. July–Sept.

15. P. montànum Michx. (of mountains). — *Stem* simple or with few long ascending branches, *glabrous; leaves glabrous, green throughout,* lance-ovate to -oblong, thin, acuminate, sharply serrate, short-petioled, the larger 6–12 cm. long and 2–4.5 cm. broad; *glomerules dense, subspherical, closely sessile in upper axils or at tips of long branches; bracteal leaves long-acuminate; calyx teeth deltoid-subulate,* the lips unequal. — Upland woods, W.Va. and w. Va. to w. S.C. and e. Tenn. June–Aug.

30. ORÍGANUM L. Wild Marjoram

Calyx hairy in the throat, striate, 5-toothed. Tube of the corolla about the length of the calyx; the upper lip rather erect and slightly notched, the lower longer, of 3 nearly equal spreading lobes. Stamens exserted, diverging. — Perennials nat. of Mediterr. reg., with nearly entire leaves, and purplish flowers crowded in cylindrical or ellipsoid spikes imbricated with colored bracts. (An ancient Greek name, composed of the Greek *oros, a mountain,* and *ganos, ornament.*)

1. O. vulgàre L. (common). — Upright, hairy, corymbose at the summit; leaves petioled, round-ovate; bracts ovate, obtuse, purplish. — Roadsides, old fields and thin woods, sw. Que. and s. Ont., s. to N.C. June–Oct. (Natzd. from Eu.)

31. THỲMUS L. Thyme

Calyx 13-nerved, hairy in the throat; the upper lip 3-toothed, spreading; the lower 2-cleft, with the subulate divisions ciliate. Corolla short; the upper lip straight and flattish, notched at the apex, the lower 3-cleft. Stamens 4, straight and distant, usually exserted. — Low perennials chiefly of Old World, with small and entire strongly veined leaves, and purplish or whitish flowers. (The ancient Greek name of the Thyme, probably from *thyein, to burn perfume,* because it was used for incense.)

1. T. Serpýllum L. (old generic name), Creeping T., Serpolet (Que.). — Prostrate; leaves green, flat, ovate, entire, short-petioled; flowers crowded at the ends of the branches. — Old fields, etc., Que. and Ont., s. to N.S., s. N.E., N.C., O. and Ind. July, Aug. (Natzd. from Eu.)

32. CUNÌLA L. Dittany

Calyx ovoid-tubular, equally 5-toothed, very hairy in the throat. Upper lip of corolla erect, flattish, mostly notched; the lower spreading, 3-cleft. Stamens 2, erect, exserted; sterile filaments short, minute. — Am. perennials, with small white or purplish flowers in corymbed cymes or clusters. (An ancient Latin name for some fragrant plant, transferred to this Am. genus.) Mappia Heist.

1. C. origanoìdes (L.) Britt. (like *Origanum*), Common D. — Stems tufted, corymbosely much branched, 2–4 dm. high; leaves smooth, ovate, serrate, rounded or cordate at base, nearly sessile, dotted, 1.5–4 cm. long; cymes peduncled; calyx striate. (*C. Mariana* L.; *Mappia origanoides* (L.) House) — Dry open woods, thickets and clearings, Fla. to Tex., n. to se. N.Y., e. Pa., W.Va., O., Ind., Ill., Mo. and Okla. July–Oct.

33. LÝCOPUS L. Water-Horehound. Bugleweed

Calyx campanulate to ovoid, 4–5-toothed, naked in the throat. Corolla more or less campanulate. Stamens 2, distant, the upper pair either sterile rudiments or wanting. Nutlets with thickened margins. — Perennial mostly stoloniferous herbs of N. Hemisph., resembling *Mentha* but not fragrant, with mostly toothed or pinnatifid leaves, the floral bracts similar and much longer than the dense axillary whorls of small mostly white flowers; in summer. (Name compounded of the Greek *lycos, a wolf,* and *pous, foot,* from some fancied likeness in the leaves.)*

---

* Nutlet-characters taken chiefly from study by Frederick J. Hermann in Rhodora, xxxviii. 373–375 (1936).

*a.*Calyx-teeth lanceolate to triangular, blunt or barely acutish, shorter than to about equaling mature nutlets.

Stem usually puberulent, rising from a slender base, with elongate mostly non-tuberiferous stolons; leaves dark green to purple; calyx ovoid-cylindric; corolla tubular, with erect lobes; stamens and style included. . . 1. *L. virginicus.*

Stem glabrate, from a tuberous base, the stolons bearing whitish tubers; leaves light green; calyx campanulate; corolla with flaring lobes; stamens and style evident, often exserted. . . 2. *L. uniflorus.*

*a.*Calyx-teeth subulate- or cuspidate-tipped, overtopping the nutlets. . . *b.*

*b.*Lower and median leaf-blades tapering to petioles or to subpetiolar bases; tubers rarely developed. . . *c.*

*c.*Upper surface of leaf strigose; nutlets 1.5–2 mm. long, 1–1.3 mm. broad; introd. weed. . . . . . . 3. *L. europaeus.*

*c.*Upper surface of leaf glabrous or at most minutely puncticulate when dry; nutlets 1–1.6 mm. long, 0.6–1 mm. broad; native.

Lower and median primary leaves (often wanting at maturity of fruit) commonly incised or pinnatifid at least at base; calyx-teeth with rigid needle-like tips; sterile filaments usually exserted; corolla-tube scarcely or but slightly exceeding calyx-teeth; summit of nutlet even and entire.

Stem with low blunt angles; leaves mostly incised toward base; filaments exserted; nutlets with corky ridge; wide-ranging in fresh soils. . . . . 4. *L. americanus.*

Stem with thin-winged angles; leaves rarely incised, merely sharp-dentate; filaments included; ridge of nutlet not corky; plant of tidal estuary of St. Lawrence R. . . . . 5. *L. laurentianus.*

Lower and median leaves merely serrate; calyx-teeth sharp but not needle-tipped; corolla-tube twice as long as calyx; sterile filaments scarcely exserted; dorsal low and rounded face of nutlet firm, not darkened; summit of nutlet undulate. . . . . . . 6. *L. rubellus.*

*b.*Lower and median leaf-blades sessile; plant tuberiferous.

Stem slender, glabrous to minutely puberulent; leaves smooth or smoothish, pale green, the larger ones with 3–7 teeth on each margin; bracts minute, nearly filiform, much shorter than glomerule; calyx-teeth subulate; plant of Coastal Plain and southern Great Lakes reg. . . . 7. *L. amplectens.*

Stem stout, bearing multicellular hairs; leaves scabrous, dark green, the larger ones with 6–12 teeth on each margin; bracts nearly equaling glomerule, dilated; calyx-teeth triangular to ovate; western. . . . . 8. *L. asper.*

1. L. virgínicus L. (Virginian). — *Stem* obtusely angled, *usually puberulent,* 2–8 dm. high, rising *from a slender (not conspicuously tuberous-thickened) base;* stolons filiform, mostly not tuberiferous; *leaves* dark green (or purple-tinged), *ovate or ovate-oblong,* firm, rather abruptly acuminate at both ends, coarsely toothed, 6–15 cm. long, 2–5 cm. broad; glomerules dense, often seemingly compound, in maturity 8–15 mm. broad, the mature nutlets usually concealing the calyces; *calyx ovoid-cylindric; corolla tubular, with erect lobes; stamens mostly included;* nutlets asymmetrical, their summits deeply muricate. — Rich moist soil, Ga. to e. Tex., n. to s. Me., s. N.H., Vt., N.Y., O., Ind., Wisc., Minn. and Neb. July–Oct.

2. L. uniflòrus Michx. (one-flowered; a misnomer, Michaux having small and immature specimens). — Similar, but *usually more* slender and *glabrate, from a tuberous base;* stolons basal, or in forma **flagellàris** Fern. (like a whip-lash) from middle and upper axils, finally tuberiferous; *leaves light green* (rarely purple-tinged), thinner, *lanceolate to lance-oblong,* gradually narrowed at both ends, 2–11 cm. long, 0.5–3.5 cm. broad; glomerules smaller and less dense, in maturity 4–9 mm. broad; *calyx campanulate; corolla with flaring lobes; stamens exserted.* — Low ground, Nfld. to B.C., s. to N.S., N.E., L.I., Del., Md., upland to N.C., O., Ind., Ill., Ia., Neb., w. Okla., Mont. and Oreg. Late June–Sept. (E. Asia)

3. L. europaèus L. (European). — *Stems* arising from stoutish subterranean stolons or rhizomes, *without tubers,* 0.4–1 m. high, branching, *pilose to villous above; leaves* broadly lanceolate to narrowly ovate, acuminate, sinuate-dentate, *the lower and median primary ones deeply pinnatifid or incised,* with the longer teeth or lobes 1–4 cm. long, *upper surface strigose,* lower surface slightly pubescent to glabrescent; calyx-teeth lanceolate, subulate-tipped; *corolla* white, spotted with purple, the *tube scarcely to barely exceeding calyx,* the oblong lobes short; *filaments exserted; nutlets* smoothish, 1.5–2 *mm. long,* 1–1.3 *mm. broad.* — Waste places, roadsides and neighborhoods of settlements, local, Mass. to Va. Aug.–Oct. (Adv. from Eu.)

Var. móllis (Kern.) Briq. (soft). — *Leaves oval, obtuse to subacute,* short-petioled, *with shorter blunt teeth, villous or soft-pilose beneath.* — Waste places, spring-heads, swales and shores, locally natzd., Ala. to La., n. to Mass., N.Y. and s. Ont. Aug.–Oct. (Natzd. from Eu.)

4. L. americànus Muhl. (American). — Stem erect, slender, 1.5–9 dm. high, glabrous or very sparingly appressed-pubescent with dark hairs, freely stoloniferous, without tubers; *lower primary leaves incised or pinnatifid,* especially at base, *glabrous or minutely pubescent on veins beneath,* lanceolate, narrowly ovate or oval; the upper leaves lance-attenuate, sinuate to sharply toothed, all petioled; bracts short; *calyx-teeth with long subulate tips; corolla-tube scarcely or barely longer than calyx-teeth, the filaments exserted; nutlets* 1–1.5 mm. long, 0.6–1 mm. broad, with entire or barely undulate angles, *the dorsal angular face relatively soft and dark, the summit entire.* — Low grounds, Nfld. to s. B.C., s. to N.S., N.E., L.I., Fla., Ala., Miss., Tex., N.M. and Calif. Late June–Sept.

Var. **Lóngii** Benner (for its discoverer, BAYARD LONG, 1885–). — *Upper internodes villous with long multicellular hairs;* upper *leaves* lance-linear, remotely toothed or subentire, *smooth above,* sessile or nearly so, mostly longitudinally folded. — Wet sandy or peaty swamps, wet woods, swales, etc., Coastal Plain and outer Piedmont, L.I. to Va.

Var. **scabrifòlius** Fern. (harsh-leaved). — Similar to the preceding var., but *leaves scabrous-puncticulate above.* — Similar habitats, n. O. to s. Mich. and Ill., s. to La. and e. Tex.

5. L. laurentiànus Rolland-Germain (of the River St. Lawrence). — Differing from no. 4 in having thicker and shorter subterranean stolons; *stem with narrow wing-angles;* leaves rarely incised; *stamens not exserted; nutlet with a firm smooth ridge.* — Fresh estuary of St. Lawrence R., Que.

6. L. rubéllus Moench (reddish). — Stem arising from slender stolons and rhizomes, 0.4–1.2 m. high; *leaves elliptic, elliptic-ovate or -lanceolate,* petioled, *at most sharply serrate or serrate-dentate,* acuminate; bracts minute; *calyx-teeth* acuminate and sharp-pointed, *hardly subulate-tipped; corolla* white, often with purple spots, 3.5–4 *mm. long, twice as long as calyx; sterile filaments mostly included; nutlets* 1–1.6 mm. long, about 1 mm. broad, *firm throughout, the low dorsal angle rounded and the lateral ones often undulate, the summit definitely so.* — Variable species; the following vars. recognized:

Primary leaves elliptic to elliptic-ovate or elliptic-lanceolate, 1–4 cm. wide.
 Primary leaves sharply serrate or serrate-dentate; stem glabrous or essentially so; calyx-teeth and tube subequal, glabrous, the teeth straight. . . *L. rubellus* (typical).
 Primary leaves with low and depressed teeth; younger internodes and lower surfaces of leaves pubescent; calyx-teeth often with recurving tips. . . Var. *arkansanus.*
Primary leaves narrowly lanceolate, 0.5–1(–2) cm. broad, long-acuminate to base and apex, low-dentate; stem, etc. glabrous. . . . . Var. *lanceolatus.*

L. rubéllus (typical). — Damp soils of woods, thickets, etc., Fla. to e. Tex., n. to s. N.E., N.Y., O., Mich., Ill. and Mo. July–Oct.

Var. **arkansànus** (Fresn.) Benner (of Arkansas). — Ark. to ne. Tex., n. to s. Mich. and Mo.

Var. **lanceolàtus** Benner (lanceolate). — Fla. to La., n. to Va., Tenn. and Mo.

7. L. ampléctens Raf. (clasping). — Stem rising from a tuberous base, the filiform stolons *tuberiferous,* the simple to freely branched *stem glabrous* or only *minutely puberulent,* 0.15–1.2 m. high, rather strict; *leaves sessile, glabrous or nearly so, pale green,* firm, with broad subclasping or narrowed base, *ovate, oblong, elliptic or lanceolate, with 3-7 broad teeth on each margin; bracts minute, slender; calyx* glabrous or glabrescent, *with sharp subulate teeth;* corolla twice as long as calyx; nutlets 1.2–1.5 mm. long, with undulate angles and summit. (*L. sessilifolius* Gray) — Damp sand and peat, chiefly on or near Coastal Plain, Fla. to Miss., n. to e. N.C.; mts. of w. N.C.; e. Md. to s. Ct., s. R.I. and se. Mass.; nw. Ind. Aug., Sept.

Var. **pùbens** (Britt.) Fern. (downy). — *Upper internodes positively pilose; lower surfaces of expanding leaves densely pilose.* — Miss. to Fla., n. to e. S.C.; s. N.J.; Block Island, R.I.

8. L. ásper Greene (harsh). — *Stem* strict, stoutish, simple or slightly forking, *from elongate tubers, the internodes spreading-pubescent with long multicellular hairs; leaves lanceolate to lance-oblong, sessile, dark green,* much longer than internodes, appressed- to spreading-ascending, *scabrous, the larger ones with* 6–12 *sharp teeth on each margin; bracts nearly equaling the glomerules, dilated; calyx-teeth triangular to ovate, acuminate; corolla-tube scarcely exceeding calyx; nutlets about 2 mm. long and 1.7 mm. broad,* with entire angles and low-hubbled summit. (*L. lucidus,* var. *americanus* Gray) — Marshes and wet shores, Alaska to Calif., e. to Man., Mich., Ill. and Mo.; occasionally adv. along railroads, etc., eastw. July, Aug.

34. MÉNTHA L. MINT

Calyx campanulate or tubular, the 5 teeth equal or nearly so. Corolla with a short included tube; the upper lobe slightly broader, entire or notched. Stamens 4, equal, erect, distant. —

Odorous perennial herbs of temp. reg.; the small flowers mostly in close clusters, forming axillary capitate whorls, sometimes approximate in interrupted spikes, of two sorts as to the fertility of the stamens in most species. Corolla pale purple or whitish. Species mostly adv. or natzd. from Eu., with many hybrids. (*Minthe* of Theophrastus, from a Nymph of that name, fabled to have been changed by Proserpine into Mint.)*

a. Flowers in terminal crowded or interrupted spikes or subcapitate inflorescences, with or without some clusters in the upper axils. . . *b*.

b. Spikes slender and leafless, densely crowded; leaves sessile or nearly so. . . *c*.

c. Spikes canescent-pilose to tomentulose.

Leaves mostly acute, sharply serrate, with slender veins; spike 0.8–1.2 cm. thick, the individual glomerules distinct; calyx-teeth lance-subulate. . . . . 1. *M. longifolia.*

Leaves rounded at tip, open-dentate, with veins prominent beneath; spikes 1.2–1.5 cm. thick, with crowded glomerules; calyx-teeth lance-attenuate. . . . . 2. *M. alopecuroides.*

c. Spikes with green or greenish bracts and calyces.

Plant soft-pubescent; leaves round-ovate to broadly elliptical, cordate, rugose. . . . . 3. *M. rotundifolia.*

Plant essentially glabrous; leaves oblong- or ovate-lanceolate, rounded to base, not rugose. . . . . 4. *M. spicata.*

b. Spikes more open and interrupted, with the clusters somewhat separated in the terminal spikes or the flowers aggregated into terminal capitate spikes; leaves definitely petioled. . . *d*.

d. Calyx, or at least the teeth, more or less hirsute.

Leaves simply serrate or dentate.

Plant glabrous; leaves lance- or ovate-oblong; spikes loosening, becoming elongate, with 3–12 or more distinct nodes; calyx glabrous below the teeth. . . . . 5. *M. piperita.*

Pubescent with recurved hairs; leaves round-ovate; spikes globular and terminal or with 1–3 secondary glomerules below; calyx pubescent. . . . . 6. *M. aquatica.*

Leaves lacerate and crisped, ovate or orbicular; spike elongate, with many nodes. . . . . 7. *M. crispa.*

d. Calyx glabrous; stem and ovate leaves glabrous or glabrate; flowers in dense capitate to ovoid spikes. . . . . 8. *M. citrata.*

a. Flowers in subglobose clusters, all in axils of leaves, the uppermost axils often without glomerules; leaves more or less petioled. . . *e*.

e. Upper leaves conspicuously reduced, only twice or thrice exceeding glomerules. . . . . 9. *M. cardiaca.*

e. Upper leaves scarcely reduced, many times exceeding the glomerules.

Stem glabrous or rarely with a few scattered hairs. . . . . 10. *M. gentilis.*

Stem retrorse-pubescent, at least on the angles, with fine hairs. . . . . 11. *M. arvensis.*

1. M. LONGIFÒLIA (L.) Huds. (long-leaved), HORSE-M. — *Stems* tall (up to 1.5 m. high), *canescent or finely pubescent; leaves ovate-oblong to oblong-lanceolate, acute, sharply serrate, with slender veins, glabrous or glabrate above, canescent-tomentulose beneath; spikes slender*, 0.8–1.2 *cm. thick*, elongate, *with many distinct glomerules* and linear-subulate tomentulose bracts; *calyx tomentose, with lance-subulate teeth; corolla glabrous within.* — Thickets, roadsides, damp shores, etc., Mass. to Ind., s. to Va. July–Sept. (Introd. from Eu.)

Var. MOLLÍSSIMA (Borkh.) Rouy (softest). — Similar, but *upper surfaces of leaves canescent-pubescent.* — Mass. to Ind., s. to Pa. and Mo. (Introd. from Eu.)

Var. UNDULÀTA (Willd.) Fiori & Paoletti (wavy). — *Leaves round-ovate, cordate, with jagged toothing.* — Roadsides, etc., Ind. (Introd. from Eu.)

2. M. ALOPECUROÌDES Hull (like Foxtail-grass). — Tall as in no. 1, *canescent-downy; leaves round-oval or -oblong*, sessile and *subcordate, with rounded or blunt tips*, coarsely open-dentate, *with veins prominent beneath; spikes* stouter, 1.2–1.5 cm. *thick, with crowded glomerules;* bracts

* The Old World species have crossed and recrossed until hundreds of the vegetatively propagating progeny have been described and very unsatisfactorily defined. Many of these have been introd. into this country and have found their way to waste places, roadsides, etc. Only the strongest-marked and definitely natzd. plants, without hypothetical guesses as to their derivation, are here included.

lance-linear; *calyx-teeth lance-attenuate,* tomentose. — Roadsides and waste places, Mass. to N.J. and Pa. July–Oct. (Introd. from Eu.) — Often considered a hybrid of nos. 1 and 3.

3. M. ROTUNDIFÒLIA (L.) Huds. (round-leaved). — Stems 0.3–1 m. high, *downy; leaves broadly elliptic to round-ovate, cordate, nearly sessile,* round-tipped, *rugose-reticulated,* crenate-dentate; *spike very slender,* in age almost tail-like, *compact; lower bracts ovate-lanceolate,* upper lanceolate, *greenish; calyx greenish,* velvety, *with lance-triangular teeth connivent in fruit; corolla* white or pink, *with a ring of hairs within.* — Roadsides, old fields, thickets and waste places, Me. to Mich., s. to Fla., La. and Tex. Late June–Sept. (Introd. from Eu.)

4. M. SPICÀTA L. (spiked), SPEARMINT, BAUME (Que.). — *Stem* up to 1 m. or more high, usually branched, nearly or quite *glabrous; leaves oblong- or ovate-lanceolate, unequally serrate, acute, rounded at base, subsessile or very short-petioled; spikes* very slender and elongate, the lower glomerules distinct or separated, *with green and glabrous conspicuous narrow bracts; pedicels and bases of calyx glabrous;* flowers pinkish or pale violet. — Wet places near settlements, throughout our range and beyond. Late June–Oct. (Introd. from Eu.)

5. M. PIPERÌTA L. (peppery), PEPPERMINT, MENTHE POIVRÉE (Que.). — Tall, *glabrous,* commonly branching; *leaves lance- or ovate-oblong, definitely petioled,* acute, sharply serrate; *spike* at first ovoid, *lengthening and becoming loose, the lowest glomerule then remote and subtended by reduced leaves;* pedicels elongate; *calyx* rather large, *with glabrous tube and hirsute teeth;* corolla purplish. — Brooksides, wet meadows, etc., throughout our range and beyond. Late June–Oct. (Introd. from Eu.) — By Europeans often considered a hybrid of nos. 4 and 6.

6. M. AQUÁTICA L. (aquatic), WATER-M. — *Pubescent with recurved hairs; leaves round-ovate, blunt, broadly rounded to subcordate at base,* petioled, dentate; *terminal spikes globular or subglobose, peduncled,* with or without 1–3 secondary sessile glomerules below; *calyx* large, with lanceolate teeth, it and the pedicel *pubescent.* — Wet places, local, N.S. to Del. Aug.–Oct. (Introd. from Eu.)

7. M. CRÍSPA L. (curled or crisped). — Somewhat like nos. 5 and 6, glabrous or slightly pubescent; *leaves* short-petioled, *ovate to orbicular, lacerate-dentate and crisped;* spike slender and elongate, with many slightly separated glomerules; calyx small, pubescent or glabrate below. — Roadsides, ditches, etc., Mass. to Mich., s. to N.J. and Pa. June–Sept. (Introd. from Eu.)

8. M. CITRÀTA Ehrh. (lemon-like; from the odor). — Resembling no. 6; *stem and oval or ovate* slender-petioled appressed-serrate *leaves glabrous* or nearly so; *flowers in* capitate glomerules forming *dense ovoid spikes,* the upper glomerules subtended by large leaves; *calyx glabrous.* — Damp places, N.S. to Mich., s. to Va. and Mo. July–Sept. (Introd. from Eu.) — Sometimes treated as a var. of no. 5.

9. M. CARDÌACA Baker (for the heart). — *Tall and upright, with ascending branches above;* stem more or less pubescent; *leaves* lanceolate to lance-oblong, acuminate, *sharply serrate,* slightly pubescent, *the upper leaves conspicuously reduced and twice or thrice as long as the axillary* subglobose *glomerules;* the uppermost pairs of leaves often without glomerules. — Wet meadows, shores and waste places, Nfld. to Mich., s. to N.S., N.E., Va., O., Ind. and Ill. July–Oct. (Introd. from Eu.) — In Europe treated as a hybrid of nos. 4 and 11.

10. M. GENTÌLIS L. (related). — *Stem* slender, 0.3–1 m. high, mostly branching, reddish below, *glabrous or with few scattered hairs; leaves* ovate to obovate, *coarsely and sharply serrate,* especially above, slightly pubescent to glabrate, green, or in forma VARIEGÀTA Moldenke (variegated) white-mottled; *upper leaves scarcely reduced, many times exceeding the axillary glomerules.* — Rich damp soil, Que. to w. Ont., s. to N.E., N.C., Ind., Ill., Ia. and Neb. July–Sept. (Introd. from Eu.) — Supposed to be a hybrid of nos. 4 and 11.

11. M. arvénsis L. (of cultivated ground). — *Stems* usually branching, sometimes subsimple or simple, 1.5–8 or more dm. high, *retrorse-pubescent on the angles,* either pubescent or glabrous on the sides; *leaves* ovate, oblong or lanceolate, serrate, *petioled; flowers glomerulate in the upper axils, the leafy bracts little smaller than the foliage-leaves, many times longer than the axillary glomerules;* calyx pubescent, with narrowly deltoid to subulate teeth; corolla lilac-pink or purplish (rarely white). — Circumboreal, our only indig. species, complicatedly variable. The following are our chief vars. and forms:

a. Foliage-leaves and those at bases of lower glomerules ovate to elliptic, broadest chiefly below the middle, three-sevenths to four-fifths as broad as long, rounded to short petioles. . . *b.*

b. Stem near the lowest flowering axils pubescent on both sides and angles; leaves pubescent.

Angles of stem much more pubescent than the sides. . . . . . . *M. arvensis* (typical).

Angles and sides of stem about equally pubescent.

Hairs elongate, with elongate cells, mostly implexed or spreading, 1–3.5 mm. long. . . . . *M. arv.*, forma *lanata.*

Hairs appressed, only 0.2–1.5 mm. long, with very short cells. . . . *M. arv.*, forma *puberula.*

*b.*Stem near the lowest flowering axils glabrous on the sides, only minutely pubescent on the angles; leaves glabrescent. . . . . *M. arv.*, forma *glabra.*

*a.*Foliage-leaves and those at bases of lower glomerules lanceolate or lance-oblong, broadest slightly above to near middle, one-fourth to half as broad as long, tending to be cuneate or attenuate to petiole. *c.*

*c.*Stem near the lowest flowering axils pubescent on both sides and angles; leaves pubescent.

Angles of stem obviously more pubescent than sides. . . . . Var. *villosa.*

Angles and sides of stem almost equally pubescent; the hairs elongate, with long cells, implexed or spreading, 1–3 mm. long. . . . Var. *vill.*, forma *lanigera.*

*c.*Stem near the lowest flowering axils glabrous on the sides, only minutely pubescent on the angles; leaves glabrescent. . . . . . Var. *vill.*, forma *glabrata.*

M. arvénsis (typical). — Damp open soils, shores, etc., s. Lab. to Wash., s. to N.S., N.E., Va., Ky., Neb., N.M. and Calif., both native and natzd. from Eu. July–Sept. — Forma **lanàta** (Piper) S. R. Stewart (woolly), local, mostly on gravels, Nfld. to Wash., s. to N.S., N.E., Del., Md. and Calif. — Forma **pubérula** S. R. Stewart (with minute pubescence), local, Ont. to Oreg., s. to Pa. and Calif.; forma **glàbra** (Benth.) S. R. Stewart (hairless), Nfld. and e. Que. to Wash., s. to N.S., N.E., N.Y. and Kans.

Var. **villòsa** (Benth.) S. R. Stewart (with soft hair), *M. canadensis* L., and *M. arvensis*, var. *canadensis* (L.) Briq. — Mostly common, Nfld. to Alaska, s. to N.S., N.E., Del., Pa., upland to Va., O., Ind., Ill., Neb., N.M. and Calif. — Forma **lanígera** S. R. Stewart (bearing wool), chiefly on gravels, n. N.E.; Alaska to Ida. and Calif.; forma **glabràta** (Benth.) S. R. Stewart (becoming smooth), var. *glabrata* (Benth.) Fern., Lab. to B.C., s. to N.S., N.E., Pa., Md., W.Va., O., Ill., Mo., N.M., Ariz. and Calif.

35. ELSHÓLTZIA Willd.

Calyx with equal teeth. Corolla 4-lobed, slightly 2-lipped. Stamens 4, ascending, exserted, didynamous; anther-locules divergent. — Herbs, nat. of Eurasia and n. Afr., with ovate or oblong petioled leaves and spicate small flowers. (Named for *Johann Siegesmund Elsholtz*, 1623–1688, German physician and botanist.)

1. E. ciliàta (Thunb.) Hylander (ciliate). — Smooth annual, 3–7 dm. high; bracts of the spike ovate, veiny, mucronate; calyx hirsute; corolla purplish, 2–3 mm. long. (*E. Patrini* (Lepechin) Garcke) — Roadsides, old fields, etc., local, Temiscouata Co., Que., and Aroostook Co., Me. Aug.–Oct. (Natzd. from Asia)

36. COLLINSÒNIA L. Horse-Balm

Calyx ovoid, enlarged and declined in fruit, 2-lipped; upper lip truncate and flattened, 3-toothed, the lower 2-cleft. Corolla elongated, expanded at the throat, somewhat 2-lipped, the tube with a bearded ring within; the 4 upper lobes nearly equal, but the lower much larger and longer, pendent, toothed or lacerate-fringed. Stamens 2, much exserted, diverging; anther-locules divergent. — Strong-scented e. N.Am. perennials, with large ovate leaves, and yellowish flowers on slender pedicels. (Named in honor of *Peter Collinson*, 1694–1768, English botanist.)

1. C. canadénsis L. (Canadian), Richweed, Stoneroot. — Nearly smooth, 5–10 dm. high; leaves serrate, pointed, petioled, 1–2 dm. long; panicle loose; corolla 1.5 cm. long, lemon-scented; stamens 2. — Rich moist woods, Fla. to Ark., n. to Mass., Vt., N.Y., Ont. and Wisc. July–Sept.

Var. **punctàta** (Ell.) Gray (dotted). — Leaves closely tomentulose beneath; inflorescence puberulent. (*C. serotina* Walt.; *C. punctata* Ell.) — Local, Ga. and S.C. to Ct.

37. MICHELIÉLLA Briq. Stoneroot

Similar to *Collinsonia*, but the fertile stamens 4, usually 2 ascending and 2 descending; corolla rather broad, viscid-pubescent. — N. Am. perennials with hard rhizomes, broad toothed leaves and racemes of tawny or creamy flowers. (Named for *Marc Micheli*, 1844–1902, Swiss botanist.)

1. M. verticillàta (Baldw.) Briq. (whorled). — Stem slender, 1–5 dm. high, with subapproximate, seemingly subverticillate, petioled leaves at summit; leaf-blade ovate to obovate, thin, glabrous, coarsely dentate, 0.6–2 dm. long, acuminate, peduncle slender, viscid; bracts minute; lower pedicels often in 2's or 3's; calyx-teeth all attenuate-subulate; corolla tawny or yellowish-purple. — Rich woods, Ga. to Miss., n. to s. Va. and Tenn. May.

38. PERÍLLA L.

Calyx as in *Collinsonia*. Corolla-tube included, the limb 5-cleft; lower lobe a little larger. Stamens 4, included, erect, distinct, anthers with parallel locules. Nutlets globose, reticulated. — Coarse aromatic annual, nat. of Asia, with small flowers in 1-sided racemes. (Native East Indian name.)

1. P. frutéscens (L.) Britt. (shrubby). — Erect, branching, 0.3–1 m. high; leaves ovate, blunt-dentate; flowers white. — Roadsides and waste places, occasional garden-escape, and locally a troublesome weed, Fla. to Tex., n. to s. N.E., N.Y., O., Ind., Mo. and Kans. Aug.–Oct. (Introd. from e. Asia)

Var. críspa (Benth.) Deane (crisped). — Leaves laciniate-dentate. (Var. *nankinensis* (Lour.) Britt.) — Occasional, n. to Mass., N.Y., W.Va. and Ill. (Introd. from e. Asia)

39. MÓSLA Hamilton

Calyx as in *Perilla* and *Collinsonia;* corolla-tube naked or with a ring of hairs within, the lips short, the upper notched, the lower 3-cleft. Fertile stamens 2, included, erect, the anthers with divergent locules. Nutlets subglobose, reticulated. — Annual aromatic herbs, nat. of e. Asia, with 1-sided racemes of nearly bractless minute flowers. (Aboriginal name.)

1. M. dianthèra (Hamilton) Maxim. (with two anthers). — Stiffly branching, 3–7 dm. high; leaves narrowly ovate, petioled, serrate, the blades 2–3 cm. long. — Moist slopes and floodplains, McCreary Co., Ky.; also N.C. (Natzd. from Asia)

FAM. 152. SOLANÀCEAE (Nightshade Family)

Herbs (or often shrubs, especially in trop. reg.), with colorless juice and alternate leaves, regular or nearly 5-merous and 5-androus flowers, on bractless pedicels; the corolla imbricate or valvate in the bud, and mostly plaited; the fruit a 2-locular (rarely 3–5-locular) many-seeded capsule or berry. Seeds campylotropous or amphitropous. Embryo mostly slender and curved in fleshy albumen. Calyx usually persistent. Stamens mostly equal, inserted on the corolla. Style and stigma single. Placentae in the axis, often projecting far into the locule. Foliage rank-scented and, with the fruits, mostly narcotic, often very poisonous, though some are edible. — A large family in the tropics, but sparingly indigenous in our district, shading off into *Scrophulariaceae*, from which the plaited regular corolla and 5 equal stamens generally distinguish it.

a. Corolla regular (slightly irregular in no. 6); stamens all perfect and alike; fruit indehiscent or a 4-valved or circumscissile capsule. . . *b.*

  b. Corolla rotate, 5-parted or 5-lobed, the lobes valvate and their margins incurved in bud; anthers converging and forming a tube around the style, opening by terminal pores or chinks; fruit a berry. . . . . 1. *Solanum.*

  b. Corolla not valvate in bud; anthers separate, longitudinally dehiscent. . . *c.*

    c. Fruit a berry (juicy or dryish). . . *d.*

      d. Herbs; corolla plicate in bud; berry covered by the enlarged fruiting calyx. . . *e.*

        e. Calyx only obscurely if at all angulate, herbaceous, ribless, in fruit closely investing berry; flowers solitary or fascicled in leaf-axils; corolla rotate, clear white, whitish or blue-tinged, with yellow eye, cushioned in center. . . . . 2. *Chamaesaracha.*

        e. Calyx 5-angled, often 10-ribbed, membranaceous to scarious, in fruit inflated and reticulate-veiny, loosely covering the berry; pedicels solitary, axillary or extra-axillary; corolla campanulate to campanulate-rotate.

          Calyx gamosepalous, lobed only toward summit, in fruit with rounded or cordate base; corolla rotate-campanulate, lurid, greenish or yellowish-white to deeper yellow, often with dark center; ovary 2-locular; berry juicy. . . . . 3. *Physalis.*

Calyx of distinct sagittate-based sepals; corolla open-campanulate, pale blue; ovary 3–5-locular; berry dry. . . . . 4. *Nicandra.*

*d.*Shrubs; corolla imbricated in bud, funnelform or salverform; calyx not enlarged, persisting at base of 2-locular small berry. . . . . 5. *Lycium.*

*c.*Fruit a capsule. . *f.*

*f.*Calyx urceolate, becoming strongly ribbed and reticulate; corolla slightly irregular and oblique, reticulate-veiny; capsule inclosed in calyx, circumscissile near apex. . . . . 6. *Hyoscyamus.*

*f.*Calyx cylindric or tubular-campanulate, not reticulate; corolla regular, not reticulate-veined; capsule opening by valves.

Calyx prismatic, circumscissile near base, leaving a peltate border; corolla convolute-plicate in bud; capsule free, often armed, 4-locular at least at base; seeds large. . . . . 7 *Datura.*

Calyx terete, persistent; corolla imbricate in bud; capsule inclosed in calyx, 2-locular; seeds minute. . . . . 8. *Nicotiana.*

*a.*Corolla slightly oblique and a little bilabiate, the two upper lobes external, imbricated in bud; stamens unequal, 1 much smaller than the others; capsule 2-valved; seeds minute, plump. . . . . 9. *Petunia.*

1. SOLÀNUM L. Nightshade

Calyx and rotate corolla 5-parted or 5-cleft (rarely 4–10-parted), the latter plaited in the bud, and valvate or induplicate. Stamens exserted; filaments very short; anthers converging around the style, opening at the tip by two pores or chinks. Berry usually 2-locular. Herbs or shrubs, the larger leaves often accompanied by a smaller lateral (rameal) one; the peduncles mostly lateral (or becoming lateral) and extra-axillary. — A polymorphous and largely tropical genus of nearly 1000 species. (The classical Latin name, unexplained.)*

*a.*Berries naked, not wholly covered by the calyx; stamens all alike. . . *b.*

*b.*Plant without prickles; flowers 0.5–1.5 cm. broad; anthers blunt; berries 0.5–1.5 cm. in diameter; pubescence, when present, of simple hairs. . *c.*

*c.*Woody climbing or twining shrub; corolla purple (rarely white); berries ovoid to ellipsoid, red. . . . . 1. *S. Dulcamara.*

*c.*Herbs, not twining; berries green, black, yellow or (rarely) red, globose. *d.*

*d.*Leaves pinnate, with 5–11 leaflets; perennial by subterranean tubers. . 2. *S. Jamesii.*

*d.*Leaves simple; annuals, without tubers. . . *e.*

*e.*Leaves deeply pinnatifid, oblong; flowers 1–3; berries green, 1–1.5 cm. in diameter; western. . . . . 3. *S. triflorum.*

*e.*Leaves entire or sinuate, rarely bluntly lobed at base, ovate; flowers often several; berries black, green, yellow or red, 0.5–1.3 cm. in diameter. *f.*

*f.*Stem glabrous or only very sparsely and remotely pubescent; berries black.

Leaves thin and membranaceous, translucent to transmitted light; anthers 1.3–2 mm. long; berries lustrous-black, 5–9 mm. in diameter, with seeds 1.2–1.8 mm. long; native in undisturbed (or disturbed) habitats. . . . . 4. *S. americanum.*

Leaves thickish, dense or opaque to transmitted light; anthers 1.8–2.6 mm. long; berries dull black, 8–13 mm. in diameter, with seeds 1.7–2.3 mm. long; weed of waste places and disturbed soil. . . . . 5. *S. nigrum.*

*f.*Stem copiously pubescent; berries green, yellow or red, with seeds 1.8–2.2 mm. long; adventive weeds.

Stem and leaves densely appressed-villous or -pilose; fruiting pedicel slenderly clavate; calyx not accrescent in fruit, the yellow to red berry standing above it. . . . . . . 6. *S. villosum.*

Stem and leaves hirsute with spreading hairs; fruiting pedicel enlarged upward into the base of the calyx; calyx strongly accrescent after anthesis, embracing much of the green berry. 7. *S. sarachoides.*

*b.*Plant more or less prickly, perennial, with deep creeping rootstock; flowers larger; anthers tapering upward; berries mostly larger; pubescence wholly or partly stellate. . *g.*

---

* Treatment of nos. 4–7 with aid from G. Ledyard Stebbins.

*g.*Hirsute with 4–8-rayed hairs, the central division often elongate; leaves greenish, elliptic-oblong to oval, coarsely sinuate; calyx-lobes lance-acuminate; eastern and wide-ranging. . . . . 8. *S. carolinense.*

*g.*Scurfy with closely matted canescent many-rayed short hairs; western.

Leaves linear-lanceolate to narrowly oblong, gradually tapering at base; calyx-lobes linear-subulate. . . . . 9. *S. elaeagnifolium.*

Leaves ovate, broadly rounded to cordate at base; calyx-lobes broadly ovate, with abrupt acumination. . . . . 10. *S. Torreyi.*

*a.*Berries wholly or partially covered by the prickly calyx; annuals. . . *h.*

*h.*Anthers equal; fruiting pedicels spreading; calyx loosely covering the red berry; seeds minutely reticulate-pitted. . . . . 11. *S. sisymbriifolium.*

*h.*Anthers dissimilar, the lowest much larger and longer, with an incurved beak; fruiting pedicels erect; berry wholly inclosed by the close-fitting and often adherent calyx; seeds coarsely undulate-rugose.

Pubescence all stellate; corolla yellow, nearly regular, with broadly ovate lobes. . . . . 12. *S. rostratum.*

Pubescence partly of simple gland-tipped hairs; corolla violet, with acuminate lobes. . . . . 13. *S. citrullifolium.*

1. S. DULCAMÀRA L. (old generic name, bittersweet, early and bold experimenters stating that the chewed root first tasted bitter, then sweet), BITTERSWEET, NIGHTSHADE. — *Woody climber; leaves ovate and acuminate,* with or without auricled or lobed bases, glabrous or nearly so; *peduncles terminal, becoming lateral, the cyme several-flowered; pedicels nodose-articulate at base; corolla violet or purple,* or white in forma ALBIFLÒRUM House (white-flowered); *berry red, ovoid or ellipsoid.* — Thickets and clearings near settlements, throughout. Mid-May–Sept. (Introd. and natzd. from Eu.)

Var. VILLOSÍSSIMUM Desv. (most hairy). — Branches and leaves copiously pubescent. — Nfld. to s. N.E., w. to Mich. and Ill.

S. PSEÙDO-CÁPSICUM L. (false Pepper), JERUSALEM-CHERRY, a low erect shrub with oblanceolate or oblong shining leaves, white corolla and globose scarlet or yellow berry, occasionally spreads from cult. but hardly persists. (Introd. from s. Eu.)

2. S. JÀMESII Torr. (for its discoverer, EDWIN JAMES, 1797–1861), WILD POTATO. — Small *herb* up to 3 dm. high, *perennial by small subterranean tubers;* the soft stem branching; *leaves odd-pinnate, with* 2–5 *pairs of oblong-ovate to lanceolate leaflets,* the lowest pairs reduced; cyme few–several-flowered; the deeply cleft white corolla with lanceolate lobes about 7 mm. long; berry globose. — Introd. and locally estab. in s. Ia. (Introd. from our Southwest)

S. TUBERÒSUM L. (tuberous), the cult. POTATO or PATATE (Que.), occasionally springs up in garden-waste but is hardly persistent with us. (Introd. from w. S. Am.)

3. **S. triflòrum** Nutt. (3-flowered). — *Annual,* with more or less pubescent stem commonly branching and again branching from near base; *leaves oblong, deeply pinnatifid, with rounded sinuses; peduncles* 1–3-*flowered,* the pedicels soon reflexed; corolla white, scarcely 1 cm. broad; *berry globose, green,* 1–1.5 *cm. in diameter.* — Dry plains and open woods, often as a weed of fields, roadsides, etc., B.C. to Calif., e. across the plains to Minn., Ia., Kans. and Okla., and as an adv. weed occasionally e. to Atl. states. June–Aug.

4. **S. americànum** Mill. (American). — *Slender usually divergently branched glabrous or glabrescent annual* 0.05–1 m. high; *leaves ovate to oval or ovate-lanceolate, entire or sinuate-dentate, pale green, membranaceous, translucent to transmitted light,* slender-petioled, the larger ones with blades 2.5–10 cm. long; peduncles filiform, up to 3 cm. long; the 2–4-*flowered umbels with filiform soon reflexed pedicels; calyx of unequal acutish to blunt* spreading *lobes* 0.9–1.5 *mm. long; corolla* white or purple-tinged, *its lobes* 4.5–7 *mm. long; anthers* 1.3–2 *mm. long; berry lustrous-black,* 5–9 *mm. in diameter, with seeds* 1.2–1.8 *mm. long and with* 4–8 *stone-cell concretions in most berries,* the latter quickly falling when ripe. (*S. nigrum* in large part of ed. 7) — Rocky or dry open woods, thickets, shores or openings, often spreading to cult. or waste ground, w.-centr. Me. to N.D., s. to s. N.E., n. Fla., La. and e. Tex. June–Nov.

5. S. NÌGRUM L. (black). — *Coarser than no.* 1; *leaves darker green, thicker and more opaque; inflorescence umbellate, or often corymbose or subracemose, with mostly* 5–10 *flowers on stouter pedicels; calyx-lobes with rounded tips;* corolla-lobes 5.5–7.5 mm. long; *anthers* 1.8–2.6 *mm. long; berries dull black,* 8–13 *mm. in diameter, with seeds* 1.7–2.3 *mm. long and without concretions or sometimes with* 1 *or* 2 *minute ones only.* — Waste places, roadsides and other disturbed soil (sea-beaches, etc.), N.S. to Fla. and locally westw. May–Oct. (Natzd. from Eu.)

6. S. VILLÒSUM Mill. (villous). — *Resembling small-leaved forms of no.* 5 *but the stems, branches,* *leaves, peduncles, etc., cinereous with dense appressed villosity or pilosity; the* 1–5 *umbellate flowers on slenderly clavate stiff pedicels; calyx as in no.* 5 *but densely pubescent; berries yellow or red, with seeds* 1.8–2.2 *mm. long.* — Waste places, local, Mass. southw. (Adv. from Eurasia)

S. INTÈRIUS Rydb. (of the Interior), a similar plant of the Great Plains but with racemose-corymbose inflorescences and greenish-black berries, may reach our western borders.

7. S. SARACHOÌDES Sendtner (resembling the genus *Saracha*). — *Differing from nos.* 5 *and* 6 *in the spreading-hirsute stem, branches, peduncles, etc.; racemose-corymbose inflorescence up to* 9*-flowered; pedicels after anthesis becoming thick-clavate above and merging with the base of the calyx; calyx after anthesis very accrescent and becoming* 5–8 *mm. long, its broad lobes and tube closely embracing much of the green berry.* — Weed of gardens and waste places, local, N.E. to Minn. and southw.; w. N.Am. (Adv. from S.Am.)

8. S. carolinénse L. (of Carolina), HORSE-NETTLE, BALL-NETTLE. — Prickly perennial with creeping subterranean rhizome; *stems* simple or forking, *hirsute with* 4–8*-rayed hairs, the slender prickles stramineous; leaves green, elliptic-oblong to oval, sinuate and coarsely toothed,* scabrous; *racemes* several-flowered, *soon becoming* 1*-sided; calyx-lobes lance-acuminate;* corolla violet, or white in forma **albiflòrum** Benke (white-flowered); *anthers tapering to tip;* berry yellow, naked, 1–1.5 cm. in diameter. — Sandy openings, fields and waste places, often a troublesome weed, Fla. to Tex., n. to N.E., N.Y., s. Ont., O., Ind., Ill., Ia. and Neb., northw. chiefly adv. May–Oct.

9. S. elaeagnifòlium Cav. (with leaves of *Elaeagnus*), SILVERLEAF-N., WHITE HORSE-NETTLE. — *Differing from no.* 8 *in the dense canescent tomentum of scurf-like many-rayed hairs; prickles smaller and fewer (even wanting); leaves very thick, linear-lanceolate to narrowly oblong, gradually tapering at base, entire or shallowly sinuate; calyx and outside of corolla cinereous-tomentulose, the calyx-lobes linear-subulate.* — Dry open woods, prairies, waste places and disturbed soil, sw. U. S. and adj. Mex., ne. to Mo., and as an adventive to Ind., O. and Fla.

10. S. Tórreyi Gray (for JOHN TORREY, 1796–1873). — *Differing from no.* 9 *in less cinereous pubescence; ovate leaves with broadly rounded or cordate bases and sinuate-lobed margins; cymes often forking; calyx-lobes broadly ovate, abruptly acuminate; corolla* 2–2.5 *cm. broad; berry much larger.* — Rocky open ground and dry prairies, Tex. to Kans. and sw. Mo., and adv. over much of Mo.

11. S. SISYMBRIIFÒLIUM Lam. (with leaves of *Sisymbrium*). — *Very prickly annual, the very unequal (long and short) prickles abundant on the calyx as well as stem, leaves, pedicels, etc.; stem, branches and petioles villous-viscid; leaves deeply pinnatifid with deeply cut or sinuate lobes; racemes in maturity with spreading pedicels;* corolla bluish or white; *anthers equal; fruiting calyx loosely embracing the red berry; seeds minutely reticulate-pitted.* — Roadsides, waste places, etc., Fla. to Tex., n. locally (and fortunately so) to Mass., N.Y. and O. July–Sept. (Adv. from Trop. Am.)

12. S. ROSTRÀTUM Dunal (beaked; from the odd anther), BUFFALO-BUR, KANSAS-THISTLE. — *Annual, with very abundant subulate yellowish prickles; pubescence of stem, etc. hoary, of stellate hairs; leaves once or twice pinnatifid;* racemes with *pedicels soon ascending; calyx nearly hidden by spine-like prickles; corolla yellow, nearly regular, with broadly ovate lobes; stamens* and style much declined, *the lowest anther much larger and longer than the others and with an incurved beak; berry wholly inclosed by the close-fitting* and often adherent *calyx;* seeds coarsely undulate-rugose. — Native on the Plains w. of our range, now an aggressive weed through most of the e. and n. states. July–Oct. (Natzd. from farther west)

13. S. CITRULLIFÒLIUM A. Br. (with leaves of watermelon, *Citrullus*). — *Differing from no.* 12 *in having simple gland-tipped hairs mixed with the stellate ones; corolla violet, with acuminate lobes;* 4 anthers yellow, the larger one violet-tinged; fruiting calyx rather smaller. — Roadsides and waste places, locally n. to Ia. and e. to Atl. states. (Adv. from the Southwest)

The cult. S. MELONGÈNA L. (a kind of melon), the EGGPLANT, occasionally comes up in garden-refuse but hardly persists. LYCOPÉRSICUM (wolf-peach) ESCULÉNTUM Mill. (esculent), the TOMATO, likewise will grow in garden-refuse but is scarcely hardy with us.

2. CHAMAESARÀCHA Gray

Calyx herbaceous, closely investing the globose berry (or most of it), obscurely if at all veiny. Corolla rotate, 5-angulate, plicate in the bud, white, whitish or violet-tinged, often with a pubescent ring at base. Filaments filiform; anthers separate, oblong. — Perennials or annuals of N. and S.Am. and e. Asia, with mostly entire or pinnatifid leaves tapering into margined petioles, and filiform naked pedicels solitary or fascicled in the axils, refracted or recurved in fruit. (Name from the Greek *chamai, on the ground,* and *Saracha,* name of a trop. Am. genus dedicated in 1794 to *Isadore Saracha,* a Spanish Benedictine.)

1. C. grandiflòra (Hook.) Fern. (large-flowered), Large White-flowered Ground-cherry. — Erect *annual*, 1.5–9 dm. high, *with viscid-villous stem* simple or with ascending branches; *leaves lance-ovate*, entire or nearly so, *the larger ones* 0.5–2 *dm. long; flowers fascicled; calyx long-villous; corolla white* (or violet-tinged), *with yellow eye and broad tomentulose central cushion*, 3–5 *cm. broad; fruiting calyx globular*, 1.5–2 *cm. in diameter*, with lance-attenuate lobes. (*Physalis* Hook.; *Leucophysalis* Rydb.) — Sandy or rocky shores, open woods, recent clearings and roadsides, Rimouski Co., Que., to Sask., s. to L. Champlain, Vt., s. Ont., n. Mich., n. Wisc. and n. Minn. June–Aug. — The largest-flowered member of the genus, closely related to Chinese and Andean species.

3. PHÝSALIS L. Ground-cherry

Calyx 5-lobed, reticulated and enlarging after flowering, at length much inflated and inclosing the 2-locular globular (edible) berry. Corolla between rotate and campanulate, the very short tube marked with 5 concave spots at the base; the plaited border somewhat 5-lobed or barely 5–10-toothed. Stamens 5, erect; anthers separate, opening lengthwise. — Chiefly of warmer and temp. Am., a few in Old World; ours herbs with extra-axillary peduncles; mostly flowering through the summer. (Name Greek for *a bladder*, from the inflated calyx.)*

a. Annuals, without rhizomes. . . *b*.
b. Plant obviously villous or pilose with simple viscid or glandular hairs. . . *c*.
c. Calyx-lobes during anthesis acuminate, nearly or quite as long as tube; fruiting calyx 2–4 cm. long, carinately 5-angled, the basal auricles prominent, the base depressed. . . *d*.
d. Leaves when viewed by transmitted light showing no or only obscure reticulation between the lateral nerves, the margin entire or only 1–7-toothed on each side.
Leaves pale green, thin-membranaceous, almost flaccid, very translucent, entire or with low undulations; fruiting calyx subglobose-ovoid, abruptly pyramidal above; plant weak and low, 0.5–5 dm. high. . . . 1. *P. pubescens.*
Leaves dark green, firm, subopaque, the primary ones with coarse often acute triangular teeth; fruiting calyx flask-shaped, gradually conic-pyramidal above; plant strong, up to 1 m. high. . . . 2. *P. barbadensis.*
d. Leaves dark, firm, when viewed by transmitted light showing a prominent reticulum of veinlets between the lateral nerves, the margins prominently dentate to base. . . . 3. *P. pruinosa.*
c. Calyx-lobes during anthesis triangular and subacute (not acuminate), much shorter than tube; fruiting calyx 1.5–2 cm. long, obscurely obtuse-angled, scarcely auricled, the base not depressed; leaves without evident reticulation. . . . 4. *P. missouriensis.*
b. Plant glabrous, promptly glabrate or puberulent. . . *e*.
e. Flowering peduncles 3–6 mm. long, in fruit 0.4–2 cm. long.
Leaves cuneate or subcuneate at base; corolla 1–2(–2.5) cm. broad; fruiting calyx obscurely angled, round-ovoid, with bluntish triangular teeth. . . . . 5. *P. ixocarpa.*
Leaves broadly rounded or cordate at base; corolla 5–10 mm. broad; fruiting calyx strongly angled, with long acuminate teeth. . . . 2. *P. barbadensis*, var. *glabra.*
e. Flowering peduncles 1–3 cm. long, the longer ones in fruit 1–5 cm. long; corolla 5–15 mm. broad.
Fruiting calyx 2.5–3.5 cm. long, with slender-tipped teeth; longer peduncles 1–2.5 cm. long; wide-ranging southward. . . . 6. *P. angulata.*
Fruiting calyx 2–3 cm. long, with merely acutish to blunt teeth; longer peduncles 2.5–5 cm. long; western. . . . . . 7. *P. pendula.*
a. Perennials, chiefly with deep-seated rhizomes. . . *f*.
f. Corolla yellow or yellowish, often with dark center, with shallowly angulate margin; fruiting calyx green to stramineous or brown or merely veined with purple, dropping before winter. . . *g*.
g. Stem, leaves and calyx closely canescent-tomentulose with stellate hairs; leaves elliptic, oval or oblong, blunt, entire or barely undulate; plant of coastal sands southeastward. . . . . 8. *P. maritima.*

---

* Treatment based largely on that of P. A. Rydberg in Mem. Torr. Bot. Cl. iv. no. 5 (1896).

g. Stem, leaves and calyx pubescent with simple or merely jointed or branched hairs or glabrous. . . *h*.

h. Leaves, pedicels and calyces glabrous or with only sparse and minute puberulence when young; leaves cuneate or gradually tapering to base, longer than broad. . . *i*.

i. Leaves opaque or subopaque to transmitted light; fruiting calyx 2–3.5 cm. long, finally filled and often burst by the mature berry, scarcely sunken at base.

Leaves ovate or ovate-lanceolate, thinnish, the petioles of primary blades 3–6 cm. long; calyx-lobes ovate or triangular; berry red or purple; wide-ranging. . . . . 9. *P. subglabrata.*

Leaves lanceolate, oblanceolate or linear-oblong, thick, the principal petioles 1–3 cm. long; calyx-lobes triangular; berry yellow or greenish; western. . . . . 10. *P. longifolia.*

i. Leaves very translucent to transmitted light, membranous, broad-ovate; fruiting calyx pyramidal to ovoid-conic, 3–6 cm. long, deeply depressed at base, much inflated around the small berry; chiefly southwestern. . . . . 11. *P. macrophysa.*

h. Leaves, pedicels and calyx villous, pilose or hirsute with mostly spreading pubescence. . . *j*.

j. Stem, lower surfaces of leaves, etc., with once or twice branched long hairs; leaves oblong to oblong-ovate or subrhombic, mostly entire; fruiting calyx ellipsoid-ovoid, 4–5 cm. long, only slightly depressed at base; western. . . . . 12. *P. pumila.*

j. Stem, leaves, etc., with simple (often jointed or viscid) hairs. . . *k*.

k. Leaves cuneate or gradually tapering to base, longer than broad; pubescence scattered or sparse on stem (not densely crowded), of flattened hairs.

Leaves oblanceolate, oblong or subspatulate, entire or barely undulate, thick and hard; fruiting calyx pyramidal-ovoid, not depressed at base; berry yellow or green (sometimes reddish). 13. *P. lanceolata.*

Leaves ovate or ovate-lanceolate, often sinuate-dentate, pliable and submembranaceous; fruiting calyx pyramidal-ovoid, with depressed base; berry reddish. . . . . 14. *P. virginiana.*

k. Leaves cordate to broadly rounded at base, ovate to orbicular or rhombic; pubescence very dense, often glandular or viscid. . . *l*.

l. Leaves cordate-ovate, long-acuminate, mostly 0.8–1.5 cm. long; pubescence glandless, not viscid; calyx-lobes lance-acuminate; spread from cult. . . . . 15. *P. peruviana.*

l. Leaves ovate, orbicular or rhombic, not prominently acuminate, the larger blades 2–7 (–11) cm. long; pubescence glandular or eglandular; calyx-lobes triangular; natives. . . *m*.

m. Larger leaves 6–11 cm. long, ovate, mostly cordate; wide-ranging. . . . . 16. *P. heterophylla.*

m. Larger leaves 2–4.5 cm. long, round-ovate to suborbicular or rhombic, rounded at base; western.

Calyx with long glandless white trichomes greatly exceeding the short glandular hairs. . . . . 17. *P. comata.*

Calyx covered with short (rarely elongate) gland-tipped hairs. . . . . 18. *P. rotundata.*

f. Corolla white or whitish, deeply lobed; fruiting calyx bright red or scarlet, persistent; spread from cult. . . . . 19. *P. Alkekengi.*

1. P. pubéscens L. (pubescent). — *Weak* and low (0.5–5 dm. high) *villous annual, with readily crushed stem; leaves ovate, often obliquely rounded or cordate at base,* rather abruptly tipped, *pale green, thin-membranaceous, translucent and (as seen by transmitted light) without evident reticulum between the lateral veins, entire or with few shallow dentations;* flowering peduncles 3–5 mm. long, in fruit up to 1 cm. long; *calyx-lobes at anthesis narrow and acuminate;* corolla 0.5–1 cm. broad, yellow, with dark center; anthers mostly purplish; *fruiting calyx membranaceous* and very veiny, 2–3 *cm. long, carinately angled, with prominent auricles standing above the depressed base, subglobose-ovoid, abruptly pyramidal at summit;* berry yellow, sweetish, 1.5 cm. in diam. — Damp to dry open woods, clearings, sand-dunes and disturbed soil, Fla. to Tex. and Mex., n. to Va. (reported n. to Pa.), O., Ind., Wisc., Ia., Kans. and s. Calif. July–Oct. (Trop. Am.)

2. P. barbadénsis Jacq. (of Barbadoes). — *Stouter* and taller than no. 1, *up to* 1 *m. or more high*, villous; *leaves dark green, firm, subopaque, broadly ovate to subrotund, the primary ones angulate-lobed or with prominent teeth; fruiting calyx flask-like, gradually conic-pyramidal above, of firm texture,* 3–4 *cm. long.* — Low grounds, disturbed soil and waste places, Fla. to Tex. and Mex., n., locally, to Pa., Ill. and Ia. June–Oct. (Adv. from Trop. Am.)

Var. **glàbra** (Michx.) Fern. (glabrous). — Similar but glabrous or barely puberulent on young growth. (Var. *obscura* (Michx.) Rydb.) — Less common, n. to N.C. and Mo.

3. P. pruinòsa L. (with waxy powdery surface), Strawberry-Tomato. — Habitally resembling nos. 1 and 2, *hoary with dense short viscid pubescence; leaves cordate-ovate, coarsely dentate to the often oblique base, firm, with prominent reticulation between the lateral nerves;* calyx-lobes as long as tube of flowering calyx; *fruiting calyx firm, pubescent, subglobose-ovoid, abruptly acuminate* with long narrow teeth, 2–3 *cm. long,* very short-peduncled. — Dry open (often sandy) soil, old fields and waste places, n. Fla. to Mo., n. to sw. Me., Vt., N.Y., O., Ind., Wisc., Ia. and Kans. July–Oct.

4. P. missouriénsis Mackenz. & Bush (of Missouri). — Similar to nos. 2 and 3 but *stem glandular-villous; leaves elliptic-ovate or oval, with rounded* (rarely cordate) *base, thick and firm, without evident reticulation, coarsely repand-dentate to base, blunt or subacute;* villous *flowering calyx with short triangular lobes much shorter than tube; fruiting calyx* 1.5–2 *cm. long, obscurely obtuse-angled, scarcely auricled, the base not depressed, the triangular lobes short.* — Rocky open woods and barrens, spreading to cult. ground and waste places, Mo. and Kans. to Ark., Okla. and e. Tex.; casually adv. in Mass. June–Sept.

5. P. ixocárpa Brotero (with viscid fruit), Tomatillo. — *Glabrous* or glabrate annual up to 1 m. or more high; long-petioled *leaves lance-ovate or narrowly subelliptic, cuneate or tapering at base, acuminate,* sharply dentate to entire; *flowering peduncles* 3–6 *mm. long;* flowering calyx with deltoid teeth; *corolla widely expanding,* 1–2.5 *cm. broad, bright yellow with* 5 *blackish basal spots; fruiting calyx purple-veined, obscurely angled, round-ovoid, with blunt triangular teeth, nearly filled and often burst by the viscid purplish berry.* — Spread from cult. and frequently estab. in waste places n. to s. Que. and n. states. (Introd. from the Southwest and Mex.)

6. P. angulàta L. (angled; from the stems). — *Similar to no.* 5 *but with filiform flowering peduncles* 1–1.5 *cm. long; flowering calyx with narrowly triangular to lanceolate attenuate teeth; corolla* 0.6–1.5 *cm. broad, yellow, without dark spots; fruiting calyx* green or purple-veined, 2.5–3.5 *cm. long,* 10-*angled, ovoid, with pyramidal tapering summit and long slender-tipped teeth,* loosely surrounding the *yellow berry; longer fruiting peduncles* 1–2.5 *cm. long.* — Borders of woods and thickets or in loosened soil or waste ground, Fla. to Tex., n. to Va. and Okla., and casually as a weed to Ct. and Ill. July–Sept. (W.I.; S. Am.; Asia)

7. P. péndula Rydb. (pendulous). — *Differing from no.* 6 *in its longer peduncles* (*in fruit* 2.5–5 *cm. long and more pendulous*)*; fruiting calyx roundish-ovoid, obscurely angled,* 2–3 *cm. long, with blunt or merely acutish teeth.* — Damp thickets, shores and waste places, Mo. and Kans., s. to La. and Tex.; casually as a weed to n. Ill. June–Sept.

8. P. marítima M. A. Curtis (maritime). — *Perennial* from a deep-seated heavy rhizome, *the slender and cord-like subterranean horizontal branches or caudices spreading to* 1 *m. or more;* ascending leafy simple or slightly branching *stems* 1.5–6 dm. high, *cinereous-tomentulose with crowded and appressed stellate hairs; leaves elliptic, oval or oblong, blunt, entire* or barely undulate, *closely stellate-tomentulose,* mostly 2.5–7 cm. long, *tapering to short petioles; flowering peduncle, calyx and outside of corolla cinereous-stellate;* calyx-lobes in anthesis triangular-ovate; corolla greenish-yellow, with dark center, 2 cm. broad; fruiting pedicels up to 2.5 cm. long; fruiting calyx submembranaceous, globose-ovoid, 2–3 cm. long; berry yellow or orange. (*P. viscosa* of Rydb. and of ed. 7, not L.; earlier cited names too doubtful) — Sandy shores, dunes and swales along the coast, Fla. to Tex., n. to se. Va. June.

9. P. subglabràta Mackenz. & Bush (nearly glabrous). — Fleshy and stout horizontal rhizome subterranean; *leafy stem, foliage, etc., nearly or quite glabrous;* branching stem 0.3–1.5 m. high, slenderly angled; *leaves ovate, ovate-oblong or broadly lanceolate, entire or slightly repand-dentate, thin but opaque, tapering to base, the petioles of the primary ones* 3–6 *cm. long;* flowering peduncle and calyx glabrous or at first sparsely strigose; *calyx-lobes ovate, ovate-lanceolate or triangular;* corolla yellow or yellowish-green, with purple center, 1.5–2.5 cm. broad; *fruiting calyx* usually longer than its peduncle, *subglobose, depressed at base,* 2–3 cm. long, *nearly or wholly filled by the reddish or purple berry and open at the mouth.* — Shores, meadows, fields, roadsides and waste places, Ga. to Tex., n. to Vt., N.Y., s. Ont., Mich., Ill. and Ia. June–Sept.

10. P. longifòlia Nutt. (long-leaved). — Mostly *lower than no.* 9; *leaves thick, lanceolate or oblanceolate to linear, the principal petioles only* 1–3 *cm. long;* calyx-lobes triangular; *corolla* 1–1.5 *cm. broad; fruiting calyx globose-ovoid, scarcely sunken at base, mostly* 3–3.5 *cm. long,* *loosely surrounding the glutinous yellow berry.* — Bottoms, fields and cult. ground, Wyo. to Ariz. and Mex., e. to Ia., Mo. and Ark. June–Aug.

11. P. macrophỳsa Rydb. (with large bellows). — *Differing from no. 9 in its* membranaceous and by transmitted light *translucent coarsely dentate ovate leaves;* the *strongly inflated fruiting calyx pyramidal to conic-ovoid and 3–6 cm. long, twice to thrice length of peduncle, the base deeply sunken; berry relatively small.* — Bottoms, clearings, etc., Tex. to Ark. and Ia., and casually as a weed to Mich.

12. P. pùmila Nutt. (dwarf). — *Stems* arising from cord-like rhizomes or buried caudices, simple or slightly forking, 1.5–4.5 dm. high and, *like the leaves, peduncles, etc., spreading-villous with once or twice forking trichomes; leaves oblong, oblong-obovate or narrowly subrhombic, mostly entire, subcuneate to petioles, mostly* 5–10 *cm. long; calyx villous-hirsute,* its lobes narrowly triangular; corolla yellow, with brown center, up to 2 cm. broad; *fruiting calyx ellipsoid-ovoid, 4–5 cm. long, only slightly depressed at base.* — Dry open slopes, prairies, etc., Mo. to Colo. and Tex.

13. P. lanceolàta Michx. (lanceolate). — *Stems* slender, arising from a deep subterranean rhizome, 1–4 *dm. high,* simple to divergently branched, *hispid or hirsute with short flattened simple hairs; leaves oblanceolate, oblong or subspatulate, tapering to base, entire or barely undulate, thick and hard, the primary ones* 2.5–6 *cm. long, short-hirtellous; calyx short-hispid;* corolla dull yellow, with brown center; *fruiting calyx* pyramidal-ovoid, *not sunken at base,* 3–3.5 cm. long, with lance-triangular lobes; berry yellow, green or reddish. — Dry open soil, old fields and waste places, S.C. to La., w. to Ariz., n. to N.C., Mich., Wisc., Minn., s. Man. and Wyo. May–Sept.

14. P. virginiàna Mill. (Virginian). — Differing from no. 13 in its long-villous stem 1.5–6 dm. high; *ovate to ovate-lanceolate often sinuate-dentate submembranaceous and pliable leaves; fruiting calyx deeply sunken at base; berry red.* (Incl. *P. monticola* sensu Fern., not Mohr) — Dry sandy or rocky woods, openings and clearings, n. Fla. to Tex., n. to Ct., N.Y., s. Ont., Mich., Wisc., Minn. and s. Man. May–Aug.

15. P. PERUVIÀNA L. (of Peru), CAPE-GOOSEBERRY. — Much branched, with dense *non-glandular villosity on stem, leaves, etc.; leaves ovate, cordate or rounded at base, long-acuminate,* densely pubescent, entire or dentate; the primary ones 0.8–1.5 *cm. long; calyx-lobes lance-acuminate;* anthers bluish; fruiting pedicels very short; berry yellow. — Locally spread from cult. to waste lots. (Introd. from S. Am.)

16. P. heterophỳlla Nees (variable-leaved). — Rhizome thick and fleshy, usually horizontal, deeply buried; *stem* widely branching, *densely covered with stipitate glands or viscid hairs, these sometimes overtopped by long jointed and flattened villi; leaves mostly cordate at base, ovate, the larger blades* 6–11 *cm. long;* calyx-lobes triangular; corolla yellowish, with dark center, usually 1.5–2.5 cm. broad; *fruiting calyx* globose-ovoid, *sunken at base,* 2–3.5 cm. long, *with somewhat connivent lobes;* berry yellow, relatively small. — A highly variable (heteromorphic as well as heterophyllous) species. The following vars. can be defined:

Pubescence of stem a dense mixture of stipitate glands and viscid hairs only 0.5–0.7 mm. long.
- Base of stem slender and herbaceous; leaves shallowly dentate, opaque to transmitted light, their veins obscure. . . . . *P. heterophylla* (typical).
- Base of stem heavily thickened and subligneous; leaves with acute triangular teeth, conspicuously and superficially veined, translucent. . . . . Var. *clavipes.*

Pubescence of stem with crowded and flattened jointed villi up to 2 or 3 mm. long.
- Leaves thick and opaque, commonly dentate. . . . . Var. *ambigua.*
- Leaves membranaceous, translucent, entire or sparingly toothed. . . . Var. *nyctaginea.*

P. heterophỳlla (typical). — Dry open woods, clearings, etc., N.E. to Sask. and Colo., s. to S.C., Ky., Mo., Okla. and e. Tex. June–Sept.

Var. **clàvipes** Fern. (club-footed). — Sandy woods, local, se. Va.

Var. **ambìgua** (Gray) Rydb. (doubtful). — S. Que. to Minn., s. to N.E., Ga., W.Va., O., Ind., Ill., Mo., Okla. and e. Tex.

Var. **nyctagínea** (Dunal) Rydb. (resembling *Nyctago* or *Mirabilis*). — Open woods and clearings, s. Que. and Vt. to s. Ont., s. to Va., O., Ill., Okla. and Tex.

17. P. comàta Rydb. (with a tuft of hairs). — *Much smaller than no.* 16; *stem, etc., densely covered with very short glandular hairs; leaves broadly ovate to angulate-rotund, rounded to base and apex, mostly* 3–4.5 *cm. long, very thick; calyx, petioles and young parts with long flat glandless trichomes much overtopping the glandular ones;* fruiting calyx ovoid, 2–3 cm. long. — Prairies, bluffs and fields, Neb. and e. Kans. to Colo. and Tex. June–Aug.

18. P. rotundàta Rydb. (rounded). — *Similar to no. 17 but more glandular; leaves cordate or subcordate,* nearly orbicular, 2–4 *cm. in diameter; pubescence of calyx, etc., heavily glandular, without long glandless trichomes.* — S.D. to e. Kans. and Tex., w. to Colo. and N.M.

19. P. Alkekéngi L. (Arabic name), Winter-cherry, Chinese Lantern-plant. — Subsimple stems smoothish, angled; leaves ovate or ovate-rhombic, firm, with upwardly broadened petiole; *corolla whitish, deeply lobed; fruiting calyx firm, conic-ovoid, acuminate, deeply sunken at base, bright red to scarlet, subpersistent.* — Locally spread from cult. (Introd. from Asia)

4. NICÁNDRA Adans. Apple-of-Peru

Calyx 5-parted, 5-angled, the divisions rather sagittate, enlarged and bladder-like in fruit, inclosing the 3–5-locular globular dry berry. Corolla with border nearly entire. Otherwise much like *Physalis.* — Coarse smooth annual, nat. of S.Am., with ovate sinuate-toothed or angled leaves, and solitary pale blue flowers on axillary and terminal peduncles. (Named for the poet *Nicander* of Colophon.) Physalodes Boehmer

1. N. physalòdes (L.) Pers. (resembling *Physalis*), *Physalodes* Britt. — Waste grounds, near dwellings and old gardens. July–Sept. (Introd. from Peru) — Sometimes used as a fly-poison.

5. LÝCIUM L. Matrimony-vine

Calyx 3–5-toothed or -cleft, not enlarging, persistent at the base of the berry. Corolla funnelform or salver-shaped, 5-lobed, the lobes imbricated and not plaited in the bud. Stamens 5; anthers opening lengthwise. Style slender; stigma capitate. Berry small, 2-locular. — Shrubby, often spiny plants of temp. and trop. reg., with alternate and entire small leaves, and mostly axillary small flowers. (Ancient Greek *lycion,* a prickly shrub growing in *Lycia.*)

1. L. halimifòlium Mill. (with leaves of *Halimus*), Common M. — Shrub with long sarmentose recurved-drooping branches, smooth, sparingly if at all spiny; *leaves lanceolate or lance-spatulate,* on petioles 3–10 mm. long, *gray-green,* often fascicled; flowers on slender *pedicels 1–2.3 cm. long; calyx-lobes obtuse; corolla* short-funnelform, greenish-purple, the *tube exceeding the limb;* berry ovoid to ellipsoid, orange-red or scarlet, 1–2 cm. long. — Roadsides, thickets and waste ground, spread from cult., common southw., local n. to s. Can. June–Sept. (Introd. from Eu.)

2. L. chinénse Mill. (Chinese), Chinese M. — Similar to no. 1, usually without prickles, *leaves lance- to rhombic-ovate, deep green,* less fascicled; *pedicels 3–12 mm. long; calyx-lobes acute; corolla-tube short and broad;* fruit often larger. — Roadsides and waste places, esc. from cult., Mass. to Mich., s. to Va., La. and Okla. June–Oct. (Introd. from Asia)

6. HYOSCỲAMUS L. Henbane. Jusquiame (Que.)

Calyx campanulate or urceolate, 5-lobed. Corolla funnelform, oblique, with a 5-lobed more or less unequal plaited border. Stamens declined. Capsule inclosed in the persistent calyx, 2-locular, opening transversely all around near the apex, which falls off like a lid. — Clammy-pubescent fetid narcotic herbs, nat. of Old World, with lurid flowers in the axils of angled or toothed leaves. (Name composed of the Greek *hyos, of a hog,* and *cyamos, bean,* because poisonous to swine; likewise to humans.)

1. H. nìger L. (black), Black H. — Biennial or annual; leaves clasping, sinuate-toothed and angled; flowers sessile, in one-sided leafy spikes; corolla dull yellowish, strongly reticulated with purple veins. — Roadsides and waste places, Que. and s. Ont., s., locally, to N.S., Me., Mass., N.Y. (formerly to Del.), Mich. and S.D. June–Aug. (Natzd. from Eu.) — A slimy-leaved and poisonous plant, source of the narcotic hyoscyamin.

7. DATÙRA L. Stramonium. Jimsonweed. Thorn-apple. Herbe aux sorciers (Que.)

Calyx prismatic or cylindrical, 5-toothed, separating transversely above the base in fruit, the upper part falling away. Corolla funnelform, with a large and spreading 5–10-toothed plaited border. Stigma 2-lipped. Capsule globular, usually prickly, 4-valved, 4-locular except near the 2-locular top. Seeds rather large, flat. — Rank weeds, narcotic-poisonous, with ovate leaves, and large showy flowers produced all summer and autumn on short peduncles in the forks of the branching stem. Genus of warm or trop. reg. (Altered from the Arabic name, *Tatorah* or the Hindustani *Dhatura.*)

Glabrous; leaves coarsely angulate-toothed; corolla 7–10 cm. long, 5-lobed; capsule erect. . . . . 1. *D. Stramonium.*

Cinereous-puberulent; leaves undulate or with broad low teeth; corolla 1.5–2 dm. long, 10-toothed; capsule nodding. . . . . 2. *D. innoxia.*

1. D. Stramònium L. (the old generic name, said to be from *struma* or *strama*, a swelling), Stramonium. — Annual, *glabrous; leaves* ovate, *sinuate-toothed or angled; stem green; calyx prismatic; corolla white, 7–10 cm. long, the border with 5 teeth;* lower prickles of the capsule mostly shorter. — Waste ground, abundant southw., locally northw. into s. Can. July–Oct. — An ill-scented and dangerously poisonous subcosmop. weed. (Natzd. from Asia) — Forma inérmis (Juss.) Hupka (unarmed) has all or at least the later fruits unarmed, infrequent as a weed, Mass. to O. and Pa.

Var. Tátula (L.) Torr. (Arabic name), Purple T. — Stem purple-tinged; corolla lavender or violet; prickles of capsule often subequal. (*D. Tatula* L.) — Waste ground, abundant southw., locally northw. into the N. states. (Natzd. from Asia)

2. D. innóxia Mill. (harmless). — *Cinereous-puberulent or -pruinose; leaves* ovate, rounded at base, *entire, undulate or low-dentate; corolla* white, 1.5–2 *dm. long,* 10-*toothed; capsule nodding,* on an abruptly recurved peduncle, slenderly prickly. (*D. Metel* of ed. 7, not L.) — Waste ground, infrequent, n. to N.E., N.Y. and Ill. Aug.–Oct. (Adv. from Trop. Am.)

D. féroх L. (fierce), with capsules bearing very unequal spines, the 4 terminal spines prolonged and stout, is occasional on dumps southw. (Adv. from Asia)

8. NICOTIÀNA L. Tobacco

Calyx tubular-campanulate, 5-cleft. Corolla funnelform or salverform, usually with a long tube; the plaited border 5-lobed. Stigma capitate. Capsule 2-locular, 2–4-valved from the apex. Seeds minute. — Rank acrid-narcotic Am. herbs, mostly clammy-pubescent, with ample entire leaves, and racemed or panicled flowers. (Named after *Jean Nicot,* who sent seeds of Tobacco, *N. Tabacum* L., to France in 1560.)

1. N. rústica L. (of the country), Wild T. — Annual; leaves ovate, petioled; tube of the dull greenish-yellow corolla cylindrical, two-thirds longer than the calyx, the lobes rounded. — Old fields and waste places, rare and sporadic, Fla. to N.M., n. to Mass., N.Y., s. Ont. and Minn. Sept., Oct. — Formerly cult. by Indians. (Of unknown nativity)

N. Tabácum L. (aboriginal name), Tobacco, with large ovate to lanceolate leaves and greenish funnelform corolla 5–8 cm. long with reniform acute spreading lobes changing to reddish, esc. from cult. to waste places but is scarcely persistent. (Introd. from Trop. Am.)

N. longiflòra Cav. (long-flowered), with slender corolla-tube many times exceeding the calyx, is a common garden plant, sometimes found in waste. (Introd. from Trop. Am.)

9. PETÙNIA Juss. Petunia

Calyx 5-parted to below the middle. Corolla funnelform or salverform, nearly regular, its lobes plicate in bud. Stamens 5, all perfect, inserted low down, included, one conspicuously smaller; anther-locules distinct. Hypogynous disk fleshy. Stigma dilated-capitate, unappendaged. Capsule with 2 undivided valves parallel with and separating from the placentiferous dissepiment. — Clammy Am. herbs with entire leaves, the upper ones tending to become opposite, the scattered flowers becoming lateral. (*Petun,* an aboriginal name for Tobacco.)

1. P. parviflòra Juss. (small-flowered), Seaside-P. — Much branched diffuse annual; leaves linear-oblong to spatulate, fleshy, about 1 cm. long; peduncles very short; calyx-lobes resembling the smaller leaves; corolla purple, with yellow tube, 5–8 mm. long, with slightly unequal retuse short lobes; capsule ovoid, 3–4 mm. long. — Waste and cult. ground, seabeaches, etc., Fla. to s. Calif., n. to Va., and casually to s. N.Y. June–Sept. (Natzd. from Trop. Am.)

P. axillàris (Lam.) BSP. (axillary), with the lower elliptic to ovate broad-petioled leaves up to 6 cm. long, the calyx 1.5–2 cm. long and with broad lobes, the salverform white corolla 4–6 cm. broad; and **P. violàcea** Lindl. (violet), with slender-petioled lower leaves, shorter calyx with narrow lobes, and funnelform purple corolla 3–4 cm. broad, spread from cult. to waste ground, the former scarcely persistent, the latter clearly persistent and spreading from year to year from e. Pa. southw. (Introd. from S. Am.)

FAM. 153. SCROPHULARIÀCEAE (Figwort Family)

Chiefly herbs (rarely trees), with didynamous stamens (or perfect stamens often only 2, rarely 5) inserted on the tube of the 2-lipped or more or less irregular corolla, the lobes of which are imbricated in the bud; fruit a 2-locular and usually many-seeded capsule, with the placentae in the axis; seeds anatropous or amphitropous, with a small embryo in copious albumen. Style single; stigma entire or 2-lobed. Leaves and inflorescence various, but the flowers not terminal in any genuine representatives of the family (solitary and terminal in the acaulescent genus *Limosella*). — A large family of bitterish plants, some of them narcotic-poisonous.

a. Upper lip or lobes of the corolla covering the lower in the bud (with rare exceptions); capsule usually septicidal. Subfam. I. Antirrhinoideae. . *b.*

b. Corolla rotate, nearly regular; alternate-leaved biennials, often with decurrent leaves; flowers in elongate spikes or racemes. Tribe I. Verbasceae.
Stamens 5, all perfect; 3 or more filaments bearded. . . . . . . 1. *Verbascum.*

b. Corolla tubular to subglobose, 2-lipped; stamens 2–4 (or 5), all perfect or some sterile. . . *c.*

c. Corolla spurred or saccate at base, the throat with a palate; capsule opening by chinks, pores or valves; median and upper leaves alternate. Tribe II. Antirrhineae. . . *d.*

d. Capsule symmetrical, glabrous. . . *e.*

e. Capsule opening by 2 pores; seeds ellipsoid, ovoid or rounded; broad-leaved plants with axillary flowers.
Leaves palmately lobed and veined, glabrous; stem glabrous, creeping or twining; the 2 valves of the capsule each splitting into 3; seeds with wing-like crests. . . . . . . 2. *Cymbalaria.*
Leaves pinnately veined, oval, rounded or hastate, pubescent; capsule circumscissile, the 2 pores formed by the dropping off of the opercula; seeds wingless; diffuse, not twining, pubescent. . 3. *Kickxia.*

e. Capsule opening by 4–10 valves or rupturing, mostly glabrous; seeds angular; leaves linear or linear-lanceolate, in some cult. or casual introductions broader; flowers in racemes or spikes. . . . . 4. *Linaria.*

d. Capsule asymmetrical, with the 2 carpels unequal, pubescent; narrow-leaved glandular-pubescent annual with axillary flowers. . . . 5. *Chaenorrhinum.*

c. Corolla neither spurred nor saccate at base; capsule regularly splitting into valves, not opening by pores or chinks; leaves all or nearly all opposite. . . *f.*

f. Flowers in racemes, spikes, panicles or cymes or, if single, on 2-bracteate peduncles; the 4 stamens usually accompanied by a rudiment of the fifth one. Tribe III. Cheloneae. *g.*

g. Herbs, flowering from new shoots of the season; leaves narrow or, if cordate-ovate, membranaceous and not densely felted beneath. . . *h.*

h. Weak annuals or biennials; inflorescence umbelliform, the upper flowers appearing subverticillate; corolla cleft nearly to base, its upper lip somewhat recurving. . . . . 6. *Collinsia.*

h. Strong perennials; inflorescences (at least the terminal) panicles or spikes; corolla lobed only above the middle, the lobes merely spreading or porrect. *i.*

i. Inflorescence an open panicle of peduncled cymes with lurid purple to greenish leathery globular or inflated-ellipsoid corollas with upper lip projected forward; sepals united below; sterile filament flattened, as broad as or broader than long; leaves membranaceous, ovate to lanceolate. . . . . . . 7. *Scrophularia.*

i. Inflorescence spicate to racemose-paniculate, with white, pink, purple or blue petaloid tubular corollas; sepals distinct; sterile filament filiform; leaves firm to coriaceous.
Inflorescence spiciform, simple, the nearly sessile flowers subtended by broad bracts; sterile filament shorter than the others; anthers densely woolly; seeds flattish, winged. . . 8. *Chelone.*
Inflorescence a compound panicle or raceme of peduncled cymes; flowers pedicelled; bracts narrow; sterile and fertile filaments subequal; anthers glabrous or hirsute; seeds plump, wingless. . . . . 9. *Penstemon.*

g. Tree, flowering in spring from tips of the last year's branches, the large panicle partly grown in the preceding autumn; large firm cordate leaves (up to 3 or 4 dm. broad) densely felted beneath; sterile stamen 1. . . . . 10. *Paulownia.*

*f.*Flowers solitary in the axils of leaves or bracts or in no. 12 a terminal bractless raceme; 2 or 4 stamens perfect, without a rudiment of the fifth; peduncles naked or 2-bracted at base or apex. Tribe IV. GRATIOLEAE. . . *j.*

*j.*Anther-bearing stamens 3 or 4, similar, with no sterile filaments or, if with a pair of sterile filaments (in no. 13), these short and included and the calyx 5-parted. . . *k.*

*k.*Corolla strongly irregular; anthers definitely 2-locular, with the locules distinct; plants with definite stems and several to many flowers. . . *l.*

*l.*Calyx with a campanulate tube, lobed from below or above the middle.

Calyx-tube 5-angled, longer than the ascending lobes or teeth; flowers in axils of opposite leaves or leafy bracts; corolla 1–4 cm. long; mostly native. . . . . 11. *Mimulus.*

Calyx-tube not angled, shorter than the divergent blunt segments; flowers alternate in an open and naked or minutely bracted terminal raceme; corolla about 1 cm. long; adventive weed. . . . . 12. *Mazus.*

*l.*Calyx divided nearly to base into barely united segments.

Pedicels 2-bracted at apex, close to calyx. . . . . . . 13. *Gratiola.*

Pedicels naked or in one species of no. 15 with 2 basal bracts.

Segments of calyx equal; leaves pinnately divided. . . . 14. *Conobea.*

Segments of calyx unequal, the upper one largest; leaves not pinnatifid. . . . . 15. *Bacopa.*

*k.*Corolla with nearly regular rotate white or purplish limb; locules of anther united, the anther seemingly 1-locular; dwarf acaulescent (or merely slenderly stoloniferous) tufted plants with 1 flower terminating naked scapiform soon recurving peduncles. 16. *Limosella.*

*j.*Anther-bearing stamens 2. . . *m.*

*m.*Pedicel 2-bracted at summit. . . . . 13. *Gratiola.*

*m.*Pedicel naked at summit. . . *n.*

*n.*Stamens 2, both antheriferous, without accompanying sterile filaments; flowers uniform and expanding.

Calyx-segments 5, unequal, the upper largest; capsule 2-locular, regularly dehiscent. . . . . 15. *Bacopa.*

Calyx-segments or lobes 4, equal; capsule 1-locular, rupturing. 17. *Micranthemum.*

*n.*Stamens 4, the 2 lower antheriferous, the 2 upper reduced to sterile elongate filaments; some flowers often cleistogamous and unexpanding. . . . . 18. *Lindernia.*

*a.*Lower lip or lateral lobes of the corolla external in the bud. Subfam. II. RHINANTHOIDEAE. *o.*

*o.*Corolla little if at all bilabiate, the lobes all plane. . . *p.*

*p.*Locules of anther contiguous at apex and often confluent; stamens 2 or 4; non-parasitic plants, usually not darkened in drying. Tribe V. DIGITALEAE. . . *q.*

*q.*Leaves mostly alternate; corolla campanulate or tubular; stamens 2 or 4.

Flowers pedicelled, in a long raceme; corolla reddish-purple or white, 4–5 cm. long, greatly exceeding calyx; stamens 4, included. 19. *Digitalis.*

Flowers sessile in a spike; corolla greenish-white, very small, shorter than calyx; stamens 2 (rarely 4), long-exserted. . . . 20. *Wulfenia.*

*q.*Leaves chiefly opposite or verticillate; corolla rotate or salverform; stamens 2.

Leaves mostly in whorls of 3–7; corolla salverform, its tube longer than the lobes; capsule turgid, blunt; tall, with spiciform racemes. 21. *Veronicastrum.*

Leaves opposite (or the upper alternate); corolla rotate, the tube much shorter than the spreading lobes; capsule compressed, obcordate or emarginate; inflorescences various. . . . . 22. *Veronica.*

*p.*Locules of anther distinct to the apex (1 of them wanting in no. 25); stamens 4; plants more or less root-parasitic or saprophytic, the foliage and often the flowers frequently blackened in drying; leaves opposite, or the upper alternate. Tribe VI. GERARDIEAE. . . *r.*

*r.*Corolla with campanulate to funnelform tube, ebracteolate, yellow or roseate to red-purple; anthers 2-locular.

Tube of corolla shorter than the rotate limb; stamens subequal, exserted from the throat of the corolla. . . . . . 23. *Seymeria.*

Tube of corolla longer than the limb; stamens in 2 very unequal pairs, included. . . . . 24. *Gerardia.*

*r.*Corolla salverform, with slender tube and spreading dark purple limb, with 2 basal bracteoles; anthers 1-locular. . . . . 25. *Buchnera.*

*o.*Corolla definitely 2-lipped, the narrow upper lip concave and erect or arching and hood-like, inclosing the 4 stamens; somewhat root-parasitic or saprophytic herbs, usually darkening in drying. Tribe VII. Euphrasieae. . . *s.*

*s.*Locules of anther unequal and separated; leaves alternate.

Perennials (or biennial); the showy colored bracts much larger and more conspicuous than the true flowers; calyx tubular, cleft down lower and sometimes also the upper side; upper lip of corolla elongate, the lower lip short. . . . . 26. *Castilleja.*

Annuals; bracts not brightly colored; calyx tubular-campanulate, 4-cleft; lower inflated lip of corolla broader than and nearly as long as upper lip. . . . . 27. *Orthocarpus.*

*s.*Locules of anther equal, parallel and uniform; leaves often opposite. . . *t.*

*t.*Seeds 1–4, subcylindric, 2–4 mm. long, white, drying blackish, strophiolate; annual with stramineous flowers solitary in axils of the upper often lacerate-based leaves. . . . . 28. *Melampyrum.*

*t.*Seeds numerous or abundant; perennials or biennials, if annual with brightly colored flowers and tiny or thin seeds. *u.*

*u.*Calyx-lobes or -segments 4; pedicel not bracteolate; capsule loculicidally dehiscent; seeds oblong to orbicular. . . *v.*

*v.*Capsule symmetrical, the 2 locules equally dehiscing; calyx nearly symmetrical; leaves mostly opposite. . . *w.*

*w.*Calyx not inflated, tubular to campanulate; corolla white, bluish, purple or roseate; capsule elongate; seeds not orbicular. *x.*

*x.*Annuals with branching slender stem and glabrous corollas.

Upper lip of white, bluish or purplish corolla 2-lobed, the sides folded back; leaves and bracts palmately veined; anthers cohering, hairy at base. . . . . 29. *Euphrasia.*

Upper lip of roseate corolla entire, its margins not turned back; leaves and bracts pinnately veined; anthers glabrous, not coherent. . . . . 30. *Odontites.*

*x.*Perennial, the simple low stems crowded at summit of a stout rhizome; corolla rich dark purple, densely glandular-pubescent; arctic-alpine. . . . . 31. *Bartsia.*

*w.*Calyx inflated, ovate or ovoid; corolla yellow or bronze-tinged; capsule and large winged seed suborbicular; annuals. . . . 32. *Rhinanthus.*

*v.*Capsule asymmetrical, flattened, arching or falcate, usually opening along 1 side; calyx asymmetrical (its lobes mostly united), seemingly 2-lobed; lower lip shorter than the arching galea; cauline leaves mostly alternate and pinnatifid; perennials or biennials. . . . . 33. *Pedicularis.*

*u.*Calyx-teeth 5, strongly oblique, the uppermost much the smaller; pedicels 2-bracteolate at summit (at base of calyx); capsule plump, at first septicidally (finally loculicidally) dehiscent; seeds linear, winged. . . . . 34. *Schwalbea.*

Subfam I. ANTIRRHINOÏDEAE

Tribe I. Verbásceae Benth.

1. VERBÁSCUM L. Mullein. Molène (Que.)

Calyx 5-parted. Corolla 5-lobed, open or concave, with very short tube, the lobes broad and rounded, a little unequal. Stamens 5, all fertile, 3 or all of the filaments villous. Style flattened at the apex. Capsule globular, ovoid or ellipsoid, many-seeded. — Tall and usually hairy biennial herbs, nat. of Old World; the leaves of the stem sessile, often decurrent. Flowers in large terminal spikes, racemes or panicles, ephemeral, in summer. (The ancient Latin name, unexplained.)

Tomentose, woolly or scurfy with branching non-glandular hairs; lower filaments beardless or only sparsely bearded, upper ones villous with slender yellow hairs; capsule ovoid or ellipsoid, densely tomentose.

Leaves felted on both sides, yellowish-drab; inflorescence dense or only slightly interrupted, spiciform, simple or with spiciform lower branches; pedicels much shorter than calyx; corolla 1.5–4 cm. broad; capsule about equaling calyx.
Cauline leaves decurrent down the stem from leaf to leaf; spike crowded. . 1. *V. Thapsus.*
Cauline leaves clasping or only short-decurrent; spike somewhat interrupted below. . . . . 2. *V. phlomoides.*
Leaves green above, canescent beneath with thin powdery wool; inflorescence a panicle of spiciform branches; pedicels longer than calyx; corolla about 1 cm. broad; capsule exceeding calyx. . . . . 3. *V. Lychnitis.*
Glabrous and green or with glandular simple hairs; filaments all violet-villous; capsule subglobose, minutely pubescent or glabrate.
Pedicels 1–1.5 cm. long. . . . . 4. *V. Blattaria.*
Pedicels 3–5 mm. long. . . . . 5. *V. virgatum.*

1. V. Thápsus L. (classical name, from ancient Thapsus), Common M., Flannel-plant, Tabac du diable or Bouillon blanc (Que.). — *Densely woolly throughout; stem tall and stout, simple, winged by the decurrent bases of the* oblong acute *leaves; flowers* yellow, 1.5–2.5 *cm. broad,* or white in forma cándicans House (becoming white), *in a prolonged and very dense cylindrical spike;* lower stamens usually beardless. — Fields, rocky or gravelly banks, etc., throughout our range and beyond. Late June–Sept. (Natzd. from Eu.)

2. V. phlomoìdes L. (resembling *Phlomis*). — Similar but usually taller; the *sessile leaves not at all or only slightly decurrent; spike more interrupted,* frequently branching below; *flowers* 2.5–4 *cm. broad.* — Roadsides and clearings, local, Me. to Minn., s. to N.C., Ky. and Ia. July, Aug. (Natzd. from Eu.)

3. V. Lychnìtis L. (name used by Pliny), White M. — *Clothed with thin powdery woolliness;* stem and branches angled above; leaves ovate, acute, not decurrent, greenish above; *flowers* yellow, or white in forma álbum (Mill.) House (white), *about* 1 *cm. broad, in a pyramidal panicle;* filaments with white wool, the *anthers all reniform; capsule ellipsoid, twice as long as calyx.* — Roadsides and fields, local, Mass. to Ont., s. to Va. and W.Va. June–Sept. (Adv. from Eu.)

4. V. Blattària L. (name used by Pliny, pertaining to a moth), Moth-M. — *Green and smoothish,* or somewhat glandular-pubescent above, *slender;* lower leaves petioled, oblong, doubly serrate, sometimes lyrate, the upper partly clasping; *racemes loose; the pedicels* 1–1.5 *cm. long, longer than the subglobose fruit;* corolla yellow, or white with purplish base in forma albiflòrum (Don) House (white-flowered); *filaments all bearded with violet wool.* — Roadsides and old fields, N.E. to Ont., s. and w. beyond our limits. June–Sept. (Natzd. from Eu.)

5. V. virgàtum Stokes (wand-like). — Similar to the preceding species, but somewhat more pubescent and glandular; *pedicels shorter* than the fruit. — Roadsides, local, C.B. to s. Ont., s. to S.C., O., and Tex.; w. states. (Adv. from Eu.)

Tribe II. Antirrhìneae Chavannes

2. CYMBALÀRIA Hill

Calyx 5-parted. Corolla short-spurred at base on the lower side, the prominent palate closing the throat. Capsule opening by 2 pores, the 2 valves each splitting into 3. Seeds ellipsoid, rugose or crested. — Trailing or twining herbs, nat. of Old World, with long-petioled palmately veined and lobed leaves and scattered axillary flowers. (Name from *cymbalum, a cymbal,* from the round leaves.)

1. C. muràlis Gaertn., Mey. & Schreb. (of walls), Kenilworth- or Coliseum-Ivy. — Stems filiform, glabrous; leaves reniform-orbicular, 5–9-lobed; peduncles slender, becoming recurved in fruit; corolla twice or thrice length of calyx, 7–9 mm. long, pale violet, with yellow palate; capsule globose. (*Linaria Cymbalaria* (L.) Mill.) — Waste places and roadsides, infreq., spreading from cult. May–Oct. (Introd. from Eu.)

3. KÍCKXIA Dumort. Fluellin

Calyx 5-parted. Corolla with prominent palate closing the throat and exceeding the upper lip, the spur short. Capsule circumscissile, opening by 2 pores, each formed by the dropping off of the 2 opercula. Seeds rounded or ovoid-truncate, wingless, tuberculate or faveolate. — Diffuse herbs, nat. of Old World, with short-petioled dilated pinnately veined leaves and scattered axillary flowers. (Named for *Johann Kickx,* Professor at Ghent early in the 19th century.) Elatinoides (Chev.) Wettst.

1. K. Elatìne (L.) Dumort. (classical name of some low plant), Canker-root. — Procumbent much-branched glandular and pubescent annual 2–6 dm. across; *leaves* oval, the lowest opposite and more or less dentate, *the median alternate and hastate;* the upper sagittate, much surpassed by the filiform *glabrous* deflexed *peduncles; calyx-lobes lanceolate;* corolla 0.5–1 cm. long, pale yellow, with violet upper lip; capsule globose; seeds strongly alveolate. (*Linaria* (L.) Mill.; *Elatinoides Elatine* (L.) Wettst.) — Waste places, roadsides, sandy shores, etc., locally abund., Mass. to Ind., s. to Ga., La. and Kans. June–Oct. (Natzd. from Eu.)

2. K. spùria (L.) Dumort. (illegitimate), Female F. — Similar; *leaves oval to suborbicular, not hastate; peduncles hairy; calyx-lobes accrescent, becoming ovate and subcordate;* seeds reniform, closely and finely alveolate. (*Linaria* (L.) Mill.; *Elatinoides* (L.) Wettst.) — Dry fields, roadsides and gravelly bars, local, R.I. to Fla., Ala. and Mo. June–Sept. (Natzd. from Eu.)

4. LINÀRIA Mill. Toadflax

Calyx 5-parted. Corolla with prominent palate or ridges at base of lower lip closing the throat, spurred. Capsule dehiscent by 4–10 valves at apex. Seeds angular, winged or wingless. — Perennials (rarely annuals) of the N. Hemisph., with erect stems and entire leaves, the lowest leaves and those of short basal offshoots opposite or whorled, the median and upper cauline leaves alternate; flowers racemose or spicate. (Name from *Linum*, the *Flax*, from similarity of foliage.)

Cauline leaves linear or linear-lanceolate; flower 0.5–3.3 cm. long (even smaller, tubular or conic and closed in cleistogamous form of no. 4).
- Corolla pale blue, whitish or creamy-yellow, with violet stripes; spur much shorter than tube and limb.
  - Corolla, without spur, 7–12 mm. long; spur conic, 2–3 mm. long. . . 1. *L. repens.*
  - Corolla, without spur, 1.2–2 cm. long; spur compressed, 5–7 mm. long. . 2. × *L. sepium.*
- Corolla blue, yellow or orange, without violet stripes; spur nearly as long as to longer than tube and limb.
  - Corolla yellow (sometimes creamy), 2–3 cm. long (including spur); capsule 8–12 mm. long; flowers approximate or imbricated. . . . . . . 3. *L. vulgaris.*
  - Corolla wholly or partly blue-violet or purple; capsule 2–5 mm. long; flowers becoming distant.
    - Lower lip violet, with 2 paler rounded ridges at base, the filiform spur 2–9 mm. long; seeds wingless. . . . . 4. *L. canadensis.*
    - Lower lip with a full orange palate, the slenderly conical spur 0.8–1.5 cm. long; seeds with 4–6 ring-like wings. . . . . . . 5. *L. maroccana.*

Cauline leaves ovate, ovate-lanceolate or oblong; corolla yellow, 3.5–4 cm. long. 6. *L. dalmatica.*

1. L. rèpens (L.) Mill. (creeping), Striped T. — Rootstock creeping, rather stout, sending up several ascending glabrous mostly branching stems 2–8 dm. high; leaves glaucous, numerous, linear, the lower in 4's, the others alternate; racemes spiciform, at first crowded, becoming lax; pedicels about equaling calyx; *corolla pale blue, white or creamy, striped with violet, without the spur* 7–12 *mm. long, the conic spur* 2–3 *mm. long*, palate usually with a central yellow spot; capsule exceeding calyx; seeds triquetrous, strongly reticulated. (*L. striata* DC.) — Roadsides, thickets and fields, Nfld., locally to Mass.; casual on ballast and waste to N.J. and e. Pa. Late June–Sept. (Natzd. from Eu.)

2. × L. sèpium Allman (of hedges). — Similar to nos. 1 and 3, presumably a hybrid of them; *corolla* striped, often more yellowish than in no. 1, *without spur* 1.2–2 *cm. long, the compressed spur* 5–7 *mm. long.* — Roadsides, se. Nfld. (Natzd. from Eu.)

3. L. vulgàris Hill (common), Butter-and-eggs, Gueule de lion (Que.). — Perennial, with several ascending glabrous stems up to 1.3 m. high; *leaves* pale, linear or linear-lanceolate, *very numerous*, subalternate; *raceme dense*, finally becoming more or less lax; *corolla* (*including the subulate spur*) 2–3 *cm. long, bright yellow, with rounded orange palate*, or corolla whitish in forma leucàntha Fern. (white-flowered); *capsule* ovoid, 9–12 *mm. long;* seeds flattened, with a circular wing. — Roadsides, dry fields and waste places, throughout our range and beyond. May–Oct. (Natzd. from Eu.) — Forma Pelòria (L.) Rouleau (name used by Linnaeus, who supposed this plant a distinct genus, Peloria) has the corolla regular, with 3 or 5 spurs, or spurs lacking; plant usually sterile; a highly interesting state, occurring sporadically.

4. L. canadénsis (L.) Dumont (Canadian), Old-field-T. — Very *slender annual or biennial*, 1–8 dm. high; *basal offshoots trailing, forming overwintering rosettes* with opposite or ternate linear to oblong leaves; cauline leaves scattered, narrowly linear; *raceme becoming very lax; corolla* 5–10 *mm. long*, or reduced to a mere closed cap in forma **cleistógama** Fern. (with flowers fertilized without expanding), *blue-violet*, or white in forma **albìna** Fern. (plaster-colored), *with*

2 *paler rounded ridges at base of lower lip, the filiform spur* 2–6 *mm. long;* capsule 2–3.5 mm. long, equaling or exceeding calyx; *seeds* slenderly obpyramidal, *wingless*, nearly smooth, with acutish angles. — Dry sandy or sterile soil, often a weed in sandy loam, throughout our range, northw. to N.S., n. N.E., sw. Que., s. Ont., Mich., Wisc., Minn. and S.D. Apr.–Sept. Passing into

Var. **texàna** (Scheele) Pennell (Texan). — Corolla 1–1.2 cm. long, deeper violet, the spur 5–9 mm. long; seed rougher and with more rounded angles. (*L. texana* Scheele) — Fla. to Mex. and Lower Calif., n. to se. Va., Mo., S.D. and s. B.C.

5. L. MAROCCÀNA Hook. f. (of Morocco). — Annual, up to 5 dm. high, with remote linear leaves; raceme becoming lax; *corolla violet, with full orange palate, the conical spur* 0.8–1.5 *cm. long;* capsule shorter than calyx; *seeds* obconic, *with* 4–6 *ring-like wings.* — Waste places and roadsides, local, N.E.; also cult. June–Sept. (Introd. from n. Afr.)

6. L. DALMÁTICA (L.) Mill. (Dalmatian). — Comparatively coarse, up to 1 m. high, simple or paniculately branched; *leaves ovate, ovate-lanceolate or oblong, clasping; corolla* yellow, 3.5–4 *cm. long.* — Roadsides and fields, local, C.B. to Pa. and O. June–Sept. (Adv. or introd. from se. Eu.)

Several species of the Old World have occurred casually but are hardly established: **L. PURPÙREA** (L.) Mill. (purple), **L. SUPÌNA** (L.) Desf. (weak), **L. SPÁRTEA** (L.) Hoffmgg. & Link (made of broom), **L. RETICULÀTA** Desf. (netted), and some others.

ANTIRRHÌNUM L. (a plant name used by Pliny), SNAPDRAGON, with corolla merely saccate or gibbous at base, the throat closed by the conspicuous palate, is represented in gardens by **A. MÀJUS** L. (larger), COMMON S., perennial, with lance-oblong leaves, short oblong or ovate calyx-lobes, and variously colored corolla 2–3 cm. long. This frequently spreads to waste ground and roadsides; hardly long-persistent. **A. ORÓNTIUM** L. (old Greek name for some plant), LESSER S., annual, with lanceolate leaves, prolonged linear calyx-lobes, and purple or white corolla 1–1.2 cm. long, is a casual weed; apparently not persistent. (Both from Eu.)

5. CHAENORRHÌNUM Reichenb. DWARF SNAPDRAGON

Calyx 5-parted. Corolla open at throat, the upper lip arched above the depressed palate, spur short. Capsule opening by 2 pores formed by the dropping of 2 opercula. Seeds ellipsoid-ovoid, truncate, with longitudinal crests. — Branching herbs, nat. of Old World, with flowers scattered in the axils of entire leaves or in lax terminal racemes. (Name from the Greek *chainein, to gape*, and *rhis, snout.*)

1. C. MÌNUS (L.) Lange (smaller). — Glandular and hairy annual 0.5–4 dm. high, with ascending branches; lower leaves opposite and oblong, others alternate and lanceolate or linear; inflorescences loose leafy racemes; corolla slightly longer than calyx, 5–8 mm. long, lilac, with yellow throat; capsule pubescent, subglobose, shorter than the linear calyx-segments. (*Linaria* (L.) Desf.) — Waste places, roadsides, railroad-cinders, etc., becoming freq., Que. and Ont., s. to N.S., N.E., Va., O., Ill. and Mo. June–Sept. (Natzd. from Eu.)

Tribe III. CHELÒNEAE Benth

6. COLLÍNSIA Nutt.

Calyx deeply 5-cleft. Corolla declined; upper lip 2-cleft, its lobes partly turned backward. Fifth stamen gland-like. Capsule 2–many-seeded. — Slender N. Am. annuals or biennials, with particolored flowers in umbel-like clusters, appearing whorled in the axils of the upper leaves. (Dedicated to *Zaccheus Collins*, Philadelphian botanist, 1764–1831.)

Pedicels slender, as long as to longer than corolla.
  Corolla 0.8–1.5 cm. long, the very gibbous throat much shorter than limb; upper filaments more or less bearded below; seeds rugose.
    Upper leaves somewhat clasping; stem and pedicels with lines of puberulence; corolla-lobes barely emarginate, the upper lip white, the lower one bright blue; the 2–4 seeds 1.5–3 mm. long. . . . . 1. *C. verna.*
    Upper leaves sessile; stem and pedicels uniformly puberulent; corolla violet-purple, its lobes obcordate; the 6–12 seeds 1–1.5 mm. long. . . . 2. *C. violacea.*
  Corolla 4–6 mm. long, blue or blue and white, the ellipsoid throat longer than limb; filaments all glabrous; seeds smooth. . . . 3. *C. parviflora.*
Pedicels stouter, shorter than calyx-lobes; corolla 1–2 cm. long, purple or violet and whitish, the saccate throat very oblique and as broad as long; seeds rugose-reticulate. . . . . 4. *C. heterophylla.*

1. C. vérna Nutt. (vernal), Blue-eyed Mary. — Stem 1–6 dm. high, *puberulent in lines*, often branching from base; lower *leaves* ovate, the *upper* ovate-lanceolate and *partly clasping* by the cordate base, toothed; whorls about 6-flowered; *flowers long-pedicelled; corolla blue and white, 1–1.5 cm. long*, more than twice exceeding calyx, *its lobes barely emarginate; upper filaments bearded at base;* the 2–4 seeds 1.5–3 mm. long. — Rich woods and thickets, N.Y. to s. Wisc. and e. Ia., s. to w. Va., Ky., n. Ark. and e. Kans. Apr.–June.

2. C. violàcea Nutt. (violet). — Mostly lower than no. 1; *stem uniformly puberulent; upper leaves sessile;* whorls 2–5-flowered; upper lip of corolla shorter, pale; *lower lip violet, the lobes obcordate;* the 6–12 *seeds* 1–1.5 *mm. long.* — Rich woods, s. and w. Mo. and e. Kans. to Tex. Apr.–June.

3. C. parviflòra Dougl. (small-flowered). — Stem minutely puberulent, 0.5–3 dm. high, simple to diffusely branched; lower *leaves* ovate or rounded, the upper oblong-lanceolate to linear, *mostly entire;* whorls 2–6-flowered; *corolla* 4–6 *mm. long*, the ellipsoid *throat longer than the limb*, the pale upper lip with 2 yellow crests at base, the lower lip with blue-violet lobes; *filaments all glabrous;* the 2–4 *seeds* 1–2 mm. long, *smooth.* — Gravelly bluffs and openings, w. Can. and w. U. S., e., locally, to w. S.D.; n. Mich.; e. Ont.; w. Vt. Late May–July.

4. C. heterophýlla Graham (diverse-leaved). — Stem simple or diffusely branched, 1–5 dm. high; middle and upper leaves broad-based, lanceolate; flowers on *pedicels up to* 5 *mm. long*, 2–7 in a whorl; *corolla* 1–2 cm. long, the upper lip lilac or white and nearly equaling the purple or violet lower one, *the saccate tube as broad as long;* seeds numerous, flattened, winged, 1.5–2 mm. long, rugose-reticulate. (*C. bicolor* Benth., not Raf.) — Calif. species, often cult., tending to spread to cool situations in Ill. and Ky. (Introd. from farther west)

7. SCROPHULÀRIA L. Figwort

Corolla short, the tube ventricose and globular or ellipsoid; the 2 upper lobes projected forward, the 2 lateral short and rounded. the lowest drooping and reflexed. Stamens 4, declined, usually included, the anther-locules transverse and confluent into 1. Stigma entire or emarginate. Seeds rugose, marginless. — Coarse perennial herbs of temp. reg., with 4-angled stems, mostly opposite leaves and leathery greenish, purplish or lurid flowers in loose cymes, forming a terminal panicle. (So called from the fleshy knobs on the rhizomes of some species, which, by the doctrine of signatures, were supposed to cure *scrofula* and to remove *fig-warts.*)

Principal leaves cuneate to subtruncate at base, coarsely serrate to incised, their petioles wing-margined; corollas 5–11 mm. long, lustrous; rudimentary stamen greenish-yellow. . . . . 1. *S. lanceolata.*
Principal leaves rounded to cordate at base, serrate or serrate-dentate, their petioles barely margined; corollas 5–8 mm. long, dull; rudimentary stamen brown or purplish.
  Stem with rounded angles and grooved sides; panicle pyramidal to subcylindric, with spreading-ascending branches and filiform pedicels; corolla brown. 2. *S. marilandica.*
  Stem with thin or acute angles and flat sides; panicle virgate, with stiffly ascending branches and thickish pedicels; corolla with green base and tips of lobes. . . . . 3. *S. nodosa.*

1. S. lanceolàta Pursh (lance-shaped). — *Stem* 0.5–2 m. high, *with* rounded angles and *flat sides*, usually puberulent; *principal leaves* lanceolate to ovate, *tapering or cuneate to subtruncate at base, coarsely sharp-serrate or incised, with wing-margined petioles; inflorescence virgate, with strongly ascending cymes* and stoutish pedicels; *corolla greenish-brown, lustrous,* 5–11 *mm. long*, the dorsal lobes broadly oblong and considerably longer than the others; *rudimentary stamen greenish-yellow, flabelliform, usually broader than long,* 1.3–1.8 *mm. long;* capsule slenderly ovoid, acuminate, 6–9 mm. long. (*S. leporella* Bickn.) — Thickets and borders of woods, Rimouski Co., Que. to s. B.C., s. to N.S., N.E., Va., w. S.C., Ind., Ill., Mo., Okla., N.M., Utah and n. Calif. May–July.

2. S. marilàndica L. (of Maryland), Carpenter's-square. — *Stem with rounded angles and grooved sides, glabrous; leaves* ovate, acuminate, *with strongly rounded to cordate bases, serrate or serrate-dentate, on slender petioles; panicle pyramidal to subcylindric, with spreading-ascending branches and filiform pedicels; corollas* 5–8 mm. long, *brown,* dull, the dorsal lobes suborbicular; *rudimentary stamen brown or purplish, clavate, longer than broad,* 1–1.2 mm. long; capsule globose-ovoid, 4–7 mm. long. — Rich woods and thickets, sw. Me. and sw. Que. to Minn., s. to S.C., n. Ga., Ala., La. and Okla. Late June–Oct.

3. S. nodòsa L. (knotty; from the knobby rhizome). — Similar to no. 2; *stem acutely angled, with thin marginal lines, the sides flat;* leaves less attenuate at tip; *panicle stiff and virgate, with short stiffly ascending branches and thickish pedicels; corolla with green base and tips of lobes,* otherwise suffused with brown; rudimentary stamen chocolate-brown. — Rocky or gravelly woods and thickets (chiefly calcareous), Nfld.; locally adv. from Eu. in s. N.E. and formerly on ballast to N.J. Late June–Sept. (Eu.)

8. CHELÒNE L. Turtlehead. Snakehead. Tête de Tortue or Galane (Que.)

Calyx of 5 distinct imbricated sepals. The mouth of the corolla a little open; upper lip broad and arched, keeled in the middle, notched at the apex; lower woolly-bearded in the throat, 3-lobed at the apex, middle lobe smallest. Seeds many. — Smooth N. Am. perennials, with upright branching stems, serrate leaves, and large white or purple flowers, which are nearly sessile in spikes or clusters, and closely imbricated with round-ovate concave bracts and bractlets. (Name from the Greek *chelone, a tortoise*, the corolla resembling in shape the head of a reptile.)*

Leaves narrowed to base or petioled; corolla creamy-white to purple; sterile filament white or greenish.
  Leaves tapering or rounded to winged or obscure petioles or subsessile; corolla white or creamy throughout or with pink or purple summit; sterile filament greenish; bracts with prolonged tips; sepals with distinct scarious margins, only obscurely ciliolate. . . . . 1. *C. glabra.*
  Leaves distinctly slender-petioled; corolla purple throughout; sterile filament white or whitish; bracts rounded or only short-tipped; sepals obscurely scarious-margined, variously ciliolate.
    Leaves lanceolate, tapering gradually to base and apex, the principal ones 1.4–4 cm. broad; corolla scarcely ridged above, the lower lip with pale-yellow beard and with or without faint interior lines. . . . . . 2. *C. obliqua.*
    Leaves ovate, rounded to base, the principal ones 3–10 cm. broad; corolla sharply ridged above, the lower lip with deep yellow beard and with many fine lines. . . . . 3. *C. Lyoni.*
Leaves sessile, strongly rounded to subcordate at base; corolla purple, with broad deep-colored lines on lower lip; sterile filament purple. . . . . 4. *C. Cuthbertii.*

1. C. glàbra L. (smooth), Balmony. — Stem 0.4–2 m. high; *leaves* narrowly to broadly lanceolate or narrowly ovate, *with short winged petiole or subsessile*, acuminate, serrate; spike subtended at base by well-developed leaves, at first dense, elongating and loosening in fruit; *bracts acuminate; sepals* rounded, *scarious-margined, obscurely or not at all ciliolate; corolla* 2.3–3.5 cm. long, *creamy-white below*, the summit and lobes roseate, purple or greenish-yellow, the lower lip with white to pale yellowish beard; the *sterile filament greenish, much shorter than fertile ones;* capsule globose-ovoid, 1–1.5 cm. long. — Highly variable; the following varieties and forms the best marked:

Corolla whitish with only slight blush of pink at summit or with greenish-yellow lobes.
  Leaves decreasing in size from the median to the upper pair, lanceolate to narrowly ovate, the uppermost tapering to narrow bases, those subtending the terminal spike 0.5–3(–4) cm. long.
    Leaves lanceolate to ovate, the larger 1.5–5 cm. broad; corolla creamy-white to pinkish at summit or within the lobes.
      Lower surfaces of leaves glabrous. . . . . *C. glabra* (typical).
      Lower surfaces minutely pubescent. . . . . *C. glabra*, forma *tomentosa.*
    Leaves linear-lanceolate, prolonged, the larger 0.8–2.3 cm. broad; corolla greenish-yellow at summit; midwestern.
      Lower surfaces of leaves glabrous. . . . . Var. *linifolia.*
      Lower surfaces of leaves minutely pubescent. . . . . Var. *linifolia*, forma *velutina.*
  Leaves increasing in size from the median to the upper pair; the latter oblong to ovate, rounded to subcordate at base, (1.5–)2.5–5 cm. broad; flowers as in typical form. . . . . Var. *dilatata.*
Corolla purple or deep rose-color at summit and throat; leaves broadly lanceolate to narrowly oval, relatively thin and with longer petioles, the larger blades 2–6 cm. broad. . . . . Var. *elatior.*

C. glàbra L. (typical), incl. var. *elongata* Pennell & Wherry — Low grounds, margins of streams, swales and wet thickets, Nfld. to Ont. and Minn., s. to N.S., N.E., Ga., Ala. and Mo.

* Key based largely on that of Pennell & Wherry in Bartonia, no. 10: 12–23 (1929).

Late July (northw.)–Oct. — Forma **tomentòsa** (Raf.) Pennell (with dense soft hairs), scattered through the range.

Var. **linifòlia** Coleman (with leaves of flax), *C. linifolia* (Coleman) Pennell — Bogs and swales, Ont. and e. Man., s. to O., Ind., Ill. and Ia. — Forma **velùtina** Pennell & Wherry (velvety), scattered through the range.

Var. **dilatàta** Fern. & Wieg. (expanded). — Shores, ditches and low thickets, Nfld. to James Bay, s. to N.S., N.E., N.Y., and rarely along the mts. to N.C.

Var. **elàtior** Raf. (taller), *C. glabra*, forma *rosea* Fern.; *C. montana* (Raf.) Pennell & Wherry — Shaded spring-heads, wet woods, bogs, etc., N.H. to Ind., s., mostly in the uplands, to Ga. and Ala.

2. C. oblìqua L. (oblique). — Resembling the last var.; *leaves lanceolate or lance-elliptic,* thin, the principal ones 1.4–4 *cm. broad, tapering to slender petioles* 0.5–1.5 cm. long; *bracts short-acuminate or merely acute; sepals scarcely scarious-margined, slightly ciliolate; corolla purple throughout,* 2.5–3.2 cm. long, *scarcely ridged above, the lower lip with pale yellow beard* and with or without faint slender lines within; *sterile filament white,* less than half as long as fertile ones. — Cypress-swamps and wet woods, Fla. to Miss., n. to Md. and Tenn. Late Aug.–Oct.

Var. **speciòsa** Pennell & Wherry (showy). — Bracts more obtuse; sepals copiously ciliolate; corolla 3–3.7 cm. long. — Alluvial woods and swamps, Ind. to s. Minn., s. to Ark.

3. C. Lýoni Pursh (for its discoverer, John Lyon, 17?–1818). — Differing from no. 2 in its *round-based ovate leaves on petioles up to* 4 *cm. long,* the broader *blades* 3–10 *cm. wide;* bracts suborbicular or reniform, obtuse or rounded above; *corolla sharply ridged above,* the lower lip *with deeper yellow beard and brighter purple lines.* — Cult. and esc. to shores and thickets, s. N.E. July–Oct. (Introd. and natzd. from s. of our range)

4. C. Cuthbértii Small (for its discoverer, Alfred Cuthbert, 1857–1932). — Slender, stiffly erect, 0.4–1.5 m. high; *leaves* broadly lanceolate to oblong-ovate, slenderly acuminate, sharply serrate, *sessile, with rounded or subcordate bases,* the larger 1–5 cm. broad; spike 4-sided, usually well prolonged above the upper leaves; *bracts scarious-margined;* sepals ciliolate; *corolla violet-purple, with broad lines inside on lower side,* 2–3 cm. long; *sterile filament purple, one-half to two-thirds as long as the fertile.* (*C. Grimesii* Weath.) — Boggy woods and thickets, se. Va.; upland of N.C. Aug., Sept.

9. PENSTÉMON Mitchell Beard-tongue

Calyx 5-parted. Corolla tubular, gradually or abruptly dilated in the throat, more or less 2-lipped; upper lip 2-lobed, the lower 3-cleft. Fertile stamens 4, declined at base, ascending above, the fifth sterile filament either naked or bearded. Seeds numerous. — About 300 perennials of N. Am. (1 in ne. Asia, 1 in S. Am.), with opposite entire or toothed leaves, the upper sessile and mostly clasping. Flowers chiefly showy, thyrsoid or in open racemose panicles. (Name from *pente, five,* and *stemon,* in the sense of *stamen;* the fifth stamen being present and conspicuous, although sterile.)* Name often spelled Pentstemon.

a. Upper leaves elongate, herbaceous to firmer, not strongly if at all glaucous; inflorescence pubescent (or in nos. 5 and 6 glabrescent); corolla 1–3.8 cm. long or, if longer (in no. 2), the plant with closely pubescent stem and green, pubescent and sharply toothed leaves. . . *b.*

b. Inner surface of corolla or its lobes minutely (microscopically) glandular-puberulent. . . *c.*

c. Internodes of stem densely puberulent; throat of corolla ampliate.

Leaves lance-oblong to linear, entire or nearly so; corolla 1.7–2.5 cm. long, the limb 1.5–2 cm. broad. . . . . 1. *P. albidus.*

Leaves ovate, oval, broadly lanceolate or oblong, mostly sharp-toothed; corolla 3.5–5 cm. long, nearly as broad. . . . . 2. *P. Cobaea.*

c. Internodes of stem and leaves glabrous, the latter mostly entire; corolla 1.5–2 cm. long, trumpet-shaped, with widely spreading limb. . . . 3. *P. tubaeflorus.*

b. Inner surface of corolla glandless. . . *d.*

d. Throat of corolla strongly inflated, with only moderate ridging within, the upper and lower lobes subequal; sterile filament only slightly bearded. . . *e.*

e. Anthers usually bearded on back with few stiff white hairs. . . *f.*

---

* Many statements of characters derived from F. W. Pennell's treatment in his *Scrophulariaceae of Eastern Temperate North America,* Acad. Nat. Sci. Phila. Monogr. i. (1935) and from C. C. Deam's Flora of Indiana (1940).

f. Inflorescence strongly glandular; stem lustrous, glabrous except for slender decussate lines; sepals during anthesis 5–8 mm. long, scarious-margined, caudate-tipped; corolla (2–)2.3–3 cm. long. 4. *P. Digitalis.*

f. Inflorescence glabrous or very sparsely glandular; stems dull, pubescent or glabrous; sepals during anthesis 2–5 mm. long; corolla 1.5–2.3 cm. long.

Lower cauline leaves lance-acuminate, sharply serrate; sepals after anthesis elongating to 5–9 mm., caudate-tipped, scarcely scarious-margined. . . . . 5. *P. alluviorum.*

Lower cauline leaves oblong or oval, rounded at tip, entire or slightly denticulate; sepals in maturity only 3–4 mm. long, acute, with evident scarious margin. . . . . 6. *P. Deamii.*

e. Anthers glabrous.

Cauline leaves mostly with abundant dentations; sepals lance- or linear-attenuate, 5–12 mm. long and wide-spreading during anthesis; corolla usually 2–3.5 cm. long. . . . . . . 7. *P. calycosus.*

Cauline leaves entire or only obscurely dentate; sepals lance-ovate, 3–6 mm. long, with ascending acute to caudate tips; corolla 1.5–2 (–2.2) cm. long. . . . . 8. *P. laevigatus.*

d. Throat of corolla flattened and strongly ridged within; limb bilabiate, with the lower lobes projecting well beyond the upper ones; sterile filament densely bearded. *g*.

g. Corolla with deeper color lining the throat; the lower lip projected forward, the orifice of the throat consequently open; locules of anthers longer than broad, in maturity becoming cup-like; capsule ovoid to conic. . . *h*.

h. Leaves herbaceous and membranaceous; the median and upper cauline ones ovate to broadly lanceolate or oblong, broadly rounded to cordate at base; the second pair below the bracteal ones 1.5–4.5 cm. broad, sharply dentate to entire; corolla purple or violet, dark-striped within.

Throat of corolla abruptly enlarged above the tube, 8–10 mm. in diameter at summit, the whole corolla 2–3.5 cm. long; sepals after anthesis greatly prolonged, their caudate usually arching tips equaling or exceeding the capsule; bracts at base of inflorescence like reduced foliage-leaves, 1.5–9 cm. long. . . . 9. *P. canescens.*

Throat of corolla gradually tapering into the tube, 5–8 mm. in diameter at summit, the corolla 1.5–2.6 cm. long; sepals ovate-acuminate, about half as long as capsule; bracteal leaves often smaller. . . . . 10. *P. brevisepalus.*

h. Leaves firm to coriaceous; the median and upper cauline ones lanceolate, lance-oblong or lance-linear; the second pair below the bracteal ones 0.2–2 cm. broad, entire or remotely (exceptionally abundantly) denticulate; corolla white to violet or purple, its throat nearly tubular. *i*.

i. Corolla white or whitish outside; principal branches of panicle loosely ascending.

Leaves soft-velutinous or -pilose beneath; sepals in anthesis 3–5 mm. long; corolla 1.7–2.2 cm. long, its throat much longer than the tube; wide-ranging. . . . . 11. *P. pallidus.*

Leaves glabrous or nearly so; sepals in anthesis 2–4 mm. long; corolla 1.5–1.8 cm. long, its throat and tube subequal; southwestern. . . . . 12. *P. arkansanus.*

i. Corolla with lower lip purple to blue-violet outside; principal branches of the stiff panicle erect, suberect or essentially suppressed.

Corolla with reddish-purple lower lip, in age becoming broader than high; leaves of basal rosette oval, ovate or obovate; larger cauline leaves 1–2 cm. broad; southeastern. . . . 13. *P. australis.*

Corolla with violet-blue lower lip, as high as broad; leaves of basal rosette oblanceolate; larger cauline leaves 0.2–1 cm. broad; northwestern. . . . . 14. *P. gracilis.*

g. Corolla white or with throat pinkish or pale violet, of same color within and without; lower lip arching up against the upper one, the orifice of the throat consequently closed (until forced open by bees); locules of anthers as broad as long, in maturity saucer-like.

Corolla with violet-purple throat and white lobes; sepals acuminate or attenuate, in anthesis 3–7, in fruit up to 8 mm. long, more than half length of capsule; leaves glabrate in age, the principal cauline ones sharply serrate or serrate-dentate; wide-ranging. . . . . 15. *P. hirsutus.*

Corolla white throughout; sepals acute or short-acuminate, in anthesis 2–4, in fruit 3–5 mm. long, less than half length of capsule; leaf-surfaces permanently soft-pilose, the principal cauline blades remotely repand-denticulate; southern, west of Alleghanies. . 16. *P. tenuiflorus.*

*a.*Upper leaves orbicular or round-ovate, heavy and strongly glaucous, entire, clasping or perfoliate; stout stem glabrous; ventricose lilac to blue-lavender corolla 4–5 cm. long. 17. *P. grandiflorus.*

1. P. álbidus Nutt. (whitish). — *Stems* 1.5–4 dm. high, usually in tufts from a branching caudex, *closely puberulent; leaves* lance-oblong to linear, firm, *densely pubescent; thyrse strict,* 0.5–2 dm. long; *sepals viscid-villous,* in anthesis 5–7 mm. long; *corolla* 1.7–2.5 *cm. long,* white or slightly tinged with violet, with darker purple orifice of the throat, *ampliate, the inside of the lobes minutely (microscopically) glandular-puberulent,* the limb 1.5–2 cm. broad; *capsule* 8–10 mm. long, *equaled by the sepals.* — Plains, prairies and calcareous rock, Alta. to N.M., e. to Man., Minn., Ia., Neb., Kans. and Okla. May, June. — Whole plant blackened in drying.

2. P. Cobaèa Nutt. (for the genus *Cobaea*). — Stems pubescent, up to 7.5 dm. high; *cauline leaves mostly oblong to narrowly ovate and sharp-toothed, the median and upper large ones cordate-clasping and* 2.5–6 *cm. broad;* thyrse rather short; *sepals pubescent,* 9–11 *mm. long; corolla* 3.5–5 *cm. long, nearly as broad, ventricose,* violet, purple or whitish, *the inside slightly glandular-puberulent;* capsules scarcely exceeding sepals. — Plains, prairies and calcareous rocky openings, bluffs, etc., Neb. to Ark., Okla. and Tex.; locally spread from cult. farther eastw. May, June.

3. P. tubaeflòrus Nutt. (with tubular flowers). — *Stems* solitary or few, 3–9 dm. high, *glabrous; principal cauline leaves* (above the subpetiolate lower and below the reduced upper ones) 2–4 *distant pairs, the panicle standing* 1.5–4.5 *dm. above the fully developed upper leaves; blades of principal leaves oblong or oblong-ovate, bluntish or subacute, submembranaceous,* 2–4 *cm. broad, entire or nearly so; panicle with dense* to slightly open *cymules, the lowest cymules on peduncles* 0.2–1.5 (rarely –3) *cm. long; sepals lance-ovate,* 3–5 *mm. long, appressed; corolla* 1.5–2 cm. long, *trumpet-shaped, with gradually dilated throat and widely spreading limb, white or whitish,* the *inside of the throat minutely (microscopically) glandular-puberulent.* — Open woods, fields and roadsides, Neb. to e. Tex., e. to Miss., Tenn., Ind. and Wisc., partly adv. eastw.; adv. locally to Atl. states. May–July.

Var. ACHÒREUS Fern. (without a native land). — More slender, 0.6–1 m. high; *leaves more coriaceous, the pairs less distant, the large cauline ones* 1–2.5 *cm. broad* and often subacuminate; *lowest cymules on peduncles* 1.5–8 *cm. long.* — Dry old fields, pastures, etc., N.E. to Ont., s. to Pa. June, July (Natzd. but source unknown)

4. P. Digitàlis Nutt. (for the genus *Digitalis*). — *Stems glabrous except for slender decussate lines* running down from the leaf-bases, *sublustrous,* often purplish and *slightly glaucous,* up to 1.5 m. high; rosette-leaves broadly oblanceolate to elliptic; *cauline leaves subcoriaceous, glabrous, lanceolate, oblanceolate or oblong-ovate, acuminate,* entire to short-denticulate, the larger ones mostly 5–7 pairs, upper pair usually much smaller; *thyrse glandular, open-paniculate,* with lower branches elongate; *sepals during anthesis* 5–8 (*in fruit* –10) *mm. long, scarious-margined, caudate-tipped; corolla* 2–3 *cm. long, white or purple-tinged, its throat strongly inflated, with only moderate ridging within, the limb scarcely spreading, the bases of the lobes somewhat pubescent within; sterile filament only slightly bearded; anthers usually with stiff hairs along the back; larger mature capsules* 9–12 *mm. long;* seeds 0.7–1 mm. long. (*P. laevigatus,* var. Gray) — Open woods, meadows and prairies, now spread to fields and clearings, Me. and s. Que. to S.D., s. to s. N.E., L.I., Va., Ala., La. and Tex. May–July. — Original indigenous area largely in Miss. Basin.

5. P. alluviòrum Pennell (of alluvial soils). — *Differing from no.* 4 *in its dull and minutely pubescent or glabrescent stems without bloom; leaves more or less pubescent, the lowest ones lance-acuminate and sharply serrate; inflorescence only sparsely glandular; sepals in anthesis only* 3–5 *mm. long, broader, with narrower scarious margin; corolla* 1.7–2.3 *cm. long, its upper lobes up-curving;* larger capsules 8–9 mm. long. — Low grounds, bottoms and slopes, s. Ind. and s. Ill. to Mo., s. to Miss. and Ark. Mid-May–June.

6. P. Deàmii Pennell (for its discoverer, CHARLES CLEMON DEAM, 1865–). — *Differing from no.* 4 *in* lower stature (up to 9 dm. high); *dull puberulent* or glabrate *stems; less glandular* *inflorescence; sepals in anthesis only* 2–3, *in fruit* 3–4 *mm. long, merely acute; corolla* 1.5–2.2 *cm. long; capsules* 6–8 *mm. long.* — Sterile woods and fields, s. Ind. and s. Ill. Late May, June.

7. P. calycòsus Small (with prominent calyx). — Differing at once from nos. 4–6 in its *glabrous anthers; resembling no.* 6 *but with stem often pubescent; greener, subherbaceous and usually more sharply toothed leaves; sepals lance- or linear-attenuate, up to* 1.2 *cm. long, with recurving tips; capsule* 7–8 *mm. long.* — Woods, meadows, calcareous rocky slopes, etc., O., s. Mich. and Ill., s. to Ga., Ala. and Tenn.; spread eastw. in fields to N.E., Pa. and Md. June, July.

8. P. laevigàtus Ait. (smooth). — *Very close to no.* 6 *but* smaller and *with glabrous anthers; sepals lance-ovate,* 3–6 *mm. long; corolla* 1.5–2(–2.2) *cm. long; capsules* 5–8 *mm. long.* — Meadows, bottoms, rich woods, calcareous bluffs, etc., s. N.J., Pa. and W.Va., s. to S.C., n. Fla., Ala. and Miss. May, June. — A highly conservative species, without the aggressive and weedy tendencies of nos. 4 and 7.

9. P. canéscens Britt. (canescent). — Plant 3–9 dm. high, the *stem closely cinereous-puberulent,* or in forma **villicaùlis** Fern. (with villous stem) with the internodes *villous* with elongate trichomes; *leaves membranaceous and herbaceous, full green, mostly serrulate-dentate, pubescent,* or in the local forma **Brittonòrum** (Pennell) Fern. (for Nathaniel Lord Britton, 1859–1934, and Elizabeth Gertrude Britton, 1858–1934) glabrous, the rosulate and lower cauline ones petioled; *the principal cauline ones ovate to broadly lanceolate or oblong, broadly rounded to cordate at base; the second pair below the bracteal ones* 1.5–4.5 *cm. broad; bracts at base of inflorescence like reduced foliage-leaves,* 1.5–9 *cm. long; thyrse open, with loosely ascending branches; sepals after anthesis greatly prolonged, their caudate and usually arching tips equaling or exceeding the capsule; corolla purple or violet,* with more deeply colored lines in the *inside of the flattened and strongly ridged throat,* 2–3.5 *cm. long, with the lower lobes projected forward; the throat abruptly enlarged above the tube,* 8–10 *mm. in diameter at summit; locules of anthers longer than broad, becoming cup-like in maturity; sterile filament densely bearded;* capsule 6–8 mm. long. — Sandy or loamy woods and thickets, chiefly of the upland but reaching lower levels, Pa. to s. Ind., s. to se. Va., nw. S.C., n. Ga. and n. Ala. Mid-May–early July.

10. P. brevisépalus Pennell (with short sepals). — *Differing from no.* 9 *in having the throat of the corolla gradually tapering into the tube and only* 5–8 *mm. in diameter at summit, the corolla only* 1.5–2.6 *cm. long; sepals in anthesis only* 2–4 *mm. long, becoming about half as long as capsule, their tips short;* capsule 5–7 mm. long; internodes of lower half of stem closely canescent-puberulent, or in forma **heterolásius** Fern. (variously hairy) with numerous long villi intermixed. — Open woods, rich slopes and rocky banks, se. Va. to W.Va., s. chiefly along the mts. to nw. Ga. and Tenn. Late April–early July.

11. P. pállidus Small (pale). — *Differing from no.* 10 *in its firm to coriaceous paler leaves soft-pilose on both surfaces, the median and upper cauline ones lanceolate to lance-oblong and* 1–2 *cm. broad,* entire or only remotely toothed; *panicle with loosely ascending branches; sepals in anthesis* 3–5 *mm. long, lance-attenuate; corolla white outside,* 1.7–2.2 *cm. long, its throat nearly tubular and much longer than the tube.* — Sandy or loamy woods and openings, taking to fields and roadsides, N.E. to Mich. and Ia., s. to Ga., Tenn., Ark. and Kans., eastw. as a natzd. plant only. Late April–July.

12. P. arkansànus Pennell (of Arkansas). — *Similar to no.* 11 *but the leaves glabrous or essentially so;* the diffuse panicle with very numerous small flowers; *sepals in anthesis only* 2–4 *mm. long; corolla* 1.5–1.8 *cm. long, its throat and tube subequal.* — Rocky woods and slopes, Mo. and Ark. to e. Tex. May, June.

13. P. australis Small (southern). — Stems canescent-puberulent and leaves firm as in no. 11; *radical leaves oblanceolate to obovate, entire or nearly so; principal cauline leaves lanceolate, acuminate,* 0.5–2 *cm. broad, entire or remotely few-toothed,* or in forma **odontophýllus** Fern. (with toothed leaves) the rosulate leaves regularly crenate-dentate and the cauline abundantly and somewhat doubly serrate-dentate; *panicle with stiffly erect branches; sepals* 4–8 *mm. long,* acute or acuminate; *corolla creamy, with reddish-purple lower lip, in age becoming broader than high, paler inside,* 2–2.5 *cm. long;* capsule 8–9 mm. long. — Sandy pinelands and dry oak-woods of Coastal Plain, extending into outer Piedmont and rarely to the mts., Fla. and Ala., n. to se. Va. May, June.

14. P. grácilis Nutt. (slender). — *More slender than no.* 13, 2–4 *dm. high; rosette-leaves narrowly oblanceolate; principal cauline leaves linear or linear-lanceolate,* 0.2–1 *cm. wide, mostly with minute sharp teeth, glabrous; thyrse stiff and slender, usually virgate, with suppressed branches; pedicels closely glandular-puberulent; sepals* 4–5 *mm. long, in fruit nearly equaling the capsule; corolla slender, as high as broad,* 1.8–2 *cm. long, its lower lip violet-blue.* — Dry open woods and prairies, Alta. to N.M., e. to w. Ont., Wisc., S.D., Neb., and as a weed farther eastw. June, July.

Var. **wisconsinénsis** (Pennell) Fassett (of Wisconsin). — Up to 6 dm. high; *leaves puberulent; pedicels, etc., more loosely glandular;* corolla slightly longer. — Centr. and s. Wisc.

15. P. hirsùtus (L.) Willd. (hirsute). — *Stem* 2–9 dm. high, *covered with fine spreading* whitish and often gland-tipped *hairs*, leafy nearly to the inflorescence; radical leaves lanceolate, oblanceolate, oblong or elliptic; *cauline leaves* lanceolate to oblong, *usually with numerous acute serrations or dentations*, at first pubescent, *becoming glabrate;* thyrse open, with elongate branches; *sepals acuminate or attenuate*, 3–7, or in fruit *to 8 mm. long; corolla slender, dull violet or purplish, with white lobes, bilabiate, the throat nearly closed by the upward arching villous palate; anthers as broad as long; capsule* 8–9 mm. long, *more than half covered by sepals.* — Dry or rocky ground, s. Que. and s. Ont. to Wisc., s. to s. N.E., and inland from Coastal Plain, to Va., Ky. and n. Tenn. June, July.

16. P. tenuiflòrus Pennell (slender-flowered). — *Differing from no.* 15 *in permanently pilose leaf-surfaces; margins of leaves remotely small-toothed or nearly entire; sepals in anthesis* 2–4, *in fruit* 3–5 *mm. long and less than half as long as capsule; corolla white throughout.* — Dry open woods, cedar-glades and calcareous barrens, Ky., Tenn. and Ala. Late April–early June.

17. P. grandiflòrus Nutt. (large-flowered). — *Plant glabrous and glaucous throughout*, up to 1.2 m. high; *leaves thick and fleshy, entire*, all but the obovate radical ones *clasping or perfoliate, the upper and bracteal ones orbicular or round-ovate; thyrse with sessile or subsessile distant fascicles; sepals oblong*, 7–12 *mm. long; corolla strongly ventricose-campanulate, lilac or blue-lavender*, 4–5 *cm. long; sterile filament hooked and only minutely bearded at apex; capsule subligneous*, 1.8–2.5 *cm. long.* — Prairies and plains, Wyo. to Tex., e. to Wisc., w. Ill., w. Mo. and Okla.; locally spread from cult. eastw. to N.E. May, June.

10. PAULÓWNIA Sieb. & Zucc.

Calyx 5-cleft. Corolla-tube enlarged upward, the 5 unequal lobes spreading. Stamens didynamous; sterile filament none. Seeds numerous, winged. — Tree, with large cordate pubescent petioled leaves and terminal panicles of large violet flowers formed in autumn, remaining naked over winter and expanding in spring. (Named for *Anna Paulowna*, 1795–1865, princess of the Netherlands.) — Chinese tree, approaching the *Bignoniaceae*.

1. P. TOMENTÒSA (Thunb.) Steud. (densely soft-pubescent), PRINCESS-TREE, KARRI-TREE. — Gray-barked tree up to 15 m. high; leaves entire or slightly lobed, 1.5–4 (on sprouts –5) dm. broad; calyx 1–1.5 cm. long, rusty-pubescent, with obtuse lobes; corolla about 5 cm. long, violet, with yellow stripes within, glandular on the outside; capsule 3–4 cm. long. — Roadsides, clearings and borders of woods, Fla. to s. N.Y. and W.Va. Apr., May. (Introd. and natzd. from e. Asia)

Tribe IV. GRATIÓLEAE Benth.

11. MÍMULUS L. MONKEY-FLOWER

Calyx prismatic, 5-angled, 5-toothed, the uppermost tooth largest. Upper lip of corolla erect or reflexed-spreading, 2-lobed; lower spreading, 3-lobed. Stigma 2-lobed; lobes ovate. Seeds numerous. — Am., Asiatic, Afr. and Austral. herbs, with opposite (rarely whorled) leaves, and mostly handsome flowers. (Diminutive of *mimus*, *a buffoon*, from the grinning corolla.)

Corolla violet-purple (rarely white), the throat nearly closed by the prominent palate; calyx-teeth subequal; placentae of mature capsule separating at tip; erect glabrous perennials with pinnately veined leaves and terminal leafy racemes. § EUMIMULUS.
- Leaves sessile; stem wingless; fruiting pedicels 1–6 cm. long; calyx-teeth 1.5–8 mm. long. . . . . 1. *M. ringens.*
- Leaves petioled; angles of stem somewhat winged; fruiting pedicels 0.2–1 (–1.5) cm. long; calyx-teeth 1–1.5 mm. long. . . . . 2. *M. alatus.*

Corolla yellow or yellowish, with 2 prominent ridges or lines below the lower lip; placentae completely united; spreading or loosely ascending plants.
- Viscid-villous; calyx not inflated at maturity, its teeth elongate. § PARADANTHUS. . . . . 3. *M. moschatus.*
- Glabrous or only slightly pubescent; calyx inflated in maturity and loosely surrounding capsule, its uppermost tooth longest. § SIMIOLUS.
  - Flowering calyx 0.8–1.7, fruiting 1–2.5 cm. long; corolla 1–4 cm. long, the throat nearly closed by the 2 hairy ridges. . . . . 4. *M. guttatus.*
  - Flowering calyx 0.5–0.8, fruiting 1–1.2 cm. long; corolla 0.7–2.2 cm. long, the throat open. . . . . 5. *M. glabratus*, vars.

§ Eumímulus Gray

1. M. ríngens L. (gaping). — *Stem* erect, *square* in section, 0.4–1.3 dm. high, from stoloniferous rhizome; leaves oblong, elliptical, oblanceolate or narrowly obovate, sessile, dentate or low-serrate; pedicels stout, from the upper axils, 1–6 cm. long; *calyx tubular*, strongly angled, in fruit 0.8–2 cm. long, *with slender subequal teeth; corolla blue-violet or pinkish*, rarely white, 2–4 cm. long, the throat nearly closed by the large palate; capsule closely invested and exceeded by calyx, the mature placentae separating at tip. — Three vars.:

Leaves mostly clasping or strongly rounded at base; mature calyx-teeth 1.5–5 mm. long.
  Internodes mostly 3–7 cm. long; principal leaves 5–13 cm. long; lowest pedicels 2–6 cm. long; mature calyx 1.3–2 cm. long, with teeth 3–5 mm. long. . *M. ringens* (typical).
  Internodes 1.5–2.5 cm. long; principal leaves 2.5–5 cm. long; pedicels 1–1.7 cm. long; mature calyx 0.8–1 cm. long, with teeth 1.5–2.5 mm. long. . . Var. *colpophilus.*
Leaves acuminate to base and apex; mature calyx-teeth 4–8 mm. long. . . Var. *minthodes.*

M. ríngens (typical). — Shores, meadows and wet places, C.B. to James Bay and Man., s. to Ga., Ala., La., ne. Tex. and Colo. Late June–Sept. — Forma **Péckii** House (for Charles Horton Peck, 1833–1917) with white corollas, rare.

Var. **colpóphilus** Fern. (lover of estuaries). — Estuaries of the St. Lawrence system, Que., and the Penobscot R. system, Me.

Var. **minthòdes** (Greene) Grant (resembling Mint). — Marshes and shores, w. Mass. to Ga., Ala. and Kans.

2. M. alàtus Ait. (winged). — Similar to no. 1; *stem with thin wing-angles; leaves definitely petioled*, elliptic, ovate or lanceolate; *pedicels* 0.2–1(–1.5) *cm. long; calyx-teeth* 1–1.5 *mm. long, broad-based;* corolla 2–2.5 cm. long, blue-violet or pink, or white in forma **albiflòra** House (white-flowered), the tube less exserted. — Swamps and low grounds, n. Fla. to e. Tex., n. to Ct., N.Y., s. Ont., s. Mich., Ill., s. Ia. and e. Neb. June–Sept.

§ Paradánthus Grant

3. M. moschàtus Dougl. (musky), Muskflower. — *Clammy-villous or -tomentose* perennial from stoloniferous rhizomes, the stolons often moniliform; stems loosely spreading or ascending, weak, up to 3 dm. high, rooting at lower nodes; leaves oblong-ovate, short-petioled, dentate or entire, pinnately veined; *flowers axillary*, on slender pedicels shorter than leaves; *calyx* campanulate, *not inflated*, with elongate teeth; corolla funnelform, yellow, often striped with red in throat, 1.4–2 cm. long, with 2 densely hairy lines below the lower lip; capsule included. — Springy places, margins of streams and ditches, Nfld. to Ont., s., locally, to N.S., N.B., Mass., N.C., W.Va. and Mich.; Pacific N. Am.; apparently indigenous in Nfld., M.I. and n. Mich., elsewhere with us adv. or introd. June–Sept. — Plant only rarely with musky odor.

§ Simìolus Greene

4. M. guttàtus DC. (spotted). — *Smooth*, suberect or decumbent, rooting at the lower nodes, up to 5.5 dm. high; leaves rounded-ovate to oblong or broadly lanceolate, 0.8–1.5 cm. long, coarsely dentate, longitudinally ribbed, the lower long-petioled, the upper sessile; *flowers in loose terminal racemes*, remote, long-pedicelled; *calyx* campanulate, *at first* 0.8–1.7, *in fruit* 1–2.5 *cm. long, with obtuse teeth*, the upper tooth longest; *corolla* 1–4 *cm. long*, yellow, *with* red-spotted *throat nearly closed by the hairy ridges.* (*M. Langsdorfii* Donn) — Local, in brooks and meadows, Vt., Ct. and e. N.Y. June, July. (Natzd. from w. N. Am.)

5. M. glabràtus HBK. (becoming smooth). — Similar to no. 4; stems usually more depressed and repent below; *calyx* broadly campanulate, *in flower* 0.5–0.8, *in fruit* 1–1.2 *cm. long, the teeth rather lower; corolla* 0.7–2.2 *cm. long, with open throat.* — Highly variable, the typical plant of S. Am., Mex. and sw. U. S. with undulate-dentate rounded-oval leaves and spotted corollas 0.7–1.5 cm. long, the calyx 1–1.2 cm. long. — Represented with us by

Var. **Fremóntii** (Benth.) Grant (for its discoverer, John Charles Fremont, 1813–1890). — *Leaves* mostly suborbicular, oval or cordate to reniform, *with entire or merely denticulate or undulate margin; calyx* 0.5–1 *cm. long; corolla* 0.8–1.2 *cm. long*, without or with scant spotting on lower lip. (Var. *Jamesii* (T. & G.) Gray) — Wet shores, springs, brooks, etc., Mex., ne. and n. to Ont., Mich., Wisc., Minn., s. Man. and Mont. June–Oct.

Var. **michiganénsis** (Pennell) Fassett (of Michigan). — *Leaves sharply and doubly sinuate-dentate;* mature calyx 0.8–1.2 cm. long; *corolla* 1.5–2.2 *cm. long.* — Wet shores and springy places, n. Mich. June–Sept.

12. MÀZUS Lour.

Calyx campanulate, 5-cleft, the lobes soon divaricate. Corolla with slightly bifid erect upper lip, the 3-lobed lower lip much larger, with 2 gibbosities at throat. Style with 2 ovate lobes. Fertile stamens 4, the contiguous anther-locules divaricate. Capsule globose or compressed, obtuse, with 2 entire valves. — Herbs, nat. of Asia and Australia, with alternate flowering branches, subsecund terminal racemes, minute or obsolete bracts and ebracteolate or minutely bracteolate alternate pedicels. (Name from the Greek *mazos, papilla*, from the gibbosities at throat of corolla.)

1. M. JAPÓNICUS (Thunb.) Ktze. (Japanese). — Slender, 0.5–2 dm. high; leaves obovate or cuneate-oblong, coarsely dentate, the radical 2–5 cm. long; racemes loosely 2–7-flowered; calyx-lobes ovate-lanceolate, equaling the tube; corolla scarcely twice length of calyx, blue, the yellow or whitish palate bordered by reddish-brown; capsule shorter than calyx. — Grasslands and roadsides, local, Pa. to Mo. and La.; Pacific slope. July–Oct. (Natzd. from e. Asia)

13. GRATÌOLA L. HEDGE-HYSSOP

Narrow divisions of calyx usually nearly equal. Upper lip of corolla entire or 2-cleft, lower 3-cleft. Style dilated or 2-lipped at apex. Capsule 4-valved, many-seeded. — Low herbs of most temp. regions, mostly perennials, some biennials or annuals, with sessile leaves, and usually with 2 bractlets at the base of the calyx. Flowering in spring and summer; in wet or damp places. (Diminutive of *gratia, grace* or *favor*, from supposed medicinal properties.)

a. Anthers with a broad connective, the locules transverse; calyx-segments essentially equal; corollas mostly much exceeding calyx, 0.8–1.5 cm. long; capsule ovoid to globose; soft-stemmed, glabrous or viscid-puberulent or -pilose plants, with peduncled flowers. . . *b.*

b. Sterile stamens minute or none; capsules 3–7 mm. long, nearly or quite equaling calyx; cauline leaves narrowed or hardly clasping at base; annuals or biennials with honey-colored to whitish corolla with yellow tube. § NIBORA.

Pedicels filiform, in fruit becoming 1–2.5 cm. long; corolla of earlier flowers 8–12 mm. long, bearing clavate hairs within at the throat; capsule globose-ovoid, 3–5 mm. long; seeds thick-cylindric, 0.4–0.5 mm. long. 1. *G. neglecta.*

Pedicels stouter, in fruit only 0.1–1.4 cm. long; corolla of earlier flowers 0.9–1.5 cm. long, bearing slender hairs within at throat; capsule globose, 3–7 mm. long; seeds slender-cylindric, 0.7–0.8 mm. long. . . . 2. *G. virginiana.*

b. Sterile stamens obviously represented by a pair of filiform filaments with a capitate tip; capsules 1–3 mm. long, much overtopped by the calyx-segments; cauline leaves more or less clasping; perennials with creeping rhizomes and stolons. § GRATIOLARIA. . . . *c.*

c. Leaves entire or obscurely toothed, these and sepals glandular-punctate with dark dots; corolla golden-yellow (rarely pale) throughout; capsule 3 mm. long. . . . . 3. *G. aurea.*

c. Leaves dentate or serrate, at least above middle, only obscurely dotted; corolla white or purple-tinged, the yellow tube with brown or purple lines; capsule 1–2 mm. long.

Leaves oblong or ovate, finely dentate; bractlets foliaceous, nearly equaling calyx; calyx-segments 4–7 mm. long, lanceolate to elliptic. 4. *G. viscidula.*

Leaves linear or lanceolate, coarsely serrate above middle; bractlets minute or obsolete; calyx-segments 3–5 mm. long, linear-subulate. 5. *G. ramosa.*

a. Anthers with contiguous vertical locules; flowers subsessile; calyx-segments very unequal; corolla little exceeding calyx, white or purple-tinged, 5–9 mm. long; capsule slenderly conical, firm-stemmed, hirsute. § SOPHRONANTHE. . . . . . . 6. *G. pilosa.*

§ NIBÒRA (Raf.) Pennell

1. G. neglécta Torr. (overlooked). — *Annual*, with simple or loosely branched soft stem 0.3–4 dm. high, *the upper internodes, pedicels and expanding leaves* more or less *clammy-pilose or -puberulent; leaves* thin, *rhombic-lanceolate, tapering to base and apex*, undulate-dentate or entire, 1–5.5 cm. long; *pedicels filiform, elongating to* 1–2.5 *cm. long;* bractlets foliaceous, equaling or exceeding calyx; *corollas* honey-color to creamy-white, with yellowish tube, *the earlier* 8–12 *mm. long, the inside of the throat with clavate bearding;* sterile stamens minute or none; *capsule globose-ovoid,* 3–5 *mm. long; seeds thick-cylindric,* 0.4–0.5 *mm. long.* (*G. virginiana* of ed. 7, not L.; *G. lutea* Raf., in part, name too loosely applied) — Wet or muddy places, sw. Que. and centr. Me. to s. B.C., s. and w. beyond our range. May–Oct.

Var. **glabérrima** Fern. (smoothest). — Leaves more oblong, with rounded subamplexicaul bases, these and upper internodes of stem glabrous; fresh corolla milk-white except at base. — Tidal mud, estuary of St. Lawrence R., Que. Aug.–Oct.

2. G. virginiàna L. (Virginian). — Coarser than no. 1, the fleshy base often biennial; flowering *stems* simple or with few ascending branches, 1–4.5 dm. long, *usually glabrous;* leaves lanceolate to elliptic or oblong-obovate, shallowly undulate to sharply serrate, 1.5–7 cm. long; *pedicels stoutish,* 2–14 *mm. long;* calyx 4–8 mm. long; *corolla* milk-white or pink-tinged to honey-colored, *the earlier ones* 0.9–1.5 *cm. long, with filiform hairs inside at throat; capsule globose,* slightly exceeding to shorter than calyx, 4–7 *mm. long; seeds slender-cylindric,* 0.7–0.8 *mm. long.* (*G. sphaerocarpa* Ell.) — Brooks, pools, and ditches, Fla. to Tex., n. to N.J., Md., W.Va., O., Ind., Ill., Ia. and Kans. March–Oct. — In the robust forma **acùtidens** Fern. (with acute teeth) the leaves are 6–7 cm. long, long-attenuate to tip, and with long slender teeth, the sepals overtopping the capsule.

Var. **aestuariòrum** Pennell (of estuaries). — Stem simple or with few ascending branches, 0.4–2.5 dm. high; leaves elliptic, entire or undulate-crenate; flowers subsessile or with pedicels only up to 2 mm. long; calyx 2–3.5 mm. long; corolla small, unexpanding; capsule 3–4 mm. long. — Tidal and other mud, s. N.J., Md. and Va. June–Oct.

§ GRATIOLÀRIA Benth.

3. G. aùrea Muhl. (golden), GOLDEN-PERT. — Perennial with fleshy rhizome and fleshy purplish stolons; stems obtusely 4-angled, simple to bushy-branched, 0.4–4 dm. high; *leaves linear, lanceolate, oblong or ovate,* somewhat clasping, *entire or only obscurely toothed, glandular-punctate with dark dots,* the principal ones 0.8–3 cm. long and 1.5–9 mm. broad, or in forma **pusílla** Fassett (very small) in matted submersed colonies with the stems only 1.5–4 cm. high and the crowded or subapproximate linear-lanceolate leaves only 1–5 mm. long; pedicels slender, mostly ascending, half as long as to much overtopping leaves; *corolla golden-yellow throughout,* or honey-colored in forma **helvèola** Bartlett (honey-colored), or milk-white in forma **leucántha** Bartlett (white-flowered), 0.8–1.8 cm. long; *capsule* 3 *mm. long,* overtopped by calyx-segments. (*G. lutea* Raf., only doubtfully and in part) — Sandy, gravelly or peaty shores and open swamps, se. Nfld.; N.S. to e. N.D., s. on or near Coastal Plain to n. Fla., and inland s. to centr. N.Y. and n. Ill. — Late June–Sept.

4. G. viscídula Pennell (somewhat viscid). — Perennial with creeping rhizome; flowering *stems clammy-pubescent or glandular,* slender, 1–6.5 dm. long; *leaves oblong or ovate, finely dentate;* pedicels slender, shorter than to overtopping leaves; *bractlets foliaceous, nearly equaling calyx; calyx-segments lanceolate to elliptic,* 4–7 *mm. long;* corolla creamy-white or pink-tinged, with blue-purple lines and yellow center, 0.8–1.5 cm. long; capsule 1–2 mm. long. (Incl. *G. viscidula Shortii* Pennell; *G. viscosa* Schwein., not Hornem.) — Swales, marshes, pond-margins and alluvial woods, s. N.C. to nw. Ga., n. to Del., Md. and s. O. June–Sept.

5. G. ramòsa Walt. (branching). — Slender perennial, minutely viscid-puberulent, with simple or slightly forking stems 1–3.5 dm. high; *leaves linear or lanceolate, coarsely sharp-serrate above the middle; bractlets minute or obsolete; calyx-segments linear-subulate,* 3–5 *mm. long;* corolla 1–1.4 mm. long, the lobes white, tube with brown lines, throat yellow; capsule 1–2 mm. long. — Damp sandy pinelands, Fla. to La., n. to se. N.C.; Wicomico Co., Md. June–Sept.

§ SOPHRONÁNTHE Benth.

6. G. pilòsa Michx. (soft-hairy). — *Firm-stemmed, from a subligneous crown, villous-hirsute,* the mostly tufted stems 1.5–7.5 dm. high; leaves ovate or ovate-lanceolate, crenate-serrate, sessile by broad rounded base; *flowers subsessile or short-stalked* in the axils; *calyx-segments very unequal; corolla* white or purple-tinged, 5–9 *mm. long, little exceeding calyx; anthers with contiguous vertical locules; capsule slenderly conical.* (*Sophronanthe* (Michx.) Small; *Tragiola* (Michx.) Small & Pennell) — Sandy woods, clearings and roadsides, Fla. to Tex., n. to s. N.J., s. Md., w. N.C., Ky. and Ark. July–Oct.

14. CONÒBEA Aubl.

Upper lip of corolla 2-lobed, the lower 3-parted. Anthers approximate. Stigma 2-lobed, the lobes cuneiform. Seeds numerous. — Low branching Am. herbs, with small solitary flowers. (Name unexplained.)

1. C. multífida (Michx.) Benth. (much-cleft). — Annual, diffusely spreading, much-branched, minutely pubescent; leaves petioled, pinnately parted, divisions linear-cuneate; pedicels naked; corolla greenish-white, scarcely longer than the calyx. (*Leucospora* (Michx.) Nutt.) —

Along streams and shores in loam or sand, s. Ont. to Ia. and n. Kans., s. to nw. Ga., Ala., La. and Tex. June–Oct.

15. BACÒPA Aubl. WATER-HYSSOP

Calyx 5-parted; the uppermost division broadest, the innermost often very narrow. Upper lip of the corolla entire, notched or 2-cleft, and the lower 3-lobed, or the limb almost equally 5-lobed. Style dilated or 2-lobed at the apex. Seeds numerous. — Low herbs of trop. and warm reg., flowering in summer; ours mostly rather succulent perennials. (Said to be an aboriginal South American name.) HERPESTIS Gaertn.f. BRAMIA Lam. HYDROTRIDA Small. MACUILLAMIA Raf. PAGESIA Raf.

a. Stems erect, firm, from a strong crown; leaves serrate, pinnately veined; pedicels with 2 slender basal bracts. . . . . 1. *B. acuminata.*

a. Stems depressed, creeping or floating, with soft bases, if erect fleshy and with palmate-veined leaves; leaves entire or merely crenate; pedicels without basal bracts. . . *b.*

b. Leaves rounded to broad or more or less clasping bases; pedicels mostly shorter than subtending leaves. . . *c.*

c. Leaves ovate, pellucid-punctate; ovary surrounded by a slenderly 10–12-toothed disk; upper lip of corolla obcordate. . . . . 2. *B. caroliniana.*

c. Leaves suborbicular, rounded-oblong, ovate or obovate, not obviously punctate; disk at base of ovary none or without slender teeth; upper lip of corolla merely emarginate. . . *d.*

d. Outer sepals not cordate; larger leaves 1–3.5 cm. long; fruiting pedicels 0.5–2 cm. long.

Stem copiously hirsute, lax and elongate; leaves thin and clearly nerved, the larger ones 2–3.5 cm. long and 1.5–2.5 cm. broad; pedicels slender, pubescent; corolla 6–8 mm. long, nearly as broad; plant of Miss. basin. . . . . 3. *B. rotundifolia.*

Stem glabrous, ascending or erect, abbreviated; leaves fleshy, opaque, only obscurely veined, the larger ones 1–2 cm. long and 0.6–1.5 cm. broad; pedicels thick, glabrous; corolla 4 mm. long, 2 mm. broad; plant of e. Va. . . . . 4. *B. simulans.*

d. Outer sepals deeply cordate, about equaling corolla; larger leaves 5–10 mm. long, 3.5–10 mm. broad, very fleshy; fruiting pedicels 3–6 mm. long. . . . . 5. *B. stragula.*

b. Leaves cuneate to narrow bases; pedicels much exceeding subtending leaves, conspicuously 2-bracted below the calyx. . . . . 6. *B. Monnieri.*

1. B. acuminàta (Walt.) Robins. (tapering at the end). — *Stems from a subligneous crown, erect or ascending,* 4-angled, glabrous, simple or branched, 1–6(–8.2) dm. high; *leaves pinnately veined,* cuneate- to oblong-lanceolate, firm, *serrate,* with entire base, mostly 2–4 cm. long; *pedicels* filiform, *equaling or surpassing upper leaves, with* 2 *slender bracts at base;* upper sepals oblong-lanceolate, broader than the lanceolate lower ones; corolla white (or tinged with purple), 7–10 mm. long, bearded within at base of upper lip. (*Mecardonia* Small; *Pagesia* Pennell) — Ditches, damp fields, pinelands, etc., Fla. to e. Tex., n. to Md., Tenn. and s. Mo. July–Sept. — Plant often blackened in drying.

2. B. caroliniàna (Walt.) Robins. (Carolinian). Stems from creeping rhizome, floating or ascending, lanate at summit; *leaves* in subapproximate pairs, *ovate,* obtuse, *clasping,* 1–2.5 cm. long, entire, *pellucid-punctate,* fragrant, the nerves pedately radiating from base; flowers barely exserted beyond leaves, on short pedicels; calyx often with 2 minute basal bracts, outer sepal cordate; *corolla blue,* about 1 cm. long, *pubescent within, its upper lip obcordate; ovary surrounded by a slenderly* 10–12-*toothed disk.* (*Hydrotrida* Small) — Margins of ponds and streams in pinelands, Fla. to Tex., n., locally, to se. Va. Aug., Sept. — Bruised plant lemon-scented.

3. B. rotundifòlia (Michx.) Wettst. (round-leaved). — Creeping, *with elongate and lax terminally pubescent branches* up to 6 dm. long, *forming floating mats; leaves thin, round-obovate to suborbicular,* with subcuneately narrowed but clasping bases, *clearly* palmately many-*nerved, the larger* 2–3.5 *cm. long and* 1.5–2.5 *cm. broad; pedicels twice or thrice length of calyx, slender, pubescent,* 0.8–2 *cm. long; outer sepals ovate,* 6–8 *mm. long, longer than broad; corolla exserted, campanulate,* 6–8 mm. long, *the wide-spreading limb about as broad,* white, with yellow throat; *capsule globose or subglobose,* nearly equaling sepals. (*Macuillamia* Raf.) — Muddy ponds and pools, Miss. to Tex., n. to Ind., Ill., Minn., N.D. and Mont. July–Sept.

4. B. símulans Fern. (resembling; from its superficial similarity to no. 3). — *Stem decumbent* and rooting at base *or suberect, succulent, glabrous,* simple or sparsely branched, 0.5–2 dm. high; *leaves fleshy,* round-obovate or elliptic, rounded above, *obscurely palmate-nerved,* 1–2 *cm. long and* 6–15 *mm. broad;* flowers 1 or 2 in the axils; the *thick pedicels finally* divergent or reflexed and 5–11 *mm. long;* outer sepals tightly appressed, broadly ovate, rounded above, 4–5.5 mm. long; *tubular corolla insignificant, whitish,* 4 *mm. long, about* 2 *mm. broad,* 5-lobed, the lobes nearly equaling the tube and emarginate, the throat yellow; the 4 stamens included, the anthers dark purple; capsule ellipsoid, included. — Fresh tidal mud along Chickahominy R., Va. Aug., Sept.

5. B. strágula Fern. (forming mats). — Prostrate, *forming mats* 0.5–3 dm. across; *stems succulent, glabrous, much branched, rooting at nodes,* the branches ascending; *leaves thick, opaque, round-ovate,* sessile, somewhat clasping, 5–10 *mm. long and* 3.5–10 *mm. wide,* rounded above, *obscurely* palmate-*nerved; flowers* axillary, *on ascending or spreading finally arched-recurving pedicels* 3–6 *mm. long; outer sepals cordate,* rotund-ovate, rounded at apex, 4–6 mm. long; *corolla tubular, whitish,* 4–5 *mm. long,* 5-lobed, the lobes rounded-subtruncate at tip; stamens 2, 3 or 4; capsule ovoid, 2–4 mm. long. — Fresh tidal mud of Chesapeake drainage, Md. and Va. Late July–Oct.

6. B. Monnièri (L.) Pennell (for Louis Guillaume le Monnier, 1717–1799). — Prostrate, decumbent or loosely ascending, forming mats, *fleshy,* glabrous; *leaves spatulate to cuneate-obovate,* obtuse, entire or nearly so, 0.7–2 cm. long; *pedicels soon exceeding subtending leaves; flower with* 2 *narrow bracts* at base of calyx; larger sepal lanceolate to ovate, 1.5–5 mm. broad; *corolla campanulate, only obscurely bilabiate,* white, lilac or pale blue, 8–10 mm. long, *glabrous within;* capsule slenderly conic to ovoid, shorter than calyx. (*B. Monnieria* (P. Br.) Wettst. and var. *cuneifolia* Fern.) — Coastal sands, Fla. to Tex., n. to se. Va. Mid-June–Sept. (Pantrop.)

16. LIMOSÉLLA L. Mudwort

Calyx campanulate. Anthers confluently 1-locular. Style short, clavate. Capsule globular, many-seeded; the partition thin and vanishing. — Small annuals of most trop. and temp. reg., growing in mud, often near the seashore or in subsaline soil, creeping by slender runners, without ascending stems; the entire fleshy leaves in dense clusters around the naked pedicels. Flowers small, white or purplish. (Name a diminutive of *limus, mud,* Linnaeus explaining it as "*a limo natali dicte,* called from its native mud").

1. L. subulàta Ives (awl-shaped). — *Closely matted* and creeping, *the repent stems forming turf; leaves* tufted, *filiform or subterete,* ascending or arching, 1.5–6 cm. long; *pedicels promptly recurving; calyx regular or with sepals somewhat united in 2's; corolla white; valves of capsule thickened at margin.* (*L. aquatica* var. *tenuifolia* of ed. 7, not Schübler & Martens) — Brackish sand and mud, Nfld. to the lower St. Lawrence, Que., s. along the coast to Va. Late June–Oct.

2. L. aquática L. (aquatic). — Rosulate, with *decumbent branches producing terminal rosettes of* long-petioled *oblong to elliptic leaves;* pedicels less arching; *calyx regular; corolla pink; valves of capsule not thickened at margin.* — Fresh to brackish shores and wet sands, local, Nfld. and se. Lab. Pen.; s. Ung. to n. Alta. and B.C., s. to Minn., N.M., Ariz. and Calif. July–Sept. (Eu.)

17. MICRÁNTHEMUM Michx.

Stamens anterior, the short filaments with a glandular (mostly basal) appendage; anthers 2-locular, didynamous. Style short; the stigma 2-lobed. Capsule globular, thin, with an evanescent partition, several–many-seeded. — Small smooth depressed and tufted or creeping Am. annuals, in mud or shallow water, with opposite and entire rounded or spatulate sessile leaves, and minute white or purplish flowers solitary in the axils of some of the middle leaves (usually one axil floriferous, that of the opposite leaf sterile). (Name formed of the Greek *micros, small,* and *anthemon, flower.*)

1. M. micranthemoìdes (Nutt.) Wettst. (like *Micranthemum,* Nuttall having placed it in a separate genus). — Branches ascending, 1–6 cm. high; leaves obovate-spatulate or oval; pedicels at length recurved, about the length of the *calyx,* which is campanulate, 4-toothed and *split down the anterior side to the base,* in fruit becoming pyriform; *corolla* white, *falling unopened,* cleistogamous, *its upper lip wanting,* middle lobe of lower lip oblong and nearly twice length of lateral ones; *appendage of stamens nearly as long as filament itself; stigmas subulate.* (*Hemianthus* Nutt.) — Fresh tidal mud, local, N.Y. to Va. Aug.–Oct.

2. M. umbròsum (Walt.) Blake (shaded). — Repent; leaves roundish, 4–9 mm. broad; pedicels shorter than the *deeply and equally cleft calyx; corolla* barely equaling calyx, *with upper lip developed; appendage of stamen a mere tooth; stigmas short.* — On mud or wet sand in low woods or along streams, Fla. to Tex., n. to se. Va. July–Oct. (Trop. Am.)

18. LINDÉRNIA All. False Pimpernel

Upper lip of corolla short, erect, 2-lobed; lower larger and spreading, 3-cleft. Fertile stamens 2, included, posterior; anterior pair sterile, inserted in the throat, 2-lobed; one of the lobes glandular, the other smooth, usually short and tooth-like. Stigma 2-lobed. Capsule ovoid or ellipsoid, many-seeded. — Small and smooth annuals of temp. and warm reg., the purplish flowers on filiform pedicels, or the upper racemed, produced all summer. (Named for *Franz Balthasar von Lindern*, 1682–1755.) Ilysanthes Raf.

Lower pedicels shorter than their subtending leaves; principal leaves obovate to elliptic or ovate, rounded or narrowed to base; later (sometimes all) flowers often cleistogamous and unexpanded; capsule often equaled or exceeded by calyx-lobes; seeds twice to thrice as long as thick. . . . . 1. *L. dubia.*
Lower and upper pedicels exceeding their subtending leaves; principal leaves ovate to elliptic, broadest at the rounded to cordate base; corollas all expanding; capsules mostly exceeding calyx-lobes; seeds shorter. . . . . . . 2. *L. anagallidea.*

1. L. dùbia (L.) Pennell (doubtful; when originally placed in *Gratiola*). — Simple to much branching, depressed or ascending, 0.5–3.5 dm. long; *leaves oblong or elliptic to obovate,* entire or remotely dentate, the larger ones mostly 1–3 cm. long and *gradually rounded or narrowed to base; at least the lower pedicels shorter than to barely equaling their subtending leaves; calyx-lobes* linear, *about equaling or slightly exceeding* the oblique ellipsoid *capsule; corolla* 5–10 mm. long; or small, pale *tubular and unexpanding in cleistogamous flowers;* anterior filaments without anthers; style 2.5–3.5 mm. long; seeds about 0.4 mm. long, faintly lineolate and pebbled. — Very variable; the following vars. recognized:

Bracteal leaves tapering at apex, merely blunt to acutish; pedicels becoming 0.5–2 cm. long.
Foliage-leaves and bracts nearly uniform; the bracts only slightly or scarcely smaller than the former, gradually rounded at base, all overtopping the pedicels, the upper 5–10 mm. broad; corollas all expanding. . . . . *L. dubia* (typical).
Foliage-leaves larger than the bracts; the latter, as the axis prolongs, rapidly reduced upward, lanceolate to lance-ovate or oblong, 1–6 mm. broad, narrowed to base, the upper exceeded by the pedicels; later flowers often cleistogamous and unexpanding. . . . . Var. *riparia.*
Bracteal leaves elliptic to obovate, rounded at tip, scarcely or little reduced; flowers all cleistogamous, on pedicels 1–5 (rarely –10) mm. long. . . . Var. *inundata.*

L. dùbia (typical), *Ilysanthes dubia* (L.) Barnhart; *L. dubia*, subsp. *major* Pennell. — Shores, damp open spots and disturbed soil, s. Que. to w. Ont. and Minn., s. to N.S., N.E., L.I., Ga., Ala., Miss., La., Tex. and Mex.; Wash. to Calif. Late June–Oct.

Var. **ripària** (Raf.) Fern. (of river-shores). — Centr. Me. to Minn. and Neb., s. beyond our limits.

Var. **inundàta** Pennell (inundated). — Fresh tidal shores, Que. to Va., occasionally inland in the Atl. states.

L. Pyxidària L. (old generic name; like a small box or pouch), resembling the cleistogamous states of no. 1 but with all 4 filaments antheriferous, the style 1–1.5 mm. long, the capsule symmetrical, the seeds smooth, has been reported as casual with us. (Adv. from Eu.)

2. L. anagallídea (Michx.) Pennell (resembling *Anagallis*). — More slender, diffusely branched; *principal leaves ovate or elliptic, broadest at the rounded base,* commonly entire or nearly so, 0.5–2 cm. long; *pedicels filiform,* prolonged, *all clearly exceeding the bracteal leaves;* corollas all expanding; *capsule usually exceeding calyx-lobes;* seeds slightly shorter. (*Ilysanthes* Robins.) — Damp shores, sands, etc., Fla. to Tex. and Mex., n. to s. N.H., Mass., N.Y., O., Ind., Wisc., Minn., N.D., Colo. and Wash. June–Oct.

Subfam. II. RHINANTHOÌDEAE

Tribe V. Digitàleae Benth.

19. DIGITÀLIS L. Foxglove

Calyx 5-parted; the lobes often foliaceous. Corolla with a somewhat inflated tube and short scarcely spreading limb, declined. Stamens 4, didynamous, included in the corolla. — Tall herbs nat. of Old World, with alternate or scattered leaves and showy racemose flowers. (*Digitalis*, of or belonging to the finger, as the fingers of a glove, the tubular corollas early suggesting the old English name, Folks' (or Fairies') Glove or Thimble.)

1. D. purpùrea L. (purple) — Stoutish pubescent biennial; *calyx-lobes ovate; corolla purple* to white, *spotted,* 4–5 *cm. long.* — Clearings, old fields, etc., locally abund. in Nfld. and C.B., casually as an esc. from cult. elsewhere. Summer. (Introd. and natzd. from Eu.)

2. D. lanàta Ehrh. (woolly). — *Stem and calyx densely woolly;* leaves narrowly oblong to lanceolate or oblanceolate, glabrous; *calyx-lobes linear; corolla creamy-white, with violet lines, about* 2 *cm. long,* the tube subglobose. — Waste places, groves and roadsides, spread from cult., N.E. to Ind. June–Sept. (Introd. from se. Eu.)

20. WULFÈNIA Jacq.

Stamens inserted just below the upper sinuses, occasionally with another pair from the other sinuses, exserted; anther-locules not confluent. Style slender; stigma simple. Capsule flattened, rounded, obtuse or notched, 2-locular (rarely 3-lobed and 3-locular), many-seeded, loculicidal; the valves cohering below with the axis. — Perennial herbs of temp. N.Am. and Asia, with the simple stems beset with partly clasping bractlike alternate leaves; the radical leaves rounded and petioled, crenate. (Named for *Franz Xaver, Freiherr von Wulfen,* 1728–1805.) Synthyris Benth.

1. W. Búllii (Eat.) Barnh. (originally named in 1840 for its discoverer, George Bull). — Hairy; radical leaves ovate, cordate; spike dense, 1–3 dm. long; corolla greenish-white or yellowish, not longer than the calyx, usually 2–3-parted. (*Synthyris* Heller; *Besseya* (Eat.) Rydb.) — Dry prairies, oak-barrens and bluffs, s. Mich. to Minn., s. to O., Ind., Ill. and Ia. May, early June.

21. VERONICÁSTRUM Fabricius Culver's-physic

Corolla salverform, the tube much longer than the lobes. Stamens and style much exserted, the 2 stamens inserted low on the tube. Capsule ovoid, turgid, hardly at all compressed, not at all emarginate, dehiscent at apex by all four sutures, at length more loculicidal. Seeds numerous, terete, minutely reticulated. — Tall Am. and Asiatic perennial, with leaves mostly in whorls of 3 to 7, the terminal spiciform racemes minutely bracted. (Name from *Veronica* and the suffix *astrum,* i.e., *false Veronica.*) Leptandra Nutt.

1. V. virgínicum (L.) Farw. (Virginian), Culver's-root. — Smooth or minutely pubescent, or copiously villous in forma **villòsum** (Raf.) Pennell (villous), up to 2 m. high; leaves lanceolate to lance-ovate, acuminate, sharply serrate; spikes panicled; corolla white or purplish. (*Veronica* L.; *Leptandra* Nutt.) — Rich woods, thickets, meadows and prairies, Mass. and Vt. to Man., s. to N.C., w. S.C., nw. Fla., Miss., La. and e. Tex. June–Sept. (E. Asia, as the "scarcely distinguishable" *V. sibiricum* (L.) Pennell)

22. VERÓNICA L. Speedwell

Corolla rotate, with short tube, the lateral lobes or the lowest one commonly narrower than the others. Stamens 2, one on each side of the upper lobe of the corolla, exserted; anther-locules confluent at apex. Style entire, stigma single. Capsule emarginate to obcordate, more or less flattened, 2-locular, few–many-seeded. Seeds not reticulate, more or less compressed or planoconvex or the inner face hollowed. — Herbs (rarely suffruticose) with opposite or alternate leaves, mostly of N. Hemisph., and blue to white flowers in terminal racemes or axillary. (Named for *St. Veronica,* popularly thought to be from *vera, true,* and the Greek *eicon, image;* an early Christian legend picturing St. Veronica, pitying Christ on the way to Calvary, wiping his face with her handkerchief which received a miraculous *true image* of his features.)

a. Perennials with racemes of flowers subtended by small narrow bracts. . . *b.*
- *b.* Racemes terminal; the bracteal leaves mostly alternate. . . *c.*
  - *c.* Stems 0.2–1.5 m. high, from subligneous base; leaves lanceolate, copiously toothed, distinctly petioled, the larger 2–15 cm. long; raceme spiciform, attenuate; style 2–3 times length of the plump capsule.
    - Leaves very sharply serrate, acuminate or acute; bracts and sepals not strongly villous; corolla-lobes broadly oblong; fruiting pedicels nearly as long as calyx; stems 0.3–1.5 m. high. . . . . . . 1. *V. longifolia.*
    - Leaves crenate, blunt or merely acutish; bracts and sepals conspicuously white-villous; corolla-lobes oblong-lanceolate; fruiting pedicels much shorter than calyx; stems rarely 4 dm. high. . . . . . . 2. *V. spicata.*
  - *c.* Stems (flowering) 0.5–4 dm. high, slender, from slender creeping bases; leaves suborbicular, oblong or elliptic, entire or shallowly few-toothed, mostly sessile or very short-petioled, 1–3 cm. long; raceme lax or, if

dense, thick and obtuse; style shorter than to but slightly longer than flattened capsule. *d.*

*d.*Plant with few if any prostrate sterile leafy shoots; raceme dense, in fruit ellipsoid, thick-cylindric or obovoid, with closely imbricated fruits; corolla deep dark blue; upper leaves and calyx-lobes ciliate; capsule ellipsoid or narrowly obovoid, longer than broad, slightly emarginate, pubescent. . . . . 3. *V. alpina*, vars.

*d.*Plant with repent sterile leafy branches forming mats; raceme becoming lax and slender; corolla violet-blue to white, with slender markings; leaves and calyx-lobes non-ciliate; capsule broadly obcordate, broader than long, deeply notched, glabrous or at most ciliolate.

Rachis and pedicels puberulent or minutely pilose with incurved hairs; corolla whitish or pale blue with violet lines, 2–4 mm. broad; capsules mostly 3–4 mm. wide. . . . . 4. *V. serpyllifolia.*

Rachis and pedicels pubescent with spreading viscid or gland-tipped hairs; corolla deep blue-violet, the earlier ones 5–8 mm. broad; larger capsules 4–6 mm. broad. . . . . 5. *V. tenella.*

*b.*Racemes lateral or on lateral branches; leaves all opposite. . . *e.*

*e.*Stems and at least young leaves pilose; leaves coarsely toothed; pedicels ascending, shorter than to little longer than calyx; capsule pubescent, strongly flattened; plants of dry habitats. . . *f.*

*f.*Stems extensively creeping, repent, forming mats; leaves elliptic to narrowly obovate, permanently pubescent, thick, subcoriaceous; flowers and fruits mostly imbricated, their pedicels shorter than the equal obtuse calyx-lobes; corolla about 5 mm. broad, with obtuse lobes; capsule triangular-obcordate, much exceeding calyx. . . . . . 6. *V. officinalis.*

*f.*Flowering stems erect or ascending; leaves ovate to oblong, membranaceous, glabrate; flowers and fruits becoming remote; calyx-lobes unequal, acutish; corolla 0.6–1 cm. broad; capsule shorter than to but little exceeding calyx.

Stems 3–8 dm. high; leaves sessile; racemes stiffly ascending; longer calyx-lobes twice as long as shorter ones; capsule obovoid, slightly exceeding calyx. . . . . 7. *V. latifolia.*

Stems 1–4 dm. high; leaves short-petioled; racemes flexuous; calyx-lobes only slightly unequal, longer than the obcordate capsule. . . 8. *V. Chamaedrys.*

*e.*Stems glabrous or only remotely pubescent (if sometimes pilose in no. 9 the leaves linear or linear-lanceolate and only minutely toothed); fruiting pedicels divergent or loosely ascending; capsule glabrous or glandular; plants of wet habitats. . . *g.*

*g.*Leaves linear to linear-lanceolate, entire or minutely toothed; stem glabrous or with non-glandular pilosity; racemes with flexuous or zigzag filiform rachis and filiform divergent to reflexed pedicels much exceeding flowers and fruits; capsule strongly flattened, deeply obcordate, broader than long, much exceeding calyx. . . . . . 9. *V. scutellata.*

*g.*Leaves oblong, lanceolate or ovate, fleshy, crenate or serrate; stem glabrous or minutely glandular; racemes with straight or arching rachis and divergent to spreading-ascending pedicels; capsule turgid, orbicular or nearly so, shorter than to slightly exceeding calyx. . . *h.*

*h.*Leaves of flowering stems distinctly petioled; racemes 4–30-flowered.

Principal leaves of flowering stem lanceolate to narrowly ovate, acute or acutish, serrate or dentate; lower mature filiform fruiting pedicels 6–11 mm. long. . . . . 10. *V. americana.*

Principal leaves of flowering stem oval, short-oblong or oblong-obovate, rounded at tip, shallowly crenate; fruiting pedicels thicker, 4–5 mm. long. . . . . 11. *V. Beccabunga.*

*h.*Leaves of flowering stems sessile, at least the upper cordate-clasping.

Fruiting pedicels loosely ascending; racemes mostly 20–65-flowered; corolla bluish-lilac, 4–5 mm. broad; capsule ovate to roundish, scarcely notched, shorter than to about equaling the lance-acuminate sepals. . . . . 12. *V. Anagallis-aquatica,*

Fruiting pedicels loosely and horizontally divergent; racemes mostly 5–35-flowered; corolla whitish to pale rose, smaller; capsule rounded-reniform or broadly obcordate, deeply notched, longer than the blunt or merely acutish lance-oblong to narrowly ovate sepals. . . . . 13. *V. comosa.*

*a.*Annuals with flowers from the axils of alternate leaves. . . *i.*

*i.*Flowers and fruits sessile or barely pedicelled (pedicel up to 1 mm. long), ascending in the axils of reduced simple or cleft upper leaves; seeds flat, mostly less than 1 mm. long, smooth or smoothish; stems ascending.

Fleshy, glabrous or glandular-puberulent; leaves spatulate, oblong to linear, entire or low-dentate; sepals subequal; corolla whitish; capsule shallowly emarginate, glabrous or glandular-puberulent, with minute style. . . . . 14. *V. peregrina.*

Scarcely fleshy, pilose; lower leaves rounded or oval, toothed; sepals very unequal; corolla violet-blue; capsule pilose, emarginate, style elongate.

Median and upper leaves merely toothed or entire; style equaling or longer than lobes of capsule. . . . . 15. *V. arvensis.*

Median leaves deeply cleft; style much shorter than lobes of capsule. 16. *V. verna.*

*i.*Flowers on filiform elongate pedicels in the axils of leaves scarcely unlike the lower foliage; seeds cupuliform, 1.3–3 mm. long, rugose; stems weak, often reclining, sometimes becoming repent at base. . . *j.*

*j.*Principal leaves longer than broad, ovate, dentate or serrate; sepals lanceolate to ovate, minutely ciliate, in maturity prominently ribbed; capsule slightly flattened, deeply notched, pubescent; seeds 1.3–2.5 mm. long, mostly several in each locule.

Corolla only slightly exceeding calyx; longer fruiting pedicels 0.5–1.2 cm. long; fruiting calyx 5–10 mm. broad; capsules 3.5–6 mm. broad; seeds 1.3–1.5 mm. long.

Longer sepals in maturity 5–9 mm. long, ovate or broadly lanceolate, tapering to tip, much exceeding capsule; seeds 3–8 in each locule of capsule. . . . . 17. *V. agrestis.*

Longer sepals in maturity 3–4.5 mm. long, elliptic or oval, blunt, little if at all longer than capsule; seeds 9–12 in each locule. 18. *V. polita.*

Corolla much exceeding calyx; longer fruiting pedicels 1.3–2.7 cm. long; fruiting calyx 1–1.6 cm. broad; capsules 6–9 mm. broad; seeds about 8 in each locule, 1.8–2.5 mm. long. . . . . . . 19. *V. persica.*

*j.*Principal leaves broader than long, reniform-suborbicular, 3–5(–7)-lobed; sepals cordate-ovate, long-ciliate, obscurely nerved; capsule turgid, slightly 4-lobed, glabrous; seeds 2.5–3 mm. long, 2 in each locule. 20. *V. hederaefolia.*

1. V. longifòlia L. (long-leaved). — Perennial with subligneous base; *stems* 0.3–1.5 *m. high*, canescent-puberulent; *leaves* opposite or in 3's, slender-petioled; middle and upper *lanceolate, acuminate, sharply and doubly serrate*, rounded at base, puberulent, or whole plant essentially glabrous in forma glàbra (Schrad.) Aschers. & Graebn. (smooth), mostly 0.5–1.5 dm. long; racemes elongate; *bracts and sepals subulate to linear, minutely ciliolate;* corolla blue-violet, with broad oblong lobes; fruiting pedicels nearly as long as calyx. (*V. maritima* L.) — Roadsides and thickets, Nfld. to w. Que., s. to N.S., N.E., N.J., Md. and O. July–Sept. (Introd. and natzd. from Eu.)

V. grándis Fisch. (large), *V. Bachofenii* Heuffel, with cordate-ovate leaves with broader and simpler toothing, has spread locally to roadsides in Que. (Introd. from Eurasia)

2. V. spicàta L. (spiked). — Usually lower (1.5–4 dm. high) than no. 1, cinereous; *leaves short-petioled to subsessile*, thick, *oblong-lanceolate, blunt or acutish, coarsely dentate*, the larger 2–8 cm. long; *bracts and sepals conspicuously white-villous;* corolla-lobes narrowly oblong; fruiting pedicels very short. — Roadsides and rocky banks, Que. and n. N.Y. July–Sept. (Introd. from Eu.)

3. V. alpìna L. (alpine). — Perennial with *slender caudices;* stems erect, glabrous below, pilose above, 0.5–2.5 dm. high; *leaves oval, elliptic or rounded*, sessile or short-stalked, paired, 1–3 cm. long, shallowly toothed or entire (blackened in drying); *raceme dense, in fruit becoming ellipsoid to obovoid, with strongly imbricated fruits;* backs of sepals and obovate exserted capsule glabrous; corolla dark blue. — Arct. reg., s. to n. Lab. (Eurasia) — With us as

Var. **unalaschcénsis** C. & S. (of Unalaska). — Stem 0.5–3 dm. high, loosely villous above with wide-spreading gland-tipped hairs; leaves blackened in drying, those midway on stem 4–20 mm. broad; sepals and capsules long-villous with gland-tipped hairs; capsules fuscous or greenish- or bluish-black. (*V. Wormskjoldi* R. & S.) — Greenl. to Alaska, s. to cool ravines, wet moss and alpine rocks, high mts. of e. Que., Me., N.H. and Colo. July, Aug.

Var. **térrae-nòvae** Fern. (of Newfoundland). — Only slightly blackening; stem subrigid, 1–2 dm. high, appressed- or incurved-villous and scarcely glandular above; median leaves only 3–7 mm. broad; sepals short-pilose; capsule pale brown, only sparsely villous. — Alpine brooksides, Highlands of St. John, nw. Nfld.

4. V. SERPYLLIFÒLIA L. (thyme-leaved), THYME-LEAVED S. — Much branched at the creeping base, *nearly smooth;* branches ascending and simple, 0.5–2 dm. high; *leaves ovate or oblong,* obscurely crenate, 1.5 cm. or less long, *the lowest petioled* and rounded, the upper passing into lanceolate bracts; *raceme loose, the rachis and pedicels appressed-puberulent; corolla 2–4 mm. broad, whitish or pale blue, with deeper stripes;* capsule rounded, broader than long, obtusely notched, 3–4 mm. broad. — Grassy places, roadsides, damp open woods, etc., Nfld. to Ont. and Minn., s. beyond our limits. April–July. (Natzd. from Eu.)

5. V. tenélla All. (slender). — *Stouter, 2–4 dm. high; leaves 1–2.5 cm. long; rachis and pedicels pubescent with spreading viscid or gland-tipped hairs; corolla 5–8 mm. broad, deep blue; larger capsules 4–6 mm. broad.* (*V. humifusa* Dickson) — Springy places, brooksides and damp woods, Lab. to Alaska, s. to Nfld., N.B., n. N.E., N.Y., Mich., Wisc., Minn., mts. to Mex. May–July. (Eurasia; S.Am.) — Perhaps better treated as *V. serpyllifolia* var. *borealis* Laestad.

6. V. officinàlis L. (of the shops). COMMON S., GYPSYWEED. — *Matted and trailing pubescent perennial,* the branches repent and forking; *leaves* short-petioled, *thick, pubescent, obovate-elliptical or cuneate-oblong, obtuse, serrate,* mostly 2.5–6 cm. long and 1–3 cm. broad; peduncles stoutish, ascending, shorter than to exceeding subtending leaves; *racemes densely many-flowered;* pedicels shorter than calyx; calyx-lobes obtuse, subequal; *corolla* lilac-blue or lavender, or white in forma **albiflòra** (G. Don) House (white-flowered), *about 5 mm. broad,* with obtuse lobes; *capsule triangular-obcordate,* pubescent, about as broad as long. — Dry hills, open woods and fields, Nfld. to Ont., s. to N.S., N.E., N.C., Tenn. and Wisc., both indigenous and natzd. from Eu. May–July. (Eurasia)

Var. **Tournefórtii** (Vill.) Reichenb. (for JOSEPH PITTON DE TOURNEFORT, 1656–1708). — More delicate and slender throughout; leaves 1–3 cm. long, 4–17 mm. broad; filiform peduncles much longer than subtending leaves; racemes linear-cylindric, lax and flexuous; corolla blue-violet, with narrowly oblong or elliptic lobes; capsules as broad as or broader than long. — Coniferous woods and thickets, peaty knolls and sphagnous bogs, Nfld., P.E.I. and N.S. June–Aug. (Eu.)

7. V. LATIFÒLIA L. (broad-leaved). — Perennial; *stems* pubescent, *stiff and upright,* 3–8 dm. high; *leaves sessile, oblong,* coarsely toothed, pubescent; *racemes* axillary, *stiffly ascending,* densely flowered, the flowers and fruits becoming remote; *calyx unequally 5-parted, the lobes acutish, the longer ones twice as long as the others;* corolla about 1 cm. broad, bluish; *capsule obovoid, slightly exceeding calyx.* (*V. Teucrium* L.) — Esc. from cult. to roadsides, N.E. to S.D., s. to Pa. and Ind. June, July. (Introd. from Eu.)

8. V. CHAMAÈDRYS L. (old generic name, meaning ground-oak), BIRD'S-EYE. — *Stems very slender,* 1–4 *dm. high,* from a creeping base; *leaves short-petioled, ovate or cordate, incisely crenate; racemes* axillary, *flexuous, loosely flowered;* pedicels little longer than the but slightly unequal calyx-lobes; corolla showy, blue; *capsule obcordate,* deeply notched, *shorter than or but little exceeding calyx.* — Roadsides, fields and borders of woods, Nfld. to Ont., s. to N.S., N.E., Md., W.Va., O., Mich. and n. Ill. May–July. (Natzd. from Eu.)

9. V. scutellàta L. (platter-like; from the flat corolla), MARSH-S. — Weak glabrous perennial with filiform rhizome and long stolons; stems 0.5–7.5 dm. long, simple or branching; *leaves sessile, linear to lanceolate, entire or minutely toothed,* 1.5–9 cm. long, 1–15 mm. broad; *racemes* several, *divergent, with filiform zigzag rachis and filiform divergent to reflexed pedicels much exceeding the few flowers and fruits;* corolla lilac or lavender-blue; *capsule strongly flattened, deeply obcordate* and somewhat cordate, *broader than long,* much exceeding calyx. — Wet places, shores and swamps, Lab. to Alaska, s. to Nfld., N.S., N.E., n. Va., W.Va., O., Ind., Ill., e. Ia., Colo. and Calif. May–Aug. (Eurasia)

Var. **villòsa** Schumacher (soft-hairy). — Stems (at least above) and leaves more or less villous. — Similar places, w. N.Y. to L. Sup. reg. of Ont. and Minn.; s. B.C. and Wash. (Eu.)

10. V. americàna (Raf.) Schwein. (American), AMERICAN BROOKLIME. — Fleshy and more or less succulent perennial with creeping or decumbent bases; *principal leaves* (excluding lowest) *of ascending flowering stems lanceolate to narrowly ovate, acute or acutish, serrate or dentate, distinctly petioled,* the blades 1–9 cm. long; racemes lax, with arching rachises, axillary below the prolonging tip, 4–30-flowered, *the lower mature filiform divergent pedicels 6–11 mm. long;* corolla violet to lilac; capsule turgid, orbicular or nearly so. — Shallow water, springheads, rills and swamps, Nfld. to Alaska, s. to N.S., N.E., N.C., Tenn., Mo., Neb., Mex. and s. Calif. June–Aug.

11. V. BECCABÚNGA L. (old generic name from the colloquial Germanic "*Bachbungen*"), BROOKLIME of Eu. — *Almost fleshy, prostrate* and strongly repent; *leaves oval or short-oblong,* *rounded at tip, crenate,* short-petioled; mature *fruiting pedicels* thickish, 4–5 *mm. long.* — Brooks and ditches, Que. to Mich., s. to N.J. and W.Va., local. June–Aug. (Natzd. from Eu.)

12. V. Anagállis-aquática L. (aquatic *Anagallis*), Water-S., Brook-Pimpernel. — Habitally similar to no. 10; *leaves of flowering stems sessile,* more or less cordate-clasping, lance- to obovate-oblong; summit of stem and axis of inflorescence glabrous, or more or less glandular in forma **anagallifórmis** (Boreau) G. Beck (with form of *Anagallis,* Pimpernel) = *V. glandifera* Pennell; *racemes mostly* 20–65-*flowered; the fruiting pedicels loosely ascending; corolla bluish-lilac,* 4–5 *mm. broad; capsules ovate to roundish, scarcely notched, shorter than to about equaling the lance-acuminate sepals.* — Springs, rills, ditches and shores, chiefly calcareous, N.E. to Wash., s. to N.C., Tenn., Tex., N.M. and Ariz. May–Oct. (Eurasia; S. Am.) — Apparently both indigenous and natzd. from Eu.

13. V. comòsa Richter (tufted), Water-S. — Very similar to no. 12; leaves of flowering stem lanceolate or lance-oblong, dentate; *racemes mostly* 5–35-*flowered, with loosely and horizontally divergent* more or less glandular or glabrescent *pedicels; corolla whitish or pale rose, smaller; capsule rounded-reniform or broadly obcordate, deeply notched, longer than the blunt or merely acutish lance-oblong or narrowly ovate sepals.* (*V. catenata glandulosa* (Farw.) Pennell; *V. connata* sensu Pennell, perhaps not Raf.) — Calcareous springs, rills, sloughs, ditches and shores, sw. Que. to Sask. and Wash., s. to s. Pa., Tenn., Mo., Okla., N.M., Ariz. and s. Calif., the glabrous extreme mostly westw., the glandular eastw. June–Oct. (Eurasia)

14. V. peregrìna L. (wandering), Neckweed, Purslane-S. — Simple or branching erect to depressed *rather fleshy* glabrous annual 0.5–4.5 dm. high; *leaves spatulate, oblong or linear, entire or low-dentate,* 2–6 mm. broad; the lowest more oval and petioled, the others narrower and sessile, obtuse; *the uppermost smaller, subtending the sessile or short-stalked whitish flowers; sepals subequal; capsule shallowly emarginate, glabrous, with minute style.* — Damp open soil, often becoming a roadside-weed, Que. to Minn., s. to N.E., Fla., Ala., Miss., La. and e. Tex.; Alaska to Oreg. March–Aug. (Eu.)

Var. **xalapénsis** (HBK.) St. John & Warren (growing near Xalapa, Mexico). — Stem and capsule more or less glandular-pubescent. — Moist or even wet soil, sometimes in marshes or estuaries, Que. to Alaska, s. to N.E., Ga., Ala., Tex. and Mex., common westw., more local eastw.

15. V. arvénsis L. (of cultivated ground), Corn-S. — Simple to diffusely branched annual 0.5–4 dm. high, more or less *pilose; lower leaves rounded or oval, low-toothed;* upper leaves smaller, sessile, lanceolate to linear, subtending the sessile to short-stalked *violet-blue flowers; sepals very unequal; style equaling or longer than the lobes of the pilose* deeply emarginate *capsule.* — Waste open ground, rocky or sterile pastures, shaded ledges and open woodlands, Nfld. to Minn. and s. beyond our range; also w. N. Am. March–Aug. (Natzd. from Eu.)

16. V. vérna L. (vernal). — Very similar to no. 15, the stems and branches strongly ascending, minutely incurved-pilose; *median leaves deeply cleft into linear lobes; style much shorter than lobes of capsule.* — Waste ground, Bruce Pen., Ont. May–July. (Adv. from Eu.)

17. V. agréstis L. (of fields), Field-S. — Annual with prostrate or reclining pilose stems mostly forking at base, the branches 0.5–3 dm. long; *leaves ovate or rounded-oblong, coarsely crenate-dentate, the bracteal similar and subtending long-pedicelled flowers; pedicels* 0.5–1.2 *cm. long in fruit;* the *calyx in maturity* 5–10 *mm. broad* and much exceeding the slightly flattened deeply notched pubescent capsule; *sepals lanceolate to ovate, tapering to tip,* minutely ciliolate and prominently ribbed, *the longer becoming* 5–9 *mm. long;* corolla 6–8 mm. broad, blue, with pale lower lip; capsule 3.5–6 mm. broad, *each locule with* 3–8 cupuliform rugose *seeds* 1.3–1.5 mm. long. — Waste places, cult. ground and roadsides, Nfld. to Mich., locally s. to N.E. and Pa. May–Sept. (Natzd. from Eu.)

18. V. polìta Fries (smoothed; probably from its glabrescence). — Similar to no. 17, smaller throughout; leaves oval or ovate, rounded to subcordate at base; *longer sepals in maturity* 3–4.5 *mm. long, elliptic to oval, blunt, little if at all longer than capsule; corolla* about 5 *mm. broad;* style 1 mm. long; *seeds* 9–12 *in each locule.* (*V. didyma* Ten. (1830), not Ten. (1811)) — Waste places, lawns and roadsides, N.Y. to Mich., s. and sw. beyond our limits. March–May. (Natzd. from. Eu.)

V. filifórmis Sm. (hair-like), similar to no. 18 but densely matted, the leaves reniform, pedicels elongate, corolla nearly 1 cm. broad and style 2–3 mm. long, is appearing in lawns, etc., of N.Y. and Pa. (Adv. or introd. from Eurasia)

19. V. pérsica Poir. (Persian), Bird's-eye. — Similar to no. 17, larger throughout; leaves ovate to rounded, coarsely dentate or crenate-serrate, petioled; *longer fruiting pedicels* 1.3–2.7 *cm. long; fruiting calyx* 1–1.6 *cm. broad,* with broadly lanceolate sepals; *corolla* 8–12 *mm. broad,* deep blue, with paler lower lip; *capsules* 6–9 *mm. broad,* each locule with about 8 *seeds*

1.8–2.5 *mm. long.* (*V. Tournefortii* of ed. 7, not C. C. Gmel.) — Waste places and roadsides, Nfld. to Alaska, s. beyond our limits. March–Sept. (Natzd. from Eurasia)

20. V. HEDERAEFÒLIA L. (with leaves of *Hedera*, ivy), IVY-LEAVED S. — Loosely ascending to prostrate annual; *principal leaves broader than long, reniform-suborbicular,* 3–5(–7)-*lobed;* flowers long-stalked in axils of scarcely reduced leaves; *sepals cordate-ovate, long-ciliate, only obscurely nerved,* much enlarged in fruit; corolla only 2–2.5 mm. broad, lilac to bluish; *capsule turgid, slightly* 4-*lobed,* glabrous; *seeds* 2 *in each locule,* 2.5–3 *mm. across.* — Grassy slopes, fields, woodlands and waste places, local, N.Y. and O. to N.C. March–June. (Natzd. from Eu.)

Tribe VI. GERARDÌEAE Benth. (Incl. BUCHNÈREAE Benth.)

23. SEYMÈRIA Pursh

Calyx campanulate, deeply 5-cleft. Corolla with a short and thick tube, not longer than the 5 ovate or oblong nearly equal and spreading lobes. Anthers approximate by pairs, oblong, 2-locular, the locules equal and pointless. Capsule many-seeded. — Erect branching herbs of N. Am. and Madagascar; leaves mostly dissected or pinnatifid, the uppermost alternate and bract-like. Flowers yellow, in interrupted racemes or spikes. (Named for *Henry Seymer*, 1745–1800, an English naturalist.)

1. S. macrophýlla Nutt. (large-leaved), MULLEIN-FOXGLOVE. — Rather pubescent, 1–1.5 m. high; *leaves* large; *the lower pinnately divided, with the broadly lanceolate divisions pinnatifid and incised, the upper lanceolate; tube of the corolla* incurved, *very woolly inside,* as are the filaments except at the apex; *style short, dilated* and notched at the point; capsule ovoid, pointed. (*Dasystoma* Raf.) — Rich woods and banks of streams, W.Va. to s. Wisc., Ia. and e. Neb., s. to nw. Ga., n. Ala., Miss., La. and Tex. June–Sept.

2. S. cassioìdes (Walt.) Blake (resembling *Cassia*). — *Bushy-branched* slender annual, 0.5–1 m. high; *leaves* 1–2-*pinnately parted into filiform segments;* pedicels filiform; *corolla* lemon-yellow, 6–8 mm. long, *glabrous within,* except for a line at insertion of stamens; calyx-lobes setaceous; filaments woolly at base, anthers dehiscent from apex; *style filiform, elongate;* stigma simple; capsule urceolate-acuminate, 5 mm. long; seeds furrowed. — Sandy pinelands, Bahama Ids. and Fla. to La., n. to se. Va., se. Tenn. and n. Ala. Sept., Oct. — Fresh plant pale green, blackening in drying.

24. GERÁRDIA L. GERARDIA

Calyx hemispherical, campanulate or turbinate, 5-toothed or 5-cleft. Corolla swelling above, with more or less unequal lobes, the 2 upper usually rather smaller and more united. Stamens mostly hairy; anthers connivent in pairs, 2-locular; the locules parallel, often pointed at base. Style elongated, slender or enlarged and flattened at the apex. Capsule globular or ovoid, mucronate or pointed, many-seeded. — Erect branching Am. herbs (often more or less root-parasitic); cauline leaves opposite, or the upper alternate, the uppermost reduced to bracts and subtending 1-flowered pedicels which often form a raceme or spike. Flowers showy, pink, purple or yellow, in late summer and autumn. (Dedicated to the celebrated herbalist, *John Gerarde*, 1545–1611 or '12.)* Incl. AUREOLARIA Raf. and TOMANTHERA Raf.

a. Corolla purple or pink (rarely white); anther-locules blunt to mucronate or caudate at base; stigma linguliform or linear; capsules rounded at summit, often mucronate. . . *b.*

b. Stems smooth or scabrous with ascending minute points; leaves narrowly linear, lanceolate, oblanceolate or filiform, entire (except basal ones sometimes in no. 1); anthers all alike, their locules usually mucronate or caudate at base; stigma linguliform. Subgen. I. EUGERARDIA. . . *c.*

c. Plant perennial, with elongate rhizome; raceme very slender, virgate or with virgate branches, with erect pedicels; species of Coastal Plain southw. § 1. SPARTORHIZOMA. . . . . 1. *G. linifolia.*

c. Plants annual with fibrous roots, variously branched; pedicels ascending to divergent. . . *d.*

d. Fresh plants deep green to purple-tinged, mostly blackened in drying; calyx-tube little if at all reticulate-veiny; seeds blackish or dark brown. § 2. CHYTRA. . *e.*

---

* Many details derived from the studies of the genus (as a number of segregated genera) by F. W. PENNELL. The present treatment following the generic concept of BENTHAM, GRAY, WETTSTEIN, BLAKE and others.

e. Calyx-lobes exceeding calyx-tube and capsule; pedicels very short.
Leaves and calyx-lobes subrigid, with attenuate tip; corolla 2.5–3 cm. long, with spreading lobes 6–8 mm. long; southwestern. . . . . 2. *G. heterophylla.*
Leaves and calyx-lobes soft-herbaceous, blunt to subacute; corolla 1–1.5 cm. long, its scarcely spreading lobes 3–5 mm. long; Nova Scotian. . . . . 4. *G. neoscotica.*

e. Calyx-lobes shorter than tube. . . *f.*

f. Capsule ellipsoid, subcylindric, distinctly longer than thick; fastigiately branched stem, subrigid leaves and erect pedicels harshly scabrous; northwestern. . . . . 3. *G. aspera.*

f. Capsule globose, subglobose or globose-ovoid, shorter than to but slightly longer than thick; variously spreading branches, more flexible leaves and spreading to ascending pedicels scabrous to smooth. . . *g.*

g. Upper lobes of corolla reflexed-spreading; corolla pubescent within, its outer surface pubescent. . . *h.*

h. Pedicels shorter than flower, little if at all longer than calyx and capsule; calyx shorter than capsule, its lobes or teeth 0.5–3.5 mm. long; capsule subglobose, 4.5–7 mm. long. . *i.*

i. Plant fleshy and succulent; leaves and calyx-lobes obtuse; anther-locules blunt or merely acutish at base; plant of saline habitats. . . . . 5. *G. maritima.*

i. Plants of dry texture; leaves and calyx-lobes acute or acuminate; anther-locules mucronate to caudate at base; plants of fresh (mostly acid) soil. . . *j.*

j. Stem angled, smooth or but slightly scabridulous; principal leaves without or with but slightly developed short axillary fascicles, the fascicles when present much shorter than subtending leaves; stigma 2–3 mm. long; seeds 0.6–1.5 mm. long. . . *k.*

k. Leaves linear, in drying merely arching or becoming falcate; calyx-tube two-fifths to two-thirds as long as capsule; wide-ranging species.
Calyx-tube two-fifths to half as long as capsule, the sinuses V- or narrowly U-shaped, the lobes 2–3.5 mm. long; corolla 1–2(–2.3) cm. long; lower filaments 7–9 mm. long; style 6–10 mm. long; northern, chiefly between lats. 40° and 47°. . . . 6. *G. paupercula.*
Calyx-tube one-half to two-thirds as long as capsule, the sinuses subquadrate, the lobes 0.5–2 mm. long; earlier (larger) corollas 2–3.8 cm. long; lower filaments 0.8–1.6 cm. long; style 1.5–2 cm. long; wide-ranging from Fla. to Tex., n. to northern states. . . 7. *G. purpurea.*

k. Leaves nearly filiform, in age strongly curling or spiraling; calyx-tube two-thirds to three-fourths as long as capsule; corolla 2–3.5 cm. long; lower filaments 8–9 mm. long; style 1.3–1.5 cm. long; plant of Coastal Plain. . . . . 8. *G. racemulosa.*

j. Stem subterete at base, scabrous-puberulent; principal leaves arching, subtending and nearly equaled by axillary fascicles; stigma 3–4 mm. long; seeds 0.5–0.8 mm. long; southern species, chiefly on Coastal Plain. . . . 9. *G. fasciculata.*

h. Pedicels elongate, mostly as long as or longer than flower (including corolla), 1–4 cm. long; leaves slenderly linear to filiform; calyx nearly covering capsule, its teeth only 0.2–0.5 mm. long; capsule 3–4 mm. long, globose-ovoid; seeds 0.5–0.6 mm. long. . . . . 10. *G. setacea.*

g. Upper lobes of corolla arching forward over stamens and style, much longer than lower lobes; corolla glabrous within; pedicels mostly longer than flower. . . . . 11. *G. tenuifolia.*

d. Fresh plants yellow-green or pale, mostly not blackening in drying; calyx-tube definitely reticulate-veiny; corolla pink, its lobes widely spreading; seeds yellow or yellowish. § 3. CHLOROMONE. . . *l.*

l. Pedicels exceeding subtending leaves; calyx-tube 1.5–3.5 mm. long, the lobes or teeth 0.05–1.8 mm. long; corolla 1.2–2 cm. long, deep pink; capsule globose to ellipsoid-ovoid; seeds rounded and turgid. . . *m.*

*m.*Stem striate-angled; leaves narrowly linear to nearly filiform, the principal ones 1–2.5 cm. long; upper pedicels two to four times as long as their subtending leaves, the inflorescences forming racemes; stigma 1–2 mm. long. . . *n.*

*n.*Throat of corolla with 2 conspicuous yellow lines and with red or purple spots; capsule globose or globose-ovoid. . . *o.*

*o.*Angles of stem prominent, harshly scabrous; lobes of corolla truncate; stigma 1.5–2 mm. long; capsule 4–5 mm. long; plant of the interior. . . . . 12. *G. Skinneriana.*

*o.*Angles of stem less prominent, smooth or nearly so; lobes of corolla emarginate to obcordate; stigma 1–1.5 mm. long; capsule 3–4 mm. long.

Lower and median pedicels equaling to twice length of subtending bracteal leaves; calyx-tube campanulate, its triangular-acuminate teeth 0.5–1 mm. long; seeds 0.4–0.6 mm. long; plant of s. N. E. and L. I. . . . . 13. *G. acuta.*

Lower and median pedicels mostly twice or thrice length of bracteal leaves; calyx-tube hemispherical, its teeth reduced to callosities only 0.05–0.3 mm. long; seeds 0.6–0.8 mm. long; plant of the Piedmont and upland, n. to s. Pa. . . . . 14. *G. decemloba.*

*n.*Throat of corolla with only faint or obsolete lines and spots; capsule subglobose-ellipsoid; pedicels mostly three to five times as long as bracteal leaves; plant of se. Coastal Plain. 15. *G. obtusifolia.*

*m.*Stem nearly terete, smooth or scabridulous; leaves narrowly linear, the principal ones 2–3 cm. long; upper bracteal leaves one-half to nearly as long as pedicels; inflorescence rather loose and open; stigma 2–3 mm. long; plant of the interior. . . . . . . 16. *G. Gattingeri.*

*l.*Pedicels mostly overtopped by subtending leaves; calyx-tube 3.5–4 mm. long, the lobes 1.5–2 mm. long; corolla 8–12 mm. long, pale; capsule globose-obovoid, 5–6 mm. long; seeds angled, not turgid; southwestern. . . . . 17. *G. viridis.*

*b.*Stem retrorse-hispid; leaves lanceolate to lance-ovate, the uppermost auricled at base; calyx-lobes herbaceous, broad; anthers dissimilar, the lower pair longer; their locules not pointed; stigma linear. Subgen. II. OTOPHYLLA. . . . . 18. *G. auriculata.*

*a.*Corolla yellow, or yellow with brown or reddish tinge; anthers all alike, slightly or hardly included, their locules awned at base; stigma broadened; capsules acute; leaves dilated, entire to deeply bipinnatifid. Subgen. III. PANCTENIS. . . *p.*

*p.*Perennials, non-glandular with simple and entire to coarsely bipinnatifid leaves; corolla yellow throughout, glabrous outside; capsule not glandular; seeds winged, 1.3–2.7 mm. long. . . *q.*

*q.*Capsule densely rusty-pubescent; pedicels 1–6 mm. long; east of Miss. R. 19. *G. virginica.*

*q.*Capsule glabrous; pedicels 0.1–2.5 cm. long. . . *r.*

*r.*Plant densely cinereous-pubescent; calyx-lobes dentate or entire; wings of outer side of seed one-sixth to one-fourth the diameter of the seed; western. . . . . 20. *G. grandiflora*, vars.

*r.*Plants with glabrous stems and glabrous or minutely puberulent leaves; calyx-lobes entire; wings one-fourth to two-fifths diameter of seed. . . *s.*

*s.*Stem green, not glaucous; lower leaves lanceolate, broadest near base, acuminate, entire or merely dentate or slightly lobed; pedicels in anthesis mostly shorter than calyx-tube, becoming 3–8 mm. long in fruit; corolla 3–3.5 cm. long; capsule 1–1.2 cm. long; seeds 1.5–1.7 mm. long. . . . . 21. *G. laevigata.*

*s.*Stem glaucous, often purplish; at least the lower leaves pinnately cut or lobed, if entire elliptic and broadest near middle; pedicels in anthesis mostly longer than calyx-tube, becoming 0.4–1.5 cm. long; corolla 3.5–6 cm. long; capsules 1.2–2 cm. long; seeds 1.7–2.7 mm. long.

Segments of dissected leaves lanceolate or broader; corolla pubescent within above bases of filaments; anthers with awns 1–1.5 mm. long; seeds 2–2.7 mm. long; wide-ranging, mostly e. and ne. of Miss. R. . . . . 22. *G. flava.*

Segments of dissected leaves linear or linear-lanceolate; corolla glabrous within above bases of filaments; anthers with awns 1.5–2 mm. long; seeds 1.8–2 mm. long; southwestern, w. of Miss. R. . . . . 23. *G. calycosa.*

p. Annuals, glandular or viscid, with bipinnatifid leaves; corolla yellow, often tinged with reddish-purple, glandular-pubescent outside; capsule glandular; seeds wingless, 0.8–1 mm. long.

Calyx-tube turbinate, glandular-puberulent outside; capsule ellipsoid, its inner half enclosed by the calyx-tube; wide-ranging eastward and northward. . . . . 24. *G. pedicularia.*

Calyx-tube hemispherical, hirsute to lanate outside; capsule ovoid, only its base embraced by the calyx-tube; southern and southwestern. . . . 25. *G. pectinata.*

Subgen. I. EUGERÁRDIA (Benth.) Fern. *Agalinis* Raf.

§ 1. SPARTORHIZÒMA Pennell

1. G. linifòlia Nutt. (with leaves of *Linum*, flax). — *Perennial*, the *creeping rhizome with clustered fleshy roots; stem* green or purple, blackened in drying, *terete*, glabrous, 0.8–1.5 m. high, simple or with few erect virgate branches; *leaves all opposite*, erect, *linear*, *the cauline 3–5 cm. long and up to 3 mm. broad*, glabrous on upper surface; *racemes virgate, with* short *erect clavate pedicels* becoming 1–2.5 cm. long in fruit; *calyx-tube* 4–4.5 *mm. long, with subulate teeth scarcely* 0.5 *mm. long; corolla* 3–4 *cm. long*, its straight tube 2–2.7 cm. long, *the rounded or subtruncate spreading lobes* 0.9–1.3 *cm. long;* the *throat with diffuse reddish-purple spots on lower inner surface;* upper *filaments* 1.1–1.4 cm. long, *the lower ones* 1.3–1.7 *cm. long*, all lanate; *anther-locules* 4 *mm. long, awned at base;* style pubescent, 1.5–2.5 cm. long; *stigma* 4–7 *mm. long;* capsule subglobose, 6–7 mm. long. (*Agalinis* Britt.) — Damp pinelands and pond-margins on Coastal Plain, Fla. to La., n. to se. N.C.; very local, Sussex Co., Del. Late Aug.–Oct. (Cuba)

§ 2. CHŸTRA (Gaertn.) Pennell

Sub-§ 1. HETEROPHÝLLAE Pennell

2. G. heterophýlla Nutt. (various-leaved). — Annual with 4-angled glabrous stem 0.3–1 m. or more high and stiff spreading ascending branches; *leaves mostly opposite, subrigid*, lanceolate or lance-linear, *subulate-tipped, some lower ones often 3-cleft; pedicels* very short, *suberect; calyx with stiff lanceolate lobes much longer than the tube and overtopping the capsule; corolla* 2.5–3 *cm. long, with spreading lobes* 6–8 *mm. long*, pale pink, the lower half of the throat with purplish spots and 2 yellow lines; anther-locules 3.5 mm. long, mucronate at base; capsule ellipsoid or ellipsoid-ovoid, 8–9 mm. long. (*Agalinis* Small) — Sandy prairies and dry woods, w. La. and e. Tex., n. to se. Mo. and n. Okla. Late Aug., Sept. — Plant blackened in drying.

Sub-§ 2. ÁSPERAE Pennell

3. G. áspera Dougl. (harsh). — Stiffly upright, with simple or branching coarse *scabrous stems* up to 8 dm. high; *leaves slender, linear or infolded, sharp-tipped, harshly scabrous, with axillary fascicles;* pedicels suberect; *calyx-tube* campanulate, shorter than capsule, 4–6 *mm. long, its lanceolate lobes* 1.5–3 *mm. long; corolla* 1.8–2.5 *cm. long*, its lobes 5–6 mm. long, *the 2 upper lobes ascending; anther-locules* 1.8–2.2 *mm. long, obtuse to subacute at base; style about 1 cm. long; stigma* 1.6–2.3 *mm. long; capsule ellipsoid or ellipsoid-ovoid*, 7–11 *mm. long.* (*Agalinis* Britt.) — Dry sandy or rocky slopes and prairies, N.D. to Okla., e. to se. Man., Minn., s. Wisc. and Ill. Aug., Sept.

Sub-§ 3. PURPÙREAE Pennell PURPLE GERARDIAS

4. G. neoscótica Greene (Nova Scotian). — *Small* annual *with simple or slightly short-branched glabrous 4-angled slender stem* 0.5–4 *dm. high; leaves soft-herbaceous*, opposite, *linear to linear-lanceolate or oblanceolate, the primary ones* 1–4.5 *mm. broad*, without or with few small axillary fascicles, the *margins scabrous; bracteal leaves scarcely reduced, overtopping the few* short-pedicelled ascending *flowers; calyx-tube hemispherical*, 2–3 *mm. long*, two-fifths to half as long as capsule, *with narrow U- or V-shaped sinuses; calyx-lobes* lanceolate, scabrous-margined, 3–8 *mm. long, exceeding the capsule, blunt or acutish; corolla* 1–1.5 *cm. long*, 5–7 *mm. broad at summit, its tube* 7–10 *mm. long, its rounded lobes* 3–5 *mm. long and projected forward* (scarcely spreading), pink, with deeper-colored spots at base of tube below, the center green; *upper filaments* 2–4, *lower ones* 3–5 *mm. long; anther-locules* 1 *mm. long*, cuspidate at base; *style* 4–5 *mm. long; stigma* 1 *mm. long; capsule* globose, 5–6 mm. long, *overtopped by the calyx-lobes;* seeds 1–1.5 mm. long.

(*Agalinis* Fern.) — Damp sand, peat or sphagnous pockets, Sable I. and w. N.S. Mid-July–early Oct.

5. G. marítima Raf. (maritime). — *Fleshy-leaved and succulent* green to purplish annual, the simple or branching *glabrous* obscurely 4-angled *stem* 0.5–3 (–4) *dm. high; leaves* linear, *obtuse or subacute*, the primary ones 1.5–3 cm. long and 1.5–2.5 mm. broad, often scabridulous above, *without axillary fascicles; racemes* 2–10-*flowered, not strongly elevated above the foliage-leaves, the bracts only gradually shortened*, the *mature internodes* 1–4.5 *cm. long;* pedicels ascending, glabrous, 2–12 mm. long; *calyx-tube* 3–4 *mm. long*, hemispherical-campanulate, *up to half as long as capsule, the sinuses subquadrate, the obtuse triangular lobes* 0.5–1.5 *mm. long; corolla* 1–1.7 *mm. long*, its somewhat spreading lobes 4–5 mm. long; *upper filaments* 5–6, *lower* 7–8 *mm. long; anther-locules* 1.3–1.8 *mm. long, with bluntish bases, glabrous or sparingly short-pubescent; style* 6–10 *mm. long;* stigma 2.5–3 mm. long; capsule subglobose, 5–6 mm. long; *seeds* 1–1.3 *mm. long.* (*Agalinis* Raf.) — Saline marshes along coast, Yarmouth Co., N.S.; Cumberland Co., Me. to N.C. Mid-July–Sept. — Southward passing into

Var. **grandiflòra** Benth. (large-flowered). — *Taller*, 2–6 *dm. high, with* narrower and *more acute leaves, the bracteal ones strongly reduced; racemes more elevated above foliage-leaves*, with mature internodes often longer; *corolla* 1.5–2 *cm. long; anther-locules*, 1.8–2.3 *mm. long and with long pubescence; style* 1.2–1.4 *cm. long; seeds* 0.8–1 *mm. long.* (*G. spiciflora* Engelm.) — Fla. to Tex.; ne. N.C. and se. Va. July, Aug. (W.I.; Yucatan)

6. G. paupércula (Gray) Britt. (poor; as contrasted with no. 7). — Simple or with ascending branches, *slender*, the 4-*angled stem glabrous or nearly so*, 2–8 dm. high; *leaves of dry texture, linear, acute*, smooth to scabrous above, *without or with slightly developed axillary fascicles*, the primary leaves 2–4.5 cm. long and 1–2.5 mm. wide, *in drying blackened and slightly arching; racemes* elongate, *with long bracteal leaves; pedicels ascending*, slenderly clavate, glabrous, *in anthesis* 2–4, *in fruit* 2–5 *mm. long; calyx-tube* 2.5–4 *mm. long, hemispherical, two-fifths to half as long as capsule*, 2.5–4 *mm. long, with V- or U-shaped sinuses and lance-triangular acuminate lobes* 2–3.5 *mm. long; corolla* 1.5–2 (–2.3) *cm. long*, 1–1.5 *cm. broad across the spreading lobes*, rose-purple to pink, darker-spotted in throat and with 2 yellow lines below; *upper filaments* 4–6, *lower* 7–9 *mm. long, loosely woolly; anther-locules* 1–2 *mm. long, acuminate to cuspidate at apex; style* 8–10 *mm. long;* stigma 2 mm. long; capsule globose, 4.5–6 mm. long; seeds lunate-triangular, reticulate, 1–1.5 mm. long. (*Agalinis* Britt.; *G. purpurea*, var. Gray) — Damp open ground, shores, bogs, etc., sw. N.B. to s. Wisc., s. to s. N.E., L.I., w. N.J., e. and nw. Pa., n. O., n. Ind., n. Ill. and e. Ia. Aug., Sept. — Inland largely replaced by

Var. **boreàlis** (Pennell) Deam (northern). — *Differing in its smaller corolla* (1–1.7 *cm. long, with spread at summit of only* 0.5–1 *cm.*), *the lobes more porrect; filaments with shorter hairs; style* 6–8 *mm. long.* — Sw. Que. to Minn., s. to ne. Vt., centr. and w. N.Y., nw. Pa., s. Ont., n. Ind. and ne. Ill. — Forma **albiflòra** Vict. & Rousseau (white-flowered) has white corollas.

7. G. purpùrea L. (purple). — Differing from no. 6 in greater size of most parts; stem up to 1.2 m. high; primary leaves often slightly broader; racemes 6–14-flowered; *calyx-tube one-half to two-thirds length of capsule, with sinuses nearly quadrangular and lobes only* 0.5–2 *mm. long; larger corollas* 2–3.8 *cm. long*, rose-purple, or in forma **albiflòra** Britt. (white-flowered) white; *lower filaments* 0.8–1.6 *cm. long; style* 1.5–2 *cm. long; stigma* 2–3 *mm. long;* capsule up to 7 mm. long. (*Agalinis* Pennell) — Damp, mostly acid soils, Fla. to e. Tex., n. to s. N.E., se. N.Y., Pa., n. O., s. Mich., Wisc., Minn. and Neb. Late July–Sept. (W.I.)

8. G. racemulòsa Pennell (somewhat racemose). — *Differing from no. 7 in its very narrowly linear to almost filiform leaves only* 0.5–1 *mm. wide and, when dried, strongly curling or spiraling, thus exposing the axillary fascicles; racemes often more elongate, with* 8–26 *flowers; calyx-tube two-thirds to three-fourths as long as capsule; lower filaments* 8–9 *mm. long; style* 1.3–1.5 *cm. long; stigma* 2 *mm. long; capsule* 4.5–5 *mm. long; seeds* 0.9–1.2 *mm. long.* — Damp to dryish pine-barrens and bogs of Coastal Plain, L.I. to S.C. Sept., Oct.

9. G. fasciculàta Ell. (in fascicles). — *Differing from nos. 6–8 in the nearly terete base of the scabrous-puberulent stem*, but with angled branches; *leaves curving or arching, thus exposing the strongly developed fascicles, scabrous; bracteal leaves reduced, much shorter than the flowers;* racemes elongate, 12–30-flowered; *calyx-tube one-half to two-thirds as long as capsule*, with subquadrate sinuses, the acuminate lobes 0.5–2 mm. long; corolla 2–3.5 cm. long, with rounded to truncate spreading lobes 7–10 mm. long; upper filaments 8–11, lower ones 1.1–1.5 cm. long; *anther-locules* 2.5–3.5 *mm. long*, with acute to cuspidate bases; style 1.5–2.1 cm. long; *stigma* 3–4 *mm. long;* capsule globose-ovoid, 5–6 mm. long; *seed only* 0.5–0.8 *mm. long.* (*Agalinis* Raf.) — Dry to moist siliceous or argillaceous soils, dune-hollows, tidal marshes, etc., Fla. to Tex., n. on outer Coastal Plain to S.C. and se. Va., and inland from the Gulf to w.-centr. Ga., Mo. and Ark. Aug., Sept. (W.I.)

Sub-§ 4. SETÀCEAE Pennell

10. G. setàcea (Walt.) J. F. Gmel. (bristleform). — *Very slender, mostly bushy-branched* 1.5–7 dm. high, glabrous or nearly so; *leaves spreading, narrowly linear to filiform, rarely* 1 *mm. broad*, scabrellous, the bracteal ones shorter; *pedicels filiform, mostly exceeding flowers,* 1–3, *in fruit up to* 4 *cm. long; calyx-tube short-campanulate,* 2.5–3 *mm. long, with broad open sinuses and triangular teeth less than* 0.5 *mm. long; corolla* 1.5–2.5 *cm. long*, rosy-pink, *with ampliate throat and* rounded to retuse *spreading* broad *lobes* 5–7 mm. long; *upper filaments* 4–5, *lower* 7–9 *mm. long;* anther-locules 2–2.5 mm. long, acuminate to basal awns; style 0.7–1.5 cm. long; stigma 1.5–2 mm. long; *capsule* globose-ovoid, 3–4 *mm. long, nearly equaled by calyx-tube; seeds* 0.5–0.6 *mm. long.* (*Agalinis* Raf.; *G. Holmiana* Greene) — Dry sandy woods and openings on Coastal Plain and outer Piedmont, Ga. and Ala. to L.I. Aug.–Oct.

Sub-§ 5. TENUIFÒLIAE Pennell

11. G. tenuifòlia Vahl (slender-leaved). — Similar in habit to no. 10, but with less diffuse branching; *leaves linear*, acuminate, *the primary ones* 2–5 cm. long and 1–3.5 *mm. broad, spreading to arched-ascending*, the rameal much narrower, with few or no axillary fascicles; *pedicels filiform, spreading,* 0.7–2.7 *cm. long; calyx-tube* hemispherical, 2.5–3 *mm. long, with broad sinuses and subulate teeth only* 0.2–1 *mm. long; corolla glabrous inside*, roseate, or white in forma **albiflòra** Britt. (white-flowered), 1–1.5 *cm. long*, with ampliate throat, *the upper lobes arching forward over the stamens and style and much longer than the spreading lower lobes; upper filaments* 2–3, *the lower* 4–6 *mm. long; anther-locules* 1.5–2.2 *mm. long;* style 1–1.4 cm. long; *stigma* 1–1.5 *mm. long;* capsule globose, 3–5 mm. long; seeds 0.6–0.9 mm. long. (*Agalinis* Raf.) — Dry woods, thickets, fields, etc., s.-centr. Me. to s. Mich., s. to s. N.E., L.I., s. N.J. and Del., and, chiefly inland from Coastal Plain, to Ga., Ala., Miss. and La. Aug.–Oct.

Var. **macrophýlla** Benth. (large-leaved). — Rather stouter; primary leaves linear to linear-lanceolate, 3–7 cm. long and 1–6 mm. broad; calyx-teeth 1–2 mm. long; capsules mostly 5–7 mm. long; seeds 0.8–1.3 mm. long. (*G. Besseyana* Britt.) — Often in moister habitats of low woods, prairies, peaty spots, banks of streams, etc., sw. Ct. to Minn. and Neb., s., chiefly west of the mts., to Ala., Miss., Ark. and e. Tex.

Var. **parviflòra** Nutt. (small-flowered). — Branches fastigiate-ascending; leaves stiffish and often ascending, more scabrous, often with conspicuous axillary fascicles; calyx-teeth 1–2 mm. long; anthers less heavily pubescent; capsules usually 5–7 mm. long. — Usually in damp soil, sw. Que. and nw. Vt. to s. Man. and Wyo., s. to N.Y., O., Ind., Ill., Mo., Okla. and Colo.

§ 3. CHLORÓMONE Pennell

12. G. Skinneriàna Wood (discovered by Wood in 1847 "on land belonging to Dr. A. G. Skinner"). — *Pale or yellowish-green plant rarely darkening in drying; stem* stiff, 2–6 dm. high, *with prominent harshly scabrous angles*, simple or with short ascending branches; *leaves* ascending, linear, acuminate, the primary ones 1–2 cm. long and 0.5–1.2 mm. broad, *scabrous;* racemes (1–)2–8-flowered; pedicels ascending, slenderly clavate, 0.5–2 cm. long, up to twice length of subtending bracteal leaves; *calyx-tube hemispherical,* 2.5–3.5 *mm. long, reticulate-veiny, two-thirds length of capsule, the sinuses broad, the triangular callous teeth* 0.3–0.8 *mm. long; corolla* roseate or pink, 1.2–1.5 *cm. long, its throat with* 2 *conspicuous yellow lines and red- or purple-spotted within at base, its truncate spreading lobes* 3–4 *mm. long;* upper filaments 3–4, lower 4–5 mm. long; anther-locules about 1 mm. long, cuspidate at base; style 4–5 mm. long; *stigma* 1.5–2 *mm. long; capsule globose to globose-ovoid,* 4–5 *mm. long;* seeds 0.7–0.9 mm. long. (*Agalinis* Britt.) — Dry sands, bluffs and barrens, sw. Ont. to s. Wisc., s. to n. O., s. Ind. s. Ill., n. Ark. and Okla., local. Late July–Sept.

13. G. acùta Pennell (acute; from the calyx-teeth). — *Differing from no.* 12 *in the smooth or smoothish angles of the stem and glabrous pedicels; calyx-tube campanulate,* up to three-fourths length of capsule, *its teeth not callous, triangular-acuminate* and up to 1 mm. long; *corolla* 1–1.3 cm. long, *its* spreading *lobes retuse or emarginate,* the spots at base of tube faint; *lower filaments* 5–6 *mm. long; anther-locules* 1.5–1.7 *mm. long;* style 5–6 mm. long; *stigma* 1–1.2 *mm. long; capsule ovoid, about* 4 *mm. long; seeds* 0.4–0.6 *mm. long,* sharply reticulate. (*Agalinis* Pennell) — Dry sandy soil, Cape Cod and Nantucket I., Mass. to w. L.I., inland to w. Middlesex and Worcester Cos., Mass., n. Providence Co., R.I., and Hartford Co., Ct. Aug., Sept.

14. G. decémloba Greene (ten-lobed; from the 5 emarginate corolla-lobes). — Resembling nos. 12 and 13; *angles of stem smooth or smoothish; glabrous pedicels mostly twice or thrice length of bracteal leaves; calyx-tube hemispherical, its teeth reduced to callosities only* 0.05–0.3 *mm. long; corolla-lobes emarginate to obcordate; anther-locules* 1.2–1.4 *mm. long; capsule globose-ovoid,* 3–4 *mm. long; seeds* 0.6–0.8 *mm. long.* (*Agalinis* Pennell) — Dry open soil, se. Pa. to s. Ky., e. Va., centr. and w. N.C., w. S.C. and Ala. Aug., Sept.

15. G. obtusifòlia (Raf.) Pennell (blunt-leaved). — *Similar to nos.* 12–14, distinguished by the height of its smoothish stem (3–9 dm.); *leaves* commonly *linear-oblanceolate* (*slightly broadened above*), obtuse to acutish, *the primary ones only* 0.6–1.5 *cm. long; pedicels mostly three to five times length of very short bracteal leaves,* 2–2.5 mm. long; *the subulate calyx teeth minute* (0.05–0.15 *mm. long*); *corolla-lobes rounded to slightly retuse,* 3–4 mm. long, *the throat with obsolete or faint markings within; anther-locules* 1.6–2.1 *mm. long; capsule globose-ellipsoid;* seeds 0.7–0.9 mm. long. (*G. parvifolia* (Hook.) Chapm.) — Dry to moist siliceous or argillaceous pinelands, thickets and openings of Coastal Plain, Fla. to La., n. to Del. and ne. Md. Sept., Oct.

16. G. Gattíngeri Small (for its discoverer, AUGUSTIN GATTINGER, 1825–1903). — *Resembling nos.* 12–15 *but with* smoothish nearly *terete stems with spreading branches; leaves* linear, *acuminate, the primary ones* 2–3 *cm. long* and up to 1.2 mm. broad; upper bracteal leaves one-half to nearly as long as pedicels; *calyx-tube campanulate, its triangular-lanceolate teeth* 0.5–1.8 *mm. long and scarcely callous-thickened;* corolla 1.2–1.8 cm. long, its lobes 4–6 mm. long; *stigma* 2–3 *mm. long.* (*Agalinis* Small) — Moist to dry argillaceous to siliceous slopes, open woods, barrens, etc., sw. Ont. and se. Mich. to s. Minn. and e. Neb., s. to Ala., La. and e. Tex. Aug.–early Oct.

17. G. víridis Small (green). — Suggesting no. 16 in its loosely ascending branches and subterete stem, but with shorter leaves (to 2.2 cm. long); *pedicels mostly overtopped by the leafy bracts; calyx-tube* 3.5–4 *mm. long, its lanceolate lobes* 1.5–2 *mm. long; corolla only* 8–12 *mm. long, its truncate to erose lobes only* 2–3 *mm. long; upper filaments only* 2–3 *mm. long; anther-locules* 0.9–1.3 *mm. long; style* 3–4 *mm. long; stigma* 1 *mm. long.* (*Agalinis* Pennell) — Pinelands and prairies, usually damp, w. La. and e. Tex., n. to se. and sw. Mo. Sept., early Oct.

Subgen. II. OTOPHÝLLA (Benth.) Fern. *Tomanthera* Raf.

18. G. auriculàta Michx. (eared). — *Harshly scabrous annual with retrorse-hirsute* 4-angled *stiff stem* up to 8 dm. high; *leaves* subsessile, *lanceolate or lance-ovate,* 2.5–5.5 cm. long and 0.8–2 *cm. broad, the uppermost with divergent basal auricles; flowers sessile in leafy-bracted spikes; calyx* campanulate, *with unequal ovate-lanceolate lobes* 0.9–1.2 *cm. long; corolla* about 2 cm. long, purple, *the upper lobes longer than the lower,* the throat dark-dotted; *upper filaments* 5–6, *lower about* 10 *mm. long; upper anthers about* 2 *mm. long, lower* 2.2–5 *mm. long and blunt-based; capsule* thick-ovoid, 1–1.3 *cm. long; seeds ellipsoid-ovoid,* 1.3–1.6 *mm. long.* (*Tomanthera* Raf.) — Prairies, open woods and fields, n. N.J. to s. Minn., s. to the Fall Line of n. Va., n. Ala., Tenn., nw. Ark. and e. Tex. Aug., Sept.

Subgen. III. PÁNCTENIS (Rydb.) Fern. *Gerardia* § *Dasystoma* sensu Gray, not *Dasystoma* Raf., basonym. *Aureolaria* Raf. FALSE FOXGLOVES

19. G. virgínica (L.) BSP. (Virginian), DOWNY FALSE FOXGLOVE. — *Perennial, downy nearly throughout, glandless;* stem 0.3–1.5 m. high, simple or with stiff ascending branches toward summit; *leaves petioled, lanceolate or lance-ovate, acuminate, the lower coarsely sinuate or pinnatifid, the upper more oblong-lanceolate and entire to pinnatifid,* the bracteal leaves narrower and shorter; *pedicels* stout, *in anthesis* 1–3, *in fruit* 1.5–6 *mm. long; densely pubescent calyx* with tube 4–5 mm. long, *the foliaceous lobes* 0.4–1.2 *cm. long; corolla yellow throughout,* 3–4.5 *cm. long, glabrous outside,* its throat ampliate below, its rounded lobes 0.6–1.3 cm. long; upper filaments 1.2–1.6, lower 2–2.5 cm. long; anther-locules 4–4.5 mm. long, their curved basal awns 1 mm. long; style 3–3.5 cm. long; *capsule densely rusty-pubescent,* ovoid to ovoid-subglobose, 1.2–1.5 cm. long, acute; *seeds* 1.5–1.8 *mm. long,* broadly lunate-angulate, flattish, *the outer side with several thin wings, the large wings one-fourth to one-third the breadth of the seed.* (*G. flava* of ed. 7, not L.; *Aureolaria* Pennell; *Dasystoma* Britt.) — Dry open woods, s. N.H. to s.-centr. Mich., s. to n. Fla., Ala. and La. June–Aug.

20. G. GRANDIFLÒRA Benth. (large-flowered). — *Differing from no.* 19 *in more spreading branches;* petioles 5–6 mm. long; *pedicels longer* (*up to* 1.4 *cm. long*); *calyx-lobes broadly lanceolate, dentate; anther-locules* 4–5.5 *mm. long,* the awns 1.5 mm. long; *style* 3.5–4 *cm. long; seeds* 1.7–2 *mm. long, turgid, the broadest wings one-sixth to one-fourth the diameter of the seed.* — Southwest of our range; represented with us by two vars.:

Var. **cinèrea** (Pennell) Cory (ashy). — *Lower leaves only shallowly pinnatifid; bracts serrate-dentate or slightly laciniate at base; pedicels* 5–9 *mm. long; calyx-lobes linear to narrowly lanceolate, entire to but slightly dentate;* corolla 4.5–5.5 cm. long; *capsule* 1.2–1.7 *cm. long.* — Rocky woods, Mo. and n. Ark. to e. Tex. Aug., Sept.

Var. púlchra (Pennell) Fern. (beautiful). — *Differing from preceding* var. *in* less cinereous stem; *more pinnatifid leaves, the lower ones usually cut nearly to midrib; bracts commonly pinnatifid at base; corolla up to* 5.5 *cm. long; capsule* 1.5–2.3 *cm. long.* — Oak-woods, Wisc. and se. Minn., s. to w. Ind., s. Ill. and Mo. July–Sept.

21. G. laevigàta Raf. (smooth). — *Distinguished from no.* 19 *by its glabrous green stem; leaves very short-petioled or subsessile, thinnish,* lanceolate, acuminate, *mostly entire; slender pedicels glabrous,* in anthesis 1–4, *in fruit* 3–8 *mm. long; calyx glabrous outside,* its tube 3–5 mm. long, *its lobes* 4–6 *mm. long; corolla* 3–3.5 *cm. long; anthers* 3.5–4 *mm. long; style* 2.2–2.5 *cm. long; capsule glabrous,* 1–1.2 *cm. long.* (*Aureolaria* Raf.) — Dry to moist deciduous woods, chiefly along the mts., Pa. to s. O., s. to n. Ga. and Tenn. Late July–early Oct.

22. G. flàva L. (yellow). — Resembling nos. 19–21; *stem glabrous and glaucous, often purplish,* up to 2.5 m. high; *lower leaves elliptic-ovate or ovate-lanceolate, broadest near middle, deeply lobed or incised with lanceolate, oblong or ovate lobes or segments, the middle and upper* more shallowly cleft, at first puberulent, *soon glabrate, more or less glaucous and reticulate beneath; pedicels glabrous, in anthesis* 4–7, *in fruit* 5–15 *mm. long; calyx-tube* 4–7 mm. long, *glabrous outside; its lobes* 2–5 *mm. long,* acute; *corolla* 3.5–4 *cm. long, glabrous outside, densely pubescent inside the throat; lower filaments* 1.7–2 *cm. long; anther-locules* 5–6 *mm. long, their basal awns* 1–1.5 *mm. long;* capsule glabrous; *seeds* 2–2.7 *mm. long.* (*Aureolaria* Farwell; *G. quercifolia* Pursh; *G. virginica* of ed. 7, not (L.) BSP.) — Deciduous woods, sw. Me. to s. Minn., s. to e. Md. and inland s. to Ga. and Ala. July–Sept.

Var. **macrántha** (Pennell) Fern. (large-flowered). — Leaves more permanently puberulent; *calyx-lobes* 5–14 *mm. long; corolla* 3.5–6 *cm. long; capsule* 1.5–2 *cm. long.* (*Aureolaria* Pennell) — Chiefly inland, from w. N.Y. to Ill. and s. Mo., s. to w. Md., W.Va., w. N.C., n. Ala., Miss. and La.

Var. **reticulàta** (Raf.) Cory (reticulated). — *Lower,* 1–1.6 m. high, *only slightly glaucous; lower leaves nearly entire or cut less than half-way to midrib,* the *upper leaves entire; corolla* 3.5–4.5 *cm. long.* (*Aureolaria* Pennell) — Chiefly on Coastal Plain, Fla. and Ala., n. to s. Del.

23. G. calycòsa (Mackenz. & Bush) Fern. (with prominent calyx). — *Differing from no.* 22 *in its slender and more openly branched stem only* 0.8–1.6 *m. high; segments of leaves linear or narrowly linear-lanceolate and only* 1–3 *mm. wide; bracteal leaves linear; calyx-lobes linear,* 5–10 *mm. long; corolla glabrous within above bases of filaments; lower filaments* 2–2.4 *cm. long; anther-locules* 4–5.5 *mm. long, the awns* 1.5–2 *mm. long; seeds* 1.8–2 *mm. long.* (*Dasystoma* Mackenz. & Bush; *Aureolaria* Pennell) — Dry wooded slopes and barrens, Mo. and Ark. July, Aug.

24. G. pediculària L. (resembling *Pedicularis*). — *Annual with* much branched *stem* 0.4–1.2 m. high, *glandular-villous below, puberulent and glandless above; leaves thin,* short-petioled to subsessile, lanceolate or lance-ovate, *bipinnatifid, finely pubescent; pedicels slender, stipitate-glandular,* up to 2.5 cm. long; *calyx glandular-puberulent, its tube turbinate, its pinnate lobes nearly* 1 *cm. long; corolla yellow, commonly purple-tinged,* 3–4 cm. long, glandular-pubescent outside; upper filaments 1.2, lower 1.7 cm. long; anther-locules 3–3.4 mm. long, their awns 0.8 mm. long; style 2.5–3 cm. long; *capsule slenderly ellipsoid-subcylindric, stipitate-glandular,* 9–11 mm. long, *its lower half covered by the calyx-tube; seeds wingless,* compressed, 0.8 *mm. long.* (*Aureolaria* Raf.) — Dry deciduous woods and clearings, sw. Me. to s. Minn. (locally), s. to s. N.E., L.I., N.C., W.Va., n. Ind. and n. Ill. Aug., Sept.

Var. **intercèdens** (Pennell) Fern. (standing between). — *Upper half of stem with glands scattered* among the hirsute pubescence; *capsules thick-ellipsoid,* about 1.2 cm. long. — Mass. to Minn., s. to s. N.J., upland to N.C., n. Ind. and n. Ill.

Var. **ámbigens** Fern. (ambiguous). — *Upper half of stem densely glandular and hirsute; leaves less cut; calyx-lobes lanceolate,* 5–10 *mm. long; corolla* 2.5–3.5 *cm. long.* — Sandy woods, thickets and dunes, nw. O. to se. Minn.

Var. **austromontàna** (Pennell) Fern. (of the southern mountains). — Differing from var. *ambigens* in having more deeply and sharply cut leaves; *narrower* (often *linear*) *and more deeply cut calyx-lobes* 0.8–1.6 *cm. long; corolla* 3–4 *cm. long.* — Mts. of sw. Va. and se. Ky., s. to nw. Ga.

25. G. pectinàta (Nutt.) Benth. (comb-like). — *Differing from no.* 24 in its low stems (3–6 dm. high); more sharply cut leaves; *pedicels* 3–10 *mm. long; calyx-tube hemispherical, hirsute to lanate; capsule ovoid, only its base embraced by the calyx-tube.* (*Aureolaria* Pennell and var. *ozarkensis* Pennell) — Dry woods and openings, se. Ga. and nw. Fla. to La., n. to se. N.C., s. Ky. and Mo. Aug., Sept.

25. BÙCHNERA L. Blue-hearts

Calyx obscurely nerved. Corolla with a straight or curved tube and an almost equally 5-cleft limb, the lobes oblong or cuneate-obovate, flat. Stamens included; anthers one-locular (the other locule wanting). Style clavate and entire. Capsule 2-valved, many-seeded. — Peren-

nial rough-hairy herbs (doubtless root-parasitic) of S. Hemisph., se. N. Am. and se. Asia, turning blackish in drying, with opposite leaves, or the uppermost alternate; the flowers opposite in a terminal spike, bracted and with 2 bractlets. (Named in honor of *Johann Gottfried Büchner*, 1695–1749, German botanist.)

1. B. americàna L. (American). — Rough-hairy; stem virgate, 3–8 dm. high; lower leaves obovate-oblong, the others ovate-oblong to linear-lanceolate, sparingly and coarsely toothed, veiny; spike interrupted; calyx longer than the bracts, one-third the length of the deep purple corolla (2 cm. long). — Moist sandy soil, open woods and prairies, w. Fla. to Tex., n. to N.J., w. N.Y., s. Ont., s. Mich., Ill., Mo. and se. Kans., mostly in the Appalachian and Ozarkian uplands. June–Aug. — Fresh foliage pale green, this and the flowers blackening in drying.

Tribe VII. Euphrasìeae Benth.

26. CASTILLÈJA Mutis Painted-cup

Divisions of the calyx entire or 2-lobed. Tube of the corolla included in the calyx; its upper lip (*galea*) keeled, flattened laterally. Anther-locules oblong-linear, the outer fixed by the middle, the inner pendulous. Capsule many-seeded. — Am. and ne. Asiatic herbs (somewhat root-parasitic), with alternate entire or cut-lobed leaves; the bracteal ones usually dilated, colored, and more showy than the yellow or reddish spiked flowers. (Dedicated in 1781 to *Domingo Castillejo*, a Spanish botanist.)

Annual or biennial, the usually solitary flowering stem arising from a spreading rosette of simple leaves; cauline leaves mostly cleft; calyx about equally 2-cleft into 2 broad rounded lobes; bracts bright red, yellow or white. . . . . . 1. *C. coccinea.*
Perennials with clustered flowering stems; calyx with narrowed lobes.
  Floral bracts green; corolla 4–5 cm. long, long-exserted. . . . . 2. *C. sessiliflora.*
  Floral bracts colored or white; corolla 1.5–3.5 cm. long, covered by the bracts or slightly exserted.
    Cinereous-pubescent; leaves narrowly linear, entire or deeply cleft; summits of bracts purple or violet; corolla 2.5–3.5 cm. long. . . . . 3. *C. purpurea.*
    Smoothish; leaves lanceolate, entire; floral bracts whitish, creamy or suffused with bronze or dull purple; corolla 1.5–2.5 cm. long. . . . . . 4. *C. septentrionalis.*

1. C. coccínea (L.) Spreng. (scarlet), Scarlet P. — Hairy *biennial or annual;* stem simple; rosette-leaves clustered, mostly entire, obovate or oblong; those of the stem incised; the floral bracts 3–5-cleft, bright scarlet toward the summit, or yellow in forma **lutéscens** Farw. (yellowish), or white in forma **álba** Farw. (white); *calyx* about the length of the pale yellow corolla, *equally cleft both sides; the lobes quadrate-oblong, entire or retuse.* — Peaty meadows, prairies and damp sands and gravels, s. N.H. to s. Man., s. to n. Fla., n. Miss., La. and Okla. April–Aug.

2. C. sessiliflòra Pursh (sessile-flowered), Downy P. — Perennial, 1.5–3 dm. high, very leafy, cinereous-pubescent; leaves mostly 3–5-cleft, with narrow diverging sometimes cleft lobes; *the floral bracts* similar or broader, *not at all colored; calyx green, deeper-cleft in front,* the narrow lobes deeply 2-cleft; corolla 4–5 cm. long, yellowish-white, *the short* green-backed *galea but twice as long as the slender-lobed lip.* — Dry plains, prairies and hills, se. Wisc. and nw. Ill. to Man. and Mont., s. to nw. Mo., Kans., Okla., Tex., N.M. and Ariz. Late April–July.

3. C. purpùrea (Nutt.) G. Don (purple). — *Cinereous-pubescent;* stems clustered, 1–4 dm. high; *leaves narrowly linear and entire or deeply cleft; bracts* similar or with cuneate-dilated base, *the summits and calyx purple or violet;* calyx about equally 2-cleft into narrow acute lobes; *corolla* 2.5–3.5 *cm. long,* somewhat exserted; galea much shorter than tube, only twice length of the elongate-lobed lip. — Dry prairies and calcareous bluffs, sw. Mo. to Tex. April, May.

4. C. septentrionàlis Lindl. (northern). — Perennial, smooth or sparingly hairy, at the summit woolly; *leaves mainly entire,* the lower linear, upper broader; the floral oblong or obovate, greenish-white, varying to yellowish, purple, or red; *calyx equally cleft, the lobes oblong or lanceolate, 2-cleft;* corolla 1.5–2.5 cm. long, the *galea decidedly shorter than the tube,* not more than twice or thrice as long as the lip. (*C. pallida,* var. Gray; *C. acuminata* of e. Am. auth., not Spreng.) — Rocky and gravelly (often calcareous) or peaty soils, Lab. to Alta., s. to Nfld., e. Que., N.B., n. Me., subalpine N.E., n. Mich., n. Minn., S.D., Colo. and Utah. June–Aug.

27. ORTHOCÁRPUS Nutt.

Corolla with the upper lip (*galea*) little longer and usually much narrower than the inflated 1–3-saccate lower one. Otherwise nearly as *Castilleja.* Small genus of N. and S. Am. (Name from the Greek *orthos, upright,* and *carpos, fruit.*)

1. O. lùteus Nutt. (yellow). — Annual, pubescent and hirsute, sometimes viscid, erect, 1.5–4 dm. high; leaves linear to lanceolate, occasionally 3-cleft; spike dense; bracts broader, mostly 3-cleft, about equaling the flowers, not colored; corolla golden-yellow, 1 cm. long, twice or thrice length of calyx. — Sandy soil, Man. to B.C., s. to Minn., Neb., N.M., Ariz. and Calif. June–Aug.

28. MELAMPỲRUM L. Cow-wheat

Calyx campanulate, sharply cleft. Tube of corolla cylindrical, enlarging above; upper lip compressed, straight in front; lower erect-spreading, biconvex, 3-lobed at apex. Anthers approximate, oblong, nearly vertical, hairy; the locules minutely pointed at base. Capsule 1–4-seeded. — Erect branching annuals of N. Hemisph., with opposite leaves, the lower entire, the upper mostly toothed at base. Flowers solitary in the upper axils. (Name from the Greek *melas*, *black*, and *pyros*, *wheat;* from the color of the seeds of some species.)

1. M. lineàre Desr. (linear). — Simple to densely bushy-branched, 0.5–5 dm. high; leaves linear to elliptic-ovate, the lower entire, the bracteal entire or more or less toothed at base; calyx-teeth not half the length of the slender tube of the white or pale stramineous to purplish yellow-tipped corolla (5–13 mm. long); seeds white, ripening blackish with white to brown tips, subcylindric, 2–4 mm. long. — Highly plastic, somewhat arbitrarily divided into 4 confluent vars.:

a. Principal leaves of primary axis linear to narrowly lanceolate, 1–10 mm. broad; mature internode separating the 2 lowest fruiting nodes of the primary axis 0.5–3 cm. long; lowest bracteal leaf of primary axis 1–4(–6) cm. long, 1–10(–20) mm. broad. *b.*

b. Stem simple or loosely few-branched, 0.5–2 dm. high, the simple branches only 1–10 cm. long; foliage-leaves and bracts linear, 1–4(–6) mm. broad, all entire or the uppermost bracts rarely toothed at base; mature capsule 3–5 mm. broad. *M. lineare* (typical).

b. Stem bushy-branched (exceptionally unbranched), (1–)2–5 dm. high, the branches in well-developed plants 0.2–2.5 dm. long; foliage-leaves linear to lanceolate, 2–10 mm. wide; larger bracts linear-lanceolate to lance-ovate, 3–20 mm. broad, some or all of them sharply toothed at base; mature capsule 3.5–6 mm. broad.

Branches mostly simple or with few short or flexuous branchlets; foliage-leaves 2–10 mm. broad; bracts (excluding teeth) up to 20 mm. broad, the lower ones 2.5–6 cm. long; basal teeth of the middle and upper bracts shorter than breadth of bracts. . . . . Var. *americanum.*

Branches mostly stiffly forking, the plant thus intricately branched and very leafy with linear leaves; foliage-leaves 2–6 mm. broad; bracts (excluding teeth) 1–7 mm. broad, the lowermost 1–3.5 cm. long; basal teeth of the middle and upper bracts about as long as breadth of bracts. Var. *pectinatum.*

a. Principal leaves of primary axis 0.5–3 cm. broad, lanceolate to narrowly ovate; mature internode separating 2 lowest nodes of primary axis 3–4.5 cm. long; lower foliaceous bracts of primary axis broadly lanceolate to ovate, 3–7 cm. long, 1–3 cm. broad, the bracts all toothless or the middle and upper ones with relatively short basal teeth; branches of plant few, simple or only loosely few-forked. . . . . Var. *latifolium.*

M. lineàre (typical). — Bogs, heaths and damp peaty or rocky barrens, Nfld. and s. Lab. Pen. to n. Alta. and B.C., s. to N.S., e. and n. Me., alpine areas of Me., N.H. and Vt., n. Mich., n. Wisc., nw. Mont., n. Ida. and Vancouver I. July, Aug.

Var. **americànum** (Michx.) Beauverd (American). — Dry woods, Anticosti I. to L. Mistassini, Que., and Minn., s. to N.S., N.E., L.I., Md. and mts. of S.C. and Tenn.; nw. Mont. and n. Ida. June–Aug.

Var. **pectinàtum** (Pennell) Fern. (comb-like). — Dry sandy pineland and oak-scrub, Mass. and e. N.Y. to Va.; n. Ind. Late June–Sept.

Var. **latifòlium** Bart. (broad-leaved). — Dry to moist woods, sw. Que. and n. N.Y. to s. N.E., L.I., Va., and in the uplands to Ga. May–July.

29. EUPHRÀSIA L. Eyebright

Calyx tubular or campanulate, 4-cleft. Upper lip of the corolla erect, scarcely arched, 2-lobed, with the sides folded back; lower lip spreading, 3-cleft, the lobes obtuse or notched. Anther-locules pointed at the base. Capsule flattened. — Herbs of cool and temp. reg., with opposite toothed or cut leaves. Flowers small, spiked. (Name from the Greek *euphrasia, cheerfulness,* in allusion to its anciently reputed virtue in clarifying the eyesight.)*

a. Upper lip of purple or white corolla very shallowly bilobed; the lobes very short, rounded, entire, narrowly revolute or rarely erect; lower lip scarcely flabellate and scarcely exceeding upper lip; corollas tiny, 2–4 (very rarely –4.8) mm. long. . . *b.*

b. Inflorescence capitate or subcapitate, the flowers and bracts closely crowded, only 1–rarely 4 lowest pairs of bracts slightly remote in fruit.

Corolla white, with violet lines; nodes below the head 2–5; leaves crisp-pubescent, much smaller than bracts. . . . . 1. *E. Oakesii.*

Corolla chocolate-color to purple; nodes below the head 4–9; leaves glabrous or pubescent, only gradually smaller than bracts. . . . . 2. *E. Williamsii.*

b. Inflorescence becoming loose and elongate, flowers or capsules scattered in maturity, with few–many pairs of remote bracts; corolla purple or white. 3. *E. Randii.*

a. Upper lip of white or whitish corolla prominently bilobed; the lobes reflexed from the base, truncate, undulate or coarsely 3-toothed; lower lip often flabellate, spreading, much larger than the upper; corollas 4–10 mm. long; spikes usually elongating and loosening in maturity. . . . *c.*

c. Bracts of inflorescence with obtusish to acute but not bristle-tipped teeth.

Internodes long (mostly 3 cm.); spikes becoming open and lax, with lower spreading bracts in maturity 2–5 cm. apart; corolla 4–5.5(–6) mm. long. 4. *E. disjuncta.*

Internodes short (mostly 8–15 mm. long); spikes dense, uninterrupted above, with ascending imbricated bracts, the lowest in maturity 0.2–2 cm. apart; corollas 5–8 mm. long. . . . . 5. *E. arctica.*

c. Bracts with subulate or bristle-tipped teeth. . . . *d.*

d. Spikes in maturity comprising the larger portion of the plant, the lower fruits near the bases of stems and branches. . . . *e.*

e. Corolla 5–6.5 mm. long, with lavender or bluish lines; the lower lip with lateral lobes not strongly spreading.

Bracts orbicular or broadly oval, glabrous; branches arcuate-ascending; northeastern. . . . . 6. *E. canadensis.*

Bracts oblong, pubescent; branches strongly ascending; northwestern. . . . . 7. *E. hudsoniana.*

e. Corolla 6–10 mm. long, with dark purple lines; the lower lip with wide-spreading lateral lobes.

Bracts glabrous, all appressed, cuneate at base. . . . . 8. *E. rigidula.*

Bracts strongly pubescent, the lower spreading, rounded at base. . 9. *E. tatarica.*

d. Spikes in maturity occupying only the upper half or third of the stems and branches; corolla 7–10 mm. long, with dark purple lines; lower lip with wide-spreading lateral lobes. . . . . 10. *E. americana.*

1. E. Oàkesii Wettst. (for its discoverer, WILLIAM OAKES, 1799–1848). — Low, 2–12 cm. high, usually unbranched, rarely with ascending branches from the middle axils; stem crisp-pubescent; *leaves 2–5 pairs, much smaller than the bracts,* 1.5–7 mm. long, rounded, *crisp-pubescent* on both faces; *bracts* similar but larger, *with rounded teeth; inflorescence* at first capitate, *in maturity globose to ellipsoid,* 0.5–2.5 *cm. long,* at most with 1 or 2 lowest pairs of bracts becoming slightly remote; *corolla* 2–3.3 *mm. long, whitish, with violet lines; upper lip shallowly bidentate, with narrow revolute entire margins; lower lip subequal, with oblong bidentulate lobes;* eye yellow; capsules exceeding blunt calyx-lobes. — Exposed turfy or gravelly slopes or crests, s. Lab. and n. Nfld.; sea-cliffs, Gaspé Co., Que.; alpine areas, Shickshock Mts., Que., Mt. Katahdin, Me., and White Mts., N.H. July, Aug.

2. E. Williámsii Robins. (for its discoverer, EMILE FRANCIS WILLIAMS, 1858–1929). — Similar to no. 1, simple or sparingly branched, 1.5–12 cm. high; *leaves 4–7 pairs, glabrous* to sparsely pubescent on both surfaces; *mature inflorescence cylindric,* 0.7–5 *cm. long,* usually *with only the* 1–4 *lowest pairs of bracts a little separated; bracts* similar to and *only gradually larger than leaves,* 5–10 mm. long, with obtuse teeth; *corolla* 2.5–4 *mm. long, chocolate-brown or maroon,* in lobing similar to that of no. 1. — Gravelly or turfy crests and slopes, n. Nfld.; alpine areas, Shickshock Mts., Que. and White Mts., N.H. Late July–early Sept.

Var. **vestìta** Fern. & Wieg. (clothed). — *Leaves* 5–9 *pairs, crisp-pubescent; bracts* 4–7 mm. long, *with acutish teeth;* corolla chocolate-brown to lilac-purple. — Exposed crests and summits, n. and w. Nfld. and adj. Que.

3. E. Rándii Robins. (for its discoverer, EDWARD LOTHROP RAND, 1859–1924). — Simple to

* Treatment based on that of FERNALD & WIEGAND in Rhodora, xvii. 181–201 (1915).

divergently much branched, 0.1–4 dm. high; leaves ovate-oblong to nearly orbicular, 0.2–2 cm. long, with rounded to subacute teeth; bracts similar; *spikes becoming elongate, with the pairs of bracts separated, the lower scattered, the mature central spike finally one-half to three-fourths full height of plant; corollas* 2.5–4 mm. long, from *deep purple to white,* usually with yellow throat; upper lip shallowly notched, with revolute or rarely erect entire or bidentulate lobes; lower lip scarcely flabellate, with ascending truncate or bidentulate linear or oblong lobes; capsule barely equaling the lance-attenuate calyx-lobes. — Our most variable species, of which the following varieties and forms are recognized:

Leaves and bracts glabrous or only sparingly crisp-pubescent, those of the main axis 0.5–2 cm. long; stems 0.1–4 dm. high.
  Leaves and bracts sparingly crisp-pubescent on both surfaces.
    Corollas purple. . . . . *E. Randii* (typical).
    Corollas white. . . . . Forma *albiflora.*
  Leaves and bracts glabrous on both surfaces.
    Corollas deep to pale purple, with darker lines. . . . . Var. *Reeksii.*
    Corollas with white lobes. . . . . Forma *candida.*
Leaves and bracts densely white-pubescent, mostly smaller, 2–7 mm. long, the flowers thus more exserted; stems 1–13 cm. high, simple or bushily divergent and branched.
  Corollas whitish. . . . . Var. *Farlowii.*
  Corollas purple. . . . . Forma *iodantha.*

E. Rándii (typical). — Turfy slopes and knolls, peaty crests and brackish shores, coasts of Lab. and Nfld. to Gaspé Co., Que., s. to Knox Co., Me. — Forma **albiflòra** (Fern. & Wieg.) Fern. (white-flowered) less frequent.

Var **Reèksii** Fern. (for its discoverer, Henry Reeks, 1838–1882). — Grassy or peaty banks and brackish shores, coasts of Nfld., e. Que., M.I., P.E.I., N.S. and e. N.B. — Forma **cándida** (Fern. & Wieg.) Fern. (white), less common, of similar range. July–Sept.

Var. **Farlòwii** Robins. (for its discoverer, William Gilson Farlow, 1844–1919). — Exposed crests or dry turfy or gravelly coasts, Nfld. and adj. Que. to Knox Co., Me. July–Sept. — Forma **iodántha** (Fern. & Wieg.) Fern. (purple-flowered), less frequent, Nfld. and Me.

4. E. disjúncta Fern. & Wieg. (separated). — *Slender,* 0.6–4 dm. high, *simple or with loosely ascending branches* from below the middle; *leaves* up to 16 pairs, ovate or orbicular, 2–18 mm. long, *coarsely crenate, sparingly pubescent,* somewhat remote, *the internodes mostly* 3 (rarely–7) *cm. long* (shorter in dwarfed plants); *spikes interrupted, becoming very elongate,* the central one in maturity 0.5–3 dm. long; *bracts* similar, *scarcely smaller* than leaves, *remote, spreading, coarsely acute-toothed, the lower pairs becoming* 2–5 *cm. apart; corolla* 4–5.5(–6) *mm. long,* with yellow eye; *upper lip* lavender-tinged, *slightly 2-lobed, the lobes* subtruncate, *undulate or denticulate, semireflexed; lower lip white,* with purple lines, *spreading* but scarcely flabellate, *the oblong lobes notched.* — Damp open soil, Lab. Pen. and Nfld. to n. Me.; Alta., Mackenz. and Alaska. July–Sept.

5. E. árctica Lange (arctic). — Similar to no. 4, stouter, 0.05–2(rarely–4.5) dm. high, simple or with few ascending branches; leaves 5–18 mm. long, more or less pubescent; *internodes mostly* 8–15 *mm. long; spikes dense and uninterrupted except at base,* 0.3–2(–4) dm. long; *bracts ascending and imbricated,* acute-toothed, *the lowest finally* 0.2–2 (rarely –4) *cm. apart; corollas* white, with lavender lines and often bluish upper lip, 5–8 *mm. long; lobes of lower lip nearly parallel.* — Greenl. and Lab. to Keewatin, s. to Nfld., Gaspé Pen., Que., n. Mich. and n. Minn. July–Sept. (Eu.)

6. E. canadénsis Townsend (Canadian). — Simple or bushy-branched, 0.5–2.5 dm. high; leaves glabrous, those of the main axis 0.5–1 cm. long, with acute or obtuse teeth; *spikes* becoming very elongate, *the central one finally almost the full height of the plant,* dense above, open below; *bracts spreading,* rarely imbricated, *with bristle-tipped teeth,* the lowest pairs in maturity 0.3–2.5 cm. apart; *corolla* 5–6.5 *mm. long,* white, *with lavender or bluish veins,* the violet-tinged upper lip 2-lobed; the lobes half-reflexed, truncate and shallowly toothed; *lower lip only slightly fan-shaped.* — Open sterile fields and roadsides, Que. to N.S., Me., n. N.H. and w. Mass. July–Sept.

7. E. hudsoniàna Fern. & Wieg. (of Hudson Bay). — Differing from no. 6 in its *strongly ascending branches and long lower internodes* (1.5–4 *cm. long*); *leaves pubescent, oblong,* up to 1.5 cm. long, with few very coarse acute teeth; *bracts* of the inflorescence *oblong, ascending, pubescent.* — Open ground, shores of Hudson Bay, Ung., to Alta., s. to L. Sup., ne. Minn. July–Sept.

8. E. rigídula Jord. (stiffish). — Similar to no. 6; branches ascending; *bracts all ascending,* *the upper imbricated,* glabrous, the *lowest rarely more than 1 cm. apart, cuneate at base, the teeth* very sharp and *bristle-tipped;* calyx glabrous; *corolla 6–8 mm. long, with deep violet lines;* upper lip more purplish; *lower lip flabellate, with widespreading lateral lobes; capsules much overtopped by* the sharp *calyx-lobes.* (*E. stricta* Host, not HBK.) — Dry fields and sterile grasslands, Nfld. and Que., s. to N.S., n. N.E., ne. N.Y., and ne. Pa. Late June–Oct. (Natzd. from Eu.)

9. E. tatárica Fisch. (of Tartary). — Similar to no. 8; *leaves and bracts strongly setulose; bracts* broadly ovate, *the lower spreading and rounded at base, the teeth subulate-tipped; calyx densely pubescent; corolla* 7–10 *mm. long; capsules about equaling calyx.* — Thin calcareous soil, e. Que. July–Sept. (Eurasia) — Status in our flora not wholly clear.

10. E. americàna Wettst. (American). — Simple or freely branching, 1–4.5 dm. high, the branches long and arcuate-ascending; *leaves* ovate-oblong, *glabrous,* those of the main axis 0.5–2 cm. long, with coarse acute to obtuse teeth; *spikes* rather short, *in maturity* lax and slender, 3–15 cm. long and *occupying only the upper halves of stems and branches; bracts glabrous,* conspicuous, broadly ovate, *spreading, coarsely aristate-toothed,* the lowest pairs becoming 0.5–2 cm. apart; *corolla* 7–10 *mm. long; upper lip* purple-tinged, shallowly bilobed, *the lobes reflexed and each 2–3-toothed; lower lip large,* with dark purple lines, flabellate, *with wide-spreading lateral lobes.* — Fields and roadsides, common, Nfld. to centr. and coastal Me. Late June–Sept.

30. ODONTÌTES Ludwig

Calyx equally cleft. Corolla with upper lip entire and sides not folded back. Seeds pendulous or recurving. Otherwise much as *Euphrasia.* Herbs, nat. of Eurasia, with opposite sessile leaves, and subsessile flowers in the upper axils and in a terminal leafy spike. (*Odontitis,* an ancient plant-name from the Greek *odous, odontos, tooth,* applied to some herb used for toothache.)

1. O. seròtina (Lam.) Dum. (late), Red Bartsia. — Stem 1–4 dm. high, from an annual root, branching, scabrous-pubescent; leaves oblong-lanceolate, coarsely and remotely serrate; spikes elongated, loosely-flowered; corolla small, rose-red. (*O. rubra* Gilib.) — Roadsides and fields, Nfld. to w. Que., s. to N.S., n. N.E. and n. N.Y. June–Aug. (Natzd. from Eu.)

31. BÁRTSIA L. Bartsia

Calyx tubular or campanulate, 4-cleft. Corolla with relatively long tube and 2-lipped limb; upper lip entire or emarginate, erect, concave; lower lip with 3 entire suberect lobes. Stamens 4, didynamous, under upper lip; anther-locules mucronate. Capsule ovoid, pointed; the large seeds with dorsal ribs winged, the hilum lateral. — Small Eu. and n. Afr. genus, with the following circumpolar arctic-alpine species. (Named for *John Bartsch,* 1709–38, a botanist sent by Boerhaave to Surinam where he died.)

1. B. alpìna L. (alpine), Velvet-bells, Alpine Bartsia. — Stems erect, from a subligneous rhizome, 0.5–2.5 dm. high, viscid-villous; leaves sessile, opposite, subcoriaceous, ovate, subamplexicaul, dentate, the principal ones 1–2.3 cm. long; raceme dense, spiciform; calyx purple; corolla deep purple, with metallic tone, glandular, 1.2–1.7 cm. long; anthers white-hairy, slightly exserted and strongly contrasted with color of corolla. — Greenl. and e. Arct. Am. s. to peaty limestone-barrens of n. Nfld., and to n. Ont. July. — Dried plant blackening. (N. Eurasia)

32. RHINÁNTHUS L. Yellow-rattle

Calyx membranaceous, flattened, much inflated in fruit, 4-toothed. Upper lip of corolla arched, ovate, obtuse, flattened, entire at the summit, but with a tooth or nipple on each side below the apex; lower lip 3-lobed. Anthers approximate, hairy, transverse; locules pointless. Capsule orbicular, flattened. Seeds suborbicular, winged. — Annual upright herbs of cool parts of N. Hemisph., with opposite leaves; the yellow, yellowish or bronze flowers crowded in a one-sided leafy-bracted spiciform raceme. (Name composed of the Greek *rhis, snout,* and *anthos, flower,* originally given to plants now excluded from the present genus.) Alectorolophus Moench

Bracts of the raceme with at least the lower teeth lance-attenuate and ending in slender bristle-tips.
  Nipples of upper lip of corolla broad and low, less than 1 mm. long.
    Stem and branches green; corolla yellow without dark base, the nipples of upper lip whitish or pale yellow. . . . . 1. *R. Crista-galli.*
    Stem and branches black-lineolate; corolla yellow, in age brownish, with

dark spot at base of each lateral lobe of lower lip, the nipples violet or blue. . . . . 1. *R. Crista-galli*, var. *fallax*.
Nipples of upper lip elongate, 1.5–2 mm. long. . . . . 2. *R. major*.
Bracts of raceme with the teeth deltoid, blunt or merely acute, without bristle-tips.
Stem loosely branched, black-lineolate above, bronze-tinged; leaves lance-attenuate, serrate-dentate; bracts and fruiting calyx purplish; bracts with acute teeth; lower lip of corolla brown-mottled; nipples bluish. . . . . 3. *R. stenophyllus*.
Stem simple or short-branched above, green; leaves oblong, crenate-dentate; bracts and fruiting calyx green to stramineous; bracts with broad bluntish teeth; corolla yellow throughout or with whitish nipples. . . . . . . 4. *R. borealis*.

1. R. Crísta-gálli L. (old generic name, meaning cock's comb, from the deeply toothed bracts). — *Stems* 0.1–1 m. high, simple or becoming loosely branched, *greenish; leaves lance-attenuate* or narrowly oblong-lanceolate, serrate- to crenate-dentate, *often with suppressed branches in upper axils; bracts of raceme with lower lance-attenuate and bristle-tipped teeth* up to 5 mm. long, the upper teeth shorter; calyx pale green to yellowish; *corolla yellow*, about 1.5 cm. long, *the whitish or pale nipples* of the upper lip *broad and low, less than* 1 *mm. long;* fruiting calyx round-ovate. (Incl. *R. Kyrollae* Chabert) — Fields, thickets and openings, Nfld. to reg. of Hudson Bay and of L. Superior, thence to Alta., s. to N.S., n. N.E., and in coastwise distr. to s. Me.; indig. northw., natzd. southw. Late May–Sept. (Eurasia) — Plant blackening in drying.

Var. **fállax** (Wimm. & Grab.) Druce (false). — *Stem and branches marked with black lines; corolla often fading to brownish; nipples bluish to violet; lateral lobes of lower lip with dark basal spot.* (*R. Crista-galli* of ed. 7) — Nfld. and Que., s. to N.S. and s. N.E.

2. R. májor Ehrh. (larger). — *Differing from var. of no.* 1 *in the larger corolla* (2 *cm. long*) *with elongate nipples* 1.5–2 *mm. long.* — Damp fields, Plymouth Co., Mass., perhaps now extinct (coll. over a period of half a century but not recently). (Adv. from Eu.)

3. R. stenophýllus (Schur) Schinz & Thell. (narrow-leaved). — *Plant with strong bronze or purple coloration, remotely and loosely long-branched; stem black-lineolate; leaves lance-attenuate, serrate-dentate* with appressed teeth; raceme remotely flowered; *bracts purple-tinged, with the lower pairs remote,* like the foliage-leaves, *acutely serrate-dentate;* upper bracts subapproximate, more triangular and shorter; *calyx purple-tinged* in fruit, 1.5 cm. long; *corolla* 1.5 *cm. long, canary-yellow, becoming brownish, the nipples blue-gray, the lower lip with brown markings.* — Boggy meadows and shores, e. Gaspé Co., Que. July–Sept. (Eu.) — Plant blackening in drying.

4. R. boreàlis (Sterneck) Chabert (northern). — *Stem simple or sparsely short-branched, green,* 0.3–5 dm. high; *leaves oblong* to oblong-linear or -lanceolate, *crenate-dentate; bracts* scabrous, mostly shorter than mature calyces, deltoid-ovate, green, *with deltoid blunt to acutish teeth; calyx* usually ciliate, *green to stramineous, in fruit becoming* 1.3–2 *cm. long; corolla yellow* (or with paler nipples), 1–1.2 *cm. long.* (*R. groenlandicus* Chabert; *R. oblongifolius* Fern.) — Greenl. and arct. Am., s. to meadows, slopes, shores, etc. of Nfld., C.B., e. Que. and n. N.Y., and in alpine and subalpine areas of Que., n. N.E. and n. N.Y. July, Aug.

33. PEDICULÀRIS L. Lousewort or Wood-betony

Calyx various. Corolla strongly 2-lipped; the upper lip flattened, often beaked at the apex; the lower erect at base, 2-crested above, 3-lobed; lobes commonly spreading, the lateral ones rounded and larger. Anthers transverse; the locules pointless. Capsule mostly oblique, several-seeded. — Perennial (sometimes annual or biennial) herbs of N. Hemisp., with chiefly pinnatifid leaves (the floral bract-like) and rather large flowers in a raceme or spike. (Name from *pediculus, a louse,* because of the early European belief that cattle, feeding where *P. palustris* abounded, became covered with lice.)

a. Flowers axillary and terminal, rose-pink to purple; annuals or biennials, mostly low-branching.
Stem erect, 1.5–8 dm. high, with stiffly ascending branches; central raceme less than half the height of the plant; calyx with 2 crested-toothed lobes, much shorter than capsule; corolla rose-purple with violet galea. . . . 1. *P. palustris*.
Stem simple or diffusely branched, 0.3–2 dm. high; central raceme often one-half to nearly the full height of plant; calyx subequally 4–5-toothed, the foliaceous teeth equaling or exceeding capsule; corolla flesh-pink. . . . 2. *P. sylvatica*.
a. Flowers in terminal spikes or racemes, creamy or straw-color to reddish-purple or bronze; perennials with stem simple or few-branched. . . *b.*
b. Dwarf arctic-alpine species 0.2–2 dm. high; flowering raceme 1–2 cm. thick;

larger leaves 0.5–1 cm. broad, the blade 1.5–4 cm. long; fruiting calyces long-pedicelled. 3. *P. flammea.*

*b.*Coarser species of Canadian or continental temperate range, 1–9 dm. high; flowering spikes 3–5 cm. thick; larger leaves 1–5 cm. broad, the blade 0.5–2.5 dm. long; fruiting calyx short-pedicelled to subsessile. . . *c.*

*c.*Cauline leaves opposite or subopposite, sessile; summit of stem and rachis of raceme glabrous or nearly so. . . . . 4. *P. lanceolata.*

*c.*Cauline leaves scattered and alternate, the lower long-petioled; summit of stem and rachis of raceme densely villous.

Plants 1.5–4 dm. high; the flowering stems often clustered, simple; calyx split in front, otherwise entire; capsule lance-oblong, flattened, twice as long as calyx. . . . . 5. *P. canadensis.*

Plants 4–9 dm. high; the flowering stems 1 or 2, simple or with arched-ascending branches; calyx 5-lobed; capsule ovoid, plump, barely exceeding calyx. . . . . 6. *P. Furbishiae.*

1. P. palústris L. (of marshes), SWAMP-L. — *Annual or biennial, erect,* 1.5–8 *dm. high,* usually *with stiffly ascending flowering branches;* leaves lanceolate, 2–5 cm. long, pinnately parted, with small crenate oblong segments; *central raceme less than half the height of the plant; calyx* pilose or glabrous, *with 2 crested-toothed lobes, much shorter than the capsule; corolla rose-purple, with violet galea,* or whitish in forma **ochroleùca** Laestad. (creamy), 1.8–2.5 cm. long; seeds elliptic or oval, 2.5–3 mm. long. — Marshes and wet meadows, se. Nfld., M.I., e. Que. and N.S. June–Aug. (Eu.)

2. P. sylvática L. (of woodland), SMALL L. — Much smaller than no. 1; often with a basal rosette of fleshy simple pinnatifid leaves; *stem simple or diffusely branched,* 0.3–2 *dm. high, the central raceme often one-half to nearly the entire height of the plant; calyx subequally* 4–5-*toothed, the foliaceous teeth equaling or exceeding capsule; corolla flesh-pink,* 2–3 cm. long; seeds 1.5–1.8 mm. long. — Mossy or turfy depressions, slopes and woodland-glades, se. Nfld. July, Aug. (Eu.)

3. P. flámmea L. (flame-colored). — Roots fusiform; *stem* simple, glabrous, 0.2–2 *dm. high; leaves* petioled, several at the base, 1–few cauline; the closely and deeply pinnate *blades* 1.5–4 *cm. long,* 0.5–1 *cm. broad; raceme slender,* few-flowered, 1–2 *cm. thick,* with short linear-lanceolate bracts; calyx-teeth lanceolate, much shorter than cylindric tube; *corolla very slender, erect,* citron-yellow, with purplish or bronze summit, *the galea slightly arching and much exceeding the small lip; fruiting calyx long-pedicelled;* capsule lance-falcate, twice or thrice length of calyx. — Arct. reg., s. in peaty or gravelly calcareous spots to n. Nfld. and Shickshock Mts., Que. June, July. (Eurasia)

4. P. lanceolàta Michx. (lanceolate). — Stem upright, 3–9 dm. high, nearly simple, mostly smooth; *leaves partly opposite, sessile, oblong-lanceolate, doubly cut-toothed;* spike crowded; *calyx 2-lobed,* leafy-crested; *upper lip of the* pale yellow *corolla incurved* and bearing a short truncate beak at the apex; the lower erect, so as nearly to close the throat; *capsule ovoid, scarcely longer than the calyx.* — Rich, often calcareous, meadows and shores, Mass. to s. Ont. and Man., s. to Va., w. N.C., Tenn., Mo. and Neb. Aug.–Oct.

5. P. canadénsis L. (Canadian), COMMON L., WOOD-BETONY. — Hairy; stems simple, closely clustered, 1.5–4 dm. high; *leaves scattered, the lowest pinnately parted, the others half-pinnatifid,* all or nearly all petioled; raceme dense and short in flower; *calyx split in front, otherwise almost entire,* oblique; corolla yellow or yellowish, or crimson on the back in forma **bìcolor** Farw. (two-colored) or crimson throughout in forma **praeclàra** A. H. Moore (very beautiful); *the upper lip hooded, incurved,* 2-toothed under the apex; *capsule lance-oblong, flattened, twice as long as calyx.* — Woods and clearings, centr. Me. and s. Que. to Man., s. to Fla., Miss., La., Tex. and n. Mex. Apr.–June.

Var. **Dóbbsii** Fern. (for its discoverer, RAYMOND JOSEPH DOBBS, 1883–). — Scarcely clustered, the flowering stems mostly solitary; the basal offsets prolonged, stoloniform, often repent. — N.Y. to Minn., s. to n. Fla., Ala., La. and e. Tex.

6. P. Furbíshiae S. Wats. (for its discoverer, CATHERINE (KATE) FURBISH, 1834–1931). — Tall (4–9 dm. high), pubescent or glabrate; *leaves petioled,* lanceolate, *pinnately parted and the short oblong divisions pinnatifid-incised,* or the upper simply pinnatifid and the lobes serrate, silvery-margined; bracts ovate, laciniate-dentate; *calyx 5-lobed,* the lobes rather unequal, linear-lanceolate, entire or toothed; *upper lip of corolla straight* and beakless, the truncate apex bicuspidate; the lower erect, truncately 3-lobed; capsule broadly ovoid, barely exceeding calyx. — Banks of St. John R., n. N.B. and n. Me. July, Aug.

34. SCHWÁLBEA L. Chaffseed

Calyx tubular, 10–12-ribbed, 5-toothed; the posterior tooth much the smallest, the 2 anterior united higher than the others. Upper lip of the corolla oblong, entire; the lower little shorter, erect, 2-plaited, with 3 very short and broad obtuse lobes. Anther-locules parallel. Capsule ovoid. Seeds linear, with a loose chaff-like coat. — A perennial minutely pubescent upright herb, with leafy simple stems terminated by a loose spike of rather large dull purplish-yellow flowers; leaves alternate, sessile, 3-nerved, entire, ovate or oblong, the upper gradually reduced to narrow bracts; pedicels very short, with 2 bractlets under the calyx. (Dedicated to *Christian Georg Schwalbe*, who wrote on botany in 1715.)

1. S. americàna L. (American). — Stems 3–8 dm. high; larger leaves 0.7–2 cm. broad. (Incl. *S. australis* Pennell) — Moist to dry pinelands, oak-woods or clearings, local, Fla. to La., n. to e. Mass., Ct., e. N.Y. and mts. of Ky. and Tenn. May–July. — Plant darkening when dried.

FAM. 154. BIGNONIÀCEAE (Bignonia Family)

Woody plants, gamopetalous, didynamous or diandrous, with the ovary commonly 2-locular by the meeting of the two parietal placentae or of a projection from them, many-ovuled; fruit a dry capsule, the large flat winged seeds with a flat embryo and no albumen, the broad and leaf-like cotyledons notched at both ends. Calyx 2-lipped, 5-cleft or entire. Corolla tubular or campanulate, 5-lobed, somewhat irregular or 2-lipped, deciduous; the lower lobe largest. Stamens inserted on the corolla; the fifth or posterior one, and sometimes the shorter pair also, sterile or rudimentary; anthers of 2 diverging locules. Ovary free, bearing a long style, with a 2-lipped stigma. Leaves compound or simple, opposite, rarely alternate. Flowers large and showy. — Chiefly a tropical family.

Leaves compound; high-climbing shrubs; calyx 5-toothed or truncate; capsule flattened.
Leaves without tendrils, with 7–11 leaflets; capsule flattened contrary to the partition. 1. *Campsis.*
Leaves tendril-bearing, with 2 (rarely 1) leaflets; capsule flattened parallel with the partition. 2. *Bignonia.*
Leaves simple, cordate or ovate; erect trees; calyx 2-lipped; capsule terete. . . 3. *Catalpa.*

1. CÁMPSIS Lour. Trumpet-flower

Calyx campanulate, 5-toothed. Corolla funnelform, 5-lobed, a little irregular. Stamens 4. Capsule 2-locular, with the partition at right angles to the convex valves. Seeds transversely winged. — Woody, with compound leaves, climbing by aërial rootlets, one species e. Asiatic, the other ours. (Name from the Greek *campsis, curvature*, from the curved stamens.) Tecoma of many auth., not Juss., as now restricted.

1. C. radìcans (L.) Seem. (rooting), Trumpet-creeper, Cow-itch. — Leaves pinnate; leaflets 9–11, ovate, pointed, toothed; flowers corymbed; stamens not protruded beyond the tubular-funnelform orange and scarlet corolla (6–8 cm. long); capsule oblanceolate, 1–2 dm. long. (*Bignonia* L.; *Tecoma* Juss.) — Low woods and thickets, Fla. to Tex., n. to N.J., e. Pa., W.Va., Ky., Ill. and se. Ia.; natzd. n. to Ct. and Mich.; handsome in cult. but southw. becoming an aggressive weed. July–Sept.

2. BIGNÒNIA L.

Calyx truncate or slightly 5-toothed. Corolla somewhat campanulate, 5-lobed and rather 2-lipped. Stamens 4, often with a rudiment of the fifth. Capsule linear, 2-locular. Seeds transversely winged. — Woody climber of e. N. Am. (Named for the Abbé *Jean-Paul Bignon*, 1662–1743, court-librarian at Paris and friend of Tournefort.)

1. B. capreolàta L. (twining), Cross-vine. — Smooth; leaves of 2 ovate or oblong leaflets and a branched tendril, often with a pair of accessory leaves in the axil resembling stipules; peduncles few and clustered, 1-flowered; corolla deep orange to reddish outside, paler orange and reddish within the throat; capsules 1.5–2 dm. long; seeds, with wing, 4 cm. long. (*Anisostichus* Bureau) — Rich woods and swamps, Fla. to La., n. to e. Md., W.Va., s. O., Ind., Ill. and Mo. Apr.–June. — Climbing tall trees and very handsome in flower; transverse section of the wood showing a cross.

3. CATÁLPA Scop. Catalpa. Catawba. Indian-bean

Calyx deeply 2-lipped. Corolla campanulate, swelling; the undulate 5-lobed spreading border irregular and 2-lipped. Fertile stamens 2, or sometimes 4; the 1 or 3 others sterile and rudimentary. Capsule very long and slender, nearly cylindrical, 2-locular, the partition at right angles to the valves. Seeds winged on each side, the wings cut into a fringe. — Trees of warm reg. of e. N.Am. and e. Asia, with ovate or cordate and mainly opposite leaves. (The aboriginal name.)

Leaves permanently soft-pubescent beneath; corollas 2–4 cm. broad, white, marked with yellow and purple-brown; capsules 0.8–1.5 cm. in diameter; seeds 2.5–5 mm. broad.
 Leaves long-tapering at apex; corolla about 4 cm. broad, its lower lobe notched; valves of capsules remaining semiterete after dehiscence. . . . 1. *C. speciosa.*
 Leaves abruptly short-acuminate; corolla 2–3 cm. broad, its lower lobe entire; valves of capsules flattening after dehiscence. . . . . 2. *C. bignonioides.*
Leaves glabrous or quickly glabrate beneath; corollas 1–2 cm. broad, yellow, striped with orange and spotted with violet; capsules 5–8 mm. in diameter; seeds 2–4 mm. broad. . . . . 3. *C. ovata.*

1. C. speciòsa Warder (showy), Catawba-tree, Cigar-tree. — Large tree up to 30 m. high, with *deeply fissured and ridged bark; leaves* cordate-ovate, *long-acuminate,* softly pubescent beneath, 1.5–3 dm. long, essentially inodorous when crushed; *corolla 3.6–5 cm. long, about 4 cm. broad,* nearly white, inconspicuously spotted, *the lower lobe emarginate; capsule* cylindric, 2–5 dm. long, about 1.5 cm. thick, *the valves* after dehiscence *remaining semiterete; seeds 4–5 mm. wide, the hairs of the terminal tufts not connivent.* — Damp woods and swamps, Tenn. and Ark. to Tex., n. to s. Ind., s. Ill. and se. Ia.; cult. and esc. or natzd. eastw. to O., W.Va. and in e. Va., doubtless elsewhere. May, June.

2. C. bignonioìdes Walt. (like *Bignonia*), Common C. — Lower tree, up to 15 m. high, with *thin scaly bark; leaves abruptly short-acuminate,* fetid when bruised; *corolla* 2.5–4 cm. long, 2–3 *cm. broad,* conspicuously spotted, *the lower lobe entire; capsule* 1.5–4 dm. long, 0.8–1.2 cm. thick, *its valves flattening after dehiscence; seeds 2.5–4.5 mm. wide, the hairs of the terminal tufts connivent into a point.* (*C. Catalpa* (L.) Karst.) — Cult. and frequently esc. n. to s. N.E., N.Y., O. and Mich. June, July. (Introd. from farther south)

3. C. ovàta G. Don (egg-shaped), Chinese C. — Small tree or large shrub; *leaves abruptly acuminate, glabrous or quickly glabrate,* often with sharp lobes; *corollas 1–2 cm. broad, yellow, striped with orange and spotted with violet; capsules 5–8 mm. in diameter,* thin-walled; *seeds 2–4 mm. broad,* the hairs of the terminal tufts not connivent. (*C. Kaempferi* Sieb. & Zucc.) — Cult.; esc. and natzd. from Ct. to s. Ont., O. and Md. June–Aug. (Introd. from e. Asia)

FAM. 155. MARTYNIÀCEAE (Martynia Family)

Herbs, with chiefly opposite simple leaves, and flowers as of the Lentibulariaceae, *except in structure of ovary and fruit, the former being 1-locular, the latter fleshy-drupaceous, with wingless seeds and thick entire cotyledons.* Ovary (in ours) 1-locular, with 2 parietal intruded placentae expanded into 2 broad lamellae or united into a central columella. — Chiefly tropical.

1. PROBOSCÍDEA Schmidel Unicorn-plant

Calyx unequally 5-cleft; corolla gibbous, campanulate, 5-lobed and somewhat 2-lipped. Fertile stamens 4. Flesh of fruit at length falling away in 2 valves; inner part woody, terminated by a 2-horned beak, imperfectly 5-locular. Seeds several, with a thick roughened coat. — Low Am. branching annuals, clammy-pubescent, exhaling a heavy odor; stems thickish; leaves simple, rounded; flowers racemed, large. (Greek name for the proboscis-like beak of some fruit.)

1. P. louisiànica (Mill.) Thell. (of Louisiana), Proboscis-flower, Ram's-horn. — Leaves cordate, oblique, entire or undulate, the upper alternate; corolla dull white or purplish, or spotted with yellow and purple; endocarp of the fruit crested on one side, long-beaked. (*P. Jussieui* Van Eseltine, wrongly ascribed to Schmidel; *Martynia louisiana* Mill., corrected by him to *M. louisianica* Mill.) — River-banks and waste places, Ga. to n. Mex., n. to W.Va., s. O., s. Ind., Ill. and Minn.; cult. for pickles, and spontaneous northw. June–Sept.

Sésamum (ancient name) índicum L. (of East Indies), Sesame or Bené (*S. orientale* L.), of the Pedaliàceae, an erect annual with lanceolate to oblong-ovate leaves, the lower often divided, white to roseate ventricose-campanulate corolla with bilabiate 5-parted limb, linear stigmas and oblong-quadrangular velvety capsule, is sporadic in waste ground but rarely persistent. (Adv. from Old World)

FAM. 156. OROBANCHÀCEAE (Broom-rape Family)

Herbs (root-parasites) destitute of green foliage, gamopetalous, the ovary one-locular with 2 or 4 parietal placentae; capsule very many-seeded; seeds minute, with albumen and a very minute embryo. Calyx persistent, 4–5-toothed or -parted. Corolla tubular, more or less 2-lipped, ringent, persistent and withering; upper lip entire or 2-lobed, the lower 3-lobed. Stamens 4, didynamous, inserted on the tube of the corolla; anthers 2-locular, persistent. Ovary free, ovoid, pointed with a long style; stigma large. Capsule 1-locular, 2-valved; each valve bearing on its face one placenta or a pair. Seeds very numerous, minute. — Low thick or fleshy herbs, bearing scales in place of leaves, lurid yellowish to brownish or purple throughout. Flowers solitary or spiked.

Stems slender, usually branching; flowers scattered along the branches, of two sorts, the upper sterile, with a tubular corolla; the lower fertile, cleistogamous, with small unopening corollas. . . . . . . 1. *Epifagus.*

Stems stouter, simple (rarely forking); flowers all expanding and perfect, solitary or in a dense spike.

Flowers in a thick scaly spike; calyx spathiform, the upper lip large and 4-toothed; stamens exserted. . . . . 2. *Conopholis.*

Flowers solitary and long-stalked or spicate; calyx regular or about equally 2-lipped; stamens included. . . . . 3. *Orobanche.*

1. EPIFÀGUS Nutt. Beech-drops. Cancer-root

Flowers racemose or spiked; the upper sterile, with long filaments and style; the lower fertile, with a very short corolla which is forced off from the base by the growth of the capsule; stamens and style very short. Calyx 5-toothed. Stigma capitate, a little 2-lobed. Capsule 2-valved at the apex, with 2 approximate placentae on each valve. — N. Am. herb, slender, purplish or yellowish-brown, much branched, with small scattered scales. (Name from the Greek *epi, upon,* and *phagos, the Beech,* because it grows on the roots of that tree.) Leptamnium Raf. Epiphegus Spreng.

1. E. virginiàna (L.) Bart. (Virginian). — Stems 1.5–4.5 dm. high, striped or suffused with brown-madder; corolla of the upper (sterile) flowers whitish laterally, with broad dorsal band and a narrower lower band of brown-madder, about 1 cm. long, curved, 4-toothed, or whole plant pale brown except for the white to dull pink corollas in forma **pállida** Weath. (pale). — Under beech-trees, parasitic or saprophytic, P.E.I. to Ont., s. to N.S., N.E., Fla., Ala., Miss. and La. Aug.–Oct.

2. CONÓPHOLIS Wallr. Squawroot. Cancer-root

The irregularly 4–5-toothed calyx with 2 bractlets at the base; its tube split down on the lower side. Corolla tubular, swollen at base; upper lip arched, notched at the summit; the lower shorter, 3-parted, spreading. Stigma depressed. Capsule with 4 placentae, a pair on the middle of each valve. — Upper scales forming bracts to the flowers, regularly imbricate, not unlike those of a pine-cone (whence the name, from the Greek *conos, cone,* and *pholis, scale*). A single species.

1. C. americàna (L.) Wallr. (American). — Plant forming large rounded knobs on tree-roots; the several thick subcylindric chestnut-brown to yellowish stems resembling old White Pine cones, 1–2.5 dm. long, covered with fleshy, finally dry and hard scales. — In woods, mostly under oaks, often hidden by fallen leaves, N.S. to Wisc., s. to N.E., Fla. and Ala. April–July.

3. OROBÀNCHE L. Broom-rape

Upper lip of corolla more or less spreading and 2-lobed, emarginate or entire; the lower spreading, 3-lobed. Stigma broadly 2-lipped or crateriform. Capsule with 4 placentae equidistant or contiguous in pairs. — Plants brownish, purplish or whitish, mostly Eurasian but some found in most temp. or warm reg. Flowers (blue, purple, or yellowish) and naked or bracted stems minutely glandular-pubescent. (Name from the Greek *orobos, vetch,* and *anchein, to strangle.*) Including APHYLLON (Mitchell) Gray

Flowers spicate or thyrsoid-spicate, with 1–3 bracts at base of calyx; corolla 2-lipped, the upper lip generally 2-cleft.
Each flower with 3 bracts, 1 large and 2 small, at base of calyx.
Stem simple; calyx 5-lobed. . . . . 1. *O. purpurea.*
Stem branching; calyx 4-lobed. . . . . 2. *O. ramosa.*
Each flower with 1 or 2 basal bracts.
Spike loosely flowered; calyx cleft before and behind; corolla 1–1.5 cm. long, with rounded lobes. . . . . 3. *O. minor.*
Spike dense; calyx nearly regularly 5-cleft; corolla 1.5–2 cm. long, with acute lobes. . . . . 4. *O. ludoviciana.*
Flowers solitary on long naked scapose pedicels, without bracts at base of calyx; corolla with long curved tube and spreading 5-lobed limb.
Caudex subligneous, it or its branches prolonged above ground, with 5–12 firm bracts and several axillary pedicels, the latter rarely longer than the caudices. . . . . 5. *O. fasciculata.*
Caudex soft, it or its slender branches not much prolonged, with 1–few much longer pedicels from the axils of the few basal bracts.
Corolla-lobes broad-oblong to obovate, minutely ciliolate with trichomes 0.1–0.3 mm. long; capsule ovoid, capped by the marcescent corolla; seeds thick-ellipsoid to ovoid, 0.3–0.4 mm. long. . . . . 6. *O. uniflora.*
Corolla-lobes narrowly oblong, pilose-ciliate with trichomes 0.4–0.6 mm. long; capsule lance-attenuate, readily pushing off corolla; seeds slenderly ellipsoid, 0.6–0.8 mm. long. . . . . 7. *O. terrae-novae.*

1. O. PURPÙREA Jacq. (purple). — *Stem simple,* 1–2 dm. high, *bluish- or purplish-tinged; flowers* subsessile in a spiciform raceme, *each with 1 large and 2 small bracts at base of the 5-lobed calyx; corollas deep violet,* 2 cm. long, slightly bilabiate. — Grasslands, local, Ont. June–Aug. (Adv. from Eu.)

2. O. RAMÒSA L. (branched), BRANCHED B. — Much more slender than no. 1, usually *freely branched, stramineous,* 1.5–4.5 dm. high; *calyx 4-lobed; corollas yellow and pale blue,* 1–1.5 cm. long. — Parasitic on tomato, tobacco, coleus, hemp, etc., N.Y. to Ill. and Ky. July–Sept. (and through winter in greenhouses). (Adv. or natzd. from Asia)

3. O. MÌNOR Sm. (smaller). — Resembling no. 1, pubescent, pale yellowish-brown, 1–4.5 dm. high; *spike lax, each of the* purple-tinged *flowers in the axil of* 1 *basal bract; calyx cleft before and behind almost or quite to base,* the divisions usually 2-cleft; *corolla* 1–1.5 *cm. long,* the limb bluish, *with rounded lobes,* the upper lip emarginate. — Parasitic on clover, tobacco, etc., N.J. to N.C. April–July. (Adv. or natzd. from Eu.)

4. O. ludoviciàna Nutt. (of old Louisiana). — Single or clustered, stout, viscid-puberulent, 1–2.5 dm. high; *flowers densely spicate-racemose or -thyrsoid,* only the lowest on pedicels up to 1 cm. long, each with 1 or 2 basal bracts; calyx about 1 cm. long, the 5 subequal divisions linear-lanceolate; *corolla purplish,* 1.5–2 *cm. long,* with 2 yellow bearded folds in the throat; the *lobes acute.* (*Myzorrhiza* Rydb.) — Parasitic on *Ambrosia* and other *Compositae,* plains, Alta. and B.C. to Tex., e. to Man., Minn., Ill. and e. Kans.; O. and Ind. July–Sept.

5. O. fasciculàta Nutt. (bunched). — *Caudex* usually forked, *subligneous, stout, it or its branches prolonged above ground for* 0.5–1 *dm., each branch with* 5–12 *firm bracts and* 4–10 erect crowded axillary naked *pedicels* 3–12 *cm. long; calyx with triangular lobes* shorter than to equaling the tube; *corolla purple,* its tube long and curved, 1.5–2.2 cm. long, with 5 semiorbicular spreading lobes. (*Aphyllon* Gray; *Thalesia* Britt.; *Anoplanthus* Walp.) — Parasitic on *Eriogonum, Artemisia,* etc., Mich. to Yuk., s. to n. Ind., Ill., Neb., N.M., Ariz. and Calif. April–Aug.

6. O. unifløra L. (one-flowered), ONE-FLOWERED CANCER-ROOT. — *Caudex subterranean or but slightly prolonged* above ground, *soft, each branch with 1–few scales* and 1–few erect crisp-pubescent bractless *pedicels* 0.5–2.5 *dm. high; calyx-lobes lance-subulate, longer than the tube; corolla creamy-white to lilac,* with 2 linear bearded yellow folds in the throat, delicately fragrant, its curved tube 1–2.5 cm. long, *the wide-spreading short-oblong to round-obovate lobes minutely ciliolate with trichomes* 0.1–0.3 *mm. long; capsule ovoid,* 6–12 *cm. long,* 5–8 mm. thick, usually capped by the marcescent corolla; *seeds thick-ellipsoid to ovoid,* coarsely reticulate, 0.3–0.4 *mm. long.* (*Aphyllon* Gray; *Thalesia* Britt.; *Anoplanthus* Endl.) — Parasitic on various plants, damp woods and thickets, sw. N.B. to sw. Que., w. to Mont., s. to N.S., N.E., n. Fla., Miss. and Tex. April–June.

7. O. térrae-nòvae Fern. (of Newfoundland). — Similar to no. 6; flowers not fragrant; *corolla-lobes narrowly oblong, pilose-ciliate with trichomes mostly* 0.4–0.6 *mm. long; capsule lance-attenu-*

ate, 1–2 cm. long, 3–7 mm. thick, usually pushing off corolla; *seeds slenderly ellipsoid*, more delicately reticulate, 0.6–0.8 *mm. long*. — On various plants, rich woods and thickets, chiefly calcareous, Nfld.; Anticosti I., Gaspé Pen. and Rimouski Co., Que. June–Aug.

FAM. 157. LENTIBULARIÀCEAE (Bladderwort Family)

Small herbs (growing in water or wet places), with a 2-lipped calyx, and a 2-lipped personate corolla, 2 stamens with (confluently) 1-locular anthers, and a 1-locular ovary with a free central placenta, bearing several anatropous seeds with a thick straight embryo and no albumen. Corolla deeply 2-lipped; the lower lip larger, 3-lobed and with a prominent palate, spurred at the base in front; the palate usually bearded. Ovary free; style very short or none; stigma 1–2-lipped. Capsule often bursting irregularly. Scapes 1–few-flowered. — The following are the two principal genera:

Aquatic, amphibious or terrestrial, with dissected or very fine leaves, these or specialized branches bearing traps or bladders; calyx-lobes mostly entire; upper lip of corolla usually ascending. . . . . . . 1. *Utricularia.*

Terrestrial, with rosettes of entire broad leaves with a fold near the margin; calyx with upper lip deeply 3-cleft and the lower 2-cleft; corolla-lobes spreading. . . . . 2. *Pinguicula.*

1. UTRICULÀRIA L. Bladderwort

Corolla personate, the palate on the lower lip projecting, often closing the throat. Anthers convergent. — Aquatic and immersed, with capillary dissected leaves bearing little bladders or traps, which float the plant at the time of flowering; or rooting in the mud, and sometimes with simple leaves and minute subterranean traps. Scapes 1–few-flowered. Traps furnished with a valvular lid and usually with a few flagellae at the orifice. Inclusive genus nearly cosmop. (Name from *utriculus*, a *little bladder*.)*

*a.*Leaves dissected into 3-∞ elongate segments; plants aquatic or amphibious. . . *b.*

*b.*Leaves all whorled and uniform, long-petioled; the subglobose or ovoid beakless traps at tips of filiform segments and without long flagellae; flowers purple. . . . . 1. *U. purpurea.*

*b.*Leaves mostly alternate (a whorl of inflated petioles on the scape of no. 2), sessile or only short-petioled; the traps lateral or on separate branches, the prolonged summit bearing long flagellae at the orifice; flowers yellow or yellowish (purple in no. 13). . . . *c.*

*c.*Scapes bearing a whorl of 3–10 leaves with inflated petioles forming floats; submersed leaves 4–6 times dichotomous, with capillary segments. . 2. *U. inflata.*

*c.*Scapes naked or with only small scarious bracts; submersed leaves less divided. . *d.*

*d.*Leaf-segments terete or capillary; traps not on separate branches. . . *e.*

*e.*Flowers of 2 kinds, small apetalous cleistogamous short-stalked ones borne from some axils; winter-buds 2–5 mm. thick, their component leaves not hairy. . . . . 3. *U. geminiscapa.*

*e.*Flowers uniform, none cleistogamous. . . *f.*

*f.*Coarse, the very long stems 0.5 mm. or more thick, free-floating; leaves 1.5–7 cm. long, with very numerous segments; scape stout, with 6–20 or more flowers; fruiting pedicels arched-recurving; winter-buds coarse, 1–2 cm. long, their component leaves hairy. . . . . 4. *U. vulgaris.*

*f.*Delicate, with filiform stems less than 0.5 mm. thick, creeping; leaves 0.2–2 cm. long, with few remote segments; scape filiform, 1–6-flowered; fruiting pedicels erect; winter-buds wanting or minute (less than 1 mm. thick). . . *g.*

*g.*Leaves uniform, all or nearly all trap-bearing, the traps scattered; scapes 2–16 cm. high; flowers 1–3, 6–17 mm. high, the lips projected forward, the spur oblong-conic and shorter than lower lip.

* Many characters drawn from J. H. Barnhart in Mem. N. Y. Bot. Gard. vi. 39–64 (1916) and some from G. B. Rossbach in Rhodora, xli. 113–128 (1939).

Fully developed leaves mostly with 2 segments; larger sepal 2–3 mm. long; corolla 6–12 mm. high, its spur much shorter than lower lip; capsule 2–3 mm. thick. . . . . . . 5. *U. gibba.*

Fully developed leaves with 3–several segments; larger sepal 2.5–4 mm. long; corolla 1–1.7 cm. high, its spur one-half as to nearly as long as lower lip; capsule 3.5–4 mm. thick. 6. *U. biflora.*

*g.*Leaves of two sorts, those of branches from which scapes arise trap-bearing, those of vegetative branches trapless; scapes 1–4 dm. high; flowers 2–7, 1.5–2.3 cm. high, the upper and lower lips nearly vertical; spur more slender, about equaling or longer than lower lip. . . . . 7. *U. fibrosa.*

*d.*Leaf-segments flat.

Traps borne on the leaves; margins of terminal leaf-segments entire. 8. *U. minor.*

Traps mostly borne on separate elongate branches; margins of terminal leaf-segments minutely serrulate.

Trap-bearing branches leafless. . . . . 9. *U. intermedia.*

Trap-bearing branches with small dissected leaves subtending many pedicels. . . . . 10. *U. ochroleuca.*

*a.*Leaves simple or slightly parted, very slender, borne on slender basal branches, mostly hidden in peat, mud or sand; plants of damp soils, with filiform scapes and minute subterranean traps. . . *h.*

*h.*Bracts at base of pedicel or calyx basally attached; traps borne along leaf-margins and -bases; flowers sessile or very short-pedicelled; spur of corolla divergent. . . *i.*

*i.*Flowers 1–12, subsessile or very short-pedicelled, each subtended by a free bract and 2 smaller bractlets; longer sepal acuminate; corolla yellow to whitish or purplish, with subulate spur; capsule covered by beaked calyx.

Flowers 1–3 (rarely –9), approximate, the expanding lower ones over-topping the unexpanded buds above; pedicels mostly exserted beyond the basal bracts; corolla (from tip of spur to tip of upper lobe) 1.5–2.5 cm. high. . . . . 11. *U. cornuta.*

Flowers (1–)2–12, not crowded, the expanding lower ones not reaching the unexpanded buds above; pedicels overtopped by the basal bracts; corolla 3–15 mm. high. . . . . 12. *U. juncea.*

*i.*Flower 1, on an erect pedicel (like continuation of scape) 0.3–4.2 cm. long, this subtended by a truncate or notched cup; sepals obtuse; corolla purple; capsule exceeding calyx. . . . . 13. *U. resupinata.*

*h.*Bract at base of pedicel peltate; traps borne on separate branches underground; flowers long-pedicelled, yellow, whitish or purplish, with appressed blunt spur; capsule much exceeding calyx. . . . . 14. *U. subulata.*

1. U. purpûrea Walt. (purple). — *Leaves whorled* along the long immersed free-floating slender stems, *long-petioled,* decompound, with very many capillary segments; *the subglobose or ovoid beakless traps* with short glandular hairs at orifice, *borne at the tips of some leaf-segments;* peduncles at irregular intervals along the stems, 2–5-flowered; flowers on ascending pedicels; *corolla purple or deep pink,* 1–1.3 cm. broad, the somewhat flattened spur appressed to the 3-lobed 2-saccate lower lip of the corolla and about half its length. (*Vesiculina* (Walt.) Raf.) — Ponds and sluggish streams, Fla. to La., n. to N.S., N.B., N.E., sw. Que. and N.Y.; n. Ind., s. Mich. and Wisc. June–Sept. (W.I.; Centr. Am.) Fig. 1485.

1485. U. purpurea.

2. U. inflàta Walt. (inflated). — Stems prolonged, free-floating, with alternate *submersed leaves* 4–6 *times dichotomous* into delicate capillary segments and bearing small ovoid traps; *a whorl of* 4–10 *leaves with inflated petioles borne upon the scape and acting as floats,* these leaves 4–9 cm. long, the petiole 4–8 mm. thick, the basal divisions of their pinnately dissected blades 0.5 mm. or more in diameter; upper portion of scape, from whorl of floats to lowest pedicel, 7–17 cm. long; flowers 3–14; calyx-lobes 4–6 mm. long; corolla yellow, about 2 cm. broad, up to 2.5 cm. high, the lower lip 3-lobed and twice as long as appressed spur; fruiting pedicels 1.5–3.5 cm. long, spreading or recurving. — Ditches and ponds, Fla. to Tex., n. to N.J. May–Nov. Passing into

Var. **minor** Chapm. (smaller). — More delicate; the 3–8 floats 2–5.5 cm. long, inflated petiole 2–4 mm. thick; the basal divisions of the blades at most 0.25 mm. in diameter; upper portion of scape, from whorl of floats to lowest pedicel, 1.5–5 cm. long; flowers 1–5, oftenest 3; calyx-lobes 3–4 mm. long; corolla about 1.5 cm. broad; fruiting pedicels 1–2 cm. long. (*U. radiata* Small) — Ponds and sluggish waters, Fla. to Tex., n. to N.S., Me., s. N.H., s. Vt., s. N.Y., Pa. and Tenn.; nw. Ind. (S. Am.)

3. U. geminiscàpa Benj. (with paired scapes). — Free-floating and relatively slender, the *plumes of foliage* 1–3.5 *cm. in diameter;* leaves circular or oval, thrice dichotomous into capillary segments about 0.25 mm. thick, many of them copiously trap-bearing; *flowers of 2 sorts, cleistogamous apetalous ones on short pedicels and developing into ovoid capsules borne from many axils;* petaliferous yellow flowers 2–6 on filiform erect scattered peduncles 5–10 cm. long; corolla 5–8 mm. high; the larger lower lip 3-lobed, somewhat longer than the approximate thick and blunt spur; winter-buds 2–5 mm. in diameter. (*U. clandestina* Nutt.) — Pools, ponds and sluggish streams, Va., northw. to Nfld., M.I., N.B., N.E. and N.Y., thence inland to Wisc. June–Sept. Fig. 1486.

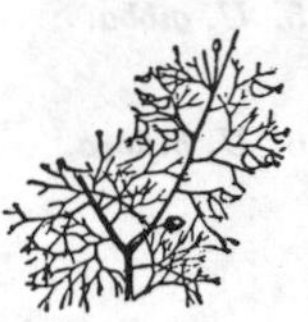
1486. U. geminiscapa.

4. U. vulgàris L. (common). — Relatively *coarse*, with prolonged free-floating stems 0.5 mm. or more thick, the *plumose branches of foliage* 3–12 *cm. in diameter;* leaves elliptic to ovate, pseudopinnately much dissected, the coarser capillary segments up to 0.75 mm. thick, the numerous traps large; *scapes* coarse, 1–8 dm. long, *with* 6–20 *or more flowers*, naked at base or producing slender small-leaved stolon-like divergent offshoots; corolla yellow, with brown vertical stripes on the large palate, 1.5–2.5 cm. high, the broad lower lip slightly 3-lobed and slightly shorter than the curved obtuse or acutish spur; *fruiting* raceme elongate, the *pedicels arched-recurving; winter-buds* ellipsoid, 1–2 *cm. long, their component crowded leaves hairy.* (Incl. var. *americana* Gray; *U. macrorhiza* LeConte) — Deep or shallow quiet waters, s. Lab. to Alaska, s. to N.S., N.E., Va., O., Ind., Mo., Tex. and Mex. May–Sept. (Eurasia) Fig. 1487.

1487. U. vulgaris.

5. U. gíbba L. (humped). — Delicate, with filiform creeping or floating branches; the *sparsely leafy plumes of foliage* 0.5–2 *cm. in diameter* and up to 3 dm. long, often intricately entangled; *leaves mostly with* 2 *filiform segments and with scattered traps;* scapes 2–9 cm. high, 1–3-flowered; *larger sepal* 2–3 *mm. long; corolla* 6–12 *mm. high*, 6–8 mm. broad, *the nearly equal lips projected forward*, the oblong-conic *spur much shorter than the lower lip;* fruiting pedicels ascending, the lower 3–9 mm. long; *capsule* 2–3 *mm. thick*, nearly equaled or exceeded by sepals. — Shallow pools, spring-heads, quaking bogs, etc., Fla. to Tex. and Mex., n. to N.S., N.E., s. Que., N.Y., s. Ont., Mich., Wisc., Minn., Okla. and Calif. Late June–Sept. (W.I.; Centr. Am.) Fig. 1488.

1488. U. gibba.

6. U. biflòra Lam. (two-flowered). — Similar to no. 5, but branches longer; *fully developed leaves with* 3 *or more segments;* scapes 0.3–1.6 dm. high; *larger sepal* 2.5–4 *mm. long; corolla* 1–1.7 *cm. high, its spur one-half as to nearly as long as lower lip;* lower pedicels 0.5–2.5 cm. long; *capsule* 3.5–4 *mm. thick*, slightly exserted. (Possibly *U. pumila* Walt.) — Shallow pools, Fla. to N.M., n. to se. N.E., N.J. and Okla. July–Oct. Fig. 1489.

1489. U. biflora.

7. U. fibròsa Walt. (fibrous). — Coarser than nos. 5 and 6; *leaves* more crowded on the submersed branches, *of two sorts, those associated with flowering scapes trap-bearing, with traps toward the base, those of vegetative branches trapless* and several times forked; *scapes* 1–4 *dm. high*, 2–7-flowered; *corolla* 1.5–2.3 *cm. high; the upper and lower lips nearly vertical*, broad and widely expanded, nearly equal; the upper undulate, concave, plaited-striate near the middle; *spur slender, about equaling or longer than lower lip;* lower fruiting pedicel 1–3 cm. long; *capsule much exceeding calyx*, 4–6 mm. in diameter. — Pineland-ponds, pools and wet peat, Fla. to Miss., n. to se. Mass., L.I. and N.J. Late May–Nov.

8. U. minor L. (smaller). — Slender, the immersed or creeping thread-like stems with 2–4 times forked small *leaves with flat* and *entire segments, the blades* 2–12 *mm. across, many of them trap-bearing* or sometimes consisting mostly of traps; scapes filiform, 0.3–2 dm. high, 2–9-flowered; *corolla* 5–8 *mm. long*, gaping, *the upper lip not extending beyond the depressed and obscure palate*, the lower lip prolonged, the *spur very short and saccate or almost wanting; fruiting pedicels arched-recurving*, the lower 4–8 mm. long; capsule exserted, subglobose, about 2 mm. in diameter; winter-buds, when formed, 1.5–5 mm. in diameter, lax, resembling miniature cabbages. — Shallow (often calcareous) pools, wet meadows, bogs and shores, Greenl. and s. Lab. to B.C., s. to Nfld., N.S., N.E., L.I., n. N.J., Pa., n. O., n. Ind., n. Ill., Ia., N.D., Colo. and Calif. May–Aug. (Eurasia) Fig. 1490.

1490. U. minor.

9. U. intermèdia Hayne (intermediate). — Slender, creeping, the *traps borne on elongate racemose or paniculate branches separate from the distichous trapless leaves; the latter* 0.5–1.5

cm. across, their thin to firm *flat segments minutely serrulate; racemes or panicles of* large pale *traps* up to 1.5 dm. long, *leafless;* scapes 0.5–3 dm. high, 2–5-flowered; corolla 1.2–2 cm. high, the *broadly deltoid upper lip much shorter than the broad* (1–1.5 cm.) *-ovate lower one,* the acute conic-subulate spur appressed to the latter and nearly as long; fruiting pedicels erect, the lower 0.5–2 cm. long; winter-buds ovoid or ellipsoid, dense, hairy, 3–10 mm. long. — Creeping (rarely free-swimming) at bottoms of shallow pools, on shores and in quagmires, Greenl. and Nfld. to Alaska, s. to N.S., N.E., L.I., n. Del., Pa., O., n. Ind., n. Ill., n. Ia. and Calif. Late May–Sept. (Eurasia) Fig. 1491.

1491. U. intermedia.

10. U. ochroleùca R. Hartman (yellowish-white). — Very similar to no. 9; tips of leaf-segments more prolonged; *many or all pedicels of the traps subtended by dissected leafy bracts;* scapes very slender, rarely 1.7 dm. high; *upper lip of corolla ovate, subtruncate* at summit, *lower lip much broader than long* but twice as long as the spur. — Shallow water, St. Paul I., N.S.; Greenl. (Eu.)

11. U. cornùta Michx. (horned). — Terrestrial, *with* delicate stems creeping underground and bearing *linear-filiform simple leaves* (seen only by collecting sods and gently washing away the soil); *traps minute, borne along leaf-margins;* scapes erect, wiry, slender, 0.3–3.5 dm. high, 1–3 (rarely –9)-flowered; *flowers subtended by an acute sessile bract and 2 smaller bractlets,* very fragrant, at first *approximate, the freshly expanding lower ones overtopping the unexpanded buds above; pedicels mostly exserted slightly beyond the bracts;* longer sepal acuminate, yellowish; *corolla,* from tip of long spur to tip of upturned upper lip, 1.5–2.5 cm. high, yellow, nearly as broad; *the larger lower lip helmet-shaped, with projecting convex center and recurved sides;* spur subulate, turned downward and outward, 7–12 mm. long; *capsule covered by the beaked calyx.* (*Stomoisia* Raf.) — Wet peaty, sandy or muddy shores or bogs, Nfld. to n. Ont. s. to N.S., N.E. (ascending to 1070 m. alt.), L.I., Del., Pa., O., n. Ind., ne. Ill., Wisc. and Minn.; se. N.C. to Fla.; e. Tex. June–Sept.

12. U. júncea Vahl (rush-like). — Similar to no. 11, the more slender scapes 1–4 dm. high, (1–)2–12-flowered, the *flowers not crowded, the expanding ones not reaching the unexpanded buds above; pedicels mostly overtopped by the bracts; corolla* 1–1.5 *cm. high,* 0.8–1 *cm. broad; the lower lip obovate, consisting mostly of the high-arched palate, without broad recurving margin; spur* 5–7 *mm. long; fruiting calyx* 5–7 *mm. long.* (*Stomoisia* Barnh.) — Wet sand, peat and pond-margins, Fla. to Miss., n. on Coastal Plain to L.I. Aug., Sept. (W.I.; S. Am.) — Passing through a transitional series into forma **virgátula** (Barnh.) Fern. (like a little wand), which is smaller, with scapes 0.2–2.8 dm. high and 1–6-flowered, the *corolla* only 3–6(–8) *mm. high* and 4–5(–7) *mm. broad, its spur* 2–3 *mm. long,* the *fruiting calyx* 3–5 *mm. long.* (*Stomoisia virgatula* Barnh.)

13. U. resupinàta B. D. Greene (turned upside down). — Terrestrial, forming insignificant mats with filiform subterranean stems and mostly hidden simple or slightly forking linear-acicular leaves bearing tiny traps; *filiform scapes* 0.25–1.7 dm. high, *terminated by a single erect pedicel* 0.3–4.2 *cm. long, this subtended by a cup-like truncate or notched bract;* flower 1; sepals obtuse; *corolla purple,* 6–12 mm. long, *transversely oblique,* the broad and round lower lip much larger than the narrowly obovate upper one; spur much shorter than lower lip, curved, its obtuse tip remote; *capsule exserted.* (*Lecticula* Barnh.) — Pond-, lake- and river-shores and -margins, local, Laurentide Mts., Que., to nw. Wisc., s. to N.S., s. N.E., N.Y. (ascending to mountain-ponds), s. Del., nw. Pa., n. Ind. and ne. Ill.; pools in pine-barrens, S.C. to Fla. July–Sept.

14. U. subulàta L. (awl-shaped). — Terrestrial, with filiform subterranean branches and slender uncut leaves, the *minute traps borne on separate underground branchlets;* scapes filiform, 0.2–2 dm. high, 1–12-flowered, the *flowers on elongate pedicels; bracts at base of pedicels centrally attached,* tapering to base and apex; sepals elliptic, blunt; corolla yellow, 3.5–12 mm. long, the rounded-ovate upper lip smaller than the 3-lobed lower one, the prominent palate 2-lobed, the compressed spur appressed to lower lip; or corolla smaller and pale yellow to whitish or purplish (consisting mostly of the enlarged saccate spur and gaping or closed throat) in forma **cleistógama** (Gray) Fern. (fertilized without the flower expanding) = *U. cleistogama* (Gray) Britt. and *Setiscapella cleistogama* (Gray) Barnh.; *capsule exserted.* (*Setiscapella* Barnh.) — Wet peat, sands and pond-shores, Fla. to Ark. and Tex., n. to L.I., se. Mass. and w. N.S. Late April–Sept. (W.I.)

2. PINGUÍCULA L. Butterwort

Upper lip of the calyx 3-cleft, the lower 2-cleft. Corolla with an open hairy or spotted palate, the lobes spreading. — Small and stemless perennials of N. and S. Am. and Eurasia, growing on damp earth or rocks, with 1-flowered scapes; the broad and entire leaves soft-fleshy, mostly greasy to the touch (whence the name, from the Latin adjective *pinguis, somewhat fat*).

1. P. vulgàris L. (common). — Leaves spatulate to elliptical, 2–6 cm. long, yellowish-green, promptly revolute on removal from ground; scape 0.4–1.5 cm. high; corolla, excluding the straight spur, 1–2 cm. long, violet, its lips very unequal, its tube funnelform. — Wet rocks and open spots, Greenl. and Lab. to Alaska, s., chiefly in calcareous areas, to Nfld., C.B. and n. N.B., and locally to n. Vt., n. and centr. N.Y., n. Mich. and n. Minn., Alta. and Wash. June–early Aug. (Eurasia)

FAM. 158. ACANTHÀCEAE (Acanthus Family)

Chiefly herbs, with opposite simple leaves, didynamous or diandrous stamens inserted on the tube of the more or less 2-lipped corolla, the lobes of which are convolute or imbricated in the bud; fruit a 2-locular and few (2–20)-seeded capsule; seeds anatropous, without albumen, usually flat and supported by hooked projections of the placentae (retinacula). Flowers commonly much bracted. Calyx 5-cleft. Style filiform; stigma simple or 2-cleft. Capsule loculicidal, usually flattened contrary to the valves and partition. Cotyledons broad and flat. — Mucilaginous and slightly bitter, not noxious. A large family in the warmer parts of the world; represented in gardens by Thunbergia, which differs from the rest by the globular capsule and seeds, the latter not on hooks.

Inflorescences spicate-racemose or -paniculate; corolla imbricated in bud, bilabiate, the upper lip erect and concave or plane; stamens 2.
  Flowers without broad involucels, in long-peduncled axillary spikes or heads; anther-sacs disjointed and oblique on a dilated connective; placentae not separating from valves of capsule. . . . . 1. *Justicia.*
  Flowers subtended by broad involucel-bracts, in subsessile or short-peduncled terminal or axillary paniculate clusters; one anther-sac nearly over the other, with narrow connective; placentae at length separating from valves of capsule. . . . . 2. *Dicliptera.*
Inflorescences of 1–few loosely disposed or glomerulate flowers; corolla sinistrorsely convolute in bud, nearly or quite regular, the 5 similar lobes spreading; stamens 4. . . . . 3. *Ruellia.*

1. JUSTÍCIA L. Water-willow

Calyx 5-parted. Upper lip of corolla notched; the lower spreading, 3-parted, external in the bud. Anthers 2-locular, the locules separated and somewhat unequal. Capsule contracted at base into a short stalk. — Perennial herbs, growing in water or wet places, with entire leaves, and purplish to white flowers in axillary peduncled spikes or heads; a large genus of the Tropics and of warm-temp. Am. and Asia. (Named for *James Justice*, Scotch horticulturist and botanist of the 18th century.) Dianthera L.

1. J. americàna (L.) Vahl (American). — Stout-based, *with* abundant thick cord-like *superficial stolons and rhizomes; stems* simple or branched, 0.2–1 m. high, *colonial; primary leaves* (or nodes) 6–14, *erect*, the well-developed ones *elongate-lanceolate or oblanceolate to lance-linear*, 0.8–2 dm. long, 0.5–2.5(–3) cm. broad, sessile or short-petioled; peduncles stiffly ascending to erect, nearly equaling to longer than leaves; *inflorescence capitate to ellipsoid, dense*, often overtopped by the developing leafy tip; *corolla pale violet to white*, the upper lip recurved; *lower lip arched-recurving, its middle lobe fiddle-shaped at base, its sides reflexed; anther-sacs* approximate, *blunt, the terminal one horizontal; seeds reniform-orbicular*, drab or pale brown, *coarsely pebbled with low flattened domes, the margin not thickened.* (*Dianthera* L.) — Margins and shallow beds of rivers and lakes, rarely in swamps, Ga. to Tex., n. to sw. Que. and nw. Vt., N.Y., s. Ont., Wisc., Mo. and Kans. June–Oct.

Var. **subcoriàcea** Fern. (somewhat leathery). — Stiff, *pale; leaves* erect, in overlapping pairs, *subcoriaceous*, oblong, lanceolate, oblanceolate or narrowly ovate, obtuse to subacute, the principal ones 4.5–15 cm. long and 1–3 cm. broad, sessile; *peduncles mostly from upper axils, the heads raised well above the upper leaves.* — Margins and beds of streams, Tex. to Kans. and Mo.

2. J. mortuiflùminis Fern. (of deadwater). — Colonial, *with deep-seated stolons and rhizomes* much prolonged and 3–7 *mm. thick;* stems 2–6 dm. high; *leaves* (or nodes) 7–11, *oblong-lanceolate to narrowly elliptic-oblong*, mostly 5–9 cm. long and 1.5–3.5 cm. broad, subacuminate to tip and petiole; *peduncles* 3.5–10 *cm. long, spreading-ascending; spikes compact, subcapitate*, in flower 1.5–3 *cm. long, the flowers densely overlapping; corolla pale violet or lilac throughout, the lower* *lip flat and projected forward, its* oblong *lobes not constricted at base, the lateral ones divergent; terminal anther-sac oblique; the lower remote, pointed at base and appressed parallel with the connective; seeds quadrate-orbicular, deep-brown, without thickened margin, minutely and subacutely pebbled.* — Wooded bottomlands and shaded margins of quiet waters, se. Va. June, July.

3. J. ovàta (Walt.) Lindau (ovate). — *Slender, from filiform creeping rhizome* only 2–4 *mm. thick;* stems 1–few, 1.5–4.5(–5) dm. high; *leaves* in *few* (5–8) distant pairs, *the larger ones rhombic-oblong to elliptic-ovate, tapering to base and apex;* peduncles filiform, spreading-ascending, 1–5 (–8.5) *cm. long; inflorescence* 1.5–5 *cm. long, loosening and elongating,* the fruits becoming subdistant; corolla much as in no. 2, thinner, *the lobes of the lower lip parallel;* anthers as in no. 2; *seeds rotund, nearly smooth, with thick margin.* (*J. humilis* Michx.) — Wooded swamps and bottomlands, Fla. to se. Va. and s. Ill. June.

2. DICLÍPTERA Juss.

Calyx deeply 5-parted. Corolla deeply bilabiate; upper lip entire or emarginate; the lower spreading, slightly if at all 3-lobed. Stamens 2. — Branched perennial herbs of trop. and warm reg., chiefly trop. and subtrop. Bractlets 2 or 4, opposite in pairs and forming a sort of involucel, the outer subequal, appressed to each other and more or less inclosing the fruit. (Name from the Greek *diclis, double-folding,* as of doors, etc., and *pteron, a wing,* alluding to the involucel.) DIAPEDIUM König

1. D. brachiàta (Pursh) Spreng. (with arm-like branches). — Erect or somewhat decumbent, 3–7 dm. high, smoothish or the upper internodes minutely pilose or puberulent; leaves membranaceous, ovate, acuminate to apex and slender petiole; flowers often cleistogamous; the well-developed corolla (when present) pink or pale purple, 1.2–1.8 cm. long; bractlets obovate. (*Diapedium* Ktze.) — Bottomlands and rich woods, Fla. to La. and Okla., n. to se. Va., s. Ind., Mo. and e. Kans. Sept.

3. RUÉLLIA L. RUELLIA

Calyx 5-parted. Corolla funnelform, with spreading ample border, convolute in bud (or in cleistogamous forms corolla small, tubular and closed). Locules of the somewhat sagittate anthers parallel and nearly equal. Capsule narrow, in ours somewhat flattened, contracted and seedless at base. Seeds with a mucilaginous coat, when wet exhibiting under the microscope innumerable tapering short spiricles, their walls marked with rings or spirals. — Large genus of perennials of trop. and warm-temp. reg., with large showy blue or purple flowers, sometimes also with small flowers precociously close-fertilized in the bud. Calyx often 2-bracteolate. (Named for the early French herbalist, *Jean Ruelle,* 1474–1537.)*

a. Main axis terminating in a prolonged flowerless leafy tip, with flowers 1–few at leafy-bracted summits of branches, or peduncles from the median or lower axils, or the cymose inflorescence loosely much branched. . . *b.*

b. Calyx-segments lanceolate or lance-linear, flat to the tip, 2–4 mm. broad; ovary and capsule glabrous; larger leaves with blades 7–18 cm. long and 3–9 cm. broad. . . . . 1. *R. strepens.*

b. Calyx-segments narrowly linear, tapering to elongate linear-acicular tips, the segments 0.5–1.2 mm. wide at base; larger leaf-blades 2–11 cm. long and 0.8–5 cm. broad. . . . *c.*

c. Stem usually divergently branching, the branches simple or forking, or stem simple to but slightly branched and with peduncles from axils of main stem or branches; calyx-segments 0.5–1 mm. wide, tapering from base to very slender tips.

Calyx, ovary and capsule pilose with slender-tipped but short spreading pubescence; plant of Ozarkian region. . . . . 2. *R. pedunculata.*

Calyx closely covered with partially imbedded and appressed elongate cystoliths; ovary and capsule glabrous; plant of southern Coastal Plain. . . . . 3. *R. pinetorum.*

c. Stem simple or with few ascending branches; peduncles 1-flowered, from 1 or 2(–3) nodes; calyx-segments linear, 0.7–1.2 mm. wide to near middle, thence tapering to apex, minutely canescent-pilose; capsules pilose; plant of Appalachian and eastern Piedmont area. . . . . . . 4. *R. Purshiana.*

a. Main axis with sessile or very short-peduncled glomerules, or flowers in the axils of the upper (sometimes median or lower) leaves, or the terminal pairs of leaves crowded and approximate to the upper fertile axils. . . *d.*

---

* Treatment condensed from that in Rhodora, xlvii (1945).

d. Calyx-segments flat, lanceolate, 2–4 mm. broad, shorter than to but slightly exceeding capsule; flowers chiefly or wholly cleistogamous, with small closed clavate corollas; stem tall, erect, simple or but slightly branched. (1). *R. strepens*, forma *cleistantha.*

d. Calyx-segments narrowly linear, with prolonged and slender to almost bristle-like tips mostly overtopping capsules; flowers rarely cleistogamous (except in form of no. 4), the showy corollas usually expanding. . *e.*

e. Upper internodes elongate, minutely pilose-puberulent; upper pairs of leaves distant; corolla usually insignificant, clavate, closed, the flowers cleistogamous; calyx-segments linear to near middle, tapering above; capsule pilose-strigose. . . . . (4). *R. Purshiana*, forma *claustroflora.*

e. Upper internodes approximate or abbreviated, villous, hirsute, puberulent or glabrescent; upper pairs of leaves approximate to crowded; corolla usually showy and expanding, cleistogamous flowers exceptional; calyx-segments linear-setaceous, attenuate from near base to slender tip.

Leaves sessile or barely petioled, nearly uniform, coriaceous, blunt, entire, 4–12 pairs on well-developed main axis equaling or longer than internodes; 4–10 nodes floriferous; stem frequently depressed or with low bushy branching. . . . . 5. *R. humilis.*

Leaves definitely petioled, the upper larger or definitely smaller than the lower and median ones, membranaceous to firm, acute or obtuse; 1–4 nodes floriferous or, if more, then the leaves usually undulate-dentate; stem remotely branched to simple, erect. . . . . . . 6. *R. caroliniensis.*

1. R. strèpens L. (rustling; from the explosive capsules). — Stem up to 1.1 m. high, simple or with few ascending branches, minutely pilose to glabrous; *principal leaves* membranaceous, ovate, rounded or tapering to slender petioles, acuminate, entire or barely undulate, *the larger ones* 7–18 *cm. long and* 3–9 *cm. broad*, short-strigillose to glabrous; *peduncles borne from* 1–3 *median nodes*, 0.2–9 cm. long, *terminated by* 2 *dilated leafy bracts which subtend* 1–3 *showy flowers; calyx-segments lanceolate to lance-linear, flat to tip,* 2–4 *mm., broad,* villous-ciliate, villous to glabrescent on back; corolla pale blue-violet, broadly expanding, 3–6 cm. long, with broadly funnelform throat; capsule (infrequent), 1–2 cm. long, glabrous, usually overtopped by calyx-segments. — Rich woods and bases of bluffs, wooded bottoms, etc., chiefly calcareous, S.C. to e. Tex., n. to n.-centr. N.J., s. Pa., centr. O., Ind., Ill., s. Mo. and e. Kans. Late May–July (rarely–Oct.). — Forma **cleistántha** (Gray) S. McCoy (with closed flowers) bears nearly sessile glomerules of small clavate tubular unexpanding pale flowers in many axils (including the upper ones) and is highly fruitful, the capsules plump and relatively short; seeds 6 or 8. (Var. *cleistantha* Gray).

2. R. pedunculàta Torr. (peduncled). — *Stem* 1–7.5 dm. high, slender, upright, *puberulent, simple and with axillary peduncles to freely branched,* the ascending to divergent *branches often much divided into a loosely cymose inflorescence; leaves ovate to lanceolate,* short-petioled, tapering from above the base, pale green, puberulent, *the larger ones* of the main axis 3–11 cm. long and 2–4.5 *cm. wide,* rameal smaller; *flowers solitary at tips of simple* 2-bracted *peduncles or loosely cymose; calyx-segments linear-filiform,* 0.5–1 *mm. wide at base, thence tapering to very slender flexuous tips,* becoming 1–3 cm. long, *cinereous with slender-tipped spreading pubescence;* corolla 2.5–5.5 cm. long, violet or purple, or in forma **Baùeri** Steyerm. (named in 1949 for its discoverer, Bill Bauer) white; *ovary and capsule* (1–2 cm. long) *cinereous-puberulent.* — Woods, bluffs, rocky slopes, barrens, etc., w. La. and e. Tex., n. to s. Ill., e. and s.-centr. Mo. and e. Okla. May–July (–Sept.).

3. R. pinetòrum Fern. (of the pines). — Habitally like no. 2, 1–3 dm. high; *leaves oblong or elliptic-lanceolate, the larger ones* 0.8–1.8 *cm. wide, subcoriaceous, minutely lineolate-puberulent or glabrate; calyx closely covered with partly immersed elongate cystoliths,* the segments 1.3–2 cm. long; corolla 3–4 cm. long; *capsule glabrous.* — Low pine-barrens, n. Fla. to La., n. to S.C. and (?) Va. June–Sept.

4. R. Purshiàna Fern. (for its discoverer, Frederick Traugott Pursh, 1774–1820). — Habitally resembling simpler plants of no. 2; *stem simple,* or with few erect short branches, *slender,* 1.5–6 dm. high, *cinereous-puberulent, with elongate internodes; leaves membranaceous,* the larger ones *elliptic- or lance-ovate,* attenuate at base, blunt to acutish, minutely hirtellous to glabrous, *usually as* 3–6 *remote pairs,* 2.5–10 cm. long and 1.5–4.3 *cm. broad,* the slender petiole 0.5–2 cm. long; *peduncles* 1-*flowered,* 0.2–3 cm. long, *from* 1 *or* 2 (rarely –4) *lower nodes, the* *upper nodes without flowers;* bracts elliptic or oval; *calyx-segments narrowly linear to near the middle,* 0.7–1.2 *mm. wide, thence slenderly attenuate,* becoming 1.6–2.8 cm. long, *minutely cinereous-pilose on back;* corolla 3–5 cm. long, the ampliate throat 0.8–1.4 cm. in diameter at summit; *capsule minutely pubescent.* — Dry or moist woods, bluffs and rich slopes, w. Md., s. along mts. and locally on the Piedmont to e. Va., centr. S.C., Ga. and Ala. Late May–July. — Forma **claustroflòra** Fern. (with closed flowers) bears glomerulate cleistogamous flowers and abundant fruits in all the upper axils; corolla 4–8 mm. long, tubular, not opening.

5. **R. hùmilis** Nutt. (low).—*Stem,* sometimes at first simple, *soon with elongate arched-ascending to horizontally divergent to reclining branches; main axis* 1–8.5 dm. high, *with* internodes villous-hirsute or canescent-pilose to glabrescent, with 4–12 *pairs of leaves longer than internodes; leaves coriaceous, entire,* often villous or hirsute on veins and margin, oblong or oblong-lanceolate to ovate, *sessile or subsessile, nearly uniform or but slightly reduced upward; flowers* few, *at* 4–10 *nodes;* calyx-segments 1.5–2.5 cm. long, linear-attenuate, mostly hirsute- to villous-ciliate; corolla lavender to bluish (rarely white), showy (small and closed in exceptional cleistogamous flowers); *capsule glabrous,* 1.2–1.5 *cm. long.* (*R. ciliosa* in part of ed. 7, not Pursh) — Polymorphic; the following vars. and forms with us:

Younger internodes of stem copiously villous-hirsute to canescent-pilose or -puberulent; larger leaves 3–8 cm. long, their veins and margins usually villous-ciliate.
- Corolla 3–4.5(–5) cm. long, its slender tube 1.2–2.5 cm. long.
  - Larger leaves of main axis elliptic-oblong to oblong-lanceolate, 1–2.5 cm. broad, obtuse to subacute.
    - Internodes of stem copiously villous-hirsute with long divergent hairs. . . . *R. humilis* (typical).
    - Internodes canescent-pilose or -puberulent, with no or very few scattered long hairs. . . . . Forma *grisea.*
  - Larger leaves of main axis ovate to oval-oblong or broadly elliptic, 2–4 cm. broad, mostly rounded at tip; stems strongly spreading-villous. . . . . Var. *frondosa.*
- Corolla 5–8 cm. long, its slender tube 3–6 cm. long; internodes of stem villous-hirsute.
  - Larger leaves of main axis elliptic-oblong to oblong-lanceolate, 1–2.5 cm. broad, obtuse to subacute. . . . . Var. *longiflora.*
  - Larger leaves of main axis ovate-oblong to broadly oval, rounded at summit, 2.5–4 cm. broad. . . . . Var. *expansa.*

Younger internodes glabrous, glabrescent or with only few scattered hairs on the angles; leaves glabrous, glabrescent or only sparsely short-hirsute on nerves beneath and margin, narrowly elliptic-oblong to oblong-lanceolate, the larger ones 1–3 cm. broad and 2–6 cm. long; corolla 2–3.5 cm. long, its tube 0.7–2.3 cm. long. . . . . Var. *calvescens.*

R. hùmilis Nutt. (typical). — Dry prairies, rocky banks, open slopes, open woods, etc., s. and e. Ia. to Tex., e. to Miss. R. and locally to s.-centr. Pa., w. Md., w. Va., Tenn. and nw. Ala. June–Aug. (–Oct.). — Plants with white corolla are forma **álba** (Steyerm.) Fern. (white). Forma **grísea** Fern. (grayish) local.

Var. **frondòsa** Fern. (with expanded leaves). — Se. Neb. to e. Tex. and La., e. to s.-centr. Pa. and w. Va.

Var. **longiflòra** (Gray) Fern. (long-flowered). — S. Ill., Mo. and e. Kans. to La. and Tex.

Var. **expánsa** Fern. (expanded; from the broad leaves). — Ia. and Neb. to Tex., e. to s. Mich., w. Ind., Ill., Mo., and locally to nw. Fla.

Var. **calvéscens** Fern. (becoming bald). — Cumberland Plateau of Ky. and Tenn., extending into s. O. and s. Ind.; mts., nw. Va. to Ala.

6. **R. caroliniénsis** (Walt.) Steud. (Carolinian). — *Erect, simple to divergently or strictly branched,* 1–9 dm. high, canescent-pilose, villous, hirsute, puberulent or rarely glabrescent, *the upper internodes greatly abbreviated,* the lower elongate, *at least the upper with crowded leaves and glomerules; middle and upper leaves distinctly petioled,* lanceolate to ovate, oval, elliptic or oblong, commonly strigose above, either strigose, hispid, pilose or glabrescent beneath; *glomerules very short-peduncled to subsessile, mostly crowded, at* 1–4 *upper nodes* (or in var. *dentata* up to 9 glomerules extending down often to base of plant); calyx-segments linear-setaceous, 1.3–2.5 cm. long, often copiously ciliate (rarely not); corolla lavender to lilac-blue, 2–5 cm. long; capsule glabrous, pilose or hirtellous, 1.2–1.7 cm. long. (*R. ciliosa,* in part, of ed. 7, not Pursh; *R. parviflora* sensu Britt. not the plant of Nees (basonym)). — Polymorphous; the following vars. and forms with us:

a. Stem simple or with mostly divergent branches from lower and median axils, these branches commonly elongate and flowering at summit; upper internodes canescent-pilose to softly villous; leaves elliptic, subrhombic, oval or narrowly ovate, membranaceous and pliable, more or less undulate (slightly crumpled when pressed); calyx-segments canescent-puberulent to villous on back; capsule pilose to glabrous.

Upper internodes copiously white-villous; leaves white-villous to canescent-pilose beneath, copiously strigose above; capsule often densely pubescent. *R. caroliniensis* (typical).

Upper internodes merely short-pilose or with only few elongate hairs; leaves glabrescent or merely sparsely hirtellous beneath; capsule usually glabrous or promptly glabrate. . . . . Var. *semicalva*.

a.Stem simple or with few ascending to suberect (very rarely strongly divergent) branches, spreading-hirsute, pilose, puberulent or glabrescent, the branches when borne from the middle and upper axils rarely prolonged or floriferous; leaves flat (not crumpled when carefully pressed), lanceolate to ovate, oval, elliptic or oblong, membranaceous and flexible to thickish, hard and firm, strigose to hispid along principal veins or quite glabrous beneath, remotely strigose to glabrescent above; capsule usually glabrous. . . *b*.

b.Leaves of upper more or less approximate nodes chiefly longer than those below, entire or not very dentate; 1 or 2 (rarely 3 or 4) nodes of the main axis floriferous, the lowest floriferous ones from closely crowded to 10 (–15) cm. apart; stem simple, only rarely branched. . . *c*.

c.Longer leaves of the uppermost nodes broadly lanceolate to oval, ovate, or elliptic, if lance-subacuminate 1.5–4.5 cm. broad; calyx-segments glabrous or nearly so on back; summit of throat of corolla 0.7–1.5 cm. broad, expanded limb 2.5–4 cm. broad. . . *d*.

d.Leaves membranaceous, pliable, oval, ovate or elliptic to broadly lanceolate. . . *e*.

e.Stem stoutish to slender, 1–5 mm. thick at base, 1–9 dm. high; larger leaves 5–12 cm. long, if obtuse mostly more than 7 cm. long. . . *f*.

f.Leaves strigose or hispid on veins beneath; calyx-segments ciliate.

Internodes of stem (or some of them) copiously spreading-hirsute. . . . . Var. *membranacea*.

Internodes puberulent to glabrescent, at most with remote elongate trichomes. . . . . Var. *membranacea*, forma *breviberbis*.

f.Leaves glabrous beneath.

Calyx-segments ciliate; internodes of stem usually pubescent; upper surfaces of leaves usually strigose. . . . . . . Var. *membranacea*, forma *hypopsila*.

Calyx-segments eciliate; internodes glabrescent or glabrous; upper surfaces of leaves glabrous or essentially so. . . . . Var. *membranacea*, forma *laevior*.

e.Stem filiform, 1–2 mm. thick at base, 1–2(–4) dm. high; leaves membranaceous, elliptic to oblong, the larger ones 2–6 cm. long, obtuse to rounded at apex.

Calyx-segments ciliate; internodes (or some of them) of stem copiously villous-hirsute with divergent hairs. . . . . . Var. *nanella*.

Calyx-segments eciliate; internodes glabrescent or merely puberulent . . . . Var. *nanella*, forma *eciliata*.

d.Leaves subcoriaceous, firm and stiff, lanceolate to lance-ovate, subacuminate; stem rigid, 4.5–9 dm. high. . . . . . . Var. *cheloniformis*.

c.Longer leaves of the uppermost nodes narrowly lanceolate to lance-linear, 0.5–1.5(–2) cm. broad; stem slender, 1.5–7 dm. high; calyx-segments canescent-pilose to glabrescent; throat of corolla 5–10 mm. broad at summit, expanded limb 2–3.5 cm. broad. . . . . Var. *salicina*.

b.Leaves of flowering summit gradually much reduced in size; 4–9 nodes of well-developed main axis bearing glomerules; all but upper internodes elongate; stem frequently with elongate floriferous ascending basal branches; median leaves oblong, oblong-lanceolate, ovate or elliptic, subequal, subcoriaceous, often undulate-dentate. . . . . Var. *dentata*.

R. caroliniénsis (typical). — Sandy woods, Fla. to s. S.C., Tenn., Ky. and e. Ark. June–Aug. Var. **semicálva** Fern. (half bald). — Fla. to e. Tex., n. to se. Va., centr. N.C., Tenn., n. Ala. and n. Miss.

Var. **membranàcea** Fern. (thin, membrane-like). — Woods, openings and clearings, S.C. to Tenn., n. to s. N.J., n. Md., W.Va., se. O. and Ky. June–Sept. — Forma **breviběrbis** Fern.

(with short beard), local throughout. Forma **hypopsìla** Fern. (smooth beneath), occasional, Va. and N.C. Forma **laèvior** Fern. (smoother), local, Va. and N.C.

Var. **nanélla** Fern. (very small) and forma **eciliàta** Fern. (without marginal hairs). — Local in woods, e. Md. to Ky., s. to S.C. and s. Miss.

Var. **cheloniiórmis** Fern. (like *Chelone*). — Woods and clearings, n.-centr. S.C. and n. Ga., n. to s. N.J., n. Md., n. Va., Ky. and s. Ind.

Var. **salicìna** Fern. (willow-like). — N. Fla. to e. Tex., n. to n. Ga., Tenn. and s. Ind.

Var. **dentàta** (Nees) Fern. (toothed). — Dry woods and clearings, S.C. to Tenn., n. to Del., se. Pa., W.Va., Ky. and s. Ind.

Dyschoríste (meaning, appropriately, *hard to separate*) oblongifòlia (Michx.) Ktze. (oblong-leaved), suggesting small-leaved erect plants of *Ruellia humilis*, with corolla slightly irregular, anthers mucronate at base and the 2–4 seeds not mucilaginous, has been reported but needs verification in our area.

FAM. 159. PHRYMÀCEAE (Lopseed Family)

Perennial herb, with slender branching stems, and coarsely toothed ovate leaves, the lower long-petioled; the small opposite flowers in elongated and slender terminal spikes, strictly reflexed in fruit. Corolla purplish or rose-color. Calyx cylindrical, 2-lipped; the upper lip of 3 bristly-subulate teeth; the lower shorter, 2-toothed. Corolla 2-lipped; upper lip notched; the lower much larger, 3-lobed. Stamens included. Style slender; stigma 2-lobed. Fruit dry, in the bottom of the calyx, oblong, 1-locular and 1-seeded. Seed orthotropous. Cotyledons convolute around their axis. A single genus:

1. PHRỲMA L. Lopseed

A single species, with characters of the family. (Derivation of the name unknown.)

1. **P. Leptostáchya** L. (early generic name, *slender-spiked*). — Plant 3–9 dm. high, smooth or slightly pubescent, with elongate internodes; leaves bright green, only sparingly pubescent, 0.5–1.5 dm. long, thin, the tip prolonged or acuminate, the marginal teeth prolonged and often double; calyx strongly ribbed and closed in fruit, the long slender teeth hooked at the tip. — Rich thickets and woods, w. N.B. to Man., s. to Fla., Ala., La. and e. Tex. July–Sept. (Asia)

Var. **confertifòlia** Fern. (with crowded leaves). — Lower, 2–4.5 dm. high, with short internodes (0.4–1.5 cm. long), cinereous-pilose; leaves grayish-puberulent beneath, the apex scarcely prolonged; the marginal teeth simple, crenate and short. — Rich woods, e. Va. to Del.

FAM. 160. PLANTAGINÀCEAE (Plantain Family)

Chiefly stemless herbs, with regular 4-merous spiked flowers, the stamens inserted on the tube of the dry and membranaceous veinless gamopetalous corolla, alternate with its lobes. — Chiefly represented by the two following genera:

Scape or peduncle with several–very many flowers in a spike; ovary and capsule 2-locular, 2–∞-seeded, usually circumscissile; terrestrial. . . . . . . 1. *Plantago*.

Scape 1–2-flowered; ovary and indehiscent fruit (achene) 1-locular, 1-seeded; aquatic. . . . . 2. *Littorella*.

1. PLANTÀGO L. Plantain. Ribwort. Queue de rat (Que.)

Calyx of 4 imbricated persistent sepals, mostly with dry membranaceous margins. Corolla salverform or rotate, withering on the capsule, the border 4-parted. Stamens 4, or rarely 2, in all or some flowers with long and weak exserted filaments, and fugacious 2-locular anthers. Ovary 2 (or in no. 6 falsely 3–4)-locular with 1–several ovules in each locule. Style and long hairy stigma single, filiform. Capsule 2-locular, 2–several-seeded, opening transversely, so that the top falls off like a lid and the loose partition (which bears the peltate seeds) falls away. Embryo straight, in fleshy albumen. — Leaves ribbed. Flowers whitish or pale, small, in a bracted spike or head, raised on a naked scape or peduncle. Genus nearly cosmop. (The Latin name, from *planta, footprint.*)

a. Leaves alternate in rosettes; scapes arising from the rosettes. Subgen. I. Euplantago. . . *b*.

b. Flowers all perfect or fertile; corolla not closed over the fruit, merely uplifted and with spreading or reflexed lobes at summit of maturing capsule; stamens 4. . . *c*.

c. Seeds plump, not boat-shaped (except for pit on inner face), 2–30; spikes slenderly cylindric or interrupted. . . *d.*

d. Tube of corolla glabrous; leaves flat, with dilated regularly toothed or entire strongly ribbed blade. . . *e.*

e. Lower 2–4 lateral ribs scattered, confluent with midrib; scapes hollow; roots few, fleshy, 0.5–1 cm. thick, one of them often becoming horizontal. 1. *P. cordata.*

e. Lower lateral ribs approximate to base of midrib; scapes solid; roots numerous, slender and fibrous.

Capsule circumscissile near the middle, 6–20-seeded; seeds reticulated, 1–1.7 mm. long. . . . . 2. *P. major.*

Capsule circumscissile toward the base, 2–9-seeded; seeds not reticulated, 1.5–3 mm. long.

Leaves thin, without dense wool at base, with slender elongate petioles; sepals sharply keeled on the back; seeds 4–9, dull, 1.5–2.5 mm. long. . . . . 3. *P. Rugelii.*

Leaves fleshy or subcoriaceous, often with dense yellowish wool at base, on stout petioles; sepals with broadly rounded back; seeds 2–4, lustrous, 2–3 mm. long. . . . . 4. *P. eriopoda.*

d. Tube of corolla pilose; leaves with linear to linear-lanceolate entire or irregularly toothed thick and fleshy obscurely ribbed blades; seeds 2–6.

Bracts or sepals or both minutely ciliolate; bracts rarely prolonged, only slightly keeled; seeds oblong to narrowly oval, 1.2–2.3 mm. long; spikes usually dense to the base. . . . . 5. *P. juncoides*, vars.

Bracts and sepals glabrous (rarely remotely ciliate); bracts often prolonged, with prominent gibbous keel; seeds linear-oblong, 2–3 mm. long; spikes mostly remotely flowered at base. . . . . . . 6. *P. oliganthos.*

c. Seeds flattened, or hollowed on inner face or boat-shaped, 2–4; spikes dense, ovoid to cylindric. . . . *f.*

f. Leaves lanceolate to ovate; flowers all perfect, proterogynous, the style first projecting from the unopened corolla, the anthers long-exserted after the opening of the corolla; inflorescence glabrous or essentially so; perennials with strong rhizomes and caudices.

Leaves ovate to elliptic or obovate; sepals distinct; flowers strongly fragrant. . . . . 7. *P. media.*

Leaves lanceolate to lance-oblong; forward (next the bract) sepals united; flowers barely fragrant.

Rootlets slender; leaves 0.2–2.3 dm. long, when fresh with the nerves impressed above; sepals 3–3.5 mm. long; seeds about 3 mm. long. . . . . 8. *P. lanceolata.*

Rootlets fleshy, thick; leaves 2–4 dm. long, when fresh with the nerves prominent above; sepals 4 mm. long; seeds 3.2–3.5 mm. long. 9. *P. altissima.*

f. Leaves linear to oblanceolate; flowers of 2 sorts (in respect to length of anthers and filaments) on different plants, mostly cleistogamous; inflorescence silky-villous or woolly; annuals or short-lived perennials with slender tap-roots.

Bracts much longer than the flowers.

Leaves and bracts dark-green, longer bracts (except in dwarf individuals) several times longer than flowers, the glabrous to short-pilose stiffly flexible linear-attenuate awns 1–5 cm. long; corolla-lobes about 2 mm. wide. . . . . 10. *P. aristata.*

Leaves and bracts pale; longer bracts 2–4 times longer than flowers, the firm and stiff bristleform long-villous awns 0.5–1 cm. long; corolla-lobes about 1 mm. wide. . . . . 11. *P. spinulosa.*

Bracts only slightly or not at all longer than flowers, mostly hidden in the densely silky-villous spike. . . . . 12. *P. Purshii.*

b. Flowers subdioecious or polygamo-cleistogamous; corolla in the fertile (or mainly fertile) plants closed over the maturing capsule and forming a beak, their anthers not exserted; sterile flowers with spreading corolla and exserted stamens; small annuals or biennials. . . . *g.*

g. Leaves spatulate or oblanceolate to obovate; corolla-lobes 1–3 mm. long; stamens 4.

Sepals rounded at tip or with the midrib barely projecting; seeds 1.2–2 mm. long, pale brown. . . . . 13. *P. virginica.*

Sepals attenuate; seeds 2.5–3 mm. long, red. . . . . 14. *P. rhodosperma.*

g. Leaves linear to filiform; corolla-lobes 0.25–0.5 mm. long; stamens 2.

Seeds 4, symmetrical, 1.25–1.8 mm. long; sepals obovate; corolla-lobes 0.5 mm. long. . . . . 15. *P. pusilla.*

Seeds 10–30 (fewer by abortion), unsymmetrical, about 0.75 mm. long; corolla-lobes 0.25 mm. long. . . . . 16. *P. heterophylla.*

a. Leaves opposite in pairs on the elongate stem and branches. Subgen. II. PSYLLIUM.

Leaves linear; peduncles from upper axils; spikes dense, ellipsoid-ovoid, their lowest bracts prolonged. . . . . 17. *P. indica.*

Subgen. I. EUPLANTÀGO Harms

1. P. cordàta Lam. (heart-shaped). — Tall *glabrous* perennial with few stout roots and heavy caudex; leaves fleshy, dark green, drying blackish, oval to cordate-ovate, 1–3 dm. long, long-petioled, *the venation pinnate, the lateral ribs confluent with the midrib; scape* 1–5 dm. long, *hollow;* spike loosely flowered, slender, up to 4.5 dm. long; bracts fleshy, round-ovate, about equaling calyx; sepals rounded, 3–3.5 mm. long, with broad keel; *capsule 2–4-seeded*, globose-ovoid, 4–5 mm. long, circumscissile near middle; *seeds* slenderly ellipsoid, dark brown, 3–3.5 *mm. long.* — In and along streams and in swampy woods, Ala. to La., n. to the lower Potomac R., Va., tidal estuary of the Hudson R., N.Y., s. Ont., Mich., Wisc. and Mo. May–July.

2. P. MÀJOR L. (larger), COMMON P., WHITEMAN'S-FOOT. — Annual or perennial; *leaves thick, usually roughish on one or both surfaces, when dry, with minute hairs,* 0.5–3 dm. long, with broad petiole and broadly elliptic to cordate-ovate, undulate, sinuate-dentate or angular-toothed, strongly ribbed blades either depressed or ascending; scapes solid, curved-ascending to erect; spike rather dense, obtuse, 0.1–5 dm. long; bracts shorter than to equaling or rarely longer than calyx, ovate, with slender keel; sepals elliptic to elliptic-rotund, 1.5–2 mm. long; *capsule* brown or purplish, *broadly conic to the rounded summit, circumscissile below the tips of the sepals; seeds* 6–15, angulate, *reticulated,* 1–1.7 *mm. long.* — Roadsides, dooryards, etc., semicosmop. weed, throughout our range and beyond. June–Oct. (Natzd. from Eu.) FIG. 1492. — Polymorphous; scores of confluent variations have been named. Forma INTERMÈDIA (Gilib.) Pilger (intermediate) is nearly or quite glabrous, smooth to touch, with round-based or subcordate coriaceous leaves. The tiniest plants, with leaves only 1–4 cm. long and spikes 1–3 cm. long are forma MICROSTÀCHYA (Hayne) Pilger (tiny-spiked). Numerous monstrous forms or sports occur: forma PANICULÀTA Domin (with a panicle), with spike changed to a panicle; forma RÒSEA (Dcne.) Prahl (having a rosette), with dilated leaves borne near summit of scape.

1492. P. major.

Var. **Pìlgeri** Domin (in honor of ROBERT PILGER, contemporary student of the genus). — *Leaves membranaceous,* nearly or quite *glabrous, elliptic-ovate,* entire or merely undulate, *tapering to slender ascending petioles;* spike relatively slender; *capsules* conic above, *circumscissile near tips of sepals;* seeds 0.7–1.5 mm. long. (Var. *asiatica* of ed. 7, not *P. asiatica* L.) — River-gravels, damp ledges, etc., Nfld. to n. B.C., s. to N.B., n. N.E., L. Sup., Ont., N.D. and Ariz. (Eu.)

Var. **scopulòrum** Fries & Broberg (of rocks). — *Fleshy-subcoriaceous,* copiously villous to sparingly pilose; *leaves* decumbent or slightly ascending, narrowly to broadly elliptic to ovate, undulate to irregularly sinuate-dentate, *with broad and short petioles;* scapes arched-ascending; spikes mostly dense, up to 2.5 dm. long; *capsule* purple or brown, *rounded-ellipsoid, with broadly rounded (scarcely conical) summit; seeds* 10–20. (Var. *intermedia* of ed. 7, not (Gilib.) Dcne.; *P. halophila* Bickn.) — Brackish or saline shores, rarely inland, P.E.I. and C.B. to Del.; James Bay; Wash. to Calif. July–Oct. (Eu.)

3. P. Rugèlii Dcne. (for its discoverer, FERDINAND RUGEL, 1806–1879). — Resembling no. 2; leaves erect or spreading, with thinnish pale smooth or slightly hirtellous blades with red petioles; *spikes* slender, dense to alternate-flowered, *attenuate to apex; calyx stipitate;* sepals oblong to lance-triangular, acutely keeled; *capsules cylindraceous, circumscissile nearly at base,* 4–9-*seeded; seeds oval,* 1.5–2.5 mm. long, *not reticulated.* — Damp shores, roadsides and waste places, sw. Que. to N.D., south beyond our range. July–Oct. FIG. 1493.

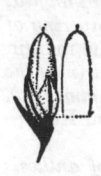

4. P. eriòpoda Torr. (woolly-based). — *Fleshy* perennial *with stout crown usually bearing a mass of yellowish wool; leaves* thick, *oblanceolate to narrowly ovate,* gradually tapering into the broad petiole; spike remotely flowered or with approximate flowers above; *bracts* about equaling calyx, broadly ovate, *with broad rounded keel; sepals* rounded-elliptic, *with rounded scarcely keeled back;* capsule conic, narrowed to the truncate apex; *seeds 2–4, lustrous,* reddish-brown, 2–3 *mm.*

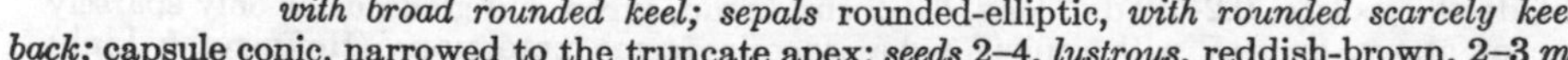

1493. P. Rugelii.

long. — Saline marshes and shores, Anticosti I. and e. Gaspé Pen. to Kamouraska Co., Que.; Man. to Mackenz., s. to Ia., Neb., Mex. and n. Calif. July, Aug.

5. P. **juncoìdes** Lam. (rush-like), Seaside-P. — Perennial with deep root and 1–∞ crowns; *leaves linear-lanceolate, fleshy,* attenuate at tip, strongly ascending; scapes mostly exceeding leaves; spike with mostly crowded spreading-ascending flowers; *bracts or calyx-segments or both minutely ciliolate; bracts rarely prolonged and with only slight keel,* broadly ovate, subreniform or suborbicular; calyx-segments broadly oblong to suborbicular, with thick or obscure keel; *corolla-tube pilose;* mature capsules ovoid to broadly conic, blunt to rounded at summit; mature seeds oblong to narrowly oval, 1.2–2.3 mm. long. — Wide-ranging mostly maritime species; typical form of Pacific N.Am. and Patagonia, with anthers 1.5–2 mm. long, their subulate tips 0.3–0.7 mm. long. — Represented with us by three vars. with anthers 1–1.5 mm. long, their subulate tips only 0.1–0.4 mm. long:

Leaves linear, erect or ascending, only rarely spreading, entire, 1–8 mm. broad, commonly shorter than scapes.
  Scapes 0.5–2.3 dm. high; longer spikes 2–10 cm. long; leaves 1–8 mm. broad. Var. *decipiens.*
  Scapes 1–7 cm. high; spikes 0.5–2 cm. long; leaves 1–2 mm. broad. . . . Var. *glauca.*
Leaves lanceolate to oblanceolate, depressed or wide-spreading, entire or often toothed, 3–15 mm. broad, commonly equaling or much exceeding the scapes. Var. *laurentiana.*

Var. **decípiens** (Barnéoud) Fern. (deceiving), *P. decipiens* Barnéoud — Headlands, cliffs and dry beaches, chiefly maritime, s. Greenl. and s. Lab. to lower St. Lawrence, Que., s. to N.J. June–Sept. (N. Eu.) In forma **vivípara** Vict. & Rousseau (sprouting on the parent plant) of the lower St. Lawrence, flowers changed to leafy tufts. — Passing insensibly into

Var. **glaùca** (Hornem.) Fern. (blue-green), *P. borealis* Lange — Similar habitats, Greenl. to Hudson Bay, s. to e. Me. (N. Eu.)

Var. **laurentiàna** Fern. (of the Gulf of St. Lawrence). — Cliffs, ledges and sands, Nfld.; Côte Nord and Anticosti, Que.; M.I.; P.E.I. and N.S.

6. P. oligánthos R. & S. (few-flowered), Seaside-P. — More fleshy than no. 5; *leaves mostly* erect or strongly ascending, 5–12 mm. broad, mostly *equaling or exceeding* the erect *scapes; spikes* 0.3–2 dm. long, *often remotely flowered at base; bracts and calyx-segments glabrous or very rarely with remote ciliation, the bracts often with prolonged tips and with thick or gibbous keel; seeds linear-oblong,* 2–3 *mm. long.* — Salt-marshes and saline or brackish shores, s. Lab. and Nfld. to lower St. Lawrence, Que., s. to N.J.; Man. and Alta. July–Sept.

Var. **fállax** Fern. (deceitful). — Leaves loosely spreading, arcuate, narrowly linear, 0.5–4 mm. broad, longer than the arcuate scapes and short (0.5–7 cm. long) denser spikes. — Lab. and Nfld. to e. Que., s. to e. Me.; Hudson B.

P. Corónopus L. (treated by early writers as a species of *Coronopus*), with thinner pinnately dissected leaves and tightly appressed flowers, occurs sporadically about ports but does not long persist. (Adv. from Eu.)

7. P. mèdia L. (intermediate), Hoary P. — Perennial with short stout caudex; *leaves* closely rosulate, usually flat on the ground, *ovate, broadly elliptic or obovate,* 0.5–1.7 dm. long, entire, undulate or coarsely dentate, canescent, 7–9-nerved; *spike dense,* cylindric, 2–10 *cm. long,* much shorter than peduncle; bracts narrowly ovate, nearly equaling calyx; *flowers fragrant; sepals distinct,* rounded-ovate to obovate, 2.2–2.5 mm. long, strongly keeled, hyaline; filaments prolonged; capsule about 3 mm. long; seeds 2–4, slightly concave or flat on the face, 2 mm. long. — Fields, lawns, etc., occasional, Que. and Ont., s. to N.E., N.Y. and Mich. June–Sept. (Adv. from Eu.)

8. P. lanceolàta L. (lance-shaped), Ribgrass, Ripplegrass, English P., Buckhorn. — Annual, becoming perennial *with strong caudex and tough slender rootlets; leaves lanceolate to lance-oblong,* ascending to spreading, tapering into the petioles, strongly ribbed, 0.5–2.3 dm. long, 0.6–4 cm. broad, *when fresh with the nerves impressed above; scape* tough, *grooved-angled,* elongating, 2–8 *dm. high; spike dense, at beginning of anthesis slenderly ovoid-conic, tapering at apex, in fruit* cylindric and obtuse, 1.5–8 *cm. long;* bracts scarious, broadly ovate, the margin erose-undulate; *flowers about* 5 *mm. broad,* not strongly fragrant; *forward sepals united* (*next the bract*), 3–3.5 mm. long; anthers long-exserted; *seeds* 1–2, *hollowed on the face, about* 3 *mm. long.* — Common and troublesome weed of grassland, throughout. May–Oct. (Natzd. from Eu.) — A common monstrosity has the spike very compound and lobate.

Var. **sphaerostáchya** Mert. & Koch (spherical-spiked). — *Spike at beginning of anthesis subglobose, rounded at apex, in fruit subglobose to cylindric and round-tipped,* 0.5–2.3 *cm. long;* leaf-blades lanceolate, 0.2–1.2 dm. long, 0.3–2 cm. broad, green and glabrous or only sparsely pubescent above, or gray with copious pubescence in forma **erióphora** (Hoffmgg. & Link)

G. Beck (bearing wool); scapes 0.3–4.5 dm. tall. — Nfld. to Mich., s. to N.S., s. N.E., N.Y. and Mo. (Natzd. from Eu.) — Forma VERNÀLIS Béguinot (vernal), with glabrous elliptic leaves, w. Nfld. and Que.

9. P. ALTÍSSIMA L. (very tall). — Stouter than no. 8, *with heavy creeping rhizome and thick fleshy roots; leaves* glabrous, up to 4 dm. long and 4 cm. broad, *when fresh with nerves prominent above;* scape stout, 0.6–1 m. high; flowers 6–7 mm. broad; *sepals 4 mm. long; seeds* 3.2–3.5 *mm. long.* — Fields, as yet local, Mass. and Ct. to Mich. June–Sept. (Adv. from Eu.)

10. P. aristàta Michx. (bearing bristles), BRACTED P. — Annual or short-lived perennial, *with dark green* loosely villous to glabrous linear entire *leaves* narrowed to margined semi-clasping petioles; scapes stiffly ascending or arcuate; *spike* 0.4–17 cm. long, *with prolonged dark green herbaceous stiffly flexible linear-attenuate smooth or sparingly pilose bracts, these in larger plants several times longer than flowers,* in dwarf individuals less prolonged; flowers mostly cleistogamous, of two sorts on different plants; sepals herbaceous, spatulate-oblong, 2.5 mm. long; *corolla-lobes* rounded-ovate, subauriculate at base, *about* 2 *mm. broad;* seeds 2, oblong, finely pitted. (Incl. subsp. *Nuttallii* (Rapin) E. L. Morris, the smallest individuals) — Dry open soil, originally from Ill. to La. and westw., now (beginning about 1869) natzd. generally eastw. to N.E. and the Atlantic states. June–Nov.

11. P. spinulòsa Dcne. (with small spines). — Similar to no. 10, *pale green;* spike loosening and becoming slightly interrupted in maturity; *bracts rigid, stiffly bristleform, long-villous, the longer two to four times longer than flowers,* 0.5–1 *cm. long; corolla-lobes about* 1 *mm. broad.* — Dry open soil, Minn. to Alta., s. to Mo., Okla. and Tex. May–July.

12. P. Púrshii R. & S. (for FREDERICK TRAUGOTT PURSH, 1774–1820). — Similar to no. 10, but *white with silky wool;* leaves linear-oblong to filiform; *spike slender-cylindric,* very dense, 0.5–15 cm. long, *silvery* (drying tawny) *-woolly; bracts hidden or barely projecting.* — Dry plains, openings and prairies, Ont. to B.C., s. to Ind., Ill., Mo., Okla., Tex., etc.; casual e. to N.E. and N.J. May–Aug.

13. P. virgínica L. (Virginian), HOARY or PALE-SEED-P. — *Hoary-pubescent* annual or biennial; leaves narrowly obovate, 2–12(–19) cm. long, permanently villous with long soft trichomes; the margin entire, undulate or shallowly dentate; scapes long-villous with flexuous hairs mostly 1–2 mm. long, up to 2.5 dm. high; spikes dense, 0.1–2(–3) dm. long; *flowers subdioecious or polygamodioeceous; sepals* 2–2.5 mm. long, *rounded at tip, the midrib barely projecting;* corollas of fertile flowers closed over maturing capsule and forming a beak, their anthers included; *corolla-lobes* 2–3 *mm. long; seeds pale brown,* 1.6–2 *mm. long.* — Dry open soil, Fla. to Tex. and s. Calif., n. to s. Me., Mass., s. and w. N.Y., O., Mich., Ill., Mo., Kans. and Oreg. Mid-Apr.–June.

Var. **viridéscens** Fern. (greenish). — *Leaves oblanceolate,* 0.5–10 cm. long, repand-dentate or entire, *green and glabrous or sparsely hirtellous and glabrate; scapes* 1–12 cm. long, *pilose-hispid with hairs* 0.2–0.5 *mm. long;* spikes 0.3–13 cm. long; sepals 1.5–2.3 mm. long; corolla-lobes 1–2.5 mm. long; *seeds* 1.2–1.5 *mm. long.* — Fla. to Tex., n. to s. N.J. and Md. March–May.

14. P. RHODOSPÉRMA Dcne. (red-seeded). — Similar to no. 13, larger; leaves up to 3 dm. long, commonly with broad horizontal remote teeth; spikes up to 2 dm. long, usually lax at base; green bracts and greenish *bristly sepals attenuate,* the latter 2.5–3 mm. long; corolla-lobes 2–3 mm. long; *seeds* 2.5–3 *mm. long, red.* — Sandy openings, roadsides and waste places, Mo. and e. Kans. May–July. (Adv. from farther west and southwest)

15. P. pusílla Nutt. (very small). — Tiny annual; leaves filiform to narrowly linear, entire (rarely laciniate), 1–5 cm. long; scapes filiform, erect or strongly ascending, 1–10 cm. high; spikes slenderly cylindric, 1–6 cm. long, the flowers closely imbricated but becoming scattered; *bracts* appressed, *equaling calyx,* fleshy, dorsally subsaccate, hardly keeled, triangular-ovate, blunt, 1.5–2 mm. long; *sepals obovate; corolla-lobes* 0.5 *mm. long,* acute; stamens 2; *capsule about one-third longer than calyx,* circumscissile below middle; *seeds* 4, *symmetrical,* 1.25 *mm. long.* (*P. elongata* of ed. 7, not Pursh) — Sandy fields and openings, Ga. to Tex., n. to se. Mass., R.I., Ct., se. N.Y., Ind., Ill., Minn., Neb. and Oreg. April–June.

Var. **màjor** Engelm. (larger). — Coarser; scapes up to 2.5 dm. high; spikes lax, 6–12 cm. long, with flowers scattered from the first; seeds 1.7–1.8 mm. long. (*P. pusilla,* subsp. *Engelmanniana* E. L. Morris) — La. and Okla. to Mo. and s. Ind.

16. P. heterophýlla Nutt. (diverse-leaved). — Resembling no. 15, depressed or ascending; *leaves* thin and flat, soft, 0.7–5 *mm. wide,* up to 1.5 dm. long, entire or remotely repand-dentate; spikes loosely-flowered, 1–12 cm. long; *bracts equaling to twice as long as calyx,* closely investing its base; sepals ovate, with slender green midrib; *corolla-lobes about* 0.25 *mm. long; capsules about twice length of calyx; seeds* 10–25 *or more, irregular,* 0.75 *mm. long.* — Sandy fields and openings, Fla. to Lower Calif., n. to e. Va. (adv. to N.J. and se. Pa.), Ky., Ill., Mo. and Calif. Apr., May.

Subgen. II. PSÝLLIUM (Juss.) Harms

17. P. ÍNDICA L. (East Indian). — Elongate-stemmed *bushy-branched* annual up to 6 dm. high; *leaves opposite*, chartaceous, straight or curving, linear, attenuate; *peduncles axillary*, stiff, 4–8 cm. long; spikes dense, ellipsoid or subglobose, 1–2 cm. long; lowest bracts with rounded blade and abrupt prolonged tip; sepals elliptic to obovate, obtuse; seeds shining, 2.5 mm. long, with thick margins. (*P. arenaria* Waldst. & Kit.) — Roadsides and waste places, becoming frequent, Que., Ont. and Minn., s. to Va., O., Ind. and Ill. July–Oct. (Natzd. from Eu.)

P. PSÝLLIUM L. (old generic name), similar to no. 17 but with narrow and gradually tapering bracts and sepals, is casual in Mass. (Adv. from Eu.)

2. LITTORÉLLA Bergius

Flowers monoecious. The staminate solitary, on a mostly simple naked scape; calyx 4-parted, longer than the cylindraceous 4-cleft corolla; anthers exserted, on very long capillary filaments. Pistillate flowers usually 2, sessile at the base of the scape; calyx of 3 or 4 unequal sepals; corolla urceolate, with a 3–4-toothed orifice. Ovary with a single locule and ovule, tipped with a long laterally stigmatic style, maturing as an achene. Small genus of e. N. Am., Eu. and s. S. Am. (Name from *litus* or *littus*, *shore*, from the place of growth.)

1. L. americàna Fern. (American). — Stoloniferous, with filiform rhizome and roots; leaves arching or straightish, 1.5–6 cm. long, linear-subulate, flattish, with base narrowly scarious-margined; peduncles of staminate flowers 0.7–4 cm. long, naked or with a bract above the middle; calyx 2.4–4 mm. long, with oblong segments; filaments 7–12 mm. long, anthers 2–4 mm. long; achenes slenderly oblong, 2 mm. long, blackish. (*L. uniflora* of ed. 7, not (L.) Aschers.) — Sandy, gravelly or muddy shores and margins of lakes and ponds, local, Nfld. to Ont., s. to N.S., Me., Vt., n. N.Y., n. Wisc. and n. Minn. July–Sept.

FAM. 161. RUBIÀCEAE (MADDER FAMILY)

Woody or herbaceous plants, with opposite entire leaves connected by interposed stipules, or in whorls without apparent stipules, the calyx adherent to the 2–4-locular ovary, the stamens as many as the lobes of the regular corolla (4–5) and inserted on its tube. Flowers perfect, but often dimorphous (as in *Mitchella* and *Houstonia*). Fruit various. Seeds anatropous or amphitropous. Embryo commonly rather large, in copious hard albumen. — A very large family, the greater part, and all its most important plants (such as the Coffee- and Peruvian-bark-trees), tropical.

a.Ovules solitary in the locules; carpels separating when ripe (or in no. 6 the fruit a double berry). Subfam. I. COFFEOIDEAE. . *b*.
- *b*.Herbs; the flowers not in large terminal naked spherical heads. . . . *c*.
  - *c*.Leaves in whorls.
    - Flowers involucrate; calyx with lanceolate lobes; corolla funnelform, with limb 4–5-lobed, blue or pink. . . . . 1. *Sherardia*.
    - Flowers exinvolucrate; calyx-teeth obsolete; corolla rotate, 4- or 3-parted, white, greenish, dull purplish or yellow. . . . . . 2. *Galium*.
  - *c*.Leaves opposite.
    - Stems ascending to erect, with elongate deciduous leaves; flowers axillary or terminal, separate; fruits dry, finally separating into the carpels.
      - Flowers crowded into whorled glomerules or dense cymes.
        - Cymes sessile in the axils as well as terminal; carpels 2, one or both dehiscent. . . . . 3. *Spermacoce*.
        - Cymes conspicuously involucrate, terminating elongate branches; carpels 2–4, closed or nearly so. . . . . . . 4. *Richardia*.
      - Flowers (and fruits) 1–3 in an axil; the 2 or 3 carpels indehiscent. . 5. *Diodia*.
    - Stems trailing, with rounded evergreen leaves; flowers terminating branchlets, twin, their ovaries united into 1; fruit a fleshy 2-eyed berry. . . . . 6. *Mitchella*.
- *b*.Shrub or tree; flowers in large terminal naked spherical heads; corolla tubular; fruits inversely pyramidal. . . . . 7. *Cephalanthus*.

a.Ovules numerous in each locule of the dehiscent capsule; leaves opposite. Subfam. II. CINCHONOIDEAE (ours all herbs).
- Corolla salverform or funnelform; seeds 4–20 in each locule, thimble-shaped, bowl-shaped or saucer-shaped. . . . . 8. *Houstonia*.
- Corolla rotate; seeds very numerous, minute, angled. . . . . . . 9. *Hedyotis*.

Subfam. I. COFFEOÌDEAE

1. SHERÁRDIA L. Field-Madder

Calyx-lobes lanceolate, persistent. Corolla funnelform, the limb 4–5-lobed. Style filiform, 2-cleft; stigmas capitate. Fruit dry, twin, of 2 indehiscent 1-seeded carpels. — A slender procumbent Old World herb, with square stems, lanceolate pungent leaves in whorls of 4–6, and small blue or pinkish flowers surrounded by a gamophyllous involucre. (Named for Dr. *William Sherard*, 1659–1728, patron of Dillenius.)

1. S. arvénsis L. (of cultivated fields). — The only species. — Fields, orchards, recent plantings and waste places, local, N.S. and sw. Que. to O., s. to N.C., Tenn. and Mo. May–Sept. (Adv. from Eu.)

2. GÀLIUM L. Bedstraw. Cleavers. Gaillet (Que.)

Calyx-teeth obsolete. Corolla rotate, valvate in the bud. Stamens 4, rarely 3, short. Styles 2. Fruit dry or fleshy, globular, twin, separating when ripe into the 2 seed-like indehiscent 1-seeded carpels. — Slender herbs, with mostly small cymose flowers, 4-angled stems, and whorled leaves, the roots often containing a red coloring matter. Genus nearly cosmop. (Name from the Greek *gala*, *milk*, which is curdled by some species of the genus.)

a. Ovary and fruit bristly. . . *b.*
b. Flowers solitary, sessile or subsessile, reflexed; leaves 4–10 mm. long. . . . 1. *G. virgatum.*
b. Flowers more numerous, definitely stalked; leaves longer. . . . *c.*
c. Principal leaves in 5's–8's; stems weak, leaning, diffuse or reclining, their angles or the leaf-margins and nerves (beneath) harsh.
Annual; stem harsh; leaves bristle-tipped, the cilia at their margins divergent or reflexed. . . . . 2. *G. Aparine.*
Perennial, stoloniferous; stem smooth; leaves cuspidate, their margins with ascending cilia. . . . . 3. *G. triflorum.*
c. Principal leaves in 4's; stems ascending, scarcely diffuse or reclining. . . *d.*
d. Peduncles loosely 3–several-flowered; flowers dull purple to greenish-white; leaves lanceolate to oblong or ovate. . . *e.*
e. Mature flowers and fruits on distinct pedicels.
Leaves firm, dull, in numerous subuniform whorls; stems firm. . 4. *G. pilosum.*
Leaves membranaceous, lucid; the upper of the 2–4 whorls largest; stems flaccid. . . . . 5. *G. kamtschaticum.*
e. Flowers and fruits mostly sessile along the loosely divergent branches of the peduncles.
Leaves oval or oblong, obtuse; corollas commonly pubescent outside. . . . . 6. *G. circaezans.*
Leaves lance-acuminate; corollas glabrous. . . . . 7. *G. lanceolatum.*
d. Flowers bright white, numerous, in a compact panicle; leaves linear to narrowly lanceolate. . . . . 10. *G. boreale.*
a. Ovary and fruit smooth or merely granulose. . . *f.*
f. Fruits dry, not succulent, if more than 2 mm. in diameter on elongate peduncles or branches; leaves not evergreen. . . *g.*
g. Erect plants, with neither stem nor leaves retrorsely scabrous. . . *h.*
h. Leaves of primary whorls in 4's, 2–6 cm. long; bases of plants chiefly without leafy perennating offshoots. . . *i.*
i. Flowers bright white, numerous, in compact erect panicles; leaves linear-lanceolate. . . . . 10. *G. boreale.*
i. Flowers dull purple to greenish-white, in loose panicled cymes.
Leaves lanceolate to lance-ovate, with prominent lateral nerves. . 8. *G. latifolium.*
Leaves linear to lanceolate, the lateral nerves obscure or wanting. 9. *G. arkansanum.*
h. Leaves of primary whorls mostly in 6's or 8's; bases of stems producing leafy autumnal offshoots. . . *j.*
j. Flowers yellow, in elongate panicles; leaves linear-acicular.
Panicle rather dense, its lower branches much exceeding the internodes. . . . . 11. *G. verum.*
Panicle loose, slender, interrupted, its lower branches shorter than the internodes. . . . . 12. *G. Wirtgenii.*

j. Flowers white, in loose leafy panicles; leaves flat, linear to lanceolate or oblanceolate. . . *k.*

k. Leaves firm, linear or oblanceolate, rarely 2 cm. long.

Flowering branches and pedicels strongly divaricate. . . . . . 13. *G. Mollugo.*

Flowering branches and pedicels mostly ascending. . . . . 14. *G. erectum.*

k. Leaves thin, lanceolate, mostly 3–5 cm. long. . . . . . . 15. *G. sylvaticum.*

g. Matted, reclining or loosely ascending plants, the stems or leaves either scabrous or smooth; leaves linear to oblanceolate or oblong, 0.3–3 cm. long; flowers white or whitish (yellow in no. 28). . . *l.*

l. Flowers white; fruits with hemispherical or semiovoid mericarps; leaves glabrous or merely with scabrous, hispid or bristly margins or nerves (beneath). . . *m.*

m. Leaves blunt or rounded at tip, without terminal bristle. . . *n.*

n. Flowers several, in small dichotomous cymes; pedicels horizontally spreading. . . . . 16. *G. palustre.*

n. Flowers solitary or in mostly simple cymes of 2–several flowers, if numerous with ascending pedicels. . . *o.*

o. Corollas greenish-white, small (1.5 mm. or less broad), commonly with 3 (sometimes 4) obtuse lobes; stems scabrous or smooth, diffuse or weakly reclining, developing prostrate matted basal autumnal offshoots. . . *p.*

p. Mature fruits 1.25–2 mm. in diameter; peduncles in maturity 0.3–3 cm. long; principal leaves 0.6–2 cm. long. . . *q.*

q. Leaves (at least when young) with retrorse-scabrous margins and midrib (beneath); upper internodes of stem often scabrous.

Leaves linear or linear-oblanceolate, chiefly in 4's; the filiform simple peduncles or the 2 or 3 filiform arcuate pedicels scabrous, 0.5–3 cm. long. . . . . . . 17. *G. trifidum.*

Leaves broadly oblanceolate to oblong-spatulate, those of main axis in 4's, 5's or 6's; the stiffer and straighter peduncles mostly shorter, the 2–4 pedicels glabrous. . 18. *G. tinctorium.*

q. Leaves (and usually the internodes of stem) glabrous, mostly in 4's.

Leaves linear or linear-oblanceolate, those of primary axis 0.7–2 cm. long; the filiform simple peduncles or the filiform pedicels mostly arcuate, 0.5–3 cm. long. . . 17. *G. trifidum*, var. *halophilum.*

Leaves broadly oblanceolate to oblong-spatulate, those of primary axis 6–12 mm. long; peduncles and pedicels straight or straightish, 3–10 mm. long. . . . . 19. *G. Brandegei.*

p. Mature fruits 0.8–1 mm. in diameter; peduncles mostly 0.5–4 mm. long; principal leaves 2–10 mm. long; stems matted, scabrous. . . . . 20. *G. brevipes.*

o. Corollas white, 2–2.5 mm. broad, commonly with 4 acute or acutish lobes; stems smooth, stiffish, erect or suberect, without matted basal offshoots; leaves mostly in 4's.

Leaves loosely spreading or ascending; cymes terminal and terminating the branches, not overtopped by lateral branches; fruits 2.5–3.5 mm. in diameter. . . . . . 21. *G. obtusum.*

Leaves soon reflexed; cymes soon overtopped by erect lateral branches; fruits 1–1.5 mm. in diameter. . . . . . 22. *G. labradoricum.*

m. Leaves sharply acute, cuspidate or bristle-tipped. . . *r.*

r. Margins of leaves smooth or upwardly scabrous; flowers in terminal panicles of cymes. . . *s.*

s. Leaves linear; floral bracts reduced or wanting; panicles very lax; stems without conspicuous leafy basal offshoots.

Leaves spinulose-margined, bristle-tipped, the principal ones 5–8 mm. long; corolla-lobes 0.2–0.4 mm. long; fruits 0.7–1 mm. high. . . . . 23. *G. parisiense.*

Leaves smoothish, subulate-tipped, the principal ones 1–2 cm. long; corolla-lobes 0.8–1 mm. long; fruits 1.5–2.3 mm. high. 24. *G. concinnum.*

s. Leaves spatulate to narrowly obovate, spinulose-margined, cuspidate; branches smooth; floral bracts foliaceous; panicle open-cylindric; stems producing prostrate leafy basal offshoots. . . . . 25. *G. saxatile.*

*r.*Margins of leaves retrorse-scabrous.

Inflorescence a leafy panicle, the cymes mostly exceeding the bracteal leaves; pedicels ascending or divergent; fruits about 2 mm. long; intricately branched perennial. . . . . . . 26. *G. asprellum.*

Inflorescences axillary, the few-flowered peduncles shorter than or barely exceeding the subtending leaves; pedicels recurving; fruits 3–4 mm. long; simple or subsimple annual. . . . . 27. *G. tricorne.*

*l.*Flowers yellow; fruit with oblong-cylindric mericarps; leaves and stem villous. . . . . 28. *G. pedemontanum.*

*f.*Fruits succulent or at least juicy when fresh; inflorescences 1–few-flowered, on short divergent axillary peduncles; leaves often evergreen.

Glabrous; leaves linear or linear-lanceolate. . . . . 29. *G. uniflorum.*

More or less hispidulous; leaves elliptic. . . . . 30. *G. hispidulum.*

1. G. virgàtum Nutt. (wand-like). — Slender erect *annual*, simple or forking below, 1–4 dm. high; stem wiry, smooth; *leaves 4–10 mm. long, mostly in 4's*, oblong, *bristly; flowers solitary, sessile or subsessile, reflexed*, white, subtended by a pair of bracteal leaves; *fruit uncinate-hispid.* — Dry barrens, glades and rocky woods, La. and Tex., n. to Tenn. and Mo. May, June. Fig. 1494.

1494. G. virgatum.

2. G. Aparìne L. (old generic name, by botanists for centuries interpreted as meaning to catch, cling or scratch), Cleavers, Spring-C., Goosegrass. — *Weak* or reclining *annual; stem retrorsely bristly*, hairy at the joints; *leaves mostly 8 in a whorl*, lanceolate, tapering at the base, chiefly 2–7 cm. long, *bristle-tipped, with coarse divergent or reflexed marginal cilia; peduncles 1–3-flowered;* corollas white; *fruits bristly*, 1.5–4 *mm. in diameter*. (Incl. var. *Vaillantii* W. D. J. Koch; *G. spurium* L.) — Rich woods, thickets, seashores and waste ground, Nfld. to Alaska, s. to N.S., N.E., Fla. and Tex., both native and introd. May–July. (Eurasia) Fig. 1495.

1495. G. Aparine.

3. G. triflòrum Michx. (three-flowered), Sweet-scented B. — *Weak perennial; stems simple or remotely forking*, 2–10 or more dm. long, *smooth; leaves mostly in 6's, thin*, elliptic-lanceolate, or in the possibly teratological forma **Rollándii** Vict. (for Louis Roland, Frère Rolland-Germain, 1881–) some or all linear and reduced nearly to midribs, the primary ones 2–8.5 cm. long, *the upper ones only slightly reduced, cuspidate, their margins with minute ascending cilia; peduncles axillary, comparatively short, terminally 3-flowered or 3-forked, the flowers all pedicelled;* corolla greenish-white; *fruit densely bristly.* — Woods and thickets, Nfld. to Alaska, s. through e. Can. and n. states and more sparingly to Va., Tenn., La., Tex. and Mex. Mid-May–Sept. (Greenl.; Eurasia) Fig. 1496. — Often sweet-scented in drying. Passing into

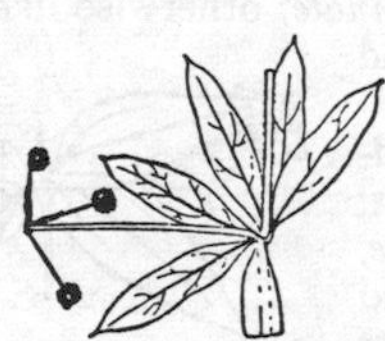

1496. G. triflorum.

Var. **asprellifórme** Fern. (with form of *G. asprellum*). — *Inflorescence paniculate, the leaves of the upper whorls strongly reduced in size; the flowering branches elongate, paniculate-branched, mostly with* 3–8 *nodes, and leaves only* 3–10 *mm. long.* — Sw. Que. and N.Y. to Fla. and Tenn. July–Sept.

4. G. pilòsum Ait. (soft-hairy). — *Stems* tufted from a perennial crown, *stiff, erect or ascending*, simple or forking, 1.5–9 dm. high, *densely to sparsely spreading-pilose* at least on the lower angles; *leaves oval, firm*, dotted, *hairy*, those of the numerous primary whorls *in 4's*, 1.5–2.5 *cm. long*, the lateral nerves obscure; inflorescence paniculate, the peduncles 2–3-forked, the greenish-white to purplish flowers on distinct pedicels; *fruits bristly.* — Dry woods and copses, s. N.H. to s. Ont. and Mich., s. to N.C., Tenn., Mo., Okla. and Tex. Late June–Aug. Fig. 1497.

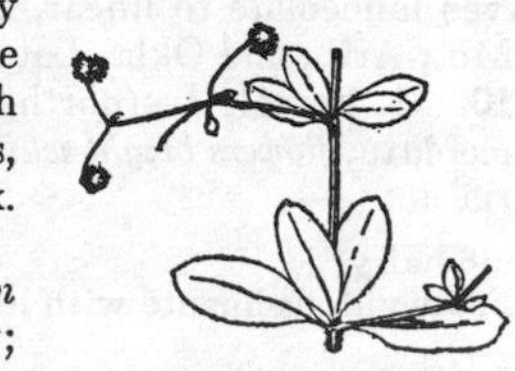

1497. G. pilosum.

Var. **puncticulòsum** (Michx.) T. & G. (with minute dots). — *Stem finely and usually densely incurved-uncinate on angles* at least below; leaves oval, their pubescence incurved-uncinate. — Fla. to Miss., n. to N.J.

Var. **laevicaùle** Weath. & Blake (smooth-stemmed). — *Stem glabrous throughout; leaves narrowly elliptic*, their sparse hairs not strongly incurved. — Fla. to e. Tex., n. to se. Va.

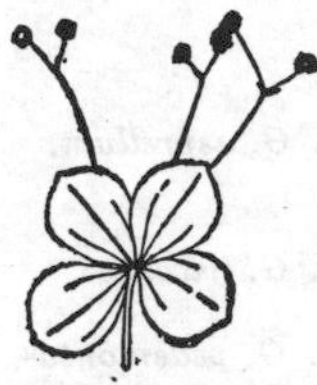
1498. G. kamtschaticum.

5. **G. kamtschâticum** Steller (of Kamtchatka). — With filiform rhizome and stolons; *stem weak, glabrous,* simple or forking at the 2–4 nodes, 0.8–2 dm. long; internodes elongate; 1 *or* 2 *lower whorls of* 4 *leaves smaller than the* 1 *or* 2 *upper whorls; the blades of the latter elliptic to obovate or ovate, thin, lustrous, distinctly 3-ribbed,* with upwardly ciliate margin, 1–3.5 *cm. long;* peduncles 1–3, terminal, smooth, each 1–3-flowered or 2–3-forked; pedicels elongate; corolla glabrous, yellowish-white, with acute lobes; fruit finely bristly. — Cool woods, thickets and brooksides, w. Nfld.; mts. of C.B., Que., n. N.E. and n. N.Y. June–Aug. (E. Asia and Aleutian Ids.) FIG. 1498.

6. **G. circaèzans** Michx. (resembling *Circaea,* Enchanter's Nightshade), WILD LICORICE. — Perennial with a short crown; *stems* wiry, very slender, smooth or *sparsely pilose on the angles,* mostly simple, 1–3.5 dm. high; whorls on main axis 4–6, of 4 *oval to ovate-oblong or ovate-lanceolate thin blunt upwardly ciliate leaves, the larger blades* 1.5–2.5 (rarely –4) *cm. long and* 0.7–1.4 (rarely –1.8) *cm. broad, glabrous beneath or sparsely short-hispid on the nerves; peduncles* usually once forked, *the flowers and fruits mostly sessile or very short-pedicelled on the elongate and widely divergent* stiff *branches; corolla* greenish, its acute or acuminate lobes usually *hairy outside;* fruits bristly. (Var. *glabrum* Britt.) — Rich woods, Fla. to Tex., n. to Va. and locally to s. N.E., N.Y., Ky. and Mich. June. FIG. 1499.

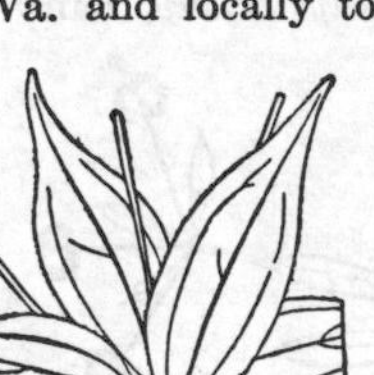
1499. G. circaezans.

Var. **hypomâlacum** Fern. (hairy beneath). — Coarser; *stems hairy on the angles,* 1.5–4.5 dm. high; *larger leaves* 2–5 *cm. long and* 1–2.5 *cm. broad,* appressed-pilose beneath, *the nerves beneath long-hirsute.* (*G. circaezans* of ed. 7) — Rich woods, s. Que. to Minn. and Neb., s. to s. N.E., Va., uplands of N.C., Ky., Mo. and Tex. June, July.

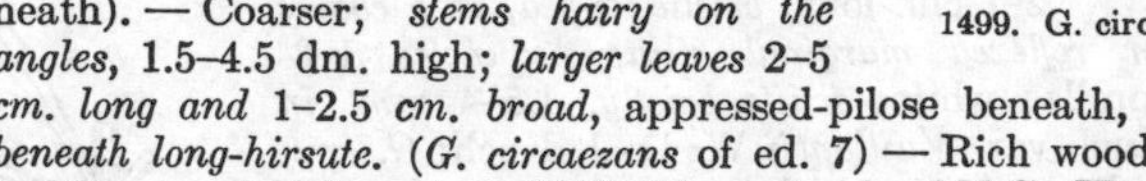
1500. G. lanceolatum.

7. **G. lanceolàtum** Torr. (lance-shaped), WILD LICORICE. — *Nearly or quite glabrous,* 2–6 dm. high; *middle and upper leaves lance-acuminate,* 3–8 *cm. long,* glabrous, or sparsely short-hispid on the nerves; *corolla glabrous,* yellowish, *turning dull purple, its lobes more acuminate;* otherwise like no. 6. — Dry woods, s. Me. and w. Que. to Minn., s. to mts. of N.C. and Tenn. June, July. FIG. 1500.

8. **G. latifòlium** Michx. (broad-leaved). — Smooth, 3–6 dm. high; *leaves lanceolate or ovate-lanceolate,* acute, 3–6 cm. long, the midrib and margins rough, *the lateral nerves prominent;* cymes panicled, loosely many-flowered, the purple *flowers on slender spreading or loosely ascending pedicels; fruit* rather fleshy, *smooth.* — Dry woods, mts. of Pa. and W.Va. to Ala. and Tenn. June, July. FIG. 1501.

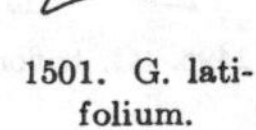
1501. G. latifolium.

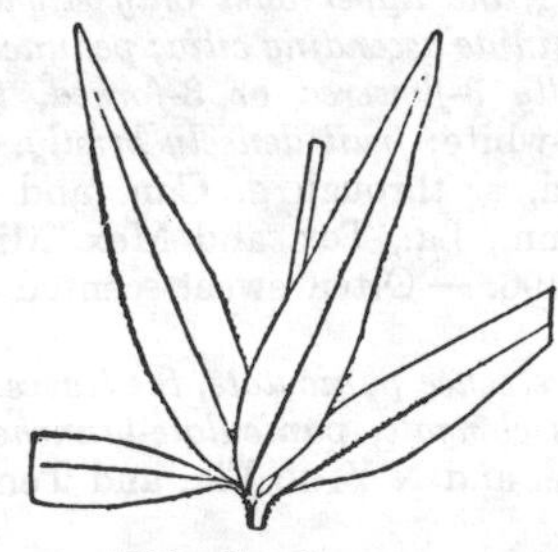
1502. G. arkansanum.

Var. **híspidum** Small (stiffly hairy). — Stems and leaves hispid. — Iron Mts., Va.

9. **G. arkansànum** Gray (of Arkansas). — Similar; leaves lanceolate to linear, 2–3.5 cm. long, *the lateral nerves obscure or none.* — Rocky woods, s. Mo., Ark., and Okla. Late May, June. FIG. 1502.

10. **G. boreàle** L. (northern), NORTHERN B. — Smooth, 3–9 dm. high; *leaves in 4's,* linear-lanceolate; *flowers bright white, in compact panicles;* fruit minutely bristly to smooth. — Highly variable:

Fruit hairy.
Fruit villous-hirsute with long hairs. . . . . *G. boreale* (typical).
Fruit covered with short appressed or incurving hairs. . . . . Var. *intermedium.*
Fruit glabrous or glabrate. . . . . Var. *hyssopifolium.*

G. boreàle (typical). — Shores and gravelly or rocky banks, Man. to Alaska, s. to W.Va. and N.M.; sparingly e. to n. N.E. June–Aug. (Eurasia)

Var. intermèdium DC. (intermediate). — N.S. to n. Ont. and westw., s. to Del., W.Va., O. and Ind. (Eurasia) FIG. 1503.

Var. hyssopifòlium (Hoffm.) DC. (with leaves of Hyssop). — Gaspé Co., Que., to N.D. and Colo., s. to n. N.J., N.Y., O., Ind. and Mo.; Pacific slope. (Eurasia)

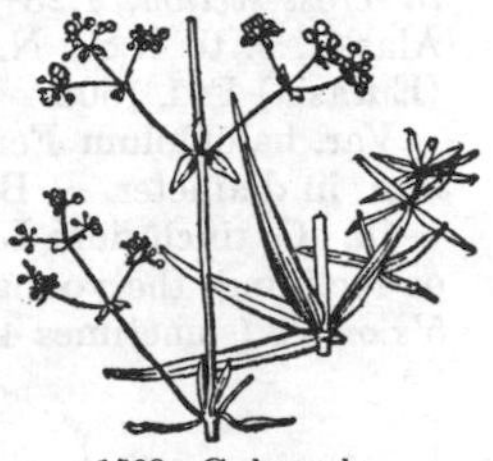
1503. G. boreale, v. intermedium.

G. RUBIOÌDES L. (resembling Madder, *Rubia*), coarser than no. 10, the scabrous elliptic-oblong leaves with 3 parallel ribs prominent beneath, 1–2 cm. broad, the flowers in a corymbiform terminal panicle, the ovaries glabrous (fruit apparently not forming here), is locally estab. in open woods near Bangor, Northampton Co., Pa. (*Mrs. Tanger*). (Introd. from Eu.)

11. G. VÈRUM L. (true; *i.e.*, the true Bedstraw, early Christian tradition claiming it to have been the Bedstraw which filled the manger at Bethlehem), YELLOW B., OUR LADY'S B. — Perennial with strong creeping base; stems stiff, erect, smooth, 3–9 dm. high; *leaves* 8 or 6 in a whorl, *often with suppressed axillary branches, linear, roughish*, ascending or divergent; *panicle dense, its lower branches much exceeding the internodes; corollas bright yellow*, or very pale in forma ÁLBIDUM (Hartm.) Lindm. (whitish), drying blackish; fruit usually smooth. — Dry fields and roadsides, locally abund., Nfld. to Ont. and N.D., s. to N.S., N.E., Va., W.Va., O., Ind., Mo. and Kans. June–Aug. (Natzd. from Eu.) FIG. 1504.

1504. G. verum.

12. G. WIRTGÈNII F. Schultz (for PHILIPP WIRTGEN, 1806–1870). — Similar to the preceding; flowers yellow, slightly larger, 3 mm. in diameter; the panicle long and interrupted, *the lower branches at anthesis shorter than or scarcely surpassing the subtending leaves.* — Fields and meadows, locally abund., Que. to Pa., w. to Mich. and Ind. June, July. (Natzd. from Eu.) FIG. 1505.

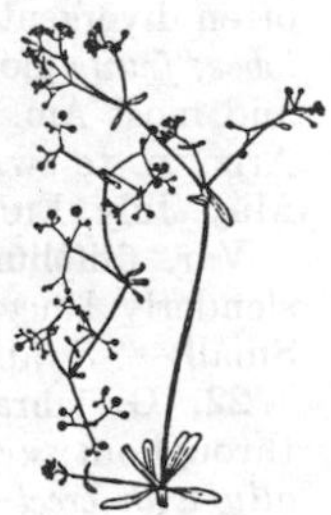
1505. G. Wirtgenii.

13. G. MOLLÙGO L. (old generic name for this and other plants with whorled leaves). — Perennial, smooth throughout or pubescent below; stems erect or diffuse, usually numerous, 3–9 dm. long; *leaves in 8's or on the branchlets in 6's*, oblanceolate to nearly linear; *flowers white, very numerous in loose* ample almost leafless *panicles; branches and pedicels mostly wide-spreading;* fruit smooth. — Roadsides and fields, Nfld. to Ont., s. to N.S., N.E., Va., W.Va. and Ind. June–Aug. (Natzd. from Eu.) FIG. 1506.

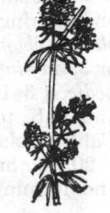

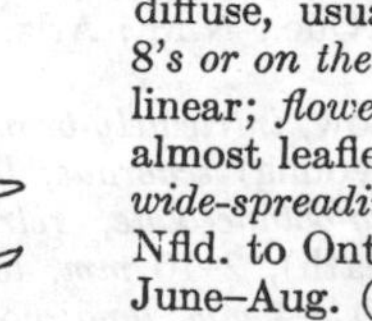
1506. G. Mollugo.

× G. OCHROLEÙCUM Wolff (yellowish-white), a hybrid of nos. 13 and 11, has been found in Ct. (Adv. from Eu.)

14. G. ERÉCTUM Huds. (erect). — Similar; stems mostly erect; flowers fewer and slightly larger; *the branches and pedicels mostly ascending.* — Fields, etc., e. Que. to N.S., N.E., N.J. and Pa. June, July. (Natzd. from Eu.)

15. G. SYLVÁTICUM L. (of the woods), SCOTCH-MIST, BABY'S-BREATH. — Stems very many, tall, suberect, shining, somewhat geniculate at base; lower *leaves* 8, upper 4 or 6 in a whorl, *acuminate, smooth*, entire, glaucous beneath; *pedicels capillary, very ascending, in loose terminal panicles;* fruit smooth. — Fields and thickets, N.E. and N.Y., esc. from cult. July–early Sept. (Introd. from Eu.)

16. G. palústre L. (of marshes). — Slender, 2–5 dm. high, slightly branched, branches solitary or opposite; leaves linear-elliptic or spatulate, thin, dull, barely 1 cm. long; *flowers numerous in terminal cymes; pedicels becoming strongly divaricate; corolla 4-parted, white or rose-tinged,* 2.5–3.3 *mm. broad; fruit* glabrous, *lunate in cross-section.* — Wet meadows and banks, Nfld. to Ont., s. to N.S., N.E., N.Y., se. Pa., Mich. and Wisc. June–Aug. (Eurasia) FIG. 1507.

1508. G. trifidum.

17. G. trífidum L. (three-cleft). — Slender and weak, often freely branched and forming dense mats; *internodes when young scabrous on angles; primary leaves* chiefly in 4's, *linear or linear-oblanceolate, retrorsely scabrous-margined,* 0.7–2 cm. long, blunt; *flowers solitary or, when terminal, in 3's, on capillary scabrous peduncles or pedicels* 0.5–3

1507. G. palustre.

cm. long; corolla whitish, 0.5 *mm. long, its* 3 (rarely 4) *lobes obtuse; fruit smooth, annular in cross-section,* 1.25–1.75 *mm. high.* — Swamps, wet shores, etc., s. Lab. and Nfld. to Alaska, s. to N.S., N.E., N.Y., O., Mich., n. Ill., Minn., Neb., Colo. and Calif. July–Sept. (Eurasia) Fig. 1508.

Var. **halóphilum** Fern. & Wieg. (of saline soils). — Fleshy, glabrous throughout; fruits 1.5–2 mm. in diameter. — Brackish or saline shores, se. Lab. and Que. to Mass.

18. G. tinctòrium L. (used for dyeing). — Firmer and usually stouter than no. 17, ascending or reclining, the younger internodes retrorse-scabrous on angles; *leaves of main axis mostly in 5's or 6's* (sometimes 4's), *broadly oblanceolate to oblong-spatulate,* firm, 0.6–2 cm. long; *peduncles stiff, straight or straightish, the longest ones* 1–1.7 cm. long *with* 3 (sometimes more) *nearly uniform smooth short* (*up to* 8 *mm. long*) *straight divergent pedicels;* flowers and fruit much as in no. 17; the mature pairs of fruit 2–3 mm. across. (*G. Claytoni* Michx.) — Swamps and damp places, Nfld. to Ont. and Neb., s. to N.S., N.E., L.I., S.C., Ky., Mo. and Tex. June–Sept. Fig. 1509.

1509. G. tinctorium (typical).

1510. G. tinctorium, v. subbiflorum.

Var. **subbiflòrum** (Wieg.) Fern. (sometimes two-flowered). — *Many flowers on* 1, 2 or 3 *separate axillary peduncles* or a few peduncles 2- or 3-forked; mature pairs of fruit 2–3 mm. across. (*G. subbiflorum* Rydb.) — Se. Lab. and Nfld. to Alaska, s. to M.I., C.B., N.B., Me., n. N.Y., Mich., Minn., N.M. and Calif. Fig. 1510. — Suggesting no. 17.

Var. **floridànum** Wieg. (of Florida). — Coarser throughout; longest peduncles up to 2.5 cm. long, frequently simple; mature pairs of fruit 3.5–5 mm. across. — Fla. to La., n. on or near Coastal Plain to se. Va.

19. G. Brandegèi Gray (for its discoverer, Townshend Stith Brandegee, 1843–1925). — *Glabrous throughout,* low and simple to matted and freely forking, the weakly ascending stems or flowering branches 0.5–1.5 dm. high; *leaves of primary axes broadly oblanceolate to oblong-spatulate,* fleshy, 6–12 *mm. long; flowers* as in no. 17, solitary or 2 or 3, *on simple or forking peduncles, these and the straightish to arching glabrous pedicels* 3–10 *mm. long; fruits* 1.5–2 *mm. in diameter.* — Low grounds, Greenl. and Lab. to Alaska, s. to n. Nfld., Shickshock Mts., Que., N.M., Ariz. and Calif. (Iceland) July, Aug. Fig. 1511.

1511. G. Brandegei.

20. G. brèvipes Fern. & Wieg. (short-stalked). — Low, *intricately branched* and forming dense mats, *the short internodes* (when young) *scabrous; leaves* mostly in 4's, *oblanceolate to oblong-oblanceolate, retrorse-scabrous* on margin and midrib (beneath), 2–10 *mm. long; flowers* axillary, *solitary or paired on peduncles finally* 0.5–4 *mm. long;* corolla minute, with 3 obtuse lobes; *fruits* globular, 0.8–1 *mm. in diameter.* — Calcareous swamps and wet shores, Gaspé Co., Que., to Minn., s. to ne. Me., sw. Vt., w. N.Y., Bruce Pen., Ont., Mich. and Minn. July–Sept. (Greenl.) Fig. 1512. — Sometimes united with no. 19.

1512. G. brevipes.

21. G. obtùsum Bigel. (obtuse). — *Stems erect,* from capillary rhizomes, simply (not intricately) branched, up to 8 dm. high, *smooth, stiffish; leaves mostly in 4's* (rarely 5's or 6's), *elliptic-oblong or lanceolate* to broadly linear, *obtuse,* those of the primary axis 1.5–3 cm. long, 2–6 *mm. broad, loosely spreading,* slightly scabrous on the margins; cymes terminating stems and branches, their several flowers grouped in 2's or 3's; *the straight peduncles and pedicels ascending in anthesis,* often divergent in fruit; *corolla* white, 2–2.5 *mm. broad,* commonly *with 4 acute lobes; fruits* globose, smooth, 2.5–3.5 *mm. in diameter.* (*G. tinctorium* of ed. 7 and most Am. auth., not L.) — Low woods, swamps and wet shores, Fla. to Ariz., n. to sw. N.S., s. and w. N.E., s. Ont., Mich., Wisc., Minn. and Neb. May–July. Fig. 1513. Southeastw. mainly as

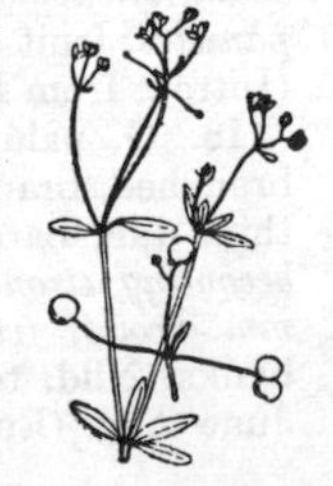

1513. G. obtusum.

Var. **filifòlium** (Wieg.) Fern. (thread-leaved). — Leaves linear-filiform to slenderly lanceolate, often strongly revolute, 1–2 mm. broad. (*G. filifolium* Small) — Swamps and wet woods, Ga. to Md. May, early June.

22. G. labradóricum Wieg. (of Labrador). — Similar to no. 21, *smaller* throughout; *stems* slender, erect or ascending, smooth, 0.5–4 dm. high, *eventually with erect branches; leaves* in 4's, oblanceolate or spatulate, 0.5–1.5 *cm. long,* scabrous beneath on margin and midrib, *loosely spreading or reflexed; cymes* terminal, *soon overtopped by elongating branches; fruits* 1–1.5 *mm. in diameter.* — Bogs and mossy thickets and woods,

1514. G. labradoricum.

s. Lab. and Nfld. to Man., s. to P.E.I., N.B., n. and w. N.E., nw. N.J., e. Pa., n. O., n. Ind., Wisc. and Minn. Late May–Aug. Fig. 1514.

23. G. parisiénse L. (Parisian). — Slender diffusely branched *annual*, the very slender stems minutely scabrous; *leaves* chiefly in 5's–7's, *linear, upwardly spinulose-margined, bristle-tipped, the principal ones* 5–8 *mm. long;* inflorescence a lax leafy panicle of small cymules, the *upper floral bracts reduced or wanting; corolla-lobes* 0.2–0.4 *mm. long, greenish-white; fruits* 0.7–1 *mm. high,* smooth or tuberculate. — Fields and roadsides, w. Va. and W.Va. to Ind., s. to N.C. and Tenn. June, July. (Natzd. from Eu.) Fig. 1515.

1515. G. parisiense.

24. G. concínnum T. & G. (elegant). — *Perennial;* the slender erect and ascending-branched stems 1.5–6 dm. high, quite smooth to minutely scabrous; *leaves in 6's, linear, smoothish,* at most minutely toothed, *subulate-tipped, the principal ones* 1–2 *cm. long, veinless;* peduncles 2–3 times forked, the terminal *panicles* delicate, *diffuse; corolla-lobes* 0.8–1 *mm. long; fruits* 1.5–2.3 *mm. high.* — Woods and thickets, N.J. to Minn., s. to Va., Ky., Ark. and Kans. June, July. Fig. 1516.

1516. G. concinnum.

25. G. saxátile L. (of rocky places), Heath-B. — *Matted perennial, with numerous trailing leafy stems* among the weakly ascending flowering ones; *branches smooth; leaves* in 4's–8's, *spatulate to narrowly obovate, upwardly spinulose-margined, cuspidate,* those of the primary flowering stem 8–15 mm. long; *inflorescence an elongate leafy panicle; floral bracts foliaceous;* cymules on short stiff peduncles, with short straight pedicels; corolla white, about 3 mm. broad, its 4 lobes acute; fruits smooth, 1.5–2 mm. high. — Peaty and rocky barrens and spring-heads, Trepassey Bay, Nfld. July, Aug. (Eu.) Fig. 1517.

1517. G. saxatile.

26. G. aspréllum Michx. (slightly rough), Rough B. — *Stem* 0.5–1.8 m. high, much-branched, *rough* backwards *with hooked prickles,* leaning on bushes; leaves in whorls of 6, or 4–5 on the branchlets, oval-lanceolate, with almost prickly margins and midrib; *inflorescence a leafy panicle, the cymes mostly exceeding the bracteal leaves; pedicels ascending or divergent;* corollas white; *fruits* smooth, *about* 2 *mm. long.* — Low grounds and damp thickets, Nfld. to w. Ont., s. to N.S., N.E., L.I., N.C., O., Ind., Ia. and Neb. July–Sept. Fig. 1518.

27. G. tricórne Stokes (three-pronged). — Resembling no. 2, rather stout, with simple branches; leaves 6 or 8, oblanceolate, cuspidate-mucronate, the margins and stem retrorsely prickly-hispid; *flowers* mostly *in clusters* of 3, dull white; *fruits* rather large, *tuberculate-granulate,* not hispid, 3–4 *mm. in diameter, on short recurving pedicels.* — Waste lands and cult. ground, sporadic and infrequent. May–July. (Adv. from Eu.) Fig. 1519.

1518. G. asprellum.

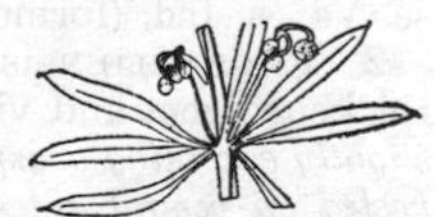
1519. G. tricorne.

28. G. pedemontànum All. (of Piedmont). — *Annual, with hairy simple stems* 0.5–3 dm. high and retrorse-bristly; *leaves* in 4's, soon reflexed, *villous,* ciliate, elliptic or oblong, 4–10 mm. long; *flowers* very small, *yellow,* in small axillary bractless cymes of 2–4; *fruit* smooth, *the small mericarps oblong-cylindric.* — Open woods, thickets and pastures, W.Va. and Ky. April–June. (Natzd. from Eu.)

29. G. uniflòrum Michx. (single-flowered). — Slender *glabrous perennial with filiform yellow rhizomes; flowering stems simple or sparingly branched,* 1–3.5 dm. high; *leaves* in 4's, *linear or linear-lanceolate,* 1–2.5 *cm. long,* subcoriaceous, often evergreen, the revolute margin scabridulous; *peduncles axillary, divergent, much shorter than leaves,* leafy-bracted, with 1–few white flowers with corolla-lobes blunt; *fruit succulent, purple-black,* smooth, 2–4 mm. in diameter. — Rich woods, Fla. to Tex., n. to e. Va. June. Fig. 1520.

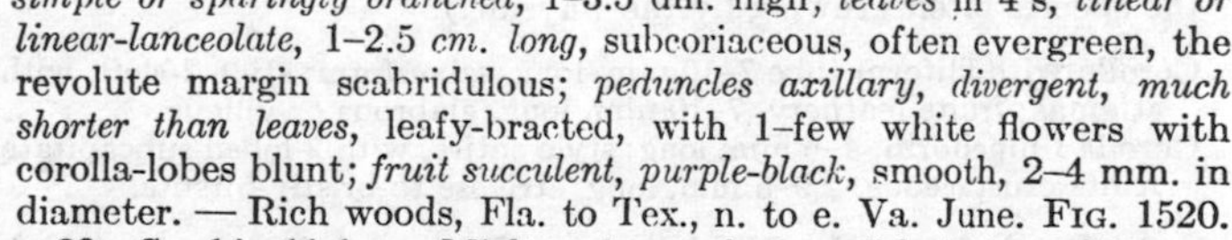
1520. G. uniflorum.

30. G. hispídulum Michx. (somewhat stiff-hairy). — Hirsute-pubescent, scabrous, or sometimes nearly smooth, 3–6 dm. high, diffusely branched; leaves oblong or oval, mucronate, 0.5–2 cm. long; pedicels solitary or commonly 2 or 3 from the small involucral whorl, all naked, or one of them bracteolate; flowers white; *berry purple, glabrate.* (*G. bermudiense* of some Am. auth., not L.) — Sandhills and dry sandy woods, Fla. to La., n. to Cape May, N.J. June, July. Fig. 1521.

1521. G. hispidulum.

Aspérula L. (slightly harsh), an Old World genus differing from *Galium* in having a definite corolla-tube, is represented in gardens by **A. odoràta** L. (fragrant), Sweet Woodruff, with slender creeping rhizomes, smooth erect stems 1–2 dm. high, their oblong-oblanceolate leaves in whorls of 6–8, and few terminal flowers, the plant fragrant in drying. It sometimes spreads to roadsides, thickets and rocky, open woods. (Introd. from Eu.)

3. SPERMACÒCE L. Buttonweed

Calyx-tube short; the limb parted into 4 teeth. Corolla funnelform or salverform, valvate in the bud. Stigma or style 2-cleft. Fruit small and dry, 2-locular, splitting when ripe into 2 carpels, one of them usually carrying with it the partition, and therefore closed, the other open on the inner face. — Small Am. herbs, the bases of the leaves or petioles connected by a bristle-bearing stipular membrane. Flowers small, whitish, crowded into sessile axillary whorled clusters or heads. (Name compounded of the Greek *sperma, seed,* and *acoce, a point;* probably from the pointed calyx-teeth on the fruit.)

1522. S. glabra.

1. S. glàbra Michx. (smooth). — Glabrous perennial; stems spreading, 2–5 dm. long; leaves oblong-lanceolate; heads many-flowered; corolla little exceeding the calyx, bearded in the throat, bearing the anthers at its base; filaments and style hardly any. — Damp shores, low woods and openings, Fla. to Tex., n. to s. O., s. Ind., s. Ill., Mo. and se. Kans. June–Oct. Fig. 1522.

4. RICHÁRDIA L.

Calyx-teeth 4–8 (commonly 5–6), lanceolate, gamophyllous at base, circumscissile-deciduous as a whole at or before the separation of the 2–4 closed or barely opening rough coriaceous carpels. Corolla funnelform, with 4–8 lobes. Stigmas 2–4, linear or spatulate. Flowers densely crowded in terminal leafy-bracted glomerules or dense cymes. — Trop. and subtrop. Am. annuals (rarely perennials), now spread to warm-temp. reg., with whitish to lavender corollas, setiferous stipules and broad short-petioled to subsessile leaves. (Named for *Richard Richardson,* 1663–1741, of London.) Richardsonia Kunth

1. R. scàbra L. (rough). — More or less loosely branching short-pubescent annual up to 8.5 dm. high; leaves elliptic-lanceolate to ovate, fleshy; glomerules depressed; *calyx-lobes united only at base, three to four times length of ovary; corolla hypocrateriform,* 5–6 mm. long, its tube exserted from the calyx, *the lobes much exceeding the stamens;* mature *fruit* 3–3.5 mm. long, *the cocci ventrally sulcate.* — Roadsides, cultivated fields and waste places, Fla. to Tex., n. to se. Va., s. Ind. (formerly) and Ark. July–Oct. (Natzd. from Trop. Am.)

2. R. brasiliénsis (Moq.) Gomez (Brazilian). — Annual, becoming perennial, with deep thickened root and villous diffusely branched or matted stems; *calyx-lobes nearly united, only slightly exceeding ovary; corolla infundibuliform, its lobes barely exceeding stamens; cocci* of *fruits keeled on ventral sides.* — Roadsides and waste places, Fla. to Tex. and Mex., n. to se. Va. July–Oct. (Natzd. from S.Am.)

5. DIÓDIA L. Buttonweed

Calyx-teeth 2–5, often unequal. Fruit 2 (rarely 3)-locular, the crustaceous carpels into which it splits all closed and indehiscent. Flowers 1–3 in each axil. — Resembling *Spermacoce.* Flowering all summer. A genus of trop. and warm reg. (Name from the Greek *diodos, a thoroughfare:* the species often growing by the wayside.)

Corolla with filiform tube 7–10 mm. long, salverform; style 2-cleft, with filiform stigmas; fruits leathery, 7–10 mm. long, glabrous or villous. . . . . . . 1. *D. virginiana.*
Corolla funnelform, 4–6 mm. long; style entire, with 2-lobed subcapitate stigma; fruits crustaceous, 2.9–5 mm. long, strigose to bristly-hirsute. . . . . . 2. *D. teres.*

1. D. virginiàna L. (Virginian). — *Weak* spreading *perennial, glabrous or sparsely villous* on the angles; the forking branches 1.5–6 dm. long; leaves lanceolate or oblong-lanceolate, sessile, only slightly tapering at tip; *corolla* white, about 1 cm. long, the *slender tube abruptly expanded into the large limb; style 2-parted; fruit ellipsoid,* 3.5–5 mm. in diameter, *strongly furrowed, crowned* mostly *with* 2 *prolonged lanceolate calyx-teeth, glabrous or sparsely villous,* or stems as well

1523. D. virginiana.

as leaves and fruits strongly villous-hirsute in forma **hirsùta** (Pursh) Fern. (hairy). — Low grounds, Fla. to Tex., n. to N.J., s. Ill. and Mo., casually to Ct. June–Sept. Fig. 1523.

Var. **attenuàta** Fern. (slenderly tapering). — Very slender, weak; *leaves* thin, narrowly lanceolate, *strongly attenuate to base and apex; fruits subcylindric,* 2.5–3.5 *mm. in diameter; calyx-lobes linear.* — Wet shores and bottoms, se. Va.

2. D. tères Walt. (with circular cross-section). — Hairy or minutely pubescent annual; stem spreading, 1–8 dm. long, nearly terete; leaves linear-lanceolate to elliptic, closely sessile, rigid; *corolla funnelform,* 4–6 *mm. long,* whitish, with short lobes, usually not exceeding the long bristles of the stipules; *style undivided; fruit* obovoid-turbinate, *not furrowed,* crowned *with* 4 *short ciliate calyx-teeth.* (*Diodella* Small) —Very variable:

Leaves without long bristle-tips.
  Stipules very much overtopping fruits; fruits (excluding calyx-lobes) 2.9–3.6 mm. long, stiffly appressed- to spreading-hirsute; leaves linear or linear-lanceolate, the primary ones 1.5–4.5 cm. long and 2–6(–9) mm. broad.
    Stem puberulent or minutely pilose, without long divergent hairs. . . . *D. teres* (typical).
    Stem more or less hirsute with long divergent hairs. . . . . Var. *hirsutior.*
  Stipules barely equaling the hispid fruit.
    Stems hispid or hirsute; leaves linear-lanceolate, the primary ones 1.5–4 cm. long and 3–8 mm. broad; bodies of fruits 3.8–5 mm. long. . . . . . Var. *hystricina.*
    Stems puberulent; leaves oblong or oblong-elliptic, the primary ones 1–2.5 cm. long and 5–8 mm. broad; bodies of fruits 3–3.5 mm. long. . . . Var. *oblongifolia.*
Leaves when young with slender apical setae; stems hirsute; stipules and fruits as in typical var. . . . . Var. *setifera.*

D. tères (typical). — Dry sandy soil, Fla. to Tex., n. to s. R.I., Ct., s. N.Y., Pa., s. O., Ill. and Mo. Late June–Oct.

Var. **hirsùtior** Fern. & Grisc. (more stiffly pubescent). — Dry sands, Fla. to Miss., n. to se. Va.

Var. **hystricìna** Fern. & Grisc. (hedgehog-like). — Dry sands, chiefly coastal, Va.

Var. **oblongifòlia** Fern. (oblong-leaved). — Loose sands, se. Va.

Var. **setífera** Fern. & Grisc. (bearing bristles). — Tex. to Wisc., Mich. and Ky.

6. MITCHÉLLA L. Partridge-berry. Pain de perdrix (Que.)

Flowers in pairs, with their ovaries united. Calyx 4-toothed. Corolla-lobes spreading, densely bearded inside, valvate in the bud. Style 1; stigmas 4, linear. Fruit a berry-like double drupe, crowned with the calyx-teeth of the two flowers, with 4 small seed-like bony nutlets to each flower. — Two smooth and trailing small evergreen herbs, ours, another in Japan, with round-ovate and shining petioled leaves, minute stipules, white fragrant flowers often tinged with purple, and scarlet (rarely whitish) edible (but nearly tasteless) berries, which remain over winter. Flowers occasionally 3–6-merous, always dimorphous; all those of some individuals having exserted stamens and included stigmas; of others, included stamens and exserted style. (Commemorating Dr. *John Mitchell*, 1676–1768, an early correspondent of Linnaeus and a botanist who resided in Virginia.)

1. M. rèpens L. (creeping), Two-eyed-berry, Running Box. — Dry or moist knolls in woods, Fla. to Tex., n. to sw. Nfld., s. Que., Ont. and Minn. June, July. — Leaves often variegated with whitish lines. The two flowers rarely completely confluent into 1 with 10-lobed corolla. Berries whitish in the rare forma **leucocárpa** Bissell (white-fruited).

7. CEPHALÁNTHUS L. Buttonbush. Bois noir (Que.)

Calyx-tube inversely pyramidal, the limb 4-toothed. Corolla-teeth imbricated in the bud. Style filiform, much protruded. Stigma capitate. Fruit small, at length splitting from the base upward into 2–4 closed 1-seeded portions. — Shrubs or small trees of temp. and trop. e. Am., Asia and Afr., with the white flowers densely aggregated in spherical peduncled heads. (Name composed of the Greek *cephale, a head,* and *anthos, a flower.*)

1. C. occidentàlis L. (western; of the Western Hemisph.). — Smooth; leaves petioled, essentially glabrous, green above, oblong-ovate, abruptly short-acuminate, 2.3–15 cm. broad, opposite or whorled in 3's or 4's, with short intervening stipules, or opposite lanceolate blades attenuate to both ends and only 1–3 cm. broad in forma **lanceolàtus** Fern. (lanceolate). — Swamps, pond-borders and margins of streams, Fla. to Mex., n. to w. N.S., sw. N.B., N.E., sw. Que. N.Y., s. Ont. and Calif. July, Aug. — Usually a shrub with us, rarely arborescent. (W.I.)

Var. **pubéscens** Raf. (hairy). — Branchlets and at least the lower leaf-surfaces soft-pubescent, upper leaf-surfaces dull and pale. — Ga. to Tex., n. to se. Va., s. O., s. Ind., s. Ill. and Okla.

Subfam. II. CINCHONOÌDEAE

8. HOUSTÒNIA L.

Calyx 4-lobed, persistent; the lobes in fruit distant. Corolla usually much longer than the calyx-lobes, the lobes valvate in the bud. Anthers linear or oblong. Style 1; stigmas 2. Ovary 2-locular. Capsule turbinate, globular, or didymous, thin, its summit or upper half free from and projecting beyond the tube of the calyx, loculicidal across the top. Seeds 4–20 in each locule, pitted. — Small N. Am. herbs, with short entire stipules connecting the petioles or narrowed bases of the leaves, and cymose or solitary and peduncled flowers; these dimorphous, in some individuals with exserted anthers and short included style; in others the anthers included and the style long, the stigmas therefore protruding. (Named for Dr. *William Houstoun*, 1695–1733, an English botanist who collected in trop. America.)

a. Delicate vernal-flowering annuals or weak perennials with creeping bases or slender rhizomes; peduncles 1-flowered, filiform; corolla salverform; seeds globular, with a deep round cavity occupying the inner face. . . *b.*

b. Plants without long repent stems; basal leaves elongate.

Corolla pale bluish or lilac to white, with yellow eye, its lobes 2.5–5 mm. broad, its tube 3–10 mm. long; annual, or matted perennial with filiform short rhizomes. . . . . 1. *H. caerulea.*

Corolla deep purple, with dark eye, its lobes 1–2.5 mm. broad, its tube 2–5 mm. long; southern annuals.

Calyx-lobes 1–2 mm. long, lance-linear, becoming subulate; corolla-tube about twice as long as calyx-lobes. . . . . 2. *H. patens.*

Calyx-lobes 2–4 mm. long, broadly lanceolate, foliaceous, nearly equaling corolla-tube. . . . . 3. *H. minima.*

b. Plants with long repent superficial stems, with suborbicular leaves; corolla similar to that of no. 1. . . . . 4. *H. serpyllifolia.*

a. Firm-stemmed mostly tufted perennials with sessile cauline leaves and with terminal corymbs, cymes or panicles; corolla funnelform, often hairy within; seeds meniscoidal, with a ridge across the hollowed inner face. . . *c.*

c. Middle and upper stipules rounded to deltoid, not bristle-tipped; capsules free from near the middle. . . *d.*

d. Capsules broader than high.

Cauline leaves broadly lanceolate to ovate, 3–5-ribbed; corymb leafy, compact; fruiting calyx 3–4 mm. long. . . . . 5. *H. purpurea.*

Cauline leaves linear, 1-ribbed; corymb loosely paniculate, open, with much reduced leaves; fruiting calyx 1.5–2.5 mm. long. . . . . . 6. *H. tenuifolia.*

d. Capsules as high as or higher than broad. . . *e.*

e. Leaves smooth.

Calyx-lobes subulate, 1.5–2.5 mm. long. . . . . 7. *H. longifolia.*

Calyx-lobes lanceolate, herbaceous, 5–9 mm. long. . . . . 8. *H. lanceolata.*

e. Leaves (at least the basal) ciliate.

Stem and calyx glabrous. . . . . 9. *H. canadensis.*

Stem and calyx grayish-scabrous. . . . . 10. *H. setiscaphia.*

c. Middle and upper stipules setiform or with bristle-tips; capsule two-thirds included in hypanthium; leaves narrowly linear, often fascicled. . . . . 11. *H. nigricans.*

1. H. caerùlea L. (blue), Bluets, Innocence, Quaker-ladies. — Delicate tufted winter-annual, or matted perennial with filiform short rhizomes connecting the tufts; *stems* erect, *filiform*, sparingly branched below, 0.5–2 *dm. high; rosette-leaves oblong-spatulate to elliptic*, 0.5–1.5 cm. long, on slender petioles, glabrous or hirtellous; cauline leaves and bracts reduced, much narrower; *peduncles* filiform, arched in bud, erect and *becoming* 1.5–7 *cm. long*, barely dilated beneath the nerveless or delicately nerved hypanthium; *corolla* salverform, *pale bluish to lilac, or* clear *white* in forma **albiflòra** Millsp. (white-flowered), *with a yellow eye, the* straight slender *long-exserted tube* 0.5–1 *cm. long*, glabrous within, the horizontally spreading *lobes* 2.5–4 *mm. wide; capsules* 2.5–4 (–4.5) *mm. broad; ripe seeds* 0.4–0.65 *mm. in diameter, the central* *depression* 0.15–0.25 *mm. across.* — Open turfy slopes and fields and (southw.) in thickets and woods, N.S. and N.B. to s. Ont. and Wisc., s. to Ga., Ala. and Mo. Apr.–June (sporadically to frost).

Var. **Faxonòrum** Pease & Moore (for Edwin and Charles Edward Faxon, 1823–1898 and 1846–1918). — *Stiffer, fleshy* (drying black); *cauline leaves and bracts hardly reduced, oblong, elliptic or oval; peduncles* stiffly erect, *broadly expanded and corrugated at base of hypanthium; corolla-tube short, gradually expanding to the white limb,* the lobes 3–5 mm. broad; *capsules* 3.5–5.5 *mm. broad; seeds* 0.75–1 *mm. in diameter, the central depression* 0.3–0.5 *mm. across.* (*H. Faxonorum* Fern.) — Alpine reg., White Mts., N.H.; borders of streams, St. P. et. Miq. June–Sept.

2. H. pàtens Ell. (spreading), Small Bluets, Star-violet. — Winter-annual, resembling no. 1, 0.2–1.2 dm. high, with capillary ascending branches and peduncles 1–4 cm. long; lower leaves spatulate to ovate, the upper smaller and narrower; *calyx-lobes* 1–2 *mm. long, lance-linear, becoming subulate; corolla deep purple, with dark eye, the tube* 3–5 *mm. long, about twice length of calyx;* the limb 6–8 mm. across, *with lobes* 1–2 *mm. broad.* (*H. pusilla* sensu Standl., perhaps of Schoepf) — Sandy or rocky fields and openings, Fla. to Tex., n. to se. Va., Ill., Mo. and Okla. March, April. Fig. 1524.

3. H. mínima Beck (smallest), Star-violet. — Similar to no. 2, diffusely branched; *calyx-lobes* 2–4 *mm. long, broadly lanceolate, foliaceous, nearly equaling the corolla-tube.* — Dry woods, openings and fields, Ill. to Ia. and Kans., s. to Ark. and Tex. April.

1524. H. patens.

4. H. serpyllifòlia Michx. (thyme-leaved). — *Filiform stems prostrate, repent, elongate, with suborbicular petioled blades* 2–7 *mm. across;* flowering branches with narrower and longer leaves and bracts; peduncles elongate and large; bluish *corolla* much as in no. 1, its *tube pubescent within.* — Damp slopes and borders of streams, mts. of Pa. and W.Va. to Ga. and Tenn. Apr.–July.

5. H. purpùrea L. (purple). — Stems 1–many from a subligneous crown, erect, 1–5 dm. high, glabrous or sparsely hirtellous, or densely pilose and leaves very pubescent in forma **pubéscens** (Britt.) Fern. (hairy); *leaves* cordate-ovate to lanceolate, 3–5-*ribbed,* glabrous or sparsely pubescent, 1.5–4.5 cm. long, 0.5–3.5 *cm. broad,* usually 2–4 pairs below the branching summit; flowers numerous, in terminal cymes; *calyx-lobes subulate to linear-lanceolate;* corolla funnelform, pale purple or lilac to white; fruiting calyx 3–4 mm. long, exceeding the *half-free broad* (*broader than long*) *capsule.* — Open sandy woods or rocky slopes, Del. to Ia., s. to Ga., Ala., La. and Okla. May, June. Fig. 1525.

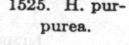

1525. H. purpurea.

6. H. tenuifòlia Nutt. (slender-leaved). — Very slender, lax, *diffusely paniculate,* 1.5–3 dm. high; basal rosette-leaves oblanceolate to spatulate; *cauline leaves linear,* 1-*ribbed,* 1–2.5 *mm. wide;* panicle with almost filiform branches and peduncles; flowers and fruit much as in no. 5; *fruiting calyx* 1.5–2.5 *mm. long.* — Dry woods and rocky slopes, Ga. to Tex. and n. Mex., n. to s. Pa., W.Va., Ky. and Mo. June–Sept.

7. H. longifòlia Gaertn. (long-leaved). — Similar to no. 5; stem 1–2.5 dm. high, mostly glabrous; cauline leaves oblong-lanceolate to linear, 1.5–2.5 cm. long; the radical oval or oblong, less rosulate, not ciliate; *calyx-lobes subulate,* 1.5–2.5 *mm. long; capsule as high as broad.* — Rocky or gravelly soil, centr. Me. to Sask., s. to Ga., Ala., Miss., Ark. and Okla. June–Sept. Fig. 1526.

1526. H. longifolia.

8. H. lanceolàta (Poir.) Britt. (lanceolate). — Coarser than no. 7, 1.5–4 dm. high; leaves broadly lanceolate, thickish, smooth; inflorescence very leafy; *calyx-lobes lanceolate, herbaceous,* 5–9 *mm. long;* capsule as long as broad, much overtopped by calyx-lobes. — Dry open woods, slopes and pastures, Ala. to Okla., n. to s. Me., W.Va., Ky., Ill. and Mo. Late May–July. Fig. 1527.

1527. H. lanceolata.

9. H. canadénsis Willd. (Canadian). — Resembling no. 7, 0.5–2 dm. high; the stems solitary or few, smooth; *basal leaves oblong or oval, hirsute-ciliate, persistent; cauline leaves remote, oblong to oblanceolate, usually ciliate,* 1–2.5 cm. long; *calyx glabrous,* its lobes longer than the ovoid capsule. (*H. ciliolata* Torr.) — Rocky or gravelly (chiefly calcareous) banks and shores, s. Ont. to Minn., s. to w. N.Y., w. Pa., Tenn. and Ill. May–Aug.

10. H. setiscáphia L. G. Carr (with bristly bowl). — Differing from no. 9 in its *oblanceolate leaves; angles of stem scabrous or hispid with grayish trichomes;* inflorescence denser; *calyx grayish-hispid;* corolla usually smaller. — Calcareous barrens and glades, Lee Co., Va.

11. H. nígricans (Lam.) Fern. (blackening). — Stems tufted, from a subligneous crown, 1.5–7.5 dm. high; *leaves* narrowly linear, 1-ribbed, *usually with short axillary fascicles*, several pairs below the stiffly branching inflorescence; *stipules bristle-tipped or the uppermost setiform;* flowers short-pedicelled; calyx-lobes usually scabrous-ciliate at base; *capsule* longer than broad, *only the upper third free.* (*H. angustifolia* Michx.) — Dry soils, Fla. to Tex. and Mex., n. to Ga., O., s. Mich., Ill., Ia. and Neb. June–Oct. — Plant blackened in drying.

9. HEDYÒTIS L.

Calyx 4-lobed, persistent. Corolla short, the limb valvate in the bud. Anthers short. Style 1 or none; stigmas 2. Capsule thin, 2-locular, opening loculicidally across the summit. — Low herbs of trop. and warm reg., with small stipules united to the petioles. (Name from the Greek *hedys, sweet,* and *ous, otos, ear,* significance not explained.) OLDENLANDIA L.

1. H. uniflòra (L.) Lam. (one-flowered; a misnomer). — *Pubescent,* loosely branched *annual,* 0.2–5 dm. tall or across; *leaves ovate to oblong; flowers in dense* sessile terminal and axillary *glomerules;* corolla nearly rotate, whitish, shorter than calyx; *calyx-lobes ovate to oblong, foliaceous, longer than* the subglobose *hirsute capsule.* (*Oldenlandia* L.) — Damp sands and open soils, Fla. to Tex., n. to L.I. July–Oct. (W.I.) FIG. 1528.

1528. H. uniflora.

2. H. Bóscii DC. (for its discoverer, LOUIS AUGUSTIN GUILLAUME BOSC, 1759–1828). — *Glabrous perennial,* diffusely branched, the principal branches 0.5–2.3 dm. long; *leaves linear,* attenuate at base; *flowers few or solitary* in the axils; *calyx-teeth broadly subulate, shorter than the smooth capsule.* (*Oldenlandia* Chapm.) — Sandy borders of ponds and streams or in ditches, Fla. to Tex., n. to se. Va., Tenn. and se. Mo. July–Oct.

FAM. 162. CAPRIFOLIÀCEAE (HONEYSUCKLE FAMILY)

Shrubs, or rarely herbs, with opposite exstipulate (*occasionally stipulate*) *leaves, the calyx-tube adherent to the 2–5-locular ovary, the stamens as many as* (*1 fewer in* Linnaea) *the lobes of the tubular or rotate corolla, and inserted on its tube.* Fruit a berry, drupe or capsule, 1–several-seeded. Seeds anatropous, with small embryo in fleshy albumen. Family chiefly of cool or temp. reg.

a. Corolla elongate, funnelform to campanulate, often more or less irregular (*i.e.*, inequilaterally saccate at base or bilabiate); style 1, elongate, with usually capitate stigma; inflorescences lateral or terminal. Tribe LONICEREAE. . . *b.*

b. Erect or climbing shrubs, with scaly winter-buds. . . . *c.*

c. Corolla funnelform to tubular; fruits many-seeded.

Calyx with slender tube and 5 subulate long lobes; ovary 2-locular, maturing into a slender long-beaked capsule with persistent slender calyx-lobes. . . . . 1. *Diervilla.*

Calyx with globular to ovoid tube and short 5-toothed or truncate limb; ovary 2- or 3-locular; fruit a berry. . . . . 2. *Lonicera.*

c. Corolla short-campanulate to salverform; ovary 4–locular; fruit a berry-like drupe, with only 2 nutlets maturing. . . . . 3. *Symphoricarpos.*

b. Creeping slender-suffruticose plants or ascending herbs.

Creeping, with delicate subligneous stems and small leaves; flowers mostly paired, drooping from tips of axillary peduncles; fruit a 1-seeded small capsule. . . . . 4. *Linnaea.*

Erect coarse herbs with large leaves and sessile flowers; fruit drupaceous, with 2–5 stones. . . . . 5. *Triosteum.*

a. Corolla rotate to open-campanulate, regular, deeply 5-lobed; stigmas 1–5, sessile or nearly so; inflorescences terminal and cymose. Tribe SAMBUCEAE.

Leaves simple; fruit a drupe with a single stone. . . . . . 6. *Viburnum.*

Leaves pinnate; fruit berry-like, with 3 or more seed-like nutlets. . . 7. *Sambucus.*

Tribe LONICÉREAE R. Br.

1. DIERVÍLLA Duham. BUSH-HONEYSUCKLE

Calyx-tube tapering at the summit; the lobes slender, subulate, persistent. Corolla 5-lobed. Capsule slender, pointed, septicidal. — Low upright N. Am. shrubs, with ovate or oblong pointed serrate leaves, and cymosely 3–several-flowered peduncles, from the upper axils or terminal. (Named in compliment to Dr. *N. Dièreville*, a French traveller who carried it from Canada to the botanist, Tournefort, in 1699.)

1. D. Lonícera Mill. (from its resemblance to Honeysuckle, *Lonicera*), HERBE BLEUE (Que.). — Leaves oblong-ovate, taper-pointed, petioled, glabrous or essentially so; peduncles mostly 3-flowered; capsule long-beaked. — Dry woods, clearings and rocky places, Nfld. to Man., s. to N.S., N.E., Del., inland Va., mts. of N.C., O., Ind., Ill. and Ia. June–Aug. — Flowers at first pale yellow, turning to deep yellow, scarlet, crimson, or even maroon.

Var. **hypomálaca** Fern. (hairy beneath). — Leaves densely pilose beneath. — Ont., n. Wisc. and n. Minn.

2. D. sessilifòlia Buckl. (sessile-leaved). — Differing from no. 1 in *quadrate branchlets; sessile leaves more attenuate; cymes several-flowered;* calyx-lobes shorter; *corolla-lobes shorter than the tube.* — Upland woods, Ga. and Ala., n. to Blue Ridge of Warren Co., Va. (*Hunnewell*) and mts. of Tenn.

WEIGÈLIA (for *Christian Ehrenfried von Weigel*, 1748–1831) FLÓRIDA (Sieb. & Zucc.) A. DC. (flowering), with large crimson to pinkish-white corollas, much cult., sometimes persists about deserted gardens. (Introd. from Asia.)

2. LONÍCERA L. HONEYSUCKLE. CHÈVREFEUILLE (Que.)

Calyx-teeth very short. Corolla tubular or funnelform, often gibbous at the base, irregularly or almost regularly 5-lobed. Berry several-seeded. — Erect or climbing shrubs. Leaves entire. Flowers often showy and fragrant. A large genus of the N. Hemisph., most abundant in Asia and long popular in cult. (Named in honor of *Adam Lonitzer*, latinized *Lonicerus*, a German herbalist of the 16th century.)

a. Leaves all distinct; peduncles axillary, single, 2-flowered at summit (or in no. 10 sometimes in terminal panicles). Subgen. I. CHAMAECERASUS. . . . *b.*
  b. Upright or ascending bushy shrubs; peduncles bearing narrow terminal bracts (except in no. 9); corollas 1–2 cm. long. . . *c.*
    c. Bracts at summit of peduncle 2, linear to lance-oblong (sometimes with 4 bractlets above); berries red or blue. . . . *d.*
      d. Peduncles shorter than to about equaling flowers, 3–15 mm. long. . . *e.*
        e. Ovaries appearing solitary; corolla yellowish-white or straw-colored; berries blue; low native species of bogs and peaty soils. . . . . 1. *L. villosa.*
        e. Ovaries obviously 2; corolla pink, or white, changing to yellow; berries red; shrubs 1–3 m. high, in dry thickets, etc. (escaped from cult.). . . *f.*
          f. Branches with solid pith, retrorsely bristly, especially near nodes; leaves 5–10 cm. long, long-acuminate, nearly evergreen, ciliate; flowers from scaly buds before expanding of leaves. . . . . 2. *L. Standishii.*
          f. Branches fistulose (hollow at center), soft-pilose to glabrous; leaves 2–6 cm. long, deciduous, obtuse or merely acute, not ciliate; flowers from leaf-axils.
            Bractlets much shorter than ovaries; filaments pubescent. . . . 3. *L. Xylosteum.*
            Bractlets about as long as ovaries; filaments glabrous.
              Leaves pubescent below; peduncles densely pilose; corolla white, becoming yellow, about 1.5 cm. long. . . . . . 4. *L. Morrowi.*
              Leaves glabrous or glabrate; peduncles sparsely pilose to glabrate; corolla pink, becoming yellow, shorter. . . . . . 5. × *L. bella.*
      d. Peduncles mostly longer than flowers, 1.5–4 cm. long. . . *g.*
        g. Branchlets fistulose (with hollow center); broad bractlets obvious at bases of ovaries; introduced species spread from cult.
          Leaves, peduncles, etc., pilose; flowers yellowish-white. . . . . 3. *L. Xylosteum.*
          Leaves and peduncles glabrous; flowers pink. . . . . 6. *L. tatarica.*
        g. Branches solid (filled by pith); bractlets at bases of ovaries minute or obsolete; native species of woods or bogs.
          Leaf-surfaces glabrous, margin and definite petiole ciliate; corolla nearly regular; ovaries distinct; berries divergent, elongate; woodland species. . . . . 7. *L. canadensis.*
          Leaf-surfaces minutely pilose beneath, margin and very short petiole not ciliate; corolla 2-lipped; ovaries united or distinct, erect; berries spherical; species of bog and swamp. . . . . 8. *L. oblongifolia.*
    c. Bracts at summit of peduncle 4, broad and leaf-like, reflexed in fruit, forming an involucre; berries purple-black. . . . . 9. *L. involucrata.*

*b.*Twining and high-climbing or trailing; bracts like the leaves, but becoming smaller; corolla 3–4 cm. long; berries black. . . . . 10. *L. japonica.*

*a.*Some of upper leaves (except in no. 17) connate; flowers in sessile whorled clusters from their axils or terminal, forming heads or interrupted spikes; twining shrubs. Subgen. II. PERICLYMENUM. . . *h.*

*h.*Key to flowering material. . . *i.*

*i.*Corolla slenderly trumpet-shaped, nearly regular, with subequal short rounded erect lobes, deep red outside, about 5 cm. long; stamens and style little exserted or included. . . . . 11. *L. sempervirens.*

*i.*Corolla ringent, the lower lip narrow, the upper broad and 4-lobed; stamens and style conspicuously exserted. . . *j.*

*j.*Corolla 1.5–3 cm. long; flowers in heads or interrupted mostly leafless spikes; uppermost leaves connate, forming disks. . . *k.*

*k.*Leaves glabrous above, glaucous beneath; corollas glabrous outside (sometimes sparsely pilose in var. of no. 12). . . *l.*

*l.*Disk of uppermost connate leaves longer than broad, green or at most slightly glaucous above; whorls of flowers 1 or 2, rarely 3. . . *m.*

*m.*Corolla gibbous on one side at base, 1.5–2.3 cm. long, yellowish or greenish to purple; the tube short, gradually expanding upward; leaves below the connate terminal ones oblong to elliptic, strongly whitish beneath. . . . . 12. *L. dioica.*

*m.*Corolla scarcely gibbous at base, 2.5–3 cm. long, yellow or orange, with slenderly cylindric tube barely enlarged up to the limb; leaves below the connate terminal ones obovate-elliptic to suborbicular, merely gray beneath.

Corolla 2–2.5 cm. long, pale yellow; the tube very hairy within, shorter than the lobes; filaments equaling corolla-lobes. . . . . 13. *L. flavida.*

Corolla 3 cm. long, orange-yellow; the tube glabrous or but sparingly pubescent within, longer than the lobes; filaments shorter than the corolla-lobes. . . . . 14. *L. flava.*

*l.*Disk of uppermost connate leaves orbicular or nearly so, glaucous above; spikes (sometimes in 3's) of 2–6 whorls, the lowest whorl sometimes subtended by a remote lower disk; leaves below the disks obovate to suborbicular. . . . . 15. *L. prolifera.*

*k.*Leaves strigose-pilose above, soft-pubescent beneath, green both sides; corolla pubescent outside. . . . . 16. *L. hirsuta.*

*j.*Corolla 4–5 cm. long; uppermost leaves distinct and petioled. . . . 17. *L. Periclymenum.*

*h.*Key to fruiting material. . *n.*

*n.*Uppermost leaves connate. . . *o.*

*o.*Leaves glabrous above, glabrous or only minutely pilose beneath (not ciliate), gray or glaucous beneath. . . *p.*

*p.*Connate upper leaves green or barely glaucous above, longer than broad; seeds 3–4.5 mm. long, 1.5–3 mm. broad. . . *q.*

*q.*Leaves strongly whitened beneath.

Fruiting spike generally elongate and interrupted, of 2–4 remote whorls; peduncle 1–5 cm. long; seeds 4–4.5 mm. long, 3–3.5 mm. broad. . . . . 11. *L. sempervirens.*

Fruiting spike dense, globose-ellipsoid, of 1–3 crowded whorls; peduncle 0–1(–3) cm. long; seeds 3–3.5 mm. long, 1.5–2.5 mm. broad. . . . . 12. *L. dioica.*

*q.*Leaves merely gray beneath; fruiting spike capitate, sessile or short-stalked.

Leaves immediately below the disks rhombic-elliptic, tapering to short petioles. . . . . 13. *L. flavida.*

Leaves immediately below the disks broadly elliptic, rounded to sessile bases. . . . . 14. *L. flava.*

*p.*Connate upper leaves glaucous above, broader than long, forming a suborbicular disk; fruiting spikes (sometimes in 3's) of 2–6 whorls, the lowest whorl frequently subtended by a remote lower disk; seeds 4.5–5 mm. long, 3.5–4.5 mm. broad. . . . . . 15. *L. prolifera.*

*o.*Leaves strigose-pilose above, soft-pubescent beneath, green both sides, villous-ciliolate. . . . . 16. *L. hirsuta.*

*n.*Uppermost leaves, subtending inflorescences, distinct and petioled; fruiting spikes dense. . . . . 17. *L. Periclymenum.*

Subgen. I. Chamaecérasus L.

1. L. villòsa (Michx.) R. & S. (soft-hairy), Mountain-Fly-H. — Low depressed or ascending shrub, rarely up to 1 m. high, *with strongly ascending branches;* winter-buds appressed or ascending, usually without accessory buds; *leaves oblong to oblong-oblanceolate, blunt,* almost sessile, thick, often pubescent; flowers on short *peduncles* (1–7 *mm. long*), with subulate terminal bracts longer than the ovaries; corolla yellowish, with subequal lobes; *berries* consisting of 2 ovaries surrounded by a *blue* fleshy cup, *edible.* (*L. caerulea* of ed. 7, not L.) — Several varieties:

*a.*Leaves densely villous-subtomentose on both surfaces; branchlets tomentose or densely short-pilose beneath the long pilosity; limb of calyx ciliate at least in anthesis; corolla villous or pilose; depressed shrub. . . . . . . *L. villosa* (typical).
*a.*Leaves pilose-hirsute to glabrous beneath, strigose to glabrous above; branchlets with puberulent to glabrous surfaces or rather sparsely pilose-hirsute; limb of calyx glabrous; corolla glabrous (rarely pilose); mostly taller. . . *b.*
*b.*Base of corolla-tube strongly gibbous on one side. . . *c.*
*c.*Surfaces of young branchlets puberulent.
Young branchlets puberulent and more or less pilose-hirsute; leaves pilose beneath, strigose to glabrate above. . . . . Var. *Solonis.*
Young branchlets merely puberulent; leaves finely pilose to glabrate beneath. . . . . Var. *calvescens.*
*c.*Surfaces of young branchlets glabrous; leaves sparingly pilose to glabrous. Var. *tonsa.*
*b.*Base of corolla-tube slender and almost regularly tapering, scarcely gibbous on either side; young branchlets glabrous. . . . . Var. *Fulleri.*

L. villòsa (typical). — Peaty or rocky barrens and bogs, s. Lab. to Hudson Bay reg., s. to Nfld., Me. and n. N.H. *Fl.* May–July; *fr.* June–Aug.

Var. **Solònis** (Eat.) Fern. (for (David) Solon (Chase Hall) Smith, who discovered it in 1815). — Bogs, peaty slopes, swamps and rocky pastures, s. Nfld. to se. Man., s. to N.S., N.E., Mich., Wisc. and Minn. *Fl.* late April–July; *fr.* late May–Aug.

Var. **calvéscens** (Fern. & Wieg.) Fern. (becoming hairless). — Peaty or swampy habitats, often subalpine, Lab. to Ont., s. to Nfld., N.S. and N.E.

Var. **tónsa** Fern. (shaved). — Nfld. to Ont., s. to N.S., N.E., Pa. and Minn.

Var. **Fúlleri** Fern. (for its discoverer, Timothy Otis Fuller, 1845–1916). — Boggy meadows, e. Mass.

2. L. Standíshii Jacques (for the recent English nurseryman, John S. Standish). — *Semi-evergreen* widely branching shrub 1–2 m. high; the *branches with solid pith, retrorsely bristly, especially near the nodes; leaves firm, veiny,* lanceolate to oblong-ovate, *long-acuminate,* rounded at base, more or less appressed-setose, *ciliate,* 5–10 *cm. long; flowers from scaly buds before leaves,* fragrant, slender-stalked; corolla bilabiate, whitish, pubescent, about 1.5 cm. long; berries red, elongate, united below. — Thickets, L.I., esc. from cult. *Fl.* Apr.; *fr.* May, June. (Introd. from Asia)

3. L. Xylósteum L. (old generic name, meaning Bonewood), Fly-H. of Europe. — Bushy shrub 1–3 m. high, with *hollow branches* soft-pilose to glabrate; *leaves* elliptic-ovate to obovate, acute, rounded to broadly cuneate at base, 3–6 *cm. long, gray-green,* glabrous or sparingly pilose above, *pale and pubescent beneath; flowers* axillary, *on filiform pubescent peduncles* 0.8–2 *cm. long;* bracts linear-subulate, spreading; *bractlets elliptic or oval, glandular, half as long as glandular ovary;* corolla bilabiate, whitish, yellowish or pinkish, pubescent, about 1 cm. long; *filaments pubescent;* berries red, distinct. — Roadsides and thickets, esc. from cult., N.E. to Mich., s. to N.J., Pa. and O. *Fl.* May, June; *fr.* June–Aug. (Introd. from Eurasia)

4. L. Mórrowi Gray (for its discoverer, James Morrow, 1820–1865). — Similar to no. 3, branches divergent; *leaves oblong or narrowly elliptic,* 2.5–5 cm. long, *grayish-tomentose beneath; peduncles densely pilose,* 0.5–1.5 cm. long; *bractlets pubescent, about equaling ovaries; calyx-teeth ciliate; corolla white, turning yellow,* about 1.5 *cm. long; filaments glabrous;* berries red or yellow. — Roadsides and thickets, esc. from cult., Me. to Mich., s. to L.I., N.J. and Pa. *Fl.* May, June; *fr.* July–Aug. (Introd. from Eurasia)

5. × L. béllA Zabel (pretty). — Midway between nos. 4 and 6; *leaves* elliptic to oblong-ovate, *glabrous or glabrate; peduncles sparsely pilose or glabrate,* 0.5–1.5 *cm. long; corolla pink,* becoming yellow, shorter than in no. 4. — Thickets, esc. from cult., Me. to N.Y., N.J. and Pa. *Fl.* May, June; *fr.* July, Aug. (Introd. from Asia)

6. L. tatárica L. (Tartarian), Tartarian H. — *Smooth* shrub 1.5–3 m. high; branches hollow, slender; *leaves* thin, *glabrous,* cordate-oval, short-petioled, 3–6 cm. long; *peduncles* filiform, 1.5–2 *cm. long, glabrous;* bractlets rounded, rarely half as long as ovaries; *corolla* bilabiate, 1.5–2 *cm. long, pink,* or white in forma ALBIFLÒRA (DC.) House (white-flowered); *the lobes subequal, widely spreading, nearly as long as tube;* berries red (or yellow), globose, united at base. — Thickets, borders of woods, shores, etc., esc. from cult., Que. and Ont., s. to N.E., N.J., Pa., Ky. and Ia. *Fl.* May, June; *fr.* late June–Aug. (Introd. from Eurasia)

7. L. canadénsis Bartr. (Canadian), FLY-H. — *Straggling* loosely branched shrub up to 1.5 m. high, *with* slender *solid* (not fistulose) *branches; leaves* oblong-ovate, often cordate, *glabrate,* the *margin and distinct petiole ciliate;* peduncles filiform, 2–3 cm. long; bractlets obsolete; *corolla* funnelform, *nearly regular,* about 2 cm. long, *yellowish-green or straw-color, the lobes much shorter than the tube; berries* separate, *elongate,* red, *divergent.* (*L. ciliata* Muhl.) — Cool woodlands, Gaspé Pen., Que., to Sask., s. to N.S., N.E., n. N.J., Pa., upland to N.C., W.Va., O., Ind., Wisc. and ne. Ia. *Fl.* April–early July; *fr.* June–Aug.

8. L. oblongifòlia (Goldie) Hook. (oblong-leaved), SWAMP-FLY-H. — Erect shrub up to 1.5 m. high, with *solid ascending minutely pilose branchlets; leaves oblong to narrowly obovate,* 3–8 cm. long, *minutely downy beneath,* becoming glabrescent, *short-petioled; peduncles* filiform, 1.5–4 *cm. long;* bracts minute or deciduous; bractlets obsolete; *corolla yellowish-white* (rarely clear white), 1–1.5 cm. long, *deeply 2-lipped; berries* orange-yellow to deep red, *subglobose, distinct or united.* — Bogs, swampy thickets and wet woods, se. Que. to Man., s. to N.B., n. and e. Me., n. and w. N.Y., w. Pa., n. O., Mich., Wisc. and Minn. *Fl.* late May, June; *fr.* July, Aug.

Var. **altíssima** (Jennings) Rehd. (very tall). — *Glabrous or essentially so from the first.* — Occasional, n. Me. to s. Ont. and nw. Pa.

9. L. involucràta (Richards.) Banks (with an involucre). — Straggling to ascending shrub 1–3 m. high, with glabrous 4-angled branchlets; *leaves* obovate, elliptic-oblong or oval, *acuminate, with stout midrib, strongly ascending,* green and glabrous or glabrate above, paler and hirsute on veins beneath, 0.5–1.5 *dm. long;* flowers on ascending *clavate peduncles one-fifth to one-third as long as leaves; involucre of* 4 green to dark purple *ascending, finally reflexed, foliaceous bracts* 1–2 *cm. long;* corollas embraced by involucre, stramineous, cylindraceous, 1–1.5 cm. long, with erect lobes, glandular outside; *berries purple-black.* — Calcareous woods, banks of streams and swamps, Gaspé Pen., Rimouski Co. and Laurentide Park, Que., and ne. N.B.; James Bay reg. to L. Superior reg. of Ont., Mich. and Wisc.; Alaska to Mex. *Fl.* and *fr.* June–Aug.

10. L. JAPÓNICA Thunb. (Japanese), JAPANESE H. — *High-twining or trailing* pubescent shrub; leaves ovate or oblong, thickish, entire (early ones often lobed), short-petioled, green; *bracts leaf-like,* becoming smaller; *corolla* 3–4 *cm. long,* white or white tinged with purple, becoming yellow, very fragrant, with pubescent tube; *berries black.* (*Nintova* Sweet) — Thickets, borders of woods and roadsides, a most pernicious and dangerous weed, overwhelming and strangling the native flora and most difficult to eradicate; extensively planted and encouraged by those who do not value the rapidly destroyed indigenous vegetation, Fla. to Tex., n. to Mass., N.Y., O., Ind., Mo. and Kans. *Fl.* late April–July (rarely –Nov.); *fr.* Sept.–Nov. (Unfortunately natzd. from Asia)

Var. CHINÉNSIS (P. W. Wats.) Baker (Chinese). — Branchlets and leaves purple, glabrous or nearly so; corolla more deeply cleft, carmine on the outside. — Similar habitats, n. to N.J. and Pa. (Introd. and natzd. from Asia)

Subgen. II. PERICLÝMENUM L.

11. L. sempérvirens L. (evergreen), TRUMPET- or CORAL-H. — Twining shrub, with glabrous or nearly glabrous branchlets; leaves glabrous on both surfaces or minutely pilose beneath, green above, white beneath, oblong to elliptic or obovate, firm, the 1 or 2 upper pairs connate into a disk; flowers in mostly 2–4 sessile remote whorls, forming interrupted spikes; *corolla slenderly trumpet-shaped, nearly regular, with subequal short rounded erect lobes, deep red* (rarely orange to yellow) *outside,* 4–5.5 *cm. long; stamens and style little exserted or included;* berries red; *seeds* 4–4.5 *mm. long,* 3–3.5 mm. broad. (*Phenianthus* Raf.) — Woods and thickets, Fla. to Tex., n. to s. Me., Mass., N.Y., O., Ia. and Neb., partly as escape from cult. northeastw. *Fl.* late March–July; *fr.* Aug.–Oct.

Var. **mìnor** Ait. (smaller). — Corolla only 2–3 cm. long. — Infrequent, N.Y. to N.C.

Var. **hirsùtula** Rehd. (with short stiffish hairs). — Branchlets hirtellous; leaves strigose-pilose above, ciliolate; corolla pubescent. — Local, Va. and N.C.

12. L. dioìca L. (dioecious). — Twining or loosely ascending or reclining shrub, *glabrous* **essentially throughout;** *leaves oblong or elliptic,* short-petioled or subsessile, 4–10 cm. long, **green above,** *whitened beneath,* often undulate, *the upper* 1–4 *pairs connate into* oblong to rhombic *at least bluntly pointed disks; peduncle* 0–1 (rarely –3) *cm. long*, the *flowers in* 1–3 subapproximate *whorls, these crowded in fruit; flowers greenish-yellow or stramineous to brick-red or purplish,* 1.5–2.3 *cm. long*, glabrous outside, hirsute within, bilabiate; *the tube gibbous on one side at base, gradually expanding upward;* style and stamens hairy; berries red; *seeds* 3–3.5 *mm. long*, 1.5–2.5 *mm. broad.* — Rocky banks, dry woods and thickets, sw. Me. and sw. Que. to Man., s. to Ga. and Mo. *Fl.* May–July; *fr.* July–Sept. — Upper leaves in whorls of 3, involucral disk 3-lobed in forma **trifòlia** Vict. & Rolland. Passing freely into

Var. **glaucéscens** (Rydb.) Butters (somewhat blue-green). — Mostly with larger leaves and smaller flowers; leaves pubescent beneath; ovary often glandular, or densely glandular and hirsute in forma **dasýgyna** (Rehd.) Deam (with hairy ovary); corolla often pubescent outside. (*L. glaucescens* Rydb.) — Sandy or rocky soils, w. Que. to B.C., s. to w. N.C., Ky., Mo. and ne. Kans.

13. L. flávida Cockerell (yellowish). — Somewhat like no. 14; *leaves* merely grayish beneath, *those immediately below the disks rhombic-elliptic, tapering to short petioles; corolla pale yellow,* 2–2.5 *cm. long; the tube very hairy within, shorter than the lobes,* not gibbous at base; *filaments equaling corolla-lobes.* — Calcareous or circumneutral rocky shores and bluffs, w. S.C., Ky., Tenn. and Mo. *Fl.* April, May.

14. L. flàva Sims (yellow), Yellow H. — Resembling no. 12; *leaves merely grayish beneath, those immediately below the disk broadly elliptic and rounded to sessile bases; corolla* 3 *cm. long, orange-yellow;* the glabrous *tube* not gibbous at base, *glabrous or but sparsely pubescent within, longer than lobes; filaments shorter than corolla-lobes;* berries crowded in a capitate sessile or only short-peduncled head. — Rocky woods and bluffs, N.C. to Mo., s. to Ga., Ala., Ark. and Okla. *Fl.* April, May.

15. L. prolífera (Kirchn.) Rehd. (proliferous), Grape-H. — Coarser than no. 12; *much whitened* with glaucous bloom, *glabrous except for slight puberulence on lower surfaces of leaves; disk of uppermost connate leaves orbicular or nearly so, glaucous above,* its ends rounded or often retuse; *spikes* (sometimes in 3's) *of* 2–6 *whorls,* the lowest whorl sometimes subtended by a remote lower disk; *leaves below disks obovate to suborbicular;* corolla 2.5–3 cm. long, only slightly gibbous at base, pale yellow; *seeds* 4.5–5 *mm. long,* 3.5–4.5 *mm. broad.* (*L. Sullivantii* Gray) — Rocky woods and banks, s. Ont. and se. Man., s. to Tenn., Ark. and e. Kans.; spread from cult. locally e. to Mass. *Fl.* May, June; *fr.* July–Oct.

16. L. hirsùta Eat. (stiffly hairy), Hairy H. — Twining and high-climbing; *branchlets glandular-villous; leaves ciliate,* deep green and more or less *appressed-setulose above, downy-hairy beneath,* veiny, dull, broadly oval, the uppermost united into 1 or 2 *disks usually acuminate at each end;* spikes (sometimes 2 or more) sessile or peduncled, the whorls approximate; *corolla* 2–2.5 cm. long, orange-yellow, *clammy*-pubescent, hairy within, the *slender tube slightly gibbous at base.* — Calcareous shores, thickets and bluffs, w. Que. and w. N.E. to Sask., s. to Pa., O., Mich., Wisc., Minn. and Neb. *Fl.* May–July.

L. Caprifòlium L. (old generic name, meaning goat-leaf), Italian Woodbine, glaucous and glabrous twiner with connate upper leaves, and white or purplish fragrant corollas 4–5 cm. long, has long been reported but seems not to be a true member of our flora.

17. L. Periclýmenum L. (the old generic name), Woodbine of Europe. — High-climbing or trailing; *leaves all distinct,* petioled, ovate to oblong, 4–6 cm. long, the uppermost often sessile; flowers in 3–5 whorls, forming close spikes; *corolla* 4–5 *cm. long,* glandular, white or purple-tinged, turning yellow; berries red. — Thickets and roadsides, esc. from cult., Nfld. and N.S. *Fl.* July, Aug. (Introd. from Eu.)

3. SYMPHORICÁRPOS Duham. Snowberry

Calyx-teeth short, persistent. Corolla campanulate, regularly 4–5-lobed, with as many short stamens inserted into the throat. Drupe 4-locular, 2-seeded. Seeds bony. — Low and branching upright shrubs, most species of N.Am., one Chinese, with oval or rounded short-petioled leaves, which are usually entire, or wavy-toothed or lobed on the young shoots. Flowers white or greenish, tinged with rose-color, in close short spikes or clusters. (Name composed of the Greek *symphorein, to bear together,* and *carpos, fruit;* from the clustered berries.)*

Corolla 5–9 mm. long, pinkish or pink and white; style glabrous (sometimes pilose near the middle in no. 2); fruit white or whitish when fresh, then 6–15 (–20) mm. long; nutlets 3.5–6 mm. long.

Flowers short-pedicelled; corolla 5–8 mm. long, its lobes 2–3 mm. long; style 2–3 mm. long, glabrous, included; anthers about equaling to longer than

*Treatment derived in part from that of G. N. Jones in Journ. Arnold Arb. xxi. 201–252 (1940).

filaments; fruit 6–15(–20) mm. long; nutlets 4–6 mm. long, 2.5–3.5 mm. broad. . . . . 1. *S. albus.*
Flowers sessile; corolla 6–9 mm. long, with lobes 3–4 mm. long; style 4–8 mm. long, pilose near middle or glabrous, exserted; anthers half as long as filaments; fruit 6–8 mm. long; nutlets 3.5 mm. long, 2–2.5 mm. broad. . . . 2. *S. occidentalis.*
Corolla 3–4 mm. long, greenish and purplish; style pilose; fruit coral-pink to purple, sessile, 5–7 mm. long; nutlets 2.5–3.5 mm. long. . . . . 3. *S. orbiculatus.*

1. S. àlbus (L.) Blake (white), Snowberry, Belluaine (Que.). — Slender loosely ascending shrub 0.1–1 m. high, with shreddy bark; branches slender, the branchlets minutely puberulent to glabrous; *leaves membranaceous;* those of flowering branches narrowly elliptic-ovate to suborbicular, 1–5 cm. long, more or less pilose beneath to quite glabrous; *flowers short-pedicelled,* commonly in small groups of 1–5 in the upper axils and frequently in terminal racemes of 5–20; *corolla pink outside,* its lobes obtuse and villous on inner side; *style* 2–3 *mm. long, glabrous, included; anthers* 1–2 mm. long, *slightly shorter than the filaments; fruits* 1–few, mostly in the axils, sometimes in terminal racemes, white, depressed-globose, 6–10(–12) mm. in diameter, *capped by the depressed calyx-lobes; nutlets* oval, 4–5 mm. long, 2.5–3.5 *mm. broad.* (*S. racemosus* Michx., not Man. ed. 7 and most authors; *S. racemosus*, var. *pauciflorus* Robbins) — Calcareous ledges, barrens and gravels, Rimouski Co., Que., to B.C., s. to w. Mass., w. Va., W.Va., O., Mich., Wisc., Minn., Neb. and Colo. *Fl.* May–July; *fr.* Aug.–Oct. (often holding into winter).

Var. laevigàtus (Fern.) Blake (smooth). — Coarser, 0.8–2 m. (in cult. –3 m.) high; branchlets usually glabrous; leaves glabrous beneath; flowers more generally in elongate terminal racemes; fruit 1.2–1.5 cm. in diameter (some 2 cm. long in cult.); nutlets 4–6 mm. long. (*S. racemosus*, var. *laevigatus* Fern.; *S. rivularis* Suksd.) — Roadsides, rocky banks, etc., spread from cult., from Que. westw. and southw. (Introd. from the Pacific slope) — Too difficult to separate on stable morphological characters from the smaller-leaved typical *S. albus*, which, even in the type-region, may be quite glabrous and with terminal racemes.

2. S. occidentàlis Hook. (western), Wolfberry. — Stoloniferous shrub 0.3–1 m. high; young twigs puberulent (rarely glabrous); *leaves* becoming *coriaceous*, elliptic or oval, up to 1 dm. or more long, sparsely pubescent along nerves beneath or glabrous; *flowers sessile* in mostly crowded terminal and axillary simple or forking clusters; *corolla* 6–9 *mm. long, somewhat broader, cleft about to middle into lobes* 3–4 *mm. long,* pale pink, densely villous within; *style* pilose near middle or glabrous, 4–8 *mm. long, usually exserted;* anthers 2 mm. long, about half as long as filaments; *fruit* greenish-white, soon darkening, *beaked by the persistent calyx-lobes;* nutlets about 3.5 mm. long. — Bluffs, dry prairies and plains, Ont. to B.C., s. to Mich., n. Ill., Mo., Kans. and N.M.; natzd. e. to N.E. and Pa. *Fl.* late May–Sept.; *fr.* Aug.–winter.

3. S. orbiculàtus Moench (round), Coralberry, Indian-currant. — Low shrub, 0.5–2 m. high; young branches villous to puberulent; leaves oval to suborbicular, pubescent beneath; *flowers sessile* and *crowded in numerous short axillary* and terminal *subcapitate spikes; corolla greenish and purplish,* 3–4 *mm. long; style pilose,* about 2 mm. long; anthers 1 mm. long, shorter than filaments; *fruit coral-pink to purple,* ellipsoid, 5–7 mm. long, 4–5 mm. thick, beaked by the calyx-lobes; *nutlets* 2–2.5 *mm. broad.* (*S. Symphoricarpos* (L.) MacM.) — Open woods, thickets and dry banks, Fla. to Tex. and Mex., n. to Pa., O., Ind., Ill., Minn., S.D. and Colo.; spread from cult. n. to N.E., N.Y., etc. *Fl.* July, Aug.; *fr.* Sept.–winter.

4. LINNAÈA Gronov. Twinflower

Calyx-teeth 5, subulate, deciduous. Corolla slender-campanulate or funnelform, almost equally 5-lobed. Stamens 4, two of them shorter, inserted toward the base of the corolla. Ovary and the small dry fruit 3-locular, but 1-seeded. — A slender creeping and trailing little boreal evergreen, somewhat hairy, with rounded-oval sparingly crenate leaves contracted at the base into short petioles, and filiform upright peduncles forking into 2 (rarely 4 or 6) pedicels at the top, each bearing a delicate and fragrant nodding flower. Corolla whitish, tinged and striped with rose-purple, hairy inside. (Dedicated, originally by Gronovius, to the immortal *Linnaeus*, 1707–1778, who first pointed out its characters, and with whom the European type of this pretty little plant was a special favorite.)

1. L. borreàlis L. (northern). — Corolla campanulate, 6–9 mm. long, flaring from well within the calyx, the tube very short. — N. Eurasia and Alaska. Represented with us by

Var. **americàna** (Forbes) Rehd. (American). — Corolla 8–15 mm. long, more funnelform, flaring from above the calyx, its tube exserted. (*L. americana* Forbes) — Cool woods and peaty knolls, w. Greenl. and Lab. to Alaska, s. to Nfld., N.S., s. N.E., L.I., ne. Pa., Md., W.Va., ne. O., n. Ind., S.D., Colo., Utah and n. Calif. June–Aug., rarely in late autumn.

5. TRIÓSTEUM L. Feverwort. Horse-gentian

Calyx-lobes linear-lanceolate, foliaceous, persistent. Corolla tubular, somewhat equally 5-lobed, scarcely longer than the calyx. Ovary mostly 3-locular, in fruit forming a dry drupe containing 3 ribbed 1-seeded bony nutlets. — Coarse upright perennial herbs, leafy to the top; the ample entire pointed leaves tapering to the base or connate around the simple stem. Flowers solitary or clustered in the axils. Small genus of Atl. N.Am. and se. Asia. (Name an abbreviation of *Triosteospermum*, alluding to the three bony nutlets.)*

Stem glabrous, crisp-puberulent, or villous with hairs at most 1.5 (in var. of no. 2, –2.5) mm. long; leaves ovate to ovate-oblong or panduriform, glabrous above or with minute pubescence (less than 1 mm. long); larger mature blades 0.5–1.5 dm. broad, 1–3 dm. long; flowers 1–4 in each axil; calyx-lobes minutely and closely puberulent; corolla crisp-puberulent.

Leaves (at least 3–5 median pairs) panduriform, the broadly connate enlarged bases of the larger 3–9 cm. broad; corolla tubular-campanulate, hardly bilabiate, firm, yellowish or greenish to dull purple, about equaling stamens; style exserted 1.5–3 mm.; fruits subglobose, dull orange-yellow. . . . . . 1. *T. perfoliatum.*

Leaves distinct (rarely 1–3 pairs with connate bases 1–2 cm. broad), tapering to base; corolla dilated above, bilabiate, petaloid, purple-red, much exceeding stamens; style usually included; fruit ellipsoid-ovoid, bright orange-red. 2. *T. aurantiacum.*

Stem setose-hispid with hairs up to 1.5–3 mm. long; leaves oblong-ovate to lanceolate or oblanceolate, strigose above with hairs 0.8–1.8 mm. long; the larger mature blades 2–6 cm. broad, 0.8–1.8 dm. long; flowers mostly solitary in the axils; calyx-lobes hispid-ciliate; corolla loosely villous, yellowish. . . . . 3. *T. angustifolium.*

1. T. perfoliàtum L. (with leaves meeting around the stem), Tinker's-weed, Wild Coffee. — Coarse, 0.5–1.2 dm. high; stem densely glandular-puberulent above; *leaves* dark green, thickish, densely velutinous-puberulent beneath, *oval to oblong-ovate; the larger with mature blades* 0.5–1.5 *dm. broad,* 1–3 *dm. long;* 3–5 *of the middle ones strongly panduriform, the connate enlarged bases of the larger* 3–9 *cm. broad;* flowers erect, mostly 3 or 4 to an axil; *calyx-lobes attenuate to acute tips,* 0.9–2 (*av.* 1.4) *mm. broad; corolla tubular-campanulate, hardly bilabiate,* 0.8–1.7 cm. long, *firm, yellowish or greenish to dull purple,* densely glandular-puberulent, *about equaling stamens; style* usually *exserted* 1.5–3 *mm.; fruits subglobose, dull orange-yellow.* — Rocky or thin soils, woods and thickets, Ga. to e. Kans., n. to Mass., e. N.Y., W.Va., Mich., Wisc., Minn. and Neb. *Fl.* mid-May–early July; *fr.* Aug.–Oct.

2. T. aurantìacum Bickn. (orange-colored), Wild Coffee. — Differing from no. 1 in stems less densely glandular-puberulent, *usually with spreading glandless hairs up to* 1.5 *mm. long; leaves all distinct and tapering to narrow bases or rarely* 1–3 *pairs with undilated connate bases* 1–2 *cm. broad,* the lower surfaces less densely pubescent; flowers 1–3 in an axil; *calyx-lobes* blunt or merely acutish, 1.5–2.8 (*av.* 2) *mm. broad; corolla dilated above, distinctly bilabiate, red-purple, petaloid, much exceeding stamens; style* usually *included; fruit ellipsoid-ovoid, bright orange-red.* (*T. perfoliatum,* var. Wieg.) — Rich woods and thickets, C.B., w. N.B. and s. Que. to w. Ont., s. to N.E., Md., upland Ga., Ky., Ill. and Ia. *Fl.* mid-May–July; *fr.* Aug.–Oct. In neighboring areas flowering later than no. 1.

Var. **glaucéscens** Wieg. (becoming blue-green). — *Leaves glabrous or nearly so.* (*T. perfoliatum,* var. Wieg.) — Centr. N.Y. and w. Pa. to Ind.

Var. **illinoénse** (Wieg.) Palmer & Steyerm. (of Illinois). — *Stems bearing glandless hairs* 1.5–2.5 *mm. long;* upper leaf-surfaces setose with hairs up to 1 mm. long; *calyx-lobes somewhat bristly-ciliate.* (*T. perfoliatum,* var. Wieg.) — O. to Ill., Mo. and Kans.

3. T. angustifòlium L. (narrow-leaved). — Slender, 3–8 dm. high; *stem retrorse-setose with glandless hairs up to* 1.5–3 *mm. long; leaves oblong-ovate to lanceolate or oblanceolate, tapering at base, strigose above with hairs* 0.8–1.8 *mm. long, glabrous or merely hispid on nerves beneath, the larger mature blades* 2–6 *cm. broad,* 0.8–1.8 *dm. long; flowers mostly solitary in axils; calyx-lobes hispid-ciliate; corolla loosely villous,* greenish- or yellowish-white; fruits strigose-hirsute. — Woods and thickets, Ala. to La., n. to centr. Ct., L.I., Pa., W.Va., Ind., Ill. and Mo. *Fl.* April–June; *fr.* July–Sept.

Var. **Eàmesii** Wieg. (for its discoverer, Edwin Hubert Eames, 1865–1948). — Leaves minutely soft-pilose beneath. — Rocky thickets and open woods, sw. Ct.

---

* Characters largely taken from treatment of Wiegand in Rhodora, xxv. 199–203 (1923).

Tribe Sambùceae HBK.

6. VIBÚRNUM L. Viburnum. Arrow-wood. Laurestinus. Viorne (Que.)

Calyx 5-toothed. Corolla spreading, deeply 5-lobed. Stamens 5. Stigmas 1–3. Fruit a 1-locular 1-seeded drupe, with soft pulp and a thin-crustaceous (flattened or tumid) stone. — Shrubs or small trees mostly of N. Hemisph., with simple leaves and white (rarely pink) flowers in compound cymes. Petioles sometimes bearing little appendages which are evidently stipules. Leaf-buds naked, or with pairs of scales. (The classical Latin name, of doubtful meaning.)

a. Cyme umbelliform, with several branches radiating from a common point. . . b.

b. Winter-buds naked; pubescence of young growth stellate.

Upright shrub or small tree; leaves oblong-ovate to oval, 0.5–1.2 dm. long; cymes peduncled, their flowers uniform, small, perfect; filaments longer than corolla-lobes; stones with 3 ventral grooves. § Lantana. . 1. *V. Lantana.*

Straggling or loosely ascending shrub; leaves round-ovate to orbicular, 1–2 dm. long; cymes sessile; their marginal flowers neutral, showy, up to 2.5 cm. broad; filaments shorter than lobes of small perfect flowers; stones with 1 ventral groove. § Pseudotinus. . . . . 2. *V. alnifolium.*

b. Winter-buds with 1 or 2 pairs of outer scales; pubescence (when present) not stellate; drupes bluish to black, or red. . . c.

c. Leaves pinnately veined, not strongly lobed (palmately veined and lobed in no. 12, with winter-buds with 2 pairs of outer scales, pubescence of young leaves fascicled, filaments longer than corolla-lobes, and drupes purple-black); pubescence (when present) scurfy or of fascicled hairs; fruit bluish, purple or black, with little juice, not strongly acid. . . d.

d. Winter-buds with 1 pair of outer scales; pubescence (when present) scurfy or tomentulose; leaves entire or finely toothed, their veins curved and anastomosing near the margin. § Lentago. . . e.

e. Cyme definitely peduncled; leaves entire, undulate, crenulate or dentate.

Winter-buds yellow-brown or golden; leaves dull above, usually with short abrupt blunt acumination, commonly dentate or crenate; anthers elevated 1–3.5 mm. above throat of corolla; pulp of drupe sweet; stone elliptic-oblong or oval; northern and upland species. . . . . 3. *V. cassinoides.*

Winter-buds reddish-brown to fuscous; leaves lustrous above, gradually tapering or rounded to tip, entire (rarely dentate); anthers elevated 4–5 mm. above throat of corolla; pulp of drupe bitter; stone round-obovate; southern species. . . . . 4. *V. nudum.*

e. Cyme sessile or essentially so; leaves serrate or serrulate with sharp teeth.

Leaves mostly sharp-acuminate; petioles undulate-margined, those subtending the cymes 1.5–3.5 cm. long. . . . . 5. *V. Lentago.*

Leaves merely acute or with obtuse to rounded tips; petioles little if at all undulate-margined, those subtending the cymes 0.5–2 cm. long.

Petioles, midribs beneath and rays of cyme without rufescent tomentum; leaves not lustrous, membranaceous to subcoriaceous. . . . . 6. *V. prunifolium.*

Petioles, midribs beneath and rays of cyme more or less rufescent-scurfy or -tomentulose; leaves lustrous above, coriaceous. 7. *V. rufidulum.*

d. Winter-buds with 2 pairs of outer scales; pubescence (when present) of fascicled hairs; leaves with straightish veins terminating in the coarse teeth. § Odontotinus. . . f.

f. Leaves pinnately veined, with unlobed margin, not evidently dotted beneath.

Leaves short-petioled or subsessile, with a pair of linear-subulate stipules; blades immediately below peduncles with petioles 0–7 mm. long; cymes 1.5–6 cm. broad. . . . . 8. *V. Rafinesquianum.*

Leaves longer-petioled, those immediately below peduncles with petioles 1–5 cm. long (if shorter, without stipules); cymes 3–10(–14) cm. broad.

Leaves deeply cordate, with 20–30 teeth on each half of blade; petioles 1.5–5 cm. long, bearing stipules at flowering time; fruit flattened, ellipsoid, about 1 cm. long. . . . . 9. *V. molle.*

Leaves narrowed, rounded or only subcordate at base, with 4–22 teeth on each half of blade; petioles 0.5–2.5 cm. long, mostly exstipulate; fruit globose- to ellipsoid-ovoid, 5–8 mm. long.

Branchlets, petioles and rays of cyme and often the leaf-surfaces closely pubescent (glabrous or glabrate in the var.); stones ellipsoid-ovoid, 5–8 mm. long, with deep furrow-like ventral groove. . . . . 10. *V. dentatum.*

Branchlets glabrous; leaves glabrous, or usually pubescent only on nerves beneath; stones globose-ovoid, 5–7 mm. long, with shallow and broad trough-like ventral groove. . . . . 11. *V. recognitum.*

*f.*Leaves palmately 3-ribbed from the rounded or subcordate base, mostly 3-lobed (maple-like), copiously dotted beneath; slender petiole often stipulate at base. . . . . 12. *V. acerifolium.*

*c.*Leaves palmately 3–5-nerved from base, mostly lobed; drupes red or orange, very juicy, acid, subglobose; stone flattened, not sulcate; winter-buds with 2 connate outer scales; foliage glabrous or with simple hairs. § Opulus.

Petioles exstipulate at base, glandless at summit or with small glands at base of leaf-tissue; blades shallowly lobed or unlobed; cymes 1.5–3.5 cm. broad, of uniform perfect flowers; stamens much shorter than corolla-lobes; straggling or loosely ascending or erect shrub. 13. *V. edule.*

Petioles with slender basal stipules, usually with glands well below the leaf-tissue; blades deeply lobed; cymes 0.4–1.5 dm. broad, their marginal flowers neutral and with greatly enlarged flat corollas; stamens of perfect flowers longer than corolla-lobes; upright shrubs or small trees.

Glands on petioles dome-shaped, columnar or clavate, with rounded summits; stipules filiform-clavate or with thickened tips. . . . 14. *V. trilobum.*

Glands on petioles forming depressed concave-topped disks; stipules filiform-attenuate, with slender tips. . . . . 15. *V. Opulus.*

*a.*Cyme a round-topped panicle with opposite branches. . . . . . . 16. *V. Sieboldii.*

§ Lantàna Spach

1. V. Lantàna L. (old name, meaning flexible or bending), Wayfaring-tree, Twistwood. — Shrub or small tree up to 5 m. high; naked winter-buds, young branchlets, lower surfaces of leaves, etc., *cinereous with minute stellate pubescence; leaves oblong-ovate to oval,* closely serrulate, 0.5–1.2 *dm. long; cymes peduncled,* 5–10 cm. broad, *with uniform small perfect flowers; filaments longer than corolla-lobes;* fruits ellipsoid, 8–10 mm. long, red, becoming black; *stone with* 3 *ventral grooves.* — Fence-rows and roadsides, spread from cult., Ct. to s. Ont. *Fl.* May, June; *fr.* Sept. (Introd. from Eu.)

§ Pseudotìnus Clarke

2. V. alnifòlium Marsh. (alder-leaved), Hobblebush, Witch-hobble, Tangle-legs, Moose-wood, Bois d'orignal (Que.). — *Straggling* shrub up to 3 m. high, branches often reclining and layering; *branchlets and large naked winter-buds scurfy; leaves round-ovate to suborbicular,* stellate-scurfy beneath (especially when young), closely serrate (midsummer leaves and those of 2nd growth after browsing often narrower, coarsely toothed, thin and glabrous), *becoming* 1–2 *dm. long and broad; cymes sessile,* 0.7–1.5 dm. broad, *flat, with broad flat neutral white,* or pink in forma **ròseum** House (rose-colored), *marginal flowers up to* 2.5 *cm. broad; perfect central flowers with stamens shorter than corolla-lobes;* drupes ellipsoid, changing through red to nearly black; *stone with* 1 *ventral groove.* (*V. lantanoides* Michx.) — Woods and cool ravines, P.E.I. to Ont., s. to N.S., N.E., n. N.J., Pa. and ne. O., and upland to Ga. and Tenn. *Fl.* May, June; *fr.* Aug.–Oct.

§ Lentàgo DC.

3. V. cassinoìdes L. (like *Ilex Cassine*), Witherod, Wild-raisin, Alisier (Que.). — Shrub 1–4 m. high; *winter-buds covered by* a single pair of *yellow or golden scurfy scales, the axillary ones often well-developed at flowering time; young shoots scurfy; leaves dull above,* oblong or lanceolate to oval or rhombic-ovate, usually *with abrupt short blunt acumination, often crenulate or dentate;* those of fertile branches 2.5–15 cm. long, 1.5–6 cm. broad, the pinnate veins curving and anastomosing near the margin; *cyme peduncled,* 3–10 cm. broad; flowers ill-scented; *anthers elevated* 1–3.5 *mm. above throat of corolla;* drupes ellipsoid to subspherical, 6–9 mm. long, changing from whitish-yellow through pink to blue-black, with heavy bloom; ripe *pulp sweet; stone* flattish, *elliptic-oblong.* — Thickets, clearings, swamps and borders of woods, Nfld. to Ont. (? se. Man.), s. to N.S., N.E., L.I., Del., Md., upland to Ala. and Tenn., O., n. Ind. and Wisc.

Fl. June–early Aug.; *fr.* Sept., Oct. — Highly variable in shape and size of foliage; hybridizes southeastw. with no. 4.

4. V. nùdum L. (naked; from the definitely peduncled cyme), POSSUM- or SWAMP-HAW. — Coarser than no. 3; trunks either tree-like, 1–2 dm. in diameter and up to 6 m. high, or loosely spreading or straggling; *winter-buds browner or fuscous, the lateral rarely developed at fruiting time; young shoots green or merely dotted; leaves lustrous above, coriaceous,* elliptic to narrowly ovate or obovate, *acute to rounded at tip (not acuminate) often appearing entire* (through recurving of crenulate margin) or usually crenulate; those of fertile branches 6–15 cm. long and 2.5–7.5 cm. broad; cyme 7–10 cm. broad; *anthers elevated 4–5 mm. above throat of corolla;* drupe subglobose; flesh often bitter; *stone round-obovate.* — Wooded swamps, wet pinelands and bogs, Fla. to e. Tex., n. to s. Ct., Ky. and Ark. *Fl.* May–early July; *fr.* July–Oct.

Var. **angustifòlium** T. & G. (narrow-leaved). — Lower, down to 4 dm. high; *mature leaves of fertile branches lanceolate to narrowly oblong,* 3.5–10 cm. long, 1.7–3 *cm. broad,* more crowded; *cymes only* 2.5–7 *cm. broad.* — Bogs, savannas and wet woods, Fla. and Ala., n. to se. Va.

5. V. Lentàgo L. (an old name, meaning flexible, for no. 1, transferred by Linnaeus to this species), SWEET VIBURNUM, SHEEPBERRY, NANNYBERRY, WILD-RAISIN, ALISIER (Que.). — Large shrub or small tree (up to 10 m. high); winter-buds drab; branchlets often a little scurfy; *leaves* ovate to elliptic-lanceolate, *caudate-acuminate, sharply serrulate, on undulate-margined petioles,* their veins anastomosing, the *petioles of leaves subtending the cymes* 1.5–3.5 *cm. long; cymes sessile,* 5–12 cm. broad; drupes ellipsoid to subglobose, blue-black, 0.8–1.5 cm. long, with sweet pulp. — Borders of woods, stream-banks, etc., w. Que. to Man., s. to N.E., N.J., Pa., upland to Ga., O., Ind., Ill., ne. Mo., S.D. and Colo. *Fl.* May, June; *fr.* Aug.–Oct.

6. V. prunifòlium L. (plum-leaved), BLACK- or SWEET-HAW, STAGBUSH. — Coarse shrub or small tree (up to 8 m. high), the gray bark with interrupted low ridges; *branchlets, petioles, lower surfaces of leaves, etc., hardly scurfy; leaves dull* (not lustrous), *membranaceous, becoming subcoriaceous,* oval or broadly ovate, *finely serrulate, acute or obtuse,* glabrous or nearly so, *the slender petioles of those below the sessile or short-peduncled cymes* 0.5–2 *cm. long;* cyme, flowers and fruit much as in no. 5. — Thickets, borders of woods, shores, etc., n. Fla. to Tex., n. to Ct., N.Y., O., Mich., Ill., Ia. and Kans. *Fl.* April, May; *fr.* Sept., Oct.

Var. **Búshii** (Ashe) Palmer & Steyerm. (in honor of BENJAMIN FRANKLIN BUSH, 1858–1937). — Shrub; leaves membranaceous to maturity, rather narrow, oblong-lanceolate, their petioles somewhat winged. — S. Ill. to Ark.

7. V. rufídulum Raf. (reddish), SOUTHERN BLACK-HAW, RUSTY NANNYBERRY. — Coarse shrub or small tree up to 10 m. high, with blackish bark broken into squarish blocks; *branchlets, petioles, lower leaf-surfaces, etc., densely scurfy (at least in areas) with rufescent minute tomentum; leaves lustrous, coriaceous,* finely serrulate, elliptical to obovate or oblanceolate, obtusely pointed or rounded above, the stout petioles rarely 2 cm. long; cymes sessile, with more or less scurfy rays; flowers and fruits much as in nos. 5 and 6. — Woods and thickets, Fla. to Tex., n. to Va., s. O., s. Ind., s. Ill. and Mo. *Fl.* April, May; *fr.* Sept., Oct.

§ ODONTOTÌNUS Rehd.

8. V. Rafinesquiànum Schultes (for its discoverer, CONSTANTINE SAMUEL RAFINESQUE-SCHMALTZ, 1783–1840), DOWNY ARROW-WOOD. — Loose, straggling or dense shrub up to 2 m. high; winter-buds with 2 pairs of outer scales; branchlets glabrous, pale; *leaves ovate to oblong-ovate,* acute or taper-pointed, *subsessile or very short-petioled; the petioles of those immediately below the peduncles* 0–7 *mm. long, with a pair of linear-subulate stipules; blades* 3–8 cm. long, *cordate or subcordate at base,* with straight veins running to the 4–10 coarse dentations on each margin, the lower surface and short petioles short-downy at least when young; cymes 1.5–6 cm. broad; fruit dark purple, ellipsoid, 7–9 mm. broad; stone slightly 2-sulcate on the faces. (*V. pubescens* of ed. 7, not (Ait.) Pursh; *V. affine* Bush, var. *hypomalacum* Blake) — Dry slopes, open woods, low grounds or barrens, sw. Que. to Man., s. to Ga., Ky. and Mo. *Fl.* May, June; *fr.* late July–Sept.

Var. **affìne** (Bush) House (related). — Leaves glabrous or pilose on nerves beneath. (*V. affine* Bush) — S. Ont. to Minn., s. to Va. and Ark.

9. V. mólle Michx. (soft). — Shrub 3–4 m. high, with *gray exfoliating bark; leaves deeply cordate,* suborbicular to broadly ovate, short-acuminate, *coarsely dentate with* 20–30 *teeth on each half of the blade,* 5–13 cm. long, slender-petioled, dark green and glabrous above, pale and soft-pubescent beneath; *petioles* 1.5–5 *cm. long, bearing short stipules at flowering time,* these deciduous; cymes 5–8 cm. broad; *fruit* ellipsoid, about 1 *cm. long, much compressed,* the stone deeply grooved ventrally. — Bluffs and rocky woods, Ind. to Mo., s. to Ky. and Ark. *Fl.* May, early June; *fr.* late June–Aug.

10. V. dentàtum L. (toothed), SOUTHERN ARROW-WOOD. — Shrub 1–3 m. high, with close gray bark; *young branchlets* often densely *pubescent*, sometimes glabrate or glabrous; *leaves* orbicular to oblong-ovate, rounded to subcordate at base, coarsely dentate with 4–22 broad teeth on each side; the blade thick, firm and corrugated beneath in exposed habitats, thinner, broader and flatter in less exposed situations, *sparsely pubescent to glabrous above, densely to sparsely stellate-pilose to glabrate beneath*, those of fertile branches 2.5–11.5 cm. long, 2.3–11 cm. broad, with the 5–11 pairs of straightish veins prominent beneath; petioles 0.5–2 cm. long, pubescent to glabrate, mostly without basal stipules; *cymes* peduncled, 3–11 cm. broad, *more or less pubescent;* drupes blue-black; *stone ellipsoid-ovoid,* 5–8 mm. long; *the ventral groove narrow, deep and furrow-like.* (Incl. *V. venosum* Britt., *V. scabrellum* (T. & G.) Chapm., *V. pubescens* (Ait.) Pursh and numerous minor varieties and forms) — Moist or dry sandy thickets and borders of woods, Fla. to Tex., n. to se. Mass., s. R.I., L.I., N.J., Pa., W.Va. and Tenn. *Fl.* June–Aug.; *fr.* late Aug.–Nov.

Var. **Deàmii** (Rehd.) Fern. (for its discoverer, CHARLES CLEMON DEAM, 1865–). — *Branchlets and petioles glabrous or essentially so; subulate stipules frequent at bases of petioles.* (*V. pubescens*, vars. *Deamii* and *indianense* Rehd.) — Similar habitats, w. Pa. to e. Mo. and Tenn.

11. V. recógnitum Fern. (restudied), ARROW-WOOD. — Strongly resembling no. 10; *branchlets glabrous; petioles glabrous,* up to 3 cm. long, mostly exstipulate; *leaf-blades glabrous or merely pilose on veins beneath,* orbicular to narrowly ovate or oblong, with veins prominent beneath, 3–9 cm. long, 2–8 cm. broad, with coarse mostly acute teeth; *cyme glabrous or essentially so; stone* 5–7 mm. long, globose-ovoid, *with shallow and broad trough-like furrow.* (*V. dentatum* of ed. 7, not L.) — Damp thickets, N.B. to s. Ont., s. to s. N.E., L.I., S.C., n. O. and Mich. *Fl.* mid-May–early July; *fr.* late July–Sept.

12. V. acerifòlium L. (maple-leaved), DOCKMACKIE, ARROW-WOOD, MAPLE-LEAVED V. — Slender-branched ascending shrub, 1–2 m. high, the young branchlets often pilose; *slender petioles often stipulate at base; leaf-blades palmately 3-ribbed from the rounded to subcordate base,* mostly suborbicular to ovate, *soft-downy and copiously dotted beneath,* 4–12 cm. broad, coarsely dentate, *with* 3 *prolonged acuminate lobes,* or unlobed or with suppressed short lobes in forma **ovàtum** Rehd. (ovate); cymes slender-peduncled, 2–9 cm. broad; flowers creamy-white, or in forma **Collínsii** Rouleau (for its discoverer, JAMES FRANKLIN COLLINS, 1863–1940) pink; stamens exserted; drupes ellipsoid, crimson, turning purple-black, or white in forma **ebúrneum** House (ivory-white); *stone lenticular, hardly sulcate.* — Dry or rocky woods, sw. Que. to Minn., s. to N.E., Ga. and Tenn. *Fl.* May–early Aug.; *fr.* late July–Oct. — Autumnal foliage pink to purple.

Var. **glabréscens** Rehd. (becoming smooth). — New branchlets glabrous; leaves glabrous or glabrescent beneath, except along veins. — Uplands of N.C., Ky., Tenn. and Ga.

§ ÓPULUS DC.

13. V. edùle (Michx.) Raf. (edible), SQUASHBERRY, MOOSEBERRY, PIMBINA. — Straggling to erect shrub 0.15–2.5 m. high, with glabrous branchlets; *leaves palmately 3–5-nerved from base,* suborbicular and *with* 3 *short lobes to* elliptic and *unlobed,* coarsely serrate, those of fertile branches 2–11 cm. broad (on vegetative sprouts to 2 dm. broad), *with or without glands above junction with the exstipulate petiole; cymes* mostly on short branches, dense, 1.5–3.5 *cm. broad, of uniform perfect* milk-white *flowers; stamens included; drupe* yellow, *becoming red or orange,* subglobose, *acid and juicy; stone flattened, not sulcate; winter-buds with* 2 *connate outer scales.* (*V. pauciflorum* La Pylaie) — Woods and thickets, Lab. to Alaska, s. to Nfld., and on mts. or cool slopes to C.B., N.B., n. N.E., n. N.Y., w. Pa., n. Mich., n. Wisc., n. Minn., Man., Colo. and Oreg. *Fl.* late May–early Aug.; *fr.* Aug.–Oct. — Fruit highly valued for jam. (Ne. Asia)

14. V. trílobum Marsh. (three-lobed), HIGHBUSH-CRANBERRY, CRANBERRY, PIMBINA, QUATRE-SAISONS DES BOIS (Que.). — Coarse shrub or small tree up to 4 m. high, with gray bark; *winter-buds with* 2 *connate outer scales; petioles with filiform-clavate or thick-tipped basal stipules* and *bearing* toward the summit *dome-shaped, columnar or clavate glands; blades* ovate, *palmately* 3–5-*ribbed,* rounded to truncate at base, *the* 3 *lobes long-acuminate* and entire in the sinuses; cymes terminating elongate branches, 0.4–1.5 dm. broad, their *marginal neutral flowers with broadly expanded flat corollas; stamens* of small perfect flowers *exserted;* drupe becoming orange to red, subglobose to ellipsoid, acid, finally juicy; stone flat. (*V. Opulus,* var. *americanum* Ait.; *V. americanum* of authors, not Mill.) — Cool woods, thickets, shores and rocky slopes, Nfld. to B.C., s. to N.S., N.E., Pa., n. O., n. Ind., n. Ill., ne. Ia., Black Hills of S.D. and Wyo., and Wash. *Fl.* late May–July; *fr.* Sept., Oct. — Fruit much used for preserves and jelly.

15. V. Ópulus L. (old generic name), Guelder-Rose. — Similar to no. 14; leaves mostly more rounded, with less prolonged lobes; *petioles with basal stipules filiform-attenuate to slender tips and with depressed concave-topped saucer-like glands;* cymes mostly smaller; fruit less pleasant to taste, often bitter. — Frequently planted and sometimes escaping northw. — The commonly cult. Snowball-tree (var. ròseum L.) has a rounded inflorescence with all the flowers enlarged and neutral. (Introd. from Eu.)

16. V. Siebòldii Miq. (for Philipp Franz von Siebold, 1796–1866). — Tree or large shrub *with cinereous-pilose branchlets; leaves lustrous, cuneate-obovate,* dentate, 1 dm. or so long, the large *round-topped paniculate cyme with opposite branches;* corollas uniform, 6–8 mm. long; stone only slightly compressed, with deep ventral furrow. — Esc. from cult., Ct. to e. Pa. (Introd. from Japan)

7. SAMBÙCUS L. Elder. Sureau (Que.)

Calyx-lobes minute or obsolete. Corolla open-urceolate, with a broadly spreading 5-cleft limb. Stamens 5. Stigmas 3. Fruit a berry-like juicy drupe, containing 3 small seed-like nutlets. — Shrubby, arborescent or even herbaceous plants, mostly of N. Hemisph., with a rank smell when bruised, pinnate leaves, serrate-pointed leaflets, and numerous small flowers in compound cymes. (The Latin name, perhaps from the Greek *sambuce,* an ancient musical instrument; because of the readily removed tubes of bark, these used for flutes and whistles.)

Cymes flat, umbelliform; fruits mostly black; pith white.
  Herb with stipule-like lower leaflets close to stem. . . . . 1. *S. Ebulus.*
  Shrub or tree with naked slender petioles.
    Stoloniferous tall shrub forming clumps; leaflets 5–11; ovary usually 4-locular; fruits purple-black, 4–5 mm. in diameter; native. . . . 2. *S. canadensis.*
    Tree or non-stoloniferous shrub; leaflets mostly 5 or 7; ovary usually 3-locular; fruits lustrous-black, 6–8 mm. in diameter; introd. . . . 3. *S. nigra.*
Cymes ovoid, paniculiform; fruits mostly red; pith brown. . . . . . 4. *S. pubens.*

1. S. Èbulus L. (old generic name), Dwarf E., Danewort. — *Herb with* subterranean perennial stem and *annual herbaceous fertile stems* 6–9 dm. high; leaves with 7–11 lanceolate finely toothed *leaflets, the lowest pair stipule-like and close to the stem;* cyme irregular, with white to roseate pleasantly fragrant flowers and black fruits. — Roadsides and waste places, spread from cult., s. Que. July–Sept. (Introd. from Eu.)

2. S. **canadénsis** L. (Canadian), Common E., Sureau blanc (Que.). — Stems scarcely woody, 1–4 m. high, with *white pith; leaflets* 5–11, *elliptic to lanceolate,* sharply serrate, glabrous or hirtellous beneath especially along the veins, the lower often 3-parted; cymes flat; flowers white, fragrant; *ovary usually 4-locular; fruit purple-black,* or bright red in forma **rùbra** Palmer & Steyerm. (red), or dark yellow or orange in forma **atrofláνula** House (dark yellow), or greenish and with leaves pale in forma **chlorocárpa** Rehd. (green-fruited), 4–5 *mm. in diameter.* — Wet, damp or rich soils, C.B. to Man., s. to N.E., Ga., La. and Okla. June, July.

Var. **submóllis** Rehd. (somewhat soft). — Leaflets canescent beneath with close puberulence. — Ill. and Ia., s. to Okla. and Tex.

3. S. nìgra L. (black), Elder of Europe. — Tree up to 10 m. high or shrub, resembling no. 2; branchlets with more lenticels; *leaflets mostly* 5–7; flowers with heavier odor; *ovary usually* 3-*locular; fruit lustrous-black,* 6–8 *mm. in diameter.* — Cult. and occasionally spread to waste ground, N.E. to Va. (Introd. from Eu.)

4. S. **pùbens** Michx. (pubescent), Red-berried or Stinking E., Sureau rouge (Que.). — Shrub up to 8 m. high, with warty bark and *brown pith;* branchlets pubescent; leaflets 5–7, ovate-lanceolate or narrower, downy underneath, or glabrous beneath and branchlets glabrous in forma **cálva** Fern. (bald), or finely dissected in forma **dissécta** (Britt.) Fern. (dissected); *cymes panicle-like, ovoid and convex or pyramidal; flowers* yellow-white, or pink in forma **rosaeflòra** (Carr.) Dansereau (rose-flowered), *ill-scented; fruit bright red,* or whitish in forma **leucocárpa** (T. & G.) Fern. (white-fruited) or yellow in forma **xanthocárpa** (Cockerell) Fern. (yellow-fruited), 4–6 mm. in diameter. (*S. racemosa* of Am. auth., not L.) — Woods and openings, often rocky, Nfld. to Alaska, s. to N.S., N.E., N.J., Pa., upland to Ga. and Tenn., W.Va., O., Ind., Ill., ne. Ia., S.D., Colo. and Oreg. *Fl.* Apr.–July; *fr.* June–Aug.

FAM. 163. ADOXÀCEAE (Moschatel Family)

Calyx-tube reaching not quite to the summit of the 3–5-locular ovary, its limb of 3 or more teeth. Corolla rotate, 4–6-cleft, bearing at each sinus a pair of separate or partly united stamens with 1-locular anthers. Style 3–5-parted; dry drupe greenish, with 3–5 cartilaginous nutlets.

A dwarf perennial, with scaly rootstock and ternately divided leaves, the cauline a single pair, the flowers capitate.

1. ADÓXA L. Moschatel

Characters of the family. (Name from the Greek *adoxos, obscure* or *insignificant.*)

1. A. Moschatéllina L. (generic name used by Tournefort, meaning little musk). — Smooth, musk-scented; radical leaves 1–3-ternate, the cauline 3-cleft or 3-parted; leaflets obovate, 3-cleft; flowers several in a close cluster on a slender peduncle, greenish or yellowish. — Mossy woods, wet rocks, etc., local, Catskill Mts., N.Y.; Wisc., Minn. and Ia.; Alaska and Yuk. to S.D. and N.M. May–July. (Eurasia)

FAM. 164. VALERIANÀCEAE (Valerian Family)

Herbs, with opposite leaves and no stipules; the calyx-tube adherent to the ovary, which has one fertile 1-ovuled locule and two abortive or empty ones; the stamens distinct, 1–3, *fewer than the lobes of the corolla and inserted on its tube.* Corolla tubular or funnelform, often irregular, mostly 5-lobed, the lobes imbricated in the bud. Style slender; stigmas 1–3. Fruit indehiscent, 1-locular (the two empty locules of the ovary disappearing), or 3-locular, two of the locules empty, the other 1-seeded. Seed suspended, anatropous, with a large embryo and no albumen. Flowers in panicled or clustered cymes. — Roots often odorous and antispasmodic.

Perennials; cauline leaves pinnate or deeply parted; calyx with 5–15 setiform lobes, tightly rolled up in flower, unrolling in fruit and developing into conspicuous plumose pappus. . . . . 1. *Valeriana.*

Annuals or biennials; leaves entire, dentate or rarely incised; calyx-lobes obsolete or very short teeth. . . . . 2. *Valerianella.*

1. VALERIÀNA L. Valerian

Calyx-limb of several plumose bristles (like a pappus) which are rolled up inward in flower, but unroll and spread as the seed-like 1-locular fruit matures. Corolla commonly gibbous near the base, the 5-lobed limb nearly regular. Stamens 3. — Nearly cosmop. group of perennial herbs, with thickened strong-scented roots, and simple or pinnate leaves. Flowers in many species imperfectly dioecious or dimorphous. (Mediaeval name, perhaps from *valere, to be strong;* said by Linnaeus to be in honor of *Publius Aurelius Licinius Valerianus*, Roman emperor from 253–260.)

a. Leaves thickish, parallel-ribbed, the basal narrowed to short winged petioles; flowers dioecious, yellowish-white; root large, fusiform, deep. . . . . . 1. *V. ciliata.*

a. Leaves thin, loosely pinnate-veined, the basal on long slender petioles; flowers perfect, often dimorphic, white, pinkish or lilac; roots fibrous. . . *b.*

b. Basal leaves all or in part simple, when pinnate with the terminal leaflet much larger than the lateral ones; cauline leaves with 3–11 leaflets or divisions, the rachis and lower surfaces glabrous. . . *c.*

c. Basal leaves oblong to ovate or spatulate, narrowed to the petioles; corollas inversely conical, 3–7 mm. long.

Leaves of basal tufts all simple; blades of the larger 2.5–7.5 cm. long, 0.8–3 cm. broad; larger cauline leaves 3–7 cm. long, 1.5–4.5 cm. broad; bractlets of inflorescence non-ciliate. . . . . 2. *V. septentrionalis.*

Leaves of basal tufts either simple or cleft; blades of the larger ones 0.5–1.5 dm. long, 2–8 cm. broad; larger cauline leaves 0.6–1.6 dm. long, 5–10 cm. broad; bractlets ciliate when young. . . . . . 3. *V. uliginosa.*

c. Basal leaves cordate-ovate· corollas very slender, the tube 1–1.5 cm. long. 4. *V. pauciflora.*

b. Basal leaves all pinnate; pinnae nearly equal, those of lowest leaves and the rachis hirsute; cauline leaves with 7–25 lance-acuminate leaflets. . . . 5. *V. officinalis.*

1. V. ciliàta T. & G. (with marginal hairs). — *Root fusiform,* large and deep; stems stout, smooth, 3–7.5 dm. high; *leaves subcoriaceous, parallel-ribbed;* the basal narrowly oblanceolate to broadly spatulate or sometimes pinnately divided, tapering to short winged petioles; cauline leaves smaller, in remote pairs, pinnate to simple; panicle slenderly thyrsiform to ellipsoid, at first compactly flowered, finally loosening; *flowers* nearly or quite *dioecious;* corolla inversely conical, whitish; fruits glabrous, 2.5–3 mm. long. (*V. edulis* of ed. 7, not Nutt.) — Prairies, sloughs or hillsides, s. Ont. to Minn., s. to O., Ind., Ill. and ne. Ia. May, June. — One of a complex and not satisfactorily worked out series, chiefly of western N. Am.; the original *V. edulis* from Oregon had pubescent fruits.

2. V. septentrionàlis Rydb. (northern). — Rhizome slender, more or less creeping; *rosette-leaves simple*, the membranous loosely veiny *oblong to lanceolate* blades tapering to the long slender petiole, *the larger* 2.5–7.5 *cm. long* and 0.8–3 *cm. broad;* stem slender, glabrous, 1.5–6 dm. high; *cauline leaves* 3 or 4 pairs, the upper very small; *the larger and lower* pinnate, 3–7 *cm. long*, 1.5–4.5 *cm. broad*, with 2–6 pairs of lanceolate or oblong *leaflets* 2–5 *mm. broad*, and a larger terminal one; *inflorescence in anthesis a corymb* 2–3.5 *cm. broad*, soon loosening and becoming an elongate interruptedly *thyrsoid fruiting panicle* 0.5–2 dm. long and 3.5–9 *cm. in diameter; bractlets of inflorescence* 2–4 mm. *long, smooth; corollas* white or pinkish, 3–4 *mm. long;* fruits thick-lanceolate to flask-shaped, 3–4 mm. long, 1–1.6 mm. thick. (*V. sylvatica* Banks, not F. W. Schmidt) — Damp calcareous ledges, gravels and meadows, Nfld.; Anticosti I. and Gaspé Pen., Que.; w. Ung.; Algoma Distr., Ont., to Yuk., s. to S.D., Wyo., Nev. and Calif. June–Aug.

3. V. uliginòsa (T. & G.) Rydb. (growing in marshes). — Coarser, 0.5–1 m. tall; *rosette-leaves either simple or cleft, the larger* 0.5–1.5 *dm. long and* 2–8 *cm. broad; larger cauline leaves* 0.6–1.6 *dm. long and* 5–10 cm. broad, *with the lateral leaflets* 0.4–1.6 *cm. broad; inflorescence in anthesis a corymb* 0.3–1.5 *dm. broad, in fruit becoming* 0.6–2 dm. long and 0.6–1.5 *dm. in diameter; bractlets* 4–11 *mm. long, ciliate when young; corollas* 5–7 *mm. long;* fruits 3.5–5 mm. long, 1.5–2 mm. thick. — Calcareous swamps and wet woods, chiefly with *Larix* and *Thuja*, Matane Co., Que., to w. Ont., s. to N.B., n. and w. N.E., s. N.Y. and Mich. Late May–July.

4. V. pauciflòra Michx. (few-flowered). — *Freely surculose*, the stolons elongate and leafy-tipped; *basal leaves cordate-ovate*, simple or with a pair of leaflets; stems slender, 0.5–1 m. high; cauline leaves with a broadly ovate to rhombic terminal leaflet and 1–3 pairs of narrower lateral ones; inflorescence at first a dense corymb, finally elongating; *corollas very slender, with a tube* 1–1.5 *cm. long*, pale pink; fruits 4–5 mm. long. — Rich woods and alluvium, Pa. to Ill., s. to Va. and Tenn. May, June.

5. V. officinàlis L. (of the shops), "Garden-Heliotrope." — Coarse, erect; *leaves all pinnate, the rachises and lower leaf-surfaces* (at least at base of plant) *hirsute; leaflets* subequal, *lance-acuminate*, those of the larger cauline leaves 5–12 *pairs;* corollas roseate, 4 mm. long. — Roadsides and thickets, Que. to Minn., s. to N.S., N.E., N.J., Pa. and O.; esc. from cult. June, July. (Introd. and natzd. from Eu.)

2. VALERIANÉLLA Mill. Corn-salad. Lamb's-lettuce

Limb of the calyx obsolete or merely toothed. Corolla funnelform or salverform, equally or unequally 5-lobed. Stamens 3, rarely 2. Fruit 3-locular, two of the locules empty and sometimes confluent into one, the other 1-seeded. — Eurasian and Am. annuals and short-lived biennials, usually smooth, with forking stems, tender and rather succulent leaves (entire or cut-lobed toward the base), and white, blue or pale cymose-clustered and bracted small flowers. Our species all have the limb of the calyx obsolete and are so much alike in aspect, flowers, etc., that the best characters are taken from the fruit. (Name a diminutive of *Valeriana.*)*

a. Bracts and bractlets of the cymules non-ciliate or with ascending mostly non-glandular spinulose ciliation; corolla funnelform, 1.5–5 mm. long, its tube shorter than to about equaling limb, with a gibbosity at the base of the throat on the ventral side. . . *b.*

b. Bracts and bractlets of cymules more or less spinulose-ciliate; corollas 1.5–2 mm. long. . . *c.*

c. Bracts all ciliate, the outer spatulate and rounded at tip; corolla-lobes bluish; stamens included or barely exserted; fruit laterally compressed, fertile locule with a very turgid corky back. . . . . 1. *V. olitoria.*

c. Bracts lanceolate to oblanceolate, acute, the outer ciliate, the inner often eciliate; corolla-lobes white or pink-tinged; stamens well exserted; fruit dorsiventrally compressed, the fertile locule without thickened corky back.

Fruits ellipsoid-oblong, two-and-a-half to three times as long as broad (2 mm. long, 0.7 mm. broad), the ventral groove extending to base, the dorsal side of fertile locule rounded. . . . . . . 7. *V. stenocarpa*, var. *parviflora.*

---

* Many characters and the figures taken from the treatment by Sarah C. Dyal in Rhodora, xl. 186–212 (1938).

Fruits ovoid, one-and-a-half to two-and-one-fourth times as long as broad (2–2.5 mm. long, 1–1.5 mm. broad), the ventral groove extending only partly to base, the dorsal side of fertile locule flattened. . 2. *V. radiata.*

b. Bracts and bractlets eciliate or essentially so (except sometimes at tips); corollas 1.5–5 mm. long. . . *d.*

d. Fruits suborbicular as viewed from front or back, 2–3 mm. broad, the inner face irregularly saucer-shaped, the fertile locule one-third to one-half as broad as both empty locules, the latter divergent and with remote tips.

Fruits 3–3.5 mm. long, 3 mm. broad, dorsiventrally flattened. . . . 3. *V. patellaria.*

Fruits 2–2.5 mm. long, 2 mm. broad, subglobose. . . . . . 4. *V. umbilicata.*

d. Fruits definitely longer than broad, 0.7–2.6 mm. broad, the inner face not saucer-shaped, the fertile locule nearly as broad as to much broader than the combined empty ones, the latter approximate or but slightly divergent. . . *e.*

e. Upper leaves acute or acutish; fruits 3–4 mm. long, 2.5–2.8 mm. broad, the fertile locule twice as broad as the united empty ones. . . . . 5. *V. chenopodifolia.*

e. Leaves all rounded at summit; fruits 1.2–2.5 mm. long, 0.7–1.2 mm. broad, the fertile locule only slightly broader than the two empty ones. . . *f.*

f. Fruits ovoid, 1.2–2 mm. long, densely uncinate-hirsute, the round-backed fertile locule much larger than the confluent empty ones, the ventral face not furrowed. . . . . 6. *V. amarella.*

f. Fruits ovoid-ellipsoid to oblong, glabrous or with straight pubescence, the depressed fertile locule about equaling the two partly separated empty ones, the ventral face furrowed.

Corolla 1.5 mm. long; fruits 0.7 mm. broad, oblong, the fertile locule gradually rounded on back. . . . . 7. *V. stenocarpa,* var. *parviflora.*

Corolla 3–5 mm. long; fruits 1.2 mm. broad, ovoid-ellipsoid, the fertile locule flattened on back. . . . . 8. *V. intermedia.*

a. Bracts and bractlets bordered by strong gland-tipped denticulations; corolla rotate, 1–1.2 cm. long, the very slender tube three or four times as long as limb and bearing a gibbosity toward the base.

Fruits laterally compressed, ovoid-ellipsoid, 1.7 mm. broad, with 3 lines of long hairs, 1 each at the margins of the prolonged keels of the locules. . 9. *V. ozarkana.*

Fruits dorsiventrally compressed, narrowly oblong, 0.7 mm. broad, glabrous or with short scattered hairs, the locules low and keelless. . . . . . 10. *V. Bushii.*

1. V. olitòria (L.) Poll. (of the kitchen-garden), Corn-salad, Lamb's-lettuce. — Angled stems slightly pubescent, 1–5 dm. high, loosely forking; basal leaves spatulate to cuneate-obovate; *cauline leaves* oblong to oblong-oblanceolate, obtuse, *more or less pubescent,* toothed at base; *cymules with ciliate bracts, the outer ones spatulate and rounded at apex; corollas* 1.5 mm. long, funnelform, white, *with bluish limb,* bearing a gibbosity at the junction of the long throat and short tube; *stamens included or barely exserted; fruit* yellowish, 2–4 mm. long, *laterally compressed,* obliquely roundish-rhomboidal, *the fertile locule with a very turgid corky back.* (*V. Locusta* Betcke) — Waste places, roadsides and old fields, N.E. to Ind., s. to N.C. and Tenn.; western states. April, May; Oct. (Introd. from Eu.) Fig. 1529. — Cult., especially in Eu., as a salad.

1529. V. olitoria.

2. V. radiàta (L.) Dufr. (having rays). — Habit of no. 1; stem 1–6 dm. high, sparingly pubescent on angles; basal leaves oblong-spatulate; *principal cauline leaves oblong,* entire or toothed at base, *broadly rounded at summit,* glabrous or sparsely pubescent, those at the lower forks often oblong-ovate; cymules compact, hemispherical, 6–13 mm. broad; *bracts lanceolate, the outer* and sometimes the inner *spinulose-ciliate; corolla* white or tinged with pink, 1.5–2 *mm. long;* stamens well exserted; *fruits dorsiventrally compressed,* quadrate in cross-section, *oblong- or ellipsoid-ovoid,* glabrous or minutely pubescent, 1.4–1.5 mm. long, 0.8 mm. broad, *thick-stipitate, the fertile locule slightly broader than the pair of* slightly divergent rounded *empty ones, back of fertile locule flattened, the ventral groove ending* 0.4 *mm. above base of fruit.* — Damp to dry woods, meadows, fields and roadsides, Fla. to Tex., n. to N.J., Pa., O., Ill., Mo. and Kans. April, May. Fig. 1530.

1530. V. radiata (typical).

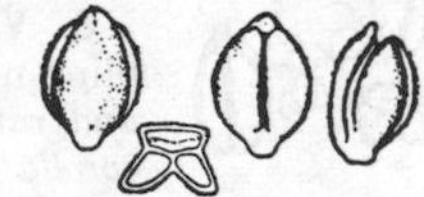

1531. V. radiata, v. missouriensis.

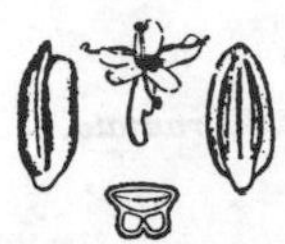

Var. **missouriénsis** Dyal (of Missouri). — Fruits ellipsoid-ovoid, 1.7–1.8 mm. long, 1.2 mm. broad; the empty locules divergent and in cross-section with prolonged backs, the latter in cross-section about as long as breadth of fertile locule. — Mo. and Ark. Fig. 1531.

1532. V. radiata, v. Fernaldii.

Var. **Fernáldii** Dyal (for one of its collectors, 1873–). — Often looser, up to 7.5 dm. high; corolla slightly larger; fruit oblong-ovoid, 1.7–1.8 mm. long, 0.7–0.8 mm. broad, the fertile locule broader than the pair of closely contiguous slender empty ones. — Woods, swamps and fields, Fla. to Ct.; Mo. and Kans. Fig. 1532.

3. V. patellària (Sulliv.) Wood (saucer-like; from the flattened fruit). — In habit and foliage similar to no. 2, 0.3–1 m. high; upper leaves lance-oblong; cymules looser; *bracts glabrous; corolla* 3–5 *mm. long; fruits orbicular as viewed from front or back*, 3–3.5 *mm. long*, 3 *mm. broad, dorsiventrally flattened*, glabrous or minutely pubescent, *the inner face irregularly saucer-like (with margins incurved), the fertile locule* nearly a third *as broad as the pair of widely divergent empty ones and protruding at summit as a slender beak, the 2 empty locules wing-like and forming a crescent 2 mm. across.* (*V. Woodsiana*, var. *patellaria* (Sulliv.) Gray) — Meadows and damp roadsides, N.J. to Ill., s. to N.C. and Tenn. May, early June. Fig. 1533.

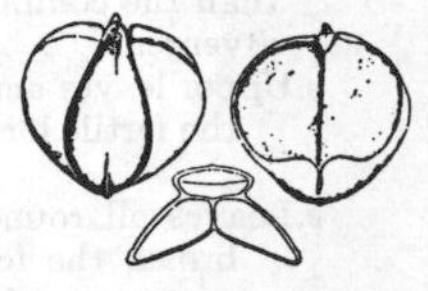

1533. V. patellaria.

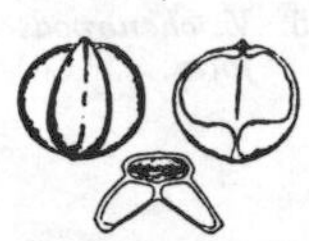

1534. V. umbilicata.

4. V. umbilicàta (Sulliv.) Wood (navel-shaped; from the fruit). — Closely resembling no. 3 but bracts sometimes with weak fimbriation at tip; *fruits globular or nearly so*, 2–2.5 *mm. long, about* 2 *mm. thick, the inflated empty locules with rounded edges about* 1 *mm. apart.* (*V. Woodsiana*, var. *umbilicata* (Sulliv.) Gray) — Low grounds, se. N.Y. to Ill. and Tenn. May. Fig. 1534.

5. V. chenopodifòlia (Pursh) DC. (with leaves of *Chenopodium*). — With habit of nos. 2–4; *upper and rameal leaves lanceolate and acute or acutish; cymules* 0.8–1.7 cm. broad, *mostly overtopped by the broadly lanceolate glabrous outer bracts;* corolla 3–4 mm. long; *fruit* trigonous-ovoid, with ventral angle rounded, 3–4 *mm. long*, 2.5–2.8 *mm. broad*, glabrous or minutely pubescent, *dorsiventrally compressed, the fertile locule twice as broad as the united empty ones*, the latter separated by a very shallow groove. — Meadows, bottoms and fields, N.Y. and Pa. to s. Ont. and Ind. May, June. Fig. 1535.

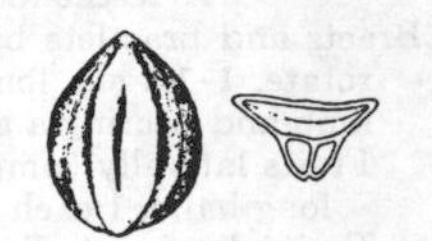

1535. V. chenopodifolia.

1536. V. amarella.

6. V. amarélla (Lindheimer) Krok (somewhat bitter). — Stiffly erect, 1–3 dm. high, *fastigiately branched, with* few to very numerous *glomerules in a broad corymb;* leaves all rounded at tip; cymules subspherical, 0.5–1 cm. broad, mostly short-stalked; bracts glabrous, lanceolate to lance-ovate; corolla 1.5–3 mm. long; *fruits trigonous-ovoid*, 1.2–2 *mm. long, densely uncinate-hirsute with white hairs, the round-backed fertile locule much larger than the confluent filiform empty ones, the ventral face not furrowed.* — Open calcareous barrens and hills, e. Kans., Okla. and Tex. April, May. Fig. 1536.

7. V. stenocárpa (Engelm.) Krok (slender-fruited). — Differing from no. 6 in its looser habit; bracts often fimbriate-tipped; cymules looser; corolla 2–3 mm. long; *fruits tetragonal-oblong, glabrous or with straight pubescence*, 1.5 mm. long, 0.7 *mm. broad, the round-backed depressed fertile locule only slightly broader than the* 2 slightly divergent and *distinct empty ones*, the groove between the latter deeply V-shaped. — Southwest of our range; represented with us by

Var. **parviflòra** Dyal (small-flowered). — Corolla only 1.5 mm. long; fruit smaller, the ventral groove more shallow. — Woods, prairies and barrens, Mo., Okla. and Tex. April, May. Fig. 1537.

1537. V. stenocarpa, v. parviflora.

1538. V. intermedia.

8. V. intermèdia Dyal (intermediate). — In habit resembling nos. 2 and 5, 1.5–7 dm. high, loosely corymbose; *upper and rameal leaves acutish;* lanceolate bracts glabrous or promptly glabrate; *corolla* 3–5 *mm. long; fruits ovoid-ellipsoid*, quadrate, 2–2.5 *mm. long*, 1.2 *mm. broad, the flat-backed fertile locule only slightly broader than the pair of empty ones and separated from them on each side by a prominent nerve*, the groove between the empty locules shallow. — Banks of streams, meadows and damp fields, centr. Mass. to Ill., s. to S.C. and Ky. May, June. FIG. 1538.

9. V. ozarkàna Dyal (of the Ozarks). — Stems 1–3 dm. high, open-corymbose above; cauline leaves oblong, broadly rounded at tip; *bracts bordered by remote strong gland-tipped denticulations,* the outer broad, the inner narrow; *corolla rotate,* 1–1.2 *cm. long, the filiform red tube three or four times as long as the wide-spreading limb, with a gibbosity near the base; fruits laterally compressed, ovoid-ellipsoid,* 3 mm. long, 1.7 *mm. broad, with* 3 *lines of long hairs, one each at the margins of the prolonged keels of the locules.* (*V. longiflora* in part of ed. 7, not (T. & G.) Walp.) — Moist calcareous open soil, sw. Mo. and nw. Ark. April, May. FIG. 1539.

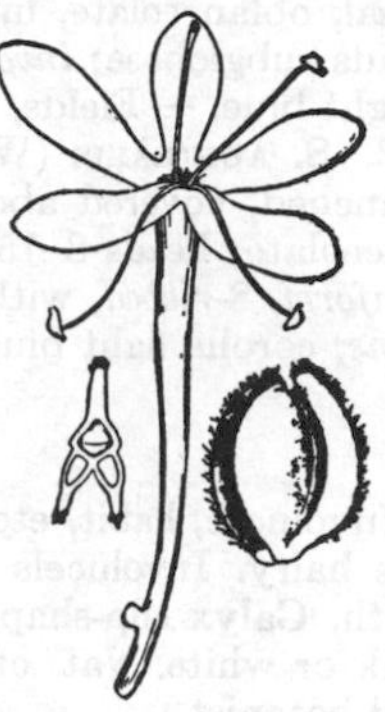

1539. V. ozarkana.

10. V. Búshii Dyal (for its discoverer, BENJAMIN FRANKLIN BUSH, 1858–1937). — Closely resembling no. 9, up to 6 dm. high; *fruit dorsiventrally compressed, narrowly oblong,* 0.7 *mm. broad, glabrous or with short scattered hairs, the locules low and keelless.* — Calcareous soil, sw. Mo. and nw. Ark. April, May.

FAM. 165. DIPSACÀCEAE (TEASEL FAMILY)

Herbs, with opposite or whorled leaves, no stipules, and the flowers in dense heads surrounded by an involucre, as in the Composite Family; *but the stamens distinct, and the suspended seed destitute of albumen.* — Represented by the following introduced genera:

Receptacle chaffy.
Chaff with long rigid points. . . . . 1. *Dipsacus.*
Chaff herbaceous, about equaling the flowers, not rigid-pointed. . . . 2. *Succisa.*
Receptacle without chaff. . . . . 3. *Knautia.*

1. DÍPSACUS L. TEASEL. CARDÈRE (Que.)

Involucre many-leaved, longer than the chaffy leafy-tipped bracts among the densely capitate flowers; each flower with a 4-leaved calyx-like involucel investing the ovary and fruit (achene). Calyx-tube adherent to the ovary, the limb cup-shaped, without a pappus. Corolla nearly regular, 4-cleft. Stamens 4, inserted on the corolla. Style slender. — Stout and coarse biennials, nat. of Old World, hairy or prickly, with large ovoid-ellipsoid heads. (Greek name of the teasel, said to be from *dipsa, thirst,* because the united cup-shaped bases of the leaves in some species (as in no. 2) hold water.)

1. D. SYLVÉSTRIS Huds. (of the woods), WILD T. — Prickly; *leaves lance-oblong, toothed and often prickly on the margin;* leaves of the involucre slender, ascending or spreading, longer than the head; bracts (chaff) tapering into a long flexible awn with a straight point. — Roadsides, old fields and pastures, w. Que. and w. N.E. to Ont. and Mich., s. to N.C., Tenn. and Mo. July–Oct. (Natzd. from Eu.)

The closely related **D. FULLÒNUM** L. (of fullers), FULLER'S T., with the bracts (chaff) stout and with recurving hooked tips, used in fulling in textile mills, is spontaneous but not persistent.

2. D. LACINIÀTUS L. (slashed). — *Leaves* coarsely *pinnatifid* or bipinnatifid, *their bases confluent and forming a cup,* conspicuously *ciliate;* leaves of involucre lance-linear, usually shorter than or barely exceeding the head. — Waste places, Mass. to Mich., local, probably esc. from cult. July–Sept. (Adv. or introd. from Eu.) — Reputed to drown insects in fluid gathered in the foliar cups.

2. SUCCÌSA Neck. DEVIL'S-BIT

Involucre many-leaved. Involucels (often called outer calyx) closely investing the ovary and fruit, 4–8-furrowed throughout their entire length, prismatic or somewhat fusiform, the limb shortly 4-lobed or -toothed, erect or spreading. Limb of the true calyx minutely 5-toothed, or of 5 awns. Corolla funnelform or campanulate, 4(–5)-lobed. Stamens 4, borne on the corolla. Small genus nat. of the Mediterr. reg. and Eu. (Name from *succidere,* to bite off, from the praemorse rootstock, once supposed to have been bitten off by Satan.) Often united with SCABIOSA L.

1. S. PRATÉNSIS Moench (of meadows). — Smoothish or hairy, 4–9 dm. high; leaves chiefly basal, oblanceolate, undulate or entire, the cauline mostly 2 pairs and considerably reduced; heads subglobose; *involucels somewhat 4-angled, villous,* 4-toothed; *calyx-limb 4-awned;* corolla bright blue. — Fields, C.B. to Mass., local. Aug., Sept. (Adv. from Eu.)

2. S. AUSTRÀLIS (Wulf.) Reichenb. (southern). — Tall, loosely and mostly trichotomously branched, covered above with minute crisped mostly reflexed gray hairs; leaves elongated-lanceolate; heads 8–15 mm. in diameter, at length subcylindric; *involucels glabrous, somewhat fusiform, 8-ribbed,* with small crenate-lobed spreading border; *calyx* shortly 5-toothed, *without awns;* corolla light blue. — Low fields, w. N.E., N.Y. and Pa. July–Oct. (Natzd. from Eu.)

3. KNAÙTIA L.

Involucre, habit, etc., much as in the preceding. Chaff wanting, but the receptacle more or less hairy. Involucels strongly compressed, the limb obscure, at most indicated by minute teeth. Calyx cup-shaped, the limb mostly 8-awned. Corolla light blue to lilac-purple, rarely pink or white. Nat. of Eurasia. (Dedicated to *Christian Knaut,* 1654–1716, Saxon physician and botanist.)

1. K. ARVÉNSIS (L.) Duby (of cultivated ground). — Pubescent, 4–9 dm. high, few-branched; some or all of the leaves deeply pinnatifid or bipinnatifid; heads depressed-hemispherical; corollas lilac. (*Scabiosa* L.) — Dry pastures, fields and waste places, Nfld. to N.D., s. to N.E. and Pa. June–Aug. (Natzd. from Eu.)

Numerous species of *Scabiosa,* with chaff of receptacle reduced to hairs, are cultivated and casually escape.

FAM. 166. CUCURBITÀCEAE (GOURD FAMILY)

Mostly succulent herbs with tendrils, dioecious or monoecious flowers, the calyx-tube adhering to the 1–3-locular ovary, and the 5 or usually 2½ stamens (*i.e.,* 1 with a 1-locular and 2 with 2-locular anthers) *commonly united by their often tortuous anthers, and sometimes also by the filaments.* Fruit (pepo) *fleshy, or sometimes membranaceous.* Limb of the calyx and corolla usually more or less combined. Stigmas 2 or 3. Seeds large, usually flat, anatropous, with no albumen. Cotyledons leaf-like. Leaves alternate, palmately lobed or veined. — Mostly a tropical or subtropical family.

Ovaries and fruits smooth or merely pilose, not prickly.
- Corollas campanulate, 5 cm. or more long; fruits 7–8 cm. long; plant trailing. . . . 1. *Cucurbita.*
- Corollas rotate or short-campanulate, much smaller; berries 1–5 cm. long; plants climbing.
  - Seeds erect, 1–4 in each locule; anther-locules strongly contorted; berries subsessile. . . . . 2. *Cayaponia.*
  - Seeds horizontal, numerous; anthers straight or but slightly curved; berries on elongate peduncles.
    - Leaves 5-angled or -lobed; staminate flowers in small racemes, very small; pistillate flowers on filiform peduncles, 4–5 mm. broad, the corolla-lobes rotate; berry smooth, about 1 cm. in diameter. . . . . . . 3. *Melothria.*
    - Leaves not angled; flowers all axillary, on thick peduncles, about 3 cm. broad, the corolla-lobes strongly recurved; berry strongly ribbed and cross-ribbed, about 5 cm. long. . . . . . . 4. *Thladiantha.*

Ovaries and fruits covered with firm to soft prickles.
- Corollas 5-lobed; fruits clustered, firm, bristly, 1-seeded, indehiscent. . . . 5. *Sicyos.*
- Corollas 6-parted; fruits single, inflated, with soft prickles, 2-locular, 4-seeded, at length rupturing at summit. . . . . 6. *Echinocystis.*

1. CUCÙRBITA L. GOURD. SQUASH. PUMPKIN

Flowers monoecious. Corolla campanulate, deeply 5-lobed, the lobes with recurved tips. Anthers united, 1 of them 1-locular, the others 2-locular. Style short, with 3 lobed or divided stigmas. Fruit fleshy, covered by a firm rind, with many horizontal seeds. — Herbs of trop. and warm reg., with annual or perennial roots, large cordate angulate or lobed leaves, large flowers solitary in the axils, and large fruits. (Classical Latin name for a gourd.)

1. C. foetidíssima HBK. (most fetid), MISSOURI GOURD, FETID WILD PUMPKIN. — Stems elongate, scabrous, from a thickish fusiform root; leaves ovate, the margin somewhat angulate, denticulate, densely scabrous-pubescent, somewhat whitened beneath; corolla 5–12 cm. long; fruit smooth, subglobose, 7–8 cm. long. (*Pepo* Britt.) — Dry or sandy soil, Mo. and Neb. to Tex., Mex. and Calif.; adv. e. to Ind. May–July.

C. MÁXIMA Duchesne (Squash), C. MOSCHÀTA Duchesne (CROOKNECK-SQUASH), and C. PÉPO L. (PUMPKIN) are familiar in cult., and incline to appear spontaneously in waste places southw., as do CÙCUMIS MÈLO L. (MUSKMELON), C. SATÌVUS L. (CUCUMBER), CITRÚLLUS VULGÀRIS Schrad. (WATERMELON), and LAGENÀRIA VULGÀRIS Ser. (Gourd).

2. CAYAPÒNIA Silva Manso

Flowers monoecious or dioecious; the staminate short-campanulate or saucer-shaped with spreading lobes; the pistillate similar, subsessile or short-stalked. Anthers distinct, with sigmoid locules. Berry ellipsoid, small, with 1–4 erect ovules in each locule. — Slender high-climbing vines of Am. and Afr., resembling no. 3, with small flowers and simple tendrils. (A Brazilian name.)

1. C. grandifòlia (T. & G.) Small (large-leaved). — Slightly pubescent; leaves thin, 3-lobed with the lateral lobes sometimes again lobed, 1–2 dm. across; flowers greenish-white, barely 1 cm. broad; berry 1.5–2 cm. long. — Bottomlands, Ga. and Fla. to La., Ark. and se. Mo. June–Aug.

3. MELÒTHRIA L. MELONETTE

Flowers polygamous or monoecious; the staminate campanulate, the corolla 5-lobed; the pistillate with the calyx-tube constricted above the ovary, then campanulate. Anthers more or less united. Berry small, pulpy, filled with many flat and horizontal seeds. — Tendrils simple. Flowers very small. Genus of trop. and warm reg. (Altered from *melothron*, an ancient name for some fruiting vine.)

1. M. péndula L. (hanging down). — Slender, from a perennial root, climbing; leaves small, roundish and cordate, 5-angled or -lobed, roughish; staminate flowers few, in small racemes; pistillate solitary, greenish or yellowish; berry ovoid, green, becoming blackish, 1 cm. long, purgative. — Rich or damp thickets, Fla. to Tex. and Mex., n. to Va. (formerly to Pa.), s. Ind., s. Mo. and Okla. July–Sept.

4. THLADIÁNTHA Bunge

Flowers dioecious, peduncled from the axils of cordate-ovate leaves; the staminate and pistillate similar, with reflexed sepals and showy short-campanulate yellow corolla with recurving lobes. Anthers free, their locules straight; styles 3, dilated above. Ovary villous; fruit ellipsoid, longitudinally ribbed and with reticulate cross-ribs, with many horizontal seeds. — Climbing showy-flowered perennials nat. of Asia. (Name from the Greek *thladias, eunuch*, from Bunge's impression that the flowers are sterile, and *anthos, flower.*)

1. T. DÙBIA Bunge (doubtful; originally published with doubt as to its relationship). — Root tuberous; leaves harshly scabrous, long-acuminate; showy yellow corolla 1.5–2.5 cm. high; fruit about 5 cm. long. — Roadsides, waste places, etc., sw. Que. and N.H. to Man. June, July. (Introd. from Asia and spread from cult.)

5. SÍCYOS L. BUR-CUCUMBER

Flowers monoecious. Petals 5, united below into a campanulate or flattish corolla. Anthers cohering in a mass. Ovary 1-locular, with a single suspended ovule; style slender; stigmas 3. Fruit ovoid, dry and indehiscent, filled by the single seed, covered with barbed prickly bristles which are readily detached. — Climbing Am., Polynesian and Australian annuals, with 3-forked tendrils, and small whitish flowers; the staminate and pistillate mostly from the same axils, the former corymbed, the latter in a capitate cluster, long-peduncled. (Greek name for the Cucumber.)

1. S. angulàtus L. (angular). — Leaves roundish-cordate, 5-angled or -lobed, the lobes pointed; plant clammy-hairy. — River-banks and damp yards, s. Me. and w. Que. to Minn., s. to Fla., La. and Tex. July–Sept.

6. ECHINOCỲSTIS T. & G. WILD BALSAM-APPLE

Flowers monoecious. Petals 6, lanceolate, united at the base into an open spreading corolla. Anthers more or less united. Ovary 2-locular, with 2 erect ovules in each locule; stigma broad. Fruit fleshy, at length dry, clothed with weak prickles, bursting at the summit, 2-locular, 4-seeded, the inner part fibrous-netted. Seeds large, flat, with a thickish hard and roughened coat. — Tall climbing Am. annuals, nearly smooth, with 3-forked tendrils, thin leaves, and very numerous small greenish-white flowers; the staminate in compound racemes often 3–4 dm. long, the pistillate in small clusters or solitary, from the same axils. (Name composed of the Greek *echinos, a hedgehog*, and *cystis, a bladder*, from the prickly fruit.) MICRAMPELIS Raf.

1. E. lobàta (Michx.) T. & G. (lobed), WILD OR PRICKLY CUCUMBER, CONCOMBRE GRIMPANT (Que.). — Leaves deeply and sharply 5-lobed; fruit ovoid, 5 cm. long; seeds dark-colored. — Rich soil along streams, N.B. to Sask., s. to Fla. and Tex.; cult. for arbors and freely esc. June–Oct.

FAM. 167. CAMPANULÀCEAE (BLUEBELL FAMILY)

Herbs (outside our area sometimes shrubs or trees) with alternate exstipulate leaves and often with latex; flowers epigynous, solitary, clustered or in racemes, panicles or cymes; calyx usually herbaceous; corolla regular or zygomorphic, of united petals; anthers free or united into a tube around the brush-like summit of the style; ovary 2–5(–10)-locular, with many anatropous ovules on axile placentae; fruit a capsule; embryo straight, in fleshy albumen. — Two subfamilies:

Subfam. I. CAMPANULOÌDEAE (BLUEBELL SUBFAMILY)

Herbs, with milky juice, alternate leaves, and scattered flowers; calyx adherent to the ovary; the regular 5-lobed corolla campanulate, rarely rotate, valvate in the bud; the 5 stamens usually free from the corolla. Style 1, usually beset with collecting hairs above; stigmas 2 or more. Capsule 2–several-locular, many-seeded. Flowers generally blue and showy.

Anthers free; capsule opening by pores on the sides.
- Corolla rotate; capsule slender-cylindric or -prismatic. . . . . 1. *Specularia.*
- Corolla campanulate (rarely with flaring limb); capsule obconic or turbinate to subglobose. . . . . 2. *Campanula.*

Anthers connate at base; capsule opening by valves at apex; flowers crowded in dense involucrate heads; corolla with slender segments. . . . . 3. *Jasione.*

1. SPECULÀRIA Fabricius VENUS'S LOOKING-GLASS

Calyx 5 (or 3–4)-lobed. Corolla of perfect flowers rotate, 5-lobed; cleistogamous and greatly reduced flowers often in lower axils. Stamens 5, separate; the membranaceous hairy-based filaments shorter than the anthers. Stigmas 3. Capsule prismatic or slender-cylindric to subulate, 3-locular, opening by 3 small lateral valves above the middle. — Low annuals of Eurasia, n. Afr. and Am., with axillary blue or purplish flowers, these in American and an Old World species dimorphous, the earlier ones small and cleistogamous. (The early name of the common European *S. Speculum-Veneris, mirror of Venus*, derived from the Latin *specularius, pertaining to mirrors*)* LEGOUZIA Durande. TRIODANIS Raf., maintained by McVaugh as a separate genus.

Foliaceous bracts of inflorescence ovate to suborbicular; capsules of cleistogamous flowers straight, appressed, 2–10(–12) mm. long, ovoid, obovoid or ellipsoid to slender, dehiscent by 2 or 3 pores.
- Pores of capsule broadly elliptic to roundish, 0.5–1.5 mm. broad, borne from near middle to apex of capsule; seeds smooth or with low tubercles scattered over the surface.
  - Seeds 0.5–0.65 mm. long, plump-lenticular.
    - Leaves clasping, about as broad as long; expanded corollas usually borne at several upper nodes; capsule ellipsoid to obovoid, its pores submedian. . . . . 1. *S. perfoliata.*
    - Leaves not clasping, ovate to oblong, longer than broad: expanded corolla usually 1 and terminal; capsule usually more slender, its pores subapical. . . . . 2. *S. biflora.*
  - Seeds 0.8–1 mm. long, flattened to barely biconvex; leaves not clasping, longer than broad; expanded flowers from several upper nodes; pores near apex of capsule. . . . . 3. *S. lamprosperma.*
- Pores of capsule linear, 0.2–0.4 mm. broad, borne near middle of capsule; seeds 0.4–0.7 mm. long, with longitudinal rows of tubercles; upper leaves and bracts broadly ovate to rounded, sessile. . . . . 4. *S. Holzingeri.*

* Treatment based in part on the study of *Triodanis* by ROGERS MCVAUGH in Wrightia, i. no. 1 (1945).

Foliaceous bracts of inflorescence lanceolate to linear; capsules of cleistogamous flowers slenderly cylindric to subulate, 8–20 mm. long, inclined to curve outward, dehiscing by subapical longitudinal slits or 1 definite pore; capsule from expanded flowers up to 2.5 cm. long, sometimes with 1 or 2 pores; seeds biconvex, 0.7–1 mm. long. . . . . 5. *S. leptocarpa.*

1. S. perfoliàta (L.) A. DC. (with leaf meeting around the stem). — Stem pilose or hispid below, retrorse-hispidulous or scabrous above, simple to loosely ascending-branched, 0.5–1 m. high; *leaves and bracts roundish to ovate, clasping by the cordate base,* 0.5–2.5 cm. broad, hirsute, or the bracts glabrous, toothed; *flowers* 1–few at a node, *those at lower nodes cleistogamous* and rarely 5 mm. long, *those at upper nodes expanded* and about 1 cm. long, with *anthers* 2.5–3.5 *mm. long; capsule ellipsoid to obovoid,* 4–10 *mm. long; the 2 or 3 broadly elliptic to roundish pores* 0.5–1.5 *mm. broad, borne about midway up the capsule; seeds plump-lenticular, ellipsoid,* 0.5–0.6 *mm. long,* smooth or variously pebbled, somewhat shining. (*Legouzia* Britt.; *Triodanis* Nieuwl.) — Sterile open ground, Fla. to Tex. and n. Mex., n. to s. Me., s. N.H., Vt., s. Que., N.Y., s. Ont., Mich., Wisc., s. Minn., S.D., Mont. and s. B.C. May–Aug. (Trop. Am.)

2. S. biflòra (R. & P.) Fisch. & Mey. (two-flowered; referring to the two types of flowers). — Similar to no. 1, the stems simple or branching chiefly from below; *leaves ovate to oblong, not clasping;* bracts sessile or slightly clasping; *cleistogamous flowers at most nodes; expanded flower terminal or wanting, its anthers* 1.5–2.5 *mm. long; capsule more slender, its pores subapical; seeds smooth and lustrous.* (*Legouzia* Britt.; *Triodanis* Greene) — Dry open soil, borders of woods, etc., Fla. to e. Tex., ne. Mex. and Lower Calif., n. to se. Va., s. Ky., Mo., Kans. and sw. Oreg. April–June (S. Am.)

3. S. lamprospérma (McVaugh) Fern. (with shining seeds). — Differing from no. 1 in its *nearly sessile or very short-petioled hard leaves;* cleistogamous flowers at the lower and median nodes larger, the expanded flowers larger; *pores of capsule subapical; seeds flatter, highly lustrous,* 0.8–1 *mm. long.* (*Triodanis* McVaugh) — Rocky openings, glades or thin woods, Mo., Ark. and Okla.

4. S. Holzíngeri (McVaugh) Fern. (for John Michael Holzinger, 1853–1929, who first noted its distinctions). — Differing from the last in the *pores of capsule linear and only* 0.2–0.5 *mm. wide by* 1.4–2.5 *mm. long and borne midway from base to apex; calyx-lobes broader; lustrous seeds with longitudinal lines of low tubercles.* (*Triodanis* McVaugh) — Open prairies, thin woods and disturbed soil, Neb., Mo. and Okla. to Ariz.

5. S. leptocárpa (Nutt.) Gray (slender-fruited). — Minutely hirsute to glabrous, up to 7.5 dm. high; *leaves narrowly elliptic to oblanceolate; bracts lanceolate to linear;* upper nodes with expanded or cleistogamous flowers, lower nodes with cleistogamous flowers only; *capsules* very variable; *those of cleistogamous flowers terete,* often subulate, *outwardly arching and with straight or spiraling summits, or straight and appressed,* 0.8–2 *cm. long,* opening by longitudinal slits or by a pore; *capsules from expanded flowers straighter, erect, up to 2.5 cm. long;* subulate calyx-lobes falcate or divergent; seeds smooth, lustrous, lenticular, 0.7–1 mm. long. (*Legouzia* Britt.; *Triodanis* Nieuwl.) — Prairies, barrens, glades, open woods, etc., Ind. and Minn. to Mont., s. to Ark., Okla. and Mo. May–Aug.

S. Spéculum-Véneris (L.) Tanfani (mirror of Venus), the European type of the genus, mostly branched to summit, with narrowly ovate to lanceolate leaves, flowers all perfect and expanding, mostly in 3's at tips of stem and branches, and glabrous filaments, is casual in waste near ports but not persistent. (Adv. from Eu.)

2. CAMPÁNULA L. Bellflower

Calyx 5-cleft. Corolla generally campanulate, 5-lobed. Stamens 5, separate; the filaments broad and membranaceous at the base. Stigmas and locules of the capsule 3 in our species, the short capsule opening on the sides by as many valves or holes. — Herbs, chiefly of N. Hemisph., with terminal or axillary flowers. (A diminutive of the Latin *campana, a bell,* from the shape of the corolla.)

a. Flowers in glomerules, spikes or racemes, mostly numerous. . . *b.*

b. Flowers sessile, chiefly in glomerules or leafy-bracted heads. . . . . . . 1. *C. glomerata.*

b. Flowers distinctly pedicelled.

Calyx and outer surface of unexpanded corolla bristly-ciliate with long pale trichomes. . . . . 2. *C. Trachelium.*

Calyx and corolla smooth, or the former minutely pubescent; racemes elongate and spiciform.

Corolla campanulate; style straight; capsule campanulate-ovoid, opening by pores at the base. . . . . 3. *C. rapunculoides.*

Corolla rotate; style declined and upwardly curved; capsule obconic to subcylindric, the pores at the summit. . . . . 4. *C. americana.*

*a.*Flowers 1-∞, on slender peduncles or in loose inflorescences. . . *c.*

*c.*Style not exserted; cauline leaves entire or only shallowly toothed. . . *d.*

*d.*Stem smooth (rarely villous), erect or ascending, not retrorse-scabrous on the angles; leaves not retrorse-scabrous, gradually diminishing in size upward.

1-flowered; capsule erect, opening by subterminal pores; the mature hypanthium thick-clavate, 1–2 cm. long, with erect or connivent calyx-lobes; corolla 5–10 mm. long. . . . . 5. *C. uniflora.*

1–many-flowered; capsule nodding, opening near base; the mature hypanthium turbinate to campanulate, 3–8 mm. long, with divergent to loosely ascending calyx-lobes; corolla 1.5–2.7 cm. long. . . . . 6. *C. rotundifolia.*

*d.*Stem retrorse-scabrous on the angles; leaves retrorse-scabrous on margin or midrib beneath, of nearly uniform size; corollas 5–12 mm. long.

Peduncles mostly divergent, leafy-bracted, the naked summit 0.3–3.5 cm. long; flowering calyx 1.3–3.8 mm. long, its lobes 0.7–2 mm. long; corolla whitish, 5–8 mm. long; fruiting hypanthium 1.2–2 mm. long. 7. *C. aparinoides.*

Peduncles ascending, mostly naked, 1–8 cm. long; flowering calyx 3–6.7 mm. long, its lobes 2–4 mm. long; corolla bluish, 10–12 mm. long; fruiting hypanthium 3.2–5 mm. long. . . . . . . 8. *C. uliginosa.*

*c.*Style long-exserted; cauline leaves saliently toothed. . . . . 9. *C. divaricata.*

1. C. GLOMERÀTA L. (clustered), CLUSTERED B. — Somewhat hairy, stout and erect, 3–6 dm. high; *basal leaves lanceolate to oblong-ovate,* long-petioled; *cauline leaves* oblong or lanceolate, *cordate-clasping; flowers sessile,* clustered in the upper axils, *forming a leafy head;* corolla open-campanulate, deep purple, 2–3 cm. long. — Roadsides, old fields and pastures, Que. to Mass. and Minn. June, July. (Introd. from Eu.)

2. C. TRACHÈLIUM L. (old generic name), NETTLE-LEAVED B., THROATWORT. — Stems simple below, 3–9 dm. high, often bristly above; *basal leaves broadly cordate-ovate to reniform,* coarsely toothed; *the upper ones short-petioled or subsessile,* ovate to lanceolate; *flowers in 2's or 3's* (rarely solitary) *in* terminal and axillary *loose clusters;* calyx bristly, the oblong-lanceolate lobes somewhat foliaceous; corolla 2.5–3.5 cm. long. — Roadsides, thickets and waste places, Que. to Mass. and Ohio. Aug.–Oct. (Introd. from Eu.)

3. C. RAPUNCULOÌDES L. (like *C. Rapunculus,* the Rampion). — Stems slender, 6–10 dm. high, smoothish, or finely pubescent above; *lower leaves* long-petioled, *cordate-ovate; the upper ovate-lanceolate,* short-petioled to sessile, irregularly serrate-dentate, *hispidulous beneath; flowers* nodding, single in the axils of bracts, *forming racemes;* calyx and capsule scabrous-puberulent; *corolla campanulate,* 2–3 cm. long; *capsule opening by pores at base.* — Roadsides, thickets, etc., Nfld. to N.D., s. to Del., Md., W.Va., O., Ind., Ill. and Mo. July–Sept. (Natzd. from Eu.)

Var. UCRÀNICA (Bess.) K. Koch **(Ukrainian)**. — Smoother; hypanthium and calyx-lobes glabrous. — Occasional, Que., N.E. and N.Y. (Natzd. from Eu.)

Several cult. Old World perennials are sporadic on rubbish or about old homesteads. These include **C. PUNCTÀTA** Lam. (dotted), suggesting no. 3, with few drooping flowers 5–7 cm. long, the calyx with basal appendages; **C. BONONIÉNSIS** L. (of Bologna), resembling no. 3, but non-stoloniferous, the stem and leaves tomentulose, upper leaves clasping, corollas short (1–2 cm. long); **C. PERSICIFÒLIA** L. (peach-leaved), smooth, with spatulate to linear-lanceolate leaves and few erect flowers with unappendaged calyx and corolla about 3 cm. high; **C. MÈDIUM** L. (name used by classical writers for this or a similar plant of Media), CANTERBURY-BELLS, hispid, with narrowly oblong leaves and ascending flowers, 1 or 2 together, with calyx basally appendaged, corolla 4–5 cm. long and stigmas 5; and **C. LATIFÒLIA** L. (broad-leaved), similar to no. 2, with leaves all ovate-lanceolate, peduncles 1-flowered, corolla 4–5 cm. long.

4. C. americàna L. (American), TALL B. — *Annual,* with mostly simple stem 0.5–2 m. high; leaves narrowly ovate to oblong-lanceolate, serrate, the lower broader, all acuminate at both ends, the lower and median long-tapering to the petiole; spiciform raceme 3–6 dm. long, the flowers solitary or clustered in axils of the upper leaves and bracts; *corolla rotate,* light blue, 2–2.5 cm. broad, or cylindric-tubuliform and narrowed to the closed apex in forma **tubuliflòra** Fern. (tubular-flowered); *style declined and upwardly curved; capsule obconic to subcylindric,* glabrous, *opening by pores at summit.* (*Campanulastrum* Small) — Rich moist soil, s. Ont. and N.Y. to Minn., s. to Fla., Ala. and Mo. June–Aug.

Var. **illinoénsis** (Fresn.) Farw. (of Illinois). — Lower and middle leaves broader, ovate, abruptly contracted to the petiole. — O. to S.D., s. to Tenn., Ark. and Okla.

5. C. uniflòra L. (one-flowered). — Smooth or nearly so, with ligneous caudex; *stems* firm, *simple,* 0.5–3 dm. high; *leaves subcoriaceous,* rather crowded at the base, oblong to spatulate, the scattered upper ones linear or lanceolate; *flower solitary,* slightly nodding to erect; *corolla* 5–10 *mm. long,* pale blue, slenderly campanulate, with narrow segments; *mature hypanthium thick-clavate, erect,* 1–2 *cm. long, with erect or connivent calyx-lobes* 3–8 mm. long; *capsule opening by subterminal pores.* — Arct. reg., s. to calcareous alpine areas of Shickshock Mts., Que., and to Ung., Colo. and Utah. July. (N. Eurasia)

6. C. rotundifòlia L. (round-leaved; from the basal foliage), HAREBELL, BLUEBELL. — *Slender perennial,* simple to freely branched, 1–5 dm. high, 1–15-flowered; *basal leaves* (rarely present on the flowering stems) *round-cordate to ovate,* mostly toothed, *long-petioled,* early withering; *cauline leaves numerous, linear or narrowly lanceolate, smooth;* calyx-lobes subulate, from one-third to two-thirds the length of the purplish-blue, or white in forma **albiflòra** Rand & Redf. (white-flowered) corolla (1.5–2.7 cm. long); *capsule* nodding, short-ovoid to subcylindric, *opening by pores at base.* — Open or rocky banks, meadows, shores, etc., widely distributed in boreal reg., s. to N.S., N.E., n. N.J., Pa., W.Va., O., Ind., Ill., Mo., Neb., Tex., N.M., Ariz. and Calif. June–Sept. (Eurasia) — Extremely variable in stature, degree of branching, number and size of flowers, texture and length of cauline leaves, length and divergence of calyx-lobes, etc., characters which readily respond to slight changes of aridity, moisture, nutriment, and exposure. In meadows and other mesophytic habitats the tall many-flowered plants have thin and elongate cauline leaves (var. *intercedens* (Witasek) Farw.); in drier or exposed habitats the foliage is firm and the upper leaves reduced (var. *petiolata* (A. DC.) J. K. Henry); and in arctic-alpine areas the stems are dwarfed (0.3–2 dm. high) and 1-flowered (var. *arctica* Lange); all tendencies which quickly become altered with change of environment and which seem of little taxonomic value.

Var. **velùtina** A. DC. (velvety). — *Stems and narrow cauline leaves closely puberulent throughout.* — Cheboygan Co., Mich. (*E. J. Hill*) (Eu.)

Var. **lancifòlia** Mert. & Koch (lance-leaved). — *Cauline leaves,* or all but the uppermost, *broadly lanceolate to ovate, oblong or oblanceolate, petioled.* (Var. *alaskana* Gray) — W. Nfld.; Anticosti I. and Gaspé Pen., Que.; St. Paul's I., N.S.; Alaska and B.C. (Eurasia)

The Old World **C.** CARPÁTICA Jacq. (Carpathian), with slender matted stems, long-petioled ovate dentate cauline leaves and ascending peduncles bearing solitary large flowers with lanceolate calyx-lobes, the erect fruits with hypanthium 0.7–1.2 cm. high, tends to spread from gardens.

7. C. aparinoìdes Pursh (like *Galium Aparine,* from its adherent foliage), MARSH-.B — *Stem* simple or branched, *weak,* 2–6 dm. long, *somewhat 3-angled, rough backward on the angles, as are the* slightly toothed *edges and midrib of the lanceolate or linear-lanceolate soft dentate leaves (longer ones* 1.5–4.5 *cm. long); flowers* chiefly *terminating strongly divergent leafy branches with naked summits* 0.3–3.5 *cm. long; flowering calyx* 1.3–3.8 *mm. long, its lobes* 0.7–2 *mm. long; corolla whitish,* 5–8 *mm. long; fruiting hypanthium* 1.2–2 *mm. long.* — Meadows, swales and wet shores, Me. to Minn., s. to Ga., Ky., Ia., Neb. and Colo. June–Aug. — With the habit of a *Galium.*

8. C. uliginòsa Rydb. (growing in marshes), MARSH-B. — Stiffer than no. 7; *leaves narrowly linear to linear-lanceolate, attenuate,* entire or barely dentate, the larger 2–8 cm. long; *peduncles ascending, mostly naked,* 1–8 *cm. long; flowering calyx* 3–6.7 *mm. long, its lobes* 2–4 *mm. long; corolla bluish,* 10–12 *mm. long; fruiting hypanthium* 3.2–5 *mm. long.* — Swales and wet thickets, Que. to James Bay and Sask., s. to N.B., Me., Mass., e. Pa., N.Y., O., Ind., Ill., Ia. and Neb. June–Aug.

9. C. divaricàta Michx. (widely divergent), SOUTHERN HAREBELL. — Very smooth; stem loosely branched, 3–9 dm. high; *leaves lanceolate to ovate, pointed at both ends, coarsely and sharply toothed; flowers numerous* in a large compound panicle; *calyx-lobes subulate,* about half the length of the small pale blue corolla (6–8 mm. long); *style exserted.* (*C. flexuosa* Michx.) — Dry woods and rocky slopes, w. Md., W.Va. and Ky., s. to Ga. and Ala.; rarely adv. n. to N.H. — July–Sept.

3. JASIÒNE L. SHEEP'S-BIT

Calyx 5-lobed. Corolla with 5 very narrow lobes. Anthers united at base into a ring about the style. Capsule opening by 2 valves. — Herbs, nat. of Old World, with small flowers in involucrate heads. (Name used by Theophrastus, from the Greek *iásis,* healing.)

1. **J.** MONTÀNA L. (of mountains). — Annual or biennial, 2–5 dm. high, simple or branching; leaves linear or lanceolate; flowers blue, in heads 1–2 cm. broad. — Fields and roadsides, e. Mass. to N.J. July–Sept. (Natzd. from Eu.)

Subfam. II. LOBELIOÌDEAE (LOBELIA SUBFAMILY)

Herbs, with acrid milky juice, alternate leaves, and scattered flowers, an irregular gamopetalous 5-lobed corolla, the 5 stamens free from the corolla and united into a tube, commonly by their filaments and always by their anthers. Style 1; stigma often fringed. — A small genus with us.

4. LOBÈLIA L. Lobelia

Calyx 5-cleft, with a short tube. Corolla with a straight tube split down on the (apparently) upper side, somewhat 2-lipped; the upper lip of 2 rather erect lobes, the lower lip spreading and 3-cleft. Two of the anthers in our species (all 5 in one) bearded at the top. Capsule 2-locular, many-seeded, opening at the top. — Flowers axillary or chiefly in bracted racemes; in summer and early autumn. Genus semicosmop. (Dedicated to *Matthias de l'Obel*, 1538–1616, Flemish herbalist.)*

a. Flowers (calyx and corolla), straightened out, 1.8–4.5 cm. long; corolla-tube fenestrate (with lateral openings). . . *b.*

b. Flowers 3–4.5 cm. long; corolla vermilion; filament-tube 1.9–3.3 cm. long.

Leaves lanceolate to ovate-lanceolate, the broader about thrice as long as broad; filament-tube 2.4–3.3 cm. long, much exceeding corolla-tube; anther-tube 4–5.5 mm. long. . . . . 1. *L. Cardinalis.*

Leaves narrowly lanceolate to linear, the broader about seven times as long as broad; filament-tube 2–2.4 cm. long, barely exserted; anther-tube 3–4 mm. long. . . . . 2. *L. splendens.*

b. Flowers 1.5–3.3 cm. long; corolla purple to blue; filament-tube 6–15 mm. long. . . *c.*

c. Calyx-lobes with foliaceous basal auricles 2–5 mm. long; pedicels with bracteoles at or above the middle; filament-tube 1.2–1.5 cm. long. . . . 3. *L. siphilitica.*

c. Calyx-lobes exauriculate or with very small auricles; bracteoles at or near bases of pedicels; filament-tube 6–11.5 mm. long. *d.*

d. Plant glabrous, or only sparsely hirsute at base; flowers 2–3.3 cm. long, in slender racemes; calyx-teeth linear or with linear to subulate tips. . . *e.*

e. Pedicels arching to one side in fruit, rough; calyx-teeth entire; leaves narrowly lanceolate, tapering to both ends. . . . . . 4. *L. elongata.*

e. Pedicels erect or ascending; calyx-teeth usually prominently toothed.

Cauline leaves elliptic or short-ovate; pedicels smooth; flowers 2–2.6 cm. long; fruiting calyx strongly ribbed, its teeth gland-tipped. . . . . 5. *L. georgiana.*

Cauline leaves narrowly lanceolate to linear; pedicels hairy; flowers 2–3.3 cm. long; calyx-teeth callus-tipped. . . . . . . 6. *L. glandulosa.*

d. Plant closely short-pubescent throughout; flowers 1.5–2.4 cm. long, in usually dense racemes; calyx-teeth lanceolate to ovate, cordate-based. . . . . 7. *L. puberula.*

a. Flowers 0.7–1.6(–2.2 in the scapose aquatic no. 15) cm. long; corolla-tube not fenestrate. *f.*

f. Leafy-stemmed, with mostly flat leaves; only the two smaller anthers hairy-tufted at tip; capsules ascending to pendent. . . *g.*

g. Lower lip of corolla densely hairy at base; cauline leaves lanceolate, oblong or ovate (linear or filiform only in the southern nos. 11 and 12). . . *h.*

h. Cauline leaves lanceolate, oblong or ovate, mostly 1 cm. wide or wider. . . *i.*

i. Stems not long-hirsute, simple or with few upright branches; flowers 0.7–1.5 cm. long on curved pedicels, in elongate spiciform racemes; capsules not inflated, partly exserted.

Stem densely short-pubescent at base; cauline leaves narrowed at base; flowers 7–12 mm. long. . . . . 8. *L. spicata.*

Stem smooth or nearly so; cauline leaves broad-based; flowers 1–1.5 cm. long. . . . . 9. *L. appendiculata.*

i. Stems long-hirsute, at least below, usually divergently branched; flowers 7–10 mm. long, on ascending pedicels, in open short racemes terminating stem and branches; capsules becoming inflated, included within the calyx. . . . . 10. *L. inflata.*

h. Cauline leaves linear-lanceolate to filiform, 0.5–4 mm. broad.

Fibrous-rooted annual; leaves linear-lanceolate or linear; pedicels with basal bracteoles; marsh-plant. . . . . 11. *L. Canbyi.*

Rhizomatous perennial; leaves filiform; pedicels without basal bracteoles; aquatic. . . . . 12. *L. Boykinii.*

*** Treatment based partly on that of Rogers McVaugh in Rhodora, xxxviii (1936).**

*g.*Lower lip of corolla smooth at base; middle and upper cauline leaves linear to narrowly lanceolate; plants slender, often with filiform branches.
Pedicels with basal bracteoles; flowers 8–11 mm. long; calyx short-hemispheric; capsule subglobose; southern species. . . . . 13. *L. Nuttallii.*
Pedicels with median bracteoles; flowers 7–16 mm. long; calyx elongate; capsule ovoid; northern species. . . . . 14. *L. Kalmii.*
*f.*Scapose (or with minute bracts), with a rosette of linear fleshy hollow curving basal leaves; all 5 anthers bearded at tip; flowers 1.2–2.2 cm. long; fruit pendent; aquatic. . . . . 15. *L. Dortmanna.*

1. L. Cardinàlis L. (pre-Linnaean generic name for *Lobelia*, meaning of the Cardinal; from color and shape of the corolla, like a miter), CARDINAL-FLOWER. — Perennial by short basal offshoots, smoothish; stem simple, coarse, 0.4–1.8 m. high; *leaves* numerous, thin, glabrous or hirtellous, *lanceolate to lance-ovate,* irregularly serrate; raceme terminal, simple, 1–5 dm. long, somewhat 1-sided; pedicels much shorter than the leafy bracts; *flowers* 3–4.5 *cm. long; corolla vermilion or deep red,* or roseate in forma **ròsea** St. John (roseate), or white in forma **álba** (Eat.) St. John (white), its tube fenestrate; *filament-tube* 2.4–3.3 *cm. long, long-exserted; anther-tube* 4–5.5 *mm. long.* — Damp shores, meadows and swamps, Fla. to e. Tex., n. to ne. N.B., s. Que., s. Ont., Mich., Wisc. and Minn. Late July–Sept.

2. L. spléndens Willd. (brilliant). — Similar, more slender; *leaves narrowly lanceolate to linear; filament-tube* 2–2.4 *cm. long, barely exserted; anther-tube* 3–4 *mm. long.* — Low grounds, Mex. and Calif., east to w. Mo., Okla. and Tex.

3. L. siphilítica L. (from formerly reputed curative properties), GREAT L., BLUE CARDINAL-FLOWER. — Perennial by basal offshoots; stem simple, coarse, smooth or sparsely hirsute on angles, 0.2–1.3 m. high; *leaves* thin, *strigose above,* ovate, oblong or broadly lanceolate, irregularly serrate, *acute at both ends,* the upper merging into the floral bracts; raceme dense, 1–5 dm. long, scarcely secund; *flowers* 2.3–3.3 *cm. long; bracts of pedicels mostly median; calyx hirsute, with* broad foliaceous lobes and *broad basal auricles* 2–5 *mm. long;* corolla blue, or white in forma **albiflòra** Britt. (white-flowered); base of lower lip white, with 2 tubercles; the tips of corolla-lobes pubescent, or in forma **laevícalyx** Fern. (with smooth calyx) leaves as well as calyx and corolla-lobes glabrous; filament-tube 1.2–1.5 cm. long; anther-tube 4–5.5 mm. long; capsule partly exserted. — Rich low woods and swamps, Me. (rare) and w. N.E. to Minn. and S.D., s. to e. Va., upland of N.C. and Ala., Miss., Mo. and e. Kans. Aug., Sept. Passing into

Var. **ludoviciàna** A. DC. (of Louisiana). — *Smoother and lower* (up to 6 dm. high); *leaves* lanceolate, *glabrous,* less toothed or subentire; inflorescence few (6–20)-flowered; *calyx glabrous or only sparsely hirsute;* auricles narrower. — Wisc. to s. Man., s. to La. and Tex.

4. L. elongàta Small (elongate). — Perennial by basal offshoots; stem usually simple, glabrous, 0.3–1.5 m. high; cauline *leaves* fleshy, *lanceolate, acute at each end,* sharply dentate; raceme strongly 1-sided, rather open, 0.4–3 dm. long; *pedicels rough, with basal bractlets,* ascending, *finally arching;* flowers 2–2.5 cm. long; *calyx* hemispherical, becoming subglobose in fruit; *its teeth lance- or linear-subulate, entire,* often revolute; corolla deep blue; filament-tube 8–11.5 mm. long; anther-tube 4 mm. long. (*L. amoena* of ed. 7, not Michx.) — Fresh to brackish tidal marshes and swamps, Del. and Md. to Ga. Aug.–Oct.

5. L. georgiàna McVaugh (of Georgia). — *More slender* than no. 4; stem smooth or hirtellous, 0.3–1.2 m. high; *leaves elliptic to oval,* firm, with parchment-like pale lower surface; raceme remotely few (1–30)-flowered, 0.3–2.5 dm. long; *pedicels smooth, erect;* flowers 2–2.6 cm. long; *calyx strongly ribbed, its teeth usually with* 1–6 *prominent glandular teeth;* corolla pale blue; filament-tube 6.5–8.5 mm. long; anther-tube 3–3.5 mm. long. (*L. amoena,* var. *glandulifera* Gray; *L. glandulifera* (Gray) Small, not Ktze.) — Meadows, peaty depressions and wet woods, e. Va. to e. Ky., s. to nw. Fla. Sept.–Nov.

6. L. glandulòsa Walt. (glandular). — Similar to nos. 4 and 5; stem slightly flexuous, slender, usually 0.6–1.4 m. high; *leaves narrowly lanceolate to linear, not much narrowed below,* callous-denticulate to subentire; *flowers* remote, 2–3.3 *cm. long; pedicels* ascending, *rough-pubescent,* basally bracted; *calyx* hirsute to glabrous, *its teeth gland-tipped; corolla* blue with white eye, *base of lower lip villous on upper surface;* filament-tube 7.5–10 mm. long; anther-tube 3.5–4 mm. long. — Damp pinelands and swamps, Fla. to se. Va. (rare). Sept., Oct.

7. L. pubérula Michx. (puberulent). — *Densely short-hirsute or puberulent throughout,* rarely glabrate; stem usually simple, 0.3–1.6 m. high; *leaves oblong to lanceolate or narrowly obovate,* with callous-tipped teeth or subentire, the upper passing gradually into the floral bracts; raceme somewhat interrupted or dense, 0.4–5 dm. long, commonly 1-sided; bractlets of pedicels basal or nearly so; flower 1.5–2.4 cm. long; *calyx-lobes lanceolate to narrowly deltoid-acuminate,* revolute; corolla blue to purple or rarely white, with white eye, pubescent. — **With pronounced geographic vars. as follows:**

Stem densely short-hirsute; calyx long-hirsute or villous, its undulate or crisped-margined lobes 2.5–5 mm. broad at base and with large and broadly rounded basal auricles; leaves ascending to appressed, the lower and median ones obovate, with rounded bases, their teeth not strongly callous-tipped; lower bracts ovate-lanceolate to oblong, broad-based; plant of Coastal Plain and outer Piedmont. . . . . *L. puberula* (typical).

Stem minutely puberulent; calyx sparsely short-pubescent to glabrous, its mostly entire lobes 1.2–2 mm. broad; leaves with callous-tipped teeth; lowest bracts lanceolate to linear, often tapering at base.

Leaves spreading to loosely ascending, glabrous or merely scabrous; lower bracts linear-lanceolate, 1–1.5 cm. long; sinuses of calyx with small auricles; plant chiefly of Appalachian Upland. . . . . Var. *simulans*.

Leaves loosely to strongly ascending, often short-pilose beneath; lower bracts lanceolate, 1.5–3 cm. long; sinuses of calyx with prominent auricles; southwestern. Var. *mineolana*.

L. pubérula (typical). — Woods and openings, Coastal Plain and Piedmont, Ga. and Ala., n. to N.J. and se. Pa. Aug.–Oct. — The rare forma **cándida** Fern. (white) is an albino.

Var. **símulans** Fern. (simulating). — Chiefly on Appalachian Upland, slightly on Coastal Plain, Va. and W.Va. to Ill., s. to n. Fla., Ala., Miss. and La.

Var. **mineolàna** E. Wimm. (for Mineola, Tex.). — Ala. to e. Tex., n. to se. Mo., Ark. and Okla.

8. L. spicàta Lam. (spicate), "Highbelia", Pale-spike-L. — *Stem* simple or with few branches, erect, 0.2–1.1 m. high, *densely short-pubescent at base, smooth above;* lower leaves oblanceolate, obovate or oblong, with short margined petioles, strigose; *middle and upper leaves lanceolate,* sessile, *with narrow* decurrent *bases; raceme virgate, slender, dense, 2–6 dm. long,* with smooth bracts; pedicels puberulent, inconspicuously bracted at base; flowers 9–12 mm. long; *calyx in anthesis flattish,* its teeth usually *with short* (*mostly less than* 1 *mm. long*) *basal auricles; corolla light blue to white,* the lower lip pubescent at base within; *anther-tube* 1.7–2 mm. long, *pale bluish-gray; capsules short-hemispherical,* partly exserted. — Rich meadows, field and thickets, often a weed, s. N.B. to Minn., s. to Ga. and Ark. June–Aug.

Var. **campanulàta** McVaugh (bell-shaped). — *Raceme few* (10–30)*-flowered; flowers* **7–9** *mm. long; calyx in anthesis short-campanulate; corolla deep purplish-blue; anthers white; capsule globose.* — Locally, s. Me. to Minn., s. to N.J., Del., W.Va., s. Ind. and Ill. Late May–July.

Var. **hirtélla** Gray (stiffly hairy). — Strongly *hirtellous throughout, low* (*2–5 dm.*), *with leaves mostly on lower part of stem;* lower floral bracts elongate; flowers as in typical form. — Prairies and dry fields, Gaspé Co., Que., to Alta., s. to L.I., w. N.Y., Mich., Ind., Ill., Mo. and Kans., commoner westw.

Var. **scapòsa** McVaugh (scapose). — *Stem subscapose,* up to 1.1 m. high; *cauline leaves* 1–6, *all below middle of stem, inconspicuous, 1–2 cm. long,* lanceolate; *basal leaves obovate to rotund,* conspicuous; *flowers* 7–10.5 mm. long, *in a loose apparently naked raceme* 1–6 dm. long; auricles of calyx usually conspicuous; corolla white or pale blue. — Open woods and slopes, s. Pa. to Ky. and n. Miss., s. to N.C., S.C. and La.

Var. **leptóstachys** (A. DC.) Mackenz. & Bush (slender-spiked). — Similar to typical var., tall (0.3–1.2 m.), leafy, *the leaves mostly appressed to stem, oblong;* raceme dense, 2–5 dm. long; *auricles of calyx deflexed, filiform, 1–5 mm. long.* (*L. leptostachys* A. DC.) — Dry sandy woods and openings, W.Va. to Ill. and e. Kans., s. to Ga., Ala., Miss. and Ark.

9. L. appendiculàta A. DC. (appendaged). — Stem erect, 2.5–9 dm. high, *glabrous* except for sparse chaffy basal hairs; *cauline leaves very thin,* essentially glabrous, oblong or ovate, *sessile, with broad to clasping bases,* 2–7 cm. long; *raceme 1-sided,* 2–3 dm. long, with linear or narrowly lanceolate callous-denticulate bracts; pedicels puberulent; *flowers 1–1.5 cm. long; calyx* long-campanulate in fruit, subinflated, its lobes bristly-ciliate, *its conspicuous lanceolate auricles 1–3 mm. long;* corolla lilac or violet, pubescent at base of lip inside; filament-tube 3–4 mm. long; anther-tube 2 mm. long, bluish-gray; capsule partially exserted, horizontal or nodding. — Prairies and open woods, Ala. to e. Tex., n. to Ill., Mo. and Kans. May, June.

10. L. inflàta L. (inflated), Indian-tobacco. — *Stem commonly* paniculately much *branched,* from an annual root, 0.15–1 m. high (in sterile soil dwarfed and simple), *villous or long-hirsute* at least below with flat hairs; cauline *leaves* obovate to ovate or ovate-oblong, dentate or serrate, usually somewhat *hairy beneath,* sessile, or the lower subpetiolate; *racemes terminal and at tips of branches,* the central one 0.5–3 dm. long, not secund; pedicels ascending, minutely bristly, with very inconspicuous basal bracts; *flowers* rather inconspicuous, 7–10 *mm. long; calyx* campanulate, smooth, *in fruit covering the capsule and becoming inflated* and ovoid to subglobose; auricles 0; corolla pale violet to whitish or pinkish, the lower lip pubescent at base. — Fields, roadsides, waste places and open woods, C.B. to Sask., s. to Ga., Ala., Miss., Ark. and e. Kans. June–Oct.

11. L. Cánbyi Gray (for its discoverer, WILLIAM MARRIOTT CANBY, 1831–1904). — *Annual;* stem simple or racemosely branched, smooth or nearly so, 3–10 dm. high; *leaves linear or linear-lanceolate,* 1–5 cm. long, *thin,* callous-denticulate, often appressed; *raceme lax,* 1–3 dm. long; *pedicels* ascending, upwardly ciliate, *basal bractlets inconspicuous; flowers* 0.9–1.4 *cm. long; calyx-tube top-shaped and acute at base, in fruit becoming ellipsoid* and nearly covering the capsule, *auricles none; corolla* deep blue, *bearded in the throat;* filament-tube 3–4 mm. long; anther-tube 2 mm. long, light bluish-gray. — Sandy swamps and wet pinelands, Coastal Plain of Ga. to N.C. and upland of N.C. and Tenn.; Sussex Co., Del. and s. N.J. July–Sept.

12. L. Boykínii T. & G. (for its discoverer, SAMUEL BOYKIN, 1786–1848). — Aquatic *perennial with thick creeping rhizomes;* stem simple or racemosely branched, fistulose, 5–8.5 dm. high; *leaves filiform,* 0.5–2.5 cm. long, often deciduous; *raceme* lax, 1-*sided,* 0.7–2 dm. long, with very short filiform bracts; *pedicels spreading,* glabrous, *bractless;* flowers 1–1.3 cm. long; calyx flattish, round-based in anthesis, becoming hemispherical in fruit; corolla blue, with white eye; capsule partly exserted. — Cypress-ponds and wet pinelands, Sussex Co., Del.; S.C. to nw. Fla. May, June.

L. PALUDÒSA Nutt. (of bogs), *with mostly basal oblanceolate leaves* about 1 dm. long, the *stem subscapose* but bracteose, *flowers* 1–1.6 *cm. long, corolla-tube fenestrate,* was reported, perhaps erroneously, by Nuttall from Del.; now known only in Fla. and se. Ga.

13. L. Nuttállii R. & S. (for its discoverer, THOMAS NUTTALL, 1786–1859). — Very slender, 2–7.5 dm. high, simple or with filiform ascending branches; lower leaves ovate to oblanceolate, the lowest petioled, middle and upper ones linear; racemes lax; *pedicels* ascending, flexuous, *bracteolate at base; flowers* 8–11 *mm. long; calyx-tube flattish, in fruit becoming short-hemispherical,* without auricles; corolla blue, with a white eye and 2 greenish tubercles at base of lower lip. — Damp to dry sandy or argillaceous soil, nw. Fla. and Ala., n. to L.I., N.J., e. Pa. and Ky. July–Oct.

14. L. Kálmii L. (for its discoverer, PEHR KALM, 1715–1779). — Stems slender, simple to diffusely branched, 1.5–6 dm. high; basal leaves spatulate to obovate, petioled, pubescent; cauline linear or linear-lanceolate to oblanceolate; racemes lax, often 1-sided; *pedicels* lax to appressed, *with median bracteoles;* flowers 7–16 mm. long; *calyx campanulate or conical, in fruit becoming ellipsoid to subglobose;* corolla blue, with conspicuous white eye, or rarely white throughout in forma **leucántha** Rouleau (white-flowered). — Damp calcareous or sweet soils, wet ledges, shores, meadows or bogs, Nfld. to Mackenz., s. to C.B., s. N.B., centr. Me., w. N.E., n. N.J., se. Pa., s. O., Ind., Ill. s. Ia., S.D. and Colo. July–Sept.

15. L. Dortmánna L. (generic name given by Rudbeck in 1720, after an early Dutch apothecary, DORTMANN), WATER-L., WATER-GLADIOLE. — Smooth; *stem scapose* or with few small bracts, simple, rarely with 1 or 2 branches, varying with depth of water from 0.5–10 dm. high, *hollow; leaves numerous in a basal rosette, linear, fleshy,* curving, *hollow;* raceme lax, of 1–11 flowers; *pedicels* finally curving or horizontal, *bractless; flowers* 1.2–2.2 *cm. long; calyx* conic, *in fruit becoming obconic to cylindric, with the fruit pendent;* corollas pale violet to white; *all* 5 *anthers bearded at tip.* — Sandy or gravelly margins of ponds, emersed to deeply submersed, flowering above water, Nfld. to Minn., s. to N.S. (incl. Sable I.), N.E., L.I., n. N.J., ne. Pa., n. Mich. and Wisc.; also sw. B.C. to Oreg. July–Oct. (W. Eu.)

FAM. 168. COMPÓSITAE (COMPOSITE FAMILY)

Flowers in a close head (the compound flower of the older botanists), *on a common receptacle, surrounded by an involucre, with* 5 (*rarely* 4) *stamens inserted on the corolla, their anthers united in a tube* (*syngenesious*). Calyx-tube united with the 1-locular ovary, the limb (called a *pappus*) crowning its summit in the form of bristles, awns, scales, teeth, etc., or cup-shaped, or else entirely absent. Corolla either strap-shaped (*ligulate*) or tubular; in the latter chiefly 5-lobed, valvate in the bud, the veins bordering the margins of the lobes. Style usually 2-cleft at the apex (in sterile flowers usually entire). Fruit seed-like (*achene*), formed by fusion of carpel to calyx-tube, dry, containing a single erect anatropous seed, with no albumen. — An immense family, in temp. reg. chiefly herbs, without stipules, with perfect, polygamous, monoecious, or dioecious flowers. The flowers with a strap-shaped (*ligulate*) corolla are called *rays* or *ray-flowers;* the head which presents such flowers, either throughout or at the margin, is *radiate.* The tubular flowers compose the *disk;* and a head which has no ray-flowers is said to be *discoid.* When the head contains two sorts of flowers it is said to be *heterogamous;* when only one sort, *homogamous.* The leaves of the involucre, of whatever form or texture, are termed *phyllaries* (formerly called bracts). The bracts or scales, which often grow on the receptacle among the flowers, are called the *chaff* or *pales;* when these are wanting, the receptacle is said to be *naked.* The largest family of flowering plants. The genera are divided by the characters of the corolla into three subfamilies, only two of which are represented in our region. The first is much the larger.

KEY TO THE SUBFAMILIES AND TRIBES

*a.*Corollas tubular in all the perfect discoid flowers, regularly 5 (rarely 3 or 4)-lobed; ligules only in the marginal or ray-flowers which, when present, are either pistillate only or neutral (with neither stamens nor pistil). Subfam. I. TUBULIFLORAE. . . *b.*

*b.*Heads alike (homogamous) and discoid; flowers all perfect and rarely yellow; anthers not caudate at base; stigmatic lines obscure, only below middle of style-branch.

Branches of style long and slender, terete, thread-like, minutely bristly all over; stigmatic lines only near base; leaves alternate, scattered or rosulate. . . . . Tribe I. VERNONIEAE (p. 1358).

Branches of style thickened upward or clavate, obtuse, very minutely and uniformly pubescent; stigmatic lines indistinct below middle; leaves opposite, alternate or in whorls. . . . . Tribe II. EUPATORIEAE (p. 1361).

*b.*Heads homogamous or heterogamous (of two kinds), discoid or radiate, some of them often yellow (or not); style-branches of the perfect or pistillate flowers with stigmatic lines mostly prominent and extending nearly or quite to the naked summit or to a pubescent or hispidulous tip or appendage. . . *c.*

*c.*Flowers discoid or discoid and ligulate. . . *d.*

*d.*Style-branches of the perfect flowers flattened and smooth, extended into lanceolate or elongate-deltoid hairy appendages; disk-corollas mostly yellow; receptacle naked (in ours); anthers obtuse and entire or barely notched at base. . . . . Tribe III. ASTEREAE (p. 1376).

*d.*Style-branches of the disk-flowers not appendaged at tip or with a very short or slender appendage. . . *e.*

*e.*Disk-corollas mostly filiform and truncate (more tubular and lobed in no. 30); anthers sagittate, with caudate bases; style-branches unappendaged at tip; pappus capillary or none (of scales only in no. 29); heads discoid (radiate only in nos. 28 and 29); leaves mostly alternate. . . . . . . Tribe IV. INULEAE (p. 1448).

*e.*Disk-corollas tubular to goblet-shaped, mostly lobed at summit; anthers not caudate at base. . . *f.*

*f.*Pappus not soft-capillary, of scales, bristles, awns, etc., or none. . . *g.*

*g.*Phyllaries (involucral bracts) or at least the outer ones herbaceous or membranaceous, not scarious; leaves (especially the lower) often opposite.

Receptacle chaffy, the pales or bracts subtending some or all of the disk-flowers; style branched or uncleft. Tribe V. HELIANTHEAE (p. 1465).

Receptacle naked, without chaff; styles forking. Tribe VI. HELENIEAE (p. 1511).

*g.*Phyllaries scarious or papery; pappus none or a low crown; style-tips truncate; alternate-leaved, mostly strong-scented. . . Tribe VII. ANTHEMIDEAE (p. 1514).

*f.*Pappus of soft very slender hairs; involucre herbaceous, little or not at all imbricated; receptacle naked; cauline leaves mostly alternate. . . . . . . Tribe VIII. SENECIONEAE (p. 1524).

*c.*Flowers all with equally or subequally cleft corollas with elongate lobes (simulating deeply cleft ligules); involucre strongly imbricated with prickly or fringed firm phyllaries; anthers caudate, long-appendaged at tip; style-branches short or united, unappendaged; pappus mostly bristly; cauline leaves alternate. . . . . . Tribe IX. CYNAREAE (p. 1537).

*a.*Corollas of all flowers ligulate, the flowers perfect; herbs or half-shrubs with milky juice (latex); cauline leaves alternate. Subfam. II. LIGULIFLORAE, With a single tribe. . . . . Tribe X. CICHORIEAE (p. 1546).

Subfam. I. TUBULIFLÒRAE (see above)

Tribe I. VERNONÌEAE Cass. (see above)

Heads not glomerulate, several–many-flowered; phyllaries uniform, strongly imbricated; corolla regularly 5-cleft; pappus double, the inner of capillary bristles, the outer shorter and stouter; leaves in ours not rosulate. . . . . 1. *Vernonia.*

Heads in glomerules, 2–5-flowered; phyllaries 8, alternately flat and conduplicate; corolla slightly irregular, more deeply cleft on inner side; pappus a single row of chaffy bristles; large basal leaves often rosulate. . . . 2. *Elephantopus*.

1. VERNÒNIA Schreb. IRONWEED

Heads discoid, 15–many-flowered, in corymbose cymes; flowers perfect; involucre shorter than the flowers, of much imbricated phyllaries. Achenes cylindrical, ribbed; pappus double, the outer of minute scale-like bristles, the inner of copious capillary bristles. — Perennial herbs of N. and S. Am., Afr. and s. Asia, with leafy stems, alternate acuminate or very acute serrate leaves and mostly purple (rarely white) flowers. (Named for *William Vernon*, 16?–1711, an English botanist who traveled in North America.)* — Species freely hybridizing.

*a.*Phyllaries mostly long-acuminate or with a filiform appendage. . *b.*
*b.*Involucres 1.2–1.8 cm. high, 1.2–2 cm. in diameter; heads 55–90-flowered. 1. *V. crinita.*
*b.*Involucres 6–12 mm. high, 5–12 mm. in diameter; heads 30–50-flowered.
Pappus purplish or purple-tinged (exceptionally pale); involucres 7–12 mm. high. . . . . 2. *V. noveboracensis.*
Pappus creamy or stramineous; involucres 6–8 mm. high. . . . . 3. *V. glauca.*
*a.*Phyllaries obtuse to acute or abruptly short-acuminate. . . *c.*
*c.*Lower surfaces of leaves glabrous or only minutely scabrous-hirtellous or puberulent.
Leaves essentially glabrous and (when dry) puncticulate beneath; achenes glabrous or essentially so. . . . . 4. *V. fasciculata.*
Leaves minutely puberulent to hirtellous and scarcely puncticulate beneath; ribs of achenes hispidulous. . . . . . . 5. *V. altissima.*
*c.*Lower surfaces of leaves tomentulose.
Heads 34–55-flowered; phyllaries blunt or merely mucronate-tipped, the midribs not prominent; achenes 4–4.5 mm. long. . . . . 6. *V. missurica.*
Heads 18–34-flowered; phyllaries with attenuate to subulate tips, the midribs mostly prominent; achenes 3 mm. long. . . . . 7. *V. Baldwini.*

1. V. crinita Raf. (long-haired). — Nearly glabrous, glaucous, 1–3 m. high; leaves linear or linear-lanceolate, 0.5–2 cm. broad, entire or finely serrate-denticulate; cyme open, with very large 55–90-*flowered heads; involucres* 1.2–1.8 *cm. high,* 1.2–2 *cm. in diameter, of loosely imbricated phyllaries with flexuous linear-filiform prolonged appendages;* pappus brownish or purplish; achenes 5–6 mm. long. — Rich low grounds and open woods, Mo., Kans., Ark. and Okla.; adv. in O. Late July–Sept. FIG. 1540. — Hybridizes with nos. 4–7.

1540. V. crinita.

2. V. noveboracénsis (L.) Michx. (of New York). — Stems 1–2 m. high, glabrous or sparsely pubescent; leaves long-lanceolate to lance-oblong, 1.5–4.7 cm. broad, gradually narrowed to base, the lower face glabrous or sparsely puberulent; cyme open, 1–3.5 dm. broad; heads 30–50-flowered; *involucres 7–12 mm. high, about as broad,* their *phyllaries purplish, or green and with corollas whitish* and pappus pale in forma **albiflòra** Britt. (white-flowered), ovate or lance-ovate, *with prolonged loosely ascending to recurved-spreading filiform tips;* pappus purplish or purple-tinged; achenes 4–4.5 mm. long. — Low grounds, margins of streams, etc., Ga. to Miss., n. to Mass., se. N.Y., W.Va. and s. O. Aug.–Oct. FIG. 1541.

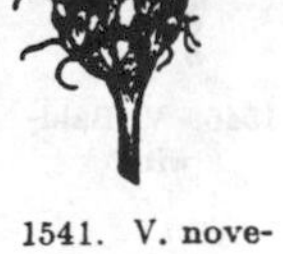

1541. V. noveboracensis.

Var. **tomentòsa** (Walt.) Britt. (densely soft-hairy). — Leaves tomentulose on the lower faces. — Swamps of the Coastal Plain, S.C. to s. N.J.

3. V. glaùca (L.) Willd. (blue-green). — Stem slender, glaucous, glabrous or essentially so, 0.7–1.6 m. high; *leaves ovate-lanceolate, acuminate to both ends,* sharply serrate, the larger 3–10 cm. broad; cyme open; *involucres* fuscous, 6–8 *mm. high, their* closely imbricated lanceolate to deltoid *phyllaries* without or *with filiform tips* 1–4 *mm. long,* or all phyllaries prolonged into awns 4–6 mm. long in forma **longiaristàta** Fern. (long-awned); *pappus creamy-white to stramineous;* achenes 3–3.5 mm. long, with smooth ribs. — Rich woods, N.J. and Pa. to Ga. and Ala. July–Oct. FIG. 1542.

1542. V. glauca.

* Treatment prepared with assistance of BERNICE G. SCHUBERT.

4. V. fasciculàta Michx. (bunched). — Stem 0.7–2 m. high, glabrous; *leaves* very numerous, *ascending,* linear to oblong-lanceolate, acuminate, spinulose-denticulate, *glabrous and when dry punctate beneath,* the larger 1–3 cm. broad; *cyme dense, with fastigiate branches,* turbinate or flattened, 0.5–2 dm. broad; *heads slenderly campanulate,* mostly sessile or short-stalked, 20–30-flowered; *involucres* 5–8 mm. high, *their closely imbricated round-tipped to subacute phyllaries with exposed tips* 1.6–2 *mm. broad;* pappus purplish; *achenes* about 3 mm. long, *glabrous or essentially so.* — Rich low ground and prairies, O. to N.D., s. to Mo., Okla. and e. Tex.; casual northeastw. July–Sept. Fig. 1543. — Hybridizes with nos. 1 and 7. Passing into

1543. V. fasciculata.

Var. **corymbòsa** (Schwein.) Schub. (corymbed). — Lower, 0.4–1 m. high; leaves lanceolate to lance-ovate; involucres 8–9 mm. high, their exposed phyllary-tips 2–3 mm. broad. (*V. corymbosa* Schwein.) — Prairies and bottoms, Minn. and Man. to centr. Kans. July–Sept.

5. V. altíssima Nutt. (very tall). — Usually tall (1–2 m. high); *leaves* loosely ascending to spreading, lanceolate, lance-oblong or narrowly ovate, sharply toothed, mostly 3–8 cm. broad; *the lower surfaces minutely puberulent, rarely punctate,* with tomentulose midrib; cyme 1–5 dm. broad, open (rarely dense); heads campanulate, 13–30-flowered; involucres 4–6 mm. high, their appressed ovate phyllaries obtuse, acute or short-cuspidate; *achenes with hispidulous ribs.* — Damp rich soil, Ga. to La., n. to N.Y., O., Ind., Ill. and Mo. Aug.–Oct. Fig. 1544. — Hybridizes with nos. 1 and 6.

1544. V. altissima.

Var. **taeniótricha** Blake (ribbon-haired). — Stems, peduncles and veins of lower leaf-surfaces hirtellous with multicellular hairs. — W. Pa. to s. Mich., Ill. and Mo., s. to Miss.

6. V. missùrica Raf. (of Missouri). — Similar in habit to no. 5; *leaves tomentulose beneath; heads 34–55-flowered;* involucres campanulate or ovoid, with *obscurely ribbed blunt or merely mucronate-tipped phyllaries;* pappus tawny to purple; achenes 4–4.5 mm. long. (Incl. *V. illinoensis* Gleason) — Rich low ground and prairies, s. Ont. and O. to Ia., s. to Ala., Miss., Ark., Okla. and Tex. July–Sept. Fig. 1545. — Hybridizes with nos. 1, 5 and 7.

1545. V. missurica.

7. V. Báldwini Torr. (for its discoverer, William Baldwin, 1779–1819). — Habit of nos. 5 and 6; leaves tomentulose beneath; *heads 18–34-flowered;* involucre thick-cylindric to campanulate, its *attenuate to subulate-tipped phyllaries squarrose or recurving at tip, their midribs prominent; achenes about 3 mm. long.* — Prairies and open ground, Ill. to Ark. and Okla. July–Sept. Fig. 1546. — Hybridizes with no. 1.

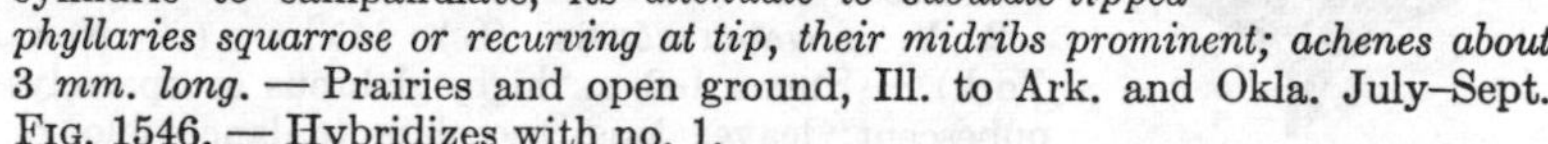

1546. V. Baldwini.

Var. **intèrior** (Small) Schub. (inland). — Phyllaries mostly appressed. (*V. interior* Small) — Ill. to Minn. and Neb., s. to Ark. and Tex. — Hybridizes with nos. 1, 4 and 6.

2. ELEPHÁNTOPUS L. Elephant's-foot

Heads discoid, 2–5-flowered, several together clustered into a compound pedunculate leafy-bracted glomerule; flowers perfect. Inner involucre narrow, flattened, of 8 oblong dry phyllaries. Achenes 10-ribbed; pappus of stout bristles, chaffy-dilated at the base. — Perennials of trop. and warm reg., with alternate or rosulate leaves and purplish flowers. (Name composed of the Greek *elephas, elephant,* and *pous, foot,* translation of aboriginal name.)

Stems leafy, without definite basal rosette; the 3–6 scattered lower cauline leaves essentially of uniform size, rhombic-oval, tapering about equally from the middle to base and apex. . . . . 1. *E. carolinianus.*

Stems scapose or with the few leaves much smaller than those of the definite basal rosette, the latter obovate to oblanceolate or spatulate.

Leaves velvety to touch, the lower surfaces soft-pubescent; foliaceous outer phyllaries broadly ovate or rounded, villous. . . . . 2. *E. tomentosus.*

Leaves not velvety, the lower surface strigose-hirsute along the principal veins; foliaceous outer phyllaries ovate-acuminate to lanceolate, strigose-hispid. 3. *E. nudatus.*

1. E. caroliniànus Willd. (of Carolina). — Stem 2–9 dm. high, simple or forking, only the lower internodes with divergent villi, the upper internodes and branches merely with somewhat scattered appressed strigae, with scattered alternate large rhombic-ovate essentially uniform crenate thin leaves, loosely corymbose-branching above; outer foliaceous phyllaries unequal, the larger ovate; chaffy base of pappus lance-subulate, much longer than the diameter of the achene and gradually attenuate to the awn. — Open woods and thickets, Fla. to Tex., n. to s. N.J., Pa., W.Va., O., Ind., Ill., Mo. and Kans. Aug.–Oct. — Forma **vestìtus** Fern. (clothed), is cinereous, with short and close pubescence to summit, the lower and middle internodes heavily tomentulose, the upper with close tomentum mixed with spreading hairs; in wooded swamps, se. Va.

2. E. tomentòsus L. (with dense pubescence), Devil's-grandmother, Tobaccoweed. — Principal leaves rosulate, velvety on both surfaces, narrowly to broadly obovate, tapering at base, once-and-a-half to thrice as long as broad, the bracteal leaves narrow and small, or rosette-leaves round-obovate or -oval and rounded at base with the lower cauline large and similar in forma **rotundàtus** Fern. (rounded); stems scapose or subscapose, 1.5–5 dm. high, pubescent; foliaceous phyllaries broadly ovate or rounded, villous; chaffy base of pappus triangular-subulate, about as long as breadth of achene, tapering to the bristle-tip. — Woodlands, Fla. to Tex., n. to Md., W.Va. and Ky. Late July–Oct.

3. E. nudàtus Gray (naked). — Similar to no. 2; rosette-leaves oblanceolate or spatulate to narrowly obovate, glabrous or strigose-pubescent, the lower surface strigose-hirsute along the nerves; foliaceous phyllaries ovate-acuminate to broadly lanceolate, strigose-hispid; scales of pappus broadly deltoid, abruptly terminated by the awn. — Woods and sandy openings, Fla. to La., n. to Del. and Ark. Late Aug.–Oct.

Tribe II. Eupatorìeae Cass. (see p. 1358).

Pappus a row of hard oval obtuse scales; slender aquatic or subaquatic plant with whorled leaves. 3. *Sclerolepis*.
Pappus of slender bristles; plant not aquatic.
  Achenes 5-angled, without intertwining ribs.
    Involucre of more than 4 phyllaries; flowers 3–40; plant erect, not twining. 4. *Eupatorium*.
    Involucre of 4 phyllaries; flowers 4; plant slender and twining. . . . . 5. *Mikania*.
  Achenes 10-ribbed.
    Phyllaries dry, not herbaceous nor with colored margins, obviously striate-nerved; flowers greenish, yellowish or whitish.
      Phyllaries in several series; pappus merely scabrous. . . . . . 6. *Brickellia*.
      Phyllaries few; pappus strongly plumose. . . . . . 7. *Kuhnia*.
    Phyllaries herbaceous, often with roseate or colored margins, only faintly if at all nerved; flowers (except in albinos) roseate or purplish; involucre strongly imbricated, few–many-flowered.
      Inflorescences simple or panicled spikes or racemes; heads 4–many-flowered; receptacle naked; corolla-lobes narrowly lanceolate or linear. . 8. *Liatris*.
      Inflorescence an open corymb or cyme; heads many-flowered; receptacle with deciduous chaff; corolla-lobes ovate or short-lanceolate. . . . 9. *Carphephorus*.

3. SCLERÓLEPIS Cass.

Head discoid, many-flowered; flowers perfect. Phyllaries linear, equal, in 1 or 2 rows. Receptacle naked. Corolla 5-toothed. Achenes 5-angled; pappus a single row of 5 almost horny oval and obtuse scales. — Smooth perennial, with simple or slightly forking stems, rooting at the base, linear entire leaves in whorls of 4–6, and a terminal head of flesh-colored flowers in summer. (Name composed of the Greek *scleros, hard,* and *lepis, a scale,* from the pappus.)

1. S. uniflòra (Walt.) BSP. (one-flowered). — Extensively creeping; flowering stems erect, slender, suggesting those of *Hippuris vulgaris,* 1–3 (in water –6) dm. high; leaves firm, callus-tipped, mostly 1–2 cm. long (in water flaccid and elongated); head 0.8–1.5 cm. broad. — Sandy or peaty swamps, shores and shallow water, Ala. and Fla. to N.C.; e. Md. to s. N.J.; Wallum P., R.I. and Mass.; L. Massasecum, Merrimack Co., N.H. July–Nov.

4. EUPATÒRIUM L. Thoroughwort

Heads discoid, 3–many-flowered; flowers perfect. Involucre cylindrical or campanulate, of more than 4 phyllaries. Receptacle flat or conical, naked. Corolla 5-toothed. Achenes 5-angled; pappus a single row of slender capillary barely roughish bristles. — Perennial (sometimes annual) herbs or shrubs, chiefly of trop. and temp. Am. (few in Eurasia and Afr.), often sprinkled with bitter resinous dots, with generally corymbose heads of white, bluish or purple flowers appearing near the close of summer. (Dedicated to *Mithridates Eupator*, 132–63 B.C., who is said to have used a species of the genus in medicine.)*

*a.*Receptacle flat; flowers white to lilac or purple; base of stem without repent superficial stolons. § EUPATORIUM proper. . . *b.*

*b.*Involucre cylindric; the exterior closely imbricated phyllaries successively shorter in several series, purple to creamy-white; flowers purple to pale pink (except in albino forms); leaves whorled or opposite. . . *c.*

*c.*Lower leaf-surfaces and summit of stem dotted with viscid lustrous atoms; one of the lower pairs of lateral veins more prominent and more prolonged than the others, the leaves thus subpalmately 3-ribbed; leaf-blade ovate to ovate-lanceolate, abruptly contracted to petiole; stem speckled with purple dots, viscid above; inflorescence with rounded summit; involucre purple; florets 5–12, purple. . . . . . . 1. *E. dubium.*

*c.*Lower leaf-surfaces and stems usually without viscid atoms; leaves pinnately veined (if somewhat palmately 3-veined in no. 2, the inflorescence flat-topped and the florets 8–20). . . *d.*

*d.*Stem deep purple or purple-spotted, not glaucous; inflorescence or its divisions flattish-topped; florets 8–20. . . . . 2. *E. maculatum.*

*d.*Stem glaucous, rarely much dotted with purple; inflorescence with rounded or strongly convex summit; florets 3–8.

Stem hollow, purplish, strongly glaucous; leaves scabrous-puberulent to glabrate beneath, blunt-toothed; involucres and flowers purple; corollas 3.5–4.8 mm. long, scarcely exserted; plant not vanilla-scented. . . . . 3. *E. fistulosum.*

Stem solid, green, with dark purple nodes, faintly glaucous; leaves pilose, villous or glabrescent beneath, often acutely toothed; involucres creamy-white to pale purple; corollas 5.5–7.5 mm. long, strongly exserted, pale; bruised fresh plant strongly vanilla-scented. . . . . 4. *E. purpureum.*

*b.*Involucre relatively lax, the green or greenish phyllaries not tightly imbricated, of 2 or 3 lengths to subequal; flowers white to lilac; leaves only exceptionally whorled. . . *e.*

*e.*Leaves dissected into many capillary divisions and segments; annual with elongate paniculate-thyrsoid inflorescence. . . . . 5. *E. capillifolium.*

*e.*Leaves merely entire, toothed or lobulate; inflorescence more or less corymbose; perennials. . . *f.*

*f.*Leaves sessile or very short-petioled; phyllaries mostly in 2 or 3 (sometimes more) lengths. . . *g.*

*g.*Heads (3–)5–7-flowered; involucre subcylindric. . . *h.*

*h.*Leaves tapering to base, linear to lanceolate, oblanceolate or oblong, the principal ones 0.5–3.5(–6) cm. broad. . . *i.*

*i.*Phyllaries with conspicuously white-scarious acute tips, the inner equaling the flowers.

Phyllaries glabrous (except sometimes at very base); leaves flat, their surfaces not cinereous-puberulent. . . . . 6. *E. album.*

Phyllaries, leaves, etc., cinereous-puberulent, the leaves often longitudinally folded. . . . . 7. *E. leucolepis.*

*i.*Phyllaries obscurely, if at all, white-scarious, mostly obtuse (acutish in no. 10), soon overtopped by the flowers. . *j.*

*j.*Principal leaves prominently 3- or 5-nerved beneath with straight parallel ribs; 20–30 or more pairs of long-acuminate blades 1–3 cm. broad below the inflorescence; inner phyllaries broadly oblong to elliptic, broadly rounded at summit; chiefly inland. . . . . 8. *E. altissimum.*

*j.*Principal leaves not at all or only faintly triple-nerved, mostly at fewer nodes below the inflorescence, the rarely long-acuminate blades 0.5–3 cm. broad; inner phyllaries narrowly oblong to linear or lanceolate; coastwise and eastern species. . . *k.*

*k.*Involucres 3–4 mm. long, of essentially 2 series of oblong broadly round-tipped phyllaries, the inner series much

---

* In the treatment of the first four species the diagnostic characters are largely copied from the treatment by WIEGAND in Rhodora, xxii. 62–70 (1920). In species 6–26 exceptional plants may have leaves in 3's.

longer than the outer; leaves cuneate-oblanceolate to oblong, regularly toothed toward tip, the principal ones 0.5–3 cm. broad; fresh flowers sordid or brownish-white. 9. *E. cuneifolium.*

k. Involucre 4–7 mm. long, of 2 or 3 (or more in no. 11) more imbricated series of phyllaries; leaves linear to lanceolate or oblanceolate, entire or, if toothed, with entire tip, the principal ones 0.05–2.5 cm. broad; fresh flowers clearer white. . . *l.*

l. Leaves soon recurving or ascending and twisted at base, entire or bluntly few-toothed, the principal ones 2–5.5 cm. long, blunt.

Leaves linear-oblanceolate, ascending or erect, entire, the slender base spirally twisted; involucre 5.5–7 mm. long, the mostly 2-ranked phyllaries lance-linear, pilose, acutish; base without tubers. . . . 10. *E. linearifolium.*

Leaves lanceolate, arched-recurving, not spirally twisted, the middle of each margin with 1–5 low teeth; involucre 4–6 mm. long, the longer linear blunt phyllaries glandular-pruinose, the short calyculate outer ones extending as several bracteoles down the pedicels; base tuberous. . . . 11. *E. recurvans.*

l. Principal leaves divergent or ascending, the larger ones 3–11 cm. long, entire or with few–many sharp teeth.

Axils of leaves bearing crowded many-leaved sessile or subsessile fascicles; larger primary leaves linear and entire to lanceolate and serrate to lacerate, 0.05–1.7 cm. broad; involucre 4.5–5.5 mm. long. . . . . 12. *E. hyssopifolium.*

Axils naked or with solitary elongating axillary branches; primary leaves lance-acuminate, regularly serrate, 1.3–2.5 cm. broad; involucre 6–7 mm. long. 13. *E. saltuense.*

h. Leaves broadly subtruncate, rounded or cordate at base, the principal blades broadly oblong to lanceolate, ovate or suborbicular, 2–7 cm. broad, regularly toothed from near base to apex. . . *m.*

m. Involucres 10–12 mm. long; phyllaries nearly or quite glabrous, scarious above the middle, attenuate to sharp tips, longer than the flowers. . . . . 6. *E. album.*

m. Involucres 4.5–6.5 mm. long; phyllaries pubescent, firm, except for narrow or short scarious margin and tip, acute or obtuse, shorter than the flowers. . . *n.*

n. Leaves oblong-lanceolate to ovate or suborbicular, hirsute at least on veins beneath, blunt to subacute; the lower lateral veins more prominent than the others, ascending from near base of slender midrib at angle of 45° or less; inner phyllaries acute, acutish or abruptly pointed. . . *o.*

o. Median cauline leaves oblong to oblong-lanceolate or -ovate, blunt to subacute, submembranaceous, each margin with 3–12 coarse often unequal rounded to subacute teeth or sublacerate at base; upper leaves frequently alternate, narrowly lanceolate to lance-linear, entire or nearly so; branches of open corymb often alternate; involucre 4.5–5.5 mm. long. . . . . 14. *E. pilosum.*

o. Median cauline leaves elliptic-oblong or ovate to suborbicular, rounded to short often subacute apex, each margin with 8–35 uniform or nearly uniform subacute to merely blunt teeth; upper leaves mostly opposite, ovate, regularly toothed; branches of corymb mostly opposite; involucre 5.5–6.5 mm. long. . . *p.*

p. Principal cauline leaves gradually rounded to cordate at the toothed base, each margin with 12–35 teeth; one or more pairs of ascending prolonged lateral nerves united with base of midrib and leaving it 2–10 mm. or more above the base.

Median cauline leaves oblong to ovate, gradually rounded to base, one-half to seven-tenths as broad as long, each margin 12–25-toothed; inner phyllaries merely tapering to attenuate tips. . . . . 15. *E. pubescens.*

Median cauline leaves cordate, four-fifths to seven-eighths as broad as long, each margin with 20–35 teeth; inner phyllaries prolonged into slender curving appendages. . . . . 16. *E. cordigerum.*

*p.*Principal cauline leaves with subtruncate to broadly cuneate straight entire bases, each margin 8–20-toothed; ascending prolonged lateral veins arising from base of midrib. . . . . 17. *E. rotundifolium.*

*n.*Leaves lanceolate to ovate, acuminate to long-attenuate to sharp tip, rounded to sessile base, glabrous or only minutely pilose beneath, pinnately veined, the fine veins inconspicuous, the rounded quill-like midrib prominent; phyllaries blunt. . . . . 18. *E. sessilifolium.*

*g.*Heads 10–40-flowered. . . *q.*

*q.*Leaves lanceolate, with long acuminate tips, distinct or perfoliate, when perfoliate often with dilated bases; phyllaries definitely of 2 or 3 series; summits of flower-buds and backs of corolla-lobes glabrous.

Leaf-bases united (perfoliate) or, if distinct, the leaves rugose-veiny, with the midrib beneath villous, the stem villous; phyllaries narrowly lanceolate, the inner ones acutish; wide-ranging species. . . . . 19. *E. perfoliatum.*

Leaves narrowed to distinct clasping bases, with inconspicuous lateral veins, the lower surfaces and the upper internodes of the stem puberulent or very minutely pilose; phyllaries oblong, rounded at tip; local plant of N. J., Del. and central Ky. 20. *E. resinosum.*

*q.*Leaves ovate, acute to blunt, not long-tapering, essentially glabrous; phyllaries subequal; summits of buds and backs of corolla-lobes hairy. . . . . 24. *E. aromaticum.*

*f.*Leaves (at least the larger ones) definitely petioled, the median ones lance- to broadly cordate-ovate; heads 8–40-flowered. . . *r.*

*r.*Involucre of 2 or 3 definitely imbricated series of phyllaries; tips of flower-buds and backs of corolla-lobes glabrous; tall stiffly erect plant, with cinereous-scabrous lanceolate to narrowly ovate coarsely toothed long-pointed leaves subtending suppressed axillary fascicles. . . . . 21. *E. serotinum.*

*r.*Involucre of essentially equal phyllaries, exceptionally with 1–few basal ones; tips of flower-buds and backs of corolla-lobes hairy; leaves very broadly to narrowly ovate, ovate-lanceolate or triangular, scarcely cinereous, without or sometimes with axillary fascicles. . . *s.*

*s.*Principal leaves ovate to rounded, if lance-ovate with cuneate to subtruncate bases; corymb terminal; flowers white. . . *t.*

*t.*Petioles elongate, those of larger primary leaves 2–8 cm. long; the larger blades 3–12 cm. broad, acuminate or short-tipped, with 9–25 teeth on each margin.

Leaves ovate, longer than broad; involucres 4–6 mm. long; phyllaries merely acute or blunt; wide-ranging species. . 22. *E. rugosum.*

Leaves broadly cordate-deltoid, as broad as long; involucre 4 mm. long; phyllaries mucronate or caudate-tipped; local plant of Ky. . . . . 23. *E. Luciae-Brauniae.*

*t.*Petioles short, those of larger leaves 0–1.5 (rarely –2) cm. long; the larger blades 2–5.5 cm. broad, with 7–12 well-developed teeth on each margin. . . . . 24. *E. aromaticum.*

*s.*Principal leaves narrowly deltoid-ovate, chiefly with cordate or subcordate bases, the larger long-petioled ones 2–5 cm. broad; inflorescence open and diffuse, with divergent slender few-headed branches borne from one-third to two-thirds of the axils; flowers lilac or pink. . . . . 25. *E. incarnatum.*

*a.*Receptacle conical, § Conoclinium; ours with deltoid-ovate leaves, slender superficial stolons, campanulate involucre of linear subequal phyllaries, and bluish or purple flowers. 26. *E. coelestinum.*

§ Eupatòrium (L.) Gray

1. E. dùbium Willd. (doubtful), Joe-Pye-weed. — Coarse perennial; *stems* 0.5–1.5 m. or more high, *finely dotted with purple,* often *purple-suffused, not glaucous,* solid or hollow, scabrous-puberulent and *viscid toward summit; leaves* in 3's or 4's (sometimes 2's or 5's), broadly to narrowly ovate, subacuminate, *abruptly narrowed to petiole,* coarsely serrate or serrate-dentate, usually scabrous above, *bearing beneath shining viscid atoms; one pair of lower veins coarser and more prolonged than the others, the leaf thus appearing palmately 3-veined from near base; corymb* dense, *convex at summit,* 0.5–3 dm. or more broad, when fully developed elongate and hemispherical to subcylindric; *heads 5–12-flowered,* deep purple, rarely whitish in forma **elùtum** Fern. (washed out); *involucre cylindric, the oblong phyllaries tightly appressed, in several lengths, the outer ones blunt, the inner acutish;* corollas 4.5–5.5 mm. long, slightly exserted; achenes 3.5–4.5 mm. long. (*E. punctatum* Willd., not Lam.; *E. purpureum* in part of ed. 7, not L.; *E. verticillatum* Lam. in part and sensu Wieg.) — Swamps, damp thickets and shores, in acid soils, N.S. and sw. Me. to e. N.Y., s. along Coastal Plain, outer Piedmont and mts. to S.C. Mid-July (northw.)–Oct. (southw.).

2. E. maculàtum L. (mottled), Joe-Pye-weed. — Similar to no 1; *stem* often taller, *usually solid, viscid-puberulent above* (especially about the inflorescence); *leaves* in (3's) 4's or 5's (rarely 6's), *oblong-lanceolate to narrowly ovate-lanceolate, tapering to base and apex,* petioled to subsessile, sharply irregularly serrate to crenate, *pinnately veined,* glabrous but slightly scabrous above, minutely crisp-puberulent to glabrate and scabrous beneath, rarely atomiferous; *corymb or its divisions flattish-topped,* standing above the reduced upper leaves (whorled leaves of node next below corymb not as long as height of corymb, 3–18 cm. long by 0.5–5.5 cm. broad); *heads thick-cylindric,* purple, rarely white in forma **Fáxoni** Fern. (for its discoverer, Charles Edward Faxon, 1846–1918), 8–20-*flowered; phyllaries broadly oblong,* obtuse; corollas about 5 mm. long, their tips scarcely or barely exserted; achenes 3.4–4.2 mm. long. (*E. purpureum,* var. Darl.) — Damp thickets, meadows, shores, etc., usually in rich or calcareous soils, Nfld. and Côte Nord, Que., to B.C., s. to N.S., n. and w. N.E., Pa., mts. to N.C., Mich., n. Ind., Ill., Ia., Neb., N.M. and Wash. Mid-July–early Sept. Passing into

Var. **foliòsum** (Fern.) Wieg. (leafy). — Leaves very large, with tendency to 3-ribbed base, the upper ones (below the corymb) mostly longer than height of corymb (0.8–3 dm. long by 3–8 cm. broad). (*E. purpureum,* var. Fern.) — Nfld. and Côte Nord, Que., to n. Ont., s. to n. N.S., n. N.B., n. and w. N.E., n. N.Y. and n. Mich.

3. E. fistulòsum Barratt (tubular), Joe-Pye-weed, Trumpet-weed. — Similar to nos. 1 and 2; *stems strongly glaucous* (when fresh), suffused with purple, *unspotted or rarely spotted,* glabrous, *hollow; leaves* in 4's–7's, narrowly to broadly *lanceolate,* attenuate to base and apex, *regularly crenate or crenate-serrate,* sparingly fine-pubescent to glabrous beneath, pinnately veined; *corymb round-topped or doming,* when fully developed hemispherical to thick-subcylindric, up to 1–5 dm. high and 1–3 dm. broad, the lower branches divergent; *heads purple or lilac-pink, 5–8-flowered; corolla* 3.5–4.8 *mm. long, barely or not at all exserted;* achenes 3.2–4.5 mm. long. (*E. purpureum* in part of ed. 7, not L.) — Damp thickets, meadows, etc., Fla. to e. Tex., n. to sw. Me., sw. Que., N.Y., O., Ind., Ill., Ia. and Okla. Mid-July–Sept.

4. E. purpùreum L. (purple), Sweet or Green-stemmed Joe-Pye-weed. — *Fresh and bruised or drying plant sweetly vanilla-scented; stem* rather slender, *green or suffused with purple, dark purple at nodes, usually unspotted, slightly glaucous* when fresh, nearly or quite glabrous, *solid; leaves* in 2's–5's, thinnish, *lanceolate to ovate,* acuminate, *gradually tapering to abruptly cuneate at base,* sharply serrate (rarely crenate-dentate), glabrous above, *pilose to glabrous beneath; paniculate corymb rounded or doming at summit,* rather open, its slender branches minutely soft-pilose; *heads slenderly cylindrical, creamy-white to pale pink or pale lilac, 3–7-flowered;* phyllaries narrowly oblong, the inner acutish; *corollas* (5–)5.5–7.5 *mm. long, their summits prominently exserted;* achenes 3–5 mm. long. (Var. *amoenum* (Pursh) Gray; *E. falcatum* Michx.) — Rich, dry to moist, chiefly calcareous, woods, s. N.H. to Minn. and Neb., s. to n. Fla., Tenn., Ark. and Okla. Mid-July–mid-Sept.

5. E. capillifòlium (Lam.) Small (hair-leaved), Dog-fennel. — Slender upright *annual* up to 3 m. high, the *stem scabrous-hirsute; leaves crowded,* mostly alternate, once or twice *pinnately divided into filiform segments; inflorescence a tall flexuous much-branched wand-like or thyrsiform leafy panicle* with innumerable green to bronze small 3–5-flowered heads 2–3 mm. long. — Borders of woods, clearings, fields and roadsides, aggressive and weedy, Fla. to Tex., n. to

N.J., n. Va., and Tenn.; casual to Mass. Mid-Sept.–Nov. — Superficially suggesting *Artemisia.*

6. E. álbum L. (white). — Stem arched-ascending or erect, villous-hirsute to scabrous-pilose, especially below, 2–9 dm. high; *leaves spatulate to oblong, narrowly obovate or oblong-lanceolate, veiny,* often *coarsely toothed, sessile or nearly so;* corymb with stiffly spreading-ascending branches with rather crowded secondary corymbs; *involucres* cylindric-turbinate, 10–12 *mm. long; phyllaries lance-acuminate, glabrous or nearly so, thin and whitish-scarious above the middle, mostly exceeding the* white *flowers.* — Four vars.:

Principal leaves spatulate, oblanceolate or narrowly obovate, obtuse, narrowed to base, they and the stem villous or strongly pilose.
- Involucre glandless or essentially so. . . . . *E. album* (typical).
- Involucre copiously dark-glandular. . . . . Var. *glandulosum.*

Principal leaves oblong, oblong-lanceolate or narrowly oblong-ovate, acute, the broad bases more rounded, their pubescence short and sparse or wanting.
- Leaves firm, the larger ones with 10–20 prominent coarse teeth on each margin, pinnately veined. . . . . Var. *monardifolium.*
- Leaves submembranaceous to firm, the larger with 3–10 low teeth on each margin or entire, triple-nerved from the base. . . . . Var. *subvenosum.*

E. álbum (typical). — Dry or sandy woods, thickets and clearings, n. Fla. to La., n. to mts. of S.C. and on or near Coastal Plain to N.J. Mid-July–Oct.

Var. **glandulòsum** (Michx.) DC. (glandular). — N. Fla. to La., n. to Md., w. Va., Ky. and Ark.

Var. **monardifòlium** Fern. (with leaves of *Monarda*). — So. N.J. and Del. to n. Md., s. to mts. of N.C.

Var. **subvenòsum** Gray (somewhat veiny). — Pine-barrens, sandy shores and openings, L.I., N.J. and Del.

7. E. leucólepis (DC.) T. & G. (white-scaled). — Stiffly erect, the *slender minutely gray-puberulent stem* 0.4–1 m. high; *leaves* thick, sessile, *often plicate, linear or lanceolate, obtuse, cinereous-puberulent, the larger lower and median ones* 3–8 cm. long and 3–10 *mm. broad,* 1*-nerved or with 2 faint lateral nerves arising from base;* corymb hoary-puberulent; involucre as long as or exceeding white flowers, the *linear phyllaries tomentulose.* — Damp sands and peats of Coastal Plain, Fla. to La., n. to L.I. Aug.–Oct.

Var. **nòvae-ángliae** Fern. (of New England). — Upper internodes pilose; leaves flat, acutish or subacuminate, more sharply toothed, merely pilose or hirtellous beneath, the lateral veins elevated beneath and joining the midrib above its base, the larger blades 0.8–2 cm. broad. — Sandy and peaty pond-shores, Plymouth Co., Mass., and Washington Co., R.I., local. Aug., Sept.

8. E. altíssimum L. (tallest). — Stem erect, 0.6–2 *m. high, hoary-downy, very leafy; the* 20–30 *or more pairs of* ascending sessile or barely petioled *firm lanceolate to elliptic-oblong acuminate leaves tapering to base,* entire or sharply serrate above the middle, 1–3 *cm. broad, with 3 or 5 long parallel veins very prominent beneath;* corymb dense; involucres cylindric-campanulate, 5 or 6 mm. long; *phyllaries downy, oblong to elliptic, the longer inner ones with broadly rounded* narrowly scarious-margined *summits* exceeded by the white flowers. — Calcareous shores, bluffs, open woods, clearings, prairies, etc., Pa. to Minn. and Neb., s. to N.C., Ala., Mo. and Tex. Aug.–early Oct.

9. E. cuneifòlium Willd. (with wedge-shaped leaves). — *Stem* stiffly ascending, *closely almost velvety-pubescent above,* 4.5–10 dm. high; *leaves* opposite, or alternate above, subtending short or finally elongating and floriferous branches, *narrowly cuneate-obovate, subspatulate or oblanceolate* (or uppermost smaller and narrower), *thickish, very short-petioled; larger blades* 3–4 *cm. long and* 0.5–1.5 *cm. broad, blunt or rounded at apex, crenate-serrate above middle, the prolonged lateral nerves often evident beneath;* corymb round-topped or doming, made up of several to many secondary small corymbs of few approximate heads; *involucres* 3–4 *mm. long, of essentially 2 series of* soft-pubescent *oblong* narrowly scarious-margined *round-tipped phyllaries, the outer phyllaries very short; fresh flowers sordid- or brownish-white,* overtopping involucre. — Dry to moist siliceous or argillaceous pinelands, open woods, clearings, etc., Fla. and Ala., n., locally, to se. Va. Sept. Passing with us into the commoner

Var. **semiserràtum** (DC.) Fern. & Grisc. (half-serrate). — Often larger; *leaves* thinner, *oblong to lanceolate or oblanceolate,* less evidently triple-nerved, more sharply toothed, *acute or subacute, the principal ones* 4–8 *cm. long and* 1–3 *cm. broad.* (*E. semiserratum* DC.) — Dry to wet open woods, shores, prairies, clearings, etc., Fla. to e. Tex., n. to Va., Tenn. and se. Mo. Aug., Sept.

10. E. linearifòlium Walt. (linear-leaved). — *Stem loosely bushy-branched from near base,* not tuberous, 0.4–1 m. high, *pilose* above; *leaves narrowly oblanceolate to linear-spatulate,* tapering to slender base, *obtuse, entire,* minutely pilose, *inclined to be thrown into ascending position by a spiral twist at base, the larger ones 2–5 cm. long and 3–8 mm. wide,* usually subtending few-leaved suppressed branchlets; corymb open, often diffuse; *involucre 5.5–7 mm. long, the mostly 2-ranked phyllaries pilose, the lance-linear inner ones acutish.* (*E. tortifolium* Chapm.) — Dry sands and pine-barrens, Fla. to e. Tex., n. to se. Va. Aug.–Oct.

11. E. recúrvans Small (recurving). — Stem erect, up to 1 m. high, with 1–3 *thick tuberous roots,* stiffly corymbose at summit, minutely pilose; *leaves lanceolate, acutish, arched-recurving, minutely rugulose, the larger ones 2–5.5 cm. long and 3–10 mm. broad, with 1–5 low teeth near middle of each margin,* subtending suppressed branches; corymb terminal; *involucre 4–6 mm. long,* subcylindric; the *longer phyllaries linear, blunt, glandular-pruinose, the short calyculate ones numerous and passing insensibly to the upper bracteoles of the pedicels.* — Wet pinelands or peaty or sandy swamps, thickets and shores, Fla. to Miss., n. to se. Va. Sept., Oct.

12. E. hyssopifòlium L. (hyssop-leaved). — Stem or stems erect, 0.3—1.4 m. high, *with rather dense fastigiate-branched flattish-topped corymb; leaves slenderly linear and entire to lanceolate and serrate to lacerate, the larger ones* 0.05–1.7 *cm. wide,* all but the lowest *subtending dense fascicles of sessile or subsessile suppressed branches; involucre* 4.5–5.5 *mm. long,* subcylindric, canescent-pilose, the inner narrowly oblong round-tipped or blunt phyllaries with paler scarious margins; flowers white, exceeding involucre; leaves frequently in 3's. — Highly variable; the following vars. confluent:

Principal leaves entire or with appressed teeth, 0.5–10 mm. broad; stems 0.3–1 m. high.
- Primary leaves narrowly lanceolate to oblanceolate, entire or with appressed teeth; the larger blades 5–10 cm. long and 5–10 mm. wide, much longer than the axillary fascicles; stems solitary or few. . . . . *E. hyssopifolium* (typical).
- Primary leaves narrowly linear to linear-oblanceolate, mostly quite entire or with revolute margins, 3.5–6 cm. long and 0.5–5 mm. wide, rarely twice length of axillary fascicles; stems often clustered. . . . . Var. *calcaratum.*

Principal leaves sharply serrate to almost lacerate, lance-linear to lanceolate; the larger ones 6–11 cm. long and 5–17 mm. broad, much longer than the relatively sparse fascicles; stems mostly solitary, 0.75–1.4 m. high. . . . . . Var. *laciniatum.*

E. hyssopifòlium (typical). — Dry to moist siliceous or argillaceous open woods, clearings, fields, etc., Fla. to e. Tex., n. to s. R.I., L.I., N.J. and Md. Aug.–Oct.

Var. **calcaràtum** Fern. & Schub. (bearing spurs; from the axillary clusters of leaves), *E. hyssopifolium* of ed. 7 and var. *linearifolium* sensu Fern., not *E. linearifolium* Walt., basonym. — Dry open woods and clearings, on or near Coastal Plain, Ga. to Tex., n. to se. Mass., R.I., Ct., se. N.Y., N.J., Pa. and Md. Aug.–Nov.

Var. **laciniàtum** Gray (slashed), *E. Torreyanum* Short — Sandy or peaty dry or damp open woods, shores, clearings, etc., Fla. to La., n. to Pa., Md. and se. O. Aug., Sept.

13. E. saltuénse Fern. (of bushy pastures). — Stem 0.9–1.4 m. high, minutely hirtellous above, with elongate internodes; *leaves* 12–14 pairs, spreading or slightly ascending, glabrous or minutely hirtellous beneath, *without well-developed axillary fascicles* or merely with solitary elongate branches; *primary* sessile or subsessile *blades lanceolate, attenuate-acuminate,* 6.5–11 cm. long by 1.3–2.5 *cm. broad, with* 12–15 *sharp regular teeth on each margin;* corymb terminal, open; *involucre 6–7 mm. long;* the 3-seriate phyllaries puberulent, obtuse; the inner ones oblong, with scarious margins, shorter than white flowers. — Rich woods, thickets and clearings, se. Va. Aug., Sept.

14. E. pilòsum Walt. (hairy). — Scabrous or harshly pubescent, 0.5–1.5 m. high; *median cauline leaves oblong to oblong-lanceolate or -ovate, submembranaceous, blunt to subacute,* rounded to base, *each margin with 3–12 coarse rounded or blunt teeth, sometimes incised toward base, the blades mostly* 3–10 *cm. long and* 1–4 *cm. broad,* the lower veins prolonged and often united with base of midrib and leaving it at an acute angle above its base; *upper leaves* (below corymb) *greatly reduced, often alternate, lanceolate or lance-linear, mostly entire; corymb open, the lower distant ascending branches frequently alternate; involucre* pubescent, 4.5–5.5 *mm. long,* about 5-flowered, *the inner phyllaries rather abruptly sharp-tipped;* flowers white. (*E. rotundifolium,* var. *lanceolatum* (Muhl.) Fern. & Grisc.; *E. verbenaefolium* Michx.; *E. lanceolatum* Muhl.) — Bogs, damp sands, low or dry thickets or wet or dry openings, in acid soils, Fla. to La., n., especially in coastwise reg., to s. N.E., se. N.Y., N.J., e. Pa., Md. and upland of Va. and Ky. Aug., Sept. — Hybridizes with no. 15.

15. E. pubéscens Muhl. (hairy). — Similar to no. 14; *stem villous,* 0.5–1.2 m. high; *median cauline leaves oblong to ovate, acutish, gradually rounded and toothed to* subsessile *base, one-half to seven-tenths as broad as long,* mostly 4–11 cm. long and 2–6 cm. broad, *each margin with* 12–25 *subequal subacute to bluntish* crenate-serrate *teeth,* the surfaces harsh-pilose, the *veins beneath mostly villous or long-hirsute,* the elongate lower veins united at base with the midrib; *upper leaves mostly opposite, ovate, regularly toothed; branches of corymb most often opposite; involucre* 5.5–6.5 *mm. long,* 5–8-flowered, *the inner phyllaries with acuminate tips.* (*E. rotundifolium,* var. *ovatum* (Bigel.) Torr.) — Moist to dry woods, thickets and clearings, Fla. to La., n. to sw. Me., Mass., se. N.Y., N.J., Pa., W.Va. and se. O. Mid-July–Sept. — Hybridizes with nos. 14 and 17.

16. E. cordígerum Fern. (bearing hearts). — Differing from no. 15 in coarser habit; the stout heavily white-pilose to tomentulose stems 1–1.3 m. high; *principal leaves strongly cordate-amplexicaul, broadly ovate, four-fifths to seven-eighths as broad as long,* 5–8.5 cm. long by 4–7 cm. broad, densely pilose and glandular-dotted beneath, *each margin with* 20–35 *teeth; inner phyllaries prolonged into long arching linear scarious appendages.* — River-marshes, swales and boggy thickets, se. Va. and e. N.C. Aug., Sept.

17. E. rotundifòlium L. (round-leaved). — Stems stiffly ascending, closely pilose above, rather slender, 0.3–1.2 m. high; principal *leaves* all or nearly all opposite, *broadly deltoid-ovate to suborbicular, with straight entire subtruncate to broadly cuneate bases, each margin with* 8–20 *teeth,* the blades subcoriaceous to very firm, hirtellous and usually scabrous above, pilose and *rugose-veiny beneath,* 1.3–7 cm. long by 1–6.5 cm. broad, *the* 1 *or more pairs of ascending and prolonged basal veins joining the midrib at its base;* corymb with opposite branches; involucre 5.5–6.5 mm. long, about 5-flowered, the linear-lanceolate inner *phyllaries with tapering slender-pointed tips.* — Dry to wet siliceous, argillaceous or peaty acid soils, open woods, clearings, savannas, etc., Fla. to Tex., n. to L.I., N.J., se. Pa., Md., D.C., Tenn. and Ark. July–Sept.— Hybridizes with no. 15.

18. E. sessilifòlium L. (sessile-leaved), Upland Boneset. — *Stem* slender, firm, *glabrous,* 0.5–2 m. high; *leaves* opposite (rarely in 3's), *lance-attenuate, with prolonged very slender acumination and rounded sessile base, thinnish, very smooth,* bright green, *with the quill-like midrib prominent beneath, the delicate venation pinnate; larger median blades* 0.9–1.8 *dm. long* by 2–4 *cm. wide, averaging five times as long as broad, regularly appressed-serrate;* corymb with dense clusters of heads; *involucres canescent-pilose,* about 5-flowered, 5–6 mm. long, *with blunt or round-tipped phyllaries.* — Rich, often calcareous, woods, thickets, and bluffs, Mass. to Ind., s. to Ga., Ala., and Mo. Late July–Sept.

Var. **Brittoniànum** Porter (in honor of Nathaniel Lord Britton, 1859–1934). — Glabrous except in inflorescence; *leaves firm* to subcoriaceous, finely toothed, *ovate-lanceolate to ovate,* long-acuminate; *the larger ones* 0.8–1.8 dm. long by 3–6 *cm. wide, averaging three times as long as broad.* — Commoner northw., s. N.H. to se. Minn., s. to s. N.E., n. N.J., Pa., Md., upland to N.C., Ky. and Mo.

Var. **Vàseyi** (Porter) Fern. & Grisc. (named in 1902 for its discoverer, George Richard Vasey). — *Upper half of stem puberulent or minutely pilose,* sometimes glutinous; *leaves dull green,* oblong-ovate, *merely acuminate or acute, scabrous beneath with minute puberulence,* the larger median blades 4–12 cm. long by 2.3–6 cm. broad. (*E. Vaseyi* Porter) — Dry siliceous or argillaceous open woods, thickets and clearings, se. Pa. and e. Md. to W.Va., s. to e. Va., N.C. and Tenn.

19. E. perfoliàtum L. (through the leaf), Thoroughwort (originally Throughwort), Boneset. — Coarse, *villous or hairy,* the stems 0.3–1.5 m. high; *leaves* opposite (rarely ternate), *lanceolate, tapering to a prolonged acuminate tip,* serrate or crenate, *rugose, the veins prominent beneath,* the *midrib beneath* and the lower surface *villous or pilose,* the often broadened *bases united around the stem,* or bases free and truncate or broadly rounded in forma **truncàtum** (Gray) Fassett (truncate), the larger blades 0.5–2 dm. long by 1.2–7 cm. broad at base; corymb flattish-topped, its branches usually with crowded heads; *involucre* 10–40-*flowered; phyllaries* 2- *or* 3-*ranked, lanceolate, tomentulose, with acutish scarious tips;* flowers whitish, or purple-tinged in forma **purpùreum** Britt. (purple). — Low woods or thickets, swales, wet shores, prairies, etc., usually common, Que. to se. Man., s. to N.S., N.E., L.I., Fla., Ala., La. and Tex. Late July–Oct.

Var. **cuneàtum** Engelm. (wedge-shaped). — Leaves distinct, gradually tapering at base, the larger ones often with a prolonged entire base; pubescence of lower surfaces and of stems short-hirtellous. (*E. serotinum,* var. *polyneuron* F. J. Herm.) — Local, s. Ind. to Mo. and La.

Var. **colpóphilum** Fern. & Grisc. (lover of estuaries). — Stem slender, puberulent to glabrate; leaves and involucres glabrous or promptly glabrate. — Fresh to brackish tidal shores, Que. to Me.

20. E. resinòsum Torr. (resinous). — *Stem* slender, 6–9 dm. high, *minutely soft-puberulent; leaves narrowly lanceolate, narrowed to sessile clasping bases,* tapering to tip, undulate, crenate or serrate, *faintly veiny and canescent-puberulent and sometimes resinous-atomiferous beneath,* the *larger* opposite *blades 5–12 cm. long by 0.5–1.7 cm. broad;* corymb stiffly branched; involucre much as in no. 19, but *minutely puberulent oblong phyllaries blunt.* — Bogs in pine-barrens, N.J. and Del. Mid-Aug.–Sept.

Var. **kentuckiénse** Fern. (of Kentucky). — Stems, branches of corymb and prominent veins of lower side of leaf *with longer curving pilosity.* — Marsh near Bean's L., Nelson Co., Ky. (*Sister Rose-Agnes*).

21. E. seròtinum Michx. (late-flowering). — Erect, 0.9–2 m. high, *pulverulent-pilose and bushy-branched above; leaves broadly lanceolate to narrowly ovate,* coarsely toothed, *long-petioled, cinereous-scabrous, subtending suppressed axillary fascicles;* the larger blades triple-nerved, 0.6–1.6 dm. long by 2–9 cm. broad; corymb with stiffly ascending cinereous branches; *involucre* gray-green, 12–15-*flowered, imbricated, the longer oblong phyllaries with strongly pubescent whitish obtuse or rounded tips;* corollas white or pale lilac. — Damp to dry thickets, borders of woods, clearings, fallow fields, etc., somewhat aggressive and weedy, Fla. to Tex. and ne. Mex., n. to N.J. (adv. to Mass.), W.Va., O., Ind., Wisc. and Kans. Late Aug.–Oct.

22. E. rugòsum Houtt. (wrinkled; name rather inept), White Snakeroot. — Stems firm, solitary or clustered from a knotty tough rhizome, 0.2–1.5 m. high; *leaves ovate, longer than broad, the larger membranaceous plane blades* 0.5–1.8 *dm. long by* 3–11 *cm. broad,* coarsely and often sharply 9–25-*toothed on each margin, gradually tapering from above the base to a long acuminate tip;* corymb in well-developed plants open with loosely ascending branches from the upper axils, in small plants more compact and terminal; *involucre* 15–30-*flowered,* 4–6 mm. long; the narrow acute or blunt ***phyllaries subequal,*** with or without 1 or more short basal ones; *flowers bright white.* (*E. urticaefolium* Reichard) — Very variable widespread species:

Phyllaries scarious, linear-attenuate, acute; larger leaves 5–18 cm. long, broadly subcuneate to rounded, rarely cordate, at base.
  Stem, petioles and leaf-blades glabrous or nearly so. . . . . *E. rugosum* (typical).
  Stem and petioles pubescent.
    Pubescence of stem and petioles loosely villous. . . . . Forma *villicaule.*
    Pubescence of stem and petioles a minute tomentum. . . . . . . Var. *tomentellum.*
Phyllaries herbaceous or with broad herbaceous backs, oblanceolate or linear-oblong; leaves with broadly rounded or cordate bases.
  Leaves glabrous, smooth, the larger blades 8–15 cm. long; involucre 5–6 mm. long; phyllaries oblanceolate, upwardly dilated, scarious-margined, the herbaceous back not prominently corrugated. . . . . Var. *roanense.*
  Leaves strigose-setulose, the larger ones 5–10 cm. long; involucre 4–5.5 mm. long; phyllaries broadly linear or narrowly oblong, herbaceous except at tip, strongly corrugated. . . . . Var. *chlorolepis.*

E. rugòsum (typical). — Rich woods, thickets and clearings, chiefly in basic soils, Gaspé Pen., Que., to se. Sask., s. to N.B., N.E., Va., upland to Ga., Ala., n. La. and ne. Tex. Late July–Oct. Forma **villicaùle** Fern. (with villous stem), var. *villicaule* Fern., scattered through range.

Var. **tomentéllum** (Robins.) Blake (finely tomentose). — Woods and thickets, Mich., Ind., Wisc. and Ill.

Var. **roanénse** (Small) Fern. (for Roan Mt.). — Woods along the mts., w. Va. and W.Va. to nw. Ga.

Var. **chlorólepis** Fern. (green-scaled). — Stem slender, weak, often reclining, pilose above; inflorescence very lax, with widely divergent branches often extending low on the plant. — Wooded marl-bluffs of James R., Surry Co., Va.; calcareous cliffs, nw. Fla.

23. E. Lùciae-Braùniae Fern. (for its discoverer, Emma Lucy Braun, 1889–). — Differing from no. 22 in its slender and softer (flattened in pressing) *stem* 4–6 dm. high, ***with** only 5 remote pairs of well-developed leaves; leaf-blade* extremely *thin and flaccid, broadly cordate-deltoid,* as broad as long; the *petioles as long as or longer than the blades,* passing to the blade through a prominent cuneate extension of the latter; teeth very coarse; *corymb frail, overtopped by upper leaves, with very short filiform divergent branches from all but lowest axils;* heads few, 15-flowered; *involucre 4 mm. long,* the costate lanceolate green *phyllaries with mucronate or caudate tips.* — Moist sheltered sandstone-rockhouses, se. Ky. Sept.

24. E. aromáticum L. (aromatic; a rather inappropriate name). — Somewhat resembling slender states of no. 22; stem slender, glabrous or nearly so, simple or often with loosely ascending long branches above; *leaves subcoriaceous or thickish, full green; the larger ones sessile or on petioles* 0.1–1.5 (rarely –2) *cm. long;* the ovate subtruncate to round-based *blades* merely acutish or blunt, 3–8 cm. long by 2–5.5 *cm. broad, with* 7–12 well-developed *blunt teeth on each margin;* corymb rather open, when well-developed with long loosely ascending branches from the axils of reduced upper leaves; *involucre* 4–4.5 *mm. long;* flowers bright white. — Dry woods, thickets and clearings, Fla. to La., n., chiefly on or near Coastal Plain, to Mass., se. N.Y., N.J., Md., W.Va. and se. O. Late Aug.–Oct. Southward passing into

Var. **incisum** Gray (sharply cut). — *Leaves paler and thinner, rhombic- or triangular-ovate, cuneate at base, coarsely and sharply toothed or incised toward base,* acuminate, on slender petioles 0.5–2 cm. long. — Moist or dry pine- or oak-woods and clearings, Fla. to se. Va.

25. E. incarnàtum Walt. (flesh-colored). — Slender, *loosely branched,* the often straggling *minutely pilose stem* 0.5–2 m. long; *leaves membranaceous, deltoid-ovate to slightly hastate, with truncate, rounded or cordate bases,* blunt-toothed, acuminate, minutely pilose; *the larger ones* 4–8 *cm. long and* 2–5 *cm. broad, on slender petioles* 1.5–4 *cm. long; corymb* very irregular and lax, *with slender divergent few-headed long branches from many axils;* heads about 20-flowered; *involucre of subequal linear scarious 2-nerved phyllaries* and a few scattered very pubescent basal ones; *flowers pink or lilac,* rarely white. — Rich woods and wooded swamps, Fla. to s. Ariz. and Mex., n. to se. Va., w. W.Va., s. O., s. Ind., s. Ill. and se. Mo. Late Aug.–Oct. — Drying plant vanilla-scented.

§ Conoclínium (L.) Benth.

26. E. coelestìnum L. (heavenly), Mistflower, Blue Boneset, "Ageratum". — Stem arising from a cord-like rhizome, *with slender scaly superficial stolons,* pubescent, 0.3–1 m. high; *leaves triangular-ovate or subcordate,* petioled, often with axillary fascicles, thinnish, *rugulose, hirtellous,* blunt or bluntish, with blunt teeth; *corymbs compact,* terminal or terminating upper branches; *involucre many-flowered, campanulate, the linear-attenuate slenderly ribbed phyllaries subequal; receptacle conical; flowers bluish-violet,* or in forma **illinoénse** Benke (of Illinois) red-purple, or in forma **álbum** E. J. Alex. (white) white. (*Conoclinium* DC.) — Low woods, damp thickets or clearings, borders of streams, etc., Fla. to Tex., n. to N.J., se. Pa., Md., W.Va., O., Ind., Ill., Mo. and e. Kans. Late July–early Oct.

5. MIKÀNIA Willd. Climbing Hempweed

Heads discoid, 4-flowered. Involucre of 4 phyllaries. Receptacle small. Flowers, achenes, etc., as in *Eupatorium.* — Twining perennials of trop. and warmer areas of Am., with opposite commonly cordate and petioled leaves, and corymbose-panicled flesh-colored to white flowers. (Named for *Joseph Gottfried Mikan,* 1743–1814, professor in the University of Prague.) Willugbaeya Neck.

1. M. scándens (L.) Willd. (climbing). — Nearly smooth; leaves somewhat triangular-cordate or hastate, pointed, toothed at the base, submembranaceous, usually prolonged to a caudate tip; flowers lilac or purplish. — Thickets, swamps and banks of streams, Fla. to Tex., n. to s. Me., s. N.H., Mass., N.Y. and s. Ont. July–Oct.

Var. **pubéscens** (Muhl.) T. & G. (hairy). — Often pubescent; leaves fleshy, more scabrous and more rounded; flowers milk-white. — Fla. to Tex., n., near the coast, to N.J.

6. BRICKÉLLIA Ell.

Characters much as in *Kuhnia;* phyllaries more numerous. Bristles of the pappus merely scabrous or at most barbellate or subplumose. Leaves often all opposite. Plants of warmer reg. of Am. (In memory of Dr. *John Brickell,* 1749–1809, of Savannah, Ga., amateur botanist and helpful correspondent of Muhlenberg, Fraser and others.)

1. B. grandiflòra (Hook.) Nutt. (large-flowered). — Herbaceous, nearly glabrous, 3–7 dm. high; leaves mostly alternate, long-petioled, deltoid-ovate or -lanceolate, 3–11 cm. long, coarsely crenate; heads slender-pedicelled and subumbellate at the tips of the panicled branches, 20–38-flowered; flowers greenish- or yellowish-white; achenes 5 mm. long; pappus of 19–26 white caducous bristles. — Dry or rocky calcareous soil, Mo. and Neb. to Wash., s. to Ark., Tex., N.M., Ariz. and Calif. July–Oct.

7. KÙHNIA L. False Boneset

Heads discoid, 10–25-flowered; flowers perfect. Phyllaries thin, few and loosely imbricated, narrow, striate-nerved. Corolla slender, 5-toothed. Achenes cylindrical, 10-striate; pappus a single row of very plumose bristles. — Am. perennials, resinous-dotted, with mostly alternate leaves, and paniculate-corymbose heads of cream-colored flowers. (Dedicated to Dr. *Adam Kuhn* of Philadelphia, who carried the living plant to Linnaeus.)

1. K. eupatorioìdes L. (resembling *Eupatorium*). — Stems 0.3–1.5 dm. high, minutely pubescent above; *leaves* very variable, from linear to narrowly ovate, membranaceous, entire or variously toothed, glabrous to slightly pubescent, *the principal ones* 4–10 *cm. long and* 0.3–2.5 *cm. wide; heads mostly on pedicels their own to half their length; involucre minutely hispid to glabrate,* 7–10 *mm. long; the inner prolonged linear phyllaries mostly* 2–4*-costate,* obtuse to attenuate; pappus sordid or brown. — Open woods, thickets, clearings and rocky slopes, Fla. to Tex., n. to N.J., Pa., O., Ind. s. Ill. and Mo. Late July–Oct. Westw. and southwestw. passing by various transitions to

Var. **corymbulòsa** T. & G. (with small corymbs). — Coarser, the *leaves* usually *firmer and puberulent; peduncles, pedicels and involucre commonly pilose or tomentulose; involucres mostly much longer than their pedicels,* 8–12 *mm. long, the broader phyllaries mostly* 4–6*-costate.* (*K. glutinosa* Ell.; *K. suaveolens* Fresn.) — Dry plains, prairies and open woods, Ala. to Tex., n. to O., Mich., Wisc., Minn., N.D. and Mont. Aug.–Oct.

Var. **angustifòlia** Raf. (narrow-leaved). — *Leaves narrowly linear,* 1–3.5 *cm. long,* 1–2 *mm. broad; peduncles, etc., puberulent and gummy;* pedicels much longer than involucres. — Dry soil, Fla. to Tex. and Mo.

8. LIÀTRIS Schreb. BUTTON-SNAKEROOT. BLAZING-STAR

Heads discoid, few–many-flowered; flowers perfect. Phyllaries well imbricated, appressed, at least at base. Receptacle naked. Corolla 5-lobed, the lobes long and slender. Achenes slender, tapering to the base, 10-ribbed. Pappus of 15–40 capillary plumose or barbellate bristles. — Perennial N. Am. herbs, often resinous-dotted, with simple stems chiefly from a roundish corm or tuber, alternate narrow entire leaves (sometimes twisted so as to become vertical), and spicate, racemose or cymose handsome rose-purple (or exceptionally white) flowers in summer or in autumn. (Derivation of name unknown.)* LACINARIA Hill. LACINIARIA Hill

a. Pappus barbellate, the lateral cilia only 3–6 times the diameter of the bristle. § 1. SUPRAGO. . . *b.*

b. Heads cylindric to cylindric-turbinate, 3–20-flowered, the phyllaries mostly erect. . . *c.*

c. Corolla-tube not pilose within; inflorescence spiciform or racemose, the heads sessile or pedicelled; leaves gradually reduced upward to the bracts subtending the heads. . . *d.*

d. Phyllaries mostly obtuse or nearly so, with appressed tips. Ser. 1. SPICATAE.

Heads mostly sessile; involucre cylindric-campanulate, 3–10 mm. thick; flowers 4–18; pappus 5–7 mm. long. . . . 1. *L. spicata.*

Heads pedicelled; involucre slenderly cylindric, 2–4 mm. thick; flowers 4–6; pappus 3.5–4 mm. long. . . . . 2. *L. microcephala.*

d. Phyllaries acuminate or acute, with squarrose tips; sessile heads densely crowded. Ser. 2. PYCNOSTACHYAE. . . . . 3. *L. pycnostachya.*

c. Corolla-tube pilose within; inflorescence spicate, racemose or paniculate. Ser. 3. GRAMINIFOLIAE. . . *e.*

e. Phyllaries without prominent keel, mostly ciliate and obtuse; heads 1–1.5 cm. long; corollas 6–9 mm. long; pappus 5–7 mm. long.

Heads 5–15-flowered; outer phyllaries ovate-lanceolate; corollas 6–8 mm. long; pappus 5–6 mm. long; achenes 3–4 mm. long. . . . 4. *L. graminifolia.*

Heads 9–20-flowered; outer phyllaries ovate-suborbicular; corollas 8–9 mm. long; pappus 6–7 mm. long; achenes 4–5 mm. long. . . 5. *L. turgida.*

e. Phyllaries prominently keeled, eciliate, mostly acute or acuminate; involucres 1 cm. long, 6–12-flowered; corollas 7 mm. long; pappus 3.5–4 mm. long. . . . . 6. *L. regiomontis.*

b. Heads spherical, hemispherical, short-cylindric or campanulate, 15–70-flowered; phyllaries broad, often squarrose or bullate; corolla-tube pilose within (except in no. 12). Ser. 4. SCARIOSAE. . . *f.*

f. Phyllaries ciliate; corolla-tube pilose within. . . *g.*

---

* Treatment based in part on that of L. O. GAISER in Rhodora, xlviii (1946); measurements made from dried material.

*g.*Basal leaves oblanceolate, lanceolate, oblong or narrowly obovate, 1–5 cm. broad.
Basal leaves 2–5 cm. broad; heads subglobose; phyllaries mostly with recurved or loosely ascending tips; corolla-tube 7–11 mm. long; montane species. . . . . . 7. *L. scariosa.*
Basal leaves 1–2.5 cm. broad; heads subcylindric to turbinate; inner phyllaries appressed; corolla-tube 1–1.5 cm. long; campestrian species. . . . . 8. *L. scabra.*
*g.*Basal leaves lanceolate, 0.5–3 cm. broad.
Heads 35–60-flowered, broadly campanulate to hemispherical; inner phyllaries pink or purple; corolla-tube 9–10 mm. long; pappus 8 mm. long; achenes 6–7 mm. long; northeastern. . . . 9. *L. borealis.*
Heads 15–25-flowered, turbinate or slenderly campanulate; inner phyllaries green or with narrow petaloid margin; corolla-tube 7–9 mm. long; pappus 6 mm. long; achenes 3–4 mm. long; southern and midwestern. . . . . 10. *L. Earlei.*
*f.*Phyllaries eciliate, or ciliate in no. 12 with corolla glabrous within.
Leaves eciliate; heads subglobose, 1.5–2.5 cm. in diameter, with eciliate phyllaries; corolla-tube pilose within; pappus 7–8 mm. long; southern and midwestern. . . . . 11. *L. aspera.*
Leaves ciliate; heads campanulate or hemispherical, 2–3 cm. thick, with ciliate phyllaries; corolla-tube glabrous within; pappus 8–10 mm. long; northwestern. . . . . 12. *L. ligulistylis.*
*a.*Pappus plumose, the lateral cilia 15 or more times the diameter of the bristle. § 2. Euliatris. . *h.*
*h.*Heads 3–8-flowered, slenderly cylindrical, definitely longer than thick; inflorescence spicate or racemose; western. Ser. 5. Punctatae. . . *i.*
*i.*Stems mostly arising from the crowns of an elongate subterranean trunk; leaves strongly punctate; inflorescence 0.3–4.5 dm. long; pappus 9–11 mm. long.
Subterranean trunk vertical; leaves rigid, the basal ones 3–5 mm. broad, often ciliate; heads 1.5–2 cm. long; achenes 6–7 mm. long; wide-ranging westward. . . . . 13. *L. punctata.*
Subterranean trunk horizontal; leaves flexible and soft, the basal ones 1–2 mm. wide, eciliate; heads 0.8–1.2 cm. long; achenes 8 mm. long; northwestern. . . . . 14. *L. densispicata.*
*i.*Stems arising from nearly globose corms 2–4 cm. thick; leaves scarcely or but slightly punctate; inflorescence 2–6 dm. long; pappus about 7 mm. long; western.
Leaves hard and stiff, punctate; spike rarely 3 dm. long; phyllaries firm, ovate or broadly oblong, abruptly cuspidate or mucronate. . . . 15. *L. mucronata.*
Leaves softer and epunctate; spikes 2–6 dm. long; phyllaries submembranaceous, lance-acuminate. . . . . 16. *L. angustifolia.*
*h.*Heads 15–60-flowered, thick-cylindric, nearly as thick as long; inflorescence loosely racemose to cymose.
Phyllaries appressed, obtuse or mucronate. Ser. 6. Cylindraceae, represented with us by . . . . 17. *L. cylindracea.*
Phyllaries loosely spreading, the median with acute slender divergent or squarrose tips. Ser. 7. Squarrosae. . . . . 18. *L. squarrosa.*

§ 1. Supràgo (Cass.) DC.

Ser. 1. Spicàtae (E. J. Alex.) Gaiser

1. L. spicàta (L.) Willd. (spicate). — *Stems* stiffly erect from fibrous-coated cormiform rhizomes, 0.3–1.8 m. high, *glabrous* or rarely hirsute; *leaves very numerous, glabrous, linear to linear-lanceolate;* the lower ones widest, 1–4 dm. long and 0.5–2 cm. broad; *inflorescence a usually dense spike* 0.6–7 dm. long, *the lower heads overtopped by or overtopping the bracteal leaves, the upper all exceeding them; heads cylindric-campanulate; involucres* 8–10 mm. long, about half as thick, *with appressed oblong mostly round-tipped viscid phyllaries* green to purple-tinged; *flowers* 10–18; *corolla* rose-purple, or white in forma **albiflòra** Britt. (white-flowered), *glabrous within,* 6.5–9 mm. long; *achenes 4–6 mm. long; pappus 5–7 mm. long.* — Meadows, borders of marshes, savannas, damp slopes, etc., Fla. to La., n. to L.I., n. N.J., e. Pa., W.Va., sw. Ont., O., Mich. and Wisc.; cult. and locally natzd. in s. N.E. and N.Y. July–Sept. — Hybridizes with no. 11.

Var. **resinòsa** (Nutt.) Gaiser (resinous). — More slender, up to 1.2 m. high; lowest leaves mostly narrowly linear and 1.5–5(–8) mm. broad; spike more open, with many or all internodes elongate; involucres usually dark purple, 6–10 mm. long; flowers 4–7(–12). (*L. resinosa* Nutt.) — Savannas, bogs and damp to dry pinelands and open woods, Fla. to Miss., n. to se. Va. and Tenn.; recorded northeastw. to e. Pa. Late July–Sept.

2. L. microcéphala (Small) K. Schum. (tiny-headed). — *Differing from the var. of no.* 1 *in its slender and less rigid stem* 3–12 dm. high; *basal leaves only* 0.5–1 *dm. long; heads in a loose raceme, on definite pedicels* about their own length; *involucres slenderly cylindric,* 2–4 *mm. thick; phyllaries green, the longer* 5–6 *mm. long; flowers* 4–6; *tube of corolla* 5–6 *mm. long; achenes* 3–4 *mm. long; pappus* 3.5–4 *mm. long.* — Dry granitic or siliceous soil, oak-barrens, etc., Ga. and Ala., n. to w. N.C. and Ky. Aug., Sept.

Ser. 2. Pycnostáchyae (E. J. Alex.) Gaiser

3. L. pycnostáchya Michx. (with crowded spike). — As coarse and tall as typical no. 1, *hirsute* (rarely glabrous); the linear punctate *leaves very crowded, the lower ones up to* 1 *dm. long and mostly* 4–12 *mm. broad; spike very dense,* 2–3.5 *cm. thick,* 0.35–3 dm. long; *the rachis usually hirsute; heads cylindric, with acute or acuminate spreading- to squarrose-tipped phyllaries; flowers* 5–12; corollas rose-purple, or rarely white in forma **Hubríchti** E. Anders. (named in 1937 for its discoverer, Leslie Hubricht), 7–9 mm. long, glabrous or sparsely hairy within; *achenes* 4–7 *mm. long; pappus* 6–7 *mm. long.* — Damp prairies, Wisc. to Minn. and S.D., s. to Ky., La., Okla. and Tex. Mid-July–Oct. — Hybridizes with no. 11; and with no. 18, producing × **L. Ridgwàyi** Standl. (for Robert Ridgway, 1862–1938).

Ser. 3. Graminifòliae (E. J. Alex.) Gaiser

4. L. graminifòlia (Walt.) Willd. (grass-leaved). — *Stems* 1–few from globose corms, 3–7 (–10) *dm. high, usually more or less hirsute* and striate; *basal leaves* linear or linear-lanceolate, *either ciliate or with long hairs along the winged petiole;* inflorescence a slender spike, spiciform raceme or panicle; *heads slenderly turbinate,* 5–15-*flowered,* 1–1.5 cm. high; *phyllaries* glabrous or hirsute; *the longer ones oblong to linear, obtuse* and 7–9 mm. long, *ciliate; corolla* 6–8 mm. long, *its tube pilose within;* achenes 3–4 mm. long; pappus 5–6 mm. long. — A highly variable eastern species; the following vars. are recognized:

*a.*Inflorescence spicate or spiciform and simple or very short-branched below, the numerous heads sessile or on pedicels rarely up to 1 cm. long. . . *b.*

*b.*Heads mostly approximate or overlapping, the greater part of the inflorescence rather dense. . . *c.*

*c.*Stem and pedicels glabrous or with very remote trichomes.

Basal leaves mostly 2–4 mm. broad, with ciliate blades; heads 5–9-flowered, about 1 cm. high and 5–6 mm. broad across summit; principal phyllaries linear, about 1 mm. broad; corolla-tube 6–7 mm. long; achenes about 3 mm. long; pappus 4–5 mm. long. . . . . . *L. graminifolia* (typical).

Basal leaves 3–10 mm. broad, eciliate except for long trichomes at base or on petiole; heads 10–15-flowered, 1.2–1.5 cm. high, about 1 cm. broad across summit; principal phyllaries oblong, about 2 mm. broad, ciliolate; corolla-tube 6–8 mm. long; achenes 4 mm. long; pappus about 5 mm. long. . . . . Var. *racemosa.*

*c.*Stem and pedicels copiously hirsute; leaves long-ciliate; phyllaries coarsely ciliate; heads, etc., otherwise much as in var. *racemosa;* plant of N.J. and Del. . . . . Var. *lasia.*

*b.*Heads remote and few (4–25), many of them 1–2 cm. apart; slender stem glabrous; leaves few; the lower narrowly lanceolate, prolonged to one-third or half the height of plant, up to 1 cm. broad; involucre 8–9 mm. high, with broad oblong phyllaries; montane southw. . . . . Var. *Smallii.*

*a.*Inflorescence loosely paniculate below, with 1–few-headed slender ascending branches or peduncles 0.4–1.5 dm. long; heads much as in var. *racemosa.* . Var. *virgata.*

L. graminifòlia (typical). — Dry to moist pinelands, oak-woods and clearings, chiefly on Coastal Plain, N.J. to S.C. and very rarely to Ala. Sept., Oct.

Var. **racemòsa** (DC.) Venard (with a raceme). — Much coarser. (Var. *dubia* (Bart.) Gray) — Dry sandy woods and barrens, Coastal Plain and outer Piedmont, southw. back to Blue Ridge, N.J. to S.C. Sept., Oct.

Var. **lásia** Fern. & Grisc. (hairy). — N.J. and Del.

Var. **Smállii** (Britt.) Fern. & Grisc. (for John Kunkel Small, 1869–1938). — Dry woods, mts. of Va. and N.C. Aug., Sept. — Anomalous plant, perhaps with admixture of no. 5 or 6.

Var. **virgàta** (Nutt.) Fern. (twiggy; from the many ascending branches). — Pinelands, local, N.J. to Ga. — More extreme than slightly paniculate states of other vars.

5. L. túrgida Gaiser (turgid). — Differing from var. *racemosa* of no. 4 in the *curving pilosity often present at base of stem and on lower leaves; leaves broader*, the *lanceolate basal ones* 0.5–2 *cm. broad, usually pilose beneath; heads more distant,* 9–20-*flowered, subglobose-turbinate; the usually purple involucre* 0.9–1.5 *cm. high and broad; outer phyllaries ovate-suborbicular*, the others broadly oblong, *glabrous or minutely ciliolate; corollas* 8–9 *mm. long; achenes* 4–5 *mm. long; pappus* 6–7 *mm. long.* — Dry upland woods, openings and barrens, mts. of Va., W.Va. and N.C. Late July–early Sept.

6. L. regiomóntis (Small) K. Schum. (of King's Mt., N.C.). — *Differing from no.* 5 *in its more slender glabrous sometimes paniculate-branched stem* 0.3–1 *m. high; leaves narrower, the basal mostly less than* 5 *mm. broad; spikes or panicles of slender elongate spikes* (except in obvious dwarfs) *occupying one-half to three-fourths height of plant*, often secund; *heads slenderly cylindric-turbinate,* 6–12-*flowered; involucre* green to purplish, *about* 1 *cm. long; phyllaries prominently keeled, eciliate, the median and upper lanceolate ones with loosely ascending acute or acuminate tips;* corolla 7 mm. long, appressed-pubescent at base of throat and in tube; achenes 3–3.5 mm. long; pappus 3.5–4 mm. long. — Sands and rocky slopes from mts. to Coastal Plain, sw. Va. to Ga. Late Aug., Sept.

Ser. 4. Scariòsae (E. J. Alex.) Gaiser

7. L. scariòsa (L.) Willd. (scarious). — *Stems* arising from a subglobose corm, 0.3–1.5 m. high, *mostly subappressed-pubescent; leaves pubescent to scabrous; the basal ones narrowly ovate or obovate to lanceolate,* 2–5 *cm. broad;* median and upper from oblanceolate to lanceolate or nearly linear; *inflorescence a raceme, spike or cylindric panicle, mostly open and with divergent pedicels or branches; heads subglobose,* 1.5–2.5 *cm. in diameter,* 25–60-*flowered; phyllaries green, with* a narrow thin *ciliolate margin, the outer ones ovate and more or less squarrose, the oblong to narrowly obovate subcoriaceous round-tipped upper ones with loosely ascending to slightly recurving tips; corolla pilose at inner base of tube, the tube* 7–11 *mm. long;* achenes 4–5 mm. long; *pappus* 6–9 *mm. long.* (Incl. var. *virginiana* (Lunell) Gaiser) — Dry woods and openings among the mts. Pa. and W.Va., s. to S.C. Aug., Sept.

8. L. scàbra (Greene) K. Schum. (harsh). — *Differing from no.* 7 *in its cinereous-puberulent scabrous stem; basal leaves only* 1–2.5 *cm. broad; heads thick-cylindric to subturbinate,* 1.5–2 *cm. thick,* 20–35-*flowered; phyllaries heavily cinereous-puberulent, the outer ones with acute or acutish squarrose tips, the inner appressed and with almost orbicular tips; corolla-tube* 1–1.5 *cm. long;* pappus 8–10 mm. long. — Dry open woods, ridges and barrens, Ala. to La. and Okla., n. to O., Ind. and Ill. Aug.–Oct.

9. L. boreàlis Nutt. (northern). — *Resembling no.* 7, *with more ascending pedicels; stems* 0.25–1 m. high, *glabrous or sparsely appressed-pubescent; leaves numerous, often crowded, glabrous, or sparsely pubescent on midrib beneath, the lower lanceolate ones* 0.5–3 *cm. broad,* the upper greatly reduced; *rachis of raceme* 0.3–3.5 dm. long, *pilose, bearing* 3–30 *hemispherical* 35–60-*flowered heads* 1.8–3 *cm. broad; phyllaries* commonly *dark reddish, appressed-ascending,* with broadly rounded tips and *with ciliolate or petaloid narrow dark margin;* corolla rose-purple, or white and with green phyllaries in forma **albiflòra** (Shinners) Fern. (white-flowered), its tube scantily pilose within and 9–10 mm. long; *achenes* 6–7 *mm. long;* pappus about 8 mm. long. (*L. novae-angliae* Shinners) — Dry argillaceous to siliceous open woods, thickets and clearings, sw. Me. to e. N.Y., s. to s. N.E., N.J. and centr. Pa. Aug., Sept.

10. L. Eàrlei (Greene) K. Schum. (for its discoverer, Franklin Sumner Earle, 1857–1929). — *Stem densely pilose* to glabrescent; leaves smooth or scabrous, much like those of no. 9; pedicels pilose; *heads subturbinate-campanulate,* 15–25-*flowered; phyllaries green, soft-pilose, the basal ones lance-attenuate and merging into bracts at summit of pedicel, the middle and upper ones oblong, with dilated blunt to abruptly tipped erose-margined summits; corolla-tube* 7–9 *mm. long; achenes* 3–4 *mm. long; pappus* 6 *mm. long.* — Dry soil, Fla. to e. Tex., n. to w. N.C., Ky. and s. Ind.

11. L. áspera Michx. (harsh). — *Stiffly erect stem* (at least *above basal third*) *puberulent or substrigose with grayish hairs; leaves* lanceolate or *mostly nearly linear, scabrous, eciliate, glabrous,* the lower chiefly 1–2 cm. broad; *inflorescence an elongate stiff spike* or virgate panicle of spikes, mostly 1.5–7.5 *dm. long; heads* 20–150 or more, *subglobose,* 1.5–2.5 *cm. in diameter,* usually closely sessile, 25–40-*flowered; phyllaries glabrous, the lower roundish; the middle and upper broadly oblong, with* broad rounded *scarious-margined eciliate summits; corollas* purple, or white in forma **Bénkii** (Macbr.) Fern. (for Herman C. Benke, 1870–1947), the *tube* 8–10 mm. long and *pilose within;* achene 4–6 mm. long; *pappus* 7–8 *mm. long.* — Dry often sandy soil, O. to N.D., s. to La. and e. Tex. Aug., Sept. — Hybridizes with no. 1, producing × **L. Steèlei** Gaiser (for Edward Strieby Steele, 1850–1942).

Var. **intermèdia** (Lunell) Gaiser (intermediate). — *Stem,* except at summit, *glabrous or only sparsely pubescent; leaves smooth,* often a little broader. — N. Fla. to e. Tex., n. to w.

N.C., W.Va., sw. Ont., O., Mich., Wisc., Mo. and Okla. — Said to hybridize with some other species, producing × **L. sphaeroìdea** Michx. (spheroidal).

12. L. ligulistỳlis (Nels.) K. Schum. (with strap-like styles). — *Differing* from no. 11 *in stem* (1–6 *dm. high*); *leaves ciliate*, densely scabrous-pubescent to glabrescent but harsh; *inflorescence a raceme of* 2–30 *hemispherical* 40–70-*flowered heads* 2–3 *cm. thick* (or terminal one much larger); glabrous *phyllaries dark purple above, ciliate, the middle and upper ones with deeply colored erose or variously toothed scarious upper margins; corolla-tube glabrous within*, 9–11 mm. long; achenes 5–6 mm. long; pappus smoky-purple, 8–10 mm. long. — Dry to moist often sterile soil, Alta. to N.M., e. across the plains to s. Man. n. Wisc., Minn. and S. D. Late June–Aug. — The anomalous × **L. Nieuwlándii** (Lunell) Gaiser (for Julius Arthur Nieuwland, 1878–1936), considered by Gaiser, from cytological evidence, a hybrid of no. 12 with a second but unidentified species (although occurring almost wholly outside the range of the western inferred parent, *i.e.*, from Mich. and Wisc. s. into O., Ind., Ill. and Mo., and flowering one to two months later, *i.e.*, mid-Aug.–Oct.), needs further consideration.

§ 2. Euliàtris DC.

Ser. 5. Punctàtae (E. J. Alex.) Gaiser

13. L. punctàta Hook. (with translucent dots). — *Stems several from crowns of an elongate* (sometimes corm-like) *deep vertical subterranean trunk*, glabrous, 1.5–8 dm. high; *leaves very numerous and overlapping, strongly ascending, linear, firm, glabrous but ciliate, punctate, the basal* 3–5 *mm. broad*, the others narrower; *inflorescence a dense spike* 0.3–4.5 *dm. long*, with long leafy bracts below; *heads cylindric or subcylindric*, 1.5–2 cm. long, 4–8-*flowered; phyllaries long-ciliate, narrow, with long acute or acuminate tips; corollas* 9–12 *mm. long, the inside of the tube and the filaments pilose; achenes* 6–7 *mm. long; pappus long-plumose*, 9–11 *mm. long.* — Dry prairies and plains, Alta. to N.M. and Trans-Pecos Tex., e. to Man., Minn., w. Ia. and Neb.

Var. **nebraskàna** Gaiser (of Nebraska). — Leaves flexible, scarcely or barely ciliolate; heads about 1.5 cm. long; phyllaries submembranaceous. — E. to Mich., Ill., Mo. and Ark.

14. L. densispicàta (Bush) Gaiser (dense-spiked). — *Differing from no.* 13 *in its elongate horizontal rootstock; softer and narrower eciliate leaves; heads* 8–12 *mm. long.* — Sands, local, Minn. — Perhaps better as an extreme of no. 13.

15. L. mucronàta DC. (mucronate). — *Differing from no.* 13 *in its nearly superficial subglobose corm* 2–4 *cm. in diameter; leaves mostly eciliate; inflorescence rarely more than* 3 *dm. long; heads* 3–6-*flowered*, 1.2–1.8 cm. long; *phyllaries broader, mostly ovate to broadly oblong, abruptly mucronate or cuspidate, sparsely ciliolate or merely membranous-margined; corollas* 9–10 *mm. long; pappus* 6–7 *mm. long.* — Dry open soil, Tex., ne. to Kans. and Mo.

16. L. angustifòlia (Bush) Gaiser (narrow-leaved). — *Differing from no.* 15 *in relatively softer and epunctate or barely punctate narrower leaves; inflorescence* 2–6 *dm. long; heads* 8–16 *mm. long; phyllaries thinner, lance-acuminate.* — Dry prairies and rocky slopes, Tex. to Neb., e. to Mo. and Ark.

Ser. 6. Cylindràceae Gaiser

17. L. cylindràcea Michx. (cylindric). — Stems slender, arising from a roundish corm, 2–7 dm. high, glabrous or sparingly hirsute; leaves linear, rigid, punctate, often lustrous, glabrous or sometimes with some trichomes beneath; *raceme of* (1–)4–20 *thick-cylindric or subturbinate* 30–60-*flowered heads*, the larger heads 2–3 cm. long and 1–1.5 cm. thick; *involucre of coriaceous sublustrous tightly appressed ovate round-tipped mucronate or abruptly acuminate phyllaries;* corolla 1.2–1.4 cm. long, the inner surface of the spreading lobes pubescent; achenes 5–6 mm. long; pappus plumose, about 1 cm. long. — Dry calcareous or siliceous soil, s. Ont. to Minn., s. to w. N.Y., O., Ind., Ill. and Mo. July–Sept.

Ser. 7. Squarròsae (E. J. Alex.) Gaiser

18. L. squarròsa (L.) Michx. (with spreading tips). — Stems from a roundish corm, 2–7.5 dm. high, pubescent to glabrous; *leaves rigid, punctate*, lanceolate to linear, *the basal usually much elongate; inflorescence a spike, raceme or cyme of remote mostly* 25–40 (or the terminal –60 or more)-*flowered heads about as thick as long* (1.5–3 cm.); *phyllaries foliaceous*, with callous or ciliate margins, *lanceolate, with prolonged acuminate wide-spreading or recurved firm tips, the longer ones* 1.5–2 *cm. long;* corolla-tube 0.9–1.5 cm. long, the inner surface of the lobes very hairy, the *outer corollas abruptly spreading;* achenes 5–6 mm. long; pappus strongly plumose, 7–12 mm. long. — Dry open woods, clearings and fields, chiefly argillaceous, n. Fla. to Ala., n. to Del., e. Md., W.Va., O., Ind., s. Ill. and se. Mo. July–Sept.

Var. **gracilènta** Gaiser (becoming slender). — More slender, usually soft-pubescent; leaves very narrow, the lower 4 mm. wide; heads sessile or subsessile, usually only 1–5; involucre elongate-cylindric, mostly about 2 cm. long and 1.2 cm. thick; tips of phyllaries spreading but not recurved. — Fla. to La., n. to w. Va. and s. Mo.

Var. **hirsùta** Rydb. (hirsute). — Plant strongly hirsute; leaves narrowly linear; outer phyllaries oblong, abruptly pointed and reflexed; heads about 20-flowered and 1.3 cm. long; pappus 8 mm. long. — Miss. and La., n. to Ky., Ia. and Neb.

9. CARPHÉPHORUS Cass.

Similar to *Liatris*. Heads many-flowered; flowers perfect. Involucre campanulate, with appressed imbricated phyllaries. Receptacle chaffy; the chaff subtending the outer flowers, thin, deciduous with the fruit. Corolla 5-lobed, the lobes ovate or short-lanceolate. Achenes as in *Liatris*. Pappus of 1 or more series of plumose bristles. — Perennials (often forming large clumps) of se. U.S., with caudices not bulbous- nor tuberous-thickened, and corymbose heads of roseate to white flowers. (Name from the Greek *carphe, chaff*, and *phoros, bearing.*)

1. C. tomentòsus (Michx.) T. & G. (densely soft-pubescent). — *Stems* mostly solitary, 2.5–7.5 dm. high, *villous-hirsute below, tomentose above*, leafy to summit *with* alternate *subapproximate and appressed* oblong to lanceolate *sessile small* (0.5–5 *cm. long) leaves;* leaves of basal rosettes oblanceolate or spatulate, petioled, 5–15 cm. long, villous; corymb short, with 1–40 heads; *involucres* 8–13 mm. high, *the acute tomentulose and villous-ciliate* herbaceous *phyllaries* 3–5-seriate, with free tips; corollas rosy-pink. — Dry or damp pine-barrens and sandy woods of Coastal Plain, rare, Fla. to se. Va. Aug.–Oct.

Var. **Wálteri** (Ell.) Fern. (for THOMAS WALTER, (?) 1740–1789). — Rosette-leaves glabrous; stems only sparsely pubescent. — Sandy pine-barrens and bogs, e. Ga. to se. Va.

2. C. bellidifòlius (Michx.) T. & G. (with leaves of *Bellis*, English Daisy). — Coarser; the *stems* commonly clustered, 2–6 dm. high, *glabrous or essentially so; cauline leaves scattered, spreading; the lower and median spatulate, petioled,* 0.5–1.6 *dm. long, glabrous*, punctate; rosette-leaves similar, 1–2 dm. long; corymb elongate, with many (up to 100 or more) heads; *involucres with closely appressed round-tipped glabrous minutely ciliolate phyllaries.* — Dry sandy barrens and woods of Coastal Plain, Fla. to se. Va. Aug.–Oct.

Tribe III. ASTÈREAE Cass. (see p. 1358)

a. Heads uniformly with perfect disk-flowers with tubular corollas; ours all herbaceous or barely suffruticose, commonly with ray-flowers. . . *b.*

b. Ray-flowers yellow (creamy-white in two species of no. 14), only exceptionally wanting; disk-corollas yellow. . . *c.*

c. Pappus of disk-flowers of 2–9 chaffy scales or awns; involucre more or less resin-coated, of firm green-tipped phyllaries.

Heads large, terminating leafy branches, many-flowered, the disk 0.8–2 cm. broad; pappus of 2–8 caducous awns. . . . . . . 10. *Grindelia.*

Heads small, in fastigiate or paniculate cymes, the disk-flowers 3–20; pappus a crown of scales. . . . . 11. *Gutierrezia.*

c. Pappus (at least of the disk) of copious slender or capillary bristles. . . *d.*

d. Pappus of disk-flowers double (of long inner bristles and short outer ones).

Ray-flowers without pappus or merely with an obsolete crown. . 12. *Heterotheca.*

Ray- and disk-flowers all with abundant bristleform pappus. . 13. *Chrysopsis.*

d. Pappus of all flowers simple.

Perennials; heads relatively small, in thyrsiform, racemose or secund close panicles or in corymbs or dense cymes; leaves not spinulose-pectinate; pappus of equal bristles. . . . . 14. *Solidago.*

Annuals or biennials or, if suffruticose-perennial, with spinulose-pectinate leaves; heads solitary at tips of leafy branchlets or long-stalked in open panicles; involucre in perennial species up to 2.5 cm. broad; pappus-bristles unequal. . . . . 15. *Haplopappus.*

b. Ray-flowers white, pink, purple or blue. . . *e.*

e. Pappus none or very short, with or without a few awns.

Receptacle conical; pappus none; phyllaries equal; annual. . 16. *Astranthium.*

Receptacle flat to low-conical; pappus present; phyllaries imbricated.

Low annual with tiny slender heads; achenes fusiform, pubescent. . 17. *Chaetopappa.*

Tall perennials with broad heads; achenes flat, glabrous. . . . 18. *Boltonia.*

e. Pappus of numerous long capillary bristles; receptacle flat. . . *f.*

f. Involucre saucer-shaped or hemispherical to campanulate or turbinate, its phyllaries herbaceous to scarious or firm; ligules white, pink, pur-

ple or bluish; disk-flowers yellow, brownish or purplish; achenes glabrous or pubescent, scarcely silky.
Heads on mostly leafy branchlets, variously paniculate, corymbed or umbellate; phyllaries definitely imbricated or with 1 or more foliaceous outer series nearly or quite as long as the inner; style-appendages slender-subulate, lanceolate or acutely ovate. . . . . . 19. *Aster*.
Heads on naked peduncles or scapes, solitary or in corymbs or panicles; phyllaries a single equal series, or a long inner series and a very short basal one; style-appendages roundish or obtuse. . . . 20. *Erigeron*.
f. Involucre slenderly cylindric or ovoid, its phyllaries coriaceous or cartilaginous, with green tips; ligules and disk-flowers milk-white or creamy; style-appendages lance-subulate; achenes silky; inflorescence corymbose. . . . . 21. *Sericocarpus*.
a. Heads dioecious and discoid; flowers of staminate heads with tubular-funnelform 5-lobed corolla and style-branches without stigmatic lines; flowers of pistillate heads with truncate corollas; shrubs with firm leaves. . . . . 22. *Baccharis*.

10. GRINDÈLIA Willd. GUMWEED. TARWEED. STICKY-HEADS

Heads many-flowered, radiate (or rayless); rays pistillate. Phyllaries of the hemispherical involucre imbricated in several series, with slender more or less spreading green tips. Achenes short and thick, compressed or turgid, truncate, glabrous; pappus of 2–8 caducous awns. — Coarse Am. perennial or biennial herbs, often resinous-viscid, ours glabrous and leafy with sessile or clasping alternate and entire, serrate or laciniate rigid leaves, and large heads terminating leafy branches. Disk and rays yellow. (Named for Prof. *David Hieronymus Grindel*, 1776–1836, a Russian botanist.)*

Involucre heavily glutinous throughout, its phyllaries with strongly recurving tips. . . . . 1. *G. squarrosa*.
Involucre barely glutinous, its phyllaries loosely spreading or ascending. . . . 2. *G. lanceolata*.

1. G. SQUARRÒSA (Pursh) Dunal (having parts (phyllaries) with recurved tips), CURLYCUP-G. — Stem stoutish, corymbosely branched, 0.1–1 m. high; *leaves strongly punctate, crenate to serrate* or sometimes entire, linear-oblong to ovate- or narrowly obovate-oblong, more or less clasping, the principal ones 3–7 cm. long; heads rather large, the disk 0.8–2 cm. broad; *involucre heavily glutinous, with strongly recurving phyllary-tips; achenes* oblong, *2–3.3 mm. long, 0.7–1.5 mm. broad*, striate; *awns* 2–8, subentire or mostly serrulate, *3–5 mm. long*. Highly variable:

Phyllaries in 5 or 6 series, the upper half or two-thirds loosely squarrose or recurving; leaves copiously serrate; plants often 4–10 dm. high.
Rays present; leaves strongly clasping.
Principal leaves oblong to narrowly obovate; rays oblong-elliptic to oblanceolate, 8–10 mm. long. . . . . *G. squarrosa* (typical).
Principal leaves narrowly linear-oblong to oblanceolate; rays spatulate-oblong to oblanceolate, 10–14 mm. long. Var. *serrulata*.
Rays wanting; leaves less clasping. Var. *nuda*.
Phyllaries in 4 or 5 series, the upper third or fourth strongly recurved; leaves entire to serrate; plant rarely more than 4 dm. high. . . . . Var. *quasiperennis*.

G. SQUARRÒSA (typical). — Prairies and plains, indigenous w. of our range; more or less adv. or natzd. in dry soil eastw. to the Atl. states. July–Sept. — A compass-plant.

Var. SERRULÀTA (Rydb.) Steyerm. (finely toothed). — Native w. of our range; locally in waste ground e. to Ont., N.Y., Ct. and Pa.

Var. NÙDA (Wood) Gray (naked). — Native sw. of our range; adv. e. to Mo. and locally in N. J.

Var. **quasiperénnis** Lunell (almost perennial), *G. perennis* Nels. — Alta. to Minn., s. to Utah, Colo. and S.D.; adv. e. to Que.

2. G. lanceolàta Nutt. (lance-shaped). — Similar to no 1; *leaves scarcely punctate*, linear to lance-oblong, entire or serrulate; *involucre scarcely or only slightly viscid, its phyllaries loosely ascending or spreading*, not recurved; *achenes 4–6 mm. long, 2–2.5 mm. broad*, smooth and lustrous; *awns 2, entire, 4–7 mm. long*. — Rocky prairies and calcareous barrens and glades, La. and Tex., n. to Tenn., Mo. and se. Kans. Aug.–Oct.

* Treatment based on that of J. A. STEYERMARK in Ann. Mo. Bot. Gard. xxi. 433–608 (1934).

11. GUTIERRÈZIA Lag. Broom-Snakeroot

Heads few–several-flowered, radiate; rays 1–10, pistillate or perfect but infertile. Involucre cylindric-clavate to hemispherical; phyllaries coriaceous, with green tips, closely imbricated, the outer shorter. Receptacle small, naked. Achenes short, terete; pappus of 5–9 chaffy or bristleform scales, shorter in the ray-flowers. — Plants glabrous and often glutinous, much branched, with narrowly linear entire alternate leaves, and small heads of yellow flowers in fastigiate or paniculate cymes, of N. and S. Am. (Named in 1816 for *Pedro Gutierrez*, correspondent of the botanical garden of Madrid.)

Suffruticose, with crowded sessile cylindric heads, with disk- and ray-flowers 3 or 4 each. . . . . 1. *G. Sarothrae.*
Annual, with pedicelled hemispherical heads, with 10–20 disk-flowers and half as many rays. . . . . 2. *G. dracunculoides.*

1. G. Saròthrae (Pursh) Britt. & Rusby (thought by Pursh to resemble *Hypericum Sarothra*). — *Suffruticose*, 1–3 dm. high, with woody-based stiff stems; *heads crowded in small clusters, cylindrical*, 4–5 mm. high; *disk-flowers 3 or 4, with pappus of about 9 chaffy scales; ray-flowers 3 or 4, pistillate*, with shorter pappus. — Dry plains, Man. to Mont., s. to Tex., N.M., Ariz. and Calif.; rarely adv. eastw. July–Sept.

2. G. dracunculoìdes (DC.) Blake (resembling *Dracunculus*). — *Annual*, 1.5–9 dm. high, *with* numerous *pedicelled hemispherical heads; disk-flowers perfect but infertile*, 10–20, *with pappus of 5–8 bristle-like scales united at base and slightly dilated upward; rays 5–10, the pappus a minute crown.* (*Amphiachyris* Nutt.) — Prairies, rocky slopes and roadsides, Ala. to Tex., n. to w. Mo. and Kans. July–Oct. (Mex.)

12. HETEROTHÈCA Cass.

Characters as in *Chrysopsis*, but the achenes of the ray thickish or triangular, without pappus or obscurely crowned; and those of the disk compressed, with a double pappus, the inner of numerous long bristles, the outer of many short and stout bristles. — Small Am. genus. (Name from the Greek *heteros, different*, and *thece, case*, alluding to the unlike achenes.)

1. H. subaxillàris (Lam.) Britt. & Rusby (subaxillary), Camphorweed. — Annual or biennial, 3–9 dm. high, bearing numerous small heads; leaves oval or oblong, the lower with petioles auricled at base, the upper mostly subcordate-clasping. — Sandy soil, Fla. to Ariz. and Mex., n., chiefly as an adv. weed, to Del. and s. N.J., Ill., nw. Mo. and Kans. Aug.–Nov. (Trop. Am.)

13. CHRYSÓPSIS Ell. Golden Aster

Heads many-flowered, radiate; the rays numerous, pistillate. Phyllaries linear, imbricated, without herbaceous tips. Receptacle flat. Achenes obovate to linear, flattened, hairy; pappus in all the flowers double, the outer of very short and somewhat chaffy bristles, the inner of long capillary bristles. — Chiefly perennial low herbs of N. Am., woolly, hairy or glutinous, with rather large often corymbose heads terminating the branches. Disk- and ray-flowers yellow. (Name composed of the Greek *chrysos, gold*, and *opsis, aspect*, from the golden inflorescences.)*

a. Leaves all prolonged, linear to lanceolate and silvery, with parallel nerves or ribs; involucre turbinate; achenes compressed-fusiform or nearly linear.
Leaves erect or ascending, the basal ones prolonged and resembling leaves of *Luzula*, the cauline ones rapidly shorter and appressed-ascending.
Plant cespitose, forming small tufts with new basal offshoots close to or only slightly distant from flowering stems; involucre glabrous, the backs of the phyllaries stipitate-glandular. . . . . 1. *C. graminifolia.*
Plant producing slender elongate lash-like or cord-like prolonged stolons and slender rhizomes; involucre silky-webbed at least at base, the phyllaries not stipitate-glandular. . . . . 2. *C. nervosa.*
Leaves divergent, spreading-recurving, along the slightly zigzag stem. . 3. *C. falcata.*
a. Leaves from linear to oblanceolate, oblong or obovate, not parallel-ribbed; involucre hemispherical; achenes obovate, more or less flattened. . . *b*.

* The specific names here used may have to be altered when the types of some species can be examined.

*b.*Pubescence woolly-tomentose or -arachnoid, often deciduous in flocculent tufts.

Leaves, involucres and sometimes the stems becoming green and glabrescent; phyllaries with ascending tips, minutely glandular. 4. *C. mariana.*

Leaves, stems and involucres densely and permanently silvery-lanate; phyllaries with recurving tips. . . . . 5. *C. Longii.*

*b.*Pubescence of distinct hairs, from villous or pilose to silky or hirtellous, not cobwebby nor deciduous. . . *c.*

*c.*Annual with soft tap-root; leaves of primary axis oblong or oblong-lanceolate, with broad sessile bases, the lower often toothed; achenes obscurely 10-nerved; outer pappus pale-like, inner of about 20 bristles; receptacle not alveolate. . . . . 6. *C. pilosa.*

*c.*Perennials with subligneous bases; leaves linear to oblanceolate, narrow-based, rarely toothed; achenes 3–5-nerved; outer pappus of minute bristles, inner of 25–50 bristles; receptacle deeply alveolate, the broad alveolae 4–5-angled, with thin walls. . . *d.*

*d.*Involucres and peduncles whitish-pubescent (rarely glabrescent), not obviously glutinous; leaves subpersistent. . . *e.*

*e.*Foliage greenish or merely griseous, not silvery. . . *f.*

*f.*Leaves oblanceolate to spatulate, the larger ones of the main axis 0.5–2 cm. broad. . *g.*

*g.*Pubescence of phyllaries closely appressed to almost wanting; peduncles with appressed to obliquely ascending short pilosity.

Leaves oblanceolate, acute to acuminate, the larger ones 5–10 cm. long and 0.8–2 cm. broad and frequently toothed; involucres of fully developed heads 7–11 mm. high, their phyllaries sparsely pilose to glabrescent; prairie-species. . . 7. *C. camporum.*

Leaves spatulate, obtuse or merely mucronate, the larger ones 2–4 cm. long and 0.5–1 cm. broad, entire; involucre 6–8 mm. high, their phyllaries canescent-strigose; species of the Great Plains. . . . . 8. *C. Bakeri.*

*g.*Pubescence of phyllaries spreading; peduncles with long divergent villosity. . . . . . . 9. *C. Ballardi.*

*f.*Leaves linear-oblanceolate, acuminate, often toothed, the larger ones 2.5–5 mm. broad. . . . . . 10. *C. angustifolia.*

*e.*Foliage and phyllaries silvery-sericeous; leaves linear-oblanceolate, acute, 2–5 mm. broad. . . . . . . 11. *C. Berlandieri.*

*d.*Involucre and peduncles glabrous except for glands and resinous atoms, glutinous; leaves linear-oblanceolate, acuminate, 1–6 mm. broad, green, soon deciduous. . . . . 12. *C. stenophylla.*

1. C. graminifòlia (Michx.) Ell. (grass-leaved), SILKGRASS. — *Cespitose or subcespitose* perennial, *with erect leafy offsets close or approximate to the* 1–10 erect *flowering stems; the latter* 3–7.5 dm. high, *cottony-sericeous* with somewhat flocculent white pubescence, *covered by numerous erect linear or linear-lanceolate silky-arachnoid leaves; basal and lower cauline leaves linear to elongate-lanceolate,* 1–3 *dm. long,* 0.2–1.2 cm. broad, *with midnerve and parallel lateral nerves prominent beneath;* inflorescence corymbiform-paniculate, with ascending branches; *peduncles remotely bracted, glutinous or stipitate-glandular;* involucre turbinate, 7–9 mm. high, with several series of *narrowly linear glabrous but stipitate-glandular or glutinous phyllaries.* (Var. *aspera* (Shuttlew.) Gray; *C. aspera* Shuttlew.; *Pityopsis aspera* (Shuttlew.) Small) — Siliceous or argillaceous dry pine- or oak-woods and openings, n. Fla. to La., n. to Va. July–Sept.

2. C. nervòsa (Willd.) Fern. (nerved), SILKGRASS. — Similar to no. 1; *producing elongate lash-like slender horizontal stolons* up to 3 dm. long, these tipped by erect linear to lanceolate parallel-ribbed white-silky leaves; stems few (1–4), 0.15–1.2 m. high; panicle open, with elongate branches (rarely with 1 head); *peduncles closely covered with imbricated bracts, these merging into the lower phyllaries and, like them, arachnoid-lanate; involucre 8–13 mm. high, silky-pilose to -lanate,* without evident glands; longer phyllaries scario-chartaceous except for slender midrib and tip, linear, linear-lanceolate or narrowly oblong, 1–1.4 mm. broad. (*C. graminifolia* sensu most auth., not (Michx.) Ell.; *C. argentea* (Pers.) Ell.) — Dry sands of the Coastal Plain, Fla. to e. Tex., n. to Del., centr. Ala., s. Ark. and se. Okla.; siliceous rock and sands, W.Va., Ky. and Tenn. Late July–Oct.

Var. **virgàta** Fern. (wand-like). — Inflorescence slenderly cylindric-thyrsiform, with very short branches; involucre as in typical *C. nervosa.* — Sands of se. Va. and e. N.C.

Var. **stenòlepis** Fern. (with narrow scales). — Habit of typical *C. nervosa; phyllaries herba-*

ceous except for very narrow margins, *linear-attenuate, only* 0.5–1 *mm. broad.* — Sandy pine-barrens, e. Md. to S.C.

3. C. falcàta (Pursh) Ell. (sickle-shaped). — *Stems* few–many from a loosely branching crown, slender, often branching, *slightly zigzag, cottony,* 1–4 dm. high; *leaves subuniform, crowded along the stems,* linear-attenuate to linear-oblong, *loosely ascending to widely spreading, sessile, often falcate;* heads corymbose, numerous, with turbinate-campanulate involucres. — Dry sandy soil of Coastal Plain, se. Mass. to N.J. July–Oct.

4. C. mariàna (L.) Ell. (of Maryland). — Perennial, with 1–several erect or arched-ascending stems from a thick crown; *stems silky with long and weak arachnoid and floccose hairs, becoming glabrate,* 2–7 dm. high; lower *leaves* spatulate to narrowly obovate, netted-veined, the upper smaller and oblong to lanceolate, *all becoming green and glabrate;* heads rather large, in corymbs; *peduncles and hemispherical involucre bearing minute pale sessile to short-stipitate glands; phyllaries green, merely glandular, with ascending tips.* — Dry sandy to argillaceous or rocky woods and openings, Fla. to e. Tex., n. to se. N.Y., Pa., W.Va. and s. O. Aug.–Oct. — Forma **efúlgens** Fern. (without glitter) has rayless heads.

Var. **macradènia** Fern. (with large glands). — Glands of peduncles and involucres sordid or fuscous and largely long-stipitate. — E. Va. tc W.Va. and mts. of N.C.

5. C. Lóngii Fern. (for one of its discoverers, Bayard Long, 1885–), Long's Flannel-weed. — Biennial or short-lived perennial with new rosettes on short basal offshoots; leaves, stems, peduncles and involucres *densely and permanently silvery-lanate;* stems 1–10, arched-ascending or decumbent, 4.5–7.5 dm. high, loosely corymbose-branched; basal leaves oblanceolate, blunt; *cauline leaves* numerous, from *narrowly obovate to oblong, round-tipped, sessile,* broad-based; *involucre hemispherical,* 1–1.5 cm. high, the 5–6 series of *attenuate linear-lanceolate phyllaries with recurving tips;* lobes of disk-corollas long-villous; achenes oblanceolate, stipitate. — Dry sandy pine-barrens, s. Southampton County, Va. Mid-July–early Oct.

6. C. pilòsa Nutt. (softly pubescent). — *Annual,* with simple or branching erect *villous* and often somewhat viscid *stem* 0.5–1.2 m. high; *leaves of main stem oblong or oblong-lanceolate, broad-based, sessile,* pilose, the lower often toothed; heads few–several, corymbose; peduncles and hemispherical involucre viscid; *phyllaries* attenuate, *the herbaceous back* more or less *villous; achenes obscurely* 10-*nerved; outer pappus pale-like,* the inner of about 20 bristles; *receptacle not alveolate,* the small pits roundish. (*C. Nuttallii* Britt.) — Sandy or rocky woods and openings, s. Mo. and Kans. to La. and Tex. July–Oct.

7. C. campòrum Greene (of the prairies). — Simple to slightly branched *stoloniferous* perennial, with more or less villous stem 0.3–1.1 m. high; *leaves oblanceolate, acute to acuminate, the larger ones* 5–10 *cm. long and* 0.8–2 *cm. broad, often toothed,* strigose-pilose but greenish; corymb of 1–several large peduncled heads; peduncles with close short subappressed pilosity; hemispherical *involucre* 7–11 *mm. high, with minutely appressed-pilose to glabrous* linear-lanceolate acuminate *phyllaries;* outer pappus of short bristles, inner of 40–50 long bristles. (*C. villosa* in part of ed. 7, not (Pursh) Nutt.) — Sandy barrens, prairies and hills, Wisc. to w. Ind. and e. Mo. July–Oct. .

8. C. Bàkeri Greene (for its discoverer, Charles Fuller Baker, 1872–1927). — Stems *cespitose, from a heavy crown,* 1–3 dm. high, mostly arched-ascending, puberulent and loosely hirsute-villous, mostly branching above middle; *leaves spatulate, obtuse,* the lower petioled, *the longer ones* 2–4 *cm. long and* 0.5–1 *cm. broad, entire,* strigose-pilose; *peduncles short-pilose with appressed or ascending hairs; involucres* hemispherical, 6–8 *mm. high; the phyllaries* about 4-ranked, *canescent-strigose.* (*C. villosa,* in part, of ed. 7, not (Pursh) Nutt.) — Dry plains and hills, Minn. to Ida., s. to Okla. and N.M. July–Sept.

9. C. Bállardi Rydb. (for its discoverer, Caswell Aden Ballard, 1867–). — Similar to no. 8; *stems* up to 6 dm. high, *copiously and divergently long-villous;* leaves 3–7 cm. long, up to 1.5 cm. broad, apiculate; *peduncles and phyllaries spreading-villous; involucre* 1–1.2 *cm. high.* — Siliceous soils, woods, hillsides and prairies, Wisc. to Sask., s. to Ill. and S.D. June–Aug.

10. C. angustifòlia Rydb. (narrow-leaved). — Habit of nos. 8 and 9, the caudex more slender; *stems* very stiff, subsimple to much branched, *harsh, hispid-villous; leaves linear-oblanceolate, acuminate,* often crowded, *frequently toothed, the larger ones* 2.5–5 *mm. broad, hispid and strongly bullate,* bristly-ciliate; *peduncles densely short-pilose;* involucres campanulate-hemispherical, 9–10 mm. high; phyllaries pilose on back. — Dry sands and rocky slopes, Minn. to Alta., s. to Mo. and Kans. July–Sept.

11. C. Berlandièri Greene (for its discoverer, Jean Louis Berlandier, 1805–1851). — Habitally like nos. 8–10; *stems, leaves and phyllaries very densely and closely silvery-silky,* the upper internodes often also spreading-hirsute; *leaves narrowly oblanceolate, acute,* 1–4 cm. long, 2–5 *mm. broad,* with long basal cilia; involucre campanulate, 6–10 mm. high, often subtended by leafy bracts. — Sands and dry rocks, nw. Mo. and Kans. to Tex. and N.M. July–Oct.

12. C. stenophýlla (Gray) Greene (slender-leaved). — Cespitose, the stiffly erect stems 1.5–4 dm. high, hispid and *glutinous; leaves linear-oblanceolate, acuminate,* 1–6 *mm. broad,* rigid, stiffly ascending, *coarsely bullate, hispid, glutinous, soon deciduous; peduncles glabrous except for glands and resinous atoms; involucres glandular and resinous,* about 8 *mm. high.* — Hillsides and dry prairies, Kans. and Colo. to Ark. and Tex. June–Sept. (Possibly not in our range)

14. SOLIDÀGO L. Goldenrod. Verge d'Or (Que.)

Heads few–many-flowered, mostly radiate; rays 1–20 or so, pistillate. Phyllaries appressed, destitute of herbaceous tips (except nos. 4–6). Receptacle small, not chaffy. Achenes many-ribbed, nearly terete; pappus simple, of equal capillary bristles. — Perennial herbs, chiefly of N. Am. (few in S. Am. and Eurasia, 1 on the Azores), with mostly virgate stems and sessile or nearly sessile rarely cordate cauline leaves. Heads small, racemed or clustered; flowers both of the disk and ray yellow (cream-color in nos. 9 and 13 and rare albinos). Closely related species tending to hybridize freely. (Name from the Latin *solidus, whole,* with the suffix *-ago,* probably in allusion to reputed vulnerary qualities.)*

Artificial Key

A. Inflorescences in the upper axils or terminal and with elongate axes, forming racemose, thyrsoid or spreading panicles or, if corymbiform, with the heads racemosely arranged on the branches. § Virgaurea. . . B.

B. Inflorescence of small axillary clusters or forming a terminal panicle or thyrse, when paniculate or corymbiform with the heads spirally arranged and not in strongly 1-sided racemiform branches. . . C.

C. Cauline leaves nearly uniform or only gradually decreasing in size upward, including the longer basal bracts of the panicle and the leaf-scars (on lower part of stem) 20–90 or more (only 7–20 in no. 3, with zigzag angled stems and essentially uniform ovate saliently toothed leaves abruptly narrowed to winged petioles); separate basal rosettes wanting. . . D.

D. Leaves pinnately veined. . . E.

E. Stems glabrous (rarely pilose, or in rare form of no. 3 copiously pilose); leaves membranaceous, sharply serrate, the upper elongate and subtending remote axillary clusters of heads or distant loosely flowering branches; achenes pubescent.

Stem terete, when fresh covered with bloom; primary leaves lanceolate to narrowly oblong. . . . . 1. *S. caesia.*

Stem angled, with decurrent lines below the leaf-bases, not glaucous.

Leaves oblong to elongate-lanceolate, attenuate to the entire sessile base. . . . . 2. *S. Curtisii.*

Leaves ovate, abruptly contracted to a winged petiole. . . . 3. *S. flexicaulis.*

E. Stems puberulent or scabrous-pilose; leaves firm, at least the middle and upper entire or shallowly dentate; inflorescences thyrsoid to corymbose, terminal, their lower bracts not much prolonged; achenes glabrous.

Leaves elliptic-oblong or narrowly obovate to broadly lanceolate; involucres 5.5–8.5 mm. high, their phyllaries oblong.

Margins of the firm to coriaceous leaves scabrous-ciliolate with short incurving hairs; uppermost leaves (below inflorescence) 1–6 cm. long; inner phyllaries lance-attenuate. . . . . . 4. *S. petiolaris.*

Margins of the membranaceous leaves with long divergent cilia; uppermost leaves 3–14 cm. long; inner phyllaries oblong, obtuse to acute. . . . . 5. *S. Buckleyi.*

Leaves (cauline) narrowly lanceolate; involucres 3–4.5(–5) mm. high, their phyllaries linear-attenuate. . . . . . . 11. *S. puberula.*

D. Leaves with 2 of the nerves much prolonged parallel to the midrib ("triple-nerved"); inflorescence a terminal thyrse or panicle. . . Species under the second "B" with compact panicles might be sought here.

---

* Treatment of many species based on that in Rhodora, xxxviii. 201–229 (1936); that of § *Euthamia* with the aid of Stuart K. Harris. Solidago, like Aster, is one of our most difficult genera. Natural hybridization frequently occurs and the species are also highly plastic. For proper study full specimens, showing subterranean parts and basal leaves as well as the whole flowering stem, are essential. Identification of fragmentary specimens is safe only after long familiarity with the group. In the figures (in general) the habit (in outline) is × 1/6; separate leaves, × 1/4; leaf-margin or back, × 2; section of stem, × 2; involucre, × 4.

C. Cauline leaves very unequal, the lower conspicuously larger than the middle and upper, 2–30, rarely –60 or more; basal offshoots usually producing rosettes of large leaves. . . F.

F. Phyllaries (especially the 2–4 outer series) with strongly recurving green tips; involucres 7–9 mm. high. . . . . 6. *S. squarrosa.*

F. Phyllaries with appressed-ascending to but slightly spreading tips. . . G.

G. Lower and principal cauline leaves tapering to long acuminate tips, membranaceous, coarsely and saliently serrate; stems (except in alpine dwarfs) 0.5–1.5 m. high; northern species.

All but uppermost leaves elliptic to ovate, abruptly contracted to petioles; involucres 8–12 mm. high; achenes 3–4 mm. long, glabrous. . . . . 7. *S. macrophylla.*

All or all but uppermost leaves elliptic-lanceolate to narrowly ovate, tapering about equally to base and elongate caudate tip; involucres 4.5–7 mm. high; achenes 1–2 mm. long, pilose. . . 8. *S. calcicola.*

G. Lower and principal cauline leaves obtuse to acute, if rarely long-acuminate not saliently serrate and the lowest tapering into petioles. . . H.

H. Leaves pilose or setose on one or both surfaces.

Ligules cream-color to nearly white; outer phyllaries usually with strongly contrasting green tips and whitish to stramineous bases and margins. . . . . 9. *S. bicolor.*

Ligules orange-yellow; outer phyllaries usually with less contrasted coloring. . . . . 10. *S. hispida.*

H. Leaves (cauline and commonly the rosette-leaves) glabrous or at most minutely puberulent on both surfaces or merely with the principal nerves minutely setulose beneath and the margins ciliate. . . I.

I. Stems and branches of thyrse covered with minute dust-like puberulence; cauline leaves numerous (12–60), approximate, the upper often subtending small axillary tufts; involucres 3–4.5 mm. high, with linear-attenuate phyllaries. . . . . . 11. *S. puberula.*

I. Stems and branches of thyrse glabrous or pilose (sometimes puberulent in nos. 19 and 22 with involucres 5–9 mm. high); cauline leaves 3–40 (rarely more), mostly remote or scattered; involucres often longer, with broader blunt phyllaries or, if these attenuate, the heads larger and plants low. . . J.

J. Cauline leaves 2–20 (rarely –60 or more), membranaceous to firm, mostly with evident primary veins; the basal narrowly oblanceolate to obovate or elliptic, oftenest toothed, 0.5–7 cm. wide, tapering to subsessile or wing-petioled bases; plants rather low, 0.2–rarely 10 dm. high, with campanulate to hemispherical mostly many-flowered heads. . . K.

K. Well-developed heads 15–50-flowered; flowering stems single or often tufted; basal and rosette-leaves 0.2–2 dm. long, the blade membranaceous to coriaceous, not fleshy, usually much longer than the petiole or petiolar base; plants often gummy or glutinous, chiefly of dry, rocky or alpine habitats. . . L.

L. Achenes glabrous. . . M.

M. Leaves dilated, flat, rather thin, the lower cauline distinctly toothed; the basal mostly 1–5 cm. wide; involucres (3–)3.5–6 mm. high.

Principal cauline leaves oblanceolate to cuneate-spatulate; rachis and branches of thyrse pilose with elongate flexuous hairs; plant of the Northeast. . 10. *S. hispida,* var. *tonsa.*

Principal cauline leaves elliptic-oval or -lanceolate; rachis and branches of thyrse hispid with short stiff hairs; southeastern species.

Phyllaries linear to linear-attenuate, mostly slenderly 3-nerved. . . . . 12. *S. roanensis.*

Phyllaries oblong, rounded at tip, coarsely 1-nerved. . . . . 13. *S. erecta.*

M. Leaves (especially the upper) subrigid, with inrolled subulate tips, entire, linear to narrowly oblanceolate, the basal less than 1 cm. wide; involucres 2.5–3.5

mm. high, with oblong blunt firm phyllaries; plant of Great Lakes reg. . . . . 14. *S. jejunifolia.*

L. Achenes pubescent. . . N.

N. Cauline leaves elliptic-ovate to -obovate, membranaceous, distinctly many-toothed, the median 2–4 cm. broad. . . . . 15. *S. sciaphila.*

N. Cauline leaves oblanceolate, lanceolate or spatulate, firm to submembranaceous, entire or with few small teeth. . O.

O. Inflorescence corymbiform, its branches and pedicels white-villous; margins of lower leaves (at least toward base) densely villous with long soft curling and mostly entangling hairs; arctic-alpine species. 16. *S. multiradiata.*

O. Inflorescence racemiform, thyrsiform or rarely corymbiform, its branches and pedicels glabrous or setose to hirsute or pilose, hardly villous; margins of lower leaves smooth or scabrous, if ciliate rarely with long entangling hairs. . . P.

P. Pedicels and branches of inflorescence pubescent; flowering stems in tufts or solitary, the base not producing stolons but often with rosettes of leaves. . Q.

Q. Heads mostly only short-pedicelled (pedicels rarely more than 4 mm. long) or sessile, forming a dense thyrse; midrib of basal leaves broad below. . R.

R. Achenes 3–3.5 mm. long; heads with 30–60 or more flowers; phyllaries herbaceous, mostly 1–2 mm. wide; lobes of disk-corolla 1.2–2 mm. long; anthers 2–3 mm. long; alpine, of N. E. and N. Y. . . . . . 18. *S. Cutleri.*

R. Achenes 1.8–2.8 mm. long; heads with 15–30 (rarely –50) flowers; phyllaries mostly chartaceous. . . S.

S. Lower and median cauline leaves oblanceolate, obtuse or merely subacute; principal phyllaries oblong to spatulate-oblong.

Leaves dark green, thinnish; involucre 5.5–8 mm. high; phyllaries with prominent deep green glutinous midrib and tip; lobes of disk-corolla 1–1.8 mm. long; plants of Anticosti Island.

Involucre 5.5–8 mm. high, inner phyllaries oblong and 1–1.7 mm. wide; lobes of disk-corolla 1.3–1.8 mm. long. . . . . . . 19. *S. anticostensis.*

Involucre 5.5–6 mm. high, inner phyllaries spatulate and barely 1 mm. wide; lobes of disk-corolla 1 mm. long. . . . . . . 20. *S. Victorinii.*

Leaves pale green, coriaceous; involucre 4–5 mm. high, its stramineous phyllaries with pale green midrib not conspicuous; lobes of disk-corolla 0.7–1 mm. long; western. . . . 21. *S. decumbens,* var. *oreophila.*

S. Lower and median cauline leaves lanceolate to narrowly elliptic, acute; plants of e. U. S. and adj. Can. . . T.

T. Leaves submembranaceous to membranaceous, deep green, the basal narrowly obovate, oblanceolate or lanceolate; involucres 4.5–6 mm. high; lobes of disk-corollas 0.8–1.5 mm. long; anthers 1.5–2 mm. long.

Phyllaries lanceolate to linear-oblong,

mostly acute or subacute; achenes strigose with long slender hairs. . . 22. *S. Randii.*

Phyllaries oblong, rounded at tip; achenes sparsely hispid with short divergent hairs. . . . . . 23. *S. Maxoni.*

T. Leaves coriaceous, pale green, the basal rounded-obovate; involucres 6–9 mm. high, with oblong obtuse phyllaries; lobes of disk-corollas 2 mm. long; anthers 2.7–3 mm. long. . . . . 24. *S. Deamii.*

Q. Heads mostly long-pedicelled (pedicels chiefly 5–15 mm. long), forming a loose raceme or racemiform thyrse; midrib of basal leaves very slender.

Leaves mostly acute or subacute, the cauline 6–40; involucres 5–10 mm. high.

Foliage membranaceous; basal leaves elliptic, long-petioled; involucre hemispherical, its herbaceous phyllaries lance- or linear-attenuate. . . . . . 17. *S. mensalis.*

Foliage firm; basal leaves oblanceolate, short-petioled; involucre cylindric-campanulate; its chartaceous phyllaries linear, obtuse to acutish. . . . . . 25. *S. racemosa.*

Leaves rounded or subtruncate at apex, spatulate-oblanceolate, firm, the cauline 3–8; involucres 3–4 mm. high. . . . . . . 26. *S. chlorolepis.*

P. Pedicels, etc., glabrous or nearly so; flowering stems solitary; rhizome bearing flagelliform stolons.

Leaves scabrous-ciliolate, gradually reduced upward, spreading; panicle one-half to as broad as high; inland and western species. . . . 31. *S. missouriensis.*

Leaves glabrous throughout; the cauline very numerous, abruptly reduced in size above the lower third or half of stem, appressed; panicle spiciform, much longer than thick; species of Coastal Plain. . . . . . . 29. *S. stricta.*

K. Well-developed heads 9–15-flowered; flowering stems usually single; basal and rosette-leaves lanceolate to oblanceolate, 1–4.5 dm. long, the fleshy blade about equaled by the long petiole; plants not glutinous, of peaty or wet habitats; achenes glabrous or nearly so. . . . . 27. *S. Purshii.*

J. Cauline leaves 18–40, coriaceous, finely reticulate but with lateral veins obscure above; basal leaves ovate to oblong-lanceolate, mostly entire, 2–10 cm. wide, contracted to elongate petioles; plant stiff and tall (0.6–2 m.), with heads sessile in stiff racemes forming a dense thyrse; involucre cylindric-campanulate, few-flowered, with chartaceous broad-oblong blunt phyllaries; achenes glabrous. . . . . 28. *S. speciosa.*

B. Inflorescence a terminal panicle or thyrse, the heads borne on the upper side of the racemose or spicate branches (secund). . . U.

U. Basal leaves much larger than the middle and upper, often forming rosettes; cauline leaves 5–40 (rarely –70), decreasing rapidly in size up to the inflorescence; flowering stems solitary or few. . . V.

V. Stems and leaves glabrous or merely pilose or hirsute, not closely cinereous-puberulent throughout; rays mostly orange- or deep yellow. . . W.

W. Pedicels and branchlets of inflorescence glabrous. . . X.

X. Panicle slender and spiciform, 1.5–3.5 cm. thick, many times longer than broad, or with slender wand-like erect branches; basal leaves spatulate, obtuse; upper half or two-thirds of the solitary stem with numerous ascending or appressed callus-tipped abruptly reduced small entire leaves; rhizome cord-like, tardily producing horizontal stolons; Coastal Plain species. . . . . 29. *S. stricta.*

X. Panicle broader, usually with widely spreading to arched-recurving branches; if spiciform the plant nonstoloniferous, with acute basal leaves and only gradually decreasing cauline ones. . . Y.

Y. Leaves fleshy, entire, smooth or smoothish; the lower and median cauline obtuse or merely acute, 1–4 dm. long, their bases clasping half or three-fourths of the circumference of the stem; heads hemispherical, 20–50-flowered; involucre 4–7 mm. high, with linear- to oblong-lanceolate acute or actuish subherbaceous phyllaries; pappus 3.5–5.5 mm. long; disk-corollas 4–5.5 mm. long; mature achenes 2.2–3.5 mm. long; maritime species. . . . . 30. *S. sempervirens.*

Y. Leaves scarcely fleshy, serrate or entire, mostly scabrous-ciliolate; the lower and median cauline blunt to sharply acute, their bases rarely covering half the circumference of the stem; heads cylindric-campanulate, 5–25-flowered; involucres 2.5–6 mm. high, with oblong or narrowly ovate mostly obtuse and firm phyllaries; pappus 2–3.5 mm. long; disk-corollas 2.5–4 mm. long; mature achenes 1–2.2 mm. long. . . Z.

Z. Heads 5–12-flowered; involucres 3.5–5 mm. high, with linear-oblong subherbaceous or thin-chartaceous phyllaries; plant of bogs and swamps. . . . . 37. *S. uliginosa*, var. *levipes.*

Z. Heads 15–25-flowered; involucres with broader and firmer chartaceous phyllaries; plants mostly of dry habitats. . . *a.*

a. Base of plant freely producing slender stolons; principal leaves clearly triple-nerved (2 of the lateral nerves prolonged parallel with the midrib), scabrous-ciliolate; inland and western species. . . . . 31. *S. missouriensis.*

a. Base without slender stolons or rarely producing them (in no. 33); leaves not at all or only delicately triple-nerved. . . *b.*

b. Middle and upper cauline leaves very small, oblong, blunt, reduced below inflorescence to only 0.2–2 cm. long; panicle subcorymbiform, with ascending scarcely recurved branches; achenes coarsely corrugate-ribbed; inland species. . . . . 32. *S. Gattingeri.*

b. Middle and upper cauline leaves lanceolate or linear, attenuate, acute or subacute, those just below inflorescence 0.5–10 cm. long; panicle recurving at tip or with mostly arched-recurving chiefly secund-racemiform branches; achenes delicately slender-ribbed. . . *c.*

c. Leaves glabrous throughout, many of them with 2 of the lateral nerves prolonged parallel with the midrib, the basal leaves only 5–17 mm. broad; achenes glabrous or glabrate; southeastern species. . . . . 33. *S. pinetorum.*

c. Leaves with ciliolate-scabrous margins, none of them clearly triple-nerved; achenes pubescent. . . *d.*

d. Inflorescence broadly pyramidal or rhombic, nearly or quite as broad as long; involucres 3–4.5 mm. high, hemispherical, with usually pale phyllaries; petioles of rosette-leaves wing-margined, long-ciliate at base; uppermost cauline leaves sharply pointed, 2–10 cm. long; chiefly northern and inland species. . . . . 34. *S. juncea.*

d. Inflorescence elongate, two to fifteen times as long as thick; involucres 4–6 mm. high, campanulate, their phyllaries with deep green median bands; petioles of rosette-leaves essentially wingless, eciliate; uppermost cauline leaves blunt or merely acute, 0.5–3 cm. long; southeastern species.

Blades of basal leaves 0.7–1.5 dm. long, 1–2.5 cm. broad; the midrib whitish, wire-like, rounded on back, the reticulum fine and inconspicuous; involucres 4–5 mm. high; ligules 1–1.5 mm. wide; achenes 3–3.6 mm. long. . . . . 35. *S. austrina.*

Blades of basal leaves 1–2.5 dm. long, 1.5–5 cm. broad; the midrib greenish and broad, acutely angled, the reticulum coarse and conspicuous; involucres 5–6 mm. high; ligules 1.7–2 mm. wide; achenes 1.4–1.8 mm. long. . . . . . 36. *S. perlonga.*

W. Pedicels and branchlets of inflorescence hirtellous or pilose. . . *e.*

e. Leaves fleshy, entire, smooth (or scabrous-margined), the bases of the lower and median cauline clasping half to three-fourths the circumference of the stem; maritime species. . . . . . . 30. *S. sempervirens.*

e. Leaves not fleshy, at least the basal toothed, the bases of the lower and median cauline rarely (in nos. 37 and 43) covering half the circumference of the stem or, if broadly clasping the stem, the leaves ciliate or scabrous. . . *f.*

f. Lower faces of the leaves all glabrous or at most scabrous, not hirsute. . . *g.*

g. Upper half of stem not obviously quadrangular; leaf-blades smooth or scabrous, not conspicuously papillate, the margins usually ciliolate. . . *h.*

h. Bases of lower cauline leaves clasping half to three-fourths of the circumference of the stem, their blades thickish, not very veiny, elongate, tapering into the petioles; heads 7–15-flowered; marsh- or bog-plant. . . . . . 37. *S. uliginosa.*

h. Bases of lower cauline leaves hardly clasping, covering less than half the circumference of stem; heads 15–20-flowered; plants of dry or upland habitats. . . *i.*

i. Achenes glabrous or glabrate.

Basal leaf-blades tapering gradually to petiole; upper leaves entire or remotely few-toothed; branches of panicle only sparsely hirtellous. . . . . . . 34. *S. juncea.*

Basal leaf-blades rather abruptly contracted to petiole; upper leaves closely serrate; branches of panicle copiously pilose. . . . . . . 38. *S. arguta.*

i. Achenes definitely pubescent. . . *j.*

j. Disk-corollas 2.8–3.3 mm. long; pappus 2–3 (rarely –3.5) mm. long; involucres 3–5 mm. high.

Leaves coriaceous, scabrous above; the basal broadly ovate, abruptly contracted to the petioles, 3–10 cm. broad; phyllaries with broad green midrib. . 39. *S. Harrisii.*

Leaves membranaceous, smooth or barely scabrous; the basal lanceolate to oblong-ovate, tapering to the petioles, 1–3.5 cm. broad; midrib of phyllaries slender. . . . . 40. *S. Boottii.*

j. Disk-corollas 4–5.5 mm. long; pappus 3–5.5 mm. long; involucres 5–8 mm. high.

Rhizome thick, without slender stolons; pappus 3–4 mm. long; blade of basal leaf tapering to petiole; achenes 1.5–2.2 mm. long. . . . . . . 41. *S. yadkinensis.*

Rhizome slender, elongate, often producing cord-like stolons; blade of basal leaf broadly rounded above the petiole; pappus 4–5 mm. long; achenes 2–3 mm. long. . . . . 42. *S. ludoviciana.*

g. Upper half of stem quadrangular, the angles narrowly winged; leaf-blades papillate-scabrous above, the margins eciliate.

Cauline leaves 12–23, the lower and median ones sharply toothed; heads mostly in short racemes borne subspirally along the closely pilose panicle-branches; pedicels mostly much shorter than the heads; northern and upland calcicolous species. . . . . 43. *S. patula.*

Cauline leaves 20–45, the lower and median ones crenate-serrate; heads solitary or few in fascicles, borne strictly along upper sides of the sparsely pilose or glabrescent panicle-branches; pedicels mostly equaling or longer than heads; southeastern species of bogs. . . . . 44. *S. salicina.*

f. Lower faces of the basal or of some or all of the leaves hispid, hirsute or pilose, especially on the nerves. . . *k.*

k. Basal leaves often in definite rosettes, ovate or rounded, with rounded to cordate bases, 3.5–10 cm. broad.

Heads definitely pedicelled, scattered or loosely fascicled along the branches; pappus 2–4 mm. long, as long as or longer than the achene.

Inner phyllaries oblong, obtuse, about 1 mm. wide, suffused with green, without prominent keel; disk-corollas

3.5–4 mm. long; pappus 2.5–4 mm. long; mature achenes 2.5–3.5 mm. long. . . . . 45. *S. strigosa.*

Inner phyllaries linear-lanceolate, attenuate, 0.5 mm. wide, pale, with prominent yellow-brown keel; disk-corollas 3–3.5 mm. long; pappus 2–2.5 mm. long; mature achenes 1.3–1.5 mm. long. . . . . . 46. *S. neurolepis.*

Heads sessile or subsessile in rounded glomerules along the branches; pappus 0.5–1 mm. long, much shorter than to barely equaling achene. . . . . 47. *S. sphacelata.*

*k.*Basal leaves rarely in definite rosettes, lanceolate or oblanceolate to narrowly obovate, tapering gradually into the petiole, 1.5–7.5 cm. broad; inner phyllaries lance- or linear-oblong, not strongly keeled. . . . . 55. *S. ulmifolia.*

V.Stems and often the leaves closely puberulent or very minutely hispidulous; rays mostly bright pale yellow. . . *l.*

*l.*Leaves cinereous with minute close pubescence.

Leaves narrowly obovate, spatulate or oblanceolate to linear, not prominently 3-ribbed, mostly petioled or with subpetiolar bases, the upper usually subtending small axillary tufts; panicle 1-sided, with mostly spreading branches; base without slender stolons. 48. *S. nemoralis.*

Leaves elliptic or oval to narrowly obovate, mostly sessile, definitely 3-ribbed, the upper usually without axillary tufts; panicle a dense thyrse with ascending branches; base with slender stolons. . . . . 49. *S. mollis.*

*l.*Leaves green, pustular-scabrous, coriaceous, elliptic or oblong to obovate or oblanceolate, mostly sessile, the upper usually without axillary tufts; panicle with spreading or recurving branches; base without slender stolons. . . . . 50. *S. radula.*

U.Basal leaves not much larger than the cauline and rarely forming rosettes (except on late basal offshoots); cauline 20–100 or more, approximate or subapproximate (rarely very distant), nearly uniform or decreasing very gradually in size nearly to the inflorescence; flowering stems few or clustered. . . *m.*

*m.*Leaves pinnately veined or with the lateral veins not obvious, not triple-nerved, *i.e.*, without 2 lateral nerves obviously prolonged parallel with the midrib. . . *n.*

*n.*Leaves sessile by broad or rounded bases, entire or only sparingly toothed, linear to oblong, the midrib prominent beneath, the lateral nerves faint.

Foliage with punctate dots (by transmitted light); the sessile linear-lanceolate blades entire, glabrous, divergent; stem glabrous or puberulent only in lines, without slender stolons. . . . 51. *S. odora.*

Foliage not punctate; blades linear to oblong-oblanceolate, shallowly toothed, puberulent or pubescent; stem pubescent, with slender stolons.

Leaves linear or linear-oblanceolate, scabrous-puberulent, 1.5–8 mm. wide, spreading; stem minutely puberulent; involucres 2.5–4 mm. high. . . . . 52. *S. tortifolia.*

Leaves oblong or oblong-oblanceolate, usually hirsute beneath, the principal ones 1–2.5 cm. wide, ascending; stem commonly hirsute above; involucres 4–5 mm. high. . . . . . . 53. *S. fistulosa.*

*n.*Leaves gradually narrowed to base (if strongly rounded prominently veiny), coarsely toothed, oblanceolate or oblong to elliptic or ovate, their lateral nerves usually evident beneath. . . *o.*

*o.*Stem glabrous throughout or to near the inflorescence.

Leaves glabrous beneath.

Involucres 3.5–6.5 mm. high, with oblong obtuse median phyllaries; disk-corollas 4–5.5 mm. long; pappus 3–5 mm. long. . . . . 54. *S. Elliottii.*

Involucres 3–4 mm. high, with linear-lanceolate to linear-oblong median phyllaries; disk-corollas 2.5–3.5 mm. long; pappus 2–3 mm. long. . . . . 56. *S. rugosa*, var. *sphagnophila.*

Leaves hispid to villous on the nerves beneath. . . . . . 55. *S. ulmifolia.*

*o.*Stem either villous, scabrous-hirsute or scabrous-puberulent; disk-corollas 2.5–3.5 mm. long; pappus 2–3 mm. long. . . . 56. *S. rugosa.*

*m.*Leaves triple-nerved, *i.e.* with two lateral nerves obviously prolonged parallel with the midrib. . . *p.*

*p.*Summit of stem below the inflorescence densely pilose. . . *q.*

q. Principal leaves ovate or broadly oval, puberulent beneath; southwestern species. . . . . 57. *S. Drummondii.*

q. Principal leaves linear- to oblong-lanceolate, acuminate. . . r.

r. Involucres 2–2.8 mm. high; disk-corollas 2.4–2.8 mm. long; branches of panicle strongly divergent or recurving; leaves mostly serrate. . . . . 58. *S. canadensis.*

r. Involucres 3–5 mm. high; disk-corollas 3–4 mm. long.

Branches of pyramidal panicle strongly divergent or recurved; heads slender, 12–18-flowered; phyllaries scarious or scario-herbaceous, with midrib conspicuous; leaves cinereous-puberulent; species of U.S. and s. Can. . . . . 59. *S. altissima.*

Branches of thyrsiform to corymbiform-pyramidal panicle strongly ascending, tips rarely recurved; heads campanulate to hemispheric, 20–40-flowered; phyllaries subherbaceous, green, midrib not conspicuous; leaves rarely cinereous; boreal species. . . . . 60. *S. lepida.*

p. Summit of stem below inflorescence glabrous or only sparsely pilose. . . s.

s. Branchlets of panicle and pedicels glabrous; phyllaries oblong; leaves mostly subtending axillary fascicles; stems solitary, freely stoloniferous. . . . . 31. *S. missouriensis.*

s. Branchlets of panicle and pedicels pilose; phyllaries linear, linear-lanceolate or linear-oblong; leaves mostly without axillary fascicles; plants only tardily stoloniferous. . . t.

t. Involucres 2.5–3 mm. high, their herbaceous bracts linear-attenuate and at most 0.6 mm. wide.

Panicle loosely pyramidal, with elongate secund divergent branches; phyllaries in 3 or 4 series; species of the southern Appalachians. . . . . 61. *S. rupestris.*

Panicle thyrsiform or corymbiform, with suberect branches; phyllaries in 1–2 series; Newfoundland species. . . . . 62. *S. Bartramiana.*

t. Involucres 3.2–6 mm. high.

Branchlets of panicle minutely hirtellous; heads slender, subcylindric, 10–14-flowered, on long pedicels; phyllaries narrowly oblong to lanceolate, coriaceous. . . . . . 63. *S. Shortii.*

Branchlets soft-pilose; heads broadly campanulate to hemispherical, many-flowered, short-pedicelled; phyllaries linear to linear-lanceolate, subherbaceous. . . . . . 64. *S. gigantea.*

A. Inflorescence a compound corymb with flat or flattish top, the heads not racemosely arranged. . u.

u. Leaves of different sizes; the basal petioled, usually much larger than the upper, 0.5–3 dm. long; heads pedicelled; involucres 4.5–8 mm. high, with thick round-tipped phyllaries; rays 6–10, fewer than the disk-flowers; receptacle not fimbrillate. . v.

v. Stems densely puberulent; leaves scabrous-hispid, the basal oval or oblong, crenate-dentate, 3–10 cm. broad. . . . . 65. *S. rigida.*

v. Stems glabrous or nearly so; leaves glabrous at least beneath, entire or serrate. . . w.

w. Basal and lower leaves obtuse or subacute, flat, often serrate above the middle, strictly 1-nerved.

Lower leaves 3.5–9 cm. broad, scabrous above; pedicels pubescent; heads 30–40-flowered; involucres 6–8 mm. high. . . 66. *S. Jacksonii.*

Lower leaves 1–3.5 cm. broad, smooth above; pedicels glabrous; heads 16–20-flowered; involucres 4.5–6 mm. high. . . . . 67. *S. ohioensis.*

w. Basal and lower leaves acute, usually falcate, entire, 1–3-nerved; pedicels pubescent; involucres 5–8 mm. high, 20–30-flowered.

Stout, 0.5–1 m. high, very leafy; cauline leaves elongate, often longitudinally folded, conspicuously overlapping, commonly with divergent or recurved tips, the basal 1–2 cm. broad; branches of dense corymb very pubescent. . . . . 68. *S. Riddellii.*

Slender, 3–6 dm. high; leaves flat, remote, erect, the basal 0.3–1 cm. broad; branches of rather open corymb sparingly puberulent. 69. *S. Houghtonii.*

u. Leaves uniform, sessile, linear to lanceolate or narrowly oblong, very numerous; heads often in glomerules forming the corymb; involucres 3–6 mm. high, with appressed often glutinous phyllaries; rays 6–20, usually more numerous than disk-flowers; receptacle fimbrillate. § EUTHAMIA. . x.

x. Primary cauline leaves seldom subtending well-developed axillary leafy fascicles; the principal leaves 2–12 mm. broad, mostly with 3–5 distinct nerves. . . *y*.

y. Glabrous plants with heads 9–22-flowered; involucre cylindric to turbinate, 4.5–6 mm. high, the outer phyllaries passing gradually to the inner linear and obtuse series; plants of the central and Gulf states.

Leaves thin, linear-lanceolate, 4–8 mm. wide, with inconspicuous translucent punctation, the surfaces often pustular; involucre not viscid, 10–15(–19)-flowered; southern species. . . . . 70. *S. leptocephala.*

Leaves thickish, narrowly linear, 2–5 mm. wide, with conspicuous dark punctation, the surfaces rarely pustular; involucres usually viscid, 14–22-flowered; chiefly a mid-western species. . . . . . 71. *S. gymnospermoides.*

y. Glabrous or pubescent; heads 12–50-flowered; involucre cylindric to turbinate, 3–5 mm. high, with the inner phyllaries oblong to lanceolate; eastern or transcontinental species.

Stems and leaves essentially glabrous to pubescent, the angles of branchlets and pedicels usually scabrous or hirtellous; leaves of primary axis usually with 2 or 4 lateral veins evident on drying, punctate, 2.5–15 cm. long, (2–)4–12 mm. broad, attenuate or acuminate, if blunt minutely pubescent; corymb compact to open, 0.1–3 dm. or more broad. . . . . 72. *S. graminifolia.*

Stems, leaves and branchlets glabrous, the angles of the latter smooth; leaves of primary axis 1-nerved, without evident laterals, scarcely punctate, 2–6 cm. long, 2–6 mm. broad, obtuse or only subacute; corymb compact, 1–10 cm. broad; plant of N. S. and N. E. . . . . (73). *S. tenuifolia*, var. *pycnocephala.*

x. Primary cauline leaves subtending well-developed leafy fascicles or sterile leafy branches; the principal leaves 1–6 mm. broad, 1-nerved or in nos. 73 and 74 faintly 3-nerved; involucre slenderly campanulate or cylindric to slenderly turbinate, 3–5 mm. high, 9–30(–50)-flowered. . . *z*.

z. Involucre campanulate to turbinate; phyllaries firm and thickish, the outer ones ovate and round-tipped, the inner broadly oblong and obtuse; leaves flat or with incurved margin, linear-lanceolate, the primary ones 2–3.5 mm. wide; heads often in glomerules. . . . 73. *S. tenuifolia.*

z. Involucre slenderly cylindric, after pressing becoming slenderly turbinate; phyllaries thin, pale, the outer ones narrowly ovate or deltoid and tapering above, the inner narrowly oblong to linear-lanceolate, subacute to acute; many of the heads distinctly pedicelled.

Leaves flat, the primary ones 2–4.5 mm. broad, darkened in drying, often 3-nerved, axillary fascicles few; in well-developed inflorescences the lower branches spreading-ascending, with their secondary corymbs nearly reaching to overtopping the central ones, the open corymb thus nearly flat-topped; transition from outer to longer inner phyllaries abrupt; plant of Great Lakes area. . . . 74. *S. remota.*

Leaves soon strongly folded or subacicular; the lower and median primary ones 1–2.5 mm. wide, pale green, rarely with any trace of lateral nerves; axillary fascicles abundant; in well-developed inflorescences the lower and succeeding suberect or strongly ascending branches with their secondary corymbs failing to reach the summits of the central ones, the compact corymb dome-like; transition from outer to inner phyllaries gradual; plant of southeastern states. 75. *S. microcephala.*

§ Virgaùrea Endl.

1. S. caèsia L. (bluish-gray), Blue-stem G. — Slender and glabrous, the purplish or green *terete* simple to paniculately branched *stem* usually *covered with bloom*, 0.2–1 m. high, from a stout crown, tardily producing stolons; *leaves nearly uniform*, those of the primary axis 20–60, divergent or loosely ascending, *elongate-lanceolate to narrowly oblong*, thinnish and smooth, sharply serrate, *narrowed to sessile base and acuminate apex, the principal ones* 4–12 *cm. long and* 1–3.5 *cm. broad;* heads small, forming loosely paniculate leafy inflorescences, or more often simply in loose axillary glomerules on a simple stem in forma **axillàris** (Pursh) House (axillary); *involucres* 3–5 *mm. high*, with obtuse phyllaries; rays 3 or 4; achenes pubescent. (Var. *paniculata* Gray; *S. latissimifolia* Mill.) — Rich deciduous or open woods, thickets and clearings, Fla. to Tex., n. to Wisc., s. Ont., sw. Que., centr. Me. and N.S. Aug.–Oct. Fig. 1547. — Hybridizes with nos. 3, 9, 55, 56 and 58.

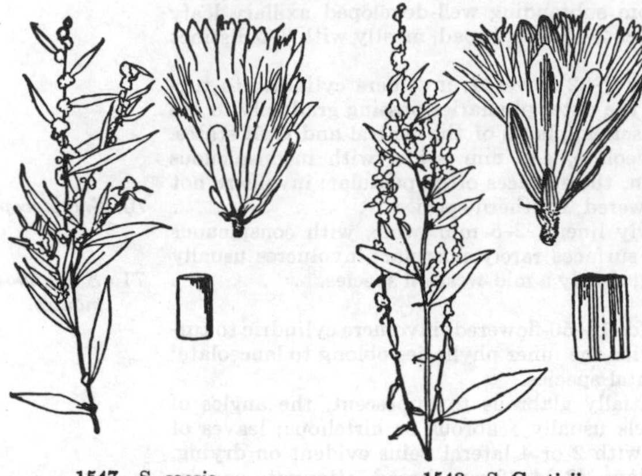
1547. S. caesia. 1548. S. Curtisii.

× **S. ulmicaèsia** Friesner (name from those of its probable parents), is an apparent hybrid of no. 1 and *S. ulmifolia.*

2. S. Curtísii T. & G. (for its discoverer, MOSES ASHLEY CURTIS, 1808–1872). — Resembling no. 1, but *stems not glaucous, striate-angled,* 0.4–1.5 m. high; *leaves oblong to elongate-lanceolate, tapering to the subsessile entire base,* long-acuminate, 1–5 cm. wide; *involucres* 3–4.5 *mm. high,* with oblong phyllaries; rays 4–7. — Rich open woods, chiefly in the uplands, Va., W.Va., and Ky., s. to Ga. and Ala. Aug.–Oct. FIG. 1548.

Var. pùbens (M. A. Curtis) Gray (short-pubescent). — Stems and often the lower leaf-surfaces pilose to tomentulose; leaves usually broad (2–6 cm.). (*S. pubens* M. A. Curtis) — Va. to Ga. and Tenn.

3. S. flexicaùlis L. (flexuous-stemmed). — *Stem* slender, *zigzag,* erect or strongly ascending, *somewhat angled,* green, simple or paniculately branched, 0.2–1 m. high, *glabrous or nearly so,* except somewhat pilose on upper internodes, the slender rhizome freely stoloniferous; *cauline leaves* 7–20, *nearly uniform, ovate,* 3–8 *cm. wide, abruptly narrowed to a short winged petiole,* acuminate, thin, *divergently sharp-serrate,* glabrescent to hirtellous-pilose beneath; the upper and bracteal leaves narrower and less toothed or entire; *heads clustered in the upper axils or a terminal leafy-bracted thyrse; involucres* 4–6 *mm. high,* their phyllaries obtuse; rays 3–4; achenes pubescent. (*S. latifolia* L.) — Rich woods, thickets and cool slopes, Gaspé Co., Que., to N.D., s. to N.S., N.E., N.C., Tenn., Ia., S.D. and ne. Kans. July–Oct. FIG. 1549. — Hybridizes with no. 1.

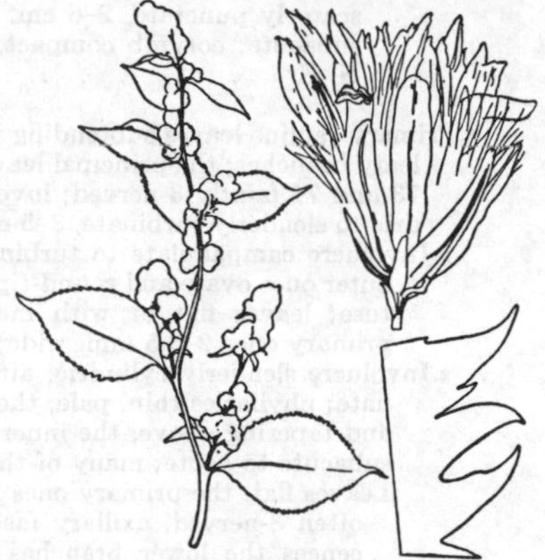
1549. S. flexicaulis.

S. albopilòsa E. L. Br. (white-pilose) has weak reclining white-pilose stem, rather small pilose leaves and small heads. — Found only in and about sandstone-rockhouses of Menifee and Powell Cos., Ky., it seems like an ecological development of no. 3, into which it passes or with which it hybridizes.

4. S. petiolàris Ait. (with foot-stalks). — *Strict,* 0.5–1.5 m. high; stem glabrous below, scabrous-puberulent or hirtellous above; *leaves* 40–60 or more, overlapping, ascending, *firm or subcoriaceous,* oblong, *entire* or somewhat dentate, opaque or sublustrous, more or less *scabrous-puberulent on both surfaces, the margins minutely ciliolate with short incurving hairs,* the very short petioles soft-pilose; lower *blades* 3–10 cm. long, *uppermost* 1–6 *cm. long; thyrse strict,* elongate (0.7–3.5 dm. long), rarely corymbiform, simple and *crowded* or with erect branches; *involucres* 5.5–8.5 *mm. high, the firm lance-attenuate often scabrous-pubescent phyllaries frequently with recurving green tips.* (*S. Milleriana* Mackenz.) — Dry woods, thickets and slopes, Fla. to Tex., n. to N.C., Mo. and Neb. Late Aug.–Oct. FIG. 1550. — Name misleading.

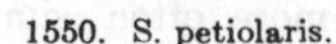
1550. S. petiolaris.

Var. Wárdii (Britt.) Fern. (for its discoverer, LESTER FRANK WARD, 1841–1913). — *Leaves*

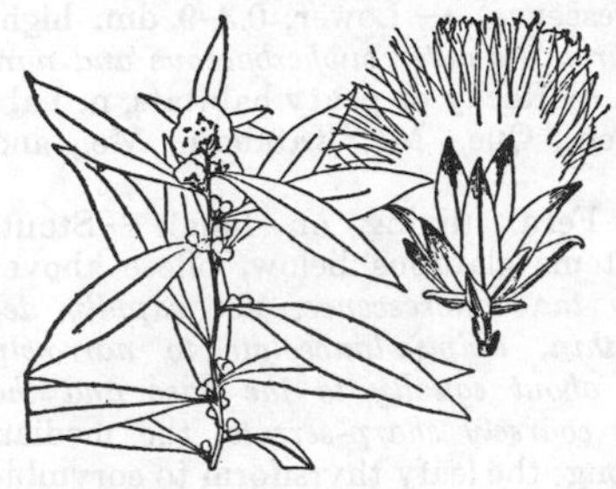
1551. S. Buckleyi.

very firm, strongly *glutinous, lustrous,* commonly less pubescent. (*S. Wardii* Britt.) — La. and Tex., n. to Mo. and Kans.

5. S. Búckleyi T. & G. (for its discoverer, SAMUEL BOTSFORD BUCKLEY, 1809–1884). — Stem 0.5–1.5 m. high, pilose at least above, simple or subcorymbosely branched; *leaves* 20–40 *on the primary axis, membranaceous, spreading,* elliptic to narrowly obovate, *hispid or hirsute on the veins beneath and hispid-ciliate with divergent* to incurving *hairs,* the lower and median 5–17 cm. long, *the uppermost and rameal* 3–14 *cm. long; thyrse loose and interrupted,* leafy-bracted; *involucres* much as in no. 4, but *the inner phyllaries oblong, obtuse to acute,* but hardly attenuate. — Woods and bluffs, W.Va. to s. Ill. and Mo., s. to Ala. Sept., Oct. FIG. 1551.

6. S. squarròsa Muhl. (from the abruptly recurved phyllaries). — *Stout* and usually tall, 0.2–1.7 m. high, *with stout rhizome producing conspicuous basal rosettes;* cauline leaves 12–25; *the lowest very large and like the rosette-leaves,* narrowly oval, *coarsely serrate,* 0.8–3.5 *dm. long,* 3–10 *cm. wide;* the middle and upper leaves much smaller, entire or subentire; *thyrse wand-like,* interrupted, or forking in forma **ramòsa** (Peck) House (branching), 1–8 dm. long (in dwarfed plants down to 4 cm.); *involucres* 7–9 *mm. high,* rather crowded, *with green tips of the* 2–4 *outer series of* lanceolate *stiffish phyllaries strongly recurved;* rays 12–16; *achenes glabrous.* — Rich dry or rocky open woods, thickets and clearings, e. Que. to s. Ont., s. to N.B., N.E., Del., Md., upland of N.C., W.Va. and Ky. Aug.–early Oct. FIG. 1552. — Hybridizes with no. 7.

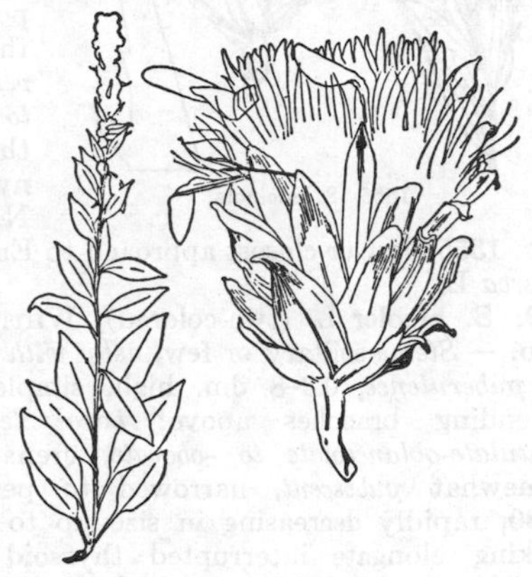
1552. S. squarrosa.

7. S. macrophýlla Pursh (large-leaved). — Stoutish, the soemwhat flexuous stem 0.1–1.5 m. high, glabrous below, pilose at summit; *leaves thin,* glabrous or essentially so except on nerves and margin below, *acuminate,* 2–20, *all but the uppermost elliptic to ovate, coarsely and sharply serrate and abruptly contracted to petioles;* the lowest blades 3–15 cm. long, 1–7.5 cm. wide; the uppermost smaller, subsessile, slightly serrate to entire; inflorescence a *loose leafy* simple or forking *raceme or thyrse* 0.3–8 dm. long; heads on naked or few-bracted pedicels, the terminal ones 15–30-flowered; *involucres slender-campanulate to thick-cylindric,* 8–12 *mm. high; their phyllaries thin, scarious, linear-attenuate;* rays 8–10; *achenes glabrous,* 3–4 *mm. long.* — Cool damp woods and thickets, s. Lab. and Nfld. to L. Superior, Ont., s. to N.S., n. N.E., w. Mass. and N.Y.; ascending to subalpine areas. Mid-July–mid-Sept. FIG. 1553. — Hybridizes with nos. 6, 56 and 58.

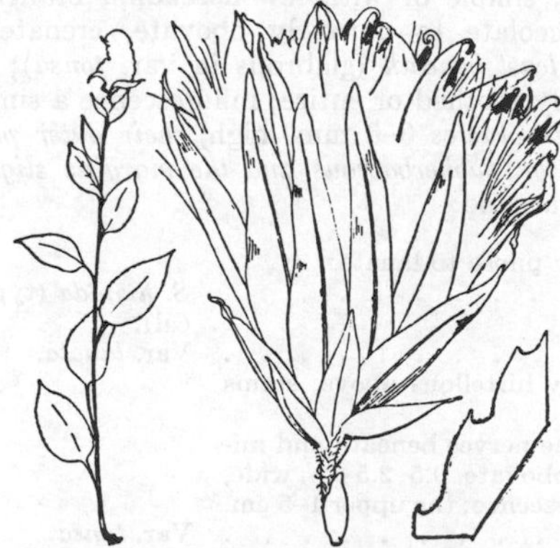
1553. S. macrophylla (typical).

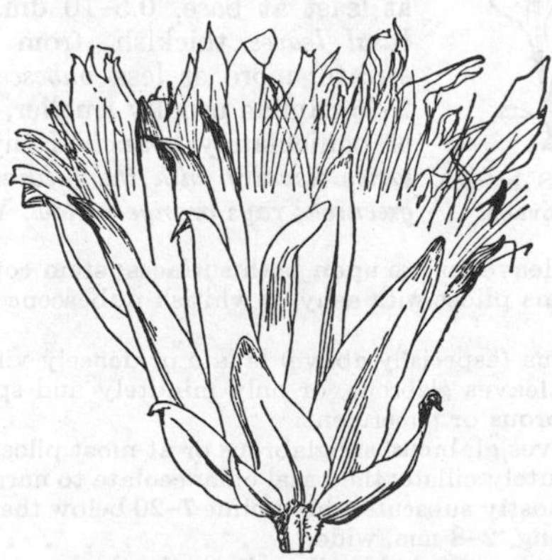
1554. S. macrophylla, v. thyrsoidea.

Var. thyrsoidea (Mey.) Fern. (with thyrse-like inflorescence). — Lower, 0.3–9 dm. high; *involucres subglobose to urceolate;* their *phyllaries* rather *firm, the outer subherbaceous and narrowly deltoid to lanceolate;* terminal heads 30–100-flowered. — Rocky or peaty habitats, n. Lab. and Ung. to exposed or alpine areas of Nfld., Gaspé Pen., Que., Mt. Katahdin, Me., and White Mts., N.H. Late July, Aug. Fig. 1554.

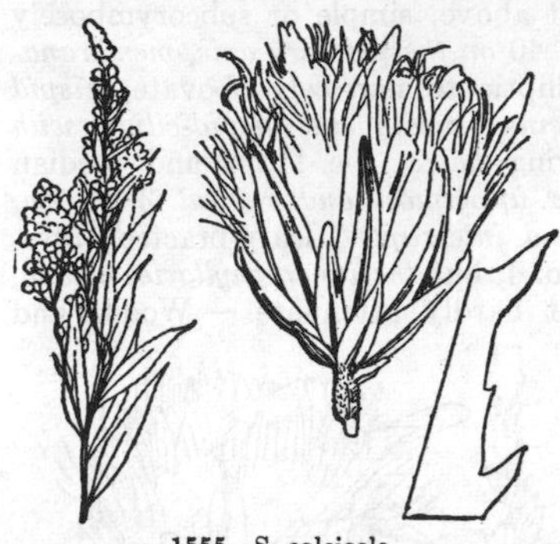

1555. S. calcicola.

8. S. calcícola Fern. (living on lime). — Stout, 0.6–1.3 m. high; stems glabrous below, pilose above; *leaves* 21–35 *below the inflorescence, not rapidly decreasing in size, thin, elliptic-lanceolate to narrowly ovate, all tapering about equally to the base and the long-acuminate tip, coarsely sharp-serrate,* the median cauline 1–1.5 dm. long; the leafy thyrsiform to corymbiform panicle 1–3.5 dm. long, with loose racemiform branches; heads mostly equaled or exceeded by the pedicels; *involucres* campanulate, 4.5–7 *mm. high,* their *linear* acute or acutish *green phyllaries in* 3–4 *very unequal series;* rays 12–15; *achenes* 1–2 *mm. long, pilose.* — Rich woods and rocky or gravelly thickets, Nfld. to nw. Que. and n. N.E. Aug., Sept. Fig. 1555. — Our closest approach to Eurasian *S. Virgaurea* L.

9. S. bícolor L. (two-colored), White G., Silver-rod. — *Stems* solitary or few, *ashy with soft pubescence or puberulence,* 0.2–8 dm. high, simple or with few ascending branches above; *basal leaves* thickish, *spatulate-oblanceolate to -obovate,* crenate or serrate, somewhat *pubescent,* narrowed to petioles; cauline 3–30, rapidly decreasing in size up to the simple or forking elongate interrupted thyrsoid or spike-like panicle; *involucres* 3–6 mm. high, *their outer phyllaries usually with strongly contrasting green tips and whitish to stramineous bases and margins; rays 7–9, whitish to cream-color,* paler than the yellow disk-flowers; achenes glabrous (rarely hirtellous), whitish-brown, 3.5–5 mm. long. — Dry, sterile open soil or thin woods, Ga. to Ark., n. to C.B., M.I., N.B., Me., sw. Que. and s. Ont. July–Oct. Fig. 1556. — Hybridizes with nos. 1, 10, 11, 48, 51 and 56.

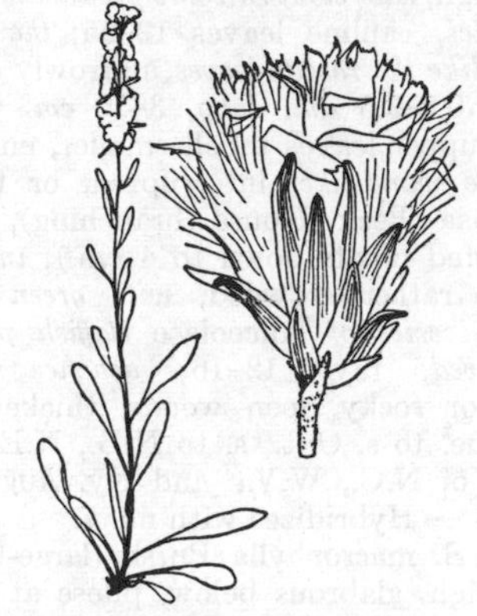

1556. S. bicolor (typical).

1557. S. bicolor, v. ovalis.

Var. ovàlis Farw. (oval). — *Basal and lower leaves broadly oval or rhombic-elliptic;* cauline leaves relatively broad. — N.J. to s. Mich., s. to Va., Ky. and Tenn. Fig. 1557.

10. S. híspida Muhl. (stiffly hairy). — Habit of no. 9; stems pubescent, at least at base, 0.5–10 dm. high, simple or with few ascending branches; *basal leaves* thickish, from oblanceolate to narrowly obovate, crenate or serrate, more or less *pubescent at least beneath* (glabrous in var. *tonsa*); the 2–25 cauline usually smaller, slightly toothed or entire; inflorescence a simple or paniculately branched thyrse; *involucres* 3–6 mm. high, *their outer phyllaries usually with the median portion subherbaceous and the margins slightly greenish; rays orange-yellow.* Very variable:

Basal leaves pilose upon both surfaces; stem copiously pilose to lanate.
  Stems pilose with ashy or whitish pubescence. . . . . *S. hispida* (typical).
  Stems (especially above) lanate or densely villous. . . . . Var. *lanata.*
Basal leaves glabrous, or only minutely and sparingly hirtellous above; stems glabrous or puberulent.
  Leaves glabrous, subglabrous or at most pilose on the nerves beneath and minutely ciliate; the basal oblanceolate to narrowly obovate, 0.5–2.5 cm. wide, mostly subacute; the cauline 7–20 below the inflorescence; the upper 1–5 cm. long, 2–8 mm. wide. . . . . Var. *tonsa.*
  Leaves sparsely hirtellous beneath; the basal oblong-obovate, 2.5–6 cm. wide,

rounded at apex; cauline 5–9 below the inflorescence; the upper 5–10 cm. long, 2–3 cm. wide. . . . . Var. *arnoglossa.*

S. híspida (typical), *S. bicolor*, var. *concolor* T. & G. — Dry or moist rocky places, chiefly calcareous, Nfld. to Man., s. to N.S., n. and w. N.E., L.I., Ga., Tenn. and Ark. July–Oct. Fig. 1558. — Simulated by hybrids of nos. 9 and 11.

Var. **lanàta** (Hook.) Fern. (woolly). — Calcareous gravels and ledges, w. Nfld.; Gaspé Pen. and L. Mistassini, Que.; n. Aroostook Co., Me.; plains of the Saskatchewan.

Var. **tónsa** Fern. (shaved). — Dry rocky or peaty soils, w. Nfld. to Algoma Distr., Ont., s. to ne. N.B., Bruce Co., Ont., and n. Mich. — Simulates the southern nos. 12 and 13.

Var. **arnoglóssa** Fern. (with a lamb's tongue). — Sea-cliffs and rocky woods, w. Nfld.; ne. Minn.(?) Fig. 1559.

1558. S. hispida (typical).

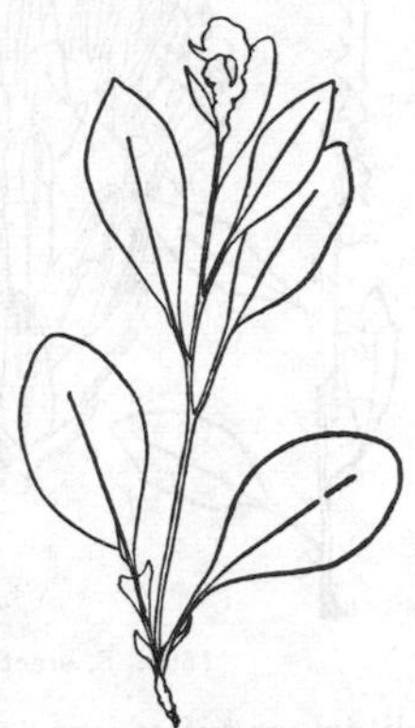

1559. S. hispida, v. arnoglossa.

11. S. pubérula Nutt. (minutely short-pubescent). — *Stems* often purplish, 0.2–1 m. high, *very minutely cinereous-puberulent* (under magnification); *leaves numerous, minutely puberulent; the lower spatulate-oblanceolate to -obovate,* mostly serrate-dentate, 0.5–2 dm. long, 1.5–5 cm. wide; *the cauline lanceolate, acutish,* narrowed at base, *very gradually decreasing in size, the upper* 1.5–5 *cm. long* and mostly subtending small axillary tufts; inflorescence a *virgate* crowded *thyrse* or pyramidal (rarely corymbiform) *panicle of wand-like racemes; involucres* 3–4.5(–5) *mm. high, with linear-attenuate phyllaries;* rays about 10, orange-yellow. — Dry or peaty sterile soils, sands, rocky barrens, etc., nw. Fla. to Miss., n. to C.B., M.I., P.E.I., n. N.E., sw. Que., N.Y. and W.Va. Late July–Oct. Fig. 1560. — Hybridizes with nos. 9 and 22.

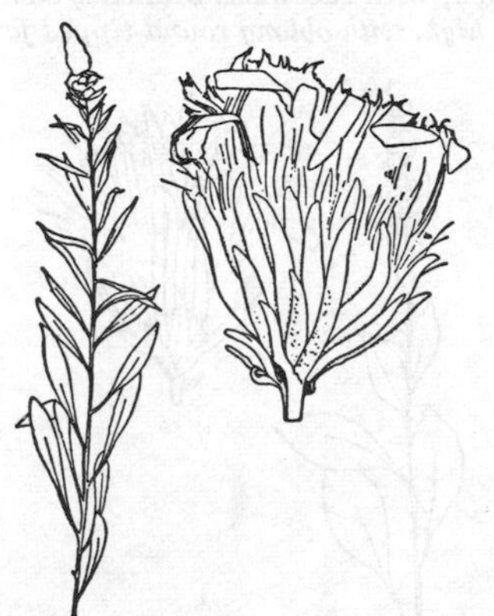

1560. S. puberula.

Var. **pulverulénta** (Nutt.) Chapm. (powdered). — *Leaves of upper two-thirds of stem abruptly reduced, elliptic or oblong, bluntish, the uppermost only* 1–1.5 *cm. long;* pubescence of thyrse coarser, a short hispidity. (*S. pulverulenta* Nutt.) — Pinelands and damp woods, Coastal Plain, Fla. to e. Va. Sept.–Nov.

12. S. roanénsis Porter (of Roan Mt.). — Stems glabrous or *short-hispid* above and *in the inflorescence,* 4–8 dm. high; *leaves membranous, glabrous,* 14–18 below the inflorescence, rather distant; *the lowest and the rosulate oblong or oblong-obovate, acute, tapering to winged petioles, serrate or serrate-dentate;* middle and upper leaves much smaller, lanceolate, less toothed or entire; *thyrse dense above,* with crowded dense glomerules or short racemes, 0.6–3.5 dm. long; *involucres* campanulate, 5–5.5 *mm. high, with herbaceous to subchartaceous linear or linear-lanceolate acute to bluntish phyllaries* 0.5–1 *mm. wide, all but the outermost usually delicately* 3*-ribbed below the middle;* disk-flowers about 10, rays about 7, deep yellow; achenes glabrous. — Mountain-woods, w. Va. and Tenn. to Ga. Aug., Sept. Fig. 1561.

Var. **montícola** (T. & G.) Fern. (mountain-dweller). — Smaller; stem simple or paniculately branched, 1–8 dm. high; *glomerules mostly* rounded, *scattered* on the lower half of the slender thyrse or racemose branches; *involucres* 4–5 *mm. high,* paler; ligules pale yellow

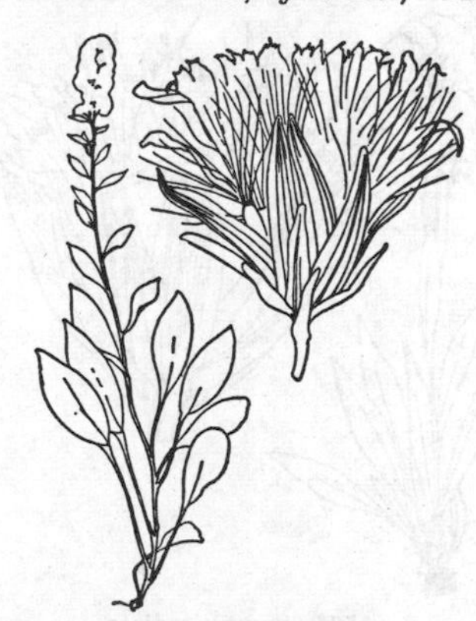

1561. S. roanensis.

or whitish. (*S. monticola* T. & G.) — Montane woods, Md. to Ky., s. to Ga. and Ala. Late July–Sept.

13. S. erécta Pursh (erect).— Stems 0.3–1.3 m. high, glabrous, or slightly hispid above; *leaves glabrous, rather firm; the lower* and those of the rosettes oblong-obovate to -oblanceolate, *crenate,* narrowed to winged petioles; *middle and upper leaves* much smaller, narrower, entire, *acute;* inflorescence a *simple wand-like interrupted thyrse or a panicle of virgate branches,* suggesting that of no. 9; *involucres* 4–5 mm. high, *of* very unequal *firm pale oblong round-tipped phyllaries with broad midrib;* rays pale yellow to whitish; *achenes glabrous.* — Dry or moist woods and thickets, Ga. to Miss., n. to s. N.E., L.I., N.J., Pa., W.Va., s. O. and Ind. Aug.–Oct. FIG. 1562.

1562. S. erecta.

1563. S. jejunifolia.

14. S. jejunifòlia Steele (insignificant-leaved). — Stems solitary or tufted, erect, 4–8 dm. high, glabrous; *basal leaves oblanceolate, acute, barely* 1 *cm. wide, entire,* short-petioled, *firm; cauline leaves* 15–20, oblanceolate, lanceolate or linear, glabrous, *firm, inrolled at the subulate tip;* thyrse rigid, 0.8–2 dm. long, 1.5–6 cm. thick, with racemose branches short-hispid; *involucres* short-campanulate, glutinous, 2.5–3.5 *mm. high, with oblong round-tipped firm phyllaries;* ligules orange, 5–10; *achenes glabrous.* (*S. Klughii* Fern.) — Open sandy woods, Bruce Pen., Ont., and n. Mich. Late July, Aug. FIG. 1563.

15. S. sciáphila Steele (shade-loving). — Resembling no. 10; stem 4–8 dm. high, *glabrous; leaves membranous;* basal and rosette-leaves elliptic to obovate, serrate or dentate; *cauline leaves glabrous, elliptic-ovate to -obovate, mostly toothed, acute, the median* 2–4 *cm. broad;* thyrse lax, 1–2.5 dm. long, with foliaceous bracts nearly equaling to exceeding the short branches; involucres cylindric-campanulate, 4–5 mm. high, with *oblong round-tipped scarious-herbaceous phyllaries* in 3–4 series; flowers 15–20, the yellow rays 5–10; *achenes strigose.* — Ledges, cliffs and sands, w. Ont. and Mich. to Minn., s. to n. Ill. and ne. Ia. Aug., Sept. FIG. 1564.

1564. S. sciaphila.

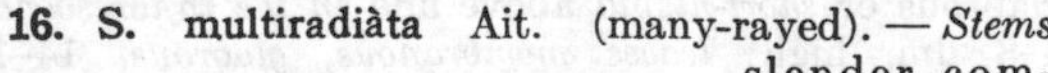
16. S. multiradiàta Ait. (many-rayed). — *Stems* slender, commonly tufted, arcuate at base, 0.3–4.5 dm. high, *villous* at least at summit; *leaves oblanceolate,* with glabrous surfaces, *the reticulate veins conspicuous* (but not elevated) *beneath; the basal* crenate to serrate, *with* broad *villous-margined petiolar base;* the cauline much shorter, sessile, entire or subentire; *inflorescence* rather dense, *corymbiform,* 2–15 cm. high, *its branches and pedicels white-villous; involucres* hemispherical, 5–7 *mm. high,* with 20–30 2–3-seriate often subequal lanceolate to linear subacute to attenuate *herbaceous phyllaries;* rays 12–18; *achenes 2 mm. long,* pilose. — Rocky and peaty soil, chiefly calcareous, argillaceous or magnesian, n. Lab. and Ung., s. to Nfld., Gaspé Pen., Que., and St. Paul I., N.S., and in minor vars. w. to Alaska and down the Rocky Mts. July, Aug. FIG. 1565.

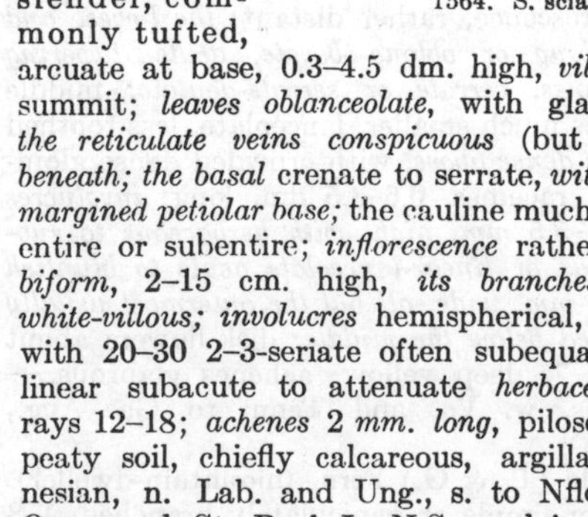
1565. S. multiradiata.

Var. párviceps Fern. (small-headed). — *Involucre with about* 15 *phyllaries,* 3–5 *mm. high.* — Local, Gaspé Pen., Que.

17. S. mensàlis Fern. (of a tableland). — Stem 2–3 dm. high, glabrous except for pilose summit; *leaves thin, glabrous, ciliolate,* deep green above, *pale and reticulate-veiny beneath; the basal narrowly elliptic, long-petioled, with blade* 3–4.5 *cm. long,* 1–2 *cm. broad, acute, serrate-dentate; cauline leaves* 9–13, *cuneate-oblanceolate, short-acuminate,* the lower subpetiolate and serrate, the upper sessile and entire; *inflorescence a loose subsimple raceme* 3–9 cm. long, *with pilose pedicels* 1–2 *cm. long; involucres campanulate-hemispherical,* 7–10 *mm. high, with green* subequal 3–4-seriate *lance-attenuate to linear-acuminate* ciliolate *phyllaries;* flowers about 50, rays 15; *achenes* 4 *mm. long,* strigose. — Calcareous alpine cliffs, Table-top Mt., Gaspé Co., Que. Aug. Fig. 1566.

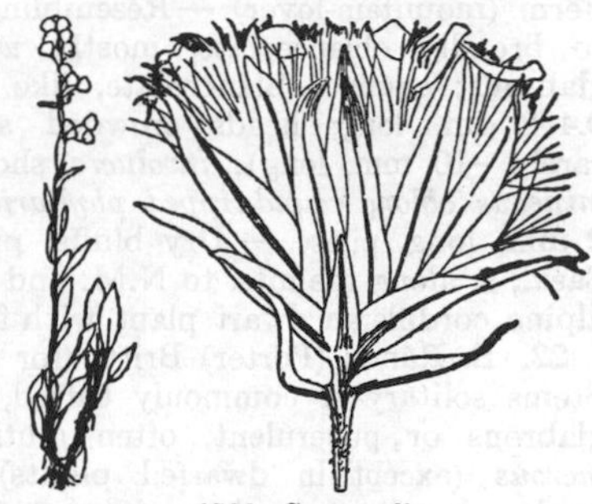

1566. S. mensalis.

18. S. Cútleri Fern. (in honor of Manasseh Cutler, 1742–1823). — Stems solitary or commonly tufted, 0.2–3.5 dm. high, glabrous or pilose; basal *leaves* oblanceolate to spatulate-obovate, obtuse or acute, crenate to jagged-serrate, narrowed to a broad petiolar base; *cauline* similar, smaller, *of nearly uniform size,* less toothed or entire, 2–7 (–15) *below the inflorescence,* the upper sessile; the simple or loosely branched *racemiform to corymbiform* often leafy *thyrse or panicle* 0.2–2.5 dm. high; heads 1–few in the axillary and terminal glomerules; *involucres* campanulate to hemispherical, 30–60 (or more)-*flowered,* 6–9 *mm. high; the green and herbaceous oblong to lanceolate* obtuse to acutish *phyllaries mostly* 1–2 *mm. broad; lobes of disk-corollas* 1.2–2 *mm. long;* rays 8–12; *anthers* 2–3 *mm. long; achenes* 3–3.5 *mm. long,* hirsute. — Granitic alpine areas, higher mts., Me. to n. N.Y. Early July–early Sept. Fig. 1567. — Hybridizes with no. 22.

1567. S. Cutleri.

19. S. anticosténsis Fern. (of Anticosti). — Resembling no. 18, glutinous, glabrous, or stems minutely setulose above; basal leaves elliptic-ovate to spatulate-obovate; *cauline leaves* subdistant, 8–15, *narrowly obovate or oblanceolate;* thyrse dense, 2–6 cm. long, the heads subsessile or on sparsely setulose pedicels up to 5 mm. long; *involucres* broadly campanulate, *about* 30-*flowered,* 5.5–8 *mm. high, with* 5-seriate *chartaceous glutinous* mostly *oblong and obtuse phyllaries* 1–1.7 *mm. broad; lobes of disk-corollas* 1.3–1.8 *mm. long;* rays 10; *anthers* 2 *mm. long; achenes* 2 *mm. long,* strigose. — Argillaceous or calcareous rocks, Anticosti I., Que. July, Aug. Fig. 1568.

1568. S. anticostensis.

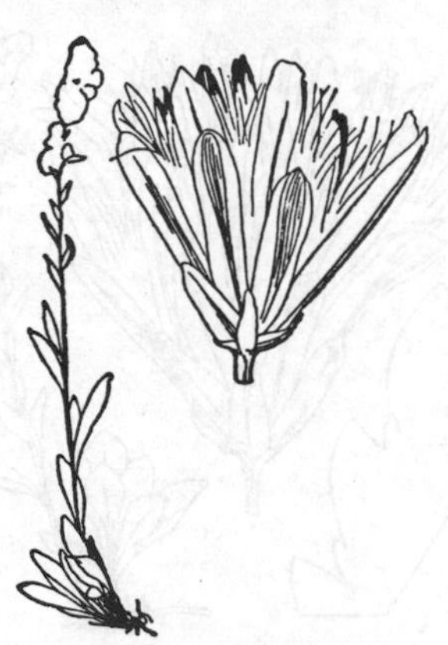

1569. S. Victorinii

20. S. Victorínii Fern. (for its discoverer, Joseph-Louis-Conrad Kirouac, Frère Marie-Victorin, 1885–1944). — Similar to no. 19 and possibly a variation of it; *involucre* cylindric-campanulate, 5.5–6 *mm. high,* with 3 series of coriaceous *dorsally pruinose phyllaries;* the outer ones linear-oblong, *the inner spatulate,* round-tipped and *only* 1 *mm. broad; flowers* 15–20; *lobes of disk-corolla* 1 *mm. long;* rays 7; immature achenes hirsute. — Calcareous ledges, Anticosti I., Que. Aug., Sept. Fig. 1569.

21. S. decúmbens Greene (reclining, with upturned tips), var. **oreóphila** (Rydb.) Fern. (mountain-lover). — Resembling no. 22; *basal leaves pale green, coriaceous,* spatulate to broadly oblanceolate, mostly *with strongly rounded summit,* crenate, 1–2 *cm.* wide, glabrous; cauline oblanceolate, like the basal or subacute; inflorescence a dense thyrse 0.4–2 dm. long; heads crowded and subsessile to short-pedicelled (pedicels up to 5, rarely –10 mm. long); *involucre* short-campanulate, 4–5 *mm. high, with chartaceous stramineous oblong round-tipped phyllaries; lobes of disk-corolla* 0.7–1 *mm. long;* achenes barely 2 mm. long, pilose. — Dry bluffs, prairies, sands and open woods, Mackenz. and Yuk. to Sask., s. along the mts. to N.M. and Ariz.; n. Mich. July–Sept. — Typical *S. decumbens* an alpine cordilleran dwarf plant with few larger subcorymbose heads.

22. S. Rándii (Porter) Britt. (for its discoverer, Edward Lothrop Rand, 1859–1924). — Stems solitary or commonly tufted, often depressed, glabrous or puberulent, often glutinous; *leaves numerous* (except in dwarfed plants); *the basal* narrowly obovate to oblanceolate, 1–5 cm. wide, glabrous, entire, crenate or serrate (sometimes jagged), *with the midrib* (beneath) *about as broad as the short petiolar base; cauline lanceolate to elliptic,* the upper entire, 1–7 cm. long, 0.2–1.5 cm. wide; *inflorescence* 0.2–4.5 dm. high, a *dense* thyrse or ample panicle with ascending thyrsoid branches; *heads crowded, sessile or short-pedicelled* (pedicels 1–4, rarely –10 mm. long); *involucres* campanulate, 5–6 *mm. high, with linear or linear-lanceolate* obtuse or acutish 3-seriatc mostly *chartaceous phyllaries* 0.5–1.5 *mm. wide; lobes of disk-corollas* 0.8–1.4 *mm. long; anthers* about 2 *mm. long;* rays 8–12; *achenes* 2–2.6 *mm. long,* appressed-setulose. (Including var. *monticola* (Porter) Fern., the smallest state) — Granitic, siliceous or magnesian rocks and gravels, local, N.S. and se. Me. to sw. Que. and n. N.Y.; near L. Superior, Mich. to Minn. July–Sept. Fig. 1570. — Hybridizes with nos. 11, 18 and 58.

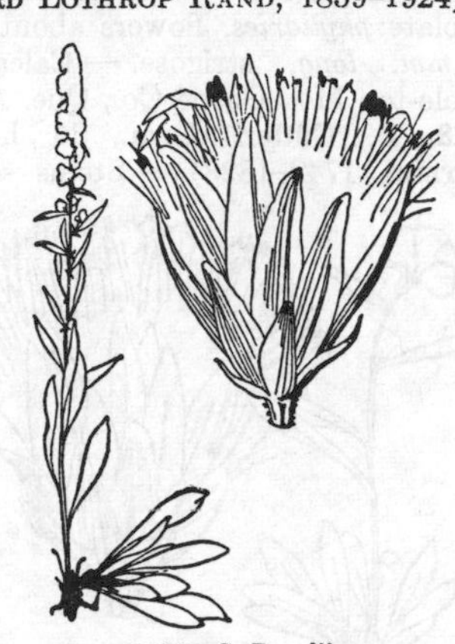

1570. S. Randii.

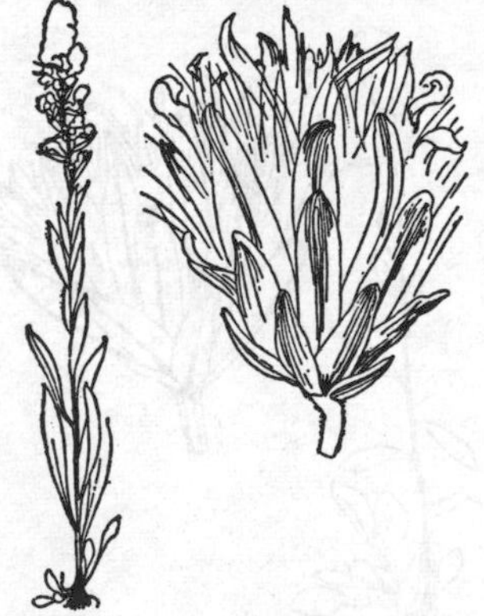

1571. S. Maxoni.

23. S. Máxoni Pollard (for its discoverer, William Ralph Maxon, 1877–1948). — Resembling no. 22; stems erect, 6–7.5 cm. high, glabrous; *leaves submembranaceous, glabrous; the basal* oblanceolate, *acutish,* dentate-serrate; *cauline leaves* 15–25, distant, *acute,* the lower resembling the basal; the median and upper lanceolate, entire or subentire and only 4–8 cm. long and 1–2 cm. wide; *thyrse* dense or interrupted, 1.5–3.5 dm. long, 3–7 cm. thick, *its branches and short* (1–5 *mm. long) pedicels minutely setulose, viscid; involucre* campanulate, 4.5–6 *mm. high, with* 3–4-seriate *glutinous phyllaries;* the outer green and oblong-lanceolate, *the intermediate oblong, obluse and with dilated green midrib;* the inner chartaceous, stramineous and with viscid midrib; flowers 15–20, the ligules yellow; anthers 2 mm. long; *achenes* 3 mm. long, *sparsely divergent-hirtellous* to subglabrous. — Ledges and cliffs of mts., Va. Aug., Sept. Fig. 1571.

24. S. Deàmii Fern. (for its discoverer, Charles Clemon Deam, 1865–). — Resembling no. 22; stems 4–5 dm. high, minutely pilose above; *leaves*

1572. S. Deamii.

coriaceous, pale green; those of the rosettes obovate, coarsely serrate-dentate, narrowed to short winged petioles, the blades 3.5–5.5 cm. long, 2–3.5 cm. broad, *rounded at summit;* cauline leaves 35–40, the lower oblanceolate and serrate; the upper much smaller, entire, acute; thyrse ellipsoid, dense, 1 dm. long, 4 cm. broad, its branches strigose-pilose; heads sessile or on bracted pedicels up to 8 mm. long; *involucre* cylindric-campanulate, 6–9 *mm. high; the pale chartaceous oblong phyllaries obtuse,* in about 4 series, the outer with dilated green backs; flowers about 20, the 8 ligules pale yellow; *lobes of disk-corollas 2 mm. long; anthers* 2.7–3 *mm. long;* achenes strigose-pilose. — Dunes of L. Michigan, n. Ind. Sept. Fig. 1572.

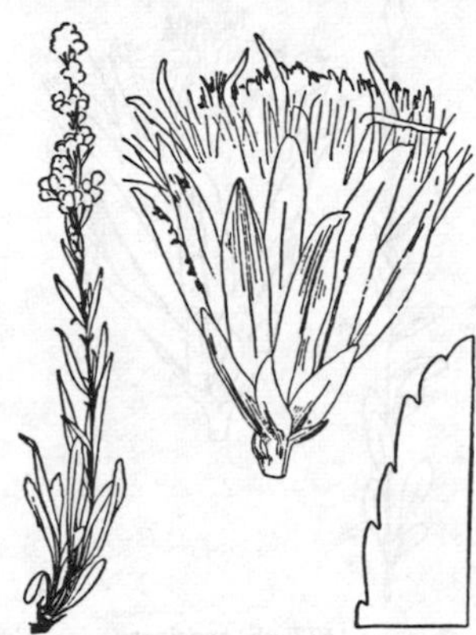
1573. S. racemosa (typical).

25. S. racemòsa Greene (racemose). — Stems commonly tufted, slender, 1–6 dm. high, glabrous or sparingly puberulent, *glutinous; leaves* glabrous, numerous; *the basal and lower oblanceolate, acute or subacute,* entire or serrate-dentate above the middle, 3–15 cm. long, 0.5–1.5 cm. wide, *the midrib very slender; cauline leaves* 5–40, rapidly decreasing in size upward, linear or oblanceolate, *mostly subtending axillary tufts;* inflorescence a loose raceme, racemiform thyrse or panicle of upright racemes, 0.2–3 dm. long; *many* of the *heads on upright pedicels* 0.5–1.5 *cm. long; involucres slenderly campanulate,* 5–8 *mm. high,* with about 3 series of glutinous chartaceous *linear* obtuse to acutish *phyllaries;* rays about 10, yellow, or white in forma **leucántha** Fern. (white-flowered); achenes 2–3 mm. long, appressed-setulose. — Dry, often calcareous, ledges and rocky banks, Anticosti I., Que., to s. Ont., s. to n. N.E. and N.Y., locally to the Susquehanna R., Pa., the Potomac, Va., Cumberland R., Ky., Mich., n. Ind. and Wisc. Late June–Oct. Fig. 1573. — Passing to

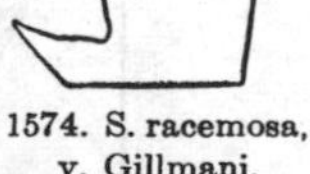
1574. S. racemosa, v. Gillmani.

Var. **Gillmáni** (Gray) Fern. (named in 1882 for its discoverer, Henry Gillman). — *Coarser,* 0.45–1 m. or more high; *basal leaves* 1–3 *dm. long, up to 4 cm. broad, coarsely toothed;* inflorescence coarser; *heads fuller.* (*S. Gillmani* Steele) — Gravels of St. John R., N.B.; dunes and rocky shores of Lakes Huron, Superior and Michigan, Ont., Mich., Wisc. and n. Ind. Fig. 1574.

26. S. chlorólepis Fern. (with green phyllaries). — Stems tufted, slender, 1–3 dm. high; *leaves* coriaceous, *the basal and rosulate spatulate-oblanceolate, rounded or subtruncate and crenate-serrate at apex,* cuneate-petiolate, 2–6 cm. long, 0.6–1.3 cm. wide; *cauline remote,* 3–8, similar, *with naked axils;* panicle loosely cylindric, racemiform, subsimple, 3.5–20 cm. long; *pedicels remote,* 1–4-headed, strongly ascending, 1–1.5 *cm. long; involucres hemispherical,* 3–4 *mm. high;* the viscid phyllaries 4-seriate, the outer thick, lanceolate, green; the median narrowly oblong, obtuse with green center, the inner stramineous with green tip; rays 15–20, 2 mm. long; achenes hispid. (*S. decumbens,* in part, of ed. 7, not Greene) — Serpentine alpine area, Mt. Albert, Gaspé Co., Que. Mid-July, Aug. Fig. 1575.

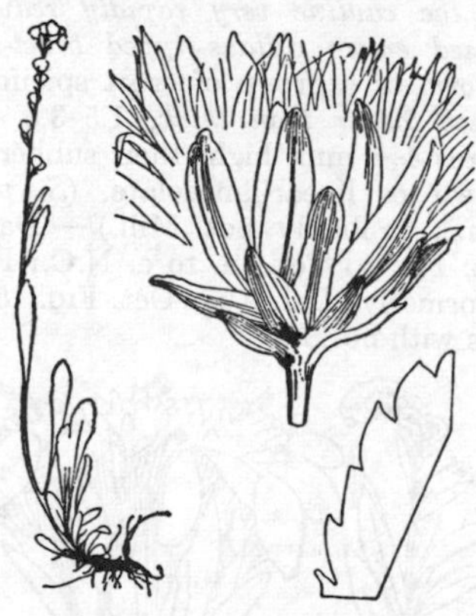
1575. S. chlorolepis.

27. S. Púrshii Porter (for Frederic Traugott Pursh, 1774–1820, who described it under an untenable name). — *Stems solitary* or few, *glabrous,* 2–10 dm. high; *leaves* lanceolate to oblanceolate, *glabrous* and *rather fleshy; the lower and rosulate* 1–4.5 *dm.*

1576. S. Purshii

long, 2–7 cm. wide, obtuse to acute; *blade about equaled by the petiole*, entire or shallowly toothed; cauline rather remote, the upper small, entire, 2–8 cm. long; *thyrse* elongate, 0.3–4.5 dm. long, *with strongly ascending cylindric (non-secund) racemose branches;* pedicels 0.1–2 cm. long; *involucres* campanulate, 4–6 mm. high, *with 2–3 series of few broad* lanceolate to oblong blunt to acute *phyllaries;* disk-flowers 9–15, rays 4–6; *tube of disk-corolla very slender, about equaling throat and limb;* achenes usually glabrous. (*S. uliginosa* of ed. 7, not Nutt.; *S. chrysolepis* Fern.) — Peaty or damp habitats, often calcareous, n. Lab. and n. Ung. to Man., s. to Nfld., C.B., N.B., n. and w. N.E., N.Y., Pa., W.Va., O., n. Ind., Wisc. and Minn. Mid-July–Sept. FIG. 1576. — Sometimes hybridizes with no. 37, producing ×**S. Farwéllii** Fern. (for its discoverer, OLIVER ATKINS FARWELL, 1868–1944).

28. S. speciòsa Nutt. (showy). — Stoutish, 0.6–2 m. high, the stem glabrous except at summit; *leaves firm, finely reticulate but with the lateral veins not prominent; the basal ovate to oblong or obovate*, 4.5–10 *cm. wide*, entire or crenate to serrate, *contracted to a long margined petiole; cauline leaves* 18–40, oblong to lanceolate, entire, all but the much smaller uppermost 2.5–5 *cm. wide; inflorescence a dense* pyramidal to thyrsiform *panicle* 0.3–4 dm. long, *with* usually *crowded strongly ascending spiciform racemes or glomerules*, its branches and pedicels hirsute; *involucres cylindric, few-flowered*, 4.5–6 mm. high, *with pale firm oblong obtuse phyllaries;* rays about 5, large; *achenes glabrous.* (*S. conferta* Mackenz., not Mill.) — Dry to moist thickets, open woods and prairies, N.C. to La. and Okla., n. to Mass., N.Y., O., Mich., Ill. and s. Minn. Late Aug.–Oct. FIG. 1577.

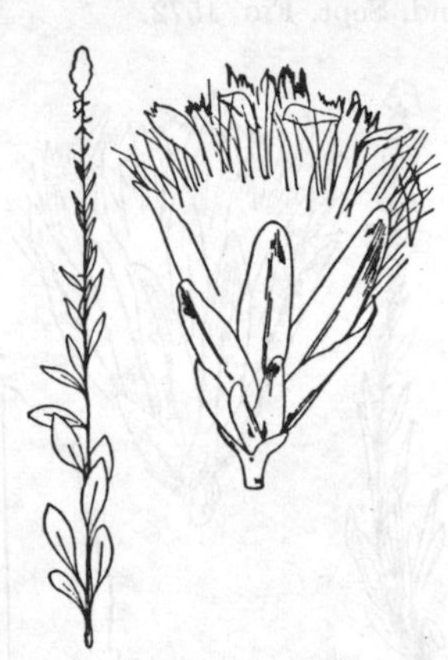

1577. S. speciosa.

Var. **angustàta** T. & G. (very narrow). — *Smaller*, 3–10 dm. high; *basal leaves spatulate-oblong to oblanceolate*, 1.5–3.5 *cm. broad*, mostly entire, *the median cauline* 0.8–3 *cm. broad;* involucres slightly smaller. (*S. rigidiuscula* Porter) — Prairies and dry thickets, Ga. to Tex., n. to s. Ont., Mich., Wisc., Minn. and Sask.

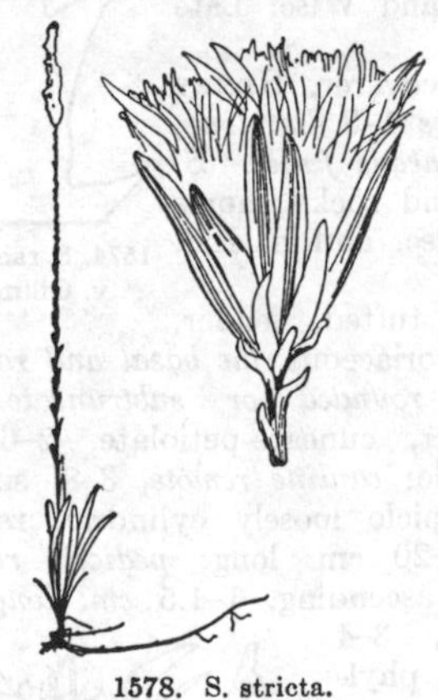

1578. S. stricta.

29. S. strícta Ait. (erect). — *Glabrous throughout; rhizome* horizontal or ascending, *finally producing slender stolons; stems* solitary, strict and simple, *wand-like*, 0.3–2.5 m. high; *basal leaves spatulate or narrowly oblanceolate, obtuse, the cauline very rapidly reduced to numerous appressed entire callous-tipped bract-like leaves; panicle spiciform* or a group of erect spiciform branches, *many times longer than thick*, 1.5–3.5 *cm. in diameter;* involucres 4–5 mm. high, their subherbaceous phyllaries linear or linear-lanceolate. (*S. petiolata* sensu Mackenz., probably not Mill.) — Damp sands and pinelands, Fla. to Tex., n. to e. N.C.; Pine Barrens, N.J. and (formerly) Del. Aug.–Oct. FIG. 1578. (W.I.) — Hybridizes with no. 37.

30. S. sempérvirens L. (evergreen), SEASIDE G. — Stout, the 1–few glabrous stems 0.2–2.5 m. high; *leaves thick*, commonly *fleshy, smooth, entire; the basal* oblanceolate to spatulate-oblong, obtuse or acutish, *broad-petioled*, 1–4 dm. long, 1.2–7 cm. broad; *cauline leaves* 10–35, oblong-oblanceolate to lanceolate, obtuse to acutish, sessile, *the lower clasping one-half to three-fourths the circumference of the stem*, the uppermost 0.3–1.5 dm. long; *panicle* loosely corymbiform to thyrsiform, 0.4–4.5 dm. high, *the secund branches glabrous or only sparingly hirtellous*, the large heads mostly pedicelled; *involucres hemispherical*, 4–7 *mm. high, with* 3–4 series of *linear- to oblong-lanceolate acute or acutish sub-*

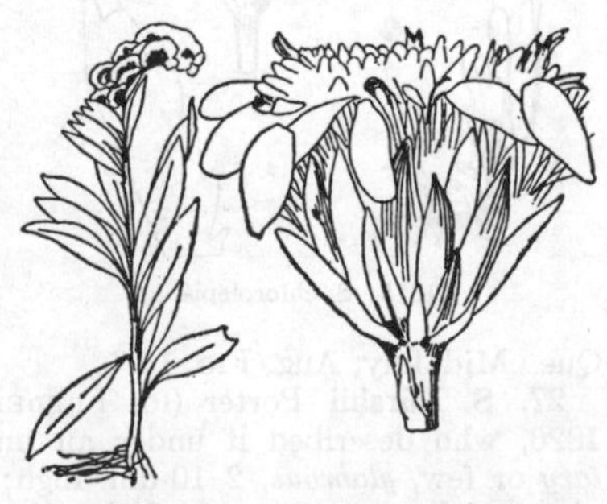

1579. S. sempervirens (typical).

1580. S. sempervirens, v. mexicana.

herbaceous phyllaries; disk-flowers 12–40; rays 8–10; *pappus* 3.5–5.5 *mm. long; disk-corollas* 4–5.5 *mm. long; mature achenes* 2.2–3.5 *mm. long,* sparingly pubescent. — Saline, brackish or even fresh habitats near the coast, w. Nfld., M.I. and lower St. Lawrence, Que., to N.J. and locally to Va. July–Nov. (occasionally to Dec.). Fig. 1579. — Hybridizes readily with no. 56, producing the frequent × **S. aspérula** Desf.; also with no. 34.

Var. **méxicana** (L.) Fern. (Mexican). — *Leaves narrower;* the lower linear- to oblong-lanceolate, 1–3 cm. broad; *the cauline linear-lanceolate,* obtuse to acute, *frequently ciliolate;* panicle and heads slightly smaller, *the branches usually hirtellous.* (*S. mexicana* L.) — Similar habitats, Fla. to Tex. and Mex., n. to s. N.Y. and se. Mass. Sept.–Nov. (W.I.) Fig. 1580.

31. S. missouriénsis Nutt. (of Missouri). — *Rhizome cord-like, horizontal, producing abundant slender stolons;* stems solitary, 1–8 dm. high, glabrous; *leaves* mostly *triple-nerved* (with 2 nerves elongated parallel with the midrib), glabrous except for *scabrous-ciliolate* margins; *the lower oblanceolate,* petioled, entire or dentate, *the cauline rapidly reduced upward,* becoming lanceolate or linear, the upper often subtending small fascicles; *panicle* 2–12 cm. broad, *usually with strongly ascending glabrous* (sparingly hirtellous w. of our range) *branches* bearing scarcely or only slightly recurving 1-sided racemes; *involucres* 3.5–5 mm. high, *with linear-oblong subchartaceous* blunt or acutish *phyllaries;* flowers 15–25, the disk-corollas 3–4 mm. long; *achenes linear-cylindric or -fusiform, strigose-pilose or hirsute,* 1.3–2.2 *mm. long.* (Var. *montana* Gray) — Dry prairies, gravels and rocky slopes, s. B.C., to Wisc., s. to Oreg., Ariz., N.M., Okla. and Mo.; natzd. along railroad, Burlington Co., N.J. July–Sept. Fig. 1581. — Passing insensibly to

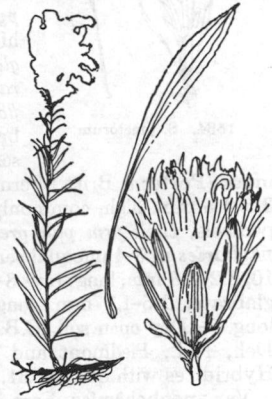

1581. S. missouriensis (typical).

1582. S. missouriensis, v. fasciculata.

Var. **fasciculàta** Holzinger (bunched). — Mostly *taller,* up to **1 m.** high; *panicle* usually larger, *up to 2 dm. broad,* usually *with arched-recurving branches;* involucres 3–4.5 mm. high; disk-corollas 2–3.5 mm. long; *achenes oblong to slender-cuneate, glabrous or only sparsely short-hispid,* 1–1.3 *mm. long.* (*S. missouriensis* Gray, not Nutt.; *S. glaberrima* Martens) — Similar habitats, Ida. to N.M., e. to Mich., Ind. and Tenn. Fig. 1582. — Hybridizes with no. 34.

32. S. Gattíngeri Chapm. (for its discoverer, Augustin Gattinger, 1825–1903). — Slender and strict, 4–10 dm. high; stems smooth throughout; *leaves* ciliolate, smooth beneath, harsh above; the lowest 0.7–1.5 dm. long, lance-spatulate, appressed-serrulate or subentire, *the upper reduced rapidly to minute entire bracts with naked axils; branches of the naked subcorymbiform panicle elongate and ascending, hardly recurved;* involucre 3–4 mm. long, its oblong phyllaries obtuse; rays 6–10; *achenes appressed-puberulent,* or glabrate throughout, *coarsely corrugated.* — Limestone-knobs, barrens and glades, Tenn. and Mo. July–Sept. Fig. 1583.

1583. S. Gattingeri.

33. S. pinetòrum Small (of pine-woods). — Stems solitary or clustered from a *non-stoloniferous* base, glabrous, 0.3–1.5 m. high, slender; *leaves* glabrous, *non-ciliate,* most of them *with 2 lateral nerves prolonged parallel with the midrib; basal leaves* petioled, *narrowly oblanceolate,* 0.5–1.7 *cm. broad,* entire or sparingly sharp-serrate; *cauline leaves* rapidly reduced upward, *linear, attenuate,* sessile, often with axillary tufts; *panicle with* spreading or *arched-recurving 1-sided racemes;* heads about 15-flowered; *involucres* 3–4.5 mm. high, *with yellowish chartaceous oblong blunt phyllaries;* disk-corollas 3 mm. long; pappus 2 mm. long; *achenes* 1–1.2 mm. long, *glabrous or glabrate, delicately slender-ribbed.* — Dry open soil or woodlands, e. Va. to S.C. Late June–Sept. Fig. 1584.

1584. S. pinetorum.

34. S. júncea Ait. (stiff, like a rush). — *Glabrous* throughout *except for the ciliolate margins of petioles and blades*, or leaves scabrous and branches of panicle often sparingly hirtellous in forma **scabrélla** (T. & G.) Fern. (slightly harsh); stems solitary or few, 3–12 dm. high, firm, nearly terete; *basal leaves firm*, borne on the caudex or occasionally on slender stolons, *oblanceolate to narrowly oval, long-acuminate*, 1–4 dm. long, 1.5–8 *cm. broad, sharply serrate, tapering gradually to the winged ciliate petiole; cauline leaves* remote, decreasing rapidly in size upward; *the uppermost lanceolate*, entire, often subtending axillary tufts; *panicle pyramidal or loosely rhomboid*, 1–4 dm. high, 0.3–3 dm. broad, *with essentially glabrous 1-sided more or less recurving racemes; the lowest branches ascending, floriferous chiefly above the middle*, or branches and racemes all erect and scarcely recurving at tip in forma **ramòsa** (Porter & Britt.) Fern. (branching); involucres hemispherical, 3.5–4.5 mm. high, commonly a little exceeding the mostly glabrous pedicels, their *firm pale green or stramineous oblong blunt* ciliolate *phyllaries* in 3–4 unequal series, often glutinous; disk-corollas about 10, 3.2–3.8 mm. long; rays 8–12, 4–5 mm. long; achenes pubescent or glabrate, 1.5–1.7 mm. long; longer pappus-bristles 2.6–3.3 mm. long. — Dry open soil, C.B. and N.B. to Sask., s. to s. N.E., L.I., Del., Md., Piedmont and upland to Ga., Tenn. and Mo. Late June–Oct. Fig. 1585. — Hybridizes with nos. 30, 31, 37, 38 and 48.

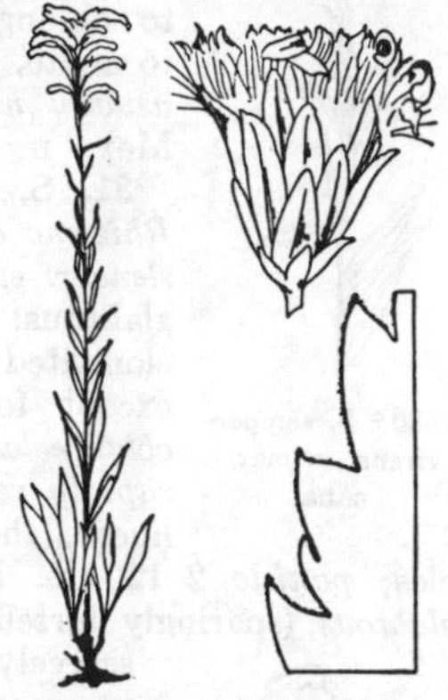
1585. S. juncea.

Var. **neobohèmica** Fern. (for the type-region, New Bohemia, Va.). — Smaller and narrow-leaved; involucre 3–4 *mm. high; phyllaries vividly green along midrib; rays* 3.5–3.8 *mm. long; disk-corollas* 3–3.4 *mm. long;* longest *pappus* 2–2.5 *mm. long; achenes* 1.4 *mm. long.* — Dry woods and clearings, se. Va. Late May, June.

1586. S. austrina.

35. S. austrìna Small (southern). — *Glabrous* throughout except for the scabrous foliage; *stems* solitary or few from the caudex, 0.7–1.5 m. high, *somewhat virgate; leaves coriaceous, scabrous at least on the margin, the basal narrowly oblanceolate; entire or appressed-dentate, tapering to slender eciliate petioles*, the blades 0.7–1.5 dm. long, 1–2.5 *cm. broad, with wire-like whitish round-backed midribs and fine inconspicuous reticulum;* cauline leaves rather numerous, lanceolate to narrowly oblong, the lower similar to the basal, *the uppermost blunt or bluntish and only* 0.5–2 *cm. long; panicle slender and elongate, two to ten times as long as thick*, 1–3 dm. long, *with very short* (rarely elongate) spreading few-headed *secund glabrous branches floriferous nearly to base;* involucres campanulate, 4–5 mm. high, *the firm green oblong obtuse phyllaries continuing indistinguishably down the short pedicels; ripe achenes* 3–3.6 *mm. long, pilose.* — Meadows and swamps, Va. (acc. to *Small*) to Ga. and Tenn. Aug., Sept. Fig. 1586.

36. S. perlónga Fern. (very elongate). — Coarser than no. 35; *rosette-leaves with* broadly oblanceolate to narrowly ovate-lanceolate acute or subacute serrate or crenate *blades* 1–2.5 *dm. long,* 1.5–5 *cm. broad; their midribs broad, greenish and acutely angled; the reticulum loose, coarse and conspicuous;* cauline leaves much reduced, the uppermost 1–3 cm. long; *inflorescence slender, secund, virgate or with erect virgate branches,* 2–4 *dm. long, the simple racemose panicles or their branches* 2–5 *cm. in diameter; pedicels up to* 1 *cm. long,* bracted; *involucres* 5–6 *mm. high; achenes* 1.4–1.8 *mm. long.* — Wet or peaty woods and clearings, se. Va. Late June–Oct.

37. S. uliginòsa Nutt. (of marshes). — Stems solitary or few, glabrous, 0.4–15 dm. high; *leaves thickish, glabrous* (rarely scabrous), not strongly veiny; *the lower from narrowly oblanceolate or lanceolate to ovate-lanceolate or elliptic, serrate, tapering at the base*, the winged *petioles clasping one-half to three-fourths the circumference of the stem; cauline leaves* strongly reduced upward, *the upper entire;* panicle with 1-sided branches recurved-ascending or spreading; *involucres* 3–5 *mm. high, with* subherbaceous to stramineous and chartaceous *oblong or linear-*

oblong blunt phyllaries in 2 or 3 series; *disk-flowers* 4–8, rays 1–8; achenes glabrous or nearly so. — Very variable species of bogs and peat. The following the most pronounced vars.:

Stoutish, 0.6–1.5 m. high; cauline leaves 20–40, the upper oblong-lanceolate; the lower ovate-lanceolate to oblong and 3–8 cm. broad; panicle elongate-pyramidal to ellipsoid, 1–4.5 dm. long, 0.3–2.5 dm. broad. . . . . *S. uliginosa* (typical).

Slender, 0.2–9 dm. high; cauline leaves 5–20 (rarely –30), linear or lanceolate; lower narrowly lanceolate or oblanceolate, 0.7–3 cm. broad.
- Panicle elongate-pyramidal to cylindric-ellipsoid, commonly 1-sided, 0.2–2.5 dm. long, 1–10 cm. thick.
  - Branches of panicle and pedicels conspicuously hirtellous. . . . . Var. *linoides*.
  - Branches of panicle and pedicels glabrous or glabrate and glutinous. . . Var. *levipes*.
- Panicle corymbiform to rhomboid or broadly pyramidal, 2.5–14 cm. broad. . Var. *terrae-novae*.

S. uliginòsa (typical), *S. neglecta* T. & G. — Acid swamps, meadows and moist to dryish thickets, Del. and Md. and upland of N.C., n. to N.S., se. and s. Me., s. N.H., s. Vt., N.Y., O., Lambton Co., Ont., s. Mich. and s. Wisc. Late July–Oct. Fig. 1587.

Var. **linoìdes** (T. & G.) Fern. (resembling flax), *S. uniligulata* (DC.) Porter. — N.J. and e. Pa., n. to centr. Nfld., M.I., s. N.B., centr. Me., N.H., Vt., N.Y., s. Ont. and s. Mich. Fig. 1588. — Hybridizes with nos. 27, 29, 34 and 58.

Var. **lèvipes** Fern. (smooth-stalked). — Bogs and wet sands, centr. N.Y. to s. Ont.

Var. **térrae-nòvae** (T. & G.) Fern. (of Newfoundland). — Peaty barrens, tundra and acid rock, s. Nfld. and M.I. to s. N.H. Fig. 1589.

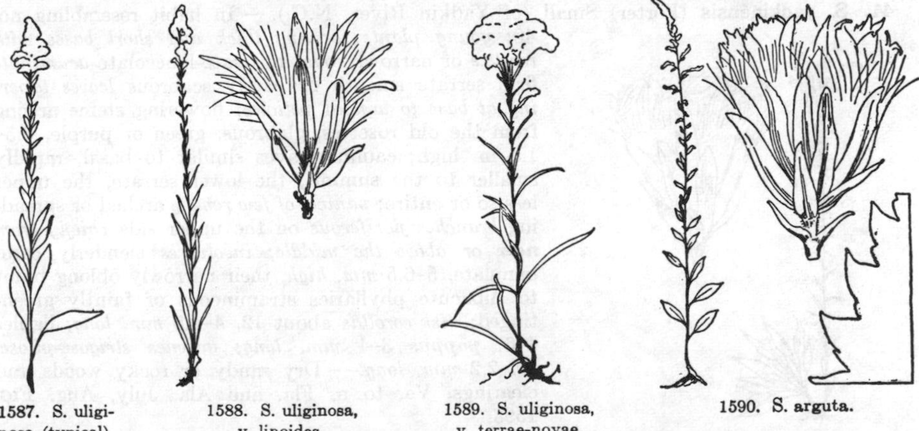

1587. S. uliginosa (typical). 1588. S. uliginosa, v. linoides. 1589. S. uliginosa, v. terrae-novae. 1590. S. arguta.

38. S. argùta Ait. (sharp; referring to teeth of leaves). — Stems mostly solitary, glabrous or essentially so, terete or obscurely angled, 0.5–2 m. high; *leaves thinnish*, smooth to scabrous, *mostly serrate; the basal lanceolate to ovate, acuminate* (rarely obtuse), *obviously contracted to the ciliate petiole*, 1–4.5 dm. long *sharply double-serrate;* the *cauline remote, the upper* reduced, lanceolate, sessile, *finely serrate*, or obovate and incised in forma **tomophýlla** Fern. (incised-leaved); panicle open, its 1-*sided spreading racemes with pilose rachises and pedicels;* heads campanulate-hemispherical; involucres 3.5–5 mm. high, their *thinnish* linear-oblong obtuse *greenish phyllaries* 3-seriate; disk-flowers about 12, rays 5–8; achenes glabrous or nearly so. — Open woods, thickets and clearings, sw. Me. to s. Ont., s. to N.C., n. Ala. and Ill. July–Sept. Fig. 1590. — Hybridizes with nos. 34 and 40.

39. S. Harrísii Steele (named in 1911 for its discoverer, Edward Harris). — Resembling no. 38, darker green, 0.4–1.2 m. high; *leaves coriaceous, scabrous above; the basal broadly ovate, strongly rounded to the* winged *petiole*, 3–10 *cm. broad*, sharply and closely double-serrate; cauline strongly reduced in size upward, narrower; panicle open, its remote branches spreading or arched-recurving, floriferous on the upper side mostly above the middle; *involucres* cylindric-campanulate, 4–5 *mm. high*, their firm oblong to lanceolate bluntish *phyllaries with broad green midribs; disk-flowers* 9–15, *with corollas* 3–3.3 *mm. long*, rays about 5; *pappus* 2–2.7 *mm. long; achenes strigose*, 2.3–2.5 mm. long. — Dry shales, w. Md., w. Va., e. W.Va. and Ky. Late May–Sept. Fig. 1591.

1591. S. Harrisii.

40. S. Boóttii Hook. (in honor of FRANCIS BOOTT, 1792–1863). — Glabrous or nearly so, slender, 0.5–1.5 m. high; *leaves membranaceous, smooth or barely scabrous above,* finely serrate; *the basal lanceolate to oblong-ovate, tapering to the petioles* and to the sharp tips, 1–3.5 cm. broad; *rosette-leaves* formed *on young plants, rarely* developed *from basal offshoots* of old plants; cauline leaves much reduced upward, often entire; panicle loose and open, the remote divergent or loosely ascending long branches floriferous on the upper side, chiefly above the middle; *involucres* campanulate, 3–5 mm. high, *the oblong-linear* blunt or acutish *phyllaries with slender midribs; disk-corollas* 2.8–3.2 *mm. long;* pappus 2–3.5 mm. long; *achenes pubescent,* 1.8–2.5 mm. long. — Woods and rocky slopes, Fla. to Tex., n. to Va., ("Md."), W.Va. and Ky. Late July–early Sept. FIG. 1592.

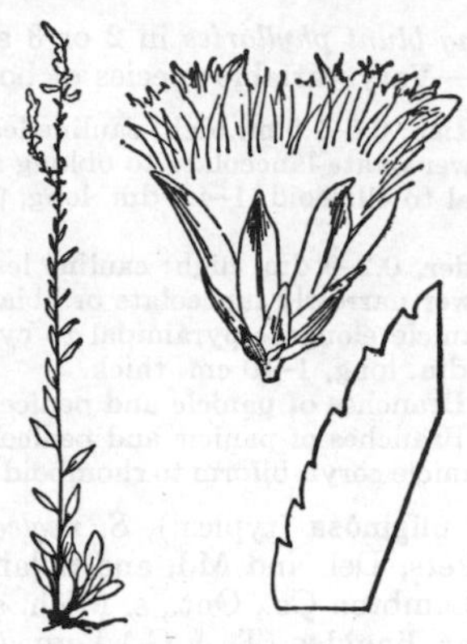

1592. S. Boottii.

41. S. yadkinénsis (Porter) Small (of Yadkin River, N.C.). — In habit resembling no. 40; *young plants forming thick and short bases with rosettes* of narrowly oval to ovate-lanceolate *acuminate* firm serrate smooth or barely scabrous *leaves tapering at base to winged petioles;* flowering stems arising from the old rosettes, glabrous, green or purple, 0.5–1.5 m. high; cauline leaves similar to basal, rapidly smaller to the summit, the lower serrate, the upper less so or entire; *panicle of few remote* arched or spreading *branches floriferous* on the upper side *chiefly from near or above the middle; involucres* slenderly campanulate, 5–6.5 *mm. high,* their narrowly oblong blunt to subacute phyllaries stramineous or faintly green-tinged; *disk-corollas* about 12, 4–4.5 *mm. long;* ligules 4–6; *pappus* 3–4 *mm. long; achenes strigose-pilose,* 1.5–2.2 *mm. long.* — Dry sandy or rocky woods and clearings, Va. to n. Fla. and Ala. July, Aug. FIG. 1593.

1593. S. yadkinensis.

42. S. ludoviciàna (Gray) Small (of Louisiana). — *Rhizome slender, cord-like, elongate,* late in the autumn *producing slender to filiform stolons terminated by rosettes of broadly ovate, elliptical or obovate* mostly *round-based subacute to short-acuminate serrate leaves;* flowering stem 0.5–1.5 m. high, smooth; cauline leaves scattered, rapidly decreasing in size from the large and serrate lower to the slightly toothed to entire and small upper ones; *panicle* elongate, *with* subascending to somewhat spreading *1-sided branches mostly closely floriferous nearly to base; involucres* campanulate, 5–8 *mm. high;* phyllaries oblong, obtuse, the outer with dilated greenish tips; *disk-corollas* 8–10, 4.5–5.5 *mm. long;* ligules nearly as numerous, oblong, rather large; *pappus* 4–5 *mm. long; achenes* pilose, 2–3 *mm. long.* (*S. tarda* Mackenz.) — Sandy pine-woods (where plants mostly sterile) and clearings, Ga. to Tex., n., locally, to s. N.J. and Ark. Sept., Oct. FIG. 1594.

1594. S. ludoviciana.

43. S. pátula Muhl. (spreading). — *Stem quadrangular above, the angles narrowly winged,* smooth, 0.5–2 m.

high; *leaves thick, harshly papillate-scabrous above,* smooth and veiny beneath; *the lower elliptical,* acuminate, closely serrate, *rather abruptly narrowed to the non-ciliate broad petiole;* the *cauline* 12–23; the upper much smaller, sessile and lanceolate; panicle rather stiff, with ascending to arched-recurving elongate closely white-pilose branches closely floriferous on and about the upper side nearly to base, the *spikelets often in subspirally arranged short racemes; pedicels mostly shorter than heads; involucres* hemispherical, 3–4.5 *mm. high, with oblong obtuse green* ciliolate *phyllaries;* disk-flowers 8–10, rays 6–8; achenes minutely pubescent. (*S. rigida* sensu Mackenz., not L.) — Swamps, meadows and wet woods or even bare ledges, oftenest calcareous to argillaceous, Vt. to s. Ont. and Minn., s. to Del., Md., upland of N.C., O., Ind., Ill. and Mo. Aug., Sept. Fig. 1595.

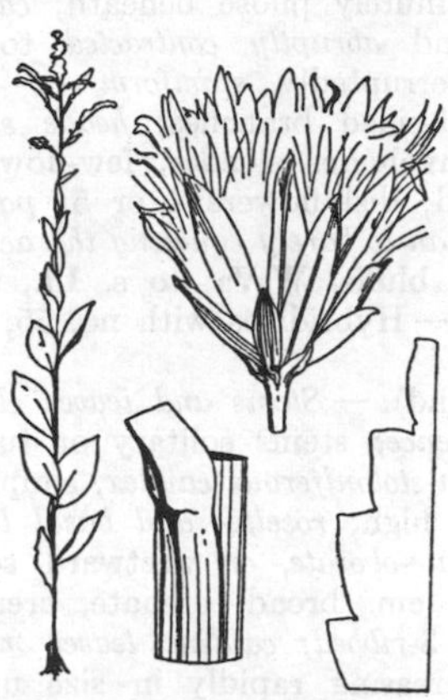

1595. S. patula.

Var. **stríctula** T. & G. (erectish). — Panicle slender and elongate, with the 1-sided subapproximate densely flowered branches only 1–3 cm. long. — Mts. of Va. to n. O., s. to w. N.C. and La.

44. S. salícina Ell. (willow-like). — Similar to no. 43, slender; *lower leaves* lanceolate to narrowly oblong-ovate, *crenate; cauline leaves* 20–45, *rapidly reduced upward; panicle* very open, *the elongate lower branches floriferous chiefly above the middle, sparsely pilose or glabrescent;* the *heads solitary or in fascicles of* 2–4, *borne strictly along upper side of panicle-branches; pedicels mostly equaling or longer than involucre.* — Bogs and swamps, Fla. to Tex., n. to se. Va. Sept., Oct.

45. S. strigòsa Small (with appressed straight hairs). — Resembling nos. 38–40; *basal rosettes with* the firm ovate to round-oblong coarsely toothed *blades* (8–12 cm. long) *abruptly contracted to the petiole, villous or setose on the nerves beneath; bases* late in the season *producing filiform stolons;* cauline leaves few, remote, ovate to lanceolate, some or all strigose on the nerves beneath; panicle elongate, lax, of a few ascending-recurving long 1-sided branches floriferous from below or near the middle; *involucres* 3.5–4.5 *mm. high; their inner phyllaries oblong, obtuse, about* 1 *mm. wide, suffused with green; disk-corollas* 3.5–4 *mm. long; pappus* 2.5–4 *mm. long; mature achenes* 2.5–3.5 *mm. long.* — Dry open woods, Ga. to La., Ark. and Mo. Late Aug.–Oct. Fig. 1596.

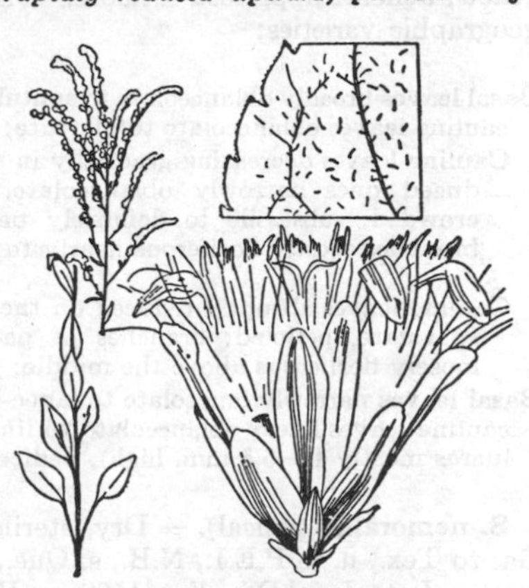

1596. S. strigosa.

46. S. neurólepis Fern. (nerve-scaled; from the glutinous-keeled phyllaries). — Similar to no. 45; stem 1–1.5 m. high; the leaves subcoriaceous, glabrous to minutely scabrous above, hispid or glabrate on the nerves beneath; *basal leaves* with long margined petioles; *the blade oblong-ovate,* 8–17 *cm. long,* coarsely serrate, *round-based, acuminate;* cauline leaves 30–40, the lower and median oblong-ovate or -obovate, subsessile and coarsely acute-serrate, the upper reduced and entire; *panicle lax,* with remote ascending or arching 1-sided branches floriferous near the tips or the upper branches floriferous to base; *involucre* cylindric, 3.5–4.5 mm. high; *its stramineous phyllaries very unequal,* the outer very short and lance-attenuate, *the intermediate lance-attenuate* and 0.5 *mm. wide and with a strong glutinous keel,* the interior linear; *disk-flowers* 4 or 5, *their corollas* 3–3.5 *mm. long; pappus* 2–2.5 *mm. long; achenes* 1.3–1.5 *mm. long,* strigose; rays 3 or 4, very small. — Dry open woods, Mo. Aug., Sept. Fig. 1597.

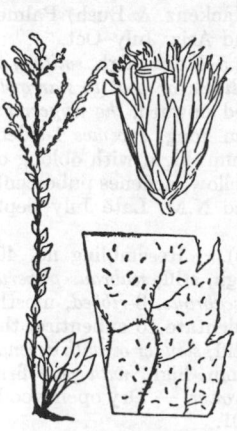

1597. S. neurolepis.

47. S. sphacelàta Raf. (blighted; application not explained).

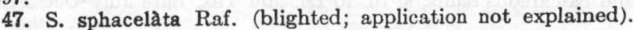

1598. S. sphacelata.

— Slender, 0.4–1.5 m. high; the rhizome elongate, bearing *rosettes of cordate coarsely toothed leaves* minutely pilose beneath; *cauline leaves* remote, *the lower* ovate and *abruptly contracted* to the petiole; *inflorescence* elongate, interruptedly *spiciform* or with spiciform ascending or spreading 1-sided branches; *heads sessile or subsessile in dense glomerules;* involucres slender, few-flowered, their oblong phyllaries green-tipped; disk-flowers 4 or 5; *pappus only* 0.5–1 *mm. long, much shorter than to barely equaling the achene.* (*Brachychaeta* Britt.) — Woods and bluffs, W.Va. to s. Ill., s. to Ga. and Ala. Aug.–Oct. Fig. 1598. — Hybridizes with no. 55, producing × **S. ovàta** Friesner.

48. S. nemoràlis Ait. (of woodland). — *Stems and leaves closely invested with minute grayish pubescence;* stems solitary or tufted, erect to depressed *from a strong non-stoloniferous caudex,* simple or corymbosely branched, 1.5–13 dm. high; *rosette- and basal leaves mostly tufted, oblanceolate to spatulate-obovate, or* westward sometimes *lance-linear, petioled,* 0.5–2.5 cm. broad, crenate, crenate-serrate or subentire, *not definitely 3-ribbed; cauline leaves mostly subtending small axillary tufts,* decreasing rapidly in size up to the inflorescence, the upper entire; *panicle 1-sided, with arched-ascending to recurving branches;* heads mostly crowded; involucres 3.5–6.5 mm. high, with firm yellowish mostly linear-oblong blunt or acute phyllaries in 3–4 very unequal series; disk-flowers 3–6; *rays 5–9, bright pale yellow;* achenes appressed-setulose. — A common and variable species, with three well-defined geographic varieties:

Basal leaves broadly oblanceolate to spatulate-obovate, crenate-serrate; principal cauline leaves oblanceolate to obovate; involucres 3.5–6 mm. high.

Cauline leaves decreasing gradually in size to the inflorescence; the upper reduced ones narrowly oblanceolate, acute, sessile or subsessile; heads crowded, subsessile to definitely pedicelled, the divergent or recurving branches mostly floriferous nearly to base. . . . . *S. nemoralis* (typical).

Cauline leaves abruptly reduced on the upper part of stem; the upper leaves spatulate, petioled; branches of panicle ascending, with recurved tips, loosely floriferous above the middle; pedicels elongate. . . . . Var. *Haleana.*

Basal leaves narrowly lanceolate to lance-linear, subentire or shallowly toothed; cauline leaves linear-oblanceolate to linear; heads comparatively large (involucres mostly 4.5–6.5 mm. high), pedicelled. . . . . Var. *decemflora.*

S. nemoràlis (typical). — Dry, sterile, siliceous or argillaceous open soil or thin dry woods, Ga. to Tex., n. to P.E.I., N.B., s. Que., s. Ont., Minn. and N.D., abundant eastw., less so westw. Late June–Dec. Fig. 1599. — Hybridizes with nos. 9, 11 and 34.

Var. **Haleàna** Fern. (for its discoverer, Josiah Hale, 1791–1856). — Dry open woods, Ga. to Tex., n. to e. Va. and less characteristically to Ky. Late Sept.–Dec.

Var. **decemflòra** (DC.) Fern. (ten-flowered), var. *longipetiolata* (Mackenz. & Bush) Palmer & Steyerm. — Dry soils, w. Ont. to n. Alta., s. to Ky., Ark., Tex. and Ariz. July–Oct.

49. S. móllis Bartl. (soft). — Stout, stiffly erect, *canescent; stems* 1.5–5 dm. high, *solitary or loosely clustered from a cord-like freely stoloniferous caudex; leaves elliptic or oval to narrowly obovate, coriaceous, definitely 3-ribbed, mostly sessile,* gradually reduced upward, *the uppermost* 1–3 cm. long and *usually without axillary fascicles,* the lower 3–8 cm. long; *racemes* 1-sided, crowded and *strongly ascending in a narrow thyrse;* involucres 4–6.5 mm. high, with oblong or narrowly ovate acutish yellow chartaceous phyllaries; ligules bright yellow; achenes pubescent. — Dry plains, Minn. and Man. to Sask. and Mont., s. to Kans. and N.M. Late July–Sept. Fig. 1600.

50. S. rádula Nutt. (a scraper; from the rigid and harsh leaves). — Resembling no. 46, greener; *stems from* a loosely branching *stoloniferous base,* 0.2–1 m. high, stiff, *scabrous-puberulent; leaves rigid, oblong to obovate or oblanceolate, greenish, pustular-scabrous, 3-ribbed,* mostly sessile, without axillary fascicles, *prominently veiny beneath,* serrate-dentate to subentire, the lower 0.5–1.5 dm. long, the upper gradually smaller; *panicle* 1-sided, stiff, *slender or with ascending* to arched-recurving *slender branches; involucres* slender, 3.5–5 mm. high, *with pale* firm narrowly oblong *obtuse phyllaries* rarely 1 *mm. wide, the inner ones narrower.* — Dry open woods and calcareous bluffs, s. Ill. to La. and Tex. Late July–Oct. Fig. 1601.

1599. S. nemoralis.

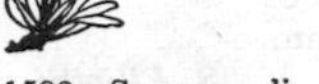

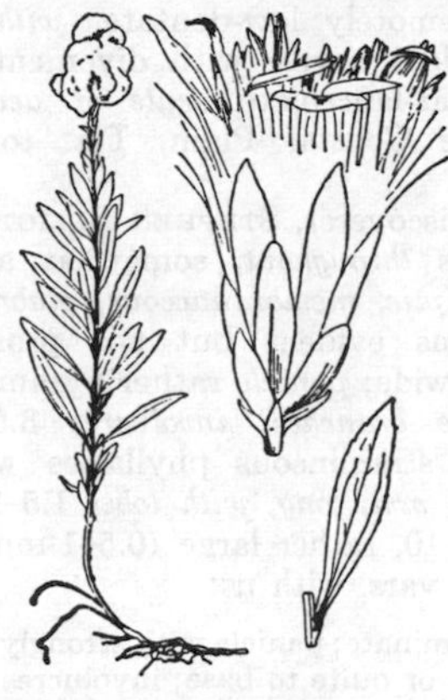

1600. S. mollis.

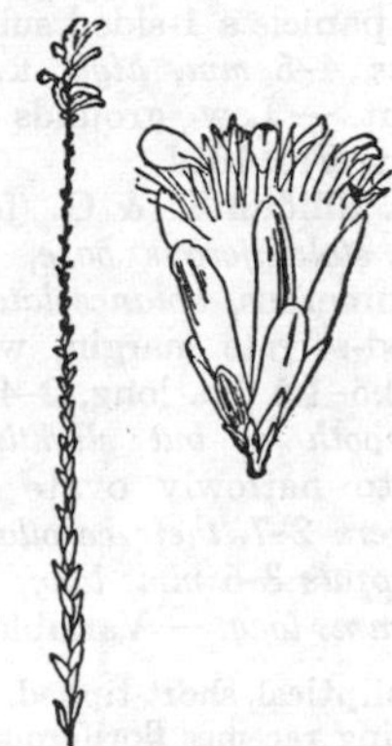

1601. S. radula.

Var. **laéta** (Greene) Fern. (gay). — *Phyllaries* ovate to broadly oblong, *the inner scarcely narrower than the median, 1–1.2 mm. wide.* — Mo. to Tex.

Var. **stenólepis** Fern. (slender-scaled). — *Phyllaries narrowly linear, 0.3–0.5 mm. wide.* — Mo. to La. and Tex.

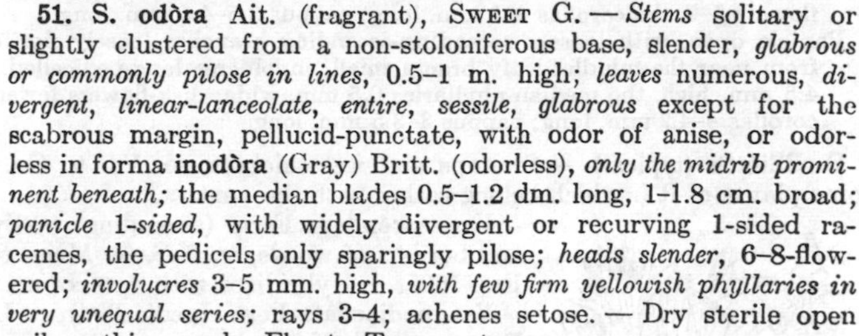

51. S. odòra Ait. (fragrant), Sweet G. *Stems* solitary or slightly clustered from a non-stoloniferous base, slender, *glabrous or commonly pilose in lines,* 0.5–1 m. high; *leaves* numerous, *divergent, linear-lanceolate, entire, sessile, glabrous* except for the scabrous margin, pellucid-punctate, with odor of anise, or odorless in forma **inodòra** (Gray) Britt. (odorless), *only the midrib prominent beneath;* the median blades 0.5–1.2 dm. long, 1–1.8 cm. broad; *panicle* 1-*sided,* with widely divergent or recurving 1-sided racemes, the pedicels only sparingly pilose; *heads slender,* 6–8-flowered; *involucres* 3–5 mm. high, *with few firm yellowish phyllaries in very unequal series;* rays 3–4; achenes setose. — Dry sterile open soil or thin woods, Fla. to Tex., n. to s. N.H., s. Vt., centr. N.Y., O., Ky., Mo. and Okla. Early July–Sept. Fig. 1602. — Hybridizes with no. 52.

1602. S. odora.

52. S. tortifòlia Ell. (with twisted leaves). — *Stems from a creeping stoloniferous base,* 0.3–1 m. high, slender, *minutely puberulent above,* very leafy; *leaves* 100–200, *linear or linear-oblanceolate, scabrous-puberulent,* 1.5–8 mm. wide, *often twisted, spreading, only the midrib prominent beneath,* the larger leaves remotely and shallowly toothed; panicle 1-sided, with many recurving 1-sided racemes; heads very slender, distinctly pedicelled; *involucres* 2.5–4 *mm. high, with* pale stramineous *obtuse phyllaries;* achenes sparingly short-pubescent. — Dry open soil and open woods, Fla. to Tex., n. to e. Va. Sept., Oct. Fig. 1603. — Hybridizes with no. 51.

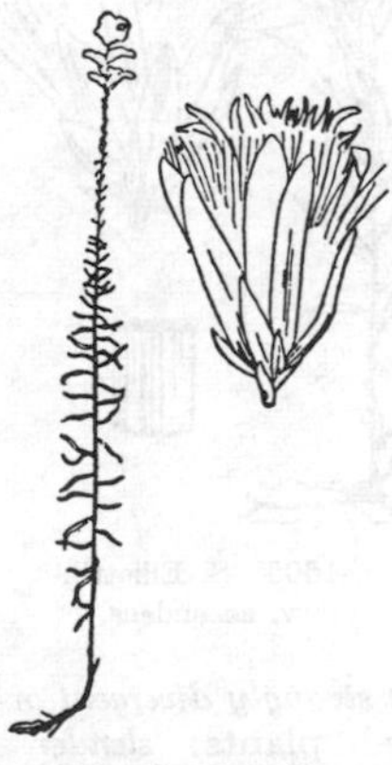

1603. S. tortifolia.

1604. S. fistulosa.

53. S. fistulòsa Mill. (hollow and cylindric). — *Stems* from a creeping stoloniferous base, 0.6–2 m. high, *hirsute* at least *above,* or in forma **épilis** Fern. (without hair) glabrous and the leaves essentially so; *leaves* very numerous, rather crowded, *ascending, oblong or oblong-oblanceolate,* sessile, often subclasping and frequently *hirsute* along the midrib *beneath, the* *principal ones* 1–2.5 *cm. broad,* remotely low-dentate, *with only the midrib prominent beneath;* panicle a 1-sided subpyramidal thyrse with divergent or recurving dense racemes; *involucres* 4–5 *mm. high, with* linear-lanceolate *acute or acutish thin phyllaries;* achenes pubescent. — Low grounds of the Coastal Plain, Fla. to La., n. to N.J. Late Aug.–early Oct. Fig. 1604.

54. S. Elliόttii T. & G. (for its discoverer, Stephen Elliott, 1771–1830). — *Stems from a creeping stoloniferous base, glabrous throughout,* somewhat angled above, 0.5–2 m. high; *leaves* numerous, *oblanceolate to elliptic, membranaceous, glabrous throughout* except for the appressed-serrate margin, with veins evident but not prominent beneath; the median blades 0.5–1.5 dm. long, 2–4.5 cm. wide; *panicle* rather pyramidal to thyrsoid, often leafy, *with smooth or but slightly pilose branches; involucres* 3.5–6.5 *mm. high,* with linear-oblong to narrowly ovate obtuse stramineous phyllaries with distinct pellucid midrib; *disk-flowers* 2–7, *their corollas* 4–5.5 *mm. long, with lobes* 1.5–2 *mm. long; anthers* 1.5–2 *mm. long; pappus* 3–5 *mm. long;* rays 8–10, rather large (0.5–1 mm. broad); *achenes* hispidulous, 1.5–1.8 *mm. long.* — Variable, three vars. with us:

Leaves elliptical, short-tipped, not acuminate; panicle with strongly divergent or recurving racemes floriferous nearly or quite to base; involucres 4.5–5.5 mm. high; median phyllaries 0.7–1 mm. broad; pappus 3.5–5 mm. long. . . . *S. Elliottii* (typical).

Leaves oblanceolate to elliptic, acuminate; panicle with strongly to slightly ascending branches.
- Panicle dense, the strongly ascending or spreading-ascending branches densely floriferous nearly to base; leafy bracts often conspicuous; involucres short-pedicelled, 4.5–6.5 mm. high, the median phyllaries 0.7–1 mm. wide; disk-flowers 5–7, the corollas 4–5 mm. long; pappus 3.5–4.5 mm. long. . . . Var. *ascendens.*
- Panicle open, with loosely spreading-ascending branches loosely floriferous from near the middle; leafy bracts small; involucres long-pedicelled, 3.5–4.5 mm. high, the median phyllaries 0.5 mm. wide; disk-flowers fewer, the corollas 4–4.2 mm. long; pappus 3–3.5 mm. long. . . . . Var. *pedicellata.*

S. Elliόttii (typical). — Swamps and wet thickets, se. Va. to Ga.; nearly approached by some plants of R.I. which belong rather with the next.

Var. **ascéndens** Fern. (ascending). — Swamps and low thickets and borders of woods, w. N.S.; e. Mass. to e. Va. Late Aug.–Oct. Fig. 1605. — Hybridizes with vars. of no. 56.

Var. **pedicellàta** Fern. (pedicelled). — Low woods and thickets, Eastern Shore, Va. Oct.

1605. S. Elliottii, v. ascendens.

55. S. ulmifòlia Muhl. (elm-leaved). — *Stem* slender, commonly solitary, 0.3–1.5 m. high, *glabrous* below the inflorescence; *leaves* thinnish and loosely veined, *long-pilose beneath,* especially along the nerves, *coarsely and unequally serrate,* often jagged; *the basal elliptic-ovate to rhombic,* sometimes rosulate, acuminate, rather abruptly narrow to a broad petiole, 0.6–2.5 dm. long, 1.5–7.5 cm. broad; *cauline leaves* 15–40 *below the inflorescence,* the lower remote and similar to basal, the median and upper distinctly smaller and less toothed; the uppermost entire, 1–7 cm. long, 0.3–2 cm. broad; *inflorescence* an open panicle 0.4–4 dm. high, *its few strongly divergent or arched-ascending* (in well-developed plants) *slender conspicuously secund branches* pilose, bearing inconspicuous bracts on the floriferous portions; *involucres slender,* 3–4 mm. high; *phyllaries* 3–4-seriate; the outer oblong or oblong-lanceolate, *stramineous; the inner lance- or linear-oblong and not strongly keeled;* disk-flowers 4–6, rays about 4; achenes minutely pubescent. — Dry rocky woods and thickets, rarely in meadows, ne. Mass. and Vt. to se. Minn., s. to Ga., Tenn., Ark., Okla. and Tex. Aug.–Nov. Fig. 1606. — Hybridizes with nos. 1, 47 and 56.

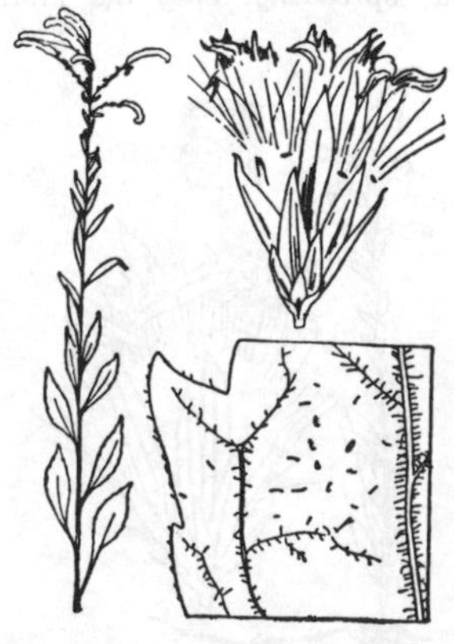

1606. S. ulmifolia.

56. S. rugòsa Ait. (wrinkled). — *Stems* in small clumps or solitary, 0.4–2 m. high, *very leafy; leaves* up to 100 or more, *rather crowded,* lanceolate to elliptic or oval, *mostly serrate or dentate, loosely veiny beneath; panicle* 0.5–4 dm. high, *its strongly divergent to recurved-ascending branches secund, pilose, with conspicuous foliaceous bracts on the floriferous portions;* heads hemispherical-campanulate, 3–5.5 mm. high, with linear to narrowly oblong green to stramineous phyllaries; *disk-flowers* 4–7, with *corollas* 2.5–3.5 *mm. long; pappus* 2–3 *mm. long;* rays 6–9; achenes pubescent. — A common and hopelessly variable species; the following the most definite and widespread vars.:

*a.*Stem and leaves copiously pubescent (at least above), with or without prominent decurrent lines running down from the leaf-bases. . . *b.*

*b.*Stems (at least near and in the panicle) sordid-villous; leaves lanceolate to narrowly oblanceolate, acuminate, gradually tapering at base, commonly coarsely sharp-serrate, usually rather thin and loosely veiny, not conspicuously rugose, more or less villous beneath; involucres 3–5 mm. high, their thin green linear or linear-lanceolate phyllaries attenuate or bluntish.

Panicle broadly pyramidal, its long curving lower branches much exceeding the comparatively small (1–7 cm. long, 0.5–1.5 cm. broad) subtending leaves. . . . . *S. rugosa* (typical).

Panicle elongate-pyramidal to cylindric, the lower lateral racemes nearly equaled to overtopped by the large (0.5–1 dm. long, 1–3.5 cm. broad) subtending leaves. Var. ***villosa.***

*b.*Stems scabrous-puberulent to short-hispid, rarely glabrous, terete or only occasionally angulate-striate; leaves oval or elliptic to lanceolate, rounded at base, subacute to short-acuminate, low-serrate to crenate (sometimes coarsely serrate), scabrous on both surfaces, thick and strongly rugose, lower surface hispid.

Panicle pyramidal, usually longer than broad, its ascending to spreading branches floriferous throughout or the lower sometimes merely leafy-bracted below, the reduced rameal leaves elliptic to lanceolate and acute; involucres 3–4 mm. high; their principal phyllaries firm, stramineous to pale green, linear-oblong and round-tipped, 0.4–0.8 mm. broad. . . . . Var. ***aspera.***

Panicle very lax, its few distant very prolonged and divergent branches (up to 4.5 dm. long) floriferous chiefly above the middle, their bracteal leaves elliptic to oval; involucres 3.5–5.5 mm. long; the inner phyllaries often prolonged, linear, obtuse or acute, often membranaceous. . . . Var. ***celtidifolia.***

*a.*Stems glabrous, commonly with prominent decurrent ridges from bases of the glabrous firm appressed-serrate lanceolate to narrowly elliptic leaves; upper leaves reduced; panicle as in typical var. but with glabrous or only sparsely pubescent branches; phyllaries linear-lanceolate to linear-oblong, obtuse or acute. . . . . Var. ***sphagnophila.***

S. rugòsa (typical), *S. altissima* sensu Mackenz. in Small, not L. — Damp open soil, thickets and borders of woods and streams, Nfld. to Ont., s. to N.S., N.E., L.I., w. Va., W.Va. and La., abund. northw. Aug.–Oct. Fig. 1607. — Hybridizes with nos. 1, 7, 9, 30, 54, 55 and 58.

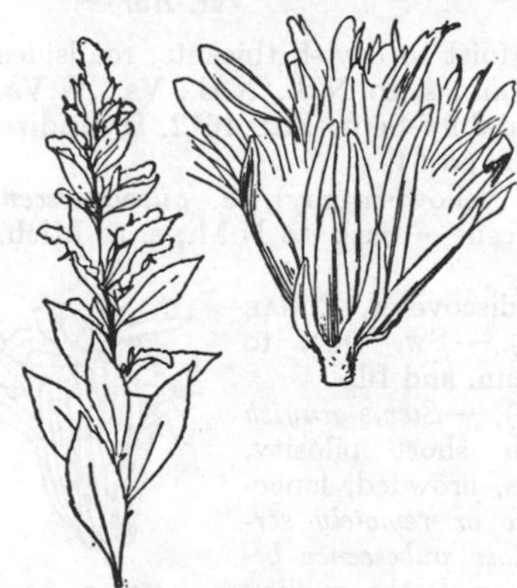

1608. S. rugosa, v. villosa.

Var. **villòsa** (Pursh) Fern. (with long soft hairs). — Low grounds, Nfld. to Ont. (often abundant), s., locally, to N.S., N.E., Va., W.Va., O. and Mich. July–Oct. Fig. 1608.

Var. **áspera** (Ait.) Fern. (rough), *S. aspera* Ait. — Dry or damp open soil or thin woods and thickets, Fla. to Tex., n. to s. Me., Mass., N.Y., O., Mich. and Mo. Mid-Aug.–Oct. Fig. 1609.

Var. **celtidifòlia** (Small) Fern. (with leaves of *Celtis,* Hackberry), *S. celtidifolia* Small — Dry or moist open

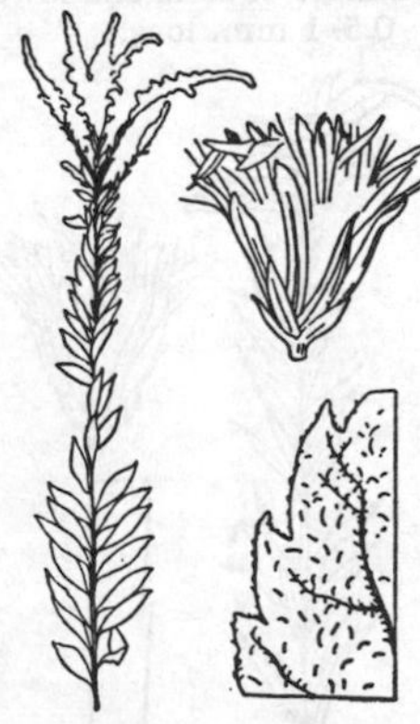

1607. S. rugosa (typical).

woods, clearings and thickets, Ga. to Tex., n. to s. N.J., s. Ind. and Ark. Sept., Oct. FIG. 1610.

Var. **sphagnóphila** Graves (*Sphagnum*-lover), *S. aestivalis* Bickn. — Swampy, often boggy, habitats, s. Me. to N.C. Aug., Sept.

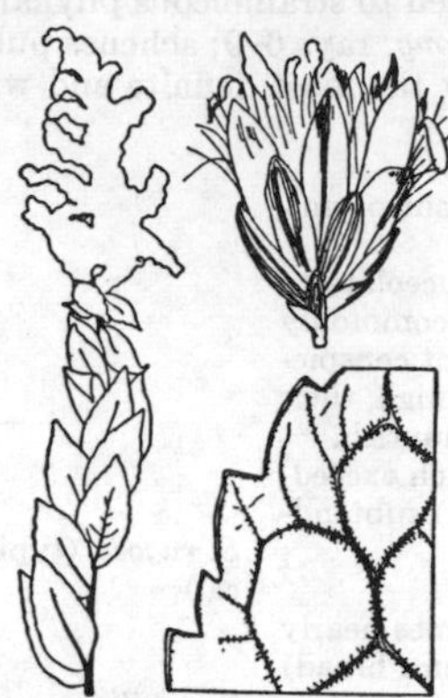
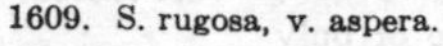
1609. S. rugosa, v. aspera.

1610. S. rugosa, v. celtidifolia.

1611. S. Drummondii.

× **S. aspérula** Desf. (roughish) or plants passing as such is a series of hybrids of various forms of nos. 30 and 56, growing near the coast and more frequently occurring than most hybrids in the genus.

57. S. Drummóndii T. & G. (for its discoverer, THOMAS DRUMMOND, 1780–1835). — *Stem* (3–9) dm. high) *and lower surface of the broadly ovate or oval somewhat triple-ribbed leaves minutely velvety-pubescent;* some of the leaves almost entire; racemes panicled; phyllaries oblong, obtuse; rays 4 or 5. — Calcareous cliffs and rocky woods, sw. Ill. and Mo. to La. Sept., Oct. FIG. 1611.

58. S. canadénsis L. (Canadian). — *Stems* clustered or solitary, *glabrous at base, closely pubescent toward summit below the lowest branches of the panicle,* 0.3–1.5 m. high; *leaves very numerous, rather crowded, lanceolate, long-attenuate, triple-nerved, mostly sharp-serrate;* the median 6–13 cm. long, 0.5–1.8 cm. wide; *broad pyramidal panicle* 0.5–4 dm. high, *its* crowded and densely floriferous *strongly secund divergent racemes with recurved tips; involucres* 2–2.8 *mm. high,* with thin linear-attenuate yellowish-green phyllaries in 3–4 series; rays minute; *disk-corollas* 2.4–2.8 *mm. long;* achenes pubescent. — Very variable, three of its vars. with us:

Summit of stem covered with minute pubescence 0.2–0.5 mm. long.
- Leaves green, glabrous or merely a little scabrous above, slightly pilose along nerves beneath. . . . . *S. canadensis* (typical).
- Leaves grayish with scabrous puberulence on both sides. . . . . . Var. *gilvocanescens.*

Summit of stem and lower surfaces of leaves pilose with loosely spreading hairs 0.5–1 mm. long. . . . . Var. *Hargeri.*

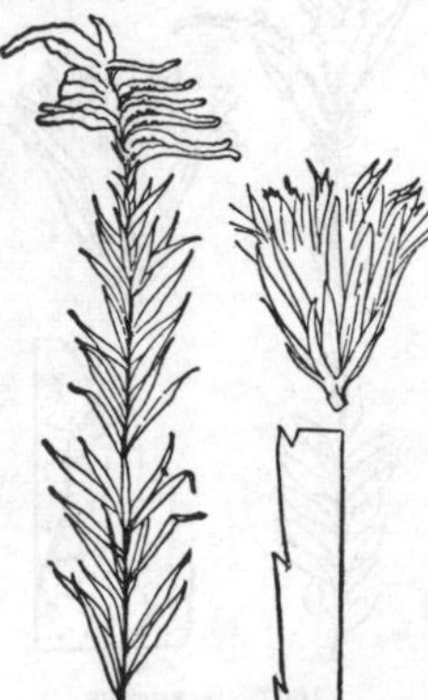
1612. S. canadensis.

S. canadénsis (typical). — Moist to dryish thickets, roadsides, clearings and slopes, Nfld. to Man., s. to N.S., N.E., Va., W.Va., Ill., Minn., S.D. and Colo. Mid-July–Sept. FIG. 1612. Hybridizes with nos. 1, 7, 22, 37, 56, 59, 60 and 64.

Var. **gilvocanéscens** Rydb. (yellowish-gray), *S. gilvocanescens* (Rydb.) Smyth; *S. pruinosa* Greene — Sask. to N.M., e. to Mich., W.Va. and n. Va.

Var. **Hárgeri** Fern. (for its discoverer, EDGAR BURTON HARGER, 1867–1946). — Sw. Que. to Mich., s. to w. N.E., w. Va., Tenn. and Ill.

59. S. altissima L. (very tall). — *Stems grayish* with a close puberulence or short pilosity, 0.7–2 m. high; *leaves* numerous, crowded, lanceolate, long-acuminate, *subentire or remotely serrate,* thickish, *cinereous with close pubescence beneath,* scabrous above, triple-nerved; the median 6–13 cm. long, 1–2 cm. broad; *panicle pyramidal,*

1613. S. altissima.

0.5–3 dm. high, *its* numerous *recurved-spreading branches cinereous; heads slender*, 12–18-*flowered; involucres* 3.2–5 *mm. high;* the linear blunt *phyllaries mostly pale, scarious-herbaceous, with prominent keel; disk-corollas* 3–4 *mm. long.* (*S. hirsutissima* Mill.) — Dry to damp thickets, roadsides and clearings, Fla. to Tex., n. to sw. Que., N.Y., sw. Ont., Mich., Wisc., Minn. and Neb. Late Aug.–Nov. Fig. 1613. Hybridizes with no. 58.

60. S. lépida DC. (elegant). — *Stem puberulent or minutely pilose* at least toward summit, 0.3–1.5 m. high; *leaves numerous*, crowded, lanceolate to oblong, acuminate, *serrate*, the median 0.6–1.5 dm. long, 1–3.5 cm. broad; *panicle dense, thyrsiform to corymbiform-subpyramidal*, 0.3–3 dm. high, *its branches strongly ascending or but slightly spreading, densely pilose; heads campanulate-hemispheric*, 20–40-*flowered; involucre* 3–5 *mm. high, with* linear-lanceolate *greenish phyllaries.* — Very variable boreal species; four vars. with us:

Inflorescence a dense pyramidal to corymbiform or thyrsiform panicle very leafy at base; upper leaves not much reduced in size, extending into the panicle as elongate bracts.
  Panicle or thyrse pyramidal or rhomboid, 0.2–1.1 dm. in diameter, 0.2–1.8 dm. high; leaves oblong or broadly lanceolate.
    Leaves green and glabrous or nearly so above, merely somewhat pilose on the nerves beneath. . . . . . *S. lepida* (typical).
    Leaves densely cinereous-puberulent on both faces. . . . . Var. *molina.*
  Panicle corymbiform or broadly subpyramidal, 0.7–2 dm. in diameter, 1–3 dm. high; leaves lanceolate, green, glabrous or sparsely pilose. . . . . Var. *fallax.*
Inflorescence an elongate scarcely leafy thyrse 2–8 cm. in diameter, 0.5–2.5 dm. long; upper leaves reduced in size below thyrse. . . . . Var. *elongata.*

S. lépida (typical). — Moist to dryish thickets, shores and banks, Lab., Nfld. and e. Que; Alaska and B.C. Aug., Sept. Fig. 1614.

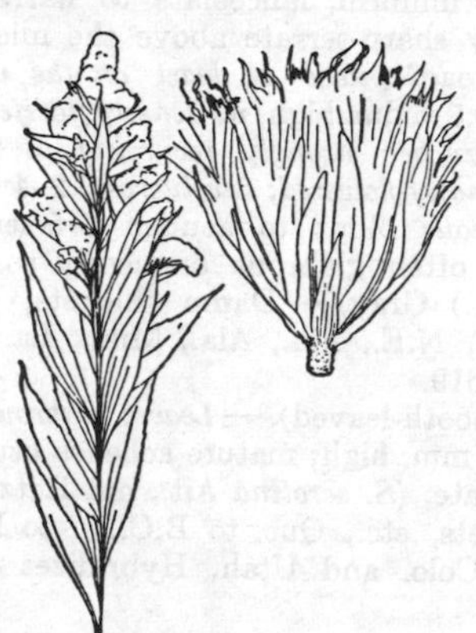

1614. S. lepida (typical).

Var. **molìna** Fern. (of a mill; from its floury aspect). — Dry gravels, Gaspé Pen., Que., and ids. of Penobscot Bay, Me.

Var. **fállax** Fern. (deceitful). — Nfld. to B.C., s. to s. Que., L. Huron, Ont., n. Mich., n. Minn. and Colo. Late July–Sept.

Var. **elongàta** (Nutt.) Fern. (elongate), *S. elongata* Nutt. — Nfld. and e. Que. to B.C., s. to L. Huron, Ont., S.D. and N.M. Fig. 1615.

× **S. leiophállax** Friesner (compounded from the names of the parents) is a probable hybrid of *S. lepida*, var. *fallax* and *S. gigantea*, var. *leiophylla.*

1615. S. lepida, v. elongata.

61. S. rupéstris Raf. (of rocks). — Resembling no. 58; *stems* slender, up to 1.2 m. high, *glabrous to base of inflorescence, the long divergent branches of the loosely pyramidal panicle minutely pilose; leaves* linear-lanceolate, glabrous, very numerous, 5–12 mm. wide, *membranaceous*, low-serrate to subentire; *involucres* 2.5–3 *mm. high, of* 3–4 *series of linear herbaceous phyllaries* 0.2–0.5 *mm. wide.* — Rocky river-banks and openings, D.C. to Ind. and Tenn. Aug., Sept. Fig. 1616.

62. S. Bartramiàna Fern. (for one of its discoverers, Edwin Bunting Bartram, 1878–). — *Stems* clustered, *glabrous to inflorescence* or very sparsely pilose, *slender*, 2–3.5 dm. high; *leaves* crowded, uniform, *lance-attenuate, thin, glabrous*, or margin scabrous, 3–6 *cm. long*, 4–8 *mm. wide, entire or obscurely serrate; inflorescence thyrsiform or corymbiform*, 3–12 cm. high, *its branches strongly ascending* and only sparsely hirtellous; *heads* few, mostly *on slender pedicels* 0.7–1.2 *cm. long; involucres* 2.5–3 *mm. high, their* few narrowly linear *thin green phyllaries in only* 1–2 *series;* rays about 10; achenes hispid. — Slaty ledges and gravel by Exploits R., Nfld. July. Fig. 1617

63. S. Shórtii T. & G. (for its discoverer, Charles Wilkins Short, 1794–1863). — *Stem* slender, 0.5–1.2 m. high, *glabrous* up to inflorescence or *minutely and sparsely hirtellous* above and *in the panicle; leaves subcoriaceous*, sublustrous, glabrous, *narrowly oblong-*

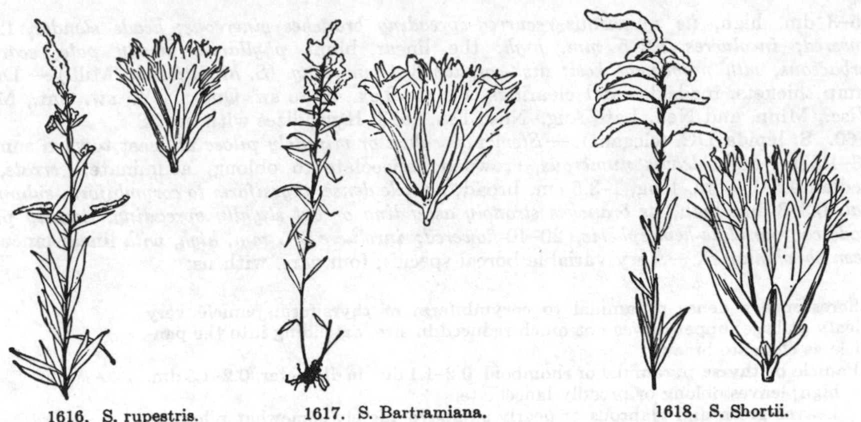

1616. S. rupestris. 1617. S. Bartramiana. 1618. S. Shortii.

lanceolate, the larger 0.5–1 dm. long, the lower finely serrate; *panicle pyramidal*, with divergent and recurving branches; *heads long-pedicelled, slenderly subcylindric*, 10–14-*flowered; involucre* 4–6 mm. high; its *phyllaries coriaceous, narrowly oblong to lanceolate, the broader* 1 *mm. wide;* achenes silky. — Rocky slopes, Ky., local. Aug., Sept. Fig. 1618.

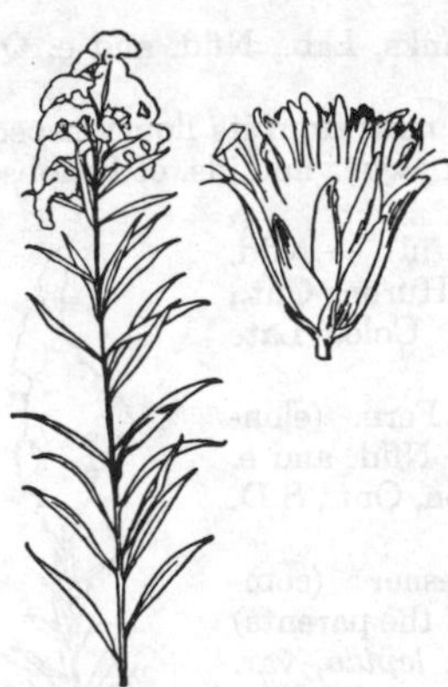

1619. S. gigantea.

64. S. gigantèa Ait. (very large). — *Stems* commonly clustered, sometimes solitary, *stoutish, glabrous* or essentially so up to the inflorescence, *often with whitish bloom*, 0.5–2.5 m. high; *leaves* numerous, essentially uniform, lanceolate to narrowly oblong, acuminate, regularly sharp-serrate above the middle, 7–14 cm. long, 1–3 cm. broad, *pilose at least on the veins beneath; panicle* pyramidal, 0.5–5 dm. high, *its densely floriferous* 1-sided *copiously pilose branches strongly recurving at tips; heads broadly campanulate, many-flowered; involucres* 3.2–4 *mm. high, with linear subherbaceous* blunt to acutish 3–5-seriate *phyllaries;* mature achenes often glabrous or nearly so. (*S. serotina*, var. *gigantea* (Ait.) Gray) — Damp thickets, etc., P.E.I. to Oreg., s. to N.S., N.E., Fla., Ala., Miss., La. and Tex. Late July–Sept. Fig. 1619.

Var. **leiophýlla** Fern. (smooth-leaved). — *Leaves glabrous on both sides;* involucres 3.5–5 mm. high; mature achenes usually pubescent, sometimes glabrate. (*S. serotina* Ait., not Retz.) — Alluvial or rich soil, thickets, etc., Que. to B.C., s. to N.S., N.E., N.C., Tenn., Tex., Colo. and Utah. Hybridizes with no. 58.

65. S. rígida L. (stiff). — *Stem* coarse, *puberulent*, 0.3–1.5 m. high; *leaves* rather numerous, thick and *harsh, grayish-pubescent, subentire or with appressed rounded teeth; lower leaves* long-petioled, *with oval or oblong blades* 1–3 dm. long, 3–10 *cm. broad*, the upper shorter, sessile or slightly clasping, ovate to oblong; *corymb* compound, 0.8–2.5 *dm. broad*, its branches densely hispid; heads showy, 30–40-flowered; *involucres* 6–8 *mm. high*, with oblong round-tipped thick greenish phyllaries; rays 7–10; achenes glabrous. (*Oligoneuron* Small; *O. grandiflorum* (Raf.) Small) — Dry or gravelly open woods, thickets and prairies, Mass. to Sask., s. to Ga., La. and Tex. Aug.–Oct. Fig. 1620.

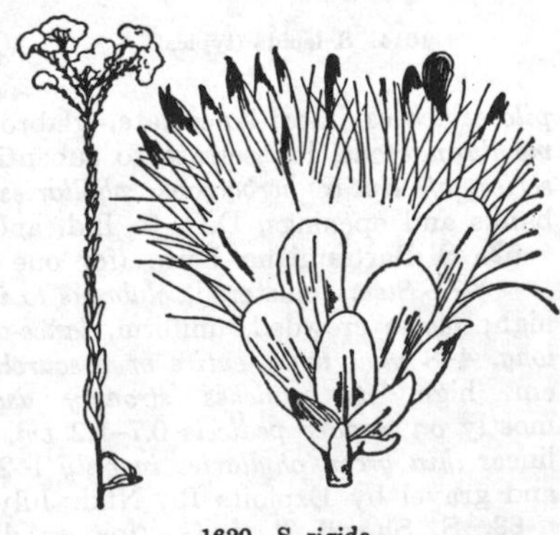

1620. S. rigida.

66. S. Jacksònii (Ktze.) Fern. (named in 1891 for its discoverer, "Mr. Jackson" of Louisville). — Resembling no. 64, smaller; *stem glabrous;* leaves glabrous beneath, scabrous or smoothish above, the lower commonly serrate above the middle; *corymb only* 3–10 *cm. broad,* its branches pubescent; *heads* 30–40-*flowered;* involucres 6–8 mm. high, resembling those of no. 65; achenes turgid, 10-nerved, glabrous. — Barrens and dry openings, Ga. and Ala., n., locally, to centr. O. Aug., Sept. FIG. 1621.

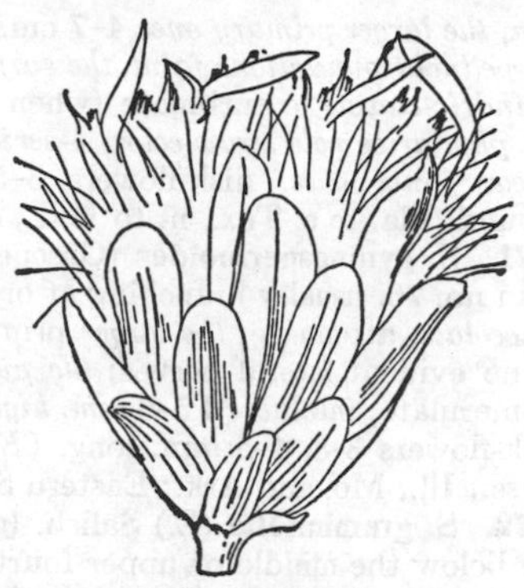
1621. S. Jacksonii.

67. S. ohioénsis Riddell (of Ohio). — *Stem glabrous* or essentially so, rather slender, 6–9 dm. high; *leaves* rather numerous, thick, *glabrous;* the *lower* long-petioled, *narrowly oblanceolate to spatulate, obtuse or obtusish,* entire or serrate above the middle; the blade 0.7–2 dm. long, 1–3.5 cm. broad; *cauline* much shorter, sessile, oblong-lanceolate, *flat,* entire, *erect; corymb* 0.7–2 dm. broad, *its slender* fastigiate *branches and* numerous *pedicels glabrous; heads* 16–20-*flowered,* cylindrical; involucres 4.5–6 mm. high, their oblong phyllaries yellowish; rays 6 or 7. — Calcareous bogs, wet prairies and sandy shores, nw. N.Y. and e. Ont. to Wisc. and Ill. Aug., Sept. FIG. 1622.

1622. S. ohioensis.

68. S. Riddéllii Frank (for its discoverer, JOHN LEONARD RIDDELL, 1807–1865). — *Smooth and stout,* 0.5–1 m. high, *very leafy,* the branches of the dense corymb and pedicels rough-pubescent; *leaves linear-lanceolate, acute, often longitudinally folded, commonly with recurved or divergent tips; the basal* 1–2 *cm. broad,* 3–5 dm. long, tapering into a long keeled petiole; *branches of the dense corymb very pubescent;* heads 20–30-flowered; involucres 5–6 mm. high; rays small, 7–9. (*Oligoneuron* Rydb.) — Wet prairies, swamps and ditches, s. Ont. to Minn., s. to O., Ind., Ill. and Mo.; Fortress Monroe, Va. Aug., Sept. FIG. 1623.

1623. S. Riddellii.

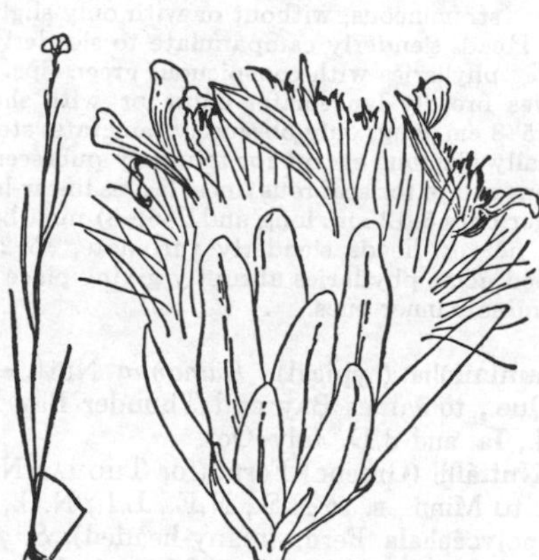
1624. S. Houghtonii.

69. S. Houghtònii T. & G. (for its discoverer, DOUGLAS HOUGHTON, 1809–1845). — *Smooth; stem rather low and slender,* 3–6 dm. high; *leaves scattered, linear-lanceolate, acutish,* rough-margined, 0.5–1.3 dm. long, 5–10 mm. wide, tapering into a narrowed slightly clasping base, 1-nerved, or the lower 3-nerved and with margined petioles; veins obscure; *heads few or several,* 20–30-flowered; involucre 6–8 mm. long, with obtuse phyllaries; rays 7–9. — Damp moss and marl-bogs, local, Genesee Co., N.Y., to Mich. and Ill. Aug. FIG. 1624.

§ EUTHÀMIA Nutt.

70. S. leptocéphala T. & G. (slender-headed). — Stem 0.4–1 m. high, simple to above the middle, then with ascending branches forming a compact to open corymb; *leaves* lanceolate, *thin, the larger primary ones* 4–7 cm. long and 4–8 *mm. wide*, with midrib and 2 lateral nerves, *the pellucid punctations faint, the surface often pustular;* some or all heads glomerulate; *involucre cylindric-clavate* to turbinate (when dry), 4.5–5 *mm. high*, 10–19-flowered, *not viscid* or barely so; *phyllaries pale straw-color, 4-seriate, the outer* ones ovate, *passing insensibly into the obtuse linear inner ones;* disk-flowers 3–5, 3–4.5 mm. long. (*Euthamia* Greene) — Damp sandy ground, Fla. to e. Tex., n. to N.C., sw. Ky. and s. Mo. Aug.–Oct.

71. S. gymnospermoìdes (Greene) Fern. (resembling the genus *Gymnosperma*). — Stronger than no. 70, usually branching at or below the middle; *leaves thickish, linear or narrowly linear-lanceolate,* attenuate; *the larger* primary ones 4–7 cm. long and 2–5 *mm. wide*, with only faint or no evident lateral nerves; *the punctation conspicuous, dark and viscid;* heads pedicelled or glomerulate; *involucre* 4.5–6 *mm. high, viscid*, 9–22-flowered; *phyllaries more numerous, 5-seriate;* disk-flowers 3–8, 4–5 mm. long. (*Euthamia* Greene) — Dry prairies, Colo. to nw. Tex., e. to Wisc., Ill., Mo. and Ark.; Eastern Shore, Va. Aug.–Oct.

72. S. graminifòlia (L.) Salisb. (grass-leaved). — *Stem* 0.3–1.5 m. high, nearly or quite simple below the middle or upper fourth, *glabrous to closely puberulent or hirtellous*, the fastigiate to spreading-ascending *branches of the corymb usually scabrous-toothed along the angles; leaves lanceolate, the lower primary ones* 0.25–1.5 dm. long and (2–)4–12 *mm. wide, usually with* 2 *or* 4 *lateral nerves, glabrous to closely short-pubescent beneath*, puncticulate; *corymb* 0.1–3 *dm. or more broad*, open to compact, the *heads commonly* all or chiefly *glomerulate; involucre slenderly campanulate to turbinate*, 2.5–5 mm. high, 12–45-flowered; disk-flowers 2.5–5 mm. long. — Polymorphous transcontinental species. We have the following varieties:

a. Glabrous or pubescent; leaves lanceolate, the larger ones 4–12 mm. broad, with 2 or 4 lateral nerves; heads 20–45-flowered, campanulate to turbinate, the outer ovate phyllaries merging gradually to the obtuse to merely acute oblong inner ones. . . *b.*

b. Leaves narrowly lanceolate, with attenuate tip, the longer ones 0.3–1.5 dm. long, clearly punctate under magnification; stem up to 1.5 m. high. . . *c.*

c. Stem, branches and leaves glabrous, except for occasionally scabrous lines. — *S. graminifolia* (typical).

c. Stems sometimes and branches and branchlets definitely puberulent or hirtellous; leaves usually pubescent.

Heads campanulate to broadly turbinate, 3–5 mm. high; phyllaries stramineous, without or with only slightly green tips. . . . . . Var. *Nuttallii*.

Heads slenderly campanulate to slenderly turbinate, 3.5–4 mm. high; phyllaries with conspicuous green tips. . . . . Var. *polycephala*.

b. Leaves broadly lanceolate, blunt or with short subacute tips, the larger 2.5–8 cm. long, only obscurely punctate; stem 3–7 dm. high; plant essentially glabrous except for minutely pubescent leaves. . . . . . . Var. *major*.

a. Glabrous except for scabrous lines; leaves linear-lanceolate, long-attenuate, the longer ones 5–12 cm. long and 2–6(–8) mm. broad, usually with only 2 lateral nerves; heads slenderly turbinate, 15–20-flowered, the outer green-tipped acute phyllaries abruptly giving place to the linear-oblong acute or acuminate inner ones. . . . . Var. *media*.

S. graminifòlia (typical), *Euthamia* Nutt. — Damp to dryish shores, thickets, etc., Côte Nord, Que., to James Bay and Thunder Bay Distr., Ont., s. to N.S., N.E., Va., W.Va., O., Ind., Ill., Ia. and S.D. July–Oct.

Var. **Nuttállii** (Greene) Fern. (for Thomas Nuttall, 1786–1859), *Euthamia Nuttallii* Greene — Nfld. to Minn., s. to N.S., N.E., L.I., N.C., Ky. and Mo. Aug.–Oct.

Var. **polycéphala** Fern. (many-headed), *S. polycephala* Fern.; *Euthamia floribunda* Greene — On or near Coastal Plain, se. Mass. to Va.; n. Ind. Aug.–Oct.

Var. **màjor** (Michx.) Fern. (larger; Michaux comparing its broad leaves with the narrower ones of no. 75), var. *tricostata* (Lunell) Harris — Nfld. and Côte Nord, Que., to James Bay, thence to n. Alta., s. to Gaspé Pen., Que., n. Mich., n. Minn., Black Hills. S.D., mts. to n. N.M., and s. B.C. July–Sept.

Var. **mèdia** (Greene) Harris (intermediate), *Euthamia media* and *camporum* Greene — Prairies, sands, rocky places, etc., Minn. to Mo., e. to Wisc. and O. Aug., Sept. — Perhaps better treated as a species, *S. media* (Greene) Bush, intermediate between nos. 71 and 74.

73. S. tenuifòlia Pursh (slender-leaved). — *Stem slender*, rather wiry, 2.5–8 dm. high, *usually with well-developed leafy fascicles or elongate branches from the axils of the primary cauline leaves,* usually corymbosely or subpaniculately branching; *leaves narrowly linear-lanceolate, acuminate, flat, thin, the larger ones 2.3–7 cm. long and 2–4 mm. wide*, subtending axillary fascicles, *with or without a pair of lateral nerves*, sparsely punctate, glabrous except for minute ciliation, *spreading,* little reduced upward; *corymb open to compact,* 0.3–3 *dm. broad, with loosely ascending branches, the lower branches commonly extending well down the stem; heads chiefly glomerulate,* 12–20-*flowered; involucre campanulate to broadly turbinate,* 3–4 (rarely –5) mm. high, commonly glutinous; *phyllaries firm,* usually 4-seriate, *the outer* ovate and obtuse ones *passing gradually to the oblong blunt to subacute inner ones;* disk-corollas 2.5–4 mm. long. — Damp to dry sandy soil, w. N.S. and s. Me. to s. N.Y., s. to Va.; s. Mich. and n. Ind. Aug.–Oct. — At the northeastern margin of its range passing to

Var. **pycnocéphala** Fern. (with crowded heads). — *Stem simple* to or nearly to summit, strict, *without or with scarcely developed axillary fascicles; leaves thick or subcoriaceous, linear-oblong, obtuse or with short acutish tips* (not long-attenuate), *strongly ascending,* the larger 2–6 cm. long and 2–6 mm. broad, *lateral nerves wanting or obscure, punctation weak; corymb compact,* 1–10 *cm. broad,* only 0.1–1 (rarely –1.5) *dm. high, its branches nearly erect; heads* 25–50-*flowered.* (*S. galetorum* (Greene) Friesner) — Sandy, peaty or gravelly shores, swales and savannas, N.S.; local, sw. Me. and centr. N.H. to Cape Cod, Mass. Mid-July–Sept.

× **S. hírtipes** Fern. (with rough pedicels) is a probable hybrid of *S. graminifolia,* var. *Nuttallii* and no. 75, robust, up to 1.5 m. high, with 3–5-nerved scabrous leaves, more or less axillary fascicles, platformed large corymb with densely hirtellous branchlets and pedicels, the glutinous cylindric heads mostly long-pedicelled. Found in Sussex Co., Va.

74. S. remòta (Greene) Friesner (remote; Greene considering it geographically remote from its nearest relative). — In habit suggesting no. 73; *leaves flat,* darkened in drying, *linear-lanceolate,* long-attenuate, *thin, the larger primary ones* 5–9 cm. long and 2–4.5 *mm. wide, usually with a few axillary fascicles,* 1- often 3-nerved; *corymb flattish-topped, up to* 2.5 *dm. broad, its lower spreading-ascending branches with their secondary corymbs nearly equaling to overtopping the central secondary corymbs; heads mostly pedicellate, slenderly cylindric* (becoming slenderly turbinate on pressing or drying), 12–20-*flowered;* involucre 3–4 mm. high; *phyllaries thin and soft,* pale, *the narrowly ovate or deltoid acutish outer ones abruptly giving way to the narrowly oblong to linear-lanceolate acutish inner ones.* (*Euthamia* Greene; *S. Moseleyi* Fern.; *S. graminifolia,* var. *remota* (Greene) Harris) — Sandy woods, openings and prairies, near Lakes Erie and Michigan, n. O. to s. Mich., Wisc. and ne. Ill. Mid-Aug., Sept.

75. S. microcéphala (Greene) Bush (tiny-headed). — Glabrous or minutely pubescent, up to 1 m. high, *with abundant and very leafy axillary fascicles or sterile branches; leaves soon folded or almost bristle-like, the lower and median primary ones* (usually falling on maturity) 1–2.5 *mm. wide, pale green, rarely with any trace of lateral nerves; well developed inflorescence a fastigiate corymb* 0.5–2.5 dm. broad and 0.6–4 dm. high, *the lower and succeeding median strongly ascending branches with their secondary corymbs failing to reach those above, the several-storied corymb thus becoming dome-like;* heads mostly pedicelled; involucre slenderly cylindric, 3.5–4.5 mm. high, 9–18-flowered, often glutinous; *phyllaries* 4-seriate, *the outer* acutish ovate ones *gradually passing into the thin acute narrowly oblong inner ones.* (*S. minor* (Michx.) Fern., not Mill.; *Euthamia caroliniana* sensu Greene and *S. caroliniana* sensu BSP., not *Erigeron carolinianus* L., basonym) — Sandy soil, chiefly on Coastal Plain, Fla. to La., n. to N.J. Sept., Oct.

15. HAPLOPÁPPUS Endl.

Heads many-flowered, radiate; rays many, pistillate. Involucre hemispherical, of many closely imbricated phyllaries in several series. Receptacle flat. Achenes short, turbinate to linear; pappus simple, of numerous unequal bristles. — Mostly herbaceous N. or S. Am. plants, with alternate rigid leaves. Ray- and disk-flowers both yellow. (From the Greek *haplous, simple,* and *pappos, pappus.*) APLOPAPPUS Cass.

Leaves merely toothed to subentire; tall annuals or biennials.
- Glabrous, with broadly oblong, oval or obovate leaves; involucre hemispherical, 2–2.5 cm. broad, with slender squarrose phyllaries in several ranks; southwestern. . . . . 1. *H. ciliatus.*
- Glandular-hirtellous, with mostly linear or linear-oblanceolate leaves; involucre turbinate, opening to 4–6 mm. broad, the few phyllaries straight and ascending; southeastern and southern. . . . . 2. *H. divaricatus.*

Leaves pinnatifid to bipinnatifid; stems from perennial subligneous caudex, 1–4 dm. high; involucre hemispherical, about 1 cm. broad, of closely imbricated phyllaries. . . . . 3. *H. spinulosus.*

1. H. ciliàtus (Nutt.) DC. (with marginal hairs). — *Annual or biennial,* glabrous, 0.5–1.5 m. high, leafy; *leaves oblong, oval* (or the lower obovate), obtuse, *dentate* with bristle-pointed teeth; heads few and clustered; *involucre* hemispheric, 1–2 *cm. high, its outer phyllaries spreading;*

achenes glabrous, the central ones abortive. (*Prionopsis* Nutt.) — Hillsides and ravines, w. Mo. to Colo., s. to Tex. and N.M. Aug., Sept.

2. H. divaricàtus (Nutt.) Gray (divergently branched). — *Annual* with tap-root; *stem* up to 1.8 dm. high, *hirtellous, often glandular,* loosely branched especially above middle; *leaves linear to linear-oblanceolate, hirtellous,* with remote spinulose teeth; heads loosely paniculate, on glandular pedicels; *involucre turbinate-hemispherical,* glandular, 5–7 *mm. high;* phyllaries ascending, finally spreading, in 3 or 4 lengths; ray-flowers 7–18. (*Isopappus* T. & G.) — Dry fields and clearings, Fla. to e. Tex., n. to se. Va., Ark. and Kans. Sept.–Nov.

3. H. spinulòsus (Pursh) DC. (with slender small spines). — *Perennial with a woody caudex,* branching, more or less white-tomentose, especially about the summit; *leaves narrow, pinnately or bipinnately parted,* the lobes and teeth bristle-tipped; heads peduncled; *involucre* 5–6 *mm. high,* the *appressed phyllaries bristle-tipped; achenes pubescent.* (*Sideranthus* Sweet) — Prairies and plains, Minn. to Alta., s. to Tex. and Mex. June–Sept.

Var. **glabérrimus** (Rydb.) Blake (very smooth). — Glabrous or sparingly glandular-puberulent. (*Sideranthus glaberrimus* Rydb.) — Dry plains, Ia. to Wyo., s. to Tex. and N.M.

16. ASTRÁNTHIUM Nutt. WESTERN DAISY

Heads many-flowered; rays numerous. Phyllaries 2–3-seriate, membranous-margined, pointed. Receptacle convex, strongly alveolate. Style-branches linear-lanceolate, acute. Achenes obovate, compressed, wingless, without pappus (or with a minute crown). — Leafy mostly Am. herbs, with alternate leaves and branches. (Greek *astron, star,* and *anthos, flower.*)

1. A. integrifòlium (Michx.) Nutt. (entire-leaved). — Loosely forking annual 1–4.5 dm. high, smoothish; leaves oblong-spatulate to lanceolate or narrowly ovate; heads slender-peduncled; involucre 0.6–1.2 cm. broad, with 2 series of lanceolate acute phyllaries; ray-flowers purplish, fertile; achenes with glochidiate hairs. (*Bellis* Michx.) — Sandy soils, nw. Ga. to Okla., n. to Ky., Mo. and Kans. May, June.

The related BÉLLIS L. (Latin name of the daisy) PERÉNNIS L. (perennial), ENGLISH DAISY, a rosulate perennial with scapose solitary heads, foliaceous blunt phyllaries nearly uniseriate, and high-conical smooth receptacle, escapes from cult. northw. and is somewhat natzd. in Nfld. Apr.–July. (Introd. from Eu.)

17. CHAETOPÁPPA DC.

Heads several-flowered, radiate; disk-flowers often sterile. Phyllaries imbricated in 2 or more rows, the outer shorter. Receptacle flat, naked. Achenes fusiform or compressed; pappus of 5 or fewer thin nerveless scales, alternating with rough bristly awns, or these wanting. — Low southwestern N. Am. branching annuals, with narrow entire leaves and solitary terminal heads; rays white or purple. (Greek *chaite, a bristle,* and *pappos, pappus.*)

1. C. asteroìdes DC. (resembling *Aster*). — Slender, 0.5–3 dm. high, pubescent; involucres slender, 4 mm. long; rays 5–12; achenes pubescent. — Dry grounds, sw. Mo., e. Kans. and Ark. to Tex. Apr.–Aug.

18. BOLTÒNIA L'Hér. BOLTONIA

Heads many-flowered; rays numerous, pistillate. Phyllaries of the hemispherical involucre imbricated somewhat in 2 or more rows, appressed, with narrow membranaceous margins. Receptacle conical or hemispherical, naked. Achenes very flat, obovate or inversely cordate, margined with a callous wing, or in the ray 3-winged, crowned with a pappus of several minute bristles and usually 2–4 longer awns. — Perennial smooth N.Am. herbs, pale green, with the aspect of *Aster;* the thickish leaves chiefly entire, often turned edgewise. Flowers autumnal; disk yellow; rays white or purplish. (Dedicated to *James Bolton,* English botanist, active the 2nd half of the 18th century.)

a. Phyllaries linear to linear-attenuate or subulate, 0.2–0.75 (rarely –1) mm. broad; disk 3–8 mm. broad; awns wanting or up to two-thirds length of achene, much shorter than disk-corolla; achenes 1.5–2 mm. long. . . *b.*

b. Phyllaries 0.5–1 mm. broad, not long-attenuate nor subulate; ligules 0.8–1.5 cm. long, usually lilac or purplish; heads few, on strongly ascending to erect naked or few-bracted peduncles; plant often spreading by elongate stolons. . . . . 1. *B. asteroides.*

b. Phyllaries 0.2–0.4 (rarely –0.5) mm. broad, long-attenuate to linear-subulate;

ligules 5–8 mm. long, white to lilac; heads numerous, more or less diffusely paniculate or corymbose. . . c.

c.*Involucre of 2–3 closely imbricated often subequal series of phyllaries, not often extending down the peduncles as bracts; peduncles 0.5–5.5 cm. long; achenes wingless or only narrowly margined. . . *d.

***d.*Diffusely panicled, the flowering branches and elongate (mostly 1.5–5.5 cm. long) peduncles spreading to loosely ascending; awns wanting or minute; achenes narrowly cuneate-obovate.**

Base without elongate stolons; leaves subtending flowering branches narrowly lanceolate to linear-oblanceolate, attenuate to both ends; ligules 5–7 mm. long, white (rarely lilac); disk 3–5 mm. broad; anthers included; awns about 0.1 mm. long. . . . . . 2. *B. caroliniana.*

Base bearing elongate stolons; leaves subtending flowering branches narrowly obovate to broadly oblanceolate; ligules 7–8 mm. long, lilac; disks 5–8 mm. broad; anthers soon exserted; awns wanting. 3. *B. Ravenelii.*

d.*Strongly corymbose, the leafy corymb with very many erect branches and short peduncles (0.5–2.5 cm. long); awns 0.7–1 mm. long, one-half to two-thirds as long as the broadly obovate achene. . . . . 5. *B. latisquama,* var. *microcephala.

c.*Involucre of 3–5 unequal series, the lower commonly merging down the peduncles with the numerous small bracts; peduncles stiff and straight, many of them 2.5–15 cm. long; disks 3–5 mm. broad; achenes mostly broadly winged. . . . . 4. *B. diffusa.

a.*Phyllaries (at least the larger) oblong to rhombic or cuneate-obovate, 0.5–2 mm. broad; disk 7–15 mm. broad; awns nearly equaling disk-corolla, mostly 1–2 mm. long; achene obovate, 2.5–3 mm. long; coarse leafy-corymbose plants; ligules lilac or white, 1–1.8 cm. long. . . . . 5. *B. latisquama.

1. B. asteroìdes (L.) L'Hér. (resembling *Aster*). — *Slender*, simple or with loosely ascending branches, 2–7 dm. high, with stolons developed in autumn; *leaves* submembranaceous, linear to oblong-lanceolate or oblanceolate, *broad-based, scarcely petioled,* the principal ones 2.5–12 cm. long and 3–13 mm. broad; *heads* 1–23, usually few, *on* loosely ascending to erect *naked or few-bracted peduncles* 2.5–11 *cm. long; involucre* of 2–3 series *of* subequal *linear phyllaries* 0.5–1 *mm. broad; ligules* lilac or purplish, 0.8–1.5 *cm. long; disk* 6–8 *mm. broad;* achenes obovate, about 2 mm. long, the *awns wanting or up to* 0.7 *mm. long.* — Gravelly shores and sandy thickets, se. N.Y. and nw. N.J. to n. O., s. to Md. and mts. of N.C. Mid-July–Sept.

Var. **glastifòlia** (Hill) Fern. (with leaves of *Glastum*). — Freely stoloniferous, up to 1.2 m. high; *leaves coriaceous,* the lower *narrowed to subpetiolar bases;* peduncles often more bracted; awns well-developed. — Muddy shores and tidal marshes, s. N.J. to Ga. and La. Aug.–Oct.

2. B. caroliniàna (Walt.) Fern. (of Carolina). — Very *tall, non-stoloniferous, the paniculate stem up to* 2.3 *m. high,* the prolonged *branches* very *diffusely divided;* principal cauline *leaves* at flowering time *submembranaceous,* lanceolate to linear-oblanceolate, *attenuate to base and apex;* peduncles filiform, 1.5–4.5 cm. long; *involucre* of 2–3 close series, *the linear-subulate phyllaries only* 0.2–0.3 *mm. broad; ligules white* (rarely lilac), 5–7 *mm. long; disks* 3–5 *mm. broad;* achenes narrowly cuneate-obovate, narrow-rimmed; *awns minute, about* 0.1 *mm. long.* — Bottomlands, rich thickets and meadows, se. Va. and e. S.C. Aug.–Oct.

3. B. Ravenélii Fern. & Grisc. (for its discoverer, Henry William Ravenel, 1814–1887). — Resembling no. 1; stems 2–8 dm. high, stoloniferous; principal *leaves thin, narrowly obovate or broadly oblanceolate,* blunt to subacute; *panicle elongate, subcylindric,* 1.5–5.5 dm. long, 1–1.8 dm. in diameter; peduncles 2–5 cm. long; *phyllaries linear-subulate,* 0.2–0.4 *mm. broad;* ligules lilac, 7–8 mm. long; disk 5–8 mm. broad; *anthers promptly exserted from the disk-corollas; achenes awnless.* — Wooded bottomlands, local, se. Va. and e. S.C. Sept., Oct.

4. B. diffùsa Ell. (loosely spreading). — Very slender, *from slender rhizomes* and stolons; *stem* 0.5–1 m. high, 1.5–3 *mm. thick at base, diffusely open-paniculate; leaves coriaceous, linear or linear-subulate, the upper much reduced; peduncles* filiform, *many-bracteolate,* straight, 2.5–15 *cm. long;* heads remote; *involucre of* 3–5 *unequal series, the lower phyllaries commonly merging down the peduncles into the numerous small bracts;* phyllaries linear-subulate, 0.2–0.5 mm. broad; ligules white or lilac, 5–8 mm. long; disk 3–5 mm. broad; *achenes broadly winged,* awns about one-fourth as long. — South of our range (Fla. and sw. Ga. to Okla. and e. Tex.); represented with us by

Var. **intèrior** Fern. & Grisc. (inland). — Less stoloniferous, *coarser; stems* 4–7 *mm. thick at base,* paniculate-ramose, the *branches mostly forking,* the peduncles thicker; *phyllaries linear-oblong,* scarcely subulate-tipped. — Woods, barrens and fields, Miss. to Okla., n. to Ky., s. Ill., and Mo. July–Sept.

5. B. latisquàma Gray (with broad scales). — *Coarse*, up to 1.5 m. or more high, *strongly corymbose-paniculate;* stem conspicuously corrugated; *leaves firm*, the primary ones *lanceolate, the numerous ones of the corymb smaller* and becoming *linear-attenuate; peduncles leafy*, 1.5–5 *cm. long; phyllaries firm, narrowly cuneate- to spatulate-obovate, short-tipped*, 1–2 *mm. broad; ligules* white or lilac, 1–1.8 *cm. long; disk* 7–14 *mm. broad; achenes* obovate, 2.5–3 *mm. long*, broad-winged; *awns* 1.5–2 *mm. long.* — Prairies and banks of streams, Mo., Kans. and Okla.; cult. and esc. eastw. to N.E. July–Oct.

Var. **recógnita** Fern. & Grisc. (restudied). — Leaves larger; corymb up to 4 dm. broad; *phyllaries narrowly oblong*, 0.5–1.3 *mm. broad.* (*B. asteroides* in large part of ed. 7) — Prairies and borders of streams, Mich. to Man. and N.D., s. to Ky., Mo. and Kans.; cult. and esc. e. to N.E. and N.J. Late July–Oct.

Var. **decúrrens** (T. & G.) Fern. & Grisc. (decurrent). — Like the last; *principal leaves decurrent as wings down the stem.* (*B. decurrens* Wood; *B. asteroides*, var. *decurrens* Engelm.) — With the last, occasional, Ill. and Mo.

Var. **microcéphala** Fern. & Grisc. (small-headed). — Similar to var. *recognita* but *phyllaries linear-subulate or -lanceolate*, only 0.3–0.4 *mm. broad; ligules* 5–8 *mm. long; disk* 5–8 *mm. broad; achenes* 1.5–2 *mm. long, the awns about* 1 *mm. long.* — Wisc., Ill., Minn. and Ia.

19. ASTER L. Aster. Starwort. Frost-flower

Heads many (rarely few)-flowered, radiate except in § 8 (small annuals); the ray-flowers in a single series, or often in 2 series in § 7, fertile. Phyllaries more or less imbricated, either herbaceous or with herbaceous or foliaceous tips, or scarious or somewhat coriaceous. Receptacle flat, alveolate. Achenes more or less flattened; pappus simple, of capillary bristles (double in § § 5 and 6). — Perennial herbs, nearly cosmop. (annual only in § § 7 and 8), with mostly corymbed, panicled or racemose heads, flowering chiefly in autumn. Ligules (rays) white (creamy in no. 58), purple, blue or pink; the disk yellow, often changing to purple. Species, especially in § 3, freely hybridizing and often with limits obscured. (Name the Greek *aster, a star*, from the radiate heads of flowers.) Doellingeria Nees. Brachyactis Ledeb. Ionactis Greene. Unamia Greene

A. Group I. Perennials with strong caudices, crowns or creeping rhizomes; ligules conspicuous. . . *B.*

B. Pappus a single series of bristles. . . *C.*

C. Pappus coarse and rigid, the stronger bristles clavate; phyllaries rigid, green, subequal; leaves rigid, linear to narrowly oblanceolate, entire. § 1. *Heleastrum*, p. 1417.

C. Pappus fine and slender, not clavate, soft to firm; phyllaries of different lengths or, if with the outer elongate and equaling or exceeding the inner, these not rigid. . . *D.*

D. Inflorescence a flat- or flattish-topped corymb; achenes linear, scarcely compressed; pappus firm or subrigid; lower leaves cordate and petioled. . . . . . § 2. *Biotia*, p. 1417.

D. Inflorescence paniculate to racemose or virgate, if (in § 4) a flattish-topped corymb, the leaves not cordate; achenes flattened, obovate-oblong to linear.

Inflorescence paniculate to racemose or virgate; phyllaries (at least the outer) with herbaceous tips or wholly herbaceous, their margins not strongly scarious; leaves linear to cordate-ovate. . . . . § 3. *Euaster*, p. 1419.

Inflorescence corymbose; phyllaries without herbaceous tips, scarious or scarious-margined; leaves narrow, not cordate. . . . . . . § 4. *Orthomeris*, p. 1438.

B. Pappus double, of an inner series of long bristles and an outer series of short ones; inflorescence more or less corymbose or heads few or solitary; phyllaries without herbaceous tips.

Leaves herbaceous, veiny, dilated; ligules white (rarely roseate); pappus distinctly of 2 series, some of the inner bristles thickened at summit. . § 5. *Doellingeria*, p. 1440.

Leaves rigid, linear, veinless (but with strong midrib); ligules pink or violet; pappus obscurely 2-seriate, the bristles all slender. . . . . § 6. *Ianthe*, p. 1441.

A. Group II. Annuals or biennials with only fibrous roots; ligules very short or wanting.

Phyllaries without herbaceous tips, linear to linear-subulate, appressed;

ligules numerous, usually more than disk-flowers and overtopping them; style-appendages lance-subulate. . . . . §7. *Oxytripolium*, p. 1441.

Phyllaries (at least the outer) herbaceous and loose, linear to oblong; ligules very short and inconspicuous or wanting; style-appendages lanceolate. §8. *Conyzopsis*, p. 1441.

Group I. Perennials

§ 1. HELEÁSTRUM (DC.) B. & H. One species with us:

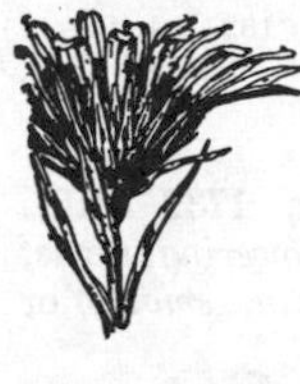

1625. A. hemisphericus.

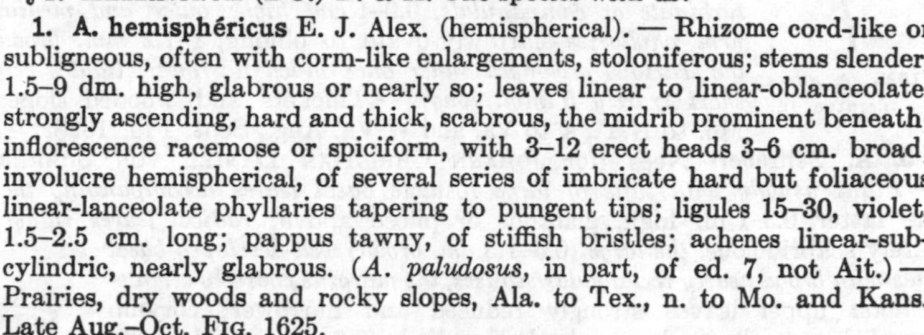

1. A. hemisphéricus E. J. Alex. (hemispherical). — Rhizome cord-like or subligneous, often with corm-like enlargements, stoloniferous; stems slender, 1.5–9 dm. high, glabrous or nearly so; leaves linear to linear-oblanceolate, strongly ascending, hard and thick, scabrous, the midrib prominent beneath; inflorescence racemose or spiciform, with 3–12 erect heads 3–6 cm. broad; involucre hemispherical, of several series of imbricate hard but foliaceous linear-lanceolate phyllaries tapering to pungent tips; ligules 15–30, violet, 1.5–2.5 cm. long; pappus tawny, of stiffish bristles; achenes linear-subcylindric, nearly glabrous. (*A. paludosus*, in part, of ed. 7, not Ait.) — Prairies, dry woods and rocky slopes, Ala. to Tex., n. to Mo. and Kans. Late Aug.–Oct. FIG. 1625.

§ 2. BIÒTIA (DC.) T. & G.

a. Ligules white (sometimes colored in age); branches of inflorescence glandless. . . *b.*

b. Involucre ovoid-campanulate or turbinate; phyllaries appressed, of gradually increasing length; plants rarely producing large tufted basal leaves. . . *c.*

c. Leaves thin, glabrous or sparsely pilose but smooth or barely scabridulous to touch; lower blades ovate, ovate-lanceolate or deltoid-ovate, taper-pointed; involucre 6–8 mm. high. . . . . 2. *A. divaricatus.*

c. Leaves thick, scabrous above; involucre 0.7–1.5 cm. high.

Lower cordate leaves narrowly ovate, long-attenuate, the blade once to several times as long as petiole; involucre 7–10 mm. high; western. . . 3. *A. furcatus.*

Lower cordate leaves broadly ovate, abruptly short-acuminate, the blade and petiole subequal; involucre 0.9–1 cm. high; eastern. . . . 4. *A. glomeratus.*

b. Involucre slenderly cylindric; its phyllaries loosely ascending, the inner ones prolonged far beyond those below; large tufted thin basal leaves abundant; eastern. . . . . 5. *A. Schreberi.*

a. Ligules violet or bluish (exceptionally whitish); branches of inflorescence glandular; sterile large-leaved basal tufts usually abundant; wide-ranging. . . . 6. *A. macrophyllus.*

2. A. divaricàtus L. (widely branching). — *Stems* arising from slender creeping rhizomes, 0.15–1 m. or more high, rather *slender, often zigzag*, glabrous or promptly glabrate; *leaves membranaceous or thinnish, glabrous or with few scattered hairs, smooth to touch or barely scabridulous; lower ones cordate, with tapering or acuminate ovate, ovate-lanceolate or deltoid blades with coarse and prominent* more or less divergent to forward-arching *teeth*, the petioles slender; *median blades* similar but *short-petioled; upper much smaller, varying from toothed to entire; inflorescence a broad flattish- to round-topped corymb of many white-rayed heads; involucre ovoid-campanulate, 6–8 mm. high; phyllaries thin, scarious, mostly obtuse and ciliate, appressed, with inconspicuous green tips, the outer* oblong to oblong-triangular and 1–1.3 *mm. broad*, the inner narrower; ligules 6–12. — Dry woods and clearings, s. Me. to O., s. to s. N.E., L.I., Ga., Ala. and Tenn. July–Oct. FIG. 1626. — Highly variable throughout much of its range, neighboring colonies readily placed in typical *A. divaricatus*, as well as in *A. tenebrosus, stilettiformis, excavatus, castaneus, chlorolepis* and a score more proposed by Burgess as species upon fluctuating degrees of leaf-toothing, outline and apex, as well as varying intensity of coloring of the disk-florets.

1626. A. divaricatus.

3. A. furcàtus Burgess (forking). — Coarser than no. 2, the stiff stem 5–9 dm. high; *leaves harshly scabrous above* and often *hispidulous beneath;* the *lower cordate blades longer than broad,* coarsely crenate-dentate, *on petioles much shorter than to about as long, attenuate; median and upper leaves scarcely reduced, narrowly ovate to oblong-lanceolate, short-petioled to subsessile; involucres* (in dried specimens) *becoming as thick as high,* 7–10 *mm. high, the firm outer and median phyllaries* 1.4–1.8 *mm. broad; ligules* 1 *cm. or more long,* white, or roseate in forma

erythráctis Benke (red-rayed). — Dry woods and shaded bluffs, w. Ind. and s. Wisc. to Mo. Aug.–Oct. Fig. 1627.

1627. A. furcatus.

4. A. glomerȧtus Bernh. (compactly clustered). — As coarse as no. 3, up to 1 m. high, usually without basal leafy tufts; *leaves thick, scabrous above, smoothish beneath; the lower cordate ones broadly ovate, abruptly short-acuminate, with fine sharp serration,* the *petioles equaling or longer than blades; median, upper and bracteal leaves rapidly reduced upward, sessile or wing-petioled; involucres green, thick-turbinate or campanulate, 0.9–1 cm. high; outer and median firm phyllaries* narrowly ovate to oblong, *1–1.2 mm. broad, the scarious prolonged inner ones much narrower; ligules very short* (*rarely 5 mm. long*). — Thickets and wooded slopes, Me. to N.Y., s. to Va. and W.Va. Aug., Sept. Fig. 1628.

1628. A. glomeratus.

5. A. Schrėberi Nees (for Johann Christian Daniel von Schreber, 1739–1810). — *Plant usually with tufts of large rosulate basal leaves accompanying the flowering stems;* the latter 0.3–1 m. high, glabrous or pilose above; *radical leaves rather thin, smooth* or barely scabridulous, *the large* (0.6–1.5 *dm. broad*) *well-developed basal ones with broad nearly rectangular sinuses, the margins coarsely serrate-dentate;* upper leaves strongly reduced and narrower; corymb roundish-topped; *involucres slenderly cylindric,* 8–9 mm. high; *the scarious phyllaries loosely ascending, about* 1 mm. *wide, the inner ones prolonged far beyond those next below; white ligules* 1 *cm. long.* (*A. curvescens* Burgess) — Damp woods and thickets, N.E. to Ill., s. to Del., Md., Pa., upland Va. and Ky. July–Sept. Fig. 1629.

1629. A. Schreberi.

6. A. macrophýllus L. (large-leaved). — Rather *coarse,* 0.2–1.5 m. high, *viscid-glandular* at least *in* the *inflorescence; sterile tufts of large ovate cordate leaves numerous;* basal leaves large, cordate; upper leaves variable, usually smaller and narrower; corymb broad or contracted; involucre slenderly campanulate, of 3 or 4 series; *phyllaries greenish,* the *outer* short-ovate, *blunt and pubescent;* the *inner elongated,* linear, *more scarious,* sometimes with roseate margins; *rays violet or pale blue* (only exceptionally white). — An extremely variable plant, characterized by the abundant basal tufts, glandular inflorescence and mostly colored ligules; the following are the most evident vars.:

a. Glands of pedicels elongate or stipitate; leaves thick or firm. . . *b.*
b. Stem simple two-thirds to five-sixths its height (up to lower branches of the broad corymb). . . *c.*
c. Stems (and usually petioles) glabrous, or only minutely pubescent toward summit. . . *d.*
d. Basal and usually the cauline leaves very harshly scabrous and thick; glands of pedicels abundant. . . . . *A. macrophyllus* (typical).
d. Basal leaves (or some of them) and usually the cauline ones smooth or merely scabridulous; glands fewer among the trichomes of the pedicels.
Stem green or colored; leaves deep green, very smooth (almost greasy to touch); the principal cauline ones broadly cordate-ovate and slenderly long-petioled. . . . . Var. *pinguifolius.*
Stem often glaucous when fresh; leaves pale green, slightly scabridulous; the principal cauline ones narrowly elliptic-ovate to oblong or lanceolate, sessile or shortly wing-petioled. . . . . Var. *excelsior.*
c. Stems and petioles villous; lower leaf-surface pilose.
Most (all but lowest) cauline leaves truncate to tapering at base. . Var. *velutinus.*
Most cauline leaves rounded or cordate at base. . . . . Var. *sejunctus.*
b. Stem forking well below the middle, the corymb therefore subpaniculate; plant pilose to glabrate; leaves small, thick and harsh. . . . . . Var. *apricensis.*
a. Glands of pedicels minute, rarely stipitate; leaves thinnish and smooth to scabridulous above. . . . . Var. *ianthinus.*

A. macrophýllus (typical). — Stem 0.3–1.5 m. high; basal cordate leaves 0.3–3 dm. broad. (Incl. many minor trends such as *A. roscidus,* etc. of Burgess) — Dry to moist open woods, thickets and clearings, Gaspé Pen., Que., to Minn., s. to N.S., N.E., Del., Md., upland to N.C. and Tenn., W.Va., O., Ind. and Ill. Aug., Sept. Fig. 1630.

1630. A. macrophyllus.

Var. **pinguifȯlius** Burgess (greasy-leaved). — Basal leaves up to 1.5 dm. broad. — More local, N.E. to W.Va. and Minn.

Var. **excėlsior** Burgess (taller). — Stem 0.2–1.5 m. high; cordate basal leaves 0.4–1.2 dm. broad. — L. St. John, Que., to Mich., s. to N.S., N.E., Pa. and W.Va.

Var. **velútinus** Burgess (velvety). — Radical leaves 0.3–1.5 dm. broad. — Gaspé Pen., Que., to Thunder Bay Distr., Ont., and n. Minn., s. to N.S., n. N.E., n. and w. N.Y., W.Va. s. Ont. and Mich.

Var. **sejúnctus** Burgess (separated). — L. St. John, Que., to Thunder Bay Distr., Ont. and Minn., s. to N.E., Pa. and n. O.

Var. **apricénsis** Burgess (of a sunny opening). — Me. to N.Y. and Pa.

Var. **iánthinus** (Burgess) Fern. (violet-colored). — The least glandular extreme of the species. (Incl. *A. ianthinus, violaris, multiformis* and *nobilis* Burgess) — Ile d'Orleans, Que., to Algoma Distr., Ont., and Wisc., s. to N.E., n. N.J., Pa., upland to Ala. and Tenn., O. and Ind.

§ 3. EUÁSTER Gray *

a. At least the basal leaves cordate and definitely petioled. . . **b.**

b. Inflorescence a flat-topped corymb; involucre of coriaceous minutely glandular closely appressed oblong phyllaries about 1 cm. high; ligules violet, 1–1.5 cm. long; most basal leaves not cordate. . . . 26. *A. Herveyi.*

b. Inflorescence paniculate or racemose; involucre of softer glandless phyllaries, shorter; ligules bluish, whitish or violet, mostly shorter; most or all basal leaves cordate. . . **c.**

c. Phyllaries with narrowly linear wide-spreading or recurving pubescent foliaceous tips; ligules 30–45, violet, 1 cm. or more long; western. . . . 7. *A. anomalus.*

c. Phyllaries appressed or ascending; ligules fewer. . . **d.**

d. Median and upper cauline leaves and petioles of lower cauline leaves not cordate-clasping. . . **e.**

e. Cauline leaves entire or merely shallowly undulate, firm; basal leaves lance-ovate to lanceolate, much longer than broad; ligules violet to blue; plants of the interior.

Leaves scabrous-hispid on both surfaces, only the lowermost definitely cordate at base; ligules deep blue (rarely purple). . . . 8. *A. azureus.*

Leaves smooth to slightly scabrous but glabrous above, pubescent to glabrous beneath, all but the uppermost cordate at base; ligules pale violet (rarely white or roseate). . . . . . . 9. *A. Shortii.*

e. Cauline leaves serrate or serrate-dentate; basal leaves broadly to narrowly cordate-ovate; ligules pale blue, blue-violet, pinkish or white. . . **f.**

f. Phyllaries closely appressed, at least the outer ones blunt or short-tipped, with spatulate, oblanceolate or narrowly subrhombic deeply colored contrasting central bands dilated near their summits; heads very abundant and short-pedicelled in panicles of racemose branches.

Principal cauline leaves on slender petioles, neither glaucous nor very smooth; wide-ranging. . . . . 10. *A. cordifolius.*

Principal cauline leaves on winged petioles, glaucous beneath, smooth; eastern. . . . . 11. *A. Lowrieanus.*

f. Phyllaries loosely ascending, acute to attenuate, the linear or oblanceolate median band not strongly broadened to summit nor strongly contrasting in color.

Heads in close panicles of rather dense racemes, mostly on short several-bracted pedicels 1–5(–10) mm. long; involucres 5–7 mm. high, their narrow-margined phyllaries about 0.5 mm. broad; wide-ranging inland from southern to northern states. 12. *A. sagittifolius.*

Heads in open panicles of short loose racemes or corymbs, mostly on elongate naked or 1- or 2-bracted slender pedicels; involucres 6–8 mm. high, their broadly scarious-margined phyllaries mostly 0.5–1 mm. wide; boreal, extending south into northernmost states. . . . . 13. *A. ciliolatus.*

d. Median and upper cauline leaves mostly cordate-clasping at base; lower cauline petioles often dilated above the clasping base; leaf-margins undulate or with low teeth; plant with pale or hoary minute pubescence. . . . . 14. *A. undulatus.*

***** The most complicated and difficult section in our flora, the specific lines, especially in the narrow-leaved series, including nos. 34–56, too often obscured by hybridization.

a. None of the leaves at once cordate and definitely petioled. . . **g.**

g. Cauline leaves with cordate- or auriculate-clasping bases. . . **h.**

h. Involucre glandular-hairy. . . **i.**

i. Leaves elliptic-lanceolate to oblong, oblong-linear or oval, blunt or merely acute or mucronate, harshly scabrous; stem cinereous-puberulent, -pilose or -hirsute; heads few and scattered, at tips of many-bracted elongate branches or peduncles; involucres turbinate to campanulate, 6–8 mm. high, with firm scarious oblong, oblanceolate or oblong-linear conspicuously green-tipped phyllaries; ligules 15–30; plants of dry soil.

Principal leaves strongly auriculate-clasping, almost as if perfoliate, the elliptic-lanceolate to oval blades 0.7–4 cm. broad. . . . . . 15. *A. patens.*

Principal leaves only half-clasping, oblong to nearly linear, 0.2–2 cm. broad. . . . . 16. *A. oblongifolius.*

i. Leaves narrowly to broadly lanceolate, acuminate, almost crowded on the tall hirsute to villous stem; heads subcorymbose, on glandular-hispid nearly bractless peduncles; involucres broadly hemispherical, 8–10 mm. high, with thin herbaceous linear-attenuate phyllaries; ligules 40–50; plants of damp soil.

Leaves entire, strongly auriculate-clasping; heads in close corymbiform clusters with short and inconspicuous peduncles; wide-ranging. . . . . 18. *A. novae-angliae.*

Leaves more or less serrate, only half-clasping; heads in loose and open small corymbs, on elongate peduncles; northwestern. . . 19. *A. modestus.*

h. Involucre not glandular. . . **j.**

j. Leaves (or at least the lower ones) abruptly contracted below the middle into a winged base, often serrate. . . **k.**

k. Winged or abruptly constricted lower third of larger leaves with dilated auriculate-clasping (appearing almost perfoliate) base, the blade sharply serrate or serrate-dentate with prolonged teeth; involucre 5–8 mm. high, of narrowly linear-attenuate phyllaries; wide-ranging s. of Canada. . . . . 20. *A. prenanthoides.*

k. Winged or abruptly constricted lower third of larger leaves only half-clasping, short-serrate to entire; boreal plants. **l.**

l. Leaves 8–15 or more below the loosely paniculate inflorescence, submembranaceous; involucre 6–8 mm. high, of narrowly linear-attenuate scarious-margined unequal phyllaries 0.2–0.5 mm. broad, the outermost short; northern continental species. . . 21. *A. tardiflorus.*

l. Leaves 3–9 below the few-headed corymbose or subspiciform inflorescence, subcoriaceous to submembranaceous; involucre of subequal phyllaries, the outer ones foliaceous and 1–2 mm. broad; plants of reg. near Gulf of St. Lawrence.

Cauline leaves coriaceous, 6–9 below the open corymb; petiole of lowest leaf shorter than blade; involucre 8–10 mm. high. 51. *A. foliaceus,* var.

Cauline leaves thin-membranaceous, 3–5 below the leafy virgate spiciform inflorescence; petiole of lowest leaf much longer than the blade; involucre 5–6 mm. high. 22. *A. subgeminatus.*

j. Leaves gradually rounded or tapering to auriculate-clasping base, not abruptly contracted below. . . **m.**

m. Stem and leaves glaucous, the leaves very smooth and leathery, ovate, elliptic, oblanceolate or oblong; phyllaries firm, with rhombic green tips, tightly appressed; achenes glabrous.

Lower leaves tapering to a winged petiole; only the median and upper ones auriculate-clasping, these narrowly to broadly lanceolate or oblanceolate to ovate or elliptic; wide-ranging. . . . 28. *A. laevis.*

Lower leaves not tapering to petiole, like the median and upper, all narrowly oblong or oblong-lanceolate and strongly auriculate-clasping; plant of the Alleghanies. . . . . 29. *A. Steeleorum.*

m. Stem and leaves not glaucous, the stem variously pubescent to glabrescent; leaves herbaceous, linear, lanceolate, lance-oblong or oblanceolate; phyllaries herbaceous; achenes pubescent.

Stem coarse; leaves strongly auriculate-clasping, lanceolate, lance-oblong or narrowly subrhombic, the primary ones mostly 1.5–4 cm. broad; involucres 7–12 mm. high, of subequal mostly linear-attenuate loose phyllaries; ligules 1.2–1.5 cm. long. . . . . . 23. *A. puniceus.*

Stem slender; leaves only half-clasping, linear to lanceolate or oblanceolate, mostly narrower; involucres and ligules mostly smaller. . . . . Various extreme forms of nos. 43–56.

g. Cauline leaves tapering to rounded or half-clasping but not wholly clasping nor auriculate bases. . **n.**

n. Leaves whitened or silvery-silky on both sides, all sessile and entire, mucronulate; involucre imbricated in 3–several rows.

Heads few or solitary on loosely corymbose-paniculate branches; phyllaries resembling reduced foliage-leaves; achenes glabrous; inland species. . . . . 24. *A. sericeus.*

Heads abundant, in virgate close spiciform raceme; phyllaries subrigid, in several short series; achenes pubescent; plant of Coastal Plain. . 25. *A. concolor.*

n. Leaves green or greenish, of various forms. . . **o.**

o. Involucre or summit of pedicel glandular; inflorescence more or less corymbiform. . . **p.**

p. Stem rigid, with elongate ascending branches and branchlets covered with many bract-like greatly reduced hard leaves 0.5–1.5 cm. long and bearing solitary terminal heads; involucre hemispherical, 1–1.5 cm. high, 1.5–2 cm. broad, with many series of squarrose foliaceous phyllaries; ligules 1.5–2.5 cm. long; southeastern. . . . 17. *A. grandiflorus.*

p. Stem simple up to the corymb, the latter with naked or few-leaved branches, the leaves large and dilated.

Lowest or tufted leaves ovate, broadly rounded or cordate at junction with petiole; median and upper leaves mostly serrate-dentate; involucres turbinate-campanulate, about 1 cm. high, the closely appressed phyllaries with short green tips; of southern N.E. and L.I. . . . . 26. *A. Herveyi.*

Lowest or tufted leaves lanceolate, oblong-spatulate or narrowly ovate, gradually tapering into petiole; median and upper leaves slightly toothed to entire; involucres campanulate-hemispherical, 0.8–1.6 cm. high, the phyllaries with broad and long foliaceous and often squarrose tips. . . . . 27. *A. spectabilis.*

o. Involucre and summit of pedicel glandless. . . **q.**

q. Phyllaries hard or firm, pale-coriaceous except for the tip. . . **r.**

r. Principal cauline leaves with cordate or half-clasping bases; plants glabrous and smooth throughout, often glaucous, scarcely stoloniferous, with panicles; ligules violet or bluish. . . . . . . 28. *A. laevis.*

r. Principal or all cauline leaves with attenuate to rounded but not cordate nor clasping bases, scabrous or scabrous-margined; inflorescences often corymbiform. . . **s.**

s. Stems simple up to the 1–few-headed corymb (or open panicle in no. 30 with involucre 1–1.5 cm. high); lower and usually the cauline leaves dilated, none with axillary fascicles or abbreviated sterile branchlets; outer phyllaries neither subulate-tipped nor bristly-ciliate. . . **t.**

t. Plants with hard and thickened subligneous bases, non-stoloniferous; involucres slenderly subcylindrical or turbinate; phyllaries pale and coriaceous except for strongly contrasting green appressed or subappressed tips; leaves mostly entire.

Tall (up to 1 m. or more) with mostly solitary stems and open panicle; involucre 1–1.5 cm. high, its short outer phyllaries passing insensibly into the tiny bractlets of the long pedicels; western species. . . . . 30. *A. turbinellus.*

Low (2–7 dm. high), with clustered stiff erect stems and corymbs; involucres 5–10 mm. high, not merging into the 0–few foliaceous bractlets of the short pedicels; eastern species. . . . . 31. *A. gracilis.*

t. Plants with filiform to slenderly cord-like elongate rhizomes and stolons; heads 1–rather few in corymbs; involucres hemispherical or hemispherical-turbinate, their phyllaries mostly with prolonged broad foliaceous summits often squarrose.

Basal leaves long-petioled, often in tufts; cauline leaves linear-lanceolate to oblanceolate, entire or nearly so; involucre 9–13 mm. high, the phyllaries of 5–7 lengths; southern. . . . . 32. *A. surculosus.*

Basal leaves not petioled nor tufted, like the cauline sessile, oblong and serrate or dentate; involucres 5–9 mm. high, the phyllaries of fewer lengths to subequal; northern. 33. *A. radula.*

s. Stems more or less branching into panicles or panicled racemes of usually abundant heads; leaves mostly narrow and hard, many of them subtending reduced and sterile axillary fascicles or shortened tufts of smaller leaves; phyllaries all subulate-tipped or bristly-ciliate; plants not stoloniferous. . . **u.**

u. Stems glabrous or merely pilose; upper leaves and the primary ones tapering to base; phyllaries entire, glabrous, tapering to subulate tips. . . **v.**

v. Involucres hemispherical or campanulate, as broad as high, 4.5–8 mm. high, with 20–40 or more disk-florets. . . 34. *A. pilosus.*

v. Involucres slenderly turbinate, 3.5–4.5 mm. high, 2–3 mm. thick, with 6–12 disk-florets.

Plant pilose; primary leaves 4–10 cm. long and 4–10 mm. broad; flowering branchlets or pedicels mostly approximate, 0.1–2 cm. long; plant of woods and prairies westward. . . . . 35. *A. parviceps.*

Plant glabrous; primary leaves 2–4 cm. long, 1–4 mm. broad; flowering branchlets or pedicels remote, mostly 1–5 cm. long; on serpentine-barrens eastw. . . 36. *A. depauperatus.*

u. Stems copiously hirsute to hoary-puberulent; herbaceous tips of the phyllaries loosely ascending or squarrose; upper bracteal leaves and at least the outer phyllaries bristly-ciliate; primary leaves scarcely narrowed to the sessile base; small heads mostly racemose or racemose-paniculate. . . **w.**

w. Phyllaries acute, loosely ascending, scarcely squarrose; ligules 7–10 mm. long, usually blue to purple; plant of moist habitats. . . . . 37. × *A. amethystinus.*

w. Phyllaries with broad obtuse squarrose tips; ligules 3–6 mm. long, usually white; plants of dry soil.

Involucre 3.5–5 mm. high, its phyllaries in 3 or 4 lengths, the outer much shorter than the inner; disk-florets 10–20; wide-ranging. . . . . 38. *A. ericoides.*

Involucre 5–8 mm. high, its phyllaries strongly overlapping and not in clear series, the outer nearly as long as the inner; disk-florets 20–30; western. . . . . . 39. *A. commutatus.*

q. Phyllaries thin and scarious or foliaceous, pliable and not strongly coriaceous below. . . **x.**

x. Involucres 3–7.8 mm. high, their phyllaries not conspicuously foliaceous and in several evident lengths; inflorescence paniculate or with racemose branches; ligules white or pale, if violet or bluish then with the abundant heads in branching panicles. . . **y.**

y. Slender- to medium-stemmed and often rather low plants with racemose-paniculate heads or these scattered on minutely foliose branchlets; branches (except in no. 44) mostly divergent; phyllaries appressed-ascending and with rather conspicuous green or purple upwardly dilated midrib. . . **z.**

z. Leaves glabrous, or pubescent only on the midrib beneath. . . ***a.***

a. Lobes of disk-corollas 0.4–0.8 mm. long, one-to two-fifths total length of throat, erect or spreading, the limb usually funnelform; ligules 9–30; midrib of leaf glabrous beneath.

Involucres 4.5–6 mm. high, their phyllaries oftenest with rhombic or dilated tips, the inner phyllaries 3–4.8 mm. long; ligules 5.2–9 mm. long; heads scarcely racemose or unilateral on the branches; rameal leaves uniform or nearly so. . . . . 40. *A. dumosus.*

Involucres 3–3.6 mm. high, their phyllaries mostly narrowly linear, the inner phyllaries 2–3 mm. long; ligules 3–5.8 mm. long; heads unilaterally subracemose or more rarely scattered; rameal leaves very unequal. . 41. *A. vimineus.*

*a.*Lobes of disk-corollas 1–1.6 mm. long, one-half to three-fourths total length of throat, usually becoming revolute, the limb goblet-like; ligules 7–14; involucres 4–6.6 mm. high; midrib of leaf pubescent or glabrous beneath. . . 42. *A. lateriflorus.*

z.Leaves closely pubescent on one or both surfaces; ligules 15–26. . . *b.*

*b.*Involucres 5–6 mm. high; phyllaries with somewhat deltoid or rhombic tips, the inner phyllaries 4–4.8 mm. long; lobes of disk-corolla 0.4–0.8 mm. long, one-fifth to one-third total length of throat. . . . . . 40. *A. dumosus,* var.

*b.*Involucres 3–4.6 mm. high; phyllaries with slender at most linear-spatulate upper green band, the inner phyllaries 2.8–4 mm. long; lobes of disk-corolla 0.7–1.2 mm. long, two-fifths to two-thirds total length of throat.

Heads in loosely ascending or divergent panicles of loose leafy-bracted racemes or in diffuse panicles, with definite pedicels; stem merely striate, soft-pilose or -villous; leaves soft-pilose beneath, flat; ligules white; inland plant. . . . . 43. *A. ontarionis.*

Heads in close spiciform and stiff nearly erect racemes, almost sessile; stem strongly angled by decurrent ridges, scabrous with dense cinereous pubescence; leaves harsh on both faces, their margins revolute; ligules purple; southeastern. . . . . 44. *A. racemosus.*

y.Stoutish to slender-stemmed, with ascending or ascending-spreading paniculate branching; phyllaries loosely ascending, linear, lanceolate or lance-oblong, their central band not upwardly dilated (narrowly rhombic in no. 45 with blue or violet ligules and subcoriaceous leaves); leaves linear or lanceolate, often with sterile axillary fascicles; ligules white (or blue or violet in no. 45). . . *c.*

*c.*Veinlets of subcoriaceous leaves forming an evident reticulum of isodiametric areolae; involucres 5–7 mm. high, of oblong- or linear-lanceolate firm phyllaries with narrowly rhombic green tips; ligules blue-violet or purple. . . . . . . 45. *A. praealtus.*

*c.*Veinlets barely evident or, if evident, of chiefly elongate areolae; leaves merely firm to membranaceous; involucres 3–6 mm. high, of thin or flexible linear or linear-lanceolate phyllaries with slender midrib.

Stoutish; the stems 3–6 mm. in diameter toward base, 0.5–2 m. high; longer (lower) primary leaves 1–2 dm. long; panicle several–very many-headed, its branches simple or forking; involucres 3–6 mm. high; ligules 20–40; pappus 2.8–7 mm. long; wide-ranging. . . . . 46. *A. simplex.*

Slender, the stems 1–2.5(–3) mm. in diameter near base, 0.5–6 dm. high; longer primary leaves 2–10 cm. long; panicle with simple 1–10-headed branches; involucre 3.3–4.5(–5) mm. high; ligules 14–27; pappus 2.5–3.5 mm. long; northeastern. . . . . 47. *A. Tradescanti.*

x.Involucres 5–15 mm. high, their phyllaries unequal or nearly equal or with spreading tips, at least the outer ones more or less foliaceous; ligules mostly 1–2 cm. long, usually blue, violet or purple; plants mostly spreading by elongate rootstocks and stolons, often colonial. . . *d.*

*d.*Basal leaves on broadly winged erect petioles 1.5–3 dm. long, these gradually expanding into a narrowly obovate very harshly scabrous blade 2–3.5 dm. long and 0.5–1.5 dm. broad; cauline leaves rapidly reduced up to the corymbiform panicle; coarse stem deeply angulate-sulcate, stiffly erect and harsh; introd. . . . . 48. *A. tataricus.*

*d.*Basal or lower cauline leaves much smaller, smooth or but slightly scabrous; stem not angulate-sulcate, smooth or soft-pubescent; natives. . . *e.*

*e.*Phyllaries in 3–5 definite lengths, the outer much shorter than the others; involucres 5–8 mm. high; leaves very narrow (mostly 2–10 mm. wide), entire.

Stems solitary from nearly filiform elongate stolons 0.8–2 mm. thick; principal leaves 0.5–1.5 dm. long, harshly scabrous on the margins, long-attenuate, the lower surface with clearly marked reticulation; inner phyllaries linear-attenuate, 0.5–1 mm. broad, with slender midrib. 49. *A. junciformis.*

Stems approximate on strong branches of coarse subligneous rhizome; principal leaves blunter, mostly shorter, barely scabridulous-margined, reticulum obscure or not evident; phyllaries linear-oblong to -spatulate, bluntish, all with rhombic or subrhombic foliaceous summits. . 50. *A. adscendens.*

*e.*Phyllaries nearly equal, the outer not conspicuously shorter or, if so, usually somewhat squarrose or with arched-recurving tips; involucres larger; leaves mostly broader. . . *f.*

*f.*Outer phyllaries broadly foliaceous, 1.5–5 mm. broad, closely simulating reduced upper leaves, herbaceous to the margin and usually with a clear (under lens) reticulum of areolae; northern.

Leaves entire or remotely serrate; the midrib beneath glabrous or only sparingly pubescent. . . . . 51. *A. foliaceus.*

Leaves with many crenations; midrib beneath copiously pilose. . . . . 52. *A. crenifolius.*

*f.*Outer phyllaries mostly narrower, merely herbaceous, if up to 2 mm. broad with scarious margins or bases, the reticulum not evident. . . *g.*

*g.*Leaves (except the uppermost) tapering and gradually narrowed from near the middle to the prolonged base; phyllaries rarely 1 mm. broad, their loosely spreading tips acicular; tube and throat of disk-corollas subequal; plant of se. Coastal Plain. . . . . 53. *A. Elliottii.*

*g.*Leaves, if tapering below, sloping from well below middle, commonly without prolonged bases; outer phyllaries broader; tube much shorter than throat of disk-corollas; northern or wide-ranging. . *h.*

*h.*Phyllaries strongly appressed-ascending, their tips not recurving (except through mechanical injury); disk-florets and pappus 6–8 mm. long; northern.

Larger cauline leaves one-tenth to one-third as broad as long, membranaceous or submembranaceous; phyllaries herbaceous nearly to base, thin and pliable, mostly 1–1.5 mm. broad; wide-ranging northeastward. . . . . 54. *A. johannensis.*

Larger cauline leaves one-eighteenth to one-twelfth as broad as long, coriaceous; phyllaries herbaceous only toward summit, hard and coriaceous below, 1.5–2 mm. broad; local plant of Jupiter R., Anticosti. . . . . 55. *A. anticostensis.*

*h.*Phyllaries or many of them (at least the outer) squarrose or recurving and narrowly herbaceous; disk-florets and pappus 4–6 mm. long; wide-ranging eastw. . 56. *A. novi-belgii.*

7. A. anómalus Engelm. (anomalous). — Stiffly erect *cinereous-scabrous stem* 0.3–1.5 m. high, *simple or racemose-branched above; foliage scabrous-hirtellous;* lowest leaves slender-petioled, with entire or nearly entire ovate to ovate-lanceolate pointed blades, the cauline progressively shorter-petioled to the leafy-bracted inflorescence; *inflorescence elongate, made up of dense and short glomerules to open and elongate spiciform racemes or 1-headed divergent branches; involucre* turbinate-hemispherical, 1 cm. or more broad, *of several closely imbricated pubescent narrow phyllaries with strongly squarrose linear-attenuate foliaceous tips; ligules* 30–45, violet, 1 *cm. or more long.* — Calcareous cliffs, open rocky woods and bluffs, Ill. to Ark. and Kans. Sept., Oct. Fig. 1631.

1631. A. anomalus.

8. A. azûreus Lindl. (bright blue). — *Stems stiffly upright,* 0.3–1 m. or more high, *scabrous-puberulent* from above base, or in forma **laevicaûlis** Fern. (smooth-stemmed) *smooth and polished; leaves scabrous-hispid on both surfaces, of dry but thick texture; lower slender-petioled blades ovate-lanceolate to oblong, many of them cordate at base,* 2–5 cm. broad, often crenate-serrate; *prin-*

cipal cauline leaves entire or rarely undulate, gradually narrowed to summit of stem, *the larger short-petioled lanceolate ones mostly* 1–3 *cm. broad; panicle of stiffly ascending racemes with tiny subulate-tipped bracts; involucres* campanulate-obconic or turbinate, *of firm appressed scarious oblong phyllaries with conspicuous rhombic green tips; ligules* 10–20, *deep blue or blue-violet*, or in forma **incarnàtus** Farw. (flesh-colored) reddish-purple, or white in forma **álbidus** Steyerm. (white). (Var. *scabrior* Engelm.) — Dry open woods, thickets and prairies, w. N.Y. and adj. Ont. to Minn., s. to Ga., Ala., La. and e. Tex. Late Aug.–Oct. FIG. 1632. — Passing southwestw. into

1632. A. azureus.

Var. **poàceus** (Burgess) Fern. (like a grass, *Poa*). —*Slender, more diffuse, the stems smooth; upper leaves linear-lanceolate or linear, mostly* 2–10 *mm. broad; panicle diffusely wide-branching, the heads mostly solitary at tips of* slender *elongate* subulate-bracted *branches or branchlets;* heads slightly smaller. (*A. poaceus* Burgess) — Open woods, local, Mo. and Ark. to e. Tex.

× **A. Wòldeni** Rydb. (for its discoverer, BERNT OLAF WOLDEN, 1886–), an evident hybrid, combines in varying degrees the characters of nos. 8 and 9 and the var. of no. 12.

9. A. Shórtii Lindl. (for its discoverer, CHARLES WILKINS SHORT, 1794–1863). — *Differing from no.* 8 *in more numerous leaves glabrous* (though *either smooth or scabrous*) *above* and glabrous to pubescent beneath, *all but the uppermost blades cordate; basal leaves ovate to ovate-lanceolate, deeply cordate, membranaceous to firm; panicle more leafy-bracted; phyllaries pubescent on back; ligules pale violet*, or in forma **Gronemánni** Benke (named in 1929 for CARL F. GRONEMANN) roseate, or white in forma **cándidus** Benke (white). — Open woods, thickets and rocky slopes, w. Pa. and W.Va. to Wisc. and e. Ia., s. to Ga., Ala. and Tenn. Aug.–Oct. FIG. 1633.

1633. A. Shortii.

× **A. Fínkii** Rydb. (for its discoverer, BRUCE FINK, 1861–1927), combining the characters of nos. 9 and 10, is evidently a hybrid of them.

10. A. cordifòlius L. (heart-leaved). — *Stems* single or several *from* crowns of *creeping rhizome, slender*, up to 2 m., *glabrous or nearly so* (except in one var.); *leaves* thin to firm, *glabrous or minutely appressed-hispidulous above* (rarely villous); *the lower all cordate-ovate, on mostly slender petioles*, conspicuously serrate, the cauline similar or the short-petioled and less toothed upper ones truncate or tapering at base; *panicle of subsimple divergent racemes or these intricately forking*, with small leafy bracts; *involucres obconic*, 3.5–6.5 *mm. high, closely imbricated; phyllaries linear-oblong to linear-attenuate, with usually oblanceolate dark central and terminal band*, the outermost 0.2–1 mm. broad; ligules 10–20, pale blue-violet, pale roseate or whitish, 5–8 mm. long. — A polymorphic species; the following geographic vars. the best defined:

Phyllaries closely appressed, obtuse to subacute, the outer ones 0.4–1 mm. broad, the deeper-colored median band broadened upward.
- Panicle much forking, its many primary branches distinctly once to several times divided; involucres 3.5–5 mm. high; wide-ranging.
  - Upper cauline leaves cordate. . . . . *A. cordifolius* (typical).
  - Upper cauline leaves with truncate or tapering bases. . . . . . Var. *polycephalus*.
- Panicle of remote simple or subsimple short racemes or short-forking racemiform branches; leaves all cordate; involucres 5–6.5 mm. high; northern and montane.
  - Stems and leaves glabrous or only sparsely pubescent. . . . . . . Var. *racemiflorus*.
  - Stems and petioles densely white-pilose or villous; lower leaf-surfaces somewhat so. . . . . Var. *Furbishiae*.

Phyllaries linear-attenuate, the median band nearly linear, the outer phyllaries 0.2–0.4 mm. broad; western. . . . . Var. *moratus*.

A. cordifòlius (typical). — Plants 0.25–2 m. or more tall; panicle (except in smallest plants) 1.5–7 dm. high and 1–3 dm. thick. — Open woods, thickets and clearings, usually common, Que. to Wisc. and Ia., s. to N.S., N.E., L.I., Ga., Ala. and Mo. Aug.–Oct. FIG. 1634.

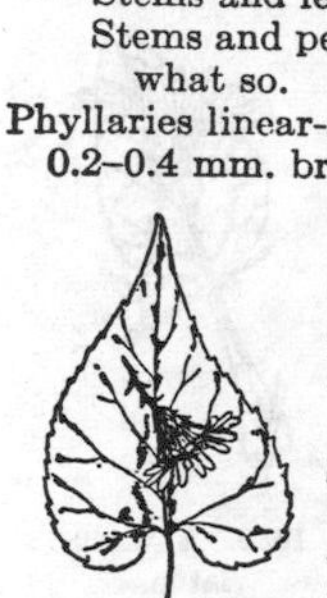
1634. A. cordifolius.

Var. **polycéphalus** Porter (many-headed). — As large as the typical var. — Sw. Que. to Ind., s. to N.E. and Ga.

Var. **racemiflòrus** Fern. (with racemose flowers). — Plant 2–8 dm. high; panicle 0.5–3 dm. high, 0.3–1.5 dm. thick, its simple or subsimple racemose branches mostly 2–12 cm. long. — Gaspé Pen. to Wright Co., Que., s. to N.S., n. and w. N.E., n. N.Y. and along the mts. of N.C. July–Sept.

Var. **Furbíshiae** Fern. (for its discoverer, CATHERINE ("KATE") FURBISH, 1834–1931). — River-gravels and banks, N.B., Me. and N.H. Aug., Sept.

Var. **moràtus** Shinners (characteristic). — Open woods and thickets, Wisc., Ill. and Ia., s. to Kans. Sept., Oct.

11. A. Lowrieànus Porter (for its discoverer, JONATHAN ROBERTS LOWRIE, 1823–1885). — *Differing from no.* 10 *in the very smooth and rather fleshy leaves glaucous or glaucescent beneath when fresh, the upper surface almost greasy to touch, the marginal teeth often nearly suppressed, the principal cauline blades on broadly winged petioles;* lower leaves ovate, cordate. (*A. schistosus* Steele) — Open woods and thickets, Ct. to sw. Ont., s. to w. N.C. and Ga. Aug.–Oct.*

Var. **lanceolàtus** Porter (lanceolate). — Lower leaves narrowly ovate to lanceolate, gradually rounded at base; upper leaves lanceolate. — Less common, Ct. to s. Mich., s. to Md. and W.Va.

12. A. sagittifòlius Wedemeyer (arrow-leaved). — *Stiffly erect,* up to 2 m. high; stem glabrous or slightly pubescent in lines; *lower leaves lance-ovate, cordate, on winged petioles,* serrate or dentate, *smooth, or in the commoner* forma **hirtéllus** (Lindl.) Shinners (somewhat harsh) *scabrous above; upper leaves narrower; panicle tall, with ascending to nearly erect branches bearing abundant often crowded racemose heads* on pedicels almost wanting or up to 7 mm. long; *involucre* slenderly ovoid or turbinate, 5–7 mm. high, *of loosely ascending linear-attenuate or subulate phyllaries with slender green median line;* ligules 10–15, pale blue, pinkish or white, 6–9 mm. long. — Dry open woods, thickets and clearings, Vt. to N.D., s. to Ga., Ala., Tenn. and Mo. Aug.–Oct. FIG. 1636.

1636. A. sagittifolius.

Var. **Drummóndii** (Lindl.) Shinners (for its discoverer, THOMAS DRUMMOND, 1780–1835). — Stem, at least above, and lower surfaces of leaves closely cinereous-pubescent. (*A. Drummondii* Lindl.) — O. to Minn., s. to Tenn., La. and e. Tex.

13. A. ciliolàtus Lindl. (slightly ciliate; from the finely ciliate young leaves). — *Stem* smooth or slightly pubescent in lines, 0.2–1 m. high, *remotely leafy; basal and lower leaves shallowly cordate or rounded at base,* on winged to slender petioles, *narrowly ovate to broadly lanceolate, acuminate, firm and smooth,* or in forma **comàtus** Fern. (with tufts of hair) the upper part of the stem and the petioles and lower leaf-surfaces pilose, *appressed-serrate; upper leaves much narrower, the median with short winged petioles,* the uppermost sessile and often narrowly lanceolate to linear; *panicle loosely thyrsiform or subcorymbiform, the branches with short racemes or corymbs of few large heads on naked or 1- or 2-bracted peduncles; involucres 6–8 mm. high, of loosely ascending linear-attenuate subfoliaceous phyllaries broadly scarious-margined and mostly 0.5–1 mm. wide;* ligules pale blue, rarely pinkish, 0.8–1.5 cm. long. (*A. Lindleyanus* T. & G.) — Dry to moist thickets, shores and clearings, Anticosti I. to Hudson Bay reg., nw. to n. B.C., s. to N.S., n. N.E., n. N.Y., Mich., Wisc., Minn., S.D., Wyo., etc. FIG. 1637. — *A. Lindleyanus,* var. *eximius* Burgess, a stiffer plant with scabrous foliage, found s. of range of no. 13, suggests a hybrid or local phase of *A. sagittifolius,* var. *Drummondii.*

1637. A. ciliolatus.

14. A. undulàtus L. (undulate; from the leaf-margin). — *Pale or somewhat hoary with dense minute pubescence;* stem 0.2–1 m. or more high, rather stiffly ascending; *leaves scabrous above, soft-pilose beneath; the basal cordate-ovate, their petioles with clasping dilated bases, the margins wavy or with low dentation; lower cauline leaves ovate to deltoid, their short petioles strongly dilated at clasping base; middle and upper cauline leaves ovate to oblong, sessile or nearly so; inflorescence stiffly branching, varying from an open panicle* (*usually higher than broad*) *of racemes to a thyrsiform raceme,* the numerous foliaceous bracts and bractlets greatly reduced; *involucre* obovoid, 3.5–5 mm. high, *of firm pubescent blunt to acutish green-tipped phyllaries;* ligules 8–15, pale violet or pale blue, 5–10 mm. long. — Dry open woods, thickets and clearings, N.S. and N.E. to s. Ont. and Minn., s. to Fla., Ala., La. and Ark. Aug.–Nov. FIG. 1638. — Very variable in outline of leaves but without seeming geographic consistency in its numerous phases ("species" of Burgess); the two following are the most definite vars.:

Var. **lorifórmis** Burgess (strap-shaped). — *Principal cauline leaves narrowly lanceolate to lance-linear;* plant otherwise not differing. (*A. loriformis* Burgess) — N.H. to N.Y., s. to Ala. and Ky.

1638. A. undulatus.

---

* There is no fig. 1635.

Var. **diversifòlius** (Michx.) Gray (variable-leaved). — *Differing in its very diffuse and less rigid panicle usually as broad as or broader than high, the axes of the subflexuous or upward-arching branches filiform* and *covered by linear-subulate bractlets; heads mostly scattered and solitary* at tips of branchlets. (*A. diversifolius* Michx.; *A. asperifolius, linguiformis, Mohrii,* etc., of Burgess) — Chiefly on Coastal Plain, Fla. to La., n. to se. Va. Late Aug.–Oct.

15. A. pàtens Ait. (spreading; from the divergent branches of the panicle). — *Scabrous-puberulent*, with firm slender stems 3–9 dm. high, *usually simple up to the inflorescence; leaves of primary stem divergent, with cordate-auriculate clasping bases almost meeting around the stem, oblong to oval, blunt* (or merely mucronate-tipped), entire, 2–7 cm. long, 1–3 cm. broad; *inflorescence of divergent to subascending branches with much reduced bracteal leaves, the bracts of the mostly single-headed flowering branches, branchlets or slender stiff peduncles flat and like reduced leaves; involucre puberulent and glandular, turbinate-campanulate, 6–8 mm. high, of* 3 *or* 4 *series of firm scarious oblong or oblong-linear imbricated phyllaries with spreading green tips;* ligules 20–30, blue-purple or violet, or roseate in forma **roseus** Svenson (rose-color), about 1 cm. long. — Dry open woods, clearings and fields, centr. Me. to Minn., s. to Fla., Ala., Miss., La. and Tex. Aug.–Oct. Fig. 1639.

1639. A. patens.

Var. **phlogifòlius** (Muhl.) Nees (Phlox-leaved). — *Leaves thinner or membranaceous,* only scabridulous; *the larger blades slightly constricted above the clasping base or panduriform,* 0.6–1.7 dm. long and 2–4 cm. broad. (*A. phlogifolius* Muhl.) — Ga. and Tenn., n. to Mass. (rare), N.Y. and O.

Var. **gràcilis** Hook. (slender). — *Leaves* as in typical var. but *smaller, thicker and harder, the principal cauline ones* 1.5–7 cm. long and 0.7–2 *cm. broad; inflorescence diffusely long-branching, with greatly reduced bracts, those of the ultimate very long flowering branchlets or peduncles becoming linear-subulate or almost setiform tiny appressed bractlets.* (*A. tenuicaulis* (Mohr) Burgess) — Ga. to Tex., n. to N.J., Md., Ky., Mo. and Kans.

16. A. oblongifòlius Nutt. (oblong-leaved). — *Stiffly erect and usually bushy-branched, with cinereous-pilose or -hispid stem* 1.5–7.5 dm. high; *leaves oblong to nearly linear, hard and harshly scabrous, half-clasping, entire,* divergent; the *primary blades* 2–10 *cm. long and* 0.2–2 *cm. broad, the crowded and very abundant rameal ones smaller; heads* loosely corymbose, *each at the end of a very leafy branchlet; involucre hemispherical-campanulate, about* 6 *mm. high, of imbricated squarrose linear or linear-oblanceolate glandular phyllaries;* ligules 20–30, up to 1 cm. long, violet-purple, or pinker in forma **roseoligulàtus** (Benke) Shinners (with roseate rays). (Incl. *A. Kumleini* Fries) — Dry prairies, calcareous rocky banks, etc., Pa. to Wisc., Minn. and N.D., s. to w. N.C., Ala., Ark., Okla. and Tex. Sept., Oct. Fig. 1640.

1640. A. oblongifolius.

17. A. grandiflòrus L. (large-flowered). — *Stem* rather *slender, hard, hispid, with long ascending branches,* 0.6–1.5 m. high; *leaves hard, scabrous, ciliate;* the lowest (soon shrivelled) ones narrowly oblong to oblanceolate, 3–7 cm. long; *the upper and rameal ones very numerous and bracteiform, sessile, oblong to linear and only* 1.5 *down to* 0.5 *cm. long; heads solitary at tips of elongate leafy-bracted branchlets; involucre hemispherical, glandular,* 1–1.5 *cm. high,* 1.5–2 *cm. broad, of many series of squarrose foliaceous phyllaries,* these somewhat transitional into the bracteal leaves; *ligules purple-violet,* very many, 1.5–2.5 *cm. long;* pappus pale brown; achenes strigose-villous. — Dry or dryish siliceous or argillaceous open pine- or oak-woods, thickets and clearings, on or near Coastal Plain, Fla. to e. Va. (July–) late Sept.–early Nov. Fig. 1641.

1641. A. grandiflorus.

18. A. nòvae-ángliae L. (of New England), New England A. — *Stems* rather stout, *spreading-villous,* up to 2.5 m. high; *cauline leaves lanceolate, acuminate, entire, cordate-auriculate and clasping,* of dryish texture, *pilose or hispidulous on both surfaces, the principal ones crowded* and mostly 0.5–1 dm. long; *heads in close corymbiform clusters at the tips of the leafy branches of the broad corymb, their peduncles usually shorter than the involucre and, like it, with numerous glands; involucres broadly hemispherical,* 8–10 *mm. high, of thin herbaceous linear-attenuate* often purple-tinged *recurving viscid phyllaries; ligules* 40–50, rather long, violet-purple, or

1642. A. novae-angliae.

roseate in forma **ròseus** (Desf.) Britt. (rose-color), or white in forma **geneseénsis** House (of Genesee Co., N.Y.). — Damp thickets, shores, meadows, etc., sw. Que. to s. Alta., s. to centr. Me., s. N.E., Del., Md., upland to w. N.C., Ky., Ark., Kans. and Colo.; much cult. and esc. elsewhere. Aug.–Oct. FIG. 1642.

19. A. modéstus Lindl. (modest). — Similar to no. 18, *more slender, with serrate or subentire only half-clasping leaves* less crowded on the villous to glabrescent stem; *heads few, at the tips of elongate peduncles,* somewhat smaller; *ligules darker violet.* (*A. major* (Hook.) Porter) — Low rich soil, Alaska to Oreg., e. to James Bay and Thunder Bay Distr., Ont., and n. Mich. Aug., Sept. FIG. 1643.

1643. A. modestus.

20. A. prenanthoìdes Muhl. (suggesting *Prenanthes;* at least to Muhlenberg). — Stem 0.3–1 m. high, glabrous or pubescent in lines, *corymbose-paniculate above; leaves* scabrous above, smooth beneath, *narrowly ovate to lanceolate, long-acuminate, strongly contracted to a long broadly winged portion with an abruptly enlarged and auriculate-clasping (almost perfoliate) base, the middle of the blade with many elongate serrations; leafy panicle broad and open,* with peduncled heads; *involucre* turbinate-campanulate, 5–8 *mm. high; the narrowly linear phyllaries* scarious, *with narrow green central band;* ligules 20–30, blue-violet, or white in forma **milwaukeénsis** Benke (of Milwaukee). — Damp thickets, rich woods, shores and bottoms, w. Mass. to Minn., s. to Del., interior Va., Ky., and Ia. Aug.–Oct. FIG. 1644.

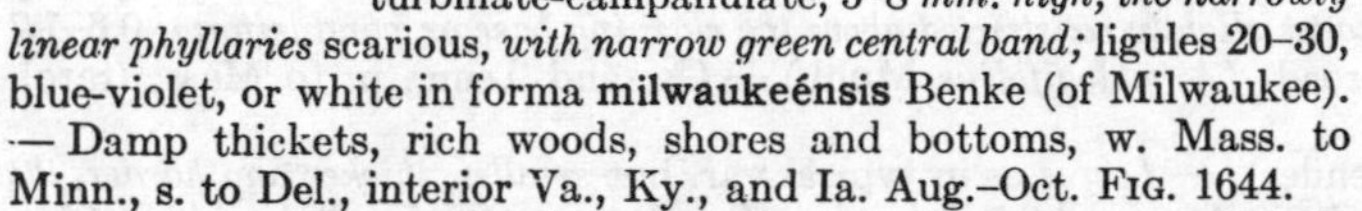

1644. A. prenanthoides.

21. A. tardiflòrus L. (late-flowering). — Glabrous or slightly pubescent, or in forma **vestìtus** Fern. (clothed) densely pilose, 0.3–1.5 m. high; *leaves submembranaceous,* 8–15 *or more below the loosely paniculate inflorescence, the cauline ones suggesting those of no.* 20, *the narrowly ovate or broadly lanceolate long-acuminate blade* short-serrate to subentire and *abruptly contracted to a winged but only half-clasping base; panicle often large, rather open; involucres* 6–8 *mm. high, of linear-attenuate thin scarious-margined unequal phyllaries* 0.2–0.5 *mm. broad, the outermost ones much shorter than the others;* ligules light blue or blue-violet. — Dry to moist thickets, borders of woods and clearings, n. N.E., s. Que. and e. Ont., s. to w. Ct., N.J. and e. Pa. Aug.–Oct. FIG. 1645. — Often seeming like an unusually abundant hybrid of nos. 10 and 23.

1645. A. tardiflorus.

22. A. subgeminàtus (Fern.) Boivin (almost paired; from the often twinned heads). — In aspect suggesting no. 21, very slender, 2.5–5 dm. high, glabrous; *cauline leaves only* 3–5 *below the inflorescence, thin-membranaceous, the narrowly ovate to lanceolate low-toothed blades acuminate, constricted below to winged half-clasping bases, the narrow bases of the lowest leaves up to twice the length of the blade; inflorescence interruptedly spiciform,* 1 *or* 2 *heads borne on short pedicels in axils of the uppermost* 2–5 *elongate and much overtopping leaves; involucre* 5–6 *mm. high, with subequal phyllaries, the outer ones foliaceous* and about 1 mm. broad; ligules 10–15, blue-violet, 1–1.3 cm. long. — Damp calcareous ravines, subalpine, sw. Nfld. Aug.

23. A. punìceus L. (red or purple; from the usual color of the stem). — *Stem usually stout,* 0.2–2.5 m. high, *from hirsute with often tuberculate-based hairs* or densely pilose above *to glabrous; leaves of dry texture, narrowly to broadly lanceolate to narrowly subrhombic, up to* 2 *dm. or more long; the primary ones mostly* 1–4 *cm. broad, strongly auriculate-clasping,* serrate, dentate or entire, harsh to smooth; *inflorescence an open to dense corymbiform panicle,* the branches longer than to much shorter than the subtending leaves; *involucres hemispherical,* 7–12 *mm. high, of* 2 *series of loose subequal linear-attenuate scarious-margined phyllaries* about 0.5 *mm. broad or with outer foliaceous and broader ones; disk-flowers* 35–50 *or more; ligules* 20–40, 1.2–1.5 *cm. long,* blue-violet to lilac, pink or, rarely, white. — One of our most intricately variable and wide-ranging species, too freely crossing with others. Our best marked vars. are as follows:

a. Inner phyllaries evident, linear-attenuate, scarious-margined, with slender green median band; only a few outer foliaceous ones present, these less than 1 mm. broad; pedicels short, rarely much prolonged beyond the median small bractlets; inflorescence with usually several to very many heads. . . *b.*

*b.*Stem harshly hispid or hirsute on median internodes, either densely hirsute or densely villous above and in the inflorescence.
Leaves lanceolate to lance-oblong, the median ones one-seventh to one-fourth as broad as long, hispidulous along midrib beneath; inflorescence with branches much longer to shorter than subtending leaves; wide-ranging. . . . . *A. puniceus* (typical).
Leaves subrhombic, the median ones one-fifth to one-third as broad as long, the midrib glabrous or nearly so beneath; inflorescence compact, its branches overtopped by the bracteal leaves; chiefly coastwise and eastward. . . . . Var. *compactus.*
*b.*Stem glabrous up to the middle, glabrous or only sparsely hispid above; leaves glabrous beneath, lanceolate. . . . . Var. *firmus.*
*a.*Inner phyllaries mostly hidden by the lanceolate or lance-linear foliaceous outer phyllaries, these 1–3 mm. broad; some pedicels often prolonged 2–5 cm. up to the naked heads or to their subapproximate subtending large foliaceous bracts; heads few.
Stem stout, 5–7 mm. thick near base; leaves broadly lance-oblong, 2–4 cm. broad, ascending and overlapping, harshly scabrous above; inflorescence compact, with approximate mostly ascending leafy bracts. . . . . . Var. *oligocephalus.*
Stem slender, 3–4 mm. thick at base; leaves linear-lanceolate or -oblanceolate, 1.2–2.2 mm. broad, glabrous and lustrous, widely divergent; inflorescence open, with distant leafy bracts and divergent branches. . . . Var. *perlongus.*

A. puníceus (typical). — Coarse and tall, up to 2.5 m. high; inflorescence open and broad to interruptedly thyrsoid, its lower and often the upper branches usually exceeding the subtending leaves; branches of inflorescence and pedicels hispid or hirsute with loosely divergent trichomes, or pedicels densely villous or villous-tomentose with incurving implexed soft hairs in forma **blándus** (Pursh) Fern. (pleasing); ligules blue-violet, or in forma **Cȯlbyi** (Benke) Shinners (named in 1935 for Carl Colby) roseate, or white in forma **cándidus** Fern. (white). — Damp thickets, meadows, shores, etc., Nfld. to s. Man., s. to N.S., N.E., L.I., Ga., Ala. and Ia. Aug.–Oct. Fig. 1646. — Forma **demíssus** (Lindl.) Fern. (unassuming) has the interruptedly thyrsiform panicle usually longer than broad, the branches much shorter than the subtending leafy bracts.

1646. A. puniceus.

Var. **compáctus** Fern. (compact). — Stout stem 3.5–8 dm. high; the narrowly rhombic leaves very heavy. — Marshes, often brackish or saline, along the coast, e. Mass. to Va., sometimes inland, as in Fauquier Co., Va., and Parke Co., Ind. Mid-Sept.– early Nov.

Var. **firmus** (Nees) T. & G. (firm). — Plant usually tall, its foliage mostly paler than in the typical var., the leaves more or less serrate, or in forma **lucídulus** (Gray) Fern. (somewhat shining) entire. (Incl. *A. firmus* Nees; *A. lucidulus* (Gray) Wieg.; *A. puniceus*, vars. *laevicaulis* and *lucidulus* Gray) — Nfld. to Thunder Bay Distr., Ont. and Minn., s. to N.S., N.E., w. N.C., Tenn. and Ia. — Forma **etiamálbus** Venard (likewise white) has white ligules; these are reddish in forma **ruféscens** Fassett (reddish).

Var. **oligocéphalus** Fern. (few-headed). — Lab. to Mackenzie Distr., s. to Nfld. and Que. (ascending to alpine areas), higher mts. of n. N.E. and ne. N.Y., and Sask. Late July–Sept.

Var. **perlóngus** Fern. (very long; from the leaves). — Membranaceous smooth and lustrous leaves eight to fifteen times as long as broad. — Known only from coniferous forest of Gaspé Co., Que. Aug.

24. A. seríceus Vent. (silky). — *Stems* slender, erect, *glabrous*, mostly *with stiff ascending branches*, 2–6 dm. high, *bearing few heads at tips of branches of open corymbiform panicles; leaves* hard, *lanceolate to oblong*, mucronate, entire, *silvery-silky*, 1–4 cm. long, *those of the flowering branches greatly reduced and passing insensibly into the squarrose phyllaries of the globular involucre* (about 1 cm. in diameter); ligules 15–30, purple-violet, changing to bluish, or white in forma **albiligulàtus** Fassett (white-rayed), 1–1.5 cm. long; pappus brownish; *achenes glabrous.* — Dry open woods, bluffs and prairies, n. Mich. to s. Man. and N.D., s. to Tenn., Mo. and e. Tex. Aug.–Oct. Fig. 1647.

1647. A. sericeus.

25. A. cȯncolor L. (of uniform color; *i.e.*, the leaves). — *Differing from no. 24 in its usually simple minutely cinereous-puberulent stem,*

1648. A. concolor.

or stem and leaves spreading-villous in forma **lasiocaùlis** Fern. (hairy-stemmed); *inflorescence an elongate subsimple (or sparsely divergent-branched) virgate more or less spiciform raceme; leaves* rather crowded, elliptic-oblong to lanceolate, *minutely cinereous-puberulent,* greatly reduced above; *involucre obovoid or turbinate, with several series of firm green-tipped tightly appressed phyllaries;* ligules 10–15, less than 1 cm. long; *achenes villous.* — Dry sandy open woods and barrens (forma *lasiocaulis* in damper soil), Fla. to La. and Ky., n. on or near Coastal Plain to L.I., and Martha's Vineyard and Nantucket Ids., Mass. Late Aug.–Nov. Fig. 1648.

26. A. Hérveyi Gray (for its discoverer, Eliphalet Williams Hervey, 1834–1925). — *Flowering plants usually accompanied by conspicuous leafy tufts; radical and lowest cauline leaves ovate,* dark to light green and *long-petioled; the blade rounded to cordate at base,* serrate-dentate; *stem* usually *simple* and solitary, slender, smooth to scabrous, 3–9 dm. high, *the corymbose summit and pedicels glandular-puberulent;* median and upper leaves sessile or short-petioled, elliptic-oval to oblong-lanceolate, scabrous; *heads rather few in an open corymb; involucres glandular, turbinate-campanulate, about* 1 *cm. high, the firm closely appressed phyllaries with short green subrhombic tips; ligules violet,* 1–1.5 cm. long. — Open woods and clearings, local, e. Mass. to L.I. Aug.–Oct. Fig. 1649. — Anomalous plant, presumably originally a hybrid of nos. 6 and 27.

1649. A. Herveyi.

27. A. spectábilis Ait. (showy). — *In habit much like no. 26; foliage light green; lowest petioled leaves lanceolate, oblong-spatulate or narrowly ovate, gradually tapering to petiole; involucre campanulate-hemispherical,* 0.8–1.4 *cm. high; phyllaries in about* 6 *series, the outer ones with oblong or oblanceolate herbaceous tips ascending to slightly squarrose;* ligules about 20, bright violet, 1.5–2 cm. long. — Dry sandy open woods and clearings, e. Mass. to Del. and Md.; w. N.C. Aug.–Oct. Fig. 1650.

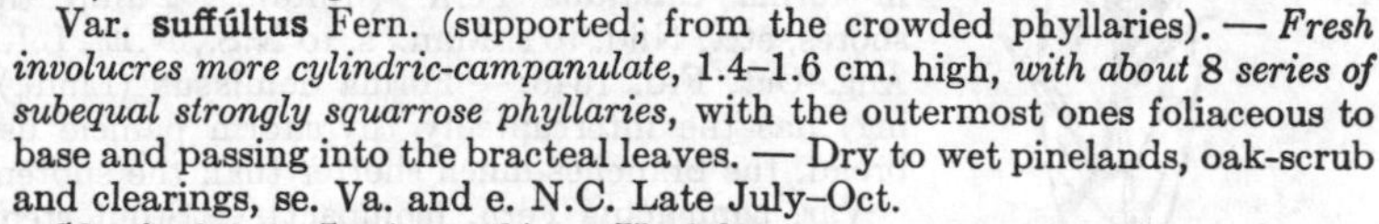

Var. **suffúltus** Fern. (supported; from the crowded phyllaries). — *Fresh involucres more cylindric-campanulate,* 1.4–1.6 cm. high, *with about* 8 *series of subequal strongly squarrose phyllaries,* with the outermost ones foliaceous to base and passing into the bracteal leaves. — Dry to wet pinelands, oak-scrub and clearings, se. Va. and e. N.C. Late July–Oct.

1650. A. spectabilis.

28. A. laèvis L. (smooth). — Upright, *more or less glaucous* and *quite glabrous,* 0.3–1 m. or more high; *leaves thick, mostly entire,* narrowly lanceolate, oblanceolate, elliptic or ovate, *the basal ones tapering to winged petioles, the middle and upper ones sessile and more or less clasping; inflorescence* variable, stiffly racemose-paniculate to subdensely thyrsoid or open-paniculate, *with greatly reduced bracteal leaves; involucre short-obovoid or campanulate,* 5–8 *mm. high, of several series of closely appressed subcoriaceous yellowish-white linear to linear-oblong phyllaries with narrowly rhombic or oblanceolate green summits; ligules* 15–30, *violet,* or white in forma **Beckwíthiae** House (for Florence E. Beckwith, 1843–1929), *about* 1 *cm. long; pappus* 4–5.5 *mm. long.* (Incl. *A. concinnus* of ed. 7, not Willd.; *A. virginiensis* Shinners) — Dry open habitats, borders of woods, etc., Me. to Sask. (and in one var. farther west), s. to s. N.E., L.I., Ga., Ala., La. and Kans. Aug.–Oct. Fig. 1651. — A highly variable species, sometimes split into several quite inconstant segregates. The most striking form is forma **latifòlius** (Porter) Shinners (broad-leaved) = forma *amplifolius* (Porter) Fern., with ovate cauline leaves hardly reduced upward.

1651. A. laevis.

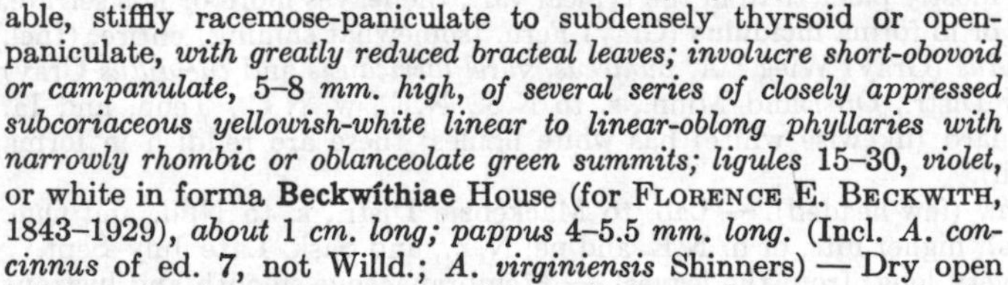

29. A. Steeleòrum Shinners (for Edward Strieby Steele, 1850–1942, and Mrs. Steele). — Stems glabrous, 1–1.3 m. high, with 16–25 *leaves* below the inflorescence; *these all sessile and deeply auriculate-clasping* (as in no. 23), *narrowly oblong to oblong-lanceolate, the median ones* 1–1.5 *dm. long and* 1–1.4 *cm. broad,* entire or nearly so or undulate, *quite glabrous;* inflorescence racemose-paniculate, with short ascending branches; involucres 6–8 mm. high, campanulate, the *phyllaries much as in no.* 28; *pappus* 5.2–6.2 *mm. long.* — Allegheny Mts. and Plateau, w. Md., w. Va. and e. W.Va. Aug., Sept.

30. A. turbinéllus Lindl. (somewhat top-shaped). — *Stem slender, usually solitary from a short subligneous base,* 0.5–1 *m. or more high,* glabrous, or minutely puberulent above, *open-*

1652. A. turbinellus.

paniculate at summit; *cauline leaves lanceolate or lance-oblong, tapering to both ends, entire,* mostly 6–12 cm. long, *smooth except for scabrous margins, the lowest petioled; panicle up to 6 dm. high, usually very open, with* ascending long branches and *long slender branchlets covered with subulate small bractlets; heads* mostly *solitary at tips of branchlets; involucres slenderly subcylindric* (under pressure turbinate), 1–1.5 *cm. high, with coriaceous stramineous linear bluntish appressed phyllaries definitely in several lengths, the outer ones passing into the bractlets below, the others with conspicuous spatulate green tips;* ligules 15–25, violet, 1–2 cm. long. — Dry prairies, open woods, rocky bluffs, etc., Ill. to Neb., s. to La., Ark. and Kans. Sept., Oct. Fig. 1652.

31. A. grácilis Nutt. (slender). — *Stems usually several from a subglobose or ovoid corm-like woody base* (the enlargements sometimes becoming short-moniliform), slender, 2–7 *dm. high,* scabridulous and stiffly erect; *lowest* (sometimes tufted) *leaves narrowly elliptic to oval, tapering to erect petioles,* mostly scabrous; *cauline leaves narrower,* ascending, scabrous to smooth, *entire or nearly so,* 2–7 cm. long; *inflorescence a corymb with* ascending branches and *slender short branchlets with few or no dilated leafy bracts; involucre subcylindric-turbinate,* 5–10 *mm. high, its strongly coriaceous short outer phyllaries with short and broad green tips, the narrower inner phyllaries tapering to subulate tips; ligules* 9–12, violet or bluish, 0.5–1 *cm. long.* — Dry to moist pinelands, oak-thickets and bogs or barrens, Coastal Plain and outer Piedmont, N.J. to S.C.; described by Nuttall from "Savannahs of Kentucky and Tennessee," perhaps through confusion of notes. Mid-July–Sept. Fig. 1653.

1653. A. gracilis.

32. A. surculòsus Michx. (producing suckers). — *Stem arising from cord-like rhizomes, producing slender elongate stolons; otherwise differing from no.* 31 *in the larger heads* (solitary to numerous) *with hemispherical involucres* 9–13 *mm. high* and *with the spreading or squarrose phyllaries thick-foliaceous except at base,* the inner ones with dilated green tips; ligules 15–30. — Dry sandy or rocky siliceous slopes, woods and clearings or in peaty savannas, Ga. and Ala., n. both on Coastal Plain and in mts. to N.C. and Ky. Aug.–Oct. Fig. 1654.

1654. A. surculosus.

33. A. rádula Ait. (a scraper; from the slightly squarrose firm phyllaries). — *Stems* single, *from cord-like or nearly filiform rhizomes, producing long subterranean stolons,* slender, *simple or with terminal corymb,* 0.2–1.2 m. high, smooth or sparsely pubescent, *very leafy; leaves all sessile, oblong-lanceolate, acute, sharply serrate or crenate in the middle, scabrous on both faces, rugose-veiny,* the larger 5–8 cm. long, *only slightly if at all reduced upward;* heads 1–40; *involucres campanulate to hemispherical,* 5–9 *mm. high; phyllaries oblong, obtuse, thick-herbaceous above the base,* scarious-margined, *appressed or with spreading ciliate tips,* of 3–5 lengths; ligules 20–30, pale violet, about 1 cm. long. — Low woods, swamps and bogs, Nfld. and Que., s. to N.S., N.E., Del., Pa. and upland of Va. and W.Va. Fig. 1655.

1655. A. radula.

Var. **stríctus** (Pursh) Gray (strict). — Slender and usually smaller, with 1–8 heads; *phyllaries subequal, herbaceous nearly throughout, acute or acutish.* — Lab. Pen., s. to Nfld., C.B. and n. N.E.

34. A. pilòsus Willd. (pilose; largely a misnomer). — *Stem pilose or more often glabrous,* 0.2–1.5 m. high, *with few heads on stiff ascending to divergent branches or more often with heads racemose along the upper side of the more or less virgate stiffish spreading branches;* lowest leaves narrow to dilated; *cauline leaves subrigid, the uppermost subulate-tipped,* linear or linear-subulate to oblong-lanceolate, *the principal ones often subtending small sterile tufts; inflorescence of racemes* or subcorymbose and with more scattered heads, short to more than half the height of the stem; *involucres hemispherical or campanulate,* 4.5–8 *mm. high, equally broad, of subrigid smooth lance-subulate phyllaries* either subequal or with the outer ones much shorter; *disk-florets* 20–40 *or more;* ligules white, sometimes purple, 5–10 mm. long. — A highly variable but discouragingly intergradient series of variations and forms, strikingly dissimilar in their extremes but surely not several definite species. The following are often distinguishable:

Stem, branches and often the leaves pilose-hirsute or villous.
Stem, etc., pilose-hirsute; principal leaves linear or linear-lanceolate. . . . *A. pilosus* (typical).

Stem densely white-villous; principal leaves lanceolate or lance-oblong. . . . Var. *platyphyllus.*
Stem and leaves glabrous.
Inflorescence racemose-paniculate, with loosely ascending to widely divergent racemose branches; involucre 4.5–6(–7) mm. high. . . . . Var. *demotus.*
Inflorescence subcorymbose-paniculate or thyrsiform, with stiffly ascending rather short branches with scattered or barely racemose heads; involucres 5–8 mm. high. . . . . Var. *Pringlei.*

A. pilôsus (typical). — Stems 0.3–1.5 m. high. (*A. ericoides*, var. Porter) — Dry thickets, clearings, fields and roadsides, centr. Me. to Ont. and Minn., s. to s. N.E., Ga., Ala., Miss. and Ark. Aug.–Oct. — Ligules ordinarily white; roseate in forma **pulchéllus** Benke (beautiful).

Var. **platyphýllus** (T. & G.) Blake (broad-leaved). — Similar habitats, N.J. to Wisc. and Ia., s. to N.C., Tenn. and Mo.

Var. **demôtus** Blake (removed; from *A. ericoides* L., with which it was erroneously placed). — As tall as the typical var., differing in lack of much pubescence. (*A. ericoides* of ed. 7, not L.; *A. ramosissimus* sensu E. J. Alex., not Mill.) — The commonest var.; centr. Me. to Minn., s. beyond our limits. Fig. 1656.

1656. A. pilosus, v. demotus.

Var. **Prínglei** (Gray) Blake (for Cyrus Guernsey Pringle, 1838–1911). — Stem 2–8 dm. high; thyrsiform or subcorymbiform panicle 0.2–1.5(–2) dm. in diameter. (Incl. var. *Reevesii* (Gray) Blake; *A. polyphyllus* Willd., not Moench; and *A. Faxoni* Porter) — Rocky, gravelly or sandy soil, se. Me. to Ont. and Wisc., s. to s. N.E., Pa., D.C. and along the mts. to Ala. and Tenn. Fig. 1657. — The anomalous *A. Priceae* Britt. is presumably a pink- or purple-rayed form of this var. or a hybrid of it with another species.

1657. A. pilosus, v. Pringlei.

35. A. párviceps (Burgess) Mackenz. & Bush (tiny-headed). — *Stems pilose,* 3–7.5 dm. high, *paniculate-branched;* basal shoots with spatulate entire blunt petioled leaves becoming 1–1.5 cm. broad; *cauline leaves linear or lanceolate, entire, the primary ones* 4–10 *cm. long and* 4–10 *mm. broad,* often with axillary tufts, *the lower blades tapering to petioles; panicle* one-fourth to three-fourths height of plant, 0.6–3 dm. broad, *its stiffly spreading branches racemose; heads subapproximate, on subulate-bracted branchlets or pedicels mostly* 0.1–2 *cm. long; involucre slenderly turbinate,* 4–4.5 *mm. high and* 2–3 *mm. thick, its narrowly linear-subulate coriaceous pale phyllaries with slenderly linear-oblanceolate upper green band; disk-corollas* 6–12; ligules 10–12, white, 4 or 5 mm. long. (*A. depauperatus*, var. *parviceps* Fern.) — Dry open woods and prairies, Ill., Ia. and Mo. Mid-Aug.–Oct.

36. A. depauperâtus (Porter) Fern. (depauperate). — *Differing from no.* 35 *in its* more slender *glabrous stem; leaves of basal shoots* 3–5 *mm. wide,* often acute; *cauline leaves smaller; the primary ones only* 2–4 *cm. long and* 1–4 *mm. broad, soon deciduous; panicle more diffuse, the flowering branchlets or pedicels remote and mostly* 1–5 *cm. long;* involucre 3.5–4 mm. high; disk-florets 6–10; ligules white, often drying pink. — Serpentine-barrens, Pa., Del., Md. and W.Va. Mid-July–Oct. Fig. 1658.

1658. A. depauperatus.

37. × A. amethýstinus Nutt. (amethyst-colored). — *Stems* 0.5–1.5 m. high, *densely puberulent to hirsute,* much branched and subcorymbose to paniculate above; *leaves* submembranaceous, *crowded, linear to oblong, sessile or half-clasping, harsh on both faces,* often nearly subulate at tip, the primary ones 2–5 cm. long and 3–6 mm. wide; *involucres broadly turbinate, with narrow acute loosely ascending to barely spreading phyllaries; ligules blue, purple or rarely white,* 7–10 *mm. long.* — Moist thickets and meadows, local, e. Mass. and Vt. to Ia. and Neb., s. to Pa., Ky. and Mo. Fig. 1659. — Pretty clearly a recurrent hybrid of nos. 38 and 18. — The plants with pink ligules are forma **leucérythros** Bemis (whitish-red); those with them white are forma **leûcos** Bemis (white).

1659. × A. amethystinus.

38. A. ericoìdes L. (resembling *Erica;* in its slender branchlets and bracteal leaves). — *Stems from a slender rhizome producing slender subterranean stolons, stiffly erect,* or prostrate in forma **prostrâtus** (Ktze.) Fern. (reclining) of very bleak habitats, *densely cinereous-hirsute or -pilose* with appressed, incurving, recurving or divergent hairs, 0.2–2 m. high, *bushy-branched, with divergent to ascending mostly crowded racemose branches; leaves almost rigid, crowded, spreading, with harsh or hispidulous-ciliate margins, linear, entire, scarcely narrowed to sessile bases;* the *uppermost* reduced and crowded, *gradually passing into the phyllaries;*

heads in elongate 1-*sided dense racemes* forming panicles, or heads mostly solitary at the tips of the branches in forma **exíguus** Fern. (petty or scanty); *involucres turbinate,* 3.5–5 *mm. high, their spatulate-oblong mostly obtuse hard and partly bristly-ciliate phyllaries with green spreading tips and in* 3 *or* 4 *different lengths, the outermost ones much shorter than the others;* disk-florets 10–20, white, rarely blue or violet in forma **caerùleus** (Benke) Blake (sky-blue), or roseate in forma **Grámsii** Benke (named in 1932 for WILLIAM F. C. GRAMS), 3–4 mm. long. (*A. multiflorus* Ait.; *A. exiguus* (Fern.) Rydb. as to name, not as to type; *A. mult.*, var. *pansus* Blake; *A. eric.*, var. *prostratus* Blake, as to plant, not as to original description of name-bringing synonym, *A. mult.*, var. *prostratus* Ktze.) — Dry open soil or in thickets, Me. to s. B.C., s. to s. N.E., Ga., Ala., Miss., Ark., Okla., Tex. and Ariz. July–Oct. FIG. 1660.

1660. A. ericoides.

39. A. commutàtus (T. & G.) Gray (changed). — *Differing from no.* 38 *in usually less crowded leaves; heads fewer, either scattered or in small clusters; involucres* 5–8 *mm. high, of strongly overlapping phyllaries, not so clearly of* 3 *or* 4 *lengths, the outer ones nearly as long as the inner; disk-florets* 20–30, *up to* 8 *mm. long.* — Alta. to N.M., e. to Man., Wisc., Ill. and Mo.

40. A. dumòsus L. (bushy). — *Stems* slender, up to 9 dm. high, *mostly puberulent,* with spreading to ascending *branches, these or the branchlets with several to many approximate or subapproximate firm small bractlets; primary leaves linear to narrowly oblong, bluntish or abruptly mucronate,* spreading to reflexed; *rameal leaves much reduced, nearly uniform; heads scattered* (rarely unilaterally racemose) *on copiously small-bracteolate branchlets or pedicels,* 5–9 mm. broad when expanded; *involucres* (4.5–)5–6 *mm. high, the strongly imbricated phyllaries with conspicuous more or less deltoid or rhombic green tips,* the *inner phyllaries* 3–4.8 *mm. long, the outer mostly oblong ones merging with the upper bracteal leaves; ligules* 13–26, 5–9 mm. long, 0.7–1.4 mm. broad, *pale lavender or bluish* (rarely white); *disk-flowers* 15–30, pale yellow or brown, 3–4 mm. long, *the throat funnelform; lobes erect or slightly spreading,* 0.4–0.8 *mm. long, one-fifth to one-third total length of throat.* — Polymorphous; the following vars. often well-marked:

a. Inflorescence usually one-third to two-thirds height of stem, its branches spreading and commonly forking; stems puberulent.
- Primary cauline leaves ascending or spreading (rarely reflexed), linear to oblanceolate or narrowly oblong, mostly 3–5 cm. long and 4–8 mm. broad; branches of panicle rather stiffly spreading; bractlets flexible; ligules 19–26, 0.7–1.4 mm. wide. . . . . *A. dumosus* (typical).
- Primary cauline leaves spreading or reflexed, narrowly linear, 3–9 cm. long and 1–4 mm. broad; panicle very diffuse; bractlets hard and firm; ligules 9–19, up to 1.9 mm. broad. . . . . Var. *coridifolius.*

a. Inflorescence occupying one-tenth to half the height of the stem, its simple or but slightly forking branches stiffly ascending; primary cauline leaves ascending, 5–12 cm. long, 2.5–11 mm. broad. . . . *b.*

b. Bracteal leaves along branchlets or pedicels sharply subulate-tipped, linear; dark band of phyllaries narrowly oblanceolate or linear-spatulate; ligules 20–25; plant of Coastal Plain. . . . . Var. *subulaefolius.*

b. Bracteal leaves blunt or merely acutish, linear-lanceolate to oblong; dark band of phyllaries narrowly rhombic; inland and northern plants.
- Stem glabrous or sparingly and minutely puberulent; leaves smooth or smoothish; ligules 13–20. . . . . Var. *strictior.*
- Stem densely cinereous-hispidulous; leaves harshly scabrous on both faces; ligules 18–24. . . . . Var. *Dodgei.*

A. dumòsus (typical). — Dry to wet siliceous and argillaceous open ground, thickets and shores, sometimes in bogs, sw. Me. to W.Va., s. to Ga. Aug.–Oct. FIG. 1661.

Var. **coridifòlius** (Michx.) T. & G. (leathery-leaved). — Dry soil, Fla. to La., n. to L.I., N.J., e. Pa., Ind. and Ill.

Var. **subulaefòlius** T. & G. (subulate-leaved). — Coastal Plain, Fla. to e. Tex., n. to se. Mass.

Var. **stríctior** T. & G. (more erect). — W. N.Y. and adj. Ont. to Mich., s. to upland N.C., Ind. and Ill.

Var. **Dódgei** Fern. (for its discoverer, CHARLES KEENE DODGE, 1844–1918). — Marshy ground, se. Mich. and n. O.

1661. A. dumosus.

41. A. vimíneus Lam. (like osier; "its branches like a sprig of osier" —

Lam.). — *Less rigid than no.* 40, *the* usually purple-tinged *slender stem* up to 1.5 m. high, *glabrous or with puberulent lines, commonly arching; the usually long and often much-forking branches arched-recurving, horizontally divergent or ascending, gradually reduced in length* and branching upward; *leaves linear or lanceolate, flexible,* entire or the lower ones serrate, *acute at both ends,* up to 1 dm. or more long and the broader ones 5–10 mm. broad, *glabrous beneath,* scabrous-margined, smooth to scabrous above; *heads unilaterally spicate or racemose* (*or* in the var. often *scattered*); *involucres* 3–3.6 *mm. high; phyllaries thin and flexible, narrowly linear, acute,* with linear-oblanceolate green midrib; *ligules* 15–30, white (sometimes changing to pink), 3–5.8 *mm. long and* 0.5–0.9 *mm. broad; disk-corollas* 2.5–3.5 *mm. long, their lobes* 0.6–0.8 *mm. long* (*about two-fifths length of throat*), *erect or ascending.* — Two confluent vars. recognized:

Heads mostly densely crowded in spiciform or close racemes, on short bracteate branchlets or pedicels mostly less than 7 mm. long. . . . . *A. vimineus* (typical).

Heads loosely racemose or scattered on diffuse branchlets or pedicels mostly 0.5–3 cm. long. . . . . Var. *subdumosus.*

A. vimíneus (typical), including var. *dubius* Wieg., individuals with more or less ascending branches. — Dry to moist fields, meadows, shores, etc., s. Me. to Simcoe Co., Ont., and Mich., s. to s. N.E., N.C. and W.Va. Mid-Aug.–Oct. Fig. 1662.

1662. A. vimineus.

Var. **subdumòsus** Wieg. (somewhat like *A. dumosus*), incl. var. *foliolosus* sensu Gray, probably not *A. foliolosus* Ait. — Sw. Que. to Mich., s. to s. N.E., Ga., Ala., La. and e. Tex.

42. A. lateriflòrus (L.) Britt. (with one-sided flower-clusters). — *Stems* slender, purplish or green, *glabrous to densely pilose,* up to 1 m. high, *usually with wide-spreading or arched-ascending branches* from near or below the middle; *leaves* narrowly *linear-lanceolate to broadly lanceolate, lance-oval or subrhombic,* thin and flexible; the primary ones 0.5–1.5 dm. long and 0.5–3 cm. broad, tapering to base and to acute tip, serrate, dentate or entire, *the midrib beneath more or less pubescent or sometimes glabrous; inflorescence racemose-paniculate, rarely with heads simply paniculate,* the divergent to subascending branches simple to much branched; *involucres* 4–6.6 *mm. high; phyllaries glabrous, in* 3 *or* 4 *series, firm, linear-oblong,* blunt or acute, *with conspicuous subspatulate green or purplish midrib; ligules* 9–15, *white,* rarely pinkish, 4.5–7.5 *mm. long and* 0.7–2 *mm. broad; disk-corollas* purplish, 3–5 mm. long, *the lobes* 1–1.6 *mm. long* or *one-half to three-fourths the full length of the goblet-shaped throat and soon recurving.* — Three vars. recognized:

Heads sessile or on bracted branchlets or pedicels mostly shorter than involucre, chiefly secund-spicate or -racemose; midrib of leaf usually pilose beneath; wide-ranging. . . . . *A. lateriflorus* (typical).

Heads on slender branchlets or pedicels once to four times length of involucre.
- Heads on long open racemose branches; involucres 4–5 mm. high; midrib of leaf pilose beneath; southern. . . . . Var. *pendulus.*
- Heads loosely paniculate or in few-headed short racemes; involucres (4–) 4.5–6 mm. high; midrib usually glabrous; northern. . . . . . Var. *tenuipes.*

A. lateriflòrus (typical), including var. *pendulus* sensu Wieg., not (Ait.) Burgess; var. *hirsuticaulis* (L.) Porter and var. *angustifolius* Wieg. — Dry to moist fields, clearings, thickets, shores, etc., M.I. to Algoma Distr., Ont. and Minn., s. to N.S., N.E., L.I., Ga., Tenn. and Ark. Aug.–Oct. Fig. 1663.

1663. A. lateriflorus.

Var. **péndulus** (Ait.) Burgess (pendulous; from the long flowering branchlets). — Rather local, Mass. and N.Y. to N.C.

Var. **tenùipes** Wieg. (with slender footstalks), including *A. tenuipes* (Wieg.) Shinners, not Makino, and *A. acadiensis* Shinners — Low woods, swamps and shores, M.I. to Thunder Bay Distr., Ont., s. to N.S., n. N.E., n. N.Y. and Mich. — Strikingly distinct in its extreme but passing freely into typical *A. lateriflorus.*

43. A. ontariònis Wieg. (of Ontario). — Somewhat *resembling no. 42, but more colonial and spreading by more elongate rhizomes and numerous stolons; stems coarser,* mostly 0.5–2 *m. or more high,* with abundant spreading to subascending mostly forking branches and racemose-paniculate to diffuse inflorescences; *leaves closely pilose beneath and sometimes above; flowering branchlets or pedicels elongate, less secund than in no.* 42, *or not at all secund;* involucres 3–4.6 mm. high; *phyllaries narrowly linear to linear-oblanceolate, acute, commonly puberulent on back, the green midrib slender; ligules* 15–26, white, 4.8–7 mm. long, 0.6–1 *mm. broad; disk-corollas* creamy to purplish, 3–4.3 mm. long, *their spreading or ascending lobes* 0.7–1.2 *mm. long or two-fifths to two-thirds length of funnelform throat.* (Incl. *A. missouriensis* Britt., not (Nutt.) Ktze. and *A. pantotrichus* Blake) — Thickets, meadows, bottoms, etc., oftenest calcareous, sw. Que. to Minn. and e. S.D., s. to interior N.C., n. Ala., n. Miss., Ark., Okla. and e. Tex. Late Aug.–Oct. Fig. 1664.

1664. A. ontarionis.

44. A. racemôsus Ell. (racemose). — Somewhat suggesting no. 42 but *with stiffly erect densely and harshly cinereous-puberulent stems prominently angled with decussate ridges, up to* 1.5 *m. high; leaves* (soon wilting and dropping) *harshly and very densely pilose on both faces, their margins revolute; spiciform stiff racemes on very long* (*up to* 3 *dm. or more*) *erect branches;* heads sessile or short-stalked; *involucres* 3–4.5 *mm. high; phyllaries narrowly linear or linear-oblanceolate, only* 0.2–0.3 *mm. broad,* the prolonged *inner ones acute; ligules purple or lavender,* 3–4 *mm. long.* — Dry open ground and borders of woods, very rare, Fla. to e. Tex. n. to e. S.C.; Norfolk Co., Va. Oct. — Adequate material with foliage and bases much needed.

45. A. praeâltus Poir. (very tall). — *Stems* up to 2 m. or more high, 3–6 *mm. thick at base,* glabrous or pubescent in lines, *with rather crowded paniculate ascending to spreading very leafy branches; leaves subcoriaceous or firm, with slightly revolute margins,* lustrous and either smooth or scabrous above, *entire; veinlets forming an evident reticulum beneath with small mostly isodiametric areolae,* the primary veins from the midrib usually not evident; *cauline leaves lanceolate to lance-elliptic, attenuate to acute tips,* mostly 0.7–1.3 dm. long and 0.8–1.8 cm. broad, *six to ten times as long as broad; rameal leaves somewhat smaller; panicle* broad and long, *slightly racemose, the heads rather large, on leafy short pedunculate branchlets;* disk 7–10 mm. broad; *involucres* 5–7 *mm. high; phyllaries* in 3 or 4 successively shorter series, *oblong- or linear-lanceolate,* 0.6–1 *mm. broad,* acute, *with narrowly subrhombic green midrib; ligules* 20–35, *blue-violet or -purple* (rarely white), 6–15 *mm. long and* 0.8–1.5 *mm. broad; slenderly funnelform or tubular disk-corollas with short ascending lobes* 0.5–0.9 mm. long, *one-sixth to one-fourth the length of the throat; pappus* of 40–60 hairs, the longer ones *up to* 7 *mm. long.* (*A. salicifolius* Ait., not Lam.) — Low thickets, meadows and prairies, Mass. to Man., s. to Md., W.Va., Ky., Mo., Okla. and e. Tex. Aug.–Oct. Fig. 1665. — Not *A. coerulescens* sensu Wieg., which is a more western species, *A. hesperius* Gray.

1665. A. praealtus.

Var. **angústior** Wieg. (narrower). — *Leaves linear or linear-lanceolate, rather thin and loosely spreading, eleven or more times as long as broad.* — Low woods, thickets and clearings, s. Que. to Algoma Distr., Ont., w. to Ia., s. to Mass., D.C., O., Ind., Ill. and Mo.

Var. **subásper** (Lindl.) Wieg. (somewhat harsh). — Rigid, with harshly scabrous appressed or subappressed thick leaves mostly three to six times as long as broad; heads more crowded; phyllaries 1–1.5 mm. wide. — Ind. and Ill. to e. Tex.

46. A. símplex Willd. (simple). — Relatively stout, *from strong extensively creeping rootstocks* 2–6 *mm. in diameter; differing from no.* 45 *in the more obscure reticulum of the lower leaf-surface made up largely of elongate areolae,* the *blades thinner and smoother; the larger ones more or less serrate or entire, oblong-lanceolate to oblanceolate, mostly* 1–2 dm. long and 2–4 *cm. wide, one-eleventh to one-fifth as broad as long; panicle* usually large and *with forking branches* (simple on reduced or second-growth specimens); *involucres* 4.5–6 *mm. high; phyllaries narrowly linear or linear-oblong, the attenuate tips with slender green band; ligules* 20–40, *white* (rarely blue-tinged), 6–11 mm. long; *lobes of disk-corolla one-third to half as long as whole throat, spreading; pappus* 3.6–7 *mm. long.* (*A. paniculatus,* var. *simplex* (Willd.) Burgess) — Damp thickets, meadows and shores, w. Nfld. to Sask., s. to N.E., upland N.C., W.Va., Ky., Mo. and Kans. Aug.–Oct. — Passing gradually into the two vars.:

Var. **ramosíssimus** (T. & G.) Cronq. (most branched). — Often more freely branching; primary leaves linear or linear-lanceolate, mostly 0.8–1.5 dm. long and 3–12 mm. wide, one twenty-fifth to one-twelfth as broad as long. (*A. paniculatus* Lam., not Mill.; *A. Lamarckianus* Nees; *A. "longifolius"* in part of ed. 7, with bluish ligules, not Lam.; *A. panic.,* var. *bellidiflorus* (Willd.) Burgess and var. *acutidens* Burgess) — S. Que. to Minn., s. to N.S., N.E., Va., O., Ind., Ill., Mo. and Ia. Fig. 1666.

Var. intèrior (Wieg.) Cronq. (inland). — Differing from the preceding var. in its firmer leaves; *involucre only* 3–4 *mm. high; ligules often lavender* (sometimes white); *pappus shorter,* 2.8–3.5 *mm. long.* (*A. interior* Wieg.; *A. Tradescanti* in part of ed. 7, not L.) — Gravelly or rocky (oftenest calcareous) shores, prairies, etc., n. N.Y. to Algoma Distr., Ont., and Wisc., s. to O., Ind. and Ill. Fig. 1667.

1667. A. simplex, v. interior.

1666. A. simplex, v. ramosissimus.

47. A. Tradescánti L. (for John Tradescant, 1608–1662, who had the plant growing at Lambeth). — *More slender and lower than no.* 46, *with slender rootstocks and stolons* 1.5–3 *mm. thick; stems* 1–2.5(–3) *mm. in diameter near base,* 0.5–6 *dm. high; longer primary leaves* 2–10 *cm. long,* linear-lanceolate or oblanceolate to oblong; *panicle of* mostly *simple* 1–10-*headed ascending branches with scattered long-stalked heads; involucres* 3.3–4.5(–5) *mm. high; phyllaries narrowly linear; ligules* 14–22, white, 4.3–8.2 *mm. long; lobes of disk-corollas one-third to nearly half full length of throat; pappus* 2.5–3.5 *cm. long.* (*A. vimineus,* var. *saxatilis* and *A. saxatilis* Fern.) — Damp rocky or gravelly shores, etc., s. Nfld. to L. St. John, Que., w. to n. Mich., s. to N.S., n. N.E., n. N.Y. and Bruce Pen., Ont. July–Sept.

48. A. tatáricus L. f. (Tartarian), Tartarian A. — Very *coarse and harsh,* forming clumps; *erect and stout stem deeply corrugate-angulate,* up to 2 m. or more high, *corymbose-paniculate above; radical leaves on erect broadly winged petioles* 1.5–3 *dm. long; these expanding into narrowly obovate coarsely toothed blades* 2–3.5 *dm. long and* 0.5–1.5 *dm. broad, all spinulose-scabrous;* cauline leaves much smaller, entire, narrow, subulate-tipped; *branches of inflorescence stiffly ascending; campanulate involucre* 1–1.5 *cm. high, of lance-oblong ascending foliaceous phyllaries in* 2–3 *lengths;* ligules blue-violet, 1–1.5 cm. long; *receptacle deeply pitted.* — Roadsides, thickets and borders of woods, spread from cult., N.E. to Ia., s. to Va. Oct., Nov. (Introd. from ne. Asia)

49. A. junciformis Rydb. (with form of *A. junceus*). — *Very slender, the solitary stems arising from nearly filiform and stoloniferous elongate rootstocks* 0.8–2 *mm. thick;* the glabrous or nearly glabrous stems 2–9 dm. high; *leaves very narrow, linear or nearly so,* 0.5–1.5 *dm. long,* 2–10 *mm. broad, entire, with harshly scabrous margins, long-attenuate, the lower surface with clearly marked reticulation;* heads few, on strongly ascending branches; *involucres* turbinate-hemispherical, 5–8 *mm. high; phyllaries in* 3–5 *series, the outer much the shorter, linear-attenuate, thin and flexible,* 0.5–1 *mm. wide, the green midrib slender;* ligules purple, roseate or white, about 1 cm. long. (*A. junceus* of ed. 7 and most auth., not Ait.) — Calcareous bogs, swamps, wet gravels and shores, Anticosti I. and Gaspé Pen., Que., to n. Alaska, s. to P.E.I., N.B., n. and w. N.E., n. N.J., n. Pa., O., n. Ind., Wisc., Ia., Colo. and Wash. June–Sept. Fig. 1668.

1668. A. junciformis.

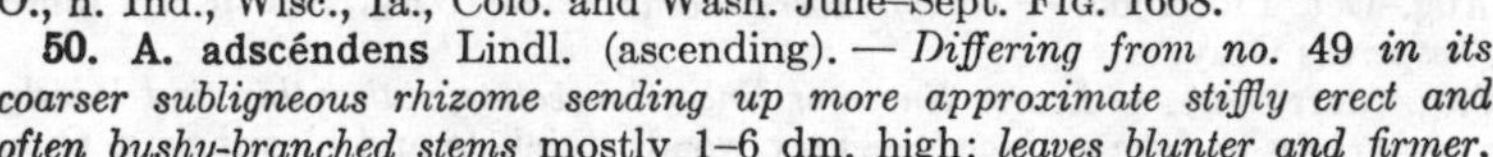

50. A. adscéndens Lindl. (ascending). — *Differing from no.* 49 *in its coarser subligneous rhizome sending up more approximate stiffly erect and often bushy-branched stems* mostly 1–6 dm. high; *leaves blunter and firmer, mostly shorter* (primary ones 2–12 cm. long), *barely scabridulous on margins, the reticulum obscure or not evident; involucre more hemispherical; the linear-oblong to -spatulate phyllaries bluntish and with conspicuous rhombic or subrhombic foliaceous summits.* (*A. chilensis* Nees, subsp. Cronq.) — Damp rocks, shores and slopes, often subalpine, w. Nfld., and Gaspé Pen., Que., to Manitoulin Distr., Ont.; Sask. to Alta., s. to N.M. and Calif. July, Aug.

51. A. foliàceus L. (leaf-like; from the thin and broad phyllaries). — *Stems* from extensively creeping and forking rootstocks, 1–*several,* erect or ascending, 0.1–1 m. high, with 6–15 leaves or nodes below the inflorescence; *leaves* subrhombic, elliptic, oblanceolate or lanceolate, *sessile, entire or with few serrations,* or the lowest narrowed to margined short petioles, smooth or scabrous, *glabrous or with midrib beneath sometimes sparsely pilose; inflorescence with* 1–30 *corymbose-paniculate heads mostly terminating long few-leaved branchlets or peduncles; involucre broadly campanulate-hemispherical,* 0.8–1.5 *cm. high, of strongly overlapping ascending phyllaries; outer phyllaries mostly covering the inner, broadly foliaceous,* 1.5–5 *mm. broad, closely simulating bracteal leaves, thin and leaf-like, green to margin, usually with well defined reticulum on the back;* ligules 30–50, bluish or violet, 1–1.5 cm. long; *disk-corollas with slenderly cylindric-campanulate throat longer than tube.* — A polymorphous boreal and montane species of which we have the following vars.:

Stem stiffly erect, 1–4(–6) dm. high, mostly 2–5 mm. in diameter; median cauline internodes mostly 1–4 cm. long; leaves firm or subcoriaceous, broadly oblanceolate or lanceolate to elliptic or rhombic, mostly 1–4 cm. broad; heads 1–10

(rarely –23), on stiffly ascending branchlets or peduncles rarely more than 5 cm. long.
Leaves little or not at all prolonged at base. . . . . *A. foliaceus* (typical).
Leaves tapering to prolonged broadly subpetiolate bases. . . . . Var. *subpetiolatus.*
Stem loosely ascending or subgeniculate, up to 1 m. high, mostly 1–2(–3) mm. in diameter; median cauline internodes mostly 2–6(–10) cm. long; leaves thin, submembranaceous, narrowly ovate- to linear-lanceolate, 0.5–2.5 (–3.5) cm. broad; inflorescence diffuse, with (6–)10–30 heads, some of them on nearly naked flexuous branchlets or peduncles 3–10 cm. long. . . . . Var. *arcuans.*

A. foliàceus (typical), *A. crenifolius* in large part, sensu Cronq., not as to type. — Meadows, shores, thickets, rocky slopes, etc., s. Lab. Pen. and e. Que., often ascending to subalpine areas; subalpine and alpine areas of n. N.E.; Alaska, etc. Late July–Sept.

Var. **subpetiolàtus** Fern. (somewhat petioled). — Shores, etc., Anticosti and Gaspé Pen., Que.; C.B.

Var. **árcuans** Fern. (arching), *A. crenifolius*, var. Cronq. — Shores, etc., Nfld. and Côte Nord to the Laurentides, Que., s. to N.S., se. N.B. and n. N.E.

52. A. crenifòlius (Fern.) Cronq. (crenate-leaved). — Similar to stiffly erect and stoutish plants of no. 51 but the *oval to subrhombic leaves with regularly crenate margins, the midrib beneath densely pilose.* (*A. foliaceus*, var. Fern.) — Recent clearing on rapidly disintegrating friable sandstone-bluff, Grand River, Gaspé Co., Que., probably now extinct. — A perplexing plant, perhaps a hybrid of nos. 23 and 51.

53. A. Ellióttii T. & G. (for its discoverer, Stephen Elliott, 1771–1830). — *Stout and stiffly erect, suggesting no.* 23; *stem* 0.5–1.5 m. high, *glabrous* or with fine lines of pubescence; *leaves stiff, subcoriaceous,* scabrous above, *glabrous beneath* (or midrib sometimes pilose); *the lowest blades* (rarely collected and often dry at flowering time) *oblanceolate to elliptic, 2–3 dm. long, long-petioled, crenate; lower and median* and often all but the upper *blades narrowed from near middle to broadly winged subpetiolar base; upper leaves sessile;* inflorescence corymbose-paniculate or virgate; *involucres turbinate-campanulate, 9–10 mm. high; phyllaries thin, narrowly linear-attenuate, rarely 1 mm. wide, subequal, the loosely spreading to squarrose tips bristle-like;* ligules lilac or purple, 50 or more, very narrow, 1–1.5 cm. long; *disk-florets* 6–8 mm. long, *the tube and slender throat subequal.* — Fla. to La., n. on outer Coastal Plain to fresh tidal reed-marshes of se. Va. Oct. Fig. 1669.

1669. A. Elliottii.

54. A. johannénsis Fern. (of Lake St. John, Que.). — Habitally much like nos. 51 and 56; *stem simple or with corymbose-paniculate branching,* 1–9 dm. high, glabrous or sparsely pubescent, from extensively creeping and forking rootstocks; *leaves* linear-lanceolate to narrowly oval, *entire or shallowly toothed,* mostly sessile or half-clasping, *membranaceous or submembranaceous* to slightly firm, *the principal ones one-tenth to one-third as broad as long; heads* 1–several *at tips of ascending branchlets, the longer simple and naked or few-bracted branchlets or peduncles 1–9 cm. long; involucre* turbinate-hemispherical, 8–12 *mm. high* and (in dried material) often broader; *phyllaries subequal, appressed-ascending, lanceolate or linear-oblong, 1–2 mm. broad, herbaceous throughout or except at base and very narrow margin;* ligules violet or bluish, 1–1.5(–2) cm. long; *disk-corollas and pappus 6–7 mm. long.* (*A. longifolius* in large part of Am. auth., not Lam.; *A. gaspensis* Vict.) — Shores, damp clearings, etc., s. Lab. to Sask., s. to Nfld., P.E.I., N.B., n. N.E., ne., centr. and w. Mass., n. N.J., N.Y. and s. Ont. Late July–Oct. Fig. 1670.

1670. A. johannensis.

Var. **villicaùlis** (Gray) Fern. (villous-stemmed). — Stem very densely villous; leaves hard and firm, densely villous along midrib beneath. (*A. longifolius*, var. Gray). — River-gravels and -ledges, e. Que., N.B. and n. N.E.

55. A. anticosténsis Fern. (of Anticosti). — Differing from no. 54 in its greatly prolonged and *almost erect overlapping hard-coriaceous* linear-lanceolate *leaves* many times longer than broad (*one-eighteenth to one-twelfth as broad as long*), the median ones 1.2–1.6 dm. long and 0.9–1 cm. broad; *phyllaries pale and coriaceous below, only the upper third or half foliaceous, 1.5–2 mm. broad; disk-corollas and pappus 7–8 mm. long.* — Banks and slopes along Jupiter R., Anticosti I., Que. July. Fig. 1671. — A unique and highly localized plant unlike anything known elsewhere in the East. *Further material needed from* Jupiter River, the plants of other Anticosti valleys not matching it.

1671. A. anticostensis.

56. A. nòvi-bélgii L. (of New Belgium or New Netherland, early name of New York). — Rather *slender*, 0.2–1 m. high, *mostly branching; leaves oblong to linear-lanceolate, entire or sparingly serrate*, the upper ones partly clasping at base; *inflorescence corymbose-paniculate*, much as in no. 54; *ultimate flowering branchlets or peduncles with several arched-recurving foliaceous bractlets; involucre with at least the outer and often all the narrow herbaceous phyllaries with loosely spreading or squarrose tips;* ligules blue-violet, or roseate in forma **ròseus** Rand & Redf. (roseate), or white in forma **albiflòrus** Rand & Redf. (white); disk-florets and pappus 4–6 mm. long. (Incl. *A. Rolandii* Shinners) — Damp thickets, meadows, shores, etc., mostly within 100 miles of the sea, Nfld. to s. Que., s. to N.S., s. N.E., L.I. and Ga. Late July–Oct. Fig. 1672. — Highly variable, with many minor trends proposed as vars. The following seem to be the best defined:

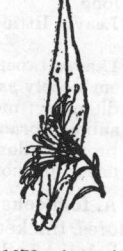

1672. A. novi-belgii (typical).

Var. **elòdes** (T. & G.) Gray (of marshes). — Leaves narrowly linear-lanceolate, tapering and hardly clasping at base. (*A. elodes* T. & G.) — N.B. to N.C.

1673. A. novi-belgii, v. litoreus.

Var. **litòreus** Gray (of seashore). — Compact, with large oblong or broadly lanceolate heavy and often smooth broad-based leaves; phyllaries relatively broad and less squarrose. — Seashores and borders of saline marshes, N.S. to Del. Fig. 1673. — A puzzling series, perhaps a group of hybrids.

§ 4. Orthómeris T. & G.

a.Plants not halophytic; leaves numerous, not fleshy, linear or broader. . . *b*.

b.Non-stoloniferous, forming tussocks from the subligneous crowns and holding marcescent basal leaves; leaves coriaceous, linear to broadly lanceolate or oblanceolate, entire or nearly so; heads several in a flat-topped corymb; involucres 4–8 mm. high, of coriaceous phyllaries; ligules white or yellowish, 5–8 mm. long; xerophytic.

Leaves linear or linear-oblanceolate, with 2 or 4 elongate veins (evident beneath) running from near base parallel with the midrib; involucres 4–6 mm. high; phyllaries lanceolate; ligules white. . . . . 57. *A. ptarmicoides*.

Leaves linear to broadly lanceolate or oblanceolate, with pinnate venation; involucres 5–8 mm. high, the inner phyllaries oblong; ligules yellowish. 58. *A. lutescens*.

b.Stoloniferous, the stems rising singly from tips of last year's stolons; leaves membranaceous or submembranaceous, pinnately veined; heads single or several and paniculate-corymbose; involucres of thin and flexible phyllaries; ligules longer, lilac-pink to white; plants of woodland or bog. . . *c*.

c.Leaves essentially uniform, very numerous or almost crowded, linear or lanceolate to oblong, entire or shallowly toothed, 0.1–2.5 cm. broad, blunt or acute; ligules lilac-purple.

Stem 1–2 mm. thick; cauline leaves 30–100, rather crowded, 1–8(–10) mm. apart at base, linear to narrowly lanceolate, entire, with scabrous margin revolute, blunt or acute, the larger blades 1–6 cm. long and 1–8 mm. broad; plant of bogs and peaty openings. . . . . 59. *A. nemoralis*.

Stem 1.5–4 mm. thick; cauline leaves 20–40, subdistant, 0.5–3 cm. apart, lanceolate to oblong, with flat often toothed margin, acute or short-acuminate, the larger blades 4–10 cm. long and 0.5–2.5 cm. broad; plant of wet thickets, woods and shores. . . . . . . 60. *A. Blakei*.

c.Leaves unequal, oblong-lanceolate or oblanceolate to oval, coarsely toothed, mostly long-acuminate, cuneate and entire at base, the larger ones mostly 1.5–6 cm. broad; ligules white or merely roseate-tinged; plant of woods and clearings. . . . . 61. *A. acuminatus*.

a.Plants halophytic; leaves few, remote, very narrowly linear or linear-lanceolate, fleshy, entire, long-attenuate to both ends; heads 1–several, loosely paniculate-corymbose. . . . . 62. *A. tenuifolius*.

57. A. ptarmicòides (Nees) T. & G. (resembling *Achillea Ptarmica*). — *In tussocks from the crowns of a subligneous short rhizome; stems firm, erect*, slender, glabrous, or hirtellous at summit,

1–6 dm. high; *leaves coriaceous and rigid*, scabrous-margined, *entire* or nearly so, *linear to linear-oblanceolate, erect or nearly so, the larger ones* 3–10 *mm. wide; midrib prominent beneath, with* 1 *or* 2 *nerves running parallel with it from near the base but anastomosing at apex; inflorescence a flat-topped corymb* of 2–∞ heads, *the branches strongly ascending; involucre nearly glabrous*, turbinate-campanulate, 4–6 *mm. high; phyllaries coriaceous*, in 3 or 4 series, *the outer lance-subulate*, the inner lance-oblong; *ligules pure white*, 5–8 *mm. long; marginal flowers with well-developed achenes.* (*Unamia alba* (Nutt.) Rydb.) — Dry, mostly calcareous, rocks, bluffs and sands, w. Que. to Sask., s. to w. Vt., N.Y., s. Ont., n. O., n. Ind., Ill., Mo., S.D. and Colo. June–Sept. Fig. 1674. — A more slender var. on mts. from N.C. to Ga.

1674. A. ptarmicoides.

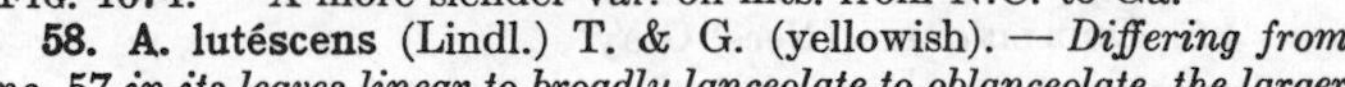

58. A. lutéscens (Lindl.) T. & G. (yellowish). — *Differing from no.* 57 *in its leaves linear to broadly lanceolate to oblanceolate, the larger ones* 0.6–2 *cm. broad; venation pinnate; heads mostly larger; involucres* 5–8 *mm. high, their phyllaries more oblong; ligules yellowish or ochroleucous.* (*A. ptarmicoides*, var. Gray; *Unamia* Rydb.) — Prairies, meadows, sands and bluffs, Bruce Pen., Ont., to Sask., s. to Ill. and S.D. Aug.–Sept.

59. A. nemoràlis Ait. (of woodland; a misnomer), Bog-A. — *Extensively creeping by long nearly filiform rhizomes and stolons; the slender stems rising singly from their tips,* 1–2 *mm. thick,* 0.5–9 dm. high, *simple or corymbosely branched at summit; leaves nearly uniform, linear to narrowly lanceolate, very abundant* (40–100), *entire*, 1–8 (–10) mm. apart, *the scabrous margins revolute*, the lower face minutely pilose, *blunt or acutish, the larger blades* 1–6 *cm. long and* 1–8 *mm. broad; heads* 1 *or* corymbed and 2–15 (exceptionally more), *on filiform small-bracted long peduncles; involucre* turbinate-campanulate, 5–10 mm. high, *of very narrowly linear or linear-lanceolate attenuate flexible often purple-tinged phyllaries*, the outer phyllaries subulate; *ligules lilac-purple*, or white in the rare forma **albiflòrus** Fern. (white-flowered), 1.2–1.8 *cm. long;* disk-corollas 6–7.5 mm. long. — Sphagnous bogs, peaty shores, pond-margins, etc., Nfld. and Côte Nord, Que., to Algoma Distr., Ont., s. to N.S., N.E., n. Del. (formerly) and Mich. July–Sept. Fig. 1675.

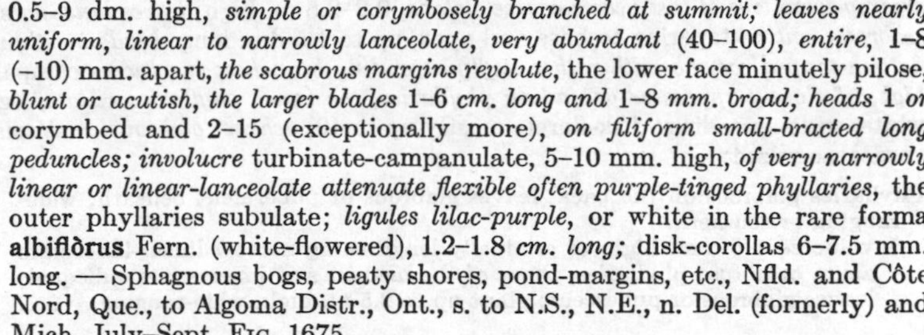

1675. A. nemoralis.

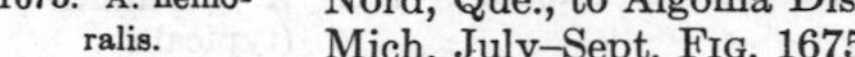

60. A. Blàkei (Porter) House (for its discoverer, Joseph Blake, 1814–1888). — Coarser than no. 59; stem 1.5–4 mm. thick; cauline *leaves* 20–40, *subdistant*, 0.5–3 *cm. apart, thin and brittle, lanceolate to oblong, with flat often low-serrate margin, acute or short-acuminate, the larger blades* 4–10 *cm. long and* 0.5–2.5 *cm. broad;* heads 1–50 (rarely –120); disk-corollas 7–8 mm. long. (*A. nemoralis*, var. *major* Peck and var. *Blakei* Porter) — Damp thickets, low woods, shores and (rarely) bogs, Cape Ray, Nfld., and C.B. to n. N.Y., s. to N.S., N.E. and N.J. Late July–Oct. — Sometimes called a hybrid of nos. 59 and 61; rarely associated with either or both, very fertile and uniform and unknown from much of their coincident areas.

61. A. acuminàtus Michx. (acuminate). — Stems slightly flexuous, arising from the often thickened tips of slender elongate stolons, 0.2–1 m. high, corymbose-paniculate above; *leaves oblong-lanceolate to -oblanceolate or narrowly subelliptic, long-acuminate and serrate or serrate-dentate above, cuneate and entire toward base,* membranaceous, *almost uniformly scattered up the stem*, the lower only slightly smaller than the median and upper, with the *upper internodes mostly* 1–3 *cm. long; principal blades* 0.5–1 dm. long and 1–6 *cm. broad; corymbiform panicle lax, with mostly naked filiform peduncles;* involucre 0.5–1 cm. high, of few loosely ascending and imbricated thin linear or linear-lanceolate phyllaries; *ligules* 12–18, *white or purple-tinged*, loosely spreading, about 1.5 cm. long. — Forma **subverticillàtus** Fern. (almost whorled) has the lower leaves much smaller than the upper; the latter approximate or subapproximate at summit of stem, with internodes 1–20 mm. long, the latter approximate or subapproximate at summit of stem, the blades 0.6–1.5 dm. long. — Dry or moist woods and clearings, sw. Nfld. (rare) and w. Saguenay Co., Que., to e. Ont., s. to N.S., N.E., N.J., Pa., W.Va. and along the mts. (up to 1930 m.) to Ga. and e. Tenn. (plants in alpine areas of N.E. very small, with leaves down to 3 or 4 cm. long). July–Sept. Fig. 1676. — A monstrous (presumably pathological) state with a chaffy receptacle and flowers turned to tufts of chaffy pales has been called forma *discoidcus* Ktze. or forma *virescens* Vict. & Rousseau.

1676. A. acuminatus.

1677. A. tenuifolius.

Var. **magdalenénsis** Fern. (of the Magdalen Islands). — Leaves regularly scattered, broadly oval and nearly half as broad as long, merely acute or short-acuminate. — M.I. July, Aug.

62. A. tenuifòlius L. (slender-leaved). — *Stoloniferous; the glabrous simple or forking firm stem arising from tips of last year's stolons,* 1.5–6 dm. high, *with prolonged internodes between the few remote fleshy narrowly linear or linear-lanceolate attenuate entire leaves; subulate-tipped bracts of the* 1–several *scattered peduncles merging with the outer phyllaries; involucres* turbinate, the larger ones about 1 *cm. high; phyllaries firm, subulate-lanceolate,* acute, appressed *in 4 or 5 lengths; ligules pale purple, rarely white.* — Saline or brackish shores and marshes, Fla. to Miss., n. along coast to N.H. Late Aug.–Oct. Fig. 1677.

§ 5. Doellingèria (Nees) Gray

Stem from a stoloniferous base; leaves lanceolate to ovate; inflorescence a usually compact corymb with stiff ascending branches; involucres 2.5–5 mm. high; inner phyllaries ciliate; plant of low grounds. . . . . 63. *A. umbellatus.*

Stem from non-stoloniferous base; leaves obovate to subelliptic; inflorescence an open corymbiform panicle with spreading-ascending slender branches; involucres 5–7 mm. high; phyllaries eciliate; plant of dry woods. . . . . . 64. *A. infirmus.*

63. A. umbellàtus Mill. (umbellate). — Smooth or nearly so, *leafy to the top, the stems arising from tips of last year's coarse stolons,* 0.2–2.5 m. high; *leaves lanceolate to ovate, smooth to scabrous, entire,* tapering to base and apex, up to 1.5 dm. long; *heads mostly very numerous in a flattish-topped corymb with rather stiffly ascending branches; involucres* campanulate, 2.5–5 *mm. high, of closely appressed-ascending phyllaries, the inner phyllaries ciliate;* ligules 2–15, mostly white, wide-spreading; disk-florets mostly up to 40; *achenes glabrous.* — Variable, the following vars. recognized:

Phyllaries glabrous on the back; leaves glabrous or puberulent beneath; wide-ranging or southern.

Involucres 4–5 mm. high, of gradually lengthening narrow linear-lanceolate acute or obtuse phyllaries, those of the median and inner series subequal; leaves glabrous or pubescent; plant up to 2.5 m. high; wide-ranging. . . *A. umbellatus* (typical).

Involucres 2.5–4 mm. high, of very unequal 4–5-seriate phyllaries, the median phyllaries oblong and obtuse; leaves glabrous; plant up to 1.4 m. high; southeastern. . . . . Var. *brevisquamis.*

Phyllaries pubescent on the back, relatively few, acutish; involucres 3.5–5 mm. high, slightly turbinate; florets relatively few; leaves more or less tomentulose beneath, these and stem usually scabrous above; northwestern. . . . . . Var. *pubens.*

A. umbellàtus (typical). — Leaves glabrous or essentially so, or in forma **intercèdens** Fern. (standing between) more or less pilose on the surface and midrib beneath; ligules 7–10 mm. long, or in forma **discoìdeus** Vict. (discoid) wanting, otherwise as in key. (*Doellingeria* Nees) — Damp to dryish thickets, borders of woods and meadows, Nfld. to Algoma Distr., Ont., and Minn., s. to N.S., N.E., L.I., N.C., upland Ga., and Ky. Aug., Sept. Fig. 1678. — A very slender extreme with remote linear-lanceolate thin and long-attenuate glabrous leaves only 1–1.3 cm. broad and very few scattered heads, is forma **flexicaùlis** (House) Fern. (with flexuous stem).

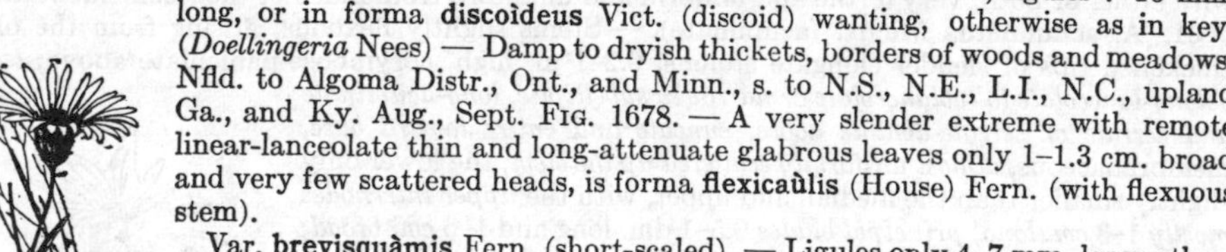

1678. A. umbellatus.

Var. **brevisquàmis** Fern. (short-scaled). — Ligules only 4–7 mm. long; otherwise as in key. — Bogs and swamps, D.C. to N.C. Sept., Oct.

Var. **pùbens** Gray (pubescent), incl. *A. pubentior* Cronq. — Thunder Bay Distr., Ont., to s. Alta., s. to Mich., Wisc., Ia. and Neb. Late July–Sept. — Well-defined in more extreme colonies but too often passing into the typical var.

64. A. infírmus Michx. (weak; only as contrasted with some stouter species). — *Non-stoloniferous;* the slender and flexuous stem 1 m. or less high; *leaves obovate to subelliptic,* tapering to base and apex, glabrous except for hispidity on veins beneath; *inflorescence an open corymbiform panicle with scattered heads; involucre 5–7 mm. high; phyllaries* strongly imbricated, *eciliate, obtuse, rather firm;* ligules white or creamy; *pappus subrigid; achenes glabrous.* (*Doellingeria* Greene) — Deciduous woods, thickets and slopes, mostly

1679. A. infirmus.

inland from Coastal Plain, Mass. to O., s. to Ga., Ala. and Tenn. July–early Sept. FIG. 1679.

§ 6. IÁNTHE Gray

65. A. linariifòlius L. (with leaves of *Linaria*, Toadflax). — *Stems* rather *rigid, in dense tussocks* from a subligneous base, 1–6 dm. high, *rather closely covered with spreading or ascending hard and dryish rough-margined 1-nerved veinless linear to linear-spatulate subulate-tipped leaves; larger leaves 2–4 cm. long,* the upper ones reduced to crowded rigid bracts on the peduncles; *heads* (1–)2–several, solitary *at the tips of erect or ascending bracted peduncles, mostly corymbose* (sometimes scattered); *involucre 1–1.2 cm. high; phyllaries firm, tightly imbricated, at least the outer ones linear-lanceolate and tapering to acute tips;* ligules blue-violet, or white in forma **leucáctis** Benke (white-rayed), 0.7–1.5 cm. long. (*Ionactis* Greene) — Dry open soil, ledges and rocky banks, Northumberland Co., N.B. to s. Minn., s. to s. N.E., Ga., nw. Fla., Ala., Miss., La. and e. Tex. Aug.–Oct. FIG. 1680.

1680. A. linariifolius.

Var. **Victorínii** Fern. (for its discoverer, JOSEPH-LOUIS-CONRAD KIROUAC, FRÈRE MARIE-VICTORIN, 1885–1944). — Low (1–1.6 dm. high); the crowded ascending *greener and subherbaceous leaves rounded at tip, the larger ones 1.3–1.8 cm. long;* involucres 8–10 mm. high; *phyllaries more herbaceous, oblong-linear, obtuse.* — Rocky river-banks, Portneuf Co., and less characteristically in St. Maurice Co., Que.

Group II. Annuals or biennials.

§ 7. OXYTRIPÒLIUM (DC.) T. & G.

66. A. subulàtus Michx. (awl-shaped). — *Annual, with more or less fleshy stem and leaves; stem glabrous,* 0.1–1 m. high, subsimple to strongly bushy-branched; *leaves narrowly linear to lanceolate, oblanceolate or spatulate, entire, the primary ones remote;* heads racemose, glomerulate or solitary on the ascending branches or branchlets; *involucre subcylindric* (slenderly turbinate in drying), 5–8 *mm. high; phyllaries linear-subulate or attenuate; ligules barely or but slightly exserted, purplish; disk-corollas* with prolonged slender tube, *mostly fewer than ligulate flowers; pappus very soft and fine, simple;* achenes pubescent. — Three vars.:

Primary leaves narrowly linear to lanceolate, long-attenuate; phyllaries in 3 or 4 lengths; ligules 20–30, mostly in 2 series, barely or scarcely exceeding the pappus.
Plant often suffused with purple; heads 6–8 mm. high, tending to become crowded and subsessile; phyllaries herbaceous, lance-linear; basal and primary cauline leaves mostly lanceolate; plant of saline marshes n. to s. Me. . . . *A. subulatus* (typical).
Plant usually greener; heads 5–7 mm. high, mostly scattered and long-stalked; phyllaries scarious, linear-attenuate; leaves mostly narrowly linear; plant of brackish to fresh soil southward. . . . Var. *euroauster.*
Primary leaves spatulate-oblanceolate, with obtuse or rounded tips; heads few, terminating erect branchlets; phyllaries subequal, the outer herbaceous; ligules in 1 row, 4–5 mm. long, much exceeding pappus; plant of northeastern N. B. . . . Var. *obtusifolius.*

A. subulàtus (typical). — Saline marshes, s. Me. to Del.; s. Mich. Late July–Oct. FIG. 1681.

Var. **euroaùster** Fern. & Grisc. (southern). — Brackish to fresh tidal marshes or in fresh thickets, borders of woods and clearings inland to the Fall-line, Ct. to Fla. and La.; Onondaga Co., N.Y.

Var. **obtusifòlius** Fern. (blunt-leaved). — Saline marshes about Nepisiguit Bay, ne. N.B. Late July–Sept.

A. exilis Ell. (meagre), very similar to no. 66 but with more hemispheric-campanulate heads, the *ligules well surpassing the pappus and mostly fewer than the disk-florets,* has recently (1948) been added to our flora. — Fla. to Ariz. and s. Calif., n. to S.C., se. Mo., Kans., etc. (Trop. Am.)

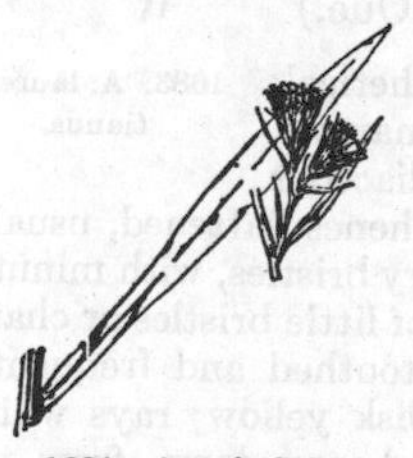

1681. A. subulatus.

§ 8. CONYZÓPSIS T. & G.

Leaves linear-attenuate, more or less ciliate near base, dry and scabrous; phyllaries uniform, linear-attenuate, often ciliate at base. . . . . 67. *A. Brachyactis.*

Leaves linear-oblong to spatulate-oblanceolate, eciliate, fleshy and glabrous; outer phyllaries foliaceous, broader and longer than the inner ones. . . . 68. *A. laurentianus.*

67. A. Brachyáctis Blake (old generic name; short-rayed). — *Annual*, simple to bushy-branched, 1–7 dm. high, nearly glabrous; *leaves linear-attenuate*, entire, *mostly ciliate at base, dry and scabrous;* heads solitary or more often spicate-racemose along the branches and branchlets; *involucres* campanulate, 6–8 *mm. high; phyllaries in* 2 *nearly equal series, linear-attenuate, somewhat scarious, essentially uniform, the outer ones often ciliate at base;* marginal flowers rayless, reduced to a tube much shorter than the elongate style, or the ligule a mere rudiment. (*A. angustus* (Lindl.) T. & G., not Nees; *Brachyactis angustus* (Lindl.) Britt.) — Saline or subsaline shores or locally natzd. on roadsides or in waste ground, Temiscouata Co. to reg. of Montreal, Que.; Onondaga Co., N.Y. and reg. of Great Lakes; saline plains of Minn. and Man. to Alta., s. to nw. Mo., S.D., Colo. and Utah. Late July–Sept. (Centr. Asia) Fig. 1682.

1682. A. Brachyactis.

68. A. laurentiànus Fern. (of Gulf of St. Lawrence). — *Annual, glabrous throughout, fleshy; stem* 0.2–2.5 dm. high, *subsimple or with racemose axillary* 1-*headed branches; leaves very soft and fleshy, quite smooth, eciliate, linear-lanceolate to spatulate;* involucre hemispherical-campanulate; *outer phyllaries foliaceous and green, equaling or overtopping the inner; the latter* in 1 or 2 series, mostly *foliaceous, lanceolate, oblong or spatulate,* mostly subequal and 5–8 mm. long; *outer flowers filiform, without ligules,* shorter than style; central flowers few, filiform, with a 4–5-toothed yellow campanulate limb, about equaling the purplish stigmas. Three vars.:

Stem 0.3–2.5 cm. high; principal leaves linear-spatulate or -lanceolate, obtuse or acutish; outer 3–5 phyllaries linear-lanceolate, acute, greatly elongated, 8–18 mm. long and 1–2 mm. wide. . . . . *A. laurentianus* (typical).

Stem 2–13 cm. high; leaves spatulate, with rounded or obtuse tips; outer 5–11 phyllaries barely elongate.

Outer phyllaries spatulate, oblong or narrowly elliptic, mostly obtuse, 8–11 mm. long, 2–4 mm. broad. . . . . Var. *magdalenensis.*

Outer phyllaries linear-oblong, acute or acutish, 5–10 mm. long, 1–2 mm. broad. . . . . Var. *contiguus.*

A. laurentiànus (typical). — Brackish sands or mud, P.E.I. and w. M.I. Aug., Sept. Fig. 1683.

Var. **magdalenénsis** Fern. (of the Magdalen Islands). — Brackish sands, M.I.

Var. **contíguus** Fern. (neighboring). — Dryish spots in saline marsh, e. Gloucester Co., N.B.

A. frondòsus (Nutt.) T. & G. (leafy), superficially suggesting no. 68 but with more regularly imbricated phyllaries, the outer series much shorter than the inner, the marginal flowers with pinkish ligules, is casually adv. from the West about wool-waste. Fig. 1684.

1684. A. frondosus.

1683. A. laurentianus.

20. ERÍGERON L. Fleabane. Vergerette (Que.)

Heads many-flowered, radiate, mostly flat or hemispherical; the narrow rays very numerous, pistillate. Phyllaries narrow, equal, and little imbricated, never coriaceous, neither foliaceous nor green-tipped. Receptacle flat or convex, naked. Achenes flattened, usually pubescent and 2-nerved; pappus a single row of capillary bristles, with minuter ones intermixed, or with a distinct short outer pappus of little bristles or chaffy scales. — Herbs, largely Am. but genus semicosmop., with entire or toothed and frequently sessile leaves, and solitary or corymbed naked-pedunculate heads. Disk yellow; rays white, pink, or purple. (The ancient name of an early-flowering plant with hoary down, from the Greek *eri, early,* and *geron, old man.*) *

a. Involucres saucer-shaped to hemispheric, 0.6–2 cm. broad; disks 0.3–2 cm. broad; ligules exceeding tube and pappus of marginal flowers. . . *b.*

b. Ligules conspicuous, flat (wanting only in exceptional forms); disk-flowers

* In the figures the heads are × 1½; the separate leaves × ½.

perfect, with no pistillate more slender ones next the ligulate flowers. §EUERIGERON. . . *c*.

c.Radical leaves dissected into 3–5 deeply cleft small divisions; densely cespitose low plant with subscapose 1-headed flowering stems only 0.3–1.3 dm. high. . . . . 1. *E. compositus*, var. *discoideus*.

c.Radical leaves simple, entire, toothed or at most lacerate. *d*.

d.Flowering stem scapose or subscapose, with only small bracteiform leaves; rosette-leaves obovate to spatulate, glabrous; heads corymbose (rarely 1), with only 20–30 ligules. . . . . 2. *E. vernus*.

d.Flowering stem definitely leafy; ligules (except in no. 3) more numerous. . . *e*.

e.Cauline leaves very numerous, linear to linear-oblanceolate, 1–4 mm. broad; heads solitary at tips of stems or branches; ligules 20–30; subfiliform stems mostly densely tufted and bearing very leafy sterile axillary branches. . . . . 3. *E. hyssopifolius*.

e.Cauline leaves relatively few, broader; heads in corymbs (or sometimes solitary); ligules 40–100 or more; stems single or few, rarely with axillary sterile shoots. . . *f*.

f.Perennial with ligneous caudex; leaves entire, coriaceous; pappus double, with a ring of very short bristles outside the long ones; heads 1–9, with more than 100 ligules; northwestern species. . 4. *E. glabellus*, var. *pubescens*.

f.Annuals with tap-roots, biennials with flowering stem arising from rosette of preceding year, or soft-leaved rosulate perennials; lower leaves toothed to entire, membranaceous or firm; wide-ranging or southwestern species. . . *g*.

g.Stems soft, usually flattened in pressing; leaves membranaceous; the cauline sessile, with rounded or subamplexicaul bases; disk-corollas 2.5–4.5 mm. long; pappus simple, of slender bristles.

With flagelliform elongate superficial stolons; cauline leaves subamplexicaul; heads 1–7, when expanded 2.5–4 cm. broad; ligules about 50; disk-corollas 3.5–4.5 mm. long. . 5. *E. pulchellus*.

Without elongate superficial stolons; cauline leaves sessile or amplexicaul; heads mostly more numerous, smaller; ligules 100 or more; disk-corollas 2–3.5 mm. long.

Stem long-pilose below; heads 1.5–2.5 cm. broad when expanded, the flesh-pink to whitish ligules 5–10 mm. long; disk-corollas 2.5–3.5 mm. long. . . 6. *E. philadelphicus*.

Stem short-pilose to cinereous-hirtellous; heads 1–1.5 cm. broad when expanded, the violet or bluish ligules 3–5 mm. long; disk-corollas 2–2.5 mm. long. . . 7. *E. quercifolius*.

g.Stems firm, usually terete after pressing; cauline leaves tapering at base; disk-corollas 1.5–2.5 mm. long; pappus of disk-achenes more or less double, of slender inner bristles and an outer crown of short squamellae; annuals or biennials. . . *h*.

h.Stem simple up to the corymbose flowering summit, 0.3–1.5 m. high; larger lower leaf-blades 0.25–1.5 dm. (or more) long; pappus of ligulate flowers wanting or of few fragile bristles, of disk-flowers double; wide-ranging species.

Leaves membranaceous; the principal cauline ovate to lanceolate, coarsely toothed or cleft, bristly-ciliate; ligules longer than breadth of disk. . . . . 8. *E. annuus*.

Leaves firmer; the principal cauline narrowly to broadly oblanceolate to linear-lanceolate, entire, only minutely strigose to glabrous; ligules shorter than to but slightly longer than breadth of disk. . . . . 9. *E. strigosus*.

h.Stem simple or with erect branches, 1–4 dm. high; larger lower leaves with blades 1–3 cm. long; pappus of all flowers double; southwestern species. . . . . 10. *E. tenuis*.

b.Ligules slender, becoming involute and often quite inconspicuous on drying; a ring of filiform pistillate flowers between the ligulate flowers and the perfect disk-flowers (except in no. 12). §TRIMORPHAEA. *i*.

i.Peduncles glandular-pruinose or viscid; phyllaries glabrous or nearly so, glandular or viscid; inflorescence an elongate raceme or thyrsiform to corymbiform panicle of (4–)7–80 heads on stiffly spreading-ascending peduncles; involucres 5–8 mm. high; stems 2–8 dm. high. . . . 11. *E. angulosus*, var. *kamtschaticus*.

i.Summits of peduncles and phyllaries copiously hispid or villous, glandless; stems 0.2–4.5 dm. high.

Cauline leaves narrowly linear; heads 1–70, on erect peduncles; no filiform pistillate flowers between the marginal ligulate and the perfect central flowers. . . . . 12. *E. lonchophyllus*.

Cauline leaves linear-oblong, lanceolate or oblanceolate; heads 1–8, on elongate flexuous or arching peduncles; filiform pistillate flowers forming a ring between the ligulate and the perfect tubular central flowers. . . . . 13. *E. elatus*.

a.Involucre campanulate, 1.5–4 mm. (when dry spreading to 10 mm.) broad; disks in fresh heads 1–4 mm. broad; ligules minute, in several rows, shorter than corolla-tube and pappus, or wanting; annuals. § CAENOTUS (*Leptilon*). . *j*.

j.Involucre cinereous-hispid, 4–6 mm. long; flowers 80–100 or more; denuded receptacle 2–4 mm. broad. . . . . 14. *E. bonariensis*.

j.Involucre glabrous or nearly so, 2.5–5 mm. long; flowers fewer; denuded receptacle 0.5–2.5 mm. broad. . . *k*.

k.Stems erect, spreading-hirsute or glabrous, with abundant cymose-paniculate to racemose-corymbose heads; lower leaves oblanceolate to spatulate or oblong, 3–13 mm. wide; denuded receptacle 1.2–2.5 mm. broad.

Stem hirsute to summit; leaves bristly-ciliate nearly or quite to tip, more or less hispid; phyllaries without dark apical spot. . . . 15. *E. canadensis*.

Stem glabrous (except sometimes at base); leaves bristly-ciliate only at base, otherwise glabrous; some or all phyllaries with a dark purple spot at tip. . . . . 16. *E. pusillus*.

k.Stems broadly or diffusely bushy-branched, cinereous-strigose; leaves narrowly linear, the lowest 1–2.5 mm. wide; heads solitary or few, scattered; denuded receptacle 0.5–1 mm. broad. . . . . 17. *E. divaricatus*.

§ EUERÍGERON Gray

1. E. compósitus Pursh (compound). — Dwarf *densely cespitose* perennial with multicipital caudex; *leaves* crowded at base, *the well-developed ones three to four times ternately divided* into narrowly linear segments; flowering stem scapose or with few slender bracts, 1-headed; involucre 6–8 mm. high, of linear-attenuate hirsute phyllaries; ligules 40–60, white to purplish. — Polymorphous species, chiefly of western N. Am. Represented with us by

Var. **discoìdeus** Gray (discoid; an unfortunate name for a rare extreme). — *Leaves 3-cleft or the divisions again 3-cleft;* scapes naked or rarely with 1 or 2 bracts, 0.3–1.3 dm. high; ligules wanting, or about 1 cm. long in the more usual forma **trífidus** (Hook.) Fern. (three-cleft), FIG. 1685, *E. trifidus* Hook. — Dry calcareous rock and gravel, Gaspé Co. to Rimouski Co., Que.; Greenl. and Arct. Am., s. to S.D., Colo. and Calif. June, early July.

1685. E. compositus, v. discoideus, f. trifidus.

2. E. vérnus (L.) T. & G. (vernal). — Perennial with short thick rhizome and *rosette of coriaceous glabrous* obovate to oblanceolate or spatulate entire or obscurely dentate *leaves; flowering stem* 2–4.5 dm. high, *scapose or subscapose, with 1–4 very small and remote bracteiform leaves;* heads 1 or 2–12 in a loose corymb; *involucre with shorter outer phyllaries*, inner linear-attenuate phyllaries 3.5–6 mm. long; ligules 20–30, white to lilac. — Damp sands, peats and pinelands of Coastal Plain, Fla. to La., n. to se. Va. April–Aug. FIG. 1686.

1686. E. vernus.

3. E. hyssopifòlius Michx. (hyssop-leaved). — *Densely cespitose* perennial with crowded slender caudices; *stems subfiliform*, glabrous or sparingly pilose, erect or ascending, 0.5–3 dm. high, *with abundant linear, linear-lanceolate or oblanceolate membranaceous entire acute* short-ciliate *leaves 1–4 mm. broad, commonly bearing sterile axillary very leafy branches; heads solitary* on erect naked or sparsely bracted filiform scapiform peduncles one-fourth to one-half the height of the plant; linear phyllaries with long acicular tips; *ligules* 20–30, lilac-purple to white. — Calcareous rocks, talus and gravels, Nfld. to Mackenz., s. to N.S., N.B., centr. Me., n. Vt., n. N.Y., Ont. and n. Mich. June–Aug. FIG. 1687.

Var. **villicaùlis** Fern. (villous-stemmed). — More pubescent; stems 1-few, depressed or decumbent, the leaves crowded at base (leafy portion of stem only 2–8 cm. long), mostly oblanceolate; scapiform peduncles nearly equaling to exceeding length of leafy base. — W. Nfld. and Anticosti I.

1687. E. hyssopifolius.

4. E. glabéllus Nutt. (smoothish), var. **pubéscens** Hook. (hairy). — *Perennial with woody caudex* and 1-few arched-ascending stiff slightly pubescent flowering stems 1–4.5 dm. high, the pubescence spreading, the leafless summit bearing 1–9 large heads; *leaves nearly glabrous,* except the margins, *entire;* the upper oblong-lanceolate and pointed, closely sessile or partly clasping, the lower spatulate and petioled; *rays more than* 100, purple, more than twice the length of the hoary-hispid involucre; *outer pappus of minute bristles.* (*E. asper* Nutt.) — Dry barrens, plains and hills, Man. to Alaska, s. to Wisc., Minn., N.D. and Colo. Late May–July. — Typical *E. glabellus,* with appressed pubescence, is more western.

5. E. pulchéllus Michx. (handsome), Robin's-plantain. — *Villous,* especially below, *producing elongate superficial flagelliform stolons terminated by rosettes;* rosulate and lower cauline leaves obovate to spatulate, membranaceous, crenate or dentate; *cauline leaves* remote, 3–13, few-toothed to entire, rapidly reduced upward, *subamplexicaul; flowering stem soft, hollow,* 1.5–4 (elongating to 7.5) dm. high, *with a loose corymb of* (1–)2–5(–7) *mostly long-peduncled heads; expanded head* 2.5–4 *cm. broad, with about* 50 *broad* blue-purple to whitish *ligules; disk-corollas* 3.5–4.5 *mm. long;* achenes glabrous or nearly so. — Copses, open woods and meadows, s. Me. to s. Que. and s. Ont. and Minn., s. to Fla., Ala., La. and se. Kans. Early April–early July.

Var. **Braùniae** Fern. (for its discoverer, Emma Lucy Braun, 1889–). — Leaves glabrous on both surfaces, the basal and lower entire or merely undulate; stem glabrous or only very sparsely pilose. — Sandy open woods, Ky.

6. E. philadélphicus L. (Philadelphian). — *Villous* short-lived perennial with short basal offshoots; *rosulate and lower membranaceous leaves oblong or narrowly obovate,* crenate or dentate; *principal cauline leaves oblong, sessile, round-based or subamplexicaul; stem soft, flattened under slight pressure,* 0.1–1 m. high; *heads* few to very many, *nodding in bud,* 1.5–2.5 *cm. broad when expanded, with* 100 *or more narrow* flesh-pink to roseate or whitish *ligules; disk-corollas* 2.5–3.5 *mm. long.* — Rich thickets, alluvium, shores and springy slopes, Nfld. to B.C., s. to N.S., N.E., Fla., La. and Tex. Apr.–Aug.

Var. **scaturícola** Fern. (growing about gushing springs). — Perennial by *multicipital elongate caudices; leaves fleshy and brittle,* the basal ones very large (up to 3.8 dm. long and to 8 cm. broad); cauline *leaves* strongly clasping, *the bracteal ones* 1.5–6 *cm. broad; young unexpanded heads erect;* ligules white or promptly changing to white. — Seeping calcareous bluffs of lower James R., Va.

E. Provâncheri Vict. & Rousseau (in honor of Léon Provancher, 1820–1892) of damp rocks at tidal limit in Bellechasse Co., Que., with smoother stem and nearly or quite glabrous leaves, seems like an extreme of no. 6.

7. E. quercifòlius Lam. (oak-leaved). — Similar to no. 6, lower, rarely 5 dm. high, *cinereous-hirtellous or short-pilose; heads* rather few, *only* 1–1.5 *cm. broad when expanded; ligules violet or bluish,* 3–5 *mm. long; disk-corollas only* 2–2.5 *mm. long.* — Fields and roadsides, Fla. to Tex., n. to se. Va. and Tenn. April, May (rarely –Aug.)

8. E. ánnuus (L.) Pers. (annual), Daisy-F., Sweet-scabious, White-top. — Coarse annual or biennial, with sparsely spreading-hispid stem 0.3–1.5 m. high; *leaves coarsely and sharply toothed, membranaceous; the lowest ovate,* up to 1.5 dm. or more long, the blade tapering into a margined petiole; *the upper ovate to lanceolate, sharply toothed or cleft,* acute and entire at both ends, *bristly-ciliate* near base and often on midrib beneath; upper branches few to numerous, forming a many-headed corymb; *ligules* white to lavender, *longer than breadth of disk,* or wanting in forma **discoìdeus** Vict. & Rousseau (discoid); *pappus of ligulate flowers a short crown, with or without slender bristles, of the disk-flowers both a short crown and slender caducous bristles.* — Fields and waste places, throughout our range (except extreme north) and beyond, a semicosmop. weed. June–Oct. Fig. 1688.

1688. E. annuus.

9. E. strigòsus Muhl. (with straight appressed hairs), Daisy-F., White-top. — Similar; stem 0.3–1 m. high, minutely cinereous-strigose above; *leaves firm, entire or nearly so;* basal narrowly oblanceolate to spatulate or oblong, tapering into a slender petiole; *upper narrowly to broadly oblanceolate to linear-lanceolate, minutely strigose to glabrous, entire; ligules* white or purple, *shorter than to but slightly longer than breadth of disk.* (*E. ramosus* (Walt.) BSP., probably not Raf.) — A variable species:

Stems and leaves minutely cinereous-strigose.
- Larger basal leaves broadly oblanceolate or oblong, 0.7–3 (–5) cm. broad; lower cauline leaves broadly oblanceolate; disks of leading heads 0.6–1.2 cm. broad.
  - Ligules elongate, white, rarely purplish. . . . . *E. strigosus* (typical).
  - Ligules wanting or minute. . . . . Forma *discoideus.*
- Larger basal leaves narrowly oblanceolate, 0.5–1.5 cm. broad; lower cauline leaves nearly linear; disks of leading heads 3–8 mm. broad; ligules purple or white. . . . . Var. *Beyrichii.*

Stems (at least near base) and lower broadly oblanceolate or oblong leaves glabrous or sparsely spreading-hispid; heads as in typical form. . . . . Var. *septentrionalis.*

E. strigòsus (typical). — Dry open soil, rocky slopes, pastures and fields, P.E.I. to Man., s. to N.S., N.E., L.I., S.C., O., Ind., Mo. and Okla. Late May–Aug. (–Sept.). — Forma **discoideus** (Robbins) Fern. (disc-like), local, N.E. to Mo. and Va. (incl. vars. *discoideus* Robbins and *eligulatus* Cronq.).

Var. **Beyrìchii** (Fisch. & Mey.) T. & G. (for its discoverer, Heinrich Carl Beyrich, 1796–1834). — Fla. to Tex., n. to R.I., Wisc., Minn., N.D., Mont., Ida. and Wash. June–Sept.

Var. **septentrionàlis** (Fern. & Wieg.) Fern. (northern). — Nfld. to B.C., s. to N.S., n. and w. N.E., N.Y., n. Mich., Wisc., Minn., Ida. and Calif. Late June–Aug.

10. E. ténuis T. & G. (slender). — Annual or winter-annual, *simple or with few erect slender branches, green and sparingly pubescent,* 1–4 *dm. high; leaves thin; the lowest* oblanceolate, entire or with few teeth, 1–3 *cm. long;* heads few, long-peduncled, in a loose corymb; *involucre green, nearly glabrous,* 3–6 mm. high; ligules about 100, pink, purple or white; *pappus of ligulate as well as disk-flowers double.* — Sandy prairies, barrens, fields and waste places, Mo. to La. and Tex. May, June.

§ Trimorphaèa (Cass.) Gray

11. E. angulòsus Gaudin (full of angles), var. **kamtscháticus** (DC.) Hara (Kamtchatkan). — Biennial or short-lived perennial, with erect *rather stout* leafy *stem* 2–8 *dm. high, with elongate raceme or thyrsiform to corymbiform panicle of* (4–)7–80 *heads on stiffly spreading-ascending glandular- or viscid-pruinose peduncles;* basal leaves oblanceolate; cauline narrowly oblanceolate to linear-lanceolate, smoothish, ciliolate; *involucres* hemispheric (spreading when dried), 5–8 *mm. high, with glabrous* or very sparsely hispid slenderly linear *viscid phyllaries;* ligules lilac, becoming filiform and insignificant in drying; a ring of filiform pistillate flowers between the ligulate and the perfect central ones. (*E. acris,* var. *asteroides* of ed. 7, not (Andrz.) DC.; *E. elongatus* Ledeb., not Moench) — Damp banks, open sands, thickets and clearings, Lab. to Alaska, s. to Anticosti I. and Gaspé Pen., Que., n. N.B., n. Me., n. Mich., Minn., Sask., Colo., Utah and Oreg. July, Aug. (E. Asia) Fig. 1689. — Typical *E. angulosus* of Eu. has more hairy stem and narrower leaves.

1689. E. angulosus, v. kamtschaticus.

12. E. lonchophýllus Hook. (spear-leaved). — Similar, 1–4.5 dm. high; *cauline leaves narrowly linear,* bristly-ciliate, *erect, extending up the stem into the spiciform or slenderly racemose inflorescence, equaling or overtopping some of the* 3–70 *hispid erect peduncles; involucre hispid,* 7–10 mm. high, with lance-attenuate phyllaries; *no filiform pistillate flowers between the ligulate and the perfect central ones.* — Calcareous gravels, shores and meadows, Anticosti I. and Mingan Ids., Que.; Yuk. and Alaska to S.D., N.M. and s. Calif. Late July, Aug. (Asia) Fig. 1690.

Var. **laurentiànus** Vict. (of the Gulf of St. Lawrence). — Stems 0.2–1.3 dm. high; leaves crowded at base, the cauline reduced to tiny linear bracts; heads 1 or 2 (or 3), terminating the scapiform stem. — Anticosti I., Mingan Ids. and adj. mainland, Saguenay Co., Que. Late July, Aug.

1690. E. lonchophyllus.

13. E. elàtus Greene (tall). — Slender perennial, with 1–few *flexuous remotely leafy slender sparsely hirsute stems* 0.5–3 *dm. high;*

rosette-leaves oblanceolate, thin; *cauline leaves* 3–8, remote, *linear-oblong or oblanceolate; heads* 1–8 *on elongate erect, flexuous or arching long hirsute peduncles; involucres* 5–9 mm. high, *of* 20–36 linear-attenuate *hirsute phyllaries,* the outer phyllaries one-half to nearly equaling the inner; disks of leading heads 0.8–1.5 cm. broad; *a ring of filiform pistillate flowers below the marginal ligulate and the central perfect tubular ones.* — Calcareous turf, gravel and shores, R. Ste. Anne des Monts, Gaspé Co., Que.; Ung. to Mackenz., s. to s. Alta. and s. B.C. July. Fig. 1691.

1691. E. elatus.

Var. **oligocéphalus** (Fern. & Wieg.) Fern. (few-headed). — Coarser; cauline leaves less remote, up to 18 in number, linear-oblong to lanceolate; involucres 7–12 mm. high, of 36–60 phyllaries; disks of leading heads 1.5–2 cm. broad. — Calcareous slopes, Straits of Belle I., Lab. Pen. and Nfld.; Can. Rocky Mts. Aug., Sept.

§ Caenòtus Nutt.

14. E. bonariénsis L. (of Buenos Aires). — *Cinereous-hispid* or strigose-hirsute annual 1–9 dm. high, erect; lower *leaves* oblanceolate, entire or often incised or coarsely toothed, the upper linear, *many of them subtending leafy branches;* heads racemose, usually in panicles, long-peduncled; *involucre cinereous-hispid,* campanulate, 4–6 *mm. long, with* 80–100 *or more flowers;* ligules white to lilac, numerous, very short, much shorter than corolla-tube and tawny pappus; denuded old receptacle 2–4 mm. broad. (*Leptilon* Small) — Roadsides and waste places, Fla. to Mex. and Calif., n. to se. Va. (casual and non-persistent to Mass.). June–Aug. (Pantropical weed) Fig. 1692.

1692. E. bonariensis.

15. E. canadénsis L. (Canadian), Horse-weed, Hog-weed, Butter-weed. — Stem erect, simple below, 0.1–2 m. high, *remotely spreading-hirsute to summit; leaves bristly-ciliate;* the lowest oblanceolate and toothed or entire, 3–13 mm. broad; the abundant cauline ones narrowly oblanceolate to linear; *heads racemose or cymose on the branches, forming elongate panicles; involucres* slenderly campanulate, 2.5–5 *mm. long, the linear-attenuate phyllaries with pale tips;* flowers few; ligules relatively numerous, white to lavender; denuded old receptacle 1.2–2.5 mm. broad. (*Leptilon* Britt.) — Waste places, cult. fields, etc., throughout our range and far westw. and southw. July–Nov. (Semicosmop. weed) Fig. 1693.

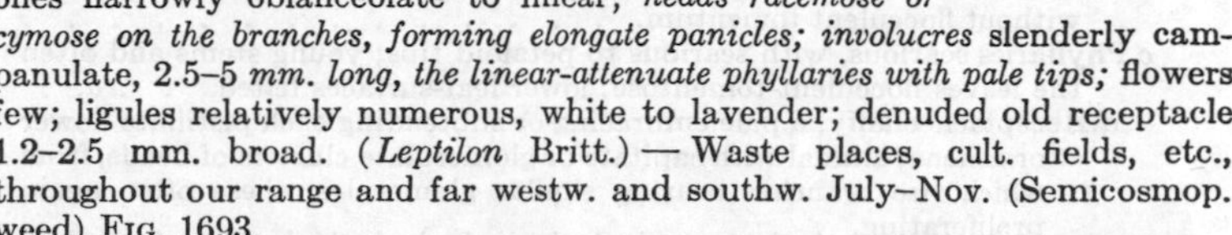

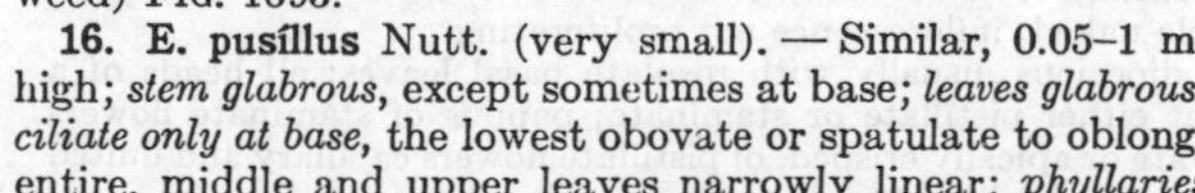

1693. E. canadensis.

16. E. pusillus Nutt. (very small). — Similar, 0.05–1 m. high; *stem glabrous,* except sometimes at base; *leaves glabrous, ciliate only at base,* the lowest obovate or spatulate to oblong, entire, middle and upper leaves narrowly linear; *phyllaries* bluntish, some or all *with a purple spot at or just below apex.* (*Leptilon* Britt.) — Dry sandy soil, Fla. to Tex., n. on Coastal Plain to se. Mass., and inland to s. Ind. (and ? s. Mich.). Late June (rarely May)–Oct. (Trop. Am.)

17. E. divaricàtus Michx. (divergent). — *Diffusely branched, depressed or bushy annual* 1–3 dm. high, *with cinereous-strigose slender branches and slenderly linear to almost acicular cinereous leaves; heads few in small corymbs or solitary;* involucre 2–4 mm. long, with few purple ligules; *denuded old receptacle* 0.5–1 *mm. broad.* (*Leptilon* Raf.) — Dry fields, prairies and waste places, Ala. to Tex. n. to O., Ind., Wisc., Minn. and N.D. Late June–Sept. Fig. 1694.

1694. E. divaricatus.

21. SERICOCÁRPUS Nees White-topped Aster

Heads 12–20-flowered, radiate; rays about 5, fertile, white. Involucre somewhat cylindrical or clavate; the phyllaries closely imbricated in several rows, cartilaginous and whitish, appressed, with short and abrupt often spreading green tips. Receptacle alveolate-toothed. Achenes short, inversely pyramidal, very silky; pappus simple, of numerous capillary bristles. — Perennial tufted N. Am. herbs, 2–7 dm. high, with sessile somewhat 3-nerved leaves, and small heads mostly in little clusters, disposed in a flat corymb. Disk-flowers pale yellow. (Name from the Greek *sericos, silky,* and *carpos, fruit.*)

1. S. asteroìdes (L.) BSP. (Aster-like). — Somewhat pubescent; *leaves oblong-lanceolate or the lower spatulate, few, veiny, ciliate and more or less toothed; heads loosely corymbed, obconical,* *the involucre 5–9 mm. high; pappus sordid to rufescent*, at least in age. — Dry woods and clearings, s. Me. to Mich., s. to Fla., Ala. and Miss. June–Aug.

2. S. linifòlius (L.) BSP. (*Linum* (flax)-leaved). — *Glabrous*, slender; *leaves linear* and firm, *numerous*, only obscurely veined, *entire*, scabrous-margined; corymb denser, the *heads slender; involucre 4–7 mm. high; pappus whitish.* — Dry woods, thickets and clearings, s. N.H. to O., s. to Ga., Ala., Miss. and La. June–Sept.

22. BÁCCHARIS L. GROUNDSEL-TREE

Heads many-flowered; flowers all tubular, dioecious, *i.e.*, the pistillate and staminate borne by different plants. Involucre imbricated. Corolla of the pistillate flowers very slender and filiform; of the staminate larger and 5-lobed. Anthers tailless. Achenes ribbed; pappus of capillary bristles, in the staminate plant scanty and tortuous, in the pistillate very long and copious. — Shrubs, chiefly of trop. Am., commonly smooth and resinous or glutinous. Flowers whitish or yellow. (Name of some shrub, *Baccharis*, anciently dedicated to *Bacchus*, the name transferred to this genus.)

1. B. halimifòlia L. (with leaves of *Halimus*), SEA-MYRTLE, CONSUMPTION-WEED. — Glabrous but somewhat scurfy, 1–3 m. high; branches angled; leaves obovate and cuneiform-petiolate, coarsely toothed, or the upper entire; *heads scattered at the ends of the branches, forming pyramidal panicles;* involucre 5–6 mm. high; phyllaries acutish. — Open woods, thickets and borders of marshes near the coast, Fla. to Tex. and Mex., n. to coast of Mass. Late Aug.–Oct. — The pistillate shrub conspicuous in autumn, with its white pappus 1–1.5 cm. long. (Trop. Am.)

Tribe IV. INÙLEAE Cass. (see p. 1358)

*a.*Corollas, at least of pistillate or perfect flowers, mostly filiform and truncate; pappus present. . *b.*
*b.*Heads discoid, small; leaves not clasping. . . *c.*
*c.*Phyllaries hardly scarious; most of the flowers pistillate, the few central ones hermaphrodite but sterile; rank-smelling mostly glutinous plants without flocculent tomentum. . . . . 23. *Pluchea.*
*c.*Phyllaries scarious, with scarious to petaloid tips; young stems and often the leaves flocculent-tomentose; lower leaf-surfaces felted. . . *d.*
*d.*Receptacle chaffy, a pale embracing or subtending each pistillate flower or achene; annual with capitate or glomerulate clusters of heads, from which rise branches bearing similar glomerules, these often again proliferating. . . . . 24. *Filago.*
*d.*Receptacle naked; inflorescence not proliferating. . . *e.*
*e.*Plants dioecious, usually with rosulate basal leaves; all heads of a plant either pistillate or staminate; pappus of staminate flowers clavate or apically crisped, of pistillate flowers capillary and united at base and falling together. . . . . 25. *Antennaria.*
*e.*Plants dioecious or with perfect flowers, usually not rosulate-leaved, except when young; pappus capillary, of distinct bristles.
Heads dioecious or nearly so, usually a few sterile hermaphrodite flowers in center of pistillate heads; upright leafy-stemmed perennials with corymbose inflorescences and papery whitish involucres. . . . . 26. *Anaphalis.*
Heads with perfect flowers; annual, biennial or perennial with glomerulate, spicate, racemose, panicled or corymbose inflorescences; involucres whitish to fuscous, brown or purple. . . 27. *Gnaphalium.*
*b.*Heads with yellow rays, relatively large, the outer phyllaries foliaceous; cauline leaves clasping.
Receptacle naked; anthers with caudate bases; pappus simple, of capillary bristles. . . . . 28. *Inula.*
Receptacle alveolate; anthers merely sagittate; pappus double, the outer short pales forming a cup, the inner series of caducous bristles. 29. *Pulicaria.*
*a.*Corollas all broadly tubular and lobed; the outer flowers pistillate and fertile, the inner hermaphrodite but sterile; phyllaries equal, in 1 series, not scarious; pappus none. . . . . 30. *Adenocaulon.*

23. PLÙCHEA Cass. MARSH-FLEABANE. STINKWEED

Heads many-flowered; the flowers all tubular; the central perfect but sterile, few, with a 5-cleft corolla; all the others with a filiform truncate corolla, pistillate and fertile. Involucre imbricated. Receptacle flat, naked. Anthers with tails. Achenes grooved; pappus in a single row. — Herbs of warm or trop. reg., somewhat glandular, emitting a strong or camphoric odor, the heads cymosely clustered. Flowers purplish to creamy. (Dedicated to the Abbé *Nat. Ant. Pluche,* French naturalist of the 18th century.)

Leaves sessile or broadly half-clasping, flowers creamy. . . . . 1. *P. foetida.*
Leaves petioled or narrowed to base; flowers pink or purple.
  Phyllaries pubescent on the back; leaves sessile or nearly so; annual.
    Leaves firm, lanceolate to elliptic, canescent-pilose beneath; involucre 4–5 mm. high, its outer phyllaries ovate. . . . . 2. *P. purpurascens.*
    Leaves succulent, ovate-lanceolate to ovate or obovate, glabrous or only sparsely pubescent beneath; involucre 5.5–9 mm. high, its outer phyllaries elliptic or oblong-ovate. . . . . 2a. *P. purpurascens,* var. *succulenta.*
  Phyllaries glabrous or merely granular on back; leaves distinctly petioled; annual or perennial. . . . . 3. *P. camphorata.*

1. P. foètida (L.) DC. (foul-smelling), Stinking Fleabane. — *Perennial,* 5–9 dm. high; *leaves closely sessile or half-clasping,* oblong to lanceolate, sharply denticulate, veiny, only 5–10 cm. long; heads clustered in a corymb; phyllaries lanceolate; *flowers creamy.* — Wet sand, ditches and swamps, Fla. to Tex. and Mex., n. to s. N.J. and se. Mo. Late July–Sept. (W.I.)

2. P. purpuráscens (Sw.) DC. (purplish). — Stiffish annual up to 1.5 m. high; *leaves sessile,* lanceolate to elliptic, dentate, *firm, canescent-pilose beneath;* inflorescence a leafy flat-topped corymb; *involucres* purplish to green, 4–5 *mm. long,* strongly pilose; *outer phyllaries ovate, acuminate;* flowers pink or purple. — Saline or brackish marshes, Fla. to Tex. and Mex., n. to se. Va., Kans. and Calif.; adv. about salt-springs in Mich. July–Sept. (Trop. Am.)

Var. **succulénta** Fern. (succulent), Salt-marsh-Fleabane. — *Annual, pale,* 3–15 dm. high; *leaves sessile or slightly petioled,* rhombic- to oblong-ovate or lanceolate, tapering to tip, or in the rare forma **obovàta** Fern. (obovate) the blades broadly obovate and rounder or notched at summit, *fleshy,* obscurely veiny, subentire or serrate, *glabrous or nearly so;* corymb flat; *heads* 5.5–9 *mm. high; outer phyllaries elliptic to oblong-ovate,* slightly *pubescent;* flowers pink or purple. (*P. camphorata* of ed. 7, not (L.) DC.) — Saline (rarely fresh) marshes, Fla. to s. Me.; reported in w. N.Y. Aug., Sept.

3. P. camphoràta (L.) DC. (camphor-like), Camphorweed, Stinkweed. — *Annual or perennial,* resembling no. 2, greener and smoother; leaves slenderly petioled, more finely and sharply serrate; *inflorescence a round-topped elongate to corymbiform panicle;* heads smaller; *phyllaries smooth or merely granular on the back.* (*P. petiolata* Cass.) — Fresh to brackish marshes, shores and ditches, Fla. to Tex., n. to Del., O., Ind., Ill., s. Mo. and Kans. Aug., Sept.

24. FILÀGO L. Cotton-rose

Heads rather many-flowered, discoid; flowers as in *Pluchea,* the central usually sterile. Receptacle elongated or turbinate; the chaff resembling the proper involucral phyllaries, each scale covering a single pistillate flower. Achenes terete; pappus of the central flowers capillary, of the outer ones mostly none. — Annuals of N. and S. Am., Eurasia and n. Afr., with entire leaves, and small heads in capitate clusters. (Name from *filum, a thread,* alluding to the cottony hairs of the plants.) Gifola Cass.

1. F. germánica (L.) Huds. (German), Herba Impia. — Stem erect, short, clothed with lanceolate upright crowded leaves, and producing a capitate cluster of woolly heads, from which rise one or more branches, each terminated by a similar head, and so on — hence the common name applied to it by the old botanists, as if the offspring were undutifully exalting themselves above the parent. (*Gifola* Dumort.) — Dry fields, local, s. N.Y. to O., W.Va. and Ga. June–Oct. (Natzd. from Eu.)

25. ANTENNÀRIA Gaertn. Everlasting.

Ladies'-tobacco. Pussy's-toes. Immortelle (Que.)

Heads many-flowered, dioecious; flowers all tubular; pistillate corollas very slender. Involucre dry and scarious, white or colored, imbricated. Receptacle convex or flat, not chaffy. Anthers caudate. Achenes terete or flattish; pappus a single row of bristles, in the fertilé flowers capillary, united at the base so as to fall in a ring, and in the sterile thickened and clavate or barbellate at the summit. — Perennial mostly woolly or silky herbs, nearly world-wide but chiefly of N. Am., with entire leaves and corymbose or racemose (rarely single) heads. Staminate plants smaller than the pistillate, abundant only in nos. 1, 2, 18, 19, a var. of 22, 23, 24, 31 and 32, though occasionally found in nos. 28 and 29, very rarely in nos. 17, 20, 25 and 27, but unknown in most species; these parthenogenetic or apogamous. Phyllaries of the staminate heads with broad petaloid tips. (Name from the resemblance of the pappus of the staminate flowers to the *antennae* of certain insects.) *

a. Basal leaves erect, oblanceolate to elliptic-acuminate, 2–18 cm. long, similar to the cauline ones; involucres of pistillate heads deep brown to blackish; achenes glabrous, not papillate.

Flowering stem with 4–10 leaves; pistillate corollas 3–4.3 mm. long; exserted tip of mature style three-fourths as long as corolla; longer pappus-bristles 6–7 mm. long; phyllaries of staminate heads fuscous; staminate corollas 3–4 mm. long. . . . . 1. *A. eucosma.*

Flowering stem with 6–12 leaves; pistillate corollas 4.6–5.2 mm. long; exserted tip of mature style at most one-third length of corolla; longer pappus 9–11 mm. long; phyllaries of staminate heads whitish; staminate corollas 4–5 mm. long. . . . . 2. *A. pulcherrima.*

a. Basal leaves spreading, forming depressed rosettes, strongly contrasting in outline with the cauline leaves; plants humifuse to stoloniferous. . . *b.*

b. Larger basal leaves only 1–5 mm. wide, blunt or only obscurely mucronulate, whitened above; flowering stems only 0.05–1.8 dm. high; only pistillate plants known, their involucres deep brown to blackish or, if pale, at most 7 mm. high. *c.*

c. Cauline leaves 15–28, very crowded (except in old individuals), the upper 7–20 with twisted scarious tips 2–3 mm. long; taller stems up to 4 (rarely –6) cm. high; involucre with 3–4 very unequal series of conspicuously imbricated phyllaries; achenes glabrous. . . . . 3. *A. columnaris.*

c. Cauline leaves 4–16, only the upper 1–7 with scarious tips; flowering stems mostly 4–18 cm. high; involucres with phyllaries subequal or in 2 or 3 unequal series (or in 4–6 series in no. 10, which has only 8–10 remote cauline leaves, with 3–5 appendaged, and pale involucres). . . *d.*

d. Involucres deep brown to blackish; phyllaries subequal or in 2–3 unequal series. . . *e.*

e. Involucres with the lower half prolonged, green and viscid, the phyllaries closely and firmly appressed or agglutinated to form an ellipsoid-campanulate falsely gamophyllous cup 7–9 mm. high; corollas 5–5.5 mm. long; achenes glabrous. . . . . 4. *A. Foggii.*

e. Involucres with loose and distinct phyllaries; corollas 3–5 mm. long. . . *f.*

f. Phyllaries conspicuously unequal, in 3 series, the outer about half as long as the inner; corollas 4–5 mm. long.

Cauline leaves 6–9, the 2 or 3 upper with unguiculate subulate tips 0.6–1.5 mm. long; involucres 6–7 mm. high, their outer and median phyllaries with scarious tips 1.2–2 mm. wide; style included or nearly so, subentire; achenes papillate, 1.1–1.4 mm. long. . . . . 5. *A. Bayardi.*

Cauline leaves 8–14, the 4 or 5 upper with flat scarious tips 2–3.5 mm. long; involucres 7.5–9 mm. high, their outer and median phyllaries with scarious tips 2–2.5 mm. broad; style exserted, 2-cleft; achenes glabrous, 1.8 mm. long. . . . . 6. *A. brunnescens.*

f. Phyllaries subequal, in 2 or 3 series, the outer nearly as long as the inner; corollas 3–4.5 mm. long; achenes glabrous. . . *g.*

g. Flowering stems at most 1.2 dm. high, with 5–8 leaves; the 3–6 upper leaves with flag-like oblong-lanceolate flat tips.

Involucres 7–10 mm. high, with squarrose pale brown phyllary-tips 1.3–2 mm. broad; upper 3 or 4 cauline leaves appendaged; corollas 4–4.5 mm. long; achenes 1.2–1.4 mm. long. . . . . 7. *A. cana.*

---

* The pistillate involucres only are intended in the measurements (unless otherwise specified); the counts and descriptions of *upper* cauline leaves refer to those *below* the inflorescence and exclude the bracteal leaves in the corymb. Treatment based largely on those in Rhodora, xxxv. 327–346 (1933) and xlvii. 221–239 (1945). In the figures the outline of a flowering plant (or portion of plant) is shown, × ¼; separate rosette-leaf (in some), × ¾; summit of stem (in nos. 29 and 30), × 5; upper cauline leaf and its tip, × 3; pistillate head, × 3; receptacle in nos. 20–22, × 3; achene (when shown), × 10.

Involucres 6–7 mm. high, with ascending fulvous phyllary-tips rarely more than 1 mm. broad; upper 4–6 cauline leaves appendaged; corollas 3–4 mm. long; achenes 1.6–1.8 mm. long. . . . . 8. *A. vexillifera.*

*g.*Flowering stems slender, up to 18 cm. high, their 8–15 leaves mostly subulate-tipped, only the uppermost with lance- or linear-involute scarious tips; involucres 5–6 mm. high, the outer ascending tips 1–1.7 mm. broad. . . . . . . 9. *A. confusa.*

*d.*Involucres milk-white or ochroleucous or pale brown; achenes glabrous (papillate in no. 12). . . *h.*

*h.*Cauline leaves 5–10; involucres pale brown or stramineous, sometimes roseate or greenish, 5–7 mm. high; corollas 3.7–5 mm. long; mature pappus 4.5–5 mm. long. . . *i.*

*i.*Involucres of 4–6 series of conspicuously unequal phyllaries, stramineous to pale brown. . . . . 10. *A. straminea.*

*i.*Involucres of 2 or 3 series of subequal phyllaries.

Flowering stems 3–7 cm. high, not glandular; upper cauline leaves with oblong-lanceolate scarious appendages 2–3 mm. long; involucres not glandular; achenes glabrous. . . . 11. *A. Peasei.*

Flowering stems 5–15 cm. high, glandular above; upper cauline leaves with subulate or involute tips; involucres glandular-viscid at base; achenes papillate. . . . . . . 12. *A. subviscosa.*

*h.*Cauline leaves 9–15; involucres milk-white or ochroleucous, 4.5–6 mm. high; corollas 3–3.3 mm. long; mature pappus 4–4.3 mm. long. . 13. *A. albicans.*

*b.*Larger basal leaves mostly wider, 0.2–5.5 cm. broad, usually distinctly apiculate or mucronate, green and glabrous to white-tomentose above; flowering stems 0.4–5 dm. high; involucres whitish, greenish, pale brown or fulvous, 5–11 mm. high; staminate plants of some species known. . . *j.*

*j.*Rosette-leaves comparatively small, 0.2–2.1 cm. wide, with only the midrib prominent to the tip beneath, the lateral ribs short and evanescent. . . *k.*

*k.*Middle and upper cauline leaves of pistillate plants terminated by a flat or merely inrolled scarious appendage. . . *l.*

*l.*New rosette-leaves bright green and glabrous or promptly glabrate on the upper face. . . *m.*

*m.*Rosette-leaves spatulate to cuneate-oblanceolate or narrowly cuneate-obovate, scarcely petioled, rounded at tip or only subacute, terminated by a mucro less than 0.5 mm. long; heads 1–6; plants of Nfld. and e. Que. . . *n.*

*n.*Basal offshoots crowded, scarcely elongate; basal leaves 6–13 mm. long, 2–4 mm. wide; flowering stem 5–13 cm. high, its longer leaves barely 1 cm. long; involucres brown, 6–8 mm. high; corollas 4.5 mm. long; achenes glabrous. . . . 14. *A. Wiegandii.*

*n.*Basal offshoots elongate, when well-developed cord-like; basal leaves 1–3.5 cm. long, 4–12 mm. wide; flowering stem 0.5–3 dm. high, its longer leaves 1–2.5 cm. long; involucres whitish, 7–11 mm. high; corollas 4.8–6 mm. long; achenes papillate.

Rosette-leaves broadly rounded at summit; lower and median cauline leaves obtuse, merely mucronate-tipped, the upper 1–3 with scarious appendages. . . . . 15. *A. spathulata.*

Rosette-leaves subacute to round-tipped; all or all but the very lowermost cauline leaves ending in prolonged scarious appendages. . . . . 16. *A. appendiculata.*

*m.*Rosette-leaves oblanceolate to narrowly obovate, somewhat narrowed to acute tip and tapering to a subpetiolar base, tipped by a sharp mucro 0.5–1.5 mm. long; heads 3–18; plant of broad continental range; achenes (at least when young) papillate. . 17. *A. canadensis.*

*l.*New rosette-leaves grayish- or whitish-tomentulose or -sericeous on upper face; achenes (at least when young) papillate. . . *o.*

*o.*Rosette-leaves cuneate-oblanceolate, -obovate or -spatulate, mostly 1.5–6.5 cm. long, 0.5–1.8 cm. broad, the old ones becoming green with weathering; stolons filiform, lash-like or cord-like, elongate; heads crowded in a glomerule or becoming loosely racemose; corollas 5–6.2 mm. long; continental plants, both staminate and pistillate.

Stolons flexuous and lash-like; rosette-leaves cuneate-oblanceolate to -spatulate, narrowed to a subpetiolar base, 0.5–1.3 cm.

broad; stems of pistillate plants elongating in fruit to 1.5–4 dm. high; pistillate heads in maturity becoming spicate to racemose; staminate involucres 4–6 mm. high; eastern species. 18. *A. neglecta.*

Stolons stiffer, cord-like; rosette-leaves narrowly cuneate-obovate, broad-based, 0.6–1.8 cm. broad; stems of pistillate plants 0.4–2 dm. high; pistillate heads closely glomerulate; staminate involucres 6–8 mm. high; Great Plains species. . . 19. *A. campestris.*

*o.*Rosette-leaves oblanceolate or narrowly obovate, acute or acutish, 0.5–2 cm. long, 2–5 mm. broad, strongly whitened; basal offshoots crowded, very short and assurgent, consisting chiefly of depressed rosettes; heads loosely corymbose (or solitary); corollas 4–4.5 mm. long; staminate plant unknown; species of e. Que. and Nfld. . . . . 20. *A. gaspensis.*

*k.*Middle and upper cauline leaves of pistillate plants blunt or with subulate or subulate-aristate tips (only those about the corymb with flat scarious appendages); achenes (at least when young) papillate. . . *p.*

*p.*Stolons and basal offshoots short and assurgent, ending in depressed rosettes; rosette-leaves oblanceolate to broadly obovate, often with definite petioles. . . *q.*

*q.*Rosette-leaves and cauline leaves all or nearly all with a naked terminal mucro or subulus; eastern and northeastern species.

Basal leaves oblanceolate or narrowly spatulate-obovate, acute or subacute, scarcely petioled; flowering stem stiff, its 8–18 leaves subapproximate and evenly spaced; corymb compact; involucres greenish or light brown only at base, with firm chartaceous milk-white to creamy-brown blunt phyllaries mostly 1.4–2 mm. broad; pits of denuded receptacle narrower than the ridges. . . . . 21. *A. rupicola.*

Basal leaves narrowly to broadly obovate, mostly rounded to tip, petioled; flowering stem usually becoming flexuous upon elongation, its 4–10(–14) leaves often becoming remote; pistillate involucres greenish-, purplish- or brown-tinged, their thin and usually scarious pale-tipped phyllaries 0.4–1.4 mm. broad; pits of old receptacle broader than ridges.

Plants all or chiefly pistillate, these with stems 0.5–5 dm. high; rosette-leaves 1–5.5 cm. long, 0.3–2 cm. wide; lower cauline leaves 1.5–4 cm. long, 2–6 mm. broad; pistillate involucres 6–9 mm. high, with 30–60 or more phyllaries; florets 50–140; corollas 3.2–6 mm. long; pits of receptacle deep; staminate involucre 3.8–6.5 mm. high; wide-ranging from Nfld. to n. Ont., s. to N. S., N. E., L. I., Va., Ind., Wisc. and Minn. . . . . 22. *A. neodioica.*

Plants about equally staminate and pistillate; the pistillate ones with very slender stems 0.6–2 dm. high; rosette-leaves 1–2.5 cm. long, 3–8 mm. wide; lower cauline leaves 1–1.4 cm. long, 1–2 mm. wide; pistillate involucre 4.5–7 mm. high, with 25–35 phyllaries; florets 40–70; corollas 3.2–4.5 mm. long; pits of receptacle shallow; staminate involucre 3.8–5 mm. high; plant of Appalachian Upland, Va., W. Va. and Pa., very locally to w. Vt. . . . . 23. *A. virginica.*

*q.*Rosette-leaves and all or nearly all the cauline leaves blunt, without an evident subulus; Great Plains species. . . . . 24. *A. aprica.*

*p.*Stolons elongate, lash- or cord-like, only tardily developing terminal rosettes; rosette-leaves cuneate-oblanceolate to spatulate-obovate. 25. *A. petaloidea.*

*j.*Rosette-leaves comparatively large, 0.7–5.5 cm. broad, with 3–7 nerves somewhat prominent and prolonged beneath; achenes (at least when young) papillate. . . *r.*

*r.*Heads in corymbs; stolons assurgent and leafy or with assurgent leafy tips. . . *s.*

*s.*Pistillate involucres 7–11 (rarely only 6 in no. 30) mm. high; central corollas 4.5–7 mm. long; achenes 1.3–2.2 mm. long; mature pappus (longer bristles) 5.5–9 mm. long. . . *t.*

*t.*Rosette-leaves mostly subtruncate, abruptly constricted above the middle to a concave curve arching to the prolonged slenderly cuneate base; mature pappus 5.5–6 mm. long; local species of upper Great Lakes. . . . . 26. *A. Farwellii.*

*t.*Rosette-leaves acutish or gradually rounded at summit, more gradually narrowed to the base; mature pappus 6–9 mm. long. . . *u.*

*u.*Blades of larger rosette-leaves 2.5–8 cm. long above the petiolar base, minutely canescent, or glabrous or glabrate; mature flowering stems (pistillate) 1.5–5 dm. high; involucres 7–11 mm. high; corollas 5–7 mm. long; longer pappus 6–9 mm. long.

Rosette-leaves spatulate to narrowly spatulate-obovate, strongly rounded at summit, closely canescent-tomentose above; heads glomerulate or densely crowded; longer mature pappus 8–9 mm. long; stems glandless. . . . . . . 27. *A. munda.*

Rosette-leaves broadly obovate-spatulate to obovate and subacute to suborbicular and rounded above; heads densely to loosely corymbed; longer mature pappus 6–8.5 mm. long; stems frequently glandular above.

New rosette-leaves permanently and closely canescent-tomentulose above; stems and involucres rarely purple; mature achenes 1.3–1.6 mm. long, densely papillose, their longer pappus 6–8 mm. long; summit of plant glandless or glandular. . . . . 28. *A. fallax.*

New rosette-leaves bright green and glabrous or only lightly canescent and glabrate; stems and involucres frequently purplish; mature achenes 1.6–2.2 mm. long, smooth or slightly papillose, their longer pappus 7.5–8.5 mm. long; summit of stem usually glandular. . . . . 29. *A. Parlinii.*

u. Blades of larger rosette-leaves 1.5–3.5 cm. long, strongly rounded, loosely tomentose above; mature flowering stem (pistillate) 1.3–3.5 dm. high, with stipitate glands at summit and in the corymb; involucres 6–8.5 mm. high; corollas 4.5–5.5 mm. long; longer pappus 6–7 mm. long. . . . . 30. *A. Brainerdii.*

s. Pistillate involucres 5–7 mm. high; central corollas 2.5–4.3 mm. long; achenes 1–1.5 mm. long; mature pappus 4–5.5 mm. long; rosette-leaves oblanceolate, obovate or suborbicular, minutely canescent above. . . . . 31. *A. plantaginifolia.*

r. Heads solitary; stolons filiform, lash-like, tardily producing terminal rosettes of sessile to broad-petioled obovate- to broadly oblong-spatulate leaves. . . . . 32. *A. solitaria.*

1. A. eucósma Fern. & Wieg. (elegant). — Pistillate plant: stems usually solitary from a slender subterranean caudex or stoloniferous rhizome, the stolons cord-like; *leaves* silky-tomentose, acute or short-acuminate; *the basal and those of the offshoots erect, lanceolate to oblanceolate,* rarely elliptical, tapering to the petiole, 3-5-nerved, 0.2–1.7 dm. long, 0.5–1.8 cm. broad; *flowering stem* 0.8–2.5 *dm. high; the* 4–10 *cauline leaves* rapidly diminishing in size, *the lower like those of the basal tufts,* callus-tipped; *the uppermost* linear-attenuate, 0.7–3 cm. long, *with a slender scarious appendage;* corymb dense or glomerulate, with 3–15 heads; *involucres chestnut to blackish,* 7–10 mm. high, lanate at base; their outer phyllaries ovate to oblong, the inner lance-linear, all with conspicuous brown, castaneous or drab tips; *corollas* 3–4.3 *mm. long; exserted tips of mature dark brown or purple style three-fourths as long as corolla; achenes glabrous,* 1.2–1.5 *mm. long;* longer pappus-bristles 6–7 mm. long. Staminate plant: similar, smaller; inflorescence looser; involucres about 6 mm. high, their *phyllaries with conspicuous fuscous oblong tips; corollas* 3–4 *mm. long;* pappus-bristles slightly thickened upward, serrulate to tip. — Turf and humus over limestone, Straits of Belle Isle to Bay St. George, Nfld. Mid-July, Aug. Fig. 1695.

1695. A. eucosma.

2. A. pulchérrima (Hook.) Greene (very handsome). — Pistillate plant: similar to no. 1, scarcely stoloniferous, *taller,* 2–5 *dm. high, with* 6–12 *leaves, corymb looser,* with 5–20 heads; *involucres pale brown to chestnut; corollas* 4.6–5.2 *mm. long; exserted tip of style one-sixth to one-third length of corolla; achenes* 0.9–1.3 *mm. long; longer pappus* 9–11 *mm. long.* Staminate plant: *phyllaries of involucre with whitish tips; corollas* 4–5 *mm. long.* — Calcareous gravels and turf,

Anticosti I., Que.; shores of Hudson B. to n. Alta., s. in Rocky Mts. Late June–early Aug. FIG. 1696.

1696. A. pulcherrima.

3. **A. columnàris** Fern. (columnar). — Dwarf, forming dense mats 2–10 cm. broad; basal leaves and those terminating very short offshoots oblanceolate, scarcely petioled, 5–9 mm. long, 1.5–4 mm. wide, barely mucronulate, with felt-like tomentum above; *flowering stem* 0.5–4(–6) *cm. high; cauline leaves* 15–28, *crowded*, or the upper finally subdistant, *linear*, *erect*, the lower obtuse, the median attenuate and mucronate, *the upper 7–20 with a terminal scarious linear-lanceolate to -acicular appendage* 2–3 *mm. long;* heads (pistillate) 1–6 in a close corymb, subcylindric (in drying becoming turbinate-campanulate); *involucre* 5.5–8.5 mm. high, *its* 30–50 3–4-*seriate strongly imbricated phyllaries* fulvous or fuscous, the outer linear-oblong and obtuse, the inner narrower and acuminate; corollas 4 mm. long, purple-tipped; style purple, exserted, 2-cleft; achenes glabrous, 1.2 mm. long; staminate plant unknown. — Peaty and turfy limestone-barrens near St. John and Ingornachoix Bays, Nfld. July, early Aug. FIG. 1697.

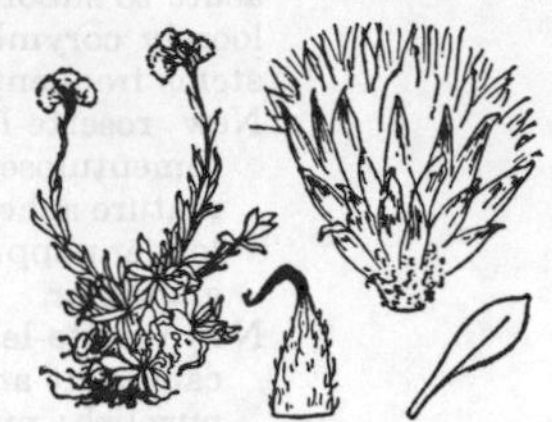
1697. A. columnaris.

4. **A. Fóggii** Fern. (for one of its discoverers, JOHN MILTON FOGG, 1898–). — Forming mats 0.3–2 dm. across, with leafy offshoots crowded and 1–3 cm. long; rosette-leaves cuneate-oblanceolate or narrowly obovate, scarcely petioled, 4–16 mm. long, 2–5 mm. broad, round-tipped to subacute, heavily white-felted; flowering stems 4–13 cm. high, floccose-tomentose; *cauline leaves* 8–16, *subdistant*, floccose, linear, the lower obtuse, the median acute or subulate-tipped, the 5–7 *upper with lanceolate scarious appendages* 1.5–4 *mm. long;* heads (pistillate) 1–5 in a corymb, ellipsoid-campanulate (when dry broadly campanulate); *involucre 7–9 mm. high, viscid-lanate, green and prolonged below the middle and appearing gamophyllous, the lower phyllaries appressed or agglutinated, the upper with* 3 *series of rosulate oblong brown or fuscous tips* 1.5–2 *mm. broad;* corollas 5–5.5 *mm. long*, with purple tips; style exserted, purple, 2-cleft; achenes glabrous, 1.7 mm. long; staminate plant unknown. — Limestone-barrens bordering St. John and Ingornachoix Bays, Nfld. July, early Aug. FIG. 1698.

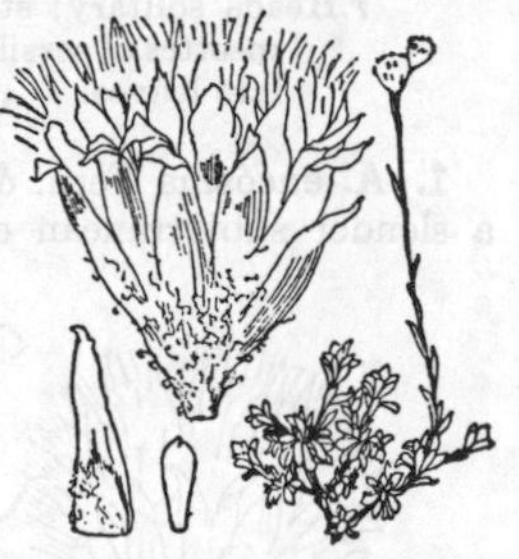
1698. A. Foggii.

5. **A. Bayárdi** Fern. (for one of its discoverers, BAYARD LONG, 1885–). — Resembling no. 4; rosette-leaves cuneate-obovate, subrhombic, subacute or mucronate, 5–10 mm. long, 2–5 mm. broad, minutely tomentose; *cauline leaves* 6–9, *remote*, linear-oblong, loosely tomentose, lower obtuse, median subacute and mucronate, *the 2 or 3 upper with a lance-subulate and claw-like tip* 0.6–1.5 *mm. long;* heads (pistillate) 1–6, on unequal peduncles, subcylindric-urceolate (subturbinate when dry); *involucre* 6–7 *mm. high*, its 20–25 very unequal 3-seriate *brown or fuscous oblong outer and median phyllaries* 1.2–2 *mm. broad*, the inner phyllaries slender and acuminate; corollas 4–4.8 mm. long, with purplish tips; *style included or barely exserted, subentire; achenes hirtellous*, 1.1–1.4 *mm. long;* staminate plant unknown.— Dry humus on limestone, Bonne Bay and Bay of Islands, Nfld. July, early Aug. FIG. 1699.

1699. A. Bayardi.

6. **A. brunnéscens** Fern. (brownish). — Similar to no. 5; rosette-

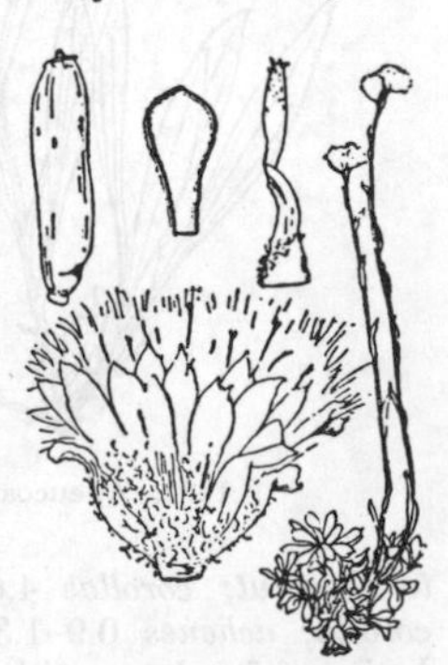
1700. A. brunnescens.

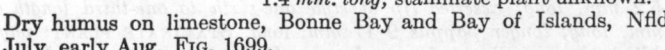

leaves spatulate or narrowly cuneate-obovate, 5–14 mm. long, 2–4 mm. wide, white with minute tomentum; *cauline leaves* 8–14, subdistant, linear, strict, glabrate, the median subulate-tipped, *the upper* 4 *or* 5 *with a scarious lanceolate or oblong flat appendage* 2–3.5 *mm. long;* heads (pistillate) 1–4 in a close corymb; *involucre* 7.5–9 *mm. high; its* 25–35 3-seriate thin *phyllaries very unequal,* oblong-lanceolate, brown, *the outer* 2–2.5 *mm. broad;* corolla 4.8–5 mm. long, purple-tipped; *style exserted,* purple, 2-*cleft; achenes glabrous,* 1.8 *mm. long;* staminate plant unknown. — Turfy limestone-crest, alt. 670 m., Killdevil, Bonne Bay, Nfld. Aug. Fig. 1700.

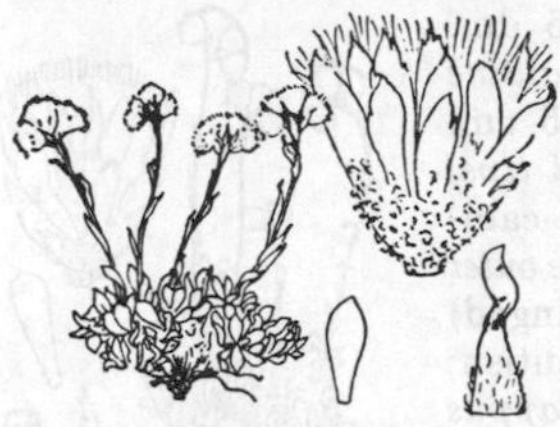
1701. A. cana.

7. A. càna (Fern. & Wieg.) Fern. (grayish-white). — Forming mats up to 1 dm. broad; rosette-leaves broadly spatulate or cuneate-obovate, rounded at summit, 5–11 mm. long, 2.5–4 mm. broad, densely white-tomentose; flowering stems 4–10 cm. high; *cauline leaves* 4–8, subdistant; the median linear, subulate-mucronate, *the* 3 *or* 4 *upper with an oblong-lanceolate scarious tip* 1.5–2 *mm. long;* heads (pistillate) 1–6, turbinate-campanulate, loosely corymbose; *involucre* 7–10 *mm. high, its thin subequal fulvous or castaneous phyllaries* 2–3-seriate, *the pale brown oblong tips* 1.3–2 *mm. broad; corollas* 4–4.5 *mm. long;* style yellowish or brown, exserted, 2-cleft; *achenes* glabrous, 1.2–1.4 *mm. long;* staminate plant unknown. — On dry limestone or trap (up to 760 m. alt.), Straits of Belle Isle to Bonne Bay, Nfld.; n. B.C. July, early Aug. Fig. 1701.

8. A. vexillífera Fern. (bearing pennants). — Similar to no. 7; pubescence of rosette-leaves minute; flowering stems more slender; 4–6 *of the cauline leaves with broad appendages* 1.5–3 *mm. long;* heads campanulate, rounded at base; *involucre* 6–7 *mm. high, its ascending* fulvous *phyllaries narrower, the principal ones rarely* 1 *mm. broad; corollas* 3–4 *mm. long; achenes* 1.6–1.8 *mm. long;* staminate plant unknown. — Dry calcareous gravels, nw. Nfld.; Shickshock Mts. (at 1070 m. alt.), Que. July. Fig. 1702.

1702. A. vexillifera.

9. A. confùsa Fern. (confused). — Similar to nos. 7 and 8; rosette-leaves oblanceolate or narrowly cuneate-obovate; *flowering stems slender, up to* 1.8 *dm. high; cauline leaves* 8–15, *mostly subulate-tipped, only the* 3 *or* 4 *upper with scarious lance- or linear-involute appendages; involucre* 5–6 *mm. high,* somewhat viscid at base, *its outer* ascending fulvous or fuscous *phyllaries oblong-linear,* 1–1.7 *mm. wide;* corollas 3–3.8 mm. long, purple-tipped; *style purple, exserted, bifid; achenes* 1.5 *mm. long;* staminate plant unknown. — Calcareous gravel and humus, w. Nfld. July, Aug. Fig. 1703.

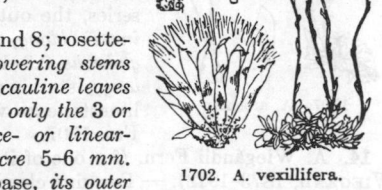

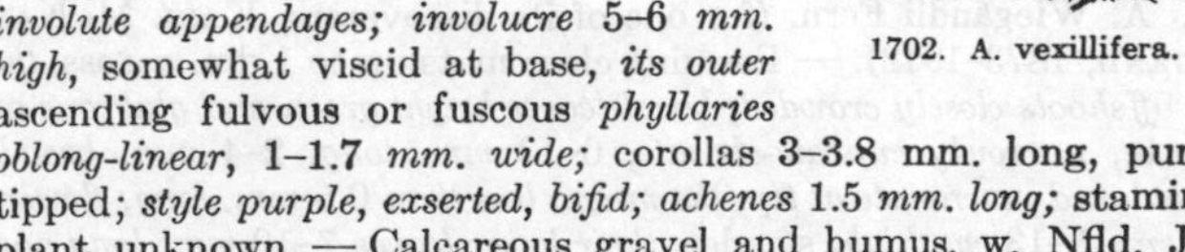

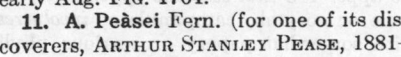

1703. A. confusa.

10. A. stramínea Fern. (straw-colored). — Forming dense to loose mats up to 1.5 dm. broad; *rosette-leaves spatulate,* 5–12 mm. long, 2–4 mm. broad, minutely white-tomentulose; flowering stems slender, 3–14 cm. high; *cauline leaves remote,* 8–10, linear, the median attenuate to a dark subulate tip, *the upper* 3–5 *with a linear scarious appendage;* heads (pistillate) 1–7, mostly in a close corymb, hemispheric-campanulate; *involucre* 5.5–7 mm. high, *with* 4–6 *series of very distinctly unequal phyllaries,* the outer ones ovate or oblong and brown, *with stramineous tips,* the inner narrower and paler; corollas 3.7–4.2 mm. long; achenes glabrous, 1.4 mm. long; longer pappus-bristles 4.5–5 mm. long; staminate plant unknown. — Dry calcareous gravel or humus, n. and w. Nfld. July, early Aug. Fig. 1704.

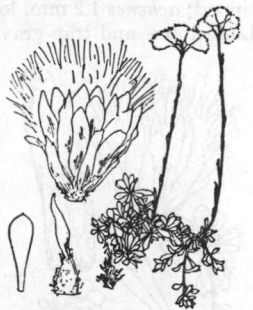

1704. A. straminea.

1705. A. Peasei.

11. A. Peàsei Fern. (for one of its discoverers, Arthur Stanley Pease, 1881–). — Forming mats 2–8 cm. across; rosette-leaves broadly oblanceolate or narrowly cuneate-obovate, 7–12 mm. long, 2–4.5 mm. wide, mucronate, minutely white-tomentose above; *flowering stems* 3–7 *cm. high,* slender; *cauline leaves* 5–7, linear-lanceolate, the lower and median subulate-mucronate, *the upper* 2 *or* 3 *with an oblong-lanceolate scarious appendage* 2–3 *mm. long;* heads (pistillate) 2–5 in a dense corymb, campanulate; *involucre* 6–7 *mm. high, with* 2–3 *series of subequal subpetaloid phyllaries; the outer* oblong, *pale-brown, the inner* similar, acuminate, *whitish; corollas* 3.8–5 *mm. long;* style purple; *achenes glabrous;* staminate plant unknown. — Alpine gravels, Shickshock Mts., Que. July. Fig. 1705.

12. A. subviscôsa Fern. (somewhat sticky). — Forming mats with crowded *trailing subligneous branches* 1–5 *dm. long,* the offshoots short and crowded; *rosette-leaves spatulate,* obtusish, scarcely or barely mucronate, 0.5–1.5 cm. long, 2–5 mm. wide, white-tomentose, *marcescent for many years; flowering stems* 0.5–1.5 dm. high, white-tomentose, *glandular-hirsute above, marcescent; cauline leaves* 7–10; the lower linear-oblanceolate, mucronate, 1.5–2.5 cm. long, *the upper with subulate or involute subscarious pubescent tips;* heads (pistillate) 3–9, densely to loosely corymbose, turbinate-campanulate; *involucre* 5–6.5 *mm. high, with phyllaries* 2–3-*seriate; the outer glandular-viscid,* greenish or stramineous (sometimes pink-tinged) *with creamy or roseate thin obtuse tips,* the inner narrow, acutish; *corollas* 3.8–4.3 *mm. long; achenes papillose,* 1–1.2 mm. long; *pappus* 4.5–5 *mm. long;* staminate plant unknown. — Calcareous sea-cliffs, Rimouski Co. to n. Gaspé Co., Que. Late June, July. Fig. 1706.

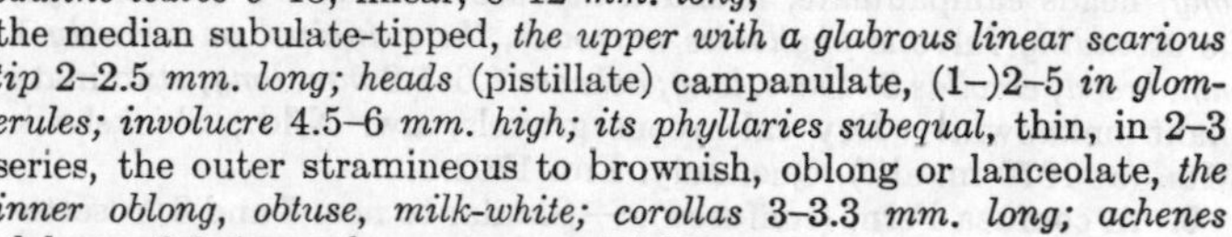

1706. A. subviscosa.

13. A. álbicans Fern. (whitish). — Similar to no. 12, much smaller, the mats only 2–8 cm. across; *rosette leaves* 3–8 *mm. long,* 2–3 mm. wide, *covered above with sublustrous minute pubescence;* flowering stems 4.5–9 cm. high; *cauline leaves* 9–15, linear, 6–12 *mm. long,* the median subulate-tipped, *the upper with a glabrous linear scarious tip* 2–2.5 *mm. long; heads* (pistillate) campanulate, (1–)2–5 *in glomerules; involucre* 4.5–6 *mm. high; its phyllaries subequal,* thin, in 2–3 series, the outer stramineous to brownish, oblong or lanceolate, *the inner oblong, obtuse, milk-white; corollas* 3–3.3 *mm. long; achenes glabrous,* 0.8–1 mm. long; *pappus* 4–4.3 *mm. long;* staminate plant unknown. — Dry limestone-gravels, w. Nfld. July, early Aug. Fig. 1707.

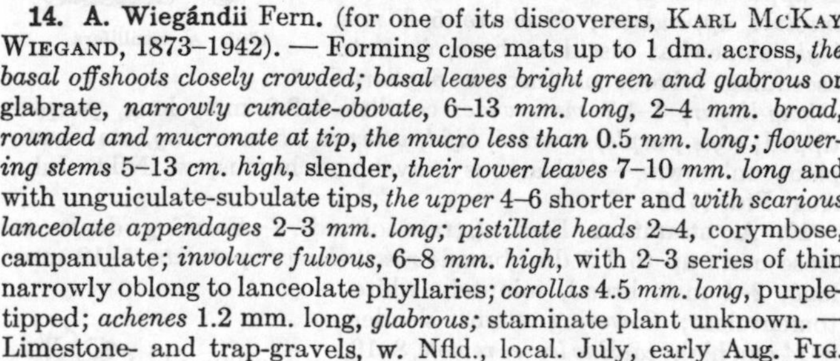

1707. A. albicans.

14. A. Wiegándii Fern. (for one of its discoverers, Karl McKay Wiegand, 1873–1942). — Forming close mats up to 1 dm. across, *the basal offshoots closely crowded; basal leaves bright green and glabrous* or glabrate, *narrowly cuneate-obovate,* 6–13 *mm. long,* 2–4 *mm. broad, rounded and mucronate at tip, the mucro less than* 0.5 *mm. long; flowering stems* 5–13 *cm. high,* slender, *their lower leaves* 7–10 *mm. long* and with unguiculate-subulate tips, *the upper* 4–6 shorter and *with scarious lanceolate appendages* 2–3 *mm. long; pistillate heads* 2–4, corymbose, campanulate; *involucre fulvous,* 6–8 *mm. high,* with 2–3 series of thin narrowly oblong to lanceolate phyllaries; *corollas* 4.5 *mm. long,* purple-tipped; *achenes* 1.2 mm. long, *glabrous;* staminate plant unknown. — Limestone- and trap-gravels, w. Nfld., local. July, early Aug. Fig. 1708.

1708. A. Wiegandii.

1709. A. spathulata.

15. A. spathulàta Fern. (spatula-shaped). — *Basal offshoots elongate and cord-like,* only tardily developing rosettes, forming loose mats up to 1.5 dm. across; *rosette-leaves spatulate to cuneate-oblanceolate, rounded at summit,* 1–2.5 cm. long, 4–10 mm. wide, *bright green and glabrous above,* tipped by a very short (less than 0.5 mm. long) mucro; flowering stems 0.5–2 dm. high; their *lower and median leaves* 1–2 *cm. long, obtuse, merely mucronate-tipped, the upper* 1–3 *with scarious appendages;* heads 1–6 in a close glomerule or corymb, campanulate; *involucres whitish,* 7–11 *mm. high,* with somewhat papery or scarious phyllaries; *corollas* 4.8–6 *mm. long; achenes papillate,* 1.3–1.5 mm. long; staminate plant unknown. — Turfy and peaty soil, Nfld. and St. P. et Miq.; Côte Nord, Anticosti I., and L. Mistassini, Que. Late June–early Aug. Fig. 1709.

16. A. appendiculàta Fern. (bearing appendages). — Similar to no. 15, coarser; *rosette-leaves subacute to round-tipped,* 1.5–4 cm.

1710. A. appendiculata.

long; *flowering stems* 1–3 dm. high, *all or nearly all of their leaves with long scarious appendages.* — Peaty or turfy calcareous soil, nw. Nfld.; Anticosti I. and Gaspé Co., Que., and reg. of James Bay. Late June–early Aug. FIG. 1710.

17. A. canadénsis Greene (Canadian). — Mats often several dm. across, the assurgent offshoots terminating in rosettes; *rosette-leaves oblanceolate to narrowly obovate,* narrowed to an *acutish tip with sharp mucro* 0.5–1.5 *mm. long, tapering to a petiolar base, bright green and glabrous* or only sparsely arachnoid above; *flowering stems slender,* of the pistillate plants 1–4.5, of the very rare staminate plants 0.7–2.3 dm. high; *cauline leaves* 5–11, linear-attenuate, the lower and median subulate-tipped, *the* 2–4 *upper with a slender soft scarious appendage, becoming very remote by elongation of stem; pistillate heads* 3–18, *loosely corymbed; involucres* campanulate, 8–10 mm. high, with 3–4-seriate phyllaries; outer *phyllaries* oblong to lanceolate, *with pale green to brown bases and pale scarious to petaloid tips,* inner phyllaries narrower; corollas 4.3–6 mm. long; achenes 1.2–1.5 mm. long, smooth or sparingly papillate; staminate corymb 1.5–3.5 cm. broad, the small heads with blunt oblong spreading white phyllaries. — Open woods, pastures, fields and rocky barrens, Gaspé Co., Que., to n. Man., s. to N.S., N.E., N.Y., mts. of Pa. and Va., n. Ind., Mich. and Minn. May–early July (–Aug. in n. mts.) FIG. 1711.

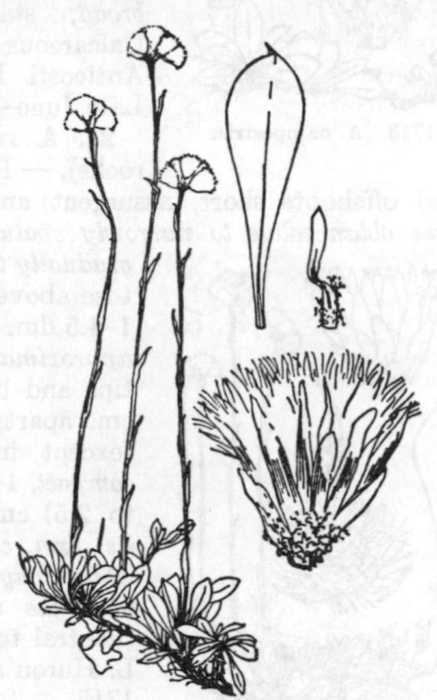
1711. A. canadensis.

18. A. neglécta Greene (overlooked). — Loosely spreading by *slender filiform creeping branches and elongate flexuous lash-like stolons* which tardily develop terminal rosettes; *rosette-leaves cuneate-oblanceolate to spatulate,* prominently 1-nerved beneath, 1.5–6.5 cm. long, 0.5–1.3 cm. broad, *narrowed to a subpetiolar base,* the new ones *canescent above; flowering stems* slender, of the pistillate plant 0.5–2.5 dm., *elongating in fruit to* 1.5–4 *dm. high,* of the smaller soon shriveling staminate plants 0.2–1.3 dm. high; *cauline leaves* 3–8, linear, *the upper terminated by a lance-attenuate flat or involute scarious appendage; pistillate heads* 2–8, or 1 and erect in forma **símplex** (Peck) Fern. (simple), at first crowded and glomerulate, later *becoming by elongation of the rachis* (up to 1.5 dm. long) *spicate or racemose; involucres of pistillate heads* 6.5–9 mm. high, *with* 3–4 series of obtuse to acutish white- or scarious-tipped *usually purple-based phyllaries;* corollas 5–6.2 mm. long; *styles usually crimson* (sometimes yellowish), *their branches less than* 0.3 *mm. long;* achenes 1–1.4 mm. long; the longer pappus-bristles 6–9 mm. long; *staminate* inflorescence a dense hemispherical glomerule (rarely a loose corymb) 1–2.5 cm. broad; the *involucres* 4–6 *mm. high,* with conspicuous white petaloid blunt tips. (*A. longifolia* Greene) — Sterile fields, pastures, plains or open woods, one of the usually abundant species (forma *simplex* local), N.S. and Me. to s. Ont. and Minn., s. to Va., W.Va., O., Ind., Mo. and Kans. Late April–July. FIG. 1712.

1712. A. neglecta.

19. A. campéstris Rydb. (of the Plains). — Resembling no. 18; stiffer and lower; the *stolons less flexuous; rosette-leaves narrowly cuneate-obovate, broad-based,* 0.6–1.8 cm. broad; *stems of pistillate plants only* 0.4–2 *dm. high;* pistillate *heads remaining glomerulate, the rachis not elongating; staminate involucres* 6–8 *mm. high.* — Plains and prairies, n. B.C. to Man., s. to Okla., e. to Mich. and Mo. Late Apr.–June. FIG. 1713.

20. A. gaspénsis Fern. (of the Gaspé Peninsula). — *Basal offshoots very short, crowded and assurgent, consisting chiefly of rosettes; rosette-leaves oblanceolate or narrowly obovate, acute or acutish,* 0.5–2 *cm. long,* 2–5 *mm. broad, strongly whitened;* flowering stems slender,

1713. A. campestris.

0.5–2.8 dm. high; *cauline leaves scattered, the median with an unguiculate-subulate tip, the 3–5 upper with a pale scarious linear-oblong appendage;* heads 2–7 (or 1), loosely corymbed, cylindric (turbinate in drying); *involucres* 7–10 mm. high, *of 25–35 linear-oblong whitish or light brown phyllaries; corollas 4–4.5 mm. long; denuded receptacle ovoid, as high as or higher than broad;* staminate plants unknown. — Calcareous gravels and ledges, w. Nfld.; Anticosti I. and e. Gaspé Pen., Que. Late June–Aug. FIG. 1714.

1714. A. gaspensis.

21. A. rupícola Fern. (growing among rocks). — Resembling nos. 20 and 22; the basal offshoots short, assurgent and ending in rosettes; *rosette-leaves oblanceolate to narrowly spatulate-obovate, acute or subacute, gradually tapering to base,* grayish-tomentose above; flowering stems stiffly erect, 1–4.5 dm. high; *cauline leaves* 8–18, *subapproximate or almost evenly spaced,* the tips and bases of successive ones rarely becoming more than 1–4 cm. apart; *the tips subulate, the upper subulate-aristate; corymb* (except in aberrant individuals with extremely long branches) *compact, 1–4 cm. broad;* heads subsessile or on rays up to 1 (rarely to 2.5) cm. long; *involucre greenish or light brown at base only, its firm chartaceous blunt lemon-white or whitish-brown to milk-white phyllaries* 1.4–2 *mm. broad; pits of receptacle narrower than the ridges;* staminate plant unknown. — Rocks and talus (neutral to calcareous), local, Nfld. and Anticosti I. to ne. Me.; n. L. Huron and L. Sup. region of Ont. and n. Mich. June, July. FIG. 1715.

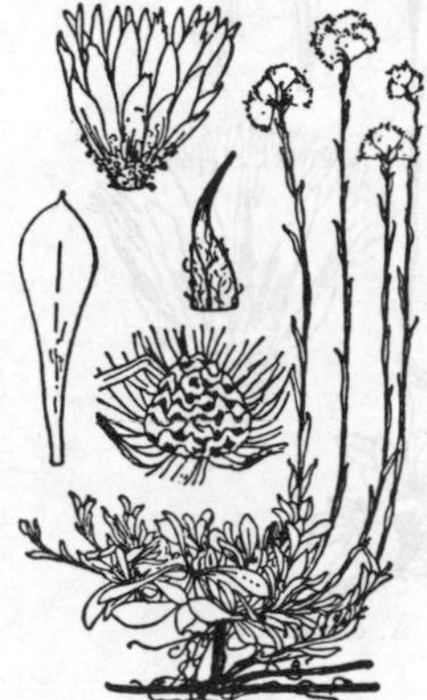

1715. A. rupicola.

22. A. neodioìca Greene (the new *A. dioica*). — Forming dense to loose mats, often several dm. across; *basal offshoots assurgent, soon producing terminal rosettes; rosette-leaves* narrowly to broadly *obovate,* 3–20 mm. broad, *mostly rounded at summit and tipped by a prominent mucro,* commonly *petioled, gray-tomentose* to green and glabrous *above; flowering stems* slender, *usually becoming flexuous upon elongation; cauline leaves* 5–14, *often becoming remote and unequally spaced* (up to 3–14 cm. apart) *by elongation of stem, subulate-tipped* in pistillate plant, sometimes scarious-tipped in the staminate; pistillate corymbs loose and up to 6 cm. broad or dense, the branches sometimes up to 4 or even 10 cm. long; *involucres* 5–9 mm. high, *greenish-, purplish- or brown-tinged, with paler phyllary-tips, the thin and usually scarious* (sometimes petaloid) *phyllaries mostly* 0.7–1.4 *mm. broad;* corollas 3.2–6 mm. long; *pits of receptacle broader than the ridges;* staminate plants rare or unknown in most vars., common only in one, small, their involucres 3.8–6.5 mm. high, with spreading white to roseate phyllaries. — Our most variable species, with locally distinct but broadly confluent varieties:

*a.*Rosette-leaves 1–4 cm. long, 3–17 mm. broad; lower cauline leaves 1–3 cm. long, 1–5 mm. broad; involucres of pistillate heads greenish-, brown-, or purple-tinged, with scarious tips; corollas 3.2–5 mm. long. . . *b.*

*b.*Rosette-leaves gray- or grayish-tomentose above. . . *c.*

*c.*Flowering stems flexuous upon elongation, 1–5 dm. high; the 5–14 leaves becoming distant; rosette-leaves obovate, petioled; corymb open.

Outer and middle phyllaries of pistillate involucre obtuse to subacute. *A. neodioica* (typical).

Outer and middle phyllaries lance- to linear-attenuate. . . . . . Var. *attenuata.*

*c.*Flowering stems stiffly erect, 0.5–2.5 dm. high, the 8–14 leaves subapproximate or imbricated; rosette-leaves oblanceolate or narrowly obovate, acutish, cuneate at base; corymb subglomerulate, globose or hemispherical. . . . . Var. *interjecta.*

*b.*Rosette-leaves green and glabrous above, 3–13 mm. wide; flowering stem 1.5–5 dm. high. . . . . Var. *chlorophylla.*

*a.*Rosette-leaves 2–5.5 cm. long, 0.7–2 cm. broad; lower cauline leaves 2–4 cm.

long, 3–6 mm. broad, often overlapping; flowering stems stoutish, 1.5–4.5 dm. high; corymb loose; heads very full; involucres (except in shade) reddish, the phyllaries with white petaloid tips; corollas 4.8–6 mm. long. . . Var. *grandis*.

A. neodioica (typical). — Dry fields, pastures, open woods and rocky barrens, Nfld. to Ont., s. to N.S., N.E., Va., Ind., Wisc. and Minn. May–early July. Fig. 1716. — Staminate plant very rare and local.

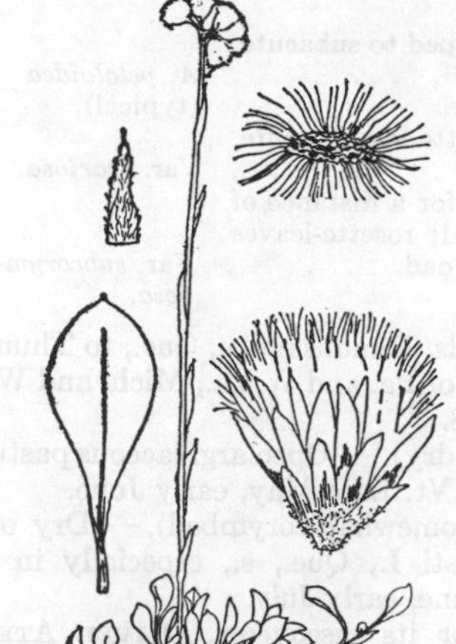
1716. A. neodioica.

Var. **attenuàta** Fern. (slenderly tapering). — Similar range. May–July (–Aug. northw.) — Staminate plant almost unknown.

Var. **interjécta** Fern. (thrown in between). — Dry open woods and gravels, Gaspé and Rimouski Cos., Que. to n. shore of L. Sup., Ont., s. to s. Ont. and e. Wisc. Mid-June–mid-July. — Staminate plant unknown.

Var. **chlorophýlla** Fern. (green-leaved). — Open woods, pastures and fields, Nfld. to Wisc., s. to N.S., N.E. and N.Y. May–July (–Aug. northw.). — Staminate plant unknown.

Var. **grándis** Fern. (large), *A. grandis* (Fern.) House. — Woods and clearings, N.S. and s. N.B. to s. Que., w. to Mich., s. to Mass. and N.Y. Late May–July. — Staminate plant unknown.

23. A. virgínica Stebbins (Virginian). — Resembling smallest extremes of no. 22; *staminate plants as abundant as pistillate; pistillate plant with slender stems* 0.6–2 *dm. high; rosette-leaves* 1–2.5 *cm. long,* 3–7 *mm. wide, canescent,* cuneate-oblanceolate, acute; *cauline leaves* 4–7, *the lower only* 1–1.4 *cm. long and* 1–2 *mm. wide; pistillate involucre* 4.5–6.5 *mm. high, of* 25–35 oblong to lanceolate *phyllaries;* outer phyllaries obtuse to merely acute, 0.8–0.9 mm. wide, the inner with acuminate milk-white to stramineous erose-margined tips; *corollas* 3.2–4.2 *mm. long; receptacle with shallow pits; staminate plants with scarious-tipped upper leaves;* the 1–5 heads with *involucre* 3.8–5 *mm. high; corollas* 2.5–3.5 *mm. long.* — Dry woods and openings, Appalachian Upland of Va., W.Va. and Pa., rarely n. to w. Vt. April–June.

Var. **argillícola** Stebbins (growing on shale). — Basal leaves obovate, 6–8 mm. wide, obtuse; pistillate involucres 5.5–7 mm. high, the outer phyllaries acute, the inner attenuate; phyllaries of staminate plants white to roseate, entire. — Shale-barrens and dry slopes, w. Pa., e. W.Va. and w. Va.

24. A. aprìca Greene (of sunny places). — Low, matted, the basal offshoots crowded and assurgent; *basal leaves* cuneate- or spatulate-obovate, *rounded to the obscure pubescent point,* densely gray-tomentose, 1–2 cm. long, 0.5–1 cm. broad; flowering stems 0.3–1.5 dm. high, densely pubescent; the narrowly oblanceolate to linear-oblong canescent *cauline leaves blunt;* pistillate corymb dense, with 2–6 subsessile to short-pedicelled heads; involucre 8–10 mm. high, with several series of whitish to creamy long tips; staminate plant with smaller heads and broad white phyllaries. — Plains, rocky bluffs and open woods, e. Man. to B.C., s. to w. Minn., Neb., N.M. and n. Mex. May–July. Fig. 1717.

1717. A. aprica.

A. microphýlla Rydb. (tiny-leaved), discovered in our area since this treatment went into type, *differs from no.* 24 *in smaller rosette-leaves* (0.5–1.5 *cm. long,* 2–5 *mm. broad*), *slender flowering stems up to* 3.5 *dm. high* and with more numerous and narrower cauline leaves, *pistillate corymb with* 5–30 *heads, involucres only* 5–7 *mm. high;* extending from Rocky Mts. across the plains to Man., w. Minn., S.D. and Neb.

25. A. petaloìdea Fern. (petal-like). — Loosely spreading by slender stiffish prostrate branches and *slender cord-like stolons* and procumbent branches; *rosette-leaves cuneate-oblanceolate to spatulate or spatulate-obovate,* 1.5–6.5 *cm. long,* 0.5–2.1 *cm. broad, the new ones canescent above;* flowering stems of the pistillate plant 0.4–5 dm. high; *cauline leaves* 5–13, linear, *the middle and upper with a long colored subulate-aristate tip; pistillate heads* (1–)5–15, *corymbose,* from sessile to long-pedicelled; involucre 7–11 mm. high, with 3–4 series of green- or brown-based phyllaries; corollas 4–6.5 mm. long; *styles commonly yellowish or brownish, with long branches* (0.3–0.6 *mm. long*); achenes 1.2–1.7 mm. long; *longer pappus-bristles* 5.5–7.5 *mm. long;* staminate plant (very rare), smaller; its few involucres 6–7 mm. high, with broad spreading white phyllary-tips. — Very variable:

Flowering stems 1–3 (–4) dm. high during anthesis, their leaves mostly at nearly regularly decreasing intervals up to the inflorescence; rosette-leaves 1.5–5 cm. long, 0.5–1.7 cm. broad, rounded at summit to acute.
Phyllaries with whitish petaloid tips; rosette-leaves round-tipped to subacute, up to 1.7 cm. broad. . . . . *A. petaloidea* (typical).
Phyllaries scarious, lustrous, yellowish, long-attenuate; rosette-leaves acute, up to 1.2 cm. broad. . . . . Var. *scariosa.*
Flowering stems 2–5 dm. high, nearly or quite without leaves for a distance of 0.7–1.7 dm. below the inflorescence; phyllaries white-tipped; rosette-leaves acute or acutish, mostly 3–6.5 cm. long and up to 2.1 cm. broad. . . . . Var. *subcorymbosa.*

A. petaloìdea (typical). — Fields, pastures and open woods, Rimouski Co., Que., to Thunder Bay Distr., Ont., and Minn., s. to s. N.B., N.E., N.Y., mts. of Pa. and W.Va., Mich. and Wisc. Mid-May–early July. Fig. 1718.

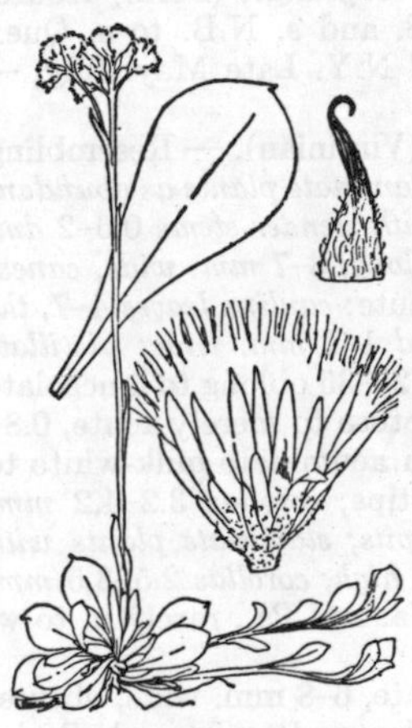
1718. A. petaloidea.

Var. **scariòsa** Fern. (thin and dry). — Open argillaceous pastures and slopes, local, centr. Me. to Vt. Late May, early June.

Var. **subcorymbòsa** Fern. (somewhat corymbed). — Dry open habitats, e. Nfld. and Anticosti I., Que., s., especially in the coastwise areas, to se. Mass. June, early July.

26. A. Farwéllii Greene (for its discoverer, Oliver Atkins Farwell, 1868–1944). — Basal offshoots rather crowded, very short and assurgent, consisting practically of basal-rosettes; *rosette-leaves mostly subtruncate, abruptly constricted above the middle to a concave curve arching to the prolonged narrowly cuneate base,* 2.5–6.5 cm. long, 1–2.2 cm. broad, cinereous-tomentose above, clearly 3-nerved; flowering stems slender, 1–3 dm. high, with 7–10 leaves, *the middle and upper leaves tapering to a dark subulate tip;* corymb becoming lax; involucres 8–10 mm. high, their linear and lustrous phyllaries with thin whitish tips; corollas 5–6 mm. long; achenes 1.5–1.8 mm. long; *pappus* 5.5–6 *mm. long.* — Dry gravelly banks and rocky bluffs (calcareous or basic), very local, Bruce Pen., Ont., and Keweenaw Co., Mich. June, early July. Fig. 1719. — Staminate plant unknown.

1719. A. Farwellii.

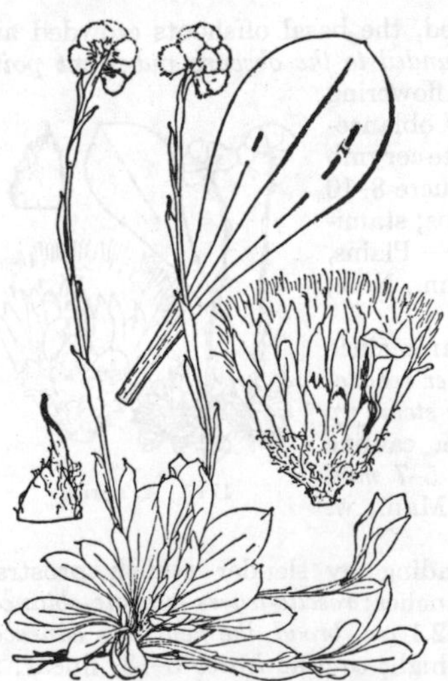
1720. A. munda.

27. A. múnda Fern. (neat). — Forming close to open mats, with basal offshoots short and assurgent or prolonged and decumbent and ending in rosettes; *rosette-leaves spatulate to narrowly spatulate-obovate, strongly rounded at summit;* the blade 2–6 cm. long, 1.3–5 cm. broad, 3–5-nerved, *canescent above;* flowering stem of pistillate plant stoutish, 1–4 dm. high; cauline leaves 6–15, the lower oblanceolate to narrowly oblong and 4.5–8 mm. broad, the middle and upper tapering to a slender subulus 0.6–1.4 mm. long; *heads* 5–20, *in a* *subglobose glomerule or dense rounded corymb* 2–4.5 cm. in diameter; involucres 8–10 mm. high, their phyllaries purplish at base; the outer narrowly oblong, with blunt erose white tips, the inner narrower and acute; pistillate corollas 5.5–7 mm. long; style reddish; achenes 1.5–1.8 mm. long; *longer pappus* 8–9 *mm. long;* staminate plant smaller, with stem 1 dm. high and dense corymb 1.5–2.7 cm. broad, the phyllaries broad and white. (*A. occidentalis* of ed. 7, not Greene) — Dry sandy, gravelly or rocky ground, centr. Me. to Thunder Bay Distr., Ont., s. to Mass., Ct., N.Y., e. Va., W.Va., n. Ind., Wisc. and Minn. Mid-May–mid-June. Fig. 1720. — Staminate plant only once found.

28. A. fállax Greene (deceitful). — Pistillate plant: similar to no. 27; *rosette-leaves obovate or rhombic-obovate* to rounded, minutely *cinereous-tomentose above, gradually tapering to the subacute or blunt but mucronate tip* and gradually to the petiole, 1.5–4 *cm. broad;* flowering stems at first stoutish, later prolonged and more slender, 1.5–5 dm. high; lower cauline leaves oblong-lanceolate, 5–12 mm. wide, rather crowded, the upper becoming remote and narrower, subulate-tipped; corymb dense to lax, with densely tomentose (occasionally glandular) branches; *involucres* 8–11 *mm. high,* with scarious or somewhat petaloid phyllaries; *corollas* 5–6 *mm. long; styles yellowish or brown* (rarely purple); *achenes columnar,* 1.3–1.6 *mm. long,* densely papillose; *longer pappus* 6–8 *mm. long.* Staminate plant: lower; stem 0.5–2.5 dm. high, with a hemispherical to flattish simple or proliferating umbel; the involucres with broad white tips. — Clearings, fields and open woods, centr. Me. to s. Ont. and Minn., s. to Va., Tenn., Ark. and e. Tex. Apr.–June. Fig. 1721. — Staminate plants frequent southw. and westw., rare northeastw.

1721. A. fallax.

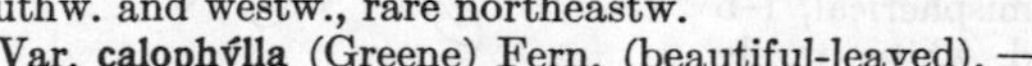

Var. **calophýlla** (Greene) Fern. (beautiful-leaved). — *Basal leaves rounded-ovate* to suborbicular, *broadly rounded at summit.* (*A. calophylla* Greene) — Open woods, fields and rocky slopes, Ga. to Tex., n. to Va., s. Mich., Ill. and Mo. Apr., May.

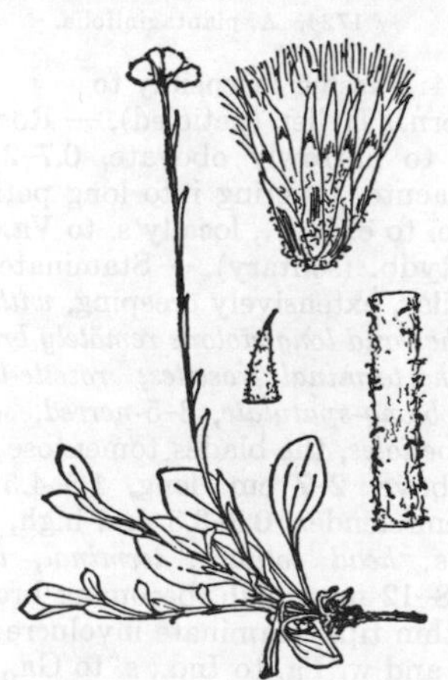

1722. A. Parlinii.

29. A. Parlínii Fern. (for its discoverer, John Crawford Parlin, 1863–1948). — Pistillate plant: similar to no. 28; *rosette-leaves bright green and glabrous* or soon glabrate *above,* broadly obovate-spatulate to obovate, petioled, the blades 2–7 cm. long, 1–4 cm. broad, with a prominent terminal mucro, 3–5-nerved; *stems* at first stout, becoming slender by elongation, purplish or green, 1.5–5 dm. high, *often purple-glandular,* especially at summit; involucres as large as in no. 28; the phyllaries lance-attenuate, scarious; style often purple; *achenes commonly slenderly fusiform,* 1.6–2.2 *mm. long,* glabrous or only sparingly papillate; mature pappus 7.5–8.5 mm. long. Staminate plant: similar, lower, the stems 1–2.5 dm. high, the involucres with lustrous rounded spreading white phyllaries. — Dry open woods and fields, w. N.S. and se. Me. to s. Ont., s. to Ga., O., Ind., Ill. and Ia. Apr.–early June. Fig. 1722. — Staminate plant freq. southw. and westw., rare northeastw.

Var. **arnoglóssa** (Greene) Fern. (with a lamb's tongue). — At least the outer phyllaries of the pistillate involucres with oblong white petaloid tips. — Similar range, s. to N.C., Tenn. and Mo. Late Mar.–early June.

30. A. Brainérdii Fern. (for its discoverer, Ezra Brainerd, 1844–1924). — Resembling *A. neodioica* var. *grandis,* but *the new basal offshoots, summit of stem and the inflorescence commonly with purple glands; rosette-leaves rounded- or somewhat rhombic-obovate; the mature blade* 1.5–3.5 *cm. long,* 0.7–2.5 *cm. broad, 3-nerved, loosely tomentose above;* pistillate flowering stem 1.3–3.5 dm. high, purple-tinged; cauline leaves 7–11, often glandular, becoming remote, the lower linear-oblong and 3–5 mm. broad, the middle and upper attenuate to purple subulate tips; pistillate heads 3–12, corymbed; *involucres* 6–8.5 *mm. high*, purple or brown at base; the outer phyllaries oblong, with erose white tips, the inner slender; *corollas purple*, 4.5–5.5 *mm. long;* mature achenes 1.4–1.8 mm. long, tapering to both ends; *longer pappus* 6–7 *mm. long.* — Argillaceous banks, rich open woods, etc., w.-centr. Me. to Ont., s. to Ct., N.Y., mts. of Va., and n. Mich. Mid-May–mid-June. FIG. 1723. — Staminate plant unknown.

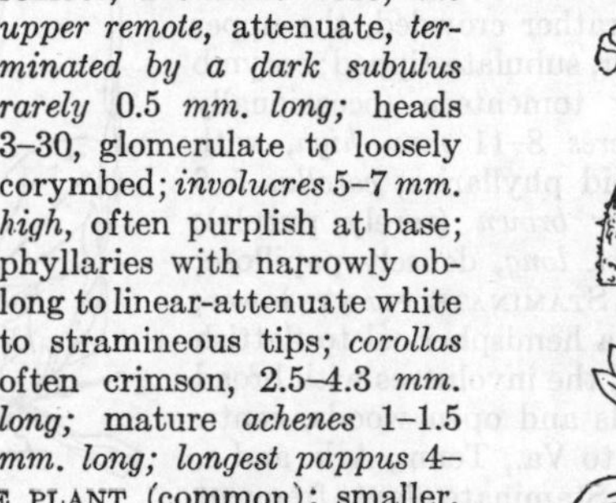

1723. A. Brainerdii.

31. A. plantaginifòlia (L.) Hook. (plantain-leaved). — PISTILLATE PLANT: with habit of no. 28, smaller throughout; *rosette-leaves* obovate to rounded-oblong or suborbicular, strongly rounded at summit, petioled, minutely *canescent* above, 3–7-*nerved*, the mature blades 1.4–6 cm. long, 1.5–4 cm. broad, *with a minute* (0.1–0.5 *mm. long*) *terminal mucro;* flowering stem slender, white-tomentose, 0.5–3 dm. high; *cauline leaves* 5–12, the lower at first crowded, becoming remote, 2–7 mm. wide; *the upper remote,* attenuate, *terminated by a dark subulus rarely* 0.5 *mm. long;* heads 3–30, glomerulate to loosely corymbed; *involucres* 5–7 *mm. high,* often purplish at base; phyllaries with narrowly oblong to linear-attenuate white to stramineous tips; *corollas* often crimson, 2.5–4.3 *mm. long;* mature *achenes* 1–1.5 *mm. long; longest pappus* 4–5.5 *mm. long.* STAMINATE PLANT (common): smaller, 0.2–1.8 dm. high; corymb loose or hemispherical, 1–5 cm. broad; phyllaries spreading, broad, white (rarely pink)-tipped. — Dry open woods, fields, pastures and rocky banks, sw. Me. to Minn., s. to Ga., Ala. and Mo. Apr.–June. FIG. 1724. Passing insensibly to

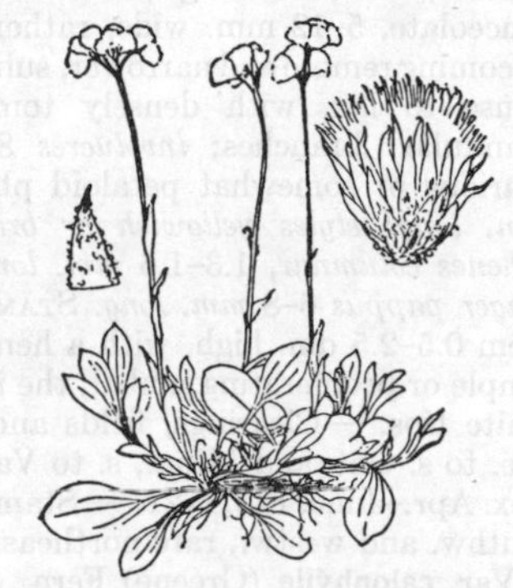

1724. A. plantaginifolia.

Var. **petiolàta** (Fern.) Heller (petioled). — Rosette-leaves oblanceolate to narrowly obovate, 0.7–2 cm. broad, acute or subacute, tapering into long petioles. — Dry soils, sw. Me. to e. N.Y., locally s. to Va.

32. A. solitària Rydb. (solitary). — Staminate and pistillate plants similar, extensively creeping, *with filiform prostrate branches and long stolons remotely bracted and ultimately with terminal rosettes; rosette-leaves obovate- to broadly oblong-spatulate,* 3–5-*nerved, sessile or with short broad petioles,* the blades tomentose (but floccose-glabrate) above, 2–7 cm. long, 1.5–4.5 cm. broad; flowering stem slender, 0.3–3.5 dm. high, with very narrow leaves; *head solitary, terminal, large;* pistillate involucre 8–12 mm. high, becoming broader in drying, brown to purple, its linear phyllaries with paler thin tips; staminate involucre with broader blunt phyllaries. — Rich woods and clearings, Md. and w. Pa. to Ind., s. to Ga., Ala. and La. Apr., May. FIG. 1725.

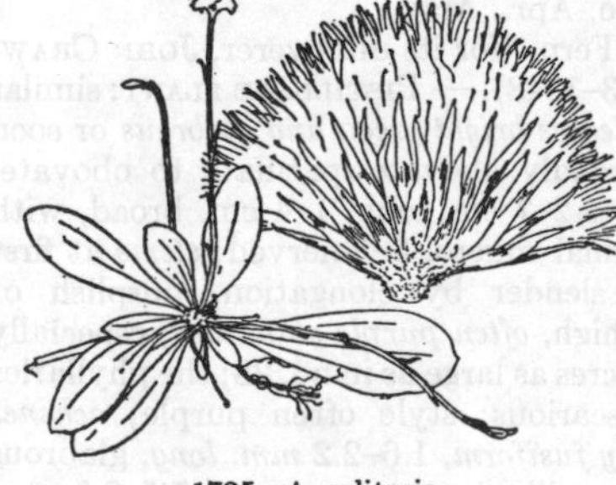

1725. A. solitaria.

26. ANÁPHALIS DC. EVERLASTING. LIFE-EVERLASTING. IMMORTELLE (Que.)

Characters of *Antennaria*, but the pappus in the sterile flowers not thickened at the summit or scarcely so, and that of the fertile flowers not at all united at base; fertile heads usually with a few hermaphrodite but sterile flowers in the center. Small genus of N. Am. and Asia. (Name an anagram of *Gnaphalium.*)

1. A. margaritàcea (L.) C. B. Clarke (pearly), PEARLY E. — Stem erect, 1–9 dm. high, very leafy; the leaves linear to lanceolate, divergent, white-tomentose beneath and often above; corymb somewhat leafy, with very numerous heads; phyllaries pearly-white, very numerous, obtuse or rounded, radiating, at least in age. — Very variable:

Leaves rather few, 15–48 (av. 36), broadly linear to narrowly oblong, bluntish to acute or only slightly attenuate, only slightly if at all reduced in size below the inflorescence.
Stems 3–8 dm. high; leaves bright green and glabrous or promptly glabrate above, those below the inflorescence 5–12 cm. long and 5–20 mm. wide. . *A. margaritacea* (typical).
Stems 1–5 dm. high; leaves cinereous-tomentose above, those below the inflorescence 2–7 cm. long and 3–15 mm. wide. . . . . . . Var. *subalpina.*
Leaves very numerous, 38–66 (av. 51), linear to linear-lanceolate, attenuate to tip; those below the inflorescence usually much reduced (1.5–6.5 cm. long, 1–3 mm. wide).
Leaves bright green and glabrous above. . . . . . Var. *angustior.*
Leaves white-tomentose or flocculent-arachnoid above. . . . Var. *intercedens.*

A. margaritàcea (typical), var. *occidentalis* Greene. — Gravelly or sandy soils and along streams, Nfld. to Alaska, s. to N.S., n. N.E., Cape Cod, Mass., w. N.Y., Minn., S.D., Colo. and Calif., ascending to subalpine areas. July–Sept. (E. Asia; early introd. into Eu.)

Var. **subalpina** Gray (subalpine). — Dry slopes and pastures, Nfld. to Alaska, s. to N.S., Wisc., Minn., S.D., N.M. and s. Calif. July–Sept.

Var. **angústior** (Miquel) Nakai (narrower), var. *revoluta* Suksd. — Clearings, thickets and shores, Nfld. to B.C., s. to N.S., N.E., W.Va., S.D., N.M. and Calif. (Asia)

Var. **intercèdens** Hara (standing between). — Nfld. to Alaska, s. to N.S., N.E., N.C., W.Va., O., Mich., Wisc., Minn., Mo. (adv.), S.D. and N.M., the commonest var. eastward. July–Oct. (Japan)

27. GNAPHÀLIUM L. Cudweed. Everlasting

Heads many-flowered; flowers all tubular, the outer pistillate and very slender, the central perfect. Phyllaries dry and scarious, white or colored, imbricated in several rows. Receptacle flat. Achenes terete or flattish; pappus a single row of rough bristles. — Woolly herbs, with sessile or decurrent leaves, and clustered or corymbed heads; fl. in summer and autumn. Corolla whitish or yellowish. Genus of wide range. (Ancient Greek name of some downy plant, derived from *gnaphallon, lock of wool.*)

a. Bristles of pappus all distinct, not united at base. § Eugnaphalium. . . *b.*
b. Involucres white or whitish, woolly only at base; achenes smooth; erect annuals or biennials with terminal corymbs or panicles of corymbs.
Bases of the leaves decurrent down the stem as thin wings. 1. *G. Macounii.*
Bases of leaves not decurrent.
Stem closely pannose- or appressed-lanate, glandular-puberulent or glandular-villous; phyllaries in 3–5 lengths. . . . . . . 2. *G. obtusifolium*
Stem loosely spreading-tomentose or -villous, not glandular; phyllaries in 1 or 2 lengths. . . . . 3. *G. saxicola.*
b. Involucres brown, the small heads in capitate leafy-bracted clusters or in a small spike; achenes scabrous, strigose or smooth; plants low.
Annual of low grounds, commonly diffusely-branched; heads densely involved in wool, forming close glomerules with long leafy bracts. 4. *G. uliginosum.*
Tufted alpine perennial with leaves mostly in basal tufts; heads smooth except at base, few in terminal spike or solitary. . . . . 5. *G. supinum.*
a. Bristles of pappus united in a ring at base and deciduous together; heads in spikes (continuous or interrupted), with brown to purplish involucres. § Gamochaeta. . . *c.*
c. Northern perennials with stout caudex bearing tufts of acute leaves.
Lower leaves 2–10 mm. broad, 1-ribbed; upper cauline leaves narrowly linear; species of low altitudes. . . . . 6. *G. sylvaticum.*
Lower leaves 0.6–3 cm. broad, 3-ribbed; upper cauline leaves oblanceolate; alpine species. . . . . 7. *G. norvegicum.*
c. Southern annuals or biennials; leaves of first year's rosette and of base of flowering stem obtuse.
Upper cauline leaves spatulate-oblong, round-tipped, 2–15 mm. broad; involucres 3–6 mm. high, enmeshed in tomentum below.
Lower leaf-surface closely white-pannose, the subappressed hairs tightly enmeshed; involucres densely woolly only at base, 4–6 mm. high. 8. *G. purpureum.*
Lower leaf-surface loosely villous-lanate; involucres nearly buried in wool, 3–4 mm. high. . . . . 9. *G. peregrinum.*
Upper cauline leaves linear or linear-oblanceolate, acute, 1.5–3 mm. broad; involucres 3–4 mm. high, glabrous except at very base. . . 10. *G. calviceps.*

§ EUGNAPHÀLIUM B. & H.

1. G. Macoùnii Greene (in honor of JOHN MACOUN, 1831–1920). — Biennial with a first-year rosette of blunt oblanceolate tomentose leaves; flowering stem 2–9 dm. high, heavily glandular below and sometimes to the more tomentose-lanate summit; *leaves* linear-lanceolate, partly clasping and *decurrent as wings down the stem;* corymb, when well-developed, with stiffly ascending branches, the heads densely clustered; phyllaries yellowish-white, oval to oblong, acutish. (*G. decurrens* Ives, not L.) — Clearings, pastures and borders of woods, Que. to s. B.C., s. to N.S., N.E., Pa., W.Va., O., Ind., Wisc., Minn., S.D., N.M. and Ariz. July–Oct.

2. G. obtusifòlium L. (obtuse-leaved), CATFOOT. — Biennial with first-year rosette-leaves oblanceolate or spatulate, tomentose; flowering stems 0.1–1.5 m. high, with heads densely corymbed or glomerulate at tips of the branches or in a terminal corymb; *leaves narrowed to base, sessile but without decurrent bases;* involucres with *phyllaries in* 3–5 *lengths,* the outer rounded or oblong, the inner narrower, all whitish. Very variable:

Stems covered with a close white felt-like pubescence.
Inflorescence corymbose, the lower branches in well-developed plants elongating and often forking; leaves frequently papillate or glandular above. . . *G. obtusifolium* (typical).
Inflorescence cylindric or thyrsoid, the branches all short and subuniform; leaves glabrous above. . . . . Var. *praecox.*
Stem glandular, not densely felted.
Stem glandular-puberulent, 1–6 dm. high; cauline leaves 1.5–10 mm. wide. . Var. *micradenium.*
Stem glandular-villous, 3–9 dm. high; cauline leaves 0.5–2 cm. wide. . . . Var. *Helleri.*

G. obtusifòlium (typical). — Dry clearings, borders of woods and fields, P.E.I. to Ont., s. to N.S., N.E., Fla., Ala., La. and Tex. Aug.–Nov.

Var. **praècox** Fern. (early), *G. polycephalum* Michx. — Pine-woods and clearings, s. N.J. to Ga. and Ala. July, early Aug.

Var. **micradénium** Weath. (with small glands). — Sandy woods, thickets and clearings, s. Me. to Mich., s. to S.C., Ky. and e. Mo. Aug.–Oct.

Var. **Hélleri** (Britt.) Blake (for its discoverer, AMOS ARTHUR HELLER, 1867–1944), *G. Helleri* Britt. — Sandy woods, Va. to Fla. Sept.–Nov.

3. G. saxícola Fassett (dwelling among rocks). — Resembling small plants of no. 2, 0.5–3 dm. high; *stem loosely floccose-villous;* leaves smooth above; *involucre with the narrow acuminate phyllaries in* 1 *or* 2 *lengths.* — Siliceous ledges and cliffs, Wisc. Aug., Sept.

4. G. uliginòsum L. (growing in marshes), LOW C. — Diffusely branched or subsimple appressed-woolly annual, 0.5–3 dm. high; leaves spatulate-oblanceolate or linear, not decurrent; *heads small, in terminal sessile capitate clusters* subtended by leaves; *involucres* 3–4 *mm. high,* their *phyllaries* brownish, *only slightly imbricated.* — Ditches, damp clearings and waste places, Nfld. to B.C., s. to N.S., N.E., Va., W.Va., O., Ind., Wisc., Minn., ne. Kans., Colo., Utah and Oreg.; a troublesome garden-weed. July–Oct. (Eurasia)

5. G. supìnum L. (lying on its back), ALPINE C. — Dwarf *tufted perennial with crowded linear-oblanceolate or linear leaves in tufts; flowering stems* 0.5–10 *cm. high, simple, with* few linear leaves and 1 *terminal head or usually* 2–5 *in a terminal spike; involucres* 5–7 *mm. high,* with oblong to lanceolate smooth or smoothish brown phyllaries. — Greenl. and Lab. to granitic alpine areas of Tabletop Mt., Que., Mt. Katahdin, Me., and Mt. Washington, N.H. July–Sept. (Eu.)

§ GAMOCHAÈTA (Weddell) B. & H.

6. G. sylváticum L. (of woodland). — *Silvery-silky perennial with* 1-*ribbed linear-oblanceolate basal leaves* 2–10 *mm. broad; flowering stems* stiffly erect, 1–7 dm. high, *with numerous linear-attenuate ascending leaves; inflorescence an elongate leafy interrupted spike,* simple, or compound at base, *occupying one-third to four-fifths the entire height of the plant;* involucres 5–7 mm. high, their pale brown oblong *phyllaries with a dark spot below the hyaline tip.* — Clearings, rocky slopes, borders of woods and fields, Nfld. to w. Que., s. to N.S. and n. N.E. Late July–early Sept. (Eu.)

7. G. norvégicum Gunn. (Norwegian). — Coarser than no. 6, often greener; *basal leaves oblanceolate,* 3-*ribbed,* 0.6–3 *cm. broad; flowering stem* 2–5 dm. high, *with few oblanceolate dilated leaves; spike* denser and thicker, *occupying one-sixth to one-half the height of the plant; phyllaries castaneous or brown with a narrow pale center and base.* — Greenl. and Lab. Pen., s. to wet or peaty slopes and alpine meadows of n. Nfld. and Shickshock Mts., Que. Late July–Sept. (Eu.)

8. G. purpùreum L. (purple), PURPLE C. — *Annual or biennial;* leaves of young rosette and of base of stem spatulate, round-tipped; *flowering stem* or stems 0.5–5 dm. high, *simple* or sparingly forked, *with remote round-tipped or blunt spatulate or oblanceolate leaves closely white-*

pannose with tightly enmeshed subappressed hairs beneath; spike terminal, dense, or interrupted at base, leafy-bracted in the lower half; *involucres* crowded, in glomerules, *lanate only at base, 4–6 mm. long, brown to chestnut or purple.* — Dry siliceous or argillaceous soil, Fla. to Tex. and Calif. n. to s. N.E., N.Y., O., Ind., Ill., Mo., Kans. and Oreg. Apr.–June (–Oct.) (Trop. Am.)

9. G. peregrìnum Fern. (wandering). — Resembling no. 8, simple or loosely branching; *leaves loosely villous-lanate and cinereous beneath; heads nearly buried in dense gray wool, the involucres 3–4 mm. high.* — Old fields, roadsides and waste places, Fla. to Tex. and s. Calif., n. to se. Pa. May–July. (S. Am.)

10. G. cálviceps Fern. (bald-headed). — More slender than no. 8; *stem* subsimple or *commonly very much branched,* 1–5.2 dm. high; *cauline leaves linear-oblanceolate, falcate, the middle and upper acute or attenuate to a terminal mucro;* spike more interrupted and slender; *involucres 3–4 mm. long, stramineous, glabrous or nearly so* except at very base. — Dry open soil and pine-woods, Va., N.C. and Tenn., inclined to be a weed. June–Aug. (S. Am.)

28. ÍNULA L.

Heads many-flowered, radiate; disk-flowers perfect and fertile. Involucre imbricated, hemispherical, the outer phyllaries herbaceous or leaf-like. Receptacle naked. Anthers caudate. Achenes more or less 4–5-ribbed; pappus simple, of capillary bristles. — Coarse herbs, nat. of Eurasia and Afr., not floccose-woolly, with alternate simple leaves and large yellow heads. (The ancient Latin name.)

1. I. Helénium L. (old generic name for this plant; see genus 65), Elecampane. — Stout perennial, 1–1.5 m. high; leaves large, woolly beneath; those from the thick root ovate, petioled, the others partly clasping; rays very many, narrow. — Roadsides, fencerows and rich clearings, e. Que. to s. Ont., s. beyond our range. July–Sept. — Heads very large. Root mucilaginous. (Natzd. from Eu.)

29. PULICÀRIA Gaertn. Fleabane

Heads many-flowered, with yellow ligules in 1 or 2 series, or discoid; disk-florets perfect, ligules pistillate. Involucre hemispherical, with narrow few-seriate phyllaries, the outer phyllaries herbaceous; receptacle alveolate. Anthers sagittate, with caudate auricles. Achenes terete or ribbed; pappus double, the outer short pales more or less connate into a cup, the inner of elongate capillary caducous bristles. — Annual or perennial mostly villous or lanate herbs with amplexicaul leaves and small solitary heads terminating the stem and branches. Nat. of Eurasia and Afr. (Ancient Latin name from *pulex, flea,* meaning *pertaining to fleas.*)

1. P. dysentérica (L.) Gaertn. (curing dysentery). — Perennial, with horizontal rootstock; stem 3–8 dm. high, with loosely ascending villous-lanate branches and downy oblong auriculate-clasping leaves 3–7 cm. long; disks about 1 cm. broad; ligules about as long or shorter. — Shores of lower Potomac, Charles Co., Md. July, Aug. (Natzd. from Eu.)

30. ADENOCAÙLON Hook.

Heads 5–10-flowered; the flowers all tubular and with similar corollas; the marginal flowers pistillate, fertile, the others perfect but sterile. Phyllaries equal, in 1 row. Receptacle flat, naked. Anthers caudate. Achenes elongated at maturity, clavate, beset with stalked glands above; pappus none. — Slender Am. and Asiatic perennials, with alternate thin petioled leaves smooth and green above, white-woolly beneath, and few small (whitish) heads in a loose panicle, beset with glands (whence the name, from the Greek *aden, a gland,* and *caulos, a stem*).

1. A. bícolor Hook. (two-colored). — Stem 3–9 dm. high; leaves triangular, rather cordate, with angular-toothed margins; petioles margined. — Rich woods of the L. Huron and L. Sup. region, Bruce Pen., Ont., to n. Mich.; Black Hills, S.D.; s. B.C. to nw. Mont. and Calif. Late June, July.

Tribe V. Heliántheae Cass. (see p. 1358)

a. **Heads without ligulate flowers; the pistillate flowers only 1–5, without corolla or with the corolla reduced to a tube or ring around the base of the 2-parted style; staminate flowers with 4- or 5-lobed corolla, anthers only slightly united, their terminal appendage inflexed, the abortive style with enlarged and depressed summit; pappus none.** *b.*

b. **Heads all alike, the few pistillate flowers marginal, the many central flowers staminate; involucre of few roundish phyllaries.** . . . . 31. *Iva.*

b.Heads of two kinds, the pistillate with a tuberculate or bur-like involucre. *c.*

c.Staminate involucre with united phyllaries.

Pistillate involucre with a single row of tubercles near summit; fruiting head 1-seeded. . . . . 32. *Ambrosia.*

Pistillate involucre with more than 1 series of spines, forming a 2–4-seeded bur. . . . . 33. *Franseria.*

c.Staminate involucre of distinct phyllaries; pistillate head 2-locular and 2-flowered, maturing into a very prickly hard 2-beaked bur. . . . 34. *Xanthium.*

a.Heads ligulate (sometimes discoid), the corolla regularly developed; anthers united nearly their whole length; fruit not bur-like, except in no. 36. . . *d.*

d.Outer phyllaries not inclosing the opposite flower or achene (if achenes inclosed by inner phyllaries the plants strong and showy-flowered perennials or with bur-fruits, and not clammy-viscid). . *e.*

e.Disk-flowers sterile; ray- or marginal flowers fertile. . . *f.*

f.Achenes thick, scarcely flattened.

Achenes partially enveloped by the inner unarmed phyllaries; tall perennials with large dilated leaves; the showy large heads paniculate-corymbose. . . . . 35. *Polymnia.*

Achenes closely embraced by the prickly inner phyllaries, thus forming a bur; small diffuse annuals with axillary flowers. 36. *Acanthospermum.*

f.Achenes obcompressed, *i.e.*, strongly flattened parallel to the phyllaries. . *g.*

g.Achenes broadly winged, in 2 or 3 series; pappus none or of 2 teeth confluent with the wings; coarse and tall paniculate branched perennials. . . . . 37. *Silphium.*

g.Achenes wingless or very narrowly margined, in 1 series; phyllaries in 2 series.

Showy ligules deciduous from the wingless achene.

Achenes 5–12, without pappus; outer foliaceous phyllaries smaller than the inner; southwestern. . . . . 38. *Berlandiera.*

Achenes about 5, the pappus a one-sided toothed crown; outer foliaceous phyllaries larger than the inner chaff-like ones; southeastern. 39. *Chrysogonum.*

Insignificant ligules persisting at summit of the 5 callous-margined achenes; roundish phyllaries subequal. 40. *Parthenium.*

e.Disk-flowers fertile; marginal or ray-flowers fertile or sterile. . . *h.*

h.Achenes flat, 4-angled or, if 5-angled, subterete and slender. . *i.*

j.Ligules marcescent, rather chartaceous, persistent at summit of the 4-sided marginal achenes; leaves opposite. . . . . 41. *Heliopsis.*

j.Ligules petaloid and promptly deciduous, or none. . . *k.*

k.Chaff of receptacle subtending and often wrapped around base of disk-flowers and achenes; achenes not obcompressed. . *l.*

l.Chaff of the flat receptacle bristleform; achene 3–4-angled; small heads with short white ligules about equaling the 10 or 12 phyllaries; small weak annual with short-stalked axillary heads. 42. *Eclipta.*

l.Chaff broader, embracing or subtending achenes; strong perennials, biennials or sometimes annuals with showy mostly peduncled or terminal large heads. . . *m.*

m.Involucre double, the outer one of 4 large foliaceous phyllaries united into an angled cup; inner small chaffy phyllaries clasping the marginal obovoid achenes; pappus none. . . . . 43. *Tetragonotheca.*

m.Involucre of 1 or more series of free phyllaries. . . *n.*

n.Receptacle high, conical to columnar in fruit; pappus none or a short crown or awn. . . . *o.*

o.Achenes 4-sided.

Ray-flowers neutral, yellow (or mottled with brown); phyllaries in about 2 rows; chaff concave, not rigid. 44. *Rudbeckia.*

Ray-flowers hermaphrodite but sterile, pink or purple (in one species yellow); phyllaries imbricated; chaff folded, keeled, spinose-tipped. . . . . . . 45. *Echinacea.*

o.Achenes flattened or 3-sided.

Marginal flowers neutral, ligulate; achenes laterally flattened; leaves alternate, pinnately dissected. 46. *Ratibida.*

Marginal flowers fertile, with or without ligules; their achenes 3-sided or dorsiventrally flattened; leaves opposite, not deeply dissected. . . . . 47. *Spilanthes*.

n.Receptacle flat to convex (rarely conical in no. 51). . . *p*.

p.Leaves not decurrent; achenes 3–4-angled, wingless; pappus a 4-toothed crown or of 2 or 4 scales.

Ray-flowers fertile; achenes plump, with a persistent 4-toothed crown; chaff soon deciduous; low and shrubby halophyte with opposite coriaceous leaves. . . . . 48. *Borrichia*.

Ray-flowers neutral; achenes laterally compressed, the pappus of 2 or 4 caducous scales; tall, herbaceous, with thin to firm opposite or alternate leaves; not halophytic. . . . . 49. *Helianthus*.

p.Leaves decurrent down the stem; achenes flat; pappus of 2 or 3 persistent awns.

Phyllaries few, soon deflexed; achenes squarrosely spreading. . . . . 50. *Actinomeris*.

Phyllaries numerous, imbricated; achenes ascending. 51. *Verbesina*.

k.Chaff of receptacle flattish, scarcely enveloping achene; achenes obcompressed (flattened parallel with the phyllaries) or in the aquatic no. 57 terete; involucre double, the outer one more foliaceous than the inner. *q*.

q.Ray-flowers pistillate; receptacle conical; 3-angled marginal achenes awnless; leaves somewhat clasping; adventive weed. 52. *Guizotia*.

q.Ray-flowers neutral or wanting; receptacle flat; achenes awned or awnless; leaves mostly not clasping. . . *r*.

r.Terrestrial or merely paludal, with firm stems; the leaves, if dissected, not flaccid; heads few–very many; awns (when present) or teeth shorter than the flat or angled (terete and beakless in no. 53) achenes. . . *s*.

s.Inner involucre cup-like, its phyllaries united to or above middle; achenes terete, beakless. . . . . . 53. *Thelesperma*.

s.Inner involucre with phyllaries free nearly to base; achenes flat or angled. . . *t*.

t.Achenes flat, orbicular or subquadrate to elliptic or oblong, mostly winged, with smooth or fimbriate margins; pappus 2 barbless teeth, a mere crown or none. 54. *Coreopsis*.

t.Achenes flat, angled and thick above or subterete, oblong-ovate, cuneate, linear or fusiform, ciliate or hispid at least on margins and angles; pappus mostly of retrorsely or antrorsely barbed or hispid awns or teeth.

Achenes oblong, cuneate, linear or subterete, beakless, ciliate or hirtellous only along the margins or angles; ligules yellow or whitish; native and introduced. . . . . 55. *Bidens*.

Achenes slenderly fusiform, 5-angled and subterete, long-beaked, hispid over entire surface; ligules pink, red, white or yellow; garden-escapes. . . 56. *Cosmos*.

r.Aquatic with flaccid stem; submersed leaves whorled, very finely dissected into flaccid segments; head usually solitary, terminal; achenes nearly terete, truncate, with 3–6 much longer rigid divergent apically retrorse-barbed awns. . . . . 57. *Megalodonta*.

h.Achenes turbinate, 5-angled; pappus of several scales or wanting.

Leaves alternate, entire; rays none; disk purple; native perennials with long-peduncled heads terminating the simple stems and branches. . . . . 58. *Marshallia*.

Leaves opposite, toothed; rays small, white or purple; disk yellow; introd. and weedy branching annuals. . . . . 59. *Galinsoga*.

d.Outer phyllaries (or the larger ones when not uniseriate) laterally compressed and infolding the laterally compressed achene; disk-flowers inclosed by the pales of the receptacle; viscid and heavy-scented homely plants (mostly casual weeds). . . . . 60. *Madia*.

31. ÌVA L. Marsh-Elder. Highwater-shrub

Heads several-flowered, not radiate; pistillate and staminate flowers in the same heads, the former few and marginal. Anthers nearly separate. Phyllaries few, roundish. Receptacle small, with narrow chaff among the flowers. Achenes obovoid or lenticular; pappus none. — Herbaceous or shrubby coarse Am. plants, with thickish leaves (the lower opposite) and small nodding greenish-white heads of flowers; in summer and autumn. (Old name of some medicinal plant.)

Heads spicate or spicate-racemose in the axils of leaves or leaf-like bracts; pistillate flowers with hyaline truncate tubes; foliage-leaves oblong or lanceolate to narrowly ovate or narrowly obovate. § Euiva.
Shrubby or suffruticose smooth or strigose maritime species with sessile, or only short-petioled, oblong to narrowly obovate leaves; bracts of raceme similar but smaller.
Larger leaves 6–15-toothed on each margin, opposite; involucres with a single series of 4–6 phyllaries, 4–6 mm. broad. . . . . 1. *I. frutescens.*
Larger leaves entire or 1–5-toothed on each margin, alternate; involucres with 2 rows of 6–9 phyllaries, 6–10 mm. broad. . . . . . . 2. *I. imbricata.*
Annual hirsute inland species with long-petioled ovate or lanceolate acuminate leaves; bracts of raceme quite unlike principal leaves. . . . 3. *I. ciliata.*
Heads in panicled spikes, without foliaceous bracts; pistillate flowers with obsolete corollas; lower leaves broad-ovate to suborbicular, long-petioled. § Cyclachaena. . . . . 4. *I. xanthifolia.*

§ Euìva Hoffm.

1. I. frutéscens L. (shrubby). — Shrubby at base, 1.5–3.5 *m. high,* smooth or minutely strigose; *leaves opposite,* lanceolate, fleshy, *those of the main stems lanceolate, sharply* 6–15-*toothed on each side,* 1–3 cm. broad, the rameal narrower; heads nodding in the axils of linear-lanceolate to linear subentire leafy bracts; *involucres* hemispheric, 4–5 *mm. broad, of* 4 *or* 5 *phyllaries in a single series;* pistillate flowers 4 or 5; achenes 2–2.6 mm. long. — Saline marshes and shores, Fla. to Tex., n. to Va. and less characteristically to s. N.J. Aug.–Oct.

Var. **orària** (Bartlett) Fern. & Grisc. (of the coast). — Lower (0.5–2 m. high); larger leaves elliptic, oval or broadly lanceolate, 2–5 cm. broad; heads 5–6 mm. broad, with 5 or 6 phyllaries; pistillate flowers 5 or 6; achenes 2.6–3.3 mm. long. (*I. oraria* Bartlett) — Saline marshes, Va. to s. N.H.; w. N.S. Aug.–Oct.

2. I. imbricàta Walt. (imbricated). — *Suffruticose,* glabrous, simple, 3–10 dm. high; *leaves* alternate, fleshy, *spatulate-oblong or lanceolate, entire* or slightly serrate; *heads* 6–8 *mm. high; hemispherical involucre of* 6–9 *phyllaries,* the outer orbicular. — Coastal sands, Fla. to La. and Va. Aug.–Oct. (W.I.)

3. I. ciliàta Willd. (with marginal hairs), Sumpweed. — *Annual,* 0.3–2 m. high, *rough and hairy; leaves* ovate, pointed, coarsely toothed, *downy beneath, on slender ciliate petioles;* heads in dense spikes, with conspicuous ovate- to linear-lanceolate rough-ciliate phyllaries; these and the pistillate flowers 3–5. (*I. caudata* Small) — Alluvial or moist soils, Ind. to Neb., s. to La., Tex. and N.M.; adv. e. to N.E. July–Oct.

§ Cyclachaèna (Fresn.) Gray

4. I. xanthifòlia Nutt. (with leaves of *Xanthium,* Clotbur). — Annual, tall, roughish; leaves nearly all opposite, hoary with minute down, ovate, rhombic, or the lowest cordate, doubly or incisely toothed, or obscurely lobed; heads small, crowded, in axillary and *terminal panicles of leafless spiciform racemes; corolla of the 5 pistillate flowers a mere rudiment or none.* (*Cyclachaena* (Nutt.) Fresn.) — Rich or alluvial soil, Wisc. and Man. to Mont., s. to Mo., Okla., Tex. and N.M.; adv. e. to Que., N.B., N.E. and N.J. July–Oct.

32. AMBRÒSIA L. Ragweed

Pistillate heads 1–few together, sessile in axils of leaves or bracts, at the bases of racemes or spikes of staminate heads. Staminate involucres flattish or turbinate, of 7–12 united phyllaries, containing 5–20 flowers, with or without slender chaff intermixed. Anthers almost separate. Fruiting involucre (fruit) ellipsoid, obovoid or turbinate, closed, pointed, resembling an achene and inclosing a single flower; elongated style-branches protruding. Achenes ovoid. — Coarse homely Am. and Afr. herbs, with opposite or alternate often lobed or dissected leaves, and inconspicuous greenish flowers, in late summer and autumn; ours annuals, except the last.

(The Greek and later Latin name of several plants, as well as of the food of the gods; name most inappropriate since the pollen of these and related bitter plants is the worst cause of hay-fever.)

Staminate heads sessile in solitary terminal spikes; the involucres extended on the upper side into a large lanceolate hooded bristly tooth or appendage; pistillate involucre ellipsoid or ovoid, 4-angled. § CERCOMERIS. . . . . . . 1. *A. bidentata.*

Staminate heads pedicelled in 1–several slender racemes, the involucres nearly or quite regular. § EUAMBROSIA.

Leaves entire or palmately cleft, opposite; staminate involucres unilaterally 3-ribbed; fruits 3–12 mm. long. . . . . 2. *A. trifida.*

Leaves pinnately lobed (rarely entire) to bi- or tripinnatifid, opposite and alternate; staminate involucres indistinctly radiate-veined; fruits mostly smaller.

Annual; leaves not harsh above; staminate involucres glabrous or pilose; fruits with 4–7 acute teeth or small tubercles. . . . . 3. *A. artemisiifolia.*

Perennial by creeping roots and rootstocks or annual; leaves harsh above; staminate involucres strigose-hispid; fruits unarmed or with 4 short blunt tubercles. . . . . 4. *A. psilostachya.*

§ CERCÓMERIS T. & G.

1. A. bidentàta Michx. (twice-toothed). — Annual, stiffly hirsute, up to 1 m. high, with erect branches, very leafy; *leaves mostly alternate,* lanceolate, *partly clasping,* nearly entire or hastately toothed at base; *staminate* spikes solitary, with crowded *sessile heads, the turbinate involucre extended at summit into a lanceolate hooded tooth;* fruits ellipsoid or ovoid, 4-angled, with 4 stout spines. — Dry open prairies and roadsides, Ind. to Minn. and Neb., s. to La. and Tex.; adv. in D.C. Aug.–Oct. — A hybrid with no. 2 occurs in Ill. and Mo.

§ EUAMBRÒSIA T. & G.

2. A. trífida L. (three-cleft), GREAT R., BUFFALO-WEED. — Stout annual 0.5–6 m. high, rough-hairy; *leaves all opposite, some* or all *deeply and palmately* 3 (or the lower 5) *-cleft into ovate-lanceolate serrate lobes,* or leaves all unlobed in forma **integrifòlia** (Muhl.) Fern. (entire-leaved), *the petioles slightly margined;* staminate racemes 1–several; *involucres* slender-pedicelled, regular, shallowly scalloped, *with* 3 *strong ribs on one side; fruit* 6–12 *mm. long, with acute conical beak* 2–4 *mm. long and* 4–10 *acute tubercles* terminating the ribs. (*A. striata* Rydb.) —Alluvium, rich openings and waste places, sw. Que. to B.C., s. to Fla., Ala., La., Tex. and Ariz. Late June–Sept. (Adv. in Eu.)

Var. TEXÀNA Scheele (Texan). — Blades 3- or 5-cleft, their *petioles slender* and scarcely margined; *fruits* 3–7 *mm. long, with broader beak* 0.5–1 *mm. long, without or with short obtuse tubercles.* (Var. *aptera* (DC.) Ktze.; *A. aptera* DC.) — Waste places, N.E. and s. Ill., doubtless elsewhere. (Adv. from Tex. or Mex.)

× **A. Hélenae** Rouleau (for HÉLÈNE BOULÉ, who became the wife of SAMUEL DE CHAMPLAIN, 1567–1635) is an obvious hybrid of nos. 2 and 3, on Ile Sainte-Hélène, near Montreal.

3. A. artemisiifòlia L. (with leaves of *Artemisia,* wormwood), COMMON R., ROMAN WORMWOOD, HOG-WEED, BITTER-WEED, HERBE À POUX and SARRIETTE (Que.). — Simple to much-branched glabrous or hairy monoecious or dioecious annual 0.2–2.5 m. high; *leaves both opposite and alternate,* comparatively thin, *smoothish above, pinnately lobed,* rarely unlobed, *to bi-* or *tri-pinnatifid* with small segments; staminate racemes 1–several, slender, becoming lax; *involucres saucer-shaped,* crenate, *glabrous or pilose, indistinctly radiate-veined;* fruits 4–5 mm. long; the subulate beak 1–2 mm. long, much exceeding the usually 4–7 sharp tubercles. — A polymorphic and despised weed:

Leaves simple, coarsely pinnatifid or rarely bipinnatifid; staminate involucres 3–7 mm. broad. . . . . *A. artemisiifolia* (typical).

Leaves bi- to tripinnatifid with small segments; staminate involucres 1.5–5 mm. broad.

Staminate involucres 2.5–5 mm. broad.

Plant glabrous or appressed-pubescent. . . . . Var. *elatior.*

Plant spreading-villous. . . . . Forma *villosa.*

Staminate involucres 1.5–2.5 mm. broad. . . . . Var. *paniculata.*

A. artemisiifòlia (typical). — Sea-beaches and cult. or waste land, Nfld. and M.I. to D.C., freq. northw., local southw. July–Oct.

Var. **elàtior** (L.) Descourtils (taller), *A. elatior* L.; *A. diversifolia* (Piper) Rydb. — Roadsides, cult. land and waste places, a pernicious weed, Fla. to Tex. and Mex., n. to s. Canada. July–Oct. (W.I.; natzd. in Eu.) — Forma villòsa Fern. & Grisc. (with long soft hair), similar range.

Var. **paniculàta** (Michx.) Blankinship (paniculate), *A. monophylla* Walt.; *A. paniculata* Michx. — Se. U.S., n. to Va., and locally as a weed to N.Y. and N.E.

4. A. PSILOSTÁCHYA DC. (naked-spiked). — *Perennial by creeping roots and rootstocks* or annual, paniculately branched, 0.3–1.5 m. high, harsh to touch and somewhat hoary with stiffish short hairs; *leaves once pinnatifid*, thick, nearly or quite sessile, *scabrous above, the lobes acute*, sometimes incised; *staminate involucres minutely scabrous-hirtellous or puberulent;* fruits obovoid, unarmed or with 4 short blunt tubercles. — The typical form local, ne. Mex. and se. Tex. More generally represented in the Southwest by var. *Lindheimeriana* (Scheele) Blankinship, with the involucre strigose-hispid with pustular-based hairs. Represented with us by:

Var. **coronopifòlia** (T. & G.) Farw. (with leaves of *Coronopus*). — Staminate involucres strigose-hispid to pilose with elongate slender hairs. (*A. coronopifolia* T. & G.; *A. media* Rydb.) — Dry prairies, barrens, sands and openings, Mich. to Sask. and Mont., s. to La., Tex. and Mex.; adv. e. to Que., N.S. and N.E. Aug.–Oct. (Adv. in Eu.)

33. FRANSÈRIA Cav.

Staminate and pistillate heads separate as in *Ambrosia*, or sometimes mixed in the inflorescence. Pistillate involucre 1–4-locular, with a single pistil in each locule, armed with spines in more than 1 series, bur-like. — Herbs (with us) or shrubs of N. and S. Am., with mostly alternate leaves, flowering in late summer and autumn. (Named in 1793 for *Antonio Franseri*, Spanish botanist and contemporary of Cavanilles.) GAERTNERIA Medic.

Perennial, with creeping subterranean rootstock; leaves tomentose beneath; staminate involucres tomentose, 3.5–5 mm. broad; spines of fruit subulate-conic, terete.
  Staminate raceme dense, spiciform, the involucres with short rounded lobes; fruits canescent-tomentulose, with straight spines. . . . . . . 1. *F. discolor*.
  Staminate racemes lax, the long-pedicelled involucres with long acuminate lobes; fruits glandular-puberulent, with curved spines. . . . . . . 2. *F. tomentosa*.
Annual, with small tap-root; leaves strigose; staminate involucres glabrous or merely strigose, 2–4 mm. broad; spines of hispid or glabrate fruit lance-attenuate, flattened. . . . . 3. *F. acanthicarpa*.

1. F. díscolor Nutt. (of different colors). — Perennial, with deep creeping rootstock; the flowering stems 1–3 dm. high; leaves interruptedly bipinnatifid, canescently tomentose beneath, glabrate above; *staminate heads crowded in a spiciform raceme, short-pedicelled or subsessile*, the *shallowly scalloped* tomentulose *involucre* 4–5 mm. broad; *fruits* 5–6 mm. long, *tomentulose, with straight* subulate-conic *spines* and 2 straight beaks. — Dry sands and plains, Wyo. to Ariz., locally e. to Ill. July–Sept.

2. F. tomentòsa Gray (with dense soft hairs). — Similar to no. 1, taller, the flowering stems 3–6 dm. high; *leaves canescent on both sides; staminate heads long-pedicelled, in lax racemes, their involucres deeply cleft into acuminate lobes; fruits* up to 8 mm. long, *glandular-puberulent, with curved or uncinate spines.* — Rich prairies and low grounds, Colo. to Tex., e. to e. Neb. and e. Kans. Aug.–Oct.

3. F. acanthicárpa (Hook.) Coville (prickly-fruited). — *Annual*, 1.5–6 dm. high, *strigose-hispid* to hirsute; *leaves deeply bipinnatifid, strigose; staminate heads* slender-pedicelled, the *green and glabrous or sparsely strigose involucres 2–4 mm. broad; fruits hispid or glabrate*, 6–10 mm. long, *with straight flat lance-attenuate spines.* — Dry sands and river-draws, Alta. to s. Calif., e. to Man., Minn., Neb., Kans. and Tex. Aug.–Oct.

34. XÁNTHIUM L. COCKLEBUR. CLOTBUR. LAMPOURDE (Que.)

Staminate and pistillate flowers in different heads, the latter clustered below, the former in short spikes or racemes above. Staminate involucres and flowers as in *Ambrosia*, but the phyllaries separate and receptacle cylindrical. Pistillate involucre coriaceous, ovoid or ellipsoid, clothed with hooked prickles so as to form a rough bur, 2-locular, 2-flowered; the flower consisting of a pistil and slender filiform corolla. Achenes oblong, flat. — Coarse weedy annuals, some originally in most warm and temp. reg., many now spread far from their nat. reg., with branching or simple stems, and alternate toothed or lobed petioled leaves; flowering in summer and autumn. (Greek name of some plant used to dye the hair; from *xanthos*, *yellow*.)*

*a.*Leaves attenuate to both ends, usually with triple spines at base; fruiting bur blunt or with 1 or 2 inconspicuous beaks. § ACANTHOXANTHIUM. . . . 1. *X. spinosum.*

*a.*Leaves cordate, rounded, subdeltoid or ovate, unarmed in axils; fruiting bur with usually 2(–3) strong porrect beaks. § EUXANTHIUM. . . *b.*

*b.*Surface of body of bur or bases of prickles visible, the prickles few to very many; bur (including beaks and prickles) 0.9–3 (rarely–4) cm. long and 0.6–2 (rarely –4) cm. thick; beaks 1–7 (rarely –8) mm. long; prickles 2–7 (rarely –10) mm. long. . . *c.*

*c.*Body of bur glabrous, glabrescent, short-glandular or merely short-hirtellous or -pilose. . . *d.*

*d.*Prickles numerous, 30–200± visible on one face of bur, slenderly linear-subulate to bristleform, straight or straightish or merely uncinate at tip. . . *e.*

*e.*Many of the prickles of mature burs 4–7(–10) mm. long. . . *f.*

*f.*Prickles visible on one face (about two-thirds the whole circumference) 50–80, remote; southwestern.

Mature bur 1.3–2 cm. long by 1–1.3 cm. thick, its body 4.5–7 mm. thick; beak 3–4 mm. long; prickles 4–6 mm. long. 2. *X. globosum.*

Mature fulvous or yellowish bur 2–2.5 cm. long by 1.5–2 cm. thick, its body 6–8 mm. thick; beaks 5–7 mm. long, strongly incurved above the hispid base; prickles 4.5–10 mm. long. 3. *X. inflexum.*

*f.*Prickles visible on one face 100–200 or more, remote or subapproximate, 2–7 mm. long; mature bur 1.7–2.5 cm. long, 0.8–1.5 cm. thick; widely dispersed.

Body of mature fruit lustrous, glabrous or essentially so, the interspaces much broader than bases of prickles; prickles 100–150 visible on one face, some straight, some hooked, smooth or only remotely glandular. 4. *X. chinense.*

Body of mature fruit dull to sublustrous, glabrescent to short-pilose with pale pubescence, the interspaces hardly to but slightly broader than bases of prickles; prickles 200 or more visible on one face, mostly glandular-hispid below. . . 5. *X. pensylvanicum.*

*e.*Prickles of mature burs 2–3.5 mm. long, bristleform; bur 6–13 mm. thick; beak 1–5 mm. long.

Petioles minutely soft-pilose; leaf-blades submembranaceous and barely scabridulous above; body of bur 5–7 mm. thick, closely short-pilose, its surface with interspaces much broader than bases of remote prickles, 15–50 of these visible on one face; local weed. 6. *X. strumarium.*

Petioles scabrous-hispid; leaf-blades thick, harsh above; body of bur 6–9 mm. thick, glabrous, its surface with interspaces scarcely broader than bases of crowded prickles, 100–200 of these visible on one face; plant of Illinois Valley. 7. *X. Chasei.*

*d.*Prickles 30–50± visible on one face of bur, conic-subulate, mostly 0.8–1 mm. thick at base, strongly arching to the subspiral terminal hook; bur minutely glandular-punctate or glabrate, 1.8–2.2 cm. long, its cylindric-fusiform body 3.5–5 mm. thick; plant of shores of L. Champlain. . . . . . 8. *X. curvescens.*

*c.*Body of fruit hispid, hirsute, villous or pilose. . . *g.*

*g.*Prickles and beak of bur much more slender than the length of the elongate basal villi. . . *h.*

*h.*Beaks of mature bur subulate, the gradually tapering base one-fourth to one-third as thick as the length (5–10 mm. long); ripe burs warm brown to ferruginous; prickles mostly 4–10 mm. long; leaf-margins subacutely dentate.

Mature burs 1.7–3 cm. long, 1.2–2.2 cm. thick, the body 1.3–1.8 cm. long and 6–8 mm. thick; beaks 5–7 mm. long, 1.2–2 mm. thick at base; prickles mostly 4–7 mm. long; transcontinental. . 9. *X. italicum.*

---

* Treatment based largely on the two monographs: MILLSPAUGH & SHERFF, Field Mus. Nat. Hist. Bot. Ser. iv. no. 2 (1919); and WIDDER in Fedde, Repert. Sp. Nov. Beih. xx (1923). Fully grown fruiting burs are essential for identification; detailed photographs of many of them will be found in Rhodora, xlviii, plates 1013–1019 (April, 1946).

Mature burs 3–4 cm. long, 2.5–4 cm. thick, the body 2–2.5 cm. long and 1.2–2 cm. thick; beaks 7–10 mm. long, the base 2–3.5 mm. thick; prickles mostly 7–10 mm. long; western, adv. eastw. . 10. *X. oviforme.*

h. Beaks of mature bur with stout bases 2–3 mm. thick, two- to three-fifths as thick as length of beak (3–6 mm. long); ripe bur drab to pale brown, plump-ovoid or -ellipsoid, 1.8–3 cm. long by 1.2–1.8 cm. thick; prickles mostly 2.5–5 mm. long.

Margins of leaves shallowly undulate, the low undulations much longer than high or deep; beaks soon strongly incurved, eventually approximate or their tips crossing; plant of Atlantic sands, beaches and shores. . . . . 11. *X. echinatum.*

Margins of leaves with prominent dentation, the deltoid teeth nearly as high as to higher than broad; beaks erect or nearly so, straightish or hamate; western or inland. . . . . . 12. *X. varians.*

g. Prickles and beaks of bur as thick as or thicker than length of their superficial trichomes or glabrescent. . . *i.*

i. Bur green, 1.3–1.7 cm. long by 1–1.2 cm. thick, short-pilose with white pubescence, the 15–50 very remote bristle-like smooth prickles 2–3 mm. long; beaks 1–2 mm. long; local adventive. 6. *X. strumarium.*

i. Burs yellow-brown, stramineous or fulvous, mostly longer or thicker; the prickles and beak longer and stouter. . . *j.*

j. Burs densely prickly, about 200 slenderly subulate usually crowded prickles visible on one surface of bur, these 3–7 mm. long; beak 4–6 mm. long; transcontinental. . . . . 5. *X. pensylvanicum.*

j. Burs remotely prickly, few–100 mostly stout prickles visible on one face of bur.

12–30 coarse straight or apically slightly hooked glandular-pubescent prickles visible on each face of bur; the stout porrect beaks distant from the upper prickles. . . . . 13. *X. Wootoni.*

36–100 coarse strongly arching and strongly hooked smoothish or basally glandular prickles visible on each face, the uppermost prickles approximate to the strongly incurving beaks. . 14. *X. orientale.*

b. Surface of body of bur and bases of filiform-bristle-like densely crowded prickles hidden; mature bur (including beaks and prickles) 3–4 cm. long and 2–3 cm. thick; beaks 6–11 mm. long; hispid prickles 7–9 mm. long. 15. *X. speciosum.*

§ Acanthoxánthium DC.

1. X. spinòsum L. (spiny), Spiny C. — Hoary-pubescent, armed at the axils with triple spines; stems slender; *leaves lanceolate* or ovate-lanceolate, short-petiolate, *white-downy beneath,* often 2–3-lobed or -cut; fruit about 1 cm. long, with a single short beak or beakless. — Waste places, Me. to Ont., westw. and southw. *Fr.* Aug.–Nov. (Natzd. from Eu.)

Var. inérme Bel (unarmed), without axillary spines has been found near Philadelphia. (Adv. from Eu.)

X. ambrosioìdes H. & A. (like *Ambrosia*), an Argentinian species differing from no. 1 in its procumbent habit and bipinnate blunt-lobed leaves, is casual near woolen-mills. (Adv. from S.Am.)

§ Euxánthium DC.

2. X. globòsum Shull (globose). — Stem up to 1 m. high, usually with long branches, the lower branches up to 1 m. long, purple-lineolate; leaves cordate-deltoid, shallowly lobed, subacutely dentate, not very thick, scabrous on both sides, about equaled by petiole; glomerules of pistillate burs mostly axillary and with foliaceous bracts; *ripe bur glabrous but minutely glandular, ovoid, thick-ellipsoid or subglobose, including beak and prickles* 1.3–2 *cm. long and* 1–1.3 *cm. thick; the body green, in age light brown,* 4.5–7 *mm. thick; beaks* 3–4 *mm. long,* erect or arching; *prickles* subulate, smooth, 4–6 *mm. long, their bases about as broad as the open interspaces* of the body, *green or light brown without red coloring,* 50–80 *of them visible on one face;* seeds oblong-ellipsoid, 6–6.5 mm. long, with dark brown testa. — Bottoms, old fields and waste places, Ill. and Ky. to Kans. *Fr.* Oct.

3. X. infléxum Mackenz. & Bush (incurved). — Differing from no. 2 in its coarser habit; the glabrate or toward summit white-papillate stem up to 1.5 m. high; *cordate* harsh *leaves with long acuminate tip; burs* crowded in elongate clusters, *stramineous to fulvous,* 2–2.5 *cm. long* and 1.5–2 *cm. thick, the oblong glabrous or short-pubescent or glandular-hirtellous body* 1.3–2 *cm. long and* 6–8 *mm. thick; beaks* hispid at base, 5–7 mm. long, *at maturity* abruptly bent near middle, *inflexed, and finally strongly overlapping; prickles slender-subulate, with glandular short pubescence or glabrate at base, 4.5–10 mm. long,* commonly hooked at tip. — Bottomlands, s. Ill., Mo. and Ark. *Fr.* Sept.–Nov.

4. X. chinénse Mill. (Chinese, a misnomer; plant at first said to be from China, this later corrected to Vera Cruz). — Scabrous stem 0.3–2 m. high, usually branched; leaves cordate-ovate to subreniform, not very thick; *fruiting burs ovoid to fusiform,* including beaks and prickles 2–2.5 *cm. long by* 1–1.8 *cm. thick, somewhat fulvescent,* the *glabrous or minutely glandular hirtellous* usually *lustrous body* 0.9–1.5 cm. long, the *interspaces much broader than bases of prickles; prickles* somewhat reddish-tinged when ripe, 100–150 *visible on one face of fruit,* slenderly subulate, straight or arched, *glabrous or only remotely glandular at base,* mostly 4–7 mm. long; beaks porrect or arched, with inflexed or hooked tips, glabrous or short-pubescent at base, 3–6 mm. long; seed lanceolate, 7–10 mm. long, its testa yellow-brown. (*X. canadense* sensu ed. 7 and recent auth., not Mill.; *X. pungens* Wallr.; *X. glabratum* Britt.) — Low grounds, waste or cult. land, roadsides, etc., W.I. and Fla. to Mex., n. to s. Que., s. Ont., Mich., Wisc., Minn., Neb., etc. *Fr.* Aug.–Oct.

5. X. pensylvánicum Wallr. (Pennsylvanian). — Similar to no. 6, mostly low (2–9 dm.); *leaves thicker and harsher,* cordate to subtruncate or broadly cuneate at base; fruiting *burs paler brown or drab-brown,* only slightly fulvous in age, *dull or barely lustrous, glabrescent or with body closely covered with dense pale pilosity; interspaces rarely broader than bases of the* 200 *or more prickles visible on one face, these mostly glandular-hispid below* (*X. saccharatum* Wallr.) — Bottomlands, low grounds, cult. or waste places, etc., Fla. to Tex. and s. Calif., n. to s. Que., Ont., Mich., Wisc., Minn., N.D., etc. *Fr.* late Aug.–Oct.

6. X. STRUMÀRIUM L. (the root anciently supposed to cure scrofulous tumors or strumae). — Erect or ascending, up to 1.5 m. high, usually branching; *leaves* cordate-ovate, lobulate and subacutely dentate, long-petioled, *the petioles minutely soft-pilose, the submembranaceous blades barely scabridulous; pistillate burs in subthyrsiform terminal and smaller axillary leafy-bracted clusters; mature burs thick-ovoid to ellipsoid, green, finely pilose,* often glandular, 8–14 *mm. long, the body* 5–7 *mm. thick; beaks* 1–2 *mm. long,* thick, straight, rarely arching; *prickles bristle-like, scarcely thickened at base,* 2–3 *mm. long,* very remote, *leaving broad interspaces,* 15–50 *visible on one face.* — Waste ground, casual and local, e. Mass. to e. Pa. *Fr.* Oct., Nov. (Adv. from Eu.)

7. X. Chàsei Fern. (for its discoverer, VIRGINIUS HEBER CHASE, 1876–). — Somewhat resembling nos. 2 and 6; *stem* simple or nearly so, 6 dm. high, *scabrous; thick petioles and very thick* cordate-deltoid dentate *leaf-blades harshly scabrous; fruiting burs sessile* in terminal and axillary clusters, *subglobose to ellipsoid-ovoid, the body glabrous or minutely puncticulate* and 6–9 *mm. thick and* 1.3–1.6 *cm. long, with interspaces scarcely broader than bases of crowded prickles; these slenderly subulate, straight or barely forward-arching, glabrous,* or sparsely sessile-glandular at base, 2–3.5 *mm. long; beaks* 4–5 *mm. long, very thick, remote, suberect, straight,* glandular-puberulent. — Bottomlands of Illinois R., Ill. (*V. H. Chase*). *Fr.* Sept., Oct. — Little known.

8. X. curvéscens Millsp. & Sherff (curving). — Stem 2–3 dm. high, subsimple or slightly branched, fulvous; leaves broadly deltoid-ovate, scabrous, with cordate to subtruncate base; *burs* 1–few in axils, *reddish-brown,* 1.8–2.2 *cm. long and* 1–1.5 *cm. thick; body glabrate or minutely glandular-punctate,* 1.3–1.6 cm. long, 3.5–5 *mm. thick, with broad interspaces between the prickles; these conic-subulate, mostly* 0.8–1 *mm. thick at base, strongly arching to the subspiral terminal hook,* 30–50± *visible on each face.* — Shores of L. Champlain, Vt. *Fr.* Sept., Oct. — Probably a local hybrid of no. 14 and the glabrous-fruited no. 4.

9. X. itálicum Moretti (Italian). — Mostly *tall* (*up to* 1.5 *m.*) and branched, the scabrous stem and branches often dark-lineolate; *leaves* on harsh petioles, *the cordate-ovate blades thick and scabrous, coarsely dentate,* the larger ones lobed; *burs* usually abundant in axillary and subterminal masses, *fulvous, in maturity* 1.7–3 *cm. long and* 1.2–2.2 *cm. thick, the villous-hirsute body* 1.3–1.8 *cm. long and* 6–8 *mm. thick; subulate beaks* 5–7 *mm. long,* 1.2–2 *mm. thick at base, densely long-hirsute below the middle; prickles crowded, slender, mostly* 4–7 *mm. long, straightish or apically hooked, divergently long-villous except at tip.* (*X. commune* Britt.; *X. saccharatum* sensu Widder, not Wallr.) — Low grounds, stream-banks, waste places, cult. land, etc., often a weed, W.I. and Fla. to Mex., n. to Que., Ont., Man., Sask., etc. *Fr.* Sept.–Nov. (S. Am.; Hawaii; s. Eu.)

10. X. OVIFÓRME Wallr. (egg-shaped). — Similar to no. 9, *usually* lower (3–7 dm. high); *simple* or short-branched *stem smooth except toward summit; mature burs* 3–4 *cm. long and* 2.5–4 *cm. thick, the body* 2–2.5 *cm. long* and 1.2–2 *cm. thick; beaks stouter,* 2–3.5 *mm. thick at base,* 7–10 *mm. long; prickles mostly* 7–10 *mm. long.* — Shores and waste places, s. Que. and Vt. to Mich., s. to e. Pa. *Fr.* Sept., Oct. (Adv. from w. U.S.)

11. X. echinàtum Murr. (hedgehog-like), SEA-BURDOCK. — *Purple-mottled* or -lineolate *stem simple or short-branched,* 1.5–6 dm. high, whitened above with harsh short hispidity; later *leaves* broadly cordate- or subcordate-ovate, harsh, *slightly lobed or unlobed, the margins shallowly undulate, with undulations mostly longer than high or deep; burs olive-shaped, drab or pale brown,* 1.8–3 *cm. long by* 1.2–1.8 *cm. thick,* the body *long-hirsute; beaks with stout* hispid *bases* 2–3 *mm. thick,* 3–6 *mm. long, soon strongly incurved and eventually approximate or with their tips crossing; prickles* 2.5–5 *mm. long, remote (with broad interspaces obvious on body), hirsute-villous below middle,* mostly hooked at tip. — Beaches, dune-hollows and borders of saline marshes along the coast, N.S. to Va. *Fr.* late Aug.–Oct.

12. X. vàrians Greene (variable). — Differing from no. 11 in *greener stem with sparse pubescence; leaves subacutely dentate with teeth nearly or quite as high as broad,* the blade often rhombic or cuneate-based; *beaks erect or nearly so, straight or merely hooked.* — Shores, s. Que. to n. Alta., s. to n. N.Y., Ia., N.D., Sask. and Oreg.; adv. in Mo. and perhaps elsewhere. *Fr.* late July–Oct.

13. X. WOÒTONI Cockerell (for its discoverer, ELMER OTTIS WOOTON, 1865–). — *Stem simple* or nearly so, dark-lineolate, sparsely strigose, flexuous above; *leaves* cordate- or subcordate-ovate, scabrous, *coarsely dentate; burs* 1–few in most axils, 1.8–3 *cm. long by* 1–1.8 *cm. broad, the brownish to fulvous glandular-pubescent oblong body* 1–1.6 cm. long and 4–7 *mm. thick, remotely prickly, with broad interspaces and prolonged naked summit below the beaks; beaks stout, porrect, glandular-hirsute; prickles coarse, straight or apically slightly hooked, glandular-pubescent,* 3.5–6 *mm. long,* 12–30 *of them visible on one face of bur.* — Waste places, etc., Mass., Minn., etc. *Fr.* late Sept., Oct. (Natzd. from farther west)

14. X. ORIENTÀLE L. (eastern). — Differing from no. 13 in having the 36–100 *prickles visible on one side of the bur strongly arched and hooked, smoothish or basally glandular, the uppermost prickles approximate to the strongly incurving beaks.* (Including *X. leptocarpum* Millsp. & Sherff) — Shores and waste places, sw. Que. and nw. Vt. *Fr.* Sept., Oct. (Natzd. from Eurasia) — The relatively slender-bodied burs of the type of *X. leptocarpum* matched by the more slender fruits of the Old World plant; the L. Champlain series, including some apparently typical *X. orientale, X. curvescens* (no. 8) and extreme *X. leptocarpum,* probably a mixture of the recently arrived no. 14 and more stabilized glabrous-fruited no. 4.

15. X. speciòsum Kearney (showy). — Differing from our other species in the *hundreds of densely crowded filiform-bristle-like fulvous short-hispid prickles* 7–9 *mm. long, these so crowded that their slender bases are hidden from view; mature bur* 3–4 *cm. long and* 2–3 *cm. thick; beaks* 6–11 *mm. long,* hispid, straight or curving, slender. — Low ground, prairies, roadsides and waste places, Minn. to S.D., s. to Mo., Okla., Tex. and Mex.; natzd. eastw. along streams below woolen-mills, etc., in Me., Vt. and N.Y. *Fr.* Sept., Oct.

35. POLÝMNIA L. LEAFCUP

Heads broad, many-flowered; rays several (rarely abortive), pistillate; disk-flowers hermaphrodite but sterile. Phyllaries in two rows; the outer large, spreading; the inner membranaceous, partly embracing the thick achenes. Receptacle flat, membranous-chaffy. Pappus none. — Tall branching N. and S. Am. perennials, viscid-hairy, exhaling a heavy odor. Leaves large, thin, opposite, or the uppermost alternate, lobed, with dilated appendages at the base. Heads in panicled corymbs. Flowers light yellow to whitish, in summer and autumn. (Dedicated to the Muse, *Polyhymnia.*)

1. P. canadénsis L. (Canadian). — *Clammy-hairy,* 0.5–1.5 m. high; lower leaves deeply pinnatifid, the uppermost triangular-ovate and 3–5-lobed or -angled, petioled; heads small; *rays* 5, *obovate or wedgeform, shorter than the involucre,* usually minute or abortive, whitish-yellow, but in forma **radiàta** (Gray) Fassett (var. *radiata* Gray; *P. radiata* Small) 3-lobed and 1 cm. long and whitish; achenes 3-costate, not striate. (*Osteospermum* House) — Moist calcareous woods, ravines and bases of cliffs, w. Vt. to Ont., s. to Ga., Tenn., La. and Okla. Late June–Oct.

2. P. Uvedàlia L. (for ROBERT UVEDALE, 1642–1722, who had the plant in his English garden), BEARSFOOT. — *Roughish-hairy, stout,* 1–3 m. high; leaves broadly ovate, angled and toothed, nearly sessile; the lower palmately lobed, abruptly narrowed into a winged petiole; outer phyllaries very large; *rays* 10–15, *linear-oblong, much longer than the inner phyllaries of the involucre,* yellow; achenes strongly striate. (*Smallanthus* Mackenz.) — Three vars.:

Pedicels and branches of inflorescence densely glandular; slender (hispid-pilose) hairs few or none. . . . . *P. Uvedalia* **(typical).**

Pedicels and branches with many slender (hispid-pilose) hairs more or less concealing the glands.
Glands evident beneath the scattered hairs. . . . . Var. *floridana.*
Glands mostly hidden by the dense canescent pilosity. . . . . . Var. *densipilis.*

P. Uvedàlia (typical). — Rich woods and thickets, N.Y. to Ill., s. to Ga. and Tenn. July–Sept.
Var. **floridàna** Blake (of Florida). — Del. to Fla. and Ala.
Var. **densípilis** Blake (with dense hair). — Mo. to La. and Tex.; se. Va.

36. ACANTHOSPÉRMUM Schrank

Heads small, axillary or subsessile in the forks of the stem. Ray-flowers few, fertile; the ligules small, yellow, usually 3-dentate; the disk-flowers with campanulate yellow 5-toothed corolla, sterile. Involucre double, the outer phyllaries herbaceous; the inner more or less strongly modified, closely enveloping the fertile ray-achenes, muricate or prickly. — Diffuse chiefly trop. Am. annuals with opposite toothed or lobed leaves. (Name from the Greek *acantha, thorn,* and *sperma, seed,* from the prickly fruit formed by the achene and its investing phyllary.)

1. A. AUSTRÀLE (Loefl.) Ktze. (southern). — Sordid-pubescent; leaves ovate or obovate, 1.5–2.5 cm. long, toothed above the middle, and cuneately narrowed at the entire base to a short but slender petiole; bristly fruits 5, stellate-divaricate, 8–10 mm. long. — Waste places, roadsides, etc., Fla. to La., n. to Va., and casually to Mass. July–Oct. (Adv. from Trop. Am.)

37. SÍLPHIUM L. Rosinweed

Heads many-flowered; rays numerous, pistillate and fertile, their broad flat ovaries imbricated in 2–3 rows; disk-flowers apparently perfect but with entire style and sterile. Phyllaries of the broad and flattish involucre broad and with loose leaf-like summits, except the innermost, which resemble the linear chaff of the flat receptacle. Achenes broad and flat, dorsally compressed, surrounded by a wing, usually notched at the top, without pappus or with 2 teeth confluent with the winged margins, the achene and its subtending chaff usually falling together; achenes of the disk sterile and stalk-like. — Coarse and tall perennial N. Am. herbs, with copious resinous juice, and large corymbose-panicled yellow-flowered heads. (*Silphion,* the ancient Greek name of some resinous plant, transferred by Linnaeus to this genus.)*

*a.*Stem conspicuously 4-angled; leaves or their petiolar bases cupulate-connate or -perfoliate.
Stem and phyllaries glabrous; leaves (except the uppermost) contracted to narrowed connate bases. . . . . 1. *S. perfoliatum.*
Stem divergently or retrorsely hirsute; involucre pubescent; leaves (except the basal) not narrowed at the connate bases. . . . . . 2. *S. connatum.*
*a.*Stem terete or only obtusely angled; leaves not connate-perfoliate. . . *b.*
*b.*Leaves entire or variously lobed or toothed, if pinnate or bipinnatifid restricted to lower part of stem; phyllaries foliaceous or, if becoming coriaceous in age, not long-acuminate. . . *c.*
*c.*Well-developed leaves ascending high on the stem, mostly only 1–10 cm. broad, sessile or short-petioled; phyllaries not at all coriaceous, more or less spreading and usually squarrose at tips. . . *d.*
*d.*Wing of achene relatively broad, 2 mm. or more wide at base of the deep (2–4 mm.) sinus, tapering to a margin 1 mm. or more wide. . . *e.*
*e.*Phyllaries loosely spreading, with reflexed tips, inner phyllaries broadly ovate.
Green; stem, peduncles and exposed surfaces of phyllaries rough-hispid. . . . . 3. *S. asperrimum.*
Glaucous; stem, peduncles and phyllaries glabrous. . . . . . 4. *S. speciosum.*
*e.*Phyllaries slightly spreading, with reflexed tips but more approximate, scabrous-pubescent, inner phyllaries narrowly ovate to oblong. 5. *S. integrifolium.*
*d.*Wing of achene narrow, rarely 2 mm. wide at base of shallow (at most 2 mm.) sinus, quickly narrowing to a margin at most 1 mm. wide. . . *f.*
*f.*Peduncles and phyllaries hirsute-hispidulous; tips of chaff glandular. 6. *S. Gatesii.*

---

* Treatment based on that of L. M. Perry in Rhodora, xxxix, 281–297 (1937). In many species fully mature achenes are essential for proper identification.

f. Peduncles and phyllaries glabrous or only finely pubescent. . . *g.*

g. Tips of chaff minutely pubescent, not glandular; outer phyllaries with spreading to reflexed tips. . . *h.*

h. Peduncles and phyllaries glabrous. . . *i.*

i. Principal leaves hastate- or deltoid-lanceolate, with deep sinuses; achenes suborbicular. . . . . 7. *S. brachiatum.*

i. Principal leaves lanceolate to lance-ovate, tapering to base; achenes elliptic or elliptic-obovate.

Petioles long-hirsute; achenes elliptic, excluding wing 9–9.5 mm. long. . . . . 8. *S. atropurpureum.*

Petioles short-ciliate or hispidulous; achenes elliptic-obovate, excluding wing 6–9 mm. long. . . . . . . 9. *S. trifoliatum.*

h. Peduncles and phyllaries finely pubescent. . . . . . . 10. *S. Asteriscus.*

g. Tips of chaff minutely glandular-pubescent; outer phyllaries usually with strongly reflexed tips. . . . . 11. *S. dentatum.*

c. Well-developed leaves chiefly confined to lower part of stem, the blades 1–3 dm. broad, on petioles 1–5 dm. long; phyllaries becoming coriaceous in age, spreading-ascending, rounded at tips.

Basal leaves cordate- or ovate-oblong, regularly toothed; involucre 2.5–4 cm. broad, its phyllaries in about 3 series, the outer only slightly shorter than the 2nd and half as long as the inner. . 12. *S. terebinthinaceum.*

Basal leaves broadly cordate-ovate to reniform, variously divided or lobed or merely coarsely sinuate-dentate; involucre 1–2 cm. broad, its phyllaries in 3–4 series, the inner gradually longer than the outer. 13. *S. compositum.*

b. Leaves deeply pinnatifid or bipinnatifid, extending well up the stem; phyllaries coriaceous in age, long-acuminate. . . . . 14. *S. laciniatum.*

1. S. perfoliàtum L. (with leaf-base surrounding stem), CUP-PLANT. — *Stem* stout, *4-angled, glabrous,* 1–2.5 m. high, often branched above; *leaves* scabrous, ovate, coarsely toothed, 1.5–3.5 dm. long, *the upper united by their connate mostly narrowed bases* and forming a cup, *the lower with connate winged petioles;* heads corymbose; *phyllaries* ovate, *glabrous;* achenes winged and variously notched. — Rich woods, thickets, river-banks and prairies, Ont. to S.D., s. to Ga., Miss., Mo. and Okla.; locally natzd. e. to s. N.E. and Pa. July–Sept.

2. S. connàtum L. (with leaf-bases united). — Similar to no. 1, but *stem divergently or retrorsely hirsute;* only the basal leaves petioled; *cauline leaves not narrowed at the connate bases; phyllaries finely pubescent.* — Alluvial or rich woods, w. Va., W.Va. and w. N.C. June–Aug.

3. S. aspérrimum Hook. (very harsh). — *Stem* terete but corrugated (when dry), *rough-hispid,* 0.6–2 m. high, simple up to the inflorescence; leaves mostly opposite, extending up to inflorescence, entire or slightly dentate, firm, sessile, oblong-ovate, mostly 3–10 cm. broad, rounded or subcordate at base, harsh; *involucres with scabrous loosely spreading herbaceous phyllaries,* the outer narrowly, *the inner broadly ovate; wing of achene 2 mm. or more broad at base of the deep sinus.* — Low sandy ground, Mo., Okla. and Tex. June–Sept.

4. S. speciòsum Nutt. (showy). — Resembling no. 3, *glaucous; stem, peduncles and involucres glabrous;* leaves smooth beneath. — Open or rocky woods and prairies, Neb. to Mo. and Okla. July–Sept.

5. S. integrifòlium Michx. (entire-leaved). — Similar to no. 3; stem nearly or quite glabrous above; leaves essentially all opposite, scabrous on both surfaces; corymb broad; *involucres with rather crowded scabrous phyllaries with reflexed tips, the inner ones narrowly ovate or oblong.* — Dry or sandy prairies, etc., Ind. to e. Kans. and Miss. July–Sept.

Var. **Deàmii** Perry (for CHARLES CLEMON DEAM, 1865–). — *Involucre glandular-pubescent.* — Woods, thickets and prairies, Ind., s. Wisc. and Ia. to Ala., Miss. and Ark.

6. S. Gàtesii Mohr (in honor of HEZEKIAH GATES, ? –1850?). — In habit similar to no. 3; *stem hirsute; leaves alternate* (or some opposite), *narrowly oblong to narrowly ovate,* sessile (or the lower narrow and petioled), 1–5 cm. broad, scabrous-hispid on both surfaces; corymb broad; *peduncles and involucres hirsute-hispidulous;* phyllaries ovate, crowded, spreading-ascending, with squarrose acuminate tips; *pales of receptacle glandular at tip; achenes with very shallow (about 0.2 mm.) sinus, the wing there less than 2 mm. broad and quickly narrowing to less than 1 mm.* — Dry open woods and rocky slopes, Ga. to Mo. and Okla. Late June–Oct.

7. S. brachiàtum Gattinger (with arm-like branches). — *Stem glabrous,* 1–1.5 m. high, with numerous slender branches; *cauline leaves* scabrous above, *hastate- or deltoid-lanceolate, with deep sinuses,* petioled, mostly 1.2–2 dm. long; involucre 0.9–1.5 cm. high, with ovate phyllaries; *achenes suborbicular.* — Limestone-slopes, Cumberland Plateau, Ky., to Ala. July–Sept.

8. S. atropurpùreum Retz. (dark purple). — Habitally like species 2–6; stem 0.8–1.6 m. high, glabrous, or hispid in forma **hirticaùle** Fern. (harsh-stemmed); *leaves mostly in whorls* of 3–5 or opposite (or the uppermost alternate), *with slender villous-hirsute petioles*, the lance-acuminate serrate blades scabrous above and 1–5 cm. broad; corymb lax, with *glabrous peduncles; involucre* 1–2.5 cm. broad, *of glabrous* spreading-ascending lance-oblong outer and narrowly ovate inner *phyllaries;* pales of receptacle not glandular; *achenes elliptic*, excluding the narrow wing 9–9.5 *mm. long.* — Rich woods and banks of streams, local, W.Va. and Va. Late June–Sept.

9. S. trifoliàtum L. (three-leaved). — Leafy to summit, 0.7–3 m. high; stem terete, slender, smooth, simple up to the corymbiform panicle; *leaves in whorls* of 3–5 *or opposite or alternate,* lanceolate to oblong-ovate, *scabrous-hispid on one or both faces, their short petioles glabrous, ciliate or hispid;* peduncles slender, glabrous; involucre much as in no. 7, glabrous, the outer loosely divergent to ascending phyllaries narrowly ovate, the inner erect ones broadly elliptic; *achenes*, excluding wing, 6–9 *mm. long*, the wing 0.5–2 mm. broad, the sinus definite, or sinus wanting and wing less than 0.5 mm. broad in forma **praecìsum** Perry (cut-off). — Woods, thickets and openings, Pa. to Ind., s. to N.C. and (?) Tenn. Late June–Sept.

Var. **latifòlium** Gray (broad-leaved). — *Leaves glabrous*, all or mostly opposite. — O. and Ind. to Ala.; se. S.C.

10. S. Asterìscus L. (old generic name). — Resembling no. 9; leaves essentially all opposite, oblong-lanceolate to -ovate, sharply toothed, thick, harsh on both faces; *peduncles and involucres hispid;* heads rather larger; wings of achene prolonged into acuminate teeth. — Thickets and borders of woods, N.C., (?Va.) to Mo., s. to Ala., Ark. and Okla. June–Sept.

11. S. dentàtum Ell. (toothed). — Resembling nos. 9 and 10; stem glabrous at least below; leaves narrowly ovate, opposite or alternate, entire to coarsely dentate; *phyllaries* glabrous, *the outer* usually *with strongly reflexed tips; tips of chaff minutely glandular-puberulent;* achenes with rounded wing-tips, the sinus 1–2.5 mm. deep and up to 3 mm. broad. (*S. incisum* Greene) — Sandy woods and thickets, Ga. and Ala., n. to N.C. and Ky. Aug., Sept.

12. S. terebinthinàceum Jacq. (like turpentine), Prairie-Dock. — *Stem smooth, slender,* 1–3 m. high, panicled at the summit and bearing several to many large heads, leafless except toward the base; *leaves ovate* and ovate-oblong, somewhat cordate, *serrate-toothed,* thick, rough especially beneath, 3–6 dm. long, on slender petioles; *involucre 2.5–4 cm. broad; its phyllaries in about 3 series, the outer only slightly shorter than the 2nd and half as long as the inner, becoming coriaceous in age, rounded at tip;* achenes narrowly winged, slightly notched and 2-toothed, about 1 cm. long. — Prairies and openings, s. Ont., s. to Tenn. and Mo. Aug.–Oct.

Var. **pinnatifidum** (Ell.) Gray (pinnately cut). — Leaves pinnatifid or lobed. — Ga. and Ala. to O., Ind. and w. Tenn.

13. S. compósitum Michx. (compound). — *Glaucous and wand-like* perennial 1–3 m. high; *basal leaves long-petioled, their* thick *cut-lobed to deeply divided or pinnatifid blades cordate-ovate and* 1–3 *dm. broad;* cauline leaves rapidly reduced up to the loosely forking divergently branched large panicle; *involucre 1–2 cm. broad; its round-tipped* finally coriaceous *phyllaries in 3–4 series, the inner gradually longer than the outer;* achenes roundish-obovate, 6–8 mm. long, the broad wing terminating in a tooth each side of the broad rounded sinus. (Incl. *S. lapsuum* Small, the plants with deeply divided blades like the type of the species) — Dry open woods and clearings, Ga. and e. Tenn., ne. to se. Va. Late June–Sept.

Var. **renifórme** (Raf.) T. & G. (kidney-shaped). — Blades of lower leaves reniform or round-ovate, merely dentate. (*S. reniforme* Raf.) — Local, upland of Va. and N.C.; Coastal Plain, se. Va.

14. S. laciniàtum L. (slashed), Rosinweed, Compass-plant. — *Rough-bristly throughout;* stem stout, 1–3.5 m. high, leafy; *leaves alternate, extending well up the stem, pinnately parted,* petioled but dilated and clasping at the base; *their divisions lanceolate or linear*, acute, *cut-lobed or pinnatifid*, rarely entire; heads few, 0.5–1 dm. broad, sessile or short-peduncled along the naked summit; *phyllaries ovate, tapering into long and spreading rigid points;* achenes broadly winged and deeply notched, 1–4 cm. long. — Prairies, Mich. to N.D., s. to Okla. and Tex. July–Sept. — Lower leaves vertical, 3–9 dm. long, ovate in outline, on the wide open prairies disposed to present their edges north and south; hence the name Compass-plant.

Var. **Robinsònii** Perry (in honor of Benjamin Lincoln Robinson, 1864–1935). — Peduncles and involucres glandular as well as pubescent. — Ky. and w. Ind. to Ala., Miss., La. and Okla.

38. BERLANDIÈRA DC.

With the characters of *Silphium*, but the 5–12 fertile ray-flowers in a single series. Phyllaries in about 3 series, thinner; the inner dilated, obovate, exceeding the disk; the outer smaller and more foliaceous. Achenes without pappus, obovate, neither winged nor notched at the apex, deciduous with the subtending bract and 2–3 scales of the chaff. — Alternate-leaved perennials of the s. and sw. states; heads pedunculate. (Named for *Jean Louis Berlandier*, 1805–1855, a Swiss botanist who collected in Texas and Mexico.)

1. B. texàna DC. (Texan). — Hirsute-tomentose or villous, 6–9 dm. high, very leafy; leaves crenate, the radical oblong, petiolate, the cauline oblong-cordate to subcordate-lanceolate, the upper closely sessile; heads somewhat cymose, 3–4 cm. broad. — Rocky and gravelly woods, thickets and plains, s. Mo., Kans., Okla. and Tex. Late June–Oct.

39. CHRYSÓGONUM L.

Heads many-flowered, radiate; rays about 5, pistillate and fertile; the disk-flowers hermaphrodite but sterile. Involucre of about 5 outer leaf-like oblong phyllaries which exceed the disk, and as many interior shorter and chaff-like concave scales. Receptacle flat, with a linear scale to each disk-flower. Achene obovate, obcompressed, 4-angled, partly inclosed by the short subtending phyllary; pappus a small chaffy 2–3-toothed crown. — Hairy perennial N. Am. herbs with opposite long-petioled leaves, and solitary long-peduncled heads of yellow flowers, nearly stemless when beginning to flower, the stem later elongating in our species. (The Greek name of some plant, composed of *chrysos, golden*, and *gonu, knee*, from the yellow rays and geniculate stem.)

1. C. virginiànum L. (Virginian). — Usually low, at first acaulescent or very short-stemmed, the stem often elongating to 3–6 dm. high; leaves ovate to oblong, crenate-dentate, rarely serrate; rays 1–1.5 cm. long. — Rich woods and shaded rocks, Pa. and W.Va. to Fla. and La. Apr.–June.

40. PARTHÉNIUM L.

Heads many-flowered, inconspicuously radiate; ray-flowers 5, with very short and broad obcordate ligules not projecting beyond the woolly disk, pistillate and fertile; disk-flowers staminate, with imperfect styles, sterile. Involucre hemispherical, of 2 ranks of short ovate or roundish phyllaries. Receptacle conical, chaffy. Achenes only in the rays, surrounded by a slender callous margin, crowned with the persistent ray-corolla. — Leaves alternate. Heads small, corymbed; the flowers whitish. Small N. Am. genus. (An ancient name of some plant from the Greek *parthenos, virgin;* only the pistillate flowers fertile.)

Leaves membranous, pinnately dissected into pinnatifid lobes; annual. . . . . 1. *P. Hysterophorus.*
Leaves firm, with toothed but simple coarse leaves; perennials.
  Stem glabrous or minutely pubescent; leaves glabrous or minutely pubescent beneath. . . . . 2. *P. integrifolium.*
  Stem spreading-hirsute, at least below; lower leaves long-hirsute on veins beneath. . . . . 3. *P. hispidum.*

1. P. HYSTERÓPHORUS L. (old generic name, of meaning similar to that of the generic name), SANTA MARIA. — *Annual*, 3–9 dm. high; *leaves membranaceous, bipinnatifid;* inflorescence a loose and open corymbiform cyme, with slender-stalked heads; pappus of 2 large roundish scales. — Waste ground, Fla. to Tex., etc., locally n. to Mass., Pa., O., Mich., Ill., Mo. and Kans. Aug.–Oct. (Adv. from Trop. Am.)

2. P. integrifòlium L. (entire-leaved), WILD-QUININE. — Perennial, 0.4–1.2 m. high, from a thickened root, non-stoloniferous; *stem glabrous to minutely pubescent;* basal *leaves* long-petioled, their thick blades *harsh above and often beneath with short hairs,* oblong, ovate or obovate, crenate or serrate-dentate, 0.8–3 dm. long; cauline leaves remote, lanceolate to ovate, the upper sessile, all toothed and scabrous; corymb 0.5–2.5 dm. broad, of many erect crowded hemispherical heads; *involucres* 3.5–6 *mm. high,* with canescent ovate phyllaries; pappus a pair of small chaffy teeth or scales. — Dry open woods, thickets and prairies, Ga. to Tex., n. to se. N.Y., W.Va., O., Mich., Wisc. and Minn.; adv. in Mass. June–Sept.

3. P. híspidum Raf. (with straight spreading hairs). — Similar to no. 2, lower, 2.5–8 dm. high; the rhizome sometimes stoloniferous; *stem spreading-hirsute* below or throughout; *leaves* often more deeply, sometimes jaggedly toothed, *at least the lower and median ones long-hirsute beneath,* especially along the veins, the basal blades lanceolate to lance-ovate, the median and upper ones usually not auricled; corymb 0.3–1.5 dm. broad; *involucres* 4–7 *mm. high,* 6–10 *mm. broad.* (*P. repens* Eggert) — Calcareous ledges, barrens and openings, Mo. and Kans. to Ark. and Tex.; adv. along railroads to s. Mich. May–July.

Var. **auriculàtum** (Britt.) Rollins (eared). — Leaves frequently slightly narrowed to auriculate base; involucres 3–5 mm. high and 4–6 mm. broad. (*P. auriculatum* Britt.) — Piedmont and upland Va. and N.C.

41. HELIÓPSIS Pers. Ox-eye

Heads many-flowered; rays 10 or more, fertile. Phyllaries in 2 or 3 rows, nearly equal; the outer foliaceous and somewhat spreading, the inner shorter than the disk. Receptacle conical; chaff linear. Achenes smooth, thick, 4-angular, truncate; pappus none, or a mere border. — Perennial N. Am. herbs, resembling *Helianthus*. Heads showy, peduncled, terminal. Leaves opposite, petioled, triple-ribbed, serrate. Flowers yellow. (Name from the Greek *helios, sun,* and *opsis, appearance,* from the likeness to the sunflower.)

1. H. helianthoìdes (L.) Sweet (like *Helianthus*, sunflower). — Stem smooth or nearly so, 0.3–1.5 m. high, simple or corymbosely branched above; *leaves* ovate-lanceolate to oblong-ovate, slender-pointed, serrate or dentate, rather rounded at base, *smooth or only slightly scabrous;* outer phyllaries of central heads 3–6 mm. broad, with a rigid strongly nerved base; disk 1–1.6 cm. high and 1.7–2.5 cm. broad; ligules linear-oblong, 5–8 mm. broad, marcescent; disk-corollas 4–5 mm. long, with prolonged lobes; *achenes glabrous throughout,* 4–5 mm. long; pappus none or of 2–4 obscure teeth. — Open woods, thickets and dry banks, N.Y. and s. Ont. to Minn., s. to N.C., O., Ind. and Ill.; adv. e. to N.E. July–Sept.

Var. **solidaginoìdes** (L.) Fern. (resembling *Solidago*, goldenrod). — Leaves submembranaceous, cuneate at base; outer phyllaries of central heads 1.5–3 mm. broad; disk 0.7–1 cm. high, 1–1.5 cm. broad; disk-corollas 3–4 mm. long; achenes 3–3.5 mm. long. — Rich woods, Va. and W.Va. to Ga. and Ala. Late June–Sept.

Var. **scàbra** (Dunal) Fern. (rough). — Similar to typical var.; *leaves harshly scabrous above* and somewhat so below, rounded or subcordate at base, acute or acuminate; *achenes pubescent on the angles when young.* (*H. scabra* Dunal) — Dry thickets and openings, N.Y. and Ont. to B.C., s. to Md., Tenn., Ark., Tex. and N.M.; adv. e. to Que. and N.E. June–Sept.

Zínnia (for *Johann Gottfried Zinn*, 1727–1759) élegans Jacq. (elegant), the common Zinnia of gardens, with opposite sessile entire leaves, broad and dry phyllaries, and variously colored persistent ligules, is occasional about refuse but hardly persistent. (Introd. from Trop. Am.)

42. ECLÍPTA L.

Heads many-flowered; rays short; disk-flowers perfect, 4-toothed, all fertile. Phyllaries 10–12, in 2 rows, foliaceous, ovate-lanceolate. Receptacle flat, with almost bristleform chaff. Achenes short, 3–4-sided, or in the disk laterally flattened, roughened on the sides, hairy at summit; pappus none or an obscure denticulate crown. — Annual rough trop. and N. Am. herbs, with slender stems and opposite leaves. Heads solitary, small. Flowers white; anthers brown. (Name from the Greek *ecleipo, to be deficient,* alluding to the absence of pappus.)

1. E. álba (L.) Hassk. (white), Yerba-de-Tajo. — Rough with fine appressed hairs; stems procumbent or ascending, 2–9 dm. high; leaves lanceolate or oblong, acute at each end, mostly sessile, slightly serrate; rays equaling disk. — Alluvial or damp soils, ditches and waste places, Fla. to Tex., n. to L.I. (and as an adventive to Mass. and s. Ont.), Ind., Ill., Ia. and Neb. Aug.–Oct. — A semicosmotrop. type, possibly of more than one species.

43. TETRAGONOTHÈCA L.

Heads many-flowered, radiate; the rays 6–9, fertile. Involucre double, the outer of 4 large and leafy ovate phyllaries, united below by their margins into a 4-angled or -winged cup; the inner of small chaffy phyllaries as many as the ray-flowers and partly clasping their achenes. Receptacle convex or conical, with narrow and membranaceous chaff. Achenes very thick, obovoid, flat at the top; pappus none. — Erect N. Am. perennials, with opposite coarsely toothed sessile sometimes connate leaves, and large single heads of pale yellow flowers on terminal peduncles. (Name from the Greek *tetragonos, four-angled,* and *thece, a case,* from the shape of the involucre.)

1. T. helianthoìdes L. (resembling *Helianthus*, sunflower). — Villous and somewhat viscid, 3–7 dm. high, simple; leaves ovate or rhombic-oblong, sessile by a narrow base; phyllaries and rays 2–3 cm. long. — Sandy woods and fields, Fla. to Miss., n. to Va. and Tenn. June, July.

44. RUDBÉCKIA L. Coneflower

Heads many-flowered, radiate; the rays neutral. Phyllaries foliaceous, in about 2 rows, spreading. Receptacle conical or columnar; the short chaff concave, not rigid. Achenes 4-angled (in our species), smooth, not margined, flat at the top, with no pappus, or a minute crown-like border. — Chiefly perennial N. Am. herbs, with alternate leaves and showy terminal heads; the rays generally long, yellow, often darker at base. (Named in honor of the Professors *Rudbeck*: *Olaf*, 1630–1702, the father, and *Olaf*, 1660–1740, the son, predecessors of Linnaeus at Uppsala.)*

*a.*Hirsute or more or less pubescent to glabrous (no. 12) perennials or biennials usually developing basal leafy tufts (nos. 15 and 16 hirsute annuals without basal leafy tufts); phyllaries more than 1 cm. long (if shorter, pubescent); disk hemispherical to elongate; achenes 4-angled, with or without pappus. § Eurudbeckia. . . *b.*

*b.*Pales not attenuate to bristle-tips; style-tips blunt; achenes with a crown of pappus. . . *c.*

*c.*Pales obtuse to deltoid-tipped or mucronate or acuminate, not long-cuspidate; perennials or biennials developing leafy basal offshoots. . . *d.*

*d.*Leaves (or at least the lower) 3–7-cleft or -divided.

Stems and branches glabrous; leaves glabrous to sparsely hispid; achenes oblique at base, sublaterally attached, flattened. . . . 1. *R. laciniata.*

Stems, branches and lower leaf-surfaces densely pilose; achenes subtruncate at base, centrally attached, equilaterally quadrate. . . 2. *R. subtomentosa.*

*d.*Leaves simple or merely lobed. . . *e.*

*e.*Stems or leaves pubescent; disk hemispherical, globose or ovoid-conic, in maturity 0.5–2.5 cm. high. . . *f.*

*f.*Pales canescently puberulent above the middle; lobes of disk-corollas soon reflexed.

Barely scabridulous or smooth; disk 6–13 mm. high; ligules 1.5–2.7 cm. long, rotately spreading; pappus essentially obsolete. 3. *R. Heliopsidis.*

Harshly scabrous; disk 1.3–2.3 cm. high; ligules 4–7 cm. long, soon reflexed; pappus a conspicuous crown. . . . . . 4. *R. grandiflora.*

*f.*Pales glabrous, sparsely pilose or ciliate at summit; lobes of disk-corollas ascending; ligules 1–4 cm. long; achenes subequally quadrate. . . *g.*

*g.*Summits of all pales prominently ciliate; basal leafy offshoots erect, subsessile. . . *h.*

*h.*Basal and lower cauline leaves ovate, broadly rounded to subcordate at base, 3.5–5 cm. broad; ligules 1.5–3 cm. long. . 5. *R. umbrosa.*

*h.*Basal and lower cauline leaves lanceolate or oblanceolate to narrowly obovate or narrowly elliptic, gradually tapering at base, 0.5–4.5 cm. broad; ligules 1–2(–2.5) cm. long. . . *i.*

*i.*Lower surfaces of leaves and upper surfaces of at least some phyllaries strigose-hispid.

Middle and upper internodes and bases of leaves spreading-hirsute; basal and petioled cauline leaves 2–4.5 cm. broad; involucre 1–2.2 cm. long, its larger phyllaries 2.5–7 mm. broad. . . . . 6. *R. fulgida.*

Middle and upper internodes glabrous to strigose-hispid; bases of leaves appressed-short-strigose; basal and petioled cauline leaves 0.5–2 cm. broad; involucre 5–9 mm. long, its larger phyllaries 1–2 mm. broad. . . . 7. *R. spathulata.*

*i.*Lower surfaces of leaves and upper surfaces of phyllaries glabrous or essentially so.

Basal leaves narrowly elliptic-ovate, mostly 2–4 cm. broad; phyllaries grayish-pilose on back, not ciliate. . 8. *R. palustris.*

Basal leaves elongate-lanceolate, mostly 1–1.5 cm. broad; phyllaries bristly-ciliate, otherwise glabrous. 9. *R. acuminata.*

*g.*Summits of pales eciliate (or lowermost pales sometimes ciliate), erose or entire, the backs glabrous or pilose. . . *j.*

*j.*Leafy basal tufts terminating horizontal stolons; their blades oval to elliptic-lanceolate, pointed, dentate; cauline leaves remote, coarsely dentate, acuminate. . . *k.*

*k.*Outer pales blunt, not subulate-tipped, their backs glabrous; ligules 12–21, 2–4 cm. long.

* When the types of species long ago proposed are critically studied, some of the recent names here tentatively accepted may be changed. In the figures the leaves are shown × ca. ⅓, a disk-flower and its pale × ca. 3.

Stem sparsely hirsute; cauline leaves obviously reduced in breadth to summit of stem; upper faces of outer phyllaries glabrous or only sparingly pubescent. . . . 10. *R. speciosa.*

Stem densely and retrorsely villous-hirsute; cauline leaves scarcely reduced upward; upper faces of outer phyllaries densely pubescent. . . . . 11. *R. Deamii.*

k. Outer pales subulate-tipped, their backs pubescent; ligules 6–14, 0.5–2 cm. long; stem and lower leaf-surfaces strigose with short trichomes 0.5–1 mm. long; basal leaves elliptic-lanceolate, 1–2 cm. broad. . . . . 12. *R. tenax.*

j. Leafy basal tufts erect, subsessile; their blades linear- to lance-spatulate, long-villous, entire; cauline leaves numerous, essentially uniform, linear to narrowly oblong, obtuse or acute, not acuminate, entire. . . . . 13. *R. missouriensis.*

e. Stems and coriaceous leaves glabrous; disk columnar, in maturity 4–8 cm. long. . . . . 14. *R. maxima.*

c. Pales prolonged into a long sharp glabrous cusp; lower leaves often lobed; plant annual or biennial, without basal leafy offshoots. . . . . 15. *R. triloba.*

b. Pales attenuate, bristle-tipped and bristly-ciliate; style-tips slender-subulate; achenes without pappus.

Annual, from a tap-root, without basal tufts of leaves; cauline leaves nearly uniform, oblong to oblong-spatulate or narrowly obovate, obtuse, mostly sessile, the lower only 3–8 cm. long. . . . . 16. *R. bicolor.*

Perennial, with basal tufts of longer linear-lanceolate to ovate or obovate petioled leaves; cauline leaves decreasing in size upward, usually acute or acuminate.

Basal leaves ovate to rhombic-oval, 2.5–7 cm. broad; cauline leaves mostly with ovate blades, the lowest sessile ones (2–)2.5–6.5 cm. broad, usually coarsely toothed; indigenous plant of woods and thickets. . . . . 17. *R. hirta.*

Basal leaves oblanceolate, 1–3(–5) cm. broad; cauline leaves from linear-lanceolate to lance-elliptic or oblanceolate, the lowest sessile ones 0.5–2(–3) cm. broad, entire or nearly so; aggressive weed of woods, fields and clearings. . . . . 18. *R. serotina.*

a. Glabrous annual; leaves clasping; phyllaries 5–10 mm. long, glabrous (except for ciliate margin); disk cylindric in age; achenes terete, without pappus. § Dracopsis. . . . . 19. *R. amplexicaulis.*

§ Eurudbéckia Gray

1. R. laciniàta L. (slashed). — Coarse perennial 0.5–3 m. high; *stem glabrous,* glaucous; *leaves* smooth or harsh; *the lower* petioled, *pinnate, with 5–7-cut or 3-lobed leaflets;* the upper similar but sessile, or irregularly 3–5-parted or uncleft; heads long-peduncled; involucre of foliaceous phyllaries; ligules oblanceolate, yellow, 2–6 cm. long, soon reflexed; *disk dull greenish-yellow,* at first hemispherical to globular, 1.3–2.5 *cm. broad, in fruit becoming columnar* and 1.5–3 cm. long; pales truncate, downy at summit; *achenes flattened, oblique at base, sublaterally attached,* 5–6 mm. long. (Including *R. ampla* Nels.) — Rich low ground, w. Que. to Mont., s. to N.S., N.E., n. Fla., La., Tex., N.M. and Ariz. July–Sept. Fig. 1726.

1726. R. laciniata (typical).

Var. **horténsia** Bailey (cultivated), Golden-glow. — Ligules greatly increased in number, producing a full double head. — Spreading from cult.

Var. **digitàta** (Mill.) Fiori (digitate). — More *slender* and lower (0.7–1.5 m. high); *leaves thin, smooth or merely scabrous, the basal and lower cauline with pinnae cut into narrowly lanceolate to linear segments; disks* only 0.7–1.5 *cm. long; ligules* 1.5–3.5 *cm. long; achenes* 3.5–5 *mm. long.* — Low woods and swamps, e. Md. to Ga. Aug.–Oct.

Var. **hùmilis** Gray (humble). — Differing from the last in its thin *lower leaves ovate, all or nearly all uncleft, or but slightly cleft into* 1 or 2 *pairs of broad undivided segments; upper petioled leaves mostly simple and ovate or ovate-elliptic.* — Open woods, mts. of Va. and Ky. to Ga. and Tenn.; local on Coastal Plain, se. Va. Fig. 1727.

1727. R. laciniata, v. humilis.

2. R. subtomentòsa Pursh (somewhat tomentose), SWEET C. — Stem branching above, 0.5–1.5 m. high, *downy*, as are the petiolate ovate or ovate-lanceolate serrate leaves beneath; lower blades (sometimes all) deeply 3-cleft; involucre and receptacle anise-scented; *disk* hemispherical, becoming *globular or conical*, dull brown, or yellowish in forma **Craigii** (Sherff) Fern. (named in 1912 in honor of MOSES CRAIG); *achenes subtruncate at base, centrally attached, equilaterally quadrate.* — Prairies and low grounds, Ind., Wisc. and Ia., s. to La. and Tex. Aug., Sept.

3. R. Heliópsidis T. & G. (from resemblance to *Heliopsis*). — Rhizomes stout, with crowded branches terminating in leaf-tufts or flowering stems; *basal leaves with oblong-ovate* firm *3-ribbed* smooth or slightly scabridulous dentate *blades* 3–13 cm. long, on smoothish *petioles three to five times their length;* stems smooth except at base, or villous below; petioles villous in forma **víllipes** Fern. (hairy-stalked), 6–9 dm. high, stiffly ascending; *cauline leaves* similar to basal, *all but the upper long-petioled;* heads solitary on few long erect branches; involucre much shorter than the disk; *ligules* 1.5–2.7 *cm. long, rotately spreading; disk* dark brown, subglobose, 6–13 *mm. high,* its crowded corollas with finally recurved lobes; achenes 4-angled, about 3 mm. long, *pappus almost obsolete; chaff canescent on back.* — Pine- and oak-woods and thickets, very local, se. Va.; Ga. and Ala. Late July–early Sept.

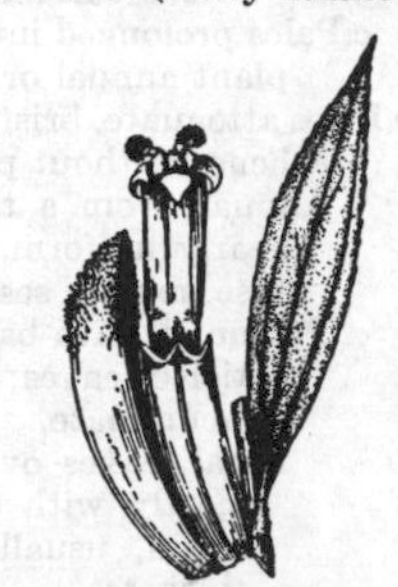
1728. R. grandiflora.

4. R. grandiflòra (D. Don) DC. (large-flowered). — *Harsh* throughout, 0.5–1 m. high; *leaves* oval, all but the uppermost *long-petioled, rigid,* 1–2.5 dm. long, distinctly 3–5-ribbed; heads mostly single, long-peduncled; involucre canescent-strigose; *ligules* yellow, 4–7 *cm. long, soon reflexed;* disk hemispherical, becoming conical, 1.3–2.3 cm. high; *pales canescently puberulent above the middle; disk-corollas* dull purple, *their lobes soon reflexed; achenes flattened,* with conspicuous crown. — Dry open ground, Mo. and Okla. to La. and Tex. July, Aug. FIG. 1728.

5. R. umbròsa C. L. Boynt. & Beadle (shady). — Perennial with short thick rhizome and ascending basal tufts; stem 0.4–1 m. high, striate-angled; *basal and lower cauline leaves* slender-petioled, *ovate, broadly rounded to subcordate at base,* 3.5–5 *cm. broad,* coarsely dentate; upper leaves much reduced; heads long-peduncled; involucre foliaceous; *ligules* 8–12, yellow or orange, 1.5–3 *cm. long;* disk dark purple, hemispheric to conic, 1–1.5 cm. broad; *pales ciliate* at the broad summit; achenes with prominent crown. — Low woods and bottoms, Va. to O., Ind. and Mo., s. to Tenn. and Ark. Aug.–Oct. FIG. 1729.

1729. R. umbrosa.

6. R. fúlgida Ait. (shining). — Similar to no. 5; *stem* more or less *villous-hirsute,* 4.5–12 dm. high; *basal and lower cauline leaves oblanceolate, narrowly obovate or narrowly elliptic, cuneate* at base, 2–4.5 *cm. broad, their lower surfaces strigose-hispid, their bases spreading-hirsute; involucres* reflexed, 1–2.2 *cm. long, their larger phyllaries* 2.5–7 mm. broad, *with inner surfaces of some or all strigose-hispid; ligules* 0.8–2 *cm. long;* disk 1.1–1.8 cm. broad; *pales glabrous* except for the ciliate margin. — Dry or moist, open or shaded places, N.J. to Ind., s. to e. Va., w. N.C. and Ky. Aug.–Oct. FIG. 1730.

1730. R. fulgida.

7. R. spathulàta Michx. (spatula-like). — Resembling no. 6, more slender, 2–10 dm. high; *stem* simple or with erect branches, *glabrous or minutely strigose-hispid; basal and lower petioled cauline leaves* oblanceolate to narrowly lance-obovate, 0.5–2 *cm. broad; upper leaves* sessile, oblanceolate to narrowly oblong, *minutely strigose on both surfaces; involucres* 5–9 *mm. long, their broader* acute *phyllaries only* 1–2 *mm. broad;* disk 1–1.4 cm. broad. — Low woods, meadows and clearings, Va. to Ga. and Ala. Aug., Sept.

8. R. palústris Eggert (of marshes). — Similar to no. 6; basal leaves narrowly elliptic-ovate, mostly 2–4 cm. broad; *lower leaf-surfaces glabrous or nearly so; upper faces of phyllaries glabrous, their backs grayish-pilose;* ligules 1.5–2.5 cm. long. — Damp or wet places, Ind. to Mo. and Tenn. July–Sept. FIG. 1731.

9. R. acuminàta C. L. Boynt. & Beadle (acuminate). — More slender than no.

8, 5–8 dm. high; stem sparsely strigose-hirsute; *lower* petioled *leaves elongate-lanceolate, mostly* 1–1.5 *cm. broad;* upper sessile leaves similar; heads 1–few; *phyllaries bristly-ciliate,* otherwise glabrous. — Woods and banks of streams, Ky. and Tenn. July–early Sept.

1731. R. palustris.

1732. R. speciosa.

10. R. speciòsa Wenderoth (showy). — *Stem sparsely hirsute,* from a slender elongate rhizome, up to 1 m. high, loosely branched; *basal leafy tufts terminating slender horizontal stolons, their blades broadly lanceolate to oval,* coarsely toothed, long-petioled; *cauline leaves remote, much reduced upward,* lanceolate, mostly tapering to both ends, coarsely toothed or lacerate; *upper faces of outer phyllaries glabrous or only sparingly pubescent; pales glabrous on the back, mostly without ciliation, the outer ones blunt; ligules* 12–21, 2–4 *cm. long;* achenes rather large, curved, equally quadrangular, with a short crown. — Woods and bottomlands, local, N.Y. to Ga., Ala. and Mo.; adv. in Ct. Late July–Sept. Fig. 1732.

Var. **Sullivántii** (C. L. Boynt. & Beadle) Robins. (for its discoverer, William Starling Sullivant, 1803–1873). — Cauline leaves ovate, less incised. (*R. Sullivantii* C. L. Boynt. & Beadle) — Swamps, damp shores, etc., O. and Mich. to Mo., s. to Ala. and Ark. Aug., Sept.

11. R. Deàmii Blake (for its discoverer, Charles Clemon Deam, 1865–). — Resembling no. 6; *stem densely and retrorsely villous-hirsute;* leaves of stolons ovate or oval, often thin; *cauline leaves scarcely reduced upward,* the lower like those of the stolons, the upper ovate and subamplexicaul; *upper faces of outer phyllaries densely pubescent.* — Wooded ridges and banks of streams, Ind. Aug.–Oct. Fig. 1733.

1734. R. tenax.

1733. R. Deamii.

12. R. tènax C. L. Boynt. & Beadle (tenacious). — *Loosely stoloniferous;* stems 3–7.5 dm. high, smooth or *strigose-hispid,* simple or sparingly forked, with long-peduncled heads; *basal* leafy tufts terminating horizontal stolons, long-petioled; their *blades lanceolate,* 1–2 *cm. broad,* dentate, prominently 3-nerved, *minutely strigose-hispid beneath with trichomes* 0.5–1 *mm. long;* cauline leaves lanceolate, not rapidly reduced upward; upper faces of phyllaries strigose-hispid; *ligules* 6–14, 0.5–2 *cm. long; pales* eciliate except at summit, *the outer subulate-tipped, pubescent on the back.* — Dry woods and clearings, Ind. and Ill., s. to Ala. July, Aug. Fig. 1734.

13. R. missouriénsis Engelm. (of Missouri). — Perennial, from a *thick rhizome, bearing erect leafy tufts;* stem 4–8 dm. high, villous-hirsute, usually branching, with several leafy peduncles; *leaves of basal tufts linear- to lance-spatulate, long-villous, entire; cauline leaves numerous, essentially uniform, linear to narrowly oblong, obtuse to acute,* entire; involucres villous; ligules 1.5–2.5 cm. long; pales glabrous, with deltoid summit; achenes with a minute crown. — Rocky prairies, glades and knobs, Mo. to La. and Tex. Late July–Oct. Fig. 1735.

1735. R. missouriensis.

14. R. máxima Nutt. (very large). — Very robust, 1–3 m. high, glaucous, *glabrous; leaves coriaceous, glabrous and glaucous,* entire or repand, ovate-lanceolate to oblong or rounded, obtuse, 0.6–3 dm. long, *the upper clasping;* heads long-peduncled; involucre of leathery oblong phyllaries; ligules 1.5–4 cm. long; *disk elongating, becoming subcylindric and* 4–8 *cm. long;* pales pubescent at summit, obtuse; achenes compressed. — Waste ground and open places, w. Mo. Late June–Aug. (Adv. from farther southwest) Fig. 1736.

15. R. trìloba L. (three-lobed). — *Annual or biennial,* usually *without leafy tufts* from the base of flowering stem, rarely short-lived perennial; stems 0.3–1.6 m. high, hispid to glabrous, mostly with leafy branches; *leaves* of young plants sometimes simple and cordate, *often divergently 3-lobed or 3-parted or even pinnately 5–7-parted;* lower cauline leaves similar to the

1736. R. maxima.

juvenile; the middle and upper ovate to lanceolate, coarsely toothed or entire, acuminate; heads short-peduncled; involucre foliaceous, soon reflexed; rays 6–12, deep yellow or with the base orange to brown; disk black-purple (rarely pale), depressed, becoming ovoid; *pales* glabrous, concave, *acuminate into a long smooth cusp.* — Highly variable; we have 3 marked extremes:

Lower cauline leaves 3-parted or 3-lobed; phyllaries lanceolate to lance-ovate, sparingly strigose-hispid on the back.
  Ligules 0.6–2 cm. long, yellow or parti-colored; mature disk 5–13 mm. broad. *R. triloba* (typical).
  Ligules 2–3.5 cm. long, orange or orange and yellow; mature disk 1.3–2 cm. broad. . . . . Var. *rupestris.*
Lower cauline leaves often 5–7-parted; phyllaries linear or linear-lanceolate, spreading-villous; ligules 1–2 cm. long; mature disk 5–13 mm. broad. . . . Var. *Beadlei.*

R. tríloba, typical. — Open woods, thickets and fields, e. N.Y. to Minn., s. to Ga., Tenn., Ark. and Okla.; adv. or natzd. in N.E. June–Oct. FIG. 1737.

Var. **rupéstris** (Chickering) Gray (of rocky places), *R. rupestris* Chickering. — Rocky slopes, mts. of N.C. and Tenn., n. in less extreme development to D.C. and Ky.

Var. **Beàdlei** (Small) Fern. (for CHAUNCEY DELOS BEADLE, 1866–). — Dry or moist soil, D.C. to Ia., s. to N.C., Ky. and Okla.

1737. R. triloba.

16. R. bícolor Nutt. (two-colored), NIGGER-TEATS. — *Annual* from a tap-root; hirsute or hispid stem 2–9 dm. high, simple or with erect branches; *leaves nearly uniform, oblong to oblong-spatulate or narrowly obovate, obtuse, mostly sessile, the lower only* 3–8 *cm. long;* phyllaries linear-oblong to lanceolate, hirsute; ligules yellow, or purple or brown at base, 1.5–2.5 cm. long; disk blackish, hemispherical, becoming conical; *pales* acuminate or attenuate, *bristle-tipped* and bristly-ciliate; *style-tips slender-subulate; pappus none.* — Dry soil, Ala. to Tex., n. to s. Ind., s. Ill., s. Mo. and Okla. June–Sept. FIG. 1738.

1738. R. bicolor.

17. R. hírta L. (rough). — Perennial with *basal leaves ovate to rhombic-oval,* 2.5–7 *cm. broad; cauline leaves mostly with ovate blades; the lowest sessile or broadly subpetiolate ones* (2–)2.5–6.5 *cm. broad, usually coarsely toothed;* stems 0.3–1 m. high, with long-peduncled heads; involucre with strongly foliaceous outer *phyllaries, the inner ones linear or linear-lanceolate;* ligules orange-yellow, 2–4 cm. long; *pales reaching summit of corolla-throat, with attenuate bristly-ciliate apex; constricted tube of corolla half as long as throat; style-tips slenderly subulate, widely divergent; pappus none.* (Var. *monticola* (Small) Fern.; *R. monticola* Small) — Open woods, thickets, barrens and fields, w. Mass. to Ill., s. to Ga. and Ala. June–Oct. FIG. 1739.

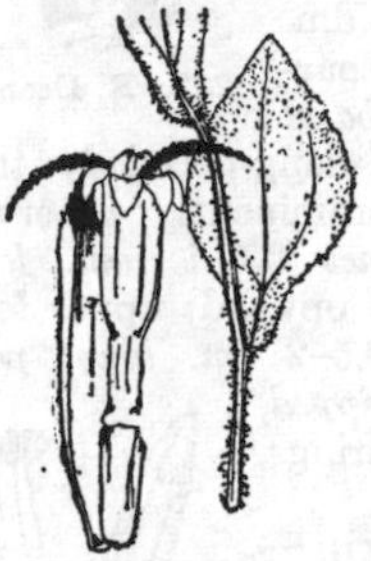

Var. **Brittònii** (Small) Fern. (for NATHANIEL LORD BRITTON, 1859–1934). — Inner phyllaries oblong. (*R. Brittonii* Small) — Md. to W.Va. and Ala.

1739. R. hirta.

18. R. SERÒTINA Nutt. (late), BLACK-EYED SUSAN, NIGGER-HEADS, YELLOW DAISY. — Differing from no. 17 in its *oblanceolate basal leaves* 1–3 (rarely –5) *cm. broad; cauline leaves linear-lanceolate to lance-elliptic or oblanceolate;* the *lower sessile ones* 0.5–2(–3) *cm. broad, entire or nearly so; heads short- to long-peduncled;* ligules orange-yellow (or variously verging to red or green forms), 1–5 cm. long; *pales shorter than throat of corolla, the bristly ciliation extending lower down; corolla-tube about one-fifth as long as throat; style-tips ascending.* (*R. hirta* sensu most auth., not L.) — A most aggressive weedy species chiefly originating west of our range and, since the opening up of the land, rapidly spreading eastw. to our extreme limits and in its newly invaded area giving rise to an indefinite number of sports and freakish forms. — The following varieties are recognized:

Heads definitely peduncled, the 1st or leading head of the primary axis on a naked peduncle becoming 6–35 cm. long (one-fifth to half the height of plant) above the uppermost well-developed leaf.
  Ligules 1 (or less)–3.5 cm. long.
    Pubescence of lower leaf-surfaces variously spreading, with broad open glabrous areas between the conspicuous green bulbous bases of the trichomes. . . . . *R. serotina* (typical).

Pubescence of both faces of leaves closely appressed (or chiefly so), the crowded hairs mostly parallel with the midrib, with minute or obscure pale enlarged bases. . . . . Var. *sericea.*
Ligules becoming 3.5–5 cm. long. . . . . Var. *lanceolata.*
Heads subcorymbose, on short erect branches, the central head on a peduncle only 1–5 cm. long (one-sixtieth to one-fifteenth height of plant); leaves as in typical var. but these and stems more villous; ligules 1–2 cm. long. . . . . Var. *corymbifera.*

R. SERÒTINA (typical), *R. flava* T. V. Moore. — Fields and clearings, nearly ubiquitous. June–Oct. (Natzd. from the Great Plains and beyond) FIG. 1740. — The following anomalous sports have been named (many others doubtless occur): forma **tubulifórmis** (Burnham) Fern. & Schub. (formed like tubules) with the orange-yellow ligules tubular, at least below; forma **rùbra** (Clute) Fern. & Schub. (red) with ligules entirely red; forma **pulchérrima** (Farw.) Fern. & Schub. (very handsome) with a red or brown spot at the base of each ligule; forma **annulàta** (Clute) Fern. & Schub. (ringed) with the orange-yellow ligules tipped with red; forma **viridiflòra** (Burnham) Fern. & Schub. (green-flowered) with ligules green; forma **homóchroma** (Steyerm.) Fern. & Schub. (with uniform coloring) with disk yellow like the ligules; forma **pleniflòra** (Moldenke) Fern. & Schub. (full of flowers) with several to many disk-flowers changed to ligules. In heads attacked by *Cecidiomyia Rudbeckiae* the disk-flowers and pales change to foliaceous green tufts.

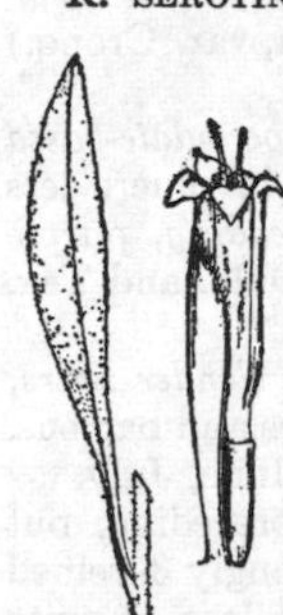
1740. R. serotina.

Var. SERÍCEA (T. V. Moore) Fern. & Schub. (silky), *R. sericea* T. V. Moore; *R. hirta,* var. Fern. — Natzd. as a weed eastw. to e. Que., N.B., N.S., N.E. and Pa.

Var. **lanceolàta** (Bisch.) Fern. & Schub. — Similar to typical var., but with *ligules* 3.5–5 *cm. long.* — Open woods and wooded bluffs, Ill. and Mo., apparently there indig.; spread eastw. to fields of Nfld., Maritime Provinces and Atlantic states.

Var. **corymbífera** (Fern.) Fern. & Schub. (bearing corymbs), *R. hirta,* var. Fern. — Fields and clearings, se. Va., seemingly a recently established but constant var.

1741. R. amplexicaulis.

§ DRACÓPSIS (Cass.) Gray

19. R. amplexicaùlis Vahl (clasping the stem). — *Glabrous annual,* glaucous, 2–8 dm. high, simple or loosely branching; *leaves* 1-ribbed, entire, sinuate or dentate, ovate or oblong, *cordate-clasping;* heads long-peduncled; *phyllaries linear,* 5–10 *mm. long, glabrous,* ciliate; ligules yellow, often with brown bases, soon reflexed, 1–2.5 cm. long; *disk becoming columnar,* 1.5–3 cm. long, brownish-olive; *achenes nearly terete,* striate, *without pappus.* (*Dracopsis* Cass.) — Low grounds, Ga. to Tex., n. to Mo. and Kans. June. FIG. 1741.

45. ECHINÀCEA Moench PURPLE CONEFLOWER

Heads many-flowered; rays mostly drooping, pistillate but sterile. Phyllaries imbricated, lanceolate, spreading. Receptacle conical, the lanceolate carinate spiny-tipped chaff longer than the disk-flowers. Achenes thick, short, 4-sided; pappus a small toothed border. — Perennial N. Am. herbs with stout and nearly simple stems naked above and terminated by a single large head; leaves chiefly alternate, 3–5-nerved. Rays rather persistent; disk purplish. (Name from the Greek *echinos, sea-urchin,* from the sharp-pointed pales.) BRAUNERIA sensu Britt., not Neck.

Leaves broadly to narrowly ovate, rounded at base; pales of receptacle linear or narrowly lanceolate, 1–1.5 mm. broad, with flexible tips; ligules purple.
Leaves scabrous above, not glaucous; pales 1–1.3 cm. long, with straight tips. 1. *E. purpurea.*
Leaves smooth and glaucous; pales 9 mm. long, with incurved tips. . . . 2. *E. laevigata.*
Leaves lanceolate to lance-linear, attenuate to base; pales lanceolate to narrowly ovate, mostly broader, tapering to rigid tips.
Hirsute; ligules purple (or white).
Plant 1–5 dm. high, with tuberculate-based hairs; ligules 2–3.5 cm. long, spreading. . . . . 3. *E. angustifolia.*
Plant 0.6–1 m. high, with slender hairs; ligules 4–9 cm. long, strongly reflexed. . . . . 4. *E. pallida.*
Glabrous or strigose-pubescent; ligules yellow. . . . . 5. *E. paradoxa.*

1. E. purpùrea (L.) Moench (purple). — Stem smooth, or in one form rough-bristly; *leaves* rough, often serrate; the *lowest ovate, 5-nerved,* veiny, long-petioled; the *others ovate-lanceolate;* involucre imbricated in 3–5 rows; ligules 15–20, dull purple (rarely whitish), 2.5–5.5 cm. long; *pales* lanceolate, 1–1.3 *cm. long,* 1–1.5 mm. broad, *with straight tips.* (*Brauneria* Britt.) — Dry open woods and prairies, Ga. to La., n. to Va., O., Mich., Ill. and Ia.; casually adv. northeastw. June–Oct.

2. E. laevigàta (C. L. Boynt. & Beadle) Blake (smooth). — Similar to no. 1, *glabrous and with quite smooth and glaucous foliage;* ligules 3.5–8 cm. long; *pales* almost linear, 9 *mm. long,* 1 mm. broad, *with incurved tips.* (*Brauneria* C. L. Boynt. & Beadle; *E. purpurea,* var. Cronq.) — Woods and fields, s. Pa. to Ga.

3. E. angustifòlia DC. (narrow-leaved). — Low, 1–5 *dm. high, hirsute with tuberculate-based hairs; leaves lanceolate and linear-lanceolate, attenuate at base,* 3-nerved, *entire;* involucre less imbricated and heads often smaller than in no. 1; *ligules* 2–3.5 *cm. long, spreading,* purple or white. (*Brauneria* Heller) — Dry prairies and barrens, Minn. to Sask., s. to Okla. and Tex. May–Aug.

4. E. pállida Nutt. (pale). — Taller than no. 3, 0.6–1 *m. high, hirsute with slender hairs; ligules slender and drooping,* 4–9 *cm. long,* 2-toothed. (*Brauneria* Britt.) — Prairies and barrens, Mich. to Neb., s. to Ala., La. and Tex.; casual in fields to the Atlantic states. June, July.

5. E. paradóxa (Norton) Britt. (unusual). — In habit similar to the two preceding, but *nearly glabrous,* 5–8 dm. high; the narrowly lance-linear somewhat rigid and strongly 3-veined leaves 1–2 dm. long, 0.8–3.5 cm. wide, scabrous on the margins, sparingly strigillose or quite smooth on the surfaces; *rays* drooping, 3–7 cm. long, *yellow.* (*Brauneria* Norton) — Prairies, glades and knobs, Mo. and s. Kans. to Tex. June.

46. RATÍBIDA Raf. PRAIRIE-CONEFLOWER

Heads many-flowered; the rays few, neutral. Phyllaries few and small, spreading. Receptacle columnar; the chaff truncate, thickened and bearded at the tip, partly embracing the flattened and margined achenes. Pappus none or of 2 teeth. — Perennial N. Am. herbs, with alternate pinnately divided leaves; the grooved stems or branches naked above, bearing single generally showy heads. Rays yellow or parti-colored, drooping; disk grayish. (Meaning, like most work of its author, not clear.) LEPACHYS Raf., a later name.

1. R. pinnàta (Vent.) Barnh. (pinnate). — Hoary with minute appressed hairs, slender, 0.5–1.5 m. high, branching; leaflets 3–7, lanceolate, acute; disk ellipsoid, much shorter than the large (5 cm. long) and drooping light yellow rays. (*Lepachys* (Vent.) T. & G.) — Dry soil, s. Ont. and w. N.Y. to Minn. and Neb., s. to Ga., Ark. and Okla.; sometimes a casual waif eastw. June–Sept. — The bruised fresh receptacle exhales an anisate odor.

2. R. columnífera (Nutt.) Wooton & Standl. (bearing columns). — Branching from base, 3–8 dm. high; leaflets 5–9, oblong to narrowly linear, entire or 2–3-cleft; disk columnar, often 3 cm. long or more; rays as long or shorter, yellow, or in forma **pulchérrima** (DC.) Fern. (very handsome) in part or wholly brown-purple. (*Lepachys columnaris* (Pursh) T. & G.; *R. columnaris* (Pursh) D. Don) — Dry plains, prairies and ravines, Alta. to Mex., e. to Man., Minn., Ill., Mo., Ark. and Tex.; adv. and occasionally natzd. e. to N.E. June–Sept.

47. SPILÁNTHES Jacq.

Heads small, many-flowered; rays, when present, fertile. Phyllaries few, loose. Receptacle elongated, columnar; chaff conduplicate, enwrapping the achenes. Ray-achenes 3-angled or obcompressed; disk-achenes somewhat compressed, with acute margins continued into setiform awns, or the pappus none. — Slender spreading or depressed herbs of trop. and warm-temp. Am., with opposite leaves and ovoid-conical pedunculate heads. Rays yellow or white. (Name from the Greek *spilos, a stain,* and *anthos, flower.*)

1. S. americàna (Mutis) Hieron. (American), var. **rèpens** (Walt.) A. H. Moore (creeping). — Pubescent or glabrous, decumbent or loosely ascending; leaves elliptic-ovate to lanceolate, 2–9 cm. long, petioled, strongly but equally toothed; peduncles 3–12 cm. long; heads 9–16 mm. in length. (*S. repens* Michx.) — Low moist places, Fla. to Tex., n. to N.C., s. Ill. and Mo. July–Oct.

48. BORRÍCHIA Adans. SEA-OX-EYE

Heads many-flowered; rays fertile. Phyllaries of the hemispherical involucre imbricated. Receptacle flat, covered with lanceolate rigid and soon deciduous chaff. Achenes somewhat cuneiform, 3–4-angled; pappus a short 4-toothed crown. — Shrubby low maritime N. and S. Am. plants, coriaceous or fleshy, with opposite nearly entire leaves, and solitary peduncled terminal heads of yellow flowers; anthers blackish. (Named for *Ole Borrich*, 1626–1690, Danish botanist.)

1. **B. frutéscens** (L.) DC. (shrubby). — Creeping by short rhizomes; stems stout, 3–6 dm. high, minutely white-silky; leaves obovate to spatulate-oblong or lanceolate, 2–10 cm. long, often toothed near the base; chaff rigidly pointed. — Borders of saline or brackish marshes, Fla. to Tex. and Mex., n. to e. Va., and Anacostia R., D.C. July–Oct. (Bermuda)

49. HELIÁNTHUS L. SUNFLOWER. SOLEIL (Que.)

Heads many-flowered; rays several or many, neutral. Involucre imbricated, herbaceous or foliaceous. Receptacle flat or convex; the persistent chaff embracing the 4-sided and laterally compressed smooth achenes, which are neither winged nor margined. Pappus very deciduous, of 2 thin chaffy scales on the principal angles, and sometimes 2 or more small intermediate scales. — Coarse and stout Am. herbs, with solitary or corymbed heads and yellow rays; flowering toward autumn. (Name from the Greek *helios, the sun*, and *anthos, a flower*.)*

a. Annuals with merely fibrous roots; leaves mostly alternate (except lowest); disk usually brown to dark purple; receptacle nearly flat; involucre spreading, its phyllaries attenuate at tip.

Phyllaries conspicuously long-ciliate, oblong-lanceolate to oval or broadly ovate; achenes whitish, dark-lineolate, glabrous except at pilose summit, 4–8 mm. wide, the base with slender "wire-edge"; coarse, erect, the stem simple or with ascending branches; the young tips villous, soon becoming scabrous; leaves scabrous. . . . . 1. *H. annuus*.

Phyllaries eciliate or with very short and close marginal hairs scarcely longer than those of the back, lance-attenuate; achenes drab or brownish, copiously pubescent, 1.2–2.5 mm. wide, with strongly developed bilobed callous base.

Pales at center of flowering disk copiously long-villous at summit; achenes long-villous with subappressed hairs, not obviously mottled; stem erect or with ascending branches, their summits densely strigose-villous; leaves pale, strigose. . . . . 2. *H. petiolaris*.

Pales merely puberulent at tip; achenes more spreading-hirsute, dark-mottled; stem rather weak, loosely long-branched from base, often decumbent, frequently dark-mottled, glabrous or sparsely hispid; leaves dark green, merely scabrous. . . . . 3. *H. debilis*.

a. Perennial, with tough overwintering base, rhizome, stolons or tuberous-thickened roots; receptacle convex to low-conical; at least the lower leaves often opposite (exceptionally ternate). . . *b*.

b. Disk-corollas with red to purplish-brown lobes. . . *c*.

c. Principal leaves linear to linear-oblong or elongate-lanceolate, with entire or barely undulate mostly revolute margins, the broader blades 0.1–1.5 cm. wide and with elongate lateral nerves faint or scarcely developed. . . *d*.

d. Phyllaries lance-linear or narrowly lance-attenuate with very slender subulate or almost bristle-like tips often as long as disk; leaves revolute-margined, mostly longer than internodes, their tips attenuate.

Leaves and stem smooth and glabrous or nearly so; leaves closely crowded, without axillary fascicles, all but the basal narrowly linear to nearly filiform, submembranaceous, 1–3 mm. wide, with scarcely narrowed sessile base; involucre smooth; achenes with rounded summit; rhizome elongate; southwestern. . . . 4. *H. salicifolius*.

Leaves and at least base of stem scabrous; leaves abundant but not crowded, often with axillary fascicles, hard, mostly 0.3–1.5 cm. wide, commonly narrowed above the base; involucre scabrous; achenes truncate; rhizome a short erect crown; wide-ranging. . . 5. *H. angustifolius*.

d. Phyllaries narrowly oblong or lanceolate, obtuse to nearly acute; leafy nodes remote, 10–12 below the inflorescence; leaves flat or flattish, blunt to subacute, mostly shorter than internodes. . . . . . 5. *H. angustifolius*, var. *planifolius*.

---

* The key to perennial species is based largely on that of C. C. DEAM in his Flora of Indiana. Species often apparently crossing.

c. Principal leaves oblong-lanceolate to rounded-ovate, toothed, much broader, with prominent elongate lateral nerves.

Stem nearly scapose above the few subbasal leafy nodes, the upper half or two-thirds with greatly reduced bracteiform blades; subbasal internodes short and villous-hirsute, the leaves there with long upwardly winged hirsute petioles and ovate to lanceolate blades villous-hirsute on veins beneath. . . . . 6. *H. atrorubens.*

Stem more uniformly leafy nearly to summit, with short scabrous pubescence or glabrescent; leaves short-pubescent and scabrous beneath.

Leaves rounded-ovate to suborbicular, abruptly slender-petioled; phyllaries broadly oval or ovate-oblong and rounded above. . . 7. *H. silphioides.*

Leaves oblong- to rhombic-lanceolate or ovate, tapering to summit and to thick-petioled or sessile base; phyllaries narrowly ovate or lanceolate, tapering above. . . . . 8. *H. laetiflorus.*

b. Disk-corollas with yellow lobes. . . *e.*

e. Involucre of series of very unequal firm unappendaged phyllaries, these tightly appressed and without long-attenuate or foliaceous spreading or recurving tips; leading heads long-peduncled.

Leaves gradually decreasing in size up the stem, the latter leafy three-fourths its height or more; petioles of lowest blades decurrent-winged nearly to base, 0.5–3 cm. long. . . . . 8. *H. laetiflorus.*

Leaves abruptly reduced in size at to far below middle of stem; petioles of lowest blades slender, winged only at summit, 3–15 cm. long. . 9. *H. occidentalis.*

e. Involucre looser, the phyllaries, or at least the spreading outer ones, with long attenuate or more or less foliaceous tips. . . *f.*

f. Principal leaves rounded to sessile or very short-petioled bases, mostly opposite, their petioles, when developed, rarely 5 mm. long. . . *g.*

g. Stem, lower leaf-surfaces and backs of phyllaries densely whitish-villous; leaves mostly cordate-clasping. . . . . 10. *H. mollis.*

g. Stems, lower leaf-surfaces and backs of phyllaries greener, glabrous, scabrous, short-pilose or hirsute; leaves rarely clasping. . . *h.*

h. Principal cauline leaves triangular-lanceolate to -ovate, broadest at base, thence long-tapering, prominently triple-nerved, the prolonged lateral nerves confluent with the midrib at or only slightly above its base.

Stem glabrous, often glaucous; leaves horizontally divergent, sessile, their long lateral ribs uniting to base of midrib. . . 11. *H. divaricatus.*

Stem scabrous to hirsute; leaves divergent to subascending, sessile or short-petioled, the green tissue usually extending slightly below junction of lateral ribs and midrib. . . . 12. *H. hirsutus.*

h. Principal cauline leaves more oblong-lanceolate to -ovate, broadest well above base, short-petioled, without elongate lateral ribs, the lowest short laterals joining the midrib 1 cm. or more above base. 17. *H. doronicoides.*

f. Principal leaves tapering to base, if strongly rounded with petioles more than 5 mm. long. . . *i.*

i. Disk 2–3 cm. broad; larger plants with heads short-peduncled in upper axils and forming spiciform to racemose inflorescences; elongate alternate leaves often longitudinally folded and arcuate-recurving. 13. *H. Maximiliani.*

i. Disk 0.4–2(–2.5) cm. broad, if 2–2.5 cm. broad the inflorescence neither spiciform nor slenderly racemose (except sometimes in no. 17) and the leaves flat. . . *j.*

j. Disk 4–9 mm. broad; leaves of main axis mostly opposite.

Leaves minutely scabrous-pubescent, slender-petioled; disk 4–6 mm. broad. . . . . 14. *H. microcephalus.*

Leaves glabrous, scarcely petioled; disk 5–9 mm. broad. . . 15. *H. laevigatus.*

j. Disk 1–2.5 cm. broad. . . *k.*

k. Disk of larger heads 2.5 cm. broad; lobes of disk-corollas and tips of pales heavily villous with long whitish hairs; involucre very lax, of filiform-tipped villous phyllaries equaling or exceeding height of disk, these squarrose or with recurved tips; leaves ovate to lance-oblong, heavily pubescent beneath, alternate, the base decurrent to base of petiole. . . . . . . 16. *H. tomentosus.*

k. Disk 1–2 cm. broad, if slightly broader then with foliage and involucre greener, not villous, and disk-corollas and pales scarcely whitened with pubescence. . . *l*.

l. Leaves of primary axis (below inflorescence) mostly opposite, triple-nerved (with the 2 lateral ribs extending submarginally from junction with midrib nearly to apex of blade) or, if not triple-nerved, very narrowly lanceolate and convolute. . . *m*.

m. Leaves ovate to broadly lanceolate, triple-nerved, flat, spreading or ascending. . . *n*.

n. Petioles of larger leaves with decurrent tissue of blade extending far down the sides; upper leaves of main axis opposite, subentire to definitely toothed. . . *o*.

o. Petioles of larger leaves rarely 1 cm. long, the leaf-tissue often extending to base of petiole; blades strongly rounded at base, thick and hard, subentire or shallowly toothed; phyllaries rather closely ascending, shorter than height of disk. . . . . 17. ***H. doronicoides.***

o. Petioles of larger leaves 1–4 or more cm. long, slender at base.

Leaves whitish or gray beneath, usually with pubescence, thick and firm, the larger ones usually subentire or shallowly dentate; petioles slender except at summit; phyllaries barely as long as height of disk, their tips not greatly prolonged. . . . . 18. ***H. strumosus.***

Leaves green both sides, the larger ones coarsely toothed; petioles winged from below or near middle upward.

Rootstocks tuber-bearing; leaves thick, hard and harshly scabrous, the lower ones ovate, with base 0.6–1.5 dm. broad and rounded to petiole; phyllaries about equaling height of disk, loosely ascending, their tips not strikingly prolonged. 21. ***H. tuberosus,*** var. ***subcanescens.***

Rootstocks not tuber-bearing; leaves membranaceous or thinnish, smoothish or but slightly scabrous, oblong-lanceolate to narrowly oblong-ovate; the larger ones with base 2–10 cm. broad and usually more gradually tapering to petiole; phyllaries mostly longer than height of disk, their loosely spreading or recurving tips much prolonged. . . . . 19. ***H. tracheliifolius.***

n. Petioles of larger ovate and coarsely toothed blades elongate; upper leaves of main axis often alternate.

Stem glabrous or essentially so; leaves membranaceous or thin, smooth or merely scabridulous, their ribs slender and inconspicuous, the petioles slender except at tips; rhizomes slender, only rarely tuber-bearing. . . . . 20. ***H. decapetalus.***

Stem scabrous or hirtellous; leaves thick, hard and scabrous; the midribs of larger ones whitish, coarse and prominent beneath; petioles broadly winged above middle; rhizomes coarse and freely tuberiferous. . . . . 21. ***H. tuberosus.***

m. Leaves narrowly lanceolate, pinnately veined, conduplicate, loosely spreading or drooping. . . . . . 22. ***H. Dalyi.***

l. Leaves of primary axis mostly alternate (occasionally opposite), pinnately veined, not triple-nerved.

Stem smooth and glaucous; ligules 1–4 cm. long.

Larger leaves linear or linear-lanceolate, 1–2 cm. broad, long-attenuate to both ends, subentire or remotely fine-serrate, recurving or drooping; lobes of disk-corollas strongly white-hairy. . . . . . . 23. ***H. Kellermani.***

Larger (lower) or median leaves lanceolate to oblong-

ovate, attenuate or rounded to petiole, 1.5–9 cm. broad, coarsely toothed, ascending; lobes of disk-corollas glabrous or glabrescent.	24. *H. grosseserratus.*
Stem scabrous or hirtellous; ligules 1.5–2 cm. long; lobes of disk-corollas pubescent.	25. *H. giganteus.*

1. H. ánnuus L. (annual), Common S., Tourne-soleil (Que.). — *Erect annual;* the simple to much branched stem *scabrous-hispid* (softer-villous on expanding tips), 0.2–3 m. or more high; *leaves mostly alternate,* from broadly cordate-ovate to elliptic-oval, harsh on both surfaces, dentate, long-petioled, the *blade* 0.2–3 *dm. or more wide; heads* large, *with* orange-yellow rays and *dark* (rarely yellow in cult.) *low-convex disk* 3–5 *or more cm. broad; involucre of* wide-spreading *oblong-lanceolate to oval or broadly ovate long-ciliate phyllaries* with pubescent to glabrescent backs, their tips attenuate; *receptacle nearly flat;* pales of receptacle 3-cleft, the longer middle lobe narrower, hispid at tip; *achene* obovate to suborbicular, *whitish or grayish,* dark-lineolate, *about* 9 *mm. long by* 4–8 *mm. broad, glabrous* except for sparsely pilose summit, *the base with slender thickened rim.* (*H. lenticularis* Dougl.) — Plains, bottoms and other rich soil, Man. and Minn. to Tex. and westw.; cult. and spread to waste places, roadsides, etc., eastw. to Que., N.B., N.S. and the Atl. states. July–Nov. — Extensively cult. and with many named cultigens.

H. argophýllus T. & G. (glistening-leaved), with stem, leaves and involucre densely covered with whitish silky wool, often cult., sometimes spreads to waste places. (Introd. from Tex.)

2. H. petiolàris Nutt. (petioled). — More *slender* than most plants of no. 1; *the stem* 0.3–3 m. high, simple or with ascending branches, these *densely strigose-villous toward summit; leaves pale green,* narrowly oblong-lanceolate to deltoid-ovate, often slightly panduriform, slender-petioled, *strigose-pilose,* entire or undulate-dentate, the larger ones 4–15 cm. long by 1–10 cm. broad; *phyllaries lanceolate, with attenuate-subulate tips, scabrous on the back, barely short-ciliolate;* ligules 1.5–3 cm. long; *disk reddish-purple,* 1–2.5 *cm. broad, low-arching; receptacle flattish; pales near center of flowering head with densely long-villous tips; achenes* drab or pale brown, 4.5–6 mm. long, 1.2–2.5 *mm. broad, long-villous with subappressed hairs, with thick bilobed callous base.* — Sandy soil, Man. and Minn. to Wash., s. to La., Tex., N.M. and Ariz.; adv. eastw. on dry roadsides, along railroads and in waste places, to N.E., N.J. and Va. June–Oct.

3. H. débilis Nutt. (weak). — *Annual with deep green glabrous or sparsely hispid decumbent, weakly ascending to suberect often purple-mottled stem usually with elongate alternate lower branches; leaves dark green,* narrowly to broadly deltoid, *merely scabrous,* slender-petioled; heads slender-peduncled as in no. 2; ligules 0.8–3 cm. long; *pales merely puberulent at tip; achenes mottled,* villous-hirsute, *the hairs subappressed to somewhat spreading.* — A species nat. farther south, with two vars. introd. with us:

Leaves entire or shallowly undulate-dentate, the teeth blunt; stem green or faintly mottled.	*H. debilis* (typical).
Leaves acutely or subacutely dentate; stem strongly mottled.	Var. *cucumerifolius.*

H. débilis (typical). — Waste places and roadsides, locally n. to Me., partly spread from cult. June–Oct. (Introd. or adv. from farther south)

Var. **cucumerifòlius** (T. & G.) Gray (cucumber-leaved). — Waste ground, chiefly spread from cult., n. to s. N.E. (Introd. or adv. from farther south)

4. H. salicifòlius A. Dietr. (willow-leaved). — *Stem glabrous, from a horizontal elongate rhizome,* 0.7–2 m. or more high; *leaves very numerous,* rather *crowded, linear to subfiliform or linear-lanceolate, glabrous,* pale green, *mostly* 1–3 *mm. wide, scarcely narrowed to sessile base, without axillary fascicles;* panicle elongate, the peduncles erect; *phyllaries linear-attenuate to filiform,* smooth, loose; ligules about 2 cm. long; disk about 1 cm. broad, purple-brown; *achenes* oblong-obovate, *rounded at summit.* (*H. orgyalis* DC.) — Calcareous prairies, glades and plains, Neb. to e. Colo., s. to w. Mo., Ark., Okla. and Tex. Aug.–Oct.

5. H. angustifòlius L. (narrow-leaved). — *Stem* ascending *from a short erect crown, scabrous at least below,* 0.5–1.7 m. high, paniculate-branched above, *often with axillary fascicles; leaves abundant* but not crowded, *narrowly lanceolate to linear, hard, scabrous, mostly narrowed above base, long-attenuate,* mostly 1–2 dm. long by 0.3–1.5(–2) *cm. wide;* peduncles slender, elongate; *phyllaries scabrous, narrowly lanceolate, attenuate, loose;* rays yellow; *disk dark purple,* about 1 cm. broad; pales entire or 3-toothed; *achene about* 4 *mm. long, truncate at summit.* — Wet or damp acid soil of bogs, thickets and pinelands, Fla. to Tex., n. to L.I., N.J., e. Pa., Ky., s. Ind. and se. Mo. Aug.–Oct.

Var. **planifòlius** Fern. (flat-leaved). — More slender and lower, the simple or subsimple *stem* 4.5–6 dm. high, *without axillary fascicles; leaves or their pairs distant, at* 10 *or* 12 *nodes below inflorescence, the flat narrowly ovate, lanceolate or oblong blades obtuse or barely acute, mostly shorter than internodes;* heads 1–few; *phyllaries narrowly oblong or lanceolate,* merely *acute.* — Bogs, low woods and damp pinelands, se. Va. to N.J. Aug.–Oct.

6. H. atrórubens L. (dark red). — *Stem solitary from a short horizontal to ascending simple rhizome* (1–4 *cm. long*), the *lower internodes copiously,* the upper somewhat *villous-hirsute* with long spreading hairs; *leaves mostly near base of stem,* dark green and scabrous above, *villous-hirsute along veins beneath; the lower opposite, with oval or ovate blunt to subacute* crenate or dentate thick *blades one-half to four-fifths as broad as long,* these mostly *abruptly contracted to a long upwardly winged villous-hirsute petiole;* median and upper leaves greatly reduced, the upper remote, alternate; peduncles elongate; *phyllaries in several appressed series, oval, obovate or broadly oblong, with rounded or abruptly short-acuminate tips,* prominently veined below; ligules yellow; disk reddish-purple; achenes narrowly cuneate-obovate, hairy at summit, sparsely setose or glabrescent below, 3.5–4 mm. long. — Dry to wet open woods, thickets, pine-barrens and clearings, Fla. to La., n. to Va. and e. Tenn. Late July–Oct.

Var. **alsòdes** Fern. (of groves). — *Lower leaves narrowly ovate to lanceolate, serrate or serrate-dentate, mostly tapering gradually to petiole and two to five times as long as broad;* uppermost sessile leaves narrowly ovate or lanceolate; phyllaries slightly narrower, often less rounded above. — Dry woods and thickets, Va., N.C., se. Ky. and e. Tenn.

7. H. silphioìdes Nutt. (like *Silphium*). — *Colonial,* with several tall (up to 3 m.) *almost uniformly leafy scabrous-puberulent stems* from subterranean branching rhizome; *leaves rounded-ovate to suborbicular,* broadly rounded above, *all slender-petioled, finely scabrous-pubescent beneath;* phyllaries oval to oblong-ovate, round-tipped, closely imbricated; rays yellow; disk red-purple; achenes cuneate-obovate, 2–2.5 mm. long, short-hairy at summit, long-strigose to glabrescent below. (*H. atrorubens,* var. *pubescens* Ktze.; *H. kentuckiensis* McFarland & Anderson) — Woods and thickets, s. Ind. to s. Minn., s. to Ala. and La. Aug., Sept.

8. H. laetiflòrus Pers. (beautifully flowered). — *Stem* stiffly ascending, *scabrous,* 0.15–2.5 m. high, simple or branched above, *leafy up to the bases of the long peduncles,* the *leaves gradually decreasing in size,* opposite, or the upper becoming alternate, *lanceolate to rhombic-ovate, scabrous on both sides, gray-green,* serrate to subentire, *slightly triple-nerved, the lower blades decurrent down the short* (0.5–3 *cm. long*) *petioles;* heads 1–several, their naked or merely bracted *peduncles* 0.3–2 *dm. long; phyllaries in several closely appressed series,* narrowly ovate or oval to lanceolate, acute to blunt, glabrate on the back; *ligules broad,* 2.5–5 *cm. long; disk* 1.5–2 *cm. broad,* its corollas with either yellow or dark purple lobes; achene oblong-obovate, about 6 mm. long; roots sometimes tuberous-thickened. — With three confluent vars.:

Disk-corollas with yellow lobes; leaves lanceolate to lance-ovate, long-acuminate, at 9–15 nodes, the larger blades 0.8–2.5 dm. long. . . . . *H. laetiflorus* (typical).

Disk-corollas with dark purple lobes.
- Stem harshly scabrous, 0.7–2.5 m. high, with 7–15 nodes; leaves oblong-lanceolate to lance-ovate, tapering to long acuminate tips, the larger and lower ones up to 3 dm. long; phyllaries narrowly ovate or lanceolate. . . Var. *rigidus.*
- Stem slender, 0.15–1.2 m. high, less harsh above, with 6–9 nodes; leaves subrhombic-ovate to -lanceolate, merely subacute to bluntish, the larger ones 0.5–1.5 dm. long; phyllaries oblong or oblong-oval. . . . . Var. *subrhomboideus.*

H. laetiflòrus (typical). — Open woods and thickets, N.Y. to Sask., s. to Ga., Mo. and Neb.; often cult. and adv. eastw. to N.E., N.J., etc. Aug.–Oct.

Var. **rígidus** (Cass.) Fern. (rigid), PRAIRIE-S. (*H. rigidus* (Cass.) Desf.; *H. scaberrimus* Ell.) — Dry prairies, etc., w. Ont. and Mich., w. nearly across the continent, s. to w. Ga., Ala., La. and Tex.; adv. in waste places, e. to Que. and N.E. to Va. — Mostly thicker-leaved than typical *H. laetiflorus.*

Var. SUBRHOMBOÌDEUS (Rydb.) Fern. (nearly rhombic), *H. subrhomboideus* Rydb. — Roadsides, waste places, railroad-yards, etc., Que. and N.E. westw. (Natzd. from farther west)

9. H. occidentàlis Riddell (western). — *Stem* slender, *rising from elongate stolons,* pilose-hirsute at base, *glabrous or glabrescent above,* 0.4–1.5 *m. high, subscapose, the leaves mostly at base of stems; basal and subbasal leaves* 3–5 *pairs,* opposite, *ovate to oblong-lanceolate,* entire or subentire, 5–15 cm. long, 1.5–7 *cm. broad,* scabrous, minutely hispidulous beneath, *tapering into slender petioles* 3–15 *cm. long; upper sessile bracteiform leaves remote and few;* heads few, long-peduncled; *phyllaries closely appressed in about* 3 *series, lanceolate to ovate,* acuminate; ligules

1–2 cm. long; *disk* 1 *cm. broad, yellow.* — Oak-barrens, dry woods, rocky banks, etc., O. to Minn., s. to Ga., nw. Fla., Tenn., Ark. and e. Tex.; sparingly natzd., N.J. Late July–Sept. (*H. illinoensis* Gleason) — Passing into

Var. **Dowelliànus** (M. A. Curtis) T. & G. (named in 1843 for SILAS MCDOWELL, botanist of N.C.). — *Stouter and taller* (0.7–2.5 *m. high*), *mostly much-branching above, with subuniform leaves extending nearly or quite half-way to summit;* radical leaves evanescent, *the lower cauline ones broadly to narrowly ovate, the larger ones* 0.7–2 *dm. long and* 5–10 *cm. broad.* (*H. Dowellianus* M. A. Curtis) — Ky. to Neb., s. to w. N.C., Tenn. and e. Kans. — Simulating no. 8, as the typical var. does no. 7.

10. H. móllis Lam. (soft). — *Stems* 1–several from creeping rhizomes, up to 1 m. high, *densely cinereous- or whitish-villous* and more or less scabrous, simple to branched above, the *branches suberect; leaves opposite, sessile, with rounded to subcordate base, narrowly ovate to ovate-lanceolate, densely cinereous-pubescent beneath,* gray and scabrous above, *triple-nerved,* the pairs *often approximate, the principal* (median) *blades* 2–6 *cm. broad; heads* terminal or *often short-stalked in spiciform racemes* (alternate); heads showy, the yellow *disk* 2–3 *cm. broad; phyllaries* narrowly lanceolate, *attenuate, loosely ascending, whitish-villous.* — Dry open soil and thin woodland, O. and Mich. to Ia., s. to Ga., Ala., La. and Tex.; adv. or spread from cult. e. to Atl. states. Aug.–Oct.

Var. **cordàtus** S. Wats. (heart-shaped). — *Principal leaves deeply cordate and clasping, broadly ovate,* 5–8 *cm. broad.* — Similar range, more frequently cult. and esc. eastw.

11. H. divaricàtus (spread asunder). — *Stem* erect, from long creeping rhizome, *glabrous, often glaucous,* 0.5–2 *mm. high,* simple or more often branched at summit, *with* 8–15 *nodes below the inflorescence; leaves opposite, horizontally divergent, triangular-ovate or -lanceolate, tapering from broad rounded sessile or subsessile base to acuminate tip, triple-nerved, the two elongate lateral ribs confluent with base of midrib,* scabrous above, sparsely hispid beneath, *in remote pairs, the larger blades* 2–7 *cm. broad and* 0.8–1.5 *dm. long; heads few* at summit or at tips of ascending branches of terminal cyme, *short-peduncled; disk about* 1 *cm. broad,* yellow; ligules rather narrow, about 2 cm. long; *involucre loose, of lance-attenuate somewhat recurving phyllaries.* — Dry openings, thin woods and thickets, w. Me. and sw. Que. to Sask., s. to s. N.E., L.I., Ga., Tenn. and Ark. Mid-July–early Oct. Passing gradually eastw. to

Var. **angustifòlius** Ktze. (narrow-leaved). — Stem very slender, 2.5–8 dm. high, with 7–12 nodes; leaves narrowly lanceolate, caudate-attenuate, the larger 0.6–2 cm. broad and 5–10 cm. long; heads 1–3, crowded. — Dry soil, Fla. to se. Mass. and nw. N.Y.

12. H. hirsùtus Raf. (stiffly hairy). — Resembling no. 11 but *with* some or all *internodes of stem hirsute or scabrous; leaves* either divaricate or subascending, *sessile or on petioles up to* 5 *mm. long,* the *lateral ribs joining midrib at or slightly above its base,* the lower surface copiously hirsute; *disk* 1–2 *cm. broad.* — Highly variable; three vars. recognized:

Internodes hirsute with long hairs; leaves ovate to lanceolate, the broader ones 2.5–9 cm. broad. . . . . *H. hirsutus* (typical).

Internodes short-hispid or merely scabrous.
- Leaves ovate to lanceolate, strongly triple-nerved, the larger ones 2–7 cm. broad and 0.8–1.8 dm. long, spreading or subascending. . . . . . . Var. *trachyphyllus.*
- Leaves narrowly lanceolate, less strongly triple-nerved, the larger ones 0.7–2 cm. broad and 5–11 cm. long, ascending. . . . . Var. *stenophyllus.*

H. hirsùtus (typical). — Open woods and thickets, w. Pa. to Minn., s. to Ga., Ala., Ark. and Okla. Mid-July–Oct.

Var. **trachyphýllus** T. & G. (rough-leaved). — W. Pa. to Wisc. and Ia., s. to Tenn. and Ark.; adv. e. to Ct.

Var. **stenophýllus** T. & G. (narrow-leaved). — La. and e. Tex., ne. and n., locally, to W.Va., s. O. and Ark.

13. H. Maximiliàni Schrad. (for its discoverer, Prinz MAXIMILIAN VON WIED-NEU WIED, 1782–1867). — Stout, from a short rhizome, the 1–several *stems scabrous-hispidulous above,* simple up to the inflorescence, 0.5–3 m. high; *leaves numerous, alternate,* lanceolate, acuminate, thick and firm, scabrous, *often infolded and with recurving tips; inflorescence soon becoming spiciform or slenderly racemiform, with short-peduncled alternate heads in upper axils; involucre* depressed, *of loosely spreading* narrowly lance-acuminate *canescent phyllaries longer than height of disk; disk* yellow, 2–3 *cm. broad; ligules concave,* 2.5–4 cm. long; achene narrowly oblong-obovate, glabrous. — Rich prairies, Minn. to Sask., s. to Mo., Okla. and Tex.; adv. or esc. from cult. e. to Atl. states. July–Oct.

14. H. microcéphalus T. & G. (tiny-headed), SMALL WOOD-S. — *Stem* erect, *slender,* from a slender rhizome, 1–2 m. high, *smooth, paniculately branched above; leaves opposite, or the upper alternate, thin and membranaceous, scabrous above, minutely pubescent beneath,* narrowly ovate to lanceolate, *taper-pointed,* serrate-dentate except at entire base, *slender-petioled; inflorescence open,* paniculate, usually *with several slender alternate* leafy and forking *branches; heads small,* the yellow *disk* 4–6 *mm. broad; involucre campanulate, of lance-acuminate* loosely ascending smoothish *phyllaries about equaling height of disk,* only the outer phyllaries slightly recurving. (Incl. *H. glaucus* Small) — Woods and thickets, centr. Pa. to Ill., s. to n. Fla., Ala., Miss. and Mo. Aug., Sept.

15. H. laevigàtus T. & G. (smooth). — Resembling no. 14; *leaves smooth and glabrous* (as well as stem), *the base decurrent down the petiole; disk 5–9 mm. broad;* rhizome coarser. — Upland woods, w. Pa., e. W.Va. and w. Va. Aug., Sept.

16. H. tomentòsus Michx. (with matted soft hairs). — Stem stout, from a subligneous base, mostly 1–3 m. high, the *upper nodes villous to hirsute; leaves* ovate to lance-oblong, scabrous above, *densely soft-pubescent beneath,* shallowly dentate, *the base extending down to base of petiole, alternate;* inflorescence open-paniculate or with racemiform branches; *heads large,* with *disk about* 2.5 *cm. broad; phyllaries loosely imbricated, long-attenuate to filiform tips, mostly exceeding height of disk, white-villous, squarrose or strongly recurved; lobes of disk-corollas and tips of pales white with dense long villi.* — Open woods and thickets, w. Va. to Mo., s. to n. Fla. and Ala. Aug., Sept.

17. H. doronicoìdes Lam. (like *Doronicum*). — *Stem* arising from creeping rhizome (often with fusiform-thickened roots), *scabrous above,* 0.7–1.5 m. high; *leaves of main axis opposite, oblong-ovate to -lanceolate, rounded to gradually narrowed at base, with tissue extending down the very short* (*less than* 1 *cm.*) *petiole nearly to its base,* the *lowest lateral ribs joining midrib* 1 *cm. or more above base,* both sides scabrous, *margin subentire or low-dentate;* heads 1–few, on peduncles up to 1 dm. or more long; *involucre of loosely ascending* pubescent subulate-tipped lanceolate to linear *phyllaries shorter than height of disk;* disk 1.5–1.8 cm. broad, its corollas with puberulent lobes; ligules 2–3 cm. long, 13–18; achenes glabrous. — Dry woods, thickets and clearings, n. N.J. to Minn., s. to s. Pa., O., Ind., Ill., Mo., Okla. and e. Tex. Late July–Oct.

18. H. strumòsus L. (scrofulous; from the thick-based hairs). — Rhizome elongate, its slender branches often slenderly tuberiferous; *stem often branching above,* slightly scabrous to smoothish, up to 3 m. high; *leaves of primary axis opposite,* ovate to broadly lanceolate, triple-nerved, *thick and hard, harsh above, pale to whitish and commonly soft-pubescent* (rarely glabrous) *beneath, entire or shallowly dentate, the lower ones* up to 2.5 dm. long and *on elongate slender-based petioles; involucre hardly as long as height of disk,* the slender *tips of* the *phyllaries not greatly prolonged;* disk 1–2 cm. broad; ligules 9–15, 2–3 cm. long; achenes glabrous. — Open woods, thickets and clearings, sw. Que. to N.D., s. to s. N.E., Ga., Ala., Ark. and Okla. Aug., Sept.

19. H. trachelifòlius Mill. (with leaves of *Trachelium*). — *Differing from no.* 18 *in non-tuberous rootstocks; leaves green both sides, thinnish to membranaceous, smoothish or merely scabridulous,* often more coarsely toothed, *their petioles winged from near middle; phyllaries mostly longer than height of disk, their prolonged tips loosely spreading or recurving.* — Woods and thickets, w. N.E. to Minn. and Neb., s. to S.C., Ky., Mo. and Kans. Late July–Sept.

20. H. decapétalus L. (with ten petals). — *Rhizome* elongate, *usually without tubers* (rarely with fusiform tubers); *stem smooth,* simple to much branched above; *lower and median leaves of primary axis opposite,* upper ones sometimes alternate, *triple-nerved with slender inconspicuous ribs, ovate, rounded, cuneate or truncate to slender petiole, thin or membranaceous,* green both sides, *glabrous or barely scabridulous,* the larger ones *coarsely serrate or serrate-dentate;* heads rather small; *involucre with loosely spreading to recurving long-tipped phyllaries longer than height of disk;* disk 1–1.5 cm. broad; *disk-corollas pubescent;* ligules 8–12 or more, 2–2.5 cm. long. — Open woods and thickets, centr. Me. and sw. Que. to Minn. and Neb., s. to s. N.E., Ga., Ky. and Mo. Aug.–Oct.

21. H. tuberòsus L. (tuberous), JERUSALEM ARTICHOKE. — *Rhizomes* elongate, *enlarging at tips into* fusiform, ovoid or subglobose *fleshy* edible *tubers; stem* mostly branching above, *coarse, scabrous or hirtellous,* up to 3 m. high; *leaves thick and hard, scabrous above, minutely puberulent beneath,* those of lower quarter of stem opposite, *the upper ones alternate; the larger ones with midvein coarse, whitish and prominent beneath;* the *petioles of the ovate triple-nerved coarsely toothed blades broadly winged* above middle, the *larger blades* 0.6–1.5 *dm. broad; phyllaries about equaling height of disk,* loosely ascending, their acuminate tips not strikingly prolonged; disk 1–1.5 cm. broad, its corollas glabrous or nearly so; ligules broad, 2.5–4 cm. long. — Rich or damp thickets, etc., Ont. and Sask., s. to Ga., Tenn. and Ark., now spread through cult. and generally natzd. throughout our range. Aug.–Oct.

Var. **subcanèscens** Gray (somewhat gray). — All or essentially all leaves opposite, more generally soft-tomentose beneath. (*H. mollissimus* E. E. Wats.) — O. to Minn., s. to Mo. and Kans.; rarely spread from cult. e. to N.E.

22. H. Dàlyi Britt. (for one of its discoverers, CHARLES PATRICK DALY, 1816–1899), JUDGE DALY'S S. — *Tuberous; stems numerous, slender, simple,* appressed-pubescent, 5–8 *dm. high; leaves mostly opposite, only the uppermost alternate,* rather crowded, *narrowly lanceolate, attenuate to both ends, drooping or loosely recurving, pinnately veined, conduplicate,* grayish-green, paler beneath, scabrous, *the larger* 1–1.3 *dm. long and* 7–14 *mm. broad;* heads 1 or 2, peduncled; phyllaries narrowly lanceolate, with long subulate tips; disk 1.5 cm. broad, its corollas with puberulent lobes; *ligules acute,* 2.5–4 cm. long. — Gravelly shores, local and little known, Plymouth Co., Mass.; L.I.; Mich. and Minn. July–Sept.

23. H. Kellermàni Britt. (for its discoverer, WILLIAM ASHBROOK KELLERMAN, 1850–1908), KELLERMAN'S S. — *Stems* 2–3 m. high, *glabrous and glaucous,* from a slender rhizome, *much branched above; leaves mostly alternate or falsely opposite,* only the lower ones definitely opposite, *very numerous, narrowly lanceolate or lance-linear, long-attenuate to both ends, pinnately veined, loosely spreading to drooping,* thinnish, scabrous above, pubescent beneath; the larger up to 2 dm. long and 1–2 cm. broad, *entire or remotely fine-serrate;* heads numerous; *phyllaries* narrowly lance-attenuate, *longer than height of disk, loosely spreading to recurving; disk* 1–1.5 cm. broad, *its corollas with white-pubescent lobes;* ligules 1–3 cm. long. — Damp or rich thickets, local, e. N.Y. to O. and Wisc. Late Aug.–Oct.

24. H. grosseserràtus Martens (coarsely serrate). — Rhizome elongate, freely branching, sending up *colonial glabrous and glaucous* coarse *stems mostly* 2–5 *m. high; leaves alternate, the larger* (lower and median) *ones lanceolate to oblong-ovate, attenuate or rounded to petiole,* 1.5–9 *cm. broad, coarsely toothed,* ascending, harshly *scabrous above,* whitened and densely pubescent beneath, with long petioles; heads few–several on erect upper branches and peduncles; *phyllaries* linear-subulate, *equaling or longer than height of disk, loosely spreading or squarrose; disk* 1–2 *cm. broad, its florets with glabrous or glabrescent lobes;* ligules showy, 2.5–4 cm. long. — Prairies and plains, O. to N.D., s. to Ark. and Tex.; cult. and spread to rich thickets, roadsides, etc., e. to Atl. states. July–Oct.

25. H. gigantèus L. (gigantic). — *Rhizome* short, *with some elongate-fusiform roots; stem* usually *scabrous or hirtellous,* up to 3 m. high, subsimple to much branched above; *leaves alternate* (or lower ones opposite), *lanceolate to lance-oblong, pinnately veined,* acuminate, narrowed to *sessile or short-petioled* base, *scabrous and green on both sides, with low serration or subentire,* mostly 1.5–3 cm. broad; heads 1–few at tips of branches; *phyllaries* linear-subulate, equaling or exceeding height of disk, *loosely spreading to recurving; disk* about 2 cm. broad, *its florets with pubescent lobes; ligules* yellow (rather pale), 1.5–2 *cm. long.* — Damp or rich thickets, swampy woods and clearings, sw. Que. to Sask., s. beyond our limits. July–Oct.

50. ACTINÓMERIS Nutt. WING-STEM

Heads many-flowered; rays neutral, few or none. Phyllaries few, herbaceous, nearly equal, soon deflexed beneath the globular disk. Receptacle small, chaffy. Achenes flat, obovate, winged or wingless, at maturity spreading in all directions; pappus of 2–3 smooth persistent awns. — Tall branching N.Am. perennials, with serrate feather-veined leaves tapering to the base and mostly decurrent on the stem. Heads corymbed; flowers chiefly yellow. (Name from the Greek *actis, a ray,* and *meris, a part;* alluding to the irregularity of the rays.)

1. A. alternifòlia (L.) DC. (alternate-leaved). — Stem somewhat hairy, usually winged above, 1–2 m. high; leaves alternate or the lower opposite (or all opposite or in 3's), oblong or ovate-lanceolate, pointed at both ends; rays 2–8, irregular. (*Ridan* Britt.) — Rich thickets and borders of woods, Fla. to La., n. to N.Y., s. Ont. and Ia. Aug., Sept.

51. VERBESÌNA L. CROWN-BEARD

Heads several–many-flowered; rays pistillate, or sometimes neutral and sterile, few or sometimes none. Phyllaries imbricated in 2 or more rows. Receptacle rather convex (conical in no. 3), the chaff concave. Achenes flat (laterally compressed), winged or wingless, 2-awned. — Mostly perennial herbs of N. and S.Am. and Afr.; the toothed leaves decurrent on the stem. Flowers mostly yellow. (Name said to be "metamorphosed" from *Verbena.*)

Tall perennials, the winged stem simple below the corymbose inflorescence;
leaves sessile or decurrent to winged petioles; involucres campanulate or hemispheric.

Heads very numerous, with slenderly campanulate involucres, cymosely paniculate; ray-flowers 1–5, pistillate, usually fertile.
Leaves opposite; ligules yellow. . . . . 1. *V. occidentalis.*
Leaves alternate; ligules white. . . . . 2. *V. virginica.*
Heads few, with broad hemispheric involucres; ray-flowers 8–15, mostly sterile. . . . . 3. *V. helianthoides.*
Loosely branching annual, with wingless stem and scattered large heads; leaves slender-petioled; involucres depressed. . . . . 4. *V. encelioides,* var. *exauriculata.*

1. V. occidentàlis (L.) Walt. (western; as contrasted with Eu.). — Stem 1–2 m. high, 4-winged; *leaves opposite,* ovate to oblong-lanceolate, triple-nerved, serrate, pointed at both ends, often pubescent beneath, large and thin; heads slender, in compound corymbs; receptacle flattish; *flowers yellow;* rays 1–5, lanceolate; achenes wingless. (*Phaethusa* Small) — Rich woods, thickets and openings, Fla. to Miss., n. to Pa., W.Va., O. and Ill. Aug.–Oct.

2. V. virgínica L. (Virginian), TICKWEED, FROSTWEED. — *Stem* narrowly or interruptedly winged below, *downy-pubescent, like the lower surface of the* ovate-lanceolate feather-veined *alternate leaves;* heads small, in compound corymbs; receptacle convex; *flowers white;* rays 3–4, oval; achenes winged. (*Phaethusa* Small) — Dry open woods, thickets and clearings, Fla. to Tex., n. to Pa., Ky., Mo. and se. Kans. Aug.–Oct.

3. V. helianthoìdes Michx. (resembling *Helianthus,* Sunflower). — *Perennial;* stem hairy, 1 m. or less high, widely winged by the decurrence of the ovate to ovate-lanceolate *sessile* alternate *leaves,* which are rough above and soft-hairy beneath; *heads broad;* involucre appressed; *rays* 8–15, pistillate or neutral, usually sterile; achenes winged, tipped with 2 fragile awns. (*Pterophyton* E. J. Alex.) — Open woods, thickets and prairies, Ga. to Tex., n. to O., Ill. and Ia. June–Oct.

4. V. ENCELIOÌDES (Cav.) B. & H. (resembling *Encelia*), var. EXAURICULÀTA Robins. & Greenm. (without auricles). — *Annual,* branching, 3–6 dm. high, cinereous; leaves alternate, ovate or cordate to deltoid-lanceolate, the *petioles destitute of the wings or auricles* (characteristic of the typical more southern var.); involucre depressed, its foliaceous spreading linear phyllaries equal; rays numerous, fertile; achenes broadly winged. (*Ximenesia exauriculata* Rydb.) — Mont. to Ariz., e. to Kans. and Tex.; adv. in Mo. and casually to N.E. June–Oct.

52. GUIZÒTIA Cass.

Heads with pistillate 3-dentate rays and fertile disk-flowers. Involucre campanulate, double, with distinct phyllaries, the outer membranous-foliaceous, the inner chaff-like. Receptacle conical or convex, with flat membranous chaff subtending the flowers. Disk-corollas with 5-lobed campanulate limb and short densely woolly tube. Achenes dorsally compressed, narrow, somewhat 4-angled, those of marginal flowers 3-angled, awnless. — Annual herbs, nat. of Afr., with mostly opposite entire or dentate leaves and peduncled yellow heads. (Named in honor of *François Pierre Guillaume Guizot,* 1787–1874, French statesman and historian.)

1. G. ABYSSÍNICA (L. f.) Cass. (of Abyssinia). — Widely branching, up to 1.3 m. high, nearly glabrous; leaves oblong-lanceolate, sessile, subamplexicaul; outer phyllaries ovate; expanded head 2–3 cm. broad. — Waste places, as yet local and somewhat transient, Ct. to Mich. and Pa. Sept., Oct. (Adv. from trop. Afr.)

53. THELESPÉRMA Less.

Heads many-flowered; rays about 8 and neutral, or none. Involucre as in *Coreopsis,* the inner phyllaries scarious-margined and united nearly to or above the middle. Receptacle flat, the whitish scarious chaff falling with the wingless terete and beakless achenes; pappus of 2 stout, subulate retrorsely hispid awns. — Smooth Am. herbs, with opposite dissected leaves and pedunculate heads of yellow flowers. (From the Greek *thele, a nipple,* and *sperma, seed,* on account of the papillose achenes.)

1. T. trífidum (Poir.) Britt. (three-cleft). — *Annual or biennial,* 3–7 dm. high, loosely branching and very leafy; leaves 2-pinnate, the lobes filiform; *outer phyllaries* 8, *subulate-linear,* hardly equaling the *inner,* which are *united only below the middle;* rays 1 cm. or more long; outer achenes conspicuously roughened on the back. — Calcareous barrens and plains, S.D. to Colo., s. to w. Mo., Okla., Tex. and N.M.; adv. eastw. to Mich. July–Oct.

2. T. grácile (Torr.) Gray (slender). — *Perennial,* rather rigid, naked above; leaves with narrow or filiform divisions, or the upper entire; *phyllaries* 4–6, *the outer very short-ovate or oblong, the inner connate to above the middle;* rays short or usually none; achenes less roughened. — Dry plains and prairies, spreading to roadsides, Neb. and Wyo., s. to w. Mo., Okla., Tex., N.M. and Ariz. Late May–July.

54. COREÓPSIS L. Coreopsis. Tickseed

Heads many-flowered, radiate; rays mostly 8, neutral, rarely wanting. Involucre double; each series of about 8 phyllaries, the outer foliaceous and somewhat spreading; the inner broader and appressed, nearly membranaceous. Receptacle flat, with membranaceous chaff deciduous with the fruit. Achenes flat, obcompressed (*i.e.*, flattened parallel with the phyllaries), often winged, not narrowed at the top, 2-toothed or 2-awned, or sometimes naked at the summit; the awns not barbed downwardly. — Herbs, chiefly of Am., Hawaii and Afr., generally with opposite leaves and yellow or parti-colored (rarely purple) rays. Too near some species of *Bidens*, but generally well-distinguished as a genus. (Name from the Greek *coris, a bug,* and *opsis, appearance;* from the form of the achene.)*

*a.*Outer involucre much shorter than inner, 1–4 (rarely –6) mm. long; style-tips truncate or obtusely short-conical; achenes oblong or elliptic. *b.*

*b.*Perennials with mostly simple leaves; ligules roseate, white or yellow; achenes nearly straight.

Leaves narrowly linear; ligules roseate to white; achenes wingless, with glabrous faces; subterranean stems creeping, stoloniferous. . . . . 1. *C. rosea.*

Leaves oblanceolate, only the reduced upper ones linear; ligules orange-yellow; achenes with fimbriate-pectinate wing, usually with papillate faces; nonstoloniferous. . . . . 2. *C. oniscicarpa.*

*b.*Annual with principal leaves opposite, the basal and lower cauline pinnately divided into narrow segments; ligules yellow, crimson-brown or parti-colored; achenes incurved, often papillose-roughened. . . . . . 3. *C. tinctoria.*

*a.*Outer involucre usually nearly as long as to equaling inner, 0.2–1(–2) cm. long; style-tips cuspidate, acutely conical or tailed; achenes oblong to orbicular; perennials (some flowering 1st year) with yellow ligules. . . *c.*

*c.*Ligules mostly palmately lobed; central and larger achenes roundish, incurved, often papillose and with a callus at base and apex of inner face. . . *d.*

*d.*Wings of central achene thin-scarious, outspread, broad when well-developed. . . *e.*

*e.*Lower leaves linear to spatulate, upper lanceolate to linear; ligules 1.3–3 cm. long. . . *f.*

*f.*Outer involucre appressed-ascending; inner involucre 8–13 mm. long; larger achenes 2.3–3 mm. long, their wings less than 1 mm. wide.

Leaves mostly on lower half of stem, chiefly entire, linear to oblanceolate; peduncles scape-like, becoming 2–4 dm. long; larger achenes 2.3–3 mm. long; pales 4–6 mm. long. . . . . . . 4. *C. lanceolata.*

Leaves extending well up the stem, the upper 3–5-parted; peduncles 1–1.5 dm. long; larger achenes 2.5 mm. long; pales 6–7 mm. long. . . . . 5. *C. grandiflora.*

*f.*Outer involucre soon reflexed; inner involucre 1.5–1.7 cm. long; larger achenes 3.5–4 mm. long, with wings 1–1.5 mm. wide. . . 6. *C. heterogyna.*

*e.*Lower leaves oval-obovate, upper oblong or oblong-lanceolate; ligules 1–2.3 cm. long; larger achenes 2.8–3 mm. long. . . . . . . 7. *C. pubescens.*

*d.*Wings of achene narrow, callous-thickened, involute; lower leaves ovate or roundish. . . . . 8. *C. auriculata.*

*c.*Ligules entire or but slightly toothed; achenes oblong to obovate, nearly straight, without callus, wingless or only very narrowly winged. . *g.*

*g.*Leaves sessile, mostly palmately divided (but leaflets sometimes pinnately parted), opposite but appearing whorled; outer phyllaries 4–9 mm. long, about equaling inner; stems mostly less than 1 m. high. . . *h.*

*h.*Leaves 3-cleft, not divided to base; achenes oblong, 1.8–2.3 mm. broad. 9. *C. palmata.*

*h.*Leaves divided to base (or uppermost and lowest simple); achenes obovate or elliptic. . . *i.*

---

* Some details derived from the revision by E. E. Sherff, Field Mus. Bot. ser. xi. no. 6 (1936).

*i.*Minutely soft-pubescent; leaf-segments lanceolate to rhombic-oval, 0.5–2.5 cm. broad; disk-corollas yellow. . . . . 10. *C. major.*

*i.*Glabrous (except sometimes at nodes); leaf-segments linear to filiform, 0.3–6 mm. broad.

Leaf-segments flat, linear, the ultimate ones 1–6 mm. wide; disk-corollas purple-brown; larger achenes 4.5–6 mm. long, 2–4 mm. broad. . . . . 11. *C. delphinifolia.*

Leaf-segments linear-filiform or linear, 0.3–1.5 mm. wide; disk-corollas yellow; larger achenes 3–5 mm. long, 1–1.7 mm. broad. 12. *C. verticillata.*

*g.*Leaves petiolate, pinnately 3–5-divided; outer phyllaries 2–3 mm. long, definitely shorter than inner; stems 1–3 m. high. . . . . . . 13. *C. tripteris.*

1. C. ròsea Nutt. (rose-colored). — Smooth perennial with *creeping subterranean stoloniferous base;* ascending leafy stem simple or much branched, 1.5–8 dm. high; leaves opposite, often with axillary fascicles, linear, 0.5–3 mm. broad, simple or rarely 2–3-parted; peduncles 2–8 cm. long; *outer involucre* of 6–10 slender phyllaries 1–3 *mm. long, inner* 4–6 *mm. high* and of oblong-ovate phyllaries connate at base; *ligules* deep to light *rose, or white* in forma **leucántha** Fern. (white-flowered), cuneate-oblong, 3-lobed at apex, 8–13 mm. long; pales linear, 2–4 mm. long; disk-flowers yellow; *style-branches abruptly and bluntly conic-tipped; achenes flat, oblong, wingless, about* 2 *mm. long.* — Damp sandy, gravelly or peaty shores and depressions, Yarmouth Co., N.S.; e. Mass.; R.I.; L.I.; s. N.J. and se. Pa. to e. Md. July–Sept.

2. C. oniscicárpa Fern. (with fruit like *Oniscus*, a sowbug). — Erect, *non-stoloniferous,* 6–9 dm. high; *leaves* opposite (rarely alternate); *the basal narrowly oblanceolate,* long-petioled, the callous-margined blades 0.5–1 *cm. broad;* upper leaves reduced to linear bracts; heads few-several, paniculate-corymbose, slender-peduncled; outer involucre irregularly biseriate, its coriaceous deltoid-lanceolate to lance-oblong phyllaries 0.7–3 mm. long; phyllaries of inner involucre dark brown or fuscous, oblong, 6 mm. long; *ligules orange-yellow, cuneate*-obovate, 0.8–1.8 cm. long, 3-lobed at tip; pales linear, 4–5 mm. long; disk-corollas purple-black, 2–3 mm. long; *achenes* 1.8–2.2 mm. long, 0.5–0.8 mm. broad, *with a conspicuous* but short *fimbriate-pectinate border and usually papillate faces*; awns 0.7–1.3 mm. long. — Wet pinelands, thickets and ditches, se. Va. and e. N.C. Late Aug.–Oct.

Var. **símulans** Fern. (simulating; the basal leaves and outer phyllaries suggesting the more southern *C. linifolia*). — *Basal leaves broadly oblanceolate to subelliptic,* 1.2–2 *cm. broad; outer phyllaries deltoid-ovate, with white-hyaline margin, up to* 5 *mm. long and* 2 *mm. wide; inner phyllaries pale; ligules* 0.5–1.2 *cm. long.* — Damp sand, Nansemond Co., Va.

3. C. tinctòria Nutt. (used in dyeing). — Glabrous *annual* 0.6–1.2 m. high, with many slightly angled branches; *leaves opposite, the basal and lower cauline pinnately divided into linear or narrowly lanceolate segments,* upper undivided; heads numerous, corymbose, slender-stalked; outer involucre 1–2-seriate, of linear-oblong to triangular mostly scarious-margined phyllaries about 2 mm. long, inner of ovate phyllaries 5–6 mm. high; *ligules yellow, or parti-colored,* or in forma **atropurpùrea** (Hook.) Fern. (dark purple) wholly purple-brown, obovate, 3-lobed at summit, 0.7–1.5 cm. long; pales nearly filiform, 4–4.5 mm. long; disk-corollas dark red; style-branches obtuse; *achenes* linear-oblong, *incurved, wingless,* awnless, *often papillose-roughened below.* — Low grounds, Minn. and Man. to Wash., s. to La., Tex., N.M., Ariz. and Calif.; generally cult. and frequently esc. e. to the Atl. states. June–Sept.

C. BASÀLIS (Dietr.) Blake (from basal branching of original specimen), *C. Drummondii* (D. Don) T. & G., cult. and sometimes spreading to waste ground, differs from no. 3 in its leaf-segments lanceolate to elliptic-oblong or orbicular; outer phyllaries slender; ligules 1.3–2.3 cm. long; achenes obovate and cartilaginous-margined.

4. C. lanceolàta L. (lance-shaped). — Glabrous perennial; the branching *stems leafy chiefly below,* 2–8 dm. high, *with scapiform erect peduncles becoming* 2–4 *dm. long; leaves mostly simple* (rarely with 1 or 2 small lateral lobes), *linear to oblanceolate,* lower slender-petioled, upper sessile; outer involucre of lanceolate to oblong-ovate phyllaries 4–8 mm. long, inner of ovate phyllaries 0.8–1.3 cm. long; ligules cuneate or obovate, bright yellow, deeply lobed at summit, 1.3–3 cm. long; *pales* 4–6 *mm. long,* oblong-linear with filiform tip; disk-corollas yellow; tips of style-branches tail-like; larger *achenes orbicular, curved, broadly thin-winged,* 2.3–3 *mm. long,* with a callus at base and apex on the inner side and wing 0.5–0.8 mm. wide; pappus 2 small fimbriolate squamulae. — Dry sandy, gravelly or rocky soils, Fla. to N.M., n. to Va., Ont., Mich., Wisc. and Mo.; cult. and frequently natzd. e. to s. N.E. and N.J. May–July.

Var. **villòsa** Michx. (villous). — Stem and leaves pubescent. — Fla. to La., n. to Va., s. Ont., Mich., Ill. and Mo.; cult. and occasionally natzd. ne. to s. N.E. May–July.

5. C. grandiflòra Hogg (large-flowered). — Similar to no. 4; *leaves extending well up the stem; the upper* 3–5-*parted*, with linear to linear-lanceolate segments mostly 1–5 mm. wide; *peduncles* 1–1.5 *dm. long;* outer phyllaries lance-subulate, the whitish margins ciliate; ligules 1.3–2.5 cm. long; *pales* 6–7 *mm. long; achenes about* 2.5 *mm. long;* pappus often wanting. — Sandy or rocky prairies and thickets, Fla. to Tex. and N.M., n. to Ga., Mo. and se. Kans.; cult., becoming natzd. n. and e. to Mich., Ind. and s. N.E. May–July.

Var. **Harveyàna** (Gray) Sherff (for its discoverer, FRANCIS LEROY HARVEY, 1850–1900). — Leaf-segments linear-filiform. — Similar habitats, Mo., Ark. and Okla.

6. C. heterógyna Fern. (fickle female; because of the variable achenes). — Similar to no. 5; stems up to 9 dm. high, pilose below; primary cauline *leaves* with petioles 0.5–1 dm. long and broadly oblanceolate *acutish* blades 2–3.5 cm. broad and pilose on both sides; peduncles 1–2.5 dm. long; *outer involucre soon reflexed,* its deltoid-lanceolate white-margined phyllaries 7–9 mm. long; *inner involucre* with ovate phyllaries 1.5–1.7 *cm. long* and about 1 cm. broad; *pales* linear-attenuate, 1–1.2 *cm. long; larger achenes* with obovate or reniform bodies 3.5–4 *mm. long,* their *wings* 1–1.5 *mm. broad.* — Alluvial woods, thickets and clearings, se. Va.; cult. and natzd. n. to Mass. Late May–July.

7. C. pubéscens Ell. (hairy). — Resembling no. 5, more or less pubescent, 0.5–1.3 m. high; *lower leaves oval-obovate,* petioled; *the upper oblong or oblong-lanceolate, uncut or with* 3–5 *segments;* peduncles 1–2 dm. long; outer phyllaries linear-lanceolate, 7–10 mm. long; *achenes* 2.8–3 *mm. long.* — Open woods, dry slopes and cliffs, Fla. to La., n. to Va., W.Va., Ky., s. Ill., Mo. and Okla. June–Sept.

Var. **robústa** Gray (stout). — Stems and branches glabrous or soon glabrate. — Va. and N.C. to Ky.; spread from cult. in s. N.E.

8. C. auriculàta L. (with basal lobes; as in the ear). — Pubescent or tardily glabrate above, 1.5–5 dm. high, subscapose, *stoloniferous; leaves mostly on lower third of plant,* petioled, membranaceous, *the lower ovate or rounded,* uncleft or with small basal lobes; *peduncles subscapiform,* 1–2.5 dm. long; outer phyllaries lanceolate or narrowly ovate, white-margined; *ligules* 1.5–2.3 *cm. long; achenes obovate or oblanceolate,* 2.2–2.9 mm. long, *with narrow callous involute wings.* — Rich calcareous hardwoods and openings, Fla. and La., n. to Va. and Ky. April–June.

9. C. palmàta Nutt. (palmate). — Erect from horizontal rootstock; the simple stems 5–9 dm. high, glabrous, or hirsute at nodes, leafy; *leaves sessile,* 3-*cleft only partly to base* into entire or irregularly lobed oblong or oblong-linear segments, thus appearing verticillate; peduncles 1–4 cm. long; *outer phyllaries* linear, 4–9 *mm. long, about equaling the inner;* ligules obtuse or acutely short-dentate, 1.5–2.7 cm. long; *achenes oblong,* 1.8–2.3 *mm. broad,* narrowly winged. — Prairies, open woods and calcareous bluffs, Wisc. to Man., s. to Ind., Ill., Mo. and Okla. June, July.

10. C. màjor Walt. (larger). — Erect, from a hard base, 5–9 dm. high, the slender *stem and branches pubescent; leaves* sessile, spreading or ascending, *pubescent, palmately divided to base into lanceolate to oval* entire obtuse *segments* 1.2–2.5 *cm. broad,* or the rameal leaves simple and narrower; heads loosely corymbose, on peduncles mostly 2–8 cm. long; *involucre hispid;* the outer phyllaries linear-oblong, 4–8 mm. long, about equaling the inner; ligules 1.5–2.5 cm. long; *disk-corollas yellow; achenes* elliptic-oblong or obovate, 4.5–6 mm. long, 2–4 *mm. wide.* — Open woods, thickets and clearings, Ga. to Miss., n. to N.C. and O. June, July.

Var. **stellàta** (Nutt.) Robins. (star-like). — Similar, but leaves glabrous. — Fla. and Ala., n. to Va., W.Va. and O. June, July.

Var. **rígida** (Nutt.) F. E. Boynt. (stiff). — Leaves firmer; their *leaflets* linear-lanceolate, *only* 5–11 *mm. broad.* — Ga. and Tenn. to Va.

11. C. delphinifòlia Lam. (with leaves of *Delphinium,* Larkspur). — Similar to no. 10; *stem glabrous,* or hispid only at nodes; *leaf-segments linear,* 1–6 mm. broad, the *ultimate ones acute or acutish,* strongly ascending; involucre glabrous; *disk-corollas reddish-brown; larger achenes* obovate, 4.5–6 *mm. long,* 2–4 *mm. broad.* — Pinelands, Ga. and Ala., n. to Va. June, July.

12. C. verticillàta L. (whorled). — Similar to nos. 10 and 11; fresh rhizome yellow; stem 1.5–11 dm. high, *glabrous,* subangulate; *leaves* palmately 3-parted, *with* linear-*filiform* to *linear slender-tipped segments only* 0.3–1.5 *mm. broad,* the leaves thus appearing strongly verticillate; involucre glabrous or minutely pubescent; outer phyllaries linear-oblong, but slightly shorter than inner; ligules entire or slightly toothed at apex, 1.2–2.5 cm. long; disk-corollas yellow; *achenes* oblong-obovate, 3–5 *mm. long,* 1–1.7 *mm. broad.* — Dry open woods and clearings, Fla. to Ala. and Ark., n. to Md. and D.C. June, July.

13. C. trípteris L. (three-winged; from ternately divided leaves), TALL C. — Stems 1–3 m. high, from a horizontal stoloniferous base; principal *leaves petiolate, pinnately* 3–5-*divided* into oblong-lanceolate to linear *glabrous* leaflets, rameal leaves often simple and sessile; peduncles

2–8 cm. long; *outer involucre of* linear-oblong *glabrous phyllaries only 2–3 mm. long, definitely shorter than the inner;* disk-corollas yellow, becoming brown; *achenes* cuneate, 5–7 mm. long, 2.8–4.2 *mm. broad.* — Thickets and borders of woods, s. Ont. to Wisc., s. to N.C., Ga., Miss., La. and Kans.; esc. from cult. ne. to Mass. Aug., Sept. — Heads with odor of anise.

Var. **Deàmii** Standl. (for its discoverer, CHARLES CLEMON DEAM, 1865–). — Lower surfaces of *leaves and outer involucres densely pubescent.* — Pa. to Mich. and Ill., s. to Ga., Ark. and Mo.

Var. **Smìthii** Sherff (for its discoverer, JOHN DONNELL SMITH, 1829–1928). — Leaves all or nearly all simple, oblong-lanceolate. — Ky. and Ark. and southw.

55. BÌDENS L. BUR-MARIGOLD. CUCKOLD. BIDENT OR FOURCHETTE (Que.)

Heads many-flowered; the rays, when present, 3–8, neutral. Involucre double, the outer commonly large and foliaceous. Receptacle flattish; chaff deciduous with the fruit. Achenes flattened parallel to the phyllaries or slender and 4-sided (rarely subterete), beakless, with wingless ciliolate margins and angles, crowned with retrorsely or antrorsely barbed or hispid awns or short teeth (these rarely naked). — Annual or perennial herbs, with opposite leaves, and mostly yellow flowers; genus semicosmop. (Latin *bidens, two-toothed.*) *

a. Achenes broadened from base to summit, from linear-oblanceolate to cuneate-obovate, the central rarely more than one-third longer than the inner involucre and the chaff. . . *b.*

b. Primary leaves simple or merely deeply cleft into 3–5 coarse lobes, but without distinct leaflets. . . *c.*

c. Achenes with a convex cartilaginous summit. . . *d.*

d. Mature disk (except in depauperate extremes) 1.3–2.8 cm. broad; fruiting heads often nodding; outer involucre reflexed, spreading or merely subascending; disk-corollas 5-toothed; anthers exserted, purple-black; achenes not conspicuously striate between margins and midribs or keels, the central 1.8–2.5 mm. broad.

Achenes straight and flat, wingless, not strongly keeled, deep-brown to purplish; the outer 6–8 mm. long, with marginal awns 2.8–4.5 mm. long; the central 8–9.5 mm. long; stem firm, usually smooth; outer involucre rarely longer than inner; chaff reddish-tipped; ligules 1.5–3 cm. long. . . . . 1. *B. laevis.*

Achenes curved, with almost wing-like pale margins and keels, olivaceous; the outer 3.3–6.3 mm. long, with marginal awns 2–2.8 mm. long; the central 4.2–7.8 mm. long; stem soft, usually hispid; outer involucre usually longer than inner; chaff yellow-tipped; ligules wanting or at most 1.7 cm. long. . . . . 2. *B. cernua.*

d. Mature disk rarely as much as 1.5 cm. broad; fruiting heads erect; outer involucre ascending; disk-corollas 4-toothed; anthers included, paler; achenes distinctly 7–15-striate on each face, the central 1.4–2.4 mm. broad, flat, olive-brown to drab. . . . . 3. *B. hyperborea.*

c. Achenes truncate or concave at summit, if convex (in no. 9) not cartilaginous. . . *e.*

e. Mature central achenes strongly 4-angled or with almost wing-like midribs. . . . . 4. *B. connata.*

e. Mature central achenes flat or flattish, with slender or obscure midribs, or slenderly linear-oblanceolate, plano-convex and many-striate. . . *f.*

f. Achenes plano-convex, uniformly and closely striate-nerved, conspicuously short-setose with ascending hairs; heads slender-cylindric; inner involucre 0.9–1.6 cm. long.

Phyllaries of outer involucre scabrous-ciliolate; disk-corollas glabrous; outer achenes 7–10, inner 9–13 mm. long; marginal awns 5.5–8.5 mm. long. . . . . 5. *B. bidentoides.*

Phyllaries of outer involucre barely ciliolate; disk-corollas pilose on tube with multicellular hairs; outer achenes 6–6.5, inner 8–10 mm. long; marginal awns 3–6 mm. long. . . . . 6. *B. mariana.*

f. Achenes flat, not striate-nerved, with distinct slender midribs, glabrous to sparsely setulose; inner involucre 6–13 mm. long. . . *g.*

---

* In the measurements only the leading or well-developed heads should be used, the small later-developed ones being uncharacteristic. Fully mature achenes are essential for proper identification

g. Central achenes 1–2 mm. broad; primary leaves slender-petioled.
Achenes truncate, with erect or ascending marginal awns 2–4.6 mm. long.
Heads campanulate-hemispherical, essentially as broad as long; inner involucre 6–9 mm. long; achenes upwardly ciliate, the outer 4–5, the central 5–8 mm. long. . . . . . 7. *B. heterodoxa.*
Heads slender-cylindric to campanulate; inner involucre 8–13 mm. long; achenes (except in a var.) retrorse-ciliate, the outer 5.5–9.8, the central 6.5–11 mm. long. . . . . . 8. *B. Eatoni.*
Achenes with convex or dome-like summit, awnless or with widely divergent awns only 0.2–0.8 mm. long. . . . . 9. *B. infirma.*
g. Central achenes 2.4–3 mm. broad; primary leaves wing-petioled.
Outer phyllaries smooth-margined or but sparingly ciliate, 2–6 cm. long; inner involucres 0.9–1.2 cm. long; anthers included; central achenes 8.5–11 mm. long, with marginal awns 5–6 mm. long. . . . . 10. *B. comosa.*
Outer phyllaries strongly hirsute-ciliate from base to apex, 0.7–3.5 cm. long; inner involucre 6–8 mm. long; anthers exserted; central achenes 6–8.5 mm. long, with marginal awns 2–3 mm. long. . . . . 11. *B. tripartita.*
b. Primary leaves pinnate, with distinct leaflets, or deeply pinnate-divided and cleft into fine lobes. . . *h.*
h. Leaves with 3–5 mostly petiolulate uncleft leaflets; disk-corollas with a cup-like or campanulate throat; ligules wanting or shorter than outer involucre. . . *i.*
i. Outer involucre of 10–20 copiously hispid-ciliate phyllaries, inner of 8–18 narrowly deltoid or oblong-ovate phyllaries 7–12 mm. long; disk-corollas 2.5–4 mm. long, pale yellow; outer achenes 6.5–11.3 mm. long, 3.5–6.3 mm. broad; awns of inner achenes 4–9.5 mm. long. . 12. *B. vulgata.*
i. Outer involucre of 2–8 sparingly ciliate or smooth-margined phyllaries, inner of 5–12 oblong phyllaries 4–9 mm. long; disk-corollas 1.5–3 mm. long, orange; outer achenes 3.5–7 mm. long, 1.5–4 mm. broad; awns of inner achenes 1.5–5 mm. long.
Outer involucre of 5–8 sparingly ciliate phyllaries; disk-corollas 2.5–3 mm. long; achenes glabrous or only sparsely hairy, flat, the outer 5.3–7 mm. long and 3–4 mm. broad, with awns 2–4 mm. long; the inner 7–10 mm. long and 2.2–3.8 mm. broad, with awns 2.5–5 mm. long. . . . . 13. *B. frondosa.*
Outer involucre of 2–5 smooth-margined phyllaries; disk-corollas 1.5–2 mm. long; achenes upwardly strigose-hirsute, thick, the outer 3.5–4.6 mm. long and 1.5–1.7 mm. broad, with awns 1–1.5 mm. long; the inner 5.5–6.5 mm. long and 1–1.5 mm. broad, with awns 1.5–2.4 mm. long. . . . . 14. *B. discoidea.*
h. Leaves chiefly 3–7-divided into lanceolate to linear pinnately incised or coarsely serrate divisions (sometimes simple in no. 16); disk-corollas with slender-cylindric throat; ligules wide-spreading, showy, conspicuously longer than outer involucre. . . *j.*
j. Achenes marginless, with glabrous or merely strigose-ciliolate borders (except toward summit).
Inner achenes 4.5–9 mm. long, strigose-ciliolate, with awns 0.6–3.7 mm. long; plant of broad continental range. . . . . . 15. *B. coronata.*
Inner achenes 2.5–4.5 mm. long, glabrous-margined, with awns 0.5–1 mm. long; plant of se. Coastal Plain. . . . . 16. *B. mitis.*
j. Achenes with a thin friable hispid-ciliate margin.
Outer involucre of 8–12 smooth or finely ciliate spreading blunt or merely acute phyllaries 4–7 mm. long; outer achenes 3.3–5.2, inner 2.5–4.6 mm. broad. . . . . 17. *B. aristosa.*
Outer involucre of 12–25 bullate- or coarsely serrate-ciliate spreading or reflexed curling attenuate phyllaries 0.7–2 cm. long; outer achenes 2.5–3.8, inner 1.8–2.8 mm. broad. . . . . 18. *B. polylepis.*
a. Achenes linear-tetragonal, gradually tapering to the 2–4-awned summit, very unequal; the inner 1.2–1.8 cm. long, twice or thrice length of involucre; primary leaves 2–3 times pinnately divided. . . . . 19. *B. bipinnata.*

1. B. laèvis (L.) BSP. (smooth). — *Stem firm and usually smooth,* simple or with floriferous ascending middle and upper branches and divergent sterile lower ones, annual and erect or perennial and with elongate rooting bases, 0.25–1 m. high; leaves sessile or nearly so, linear-lanceolate to elliptic, appressed-serrate; *heads erect in anthesis,* erect or nodding in fruit, slender-peduncled; *outer involucre* of 5–9 narrow spreading or subappressed phyllaries *mostly about equaling the disk*, inner 9–12 mm. long; *chaff reddish-tipped; ligules golden-yellow*, elliptic or oval, 1.5–3 *cm. long;* disk hemispherical, 1.5–2.8 cm. broad; disk-corollas deep yellow, 4.5–5 mm. long, with slender tube, campanulate throat and 5-lobed limb; *anthers exserted, purple-black; achenes* narrowly cuneate, *straight and flat*, deep brown or purplish, 2–4-awned; the *outer achenes* 6–8 *mm. long and* 1.8–2.3 *mm. broad, with marginal awns* 2.8–4.5 *mm. long; central achenes* 8–9.5 *mm. long, with marginal awns* 3.5–5 *mm. long.* — Marshes and margins of pools and sluggish streams, either fresh or brackish, Fla. to Mex. and Calif., n. in coastal states to s. N.H., inland in the North locally to Ind. and W.Va. Aug.–Oct. (S. Am.; Hawaii) Fig. 1742. — A complex group, needing careful attention.

1742. B. laevis.

2. B. cérnua L. (nodding), Stick-tight. — *Annual with soft usually hispid simple* or freely branching *stem* 2 cm.–1.8 m. high, the larger plants with decumbent rooting bases; leaves linear, lanceolate, oblanceolate or narrowly oblong, sessile to connate or rarely petioled; *heads* erect in anthesis, usually *strongly nodding in fruit*, discoid or radiate; *outer involucre of* 3–10 *unequal widely spreading or reflexed foliaceous phyllaries much longer than the inner involucre; chaff yellow-tipped; ligules* (when present) bright yellow, *elliptic, rarely as much as* 1.7 *cm. long;* disk-corollas and anthers as in no. 1; *achenes curved, olivaceous, with almost wing-like* coarsely retrorse-barbed *pale margins and keels*, glabrous, convex at summit, the 4 (rarely 2) erectish awns retrorsely barbed; *outer achenes* 3.3–6.3 *mm. long* and 2–2.8 mm. broad, *with marginal awns* 2–2.8 *mm. long; central ones* 4.2–7.8 *mm. long* with marginal awns 2.6–4 mm. long. — Highly variable; the best marked vars. are the following:

a. Stem stoutish, 0.25–1 cm. in diameter at base, commonly branching, 0.5–1.8 m. high; leaves sessile or at most narrowed at base, thickish, 0.2–2 dm. long; heads commonly numerous, broadly hemispherical, many-flowered, the chief ones with disks 1–2.7 cm. broad, nodding in fruit; outer involucre of 5–10 phyllaries, inner of about 8 phyllaries 6–12 mm. long. . . *b.*
  b. Leaves tapering to long acuminate-attenuate tips, the primary ones with 4–24 pairs of sharp serrations; phyllaries of outer involucre mostly linear to lanceolate, acute or acutish. . . . *c.*
    c. Leaves (at least the larger) with broad sessile or subconnate bases, only gradually narrowed below the middle.
      Leaves linear to oblanceolate, 0.3–2(rarely –3) cm. broad, with 4–17 pairs of coarse teeth 1–5 mm. high; heads radiate (rarely discoid). . *B. cernua* (typical).
      Leaves (at least the larger) linear- to lance-oblong, 0.5–2.5 (rarely –3.5) cm. broad, with 12–24 pairs of fine teeth scarcely 1 mm. high; heads radiate. . . . . Var. *integra.*
    c. Leaves conspicuously narrowed to base, elliptic-lanceolate, 1.5–4.5 cm. broad, with 5–20 pairs of coarse teeth; heads radiate (rarely discoid). . Var. *elliptica.*
  b. Leaves mostly blunt or round-tipped, the primary ones 0.6–2 cm. wide, entire or with 1–6(–10) pairs of remote teeth; phyllaries of outer involucre oblong to spatulate, with obtuse or rounded tips; heads discoid. . . . Var. *oligodonta.*
a. Stem capillary, simple or but slightly forking, 0.2–2 dm. high; leaves petioled, oblanceolate or spatulate, thin, 0.4–4 cm. long, 1–7 mm. broad; heads solitary or few, campanulate, few-flowered, with disks 1.5–10 mm. broad, scarcely nodding in fruit; outer involucre of 2–6 phyllaries, inner of 3–6 phyllaries 2–7 mm. long. . . . . Var. *minima.*

B. cérnua (typical). — Sloughs, springs, pools and shores, M.I. to B.C., s. to N.S., N.E., Del., Md., w. N.C., Tenn., Mo., S.D., Wyo., Ida. and Calif. Aug.–Oct. (Eurasia) Fig. 1743. — Typical *B. cernua* has well-developed ligules. Forma **discoìdea** (Wimm. & Grab.) Briq. & Cavill. (like a disc), without ligules, is often more abundant, especially in e. Can.

1743. B. cernua.

Var. **íntegra** Wieg. (entire; name scarcely appropriate). — Swamps and bottoms, P.E.I. to N.D., s., locally, to Mass., upland of N.C., O., Ill., Mo. and Okla., chiefly westw.

Var. **ellíptica** Wieg. (elliptic). — Similar habitats, Que. to Wash., s. to N.E., L.I., w. N.C., Ky., Kans. and Colo. — Hybridizes with no. 4.

Var. **oligodónta** Fern. & St. John (few-toothed). — A low often matted fleshy extreme of brackish or saline (rarely fresh) shores, M.I. and P.E.I. to Mass. and Minn.

Var. **mínima** (Huds.) Pursh (smallest). — Bogs and shallow pools, M.I. to Man., s. to N.S., N.E., N.Y., Ind. and Wisc.

3. B. hyperbòrea Greene (northern), ESTUARY-BEGGAR-TICKS. — Stem glabrous or sparingly hispid, simple to freely branched, erect or depressed, 0.5–7 dm. high; *leaves linear to oblanceolate, sessile or narrowed to winged petioles; heads erect, cylindric to campanulate-hemispherical,* 15–60-flowered, discoid or radiate; *outer involucre of* 2–9 *ascending phyllaries,* inner of 8–12 narrowly oblong obtuse phyllaries 7–11 mm. long; *ligules* pale yellow, 0.4–1.2 cm. long, 1.5–6 *mm. broad; disk-corollas* 3.5–4 *mm. long,* with 4 (3–5)-*toothed* orange limb; *anthers included; achenes flat, or slightly thickened at summit, linear-cuneiform, straight, very slender at the attenuate base,* conspicuously rounded at the pale cartilaginous tip, olive-brown or drab, retrorsely ciliate, 2–4-nerved, *with* 7–15 *distinct ribs on each face;* the outer 4–8.5 mm. long; *the inner* 6–10 mm. long and 1.4–2.4 *mm. broad,* with retrorsely barbed marginal awns 1.8–5 mm. long. — Highly variable species localized in fresh to brackish estuaries. The following vars. strikingly isolated:

a. Outer achenes 4–5 mm. long; inner 5–7 mm. long, with marginal awns 1.8–3 mm. long.
- Stem simple, monocephalous; leaves blunt, subentire or obscurely few-toothed, the longer 4 cm. long; phyllaries of outer involucre about 1.5 cm. long, only slightly exceeding inner involucre. . . . . . . *B. hyperborea* (typical).
- Stem branching or simple, 1–4.5 dm. high; leaves attenuate-acuminate, commonly with 3–10 pairs of sharp teeth, the primary blades 0.5–1.2 dm. long; phyllaries of outer involucre 1.5–4 cm. long. . . . . Var. *colpophila.*

a. Outer achenes 5–8.5 mm. long; inner 7.5–10 mm. long, with marginal awns 3–5 mm. long. . . *b.*
- *b.* Leaves thinnish, with midrib prominent, long-attenuate at tip; the larger 0.7–1.7 dm. long, with 3–12 pairs of sharp teeth; phyllaries of outer involucre 3–9, linear-lanceolate, acute or acuminate; marginal awns of inner achenes 3.5–5 mm. long.
  - Tall, 3–7 dm. high, with strongly ascending branches; phyllaries of outer involucre 4–9, entire, 1.5–3.5 cm. long; disk-flowers 30–60. . . . . Var. *cathancensis.*
  - Low, 2–3 dm. high, with decumbent lower branches; phyllaries of outer involucre 3–5, mostly sharp-serrate, 2–8 cm. long; disk-flowers 15–30. Var. *arcuans.*
- *b.* Leaves more fleshy, with midrib less evident or obscure, obtuse or blunt-subattenuate; the larger 1.5–11 cm. long, entire or with 1–5 pairs of mostly blunt teeth; phyllaries of outer involucre 2–6, obtuse to subacute; marginal awns of inner achenes 2.5–4 mm. long.
  - Stems 1–4.5 dm. high; leaves slightly fleshy, subattenuate; the larger 3–11 cm. long, mostly with 1–5 pairs of teeth; phyllaries of outer involucre obtuse to subacute, entire or with 1–2 pairs of teeth; inner phyllaries 3–4 mm. wide. . . . . *B. hyperborea* (typical).
  - Stems 1–2.5 dm. high; leaves very fleshy, with rounded tips; the larger 1.5–5 cm. long, entire or with 1 pair of teeth; phyllaries of outer involucre similar, inner phyllaries 2–3 mm. wide. . . . . Var. *gaspensis.*

B. hyperbòrea (typical), incl. vars. *laurentiana* and *Svensoni* Fassett — Estuary of Rupert River, James Bay; also estuaries entering the St. Lawrence, Que., and Gulf of St. Lawrence, Que. and N.B. FIG. 1744. — Based on simple and dwarf individuals, these now shown by Fassett to be not truly representative.

1744. B. hyperborea.

Var. **colpóphila** (Fern. & St. John) Fern. (lover of estuaries). — Tidal mud of estuaries, Northumberland Strait, N.B. and N.S., to Mass. Aug.–Oct.

Var. **cathancénsis** Fern. (of Cathance River, Me.). — Estuaries of Kennebec system, s. Me. — A somewhat similar plant supposed to have been collected on Hackensack marshes, N.J., more than a century ago.

Var. **árcuans** Fern. (arching). — Estuary of Miramichi R., near Newcastle, N.B.

Var. **gaspénsis** Fern. (of Gaspé Pen., Que.). — Tidal mud of St. John (Douglastown) and Dartmouth Rivers, Gaspé Co., Que.

4. B. connàta Muhl. (united), BEGGAR-TICKS, STICK-TIGHT. — Stoutish or slender, 0.1–1.5 m. high, usually with loosely ascending branches, glabrous; *leaves* more or less petioled, *undivided or the primary ones with* 2–4 *divergent decurrent* (connate) *basal lobes;* the undivided primary leaves or the terminal lobes lanceolate to elliptic and acuminate, coarsely sharp-serrate or dentate, sometimes incised, 0.25–2 dm. long; heads thick-campanulate or hemispherical, discoid, rarely radiate, on peduncles 1–8 cm. long; *outer involucre of* 2–6 spreading or loosely ascending linear to spatulate *smooth-margined or barely ciliolate phyllaries; inner involucre* of about 8 elliptic to oblong-ovate blunt brown phyllaries 6–10 *mm. long;* ligules (when developed) golden-yellow, elliptic, about 8 mm. long and 4–5 mm. broad; disk-corollas orange; *anthers exserted, blackish; achenes* slender-cuneate, *with the midribs prominent,* (3–)4–5-*awned,* glabrous or sparingly setulose, often warty, *retrorse-ciliate* (sometimes with ascending cilia), truncate at summit; *the outer* 3–4-*angled,* 2–5-awned, 4–6.5 mm. long, 1.7–2.7 mm. broad; *the inner* 5–8 mm. long, 1.6–2.6 *mm. broad, with the marginal awns* 2.2–4.7 *mm. long.* — Very variable:

a. Middle and lower primary leaves undivided or with 2–4 spreading decurrent and approximate broad basal lobes. . *b.*

b. Middle and lower (sometimes upper) primary leaves with 2–4 decurrent or confluent broad basal lobes; outer achenes 3–5.5 mm. long and 1.7–2.5 mm. broad; the inner 4.5–8 mm. long and 1.6–2 mm. broad, with marginal awns 2–3.6 mm. long. . . *c.*

c. Petioles short, broadly margined; outer achenes 4–5.3 mm. long; inner 5–6.5 mm. long.

Blades of middle cauline leaves and terminal lobes of the divided ones (except in starved specimens) with 7–20 pairs of sharp serrations; outer phyllaries linear to lanceolate or oblanceolate, the longest 1–2.5 cm. long and 1–4 mm. broad. . . . . *B. connata* (typical).

Blades or terminal lobes of middle cauline leaves coarsely dentate with 1–10 pairs of teeth; outer phyllaries oblanceolate to oblong, the larger 2.5–6 cm. long and 0.3–1.5 cm. broad. . . . . Var. *fallax.*

c. Petioles long and slender. . *d.*

d. Outer achenes 3–4 mm. long; inner 4.5–5 mm. long, their marginal awns 2–2.5 mm. long. . . . . Var. *gracilipes.*

d. Outer achenes 5–5.5 mm. long; inner 6–8 mm. long, their marginal awns 3–3.5 mm. long.

Awns retrorsely barbellate. . . . . Var. *inundata.*

Awns with mixed retrorse and antrorse barbing. . . . . . Var. *ambiversa.*

b. All (or nearly all) leaves unlobed, tapering to slender or narrowly margined petioles; achenes often larger, the outer up to 6.5 mm. long and 2.8 mm. broad; the inner up to 8 mm. long and to 2.6 mm. broad, with marginal awns up to 4.7 mm. long.

Awns retrorsely barbed. . . . . Var. *petiolata.*

Awns antrorsely barbellate. . . . . Var. *anomala.*

a. Middle and lower primary leaves pinnately 3–7-parted, with distant linear-lanceolate divisions, the terminal narrowly lanceolate division deeply cleft. Var. *pinnata.*

B. connàta (typical). — Wet shores and swamps, s. Que. and s. Ont., s. to N.S., N.E., Del., Md., O., Wisc. and Minn. Aug.–Oct.

Var. **fállax** (Warnst.) Sherff (deceitful). — Local, s. Que. to Minn., s. to Va. and Ind. (Natzd. in Eu.)

Var. **gracílipes** Fern. (slender-stalked). — Wet peaty and sandy pond-margins, s. Me. to Conn.

Var. **inundàta** Fern. (subject to inundation). — Wet or inundated places, N.S.

Var. **ambivérsa** Fassett (turned around). — Bogs, n. Wisc.

Var. **petiolàta** (Nutt.) Farw. (petioled). — Damp shores and swamps, common, s. Que., s. Ont. and Minn., s. to N.S., N.E., Va., Tenn., Mo. and Kans. Fig. 1745.

1745. B. connata, v. petiolata.

Var. **anómala** Farw. (departing from rule). — Rare and local, centr. N.Y. to Wisc., s. to Md. and O.

Var. **pinnàta** S. Wats. (pinnate), *B. Sandbergii* Rydb. — Wisc. and Minn.

5. B. bidentoìdes (Nutt.) Britt. (resembling *Bidens;* originally placed in *Coreopsis*). — Glabrous annual with purplish or green simple or paniculately branched stem 1–9 dm. high; *leaves* thin, *lanceolate, coarsely serrate, long-attenuate* to an entire acuminate tip and below *to a long slender petiole,* the primary blades 0.5–1.6 dm. long and 0.5–2.5 cm. broad; *heads erect, slender-cylindric,* 10–25-flowered; *outer involucre of* 3–5 narrowly lanceolate to oblanceolate *entire or merely scabrous-ciliolate erect foliaceous phyllaries* up to 6 cm. long; *inner involucre of* leading heads of 4–8 oblong obtuse conspicuously striate *phyllaries* 0.9–1.6 *cm. long* and 2.5–4.5 *mm. wide;* ligules wanting or inconspicuous; *disk-corollas glabrous,* yellow,

1746. B. bidentoides.

4-toothed, 3.5–4.5 *mm. long; achenes plano-convex, uniformly and closely striate-nerved, copiously short-setulose with ascending hairs, linear to linear-oblanceolate,* 2 (rarely 3–4)-*awned,* truncate at summit, *outer* 7–10 *mm. long* and 1–1.7 *mm. broad; central* 9–13 *mm. long* and 1–1.5 *mm. broad; awns* very slender, *upwardly barbellate,* the 2 *marginal nearly parallel or but slightly divergent* and 5.5–8.5 *mm. long,* the others (when present) very short. — Fresh to brackish tidal shores of the Hudson R. to Albany and Rensselaer Cos., N.Y.; tidal shores, Delaware R. system, N.J., Pa. and Del.; Maurice R., N.J. Aug.–Oct. Fig. 1746.

1747. B. mariana.

6. B. mariàna Blake (of Maryland). — Similar to no. 5; leaves more often laciniate-lobed near base; larger heads 18–32-flowered; *outer involucre* at most 3 *cm. long, its phyllaries not ciliate;* inner involucre 0.9–1.3 cm. long; *disk-corollas* 4–5.8 *mm. long,* 4–5-toothed, *the tube pilose with multicellular hairs; outer achenes* of disk 6–6.5 *mm. long; central* 8–10 *mm. long; marginal awns* 3–6 *mm. long.* (*B. bidentoides,* var. *mariana* (Blake) Sherff) — Tidal shores, Northeast R. and lower Susquehanna R., Md. Sept., Oct. Fig. 1747.

7. B. heterodóxa (Fern.) Fern. & St. John (contrary to the standard). — Glabrous, simple to freely branched, with strongly ascending branches, 1–9 dm. high; *leaves simple or* 3–5-*parted, slender-petioled, or the upper on margined petioles or subsessile;* blades or terminal lobes narrowly lanceolate to narrowly ovate, coarsely and *sharply serrate; heads* discoid or radiate, *campanulate-hemispherical;* outer involucre of 3–7 ascending linear-lanceolate to oblanceolate smooth-margined or but sparingly ciliate phyllaries 1–4 cm. long; *phyllaries of inner involucre* black-striped, oblong, obtuse, 6–9 *mm. long;* ligules (when developed) elliptical, orange-yellow, about 1 cm. long, 5–6 mm. broad; disk-corollas orange-yellow, 2.5–3.5 mm. long, 5-toothed; anthers blackish, exserted; *achenes flat,* strigose or glabrate, cuneate-oblong to narrowly obovate, *truncate,* 2(–4)-awned, *upwardly ciliate,* outer 4–5 mm. long and 1.8–2.3 mm. broad; central 5–8 mm. long and 1.5–2 mm. broad, with marginal awns 2–4.5 mm. long. — Highly localized, resembling variations of no. 4 but with marked achenial character:

Leaf-blades or terminal lobes narrowly lance-acuminate, slenderly sharp-toothed or incised, their teeth mostly longer than broad; branches elongate, much exceeding subtending leaves.
- Awns upwardly barbellate. . . . . *B. heterodoxa* (typical).
- Awns retrorsely barbed. . . . . Var. *orthodoxa.*

Leaf-blades or terminal lobes oblong-lanceolate to lance-ovate, their teeth mostly deltoid and broader than long; branches short, mostly exceeded by subtending leaves.
- Awns retrorsely barbed. . . . . Var. *monardaefolia.*
- Awns smooth and barbless or barely scabrous. . . . . Var. *agnostica.*

B. heterodóxa (typical). — Fresh, brackish or saline marshes, M.I. and P.E.I. Aug.–Oct. Fig. 1748.

Var. **orthodóxa** Fern. & St. John (conforming to standard). — Similar habitats, M.I.

Var. **monardaefòlia** Fern. (with leaves of *Monarda*). — Strand of Pocotopaug L., Middlesex Co., Conn.

Var. **agnóstica** Fern. (undecided; neither one nor the other). — With the last.

1748. B. heterodoxa.

8. B. Eàtoni Fern. (for its discoverer, Alvah Augustus Eaton, 1865–1908). — Slender, glabrous, purple- or green-stemmed, simple or *strictly branching,* 0.2–1.5 m. high; *leaves thin, lanceolate, long-attenuate above to an entire acuminate tip and* below *to the petiole;* primary ones 5–15 cm. long, 0.7–2.5 cm. broad, with 4–15 pairs of long sharp serrations, sometimes deeply cleft toward base; *heads erect, slender-cylindric to campanulate* (after pressure ovoid), 7–33-flowered; *outer phyllaries* 3–5, narrowly oblanceolate, *smooth-margined* or only obscurely ciliolate, 0.7–7 cm. long; *phyllaries of inner involucre* 5 or 6, conspicuously striate, 8–13 *mm. long;* ligules usually wanting; *achenes flattish* or with the slender midribs slightly prominent above, *scarcely or but faintly striate, linear-oblanceolate,* glabrous or sparingly appressed-setulose, usually with a few retrorse marginal setae, 2–4(rarely –5)-awned, truncate; *outer* 5.5–9.8 *mm. long; inner* 6.5–11 *mm. long* and 1–2 mm. broad; marginal awns 3–4.6 mm. long, about equaling the pale corolla. — Like no. 3 a locally segregated estuarine species:

*a.*Inner achenes 6.8–9 mm. long; awns 2, rarely 4. . . *b.*
*b.*Heads slenderly cylindric when fresh, 7–25-flowered.
Awns retrorsely barbed. . . . . *B. Eatoni* (typical).
Awns upwardly barbellate. . . . . Var. *fallax.*
*b.*Heads thick-cylindric to campanulate. . . *c.*
*c.*Inner phyllaries with dull yellow to dark brown striations; lower leaves often 3-cleft.
Awns 1.5–4 mm. long, retrorsely barbed. . . . . Var. *interstes.*
Awns 0.5–2 mm. long, both retrorsely and upwardly barbed. . . . Var. *mutabilis.*
*c.*Inner phyllaries with lustrous amber to purple-brown striations, lustrous black in age; leaves all simple. . . . . Var. *simulans.*
*a.*Inner achenes 9–11 mm. long; awns 4, rarely only 2.
Heads, when fresh, cylindric; leaves with slender petioles 1–3 cm. long, the lower usually cleft. . . . . Var. *kennebecensis.*
Heads campanulate; leaves with short winged petioles or subsessile, uncleft. Var. *major.*

B. Eàtoni (typical). — Tidal shores of Merrimac R., Essex Co., Mass. Aug.–Oct. Fig. 1749.

Var. **fállax** Fern. (deceitful). — With the last, Merrimac R., Essex Co., Mass.; St. Lawrence R., Quebec Co., Que.

Var. **intérstes** Fassett (standing between). — Tidal shores of St. Lawrence R., Bellechasse Co., Que., and of Kennebec R. system, Me.

Var. **mutábilis** Fassett (changeable). — Tidal shores of Kennebec R., Me.

Var. **símulans** Fassett (simulating). — Brackish marshes and tidal shores about mouth of Connecticut R., Ct.

Var. **kennebecénsis** Fern. (of the Kennebec River). — Tidal shores, estuaries of Kennebec R. system, Me.

Var. **màjor** Fassett (larger). — Tidal shores of Quinnipiac R., Ct., and of Hudson R., N.Y.

1749. B. Eatoni.

9. B. infírma Fern. (deprived of strength; *i.e.*, awnless). — *Diffusely branched,* 1–2.5 dm. high; *leaves* membranaceous, *slender-petioled, with narrowly lanceolate* coarsely serrate-dentate *blades tapering to base and apex;* heads discoid, slender-peduncled, campanulate, 7–13-flowered; *outer involucre quite glabrous,* its 2–4 linear or lanceolate phyllaries spreading or loosely ascending; phyllaries of inner involucre 5–8 mm. long; *achenes* flat, *oblanceolate,* 1-nerved on each face, strigose, *with convex or dome-like summit; the outer* 4–6.3 mm. long and 1–1.5 mm. broad, *rounded and without awns at the upper margins;* central 5.5–8 mm. long, *awnless or with short subulate divergent antrorsely barbellate awns only* 0.2–0.8 *mm. long.* Tidal flats of the St. Lawrence, Bellechasse Co., Que. Aug., Sept. Fig. 1750.

1750. B. infirma.

10. B. comòsa (Gray) Wieg. (tufted; from the leafy phyllaries), Beggar-ticks. — Stoutish, simple or with strongly ascending branches, 3–8 dm. high; *leaves undivided* or occasionally lobed at base, *elliptic-lanceolate,* acuminate, *tapering to wing-margined petioles;* primary blades regularly and coarsely serrate, 0.6–2 dm. long; heads campanulate-hemispherical to subglobose, on stout peduncles up to 8.5 cm. long; *outer phyllaries* 6–10, linear to oblanceolate, erect to slightly spreading, *smooth-margined or but sparingly ciliate,* unequal, 2–6 cm. long; *inner phyllaries* about 8, *oblong-ovate to ovate-lanceolate, subacute,* brownish, with stramineous border, 0.9–1.2 *cm. long;* ligules usually wanting; *disk-corollas pale yellow, about* 5 *mm. long,* 4-toothed; *anthers pale, included; achenes flat,* with the midrib not prominent, cuneate-oblanceolate, glabrous, *retrorse-ciliate,* with 3 (2–5) stout retrorsely barbed awns; *outer achenes* 5.5–9.5 *mm. long,* 2.7–3.3 *mm. broad; inner* 8.5–11 *mm. long* and 2.5–3 *mm. broad, with marginal awns*

1751. B. comosa.

5–6 *mm. long.* (Incl. var. *acuta* Wieg.) — Rich moist or comparatively dry soils, centr. Me. to N.D., s. to N.C., Tenn., La., N.M. and Utah. Aug.–Oct. Fig. 1751.

11. B. tripartìta L. (three-parted), Beggar-ticks of Eu. — Resembling no. 4; *leaves simple, incised-serrate, cleft or* 3–5-*lobed,* the uncleft blade or terminal lobe lanceolate to narrowly rhombic, *the petiole margined and short;* heads hemispherical; *outer phyllaries* 4–8, *strongly hirsute-ciliate from base to apex,* often serrate, 0.7–3.5 cm. long; *inner involucre* 6–8 *mm. long;* disk-corollas 3–4 mm. long; *anthers pale, exserted; achenes flat, with corky-thickened margin and midribs,* retrorse-ciliate, glabrous, 2–4-awned; *the inner* cuneate, 6–8.5 mm. long, 2.4–3 *mm. broad, with marginal awns* 2.3–3 *mm. long.* — Swampy thickets and waste places, local, Gaspé Co., Que., to N.H., and sometimes about ports. Aug.–Oct. (Adv. from Eu.) Fig. 1752.

1752. B. tripartita.

12. B. vulgàta Greene (common), Beggar-ticks, Stick-tight. — Coarse annual, up to 1.5 m. high; stem commonly reddish, furrowed, with spreading-ascending branches, glabrous; or stem as well as leaves and foliaceous bracts finely crisp-puberulent in forma **pubérula** (Wieg.) Fern. (with minute pubescence), var. *puberula* (Wieg.) Greene or *B. puberula* (Wieg.) Rydb.; *leaves slender-petioled, the principal cauline ones with* 3–5 *mostly petiolulate leaflets* or with the terminal and the lower lateral ones again cleft; leaflets lance-acuminate, coarsely serrate, cuneate; heads hemispherical, radiate or discoid, on stoutish angled peduncles up to 2 dm. long; *outer involucre of* 10–20 loosely ascending unequal *copiously hispid-ciliate phyllaries* 1–4(–10) cm. long; *inner of* 8–18 *narrowly-deltoid to oblong-ovate* olive to chestnut *phyllaries* 7–12 *mm. long,* these usually conspicuously shorter than disk; ligules (when developed) pale yellow; *disk-corollas* 2.5–4 mm. long, *pale yellow,* 4- or 5-toothed; *anthers mostly included; achenes flat,* with slender midribs, 2-awned, *pale brown to olive,* the margins upwardly ciliate nearly to the *retrorse-ciliate summit,* the erect to divaricate awns retrorsely barbed; *outer achenes obovate-cuneate,* 6.5–11.3 *mm. long* and 3.5–6.3 *mm. broad, with awns* 3–6 *mm. long; inner* oblong-cuneate, 0.9–1.7 *cm. long,* 2.8–6.2 mm. broad, with awns 4–9.5 mm. long. — Rich low ground, ditches, roadsides, waste places, etc., a disagreeable weed, s. Que. to Alta. and Wash., s. to N.S., N.E., N.C., Tenn., Mo., Kans., Wyo., Nev. and Calif. Aug.–Oct. Fig. 1753.

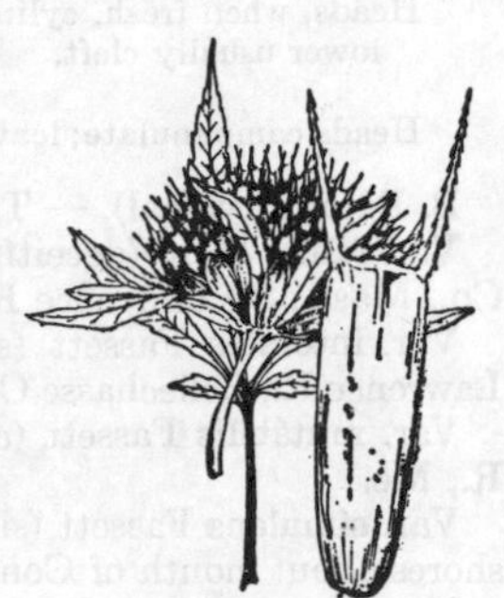

1753. B. vulgata.

Var. **schizántha** Lunell (dissected-flowered; most unfortunate name for a *cut-leaved* plant). — Principal leaves bipinnate or tripinnatisect. — L. Sup., Ont., to Sask. and N.D.

13. B. frondòsa L. (leafy; referring to the outer involucre), Beggar-ticks, Stick-tight. — More slender than no. 12, 0.1–1 m. high; terminal leaflets of principal leaves long-petiolulate, lanceolate to lance-ovate; heads hemispherical to subglobose (in dwarfed plants campanulate); *outer involucre of* 5–8 loosely ascending *entire sparingly ciliate phyllaries* 0.5–2(–6) cm. long; *inner* of 6–12 *oblong blunt* brown or lead-colored phyllaries 5–9 *mm. long,* these usually about equaling disk; ligules (when developed) golden-yellow; *disk-corollas* 2.5–3 mm. long, *orange,* 5-toothed; *stamens exserted,* purple-black; *achenes blackish, with margins upwardly ciliate to the bases of the* slightly divergent to erect *awns; the outer* obovate-cuneate, 5.3–7 *mm. long* and 3–4 *mm. broad, with awns* 2–4 *mm. long; the inner* oblong- to linear-cuneate, 7–10 *mm. long* and 2.2–3.8 *mm. broad, with awns* 2.5–5 *mm. long.*

Teeth of leaflets broadly deltoid, nearly or quite as broad as long; most phyllaries of outer involucre rarely more than twice length of inner involucre.
- Awns of achenes retrorsely barbed. . . . . *B. frondosa* (typical).
- Awns upwardly barbellate. . . . . Forma *anomala.*

Teeth of leaflets lance-attenuate; all phyllaries of outer involucre two-and-a-half to four times as long as inner. . . . . Var. *stenodonta.*

B. frondòsa (typical). — Damp open habitats or in waste or cult. ground, a bad weed, w. Nfld. to Wash., s. beyond our range. Aug.–Oct. Fig. 1754. — Forma **anómala** (Porter) Fern. (anomalous), occasional from e. Que. to w. Ont., s. to D.C. and Kans. — Var. **pállida** Wieg. (pale) a glaucous plant chiefly from shores of Cayuga L., N.Y., is probably a hybrid of no. 13 with a form of no. 4.

Var. **stenodónta** Fern. & St. John (slender-toothed). — Wet thickets, e. Nfld., M.I. and P.E.I.

1754. B. frondosa.

14. B. discoidea (T. & G.) Britt. (discoid; without rays). — Very slender, commonly red-stemmed, subsimple to intricately bushy-branched, 0.1–1.8 m. high; *leaves membranaceous*, glabrous, *the lower and middle primary ones 3-foliolate*, the upper and rameal often simple, the 3-foliolate ones *on slender marginless petioles*, with long-acuminate lanceolate obliquely cuneate coarsely toothed leaflets; terminal leaflet slender-petiolulate; *heads* mostly 15–20-flowered, campanulate, discoid, 3–6 *mm. broad*, on slender peduncles up to 4 cm. long; *outer involucre* of 2–4 (–5) loosely ascending or spreading *glabrous* narrow phyllaries 0.3–2.5 cm. long; *phyllaries of inner involucre about* 5, oblong, obtuse, 4–6.2 *mm. long; disk-corollas orange*, 1.5–2 *mm. long; achenes* cuneate-obovate to -oblanceolate, *thickish, upwardly strigose-hirsute*, truncate or notched at summit, *with 2 stout erect upwardly hairy awns; outer achenes 3.5–4.6 mm. long and 1.5–1.7 mm. broad, with awns 1–1.5 mm. long; inner achenes 5.5–6.5 mm. long, 1–1.5 mm. broad, with awns 1.5–2.4 mm. long.* — Peaty or wet sandy swamps and shores or on debris in pools, Ala. to Tex., n., somewhat locally, to N.S., s. Me., s. Que., s. Ont., Mich., Wisc. and Minn. Aug.–Oct. Fig. 1755.

1755. B. discoidea.

15. B. coronàta (L.) Britt. (crowned), Tickseed-Sunflower. — Stem slender, 0.3–1.5 m. high, smooth and firm, subsimple to freely forking; *leaves short-petioled* (principal petioles 0.5–2 cm. long), 3–7-*divided into lanceolate to linear pinnately incised or coarsely serrate divisions, or undivided;* heads few–many, on slender peduncles 0.25–1.5 dm. long, showy, radiate; *outer involucre of* 6–11 (usually 8) linear or narrowly spatulate *smooth or but slightly ciliate appressed to loosely ascending* blunt *phyllaries* 3–10 (by foliar development –18) *mm. long;* inner phyllaries narrowly oblong, subacute and 3.5–7.5 mm. long; chaff black-tipped; ligules golden-yellow, oblong, wide-spreading, 1.2–3 cm. long; disk-corolla 3.5–4.7 mm. long; anthers exserted, purple-black; *achenes marginless*, smooth or slightly hairy, flattish, upwardly ciliate; *outer achenes* cuneate-oblong or -obovate, 3–6 mm. long, 1.6–2.5 *mm. broad*, with awns 0.4–2.6 mm. long; *inner achenes* 4.5–9 mm. long, 0.9–2.4 *mm. broad, with awns* 0.6–3.7 *mm. long*. (Incl. *B. trichosperma* (Michx.) Britt.) Four geographic vars.:

Longest central achenes 6–9 mm. long.
- Awns of central achenes 2.8–3.7 mm. long; larger outer achenes 5–6 mm. long, with awns mostly 2–2.6 mm. long; larger leaf-segments 4–20 mm. broad or leaves simple; outer involucre loosely ascending to spreading. . . . . *B. coronata* (typical).
- Awns of central achenes 0.6–1.8 mm. long; larger outer achenes 3.7–6 mm. long, with awns 0.6–1 mm. long; larger leaf-segments 1–12 mm. broad; outer involucre appressed-ascending. . . . . Var. *brachyodonta*.

Largest central achenes 4.5–6 mm. long; larger leaf-segments 1–12 mm. broad.
- Ligules 1.5–2.5 cm. long; awns of central achenes 1–2 mm. long; larger outer achenes 3–5.2 mm. long, with awns 0.4–1.6 mm. long; plant of the interior. Var. *tenuiloba*.
- Ligules 1.2–1.7 cm. long; awns of central achenes 1.8–3 mm. long; larger outer achenes 4.7–6 mm. long, with awns 1.5–1.8 mm. long; plant of Atl. states. . Var. *trichosperma*.

B. coronàta (typical). — Prairies, swales and rich bottoms, Ct. R., Ct., to s. Ont. and Neb., s. to e. Va., Pa., O., Ind. and Ill. Aug.–Oct. Fig. 1756.

Var. **brachyodónta** Fern. (short-toothed). — Bogs and peaty meadows, e. Mass. to e. Va.

Var. **tenuíloba** (Gray) Sherff (slender-lobed). — Bogs and swales, s. Ont. to Minn., s. to O., Ind., Ill., Ia. and Neb.

Var. **trichospérma** (Michx.) Fern. (hairy-seeded). — Bogs and swamps, N.J. and e. Pa. to Fla.

16. B. mitis (Michx.) Sherff (ripe; the original specimens in mature fruit). — Similar to no. 15; leaves in ours finely divided and subdivided into linear segments; *achenes glabrous or nearly so, with non-ciliate margin, 2.5–4.5 mm. long, with mere obtuse marginal teeth or with awns up to 1 mm.*

1757. B. mitis.

1756. B. coronata.

long. — Brackish to fresh swamps, Fla. to La., n., very locally, to e. Md. Sept., Oct. Fig. 1757.

17. B. aristòsa (Michx.) Britt. (bearing bristles), Tickseed-Sunflower. — Resembling no. 15; *outer involucre of* 8–12 *spreading smooth or ciliate blunt or merely acute phyllaries* 4.5–7 (by foliar development –18) mm. long; *achenes with a thin friable hispid-ciliate often erose-notched margin, the outer* 4–7 mm. long and 3.3–5.2 *mm. broad, the inner* 4.8–8.2 mm. long and 2.5–4.6 *mm. broad.* — Three vars.:

Awns well-developed, commonly 2–4 mm. long.
  Awns upwardly barbellate. . . . . *B. aristosa* (typical).
  Awns retrorsely barbed. . . . . Var. *Fritcheyi.*
Awns wanting or mere rudiments. . . . . Var. *mutica.*

B. aristòsa (typical). — Low grounds, Del. to Minn., s. to Va., Ala., Miss., La. and Tex.; adv. on roadsides and waste places ne. to N.E. Aug.–Oct. Fig. 1758.

Var. **Frítcheyi** Fern. (for its discoverer in 1858, John Quincy Adams Fritchey). — Ind. and Ky. to Ill. and Mo.; adv. in waste places and on roadsides to the Atl. states.

Var. **mùtica** Gray (docked). — Range similar to that of last, eastw. to O.

18. B. polýlepis Blake (with many scales). — Similar to no. 17; *outer involucre of* 12–25 *bullate- or coarsely serrate-ciliate spreading or reflexed curling attenuate phyllaries* 0.7–2 *cm. long; achenes with a hispid-ciliate thin friable margin; the outer* 4.5–5.5 mm. long and 2.5–3.8 *mm. broad*, with smooth or upwardly barbellate awns 0.2–1 mm. long, *the inner* 5.8–6.5 mm. long and 1.8–2.8 *mm. broad.* (*B. involucrata* (Nutt.) Britt., not Phil.) — Low prairies, swales and shores, Ill. to Ia., Kans. and Colo., s. to Tenn., Mo., Okla. and Tex.; natzd. on roadsides and in low fields and marshes, e. to the Atl. states. Aug.–Oct. Fig. 1759.

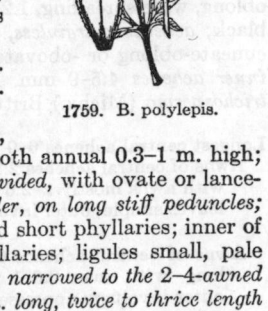

1758. B. aristosa.

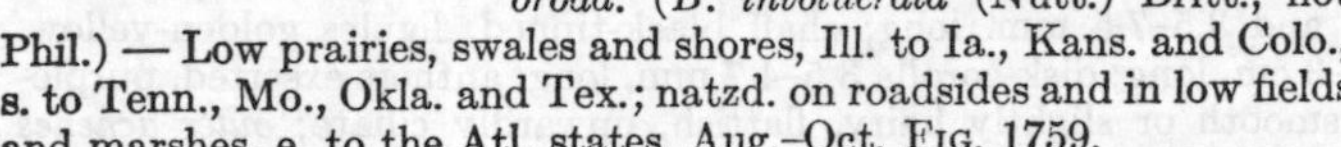

1759. B. polylepis.

Var. **retrórsa** Sherff (directed backward). — Awns retrorsely barbed. — Local, O. and Ind. to Mo.

19. B. bipinnàta L. (twice-pinnate), Spanish-needles. — *Square-stemmed* smooth annual 0.3–1 m. high; *leaves* membranous, 2–3-*times pinnately divided*, with ovate or lanceolate cuneate-based divisions; *heads slender, on long stiff peduncles;* outer involucre of about 8 linear appressed short phyllaries; inner of pale-bordered castaneous lanceolate phyllaries; ligules small, pale yellow; *achenes linear-tetragonal, gradually narrowed to the 2–4-awned summit, very unequal, the inner* 1.2–1.8 *cm. long, twice to thrice length of involucre;* awns divergent, pale, strongly retrorse-barbed. — Rocky woods, roadsides and waste places, often in sandy soil, Fla. to Mex., n. to Mass., N.Y., O., Ind., Ill., Mo., Kans., etc. Aug.–Oct. (Trop. Am. and semicosmop. weed) Fig. 1760.

1760. B. bipinnata.

B. pilòsa L. (hairy), var. radiàta Sch. Bip. (with rays), *B. leucantha* (L.) Willd., a weedy square-stemmed annual with simple ovate or 3–5-parted leaves, white to pale yellow or purplish ligules and linear-tetragonal achenes 5–9 mm. long, is a casual weed n. to Mass. (Adv. from Trop. Am.)

56. CÓSMOS Cav. Cosmos

Rather unsatisfactorily distinguished from *Bidens* by its slender subterete or somewhat 5-angled antrorsely hispid achenes tapering to slender beaks; ligules purple or red to white, in one species orange; heads on long naked peduncles; leaves (in ours) pinnatisect. — Showy annuals or perennials of the warmer reg. of N. and S.Am. (Greek *cosmos, a decoration.*)

1. C. bipinnàtus Cav. (twice pinnately divided). — Annual, with finely dissected leaves, the segments linear or narrowly lanceolate; *ligules red, pink or white;* achenes glabrous. — Roadsides and waste places, spread from cult., becoming estab. southw. July–Oct. (Introd. from Mex.)

2. C. sulphùreus Cav. (sulphur-colored). — Annual; segments of leaves lanceolate to elliptic; *ligules orange or golden-yellow;* achenes hispidulous. — Roadsides and waste places, N.J. and Pa. southw., spread from cult. Aug.–Oct. (Introd. from Mex.)

57. MEGALODÓNTA Greene WATER-MARIGOLD

As in *Bidens* but achenes almost terete, cartilaginous, truncate at both ends, with 3–6 long rigid persistent sharp awns smooth below but retrorsely barbed at summit. — An aquatic with finely dissected submersed leaves, simple emersed ones and mostly solitary heads with showy ligules. (Name from the Greek *megas, large,* and *odous, tooth.*)

1. M. Béckii (Torr.) Greene (for its discoverer, LEWIS CALEB BECK, 1798–1853). — Submersed stems elongate; the ternately multifid submersed leaves suggesting whorls; emersed leaves firm, pectinate to entire; achenes (rarely maturing) 1–1.5 cm. long, the longer divergent awns 1.3–2.5 cm. long. (*Bidens* Torr.) — Ponds and slow streams, C.B. to s. Que., w. to Minn., s. to N.S., N.E., N.J., Pa., O., Ind., Wisc. and Mo.; also Wash. and Oreg. Aug.–Oct.

58. MARSHÁLLIA Schreb. BARBARA'S-BUTTONS

Heads many-flowered; flowers all tubular and perfect; corolla-lobes slender and spreading. Phyllaries linear-lanceolate, foliaceous, erect, in 1–2 rows, nearly equal. Receptacle convex or conical, with narrowly linear rigid chaff. Achenes turbinate, 5-angled; pappus of 5 or 6 membranaceous pointed scales. — Smooth low N. Am. perennials, with alternate entire 3-nerved leaves, and long-pedunculate heads terminating the simple stem or branches. Flowers purplish; anthers blue. (Named at the request of Muhlenberg for Dr. *Moses Marshall,* 1758–1813, nephew of the more distinguished Humphrey Marshall.)

Leafy-stemmed; leaves oblanceolate to lanceolate or ovate; achenes longer than pappus.
- Leaves thin, extending without much reduction in size to summit of stem, lanceolate or ovate, strongly acuminate; achenes 2 mm. long. . . . . . . 1. *M. trinervia.*
- Leaves thick, the lower oblanceolate or narrowly obovate, round-tipped to acutish, the upper rapidly decreasing in size; achenes 3–5 mm. long.
  - Phyllaries 6–8 mm. long; pales spatulate-clavate; florets 1–1.3 cm. long, with throat 1–1.5 mm. long. . . . . 2. *M. obovata.*
  - Phyllaries 8–12 mm. long; pales linear-oblanceolate; florets 1.5–2 cm. long, with throat 3–3.5 mm. long. . . . . 3. *M. grandiflora.*

Scapose or subscapose, with crowded linear-spatulate basal leaves; achenes shorter than pappus. . . . . 4. *M. caespitosa.*

1. M. trinérvia (Walt.) Porter (three-nerved). — Leafy-stemmed, 3–7 dm. high; *leaves thin,* broadly lanceolate to ovate, *prominently acuminate,* conspicuously 3-ribbed; peduncles 0.5–2 dm. long; involucre hemispherical or broad-campanulate, with acute linear-lanceolate phyllaries 6–10 mm. long; florets 1–1.5 cm. long, pubescent, the corolla dilated at throat; *achenes 2 mm. long,* slightly *longer than the lance-acuminate entire pappus-scales.* — Pinelands and damp woods, Ala. and Miss., n., very locally, to Va. and Tenn. June, July.

2. M. OBOVÀTA (Walt.) Beadle & F. E. Boynt. (obovate). — Leafy at base only; the narrowly obovate or oblanceolate leaves 2–5 cm. long; stem scapose, naked or with 1 or 2 very small bracts, 1.5–3 dm. high; *phyllaries* oblong, *6–8 mm. long; pales of receptacle spatulate-clavate; florets 1–1.3 cm. long, with distended throat* 1–1.5 *mm. long;* achenes 3–3.5 mm. long, longer than the pappus. — Pinelands and dry woods, Fla. to Ala. and N.C. Represented with us by

Var. **platyphýlla** (M. A. Curtis) Beadle & F. E. Boynt. (broad-leaved). — Stem leafy below the middle, the larger leaves 6–12 cm. long. — Fla. to s. Mo. (acc. to *Palmer & Steyerm.*) and N.C. May–July.

3. M. grandiflòra Beadle & F. E. Boynt. (large-flowered). — Coarser, 0.3–1 m. high; *leaves thick, the lowest* narrowly obovate or oblanceolate and *blunt;* the cauline lanceolate, much reduced upward; *phyllaries* oblong-lanceolate, 8–12 *mm. long; florets* 1.5–2 *cm. long, the enlarged throat* 3–3.5 *mm. long; achenes* 4–5 *mm. long, twice or thrice length of pappus.* (*M. obovata* and var. sensu ed. 7, not Beadle & F. E. Boynt.) — Moist or dry banks of streams and openings, upland of Pa. to Ky. and N.C. June, July.

4. M. caespitòsa Nutt. (forming mats). — Cespitose, the *linear-spatulate thick leaves mostly crowded at base,* the scapiform or subscapose peduncle 1.5–4 dm. high; phyllaries linear to lanceolate, 7–9 mm. long; *achenes* 2–2.5 mm. long, *shorter than the ovate serrulate pappus-scales.* — Prairies and openings, La. and Tex., n. to s. Mo. and Okla. May, June.

59. GALINSÒGA R. & P.

Heads several-flowered, radiate; rays 4–5, small, roundish, pistillate. Involucre of 4–5 ovate thin phyllaries. Receptacle conical, with narrow chaff. Pappus of small oblong cut-fringed chaffy scales, sometimes wanting. — Annual Am. herbs, with opposite triple-nerved thin leaves, and small heads; disk yellow; rays white or reddish. (Named for *Mariano Martinez de Galinsoga*, a Spanish botanist of the 18th century.)*

Ligules white or whitish.
- Pappus of disk-flowers aristate; pappus of ligulate flowers equaling corolla-tube; marginal achenes densely hispid on inner faces. . . . . 1. *G. ciliata.*
- Pappus of disk-flowers awnless; pappus of ligulate flowers wanting; marginal achenes glabrous or merely pilose at summit. . . . . 2. *G. parviflora.*

Ligules purple.
- Disk-corollas longer than their achenes; the pappus awnless and half as long as corolla; tube of ligulate flower exceeding its pappus. . . . . 3. *G. caracasana.*
- Disk-corollas barely equaling achenes; pappus long-awned, equaling corolla; tube of ligulate flower shorter than its pappus. . . . . 4. *G. bicolorata.*

1. G. ciliàta (Raf.) Blake (with marginal hairs). — Annual, bushy-branched, 1–6 dm. high; *stems and peduncles coarsely spreading-hispid,* often glandular; leaves petioled, ovate, coarsely serrate, 2–8 cm. long; involucre 2–3 mm. high, with ciliate ovate phyllaries; *ligules* white, 1–2 mm. long, *their tubes equaled by the aristate* fimbriate *pappus; disk-flowers with aristate pappus about equaling the tube; marginal achenes densely hispid on the inner faces.* (*G. parviflora*, var. *hispida* DC.; *G. aristulata* Bickn.) — Weed of gardens, yards and waste places, reaching N.E. in 1866, now too abundant in many areas, n. to Que., Ont., Minn. and Neb. June–Nov. (–Jan.). (Natzd. from Trop. Am.)

2. G. parviflòra Cav. (small-flowered). — Similar to no. 1; *pubescence of stem wanting or minute and appressed;* leaves lance-ovate, with smaller and blunter teeth; *ligulate flowers without pappus, their achenes glabrous or merely pilose at summit; pappus of disk-flowers awnless, equaling or longer than the corollas.* — Waste places and weedy gardens, reaching Mass. in 1856, much less common than no. 1, n. to Mass., N.Y., Minn. and Oreg. June–Nov. (Natzd. from Trop. Am.)

3. G. caracasàna (DC.) Sch. Bip. (of Caracas). — Coarser than no. 1, densely villous; involucres slightly larger; *ligules purple,* oblong, 3-toothed at apex, *their tube longer than the awnless pappus; disk-corollas longer than the achene, their pappus one-half as long as the corollas;* marginal achenes glabrous, or hispidulous on one face. — Waste and cult. ground, local, Mass. to Va. July–Oct. (Adv. from Trop. Am.)

4. G. bicoloràta St. John & White (two-colored). — Like no. 3 but stem glabrate below, hispid above; *leaves* lanceolate or lance-ovate, *with prolonged tips;* the purple *ligules* cruciform, *their 2 lateral lobes widely divergent; pappus long-aristate, exceeding tubes of corollas.* — Waste places, local, e. Mass. July–Oct. (Adv. from sw. U.S. and Mex.)

60. MÀDIA Molina Tarweed

Heads few–many-flowered; ray-flowers (0–)1–20, fertile, with cuneate or oblong 3-lobed ligules, each subtended by a conduplicate-infolded laterally compressed phyllary of the mostly uniseriate involucre which partly or completely incloses the laterally compressed achene. Receptacle flat or convex, bearing a single series of bracts inclosing the disk-flowers as an inner involucre. Pappus wanting or a rudiment. — Viscid heavy-scented herbs chiefly of Pacific Am., with axillary and terminal heads, the flowers closing in bright sunshine. (From *madi*, the Chilean name.)

1. M. satìva Molina (sown). — Annual, 0.2–1.5 m. high, *viscid-villous and with long gland-tipped hairs;* stem simple or with ascending branches; leaves lanceolate to linear; *heads* scattered or in scattered glomerules, short-peduncled or sessile, *subglobose,* 6–8 mm. high; *ligules* 5–12, honey-colored, *about* 4 *mm. long;* disk-achenes cuneate-oblong, 4-angled, 1-nerved on each face; marginal achenes falcate-obovate. — Pacific Am.; with us as

Var. congésta T. & G. (crowded). — Heads crowded in 1–few glomerules at the ends of the stem and branches, 9–13 mm. long. (*M. capitata* Nutt.) — Fallow fields, new seeding, hen-yards and waste places, casual, Que. to Ind. and Del. June–Oct. (Adv. from Pacific slope)

---

* Treatment derived from that of Harold St. John & Donald White in Rhodora, xxii. 97–101 (1920).

2. M. glomeràta Hook. (clustered). — Similar to no. 1, smaller, *glandular only toward summit; leaves narrowly linear, hispid; heads* many in terminal and axillary glomerules, *slenderly campanulate,* about 6 mm. high; ligules 0–5, shorter; achenes narrower. — Pacific states, e. across the plains, locally, to Minn.; casual as an adventive e. to Que. and N.E. June–Oct.

Tribe VI. Helenìeae Cass. (see p. 1358)

*a.*Phyllaries distinct, not glandular-punctate. . . *b.*
*b.*Pappus none; phyllaries 3, keeled; narrow-leaved annual with small (2–5-flowered) glomerulate heads. . . . . 61. *Flaveria.*
*b.*Pappus of distinct scales or bristles or wanting; phyllaries 6 or more; heads solitary and long-peduncled or in corymbs or panicles. . . . *c.*
*c.*Tips of phyllaries colored or petaloid; ligules wanting.
Leaves mostly pinnately dissected; corolla-lobes ovate. . . . . . 62. *Hymenopappus.*
Leaves not dissected; corolla-lobes linear. . . . . 63. *Palafoxia.*
*c.*Tips of phyllaries not brightly colored nor petaloid; ligules present. . . *d.*
*d.*Receptacle naked; style-branches of disk-flowers dilated-truncate, penicillate at apex.
Phyllaries appressed; receptacle low-hemispherical, scapose plant with linear or narrow spatulate firm erect basal leaves and solitary head; pappus of scarious awnless scales. . . . . 64. *Actinea.*
Phyllaries reflexed; receptacle globose, ellipsoid or depressed; leafy-stemmed with 1–very many heads, pales of pappus often awn-tipped; leaves often decurrent. . . . . 65. *Helenium.*
*d.*Receptacle bearing slender fimbrillae among the disk-flowers, globose to convex; style-branches of disk-flowers with an appendage above the penicillate tuft. . . . . 66. *Gaillardia.*
*a.*Phyllaries united into a cup, glandular-punctate.
Involucre with loose phyllaries at base of the united series; receptacle bearing slender chaff; pappus of chaffy scales dissected into many bristles; low and diffuse annual. . . . . 67. *Dyssodia.*
Involucre naked at base; receptacle honeycombed; pales of pappus entire; upright annuals. . . . . 68. *Tagetes.*

61. FLAVÈRIA Juss.

Heads 3–15-flowered, usually with but 1 ray-flower; flowers all fertile. Phyllaries few, subequal or 1–2 of the outer much shorter. Receptacle small, naked or setose. Achenes oblong, 8–10-ribbed, glabrous; pappus none. — Opposite-leaved mostly trop. annuals with clustered small yellowish heads. (Name from *flavus, yellow,* the original species used as a yellow dye; Jussieu, separating it from *Milleria,* used a similar termination.)

1. F. campéstris J. R. Johnston (of the plains). — Erect and glabrous, 3–6 dm. high, branched above; leaves linear or lanceolate, 3-nerved, mostly serrulate; heads subsessile, in mostly terminal glomerules; involucre of 3 phyllaries, 2–5-flowered. — Alkaline soil, Mex. to Colo. and Kans.; adv. in w. Mo. May–Sept.

62. HYMENOPÁPPUS L'Hér.

Heads many-flowered; flowers all tubular and perfect, with large revolute corolla-lobes. Phyllaries 6–12, loose and broad, thin, the upper part petal-like, usually white. Receptacle small, naked. Achenes turbinate, with a slender base, striate; pappus of 15–20 blunt scales in a single row, very thin (whence the name of the genus, from the Greek *hymen, membrane,* and *pappos, pappus*). — Biennial or perennial N. Am. herbs, with alternate mostly dissected leaves, and corymbed small heads of usually whitish flowers.

1. H. scabiosaèus L'Hér. (like *Scabiosa*). — Somewhat flocculent-woolly when young, *leafy to the top,* 3–9 dm. high; leaves 1–2-pinnately parted into linear or oblong lobes; *phyllaries roundish,* mainly whitish, *at first surpassing the disk;* pappus-scales very small, roundish, nerveless. (*H. carolinensis* (Lam.) Porter) — Sandy woods and openings, Fla. to Tex., n. to S.C., Ind., Ill., Mo. and Kans. May, June.

2. H. corymbòsus T. & G. (corymbed), Old-plainsman. — *Leaves reduced upward,* becoming very short; *phyllaries oblong or oblong-obovate, much shorter than flowers,* their petaloid summits greenish. — Dry prairies, calcareous barrens and plains, e. Neb. to Ark., Okla. and Tex. May–Aug.

63. PALAFÓXIA Lag.

Heads few-flowered, small; flowers all tubular, deeply 5-parted. Phyllaries 8–10, herbaceous. Achenes slender-obpyramidal; pappus of short rounded pales or wanting. — Scabrous Am. herbs with narrow short-petioled mostly alternate leaves, and pedunculate loosely corymbose or paniculate small purplish heads. (Named for *José de Palafox y Melzi*, 1780–1847, a Spanish general.) POLYPTERIS Nutt.

1. P. callòsa (Nutt.) T. & G. (callous; from the tips of the phyllaries). — Slender mostly branched annual 2–8 dm. high; leaves linear; peduncles glandular; phyllaries with dry reddish tips. (*Polypteris* Gray; *Othake* Bush) — Calcareous barrens and gravels, Mo. and Ark. to Tex. Aug.–Oct.

64. ACTÍNEA Juss.

Heads many-flowered; rays several, cuneate-oblong, 3-toothed, pistillate. Phyllaries of the hemispherical involucre ovate or lanceolate, membranaceous or coriaceous, nearly equal, in 2–3 ranks, little shorter than the disk. Achenes turbinate, densely silky-villous; pappus of 5 or more ovate or lanceolate very thin chaffy scales. — Low Am. herbs, with narrow alternate leaves, dotted with resinous atoms, bitter-aromatic; the solitary heads terminating scapes or slender naked peduncles; flowers yellow. (Name from the Greek *actis, ray.*) ACTINELLA Pers. PICRADENIA Hook. TETRANEURIS Greene.

1. A. herbàcea (Greene) Robins. (herbaceous). — Perennial, cespitose; branches of the caudex rather stout, bearing numerous thickish spatulate to linear 1-nerved sparingly villous glandular-punctate leaves and scape-like peduncles (villous or lanate especially toward the summit); heads (including the showy rays) 3–4 cm. in diameter. (*Tetraneuris* Greene) — Prairies, sands and shores, very local, Manitoulin I., Ont.; Ottawa Co., O.; Will and Mason Cos., Ill. May–July. — Perhaps better as *A. acaulis* (Pursh) Spreng., var. *glabra* (Gray) K. F. Parker.

65. HELÉNIUM L. SNEEZEWEED

Heads many-flowered, usually radiate; rays several, cuneate, 3–5-cleft, fertile, rarely sterile. Involucre small, reflexed; the phyllaries linear or subulate to spatulate. Receptacle globose or ellipsoid to depressed. Achenes turbinate, ribbed; pappus of 5–8 thin 1-nerved chaffy scales, the nerve usually extended into a bristle or point. — Erect often branching Am. herbs with alternate leaves, often sprinkled with bitter aromatic resinous globules; heads yellow, rarely purple, terminal, single or corymbed. (The Greek name of some plant, said by Linnaeus to be named after *Helena* (*Helen of Troy*), wife of King Menelaus of Sparta.)

a. Leaves linear-filiform, not decurrent on the stems and branches; annual. . 1. *H. tenuifolium.*
a. Leaves linear to oblong-ovate, decurrent on the stems and branches; perennials or biennials. . . *b.*
  b. Disk ovoid- to oblate-globose; stem commonly with leafy branches and few to very many heads; foliage puberulent; pales of pappus mostly attenuate to awn-tips. . . *c.*
    c. Disk globose or globose-ovoid, brown or purplish; receptacle ovoid; ray-flowers neutral or sterile. . . . . 2. *H. nudiflorum.*
    c. Disk depressed- or oblate-globose, yellow; receptacle depressed-globose; ray-flowers fertile (rarely sterile).
      Pappus of disk-flowers one-sixth to one-third the length of the tube, keeled, attenuate to the awn; leaves lanceolate to ovate-oblong; wide-ranging species. . . . . 3. *H. autumnale.*
      Pappus one-half to two-thirds length of tube, keelless, the ovate blades abruptly long-awned; leaves narrowly oblanceolate to linear; plant of upland Va. . . . . 4. *H. virginicum.*
  b. Disk depressed, barely hemispherical, purple-brown; stem simple up to the 1–5 elongate naked or minutely bracted peduncles; pales of pappus blunt and awnless. . . . . 5. *H. brevifolium.*

1. H. tenuifòlium Nutt. (slender-leaved). — *Smooth* corymbosely branched *annual* 1.5–7 dm. high, *densely covered with linear-filiform leaves, these not decurrent* down the branches; peduncles filiform, erect, elongate; heads 1.5–2.5 cm. broad; *the globose disk yellowish,* the fertile rays wide-spreading to reflexed; pappus ovate, long-awned, about as long as achenes. — Open ground, roadsides, etc., Fla. to Tex. and Mex., n. to Va., Ky., s. Ill., Mo. and Kans., rapidly spreading as a weed northw. to Mass., N.Y., O., Mich., etc. June–Nov. (W.I.)

2. H. nudiflòrum Nutt. (naked-flowered; from the long peduncles). — *Puberulent* or minutely hirtellous *perennial* 0.2–1 m. high; the *stem* with stiffly ascending branches, *wing-angled* by the decurrent bases of the linear to lanceolate firm and often scabridulous leaves; heads slender-stalked, in open corymbs; involucre puberulent, the subulate to spatulate phyllaries soon reflexed; *disk globose or globose-ovoid, brown or purplish,* 6–10 *mm. in diameter;* ligules shorter than to longer than diameter of disk, golden-yellow or with purplish base, strongly 3-lobed at tip, neutral or sterile; *pappus of lanceolate to ovate attenuate slender-tipped or awned pales.* (Incl. *H. polyphyllum* Small) — Damp plains, openings, meadows, fields and roadsides, Ga. to Tex., n. to N.C., Ky., Mo. and Kans., rapidly spreading as a weed northw. to N.E., N.Y., O., Mich., etc. June–Oct.

3. H. autumnàle L. (autumnal). — Perennial; the wing-angled stem 0.5–1.5 m. high, simple or corymbosely branched at summit; *leaves elliptic, oblong or lanceolate,* acuminate, *membranaceous, the larger coarsely dentate or serrate ones* 0.5–1.6 dm. long and 2–5.5 *cm. broad;* heads few to many, peduncled; *disk depressed- or oblate-globose, yellow,* becoming *when fully developed* 1.6–2.3 *cm. broad; denuded receptacle depressed-globose;* ray-flowers fertile (rarely sterile); the *flat cuneate* deep yellow *ligules* soon reflexed, 1.6–2.5 *cm. long and* 7–12 *mm. broad* at the bluntly toothed or lobed summit; *pappus of disk-flowers one-sixth to one-third length of corolla-tube, keeled, attenuate to the awn.* (*H. latifolium* Mill.) — Rich thickets, meadows and shores, w. N.E. to Minn., s. to N.J., w. N.C., Ky. and Mo., northw. passing into the first, southw. into the second var.; other vars. beyond our range. Aug.–Nov.

Var. **canaliculàtum** (Lam.) T. & G. (channeled). — Often more slender, 3.5–9 dm. high, with 1-many heads; *leaves linear to lanceolate, entire or shallowly* (rarely coarsely) *denticulate, firm and subrigid, the larger ones* 3–12 cm. long and 3–18 *mm. wide; mature disk* 1–2 *cm. broad; ligules strongly narrowed and often channeled at base,* 1–2 *cm. long and* 2–7 *mm. broad* at the sharply or bluntly toothed summit. — Fresh or tidal shores, meadows and low thickets, lower St. Lawrence R., Que., to Minn. and Neb., s. to Mass. (adv.), N.Y., n. O., Ind., Ill., Mo., Tex. and Ariz.

Var. **parviflòrum** (Nutt.) Fern. (small-flowered). — *Leaves membranaceous or submembranaceous, elliptic, oblong, lanceolate or oblanceolate, coarsely toothed to subentire, the larger ones* 0.7–3.5 *cm. broad; mature disk* 0.8–1.5 *cm. broad; ligules* 3–13 *mm. long and* 3–7 *mm. broad,* often channeled at base. (*H. parviflorum* Nutt.) — Swamps, thickets and bottomlands, Fla. to Ark., n. to Ct., N.Y., Pa., Ky., Ill. and se. Ia.

4. H. virgínicum Blake (Virginian). — Resembling no. 3, 0.7–1.1 m. high, slender, more or less pilose or puberulent below, *with very open slenderly branched corymb* with nearly or only sparsely leafy branches; *cauline leaves overlapping, narrowly linear or the lowest oblanceolate,* entire or somewhat toothed, often puberulent; heads 2–20, slender-peduncled; *disk* 1–1.6 *cm. broad;* ligules 1–1.5 cm. long, 6–9 mm. broad, bluntly 3-toothed; *pales of pappus keelless, ovate, abruptly long-awned, one-half to two thirds as long as corolla-tube.* — Meadows and swamps, Augusta Co., Va. July–Sept.

5. H. brevifòlium (Nutt.) Gray (short-leaved). — *Rosulate* perennial *with simple* erect slender *stems* 2–7 dm. high; *rosette-leaves usually present,* oblanceolate, spatulate or oblong, slender, petioled; *cauline leaves* few, scattered, *rapidly decreasing upward; peduncles* 1–5, elongate, *naked or only minutely bracted;* head showy, with overlapping broad deeply 3-lobed golden ligules 1–1.5 cm. long; *disk purple-brown, depressed or barely low-hemispherical; pales of pappus blunt and awnless.* — Wet pinelands, meadows and swampy woods, Fla. to Miss., n. to se. Va. May, June.

H. Curtísii Gray (for its discoverer, MOSES ASHLEY CURTIS, 1808–1872), *differing from no.* 5 *in its less reduced cauline leaves, subulate reflexed phyllaries and ovoid disk of yellow flowers,* a vernal species (May, June) of sphagnous bogs, heretofore known only from e. N.C., has recently been found in James City Co., Va. (*J. T. Baldwin,* Jr.).

66. GAILLÁRDIA Foug. GAILLARDIA. BLANKET-FLOWER

Heads many-flowered; rays 3-cleft or -toothed, neutral or sometimes fertile, or none; *phyllaries* in 2–3 rows, the outer larger, loose and foliaceous. Receptacle convex to globose, with bristle-like or subulate or short and soft chaff. Achenes turbinate, 5-costate, villous; pappus of 5–10 long thin awn-tipped scales. — Erect alternate-leaved Am. herbs with long-peduncled showy heads of yellow or purplish fragrant flowers. (Named in 1788 after *Gaillard de Charentonneau,* a French botanical amateur.)

1. G. lùtea Greene (yellow). — Erect or nearly so, 4 dm. or more in height, *puberulent and somewhat scabrous,* leafy-stemmed, branched above; branches ascending; leaves oblong-lanceolate, somewhat amplexicaul at the broadish base, toothed or subentire; *both disk-flowers and rays yellow; fimbrillae soft and short; ligules about* 1 *cm. long.* — Moist sandy ground, s. Mo. July–Oct.

2. G. pulchélla Foug. (handsome). — *Hirsute* annual, 2–6 dm. high, with ascending branches; basal leaves oblanceolate, bluntly toothed or lobed, densely hispid beneath, sparsely long-pubescent above; upper leaves lance-oblong, sessile, often clasping, acute; *disk-corollas purple-tipped; ligules yellow and purple or purple throughout,* 1–2 *cm. long; fimbrillae firm.* (*G. aristosa* of ed. 7, not Pursh) — Dry sandy prairies and openings, Colo. and N.M., e. to Minn., Neb., Mo. and La.; casually as a migrant or as a garden-escape e. to Atl. states. June, July.

67. DYSSÒDIA Cav. FETID MARIGOLD

Heads many-flowered, usually radiate; rays pistillate. Involucre of 1 row of phyllaries united into a firm cup, with a few loose bracts at the base. Receptacle flat, beset with short chaffy bristles. Achenes slender, 4-angled; pappus a row of chaffy scales, dissected into numerous rough bristles. — Am. herbs, mostly annuals or biennials, dotted with large pellucid glands, which give a strong odor; heads terminating the branches; flowers yellow. (Modification of the Greek *dysodia, an ill smell,* which the plants exemplify.) BOEBERA Willd. DYSODIA Spreng.

1. D. pappòsa (Vent.) Hitchc. (named for the conspicuous pappus). — Nearly smooth, diffusely branched, 1–5 dm. high; leaves opposite, pinnately parted, the narrow lobes bristly-toothed or cut; rays few, scarcely exceeding the involucre. (*Boebera* Rydb.) — Alluvium, roadsides and waste places, La. to Mex. and Ariz., n. to s. Ont., Minn., N.D. and Mont.; adv. e. to N.E. July–Oct.

68. TAGÈTES L. MARIGOLD

Heads radiate. Involucres naked at base, the united phyllaries forming a cylindric or elongate tube or cup. Receptacle favose. Pappus of 3–6 firm often unequal entire pales, one or more of them often awned. Achenes angulate or compressed, hardly striate. — Am. herbs, ours annual, strong-scented, with mostly pinnate or pinnatifid leaves with copious oil-glands, and yellow, orange or reddish flowers. (Named for *Tages,* mythological and beautiful Etruscan divinity.)

1. T. MINÙTA L. (minute), MUSTER JOHN HENRY. — Annual 0.4–1 m. high, branching above; leaves opposite or alternate, pinnate, with 11–17 sharply serrate linear-lanceolate leaflets and pectinate base; heads slenderly cylindric in dense corymbs, about 1 cm. high, with 5 united phyllaries rounded at the free summits; ligules lemon-color, 1–2 mm. long; achenes strigillose. — Waste places and roadsides, locally abundant, S.C. to se. Va. (where cult. for seasoning); occasionally n. to Mass. Aug.–Oct. (Introd. from S. Am.)

T. ERÉCTA L. (erect), AFRICAN MARIGOLD, with lanceolate leaflets and clavate-peduncled heads and cup-like involucre 1.5–2 cm. high, and large orange or yellow broad ligules, escapes to roadsides from cult. (Introd. from Mex.)

Tribe VII. ANTHEMÍDEAE Cass. (see p. 1358)

a. Receptacle chaffy, at least toward summit or center; heads radiate.
- Receptacle flattish; achenes oblong, flattened; heads many in close corymbs. 69. *Achillea.*
- Receptacle conical; achenes terete or angled or ribbed; heads solitary at tips of branches and branchlets. . . . . 70. *Anthemis.*

a. Receptacle naked. . . *b.*
- *b.* Heads solitary or in corymbs. . . *c.*
  - *c.* Receptacle conical, at least as it matures; heads peduncled, not in corymbs; pappus a crown, or wanting in species with sweet fragrance. . 71. *Matricaria.*
  - *c.* Receptacle flattish or merely low-convex. . . *d.*
    - *d.* Corollas of perfect flowers 5-toothed; marginal flowers with a corolla; achenes sessile.
      - Heads single at tips of stems and branches or in corymbs, radiate (rarely discoid); achenes 5–10-nerved, pappus none. . . . . 72. *Chrysanthemum.*
      - Heads closely or loosely corymbose (rarely solitary), without ligules; achenes 3–5-angled, with a short crown. . . . . . 73. *Tanacetum.*
    - *d.* Corollas of perfect flowers 4-toothed; marginal flowers apetalous; mature achenes raised on pedicels which remain attached to the receptacle. . . . . 74. *Cotula.*
- *b.* Heads in spiciform, racemose or paniculate inflorescences, discoid; flowers all perfect; pappus none. . . . . 75. *Artemisia.*

69. ACHILLÈA L. YARROW. HERBE-À-DINDES (Que.)

Heads many-flowered, radiate; the rays few, fertile. Phyllaries imbricated, with scarious margins. Receptacle chaffy, flattish. Achenes oblong, flattened, margined; pappus none. —

Perennial herbs of N. Hemisph., with small corymbose heads. (So named because its *healing* powers are said to have been discovered by *Achilles*.)

Leaves simple; involucre campanulate; receptacle low.
  Leaves simply serrate; ligules 4–5 mm. long, nearly as broad; adventive. . . . 1. *A. Ptarmica*.
  Leaves pectinate; ligules about 1 mm. long and 2 mm. broad; indigenous boreal species. 2. *A. sibirica*.
Leaves bipinnately dissected; involucre slender-cylindric; receptacle elongating after anthesis.
  Phyllaries with dark brown to blackish margins; ligules comparatively showy, their blades half to two-thirds as long as involucre; cauline leaves 4–9; boreal and alpine. 3. *A. borealis*.
  Phyllaries pale, the uppermost rarely dark-margined; ligules small, their blades rarely half as long as the involucre; cauline leaves 5–20 or more.
    Stem cobwebby to smoothish, up to 1 m. high; corymb flattish-topped, 0.6–3 dm. broad. . . . . 4. *A. Millefolium*.
    Stem heavily woolly, 1–6 dm. high; corymb strongly convex, 2–10 cm. broad. . . . . 5. *A. lanulosa*.

1. A. Ptármica L. (old generic name from the Greek, *causing to sneeze*), Sneezeweed. — Stem rather rigid, smooth or slightly pubescent; *leaves lance-linear*, finely appressed-serrate; corymb loose, the heads long-pediceled; *rays* 8–12, white, *much longer than the involucre*. — Damp fields, roadsides, etc., s. Lab. to Ont., s. to Nfld., N.S., N.E., N.Y. and Mich., originally spread from cult. Aug., Sept. (Introd. and natzd. from Eu.)

2. A. sibírica Ledeb. (Siberian). — Similar to no. 1, villous to glabrescent; *leaves pectinate-pinnatifid* into denticulate lance-subulate fine lobes; corymb close; *ligules about* 1 *mm. long* and 2 mm. wide. (*A. multiflora* Hook.) — Damp woods, meadows and shores, Gaspé Co., Que.; Alaska to Man., Sask., etc. July, Aug. (Siberia)

3. A. boreàlis Bong. (northern). — Stem erect, 0.5–4 dm. high, more or less lanate; *cauline leaves few* (4–9), silky-lanate especially beneath; *corymbs* 2–6 *cm. broad, very convex;* involucre 4–6 mm. long, its *phyllaries all dark-margined; rays* 10–20, white (rarely pink), short-oblong or suborbicular, 2.5–4 *mm. long*. (Incl. *A. nigrescens* (E. Mey.) Rydb.) — Wet rocks, cool slopes, shores, etc., Greenl. and Lab. to Alaska, s. to Nfld., M.I., alpine areas of N.H., n. Mich., Rocky Mts., etc. June–Aug. (Eurasia)

4. A. Millefòlium L. (old generic name, *thousand-leaved*), Common Y., Milfoil. — *Stem* simple or somewhat forking above, 3–10 dm. high, *arachnoid to glabrescent; cauline leaves numerous* (mostly 8–20 or more), smooth to loosely pubescent, bipinnately parted and dissected into fine segments; *corymbs* very compound, *flattish-topped*, 0.6–3 dm. broad, with stiffly ascending branches; involucre slenderly cylindric, its phyllaries pale, or in exposed situations the uppermost becoming dark-margined; ligules whitish, or often passing to shades of pink to deep rose-purple in forma ròsea Rand & Redf. (roseate), short-oblong, 1.5–2.5 mm. long; receptacle greatly prolonged in fruit, the mature disk-flowers becoming exserted. — Fields and roadsides, common throughout. June–Sept. (Natzd. from Eu.) — A polymorphous species, sadly in need of well-balanced study, its intricacies not properly understood. Several reputed segregates reported as in our flora need checking.

5. A. lanulòsa Nutt. (woolly). — Similar to no. 4; *stem* 1–6 dm. high, *densely woolly;* cauline leaves silky-lanate, few, narrow; *corymbs* 2–10 cm. broad, *very convex; involucre* pubescent, *none of its phyllaries dark-margined;* ligules 1–2 mm. long, narrow. — Gravelly or sandy shores and open ground, Nfld. to B.C., s. along coast to s. N.E., and inland s. to Mich., Mo., Okla. and Mex. June–Aug.

70. ANTHEMIS L. Chamomile

Heads many-flowered, radiate; rays pistillate or (in no. 1) neutral. Involucre hemispherical, of many small imbricated dry and scarious phyllaries shorter than the disk. Receptacle conical, usually with slender chaff at least near the summit. Achenes terete or ribbed, glabrous, truncate; pappus none or a minute crown. — Branching often strong-scented herbs, nat. of Eurasia and n. Afr., with pinnately dissected leaves and solitary terminal heads; rays white or yellow (rarely wanting); disk yellow. (The ancient name of the Chamomile.)

Ligules white (in no. 3 white with yellow base); disks 5–12 mm. broad, their corolla-tubes cylindric.
  Ill-scented; ray-flowers sterile; pales of receptacle subtending only the central flowers of the disk. . . . . 1. *A. Cotula*.
  Odorless or pleasantly fragrant; ray-flowers fertile; pales subtending all disk-flowers.

Odorless; leaves mostly bipinnatifid; ligules white, 6–12 mm. long; disks 7–12 mm. broad; disk-corollas unappendaged. . . . . 2. *A. arvensis.*
Fragrant; leaves mostly simply pinnate; ligules white, with yellow base, 5–8 mm. long; disks 5–7 mm. broad; disk-corollas with a spur extending from base of tube down 1 side of achene. . . . . 3. *A. mixta.*
Ligules yellow; disks 1–2 cm. broad, their corolla-tubes compressed. . . . . 4. *A. tinctoria.*

1. A. Cótula L. (early united with the genus *Cotula*), Mayweed, Stinking C., Dog-fennel. — Annual, simple or bushy-branched, *ill-scented;* leaves finely 3-pinnately dissected; *ray-flowers sterile,* with white rays 6–10 mm. long; *pales of receptacle* subulate, stiff, *subtending only the central flowers of the disk;* pappus none; *achenes rough-tuberculate.* (*Maruta* DC.) — Roadsides and waste places, Nfld. to Alaska, s. beyond our limits. June–Oct. (Natzd. from Eu.) Fig. 1761.

1761. A. Cotula.

2. A. arvénsis L. (of cultivated fields), Corn-C. — Pubescent annual or biennial, resembling Mayweed, but *not ill-scented;* leaves less finely 1–2-pinnately parted; *ray-flowers fertile, with white ligules 6–12 mm. long; disks 7–12 mm. broad; pales of receptacle lanceolate,* pointed, *subtending all disk-flowers* and distinctly exceeding them; *achenes smooth on the sides,* 10-nerved; pappus a minute border. — Roadsides and waste places, occasional, Me. to w. N.Y., s. to Ga. May–Aug. (Natzd. from Eu.) — More generally represented by

Var. agréstis (Wallr.) DC. (of fields). — Chaff shorter than disk-flowers. — Fields, roadsides and waste, Que. to Mich. and Va. March–Aug. (Natzd. from Eu.) Fig. 1762.

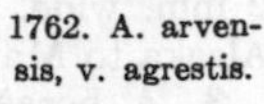
1762. A. arvensis, v. agrestis.

A. secundiràmea Biv. (with one-sided branching), resembling stiffly branched extremes of no. 2, with mostly *unilateral branching, puncticulate leaves,* small heads and *tuberculate achenes,* has appeared in railroad-ballast in se. Va. (Adv. from Mediterr. reg.)

3. A. míxta L. (mixed). — Smaller than no. 2, *pleasantly fragrant; leaves mostly simply pinnatifid;* ray-flowers fertile, the *ligules white, with yellow base and only 5–8 mm. long; disks 5–7 mm. broad; tubes of disk-corollas with a long basal spur projecting back over 1 side of the achene.* (*Ormenis* Dumort.) — Waste places, casual about towns, N.Y. to Fla. June–Oct. (Adv. from Eu.)

A. nòbilis L. (well known), Garden-C., a pleasantly strong-scented downy perennial with creeping basal shoots, pales of receptacle blunt, is sometimes spontaneous about old gardens. (Introd. from Eu.)

4. A. tinctòria L. (used in dyeing), Yellow C. — Whitish-pubescent *stoloniferous perennial* up to 8 dm. high; leaves pinnately divided; heads long-peduncled, 3–4 cm. broad; *ligules* numerous, *yellow; disks 1–2 cm. broad, their corolla-tubes compressed;* pales lanceolate; pappus a short crown. (*Cota* J. Gay) — Fields and roadsides, occasional, Que. to Minn. and N.J. June–Sept. (Natzd. from Eu.)

Santolìna (name of doubtful origin) Chamaecyparíssus L. (dwarf Cypress), Lavender-Cotton, a low shrubby tomentose evergreen with small pectinate gray leaves and long-peduncled globular yellow heads 1–1.5 cm. in diameter, sometimes spreads from cult. to dry banks and roadsides. (Introd. from Eu.)

71. MATRICÀRIA L. Wild Chamomile

Heads many-flowered; rays pistillate or wanting. Phyllaries imbricated, with scarious margins. Receptacle conical (at least in fruit), naked. Achenes 3–5-ribbed, wingless; pappus a membranaceous crown or border, or none. — Smooth and branching chiefly Old World herbs (ours annuals or biennials) with finely divided leaves and single or corymbed heads. Rays white or none; disk yellow. (Name from *matrix, womb,* for reputed medicinal virtues.)

Heads ligulate; disk-corollas 5-lobed.
Odorless; receptacle hemispherical; pappus a short crown. . . . . 1. *M. maritima.*
Pleasantly odorous; receptacle conical; pappus none. . . . . 2. *M. Chamomilla.*
Heads discoid; receptacle conical; disk-corollas 4-lobed; bruised plant with fragrance of pineapple; pappus none. . . . . 3. *M. matricarioides.*

1. M. marítima L. (of the sea-shore). — Low annual with depressed or horizontally spreading branches; leaves 2–4 cm. long, bipinnatifid, with fleshy linear segments 1.5–5 mm. long;

1763. M. maritima, v. agrestis.

heads 1.5–3 cm. broad, *on naked peduncles* terminating the branches; ligules white, 7–12 mm. long; *disk hemispherical, broadly rounded;* disk-corollas 5-lobed; achenes strongly 3-ribbed; pappus a short crown or border, more or less toothed. (*M. inodora*, var. *salina* Bab.; *Chamomilla* Rydb.) — Waste ground near ports, local, Gaspé Co., Que., to Pa. Aug.–Oct. (Adv. from Eu.) — Represented with us chiefly by

Var. AGRÉSTIS (Knaf) Wilmott (of fields). — Ascending or erect, with ascending branches; leaves larger, with almost filiform segments 0.4–2 cm. long; heads 3–4 cm. broad; ligules 1–2 cm. long. (*M. inodora* L.; *Chamomilla inodora* (L.) Gilib.) — Fields and waste places, Nfld. to Ont., s. to N.S., N.E., Pa., Ky. and Kans., abundant northw. Late June–Sept. (Natzd. from Eu.) FIG. 1763.

2. M. CHAMOMÍLLA L. (old generic name). — Similar to no. 1, but *with fragrance of pineapple; heads* smaller, *about 2 cm. broad;* rays shorter; receptacle more convex; achenes less distinctly ribbed; *pappus obsolete.* — Roadsides and waste places, locally abundant, Nfld. to Minn. and Pa. May–Oct. (Natzd. from Eu.)

Var. CORONÀTA (J. Gay) Coss. & Germ. (crowned). — Achenes with short crown of pappus. — Roadsides and waste, sw. N.B. and se. Me.; Mich. (Natzd. from Eu.)

3. M. MATRICARIOÌDES (Less.) Porter (like *Matricaria;* Lessing considering it a species of *Artemisia*), PINEAPPLE-WEED. — Low; leaves 2–3-pinnately parted into short linear lobes; heads rayless, short-peduncled; phyllaries oval, with broad margins, much shorter than the *conical disk;* achenes more terete; pappus obsolete; *odor of the bruised plant suggesting pineapple.* (*M. suaveolens* (Pursh) Buchenau, not L.) — Roadsides and old fields, Nfld. to Man., s. to N.S., N.E., Del., Pa., O., Ind., Mo., etc., rapidly spreading. June–Oct. (Natzd. from Pacific states)

72. CHRYSÁNTHEMUM L. CHRYSANTHEMUM

Heads many-flowered; rays numerous or rarely wanting, fertile. Phyllaries of the broad and flat involucre imbricated, with scarious margins. Receptacle flat or convex, naked. Disk-corollas with a flattened tube. Achenes of disk and ray similar, striate. — Annual or perennial herbs of N. Hemisph., with toothed, pinnatifid or divided leaves, and single or corymbed heads. Rays white or yellow (rarely wanting); disk yellow. (Old Greek name, *chrysanthemon, i.e.*, golden flower.)

Heads 2–6 cm. broad, solitary at tips of stem and long branches.
  Ligules white; phyllaries lanceolate to narrowly oblong; achenes merely angled and ribbed; perennial. . . . . 1. *C. Leucanthemum.*
  Ligules yellow; phyllaries elliptic or ovate; marginal achenes winged; annual. 2. *C. segetum.*
Heads 0.5–2 cm. broad, very numerous, in corymbs.
  Leaves pinnatifid, with dissected segments; ligules white, oblong, 3–8 mm. long. . . . . 3. *C. Parthenium.*
  Leaves mostly uncleft, merely crenate; heads discoid or with tiny ligules. . . 4. *C. Balsamita.*

1. C. LEUCÁNTHEMUM L. (old generic name, *white flower*), OX-EYE- or WHITE DAISY, MARGUERITE, WHITEWEED. — Stem erect, 0.3–1 m. high, simple or forked toward the summit; basal leaves spatulate-obovate, on long slender petioles, the blades crenate-dentate; middle and upper cauline leaves oblong or oblanceolate, coarsely and regularly crenate or dentate above, with larger spreading teeth at base; heads 4–6 cm. broad; *phyllaries narrow, brown-margined; rays white* (rarely tubular, laciniate or deformed). (*Leucanthemum* Rydb.; *L. vulgare* Lam.) — Fields, etc., Nfld. and e. Que. to N.Y. and N.J., rare southw. June–Aug. (Natzd. from Eu.) FIG. 1764.

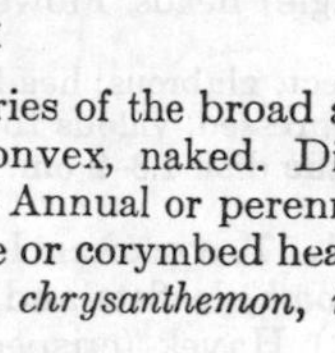

1764. C. Leucanthemum (typical).

Var. PINNATÍFIDUM Lecoq & Lamotte (pinnately divided). — Basal leaves pinnatifid, subpinnatifid or coarsely and irregularly toothed; middle and upper cauline leaves narrowly oblong or oblanceolate, conspicuously subpinnatifid at base; heads usually smaller than in the typical var. (Var. *laciniatum* Vis., published in same year, possibly earlier) — Fields, meadows, roadsides, etc., a pernicious but beautiful weed, Lab. to B.C., s. to Fla., Ky., Mo., Kans., etc. June–Aug. (Natzd. from Eu.) FIG. 1765. — Forma TUBULIFLÒRUM Henrikssen (tubular-flowered) has tubular ligules.

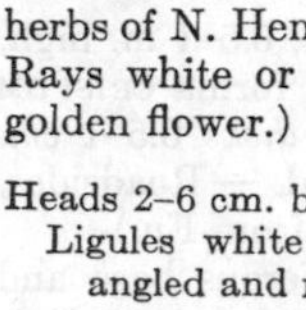

1765. C. Leucanthemum, v. pinnatifidum.

2. C. SÉGETUM L. (weed among crops), CORN-MARIGOLD. — Similar to no. 1, annual; leaves oblong, somewhat clasping, coarsely toothed or pinnatifid; *rays golden-yellow; phyllaries broad and scarious; achenes 2-winged.* — Fields, roadsides and waste places, infrequent, Nfld. to Ont., s. to Fla. June–Oct. (Adv. from Eu.)

C. CORONÀRIUM L. (used in garlands), GARLAND-C., similar to no. 2 but with bipinnatifid leaves and 3-winged achenes, occasionally spreads from gardens. (Introd. from Eu.)

3. C. PARTHÉNIUM (L.) Bernh. (old generic name), FEVERFEW. — Bushy-branched perennial 0.3–1 m. high; *leaves bipinnately divided, the divisions* ovate, *cut;* heads 1–2 cm. broad, very numerous, corymbose; ligules white, oblong, 3–8 mm. long. (*Matricaria* L.) — Waste places, roadsides and borders of woods, Que. to O., s. to Md. and Mo., esc. from cult. June–Sept. (Introd. from Eu.)

4. C. BALSÁMITA L. (old generic name), COSTMARY, MINT-GERANIUM. — *Leaves oblong, crenate,* the upper sessile, the lower petioled, often with 2 small lateral lobes at base; heads rayless or with minute white rays. (Incl. var. *tanacetoides* Boiss.; *Balsamita* Rydb.) — Roadsides, Que. and Ont., s. to N.E., Del., Pa., O., Ind. and Kans., esc. from cult. Aug.–Oct. (Introd. from Eu.)

73. TANACÈTUM L. TANSY. TANAISIE (Que.)

Heads many-flowered, nearly discoid; flowers all fertile, the marginal chiefly pistillate and 3–5-toothed. Involucre imbricated, dry. Receptacle convex, naked. Achenes angled or ribbed, with a large flat top; pappus a short crown. — Bitter and acrid mostly strong-scented herbs (ours perennial) of N. Hemisph., with 1–3-pinnately dissected leaves, and corymbed (rarely single) heads. Flowers yellow, in summer. (Name of uncertain derivation.)

Erect, glabrous; heads many in a flat-topped corymb, the disk 0.5–1 cm. broad. 1. *T. vulgare.*
Depressed, villous to woolly, at least when young; heads 1–30, loosely disposed, the disk 1.3–2 cm. broad. . . . . 2. *T. huronense.*

1. T. VULGÀRE L. (common), COMMON T., GOLDEN-BUTTONS. — Stem erect, 0.5–1 m. high, *smooth;* leaflets and wings of petiole cut-toothed, or deeply cut and crisped in forma CRÍSPUM (L.) Hayek (crisped); *corymb dense, flat-topped,* the very many heads with *disks* 0.5–1 *cm. broad;* pistillate flowers terete, with oblique 3-toothed limb; pappus 5-lobed. — Roadsides, borders of fields, etc., Nfld. to B.C., s. beyond our limits. July–Sept. (Introd. from Eu.)

2. T. huronénse Nutt. (of L. Huron). — *Depressed,* with loosely stoloniferous base and many feathery sterile tufts, *villous to lanate when young; heads* 1–30, *long-peduncled, irregularly corymbed, the disks* 1.3–2 *cm. broad;* pistillate flowers flattened, 3–5-cleft; pappus irregularly toothed. — Four geographic vars.:

Flowering stem 4–8 dm. high, with 7–15 green leaves; corymb with 6–30 heads; leaves oblong, with oblong to oblanceolate pinnae. . . . . *T. huronense* (typical).
Flowering stem 0.4–4.5 dm. high; corymb with 1–6 heads.
  Leaves oblong to narrowly elliptic, the larger 4–10 cm. long and 2–5 cm. broad; pinnae oblong to oblong-lanceolate, blunt, with approximate to imbricated pinnules.
    Flowering stem 1.2–3 dm. high, glabrous to sparingly pilose except at summit, with 4–10 green and sparsely pilose leaves above the basal rosette; heads 2–6. . . . . Var. *bifarium.*
    Flowering stem 0.4–2 dm. high, copiously lanate, with 1–4 white-lanate leaves above the basal rosette; heads 1 or 2 (rarely –4). . . . . Var. *terrae-novae.*
  Leaves (especially the basal and median) elliptic, the larger 1–3 dm. long and 0.35–1.4 dm. broad; pinnae remote, with distant simple to much dissected acute pinnules; flowering stem 2–4.5 dm. high, with 5–10 leaves and 1–5 heads. . . . . Var. *johannense.*

T. huronénse (typical). — Sands and gravels of Lakes Huron, Mich. and Sup. June–Aug.

Var. **bifàrium** Fern. (twofold). — Calcareous sands and gravels, Anticosti I., Que., and shores of Hudson Bay. Late June–Aug.

Var. **térrae-nòvae** Fern. (of Newfoundland). — Peaty and gravelly limestone-barrens, w. Nfld. July–Sept.

Var. **johannénse** Fern. (of St. John R.). — Gravels and sands of St. John and Restigouche Rivers and tributaries, Que., N.B. and Me. June–Aug.

74. CÓTULA L.

Heads hemispherical to globose, many-flowered, discoid; the marginal flowers (reduced rays) pistillate and fertile, nearly or quite apetalous; disk-flowers tubular, 4-toothed, fertile. Chaff none. Achenes at maturity raised on pedicels, which remain attached to the flat or moderately convex receptacle. Pappus obsolete or none. — Low mostly diffuse or creeping strong-scented herbs, chiefly of S. Hemisph. with alternate toothed, lobed or dissected leaves, and pedunculate heads of yellow flowers. (Name from the Latin *cotula,* after the Greek *cotyle, a small cup;* alluding to the hollow at the base of the amplexicaul leaves.)

1. C. coronopifòlia L. (with leaves of *Coronopus*). — Branched from the base, decumbent and often somewhat repent, slightly fleshy, nearly or quite smooth; leaves linear-oblong, irregular, 3–many-toothed; heads at length subglobose, about 1 cm. in diameter, on slender terminal peduncles. — Brackish mud, Bonaventure Co., Que., P.E.I. and Grand Manan I., N.B.; also on Pacific slope. July–Sept. (Semicosmop.)

CENTÍPEDA (hundred-footed) MÍNIMA (L.) A. Br. & Aschers. (tiny), a small depressed and branching annual with cuneate-oblanceolate toothed leaves and tiny discoid Tansy-like yellow axillary heads, has appeared as a local weed in e. Mass. and e. Pa. (Adv. from Asia)

75. ARTEMÍSIA L. WORMWOOD. ARMOISE (Que.)

Heads discoid, few–many-flowered; flowers all tubular, some or all hermaphrodite and in some sections fertile, in others sterile. Involucre imbricated, scarious, often lanate or tomentose. Receptacle small and flat. Achenes obovoid to oblong, with small summit and no pappus. — Herbs or shrubby plants chiefly of arid regions of N. Hemisph. (a few in w. S. Am.), bitter and aromatic, with small commonly nodding and panicled or spicate-racemose heads; flowering in summer and autumn. Corolla yellow to purplish. (Ancient name of the Mugwort, in memory of *Artemisia,* wife of Mausolus.)

a. Receptacle smooth, not hairy. . . *b.*

b. Disk-flowers with both stamens and pistil but sterile, the ovary abortive and soon shrinking; style entire or nearly so. § DRACUNCULUS. . . *c.*

c. Principal leaves dissected. . . *d.*

d. Involucres 2–3 mm. long, 2–3 mm. broad, with 14–25 flowers; disk-corollas 1.4–2 mm. long; biennials or short-lived perennials 0.3–1.5 m. high, with panicles 1.5–7.5 dm. high; plants of low altitudes and temp. reg. . . . . 1. ***A. caudata.***

d. Involucres 3–4 mm. long, 3.5–6 mm. broad, with 23–45 or more flowers; disk-corollas 2.2–3 mm. long; perennials 0.2–7.5 (very exceptionally –10) dm. high, the spiciform, racemose or paniculate inflorescence 0.2–6 dm. high; boreal or alpine species.

Broader segments of basal leaves rarely more than 1–2 mm. wide; panicle mostly with many elongate branches, 1–6 dm. high, with some heads shorter than their pedicels; involucre green and whitish, glabrous, the broad scarious white margins of the phyllaries as wide as or wider than the green center; flowers yellow; stems 0.2–1 m. high. . . . . 2. ***A. canadensis.***

Broader segments of basal leaves 1–4 mm. wide; panicle spiciform or racemiform or, if of elongate branches, with the heads sessile or much longer than their pedicels; involucre purplish to green, villous to glabrous, the herbaceous phyllaries with very narrow whitish margins; flowers orange or red; stem 0.4–3.7 dm. high. . . 3. ***A. borealis.***

c. Principal leaves entire or only the basal ones 3-cleft. . . . . . . 4. ***A. glauca.***

b. Disk-flowers with both stamens and pistil (or some merely pistillate), fertile, the achenes developed; style 2-cleft. § ABROTANUM. . . *e.*

e. Annuals or biennials, with tap-roots, glabrous throughout.

Panicle strict, leafy, spiciform or with spiciform or glomerulate erect branches; leaves bipinnatisect or the upper bipinnatifid; heads erect, sessile or nearly so; plant not sweet-scented. . . . . 5. ***A. biennis.***

Panicle loose and open, with reduced leaves; principal leaves tripinnatifid; heads nodding, pedicelled; plant sweet-scented. . . . 6. ***A. annua.***

e. Perennials with tough or ligneous bases or crowns. . . *f.*

f. Leaves green, glabrous or nearly so above, only sparsely pubescent beneath, divided into linear-filiform revolute segments; shrub up to 2 m. high. . . . . 7. ***A. Abrotanum.***

f. Leaves whitened beneath with close tomentum, simple to variously dissected, the segments not filiform; flowering stems arising annually from crowns or rhizomes, half-shrubby and persistent only in no. 8. . . *g*.

g. Half-shrubby, the stiffly erect stems subligneous at base and somewhat persistent; leaves 1–3 cm. long, with linear lobes and narrowly linear short blunt ultimate segments; heads broadly hemispherical, scattered, long-pedicelled; garden-escape. . . . . . 8. *A. pontica.*

g. Herbs with annual flowering stems; leaves mostly larger and broader or with broader lobes and segments or, if linear-lobed, with acute segments and crowded narrow heads. . . *h*.

h. Inflorescence spicate-racemose; involucres 6–7 mm. high, 6–9 mm. broad; disk-corollas 3.5–4 mm. long; heavily white-lanate plant with decumbent leafy basal offshoots of blunt-lobed obovate leaves. . . . . 9. *A. Stelleriana.*

h. Inflorescence paniculate; involucres 2.5–5 mm. high, 2–4 mm. broad; disk-corollas 1.5–3 mm. long. . . *i*.

i. Leaves of primary axis (below inflorescence) roundish-ovate to -obovate, 4–12 cm. long, cleft, often nearly to midrib, into obovate to oblanceolate frequently incised or toothed divisions; base of petiole with conspicuous leaflet- or stipule-like appendages; tall, in clumps 0.5–2 m. high, from strong branching rhizome without slender stolons. . . . . 10. *A. vulgaris.*

i. Leaves linear or lanceolate to oblong or oblanceolate, entire or once lobulate or, if broader and deeply divided, the petioles without basal appendages and the slender rhizome producing filiform stolons. . . *j*.

j. Stems clustered from stout ligneous crowns 0.8–3 cm. thick, without stolons, 0.4–3 m. high; principal leaves entire or finely serrate, lance-attenuate. . *k*.

k. Leaves bright green and glabrous above, the longer ones 0.9–2 dm. long.

Principal leaves below the panicle finely serrate, 1–3 cm. broad, the base bearing lance-subulate appendages; involucre campanulate, 2.5–3 mm. high. . . . . . 11. *A. serrata.*

Principal leaves entire, 0.5–1.5 cm. broad, without basal appendages; involucre subcylindric, 4–5 mm. long. . . 12. *A. Herriotii.*

k. Leaves canescent above with permanent close felt, linear-attenuate, the longer ones 3–10 cm. long and 2–10 mm. wide; involucre 4–5 mm. long. . . . . 13. *A. longifolia.*

j. Stems scattered or few together, in loose colonies, 2–10 dm. high, from slender soon freely stoloniferous rhizomes, the stolons elongate and slenderly cord-like; leaves entire or variously lobed or cleft; involucre 2.5–5 mm. long.

Principal leaves lance-linear or lanceolate to lance-oblong or oblanceolate, 2.5–10 cm. long, entire or with few forward-arching lobes or coarse teeth; stems 3–10 dm. high. 14. *A. ludoviciana.*

Principal leaves oblanceolate to elliptic, 1–5 cm. long, pinnately dissected essentially to midrib into divergent revolute-margined narrowly linear or linear-filiform lobes only 3–10 mm. long and 0.5–1 mm. wide; stems 2–7.5 dm. high. . . . . 15. *A. Carruthii.*

a. Receptacle beset with long woolly hairs; flowers all fertile, the marginal ones pistillate. § Absinthium.

Stems coarse, herbaceous, 0.3–1 m. high; principal leaves roundish, 3–10 cm. long, their ultimate segments 1.5–4 mm. broad. . . . . . . 16. *A. Absinthium.*

Stems slender, woody at base, 1–4 dm. high; principal leaves 0.5–1 cm. long, their ultimate segments 0.5–1 mm. broad. . . . . 17. *A. frigida.*

§ Dracúnculus Bess.

1. A. caudàta Michx. (tailed; evidently from the prolonged tip of the panicle). — *Biennial,* only *occasionally perennial; first-year plants* a crown *of numerous twice- or thrice-pinnately divided* long petioled *green and glabrous or nearly glabrous leaves with filiform-linear segments about* 0.5 *mm. wide; flowering stems very leafy,* arising from first year's rosettes, exceptionally producing from the base new and short-lived leafy tufts, 0.3–1.5 *m. high; panicle prolonged,* green to bronze, leafy, with many erect or strongly ascending crowded branches, 1.5–7.5 *dm. long,* with tapering apex; *heads rather crowded,* short-pedicelled, erect, or *finally divergent or nodding; involucre 2–3 mm. long and broad,* glabrous, with oblong narrowly scarious-margined green phyllaries; *flowers 14–25, the disk-corollas 1.4–2 mm. long.* (*A. canescens,* subsp. *caudata* Hall & Clements) — Sands, Fla. to Tex., n. near coast to s. Me.; inland from sw. Que. to Man., s. to Vt., n. N.Y., n. O., n. Ind., Ill., Mo., Okla. and Tex. July–Oct.

Var. **cálvens** Lunell (becoming bald). — *Basal* (or rosulate) *leaves grayish-pubescent;* stem canescent-pilose, inclined to be glabrate; heads more often not drooping. (*A. Forwoodii* S. Wats.) — Dry plains, prairies and sands, Sask. to Wash., s. to Ariz., e. to Man., Minn., S.D. and Neb. and along the Great Lakes to e. Ont., w. N.Y. and nw. Pa. Aug., Sept.

2. A. canadénsis Michx. (Canadian). — Subrosulate, with long *perennial* (*rarely only biennial*) *tap-root and* usually persistent *rosettes of basal leaves;* the latter glabrous or sparsely silky and glabrate, *with* very narrow *linear segments* 0.5–2 *mm. wide; flowering stems* 1–several, decumbent to erect, 0.2–7 (rarely –10) *dm. high; panicle* (except in obviously stunted plants) mostly *with strongly ascending elongate branches,* 1–6 *dm. high; the heads* suberect to nodding, *some of them on pedicels longer than the involucre; involucre* hemispherical, 3–4 *mm. long* and 3.5–5 *mm. broad; principal phyllaries* round-elliptic, *with broad pale scarious sides and narrow green backs,* the inner ones ovate and mostly scarious; *disk-corollas yellow,* 2.2–3 *mm. long.* (Included in *A. campestris,* subsp. *borealis* by Hall & Clements) — Gravels, sands, calcareous rocks and cliffs, s. Lab. to B.C., s. to Nfld., N.B., n. Me., n. Vt., Ont., n. Mich., Colo., etc. July–Sept.

3. A. boreàlis Pall. (northern). — Mostly (except in most favorable conditions) lower than no. 2, always perennial; *basal leaves with linear-lanceolate to oblanceolate* mostly short *segments* 0.5–2 *mm. wide, minutely silky, glabrate or glabrous; flowering stems* 0.4–2 (rarely –3.7) *dm. high; panicle spiciform, racemiform, or* in larger plants *with elongate spiciform branches,* the *heads sessile or nearly so,* the middle and upper leafy bracts entire; *involucre* hemispherical, 3–4 mm. long by 4–6 *mm. broad, purplish to green, glabrescent or glabrous, the herbaceous* broadly elliptic *phyllaries with narrow scarious border; disk-corollas orange to red.* (*A. campestris,* subsp. *borealis* Hall & Clements, in part) — Arct. reg., s. to calcareous and magnesian barrens and cliffs of Nfld., Gaspé Pen., Que., n. Mich., Colo., etc. July, Aug. (Eurasia)

Var. **latiśecta** Fern. (with broad segments). — As in the preceding but with *thicker rosulate leaves, their oblong or oblanceolate segments often* 3 *or* 4 *mm. broad.* — On calcareous cliffs and ledges, Lab., w. Nfld. and Anticosti I.

Var. **Púrshii** Bess. (for Frederick Traugott Pursh, 1774–1820). — *Leaves* and stem *densely* and more or less *permanently silky; involucres and axis of inflorescence* usually *villous.* (*A. campestris,* subsp. *spithamaea* Hall & Clements) — Greenl. to Alaska and Kamtchatka, s. to serpentine- and limestone-barrens of Nfld., Shickshock Mts., Gaspé Pen., Que., n. Mich., Colo. and Oreg.

4. A. glaùca Pall. (blue-green; name often inappropriate). — Mostly inodorous perennial with hard horizontal rhizome; stem erect, 0.3–1.5 m. high, glabrous or grayish-pubescent and glabrate; *leaves linear,* 3–8 cm. long, 1–4 mm. wide, glabrous and green to grayish-pubescent and glabrate, *all entire or the lowest* 3-*cleft at apex,* firm; panicle with elongate leafy ascending branches; *heads soon spreading or nodding on slender pedicels* up to 2 mm. long; *involucres subglobose,* 2–3 *mm. broad,* the larger of the 7–12 scarious-sided phyllaries broadly elliptical; flowers about 20–30; disk-corollas 1.5–2 mm. long. (Including *A. dracunculoides* Pursh; *A. Dracunculus,* subsp. *glauca* Hall & Clements) — Prairies, plains and dry slopes, B.C. to n. Mex., e. to Man., Wisc., Ill., Mo., Okla. and Tex.; adv. eastw. to N.E. July–Oct. (Siberia)

Var. **dracunculìna** (S. Wats.) Fern. (like *A. Dracunculus*). — *Inflorescence very loose,* the *nodding heads on drooping pedicels mostly* 3–10 *mm. long or on filiform arching or drooping minutely bracted branchlets up to* 3.5 *cm. long.* (*A. Dracunculus,* subsp. *dracunculina* Hall & Clements) — Minn. to Utah, s. to Mo., Kans., Tex. and n. Mex.; casually adv. eastw. to N.E.

§ Abrótanum Bess.

5. A. biénnis Willd. (biennial). — Strict *glabrous annual or* biennial with tap-root and no basal offshoots, 0.1–3 m. high; *leaves 2-pinnately parted or the upper pinnatifid; their lobes* linear, *acute, all but uppermost cut-toothed; panicle strict,* with long leafy bracts, *spiciform or with spiciform or glomerulate erect branches; heads subglobose, crowded, erect, sessile or nearly so;* involucre 2–3 mm. high, the larger glabrous phyllaries roundish and with broad scarious sides and green midrib. — Open ground, clearings, burns, roadsides and waste places, Gaspé Pen., Que., to B.C., s. to N.S., N.E., N.J., Pa., O., Ind., Ill., Mo., etc.; an aggressive weed, presumably originating in our cordilleran reg. Aug.–Nov.

6. A. ánnua L. (annual). — Slender glabrous *sweet-scented annual* up to 3 m. high, *loosely bushy-branched; lower and median leaves 3-pinnatifid into oblong-lanceolate* sharply toothed or cleft *segments; panicle loose and open, with reduced bracteal leaves, the pedicelled heads spreading or nodding; involucre* hemispherical, 1–1.5 *mm. long,* its phyllaries scarious except for green midrib. — Weed of waste places, roadsides, fallow fields, neglected gardens, etc., P.E.I. to Ont. and sparingly westw., s. to N.E., L.I., Va., Ala., Tenn., Ark., etc. Aug.–Oct. (Natzd. from Eurasia)

7. A. Abrótanum L. (Greek *abrotonon,* an aromatic plant), Southernwood. — *Shrub* up to 2 m. high, fragrant, *the slightly woody glabrous stems much branched; leaves* numerous, *dissected into revolute linear-filiform segments, green, glabrous or nearly so above,* sparsely pubescent beneath, the petiole naked; panicle elongate, with short-stalked finally nodding subglobose heads 2.5–3 mm. broad; phyllaries somewhat pubescent. — Roadsides and waste places, spread from old gardens. Aug.–Oct. (Introd. from s. Eu.)

8. A. póntica L. (from the Black Sea), Roman W. — *Half-shrubby;* the *slightly woody gray-puberulent stems simple below,* 0.15–1 m. high, with somewhat persistent bases; *leaves* numerous, *crowded, grayish,* 1–3 *cm. long,* 1–2 *cm. broad, twice pinnatifid into linear often narrow-toothed blunt ultimate segments, gray-tomentulose beneath; heads hemispherical* (rather rare with us), *nodding on slender pedicels,* scattered along the racemes or racemose slender branches of the panicle; *involucre* 3–4 mm. broad, *densely pubescent, the phyllaries herbaceous.* — Waste places, roadsides, old pastures, etc., N.S. to Ont. and Wisc., s. to s. N.E., Del., Pa. and O. July–Oct., rarely flowering. (Introd. from se. Eu.)

9. A. Stelleriàna Bess. (for its discoverer, Georg Wilhelm Steller, 1709–1746), Beach-W., Dusty Miller, Old Woman. — Matted coarse *densely white-lanate* perennial with extensively creeping and forking rhizome; *leafy decumbent basal stems frequent,* these *with broad-petioled obovate bluntly lobed leaves;* flowering stems arising from old leafy offshoots, decumbent or erect, 2–7.5 dm. high, with obovate to spatulate coarsely few-lobed leaves; *inflorescence strict, spicate-racemose,* with hemispherical erect or ascending heads; *involucre* hemispherical, heavily lanate, 6–7 *mm. high* and 6–9 *mm. broad; disk-corollas* 3.5–4 *mm. long;* achenes slenderly oblanceolate. — Sandy beaches and dunes along the coast from Gulf and lower R. St. Lawrence, Que., to Va.; inland on lake-shores from e. Ont. and N.Y. to Minn., originally spread from cult. about 1880. Mid-May (southw.)–Aug. (northw.). (Introd. and natzd. from ne. Asia)

10. A. vulgàris L. (common), Common Mugwort. — Perennial with stout horizontal forking rhizome, *without slender elongate stolons,* the 1–several *erect* corrugated greenish to purplish glabrescent *stems* 0.5–2 *m. high; principal leaves of axis* (below inflorescence) *roundish-ovate to -obovate,* petioled, *base of petiole with* 4–8 *obvious leaflet- or stipule-like appendages;* the *blades glabrous* or promptly glabrate *and green above,* white-tomentulose beneath, 4–12 cm. long, *cleft nearly to midrib into lacerate or cut-toothed lance-acuminate forward-pointing divisions;* panicle elongate, composed of ascending subspiciform branches with crowded heads in glomerules; heads sessile, ovoid, finally spreading; involucre gray-tomentulose, 3.5–4 mm. long, its outer phyllaries linear-lanceolate, the inner ones oblong; disk-corollas 2–3 mm. long. — Thickets, roadsides, stream-banks and waste places, Nfld. to Ont., s. to N.S., N.E., Del., Pa., casually to Ga., Mich., Wisc. and Minn. July–Sept. (Natzd. from Eu.)

Var. glàbra Ledeb. (smooth). — Similar; the *narrowly lance-acuminate divisions of the primary leaves entire.* — Local, w. N.E. to Ont., s. to Ct., O., Mich. and Kans. (Natzd. from Asia)

Var. latíloba Ledeb. (broad-lobed). — *Leaves thinner, the principal ones less deeply cleft, the broadly obovate or rhombic terminal division* and the oblanceolate to oblong lower ones and their few teeth *blunt or merely acutish,* not acuminate; panicle usually less developed. — Que. to s. N.E and Ill. (Natzd. from Eurasia)

11. A. serràta Nutt. (saw-toothed). — *Stem* erect, *from a stout forking ligneous rhizome,* 0.7–3 *m. high,* minutely cinereous-tomentulose to glabrate; *leaves lance-acuminate, bright green and glabrous above,* white-tomentulose beneath, *the longer primary blades* 0.9–1.5 *cm. long and* 1–3 *cm. broad, sharply fine-serrate, their bases with nearly subulate short appendages,* upper leaves shorter and entire; panicle elongate, leafy, with strongly ascending spiciform-glomerulate branches; *heads crowded, sessile or subsessile, campanulate; involucre* 2.5–3 *mm. high,* with round-tipped elliptic inner phyllaries, 16–24-flowered. (*A. vulgaris,* subsp. *serrata* Hall & Clements) — Bottoms, rich thickets and prairies, Wisc. to N.D., s. to Ill., Mo. and Kans.; sparingly adv. eastw. July–Sept.

12. A. Herriótii Rydb. (for one of its discoverers, William Herriot, 1870–1930). — Similar to no. 11, 0.4–1.2 m. high; *leaves entire* or only sparingly toothed, *without basal auricles, the longer ones* (0.5–)1–2 *dm. long and* 0.5–1.5 *cm. wide,* very long-attenuate; panicle slender, the *often pedicelled heads subcylindric; involucre 4–5 mm. long,* its oblong to ovate phyllaries narrowed to blunt tips. (Referred erroneously by Hall & Clements to the more western and very different *A. Douglasiana* Bess., their *A. vulgaris*, subsp. *heterophylla*) — Plains, dry ridges and gravelly shores, Minn. to n. Alta. and S.D. July–Sept.

13. A. longifòlia Nutt. (long-leaved). — *Rhizome woody, stout, branching into crowns* 0.6–2 *cm. thick; stems clustered,* erect, 0.3–1 m. high, closely white-pannose; *leaves* closely puberulent, *gray above, white beneath, linear-attenuate, entire* or the lowest rarely toothed, *revolute-margined,* crowded; *the principal ones* 3–10 *cm. long,* 2–5 (rarely –10) *mm. broad;* panicle slender or with slender erect racemiform branches; heads erect, sessile, slenderly campanulate; involucres lanate, 4–5 mm. long, 20–30-flowered. (*A. vulgaris*, subsp. *longifolia* Hall & Clements) — Dry plains and alkaline areas, w. Ont. and Mich. to Mont. and Ida., s. to Colo. and Wyo. June–Sept.

14. A. ludovicìana Nutt. (of St. Louis), Western Mugwort, White Sage. — *Rhizome slender and cord-like, soon producing lash-like or slenderly cord-like more or less flagelliform stolons* and becoming loosely colonial; stems 0.3–1 m. high, grayish-lanate or tomentulose, rarely glabrate; *leaves covered beneath with white felt,* lance-linear or lanceolate to lance-oblong, oblong-elliptic or oblanceolate, the principal ones 2.5–10 cm. long, *entire or with few entire forward-arching lobes or coarse teeth;* panicle dense to loose; heads globose-campanulate or cylindric, with pale phyllaries, the involucre 3–5 mm. high. (*A. vulgaris*, subspp. *ludoviciana* and *gnaphalodes* Hall & Clements) — Excessively variable and mostly aggressive and weedy plants. The following vars. occur with us:

a. Young leaves tomentose or lanate on upper as well as lower surface. . . *b.*
b. Pubescence of upper surfaces of primary and often of rameal leaves loosening and rather deciduous, the older leaves becoming glabrate and bright green; leaves entire or some of the lower and median ones with falcate-lobed margins. . . . . *A. ludoviciana* (typical).
b. Pubescence of upper as well as lower surfaces persistent, the upper surface remaining whitish or gray. . *c.*
c. Principal leaves lance-linear, lanceolate, oblong or oblanceolate, entire or with marginal falcate teeth or divisions; the blades soft and pliable, heavily tomentose, many times longer than broad. . *d.*
d. Leaves flat, mostly straight, ascending or spreading.
Blades lanceolate, acute or attenuate, longer ones 5–10 cm. long, ascending; stems usually simple below or to summit, without or more often with short suppressed axillary branches. . . . . Var. *gnaphalodes.*
Blades oblong, oblong-elliptic or oblong-oblanceolate, blunt or merely acutish, 2.5–7 cm. long, loosely ascending or spreading; stems frequently with loosely spreading or divergent elongate branches. . Var. *latifolia.*
d. Leaves mostly plicate, widely spreading or recurving, often twisted, the longer ones 2.5–5 cm. long; stem or its erect basal branches with suppressed axillary fascicles. . . . . Var. *pabularis.*
c. Principal leaves broadly oblong, with few coarse teeth around the summit, firm and thick, rather hard, only two to four times as long as broad; axillary branches short and suppressed. . . . . Var. *Brittonii.*
a. Young leaves glabrous (or only obscurely puberulent) and bright green above; blades linear to lanceolate, entire or with long falcate lobes, they or the lobes attenuate.
Panicle open and leafy, virgate or with virgate branches; involucre globose-hemispherical; stem usually covered by dense continuous felt; mostly northern. . . . . Var. *americana.*
Panicle pyramidal, dense; involucre cylindric or cylindric-ovoid; stem thinly tomentulose to puberulent, often glabrescent; southwestern. . . . Var. *mexicana.*

A. ludovicìana (typical). — Native of prairies and dry open soils or thin woodland, Mich. to Wash., s. to Ill., Ark., Tex. and Mex.; now spread along railroads, roadsides and in waste ground, dooryards, cemeteries, etc., eastw. to N.E., N.Y., N.J., Va., etc. July–Sept. — Stem 4–10 dm. high.

Var. **gnaphalòdes** (Nutt.) T. & G. (resembling *Gnaphalium*), *A. gnaphalodes* Nutt. — Native of prairies, etc., s. Ont. and Mich. to s. B.C., s. to Ark., Okla., Tex., etc.; natzd. eastw. to Que., N.E., N.J. and Del.

Var. **latifòlia** (Bess.) T. & G. (broad-leaved). — Man. and Minn. to s. B.C., s. to Ia., Kans. and N.M.; natzd. eastw. to Que., N.B. and N.E.

Var. **pabulàris** (Nels.) Fern. (fit for forage), *A. pabularis* (Nels.) Rydb. — Man. and Minn. to Oreg., s. to Ia., Neb. and Colo.; adv. eastw. to Mich. — Stem 2–7.5 dm. high.

Var. Brittònii (Rydb.) Fern. (in honor of Nathaniel Lord Britton, 1859–1934), *A. Brittonii* Rydb. — Mont. to Colo., etc.; casually adv. in N.E. — Stem 2–6 dm. high.

Var. americàna (Bess.) Fern. (American). — Alta. to Tex. and n. Mex.; casually adv. e. to N.E. — Stem 0.45–1 m. high.

Var. **mexicàna** (Willd.) Fern. (Mexican), *A. mexicana* Willd.; *A. vulgaris*, subsp. *mexicana* Hall & Clements — Mex. and Tex., ne. to barrens and sands of Mo. — Stem up to 1 m. high.

15. A. Carrùthii Wood (for its discoverer, James H. Carruth, 1807–1896). — *Crown bearing prolonged often leafy-tipped stolons;* the upright stems forking at base or clustered, 2–7.5 dm. high, mostly with suppressed axillary branches; *leaves* 1–5 *cm. long, of oblanceolate to elliptic outline, all or essentially all* (except in panicle) *pinnately dissected to midrib into divergent narrowly linear or linear-filiform lobes only* 3–10 *mm. long and* 0.5–1 *mm. wide,* white beneath, minutely canescent-pilose above; panicle slenderly pyramidal; heads slenderly campanulate; involucre canescent, 2.5–3 mm. high. (*A. kansana* Britton; *A. vulgaris*, subsp. *Wrightii* Hall & Clements, in part only) — Plains and dry scrub, w. Kans. and Colo. to w. Tex., N.M. and Ariz.; adv. along railroads, etc., in Mo., Ind., N.Y., R.I. and presumably elsewhere. Aug.–Oct.

A. tridentàta Nutt. (three-toothed), a western Sagebrush with shrubby stem, cuneate firm silvery leaves 3 (–7)-toothed across the summit, and heads with perfect flowers, has appeared casually in an old field in e. Mass. (Adv. from the West)

§ Absínthium Bess.

16. A. Absínthium L. (Greek and Latin name for Wormwood), Wormwood, Absinthium, Absinthe, Armoise (Que.). — *Coarse* perennial *herb,* the canescent *stems* rising 0.3–1 *m. high* from a branching rhizome; *lower leaves long-petioled, middle and upper ones* short-petioled to sessile, *silvery-silky; the larger roundish blades* 3–10 *cm. long,* deeply twice or thrice pinnately dissected, the *ultimate linear to oblong or oblanceolate segments* 1.5–4 *mm. broad;* panicle leafy, 1–4 dm. long, with erect branches; heads nodding; involucre hemispherical, 3–4 mm. broad; receptacles covered with long white hairs. — Roadsides, pastures, etc., Nfld. to Man., s. to N.S., N.E., N.J., Pa., rarely to N.C., O., Ind., Wisc., Minn., etc., spread from cult. July–Sept. (Introd. from Eu.)

17. A. frígida Willd. (of cold regions), Prairie-Sagewort. — *Suffruticose* from a woody base forking into many crowns, *with many closely leafy basal branches; flowering stems slender,* 1–4 *dm. high,* densely leafy; *leaves* silvery-silky, roundish, 0.5–1 *cm. long,* divided into many linear *segments* 0.5–1 *mm. broad;* panicle slender, subsimple or with erect branches, 0.5–3 dm. long; heads nodding; involucre long-villous. — Dry prairies, plains and rocks, Sask. to Alaska, s. to Wisc., Minn., Kans., Tex., Ariz., etc.; natzd. along roadsides, railroads, in dry fields, etc., eastw. to Que., N.B. and N.E. July–Sept. (Asia)

Tribe VIII. Senecìoneae Cass. (see p. 1358)

Heads monoecious or subdioecious, the hermaphrodite flowers mostly sterile, the pistillate ones fruiting; the marginal flowers (at least in the pistillate heads) in more than 1 row; style-branches obtuse, neither appendaged nor hispid; dilated leaves chiefly radical.
- Heads solitary, yellow, monoecious; marginal ligules in several series. . . 76. *Tussilago.*
- Heads in corymbs, creamy-white to purple, dioecious or subdioecious; staminate heads with a single row of short-rayed pistillate flowers; pistillate heads without or with short ligules. . . . . 77. *Petasites.*

Heads with all or nearly all flowers fertile; style-branches truncate, hirsute or capitellate at tip, often with a conical or cuspidate appendage.
- Cauline leaves opposite; heads 1–few, single at tips of stem and branches, showy; pappus stiffish, scabrous to plumose. . . . . . . 78. *Arnica.*
- Cauline leaves alternate; pappus of soft capillary bristles.
  - Heads discoid; corollas whitish or creamy, rarely pinkish.
    - Annuals; outer flowers pistillate, filiform, with 2–4-toothed limb; receptacle naked. . . . . 79. *Erechtites.*
    - Perennial; flowers all hermaphrodite; corollas tubular, 5-cleft; receptacle thickish-fibrillate or naked. . . . . 80. *Cacalia.*
  - Heads mostly radiate, sometimes discoid; corollas yellow or yellow with red or purplish tips, 5-toothed; receptacle naked; mostly perennials (some annuals). . . . . 81. *Senecio.*

76. TUSSILÀGO L. Coltsfoot. Pâs-d'âne (Que.)

Head many-flowered; ray-flowers in several rows, narrowly ligulate, pistillate, fertile; disk-flowers with undivided style, sterile. Involucre nearly simple. Receptacle flat. Achenes slender-cylindric or prismatic; pappus copious, soft and capillary. — Low perennial, nat. of Eurasia and n. Afr., with horizontal creeping rootstocks, sending up scaly scapes in early spring, these bearing a single head, and producing rounded-cordate angled or toothed leaves later in the season, these woolly when young. Flowers yellow. (Name from *tussis*, a cough, for which the plant has long been a reputed remedy.)

1. T. Fárfara L. (*Farfarus*, the Latin name for coltsfoot). — Damp argillaceous soils, brooksides, etc., Nfld. to Minn., s. to N.S., N.E., N.J., Pa. and O. March–June. (Introd. from Eu.)

77. PETASÌTES Mill. Sweet Coltsfoot

Heads many-flowered, somewhat dioecious; in the substerile plant with a single row of ligulate pistillate ray-flowers, and many tubular sterile ones in the disk; in the fertile plant wholly or chiefly of pistillate flowers, tubular or distinctly ligulate. Otherwise as *Tussilago*. — Perennial woolly boreal herbs, the leaves all from the rhizome, the scape with sheathing scaly bracts, bearing heads of purplish or whitish fragrant flowers in a corymb or racemose panicle. (The Greek name for the Coltsfoot, from *petasos*, a broad-brimmed hat; on account of its large leaves.)

Heads corymbose or corymbose-racemose, with elongating and ascending pedicels and foliaceous (at least lower) bracts; flowers whitish, the marginal ligulate; leaves lobed, or unlobed and deltoid or hastate, their petioles not broadly dilated at base; indigenous species.
  Leaves lobed, green and glabrous or essentially so above, reniform, orbicular or cordate-deltoid.
    Leaves reniform or suborbicular, deeply cleft (more than two-thirds to base) into 5–7 cut and cleft lobes. . . . . 1. *P. palmatus.*
    Leaves reniform to cordate-deltoid, cut about half-way to midrib into numerous coarse lobes. . . . . 2. *P. vitifolius.*
  Leaves unlobed, deltoid-oblong to reniform-hastate, floccose above, heavily white-tomentose beneath. . . . . 3. *P. sagittatus.*
Heads closely racemose-paniculate, with short divergent pedicels subtended by linear-acicular bracts; flowers purplish, all tubular; leaves round-cordate, merely angulate-dentate, the petioles with dilated sheathing bases. . . 4. *P. hybridus.*

1. P. palmàtus (Ait.) Gray (palmate). — Extensively creeping, with slender cord-like rhizomes and stolons; flowering stems stout, preceding the leaves, finally becoming 0.15–1 m. high, the racemose *corymb* of the staminate plants soon shriveling; that of the pistillate elongating, *with* the lowest *ascending* fruiting *pedicels* becoming 3–10 cm. long; involucres 5–10 mm. high in anthesis, enlarging in fruit; *flowers creamy-white, the marginal usually ligulate; leaves reniform or suborbicular, deeply cleft (more than two-thirds to base) into* 5–7 sharply toothed and cleft *lobes, green and glabrous above,* thinly white-tomentose beneath, becoming 0.5–2.5 dm. broad, on slender petioles; *mature fruiting heads (including pappus)* 1–2 *cm. long; mature achenes* 1.5–2.5 *mm. long.* — Low woods, glades and damp clearings, Lab. to ne. B.C., s. to Nfld., N.S., Me., ne. and w. Mass., w. Ct., N.Y., Mich., Wisc. and Minn. Apr.–June (–July in subalpine areas).

2. P. vitifòlius Greene (with grape-leaves). — Resembling no. 1; *leaves reniform to cordate-deltoid,* becoming 0.5–2.5 dm. broad, *cut about half-way to midrib into coarse divergent lobes, white-tomentose beneath;* flowering stems elongating to 1.2–6 dm. high; *mature fruiting heads (including pappus)* 1–1.5 *cm. long; achenes about* 1 *mm. long.* (Incl. *P. trigonophyllus* Greene) — Meadows, swales and boggy woods, Lab. Pen.; Anticosti and subalpine areas of Shickshock Mts., Que.; n. Minn. and Man. to Alta. May–Aug.

3. P. sagittàtus (Pursh) Gray (arrow-shaped). — Similar to nos. 1 and 2; *leaves unlobed, deltoid-oblong to reniform-hastate, floccose above,* densely white-tomentose beneath, the margin dentate, becoming 0.8–2 dm. broad; fruiting stems up to 1 m. high; *fruiting heads (including pappus)* 2.2–3.5 *cm. long; ripe achenes* 3–3.5 *mm. long.* — Meadows and bogs, Lab. to n. B.C., s. to Wisc. and Minn., Colo. and Wash. May, June.

4. P. hýbridus (L.) Gaertn., Mey. & Scherb. (hybrid; from an early misconception), Butterbur. — Coarser, with stouter rhizomes; *heads in a compact ellipsoid racemose panicle, on short horizontal pedicels with linear-acicular basal bracts; flowers purplish, all tubular; leaf round-cordate,* merely dentate-angulate, becoming 6 dm. broad, *the base of the petiole dilated.* (*P. vulgaris* Hill) — Waste places, Mass. to Pa. Apr., May. — A particularly rank substitute for tobacco. (Introd. from Eu.)

78. ÁRNICA L. Arnica

Heads many-flowered; rays pistillate. Phyllaries of the campanulate involucre narrow, equal, somewhat in 2 rows. Receptacle flat, fimbrillate. Achenes slender or fusiform; pappus a single row of rather rigid and strongly roughened-denticulate (rarely subplumose) bristles. — Perennial herbs (chiefly of mountains and cold northern regions), with simple stems, bearing single or corymbed large heads and opposite leaves. Flowers yellow. (Origin of name, like limits of species, disputed.)*

a. Basal leaves petioled or attenuate to base, 1–3(–5)-ribbed, or cordate and pinnately veined; achenes pubescent at least above; boreal species. . . *b.*
- *b.* Pappus white or whitish, its bristles merely barbellate; upper leaves much reduced, mostly entire. *c.*
  - *c.* Basal leaves attenuate or gradually tapering below, 0.5–3.3 cm. broad, parallel-ribbed from near base. . *d.*
    - *d.* Lower leaves entire or with only rare and irregular teeth. . . *e.*
      - *e.* Base of involucre and summit of peduncle densely villous-lanate; phyllaries villous to tip; lower leaves linear to narrowly lanceolate or oblanceolate, 0.5–1.7 cm. broad; teeth at tip of ligule acute, the longer 3–8 mm. long.
        - Basal leaves glabrous or glabrate, often minutely glandular, 3-ribbed; cauline leaves (1–)3–5 pairs; phyllaries 13–20, narrowly lanceolate. . . . 1. *A. terrae-novae.*
        - Basal leaves villous-tomentose on both sides, 1-nerved; cauline leaves 1–2 pairs; phyllaries 8–10, rhombic-lanceolate. . . . 2. *A. tomentosa.*
      - *e.* Base of involucre and summit of peduncle glandular-puberulent to -villous, not lanate; phyllaries only sparingly pilose to glabrate above base; lower leaves oblong to broadly lanceolate, 1.2–2.5 cm. broad, 3-ribbed, glabrous or nearly so; teeth of ligule obtuse, 1–2 mm. long. . . . . 3. *A plantaginea.*
    - *d.* Lower leaves regularly dentate. . . *f.*
      - *f.* Rhizomatous, with single flowering stems mostly 2–5 dm. high; basal leaves slender-petioled, strongly 3-ribbed from base; phyllaries pubescent; densely pilose portion of tube of disk-corollas 3.5–8 mm. long; achenes densely pubescent from base to summit, 4–7 mm. long.
        - Disk-corollas 6–8 mm. long, with glabrous or only sparsely setose lobes; densely pilose portion of tube and throat 3.5–6 mm. long; pubescence much shorter than diameter of achene; mature pappus 5–8 mm. long, bright white. . . . 4. *A. chionopappa.*
        - Disk-corollas 9–10 mm. long, lobes copiously setose; densely pilose portion of the tube 7–8 mm. long; longer hairs nearly equaling diameter of achene; mature pappus 9–10 mm. long, creamy. . 5. *A. gaspensis.*
      - *f.* Stoloniferous, with 1–several flowering stems 0.5–2.5 dm. high; basal leaves with short margined petioles; lateral ribs faint, joining midrib well above base; phyllaries glabrous except at base; densely pilose portion of corolla-tube 2.5–3.5 mm. long; achenes glabrous or only remotely setulose below middle, 3–3.5 mm. long. . . 6. *A. louiseana.*
  - *c.* Basal leaves cordate, 4.5–8 cm. broad, pinnately veined. . . . . . 7. *A. Whitneyi.*
- *b.* Pappus yellowish-brown or olive-tinged, its bristles subplumose; cauline leaves 3–5 pairs, not greatly reduced, all evenly dentate. . . . . 8. *A. mollis.*

a. Basal leaves rhombic-oval, sessile or with short broad petiole, 5–7 ribbed; achenes glabrous; southern species. . . . . 9. *A. acaulis.*

1. A. térrae-nòvae Fern. (of Newfoundland). — Stem erect, simple or with 2 or 3 peduncles, 1–4 dm. high, sparsely pilose, *summit lanate; basal leaves broadly linear to narrowly oblanceolate,* 0.5–1.7 *cm. broad,* 3 (rarely 5)-*ribbed, entire, glabrous or* sparsely pilose and *glabrate* and glandular; cauline leaves 2–5 pairs; the lower like the basal, petioled, papillose-glandular, the upper much reduced and callus-tipped; peduncles elongate, naked or bracted; heads 4.5–8 cm. broad; *involucre* 1.3–1.7 cm. high, *white-lanate at base, with* 13–20 *loosely villous lance-acuminate phyllaries; ligules* 9–13, 1.7–3 cm. long, *with acute terminal teeth, the longer* teeth 3–8 *mm. long;* achenes 5.5–6 mm. long, black, white-hirsute; pappus white, barbellate, 7.5–8.5 mm. long. — N. Lab. to Keewatin and Man.; calcareous gravel and turf, w. Nfld. July, early Aug.

---

* Treatment based largely on that in Rhodora, xxxv. 364–369 (1933).

2. A. tomentòsa J. M. Macoun (tomentose). — *Stem* simple, 1–3 dm. high, *densely villous with white hairs* 1–2 *mm. long* mixed with minute glands; *leaves densely white-villous on both sides,* 1-*nerved;* the basal lanceolate or oblanceolate, 0.5–1.5 cm. broad, entire or remotely dentate; cauline leaves 1 or 2 pairs, entire; *involucre densely villous-lanate at base,* of 8–10 glandular and *loosely villous rhombic-lanceolate phyllaries* 1.5 cm. long; *ligules* 2–2.5 cm. long, 0.5–1 cm. broad, *sharply toothed, the longer teeth* 3–6 *mm. long;* achenes white-hirsute; pappus white, barbellate. — Dry calcareous gravels, w. Nfld.; Rocky Mts., Yuk. and Alaska to Mont. July.

3. A. plantagínea Pursh (plantain-like). — *Stem* 1–4.5 dm. high, with 1–3 peduncles, *glabrous or only very sparingly villous-hispid below, glandular at summit; leaves* entire, *glabrous or minutely pilose on nerves and margins; the basal oblong to lanceolate,* 1–2.5 *cm. broad,* 3-*ribbed,* the cauline 3 or 4 pairs, the upper much reduced; heads 1–3, 4–6 cm. broad; *involucre glandular-villous at base, its* 10–12 oblanceolate *phyllaries* 1–1.5 cm. long, *sparingly pilose or glabrate above base; ligules* 10–15, 1–2 cm. long, *with blunt teeth* 1–2 *mm. long;* achenes hispid; pappus white, barbellate. — Brooksides and cool slopes, Lab. and Ung.; n. Nfld. July.

4. A. chionopáppa Fern. (with snowy pappus). — *Stem* 0.7–5.5 dm. high, *loosely or sparingly villous; leaves sparingly villous or glabrate; the basal* lanceolate or narrowly ovate, *regularly dentate,* slender-petioled, *strongly* 3-*ribbed,* 1–3.5 cm. broad; cauline 2–4 pairs, entire, the upper reduced; heads 1–4, 3–4 cm. broad; *involucre villous throughout, the* 10–15 *linear- to lance-attenuate phyllaries* 7–13 *mm. long;* ligules 10–15, with short bluntish teeth; *disk-corollas* 6–8 *mm. long,* with glabrous or only sparsely setulose lobes, *the densely pilose portion* 3.5–6 *mm. long; achenes* setulose *with white hairs about one-half as long as diameter of achene; pappus bright white,* 5–8 *mm. long.* — Calcareous gravel, ledges and cliffs, w. Nfld. to Rimouski Co., Que., and n. N.B.; ne. Minn.; Mackenz. to Alta. Late June–early Aug.

5. A. gaspénsis Fern. (of Gaspé Pen., Que.). — Similar to no. 4; *stem glandular-puberulent above;* leaves sparingly pilose on both surfaces; *involucre glandular-puberulent and sparingly pilose;* its phyllaries oblong, with triangular tips; *ligules with sharp teeth; disk-corollas* more slender, 9–10 *mm. long, with conspicuously setose lobes, the densely pilose portion* 7–8 *mm. long; longer hairs of achene nearly equaling its diameter; pappus creamy,* 9–10 *mm. long.* — Calcareous ledges and cliffs, Cap Tourelle, Gaspé Co., Que. July.

6. A. louiseàna Farr (of Lake Louise). — *Stoloniferous, with* 1-*several* sparsely hirsute flowering *stems* 0.5–2.5 *dm. high; leaves fleshy, glabrous* or quickly glabrate, *the basal oblong to lance-ovate, short-petioled to subsessile,* callous-serrate, *with faint lateral ribs joining the midrib well above the base,* 0.5–2.5 cm. broad; cauline leaves 0–2 pairs, the lower like the basal, the upper reduced; peduncles 1–3; heads 2.5–5 cm. broad; involucre 1–1.5 cm. high, minutely pilose at base, glabrous above, its 8–10 narrowly rhombic phyllaries acuminate; ligules deep yellow, 1–2 cm. long, with teeth about 1 mm. long; *disk-corollas* 6–7 mm. long, *the villous portion* 2.5–3.5 *mm. long; achenes* 3–3.5 mm. long, *glabrous below* the sparsely short-hirsute summit. — Turfy and gravelly calcareous slopes, nw. Nfld.; Shickshock Mts., Que.; Rocky Mts., Alta. July, early Aug.

7. A. Whítneyi Fern. (for its discoverer, William Dwight Whitney, 1827–1894). — Stoloniferous, with very slender and elongate rhizomes; stem 1.7–4 dm. high, sparsely villous, with admixture of minute glands; *basal leaves broadly ovate to suborbicular, deeply cordate,* with narrow sinus, *membranaceous,* 4.5–8 cm. wide, slender-petioled; cauline leaves 2–3 pairs, the lower long-petioled and resembling the basal, the median similar but short-petioled; peduncles 1–3; heads 2.5–5 cm. broad; involucre 1.2–1.8 cm. high, densely white-villous at base, glandular above; phyllaries 8–12, linear-oblanceolate; ligules 1.2–1.8 cm. long, with terminal teeth about 1 mm. long; *achenes strigose-hirsute;* pappus white, barbellate. (*A. cordifolia* of ed. 7, not Hook.) — Dry woods, Keweenaw Co., Mich. Late June, July.

8. A. móllis Hook. (soft). — Stem 0.5–6 dm. high, crisp-villous, somewhat glandular above; *leaves all regularly dentate,* sparingly pilose; the basal oblanceolate to narrowly rhombic, petioled, 1–3.5 cm. broad; *cauline* 3–5 *pairs,* oblanceolate to ovate, all but the lowest sessile and rounded at base, the median 1–3.7 cm. broad; heads 1–10, 4–6 cm. broad; involucre glandular and villous throughout; phyllaries lance-attenuate, 1–1.6 cm. long; ligules 10–15; achenes hirsute; *pappus yellowish-brown or olive-tinged, plumose.* (*A. lanceolata* Nutt.) — Ledgy or gravelly shores or wet cliffs, often subalpine, Gaspé Pen. to Megantic Co., Que., s. to mts. of Me. and n. N.H.; Alta. and B.C. to Colo. and Calif. July–early Sept.

Var. **petiolàris** Fern. (petioled). — Cauline leaves oblanceolate, all but the upper pairs narrowed to slender petioles, the median 0.6–2 cm. broad; heads 3–4 cm. broad; phyllaries 7–9 mm. long. — Chiefly at lower altitudes, local, Bonaventure Co., Que., to n. N.H.

9. A. acaùlis (Walt.) BSP. (stemless), Leopard's-bane. — Stem 2.5–9 dm. high, pilose and glandular, stoutish; *rosette-leaves rhombic-oval, coriaceous, sessile or short-based,* 2.5–7 cm. broad, *with* 5–7 *ribs prominent beneath;* cauline leaves reduced, remote; heads 2–16, corymbed; involucre pilose; *achenes glabrous;* pappus whitish, barbellate. — Open sandy woods and thickets, se. Pa. and Del. to Fla. Late Apr.–early June.

79. ERECHTÌTES Raf. FIREWEED. CRÈVE-À-YEUX (Que.)

Heads many-flowered; the flowers all tubular and fertile; the marginal pistillate, with a slender corolla. Phyllaries of the cylindrical involucre in a single row, linear, acute, with a few small bractlets at the base. Receptacle naked. Achenes oblong, tapering at the end; pappus copious, of very fine and white soft hairs. — Erect and coarse annuals of Am., Australia and N.Z., of rank smell, with alternate simple leaves, and paniculate-corymbed heads of whitish flowers. (An ancient name, *Erechthites,* used by Dioscorides for some species of Groundsel.)

1. E. hieracifòlia (L.) Raf. (with leaves of Hawkweed, *Hieracium*), PILEWORT. — Often hairy; stem grooved, 0.2–3 m. high; leaves lanceolate to oblong, acute, often cut-toothed; *heads subcylindric, only slightly gibbous at base; phyllaries linear, with a slender attenuate tip,* 0.5–1.5 mm. broad at the *scarcely dilated base; corolla* of perfect flowers not brown-lineate, the *tube pale stramineous; achenes* 2–3 *mm. long, not inflated, with* 10–12 *pale ribs* and usually strigose brown furrows, the terminal annulus (formed by bases of pappus-bristles) usually without a projecting beak (persistent style-base) protruding from center; *denuded receptacles* 5–8.5 *mm. in diameter.* — Very variable:

Upper leaves with broad sessile or somewhat clasping bases.
  Leaves scarcely decreasing in size into the inflorescence. . . . . *E. hieracifolia* (typical).
  Upper leaves rapidly reduced to bracts below the inflorescence. . . . . . Var. *intermedia.*
Upper leaves attenuate to base or petioled. . . . . Var. *praealta.*

E. hieracifòlia (typical). — Damp thickets, clearings and shores, Me. to Minn., s. to Va. and Ill. July–Oct.

Var. **intermèdia** Fern. (intermediate). — Clearings, damp woods and burns, P.E.I. to w. Ont., s. to Fla., La. and Tex.

Var. **praeálta** (Raf.) Fern. (very tall). — Que. to Wisc., s. to Fla. and Tenn.

2. E. megalocárpa Fern. (large-fruited). — Resembling typical no. 1, *fleshy; heads ovoid, abruptly acuminate* (when fresh); *phyllaries lanceolate, bluntish,* 1–3 mm. broad *at the dilated base; corolla of perfect flowers with brown margins and midribs, the tube green; achenes* 4–5.5 *mm. long, inflated, with* 16–20 *ribs,* glabrous, or slightly strigose in furrows, the annulus with distinct central beak; *denuded receptacles* 1–1.2 *cm. broad.* — Sandy seashores, se. Mass. to L.I. Sept., Oct.

80. CACÀLIA L. INDIAN-PLANTAIN

Heads 5–many-flowered; the flowers all tubular and perfect. Phyllaries in a single row, erect, connivent, with or without a few bractlets at the base. Receptacle naked or thickish-fibrillate at center. Corolla deeply 5-cleft. Achenes oblong or slender-cylindric, smooth; pappus of numerous soft capillary bristles. — Smooth and tall Am. and Asiatic perennial herbs, with alternate often petioled leaves, and rather large heads in flat corymbs. Flowers white or whitish. (An ancient name, of uncertain meaning.)

Involucre with several primary phyllaries and many linear-attenuate basal (calyculate) ones, 25–30-flowered; corolla-lobes shorter than throat; receptacle plane; leaves hastate. § EUCACALIA . . . . 1. *C. suaveolens.*
Involucre with about 5 phyllaries, without elongate basal ones, about 5-flowered; corolla-lobes longer than throat; receptacle commonly with thickish fibrillae in the center; leaves not hastate. § CONOPHORA.
  Lower leaves reniform, roundish or deltoid, lobed or coarsely angulate-dentate.
    Not glaucous; lower leaves round-reniform; stem angled or sulcate. . . 2. *C. Muhlenbergii.*
    Glaucous or strongly whitened; lower leaves triangular-ovate or angulate-cordate; stem terete. . . . . 3. *C. atriplicifolia.*
  Lower leaves lance-ovate or oval, entire or shallowly crenate or dentate, green. . . . . 4. *C. tuberosa.*

§ EUCACÀLIA DC.

1. C. suavèolens L. (sweet-smelling). — Nearly glaucous; stem sulcate, 0.5–1.6 m. high, subuniformly leafy; *leaves triangular, hastate,* long-tapering, doubly serrate-dentate, the cauline wing-petioled; *involucre 25–30-flowered, with several linear-attenuate basal phyllaries;* corolla-lobes shorter than the throat; receptacle plane. (*Synosma* Raf.) — Woods, thickets and clearings, Ct. to Ia., s. to Fla., Tenn. and Mo.; natzd. in e. Mass. Late July–Sept.

§ Conóphora DC.

2. C. Muhlenbérgii (Sch. Bip.) Fern. (for its discoverer, Gotthilf Henry Ernest Muhlenberg, 1753–1815), Great Indian-p. — Not glaucous; stem 1–3 m. high, grooved and angled; *leaves green on both sides, dilated-flabelliform, or the lowest reniform,* 3–6 dm. broad, *repand-toothed* and angled, palmately veined, petioled, the teeth pointed; corymbs large; involucre of about 5 phyllaries, and 5-flowered, its basal phyllaries none or minute; receptacle often with a scale-like pointed appendage in the center; corolla-lobes longer than throat. (*C. reniformis* Muhl., not Lam.; *Mesadenia* Rydb.) — Rich woods, thickets and clearings, w. N.J. to Minn., s. to Ga., Ala. and Mo. Late June–Aug.

3. C. atriplicifòlia L. (with leaves of Orach, *Atriplex*), Pale Indian-p. — Glaucous; stem terete, 1–2 m. high; *leaves palmately veined and angulate-lobed;* the lower triangular-reniform or slightly cordate; the upper rhombic or cuneate, *toothed.* (*Mesadenia* Raf.) — Dry open woods, thickets and openings, N.Y. to Minn., s. to Fla., Ala. and Okla. Late June–Sept.

4. C. tuberòsa Nutt. (tuberous). — Stem angled and grooved, 6–20 dm. high, from a thick tuberous root; *leaves green on both sides,* thick, strongly 5–7-*nerved;* the lower *lance-ovate or oval,* nearly entire, tapering into long petioles; the upper on short margined petioles, sometimes toothed at apex. (*Mesadenia* Britt.) — Prairies, marly bogs and damp fields, s. Ont. to Minn., s. to Ala., Miss., La. and Tex. June–Aug.

81. SENÉCIO L. Groundsel. Ragwort. Squaw-weed. Séneçon (Que.)

Heads many-flowered; rays pistillate or none; involucre cylindrical to campanulate, simple or with a few bractlets at the base, the phyllaries erect-connivent. Receptacle flat, naked. Pappus of numerous very soft and capillary bristles. — Tremendous almost cosmop. genus; ours herbs, with alternate leaves and solitary or corymbed heads. Flowers chiefly yellow. (Latin name of a plant, from *senex,* an old man; alluding to the hoariness of many species, or to the white hairs of the pappus.)

*a.*Annuals or short-lived biennials without branching rhizomes or basal tufts arising from rhizomes or crowns; inflorescences glabrous or variously hispid to arachnoid or long-villous. . . *b.*

*b.*Leaves pinnatifid, lobed or dissected; inflorescence glabrous or short-pubescent; involucre calyculate at base; pappus twice or thrice length of achene. . . *c.*

*c.*Ligules 0 or up to 3 mm. long; leaves undulate-pinnatifid, deeply pinnatifid or, if pinnatisect, hirtellous; annual weeds. . . *d.*

*d.*Calyculate bracteoles many times shorter than phyllaries; achenes pubescent.

Stem glabrous or promptly glabrate, not glutinous; leaves undulate to pinnatifid; bracteoles and tips of glabrous phyllaries blackish; rays 0. . . . . 1. *S. vulgaris.*

Stem viscid or minutely glandular-hirtellous; leaves pinnatifid to deeply pinnatisect; bracteoles and minutely pubescent phyllaries not black-tipped; rays short. . . . . 2. *S. sylvaticus.*

*d.*Calyculate bracteoles about half as long as the heavily glandular-hispid phyllaries; whole plant heavily glandular and strong-smelling; rays obvious; achenes glabrous. . . . . 3. *S. viscosus.*

*c.*Ligules conspicuous, 6–10 mm. long; leaves, or many of them, pinnately divided.

Principal leaves finely and uniformly 2–3-pinnatisect; coarse hard-stemmed northeastern field-weed; the stem, lower surfaces of leaves and inflorescence cobwebby. . . . . 4. *S. Jacobaea.*

Principal leaves lyrate-pinnatifid, the terminal division large and rounded; succulent southern paludal plant with leaves and inflorescence promptly glabrate. . . . . 5. *S. glabellus.*

*b.*Leaves undulate, dentate or shallowly pinnatifid; corymb dense to open, more or less long-villous and woolly with jointed hairs with simple non-calyculate involucres; pappus four or five times length of achene; coarse hollow- and soft-stemmed northern paludal plant. . . . . 6. *S. congestus,* vars.

*a.*Perennials, either with stout base or with rhizomes and perennating crowns bearing leafy tufts distinct from the flowering stems; pubescence, when present, usually woolly or arachnoid, rarely hispid to villous. . . *e.*

*e.*Disk 2.5–5 cm. broad; involucre 2–4 cm. high, densely cottony, calyculate by slender-subulate prolonged bracteoles extending up from the heavy clavate cottony peduncle; rhizome vertical, very stout, prolonged, praemorse; leaves very fleshy, lingulate to elliptic or obovate, lustrous-green above, white-felted beneath, crowded along the heavy upright stem; northern halophyte. . . . . **7. *S. Pseudo-Arnica.***

*e.*Disk and involucre much smaller; stems more slender, from relatively short crowns or slender rhizomes. . . *f.*

*f.*All the leaves finely, the basal bi- or tripinnately dissected, the ultimate segments 0.5–2 mm. wide; endemic in southeastern mts. . . . . **8. *S. Millefolium.***

*f.*At least the basal leaves undivided or only pinnately lobed or pinnatifid. . . *g.*

*g.*Rhizome short, erect, forming a tough crown without tufted leafy offsets; leaves entire or merely denticulate, the basal much larger than the rapidly decreasing and slender upper ones; plant at first sparsely crisp-villous with jointed hairs, soon glabrate; northwestern species. **9. *S. integerrimus.***

*g.*Rhizome usually prolonged, bearing (ordinarily) tufts of leaves distinct from the flowering stems; pubescence, when present, a tomentum, or more or less arachnoid and persistent as flocculent tufts. . . *h.*

*h.*Lower leaf-surfaces densely felted, upper surfaces of new leaves pubescent; involucre tomentulose nearly or quite to summit.

Larger basal leaf-blades 0.4–1.8 dm. long, lanceolate, oblong or oval, undulate-dentate, on petioles 0.4–2.3 dm. long; achenes columnar, drab, heavily hirtellous on the prominent rounded ridges; plant of Coastal Plain and adjacent area. . . . . . **10. *S. tomentosus.***

Larger basal leaves 1.5–4 cm. long, obovate to elliptic, more sharply dentate toward summit, tapering to petioles 1–4 cm. long; achenes subfusiform, brown, sparsely hirtellous on the obscure acutish ridges; mountain-species. . . . . . **11. *S. antennariifolius.***

*h.*Lower leaf-surfaces glabrous or, if tomentulose (in nos. 10 and 12), glabrescent; upper surfaces glabrous; involucre glabrous, or tomentose only at base. . . *i.*

*i.*Petioles, summit of stem, pedicels or base of involucre or all of them permanently tomentose; basal leaves either tomentose beneath or glabrescent; achenes usually hirtellous. . . . . . . **12. *S. plattensis.***

*i.*Petioles, summit of stem, pedicels and involucre usually glabrous or promptly glabrate; tomentum, if present, soon becoming scattered floccose tufts (persistent only in a local var. of no. 13); achenes glabrous or hirtellous. . . *j.*

*j.*Phyllaries linear or linear-lanceolate, 0.5–1.5 mm. wide, green, greenish or merely purple-tinged; disk-corollas uniformly pale to deep yellow; achenes 1.5–3.5 mm. long; heads few–very numerous; species of temperate areas. . . *k.*

*k.*Undivided basal leaves slender-petioled, the oblanceolate or spatulate to oblong, elliptic, ovate or subrotund blade not broadly decurrent into the petiole; plants without filiform stolons; phyllaries gradually tapering to apex from below or from above the middle. . . *l.*

*l.*Undivided basal leaf-blades narrowly oblanceolate or spatulate to lanceolate, oblong, elliptic or rounded, tapering to subtruncate at base, not cordate or, if slightly cordate (in no. 17), the membranaceous sharply-toothed larger blades less than half as broad as long. . . *m.*

*m.*Heads ordinarily radiate (only rarely discoid); disk-corollas and ligules deep yellow to pale orange; receptacle smooth. . . *n.*

*n.*Undivided basal leaves tapering or gradually rounded to petiole (not cordate), crenate or merely dentate-serrate, firm to subcoriaceous, oblanceolate to elliptic or oval, blunt; some of the cauline leaves often incised or pinnatifid to tip; flowers deep yellow. . . *o.*

*o.*Phyllaries linear-attenuate or linear-lanceolate and tapering gradually from near base. . . *p.*

*p.*Achenes hirtellous; stem coarse, fistulous, soon fragile and brittle; radical leaves lanceolate or lance-oblong to oblong-ovate, mostly 3–8 cm. broad; plant of se. Coastal Plain. . . . . . 10. *S. tomentosus.*

*p.*Achenes glabrous (very rarely hirtellous); stem slender, firm, long-persistent; radical leaves oblanceolate or spatulate to oblong-elliptic, 0.5–4 cm. broad; northern or inland species.

Rhizome elongated, often branching; undivided lower leaf-blades 0.5–2 cm. broad, longer than to rarely slightly shorter than the petiole; stems 0.5–4 dm. high, if rarely –6 dm. high then with 10–40 or more heads; transcontinental northern and upland species. . . 13. *S. pauperculus.*

Rhizome mostly reduced to a small erect praemorse caudex; undivided lower leaf-blades 1.3–4 cm. broad, mostly only one-fifth to one-third as long as petiole; plants 4–6 dm. high, with 2–8 heads; southern species. 14. *S. Crawfordii.*

*o.*Phyllaries linear-oblong, with parallel sides, tapering only from above the middle.

Undivided basal leaves elliptic, oval or oblong-oblanceolate; well developed cauline leaves 1–5, the lower and sometimes all with shallowly pinnatifid to uncleft blades; heads 1–25 (more in exceptional individuals); disk 6–12 mm. broad; achenes glabrous, 2.4–3 mm. long; plant of Nfld., Anticosti and Gaspé. . . 15. *S. gaspensis.*

Undivided basal leaves narrowly spatulate-oblanceolate to lance-obovate; cauline leaves 4–12, mostly pectinate-pinnatifid; heads 10–150; disk 5–8 mm. broad; achenes copiously hispidulous, 1.5–2.2 mm. long; southern species. 16. *S. Smallii.*

*n.*Undivided basal leaves or some of them shallowly cordate, sharp-serrate to lacerate below the middle, membranaceous, lanceolate to ovate and acutish; cauline leaves similar or with pectinate bases; northeastern. . . . . 17. *S. Robbinsii.*

*m.*Heads discoid (only rarely radiate); flowers lemon-color; receptacle fimbrillate; northern plant with elliptic or oval to suborbicular membranaceous lower leaves. 18. *S. indecorus.*

*l.*Undivided basal leaf-blades cordate, suborbicular or ovate, one-half to quite as broad as long, long-petioled; flowers deep yellow. . . . . 19. *S. aureus.*

*k.*Undivided basal or rosette-leaves with the short to elongate petiole broadened upward to the obovate to subrotund often cuneate-based crenate blade; filiform stolons soon developed in early summer; phyllaries abruptly acuminate from above the middle. . . . . . 20. *S. obovatus.*

*j.*Phyllaries linear-oblong, 1–2 mm. broad, purple to green and purplish; unexpanded disk-corollas and outsides of corolla-lobes scarlet to deep orange; mature achenes 3–4 mm. long; heads 1–8 (rarely –15); alpine or subalpine species of Nfld. and Gaspé.

Fleshy, 1–5.5 dm. high; corymb of (1–)2–15 heads on pedicels 0.2–6 cm. long; receptacle smooth; disk (in fresh heads) less than 1 cm. broad, after pressure 1–1.5 cm. broad, 7–10 mm. high. . . . . 21. *S. pauciflorus.*

Not fleshy, 0.3–2.5 dm. high; heads 1, if rarely 2–4 then on pedicels mostly 0.5–2 dm. long; receptacle fimbrillate; disk (after pressing) 1.5–2 cm. broad, 1–1.5 cm. high. . . 22. *S. resedifolius.*

1. S. vulgàris L. (common), Common G. — Quick-growing *annual* 1–6 dm. high, corymbosely branched above and often below, *at first more or less cobwebby, soon glabrate; leaves* soft, rather *fleshy, undulate to pinnatifid-lobed*, the divergent often toothed lobes subequal; lower

blades petioled, the median and upper sessile and more or less clasping; heads slender-cylindric; involucre 8–10 mm. high, calyculate, the *calyculate basal bracteoles black-tipped and many times shorter than the* narrowly linear acutely black-tipped *glabrous phyllaries; flowers* golden-yellow, usually *all tubular; achene* slender, *appressed-pubescent,* 1.5–2.5 mm. long. — Weed of cult. and waste land, Lab. to Alaska, s. to Nfld., N.S., N.E., N.C., O., Ind., Wisc., Tex., N.M. and Calif. March–Oct. (Natzd. from Eu.)

2. S. SYLVÁTICUS L. (of the woods). — Similar to no. 1, 0.1–1 m. high, the *firm stem, leaves and involucre viscid-hirtellous, with heavy oily odor; leaves* simply *pinnatifid to deeply pinnati-parted* with toothed or incised divisions; *heads in open terminal corymbs, on naked filiform pedicels; involucre minutely pubescent, pale; rays very short and recurved.* — Clearings, open woods, shores, rocky slopes and waste places, often appearing as if indigenous, freq. near coast, Nfld. and Gaspé Pen., Que., to Me.; locally, n. N.J., O. and Wisc.; Pacific coast. July–Sept. (Natzd. from Eu.)

3. S. VISCÒSUS L. (sticky). — Coarser than no. 2, 1–5 dm. high, *heavily viscid-glandular, fetid;* stem firm; leaves deeply pinnate-lobed or pinnately dissected; *heads on thick bracted pedicels,* larger than in nos. 1 and 2, *glandular-hirsute; the calyculate basal bracteoles lax and slender, about half as long as the phyllaries; ligules obvious,* recurving; *achenes glabrous.* — Waste places, railroad-yards, etc., Nfld. to s. N.Y., and casual about ports southw. July–Sept. (Natzd. from Eu.)

4. S. JACOBAÈA L. (of St. James; an old name, presumably from the midsummer blooming being associated with St. James's Day, July 25), STINKING WILLIE. — *Coarse biennial or winter-annual* (said to be sometimes perennial, but rarely so with us); *the tough* erect *stem* 0.3–1.2 m. or more high, *minutely arachnoid, quickly glabrate,* with a showy terminal broad corymb; *leaves* running nearly uniformly to summit, *deeply and uniformly bi- or tripinnatifid, arachnoid beneath,* the larger ones 0.5–2.3 dm. long and 3–11 cm. broad; *branches of corymb arachnoid; involucre hemispherical, with* 1 or 2 short calyculate bracteoles and *scarious-margined rhombic-oblong phyllaries; rays conspicuous,* 6–10 mm. long; marginal achenes glabrous or glabrescent, those of the disk pubescent. — Monopolistic weed of fields, pastures, roadsides, etc. (avoided by grazing animals), Nfld. and Gaspé Pen., Que., throughout much of the Maritime Provinces, s., locally, to e. Mass.; casual in Ont. and elsewhere; Pacific coast. July–Oct. (Natzd. from Eu.)

5. S. glabéllus Poir. (almost smooth), BUTTERWEED. — *Succulent annual or biennial with* erect *glabrous soft and hollow stems* 0.2–1 m. high, sometimes arachnoid in the leaf-axils; *leaves* rapidly diminishing upward; *the basal and lower petioled, lyrate-pinnatifid, with flabelliform bluntly notched lateral divisions and a much larger and rounder terminal one;* corymb arachnoid when young, quickly glabrate, with filiform pedicels; involucre about 6 mm. high, sparsely and slenderly calcyulate below the linear-lanceolate phyllaries; rays showy. — Wet woods, swamps and ditches, Fla. to e. Tex., n. to N.C., sw. O., s.-centr. Ind., s. Ill., Mo. and Okla. April–June.

6. S. CONGÉSTUS (R. Br.) DC. (crowded), MARSH-FLEABANE. — *Annual or biennial with hollow and soft (easily compressed) stout simple stems* 1.5–3 dm. high; leaves linear- to oblong-lanceolate, ascending, undulate, dentate or shallowly pinnatifid; *corymb very dense, its branches, pedicels and involucres densely blanketed with long-villous and -woolly many-jointed hairs;* the *involucres non-calyculate;* rays pale yellow, short; *mature pappus four or five times as long as the* smooth slender *achenes.* — Arct. reg., s. to coast of se. Lab. and Hudson Bay. (Eurasia) — Represented with us by

Var. **palústris** (L.) Fern. (of marshes). — Stem up to 8 dm. high; corymb more open, more or less villous-lanate. (*S. palustris* (L.) Hook., not Velloso) — Marshes, wet shores and damp slopes, Lab. to Alaska, s. to Côte Nord, Que., James Bay and n. Ia. June–Aug. (Eurasia)

Var. **tônsus** Fern. (sheared). — Still smoother; the open corymb without or essentially without long villi, the branches and pedicels merely hirtellous or short-pilose or with very few long villi. — Alta. to Wisc., Minn., N.D. and Man.

7. S. Pseùdo-Árnica Less. (false *Arnica*). — *Stem stout, from a deep vertical praemorse heavy rhizome, white-woolly,* 0.15–1 m. or more high; leaves numerous, often overlapping, *lingulate to elliptic or obovate, fleshy, lustrous-green above, white-felted beneath, repand-dentate to subentire,* the larger ones 0.8–2 dm. long; heads 1–20 in a leafy corymb; *involucre cottony, 2–4 cm. high,* hemispherical, *calyculate by extension up from the thick-clavate peduncle of filiform bracteoles; disk 2.5–5 cm. broad;* disk-corollas deeply 5-toothed; ligules showy, 1–2 cm. long, or in forma **Rollándii** (Vict.) Fern. (for its discoverer, LOUIS ROLAND, FRÈRE ROLLAND-GERMAIN, 1881–) rayless; pappus dull white; achenes thin-angled. — Sandy or gravelly seashores and upper beaches, Lab., Nfld., and Gulf of and Lower R. St. Lawrence, Que., s. to Sable I., N.S. and sw. N.B.; Alaska and B.C. Mid-July–Sept. (Ne. Asia)

8. S. Millefòlium T. & G. (thousand-leaved, like *Achillea Millefolium*). — Erect slender perennial from a small caudex; *stem glabrate*, or with white tomentum at base or in tufts, 2.5–7 dm. high; *basal leaves long-petioled, the finely bi- or tripinnatisect blades* 0.5–2 *dm. long*, suggesting leaves of carrot or caraway; *upper leaves smaller, all finely dissected;* corymb broad, with arched-ascending rays and 7–20 or more heads with campanulate *scarcely calyculate involucres* 4–6 mm. high; *achenes hirtellous.* — Wet or dry rocks, mts. of sw. Va. to S.C. May, early June.

9. S. integérrimus Nutt. (most entire). — Perennial *with a short tough simple caudex; stem* stoutish, *at first sparsely crisp-villous with jointed hairs, soon glabrate,* 0.3–1 m. high; *lower leaves* oblong, narrowly obovate or lanceolate, thickish, *entire, denticulate or sinuate, petioled,* 0.6–1.5 *dm. long; middle and upper leaves greatly reduced and sessile, with long slender tips;* corymb of several heads on naked elongate pedicels; *heads* 10–12 *mm. high,* the phyllaries linear-attenuate, often dark-tipped; rays numerous, yellow, 1 cm. long; achenes glabrous. — Wet prairies, rocky slopes and gravels, Man. to B.C., s. to Ia., Neb., Colo., etc. May–July.

10. S. tomentòsus Michx. (covered with implexed hairs). — *Tufted* from a thick caudex or short rhizome; *stem stoutish, white-tomentose* or glabrate, erect, 1.5–6 *dm. high; basal leaves lanceolate to oblong or oval, white-tomentose beneath, arachnoid* and often glabrate *above, undulate-dentate with low rounded cartilaginous teeth, the larger ones* 0.4–1.8 *dm. long* and 1.5–8.5 cm. wide, *on erect petioles* 0.4–2.3 *dm. long;* cauline leaves few, greatly reduced; corymb at first tomentose, becoming glabrescent; pedicels and erect branches prolonged; involucre tomentulose to glabrescent, 8–10 mm. high, the phyllaries linear-lanceolate; rays conspicuous, yellow; *achenes columnar, drab, heavily hirtellous on the prominent rounded ridges.* — Sandy or rocky open woods, clearings and fields on or near the Coastal Plain, Fla. to La., n. to s. N.J. and Ark. April–June. Hybridizes with no. 19. — Relatively uncommon plants have the leaf-blades glabrous from the first or promptly glabrate and the upper half of stem and the involucres glabrous; these are forma **alabaménsis** (Britt.) Fern. (of Alabama).

11. S. antennariifòlius Britt. (with leaves of *Antennaria*). — Smaller, from prolonged branching crowns; the slender white-lanate stems 1–4 dm. high; *basal leaves crowded, obovate to elliptic, prominently toothed at the broad rounded summit, white-pannose beneath* and minutely tomentulose above, *the larger* 1.5–4 *cm. long and* 0.8–2.5 *cm. broad, tapering into petioles* 1–4 *cm. long;* cauline leaves few and narrow; involucre white-lanate, 5–8 mm. high, with linear-lanceolate phyllaries; rays conspicuous, yellow; *achenes fusiform-columnar, brown, sparsely hirtellous on the obscure acutish ridges.* — Dry shaly soils, Blue Ridge, Pa. (*Pursh*) and w. Md., e. W.Va. and w. Va. Mid-April–mid-May.

12. S. platténsis Nutt. (of the Platte River region). — *Stems* 1–several from a forking rhizome or crown, *permanently tomentose-lanate below and often at summit,* 1–4 dm. high; *basal leaves* tufted, *those of the new offshoots and many at base of flowering stem unlobed,* oblong-ovate to lanceolate or oblanceolate, tapering to petiole or with rounded to subcordate bases, *the principal blades* 2–8 *cm. long* and 1–5 *cm. broad, firm and thickish, usually finely tomentulose beneath,* sometimes glabrate, crenate to serrate-dentate; other basal often, and the lower cauline leaves generally, more or less pinnatifid at base, the *petioles usually lanate; corymb usually lanate-tomentose,* sometimes glabrate; pedicels elongate, strongly ascending, tomentose to glabrate; *involucre usually tomentose at base,* 5–6 mm. high, the linear-lanceolate phyllaries slender-tipped; rays showy, yellow; *achenes usually hirtellous on the angles.* — Dry calcareous rocks, bluffs and prairies, w. Vt. to s. Ont. and w. to Sask., s. to w. Va., O., Ind., La. and Tex. May–early July. In its smoother state too close to forms of no. 13.

13. S. paupérculus Michx. (poor). — Perennial *with usually slender and often branching rhizome; stems* 0.5–6 *dm. high, glabrous or glabrescent and frequently with flocculent tufts of white wool; undivided basal leaves* either in separate tufts or at bases of flowering stems, *with firm oblanceolate, spatulate or oblong-elliptic blades longer than to slightly shorter than the petioles,* 1–10 cm. long and 0.5–2 *cm. broad,* crenate or crenate-dentate; *lower cauline leaves with* undivided or *usually pinnatifid blades,* the upper similar or reduced and bracteiform; corymb with (1–)2–40 or more heads, these mostly long-pedicelled and slightly calyculate at base; the *heads in anthesis* 5–9 *mm. high; involucre* campanulate, 4–8 mm. high, *glabrous or glabrate* (permanently tomentose in one var.), the *linear- or linear-lanceolate greenish* (or purple-tipped) *phyllaries tapering from base to apex,* 0.8–1.4 mm. wide; flowers deep yellow, the ligules usually showy; *achenes glabrous* (very rarely hispid), 1.6–3.5 *mm. long.* — Very variable; the following vars. recognized:

Basal undivided leaf-blades 1–4 cm. long, 0.5–1.5(–2) cm. broad; lower and median cauline blades 1–3.5 cm. long, 0.1–6(–10) mm. broad; the upper very short and linear to subulate, mostly entire; heads 1–8 (rarely –20); involucre 4–6 mm. high, glabrous or glabrescent; achenes 1.6–3.5 mm. long. . . . . *S. pauperculus* (typical).

Basal undivided blades (1.5–)3–10 cm. long, (0.7–)1–2 cm. broad; lower and median cauline blades well-developed, 3–12 cm. long, (0.7–)1–3 cm. broad, mostly pinnatifid; heads 2–40; involucre 4–8 mm. high; achenes 2–3.5 mm. long.

Undivided basal blades 1.5–8 cm. long; lower and median cauline blades 3–9 cm. long, the broadest 1–2.5 cm. broad; upper blades greatly reduced; achenes 2–3.5 mm. long.

Involucre glabrous or promptly glabrate, 5–8 mm. high; achenes glabrous (rarely hirtellous), 2.2–3.5 mm. long; stem 1.5–6 dm. high; heads 4–40; wide-ranging plant. . . . . Var. *Balsamitae.*

Involucre densely tomentose at base or to summit, 4–5 mm. high; achenes glabrous, 2 mm. long; stem 1–4 dm. high; heads 2–25; plant of southeastern Canada. . . . . Var. *neoscoticus.*

Undivided basal blades 3–10 cm. long; large pinnatifid cauline leaves extending nearly or quite to summit of rather coarse stem (4–6 dm. high), the larger blades 6–12 cm. long and 1.3–3 cm. broad; heads 6–40; involucre glabrous or glabrescent, 5–7 mm. high; achenes 2–2.5 mm. long, glabrous. . . . . Var. *praelongus.*

S. paupérculus (typical). — Basic or calcareous gravels, wet rocks, peats and bogs, Lab. to Alaska, s. to Nfld., n. N.B., n. N.E., e. Pa., upland to Va., n. O., Mich., n. Ill., Wyo. and B.C. Late May–Aug. — The slenderest extreme of the species, with greatly reduced cauline leaves. The rare forma **inornàtus** Fern. (without ornament) has the heads discoid and without ligules.

Var. Balsámitae (Muhl.) Fern. (of a Balsam; from the showy inflorescence), *S. Balsamitae* Muhl. — Neutral, basic or calcareous ledges, gravels, cliffs, etc., s. Lab. to Alta., s. to Nfld., N.B., Me., Mass., Ct., Del., Md., upland to nw. Ga. and n. Ala., O., n. Ind., Ill., Ia., Neb. and Ida. Late May–Aug. — A rare form with wholly discoid heads is forma **inchoàtus** Fern. (incomplete).

Var. neoscóticus Fern. (Nova Scotian). — Gypsiferous and other calcareous rocks, N.S. and Gaspé Pen., Que. July, Aug. — In its permanently tomentose involucres simulating no. 12; in aspect like the preceding var. but with smaller heads and achenes.

Var. praelóngus (Greenm.) House (very long), *S. Balsamitae* var. Greenm. — Rocky open woods, slopes and crests, often on less basic rock than the other vars., e. Mass. to Mich., s. to Ct., Md. and the upland to Va. — Simulating no. 16.

14. S. Crawfórdii Britt. (for its discoverer, JOSEPH CRAWFORD, 1858–1936). — Glabrous or glabrescent perennial *with praemorse ascending simple rhizome only* 0.5–1.5 *cm. long;* stem 3–6 dm. high; *radical leaves erect,* long-petioled, *the oblanceolate to oblong or elliptic thick coarsely serrate-dentate blunt blades* 2.5–7 cm. long and 1.3–4 *cm. broad, tapering to the very long petiole (usually three to five times as long)*; cauline leaves 3–8, the upper greatly reduced, the lower or median shallowly pinnatifid at least below the middle; *heads* 2–8, *on mostly simple slender pedicels* 2–10 *cm. long; involucre* 7–9 *mm. high, with linear-attenuate phyllaries* 1–1.5 mm. broad at base; flowers bright yellow; ligules 8–12 mm. long; *achenes glabrous,* about 2.5 mm. long. (Including *S. obovatus,* var. *umbratilis* Greenm., in part) — Bogs, wet meadows and peaty pinelands, N.C. and Tenn., n. to w. N.J., e. Pa. and Ind. Late April–early June.

15. S. gaspénsis Greenm. (of the Gaspé Peninsula). — Resembling larger vars. of no. 13; *horizontal rhizome elongate,* often forking; stems stoutish to slender, 2–6 dm. high; *undivided basal leaf-blades elliptic, oval or oblong-oblanceolate, thick, the larger ones* 3.5–8 cm. long and 2–4.5 *cm. broad, round-tipped or obtuse,* dentate, *half as long as to about equaling the erect* slender *petioles; cauline leaves well developed,* 3–5, the larger 5–12 cm. long and 1–3 cm. broad, *shallowly pinnatifid to uncleft;* heads 1–25 (exceptionally more), on mostly unforking erect pedicels mostly 2–10 cm. long, 8–10 mm. high in anthesis, the *disk* 6–12 *mm. broad; involucre* 6–9 *mm. high, the oblong-linear phyllaries* 1–1.5 mm. wide, *tapering from above the middle;* flowers bright yellow; ligules showy, or wanting in the local forma **verecúndus** Fern. (bashful); achenes glabrous, 2.5–3 mm. long. — Bottomlands, wet shores, swamps and calcareous cliffs, Anticosti I. and Gaspé Pen., Que., to n. Me. (where crossing with no. 13). July, early Aug. Passing into

Var. firmifòlius (Greenm.) Fern. (firm-leaved). — Plant *smaller; stems* 0.2–2 *dm. high, stiffly erect,* often much tufted; *undivided basal leaf-blades* 1–6 *cm. long and* 0.7–2.5 *cm. broad; flowering stem nearly naked except at base or with greatly reduced leaves only* 1–3.5 *cm. long and rarely* 1 *cm. broad; pedicels* 0.2–5 *cm. long;* achenes 2.4–2.6 mm. long. — Dry calcareous rock, shingle and talus, w. Nfld.; Côte Nord, Anticosti I. and Gaspé Pen., Que. Mid-July–early Sept.

16. S. Smállii Britt. (in honor of JOHN KUNKEL SMALL, 1869–1938). — *Rhizome* simple or *with several crowded ascending crowns; stems* 1–many, 2–8 dm. high, *tomentose at base; undivided basal leaves* firm, narrowly spatulate-oblanceolate to lance-obovate, *the larger* 3–10 *cm. long* and 0.8–4 cm. broad, closely dentate; *cauline leaves* 4–12, *mostly pectinate-pinnatifid,* the lower often as large as the basal; *heads* (10–)30–150, slender-pedicelled, 7–10 *mm. high, the disk*

5–8 *mm. broad;* involucre 5–6 mm. high, glabrous or glabrate, *the linear-oblong phyllaries with parallel sides but tapering at apex;* flowers bright yellow; ligules spreading, or wanting in forma **trístis** Fern. (sad); *achenes* usually *copiously hispidulous,* 1.5–2.2 *mm. long.* — Open woods, thickets, clearings and fields, Fla. and Ala., n. to N.J., Pa., Ky. and Ind. Mid-May, June. Northward apparently crossing with vars. of no. 13.

17. S. Robbínsii Oakes (for its discoverer, James Watson Robbins, 1801–1879). — Perennial from an oblique to prostrate rhizome; flowering stems 1–several, 0.3–1 m. high; *basal undivided leaves with membranaceous lanceolate to ovate acutish* to blunt *sharply serrate cordate to round-based blades* mostly 4–10 cm. long and 1.5–5 cm. broad, the *slender petioles elongate; lower cauline leaves similar, petioled, undivided or with pectinate bases,* the upper sessile and reduced; heads few–many on slender erect pedicels, 8–10 mm. high, radiate, or discoid in forma **invenústus** Fern. (not beautiful), the yellow flowers often paler than in no. 19; achenes glabrous. — Peaty meadows, swales, thickets and fields, sometimes ascending to high alts., w. Gaspé Co. to Portneuf Co., Que., s. to N.S., n. N.E. and ne. N.Y.; summit of Roan Mt., N.C. and Tenn. June–early Aug.

18. S. indecòrus Greene (unsightly). — Tufted perennial, with 1–several fistulous erect glabrous or glabrescent stems 0.17–1 m. high from erect crowns; *leaves membranaceous; the basal oblong, elliptic or rounded, blunt, dentate to lacerate-based blades* 2–7.5 cm. long and 1–5 cm. broad, *on long slender petioles; cauline leaves* 4–13, *the lower petioled and like the basal or more lacerate at base, the smaller median and much smaller upper sessile and lacerate-pinnatifid; heads* mostly 6–40, *slenderly cylindric-urceolate; the narrowly linear phyllaries* 0.5–1 *mm. broad, green* or with purple tips; *flowers pale yellow or lemon-color,* all *discoid,* or the marginal ligulate in forma **Búrkei** (Greenm.) Fern. (for its discoverer, Joseph Burke, who made collections in our Northwest, beginning in 1844); *filiform tube of corolla* 3.6–6 *mm. long; achenes drab or gray-brown, slender, strongly costate,* 2–3 mm. long; *denuded receptacle* convex, conspicuously alveolate, *the walls of the pits thin and jagged or fimbriate.* (*S. discoideus* sensu Hook. in part, not as to type; *S. idahoensis* Greenm.) — Rich thickets, swales and calcareous rocks or slopes, Gaspé Pen., Que., to Alaska, s. to Rimouski Co., Que., n. Mich., Mont., Ida. and n. Calif. July, early Aug.

19. S. aûreus L. (golden), Golden R., Squaw-weed. — *Rhizome and basal offshoots horizontally creeping,* becoming 0.5–3 dm. long; *flowering stems* 1–several, erect, 0.2–1.2 m. high, *glabrous* or with some flocculent and promptly deciduous tomentum when expanding, *basal leaves* slender- and long-petioled, *cordate, suborbicular to ovate;* cauline leaves variously pinnatifid; *heads* few–many, showy, *with golden yellow ligules* (flowers rarely all discoid); involucre 5–10 mm. high; *phyllaries narrowly linear,* glabrous or promptly glabrate; *achenes glabrous.* — Very variable; the following vars. recognized:

Radical leaves dentate with rounded to blunt teeth.
- Undivided basal leaves suborbicular to round-ovate, broadly rounded at summit, with a well-defined cordate base.
  - Basal offshoots at flowering time 4–10 mm. thick, deep red or purple; margins of basal sheaths, many petioles, surfaces of unexpanded leaves and unexpanded corymbs with long dense flocculent tomentum; involucres 8–11 mm. high; disk-corollas 6–10 mm. long; achenes 3.5–4 mm. long. . . *S. aureus* (typical).
  - Basal offshoots at flowering time 1–5 mm. thick, rarely purple; margins of basal sheaths, petioles, surfaces of unexpanded leaves and unexpanded corymbs glabrous or sparsely short-tomentose and glabrescent; involucres 5–8(–9) mm. high; disk-corollas 5–8 mm. long; achenes 2–3.5 mm. long.
    - Stems 3–9 dm. high, often 2 or more; basal offshoots mostly 2–10 cm. long; radical leaves becoming firm and subcoriaceous, the larger ones 6–15 cm. long and 3–12 cm. broad, often purplish beneath, the longer petioles 0.5–3 dm. long; involucres 5–9 mm. high. . . . . . . Var. *intercursus.*
    - Stems slender, mostly solitary, 2–6 dm. high; basal offshoots rarely 6 cm. long; radical leaves membranaceous, 1–3.5 cm. long, 1–2.8 cm. broad, green both sides, the longer petioles 2–10 cm. long; involucre 5–7 mm. high. . . . . Var. *gracilis.*
- Undivided basal leaves oblong-oval, rounded to base or some of them barely subcordate, the larger blades 2–6.5 cm. long and 1.5–4.3 cm. broad, submembranaceous. . . . . Var. *semicordatus.*

Radical leaves sharply serrate or dentate, at least at base, to acutely or subacutely lacerate, ovate or oval to rotund, deeply cordate, acute to rounded at tip, membranaceous; basal offshoots slender. . . . . Var. *aquilonius.*

S. aùreus (typical). — Rather stout, the thick basal offshoots 3–15 cm. long; stems 3–7.5 dm. high; radical leaves round- to oblong-ovate, becoming coriaceous, in maturity up to 2.2 dm. long and 1.8 dm. broad. (Var. *Ashei* Greenm.) — Rich calcareous woods and bottoms or in upland swamps, Fla. to Ark., n. to Susquehanna R., Md., Ky. and Mo. April, May. — Crosses with no. 10.

Var. **intercúrsus** Fern. (lying between; in this case the typical and the next var.). — Meadows, boggy swales, swamps and low woods, Que. to Ung., Mich. and n. Ill., s. to N.S., N.E., Pa., upland to Ala., Ky. and Mo. Late April (southw.)–Aug. (northw.).

Var. **grácilis** (Pursh) Wood (slender). — Meadows, swamps and bogs, Mass. to N.D., s. to N.C. and Ark. April–June.

Var. **semicordàtus** (Mackenz. & Bush) Greenm. (half-cordate). — Calcareous thickets, shores, meadows, etc., locally abundant, w. Nfld.; Gaspé Pen., Que., to Ung. and s. Ont., Mich. and Ill., s. to n. N.B., E. Townships, Que., and Mo. — Possibly a series of persistent hybrids.

Var. **aquilònius** Fern. (northern). — Cool woods, damp thickets, calcareous shores, subalpine meadows, etc., Nfld. and Anticosti I., Que., to Algoma Distr., Ont., s. to C.B., P.E.I., n. N.B., n. N.E., n. N.Y., nw. Pa., n. O., n. Ind. and Wisc. — Forma **ecoronàtus** Fern. (crownless) has discoid heads. May–Aug.

20. S. obovàtus Muhl. (obovate; from the leaf-outline). — Glabrous or promptly glabrate *rosette-forming* perennial *with*, normally, *several obovate leaves in a basal rosette and flagelliform elongate stolons terminated by similar new rosettes; radical leaf-blades* membranaceous, narrowly to broadly *obovate*, 2–10 cm. long and 1–5.5 cm. broad, crenate to serrate-dentate, *cuneate or rounded to the upwardly dilated petiole*, the latter much shorter than to but slightly exceeding the blade, the *dilated summit* 1–3(–6) *cm. long;* the flowering stem slender, usually glabrous from the first, 1.5–5.5 dm. high, soft or weak; lower cauline leaves pectinate-pinnatifid, the upper similar or usually much reduced; corymb few–many-headed; the heads mostly on elongate slender pedicels, 8–10 mm. high; *phyllaries abruptly acuminate from above the middle;* ligules conspicuous, or wanting in forma **elongàtus** (Pursh) Fern. (elongate); achenes glabrous or slightly hirtellous on angles. — Calcareous rocks and slopes and rich wooded banks, s. N.H. to s. Ont. and Mich., s. to s. N.E., S.C. and Ala. Apr.–June (rarely –Aug.).

Var. **Ellióttii** (T. & G.) Fern. (for Stephen Elliott, 1771–1830). — *Stouter, the* rather *coarse stem, petioles,* often the lower leaf-surfaces *and short pedicels floccose-tomentose* when young; *petioles of* strongly decurrent *subcoriaceous radical leaves short* to nearly obsolete. (*S. Elliottii* T. & G.) — Calcareous slopes or ravines, e. W.Va. to nw. Fla. and Ala. April.

Var. **rotúndus** Britt. (rounded). — Glabrous or glabrescent, slender, with long-pedicelled heads; *leaves of basal rosette with suborbicular to round-obovate thin blades* 1–7.5 cm. long and 1–6.5 cm. broad, *these rounded to the slender petioles, the latter as long as to thrice the length of the blade, their dilated summits only* 0.3–1.5 *cm. long.* (*S. rotundus* (Britt.) Small) — Ga. to Tex., n. to O., s. Mich., Mo. and Kans. Late March–May.

21. S. pauciflòrus Pursh (few-flowered). — Perennial *with* stout rhizome usually forking into *erect crowns; stems* 1–few, glabrous, 1–5.5 *dm. high; leaves thick and fleshy;* the basal ones long-petioled, *elliptic to reniform, blunt,* cuneate to cordate at base, *coarsely dentate,* 1.5–4 cm. long; *median and upper cauline leaves* sessile; the 2–5 below inflorescence lanceolate, oblong or oblanceolate, *with mostly obtuse pinnatifid lobing* and coarse teeth; *heads* 1–6 (rarely –15), *corymbose, on pedicels* 0.2–6 *cm. long,* broadly campanulate, *discoid,* or radiate in forma **fállax** (Greenm.) Fern. (deceitful); *phyllaries usually purple,* blunt to acute, 1–2 *mm. broad; disk* 7–10 *mm. high, less than* 1 *cm. broad, after pressure* 1–1.5 *cm. broad; disk-corollas* with filiform tube 3–4 mm. long, *the lobes deep red to red-orange;* achenes plump, dark red-brown, glabrous; *denuded receptacle smooth.* — Calcareous meadows, alpine slopes and wet rocks, Lab. to n. Nfld. and Shickshock Mts., Gaspé Pen., Que.; n. Man. to Alaska, s. to Wyo. and Calif. July, Aug.

22. S. resedifòlius Less. (with leaves of *Reseda,* Mignonette). — *Stems* 1–few from strong rhizome, 0.3–2.5 *dm. high, firm,* slender, *simple and with* 1 *head or loosely forking from near or below the middle and with* 2–4 *long-stalked heads; basal leaves coriaceous;* the ovate, elliptic or reniform *blades* 0.5–2.5 *cm. long,* crenate to sharply dentate; cauline leaves few, the lower variously cleft, the upper simpler and reduced; *involucre* campanulate, purple, purplish or green; the heads *discoid,* or in forma **columbiénsis** (Gray) Fern. (of Columbia, the original material supposed to come from northern B.C.) with long rays deep yellow or with their backs reddish; *disk* 1–1.5 *cm. high, after pressure* 1.5–2 *cm. broad; disk-corollas deep orange or with red backs of lobes; receptacle fimbrillate;* achenes reddish, smooth, about 4 mm. long. — Calcareous shingle, talus and slopes, mts. of w. Nfld. and Gaspé Pen., Que.; Alaska. July, Aug. (Asia)

Tribe IX. Cynàreae Spreng. (see p. 1358)

Flowers all alike; achenes attached by the base.
Leaves not prickly, the lower broadly ovate and mostly cordate; phyllaries with hooked tips; style-branches partly distinct. . . . . . . 82. *Arctium.*
Leaves prickly, relatively narrow; style-branches coherent, usually a pubescent ring below.
Involucres 1-flowered, the very many involucrate flowers in a globose head with a reflexed and hidden outer involucre. . . . . 83. *Echinops.*
Involucres many-flowered; heads not globose.
Receptacle densely bristly.
Pappus-bristles not plumose. . . . . 84. *Carduus.*
Pappus-bristles plumose. . . . . 85. *Cirsium.*
Receptacle deeply honeycombed, scarcely or not at all bristly. 86. *Onopordum.*
Marginal flowers often enlarged and somewhat ray-like; achenes obliquely attached.
Pappus of several series of short scales or bristles, or none; flowers red, purple, blue, white or pale yellow. . . . . 87. *Centaurea.*
Pappus of 10 short horny teeth, 10 long bristles, and 10 shorter ones; flowers yellow. 88. *Cnicus.*

82. ÁRCTIUM L. Burdock. Clotbur. Bardane (Que.)

Heads many-flowered; flowers all tubular, perfect, similar. Involucre globular; the imbricated phyllaries coriaceous and appressed at base, attenuate to long stiff points with hooked tips. Receptacle bristly. Achenes oblong, flattened, wrinkled transversely; pappus short, of numerous rough bristles, separate and deciduous. — Coarse biennial weeds, nat. of Old World, with large unarmed petioled roundish or ovate mostly cordate leaves floccose-tomentose beneath, and small solitary or clustered heads; flowers purple, rarely white. (Greek plant-name, *arction*, from *arctos*, *a bear*; from the rough involucre.)*

Heads corymbose, long-peduncled; larger leaf-blades round-ovate, broadly rounded at summit; petioles strongly angled.
Petioles solid; heads 3–4.5 cm. in diameter; involucre glabrous, its middle and inner phyllaries subequal and exceeding corollas. . . . . 1. *A. Lappa.*
Petioles hollow; heads 2–2.7 cm. in diameter; involucre often arachnoid, its middle and inner phyllaries successively longer, mostly shorter than corollas. 2. *A. tomentosum.*
Heads racemose or subracemose, sessile or peduncled; larger leaf-blades ovate-oblong, tapering at summit; petioles only slightly angled, hollow.
Heads 2.5–3.5 cm. broad; middle and inner phyllaries equaling or exceeding corollas; achenes 5.5–7.5 mm. long, dark brown or light-mottled. . . . 3. *A. nemorosum.*
Heads 1.5–2.5 cm. broad; middle and inner phyllaries shorter than corollas; achenes 5–6 mm. long, gray or ashy-brown, usually dark-mottled. . . . 4. *A. minus.*

1. A. Láppa L. (old generic name, a bur), Great B. — *Petioles solid, strongly angled,* deeply furrowed; *larger leaf-blades rounded-ovate;* inflorescence a broad *corymb with long-peduncled heads 3–4.5 cm. in diameter; involucre glabrous* and green, or purple in forma purpuráscens (Le Grand) Murb. (becoming purple); *outermost phyllaries about* 1 *cm. long; middle and inner ones subequal and exceeding corollas; innermost with pale firm slender gradually attenuate tips;* broad upper portion of corolla shorter than slender tube; *achenes* 6–7 mm. long, fawn-color, often dark-mottled, *slightly rugose below the crenate-bordered summit.* — Waste places, roadsides, etc., chiefly in calcareous soil, Que. to Mich., s. to N.B., N.E., Pa. and Ill. July–Oct. (Natzd. from. Eu.)

2. A. tomentòsum Mill. (tomentose). — Similar to no. 1; *petioles hollow; heads smaller, 2–2.7 cm. broad; involucre* more or less *arachnoid,* or smooth in forma cálva Fisch. (bald), *outermost phyllaries 4–7 mm. long; middle and inner successively longer,* all or all but the inner *shorter than corollas; innermost with membranous more or less colored blunt or cuspidate tips;* broad upper portion of corolla as long as tube; *achenes* 4.5–7 mm. long, gray, usually dark-mottled, *rugulose from base to entire summit.* — Waste ground, infrequent, Que. and N.S. to Pa., O. and Mo. Late June–Sept. (Natzd. from Eu.)

3. A. nemoròsum Lej. & Court. (of woodland). — Petioles hollow, only slightly angled; *larger leaf-blades tapering to summit; inflorescences racemose to subcorymbose,* the *heads sessile to long-peduncled,* 2.5–3.5 *cm. broad;* involucre glabrous or somewhat arachnoid, green to purple;

* Treatment largely based on that of Fernald & Wiegand in Rhodora, xii. 43–47 (1910).

outermost *phyllaries* short, 5–7 mm. long; *middle and inner* successively much longer, *nearly equaling to exceeding corollas;* innermost long-attenuate, firm and aristate; *achenes* 5.5–7.5 *mm. long, dark brown,* or light-mottled, scarcely rugulose, with entire summit. (*A. vulgare* A. H. Evans, not *Lappa vulgaris* Hill; *A. minus,* var. *corymbosum* Wieg.) — Waste ground, Nfld. to Ont., s. to N.S., N.E., Va. and Kans. July–Oct. (Natzd. from Eu.) — A mixed series, generally treated in Europe as a species, sometimes as hybrids of nos. 1 or 2 and 4 (a propinquity rare with us), sometimes as a var. of no. 4.

4. A. MÌNUS (Hill) Bernh. (smaller), COMMON B., CIBOURROCHE (Que.). — Similar to no. 3; *heads* smaller, 1.5–2.5 cm. broad, *sessile or very short-peduncled,* often crowded; *middle and inner phyllaries definitely shorter than corollas;* corollas pale pink, or deep-purple in forma PURPÙREUM (Blytt) A. H. Evans (purple), or white in forma PÁLLIDUM Farw. (pale); *achenes* 5–6 *mm. long, gray or ashy, usually dark-mottled,* smooth or faintly rugulose. — Waste land, common, Nfld. to B.C., s. to N.S., N.E., Va., W.Va., Mo., Kans. and Calif. July–Oct. (Natzd. from Eu.) — Forma LACINIÀTUM Clute (sharply divided) is an extraordinary form with some or all leaves laciniate or reduced to narrow blades, the flowers sterile. Local or infrequent, Mass. to Pa. and Ill.

83. ECHÌNOPS L. GLOBE-THISTLE

Involucres 1-flowered, very many, aggregated in dense globular capitate clusters, the common involucre of the inflorescence of small reflexed bracts. Proper involucres cylindrical, of several series of unequal imbricated spinescent paleaceous phyllaries; corollas with slender tube and cylindric 5-parted limb. Filaments glabrous. Achenes cylindrical or somewhat tetragonal; pappus coroniform or of many short distinct or connate subpaleaceous bristles. — Stately thistle-like herbs, nat. of Old World, with alternate spinose pinnatifid or dentate leaves, and large globose terminal (compound) heads of whitish or bluish flowers. (Name from the Greek *echinos, hedgehog,* and *ops, appearance,* from the bristly nature of the armed foliage or perhaps of the spreading individual involucres in the dense spherical glomerules.)

1. E. SPHAEROCÉPHALUS L. (spherical-headed). — Tall, 1–2 m. high, grayish- or white-arachnoid on the stem and lower surface of leaves. — Frequent in cult. and occasional as an escape upon waste-heaps in fields, etc., s. Que. to Ia. and Va. July–Oct. (Introd. from Eu.)

CARLÌNA L. (said by Linnaeus to have been named for Emperor *Charles V.,* 1500–1558), a Eurasian genus of thistles with spinose leaves, the outer series of phyllaries very prickly, the inner linear ones colored or lustrous and spreading like the rays of a star, the receptacle bearing dissected chaff, the achenes silky, the pappus plumose, is locally represented by **C.** VULGÀRIS L. (common), CARLINE THISTLE, a biennial up to 2 dm. high with hemispherical heads about 2 cm. in diameter, the entire very narrow inner phyllaries with horizontally spreading tips. — Beginning to spread in w.-centr. N.Y. (Adv. from Eu.)

84. CÁRDUUS L. PLUMELESS THISTLE

Bristles of the pappus naked (not plumose), merely rough or denticulate. — Leaves conspicuously decurrent, spiny; wings of stem spiny. Otherwise as in *Cirsium.* Large genus, nat. of Eurasia and n. Afr. (The ancient Latin name.)

Heads nodding, solitary; involucre 3–4 cm. in diameter. . . . 1. *C. nutans.*
Heads ascending, often clustered; involucre 1–2.5 cm. in diameter.
  Involucre hemispherical, 1.5–2.5 cm. broad; outer phyllaries somewhat herbaceous and spreading. . . . 2. *C. acanthoides.*
  Involucre ovoid, 1–1.3 cm. broad; outer phyllaries rigid, hardly spreading. . 3. *C. crispus.*

1. C. NÙTANS L. (nodding), MUSK-THISTLE. — Biennial, 0.4–1 m. high; *heads solitary, hemispherical, nodding,* 3–5 *cm. broad,* showy, *on long partly naked peduncles; phyllaries lanceolate, the outer reflexed.* — Fields and waste places, local, St. P. et Miq. to Ia., s. to N.S., N.E., Md., D.C. and Mo. June–Oct. (Adv. from Eu.)

2. C. ACANTHOÌDES L. (resembling *Acanthus*). — Annual or biennial; *heads erect,* single or clustered *at tips of spiny-winged branches; involucre hemispherical,* 1.5–2.5 *cm. in diameter; phyllaries* linear, *the outer more or less herbaceous and spreading;* corolla about 18 mm. long. — Waste places, fallow fields and roadsides, rather local, N.S. to Neb., s. to Va. and O. June–Sept. (Adv. from Eu.)

3. C. CRÍSPUS L. (crisped), WELTED THISTLE. — Similar to no. 2; heads more clustered; *involucre* 1–1.3 *cm. broad; outer phyllaries rather rigid, hardly spreading;* corolla about 14 mm. long. — Roadsides, waste places and fallow fields, local, N.S. to Minn., s. to Ct., N.Y., and Mo. June–Oct. (Adv. from Eu.)

85. CÍRSIUM Mill. COMMON OR PLUMED THISTLE. CHARDON (Que.)

Heads many-flowered; flowers all tubular, perfect and similar, rarely imperfectly dioecious; phyllaries of the ovoid, subcylindric or spherical involucre imbricated in many rows, tipped with a point or prickle. Receptacle thickly clothed with soft bristles or hairs. Achenes oblong, flattish, not ribbed; pappus of numerous bristles united into a ring at the base, plumose to the middle, deciduous. — Herbs, nat. of the N. Hemisph., mostly biennial or perennial; the sessile alternate leaves often pinnatifid, prickly. Heads usually large, terminal. Flowers reddish-purple, rarely white or yellowish; in summer or autumn. (Name *Cirsion*, used by Dioscorides, from *cirsos, a swollen vein*, for which the Thistle was a reputed remedy.) By some recent Am. auth. included in CARDUUS.

a. Flowering stem arising from center of last year's basal rosette; root biennial or perennial; flowers perfect. . . b.

b. Upper half of stem and branches with prickly pinnatifid or undulate broad wings running down from bases of leaves nearly or quite to node below.

Heads mostly scattered, at tips of short branches; involucre 3–4 cm. high; nearly all the phyllaries attenuate to rigid subulate-acerose spreading prickles. . . . . 1. *C. vulgare*.

Heads in glomerules at summit of stem and branches; involucre 1–1.5 cm. high; inner phyllaries with flat blunt tips. . . . . 2. *C. palustre*.

b. Upper half of stem and branches not long-winged, at most with leaves decurrent only at summits of internodes. . . c.

c. Phyllaries of true involucre all with weak flat tips; flowers yellowish or purple; heads 4–8 cm. broad, subtended and often overtopped by a false involucre of large very spinescent appressed bracts. . . . . . 3. *C. horridulum*.

c. At least the outer phyllaries prickle-tipped or with firm points; heads naked at base or leafy-bracted. . d.

d. Leaves blanketed beneath with dense white tomentum or close felt. . . e.

e. Leaves white- or gray-tomentose above with a dense web. . . f.

f. Principal leaves pinnate, with linear and mostly entire divisions; flowers cream-colored; species of shores of Great Lakes. . . . 4. *C. Pitcheri*.

f. Principal leaves pinnatifid or pinnately parted, the lobes linear, lanceolate or triangular; flowers reddish-purple (except in albinos); plants of prairie and plains.

Principal cauline leaves with narrowly lanceolate acerose lobes, the sinuses extending nearly to median blade; involucre 2–2.5 cm. high; deep-rooted perennial. . . . . 5. *C. Flodmani*.

Principal cauline leaves with broadly lanceolate to broadly triangular short lobes; involucre 2.5–5 cm. high; biennial. . . . . 6. *C. undulatum*.

e. Leaves green and glabrous or promptly glabrate or merely hirtellous above. . . g.

g. Heads subtended by spinescent leafy bracts, the short peduncles leafy; involucre 2–3.5 cm. high. . . h.

h. Principal cauline leaves deeply pinnatifid into linear-lanceolate lobes; inner phyllaries with long-attenuate tips. . . . . 7. *C. discolor*.

h. Principal cauline leaves entire or shallowly sinuate-lobed (only the basal sometimes deeply pinnatifid).

Involucre 2–3 cm. high; inner phyllaries with lanceolate to deltoid usually serrulate tips; wide-ranging species. . . . . 8. *C. altissimum*.

Involucre 3–3.5 cm. high; inner phyllaries with prolonged lance- or linear-attenuate tips; western prairie-species. . . . 9. *C. iowense*.

g. Heads naked at base, on elongate peduncle-like branches with greatly reduced and scattered bracteal leaves; involucre slender, 1.5–2.8 cm. high; southern species.

Leaves coriaceous, subrigid; peduncles with reduced upper leaves numerous and extending nearly to summit; involucre 2–2.8 cm. high. . . . . 10. *C. virginianum*.

Leaves membranaceous and pliable; peduncles naked or with only 1 or 2 remote bracts; involucre 1.5–2 cm. high. . . . 11. *C. carolinianum*.

d. Leaves green both sides, or with only loose cobwebby deciduous gray hairs beneath. . . i.

i. Inner phyllaries with the narrow or tapering tips not dilated and crisped; involucre 1.7–3.3 cm. high. . . j.

j.Outer phyllaries with prickles at tip; tips of inner ones with linear-attenuate curved appendages; southeastern species.

Tall (0.6–4 m. high) and slender, with prolonged ascending flowering branches, glabrescent; stem winged below by decurrent prickly prolonged bases of the deeply pinnatifid prickly leaves; long-stalked slenderly cylindric involucres 1.8–2 cm. high; divergent bristle-like prickles of small outer and median phyllaries 0.5–1.5 mm. long. . . . . 12. *C. Nuttallii.*

Low (1.5–6 dm. high), stoutish, with short or nearly obsolete peduncles, arachnoid-lanate; stem wingless; leaves oblanceolate or narrowly oblong, shallowly undulate-lobulate; involucre oblong-cylindric, 2–3.3 cm. high, the coarse outer and median phyllaries with strong subulate ascending prickles 2–3 mm. long. . . . . 13. *C. repandum.*

j.Outer ovate phyllaries barely pointed, scarcely prickle-tipped; tips of inner ones broadly lanceolate; wide-ranging continental species. . . . . 14. *C. muticum.*

i.Inner phyllaries with oblong to elliptic scarious and crisped appendage below the slender tip; involucre broadly ovoid, 4–5.5 cm. high.

Coarse terminal prickles of outer phyllaries 4–6 mm. long, the back of the phyllary rarely with dark glutinous band; eastern species. . . . . 15. *C. pumilum.*

Slender terminal prickles of outer phyllaries 1.5–3 mm. long, the back of phyllary with dark glutinous band; inland species. . . 16. *C. Hillii.*

a.Flowering stems arising from sprouts from widely creeping subterranean perennial root; flowers dioecious; heads numerous and relatively small; involucre 1–2 cm. high; ubiquitous weed. . . . . 17. *C. arvense.*

1. C. vulgàre (Savi) Tenore (common), Bull- or Common T., Gros Chardon (Que.). — Biennial with oblanceolate to elliptic coarsely toothed first year's leaves forming rosettes; *flowering stem* 0.5–2 m. high, *with prickly lobed wings; leaves* pinnatifid or pinnatiparted, pale or woolly or webbed beneath, *hirtellous* and green *above,* the divaricate lobes and lance-acuminate tip fiercely armed with long prickles; *heads solitary or few at tips of short prickly-winged branches; involucre* ovoid to subglobose, 3–4 *cm. high, with nearly all the lanceolate to linear* crowded *phyllaries attenuate to rigid subulate-acerose* spreading *prickles;* flowers purple. (*C. lanceolatum* Scop. and most auth., not Hill) — Clearings, pastures and roadsides, an aggressive weed, Nfld. to B.C., s. beyond our range. June–Sept. (Natzd. from Eu.)

2. C. palústre (L.) Scop. (of marshes). — Biennial, 0.5–1.2 m. high, the *stem and branches strongly winged with undulate-pinnatifid very prickly decurrent bands* from the leaf-bases; leaves abundant in basal rosette and at base of stem, becoming more scattered upward, narrowly lanceolate, green and hirsute above, paler and pubescent beneath, pinnatifid, the prickly-toothed divergent lobes separated by round-quadrate sinuses; *heads clustered or glomerulate at summit and at tips of ascending branches,* slenderly ovoid; *involucre* 1–1.5 *cm. high;* the relatively soft ovate-lanceolate outer phyllaries pointed by a slender prickle about 1 mm. long; *inner phyllaries with soft purple linear-attenuate tips;* flowers purple. — Damp or springy woods, thickets and clearings, Nfld. (apparently indigenous); and local (partly adv.), N.S. to n. Mich., and in waste places s. to N.Y. Late June–Sept.

3. C. horrídulum Michx. (somewhat bristly), Yellow T. — Biennial; leaves of 1st year's rosette soft, oblong-oblanceolate, hirsute above, lanate beneath, with broad quadrate prickly-margined short lobes; *stem stout,* 0.2–1.5 m. high, *webbed-hairy* when young; cauline leaves thinnish, partly clasping, soon smooth, lanceolate, the short and broad toothed and cut lobes very spiny with yellowish prickles; *heads* 4–8 cm. broad, *surrounded by and equaled or overtopped by very prickly bract-like erect leaves; true involucre with narrow soft-tipped phyllaries; flowers pale yellowish,* or pale purple in forma **Ellióttii** (T. & G.) Fern. (for Stephen Elliott, 1771–1830); anthers usually red. (*C. spinosissimum* of recent Am. auth., not (L.) Scop.) — Sandy or peaty fields, meadows, shores and borders of woods, Fla. to Tex., n. on or near Coastal Plain to s. Me. May–Aug.

4. C. Pítcheri (Torr.) T. & G. (for its discoverer, Zina Pitcher, 1797–1872). — White-woolly throughout, the tap-root up to 2 m. long; *rosettes with masses of* canescent *long-petioled leaves with blade cleft to midrib into soft elongate-linear to narrowly oblong mostly entire divisions;* flowering stem up to 8 dm. high, branched at summit; *cauline leaves similar to basal;* heads short-peduncled, solitary or crowded; *involucre* 2–3 cm. high, drab, *the outer ovate phyllaries prickle-tipped, the inner* longer ones *acuminate to a slender and weak bristle; flowers cream-colored.* — Sandy shores and dunes of Lakes Michigan, Huron and Superior in Ont., Mich., Ind., Ill. and Wisc. May–Sept.

5. C. Flodmáni (Rydb.) Arthur (named in 1896 for its discoverer, J. H. Flodman). — Biennial; 1st year's rosette with coarsely pinnatifid to subentire oblanceolate *leaves white-felted beneath and gray-tomentose above;* flowering stem slender, 2–7.5 dm. high, white-felted, usually few-branched at summit; *cauline leaves* like the basal but all *deeply pinnatifid with narrowly lanceolate acerose lobes, the median blade often reduced nearly to the midrib; involucre* 2–2.5 *cm. high,* its outer flocculent ovate phyllaries with spreading prickles 2–4.5 mm. long, glutinous on back; inner phyllaries lanceolate and attenuate to weak tips; flowers purple. (*C. canescens* of ed. 7, not Nutt.) — Clearings, shores, fields and meadows, n. Vt. and n. N.Y. (adv.), to Sask., s. to Ia., Neb. and Colo. July–Sept.

6. C. undulàtum (Nutt.) Spreng. (undulate; from wavy leaf-margin). — Resembling no. 5, coarser; *principal cauline leaves with broadly lanceolate to broadly triangular short lobes and broad median band; involucre* 2.5–3 *cm. high;* outer phyllaries ending in prickles 3.5–5.5 mm. long. — Dry open soil, B.C. to Ariz., e. to Mich., Ia., Kans. and Tex. June–Oct. Passing freely into

Var. **megacéphalum** (Gray) Fern. (large-headed). — *Involucre* 3–5 *cm. high;* prickles 5–7 mm. long. (*C. megacephalum* (Gray) Cockerell) — Eastw. to Man., N.D., Neb., Mo. and Okla.

7. C. díscolor (Muhl.) Spreng. (two-colored). — Biennial or sometimes perennial; 1st year's *rosette-leaves petioled, white-felted beneath, green and glabrous or merely hirtellous above, divided nearly to the midrib into linear-lanceolate to narrowly oblong distant divergent simple or forking thin bristly-serrulate lobes;* flowering stem strongly furrowed, greenish, hirtellous or glabrate, 1–3 m. high, usually branching at summit; cauline leaves similar to basal, very numerous, strongly reduced upward; heads mostly solitary on leafy branches, *the uppermost sessile leaves forming a false involucre;* true *involucre* 2.5–3 *cm. high;* the outer appressed *phyllaries* with a weak recurved prickle, *the inner ones linear- or lance-attenuate and with a long colorless entire appendage;* flowers purple, or white in forma **albiflòrum** (Britt.) House (white-flowered). — Thickets, river-banks, prairies, etc., sw. Que. and sw. Me. to Man., s. to s. N.E., L.I., Ga., Tenn. and Mo. July–Oct.

8. C. altíssimum (L.) Spreng. (very tall). — Similar to no. 7; the pilose stem 1–4 m. high, usually branching lower down; *leaves undivided, oblong-ovate to narrowly lanceolate, sinuate-toothed or undulate-pinnatifid, with weak marginal and terminal prickles; involucre* 2–3 *cm. high; outer phyllaries with short dark glandular line on back,* abruptly tipped by a setiform spreading prickle; *inner phyllaries with a lanceolate or deltoid usually serrulate tip;* flowers rosy-purple, or white in forma **Moòrei** Steyerm. (named in 1939 for George Moore of Leclede Co., Mo.) — Rich thickets, river-banks, woods and clearings, n. Fla. to Tex., n. to s. N.Y., O., Mich., Wisc., Minn. and N.D. July–Oct.

9. C. iowénse (Pammel) Fern. (of Iowa). — Similar to no. 8; leaves entire or coarsely angulate-lobed; *involucre* 3–3.5 *cm. high; inner phyllaries with prolonged linear- or lance-attenuate tips.* — Prairies, low grounds and wet rocky banks, Minn., Ia. and S.D. to Kans. July–Sept.

10. C. virginiànum (L.) Michx. (Virginian). — Biennial, *with tuberous-thickened* 1st-year *roots;* leaves of 1st year's rosette few, long-petioled, narrowly oblanceolate, relatively thin, with fine bristle-like prickles on the entire margin, white-felted beneath, bright green above; *flowering stem slender,* simple or loosely branching, more or less *arachnoid,* 0.3–1.4 m. high; *cauline leaves very numerous* (40–70 or more), *firm or coriaceous, white-felted beneath, bright green above, lance-acuminate, the lower and median* 0.5–1.5 *dm. long and* 0.5–1.5 *cm. wide, the upper rapidly reduced and bracteiform, all entire* and prickle-tipped, *or the lower and median ones with remote divergent short subulate prickles;* flowering branches and summit peduncle-like, with several bracteiform leaves; *involucre* slender, 2–2.8 *cm. high, its pale appressed outer small phyllaries with divergent subulate prickles* 1.5–3 *mm. long;* inner phyllaries with lance-linear prolonged tips; flowers purplish. — Wet pineland or sphagnous or peaty bogs, swales or clearings, mostly on or near Coastal Plain, Fla. to N.J. — Mid-Aug.–Oct. — In forma **revolùtum** (Small) Fern. (revolute; from the leaf-margin) the lower and median cauline leaves are 1.5–4 cm. broad, deeply pinnatifid with lanceolate to narrowly triangular simple or forked or toothed long-pointed lobes. (*C. revolutum* Small) — Often in extensive areas to the exclusion of typical *C. virginianum.*

11. C. caroliniànum (Walt.) Fern. & Schub. (of Carolina). — Similar to no. 10; *roots slender; cauline leaves few* (10–20) *and remote, thin and flexible,* the lower and median ones coarsely repand-pinnate, 0.8–1.8 dm. long and 1.5–5 cm. broad; *peduncles naked or with only* 1 *or* 2 *remote bracts;* involucre 1.5–2 cm. high. (*C. flaccidum* Small) — Dry acid woods, thickets and ravines, n. Ga. to Tex., n. to N.C., s. O., s. Ind., s. Ill. and s. Mo. May, June.

12. C. Nuttállii DC. (for Thomas Nuttall, 1786–1859). — Biennial with shallow roots; *stem slender,* glabrescent, 0.6–4 *m. high, somewhat winged below by decurrent prickly ridges* running down from leaf-bases; *leaves green on both sides, deeply pinnatifid, with* distant *prickly lanceo-*

late often toothed lobes; heads on long nearly naked peduncles; involucre slenderly cylindric, 1.8–2 *cm. high; small* ovate appressed *outer phyllaries* with dark glutinous median band, *tipped by bristle-like prickle* 0.5–1.5 *mm. long;* inner phyllaries narrow, with linear-attenuate curved soft tip. — Sandy thickets and borders of woods, Fla. to La., n. on Coastal Plain to se. Va. Late June–Aug.

13. C. repándum Michx. (sinuous). — Biennial with deep tap-root; *stem stoutish,* 1.5–6 *dm. high,* simple or few-branched at summit, *arachnoid-lanate; leaves* very many, subapproximate, divergent or reflexed, *oblanceolate or narrowly oblong, shallowly undulate-lobulate, green but more or less villous on both faces,* the larger ones 7–16 cm. long and 1–3.5 cm. broad; *heads short-stalked,* terminal; *involucre* oblong-cylindric, 2–3.3 *cm. high; coarse outer and median phyllaries with strong subulate prickles* 2–3 *mm. long,* with broad glutinous band; flowers purple. — Sandy pinelands and sandhills of Coastal Plain, Fla. to se. Va. June (rarely –Sept.).

14. C. mûticum Michx. (without points), SWAMP-T. — Biennial with small forking subnapiform root; *leaves submembranaceous,* those of basal rosette long-petioled, ovate to elliptic, deeply lobed; *flowering stem soft and hollow,* 0.45–3 m. high, usually branching above; *cauline leaves thinnish, sessile, green above, paler and more or less webby beneath, the principal ones deeply pinnatifid,* cleft two-thirds to or practically to the midrib into lanceolate to oblong often toothed or forking slightly prickly-margined lobes, or margin merely undulate-lobate and with few coarse and short lobes in forma **subpinnatífidum** (Britt.) Fern. (somewhat pinnatifid); upper leaves much reduced; heads on elongate peduncle-like branches or clustered; *involucre cobwebby,* ovoid-cylindric, 2–3.5 cm. high; *outer broad phyllaries blunt and merely subulate-mucronate;* inner ones with scarious dilated appendages, the innermost often attenuate to slender tip; flowers purple, or whitish in forma **lactiflòrum** Fern. (milky-flowered). — Low woods, thickets and swamps, Nfld. and Côte Nord, Que., to Sask., s. to N.S., N.E., Del., Md., upland to N.C. and Tenn., and n. La. July–Sept.

Var. **montícola** Fern. (growing on mountains). — *Stem* 1–7.5 (rarely –15) *dm. high; heads* mostly *in a dense glomerule or very short-peduncled; involucre often glabrous from the first; phyllaries glutinous, with broad brown or purple blades.* — Exposed or alpine areas, s. Lab., Nfld. and Shickshock Mts., Que.

15. C. pùmilum (Nutt.) Spreng. (dwarf), PASTURE- or BULL-T. — *Biennial* with deep taproot; *rosette of* 1*st year with* numerous *short-petioled or subsessile hard* oblanceolate *leaves green on both sides* and often hirtellous above, pinnatifid *with rounded-quadrate sinuses and short-oblong* prickly-margined *lobes; stem stout, hairy,* 2–9 dm. high, simple or with several upright branches; *cauline leaves* lanceolate-oblong, *green, with very prickly* oblong to lance-deltoid *lobes; heads large, short-stalked or subsessile,* without or with approximate leafy bracts, or subtended by a score or more of crowded overtopping leafy rosulate bracts in forma **fúltius** Fern. (supported); *involucre thick-ovoid,* 4–5.5 *cm. high; phyllaries* usually without glutinous central band, *the outer ones terminated by a coarse prickle* 4–6 *mm. long, the inner with an oblong to elliptic scarious crisped appendage;* flowers purple, or white in forma **cándidum** Fern. (white). (*C. odoratum* Bart.) — Dry open soil, s. Me. to O., s. to Del., Md. and interior N.C. June–Sept.

16. C. Híllii (Canby) Fern. (for its discoverer, ELLSWORTH JEROME HILL, 1833–1917). — Resembling no. 15, *with deep perennial root;* leaves less deeply lobed, with finer and shorter prickles, *the* 1*st year's rosette-leaves with repand-undulate margins; phyllaries with dark glutinous backs, the outer ones terminated by a slender prickle only* 1.5–3 *mm. long.* — Sandy open soil, Ont. and Man., s. to w. Pa., Ind., Ill. and Ia. June–Aug.

17. C. ARVÉNSE (L.) Scop. (of fields), CANADA T. — *Perennial by extensively creeping and too freely sprouting roots;* stem 3–9 dm. high, very leafy; *leaves subrigid, oblong or lanceolate, smooth,* or slightly woolly beneath, finally *green on both sides, strongly sinuate-pinnatifid, very prickly-margined,* the upper sessile but scarcely decurrent; *heads imperfectly dioecious; involucres slenderly ovoid-cylindric,* very numerous, often clustered, 1–2 *cm. high;* the appressed *outer* ovate *phyllaries subulate-tipped,* the inner with thin lance-attenuate tips; flowers pinkish-purple, or white in forma ALBIFLÒRUM (Rand & Redf.) R. Hoffm. (white-flowered). — Ubiquitous weed of cult. and waste ground. June–Oct. (Natzd. from Eu. — NOT from Canada)

Var. MÌTE Wimm. & Grab. (mild or gentle). — *Leaves thinner,* green, *the primary ones merely undulate-lobulate and with few fine and short prickles.* — Locally from N.E. westward, s. to Pa., Ia., S.D. and Colo. (Natzd. from Eu.)

Var. INTEGRIFÒLIUM Wimm. & Grab. (entire-leaved). — Green; *leaves* oblong, flat, *entire* or barely crenate, almost without prickles. — Local, Que. to B.C., s. to N.E., Del., Md. and Mich. (Natzd. from Eu.)

Var. VESTÌTUM Wimm. & Grab. (clothed). — *Stem canescent-tomentose; leaves* entire or merely undulate, nearly prickleless, *blanketed beneath with white tomentum.* — Local, N.E. to Ill. (Natzd. from Eu.)

86. ONOPÓRDUM L. Cotton- or Scotch Thistle

Receptacle deeply alveolate, not setose. Pappus not plumose. Otherwise as *Cirsium.* — Coarse branching annuals or biennials, nat. of Old World, with the stems winged by the decurrent bases of the lobed and toothed somewhat prickly leaves. Heads large; flowers purple. (Latinized from the ancient Greek name of the plant, from *onos, donkey,* and *porde, flatulence;* Pliny stating that it produced flatulence in donkeys.)

1. O. Acánthium L. (old generic name, prickly). — Stem (1–3 m. high) and leaves cotton-woolly; phyllaries linear-subulate. — Roadsides and waste places, N.B. to Ont., s. to Ala. and se. Mo., local. July–Sept. (Natzd. from Eu.)

Milk-Thistle, Sílybum Mariánum (L.) Gaertn. (of the Virgin Mary, from the milky spots on the leaves), a tall plant with white-mottled large clasping leaves and large heads of purple flowers, the phyllaries spreading or reflexed and ending in a stout spine, the receptacle densely bristly and the filaments smooth and united into a tube, occasionally spreads from gardens. (Introd. from Eu.)

87. CENTAURÈA L. Star-Thistle

Heads many-flowered; flowers all tubular, the marginal often much larger (as if radiate) and sterile. Receptacle bristly. Involucre ovoid or globose, imbricated; phyllaries commonly margined or appendaged. Achenes obovoid or oblong, compressed or 4-angled, attached obliquely or laterally at or near the base; pappus setose or partly chaffy, or none. — Herbs, largely nat. of the Mediterr. reg., with alternate leaves; the single heads rarely yellow. (*Centaurie,* an ancient Greek plant-name, poetically associated with Chiron, the *Centaur,* but without wholly satisfactory explanation.)

a. Phyllaries, at least the outer ones, terminated by a definite rigid spine; annuals or biennials. . . *b.*
- *b.* Spines spinulose only at base, the divergent to reflexed terete primary spine 1–2.5 cm. long.
  - Stems wingless; heads sessile at leafy tips and in forks of branches. . . 1. *C. Calcitrapa.*
  - Stems winged; heads terminal, not leafy at base. . . 2. *C. solstitialis.*
- *b.* Spines pinnate on the flat lower half, ascending or spreading, the flattened primary spine 2–3 mm. long. . . 3. *C. diffusa.*

a. Phyllaries not spine-tipped, or merely with short firm tip hardly longer than marginal teeth or fringe. . . *c.*
- *c.* Involucres 0.8–2.5 cm. high, rarely as broad; peduncles slender or, if thickened upward, rarely 4 mm. thick; scar or insertion of achene lateral; introd. species. . . *d.*
  - *d.* Principal cauline leaves simple or merely coarsely dentate or lyrate. . . *e.*
    - *e.* Annual with white flocculent pubescence and simple entire linear or lanceolate leaves. . . 4. *C. Cyanus.*
    - *e.* Perennial or biennial, green; at least the 1st year's or basal leaves often pinnatifid or lobed. . . *f.*
      - *f.* Involucre whitish-green, about 1 cm. high; the outer phyllaries entire, inner merely serrulate; pappus deciduous from slender achene. . . 5. *C. repens.*
      - *f.* Involucre brownish to nearly black, at least at tips, larger; phyllaries with scarious and entire, lacerate or pectinate appendages; pappus persistent or deciduous. . . *g.*
        - *g.* Phyllaries entire or merely with irregularly denticulate to lacerated (not regularly toothed or pectinate) margins. . . 6. *C. Jacea.*
        - *g.* Phyllaries, or many of them, with regularly pectinate or toothed margins. . . *h.*
          - *h.* Middle and outer phyllaries pectinate or fringed except at the hidden base.
            - Fringe of the median phyllaries twice or thrice as long as breadth of blade; flowering branches with short pointed leaves. . . . . 7. *C. nigra.*
            - Fringe of median phyllaries only about as long as breadth of blade; leaves of flowering branches blunt. . . . . 8. *C. nigrescens.*
          - *h.* Middle and outer phyllaries with conspicuous greenish blades, with dark pectinate tip about one-third as long as blade. . . 9. *C. vochinensis.*
  - *d.* Principal cauline leaves deeply pinnatifid into long narrow segments.
    - Heads 5–6 cm. broad; involucre 2–2.5 cm. high, its arachnoid outer and middle phyllaries with dark pectinate tips 4–6 mm. long. . . 10. *C. Scabiosa.*

Heads 1.5–2.5 cm. broad; involucre 1–1.4 cm. high, its smooth strongly ribbed outer and middle phyllaries with dark pectinate tips 1–2 mm. long. . . . . 11. *C. maculosa.*

c.Involucre 3.5–4.5 cm. high; phyllaries pale brown or cinnamon-color, all fringed; peduncles thickened upward, 0.6–1 cm. thick at summit; scar or insertion of achene obliquely basal; nat. of w. Am. . . . . 12. *C. americana.*

1. C. Calcítrapa L. (old name, *a caltrop*), Star-Thistle, Caltrops. — Divaricately branched annual or biennial 2–5 dm. high, with pale stem and branches; *leaves* pubescent; the basal rosulate, *pinnately divided* to the dentate rachis *into linear-lanceolate* toothed or incised *segments;* cauline leaves similar but reduced upward to simple or subsimple sessile blades; *heads sessile or subsessile at leafy tips or at leafy forks of stem; involucre ovoid,* smooth, 5–7 *of the coriaceous phyllaries tipped by terete rigid stramineous spinulose-based divergent to reflexed spines* 1–2.5 *cm. long;* corollas purplish, all tubular. — Waste places, roadsides, etc., se. states, n., locally, to N.Y. and s. Ont. June–Oct. (Natzd. from Mediterr. reg.)

2. C. solstitiàlis L. (of the summer solstice), Yellow Star-Thistle, Barnaby's Thistle. — Differing from no. 1 in having *winged ascending stem and branches; leaves tomentose,* the basal lyrate-pinnatifid, *the middle and upper entire* and with decurrent bases; *heads* terminal, *without leafy bracts; involucre globose, lanate.* — Waste places, roadsides and fallow fields, local, Fla. to Calif., n. to Mass., N.Y., s. Ont. and Ia. July–Oct. (Natzd. from Mediterr. reg.)

C. meliténsis L. (of Malta), with slender purplish spines only 6–9 mm. long, has occasionally appeared on ballast; not persistent.

3. C. diffùsa Lam. (diffuse). — Intricately branched annual or biennial *with angled* many-headed *paniculate-corymbose rigid stem,* radical *leaves* bipinnately divided, *the cauline* shorter and *with fewer divisions or the uppermost entire; heads terminating short leafy-bracted branchlets; involucre* ellipsoid-cylindric, *about* 1 *cm. high,* minutely glandular-pruinose; the coriaceous *pale lanceolate phyllaries stiffly spinulose-ciliate, ending in a flat lance-attenuate ascending or spreading spine* 2–3 *mm. long;* flowers roseate, white or creamy, not radiate; pappus deciduous. — Fields, roadsides, etc., locally but rapidly spreading, Mass.; Mich.; Ia. July–Sept. (Natzd. from se. Eu.)

4. C. Cỳanus L. (the ancient name), Bluebottle, Bachelor's-button, Cornflower. — *Slender annual,* with long ascending branches terminated by solitary heads; *involucre ovoid, of about* 4 *very unequal series of pale phyllaries; the outermost* narrowly deltoid, *and with the longer middle ones bordered by a white or silvery sharp-toothed scarious band;* the innermost elongate-lanceolate, with more or less erose and colored tips; *marginal flowers large and ray-like,* blue, violet, pink or white. — Esc. from cult. to roadsides, etc. July–Sept. (Introd. from Eu.)

C. montàna L. (of mountains), Mountain-Bluet, a perennial with subterranean stolons, the simple erect stems with broad lanceolate to elliptic decurrent leaves, the terminal large radiate head with the greenish phyllaries bordered by a dark lacerate rim, tends to spread from cult. in Nfld. (Introd. from Eu.)

5. C. rèpens L. (creeping). — *Repent perennial with* canescent or *pale stiffly and virgately branched* corymbose *stems* 2–6 dm. high; *leaves* firm, ascending, *linear-lanceolate to narrowly oblong, entire, dentate or slightly pinnatifid;* heads terminating leafy-bracted branchlets; *involucre slenderly ovoid, pale, about* 1 *cm. high; the outer* rounded-ovate *phyllaries with hyaline entire border,* the inner ones oblong-acuminate and serrulate; flowers roseate; *achene slenderly obovoid, its pappus deciduous.* (*C. Picris* Pall.) — Fields, roadsides and waste places, rapidly spreading, Mich. to Wash., s. to Mo., Tex. and s. Calif. June–Aug. (Natzd. from the Caucasus)

6. C. Jàcea L. (old generic name). — Perennial, 0.3–1 m. high, simple or with ascending branches; rosette-leaves petioled, lanceolate, entire, sinuate or pinnatifid; *cauline leaves hard, lanceolate,* scattered; heads rather showy, terminating branches; *involucre* subglobose, 1.5–2.5 *cm. high; phyllaries* imbricated, *their bases hidden by the brown concave scarious entire, erose or lacerate broad pale-margined appendages;* flowers rose-purple, the outer falsely radiate. — Roadsides, fields and waste places, Que. to Ia., s. to N.E., Va., O. and Ill. June–Sept. (Natzd. from Eu.) Fig. 1766.

1766. C. Jacea.

C. Moschàta (old generic name, musky), Sweet Sultan, an annual with thin dentate or pinnatifid leaves, heads on long naked peduncles, the ovoid-globose involucre with coriaceous yellowish entire obtuse phyllaries without scarious margin, and showy yellow, purple or white flowers, sometimes spreads from cult. (Introd. from sw. Asia)

7. C. nìgra L. (black), Knapweed, Spanish-buttons. — *Coarse* perennial with harsh somewhat branching stems up to 8 dm. high; rosette-leaves oval or elliptic-lanceolate, more or less sinuate; *cauline leaves* oblong-lanceolate, *the uppermost and bracteal ones reduced, narrow and pointed, extending nearly or quite to the terminal heads; involucre* subglobose, 1.5–2 *cm. high; the outer and median phyllaries with blades completely hidden by the deep brown to blackish long-pectinate appendages, the fringe of the median appendages two or three times as long as the breadth of the elliptic to lance-ovate blade;* innermost appendages paler, merely lacerate; corollas rose-purple, all tubular. — Fields, roadsides, etc., Nfld. to Ont., s. to N.S., N.E., Del., Md., D.C., O. and Mich. July–Sept. (Natzd. from Eu.) Fig. 1767.

1767. C. nigra.

Var. RADIÀTA DC. (having rays). — Involucre tawny (rarely blackish); marginal flowers falsely radiate. — Nfld. to Ont., s. to Va., O. and Mich. (Natzd. from Eu.)

8. C. NIGRÉSCENS Willd. (blackish). — Similar to no. 7; cauline *leaves oblong to elliptic, the uppermost blunt ones closely subtending the heads; appendages of lower and median phyllaries with fringe only about as long as the breadth of the blade.* — Fields and roadsides, St. P. et Miq. and N.S. to N.Y., s. to Pa. June–Sept. (Natzd. from Eu.)

C. NERVÒSA Willd. (nerve-like), similar to nos. 7 and 8, but with the appendages of the phyllaries elongate, arcuate-recurving, reduced to a long thread-like blade with filiform fringe, has once been collected in Washington Co., N.Y. (Adv. from Eu.)

9. C. VOCHINÉNSIS Bernh. (of the Wocheiner See region). — *Harsh,* somewhat *resembling no. 7; involucre of several very unequal series;* the *outermost phyllaries* deltoid or ovate, *short, the dark pointed tip bearing 5–7 pairs of long cilia; middle phyllaries* elongate-lanceolate, *terminated by a dilated* ovate or orbicular *dark pectinate appendage;* innermost phyllaries elongate, with dark or brightly colored erose or lacerate appendage; corollas rose-purple, the marginal falsely radiate. — Fields and roadsides, Me. to Ont., s. to Va. and Mo. June–Sept. (Natzd. from Eu.) Fig. 1768.

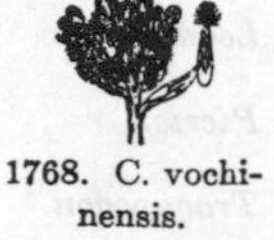
1768. C. vochinensis.

10. C. SCABIÒSA L. (originally placed with *Scabiosa;* so named from its scurfiness). — Perennial, with angulate *stems rather densely leafy at base,* forking above; *leaves,* except the reduced and few upper, *pinnately divided into linear entire or pinnatifid segments; heads* showy, 1–few, *on long erect peduncles, 5–6 cm. broad; involucre* globose-ovoid, 2–2.5 *cm. high;* the closely imbricated *outer and median arachnoid pale ovate-deltoid phyllaries with dark marginal and terminal pectinate band 4–6 mm. long;* corollas purplish, the marginal falsely radiate. — Pastures, fields and roadsides, local, Bonaventure Co., Que., to s. Ont., s. to O. and Ia. July, Aug. (Natzd. from Eu.)

11. C. MACULÒSA Lam. (spotted). — *Biennial with* slender *bushy-branched wiry stems* 0.3–1 m. high; *leaves canescent, the basal and lower cauline pinnate or pinnatifid* with linear or lanceolate entire or pinnatifid segments, upper leaves simpler; *heads* terminating branchlets, 1.5–2.5 *cm. broad; involucre pale,* 1–1.4 *cm. high, its smooth strongly ribbed outer and median ovate phyllaries with firm points and 5–7 pairs of cilia, the dark tip 1–2 mm. long;* innermost phyllaries entire or lacerate; corollas whitish, roseate or purple, the marginal falsely radiate. — Fields, roadsides and waste places, rapidly spreading, Que. to B.C., s. to N.S., N.E., Va., Tenn., Mo., Kans., etc. June–Aug. (Natzd. from Eu.)

12. C. **americàna** Nutt. (American). — *Tall* and smoothish (0.5–1.5 m. high); *stems conspicuously thickened below the showy heads; leaves oblong-lanceolate, mostly entire; phyllaries all with conspicuously fringed scarious appendages, the outermost and middle with spreading,* the elongate *innermost with ascending, teeth;* corollas rose-purple to flesh-pink, the outer conspicuously enlarged. — Plains, Mo., southw. and southwestw.; rarely adv. northeastw. and eastw. May–Aug.

88. CNÌCUS L. Blessed Thistle

Heads many-flowered; flowers all tubular, the marginal sterile, shorter than the others, which are perfect and fertile. Phyllaries of the ovoid involucre coriaceous, appressed, extended into a long and rigid spinous appendage. Receptacle clothed with capillary bristles. Achenes terete, short, strongly many-striate, crowned with 10 short and horny teeth and bearing 10 elongated rigid bristles, also 10 shorter alternating ones in an inner row. — Annual pubescent herb, nat. of Eu. and the Orient, with scarcely pinnatifid-cut but spinescent leaves and large leafy-bracted heads of yellow flowers. (Latin name of the Safflower, from the Greek *cnecos.*)

1. C. BENEDÍCTUS L. (blessed, the plant once supposed to prevent evil). — Roadsides and waste places, rare, N.B. to Ill., and southw. April–Sept. (Adv. from Eu.)

Subfam. II. LIGULIFLÒRAE (see p. 1358)

Tribe X. CICHORÌEAE Spreng. (see p. 1358)

a.Pappus none; annuals.
Leafy-stemmed; stem not strongly thickened upward, 3 dm. or more high.
Involucre calyculate-bracteate at base. . . . . 89. *Lapsana.*
Involucre naked at base. . . . . 90. *Serinia.*
Leaves all basal; low stems and branches upwardly thickened; involucre 1-seriate. . . . . 91. *Arnoseris.*
a.Pappus present. . *b.*
b.Pappus chaffy or of both chaff and bristles.
Involucre double (calyculate); pappus of uniform chaffy scales; flowers bright blue, pink or white. . . . . 92. *Cichorium.*
Involucre simple; pappus of both chaff and bristles; flowers yellow. . 93. *Krigia.*
b.Pappus of many bristles. . *c.*
c.Pappus plumose. . *d.*
d.Involucre calyculate (with bractlets at base). . *e.*
e.Calyculate bractlets appressed; achenes fusiform; flowering stem scapose or scapiform.
Inner achenes produced into long slender beaks; receptacle chaffy. 94. *Hypochoeris.*
Inner and outer achenes uniform, not slender-beaked; receptacle naked. . . . . 95. *Leontodon.*
e.Calyculate bractlets or outer phyllaries elongate and loosely spreading; achenes terete, beakless. . . . . 96. *Picris.*
d.Involucre without calyculate base; slenderly fusiform achenes long-beaked; tall, leafy-stemmed, with large terminal heads. . . . 97. *Tragopogon.*
c.Pappus not plumose. . . *f.*
f.Achenes strongly muricate or spinulose above.
Leafy, branching, with firm stems; heads small; achenes terete. 98. *Chondrilla.*
Scapose, the weak scapes hollow; heads broad; achenes fusiform. 99. *Taraxacum.*
f.Achenes not muricate above. . . *g.*
g.Achenes flat or flattish; involucre imbricated; mostly leafy-stemmed and chiefly with panicled heads.
Achenes beakless; flowers 50 or more in each finally tumid-based involucre, yellow. . . . . 100. *Sonchus.*
Achenes tapering to summit or beaked; flowers 6–30 in each head (the latter not tumid at base), blue, purple or yellow. 101. *Lactuca.*
g.Achenes columnar or nearly so. . . *h.*
h.Achenes filiform, tapering to summit but not to base; involucre slenderly cylindric; flowers 5, rose-purple, on rush-like often naked branches; western. . . . . 102. *Lygodesmia.*
h.Achenes narrowed at base; leaves well developed; flowers white, yellow or orange, rarely pink. . . *i.*
i.Achenes beaked.
Scapose; pappus white; achenes 10-ribbed; rosulate perennials. . . . . 103. *Agoseris.*
Leafy-stemmed branching annual or perennial; pappus reddish or rusty; achenes 4- or 5-ribbed. . . . . . 104. *Pyrrhopappus.*
i.Achenes beakless.
Phyllaries in a single series. . . . . 105. *Crepis.*
Phyllaries of 2 or more series (calyculate or imbricated).
Inflorescence a branching raceme or panicle of slenderly cylindrical drooping heads with whitish, creamy or pink flowers; lower leaves often cleft or lobed; root tuberous, fusiform to subglobose. . . . . 106. *Prenanthes.*
Inflorescence with 1–very many erect or ascending heads, mostly in corymbs or panicles; involucre campanulate or broader; flowers yellow, sometimes orange-scarlet; leaves simple, not cleft; root not tuberous. . . . . . 107. *Hieracium.*

89. LÁPSANA L. NIPPLEWORT

Heads 8–12-flowered. Phyllaries of the cylindrical involucre 8, erect; a short outer series also present. Receptacle naked. Achenes oblong; pappus none. — Slender branching annuals, nat. of Eurasia, with angled or toothed leaves, and loosely panicled small heads; flowers yellow. (Name used by Dioscorides for some plant.) LAMPSANA Hill

1. L. COMMÙNIS L. (common). — Nearly smooth, 3–8 dm. high; lower leaves ovate, sometimes lyrate. — Roadsides and waste places, Que. and Ont., s. to N.S., N.E., Va., W.Va. and Mo., locally abundant. June–Sept. (Natzd. from Eu.)

90. SERÍNIA Raf.

Heads as in *Lapsana*, but the involucre not calyculate; tips of the phyllaries somewhat connivent in fruit. Achenes obovoid, 10-costate, obscurely and transversely scabrous-lineolate. — Low glaucescent glabrous N. Am. annuals, with oblong to lanceolate entire or repand-dentate leaves, the upper sessile; and small scattered heads on slender peduncles. (Name intended as a diminutive of the Greek *seris, chicory.*) APOGON Ell.

1. S. oppositifòlia (Raf.) Ktze. (opposite-leaved). — Peduncles naked or glandular-bristly below the small (5 mm. high) heads. — Damp sandy fields, thickets and waste places, Fla. to Tex., n. to se. Va., s. Ill., Mo. and Kans. Apr.–June. — By Shinners placed in *Krigia.*

91. ARNÓSERIS Gaertn.

Heads many-flowered; flowers all ligulate and fertile. Involucre campanulate; phyllaries narrow, equal, in 1 series, acuminate, after anthesis thickened toward the base. — Small annual, nat. of Old World, with rosulate leaves and branched upwardly thickened scapes bearing rather small heads of yellow flowers. (Name from the Greek *arnos, lamb,* and *seris, chicory.*)

1. A. MÍNIMA (L.) Schweigger & Koerte (smallest), LAMB-SUCCORY. — Puberulent, 1.5–3 dm. high; leaves oblanceolate or spatulate, mostly toothed. — Old fields and roadsides, local, N.S. to Mich. June–Sept. (Adv. from Eu.)

92. CICHÒRIUM L. SUCCORY OR CHICORY. CHICORÉE (Que.)

Heads several-flowered. Involucre double, herbaceous, the inner of 8–10, the outer of 5 short and spreading phyllaries. Achenes striate; pappus of numerous small chaffy scales, forming a short crown. — Branching perennials or biennials, nat. of Old World, with deep roots; the sessile heads 2 or 3 together, axillary and terminal, or solitary on short thickened branches. Flowers showy. (Altered from the Arabian name of the plant.)

1. C. ÍNTYBUS L. (old generic name), COMMON C., BLUE SAILORS. — Cauline leaves oblong or lanceolate, partly clasping, the lowest often runcinate, those of the rigid flowering branches minute; flowers bright blue, or pink in forma RÒSEUM Neum. (rose-colored), or white in forma ÁLBUM Neum. (white) — Fields and roadsides, Nfld. to B.C., s. beyond our limits. June–Oct. (Natzd. from Eu.)

93. KRÍGIA Schreb. DWARF DANDELION

Heads several–many-flowered. Phyllaries several, in about 2 rows, thin. Achenes short and truncate, turbinate or columnar, terete or angled; pappus double, the outer of thin pointless chaffy scales, the inner of delicate bristles. — Small N. Am. herbs, branched from the base; the leaves chiefly radical, lyrate or toothed; the small heads terminating the naked scapes or branches. Flowers yellow. (Named in 1791 for *David Krig* or *Krieg,* a German physician, who was among the first to collect plants in Maryland.)

Phyllaries 5–8, remaining erect; dwarf annual. § CYMBIA. . . . . 1. *K. occidentalis.*
Phyllaries 9–18, reflexed in age.
  Annual; achenes 5-angled; pappus 5–7 short roundish scales alternating with as many bristles. § EUKRIGIA. . . . . 2. *K. virginica.*
  Perennial; achenes 15–20-ribbed; pappus 10–15 small scales and 15–20 bristles. § EUCYNTHIA.
    Root tuberiferous; scape simple, leafless. . . . . 3. *K. Dandelion.*
    Root fibrous; stem forking, with 1–3 clasping leaves. . . . . 4. *K. biflora.*

§ CÝMBIA T. & G.

1. K. occidentàlis Nutt. (western). — Scapes tufted, 1.5 dm. or less high, glandular-hispid, at least above; leaves obovate to lanceolate, entire, lyrately lobed or pinnatifid; *phyllaries 5–8, remaining erect;* achenes turbinate; *pappus 5 obovate scales and usually 5 alternating bristles.* (*Cymbia* Standl.) — Prairies and open sandy soil, Tex. to Ark. and Mo. April, May.

§ Eukrígia T. & G.

2. K. virgínica (L.) Willd. (Virginian). — Scapes several, 0.5–3 dm. high, becoming branched and leafy; earlier leaves roundish and entire, the others narrower and often pinnatifid; *phyllaries 9–18, reflexed in age; achenes turbinate, 5-angled; pappus of 5–7 short scales alternating with bristles.* (*Adopogon carolinianum* (Walt.) Britt.) — Dry sterile soil, Fla. to Tex., n. to s. Me., s. N.H., Mass., N.Y., s. Ont., Mich., Wisc. and Ia. Apr.–Aug.

§ Eucýnthia (DC.) Shinners

3. K. Dandèlion (L.) Nutt. (from its resemblance to the common weed), Potato-Dandelion. — *Roots* slender, *tuberiferous; scapes leafless,* 1.5–5 dm. high; leaves varying from spatulate-oblong to linear-lanceolate, entire or few-lobed; *involucre of numerous phyllaries, reflexed in age; achenes* 15–20-*ribbed; pappus* 10–15 *small scales and* 15–20 *bristles.* (*Adopogon* Ktze.; *Cynthia* DC.) — Prairies, meadows and borders of woods, Fla. to Tex., n. to s. N.J., Ky., s. Ill., Mo. and se. Kans. April–June.

4. K. biflòra (Walt.) Blake (two-flowered). — Roots fibrous; *stem* 2–7 dm. high, *forking above,* glabrous; *cauline leaves* 1–3, oblong or oval, *clasping,* mostly entire; radical leaves wing-petioled, often toothed, sometimes pinnatifid; peduncles 2–7, glabrous, or glandular in forma **glandulífera** Fern. (bearing glands); involucre and pappus similar to those of no. 3. (*K. amplexicaulis* Nutt.; *Adopogon virginicum* (L.) Ktze.; *Cynthia virginica* (L.) Don) — Open woods, prairies and meadows, s. N.E. to Man., s. to Ga., Tenn., Mo., N.M. and Ariz. May–Aug.

94. HYPOCHOÈRIS L. Cat's-ear

Similar to *Leontodon,* but at least the inner achenes extended into long slender beaks. Receptacle chaffy. — Old World and S. Am. herbs, with bracteate slightly branching scapes and yellow flowers. (A name used by Theophrastus for this or a related genus.)

1. H. radicàta L. (having roots; probably from its firm hold in the ground). — Perennial; *leaves hirsute;* scapes 2–4 dm. high, *stout;* heads 2.5–4 cm. broad; involucre 1.5–2.5 cm. long; phyllaries glabrous or strigose; achenes all beaked. — Fields and waste places, Nfld. to Ont., s. to N.S., N.E., N.C., W.Va., O., Ind. and Ill.; Pacific slope. May–Aug. (Natzd. from Eu.)

95. LEÓNTODON L. Hawkbit

Heads many-flowered. Involucre scarcely imbricated, but with several bractlets at the base. Achenes fusiform, striate, all alike; pappus persistent, composed of plumose bristles which are enlarged and flattened toward the base, with sometimes an outer paleaceous crown. Receptacle not chaffy. — Low and stemless perennials, nat. of Old World, with toothed or pinnatifid basal leaves, and simple or forking scapes bearing one or more yellow heads. (Name from the Greek *leon, a lion,* and *odous, a tooth,* in allusion to the toothed leaves.)

Scape scaly-bracted above; heads before anthesis erect; pappus a single row of plumose bristles. . . . . 1. *L. autumnalis.*
Scape naked; heads before anthesis nodding; pappus of 2 kinds.
  Pappus similar in all flowers, the outer a few short setae, the inner a row of plumose bristles with dilated bases. . . . . 2. *L. hastilis.*
  Pappus of marginal flowers a crown of short scales, of inner flowers setiform and plumose. . . . . 3. *L. Leysseri.*

1. L. autumnàlis L. (autumnal), Fall-Dandelion, "Arnica". — Leaves laciniate-toothed or pinnatifid, somewhat pubescent; scape commonly forking, 1–6 dm. high; *peduncles* thickened at summit, *scaly-bracteate* (rarely leafy); involucre glabrous or slightly pubescent; tawny *pappus a row of equal bristles.* — Fields and roadsides, Nfld. to w. Ont., Mich., and Pa. Late May–Nov. — A frequent teratological or perhaps pathological form, forma ochroleùca Sylvén (pale yellow), has greenish-yellow heads without ligules. (Natzd. from Eu.)

Var. praténsis (Link) W. D. J. Koch (of meadows). — Usually larger; involucre and tips of peduncles densely soft-pubescent with blackish hairs. — Often more abundant, Nfld. to w. N.E. (Natzd. from Eu.)

2. L. hastìlis L. (like a spear-head). — Rather stout, glabrous throughout; leaves oblong-lanceolate, dentate or pinnatifid; scape 2–7 dm. high, thickened upward; heads rather large; involucre 11–14 mm. long, its smooth lanceolate phyllaries lead-colored; *pappus of short setae and an inner row of plumose broad-based bristles.* — Fields, local, Ct. to O. (Adv. from Eu.)

Var. vulgàris W. D. J. Koch (common). — Leaves, scape, and involucre bristly-hispid. (*L. hispidus* L.) — Fields and waste places, R.I. to Ont. and s. N.Y. (Adv. from Eu.)

3. L. Leỳsseri (Wallr.) G. Beck (in honor of Friedrich Wilhelm von Leysser, 1731–1815). — Slender; leaves hispid; scape filiform, 1–2 dm. high; involucre 7–10 mm. long, its lanceolate phyllaries hirsute or glabrous; *pappus of marginal flowers short scales, of inner flowers setiform and plumose.* (*L. nudicaulis* sensu ed. 7, not Banks) — Pastures, fields and waste places, local, Vt. to O. and N.J. July–Sept. (Adv. from Eu.)

96. PÌCRIS L.

Heads many-flowered, terminating leafy stems. Outer phyllaries loose or spreading. Achenes with 5–10 rugose ribs; pappus of 1 or 2 rows of plumose bristles. — Coarse rough-bristly annuals or biennials, nat. of Old World, with yellow flowers. (The Greek name of some allied bitter herb, from *picros, bitter.*)

1. P. hieracioìdes L. (like *Hieracium*). — Rather tall, corymbosely branched, the bristles somewhat barbed at tip; *leaves* lanceolate or broader, clasping, *irregularly toothed; outer phyllaries narrow;* achenes oblong, with little or no beak; *pappus sparsely plumose.* — Fields, roadsides and waste places, local, N.E. to Mich., s. to N.J., Pa. and Ill. June–Oct. (Adv. from Eu.)

2. P. echioìdes L. (like *Echium*), Ox-tongue. — Similar, but *leaves and phyllaries spinescent; the outer phyllaries ovate, subcordate,* the narrow inner ones becoming thickened below; achenes beaked; *pappus densely plumose.* — Similar situations, local, Me. to s. Ont., s. to N.J., Pa. and O. July–Oct. (Adv. from Eu.)

97. TRAGOPÒGON L. Goat's-beard

Heads many-flowered. Involucre simple, of several erect lanceolate attenuate equal phyllaries. Achenes slender, fusiform, 5–10-ribbed, long-beaked; pappus of numerous long plumose bristles. — Stout glabrous biennials or perennials, nat. of Old World, with entire grass-like clasping leaves and large solitary heads of yellow or purple flowers. (Name from the Greek *tragos, goat,* and *pogon, beard.*)

1. T. porrifòlius L. (with leaves of Leek, *Allium Porrum*), Salsify, Oyster-plant. — Stem 1 m. or less high; peduncle thickened and fistulous below the head; *flowers purple;* achenes and pappus 5–6 cm. long. — Roadsides and fields, esc. from cult., N.S. to Ont., s. to Ga., Mo. and Kans. May–July. (Introd. from Eu.)

2. T. praténsis L. (of meadows), Goat's-beard. — Very similar; leaves somewhat broader at base; peduncle *slenderly cylindric; involucre 2–3 cm. high; flowers yellow; achenes* 1.5–2.5 *cm. long.* — Fields, rocky banks, etc., Que. and Ont., s. to N.S., N.E., Ga., Tenn., Mo. and Kans. May–Aug. (Natzd. from Eu.)

3. T. màjor Jacq. (larger). — Like no. 2, but *peduncle upwardly enlarged below the head; involucre 5–7 cm. high; achenes 2–4 cm. long.* (*T. dubius* of Am. auth., not Scop.) — Fields and roadsides, N.Y. to Wash., s. to Va., Ill., Tex. and Calif. May–July. (Natzd. from Eu.)

98. CHONDRÍLLA L. Gum-Succory

Heads few-flowered. Involucre cylindrical, of several narrow linear equal phyllaries and a row of small bractlets at base. Achenes terete, several-ribbed, smooth below, roughened at the summit by little scaly projections, from among which springs an abrupt slender beak; pappus copious, of very fine and soft capillary bristles, bright white. — Herbs, nat. of the Old World, with virgate branching stems, and small heads of yellow flowers. (A name used by Dioscorides for some plant which exudes a gum.)

1. C. júncea L. (rush-like), Skeleton-weed. — Biennial, bristly-hairy below, smooth above, 1 m. or less high; basal leaves runcinate; cauline leaves few and small, linear; heads scattered on nearly leafless branches, 1–1.5 cm. long. — Fields and roadsides, N.Y. to Va., W.Va. and Mich. July–Sept. (Natzd. from Eu.)

99. TARÁXACUM Zinn Dandelion. Blowballs. Pissenlit (Que.). Dumble-dor (Nfld.)

Heads many-flowered, large, solitary on a slender hollow scape. Involucre double, the outer of short phyllaries; the inner of long linear erect phyllaries in a single row. Achenes oblong-ovate to fusiform, 4–5-ribbed, the ribs roughened; the apex prolonged into a very slender beak, bearing the copious soft pale capillary pappus. After flowering, the inner involucre closing, and the slender beak elongating and raising up the pappus while the fruit is maturing; the whole involucre then reflexing, exposing to the wind the globular head of fruits. — Perennial or biennial herbs of cold or temp. reg. of N. Hemisph. and of more southern reg. of S.

Hemisph., with runcinate or pinnatifid to subentire leaves and yellow ligules; fruit largely produced parthenogenetically or apogamously. (Name a modification of the Arabic *Tharakhchakon.*)*

a. **Mature achenes mostly tuberculate nearly or quite to the base, the tubercles (at least of the upper half) closely crowded; surface of achene without broad and plane intermediate areas. . . *b.***

b. **Beak of achene only 2.5–5 mm. long; involucre blackish, 1–1.4 cm. high, with 10–12 inner phyllaries; leaves 2–8 cm. long, 3–8 mm. wide; scapes only 2–9 cm. long. . . . . 1. *T. phymatocarpum.***

b. **Beak 6–17 mm. long; involucre green, purplish or livid, 1.2–3 cm. high, with 10–25 inner phyllaries; leaves mostly larger; scapes mostly taller. . . *c.***

c. **Many phyllaries with horn-like appendage or knob below tip; outer phyllaries short and appressed or, if elongate and recurving, definitely broader than the inner.**

Inner phyllaries 1.2–2 cm. long, their tips unappendaged or with appendages mostly not exceeding phyllary-tips; outer calyculate phyllaries tightly appressed, short, firm, usually with conspicuous white border; beak of fruit 6–14 mm. long. . . . . . . 2. *T. ceratophorum.*

Inner phyllaries 2–2.6 cm. long, mostly with coarse appendages much longer than phyllary-tips; outer phyllaries two-thirds as long, finally loosely spreading or recurving, herbaceous, without broad margin; beak 13–17 mm. long. . . . . 3. *T. laurentianum.*

c. **Most or all phyllaries with plane unappendaged tips; outer phyllaries thin and herbaceous, when strongly recurving scarcely or but slightly broader than the inner.**

All or nearly all leaves with broad bases, their lobes and teeth mostly entire; phyllaries of inner involucre during anthesis separated to within 1 or 2 mm. of base, their herbaceous central band with white scarious margin 0.2–1 mm. wide at base; achenes pale brown or reddish, narrowly oblanceolate, 3.8–4.5 mm. long (excluding pyramid); beak 6–9 mm. long. . . . . 4. *T. ambigens.*

All or nearly all leaves slender-petioled, their lobes and teeth themselves sharply toothed; inner phyllaries during anthesis united 3–6 mm. at base, herbaceous central band with white scarious margin 1.3–1.5 mm. wide at base; achenes olivaceous, gray or fuscous, cuneate, 2.8–3.8 mm. long; beak 9–12 mm. long. . . . . 5. *T. latilobum.*

a. **Mature achenes tuberculate only above the middle or, if occasionally below, the tubercules remote and with broad plane intermediate areas or bands. . . *d.***

d. **All or at least many phyllaries corniculate-appendaged. . . *e.***

e. **Achenes grayish, drab, pale brown or olivaceous; pappus bright white; upper faces of ligules orange-yellow; leaves dentate, shallowly sinuate or deeply lobed. . . *f.***

f. **Leaves shallowly toothed or, if with teeth prolonged these mostly entire; pyramid of fruit stout, one-half as long to as long as thick.**

Outer phyllaries subscarious, stramineous to whitish-brown, prolonged-lanceolate; leaves broadly oblanceolate. . . . . . . 6. *T. dumetorum.*

Outer phyllaries herbaceous, fuscous, ovate or short-lanceolate; leaves linear-oblanceolate. . . . . 7. *T. lacerum.*

f. **Leaves deeply sinuate, the linear-caudate lobes lacerate at the deltoid base and with intermediate narrow lobes; pyramid slender, 3–4 times as long as broad. . . . . 8. *T. Longii.***

e. **Achenes red or reddish-purple; pappus creamy or sordid; upper face of ligule sulphur-yellow; leaves mostly dissected nearly or quite to midrib. 9. *T. erythrospermum.***

d. **All or nearly all phyllaries unappendaged.**

Leaves usually narrowed to slender scarcely winged petiolar bases, at least the lower and longer marginal lobes toothed and with frequent intermediate small teeth; achenes (excluding pyramid) 2–4 mm. long, drab or olivaceous. . . . . 10. *T. officinale.*

Leaves mostly broad at base or with broadly winged petiole; marginal lobes entire or with few teeth, intermediate teeth few or none; achenes 3.5–5 mm. long, olivaceous to fulvous. . . . . 11. *T. lapponicum.*

*** Treatment based on that in Rhodora, xxxv. 369–386 (1933).**

1. T. phymatocárpum Vahl (tumor-fruited; from the tuberculate achenes). — Dwarf; *scapes 2–9 cm. high; leaves* oblanceolate, entire, dentate or sinuate, 2–8 *cm. long,* 3–8 *mm. wide,* with slender petiolar bases; *involucre blackish,* 1–1.4 *cm. high, with* 10–12 *inner phyllaries* and short ovate to lanceolate outer ones; *achenes broadly cuneate-oblanceolate, strongly tuberculate* on the several ridges, 3–3.5 mm. long, with short broad pyramid, *fuscous or blackish; beak* 2.5–5 *mm. long,* about as long as the creamy pappus. — Greenl. and Arct. Am.; limestone-crevices, Pistolet Bay, Nfld. July. (Arct. Asia) Fig. 1769.

1769. T. phymatocarpum.

2. T. ceratóphorum (Ledeb.) DC. (horn-bearing; from the appendaged outer phyllaries). — Scapes 0.6–5 dm. high; *leaves* narrowly oblanceolate, 0.4–3 dm. long, 0.6–3 cm. broad, firm, shallowly dentate, or *with* horizontally divergent to slightly recurving mostly *entire deltoid long teeth; involucre* 1.2–2 *cm. high, the tips of the inner phyllaries unappendaged or with appendages rarely if ever much exceeding or overtopping the phyllary-tips; outer calyculate phyllaries tightly appressed* to and much *shorter than inner, firm, corniculate-appendaged, often with a conspicuous white scarious border; achenes* oblanceolate, *heavily pebbled-tuberculate to base; beak 6–14 mm. long;* pappus creamy. — Lab. to B.C., s., oftenest on calcareous meadows, ledges and cliffs, to w. Nfld., Gaspé and Rimouski Cos., Que., and Ung. — June–Aug. (Asia) Fig. 1770.

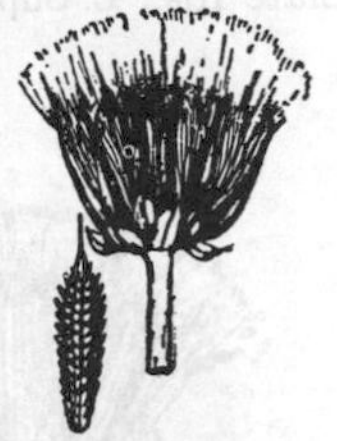
1770. T. ceratophorum.

3. T. laurentiànum Fern. (of the Gulf of St. Lawrence). — Coarse; scapes 1–3.5 dm. high; leaves membranaceous, oblanceolate, 1–3 dm. long, 2–4 cm. broad, mostly with dilated bases; margin doubly dentate, the longer teeth narrowly deltoid; *involucre* 2–2.6 *cm. high, the* 18–25 *inner phyllaries conspicuously corniculate-appendaged, the ovate horn much exceeding the bract-tip; outer phyllaries herbaceous, greenish or purplish,* broadly lanceolate to narrowly ovate, 1.5–2 *cm. long,* ascending, finally spreading or recurving, with plane or coarsely appendaged tips; *achenes* grayish-olivaceous, cuneate-oblong, 3.2–4 mm. long, obtusely *muricate below,* sharply so above, the pyramid 1.2–2 mm. long; *beak* 13–17 *mm. long;* pappus creamy. — Calcareous meadows, shores and ledges, w. Nfld.; and Mingan Ids. and Anticosti I., Que. July, Aug. Fig. 1771.

1771. T. laurentianum.

4. T. ámbigens Fern. (doubtful). — Slender scapes 0.4–6 dm. high; *leaves* membranaceous, oblanceolate, 0.3–3 dm. long, 1–3(–6) cm. broad, *with dilated bases,* the margin dentate or sinuate *with divergent mostly entire triangular teeth;* heads small; *involucres* 1.3–2 *cm. high, their* 10–20 *inner* narrowly lanceolate phyllaries mostly *with unappendaged tips, during anthesis separated to within* 1 *or* 2 *mm. of the base, their herbaceous central band with white scarious margin* 0.2–1 *mm. wide at base; outer phyllaries* linear-lanceolate, 1–1.5 cm. long, 1.5–3 mm. wide, recurved, *with unappendaged tips; achenes pale brown or reddish, narrowly oblanceolate,* pebbled to base, 3.8–4.5 *mm. long,* with stout pyramid 0.5–1 mm. long; *beak* 6–9 *mm. long;* pappus white. — Calcareous meadows, shores and ledges, se. Lab. to Ung., s. to Nfld. and Gaspé Pen., Que. June–early Aug. Fig. 1772.

Var. **fúltius** Fern. (supported). — Outer phyllaries ovate or ovate-lanceolate, 5–8 mm. long, 2.5–4 mm. wide, erect. — W. Nfld., and Shickshock Mts., Que.

1772. T. ambigens.

5. T. latílobum DC. (broad-lobed). — Coarse, up to 7.5 dm. high; *leaves broadly oblanceolate to narrowly obovate,* mostly 1.5–4.5 dm. long, 3.5–9 cm. broad, *with slender petiolar bases,* the upper half broadly dilated, the margin *with broad* obliquely deltoid and *sharply toothed divergent broad lobes; involucres* fuscous, 2–2.5 cm. high; *their inner phyllaries unappendaged, during anthesis united* 3–6 *mm. at base and with scarious margin* 1.3–1.5 *mm. wide at base;* outer phyllaries lanceolate, one-half to two-thirds length of inner, unappendaged, recurving; *achenes olivaceous to drab or fuscous, pebbled to base,* cuneate, 2.8–3.8 *mm. long;* pyramid slender, 1 mm. long; *beak* 9–12 *mm. long;* pappus milk-white. — Rich, often calcareous, slopes and talus, Nfld. to e. Me.; adv. to w. N.E. May–July. Fig. 1773.

6. T. dumetòrum Greene (of thickets). — Coarse, up to 4.5 dm. high; *leaves broadly oblanceolate, chiefly with dilated bases,* 1–2.5 dm. long, 2–4 cm. broad, *shallowly dentate or with* mostly *entire divergent falcate-lanceolate lobes; involucres pale,* 1.5–2.5 dm. high, the *phyllaries* mostly *with large ovate appendages; the outer* phyllaries *subscarious, stramineous to whitish-brown, prolonged-lanceolate; achenes drab or olivaceous,* cuneate-oblanceolate, 3.5 mm. long, *smooth or smoothish below* the muricate *summit; pyramid broad and short;* beak 7–11 mm. long. — Meadows and calcareous ledges, B.C. and Alta. to Colo., e. to Sask.; Slate Ids., L. Sup., Ont. June, early July. Fig. 1774.

1773. T. latilobum.

1774. T. dumetorum.

1775. T. lacerum.

7. T. lácerum Greene (lacerated). — Rather small; scapes up to 3 dm. high; *leaves linear-oblanceolate, tapering to petiolar bases,* 0.5–2 dm. long, 1–3 *cm. broad,* subentire, dentate or, *if sinuate or dissected, with mostly entire deltoid* divergent *lobes; involucre blackish or fuscous,* 1.5–2 dm. high; *phyllaries herbaceous,* many of them *corniculate-appendaged; the outer ovate or short-lanceolate,* closely appressed or barely spreading; *achenes* gray to olive or pale brown, *smooth* below the muriculate summit, oblong-oblanceolate, 3.5–4.5 *mm. long, with stout pyramid scarcely longer than broad; beak* 5.5–8.5 *mm. long;* pappus creamy. — Greenl. to Yuk., s. to nw. Nfld., Mingan Ids., Que., Ung. and Rocky Mts. of n. B.C. June, July. Fig. 1775.

8. T. Lóngii Fern. (for one of its discoverers, Bayard Long, 1885–). — Rather slender; scapes up to 2 dm. high; *leaves narrowly obovate-lanceolate or broad-oblanceolate, deeply laciniate,* 1–1.5 dm. long, 3–5 cm. broad, *the linear-caudate* divergent *lobes lacerate at their deltoid bases* and with shorter slender intermediate lobes; involucre 1.6–2.4 cm. high, purple-fuscous; *phyllaries* herbaceous, many of them *corniculate-appendaged;* the *outer ovate or oblong,* erect, finally spreading or recurving; *achenes* olive-gray to pale brown, oblanceolate, 3.5–4 mm. long, *smooth below* the muriculate *summit; pyramid subulate,* 1.2–1.5 *mm. long,* 3–4 *times as long as thick; beak* 1–1.3 *cm. long;* pappus milk-white. — Calcareous turf and gravel, nw. Nfld. and Gaspé Co., Que. Late June–early Aug. Fig. 1776.

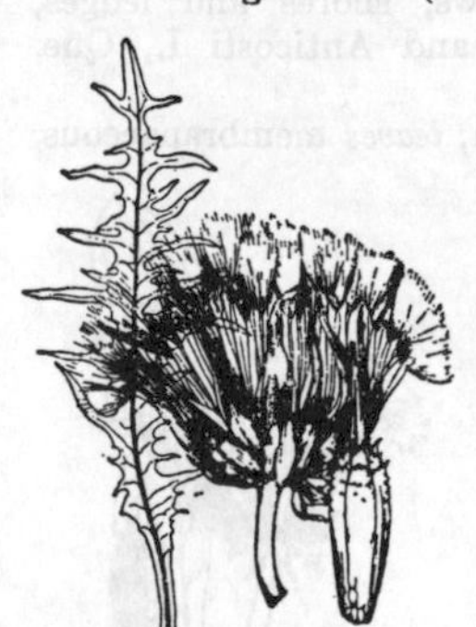
1776. T. Longii.

9. T. erythrospérmum Andrz. (red-seeded), Red-seeded D. — Slender; scapes 0.2–2 dm. high; *leaves* narrowly oblanceolate, *slender-petioled,* 0.5–2.5 dm. long, 1–5 cm. broad, *cleft nearly or quite to midrib into remote divergent to reflexed narrow lobes with numerous smaller intermediate short and slender lobes;* involucre 1.3–2.5 cm. high, its inner *phyllaries mostly corniculate-appendaged; the outer phyllaries short, lanceolate, spreading or ascending; ligules sulphur-yellow above; achenes slenderly oblanceolate, red or reddish-purple, smooth below,* sharply muriculate above, 2.8–3.5 mm. long, with subulate pyramid; beak 6–8 mm. long; *pappus sordid.* (*T. laevigatum* of many auth., not (Willd.) DC.) — Thin dry soil, s. Que. to s. B.C., s. to N.S., N.E., Va., Mo., Kans. and N.M. Apr.–July (Natzd. from Eu.) Fig. 1777.

1777. T. erythrospermum.

10. T. officinàle Weber (of the shops; from the fame of the plant in medicine), Common D. — *Leaves* usually *narrowed to slender* scarcely winged *petiolar bases,* coarsely and variously pinnatifid to sinuate-dentate or even subentire, *at least the lower and longer marginal lobes toothed and with frequent intermediate small teeth;* heads 2–5 cm. broad, orange-yellow; *phyllaries green to fuscous, all or nearly all unappendaged,* linear, the outer soon strongly recurving; *achenes* (excluding pyramid) 2–4 *mm. long, drab or olivaceous, tuberculate at summit,* the slender beak and whitish pappus elongate. — Nearly ubiquitous weed of lawns, grasslands and open grounds. March–Sept., and more or less through winter. (Natzd. from Eu.) — Polymorphous; by some Europeans split inordinately upon minor variations of foliage and involucres into hundreds of microspecies or apomicts, because parthenogenetically reproduced, but seemingly impossible to key on any usable basis.

1778. T. officinale, v. palustre.

Var. palústre (Sm.) Blytt (of marshes). — Outer phyllaries lanceolate to deltoid-ovate, often glaucous, ascending or but slightly spreading. — In damper soil, Nfld. and Que. to s. N.E. (Natzd. from Eu.)

Fig. 1778. — The exact name still in doubt.

11. T. lappónicum Kihlm. (of Lapland). — Rather slender; scapes up to 6 dm. high; *leaves* oblanceolate, mostly *with dilated bases,* 0.6–3 dm. long, 1–8 cm. broad, *the short marginal lobes entire or with few teeth, intermediate teeth few or none;* involucre 1.5–2 cm. high, with *thin herbaceous unappendaged phyllaries;* outer phyllaries shorter, ovate- to linear-lanceolate, spreading or finally recurving; *achenes olivaceous to fulvous,* 3.5–5 *mm. long, smooth below* the muriculate summit, with slender pyramid; beak 6–9 mm. long, pappus milk-white. — Meadows, damp ledges and shores, Greenl. to Alaska, s. to Nfld., Gaspé Pen., Que., Ung. and along Rocky Mts. June–Aug. (Eurasia) Fig. 1779.

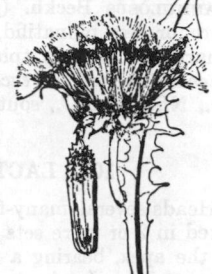

1779. T. lapponicum.

100. SÓNCHUS L. Sow-Thistle. Milk-Thistle. Laiteron (Que.)

Heads many-flowered, becoming tumid at base. Involucre more or less imbricated. Achenes obcompressed, ribbed or striate, not beaked; pappus copious, of very white exceedingly soft and fine bristles mainly falling together. — Leafy-stemmed coarse weeds, nat. of Old World, chiefly smooth and glaucous, with corymbed or umbellate heads of yellow flowers produced in summer and autumn. (The ancient Greek name.)

Perennials, with long creeping rootstock; flowers bright yellow to orange, in heads 4–5 cm. broad.
  Involucre broadly campanulate or hemispherical; the deep green to lead-colored phyllaries uniformly colored, heavily glandular to glabrous. . . . 1. *S. arvensis.*
  Involucre slenderly turbinate to cylindric; the pale phyllaries mostly with white margins, glabrous. . . . . 2. *S. uliginosus.*
Annuals, with tap-roots; flowers pale yellow, in smaller heads.
  Basal auricles of the leaves acute; achenes striate, transversely wrinkled. . . 3. *S. oleraceus.*
  Basal auricles rounded; achenes 3-nerved on each side, otherwise smooth. . 4. *S. asper.*

1. S. arvénsis L. (of cultivated ground), Field-S. — *Extensively creeping by subterranean rootstocks;* flowering stems 0.4–1.2 m. high, with many runcinate-pinnatifid to subentire spiny-toothed leaves clasping by a cordate base; *heads* few to very many, *on long glandular-hispid peduncles; involucres* 1.5–2.5 cm. high, *broadly campanulate or hemispheric, with dark green to lead-colored glandular phyllaries* in about 3 lengths; ligules long, the expanded heads 4–5 cm. broad; achenes transversely wrinkled on the ribs. — Fields, roadsides, gravelly shores and waste places, often a noxious weed, Nfld. to Alaska, s. to N.S., N.E., Del., Md., O., Ind., Ill., Mo., Id., Utah and Oreg. July–Oct. (Natzd. from Eu.)

Var. glabréscens Guenth., Grab. & Wimm. (smoothish). — Peduncles glabrous; the lead-colored involucres glabrous or nearly so. — Locally abund., e. Que. to Minn., s. to N.S., Ct. and Ind. (Natzd. from Eu.)

2. S. uliginòsus Bieb. (of marshes). — Similar, glabrous, paler; the *slenderly turbinate*

1780. S. oleraceus.

to cylindric pale involucre 1.2–2 cm. high, *its glabrous phyllaries white-margined.* — Locally abund., N.B. to Mich., s. to s. N.E., Del., Md., O., Ind. and Ill., and probably of wider range. July–Oct. (Natzd. from Eu.)

3. S. oleràceus L. (fit to be a vegetable), Common S. — *Annual; cauline-leaves* runcinate-pinnatifid, or undivided in forma integrifòlius (Wallr.) G. Beck (entire-leaved), *with* soft spiny teeth, clasping by a cordate base, *the basal auricles acute,* the terminal half much larger than the lateral lobes, or lobes and summit of leaf all narrow and subequal in forma lácerus (Willd.) G. Beck (torn); flowers pale yellow; *achenes striate, also wrinkled transversely.* — Waste places, cult. fields, etc., Nfld. to B.C., s. beyond our limits. July–Oct. (Natzd. from Eu.) Fig. 1780.

4. S. ásper (L.) Hill (rough), Spiny-leaved S., Chaudronnet (Que.). — Similar to no. 3; stem and rachis of inflorescence glabrous, or in forma glandulòsus Beckh. (glandular) with stipitate red glands; *cauline leaves* more or less pinnatifid, or undivided in forma inérmis (Bisch.) G. Beck (unarmed), strongly spiny-toothed, *the auricles* of the clasping base *rounded; achenes 3-nerved on each side, otherwise smooth.* — Waste places, roadsides, etc., Nfld. to B.C., south beyond our limits. June–Oct. (Natzd. from Eu.) Fig. 1781.

1781. S. asper.

101. LACTÙCA L. Lettuce. "Milkweed." Laitue (Que.)

Heads several–many-flowered. Involucre cylindrical, or in fruit conical; phyllaries imbricated in 2 or more sets of unequal lengths. Achenes contracted into a beak, which is dilated at the apex, bearing a copious and fugacious very soft capillary pappus, its bristles falling separately. — Leafy-stemmed herbs, with mostly panicled (rarely solitary) heads; flowers of variable color, produced in summer and autumn. Genus semicosmop. (The ancient name of the Lettuce, *L. sativa* L.; from *lac,* milk, in allusion to the milky juice.)

a. Involucre slenderly cylindric, of 3–5 elongate phyllaries and very short basal calyculate ones; flowers 5; achenes 5–7-ribbed on each face; leaves membranaceous, lyrate, the terminal segment very large. § Mycelis. . . . 1. *L. muralis.*

a. Involucre cylindric to urceolate, of several–many phyllaries of varying lengths; flowers 5–20 or more; leaves firmer. . . *b.*

b. Achenes very flat, with a distinct soft filiform beak; plants annual or biennial; flowers yellow, sometimes becoming purplish (blue only in no. 7). § Scariola. *c.*

c. Heads 5–8(–12)-flowered; achenes 5–7-ribbed on each face, oblanceolate to narrowly obovate.

Leaves copiously spinulose-denticulate, with midrib usually spinulose beneath, oblong to oblanceolate; panicle open, pyramidal. . . . 2. *L. Scariola.*

Leaves with smooth or only sparingly toothed margin, with smooth midrib, linear to oblong; panicle virgate. . . . . 3. *L. saligna.*

c. Heads 12–20-flowered; achenes 1–3-ribbed on each face, oblong, elliptic or oval. . . *d.*

d. Leaves not spinulose-margined nor strongly glaucous; larger outer phyllaries lanceolate.

Involucre 1–1.4 cm. long; mature achenes and beak 5–6 mm. long; pappus 5–7 mm. long; lower cauline leaves uncleft or with linear to lanceolate or oblong segments usually broadest at base. . . . 4. *L. canadensis.*

Involucre 1.5–2.2 cm. long; mature achenes (including beak) 7–10 mm. long; pappus 9–12 mm. long.

Pinnatifid leaves borne one-half to two-thirds up the stem, their lateral lobes oblong-obovate, not strongly recurving. . . . . 5. *L. hirsuta.*

Pinnatifid leaves basal (or wanting), their lateral lobes linear-lanceolate to oblanceolate and strongly falcate-recurving; cauline leaves numerous, much reduced, linear or linear-lanceolate, unlobed. . . . . 6. *L. graminifolia.*

d. Leaves spinulose-margined, strongly glaucous; larger outer phyllaries ovate; species of the prairie- and plains-region. . . . . . 7. *L. ludoviciana.*

b. Achenes flat or thickened, oblong, fusiform or falcate, tapering to a thick summit or stout or firm beak, 3-ribbed on each face; flowers blue, bluish or cream-color (sometimes yellow in no. 11). . . *e.*

e. Perennial with long rhizomes deep in the ground; flowers blue, the head in anthesis 2–3 cm. broad; involucre 1.4–1.8 cm. long; achenes thin, equilateral, with slender firm beak. § LACTUCASTRUM. . . . . . 8. *L. pulchella.*

c. Biennial with basal rosette and tap-root; flowers and head smaller; involucre shorter; achenes thick, usually curved or inequilateral, beakless or with a stout beak. § MULGEDIUM. . *f.*

f. Pappus white.

Achenes orange or orange-brown, not mottled, all tapering to a distinct beak. . . . . 9. *L. terrae-novae.*

Achenes gray-black or fuscous, mottled, all or at least the central beakless. . . . . 10. *L. floridana.*

f. Pappus sordid, olivaceous or tawny; achenes drab or fuscous, mottled, beakless or essentially so. . . . . 11. *L. biennis.*

§ MYCÈLIS (Cass.) DC.

1. L. MURÀLIS (L.) Gaertn. (growing on walls), WALL-L. — Slender, glabrous, 4–9 dm. high; *leaves* very thin and *membranaceous, runcinate-lyrate, with very large angulate cordate-based terminal segment;* panicle open; *involucre slenderly cylindrical,* 6–9 mm. long, *with* 3–5 *elongate phyllaries and very short calyculate ones; flowers* 5, bright yellow; *achenes* black or dark red, oblanceolate, about 0.5 mm. wide, 5–7-*ribbed on each face,* the short white beak with dilated summit. — Roadsides and waste places, local, w. Que. and e. N.Y. to Mich. July–Sept. (Adv. from Eu.) FIG. 1782.

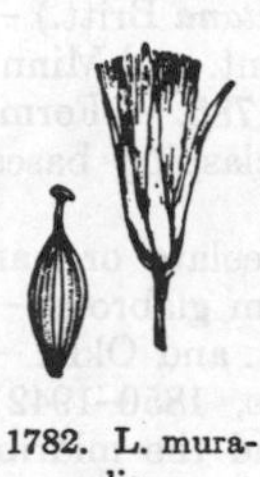
1782. L. muralis.

L. STOLONÍFERA (Gray) Maxim. (bearing stolons), *Ixeris stolonifera* Gray, a stoloniferous low species, with long-petioled small rounded to oval leaves and 1–few heads with few flowers, is a local garden-weed on L.I. and in e. Pa. (Adv. from e. Asia)

§ SCARÌOLA DC.

2. L. SCARÌOLA L. (Arabic name), PRICKLY L. — Annual or biennial; stem sparsely prickly-bristly below; *leaves* pinnatifid, or unlobed or only the lowest pinnatifid in forma INTEGRIFÒLIA (Bogenh.) G. Beck (entire-leaved) = var. *integrifolia* Bogenh. = var. *integrata* Gren. & Godr. = *L. integrata* (Gren. & Godr.) Nels. = *L. virosa* sensu Rydb., not L., *oblong or oblanceolate,* sagittate-clasping, *spinulose-denticulate* and *usually spinulose-setose on the midrib beneath,* tending to turn in a vertical position (*i.e.* with one edge up); *panicle loose, pyramidal,* with spreading branches; involucres slender, 1–1.3 cm. long, with phyllaries in 4 lengths, 5–12-flowered; flowers yellow, often drying purple or bluish; *achenes oblanceolate, with muriculate summit,* 5–7-*ribbed on each face,* with filiform beak and white pappus. (*L. Serriola* L., original form of name, later corrected by Linnaeus to the form here used.) — Roadsides and waste grounds, N.E. and w. Que. to B.C., s. beyond our range, more often as forma *integrifolia.* June–Oct. (Natzd. from Eu.) FIG. 1783.

1783. L. Scariola.

L. SATÌVA L. (of the garden), GARDEN-L., with many horticultural forms, is frequent on town-waste and garden-refuse, but hardly natzd. (Introd. from Eurasia)

3. L. SALÍGNA L. (like willow), WILLOW-LEAVED L. — More slender than no. 2; *leaves* oblong and runcinate-pinnatifid, or in forma RUPPIÀNA (Wallr.) G. Beck (for HEINRICH BERNHARD RUPPIUS, 1688–1719) lanceolate to linear and entire, *the margin and the prominent white midrib smooth or nearly so; inflorescence virgate or with erect virgate branches.* — Roadsides and waste places, becoming frequent, Pa. to Mich., s. to Va., Ky. and Mo. July–Oct. (Natzd. from Eu.) FIG. 1784.

1784. L. saligna.

4. L. canadénsis L. (Canadian). — Biennial with tap-root and basal rosette of leaves, 0.5–3.5 m. high, glaucous; *leaves* glabrous or rarely villous on the nerves beneath, *unlobed or sinuate-pinnatifid with linear to lanceolate or oblong segments usually broadest at base,* the uppermost usually lanceolate and entire; panicle elongate, open; *involucre green* or tinged with purple, urceolate, 1–1.4 *cm. long; flowers* 12–20, *yellow,* sometimes becoming purplish in age; *achenes* oblong or elliptic, rounded at both ends, blackish, cross-rugulose, 1–3-nerved on each face, *including the soft filiform pale beak* 5–6 *mm. long; pappus* white, 5–7 *mm. long.* — Our most variable species; the following variations are recognized, with full realization that they need further study in the field.

All or all but the lowermost leaves unlobed.
  Principal cauline leaves lanceolate to lance-ovate, the median entire or rarely toothed. . . . . *L. canadensis* (typical).
  Principal cauline leaves oblanceolate or narrowly obovate, usually denticulate. Var. *obovata*.
All or all but the uppermost leaves sinuate-pinnatifid, with definite lobes.
  Lobes linear-falcate, usually entire; unlobed upper leaves (when present) linear-lanceolate to linear. . . . . Var. *longifolia*.
  Lobes broadly falcate, or obovate and obliquely truncate, entire or toothed; upper leaves similar or, if unlobed, lanceolate or oblanceolate to narrowly obovate. . . . . Var. *latifolia*.

L. canadênsis, typical. — Principal leaves lanceolate to lance-ovate; the median entire or rarely toothed, their bases sagittate- or auriculate-clasping. (*L. integrifolia* Bigel.; *L. sagittifolia* Ell.; *L. canadensis*, var. *integrifolia* (Bigel.) Gray and var. *montana* Britt.) — Thickets, borders of woods and clearings, C.B. and P.E.I. to Ont. and Minn., s. to s. N.E., N.C., Ind., Ill., Mo. and Okla. July–Sept. Fig. 1785. — Forma **angustàta** Wieg. (narrowed), with leaves tapering to non-clasping bases, occasional.

1785. L. canadensis.

Var. **obovàta** Wieg. (obovate). — Principal leaves oblanceolate or narrowly obovate, usually denticulate, their bases sagittate; stem glabrous. — Similar habitats, w. Me. to Minn. and Neb., s. to s. N.E., Va. and Okla. — Forma **Steèlei** (Britt.) Fern. (for Edward Strieby Steele, 1850–1942), *L. Steelei* Britt., of D.C., Del. and e. Va., has the stems and the midribs beneath villous; forma **stenópoda** Wieg. (slender-stalked), with glabrous stems and leaves, the latter tapering to non-sagittate bases, scattered through the range.

Var. **longifòlia** (Michx.) Farw. (long-leaved). — Principal leaves sagittate-clasping, their linear-falcate lobes mostly entire. (Var. *typica* sensu Wieg., not the plant of L.) — Similar habitats, Que. to Sask., s. to N.S., N.E., N.C., O., Ind., Mo. and Colo.; B.C. to Calif. — Forma **angústipes** Wieg. (narrow-based), with leaves tapering to non-sagittate bases, occasional.

Var. **latifòlia** Ktze. (broad-leaved). — Principal leaves sagittate-clasping, their broadly falcate or obovate and obliquely truncate lobes entire or toothed. — Similar habitats, Que. and Ont., s. to N.S., N.E., n. Fla., Tenn., Mo., Okla. and Tex.; Wash. — Forma **villicaùlis** Fern. (villous-stemmed), with stems villous, and forma **exauriculàta** Wieg. (without basal lobes), with leaves narrowed to non-sagittate bases, occasional. — Plants of var. *latifolia* with spinulose-hirsute under sides of midribs of lower leaves have been separated as var. *spinulosa* Jennings.

× **L. Mórssii** Robins. (for its discoverer, Charles Henry Morss, 1857–1930), evidently a hybrid of nos. 4 and 11, has the habit, hirsute lower internodes, foliage and blue flowers of the latter, but the achenes intermediate in form or resembling either parent but with a short filiform beak and cream-colored pappus. — Rich thickets and low grounds, Me. to Mich., s. to Va. and Ill.

5. L. hirsùta Muhl. (hirsute). — *Purplish* (or greenish in shade); lower internodes of stem densely villous; *leaves chiefly on lower half or two-thirds of stem*, much reduced upward; the lower and median copiously pilose on both surfaces; *their lateral lobes oblong-obovate, commonly broadest above the base*, often more or less truncate, usually dentate; panicle cylindric and racemiform to broadly corymbiform; *involucre purple* (rarely green), when fully developed slenderly urceolate, 1.6–2.2 *cm. long;* flowers yellow or coppery, sometimes drying purplish; *achenes* as in no. 4 but (*including beak*) 7–9 *mm. long, the body and beak subequal; pappus* 9–12 *mm. long.* — Very local and little known, Pa. to Va. and La. July–Oct.

1786. L. hirsuta, v. sanguinea.

Var. **sanguínea** (Bigel.) Fern. (blood-red; from color of stem). — Stem glabrous or essentially so, 0.4–1.2 m. high, with panicle mostly corymbiform; leaves as above, or oblong or rounded and undivided in forma **indivisa** Fern. (undivided), glabrous or essentially so, except for the midribs villous beneath, or midribs glabrous in forma **calvéscens** Fern. (becoming bald). — Dry open woods and clearings, locally abundant, P.E.I. to s. Ont., s. to N.S., N.E., Va., La. and Tex. Late June–Sept. Fig. 1786.

6. L. graminifòlia Michx. (grass-leaved). — Related to no. 5; stems up to 1.5 m. high, *glabrous; leaves crowded at base, linear-attenuate and entire or with few falcate-recurving linear-lanceolate to oblanceolate lobes; cauline leaves crowded, linear-attenuate to linear-lanceolate, much reduced;* heads much as in no. 5; *bodies of achenes often longer than beak.* — Sandy woods and clearings, Fla. to Ariz., n. to se. N.C.; pine-barrens of N.J. July–Sept.

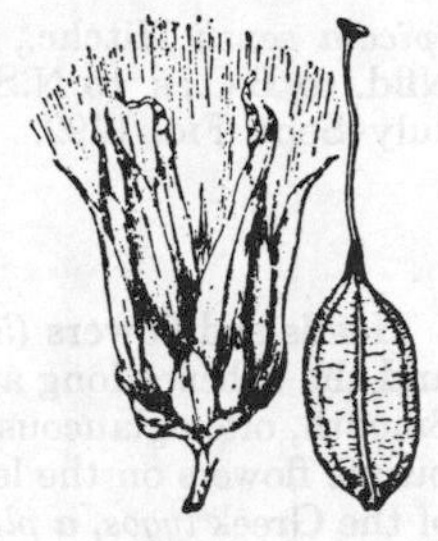
1787. L. ludoviciana.

7. L. ludoviciàna (Nutt.) Riddell (of Louisiana). — Stout, 0.6–2 m. high, rather *uniformly leafy* to summit; *leaves firm, strongly glaucous,* oblong or obovate, sinuate or sinuate-pinnatifid, *spinulose-dentate;* lower bracts of panicle leafy; *involucre* much imbricated, 1.7–2.3 cm. long, *its larger outer phyllaries ovate;* flowers yellow, sometimes drying purplish, or blue or purple from the first in forma **campéstris** (Greene) Fern. (of the plains); achenes much as in no. 4, including beak 7 mm. long; pappus bright white, 7–8 mm. long. — Prairies, low grounds and roadsides, Pacific slope, e. to Man., Wisc., Ill., w. Mo., Ark. and Tex. June–Aug. Fig. 1787.

§ Lactucástrum Gray

8. L. pulchélla (Pursh) DC. (beautiful), Blue L. — Pale or glaucous *perennial with long subterranean rhizome;* stems stoutish, 0.2–1 m. high; *leaves* sessile, glabrous, *oblong- or linear-lanceolate, entire,* or the lower runcinate-pinnatifid; *heads* few and large, *with blue or blue-purple flowers, in anthesis* 2–3 *cm. broad; involucre 1.4–1.8 cm. long,* with phyllaries 3–4-ranked, on scaly-bracted peduncles; *achenes thin, flat,* lance-oblong, marginless, *with slender but short firm beak.* — Prairies, river-banks and openings, James Bay reg., Que., to Alaska, s. to Mich., Wisc., Mo., Okla., N.M., Ariz. and Calif.; rarely adv. e. to N.E. July–Sept. Fig. 1788.

1788. L. pulchella.

1789. L. terrae-novae.

§ Mulgèdium (Cass.) Gray

9. L. térrae-nòvae Fern. (of Newfoundland). — In habit, foliage and involucres closely resembling no. 11; leaf-tips caudate; involucres urceolate-campanulate, 1.2–1.5 cm. high; *corollas bluish; pappus bright white; achenes orange or orange-brown, not mottled, obliquely lance-oblong,* the bodies 5–6 mm. long, *all attenuate to a firm beak 1 mm. long.* — Calcareous subalpine meadows, Highlands of St. John, nw. Nfld. Aug., Sept. Fig. 1789.

10. L. floridàna (L.) Gaertn. (of Florida). — Tall biennial, 0.7–2 m. high, with a large loose panicle of small heads on divergent peduncles; *leaves lyrate or runcinate-pinnatifid,* sometimes unlobed; *flowers bluish,* or white in forma **leucántha** Fern. (white-flowered); *pappus bright white; achenes* thickish, lance-ovate or oblong, *gray-black or fuscous, mottled, the outer often distinctly thick-beaked, the inner beakless.* (*Mulgedium* DC.) — Rich woods, thickets and openings, Fla. to Tex., n. to se. Mass., s. N.Y., O., Ind., Ill., Minn. and Neb. Aug.–Oct. Fig. 1790.

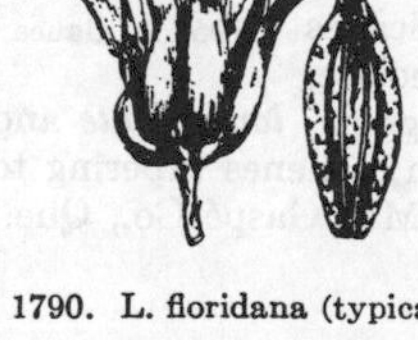
1790. L. floridana (typical).

Var. **villòsa** (Jacq.) Cronq. (soft-hairy). — Leaves ovate to oblong-lanceolate, long-pointed, uncleft; *achenes all or nearly all beakless.* (*L. villosa* Jacq.; *Mulgedium villosum* Small) — Open woods and thickets, Fla. to La., n. to N.Y., O., Ind., Ill., Ia. and Neb. Aug.–Oct. Fig. 1791.

1791. L. floridana, v. villosa.

11. L. biénnis (Moench) Fern. (biennial). — Nearly smooth biennial, coarse, 0.5–5 m. high, very leafy; leaves irregularly pinnatifid, sometimes runcinate, coarsely toothed, the upper cauline sessile and auriculate; heads in large and dense compound panicle; *flowers pale bluish to dirty blue or creamy,* or yellow and with the leaves pinnatifid in forma **aùrea** (Jennings) Fern. (golden), or yellow or dirty white and with the leaves unlobed in forma **integrifòlia** (T. & G.) Fern. (entire-leaved); *achenes* oblong or oblong-lanceolate, *thickish, drab or fuscous, mottled, beakless or essentially so; pappus sordid to olivaceous or tawny.* (*L.*

1792. L. biennis.

spicata sensu Hitchc., not *Sonchus spicatus* Lam.) — Rich or damp thickets and openings, Nfld. to B.C., s. to N.S., N.E., Del., Md., mts. of N.C. and Tenn. and to Ia., Colo. and Oreg. July–Sept. Fig. 1792.

102. LYGODÉSMIA D. Don

Heads and flowers (5–10) nearly as in *Prenanthes*, the cylindrical involucre more elongated, and the achenes long and slender, tapering at the summit; pappus whitish. — Smooth, often glaucous N. Am. herbs, with terminal or scattered heads of rose-purple flowers on the leafless or rush-like stems or branches. (Name composed of the Greek *lygos, a pliant twig,* and *desme, a bundle,* from the fascicled twiggy or rush-like stems.)

1793. L. juncea.

1. L. júncea (Pursh) D. Don (stiff, like a rush). — Perennial, 2–4 dm. high, tufted, rigid, much branched; *lower leaves* lance-linear, 2.5–5 *cm. long, rigid, the upper subulate and minute; heads* 5-flowered, *terminal; achene filiform, only slightly attenuate above,* 4–5 *mm. long.* — Prairies and plains, Wisc. to Alta., s. to Mo., Okla., Tex., N.M. and Ariz. June–Aug. Fig. 1793.

103. AGÓSERIS Raf.

Head large, solitary, many-flowered. Phyllaries of the campanulate involucre ovate or lanceolate, pointed, loosely imbricated in 2 or 3 rows. Achenes smooth, 10-ribbed, with or without a distinct beak; pappus usually longer than the achene, white, of copious and unequal rigid capillary bristles. — Perennial scapose Am. herbs, with elongated tufted basal leaves, and yellow flowers. (Name from the Greek *aix, goat,* and *seris, chicory.*) Troximon Nutt.

Achenes beakless. . . . . 1. *A. cuspidata.*
Achenes tapering to a definite firm beak.
  Scapes glabrous or essentially so; outer phyllaries not strongly ciliate. . . . 2. *A. glauca.*
  Scapes permanently villous at summit; outer phyllaries long-ciliate. . . . 3. *A. gaspensis.*

1. A. cuspidàta (Pursh) Raf. (with a cusp). — Scape 3 dm. high, from a thickened caudex; leaves lanceolate, elongated, tapering to a sharp point, entire, woolly on the margins; phyllaries lanceolate, sharp-pointed; *achenes beakless.* (*Troximon* Pursh; *Nothocalais* Greene) — Prairies and stony hills, Wisc. to Mont., s. to n. Ill., Mo., Okla. and N.M. Apr., May. Fig. 1794.

1794. A. cuspidata.

2. A. glaùca (Pursh) Raf. (blue-green). — *Scape* 3–6 dm. high, *glabrous or nearly so;* leaves linear to lanceolate, dentate or laciniate; *involucre glabrous or villous, its outer phyllaries scarcely ciliate; the inner broadly lanceolate, with a subcuspidate tip;* achenes long-beaked. (*Troximon* Pursh) — Prairies and meadows, w. Ont. and Minn. to B.C., s. to S.D., Colo. and Utah. May–July. Fig. 1795.

1795. A. glauca.

3. A. gaspénsis Fern. (of Gaspé Pen., Que.). — Scapes permanently white-villous at summit, 2–4 dm. high; leaves linear-lanceolate, entire or remotely pinnate-incised; *phyllaries lance-attenuate, the outer* green, *long-ciliate* and villous, *the inner scarious and purple-spotted;* corollas bronze to purplish; achenes tapering to a firm beak. — Rocky subalpine meadows, Tabletop Mt., Gaspé Co., Que.; mts. of n. B.C. July–Sept.

104. PYRRHOPÁPPUS DC. False Dandelion

1796. P. carolinianus.

Heads, etc., nearly as in *Taraxacum,* but the soft pappus reddish or rust-color, and surrounded at base by a soft-villous ring. — Mostly annual or biennial (sometimes perennial) Am. herbs, scapose or often branching and leafy below. Heads solitary or terminating the branches. Flowers deep yellow. (Name composed of the Greek *pyrros, flame-colored,* and *pappos, pappus.*) Sitilias Raf.

1. P. caroliniànus (Walt.) DC. (of Carolina). — *Annual or biennial, branching,* 1–9 dm. high; leaves oblong or lanceolate, entire, cut or pinnatifid, the cauline leaves partly clasping. (*Sitilias* Raf.) — Dry fields and clearings, Fla. to Tex., n. to Del., s. Ind., Mo. and Kans. June–Oct. Fig. 1796.

105. CRÈPIS L. Hawk's-beard

Involucre few–many-flowered, commonly of a single row of equal phyllaries which often become thickened at base. Pappus copious, white, soft. — Annuals, biennials or perennials, not pilose. Otherwise as *Hieracium;* genus semicosmop. (The Greek name of some plant, from *crepis, a boot.*)

Depressed tufted perennial, forming small mats; leaf-blades 0.5–1.5 cm. long, long-petioled; heads crowded among the chiefly basal leaves, the slender involucre glabrous. . . . . 1. *C. nana.*
Ascending, with simple or forking stems many times longer than basal leaves; leaves larger, short-petioled to sessile; heads terminating the branches.
Scapose or subscapose perennial, with basal rosette of obovate- to spatulate-oblong leaves; involucre 1–1.3 cm. high, its phyllaries scarcely thickened after flowering. . . . . 2. *C. runcinata.*
Leafy-stemmed to subscapose annuals or biennials, with lanceolate or oblanceolate laciniate- or runcinate-pinnatifid to dentate leaves; phyllaries finally becoming thickened and firm at base.
Involucre pubescent, 4–7 mm. high.
Cauline leaves sagittate-auriculate at base; involucres 6–7 mm. high. . . 3. *C. capillaris.*
Cauline leaves tapering to sessile or petioled bases; involucres 4–5 mm. high. . . . . 4. *C. japonica.*
Involucre glabrous, 9–11 mm. high. . . . . . . 5. *C. pulchra.*

1. C. nàna Richards. (dwarf). — *Depressed tufted perennial,* forming rounded mats 3–10 cm. across and 2–6 cm. high; *leaves chiefly basal, slender-petioled,* with elliptic to rounded entire to runcinate subfleshy *blades* 0.5–1.5 *cm. long; heads crowded among the leaves* or on very short branches; involucre slenderly cylindric, 7–10 mm. long, of few purplish-green or livid linear glabrous obtuse phyllaries; *achenes* linear-cylindric, 10-striate, *with a terminal disk* beneath the pappus. (*Youngia* Rydb.) — Arct. Am., s., very locally, to calcareous gravels of nw. Nfld. and mts. of Mont.; related plants further s. and w. July, Aug.

2. C. runcinàta (James) T. & G. (sharply incised). — *Perennial; scape* 3–9 m. high, *glabrous* or somewhat glandular-hispid; *leaves rosulate, obovate-oblong or oblong-spatulate,* repand or runcinate-dentate, glabrous or slightly hispidulous; heads loosely corymbose, 2 cm. broad; *involucre* hirsute, 1 *cm. or so high.* — On saline soil, Man. to Ia. and westw. June, July.

3. C. capillàris (L.) Wallr. (hair-like). — *Annual or biennial,* ascending, 3–6 dm. high, often branching below; *leaves lance-spatulate, laciniate-pinnatifid or merely dentate, the cauline with sagittate-auriculate bases;* heads small, *involucre* 6–7 *mm. high; achenes smooth,* 10*-nerved, slightly narrowed at both ends.* (*C. virens* L.) — Fields and waste places, becoming frequent. (Adv. from Eu.)

4. C. japónica (L.) Benth. (Japanese). — Similar to no. 3, 0.2–1 m. high, scapose or few-leaved above, smooth; *rosette-leaves* coarsely runcinate-pinnatifid, *broadly oblanceolate, cauline reduced and tapering or somewhat slender-petioled at base; involucres* 4–5 *mm. high,* with minute basal phyllaries. — Fields, locally abundant, Pa. to Va. Apr.–June (Adv. from Asia)

5. C. púlchra L. (handsome). — Coarse biennial with habit of no. 3; stem glandular and puberulent below, glabrous above; leaves pubescent; flowering stem corymbosely branched; *involucre glabrous,* 9–11 *mm. high;* achenes straight, scarcely narrowed at ends. — Roadsides and wooded slopes, local, Va. and Ind. May–July. (Natzd. from Eu.)

Several European annuals or biennials, casually introduced in impure seed (alfalfa, clover and hay), appear sporadically but do not long persist. These include C. tectòrum L. (of roofs), with revolute cauline leaves, glandular-pilose involucres 6–8 mm. long and fusiform achenes scabrous on the ribs; C. biénnis L. (biennial), coarse and scabrous, with involucres 1–1.3 cm. long, their phyllaries pubescent within, and with 13-ribbed beakless achenes 4–5 mm. long; C. nicaeénsis Balb. (of Nicaea in Asia Minor), simple or subsimple, with involucres 7–9 mm. long, their phyllaries glabrous within, and the 10-ribbed achenes 3–3.5 mm. long; and C. setòsa Haller f. (bristly), long-hirsute, with conspicuously bristly involucre.

106. PRENÁNTHES L. Rattlesnake-root

Heads 5–35-flowered. Involucre cylindrical, of 5–14 linear phyllaries in a single row, and a few small bractlets at base. Achenes short, linear-oblong, striate or grooved, not contracted at the apex. Pappus of copious whitish, stramineous or brownish rough capillary bristles. — Perennial N. Am. and Eurasian herbs, with upright leafy stems arising from fusiform to rounded tuberous roots (intensely bitter), very variable leaves, and mostly racemose-panicled nodding heads. Flowers in late summer and autumn. (Name from the Greek *prenes, drooping,* and *anthe, flower.*) NABALUS Cass.

a.Heads 20–35-flowered, in a corymbose panicle; cauline leaves petioled. . . . 1. *P. crepidinea.*
a.Heads 5–18-flowered, racemose, paniculate or thyrsoid. . . . *b.*
b.Cauline leaves sessile, the basal tapering to winged petioles; heads in a racemose panicle or thyrse; pappus creamy or straw-color. . . . *c.*
c.Inflorescence pubescent; heads ascending, 12–15-flowered.
Stem and leaves glabrous; upper cauline leaves partly clasping; flowers pink. 2. *P. racemosa.*
Stem and leaves pubescent; upper leaves not clasping; flowers creamy. 3. *P. aspera.*
c.Inflorescence glabrous; heads nodding, 8–12-flowered.
Upper leaves partly clasping, lanceolate to ovate, extending without strong reduction into the interrupted thyrse. . . . . 5. × *P. mainensis.*
Upper leaves not clasping, reduced to linear-lanceolate to subulate small bracts below the slender spiciform raceme. . . . . 4. *P. autumnalis.*
b.Cauline leaves (at least the lower) petioled; the basal slender-petioled, with cordate, truncate or hastate bases; heads racemose to corymbose-paniculate, drooping or spreading. . . . *d.*
d.Involucre slender-cylindrical (at least below the middle); phyllaries scarious-margined; the outer numerous, short, appressed. . . . *e.*
e.Involucre with 6–8 long primary phyllaries, 8–12-flowered.
Pappus deep reddish-brown. . . . . 6. *P. alba.*
Pappus whitish or brownish-white.
Primary (longer) phyllaries scarcely as long as pappus, the small outer ones lanceolate. . . . . 7. *P. Serpentaria.*
Primary phyllaries as long as pappus, the outer ones deltoid to ovate. 8. *P. trifoliolata.*
e.Involucre with 5 primary phyllaries, 5–6-flowered. . . . . 9. *P. altissima.*
d.Involucre slender-campanulate; phyllaries scarcely scarious-margined, the 2 or 3 outer ones linear and loose. . . . . 10. *P. Boottii.*

1. P. crepidínea Michx. (like *Crepis*). — Somewhat smooth; stem stout, 1.5–2.7 m. high, bearing numerous nodding 25–35-*flowered heads* in loose clusters *in a corymbose panicle; leaves* large (1.5–3 dm. long), broadly triangular-ovate or hastate, strongly toothed, *contracted into winged petioles;* flowers cream-color; pappus brown. (*Nabalus* DC.) — Rich thickets and openings, w. N.Y. to Minn., s. to Tenn. and Mo. Aug.–Oct. FIG. 1797.

1797. P. crepidinea.

2. P. racemòsa Michx. (racemose). — *Stem* 0.2–1.5 m. high, *smooth and glaucous, as well as the oval* or oblong-lanceolate denticulate *leaves;* the lower leaves tapering into winged petioles (rarely cut-pinnatifid), *the upper partly clasping; heads ascending,* in crowded clusters forming a dense or interrupted leafy thyrse; *involucres hirsute,* 12–15-*flowered; flowers pink,* or white in forma **Rollándii** Vict. & Rousseau (for LOUIS ROLAND, Frère ROLLAND-GERMAIN, 1881–); *pappus yellowish.* (*Nabalus* Hook.) — Calcareous river-banks, shores and damp prairies, Saguenay Co., Que., to James Bay, w. to Alta., s. to C.B., N.B., n. Me., n. Vt., O., Ill., Mo., S.D. and Colo.; se. N.Y. and n. N.J. Aug., Sept. FIG. 1798.

1798. P. racemosa.

3. P. áspera Michx. (harsh). — Resembling no. 2; *stem* 0.5–1.3 m. high, *rough-pubescent, as are the* oval-oblong or broadly lanceolate toothed *leaves; upper leaves not clasping;* heads in small clusters; flowers larger, cream-color. (*Nabalus* T. & G.) — Dry prairies and barrens, O. to Minn. and S.D., s. to Tenn., La. and Okla. Aug., Sept. FIG. 1799.

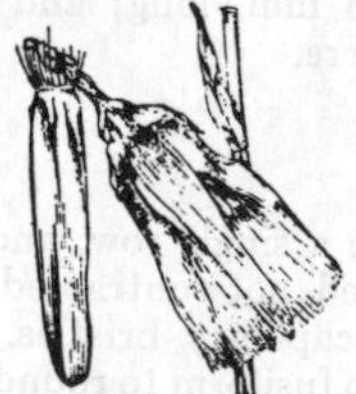
1800. P. autumnalis.

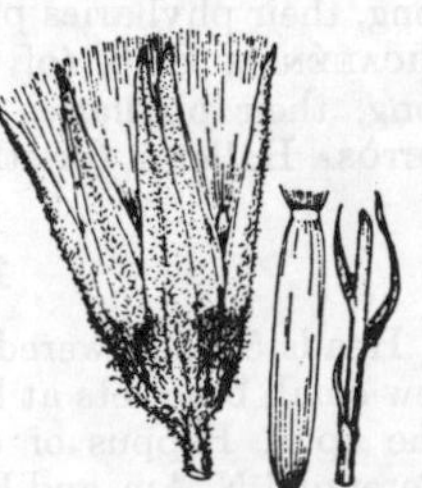
1799. P. aspera.

4. P. autumnàlis Walt. (of autumn), SLENDER R. — Slightly glaucous, *glabrous;* stem 6–12 dm. high, prolonged into a *naked* *and slender spiciform raceme* (4–6 dm. long), rarely with several erect branches; *heads nodding*, clustered and mostly unilateral; *leaves lanceolate*, acute, *closely sessile, the upper reduced to bracts*, the lower toothed or pinnatifid; involucre (purplish) of about 8 phyllaries, with about same number of whitish or pale pink flowers; pappus sordid. (*P. virgata* Michx.; *Nabalus virgatus* DC.) — Sandy pinelands and peats, Fla. to Miss., n. to s. N.J. Sept., Oct. Fig. 1800. — Hybridizes with no. 7.

5. × P. mainénsis Gray (of Maine). — Stem 0.5–1 m. high, leafy; *leaves* as in no. 2, but *the radical ovate* and more *abruptly narrowed to the short petiole; heads* persistently drooping on slender pedicels, *glabrous.* — Local, n. N.B., n. Me. and e. Que. Aug., Sept. — Apparently a hybrid of nos. 2 and 8. Fig. 1801.

1801. P. mainensis.

6. P. álba L. (white), White Lettuce, Rattlesnake-root. — Smooth and glaucous; stem stout, usually purplish, 0.5–1.5 m. high, corymbose-panicled at summit; leaves angulate or triangular-hastate, sinuate-toothed or 3–5-cleft, mostly petioled, the uppermost oblong and undivided, petioled; involucres glaucous, whitish-green and purplish, cylindric, with 6–8 scarious-margined primary phyllaries, 8–12-flowered; flowers whitish; *pappus reddish-brown.* (*Nabalus* Hook.) — Rich woods and thickets, s. Ung. and w. Que. to Sask., s. to N.E., Ga., Tenn., Mo. and S.D. Fig. 1802. — Basal leaves from simple and cordate or hastate to deeply cleft; these foliage-forms wrongly treated by some as vars. or even as species.

1802. P. alba.

7. P. Serpentària Pursh (old name for various plants used to cure snake-bite), Lion's-foot, Gall-of-the-earth. — Stem smooth, 3–12 dm. high, usually purplish, *corymbose-panicled at summit;* leaves thick, ovate, variously lobed, often pinnatifid with blunt or rounded lobes, or in forma **simplicifòlia** Fern. (simple-leaved) oblong or elliptic-oblong and unlobed, the lower on margined petioles; *heads chiefly clustered at the tips of elongate branches; involucre funnelform, cylindric below,* its subherbaceous green or purplish frequently *setulose phyllaries abruptly spreading above the middle, scarcely as long as pappus, the short outer phyllaries lanceolate;* flowers pink, cream-color or whitish; *pappus creamy.* (*Nabalus* Hook.) — Dry barrens, thickets and open woods, Fla. to Miss., n. to Mass., N.Y. and O. Sept., Oct. Fig. 1803. — Basal leaves highly variable; see note under no. 6.

1803. P. Serpentaria.

8. P. trifoliolàta (Cass.) Fern. (with 3 leaflets), Gall-of-the-earth. — Glabrous, 1.5–15 dm. high; *leaves nearly all petioled;* the lower mostly 3-divided or angulate, occasionally uncleft or with the divisions finely dissected; inflorescence an elongate to corymbiform panicle, the heads clustered at the tips of comparatively short ascending branches or in the upper axils; *involucres cylindric,* glaucous; *their* 6–8 *pale green to purple-tinged primary phyllaries linear-lanceolate,* acutish, *as long as the* white to creamy *pappus,* the inner with broad scarious margins; *outer calyculate phyllaries lance-deltoid,* rather firm, with pale hyaline margins, *regularly imbricated, the longest* 1.5–2.5 *mm. long.* (*Nabalus* Cass.) — Dry slopes, clearings and thickets, Nfld. to O., s. to N.S., N.E., N.C. and Tenn. Aug., Sept. — Lower leaves polymorphic; see note under no. 6. Fig. 1804.

1804. P. trifoliolata.

Var. **nàna** (Bigel.) Fern. (dwarf). — Lower (0.5–7.5 dm. high); *involucres thick-cylindric, with lead-colored or blackish phyllaries; the primary ones linear- to lance-oblong,* with narrow scarious margins; *outer calyculate ones ovate to ovate-lanceolate, blackish-green,* herbaceous or fleshy, *very unequal, the longest* 3–6 *mm. long.* (*P. nana* (Bigel.) Torr.; *Nabalus nanus* DC.) — Rocky or mossy exposed places, s. Lab. and Nfld. to e. Me. and higher mts. of n. N.E. and n. N.Y. July–Sept.

9. P. altíssima L. (very tall). — Similar to nos. 7 and 8; slender, 0.5–2 m. high, glabrous, or villous-stemmed with leaves hispidulous beneath in forma **hispídula** Fern. (with short stiff hairs); heads in small axillary and terminal loose clusters forming a long and virgate leafy panicle; leaves membranaceous, all petioled, ovate, cordate or triangular, and merely toothed or cleft, with naked or winged petioles, or frequently 3–5-parted, with the divisions entire

1805. P. altissima.

or again cleft; *involucre very slender*, greenish, *with* 5 *primary phyllaries;* flowers 5 or 6, greenish-white; pappus creamy-white. (*Nabalus* Hook.) — Moist woods, Que. to Man., s. to N.S., N.E., Ga. and Tenn. July–Oct. Fig. 1805. — Basal leaves highly variable; see note under no. 6; the type has trifoliolate basal leaves. Plants with uncleft basal leaves are forma **íntegra** Rousseau (entire).

Var. **cinnamòmea** Fern. (cinnamon-colored). — Pappus cinnamon-color or deep brown. — Ind. to Mo. and La.

1806. P. Boottii.

10. **P. Boòttii** (DC.) Gray (named for its discoverer, John Wright Boott, 1788–1845). — Simple *dwarf*, 1–3 dm. high, pubescent at the summit; the *heads in an almost simple raceme;* lowest leaves hastate or cordate, the middle oblong, the upper lanceolate, nearly entire, tapering into a margined petiole; *involucre* livid, 10–18-*flowered;* the *primary phyllaries* 10–15, very obtuse; pappus pale straw-color. — Alpine regions, n. N.E. and n. N.Y. July–Sept. Fig. 1806.

107. HIERÀCIUM L. Hawkweed. Épervière (Que.)

Heads 12–many-flowered. Involucre scarcely to strongly imbricated or at least calyculate, often glandular. Achenes slender, columnar or with summit tapering but not strongly beaked, 10-striate; pappus of a single equal series of long or of unequal sordid to tawny fragile capillary rough bristles. — Usually pubescent and often glandular perennials of temp. and cold reg., with simple entire to lacerate leaves and single, panicled or corymbed heads of mostly yellow (sometimes scarlet) flowers. (Name from the Greek *hierax, a hawk;* the Ancients, as recorded by Pliny and others, supposing that hawks used the plant to strengthen their eyesight.)*

a. Stem soft, scapose or nearly so, arising from basal rosette of numerous spatulate, oblanceolate or lanceolate entire or merely denticulate leaves 0.5–3.5 cm. broad; inflorescence a corymbiform cyme or of 1–few heads; involucre of nearly equal phyllaries, not strongly imbricate nor calyculate; weeds of European origin. § Pilosella. . . *b*.

b. Scape 0.4–4.5 dm. high, with 1 head or 2–4 long-peduncled heads; base freely stoloniferous.

Leaves (especially the new ones) pale beneath with close minute tomentum; mature blades 1.5–8 cm. long and 5–15 mm. broad; scape filiform, 0.4–2.5 dm. high, with 1 head; involucre 8–10(–12) mm. high. . . 1. *H. Pilosella.*

Leaves green beneath, mature blades 4–15 cm. long and 0.8–3 cm. broad; scapes stoutish, 1.2–4.5 dm. high, often forking, with (1–)2–4 heads; involucre 1–1.3 cm. high. . . . . 2. *H. flagellare.*

b. Scape (1–)2–10 dm. high, with corymbs of 7–very many heads; base stoloniferous or non-stoloniferous. . *c*.

c. Rhizome slender and cord-like, prolonged; plant at or after anthesis developing prostrate and elongate soon rooting basal stolons. . . *d*.

d. Flowers orange-red. . . . . 3. *H. aurantiacum.*

d. Flowers yellow.

Glaucous, freely and early stoloniferous; upper surface of basal leaves glabrous or nearly so. . . . . 4. *H. floribundum.*

Green, more tardily stoloniferous; upper surface of leaves abundantly setose. . . . . 5. *H. pratense.*

c. Rhizome short and thick, praemorse; prostrate repent stolons wanting (loosely ascending or divergent non-repent basal offshoots in no. 6).

Sterile ascending or spreading slender elongate branches arising from among basal leaves; the latter finely stellate-pubescent beneath. . 6. *H. praealtum,* var. *decipiens.*

* The genus, especially in Europe, broken by technical specialists, with eyesight stimulated beyond that of the ancient hawks, into thousands of so-called species, subspecies, varieties and forms, largely on degree and character of pubescence. These *apomicts* will be found unlucidly elaborated beyond practical utility in the 1700 pages of Zahn (in Engler, Das Pflanzenreich iv[280] (1921–1923)), who in our no. 1 alone recognizes but does not define 624 "subspecies". Until a rational presentation of European species is available the identification of our adventives from Europe must be tentative.

Sterile branches not developed; leaves glabrous or nearly so. . . . 7. *H. florentinum.*

*a.*Stem firm, leafy or, if scapose, with the inflorescence an open panicle or loose paniculate corymb; involucre clearly imbricated, with phyllaries of different lengths, or with short calyculate bracts at base. . . *e.*

*e.*Involucre of more or less imbricated phyllaries; pappus of numerous bristles of unequal lengths; heads large, 2.5–4.5 cm. broad. § ARCHIERACIUM. . . *f.*

*f.*Stem scapose or more or less scapiform, with long-petioled often rosulate basal leaves frequently mottled with purple, the remote cauline ones 0–7(–12), the lower similar to the large basal ones; exposed ligules of outer flowers as long as to twice as long as involucre. . . *g.*

*g.*Stem truly scapiform or subscapiform, naked or with 1 or 2 leaves borne near base; lowest leaves with rounded or cordate bases. . . . . 8. *H. murorum.*

*g.*Stem with 2–12 leaves scattered to near inflorescence; lower leaves attenuate to petioles. . . *h.*

*h.*Involucre and pedicels stipitate-glandular, with no or but few glandless villi overtopping the glands; inflorescence paniculate-corymbose, with arched- or divergent-ascending branches and pedicels; heads few–30 or more. . . . . 9. *H. vulgatum.*

*h.*Involucre and pedicels glandless or only very minutely glandular, copiously long-pilose or villous with glandless hairs; inflorescence strict, with 1–10(–20) heads on erect branches or peduncles.

Larger (lower) leaves 1–5 cm. broad, membranaceous, villous-ciliate, their petioles and midribs of lower surfaces copiously villous; heads (1–)4–20; phyllaries blunt or acuminate. . . . . 10. *H. groenlandicum.*

Larger (lower) leaves 0.7–2 cm. broad, coriaceous, eciliate or nearly so, their petioles and midribs of lower surfaces sparsely pilose to glabrescent; heads 1–5(–8); phyllaries caudate-attenuate. . . 11. *H. Robinsonii.*

*f.*Stems leafy nearly or quite to summit; lower leaves not rosulate nor long-petioled nor usually mottled, the cauline numerous; exposed ligules of outer flowers about as long as involucre. . . *i.*

*i.*Leaves lanceolate or lance-oblong to narrowly ovate, the middle and upper ones rounded to subcordate at base, usually toothed, flat, smooth or smoothish; inflorescence paniculate or paniculate-corymbose, most of its branches not definitely umbellate at summit; pits of denuded receptacle with shallowly dentate margin.

Panicle usually one-tenth to one-third height of plant, its uppermost branches often subumbellately crowded; involucre usually bearing a few glandular setae; phyllaries attenuate, all appressed; hairs (when present) of lower internodes and under surfaces of lower leaves slender; wide-ranging native species. . . . . 12. *H. canadense.*

Panicle one-third to two-thirds height of plant, loosely paniculate, not umbelliform; involucre glabrous or with few non-glandular setae; phyllaries round-tipped, the lowest spreading; hairs of lower internodes and leaf-surfaces bulbous-based; locally introd. weed. 13. *H. subaudum.*

*i.*Leaves linear to narrowly lanceolate, tapering to base, scabrous, with revolute entire or but slightly toothed margins; inflorescence with many rays umbellately approximate at summit; involucre glandless; outer phyllaries spreading; pits of denuded receptacle with fimbriate-dentate margins. . . . . 14. *H. umbellatum.*

*e.*Involucre of a single long series of equal phyllaries and small or minute calyculate or basal ones; pappus of equal and rather scanty bristles; heads small, 1–2.5 cm. broad. § STENOTHECA. . . *j.*

*j.*Inflorescence an open panicle or paniculate corymb, if subcylindric (in no. 17) leafy-bracted. . . *k.*

*k.*Leaves chiefly basal or 1–6 scattered ones extending up the stem; inflorescence an open corymbiform panicle, with minute bracts and with filiform pedicels equaling to much exceeding heads. . . . 15. *H. venosum.*

*k.*Leaves regularly extending up the stem; at least the lower bracts of the inflorescence foliaceous.

Stem slender, glabrous except at villous base; panicle lax and open; heads 12–20-flowered, on filiform pedicels; leaves lanceolate, thin, glaucous beneath. . . . . 16. *H. paniculatum.*

Stem coarse, densely pubescent at least above; corymb stiff; heads

40–50-flowered, on stout densely hairy pedicels; leaves elliptic to spatulate-obovate, thick, not glaucous. . . . . 17. *H. scabrum.*

j. Inflorescence cylindric to subcylindric, thyrsoid, three to ten times as long as broad, not leafy-bracted.

Pubescence of base of stem and lower leaves villous or short-setose; basal and lower cauline leaves obovate to oval or oblong, rounded or obtuse at summit, the larger rarely 2 dm. long; involucre 8–10 mm. high, glandless, or glandular below the middle, 15–35-flowered; wide-ranging species. . . . . 18. *H. Gronovii.*

Pubescence setose, shaggy, 1–2 cm. long; basal leaves oblanceolate to spatulate-oblong, acute to tapering at tip, the longer ones 1.5–3 dm. long; involucre 1–1.3 cm. high, heavily stipitate-glandular to summit, 40–60-flowered; prairie-species. . . . . 19. *H. longipilum.*

§ Pilosélla Fries

1. H. Pilosélla L. (old generic name; a little hairy), Mouse-ear. — *Dwarf, forming stoloniferous carpets; leaves of basal rosette* oblong-lanceolate to spatulate, becoming 1.5–8 *cm. long and* 5–15 *mm. broad,* setose on both surfaces, green above, *whitened beneath when young with close stellate tomentum;* stolons filiform, numerous, leafy; *scape filiform,* 0.4–2.5 *dm. high,* with a single head; *involucre* 8–10 (rarely –12) *mm. high,* the expanded head 2.5–3 cm. broad. — Pastures and fields, Nfld. to Minn., s. to N.S., N.E., L.I., N.C. and O., very troublesome northw. June–Sept. (Natzd. from Eu.)

Var. níveum Muell. Arg. (snowy). — Leaves permanently white-pannose beneath. — Fields, e. Me. (Natzd. from Eu.)

2. H. flagellàre Willd. (like a whip-lash; from the stolons). — Much *coarser and greener* than no. 1; *leaves green beneath, the mature blades* 4–15 *cm. long and* 0.8–3 *cm. broad; scapes* stoutish, 1.2–4.5 *dm. high, often forking* near or above the middle, *with* (1–)2–4 *heads; involucre* dark-pilose, 1–1.3 *cm. high.* (*H. Pilosella,* var. *viride* Ser.) — Becoming a very insistent and obnoxious weed, P.E.I. to Mich., s. to N.E. and Va. May–July. (Natzd. from Eu.)

H. Aurícula L. (the rosulate habit suggesting *Auricula,* an old name for *Primula*), much smaller and more slender than no. 2, the *thin narrowly oblanceolate green* (*both sides*) *leaves only* 0.5–1.5 *cm. broad,* the slender *scape with 2–several heads with involucres only* 6–8 *mm. high,* has recently appeared in pastures of N.S. (Adv., probably soon natzd., from Eu.)

3. H. aurantìacum L. (orange-colored), Orange H., Devil's Paint-brush, Marguerite rouge (Que.), Bouquets rouges (Que.). — *Long-hirsute* rank-growing perennial with many-leaved basal rosettes and coarse, rapidly spreading and rooting stolons; *leaves* oblanceolate to narrowly elliptic-obovate, green on both sides, 0.4–2 *dm. long; scapes* naked or with 1 or 2 leaves, 2–7 dm. high, *with a corymb of orange-red heads* about 2 cm. broad; involucre dark-villous. (Including *H. brunneocroceum* Pugsley, colonies with broadest leaves, quite inseparable with us) — Clearings and fields, mostly too abundant and troublesome, Nfld. to Minn., s. to N.S., N.E., Va., O., n. Ind., Ill. and Ia. June–Aug. (–Oct.). (Natzd. from Eu.)

4. H. floribúndum Wimm. & Grab. (abounding in flowers), King Devil. — In habit somewhat like no. 3, *glaucous; rhizome slender and cord-like; the prostrate* leafy *stolons abundant and early produced; leaves narrowly oblanceolate, subacute,* 5–15 cm. long, *glabrous or nearly so above* but setose on margin, midrib and sometimes sparingly so on lower surface; primary scapes 0.3–1 m. high, naked or with 1–3 small usually basal bracts; involucre nigrescent; *flowers yellow.* — Fields and clearings, Nfld. to Ct., w. to O., often very aggressive. June–Aug. (Natzd. from Eu.) — Luxuriant plants developing leafy decumbent secondary flowering axes as well as prostrate and repent stolons.

5. H. praténse Tausch (of meadows), King Devil. — *Green,* not glaucous; *rhizome slender and cord-like; repent stolons fewer and later* developed than in no. 4; *basal leaves* narrowly oblong to oblanceolate, 0.6–2.5 dm. long, *abundantly setose on both surfaces; scape* 0.15–1 m. high, commonly *with* 1–3 *well developed leaves;* otherwise similar to no. 4. — Clearings, pastures and grasslands, very aggressive, Gaspé Pen., Que., to Ont., s. to N.S., N.E., L.I., N.C. and Tenn. May–Aug. (Natzd. from Eu.)

6. H. praeáltum Gochnat (very tall), var. decípiens W. D. J. Koch (deceptive), King Devil. — *Rhizome short and thick, praemorse* or not cord-like; *plant* somewhat *glaucous, with numerous slender elongated leafy ascending or spreading sterile* (or finally flowering) *branches arising from among the* narrowly oblanceolate to somewhat hispid (on both surfaces) and *on the lower surface finely stellate-pubescent basal leaves;* scape 0.4–1 m. high, setose, often leafy below; flowers yellow. — Locally abundant in grasslands and pastures, Nfld. to N.Y. June, July. (Natzd. from Eu.)

7. H. florentìnum All. (of Florence), King Devil. — Resembling nos. 4 and 5; *rhizome short and stout, praemorse; plant* smoothish, usually *without prostrate stolons or slender elongate leafy branches; basal leaves* oblanceolate to spatulate, *thickish*, pale green, only sparingly setose, more *commonly glabrous;* scape 0.15–1 m. high, smooth or but sparingly setose. — Fields, clearings, roadsides, etc., very aggressive, Nfld. to Ont., s. to N.S., N.E., Va., O., Mich. and Ia. May–Aug. (Natzd. from Eu.)

§ Archieràcium Fries

8. H. muròrum L. (of walls), Golden Lungwort. — Rhizome short and thick; *basal leaves rosulate*, green or mottled, elliptic or oval, *rounded or cordate at base*, coarsely or shallowly dentate, *the basal teeth divergent or reflexed*, on villous petioles; *scape naked or with* 1 *or* 2 *leaves*, 2–7 dm. high; heads few, 3–4 cm. broad; *involucre densely stipitate-glandular*, the thin imbricated phyllaries linear-attenuate. — Fields, roadsides, thickets and open groves, local, Nfld. to Mich., s. to N.S., w. N.E., n. N.J. and Pa. May, June. (Adv. from Eu.)

9. H. vulgàtum Fries (common). — Rhizome short and stout; basal *leaves* more or less rosulate, *often mottled with purple or bronze*, narrowly ovate, elliptic or broadly lanceolate, *dentate with ascending or spreading teeth to subentire or lacerate*, membranaceous, setose on both surfaces, *tapering to the villous petiole; flowering stem* 0.2–1 m. high, *with* 2–12 *scattered leaves, the uppermost near the base of the paniculate-corymbose inflorescence; heads* few–30, *on arched- or divergent-ascending branches or peduncles*, about 4 cm. broad; *involucre bearing many dark gland-tipped trichomes* and few or no glandless villi. — Roadsides, thickets, open groves, etc., locally abundant, Nfld. and e. Que. to Mich., s. to N.S., N.E., L.I. and Pa. June–Sept. (Natzd. from Eu.) — A heteromorphic and aggressive species with endless formal but scarcely definable tendencies.

10. H. groenlàndicum Arv.-Touv. (of Greenland). — Habitally resembling no. 9; *basal leaves* less rosulate, *long-villous on both surfaces, as well as petioles, membranaceous, closely villous-ciliate,* the *blades* 5–15 cm. long and 1–5 *cm. broad; flowering stems* 1–3, 2–8 dm. high, *with* 2–4 *scattered leaves,* the larger nearly as large as the basal; *inflorescence strict, with* (1–)4–20 *heads* 4.5 cm. broad, *on erect branches; involucre and pedicels* glandless or only minutely glandular, *copiously long-villous with glandless hairs; phyllaries blunt to acuminate,* 1.7–2 *mm. broad,* dark on the back. — Thickets, calcareous rocky banks, river-banks, etc., Lab., n. Nfld. and Anticosti I., Que. Late July–Sept. (Greenl.)

11. H. Robinsònii (Zahn) Fern. (for one of its discoverers, Benjamin Lincoln Robinson, 1864–1935). — Much smaller than no. 10; *basal leaves lanceolate to narrowly oblong, coriaceous or firm, eciliate or nearly so, glabrous above, glabrous, glabrescent or merely pilose along midrib beneath, the larger* 2–8 cm. long and 0.7–2 *cm. broad; cauline leaves* (2–)4–10, gradually decreasing in size upward, *the lower often sharp-toothed; heads* 1–5 (rarely –8), *erect,* 3.5–4 cm. broad; *phyllaries pilose-hirtellous, lance-attenuate with caudate tips,* the dark bases 1–2 mm. broad. — Crevices of ledges and rocky (oftenest argillaceous) shores, Nfld. and e. Que., s. to C.B., n.-centr. Me. and n. N.H. Late June–Sept.

12. H. canadénse Michx. (Canadian). — *Stems* 1–several from a knotty base, erect or ascending, mostly 0.2–1.5 m. high, *leafy from base to summit, more or less pubescent at base with slender hairs; leaves lanceolate or lance-oblong to narrowly ovate,* dentate, lacerate or subentire, *flat,* smooth or but slightly scabrous, *the upper ones with rounded to subcordate bases; inflorescence paniculate or paniculate-corymbose,* usually one-tenth to one-third the height of the plant; *involucre, at least when young, usually bearing a few glands; phyllaries attenuate to tip, all appressed; pits of denuded receptacle with shallowly dentate margins.* — Three vars.:

Involucre 8–13 mm. high; upper internodes of stem and branches without or with only a few scattered setae.

Involucre dark fuscous to blackish; median phyllaries oblong-lanceolate, 1.5–2 mm. broad, the inner ones fuscous throughout or with broad dark central band; stems 1.5–10 dm. high; leaves (5–)8–20(–30), mostly remote, thin or membranaceous to firm, oblong to narrowly ovate, lanceolate or oblanceolate; heads 1–many, on arched-ascending pedicels mostly 2–10 cm. long in an open or divaricately corymbiform panicle. . . . . . . *H. canadense* (typical).

Involucre pale olive; median phyllaries narrowly lance-attenuate, 1–2 mm. broad at base, the inner ones pale throughout or merely with slender greenish midrib and green tip; stems mostly 0.5–1.5 m. high, stiffly erect; leaves 25–50 or more, often subapproximate, lanceolate to narrowly oblong-ovate, thick to coriaceous; heads many, often subvirgately fascicled or subumbellately corymbose, on stiff pedicels mostly 0.5–4 cm. long. . . . . . . Var. *fasciculatum.*

Involucres 5–10 mm. high; panicle more or less virgate-fastigiate to umbelliform, with slender and erect elongate pedicels; stem slender, 0.5–9 dm. high, its middle and upper internodes and the pedicels copiously villous-hirsute with hairs 1.5–3 mm. long; leaves as in typical var. or the lower more strongly ciliate. Var. *hirtirameum.*

H. canadénse (typical), *H. columbianum* Rydb. — Thickets, borders of woods, rocky slopes and openings, clearings, etc., s. Lab. to n. B.C., s. to Nfld., C.B., P.E.I., n. Me., n. N.H., Ont., Mich., Wisc., ne. Ia. (adv.), S.D., Mont., Ida. and Oreg. Late July–Sept.

Var. **fasciculàtum** (Pursh) Fern. (clustered). — Thickets, clearings and shores, s. Que. to sw. Ont., s. to N.S., N.E., n. N.J., Pa., O., Ind., Ill. and Ia. Late July–Oct.

Var. **hirtiràmeum** Fern. (with hairy branches). — Ledges, shores, cliffs and clearings, Nfld. to Wisc., s. to N.S., N.E. and ne. Pa. Early July–early Sept.

13. H. SABAÙDUM L. (of Savoy). — Similar to no. 12; *base of stem and surfaces of leaves bearing pustular-based hairs;* leaves mostly on lower half of plant; *inflorescence an open panicle occupying one-third to two-thirds height of plant; involucre glabrous or with few non-glandular setae; phyllaries fuscous-green,* broadly lanceolate, *obtuse, the outermost spreading.* — Grasslands and waste open ground, locally too abundant, e. Mass. to e. Pa., rapidly spreading. Sept., Oct. (Natzd. from Eu.)

14. H. umbellàtum L. (umbelled). — Resembling stiffer and more leafy extremes of no. 12; *stems not hairy at base; leaves linear to lanceolate or narrowly oblong, attenuate to base, revolute, harshly scabrous,* entire or toothed; *inflorescence subumbellate,* the heads and branches approximate at summit; *involucre* fuscous, *glandless; outer phyllaries spreading; pits of denuded receptacle with fimbriate borders.* (*H. scabriusculum* Schwein.) — Sandy or rocky shores, thickets and slopes, Ont. to Alaska, s. to n. Mich., n. Wisc., Minn., S.D., Mont. and Oreg. July, Aug. (Eurasia)

§ STENOTHÈCA T. & G.

15. H. venòsum L. (veiny), RATTLESNAKE-WEED, POOR ROBIN'S PLANTAIN. — *Basal leaves rosulate,* elliptic-oblong to narrowly obovate, *nearly entire to undulate-dentate, short-petioled or subsessile, green or purple-veined or mottled* and some or all remotely *long-setose above,* glaucous to purplish and more or less villous beneath; *scape slender,* 2–7 dm. high, naked or with 1–6 scattered oblanceolate to narrowly obovate leaves, forking into an *open corymbiform panicle with* minute bracts at bases of the *filiform branches,* the filiform pedicels equaling to much longer than the small heads; pedicels and involucres more or less glandular at base, rarely glandless, the slender involucres 6–7.5 mm. high. — Open woods and clearings, Fla. to La., n. to Va., Tenn. and Mo., and sparingly to se. Mass. May–Sept.

Var. **nudicaùle** (Michx.) Farw. (naked-stemmed). — Rosette-leaves (except sometimes the very lowest) glabrous above. (Incl. vars. *subcaulescens* T. & G. and *Blombergii* Zahn) — Similar habitats, s. Me. to s. Ont., s. to L.I., Del., n. Va. and upland to n. Ga. and Tenn. May–Sept. Hybridizes with nos. 17 and 18.

16. H. paniculàtum L. (panicled). — *Stems slender,* 0.2–12 dm. high, *glabrous* except at the villous base, *leafy up to the lax and open capillary-branched panicle; leaves thin,* the middle and upper ones lanceolate, remotely toothed, acute, *glabrous, glaucous beneath;* lower bracts of inflorescence foliaceous; *heads* small, 12–20-*flowered,* the filiform pedicels smooth, or in forma **glandulòsum** R. Hoffm. (glandular) stipitate-glandular. — Open woods, Ga. and Ala., n. to N.S., centr. Me., s. Que., s. Ont. and Mich. July–Sept. — Hybridizes with no. 17.

17. H. scàbrum Michx. (harsh). — Stoutish and *coarse,* 1–15 dm. high, *densely pubescent at least above, commonly glandular; leaves* usually *extending to inflorescence,* the larger ones elliptic to spatulate-obovate, obtuse, *thickish,* subentire, *commonly hairy* on one or both surfaces; *panicle stiff, corymbiform to short-subcylindric, with foliaceous bracts,* its axis and branches white-tomentose, often dark-glandular; *heads 40–50-flowered, on thick pedicels.* — Four vars.:

Lower internodes clothed with slender often sordid (but not gland-tipped) trichomes 2–5 mm. long; lower surfaces of leaves with midrib villous with hairs 2–5 mm. long.

Trichomes of lower internodes, petioles and midribs beneath 2–3 mm. long; upper leaf-surfaces with scattered trichomes mostly 0.5–2 mm. long. . . *H. scabrum* (typical).

Trichomes of lower internodes and of both leaf-surfaces 3–5 mm. long. . . Var. *intonsum.*

Lower internodes tomentose (not long-villous), glabrescent or hispidulous with trichomes less than 0.7 mm. long; midribs beneath with trichomes barely 0.5 mm. long or glabrescent.

Stem with 15–25 crowded leaves, densely white-tomentose to base, the tomentum mixed with dark glands; both leaf-surfaces minutely glandular-pilose. Var. *leucocaule.*

Stem subscapose or with the 4–11 leaves scattered, glabrescent or hispidulous below with hairs only 0.2–0.6 mm. long, viscid; both leaf-surfaces glabrous or the upper sparingly setulose. . . . . Var. *tonsum.*

H. scàbrum (typical). — Stem 0.2–1.5 m. high, with few–many leaves; longer (lower) leaves 0.4–2 dm. long. — Dry open woods, pastures and clearings, usually one of the commonest species, Gaspé Pen., Que., to Ont. and Minn., s. to N.S., N.E., L.I., Del., Md., upland to Ga. and e. Tenn., Ky., s. Ill. and Mo. June–Sept.

Var. **intónsum** Fern. & St. John (unshaved). — Woods and dry openings, Ill. and Ia.

Var. **leucocaùle** Fern. & St. John (white-stemmed). — Compact, 1–2.5 dm. high; longest leaves 4–5 cm. long. — Dunes and heaths, Sable I., N.S. Sept.

Var. **tónsum** Fern. & St. John (shaved). — Stem 2–6 dm. high; longest leaves 4–11 cm. long. — Gaspé Pen., Que., to Algoma Distr., Ont., s. to M.I., n.-centr. Me., n. Mich. and n. Wisc. July–early Sept.

H. mariànum Willd. (of Maryland), combining traits of nos. 15 with those of nos. **17** or, northward, of no. 16, is a heteromorphic series, presumably of hybrid origin and somewhat apomictic. Its components are not easily defined nor consistently stable. **H. Traìllii** Greene (named in 1900 for Traill Green) or *H. Greenii* Porter & Britt. is similarly baffling, as are some apparent blends from farther south.

18. H. Gronòvii L. (in honor of Jan Fredrik Gronovius, 1690–1762). — Strict, 0.25–1.5 m. high, villous and papillate-setose at base; *leaves* 2–10(–30), *mostly borne below middle of stem; the basal and lower cauline obovate to oval or oblong, rounded or blunt at tip*, setose chiefly above, minutely stellate-pubescent beneath; *inflorescence a virgate thyrsiform cylindric panicle three to ten times as long as broad; bracts small and subulate; heads* mostly on glandular slender pedicels, 15–35-*flowered; involucre* 8–10 *mm. high, glabrous, or stipitate-glandular below the middle;* achenes tapering at summit. — Dry open woods, thickets and openings, Fla. to Tex., n. to Mass., N.Y., O., s. Mich., Ill. and Kans. Late July–Oct.

19. H. longípilum Torr. (with long hairs). — Stiff and virgate, 0.6–1.5 m. high, *densely covered below and sometimes to summit with long* (1–2 cm.) finally *rusty hairs; leaves erect, chiefly crowded at base of plant, oblanceolate to spatulate-oblong, acute to tapering at tip, long-hairy, the longest ones* 1.5–3 *dm. long;* reduced and slender bracteal leaves extending up the stem; *inflorescence* much as in no. 18 or more slender, *heavily stipitate-glandular to tip of phyllaries; heads* 40–60-*flowered; involucres* 1–1.3 *cm. high.* — Dry prairies and open sands, La. and e. Tex., n. to sw. Ont., Mich., Wisc., Minn. and Neb. July, Aug.

Glossary

A, privative, or *an*. A prefix meaning without.

Abaxial. On the side of an organ away from the axis (= dorsal).

Aberrant. Departing from the normal.

Abortion. Imperfect development, or non-development of an organ.

Abortive. Defective or barren.

Abruptly pinnate. Pinnate without a single terminal leaflet.

Acaulescent. Stemless or apparently so, or with stem subterranean.

Accrescent. Enlarging with age, as the bud-scales of some hickories or the sepals of some flowers.

Accumbent (cotyledon). Having the edges against the radicle.

Aceriform. Like a maple-leaf.

Acerose. Needle-shaped.

Achene. A small dry and hard 1-locular 1-seeded indehiscent fruit, like the individual fruits of *Ranunculus*.

Achenial. Pertaining to an achene.

Achlamydeous. Without calyx or corolla.

Acicular. Slenderly needle-shaped.

Aculeate. Prickly; beset with prickles.

Aculeolate. Provided with diminutive prickles.

Acuminate. Tapering at the end to a gradual point.

Acumination. A gradually tapering point.

Actinomorphic or *-ous*. Capable of bisection through two or more planes into similar halves, as is a regular or radially symmetrical flower.

Acute. Terminating in a sharp or well-defined angle.

Adaxial. Toward the axis (= ventral).

Adnate. United to an organ of a different cycle or kind, as the inferior ovary with the calyx-tube. *Adnate anther*, one attached for its whole length to the inner or outer face of the filament.

Adventitious. Used of buds, roots, etc., which develop in an irregular or unusual position.

Adventive. Imperfectly naturalized.

Aestival. Appearing in summer.

Aestivation. The arrangement of the parts of the flower in the bud.

Agglomerate. Heaped or crowded into a dense cluster, but not cohering.

Aggregate. Heaped or crowded into a dense cluster, but not cohering; cf. *agglomerate*.

Ala (pl. *-ae*). A wing; also the lateral petals of a papilionaceous corolla.

Alate. Winged.

Albumen. Any deposit of nutritive material accompanying the embryo.

Albuminous. Having albumen or of the nature of albumen.

Alliaceous. Having the smell or taste of garlic.

Alleghanian. Belonging to or pertaining to the Alleghany and parallel mountains of the Appalachian system.

Alpine. Used of regions above tree-line.

Alternate (of leaves, etc.). Not opposite to each other on the axis, but borne at regular intervals at different levels.

Alveolate. Honeycombed; having angular depressions separated by thin partitions.

Alveolation. A honeycombed condition.

Alveoli. Pits or depressions suggesting those of honeycomb.

Ament. A catkin, or dry scaly spike, usually unisexual, such as the inflorescence of willows, birches, etc., and at least the staminate inflorescence in hickories and some other genera.

Amentiferous. Bearing aments.

Amorphous. Shapeless or of indefinite form.

Amphicarpous. Producing two kinds of fruit.

Amphigean. Native of both the Old and New Worlds.

Amphitropous (ovule or seed). Half-inverted and straight, with the hilum lateral.

Amplexicaul. Clasping the stem.

Ampliate. Enlarged.

Anastomosing. Connecting by cross-veins and forming a network.

Anatropous (ovule). Inverted and straight, with the micropyle next the hilum and the radicle consequently inferior.

Ancipital. Two-edged.

Androecium. A collective term for the stamens.

Androgynous (inflorescence). Composed of both staminate and pistillate flowers, the staminate at the apex; used chiefly in *Cyperaceae*.

Androphore. A support or column on which stamens are raised.

-androus. In composition, having stamens.

Anemophilous. With wind-borne pollen.

Angiospermous. Having the seeds borne within a pericarp.

Annotinous. A year old, or in yearly growths.
Annual. Of only one year's duration. *Winter-annual*, a plant from autumn-sown seed which blooms and fruits in the following spring.
Annular. In the form of a ring.
Annulus. A ring, such as surrounds the sporangium in some Ferns.
Anterior. On the front side and away from the axis.
Anthelate. With elongate flower-bearing branches as in some *Junci.*
Anther. The polliniferous part of a stamen.
Antheridium. In Cryptogams the organ corresponding to an anther.
Antheriferous. Anther-bearing.
Antherozoid. One of the minute organs developed in an antheridium.
Anthesis. The expansion or the time of expansion of a flower.
Antrorse. Directed upward or forward.
Apetalous. Having no petals.
Aphyllous. Leafless.
Apical. Relating to the apex or tip.
Apiculate. Ending in an abrupt short pointed tip.
Apiculation. A sharp and short, but not stiff point, in which a leaf, petal or other organ may end.
Apocarpous. With separate carpels.
Apogamous. Developed without fertilization; parthenogenetic.
Apophysis. An enlargement or swelling of the surface of an organ.
Appendiculate. Furnished with an appendage.
Appressed. Lying close and flat against.
Apterous. Wingless.
Aquatic. Living in water.
Arachnoid. Cobwebby; of slender entangled hairs.
Araneose. Like a spider-web.
Arboreous. Tree-like or pertaining to trees.
Arborescent. Approaching the size and habit of a tree.
Archegonium. The organ in the higher Cryptogams corresponding to a pistil in the Flowering Plants.
Arctic-alpine. Used for plants of arctic and alpine distribution but found only south of the Arctic zone.
Arcuate. Moderately curved; arching.
Areolate. Marked out into small spaces; reticulate.
Areole. A small space marked out upon or beneath a surface, as in many leaves, petals, etc.
Argillaceous. Of clayey quality.
Aril. An appendage growing at or about the hilum of a seed.
Arillate. Having an aril.
Aristate. Awned; provided with stiffish bristle-shaped appendages.
Article. The portion of a fruit (especially in *Leguminosae*) separated from others by a constriction or joint, as in *Desmodium.*
Articulate. Jointed; having a node or joint.
Ascending. Rising somewhat obliquely, or curving upward. *Ascending ovule*, one that is attached above the base of the ovary and is directed upward.
Asperous. Rough or harsh to the touch.
Assurgent. Ascending.
Attenuate. Slenderly tapering; gradually becoming very narrow or slender.
Auricle. An ear-shaped appendage or lobe.
Auriculate. Furnished with auricles.
Austral. Southern.
Awl-shaped. Tapering gradually upward from the base to a slender or rigid point.
Awn. A bristle-shaped appendage.
Awned. Having awns.
Axil. The angle formed between any two organs.
Axile. In the axis; used especially of the placentae in the ovary.
Axillary. In or related to the axis.
Axis (of the stem, inflorescence, etc.). The central part of a longitudinal support on which organs or parts are arranged.

Baccate. Berry-like; pulpy throughout.
Banner. The vexillum, standard or upper petal of a papilionaceous corolla.
Barbed. Furnished with rigid points or short bristles, usually reflexed like the barb of a fish-hook.
Barbellate. Finely barbed.
Barbulate. Finely bearded.
Basifixed. Attached by the base.
Basonym. The original epithet, retained when transferred to a new position.
Bast. The fibrous portion of the inner bark.
Beaked. Ending in a firm prolonged slender tip.
Bearded. Bearing or furnished with long or stiff hairs.
Berry. A pulpy fruit with immersed seeds, as the grape, cranberry, etc.; loosely extended to cover other fleshy or pulpy fruits, such as the strawberry (ripened receptacle), raspberry (coherent drupelets), etc.
Bi- or *bis-*. A Latin prefix signifying two, twice, or doubly.
Bidentate. Having two teeth.
Biennial. Of two years' duration.
Bifid. Two-cleft.
Bilabiate. Two-lipped.
Bilocular. With two cavities.
Bilateral. Arranged on opposite sides.
Bipinnate. Doubly or twice pinnate.
Bipinnatifid. Twice or doubly pinnatifid.

Bisexual. Having both stamens and pistils.
Bivalvular. With two valves.
Blade. The expanded portion of a leaf, etc.
Bloom. A whitish powdery and glaucous covering of the surface, often of a waxy nature.
Boreal. Northern.
Boss. A knob-like or rounded protuberance.
Bossed. With a rounded surface having a projection in its center.
Brachiate. With spreading branches suggesting arms.
Bract. A more or less modified leaf subtending a flower or belonging to an inflorescence, or sometimes cauline.
Bracteal. Of or pertaining to bracts.
Bracteate. Having bracts.
Bracteolate. Having bractlets.
Bracteole. Bractlet.
Bracteose. With numerous or conspicuous bracts.
Bractlet. A secondary bract, as one upon the pedicel of a flower.
Branchlet. Ultimate divisions of a branch.
Bristle. A stiff hair, or any slender body which may be likened to a hog's bristle.
Bristly. Provided with bristles.
Bud. The rudimentary state of a stem or branch; an unexpanded flower.
Bulb. A subterranean leaf-bud with fleshy scales or coats.
Bulbiferous. Bearing bulbs.
Bulbil or *bulblet.* A small bulb; a bulb-like body, especially one borne upon the stem.
Bulbous. Having the character of a bulb.
Bullate. Blistered or puckered.
Bulliform. Applied to large thin-walled epidermal cells of most *Gramineae* and *Cyperaceae.*
Bursicle. A pouch-like receptacle.
Bursicule, bursicula. The pouch-like expansion of the stigma into which the caudicle of some Orchids is inserted.

Caducous. Falling off very early.
Caespitose. See *Cespitose.*
Calathiform. Cup-shaped.
Calcarate. Produced into or having a spur.
Calcareous. Limy.
Callosity. A hardened thickening.
Callous. Having the texture of a callus.
Callus. A hard protuberance or callosity; in the Grasses the tough often hairy swelling at the base or point of insertion of the lemma or palea.
Calyculate. Having bracts around the calyx or involucre imitating an outer calyx.
Calyx. The outer perianth of the flower.
Cambium. A layer, usually regarded as one cell thick, of persistent meristematic tissue (referring to vascular and cork cambia); or a persistent meristematic layer which gives rise to secondary wood and secondary phloem (vascular cambium).
Campanulate. Bell-shaped; cup-shaped with a broadened rim.
Campestrian. Of plains or open country; specifically of the Great Plains of North America.
Campylotropous (ovule or seed). So curved as to bring the apex and base nearly together.
Canaliculate. Longitudinally channeled.
Cancellate. Latticed; resembling latticework.
Canescent. Hoary with gray pubescence.
Capillary. Hair-like.
Capitate. Shaped like a head; collected into a head or dense cluster.
Capsular. Belonging to or of the nature of a capsule.
Capsule. A dry dehiscent fruit composed of more than one carpel.
Carina. A keel; used either for the two combined lower petals of a papilionaceous corolla or for a salient longitudinal projection on the center of the lower face of an organ, as on the lemmas of many Grasses.
Carinal. On or having relation to a ridge or keel. For special use see *Equisetum.*
Carinate. Keeled.
Carpel. A simple pistil, or one member of a compound pistil.
Carpellate. Possessing carpels.
Carpophore. The slender prolongation of the floral axis which in the *Umbelliferae* supports the pendulous ripe carpels.
Cartilaginous. Firm and tough but flexible, like cartilage or the core of a pear.
Caruncle. An excrescence or appendage at or about the hilum of a seed.
Caruncalate. Having a caruncle.
Caryopsis. A grain, as of Grasses; a seed-like fruit with a thin pericarp adnate to the contained seed.
Castaneous. Of a chestnut-color; dark brown.
Catkin. An ament.
Caudate. Having a slender tail-like appendage.
Caudex. The persistent base of an otherwise annual herbaceous stem.
Caudicle. The thread-like or strap-shaped stalk of a pollinium.
Caulescent. Having a manifest stem above ground.
Cauline. Belonging to the stem.
Cell. One of the minute vesicles, of very various forms, of which organisms are composed.
Centrifugal. Developing from the center outward.
Centripetal. Developing from the outside toward the center.
Centrum. The central portion; here used spe-

cifically for the large central air-space in hollow stems such as those of *Equisetum.*

Cernuous. Nodding.

Cespitose (or *caespitose*). Growing in tufts; forming mats or turf.

Chaff. A small thin scale or bract, becoming dry and membranous.

Chaffy. Having or resembling chaff.

Channeled. Deeply grooved longitudinally.

Chartaceous. Having the texture of writing-paper.

Chlorophyll. The green coloring-matter within the cells of plants.

Chlorosis. Becoming green when usually of some other color.

Chorisis. Separation of an organ (leaf, petal, stamen, etc.) into more than one.

Cilia. Marginal hairs.

Ciliate. Marginally fringed with hairs.

Ciliolate. Minutely ciliate.

Cinereous. Ash-colored.

Circinate. Coiled from the tip downward, as the young frond of a fern.

Circumscissile. Dehiscing by a regular transverse circular line of division, as in the fruit of *Plantago.*

Cladophyll. A branch functioning as a leaf.

Clathrate. Latticed.

Clavate. Club-shaped; gradually thickened upward.

Clavellate. Diminutive of clavate.

Claw. The narrowed base or stalk of some petals, etc.

Cleft. Deeply cut.

Cleistogamous. Fertilized in the bud, without the opening of the flower.

Clinandrium. The anther-bed in Orchids, that part of the column in which the anther is concealed.

Coaetaneous. Of the same age; flowering as the leaves expand.

Coalescence. The union of parts or organs of the same kind.

Coccus (pl. *cocci*). One of the parts into which a lobed fruit with 1-seeded locules splits.

Cochleate. Spiral, like a snail-shell.

Colonial. Forming colonies; used chiefly for plants with subterranean connections.

Columella. Persistent axis of some capsules.

Column. A group of united filaments in *Malvaceae* and some other groups; in the Orchids the coalesced style and filaments.

Columnar. Column-shaped or pillar-shaped.

Coma. A tuft of hairs, especially at the tips of seeds, as in *Epilobium.*

Commissure. The surface by which one carpel joins another, as in the *Umbelliferae.*

Comose. Furnished with or resembling a tuft of hairs or coma.

Complicate. Folded upon itself.

Compound. Composed of 2 or more similar parts united into one whole. *Compound leaf,* one divided into separate leaflets.

Compressed. Flattened, especially laterally.

Conduplicate. Folded together lengthwise.

Conglomerate. Densely clustered or heaped together.

Conjugate. Coupled or in pairs.

Coniferous. Cone-bearing.

Connate. United; used especially of like structures, such as leaves, joined from the start.

Connective. The portion of a stamen which connects the two locules of the anther.

Connivent. Coming into contact; converging.

Conoidal. Nearly conical.

Contorted. Twisted; or bent or twisted on itself.

Convolute. Rolled up longitudinally.

Coralloid. Coral-like.

Cordate. Heart-shaped with the point apical.

Cordiform. Shaped like a heart.

Cordilleran. Belonging or related to the Cordillera, the north-and-south mountain-ranges of western America.

Coriaceous. Leathery in texture.

Corm. The enlarged fleshy base of a stem, bulb-like but solid; or a solid bulb.

Cormatose. Producing corms.

Corneous. Horny.

Corniculate. Furnished with a little horn or horns.

Corolla. The inner perianth, of distinct or connate petals.

Corolline. Seated on a corolla; corolla-like; petaloid or belonging to a corolla.

Corolloid. Corolline; corolla-like; petaloid.

Corona. A crown or inner petal-like appendage, as in *Narcissus.*

Coroniform. Shaped like a crown.

Corrugate or *corrugated.* Wrinkled or in folds.

Cortex. Rind or bark.

Cortical. Relating to bark.

Corymb. A flat-topped or convex open flower-cluster; in the stricter use of the word equivalent to a contracted raceme, progressing in its flowering from the margin inward.

Corymbose. In corymbs or corymb-like.

Corymbiform. Shaped like a corymb.

Costa. A rib; a midrib or mid-nerve.

Costate. Ribbed; having one or more longitudinal ribs or nerves.

Cotyledon. The foliar portion or first leaves (one, two, or more) of the embryo as found in the seed.

Crateriform. In the shape of a saucer or cup; hemispherical or more shallow.

Creeping. Running along at or near the surface of the ground and rooting (= repent).

Crenate. Dentate with the teeth much rounded.

Crenulate. Finely crenate.

Crested. Bearing elevated ridges on projections on the surface, especially on petals, etc.

Crimped. Puckered into small waves.

Cristate. Bearing an elevated appendage resembling a crest.

Cristulate. Inclined to be crested or with tiny crests.

Crown. An inner appendage to a petal, or to the throat of a corolla.

Cruciate or *cruciform.* Cross-shaped.

Crustaceous. Of hard and brittle texture.

Cryptogams. Plants destitute of stamens, pistils, and true seeds, but often reproduced as the result of a sexual act.

Cucullate. Hooded or hood-shaped; cowled.

Culm. The peculiar stem of Grasses and Sedges.

Cultigen. A plant which has arisen through cultivation.

Cuneate. Wedge-shaped; narrowly triangular with the acute angle downward.

Cupule. The cup (involucre) of the acorn.

Cupuliform. Cupola-shaped.

Cuspidate. Tipped with a cusp or sharp and firm point.

Cuticle. The outer film of epidermal cells.

Cyathium. The ultimate inflorescence of *Euphorbia*, consisting of a cup-like involucre bearing the flowers from its base.

Cyathiform. Cup-shaped.

Cycle. Circle; used for the series of similar parts in a flower.

Cyme. A usually broad and flattish determinate inflorescence, *i.e.*, with its central or terminal flowers blooming earliest.

Cymose. Bearing cymes or cyme-like.

Cymule. A diminutive cyme or portion of one.

Cypsela. An achene invested by an adnate calyx as in the fruit of most *Compositae.*

Cystolith. A stone-like concretion in the epidermis of some plants, as the *Urticaceae* and *Acanthaceae*.

Deciduous. Not persistent; not evergreen.

Declinate. Bent or curved downward or forward.

Decompound. More than once compound or divided.

Decumbent. Reclining, but with the summit ascending.

Decurrent. Extending downward from the point of insertion.

Decussate. Alternating in pairs at successive levels at right angles.

Deflexed. Bent or turned abruptly downward.

Dehiscence. Method of opening of fruits, anthers, etc., at maturity.

Dehiscent. Opening regularly by valves, slits, etc., as a capsule or anther.

Deliquescent. Softening or wasting away, as in some perianths; branching without a continuous main axis.

Deltoid. Shaped like the Greek letter Δ.

Dentate. Toothed, usually with the teeth directed outward.

Denticle. A minute tooth.

Denticulate. Minutely dentate.

Depauperate. Impoverished, starved or stunted.

Depressed. Somewhat flattened from above.

Determinate. With a definite number; in an inflorescence with a single terminal flower opening before those below, as in a cyme.

Dextrorse. Toward the right hand.

Di-, dis-. A Greek prefix signifying two or twice.

Diadelphous (stamens). Combined in two often unequal sets.

Diandrous. Having two stamens.

Diaphragm. Dividing membrane or partition, as in the pith of various trees.

Dicarpellary. Composed of two carpels.

Dichasium. A cyme with two lateral axes.

Dichotomous. Forking regularly by pairs. *Dichotomous Key.* An analytical key to groups, divided into two contrasting divisions and contrasting pairs of subdivisions.

Dicotyledonous. Having two cotyledons.

Didymous. Twin; found in pairs.

Didynamous (stamens). In two pairs of unequal length.

Diffuse. Widely or loosely spreading.

Digitate. Compound, with the members arising together at the apex of the support, suggesting the fingers of the hand.

Dimerous (flower). Having all the parts in twos. *Dimerous regularity.* Regular in two different planes, as in *Dicentra.*

Dimidiate. Halved, or as if one half is wanting.

Dimorphous. Occurring in two forms.

Dioecious. Unisexual, with the two kinds of flowers on separate plants or in separate parts of the inflorescence.

Dipterous. Two-winged.

Disciform. Depressed and circular like a disk.

Discoid. Resembling a disk. *Discoid head*, in *Compositae*, one without ray-flowers.

Discrete. Separate; not coalescent.

Disc or *disk.* A development of the receptacle at or around the base of the pistil; in the *Compositae* the central portion of the flowering head, bearing tubular flowers.

Disk-flowers. In *Compositae*, the tubular flowers of the head as distinct from the ray-flowers.

Dissected. Cut or divided into narrow segments.

Dissepiment. A partition in an ovary or fruit.

Distichous. In two vertical ranks.

Distinct. Separate; not united; evident.

Diurnal. Daily; occurring in the day.

Divaricate. Widely divergent.

Divergent. Inclining away from each other.

Divided. Lobed or separated to the base.

Dolabriform. Axe-shaped or hatchet-shaped.

Dorsal (in the morphological sense). Upon or relating to the back or outer surface of an organ (abaxial).

Dorsiventral. With distinction of back and front, or placed with reference to the back or front.

Downy. Pubescent with fine and soft hairs.

Drupaceous. Resembling or of the nature of a drupe.

Drupe. A fleshy or pulpy fruit with the inner portion of the pericarp (1-locular and 1-seeded, or sometimes several-locular) hard or stony.

Drupelet. A diminutive drupe, as in a raspberry or blackberry.

Duct. Tube or canal, usually in the wood, which carries resin, latex, or oil.

E- or *ex-*. A Latin prefix having often a privative signification, as *ebracteate*, without bracts.

Echinate. Provided with prickles.

Effuse. Very loosely spreading.

Ellipsoid. Solid but with an elliptical outline.

Elliptical. With the form of an ellipse, rounded about equally to both ends.

Emarginate. Having a shallow notch at the extremity.

Embryo. The rudimentary plantlet within the seed.

Endemic. Confined geographically to a single area.

Endocarp. The inner layer of a pericarp.

Endogenous. Growing throughout the substance of the stem, instead of by superficial layers.

Endosperm. The reserve food stored around the embryo.

Ensiform. Sword-shaped, as in the leaf of *Iris*.

Entire. Without toothing or division.

Entomophilous. Pollinated by insects.

Ephemeral. Lasting for one day or less.

Epi-. Greek prefix meaning upon.

Epicalyx. A series of bracts close to and resembling the calyx.

Epicarp. The outer layer of the pericarp or matured ovary.

Epidermis. The superficial layer of cells.

Epigynous. Growing on the summit of the ovary, or apparently so.

Epiphyte. A plant growing attached to another plant but not parasitic; an air-plant.

Equitant. Astride; used of conduplicate leaves which enfold each other in two ranks, as in *Iris*.

Erose. With the margin as if gnawed.

Etiolated. Blanched.

Ex-. Prefix meaning without or destitute of.

Exasperate. Rough with hard projecting points.

Excentric. Off-center or one-sided.

Excurrent. Running out, as a nerve of a leaf projecting beyond the margin.

Exfoliating. Cleaving off in thin layers.

Exocarp. The outer layer of a pericarp.

Exogenous. Growing by annular layers near the surface.

Explanate. Spread out flat.

Exserted. Projecting beyond an envelope, as stamens from a corolla.

Exsiccated. Dried; used especially for soils which have lost their evident moisture.

Extrorse. Facing outward.

Falcate. Scythe-shaped; curved and flat, tapering gradually.

Farinaceous. Containing starch; starch-like.

Farinose. Covered with a meal-like powder.

Fasciate. Monstrous expansion of stems, branches, etc., these appearing to have coalesced.

Fascicle. A close bundle or cluster.

Fascicled. Bearing or in fascicles.

Fasciculate. In close bundles or clusters.

Fastigiate (branches). Erect and near together.

Faucal. Pertaining to the throat (of a perianth, etc.)

Faveolate or *favose.* Honey-combed.

Feather-veined. With veins all proceeding from the sides of a midrib.

Fenestrate. Pierced with holes, as the septum in some *Cruciferae*.

Ferruginous. Rust-colored.

Fertile. Capable of producing fruit; or productive.

Fibril or *fibrilla.* A diminutive fiber.

Fibrillose. Furnished or abounding with fine fibers.

Fibrous. Composed of, or resembling, fibers. *Fibrous tissue,* a tissue formed of elongated thick-walled cells.

Fibro-vascular. Composed of woody fibers and ducts.

Filament. The part of a stamen which supports the anther; any thread-like body.

Filamentous. Composed of threads.

Filiform. Thread-shaped; long, slender, and terete.

Fimbriate. Fringed.

Fimbrillate. Having a minute fringe.

Fimbriolate. Very finely fimbriate.

Fissile. Friable or splitting.

Fistular or *fistulose.* Hollow and cylindrical.

Flabellate or *flabelliform.* Fan-shaped or broadly wedge-shaped.

Flaccid. Without rigidity; lax and weak.
Flagelliform. Resembling a runner; or lash-like.
Flexuous. Curved alternately in opposite directions.
Floccose. Clothed with locks or flocks of soft hair or wool.
Flocculent. Diminutive of floccose.
Floret. A small flower, usually one of a dense cluster.
Floricane. The flowering cane, usually the second year's development of the *primocane,* in *Rubus* and other such genera.
Floriferous. Flower-bearing.
Foliaceous. Leaf-like in texture or appearance.
Foliar. Relating to a leaf.
-foliate. -leaved; having leaves.
-foliolate. Having leaflets.
Foliose. Bearing numerous or crowded leaves.
Follicle. A fruit consisting of a single carpel, dehiscing by the ventral suture.
Follicular. Like a follicle.
Forked. Divided into nearly equal branches.
Fornicate. Provided with scale-like appendages in the corolla-tube, as in *Myosotis.*
Foveolate. Pitted.
Free. Not adnate to other organs.
Frond. The expanded leaf-like portion of Ferns and some other Cryptogams; in *Lemnaceae,* etc., the thallus-like stem which functions as foliage.
Frondose. Leafy; frond-like or bearing fronds.
Fructification. The act or organs of fruiting.
Fruit. The seed-bearing product of a plant, simple, compound, or aggregated, of whatever form.
Frutescent. Shrubby or becoming so.
Fruticose. Shrubby or relating to shrubs.
Fugacious. Falling or fading very early.
Fulvous. Tawny.
Funicle. The free stalk of an ovule or seed.
Funnelform. Infundibuliform.
Furcate. Forked; or divergently parted.
Fuscous. Grayish-brown.
Fusiform. Spindle-shaped; swollen in the middle and narrowing gradually toward each end.

Galea. A hooded or helmet-shaped portion of a perianth, as the upper sepal of *Aconitum* and the upper lip of some bilabiate corollas.
Galeate. Helmet-shaped; having a galea.
Gamopetalous. Having the petals more or less united.
Gamophyllous. Composed of coalescent leaves or leaf-like organs.
Gamosepalous. Having the sepals united.
Geminate. Equal, in pairs.
Gemma. A bud or body analogous to a bud by which some plants propagate themselves.
Gemmiparous. Producing gemmae.
Geniculate. Bent abruptly, like a knee.
Gibbosity. A swelling of moderate extent and asymmetrical character, at or near the base of an organ.
Gibbous. Protuberant or swollen on one side.
Glabrate or *glabrescent.* Becoming glabrous.
Glabrous. Smooth, especially not pubescent nor hairy.
Gladiate. Sword-shaped; in the form of a sword-blade, either straight or somewhat curved.
Gland. A secreting surface or structure; any protuberance or appendage having the appearance of such an organ.
Glandular. Bearing glands or of the nature of a gland.
Glanduliferous. Bearing glands.
Glaucescent. Tending to be glaucous.
Glaucous. Covered or whitened with a bloom.
Glochid. A barbed hair or bristle.
Glochidiate. Barbed at the tip.
Glomerate. Compactly clustered.
Glomerulate. In small compact clusters.
Glumaceous. Furnished with or resembling glumes.
Glume. A chaff-like bract; specifically one of the two empty chaffy bracts at the base of the spikelet in the Grasses.
Glutinous. Covered with a sticky exudation.
Granulose. Composed of, or appearing as if covered by, minute grains.
Gregarious. Growing in large colonies, associated but not matted. Colonial.
Gymnospermous. Bearing naked seeds, without an ovary.
Gynandrous. With the stamens borne on or adnate to the pistil.
Gynecandrous. Having staminate and pistillate flowers in the same spike, the pistillate at the apex; used chiefly in the *Cyperaceae.*
Gynobase. An enlargement or prolongation of the receptacle bearing the ovary.
Gynoecium. The pistil or collective pistils of a flower; the female portion of a flower as a whole.
Gynophore. The stipe of a pistil.
Gynostemium. The compound structure resulting from the union of the stamens and pistil in the *Orchidaceae.*

Habit. The general appearance of a plant.
Habitat. The kind of locality in which a plant grows.
Halberd-shaped. The same as hastate.
Halophyte. A plant of saline soil.
Halophytic. Growing in saline soil.
Hamate. Hooked at the tip.
Hastate. Like an arrow-head, but with the basal lobes pointing outward at wide angles.

Haustoria. Root-like suckers of parasitic plants like *Cuscuta*.

Head. A dense cluster of sessile or nearly sessile flowers or fruits on a very short axis or receptacle.

Helicoid. Curved or spiralled like a snail-shell.

Herb. A plant with no persistent woody stem above ground; also plants used in seasoning or in medicine.

Herbaceous. Having the characters of an herb; leaf-like in color and texture.

Hermaphrodite. With the stamens and pistils in the same flower.

Heterocarpous. Producing more than one kind of fruit.

Heterogamous. Bearing two kinds of flowers.

Heterogeneous. Not uniform in kind.

Heterostyly. Change of function of staminate and pistillate flowers or of the relation of their parts.

Hexa-. Greek for six; used in compounds.

Hibernaculum. Winter-bud.

Hilum. The scar or point of attachment of the seed.

Hippocrepiform. Horseshoe-shaped.

Hirsute. Pubescent with rather coarse or stiff hairs.

Hirsutulous. Slightly hirsute.

Hirtellous. Minutely hirsute.

Hispid. Provided with rigid or bristly hairs or with bristles.

Hispidulous. Minutely hispid.

Hoary. Grayish-white with a fine close pubescence.

Homogamous. Bearing but one kind of flowers.

Homogeneous. All of one kind.

Humifuse. Spreading over the ground.

Hyaline. Transparent or translucent.

Hybrid. A cross-breed of two species.

Hydrophyte. An aquatic plant.

Hydrophytic. Pertaining to aquatic plants.

Hygroscopic. Altering form or position through changes of moisture.

Hypanthium. An enlargement or development of the torus under the calyx; a syconium.

Hypochil. The (often fleshy or otherwise modified) basal portion of the labellum or lip in *Orchidaceae*.

Hypocrateriform. Salver-shaped.

Hypogaeous. Growing or remaining underground.

Hypogynium. The perianth-like structure subtending the ovary in *Scleria* and some other *Cyperaceae*.

Hypogynous. Situated on the receptacle beneath the ovary and free from it and from the calyx; having the petals and stamens so situated.

Imbricate. Overlapping, either vertically or spirally, where the lower piece covers the base of the next higher; or laterally, as in the aestivation of a calyx or corolla, where at least one piece must be wholly external and one internal.

Immersed. Growing wholly under water.

Imparipinnate (odd-pinnate). Pinnate with an odd terminal leaflet.

Incrassate. Thickened.

Implexed. Entangled, interlaced.

Implicated. Entangled, woven in.

Impressed. Bent inward, hollowed or furrowed as if by pressure.

Incised. Cut sharply and irregularly, more or less deeply.

Included. Not at all protruded from the surrounding envelope.

Incumbent (cotyledons). Lying with the back of one against the radicle.

Indehiscent. Not opening by valves, etc.; remaining persistently closed.

Indeterminate. Of indefinite growth, as a theoretical racemose inflorescence.

Indigenous. Native and original to the region.

Indument. Any hairy covering or pubescence.

Induplicate. With the edges folded or turned in.

Indurated. Hardened.

Indusiate. Provided with an indusium.

Indusium. The proper (often shield-shaped) covering of the sorus or fruit-dot in Ferns.

Inferior. Lower or below; outer or anterior. *Inferior ovary*, one that is adnate to the calyx.

Inflated. Bladdery.

Inflexed. Turned abruptly or bent inward; incurved.

Inflorescence. The flowering part of a plant, and especially the mode of its arrangement.

Infra-. In composition, below; as *infra-axillary*, below the axil.

Infundibuliform. Funnelform.

Innate. Borne at the apex of the supporting part, as some anthers.

Innovation. An offshoot from the stem.

Inserted. Attached to or growing out of.

Inter- or *intra-*. In compounds, between.

Interfoliaceous. Between the leaves of a pair, as the stipules of many *Rubiaceae*.

Internode. The portion of a stem or other structure between two nodes.

Interruptedly pinnate. Pinnate with a pair of terminal leaflets; having small leaflets interposed with those of larger size.

Intramarginal. Within and near the margin.

Introduced. Brought intentionally from another region, as for purposes of cultivation.

Introrse. Turned inward or toward the axis.

Involucel. A secondary involucre, as that of an umbellet in *Umbelliferae.*
Involucellate. Having an involucel.
Involucral. Belonging to an involucre.
Involucrate. Having an involucre.
Involucre. A circle or collection of bracts surrounding a flower-cluster or head or a single flower.
Involute. Rolled inward.
Irregular (flower). Showing inequality in the size, form, or union of its similar parts.
Isomerous. With the members of successive circles of equal number.

Keel. A central dorsal ridge, like the keel of a boat; the two anterior united petals of a papilionaceous flower.
Key. See *samara;* an abbreviated statement of contrasting diagnostic characters.

Labellum. Lip; the characteristic upper (but by a twist of the pedicel apparently lower) petal of the *Orchidaceae.*
Labiate. Lipped; belonging to the *Labiatae.*
Labyrinthiform. With complicated sinuous lines or winding passages.
Lacerate. Irregularly cleft as if torn.
Lacinia (pl. *-ae*). A slender, elongate lobe.
Laciniate. Slashed; cut into narrow pointed lobes.
Lactescent. Yielding milky juice.
Lacuna. An air-space in the midst of tissue.
Lacunose. Covered with depressions or perforated with holes.
Lacustrine. Belonging to or living in lakes.
Lamella. A thin flat plate or laterally flattened ridge.
Lamellate. Made up of thin plates.
Lamina. The blade or expanded part of a leaf, petal, etc.
Lanate. Bearing long implexed hairs or wool.
Lanceolate. Shaped like a lance-head, several times longer than wide, broadest toward the base and narrowed to the apex.
Lanuginous. Cottony or woolly (see *lanate*).
Lateral. Belonging to or borne on the sides.
Latex. The milky juice of some plants.
Leaflet. A single division of a compound leaf.
Legume. The fruit of the *Leguminosae,* bilaterally symmetrical and produced from a unilocular ovary, 1–many-seeded, variously dehiscent or indehiscent and in some groups articulated into one to several 1-seeded portions (cf. *loment*).
Leguminous. Pertaining to a legume or to the *Leguminosae.*
Lemma. The lower of the two bracts inclosing the flower in the Grasses, formerly called the flowering glume.
Lenticel. Corky spots on young bark, corresponding to epidermal stomata.
Lenticular. Lens-shaped; of the shape of a biconvex lens.
Lepidote. Provided with small scurfy scales.
Ligneous. Woody.
Ligulate. Furnished with a ligule.
Ligule. The flattened spreading limb of some marginal or ray-flowers of *Compositae;* a projection from the summit of the sheath in Grasses, etc.
Liguliform. Strap-shaped.
Limb. The expanded portion of a gamopetalous corolla above the throat; the expanded portion of any petal or leaf.
Linear. Long and narrow, with parallel margins.
Lineolate. Marked with fine lines.
Lingulate. Tongue-shaped.
Lip. Each of the upper and lower divisions of a bilabiate corolla or calyx; the characteristic upper (but by a twist of the pedicel apparently lower) petal in Orchids.
Litoral or *littoral.* Growing on shores or pertaining to shores.
Livid. Pale lead-colored.
Lobe. Any segment of an organ, especially if rounded.
Lobed or *lobate.* Divided into or bearing lobes.
Lobulate. Divided into small lobes.
-locular. In composition, having locules.
Locule. The cavity of an ovary or anther.
Loculicidal. Dehiscent into the cavity of a locule of an ovary through the dorsal suture.
Lodicule. One of the small scales or processes in the base of the flower of a grass.
Loment. A legume which is composed of 1-seeded articles.
Lunate. Of the shape of a half-moon or crescent.
Lurid. Dingy.
Lyrate. Pinnatifid with a large and rounded terminal lobe and with the lower lobes small.

Macrosporangium. The receptacle in which macrospores are developed.
Macrospore. The larger kind of spore in *Selaginellaceae, Isoëtes,* etc.
Maculate. Blotched or mottled.
Malpighiaceous hairs. Hairs which are straight and attached by the middle.
Mammiform. Breast-shaped, conical with rounded apex.
Mamillate. Bearing teat-shaped processes.
Marbled. Stained with irregular streaks of color.
Marcescent. Withering but persistent.
Marginate. With a distinct margin.
Maritime. Pertaining to the sea.
Median. Belonging to the middle.
Membranaceous, membranous. Thin, rather

soft, and more or less translucent and pliable.

Meniscoidal. Thin and concavo-convex.

Mericarp. A portion of a fruit which splits away as a seemingly separate fruit, as the two carpels in the *Umbelliferae.*

Meristem. Nascent tissue, capable of being transformed into special forms, as cambium, etc.

Meristematic. Pertaining to the meristem.

-merous. In composition, having parts; as 2-merous, having two parts of each kind.

Mesocarp. The middle layer of a pericarp.

Mesophyte. Plants of medium or favorable conditions as to moisture and light.

Mesophytic. Growing under medium conditions.

Micropyle. The opening in the integument leading into the nucellus of an ovule.

Microsporangium. The receptacle in which microspores are developed.

Microspore. The smaller kind of spore in *Selaginellaceae, Isoëtes,* etc.

Midrib. The central or main rib of a leaf or similar structure.

Mitriform. Shaped like a mitre or cap.

Monadelphous (stamens). United by their filaments into a tube or column.

Moniliform. Resembling a string of beads; cylindrical, with contractions at regular intervals.

Mono-. Prefix meaning one, or of one.

Monocotyledonous. Having but one cotyledon.

Monoecious. With stamens and pistils in separate flowers on the same plant.

Montane. Of or pertaining to mountains.

Mucilaginous. Slimy or mucilage-like.

Mucro. A short and small abrupt tip.

Mucronate. Tipped with a mucro.

Multicipital. With many heads, referring to the crown of a single root or to several caudices.

Multifid. Cleft into many lobes or segments.

Muricate. Rough with short hard points.

Muriculate. Very finely muricate.

Muticous. Pointless or blunt.

Napiform. Turnip-shaped or -rooted.

Nascent. In the act of being formed. *Nascent tissue.* Meristem.

Naturalized. Thoroughly established, originally coming from a foreign area.

Navicular. Boat-shaped.

Nectary. Any place or organ where nectar is secreted.

Nectariferous. Producing or having nectar.

Nervation. Same as venation.

Nerve. A simple or unbranched vein or slender rib.

Netted. Same as reticulated.

Neuter, neutral. Without functional stamens or pistils.

Nigrescent. Blackish.

Niveus. Snowy-white.

Node. The place upon a stem which normally bears a leaf or whorl of leaves; the solid constriction in the culm of a grass; a knot-like or knob-like enlargement.

Nodose. Knotty or knobby.

Nodule. Diminutive of node.

Nodulose. Provided with little knots or knobs.

Nut. A hard indehiscent 1-seeded fruit.

Nutlet. A diminutive nut.

Ob-. A Latin prefix, usually carrying the idea of inversion.

Obcompressed. Compressed dorso-ventrally instead of laterally.

Obconical. Inversely conical, having the attachment at the apex.

Obcordate. Inverted heart-shaped, with the point basal.

Oblanceolate. Lanceolate with the broadest part toward the apex.

Oblate. Flattened at the poles, as a tangerine-orange.

Oblique. Unequal-sided or slanting.

Oblong. Two to three times longer than broad and with nearly parallel sides.

Obovate. Inverted ovate.

Obovoid. Having the form of an egg with the broad end apical.

Obsolescent. Becoming rudimentary or extinct.

Obsolete. Not evident; rudimentary; extinct.

Obtuse. Blunt or rounded at the end.

Ochroleucous. Yellowish-white.

Ocrea. A tubular stipule or confluent pair of elongate stipules.

Ocreate. Having sheathing stipules.

Ocreolae. The smaller or secondary sheaths, as in the inflorescences of *Polygonum.*

Olivaceous. Olive-green.

Opaque. Dull; neither shining nor translucent.

Operculate. Furnished with a lid.

Operculum. A lid; the upper portion of a circumscissile capsule.

Orbicular. Circular.

Orifice. An opening by which spores, etc., escape; any opening.

Orthotropous (ovule or seed). Erect, with the orifice or micropyle at the apex.

Oval. Broadly elliptical.

Ovary. The part of the pistil which contains the ovules.

Ovate. Egg-shaped; having an outline like that of an egg, with the broader end basal.

Overwintering buds. Especially the axillary and terminal buds of trees and shrubs.

Ovoid. A solid with an ovate outline.

Ovulate. Bearing ovules.
Ovule. The body which after fertilization becomes the seed.
Ovuliferous. Bearing ovules.

Palate. A rounded projection of the lower lip of a personate corolla, closing the throat.
Pale. A chaffy scale such as often subtends the fruit in *Compositae.*
Palea. The tiny upper bract which with the lemma incloses the flower in Grasses.
Paleaceous. Chaffy.
Palmate (leaf). Radiately lobed or divided.
Palmately. In a palmate manner.
Paludal or *paludose.* Of or growing in marshes.
Pandurate, panduriform. Fiddle-shaped.
Panicle. A loose irregularly compound inflorescence with pedicellate flowers, such as a branched raceme or corymb.
Panicled, paniculate. Borne in a panicle; resembling a panicle.
Pannose. Having the appearance or texture of felt or woollen cloth of very close texture.
Papilionaceous (corolla). Having a standard, wings, and keel, as in the corolla of many *Leguminosae.*
Papillose. Bearing minute nipple-shaped projections.
Pappus. The modified calyx-limb in *Compositae*, etc., forming a crown of very various character at the summit of the achene.
Parasitic. Growing on and deriving nourishment from another plant.
Parenchyma. Soft tissue of cells with unthickened walls.
Parietal. Borne on or pertaining to the wall or inner surface of a capsule.
Paripinnate (abruptly or even-pinnate). Pinnate with an equal number of leaflets, *i.e.*, without a terminal one.
Parted, -partite. Cleft nearly but not quite to the base.
Parthenogenetic. Developing without fertilization; apogamous.
Partial. Of secondary rank.
Patelliform. Disk-shaped.
Pathological. Diseased.
Pectinate. Pinnatifid with narrow closely set segments; comb-like.
Pedate. Palmately divided or parted, with the lateral segments 2-cleft.
Pedicel. The support of a single flower.
Pedicellate. Borne on a pedicel.
Peduncle. A primary flower-stalk, supporting either a cluster or a solitary flower.
Pedunculate. Borne upon a peduncle.
Pellicle. An evident superficial membrane or epidermis.
Pellucid. Clear, transparent.
Peltate. Shield-shaped and attached to the support by the lower surface.
Pendulous. More or less hanging or declined.
Pendulous ovule, one that hangs from the side of the locule.
Penicillate. Brush-like.
Penta-. Greek for five, used in compounds.
Perennial. Lasting year after year.
Perfect (flower). Having both functional pistil and stamens.
Perfoliate (leaf). Having the stem apparently passing through it.
Peri-. Greek for around, used in compounds.
Perianth. The floral envelope, consisting of the calyx and corolla (when present), whatever their form.
Pericarp. The wall of the matured ovary.
Perigynium. The inflated sac which incloses the ovary in *Carex;* sometimes used for the reduced perianth in other *Cyperaceae.*
Perigynous. Adnate to the perianth and therefore around the ovary and not at its base.
Peripheral. On or near the margin.
Persistent. Long-continuous, as a calyx upon the fruit, leaves through winter, etc.
Personate (corolla). Bilabiate and with the throat closed by a prominent palate.
Petal. A division of the corolla.
Petaloid. Colored and resembling a petal.
Petiolate. Having a petiole.
Petiole. The support (foot-stalk) of a leaf.
Petioled. Having a petiole.
Petiolulate. Having a petiolule.
Petiolule. The foot-stalk (petiole) of a leaflet.
Phaenogamous (= *Phanerogamous*). Having flowers with stamens and pistils and producing seeds.
Phanerogam. Equals *Spermatophyte.*
Phyllary. An involucral bract in the *Compositae.*
Phyllodium (pl. *phyllodia*). A somewhat dilated petiole having the form of and serving as a leaf-blade.
Phyllotaxy. The manner in which the leaves are arranged with regard to the axis.
Phytological. Relating to the study of plants.
Pilose. Hairy, especially with soft hairs.
Pinna (pl. *pinnae*). One of the primary divisions of a pinnate or compoundly pinnate frond or leaf.
Pinnate (leaf). Compound, with the leaflets arranged on each side of a common axis.
Pinnatifid. Pinnately cleft.
Pinnule. A secondary pinna; one of the pinnately disposed divisions of a pinna.
Pisiform. Pea-shaped.
Pistil. The seed-bearing organ of the flower, consisting of the ovary, stigma and style (when present).
Pistillate. Provided with pistils, and, in its more proper sense, without stamens.

Pith. The spongy center of an exogenous stem, chiefly consisting of parenchyma.
Pitted. Marked with small depressions or pits.
Placenta. Any part of the interior of the ovary which bears ovules.
Plaited. Equals plicate (with folds, usually lengthwise).
Plane. With flat surface.
Plicate. Folded into plaits, usually lengthwise.
Plumose. Having fine and elongate hairs, like the plume of a feather, as the pappus-bristles of some Thistles (especially *Cirsium*).
Plumule. The growing point of the embryo.
Pollen. The fecundating grains contained in the anther.
Pollen-tube. The tube emitted by a pollen-grain, passing down from the stigma to the ovary and ovules.
Polliniferous. Bearing pollen.
Pollinium (pl. *pollinia*). A mass of waxy pollen or of coherent pollen-grains, as in *Asclepias* and *Orchidaceae*.
Poly-. In Greek compounds, denoting many.
Polyadelphous. With the stamens grouped into several bundles.
Polyandrous. Having an indefinite number of stamens.
Polyembryony. The production of more than a single embryo in an ovule.
Polygamodioecious. Polygamous but chiefly dioecious.
Polygamomonoecious. Polygamous but chiefly monoecious.
Polygamous. With hermaphrodite and unisexual flowers on the same or on different individuals of the same species.
Polymorphous. With several or various forms; variable as to habit.
Polypetalous. Having separate petals.
Pome. A fleshy fruit of which the apple is the type, formed from an inferior ovary with several locules.
Porose or *porous.* Pierced with small holes or pores.
Porrect. Directed outward and forward.
Posterior. Next or close to the main axis; reverse of anterior.
Praemorse. Appearing as if bitten off.
Precocious. Appearing or developing very early; in *Salix*, etc., with the aments expanding before the leaves.
Prehensile. Clasping or grasping, as in tendrils.
Prickle. A small and more or less slender sharp outgrowth from the epidermis.
Primary. Used of the part first developed; the main divisions of a leaf, etc.
Primocane. The first year's cane (usually without flowers) of *Rubus* and similar genera.
Prismatic. Of the shape of a prism, angulate, with flat sides.
Procumbent. Lying on the ground or trailing but without rooting at the nodes.
Proliferating or *proliferous.* Producing buds and plantlets from leaves or as other offshoots.
Prophyllum. The bracteole at the base of an individual flower, as in *Juncus*.
Prostrate. Lying flat upon the ground.
Proterandrous. With the anthers ripe before the pistils in the same flower.
Proterogynous. Having the stigma ripe for the pollen before the maturity of the anthers of the same flower.
Prothallus. A cellular usually flat and thallus-like growth, resulting from the germination of a spore, upon which are developed sexual organs or new plants.
Pruinose. Having a waxy powdery secretion on the surface, a "bloom."
Psammophilous. Sand-loving, as the vegetation of dunes.
Pseudo- or *pseudos-.* Greek for false, used as a prefix.
Puberulent. Minutely pubescent with scarcely elongate hairs.
Pubescent. Covered with hairs, especially if short, soft and down-like.
Pulverulent. Powdered; appearing as if covered by minute grains of dust.
Pulvinus. An enlargement close under the insertion of a leaf or of a branch of an inflorescence; the swollen base of the petiole.
Punctate. Dotted with depressions or with translucent internal glands or colored dots.
Puncticulate. Minutely punctate.
Pungent. Terminating in a rigid sharp point; acrid.
Pustular, pustulate. Having slight elevations like blisters.
Putamen. The shell of a nut; the bony part of a stone-fruit.
Pyramidal. Pyramid-shaped.
Pyriform. Pear-shaped.
Pyxidate. Furnished with a lid, as some capsules.
Pyxis. A capsule with circumscissile dehiscence, the upper portion acting as a lid, as in *Portulaca*, *Plantago*, etc.

Quadrate. Nearly square in form.
Quilled. Applied to normally ligulate florets which have become tubular.
Quinary. In fives.
Quinate. Growing together in fives, as leaflets from the same point.

Raceme. A simple inflorescence of pedicelled

flowers upon a common more or less elongated axis.

Racemiform. In the form of a raceme.

Racemose. In racemes; or resembling a raceme.

Rachilla. A secondary axis; specifically in the Grasses and Sedges the floral axis as opposed to that of the spike or spikelet.

Rachis. The axis of an inflorescence or of a compound leaf.

Radiant. Diverging from a central point.

Radiate. Spreading from or arranged around a common center; bearing ray-flowers.

Radicant. Rooting, usually applied to stems or leaves.

Radicle. The portion of the embryo below the cotyledons, more properly called the caudicle.

Rameal. Belonging to a branch.

Ramification. Branching.

Ramose. Branching; having many branches.

Ramulose. Having many branchlets.

Raphe. The ridge or adnate funicle which in an anatropous ovule connects the two ends.

Raphid (pl. *raphides*). Needle-shaped crystals in the tissues of plants, as in the *Araceae.*

Ray. The branch of an umbel or similar inflorescence; the ligule or strap-like marginal flower of many *Compositae,* when differentiated from the disk-flowers.

Receptacle. The more or less expanded or produced portion of an axis which bears the organs of a flower (the *torus)* or the collected flowers of a head; any similar structure in Cryptogams.

Reclinate. Turned or bent downward.

Recurved. Curved downward or backward.

Reflexed. Abruptly bent or turned downward.

Regular. Uniform in shape or structure.

Relic or *relict.* Used for localized plants which are evidently left over from past geological epochs.

Reniform. Kidney-shaped.

Repand. With a slightly uneven and somewhat sinuate margin.

Repent. Creeping; prostrate and rooting at the nodes.

Replicate. Folded backward.

Replum. A frame-like placenta from which the valves fall away in dehiscence, as in many *Cruciferae.*

Resiniferous. Producing resin.

Resupinate. Turned upside down.

Reticulate. In the form of network; netveined.

Reticulum. A network of fibers.

Retinaculum. The gland to which one or more pollinia are attached in Orchids; in Asclepiads a horny elastic body to which the pollen-masses are fixed; the funicle in the fruit in most *Acanthaceae,* which is curved like a hook and retains the seed until mature.

Retrorse. Directed back or downward.

Retuse. With a shallow notch at a rounded apex.

Reversion. A change backward, as to an earlier condition.

Reversionary. With the characteristics of a reversion.

Revolute. Rolled backward from the margins or apex.

Rhizomatose. Having the character of a rhizome.

Rhizome. Any prostrate or subterranean stem, usually rooting at the nodes and becoming upcurved at the apex.

Rhombic. With the outline of an equilateral oblique-angled figure.

Rhomboid, rhomboidal. A solid with a rhombic outline.

Rib. A primary or prominent vein of a leaf.

Ribbed. With prominent ribs.

Rimose, rimous. With chinks or cracks, as in old bark.

Ringent. Gaping, as the mouth of an open bilabiate corolla.

Riparian. Growing by rivers or streams.

Root. The descending axis of the plant, usually growing in the opposite direction from the stem, without nodes and internodes, mostly developing underground and absorbing moisture from the soil.

Root-hairs. Unicellular absorptive hairs on the young roots and rootlets.

Rootstock. Equals rhizome; or sometimes used for elongate, unmodified rooting subterranean offshoots.

Root-tipping. Used for the tips of arching or trailing stems which root on touching the ground.

Rosette. A cluster of leaves or other organs in a circular form.

Rostellate. The diminutive of rostrate, somewhat beaked.

Rostellum. A little beak; a slender extension from the upper edge of the stigma in Orchids.

Rostrate. Having a beak.

Rosula. A rosette.

Rosulate. In the form of a rosette.

Rotate (corolla). Wheel-shaped; flat and circular in outline.

Rotund. Rounded in outline.

Ruderal. Growing in waste places or among rubbish.

Rudiment. An imperfectly developed and functionally useless organ; a vestige.

Rufous or *rufescent.* Reddish-brown.

Rugose. Wrinkled.

Ruminate. Looking as though chewed.

Runcinate. Sharply incised, with the segments directed backward.
Runner. A filiform or very slender stolon.
Rupturing. Bursting irregularly.

Saccate. Sac-shaped.
Sagittate. Shaped like an arrow-head, the basal lobes directed downward or backward.
Salverform or *salver-shaped* (corolla). Having a slender tube abruptly expanded into a flat limb, as in the primrose or phlox.
Samara. An indehiscent winged fruit.
Sarmentose. Producing long and lithe runners.
Scabridulous. Slightly rough.
Scabrous. Rough to the touch.
Scalariform. Having markings suggestive of a ladder.
Scale. Any thin scarious body, usually a degenerate leaf, sometimes of epidermal origin; a trichome, if disc-like or flattened; sometimes used for *glume.*
Scandent. Climbing.
Scape. A naked flowering stem rising from the ground.
Scapiform. Resembling a scape.
Scapose. Bearing or resembling a scape.
Scarious. Thin, dry, and membranaceous, not green.
Schizocarp. A pericarp which splits into one-seeded portions, mericarps, or "split-fruits".
Sclerenchymatous. Of sclerenchyma, *i.e.,* of tissue composed of cells with thickened and hardened walls.
Scobiform. Having the appearance of sawdust.
Scorpioid (inflorescence). Circinately coiled while in bud, as in the forget-me-not.
Scrobiculate. Marked by minute or shallow depressions.
Scrotiform. Pouch-shaped.
Scurfy. With scale-like or bran-like particles.
Scutelliform. Platter-shaped.
Secund. Directed to one side only, usually by torsion.
Seed. The ripened ovule, consisting of the embryo and its proper coats.
Segment. One of the parts of a leaf or other like organ that is cleft or divided.
Semi-. Latin for half, used as a prefix.
Sepal. A division of a calyx.
Sepaloid. Of the texture of or resembling a sepal.
Septate. Divided by partitions.
Septicidal (capsule). Dehiscing through the partitions and between the locules.
Septum. Any kind of partition.
Sericeous. Silky.
Serotinous. Produced late in the season; in *Salix,* etc., with aments expanding after the leaves
Serrate. Having sharp teeth pointing forward.
Serrulate. Finely serrate.
Sessile. Without stalk of any kind.
Seta. A bristle.
Setaceous or *setiform.* Bristle-shaped.
Setiferous. Bearing bristles.
Setose. Beset with bristles.
Setulose. Having minute bristles.
Sheath. A tubular envelope, as the lower part of the leaf in Grasses.
Sheathing. Inclosing, as by a sheath.
Shield-shaped. Equals peltate.
Shrub. A woody perennial, smaller than a tree, usually with several stems.
Sigmoid. Doubly curved like the letter S.
Siliceous or *silicious.* Composed of or abounding in silica.
Silicle. A short silique.
Silique. A specialized capsule in which a (characteristic) replum separates the valves, notably in the *Cruciferae.*
Silky. Covered with close-pressed soft and straight pubescence.
Sinistrorse. Turned to the left.
Sinuate or *sinuous.* With the outline of the margin strongly wavy.
Sinus. The cleft or recess between two lobes.
Smooth. Without roughness; rather inaccurately used to signify without pubescence.
Sobole. A shoot, especially from the base.
Soboliferous. Bearing vigorous basal shoots.
Sordid. Dirty in tint, chiefly applied to pappus when of an impure white.
Sorus (pl. *sori*). A heap or cluster, applied to the fruit-dots of Ferns.
Spadix. A spike with a fleshy axis; as in Aroids.
Spathe. A large bract inclosing an inflorescence.
Spatulate. Oblong, with the basal end attenuated like a druggist's spatula.
Spermatophyte. A Phanerogam, a plant with true seeds.
Spermatozoid. A motile ciliated male reproductive cell.
Spicate. Arranged in or resembling a spike.
Spiciform. Spike-like.
Spiculose. With a surface covered with fine points.
Spike. A form of simple inflorescence with the flowers sessile or nearly so upon a more or less elongated common axis.
Spikelet. A small or secondary spike; especially in Grasses and Sedges.
Spindle-shaped. Same as *fusiform.*
Spine. A sharp woody or rigid outgrowth from the stem.

Spinescent. Ending in a spine or bearing a spine.
Spinose. Spine-like, or having spines.
Spinous. Spiny, having spines.
Spinule. A little spine or spine-like process.
Spiricle. A delicate coiled thread in the superficial cells of certain seeds and achenes which uncoils when moistened, as in *Ruellia.*
Sporadic. Occurring here and there, without continuous range.
Sporangium. A spore-case.
Spore. The reproductive organ in Cryptogams, which in function corresponds to a seed but possesses no embryo.
Sporocarp. The fruit-cases of certain Cryptogams, containing sporangia or spores.
Sporophyll. A leaf bearing spores.
Sport. A bud- or seed-variation from the normal.
Spray. The small branches or branchlets of trees, with their foliage.
Spur. A hollow sac-like or tubular extension of some part of a flower, usually nectariferous.
Squamellate. Having tiny or secondary scales.
Squamulose. Provided with small scales.
Squarrose. Having its parts or processes (such as the tips of phyllaries) spreading or recurved at the end.
Squarrulose. Diminutively squarrose.
Stamen. One of the pollen-bearing organs of the flower.
Stamineal. Relating to or consisting of stamens.
Staminode or *staminodium.* A sterile stamen, or any structure without anther corresponding to a stamen.
Standard. The upper dilated petal of a papilionaceous corolla.
Station. A particular locality for a given plant.
Stellate, stelliform. Star-shaped.
Sterile. Unproductive, as a flower without pistil or a stamen without an anther; infertile.
Stigma. That part of a pistil or style which receives the pollen for effective fertilization; usually distinguished by bearing minute papillae or having a viscid surface or specialized lines or other modifications.
Stigmatic. Belonging to or characteristic of the stigma.
Stipe. The stalk-like support of a pistil; the "leaf-stalk" of a fern-frond.
Stipel or *stipellum.* An appendage of a leaflet analogous to a stipule.
Stipellate. Furnished with stipellae.
Stipitate. Having a stipe; *Stipitate glands,* glands with stalk-like bases.
Stipular. Belonging to stipules.
Stipulate. Having stipules.
Stipule. An appendage at the base of a petiole or leaf or on each side of its insertion.
Stolon. A runner, or any basal branch that is inclined to root.
Stoloniferous. Producing stolons.
Stoma (pl. *stomata*). A minute orifice or mouth-like opening between two guard-cells in the epidermis, particularly on the lower surface of the leaves, through which gaseous interchange between the atmosphere and the intercellular spaces of the parenchyma is effected.
Stramineous. Straw-colored.
Striate. Marked with fine longitudinal lines or streaks.
Strict. Very straight and upright.
Striga. A small straight hair-like scale.
Strigose. Provided with appressed sharp straight and stiff hairs.
Strobile. An inflorescence marked by imbricated bracts or scales, as in the Hop and Pine-cone.
Strophiole. An appendage at the hilum of certain seeds.
Style. The usually attenuated portion of the pistil connecting the stigma and ovary.
Stylopodium. A disk-like expansion at the base of a style, as in *Umbelliferae.*
Sub-. A Latin prefix, usually signifying "somewhat" or "slightly".
Suberose. Corky in texture.
Subula. A fine abrupt sharp point.
Subulate. Awl-shaped.
Succulent. Juicy; fleshy.
Sucker. A vegetative shoot of subterranean origin.
Suffrutescent. Slightly or obscurely shrubby.
Suffruticose. Very low and woody; diminutively shrubby.
Sulcate. Grooved or furrowed.
Super or *supra.* Latin for above, used as a prefix.
Superior ovary. An ovary with the perianth inserted below it on the torus.
Supra-axillary. Borne above the axil.
Surculose. Producing suckers.
Suspended (ovule). Hanging from the apex of the locule.
Suture. A junction or seam of union; a line of opening or dehiscence.
Syconium. A multiple hollow fruit, as that of the Fig.
Symbiosis. The living together of dissimilar organisms, with benefit to one only, or to both.
Symbiotic. Relating to *symbiosis.*
Symmetrical (flower). Regular as to the number of its parts; having the same number of parts in each circle.
Sympetalous. Equals gamopetalous.

Syncarp. A multiple or fleshy aggregate fruit, as the Mulberry or Magnolia.
Syngenesious. With anthers cohering in a ring.
Synonym. A superceded or unused name.
Synonymy. The series of discarded names for identical objects.
Synonymous. Having the same meaning.

Tactile. Sensitive to touch.
Tap-root. Primary descending root.
Taxonomy. Classification.
Tendril. A slender clasping or twining cauline or foliar outgrowth.
Tepal. Used for sepals and petals of similar form and not readily differentiated.
Teratological. Monstrous; relating to a monstrosity.
Terete. Having a circular transverse section.
Ternary, ternate. In threes, trimerous.
Tessellate, -ed. Checkered.
Testa. The outer commonly hard and brittle seed-coat.
Tetrad. A group of four objects; a body formed of four cells, as in the formation of pollen in the pollen-mother-cells.
Tetradynamous. Having four long and two shorter stamens.
Tetragonal. Four-angled.
Tetramerous. Of four members.
Thalloid, thallose. Resembling a thallus.
Thallus. A vegetative body without differentiation into stem and leaf.
Thorn. Same as *spine.*
Throat. The orifice of a gamopetalous corolla or calyx; the part between the proper tube and the limb.
Thyrse. A contracted cylindrical or ovoid and usually compact panicle as a cluster of grapes.
Thyrsoid. Resembling a thyrse.
Tomentose. Densely pubescent with matted wool.
Tomentulose. Finely tomentose.
Tomentum. Closely implexed woolly hair.
Torose. Cylindrical with contractions or swellings at intervals.
Torulose. Diminutive of torose.
Torus. The receptacle of a flower; or of the flowers of a head in the *Compositae.*
Trabeculate. Cross-barred.
Trailing. Prostrate but not rooting.
Trapeziform. Unsymmetrically 4-sided.
Tree. Perennial woody plant with an evident trunk.
Tri-. In composition, three or thrice.
Triad. Group of three objects.
Triandrous. Having three stamens.
Trichome. Any hair-like outgrowth of the epidermis.
Trifoliolate. Having three leaflets.
Trigonous. Three-angled.
Trimorphous. Occurring under three forms.
Triquetrous. Having three salient angles.
Truncate. Ending abruptly, as if cut off transversely.
Tuber. A thickened and short subterranean branch having numerous buds or eyes.
Tubercle. A small tuber or tuber-like (not necessarily subterranean) body, often due to symbiotic relation of organisms.
Tuberculate. Bearing small processes or tubercles.
Tuberiferous. Bearing tubers.
Tuberoid. A fleshy-thickened root, resembling a tuber, as in many terrestrial Orchids.
Tuberous. Having the character of a tuber; tuber-like in appearance.
Tumid. Swollen.
Tunicated. Having concentric coats, as an onion.
Turbinate. Top-shaped; inversely conical.
Turgid. Swollen, or tightly drawn, said of a membrane or covering expanded by pressure from within.
Type. In taxonomy, the specimen from which the original description was made.
Typical. In classification referring to the originally described or the ordinary variation.

Ubiquitous. Occurring everywhere.
Umbel. An inflorescence in which the peduncles or pedicels of a cluster spring from the same level.
Umbellate. In or like an umbel.
Umbellet. A secondary umbel.
Umbelliform. In the shape of an umbel.
Umbellule. An umbellet.
Umbilicate. Depressed in the center.
Umbonate. Bearing a stout projection in the center; bossed.
Undulate. With a wavy surface or margin; repand.
Unguiculate. Contracted at base into a claw.
Uni-. In composition, one.
Unisexual. Of one sex, either staminate or pistillate only.
Urceolate. Hollow and cylindrical or ovoid, and contracted at or below the mouth, like an urn.
Utricular. Inflated, bladder-like.
Utricle. A small bladdery 1-seeded fruit; any small bladder-like body.

Vaginate. Sheathed.
Vallecula. Applied to the grooves in the intervals between the ridges, as in the fruit of *Umbelliferae. Vallecular.* Pertaining to such grooves.
Valvate. Opening by valves, as a capsule; in

aestivation, meeting by the edges without overlapping.

Valve. One of the pieces into which a capsule splits; the partially detached lid of an anther.

Vascular. Furnished with vessels or ducts.

Veins. Threads of fibrovascular tissue in a leaf or other organ, especially those which branch (as distinguished from nerves).

Velum. The membranous indusium in *Isoëtes.*

Velutinous. Velvety.

Venation. The character of the veining.

Ventral. Belonging to the anterior or inner face of an organ, as opposed to dorsal (adaxial).

Ventricose. Swelling unequally, or inflated on one side.

Vermiform. Worm-shaped.

Vernal. Appearing in spring.

Vernation. The arrangement of leaves in the bud.

Vernicose. Shiny, as though varnished.

Verrucose. Covered with wart-like elevations.

Versatile (anther). Attached near the middle and turning freely on its support.

Verticil. A whorl.

Verticillastrate. Bearing or arranged in clusters resembling whorls.

Verticillate. Disposed in a whorl.

Vesicle. A small bladder or air-cavity.

Vesicular, vesiculose. Composed of or covered with vesicles.

Vespertine. Appearing or expanding in the evening.

Vestige. Rudiment.

Vestigial. Rudimentary.

Vexillum. The standard or large posterior petal of a papilionaceous flower.

Viatical. Growing by roadsides or paths.

Villous. Bearing long and soft (not interwoven) hairs.

Virgate. Wand-shaped; slender, straight and erect.

Viscid. Glutinous; sticky.

Vitta. An oil-tube; a structure commonly present in the pericarp of the *Umbelliferae.*

Viviparous. Sprouting or germinating on the parent plant.

Weed. A troublesome or aggressive plant which intrudes where not wanted.

Weedy. With the attributes of a weed.

Whorl. An arrangement of leaves, etc., in a circle around the stem.

Wing. Any membranous or thin expansion bordering or surrounding an organ; the lateral petal of a papilionaceous corolla.

Winter-bud. Shortened and crowded vegetative shoots found in *Potamogeton* and some other genera.

Woolly. Clothed with long and tortuous or matted hairs.

Xerophyte. A plant of arid habitats.

Zoöspore. A free-moving spore (of the lower Cryptogams), an asexual reproductive cell with cilia.

Zygomorphic. Capable of division in only one plane of symmetry.

Index to Latin Names

OF FAMILIES, GENERA AND SPECIES

Roman type is used for maintained specific epithets and for family-names (which appear in CAPITALS).

Italic type is used to indicate synonyms as well as incidental references to maintained elements.

Index to English, French-Canadian and Colloquial Names